Aquatic Dicotyledons of North America

Aquatic Dicotyledons of North America
Ecology, Life History, and Systematics

Donald H. Les

CRC Press
Taylor & Francis Group
Boca Raton London New York

CRC Press is an imprint of the
Taylor & Francis Group, an **informa** business

CRC Press
Taylor & Francis Group
6000 Broken Sound Parkway NW, Suite 300
Boca Raton, FL 33487-2742

© 2018 by Taylor & Francis Group, LLC
CRC Press is an imprint of Taylor & Francis Group, an Informa business

No claim to original U.S. Government works

Printed on acid-free paper

International Standard Book Number-13: 978-1-4822-2502-0 (Hardback)

Library of Congress Cataloging-in-Publication Data

Names: Les, Donald H., author.
Title: Aquatic dicotyledons of North America : ecology, life history, and systematics / author: Donald H. Les.
Description: Boca Raton, FL : Taylor & Francis, 2018. | Includes
bibliographical references and index.
Identifiers: LCCN 2016053889 | ISBN 9781482225020 (hardback : alk. paper)
Subjects: LCSH: Dicotyledons--North America. | Aquatic plants--North America.
Classification: LCC QK110 .L55 2018 | DDC 583--dc23
LC record available at https://lccn.loc.gov/2016053889

Visit the Taylor & Francis Web site at
http://www.taylorandfrancis.com

and the CRC Press Web site at
http://www.crcpress.com

Printed and bound in the United States of America by Sheridan Books, Inc. (a Sheridan Group Company).

Contents

Acknowledgments

I extend my sincerest thanks to everyone who helped me during the preparation of this volume by providing contacts, information, unpublished results, and technical support (specific names are not mentioned lest I unintentionally exclude some). Special thanks go to the technical support staff for the Integrated Taxonomic Information System (ITIS) website, who promptly responded to my frequent and frantic requests for assistance. I also thank the various herbaria whose specimen data are hosted on public websites for providing access to that rich and incredibly useful body of information. Above all, I cannot adequately credit my wife, Jane, who patiently tolerated me during the 13 years of this book's preparation.

Introduction

AQUATIC PLANTS—AN ELUSIVE DEFINITION

It is difficult to agree on a mutually acceptable definition of "aquatic plant" because this term has been applied to such a broad spectrum of species representing an assortment of habitats characterized by various degrees of moisture. Also, many plants are adapted flexibly to diverse habitat conditions, and it is not unusual for a single species to thrive in sites, which differ diametrically by their extent of soil moisture. Although most would agree that certain plants like *Nymphaea* (water lilies) should be classified as aquatics, it is much more difficult to categorize species such as *Persicaria lapathifolia* L. (nodding smartweed), which occurs in habitats ranging from swamps to cultivated fields (Fernald, 1950). Whether or not such species are regarded as aquatic plants has more than theoretical importance, because precise wetland delineation is necessary for satisfying mandates imposed by various wetland conservation laws (Tiner, 1999).

Life forms (Sculthorpe, 1967) are useful in determining whether certain plants are aquatic. Submersed species, floating-leaved species, free-floating species, and suspended species all can be categorized reliably as aquatics, given their constant association with aqueous conditions throughout their life cycle. However, efforts to define aquatic plants are complicated by wetlands, which are transitional, and represent a continuum from terrestrial to aquatic habitats (Cronk & Fennessy, 2001). A distinction between the terms "aquatic" and "wetland" is evident in the published literature. Some titles refer only to "aquatic plants" (e.g., Gruchy, 1938; Muenscher, 1944; Fassett, 1957; Stanton & Smith, 1957; Eyles & Robertson Jr., 1963; Steward et al., 1963; Carlson & Moyle, 1968; Ogden et al., 1976; Winterringer & Lopinot, 1977; Nelson & Couch, 1985; Schloesser, 1986; Klussmann & Davis, 1988; Cho, 2009; Skawinski, 2010; Block and Rhoads, 2011), others only to "wetland plants" (e.g., Niering & Goodwin, 1973; Silberhorn, 1976; Montz, 1977; Pierce, 1977; Lazarine, 1980; Crawford, 1981; Faber, 1982; Weinmann et al., 1984; Reed, 1986; Tiner, 1987; 1993b; Cooper, 1989; Guard, 1995; Cooke, 1997; Newmaster et al., 1997; Erickson & Puttock, 2006), yet others to "aquatic and wetland plants" (e.g., Correll & Correll, 1975; Dennis et al., 1977; Taylor, 1977; Tarver et al., 1978; Godfrey & Wooten, 1979, 1981; Beal & Thieret, 1986; Venable, 1986; Aulbach-Smith & de Kozlowski, 1990; Combs & Drobney, 1991; Cook, 1996a; Stutzenbaker, 1999; Crow & Hellquist, 2000a,b). Often, many of the same species are treated in all of these books. So how does one distinguish between what is aquatic and what is not?

My definition of aquatic plants assumes that such species possess adequate adaptations necessary to survive in either "aquatic" or "wetland" habitats. Thus, such plants must not simply occur at a wet site, but must persist and reproduce under those conditions. Consequently, those species fulfilling the latter requirements should be those that are most commonly recognized by experts as being adapted to an aquatic habitat. To define the aquatic habitat, I have emphasized the occurrence of hydric soils, which are defined elsewhere (e.g., Anon, 1985, 1987; Fletcher et al., 1998) as a major factor.

From this line of reasoning, I define aquatic plants as "*indicative* taxa capable of *perpetuating* their life cycles and *continuing* their existence in still or flowing *standing* water or upon inundated or non-inundated *hydric* soils." This definition has several components. The primary focus of this book is to provide information on the indicative taxa, which are those flowering plant species most readily recognized as adapted to aquatic habitats.

There are several ways to determine whether a species is indicative. One need to only look in any of the books mentioned earlier to see what the experts have regarded as comprising aquatic or wetland species. Species that occur in these books are indicative, at least presuming that the various authors have used similar criteria in assessing whether the plants were "aquatic enough" to be included in their treatments. Corn plants (*Zea mays*) often are observed growing in temporarily flooded fields, but to my knowledge, corn is not included in any book pertaining to aquatic or wetland species. Consequently, corn would not be regarded as indicative. Conversely, plants like cattails (*Typha*) almost always are observed to grow in sites that would be characterized as wet, and this genus is listed in nearly all of the books cited earlier. *Typha* then is indicative. Another way to look at this comparison is that corn almost never grows under wet conditions, whereas cattails almost always do.

Fortunately, many specialists have combined their expertise over the years to develop a listing and classification of hydrophytes for the United States using a similar line of reasoning (USACE, 2016). This national list of wetland plants has undergone several revisions since its inception (Reed, 1988) including a number of revisions issued during the preparation of this book. Although this project began when the 1996 list was in use, the list issued in 2013 has been followed here as the primary reference. The 2013 list categorized plants as belonging to one of five major groups of wetland indicators based essentially on their relative occurrence in wet versus dry environments (Lichvar et al., 2012). These categories are delineated as

1. Obligate wetland (OBL): Occur in wetlands >99% of the time
2. Facultative wetland (FACW): Occur in wetlands 67%–99% of the time
3. Facultative (FAC): Occur in wetlands 34%–66% of the time
4. Facultative upland (FACU): Found in wetlands 1%–33% of the time

5. Upland (UPL): Occur in wetlands in some regions, but occur almost always in nonwetlands.

This book treats only those species designated as OBL species in at least one region. Consequently, only those species most likely to occur in aquatic and wetland habitats are included. These obligate aquatic taxa are abbreviated throughout the text as OBL. It is important to emphasize that the OBL category can include species that require water to complete an essential part of their life history, even though they might often appear to be growing on dry land after the water table subsides. More than 1500 North American plant species have been categorized as OBL, with 1016 species (including some categorized formerly as OBL) treated here within the dicotyledons. The comprehensive life-history information summarized for these OBL taxa throughout this book should facilitate their evaluation as wetland indicators with respect to the categories defined earlier. In some cases, a species originally categorized as OBL in 1996, but later reclassified as a lesser indicator, has been included in this treatment anyway. These relatively few instances reflect those species whose treatments had already been completed prior to the release of the 2013 list.

This book focuses strictly on aquatic dicotyledonous flowering plants but is not a taxonomic guide for species identification. Rather, it is a reference, which contains a compilation of information on aquatic plant species including various aspects of their ecology, life history, and systematics. Consequently, one must first identify a species in order to effectively apply this information, using one of the several available manuals (see those referenced earlier) to facilitate aquatic angiosperm species identification. The geographical coverage of this book is restricted to North America, which is defined here as comprising the lower 48 United States, Alaska, and Canada. However, many North American aquatics also are found throughout the world, and this reference should be useful in many other regions as well.

The entries in this book are first organized essentially following the informal angiosperm groups summarized by Judd et al. (2016):

DICOTYLEDONS
  DICOTYLEDONS I (monosulcates)
    A GROUP OF UNCERTAIN PHYLOGENETIC AFFINITY
    THE ANA GRADE
    MAGNOLIID DICOTYLEDONS
  EUDICOTS (tricolpate dicots; Eudicotyledoneae)
    DICOTYLEDONS II (basal tricolpates)
    CORE EUDICOTS
      DICOTYLEDONS III ("caryophyllid" tricolpates)
      DICOTYLEDONS IV ("rosid" tricolpates)
      DICOTYLEDONS V ("asterid" tricolpates)

The entries are grouped secondarily by their respective orders (arranged alphabetically), which closely follows the APG III

(2009) classification, but with deviations as noted in the text. A bracketed number follows each ordinal name to indicate the total number of families contained. A systematic overview (with text citations) follows each order and indicates specifically those families that contain OBL indicator taxa. An overview of each family containing OBL taxa follows sequentially in alphabetical order. Family names are followed by a bracketed estimate of the total number of genera contained. An overview of each family follows (again with cited references) to summarize information on the systematics and economic importance. Each family treatment ends with an alphabetical list of genera that contain OBL species, listing the taxonomic authority for each genus. Phylogenetic trees describing ordinal and familial relationships are included whenever available. However, it is important to emphasize that the putative relationships indicated by the organization of this book or by the phylogenetic trees included, *all represent hypotheses rather than fact*, and that no assurance can be given regarding their accuracy beyond the conclusions reached by the relevant studies cited.

The main part of the text focuses on genera and species. Each genus name is arranged alphabetically within a family and is followed (in parentheses) by a common name or names in English and French (when available), the latter separated by a semicolon. Generic treatments follow the same format, listing first the etymology of the name, major synonyms, and the global and North American distribution and species numbers. "In part" is used for synonyms that do not apply to the taxon in general. Asterisks designate nonindigenous distributional regions. Next is a list of any species ranked as OBL by the USF&WS system (http://rsgisias.crrel.usace.army.mil/NWPL/doc/historic_lists/NWI_1996/v3_96_intro.html), along with any alternative rankings applied to these species. Generic and species names are those accepted by Integrated Taxonomic Information System (ITIS) (http://www.itis.gov/), unless given otherwise as explained in the accompanying "Systematics" section. After that listing is a description of habitat, which indicates salinity (freshwater to marine) and general community type. The latter includes four categories: lacustrine, palustrine, riverine, and oceanic. Lacustrine refers to those species found primarily in the deeper waters of lakes, riverine to those species preferentially occupying watercourses, and palustrine to "wetland" communities, often including those that occur along the shorelines of rivers and lakes. The pH value given next represents the complete range of values reported among the individual species, or is listed as "unknown" if reports were not found. Depth also summarizes the full range of water depth associated with all of the OBL species in the genus.

Life forms also are summarized for all of the OBL species discussed. "Emergent" refers to those species where substantial vegetative growth (shoot and leaf production) occurs above the water surface; "submersed" plants grow with their shoots and leaves underwater (excluding flowering portions in some cases); "suspended" plants are those whose shoots and leaves remain buoyant just beneath the water surface without any attachment to the substrate; "floating-leaved" plants produce leaves on the surface of water bodies and are rooted in the substrate; all structures of "free-floating" plants float on the water surface and do not attach to the substrate unless stranded in very shallow water or along shorelines. Herbs are herbaceous species where the leaves are scattered along the stem (vittate) or extend from one common point (rosulate). Woody plants include trees (strong ascending trunks exceeding 6 m in height) or shrubs (multiply branched, usually less than 6 m in height).

The section on "Key morphology" is somewhat nontraditional, in that it describes the genus with respect to the salient characteristics summarized for all of the OBL species. Consequently, a measurement such as "to 50 cm" indicates that at least one OBL species in the genus is characterized by this value, even though many might be smaller in stature. Where several discrete ranges occur among the OBL species, they are separated by a comma, e.g., "to 5–10, 50–60 mm." The values provided indicate the longest axis of a structure, which is not necessarily its length. For monotypic genera or where a single OBL species exists, the description provided is of that species.

The "Life history" section summarizes information for all OBL species in the genus. For duration, the relevant structure is indicated in parentheses; for fruits, an indication of frequency is indicated parenthetically as well as presumed vectors associated with the structures involved in local and long-distance dispersal (specific details are provided within the included species treatments).

The imperilment status is summarized for each species using the data reported by NatureServe (http://www.nature-serve.org/). The global [G] rank and regional ranks (where applicable) are listed for each OBL species using standard abbreviations for the US states and Canadian provinces; the latter are underlined for clarification. Taxa that are secure globally will include only the global [G] ranking. The rankings listed reflect the status when each treatment was prepared and always should be verified because they change continually.

A general ecological overview of the genus is provided, which indicates the frequency of OBL taxa, and summarizes basic common features such as reproductive modes, pollinators, and seed ecology. When a genus is monotypic, information is excluded from this section and is simply summarized under the species treatment.

Each OBL species within the genus is treated successively in alphabetical order. Species names are provided with authority names and are highlighted in bold. Habitat types reported in literature accounts or from specimen label data are listed alphabetically. Some designations (e.g., "dunes") might appear unusual for aquatic plants, but refer to wet areas (pools, swales, etc.) occurring within these landforms. Similarly, the designation "prairie" would infer wet prairie specifically. Elevation data have been determined from literature reports or from specimen databases and represent the highest documented values. Thus, a value of "to 3489 m" does not mean that plants do not occur above this specific elevation, but only that the value was the highest found in the literature and database searches. All available life-history information that could be reasonably extracted from literature accounts and specimen database records is summarized. A conscientious effort has been made to locate information pertaining to reproductive ecology, pollination biology, seed germination and dispersal, and other important life-history information for each species. Regrettably, such information often is incomplete or essentially absent for a species, which indicates the need for additional research. Where specific organisms are mentioned by common name (e.g., "honey bees" as pollinators), additional taxonomic information (e.g., "Insecta": Hymenoptera: Apidae: *Apis*) is provided parenthetically to enable the reader to more specifically identify the intended taxon. Although it might seem obvious and/or redundant to constantly repeat groups like Aves (birds) and Insecta (insects), this convention has been followed for two reasons. First, a designation simply of "insects (Insecta)" indicates that no additional taxonomic information (e.g., order, family, etc.) is available for that particular entry. Second, the use of this book by non-English-speaking readers hopefully will be facilitated by including the scientific names along with the English common names, with which they might be unfamiliar.

To conserve space, none of the information summarized in the species treatments contains any literature citations; however, all of the literature from which any information has been obtained is cited at the end of each generic treatment under "References." Admittedly, although more space efficient, this convention does make it somewhat difficult to track down the original source of a specific entry, especially where lengthy reference lists occur.

Comprehensive lists of "Reported associates" are provided for each species, relative to the information available. The names presented here are those accepted by ITIS, except for a relatively few instances of disagreement at the author's discretion. These lists reflect the reported ecological association of any plant genus or species (including nonangiosperms) with the subject species in a variety of resources including published books and papers as well as various data repositories including but not limited to

Federal Register (https://www.federalregister.gov/endangered-threatened-species)

FEIS (http://www.fs.fed.us/database/feis/)

NatureServe (http://www.natureserve.org/)

VegBank (http://www.vegbank.org)

Alabama Herbarium Consortium (http://www.floraofalabama.org/Specimen.aspx)

Arctos (http://arctos.database.museum/)

CONN Herbarium (http://bgbaseserver.eeb.uconn.edu/)

Consortium of California Herbaria (http://ucjeps.berkeley.edu/consortium/)

Consortium of Pacific Northwest Herbaria (http://www.pnwherbaria.org/)

FLAS Herbarium (http://www.flmnh.ufl.edu/natsci/herbarium/)

FSU Herbarium (http://herbarium.bio.fsu.edu/)

LSU Herbarium (http://data.cyberfloralouisiana.com/lsu/)

MISS Herbarium (http://www.herbarium.olemiss.edu/bigsearch.html)
SEINet (http://swbiodiversity.org/seinet/)
VPlants (http://www.vplants.org/search.html)
WIS Herbarium (http://www.botany.wisc.edu/herb/)

Because these lists summarize data throughout the range of a species, they may or may not provide an accurate account of possible associations at a particular locality, especially for widespread species. Also, there are many factors that influence the inclusion of a species in these lists. Records of imperiled taxa often exclude associated species, even when reported on database records, if those records have been "masked" to ensure site confidentiality. Herbarium records for very weedy species often do not mention many associated species because such plants frequently grow in dense monocultures. Sometimes the associated reports reflect a broader community that includes the habitat of a species, and it often is difficult to distinguish such reports from those where species might be growing closely together in one portion of a habitat. Consequently, these lists should be viewed as identifying those plants most likely to be found in habitats similar to that of the subject species. At the discretion of the author, highly improbable associations have been omitted, but these occurrences represented relatively few of the records evaluated.

It was impossible to keep the nomenclature completely up-to-date for the lists of associated species, as numerous changes occurred during the 13-year preparation of this book. In many instances (depending on the date of completion for a treatment) an associated species may be listed under different synonyms. Some attempt was made to minimize such discrepancies, but many yet remain. Consequently, it is highly recommended that the ITIS website be checked for all currently accepted nomenclature. Hopefully, the nomenclature applied to the main species treatments has been applied correctly, given the few intentional deviations made at the discretion of the author.

Infraspecific taxa are not reported and the accuracy of species identities depends entirely on the source from which the data were obtained. In some cases, there will be inaccuracies. For instance, if only "*Eryngium vaseyi*" is given as an associated species, it is impossible to tell whether the author of the record intended reference to *Eryngium vaseyi* J.M. Coult. & Rose, rather than an infraspecific taxon such as *E. vaseyi* var. *castrense* (Jeps.) Hoover ex Mathias & Constance, which in this case would refer to an entirely different species, *Eryngium castrense* Jeps. Such errors are unavoidable but hopefully represent only a small fraction of those reported.

The listing of ecological associates provides another means for evaluating the overall wetland indicator status of a particular species. By enumerating the wetland designations of all the associated species (not done here), one might obtain a more accurate perspective of the typical wetland affinity of a single target species.

Following the species accounts is a section on their use by wildlife. The species are reported alphabetically as consistently as possible. Scientific names (at various ranks

determined to provide adequate categorization) are provided for all associated organisms mentioned. In addition to published literature accounts, this information also relied on several websites including

HOSTS (http://www.nhm.ac.uk/research-curation/research/projects/hostplants/)
Insect food plant database (http://www.brc.ac.uk/DBIF/hosts.aspx)
Mycobank (http://www.mycobank.org/Biolomics.aspx?Table=Mycobank&Page=200&ViewMode=Basic)
Pacific Northwest Fungi database (http://pnwfungi.wsu.edu)
USDA fungal hosts (http://nt.ars-grin.gov/fungaldatabases/fungushost/fungushost.cfm) along with numerous others.

The economic importance of species in each genus was summarized primarily from literature accounts and information provided by The Native American Ethnobotany website (http://naeb.brit.org/). **CAUTION**: *The information regarding food and medicinal uses of plant species provided in this book does not represent an endorsement of their edibility, medical efficacy, or toxicity; but simply reports information that exists in the literature.* Consequently, no wild plant should be used for food or medicinal purposes based solely on the information reported here.

The Royal Horticultural Society (RHS) horticultural database (http://apps.rhs.org.uk/horticulturaldatabase/) was routinely consulted for information on species that are maintained regularly under cultivation. To provide additional information on weeds, the database on herbicide resistant weeds (http://www.weedscience.org/summary/home.aspx) was checked.

The "Systematics" section was compiled with the objective of evaluating the most recent literature on each included genus to provide reasonable syntheses of appropriate classifications, overviews of phylogenetic relationships, cytological information, hybridization, and other systematically important topics. Although infraspecific taxa are not included in the individual OBL species accounts, they are discussed in this section whenever deemed relevant. Phylogenetic trees have been included whenever available and are provided to convey current (or at least recent) hypotheses regarding evolutionary relationships among the pertinent groups shown. These trees have been simplified without any indication of branch lengths, internal branch support, or other statistics, which usually can be found in the original source that is referenced in the figure caption. Because the trees depicted here may differ in some respects from those presented in the original citation, the original source always should be consulted for a definitive interpretation. Throughout the book the convention is followed to indicate any taxon (any rank) that contains at least one OBL species, to be set in bold type when appearing on a diagram. Non-OBL taxa appearing on diagrams typically appear in lighter-shaded (50% gray) type and also have their wetland indicator status indicated where applicable. In many cases, this representation provides a fair estimate of how many independent origins of the OBL habit have occurred in the group shown (depending on the accuracy of the given phylogenetic tree, of course).

The "Comments" section primarily summarizes the geographical distributions of each species in North America and elsewhere, occasionally adding additional important information not covered in other sections.

Finally, each genus treatment ends with a list of the primary references from which any specific information conveyed in the preceding account was obtained. A number of "general" references (e.g., Sculthorpe, 1967) were consulted throughout the book and are not listed repetitively in each reference section.

Dicotyledons

Traditionally, the term "dicotyledon" (or "dicot") was used to refer to those flowering plants that possessed (among other features) two embryonic seed leaves or cotyledons. Although this designation has long been used to distinguish the group from angiosperms having a single cotyledon (i.e., "monocotyledons"), the results of numerous molecular phylogenetic analyses have not resolved the two groups as clades (Judd et al., 2016). Rather, a more fundamental distinction has been found with respect to pollen characteristics, which separate a large group of dicotyledons by their tricolpate pollen structure ("eudicots") from a smaller group of dicotyledons, as well as all monocotyledons, which possess monosulcate pollen (Judd et al., 2016). In most analyses, the monocotyledons resolve as a clade, but the monosulcate dicotyledons as a paraphyletic grade, and have been subdivided into two major groups known as the "ANA grade" and "magnoliids"; Ceratophyllales are not resolved consistently in the same topological position but are usually depicted as a sister group to eudicots (Figure 1). The term "mesangiosperms" is sometimes used in reference to all extant angiosperms in exclusion of the ANA grade (Judd et al., 2016).

As widely accepted as the preceding scheme has become, some of the proposed phylogenetic relationships are far from resolved, and conflicting topologies abound among studies incorporating different data sets or taxon-sampling strategies. In particular, the phylogenetic affinities of two families (Ceratophyllaceae and Chloranthaceae) have proven to be exceptionally difficult to elucidate and remain uncertain even after years of study (Les, 2015). The extremely long branches rendered in phylogenetic trees for these and other early diverging groups have proven to be particularly problematic analytically, resulting in quite different topological placements for taxa when using different molecular data sets and even when using data from different genes located on the same (e.g., chloroplast DNA) molecule. The typical response to such anomalies has been to amass greater quantities of data, which are then analyzed in a combined fashion, rather than to seek a precise explanation for the inconsistencies observed, hoping that the correct result somehow will materialize. At present, we are left to rely on the results of such analyses. However, one should keep in mind that even a simple rooting error of an otherwise fairly consistently resolved tree like that shown in Figure 1 could have substantial repercussions, and this would not be unexpected given the very large branches that separate angiosperms from their gymnospermous outgroup taxa in every such analysis conducted.

Because the primary focus of this book is not to debate such phylogenetic hypotheses, the general arrangement of groups depicted in Figure 1 has been followed here with one exception: Although Ceratophyllaceae (which contains obligate wetland [OBL] indicators) have been assigned to the magnoliid mesoangiosperms (Judd et al., 2016), this disposition is highly tentative given their ambiguous phylogenetic affinity. Instead, Ceratophyllaceae are treated here first, as a group of uncertain affinity. The remaining groups that contain OBL indicators are organized sequentially as the ANA grade (here only Austrobaileyales and Nymphaeales), the magnoliids (Laurales, Magnoliales, and Piperales), and finally the eudicots.

A GROUP OF UNCERTAIN PHYLOGENETIC AFFINITY

ORDER 1: CERATOPHYLLALES [1]

1. Ceratophyllaceae Gray

Family 1: Ceratophyllaceae [1]

Because the order Ceratophyllales contains only the single extant family Ceratophyllaceae, the two groups are discussed here together. Ceratophyllaceae are intriguing aquatic plants in many respects. Phylogenetically, they represent one of only

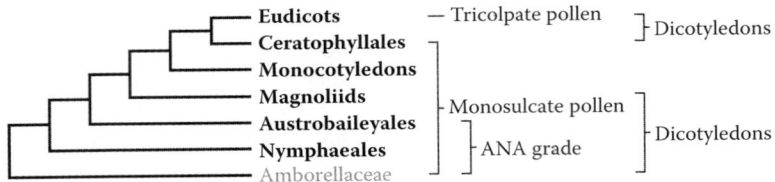

FIGURE 1 Hypothetical phylogenetic relationships among major angiosperm groups, as summarized from the results of numerous molecular systematic studies. Taxa containing OBL North American indicators are highlighted in bold. (Adapted from Judd, W. S. et al., *Plant Systematics: A Phylogenetic Approach*, Sinauer Associates, Inc., Sunderland, MA, 2016.)

a few angiosperm groups (with Boraginales, Chloranthales, and Dilleniaceae) whose placement remains designated as "uncertain" by contemporary plant systematists (Judd et al., 2016). The family was long considered to be a highly derived offshoot of the Nymphaeales that has adapted to submersed conditions (Cronquist, 1981). However, a morphological systematic reassessment of the group later found no evidence of any relationship to Nymphaeales, suggesting instead that Ceratophyllaceae might represent one of the earliest angiosperm lineages that originated even before the fundamental divergence of dicotyledons and monocotyledons (Les, 1988).

The remote relationship between Ceratophyllaceae and Nymphaeales was corroborated by an early study employing *rbcL* sequence data, which coincidentally resolved the former as the sister group to the small number of angiosperm taxa sampled, in a position distant from members of Nymphaeales (Les et al., 1991). In the first comprehensive assessment of angiosperm phylogeny (again using *rbcL* data), the placement of Ceratophyllaceae as the sister group to all other angiosperms was maintained, as well as its distinction from Nymphaeales (Chase et al., 1993). Subsequent studies consistently showed Ceratophyllaceae not to be closely related to Nymphaeales; however, they also did not consistently replicate its placement as the sister group of the angiosperms. Moreover, they did not consistently place the family elsewhere. Sampling of various DNA sequence data and taxon assemblages have resolved the family as the sister group of Piperales or as distant to the Piperales, and variously as the sister group of Chloranthales, monocotyledons, and eudicots, as well as all angiosperms (Goloboff et al., 2009; Morton, 2011; Maia et al., 2014; Ruhfel et al., 2014; Zeng et al., 2014; Les, 2015). Consequently, it is understandable why the precise phylogenetic placement of the family remains enigmatic.

Ceratophyllaceae are distinct morphologically. All of the species are perennial aquatic herbs that occur as completely submersed or suspended life-forms (Les, 1993, 1997). The plants lack roots and xylem entirely, and possess whorls of dichotomously divided leaves apparently derived from a decussate phyllotaxy (Les, 1993; Schneider & Carlquist, 1996b; Iwamoto et al., 2015). The flowers are extra-axillary, unisexual pseudanthia, which actually represent reduced cymose inflorescences that are arranged in a monoecious configuration (Les, 1988, 1993, 1997). Pollination occurs entirely underwater (Les, 1988b; Philbrick & Les, 1996). The pollen tubes branch, and the seeds have only a single integument and

lack endosperm. Overall, Ceratophyllaceae can be regarded as representing some of the angiosperms best adapted to life under water.

The group is not particularly important economically, although the species sometimes are distributed as aquarium plants. They have been reported as weeds in some localities.

Although several fossil groups have been associated with the family (Les, 1988; Dilcher & Wang, 2009; Gomez et al., 2015), only one extant genus (*Ceratophyllum*) remains; it also contains OBL indicators:

1. *Ceratophyllum* L.

1. *Ceratophyllum*
Hornwort; cornifle
Etymology: from the Greek *ceratos phyllon* ("horn leaf") in reference to the antler-like leaves
Synonyms: none
Distribution: global: cosmopolitan; **North America:** widespread
Diversity: global: 6 species; **North America:** 3 species
Indicators (USA): OBL: *Ceratophyllum demersum, C. echinatum, C. muricatum*
Habitat: brackish, freshwater; lacustrine, palustrine, riverine; **pH:** 4.9–9.9; **depth:** 0.1–4.5 m; **life-form(s):** submersed, suspended (vittate)
Key morphology: shoots (to 3 m) submersed or suspended, rootless but sometimes anchored in the substrate; leaves cauline, whorled (3–11), often crowded apically, the blades (to 3 cm) dissected dichotomously into 2–10 linear, denticulate segments; flowers (inflorescences) unisexual, naked; stamens numerous (♂ flowers); carpel 1, with a stigmatic pore at the base of a spinelike style (♀ flowers); achenes (to 6 mm) ellipsoid, slightly compressed, with a single apical spine (to 14 mm) and a pair of basal spines (to 12 mm), the margins wingless or winged by confluent bases of 2–20 spines (to 6.5 mm); first plumule leaves of embryo simple or forked
Life history: duration: perennial (dormant apices, winter buds, whole plants); **asexual reproduction:** shoot fragments, winter buds; **pollination:** water (subsurface); **sexual condition:** monoecious; **fruit:** achenes (infrequent); **local dispersal:** achenes, shoot fragments (water); **long-distance dispersal:** achenes (waterfowl [Aves: Anatidae])
Imperilment: (1) *Ceratophyllum demersum* [G5]; S1 (AK); S2 (UT, WY, <u>YT</u>); S3 (<u>AB</u>, NC, <u>QC</u>, WV); (2) *C. echinatum*

[G4]; SH (OR); S1 (KS, MB, MD, MO, NB, NJ, TN, WV); S2 (IN, KY, ME, NC, NS, VT); S3 (BC, IL, NY, ON, QC); (3) *C. muricatum* [G5]; S1 (NC)

Ecology: general: All *Ceratophyllum* species are obligate, perennial aquatics that retain all of their foliage and reproductive structures beneath the water surface throughout their life cycle (except for the fruits during dispersal). Because they lack even embryonic roots, the shoots grow unattached and ordinarily become suspended just below the surface layer, or become submersed in deeper water if partially buried (and anchored) by the sediment. The shoots are brittle and fragment easily, which allows for prolific local dispersal of vegetative propagules. Highly clonal populations are common. The species grow primarily in freshwaters, but some can also tolerate slightly brackish conditions. They occur from sea level to elevations exceeding 2000 m, and in waters ranging from strongly acidic to highly alkaline conditions. Their vegetative morphology is highly plastic depending on the ambient environment, which can span an extremely broad range of conditions. The unisexual flowers are monoecious in all of the species; they are self-compatible (enabling geitonogamous pollinations). Continuous illumination of the plants will induce the formation of staminate, but not pistillate flowers. Pollination is three-dimensional and occurs entirely under water (i.e., hypohydrophily). Pollen is released while the anthers remain attached, or from detached anthers that float near the surface. The pollen tubes often germinate precociously, forming large, tangled masses; the grains can remain viable for up to eight days. A pollen grain (or tube) must first find its way to a small opening in the carpel (the stigmatic pocket) to subsequently achieve fertilization. The group is known for its poor seed production and its predominantly vegetative mode of reproduction by shoot fragmentation; however, numerous seeds can be found where conditions are optimal. A combination of warm, shallow, and stagnant water is necessary for optimal seed production because the waterborne pollen is carried away from the plants by any currents. The achenes sink when released from the plants and are dispersed locally by water currents. They retain viability when ingested by waterfowl (Aves: Anatidae), which facilitates their long-distance dispersal by endzoic transport. Seeds of the North American species are dormant and require a period of cold stratification (or mechanical scarification) to induce germination. However, seedlings are rarely observed in the field, even at sites where seed production is evident.

Ceratophyllum demersum L. grows in standing waters (to 4.5 m deep) of backwaters, bayous, bogs, canals, channels, ditches, drains, fens, floodplains, gravel pits, hot springs, lagoons, lakes, marshes, oxbows, ponds, pools, potholes, reservoirs, rivers, sloughs, streams, and swamps at elevations of up to 2699 m. This species tolerates an extraordinary range of environmental conditions including fully open to shaded exposures, and extremely shallow to deep, fresh to brackish, eutrophic to mesotrophic, and clear to turbid waters of broad ecological amplitude (pH: 5.4–9.9; mean pH: 7.4; conductivity: 10–690 μmhos/cm; total CaCO$_3$ alkalinity: 10–310 mg/L). This wide range of tolerance helps to explain why this is one of the most commonly encountered submersed aquatic plants in the region. The recorded substrates are equally diverse and include gravel, marly sand, muck, mud (calcareous), muddy clay, peat, pumice, rock, rocky sand, rocky sandy gravel, sand, sandy clay, sandy muck, sandy silt, silt (alluvial), silty muck, and silty sand. Flowering occurs sporadically from May to September, and fruiting from June to October. The fruits will begin to develop in successfully pollinated pistillate flowers within 2 weeks. They are dormant and require a 150–210-day period of cold stratification (1–3°C) before germinating under a 19/15°C temperature regime. The seeds are dispersed locally by water currents and over longer distances by endozoic transport in waterfowl (Aves: Anatidae). Germination after 90 days of cold stratification can be achieved by cutting a slit through the thick pericarp. A transient seed bank can be produced, but is of little significance because of infrequent fruit production. Vegetative reproduction is prolific and occurs by fragmentation of the brittle stems. Genetic studies (using allozymes) have confirmed that populations are often highly clonal and lack evidence of sexual recombination. The clustered apical foliage develops into insulative winter buds, enabling the plants to overwinter and also serve as propagules. A vegetative propagule bank can develop, with average densities of 91.7 propagules/m^2 reported in one study. Whole plants also have been observed to overwinter under ice cover. **Reported associates:** *Acorus calamus, Alisma, Alopecurus aequalis, Alternanthera philoxeroides, Azolla filiculoides, Beckmannia syzigachne, Bidens beckii, Brasenia schreberi, Butomus umbellatus, Cabomba caroliniana, Callitriche palustris, Carex rostrata, C. utriculata, C. vesicaria, Cephalanthus occidentalis, Ceratophyllum echinatum, Chara globularis, Cicuta, Crypsis schoenoides, Dulichium arundinaceum, Echinochloa walteri, Echinodorus cordifolius, Egeria densa, Eichhornia crassipes, Eleocharis acicularis, E. macrostachya, E. palustris, Elodea canadensis, E. nuttallii, Equisetum fluviatile, Glossostigma cleistanthum, Glyceria grandis, Heteranthera dubia, Hippuris vulgaris, Hydrilla verticillata, Hydrocotyle ranunculoides, Hydrodictyon reticulatum, Hydrolea uniflora, Ilex cassine, Isoetes tuckermanii, Juncus marginatus, Justicia americana, Lemna gibba, L. minor, L. minuta, L. trisulca, Limnobium spongia, Limosella, Lobelia cardinalis, Ludwigia octovalvis, L. palustris, L. peploides, L. sphaerocarpa, Lythrum salicaria, Marsilea vestita, Mimulus glabratus, Myosotis laxa, Myriophyllum heterophyllum, M. quitense, M. sibiricum, M. spicatum, M. verticillatum, Najas canadensis, N. flexilis, N. gracillima, N. guadalupensis, N. marina, N. minor, Nasturtium officinale, Nelumbo lutea, Nitella allenii, Nuphar advena, N. polysepala, N. variegata, Nymphaea mexicana, N. odorata, Nymphoides peltata, Nyssa biflora, Persicaria amphibia, P. hydropiper, P. punctata, Phalaris arundinacea, Phyla lanceolata, Populus fremontii, Potamogeton alpinus, P. amplifolius, P. berchtoldii, P. bicupulatus, P. crispus, P. diversifolius, P. epihydrus, P. foliosus, P. friesii, P. gramineus, P. illinoensis, P. natans, P. nodosus, P. perfoliatus, P. praelongus, P. pulcher, P. pusillus, P. richardsonii, P. robbinsii, P. spirillus, P. strictifolius, P. vaseyi, P. zosteriformis, Ranunculus aquatilis, R. flammula,*

R. longirostris, R. macounii, R. subrigidus, Riccia fluitans, Rorippa palustris, Sagittaria kurziana, S. latifolia, Salix gooddingii, S. interior, Saururus cernuus, Schoenoplectus acutus, S. americanus, S. lacustris, S. subterminalis, S. torreyi, Sium suave, Sparganium eurycarpum, S. natans, Sphagnum, Spirodela polyrhiza, Stuckenia filiformis, S. pectinata, S. vaginata, Taxodium distichum, Typha latifolia, Utricularia geminiscapa, U. gibba, U. macrorhiza, U. minor, U. purpurea, U. radiata, Vallisneria americana, Veronica anagallis-aquatica, V. beccabunga, Wolffia columbiana, Wolffiella gladiata, Zannichellia palustris, Zizania aquatica.

***Ceratophyllum echinatum* A. Gray** inhabits standing waters (0.2–2.5 m deep) of bogs, bottomlands, canals, depressions, ditches, lakes, marshes, ponds, streams, and swamps at elevations to 410 m. The plants are much less common and have more restricted habitat requirements than the previous species. They are more frequent in oligotrophic to mesotrophic sites and are most often seen in shaded exposures and in waters that are primarily acidic and are characterized by low alkalinity and low conductivity (pH: 4.8–7.9; mean pH: 6.6; total $CaCO_3$ alkalinity: 5–70 mg/L; 20–190 μmhos/cm). The substrates include gravel, muck, peat, sand, silt, and rock. Flowering occurs from April to July, with fruiting observed from May to December. The water pollination system is typical of the genus. The fairly large fruits (with long marginal spines) are dispersed locally in the water and are transported to greater distances endozoically by waterfowl (Aves: Anatidae). Seed germination requirements are similar to those of the previous species. Although somewhat more sexual than the preceding species, the plants still propagate extensively by vegetative fragmentation. Genetic (allozyme) analyses have documented that low levels of genetic variation exist within the typically clonal populations. **Reported associates:** *Acer rubrum, Aldrovanda vesiculosa, Bidens beckii, Brasenia schreberi, Cabomba caroliniana, Callitriche heterophylla, Carex aquatilis, C. lacustris, C. obnupta, C. rostrata, C. utriculata, Cephalanthus occidentalis, Ceratophyllum demersum, Chara vulgaris, C. zeylanica, Decodon verticillatus, Dulichium arundinaceum, Eichhornia crassipes, Elatine, Eleocharis ovata, E. palustris, Elodea canadensis, E. nuttallii, Equisetum fluviatile, Galium triflorum, Glossostigma cleistanthum, Glyceria borealis, Gratiola aurea, Heteranthera dubia, H. multiflora, Iris, Isoetes echinospora, Juncus pelocarpus, J. supiniformis, Leersia oryzoides, Leitneria floridana, Lemna minor, L. trisulca, Limnobium spongia, Ludwigia palustris, Mentha arvensis, Myrica gale, Myriophyllum farwellii, M. heterophyllum, M. heterophyllum × M. laxum, M. sibiricum, M. tenellum, M. ussuriense, M. verticillatum, Najas flexilis, N. gracillima, N. guadalupensis, Nelumbo lutea, Nitella, Nuphar advena, N. polysepala, N. variegata, Nymphaea odorata, Oenanthe sarmentosa, Panicum, Persicaria amphibia, P. coccinea, P. hydropiper, P. hydropiperoides, Phalaris arundinacea, Pistia stratiotes, Pontederia cordata, Potamogeton amplifolius, P. epihydrus, P. foliosus, P. gramineus, P. natans, P. nodosus, P. obtusifolius, P. pulcher, P. pusillus, P. richardsonii, P. robbinsii, P. spirillus, P. vaseyi, P. zosteriformis,*

Ranunculus aquatilis, R. longirostris, R. trichophyllus, Riccia fluitans, Rorippa aquatica, Sagittaria cristata, S. rigida, Salix fragilis, Schoenoplectus acutus, S. lacustris, S. torreyi, Scirpus atrocinctus, S. cyperinus, S. microcarpus, Sium suave, Sparganium emersum, S. eurycarpum, S. fluctuans, S. natans, Spirodela polyrhiza, Stuckenia filiformis, S. pectinata, Taxodium distichum, Trapa natans, Typha domingensis, T. latifolia, Utricularia gibba, U. intermedia, U. macrorhiza, U. purpurea, U. radiata, U. striata, Vallisneria americana, Wolffia brasiliensis, W. columbiana, Wolffiella.

***Ceratophyllum muricatum* Cham.** grows in brackish to freshwater ditches, hydric hammocks, and ponds near coastal areas at elevations <50 m. Exposures range from full sunlight to shade. There is little life history information available for this species due to its scarcity in North America. Flowering and fruiting occur throughout the year in the southern (Central and South American) portions of its distributional range. The reproductive period in North America presumably is similar, with fruiting being observed from April to February. Unlike its aforementioned congeners, sexual reproduction is much more common in this species, and the plants typically produce numerous fruits. The North American habitats are also shallow, temporary sites that likely undergo seasonal periods of desiccation similar to its habitats in more southerly localities (e.g., Costa Rica). These sites also occur similarly in areas subject to high levels of storm disturbance (e.g., on coastal barrier islands). Accordingly, the life history of this species is perhaps more similar to that of an annual, by producing abundant fruits that are capable, in this case, of withstanding the ephemerality of the habitats. The extent of vegetative reproduction is not known. **Reported associates:** *Azolla, Lemna, Limnobium, Spirodela, Utricularia.*

Use by wildlife: *Ceratophyllum demersum* is an important cover plant for fish (Vertebrata: Osteichthyes) and muskrats (Mammalia: Cricetidae: *Ondatra zibethicus*) and for invertebrates such as amphipods (Crustacea: Amphipoda: Gammaridae: *Gammarus*; Hyalellidae: *Hyalella azteca*) and insects (Insecta: Odonata: Coenagrionidae: *Enallagma*). The foliage and achenes are relished and eaten by numerous waterfowl (Aves: Anatidae) including American wigeon (*Anas americana*), black duck (*Anas rubripes*), blue-winged teal (*Anas discors*), bufflehead (*Bucephala albeola*), Canada geese (*Branta canadensis*), canvasback (*Aythya valisineria*), cinnamon teal (*Anas cyanoptera*), gadwall (*Anas strepera*), goldeneye (*Bucephala clangula*), green-winged teal (*Anas crecca*), mallard (*Anas platyrhynchos*), merganser (*Mergus merganser*), mute swan (*Cygnus olor*), northern pintail (*Anas acuta*), oldsquaw (*Clangula hyemalis*), redhead (*Aythya americana*), ring-necked duck (*Aythya collaris*), ruddy duck (*Oxyura jamaicensis*), scaup (*Aythya affinis*; *A. marila*), surf scoter (*Melanitta perspicillata*), whistling swan (*Cygnus columbianus*), white-winged scoter (*Melanitta fusca*), wood duck (*Aix sponsa*), and other birds like coots (Rallidae: *Fulica*). The plants also are eaten by a variety of other animals including crayfish (Crustacea: Decapoda: Cambaridae: *Procambarus clarkii*), fish (Vertebrata: Osteichthyes: Cyprinidae: *Ctenopharyngodon idella*; Cyprinodontidae: *Cyprinodon* *nevadensis nevadensis*), frogs (Amphibia: Ranidae: *Rana pipiens, R. septentrionalis*), larval moths (Insecta: Lepidoptera: Crambidae: *Acentria nivea, Parapoynx diminutalis*), manatees (Mammalia: Sirenia: Trichechidae: *Trichechus manatus*), and snails (Mollusca: Gastropoda: Ampullariidae: *Marisa cornuarietis*). The plants contain about 16.6% crude protein and have also been used as fodder for livestock.

Economic importance: food: *Ceratophyllum* is not used as food; **medicinal:** *Ceratophyllum demersum* has been used in traditional medicines for treating biliousness, dermatitis, fevers, jaundice, scorpion stings, and sunburn. The aqueous and methanolic extracts have been shown to exhibit strong analgesic, antidiarrheal, anti-inflammatory, antipyretic, and wound healing activity in laboratory animals. Acetone extracts from the plants are inhibitory to Cyanobacteria; **cultivation:** *Ceratophyllum demersum* is distributed as an aquarium plant; **misc. products:** Due to its metal tolerances and absolute foliar nutrient uptake, *C. demersum* has been used extensively in the phytoremediation of waters polluted by wastewater and various heavy metals; e.g., the plants can tolerate concentrations of waterborne arsenic and lead up to 40 μM; **weeds:** Excessive growth of *C. demersum* has interfered with the operation of hydroelectric plants; **nonindigenous species:** none in North America

Systematics: As discussed earlier, the precise phylogenetic placement of *Ceratophyllum* (Ceratophyllales; Ceratophyllaceae) is uncertain, but the genus certainly is situated somewhere among the early diverging groups of monosulcate dicotyledons. The monophyly of the genus is evidenced by its unique combination of morphological features, which are shared by no other living group of plants, and also by preliminary molecular investigations. The interspecific relationships in *Ceratophyllum* have been explored phenetically (Figure 2); however, a comprehensive phylogenetic study of the group has not yet materialized. Although morphological and flavonoid data have suggested the recognition of three sections, the proposed groups are not supported by preliminary molecular data. In particular, the phenetically similar *C. echinatum* and *C. submersum* are quite divergent at the molecular level. Confusion exists in the literature regarding the distinction between *C. echinatum* and *C. muricatum*. These taxa were merged by some authors, but they are readily separable morphologically as well as by their biochemical flavonoid profiles (Figure 2). *Ceratophyllum muricatum* is widespread throughout the tropical regions, with the North

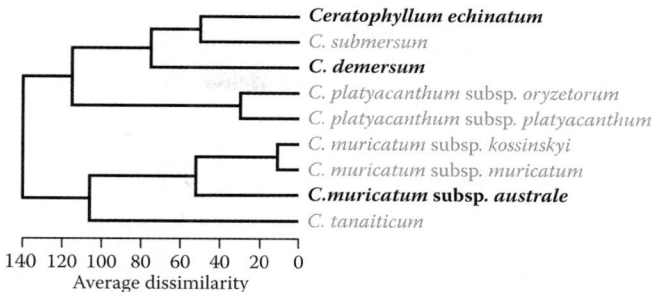

FIGURE 2 Phenetic relationships among *Ceratophyllum* species as indicated by a UPGMA cluster analysis of vegetative and reproductive morphological data. The three OBL North American indicators are highlighted in bold. (Adapted from Les, D. H., *Syst. Bot.*, 14, 254–262, 1989.)

American (and all New World) populations assigned to *C. muricatum* subsp. *australe*. The basic chromosome number of *Ceratophyllum* is *x* = 12. *Ceratophyllum demersum* and *C. echinatum* are diploids (*2n* = 24). Counts are not available for *C. muricatum*. Although naturally occurring hybrids have not been reported in the literature, successful synthetic hybrids have been made between *C. demersum* and *C. echinatum*, despite their extremely low allozymically determined genetic identity (0.17).

Comments: *Ceratophyllum demersum* is widespread in North America and has a cosmopolitain global distribution. *Ceratophyllum echinatum* occurs primarily in eastern North America, with disjunct populations in western North America. *Ceratophyllum muricatum* is restricted to coastal regions of the southeastern United States, but extends into South America.

References: Abdallah, 2012; Bailey et al., 2008; Bolser et al., 1998; Borman et al., 1997; Boyd, 1968; Campbell & Irvine, 1977; Chakraborty et al., 2016; Chapman et al., 1974; Chen et al., 2001, 2015; Chilton, 1990; Cottam, 1939; Craven & Hunt, 1984b; Crow et al., 1987; Fassett, 1953a, 1957; Hajra, 1987; Hedeen, 1972; Karale et al., 2013; Kasselmann, 2003; Khang et al., 2012; Lamont et al., 2013; Les, 1985, 1986a,b,c, 1988a,b,c,d, 1989, 1991, 1993; Lowden, 1981; Mabbott, 1920; Martin & Uhler, 1951; McAtee, 1939; Montz, 1978; Muenscher, 1936, 1940; Naiman, 1979; Nichols, 1999; Scribailo & Alix, 2002; Seaman & Porterfield, 1964; Taranhalli et al., 2011; Yi et al., 2012; Yocum, 1951.

1 Dicotyledons I
The ANA Grade and "Magnoliid" Monosulcates

THE ANA GRADE

"ANA" is derived from the acronym for Amborellales, Nymphaeales, and Austrobaileyales (Judd et al., 2016), which resolve as a paraphyletic grade in many molecular phylogenetic analyses (e.g., Figure 1.1). Two of the orders (Austrobaileyales and Nymphaeales) contain obligate wetland (OBL) indicators.

ORDER 1: AUSTROBAILEYALES [3]

This small order contains a total of only about 100 species distributed among three families (Austrobaileyaceae, Schisandraceae, and Trimeniaceae); Illiciaceae (formerly included in the order) has since been merged with Schisandraceae (Judd et al., 2016). The latter is the most specialized phylogenetically (Figure 1.1) and is the only family within the order containing OBL indicators:

1. **Schisandraceae** Blume

Family 1: Schisandraceae [3]

This group of about 90 species is almost entirely terrestrial. *Illicium*, the only member to contain an OBL North American indicator species, occurs as the sister genus to the remainder of the family (Figure 1.1). The species include lianas, shrubs, and trees with alternate, spiral leaves marked by pellucid dots, branched sclereids, and pollen that is tri- or hexacolpate, but differing structurally from the type found in the eudicots (Judd et al., 2016). The flowers are bisexual or unisexual, self-incompatible, nectariferous, and pollinated by flies (Insecta: Diptera) or other small insects (Judd et al., 2016). Their structure is highly unspecialized, with perianths consisting of five to numerous tepals, stamens with poorly differentiated anthers, and seven to numerous pistils (Judd et al., 2016). The flowers of some species are thermogenic (Luo et al., 2010). The seeds are dispersed abiotically (but over short distances) by their forceful ejection from the elastically dehiscent follicles (Roberts & Haynes, 1983). The seeds have rudimentary embryos and typically are morphophysiologically dormant (Baskin & Baskin, 1998). A persistent seed bank can develop in some species. Many of the species reproduce vegetatively by rhizome production (Judd et al., 2016).

Ornamental horticultural specimens occur in *Illicium*, *Kadsura*, and *Schisandra*. *Illicium* is the source of star anise (*I. verum*), a popular culinary spice. Oils extracted from the seeds are used in a variety of cosmetic and dermatological products. The plants also are rich in shikimic acid, which is used to synthesize the anti-influenza medicine known as oseltamivir or Tamiflu (Wang et al., 2011).

Only one genus contains OBL indicators in North America:

1. ***Illicium*** L.

1. *Illicium*
Swamp star anise, yellow anise tree
Etymology: from the Latin *illicere*, to seduce
Synonyms: *Badianifera*
Distribution: global: Asia; New World; **North America:** southeastern
Diversity: global: 42 species; **North America:** 2 species
Indicators (USA): OBL: *Illicium parviflorum*
Habitat: freshwater; palustrine; **pH:** unknown; **depth:** <1 m; **life-form(s):** emergent (shrub, tree)
Key morphology: shrubs or small trees (to 6 m); leaves (to 21 cm) aromatic, petioled (to 16 mm), alternate, simple, entire, narrowly elliptic, with blunt tips; flowers (to 1.2 cm) star-shaped, clustered near branch tips, peduncled (to 2.4 cm); tepals (11–16) yellow-green; stamens 6–7; pistils numerous; the fruit a star-like aggregate (to 3.5 cm) of 10–13 follicles
Life history: duration: perennial (buds); **asexual reproduction:** suckering basal shoots; **pollination:** insect; **sexual condition:** hermaphroditic; **fruit:** aggregates (follicles) (infrequent); **local dispersal:** ballistic, water (seeds), root suckers; **long-distance dispersal:** unknown
Imperilment: (1) *Illicium parviflorum* [G2]; SH (GA); S2 (FL)
Ecology: general: *Illicium* is primarily an eastern Asian genus of woody, terrestrial plants, which occur in forests, ravines, and thickets. A few species grow in wetter sites such as the areas along the margins of rivers. Both North American species are wetland indicators, with only one designated as OBL. The flowers are pollinated by insects (Insecta: Coleoptera; Diptera) with the seeds dispersed ballistically. Seed dormancy is morphophysiological.

Illicium parviflorum **Michx. Ex Vent.** occurs in bayheads, bottomlands, hammocks (hydric), ravines, seeps, swamps, woodlands, and along the margins of spring-fed streams at elevations to 70 m. The substrates have been described as acidic or calcareous and include loam, peat, samsula (terric medisaprists), and sand (organic). Sunny exposures are tolerated but the plants thrive in partial shade. The flowers (occurring from March to June) are self-compatible but dichogamous (protogynous) and remain open for 2–3 days. They are

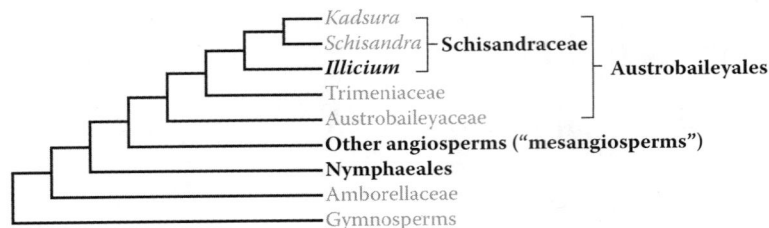

FIGURE 1.1 Interfamilial relationships in Austrobaileyales as indicated by phylogenetic analysis of combined DNA sequence data. Some authors refer to Amborellales, Austrobaileyales, and Nymphaeales as the "ANA grade." Groups containing OBL North American indicators are highlighted in bold. The general topology is consistent with results from an analysis of complete chloroplast genomes (Hansen et al., 2007). (Adapted from Qiu, Y.-L. et al., *Int. J. Plant Sci.*, 161, S3–S27, 2000.)

capable of self pollination, but have a fairly high pollen:ovule ratio (mean: 511) and exhibit symptoms of inbreeding depression when selfed. A study using AFLP markers indicated that less than 10% of self-pollination occurred in one population surveyed. Outcrossing is facilitated by small, pollen-eating insects (see *Use by wildlife*). Fruiting extends from June to October but the plants typically produce few fruits. The seeds are dispersed ballistically, but only over short distances. They have an oily seed coat, which also enables them to float for up to 5 days. The seeds will germinate fairly well (~30%) without pretreatment, but germination rates increase substantially (to 81%) following 90 days of cold (5°C), moist stratification and incubation under a 21°C/15.5°C day/night temperature regime. The plants are highly clonal and form numerous root sprouts, which can generate dense thickets. Genetic surveys using ISSR markers indicate that populations are maintained primarily by clonal reproduction. The roots are fairly sensitive to heat and can be damaged by fires.

Reported associates: *Acer rubrum, Ardisia escallonioides, Asimina parviflora, Bidens bipinnatus, Callicarpa americana, Callisia repens, Carex, Carya glabra, Celtis laevigata, Chiococca alba, Citrus ×aurantium, Conopholis, Cyperus tetragonus, Decumaria barbara, Dichanthelium commutatum, Diodia teres, Epidendrum magnoliae, Gordonia lasianthus, Habenaria floribunda, Hamelia patens, Ilex cassine, I. coriacea, Iresine diffusa, Juncus, Liquidambar styraciflua, Lyonia ferruginea, Magnolia grandiflora, M. virginiana, Mitchella repens, Morus rubra, Myrica cerifera, Myrsine floridana, Osmundastrum cinnamomeum, Persea palustris, Pharus lappulaceus, Pinus elliottii, P. serotina, P. taeda, Pleopeltis polypodioides, Psychotria nervosa, Quercus laurifolia, Q. nigra, Q. virginiana, Rhamnus caroliniana, Rhapidophyllum hystrix, Rhododendron viscosum, Sabal palmetto, Scleria triglomerata, Serenoa repens, Smilax auriculata, Solidago leavenworthii, Thelypteris kunthii, Tillandsia, Toxicodendron radicans, T. vernix, Vaccinium corymbosum, Verbesina virginica, Vitis rotundifolia, Woodwardia virginica, Zanthoxylum clava-herculis.*

Use by wildlife: The foliage and fruits of many *Illicium* species are poisonous to cattle (Mammalia: Bovidae: *Bos taurus*). The flowers of *I. parviflorum* are visited by various insects (Insecta), which forage for the pollen. These potentially include flies (Diptera: Cecidomyiidae: *Clinodiplosis,*

Giardomyia, Lestodiplosis; Ceratopogonidae: *Forcipomyia fuliginosus*; Psychodidae: *Threticus*), bugs (Hemiptera: Cixiidae: *Bothriocera datuna*), lacewings (Neuroptera: Sisyridae: *Sisyra apicalis*), and crickets (Orthoptera: Gryllidae: *Cycloptilium*).

Economic importance: food: The leaves of *I. parviflorum* reportedly have been used as a spice, but possibly are poisonous to humans; **medicinal:** The foliage of *I. parviflorum* contains the sesquiterpene lactone seco-prezizaane, which potentially affects the human central nervous system; **cultivation:** *Illicium parviflorum* is commonly cultivated as a nursery plant (often mistakenly under the name *I. anisatum*); **misc. products:** the leaves of *I. parviflorum* are a source of incense; **weeds:** none in North America; **nonindigenous species:** *Illicium parviflorum* reportedly has escaped from cultivation in Thomas County, Georgia.

Systematics: Phylogenetic analyses of DNA sequence data support the inclusion of *Illicium* as a monophyletic genus within the Austrobaileyales clade, where it occurs along with Austrobaileyaceae, Schisandraceae, and Trimeniaceae. Once segregated within a distinct family (Illiciaceae), many authors now include *Illicium* within Schisandraceae (e.g., Figure 1.2), which has been the convention followed here. Some phylogenetic analyses of morphological and molecular (nrITS) data place the New World *Illicium* species in a clade distinct from the Old World species and resolve *I. parviflorum* as being most closely related to *Illicium floridanum*. While retaining the distinction of New and Old World species, additional taxon sampling (roughly half of the recognized *Illicium* species) resolves *I. parviflorum* somewhat differently, placing it within a clade containing *I. cubense, I. ekmanii,* and *I. hottense,* which is the sister group to a clade containing *I. floridanum* and *I. mexicanum* (Figure 1.2). Phylogenetically, *I. parviflorum* also groups with congeners that share trisyncolpate pollen and a "type B" pollen aperture. The basic chromosome number of *Illicium* is $x = 14$ (or $x = 13$ by aneuploid reduction). *Illicium parviflorum* ($2n = 28$) is a diploid. No natural hybrids have been reported that involve *I. parviflorum*.

Comments: All New World species of this predominantly Asian genus are narrow endemics, with *I. parviflorum* restricted to small areas of Florida and Georgia.

References: Adams et al., 2010; Buckley, 2012; Hao et al., 2000; Ingram et al., 1986; Koehl et al., 2004; Morris et al.,

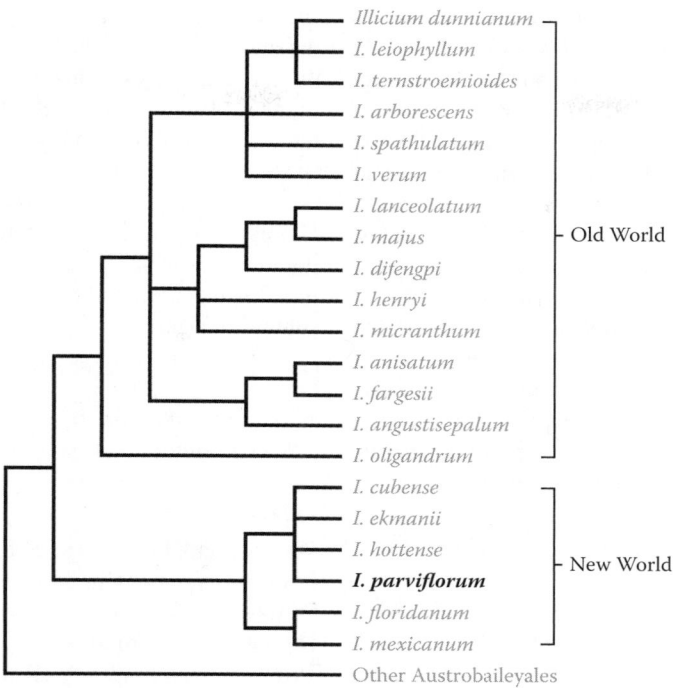

FIGURE 1.2 Interspecific relationships in *Illicium*, as indicated by analysis of nrITS sequence data. The New and Old World species resolve in distinct clades, with *I. parviflorum*, the only OBL indicator (in bold), placed within the former. (Adapted from Morris, A.B. et al., *Syst. Bot.*, 32, 236–249, 2007.)

2007; Newell & Morris, 2010; Oh et al., 2003; Olsen & Ruter, 2001; Roberts & Haynes, 1983; Schmidt, 1999; Stone & Freeman, 1968; USFWS, 2016b; Vincent, 1997; Wang et al., 2010; White & Thien, 1985.

ORDER 2: NYMPHAEALES [2]

The Nymphaeales (water lilies) are typically regarded as primitive flowering plants (e.g., Cronquist, 1981), and phylogenetic studies have regularly shown the order to occupy various phylogenetic positions situated among some of the earliest diverging angiosperm lineages. The group has resolved consistently in association with the anomalous genus *Amborella*, but in various configurations, which include such variations as a fairly derived clade of early diverging angiosperms known formerly as "paleoherbs" (e.g., Chase et al., 1993; Goremykin et al., 2004), a clade with *Amborella* that is sister to all of the remaining angiosperms (e.g., Nickerson & Drouin, 2004), a sister clade to all angiosperms in exclusion of *Amborella*, which then resolves as the basal sister group (e.g., Moore et al., 2010; Soltis et al., 2011), or in the previous configuration with *Ceratophyllum* (not *Amborella*) as the sister group of angiosperms (Morton, 2011). Although each hypothesis has received support, these inconsistencies (which occur even among different genes within the same, e.g., plastid, genomes) should caution against accepting any of the results as factual, especially given that each of the groups in question is characterized by long branches in most analyses, and these branches are associated with key nodes that lack compelling

internal support, even after incorporating massive amounts of DNA sequence data. Until it becomes possible to better understand these inconsistencies, a safe evaluation is simply to regard Nymphaeales as a relatively early diverging group of monosulcate angiosperms, which is a conclusion uncontested in all earlier studies. Studies of genome size (Pellicer et al., 2013) have shown an interesting trend in Nymphaeales, where increased chromosome numbers correspond to the largest genome sizes by virtue of smaller chromosomes, which yield lower C-values. A phylogenetic trend towards reduction in genome size is apparent in the order.

Over the past several decades, several major modifications to the circumscription of the water lilies have been implemented in order to define the group as monophyletic. Among these was the exclusion of the families Ceratophyllaceae and Nelumbonaceae, which were found not to associate closely with any of the core water lily genera (i.e., *Barclaya*, *Brasenia*, *Cabomba*, *Euryale*, *Nuphar*, *Nymphaea*, *Victoria*), either morphologically (Les, 1988a) or phylogenetically (Les et al., 1991). The distinction of both Ceratophyllaceae and Nelumbonaceae from the Nymphaeales has been verified repeatedly by subsequent phylogenetic analyses (e.g., Chase et al., 1993; Mathews & Donoghue, 1999; Saarela et al., 2007; Soltis et al., 2000, 2011). Molecular phylogenetic analyses have also indicated that the former genus *Ondinea* should be merged with *Nymphaea* because of its nested placement within that genus (Borsch et al., 2007; Löhne et al., 2009).

Another proposed revision adopted by some authors has been the inclusion of the unusual family Hydatellaceae within the Nymphaeales, because of its resolution as its sister group in molecular phylogenetic analyses (Saarela et al., 2007; Judd et al., 2016). However, this convention has not been followed here because the extremely high degree of morphological divergence that distinguishes Hydatellaceae and Nymphaeales would argue for their retention as separate orders, regardless of whether their proposed sister clade relationship is accurate. Presently, there is no reason for adopting one scheme over the other, given that each is compatible phylogenetically and simply reflects a matter of taste. Consequently, the order Nymphaeales is considered here to include seven genera distributed among two distinct clades, which represent the families Cabombaceae and Nymphaeaceae (Figure 1.3). The entire order comprises plan-mergent aquatic plants (i.e., all of the genera produce floating leaves of some sort), and OBL North American indicators are present in both families:

1. **Cabombaceae** A. Richard
2. **Nymphaeaceae** Salisbury

Family 1: Cabombaceae [2]

This small family of only two genera and six species consists entirely of aquatic herbaceous perennials with all of the species occurring normally in standing waters. Numerous phylogenetic studies have established the monophyly of the family and its resolution as the sister group to the much larger and more diverse Nymphaeaceae (e.g., Figure 1.3). Although both genera possess peltate floating leaves, only *Cabomba* is heterophyllous and grows most often as a submersed species, which is characterized by highly dissected underwater foliage and produces highly reduced floating leaves only during anthesis.

To undergo sexual reproduction, the protogynous flowers of both *Brasenia* and *Cabomba* extend from elongate peduncles above the water surface where they are pollinated by the wind or by insects (Insecta). They are solitary, bisexual, and consist of a tetramerous or trimerous perianth, 3–36 stamens, and an apocarpous gynoecium of 2–18 pistils each containing 2–5 ovules with laminar placentation (Cronquist, 1981; Wiersema, 1997). The fruits are leathery, indehiscent nutlets, containing seeds with a small embryo, little endosperm, and abundant perisperm (Cronquist, 1981; Wiersema, 1997). The

seeds are dispersed locally by water currents and over greater distances by waterfowl (Aves: Anatidae). Both genera reproduce vegetatively by rhizomes, which can result in the development of extensive clonal populations. Weedy infestations have been attributed to both *Brasenia* and *Cabomba*.

Although extending into temperate regions, *Cabomba* occurs primarily in tropical portions of the Western Hemisphere and is important economically as a source of popular aquarium plants. *Brasenia* is primarily temperate in its distribution and is grown as an ornamental curiosity in some water gardens.

Both genera of Cabombaceae contain OBL indicators:

1. ***Brasenia*** Schreb.
2. ***Cabomba*** Aubl.

1. Brasenia
Water-shield; brasénie de Schreber
Etymology: after Christopher Brasen (1738–1774)
Synonyms: *Cabomba* (in part); *Hydropeltis*
Distribution: global: Africa, Asia, Australia, New World;
North America: East/West disjunct
Diversity: global: 1 species; **North America:** 1 species
Indicators (USA): OBL: *Brasenia schreberi*
Habitat: freshwater; lacustrine, riverine; **pH:** 4.7–9.5; **depth:** 0–3 m; **life-form(s):** floating-leaved
Key morphology: shoots (to 20 dm) highly mucilaginous; leaves (to 13.5 cm) alternate, floating, peltate, broadly elliptical, highly mucilaginous, dichotomously veined, long petioled (to 61 cm), arising along an upright shoot; flowers emergent; flowers (to 2 cm) axillary, solitary, peduncled (to 15 cm); perianth segments (to 20 mm) purplish; stamens numerous; pistils (4–18) unilocular, with 1–2 ovules; fruits (to 10 mm) indehiscent, leathery; seeds (to 4 mm) 1–2, ovoid
Life history: duration: perennial (rhizomes, winter buds); **asexual reproduction:** rhizomes, winter buds; **pollination:** wind; **sexual condition:** hermaphroditic; **fruit:** achene-like (1–2 seeded) (common); **local dispersal:** rhizomes (1–2 year survival), seeds, winter buds (water currents); **long-distance dispersal:** fruits, seeds (waterfowl)
Imperilment: (1) *Brasenia schreberi* [G5]; S1 (<u>AB</u>, AK, IL, MT, <u>NF</u>, OK, PE); S2 (CA, IA, MB, NE); S3 (KY, NC)
Ecology: general: *Brasenia* is monotypic.

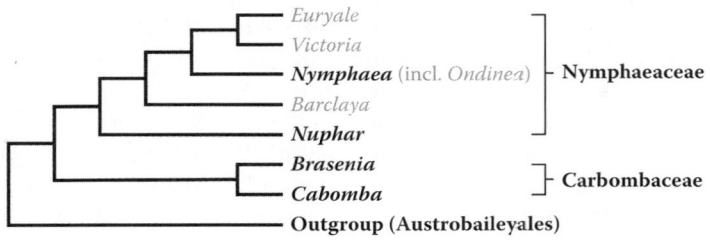

FIGURE 1.3 Intergeneric relationships in the order Nymphaeales as inferred by a phylogenetic analysis of *trnT-trnF* sequence data. The results are consistent with those based on analyses of combined molecular and nonmolecular data (Les et al., 1999). Although the entire order is aquatic, the North American taxa containing OBL indicators are highlighted in bold. (Adapted from Borsch, T. et al., *Int. J. Plant Sci.*, 168, 639–671, 2007.)

Brasenia schreberi **J. F. Gmel.** grows in still or sluggish standing waters (2.5–3.0 m deep) of bogs, depressions (deep), ditches, lakes, marshes, oxbows, ponds, potholes, reservoirs, rivers, sinkholes, and streams at elevations to 2194 m. It occupies a wide range of oligotrophic–mesotrophic habitats, which usually are less than 2 m deep and are characterized by low alkalinity (6.5–140 mg/L CaCO$_3$) and relatively low conductivity (20–250 μmhos/cm). The elastic shoots and petioles enables the plants to resist wave action, making them a conspicuous element of the floating-leaved zone. The plants are turbidity-tolerant and possess phytotoxic properties, factors that may contribute to their frequently dominant growth pattern manifested as expansive monocultures in small lakes and ponds. Plants grow in full sunlight and in substrates spanning a broad range of acidity (pH: 4.7–9.5; mean: 7.1), including clay loam, Hyde silt loam, muck, mud, muddy silt, sand, sandy clay, and silt. Flowering and fruiting occur from June to October. The protogynous flowers open over a 2-day cycle, being functionally pistillate during the first day and functionally staminate during the second day. Pollination is facilitated primarily by the wind. Occasional pollination by insects (Insecta) can also occur, but normally to an insignificant extent. High pollen:ovule ratios (e.g., 14,947:1) indicate that the species is adapted for obligate outcrossing; however, self-compatibility also allows for infrequent geitonogamous pollination to occur between different flowers of the same individual. The seeds are dispersed in the water or are consumed and dispersed endozoically by waterfowl (Aves: Anatidae). The seeds require a period of cold (4°C) stratification and incubation in the light at 25°C for optimal germination. Germination is negligible under dark conditions. Post-stratification soaking of the seeds for 6 h in a 500 mg/L solution of gibberellic acid (GA3) with a subsequent incubation in hot water (40°C) for 50 min also reportedly improves germination rates. The plants reproduce vegetatively by rhizome production and by the formation of gelatinous winter buds, which also enable the plants to overwinter. About 60% of the rhizomes die off by the end of the growing season and have an average life expectancy of 1.5 years. Genetic studies using nuclear and plastid markers have found substantial levels of variability and geographical subdivision to occur among some Asian populations. The unusual mucilaginous coating, which normally envelops the vegetative structures, has been found to reduce the level of herbivory on the plants. Large patches of the plants occur often in exclusion of other associated species; thus, the following list might not seem as extensive as anticipated for such a widespread species. **Reported associates:** *Cabomba caroliniana, Callitriche trochlearis, Carex gynandra, Ceratophyllum demersum, C. echinatum, Dulichium arundinaceum, Elatine heterandra, E. minima, Eleocharis acicularis, Elodea canadensis, E. nuttallii, Eriocaulon aquaticum, Gratiola aurea, Heteranthera dubia, Hydrilla verticillata, Imperata cylindrica, Juncus effusus, J. marginatus, J. scirpoides, Leersia hexandra, Lemna minor, Ludwigia palustris, Menyanthes trifoliata, Myriophyllum farwellii, M. heterophyllum, M. hippuroides, M. sibiricum, M. spicatum, Najas gracillima, N. flexilis, N. minor, Nymphaea odorata, N. tuberosa, Nuphar polysepala, N. variegata, Nymphoides cordata, Oenanthe sarmentosa, Persicaria amphibia, Pontederia cordata, Potamogeton amplifolius, P. bicupulatus, P. diversifolius, P. epihydrus, P. foliosus, P. gramineus, P. illinoensis, P. natans, P. nodosus, P. oaksianus, P. praelongus, P. pulcher, P. pusillus, P. robbinsii, P. spirillus, P. vaseyi, P. zosteriformis, Proserpinaca palustris, Sagittaria rigida, Schoenoplectus acutus, S. subterminalis, Sparganium emersum, Spirodela polyrhiza, Stuckenia pectinata, Taxodium distichum, Typha, Utricularia gibba, U. intermedia, U. macrorhiza, U. minor, U. purpurea, U. radiata, Vallisneria americana.*

Use by wildlife: *Brasenia schreberi* provides habitat (shade and shelter) for invertebrates, fish (Vertebrata: Osteichthyes), and waterfowl (Aves: Anatidae). The shoots and seeds are eaten by 22 species of ducks (Aves: Anatidae) including mallards (*Anas platyrhynchos*), pintails (*Anas acuta*), redheads (*Aythya americana*), wood ducks (*Aix sponsa*), and especially ring-necked ducks (*Aythya collaris*). The foliage is eaten by moose (Mammalia: Cervidae: *Alces americanus*). The leaves and roots are damaged by herbivorous insects (Insecta) including beetles (Coleoptera: Chrysomelidae: *Donacia cincticornis, Galerucella nymphaea*; Curculionidae: *Bagous cavifrons*), flies (Diptera: Chironomidae: *Polypedilum braseniae*), and snout moths (Lepidoptera: Crambidae: *Munroessa gyralis, M. icciusalis, Paraponyx allionealis, P. maculalis, P. seminealis, Synclita obliteralis, S. tinealis*). The flowers are visited by several insects (Insecta) including bees (Hymenoptera: Apidae: *Apis mellifera*), beetles (Coleoptera: Chrysomelidae: *Donacia cincticornis*; Curculionidae: *Perigaster cretura*), flies (Diptera: Ephydridae: *Notiphila cressoni*), and odonates (Odonata).

Economic importance: food: The shoots and young leaves of *B. schreberi* are eaten in Asia. The tuberous roots were ground into flour and eaten by native North Americans; **medicinal:** *Brasenia schreberi* has been used as an astringent (leaves) and as a treatment for dysentery and tuberculosis (rhizomes, seeds). The plants contain gallic acid and quercetin-7-*O*-β-D glucopyranoside, which exhibit strong antioxidant properties. The foliar extracts are alleopathic and inhibit the growth of bacteria (Bacteria) and algae (Chlorophyta: Oocystaceae: *Chlorella pyrenoidosa*; Cyanobacteria: Nostocaceae: *Anabaena flosaquae*); **cultivation:** *Brasenia schreberi* is of minor importance as an aquarium and water garden ornamental; **misc. products:** The mucilage of *B. schreberi* is said to have excellent lubrication properties, which could have applications as a coating for medicinal pills; **weeds:** The broadly adapted *B. schreberi* can become aggressive and is capable of crowding or displacing other species, especially in small ponds. It is regarded as a serious pest in parts of Missouri; **nonindigenous species:** none in North America

Systematics: *Brasenia* is monotypic. Phylogenetic studies (e.g., Figure 1.3) consistently have resolved the genus as the sister group of *Cabomba*. Although *Brasenia* appears to be uniform morphologically worldwide, it would be interesting to determine the extent of genetic variation in populations surveyed from across its range. The basic chromosome number of *Brasenia* is uncertain (it is $x = 13$ in *Cabomba*). *Brasenia schreberi* is relatively uniform cytologically ($2n = 80$), but counts of $2n = 72$ also has been reported.

Comments: *Brasenia schreberi* is disjunct in eastern and western North America, a pattern that probably resulted from the eradication of contiguous central populations during glaciation.

References: Adams, 1969; Bassett et al., 1993; Chrysler, 1938; Elakovich & Wooten, 1987; Harms & Grodowitz, 2009; Kim et al., 2012; Kunii, 1993; Legault et al., 2011; Les, 1986a; Les et al., 1999; Li et al., 2011, 2012, 2013; Nichols, 1999a; Osborn & Schneider, 1988; Podoplelova & Ryzhakov, 2005; Thompson et al., 2014; Wei et al., 1994; Wiersema, 1997.

2. *Cabomba*

Fanwort

Etymology: the name used by indigenous people of Guyana

Synonyms: *Nectris*

Distribution: global: New World; Asia*; Australia*; Europe*; **North America:** East/West disjunct

Diversity: global: 5 species; **North America:** 1 species

Indicators (USA): OBL: *Cabomba caroliniana*

Habitat: freshwater; lacustrine, palustrine, riverine; **pH:** 5.3–9.1; **depth:** 0–3 m; **life-form(s):** floating-leaved, submersed (vittate), suspended

Key morphology: shoots (1–2 m; up to 10 m) heterophyllous, flexuous, somewhat gelatinous, with adventitious roots (to 24 cm); submersed leaves (to 6 cm) opposite, long-petiolate (to 4 cm), fan-like, palmate/dichotomously dissected into 3–200 linear segments; floating leaves (to 3 cm) alternate, peltate, broadly to narrowly elliptical, rhomboidal, or sagittate; flowers (to 15 mm), 3-merous, solitary, emergent, peduncled (to 12 mm); sepals (to 12 mm) petaloid; petals (to 12 mm) 3, pinkish, purple, or white, clawed, with paired nectaries; stamens 6; pistils 2–4; fruits (to 7 mm) leathery, nutlet-like, 3-seeded; seeds (to 3 mm) with 4 rows of tubercles

Life history: duration: perennial (rhizomes); **asexual reproduction:** shoot fragments, rhizomes; **pollination:** insect; **sexual condition:** hermaphroditic; **fruit:** nutlets (infrequent); **local dispersal:** fragments (water currents, boating equipment); **long-distance dispersal:** fruits (waterfowl); fragments (aquarium disposal)

Imperilment: (1) *Cabomba caroliniana* [G3/G5]; SX (IN); S1 (IL, OK); S2 (KY, VA); S3 (NC)

Ecology: general: All *Cabomba* species are obligate aquatic perennials, which grow in standing waters up to several meters in depth. Most of the species inhabit warmer (25°C–33°C), circumneutral (pH: 6–8) waters in regions characterized by a brief winter dry period and annual rainfall amount exceeding 100 mm. The temperate species occur under cooler conditions (6°C–18°C average annual temperature) where the water temperatures remain below 27°C. *Cabomba* species inhabit floodplains, lakes, streams, and swamps where sufficient light is present. Significant annual water level fluctuations can result in reduced "mudflat" forms, which develop on exposed sediments where waters have receded. Although heterophyllous, the plants occur primarily as submersed or suspended growth forms unless they are undergoing sexual reproduction. The flowers of

all *Cabomba* species are bisexual and extend above the water surface during anthesis (their support facilitated by small, floating leaves) and are pollinated by insects (Insecta). All of the species also are capable of vigorous clonal vegetative growth.

***Cabomba caroliniana* A. Gray** grows in standing waters (6–600 cm deep) associated with bayous, bogs, brakes, canals, ditches, floodplains, lakes, ponds, pools (marshes), reservoirs, rivers, sloughs, streams, and swamps at elevations of up to 300 m. The plants are adapted to a broad range of environmental conditions, but usually occur in water depths less than 3.5 m. Biomass production is proportional to water clarity and is highest at 2–3 m depths. The plants usually occur without floating leaves unless they are rooted in shallow, quiet localities, where flowering is initiated. The plants generally grow under full sunlight; however, the plants are fairly shade-tolerant, with 99% shading found to be necessary for reducing their biomass to zero over a 4-month period. The plants thrive in clear, soft waters with low specific conductance. The waters span a wide range of acidity (pH: 5.3–9.1), but the plants grow best at pH values less than 7.0. The substrates generally are high in organic matter and nutrients and are not often cobble, sand, or similarly loose textures. They are described most often as marl, muck, mud, sandy mud, and silt. Flowering and fruiting occur from April to December. The flowers are protogynous and open (from 10:00 am to 4:30 pm) for two consecutive days, after which they are retracted underwater. The stigmas are fertile during the first day, but wither before the pollen is shed during the second day, thereby precluding self-pollination. Pollination by insects (see *Use by wildlife*) is indicated by the relatively low pollen production (averaging 560 grains/flower) and low average pollen:ovule ratios of 62. Seed set is characteristically low. The seeds develop within 2–4 weeks and usually germinate within 10 weeks after fertilization occurs; no after-ripening period appears to be necessary. The ripe fruits detach and sink to the bottom, or are dispersed in the water while attached to the buoyant shoots, which can be transported for distances up to 3 km. Once released from the decomposing fruits, the seeds can be dispersed over greater distances by adhering to the plumage or muddy feet of waterfowl (Aves: Anatidae). The seeds can remain viable in the sediment for at least 2 years. Their germination rate is higher (85%) when dried compared to seeds kept continuously wet (25%). The plants are highly clonal and are capable of prolonged survival (as suspended forms) if uprooted or fragmented. Asexual reproduction occurs locally by layering. Vegetative reproduction by fragmentation can be prolific, with only a single node and leaf necessary for regeneration. The plants overwinter as whole shoots, or by the production of turion-like propagules. They are resistant to desiccation and are not effectively controlled by short periods of drawdown (unless the substrate thoroughly dries out). Vegetative fragments that are exposed to the air can remain viable for up to 42 h, and for up to 3 h when exposed to wind during vehicular transport. Introductions to northern portions of North America have been shown to reduce light penetration to other native macrophytes, increase populations of epiphytic algae, and alter the composition of macroinvertebrate communities. Studies using genetic markers have confirmed that introduced populations can originate by dispersal from other introduced populations. **Reported associates:** *Alternanthera philoxeroides, Bacopa caroliniana, B. rotundifolia, Bidens beckii, Brasenia schreberi, Callitriche, Carya, Cephalanthus occidentalis, Ceratophyllum demersum, Chara vulgaris, Colocasia esculenta, Echinodorus cordifolius, Egeria densa, Eichhornia crassipes, Elatine minima, Eleocharis acicularis, E. robbinsii, Elodea canadensis, E. nuttallii, Eriocaulon aquaticum, Gratiola aurea, Heteranthera dubia, Hydrilla verticillata, Hygrophila lacustris, Isoetes echinospora, Lemna minor, L. minuta, Limnobium spongia, Ludwigia hexapetala, L. octovalvis, L. repens, Luziola fluitans, Mayaca fluviatilis, Myriophyllum alterniflorum, M. heterophyllum, M. heterophyllum × M. laxum, M. sibiricum, M. spicatum, M. tenellum, M. verticillatum, Najas flexilis, N. guadalupensis, Nelumbo lurea, Nuphar advena, N. variegata, Nymphaea mexicana, N. odorata, Nymphoides cordata, Nyssa, Ottelia alismoides, Persicaria amphibia, Pistia stratiotes, Pontederia cordata, Potamogeton amplifolius, P. bicupulatus, P. crispus, P. epihydrus, P. gramineus, P. natans, P. perfoliatus, P. praelongus, P. pulcher, P. pusillus, P. richardsonii, P. robbinsii, P. spirillus, P. zosteriformis, Proserpinaca, Quercus, Sagittaria graminea, S. subulata, Salvinia, Schoenoplectus subterminalis, Sparganium, Spirodela polyrhiza, Stuckenia pectinata, Taxodium distichum, Utricularia gibba, U. inflata, U. intermedia, U. macrorhiza, U. purpurea, Vallisneria americana, Wolffia brasiliensis, W. columbiana.*

Use by wildlife: *Cabomba caroliniana* provides habitat and shelter for invertebrates and fish (Vertebrata: Osteichthyes). The foliage is eaten preferentially by turtles (Reptilia: Emydidae: *Pseudemys texana*) and also by wood ducks (Aves: Anatidae: *Aix sponsa*). The seeds serve as a minor food for some waterfowl (Aves: Anatidae) such as ruddy ducks (*Oxyura jamaicensis*) and mallard ducks (*Anas platyrhynchos*). The plants are host to larval caddisflies (Trichoptera: Leptoceridae: *Nectopsyche tavara*) and to several larval moths (Insecta: Lepidoptera: Crambidae: *Paraponyx diminulatis*; Noctuidae: *Feltia jaculifera*). The flowers are visited (and pollinated) by various insects (Insecta) including bees (Hymenoptera: Apidae: *Apis mellifera*; Halictidae: *Lasioglossum*), flies (Diptera: Ephydridae: *Hydrellia bilobifera, Notiphila cressoni*), and wasps (Braconidae: *Ademon, Chaenusa*). They are visited incidentally by damselflies (Insecta: Odonata: Zygoptera) and weevils (Insecta: Coleoptera: Curculionidae). The plants also are associated with several nematodes (Nematoda) including free-living species (Dorylaimidae: *Dorylaimus, Mesodorylaimus, Michonchus*; Mononchidae: *Mononchus*; Rhabditidae: *Rhabditis*) and phytopathogens (Aphelenchoididae: *Aphelenchoides fragariae*; Criconematidae: *Criconemoides, Hemicriconemoides*; Dolichodoridae: *Dolichodorus*; Pratylenchidae: *Hirschmanniella caudacrena*; Tylenchidae: *Tylenchus*).

Economic importance: food: not edible; **medicinal:** no uses reported (the plants contain very low amounts of alkaloids); **cultivation:** *Cabomba caroliniana* is a major aquarium plant that is sold worldwide; **misc. products:** none; **weeds:** *Cabomba caroliniana* is highly invasive in northeastern North America where it has been introduced; it is listed as a noxious aquatic weed in California, Maine, New Hampshire, Vermont, and Washington. Dense monocultures can displace native species and interfere with water recreation. The introductions have occurred primarily by the careless disposal of cultivated aquarium plants; **nonindigenous species:** *Cabomba caroliniana* presumably is native to the southeastern United States but has been introduced to Oregon, Washington, the northeastern United States, and to southern Ontario. It was introduced to New England about 1920 (Massachusetts), to New York (Hudson River) around 1955, and to Michigan about 1890. *Cabomba* was first observed in Canada in 1992 from Kasshabog Lake, Ontario. All Old World occurrences of *Cabomba* (e.g., Belgium, China, England, Germany, Hungary, Japan, Netherlands, Scotland, and Sweden) are due to introductions.

Systematics: *Cabomba* is related closely to *Brasenia* as demonstrated by cladistic analysis of morphological and molecular data (Figure 1.3). Although presumably monophyletic, a phylogenetic analysis of *Cabomba* species has not yet been conducted. Ørgaard recognized three varieties of *C. caroliniana* (*C. caroliniana* var. *caroliniana*, *C. caroliniana* var. *pulcherrima*, *C. caroliniana* var. *flavida*) distinguished primarily by petal color (white, purple, or yellow, respectively). Allozyme analysis has shown very high genetic identity (0.999–1.0) between *C. caroliniana* var. *caroliniana* and *C. caroliniana* var. *pulcherrima*. The base chromosome number of *Cabomba* is $x = 13$. *Cabomba caroliniana* is variable cytologically ($2n = 24$, c. 78, c. 96, c. 104) and reportedly has reduced pollen fertility compared to the other species.

Comments: *Cabomba caroliniana* occurs in eastern North America but is disjunct in the Pacific Northwest (as a result of introductions). Disjunct populations of *C. caroliniana* in South America indicate its possible introduction into North America or vice-versa.

References: Barnes et al., 2013; Beal, 1900; Bickel, 2012, 2015; Dugdale et al., 2013; Fassett, 1953b; Fields et al., 2003; Harms & Grodowitz, 2009; Hogsden et al., 2007; Hussner, 2012; Jacobs & Macisaac, 2009; Les, 2004; Les & Mehrhoff, 1999; Les et al., 1999; McCracken et al., 2013; Ørgaard, 1990; Ørgaard et al., 1992; Osborn et al., 1991; Schneider & Jeter, 1982; Schooler, 2008; Schooler et al., 2009; Wain et al., 1983; Wiersema, 1997; Wilson et al., 2007.

Family 2: Nymphaeaceae [5]

The "water lilies" comprise a well-supported clade (Figure 1.3) of about 60 perennial, aquatic species that are dispersed throughout most parts of the world. All of the species are obligate hydrophytes and are recognized by prominent, floating leaves (often referred to as "lily pads") that in some instances (e.g., *Victoria*) can grow to well over a meter in diameter. The leaves of some species emerge from the water surface as opposed to being strictly floating. Yet others are heterophyllous and produce membranous submersed leaves in addition to coriaceous floating leaves. The foliage typically contains laticifers, sesquiterpene alkaloids, and stellate sclereids, which provide support for an extensive system of lacunae (Cronquist, 1981).

Water lily flowers are solitary and extend from the surface on elongate, rigid peduncles (tropical species) or float on the water surface (temperate species), attached to the plants by long, flexuous peduncles that resist the stressful forces of wave action. Typically, they are large and showy, bisexual, and consist of a colorful perianth (often tepaloid or intergrading) comprising 4–14 sepals and numerous petals. The stamens are numerous and laminar (often petaloid). The gynoecium consists of 5-many carpels that are joined (at least by their outer margins) into a large, compound, multichambered ovary that develops into a many-seeded, berrylike fruit (Cronquist, 1981). The flowers are typically pollinated by beetles (Insecta: Coleoptera) and also by bees (Hymenoptera) and flies (Diptera). Self-pollination can occur in some species. The seeds are dispersed primarily by water, often while still attached to the buoyant fruits or their parts. A transient seed bank develops in some species (Padgett, 2007). Perennation and vegetative reproduction occur by means of corms, rhizomes, or tubers.

Many water lilies are grown as ornamental plants for water gardens, with the most notable genera in this respect being *Nymphaea* and *Victoria*. Several hundred cultivars have been developed in *Nymphaea* alone. Although it contains only two species, *Victoria* is arguably one of the most popular specimens displayed in botanic gardens and conservatories, due to its highly exotic appearance and enormous floating leaves, which often are pictured (facilitated by means of some hidden support) with people or various domestic animals standing on their surface. *Victoria*-like plants also appeared in one of the first (and Technicolor) views of "Munchkinland" in the 1939 film *The Wizard of Oz*. Water lilies have symbolic or religious significance in some cultures and have inspired many artists (e.g., Claude Monet) and architectural designs (e.g., Joseph Paxton's *Crystal Palace*, formerly of Hyde Park, London). Some species are of minor importance as medicinal or food plants.

Two North American genera contain OBL indicators:

    1. ***Nuphar*** Sm.
    2. ***Nymphaea*** L.

1. *Nuphar*

Brandy bottle, cow lily, pond lily, spatterdock; nénuphar
Etymology: from *nouphar*, the Greek translation of an ancient Arabic name for the plants
Synonyms: *Clairvillea*; *Nenuphar*; *Nufar*; *Nymphaea* [in part]; *Nymphanthus*; *Nymphona*; *Nymphosanthus*; *Nymphozanthus*; *Ropalon*
Distribution: global: pantemperate; **North America:** widespread
Diversity: global: 8 species; **North America:** 5 species
Indicators (USA): OBL: *Nuphar advena*, *N. microphylla*, *N. polysepala*, *N. sagittifolia*, *N. variegata*

Habitat: freshwater; lacustrine, palustrine, riverine; **pH:** 4.1–9.8; **depth:** 0–2 m; **life-form(s):** floating-leaved
Key morphology: stems (to 1.5 m) rhizomatous, stout, thick (to 10 cm), branching; aerial/floating leaves (to 50 cm) broad, coriaceous (submersed leaves membranous, the margins usually crisped), orbicular to lanceolate, pinnate-dichotomous veined, basal lobes rounded, petioles round or flattened/winged; flowers (to 12 cm) spherical, floating or emergent from thick (to 15 mm), rigid peduncles (to 2.7 m long); sepals (to 6 cm) 5–14, showy, yellow, sometimes reddish or greenish at base; petals numerous, reduced, scale-like, yellow or reddish, with a nectary; stamens numerous, laminar, reddish or yellowish; carpels numerous, adnate, terminating apically as a flat, stigmatic disc (to 33 mm), constricted or not beneath; fruit (to 9 cm) a leathery, capsular berry (dehiscing by deterioration); seeds (to 6.5 mm) brown, numerous (43–1400/fruit)
Life history: duration: perennial (rhizomes); **asexual reproduction:** rhizomes; **pollination:** insect; **sexual condition:** hermaphroditic; **fruit:** capsular berries (common); **local dispersal:** fruit, seeds, rhizomes (water currents, wildlife); **long-distance dispersal:** seeds (turtles)
Imperilment: (1) *Nuphar advena* [G5]; SH (CT); S1 (AL, KS, WI); S2 (FL, ME); S3 (GA, IL); (2) *N. microphylla* [G5]; SH (NH, NJ); S1 (MA, MI, PA); S3 (<u>NB</u>, <u>ON</u>); (3) *N. polysepala* [G5]; S1 (AZ); S2 (UT); S3 (WY); (4) *N. sagittifolia* [G5]; S1 (VA); S2 (NC); (5) *N. variegata* [G5]; S1 (IL, OH); S2 (NE); (<u>BC</u>)
Ecology: general: *Nuphar* species occur in a wide range of freshwater acidic to alkaline habitats (ponds, lakes, slow streams, swamps, ditches) but grow most often on soft substrates in circumneutral waters that are between 0.5 and 2.0 m in depth and are characterized by slow or negligible currents. Some species can extend to elevations beyond 3000 m. All of the species are an integral component of the floating-leaved zone of aquatic vegetation. The coarser, stronger leaves of *Nuphar* plants enable them to grow in harsher, more exposed areas than *Nymphaea* species, but still generally in shallower water. All *Nuphar* species are tolerant to turbidity due to their ability to rapidly produce new leaves at the water surface where light is plentiful. The plants are only slightly shade tolerant. Development of floral primordia is delayed over a 2–3-year period so that the flowers do not all mature in a single season. After emerging from the water, the protogynous flowers open for 4–5 days (but close each night), with receptive stigmas but no pollen released during the first day. The stigmas wither and lose receptivity by the end of the second day, as mature pollen is shed. Pollen production extends throughout the remainder of anthesis. The flowers are pollinated by insects (Insecta). The most effective agents are beetles (Coleoptera: Chrysomelidae: *Donacia*), which can become entrapped within the flowers during their nocturnal closure. Other insect pollinators include aphids (Hemiptera: Aphididae: *Rhopalosiphum*), bees (Hymenoptera: Apidae: *Apis*, *Bombus*; Halictidae: *Halictus*), and flies (Diptera: Empididae: *Hilara*; Ephydridae: *Notiphila*; Scathophagidae: *Hydromyza*; Syrphidae: *Eristalis*, *Helophilus*). The extent of

autogamy varies and can be facilitated by the movement of entrapped pollinators. The fruits mature above the water, but the seeds are dispersed primarily by water. Although individual seeds will sink, they can be dispersed within the buoyant fruits (or portions thereof), which remain afloat for up to 72 h and have been observed to travel on the water surface at rates of up to 80 m/h. The seeds do not tolerate dessication and are not adapted for endozoic transport by most animals. However, they are passed intact through the digestive system of some turtles (Reptilia: Emydidae), which have been implicated as potential dispersal agents. The seedlings also float and can be dispersed within a water body. The seeds are typically physiologically dormant and require a period of cold stratification to induce germination. Those of some species have germinated well after 45 days of stratification (4°C) followed by incubation at 25°C; poor germination has been reported at temperatures below 13°C. The seedlings thrive and develop optimally under low light conditions. Populations are often highly clonal and develop by the proliferation of the branching rhizomes. Networks of the intertwined rhizomes (which occur on or beneath the sediment surface) can substantially increase silt deposition. Most *Nuphar* rhizomes release accumulating alcohol (a result of anaerobic conditions) from the flowers and fruits, hence the common generic name of "brandy bottle."

Nuphar advena **(Aiton) W.T. Aiton** occurs in nontidal or tidal waters (0.1–2.1 m deep) associated with bayous, bogs, borrow pits, bottomlands, canals, ditches, flatwoods, floodplains, lakes, oxbows, ponds, pools, potholes, prairies, rivers, sloughs, streams, and swamps at elevations to 600 m. The plants typically mimic emergent species by extending their leaf blades above the water surface. The floating leaves live for 31 days on average, while the average lifespan of submersed leaves is only slightly longer (33 days). The habitats range broadly from acidic to calcareous (pH: 4.3–9.4 [mean: 6.5–7.0]; conductivity <340 µmhos/cm; alkalinity <195 mg/L CaCO$_3$) and are characterized by substrates that are described as muck, mud, peat, sand, sandy alluvium, sandy shells, silt, and silty muck. Flowering occurs from March to November, with fruiting extending from April to November. The flowers are neither apomictic nor spontaneously autogamous and are outcrossed by insect pollinators, utilizing those species present in the highest abundance (see *Use by wildlife*); geitonogamous pollination can also occur. The seeds are dormant when fresh, and unstratified seeds have been observed to germinate slowly (at room temperature). Precise germination requirements have not been determined, but the seeds reportedly germinate in or on muddy sediments, which become exposed above the water surface. A persistent seedbank apparently does not develop, and seedlings are rarely observed in the field. **Reported associates:** *Alternanthera philoxeroides, Amaranthus cannabinus, Asclepias incarnata, Bidens laevis, Boehmeria cylindrica, Brasenia schreberi, Callitriche heterophylla, Carex lupulina, Ceratophyllum demersum, Chelone glabra, Commelina virginica, Decodon verticillatus, Eleocharis quadrangulata, Hibiscus laevis, Hydrocotyle verticillata, Hydrolea quadrivalvis, Hypericum fasciculatum, Itea virginica, Leersia oryzoides, Lemna minor, Ludwigia, Myriophyllum laxum, Najas*

flexilis, N. gracillima, N. guadalupensis, Nuphar variegata, Nymphaea odorata, Nyssa, Orontium aquaticum, Peltandra virginica, Persicaria arifolia, P. hydropiper, P. punctata, Pontederia cordata, Ranunculus flabellaris, Sagittaria latifolia, Samolus valerandi, Saururus cernuus, Schoenoplectus heterochaetus, Taxodium, Typha latifolia, Vallisneria americana, Zizania aquatica.

Nuphar microphylla **(Pers.) Fernald** grows in the shallow waters (e.g., 0.5–1 m deep) of backwaters, lagoons, lakes, ponds, sloughs, and slow-moving streams at elevations to 300 m. The waters are of low alkalinity and low conductivity and are usually circumneutral in pH. The substrates are described as mud. Flowering occurs from June to September. Little additional life history information is available for this species. **Reported associates:** *Alisma, Callitriche palustris, Carex, Ceratophyllum demersum, C. echinatum, Chara, Eleocharis palustris, Elodea canadensis, Equisetum arvense, Eutrochium maculatum, Glyceria striata, Heteranthera dubia, Lemna trisulca, Lycopus americanus, Myriophyllum sibiricum, Najas flexilis, Nitella, Nuphar variegata, Nymphaea odorata, Potamogeton amplifolius, P. crispus, P. epihydrus, P. natans, P. nodosus, P. perfoliatus, P. praelongus, P. richardsonii, P. robbinsii, P. zosteriformis, Ranunculus longirostris, R. trichophyllus, Ruppia maritima, Sagittaria graminea, S. latifolia, Scirpus cyperinus, Sium suave, Sparganium americanum, S. emersum, Spiraea alba, Stuckenia pectinata, Vallisneria americana, Zizania aquatica.*

Nuphar polysepala **Engelm.** inhabits fresh (or very slightly brackish), standing waters (0.2–4.0 m deep) in bogs, channels, depressions, ditches, fens (poor), floodplains, lakes (kettle), marshes, muskeg, oxbows, ponds, pools, sinkholes, sloughs, and streams at elevations to 3482 m. Although the plants grow primarily in full sunlight, they occasionally will tolerate partially shaded sites. This is the largest of the *Nuphar* species, but also often lacks submersed leaves. The floating leaves occasionally extend from the water surface as in *N. advena*. The waters are soft and mainly acidic (e.g., pH: 5.3–7.6; conductivity: 16–97 mmhos; CaCO$_3$ hardness: 23.8–42.8 mg/L; total alkalinity: 4–45 mg/L). The substrates are also characterized as acidic (e.g., pH: 4.1–4.5) and include clay, Coconino sandstone, gravel, humus, loam, muck (organic), mucky ooze, mud, muddy sand, peat, quartzite, sand, sandy loam, sandy silt, and silt. Flowering and fruiting occur from April to November. More specific details on the reproductive ecology of this species are lacking. Vegetative reproduction by rhizomes is prolific and can result in the development of monocultures, especially in smaller ponds. Depauperate mudflat forms can persist when water levels recede excessively. **Reported associates:** *Azolla, Bidens cernuus, B. frondosus, Brasenia schreberi, Cabomba caroliniana, Callitriche palustris, Carex aquatilis, C. canescens, C. diandra, C. limosa, C. rostrata, C. utriculata, C. vesicaria, Ceratophyllum demersum, Chara, Cicuta douglasii, Comarum palustre, Drosera anglica, D. rotundifolia, Dulichium arundinaceum, Egeria densa, Eleocharis palustris, Elodea canadensis, Equisetum fluviatile, Eriophorum gracile, Glyceria borealis, G. grandis, Hippuris vulgaris, Isoetes bolanderi, I. occidentalis, Juncus acuminatus, Kalmia microphylla,*

Lemna minor, L. trisulca, Mentha arvensis, Menyanthes trifoliata, Micranthes oregana, Mimulus primuloides, Myosotis laxa, Myriophyllum heterophyllum, M. sibiricum, M. spicatum, M. verticillatum, Najas flexilis, Nitella, Nuphar variegata, Nymphaea odorata, N. tetragona, Oenanthe sarmentosa, Persicaria amphibia, Phalaris arundinacea, Potamogeton alpinus, P. amplifolius, P. epihydrus, P. gramineus, P. natans, P. praelongus, P. pusillus, P. richardsonii, P. subsibiricus, P. zosteriformis, Primula tetrandra, Ranunculus aquatilis, R. flabellaris, R. trichophyllus, Rhynchospora alba, Rorippa curvisiliqua, Ruppia maritima, Sagittaria latifolia, Salix eastwoodiae, Scheuchzeria palustris, Schoenoplectus acutus, Sericocarpus oregonensis, Sparganium angustifolium, S. emersum, S. eurycarpum, S. hyperboreum, Sphagnum angustifolium, S. centrale, S. teres, Spiraea douglasii, Stuckenia pectinata, Typha latifolia, Utricularia macrorhiza, Vallisneria americana, Zannichellia palustris.

Nuphar sagittifolia (**Walter**) **Pursh** resides in standing (to 2.1 m deep), nontidal or tidal waters of backwaters, bayous, channels, ditches, estuaries, lakes, ponds, rivers, sloughs, and streams at elevations to 50 m. The plants require open exposures and are intolerant of turbidity. The habitats are characterized by acidic waters (pH: 4.2–6.7) and substrates of mud, sand, or silt. Flowering occurs from April to October, with fruiting commencing by June. Specific details on the pollination biology are not available, but probably are similar to that of the other species (i.e., protogyny and prevalent outcrossing by insects). The seeds are physiologically dormant and require a period of cold stratification to induce germiantion. Stratification at 4°C for 45 days resulted in >90% germination for seeds incubated at room temperature. High germination rates have also been observed after 35 days of cold stratification at 1.7°C. Their dispersal mechanism presumably relies on the movement of the seed-bearing fruits by water. Unlike its congeners, this species occurs more frequently in flowing waters, with submersed leaves being more prevalent than floating leaves. **Reported associates:** *Ampelopsis brevipedunculata, Asclepias perennis, Bacopa innominata, Brasenia schreberi, Carex intumescens, Carya aquatica, Ceratophyllum, Decumaria barbara, Diospyros virginiana, Eubotrys racemosa, Fraxinus caroliniana, Hydrocotyle ranunculoides, Ilex decidua, Leersia, Micranthemum, Mikania scandens, Mitchella repens, Murdannia keisak, Myriophyllum, Nymphaea odorata, Nyssa aquatica, Persea palustris, Persicaria amphibia, Quercus laurifolia, Q. lyrata, Rosa palustris, Sagittaria subulata, Solidago, Sparganium americanum, Styrax americanus, Taxodium ascendens, T. distichum, Typha, Ulmus alata, Utricularia, Woodwardia areolata, Zizaniopsis miliacea.*

Nuphar variegata **Durand** grows in standing waters (0.4–2.2 m deep) of bogs, ditches, floodplains, lakes, oxbows, ponds, pools, rivers, sloughs, and streams at elevations to 1982 m. The plants are restricted to boreal regions and occur almost exclusively north of the last glacial boundary. The habitats are characterized by waters that are clear and acidic to alkaline (pH: 5.0–9.4; conductivity: 5–340 µmhos/cm; $CaCO_3$ hardness: 25.8–109.8 mg/L; total alkalinity: 5–195 mg/L). The

substrates are described as clay, gravel, muck, mud, peat, sand, sandy silt, and silt. Flowering occurs from May to September, with fruiting observed from July to August. Additional details on the reproductive biology and seedling ecology are scarce. The seeds maintain good viability (>80%) after 5 years of storage in water at 1°C–3°C. Seed dispersal does not appear to be very effective, resulting in dispersal limitation relative to many other hydrophytes. The biomass investment in floating leaves can be substantially lower (e.g., by 50%) when growing at shallower depths. The biomass of submersed (but not floating) leaves increases with higher CO_2 levels, which enables the plants to better exploit the CO_2 present in the water column. Herbivores can consume up to nearly 1.5% of the leaf area per day. **Reported associates:** *Acorus calamus, Alisma, Alopecurus aequalis, Beckmannia syzigachne, Bidens beckii, Brasenia schreberi, Cabomba caroliniana, Callitriche hermaphroditica, C. heterophylla, Carex limosa, Cephalanthus occidentalis, Ceratophyllum demersum, C. echinatum, Chara vulgaris, Decodon verticillatus, Dulichium arundinaceum, Eleocharis acicularis, E. obtusa, E. palustris, Elodea canadensis, E. nuttallii, Eriocaulon aquaticum, Fontinalis antipyretica, Glyceria, Gratiola aurea, Heteranthera dubia, Hippuris vulgaris, Hydrocharis morsus-ranae, Hypericum boreale, Isoetes echinospora, Juncus brevicaudatus, J. canadensis, J. pelocarpus, Lemna minor, L. trisulca, Lobelia dortmanna, Myosotis laxa, Myriophyllum farwellii, M. heterophyllum, M. humile, M. sibiricum, M. spicatum, M. tenellum, Najas flexilis, N. gracillima, N. guadalupensis, N. marina, Nitellopsis, Nuphar advena, N. microphylla, N. polysepala, Nymphaea odorata, Nymphoides peltata, Persicaria amphibia, Pontederia cordata, Potamogeton amplifolius, P. berchtoldii, P. bicupulatus, P. confervoides, P. epihydrus, P. friesii, P. gramineus, P. illinoensis, P. natans, P. nodosus, P. praelongus, P. pulcher, P. pusillus, P. richardsonii, P. robbinsii, P. strictifolius, P. zosteriformis, Ranunculus longirostris, Sagittaria cuneata, S. latifolia, Schoenoplectus acutus, S. subterminalis, S. tabernaemontani, Sium suave, Sparganium americanum, S. angustifolium, S. fluctuans, Sphagnum, Spirodela, Stratiotes aloides, Stuckenia pectinata, Typha latifolia, Utricularia gibba, U. intermedia, U. macrorhiza, U. minor, U. purpurea, U. radiata, U. resupinata, Vallisneria americana, Zannichellia palustris, Zizania palustris.*

Use by wildlife: The large floating leaves of all *Nuphar* species provide habitat for many algae and aquatic invertebrates, as well as cover for amphibians (Vertebrata: Amphibia) and fish (Vertebrata: Osteichthyes); the roots are used by fish as a spawning substrate. Many of the associated invertebrates are important food for bluegills (Vertebrata: Osteichthyes: Centrarchidae: *Lepomis macrochirus*) and redear sunfish (Vertebrata: Osteichthyes: Centrarchidae: *Lepomis microlophus*). The roots and foliage of various *Nuphar* species are eaten by herbivorous mammals (Mammalia) such as beavers (Castoridae: *Castor canadensis*), deer (Cervidae: *Odocoileus*), moose (Cervidae: *Alces*), muskrats (Cricetidae: *Ondatra zibethicus*), and porcupines (Rodentia: Erethizontidae: *Erethizon dorsatus*). Because *Nuphar* fruits mature above the water, they provide wildlife with easy access to the seeds. The seeds are consumed by at least 12 species of ducks (Aves: Anatidae) and other marsh birds such as bitterns (Aves: Ardeidae), and soras (Aves: Rallidae: *Porzana*). *Nuphar* leaves are damaged by a number of insects (Insecta) including beetles (Coleoptera: Chrysomelidae: *Donacia, Galerucella*), caterpillars (Crambidae: *Nymphula nitidulata, Ostrinia penitalis*; Noctuidae: *Bellura gortynoides, Homophoberia cristata, Thurberiphaga diffusa*), and mining fly larvae (Diptera: Scathophagidae: *Hydromyza*), as well as snails (Mollusca: Gastropoda: Lymnaeidae: *Lymnaea*), and fungi (Ascomycota: incertae sedis: *Dichotomophthoropsis*; Basidiomycota: Typhulaceae: *Sclerotium*; Oomycota: Pythiaceae: *Pythium*). Herbivory is often greater on the submersed rather than floating leaves. Excessive growth of *Nuphar* can reduce water oxygen levels, thereby decreasing available fish habitat. The leaves, petioles, rhizomes, and roots of *N. advena* contain a fair amount of crude protein (24.8%) and are eaten by moose (Mammalia: Cervidae: *Alces americanus*). The seeds are a major food of ring-necked ducks (Aves: Anatidae: *Aythya collaris*). The plants host various insects (Insecta) including aphids (Hemiptera: Aphididae: *Rhopalosiphum nymphaeae*), beetles (Chrysomelidae: *Galerucella nymphaeae*), bugs (Hemiptera: Delphacidae: *Megamelus davisi*), caterpillars (Lepidoptera: Crambidae: *Munroessa gyralis*; Noctuidae: *Bellura gortynoides, Homophoberia cristata*), midges (Diptera: Chironomidae: *Polypedilum fallax*), and weevils (Coleoptera: Erirhinidae: *Onychylis nigrirostris*). The flowers are visited (and pollinated) by several insects (Insecta) including bees (Hymenoptera: Halictidae: *Lasioglossum nelumbonis, L. pectorale*), beetles (Coleoptera: Chrysomelidae: *Donacia piscatrix, D. texana*), and flies (Empididae: *Hilara bella*; Ephydridae: *Hydrellia cruralis, Notiphila*; Scathophagidae: *Hydromyza confluens*; Syrphidae: *Chalcosyrphus metallicus, Paragus, Toxomerus geminatus*). The fruits of *N. polysepala* are eaten by Pacific pond turtles (Reptilia: Emydidae: *Actinemys marmorata*). Plants of *N. sagittifolia* are hosts to larval caddisflies (Insecta: Trichoptera: Leptoceridae: *Nectopsyche waccamawensis*). The leaves of *N. variegata* are eaten by several insects (Insecta) including beetles (Coleoptera: Chrysomelidae: *Donacia cincticornis, Galerucella nymphaeae*), caddisflies (Trichoptera: Limnephilidae: *Limnephilus infernalis*), midges (Diptera: Chironomidae: *Hyporhygma quadripunctatus*), and weevils (Coleoptera: Curculionidae: *Bagous americanus*). The stems support larval zebra mussels (Mollusca: Dreissenidae: *Dreissena polymorpha*). The flowers, fruits, and seeds are eaten by eastern painted turtles (Reptilia: Emydidae: *Chrysemys picta*) and snapping turtles (Reptilia: Chelydridae: *Chelydra serpentina*).

Economic importance: food: Although usually regarded as inedible because of their alkaloid content, which can impart a bitter taste, *Nuphar* rhizomes and roots (roasted or boiled) and seeds (ground into flour or popped like popcorn) were eaten by a number of native North American tribes. The Comanche, Lakota, and Menominee people ate the boiled roots or rootstocks of *N. advena*. The Montana Indians ate

the fruits, seeds (parched and eaten like popcorn or ground into meal for soups), and rhizomes. The seeds were cooked and eaten by the Pawnee. The seeds of *N. polysepala* provided a staple grain (known as "wokas") to the Klamath, and the fruits (known as "spokwas") were stored as an important winter food. The Cheyenne ate the raw or roasted seeds, or dried them for use in porridge or as flour for bread. They also ate the raw or boiled roots. Native people in Alaska boiled or roasted the rootstocks, which were eaten as a vegetable. The fruits of *N. variegata* were eaten by the indigenous people of eastern Canada; **medicinal:** Alkaloids contained in *Nuphar* plants have exhibited antitumor activity in some laboratory animals. The seeds and pulverized roots of several *Nuphar* species have been used to treat diarrhea, external bleeding, and leucorrhea. *Nuphar advena* found wide medicinal use by many tribes, including the Iroquois (as an analgesic and for treating blood disease, epilepsy, fevers, heart problems, intestinal problems, smallpox, and swollen lungs), Micmac and Ojibwa (as a dermatological aid, to reduce bruises and swellings), Penobscot, Potawatomi, and Rappahannock (to reduce inflammation and swelling), Rappahannock (to treat boils, sores, and to reduce fever), Sioux (as a styptic), and the Thompson people (as an analgesic and to treat sores or wounds). The Bella Coola, Haisla, and Hanaksiala tribes made a decoction from the boiled roots of *N. polysepala* for treating internal pains related to gonorrhea, heart disease, rheumatism, and tuberculosis. The plants were used as a general medicinal by the Hesquiat and as a gynecological aid by the Kitasoo. The Gitksan tribe prepared an infusion from the roasted root scrapings, which was taken as a contraceptive or for alleviating lung hemorrhages. The Abenaki, Algonquin, Iroquois, and Micmac tribes ate plants of *N. variegata* to relieve symptoms of diabetes. Extracts from *N. variegata* are effective against several human pathogenic bacteria; **cultivation:** *Nuphar* species have minor value as aquarium or water garden ornamentals; *N. sagittifolia* is the "Cape Fear spatterdock" of commerce; **misc. products:** The tanniniferous leaves of *N. polysepala* and *N. advena* have been used for dying and tanning by several native North American cultures. The distinctive foliar sclereids of *Nuphar* have been used in paleoenvironmental research; *Nuphar* alkaloids exhibit some insecticidal activity; **weeds:** *Nuphar* species can grow excessively along lake and river margins or in shallow ponds; the large plants can reduce water flow and increase sedimentation in canals; *Nuphar polysepala* reportedly is problematic in parts of British Columbia; **nonindigenous species:** none in North America

Systematics: morphological and molecular data resolve *Nuphar* as the sister genus to the remainder of Nymphaeaceae, closest to *Barclaya* (Figure 1.3). Phylogenetic analyses of *Nuphar* (Figure 1.4) also indicate two fundamental clades (recognized taxonomically as sections), corresponding essentially to taxa distributed in the Old World (section *Nuphar*) or New World (section *Astylus*). One exception is *N. microphylla*, a New World species derived from the otherwise Old World clade. *Nuphar* taxonomy has been applied inconsistently. Earlier taxonomic schemes treated all North American taxa

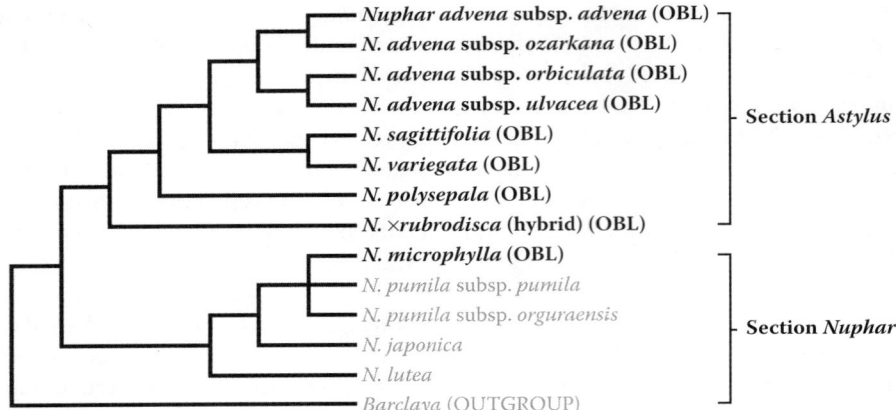

FIGURE 1.4   Interspecific relationships in *Nuphar* as evidenced by phylogenetic analysis of combined molecular and morphological data. All *Nuphar* species are aquatic plants, with OBL North American indicators (highlighted in bold) occurring within two different sections. *Nuphar microphylla* is the only North American representative of the otherwise Old World section *Nuphar*. (Adapted from Padgett, D.J. et al., *Amer. J. Bot.*, 86, 1316–1324, 1999.)

as subspecies of the Old World *N. lutea*, which is inconsistent with the results of phylogenetic analyses. Some treatments recognize *N. orbiculata*, *N. ozarkana,* and *N. ulvacea* (regarded here as subspecies of *N. advena*) as distinct species. *Nuphar microphylla* has been treated as a subspecies of *N. pumila*; however, numerical analyses have shown these taxa to be similar but distinct. The system followed here recognizes five species, four subspecies of *N. advena*, and one spontaneous hybrid in North America. The base chromosome number of *Nuphar* is $x = 17$. The species are uniform cytologically, with *N. advena*, *N. microphylla*, *N. polysepala*, and *N. variegata* ($2n = 34$) all diploid. Clinal variation in the leaf shape of *N. sagittifolia* and *N. advena* is maintained through selection on the duration of seed vernalization (cold treatment). Molecular data indicate that *Nuphar microphylla* originated recently in the New World, where it hybridizes with *N. variegata* to produce the partially fertile taxon known as *N. ×rubrodisca* Morong (also an OBL indicator), which occurs within the parental zone of sympatry. *Nuphar ×interfluitans* Fernald is a putative natural hybrid between *N. advena* and *N. sagittifolia*. A complete chloroplast genome sequence has been published for *Nuphar advena*.

Comments: In North America, *N. variegata* is northern, *N. microphylla* is northeastern, *N. advena* is southeastern, and *N. polysepala* is western, in distribution; *N. sagittifolia* is restricted to the coastal plain of Virginia and the Carolinas.

References: Beal, 1956; Beal & Southall, 1977; Bodamer & Ostrofsky, 2010; Borkholder et al., 2002; Boyle et al., 2009; Capers et al., 2010; Carr et al., 1986; Catling et al., 2003; Christy, 2013; Coville, 1902; Cronin et al., 1998; DePoe & Beal, 1969; De Vos, 1958; Dorn et al., 2001; Eichler & Boylen, 2014; Glover & Floyd, 2004; Hart & Cox, 1995; Irvine & Trickett, 1953; Jeske et al., 1993; Judd, 1964b; Leck & Simpson, 1993; Les et al., 1999, 2003; Lippok et al., 2000; Marrotte et al., 2012; McCune, 2012; Miller & Standley, 1912; Moldowan et al., 2016; Muenscher, 1944; Newberry, 1887; Nichols, 1999; Nieuwland, 1916; Padgett, 1998, 1999,

2003, 2007; Padgett et al., 1998, 1999, 2010; Perleberg & Brown, 2004; Raubeson et al., 2007; Reddoch & Reddoch, 2005; Riemer, 1985; Roley, 2005; Schneider & Moore, 1977; Smith, 1929; Snyder et al., 2016; Strecker et al., 2014; Taylor, 2008; Titus & Sullivan, 2001; Tsuchiya, 1991; Wiersema & Hellquist, 1997; Wilson & McPherson, 1981; Yanovsky, 1936.

2. *Nymphaea*

Water lily; lis-de'eau, nénuphar blanc

Etymology: from the Greek *nymphe*, a mythical goddess of waters

Synonyms: *Castalia*; *Leuconymphaea*

Distribution: global: cosmopolitan; **North America:** widespread

Diversity: global: 35–40 species; **North America:** 10 species

Indicators (USA): OBL: *Nymphaea alba, N. ampla, N. capensis, N. elegans, N. jamesoniana, N. leibergii, N. lotus, N. mexicana, N. odorata, N. tetragona*

Habitat: freshwater; lacustrine, palustrine, riverine; **pH:** 4.5–9.9; **depth:** 0–3 m; **life-form(s):** floating-leaved

Key morphology: Rhizomes thick (to 8 cm), branched or unbranched, erect (to 7 cm) or horizontal (to 120 cm), stolons (to 1 m) present or absent; leaves floating, the blades (to 45 cm) orbicular to elliptic, basally lobed, the margins entire to dentate, palmately dichotomous veined; flowers solitary, showy (to 25 cm), diurnal or nocturnal, floating or emergent (to 18 cm) from elongate peduncles (to 15 dm); sepals (to 9.4 cm) 4, greenish; petals (to 8.4 cm) 8-numerous, blue, pink, white, or yellow, grading into the stamens; stamens (to 4.2 cm) numerous; ovary (to 33-locular) with prominent stigmatic disk having marginal, upwardly incurved appendages (to 12 mm); fruits (to 8 cm) globose, attached to long, curved or coiled peduncles; seeds (to 6 mm) ellipsoid, smooth or papillate (to 220 μm), with a spongy aril

Life history: duration: perennial (rhizomes, stolons); **asexual reproduction:** stolons, rhizomes, tubers, adventitious plantlets; **pollination:** insect, self; **sexual condition:**

hermaphroditic; **fruit:** berries (common); **local dispersal:** tubers, seeds (water), stolons; **long-distance dispersal:** seeds (waterfowl [Aves: Anatidae])

Imperilment: (1) *Nymphaea alba* [GNR]; (2) *N. ampla* [G5]; (3) *N. capensis* [GNR]; (4) *N. elegans* [G4]; S3 (LA); (5) *N. jamesoniana* [G5]; S2 (FL); (6) *N. leibergii* [G5]; SH (ID); S1 (AB, ME, MI, MT, VT); S2 (BC, MN, SK); S3 (QC); (7) *N. lotus* [GNR]; (8) *N. mexicana* [G3/G4]; SH (MS); S3 (LA); (9) *N. odorata* [G5]; SH (SD); S1 (AK, KS, MB, PE); S2 (IL, KS, MB, NE, NJ); S3 (IL, QC); (10) *N. tetragona* [G5]; SH (ID, WA); S1 (AB, YT); S2 (BC, MB); S3 (AK)

Ecology: general: All water lily species are obligate, aquatic perennials that comprise a dominant element of the floating-leaf zone, which occurs typically between the emergent and submersed zones in aquatic communities. They tolerate a wide range of habitats (tidal and nontidal margins of ponds and lakes, pools in swamps and marshes, canals, ditches, slow rivers, streams, and warm springs), but occur mainly in still or slow-moving water. Shading by dense stands of water lilies reduces algal growth and provides cover for bass and sunfish (Osteichthyes: Perciformes: Centrarchidae); dense masses of their floating leaves also effectively ameliorate habitats by dampening wave action. The solitary and showy flowers either float on the water surface or are raised above it on elongate peduncles. The flowers are self compatible, but typically are protogynous and predominantly outcrossed by insects (Insecta) including bees (Hymenoptera), beetles, (Coleoptera), and flies (Diptera: Syrphidae). Beetles are attracted to some night-blooming species by their fleshy appendages, which are relished as a food item. Floral cycles (diurnal or nocturnal) last from 2 to 7 days depending on the species. Stigmatic receptivity is indicated by the accumulation of fluid in the cuplike receptive surface of first-day flowers. The fluid is acidic (pH: 4.5–5.0), contains about 1%–1.5% sugars in tropical species, and up to 3% sugars in some temperate species. It retains full functionality after 60 days of refrigeration, or when diluted in water by 50%. Because the stigmatic fluid is highly similar to ordinary water, the pollination mechanism could be regarded as having a hydrophilous phase. Some autogamous pollination can occur in first-day flowers (normally in female-phase) if the anthers dehisce (homogamous species), or if foraging pollinators cause a premature release of pollen. Anther dehiscence normally occurs only in second-day and subsequent flowers, which are functionally staminate in most species and do not contribute to fruit production unless they were pollinated previously. The fruits are retracted (by bent or highly coiled peduncles) below the water surface where they mature. The fruits are fleshy berries, but "dehisce" by rotting or deterioration. As the seeds are released, they quickly float to the surface by means of a spongy, air-trapping aril, which confers buoyancy for one to several days. Subsequent dispersal occurs mainly on the water surface. Longer distance dispersal presumably occurs by the exozoic transport of seeds on the plumage or muddy feet of waterfowl (Aves: Anatidae) and other birds. Endozoic dispersal by fish or waterfowl has been reported in some species. The seeds of most tropical day- or night-blooming species can withstand drying, with some remaining viable after 16 years of open storage; whereas, those of many temperate species are typically highly susceptible to desiccation and will lose their ability of germination unless they are stored in water (e.g., at 3°C). All of the species arise from a rhizome that extends vertically or horizontally, and in some cases produces tubers. For species with horizontally disposed rhizomes, vegetative reproduction enables the plants to spread quickly within the floating-leaf zone where large, clonal populations can develop.

Nymphaea alba **L.** is a nonindigenous species that occurs in standing waters (0.5–2.0 m deep) of backwaters, lakes, and ponds at elevations to 457 m. Because of the few documented North American localities where this species is seen, relevant ecological information is quite limited. Although the plants have been found in acidic lakes, they can tolerate a fairly broad range of acidity (pH: 5.0–8.5) in their native range. They are intolerant of disturbance or turbidity and preferentially colonize fine-grained substrates including clay, marl, mud, peat, sand, and silt. Flowering (North America) occurs from May to October. The flowers open diurnally. They are weakly protogynous and capable of self-pollination or are outcrossed by insects (Insecta). The fruits mature underwater and then release the buoyant, arillate seeds that can remain afloat for up to 3 days while they are dispersed by water currents. The seeds are not adapted for endozoic dispersal and do not tolerate desiccation. They are recalcitrant and require a period of cold stratification (30 days) to induce germination (achieved at 20°C). Vegetative reproduction occurs by means of the elongate, horizontal rhizomes. Populations in the indigenous range tend to be fairly species poor, with no constant associates. **Reported associates (North America):** *Potamogeton confervoides.*

Nymphaea ampla **(Salisb.) DC.** grows in shallower sites (to 1 m deep) in nontidal or tidal, freshwater canals, ditches, and ponds at elevations to 350 m. Flowering and fruiting occur throughout the year. The flowers are raised above the water on rigid peduncles during diurnal anthesis (7:00 am to 4:00 pm), which repeats over 3–4 consecutive days. The flowers are not protogynous, have a low pollen:ovule ratio (e.g., 195:8), and are primarily autogamous. The anthers dehisce while in bud, and the stigmas are receptive from the first day of anthesis, which promotes self-pollination facilitated by foraging bees (e.g., Insecta: Hymeoptera: Apidae: *Apis mellifera*). Fruit set is high (>96%; with >98% seed set), but is reduced in outcrossed plants, indicating the possibility of outbreeding depression. The fresh seeds are not dormant, germinate readily, and do not persist in a seed bank. They possess a large, buoyant aril and are dispersed primarily by water. **Reported associates:** *Hydrilla verticillata, Typha.*

Nymphaea capensis **Thunb.** is a nonindigenous species that occurs in shallow waters (e.g., 0.6 m deep) of ditches, ponds, roadsides, and along the margins of canals and swamps at elevations to 100 m. In North America, the plants usually occur on sandy substrates. Flowering (North America) has been observed nearly year-round (May to February). The flowers are fragrant (evocative of raspberries or strawberries) and extend above the water surface during anthesis. Unlike many of its congeners, they are not protogynous. The flowers are self-compatible and primarily autogamous. Anthesis is diurnal and extends over three consecutive days. In first day flowers, the outer anthers dehisce while the stigmas are receptive. The stamens turn outward in second-day flowers as the stigmatic surfaces become dry. All stamens dehisce in the third-day flowers. On the fourth day, the developing fruits are retracted underwater by the coiling of the peduncle. Seed production is prolific, with mature fruits containing an average of 1424 seeds; however, these are produced only by the flowers pollinated on the first day of anthesis (the stigma no longer being receptive by the second day). The fresh, mature seeds germinate immediately or after a period of drying. **Reported associates (North America):** *Sagittaria montevidensis, Taxodium, Typha.*

Nymphaea elegans **Hook.** grows in clear, shallow (to 1.2 m deep), standing, freshwater (salinity <3 ppt) sites associated with borrow pits, canals, culverts, ditches, glades, hammocks, lakes, marshes (depression), ponds, pools, prairies, sloughs, streams (quiet), swales, and swamps at elevations to 150 m. Flowering occurs from March to October, with fruiting observed from July to September. The flowering cycle extends for three consecutive days, with repeated diurnal opening and closure. The first-day flowers (opening from 9:00 am to 2:00 pm) are protogynous, with their receptive stigmatic surface secreting a fluid into a cuplike area. Potential pollinators entering the flowers land on the flexible (but undehisced) inner stamens, which bend, causing the pollinators to drop into the fluid (drowning many in the process) where any foreign pollen is washed off. Compatible pollination occurs at this time. The stigmas dry and lose receptivity before the second-day opening (~8:30 am) of the flowers (no longer a threat to floral visitors), which are functionally staminate during the remainder of the cycle. The seeds will germinate immediately or after a period of drying. The shoots arise from an unbranched, vertically oriented rhizome, and most reproduction appears to occur by seed. The plants have a low carbon-to-nitrogen ratio, and their floating leaves decompose rapidly, at a rate faster than that observed for the foliar decomposition of other hydrophytes. **Reported associates:** *Azolla caroliniana, Bacopa caroliniana, Bidens, Brasenia schreberi, Cabomba caroliniana, Cephalanthus occidentalis, Ceratophyllum demersum, C. echinatum, C. muricatum, Chara, Chloracantha spinosa, Cornus, Crinum americanum, Cyperus articulatus, Echinodorus berteroi, Eichhornia crassipes, Eleocharis macrostachya, Eupatorium, Heteranthera dubia, Hibiscus moscheutos, Hydrocotyle verticillata, Iris virginica, Landoltia punctata, Leersia hexandra, Lemna minor, Limnobium spongia, Ludwigia decurrens, L. peploides, Luziola, Myrica, Myriophyllum aquaticum, M. spicatum, Najas guadalupensis, Nelumbo lutea, Neptunia plena, Nuphar advena, Nymphaea lotus, N. mexicana, N. odorata, Nymphoides aquatica, Ottelia alismoides, Panicum hemitomon, Paspalidium geminatum, Persicaria hydropiperoides, P. punctata, Phragmites australis, Physalis pubescens, Pontederia cordata, Prosopis glandulosa, Sabal palmetto, Sagittaria lancifolia, S. latifolia, S. longiloba, S. subulata,* *Salix, Salvinia minima, Schoenoplectus californicus, Sesbania drummondii, Sideroxylon, Taxodium, Typha domingensis, T. latifolia, Utricularia foliosa, Wolffiella lingulata.*

Nymphaea jamesoniana **Planch.** inhabits shallow (e.g., 0.5 m deep), still, temporary waters associated with canals, ditches, flatwoods, marshes, ponds (retention), and sloughs at elevations to 100 m. Flowering occurs from August to November. The flowers float on the water surface and open nocturnally from 7:00 pm to 1:00 am. Although protogynous (receptive stigmas but no pollen dehiscence occurring in first day flowers), the stigmas remain receptive during the second day, when self pollen is released. However, the flowers often fail to open during the first day of the floral cycle, resulting essentially in a homogamous condition during the second day (receptive stigmas and dehisced pollen). These features are interpreted as adaptations for autogamy. Details on the seed ecology are unavailable. The arillate seeds are dispersed by water. The plants reproduce vegetatively by the rhizomes, which produce tubers that are buried 5–10 cm below the substrate surface. **Reported associates:** *Bacopa monnieri, Hydrocotyle umbellata, Panicum repens, Persicaria punctata, Pontederia, Symphyotrichum subulatum.*

Nymphaea leibergii **Morong** grows in clear, cool standing waters (to 2.5 m deep) along the shores of lakes, ponds, rivers (bays), sloughs, and streams (slow-moving) at elevations to 1000 m. The plants often occur near inlets in soft water, acidic to neutral sites (average pH: 7.1; total alkalinity: 20 mg/L $CaCO_3$) on substrates described as mud, peat, and sand. Flowering and fruiting have been observed from July to August. The floral cycle is long, lasting from five to seven consecutive days. Flowering each day is diurnal, with the buds opening early in the afternoon and closing by early evening (e.g., 6:00 pm). The flowers are protogynous and are incapable of spontaneous autogamy. The stigmas of first-day flowers exude a fluid and are receptive while the stamens remain undehisced. Pollen production begins by the third day of the floral cycle. Pollination is carried out by flies (Insecta: Diptera: Ephydridae: *Notiphila shewelli*) and other insects. The developing fruits are pulled beneath the water surface to ripen by the coiling of the peduncle; the fruits mature in 3–4 weeks. The fruits rupture when ripe and release the buoyant seeds that float on the water surface assisted by an encircling aril. Once the aril deteriorates, the large seeds sink to the bottom. The mechanism of long-distance dispersal is uncertain. Propagation relies exclusively on the seed, which is deposited in the vicinity of the maternal plants. The seeds are dormant and require a period of cold stratification to induce germination. The plants produce erect tubers, but rarely any offsets, and do not exhibit clonal reproduction. **Reported associates:** *Acorus calamus, Ceratophyllum demersum, Equisetum, Eriocaulon aquaticum, Isoetes echinospora, Lobelia dortmanna, Myriophyllum spicatum, M. verticillatum, Nuphar variegata, Nymphaea odorata, Potamogeton crispus, P. epihydrus, P. natans, P. richardsonii, P. spirillus, Sagittaria cuneata, Schoenoplectus subterminalis, Sparganium fluctuans, Utricularia macrorhiza, Zizania aquatica.*

Nymphaea lotus **L.** is a nonindigenous species, which occurs in canals, ditches, ponds, and swamps at elevations to 160 m. Little information exists on the habitats occupied in the nonindigenous North American portion of its range. Flowering (North America) has been observed from September to October. The flowers are self-compatible but protogynous and incapable of spontaneous autogamy. They open nocturnally (opening in the evening and closing by morning) each day over a 4–5-day cycle. The following has been summarized from observations made in the native (African) range of the species. Only about half of the flowers that completely emerge above the water surface open during the first day. The first-day flowers open around 8:22 pm but delay their opening until 10:00 pm on subsequent days. They open less widely, close earlier, and remain closed for a longer period (2–8 h vs. 1.5–6 h) during the first day. The first-day flowers secrete a fluid on the receptive stigmatic disk, while the stamens remain undehisced. The stigmas remain receptive for 15–17 h (as the fluid dries up), with the male phase (release of pollen) beginning about an hour afterwards on the second day. The styles cover the stigmatic disk thoroughly during the second day of anthesis. The thermogenic flowers attract beetles (Insecta: Coleoptera: Dynastidae: *Ruteloryctes morio*), which feed on the floral parts. However, the beetles are less efficient pollinators than bees (Insecta: Hymenoptera: Apidae; Halictidae), which visit the flowers during the early morning hours (5:40–7:00 am). The pollen-laden insects often slip from the stamens and into the stigmatic fluid, where their pollen is washed off and deposited. Some of the insects drown during this activity. Fruit production from pollinated first-day flowers is virtualy 100%. The fruits mature within 25–30 days, either on or just below the water surface. The deterioration of mature fruits eventually releases the seeds, which float on the water surface for 12–36 h by means of their buoyant arils. Freshly collected seeds reportedly will germinate without pretreatment when placed on moist filter paper, or in beakers under 10 cm of water; germination begins within 20–25 days after the mucilaginous seed coats have swelled with imbibed water. The seeds also germinate well following a period of drying or after wet and dry or dry pretreatments. They are known to persist in the seedbank. The plants reproduce vegetatively by branched or unbranched rhizomes and slender stolons. Ethanolic extracts of the rhizomes inhibit seed germination and seedling growth in cultivated rice (*Oryza*) and might reduce interspecific competition through such allelopathic effects. **Reported associates (North America):** *Eichhornia crassipes, Nuphar, Nymphaea elegans, N. odorata.*

Nymphaea mexicana **Zucc.** inhabits standing waters (to 1.8 m deep) in backwaters, canals, ditches, gravel pits, lagoons, lakes, mudflats, ponds, pools, prairies, rivers, sloughs, springheads, springs, and streams (sluggish) at elevations to 2000 m. The waters range from slightly acidic to alkaline (pH: 6.0–8.7), and the substrates have been described as marl and mud. Flowering extends from March to September. Anthesis is diurnal over a cycle of two consecutive days. The species is protogynous and is pollinated by insects (Insecta). The first-day flowers open from 11:00 am to 4:00 pm; the second-day flowers open somewhat earlier (10:30 am). The first-day flowers are functionally female and secrete stigmatic fluid (3%–4% total dissolved solids) upon the receptive stigmatic disk. The flowers are visited by several potential insect pollinators including bees (Hymenoptera: Halictidae: *Lasioglossum*), beetles (Coleoptera), and flies (Diptera), which land on the wet stigmas where the incoming pollen is washed off (including any conspecific grains), thereby enabling cross-pollination to occur. In second-day flowers, the stigmatic fluid dries up, the stigmas lose receptivity, and the anthers begin to dehisce and release their pollen. At the end of the floral cycle, the flowers close and are pulled beneath the water surface where the fruits mature. When fully mature, the fruits burst open to release up to 40 buoyant seeds, which float on the surface by virtue of their air-trapping arils. The requirements (if any) for seed germination have not been established. Vegetative reproduction occurs by means of banana-like tubers, which form at the tips of elongate, overwintering stolons. The structures can overwinter while buried in up to 20 cm of mud. **Reported associates:** *Cabomba caroliniana, Cephalanthus occidentalis, Ceratophyllum demersum, Chara, Cladophora, Egeria densa, Eichhornia crassipes, Enteromorpha compressa, Fraxinus caroliniana, Heteranthera, Hydrilla verticillata, Hygrophila, Isoetes flaccida, Limnobium, Ludwigia repens, Myriophyllum aquaticum, M. spicatum, Najas, N. guadalupensis, Nelumbo lutea, Paspalum vaginatum, Phyla lanceolata, Pistia stratiotes, Quercus virginiana, Ruppia maritima, Sabal palmetto, Sagittaria kurziana, S. subulata, Schoenoplectus americanus, S. californicus, Styrax, Taxodium, Typha domingensis, T. latifolia, Vallisneria americana, Zannichellia palustris.*

Nymphaea odorata **Aiton** occurs in fresh to slightly brackish, standing waters (to 2.9 m deep) of bayous, bogs, borrow pits, canals, Carolina bays, clay pits, depressions, ditches, flatwoods, floodplains, lagoons, lakes, marshes, ponds, pools, prairie fens, prairies, reservoirs, sand prairies, sinks (limestone), sloughs, streams, swamps, and along the margins of rivers at elevations of up to 2438 m. The plants are found most often in open areas receiving full sunlight but occur occasionally in partial shade. The waters span a wide range of environmental conditions (pH: 4.5–9.9; alkalinity: 2.5–277.5 mg/L; conductivity: 4–450 µmhos/cm). The sediments are often characterized as organic and include clay, Dorovan soils (typic medisaprists), gravel, humus, loam, marl, muck, mud, ooze, peat, Rutlege loamy sand, Samsula–Myakka soils, sand, silt, and silty sand. Flowering occurs from March to October. The flowers are protogynous and open diurnally over a successive 3-day cycle. Flower buds newly emerging from the water secrete a fluid that covers the stigmatic disk a day before the flowers open. The fluid consists of about 3% sugars, mainly sucrose, with lesser but equal parts of glucose and fructose. The first-day flowers open from 7:00 am to 1:00 pm with a receptive stigmatic surface but undehisced anthers. Pollinators visiting during this phase often fall from the flexible stamens into the stigmatic fluid (some drowning), which washes off any adhering pollen. The containment of conspecific pollen by the fluid facilitates outcrossing by bees

(Insecta: Hymenoptera: *Lasioglossum versatum*) and other insects (see *Use by wildlife*). In second-day flowers (opening from 6:30 am to 4:00 pm), the stigmatic fluid disappears, the stigmas are no longer receptive, and the anthers subsequently dehisce and release self-pollen. Spontaneous self-pollination does not occur. The flowers remain closed and submerge after the third day of anthesis by coiling of the peduncle, enabling the fruits to mature underwater. The peduncles remain uncoiled in unpollinated flowers. The fruits enlarge within a week after pollination. When mature, the stigmatic area detaches to release the seeds, which float as a consequence of their buoyant, air-filled arils. The liberated seeds can remain afloat for hours as they are dispersed on the surface by water and wind currents. They eventually lose buoyancy and sink to the bottom within a few days. The seeds are also reportedly dispersed endozoically by waterfowl (Aves: Anatidae) including gadwalls (*Anas strepera*) and wood ducks (*Aix sponsa*). Fresh seeds can germinate immediately, but quickly become dormant when subjected to cold water, drying, or freezing. They then require a period of cold stratification (150–210 days at 4.4°C) for germination at room temperature or under a 19°C/15°C day/night temperature regime. Germination is also reportedly enhanced in the light or when the seeds are crowded. The seeds are intolerant of drying, with their germination rates declining after 3 h of dry storage and becoming negligible after 24 h of drying. However, viable seeds can be preserved by thoroughly air-drying the fruits before they dehisce, and then storing them out of direct sunlight. The seeds can reach densities of up to 400/m^2 and are able to remain viable in the sediments for at least 2.5 years. They retain good viability (>90%) after 3 years of storage in water at 1°C–3°C, but their viability declines by their fifth year of storage. Germination is reportedly higher on saturated rather than inundated sediments; however, other studies have found that the seedlings emerge most abundantly from continuously inundated substrates, but germinate at lower levels in moist sites without standing water, or at sites where fluctuating noninundated/inundated cycles occur. The plants reproduce vegetatively by the production of elongate, horizontal rhizomes. Some populations also develop tubers, whose detachment facilitates asexual reproduction and local dispersal. Plants grown in shallow water (30 cm) produce many, small, short-lived leaves compared to those that grow at greater depths (60–90 cm). Deeper water plants allocate more bioamss to their leaves and roots; whereas, shallow water plants allocate greater biomass to their rhizomes. In some populations, the leaves have been observed to occur on the water surface early in the growing season, change to aerial leaves during midseason, and then revert back to floating leaves at the end of the season, a cycle interpreted as a response to varying light regimes. **Reported associates:** *Alisma, Alopecurus, Alternanthera philoxeroides, Amaranthus australis, Azolla, Bacopa monnieri, Beckmannia, Bidens mitis, Brasenia schreberi, Cabomba caroliniana, Carex aquatilis, C. crinita, C. cusickii, C. lasiocarpa, C. longii, C. rostrata, C. stipata, Carya, Centella asiatica, Cephalanthus occidentalis, Ceratophyllum demersum, C. echinatum, C. muricatum,*

Chara, Cicuta douglasii, Cladium jamaicense, Comarum palustre, Cyperus odoratus, Cyrilla racemiflora, Diodella teres, Dulichium arundinaceum, Echinodorus berteroi, Egeria densa, Eichhornia crassipes, Eleocharis acicularis, E. palustris, Elodea canadensis, E. nuttallii, Equisetum fluviatile, Eriocaulon aquaticum, Fraxinus pennsylvanica, Fuirena scirpoidea, Heteranthera dubia, H. reniformis, Hippuris vulgaris, Hydrilla verticillata, Hydrocotyle umbellata, H. verticillata, Ipomoea sagittata, Juncus balticus, J. roemerianus, Leersia, Lemna minor, L. trisulca, Liquidambar styraciflua, Lobelia dortmanna, Ludwigia, Luziola, Lythrum lineare, Magnolia virginiana, Menyanthes trifoliata, Mikania scandens, Myrica cerifera, Myriophyllum heterophyllum, M. humile, M. sibiricum, M. spicatum, M. tenellum, M. verticillatum, Najas flexilis, N. guadalupensis, N. minor, Nelumbo lutea, Nitella, Nuphar polysepala. N. variegata, Nymphoides aquatica, N. cordata, Nyssa, Orontium aquaticum, Panicum dichotomiflorum, P. hemitomon, P. virgatum, Peltandra virginica, Persicaria amphibia, P. hydropiperoides, P. punctata, Phalaris arundinacea, Phragmites australis, Pinus taeda, Polypogon monspeliensis, Pontederia cordata, Potamogeton amplifolius, P. bicupulatus, P. confervoides, P. diversifolius, P. epihydrus, P. foliosus, P. gramineus, P. natans, P. perfoliatus, P. praelongus, P. pulcher, P. pusillus, P. robbinsii, P. zosteriformis, Proserpinaca palustris, Quercus, Ranunculus repens, Rhynchospora elliottii, R. fascicularis, R. inundata, R. tracyi, Rubus argutus, Rumex crispus, Sacciolepis striata, Sagittaria brevirostra, S. lancifolia, S. latifolia, S. subulata, Salix exigua, Schoenoplectus acutus, S. subterminalis, S. tabernaemontani, Sisyrinchium atlanticum, Sium suave, Smilax laurifolia, S. walteri, Sparganium angustifolium, S. emersum, S. eurycarpum, Sphagnum macrophyllum, Spirodela polyrhiza, Stuckenia pectinata, Taxodium distichum, Typha angustifolia, T. domingensis, T. latifolia, Urtica dioica, Utricularia gibba, U. intermedia, U. macrorhiza, U. purpurea, U. radiata, Vallisneria americana, Veronica, Vicia sativa, Viola lanceolata, Wolffia brasiliensis, Woodwardia virginica.

Nymphaea tetragona **Georgi** inhabits fresh, standing waters (to 2.5 m deep) in bogs, depressions, lakes, marshes, muskeg, oxbows, ponds, sloughs, and streams at elevations to 1067 m. The waters are characterized as moderately alkaline (pH: 7.9–8.5; total CaCO$_3$ alkalinity: 60–225 mg/L; conductivity: 125–510 µmhos/cm). The substrates include gravel, gravelly rock, mud, sand, sandy gravelly silt, and sandy silty muck. Flowering occurs from July to August, with fruiting in August. The flowers are not fragrant and open from 11:00 am to 5:00 pm, which is later than most of the diurnal congeners. They are pollinated by flies (Insecta: Diptera: Epihydridae: *Notiphila shewelli*). Genetic studies in Eurasia indicate low levels of genetic diversity within populations and high genetic differentiation among populations. Although insect-pollinated, the plants experience high levels of biparental inbreeding due to their depauperate gene pool. The fruits develop while attached to the maternal plant. As they decay, the seeds float along the water surface by means of their buoyant, spongy arils; they remain afloat for up to a day. The seeds are eaten by fish (Vertebrata: Osteichthyes), which have been suggested as potential long-distance dispersal agents. The seeds will germinate only while underwater and are intolerant of desiccation. The leaves live for 31 days on average. The maximum seasonal foliar biomass has been estmated at 173.2 g dry wt./m². The roots and rhizomes comprise about 60% of the summer biomass. They do not branch or proliferate, but can live for more than 5 years. In some areas (especially in shallow water sites), the plants behave like annuals (having a life cycle of 1 year). **Reported associates:** *Artemisia tilesii, Azolla cristata, Brasenia scherberi, Calla palustris, Callitriche, Caltha, Carex canescens, C. rostrata, C. utriculata, Chara, Cicuta virosa, Comarum palustre, Eleocharis palustris, Equisetum arvense, Glyceria, Isoetes echinospora, Lemna minor, L. trisulca, Lupinus polyphyllus, Myrica gale, Myriophyllum spicatum, M. verticillatum, Najas flexilis, Nuphar polysepala, Persicaria amphibia, Potamogeton epihydrus, P. gramineus, P. natans, P. obtusifolius, P. perfoliatus, P. praelongus, P. pusillus, P. richardsonii, P. robbinsii, P. zosteriformis, Ranunculus aquatilis, Ruppia maritima, Sparganium angustifolium, S. hyperboreum, Sphagnum, Spirodela polyrhiza, Stuckenia pectinata, S. vaginata, Subularia aquatica, Typha, Utricularia intermedia, U. macrorhiza, Zannichellia palustris.*

Use by wildlife: Water lily leaves ("lily pads"), especially their undersides, harbor numerous organisms including bristle-worms (Annelida: Polychaeta), bryozoa (Bryozoa), clams (Mollusca: Bivalvia), flatworms (Platyhelminthes: Planariidae: *Planaria*), frogs (Amphibia: Anura), hydras (Cnidaria: Hydrozoa: Hydridae), insects (Arthropoda: Insecta), mites (Arthropoda: Arachnida: Acari), ostracods (Arthropoda: Crustacea: Ostracoda), protozoans (Protozoa), rotifers (Rotifera), snails (Mollusca: Gastropoda), and freshwater sponges (Porifera: Spongillina). *Nymphaea* fruits mature under water where the seeds are released or sought by diving birds. Water lily seeds are eaten by 17 species of ducks (Aves: Anatidae), including blue-winged teal (*Anas discors*), scaup (*Aythya*), and wood ducks (*Aix sponsa*), as well as by other animals. The tubers are eaten by wild hogs (Mammalia: Suidae: *Sus scrofa*) and deer (Mammalia: Cervidae: *Odocoileus*). Muskrats (Mammalia: Cricetidae: *Ondatra zibethicus*) feed on water lily roots and tubers and use various parts of the plants for constructing shelters. The leaves and roots are also eaten by beavers (Mammalia: Castoridae: *Castor canadensis*), moose (Mammalia: Cervidae: *Alces americanus*), porcupines (Mammalia: Rodentia: Erethizontidae: *Erethizon dorsatus*), and deer (Mammalia: Cervidae: *Odocoileus*); snails (Mollusca: Gastropoda: *Lymnaea, Physa*) and tadpoles (Amphibia: Anura) graze on the algae that cover the foliage. The leaves are damaged by various insects (Insecta) including aphids (Hemiptera: Aphididae: *Rhopalosiphum*), beetles (Coleoptera: Chrysomelidae: *Donacia, Galerucella, Neohaemonia, Pyrrhalta*; Curculionidae: *Bagous, Sitophilus*; Erirhinidae: *Lissorhoptrus*; Haliplidae: *Apteraliplus, Brychius, Haliplus, Peltodytes*; Hydrophilidae: *Berosus*), caddisflies (Trichoptera: Hydroptilidae: *Oxyethira*; Leptoceridae:

Triaenodes; Limnephilidae: *Halesus, Limnephilus*; Phryganeidae: *Phryganea*; Polycentropodidae: *Holocentropus*), caterpillars (Lepidoptera: Crambidae: *Munroessa gyralis, Nymphula, Paraponyx maculalis, Synclita occidentalis*; Noctuidae: *Bellura gortynoides*), mining fly larvae (Insecta: Diptera: Scathophagidae: *Hydromyza*), and also by coccus bacteria (Bacteria), fungi (Ascomycota: Tubeufiaceae: *Helicosporium*; Basidiomycota: Pucciniaceae: *Puccinia*; Uropyxidaceae: *Aecidium*; Oomycota: Pythiaceae: *Pythium*), and snails (Mollusca: Gastropoda: Lymnaeidae: *Lymnaea*). Several birds (Aves: Rallidae) including the American coot (*Fulica americana*), purple gallinule (*Porphyrio martinica*), and moorhen (*Gallinula*) often are seen walking on the floating leaves of water-lilies, which is a tactic used to escape predators. *Nymphaea elegans* is eaten by nutria (Mammalia: Mycocastoridae: *Myocastor coypus*) and sandhill cranes (Aves: Gruidae: *Grus canadensis*); its seeds are eaten by several ducks (Aves: Anatidae). The tubers of *N. mexicana* are an important food for canvasback ducks (Aves: Anatidae: *Aythya valisineria*) and ring-necked ducks (*Aythya collaris*). The seeds are eaten by lesser scaups (Aves: Anatidae: *Aythya affinis*). The leaves are damaged or eaten by aphids (Insecta: Hemiptera: Aphididae: *Rhopalosiphum*), beetles (Insecta: Coleoptera: Chrysomelidae: *Donacia cincticornis*), and caterpillars (Insecta: Lepidoptera: Crambidae: *Synclita obliteralis*). The foliage is used for egg deposition by odonates (Insecta: Odonata). *Nymphaea odorata* is a preferred food of beavers (Mammalia: Castoridae: *Castor canadensis*). The leaves host more than 30 species of fungi. They are eaten by apple snails (Mollusca: Gastropoda: Ampullariidae: *Marisa cornuarietis, Pomacea canaliculata, P. diffusa, P. haustrum, P. insularum, P. paludosa*) and by various insects (Insecta) including aphids (Hemiptera: Aphididae: *Rhopalosiphum nymphaeae*), beetles (Coleoptera: Chrysomelidae: *Donacia cincticornis, D. pubescens, Galerucella nymphaea*; Coccinellidae: *Coleomegilla maculata, Hippodamia tredecimpunctata tibialis*; Curculionidae: *Bagous americanus, B. magister, B. tanneri, Onychylis nigrirostris*), caddisflies (Trichoptera: Leptoceridae: *Mystacides longicornis, Oecetis, Triaenodes abus, T. injustus, T. marginatus*; Polycentropodidae: *Plectrocnemia remota*), flies (Diptera: Chironomidae: *Polypedilum braseniae, P. illinoense, Hyporhygma quadripunctatus*; Ephydridae: *Notiphila loewi*), larval moths (Lepidoptera: Crambidae: *Elophila gyralis, Langessa nomophilalis, Munroessa gyralis, M. icciusalis, Paraponyx allionealis, P. badiusalis, P. maculalis, P. obscuralis, P. seminealis, Synclita obliteralis*; Noctuidae: *Bellura vulnifica*), and plant hoppers (Hemiptera: Delphacidae: *Megamelus davisi*). The flowers are visited by several insects (Insecta) including bees (Hymenoptera: Apidae: *Apis mellifera*; Halictidae: *Lasioglossum versatum*), beetles (Coleoptera: Chrysomelidae: *Diabrotic undecimpunctata howardi, Donacia piscatrix*; Cucurlionidae: *Bagous texanus*; Scarabaeidae: *Strigoderma arbicola*), and flies (Diptera: Syrphidae: *Allograpta*). The seeds are eaten by many waterfowl (Aves: Anatidae) including gadwalls

(*Anas strepera*), ring-necked ducks (*Aythya collaris*), ruddy ducks (*Oxyura jamaicensis*), and wood ducks (*Aix sponsa*). The leaves of *N. tetragona* are eaten by leaf beetles (Insecta: Coleoptera: Chrysomelidae: *Galerucella nymphaeae*); the rhizomes are eaten by muskrats (Mammalia: Cricetidae: *Ondatra zibethicus*). The plants are fed to cattle (Mammalia: Bovidae: *Bos taurus*) and sheep (Mammalia: Bovidae: *Ovis aries*) in India.

Economic importance: food: The rootstocks of *N. capensis* are eaten as a food in Madagascar, and purportedly are delicious in taste. The boiled or raw inner rhizomes of *N. jamesoniana* were eaten by the Chorote Indians of Argentina. The rhizomes and tubers of *N. lotus* reputedly are sweet in taste and are eaten in Africa and Asia. The Egyptians ate the fruits in salads and ground the seeds into meal for breadmaking. Native North Americans consumed many parts of *N. odorata* including the young leaves, unopened flower buds (Ojibwa), seeds (whole or ground into flour), and tubers. The leaves are fairly high in tannins, but contain about 40% crude protein and are capable of yielding from 121 to 197 kg of crude protein per hectare. The leaf buds and seeds of *N. tetragona* are eaten in Japan; **medicinal:** The ethanolic extracts of *N. alba* flowers exhibit significant anti-inflammatory activity in laboratory animals. The rhizomes contain tannins and other phenolic compounds that are believed to be responsible for its antioxidant activity and analgesic properties. The flowers and other parts of *N. ampla* contain aporphine and quinolizidine alkaloids, which possess narcotic properties. Methanolic extracts of *N. capensis* leaves are cytotoxic and have sedative effects on laboratory animals. *Nymphaea jamesoniana* has been used as an astringent and to treat dysentery, irritated eyes, and skin lesions. Its flowers are allegedly narcotic. *Nymphaea lotus* has been used in traditional medicine as a source of analgesics, anti-inflammatory agents, astringents, and sedatives. The plants contain various constituents that exhibit strong antioxidant properties. The ethanolic extracts are highly inhibitory to several gram-negative and gram-positive bacteria (Enterobacteriaceae: *Escherichia coli*; Staphylococcaceae: *Staphylococcus aureus*; Streptococcaceae: *Streptococcus pyogenes*). Aqueous extracts contain potentially therapeutic ingredients that are protective against gastric ulcers in laboratory animals. The flowers of *N. mexicana* contain naringenin, which has been shown to possess anti-inflammatory properties. The plant extracts are antioxidants and exhibit DNA protective properties and antiproliferative activity. Leaf extracts of *N. odorata* have exhibited antidiabetic effects in diabetic laboratory animals. The plants were used in various medical applications by several indigenous North American tribes including the Chippewa (for mouth sores), Micmac (to remedy colds, coughs, grippe, and swelling), Ojibwa (to treat coughs and tuberculosis), Okanagan–Colville (toothache remedy), Penobscot (for swollen limbs), and Potawatomi (as a general medicinal). A tea made from root extracts has been used to treat coughs, diarrhea, and sore throats. The leaves and roots of *N. tetragona* contain geraniin, a hydrolyzable tannin, which is inhibitory to fish (Vertebrata: Osteichthyes) and microbes (Bacteria: Proteobacteria: Aeromonadaceae:

Aeromonas salmonicida; Pseudomonadaceae: *Pseudomonas fluorescens*). The plant extracts also are potentially effective against infections of drug-resistant *Salmonella* (Bacteria: Proteobacteria: Enterobacteriaceae). All parts of the plants are used in the traditional medicine of China and Siberia as analgesics, anti-inflammatories, antipyretics, and to treat urinogenital and respiratory diseases; **cultivation:** Hundreds of cultivars (too many to enumerate here) have been developed among the various *Nymphaea* species (see Knotts & Sacher, 2000), and these have been of major economic importance as water garden ornamentals for centuries. Archaeological evidence suggests that *Nymphaea* was grown in water gardens by the Egyptians more than 5000 years ago for use in various religious and ceremonial activities. In the New World, water lily flowers (e.g., *N. ampla*) were an important artistic motif used by the Mayan culture more than 2500 years ago. Europeans began to cultivate water lilies early in the 18th century. *Nymphaea odorata* was introduced into cultivation in Europe in 1786; **misc. products:** The Ojibway people used *Nymphaea* leaves to wrap their foods for cooking. *Nymphaea capensis* has been used in Australia as a biomonitor of acid sulfate soil drainage waters. Several water lily species have been considered for use in the remediation of variously polluted waters; **weeds:** *Nymphaea mexicana* is problematic in waterways and is listed as a noxious weed in California; *N. odorata* is troublesome in parts of the Pacific Northwestern United States, especially in shallow waters. *Nymphaea lotus* is a serious weed of African ricefields. *Nymphaea* ×*thiona*, a sterile hybrid between *N. mexicana* and *N. odorata*, is vigorous and forms extensive clones in Kentucky and Nevada; **nonindigenous species:** Many water lilies have been introduced as escapes from cultivation of horticultural specimens in water gardens. The European *Nymphaea alba* was introduced to British Columbia and California before 1960, to New Hampshire before 1972, and to Rhode Island; it also has been introduced to Luxembourg and Portugal. *Nymphaea capensis* and *N. lotus* were introduced to Florida and Louisiana; *N. lotus* also was introduced to Hungary and Romania. *Nymphaea mexicana* is nonindigenous to British Columbia and North Carolina and was introduced to New Zealand before 1982 and to Spain before 1993. Cultivated specimens of *N. odorata* were planted intentionally in Newman Lake, Washington, around 1931. The western localities (Alaska, Arizona, British Columbia, California, Colorado, Idaho, Montana, Nevada, New Mexico, Oregon, Utah, Washington) are all believed to represent introductions, as well as some eastern sites (Massachusetts, West Virginia).

Systematics: Cladistic analyses of morphological and molecular data show *Nymphaea* to comprise a relatively derived clade in Nymphaeaceae, which resolves as the sister group to a clade containing *Euryale* and *Victoria* (Figures 1.3 and 1.5). Broad phylogenetic surveys of the genus have provided sufficient evidence to merge the former genus *Ondinea* within subgenus *Anecphya* of *Nymphaea* because of its nested placement within the latter group (Figure 1.5). Traditionally, *Nymphaea* has been subdivided into five subgenera (*Anecphya, Brachyceras, Hydrocallis, Lotos, Nymphaea*).

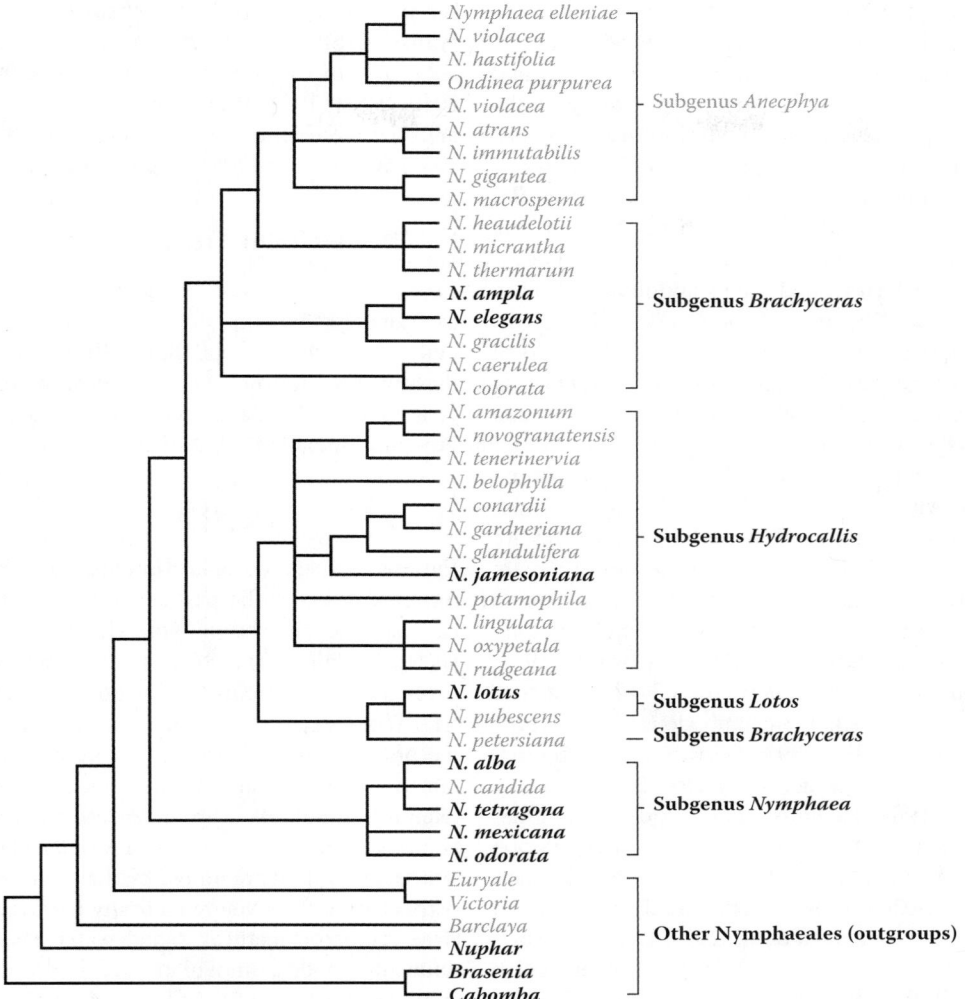

FIGURE 1.5 Interspecific relationships in *Nymphaea* as indicated by analysis of *trnT-trnF* sequence data. All subgenera of *Nymphaea* resolve as distinct clades except for *Brachyceras*. The nested placement of the former genus *Ondinea* within an Australian cluster of *Nymphaea* species (subgenus *Anecphya*) supports its merger with the latter. Although the entire genus is aquatic, the taxa containing OBL North American indicators have been highlighted in bold. (Adapted from Borsch, T. et al., *Int. J. Plant Sci.*, 168, 639–671, 2007.)

The retention of these widely recognized subgenera as monophyletic groups is supported phylogenetically (Figure 1.5) except for subgenus *Brachyceras*, which includes misplaced species and does not resolve as a clade that is distinct from subgenus *Anecphya* in some analyses. *Nymphaea ampla* has been subdivided into two species (*N. ampla*, *N. pulchella*), with the North American populations retained within the former. The long-standing species *N. tuberosa* has most recently been reclassified as a subspecies of *N. odorata* (*N. odorata* subsp. *tuberosa*) as the result of several molecular studies that have failed to resolve the taxa as discrete clades. Pink to rose-colored flowers occur occasionally in both subspecies of the normally white-flowered *N. odorata*. DNA sequence analyses show *N. odorata* to be related closely to the similar *N. mexicana* and *N. alba*, but also to the "dwarf" species *N. candida* and *N. tetragona* (Figure 1.5), the latter association most likely being the consequence of allopolyploidy.

Studies using various DNA markers have shown *N. candida* to be an allopolyploid that has originated several times from hybridization involving *N. alba* and *N. tetragona*. The basic chromosome number of *Nymphaea* is $x = 14$. The genus is diverse cytologically, with the North American taxa including $2n = 28$ diploids (*N. ampla*, *N. jamesoniana*), $2n = 56$ tetraploids (*N. mexicana*, *N. odorata*, *N. lotus*), $2n = 84$ hexaploids (*N. alba*, *N. odorata*), and $2n = 112$ octoploids (*N. odorata*, *N. tetragona*); aneuploidy is also common in several species. Natural hybrids (known as *N. ×thiona*) occur between *N. mexicana* and *N. odorata*. The name *Nymphaea loriana* has been applied to fertile populations of the otherwise sterile hybrids involving *N. leibergii* and *N. odorata*. A hybrid cultivar (*N. ×daubenyana*), which has been introduced to parts of Florida, develops adventitious plantlets capable of asexual propagation at the upper junction of the blade and petiole (such species often are erroneously said to be "viviparous").

A synthetic intersubgeneric hybrid has been documented in the origin of the cultivar *Nymphaea* 'William Phillips.'

Comments: *Nymphaea alba* occurs in British Columbia and California. Several *Nymphaea* species are confined to the southern United States including *N. ampla* (Florida and Texas, extending into South America); *N. capensis* and *N. jamesoniana* (Florida; the latter extending into South America); *N. elegans* (Florida, Louisiana, and Texas, extending into Mexico); and *N. lotus* (Florida, Louisiana). *Nymphaea mexicana* occurs in the coastal southeastern United States and more inland in Arizona and California. *Nymphaea liebergii* occurs in north-central North America, and *N. tetragona* in northern North America; the latter extends throughout Eurasia. *Nymphaea odorata* is widespread in eastern North America and has been introduced in much of western North America.

References: Afolayan et al., 2013; Akinjogunla et al., 2009; Alam et al., 2016; Alexander, 1987; Arenas & Scarpa, 2007; Barrios & Ramírez, 2008; Bonilla-Barbosa et al., 2000; Borsch et al., 2007, 2014; Bose et al., 2012; Bowerman & Goos, 1991; Boyd, 1968; Brock & Rogers, 1998; Capperino & Schneider, 1985; Catling et al., 1986; Cely, 1979; Chen et al., 2016; Conard, 1905; Conard & Hus, 1914; Cowgill, 1973; DiTomaso & Healy, 2003; Dodamani et al., 2012; Doucet & Fryxell, 1993; Else & Riemer, 1984; Emboden, 1983; Estevez et al., 2000; Garcia-Murillo, 1993; Gerritsen & Greening, 1989; Guthery, 1975; Harms & Grodowitz, 2009; Hegazy et al., 2001; Heins, 1984; Hellquist, 1972, 2003; Hill, 1988; Hirthe & Porembski, 2003; Holm Jr. et al., 2011; Hoppe et al., 1986; Hossain et al., 2014; Hsu et al., 2013; Hussner, 2012; Ikusima & Gentil, 1996; Irvine & Trickett, 1953; Jesurun et al., 2013; John-Africa et al., 2012; Johnson, 2004; Johnson et al., 1997; Johnstone, 1982; Kaufmann, 1970; Khedr & Hegazy, 1998; Knotts & Sacher, 2000; Kunii, 1993; Kunii & Aramaki, 1992; Kurihara et al., 1993; Landers et al., 1977; Les et al., 1999, 2004; Masters, 1974; McGaha, 1954; Meeuse & Schneider, 1979; Morrison & Hay, 2011; Moyle, 1945; Muenscher, 1944; Murphy et al., 1984; Nachtrieb et al., 2011; Neyland, 2007; Orban & Bouharmont, 1995; Padgett & Les, 2004; Perry & Uhler, 1982; Pfauth & Sytsma, 2005; Philomena & Shah, 1985; Pojar, 1991; Preston & Croft, 1997; Raja et al., 2010; Richards et al., 2011; Richardson & King, 2008; Sacher, 2005; Schneider, 1982; Schneider & Chaney, 1981; Shah et al., 2014; Shi et al., 1992; St. John, 1942; Stroud & Collins, 2014; Stutzenbaker, 1999; Swindells, 1983; Tomar & Sharma, 2002; Van der Valk & Rosburg, 1997; Villani & Etnier, 2008; Vincent et al., 2003; Volkova et al., 2010; Ward, 1977; Wiersema, 1982, 1987, 1988, 1996, 1997; Wiersema & Hellquist, 1994; Wiersema et al., 2008; Ward, 1977; Woods et al., 2005; Wunderlin & Les, 1980; Ya, 2015.

MAGNOLIID DICOTYLEDONS

The "magnoliid" dicotyledons correspond roughly with the subclass Magnoliidae (Cronquist, 1981), but is modified to exclude Ceratophyllales, Illiciales, Ranunculales, and Papaverales and to elevate Canellales to ordinal status (Judd et al., 2016). Three of the orders (Laurales, Magnoliales, Piperales) include OBL indicators (Figure 1.6).

ORDER 1: LAURALES [7]

Various revised circumscriptions have been provided for the Laurales over the past several decades (Cronquist, 1981; Renner, 1999), but the outcome of numerous phylogenetic studies (e.g., Figure 1.7) resolves the group as a clade comprising seven families in exclusion of Amborellaceae, Chloranthaceae, and Trimeniaceae, which have now been placed outside of the order. Some authors retain Idiospermaceae as an eighth family, whereas others have merged it with Calycanthaceae (Renner, 1999). As currently defined, Laurales consist of approximately 2400 species of woody plants (trees, shrubs, and vines), which are united by their perigynous flowers with carpels often embedded in a fleshy receptacle (Renner, 1999). Other common features include opposite leaves, stems with unilacunar nodes, uniovulate carpels, and inaperturate pollen (Renner, 1999). The flowers are typically pollinated by insects (Insecta).

In general, Laurales represent a clade of predominantly terrestrial plants; OBL indicators occur only with a single family:

    1. **Lauraceae** Jussieu

Family 1: Lauraceae [50]

The "laurel" family is a group of about 2500 woody species that are found mainly in terrestrial habitats throughout

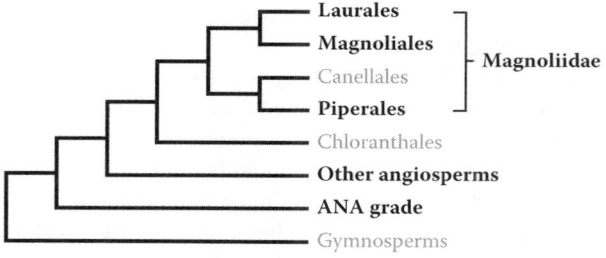

FIGURE 1.6 The phylogenetic placement and composition of magnoliid dicotyledons (subclass Magnoliidae) as indicated by combined analysis of DNA sequences from 17 nuclear, mitochondrial, and plastid genes. Taxa containing OBL North American indicators are highlighted in bold. The topology of Magnoliidae also is consistent with results from phylogenetic analysis of nuclear *Xdh* sequences (Morton, 2011). (Adapted from Soltis, D.E. et al., *Amer. J. Bot.*, 98, 704–730, 2011.)

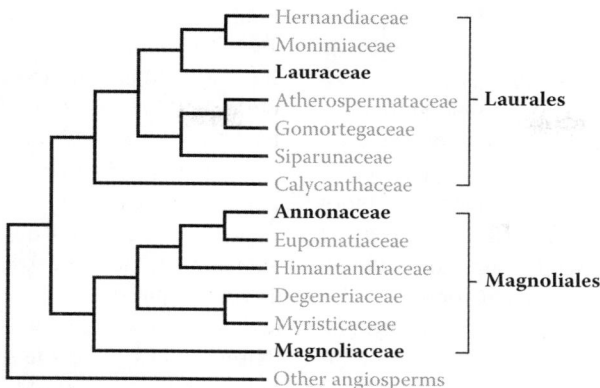

FIGURE 1.7 Phylogenetic structure of the orders Laurales and Magnoliales as indicated by combined analysis of mitochondrial, nuclear, and plastid sequence data. These relationships indicate several independent origins of OBL indicators in the groups (taxa containing OBL indicators are highlighted in bold). (Adapted from Soltis, D.E. et al., *Amer. J. Bot.*, 98, 704–730, 2011.)

tropical and subtropical regions (Judd et al., 2016). The group is characterized by the presence of benzyl-isoquinoline and aporphine alkaloids, alternate, spiral leaves with pellucid dots, no stipules, and radially symmetric flowers with 3-merous tepals, 3–12 stamens having broadly spaced sporangia, and a single, uniovulate carpel (Judd et al., 2016). The fruits are drupes (rarely berries), which often retain a persistent portion of the receptacle known as the "cupule." The seeds often contain lauric acid (Cronquist, 1981). The flowers are mainly outcrossed and are pollinated by insects (Insecta: Diptera; Hymenoptera). The drupes are dispersed by birds (Aves) or mammals (Mammalia).

Laurels are important economically as the source of avocados (*Persea americana*) and a number of spices including bay leaves (*Laurus nobilis*), camphor (*Cinnamomum camphora*), cinnamon (*Cinnamomum verum*), and oil of sassafras (*Sassafras albidum*). The laurel wreath, a symbol of victory in ancient Greece and Rome, has become a universal symbol of success and accomplishment. Ornamental cultivars have been developed in *Cinnamomum*, *Laurus*, *Lindera*, and *Persea*.

Two genera of Lauraceae contain OBL indicators:

1. *Lindera* Thunb.
2. *Litsea* Lam.

1. *Lindera*

Spicebush, bog spicebush, pondberry
Etymology: for John Linder (1676–1723)
Synonyms: *Benzoin*; *Daphnidium*; *Iteadaphne*; *Parabenzoin*; *Sinosassafras*
Distribution: global: Asia; North America; **North America:** eastern
Diversity: global: 100 species; **North America:** 3 species
Indicators (USA): OBL: *Lindera melissifolia*, *L. subcoriacea*; **FACW:** *L. melissifolia*
Habitat: freshwater; palustrine; **pH:** acidic; **depth:** <1 m; **life-form(s):** emergent (shrub)

Key morphology: shrubs (to 4.2 m) deciduous, stems 2–4, the branchlets straight; leaves large (4–16 cm), alternate, aromatic, petioles (to 10 mm) pubescent; inflorescences axillary, subsessile, near umbellate, appearing before the leaves; ♂ flowers (to 4.5 mm) with 6 yellow tepals (to 2.8 mm); stamens (to 2.9 mm) 9; ♀ flowers (to 2.6 mm) with 6 yellow tepals (to 2.2 mm); carpel 1, uniovulate; drupes (to 12 mm) globose to ellipsoidal, bright red, pedicelled (to 12 mm)
Life history: duration: perennial (buds); **asexual reproduction:** stolons; **pollination:** insect; **sexual condition:** dioecious; **fruit:** drupes (infrequent); **local dispersal:** fruits; **long-distance dispersal:** fruits (birds, mammals)
Imperilment: (1) *Lindera melissifolia* [G2/G3]; SX (FL); SH (LA); S1 (AL, MO, NC); S2 (AR, GA, MS, SC); (2) *L. subcoriacea* [G3]; S1 (AL, FL, GA, LA); S2 (MS, NC); S3 (SC)
Ecology: general: Most of the 100 or so *Lindera* species grow in more upland conditions (typically in forests), but often extend into wetter sites especially along watercourses. All three of the North American species are wetland indicators, with two designated as OBL. *Lindera* flowers are unisexual and are borne in a dioecious arrangement (a few bisexual flowers occur rarely). They are pollinated by several generalist insects (Insecta: Diptera; Hymenoptera; Lepidoptera). The fruits (drupes) are bright red and are dispersed primarily by birds (Aves). The seeds of some species are dormant physiologically and require a period of cold stratification to trigger their germination. *Lindera* habitats are threatened by drainage and by conversion to other land uses; some populations are threatened by livestock grazing and timber harvesting.

Lindera melissifolia (**Walter**) **Blume** is found in bogs, bottomlands, copses, depressions, hammocks, woodlands, and along the margins of ponds, swamps, and sinks at elevations of up to 100 m. The plants grow optimally in shaded areas that receive less than 40% sunlight. Photosynthesis increases with increased light (up to 42% sunlight), but declines at 100% sunlight. Plants grown under 100% light also exhibit reduced biomass (mainly in the roots). The substrates are described as Bosket series (mollic hapludalfs), clay, Foley series (albic glossic natraqualfs), loam, muck, sand,

Tuckerman series (typic endoaqualfs), vertic haplaquepts, and Wardell series (mollic epiaqualfs). Flowering begins in early spring. The sex of individual plants is not fixed genetically, as evidenced by a change in sex (from male to female) observed for one plant growing in cultivation at the Missouri Botanical Gardens. Sex ratios recorded in natural populations are strongly male-biased (7:1 to 19:1). The biased sex ratios are believed to result from the better growth characteristics of the male plants, providing them with a superior competitive ability and making them better adapted for colonizing suitable habitats as they become available. In one study, percent cover (which is related to light) explained 52% of the variation in the number of male flowering stems/plot and 14% of the variation in female stems/plot. Male plants produce an average of 10.9–15.0 flower clusters/stem and 46 flowers/cluster, with females producing 6.4–12.3 clusters/stem and 4 flowers/cluster. Fruiting has been observed to occur from September to October. It takes about 90 days from anthesis before the embryos fully mature. Early developing fruits contain myristic, palmitic, steric, oleic, linoleic, and linolenic fatty acids, with lauric acid becoming dominant in mature fruits. Seed production is erratic (9.8–27.8 fruit set), and the seedlings are often not observed at sites where fruiting occurs. The low seed set is not due to pollen limitation, as it is not higher in hand-pollinated flowers. The seeds have nondeep physiological dormancy. The recommended germination procedure is to cold-stratify fresh seeds at 4°C for 6 months followed by incubation under a 30°C/35°C, 16/8 h temperature and light regime. Seeds stratified or immersed in cold water (5°C) for 6–12 weeks germinated well (>63%) when incubated in light at 35°C/20°C or 30°C/20°C day/night temperature regimes. The best germination rate (100%) has been observed for seeds incubated in darkness following 6 weeks of cold stratification. The seeds tolerate submergence, but are not dispersed by water. They are dispersed primarily by birds (Aves: see *Use by wildlife*). Although seed viability can exceed 50% after a year of burial, a persistent seedbank does not appear to develop. Hydrated seeds have remained viable following 16 months of storage at −2°C to 4°C. Although viability is reduced by drying, seeds dried for 24 h to a moisture content of 8.6% can be stored successfully in liquid nitrogen. Within a year, over 70% of seeds that are left on the ground either rot or disappear. Experimentally planted populations showed a seedling survival rate of only 5%. The plants respond to light with various degrees of phenotypic plasticity. Seedlings grown under low light were 76% taller than those grown under bright light and also had larger leaves. Seedlings grown under bright light distributed a greater proportion of their biomass in the roots, possibly in response to water stress. The highest stem growth and survival rates occurred under 37% light, with stem diameter greatest below 70% light. Ramet production by female clones also increases with light availability. Overall, ramet production has been shown to increase substantially (up to 274%) when plants are grown under higher light (37%–70%) than when grown under very low light (5%). Electrophoretic (allozyme) data indicate low levels of genetic variation, owing to a combination of factors including a historical bottleneck with further erosion due to recent population losses, population declines, and clonal growth. Populations contain only one or a few genets (average of 4.5 multilocus genotypes/population) and often are dominated by male clones. Studies employing microsatellite markers have found evidence of outcrossing in at least one population, which was characterized by 2–11 alleles per locus and observed heterozygosity values ranging from 0.07 to 0.91. Additional studies using microsatellite markers identified only 67 genets out of 508 stems sampled (13%) across 11 sites (which contained from 1 to 16 genets). Of those, 94% were site specific and 39% represented single stems. The plants reproduce vegetatively from rhizomes and are highly clonal with clusters of stems capable of living for 6–7 years. Species of *Smilax* and *Vitis* have been implicated as their most likely competitors. **Reported associates:** *Acer negundo, A. rubrum, Ampelopsis arborea, Andropogon glomeratus, Aralia spinosa, Aristolochia serpentaria, Asimina triloba, Asplenium platyneuron, Berchemia scandens, Betula nigra, Bignonia capreolata, Boehmeria cylindrica, Botrychium biternatum, Brunnichia ovata, Callicarpa americana, Campsis radicans, Carex crus-corvi, C. louisianica, C. tribuloides, Carpinus caroliniana, Carya aquatica, C. glabra, C. illinoinensis, C. laciniosa, C. ovata, Celtis laevigata, Cephalanthus occidentalis, Clematis crispa, Cocculus carolinus, Commelina virginica, Cornus amomum, C. foemina, Cynosciadium digitatum, Desmodium, Dioclea multiflora, Diospyros virginiana, Eleocharis tuberculosa, Erechtites hieracifolia, Eupatorium, Forestiera acuminata, Fraxinus pennsylvanica, F. profunda, Gleditsia triacanthos, Gonolobus suberosus, Ilex decidua, I. myrtifolia, Juncus repens, Justicia ovata, Leersia virginica, Liquidambar styraciflua, Litsea aestivalis, Ludwigia glandulosa, Melothria pendula, Morus rubra, Myrica cerifera, Nyssa biflora, N. sylvatica, Parthenocissus quinquefolia, Passiflora lutea, Persea palustris, Persicaria virginiana, Phytolacca americana, Pinus elliottii, Planera aquatica, Platanus occidentalis, Populus heterophylla, Prunus serotina, Quercus alba, Q. falcata, Q. laurifolia, Q. lyrata, Q. michauxii, Q. nigra, Q. texana, Q. pagoda, Q. phellos, Q. velutina, Rhexia mariana, R. virginica, Rhynchospora cephalantha, R. fascicularis, R. wrightiana, Rubus trivialis, Sabal minor, Salix nigra, Sanicula canadensis, S. odorata, Sassafras albidum, Saururus cernuus, Sideroxylon lycioides, Smilax bona-nox, S. glauca, S. rotundifolia, S. tamnoides, Spiranthes ovalis, Styrax americanus, Taxodium ascendens, T. distichum, Thyrsanthella difformis, Toxicodendron radicans, Ulmus alata, U. americana, U. crassifolia, Vaccinium arboreum, Viola, Vitis aestivalis, V. palmata, V. rotundifolia, Woodwardia virginiana, Xyris elliottii, X. fimbriata.*

***Lindera subcoriacea* Wofford** occurs in bogs, copses, depressions, floodplains, hammocks, pocosins, seeps, swamps, and thickets at elevations to 200 m. The plants have been observed in bright light areas but occur most often in partial to deep shade. The substrates are acidic, high in organic matter, and described as peat or peaty muck. Flowering begins in March before the leaves expand. The plants display pronounced sexual dimorphism with the leaves of male plants being much longer (>9 cm) than those of the females (<7 cm). The sex ratios are consistently male-biased (58%) across all populations. About 15% of inventoried populations are entirely male. The flowers are pollinated by butterflies (Insecta: Lepidoptera: Papillionidae: *Papilio troilus*) and other insects. Fruiting proceeds from July to August. Dispersal of the bright red drupes presumably occurs by birds (Aves). Populations are small (averaging 7.9 individuals). The plants are weakly clonal (stems extending to 0.5 m), with stem clusters living for up to 6–7 years. **Reported associates:** *Acer rubrum, Agalinis aphylla, Alnus serrulata, Aronia arbutifolia, Arundinaria tecta, Bartonia paniculata, Berchemia scandens, Calamovilfa brevipilis, Calopogon barbatus, Carex atlantica, C. collinsii, C. lonchocarpa, C. turgescens, Chamaecyparis thyoides, Cladium mariscoides, Clethra alnifolia, Cliftonia monophylla, Cyrilla racemiflora, Dichanthelium scabriusculum, Dryopteris ludoviciana, Eleocharis tuberculosa, Eriocaulon aquaticum, Eubotrys racemosa, Eupatorium pilosum, E. resinosum, Gaylussacia mosieri, Glyceria obtusa, Gordonia lasianthus, Gymnadeniopsis clavellata, Ilex amelanchier, I. coriacea, I. glabra, I. laevigata, I. opaca, Illicium floridanum, Isoetes louisianensis, Itea virginica, Juncus gymnocarpus, Kalmia cuneata, Lachnocaulon digynum, Lilium iridollae, Lindernia dubia, Liriodendron tulipifera, Litsea aestivalis, Lobelia batsonii, Lycopus cokeri, Lyonia lucida, L. ligustrina, Lysimachia asperulifolia, Macranthera flammea, Magnolia virginiana, Myrica heterophylla, M. inodora, Nyssa biflora, Osmundastrum cinnamomeum, Oxypolis denticulata, Peltandra sagittifolia, Persea palustris, Pinguicula planifolia, P. primuliflora, Pinus palustris, P. serotina, P. taeda, Platanthera cristata, Rhododendron viscosum, Rhynchospora capitellata, R. crinipes, R. macra, R. miliacea, R. pallida, R. stenophylla, Rubus pensilvanicus, Sagittaria macrocarpa, Salix floridana, Sarracenia purpurea, Schoenoplectus etuberculatus, Smilax glauca, S. laurifolia, Sphagnum, Sporobolus, Thelypteris kunthii, Tofieldia glabra, Toxicodendron vernix, Triadenum virginicum, Utricularia purpurea, Vaccinium corymbosum, Veratrum virginicum, Viburnum nudum, Woodwardia virginica, Xyris drummondii, X. scabrifolia.*

Use by wildlife: The seedlings of *Lindera melissifolia* are eaten or cut by swamp rabbits (Mammalia: Leporidae: *Sylvilagus aquaticus*) and wood rats (Mammalia: Cricetidae: *Neotoma floridana*). The fruits are eaten by several mammals (Mammalia) including gray squirrels (Sciuridae: *Sciurus carolinensis*), nine-banded armadillo (Dasypodidae: *Dasypus novemcinctus*), swamp rabbits (Leporidae: *Sylvilagus aquaticus*), and white-tailed deer (Cervidae: *Odocoileus virginianus*), as well as by various birds (Aves) including cardinals (Cardinalidae: *Cardinalis cardinalis*), tufted titmouse (Paridae: *Baeolophus bicolor*), and white-throated sparrows (Emberizidae: *Zonotrichia albicollis*); the fruits are eaten and dispersed by hermit thrushes (Turdidae: *Catharus guttatus*), and brown thrashers (Mimidae: *Toxostoma rufum*). The leaves are used for nests by leafcutter bees (Insecta: Hymenoptera: Megachilidae: *Megachile texana*) and are host to several caterpillars (Insecta: Lepidoptera: Geometridae; Papilionidae: *Papilio palamedes, P. troilus*; Tortricidae: *Choristoneura rosaceana*). The stems are host to at least one species of weevil (Insecta: Curculionidae) and to several fungi (Fungi: Ascomycota: Aspergillaceae: *Aspergillus*; Diaporthaceae: *Diaporthe, Phomopsis*; Glomerellaceae: *Colletotrichum*; Mycosphaerellaceae: *Cercospora*; Nectriaceae: *Fusarium*; Pleosporaceae: *Alternaria*; Zygomycota: Mucoraceae: *Mucor*). Although the plant pathogenic laurel wilt (Fungi: Ascomycota: Ophiostomataceae: *Raffaelea lauricola*) has been documented only from *L. melissifolia*, it is presumed that *L. subcoriacea* is also potentially susceptible to infection. The fungus is transmitted by the redbay ambrosia beetle (Insecta: Coleoptera: Curculionidae: *Xyleborus glabratus*).

Economic importance: food: not edible; **medicinal:** Although several *Lindera* species have been used medicinally, there are no specific reports for either of the OBL taxa; **cultivation:** not cultivated; **misc. products:** Volatile extracts from *Lindera melissifolia* are effective at repelling ticks (Arthropoda: Arachnida: Acari) and mosquitoes (Insecta: Diptera: Culicidae); **weeds:** none; **nonindigenous species:** none in North America

Systematics: The Lauraceae, the only family of Laurales to contain aquatic taxa, are related phylogenetically to Hernandiaceae and Monimiaceae (Figure 1.7). Results from several molecular phylogenetic analyses indicate that *Lindera* is not monophyletic, but place those *Lindera* species sampled near to *Actinodaphne*, *Litsea*, and other members of the relatively derived tribe Laureae. *Lindera* and *Litsea* are the only genera in the family with OBL indicators. Unfortunately, phylogenetic studies of Lauraceae have not yet included either *Lindera melissifolia* or *L. subcoriacea*, leaving questions of their relationships in the genus largely unsettled. Further clarification of relationships among these species will require a more inclusive phylogenetic survey of species in *Lindera* and *Litsea*. As for all Lauraceae, the chromosomal base number for *Lindera* is $x = 12$; Although the chromosome number has not been determined for either of the OBL species, all other counts obtained for the genus have uniformly indicated diploids ($2n = 24$).

Comments: Both *Lindera melissifolia* and *L. subcoriacea* occur primarily along the southeastern coastal plain, with the former extending upward in the Mississippi embayment region.

References: Adams et al., 2010; Aleric & Kirkman, 2005; Anderson, 1999; Bryson et al., 1988; Cao et al., 2016; Carl III et al., 2004; Chanderbali et al., 2001; Connor et al., 2007, 2012; Devall, 2013; Devall et al., 2001; Echt et al., 2006, 2011; Fijridiyanto & Murakami, 2009a, 2009b; Godt & Hamrick, 1996a; Gramling, 2010; Gustafson et al., 2013; Hawkins et al., 2009a,b, 2011; Koch & Smith, 2008; Li et al., 2004; Lockhart et al., 2012, 2013, 2015; Martins et al., 2015; Morgan, 1983; Nie et al., 2007; Oh et al., 2012; Renner, 1999; Smith, 2003; Smith III et al., 2004; Sorrie et al., 2006; Steyermark, 1949; Stucky & Coxe, 1999; Unks et al., 2014; Wall et al., 2013; Wofford, 1997; Wright, 1989, 1990; Wright & Conway, 1994; Yager & Leonard, 2005.

2. Litsea
Pondspice

Etymology: from the Cantonese *leĭ tsaí* meaning "small plum" in reference to the fruit

Synonyms: *Glabraria*; *Hexanthus*; *Iozoste*; *Malapoënna*; *Pseudolitsea*; *Tetranthera*

Distribution: global: Asia; Neotropics; North America; **North America:** southeastern

Diversity: global: 400 species; **North America:** 1 species

Indicators (USA): OBL: *Litsea aestivalis*

Habitat: freshwater; palustrine; **pH:** alkaline; **depth:** <1 m; **life-form(s):** emergent (shrub)

Key morphology: deciduous shrubs (to 3 m), the branchlets zig-zagged; leaves (to 3 cm) simple, alternate, elliptic, aromatic; flowers unisexual (the plants dioecious); perianth yellow, 6-parted; ♂ with 9 or 12 stamens; ♀ with 9 or 12 staminodes and 1 carpel; fruit (to 10 mm) a round red drupe

Life history: duration: perennial (buds); **asexual reproduction:** none; **pollination:** insect; **sexual condition:** dioecious; **fruit:** drupes (common); **local dispersal:** fruits; **long-distance dispersal:** fruits (animals)

Imperilment: (1) *Litsea aestivalis* [G3]; S1 (MD, VA); S2 (FL, GA, NC); S3 (SC)

Ecology: general: It is difficult to summarize the ecology of this genus due to its likely polyphyletic nature (see *Systematics*). *Litsea* (at least in the New World) is primarily a terrestrial group of evergreen trees and shrubs that grow in cloud forests, pine-oak forests, and occasionally near the boundary of tropical dry forests, at elevations from 900 to 3000 m. Although deciduous, the Asian species grow in similar habitats (dry montane forests), but occasionally are found near the margins of watercourses. *Litsea aestivalis*, the only OBL indicator (and only North American species), occurs in low elevation wetland habitats, which are anomalous for the genus, and likely are associated with its distinctive morphology (see *Systematics*). The flowers are unisexual throughout the genus and are borne in a dioecious arrangement. Those of most *Litsea* species appear to be pollinated by small insects (Insecta) such as flies (Diptera) and bees (Hymenoptera). The fleshy fruits are dispersed primarily by frugivorous animals. Physiological dormancy has been reported for the seeds of some species.

Litsea aestivalis **(L.) Fernald** occurs in bogs, depressions, ditches, flatwoods, limesinks, roadsides, and along the margins of bayheads, dome swamps, pocosins, ponds, sinks, swamps, and woodlands at elevations of up to 200 m. The plants are regarded by some as calciphiles and occur on substrates described as Ellebelle loamy sand, Leon sand, limestone, Murville mucky loamy sand, peat, and sand. Flowering occurs from February to April, with fruiting reported from July to September. The flowers are precocious (appearing before the foliage) and are pollinated by flies (Insecta: Diptera). Some sites are comprised entirely of male plants and as a consequence are devoid of seedling recruitment. The fleshy red fruits are likely to be dispersed primarily by birds (Aves). Other life history information for this species is scarce.

Reported associates: *Acer rubrum, Amorpha fruticosa, Ampelopsis arborea, Andropogon brachystachyus, A. virginicus, Arundinaria gigantea, Berchemia scandens, Boehmeria cylindrica, Campsis radicans, Carex albolutescens, C. flacca, C. folliculata, C. glaucescens, C. striata, C. verrucosa, Centella asiatica, Cephalanthus occidentalis, Clematis crispa, Clethra alnifolia, Crataegus aestivalis, Cyrilla racemiflora, Diospyros virginiana, Ditrysinia fruticosa, Drosera intermedia, Dulichium arundinaceum, Eleocharis melanocarpa, Erianthus, Eriocaulon compressum, Eubotrys racemosa, Eupatorium mohrii, Fimbristylis perpusilla, Fraxinus caroliniana, Gleditsia aquatica, Gordonia lasianthus, Hypericum brachyphyllum, H. fasciculatum, H. myrtifolium, Hyptis alata, Ilex decidua, I. glabra, I. myrtifolia, I. vomitoria, Itea virginica, Juncus effusus, Lachnanthes caroliniana, Lindera melissifolia, L. subcoriacea, Liquidambar styraciflua, Ludwigia pilosa, L. suffruticosa, Lycopodiella alopecuroides, Lycopus, Lyonia ligustrina, L. lucida, L. mariana, Magnolia virginiana, Mitchella repens, Myrica cerifera, Nymphoides, Nyssa biflora, N. ogeche, Osmundastrum cinnamomeum, Panicum hemitomon, P. verrucosum, Persea palustris, Pinus clausa, P. elliottii, P. palustris, P. serotina, P. taeda, Planera aquatica, Pluchea baccharis, Polygala cymosa, Pontederia cordata, Quercus laurifolia, Q. lyrata, Q. margarettae, Q. myrtifolia, Q. nigra, Rhexia mariana, R. virginica, Rhus copallinum, Rhynchospora fascicularis, R. macrostachya, Rosa palustris, Rudbeckia mohrii, Sabal minor, Sagittaria filiformis, Salix caroliniana, S. nigra, Sarracenia minor, Saururus cernuus, Scirpus cyperinus, Serenoa repens, Smilax bona-nox, S. laurifolia, S. rotundifolia, S. walteri, Sphagnum, Styrax americanus, Taxodium ascendens, T. distichum, Thelypteris palustris, Triadica sebifera, Ulmus americana, Utricularia inflata, U. purpurea, Vaccinium corymbosum, Viburnum obovatum, Vitis rotundifolia, Woodwardia areolata, W. virginica, Xyris elliottii, Zenobia pulverulenta.*

Use by wildlife: *Litsea aestivalis* is a larval food plant for several butterflies (Insecta: Lepidoptera) including the Palamedes swallowtail (*Papilio palamedes*) and spicebush swallowtail (*Papilio troilus*). Its fruits are eaten by birds (Aves) and other wildlife. The stems and leaves are susceptible to damage by Florida wax scale (Insecta: Hemiptera: Coccidae: *Ceroplastes floridensis*). The plants can become severely infected by laurel wilt (Fungi: Ascomycota: Ophiostomataceae: *Raffaelea lauricola*), which is transmitted by the redbay ambrosia beetle (Insecta: Coleoptera: Curculionidae: *Xyleborus glabratus*).

Economic importance: food: not edible; **medicinal:** Although several *Litsea* species have been used medicinally, *L. aestivalis* has not; **cultivation:** *Litsea* rarely is cultivated as a plant to attract wildlife; **misc. products:** *Litsea aestivalis* has no other reported uses; however, some species have been used as a fragrance component or as a flavoring agent; **weeds:** none in North America; **nonindigenous species:** none in North America

Systematics: This large genus is similar to *Lindera* in that it certainly is polyphyletic as currently circumscribed and badly in need of revision. Although anther locule number has been used to separate *Lindera* (bilocular) from *Litsea* (tetralocular), cladistic analysis of morphological and leaf cuticle data was unable to effectively distinguish these genera and it seems likely that at least some members of these genera eventually will be merged with one or the other. Phylogenetic analyses of DNA sequence data resolve different *Litsea* species with those of *Dodecadenia* and *Lindera*; the genus also is relatively closely related to *Actinodaphne*, *Laurus*, *Neolitsea*, and *Parasassafras* and is placed together with them within tribe *Laureae* of Lauraceae. *Litsea aestivalis* has been excluded from molecular phylogenetic studies thus far. More comprehensive systematic surveys of *Litsea* and *Lindera* will be necessary in order to improve the circumscriptions for these genera and also to provide a better understanding of interspecific relationships within each. In addition to its anomalous ecology, *Litsea aestivalis* is also distinct morphologically from all other New World species in the genus owing to its deciduous leaves, which have trichomes only near the base of the abaxial midvein. The basic chromosome number of *Litsea* is $x = 12$. The chromosome number for *Litsea aestivalis* is unknown. No natural hybrids involving *L. aestivalis* have been reported.

Comments: *Litsea aestivalis* occurs along the outer southeastern coastal plain of the United States.

References: Boyle et al., 2007; Chanderbali et al., 2001; Devy & Davidar, 2003; Elam, 2007; Fijridiyanto & Murakami, 2009a,b; Godfrey, 1988; Hill, 1992; Hughes et al., 2014; Jiménez-Pérez & Lorea-Hernández, 2009; Li & Christophel, 2000; Renner, 1999; Rohwer, 1993, 2000; van der Werf, 1997.

Order 2: Magnoliales [6]

The Magnoliales include six families with a total of roughly 2840 species (Judd et al., 2016). Analyses of various DNA sequence data indicate that the group is monophyletic and represents the sister clade of Laurales (Soltis et al., 2011; Judd et al., 2016). Various analyses have placed either Myristicaceae (most studies) or Magnoliaceae (e.g., Figure 1.7) as the sister group to the remainder of the order (Soltis et al., 2011). In any case, the two families that contain OBL indicators (Annonaceae and Magnoliaceae) are not closely related (Figure 1.7), which indicates an independent derivation of the habit in each group.

The Magnoliales share a suite of features including benzylisoquinoline or aporphine alkaloids, woody habit, 2-ranked leaves, trilacunar nodes, reduced pit borders, stratified phloem, leaf mesophyll asterosclereids, and ruminate endosperm (Cronquist, 1981; Judd et al., 2016). Most of the species retain an unspecialized hypogynous floral morphology consisting of undifferentiated tepals; numerous, spirally arranged stamens; an apocarpous gynoecium with laminar or marginal placentation; and seeds having tiny embryos that are richly provided with endosperm (Cronquist, 1981; Judd et al., 2016).

OBL indicators occur in two North American families:

1. **Annonaceae** Jussieu
2. **Magnoliaceae** Jussieu

Family 1: Annonaceae [128]

The "custard apple" family is the largest group of Magnoliales with approximately 2300 species that are distributed mainly throughout tropical and subtropical regions worldwide (Cronquist, 1981; Judd et al., 2016). Various phylogenetic analyses, including a combined analysis of eight plastid gene sequences (Chatrou et al., 2012) indicate that the family is monophyletic and can be subdivided into four subfamilies (Ambavioideae, Anaxagoreoideae, Annonoideae and Malmeoideae), which represent the major clades; OBL indicators occur only within the relatively derived subfamily Annonoideae (Figure 1.8).

These are woody plants (trees, shrubs, and lianas) that produce benzylisoquinoline alkaloids and contain vessels with simple perforation pits, flowers that gradually increase in size after opening, and anthers vascularized by a single trace (Cronquist, 1981; Judd et al., 2016). The flowers are trimerous, typically with three sepals, six petals, and numerous stamens and carpels, the latter containing one to many seeds. They are protogynous, sometimes thermogenic, and pollinated by insects (Insecta), primarily beetles (Coleoptera), but also include bees (Hymenoptera), flies (Diptera), or thrips (Thysanoptera) (Judd et al., 2016). The fruits are usually aggregates of fleshy berries (dispersed by animals), but are sometimes explosively dehiscent follicles (Judd et al., 2016). The characteristically underdeveloped embryos result in seeds having either morphological or morphophysiological dormancy requirements (Baskin & Baskin, 1998). Persistent seed banks occur in some species.

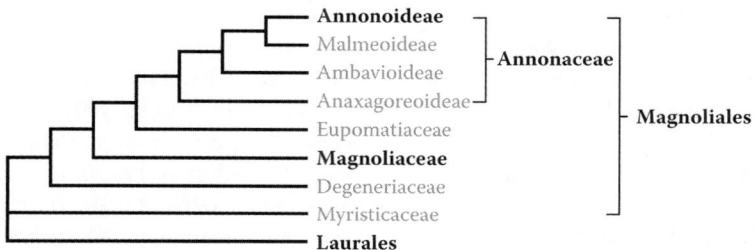

FIGURE 1.8 Subfamilial relationships in Annonaceae as indicated by phylogenetic analysis of combined plastid sequence data. Taxa containing OBL indicators are highlighted in bold. In Annonaceae, the OBL indicators occur within the relatively derived subfamily Annonoideae. (Adapted from Chatrou, L.W. et al., *Bot. J. Linn. Soc.*, 169, 5–40, 2012.)

Economically important species include the edible fruits of cherimoya (*Annona cherimola*), custard apples (*Annona reticulata*), and pawpaw (*Asimina triloba*). Some perfumes contain a fragrance ingredient derived from the flowers of *Cananga odorata* (Judd et al., 2016). Fruits of *Monodora myristica* have been used as a nutmeg substitute (Judd et al., 2016). Ornamental cultivars have been developed in *Asimina* and *Polyalthia*.

Only one North American genus of Annonaceae contains OBL indicators:

1. ***Annona*** L.

1. *Annona*

Custard apple, pond apple
Etymology: from *anón*, the Taino (Hispaniolan) name of the fruit
Synonyms: *Pseudannona*; *Rollinia*
Distribution: global: Africa; Asia*; Australia*; Neotropics; North America; **North America:** southeastern (S. Florida)
Diversity: global: 110 species; **North America:** 2 species
Indicators (USA): OBL: *Annona glabra*
Habitat: brackish, freshwater; palustrine; **pH:** 5.1–8.5; **depth:** <1 m; **life-form(s):** emergent (shrub, tree)
Key morphology: trees or shrubs (to 15 m) with trunk buttresses; leaves (to 15 cm) deciduous, simple, ovate to elliptic, distichous, leathery, petioled (to 20 mm); flowers solitary, axillary, showy, trimerous; sepals (to 6 mm) 3; petals (to 3 cm) 6, in two whorls of three, cream-white (the innermost purple at base); stamens (to 4 mm) numerous, linear; fruits (to 12 cm) pendulous, apple-like syncarps, fleshy, dull yellow with brown blotches; seeds (to 1.5 cm) 1 per carpel
Life history: duration: perennial (buds); **asexual reproduction:** suckers; sprouting cut branches; **pollination:** insect; **sexual condition:** hermaphroditic; **fruit:** berry-like syncarp (common); **local dispersal:** water (floating fruits, seeds); **long-distance dispersal:** animals, water (seeds)
Imperilment: (1) *Annona glabra* [G5]
Ecology: general: *Annona* is principally a genus of terrestrial tropical trees and shrubs, with only one North American species designated as a wetland indicator. The flowers of most species are pollinated by beetles (Insecta: Coleoptera). The fruits develop from multiple pistils into an aggregate of fleshy "syncarps" and bear a superficial resemblance to apples (thus the common names). The fruits are dispersed endozoically by birds (Aves) and other animals or passively by water. The seeds of some species contain the neurotoxin annonacin. They often contain underdeveloped embryos, which confer morphological dormancy.

Annona glabra **L.** is a mangrove-like plant that inhabits brackish to freshwater tidal and nontidally influenced depressions, ditches, hammocks, roadsides, sloughs, swamps, and the margins of canals, lakes, ponds, and streams at elevations to 50 m. The plants typically occur in shaded areas as understory trees. The substrates can span a broad range of acidity (pH: 5.1–8.5) but are often alkaline and have been described as muck and oolite. The plants are highly flood-tolerant and drought-, shade-, and salt-tolerant, but fire-intolerant. They can reach reproductive maturity within 2 years. Flowering occurs from April to July, with fruiting extending through January. The flowers are strongly protogynous and produce a strong, fruity fragrance (dominated by 3-pentanyl acetate and 1,8-cineole) that attracts weevils (Insecta: Chrysomelidae, Curculionidae) and other small beetles as pollinators. Although outcrossing would be expected, some studies have reported a lack of detectable genetic (isozyme) variation within and between populations. Anthesis lasts for about 40 h. The fruits and seeds are dispersed readily by water. The ripe fruits are buoyant, but deteriorate fairly rapidly to release their seeds, which also are buoyant and can remain afloat for more than 12 months in either fresh or salt water, while retaining fairly high germinability (e.g., 38%). The seeds are also dispersed endozoically by turtles (Reptilia: Testudines) and other vertebrates. Allegedly, they can be dispersed by alligators (Reptilia: Alligatoridae: *Alligator mississippiensis*), whose stomach contents have been found to contain up to 1286 seeds. However, some experiments have shown that seeds ingested by alligators become inviable. Undigested seeds can remain viable in the sediment for up to 3 years. The seeds require high amounts of moisture and temperatures above 25°C for germination. The seedlings exhibit high survivorship under low or high water conditions. Multiple stems can result from the germination of adjacent seeds. Damaged roots or trunks can regenerate vegetatively by sucker production, which can lead to the development of clonal thickets if extensive. Cut branches also can sprout when in contact with wet soil. The root cortical cells are colonized by arbuscular mycorrhizal (AM) fungi. Populations of *A. glabra* have declined substantially due to cutting and clearing of vegetation for sugarcane production. The plants provide a substrate for some epiphytes such as bromeliads (e.g., Bromeliaceae: *Guzmania monostachia*) and lichens (e.g., Fungi: Ascomycota: Parmeliaceae: *Parmotrema endosulphureum*). **Reported associates:** *Acer rubrum, Acrostichum, Baccharis, Blechnum serrulatum, Bumelia reclinata, Byrsonima lucida, Callicarpa americana, Celtis laevigata, Cephalanthus occidentalis, Chrysobalanus icaco, Cladium jamaicense, Croton linearis, Cucurbita okeechobeensis, Distichlis, Dodonaea viscosa, Ficus aurea, Fraxinus caroliniana, Gordonia lasianthus, Guzmania monostachya, Ilex cassine, Itea virginica, Ludwigia, Lyonia fruticosa, L. lucida, Lysiloma latisiliquum, Magnolia virginiana, Metopium toxiferum, Myrica cerifera, Nyssa sylvatica, Osmunda regalis, Osmundastrum cinnamomeum, Persea borbonia, P. palustris, Pinus ellottii, Quercus laurifolia, Q. minima, Q. pumila. Q. virginiana, Rubus argutus, Sabal palmetto, Salix caroliniana, Sambucus nigra, Taxodium ascendens, T. distichum, Vaccinium myrsinites, Viburnum nudum, Woodwardia areolata, W. virginica.*
Use by wildlife: *Annona glabra* is an important cover plant for wildlife in the Florida everglades. It has been identified as a component of the mixed-swamp forests that provide habitat for the endangered Florida panther (Mammalia: Felidae: *Puma concolor couguar*). The plants provide nesting sites for wood storks (Aves: Ciconiidae: *Mycteria americana*) and

habitat for the Florida Everglades kite (Aves: Accipitridae: *Rostrhamus sociabilis plumbeus*). The foliage is high in nitrogen (2.25%) and phosphorous (0.0832%) and provides food for several butterfly and moth larvae (Insecta: Lepidoptera: Erebidae: *Gonodonta nutrix*; Geometridae: *Orthonama obstipata*; Noctuidae: *Peridroma saucia*; Papilionidae: *Eurytides marcellus*, *Mimoides clusoculis*; Psychidae: *Oiketicus abbotii*, *Thyridopteryx ephemeraeformis*; Sphingidae: *Cocytius antaeus*, *C. duponchel*; Tortricidae: *Argyrotaenia amatana*, *Tsinilla lineana*). The plants host several bugs (Insecta: Hemiptera) including lobate lac scale (Kerriidae: *Paratachardina lobata*) and whiteflies (Aleyrodidae: *Paraleyrodes minei*, *P. pseudonaranjae*, *Tetraleurodes ursorum*). The fruits/seeds of *A. glabra* are eaten by alligators (Reptilia: Alligatoridae: *Alligator mississippiensis*), birds (Aves), Florida box turtles (Reptilia: Emydidae: *Terrapene bauri*), and lizards (Reptilia: Squamata).

Economic importance: food: The fruits of *Annona glabra* are eaten raw (Florida Seminoles) or can be made into custard, jellies, and wine. Though edible, the raw fruits are not palatable to most people; **medicinal:** Decoctions or infusions made from the leaves, flowers, twigs, or fruit rind of *Annona glabra* have been used to treat coughs, diarrhea, jaundice, rheumatism, and stomach ache and have been used as a vermifuge or antiparasitic. Tea made from the flowers has been used to treat kidney ailments. Ethanolic extracts of the leaves, pulp, and seeds contain acetogenins, which have been shown to exhibit anticancer properties; **cultivation:** *Annona glabra* is cultivated as a medicinal plant in parts of the West Indies;

misc. products: *Annona glabra* was used by the Seminoles to make spoons and as a cleaning agent (lye). The light wood is sometimes made into fishing floats. The species is used as grafting stock (to improve flood tolerance) for several species including *Annona squamosa*, *A. reticulata,* and the commercial Atemoya custard apple (a hybrid derived from *Annona squamosa* and *A. cherimola*) in Australia. The plants have been used in projects for coastal wetland restoration in Florida; **weeds:** *Annona glabra* is a seriously invasive weed in Queensland, Australia, where it was introduced in 1886. It is also a weed in Vietnam; **nonindigenous species:** None in North America; however, *A. glabra* has been introduced to Australia, Fiji, Polynesia, Sri Lanka, and Southeast Asia. The plants were introduced to Australia in 1912 for use as grafting stock (see earlier).

Systematics: Although cladistic analyses of morphological data indicate a close relationship of *Annona* to *Rollinia*, molecular data resolve the latter group as nested within *Annona* (Figure 1.9). As a result, most recent studies have merged *Rollinia* with *Annona*. In that modified circumscription, *Annona* appears to be most closely related to *Asimina* (e.g., Figure 1.9). *Annona* has been subdivided into 17 sections, but that generic classification remains to be evaluated phylogenetically. Molecular phylogenetic studies including approximately one-eighth of the recognized *Annona* species have indicated a close relationship between *A. glabra* and the African *A. senegalensis* (Figure 1.9), which characteristically inhabits semiarid coastal regions, but also extends into swampy habitats. *Annona* is one of four Annonaceae genera

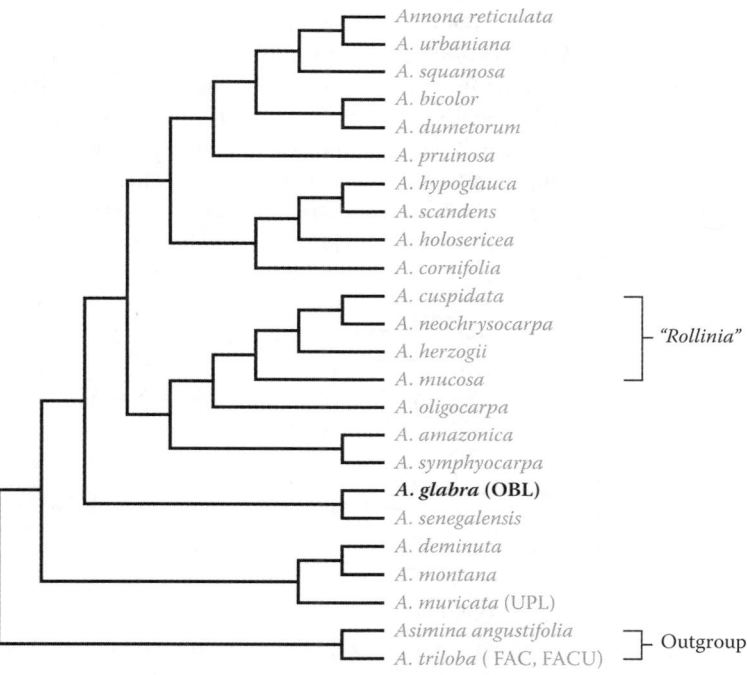

FIGURE 1.9 Phylogenetic relationships in *Annona* as indicated by the analysis of combined DNA sequence data. In this sample of taxa, the species formerly assigned to *Rollinia* are nested deeply within *Annona* and the OBL *Annona glabra* resolves as the sister species of *A. senegalensis*. The wetland indicator status is provided in parentheses for North American species; OBL indicators are highlighted in bold. (Adapted from Chatrou, L.W. et al., *Mol. Phylogen. Evol.*, 53, 726–733, 2009.)

with a base chromosome number of $x = 7$. Most *Annona* species are diploid ($2n = 14, 16$) except for *A. glabra*, which is tetraploid ($2n = 28$). No hybrids involving *A. glabra* have been reported.

Comments: *Annona glabra* is restricted to southern Florida but extends into South America, and there is disjunct between the Neotropics and Western Africa.

References: ARMC, 2001; Chatrou et al., 2009, 2012; de Queiroz Pinto et al., 2002; dos Santos & Sant'Ana, 2001; Evans, 2007; Fernando & Rupasinghe, 2013; Fries, 1959; Goodrich & Raguso, 2009; Gottsberger, 1999; Handayani & Nugraha, 2016; Jayachandran & Fisher, 2008; Johnson, 2003; Kessler, 1993; Kral, 1997; Land Protection, 2004; Platt et al., 2013; Rosenblatt et al., 2014; Saha et al., 2009; Samuel et al., 1991; Schaffer, 1998; Setter et al., 2008; Van der Valk et al., 2008; Van Riemsdijk, 2014.

Family 2: Magnoliaceae [2]

The magnolia family contains 220 species of deciduous or evergreen trees and shrubs that occur primarily in the moist forests of temperate America, Asia, and tropical South America (Judd et al., 2016). Even though as many as 12 genera have been recognized in the family, various phylogenetic studies (e.g., Azuma, 2001; Nie et al., 2008; Kim & Suh, 2013) have not found them to represent clades that are discrete from *Magnolia* (Figure 1.10). Although no real consensus has been reached, most of the species now have been transferred to the genus *Magnolia*, with two retained in the sister genus *Liriodendron* (e.g., Judd et al., 2016).

The group is recognized by a combination of the following factors: benzylisoquinoline or aporphine alkaloids, multilacunar nodes, septate pith, stipules, and solitary flowers with elongate receptacles (Cronquist, 1981; Judd et al., 2016). The flowers are protogynous, self-compatible or self-incompatible, showy, and pollinated by bees (Insecta: Hymenoptera) or beetles (Insecta: Coleoptera), which often become trapped in the flowers as they consume the pollen. The perianth is 3-merous and undifferentiated, consisting of 6–18 tepals (Cronquist, 1981). The numerous stamens

are laminar and are not clearly differentiated into an anther and filament (Cronquist, 1981). The few to numerous pistils usually contain two ovules and mature into an aggregate of dehiscent follicles or become berrylike or a fleshy syncarp. In species with follicles, the seed coats can be colored a bright red, which facilitates their dispersal by birds (Aves) once they are released from the fruits. The syncarps also often are bird-dispersed.

There are numerous ornamental cultivars in both *Liriodendron* and *Magnolia*. Species in both genera have been used for lumber. Some *Magnolia* species are regarded as medicinal plants. Only one genus contains OBL indicators:

1. *Magnolia* L.

1. *Magnolia*

Magnolia, swamp bay, sweet bay; laurier doux
Etymology: after Pierre Magnol (1638–1715)
Synonyms: *Alcimandra*; *Aromadendron*; *Buergeria*; *Dugandiodendron*; *Elmeria*; *Kmeria*; *Kobus*; *Manglietia*; *Michelia*; *Pachylarnax*; *Parakmeria*; *Talauma*; *Tulipastrum*
Distribution: global: Asia; New World; **North America:** eastern
Diversity: global: 120 species; **North America:** 8 species
Indicators (USA): OBL; FACW: *Magnolia virginiana*
Habitat: freshwater; palustrine; **pH:** 2.5–6.7; **depth:** <1 m; **life-form(s):** emergent (shrub, tree)
Key morphology: deciduous or evergreen, multitrunked shrub (to 10 m) or single-trunked tree (to 28 m) with septate pith; leaves (to 22 cm) alternate, simple, entire, white pubescent beneath, stipulate (to 6 cm); flowers showy (to 8 cm), trimerous, with 6–9 creamy white tepals (to 6 cm); stamens and pistils numerous; fruit a conelike aggregate (to 5.5 cm) of follicles; seeds (to 5 mm) with a red aril
Life history: duration: perennial (buds); **asexual reproduction:** stump sprouts; **pollination:** insect; **sexual condition:** hermaphroditic; **fruit:** aggregates (follicles) (common); **local dispersal:** wind, water, birds (seeds); **long-distance dispersal:** birds (seeds)

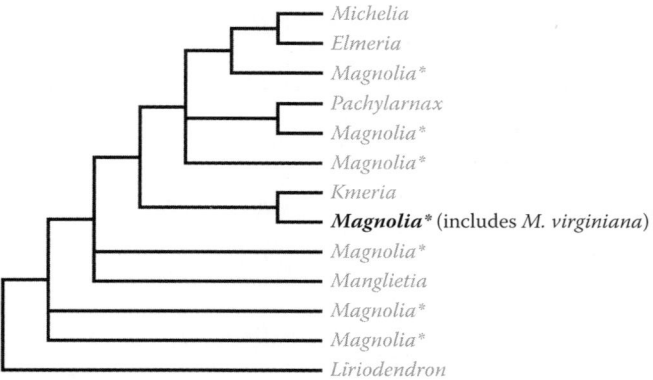

FIGURE 1.10 Intergeneric relationships in Magnoliaceae as indicated by the analysis of combined DNA sequence data. The major clades depicted by this result render those containing *Magnolia* species (denoted with an asterisk) as a highly polyphyletic group, and present a persuasive argument for the recognition of only two genera in the family (*Liriodendron* and *Magnolia*). The clade that contains the only OBL indicator in the family (*Magnolia virginiana*) is highlighted in bold. (Adapted from Kim, S. & Suh, Y., *J. Plant Biol.*, 56, 290–305, 2013.)

Imperilment: (1) *Magnolia virginiana* [G5]; S1 (MA, NY); S2 (PA, TN)

Ecology: general: Magnolias occur frequently in alluvial woods, bottomlands, and rich woods. Six of the North American species (75%) are wetland indicators, but only one is designated as OBL. The group is similar ecologically, with the North American species growing sympatrically throughout the southeastern United States. The flowers are pollinated by beetles (Insecta: Coleoptera), and the arillate seeds typically are dispersed by birds (Aves). The seeds generally contain underdeveloped embryos and are characterized by morphophysiological dormancy.

Magnolia virginiana L. inhabits barrens (pine), baygalls, bayheads, bays (Carolina), bogs, bottomlands, brakes, ditches, drains, draws, dunes, flatwoods, floodplains, hammocks, marshes, ravines, roadsides, savannas, seeps, slopes, sloughs, streamheads, swales, swamps, woodlands, and the margins of bayous, beaches, brooks, pocosins, ponds, rivers, streams, and thickets at elevations of up to 510 m. This is a late successional species with intermediate shade and flood tolerance. The plants are highly susceptible to drought. The substrates are acidic (pH: 2.5–6.7) and are characterized as Bayboro, Bayou sandy loam, Belhaven, Bibb, clay loam, Coastal Plain histosols, Corolla–Duckson sands, Greenville clay loam, Harleston fine sandy loam, Lakeland sand, mud, Myatt, Ponzer, Portsmouth, sand, sandy loam, sandy silt, Scuppernong, silty loam, spodosols, and ultisols. Flowering occurs from March to October, with fruiting extending from June to November. On the first day of anthesis, the self-compatible, protogynous flowers open briefly during the afternoon (2–5 pm) and then close. The stigmas are receptive during this phase. On the following afternoon, the stigmas wither, and the flowers reopen to release their pollen. A few flowers will close once again and reopen permanently during the afternoon. The flowers are pollinated by bees (Insecta: Hymenoptera) or beetles (Insecta: Coleoptera), which enter the buds or closed flowers to feed on nectar, pollen, or other flower parts. Annual fruit production is relatively low. When the follicles mature and dehisce, the bright red seeds first emerge by hanging from fine threads. The seeds are dispersed locally by water or wind and over greater distances by birds (Aves). They are morphophysiologically dormant and reportedly germinate well at 18°C after 90 days of cold stratification (2°C). Germination rates of 32%–50% have been achieved for seeds stored at 0°C–5°C for 3–6 months. Seeds (with the pulp dried or removed) will retain high viability for several years if stored in sealed containers at 0°C–5°C. The seedlings usually establish in natural openings but are shade-tolerant during their first year of growth and develop well under shaded conditions. Although prolonged inundation during this period can cause high mortality, greenhouse-raised seedlings have survived well (88% survival rate) after 100 days of ponded conditions. The seedlings are killed quickly by frost. Continuous flooding promotes the development of adventitious roots on submerged portions of the stems. Depending on latitude, the average growing season can extend from 180 days in the north (average minimum temperature of −23°C) to 340 days in the south (average minimum temperature of 4°C). The bark is adapted moderately to fire, and the plants can survive fires by their persistent root crown or caudex; however, seedlings are killed by fire, and repeated burning will eliminate the species. The plants can invade pine and hardwood stands to become dominant in some southern mixed hardwood forests. **Reported associates:** *Acer rubrum, Aesculus pavia, Alnus serrulata, Andropogon, Annona glabra, Aralia spinosa, Aronia arbutifolia, Arundinaria gigantea, A. tecta, Asclepias rubra, Asimina parviflora, Baccharis halimifolia, Betula nigra, Boehmeria cylindrica, Callicarpa americana, Carex, Carpinus caroliniana, Carya tomentosa, Castanea dentata, Cephalanthus occidentalis, Chamaecrista nictitans, Chamaecyparis thyoides, Chasmanthium laxum, C. sessiliflorum, Chionanthus virginicus, Chrysobalanus icaco, Cicuta maculata, Clethra alnifolia, Cliftonia monophylla, Conyza canadensis, Cornus florida, C. foemina, Crataegus aestivalis, Cyperus, Cyrilla racemiflora, Decumaria barbara, Desmodium viridiflorum, Dichanthelium ensifolium, Diospyros virginiana, Echinochloa crus-galli, Eriocaulon, Eubotrys racemosa, Eupatorium capillifolium, Euthamia graminifolia, Fagus grandifolia, Fraxinus caroliniana, F. profunda, Galactia volubilis, Gordonia lasianthus, Hamamelis virginiana, Helianthus debilis, Hieracium gronovii, Hydrangea quercifolia, Hypericum brachyphyllum, Hyptis alata, Ilex cassine, I. coriacea, I. decidua, I. glabra, I. opaca, I. verticillata, I. vomitoria, Ipomoea sagittata, Itea virginica, Juncus caesariensis, J. effusus, J. scirpoides, Juniperus virginiana, Kummerowia striata, Leucothoe axillaris, Liatris, Liquidambar styraciflua, Liriodendron tulipifera, Ludwigia alternifolia, L. microcarpa, L. palustris, Lycopodium, Lyonia ligustrina, L. lucida, L. mariana, Magnolia grandiflora, Melothria pendula, Mikania scandens, Mitchella repens, Mitreola sessilifolia, Muhlenbergia capillaris, Myrica cerifera, M. inodora, Nyssa aquatica, N. biflora, N. sylvatica, Onoclea sensibilis, Osmanthus americanus, Osmunda regalis, Osmundastrum cinnamomeum, Ostrya virginiana, Oxydendrum arboreum, Panicum virgatum, Parthenocissus quinquefolia, Paspalum plicatulum, Persea borbonia, P. palustris, Phragmites australis, Phytolacca americana, Pinus echinata, P. elliottii, P. palustris, P. serotina, P. taeda, Pityopsis graminifolia, Planera aquatica, Pluchea baccharis, Pogonia ophioglossoides, Polygala cruciata, P. cymosa, P. lutea, Polypodium, Polytrichum, Prunus caroliniana, P. serotina, Pseudognaphalium obtusifolium, Quercus falcata, Q. laevis, Q. laurifolia, Q. lyrata, Q. michauxii, Q. nigra, Q. virginiana, Rhexia alifanus, R. lutea, Rhus copallinum, Rhynchospora chapmanii, R. elliottii, R. glomerata, R. plumosa, Rubus, Sabal minor, S. palmetto, Sagittaria, Salix caroliniana, Sarracenia alata, Sassafras albidum, Schoenoplectus, Scirpus cyperinus, Sesbania herbacea, Sideroxylon lanuginosum, Smilax glauca, S. laurifolia, S. pumila, S. rotundifolia, S. walteri, Sphagnum, Stenanthium densum, Strophostyles helvola, Styrax americanus, Symplocos tinctoria, Taxodium ascendens, T. distichum, Tiedemannia filiformis, Toxicodendron radicans, T. vernix, Typha latifolia, Ulmus alata, U. americana, Vaccinium* *arboreum, V. corymbosum, V. elliottii, Verbena bonariensis, Viburnum nudum, Vitis rotundifolia, Woodwardia areolata, Xyris scabrifolia.*

Use by wildlife: *Magnolia virginiana* provides perching and nesting sites for several bird (Aves) species. Its leaves and twigs are browsed by white-tailed deer (Mammalia: Cervidae: *Odocoileus virginianus*), and its seeds are eaten by gray squirrels (Mammalia: Sciuridae: *Sciurus carolinensis*), white-footed mice (Mammalia: Cricetidae: *Peromyscus leucopus*), wild turkeys (Aves: Phasianidae: *Meleagris gallopavo*), quails (Aves: Phasianidae), red-cockaded woodpeckers (Aves: Picidae: *Picoides borealis*), and song birds (Aves: Passeriformes). It is used for food and shelter by beavers (Mammalia: Castoridae: *Castor canadensis*) and also provides important forage for cattle (Mammalia: Bovidae: *Bos taurus*). Although the foliage contains neolignan compounds (magnolol and a biphenyl ether) that are toxic to many insect (Insecta) larvae, they are detoxified by some silkmoths (Lepidoptera: Saturniidae: *Callosamia securifera*), but remain toxic to closely related *C. angulifera* and *C. promethea*. Similarly, the foliage is eaten by larvae of the eastern tiger swallowtail (Lepidoptera: Papillionidae: *Papilio glaucus*), but is toxic to the related Palamedes and spicebush swallowtails (*P. palamedes* and *P. troilus*). Other reportedly hosted insects include additional Lepidoptera (Papillionidae: *Papilio polyxenes*; Saturniidae: *Automeris io*; Sphingidae: *Eumorpha fasciatus, E. vitis*; Tortricidae: *Paralobesia cyclopiana, P. liriodendrana*) and thrips (Insecta: Thysanoptera: Thripidae: *Caliothrips striatus*). The flowers also attract a variety of insects including bees (Hymenoptera: Andrenidae; Apidae: *Apis mellifera, Bombus impatiens, Xylocopa*), beetles (Coleoptera: Cerambycidae; Chrysomelidae, *Diabrotica undecimpunctata howardii, Gibbobruchus mimus*; Coccinellidae: *Coleomegilla maculata*; Curculionidae; Mordellidae: *Mordellistena*; Nitidulidae: *Epuraea erichsoni, Glischrochilus, Pocadius helvolus, Stelidota geminata*; Scarabaeidae: *Macrodactylus subspinosus, Trichiotinus bibens*), bugs (Hemiptera: Anthocoridae), butterflies (Lepidoptera), and flies (Diptera: Sphaeroceridae; Syrphidae: *Eupeodes*). The plants are also host to several fungi including leaf spot (Ascomycota: Mycosphaerellaceae: *Mycosphaerella glauca, M. milleri*), white mold (Ascomycota: Sclerotiniaceae: *Sclerotinia gracilipes*), wood rotting fungi (Ascomycota: Xylariaceae: *Hypoxylon epiphaeum*; Basidiomycota: Polyporaceae: *Fomes fasciatus*), a wood staining fungus (Ascomycota: Hypocreales: *Cephalosporium pallidum*), and a lichen (Ascomycota: Pilocarpaceae: *Byssoloma maderense*).

Economic importance: food: The leaves of *Magnolia virginiana* are used to flavor foods; **medicinal:** Decoctions derived from the leaves and twigs of *Magnolia virginiana* have been used as an alterative, cardiotonic, laxative, stimulant, sudorific, and to treat colds, dysentery, fever, gout, malaria, and rheumatism. The Houma tribe prepared decoctions from the leaves and twigs to warm the blood and to use as a cold remedy and febrifuge. The flowers contain the essential oil 2-phenylethanol, which has significant antioxidant activity in scavenging free radicals, but is ineffective against human breast and lung carcinoma cell lines. The leaves contain magnolol and other neolignans that are highly toxic to brine shrimp (Crustacea: Anostraca) and mosquito larvae (Insecta: Diptera: Culicidae) as well as being strongly antifungal and antibacterial. The plant extracts also are highly inhibitory to thrombin, an enzyme involved in blood coagulation; **cultivation:** *Magnolia virginiana* has a long history of cultivation, being introduced to England as an ornamental in 1688. The hybrid cultivar *M.* ×*thompsoniana* (the first *Magnolia* hybrid to be described) was developed in England in 1838. Other cultivars include 'Aiken County,' 'Freeman,' 'Green Shadow,' 'Havener,' 'Henry Hicks,' 'Jim Wilson,' 'Katie-O,' 'Ludoviciana,' 'Maryland,' 'Mitton,' 'Pink Halo,' 'Plena,' 'Satellite,' and 'Sweet Thing'; **misc. products:** The Rappahannock inhaled the aroma from leaves or bark of *Magnolia virginiana* as a mild hallucinogen. The wood is used as pulpwood and in various furniture products; for making boxes, flats, and baskets for produce; and for popsickle sticks, tongue depressors, broomhandles, veneer, and venetian blinds; **weeds:** none in North America; **nonindigenous species:** none in North America

Systematics: Several molecular phylogenetic studies (e.g., Figure 1.10) have indicated that *Magnolia* is not monophyletic as traditionally circumscribed and would require the merger of several segregate genera (*Elmeria, Kmeria, Manglietia, Michelia*, and *Pachylarnax*) to resolve the group as a clade. If that course is taken (as followed here), *Magnolia* then represents the sister group of the small (but substantially divergent) genus *Liriodendron* (Figure 1.10). The distinctive *Magnolia virginiana* originally was assigned to *Magnolia* section *Magnolia* as the sole representative. However, phylogenetic analyses of molecular data (cpDNA RFLP, *ndhF* data) placed *M. virginiana* close to *M. grandiflora, M. guatemalensis, M. scheidiana, M. sharpii* and *M. tamaulipana*, which were assigned to section *Theorhodon*. Subsequent molecular analyses (*trnK, psbA-trnH* & *atpB-rbcL* spacer data) indicated that section *Theorhodon* was not monophyletic and that *M. guatemalensis* and *M. sharpii* indeed were close relatives of *M. virginiana*. Although morphological cladistic analysis indicated some affinity of *M. virginiana* to *M. delavayi* (section *Gwelimia*) and *M. pterocarpa* (section *Lirianthe*), molecular (*ndhF*) data showed those species to be related more distantly to *M. virginiana*. Further studies using additional DNA sequence data from nuclear and plastid regions consistently resolved *M. virginiana* among the species assigned to section *Theorhodon*, which resulted in an expanded circumscription of section *Magnolia* (retained nomenclaturally because it represents the type section of the genus) to include the species assigned formerly to section *Theorhodon*. The 10–11 members of that group that have been included in phylogenetic analyses resolve as a clade (Figure 1.11), which is now recognized to comprise section *Magnolia*. However, the placement of *M. virginiana* within section *Magnolia* is not consistent, with nuclear markers resolving it as the sister species to the remainder of the section (Figure 1.11), while plastid markers resolve it within the section in proximity of *M. panamensis* and *M. tamaulipana*. *Magnolia virginiana* occurs as two

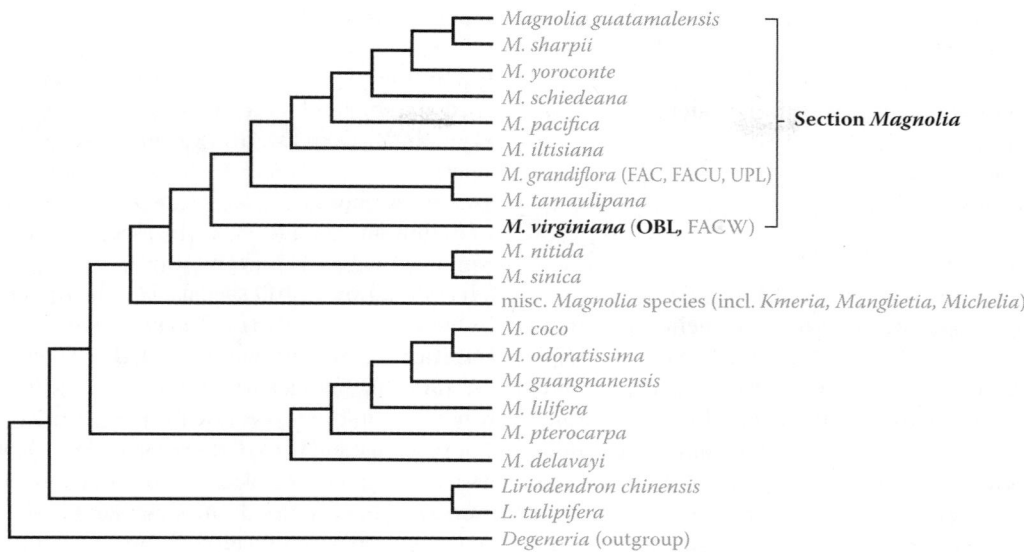

FIGURE 1.11 Interspecific relationships in *Magnolia* as indicated by analysis of combined nuclear DNA sequences. The results are similar (but not identical) to those obtained using combined plastid sequence data (Kim & Suh, 2013). Wetland indicator designations are provided in parentheses for North American species; taxa containing OBL indicators are highlighted in bold. (Adapted from Nie, Z.L. et al., *Mol. Phylogen. Evol.*, 48, 1027–1040, 2008.)

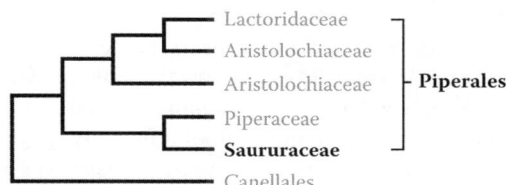

FIGURE 1.12 Interfamilial relationships within Piperales as indicated by phylogenetic analysis of combined DNA sequence data. Taxa containing OBL indicators are highlighted in bold. Although Aristolochiaceae also contain an OBL indicator, the genus (*Asarum*) was not included in the analysis. (Adapted from Soltis, D.E. et al., *Amer. J. Bot.*, 98, 704–730, 2011.)

distinct growth forms, having deciduous, multitrunked populations in the north of its range and evergreen, single-trunked populations in the south of its range. Relatively high DNA sequence divergence (0.08%) and the distribution of cpDNA haplotypes indicate a marked racial differentiation between the two distinct growth forms, which perhaps could be recognized taxonomically as subspecies. The basic chromosome number of *Magnolia* is $x = 19$. *Magnolia virginiana* ($2n = 38$) is a diploid and hybridizes with several diploids ($2n = 38$: *M. hypoleuca*, *M. insignis*, *M. macrophylla*, *M. sieboldii*, *M. tripetala*, and *M. yuyuanensis*) and with the hexaploid *M. grandiflora* ($2n = 114$). The hybrid cultivar *M.* ×*thompsoniana* ($2n = 38$) is intermediate morphologically between *M. virginiana* and *M. tripetala*.

Comments: *Magnolia virginiana* occurs along the coastal plain of the eastern United States.

References: Azuma et al., 1999, 2011; Baskin & Baskin, 1998; Breuss, 2016; Chistokhodova et al., 2002; Duberstein et al., 2014; Ducey et al., 2015; Duever & Riopelle, 1983; Elsey et al., 2015; Farag & Al-Mahdy, 2013; Forman &

Boerner, 1981; Heiser Jr., 1962; Jobes et al., 1998; Johnson, 1999; Kim & Suh, 2013; Kim et al., 2001; Li & Conran, 2003; Losada, 2014; McCormick et al., 2013; Meyer, 1997; Nie et al., 2008; Nitao et al., 1991, 1992; Outcalt, 1990; Priester, 1990; Qiu et al., 1993, 1995; Ranney & Gillooly, 2014; Slusher et al., 2014; Spongberg, 1976; Thien, 1974; Tyler-Julian et al., 2012.

ORDER 3: PIPERALES [4]

The Piperales include four families (Aristolochiaceae, Lactoridaceae, Piperaceae and Saururaceae), with OBL North American indicators occurring in two. Several authors (e.g., Judd et al., 2016) have merged Lactoridaceae within Aristolochiaceae, given its nested placement in some analyses (e.g., Figure 1.12). A close association of Piperaceae and Saururaceae is evident by their many shared morphological features (e.g., tetracytic stomata, obscurely bilateral floral symmetry, minute pollen, and perisperm) as well as by the consistent results of numerous molecular phylogenetic studies (Judd et al., 2016). The group consists primarily of herbaceous (or broadly rayed woody) plants containing benzylisoquinoline and aporphine alkaloids, with monocotyledonous sieve tube plastids and 3-merous flowers (Judd et al., 2016). Two North American families contain OBL indicators:

1. **Aristolochiaceae** Jussieu
2. **Saururaceae** E. Meyer

Family 1: Aristolochiaceae [5–7]

The "birthwort" family contains about 500 pantropical (occasionally temperate) species, with most of them represented by terrestrial members of *Aristolochia* and *Asarum*. Their habits include herbs, lianas, and shrubs, which frequently contain aristolochic acid or aporphine alkaloids (Cronquist, 1981). The bisexual, bilateral or radially symmetrical flowers have enlarged, petaloid sepals that are often fused into a curving pipe-shaped structure that provides an intricate mechanism for promoting outcrossing. Petals are reduced or absent. Although some species are self-pollinating, the modified calyx more typically attracts potential pollinators (Insecta: Diptera) by exuding nectar and a fetid or sweet odor. The peculiar shape traps the insect visitors within by a series of downward pointing hairs. During this protogynous phase, the insects can deposit pollen on the exposed stigmas, which eventually wither before the anthers dehisce and coat the insects with self-pollen. The subsequent withering of the trap hairs enables the insects to escape along with their fresh load of pollen destined for different flowers (summarized from Judd et al., 2016). The seeds are dispersed by a variety of mechanisms. Many are flattened and dispersed by wind or water, whereas others are adhesive or fleshy and dispersed by animals (endo- and exozoically). Some seeds are arillate and are dispersed by ants (Insecta: Hymenoptera: Formicidae). The seeds typically contain a rudimentary embryo and are morphophysiologically dormant (Baskin & Baskin, 1998).

Numerous species of *Aristolochia* and *Asarum* are cultivated as horticultural garden specimens or as indoor ornamentals. Because the species contain aristolochic acid (which is toxic to humans and potentially carcinogenic), they are inedible and unsuitable for medical applications (Lai et al., 2010). Herbal medicines that contain this substance should be avoided.

In North America, OBL indicators occur only within one genus (Figure 1.13):

1. *Asarum* L.

1. *Asarum*

Wild ginger; gingembre sauvage
Etymology: from the Greek *asaron*, an ancient name for the plants
Synonyms: none
Distribution: global: Asia; Europe; North America; **North America:** disjunct (East/West)
Diversity: global: 100 species; **North America:** 6 species
Indicators (USA): OBL: *Asarum lemmonii*
Habitat: freshwater; palustrine; **pH:** unknown; **depth:** <1 m; **life-form(s):** emergent (herb)
Key morphology: low, mat-forming plants (to 4 dm) arising from a horizontal rhizome; leaves (to 13 cm) paired, aromatic, partly evergreen, petioled (to 26 cm), cordate/reniform, apex rounded; flowers (to 3 cm) descending, solitary, terminal, peduncled (to 3.5 cm); calyx of three red/purple sepals (lobes to 1.5 cm), strongly reflexed, forming a tube; petals absent; stamens 12, the inner whorl longer (to 4 mm); ovary inferior; capsules (to 1.4 cm) fleshy, 6-ribbed; seeds (to 5 mm) ovoid, with fleshy appendage
Life history: duration: perennial (rhizomes); **asexual reproduction:** rhizomes; **pollination:** self; insect; **sexual condition:** hermaphroditic; **fruit:** capsules (common); **local dispersal:** ants (arillate seeds); **long-distance dispersal:** unknown
Imperilment: (1) *Asarum lemmonii* [G4]
Ecology: general: *Asarum* species occur characteristically among the understory vegetation of moist to dry forests, or sometimes on exposed, rocky slopes. They often occupy moist sites such as those along the margins of streams. Three of the North American species (50%) are wetland indicators, but only one is designated as OBL. The flowers of many species are autonomously self-pollinated but are outcrossed occasionally by flies (Insecta: Diptera). In some species, the flowers have putatively evolved morphological mimicry to resemble fungi (Basidiomycota), which tricks fungus gnats (Insecta: Diptera) into visiting them, inadvertently becoming pollinators in the act. The arillate seeds are dispersed by ants (Insecta: Hymenoptera: Formicidae) and have morphophysiological dormancy.

Asarum lemmonii S. **Watson** inhabits alluvial flats, bogs, bottomlands, ditches, meadows, roadsides, seeps, slopes (north-facing), springs, swales, and the margins of brooks,

FIGURE 1.13 Phylogenetic relationships in Aristolochiaceae as indicated by analysis of *trnL-F* sequence data. In this strict consensus tree, *Asarum* and *Saruma* resolve as sister genera. Taxa containing OBL indicators are highlighted in bold. (Adapted from Neinhuis, C. et al., *Plant Syst. Evol.*, 250, 7–26, 2005.)

rivers, and streams in coniferous woodlands at elevations of up to 1829 m. The plants are tolerant to dense shade and occur almost exclusively in shaded areas. The substrates have been described as organic or sand (granitic). Flowering occurs from May to July, with fruiting from June to August. The flowers are autonomously self-pollinated due to the close proximity of anthers and stigma surfaces. Although a protogynous phase enables outcrossing to occur (then mainly by flies), most flowers self. The seeds possess an elaiosome and are dispersed by ants (Insecta: Hymenoptera: Formicidae). The seeds presumably are morphophysiologically dormant, thereby requiring a period of cold stratification for maximum germination. Vegetative reproduction occurs by the extension of stoloniferous rhizomes, which can result in extensive, dense, clonal patches of plants. **Reported associates:** *Abies concolor, Acer circinatum, A. macrophyllum, Aconitum columbianum, Actaea rubra, Alnus rhombifolia, Asarum caudatum, A. hartwegii, Botrychium multifidum, Bromus vulgaris, Calocedrus decurrens, Carex amplifolia, Ceanothus integerrimus, Cephalanthera austiniae, Chimaphila umbellata, Circaea alpina, Cornus nuttallii, C. sericea, Darlingtonia californica, Darmera peltata, Dicentra formosa, Digitalis purpurea, Drosera rotundifolia, Equisetum hyemale, Galium triflorum, Glyceria elata, Juncus effusus, Lilium pardalinum, Lithophragma tenellum, Luzula parviflora, Neottia, Pinus lambertiana, P. ponderosa, Platanthera dilatata, Pseudotsuga menziesii, Pteridium aquilinum, Quercus, Rhamnus purshiana, Rhododendron, Rubus parvifolius, Rumex, Senecio triangularis, Symphoricarpos mollis, Taxus brevifolia, Thelypteris nevadensis, Trillium albidum, Veratrum californicum, Viola glabella.*

Use by wildlife: *Asarum lemmonii* is reportedly resistant to grazing by deer (Mammalia: Cervidae: *Odocoileus*).

Economic importance: food: The dried rootstock of *Asarum lemmonii* (boiled in water and thick syrup) has been used as a substitute for ginger; however, consumption of the plants should be avoided because of their toxic aristolochic acid content (see Aristolochiaceae); **medicinal:** none; **cultivation:**

Asarum lemmonii is sold as an ornamental, primarily for use in California native plant gardens; **misc. products:** none; **weeds:** none in North America; **nonindigenous species:** none in North America

Systematics: Various phylogenetic studies using morphological and/or DNA sequence data consistently render *Asarum* as monophyletic. All of the North American species (including *A. lemmonii*) occur in a subclade within a larger clade, which corresponds taxonomically to section *Asarum*. Phylogenetic analysis of morphological data places *A. lemmonii* close to *A. caulescens, A. himalaicum,* and *A. pulchellum*; however, molecular data (ITS sequences) or combined morphological/ITS data resolve *A. lemmonii* as the sister species of *A. caudatum*. However, analyses including larger amounts of DNA sequence data resolve *A. lemmonii* either as the sister species of *A. caudatum* (when analyzed by a Bayesian coalescence inference approach) or of *A. hartwegii* when analyzed using a maximum likelihood approach (Figure 1.14). All three species occur in western North America, with *A. lemmonii* and *A. hartwegii* sharing a nearly sympatric distribution. The basic chromosome number of *Asarum* is *x* = 13. *Asarum lemmonii* (and also *A. caudatum* and *A. hartwegii*) are diploids ($2n = 26$). There are no reports of natural hybrids involving *A. lemmonii*.

Comments: *Asarum lemmonii* is endemic to California's Sierra Nevada region.

References: Kelly, 1997, 1998, 2001; Kelly & González, 2003; Lu, 1982; Sinn et al., 2015; Whittemore et al., 1997.

Family 2: Saururaceae [4]

The "lizard tail" family is a small group of only six Asian and Central/North American species that are distributed among the following genera: *Anemopsis* (monotypic), *Gymnotheca* (2 species), *Houttuynia* (monotypic), *Saururus* (2 species). Although many authors include the monotypic *Circaeocarpus* (*C. saururoides*) in Saururaceae, phylogenetic analysis of morphological and DNA sequence data (Meng et al., 2003; Meng, 2004) consistently resolve the taxon within Piperaceae, where it is synonymous with *Zippelia begoniifolia* (Figure 1.15).

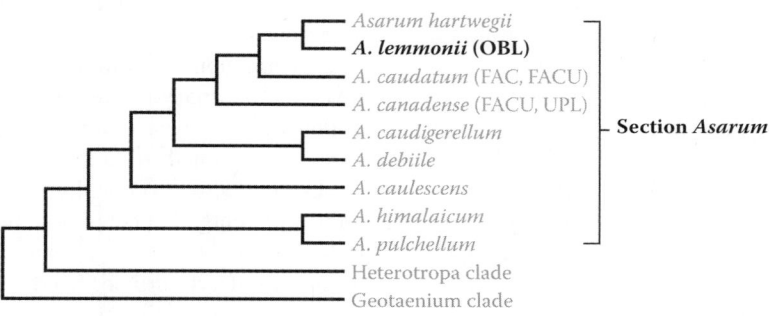

FIGURE 1.14 Phylogenetic relationships in *Asarum* as indicated by maximum likelihood (ML) analysis of combined DNA sequence data representing eight plastid and nuclear regions. *Asarum lemmonii* (OBL) is a relatively derived member of section *Asarum* and resolves as the sister species of *A. hartwegii*, another western North American species. However, analysis of the same data using Bayesian coelescence inference places *A. lemmonii* and *A. caudatum* as sister species, a result similar to that obtained by combined analysis of morphological and nrITS data (Kelly, 2001). The wetland indicator status is shown for North American species (taxa containing OBL indicators are highlighted in bold). (Redrawn from Sinn, B.T. et al., *Mol. Phylogen. Evol.*, 89, 194–204, 2015.)

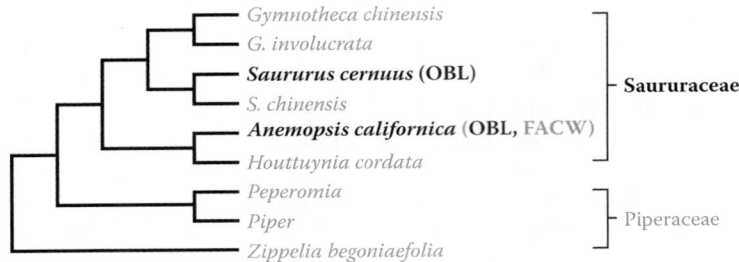

FIGURE 1.15 Phylogenetic relationships in Saururaceae as indicated by analyses of combined molecular or morphological data. All species of Saururaceae are included in this analysis, which places *Zippelia* (formerly included in Saururace as *Circaeocarpus*) within Piperaceae instead. The wetland indicator status is shown for North American species (taxa containing OBL indicators are highlighted in bold). (Adapted from Meng, S.-W., *Pract. Bioinf.*, 2004, 245–268, 2004.)

Members of Saururaceae occur commonly in wet habitats, and both of the North American genera contain OBL indicators (Figure 1.15). These are herbaceous, rhizomatous perennials with simple, alternate, stipular leaves that are often cordate in outline. The bisexual flowers lack a perianth and are arranged in dense terminal spikes or in lax spiciform racemes. The capsular or schizocarp fruits contain a minute embryo embedded within perisperm (Buddell II & Thieret, 1997). The flowers of at least some species are genetically self-incompatible (Schroeder & Weller, 1997; Pontieri & Sage, 1999). Pollination occurs by insects (Insecta: Coleoptera; Diptera; Hymenoptera) or wind (Thien et al., 1994, 2000). The seeds have no specialized dispersal mechanisms. They are nondormant, at least in *Saururus*.

Ornamental cultivars have been developed in *Houttuynia* and *Saururus*. There have been numerous medicinal properties attributed to *Houttuynia cordata* and other species, including the ability to scavenge free radicals, hepatoprotective properties, and activities against allergies, bacteria, cancer, inflammation, leukemia, mutagens, obesity, sinusitis, and viruses (Kumar et al., 2014). OBL indicators occur in two genera:

1. *Anemopsis*
2. *Saururus*

1. Anemopsis

Yerba mansa, vavish
Etymology: resembling the windflower *Anemone*
Synonyms: *Anemia*
Distribution: global: North America; **North America:** southwestern
Diversity: global: 1 species; **North America:** 1 species
Indicators (USA): OBL; FACW: *Anemopsis californica*
Habitat: brackish, freshwater; palustrine; **pH:** 5.0–9.0; **depth:** <1 m; **life-form(s):** emergent (herb)
Key morphology: shoots (to 80 cm) hollow, aromatic (spicy); principal leaves (to 20 cm) long-petioled (to 40 cm), arising from a basal cluster, the cauline leaves smaller; inflorescence a conical, flower-like spike (to 4 cm), subtended by a whorl of 4–9 pink or white petaloid bracts (to 3.5 cm); true petals absent; capsules (to 7 mm) brown; seeds (to 1.5 mm) brown, reticulate

Life history: duration: perennial (rhizomes, stolons); **asexual reproduction:** rhizomes, stolons; **pollination:** insect **sexual condition:** hermaphroditic; **fruit:** capsules (common); **local dispersal:** stolons (root at nodes); **long-distance dispersal:** seeds
Imperilment: (1) *Anemopsis californica* [G5]; S1 (OK); S2 (UT)
Ecology: general: *Anemopsis* is monotypic.

Anemopsis californica **(Nutt.) Hook. & Arn.** inhabits beaches, bottomlands, cienega, depressions, ditches, dunes, flats, floodplains, marshes, meadows, mudflats, ravines, roadsides, salt marshes, scrub, seeps, sinks, slopes (to 2%), sloughs, springs, streambeds, swales, swamps, washes, woodlands, and the margins of lagoons, lakes, ponds, pools, rivers, streams, and thickets at elevations of up to 2106 m. The plants are salt-tolerant (categorized as hydrohalophytes) and often occur in shallow standing water, typically in areas that receive full sunlight. The poorly drained substrates are characterized as alkaline or saline (pH: 5.0–9.0) and include alluvium, basalt (obsidian), clay (Willow series), clay loam, clayey Moenkopi, granite (decomposed), granodiorite, gravel, gumbo clay, Gypill-badland soil, limestone, loam, muck, mud, muscovite-biotite monzogranite, peaty adobe, Poway "clast" (clay and cobble), rock, rocky gravel, sand, sandy clay, sandy loam (fine), sandy silt, shale, silt, silty loam, and stony sand. Flowering and fruiting occur from March to September. The flowers are self-incompatibile, with substantially higher seed production resulting from outcrossed vs. selfed pollinations. The primary pollinators are presumably unspecialized pollen-eating insects (Insecta: Coleoptera; Diptera). Because of the dense clonal structure of populations (e.g., evidenced by a lack of electrophoretically detectable allozyme variation), an insufficiently diverse pool of incompatibility alleles potentially limits their seed set. The seed dispersal mechanisms have not been described. Seeds frozen for over two years at −20°C (at 15% relative humidity) have retained 89% viability. Maximum germination has been achieved for seeds placed on 1% agar supplemented with GA3 gibberellic acid (250 mg/L) and incubated at 20°C under an 8/16 light/dark regime. Vegetative reproduction occurs by means of stoloniferous rhizomes, which can lead to the development of vast swards of clonal individuals. Some populations are threatened by overcollection for herbal remedies, loss of wetland habitat, and the lowering of the water table due to irrigation activities. The plants accelerate the decay of leaf matter and produce root oils that inhibit microorganism growth by acidifying and aerating the soil. Several strains of fungal endophytes (Fungi: Ascomycota: Chaetomiaceae: *Chaetomium cupreum*) have been detected in the roots. The plants have been successfully propagated *in vitro* using nodal explants. **Reported associates:** *Abronia maritima, A. villosa, Achillea millefolium, Achnatherum hymenoides, Acleisanthes nevadensis, Acmispon glabrus, Adenostoma fasciculatum, Adiantum capillus-veneris, Agrostis stolonifera, Allenrolfea occidentalis, Almutaster pauciflorus, Alnus oblongifolia, Ambrosia chamissonis, A. monogyra, A. psilostachya, A. salsola, Amelanchier utahensis, Amorpha fruticosa, Amphipappus fremontii, Anagallis arvensis, Apium graveolens, Apocynum cannabinum, Arctostaphylos glandulosa, A. glauca, Argemone pleiacantha, Arida carnosa, Artemisia californica, A. ludoviciana, A. tridentata, Asclepias fascicularis, A. speciosa, A. subulata, A. subverticillata, Atriplex canescens, A. confertifolia, A. hymenelytra, A. lentiformis, A. patula, A. phyllostegia, A. polycarpa, A. prostrata, A. serenana, A. suberecta, A. torreyi, Avena barbata, Baccharis douglasii, B. pilularis, B. salicifolia, B. sarothroides, B. sergiloides, Bahiopsis parishii, Baileya pleniradiata, Bassia hyssopifolia, Berula erecta, Bidens laevis, B. leptocephalus, Bolboschoenus maritimus, Brachypodium distachyon, Brassica, Brickellia desertorum, Bromus catharticus, B. diandrus, B. rubens, B. tectorum, Cakile maritima, Calibrachoa parviflora, Calochortus splendens, C. striatus, Camissoniopsis cheiranthifolia, Carex pansa, C. praegracilis, C. senta, Carpobrotus edulis, Catalpa, Ceanothus crassifolius, C. greggii, Celtis reticulata, Cenchrus clandestinus, Centromadia pungens, Cephalanthus occidentalis, Cercocarpus betuloides, Chara, Chenopodium rubrum, C. watsonii, Chilopsis linearis, Chloropyron maritimum, Chorispora tenella, Chylismia walkeri, Cirsium arizonicum, C. mohavense, C. scariosum, C. vulgare, Clematis, Cleomella obtusifolia, Coleogyne ramosissima, Colocasia esculenta, Conium maculatum, Cornus sericea, Cosmos parviflorus, Cotula coronopifolia, Cressa truxillensis, Crinum, Croton setiger, C. texensis, Cryptantha, Cupressus, Cylindropuntia echinocarpa, C. ganderi, Cynodon dactylon, Cyperus esculentus, C. niger, Dasyochloa pulchella, Datura wrightii, Descurainia pinnata, Diplacus aurantiacus, Dipsacus, Distichlis spicata, Dudleya variegata, Echinocereus engelmannii, Echinochloa crusgalli, Eclipta prostrata, Elaeagnus angustifolia, Eleocharis macrostachya, E. parishii, E. rostellata, Elymus canadensis, Epilobium canum, Epipactis gigantea, Equisetum arvense, E. laevigatum, E. telmateia, Eragrostis cilianensis, E. pectinacea, Ericameria nauseosa, E. paniculata, Eriodictyon crassifolium, E. trichocalyx, Eriogonum fasciculatum, E. inflatum, Erodium cicutarium, Eucalyptus, Euphorbia albomarginata, Fallugia paradoxa, Ferocactus viridescens, Ficus carica, Foeniculum vulgare, Forestiera, Frankenia salina, Fraxinus velutina, Funastrum cynanchoides, Glycyrrhiza lepidota, Grindelia squarrosa, Gutierrezia sarothrae, Hazardia, Hedypnois cretica, Heliotropium curassavicum, Helminthotheca echioides, Heterotheca villosa, Hordeum depressum, H. jubatum, H. murinum, Hymenothrix wrightii, Isocoma acradenia, Jaumea carnosa, Juglans major, Juncus acutus, J. arcticus, J. balticus, J. bufonius, J. cooperi, J. textilis, J. xiphioides, Juniperus californica, J. monosperma, J. osteosperma, Kochia californica, K. scoparia, Krameria erecta, Lactuca serriola, Larrea tridentata, Leonurus cardiaca, Lepidium lasiocarpum, Leymus triticoides, Lotus corniculatus, Ludwigia peploides, Lycium andersonii, L. pallidum, Lythrum californicum, Malosma laurina, Marrubium vulgare, Marsilea, Melilotus albus, M. indicus, M. officinalis, Mentzelia nitens, Mimulus cardinalis, M. guttatus, Muhlenbergia asperifolia, Nasturtium officinale, Nepeta cataria, Nicotiana glauca, N. quadrivalvis, Nitrophila occidentalis, Nymphaea, Oenothera elata, Opuntia basilaris, O. chlorotica, Paspalum distichum, Penstemon centranthifolius, Peritoma arborea, P. lutea, Persicaria lapathifolia, Phragmites australis, Phyla, Pinus edulis, P. monophylla, Plantago lanceolata, P. major, Platanus racemosa, P. wrightii, Pluchea sericea, Poa pratensis, P. secunda, Polygonum aviculare, Polypogon interruptus, P. monspeliensis, Pontederia, Populus deltoides, P. fremontii, Potentilla anserina, P. gracilis, Prosopis glandulosa, P. pubescens, P. velutina, Prunus armeniaca, P. ilicifolia, Psorothamnus fremontii, Pyrrocoma racemosa, Quercus agrifolia, Q. ×alvordiana, Q. chrysolepis, Q. cornelius-mulleri, Q. dumosa, Q. lobata, Ranunculus californicus, R. cymbalaria, R. sceleratus, Rhamnus californica, Rhus integrifolia, R. ovata, Rorippa, Rosa californica, R. woodsii, Rubus ursinus, Rumex californicus, R. crispus, R. salicifolius, Salicornia depressa, Salix amygdaloides, S. exigua, S. gooddingii, S. laevigata, S. lasiolepis, Salsola, Salvia mellifera, Sambucus nigra, Samolus valerandi, Sapindus saponaria, Schedonorus arundinaceus, Schoenoplectus acutus, S. americanus, S. californicus, S. pungens, Sclerocactus johnsonii, Scrophularia californica, Senegalia greggii, Sesuvium verrucosum, Sidalcea neomexicana, Silybum marianum, Sisymbrium irio, Sisyrinchium bellum, S. demissum, Solidago spectabilis, S. velutina, Sonchus asper, S. oleraceus, Spartina gracilis, Sphaeralcea ambigua, S. fendleri, Spiranthes infernalis, Sporobolus airoides, S. wrightii, Stachys albens, Stanleya pinnata, Stephanomeria pauciflora, Suaeda nigra, Symphyotrichum subulatum, Tamarix chinensis, T. ramosissima, Taraxacum officinale, Tauschia parishii, Tiquilia latior, Toxicodendron diversilobum, Triglochin, Typha domingensis, T. latifolia, Urtica dioica, Venegasia carpesioides, Verbascum thapsus, Verbena lasiostachys, Veronica americana, V. anagallis-aquatica, Vicia benghalensis, Vitis arizonica, V. girdiana, Wislizenia palmeri, Xanthium strumarium, Yucca brevifolia, Y. schidigera, Zeltnera exaltata, Ziziphus obtusifolia.*

Use by wildlife: *Anemopsis* plants are eaten occasionally by wild turkeys (Aves: Phasianidae: *Meleagris gallopavo*). The leaves are grazed by the saltmarsh caterpillar (Insecta: Lepidoptera: Erebidae: *Estigmene acrea*). The leaves are host to a rust fungus (Fungi: Basidiomycota: Pucciniaceae).

Economic importance: food: Pulverized *Anemopsis* seeds have been cooked as mush or baked as bread; **medicinal:** *Anemopsis* is used extensively as a medicinal plant by the Chumash, Mayo, Mexican, Pima, Shoshone, Yaqui, and other tribes. The plants contain berberine, an antimicrobial, and methyleugenol, an antispasmodic, which relieves irritated stomachs. The roots are a source of furofuran lignans (asarinin and sesamin) that have been shown to inhibit the growth of nontuberculous (but still pathogenic) mycobacteria (Bacteria: Mycobacteriaceae: *Mycobacterium*). *Anemopsis* is used to treat inflammation of mucous membranes, swollen gums, toothaches, and sore throat. Root infusions are taken as a diuretic to reduce excess uric acid associated with rheumatic diseases. *Anemopsis* supposedly prevents the accumulation of uric acid, which can cause kidney stones. Its anti-inflammatory properties are useful for treating arthritis. The fungicidal properties of dried root powder are said to alleviate athlete's foot and diaper rash. Root solutions are used to treat various sores and assist with recovery following childbirth. A poultice made from the leaves has been used to relieve muscle swelling and inflammation; **cultivation:** *Anemopsis* has minor use as an ornamental ground cover in parks and around water gardens; **misc. products:** Some native tribes strung the roots onto necklaces as beads to ward off malaria; **weeds:** *Anemopsis californica* can become intrusive in saline habitats and along roadsides; **nonindigenous species:** *Anemopsis californica* was introduced recently to Nebraska along roadsides and also to Colorado.

Systematics: *Anemopsis* is monotypic, and consequently monophyletic. Phylogenetic analyses of morphological and various molecular data consistently resolve *Anemopsis* as the sister genus to *Houttuynia*, which together comprise a clade sister to the remainder of Saururaceae (Figure 1.15). The basic chromosome number of *Anemopsis* is $x = 11$. *Anemopsis californica* ($2n = 22$) is diploid. Hybrids involving *Anemopsis* have not been reported.

Comments: *Anemopsis californica* occurs commonly in the southwestern United States.

References: Aronson, 1989; Bussey III et al., 2014, 2015; Holtzman, 1990; Kay, 1996; Keener & Hampton, 1968; Lindsey, 1948; Meng et al., 2003; Moore, 1989; Rodriguez-Sahagun et al., 2012; Rolfsmeier et al., 1999; Schroeder & Weller, 1997; Sheviak, 1989; Wen et al., 2016.

2. *Saururus*

Lizard's tail; lézardelle penchée

Etymology: from the Greek *sauros oura* ("lizard tail") in reference to the inflorescence shape

Synonyms: none

Distribution: global: Asia; North America; **North America:** eastern

Diversity: global: 2 species; **North America:** 1 species

Indicators (USA): OBL: *Saururus cernuus*

Habitat: brackish, fresh; lacustrine, palustrine, riverine; **pH:** 3.0–8.7; **depth:** <1 m; **life-form(s):** emergent (herb)

Key morphology: stems (to 1.2 m) rhizomatous; leaves (to 25 cm) arising from an erect stem, cordate, petioled (to 10 cm), the tips acute; inflorescence a tapering, white, flaccid (becoming straight in fruit), "tail-like" raceme (to 35 cm), comprising numerous (up to 414) small, bisexual flowers; perianth absent; stamens 6, filaments white; schizocarps (to 3 mm) comprising 4 single-seeded, indehiscent mericarps; seeds (to 1.3 mm) brown, smooth

Life history: duration: perennial (rhizomes); **asexual reproduction:** rhizomes; **pollination:** insect, wind; **sexual condition:** hermaphroditic; **fruit:** schizocarps (common); **local dispersal:** rhizomes (floating fragments); **long-distance dispersal:** seeds

Imperilment: (1) *Saururus cernuus* [G5]; SH (MA); S1 (CT, RI); S2 (KS, QC); S3 (IL, ON)

Ecology: general: *Saururus* species are obligate aquatics and occupy similar habitats, but only the North American species has been ranked as a wetland indicator; the second species (*S. chinensis*) occurs in Asia. The plants are typically later successional species that occur in areas of low disturbance throughout slightly elevated areas of semipermanently flooded sites. The flowers are protogynous, self-incompatible, and pollinated by a combination of wind and insect (Insecta: Coleoptera; Diptera; Hymenoptera) vectors.

Saururus cernuus L. grows on wet, exposed substrates or in fresh to brackish (salinity to 1.5 ppt), nontidal or tidal standing waters (0.1–0.5 m deep) that occur within bogs, borrow pits, bottomlands, depressions, ditches, fens, flats, floodplains, hammocks, marshes, meadows, mud/sand bars, oxbows, pocosins, pools, ravines, rice fields, roadsides, seeps, shorelines, slopes, sloughs, springs, swales, swamps, terraces, thickets, woodlands (coniferous; deciduous), and along the margins of bayous, lakes, ponds, rivers, and streams at elevations of up to 305 m. The plants can tolerate a remarkable range of environmental conditions including areas ranging from full sun to dense shade and acidic to calcareous (pH: 3.0–8.7) substrates, which include alluvium (sandy), clay, gravel, gumbo soil, loam, muck, mud, Placid series (typic humaquepts), rock, Rutlege loamy sand, sand, sandy clay (with leaf litter), sandy loam, sandy peat, sandy silt, silt, silty clay, silty loam, and silty sand. The plants thrive in sites with high levels of organic matter (to 38%) and higher pH (>6.0). However, they are killed by prolonged, deep (>1 m) inundation. Flowering occurs from May to December, with fruiting extending from July to November. The flowers open along the racemes at a rate of about 1.52 cm/day. The fruits mature in about 3 weeks after flowering. The protogynous flowers are self-compatible and are pollinated primarily by the wind and also by insects (Insecta). They are also said to possess a relatively novel mechanism of insect-mediated wind pollination. This occurs when the inflorescences are jarred by large-bodied insects such as dragonflies (Insecta: Odonata), resuting in the release of a pollen cloud (spreading across several meters) that then becomes entrained by wind currents. Seed set can approach 60% in outcrossed flowers. Although self-incompatibility limits seed production in clonal populations, some colonies can produce up to 1800 seeds/m². The seeds have been reported variously as nondormant or as requiring a dormancy period. They are said to germinate optimally at 25°C. Seeds stratified (at 2°C) over the winter germinate well (e.g., 40%) at temperatures from 21°C to 23°C and also at ambient greenhouse temperatures. A seed bank of undetermined duration develops. The seedlings float and provide a means for waterborne dispersal; however, established seedlings are rarely encountered in the field where adult plants occur (perhaps because they are carried away by water to different sites). Rhizomes enable the plants to reproduce vegetatively and to grow throughout the year in some southern localities. Their peak biomass (up to 2.25 kg/m^2) is attained by the end of May. Populations may be threatened by drainage, excessive siltation, inundation, and by competition from invasive species such as *Lythrum salicaria*. **Reported associates:** *Acer rubrum, A. saccharinum, Acrostichum danaeifolium, Ageratina altissima, Alisma subcordatum, Alnus, Alternanthera philoxeroides, Amorpha fruticosa, Ampelopsis arborea, Amsonia rigida, Annona glabra, Aralia spinosa, Arundinaria gigantea, Asclepias incarnata, A. perennis, Asimina triloba, Athyrium filix-femina, Baccharis halimifolia, Bacopa monnieri, Betula alleghaniensis, B. nigra, Bidens comosus, B. mitis, Boehmeria cylindrica, Brasenia schreberi, Brunnichia ovata, Callicarpa americana, Caltha palustris, Campsis radicans, Cardamine bulbosa, Carex cherokeensis, C. comosa, C. conjuncta, C. corrugata, C. cristatella, C. crus-corvi, C. glaucescens, C. granularis, C. grayi, C. gynandra, C. hyalinolepis, C. hystricina, C. intumescens, C. lacustris, C. lasiocarpa, C. louisianica, C. lupulina, C. lurida, C. muskingumensis, C. radiata, C. rosea, C. stricta, C. tribuloides, C. typhina, Carpinus caroliniana, Carya aquatica, C. glabra, Ceanothus americanus, Celtis occidentalis, Centella asiatica, Cephalanthus occidentalis, Ceratophyllum demersum, Chasmanthium laxum, Chelone glabra, Cinna arundinacea, Cladium jamaicense, Colocasia esculenta, Cornus alternifolia, C. amomum, C. florida, C. foemina, C. racemosa, C. sericea, Crataegus marshallii, Cryptotaenia canadensis, Cuscuta gronovii, Cyperus haspan, Didiplis diandra, Diospyros virginiana, Eichhornia crassipes, Eleocharis obtusa, Elymus virginicus, Elytraria caroliniensis, Epilobium coloratum, Equisetum arvense, Eupatorium perfoliatum, E. serotinum, Eutrochium fistulosum, Fagus grandifolia, Festuca subverticillata, Ficus aurea, Fraxinus caroliniana, F. nigra, F. pennsylvanica, F. profunda, Geum canadense, Glyceria septentrionalis, G. striata, Helenium autumnale, Heliotropium indicum, Hibiscus laevis, H. moscheutos, Hydrocotyle umbellata, H. verticillata, Hydrolea, Hymenocallis, Hypericum denticulatum, Ilex cassine, I. decidua, Iris pseudacorus, I. virginica, Itea virginica, Iva annua, Impatiens capensis, Juncus effusus, J. marginatus, J. polycephalus, Juniperus, Justicia americana, J. ovata, Lachnanthes caroliniana, Laportea canadensis, Leersia lenticularis, L. oryzoides, L. virginica, Lemna minor, Lindera benzoin, Lindernia dubia, Liquidambar styraciflua, Litsea aestivalis, Lobelia cardinalis, Lonicera maackii, Ludwigia palustris, Lycopus americanus, Lyonia lucida, Lysimachia ciliata, L. nummularia, L. terrestris, Lythrum salicaria, Maianthemum stellatum, Magnolia virginiana, Mikania scandens, Mimulus ringens, Morus rubra, Myrica cerifera, Nasturtium officinale, Nymphaea odorata, Nymphoides aquatica, Nyssa aquatica, N. biflora, N. sylvatica, Oenothera gaura, Onoclea sensibilis, Osmunda regalis, Osmundastrum cinnamomeum, Peltandra virginica, Persea palustris, Persicaria amphiba, P. arifolia, P. hydropiperoides, P. maculosa, P. punctata, P. virginiana, Phanopyrum gymnocarpon, Phalaris arundinacea, Phragmites australis, Phyla nodiflora, Physostegia virginiana, Pilea pumila, Pinus echinata, P. elliottii, P. palustris, P. taeda, Planera aquatica, Platanthera flava, Platanus occidentalis, Podophyllum peltatum, Pontedaria cordata, Populus heterophylla, Proserpinaca palustris, Pteridium aquilinum, Ptilimnium capillaceum, Quercus alba, Q. bicolor, Q. laurifolia, Q. lyrata, Q. macrocarpa, Q. michauxii, Q. nigra, Q. rubra, Ranunculus hispidus, Rhododendron, Rhus copallinum, Rhynchospora, Robinia pseudoacacia, Rosa palustris, Rubus argutus, R. trivialis, Rudbeckia hirta, R. laciniata, Rumex verticillatus, Sabal minor, Sabatia, Sagittaria latifolia, Salix caroliniana, S. interior, S. nigra, Sambucus nigra, Sanicula odorata, Sarracenia, Sassafras albidum, Schoenoplectus tabernaemontani, Scutellaria lateriflora, Sesbania vesicaria, Silene nivea, Sium suave, Smilax bonanox, S. glauca, S. tamnoides, Solanum carolinense, Solidago caesia, S. patula, Sparganium eurycarpum, Sphagnum, Stachys tenuifolia, Symphyotrichum lanceolatum, S. lateriflorum, S. ontarionis, S. puniceum, Symplocarpus foetidus, Taxodium ascendens, T. distichum, Thelypteris palustris, Tilia americana, Toxicodendron radicans, T. vernix, Triadenum virginicum, Typha domingensis, T. latifolia, Ulmus alata, U. americana, U. rubra, Utricularia gibba, Verbascum thapsus, Verbesina alternifolia, Vernonia fasciculata, Veronica peregrina, Viburnum lentago, Vicia sativa, Viola lanceolata, V. sororia, Vitis riparia, V. rotundifolia, Woodwardia virginica.*

Use by wildlife: *Saururus cernuus* plants can contain up to 16% protein, with their roots comprising about 17% carbohydrate. They provide cover for beavers (Mammalia: Castoridae: *Castor canadensis*); various birds (Aves) including Canada geese (Anatidae: *Branta canadensis*), common yellowthroats (Parulidae: *Geothlypis trichas*), and wood ducks (Anatidae: *Aix sponsa*); crayfish (Crustacea: Decapoda: Cambaridae: *Cambarus*); several fish (Osteichthyes) including darters (Percidae: *Etheostoma olmstedi*), sunfish (Centrarchidae: *Lepomis macrochirus*), and topminnows (Poeciliidae: *Gambusia affinis*); frogs (Amphibia: Ranidae: *Rana catesbeiana, R. sphenocephala*); muskrats (Mammalia: Cricetidae: *Ondatra zibethicus*); northern water snakes (Reptilia: Colubridae: *Nerodia sipedon*); raccoons (Mammalia: Procyonidae: *Procyon lotor*); and spotted salamanders (Amphibia: Ambystomatidae: *Ambystoma maculatum*). They also harbor numerous invertebrates including euglenoids (Protozoa: Euglenaceae: *Euglena gracilis*); insects (Insecta) such as dragonflies (Odonata: Aeshnidae: *Anax junius*), mosquitoes (Diptera: Culicidae: *Aedes albopictus, Anopheles quadrimaculatus*), and predacious diving beetles (Coleoptera: Dytiscidae: *Dytiscus*); leeches (Annelida: Hirudinidae: *Macrobdella decora*); and snails

(Gastropoda: Lymnaeidae: *Lymnaea stagnalis*). The presence of *S. cernuus* has been found to correlate positively with relative amphibian species diversity. The plants host several stink bugs (Insecta: Hemiptera: Pentatomidae: *Euschistus ictericus*, *E. tristigmus luridus*, *Mormidea lugens*), and their pollen is collected by tarnished plant bugs (Insecta: Hemiptera: Miridae: *Lygus lineolaris*). The seeds are eaten by wood ducks (Aves: Anatidae: *Aix sponsa*) and other waterfowl. Although the leaves contain lignans and volatile oils that deter herbivory, the foliage is eaten by caterpillars (Insecta: Lepidoptera: Noctuidae: *Parapamea buffaloensis*), crane fly larvae (Diptera: Tipulidae: *Tipula*, snails (*Lymnaea stagnalis*), and especially by turtles (Reptilia: Emydidae: *Chrysemys picta picta*; Chelydridae: *Chelydra serpentina*). The stems are used as a substrate for egg deposition by insects, frogs, and salamanders. An angular leaf spot disease is caused by a fungus (Fungi: Ascomycota: Mycosphaerellaceae: *Passalora saururi*).

Economic importance: food: not edible; **medicinal:** *Saururus cernuus* has been used as an emollient and to treat irritation and inflammation of the bladder, breast (as root poultice), kidney, prostate, and urinary passages. Tea made from the leaves is drunk to relieve back and breast pain. The Cherokee and Choctaw tribes administered a poultice made from the mashed roots to treat wounds. The Ojibwa made washes from the plants to treat rheumatism and stomach problems. The Seminoles prepared treatments from the roots and other parts for treating fever, rheumatism, and spider bites. The plant extracts exhibit neuroleptic properties and have been shown experimentally to contain potent anticancer compounds. The extracts also inhibit expression of TNFα, IL-1β, COX-2, and iNOS, indicating that they are potentially useful therapeutically to treat metabolic inflammation, as has been the practice in traditional Creole medicine; **cultivation:** *Saururus cernuus* is sold as an ornamental for aquariums and garden ponds. It is a species recommended for wetland restoration and enhancement. The roots help to stabilize pond banks; **misc. products:** *Saururus cernuus* has been planted in created and restored wetlands, in storm water retention ponds, and in wetlands used to treat agricultural and domestic wastewater and coal mine drainage. It is particularly suitable for tolerating sustained soil water deficits; **weeds:** *Saururus cernuus* sometimes attains nuisance levels in reservoirs; **nonindigenous species:** *Saururus cernuus* has been introduced to Italy and other areas of Europe.

Systematics: Phylogenetic analyses of various morphological and DNA sequence data indicate that *Saururus* is monophyletic and is related most closely to *Gymnotheca* (Figure 1.15). The basic chromosome number of *Saururus* is $x = 11$. Populations of *S. cernuus* appear to be uniformly diploid ($2n = 22$) throughout its distributional range. Hybrids involving *S. cernuus* have not been reported.

Comments: *Saururus cernuus* is common throughout the eastern United States and adjacent Canada.

References: Badisa et al., 2007; Baldwin & Speese, 1949; Batcher, 2003; Boudreau et al., 2014; Boyd & Walley, 1972; Buddell & Thieret, 1997; Dybas, 2016; Hall, 1940; Hussner, 2012; Jones & Allen, 2013; Kubanek et al., 2001; Lindquist et al., 2013; Meng et al., 2003; Parker et al., 2007b; Pontieri & Sage, 1999; Sigmon et al., 2013; Swanson, 2012; Thien et al., 1994.

2 Dicotyledons II
Basal Tricolpates

The eudicots ("tricolpate dicots") have been defined as a clade on the basis of their pollen structure (triclopate- or tricolpate-derived types) as well as by their anomocytic stomata and prevalent S-type (starch bearing) sieve-element plastids (Judd et al., 2016). The flowers are structured with alternating whorls of parts and have stamens with clearly defined filaments (Judd et al., 2016).

The monophyly of the group was first demonstrated by phylogenetic analysis of *rbcL* sequence data (Chase et al., 1993), a result that has been confirmed repeatedly by the analysis of numerous other gene regions (Judd et al., 2016). Most taxa occur within a clade ("Pentapetalae") comprising the Asterid, Caryophyllid, and Rosid families, which accounts for the majority of species. A few outlying orders (Buxales, Gunnerales, Nelumbonales, Proteales, Ranunculales, Trochodendrales) represent the remainder of the tricolpate taxa (Figure 2.1). Among these outliers, Gunnerales are regarded as "core eudicots," with the other orders referred to as "basal tricolpates"; Ranunculales resolve as the sister group to the entire assemblage (Judd et al., 1996).

ORDER 1: RANUNCULALES [7]

This order represents a diverse group of about 3490 mostly herbaceous species (Judd et al., 2016). The members commonly share lobed or compound leaves, assorted alkaloids (e.g., benzylisoquinoline, berberine, or morphine), and flowers with numerous stamens (Judd et al., 2016). A number of studies place the group as the sister to the remainder of the tricolpate dicotyledons (e.g., Figure 2.1). The majority of the order comprises terrestrial species. Following the reclassification of *Corydalis aquae-gelidae* [obligate wetland (OBL) in the 1996 list] as a variety of *C. caseana* [facultative wetland (FACW)] in the 2013 indicator list, only one family in the order contains OBL indicators:

1. Ranunculaceae Jussieu

Family 1: Ranunculaceae [60]

The buttercup family (Ranunculaceae) represents a clade of about 1700 species that are primarily herbaceous plants containing alkaloids or the lactone glycoside ranunculin (Judd et al., 2016). The alkaloids impart poisonous properties to many of the species. The family is related most closely to Berberidaceae (Figure 2.2). Phylogenetic analyses of multiple DNA data sets (e.g., Cossard et al., 2016) indicate that the family is monophyletic and can be divided fairly well into five subfamilies: Coptidoideae, Glaucidioideae, Hydrastidoideae, Ranunculoideae, and Thalictroideae (Ranunculoideae may be biphyletic; Figure 2.3). Several genera with OBL species (*Coptidium*, *Halerpestes*, and *Kumlienia*) have been segregated from *Ranunculus* as the result of phylogenetic investigations (Emadzade et al., 2010; Hoffmann et al., 2010). The family principally is North Temperate in its distribution and is widespread throughout the boreal regions of the Northern Hemisphere (Judd et al., 2016). The family name is derived from the Latin *rana unculus* ("little frog"), which is a reference to the wet sites inhabited by many of the species. The group is also represented well in alpine floras.

The foliage of buttercups is often palmately compound, a feature that helps distinguish the plants from the morphologically similar Rosaceae, which typically possess pinnately compound leaves. The flowers normally are radial and bisexual (but occasionally bilateral or unisexual) and are constructed of few to numerous tepals or differentiated five-merous perianths; the stamens are numerous and the gynoecium apocarpous (often with numerous pistils). The fruits are achenes, berries, or follicles. The flowers are often fairly showy and are pollinated primarily by insects (Insecta), and also by hummingbirds (Aves: Trochilidae), or by the wind. The seeds are dispersed variously (Judd et al., 2016). Some are modified with plumes to facilitate wind dispersal, while others have various structures enabling them to attach to animal fur. The smaller, shed seeds can be dispersed abiotically by water or wind. Others are dispersed by ants (Insecta: Hymenoptera: Formicidae) or (for berries) by birds (Aves).

Buttercups contain a large number of economically important genera and include plants that are grown as popular garden ornamentals. These include *Aconitum* (monkshood), *Actaea* (doll's eyes), *Anemone* (windflower), *Aquilegia* (columbine), *Clematis* (virgin's bower), *Delphinium* (larkspur), *Ficaria* (lesser celandine), *Helleborus* (hellebore), *Hepatica* (liverleaf), *Pusatilla* (pasqueflower), *Ranunculus* (buttercup), *Thalictrum* (meadow rue), and *Trollius* (globeflower).

Twelve genera contain OBL indicators in North America:

1. *Aconitum* L.
2. *Aquilegia* L.
3. *Caltha* L.
4. *Clematis* L.
5. *Coptidium* (Prantl) Beurl. ex Rydb.
6. *Delphinium* L.
7. *Halerpestes* Greene
8. *Kumlienia* Greene
9. *Myosurus* L.
10. *Ranunculus* L.
11. *Thalictrum* L.
12. *Trollius* L.

FIGURE 2.1 Phylogenetic relationships among the major tricolpate dicotyledon groups as indicated by combined analysis of DNA sequence data from 17 multiple (cpDNA, mtDNA, nuclear DNA) gene regions. A comparable result was obtained by Zeng et al. (2014) using data from 59 low-copy nuclear genes. Taxa containing OBL North American indicators are highlighted in bold. (Adapted from Soltis, D.E. et al., *Amer. J. Bot.*, 98, 704–730, 2011.)

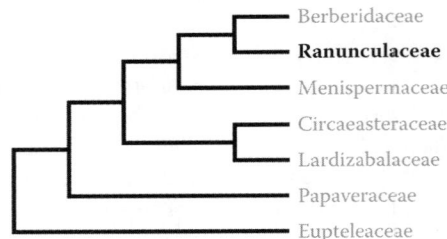

FIGURE 2.2 Interfamilial relationships in Ranunculales as indicated by phylogenetic analysis of combined DNA sequence data (*rbcL*, *matK*, *trnL-F*, 26S rDNA) and morphological characters. Ranunculaceae (in bold) are the only family containing OBL indicators in the order and occur within a relatively derived portion of the cladogram. (Adapted from Wang, W. et al., *Perspect. Plant Ecol. Evol. Syst.*, 11, 81–110, 2009.)

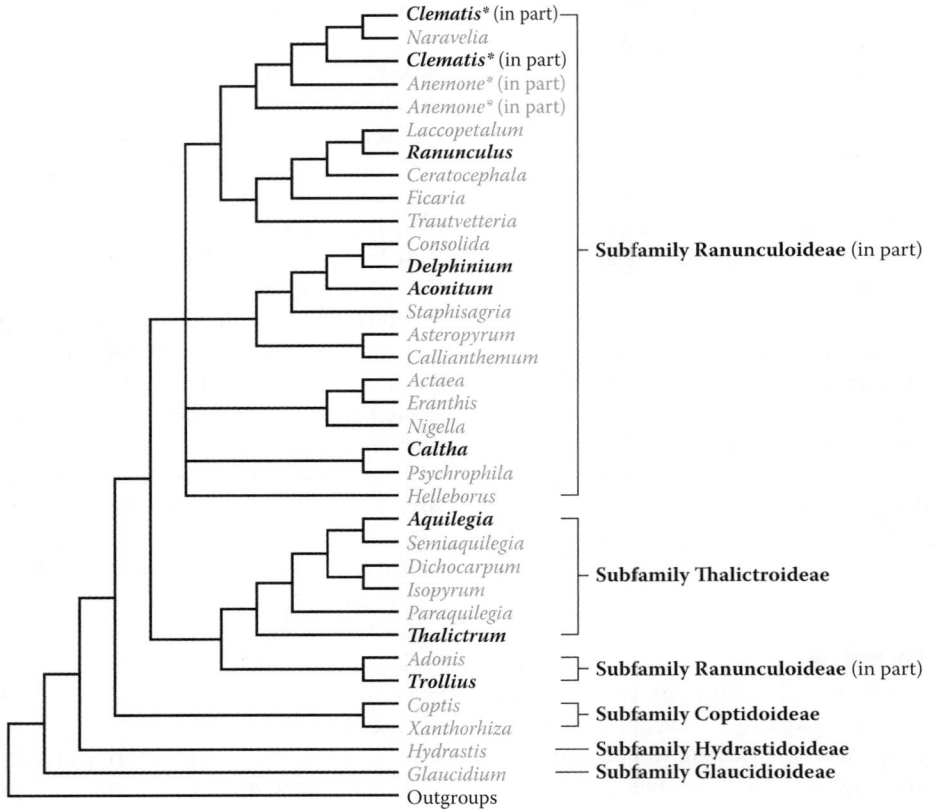

FIGURE 2.3 Intergeneric and subfamilial relationships in Ranunculaceae as indicated by analysis of combined DNA sequence data. These results reveal problems with the monophyly of *Anemone* and *Clematis* (asterisks) as well as subfamily Ranunculoideae; however, the latter discrepancy is not strongly supported in the analysis shown. Taxa containing OBL North American indicators (in bold) are dispersed broadly across the family, indicating multiple radiations of aquatic species in the group (*Myosurus* was not surveyed). (Adapted from Cossard, G. et al., *Plant Syst. Evol.*, 302, 419–431, 2016.)

1. *Aconitum*

Aconite, monkshood; aconit

Etymology: from the Greek *akóniti* ("without struggle"), in reference to the potentially debilitating properties of the plants

Synonyms: none

Distribution: global: Africa; Eurasia; North America; **North America:** widespread (except central)

Diversity: global: 100 species; **North America:** 10 species

Indicators (USA): OBL; FACW: *Aconitum infectum*

Habitat: freshwater; palustrine; **pH:** unknown; **depth:** <1 m; **life-form(s):** emergent (herb)

Key morphology: shoots tall, leafy; leaves deeply palmately dissected; flowers in racemes, the pedicels erect, ascending; flowers bisexual, bilateral, five-merous; sepals showy, dark blue to violet, the upper (to 1.5 cm) hoodlike; petals reduced; stamens numerous; gynoecium apocarpous, pistils (3–5), several seeded; follicles short-beaked

Life history: duration: perennial (tubers); **asexual reproduction:** rhizomes; **pollination:** insect; **sexual condition:** hermaphroditic; **fruit:** follicles (common); **local dispersal:** seeds; **long-distance dispersal:** seeds

Imperilment: (1) *Aconitum infectum* [G1]; S1 (AZ)

Ecology: general: *Aconitum* comprises herbaceous annual, biennial, and perennial species that grow often as colorful wildflowers not only in open meadows but also in shaded woodlands. They range from tundra to high elevation habitats, where they occur frequently in moist environments. Five of the North American species (50%) are wetland indicators, but only one is designated as OBL. The flowers are protandrous (or partially so), self-incompatible or self-compatible, and pollinated primarily by bees (Insecta: Hymenoptera: Apidae: *Bombus*) and occasionally by hummingbirds (Aves: Trochilidae). Wind pollination has also been documented, especially in some alpine species. Anthesis characteristically occurs during the late-summer period. The seeds of at least some species require 3 months of cold (e.g., 3.3°C) stratification in order to break dormancy and to induce germination.

Aconitum infectum **Greene** is a perennial that inhabits meadows, woodlands, and the margins of streams at subalpine elevations from 2743 to 3353 m. Flowering occurs from June to July, with fruiting extending into August. The plants perennate by their tuberous bases. The flowers presumably are insect pollinated, but few life history particulars are available for this species, which many authors treat as synonymous with *A. columbianum* (see *Systematics*).

Reported associates: *Abies, Picea, Pinus.*

Use by wildlife: none reported

Economic importance: food: not edible; the plants are regarded as poisonous; **medicinal:** no known uses; **cultivation:** not cultivated; **misc. products:** none; **weeds:** none; **nonindigenous species:** none.

Systematics: Although some analyses show *Aconitum* as related closely to *Delphinium* and *Consolida* (e.g., Figure 2.3), other studies incorporating much larger numbers of species indicate that *Consolida* and *Aconitella* are nested within *Delphinium* and that the monophyly of *Aconitum* is compromised by several misplaced *Delphinium* species. However, *Aconitum* itself appears to be monophyletic. The taxonomic status of *A. infectum* is uncertain. The taxon has been placed in synonymy with *A. columbianum* by many authors, and the question of its distinctness has not been resolved sufficiently. No material of *A. infectum* has yet been included in phylogenetic studies. The basic chromosome number of *Aconitum* is $x = 8$. Counts are not available for *A. infectum*.

Comments: *Aconitum infectum* is known only from the San Francisco Peaks region of Arizona

References: Bosch & Waser, 1999; Brink & Woods, 1997; Cossard et al., 2016; Dosmann, 2002; Duan et al., 2009; Jabbour & Renner, 2012; Kearney & Peebles, 1960; Quattrocchi, 2012; Utelli & Roy, 2000.

2. *Aquilegia*

Columbine; ancolie

Etymology: from the Latin *aqua legere*, to select water

Synonyms: none

Distribution: global: Northern Hemisphere; **North America:** widespread

Diversity: global: 70 species; **North America:** 21 species

Indicators (USA): OBL: *Aquilegia eximia*

Habitat: freshwater; palustrine; **pH:** alkaline; **depth:** <1 m; **life-form(s):** emergent (herb)

Key morphology: shoots (to 100 cm) with basal foliage; leaves (to 35 cm) petioled (to 30 cm), ternately compound (2–3×), leaflets (to 48 mm) slightly viscid; flowers showy, nodding; sepals (to 28 mm) red, lance-ovate; petals red, projecting backward into prominent, tubular spurs (to 35 mm); follicles (to 25 mm) beaked (to 20 mm)

Life history: duration: perennial (rhizomes); **asexual reproduction:** none; **pollination:** hummingbird, insect; **sexual condition:** hermaphroditic; **fruit:** aggregates (follicles; common); **local dispersal:** seeds (gravity); **long-distance dispersal:** seeds

Imperilment: (1) *Aquilegia eximia* [G3]

Ecology: general: *Aquilegia* is a genus of herbaceous, rhizomatous perennials that are readily recognized by their showy, distinctively spurred flowers. The spurs are rich with nectar and serve as effective attractants to pollinators. The species occur in habitats ranging from meadows and woodlands to disturbed rocky sites, at elevations extending into alpine zones (to 4000 m). They are frequently associated with moist (e.g., riparian) sites. Six of the North American species (29%) are wetland indicators, but only one is designated as OBL. The fairly large flowers are generally self-compatible but protogynous and are primarily outcrossed by insects (Insecta: Diptera: Syrphidae; Hymenoptera: Apidae: *Bombus*; Lepidoptera: Sphingidae) and hummingbirds (Aves: Trochilidae). Self-pollination occurs in some species. The seeds are large (to 2 mm) and lack any specialized features to facilitate their dispersal over long distances; most local dispersal presumably occurs by gravity or wind. Some species have morphologically dormant seeds. Some seeds require up to 4 weeks of cold stratification (at 3°C–5°C) to induce germination, whereas others are capable of germination without any pretreatment.

***Aquilegia eximia* Van Houtte ex Planch.** inhabits depressions, marshes, ravines, scrub, seeps, slopes, springs, thickets, woodlands, and the drying beds or margins of rivers and streams at elevations of up to 1523 m. The sites range from full sunlight to shade. The plants are serpentine endemics and grow on substrates described as boulders, clay, humus, limestone, loam, rock, sand, and talus. The flowers occur from May to October and are pollinated by hummingbirds (Aves: Trochilidae) or bees (Insecta: Hymenoptera: Apidae). There are no reported specialized mechanisms of seed dispersal. The seeds will germinate without stratification after 30 days of ripening and can be sown directly without any pretreatment. Germination occurs within 16–25 days and reaches its maximum within a month. The plants reproduce vegetatively by woody rhizomes. The viscid foliage protects the plants from herbivory by trapping phytophagous visitors (see *Use by wildlife*). **Reported associates:** *Acer macrophyllum, Adenostoma fasciculatum, Aesculus californica, Agrostis exarata, Alnus, Arctostaphylos glauca, A. viscida, Astragalus clevelandii, Brickellia californica, Bromus diandrus, Callitropsis sargentii, Calocedrus decurrens, Carex, Castilleja minor, Ceanothus, Elymus, Eriodictyon californicum, Eriogonum fasciculatum, Fritillaria pluriflora, Hastingsia alba, Heteromeles arbutifolia, Lilium pardalinum, Lolium multiflorum, Melica torreyana, Mimulus guttatus, Packera clevelandii, Perideridia, Pinus attenuata, P. jeffreyi, P. sabiniana, Quercus agrifolia, Q. durata, Rhamnus californica, R. ilicifolia, Rubus, Salix breweri, S. sitchensis, Schedonorus arundinaceus, Stachys albens, S. pycnantha, Toxicodendron diversilobum, Trichostema laxum, Triteleia peduncularis, Umbellularia californica, Viola.*

Use by wildlife: *Aquilegia eximia* is eaten by deer (Mammalia: Cervidae: *Odocoileus*). The seeds are eaten by small birds (Aves: Emberizidae) such as juncos (*Junco*), finches, and sparrows. The reproductive organs are consumed by cutworms (Insecta: Lepidoptera: Noctuidae: *Heliothis phloxiphaga*). The plants harbor many invertebrates including assassin bugs (Heteroptera: Reduviidae: *Pselliopus spinicollis*), crab spiders (Thomisidae: *Mecaphesa schlingeri*), mirids, (Heteroptera: Miridae: *Tupiocoris californicus*), and stilt bugs (Heteroptera: Berytidae: *Hoplinus echinatus*).

Economic importance: food: not edible; **medicinal:** *Aquilegia* species (including *A. eximia*) have been used by the Paiute, Quileute, and Shoshone tribes as an emetic (root decoction), to treat rheumatism (fresh mashed roots), and as a remedy for bee stings (root/leaf poultice), biliousness (leaf/root decoction), colds (leaf decoction), coughs (chewed leaves; root decoction), diarrhea (root decoction), dizziness (leaf/root decoction), head lice (mashed seeds), sore throat (leaves), sores (leaf poultice, root pulp), stomach ache (root decoction, seeds), and venereal diseases (whole plant decoction); **cultivation:** *Aquilegia eximia* (serpentine columbine) is distributed in commerce as a garden ornamental; **misc. products:** The Pomo, Kashaya and other tribes use the flowers of *A. eximia* in ceremonial dance wreaths; **weeds:** none; **nonindigenous species:** none.

Systematics: Molecular phylogenetic analyses resolve *Aquilegia* as a monophyletic sister genus to *Semiaquilegia*, within a clade that includes other members of Ranunculaceae subfamily Thalictroideae (e.g., Figures 2.3 and 2.4). Analyses of combined cpDNA sequence data from 21 rapidly evolving regions have provided evidence for the monophyly of the North American species and resolve *A. eximia* as closely related to *A. flavescens* and *A. formosa* [facultative (FAC), facultative upland (FACU)] (Figure 2.4). The clade of North American species is derived from within a large group of Eurasian species. The basic chromosome number of *Aquilegia* is $x = 7$. All

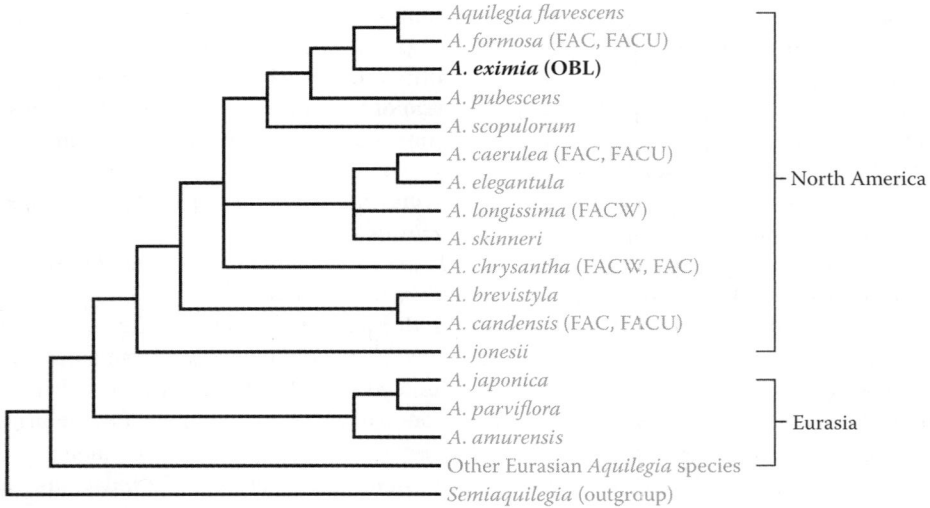

FIGURE 2.4 Phylogenetic relationships in *Aquilegia* resolved from analysis of DNA sequences from 21 rapidly evolving cpDNA regions. The North American species resolve as a clade derived from the Eurasian species. The North American wetland indicators (designations in parentheses) occur throughout the genus; *A. eximia*, the single OBL indicator (highlighted in bold) occurs within a fairly derived portion of the cladogram. (Adapted from Fior, S. et al., *New Phytol.* 198, 579–592, 2013.)

Aquilegia species (including *A. eximia*) are diploid (2n = 14). Many *Aquilegia* species are intefertile and capable of hybridization. *Aquilegia eximia* reportedly hybridizes with *A. formosa* var. *truncata* and *A. pubescens.*

Comments: *Aquilegia eximia* is restricted to western California.
References: Bastida et al., 2010; Chase & Raven, 1975; Eckert & Schaefer, 1998; Evans & San, 2004; Everett, 2012; Finnerty et al., 1992; Fior et al., 2013; Hodges & Arnold, 1994a,b; Hodges et al., 2003; Hoot, 1995; Jensen et al., 1995; Johansson, 1995; LoPresti et al., 2015; Munz, 1946; Payson, 1918; Ro & McPheron, 1997; Ro et al., 1997; Strand & Milligan, 1996; Strand et al., 1996; Whittemore, 1997a.

3. *Caltha*

Marsh marigold, cowslip; populage des marais
Etymology: from the Greek *kalathos* ("basket"), in reference to the flower shape
Synonyms: *Psychropila, Thacla*
Distribution: global: cosmopolitan; **North America:** northern
Diversity: global: 10 species; **North America:** 3 species
Indicators (USA): OBL: *Caltha leptosepala, C. natans, C. palustris*
Habitat: freshwater; palustrine; **pH:** 4.5–7.5; **depth:** 0–2 m; **life-form(s):** emergent (herb), free-floating
Key morphology: herbage arising from caudices (to 2 cm) or stolons; leaves basal or cauline, petioled or nearly sessile, the blades (to 13 cm) broadly cordate to reniform, margins crenate to dentate or entire; flowers cuplike, showy (to 45 mm), radial, solitary or (up to 6) in axillary or terminal cymes (to 22 cm); petals absent; sepals (5–12; to 23 mm) petaloid, orange, pink, white, or yellow; stamens numerous; follicles (to 20 mm) numerous, beaked (to 2 mm), arranged in aggregates; seeds (to 2.5 mm) numerous, brown
Life history: duration: perennial (caudices, stolons); **asexual reproduction:** stolons, shoot fragments; **pollination:** insect; **sexual condition:** hermaphroditic; **fruit:** aggregates (follicles; common); **local dispersal:** stoloniferous stems; **long-distance dispersal:** seeds
Imperilment: (1) *Caltha leptosepala* [G5]; S2 (AK, <u>YT</u>); S3 (<u>AB</u>, WY); (2) *C. natans* [G5]; S1 (MN, WI); S2 (<u>ON</u>, YT); S3 (<u>MB</u>); (3) *C. palustris* [G5]; SX (KY); S1 (<u>LB</u>, TN); S2 (DE, <u>NE</u>, NS); S3 (IL)
Ecology: general: All *Caltha* species are temperate aquatics or occur at least as helophytic plants. All three North American species (100%) are ranked as OBL indicators throughout their ranges. *Caltha* species flower early in the spring (as early as February) and occasionally through snow cover. The flowers of those species investigated are genetically self-incompatible, highly nectariferous, and outcrossed primarily by bees (Insecta: Hymenoptera: mainly Andrenidae, Apidae, and Halictidae) and flies (Diptera: Syrphidae); *Caltha* pollen has also been found on a variety of other insects including species of Coleoptera, Hemiptera, Lepidoptera, Plecoptera, and Thysanoptera. The seeds are dispersed locally by raindrops, which dislodge them from the follicles. In many cases, the seeds are buoyant and are known to be transported over

longer distances by water. The seeds are often morphophysiologically dormant, whereby the embryos are morphologically immature at dispersal and are also dormant physiologically.

***Caltha leptosepala* DC.** grows in bogs, bottomlands, carrs, channels, cirques, depressions (snowmelt), ditches, draws, fens, flats, floodplains, hummocks, marshes, meadows, mires, moors, ponds, pools, rills, rivulets, roadsides, scrub, seeps, shores, streams, slopes (to 40%), swales, swamps, tundra, vernal pools, and along the margins of brooks, lakes, muskeg, ponds, snowbanks, and streams at elevations of up to 4131 m. The plants are tolerant to freezing and are common at alpine or subalpine elevations on moist ground or in shallow, standing, still, or flowing water. The sites range from open to partially shaded exposures. The substrates have been categorized as acidic (e.g., pH: 4.5–4.7) or calcareous and are described as alluvium (organic), boulders, clay, clay loam, cobbles, diorite, granite, gravel, humus, limestone (Claron), loam, loamy gravel, muck, peat, rock, rocky clay, rocky sandy loam, rocky scree, rocky shale, sand, sandy loam (granitic), Sawatch quartzite, silt, silty gravel (sandstone), silty muck (organic), silty sand, talus, and till (glacial). Flowering and fruiting extend from April to September. The entomophilous flowers are pollinated mainly by butterflies (Lepidoptera), flies (Diptera), and solitary bees (Hymenoptera). Although the flowers appear to be primarily self-incompatible, evidently a low level of selfing can occur, with as many as 20 seeds produced by plants whose flowers had been "bagged" to exclude pollinators. Seed set correlates positively with temperature, which might reflect enhanced pollinator activity. The reproductive structures become transiently supercooled to prevent freezing injury. The plants will reach maturity within 2–3 years following seed germination. The seeds are morphophysiologically dormant. When shed, the embryos require an additional 4–7 months to complete their maturation, which proceeds while under snow cover. A germination rate of 70% has been reported for seeds that have been cold stratified (2.5°C) for 7 months. Unlike many alpine plants, photosynthesis is not photoinhibited, even at fairly high incident sunlight levels (>2500 μmol/m²s). The plants decline in cover in sites trampled repeatedly by livestock.

Reported associates: *Abies amabilis, A. lasiocarpa, Achillea, Aconitum columbianum, A. delphiniifolium, Agoseris aurantiaca, Agrostis exarata, A. variabilis, Allium validum, Alnus rubra, A. rugosa, A. sinuata, A. viridis, Androsace septentrionalis, Anemone occidentalis, A. parviflora, A. richardsonii, Angelica genuflexa, Antennaria corymbosa, A. lanata, Anthoxanthum nitens, Anticlea, Arnica latifolia, A. lessingii, A. mollis, Artemisia norvegica, A. scopulorum, Aruncus dioicus, Astragalus alpinus, A. australis, Athyrium, Betula glandulosa, Bistorta bistortoides, B. vivipara, Blechnum spicant, Boykinia major, Calamagrostis canadensis, C. rubescens, Callitropsis nootkatensis, Camassia quamash, Canadanthus modestus, Cardamine cordifolia, C. oligosperma, Carex anthoxanthea, C. aquatilis, C. buxbaumii, C. canescens, C. chalciolepis, C. fuliginosa, C. hoodii, C. illota, C. lenticularis, C. leptalea, C. limosa, C. luzulina, C. macrochaeta, C. mertensii, C. microchaeta, C. microptera, C. nigricans, C. nova, C. obnupta, C. podocarpa, C. praeceptorum,*

C. rostrata, C. rupestris, C. saxatilis, C. scirpoidea, C. scopulorum, C. utriculata, Cassiope mertensiana, Castilleja angustifolia, C. cusickii, C. elmeri, C. haydeniana, C. miniata, C. parviflora, Cerastium arvense, Chamaecyparis, Chamerion latifolium, Cladonia, Claytonia lanceolata, C. sarmentosa, C. sibirica, Coptis aspleniifolia, Cornus canadensis, Danthonia intermedia, Dasiphora floribunda, Delphinium, Deschampsia cespitosa, Draba sobolifera, D. streptocarpa, Drosera anglica, D. rotundifolia, Dryas octopetala, Eleocharis palustris, E. quinqueflora, Empetrum, Epilobium anagallidifolium, E. clavatum, Equisetum arvense, E. fluviatile, Eremogone fendleri, Erigeron humilis, E. peregrinus, Eriogonum, Eriophorum angustifolium, E. gracile, Eritrichium nanum, Erythronium grandiflorum, Festuca altaica, F. brachyphylla, Fragaria virginiana, Galium trifidum, Gaultheria humifusa, G. ovatifolia, G. shallon, Gentiana calycosa, G. sceptrum, Gentianella amarella, G. propinqua, Geranium, Geum calthifolium, G. macrophyllum, G. rossii, Glyceria striata, Heracleum sphondylium, Hesperochiron pumilus, Hippuris montana, Hypericum anagalloides, Iris, Juncus balticus, J. castaneus, J. drummondii, J. ensifolius, J. hallii, J. mertensianus, J. parryi, J. triglumis, Juniperus communis, Kalmia microphylla, K. procumbens, Kobresia simpliciuscula, Larix occidentalis, Leptarrhena pyrolifolia, Ligusticum canbyi, L. grayi, L. tenuifolium, Linnaea borealis, Lloydia serotina, Lonicera involucrata, Luetkea pectinata, Lupinus arcticus, L. burkei, Luzula spicata, Lysichiton americanus, Menyanthes trifoliata, Menziesia ferruginea, Mertensia ciliata, M. paniculata, Micranthes lyallii, M. nelsoniana, M. odontoloma, M. oregana, M. rhomboidea, Mimulus guttatus, M. lewisii, M. tilingii, Minuartia obtusiloba, Mitella pentandra, Moneses uniflora, Neottia cordata, Nephrophyllidium crista-galli, Osmorhiza berteroi, Oxyria digyna, Packera pseudaurea, P. streptanthifolia, P. subnuda, Parnassia fimbriata, Pedicularis bracteosa, P. groenlandica, P. ornithorhyncha, P. sudetica, Pellaea breweri, Petasites frigidus, Phleum alpinum, Phlox pulvinata, Phyllodoce empetriformis, Picea engelmannii, P. pungens, Pinus albicaulis, P. contorta, P. flexilis, Platanthera dilatata, P. hyperborea, P. stricta, Poa alpina, P. interior, P. leptocoma, P. nervosa, Podagrostis humilis, P. thurberiana, Podistera eastwoodiae, Polemonium occidentale, Polystichum lonchitis, P. munitum, Populus tremuloides, Potentilla ×diversifolia, P. flabellifolia, Primula jeffreyi, P. parryi, P. pauciflora, P. tetrandra, Pseudocymopterus montanus, Pseudotsuga, Ranunculus alismifolius, R. eschscholtzii, R. macauley, R. occidentalis, R. orthorhynchus, Rhodiola integrifolia, R. rhodantha, R. rosea, Rhododendron albiflorum, R. columbianum, Ribes lacustre, R. montigenum, Rubus spectabilis, Rumex paucifolius, Salix arctica, S. barclayi, S. bebbiana, S. brachycarpa, S. commutata, S. eastwoodiae, S. glauca, S. monticola, S. nivalis, S. planifolia, S. polaris, S. reticulata, S. wolfii, Sanguisorba officinalis, Selaginella densa, Senecio integerrimus, S. triangularis, Sibbaldia procumbens, Silene acaulis, Sisyrinchium, Solidago multiradiata, Sorbus sitchensis, Sphagnum, Spiraea douglasii, S. splendens, Spiranthes romanzoffiana, Stachys, Stellaria umbellata, Streptopus amplexifolius, Taraxacum, Tetraneuris

acaulis, Thalictrum alpinum, T. occidentale, Thuja plicata, Tiarella trifoliata, Trautvetteria caroliniensis, Triantha glutinosa, Trichophorum cespitosum, Trientalis europaea, Trifolium longipes, T. parryi, Trillium ovatum, Trollius albiflorus, T. laxus, Tsuga heterophylla, T. mertensiana, Vaccinium cespitosum, V. membranaceum, V. myrtillus, V. ovalifolium, V. parvifolium, V. scoparium, V. uliginosum, Valeriana dioica, V. sitchensis, Veratrum californicum, V. viride, Veronica americana, V. serpyllifolia, V. wormskjoldii, Viola adunca, V. macloskeyi, V. palustris, Wyethia helianthoides.

***Caltha natans* Pall.** inhabits beaches, bogs, depressions, ditches, dunes, flats (tidal), floodplains, gravel bars, gravel pits, heath, marshes, meadows, mudflats, muskeg, pools, roadsides, scrub, slopes, taiga (boreal), tundra (coastal), and the margins or shores of lakes, ponds (dessicating), sloughs, and sluggish rivers and streams at elevations of up to 1500 m. Typically, the plants are found floating in shallow waters (to at least 50 cm) along the shorelines of watercourses. They occupy often slow-flowing streams, which flood frequently, especially those sites that have been dammed by beavers (Mammalia: Castoridae: *C. canadensis*) and have an extensive cover of shrubs and *Calamagrostis*. The sites exposures usually are described as being open. The substrates are characterized as organic (mineratrophic) and include clay, gravel (fluvial), loam, muck, mud, peat, sand, silt (fluvial), silty clay, silty loess, and silty mud. Flowering and fruiting occur from June to August. The follicles lack specializations for long-distance transport and are believed to be dispersed primarily by water. Vegetative spread of this species is facilitated by its creeping, stoloniferous stems, which root at their nodes. **Reported associates:** *Alisma triviale, Alnus rugosa, Arctophila fulva, Beckmannia syzigachne, Betula papyrifera, Bidens cernuus, Calamagrostis canadensis, Callitriche palustris, Caltha palustris, Carex atherodes, C. lacustris, C. rostrata, C. stricta, C. utriculata, Cicuta virosa, Comarum palustre, Eleocharis acicularis, E. palustris, Elodea canadensis, Epilobium ciliatum, Equisetum, Glyceria grandis, G. maxima, Hypericum, Impatiens, Iris versicolor, Juncus brevicaudatus, Lemna minor, Lycopus, Lysimachia thyrsiflora, Mentha arvensis, Persicaria amphibia, P. sagittata, Phalaris arundinacea, Potamogeton epihydrus, Ranunculus gmelinii, R. pensylvanicus, Rhododendron columbianum, Rorippa palustris, Salix discolor, S. petiolaris, Scirpus cyperinus, Sium suave, Sparganium angustifolium, S. emersum, S. glomeratum, Spiraea alba, Tephroseris palustris, Typha latifolia, Utricularia macrorhiza, Vaccinium vitis-idaea, Vahlodea atropurpurea.*

***Caltha palustris* L.** occurs in brackish to fresh sites associated with beaches, bogs, channels, depressions, ditches, fens, floodplains, gravel bars, gullies, hummocks, lagoons (tidal or nontidal), marshes (tidal or nontidal), meadows, mudflats, muskeg, oxbows, pools, prairies, rivulets, roadsides, scrub (riparian), seeps, slopes, springs, streambeds, swales, swamps, thickets, tundra, woodlands, and the margins or shores of brooks, lakes, moors, ponds, pools, rivers, sloughs, and streams at elevations of up to 1847 m. The plants occur most often in open sites (but occasionally are found in partial shade), can grow in shallow (e.g., 10–15 cm) standing water, and will withstand prolonged cold temperatures down to −35°C. The substrates have been characterized as acidic and as calcareous and are described as clay, cobbly gravel, granite, gravel, humus, limestone, marl, muck (organic), mud, muddy gravel, Palms muck (terric medisaprist), peat (sedge), peaty silt, rock (marbleized carbonate; volcanic), rocky sand, sand, sandy loam, silt (deltaic; organic), and silty clay. Flowering occurs from March to August, with fruits observed from May to August. The flowers are genetically self-incompatible and are outcrossed primarily by insects (see *Use by wildlife*). The inner bases of the sepals are strongly UV reflective (a condition often lost in cultivated material), which serves to attract pollinators. The seeds weigh 0.55 mg on average and are dispersed locally by raindrops (ombrohydrochory) and over longer distance by water (nautochory). About 10% of the seed can remain afloat after 106 days in still waters and after 92 days in flowing waters. Artificial manipulations indicate that a fairly extensive seed bank potentially can develop. The seeds require 28–112 days of cold (4°C) stratification, followed by gradually rising temperatures in order to germinate; they will freeze at −7°C. Fairly high germination has also been achieved by incubating seeds at 21°C under a 16/8 h light/dark regime. The seedlings take 3 years before they are mature enough to flower. Several transplanted populations have survived for more than 10 years, indicating the potential success of restoration attempts. **Reported associates:** *Abies, Acer macrophyllum, A. rubrum, A. saccharinum, A. saccharum, Aesculus glabra, Agrimonia parviflora, Agrostis scabra, Alisma subcordatum, A. triviale, Alnus, Alopecurus magellanicus, Anemone canadensis, Anemonella thalictroides, Apios americana, Aquilegia canadensis, Arisaema triphyllum, Arnoglossum plantagineum, Asarum canadense, Asimina triloba, Barbarea vulgaris, Betula glandulosa, Bidens, Boehmeria cylindrica, Bolboschoenus fluviatilis, Botrychium crenulatum, Calamagrostis canadensis, Callitriche stagnalis, Caltha natans, Camassia scilloides, Cardamine bulbosa, C. concatenata, C. douglassii, C. pensylvanica, C. pratensis, Carex aquatilis, C. bigelowii, C. buxbaumii, C. canescens, C. conjuncta, C. corrugata, C. exsiccata, C. festucacea, C. fuliginosa, C. haydenii, C. hyalinolepis, C. hystericina, C. jamesii, C. lachenalii, C. lacustris, C. lupulina, C. lyngbyei, C. membranacea, C. muskingumensis, C. pluriflora, C. prairea, C. saxatilis, C. stipata, C. stricta, C. trichocarpa, C. utriculata, Carpinus caroliniana, Cephalanthus occidentalis, Chaerophyllum procumbens, Chrysosplenium americanum, C. tetrandrum, Cicuta douglasii, C. maculata, Cirsium muticum, Claytonia virginica, Comarum palustre, Cornus alternifolia, C. amomum, C. canadensis, C. sericea, Cryptotaenia canadensis, Cyperus, Draba ogilviensis, Dryas integrifolia, Dryopteris cristata, Eleocharis, Elymus virginicus, Equisetum arvense, E. fluviatile, E. palustre, E. telmateia, Eriophorum angustifolium, Eriophorum ×medium, E. scheuchzeri, Erythronium americanum, Euonymus obovatus, Eupatorium perfoliatum, Eutrochium maculatum, Fraxinus americana, F. nigra, F. pennsylvanica, F. profunda, F. quadrangulata, Galium triflorum, Geranium maculatum, Glechoma hederacea, Glyceria grandis, Helianthus grosseserratus, Heracleum sphondylium, Hippuris tetraphylla, H. vulgaris, Hydrophyllum virginianum, Ilex verticillata, Impatiens capensis, Iris pseudacorus, I. virginica, Juncus balticus, J. biglumis, J. tenuis, Laportea canadensis, Larix laricina, Leersia virginica, Lemna valdiviana, Leymus, Liatris spicata, Lilium michiganense, Lindera benzoin, Lindernia dubia, Liriodendron tulipifera, Lobelia siphilitica, Lonicera ×bella, Lycopus americanus, L. uniflorus, Lysichiton americanus, Lysimachia nummularia, L. quadriflora, Maianthemum canadense, Mentha arvensis, Menyanthes trifoliata, Micranthes foliolosa, M. hieracifolia, Muhlenbergia glomerata, Myosotis, Myrica gale, Napaea dioica, Nasturtium officinale, Nyssa sylvatica, Oenanthe, Onoclea sensibilis, Ostrya virginiana, Oxalis stricta, Oxypolis rigidior, Packera aurea, Parnassia glauca, Parthenocissus quinquefolia, Peltandra virginica, Persicaria virginiana, Petasites frigidus, Phalaris arundinacea, Picea glauca, P. mariana, Pilea pumila, Platanthera dilatata, Poa arctica, Podophyllum peltatum, Polymnia canadensis, Polystichum acrostichoides, Populus, Potentilla anserina, Prunus virginiana, Pycnanthemum virginianum, Quercus bicolor, Q. macrocarpa, Ranunculus aquatilis, R. flammula, R. gmelinii, R. hispidus, R. hyperboreus, R. pallasii, R. recurvatus, R. sceleratus, R. sulphureus, Rhamnus frangula, Rhodiola integrifolia, Ribes americanum, Rorippa sylvestris, Rosa, Rubus chamaemorus, Rudbeckia laciniata, Rumex, Sagittaria latifolia, Salix alaxensis, S. barclayi, S. eriocephala, S. interior, S. ovalifolia, S. planifolia, S. pulchra, S. reticulata, Sambucus nigra, Sanicula canadensis, S. odorata, Saururus cernuus, Saxifraga radiata, S. rivularis, Scirpus atrovirens, Sidalcea hendersonii, Silene nivea, Solanum dulcamara, Solidago gigantea, S. ohioensis, S. patula, Sparganium eurycarpum, S. natans, Sphagnum, Spiraea alba, Stachys pilosa, Symphyotrichum lanceolatum, S. ontarionis, S. pilosum, S. praealtum, S. puniceum, Symplocarpus foetidus, Thalictrum dioicum, Thelypteris palustris, Thuja occidentalis, Tilia americana, Toxicodendron radicans, T. vernix, Trillium grandiflorum, T. recurvatum, Tsuga canadensis, Typha latifolia, Ulmus americana, Vaccinium, Valeriana edulis, Viburnum lentago, Viola pubescens, V. sororia, Vitis riparia, Zizia aurea.*

Use by wildlife: *Caltha leptosepala* and *C. palustris* reportedly are poisonous to mammals (Mammalia) if eaten in excess. The foliage of *C. leptosepala* can contain up to 20% crude protein, 89% digestible dry matter, and yield an average of 4918 calories/g of dry weight. The plants are grazed to a small extent during summer by elk (Mammalia: Cervidae: *Cervus elaphus*) and mule deer (Mammalia: Cervidae: *Odocoileus hemionus*). The foliage is host to several fungi including leaf spot (Ascomycota: Davidiellaceae: *Cladosporium herbarum*) and rusts (Basidiomycota: Pucciniaceae: *Puccinia areolata, P. treleasiana, P. gemella*). The nectariferous flowers attract bees (Hymenoptera: Apidae) and other insects (Insecta). The plants are also a host for larval bog fritillary caterpillars (Insecta: Lepidoptera: Nymphalidae: *Boloria eunomia*). The flowers, leaves, and stems of *C. palustris* are eaten by bears

(Mammalia: Ursidae: *Ursus*), deer (Mammalia: Cervidae: *Odocoileus*), and moose (Mammalia: Cervidae: *Alces*). Fossil remains indicate that the plants were also grazed by Holocene bison (Mammalia: Bovidae: *Bison priscus*). The seeds are fed on by grouse (Aves: Phasianidae). Several beetles (Insecta: Coleoptera) feed on the nectar and pollen (Coccinellidae: *Coleomegilla maculata*; Kateretidae: *Brachypterolus pulicarius*) or bore into the stems (Chrysomelidae: *Prasocuris phellandrii*). The reproductive structures contain protoanemonin, which is believed to function as an antiherbivory agent. Thirty-nine insect (Insecta) species from 21 families have been found in association with *C. palustris*. The primary floral visitors (pollinators) are bees (Hymenoptera: Apidae), beetles (Coleoptera: Chrysomelidae; Curculionidae; Nitidulidae), flies (Diptera: Bibionidae; Empidae; Muscidae; Syrphidae; Stratiomyidae), moths (Lepidoptera), and stoneflies (Plecoptera: Perlidae). The plants are parasitized by chytrids (Fungi: Blastocladiomycota: Physodermataceae: *Physoderma johnsii*).

Economic importance: food: The raw leaves and bruised foliage of *C. palustris* are known to contain helleborin, protoanemonin, and other volatile cardiotoxic poisons that can cause gastrointestinal disorders and inflammation of the mouth and throat. Simple contact with the leaves of the plants has caused skin blisters to develop in some persons. Nevertheless, the newly emerging leaves and stems of *C. palustris* have long been eaten as spring greens by people in New England and eastern Pennsylvania. The volatile toxins are expelled by cooking. *Caltha* foliage is rich in iron. The leaves (boiled with lard) and seeds of *C. palustris* were eaten by the Abnaki as vegetables. The Chippewa, Iroquois, Menominee, Mohegan, Ojibwa, and various native Alaskan tribes also cooked the leaves, stems, and roots (sometimes with pork) as a vegetable.

Alaskan Eskimos ate the leaves and stalks either fresh or after boiling them in seal oil. The pickled flower buds of *C. palustris* have been eaten as a substitute for capers; **medicinal:** A poultice made from chewed plants of *C. leptosepala* was used by the Okanagon and Thompson tribes to treat inflamed wounds. Extracts of *C. palustris* exhibit moderate antioxidant activity. Juices from the plant have been applied to remove warts. The Chippewa used the plants as a diaphoretic (root decoction), diuretic (leaf decoction), expectorant (root decoction), and as a poultice of boiled and mashed (or powdered) roots to relieve sores. The Chippewa and Iroquois employed root decoctions as an emetic. The Western Eskimo tribes used leaf infusions to relieve constipation; **cultivation:** *Caltha leptosepala* and *C. palustris* (*C. natans* to a lesser degree) are important ornamental water garden plants. *C. leptosepala* is distributed as the cultivar 'Grandiflora.' Several double-flowered cultivars of *C. palustris* are in commerce, generally under the name 'Flore Pleno' (a Royal Horticulture Society Award of Garden Merit plant). Other cultivars of *C. palustris* include 'Aeungold,' 'Auenwald,' 'Cannington Surprise,' 'Charlie,' 'Erlenbruch,' 'Girls Eyes,' 'Golden Monarch,' 'Goldschale,' 'Himalayan Snow,' 'Honeydew,' 'Marilyn,' 'Monstrosa,' 'Monstrosa Plena,' 'Multiplex,' 'Nana Plena,' 'Pallida Plena,' 'Plena,' 'Plurisepala,' 'Professor Maatsch,' 'Purpurascens,' 'Semiplena,' 'Stagnalis,' 'Susan,' 'Tyermannii,' 'Wheatfen,' and 'Yellow Giant'; **misc. products:** The blossoms of *C. palustris* are used in winemaking and as the source of a yellow dye, which sometimes is used to dye butter; **weeds:** none; **nonindigenous species:** none.

Systematics: The phylogenetic position of *Caltha* within Ranunculaceae is not yet well resolved, but current data place the genus within subfamily Ranunculoideae near *Asteropyrum*, *Cimicifuga*, *Helleborus*, and *Souliea* (Figure 2.5). Ten species

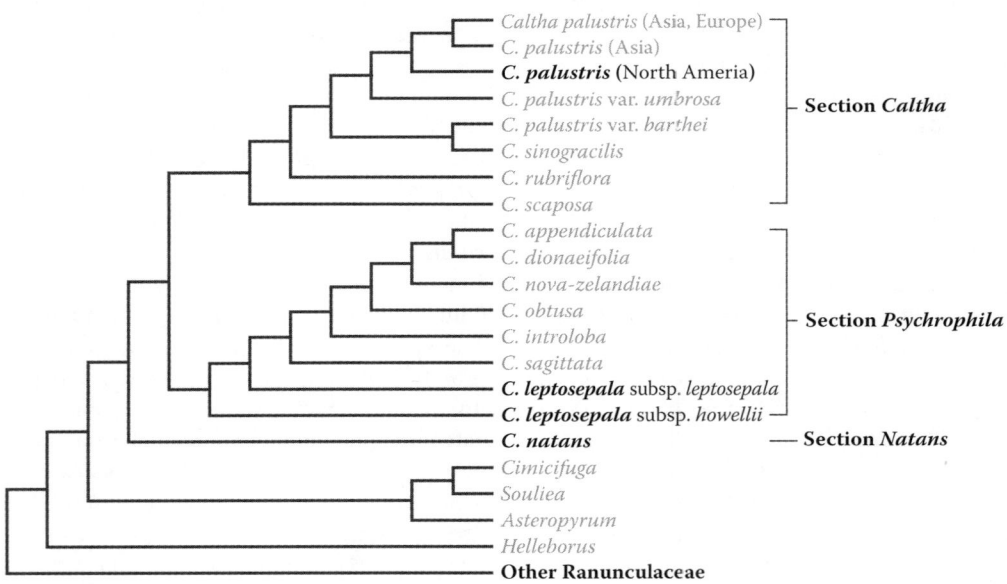

FIGURE 2.5 Interspecific relationships in *Caltha* (Ranunculaceae) based on phylogenetic analysis of combined DNA sequence data. Although the entire genus should be regarded as aquatic, the OBL North American indicator taxa (highlighted in bold) have been derived from all three clades or sections of the genus. The delimitation of specific and subspecific taxa in section *Caltha* warrants further consideration. (Adapted from Cheng, J. & Xie, L., *J. Syst. Evol.*, 52, 51–67, 2014.)

usually are accepted; however, some authors recognize two additional species (*C. rubriflora* and *C. sinogracilis*) as distinct from *C. palustris*. Cladograms constructed from multiple data sets (*atpB-rbcL* spacer, *trnL-F* region, ITS sequences, morphology) support the recognition of three groups in *Caltha*, which correspond taxonomically to sections (Figure 2.5). *Caltha natans* (section *Natans*) is uniform morphologically throughout its range and despite its derived, floating habit, is sister to the remainder of the genus (Figure 2.5). *Caltha palustris* (section *Caltha*) from North America is distinct genetically from Eurasian populations (for which many subspecies have been proposed) and is related to *C. scaposa* from the Himalayas. *Caltha leptosepala* (section *Psychrophila*) includes morphologically and genetically distinct subspecies (sometimes recognized as separate species) and is related to species in South America, Australia, and New Zealand (Figure 2.5). *Caltha* ($x = 8$) is diverse cytologically with inter- and intraspecific chromosome number variation reported in *C. leptosepala* ($2n = 48, 96$), *C. natans* ($2n = 16, 32$), and *C. palustris* ($2n = 16, 32, 48, 56, 60, 64, 72, 80$, and numerous aneuploids). Further elucidation of subspecific taxa (particularly in *C. palustris*) is needed, especially with respect to the status of "*C. rubriflora*" and "*C. sinogracilis*."

Comments: *Caltha natans* and *C. palustris* occur in northern North America and both extend into northern Eurasia; *C. leptosepala* occurs in western North America.

References: Andersen & Armitage, 1976; Anderson, 1940; Baker & Hobbs, 1982; Broek et al., 2005; Casper & Krausch, 1981; Cheng & Xie, 2014; Christy, 2004; Cole & Monz, 2002; Cook, 1996a; Cossard et al., 2016; Drayton & Primack, 2012; Forbis & Diggle, 2001; Ford, 1997; Harms, 1978; Harris & Marr, 2009; Hoot, 1995; Johnson et al., 2008; Judd, 1964; Jürgens & Dötterl, 2004; Knuth, 1908; Lundqvist, 1992; Moss, 1953; Nikolova et al., 2011; Parolin, 2006; Primack, 1982; Ro et al., 1997; Sanchez & Smith, 2015; Schuettpelz & Hoot, 2004; Sklenář, 2016; Sparrow, 1976; Straka & Starzomski, 2015; Thomson, 1980; Tymon et al., 2015; Van Geel et al., 2014; Van Leeuwen et al., 2014; Wallmo et al., 1972; Walton, 1994, 1995; Wardlow et al., 1989a,b.

4. Clematis

Clematis, leather flower; clèmatite

Etymology: from the Greek *klema*, meaning "vine"

Synonyms: *Thacla*; *Viorna*

Distribution: global: Africa; Asia; Europe; New World; **North America:** widespread

Diversity: global: 300 species; **North America:** 37 species

Indicators (USA): OBL; FACW; FAC: *Clematis crispa*

Habitat: freshwater; palustrine; **pH:** 5.5–6.5 (optimal); **depth:** <1 m; **life-form(s):** emergent (semiwoody vine)

Key morphology: vines with ribbed stems (to 3 m); leaves opposite, petiolate, once or twice pinnately compound into 4–10 leaflets (to 10 cm), a tendril-like terminal leaflet also present; flowers terminal, solitary, showy, nodding, bell-shaped; sepals (to 5 cm) 4, petaloid (rose, blue, violet, or white), spreading to recurved, thick with crisped margins and acuminate tips; petals absent; stamens and pistils numerous; achenes in aggregates, single-seeded, beaked (to 3.5 cm)

Life history: duration: perennial (buds); **asexual reproduction:** none; **pollination:** insect; **sexual condition:** hermaphroditic; **fruit:** aggregates (achenes); **local dispersal:** achenes (wind); **long-distance dispersal:** achenes

Imperilment: (1) *Clematis crispa* [G5]; S1 (IL, OK); S2 (KY); S3 (VA)

Ecology: general: *Clematis* is not characteristically a wetland genus although several species (e.g., *C. glaucophylla*, *C. ligusticifolia*, *C. terniflora*, *C. virginiana*) grow occasionally in wet sites. Ten North American species (27%) are wetland indicators, but only one species is designated as OBL. Even this species (*C. crispa*) is regarded as OBL only in the north-central portion of its range but is facultative (FACW, FAC) elsewhere. The species are perennial vines (or lianas), herbs, or subshrubs. The flowers are showy, radially symmetrical, and either unisexual (then dioecious or monoecious) or hermaphroditic. Typically, they are outcrossed by insects (Insecta) such as bees (Hymeoptera: Apidae: *Bombus*) and flies (Diptera: Muscidae); however, some of the species have evolved autogamous breeding systems. Numerous insects also visit the flowers as nectar robbers. The seeds are dispersed abiotically by gravity or by wind, the latter facilitated by the plumose styles of some species. As in many other Ranunculaceae, the seeds are often morphophysiologically dormant and require full embryo maturation as well as a period of cold stratification (e.g., 30–112 days at 4°C) in order for germination to proceed.

Clematis crispa **L.** inhabits alluvial forests, bogs, bottomlands, depressions, dikes, ditches, draws, flatwoods, floodplains, gravel pits, hammocks, ledges, marshes, meadows, ravines, roadsides, savannas, slopes, swamps, thickets, woodlands, and the margins of bayous, canals, lakes, pools, rivers, sloughs, and streams at elevations of up to 160 m. The plants are regarded as calciphiles (but are reported from acidic sites) and occur on brackish to fresh, rich substrates, which are characterized as alluvium, clay, clay loam, gravel, humus, Iredell soil, Kinston–Bibb association, limestone, loam, loamy sand, muddy clay, Ogechee series, peat, sand (Clinchfield), sandy clay loam, sandy loam (Wahee), Sharkey clay, and silty muck. This is a moderately fast-growing species that tolerates full sunlight but occurs mainly in partial to fully shaded exposures. Because of its viny habit, it relies on other vegetation for support and climbs on other plants (often shrubs) by means of its tendrils. Flowering occurs from March to September, with fruiting extending from May to October. The bisexual flowers are pollinated by large bees (Insecta: Hymenoptera: Apidae) and butterflies (Insecta: Lepidoptera). Seed germination requires 60–180 days of cold–moist stratification at 1°C–4°C. Allozyme studies indicate that the populations retain moderate levels of genetic variability (e.g., 57% polymorphic loci; expected heterozygosity = 0.279; multilocus genotypic diversity = 0.933) despite poor apparent seedling recruitment.

Reported associates: *Acer rubrum, A. saccharum, Aesculus pavia, Agrostis hyemalis, Aira elegantissima, Amorpha fruticosa, Ampelopsis arborea, Amsonia rigida, Arisaema triphyllum, Arundinaria gigantea, Baptisia alba, Berlandiera pumila, Betula nigra, Bidens bipinnatus, Bignonia capreolata, Brunnichia ovata, Campsis radicans, Cardamine hirsuta,*

Carex caroliniana, *C. cherokeensis*, *C. debilis*, *C. gigantea*, *C. glaucescens*, *C. intumescens*, *C. lonchocarpa*, *C. lurida*, *C. meadii*, *C. retroflexa*, *C. squarrosa*, *C. vulpinoidea*, *Carpinus caroliniana*, *Carya aquatica*, *Celtis laevigata*, *Cephalanthus occidentalis*, *Chasmanthium*, *Cicuta maculata*, *Cirsium*, *Cladium jamaicense*, *Clethra alnifolia*, *Commelina diffusa*, *C. erecta*, *C. virginica*, *Coreopsis lanceolata*, *Cornus drummondii*, *C. florida*, *Crataegus marshallii*, *Crepis pulchra*, *Cyperus polystachyos*, *C. surinamensis*, *Cyrilla racemiflora*, *Decumaria barbara*, *Desmodium ochroleucum*, *D. tenuifolium*, *Dichanthelium*, *Digitaria*, *Diodia virginiana*, *Diospyros virginiana*, *Ditrysinia fruticosa*, *Dyschoriste humistrata*, *Dysphania ambrosioides*, *Echinochloa*, *Erechtites hieracifolia*, *Eryngium*, *Euphorbia hyssopifolia*, *Fagopyrum esculentum*, *Fimbristylis*, *Fraxinus caroliniana*, *F. pennsylvanica*, *Fuirena*, *Galium aparine*, *G. obtusum*, *Glyceria*, *Hamamelis virginiana*, *Heliotropium indicum*, *Hibiscus moscheutos*, *Houstonia procumbens*, *Hydrocotyle bonariensis*, *Hymenocallis rotata*, *Hypericum*, *Ilex decidua*, *I. myrtifolia*, *I. opaca*, *Iris hexagona*, *Itea virginica*, *Juniperus virginiana*, *Kalmia hirsuta*, *Lindernia*, *Lipocarpha*, *Liquidambar styraciflua*, *Lobelia amoena*, *Lonicera*, *Ludwigia microcarpa*, *Lupinus perennis*, *Lyonia lucida*, *Magnolia virginiana*, *Melothria pendula*, *Micranthemum*, *Morus rubra*, *Myrica cerifera*, *Neottia bifolia*, *Nyssa aquatica*, *N. biflora*, *Ostrya*, *Panicum*, *Parthenocissus quinquefolia*, *Paspalum notatum*, *P. urvillei*, *Peltandra virginica*, *Penstemon australis*, *Penthorum sedoides*, *Pentodon pentandrus*, *Persea palustris*, *Persicaria arifolia*, *P. hydropiperoides*, *Phanopyrum gymnocarpon*, *Pinus palustris*, *Piptochaetium avenaceum*, *Planera aquatica*, *Podophyllum peltatum*, *Polypremum procumbens*, *Quercus laurifolia*, *Q. lyrata*, *Q. michauxii*, *Q. nigra*, *Q. pagoda*, *Q. virginiana*, *Ranunculus*, *Rhododendron canescens*, *Rhynchospora decurrens*, *R. mixta*, *Rubus*, *Ruellia caroliniensis*, *Rumex hastatulus*, *Sabal minor*, *Salix caroliniana*, *Salvia lyrata*, *Samolus*, *Sanguinaria canadensis*, *Saururus cernuus*, *Smilax bona-nox*, *S. rotundifolia*, *S. tamnoides*, *S. walteri*, *Stachys floridana*, *Stillingia aquatica*, *Styrax americanus*, *Taxodium distichum*, *Thelypteris interrupta*, *Thyrsanthella difformis*, *Toxicodendron radicans*, *Triadenum*

walteri, *Trillium cuneatum*, *T. pusillum*, *Ulmus alata*, *U. americana*, *Vaccinium arboreum*, *V. elliottii*, *Viburnum obovatum*, *Viola*, *Vitis cinerea*, *Xyris*, *Zephyranthes atamasco*.

Use by wildlife: *Clematis crispa* reputedly is resistant to grazing by deer (Mammalia: Cervidae: *Odocoileus virginianus*), which feed on it preferentially. The plants are fed on by blister beetles (Insecta: Coleoptera; Meloidae).

Economic importance: food: none; **medicinal:** *Clematis crispa* has been used as an alterative and purgative (root decoctions), diaphoretic, diuretic, and to treat scalp disease, syphilis, and as a wash for sores and ulcers (powdered leaves). Plants of *C. crispa* also attract blister beetles (see above), which are a source of medicinal cantharidin; **cultivation:** *Clematis crispa* (blue jasmine, swamp leather flower) is sold as an ornamental and was first introduced to Europe around 1726. Cultivars include 'Betty Corning' (presumably a hybrid between *C. crispa* and *C. viticella*), 'Burford Bell,' 'Cylindrica,' 'Daniel Deronda' (allegedly a hybrid between *C. crispa* and *C. integrifolia*), 'Garnet,' 'Mrs Harvey,' 'Odoriba,' 'Pendragon,' 'Purple Treasure,' and 'Rosea'; **misc. products:** none; **weeds:** none; **nonindigenous species:** none.

Systematics: Molecular data resolve *Clematis* as closely related phylogenetically to *Anemone*, *Hepatica*, *Knowltonia*, and *Pulsatilla*. *Clematis* is divided into four subgenera (sometimes segregated as separate genera), with *C. crispa* placed in subgenus *Viorna* (section *Viorna*, subsection *Crispae*). Interspecific relationships in *Clematis* have not been fully elucidated, but molecular phylogenetic analyses are continuing. Currently, the surveyed species fall within 10 distinct clades, with one of the groups comprising sections *Viorna* and *Viticella*. Reconstructions based on combined DNA sequence data (nrITS, *atpB-rbcL*, *psbA-trnH-trnQ*, and *rpoB-trnC* regions) for 75 species, as well as sequences from the nuclear actin I intron region for six fairly closely related species, place *C. crispa* along with other taxa assigned to section *Viorna* and subsection *Crispae*. However, varied resolution from different markers and unequal taxon sampling have made it difficult to elucidate the sister species of *C. crispa*. One possible candidate is *C. reticulata* (a species of dry habitats), which is closely related to *C. crispa* in both aforementioned analyses

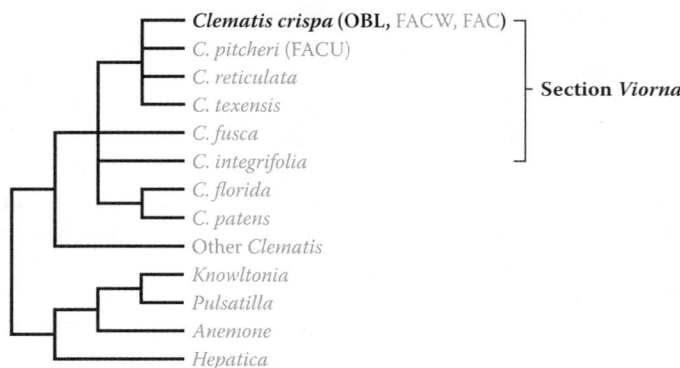

FIGURE 2.6 General overview of relationships for *Clematis crispa* as indicated by parsimony analysis of combined DNA sequence data. A low level of sequence divergence yields low resolution among several potentially closely related species within section *Viorna*. The wetland indicator status is given in parentheses (OBL taxa are highlighted in bold). (Adapted from Xie, L. et al., *Syst. Bot.*, 36, 907–921, 2011.)

(e.g., Figure 2.6). Artificial hybrids between these species have also been synthesized (see also *Cultivation*). However, low levels of DNA divergence and identical alleles shared between *C. crispa* and several potentially closely related species have made it difficult to elucidate its interspecific relationships with certainty. A better understanding of interspecific relationships in *Clematis* awaits additional sampling of taxa and genetic loci for this large and diverse genus. The leaflet width in *C. crispa* is highly variable (even within single individuals) but is not correlated geographically and does not appear to be informative taxonomically. The base chromosome number of *Clematis* is $x = 8$. The North American species of subgenus *Viorna* (including *C. crispa*) are uniformly diploid ($2n = 16$). *Clematis crispa* and *C. fusca* share a similar karyotype.

Comments: *Clematis crispa* occurs in the southeastern United States, mainly along the coastal plain and interior Mississippi embayment region.

References: Borkent & Harder, 2007; Do, 2006; Ensign et al., 2014; Godfrey & Wooten, 1981; Goertzen & Boyd, 2007; Hill, 1992; Hoot, 1995; Horrell, 2013; Jiang et al., 2010; Porcher, 1863, 1981; Pringle, 1997; Ro & McPheron, 1997; Ro et al., 1997; Sheng et al., 2014; Slomba et al., 2001, 2004; Xie et al., 2011.

5. *Coptidium*

Buttercup; renoncule

Etymology: from the Greek *koptos idium* ("cut small"), in reference to the cleft leaves

Synonyms: *Anemone* (in part); *Ranunculus* (in part)

Distribution: global: Eurasia, North America; **North America:** northern

Diversity: global: 2 species; **North America:** 2 species

Indicators (USA): OBL: *Coptidium lapponicum, C. pallasii*

Habitat: freshwater; lacustrine, palustrine; **pH:** acidic; **depth:** <1 m; **life-form(s):** emergent or floating (herb)

Key morphology: stems creeping, floating, or prostrate, rooting at nodes; leaves basal, petioled (to 75 mm), blades (to 4.3 cm) linear, obovate, or reniform, undivided or three lobed, the lobes elliptic, lanceolate, or cleft, margins crenate or entire; flowers solitary, radial, apocarpous, peduncles (to 2 dm) erect, sometimes emerging from the water; sepals (to 7 mm) spreading; petals (to 13 mm) 5–11, pink, yellow, or white; stamens and pistils numerous; achenes (to 5.2 mm) aggregated in heads (to 1.5 cm), beaked (to 2.4 mm), tip of beak hooked or straight

Life history: duration: perennial (rhizomes); **asexual reproduction:** rhizomes; **pollination:** insect; **sexual condition:** hermaphroditic; **fruit:** achenes (common); **local dispersal:** achenes (water), rhizomes; **long-distance dispersal:** achenes (animals)

Imperilment: (1) *Coptidium lapponicum* [G5]; SH (NF); S1 (MI, NB, WI); S2 (LB, ME); S3 (MN, QC); (2) *C. pallasii* [G5]; S2 (MB, ON, QC); S3 (YT)

Ecology: general: Both *Coptidium* species are OBL indicators, which occur consistently in wetland habitats. They are similar ecologically to *Ranunculus* (see later), a genus to which they previously had been assigned taxonomically.

Coptidium lapponicum (L.) Rydb. inhabits bogs, channels, depressions, drumlins, dunes, fens, floodplains, gullies, heath, hollows, hummocks, knolls, meadows, mires, mudflats, muskeg, roadsides, ruts, scrub, seeps, slopes, snowbeds, swamps, taiga, thaw pits, thickets, tundra, woodlands (boreal), and the margins of lakes, ponds, pools, rivers, and streams at elevations of up to 2100 m. The sites typically represent areas, which receive an influx of cold groundwater. Although exposures usually are reported as shaded, the plants have also been observed in open sites. The substrates are characterized as acidic and include alluvium, dioritic rock (outcrops), fluvial silt, gravel, gravelly till, mud, peat, sandy clay, and silty till. Flowering has been observed from June to July, with fruiting extending into August. The plants appear to be highly fertile, with 97% pollen stainability (the possibility of apomixis has not yet been excluded experimentally). The pollinators are unknown but the showy, nectariferous flowers indicate them to be insects (Insecta). The achenes contain spongy parenchyma apically, which probably confers some degree of buoyancy to facilitate water dispersal. The hooked tips of the achenes are also likely to attach them to animal (Mammalia) fur for longer range dispersal. The seeds are morphophysiologically dormant and difficult to germinate. In one study, the seeds did not germinate under light or dark conditions. Vegetative reproduction occurs by rhizome production, which can result in the development of clonal populations. **Reported associates:** *Alectoria, Alnus tenuifolia, Arctophila fulva, Arctous rubra, Betula glandulosa, B. pumila, Calamagrostis, Carex aquatilis, C. bigelowii, C. disperma, C. gynocrates, Cladonia, Comarum palustre, Cornus canadensis, Drosera rotundifolia, Empetrum, Eriophorum, Galium boreale, Hippuris vulgaris, Hylocomium splendens, Larix laricina, Moneses uniflora, Nephroma arctica, Petasites frigidus, Picea glauca, P. mariana, Pinguicula villosa, Populus balsamifera, Rhododendron groenlandicum, R. tomentosum, Rubus arcticus, R. chamaemorus, Salix planifolia, S. pulchra, S. richardsonii, Sphagnum, Trichophorum cespitosum, Vaccinium oxycoccos, V. uliginosum, V. vitis-idaea.*

Coptidium pallasii (Schlecht.) Tzvelev grows in brackish to fresh, shallow waters (to 1 m) associated with beach ridges, bogs, channels, depressions, ditches, fens, flats, floodplains, heath, marshes, meadows, muskeg, ponds, pools, roadsides, streams, swales, thickets, tundra, and the margins of lagoons, lakes, and ponds at elevations of up to 700 m. The plants can be found rooted in the substrate or floating on the water surface. The substrates are described as acidic (e.g., pH: 6.8) and include cobbles, muck, mud, peat, and silty peat. Flowering and fruiting extend from March to August. Little is known regarding the reproductive biology or seedling ecology of this species. The achenes are eaten by waterfowl (Aves: Anatidae) and are probably dispersed endozoically in some instances. Vegetative reproduction occurs by means of rhizomes. **Reported associates:** *Anthoxanthum arcticum, Arctagrostis latifolia, Arctophila fulva, Calliergon, Caltha, Carex aquatilis, C. chordorrhiza, C. lyngbyei, C. mackenziei, C. ramenskii, C. rariflora, C. rotundata, C. subspathacea, C. utriculata, Cetraria islandica, Cicuta virosa, Comarum palustre, Dactylina arctica, Dupontia fisheri, Empetrum, Equisetum fluviatile, Eriophorum angustifolium, E. chamissonis, E. ×medium, E. scheuchzeri, Hippuris vulgaris, Luzula* *nivalis, Menyanthes trifoliata, Pedicularis sudetica, Picea mariana, Poa arctica, Salix fuscescens, S. pulchra, S. rotundifolia, Saxifraga cernua, Sphagnum aongstroemii.*

Use by wildlife: *Coptidium pallasii* is eaten by moose (Mammalia: Cervidae: *Alces alces*). The seeds are consumed in quantity by the spectacled eider (Aves: Anatidae: *Somateria fischeri*). The nutrient rich rhizomes are eaten by barnacle and white-fronted geese (Aves: Anatidae: *Branta leucopsis, Anser albifrons,* respectively). The plants are occasional hosts of a rust (Fungi: Basidiomycota: Pucciniaceae: *Puccinia rubigo-vera*).

Economic importance: food: *Coptidium lapponicum* is regarded as toxic when eaten and is also suspected to cause contact dermatitis and photodermatitis. Nevertheless, some western Eskimo groups ate *C. lapponicum* as a dietary aid. Eskimos have also eaten the nutritious rootstocks of *C. pallasii* (known as "Kabootie") before the leaves appear, whereupon they turn bitter. The Yup'ik rake the plants from lake bottoms, boil the new shoots until tender, and eat the peppery flavored vegetable (known as "kapuukaraat," "kapukaraqor," or "uivlut") with seal oil. The flavor of older plants is too intense to be palatable; **medicinal:** unknown; **cultivation:** *Coptidium lapponicum* is sold occasionally as a garden ornamental; **misc. products:** none; **weeds:** none; **nonindigenous species:** none.

Systematics: *Coptidium* is assigned to tribe Ranunculeae of Ranunculaceae. *Coptidium lapponicum* and *C. pallasii* have long been included among the species of *Ranunculus,* in which they comprised different subgenera (*Coptidium* and *Pallasiantha,* respectively). However, phylogenetic analysis of DNA sequence data clearly resolves these species as a clade that is distinct from *Ranunculus* (Figure 2.7), a result that warrants their recognition as a separate genus. A close relationship

of both *Coptidium* species is indicated further by their ability to hybridize (see below). The base chromosome number of *Coptidium* is *x* = 8. *Coptidium lapponicum* is diploid (2*n* = 16) and *C. pallasii* (2*n* = 32) is a tetraploid. *Coptidium lapponicum* and *C. pallasii* hybridize to form a sterile (presumably triploid) hybrid, which is known as *Coptidium ×spitzbergensis.*

Comments: *Coptidium lapponicum* has a nearly circumpolar northern distribution; *C. pallasii* occurs in far northern regions of North America.

References: Ager & Ager, 1980; Aiken et al., 2007; Anderson, 1939; Bliss, 1958; Emadzade et al., 2010; Griffin, 2009; Hoffmann et al., 2010; Kistchinski & Flint, 1974; MacCracken et al., 1997; Markon & Derksen, 1994; Prescott, 1953; Rozenfeld & Sheremetiev, 2014; Savile & Savile, 1953; St. Hilaire, 2002; Villarreal et al., 2012; Whittemore, 1997a.

6. *Delphinium*

Larkspur; pied d'alouette

Etymology: from the Greek *delphinion,* the closed flowers resembling a dolphin

Synonyms: none

Distribution: global: Asia; Africa; Europe; North America; **North America:** widespread

Diversity: global: 300 species; **North America:** 61 species

Indicators (USA): OBL: *Delphinium uliginosum*

Habitat: freshwater; palustrine, riverine; **pH:** 7.6–8.3; **depth:** <1 m; **life-form(s):** emergent (herb)

Key morphology: shoots (to 70 cm) sometimes reddish at base; leaves (to 8 cm) basal or cauline, fan-shaped, palmately lobed ½ way to base, petioled (to 7 cm); flowers (5–48) in showy racemes, pedicels (to 10 cm) with bracteoles (to 5 mm); perianth bilateral; sepals (to 14 cm) dark blue, one with spur

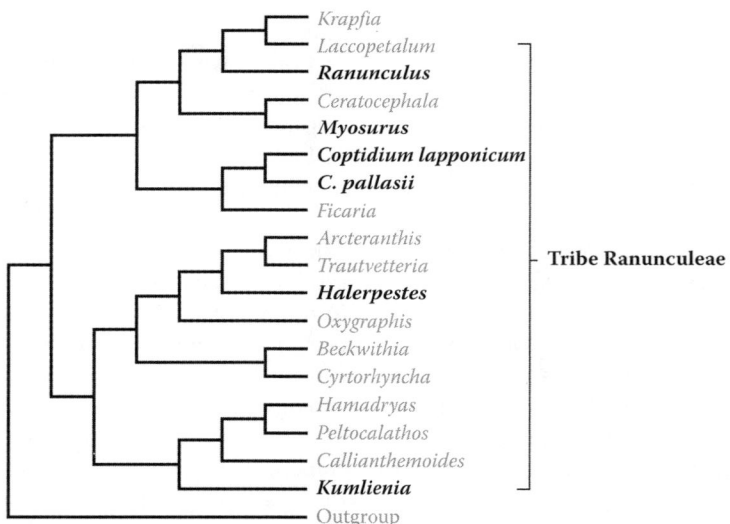

FIGURE 2.7 Intergeneric relationships in tribe Ranunculeae (Ranunculaceae subfamily Ranunculoideae) based on combined molecular data. This phylogeny provides compelling evidence to distinguish the genera *Arcteranthis* (formerly placed in *Kumlienia*), *Coptidium, Cyrtorhyncha,* and *Kumlienia* from *Ranunculus,* despite their merger with the latter by several recent authors. Although morphological data have resolved *Myosurus* as the sister group to the remainder of tribe Ranunculeae (e.g., Loconte et al., 2012), this analysis nests *Myosurus* within the tribe, a result that is consistent with other molecular studies (e.g., Lehnebach et al., 2007; Wang et al., 2009). Taxa containing OBL North American indicators are highlighted in bold. (Adapted from Emadzade, K. et al., *Taxon,* 59, 809–828, 2010.)

(to 15 mm), the laterals spreading; petals smaller, white to bluish, marginally ciliate; fruit a follicle (to 18 mm); seeds moderately winged, with wavy surfaces

Life history: duration: biennial (woody roots); **asexual reproduction:** none; **pollination:** insect; **sexual condition:** hermaphroditic; **fruit:** follicles (common); **local dispersal:** seeds; **long-distance dispersal:** seeds

Imperilment: (1) *Delphinium uliginosum* [G3]; S3 (CA)

Ecology: general: Several *Delphinium* species occur occasionally in wet habitats. Seventeen North American species (28%) are wetland indicators, but only *D. uliginosum* is designated as OBL. Most of the species are self-compatible but are protandrous and predominantly outcrossed by bumblebees (Insecta: Hymenoptera: Apidae: *Bombus*). Typically, the flowers open sequentially upward on the raceme, with the older lower flowers (in pistillate phase) producing more nectar than the newly opening upper flowers (in staminate phase). Attracted to the higher nectar volume, the pollinators move from the lower to the upper flowers, thereby reducing opportunities for self-pollination to occur. Some species are pollinated by hummingbirds (Aves: Trochilidae). Self-incompatibility and autogamy occur throughout the genus as well. The seeds are dispersed locally (over very short distances) by gravity as they are shaken free from the fruits by the wind. There is some indication that broader dispersal in some species (having somewhat modified seeds) might be achieved by wind or by animal carriage. The seeds of a few species possess an elaiosome and are dispersed by ants (Insecta: Hymenoptera: Formicidae: *Formica*). Both morphological dormancy and physiological dormancy have been reported for the seeds of various species.

Delphinium uliginosum **Curran** is a biennial or weak perennial (from woody roots) that is restricted to seeps, slopes, swales, or the margins of streams within serpentine chaparral or meadows at elevations from 190 to 853 m. Exposures are described as shaded. The substrates are categorized as serpentine and alkaline (pH: 7.6–8.3) and include gravel and sand. Flowering occurs from May to July. The flowers survive for 8.2 days on average and can reach densities averaging 10.3/m². They receive a relatively high diversity of pollinators (Insecta). The loading of stigmas by heterospecific pollen increases linearly with conspecific pollen, which limits seed set. Ground cover by moss (Bryophyta: Pottiaceae: *Didymodon tophaceus*) not only facilitates seedling emergence but also results in reduced adult plant biomass. Higher numbers of adult individuals occur at sites having moss cover. Seed dispersal is extremely limited due to the high retention rate of seeds within the moss cover. Populations typically contain from 40 to 60 individuals, but the smaller ones are subject to local extirpation. Additional life history studies are needed for this species. **Reported associates:** *Astragalus clevelandii, Carex serratodens, Cirsium douglasii, Didymodon tophaceus, Helianthus exilis, Mimulus guttatus, M. nudatus, Packera clevelandii, Toxicoscordion venenosum.*

Use by wildlife: *Delphinium* species are known to cause poisoning in mammals (Mammalia) including cattle (Bovidae: *Bos taurus*), horses (Equidae), and sheep (Bovidae: *Ovis aries*).

Economic importance: food: *Delphinium* species are not edible and potentially are quite poisonous; **medicinal:** Young plants and seeds of all *Delphinium* species reportedly are toxic and should not be taken for medicinal uses; **cultivation:** *Delphinium uliginosum* is found occasionally in cultivation but is difficult to grow due to its particular ecological requirements; **misc. products:** none; **weeds:** none; **nonindigenous species:** none.

Systematics: *Delphinium* is assigned taxonomically to Ranunculaceae tribe Delphinieae (or Cimicifugeae), wherein phylogenetic studies indicate that it is related most closely to the genera *Consolida* (sometimes placed in *Delphinium*) and *Aconitum* (Figure 2.3). Analyses including a fairly comprehensive sampling of species from all three genera indicate that *Aconitum*, *Aconitella*, and *Consolida* are nested phylogenetically within *Delphinium*, unless a small number of species are transferred to the genus *Staphisagria* (to which they had been assigned in the past). In order to maintain the monophyly of *Delphinium*, it would also be necessary to merge *Aconitella* and *Consolida* with *Delphinium* or to recognize at least one additional subordinate genus (Figure 2.8). *Delphinium uliginosum* is assigned to section *Diedropetala*, subsection *Depauperata*. It has been included in only one molecular phylogenetic analysis (using nrITS sequence data), which placed it anomalously in a weakly supported clade containing a mixture of species from three different subsections of section *Diedropetala* (Figure 2.9). Clearly the current infrageneric taxonomy of *Delphinium* requires many modifications in order to reflect strictly monophyletic taxa. The base chromosome number of *Delphinium* is $x = 8$. *Delphinium uliginosum* is diploid ($2n = 16$) and allegedly hybridizes with *D. hesperium* (subsection *Echinata*), also $2n = 16$.

Comments: *Delphinium uliginosum* is a narrow endemic restricted to the North Coast Ranges of California.

References: Arceo-Gómez et al., 2016; Baskin & Baskin, 1998; Cruden & Hermann-Parker, 1977; DeHart et al., 2014; Epling & Lewis, 1952; Freestone, 2006; Harrison et al., 2000; Hoot, 1995; Jabbour & Renner, 2012; Koontz et al., 2004; Loconte et al., 2012; Olsen et al., 1990; Ro et al., 1997; Turnbull et al., 1983; Warnock, 1997; Williams et al., 2001.

7. Halerpestes

Alkali buttercup; renoncule cymbalaire

Etymology: from the Greek *halo erpes* ("salt creeper") in reference to the habit and habitat of some species

Synonyms: *Cyrtorhyncha* (in part); *Oxygraphis*; *Ranunculus* (in part)

Distribution: global: Eurasia; Mexico; North America; South America; **North America:** northern and western

Diversity: global: 10 species; **North America:** 1 species

Indicators (USA): OBL: *Halerpestes cymbalaria*

Habitat: freshwater; palustrine; **pH:** alkaline; **depth:** <1 m; **life-form(s):** emergent (herb)

Key morphology: flowering stems erect, stolons (to 40 cm) prostrate, rooting at nodes; leaves basal, simple, petioled, the blades (to 3.8 cm) circular, cordate or oblong, base rounded or cordate, margins crenate; flowers solitary, bisexual, radial,

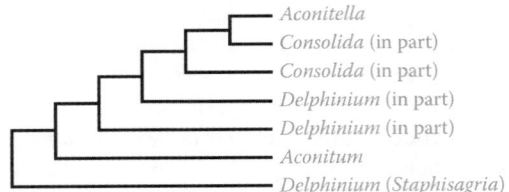

FIGURE 2.8 Intergeneric relationships within Ranunculaceae tribe Delphinieae as indicated by phylogenetic analysis of combined DNA sequence data. These results indicate that the monophyly of *Delphinium* would require either the merger of *Aconitella* and *Consolida* with *Delphinium* or the transfer of some species currently placed in *Delphinium* to *Staphisagria* and others to an additional genus. *Delphinium uliginosum*, the only OBL indicator in the genus, was excluded from the analysis. (Adapted from Jabbour, F. & Renner, S.S., *Mol. Phylogen. Evol.*, 62, 928–942, 2012.)

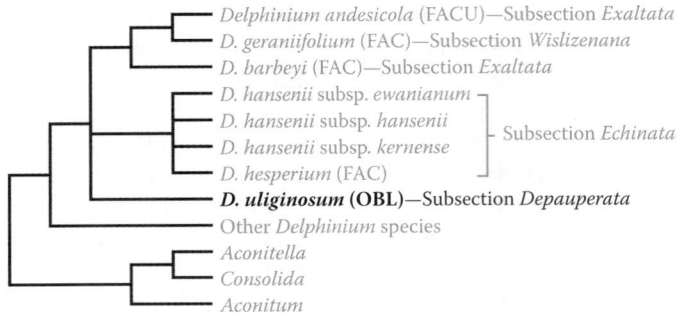

FIGURE 2.9 Interspecific relationships in *Delphinium* section *Diedropetala* as indicated by analysis of nrITS sequence data. The North American wetland indicator status is indicated in parentheses. *Delphinium uliginosum*, the only OBL indicator in the genus (highlighted in bold) occurs among a mixture of species from different subsections, which indicates the need for refinements in the infrageneric taxonomy of the group. (Adapted from Koontz, J.A. et al., *Syst. Bot.*, 29, 345–357, 2004.)

peduncled (to 15 cm); sepals (to 6 mm) 5, green, spreading; petals (to 7 mm) 5, yellow; stamens numerous; pistils numerous, aggregated in cylindrical heads (to 12 mm); achenes (to 2.2 mm), with a persistent, straight beak (to 0.2 mm) **Life history: duration:** perennial (stolons); **asexual reproduction:** stolons; **pollination:** Insect; **sexual condition:** hermaphroditic; **fruit:** achenes (common); **local dispersal:** achenes (water), stolons; **long-distance dispersal:** achenes **Imperilment:** (1) *Halerpestes cymbalaria* [G5]; SH (MO, NJ, NY, RI); S1 (CT, IL, MB, MI); S3 (WI, YT)

Ecology: general: Many *Halerpestes* species occur in wet habitats and are common especially along the margins of lacustrine and riverine communities. Most of the species are stoloniferous perennials that often grow on saline substrates. The flowers are nectariferous and presumably pollinated by insects (Insecta). The uncertain circumscription of *Halerpestes* (with most of the species currently retained in *Ranunculus*) makes it difficult to summarize other life history information for the genus.

Halerpestes cymbalaria (**Pursh**) **Greene** grows in shallow tidal or nontidal sites associated with alcoves, beaches, bogs, borrow pits (dried), bottomlands, chapparal, coasts, depressions, ditches, draws, estuaries, flats, floodplains, gravel bars, gravel pits, hanging gardens, marshes, meadows, prairies, riverbeds, river terraces, roadsides, saltmarshes, sand/ silt bars, seeps, slopes (to 10%), sloughs, springs (hot), swales, swamps, washes, and along the margins or shores of canals, coulees, lakes, ponds (ephemeral), reservoirs, rivers, springs, streams, and tidal pools at elevations of up to 3140 m. The plants can tolerate shallow (e.g., 5 cm) standing water and exposure to full sunlight but occur most often in the shade. They are considered to be somewhat salt tolerant. The substrates are acidic or alkaline/calcareous (often brackish or saline) and include alluvium, basalt rimrock, clay, clay loam, clay mud, granite, gravel, gravelly limestone, gravelly sand, gravelly silt, humus, limestone, loam, marl, muck, mud, peat, pine duff, quartzitic gravel, rock, rocky loam, sand, sandy alluvium, sandy clay, sandy gravel, sandy loam, sandy silt, shale, silt (alluvial), silty loam, stones, talus, and travertine. Flowering occurs from April to September, with fruiting extending from May to October. Information on specific pollinators is unavailable. Individual plants can produce up to 1220 seeds. The seeds (~0.073 mg) are buoyant and can be dispersed along the water surface. A seed bank of uncertain duration develops. Seeds recovered from seed bank cores have germinated following 5 weeks of cold stratification (5°C) and incubation at 25°C under ambient greenhouse conditions supplemented with 14 h of artificial light. The seeds must be exposed to light for germination to occur. Longer periods of stratification (e.g., 2 months) will increase germination rates substantially. After 2 months of cold–moist stratification (5°C), high seed germination rates (82%) have been obtained under a 14/10 h, 25°C/15°C day/night light/temperature regime. Populations are known to severely decline due to grazing by waterfowl (Aves: Anatidae). The plants reproduce vegetatively by fairly prolific stolons, which root along their nodes and enable the shoots to spread effectively across open ground. The roots are colonized by arbuscular mycorrhizal and fine endophytic fungi. **Reported associates:** Abies concolor, Acer negundo, Achillea millefolium, A. sibirica, Acmispon utahensis, Agoseris parviflora, Agropyron, Agrostis scabra, A. stolonifera, Alhagi maurorum, Alisma, Allium cernuum, A. geyeri, Almutaster pauciflorus, Alnus oblongifolia, Alopecurus aequalis, A. carolinianus, A. pratensis, Amelanchier utahensis, Amorpha californica, A. fruticosa, Amphiscirpus nevadensis, Androsace, Anemopsis californica, Antennaria, Apocynum cannabinum, Aquilegia chrysantha, Artemisia dracunculus, A. frigida, A. nova, A. tridentata, Astragalus canadensis, A. preussii, Atriplex canescens, A. patula, Baccharis salicifolia, Beckmannia syzigachne, Berberis haematocarpa, B. repens, Berula erecta, Betula occidentalis, Bolboschoenus maritimus, Brickellia desertorum, B. floribunda, Bromus frondosus, B. inermis, B. rubens, B. tectorum, Calamagrostis deschampsioides, Callitriche palustris, Calocedrus decurrens, Carex aquatilis, C. douglasii, C. emoryi, C. interior, C. lasiocarpa, C. lyngbyei, C. maritima, C. microptera, C. nebrascensis, C. pellita, C. praegracilis, C. rostrata, C. senta, C. simulata, C. subfusca, C. ursina, C. utriculata, C. xerantica, Castilleja exserta, C. miniata, Catabrosa aquatica, Ceanothus, Celtis reticulata, Centaurium pulchellum, Cercocarpus ledifolius, C. montanus, Chara, Chenopodium rubrum, Chilopsis linearis, Cichorium intybus, Cirsium arvense, C. undulatum, Claytonia perfoliata, Clematis, Coleogyne ramosissima, Conium maculatum, Conyza canadensis, Cornus, Crataegus, Cryptantha setosissima, Cylindropuntia whipplei, Cynodon dactylon, Cyperus niger, Dasiphora floribunda, Deschampsia cespitosa, Dieteria bigelovii, Distichlis spicata, Echinocereus engelmannii, E. triglochidiatus, Echinochloa crus-galli, Elaeagnus angustifolia, Eleocharis acicularis, E. bella, E. erythropoda, E. macrostachya, E. palustris, E. parishii, E. quinqueflora, E. rostellata, Elymus elymoides, E. trachycaulis, Ephedra, Epilobium ciliatum, E. palustre, Epipactis gigantea, Equisetum arvense, E. laevigatum, Eremogone eastwoodiae, Ericameria nauseosa, Erigeron divergens, E. flagellaris, E. foliosus, Eriogonum corymbosum, E. racemosum, E. umbellatum, Euphorbia glyptosperma, Euthamia occidentalis, Festuca rubra, Forestiera pubescens, Fragaria, Fraxinus pennsylvanica, F. velutina, Galium trifidum, Gaura mollis, Gentiana parryi, Geranium caespitosum, Glyceria grandis, G. striata, Glycyrrhiza lepidota, Gnaphalium, Gutierrezia sarothrae, Halogeton glomeratus, Helianthella quinquenervis, Heliomeris multiflora, Heliotropium curassavicum, Heracleum sphondylium, Heterotheca villosa, Hippuris vulgaris, Holcus lanatus, Hordeum jubatum, H. nodosum, Hosackia oblongifolia, Hydrocotyle, Hymenoclea sandersonii, Hymenoxys richardsonii, H. subintegra, Hypericum scouleri, Iris missouriensis, I. setosa, Juglans major, Juncus articulatus, J. balticus, J. bufonius, J. compressus, J. dudleyi, J. ensifolius, J. longistylis, J. nevadensis, J. orthophyllus, J. saximontanus, J. torreyi, J. xiphioides, Juniperus communis, J. deppeana, J. occidentalis, J. osteosperma, J. scopulorum, Koeleria macrantha, Lathyrus japonicus, Lemna minor, Lepidospartum squamatum, Leptochloa fusca, Leymus, Lilium philadelphicum, Limosella aquatica, Lobelia cardinalis, Lotus corniculatus, Ludwigia peploides, Lupinus hillii, Lycium pallidum, Lycopus, Lysimachia maritima, Lythrum salicaria, Medicago lupulina, M. sativa, Melilotus albus, M. officinalis, Mentha arvensis, Mimulus guttatus, M. moschatus, M. primuloides, M. tilingii, Mirabilis laevis, Monarda fistulosa, Muhlenbergia mexicana, M. montana, M. richardsonis, Myosurus minimus, M. stricta, Myriophyllum, Nasturtium officinale, Navarretia intertexta, Nolina parryi, Nuphar polysepala, Nymphaea, Oenothera curtiflora, Opuntia basilaris, O. chlorotica, O. engelmannii, Oryzopsis, Packera multilobata, Pedicularis groenlandica, Penstemon barbatus, P. pseudoputus, Perideridia parishii, Peritoma lutea, Persicaria, Phacelia lutea, Phalaris arundinacea, Phleum pratense, Phragmites australis, Picea engelmannii, P. glauca, P. pungens, P. sitchensis, Pinus contorta, P. monophylla, P. ponderosa, Plagiobothrys salsus, Plantago major, Platanus wrightii, Pluchea sericea, Poa fendleriana, P. nemoralis, P. palustris, P. pratensis, P. sandbergii, P. secunda, Polypogon monspeliensis, Populus fremontii, P. tremuloides, P. trichocarpa, Potamogeton, Potentilla anserina, P. hippiana, P. norvegica, Primula incana, Prunus virginiana, Pseudocymopterus montanus, Pseudognaphalium luteoalbum, P. stramineum, Pseudotsuga menziesii, Ptelea trifoliata, Puccinellia distans, P. nuttalliana, P. phryganodes, P. vaginata, Purshia tridentata, Quercus agrifolia, Q. chrysolepis, Q. gambelii, Q. turbinella, Ranunculus aquatilis, R. gmelinii, R. inamoenus, R. longirostris, R. macounii, R. sceleratus, Rhamnus californica, Rhaponticum repens, Rhinanthus minor, Rhus aromatica, Ribes aureum, R. cereum, Rosa woodsii, Rumex acetosella, R. crispus, R. fueginus, R. maritimus, Sagina procumbens, Sagittaria, Salix bebbiana, S. candida, S. eriocephala, S. exigua, S. geyeriana, S. gooddingii, S. laevigata, S. lasiandra, S. lasiolepis, S. lemmonii, S. lutea, S. ovalifolia, S. planifolia, Salsola, Salvia pachyphylla, Schedonorus arundinaceus, Schizachyrium scoparium, Schoenoplectus acutus, S. americanus, S. pungens, S. tabernaemontani, Scirpus microcarpus, S. pallidus, Sidalcea neomexicana, Silene involucrata, Sinapis arvensis, Sisyrinchium angustifolium, S. demissum, S. idahoense, Solidago confinis, S. multiradiata, S. nana, S. sempervirens, Sonchus arvensis, S. asper, Sorghastrum nutans, Spartina gracilis, Spergularia media, S. salina, Sphagnum, Sphenosciadium capitellatum, Spiranthes diluvialis, Spirodela, Stachys albens, Stanleya pinnata, Stellaria crassifolia, Stuckenia pectinata, Suaeda calceoliformis, Symphoricarpos rotundifolius, Symphyotrichum ascendens, S. falcatum, S. frondosum, S. spathulatum, Tamarix chinensis, T. ramosissima, Taraxacum officinale, Thalictrum fendleri, Thermopsis divaricarpa, Toxicodendron radicans, Trifolium repens, Triglochin maritimum, Trillium ovatum, T. petiolatum, Typha angustifolia, T. domingensis, T. latifolia, Urtica, Verbascum thapsus, Veronica americana, V. anagallis-aquatica, Viola adunca, V. epipsila, V. sororia, Vulpia octoflora, Wilhelmsia physodes, Xanthium strumarium, Zannichellia palustris.

Use by wildlife: *Halerpestes cymbalaria* plants are suspected of causing livestock poisoning but are grazed by Canada geese (Aves: Anatidae: *Branta canadensis*). They are host to powdery mildew fungi (Ascomycota: Erysiphaceae: *Erysiphe cruciferarum*) and rusts (Basidiomycota: Pucciniaceae: *Puccinia cinerea*, *P. clematidis*).

Economic importance: food: *Halerpestes cymbalaria* contains protoanemonin glycosides, which are regarded as poisonous. It would be advisable to avoid its consumption; however, the plants were cooked and eaten by the Gosiute of Utah; **medicinal:** *H. cymbalaria* was administered by the Navajo as an emetic and to treat venereal problems. It was used by the Kawaiisu people as a dermatological aid; **cultivation:** not cultivated; **misc. products:** The Navajo used *H. cymbalaria* as a ceremonial medicine; **weeds:** none; **nonindigenous species:** none.

Systematics: *Halerpestes* had long been merged with *Ranunculus*, until phylogenetic studies indicated that it represents a distinct, distantly related clade (Figures 2.7 and 2.10). *Halerpestes* appears to be monophyletic, given the close association of the four surveyed species (Figure 2.10). *Halerpestes cymbalaria* appears to be closely related to several species of Asian affinity (*H. salsuginosa*, *H. sarmentosa*, and *H. ruthenica*), which indicates its likely origin on that continent. The basic chromosome number of *Halerpestes* is $x = 8$; *H. cymbalaria* ($2n = 16$) is a diploid. Hybrids involving *H. cymbalaria* have not been reported.

Comments: *Halerpestes cymbalaria* is widespread throughout western and northern North America and extends into Eurasia and South America.

References: Arthur, 1909; Brotherson et al., 1980; Dawe et al., 2011; Emadzade et al., 2010; Garrett, 1921; Hoffmann et al., 2010; Hopfensperger & Baldwin, 2009; Hopfensperger et al., 2009; Khan & Weber, 2006; Pool, 1910; Staniforth et al., 1998; Steven & Franke, 1990; Stevens, 1957; Walker et al., 2010; Whittemore, 1997a.

8. *Kumlienia*
False buttercup
Etymology: after Thure Kumlien (1819–1888)

Synonyms: *Ranunculus* (in part)
Distribution: global: North America; **North America:** western
Diversity: global: 1 species; **North America:** 1 species
Indicators (USA): OBL: *Kumlienia hystricula*
Habitat: freshwater; palustrine; **pH:** unknown; **depth:** <1 m; **life-form(s):** emergent (herb)
Key morphology: stems arising from a caudex; basal leaves in a rosette, the blades (to 6.6 cm) reniform, five to seven lobed, margins crenate; cauline leaves 2 or absent, scalelike; flowers solitary from elongate peduncles (to 20 cm), radial, bisexual; sepals (to 13 mm) 5–6, spreading, showy, white; petals (to 4 mm) 8–12, inconspicuous, cuplike, green or yellow, intermixed with stamens; stamens and pistils numerous; achenes (to 4.2 mm) aggregated in heads (to 8 mm), with filiform beak (to 1.4 mm), hooked at tip
Life history: duration: perennial (caudex); **asexual reproduction:** none; **pollination:** insect; **sexual condition:** hermaphroditic; **fruit:** aggregates (achenes; common); **local dispersal:** achenes; **long-distance dispersal:** achenes
Imperilment: (1) *Kumlienia hystricula* [G3]
Ecology: general: *Kumlienia* is monotypic (see *Systematics*).

Kumlienia hystricula (A. Gray) Greene inhabits depressions, cliffs (rocky), crevices (rocky), flats, gullies, ravines (rocky), seeps, slopes, springs, and the margins of streams (commonly near waterfalls) at elevations of up to 1677 m. The site exposures are shaded. The substrates characteristically are associated with much moss (Bryophyta) or pine (*Pinus*) needle cover and include granite (decomposed), phyllite, rock, and sandy loam. Flowering and fruiting extend from February to July. The bisexual flowers resemble those of *Anemone* superficially and are pollinated by insects (Insecta), which are attracted to the showy sepals, given their prominent size with respect to the highly reduced petals. Little information exists on the reproductive or seed ecology. It is conceivable that the apically hooked achenes might be dispersed by attachment to animal fur. The plants are perennials and arise from a caudex. **Reported associates:** *Adiantum jordanii*, *Aesculus californica*, *Alnus rhombifolia*, *Arbutus menziesii*,

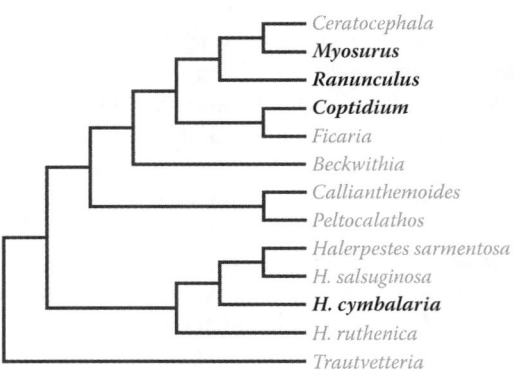

FIGURE 2.10 Phylogenetic tree constructed from combined DNA sequence data showing the monophyly of *Halerpestes* and its distinction from *Ranunuculus*, with which it has been merged by some authors. Taxa containing OBL North American indicators are highlighted in bold. The three non-North American *Halerpestes* species are of Asian provenance. (Adapted from Hoffmann, M.H. et al., *Int. J. Plant Sci.*, 171, 90–106, 2010.)

Aspidotis californica, Cystopteris fragilis, Darmera peltata, Eriodictyon californicum, Lonicera, Pinus ponderosa, Polypodium californicum, Quercus chrysolepis, Q. douglasii, Rhamnus californica, Sambucus, Selaginella hansenii, Woodwardia fimbriata.

Use by wildlife: none reported.

Economic importance: food: not edible; **medicinal:** no uses reported; **cultivation:** not cultivated; **misc. products:** none; **weeds:** none; **nonindigenous species:** none.

Systematics: The taxonomy of *Kumlienia* is confusing. *Kumlienia* should be regarded as monotypic despite the fact that many authors continue to recognize *K. hystricula* and also *Arcteranthis cooleyae* (which has also been assigned to *Kumlienia*), as species of *Ranunculus*. Phylogenetic evidence (e.g., Figure 2.7) clearly shows that neither is related closely to the latter. The distinction of *Arcteranthis* and *Kumlienia* is not surprising, given that they had been assigned to different sections (*Arcteranthis* and *Pseudaphanostemma*, respectively), even when regarded as members of *Ranunculus*. The base number of *Kumlienia* is x = 8; *K. hystricula* (2n = 16) is a diploid. There have been no reports of hybrids involving *K. hystricula*.

Comments: *Kumlienia hystricula* is endemic to the western slopes of the Sierra Nevada of California.

References: Benson, 1940, 1948; Emadzade et al., 2010; Goepfert, 1974; Whittemore, 1997c.

9. *Myosurus*

Mousetail; queue de souris

Etymology: from the Greek *mus oura* ("mouse tail"), in reference to the elongate receptacles

Synonyms: none

Distribution: global: pantemperate; **North America:** widespread

Diversity: global: 15 species; **North America:** 5 species

Indicators (USA): OBL: *Myosurus apetalus, M. minimus*; **FACW:** *M. apetalus, M. minimus*; **FAC:** *M. minimus*

Habitat: freshwater, saline (inland); palustrine; **pH:** 6.1–8.3; **depth:** <1 m; **life-form(s):** emergent (herb)

Key morphology: leaves (to 11.5 cm) basal, linear-spatulate; flowers solitary, radial, on long (to 13 cm), scapose peduncles; sepals (to 4 mm) 3–8, greenish or whitish; petals (to 2.5 mm) 0–5, white, long-clawed; stamens 5-numerous; pistils numerous, uniovulate; receptacle expanding into an elongate (to 5 cm), taillike spike bearing numerous, aggregated, spirally arranged achenes (to 2.2 mm), which are beaked (to 1.4 mm)

Life history: duration: winter annual (fruit/seeds); **asexual reproduction:** none; **pollination:** self, insect (minor); **sexual condition:** hermaphroditic; **fruit:** aggregates (achenes; common); **local dispersal:** achenes; **long-distance dispersal:** achenes (mammals)

Imperilment: (1) *Myosurus apetalus* [G5]; S1 (WY); S2 (SK, WY); S3 (AB, BC, MT); (2) *M. minimus* [G5]; S1 (MB, WV); S2 (ON, SK); S3 (BC, IA, MT, NC, WY)

Ecology: general: *Myosurus* comprises annual herbs with radial, apocarpous, bisexual flowers. Several species colonize the temporary waters of depressions, which form on clay or hardpan substrates ranging from silty clay to sandy loam with a moderate to high alkali content. Four of the North American species (80%) are wetland indicators, with two designated as OBL. The flowers are normally homogamous but occasionally protandrous. The early produced flowers are self-compatible and primarily autogamous. Outcrossing can occur to a lesser extent in later produced flowers. Although predominantly selfing, the flowers are outcrossed on occasion by bees (Insecta: Hymenoptera: Halictidae) or flies (e.g., Insecta: Diptera: Muscidae; Syrphidae). The terminally spined achenes are thought in some instances to become attached to animals for long-distance dispersal. They are also ingested by waterfowl (Aves: Anatidae) and retain germinability when excreted, providing for long-distance endozooic dispersal. The seeds of at least some species lack dormancy. All of the wetland species are threatened by habitat losses due to agricultural development.

Myosurus apetalus **Gay** occurs in or on alluvial benches, bogs, depressions, ditches, draws, dunes, flats, floodplains, gravel pits, hummocks, marshes, meadows, mudflats, playas, pools (dry), potholes, prairies, reservoir bottoms (dry), roadsides, ruts, seeps, shores, slopes (to 5%), streams (ephemeral), swales, thickets, vernal pools, washes (drying), and along the margins of channels, lakes, ponds, rivers, sloughs, springs, and streams at elevations of up to 3505 m. The plants occur in open to shaded exposures. The substrates are categorized as alkaline (pH: 6.1–8.3) and include basalt rimrock, clay, dolomite, granite, gravel, gravelly loam, lava (outcrops), loam, mud, quartz, rock (outcrops), rocky loam, sand, sandy gravel, sandy loam, shale, silt, stony clay, and stony Tuscan loam. Flowering occurs from March to July, with fruiting observed from April to July. The flowers presumably are self-pollinating, but their reproductive biology has not been studied. No information is available on seed dispersal mechanisms or germination requirements. **Reported associates:** *Abies concolor, Achnatherum thurberianum, Agave, Alisma, Allium brandegeei, Alopecurus aequalis, A. carolinianus, A. saccatus, Ambrosia, Androsace septentrionalis, Antennaria luzuloides, Aquilegia formosa, Arctostaphylos patula, A. pungens, Arenaria lanuginosa, Arnica sororia, Artemisia cana, A. tridentata, Atriplex gardneri, Balsamorhiza sagittata, Bistorta bistortoides, Blennosperma nanum, Boechera gracilipes, Bouteloua curtipendula, B. gracilis, Bromus inermis, B. japonicus, B. tectorum, Callitriche heterophylla, Camassia quamash, Cardamine oligosperma, Carex athrostachya, C. douglasii, C. filifolia, C. occidentalis, C. praegracilis, C. subfusca, Castilleja campestris, C. cusickii, C. lasiorhyncha, Ceanothus greggii, Centaurea, Cerastium fontanum, C. pumilum, Cercocarpus montanus, Chara, Chorispora tenella, Chrysothamnus viscidiflorus, Claytonia perfoliata, Clematis ligusticifolia, Collinsia parviflora, Coreopsis, Crassula aquatica, Crepis runcinata, Cylindropuntia whipplei, Deschampsia cespitosa, D. danthonioides, Descurainia pinnata, Distichlis, Downingia elegans, Draba verna, Drymocallis lactea, Echinochloa, Elatine rubella, Eleocharis acicularis, E. bella, E. macrostachya,*

E. ovata, E. palustris, E. quinqueflora, Elodea, Elymus elymoides, E. trachycaulus, Epilobium ciliatum, Ericameria nauseosa, Erigeron pumilus, Eriophorum, Erodium cicutarium, Euphorbia esula, Floerkea proserpinacoides, Galium bifolium, Garrya flavescens, Geum triflorum, Glyceria, Gnaphalium, Gratiola neglecta, Gymnosteris nudicaulis, Hordeum brachyantherum, H. jubatum, Horkelia rydbergii, Hymenopappus filifolius, Hypericum, Idahoa scapigera, Iris missouriensis, Iva axillaris, Juncus balticus, J. dudleyi, J. ensifolius, J. hemiendytus, J. kelloggii, J. nevadensis, J. orthophyllus, J. tenuis, Juniperus coahuilensis, J. deppeana, J. monosperma, J. occidentalis, J. osteosperma, J. scopulorum, Lappula occidentalis, Layia fremontii, Leptosiphon bicolor, Lewisia nevadensis, L. triphylla, Leymus cinereus, L. triticoides, Limnanthes alba, Limosella acaulis, L. aquatica, Linaria genistifolia, Lithophragma, Logfia arvensis, Lomatium bicolor, L. triternatum, Lupinus kingii, L. lepidus, Maianthemum stellatum, Marsilea vestita, Medicago sativa, Melilotus officinalis, Micranthes nidifica, Microsteris gracilis, Mimulus androsaceus, M. breweri, M. floribundus, M. primuloides, M. suksdorfii, Montia chamissoi, M. dichotoma, M. linearis, Myosotis stricta, Myosurus minimus, Navarretia leucocephala, Oenothera tanacetifolia, Opuntia macrorhiza, O. phaeacantha, Orobanche, Orthocarpus campestris, Parthenocissus vitacea, Pectocarya linearis, Phacelia exilis, P. linearis, Picea engelmannii, Picrothamnus desertorum, Pinus contorta, P. edulis, P. jeffreyi, P. monophylla, P. ponderosa, Plagiobothrys hispidulus, P. leptocladus, P. scouleri, P. scriptus, Plantago elongata, Poa arida, P. fendleriana, P. palustris, P. pratensis, P. secunda, Polemonium micranthum, Polyctenium fremontii, Polygonum aviculare, P. douglasii, P. polygaloides, Porterella carnosula, Potamogeton, Potentilla anserina, P. gracilis, P. wheeleri, Primula cusickiana, P. tetrandra, Pseudoroegneria spicata, Psilocarphus brevissimus, P. tenellus, Purshia tridentata, Quercus emoryi, Q. gambelii, Q. turbinella, Ranunculus cardiophyllus, R. flammula, R. glaberrimus, R. trichophyllus, Rhus aromatica, Ribes cereum, R. inerme, Rorippa curvisiliqua, Rosa, Rumex acetosella, Salix drummondiana, S. exigua, S. geyeriana, S. lasiolepis, S. lutea, S. scouleriana, Sarcobatus vermiculatus, Senecio crassulus, S. hydrophilus, Sphaeralcea parvifolia, Sphagnum, Stachys albens, Stellaria longipes, S. nitens, Symphoricarpos rotundifolius, Taraxacum californicum, T. officinale, Taraxia breviflora, Tetraneuris acaulis, Thalictrum fendleri, Thinopyrum ponticum, Tortula ruralis, Townsendia, Tragopogon dubius, Trifolium gymnocarpon, T. macrocephalum, T. monanthum, Triphysaria eriantha, Triteleia grandiflora, Verbascum thapsus, Veronica peregrina, Vulpia octoflora, Wyethia helianthoides.

***Myosurus minimus* L.** grows in bottomlands, cornfields, depressions, ditches, draws, flats, floodplains, lawns, marshes, meadows, mudflats, playas, prairies, ravines, roadsides, ruts, salt flats, seeps, slopes (to 1%), sloughs, springs, swales, swamps, thickets, trenches, vernal pools, washes, and along the margins (or drying bottoms) of gravel pits, lakes, oxbows, ponds, potholes, and reservoirs at elevations of up to 3352 m. The plants normally are found on wet (or drying) substrates and less often in shallow water (<5 cm deep). The plants can tolerate full sunlight to shaded exposures. The substrates are alkaline and are described as alluvium (muddy, sandy), basalt, basalt rimrock, Capay-Clear Lake soil, Capay silty clay, clay (saline, sodic), clay loam, claypan, clay silt, decayed sawdust, gravel (compacted), gravelly clay, gravelly hardpan, Hanford sandy loam, Klamath silty clay loam, loam (organic), muck, mud, obsidian rubble, Pescadero clay, rocky clay, rocky loam, sand, sandy clay, sandy clay loam, sandy loam, scree, silt (organic), silty clay, silty loam, silty sand, Solano silty clay, Stockpen soil, stony clay, Traver sandy loam, and Willows clay. Flowering and fruiting extend from March to October. The flowers are not only normally self-pollinating but also are outcrossed by insects on occasion (see *Use by wildlife*). The seeds are dispersed endozoically by waterfowl (Aves: Anatidae). They will germinate in about 2 weeks without stratification and can remain viable in the seed bank for at least 2–3 years. Much higher seed germination occurs under noninundated conditions, with little germination apparent in continually inundated sediments. Seeds that were stratified at 5°C for 56 days exhibited low germination (8%) when incubated at 20°C under a 12/12 h light/dark regime. **Reported associates:** *Acmispon parviflorus, Agoseris glauca, Aira praecox, Allium amplectens, A. bisceptrum, Alopecurus aequalis, A. carolinianus, A. saccatus, Amaranthus palmeri, Amsinckia, Antennaria plantaginifolia, Aphanes occidentalis, Arbutus, Aristida adscensionis, Artemisia arbuscula, A. cana, A. rigida, A. tridentata, A. tripartita, Astragalus argophyllus, Atriplex argentea, A. coronata, A. davidsonii, A. parishii, A. polycarpa, A. serena, A. spinifera, Avena fatua, Baccharis sarothroides, Barbarea vulgaris, Blennosperma nanum, Boechera laevigata, Bouteloua curtipendula, B. gracilis, Brodiaea filifolia, B. orcuttii, Bromus hordeaceus, B. rubens, B. tectorum, Buglossoides arvensis, Calandrinia, Calibrachoa parviflora, Callitriche longipedunculata, C. marginata, Camassia leichtlinii, C. quamash, Capsella bursa-pastoris, Cardamine parviflora, C. pensylvanica, Carex alma, C. athrostachya, C. disperma, C. hirsutella, C. nebrascensis, C. occidentalis, C. pensylvanica, C. praegracilis, C. subfusca, C. utriculata, Castilleja attenuata, Celtis reticulata, Cerastium glomeratum, C. nutans, C. semidecandrum, Chara, Cheilanthes, Chrysothamnus viscidiflorus, Claytonia gypsophiloides, C. perfoliata, C. rubra, C. virginica, Collinsia, Convolvulus arvensis, Coreopsis tinctoria, Crassula aquatica, C. connata, C. solieri, Cressa truxillensis, Crypsis schoenoides, Cryptantha, Cynodon dactylon, Cynosurus echinatus, Cystopteris, Deinandra fasciculata, Delphinium recurvatum, Deschampsia cespitosa, D. danthoides, Descurainia, Diaperia verna, Dichanthelium oligosanthes, Dichelostemma multiflorum, Disakisperma dubium, Distichlis spicata, Downingia insignis, Draba verna, Echinochloa crus-galli, Elatine brachysperma, E. californica, E. chilensis, E. rubella, Eleocharis acicularis, E. engelmannii, E. macrostachya, E. palustris, E. parishii, Elodea nuttallii, Elymus glaucus, E. villosus, Epilobium campestre, E. ciliatum, E. densiflorum, Eragrostis, Ericameria linearifolia, Erigeron divergens, E. pulchellus, E. pumilus, Eriogonum*

fasciculatum, E. umbellatum, Erodium brachycarpum, Eryngium aristulatum, E. mathiasiae, E. spinosepalum, Erysimum repandum, Euphorbia esula, Eurybia integrifolia, Festuca rubra, Floerkea proserpinacoides, Frankenia salina, Gaultheria shallon, Geranium carolinianum, Geum triflorum, Gilia flavocincta, Glandularia bipinnatifida, Gnaphalium, Grindelia hirsutula, G. integrifolia, Helenium autumnale, Helianthus divaricatus, Hemizonia pungens, Hordeum brachyantherum, H. depressum, H. intercedens, H. jubatum, H. marinum, H. murinum, H. pusillum, Hymenoxys, Ilex opaca, I. missouriensis, Ipomoea, Iris pseudacorus, Isocoma tenuisecta, Isoetes howellii, I. nuttallii, Iva axillaris, Juncus balticus, J. bufonius, J. dudleyi, J. effusus, J. ensifolius, J. hemiendytus, J. sphaerocarpus, J. tenuis, J. uncialis, Juniperus monosperma, J. osteosperma, Kochia californica, Lamium amplexicaule, L. purpureum, Lasthenia californica, L. platycarpha, Layia fremontii, Lepidium dictyotum, L. latipes, L. nitidum, L. perfoliatum, L. ruderale, L. virginicum, Lewisia, Leymus cinereus, Limnanthes alba, L. douglasii, Limosella aquatica, Lolium multiflorum, Lomatium caruifolium, L. nudicaule, Lupinus bicolor, L. kingii, Lythrum hyssopifolia, Malvella leprosa, Marsilea vestita, Matricaria discoidea, M. occidentalis, Micranthes integrifolia, M. subapetala, Microsteris gracilis, Mimosa aculeaticarpa, Mimulus breweri, M. guttatus, Monolepis nuttalliana, Montia dichotoma, M. fontana, M. howellii, M. linearis, Muhlenbergia, Musineon divaricatum, Myosurus apetalus, M. sessilis, Navarretia fossalis, N. intertexta, N. leucocephala, N. prostrata, Nemophila maculata, N. pedunculata, Neoholmgrenia andina, Oenothera elata, O. flava, Opuntia, Orcuttia californica, Oxalis violacea, Parthenocissus quinquefolia, Pascopyrum smithii, Perideridia gairdneri, Persicaria amphibia, Phalaris caroliniana, P. lemmonii, P. minor, P. paradoxa, Phleum pratense, Pilularia americana, Pinus contorta, P. edulis, P. ponderosa, Plagiobothrys acanthocarpus, P. arizonicus, P. bracteatus, P. hirtus, P. hispidulus, P. humistratus, P. leptocladus, P. nothofulvus, P. scouleri, P. stipitatus, Planodes virginica, Plantago bigelovii, P. coronopus, P. elongata, Plectritis congesta, Pleuropogon californicus, Poa annua, P. chapmaniana, P. compressa, P. fendleriana, P. interior, P. pratensis, P. secunda, Pogogyne douglasii, Polygonum aviculare, P. douglasii, P. polygaloides, Populus fremontii, Potamogeton berchtoldii, Potentilla biennis, P. gracilis, P. simplex, Primula clevelandii, P. pauciflora, Prosopis velutina, Prunella vulgaris, Pseudognaphalium stramineum, Pseudoroegneria spicata, Pseudotsuga menziesii, Psilocarphus brevissimus, P. oregonus, P. tenellus, Quercus douglasii, Q. gambelii, Ranunculus abortivus, R. aquatilis, R. cymbalaria, R. micranthus, R. repens, R. sceleratus, R. testiculatus, Rhamnus californica, Rorippa curvisiliqua, Rosa nutkana, Rumex fueginus, R. salicifolius, Sagina decumbens, S. procumbens, Salix gooddingii, S. lasiolepis, Sambucus, Sarcobatus vermiculatus, Schoenoplectus americanus, Scirpus microcarpus, Sclerochloa dura, Senecio integerrimus, Silene, Sisymbrium officinale, Sisyrinchium bellum, Sonchus, Spartina pectinata, Spergularia rubra, Sporobolus airoides, S. wrightii, Stellaria media, Suaeda,
Symphoricarpos albus, Tamarix ramosissima, Taraxacum officinale, Taraxia breviflora, Tradescantia virginiana, Trifolium beckwithii, T. depauperatum, T. fucatum, T. gymnocarpon, T. repens, T. variegatum, T. willdenovii, Triglochin maritimum, T. scilloides, Triodanis, Triteleia grandiflora, Triphysaria eriantha, Typha latifolia, Verbascum thapsus, Veronica arvensis, V. biloba, V. peregrina, V. persica, Veronicastrum virginicum, Viola adunca, V. palmata, Vitis cinerea, Vulpia, Wyethia, Yucca, Xanthocephalum gymnospermoides, Ziziphus obtusifolia, Zuloagaea bulbosa.

Use by wildlife: The seeds of *M. minimus* are eaten (and dispersed) by lesser white-fronted geese (Aves: Anatidae: *Anser erythropus*) and occasionally by horses (Mammalia: Equidae: *Equus*). The pollen is collected by short-tongued bees (Insecta: Hymenoptera: Halictidae: *Lasioglossum zephyrus*), which occasionally outcross the plants.

Economic importance: food: *Myosurus* plants are cyanogenic and consequently should never be eaten; **medicinal:** *Myosurus* foliage contains the glycoside ranunculin and when bruised, enzymatically releases the skin irritant protoanemonin. The Navajo-Ramah used *M. apetalus* as a life medicine and *M. minimus* as a remedy for ant bites; **cultivation:** not cultivated; **misc. products:** The Navaho-Ramah used *M. apetalus* to ward away evil spirits; **weeds:** *Myosurus minimus* can become weedy in some no-till crop production systems, fallow fields, or in standard agricultural fields. It has also been reported as a weed in (American) football fields; **nonindigenous species:** *Myosurus minimus* reportedly was introduced to Delaware, Illinois (common only in agricultural fields), Massachusetts, and Wisconsin. It is possible that the species was introduced to other localities in eastern North America because most records in that region are associated with cultivated sites; however the extent of its native range is difficult to determine with certainty. The seeds can be transported in horse dung (see above), and apparently, along with contaminated agricultural materials.

Systematics: *Myosurus* is assigned to tribe Ranunculeae of subfamily Ranunculoideae (Ranunculaceae). Although some phylogenetic analyses of morphological data have placed *Myosurus* as the sister group to the remainder of tribe Ranunculeae, DNA sequence data resolve the genus within the tribe, as the sister group to *Ceratocephala* (e.g., Figure 2.7). A comprehensive phylogenetic analysis of *Myosurus* species has not been carried out (molecular data currently are available only for *M. minimus*), which precludes any details of interspecific relationships at this time. The widely distributed *M. minimus* has been subdivided variously into subspecies by some authors; two varieties (differing by their sepal venation) have been recognized for *M. apetalus*. The basic chromosome number of *Myosurus* is $x = 8$. The genus is relatively uniform chromosomally ($2n = 16$), as in *M. apetalus* ($2n = 16$); multiple counts ($2n = 16, 28$) have been reported for *M. minimus*. *Myosurus minimus* and *M. sessilis* (FACW) can be hybridized artificially and are believed to produce natural hybrids, which sometimes are recognized as distinct taxa. Because *Myosurus* species mostly are self-pollinated, natural hybrids probably occur through pollen transfer by small flies (Insecta: Diptera) and other nonspecific insects, which visit the flowers incidentally. Genetic characterization of widespread *M. minimus* populations would facilitate efforts to ascertain the native range of the species and help to identify nonindigenous introductions.

Comments: *Myosurus apetalus* is distributed in western North America and is disjunct in South America; *M. minimus* is widespread in North America and extends into Africa, Eurasia, Mexico, and New Zealand.

References: Bliss & Zedler, 1998; Campbell & Gibson, 2001; Clausnitzer & Huddleston, 2002; Clausnitzer et al., 2003; Emadzade et al., 2010; Fishel et al., 2000; Godefroid et al., 2010; Hoot, 1995; Knuth, 1908; Lehnebach et al., 2007; Loconte et al., 2012; Rogers et al., 2002; Sorrie, 1990; Stone, 1959; Thorsen et al., 2009; Tóth et al., 2016; Wang et al., 2009; Whittemore, 1997b.

10. *Ranunculus*

Buttercup, crowfoot; renoncule

Etymology: from the Latin *rana unculus*, ("little frog") because of the associated wet habitats

Synonyms: *Batrachium*; *Halerpestes*; *Hecatonia*

Distribution: global: cosmopolitan; **North America:** widespread

Diversity: global: 300 species; **North America:** 71 species

Indicators (USA): OBL: *Ranunculus alismifolius R. ambigens, R. aquatilis, R. arcticus, R. bonariensis, R. confervoides, R. flabellaris, R. flammula, R. hederaceus, R. hydrocharoides, R. hyperboreus, R. laxicaulis, R. lobbii, R. longirostris, R. macounii, R. pensylvanicus, R. populago, R. pusillus, R. sceleratus, R. subrigidus, R. trichophyllus*; **FACW:** *R. alismifolius R. flammula, R. macounii, R. pensylvanicus, R. populago, R. pusillis*; **FAC:** *R. arcticus*

Habitat: freshwater; lacustrine, palustrine, riverine; **pH:** 4.2–9.6; **depth:** 0–3 m; **life-form(s):** emergent (herbs), floating, floating-leaved, submersed (vittate)

Key morphology: shoots (to 46 cm) ascending, decumbent, erect, floating, or prostrate, with or without nodal roots; roots slender throughout or thickened basally (to 1.2 mm); foliage homophyllous or heterophyllous, stipulate; basal leaves sessile or petioled, the blades (to 9.5 cm) simple and cordate or reniform, or palmately divided or lobed, the cauline leaves sessile or petioled, their blades (to 14.1 cm) simple and elliptic, lanceolate, oblanceolate, ovate, or reniform, or palmately divided or lobed, or dissected into filiform segments, the margins entire, crenate, denticulate, or serrate; flowers borne singly (1–7 per stem) or in cymes, pedicelled; sepals (to 7 mm) 2–5, spreading, reflexed, or recurved; petals (to 14 mm) 1–12, white (subgenus *Batrachium*) or yellow, nectary scales conspicuous or reduced; stamens 5-numerous; gynoecium apocarpous, pistils numerous; fruit a head-like aggregate (to 15 mm) of achenes (to 2.8 mm), pedicels straight or recurved, with curved or straight beak (to 1.8 mm), or beak absent or reduced

Life history: duration: annual (fruit/seeds), perennial (whole plants, tuberous roots, stolons); **asexual reproduction:** shoot fragments, stolons; **pollination:** insect, self; **sexual condition:** hermaphroditic; **fruit:** aggregates (achenes; common); **local dispersal:** seeds, vegetative fragments (water currents); **long-distance dispersal:** seeds (animals)

Imperilment: (1) *Ranunculus alismifolius* [G5]; S1 (BC, WY); S3 (MT, WY); (2) *R. ambigens* [G4]; SH (DC, MA, MD, ME, MI, NC, OH); S1 (CT, NH, VA); S2 (NJ); S3 (KY, PA); (3) *R. aquatilis* [G5]; S1 (WY); S3 (BC); (4) *R. arcticus* [G5]; S1 (UT); S2 (BC, LB, MB, NF, ON, SK, WY); (5) *R. bonariensis* [G4]; (6) *R. confervoides* [unranked]; (7) *R. flabellaris* [G5]; SH (WY); S1 (AL, DE, LA, MD, NC, OK, RI, UT); S2 (KS, NB, PA, TN); S3 (AR, BC, IL, MB, NJ, QC, VA, VT); (8) *R. flammula* [G5]; SH (NJ, PA); S1 (ND, WV); S2 (NS); S3 (AZ, WY); (9) *R. hederaceus* [G5]; SH (NC, VA); S1 (MD, PA); (10) *R. hydrocharoides* [G3/G5]; S1 (CA); (11) *R. hyperboreus* [G5]; S1 (MB, NF, SK, UT); S2 (ON); S3 (AB, MT, QC, WY); (12) *R. laxicaulis* [G5]; SH (DE); S1 (IN, KS, VA); S2 (NC); (13) *R. lobbii* [G4]; SH (BC); S3 (CA); (14) *R. longirostris* [G5]; SX (DE); SH (VA); S2 (AR, KY, NB, NJ); (15) *R. macounii* [G5]; S1 (MI, WV); S2 (CA, NF); S3 (QC); (16) *R. pensylvanicus* [G5]; SH (CT, DE, MD); S1 (LB, NS, WV); S2 (IL, MA, NF, NJ, VT, WY); S3 (AB, MT, PA); (17) *R. populago* [G4]; S2 (WA); S3 (MT); (18) *R. pusillus* [G5]; SX (NY); S1 (IN, OH, PA, WV); S2 (NJ); S3 (DE, IL, OK); (19) *R. sceleratus* [G5]; S1 (NB, NS, PE, RI); S2 (NC, WV); S3 (IL, KY, QC, WY); (20) *R. subrigidus* [unranked]; (21) *R. trichophyllus* [G5]; SH (WV); S1 (MD); S2 (NJ); S3 (IL)

Ecology: general: It is difficult to discuss the ecology of *Ranunculus* and its species effectively not only because of its large size but also because the taxonomy of the group remains in a state of turmoil (see *Systematics*). As a consequence, many taxa are treated under different names in the literature and in various databases, where the taxonomy of the genus can vary widely. Unfortunately, much confusion is encountered when attempting to associate literature accounts with a specific taxon, and this factor should be considered when consulting the treatment provided here. This large genus is extraordinarily variable, and the species occupy a respectable diversity of habitats. In North America, the species are most common in dry or wet meadows and woodlands, and along coastal areas at elevations from sea level to more than 3600 m. Many *Ranunculus* species are highly tolerant to cold conditions. Some of the species can become aggressive and weedy in disturbed sites. As its etymology implies, *Ranunculus* is associated commonly with wet habitats. Approximately 63 (89%) of the North American species are wetland indicators, with 22 (31%) species designated as OBL. The aquatic taxa (mostly helophytes) occupy an array of habitats ranging from bogs, ditches, pond margins, stream banks, swamps, and wet depressions to deeper, open waters with slow currents. The truly aquatic species are submersed and characterized by highly dissected foliage. Some are heterophyllous, producing dissected foliage underwater and less-divided floating or emergent leaves. Yet, others produce floating leaves but not highly dissected underwater leaves. Despite these vegetative adaptations, all of the species produce aerial flowers during sexual reproduction. However, some of the submersed species occasionally self-pollinate in closed, underwater flowers, which retain gas bubbles. Most of the species included here are perennials, but a few are annuals. In all cases, the flowers are bisexual, possess nectaries, and are pollinated by insects such as flies (Insecta: Diptera). Some members of the genus have been found to be apomictic. Like many Ranunculaceae, the seeds are often characterized by "double" (morphophysiological) dormancy, which requires a period of time sufficient to complete the embryo maturation as well as a sufficient period of cold stratification in order to induce germination. A persistent seed bank develops in many species, with some seeds remaining viable in the soil for as long as 14 years. Seed dispersal mechanisms are variable and include simple gravity, water, attachment of the achenes (often with hooked stylar appendages) to animal fur, and endozooic transport of ingested fruits.

Ranunculus alismifolius **Geyer ex Benth.** is a perennial that grows in bogs, carrs, depressions, ditches, fens, flats, floodplains, glades, hummocks, marshes, meadows, mudflats, pond bottoms (dry), prairies, roadsides, sandbars, seeps, slopes (to 15%), sloughs, swales, swamps, thickets, woodlands, and along the margins of lakes, ponds, rivers, and streams at elevations of up to 3659 m. The plants occur often in shallow standing water (to 10 cm) under exposures ranging from full sunlight to partial shade. The substrates have been characterized as acidic (e.g., pH: 4.7–5.8) or calcareous and are described as alluvium, basalt, boulders, clay, clay loam, granite (decomposed), granitic loam, gravel (coarse), gravelly clay, gravelly loam, humaquepts, humus, limestone, loam, muck, mud, peat, quartzite talus, rock, rocky loess, rocky scree, sand, sandy alluvium, sandy gravel, sandy loam, silt, typic cryaquent, and volcanic colluvium. Flowering and fruiting extend from April to November. Information is lacking on the reproductive ecology and pollination biology; presumably, the flowers are sexual (not apomictic) and pollinated by insects (Insecta). A seed bank of uncertain duration develops. The seeds have germinated well after storage for 1 year under ambient outdoor weather conditions, followed by planting in soil where high moisture was maintained. The stems reportedly can produce vegetative bulblets deep at their base. Collections have often been made in recently burned areas, which indicate at least some resistance to fire. **Reported associates:** *Abies grandis, A. lasiocarpa, Achillea millefolium, Aconitum columbianum, Agoseris monticola, Allium geyeri, A. validum, Alnus rugosa, Alopecurus, Amelanchier utahensis, Anemone parviflora, Antennaria corymbosa, A. microphylla, A. racemosa, Arctostaphylos uva-ursi, Arnica cordifolia, A. nevadensis, Artemisia cana, A. tridentata, Asclepias cordifolia, Astragalus purshii, Balsamorhiza serrata, Berberis repens, Betula glandulosa, Bistorta bistortoides, Boykinia major, Calamagrostis canadensis, C. rubescens, Caltha leptosepala, Camassia quamash, Cardamine breweri, Carex aquatilis, C. arcta, C. aurea, C. brunnescens, C. buxbaumii, C. capitata, C. geyeri, C. illota, C. luzulina, C. microptera, C. nebrascensis, C. pachystachya, C. rostrata, C. scopulorum, C. simulata, C. vesicaria, Castilleja collegiorum, C. covilleana, C. cusickii, C. linariifolia, Claytonia lanceolata,* *Collinsia parviflora, Cornus sericea, Danthonia californica, D. intermedia, Dasiphora floribunda, Delphinium nuttallianum, Deschampsia cespitosa, Dianthus armeria, Draba spectabilis, Drymocallis, Eleocharis palustris, E. quinqueflora, Epilobium alpinum, E. ciliatum, Erigeron, Eriogonum strictum, Eriophorum chamissonis, Erodium cicutarium, Erythronium grandiflorum, Festuca idahoensis, Floerkea proserpinacoides, Fragaria virginiana, Fraxinus, Fritillaria pudica, Galium trifidium, Gentiana calycosa, G. newberryi, Geranium viscosissimum, Geum, Habenaria, Helleborus, Heracleum sphondylium, Hesperochiron californicus, H. pumilus, Holcus lanatus, Hordeum brachyantherum, Hydrophyllum capitatum, Hymenoxys hoopesii, Hypericum scouleri, Isoetes bolanderi, Juncus balticus, J. drummondii, J. mertensianus, J. parryi, J. tenuis, Kalmia microphylla, Larix, Lasthenia glaberrima, Leucocrinum montanum, Lithophragma parviflorum, Lomatium bradshawii, L. cous, L. dissectum, L. grayi, L. nevadense, L. nudicaule, L. tarantuloides, Lonicera caerulea, Lotus corniculatus, Lupinus burkei, L. obtusilobus, Luzula hitchcockii, Madia glomerata, Mentha, Menziesia ferruginea, Mertensia arizonica, M. longiflora, Micranthes odontoloma, Mimulus moschatus, M. nanus, M. primuloides, Mitella breweri, M. ovalis, Muhlenbergia filiformis, M. richardsonis, Noccaea fendleri, N. parviflora, Nuphar polysepala, Oreostemma alpigenum, Packera pseudaurea, Pedicularis attollens, P. bracteosa, P. groenlandica, P. racemosa, Penstemon attenuatus, Phalaris arundinacea, Phleum alpinum, P. pratense, Phyllodoce empetriformis, Picea engelmannii, P. glauca, P. pungens, Pinus albicaulis, P. contorta, P. flexilis, P. ponderosa, Plagiobothrys figuratus, Platanthera dilatata, Poa leptocoma, P. nervosa, P. pratensis, P. reflexa, Populus tremuloides, P. trichocarpa, Potamogeton gramineus, Potentilla breweri, P. ×diversifolia, P. flabellifolia, P. gracilis, Primula cusickiana, P. jeffreyi, P. pauciflora, Prunus virginiana, Pseudocymopterus montanus, Pseudotsuga menziesii, Purshia tridentata, Pyrrocoma, Ranunculus andersonii, R. aquatilis, R. occidentalis, R. orthorhynchus, R. populago, Rhododendron columbianum, Ribes montigenum, Rumex, Rytidosperma pilosum, Salix bebbiana, S. brachycarpa, S. monticola, S. planifolia, S. wolfii, Sanguisorba canadensis, Senecio integerrimus, S. serra, S. sphaerocephalus, S. triangularis, Sibbaldia procumbens, Sidalcea oregana, Sisyrinchium, Sorbus, Sphagnum, Stellaria longifolia, S. longipes, Streptopus amplexifolius, Symphoricarpos oreophilus, Symphyotrichum foliaceum, S. spathulatum, Taraxacum officinale, Thalictrum fendleri, Trautvetteria caroliniensis, Trifolium dubium, T. kingii, T. longipes, T. repens, Triteleia hyacinthina, Trollius albiflorus, T. laxus, Vaccinium cespitosum, V. membranaceum, V. scoparium, V. uliginosum, Vahlodea atropurpurea, Valeriana acutiloba, Veratrum californicum, V. viride, Veronica serpyllifolia, Viola adunca, V. beckwithii, V. glabella, V. macloskeyi, Wyethia helianthoides, Xerophyllum tenax.*

Ranunculus ambigens **S. Watson** is a perennial that occurs in ditches, flats, floodplains, marshes, meadows, pools, streambeds, swamps, woodlands, and the margins of ponds and streams at elevations of up to 587 m. The plants are found frequently in shallow standing water (e.g., 5 cm depth). The reported substrates include clay, muck (to 2 dm deep), and mud. Flowering and fruiting extend from May to August; otherwise virtually no information exists on the reproductive biology or seed ecology of this species. The plants can spread vegetatively by means of stolons. Some sites have been described as having little other vegetation, indicating that the plants might be poor competitors. This conclusion seems reasonable given that the species is imperiled throughout much of its eastern North American range. **Reported associates:** *Acer rubrum, Alisma, Bidens discoideus, Callitriche heterophylla, Carex lupulina, Cinna arundinacea, Cyperus dentatus, Glyceria septentrionalis, G. striata, Gratiola virginiana, Isoetes engelmanni, Juncus, Leersia oryzoides, Lemna minor, Ludwigia palustris, Myosotis laxa, Nuphar advena, Orontium aquaticum, Oxypolis rigidior, Persicaria hydropiperoides, Potamogeton epihydrus, Proserpinaca palustris, Quercus bicolor, Sagittaria latifolia, Sparganium americanum, Torreyochloa pallida, Utricularia geminiscapa.*

***Ranunculus aquatilis* L.** is a perennial that inhabits backwaters, bogs, canals, coves, depressions, ditches, draws, fens, gravel bars, hummocks, lakes, marshes, meadows, mudflats, muskeg, oxbows, ponds, rivers, roadsides, seeps, sloughs, springs, stock ponds, streams, swales, swamps, thickets, vernal pools, and the margins of reservoirs and rivers at elevations of up to 3018 m. The plants grow most characteristically as submersed life-forms from 2–3 cm to 1.2 m depths in still or flowing waters, which are often high in carbonates (pH: 5.8–8.8; mean: 7.8). The plants are able to use bicarbonate ions (HCO_3^-) as a carbon source and exhibit increased bicarbonate affinity at low levels of availability. They occur along with more FAC species as they become stranded in dessicating habitats after the standing water recedes. The shoots are heterophyllous, producing dissected foliage underwater and less-divided floating or aerial leaves on emergent portions. The plants can root in the substrate or dislodge and float on the water surface (still capable of flowering). They are often pioneers of newly exposed sites but can tolerate exposures of partial sunlight. The substrates are described as clay, clay loam, colluvium, gravel, gravelly sand, loam, loamy clay, muck, mud, muddy clay, muddy gravel, ooze, sand, sandy gravel, sandy loam, sandy rock, Shingle Haverdad association, silt, silty loam, Tuscan loam, Tuscan mudflow, Worfka–Shingle–Samday complex, and Worthenton clay loam. Regardless of current velocity, growth rates are up to 3.9 times higher on mud substrates than on sand or gravel. Medium water velocities (~11 cm/s) provide the highest growth rates on gravel or sandy substrates. Flowering and fruiting extend from April to September. In submersed plants, the chasmogamous flowers extend above the water surface during sexual reproduction. Specific pollinators (Insecta) have not been elucidated. Sometimes the flowers are cleistogamous and bud pollinated while underwater. The seeds most likely are dispersed by water. They germinate readily under wet conditions throughout the year. Germination rates of 75%–86% have been reported for seeds imbibed on 1% agar for four weeks at 5°C

and then incubated under a 25°C/10°C temperature regime. The plants are freeze tolerant and reportedly remain viable even after being frozen in the ice for several months. The following list of associates excludes many dubious records from outside the distributional range of the species (e.g., east of the Great Plains). **Reported associates:** *Acer grandidentatum, Agrostis gigantea, Alisma gramineum, A. triviale, Alnus tenuifolia, Alopecurus pratensis, Ambrosia tomentosa, Artemisia arbuscula, A. tridentata, Beckmannia syzigachne, Bromus inermis, Calamagrostis canadensis, Callitriche palustris, Camassia, Carex aquatilis, C. athrostachya, C. aurea, C. disperma, C. microptera, C. nebrascensis, C. praegracilis, C. rostrata, C. subfusca, C. utriculata, Castilleja cusickii, Ceanothus fendleri, Chara, Chenopodium, Chrysothamnus viscidiflorus, Cirsium arvense, Deschampsia cespitosa, Dipsacus fullonum, Downingia bacigalupii, Eleocharis acicularis, E. bella, E. macrostachya, E. palustris, Elodea canadensis, Epilobium leptocarpum, Fontinalis, Gaultheria shallon, Geum rivale, Glyceria borealis, G. grandis, Grindelia squarrosa, Halerpestes cymbalaria, Hippuris vulgaris, Holodiscus discolor, Hordeum brachyantherum, H. jubatum, Iris missouriensis, Iva axillaris, Juncus balticus, J. effusus, J. saximontanus, J. supinus, Juniperus osteosperma, Lasthenia glaberrima, Lemna trisulca, Lepidium latifolium, Leptochloa, Lilaea scilloides, Lilium philadelphicum, Limnanthes douglasii, Limosella aquatica, Lupinus kingii, Lysichiton americanus, Marrubium vulgare, Marsilea vestita, Medicago sativa, Mentha arvensis, Menziesia ferruginea, Mimulus, Muhlenbergia rigens, Myosotis, Myosurus, Myriophyllum sibiricum, M. spicatum, Nassella viridula, Nasturtium officinale, Navarretia, Nuphar polysepala, Oenanthe, Penstemon linarioides, Persicaria amphibia, Phalaris arundinacea, Picea glauca, Pinus monophylla, P. ponderosa, P. strobiformis, Plagiobothrys cognatus, Poa fendleriana, P. palustris, P. pratensis, Polypogon, Populus tremuloides, Porterella carnosula, Potamogeton diversifolius, P. foliosus, P. gramineus, P. pusillus, P. richardsonii, Potentilla anserina, P. norvegica, Primula incana, Quercus gambelii, Q. ×undulata, Ranunculus alismifolius, R. gmelinii, Rhodiola rhodantha, Rhododendron macrophyllum, Robinia neomexicana, Rorippa teres, Rosa woodsii, Rumex triangulivalvis, Salix drummondiana, S. eriocephala, S. exigua, S. geyeriana, S. lasiandra, S. lasiolepis, Schoenoplectus acutus, S. pungens, Scirpus microcarpus, Senecio crassulus, Sidalcea oregana, Sparganium angustifolium, Spartina pectinata, Spergularia, Spiraea, Subularia aquatica, Symphoricarpos, Tamarix ramosissima, Torreyochloa pallida, Typha latifolia, Urtica dioica, Utricularia macrorhiza, Verbascum thapsus, Verbena bracteata, V. macdougalii, Veronica anagallisaquatica, V. peregrina, Zannichellia palustris.*

***Ranunculus arcticus* Richardson** inhabits alluvial fans, barrens, beaches, bogs, cliffs, depressions, ditches, drainage flats, dunes, fellfields, gullies, heath, hummocks, marshes, meadows, ridges, roadsides, seeps, shores, slopes (solifluction), stream bottoms, swales, tundra, woodlands, and the margins of lakes, ponds, rivers, and streams at elevations of up to 4069 m. This species is marginally aquatic (questionably

OBL anywhere) and usually is found in dry, moderately well-drained areas. The substrates are calcareous (sometimes highly nitrophilous) usually with a low organic content and include gravel, gravelly sand, humus, limestone, loam, mica, mudstone conglomerate, rock, rocky loam, sand, sandstone, schist, scree, silt, and talus. Flowering and fruiting occur from June to August. The reproductive ecology is poorly documented. The plants do not appear to be apomictic; they produce viable pollen and do not set seed when the flowers have been emasculated. No information is available on the seed ecology. **Reported associates:** *Androsace, Antennaria, Artemisia scopulorum, Astragalus, Betula, Botrychium pinnatum, Bupleurum americanum, Calamagrostis purpurascens, Caltha, Carex obtusata, Castilleja, Chamerion angustifolium, Chrysosplenium rosendahlii, Draba glabella, D. streptocarpa, Dryas integrifolia, Erigeron, Eriophorum vaginatum, Eritrichium nanum, Geum rossii, Kobresia simpliciuscula, Koenigia islandica, Leymus, Mertensia lanceolata, Micranthes rhomboidea, Minuartia obtusiloba, Oxytropis, Phippsia, Picea glauca, Poa, Polemonium viscosum, Populus balsamifera, P. tremuloides, P. trichocarpa, Potentilla rubricaulis, Salix arctica, S. candida, Selaginella densa, Silene williamsii, Symphyotrichum boreale, Tephroseris palustris, Tetraneuris acaulis, Thermopsis, Trisetum spicatum, Valeriana.*

***Ranunculus bonariensis* Poir.** is an annual that grows in depressions, ditches, flats, meadows, ponds, swales, vernal pools, and along the margins of ponds and streams at elevations of up to 400 m. The plants tolerate shallow water and exposures ranging from full sun to partial shade. The substrates are described as clay, granite, gravel, gravelly clay, and mud. Flowering occurs from March to May. No specific details on the reproductive biology or seed ecology are available. Experimental addition of domestic cattle dung to sites (to simulate effects of grazing) has been shown to decrease the percent cover of plants significantly. **Reported associates:** *Callitriche marginata, Cuscuta howelliana, Deschampsia danthonioides, Downingia bicornuta, Eleocharis macrostachya, Eryngium castrense, E. vaseyi, Gratiola ebracteata, Lasthenia fremontii, L. glaberrima, Legenere valdiviana, Lilaea scilloides, Limnanthes douglasii, Linanthus, Linum bienne, Montia fontana, Plagiobothrys stipitatus, P. undulatus, Polypogon monspeliensis, Psilocarphus brevissimus, Quercus douglasii, Q. wislizeni, Ranunculus muricatus, Rumex crispus.*

***Ranunculus confervoides* (Fr.) Fr.** is a taxon of highly uncertain status, which some authors have treated as an amalgam of *R. aquatilis, R. longirostris, R. subrigidus,* and *R. trichophyllus,* while others regard it as a much more narrowly distributed panarctic species (see *Systematics*). As a consequence, it is futile to attempt any ecological summary for this taxon until its status can be clarified with at least some degree of certainty, and no further discussion is provided here.

***Ranunculus flabellaris* Raf.** is a perennial that occurs in shallow standing waters (e.g., 30–100 cm deep) or on drying substrates associated with bayous, bogs, bottomlands, brakes, canal beds, channels, depressions, ditches, floodplains, lakes, marshes, meadows, oxbows, ponds, pools, potholes, prairies, roadsides, seeps, sloughs, streambeds, streams, and swamps at elevations of up to 2941 m. The plants occur in acidic to alkaline (e.g., pH: 7.5) waters in sites typically characterized by open exposures. The substrates are described as alluvium, loam, loamy sand, muck, mud, silt, and silty sand. This is a heterophyllous species, which produced dissected foliage underwater and less-divided floating or emergent leaves. The hollow stems facilitate their flotation on the water surface. Flowering and fruiting occur from April to August. The flowers are pollinated by insects such as marsh flies (Insecta: Diptera: Sciomyzidae). Additional information on the floral biology or seed ecology is not readily available. **Reported associates:** *Acer rubrum, A. saccharinum, Acorus calamus, Agrostis stolonifera, Alisma subcordatum, Amorpha fruticosa, Anemone, Asclepias incarnata, Bolboschoenus fluviatilis, Boltonia asteroides, Calamagrostis canadensis, Callitriche heterophylla, Cardamine bulbosa, Carex conjuncta, C. crinita, C. crus-corvi, C. granularis, C. grayi, C. hyalinolepis, C. lupulina, C. muskingumensis, C. projecta, C. radiata, C. rostrata, C. stipata, C. stricta, C. tribuloides, C. utriculata, Celtis occidentalis, Cephalanthus occidentalis, Ceratophyllum demersum, Cicuta maculata, Comarum palustre, Dulichium arundinaceum, Eleocharis acicularis, E. palustris, Elodea canadensis, Equisetum fluviatile, Eupatorium perfoliatum, Eutrochium fistulosum, Fontinalis, Forestiera acuminata, Fraxinus nigra, F. pennsylvanica, F. profunda, Galium palustre, Glyceria borealis, G. canadensis, G. septentrionalis, G. striata, Helenium autumnale, Heteranthera dubia, Hippuris vulgaris, Howellia aquatilis, Iris versicolor, I. virginica, Isoetes, Juncus acuminatus, J. effusus, Juniperus scopulorum, Leersia lenticularis, Lemna minor, Ludwigia palustris, Lysimachia ciliata, L. thyrsiflora, L. terrestris, Mentha arvensis, Myriophyllum sibiricum, M. spicatum, M. verticillatum, Najas flexilis, Nuphar advena, N. polysepala, N. variegata, Nymphaea tuberosa, Oemleria cerasiformis, Onoclea sensibilis, Peltandra virginica, Persicaria amphibia, P. coccinea, P. pensylvanica, Phalaris arundinacea, Pinus contorta, Populus heterophylla, Potamogeton crispus, P. gramineus, P. natans, P. richarsonii, P. zosteriformis, Proserpinaca palustris, Quercus bicolor, Q. palustris, Ranunculus hispidus, R. sceleratus, Ribes americanum, Riccia, Ricciocarpus natans, Rumex britannica, R. crispus, R. verticillatus, Sagittaria cuneata, S. latifolia, Salix discolor, S. nigra, Saururus cernuus, Schoenoplectus acutus, S. heterochaetus, S. lacustris, S. tabernaemontani, Scirpus cyperinus, Scutellaria angustifolia, S. lateriflora, Sium suave, Sparganium angustifolium, S. eurycarpum, Spirodela polyrhiza, Stuckenia vaginata, Symphyotrichum lanceolatum, S. ontarionis, Taxodium distichum, Thelypteris palustris, Torreyochloa pallida, Toxicodendron radicans, Triadenum virginicum, Typha latifolia, Ulmus americana, Utricularia intermedia, U. macrorhiza, U. minor, Verbena hastata, Viola sororia, Vitis riparia, Zizania aquatica.*

***Ranunculus flammula* L.** is a perennial that grows in shallow to deep (up to 3.2 m) standing waters or on exposed substrates associated with beaches, bogs, channels, deltas, depressions, ditches, dunes, fens, flats, floodplains, gravel bars, gravel pits, gullies, lakes, marshes, meadows, mudflats, pondbeds, ponds (ephemeral), potholes, roadsides, sandbars, seeps, shores (exposed), slopes (to 4%), sloughs (dried), springs, swales, swamps, thickets, tundra, vernal pools, woodlands, and along the margins of lagoons, lakes, ponds, pools (dried), potholes, reservoirs, rivers, sloughs, streams, and tarns at elevations of up to 3284 m. The plants are heterophyllous (especially when associated with unpredictable environments), which provides extensive plasticity and enables them to function effectively as pioneer colonists. They can also occur as floating life-forms. They tolerate a remarkable range of conditions including full sunlight to deeply shaded exposures, a broad range of acidic to alkaline water chemistry (pH: 4.2–7.9), and arguably can grow to the greatest water depths of any species in the genus. The substrates are categorized as acidic or calcareous and include alluvium, clay, clayey silty loam, cobbles, dolomite, gravel, gravelly humus, gravelly loam, gravelly sand, humus, loam, muck, mucky ooze, mud, muddy gravel, muddy sand, peat, quartz monzonite sand, rocks, sand, sandy loam, sandy mud, silt, silty gravel, silty loam, and stones. Flowering (North America) occurs from April to October, with fruits produced from May to October. Each ramet typically produces only a single flower, but only when the plants occupy exposed (not submersed) sediments. The flowers are self-incompatible and are outcrossed by various insects (see *Use by wildlife*). Studies using microsatellite markers have found populations to maintain high levels of genetic polymorphism, with two to seven alleles per locus and moderate levels of observed heterozygosity (0.261). The seeds are dispersed by water and can persist in the seed bank for at least 5 years. When kept moist, freshly collected seeds germinate sporadically under dark or light conditions. High germination rates will occur within a week if seeds are frozen (for 1 h) and thawed (for 4 h) alternately for four cycles. Seeds also germinate well under high light if first soaked for 1 h in dilute bleach (1 mL/60 mL distilled water) and then transferred to a solution of gibberellic acid (5 ppt) and incubation at 25°C in the dark. The plants reproduce vegetatively by their stoloniferous stems, which root along the nodes. **Reported associates:** *Abies concolor, A. lasiocarpa, A. procera, Achillea millefolium, Achnatherum lettermanii, Agrostis scabra, Alisma, Alnus rubra, A. tenuifolia, Alopecurus aequalis, Ambrosia, Anaphalis margaritacea, Angelica arguta, Antennaria corymbosa, A. parvifolia, Apocynum, Armeria maritima, Arnica chamissonis, Artemisia campestris, Beckmannia syzigachne, Brasenia schreberi, Bromus, Calamagrostis canadensis, Callitriche palustris, Caltha, Camassia, Carex aquatilis, C. arcta, C. atherodes, C. athrostachya, C. buxbaumii, C. canescens, C. exsiccata, C. garberi, C. jonesii, C. lasiocarpa, C. lenticularis, C. limosa, C. lyngbyei, C. obnupta, C. pellita, C. praeceptorum, C. rostrata, C. saxatilis, C. scoparia, C. scopulorum, C. siccata, C. stipata, C. utriculata, C. vesicaria, C. wootonii, Ceanothus, Ceratophyllum demersum, Chamerion latifolium, Chara, Cicuta douglasii, Cornus canadensis, C. sericea, Cryptantha setosissima, Cyperus squarrosus, Dasiphora floribunda, Delphinium nuttallianum, Deschampsia cespitosa, D. elongata, Downingia elegans, Drosera anglica, Drymocallis lactea, Dulichium arundinaceum, Eleocharis acicularis, E. bella, E. palustris, E. quinqueflora, E. rostellata, Elodea canadensis, Epilobium anagallidifolium, E. brachycarpum, E. ciliatum, Equisetum arvense, E. fluviatile, Eremogone fendleri, Erigeron speciosus, Eriogonum racemosum, Eriophorum angustifolium, Fraxinus, Galium trifidum, Gaultheria shallon, Gayophytum diffusum, Gentiana calycosa, Geranium bicknellii, G. caespitosum, Geum, Glyceria borealis, G. grandis, Gnaphalium exilifolium, G. palustre, Gratiola ebracteata, Helenium bigelovii, Hesperostipa comata, Heterotheca villosa, Hippuris vulgaris, Hordeum jubatum, Hymenoxys subintegra, Hypericum anagalloides, H. perforatum, Iris missouriensis, Isoetes echinospora, I. lacustris, Juncus alpinus, J. arcticus, J. balticus, J. bufonius, J. effusus, J. ensifolius, J. filiformis, J. laccatus, J. mertensianus, J. nevadensis, J. supiniformis, Juniperus communis, Kalmia microphylla, Ligusticum grayi, Lilium philadelphicum, Linnaea borealis, Ludwigia palustris, Lycopodiella inundata, Lycopus americanus, Lysichiton americanus, Maianthemum dilatatum, M. stellatum, Marsilea vestita, Mentha arvensis, Menyanthes trifoliata, Mertensia franciscana, Micranthes hieraciifolia, Mimulus primuloides, Moneses uniflora, Myosotis laxa, Myriophyllum sibiricum, M. spicatum, Noccaea fendleri, Nuphar polysepala, N. variegata, Oenanthe, Oenothera, Paspalum distichum, Paxistima, Pedicularis centranthera, P. groenlandica, Penstemon pseudoputus, Perideridia parishii, Persicaria amphibia, P. coccinea, Phalaris arundinacea, Phleum alpinum, P. pratense, Picea engelmannii, P. pungens, Pinus albicaulis, P. contorta, P. jeffreyi, P. ponderosa, Plagiobothrys hispidulus, Plantago australis, P. major, P. maritima, Platanthera dilata, Poa pratensis, Populus tremuloides, Porterella carnosula, Potamogeton amplifolius, P. berchtoldii, P. gramineus, P. natans, P. praelongus, Potentilla anserina, P. flabellifolia, P. hippiana, P. norvegica, Prunella vulgaris, Pseudocymopterus montanus, Pseudotsuga menziesii, Psilocarphus elatior, Quercus, Ranunculus aquatilis, R. hyperboreus, R. repens, R. sceleratus, Rhododendron columbianum, Ribes bracteosum, Rorippa sphaerocarpa, Rosa, Rubus arcticus, Rudbeckia laciniata, Rumex acetosella, R. californicus, Sagittaria cuneata, Salix bebbiana, S. geyeriana, S. lasiandra, S. lemmonii, S. sitchensis, S. wolfii, Sanguisorba, Scheuchzeria palustris, Schoenoplectus acutus, S. subterminalis, Scirpus microcarpus, S. sylvestris, Senecio triangularis, Sisyrinchium demissum, Sium suave, Sparganium emersum, Sphagnum girgensohnii, Spiraea douglasii, Spiranthes romanzoffiana, Stachys rigida, Stuckenia pectinata, Subularia aquatica, Symphyotrichum foliaceum, S. spathulatum, Taraxacum officinale, Taraxia tanacetifolia, Thuja plicata, Trianthä glutinosa, Trifolium, Triglochin maritimum, Tripolium pannonicum, Tsuga heterophylla, T. mertensiana, Typha latifolia, Utricularia, Vaccinium scoparium, V. uliginosum, Veratrum fimbriatum, Verbascum thapsus, Verbena perennis, Veronica americana, V. peregrina, V. scutellata, V. serpyllifolia, Viola.*

***Ranunculus hederaceus* L.** is a nonindigenous perennial or winter annual that occurs in brooks, deltas, depressions, marshes, pools, rills, springs, swamps, and along the margins of lakes, ponds, and streams at elevations of up to 150 m.

Floating leaves (but not highly dissected underwater leaves) occur occasionally. The plants can form compact cushions during the winter. The substrates tend to be acidic (e.g., pH: 5.0) and consist mainly of sand. Flowering (North America) occurs from April to August. The flowers are protogynous but self-compatible and normally are self-pollinating. The seeds are dispersed by water. Their germination is enhanced after being subjected to several wet and drying cycles. **Reported associates (North America):** *Bacopa monnieri, Cyperus haspan, Eleocharis albida, E. radicans, Hydrocotyle ranunculoides, Juncus megacephalus, Limosella australis, Ludwigia brevipes, Rhynchospora caduca, R. colorata.*

Ranunculus hydrocharoides **A. Gray** is a perennial that inhabits shallow water (sometimes fast-flowing) or moist substrates associated with cienega, ditches, lakes, marshes, meadows, ponds, seeps, slopes (to 5%), streambeds, streams, thickets, and the margins or shores of lakes, rivers, springs, and streams at elevations of up to 2896 m. The plants grow while rooted to the substrate or when floating on the surface of water (up to 1 m deep). They are found in exposures ranging from sun to shade. The substrates are described as alluvium, aquic fluvent soils, mud, organic loam (basalt), and silt (igneous). Flowering extends from May to September. Vegetative reproduction occurs by the formation of rhizomes or stolons. Other life-history information is unavailable for this species. **Reported associates:** *Achillea millefolium, Aconitum, Agrostis, Alnus tenuifolia, Ambrosia psilostachya, Anemopsis californica, Antennaria parvifolia, Apocynum cannabinum, Arenaria lanuginosa, Artemisia cana, Asclepias subverticillata, Baccharis sarothroides, Berula erecta, Bidens aureus, Bouteloua curtipendula, Bromus inermis, Callitriche, Carex chihuahuensis, C. occidentalis, C. pellita, C. praegracilis, C. scoparia, C. stipata, C. subfusca, C. utriculata, Clematis drummondii, Convolvulus equitans, Coreopsis, Cornus sericea, Danthonia californica, Dasiphora floribunda, Deschampsia cespitosa, Eleocharis macrostachya, E. montevidensis, E. palustris, E. parishii, E. quinqueflora, Epilobium, Equisetum arvense, E. laevigatum, Geranium richardsonii, Glyceria, Helianthella quinquenervis, Heuchera parvifolia, Hopia obtusa, Houstonia wrightii, Hymenoxys hoopesii, Hypericum formosum, H. scouleri, Juncus balticus, J. effusus, J. ensifolius, J. longistylis, J. occidentalis, Lemna, Lysimachia, Lythrum californicum, Medicago lupulina, Mentha arvensis, Mertensia franciscana, Mimulus guttatus, Monarda, Morus microphylla, Muhlenbergia asperifolia, M. montana, M. repens, M. tricholepis, Nasturtium officinale, Oenothera, Orthocarpus luteus, Packera neomexicana, Parthenocissus quinquefolia, Penstemon, Persicaria maculosa, Phalaris, Phleum pratense, Picea, Pinus ponderosa, Plantago, Poa fendleriana, P. pratensis, Polemonium foliosissimum, Populus fremontii, Potentilla hippiana, Prunella vulgaris, Pseudotsuga menziesii, Quercus gambellii, Ranunculus macranthus, Rosa woodsii, Rudbeckia laciniata, Rumex acetosella, R. crispus, Salix bebbiana, S. gooddingii, S. scouleriana, Schedonorus arundinaceus, Schoenoplectus acutus, S. americanus, Scirpus microcarpus, Scutellaria galericulata, Sidalcea neomexicana, Sisyrinchium demissum, Sonchus,* *Sorghum halepense, Thalictrum fendleri, Thermopsis montana, Tragia, Trifolium repens, T. wormskioldii, Typha domingensis, T. latifolia, Valeriana edulis.*

Ranunculus hyperboreus **Rottb.** is a perennial that grows in or on beaches, bluffs, bogs, borrow pits, channels (backwater), crevices, deltas, depressions, draws, dunes, fens, flats (estuarine), floodplains, gravel bars, heaths, hummocks, lagoons, marshes, meadows, mudflats, muskeg, ponds (ephemeral), pools, roadsides, scrub, seeps, slopes, sloughs, snowbeds, springs, streambeds, swamps, thickets, tundra (coastal), and along the margins of estuaries, lakes, ponds, pools, reservoirs, and streams at elevations of up to 3743 m. Exposures are often characterized by some amount of shade. The plants can float on the water surface, grow submersed in still or flowing shallow water (to 30 cm), or become stranded on exposed surfaces. The substrates are characterized as acidic or alkaline and include clay, clay loam, cobbles (basalt), decaying eelgrass (*Zostera marina*), gravel, gravelly sand, humus, limestone, muck, mud (organic), sand, sandy gravel, silty clay loam, silty mud, and silty sand. The plants are in flower and fruit from June to September. Details on the reproductive ecology have not been reported; however, normally developed flowers have been observed on fully submersed plants. The seeds are dispersed endozoically by animals such as the arctic fox (Mammalia: Canidae: *Vulpes lagopus*), which disperse viable propagules in their scat. The seeds have germinated well (>40%) after 6 months of cold storage (at −2°C to −14°C) followed by incubation under a 20°C/10°C day/night temperature regime. Vegetative reproduction occurs by the stoloniferous stems, which root at their nodes. The plants reach higher frequencies at sites that have been grazed by barnacle geese (Aves: Anatidae: *Branta leucopsis*), which probably remove much competing vegetation; however, grazed sites exhibit also reduced flowering frequencies (perhaps due to selective grazing of the reproductive structures). **Reported associates:** *Abies lasiocarpa, Alnus, Alopecurus aequalis, A. magellanicus, Atriplex gmelinii, Betula nana, Calamagrostis, Calliergon, Callitriche hermaphroditica, C. palustris, Caltha palustris, Cardamine pratensis, Carex aquatilis, C. brevior, C. canescens, C. disperma, C. lyngbyei, C. mackenziei, C. membranacea, C. nebrascensis, C. paysonis, C. podocarpa, C. rostrata, C. saxatilis, C. scopulorum, C. subspathacea, C. utriculata, Catabrosa aquatica, Chamerion, Chrysosplenium tetrandrum, Claytonia sarmentosa, Cochlearia officinalis, Cornus canadensis, Dasiphora floribunda, Deschampsia cespitosa, Drepanocladus capillifolius, Dryas integrifolia, Dupontia fisheri, Eleocharis quinqueflora, Elymus trachycaulus, Empetrum, Epilobium ciliatum, E. hornemannii, Equisetum arvense, E. scirpoides, Eriophorum scheuchzeri, Galium trifidum, Geum macrophyllum, Haplopappus, Heracleum sphondylium, Hippuris tetraphylla, H. vulgaris, Juncus balticus, J. ensifolius, Koenigia islandica, Lemna minor, Leymus mollis, Micranthes foliolosa, Mimulus, Montia fontana, Myrica gale, Pedicularis groenlandica, Picea engelmannii, P. glauca, P. pungens, Pinus contorta, P. pumila, Platanthera huronensis, P. obtusata, P. sparsiflora, Pleuropogon sabinei, Poa arctica, Populus tremuloides,*

Potamogeton berchtoldi, Potentilla anserina, Puccinellia, Ranunculus aquatilis, R. cymbalaria, R. flammula, R. gmelinii, Rosa acicularis, Rubus pubescens, Sagina saginoides, Salix glauca, S. planifolia, S. polaris, Saxifraga cernua, Sisyrinchium angustifolium, Sparganium hyperboreum, Stellaria humifusa, S. longifolia, Streptopus amplexifolius, Swertia perennis, Tephroseris palustris, Trifolium longipes, Triglochin, Valeriana acutiloba, Veronica peregrina, V. serpyllifolia, V. wormskjoldii.

Ranunculus laxicaulis (Torr. & A. Gray) Darby is a perennial that grows in shallow water or on exposed substrates associated with bogs, depressions, ditches, flats, floodplains, hillsides (dry), marshes, meadows, oxbows, prairies, ravines, roadsides, seeps, swales, swamps, thickets, woodlands, and the margins of ponds, pools, rivers, and streams at elevations of up to 335 m. The plants can be found in exposures ranging from open sunlight to shade. The substrates are described as clay, humus, limestone, mud, rock, sand, sandy clay, and sandy loam. Flowering (March to July) occurs earlier than many of the OBL congeners. Little additional information exists on the reproductive biology or seed ecology of this species. The plants reproduce vegetatively by means of their stoloniferous stems. The wetland status of this species should be reconsidered. Despite the consistent OBL ranking, many occurrences are described as growing on dry or "almost wet" substrates along with associates that are more typical of uplands.

Reported associates: *Acer saccharum, Antennaria plantaginifolia, Callitriche, Cardamine angustata, Carex hyalinolepis, Cephalanthus occidentalis, Cladium jamaicense, Collinsia verna, Cyperus squarrosus, Duchesnea indica, Eleocharis, Fagus grandifolia, Fragaria virginiana, Galium aparine, Geranium maculatum, Glechoma hederacea, Gratiola, Hyptis alata, Iris, Juglans nigra, Juncus, Justicia americana, Lilium, Mollugo verticillata, Panicum hemitomon, P. virgatum, Paspalum plicatulum, P. praecox, Pinus strobus, Potentilla simplex, Quercus palustris, Ranunculus abortivus, Rhododendron maximum, Rorippa, Sabal minor, Salix nigra, Thalictrum thalictroides, Triadica sebifera, Tripsacum, Veronica arvensis, Viola sororia, V. striata.*

Ranunculus lobbii (Hiern) A. Gray is an annual that inhabits shallow water or wet substrates of ditches, flats, gravel pits, marshes, meadows, ponds (ephemeral), pools, swales, swamps, and vernal pools at elevations of up to 472 m. The plants are heterophyllous, producing dissected underwater foliage and less-divided floating or emergent aerial leaves. The substrates are described as alluvium, muck, mud, and rhyolite. Flowering (March to July) begins earlier than many of the OBL congeners; fruiting occurs from April to July. Little additional information exists on the life history of this species. **Reported associates:** *Acer macrophyllum, Blennosperma bakeri, Callitriche hermaphroditica, C. heterophylla, C. palustris, Cephalanthus occidentalis, Cystopteris fragilis, Fontinalis, Geranium lucidum, Isoetes, Juncus phaeocephalus, Lasthenia burkei, Lemanea, Limnanthes douglasii, Lolium multiflorum, L. perenne, Ludwigia palustris, Marsilea, Nasturtium officinale, Persicaria hydropiperoides, Polystichum munitum, Ranunculus aquatilis,*

R. muricatus, Rosa, Rumex, Trillium albidum, T. ovatum, Veronica, Vulpia octoflora.

Ranunculus longirostris Godr. is a perennial that grows completely submersed (to 5 m depth) in still or flowing, standing waters of channels, ditches, lakes, ponds, pools, sloughs, springs, and streams at elevations of up to 2173 m. The waters generally are alkaline (pH: 6.0–9.2; mean: 8.0; conductivity: 40–570 µmhos/cm; total alkalinity to 240 mg/L CaCO$_3$). Substrates are described as gravel, peat, and sand. The foliage is highly dissected, even on plants that become stranded along shorelines. Flowering and fruiting (May to September) begin early in the spring. The flowers are produced above the water surface. The seeds are dispersed locally by water and probably over greater distances by endozoic transport mediated by feeding waterfowl (Aves: Anatidae). Additional information about the reproductive biology and seed ecology is needed. The plants are kown to overwinter vegetatively beneath the water in ice-covered ponds and rivers. The plants normally grow at water depths below 2 m and can develop into thick mats when growing conditions are optimal. In deeper waters (3 m), they reach densities of 1–16 plants/m^2; some plants can grow as deep as 5 m but only at extremely low densities (<0.1 plants/m^2). The plants can survive in waters of fairly high turbidity. Vegetative reproduction occurs by fragmentation of the stems. The following list of associates excludes many records that probably refer to *R. aquatilis, R. trichophyllus,* and other species (see *Systematics*).

Reported associates: *Agrostis, Alisma, Berula erecta, Bidens beckii, Brasenia schreberi, Callitriche palustris, Ceratophyllum demersum, Chara, Echinochloa, Elatine, Eleocharis acicularis, Elodea canadensis, E. nuttallii, Epilobium, Gratiola quartermaniae, Heteranthera dubia, Hippuris vulgaris, Juncus, Lemna minor, L. trisulca, Limosella, Lycopus, Mentha arvensis, Myosurus minimus, Myriophyllum alterniflorum, M. sibiricum, M. spicatum, Najas flexilis, Nasturtium officinale, Nitella, Nuphar advena, Nymphaea odorata, Pontederia cordata, Potamogeton alpinus, P. amplifolius, P. berchtoldii, P. crispus, P. diversifolius, P. epihydrus, P. foliosus, P. gramineus, P. illinoensis, P. natans, P. obtusifolius, P. praelongus, P. pusillus, P. richardsonii, P. robbinsii, P. zosteriformis, Ranunculus cymbalaria, R. flabellaris, R. pensylvanicus, R. sceleratus, Ricciocarpus natans, Rumex, Sagittaria latifolia, Schoenoplectus acutus, Scirpus microcarpus, Sidalcea, Sparganium angustifolium, Spirodela polyrhiza, Stuckenia pectinata, Typha, Utricularia gibba, U. macrorhiza, U. minor, Vallisneria americana, Veronica americana, V. anagallisaquatica, Wolffia columbiana, Zannichellia palustris.*

Ranunculus macounii Britton is a perennial that occurs in shallow (e.g., 10–15 cm), fresh to moderately brackish water or on exposed sediments associated with beaches (strands), bogs, bottomlands, canals, channels (river), depressions, ditches, draws, fellfields, fens, flats, floodplains, hollows, marshes, meadows, mires, mudflats, muskeg, oxbows, ponds, pools, potholes, ravines, riverbeds (dry), roadsides, sandbars, scrub, seeps, shores, slopes (to 5%), swales, swamps, terraces (alluvial), thickets, woodlands, and along the margins of lakes, ponds, reservoirs, rivers, sloughs, springs, and streams at elevations of up to 2865 m. Exposures vary from open to shaded sites, with the latter being more common. The substrates are characterized as alkaline (to moderately brackish) or calcareous and include alluvium, basalt rimrock, claron limestone, clay, clay sand, clay silt, dolomite, granite, gravel, gravelly loam, humus, loam (basalt), loess, muck, mucky clay, mud, muddy gravel, peat, peaty loam, sand, sandy clay, sandy loam, shale (marine), silt, silty clay loam, silty mud, and stones. Flowering and fruiting extend from April to September. The reproductive biology and seed ecology are not documented. The flowers are visited by butterflies (Insecta: Lepidoptera), which might function as pollinators. The roots are colonized by mycorrhizal Fungi.

Reported associates: *Abies concolor, A. grandis, Acer grandidentatum, Achillea millefolium, Achnatherum nelsonii, Aconitum columbianum, Agoseris glauca, Agropyron, Agrostis gigantea, A. scabra, A. stolonifera, Alisma, Alnus oblongifolia, A. sinuata, A. tenuifolia, A. viridis, Alopecurus aequalis, Amphiscirpus nevadensis, Amsinckia, Anemone canadensis, A. multifida, Angelica arguta, Antennaria rosea, Anticlea elegans, Apium graveolens, Arabis, Arctophila fulva, Arnica chamissonis, Artemisia tridentata, Asclepias, Baccharis sarothroides, Barbarea vulgaris, Beckmannia syzigachne, Berberis repens, Betula occidentalis, Brickellia, Bromus ciliatus, B. inermis, B. marginatus, Calamagrostis canadensis, C. stricta, Callitriche, Campanula rotundifolia, Cardamine nuttallii, Carex aquatilis, C. athrostachya, C. bolanderi, C. canescens, C. capillaris, C. deweyana, C. disperma, C. hoodii, C. hystericina, C. lasiocarpa, C. lenticularis, C. leptalea, C. mariposana, C. microptera, C. nebrascensis, C. pellita, C. praegracilis, C. rostrata, C. simulata, C. sprengelii, C. utriculata, C. vulpinoidea, Castilleja rhexiifolia, Catabrosa aquatica, Celtis reticulata, Cercocarpus, Chamerion angustifoium, Chara, Cicuta douglasii, Circaea alpina, Cirsium arvense, C. foliosum, C. vulgare, Clematis ligusticifolia, Cornus sericea, Crataegus, Cyclachaena xanthiifolia, Cypripedium parviflorum, Dactylis glomerata, Dactylorhiza viridis, Danthonia intermedia, Dasiphora floribunda, Deschampsia cespitosa, Distichlis spicata, Downingia bicornuta, Echinochloa muricata, Eleocharis acicularis, E. palustris, E. quinqueflora, Elymus trachycaulus, Epilobium ciliatum, E. palustre, Equisetum arvense, E. fluviatile, E. laevigatum, Eragrostis hypnoides, Ericameria nauseosa, Erigeron divergens, E. flagellaris, Erysimum cheiranthoides, Erythronium oregonum, Euthamia, Festuca idahoensis, Fragaria vesca, F. virginiana, Fraxinus pennsylvanica, Galium boreale, G. triflorum, Gentianella amarella, Geranium viscosissimum, Geum macrophyllum, G. rivale, Glyceria borealis, G. grandis, G. striata, G. stricta, Gnaphalium exilifolium, Hackelia floribunda, Hedeoma drummondii, Helenium, Heracleum sphondylium, Heterotheca villosa, Hilaria jamesii, Hordeum brachyantherum, H. jubatum, Hymenoxys hoopesii, Hypericum scouleri, Iris missouriensis, Juncus balticus, J. bufonius, J. castaneus, J. effusus, J. ensifolius, J. howellii, J. laccatus, J. nodosus, J. saximontanus, J. torreyi, Juniperus communis, J. monosperma, J. occidentalis, J. osteosperma, J. scopulorum, Koeleria, Lathyrus ochroleucus, Leersia oryzoides, Lemna,*

Leucanthemum vulgare, Leymus innovatus, L. mollis, Lilium philadelphicum, Limosella aquatica, Lobelia, Lonicera, Lupinus nootkatensis, Luzula campestris, L. multiflora, L. parviflora, Lysimachia maritima, Maianthemum stellatum, Medicago lupulina, Mentha arvensis, Menziesia ferruginea, Mertensia, Mimulus gutattus, M. moschatus, Minuartia rubella, Monarda, Montia chamissoi, Myosotis scorpioides, Nassella viridula, Nuphar polysepala, Oenothera nuttallii, Oncosiphon, Oplopanax horridus, Opuntia phaeacantha, Orthilia secunda, Parkinsonia aculeata, Penstemon confertus, P. linarioides, Persicaria lapathifolia, Petasites frigidus, Phacelia egena, Phalaris arundinacea, Phleum pratense, Picea engelmannii, P. glauca, P. mariana, P. pungens, Pinus contorta, P. edulis, P. monophylla, P. monticola, Plantago major, Platanthera dilatata, P. huronensis, P. stricta, Poa compressa, P. nemoralis, P. palustris, P. pratensis, Polygonum aviculare, Polypogon monspeliensis, Populus angustifolia, P. balsamifera, P. tremuloides, P. trichocarpa, Porterella carnosula, Potamogeton foliosus, P. natans, Potentilla anserina, P. gracilis, P. norvegica, Prosopis, Prunella, Prunus virginiana, Pseudotsuga menziesii, Purshia tridentata, Ranunculus aquatilis, R. cymbalaria, R. occidentalis, R. sceleratus, R. testiculatus, Rhinanthus minor, Rhododendron columbianum, Ribes aureum, Robinia neomexicana, Rorippa calycina, R. sphaerocarpa, Rosa acicularis, R. woodsii, Rubus neomexicanus, Rudbeckia laciniata, R. occidentalis, Rumex crispus, R. salicifolius, R. triangulivalvis, Sagittaria cuneata, Salix bebbiana, S. exigua, S. geyeriana, S. interior, S. irrorata, S. laevigata, S. lucida, S. monticola, S. scouleriana, Sarcobatus, Schedonorus arundinaceus, Schoenoplectus acutus, S. pungens, Scirpus microcarpus, Scutellaria, Senecio serra, S. sphaerocephalus, Setaria, Shepherdia canadensis, Sidalcea neomexicana, Sparganium emersum, Spartina gracilis, Sphagnum, Spiraea douglasii, Stachys palustris, Stellaria crassifolia, Stipa, Symphoricarpos albus, Symphyotrichum foliaceum, Taraxacum officinale, Tetradymia spinosa, Thalictrum dasycarpum, T. sparsiflorum, Thermopsis divaricarpa, T. rhombifolia, Thlaspi arvense, Toxicodendron rydbergii, Trifolium dubium, T. hybridum T. pratense, T. repens, T. wormskioldii, Triglochin palustre, Typha latifolia, Urtica dioica, Vaccinium vitis-idaea, Vahlodea atropurpurea, Veratrum californicum, Veronica americana, V. anagallisaquatica, V. peregrina, V. scutellata, Vicia americana, V. cracca, V. sativa, Viola canadensis, V. sororia.

***Ranunculus pensylvanicus* L. f.** is a perennial or FAC annual that grows on exposed substrates or in shallow waters (to 10 cm) in or on alluvial terraces, beaches, bogs, bottomlands, crevices, depressions, ditches, draws, fens, floodplains, gravel bars, gulches (dry), hummocks, lake beds (dry), marshes, meadows, roadsides, sandbars, scrub, seeps, slopes (to 5%), sloughs (dry), springs, swales, swamps, thickets, woodlands, and along the margins of brooks, channels, lakes, ponds (dry), reservoirs, rivers, sloughs, and streams at elevations of up to 2200 m. The plants occur in sunny to partially shaded exposures. The substrates are characterized as alkaline (e.g., pH: 7.6; $CaCO_3$: 42.8 mg/L) and include clay, gravel, humus, loam, loamy sand, mud, peat, rocks, sand, sandy gravel, sandy loam, shonkinite, silt, silty clay loam, and silty loam. Flowering and fruiting extend from May to October. The flowers are relatively small and produce small seeds (0.6–0.7 mg); however, single plants are capable of producing up to 11,200 seeds. The reproductive ecology and seed ecology have not been studied.

Reported associates: *Abies lasiocarpa, Acer glabrum, A. negundo, Achillea alpina, Agrimonia parviflora, Agrostis gigantea, A. stolonifera, Alisma subcordatum, Alliaria petiolata, Alnus oblongifolia, A. tenuifolia, Alopecurus, Ambrosia artemisiifolia, Amorpha fruticosa, Angelica atropurpurea, Apios americana, Apocynum sibiricum, Artemisia tilesii, Asclepias incarnata, Bacopa rotundifolia, Betula papyrifera, B. pumila, Bidens cernuus, Boehmeria cylindrica, Bolboschoenus fluviatilis, Brasenia schreberi, Calamagrostis canadensis, Callitriche heterophylla, Caltha natans, C. palustris, Carex canescens, C. cristatella, C. emoryi, C. haydenii, C. lacustris, C. lasiocarpa, C. molesta, C. nebrascensis, C. rostrata, C. simulata, C. stipata, C. stricta, C. vulpinoidea, Cicuta bulbifera, C. maculata, Cirsium arvense, Clintonia, Coleataenia longifolia, Cornus canadensis, C. sericea, Crataegus douglasii, Cyperus bipartitus, C. ferruginescens, Deschampsia cespitosa, Dichanthelium clandestinum, Dulichium arundinaceum, Echinochloa crus-galli, Eleocharis acicularis, E. erythropoda, Epilobium coloratum, E. hirsutum, Equisetum arvense, Erigeron philadelphicus, Eupatorium perfoliatum, E. serotinum, Euthamia graminifolia, Eutrochium maculatum, Fraxinus americana, F. pennsylvanica, Galium obtusum, G. trifidum, Geum canadense, G. laciniatum, Glechoma hederacea, Glyceria septentrionalis, Hordeum jubatum, Hypericum perforatum, Impatiens capensis, Iris setosa, Juncus acuminatus, J. balticus, J. dudleyi, J. interior, J. marginatus, J. torreyi, J. vaseyi, Juniperus monosperma, Larix occidentalis, Leersia oryzoides, Lemna minor, Leymus arenarius, Lobelia cardinalis, Lonicera ×bella, L. tatarica, Lupinus polyphyllus, Lycopus americanus, L. asper, L. uniflorus, Lysichiton americanus, Lysimachia ciliata, L. nummularia, Lythrum salicaria, Mentha arvensis, Mertensia, Mimulus ringens, Mitella, Myrica gale, Myriophyllum, Nasturtium officinale, Nuphar polysepala, Nymphaea tetragona, Onoclea sensibilis, Panicum capillare, P. dichotomiflorum, Parthenocissus, Pedicularis lanceolata, Penthorum sedoides, Persicaria amphibia, P. hydropiperoides, P. lapathifolia, P. pensylvanica, P. punctata, P. setacea, Phalaris arundinacea, Phragmites australis, Phyla lanceolata, Physostegia virginiana, Picea engelmannii, P. glauca, P. mariana, Pinus contorta, P. monticola, Plantago major, Poa nemoralis, P. pratensis, Polygonum aviculare, Populus angustifolia, P. deltoides, P. tremuloides, Potamogeton amplifolius, P. natans, Potentilla anserina, P. norvegica, Pseudotsuga menziesii, Ranunculus flabellaris, R. longirostris, Rhamnus cathartica, R. frangula, Ricciocarpos natans, Rorippa palustris, Rosa palustris, R. woodsii, Rubus pensilvanicus, Rudbeckia laciniata, Rumex altissimus, Sagittaria cuneata, S. latifolia, Salix alaxensis, S. alba, S. amygdaloides, S. bebbiana, S. discolor, S. drummondiana, S. exigua, S. humilis, S. interior, S. irrorata, S. lucida, S. nigra, Sambucus nigra, Schoenoplectus pungens, S. tabernaemontani, Scirpus atrovirens, S. cyperinus, S. microcarpus, Scutellaria galericulata, S. lateriflora, Sisyrinchium montanum, Sium suave, Solanum carolinense, S. dulcamara, Solidago gigantea, S. patula, Sparganium emersum, S. eurycarpum, S. glomeratum, Sphagnum, Spiraea douglasii, Sporobolus compositus, Stuckenia pectinata, Symphoricarpos albus, Symphyotrichum lanceolatum, Symplocarpus foetidus, Thalictrum dasycarpum, Thelypteris palustris, Tilia americana, Toxicodendron radicans, Trifolium repens, Typha angustifolia, T. latifolia, Ulmus americana, Urtica dioica, Verbena hastata, Vernonia gigantea, Veronica americana, Vitis riparia, Xantium strumarium.*

***Ranunculus populago* Greene** is a perennial that inhabits shallow water or exposed substrates of bogs, ditches, lakes, marshes, meadows, rivulets, seeps, slopes (to 20%), springs, streams, swamps, and the margins of lakes at elevations of up to 2440 m. Exposures vary from sun to partial shade. The substrates are described as acidic (e.g., pH: 4.4–4.5) and include alluvium (organic), granite, loamy clay, peat, sandy clay, sandy loam, and serpentine. Flowering and fruiting occur from March to August. Additional life-history information for this species is lacking. **Reported associates:** *Abies lasiocarpa, Aconitum, Alnus, Antennaria, Artemisia, Botrychium multifidum, Caltha leptosepala, Carex aquatilis, C. illota, C. mertensii, C. scopulorum, C. stipata, Circaea, Epilobium minutum, Equisetum, Erythronium grandiflorum, Gaultheria humifusa, Gentiana affinis, G. calycosa, Hieracium albiflorum, H. triste, Hypericum scouleri, Juncus drummondii, J. mertensianus, Kalmia microphylla, Lilium, Maianthemum, Mertensia paniculata, Mitella, Orogenia linearifolia, Paxistima, Pedicularis bracteosa, P. groenlandica, Penstemon, Phyllodoce empetriformis, Picea engelmannii, Pinus albicaulis, Platanthera stricta, Poa, Polemonium pulcherrimum, Potentilla flabellifolia, Primula jeffreyi, P. tetrandra, Prunus emarginata, Pseudotsuga menziesii, Ranunculus alismifolius, R. eschscholtzii, Rhododendron columbianum, Rudbeckia occidentalis, Salix scouleriana, Senecio triangularis, Sphagnum, Spiraea splendens, Triantha glutinosa, Trifolium cyathiferum, Trillium, Tsuga, Vaccinium, Veronica, Viola macloskeyi, V. palustris.*

***Ranunculus pusillus* Poir.** is a perennial that grows in shallow (to 11 cm), standing waters or on exposed substrates in borrow pits, bottomlands, depressions, ditches, fields, flats, flatwoods, floodplains, hammocks, hummocks, marshes (nontidal, tidal), meadows, mudflats, ponds, pools (shallow), prairies, puddles, ravines, ricefields, rivulets, roadsides, seeps, sinkholes, springs, streambeds, streams, swales, swamps, thickets, vernal pools, waste areas, woodlands, and along the margins of canals, impoundments, and ponds at elevations of up to 1787 m. Exposures vary from full sun to partial shade. The substrates are characterized as acidic (pH: 4.8–6.0) and are described as alluvial clay/silt, clay, clayey loam, clayey sand, Elbert soil, gravel, loam (Armenia/Iredell), muck, mud, rock (volcanic), sand, sandy loam, silty clay loam (Cape), and silty loam (Hurst). Flowering and fruiting extend from March to June. Neither the reproductive biology nor seed ecology of this species is known in any detail. **Reported associates:** *Acer rubrum, Amorpha fruticosa, Amsonia tabernaemontana, Arisaema, Baccharis salicifolia, Bacopa rotundifolia, Boehmeria cylindrica, Botrychium, Briza minor, Callitriche, Cardamine pensylvanica, Carex albicans, C. atlantica, C. bicknellii, C. caroliniana, C. crus-corvi, C. decomposita, C. gigantea, C. hyalinolepis, C. interior, C. lacustris, C. lasiocarpa, Carpinus caroliniana, Cephalanthus occidentalis, Chasmanthium, Cicuta maculata, Cornus amomum, Cynodon dactylon, Cyperus erythrorhizos, C. squarrosus, C. strigosus, Deschampsia danthonioides, Dichanthelium acuminatum, D. dichotomum, Digitaria, Downingia bella, D. concolor, Eleocharis wolfii, Eragrostis, Eryngium aristulatum, E. prostratum, Fagus grandifolia, Fontinalis sullivantii, Fraxinus americana, F. latifolia, F. pennsylvanica, Galium tinctorium, Geranium dissectum, Glyceria ×occidentalis, G. striata, Gratiola ebracteata, G. virginiana, Hedyotis, Hydrocotyle umbellata, Hypericum punctatum, Ilex decidua, Impatiens, Iris fulva, I. versicolor, Isoetes ×bruntonii, I. engelmannii, I. howellii, I. melanopoda, Itea virginica, Juncus acuminatus, J. bufonius, J. effusus, Juniperus virginiana, Lasthenia glaberrima, Leptodictyum viparium, Liparis loeselii, Lipocarpha micrantha, Liquidambar styraciflua, Lobelia siphilitica, Lolium multiflorum, Lonicera ×bella, Ludwigia palustris, Lycopus americanus, L. uniflorus, Lysimachia ciliata, L. lanceolata, Mentha pulegium, Micranthemum umbrosum, Mollugo verticillata, Murdannia keisak, Nyssa sylvatica, Onoclea sensibilis, Osmunda claytoniana, O. regalis, Packera glabella, Panicum clandestinum, Persicaria hydropiperoides, Phyla nodiflora, Pilea pumila, Pinus taeda, Plagiobothrys bracteatus, P. stipitatus, Platanus racemosa, Poa annua, Pogogyne douglasii, Populus deltoides, P. heterophylla, Proserpinaca palustris, Pteridium, Ptilimnium, Quercus bicolor, Q. laurifolia, Q. palustris, Q. shumardii, Q. virginiana, Ranunculus muricatus, R. sardous, R. sceleratus, Rhamnus cathartica, R. frangula, Rhynchospora corniculata, Rorripa palustris, R. sessiliflora, Rubus hispidus, Rumex verticillatus, Sabal minor, S. palmetto, Salix laevigata, Samolus parviflorus, Scapania nemerosa, Scirpus atrovirens, S. pendulus, Scutellaria nervosa, Sisyrinchium, Sphagnum lescurii, Styrax redivivus, Taraxacum officinale, Taxodium distichum, Triadenum walteri, Trifolium dubium, T. repens, Ulmus americana, Vaccinium, Valerianella radiata, Vicia, Viola lanceolata, Woodwardia areolata, Zizaniopsis.*

Ranunculus sceleratus **L.** is an annual that occurs in fresh to brackish shallow (to 0.5 m) standing waters or on exposed substrates in or on beaches, bogs, bottomlands, depressions, ditches, draws, fellfields, fens, flats, gravel bars, gulches, gullies, hummocks, lake beds (dry), levees, marshes (coastal), meadows, ponds (ephemeral, saline), pools, potholes, prairies, ravines, ricefields, roadsides, sandbars, seeps, slopes (to 10%), sloughs (dry, tidal), springs (hot), streambeds (dry), swales, swamps, terraces (river), thickets, and along the margins of canals, lakes, ponds (dry), reservoirs, rivers, and streams at elevations of up to 3018 m. This is an extremely adaptable and widespread species, which grows under a wide diversity of aquatic and wetland conditions. The plants can become completely submersed at times and can produce floating leaves in as little as 1 cm of water. They do not develop highly dissected underwater leaves. Exposures can be in full sunlight but typically are characterized by some degree of shade. The substrates are categorized as alkaline (or neutral) and include adobe loam, alluvium, basalt (obsidian), Broadhead, clay, clayey mud, cobbles, cobbly loam (Borvant), dolomite, granite, gravel, gravelly loam, loam, loamy sand, marl, muck, mucky gravel, mucky mud, mud, muddy silt, peat, peaty gravel, quartzite, rocks, rocky sand, rubble, sand, sandy gravel, sandy gravelly loam, sandy humus, sandy limestone, sandy loam, sandy mud, silty clay, silt, silty loam (Ritzville), and silty mud. Flowering and fruiting occur from March to October. The flowers remain open for 2 days. They are self-compatible but usually are cross-pollinated by insects (see *Use by wildlife*). If pollination by larger insects does not occur, then smaller insects can facilitatate self-pollination; hence, the flowers are described as being facultatively xenogamous. The achenes are dispersed locally by water and over greater distances via endozoic transport by birds (Aves) or mammals (Mammalia). The seeds can remain viable after passing through the guts of waterfowl (Aves: Anatidae), which often function as dispersal agents. A persistent seed bank develops. Longevity of dried seeds (when stored at 35°C) can be extended fourfold to fivefold by first "priming" them in polyethylene glycol for 1 week. Optimal seed germination occurs after 60 days of cold stratification under a 20°C/15°C temperature regime in the light. However, the plants can function as either summer or winter annuals, which respond differently to cold stratification. The efficacy of dormancy release or induction is related inversely to temperature (at 2°C–11°C). **Reported associates:** *Acer negundo, A. saccharinum, Achillea millefolium, Acmispon strigosus, Acorus calamus, Agropyron, Agrostis hyemalis, A. scabra, A. stolonfera, Alhagi maurorum, Alisma gramineum, A. subcordatum, A. triviale, Alnus rubra, A. tenuifolia, Alopecurus aequalis, A. arundinaceus, A. pratensis, Alternanthera philoxeroides, Alyssum, Amaranthus californicus, A. palmeri, A. tuberculatus, Ambrosia artemisiifolia, A. monogyra, Amelanchier, Amorpha fruticosa, Amphicarpaea bracteata, Angelica atropurpurea, Apera interrupta, Apios americana, Arabidopsis lyrata, Artemisia biennis, A. ludoviciana, A. tridentata, Asclepias incarnata, Astragalus oocarpus, Atriplex canescens, A. rosea, Axonopus fissifolius, Baccharis salicifolia, B. salicina, Beckmannia syzigachne, Berula erecta, Betula occidentalis, Bidens cernuus, Bolboschoenus maritimus, Brickellia floribunda, Briza minor, Bromus pubescens, B. tectorum, Calamagrostis canadensis, Callitriche hermaphroditica, Caltha natans, C. palustris, Campanula parryi, Cardamine bulbosa, C. hirsuta, C. parviflora, Carex amplifolia, C. aquatilis, C. blanda, C. cristatella, C. diandra, C. emoryi, C. grayi, C. grisea, C. hyalinolepis, C. hystericina, C. interior, C. jamesii, C. laeviconica, C. lasiocarpa, C. lurida, C. lyngbyei, C. microptera, C. molesta, C. nebrascensis, C. normalis, C. pellita, C. praegracilis, C. rostrata, C. simulata, C. stipata, C. utriculata, C. vesicaria, Carya ovata, Castilleja applegatei, Catabrosa aquatica, Celtis reticulata,*

Cephalanthus occidentalis, Chaerophyllum procumbens, Chamaedaphne calyculata, Chamerion latifolium, Chenopodium album, C. rubrum, Chilopsis linearis, Chorispora tenella, Cicuta bulbifera, Cirsium scariosum, C. vulgare, Clematis virginiana, Conium maculatum, Conyza canadensis, Cornus amomum, C. drummondii, C. racemosa, C. sericea, Crataegus, Crypsis, Cyperus bipartitus, C. erythrorhizos, C. ferruginescens, Delphinium, Deschampsia cespitosa, Descurainia, Diodia, Dipsacus, Draba, Drymocallis glandulosa, Echinochloa crus-galli, Eclipta prostrata, Elaeagnus angustifolia, Elatine chilensis, Eleocharis acicularis, E. erythropoda, E. macrostachya, E. palustris, E. parishii, Elodea canadensis, Elymus virginicus, Epilobium campestre, E. ciliatum, E. hirsutum, E. leptophyllum, E. palustre, Equisetum arvense, E. hyemale, E. laevigatum, E. palustre, Ericameria nauseosa, Erigeron, Eriogonum fasciculatum, Eupatorium perfoliatum, Euthamia occidentalis, Eutrochium maculatum, Festuca, Fimbristylis autumnalis, Fragaria, Galium concinnum, G. tinctorium, G. triflorum, Gaultheria shallon, Geum rivale, Glyceria grandis, G. striata, Gnaphalium palustre, Gratiola neglecta, Heliotropium curassavicum, Hesperis matronalis, Hippuris vulgaris, Hirschfeldia incana, Hordeum branchyantherum, H. jubatum, Horkelia rydbergii, Impatiens capensis, Iodanthus pinnatifidus, Juncus alpinoarticulatus, J. alpinus, J. arcticus, J. balticus, J. bufonius, J. filiformis, J. nodosus, J. saximontanus, J. torreyi, Juniperus, Justicia americana, Laennecia coulteri, Lappula occidentalis, Leersia lenticularis, L. oryzoides, L. virginica, Lemna gibba, L. minor, Lepidium thurberi, Leptochloa fusca, Leymus triticoides, Limosella acaulis, L. aquatica, Lolium perenne, Lonicera ×bella, Lotus corniculatus, Ludwigia alternifolia, L. decurrens, L. palustris, Lycopus americanus, L. asper, Lysimachia nummularia, Lythrum salicaria, Marsilea, Medicago lupulina, Melilotus albus, M. indicus, M. officinalis, Mentha arvensis, Mikania scandens, Mimulus guttatus, M. ringens, Mollugo verticillata, Montia howellii, Muhlenbergia rigens, Myosotis, Myosurus minimus, Nasturtium officinale, Nicotiana glauca, Noccaea fendleri, Onoclea sensibilis, Opuntia phaeacantha, Orobanche fasciculata, Panicum virgatum, Parthenocissus, Pedicularis lanceolata, Persicaria amphibia, P. glabra, P. hydropiper, P. hydropiperoides, P. lapathifolia, P. maculosa, P. punctata, Phalaris arundinacea, Phragmites australis, Phyla lanceolata, Physocarpus, Picea engelmannii, P. glauca, P. mariana, P. sitchensis, Pinus ponderosa, Platanthera dilatata, Poa annua, P. compressa, P. nemoralis, P. pratensis, Podophyllum peltatum, Polanisia dodecandra, Polygonum aviculare, P. ramosissimum, Polypogon monspeliensis, P. viridis, Populus balsamifera, P. fremontii, P. tremuloides, P. trichocarpa, Potamogeton foliosus, Potentilla anserina, P. gracilis, P. norvegica, P. rivalis, P. supina, Primula egaliksensis, Prosopis velutina, Prunus serotina, P. virginiana, Pseudognaphalium luteoalbum, Pseudotsuga menziesii, Ptilimnium, Puccinellia distans, P. nuttalliana, Quercus agrifolia, Q. ellipsoidalis, Q. palustris, Ranunculus aquatilis, R. californicus, R. cymbalaria, R. flabellaris, R. gmelinii, R. hispidus, R. longirostris, R. macounii, R. muricatus, R. occidentalis, R. parviflorus, R. pusillus,
R. subrigidus, Rhamnus cathartica, Rhaponticum repens, Rhinanthus minor, Rhododendron groenlandicum, Rhus copallinum, Ribes americanum, Rorippa aquatica, R. curvisiliqua, R. islandica, R. palustris, Rosa multiflora, R. nutkana, R. setigera, R. woodsii, Rotala, Rubus allegheniensis, Rudbeckia laciniata, Rumex altissimus, R. crispus, R. fueginus, R. salicifolius, R. verticillatus, Sagittaria cuneata, Salix alaxensis, S. amygdaloides, S. arbusculoides, S. eriocephala, S. exigua, S. geyeriana, S. gooddingii, S. interior, S. laevigata, S. lasiolepis, S. myricoides, S. nigra, S. pseudomonticola, S. scouleriana, Salsola, Sambucus nigra, Saponaria, Sarcobatus, Schedonorus pratensis, Schoenoplectus acutus, S. lacustris, S. pungens, Scirpus atrovirens, S. microcarpus, Scolochloa festucacea, Scutellaria lateriflora, Senecio hydrophilus, Sidalcea, Silene nivea, Silphium perfoliatum, Sium suave, Sisymbrium, Solanum dulcamara, Solidago altissima, S. canadensis, S. gigantea, S. velutina, Sonchus arvensis, Sorghum halapense, Sparganium eurycarpum, Spartina, Spergularia salina, Sphagnum, Sphenopholis obtusata, Spirodela, Sporobolus wrightii, Stellaria crassifolia, S. longifolia, Swertia, Symphoricarpos, Symphyotrichum, Symplocarpus foetidus, Tamarix chinensis, T. ramosissima, Tanacetum vulgare, Taraxacum officinale, Tephroseris palustris, Thalictrum alpinum, T. dasycarpum, T. thalictroides, Thlaspi arvense, Toxicodendron radicans, Trifolium fragiferum, T. hybridum, T. repens, T. wormskioldii, Triglochin maritimum, Typha angustifolia, T. domingensis, T. latifolia, Ulmus americana, Urtica dioica, Ventenata dubia, Verbascum thapsus, Verbena bracteata, V. hastata, Veronica americana, V. anagallis-aquatica, V. arvensis, V. biloba, V. catenata, V. peregrina, V. scutellata, Viburnum lentago, V. opulus, Vicia grandiflora, V. tetrasperma, Viola sororia, Vitis riparia, Xanthium strumarium, Xanthocephalum gymnospermoides, Zannichellia palustris.

***Ranunculus subrigidus* W.B. Drew** is a perennial that inhabits clear to slightly turbid, standing, swift-moving to still waters (5 cm to 3 m depth) associated with ditches, floodplains, gravel pits, lakes, marshes, oxbows, ponds, pools, potholes, rivers, sloughs, streams, and the margins of lakes, ponds, and streams at elevations of up to 2377 m. The waters are described as calcareous (e.g., pH: 7.6–9.6; total alkalinity: 126 mg/L) and the substrates are described as clay, gravel, muck, mud, rock, sand, sandy mud, silt, and stones. The plants grow as homophyllous, submersed life-forms with highly dissected foliage. They grow attached to the substrate by roots or can float freely on the water surface. The plants attain somewhat higher average percent cover in lakes with lower water depths. Flowering and fruiting occur from May to August. The flowers are produced above the water surface. Other details on the reproductive biology and seed ecology of this species are lacking. **Reported associates:** *Agrostis stolonifera, Alisma gramineum, Artemisia tridentata, Atriplex, Beckmannia syzigachne, Bolboschoenus maritimus, Calamagrostis canadensis, Callitriche hermaphroditica, C. palistris, Carex foenea, C. utriculata, Ceratophyllum demersum, Chara, Eleocharis acicularis, E. palustris, Elodea canadensis, Equisetum fluviatile, Fontinalis, Hippuris vulgaris, Iris missouriensis, Juncus balticus, Lemna minor,*

Myriophyllum sibiricum, M. spicatum, Nuphar variegata, Persicaria amphibia, P. coccinea, Potamogeton alpinus, P. berchtoldii, P. friesii, P. gramineus, P. natans, P. praelongus, P. pusillus, P. richardsonii, P. robbinsii, P. zosteriformis, Potentilla anserina, Ranunculus aquatilis, R. sceleratus, Ruppia maritima, Sagittaria latifolia, Salix, Schoenoplectus americanus, S. heterochaetus, S. validus, Sparganium emersum, S. eurycarpum, Stuckenia filiformis, S. pectinata, S. vaginata, Utricularia macrorhiza, Veronica americana, V. anagallis-aquatica, Zannichellia palustris.

Ranunculus trichophyllus Chaix is a perennial that grows submersed in still to flowing, clear, standing waters (to 3 m depth) in canals, channels, coves, ditches, lagoons, lakes, ponds, pools, rivers, sloughs, springs, streams, trenches, and along the margins of drying vernal pools at elevations of up to 2834 m. The waters are generally alkaline (pH: 6.4–9.5; mean: 8.2; total alkalinity: 15–340 mg/L as $CaCO_3$; conductivity: 25–750 µmhos/cm). The substrates are described as calcareous and include clay, gravel, limestone, loam, muck, mud, muddy-ooze, organic detritus, peat, rocks, sand, sandy clay, silt, and volcanics. The plants are homophyllous and possess highly divided submersed foliage. Flowering and fruiting occur from May to September. The flowers are produced on stalks above the water surface. They are self-compatible and primarily autogamous (sometimes cleistogamous and bud-pollinated underwater). The seeds are dispersed locally by water and can also be transported endozoically by waterbirds (Aves: Anatidae) or by larger animals (Mammalia), which disperse them in their dung. Effective long-distance dispersal by more vagile animal vectors (e.g., birds) has been implicated by the appearance of plants at high elevation lakes (4760 m) in Nepal. The seeds germinate readily in wet conditions throughout the year. Fairly high germination rates (86%) have been obtained for seeds placed on 1% agar and incubated under a 25°C/10°C, 8/16 h temperature regime. The plants are able to reproduce vegetatively by shoot fragmentation, but the survival rate of vegetative buds is relatively low. Fragment survival is higher in the autumn than in the spring. Overall, the fragments have high colonization rates but low regeneration ability; however, the plants recolonize open patches through peripheral propagation. Entire shoots can remain viable after being frozen in ice for several months. **Reported associates:** *Alisma gramineum, Alopecurus geniculatus, Beckmannia syzigachne, Callitriche, Carex aquatilis, C. lasiocarpa, Ceratophyllum demersum, Chara, Downingia elegans, Echinochloa, Elatine californica, E. rubella, Eleocharis acicularis, E. palustris, Elodea canadensis, E. nuttallii, Fontinalis, Glyceria borealis, Heteranthera dubia, Hippuris vulgaris, Lemna trisulca, Limosella, Marsilea, Mimulus guttatus, Myosurus apetalus, Myriophyllum sibiricum, M. spicatum, Najas flexilis, Nelumbo lutea, Nuphar polysepala, Nymphaea tuberosa, Persicaria amphibia, P. coccinea, Plagiobothrys scouleri, Potamogeton amplifolius, P. crispus, P. diversifolius, P. foliosus, P. gramineus, P. illinoensis, P. natans, P. nodosus, P. pusillus, P. richardsonii, P. zosteriformis, Ranunculus flammula, Rumex, Salix, Schoenoplectus tabernaemontani, Sium suave, Sparganium angustifolium, Stuckenia filiformis, S.*

pectinata, Utricularia macrorhiza, Vallisneria americana, Veronica peregrina, Zannichellia palustris.

Use by wildlife: *Ranunculus* species contain ranunculin, which is converted to toxic protoanemonin upon mastication of the foliage. Consequently, various *Ranunculus* species are toxic to various animals such as rabbits (Mammalia: Leporidae) and at least *R. sceleratus* is known to be toxic to cattle (Mammalia: Bovidae: *Bos taurus*). *Ranunculus longirostris* is avoided as a food source by grass carp (Osteichthys: Cyprinidae: *Ctenopharyngodon idella*), and the forced feeding of the plants has caused the fish to develop mucosal lesions. Yet, at least 10 species of ducks (Aves: Anatidae), mainly mallards (*Anas platyrhynchos*), consume the foliage and seeds of *Ranunculus* species, especially those of subgenus *Batrachium*. Also, *R. arcticus* is grazed by domestic cattle (Mammalia: Bovidae: *Bos taurus*) and naturalized mountain goats (Mammalia: Bovidae: *Oreamnos americanus*). *Ranunculus flabellaris* is a preferred food of muskrats (Mammalia: Cricetidae: *Ondatra zibethicus*) and *R. hyperboreus* is eaten by barnacle geese (Aves: Anatidae: *Branta leucopsis*) and arctic foxes (Mammalia: Canidae: *Vulpes lagopus*). Species of subgenus *Batrachium* provide important habitat for freshwater shrimp (Crustacea: Atyidae: *Syncaris*), snails (Mollusca: Gastropoda), and insects (Insecta). Submersed beds of *R. flammula* provide habitat for spawning fish (Vertebrata: Osteichthyes) and also for various invertebrates. *Ranunculus* plants are sometimes host to caterpillars (Insecta: Lepidoptera: Tortricidae: *Cnephasia longana*). The fine foliage of *R. longirostris* traps fine detritus, which attracts "gathering" invertebrates, including many that are eaten by trout (Osteichthyes: Salmonidae). Its seeds are eaten by moose (Mammalia: Cervidae *Alces alces*) and ruffed grouse (Aves: Phasianidae: *Bonasa umbellus*). The flowers of *R. flabellaris* are visited by marsh flies (Insecta: Diptera: Sciomyzidae: *Sepedon fuscipennis*), which collect the nectar and likely function as pollinators. The flowers of *R. flammula* attract various pollen- or nectar-collecting insects (Insecta), which include bees (Hymenoptera: Andrenidae; Apidae; Halictidae), beetles (Coleoptera: Carabidae; Coccinellidae; Nitidulidae; Staphylinidae), and flies (Diptera: Anthomyiidae; Calliphoridae; Cecidomyiidae; Dolichopodidae; Empididae; Muscidae; Phoridae; Syrphidae; Tachinidae). Flies are their most common pollinators. *Ranunculus macounii* is a host of several Fungi including rust (Basidiomycota: Pucciniaceae: *Uromyces alopecuri*) and smut (Basidiomycota: Entylomataceae: *Entyloma ranunculi*). The flowers are visited by butterflies (Insecta: Lepidoptera: Pieridae: *Pieris rapae*). The foliage of *R. pensylvanicus* is fed upon by stink bugs (Insecta: Hemiptera: Pentatomidae: *Holcostethus limbolarius*). Plants of *R. pusillus* are associated with a large number of actinobacteria (Bacteria: Actinomycetes) and are parasitized by chytrids (Fungi: Chytridiomycota: Physodermataceae: *Physoderma*; Synchytriaceae: *Synchytrium ranunculi*). They are also collateral hosts of the rice root-knot nematode (Nematoda: Meloidogynidae: *Meloidogyne graminicola*). The flowers of *R. sceleratus* are visited by a number of potentially pollinating nectar or pollen-gathering insects

(Insecta) including aphids (Hemiptera: Aphididae), bees (Hymenoptera: Apidae: *Ceratina calcarata*; Halictidae: *Augochlorella striata*, *Lasioglossum*; Megachilidae: *Hoplitis producta*), beetles (Coleoptera: Coccinellidae: *Coccinella*), flies (Diptera: Syrphidae: *Rhingia*, *Syrphus*, *Toxomerus*), and thrips (Thysanoptera). The achenes are eaten by waterfowl (Aves: Anatidae) such as mallard ducks (*Anas platyrhynchos*). *Ranunculus subrigidus* is eaten by dabbling waterfowl (Aves: Anatidae) including blue-winged teal (*Anas discors*), gadwall (*Anas strepera*), green-winged teal (*Anas crecca*), mallards (*Anas platyrhynchos*), redheads (*Aythya americana*), ring-necked ducks (*Aythya collaris*), and ruddy ducks (*Oxyura jamaicensis*). The seeds of *R. trichophyllus* are also eaten by various species of dabbling ducks (Aves: Anatidae).

Economic importance: food: All parts of *Ranunculus* plants contain the oily glycoside ranunculin, which is converted enzymatically to toxic protoanemonin. Consumption of large amounts may lead to secondary photosensitization and liver damage in humans. However, plants of *R. aquatilis* were cooked and eaten by the Gosiute people of Utah. When boiled, *R. sceleratus* has been used as an emergency food similar in texture to spinach; **medicinal:** Both ranunculin and protoanemonin (see *food*, earlier) have antibacterial properties. Contact with protoanemonin can cause dermatitis and its ingestion affects the human gastrointestinal tract and nervous system. *Ranunculus aquatilis* is used in Asia to treat asthma, intermittent fevers, and rheumatism. The Navajo used *R. cymbalaria* to treat venereal disease, as an emetic, and in ceremonial rituals. The Kawaiisu used it for treating dermatological disorders. *Ranunculus flabellaris* was used by the Fox tribe to treat colds and to improve respiration. *Ranunculus pensylvanicus* was used as a general medicine by the Ojibwa and as an astringent by the Potawatomi. *Ranunculus sceleratus* has been used to relieve rheumatism (leaves, root, seeds) and colds (seeds); **cultivation:** A number of terrestrial *Ranunculus* species are grown as ornamentals, but few of the OBL species are in cultivation. *Ranunculus aquatilis* is distributed as an "oxygenating plant" for coldwater ponds and aquariums. *Ranunculus flammula* and *R. hederaceus* are occasionally sold as garden ornamentals. *Ranunculus sceleratus* is used as a water garden ornamental for planting along margins. The seeds of *R. pensylvanicus* were listed in seed catalogs as early as 1804; **misc. products:** The Thompson tribes used *R. sceleratus* to make an arrow poison. Plants of *R. pensylvanicus* were used by the Potawatomi for making baskets and as an ingredient of a red or yellow dye (Ojibwa and Potawatomi); **weeds:** *Ranunculus sceleratus* is seriously invasive in marshes and along watercourses in western North America. *R. pusillus* is often referred to as weedy in the southeastern United States; **nonindigenous species:** There is some debate whether *R. sceleratus* is native to eastern North America; surely it is naturalized in the west. This issue should be clarified given that the species currently is protected in many areas. The same is true also for *R. hederaceus*, which occurs sporadically along the eastern United States but is widespread in Europe. *Ranunculus pensylvanicus* was reported as establishing near docks at Liverpool, England in 1909.

Systematics: Phylogenetically, DNA sequence data indicate that *Ranunculus* is related most closely to *Krapfia* and *Laccopetalum*, with *Ceratocephala* and *Myosurus* comprising a sister clade to that group (Figures 2.7 and 2.11). *Coptidium lapponicum*, *C. pallasii*, *Halerpestes cymbalaria*, and *Kumlienia hystricula* have been included in *Ranunculus* by some authors, but are remote phylogenetically (Figure 2.7), and must be maintained as distinct in order to preserve the monophyly of *Ranunculus*. Following current conventions, the OBL aquatic taxa can be classified tentatively within two subgenera: *Ranunculus* subgenus *Batrachium* (*R. aquatilis*, *R. confervoides*, *R. hederaceus*, *R. lobbii*, *R. longirostris*, *R. subrigidus*, *R. trichophyllus*) and subgenus *Ranunculus*: section *Chrysanthe* (*R. macounii*, *R. pensylvanicus*), section *Epirotes* (*R. arcticus*), section *Flammula* (*R. alismifolius*, *R. ambigens*, *R. bonariensis*, *R. flammula*, *R. hydrocharoides*, *R. laxicaulis*, *R. populago*, *R. pusillus*), and section *Hecatonia* (*R. flabellaris*, *R. hyperboreus*, *R. sceleratus*). Interspecific relationships indicated by a phylogenetic analysis of cpDNA (RFLP) data for 88 *Ranunculus* taxa (approximately ¼ of the species) indicated that the sections were natural but that some subgenera (e.g., subgenus *Ranunculus*) deserved further evaluation. Subgenus *Batrachium* resolves as a clade that associates closely with section *Hecatonia* (also monophyletic) of subgenus *Ranunculus*, which itself is not monophyletic (Figure 2.11). Subgenera *Batrachium* and *Ranunculus* (section *Hecatonia*) include the species that are most highly adapted to aquatic conditions; that is, those having submersed and/or floating leaves (Figure 2.11). Although the completely submersed aquatic species have been segregated as the genus *Batrachium* by some authors, they resolve as a clade nested within *Ranunculus* (e.g., Figure 2.11) and are better treated taxonomically as an infrageneric group of *Ranunculus*. Attempts to delimit species within subgenus *Batrachium* have resulted in considerable taxonomic confusion. Early studies using flavonoid and isozyme data indicated that at least some *Batrachium* species were polytypic and further study was needed to clarify the species boundaries in that group. Conflicting morphological, ecological, and genetic evidence has nevertheless inspired some authors to merge a number of species (i.e., *R. confervoides*, *R. longirostris*, *R. subrigidus*, *R. trichophyllus*) with *R. aquatilis*, which does not seem to be tenable, particularly because only *R. aquatilis* among these species is known to produce floating leaves. These taxa are represented by several ploidy levels, and a better understanding of species delimitation in this group seems necessary before such a merger can be accepted. Some of the problems are related to hybridization, which is extensive in the group and not fully elucidated. As one example, the morphologically similar *R. trichophyllus* and *R. aquatilis* are regarded as closely related (and both more distantly to *R. longirostris*) but are distinct genetically (e.g., Figure 2.11) and are also somewhat divergent ecologically (e.g., mean pH and specific conductivity). However, the results of some studies (using DNA sequence data) distribute *R. trichophyllus* accessions among five distinct genetic clusters, which indicate that this taxon is polyphyletic as currently circumscribed. This discrepancy

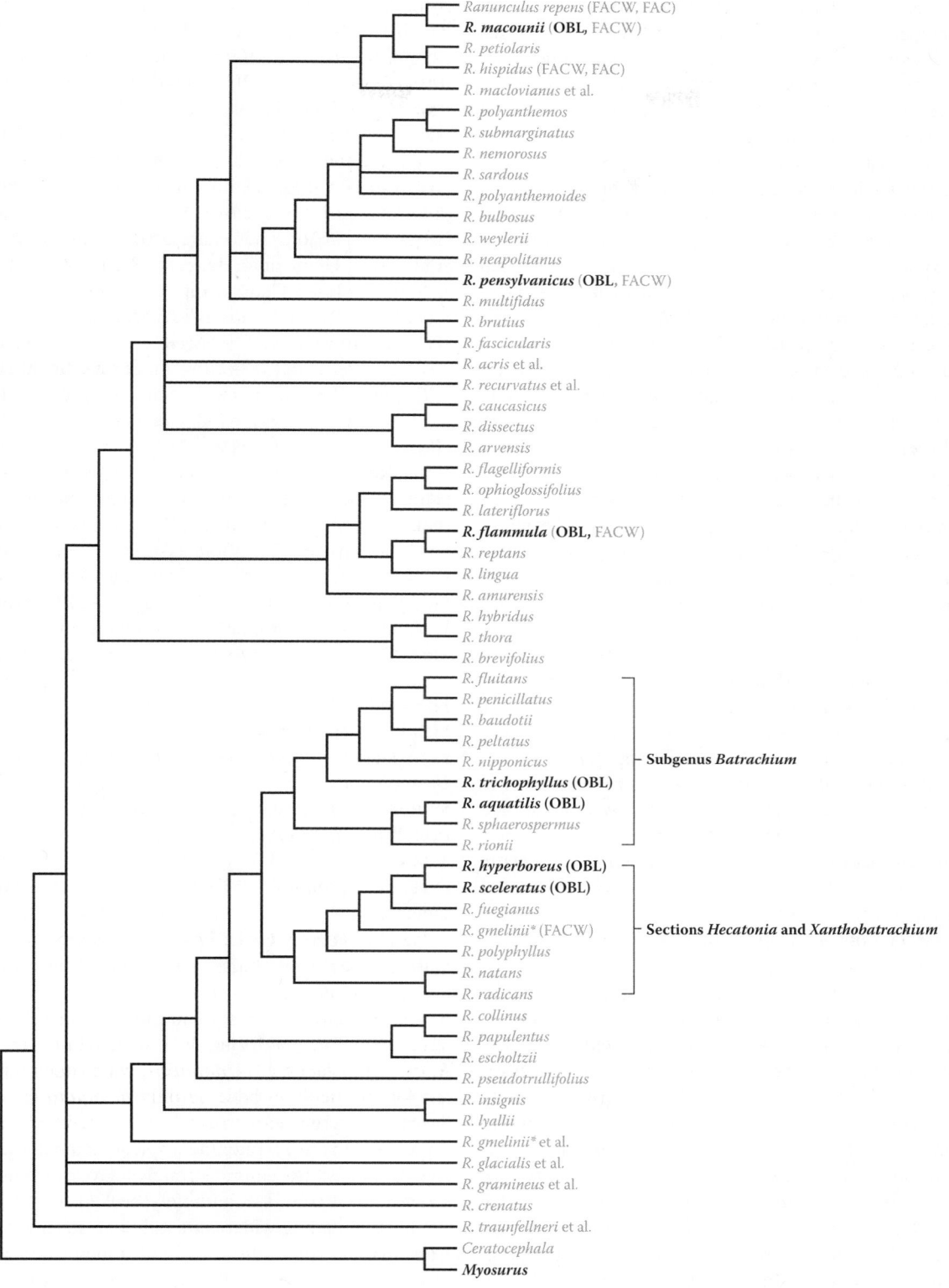

FIGURE 2.11 Interspecific relationships in *Ranunculus* based on a phylogenetic analysis of combined nrITS and *matK-trnK* DNA sequence data for 194 taxa. The OBL North American taxa (bold) are dispersed across the genus with a concentration in subgenus *Batrachium* and section *Hecatonia*, where many of the non-North American species are also aquatic. This analysis emphasizes the distinction of *R. aquatilis* and *R. trichophyllus* as well as the polyphyletic nature of *R. gmelinii* (asterisked). Because only a third of the OBL species were included, a broader sampling of species is necessary to fully evaluate the taxonomic integrity of these species as well as their interrelationships. Hybridization makes also it difficult to draw firm taxonomic conclusions at this time, pending more extensive genetic surveys of *Ranunculus*. The wetland indicator status is shown in parentheses for North American taxa. (Adapted from Hoffmann, M.H. et al., *Int. J. Plant Sci.*, 171, 90–106, 2010.)

has likely resulted from hybridization with *R. aquatilis* and other species. Although species boundaries in subgenus *Batrachium* remain tentative, analyses of various DNA sequence data have helped to clarify other taxonomic problems in the genus. Although *R. confervoides* has been placed in synonymy with *R. aquatilis* var. *diffusus*, DNA "bar codes" show it to be distinct from the latter taxon, which instead, is indistinguishable from *R. subrigidus*. Combined DNA sequence data indicate also that *R. arcticus* is closely related to, but distinct from *R. pedatifidus* var. *affinis*, with which it has been synonymized. Other close relationships evidenced by DNA sequence data include *R. flammula* with *R. reptans* and *R. macounii* with *R. repens* (e.g., Figure 2.11). Close relationships of *R. hederaceus* to *R. omiophyllus* and *R. lobbii* to *R. tripartitus* have also been hypothesized but have not yet been evaluated in a phylogenetic context. As indicated by several phylogenetic analyses (e.g., Figure 2.11), the OBL species are dispersed widely across the genus, a result consistent with their assignment to various taxonomic sections. Thus, preliminary phylogenetic analyses of *Ranunculus* indicate numerous independent origins of aquatic species in the genus, even when considering only the aquatic North American taxa. *Ranunculus* ($x = 7$, 8) contains diploids, tetraploids, hexaploids, and octoploids among the OBL taxa: $2n = 16$: *R. alismifolius*, *R. hederaceus*, *R. pensylvanicus*; $2n = 32$: *R. flabellaris*, *R. flammula*, *R. hyperboreus*, *R. macounii*; $2n = 48$: *R. aquatilis*, *R. arcticus*, *R. macounii*. Multiple counts have been reported for *R. arcticus* ($2n = 28$, 32, 48), *R. sceleratus* ($2n = 32$, 64), and *R. trichophyllus* ($2n = 16$, 32). Hybrids ($2n = 40$) between *R. aquatilis* and *R. trichophyllus* are reported in Europe and pollen sterile hybrids (*R. hyperboreus* × *R. gmelinii*) occur in North America. Artificial hybrids between *R. hederaceus* and *R. trichophyllus* are also sterile. A cpDNA restriction map of *R. sceleratus* shows a lack of structural rearrangements in the genus, with its 152 kb genome being colinear with that of tobacco (*Nicotiana*).

Comments: The distributions of the OBL *Ranunculus* species are quite variable. *Ranunculus longirostris* and *R. sceleratus* (also Eurasia) are widespread. *R. subrigidus* and *R. trichophyllus* (also Eurasia) are distributed in northern North America, *R. flammula* (also Eurasia) and *R. macounii* are distributed in northern and western North America, and *R. arcticus*, *R. confervoides*, *R. hyperboreus* (also Eurasia), and *R. pensylvanicus* are distributed in northern North America and the Rocky Mountains. *Ranunculus flabellaris* extends across central North America, *R. alismifolius*, *R. aquatilis*, and *R. lobbii* are found in western North America and *R. ambigens* is found in eastern North America. More limited distributions characterize *R. hederaceus* (eastern United States and also Europe), *R. populago* (northwestern United States), *R. laxicaulis* (southeastern United States), *R. pusillus* (southern United States), and *R. hydrocharoides* (southwestern United States but extending into Mexico and Central America). *Ranunculus bonariensis* occurs in not only California but also in South America.

References: Aiken et al., 2007; Alexander, 2016; Alsos et al., 2013; Aslam et al., 2012; Baker & Cruden, 1991; Barbour et al., 2005; Barrat-Segretain & Bornette, 2000; Baskin & Baskin, 1998; Benson, 1942; Boedeltje et al., 2003b; Boeger, 1992; Cavalli et al., 2012; Cayouette et al., 1997; Cellot et al., 1998; Chamberlin, 1911; Cook, 1966; Cook & Johnson, 1968; Cosyns et al., 2005; Croel & Kneitel, 2011b; Davis, 1993; De-Yuan, 1991; Dunn, 1905; Egger & Malaby, 2015; Emadzade et al., 2010, 2015; Engel & Nichols, 1994; Falińska, 1999; Fernald, 1935, 1940; Fiasson et al., 1997; Fraser, 1919; Fraser et al., 1980; Gleason et al., 2009; Graae et al., 2004; Green et al., 2016; Hájek et al., 2007; Hanna, 1938; Hoffmann et al., 2010; Homoya & Hedge, 1982; Hoot, 1995; Hotchkiss & Stewart, 1947; Hutchinson et al., 2007; Jain et al., 2012; Johansson, 1998; Johansson & Jansen, 1993; Johnson & Cook, 1968; Karling, 1955b, 1956; Keith, 1961; Kleyheeg et al., 2016; Kuijper et al., 2006; Lacoul & Freedman, 2006b; Lind & Cottam, 1969; Lumbreras et al., 2011, 2014; McLaughlin, 1974; M'Mahon, 1804; Monda & Ratti, 1988; Murphy et al., 2002; Nichols, 1999a; Noel et al., 2005; Probert et al., 1989, 1991; Rooke, 1984; Rossbach, 1963; Saarela et al., 2013; Schlising & Sanders, 1983; Scott, 2014; Segal, 1967; Sharp et al., 2013; Sheldon & Boylen, 1977; Steinbach & Gottsberger, 1994; Stevens, 1957; Stoffolano, Jr. et al., 2015; Stuckey et al., 1978; Svejcar & Riegel, 1998; Takos, 1947; USFWS, 2016a; Van Dijk et al., 2014; Walker et al., 2010; Walton, 1922; Webster, 1991; Whittemore, 1997a; Williams, 2015a; Wohl & McArthur, 1998; Zalewska-Gałosz et al., 2014.

11. *Thalictrum*

Meadow rue; pigamon

Etymology: from the Greek *thaliktron*, a name used by Dioscorides

Synonyms: *Anemonella*

Distribution: global: cosmopolitan; **North America:** widespread

Diversity: global: 120–200 species; **North America:** 22 species

Indicators (USA): OBL; FAC: *Thalictrum mirabile*

Habitat: freshwater; palustrine; **pH:** unknown; **depth:** <1 m; **life-form(s):** emergent (herb)

Key morphology: stems (to 30 cm) weak or reclining; leaves basal or cauline, the former long-petioled (to 2 cm), the latter shorter, the blades 1–3 times ternately compound, their leaflets (to 30 mm wide) orbicular-reniform, four to seven lobed, the margins crenate; panicles few-flowered; flowers bisexual; sepals (to 1.5 mm) white, inconspicuous; petals lacking; stamen filaments (to 3 mm) strongly clavate, white, petaloid; achenes (to 4 mm) 3–8, minutely beaked, on long stipes (to 3.5 mm), spreading widely in a radial arrangement

Life history: duration: perennial (tuberous roots); **asexual reproduction:** none; **pollination:** unknown; **sexual condition:** hermaphroditic; **fruit:** aggregates (achenes; common); **local dispersal:** achenes; **long-distance dispersal:** unknown

Imperilment: (1) *Thalictrum mirabile* [G4]; S2 (AL, TN)

Ecology: general: *Thalictrum* is primarily a northern temperate genus where most of the species occur in mesic habitats. However, the plants appear often in wetlands, with 16 of the North American species (73%) categorized as wetland indicators at some level (but only one ranked as OBL). The flowers of all *Thalictrum* species are small, inconspicuous, unspecialized, and lack nectar and petals. At least some of the species are self-compatible. They are hermaphroditic or unisexual with the latter arranged in androdioecious, dioecious, or polygamous (hermaphrodite and unisexual flowers on the same or on different individuals) sexual conditions. Pollination occurs by the wind (most unisexual flowers) or by unspecialized insects (Insecta) such as bee flies (Diptera: Bombyliidae), bees (Hymenoptera: Apidae; Halictidae), beetles (Coleoptera), flies (Diptera: Muscidae; Syrphidae), and treehoppers (Hemiptera: Membracidae). Seed dispersal appears to be primarily abiotic. Species with light but large seeds can be dispersed by the wind for as far as 30 m. The seeds of some species are buoyant (remaining afloat for at least 3.5 days) and are water dispersed. Many species produce seeds that are morphophysiologically dormant and require a sufficient afterripening time for embryo maturation as well as cold stratification in order to initiate their germination.

Thalictrum mirabile **Small** inhabits bluffs, crevices, driplines, ledges, ravines, seeps, sinks, slopes, woodlands, and margins of waterfalls at elevations of up to 1500 m. The unusual OBL/FAC indicator designation reflects the peculiarity of habitats, which do not often resemble wetlands in the traditional sense. However, they consistently represent extremely wet sites, which are characterized by dripping water, and occur in shaded exposures on limestone, rock, sand, sandstone, and shale substrates. Flowering occurs from June to early summer. The flowers are structured to indicate insect pollination, but no specific vectors have been identified. The mechanism of seed dispersal is unknown. The seeds develop nondeep simple morphophysiological dormancy, which requires a period of cold stratification to induce germination. Optimal germination (100%) has been achieved by exposing seeds to 1°C cold stratification for 6–12 weeks (12 h of light), followed by incubation at 25°C (15 h of light). Seeds that fail to germinate during the first spring season enter secondary dormancy and delay germination until the following spring. Some populations are threatened by unrestricted rock-climbing activities.

Reported associates: *Ageratina luciae-brauniae, Aquilegia canadensis, Arisaema triphyllum, Asplenium montanum, A. pinnatifidum, A. trichomanes, Dennstaedtia punctilobula, Dryopteris intermedia, D. marginalis, Heuchera parviflora, Huperzia porophila, Mitella diphylla, Oxalis montana, Rhododendron maximum, Sanguinaria canadensis, Sedum ternatum, Selaginella apoda, Silene rotundifolia, Solidago albopilosa, Tiarella cordifolia, Trichomanes intricatum, Tsuga canadensis, Vandenboschia boschiana, Viola rotundifolia, Vittaria appalachiana.*

Use by wildlife: unknown

Economic importance: food: not edible; **medicinal:** no uses are known for *T. mirabile*; however, other *Thalictrum* species contain alkaloids that may be antimicrobial, tumor surpressant, or effective in reducing hypertension; **cultivation:** *Thalictrum mirabile* is grown occasionally as a rock garden ornamental. The related (and ecologically similar) *T. clavatum* is cultivated more widely; **misc. products:** none; **weeds:** none

in North America; **nonindigenous species:** none in North America.

Systematics: Molecular phylogenetic analyses place *Thalictrum* in a relatively isolated position but close to *Aquilegia, Paraquilegia,* and *Trollius* (Figure 2.3). Various studies employing DNA sequence data indicate that *Thalictrum* is monophyletic, as long as *T. thalictroides* is retained within the genus rather than treated as the monotypic segregate *Anemonella,* which had long been the convention. *Thalictrum mirabile* is possibly not distinct taxonomically from *T. clavatum,* a FACW species found at lower elevations (to 500 m) on seepage slopes, in moist woods, and along wooded mountain stream banks. The two species are distinguished by achene characters (adaxial margin concave and longer than the stipe in the former; adaxial margin straight and equaling the stipe in the latter), which are of questionable taxonomic reliability. Otherwise they are quite similar, with *T. mirabile* simply being smaller and more delicate overall. Although approximately half of the recognized *Thalictrum* species have been included in phylogenetic analyses, *T. mirabile* has not, which makes it difficult to reconcile its taxonomic status. The basic chromosome number of *Thalictrum* is $x = 7$. Counts have not yet been reported for *T. mirabile,* but the presumably related *T. clavatum* ($2n = 14$) is diploid. Many *Thalictrum* species are polyploid.

Comments: *Thalictrum mirabile* is restricted to Alabama, Georgia, Kentucky, and Tennessee.

References: Andersson et al., 2000; Francis, 2001, 2011; Godfrey & Wooten, 1981; Hodges & Arnold, 1994b; Hoot, 1995; Kaplan & Mulcahy, 1971; Keener, 1976; Kral, 1966b; Park & Festerling, 1997; Ro & McPheron, 1997; Ro et al., 1997; Small, 1900; Soza et al., 2012, 2013; Thompson et al., 2000; Van Dorp et al., 1996; Walck et al., 1999; White & Drozda, 2006.

12. *Trollius*

Globeflower

Etymology: from the German *Trollblume* ("troll-flower"), alluding to its rotund shape

Synonyms: *Megaleranthis*

Distribution: global: Asia; Europe; North America; **North America:** eastern

Diversity: global: 30 species; **North America:** 3 species

Indicators (USA): OBL: *Trollius laxus*

Habitat: freshwater; palustrine; **pH:** 7.2–8.0; **depth:** <1 m; **life-form(s):** emergent (herb)

Key morphology: shoots (to 52 cm) with persistent thatch-like petioles; basal leaves petioled (to 30 cm), the blades palmately divided into five to seven leaflets, the margins coarsely toothed or deeply lobed, cauline leaves 2–5, their bases clasping or membranous; flowers 1–3, showy (to 5 cm), radial; sepals (5–7) greenish to bright yellow, large (to 20 mm), spreading and petal-like; petals (10–25) yellow, reduced (to 6 mm), staminodal; stamens numerous; pistils 5–12; follicles (to 12 mm) including the straight to slightly incurved beak

Life history: duration: perennial (caudex); **asexual reproduction:** vegetative offsets; **pollination:** insect; **sexual condition:** hermaphroditic; **fruit:** aggregates (follicles; common);

local dispersal: seeds (gravity, wind, rain, high water); **long-distance dispersal:** seeds (water)

Imperilment: (1) *Trollius laxus* [G5]; S1 (CT, NJ, OH, PA); S3 (NY)

Ecology: general: Although various *Trollius* species occur in moist meadows and other wet areas, only one North American species is listed as a wetland indicator. In general, *Trollius* species have large, showy flowers, which are pollinated by various insects (Insecta). The seeds are dispersed abiotically. The seeds of all species have examined intermediate complex morphophysiological dormancy, which requires only a period of cold stratification to initiate germination.

Trollius laxus Salisb. occurs in fens, meadows, swamps, and along the margins of rivers at elevations of up to 500 m. It can also be found on mossy hummocks or on moss-covered woody debris. The sites are characterized by cold, highly alkaline groundwater seepage. The plants can occur in open exposures (where water is plentiful) but generally grow better in partial shade, particularly in habitats that are exposed to periodic drying. The plants are tolerant to high light and low groundwater levels and do not tolerate extremely wet or saturated conditions for long duration. Stronger plant vigor occurs under diffuse light than direct light. The sediments are alkaline (pH: 7.2–8.0), contain high levels of lime and available calcium, and have high but variable cation-exchange capacity. This is an early- to mid-successional species, which is suppressed by competition from woody or other herbaceous plants. Many sites are flooded due to the activity of beavers (Mammalia: Castoridae: *Castor canadensis*), which reduces competition and facilitates seed dispersal. The plants emerge early and often flower (April to July) before canopy closure occurs. Flowers typically develop on plants within 2–3 years after seed germination. Because the flowers are formed at the time of snowmelt, their number in a given year likely is determined by conditions during the previous growing season. The flowers are pollinated by unspecialized insects (Insecta) including ants (Hymenoptera: Formicidae), bees (Hymenoptera), and flies (Diptera). Population genetic studies using isozyme data indicate that *T. laxus* is outcrossed with substantial interpopulational gene flow. Greater follicle production is associated with higher spring groundwater levels. The seeds ripen by mid-June and are dispersed abiotically by gravity, rain, water, and wind. They have germinated well within 2 weeks after being stratified for 90 days under cold, moist conditions (4°C) and placed subsequently under greenhouse conditions. Seed stored dry for 9 months at 4°C did not germinate when sown outdoors the following March but germinated to nearly 100% in the subsequent spring. Seeds sown outdoors within 30 days of collection (May to June) did not germinate until the subsequent spring. Seeds collected in early June, dried at room temperature for several months, stored dry in a refrigerator until November, and then sown outdoors have germinated reliably during the following spring. The habitats are threatened by inputs of agricultural nutrients, siltation, and hydrological alterations due to human or beaver activity. The

plants are also susceptible to grazing, lumbering activities, and overcollection. **Reported associates:** *Abies balsamea, Acer rubrum, Angelica atropurpurea, Calamagrostis canadensis, Caltha palustris, Carex, Carpinus caroliniana, Cicuta maculata, Cirsium muticum, Clematis virginiana, Conioselium chinense, Cornus sericea, Dasiphora floribunda, Equisetum hyemale, Fragaria virginiana, Fraxinus nigra, Geum rivale, Hamamelis virginiana, Iris versicolor, Kalmia latifolia, Micranthes pensylvanica, Mitella diphylla, M. nuda, Osmundastrum cinnamomeum, Packera aurea, Pinus strobus, Prunella vulgaris, Ranunculus recurvatus, Rhamnus alnifolia, Rubus pubescens, Solidago patula, Spiraea latifolia, Symplocarpus foetidus, Thalictrum polygamum, Thelypteris palustris, Thuja occidentalis, Toxicodendron radicans, Tsuga canadensis, Ulmus americana, Vaccinium cassinoides, V. corymbosum, V. recognitum, Viola papilionacea, Zizia aurea.*

Use by wildlife: The foliage of *T. laxus* is browsed by deer (Mammalia: Cervidae). The flowers are eaten by slugs (Mollusca: Gastropoda), which can cause enough damage to reduce seed set. The plants are a host for fritillary caterpillars (Insecta: Lepidoptera: Nymphalidae: *Boloria titania*).

Economic importance: food: *Trollius* plants are not edible. The foliage contains the glycoside ranunculin, which yields the toxic oil protoanemonin upon autolysis; **medicinal:** The Cherokee used leaf and stem infusions of *T. laxus* to treat an ailment known as "thrash"; **cultivation:** *Trollius laxus* is cultivated as an ornamental plant but not to the extent of other more common species; **misc. products:** none; **weeds:** none in North America; **nonindigenous species:** The European *T. europaeus* has escaped from cultivation in New Brunswick, but it is not aquatic.

Systematics: Various molecular sequence data show *Trollius* to be allied closely with the genus *Adonis* (e.g., Figure 2.3). However, studies using different taxon sampling (and *matK* sequences) have resolved *Megaleranthis* and *Trollius* as sister genera, with *Calathodes* and then *Adonis* as their respective sister groups. Moreover, analyses incorporating larger numbers of *Trollius* species and combined DNA sequence data place *Megalantheris* within *Trollius* but continue to resolve *Calathodes* as more closely related to *Trollius* than is *Adonis* (Figure 2.12). Consequently, it seems that the merger of *Megalantheris* with *Trollius* is necessary to maintain the monophyly of the latter genus. The western species *T. albiflorus* (unfortunately, not yet sampled in phylogenetic analyses) is recognized by some authors as a variety of *T. laxus* but is white flowered and occurs in more acidic, alpine–montane habitats. Furthermore, *T. laxus* ($2n = 32$) is isolated chromosomally from *T. albiflorus* ($2n = 16$); the basic chromosome number of *Trollius* is $x = 8$. *Trollius laxus* is the only known polyploid (i.e., a tetraploid) in the genus.

Comments: *Trollius laxus* is a rare species restricted to the northeastern United States.

References: Bai et al., 1996; Baskin & Baskin, 1998; Brumback, 1989; Ferdy et al., 2002; Jones, 2000; Jones & Klemetti, 2012; Parfitt, 1997; Ro et al., 1997; Scanga, 2011; Scanga & Leopold, 2010; Wang et al., 2009, 2010b.

ORDER 2: NELUMBONALES [1]

Although the small order Nelumbonales (water lotus) contains a single family (Nelumbonaceae) and genus (*Nelumbo*), it has experienced a convoluted taxonomic history. Throughout much of the 19th and 20th centuries, the group was treated either as an isolated magnoliid order or allied (as a family or subfamily) with the superficially similar Nymphaeales (water lilies; Les, 1988a). Serological data and conflicting alkaloid chemistry raised serious doubts regarding any close relationship between Nelumbonales and Nymphaeales (Simon, 1970; Les, 1988a), a conclusion that eventually was substantiated by the results of the first phylogenetic analysis of these groups using DNA sequence (i.e., *rbcL*) data (Les et al., 1991). More extensive *rbcL* sequence data analyses (with increased taxon sampling) resolved Nelumbonales within the tricolpate "eudicots" rather than being closely related to any of the monosulcate magnoliid groups (Chase et al., 1993; Qiu et al., 1993). In particular, those studies indicated an unexpected alliance with Platanaceae, Proteaceae, and Sabiaceae, all groups never previously associated with Nelumbonales. Nelumbonales and

Proteales cluster consistently in phylogenetic analyses using data from single genes to complete plastid genomes (Soltis et al., 2011; Wu et al., 2014a; e.g., Figure 2.1) but in many cases without reliable internal support (e.g., Kim et al., 2004; Gu et al., 2013). Genome sequencing of lotus has also proposed that the plants lack a 125-million-year-old paleohexaploid duplication that has been detected in all other sequenced eudicot genomes, including that of Platanaceae (Ming et al., 2013; but see also Zheng & Sankoff, 2014). Other peculiarities exist such as the phenylalanine ammonia-lyase gene of *Nelumbo* being more similar to those of gymnosperms than to other dicots (Wu et al., 2014b). In any case, the extremely high level of morphological divergence between Nelumbonales and any Proteales taxon provides reasonable support for continuing to recognize the group as a distinct order rather than merging it with Proteales as many authors have done. The recognition of Nelumbonales as a separate order (related at some level to Proteales) is compatible with all of the phylogenetic results thus far and has been followed here (Figure 2.13).

Biochemically, Nelumbonales are characterized by the presence of benzyl-isoquinoline and aporphine alkaloids and

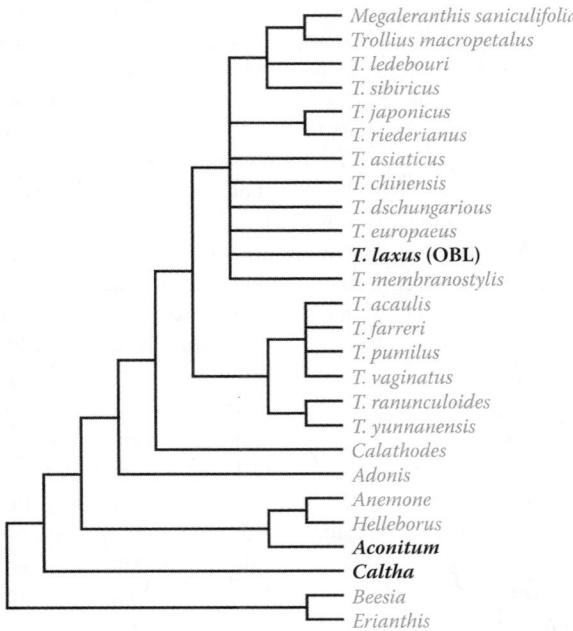

FIGURE 2.12 Phylogenetic analysis of *Trollius* and related groups based on reconstructions using combined DNA sequence data. This result argues for the inclusion of *Megaleranthis* within *Trollius*, and the recognition of *Calathodes* as its sister genus. The OBL indicator *T. laxus* is embedded centrally within the Eurasian *Trollius* species. Taxa containing OBL North American indicators are highlighted in bold. (Redrawn from Wang, W. et al., *Taxon*, 59, 1712–1720, 2010b.)

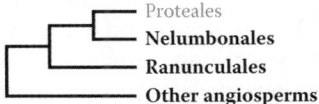

FIGURE 2.13 Phylogenetic relationships among Nelumbonales, Proteales, and Ranunculales as indicted by analysis of granule-bound starch synthase sequence data. Groups containing OBL indicator taxa are highlighted in bold. (Modified from Gu, C. et al., *Plant Mol. Biol. Rep.*, 31, 1157–1165, 2013.)

proanthocyanins but lack ellagic acid (Cronquist, 1981). The vascular bundles contain articulated lacticifers and sieve tubes with S-type plastids; vessels are confined to the rhizomes and roots (Cronquist, 1981; Schneider & Carlquist, 1996a). The leaves are large, peltate, and either floating or emergent. They are infamous for their extreme hydrophobicity and ability to shed water (the so-called "lotus effect"). The flowers are large, showy, perfect and are held above the water on large peduncles. The perianth consists of numerous tepals, the outermost being green and sepaloid and the innermost petaloid and either creamy yellow or pink to reddish in color (Cronquist, 1981). The fruits are aggregates of achenes imbedded within a large, turbinate, fleshy receptacle. The seeds are renowned for their remarkable longevity and can retain viability for more than 1000 years (Shen-Miller et al., 1995). The thermogenic flowers are pollinated by insects (Insecta), primarily flies (Diptera), beetles (Coleoptera), and bees (Hymenoptera), which they attract by the ability to elevate their temperature as high as 35°C (Schneider et al., 1990; Dieringer et al., 2014). The order contains a single family:

1. Nelumbonaceae Dumortier

Family 1: Nelumbonaceae [1]

The family Nelumbonaceae (water lotus family) contains only two species, which are distributed in temperate to tropical regions of Asia, Australia, Central America, and North America. The characteristics of the family are summarized above (see Nelumbonales).

Many Nelumbonaceae cultivars have been developed for use in ornamental water gardens. The water lotus is regarded as a sacred plant in many world cultures and is often used as an artistic motif. The rhizomes and seeds are edible and have also been used in various medicinal applications.

1. *Nelumbo*

Lotus, water chinquapin, water lotus; volée
Etymology: from the Sinhalese *nelumbu*, meaning lotus
Synonyms: *Nelumbium*, *Nymphaea* (in part)
Distribution: global: Asia; Australia; Central America; North America; **North America:** eastern
Diversity: global: 2 species; **North America:** 2 species
Indicators (USA): OBL: *Nelumbo lutea*, *N. nucifera*
Habitat: freshwater; lacustrine, palustrine, riverine; **pH:** 4.5–9.3; **depth:** 0–3 m; **life-form(s):** floating-leaved, emergent (herb)
Key morphology: leaves large (to 1 m), circular, peltate (sinus lacking), waxy (and highly water-repellent) above, flat (when floating) or concave (when emergent), long-petioled (to 2+ m); flowers solitary, showy (to 50 cm), emergent on long peduncles (to 2 m); tepals (to 13 cm) numerous, yellow, white, or pink; stamens numerous; achenes large (to 20 mm), embedded in a conical, leathery, spongy receptacle (to 1 dm)
Life history: duration: perennial (rhizomes); **asexual reproduction:** rhizomes, tubers; **pollination:** insect (first day), insect/self (second day); **sexual condition:**

hermaphroditic; **fruit:** aggregates (achenes; common); **local dispersal:** achenes, fruits, seedlings, tubers (water); **long-distance dispersal:** achenes (animals)
Imperilment: (1) *Nelumbo lutea* [G4]; SH (PA); S1 (DE, NJ); S2 (MD, MI, NC, NE, ON, WV); S3 (IL, IN, VA); (2) *N. nucifera* [G5]
Ecology: general: Both *Nelumbo* species are perennial, OBL aquatics that occur mainly in shallow water (<1 m) but occasionally can grow to depths of 3 m. The species are quite similar and share comparable life-history traits. The leaves are emergent in shallow water but float on the surface of deeper waters. The plants are turbidity tolerant but are susceptible to prolonged periods of high water. Lotus species require full sunlight and occur in habitats where the average summer (July to August) temperatures reach at least 25°C. The habitats are typically high in $CaCO_3$ alkalinity. The plants take 5–6 years to reach full flower production. The flowers are self-compatible but are protogynous and primarily are cross-pollinated by insects (Insecta), which are attracted to the thermogenic blooms. *Nelumbo* seeds can remain viable in the sediment for incredible periods of time, with some estimates ranging from 150 to 1000+ years. Their germination is enhanced by mechanical scarification followed by soaking in warm (25°C) water (changed twice daily) for 3–4 weeks. Alternatively, dormancy can be broken by mechanical abrasion of the fruit wall or by acid scarification (using concentrated sulfuric acid) of the achenes for 5 h. The individual achenes, as well as the aggregate fruits, are buoyant, which facilitate their local dispersal by water. Longer distance dispersal can occur by waterfowl (Aves: Anatidae) or other animals, which consume the achenes and transport them endozoically. The plants reproduce vegetatively by the proliferation of large rhizomes, which can grow up to 10 m in 1 year.

Nelumbo lutea Willd. occurs in backwaters, bogs, canals, ditches, floodplains, lake/pond beds (dried), lakes, levees, limesinks, marshes, oxbows, ponds, pools, rivers, roadsides, sinkholes, sloughs, streams, swamps, and along the margins of lakes, reservoirs, and rivers at elevations of up to 1200 m. The plants generally occur in shallow waters from 0.3 to 1.5 m deep and in exposures with full sunlight. Prolonged submergence is not tolerated and complete inundation of the foliage for 2 weeks is sufficient to kill the leaves. The waters tend to be alkaline in nature (e.g., pH: 6.2–9.3 mean: 8.1; total alkalinity: 25–235 mg/L $CaCO_3$; conductivity: 115–410 µmhos/cm), although the highest photosynthetic production occurs at more acidic pH values, where free CO_2 availability is high. The substrates are described as rich sediments and include limestone, marl, muck, mud, sand, and sandy loam. Flowering occurs from May to September, with fruiting observed from July to October. The flowers are protogynous and exhibit a 2-day reproductive cycle. The first-day (female-phase) flowers are outcrossed by insects (see *Use by wildlife*). If they remain unfertilized, then the second-day flowers (bisexual phase) are capable of autogamy, although insect visitation can be up to seven times higher during this phase. Only low levels of autogamy (~18%) have been recorded. Several days after pollination, the receptacles begin to reorient downward, which properly situates them for the eventual release of the ripening fruits. The ripe achenes sink immediately when shed, but within 5 h an internal air space develops, which raises them to the water surface for local dispersal by water currents. After 12 h, the fruits lose their buoyancy and sink again to the bottom where germination normally occurs. The seedlings emerge within 10 days and are also buoyant, often drifting near the margins of water bodies where they establish. Once dried, the achenes retain their viability but lose buoyancy and require mechanical or acid scarification in order to initiate germination. The large achenes occasionally are eaten (and dispersed) by waterfowl (see *Use by wildlife*) and have also been recovered from the stomach contents and feces of alligators (Reptilia: Alligatoridae: *Alligator mississippiensis*), which have been implicated as longer distance dispersal agents. Lotus plants spread mainly by vegetative reproduction resulting from rhizome extension, with the dispersal of fruits and tubers being relatively minor contributors to colony proliferation. Though often regarded as weedy (particularly in the southern portion of its range), some rare occurrences of *N. lutea* (mainly more northern populations) can be threatened by overcollecting. **Reported associates:** *Acer rubrum, Agalinis, Agrimonia pubescens, Alisma subcordatum, Alnus serrulata, Alternanthera philoxeroides, Ampelopsis arborea, Andropogon glomeratus, Azolla caroliniana, Baccharis, Berberis thunbergii, Bidens cernuus, B. trichospermus, Bolboschoenus fluviatilis, Brasenia schreberi, Cabomba caroliniana, Celtis laevigata, Cephalanthus occidentalis, Ceratophyllum demersum, Chara, Colocasia esculenta, Cornus amomum, Cyperus, Echinodorus rostratus, Egeria densa, Eichhornia crassipes, Eleocharis cellulosa, Elymus virginicus, Eupatorium, Hibiscus moscheutos, Hydrilla verticillata, Hydrocotyle ranunculoides, Hypericum, Leersia oryzoides, Lemna minor, Limnobium spongia, Lindernia dubia, Lobelia cardinalis, Ludwigia peploides, Mikania scandens, Myrica cerifera, Myriophyllum aquaticum, M. spicatum, Najas guadalupensis, N. minor, Nitella, Nuphar advena, Nymphaea mexicana, N. odorata, N. tuberosa, Oenothera, Oxycaryum cubense, Panicum hemitomon, P. repens, Paspalum repens, Persicaria amphibia, P. pensylvanica, Phragmites australis, Physocarpus opulifolius, Pistia stratiotes, Pontederia cordata, Populus deltoides, Potamogeton gramineus, P. nodosus, P. pusillus, Proserpinaca palustris, Quercus nigra, Q. palustris, Rhus copallinum, Rhynchospora, Rosa palustris, Rumex verticillatus, Sacciolepis striata, Sagittaria lancifolia, S. latifolia, S. platyphylla, Salix nigra, Salvinia minima, S. molesta, Schoenoplectus tabernaemontani, Solidago, Spartina pectinata, Spirodela polyrhiza, Stuckenia pectinata, Symphyotrichum lanceolatum, Taxodium distichum, Typha angustifolia, T. domingensis, T. latifolia, Utricularia macrorhiza, Vitis cinerea, Zizaniopsis miliacea.*

***Nelumbo nucifera* Gaertn.** is a nonindigenous species that grows in the shallow waters (0.3–1.0 m) of backwaters, ditches, drainages, marshes, pools, sloughs, and along the margins of canals, lakes, and ponds at elevations of up to 460 m. The exposures receive full sunlight. The substrates

(North America) have been described as Hyde silty loam, marl, and muck. The few North American sites were evaluated and indicate somewhat acidic water conditions (e.g., pH: 5.7–6.8). Flowering (North America) has been observed from June to September, with fruiting from July to October. The flowers are protogynous and fragrant but lack nectar. They are thermogenic during anthesis and produce an estimated 1 million pollen grains per flower. The flowers are pollinated by insects (Insecta), which primarily are bees (Hymenoptera) and flies (Diptera) in contemporary populations; however, beetles (Coleoptera) are also effective pollinators and possibly represent the primary historical pollinators. Like its congener, the dried achenes do not germinate even after imbibing water for a period of 8 months and require mechanical scarification to induce germination. Much of the life-history information available for this species pertains to its Old World distribution. However, because of its similarity to *N. lutea*, it likely shares many of the traits reported for that species in the New World. **Reported associates (North America):** *Ceratophyllum demersum, Euphorbia maculata, Hydrocotyle ranunculoides, Ipomoea lacunosa, Juncus, Lemna, Melothria pendula, Nymphaea odorata, Paspalum plicatulum, Potamogeton, Taxodium.*

Use by wildlife: *Nelumbo lutea* plants are eaten by white-tailed deer (Mammalia: Cervidae: *Odocoileus virginianus*). The large achenes are eaten occasionally by canvasback, greater scaup, ring-necked duck, mallard, and wood ducks (Aves: Anatidae: *Aythya valisineria, A. marila, A. collaris, Anas platyrhynchos, Aix sponsa*). Red-winged blackbirds (Aves: Icteridae: *Agelaius phoeniceus*) forage for insects among the flowers, often damaging the blooms in the process. The rhizomes are eaten by mammals (Mammalia) such as beavers (Castoridae: *Castor canadensis*), muskrats (Cricetidae: *Ondatra zibethicus*), and porcupines (Erethizontidae: *Erethizon dorsatus*). Large *N. lutea* populations create shelter and shade for fish and small invertebrates. They provide also a favorable habitat for mosquitos (Insecta: Diptera: Culicidae: *Anopheles quadrimaculatus*). The leaf undersides support colonies of freshwater Bryozoa (Pectinatellidae: *Pectinatella magnifica*). *Nelumbo lutea* is a host plant for caterpillars (Insecta: Lepidoptera: Noctuidae: *Bellura obliqua*; Pyralidae: *Munroessa gyralis, Paraponyx maculalis, Synclita obliteralis*). The staminal floral appendages are eaten (and damaged severely) by several grasshoppers (Insecta: Orthoptera: Acrididae) and larvae of the lotus borer (Insecta: Lepidoptera: Crambidae: *Ostrinia penitalis*). *Bellura obliqua* and *Ostrinia penitalis* are particularly serious herbivores of the leaves and petioles. More than 30 species of insects (Insecta) are associated with the flowers of *N. lutea* including various beetles (Coleoptera), bees and wasps (Hymenoptera), flies (Diptera), grasshoppers (Orthoptera), and springtails (Collembola). The thermogenic flowers of *N. lutea* volatilize the floral scent and provide warmth as an energetic reward to pollen-gathering (pollinating) insects (Insecta), which primarily are flies (Diptera: Phoridae; Syrphidae: *Eristalis tenax, Parhelophilus laetus*) and secondarily include bees (Hymenoptera: Apidae: *Apis mellifera*; Halictidae:

Agapostemon radiatus, *Augochlorella*, *Lasioglossum zonulum*) and beetles (Coleoptera: Cantharidae: *Chauliognathus*; Curculionidae; Coccinellidae: *Coleomegilla maculata*; Chrysomelidae: *Diabrotica undecimpunctata*, *D. virgifera*). Populations of *N. nucifera* that have become established in America are hosts for moth larvae (Lepidoptera: Crambidae: *Psara obscuralis*).

Economic importance: food: Lotus is eaten as a vegetable (mainly in Asia). Its rhizomes are used in stir fry and other dishes, the leaves are pickled, and the seeds are roasted or candied. Canned *Nelumbo* rhizomes are sold commercially. Boiled lotus is an ingredient in Japanese vegetable salads. The leaves are also used as wrappers to accent the presentation of other foods. In China, a paste from lotus seeds is used in sauces and cake fillings. Several native North American tribes (Comanche, Dakota, Huron, Meskwaki, Ojibwa, Omaha, Pawnee, Ponca, Potawatomi, Winnebago) ate the fruits of *N. lutea* (cooked, raw, or added to soups) and baked its banana-shaped lotus tubers, which have a texture resembling sweet potatoes. The seeds were ground and used as flour for making bread. The seeds and rhizomes have also been a traditional, sacred food (known as Tse'-wa-the) of the Osage people; **medicinal:** Lotus has been used in traditional Asian medicine for more than 2000 years. The rhizomes contain antifungicidal properties, which are effective against numerous species (Fungi: Ascomycota: Arthrodermataceae: *Trichophyton mentagrophytes*; Aspergillaceae: *Aspergillus fumigatus*, *A. niger*; Candidaceae: *Candida albicans*; Hypocreaceae: *Trichoderma viride*; Saccharomycetaceae: *Saccharomyces cerevisiae*; Trichocomaceae: *Penicillium*). Isoquinoline alkaloids in the leaves and seeds have reportedly sedative and antispasmodic properties and have been used to lower blood pressure. The seeds and stamens are said to be astringent and are used to treat diarrhea, insomnia, and palpitations. *Nelumbo lutea* contains the alkaloid anonaine, which reputedly exhibits several pharmacological properties including antibacterial, anticancer, antidepressant, antifungal, antiplasmodial, antoxidant, and vasorelaxant activity. *Nelumbo nucifera* contains various benzyl-isoquinoline alkaloids, to which are attributed numerous remedial properties. Among these are (+)-1(*R*)-coclaurine and (−)-1(*S*)-norcoclaurine, which are associated with anti-HIV properties, and the bisbenzyl-isoquinoline neferine, which exhibits cytotoxicity against hepatocellular carcinoma (HCC Hep3B cells). Quercetin extracted from the receptacle of *N. nucifera* is used in China as an antihemorrhagic. The plants reportedly contain constituents that are effective at treating numerous ailments including cholera, coughs, diarrhea, dysentery, epistaxis, fevers, hematemesis, hematuria, hemoptysis, hepatopathy hyperdipsia, hyperlipidemia, hypoglycemia, inflammations, leucoderma, metrorrhagia, obesity, pectoralgia, pharyngopathy, smallpox, spermatorrhoea, and others; **cultivation:** *Nelumbo* is an extremely important genus of water garden ornamentals with over 300 cultivars (mostly Asian) described. Lotus cultivation may extend back as far as 7000 years in Asia. Presently, both species as well as numerous inter- and intraspecific hybrids are cultivated; **misc. products:** When dried, the aggregate fruits of *Nelumbo* are used in floral arrangements and wreaths. The remarkably effective water-shedding surface of lotus leaves is densely covered by nanoscopic protuberances, which have inspired the formulation of a self-cleaning industrial paint known as Lotusan. Lotus flowers are sold as cut flowers and are used in some religious ceremonies. Lotus plants are often pictured in art and architecture. Stands of *N. lutea* were once so abundant in western Lake Erie marshes that they were identified on travel maps as scenic points of interest. Leaf extracts from *N. lutea* have been used as a topical darkening agent for skin or hair. *Nelumbo nucifera* is regarded by Buddhists as sacred; **weeds:** *Nelumbo lutea* is perceived as a weed in some areas (particularly the southeast) where it is controlled by herbicides; whereas, it is regarded as imperiled and protected in others. Extensive colonies can develop in shallow waters and can interfere with boating activity and other water recreational uses; **nonindigenous species:** *Nelumbo lutea* is native to the eastern United States, but populations that were introduced and moved by early native Americans are difficult to distinguish from unmanipulated natural occurrences. *Nelumbo nucifera* (native to eastern Asia) has been introduced throughout the eastern United States as an escape from water garden cultivation.

Systematics: Although Nelumbonaceae were long regarded as allied phylogenetically to the superficially similar Nymphaeaceae, they are distinct morphologically and have been shown to be unrelated by various molecular studies (see also discussion under Nelumbonales above). Combined molecular data sets indicate the alliance of Nelumbonaceae with Platanaceae and Proteaceae, a result that has encouraged many authors to include Nelumbonaceae within the order Proteales. However, Nelumbonaceae are disparate morphologically from either of those families, and the long branches resulting from the molecular data analyses that depict their association are not often well supported. The distinctness of Nelumbonaceae is evidenced further by results from the nearly completely sequenced (85.6%) genome of *N. nucifera*, which has been interpreted as lacking evidence of the paleohexaploid event that occurred in other eudicots (including Platanaceae) about 125 million years ago. The genome sequences of *Nelumbo* exhibit relatively slow mutation rates (e.g., 30% slower than Vitaceae) and indicate also the occurrence of a lineage-specific whole genome duplication event that occurred about 65 million years ago. Complete chloroplast and mitochondrial genome maps are also available for *N. nucifera*. The similarity of *N. lutea* and *N. nucifera* has led some authors to combine them taxonomically as a single species with two subspecies. However, molecular data (e.g., SSR and SRAP markers) have shown not only that the species are quite distinct genetically but also that populations of *N. nucifera* from China and Thailand are differentiated genetically (Figure 2.14). Genomic studies have correlated the interspecific differences in tepal color with the differential expression of genes relating to the production of anthocyanins (absent in yellow tepals), including a lack of *GST* expression and highly reduced expression of three *UFGT* genes in yellow tepals. Yellow tepals contain also two inactive forms of *MYB5*, which is functional in red tepals. Phylogenetic analysis

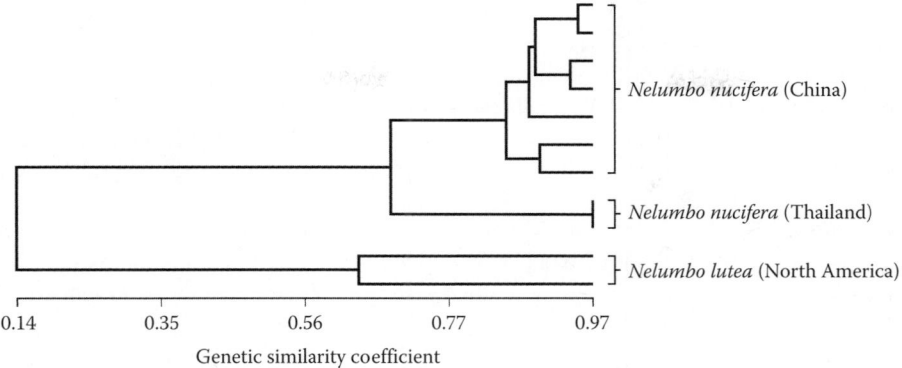

FIGURE 2.14 UPGMA dendrogram showing genetic clustering of *Nelumbo* plants surveyed from 11 Asian and North American populations (83 accessions) using simple sequence repeat (SSR) markers. The SSR data resolve *Nelumbo lutea* and *N. nucifera* (both OBL indicators) as quite distinct genetically and indicate also genetic differentiation between *N. nucifera* accessions from China and Thailand. (Adapted from Yang, M. et al., *Aquatic Bot.*, 107, 1–7, 2013.)

of plants grown from ancient Chinese seeds (estimated to be 350–1000 years old) associate them with extant *N. nucifera* accessions. *Nelumbo* (*x* = 8) is uniform cytologically with both *N. lutea* and *N. nucifera* being diploid (2*n* = 16). A triploid count (2*n* = 24) has also been reported for *N. nucifera*. Numerous hybrids have been made both between and within the two lotus species. Studies using mtDNA data indicate that cultivated lotus hybrids have involved the use of both species as their maternal parent. The similarity of lotus plants conceals their genetic differences. Even some intraspecific crosses (within *N. nucifera*) exhibit low fecundity due to embryo abortion, low stigma receptivity, and reduced pollen fertility.

Comments: *Nelumbo lutea* and *N. nucifera* occur sporadically across the eastern United States.

References: Barthlott et al., 1996; Bellrose, 1941; Borsch & Barthlott, 1994; Dieringer et al., 2014; Francko, 1986; Gu et al., 2013; Gui et al., 2016; Hall & Penfound, 1934; Han et al., 2007; Harms & Grodowitz, 2009; Hoot et al., 1999; Kanazawa et al., 1998; Kashiwada et al., 2005; Kubo et al., 2009; Les, 1988a; Les et al., 1991; Li & Huang, 2009; Li et al., 2013, 2014; Matthews & Haas, 1993; Ming et al., 2013; Mukherjee et al., 1995, 2009; Nichols, 1999a; Ohga, 1923; Platt et al., 2013; Schneider & Buchanan, 1980; Self et al., 1974; Seymour & Schultz-Motel, 1998; Slocum et al., 1996; Sohmer & Sefton, 1978; Sun et al., 2016; Swan, 2010; Teng et al., 2012; Wersal et al., 2008; Wiersema, 1997; Williamson & Schneider, 1993; Xiuwen et al., 1997; Yang et al., 2012, 2013; Yoon et al., 2013.

Eudicots (Tricolpate Dicots; Eudicotyledoneae)

3 Core Eudicots: Dicotyledons III
"Caryophyllid" Tricolpates

INTRODUCTION

The precise phylogenetic position of caryophyllids (a group equating more or less with the previously recognized subclass Caryophyllidae) remains problematic. Corroborative results from morphological analysis, amino acid sequencing, and various DNA sequencing (e.g., *atpB*, *matK*, *rbcL*, 18S sequences) leave little doubt that the group is monophyletic (Judd et al., 2016); however, whether caryophyllids are related more closely to the rosids (e.g., Goloboff et al., 2009) or asterids (e.g., Moore et al., 2010) has not yet been resolved convincingly. Some authors (e.g., Judd et al., 2016) favor the latter interpretation (related more closely to Asterids), suggesting the recognition of a "superasterid" clade comprising Berberidopsidales, Caryophyllales, Santalales, and Asteridae. Yet, support for that phylogenetic hypothesis (based primarily on the results of cpDNA analyses) is inconsistent among various different combinations of DNA sequence data that have been analyzed. Therefore, the decision has been made here to treat caryophyllids as a distinct group with no specific association with either rosids or asterids. It is evident that caryophyllids comprise two sister clades that are designated here as the orders Caryophyllales and Polygonales (Figure 3.1). Advocates of the superasterid hypothesis (e.g., Judd et al., 2016) have also abandoned this long-standing taxonomic convention in favor of recognizing a single order (Caryophyllales) with two suborders (Caryophyllineae and Polygonineae), which seems unnecessary. Morphologically, the caryophyllids are quite diverse but share characteristic features such as the presence of entire leaf margins and spinulose pollen (Judd et al., 2016).

ORDER 1: CARYOPHYLLALES [27]

The order Caryophyllales includes about 8600 species. Various phylogenetic analyses (e.g., Cuénoud et al., 2002; Figure 3.1) have indicated that several families (Molluginaceae, Portulacaceae, Phytolaccaceae) were not monophyletic as formerly circumscribed and these remain in a state of taxonomic flux. In addition, more recent studies (Nyffeler & Eggli, 2010) have provided convincing evidence to recognize Montiaceae as a distinct family, which includes several genera assigned previously to Portulacaceae. Some authors (e.g., Judd et al., 2016) have abandoned the recognition of Molluginaceae, but the family is retained here, pending additional clarification of its taxonomic status.

In North America, obligate wetland (OBL) aquatic species occur within the following five families:

1. **Aizoaceae** Martinov
2. **Amaranthaceae** Adans.

3. **Caryophyllaceae** Juss.
4. **Molluginaceae** Raf.
5. **Montiaceae** Raf.

Relative phylogenetic relationships within Caryophyllales (Figure 3.1) indicate that families with aquatic species in North America are dispersed widely across this predominantly terrestrial order, an indication that aquatic taxa have originated multiple times in the group.

Family 1: Aizoaceae [127]

The Aizoaceae family includes about 2500 species of succulent plants with opposite, entire leaves, which are adapted mainly to arid, terrestrial habitats. Molecular evidence (e.g., Figure 3.1; Crawley & Hilu, 2012) indicates that the family is monophyletic and related to Gisekiaceae, Barbeuiaceae, and Nyctaginaceae and to segregates of the polyphyletic Molluginaceae and Phytolaccaceae. The plants are unusual in possessing vasculature often arranged in concentric rings of vascular bundles. Betalains, alkaloids, and raphides are present in most species as is C_4 [often crassulacean acid metabolism (CAM)] photosynthesis. The epidermis and stem contain large, bladderlike cells (Judd et al., 2016). The fruits are capsules (some dehiscing only when wet), which contain seeds having perisperm and a curved embryo. The flowers are pollinated by insects (Insecta), which are attracted to their nectar disk, and the seeds are dispersed abiotically by wind or water. The showy floral perianth is also unusual in that it comprises numerous petal-like staminodes. The group primarily is circumtropical in coastal or arid habitats (Judd et al., 2016).

Many of the species are prized by collectors because of their unusual succulent morphology, which in some species renders the appearance of rocks and stones. The most important of these include species of aloinopsis (*Aloinopsis*), cone plant (*Conophytum*), dew plant (*Delosperma*), fig marigold (*Lampranthus*), glottiphyllum (*Glottiphyllum*), goat's horns (*Cheiridopsis*), living stones (*Lithops*), and ice plant (*Mesembryanthemum*).

Understandably, only one genus (and therein only one species) in this extensively xerophytic family contains OBL indicators in North America:

1. *Sesuvium* L.

Combined molecular data (*trnL-F*, *rps16* intron, *atpB–rbcL* spacer, *psbA–trnH* spacer sequences) place *Sesuvium* in close association with *Trianthema* and *Tribulocarpus* within subfamily *Sesuvioideae*, which resolves as the sister clade to the remainder of Aizoaceae (Figure 3.2).

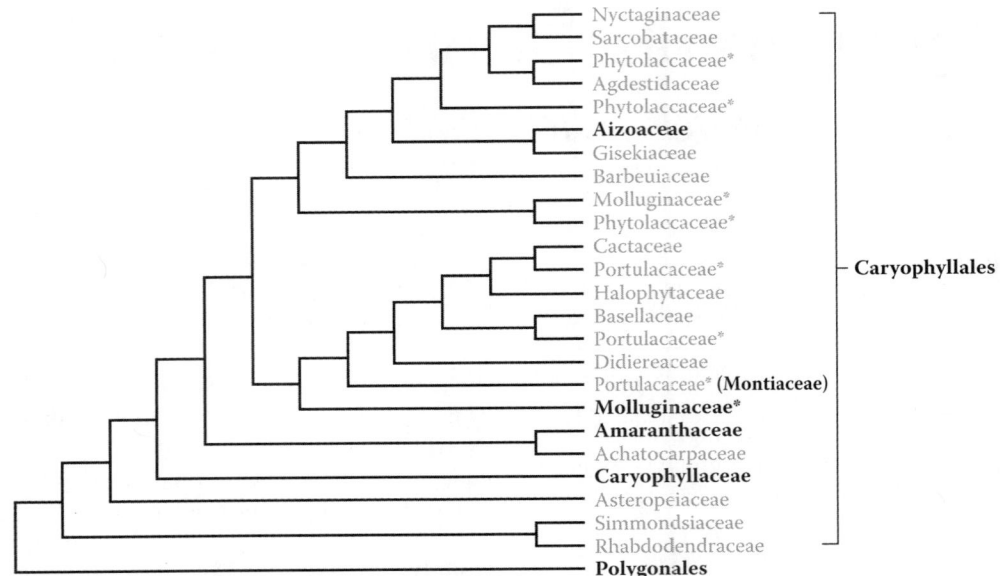

FIGURE 3.1   Phylogenetic relationships hypothesized among 19 families of Caryophyllales based on an analysis of *matK* data. The relationships are consistent with results comparing fewer families analyzed using 18S, *atpB*, and *rbcL* sequences. Groups that contain obligate aquatic taxa in North America are shown in bold type. Families resolved as polyphyletic are asterisked. For polyphyletic families, bold type identifies only those clades containing OBL North American taxa. (Adapted from Cuénoud, P. et al., *Amer. J. Bot.*, 89, 132–144, 2002.)

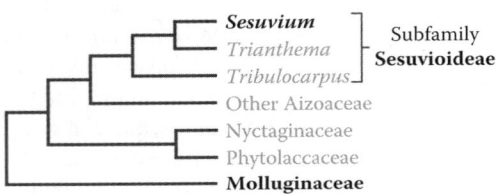

FIGURE 3.2   *Sesuvium* occurs within a sister clade to the remainder of Aizoaceae as indicated by phylogenetic analysis of four cpDNA data sets. Taxa that contain aquatic species in North America are shown in bold. *Sesuvium* is the only genus of Aizoaceae in North America to contain an OBL indicator species. More recent analyses (Bohley et al., 2015) resolve *Zaylea* as the sister group of *Sesuvium*. (Adapted from Klak, C. et al., *Amer. J. Bot.*, 90, 1433–1445, 2003.)

1. *Sesuvium*

Sea purslane

Etymology: in reference to the country of the Sesuvii, an ancient tribe

Synonyms: none

Distribution: global: New World; **North America:** southern

Diversity: global: 8 species; **North America:** 4 species

Indicators (USA): OBL: *Sesuvium trianthemoides*

Habitat: freshwater, brackish (coastal); palustrine; **pH:** unknown; **depth:** <1 m; **life-form(s):** emergent (herb)

Key morphology: plants (to 3.5 dm) succulent, the foliage with large crystalline globules; leaves (to 3 cm) fleshy, opposite, petiole bases winged, clasping (stipule-like); flowers inconspicuous (petals absent), solitary, sessile, perigynous; capsules (to 5 mm) ovoid–ellipsoid, circumscissile; seeds (to 1.5 mm) rugose, about 10 per capsule

Life history: duration: annual (fruit/seeds); **asexual reproduction:** none; **pollination:** unknown; **sexual condition:** hermaphroditic; **fruit:** capsules (common); **local dispersal:** seeds (water, wind); **long-distance dispersal:** seeds (birds)

Imperilment: (1) *Sesuvium trianthemoides* [GH]; SH (TX)

Ecology: general: *Sesuvium* species often show affinities for wet habitats, with all four North American species being either OBL or facultative wetland (FACW) indicators. *Sesuvium maritimum* is regarded as an OBL aquatic in the Caribbean region (excluded here because of geographical criteria) but only as a FACW species throughout the United States. The species include annuals or perennials. All of the North American species occur primarily in coastal areas, in habitats characterized by sandy substrates and alkaline, brackish, or saline conditions. The seeds of several species are consumed by waterfowl (Aves: Anatidae) and are probably dispersed over fairly long distances by them.

Sesuvium trianthemoides Correll is an annual species that occurs in coastal depressions, dunes, marshes, and swales in southern Texas at elevations of up to 4 m. The plants are in flower from June to August. Otherwise, virtually nothing has been reported on the ecology of this rare species, which is documented only from the vicinity of its type locality. **Reported associates:** none.

Use by wildlife: unknown

Economic importance: food: none; **medicinal:** none; **cultivation:** none; **misc. products:** none; **weeds:** none; **nonindigenous species:** none

Systematics: Phylogenetic analysis of plastid (and other) sequence data places *Sesuvium* within subfamily *Sesuvioideae* as the sister genus to *Trianthema* (Figure 3.2). Although comparable results have been obtained from analyses of morphological and nuclear (nrITS) sequence data, analysis of combined molecular data [*atpB–rbcL* intergenic spacer, *rps16* gene intron, *trnL–trnF* intergenic spacer, *petB–petD* intergenic spacer, internal transcribed spacer (ITS)] resolves *Zaylea* as the sister genus of *Sesuvium* (which is monophyletic when including *Cypselea*). Of the eight or so species worldwide, all have been included in phylogenetic analyses except *Sesuvium trianthemoides*, which is believed to be extinct. *S. trianthemoides* perhaps is (or was) related more closely to *S. maritimum*, with both species having only five stamens as opposed to the numerous stamens characteristic of other *Sesuvium* species. Yet, the two species differ by their seeds, which are conspicuously rugose in *S. trianthemoides* but smooth in *S. maritimum*. Further study is necessary to determine whether this species should be included in *S. maritimum*. The chromosome number of *S. trianthemoides* is unknown; however, the base chromosome number of subfamily *Sesuvioideae* (to which it belongs) is $x = 8$.

Comments: *Sesuvium trianthemoides* is narrowly endemic to the southern Gulf Coast of Texas, and it is uncertain whether any extant populations exist. The species was declared extinct in 2001 by the Center for Biological Diversity (Tucson, Arizona).

References: Bohley et al., 2015; Correll, 1966; Ferren, Jr., 2003; Hassan et al., 2005; Klak et al., 2003; Poole et al., 2007; Stutzenbaker, 1999.

Family 2: Amaranthaceae [169]

The "Amaranth" family includes about 2360 species and is related more closely to the family Archatocarpaceae and then to Caryophyllaceae (Figure 3.1; Crawley & Hilu, 2012; Judd et al., 2016). Historically, Chenopodiaceae and Amaranthaceae have been treated as distinct taxonomically; however, a variety of molecular studies have found these proposed families to be only weakly differentiated (e.g., Figure 3.3). Such results have inspired many authors to merge the families as an expanded Amaranthaceae as is done here. In the present treatment, the term "chenopods" makes reference to those taxa placed formerly in Chenopodiaceae; taxa placed traditionally in Amaranthaceae are indicated as "Amaranthaceae sensu stricto (s.s.)." The phylogenetic integrity of most subfamilies of Amaranthaceae is indicated by *rbcL* data; however, subfamily *Amaranthoideae* is paraphyletic (Figure 3.3).

The species of Amaranthaceae possess unusual concentric rings of vascular bundles, betalains, and P-type sieve element plastids. A large number of species have C_4 photosynthetic pathways. The floral perianth is inconspicuous and consists of papery or fleshy tepals. The pollen grains are seven-porate to polyporate. The fruits possess a single or few seeds and can develop as achenes, capsules, or utricles. The embryos are curved or spiral and are surrounded by perisperm (Judd et al., 2016). The family is a cosmopolitan element of arid or saline habitats and is often associated with disturbed sites. Many of the species are halophytes. The flowers are self-pollinating or are outcrossed by wind or insects (Insecta). Seed dispersal occurs primarily by water or wind, but to a minor extent by adhering to the fur of animals, or by their inadvertent ingestion (and subsequent excretion) by grazing animals. The family also includes "tumbleweeds" (*Salsola*) whereby entire

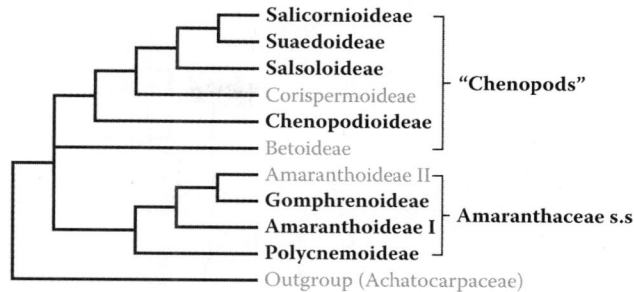

FIGURE 3.3 Phylogenetic relationships of Amaranthaceae sensu stricto (s.s.) and subfamilies assigned formerly to Chenopodiaceae as deduced from *rbcL* sequence analysis. The inability of molecular data to resolve "Chenopodiaceae" as a distinct clade (note unresolved position of subfamily *Betoideae*) has led to its merger with Amaranthaceae. Despite most Amaranthaceae being characteristic of arid, terrestrial habitats, the subfamilies containing OBL North American indicators (highlighted in bold) demonstrate that aquatic plants have evolved throughout the family. (Modified from Kadereit, G. et al., *Int. J. Plant Sci.*, 164, 959–986, 2003.)

seed-bearing plants can be dislodged and rolled across the landscape by the wind.

The family is important economically as the source of beets (*Beta vulgaris*), spinach (*Spinacia oleracea*), and grains derived from lamb's quarters or quinoa (*Chenopodium*) and pigweed (*Amaranthus*). Numerous ornamental cultivars have been derived from *Alternanthera*, *Amaranthus*, *Atriplex*, *Beta*, *Celosia*, *Chenopodium*, *Gomphrena*, *Iresine*, and *Spinacia*. Seriously weedy species include redroot pigweed (*Amaranthus retroflexus*) and alligator weed (*Alternanthera philoxeroides*).

Although *Atriplex*, *Nitrophila*, and *Suckleya* included OBL taxa in the 1996 indicator list, they no longer do and have been excluded from this treatment. The remaining OBL taxa are distributed broadly throughout Amaranthaceae (Figure 3.3). In North America, OBL indicators occur within eight largely unrelated genera:

1. ***Alternanthera*** Forssk.
2. ***Amaranthus*** L.
3. ***Arthrocnemum*** Moq.
4. ***Bassia*** All.
5. ***Chenopodium*** L.
6. ***Salicornia*** L.
7. ***Sarcocornia*** A. J. Scott
8. ***Suaeda*** Forssk. ex J. F. Gmel.

Molecular data (*rbcL* sequences) indicate that some tribal relationships and generic circumscriptions in Amaranthaceae require revision. Concerning the relationships among aquatic genera, the most problematic (i.e., not monophyletic) tribes are *Amarantheae* (subfamily Amaranthoideae) and *Chenopodieae* (subfamily Chenopodiodeae). *Amarantheae* subtribe *Aervinae* is allied more closely to tribe *Gomphreneae* (subfamily *Gomphrenoideae*) than it is to members of

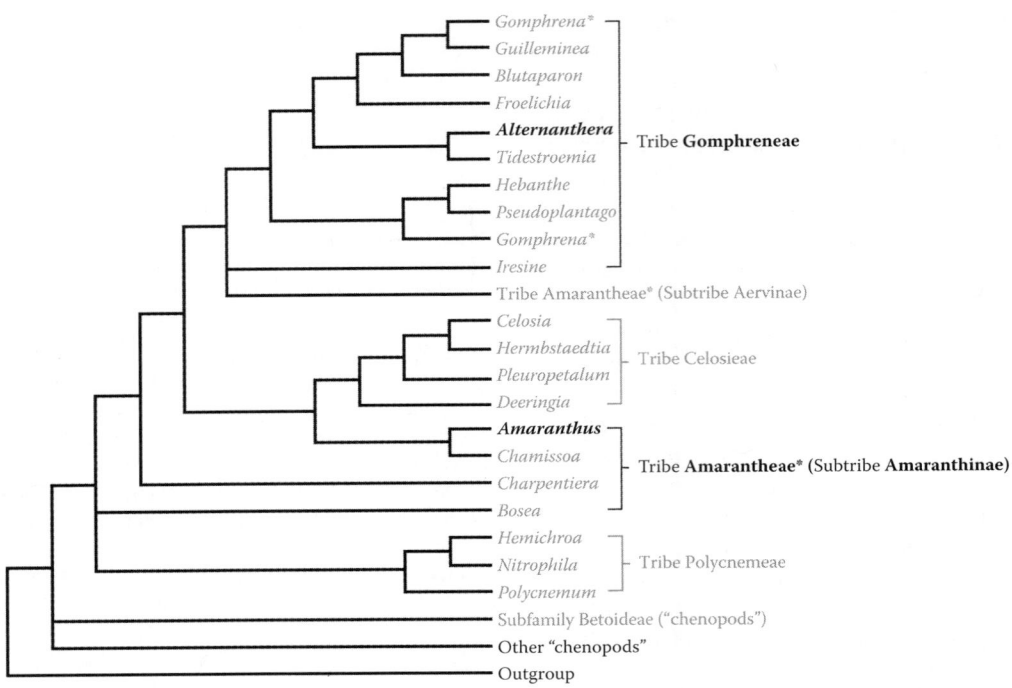

FIGURE 3.4 Intergeneric tribal relationships in Amaranthaceae sensu stricto (s.s.) based on *rbcL* sequence data. Taxa containing obligate aquatics in North America are shown in bold and polyphyletic taxa are indicated by asterisks. The wide spacing of OBL genera across the phylogeny (in different tribes) indicates independent origins of the aquatic habit in this group. (Modified from Kadereit, G. et al., *Int. J. Plant Sci.*, 164, 959–986, 2003.)

Amarantheae subtribe *Amaranthinae*. Some genera of subtribe *Amaranthinae* (*Amaranthus*, *Chamissoa*) are closer phylogenetically to tribe *Celosieae* (Figure 3.4).

Relationships among the "chenopod" genera indicate that the taxonomic circumscription of most subfamilies is also meaningful phylogenetically; that is, resolving as clades (Figure 3.5). Subfamily *Suadeoideae* is probably paraphyletic. Tribal relationships in this group (data not shown) also show relatively high phylogenetic integrity except *Chenopodieae* which is paraphyletic. With respect to the genera containing aquatic species in North America, *Bassia* and *Chenopodium* are polyphyletic (Figure 3.5) and require a thorough taxonomic reevaluation.

1. *Alternanthera*

Alligator weed, chaff flower, joyweed
Etymology: from the Latin *alternans anthera*, with reference to the alternation of staminodes and fertile stamens in the flower
Synonyms: *Achyranthes*; *Bucholzia*; *Telanthera*
Distribution: global: Africa; Asia; Australia; neotropics; **North America:** southern
Diversity: global: 80–200 species; **North America:** 9 species
Indicators (USA): OBL: *Alternanthera paronychioides*, *A. philoxeroides*, *A. sessilis*; **FACU:** *A. sessilis*
Habitat: freshwater; lacustrine, palustrine, riverine; **pH:** 4.8–8.8; **depth:** 0–2 m; **life-form(s):** emergent (herb)
Key morphology: stems (to 8 dm; to 50 cm if stoloniferous), prostrate or procumbent, glabrous to villous, often hollow, rooting at nodes; leaves (to 13 cm) opposite, sessile, blades

elliptic, lanceolate–obovate, oblanceolate, oblong or ovate, apex acute or obtuse, often tipped with a minute spine; inflorescences axillary or terminal, sessile or long stalked (to 6 cm), terminated by globose to ovoid heads (to 1.7 cm) of chaffy, inconspicuous, flowers; tepals (to 6 mm) whitish, their apex acuminate or acute; stamens 5, anthers 3–5; utricles (to 2.3 mm; sometimes absent) with lenticular seed (to 1.5 mm)
Life history: duration: perennial (rhizomes, stolons); **asexual reproduction:** shoot fragments; **pollination:** insect; **sexual condition:** hermaphroditic; **fruit:** one-seeded utricle (if present); **local dispersal:** shoot fragments (water); **long-distance dispersal:** shoot fragments
Imperilment: (1) *Alternanthera paronychioides* [G5]; *A. philoxeroides* [GNR]; *A. sessilis* [G5]
Ecology: general: *Alternanthera* includes annual and perennial species that occur in sites ranging from dry sandy conditions to wetlands. In a genus comprised mainly of terrestrial plants, three of the North American species (33%) are wetland indicators (all OBL). The plants occur often in disturbed or waste areas and include both C_3 and C_4 photosynthetic pathways. At least some of the species are self-compatible and self-pollinating; otherwise, the flowers have been observed to be pollinated by insects such as bees (Insecta: Hymenoptera). The fruits of some species possess corky cells that facilitate abiotic dispersal by enabling them to float on the surface of water.

Alternanthera paronychioides **A. St.-Hil.** is a nonindigenous perennial species that colonizes ballast sites, borrow pits, ditches, flats, roadsides, sandbars, swamps, thickets,

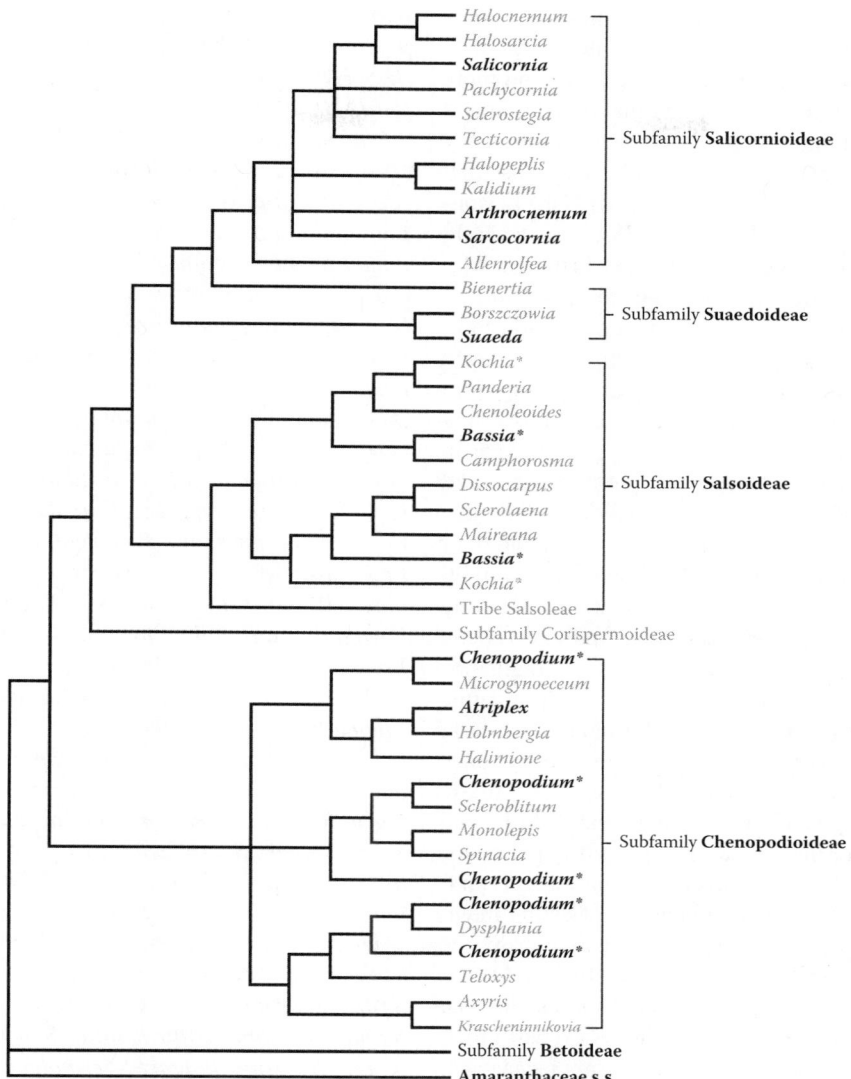

FIGURE 3.5 Relationships among "chenopod" genera of Amaranthaceae as indicated by *rbcL* sequence data. Taxa containing obligate aquatics in North America are shown in bold and polyphyletic taxa are indicated by asterisks. As in Amaranthaceae sensu stricto (s.s.), the OBL indicators are dispersed throughout the group. (Modified from Kadereit, G. et al., *Int. J. Plant Sci.*, 164, 959–986, 2003.)

woodlands, and the margins of bayous, lakes, ponds, pools, and streams at elevations of up to 10 m. Most occurrences tend to be in the proximity of saltwater. The substrates can be saline and include clay, limestone, and sand. Flowering and fruiting can persist throughout the year. **Reported associates (North America):** *Alternanthera philoxeroides, Amaranthus blitum, Ammannia latifolia, Azolla caroliniana, Bacopa monnieri, Bidens pilosus, Centella asiatica, Cephalanthus occidentalis, Colocasia esculenta, Cyperus difformis, C. polystachyos, Eclipta prostrata, Eleocharis, Emilia fosbergii, E. sonchifolia, Eragrostis hypnoides, Fimbristylis, Hibiscus moscheutos, Juncus, Leptochloa, Ludwigia decurrens, L. peruviana, Phyla nodiflora, Pluchea, Salix, Schoenoplectus, Spartina pectinata, Spilanthes, Spirodela polyrhiza, Symphyotrichum, Trifolium repens.*

***Alternanthera philoxeroides* (Mart.) Griseb.** is a perennial species that occurs in damp areas or shallow (to 1 m) standing waters of bayous, borrow pits, canals, depressions, ditches, dunes, flatwoods, lakes, levees, marshes, ponds, prairies, ricefields, roadsides, sandbars, savannahs, sink holes, sloughs, streams, swales, swamps, and the margins of canals, lagoons, rivers, and even in terrestrial sites at elevations of up to 306 m. Although the plants occur primarily where salinity is less than 0.5 ppt, they can survive short exposures to higher values; that is, up to 10% of sea strength in lentic waters and 30% in lotic waters. *Alternanthera philoxeroides* requires a minimum of 200 frost-free days for persistence. Mild frost kills the leaves, severe frosts kill the stems, but the underground portions can resist cold temperatures down to −19°C. The exposures typically receive full sunlight. The plants thrive in relatively flat areas that are protected from wave erosion and occur typically along shorelines where they form dense, floating mats over deeper water, which can reduce light penetration, crowd out native species, and promote anoxia. During low-water conditions, the plants will expand along muddy banks and into low-lying terrestrial areas. They can tolerate a wide range of water and soil conditions (pH: 4.8–8.8; mean ~7.1; alkalinity: 1.6–77 mg/L as $CaCO_3$; conductance: 43–346 µS/cm; total P: 11–125 µg/L; total N: 360–1930 µg/L; Ca: 2.5–70.5 mg/L; chloride: 8–44 mg/L; Mg: 1.2–32.4 mg/L). The substrates have been characterized as Corolla–Duckston sands, muck, mud, Newham–Corolla complex, loam, peat, sand, and sandy clay alluvium. The plants are moderately shade tolerant and fire resistant but lack long-term fire tolerance. They flourish in fertilized areas. Flowering occurs mainly from April to October (throughout the year in Florida), but mature seed production has not been observed in the introduced North American populations. Here, the plants spread vigorously by the vegetative growth of their rhizomatous/stoloniferous stems (reaching densities up to 30 plants/m²), which float and can create dense mats extending up to 70 m across the water surface. Clonal individuals can also establish from dispersal of small vegetative fragments. Despite the lack of seeds, the plants develop a propagule bank consisting of nodes, which become dormant under dark, anaerobic conditions during flooding and regenerate once flooding has subsided. Genetic studies of introduced Chinese populations found no genetic variation (using RAPD and ISSR markers) within or among seven surveyed populations, which indicated a high degree of clonal propagation in founding, nonnative populations. The maintenance of clonal integrity (i.e., connection of ramets) has been shown to improve the performance of plants, especially under stressful conditions. Terrestrial forms of *A. philoxeroides* develop solid stems that are resistant to some biological control agents. When growing with the native *Justicia americana* (Acanthaceae), *A. philoxeroides* is the superior competitor. The roots are colonized by arbuscular mycorrhizae and dark septate endophytic Fungi. **Reported associates (North America):** *Acer rubrum, Aeschynomene indica, Alternanthera sessilis, Amaranthus, Andropogon virginicus, Asclepias humistrata, Azolla caroliniana, Baccharis halimifolia, Bacopa caroliniana, B. monnieri, Bidens laevis, Boehmeria cylindrica, Boltonia asteroides, Brunnichia ovata, Cabomba caroliniana, Cakile constricta, Carex tetrastachya, C. triangularis, Cephalanthus occidentalis, Ceratophyllum demersum, Chamaecrista fasciculata, C. nictitans, Cicuta maculata, Cirsium nuttallii, Cladium jamaicense, Clethra alnifolia, Cnidoscolus urens, Coleataenia longifolia, Colocasia esculenta, Cyclospermum leptophyllum, Cynodon dactylon, Cyperus articulatus, C. haspan, C. pseudovegetus, C. virens, Dichanthelium acuminatum, Digitaria sanguinalis, Diodia teres, Dulichium arundinaceum, Echinochloa, Echinodorus cordifolius, Egeria densa, Eichhornia crassipes, Eleocharis cellulosa, E. obtusa, E. olivacea, Epidendrum magnoliae, Eragrostis hypnoides, Eryngium baldwinii, Euphorbia pinetorum, Fimbristylis autumnalis, F. littoralis, F. vahlii, Forestiera acuminata, Fraxinus caroliniana, F. pennsylvanica, F. profunda, Galium obtusum, G. tinctorium, Hibiscus moscheutos, Hordeum pusillum, Hydrilla verticillata, Hydrocotyle ranunculoides, Hymenocallis caroliniana, H. coronaria, H. crassifolia, Hypericum, Ilex vomitoria, Ipomoea, Itea virginica, Jacquemontia tamnifolia, Juncus acuminatus, J. dichotomus, J. effusus, J. repens, Justicia americana, Kalmia hirsuta, Kosteletzkya pentacarpos, Leersia lenticularis, L. oryzoides, Lemna obscura, L. trisulca, Leptochloa uninervia, Lilaeopsis chinensis, Limnobium, Lindernia dubia, Lipocarpha micrantha, Liquidambar styraciflua, Lonicera japonica, Ludwigia brevipes, L. glandulosa, L. grandiflora, L. hexapetala, L. repens, Lycopus, Lyonia lucida, Lythrum alatum, L. lineare, Marsilea, Medicago polymorpha, Micranthemum umbrosum, Mikania scandens, Mollugo verticillata, Myrica cerifera, Najas, Nelumbo lutea, Nuphar, Nuttallanthus floridanus, Nymphoides, Nyssa aquatica, N. biflora, N. sylvatica, Oenothera humifusa, Osmunda regalis, Oxalis dillenii, Panicum hemitomon, P. virgatum, Paspalum vaginatum, Peltandra virginica, Persea palustris, Persicaria hydropiperoides, P. punctata, P. setacea, Phanopyrum gymnocarpon, Phragmites australis, Phyla nodiflora, Physalis angustifolia, Physostegia correllii, P. leptophylla, Pinus elliottii, P. taeda, Pistia stratiotes, Planera aquatica, Plantago virginica, Pleopeltis polypodioides, Pluchea, Podostemum ceratophyllum, Polypremum procumbens, Pontederia cordata, Portulaca oleracea, Proserpinaca palustris, Quercus geminata, Q. nigra, Rhynchospora corniculata, Rorippa, Rotala ramosior, Sabal minor, Sacciolepis striata, Sagittaria lancifolia, S. latifolia, S. platyphylla, Salix caroliniana, S. nigra, Salvinia minima, S. molesta, Saururus cernuus, Schoenoplectus americanus, S. californicus, S. pungens, S. tabernaemontani, Scirpus cyperinus, Senna obtusifolia, Serenoa repens, Sesbania drummondii, S. herbacea, S. punicea, S. vesicaria, Sium suave, Smilax auriculata, S. walteri, Solidago sempervirens, Spartina patens, Spermolepis echinata, Sphenoclea zeylanica, Sphenopholis obtusata, Spirodela, Strophostyles helvola, Symphyotrichum lanceolatum, Taxodium distichum, Thelypteris palustris, Tillandsia usneoides, Triadica sebifera, Triadenum walteri, Typha domingensis, Vaccinium, Verbena brasiliensis, Vigna, Woodwardia areolata, Xanthium strumarium, Xyris, Zizaniopsis miliacea.*

***Alternanthera sessilis* (L.) R. Br. ex DC.** is an annual or perennial species that inhabits moist ground or shallow waters of backwaters, depressions, ditches, floodplains, hammocks, lawns, meadows, marshes, meadows, ponds, prairies, roadsides, sandbars, swamps, woodlands, and the margins of lakes, ponds, rivers, and streams at elevations of up to 122 m. Occurrences typically are associated with disturbed sites having exposures ranging from full sunlight to partial shade. The substrates have been described as alluvial sand, boulders, Grady series, Hallandale (lithic psammaquents), karst, loamy sand, marl, muck, mud, peaty sand, sand, sandy peat, and St. Johns (typic haplaquods). Loamy, alkaline, high-nitrogen soils are preferred in the native range of the species. Flowering and fruiting extend from April to December. Although the flowers are self-pollinating, the pollen is also gathered by honeybees (Insecta: Hymenoptera: Apidae: *Apis*), which likely facilitate some degree of outcrossing. A single plant produces on average 1997 light-sensitive seeds. The fruits are dispersed abiotically, either by wind or by water. **Reported associates (North America):** *Alternanthera philoxeroides, Amaranthus viridis, Anthaenantia rufa, Baccharis halimifolia, Bacopa caroliniana, Boltonia, Carex verrucosa, Centella asiatica, Clethra alnifolia, Cliftonia monophylla, Coleataenia longifolia, Commelina diffusa, Cuphea carthagenensis, Cyperus entrerianus, Cyrilla racemiflora, Dichanthelium dichotomum, Digitaria sanguinalis, Diodia virginiana, Echinochloa colona, Eclipta prostrata, Eleocharis, Eragrostis japonica, Euploca procumbens, Fimbristylis vahlii, Helianthus angustifolius, Hydrocotyle umbellata, Ilex cassine, I. coriacea, I. glabra, I. myrtifolia, Juncus, Lagerstroemia indica, Leucospora multifida, Lobelia glandulosa, Ludwigia leptocarpa, L. pilosa, L. linearis, Mitreola petiolata, Mollugo verticillata, Nyssa biflora, Oldenlandia corymbosa, Oxalis corniculata, Panicum verrucosum, Paspalum urvillei, Persea palustris, Persicaria hydropiperoides, Phyllanthus urinaria, Portulaca oleracea, Proserpinaca pectinata, Ranunculus, Rhexia mariana, Rorippa palustris, Rotala ramosior, Rubus cuneifolius, Sacciolepis striata, Sagittaria, Scutellaria racemosa, Sida rhombifolia, S. spinosa, Smilax laurifolia, Stillingia aquatica, Taxodium ascendens, Urochloa platyphylla, Woodwardia virginica, Xanthium strumarium.*

Use by wildlife: Plants of *A. philoxeroides* are highly palatable, have a high protein content, and are eaten readily by deer (Mammalia: Cervidae: *Odocoileus*), livestock (Mammalia: Bovidae), and manatees (Mammalia: Sirenia: Trichechidae: *Trichechus manatus*). Herbivorous fish (Osteichthyes: Cyprinidae: *Ctenopharyngodon idella*; Cichlidae: *Tilapia*) will eat the foliage but not preferentially. The densely tangled root system of *A. philoxeroides* provides habitat and cover for frogs (Amphibia: Ranidae), various fish species (Vertebrata: Osteichthyes), crayfish (Crustacea: Decapoda: Cambaridae), and other invertebrates including mosquitos (Insecta: Diptera: Culicidae: *Anopheles*). Various wading birds (Aves: Gruiformes: Rallidae), such as coots and gallinules, forage among the plant colonies. Ducks (Aves: Anatidae) use the dessicated plant mats as resting areas. Various insects (Insecta) have been introduced as effective biological controls of *A. philoxeroides* in the southeastern United States including flea beetles (Coleoptera: Chrysomelidae: *Agasicles hygrophila*), stem borer moths (Lepidoptera: Pyralidae: *Arcola malloi*), and alligatorweed thrips (Thysanoptera: Phlaeothripidae: *Amynothrips andersoni*); however, the former is ineffective in terrestrial stands. *Alternanthera philoxeroides* is also susceptible to a stunting virus (*Closterovirus*). The foliage hosts false celery leaftier larvae (Insecta: Lepidoptera: Pyralidae: *Udea profundalis*) and whitefly (Insecta: Hemiptera: Aleyrodidae: *Trialeurodes*).

Economic importance: food: The leaves of *A. paronychioides* are eaten as a vegetable in India. The foliage of *A. philoxeroides* is edible and is consumed as a leaf vegetable in India and other parts of Asia but not to any extent in North America. *A. sessilis* is high in protein (3.6 g/100 g) and also is eaten as a green leafy vegetable in Asia. The seed oil contains moderate levels of fatty acids including linoleic acid (25.2%), myristic acid (3.9%), oleic acid (26.0%), palmitic acid (16.9%), ricinoleic acid (22.1%), and stearic acid (5.9%); **medicinal:** Ethanolic extracts derived from *A. paronychioides* have antioxidant and other properties that protect pancreatic β cells from glucotoxicity. *Alternanthera philoxeroides* has been used medicinally in China and other regions. The plants contain various phenolics (chlorogenic acid, kaempferol, ferulic acid, salicylic acid, syringic acid), which have antimicrobial activity, antioxidant properties, and inhibit α-glucosidase activity. The extracts also exhibit antiviral properties and are effective in reducing HIV infection in laboratory cell cultures and also in treating epidemic hemorrhagic fever virus in mice. *Alternanthera sessilis* is widely used as a medicinal plant in Asia. Leaf extracts from *A. sessilis* have antimicrobial activity and facilitate the healing of wounds. Methanolic extracts have antioxidant and free-radical scavenging properties, whereas petroleum ether extracts exhibit anti-inflammatory activity. Aerial parts of the plants have been found to be analgesic and antihyperglycemic. The plants also induce hepatoprotective effects; **cultivation:** none; **misc. products:** In South America, *A. paronychioides* is administered to dogs to improve their hunting ability. The plants are used in China as a natural dye for wool. *Alternanthera philoxeroides* has been used to process wastewater in India and New Zealand but not in North America; **weeds:** All three of the OBL North American *Alternanthera* species are regarded as weedy. In particular, *A. philoxeroides* is regarded as a seriously nuisance aquatic weed throughout southern North America (notably in the southeast region) and in California, where it interferes with navigation and impedes water flow. The plants also provide breeding habitat for malaria-carrying mosquitos (Insecta: Diptera: Culicidae); **nonindigenous species:** *Alternanthera paronychioides* was first discovered as an introduction to Florida (Key West) in 1891–1892. *Alternanthera philoxeroides* is indigenous to South America and has been introduced to North America and all other continents except for Africa and Europe. It was introduced to North America around 1935, presumably as a result of ballast disposed from ships originating in South America. Similar introductions of this species via ballast discharge have been reported for New Zealand in 1906 and Australia in the 1940s, where it may also have escaped as a water-garden ornamental. *Alternanthera sessilis* was introduced to North America sometime before 1985, when it was reported from the Florida panhandle.

Systematics: *Alternanthera* is placed within Amaranthaceae subfamily *Gomphrenoideae* (tribe *Gomphreneae*), which molecular data (*rbcL* sequences) resolve as a clade with reasonable phylogenetic integrity (Figure 3.4). The same results depict *Alternanthera* and *Tidestromia* as closely related genera; however, subsequent studies (*trnL-F, rpl16*, nrITS sequences for 33 species) indicate *Pedersenia* to be the sister genus of *Alternanthera* (Figure 3.6). The study also resolved *A. chacoënsis* and *A. paronychioides* as sister species and *A. obovata* and *A. philoxeroides* as sister species (*A. sessilis* was not included). The *A. paronychioides* and *A. philoxeroides* clades are not closely related, which indicates several origins of the OBL North American taxa (Figure 3.6). Because a more

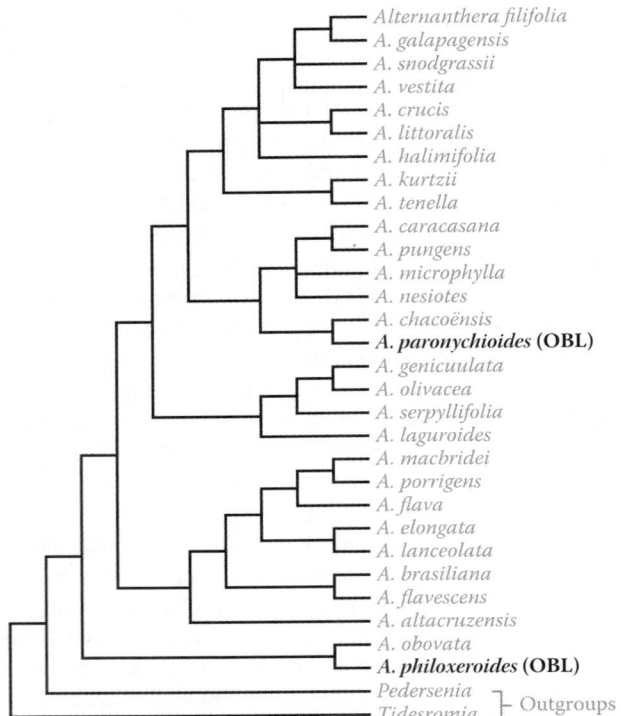

FIGURE 3.6 Phylogenetic relationships among *Alternanthera* species as reconstructed using nrITS sequence data. North American OBL indicators are shown in bold. (Adapted from Sánchez-Del Pino, I. et al., *Bot. J. Linn. Soc.* 169, 493–517, 2012.)

comprehensive phylogeny for the large genus *Alternanthera* has not yet been produced, these relationships should be regarded as tentative. Introduced, founding populations of *A. philoxeroides* often exhibit extremely limited genetic variation. It has not been determined why seeds do not mature in North American populations of *A. philoxeroides*. However, flow cytometric data indicate that US populations are either pentaploid or aneuploid, which could account, at least in part, for their sterility. A sterile polyploid origin of the species through hybridization of fertile diploids has been proposed as an explanation for the observed sterility that occurs even throughout its native range. The base chromosome number of *Alternanthera* presumably is *x* = 16. Unusually variable counts have been reported for *A. philoxeroides* (2*n* = 28, 34, 40, 96, 100), *A. paronychioides* (2*n* = 64, 100), and *A. sessilis* (2*n* = 28, 32, 34, 36, 40, 66, 100), and it is possible that at least some are due to misidentifications and/or erroneous counts. Chinese populations of *A. philoxeroides* have been determined as hexaploids (2*n* = 96). There are no reports of hybrids involving the OBL species.

Comments: *Alternanthera paronychioides* was introduced to Florida and now occurs throughout the warmer portions of the eastern and southern United States. Its range extends into Mexico and South America. *Alternanthera philoxeroides* was introduced to the Tennessee valley region and since has spread throughout the southeastern United States, to Califonia, and northward to Illinois. *Alternanthera sessilis* occurs in the southeastern United States and extends into South America, from where it was introduced.

References: Bennett & Alarcón, 2015; Bhattacherjee et al., 2014; Bhusari et al., 2005; Borah et al., 2011; Brown & Marcus, 1998; Chen et al., 1989, 2015; Clemants, 2003; Cofrancesco, 1984; Datta & Biswas, 1979; Gnanaraj et al., 2011; Guo & Hu, 2012; Holzinger, 1893; Hosamani et al., 2004; Hossain et al., 2014; Hoyer et al., 1996; Iamonico & Sánchez-Del Pino, 2016; Jalalpure et al., 2008; Kadereit et al., 2003; Kühn, 1993; Lin et al., 1994; Penfound, 1940a; Peng et al., 1997; Sánchez-Del Pino et al., 2012; Schooler, 2012; Sheela et al., 2004; Stephen et al., 2015; Stutzenbaker, 1999; Subhashini et al., 2010; Tarver et al., 1978; Townsend, 1993; Wang et al., 2005; Wu et al., 2013; Ye et al., 2003; Yu, 2015.

2. *Amaranthus*

Pigweed, water hemp; acnide, amarante

Etymology: from the Greek *amarantos* meaning unfading

Synonyms: *Acanthochiton*; *Acnida*; *Albersia*; *Amblogyna*; *Euxolus*; *Mengea*; *Sarratia*; *Scleropus*

Distribution: global: cosmopolitan; **North America:** widespread

Diversity: global: 70 species; **North America:** 40 species

Indicators (USA): OBL: *Amaranthus australis*, *A. cannabinus*, *A. floridanus*, *A. tuberculatus*; **FACW:** *A. tuberculatus*; **FAC:** *A. tuberculatus*

Habitat: freshwater, saline (coastal); palustrine, riverine; **pH:** 4.5–8.0; **depth:** <1 m; **life-form(s):** emergent (herb)

Key morphology: coarse annual, hemp-like herbs (1–9 m) with ascending branches, often tinged with red; leaves (to 30 cm) alternate, lanceolate, simple, entire, petiolate (to 20 cm); inflorescences simple to paniculate spikes, leafy bracted; flowers inconspicuous (perianth to 3 mm), unisexual; fruit a utricle (to 4 mm) containing a single shiny, reddish brown seed (to 3 mm)

Life history: duration: annual (fruit/seeds); **asexual reproduction:** none; **pollination:** wind; **sexual condition:** dioecious; **fruit:** one-seeded utricles (prolific); **local dispersal:** seeds (wind); **long-distance dispersal:** seeds (soil disturbance)

Imperilment: (1) *Amaranthus cannabinus* [G5]; S3 (PA); (2) *A. floridanus* [G3]; S3 (FL); (3) *A. tuberculatus* [G4]; SH (NE); S2 (VT); S3 (IL)

Ecology: general: *Amaranthus* species generally inhabit terrestrial sites that are often disturbed and contain high levels of nitrogen and other nutrients. A large number of species colonize sandy sites such as beaches, dunes, and semidesert areas. The group is well represented among the common agricultural weeds as well. About half of the North American species (53%) have some status as wetland indicators, but only four (10%) are designated as OBL. The flowers are unisexual and occur in dioecious and monoecious arrangements. The flowers are self-compatible, which allows for selfing (via geitonogamy) in the monoecious plants. Otherwise, the plants primarily are wind pollinated. Their reproductive output is high, with individuals often capable of producing over 100,000 seeds. The seeds are dispersed locally by gravity, wind, or water; however, they are also eaten in large quantities by various birds (Aves), which are known to excrete viable seeds (sometimes in fairly large quantities) that are capable of establishment following their endozoic transport over much longer distances. Viable seeds have been recovered as well from mammalian (Mammalia) feces, including that of deer (Cervidae: *Odocoileus virginianus*) and wild boars (Suidae: *Sus scrofa*), which represent other potential dispersal vectors. There is also some evidence to suggest that the seeds are transported along roadsides by mud that adheres to automobiles. All of the species fix carbon via a C_4 photosynthetic pathway, and nearly all are annuals.

Amaranthus australis **(A. Gray) J.D. Sauer** Sauer is an annual species that occupies fresh to brackish (oligohaline) coastal sites in backwaters, bayous, beaches, canals, ditches, estuaries, flats, glades, hammocks, tidal marshes, lakeshores, levees, marshes, roadsides, salt marshes, sinks, spoil banks, swamps, and along the margins of pools, rivers, and streams at elevations of up to 100 m. The plants can attain giant proportions (stems to 9 m high, 30 cm in diameter). They will grow in shallow standing water but tend to occur on slightly elevated sites such as levees and abandoned muskrat (Mammalia: Cricetidae: *Ondatra zibethicus*) beds. Exposures of full sunlight are most common. The substrates have been described as dredge spoil, Duckston sand, gravel over oolite, Handsboro association, Harleston fine sandy loam, muck, organic, sand, and shells. The plants are dioecious and wind pollinated. Flowering and fruiting extend from March to November. The seeds are likely dispersed primarily by water. An extensive seed bank can develop (e.g., 19,000–43,000 seeds/m²; up to 38% of total seed bank), with up to 282 seedlings/m² emerging in some germination studies. Short drawdown intervals are required to induce seed germination, which will not occur under flooded conditions or when salinity exceeds 2 ppt. This is often one of the first species to appear on spoil deposits from newly dredged sites. The roots are colonized by dark septate endophytic Fungi. **Reported associates:** *Acrostichum danaefolium, Alternanthera philoxeroides, Aralia spinosa, Atriplex, Avicennia germinans, Baccharis halimifolia, Bacopa monnieri, Batis maritima, Bidens pilosus, Bolboschoenus robustus, Borrichia frutescens, Callicarpa americana, Campsis radicans, Carex longii, Centella asiatica, Chenopodium, Colocasia esculenta, Conocarpus erectus, Crinum, Cyperus iria, C. odoratus, Dactyloctenium aegyptium, Diodia virginiana, Distichlis spicata, Echinochloa walteri, Eichhornia crassipes, Eleocharis parvula, Fimbristylis, Heliotropium curassavicum, Hydrilla verticillata, Hydrocotyle bonariensis, Hygrophila polysperma, Hypericum hypericoides, Ilex vomitoria, Iva frutescens, Juncus roemerianus, Laguncularia racemosa, Lemna, Leptochloa, Liquidambar styraciflua, Ludwigia peruviana, Magnolia virginiana, Mikania scandens, Moorochloa eruciformis, Myrica cerifera, Oenothera humifusa, Opuntia humifusa, Osmundastrum cinnamomeum, Panicum dichotomiflorum, P. virgatum, Parthenocissus quinquefolia, Paspalum repens, P. virgatum, Persicaria pensylvanica, Pinus elliottii, Pistia stratiotes, Pluchea camphorata, Pontederia cordata, Pteridium aquilinum, Quercus nigra, Q. virginiana, Rubus trivialis, Rumex crispus, R. obovatus, Sacciolepis striata, Sagittaria lancifolia, S. latifolia, Salvinia minima, Sambucus nigra, Sarcostemma clausum, Schoenoplectus americanus, Senecio glabellus, Serenoa repens, Sesbania drumondi, Sesuvium portulacastrum, Setaria magna, S. parviflora, Solidago sempervirens, Spartina alterniflora, S. patens, Sphagneticola trilobata, Suaeda linearis, Symphyotrichum tenuifolium, Taxodium, Tradescantia hirsutiflora, Triadica sebifera, Typha domingensis, Vigna luteola, Vitis rotundifolia.*

Amaranthus cannabinus **(L.) J.D. Sauer** is a relatively long-lived annual species that inhabits brackish or saline (rarely freshwater) beaches, flats (tidal), marshes (tidal), salt marshes, swamps, and the margins of sloughs and tidal rivers and streams at elevations of up to 50 m. The substrates include mud and sand. Flowering and fruiting occur from June to October. The plants are dioecious and wind pollinated. The female plants allocate more vegetative resources and experience a longer growing period than the males. Fecundity is high with individuals producing large numbers of seeds. In one study, inbreeding depression expressed as reduced percent germination, leaf size, and shoot height, occurred after two generations of inbreeding. The level of inbreeding depression is low overall and varies widely in extent within and among different populations, being more pronounced in salt-marsh populations than in freshwater marshes. The seeds float and are dispersed primarily by water; large numbers have been retrieved from floating and drift-line samples. An extensive seed bank can develop (the size increasing proportionally with elevation) with up to 3200 seeds/m² germinating from samples in some studies. However, the extent of natural germination can be quite low. Although the seeds can tolerate some degree of inundation and hypoxia, their germination is highest near the substrate surface (to 1 cm depth) and generally declines with increasing water depth. The seeds (averaging 1.6 mg) have physiological dormancy and following 180 days of cold stratification will germinate when placed under a 25°C/15°C temperature regime. Exposure to light does not increase their germinability. **Reported associates:** *Acorus calamus, Aeschynomene virginica, Alternanthera philoxeroides, Atriplex prostrata, Bidens bidentoides, B. eatonii, B. laevis, Bolboschoenus fluviatilis, B. novae-angliae, B. robustus, Brassica nigra, Cakile edentula, Chenopodium simplex, Cladium jamaicense, Colocasia esculenta, Crinium americanum, Cyperus tetragonus, Distichlis spicata, Echinochloa walteri, Eichhornia crassipes, Eleocharis parvula, Fimbristylis spadicea, Heteranthera dubia, Hibiscus moscheutos, Hydrocotyle, Impatiens capensis, Juncus gerardii, J. roemerianus, Justicia lanceolata, Leersia oryzoides, Lilaeopsis chinensis, Limonium carolinianum, Ludwigia leptocarpa, L. peploides, Lythrum salicaria, Mikania scandens, Nelumbo lutea, Nuphar variegata, Paspalum, Peltandra virginica, Persicaria arifolia, P. punctata, Phragmites australis, Pluchea odorata, Polypogon monspeliensis, Pontederia cordata, Ranunculus cymbalaria, Rumex verticillatus, Sagittaria calycina, S. falcata, S. latifolia, S. platyphylla, S. subulata, Salicornia, Samolus valerandi, Schoenoplectus americanus, S. pungens, S. smithii, S. tabernaemontani, Sium suave,*

Spartina cynosuroides, S. patens, S. pectinata, Sporobolus virginicus, Symphyotrichum subulatum, S. tenuifolium, Triglochin maritimum, T. striata, Typha angustifolia, T. latifolia, Vigna luteola, Zizania aquatica, Zizaniopsis miliacea.

***Amaranthus floridanus* (S. Watson) J.D. Sauer** grows in coastal dunes, beaches, flats, gardens, hammocks, marshes, meadows (disturbed), savannahs, and swamps at elevations of up to 10 m. Reported substrates include sand. Flowering and fruiting occur from February to October. This species is poorly known ecologically and deserves further study. **Reported associates:** *Agave, Juniperus, Sabal, Yucca.*

***Amaranthus tuberculatus* (Moq.) J.D. Sauer** is an annual species that inhabits beaches, bottoms, depressions, ditches, floodplains, gravel bars, levees, marshes, meadows, mudflats, oxbows, prairies, roadsides, sand bars, seeps, and the exposed margins of channels, lakes, ponds, rivers, sloughs, and streams at elevations of up to 1213 m. Generally, this species occurs on more mesic sites than its other OBL congeners, as is indicated by its mixed indicator status [OBL, FACW, facultative (FAC)]. However, the plants readily tolerate temporary flooding and anaerobic conditions. They are moderately $CaCO_3$ tolerant but lack salinity tolerance. The plants occur across a wide range of substrates (pH: 4.5–8.0), most often in sites (frequently disturbed) that are fairly well drained and rich in nitrogen and other nutrients. The greatest biomass is produced when the plants are grown under full sunlight exposures. The reported substrates include alluvium, ballast (railroad), clay, gravel, mud, sand, sandy gravel, sandy cinders, silt, and silty loam. Flowering and fruiting occur from May to November. These are FAC short-day plants, which initiate flowering early under short-day conditions, but grow larger, flower later, and produce more seeds when kept under long-day conditions. The plants are dioecious and wind pollinated. Sex ratios have been observed to vary between 1:1 and 2:1 female:male plants. The stigmas will remain receptive for a prolonged period, normally until fertilization is achieved. The pollen remains viable for up to 120 h, with most of the grains fertilizing plants within 50 m of their source; however, longer-distance pollen dispersal (up to 800 m) is also possible. Individual female plants can produce from 300,000 to 1,200,000 seeds under ideal conditions; however, delayed emergence can reduce seed production substantially. Both the fruits and the seeds float and are dispersed by water and to a lesser degree by the wind. Biotic dispersal vectors include various birds (Aves) and other animals along with human-mediated transport, particularly by agricultural equipment and practices (e.g., manure spreading). The seeds are long lived and retain their viability after more than 17 years of burial, resulting in an extensive and persistent seed bank, which can reach densities of more than 64,000 seeds/m². Seed germination is regulated by phytochrome and is enhanced after cold stratification at 4°C, exposure to red light, and incubation temperatures of 36°C. Minimum germination temperatures range from 10°C to 20°C. Alternation of day/night temperatures by 18°C yields optimal germination rates. Seedling emergence occurs over several months and can reach densities of 300–360 seedlings/m². Seedling emergence initiates in May to June and continues through

August. Once established, the seedlings can grow 50%–70% faster than other weedy annuals occurring in the same habitats. Localized populations (e.g., Illinois, Iowa, Missouri) have shown relatively high levels of resistance to ALS inhibiting, protoporphyrinogen oxidase inhibiting, and triazine herbicides. Genetic (microsatellite) markers indicate that North American plants are characterized by two genetic lineages, with the eastward migration of the western lineage facilitating the invasion of agricultural sites. **Reported associates:** *Abutilon theophrasti, Acalypha rhomboidea, Acer saccharinum, Actaea racemosa, Agrostis gigantea, Alliaria petiolata, Amaranthus hybridus, A. powellii, A. retroflexus, A. rudis, Ambrosia annua, A. artemisiifolia, A. trifida, Ammannia coccinea, A. robusta, Amorpha fruiticosa, Andropogon gerardii, Apocynum cannabinum, Aralia spinosa, Arenaria serpyllifolia, Asclepias incarnata, A. syriaca, Avena sativa, Bidens cernuus, B. connatus, B. frondosus, B. tripartitus, B. vulgatus, Bolboschoenus fluviatilis, Boltonia decurrens, Carex, Centaurea solstitialis, Cephalanthus occidentalis, Cerastium, Chenopodium album, Conyza canadensis, Cyperus aristatus, C. esculentus, C. niger, C. odoratus, C. squarrosus, C. strigosus, Dichanthelium depauperatum, Echinochloa crus-galli, E. muricata, E. walteri, Eclipta prostrata, Eleocharis obtusa, E. ovata, Eragrostis frankii, E. hypnoides, Erigeron annuus, Eupatorium serotinum, Euphorbia maculata, Euthamia graminifolia, Fallopia scandens, Fraxinus pennsylvanica, Grindelia, Helenium autumnale, Helianthus annuus, Hibiscus laevis, H. trionum, Iris virginica, Laportea canadensis, Leersia oryzoides, L. virginica, Leptochloa fusca, Lespedeza hirta, Leucospora multifida, Lindernia dubia, Lipocarpha micrantha, Ludwigia peploides, Lysimachia ciliata, L. nummularia, Melilotus, Mimulus ringens, Mollugo verticillata, Muhlenbergia frondosa, Onoclea sensibilis, Panax quinquefolius, Panicum capillare, P. dichotomiflorum, P. virgatum, Peltandra virginica, Pennisetum glaucum, Persicaria amphibia, P. coccinea, P. hydropiperoides, P. lapathifolia, P. maculosa, P. pensylvanica, P. punctata, Phalaris arundinacea, Phyla lanceolata, Plantago cordata, Platanus occidentalis, Populus deltoides, Potamogeton crispus, Rorippa palustris, R. sessiliflora, Rudbeckia laciniata, Rumex crispus, Sagittaria cuneata, S. latifolia, Salix amygdaloides, S. exigua, Salsola kali, Schoenoplectus acutus, S. pungens, Setaria faberi, S. viridis, Sida spinosa, Solanum carolinense, S. ptychanthum, Sorghastrum nutans, Sparganium eurycarpum, Spartina pectinata, Spermacoce glabra, Symphyotrichum lanceolatum, S. pilosum, Tamarix, Taxodium distichum, Typha angustifolia, Ulmus rubra, Verbena hastata, Vernonia fasciculata, Viburnum, Vitis riparia, Xanthium strumarium.*
Use by wildlife: Some *Amaranthus* species may be toxic to livestock. The seeds of most *Amaranthus* species are eaten by a variety of birds (Aves), insects (Insecta), and mammals (Mammalia). The wildlife food value of *A. australis* is slight, but its hollow stems provide overwintering habitat for some insects. The seeds of *A. cannabinus* are a preferred food of some waterfowl (Aves: Anatidae), especially black ducks (*Anas rubripes*). They are also consumed by songbirds (Aves: Passeriformes), small mammals, and deer (Mammalia:

Cervidae: *Odocoileus*). The seeds of *A. tuberculatus* are eaten by several insects including beetles (Coleoptera: Carabidae: *Amara aeneopolita*, *Anisodactylus rusticus*, *Harpalus pennsylvanicus*, *Stenolophus comma*) and crickets (Gryllidae: *Gryllus pennsylvanicus*). The plants also host several Fungi (Ascomycota: Botryosphaeriaceae: *Phoma macrostoma*; Mycosphaerellaceae: *Cercospora acnidae*; Sclerotiniaceae: *Phymatotrichum omnivorum*; Oomycota: Albuginaceae: *Albugo bliti*). Some Fungi (Ascomycota: Diaporthaceae: *Phomopsis amaranthicola*; Melanommataceae: *Aposphaeria amaranti*; Montagnulaceae: *Microsphaeropsis amaranthi*) are particularly pathenogenic to *A. tuberculatus* and are under investigation as biological control agents.

Economic importance: food: There are no specific accounts of any OBL *Amaranthus* species being used as food. However, many *Amaranthus* species are considered to be edible and contain vitamin A, vitamin C, calcium, and protein. Native North Americans (mainly in the southwest) used young *Amaranthus* plants (their taste is similar to spinach) as pot herbs or gathered them as greens. The seeds of various species are used in making bread flour or can be popped like popcorn; **medicinal:** The pollen of *A. tuberculatus* is strongly allergenic; **cultivation:** The aquatic species of *Amaranthus* are not found in cultivation probably because of their typically rank, unattractive habit; **misc. products:** *Amaranthus australis* has been regarded as a promising plant for biofuel production. *A. cannabinus* yields a fiber suitable for weaving but of inferior strength to many other materials; **weeds:** *Amaranthus tuberculatus* can spread to low fields, gardens, and roadsides in proximity to natural populations. Within the past decade, it has become a noxious weed of agricultural fields in the midwestern United States, exacerbated by the migration eastward of a western genotype and emergence of herbicide (e.g., glyphosate) resistant strains; **nonindigenous species:** *Amaranthus tuberculatus* has been introduced inadvertently to the Ukraine. This species has also spread to many urban areas of North America, which probably were not originally within its native distributional range.

Systematics: *Amaranthus* and *Chamissoa* (subfamily *Amaranthoideae*; tribe *Amarantheae*) are closely related. In turn, these genera associate phylogenetically as a sister group to tribe *Celosieae* (Figure 3.4). Some taxonomists have segregated the dioecious members of *Amaranthus* as the genus *Acnida* ("water hemps"); however, several studies indicate that the dioecious species are not monophyletic (e.g., Figure 3.7). Yet, substantial discrepancies exist among nearly every phylogenetic analysis available for the genus, making it difficult to ascertain interspecific relationships with a high degree of confidence. Fairly recent analyses, which evaluate concatenated nuclear data, depict much closer relationships among many of the dioecious taxa, which resolve together as a clade, albeit with inclusion of at least one monoecious taxon (*A. pumilus*). Analyses of cpDNA data produce similar results but with the dioecious species dispersed in two different clades. In any case, because the dioecious species are consistently embedded within several monoecious clades, the abandonment of generic status for *Acnida* seems justified.

Isozyme data indicate that *A. rudis* should be merged with *A. tuberculatus* and also suggest a potentially close phylogenetic relationship between *A. australis* and *A. lividus* and between *A. floridanus*, *A. crassipes*, and *A. fimbriatus*. In contrast, RAPD data denote a relationship of *A. australis* with *A. albus* and of *A. floridanus* with *A. fimbriatus*. RFLP data (Figure 3.7) similarly indicate a close relationship of *A. australis* with *A. albus* (also to *A. cannabinus* but not *A. lividus*) and indicate that *A. floridanus* is related to *A. deflexus* and *A. viridus* but is distant phylogenetically from either *A. crassipes* or *A. fimbriatus* (Figure 3.7). However, the DNA sequence data resolve quite different relationships among the species, resolving *A. australis* and *A. cannabinus* consistently as sister species, and placing *A. crassipes*, *A. deflexus*, and *A. viridis* in clades remote from any of the OBL taxa (by both nuclear and cpDNA data). Sequence data also resolve the placement of *A. albus* and *A. fimbriatus* inconsistently, either as fairly close to *A. australis* and *A. cannabinus* (cpDNA) or remote from them (nuclear DNA). A set of 14 genetic (SSR) markers have been developed, which should facilitate additional genetic studies of *Amaranthus* species. The SSR data place *A. australis* closest to *A. cannabinus*, in agreement with results from the analysis of nuclear DNA sequence data. RFLP data resolve *A. tuberculatus* in a clade with *A. wrightii* (in conflict with nuclear DNA sequence data), which, in turn, associates phylogenetically with *A. australis* and *A. cannabinus* (Figure 3.7),

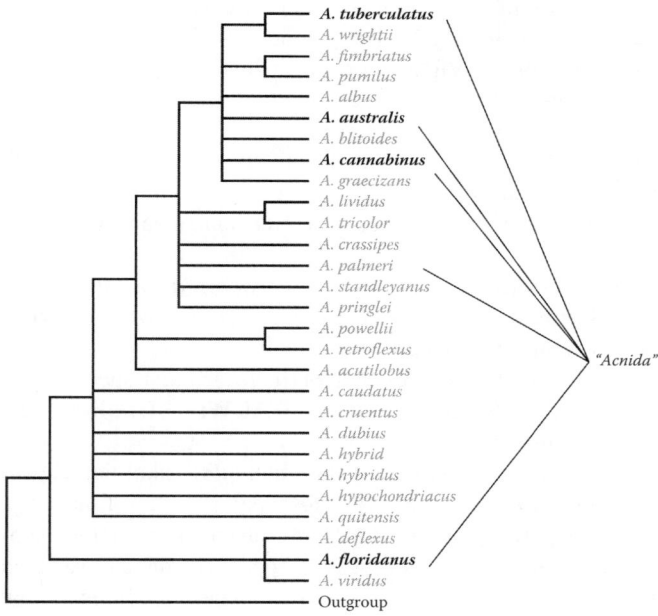

FIGURE 3.7 Phylogenetic relationships among *Amaranthus* species resolved from RFLP analysis of cpDNA and nrDNA regions. Results shown are based on a 50% majority-rule consensus tree. The OBL North American species are shown in bold. The polyphyletic distribution of dioecious taxa (formerly recognized as the genus *Acnida*) indicates that the "water hemps" are not a natural group. However, studies using nuclear DNA sequence data (Waselkov, 2013) resolve all of these dioecious species within a single clade along with the monoecious *A. pumilus*. (Adapted from Lanoue, K. Z. et al., *Theor. Appl. Genet.*, 93, 722–732, 1996.)

species which allegedly hybridize. It is evident that further investigations, which account for the influences of hybridization, etc., will be needed to better clarify the interspecific relationships in this complex genus. Differences in DNA content have proved to be useful for distinguishing natural hybrids between *A. tuberculatus* and *A. hybridus*. *Amaranthus tuberculatus* also allegedly hybridizes with *A. palmeri*. The base chromosome number of *Amaranthus* is $x = 16, 17$. The chromosome number of *A. tuberculatus* (var. *rudis*) is $2n = 34$, indicating it to be a diploid.

Comments: The OBL *Amaranthus* species are distributed primarily in eastern North America, with *A. australis* in the southeast, *A. cannabinus* mainly coastal, and *A. tuberculatus* generally throughout the region. *Amaranthus floridanus* is narrowly endemic to Florida.

References: Baldwin et al., 1996, 2010; Bram, 1998, 2002; Bram & Quinn, 2000; Buhler & Hartzler, 2001; Burnside et al., 1996; Chan & Sun, 1997; Clifford, 1959; Costea et al., 2005a; DeVlaming & Proctor, 1968; Dovrat et al., 2012; Dressler et al., 1987; Elsey-Quirk & Leck, 2015; Hartzler et al., 2004; He & Park, 2013; Hopfensperger & Baldwin, 2009; Hopfensperger & Engelhardt, 2008; Horak & Loughin, 2000; Janzen, 1984; Jeschke et al., 2003; Kadereit et al., 2003; Kandalepas et al., 2010; Khaing et al., 2013; Lanoue et al., 1996; Leck, 1996; Leon & Owen, 2003; Liu et al., 2012; Mosyakin & Robertson, 2003; Myers et al., 2004; Orłowski & Czarnecka, 2009; Peterson & Baldwin, 2004; Pratt & Clark, 2001; Schimpf, 1977; Seibert & Pearce, 1993; Stutzenbaker, 1999; Townsend, 1993; Trucco et al., 2005; Van der Valk & Rosburg, 1997; Viglasky et al., 2009; Waselkov, 2013; Waselkov & Olsen, 2014.

3. *Arthrocnemum*

Carpgrass, glasswort

Etymology: from the Greek *arthron cneme* meaning "jointed internode"

Synonyms: *Salicornia* (in part); *Sarcocornia* (in part)

Distribution: global: Africa; Asia; Europe; North America; **North America:** western

Diversity: global: 6 species; **North America:** 1 species

Indicators (USA): OBL; FACW: *Arthrocnemum subterminale*

Habitat: saline (coastal); palustrine; **pH:** alkaline; **depth:** <1 m; **life-form(s):** emergent (subshrub)

Key morphology: small (to 30 cm), succulent, evergreen subshrubs; stems articulated; leaves opposite, connate, reduced to a minute perfoliate ring (the tips sometimes indicated by small points); flowers minute, inconspicuous; in groups of three on opposite sides of succulent spikes (to 35 mm), sterile near the tip, whose apices may proliferate with branching vegetative tissue; fruit a membranous utricle containing dark brown seeds (to 1.4 mm)

Life history: duration: perennial (rhizomes); **asexual reproduction:** rhizomes; **pollination:** self, wind; **sexual condition:** hermaphroditic; **fruit:** utricles (common); **local dispersal:** seeds (water, wind), rhizome elongation; **long-distance dispersal:** seeds (birds)

Imperilment: (1) *Arthrocnemum subterminale* [G5]

Ecology: general: *Arthrocnemum* species are similar ecologically as pioneers of saline coastal habitats in temperate to tropical climates. All are characteristic of coastal salt marshes. Their flowers are unisexual or bisexual, the latter self-compatible but marginally protandrous. They are wind pollinated, with each flower capable of producing upward of 15,000 pollen grains; however, at least some degree of self-pollination seems possible. The seeds can be dispersed abiotically by water or wind. Long-distance dispersal can occur via endozoic seed transport by waterfowl (Aves: Anatidae). The genus is attributed with having heteromorphic seeds, which can exhibit different germination strategies. A persistent seed bank is known to occur in some species, but in others, the seeds can be relatively short-lived despite achieving fairly high densities (approaching 950,000 seeds/m^2). All of the species fix carbon via a C_3 photosynthetic pathway.

Arthrocnemum subterminale **(Parish) Standl.** occurs under brackish or saline (to 35 ppt NaCl) conditions within alluvial fans, beaches, berms, bluffs, depressions, ditches, dunes, estuaries, flats, lake bottoms, marine terraces, meadows, mudflats, playas, ponds, roadsides, salt marshes, seeps, sinks, strands, vernal pools, and along the margins of bays, lagoons, rivers, and streams at elevations of up to 716 m. Most occurrences occupy tidal/intertidal sites in the middle to upper coastal zones. The high tolerance to salinity provides a competitive advantage over some invasive plants in these irregularly flooded Pacific habitats. The plants are distributed in the coastal marsh from 1.9 to 3.0 m above the mean low water level. They occur in full sunlight exposures. The stems are relatively shallow rooted and drought intolerant. The substrates are described as alkaline, calcareous, or saline and have been characterized as adobe, Baywood fine sand, clay, fine sandy clay, muddy sand, Pescadero silty clay, rock, sand, sandy clay, and sandy gravel. Growth of *A. subterminale* has been shown to alter the substrate characteristics by increasing soil depth (>3 cm), decreasing soil salinity (by 27%), increasing soil moisture (by 13%), and lowering light levels. Flowering and fruiting have been observed from March to October; otherwise, the reproductive biology has not been well documented. The seeds are dispersed commonly in wrack (averaging 94 seedlings/kg in one study), which is transported by coastal waters. No seed bank is established. The seeds germinate readily in seawater, and the seedlings require saturated soils for establishment. The lack of drought tolerance in the seedlings is an important factor in the designation of this species as an obligate aquatic. This is a highly clonal, mat-forming species that can dominate the vegetation in some cases. Unusually large plants (to 1 m) have been observed when growing on *Lycium californicum* for support. The annual species *Hutchinsia procumbens* and *Parapholis incurva* are associated positively with *A. subterminale* and experience lower survival when canopies of the shrub are removed. *Spergularia salina* is associated negatively with this species, and its survival is enhanced when the shrub canopy is reduced. The plants reproduce vegetatively by means of rhizomes. The populations are sensitive to disturbance impacts attributable to trail and road traffic. **Reported**

associates: *Acacia cyclops, Acmispon, Allenrolfea occidentalis, Ambrosia psilostachya, Artemisia, Atriplex barclayana, A. parishii, A. prostrata, A. semibaccata, A. tularensis, A. watsonii, Baccharis pilularis, B. salicifolia, B. sarothroides, Bassia hyssopifolia, Batis maritima, Bolboschoenus maritimus, Bromus diandrus, B. hordeaceus, Cakile maritima, Centaurea melitensis, Centromadia pungens, Chloropyron molle, C. palmatum, Cleome sparsiflora, Conium maculatum, Cordylanthus maritimus, C. palmatus, Cressa truxillensis, Cuscuta, Deinandra fasciculata, Distichlis spicata, Downingia insignis, Dudleya nesiotica, Eryngium aristulatum, Extriplex californica, E. joaquinana, Frankenia grandifolia, F. palmeri, F. salina, Geraea canescens, Heliotropium curassavicum, Hordeum depressum, H. pusillum, Hutchinsia procumbens, Isocoma menziesii, I. veneta, Jaumea carnosa, Juncus acutus, Lasthenia californica, L. conjugens, L. glaberrima, L. glabrata, Lilaea scilloides, Limonium californicum, Lolium multiflorum, Lycium californicum, Malacothrix glabrata, Malva parviflora, Malvella leprosa, Melilotus indicus, Mesembryanthemum crystallinum, M. nodiflorum, Monanthochloe littoralis, Myosurus minimus, Oligomeris linifolia, Parapholis incurva, Pectocarya, Phalaris minor, Phyllospadix, Plagiobothrys leptocladus, P. stipitatus, Polypogon monspeliensis, Psilocarphus brevissimus, Rafinesquia neomexicana, Raphanus sativus, Rumex crispus, Ruppia maritima, Salicornia virginica, Salsola tragus, Sarcocornia pacifica, Schismus barbatus, Schoenoplectus californicus, Sesuvium verrucosum, Sonchus oleraceus, Spartina foliosa, Spergularia macrotheca, S. salina, Suaeda californica, S. esteroa, S. nigra, S. taxifolia, Symphyotrichum subulatum, Tamarix, Triglochin maritimum, Typha angustifolia.*

Use by wildlife: *Arthrocnemum subterminale* is an integral part of the salt marsh community that provides cover for numerous amphibians, fish, mammals, and shellfish and also serves as an important feeding and nesting ground for various birds (Aves).

Economic importance: food: The leaves of *A. subterminale* reportedly are edible when raw or cooked. Its seeds were ground as meal by the Cahuilla tribe of southern California; **medicinal:** none; **cultivation:** Although an unusual specimen, *A. subterminale* is not in cultivation probably because of its particular, restricted habitat requirements; **misc. products:** none; **weeds:** none; **nonindigenous species:** none in North America.

Systematics: *Arthrocnemum subterminale* was known formerly as *Salicornia subterminalis*. Although *Arthrocnemum* and *Salicornia* are related relatively closely, they are apparently not sister genera (e.g., Figure 3.5). The two genera differ principally by the presence of perisperm (absent in *Salicornia*), exserted flowers (sunken in *Salicornia*), and pubescence of the seeds (glabrous in *Antrhocnemum*). *Arthrocnemum subterminale* is believed to be closely related to the Mediterranean *A. glaucum*; however, this relationship has not been confirmed by phylogenetic analysis. A more comprehensive phylogenetic study of Amaranthaceae subfamily *Salicornioideae* is necessary to determine whether *Antrhocnemum* is monophyletic and to provide further details of its relationships. Although a fairly representative molecular survey of genera in the

subfamily has been conducted (based on nrITS, *atpB–rbcL* spacer sequence data), it included only one *Arthrocnemum* species (*A. macrostachyum*), which resolved as more closely related to *Microcnemum*. However, given the past taxonomic discrepancies, a complete survey of *Arthrocnemum* taxa should be undertaken to determine whether the genus is monophyletic as currently circumscribed. Moreover, recent morphological analyses have indicated that *A. subterminale* should be excluded from *Arthrocnemum* (which would comprise only two species) because of its numerous differences. The chromosomal base number of *Arthrocnemum* is $x = 9$. The genus includes diploids ($2n = 18$) and tetraploids ($2n = 36$); however, the chromosome number of *A. subterminale* has not been reported.

Comments: *Arthrocnemum subterminale* is restricted to the warmer portions of California and northern Mexico.

References: Aronson, 1989; Callaway, 1994; Daehler, 2003; Fernández-Illescas et al., 2011; Figuerola et al., 2003; Gul et al., 2013; James & Zedler, 2000; Kadereit et al., 2006; Khan & Gul, 1999; Morzaria-Luna & Zedler, 2007; Schroth, 1996; Standley, 1914; Sukhorukov & Nilova, 2016; Tölken, 1967.

4. Bassia

Smotherweed

Etymology: for Ferdinando Bassi (1710–1774)

Synonyms: *Chenopodium* (in part); *Echinopsilon*; *Kochia* (in part); *Spirobassia*

Distribution: global: Africa; Asia; Europe; North America*; **North America:** widespread (except central)

Diversity: global: 10 species; **North America:** 2 species

Indicators (USA): OBL: *Bassia hirsuta*

Habitat: saline; palustrine; **pH:** unknown; **depth:** <1 m; **life-form(s):** emergent (herb)

Key morphology: plants pubescent with procumbent or ascending stems (to 4 dm); leaves (to 15 mm) fleshy, linear, semicylindrical; flowers solitary, axillary; perianth of five segments, each with dorsal spine or tubercle; fruit an orbicular capsule with many seeds

Life history: duration: annual (fruit/seeds); **asexual reproduction:** none; **pollination:** wind; **sexual condition:** gynomonoecious; **fruit:** utricles (common); **local dispersal:** seeds (water); **long-distance dispersal:** seeds (birds)

Imperilment: (1) *Bassia hirsuta* [GNR]

Ecology: general: As currently circumscribed, all *Bassia* species are annual halophytes that share an ecological association with saline, coastal habitats or inland saline waste areas within Steppe and other arid regions. Both North American species are designated as OBL indicators. Reportedly, different species can exhibit either C_3 or C_4 photosynthetic pathways for carbon fixation. However, it seems futile to characterize the genus ecologically at this time because of its highly polyphyletic nature (see *Systematics*).

Bassia hirsuta (L.) Asch. inhabits ditches, eastern seaboard shores, roadsides, saline flats, salt marshes, salt pannes, strand lines, thickets, and tidal streams at elevations of up to 50 m. Typically, the plants occur in the pioneer zone of brackish or saline marshes, along mesic wracklines, and in waste areas more inland, where they grow in association with characteristic salt marsh and beach species. Exposures receive full sunlight. The substrates have been described as clay, gravel, mud, and sand. The plants prefer habitats (e.g., wrack lines) that are rich in decaying organic material. Flowering and fruiting extend at least from August to October. The flowers are wind pollinated. The seeds are dispersed by tidal waters, which move them in wrack, and by birds (Aves). Little additional information is available on the ecology of this species, especially in North America. The specific mode of carbon fixation in *B. hirsuta* is not known; however, $\delta^{13}C$ values ($-12.6‰$) indicate that it is a C_4 species, despite its lack of Kranz anatomy. **Reported associates (North America):** *Agalinis maritima, Atriplex arenaria, A. hastata, Cakile edentula, Chenopodium album, Honkenya peploides, Iva frutescens, Limonium carolinianum, Lysimachia maritima, Phragmites australis, Plantago oliganthos, Potentilla anserina, Puccinellia maritima, Salicornia europaea, S. procumbens, S. stricta, Salsola kali, Spartina alterniflora, S. patens, S. pectinata, Spergularia salina, Suaeda linearis, S. maritima, Symphyotrichum subulatum, Triglochin maritimum, Xanthium strumarium.*

Use by wildlife: none reported

Economic importance: food: none; **medicinal:** none; **cultivation:** none; **misc. products:** none; **weeds:** *B. hirsuta* is invasive in natural salt marsh communities; **nonindigenous species:** *Bassia hirsuta* is a Eurasian species, which was introduced to North America around 1900.

Systematics: The name *Bassia* has also been applied (but illegitimately) to the genus *Madhuca* in the family Sapotaceae.

Phylogenetic analyses (Figure 3.5) initially indicated that *Bassia* was polyphyletic, with *B. sedoides* related closely to the genus *Camphorosma* but *B. dasyphylla* resolved closer to members of tribe *Sclerolaeneae*. Subsequent analyses of nrITS and cpDNA sequence data (which included *B. hirsuta*) corroborated the polyphyletic nature of *Bassia* and indicated that the current taxonomic concept of the genus is highly artificial (Figure 3.8). In addition to moving a large number of *Kochia* species to *Bassia*, those results also recommended the transfer of *B. hirsuta* to a monotypic genus (*Spirobassia*), which is related more closely to *Chenolea* (Figure 3.8). Although *B. hirsuta* has been retained here in *Bassia*, it seems inevitable that the taxon will be transferred to *Spirobassia*. The basic chromosome number of *Bassia* is $x = 9$; *B. hirsuta* ($2n = 18$) is diploid.

Comments: Although found mostly throughout eastern, coastal North America, *B. hirsuta* has also been introduced to Idaho.

References: Collins & Blackwell, Jr., 1979; Dowhan & Rozsa, 1989; Kadereit & Freitag, 2011; Kühn, 1993; Mosyakin, 2003; Tutin et al., 1964; Winter, 1981.

5. *Chenopodium*
Coast-blite, goosefoot; ansérine, chénopode

Etymology: from the Greek *chen podion* meaning "goose foot"

Synonyms: *Ambrina; Blitum; Roubieva; Teloxys*

Distribution: global: cosmopolitan; **North America:** widespread

Diversity: global: 135 species; **North America:** 34 species

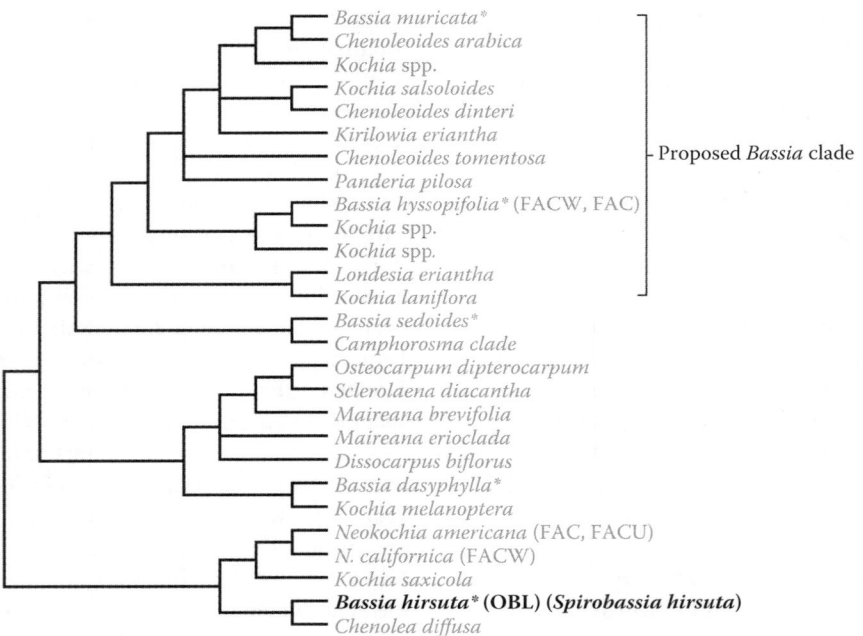

FIGURE 3.8 Phylogenetic relationships of *Bassia* species as indicated by analysis of nrITS sequence data. These results indicate that *Bassia* (denoted by asterisks) is highly polyphyletic as currently circumscribed and recommend an expansion of the genus to include many *Kochia* species, etc. and the exclusion of *B. sedoides*, *B. dasyphylla*, and *B. hirsuta* (with the latter transferred to a monotypic genus *Spirobassia*). The OBL North American *B. hirsuta* (in bold) is related more closely to the southern African genus *Chenolea* (N. American wetland indicator designations are given in parentheses). (Adapted from Kadereit, G. & Freitag, H., *Taxon*, 60, 51–78, 2011.)

Indicators (USA): OBL; FACW: *Chenopodium rubrum*
Habitat: saline (coastal/inland); palustrine; **pH:** alkaline
(>7); **depth:** <1 m; **life-form(s):** emergent (herb)
Key morphology: shoots (to 10 dm) erect, prostrate, or
spreading, fleshy, angled, reddish; leaves (to 1.5 dm) alternate,
reddish, fleshy, deltoid to rhombic, hastate or coarse toothed;
flowers inconspicuous, crowded in axillary spikes; perianth
parts (3–4; to 1 mm) reddish, fleshy, connate at base; utricles
ovoid; seeds (to 1.2 mm) vertical
Life history: duration: annual (fruit/seeds); **asexual repro-**
duction: none; **pollination:** wind; **sexual condition:** gynomonoecious; **fruit:** utricles (common); **local dispersal:** seeds;
long-distance dispersal: seeds
Imperilment: (1) *Chenopodium rubrum* [G5]; SH (CT, IA);
S1 (ME, NH, NJ, NS, PE); S2 (NB, NY); S3 (MA, WY, YT)
Ecology: general: This large genus of annuals and perennials is found nearly worldwide in a variety of habitats ranging
from desert to montane conditions. Most of the species are
terrestrial and not associated with wet habitats. About one
quarter of the North American species are FACW indicators.
The single OBL species is also ranked as FAC in some areas.
The genus is common in arid sites and in disturbed localities including agricultural systems. The flowers are wind pollinated. The seeds of at least some species require light and
will not germinate when buried at shallow depths (e.g., 3 cm).
Some seeds can germinate at salinities of 30 dS/m. Their dispersal can occur by wind, water, or birds (Aves). Several species are known to be allelopathic. All of the species fix carbon
via a C_3 photosynthetic pathway.

Chenopodium rubrum **L.** inhabits brackish to saline sites
including bare bottomlands, beaches, depressions, dessicated
pools, dikes, ditches, fens, flats, floodplains, gravel bars,
interdunal swales, marshes, meadows, mudflats, playas, potholes, roadsides, salt marshes, semipermanent lakes, shores,
slopes (to 30%), vernal pools, washes, woodlands, and the
margins of lagoons, lakes, ponds, reservoirs, rivers, sloughs,
and streams at elevations of up to 3749 m. It occurs most frequently at inland saline sites, commonly in disturbed areas
such as cultivated fields. The habitats are most often characterized by open exposures, which receive full sunlight, but
can include partially shaded sites. The substrates are highly
alkaline and include alluvium, basalt pahoehoe, clay, cobble,
decomposed granite, gravel, limestone mud, loam, marl,
muck, mud, peat, rhyolite, rock, sand, sandstone, sandy clay,
sandy pebbles, silt, silty mud, talus, and travertine. This is a
pioneer species and an indicator of high soil fertility, which
occurs frequently where high concentrations of bicarbonates,
calcium, carbonates, magnesium, sodium, and sulfates are
found. The plants are in flower from June to October, with
fruiting extending into November. These are short-day plants
with respect to photoperiod and will produce smaller seeds
when exposed to 15 h of day length than when grown under
12-h days. Pistillate flowers often appear on plants after the
maturation of the hermaphroditic ones. Seed germination
requires a 180-day period of cold stratification followed by
incubation temperatures that fluctuate from 15°C to 20°C. The
seeds are retained in the guts of waterfowl (Aves: Anatidae:

Anas platyrhynchos) for an average of 11 h. Some of the seeds
(~6%) pass through the birds intact and retain 21% viability.
They are also transported along railroads and in soil used
for landfills. Earlier germinating plants will induce flowering earlier, but plants germinating later have a shorter life
cycle, reduced vegetative growth, and fewer but larger seeds.
The larger seeds germinate more readily in the dark (70%
germination) than the smaller seeds (5% germination). The
plants are known to establish after fires, in areas where the
roots of river grass (*Scolochloa festucacea*) have been killed.
Reported associates: *Abies concolor, Achillea millefolium,
Agrostis scabra, Alopecurus arundinaceus, Amaranthus
albus, Ambrosia artemisiifolia, Anemopsis californica, Arida
carnosa, Artemisia arbuscula, A. biennis, A. frigida, A.
ludoviciana, A. tridentata, Astragalus whitneyi, Atriplex dioica, A. glabriuscula, A. micrantha, A. patula, A. prostrata, A.
rosea, A. torreyi, Baccharis, Balsamorhiza sagittata, Bassia
hyssopifolia, Berberis repens, Berula erecta, Betula, Bidens
cernuus, Bolboschoenus maritimus, B. robustus, Bromus
ciliatus, B. tectorum, Campanula rotundifolia, Carex aquatilis, C. nebrascensis, C. pellita, C. praegracilis, C. siccata,
Centromadia pungens, Chenopodium chenopodioides, C.
glaucum, C. macrospermum, C. salinum, Cirsium cymosum,
C. scariosum, C. vulgare, Clematis, Conyza, Cotula coronopifolia, Cressa truxillensis, Crypsis alopecuroides, C. vaginiflora, Cryptantha, Cynodon dactylon, Cyperus odoratus, C.
polystachyos, Dasiphora floribunda, Descurainia, Diplachne
maritima, Distichlis spicata, D. stricta, Dysphania botrys,
Echinochloa, Elaeagnus angustifolia, Eleocharis parvula,
Elymus, Equisetum arvense, Eriogonum fasciculatum, E.
umbellatum, Euthamia occidentalis, Festuca idahoensis,
Gilia, Glyceria striata, Gnaphalium palustre, Gutierrezia
microcephala, Helenium, Helianthus annuus, Heliotropium
curassavicum, Hibiscus moscheutos, Hippuris vulgaris,
Hordeum jubatum, Iva frutescens, Juncus articulatus, J.
balticus, J. bufonius, J. ranarius, J. scirpoides, Juniperus
communis, J. scopulorum, Kochia scoparia, Lepidium
appelianum, Leptochloa, Leymus cinereus, L. triticoides,
Limosella, Lomatium grayi, Lotus corniculatus, Lycopus
asper, Lysimachia maritima, Maianthemum, Malvella leprosa, Marrubium vulgare, Melilotus albus, M. indicus, M.
officinalis, Muhlenbergia asperifolia, Oenanthe sarmentosa,
Oryzopsis asperifolia, Osmorhiza berteroi, Panicum capillare, Pascopyrum smithii, Paxistima, Penstemon, Persicaria
amphibia, P. lapathifolia, Phacelia, Phalaris, Picea engelmannii, P. pungens, Pinus flexilis, P. ponderosa, Plantago
lanceolata, P. major, Pluchea odorata, Poa pratensis, P.
secunda, Polygonum ramosissimum, Polypogon, Populus fremontii, P. tremuloides, Portulaca oleracea, Potentilla gracilis, P. rivalis, P. supina, Prosopis velutina, Prunus virginiana,
Pseudognaphalium luteoalbum, Pseudoroegneria spicata,
Pseudotsuga menziesii, Pteridium aquilinum, Ptilimnium
capillaceum, Puccinellia airoides, P. distans, P. nuttalliana, Purshia, Pyrrocoma racemosa, Quercus gambelii,
Ranunculus californicus, R. sceleratus, Rorippa cusvisiliqua,
Rubus idaeus, Rumex crispus, R. fueginus, R. maritimus,
R. salicifolius, Salicornia europaea, S. rubra, Salix exigua,*

S. gooddingii, S. laevigata, S. lasiolepis, S. scouleriana, S. serissima, S. sessilifolia, Sambucus, Schoenoplectus acutus, S. americanus, S. pungens, Scolochloa festucacea, Sesuvium verrucosum, Sidalcea, Sisymbrium, Solidago lepida, S. velutina, Spartina patens, Spergularia, Spiraea, Spiranthes diluvialis, Suaeda calceoliformis, S. depressa, S. erecta, Suckleya suckleyana, Swertia, Symphoricarpos, Symphyotrichum ciliatum, S. subulatum, Tamarix ramosissima, Taraxacum officinale, Teucrium canadense, Thalictrum fendleri, Thelypodium, Trifolium fragiferum, Triglochin maritimum, Trisetum montanum, Typha angustifolia, T. domingensis T. latifolia, Urtica, Verbascum thapsus, Veronica anagallisaquatica, Vitis, Woodsia scopulina, Xanthium.

Use by wildlife: The plants are grazed by geese (Aves: Anatidae). Seeds of *C. rubrum* are eaten readily by small mammals, such as rodents (Mammalia: Rodentia), by waterfowl (Aves: Anatidae: *Anas platyrhynchos*) and other wild birds, and by domestic poultry (Aves: Galliformes). Dead stands of *C. rubrum* are used as nesting sites for eared grebes (Aves Podicipedidae: *Podiceps nigricollis*). *Chenopodium rubrum* also provides habitat for many invertebrates, especially in marshes where the water levels are manipulated.

Economic importance: food: The cooked or raw leaves of *C. rubrum* are used as a spinach substitute but should not be eaten in quantity because they contain small amounts of saponins and oxalic acid. The seeds of *C. rubrum* can be ground into flour to make bread and cakes. They were eaten extensively by the Gosiute people of Utah; **medicinal:** *Chenopodium rubrum* pollen may cause hay fever. Extracts of the plant have been used as a treatment for tumors; **cultivation:** none; **misc. products:** *Chenopodium rubrum* has been used as an experimental plant to study flowering and other physiological processes; **weeds:** *Chenopodium rubrum* is regarded as a weed around settlements and in disturbed sites throughout interior Alaska, and in anthropogenic sites in Canada. It is a minor contaminant of forage and lawn seed mixtures; **nonindigenous species:** *Chenopodium rubrum* is native to North America but is regarded as introduced to Alaska, California, and Illinois. Because it often grows in disturbed sites, it can be difficult to evaluate whether populations are indigenous or introduced.

Systematics: Phylogenetic studies demonstrate that *Chenopodium* is highly polyphyletic, with species dispersed widely among other genera of subfamily *Chenopodioideae* (e.g., Figure 3.5). *Chenopodium rubrum* is assigned to subgenus *Blitum*, which itself is polyphyletic. Phylogenetic studies incorporating DNA sequence data resolve *C. rubrum* within a clade containing *C. chenopodioides, C. glaucum,* and *C. urbicum.* However, incongruent placements result where *C. rubrum* associates strongly (100% internal support) either with *C. glaucum* (cpDNA) or with *C. chenopodioides* (nrITS data). *Chenopodium rubrum* is known to hybridize with *C. glaucum,* which could explain these results. Morphologically, *C. rubrum* is believed to be allied closely to the similar *C. humile* (often treated as a variety of *C. rubrum*), with which it intergrades. The poor circumscription of this genus at present precludes a meaningful assessment of species relationships, which will require further study involving a broader sample of species along with a better understanding of observed incongruencies. The basic chromosome number of *Chenopodium* is $x = 9$. *Chenopodium rubrum* has both diploid ($2n = 18$) and tetraploid ($2n = 36$) counts reported.

Comments: *Chenopodium rubrum* is distributed widely throughout North America except for the southeastern region. It is also a native to Eurasia.

References: Baskin & Baskin, 1998; Chamberlin, 1911; Clemants & Mosyakin, 2003; Dodd & Coupland, 1966; Eslami, 2011; Fuentes-Bazan et al., 2012; Haines, 2001; Kühn, 1993; Kunkel, 1984; Mueller & van der Valk, 2002; Rayner, 1978; USEPA, 1995; Usher, 1974; Van der Sman et al., 1988, 1992; Wahl, 1954; Ward, 1968; Wheatley & Bentz, 2002.

6. *Salicornia*

Glasswort, pickleweed, samphire; corail, salicorne

Etymology: from the Latin *sal cornu* meaning "salt horn"

Synonyms: none

Distribution: global: cosmopolitan; **North America:** widespread

Diversity: global: 10 species; **North America:** 5 species

Indicators (USA): OBL: *Salicornia bigelovii, S. depressa, S. maritima, S. rubra*

Habitat: saline (coastal, inland); palustrine; **pH:** 6.5–9.5; **depth:** <1 m; **life-form(s):** emergent (herb)

Key morphology: stems (to 4 dm) jointed, succulent, branches opposite, changing from green to red color in the Fall; leaves reduced to fleshy, perfoliate scales fused to the stem; inflorescence a fleshy spike; flowers inconspicuous, grouped in threes; sunken in the fleshy spike

Life history: duration: annual (fruit/seeds); **asexual reproduction:** none; **pollination:** wind; **sexual condition:** hermaphroditic; **fruit:** utricles (common); **local dispersal:** fruit/seeds; **long-distance dispersal:** fruit/seeds

Imperilment: (1) *Salicornia bigelovii* [G5]; S1 (DE, ME, NH); S2 (NC, NY); (2) *S. depressa* [GNR]; S3 (BC); (3) *S. maritima* [G5]; SH (ME); S3 (NC); (4) *S. rubra* [G5]; S1 (NE, KS); S2 (ID, MN); S3 (BC, WY)

Ecology: general: All *Salicornia* species occupy coastal salt marshes and inland salt flats especially on bare, exposed peat. They require abundant water and will wilt when exposed to dry air for prolonged periods. All of the species should be regarded as OBL indicators. *Salicornia* flowers are wind pollinated, but their self-compatibility facilitates self-pollination in cleistogamous flowers and between the slightly protogynous chasmogamous flowers on single individuals. Some species are protandrous and others cleistogamous. Populations of all species produce prodigious quantities of seeds, which typically remain in the upper 5 mm of sediment. The seeds fall mainly within 100 mm of the parental plants, with some reaching 400 mm and a few considerably farther. Fresh seeds sink rapidly in seawater, but dried seeds can float for up to a day. They are dispersed by tidal currents, which roll them along the substrate. Seeds of diploids are smaller than those of tetraploids and will survive for longer periods through winter. All *Salicornia* species fix carbon primarily by means of a C_3 photosynthetic pathway, although some sources represent them as C_4 plants.

Salicornia bigelovii **Torr.** forms large, dense swards in hypersaline areas of beaches, ditches, esteros, hammocks, mudflats, salt flats, salt marshes, salt meadows, salt pannes, shores, tidal flats, tidal lagoons, and the margins of bays and streams at elevations of up to 15 m. It is extremely salt tolerant and produces larger, more succulent shoots when growing in highly saline environments. Exposures receive full sunlight. The substrates include clay, heavy mud, sand, and shells. Flowering and fruiting can be observed throughout the year in some areas. The seeds germinate better in 0.3%–0.5% salt water than in distilled water and retain their viability in NaCl concentrations up to 0.06 M. Dried seeds stored at 3°C–6°C for more than a month will germinate at 5°C and 16°C. High germination rates are achieved at 15.5°C, but rates decline at 26.6°C, indicating early spring as the optimal time for seed germination. The plants are capable of increasing selenium volatilization in selenium-contaminated soils. **Reported associates:** *Acacia cyclops, Allenrolfea occidentalis, Avicennia germinans, Baccharis sarothroides, Batis maritima, Borrichia frutescens, Cressa truxillensis, Distichlis littoralis, D. spicata, Eucalyptus, Frankenia palmeri, F. salina, Heliotropium curassavicum, Jaumea carnosa, Juncus roemerianus, Limonium californicum, L. carolinianum, Polypogon monspeliensis, Quercus, Sabal palmetto, Salicornia depressa, Salix, Sarcocornia pacifica, S. perennis, Sesuvium portulacastrum, Spartina alterniflora, S. foliosa, S. spartinae, Suaeda californica, Symphyotrichum tenuifolium, Tamarix, Triglochin, Typha.*

Salicornia depressa **Standl.** occurs in coastal (or more rarely inland), brackish, or saline sites including beaches, depressions, estuaries, floodplains, marshes, mudflats, salt marshes, salt meadows, seeps, tidal flats, and the dry alkaline bottoms or margins of canals, channels, lakes, ponds, and sloughs at elevations of up to 1207 m. It commonly inhabits disturbed portions of the estuarine and salt marsh upper intertidal zone, mid to low marshes, or denuded areas of salt marshes. Exposures are characterized by full sunlight. The alkaline/calcareous substrates include clayey mud, cobble, gravel, mirabalite, mud, muddy silt, rock, sand, silt, silty clay, silty sand, and stones. Flowering and fruiting occur from June to November. The flowers are protandrous and open pollinated. The plants produce dimorphic seeds, with the larger being more salt tolerant and germinating in one season and the smaller retaining greater dormancy in the seed bank. Germination of the seeds in saline substrates is greatly enhanced after 4 weeks of cold stratification. The seeds also germinate better when covered by sea wrack (tidal litter). **Reported associates:** *Acer, Aira praecox, Allenrolfea occidentalis, Alnus, Atriplex, Bolboschoenus maritimus, Chenopodium, Chloropyron maritimum, Cressa truxillensis, Cuscuta occidentalis, C. salina, Deschampsia cespitosa, Distichlis spicata, Grindelia, Holodiscus, Isoetes nuttallii, Jaumea carnosa, Juncus gerardii, Limonium californicum, L. carolinianum, Lysimachia maritima, Micranthes integrifolia, Plantago maritima, Quercus garryana, Ranunculus, Salix, Sarcobatus vermiculatus, Sarcocornia pacifica, S. perennis, Solidago sempervirens, Spartina alterniflora, S. patens, Spergularia macrotheca, Tamarix, Triglochin maritimum.*

Salicornia maritima **S.L. Wolff & Jefferies** occurs in or on dikes, drying ponds, pools, salt flats, salt marshes, salt meadows, seashores, tidal pools, and along the margins of coastal channels, lagoons, lakes, and rivers, or (more rarely) near inland brine wells or saline springs at elevations of up to 150 m. The plants can develop into dense, monotypic stands, especially in the middle to upper zones of salt marshes. The substrates are described as brackish/saline gravel, muck, mud, muddy sand, mucky peat, sand, and stones. Flowering occurs from May to October. The flowers are cleistogamous and self-pollinating. A seed bank develops that can reach densities averaging 13 seeds/m^2 in saline substrates to 70 seeds/m^2 in brackish substrates. Field germination rates are high relative to other species, with an average seedling cover of 3.8%/0.25 m^2. **Reported associates:** *Agrostis stolonifera, Anthoxanthum hirtum, Atriplex patula, Bolboschoenus maritimus, Carex mackenziei, C. paleacea, C. salina, Cyperus strigosus, Distichlis spicata, Eleocharis parvula, Elymus repens, Epilobium ciliatum, Festuca rubra, Galium trifidum, Halerpestes cymbalaria, Hordeum jubatum, Iva frutescens, Juncus balticus, J. gerardii, Leymus mollis, Limonium carolinianum, Lysimachia maritima, Moehringia lateriflora, Phragmites australis, Plantago maritima, Potentilla anserina, Puccinellia phryganodes, Ruppia maritima, Sagina maxima, Schoenoplectus pungens, Solidago sempervirens, Spartina alterniflora, S. patens, S. pectinata, Spergularia canadensis, Stellaria humifusa, Suaeda linearis, S. maritima, Triglochin gaspensis, T. maritimum, Typha angustifolia.*

Salicornia rubra **A. Nelson** occurs in saline inland or coastal sites including beaches, bogs, depressions, ditches, draws, estuaries, flats, floodplains, marshes, meadows, mudflats, playas, potholes, riverbottoms, roadsides, salt marshes, salt plains, seeps, shores, sloughs, soda lakes, swales, wallows, and along the margins of receding lakes, ponds, springs, and streams at elevations of up to 3440 m. The growth of plants is optimal at 200 mM NaCl and declines with increased salinity. The plants frequently dominate (e.g., >65% total cover) open, periodically flooded, or seasonally wet localities where suitable substrates exist. The highly alkaline/saline substrates (e.g., pH: 7.0–9.5; 0.5%–8.0% total salts) include adobe, alluvium, basalt gravel, clay, clay silt, gleyed regosol, gravel, gypsum, loam, loamy/stony coulee, mud, peat, rock, sand, and silt. Many localities are described as having thick salty (or algal) crusts with little other vegetation present. The plants are known to readily colonize bare peat soils of hypersaline habitats that have been disturbed by geese (Aves: Anatidae: Anseriformes). Flowering occurs from June to September, with fruiting extending into October. A seed bank develops with greater densities (to 20,000/m^2) occurring in the higher than in the lower marsh sites. Seed germination is highest when incubated under a 12/12 h, 35°C/25°C day/night temperature regime, with germination inhibited at temperatures of 5°C–15°C. The germination rate decreases as salinity increases but can still occur (e.g., 15% germination) at fairly high salinity (1000 mM NaCl). Natural germination occurs in April once sites receive snowmelt water, and proceeds until early summer when evaporation raises the soil salinity. **Reported associates:** *Agropyron, Allenrolfea* *occidentalis, Almutaster pauciflorus, Alopecurus arundinaceus, Amphiscirpus nevadensis, Atriplex argentea, A. gardneri, A. micrantha, A. prostrata, A. subspicata, Bolboschoenus maritimus, Bromus inermis, B. tectorum, Calamagrostis montanensis, Chenopodium glaucum, C. rubrum, Crypsis schoenoides, Distichlis spicata, Elaeagnus angustifolia, Elymus elymoides, Ericameria albida, Grayia, Helianthus paradoxus, Hordeum jubatum, Juncus arcticus, J. balticus, Kochia, Lepidium latifolium, Leymus cinereus, Limonium californicum, Lysimachia maritima, Muhlenbergia asperifolia, Nitrophila occidentalis, Pascopyrum smithii, Peritoma multicaulis, Plantago eriopoda, P. maritima, Poa secunda, Polypogon monspeliensis, Potentilla egedii, Puccinellia nuttalliana, P. phryganodes, Ranunculus cymbalaria, Salicornia depressa, Salsola, Sarcobatus vermiculatus, Sarcocornia utahensis, Schoenoplectus pungens, Spartina gracilis, Spergularia salina, S. media, S. rubra, Sporobolus airoides, Stellaria humifusa, Suaeda calceoliformis, S. depressa, Symphyotrichum ciliatum, S. frondosum, Tamarix, Tephroseris palustris, Triglochin maritimum, Typha angustifolia, Zeltnera exaltata.*

Use by wildlife: *Salicornia* stands generally provide breeding habitat for the black salt marsh mosquito (Insecta: Culicidae: *Aedes taeniorhynchus*). *Salicornia* seeds are consumed by thrushes (Aves: Passeriformes) and various ducks (Aves: Anatidae) including gadwall (*Anas strepera*), mallard (*Anas platyrhynchos*), pintail (*Anas acuta*), scaup (*Aythya*), shoveler (*Anas clypeata*), teal (*Anas discors, A. crecca*), and wigeon (*Anas americana*). They are also eaten by mice (Mammalia: Rodentia: Cricetidae) and rabbits (Mammalia: Leporidae). *Salicornia bigelovii* marshes (especially created sites) support a diverse macrofauna including various arachnids (Arachnida), crustaceans (Peracarida), flatworms (Turbellaria), insects (Insecta), molluscs (Mollusca), ribbon worms (Nemertea), polychaetes (Polychaeta), and worms (Enchytraeidae, Naididae, Tubificidae). The stems of *S. bigelovii* and *S. depressa* are not highly palatable but are eaten by Canada and snow geese (Aves: Anatidae: *Branta canadensis, Chen caerulescens*). *Salicornia bigelovii* is the host plant for several larval butterflies (Insecta: Lepidoptera) including the eastern pygmy blue (Lycaenidae: *Brephidium exilis isophthalma*) of North America and the endemic pygmy blue of the Cayman Islands (Lycaenidae: *B. exilis thompsoni*). The roots of *S. bigelovii* are associated with a nitrogen-fixing bacterium (Bacteria: Enterobacteriaceae: *Klebsiella pneumoniae*). *Salicornia depressa* is the host plant for the eastern pygmy blue (Lycaenidae: *Brephidium exilis isophthalma*) and western pygmy blue (Lycaenidae: *B. exilis exilis*). Infection by nematodes (Nematoda: Heteroderidae: *Cactodera salina*) can reduce seed quality in *S. bigelovii*. Autumn-emerging larvae (Lepidoptera: Coleophoridae: *Coleophora caespititiella*) can consume large quantities of seeds in European *S. depressa*. Beetles (Insecta: Coleoptera: Chrysomelidae: *Erynephala maritima*) can be serious herbivores of *S. depressa*, and the larvae of *Metachroma* species (Chrysomelidae) are known to kill *Salicornia* seedlings. More than 65 species of Fungi are recorded from the roots, stems, and seeds of *S. depressa*. The stems of *S. rubra* are eaten by livestock (Mammalia:

Bovidae). Stands of *S. rubra* provide nesting sites for piping plovers (Aves: Charadriidae: *Charadrius melodus*).

Economic importance: food: Young *Salicornia* "sprouts" (sometimes called "sea asparagus") lately have become a trendy food and salad garnish, especially in the southwestern United States. *Salicornia bigelovii* is grown for fodder and as an "oilseed" crop for saline soils. The seeds contain 40%–45% protein and 30% oil. The tender, young, pickle-like shoots of *S. depressa* ("pickleweed") can be boiled in saltwater, pickled in spiced oil or vinegar, and eaten. *Salicornia maritima* is eaten in salads in the New England region. Its seeds were ground into meal for bread by the Gosiute tribe; **medicinal:** No medicinal uses are known for *Salicornia*; **cultivation:** *Salicornia bigelovii* is cultivated for oil production and as a vegetable. The special habitat requirements of *Salicornia* make these unusual plants impractical to grow as garden specimens; **misc. products:** Plants of *S. bigelovii* were used by the Seri people to line baskets and shells that they used to hold meats. The common name "glasswort" alludes to the ancient use of plants as a source of alkali for the manufacture of soda glass and soap; **weeds:** none; **nonindigenous species:** *Salicornia rubra* is nonindigenous in Michigan and Quebec.

Systematics: Molecular data (*rbcL* sequences) initially indicated the distinctness of *Salicornia* from *Arthrocnemum* and *Sarcocornia*, which have been merged with the genus in the past (Figure 3.5). Some molecular data show *Salicornia* as related closely to the Eurasian *Halocnemum* and the principally Australian *Halosarcia*, which altogether form a clade. However, nrETS sequence data (Figure 3.9) indicate that the group was likely derived "from within" the paraphyletic genus *Sarcocornia*, which creates problems with the taxonomic delimitation of the latter genus. The taxonomy of *Salicornia* is highly confused because of extensive phenotypic plasticity and morphological parallelism throughout the range of the group. Many North American authors have applied the name "*S. virginica*" (non L.) incorrectly to a perennial species now transferred to the genus *Sarcocornia* (as *S. perennis*). The name *S. virginica* L. is also treated as a synonym of *S. europaea*; however, recent chromosomal and isozyme evidence indicates that *S. europaea* (primarily diploid) is restricted to the Old World and the name has been applied inappropriately to several endemic New World species (*S. depressa*; *S. maritima*), which are distinct morphologically, chromosomally, and electrophoretically (by isozyme data). The more common of these in southern, coastal North America is *S. depressa* ($2n = 36$), a tetraploid (the chromosomal base number of *Salicornia* is $x = 9$). Molecular data (Figure 3.9) show *S. depressa* to be closely related to *S. bigelovii*, which is also tetraploid ($2n = 36$). *Salicornia borealis* and *S. rubra* are diploid ($2n = 18$) and recently have been treated as a single species (as done here), although molecular data thus far have provided no definitive answer to whether they should be retained as distinct. Morphologically, *S. depressa* and *S. maritima* appear to be closely related, but they are quite distant in phylogenetic trees derived from molecular data (e.g., Figure 3.9), which show *S. maritima* to be indistinguishable genetically from *S.*

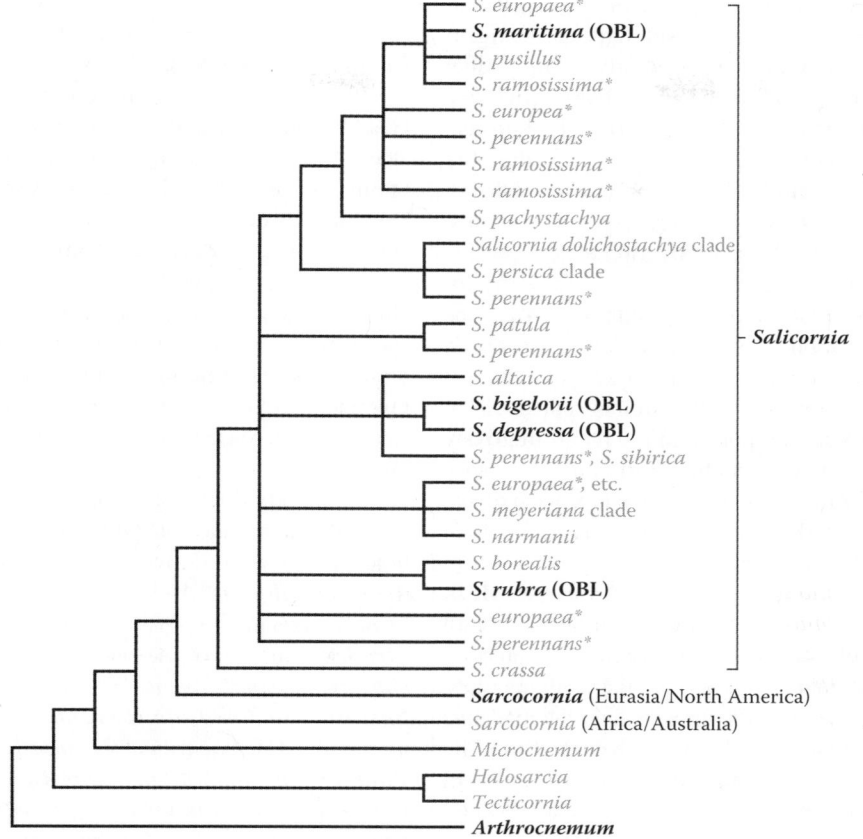

FIGURE 3.9 Phylogenetic relationships of *Salicornia* as indicated by analysis of nrETS sequence data. Extensive plasticity and convergence have created much taxonomic confusion in the genus as indicated by the wide dispersal (thus artificiality) of many taxa across the phylogenetic tree (indicated by asterisks). The OBL North American indicators (shown in bold) encompass a broad phylogenetic background. (Adapted from Kadereit, G. et al., *Taxon*, 56, 1143–1170, 2007.)

ramosissima. Like *S. ramosissima*, *S. maritima* (2*n* = 18) is diploid. It is evident that further taxonomic refinements in this genus will be forthcoming as the results of the molecular phylogenetic analyses eventually are incorporated.

Comments: *Salicornia* species mainly are coastal, although *S. depressa*, *S. rubra*, and *S. maritima* also occur in inland sites. *Salicornia maritima* occurs primarily along the northeastern coast and *S. rubra* in western North America, whereas *S. bigelovii* and *S. depressa* are found along both the east and the west coasts.

References: Ayala & O'Leary, 1995; Ball, 2003a; Buchsbaum et al., 2006, 2009; Chamberlin, 1911; Chmura et al., 2012; Crain et al., 2008; Davy et al., 2001; Desroches et al., 2013; Dodd & Coupland, 1966; Felger & Moser, 1985; Fernández-Illescas et al., 2011; Glenn et al., 1991; Gul & Weber, 2001; Haines, 2000; Handa et al., 2002; Kadereit et al., 2007; Khan et al., 2000, 2001; Lin & Terry, 1998; Parks et al., 2002; Rueda-Puente et al., 2002; Talley & Levin, 1999; Ungar, 1973; Ungar et al., 1969; Wolff & Jeffries, 1987.

7. Sarcocornia

Perennial glasswort, samphire; sarcocorne
Etymology: from the Latin *sarco cornu* meaning "flesh horn"

Synonyms: *Arthrocnemum* (in part); *Salicornia* (in part)
Distribution: global: Africa; Asia; Australia; Europe; New World; **North America:** coastal and southern
Diversity: global: 28 species; **North America:** 3 species
Indicators (USA): OBL: *Sarcocornia ambigua*, *S. pacifica*, *S. utahensis*; **FACW:** *S. utahensis*
Habitat: saline (coastal, inland); palustrine; **pH:** 6.0–8.3; **depth:** <1 m; **life-form(s):** emergent (herb, subshrub)
Key morphology: stems (to 7 dm) succulent, herbaceous or basally suffrutescent, not turning red (cf. *Salicornia*); branches articulate, opposite; leaves minute, scalelike, opposite; inflorescences fleshy, terminal spikes (to 85 mm) with up to 40 fertile segments; flowers highly reduced and adnate to spike, grouped in threes, the middle flower larger, not raised conspicuously above the laterals; the fruit a utricle containing a vertical, elliptic seed (to 1.5 mm)
Life history: duration: perennial (rhizomes, woody base); **asexual reproduction:** rhizomes; **pollination:** wind; **sexual condition:** hermaphroditic; **fruit:** utricles (common); **local dispersal:** seeds (water); rhizome fragmentation (water); **long-distance dispersal:** seeds
Imperilment: (1) *Sarcocornia ambigua* [GNR]; S1 (DE, NH); *S. pacifica* [G5]; *S. utahensis* [G4]; S3 (UT)

Ecology: general: *Sarcocornia* species occur in coastal or inland saline sites such as salt marshes, salt flats, and margins of saline lakes and shores. The species occur mainly in warmer regions but do extend fairly far northward in North America. All of the species are suffrutescent perennials. The flowers are bisexual and wind pollinated. In most species, they are arranged in a triangular fashion, with one larger central flower flanked on each side by a smaller flower, which never touch one another as in *Salicornia*. Seed production can be prolific and a seed bank generally develops. The plants are useful for stabilizing or restoring disturbed or degraded areas. All of the species reportedly fix carbon by means of a C_3 photosynthetic pathway. Extensive taxonomic problems have plagued this genus (see *Systematics*), making it extremely difficult to assign ecological information reported in the literature to any North American species with certainty. Fortunately, because the species all share similar ecological attributes, such errors should be manifest primarily by the incorrect assignments of associated species and distribution accounts.

Sarcocornia ambigua **(Michaux) M. A. Alonso & M. B. Crespo** inhabits brackish or saline, tidal sites associated with beaches, berms, ditches, estuaries, flats, floodlands (interior), mangrove swamps, mudflats, roadsides, salt flats, salt marshes, sand flats, shores (oceanic), strands, thickets, and the margins/shores of lakes and streams at elevations below 10 m. The plants can withstand mean low temperatures of −5°C and will tolerate salinities higher even than seawater (electrical conductivity up to 86.1 dS/m). The substrates are described as alkaline or saline and include coral limestone, crushed shell, marl, marly sand, mucky sand, rock, sand, sandy clay, and sandy marl. Flowering and fruiting occur from July to November. The seeds germinate well under a 12/12 h, 20°C/30°C temperature regime. The highest seed germination (81%–84%) occurs at salinities of 0–5 g NaCl/L but decreases to 41%–46% at salinities above 15 g NaCl/L. However, some seeds (3%) can germinate at even higher salinity (45 g NaCl/L). Seed viability decreases from 18% to 4% as the salinity increases. The plants perennate by means of rhizomes. **Reported associates:** *Arthrocnemum subterminale, Atriplex mucronata, A. patula, A. prostrata, A. watsonii, Avicennia germinans, Batis maritima, Borrichia frutescens, Bursera,* Cyanobacteria (mat-forming), *Dalbergia, Distichlis littoralis, D. spicata, Extriplex californica, Frankenia salina, Gouania, Iva frutescens, Juncus roemerianus, Limonium carolinianum, Lycium carolinianum, Mesembryanthemum nodiflorum, Plantago maritima, Pluchea odorata, Quercus, Salicornia bigelovii, S. depressa, Sesuvium portulacastrum, Spartina alterniflora, S. patens, S. spartinae, Spergularia salina, Suaeda californica, S. linearis, S. maritima, Swietenia, Symphyotrichum tenuifolium.*

Sarcocornia pacifica **(Standl.) A.J. Scott** inhabits brackish or saline sites (salinity: 9–46 ppt) including alluvial fans, beaches, bluffs, depressions, dikes, dunes, estuaries, floodplains, lagoons, meadows, mudflats, playas, pools, roadsides, salt flats, salt marshes, salt pans, scrub, and the margins of bays, canals, rivers, sloughs, and streams at elevations of up to 609 m. The exposures receive full sunlight. Although the substrates are usually described as alkaline, they span a wider range of acidity (pH: 6.0–7.2) and include bench gravel, clay, clay loam, gravel, muck, mud, rocky sandstone, sand, sandstone, sandy gravel, sandy loam, silt, silty loam, and silty sandy loam. Organic matter content can range from 10% to 40%. Flowering and fruiting extend from June to November. The small seeds (0.07 mg) are produced in quantity, resulting in an average of 80,755 seeds/m² in some localities. They are buoyant and dispersed by water, with 94% remaining afloat after a 24-h period. Seed viability is fairly high, up to 66% in some surveys. The shoots are rhizomatous and form clonal clumps. Vegetative reproduction can also occur by viable shoot fragments that are dispersed by water. The roots are thoroughly colonized by mycorrhizal Fungi. The plants respond favorably across all salinity levels by the addition of nitrogen, which can increase their biomass by 6- to 10-fold. They compete poorly with invasive grasses (*Polypogon monspeliensis*); however, their higher seed germination under high salinities limits the dominance of *Polypogon* to very wet years or where other factors result in reduced salinity. **Reported associates:** *Acmispon glabrus, Ambrosia psilostachya, Anemopsis californica, Artemisia californica, Arthrocnemum subterminale, Atriplex lentiformis, A. patula, A. prostrata, A. semibaccata, A. watsonii, Avena, Baccharis pilularis, B. salicifolia, Bassia hyssopifolia, Batis maritima, Beta vulgaris, Bromus diandrus, B. hordeaceus, B. madritensis, Cakile maritima, Calibrachoa parviflora, Camissoniopsis cheiranthifolia, Cenchrus clandestinus, Centaurea solstitialis, Chenopodium, Cotula coronopifolia, Cressa truxillensis, Croton setiger, Crypsis schoenoides, Cuscuta salina, Distichlis littoralis, D. spicata, Dudleya blochmaniae, Encelia, Eriogonum parvifolium, Erodium botrys, Eryngium pendletonensis, Eucalyptus, Foeniculum vulgare, Frankenia palmeri, F. salina, Glebionis coronarium, Grindelia hirsutula, Heliotropium curassavicum, Heterotheca grandiflora, Hordeum depressum, H. marinum, H. murinum, Isocoma menziesii, Jaumea carnosa, Juncus, Lactuca serriola, Lasthenia gracilis, Lepidium latifolium, Leymus triticoides, Limonium californicum, Lolium perenne, Lotus corniculatus, Melilotus albus, Mesembryanthemum crystallinum, M. nodiflorum, Myoporum laetum, Oenothera elata, Parapholis incurva, Phragmites australis, Plantago, Platanus racemosa, Poa secunda, Polypogon monspeliensis, Populus fremonti, Puccinellia, Raphanus, Rhus integrifolia, Rosa californica, Rubus ursinus, Rumex, Ruppia maritima, Salicornia bigelovii, S. depressa, S. maritima, Salix exigua, S. gooddingii, S. laevigata, S. lasiolepis, Sambucus nigra, Schoenoplectus acutus, S. americanus, S. californicus, Sesuvium verrucosum, Sidalcea malviflora, Sisyrinchium, Sonchus oleraceus, Spartina foliosa, Spergularia salina, Suaeda californica, S. esteroa, S. taxifolia, Symphyotrichum subulatum, Tamarix, Triglochin maritimum, Typha angustifolia, T. domingensis, T. latifolia.*

Sarcocornia utahensis **(Tidestr.) A.J. Scott** grows on saline beaches, bottoms, flats, marshes (coastal or inland), meadows, missle ranges, playas, woodlands, and along the margins of springs at elevations of up to 1584 m. The sites typically are characterized by open to lightly shaded exposures. The substrates are described as alkaline or saline and include clay, gypsum, sand, St. Lucie sand, and Typic Endoaquepts. A surface salt crust is often present. The plants withstand saline conditions by their ability to accumulate large numbers of Na^+ and Cl^- ions. Flowering and fruiting occur from May to August. This species is well represented in the seed bank (509–2457 seeds/m^2). The seeds not only germinate in the light at high salt concentrations (1%–5% NaCl) at 15°C–5°C but also will germinate (at reduced levels) in nonsaline conditions in the dark. Although only 65% germination has been observed in distilled water, germination generally decreases with increasing salinity, being reduced to 50% at 300 mM NaCl and less than 5% at 900 mM NaCl. The seeds are dispersed abiotically by water and wind. Vegetative reproduction occurs by means of rhizomes. **Reported associates:** *Allenrolfea occidentalis, Arthrocnemum subterminale, Atriplex, Avicennia germinans, Batis maritima, Borrichia frutescens, Chara, Cladium jamaicense, Cressa truxillensis, Distichlis littoralis, D. spicata, D. stricta, Eleocharis rostellata, Helianthus, Holosteum umbellatum, Juncus roemerianus, Lactuca, Limonium limbatum, Nuttallanthus floridanus, Phragmites australis, Pinus, Pseudoclappia arenaria, Salicornia rubra, Sarcobatus vermiculatus, Schoenoplectus americanus, Seutera angustifolia, Smilax auriculata, Spartina patens, Sporobolus airoides, Suaeda, Tamarix ramosissima.*

Use by wildlife: *Sarcocornia ambigua* is a host plant for caterpillars of the eastern pygmy blue butterfly (Insecta: Lepidoptera: Lycaenidae: *Brephidium isophthalma pseudofoea*). The species also provides habitat for the Atlantic salt marsh snake (Reptilia: Colubridae: *Nerodia clarkii taeniata*). *Sarcocornia pacifica* stands are associated with several insects (Insecta: Coleoptera: Chrysomelidae, Coccinellidae, Curculionidae; Diptera: Chironomidae, Chloropidae, Dolichopodidae, Ephyridae, Sepsidae, Tipulidae; Hemiptera: Anthocoridae, Cixicae, Cicadellidae, Lygaeidae, Miridae, Psyllidae; Thysanoptera) and spiders (Arthropoda: Arachnidae: Araneae: Araneidae, Dictynidae, Linyphiidae, Tetragnathidae).

Economic importance: food: The fleshy stems of *S. ambigua* are used as food by the Coast Salish people. The plants are also pickled or eaten as a salad by Alaskan natives; **medicinal:** *Sarcocornia ambigua* has been used to relieve pain from arthritis and rheumatism and to treat various aches and swelling; **cultivation:** *Sarcocornia* species are unsuitable for general gardening because of their high alkalinity requirement; **misc. products:** The seed oil from *S. ambigua* has been recommended for use as an animal feed and for biofuel production; **weeds:** none; **nonindigenous species:** none.

Systematics: The precise relationships of *Sarcocornia* in Amaranthaceae have remained unsettled. Results of *rbcL* data analysis resolved the genus as firmly embedded within subfamily Salicornioideae, close phylogenetically to *Arthrocnemum, Halopeplis,* and *Kalidium* but relatively distant from *Salicornia* (Figure 3.5). The same data indicated the monophyly of *Sarcocornia* with respect to the two species surveyed (*S. utahensis* and *S. blackiana*), which resolved as a clade. However, more thorough sampling of taxa and loci has depicted *Sarcocornia* as polyphyletic, with two distinct clades enveloping a third clade comprising the species assigned

to *Salicornia* (Figure 3.11). Consequently, some authors (e.g., Ball, 2012) have merged the genera *Sarcocornia* and *Salicornia*, while others continue to recognize them as distinct (as done here), pending the taxonomic disposition (i.e., a reassignment of taxa to another genus) of the "extra" clade of species. Taxonomic problems within the genus have also caused serious confusion. At one time or another, the name *S. virginica* has been applied to nearly all of the New World species; however, the type of that name actually belongs to a species of *Salicornia* rather than to any member of *Sarcocornia*. North America treatments continue to recognize the name *S. perennis* in the flora but that taxon is Eurasian and has been confused in this region with plants that actually belong to *S. ambigua* or *S. pacifica*. Therefore, it is very difficult to tell from literature accounts (especially when attributed to North American "*S. perennis*") whether the content actually refers to *S. ambigua* or to *S. pacifica*. Although some authors (e.g., Ball, 2012) have regarded *S. perennis* and *S. pacifica* as being similar morphologically, they are not particularly closely related (Figure 3.10). The base chromosome number of *Sarcocornia* is $x = 9$. *Sarcocornia ambigua* ($2n = 18$) is diploid; counts are uncertain for the other OBL species.

Comments: (*Note*: The following distributions are probably imprecise, given remaining taxonomic uncertainties in this genus.) *S. ambigua* is distributed along eastern and northwestern coastal North America; *S. pacifica* occurs along eastern and western coastal United States. *S. utahensis* occurs sporadically in western and southwestern North America.

References: Andersen, 1996; Aronson, 1989; Ball, 2003b; Buck-Diaz et al., 2012; Callaway & Zedler, 1998; Diggory & Parker, 2011; Eberl, 2011; Figueroa et al., 2003; Freitas & Costa, 2014; Gul & Khan, 2003; Gul & Weber, 2001; Gul et al., 2009; Haines, 2000; Kadereit et al., 2003, 2007; Kahn & Weber, 1986; Keammerer & Hacker, 2013; Muldavin et al., 2000; Reynolds & Boyer, 2010; Rosen & Zamirpour, 2014; Ryan & Boyer, 2012; Steffen et al., 2015; Stralberg et al., 2010; Turner & Bell, 1971.

8. *Suaeda*

Sea-blite, seepweed

Etymology: from the Arabic *suaed* meaning "black" for a dye obtained from the plants

Synonyms: *Brezia; Chenopodium* (in part); *Dondia; Salsola* (in part); *Schoberia*

Distribution: global: cosmopolitan; **North America:** widespread

Diversity: global: 115 species; **North America:** 12 species

Indicators (USA): OBL: *Suaeda conferta, S. linearis, S. maritima, S. nigra*; **FACW:** *S. maritima*; **FAC:** *S. maritima*

Habitat: saline (coastal, inland); palustrine; **pH:** 5.8–8.5; **depth:** <1 m; **life-form(s):** emergent (herb, shrub, subshrub)

Key morphology: shrubs, subshrubs, or herbs; stems (to 10 dm) branched; leaves (to 50 dm) sessile, linear or narrowly ellipsoid, subterete, often fleshy; flowers small (to 3.5 mm), sessile, 1–12 in axillary clusters (glomes); fruit a utricle, containing a flat or lenticular, black or brown seed (to 2.2 mm) disposed horizontally or vertically

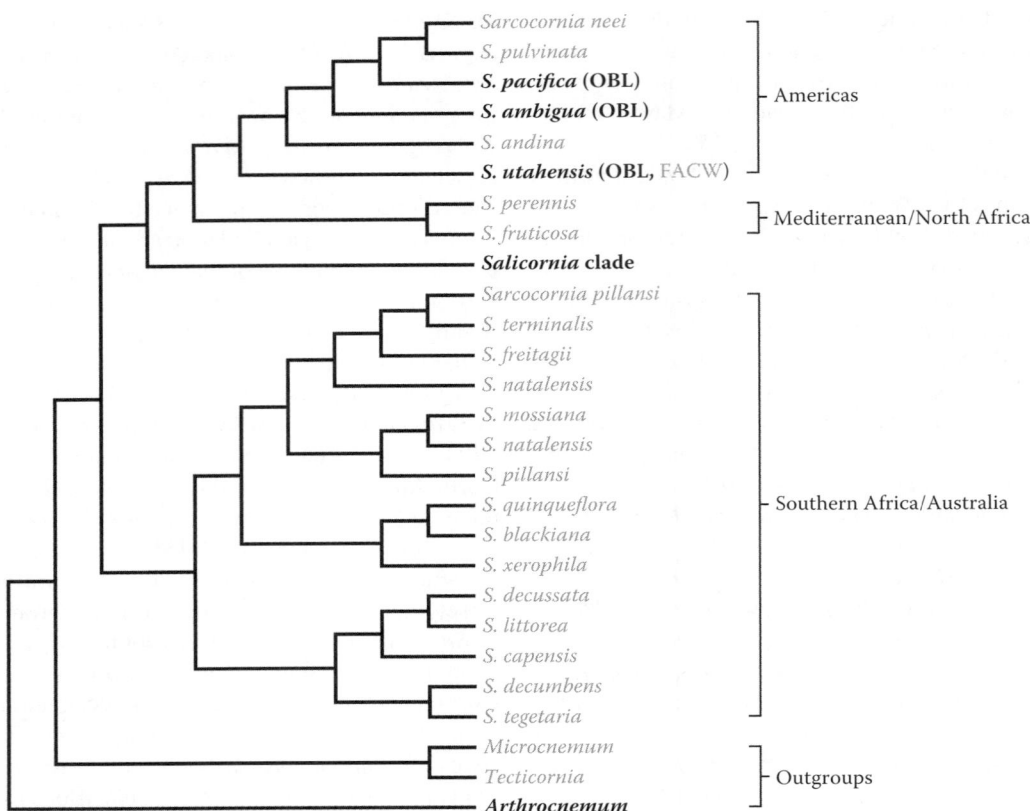

FIGURE 3.10 Phylogenetic relationships in *Sarcocornia* as indicated by analysis of combined cpDNA (*atpB–rbcL*, *rpL32–trnL* spacers) and nrETS sequences. *Sarcocornia* is polyphyletic as circumscribed, resolving in two clades, which each is distinct from *Salicornia*. The three OBL indicators (designated in bold) resolve within the clade of American species. (Adapted from Steffen, S. et al., *Ann. Bot.*, 115, 353–368, 2015.)

Life history: duration: annual, perennial (fruit/seeds, woody base); **asexual reproduction:** branches rooting at nodes, resprouting; **pollination:** wind; **sexual condition:** gynomonoecious; **fruit:** utricles (common); **local dispersal:** seeds; **long-distance dispersal:** seeds

Imperilment: (1) *Suaeda conferta* [G4/G5]; (2) *S. linearis* [G5]; S1 (NY, WV); S3 (DE, MD, VA); (3) *S. maritima* [G5]; S1 (ME, <u>NS</u>); S2 (MA); S3 (<u>NF</u>, <u>QC</u>); (4) *S. nigra* [G5]

Ecology: general: *Suaeda* species occur most often in various alkaline or saline, coastal or inland wetlands, but sometimes extend into upland sites. Ten of the North American species (83%) are wetland indicators, with four (33%) designated as OBL. The genus includes both annual and perennial life histories. These plants are able to concentrate salt in their vegetative tissues and have been considered widely for use in habitat desalinization programs. Pollination and seed dispersal are abiotic. The seeds of most *Suaeda* species examined require cold stratification followed by temperatures of 25°C–27°C to induce germination, which is reduced at NaCl concentrations above 0.17 M. They are dispersed primarily by tidal waters, which transport their buoyant fruits over relatively short distances.

Suaeda conferta (**Small**) **I.M. Johnst.** is a perennial that occurs on beaches, dunes, and saline flats at elevations of up to 200 m. The plants are usually found on bare substrates,

which can include clay, clay loam, or sand. They tend to colonize slightly elevated sites, which are inundated periodically by high tides. Flowering occurs from spring to December. Numerous, small seeds are produced, which germinate on disturbed substrates. These are dense, mat- or tuft-forming C$_4$ shrubs with prostrate, spreading branches that root at their nodes. **Reported associates:** *Amaranthus, Astrophytum asterias, Atriplex matamorensis, A. pentandra, Batis maritima, Borrichia frutescens, Castela texana, Cylindropuntia leptocaulis, Digitaria ciliaris, Distichlis littoralis, D. spicata, Gaillardia pulchella, Heliotropium curassavicum, Heterotheca subaxillaris, Iva annua, Lycium carolinianum, Parkinsonia texana, Prosopis glandulosa, Rayjacksonia phyllocephala, Salicornia bigelovii, S. depressa, Sesuvium portulacastrum, Sporobolus pyramidatus, Suaeda linearis, S. tampicensis, Uniola paniculata, Vachellia rigidula, Varilla texana, Ziziphus obtusifolia.*

Suaeda linearis (**Elliott**) **Moq.** is a FAC annual species that inhabits brackish or saline beaches, depressions, ditches, dunes, flats, marshes (coastal), meadows, mudflats, roadsides, salt marshes, sandbars, scrub (coastal), shell mounds, shores, strands, and the margins of hammocks and swamps (mangrove) at elevations of up to 10 m. The plants are very salt tolerant (electrical conductivity up to 70.1 dS/m) and usually occur in the zone from 1.0 to 1.5 m above mean high tide levels where salinity values can range from 10 to 50 ppt. However, the plants do poorly in sites that are subject to continual waterlogging. The substrates are saline and include cobble, clay, dredge spoil, mud, sand, sandy clay, sandy marl, shell fragments, and unconsolidated sand. This species has a complex life history, with annual populations occurring in the north of its range but persisting in the south of its range as perennials (from a slightly woody base). Flowering and fruiting occur from June to December. The dimorphic seeds are produced in great quantity and germinate freely under full sunlight or shaded conditions. A substantial seed bank develops. The seeds are very tolerant to salinity with a small number (~1%) able to germinate even at salt concentrations as high as 849 mM NaCl. The seedlings do not emerge at higher tidal elevations, where stands of *Spartina alterniflora* (Poaceae) increase substrate stability by reducing (up to 60%) flow-related physical disturbances. The plants fix carbon by means of a C_3 pathway. They do not compete well with other dominant salt marsh species, except on substrates comprising oyster shell deposits, which create conditions unfavorable to the other plants. **Reported associates:** *Agalinis maritima, Allowissadula lozani, Amaranthus greggii, Ambrosia artemisiifolia, A. psilostachya, Andropogon glomeratus, Atriplex cristata, A. prostrata, Avicennia germinans, Baccharis halimifolia, Bacopa monnieri, Baptisia leucophaea, Batis maritima, Blutaparon vermiculare, Bolboschoenus robustus, Borrichia frutescens, Bothriochloa laguroides, Cakile edentula, C. geniculata, Casuarina equisetifolia, Chamaecrista fasciculata, Chenopodium album, Chromolaena odorata, Citharexylum berlandieri, Commelina erecta, Conocarpus erectus, Conoclinium betonicifolium, Cressa nudicaulis, Croton punctatus, Cyperus articulatus, C. elegans, C. oxylepis, Dalea emarginata, D. scandens, Desmanthus virgatus, Dichanthium aristatum, D. portoricense, Distichlis littoralis, D. spicata, Echinocactus texensis, Echinocereus, Eleocharis, Eragrostis secundiflora, Erigeron procumbens, Euphorbia bombensis, E. mesembrianthemifolia, E. polygonifolia, Eustoma exaltatum, Fimbristylis caroliniana, F. spadicea, Forestiera angustifolia, F. segregata, Gaillardia pulchella, Havardia pallens, Heliotropium angiospermum, H. curassavicum, Heterotheca subaxillaris, Hydrocotyle bonariensis, Hypericum gentianoides, Indigofera miniata, Ipomoea imperati, I. pes-caprae, Isocoma drummondii, Iva frutescens, I. imbricata, Juniperus virginiana, Karwinskia humboldtiana, Leucophyllum frutescens, Limonium carolinianum, L. nashii, Lycium carolinianum, Lyonia, Lysimachia maritima, Lythrum alatum, Maytenus phyllanthoides, Metastelma barbigerum, Mimosa latidens, Muhlenbergia capillaris, Myrica cerifera, Neptunia pubescens, Oenothera drummondii, Opuntia engelmannii, O. pusilla, Panicum amarum, Persea borbonia, Phragmites australis, Phyla nodiflora, Physalis cinerascens, Pinus elliottii, Plantago maritima, P. rhodosperma, Pluchea odorata, Polygala alba, Polypremum procumbens, Portula pilosa, Portulaca oleracea, Prosopis reptans, Ratibida peduncularis, Rayjacksonia phyllocephala, Rhizophora, Rhynchosia americana, R. minima, R. senna, Rhynchospora colorata, Sabal palmetto, Sabatia, Salicornia bigelovii, S. depressa, Salsola kali, Samolus ebracteatus, Sarcocornia perennis, Schizachyrium scoparium, Schoenoplectus californicus, Serenoa repens, Sesuvium portulacastrum, Setaria parviflora, Sideroxylon tenax, Sisyrinchium biforme, Solanum americanum, Solidago sempervirens, Sophora tomentosa, Spartina alterniflora, S. patens, S. spartinae, Spiranthes vernalis, Sporobolus airoides, S. pyramidatus, S. virginicus, Stemodia lanata, Stenaria nigricans, Suaeda conferta, S. maritima, Tetragonia tetragonioides, Thrinax morrisii, Triglochin maritimum, Tripogandra serrulata, Uniola paniculata, Yucca treculeana, Zanthoxylum.*

***Suaeda maritima* (L.) Dumort.** is an annual species that occurs in or on coastal beaches, dunes, flats, pannes, salt marshes, shores, and strands at elevations of up to 8 m. The plants are fairly salt tolerant (electrical conductivities up to 10.0 dS/m) and typically occur in the intertidal zone, often near the high tidal limit (e.g., 6–7 m), where they are subject to frequent tidal inundation (flooded by 95% of tides annually). Optimal growth has been reported to occur at a NaCl content of 0.92%; however, Americn populations reportedly do not persist where NaCl concentrations exceed 1%. The plants can become submerged 31–564 times each year (for a total of 2–118 h) depending on whether they occur in the upper or lower intertidal regions, respectively. The substrates (pH: 5.8–8.3) are characterized as alkaline, brackish, or saline and include ballast, clay, clay silt, cobble, gravel, Kittery quartzite, peat, peaty silt, rock, rocky mud, sand, sand-pebble gravel, sandy rock, and shells. Flowering and fruiting have been observed from July to October. The flowers are weakly protandrous but readily capable of self-pollination. Individiual plants can produce upward of 1000 seeds. Seed dispersal commences in October and is facilitated by tidal waters. The buoyant seeds can remain afloat for at least 60 h. Natural germination occurs in the following spring (March). Germination rates are relatively low but are highest (to 24%) in fresh water once the seeds have been frozen at −20°C. Germination rates decline substantially (2%–6%) at NaCl concentrations from 0.5% to 1.0%. Higher germination rates occur when the seeds have been scarified, placed under cold stratification (e.g., 15 days at 4°C), or after being subjected to a 12/12 h 5°C/20°C light/temperature regime. Soil disturbance from foraging fiddler crabs (Crustacea: Ocypodidae: *Uca pugnax*) negatively impacts seedling establishment. European populations of *S. maritima* (possibly some in North America) exhibit seed dimorphism. Photosynthesis occurs by means of a C_3 pathway. **Reported associates (North America):** *Ascophyllum nodosum, Atriplex cristata, A. mucronata, A. patula, A. portulacoides, A. prostrata, Distichlis spicata, Fucus, Iva frutescens, Juncus gerardii, Limonium carolinianum, L. nashii, Lysimachia maritima, Plantago maritima, Potentilla anserina, Puccinellia maritima, Salicornia depressa, Sesuvium portulacastrum, Setaria magna, Solidago sempervirens, Spartina alterniflora, S. patens, Spergularia canadensis, S. salina, Sporobolus virginicus, Suaeda linearis, Tripolium pannonicum.*

***Suaeda nigra* J.F. Macbr.** is a perennial subshrub that inhabits bottomlands, flats, floodplains, gullies, interior lake beds (ephemeral), irrigation ditches, meadows, playas, ravines, roadsides, salt flats, saline marshes (coastal not estuarine), scrub, seeps, sinks, slopes, washes, and the margins of lakes, rivers, and vernal pools at elevations of up to 2012 m. It is often the dominant species on alkaline sinks. Exposures occur in full sun. The substrates are described as alkaline/saline and include adobe, alluvium, bentonite, boulders, Chinle clay, clay, clay shale, cobble, gravel, gypsum, loam, limestone, marl, mud, Pescadero silty clay, sand, sandy loam, shale, silt, silty clay, silty loam, silty sand, sandy clay, talus, and ustertic natrargids. Flowering and fruiting occur from May to December. Most of the seeds mature from September to October and germinate better following a period of cold stratification. The shoots resprout readily following mowing or similar disturbances. This is a fast-growing C_4 species, which is variable genetically and exhibits extensive phenotypic plasticity. **Reported associates:** *Achnatherum hymenoides, Alhagi maurorum, Alisma, Allenrolfea occidentalis, Allium fimbriatum, Ambrosia dumosa, A. salsola, Amorpha, Amsinckia tessellata, Anemopsis californica, Apocynum cannabinum, Artemisia arbuscula, A. ludoviciana, A. pedatifida, A. tridentata, Arthrocnemum subterminale, Asclepias subverticillata, Astragalus amphioxys, Atriplex argentea, A. canescens, A. confertifolia, A. elegans, A. gardneri, A. hymenelytra, A. lentiformis, A. parishii, A. phyllostegia, A. polycarpa, A. rosea, A. serenana, A. spinifera, Baccharis salicifolia, B. sarothroides, B. sergiloides, Bassia hyssopifolia, Bolboschoenus maritimus, Bouteloua barbata, Bromus rubens, B. tectorum, Calochortus striatus, Calycoseris parryi, Carex praegracilis, Carnegiea gigantea, Centromadia pungens, Chaenactis stevioides, Chenopodium leptophyllum, Chilopsis linearis, Chloracantha spinosa, Chylismia multijuga, Cleomella, Cressa truxillensis, Croton californicus, Crypsis schoenoides, Cryptantha angustifolia, Cuscuta, Cylindropuntia echinocarpa, Cynodon dactylon, Cyperus difformis, Dasyochloa pulchella, Datura wrightii, Descurainia pinnata, Distichlis spicata, D. stricta, Ditaxis serrata, Dysphania botrys, Echinocactus polycephalus, Eleocharis, Elymus elymoides, Encelia actoni, E. frutescens, E. resinifera, Ephedra nevadensis, Eragrostis lehmanniana, Eremothera boothii, Eriastrum hooveri, Ericameria nauseosa, Erigeron ochroleucus, Eriogonum brachypodum, E. brevicaule, E. inflatum, E. subreniforme, E. trichopes, Erodium cicutarium, Eschscholzia parishii, Euphorbia abramsiana, E. micromera, E. polycarpa, E. robusta, E. setiloba, Extriplex joaquinana, Ferocactus, Frankenia salina, Geraea canescens, Gutierrezia sarothrae, Halogeton glomeratus, Haplopappus, Helianthus annuus, H. nuttallii, Heliotropium curassavicum, Hemizonia, Hilaria jamesii, H. rigida, Hordeum depressum, H. marinum, H. murinum, Ipomopsis polycladon, Isocoma acradenia, I. pluriflora, I. tenuisecta, Juncus bufonius, Juniperus californica, Kochia americana, K. californica, K. scoparia, Krameria, Langloisia matthewsii, L. setosissima, Larrea tridentata, Lasthenia californica, L. glabrata, Layia, Lepidium dictyotum, L. fremontii, L. lasiocarpum, Lolium multiflorum, Lycium andersonii, L. californicum, L. fremontii, Malacothrix coulteri, Mentzelia affinis, Mirabilis multiflora, Monolepis nuttalliana, Nasturtium officinale, Nicotiana attenuata, N. glauca, N. obtusifolia, Nitrophila mohavensis, N. occidentalis, Oligomeris linifolia, Olneya tesota, Opuntia basilaris, Parkinsonia aculeata, Pascopyrum smithii, Pectis angustifolia, P. papposa, Peritoma serrulata, Phacelia crenulata, Phoradendron californicum, Phragmites australis, Pinus flexilis, Plagiobothrys leptocladus, Plantago ovata, Pluchea sericea, Poa secunda, Polypogon monspeliensis, Populus, Prenanthella exigua, Primula pauciflora, Proboscidea althaeifolia, Prosopis glandulosa, P. juliflora, P. pubescens, P. velutina, Pseudoroegneria spicata, Psorothamnus emoryi, P. fremontii, Punica granatum, Quercus john-tuckeri, Rhaponticum repens, Salazaria mexicana, Salicornia, Salix exigua, Salsola paulsenii, S. tragus, Sarcobatus vermiculatus, Schismus arabicus, Schoenoplectus acutus, S. americanus, S. pungens, Senegalia greggii, Simmondsia chinensis, Solanum elaeagnifolium, Sphaeralcea ambigua, Sporobolus airoides, Stanleya pinnata, Stephanomeria pauciflora, Stutzia dioica, Suaeda calceoliformis, Tamarix aphylla, T. chinensis, T. ramosissima, Tetradymia glabrata, Tidestromia suffruticosa, Tiquilia plicata, Trianthema portulacastrum, Typha latifolia, Valeriana edulis, Washingtonia filifera, Wislizenia refracta, Xanthisma spinulosum, Yucca brevifolia, Ziziphus obtusifolia.* **Use by wildlife:** *Suaeda conferta* provides habitat for the least tern (Aves: Laridae: *Sternula antillarum*). The seeds and foliage of *S. linearis* are eaten by seaside sparrows (Aves: Emberizidae: *Ammodramus maritimus*). The plants are host to several insects (Insecta) including larval planthoppers (Hemiptera: Acanaloniidae: *Acanalonia pumila*; Flatidae: *Cyarda*), leaf mining flies (Diptera: Ephydridae: *Clanoneurum americanum*), and plant bugs (Hemiptera: Miridae: *Creontiades signatus, Melanotrichus leviculus*). *Suaeda nigra* is a host plant for larval butterflies (Insecta: Lepidoptera: Coleophoridae: *Coleophora suaedae*; Gelechiidae: *Chionodes sistrella*; Lycaenidae: *Brephidium exilis*). The herbage of *S. nigra* is eaten occasionally by grazing animals and also is damaged by grasshoppers (Insecta: Orthoptera: Acrididae: *Aeoloplides turnbulli*).

Economic importance: food: The Cahuilla people ate *Suaeda* leaves as greens and ground the seeds (which are high in oil) into flour to make mush and cakes. The Pima and Gila ate the leaves of *S. nigra* when roasted, and the Navajo used its seeds to make porridge. The Papago and Pima tribes lined their cooking pits with leaves and stalks of *S. nigra* to impart a salty flavor to cactus; it was also added to greens and cactus fruit as a spice. *Suaeda linearis* has been used as a sweetener; **medicinal:** The Hopi and Paiute natives used *S. nigra* plants as a poultice for treating various dermatological and kidney disorders. It was used by the Shoshoni to treat bladder problems. The Kayenta and Navajo tribes used *S. nigra* to treat gasterointestinal bleeding. The roots of *S. nigra* were made into a tea for treating colds; **cultivation:** *Suaeda nigra* is found in cultivation occasionally, but most species are not suitable for gardens; **misc. products:** The Cahuilla and Seri people boiled *S. nigra* leaves to make soap and also a black dye for coloring hair, baskets, and palm mats. The dried branches of this species are used for cooking, campfires, and roofing. The Hopi used *S. nigra* plants in ceremonial baths; **weeds:** none; **nonindigenous species:** The occurrence of *Suaeda maritima* in North

America is cryptogenic and possibly represents a European introduction attributable to the disposal of shipping ballast.

Systematics: Analysis of *rbcL* sequence data (Figure 3.5) resolved *Suaeda* as being closely related to *Borszczowia* and *Bienertia* and within a paraphyletic subfamily Suaedoideae, which had also been recognized as tribe *Suaedeae* of subfamily *Salsoloideae*. However, combined DNA sequences (nrITS, *atpB–rbcL*, *matK*, *psbB-psbH*, *rbcL*, *trnL-trnF*) recover a phylogeny with three well-supported clades that correspond to the subfamilies Salicornioideae, Salsoloideae, and Suaedoideae. In that analysis, the latter clade comprises two sister genera (*Bienertia* and *Suaeda*), with *Suaeda* recircumscribed to include several species recognized formerly in *Alexandra* (now = *Suaeda* section *Alexandra*) and *Borszczowia* (now = *Suaeda* section *Borszczowia*). The two OBL species included in the phylogenetic analysis are distantly related (Figure 3.11), with *S. nigra* resolving within the perennial C$_4$ section *Limbogermen* (which also includes *S. conferta*) and *S. maritima* within the annual C$_3$ section *Brezia* (which contains nearly half of the species including *S. linearis*). *Suaeda linearis* and *S. maritima* resolve in different clades within section *Brezia* according to *rpl32-trnL* spacer and nrITS sequence analyses. However, *S. maritima* and *S. nigra* are highly polymorphic, and additional studies are necessary to clarify whether these species should be further subdivided taxonomically. The combined DNA sequence analysis resolved *S. moquinii* and *S. fruticosa* as distinct lineages (Figure 3.11), yet both are currently placed in synonymy with *S. nigra* (*S. nigra* has also been known previously as *S. suffrutescens*). Similarly, *S. prostrata* is treated as a synonym of *S. maritima* but resolves in a relatively distant clade in phylogenetic analyses (Figure 3.11). Clearly, additional taxonomic refinements will be necessary in this genus. The chromosomal base number of *Suaeda* is *x* = 9. *Suaeda nigra* is diploid (2*n* = 18), *S. maritima* is tetraploid (2*n* = 36), and *S. linearis* is hexaploid (2*n* = 54). The chromosome number of *S. conferta* has not been reported.

Comments: *Suaeda linearis* is found throughout coastal eastern North America and extends southward into Mexico and the West Indies; *S. conferta* occurs in coastal Texas and also extends into Mexico and the West Indies; *S. maritima* (possibly introduced) is not only common in northern and eastern North

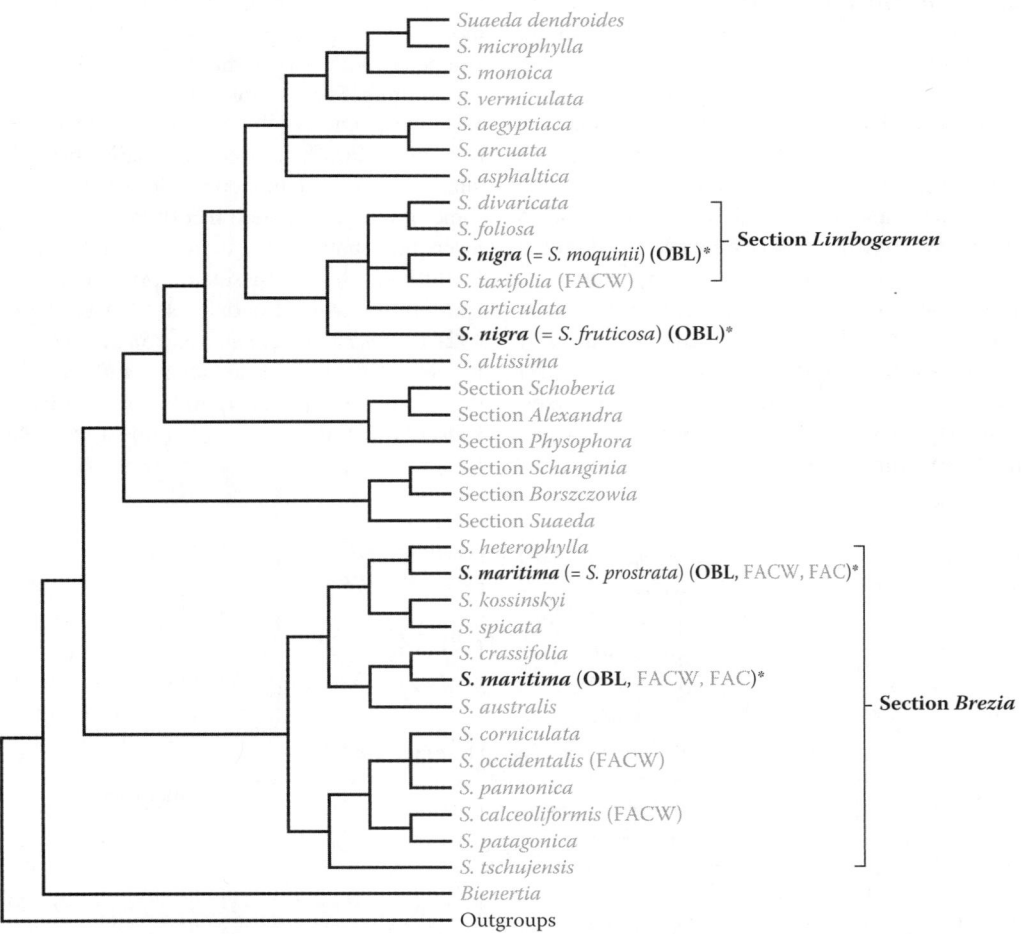

FIGURE 3.11 Phylogenetic relationships in *Suaeda* as indicated by analysis of combined nrITS and cpDNA sequence data. The North American wetland indicator status is given in parentheses, with any OBL taxa highlighted in bold. As shown here, both *S. maritima* and *S. nigra* include distinct taxa (asterisked), which merit recognition at some taxonomic level. (Modified after Kapralov, M.V. et al., *Syst. Bot.*, 31, 571–585, 2006.)

America but also occurs in Africa and Eurasia; *S. nigra* is distributed mainly in the southwestern interior of North America. **References:** Barton et al., 2016; Bartosik, 2010; Bassett & Crompton, 1978; Birnbaum et al., 2011; Boule, 1979; Brandt et al., 2015; Brewer et al., 2012; Bruno & Kennedy, 2000; Chapman, 1947; Chmura et al., 1997; Easley & Judd, 1993; Ewanchuk & Bertness, 2004; Ferren & Schenk, 2003; Fisher et al., 1997; Guo & Pennings, 2012; Henry, 1991; Javed & Urbatsch, 2014; Judd & Lonard, 2002; Judd et al., 1977; Kapralov et al., 2006; Kennedy & Bruno, 2000; Lonard et al., 2016; Perry & Atkinson, 1997; Smith & Tyrrell, 2012; Stutzenbaker, 1999; Thompson & Slack, 1982; Ungar, 1962, 1978; Wheeler, Jr., 1982; Wheeler, Jr. & Hoebeke, 1982.

Family 3: Caryophyllaceae [86]

The Caryophyllaceae contain approximately 2200 species that are found mainly throughout the temperate Northern Hemisphere but are distributed worldwide. The family is monophyletic as evidenced from both morphological and molecular synapomorphies (Judd et al., 2016). Molecular phylogenetic analyses (e.g., Figure 3.1) indicate a close relationship of the family to Achatocarpaceae and Amaranthaceae within the order Caryophyllales, a result that has been recovered consistently upon analyses of additional data (Brockington et al., 2009; Yang et al., 2015).

The family typically contains herbaceous plants with anthocyanin pigments, opposite leaves, swollen nodes, and clawed, apically notched "petals," which actually represent petaloid stamens (Rabeler & Hartman, 2005a; Judd et al., 2016). The flowers are often showy and pollinated by various insects (Insecta), including bees (Hymenoptera), butterflies and moths (Lepidoptera), and flies (Diptera). Many species are protandrous and outcrossed, but some have reduced flowers and are self-pollinated (Judd et al., 2016). The fruits are usually loculicidal capsules (opening by apical teeth or valves) with free-central (or basal) placentation. The seeds are dislodged and dispersed from the capsules by the wind or from the movements of animals (Judd et al., 2016). Some are scattered by raindrops. This group also contains "tumbleweeds" (*Gypsophila*), whereby entire plants are dislodged and rolled across the ground by the wind, thereby dispersing seeds as they rotate.

Many popular ornamental plants occur here, including baby's breath (*Gypsophila*), carnations (*Dianthus*), campions and catchflies (*Lychnis, Silene*), and soapworts (*Saponaria*). The family is also known for some weedy agricultural species, which are often referred to as chickweeds (*Cerastium, Paronychia, Stellaria*).

Most Caryophyllaceae are terrestrial with many of the species commonly inhabiting fairly arid, disturbed, or open sites. In North America, OBL indicators occur within only four genera:

1. *Arenaria* L.
2. *Honckenya* Ehrh.
3. *Spergularia* (Pers.) J. Presl & C. Presl
4. *Stellaria* L.

Three of these genera (*Arenaria, Honckenya*, and *Stellaria*) traditionally have been placed within subfamily Alsinoideae (tribe Alsineae) with *Spergularia* assigned to subfamily Paronychioideae (tribe Polycarpeae). However, analyses of molecular data (e.g., Smissen et al., 2002) began to reveal inconsistencies with the traditionally recognized subfamilial and tribal limits. Subsequent studies, which included a larger number of taxa and combined DNA sequence data (Greenberg & Donoghue, 2011), showed eventually that the recognized subfamilies were not monophyletic and the tribal placements of some of the genera were incorrect (Figure 3.12). In the light of recent information, all four of the OBL genera would be assigned to different tribes (Alsineae, Arenarieae, Sclerantheae, and Sperguleae), which are dispersed across the phylogeny (Figure 3.12). *Arenaria, Honckenya*, and *Stellaria* occur within the limits of subfamily Alsinoideae, which would be monophyletic if tribe Eremogoneae was transferred to subfamily Caryophylloideae; *Spergularia* falls within subfamily Paronychioideae, which is paraphyletic as currently recognized (Figure 3.12).

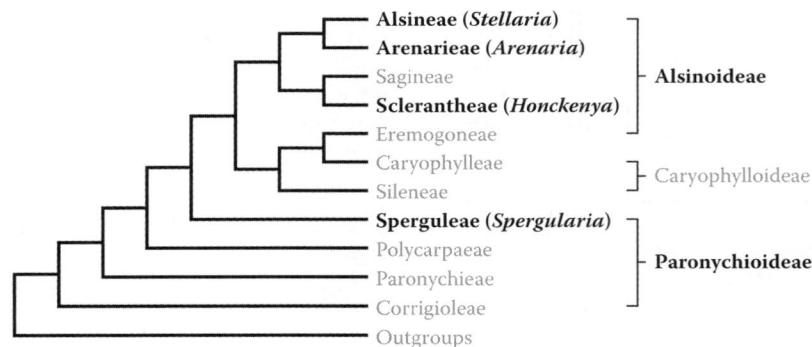

FIGURE 3.12 Phylogenetic relationships at the subfamilial and tribal levels in Caryophyllaceae as indicated by the analysis of combined DNA sequence data. Tribe Eremogoneae would have to be transferred to subfamily Caryophylloideae in order to retain that group and subfamily Alsinoideae as monophyletic. Tribe Sperguleae is recognized within subfamily Paronychioideae, which is paraphyletic. The taxa containing OBL North American indicators (highlighted in bold) are dispersed across the phylogenetic breadth of the family. (Adapted from Greenberg, A.K. & Donoghue, M.J., *Taxon*, 60, 1637–1652, 2011.)

1. *Arenaria*

Sandwort

Etymology: from the Latin *harena* meaning "sand"

Synonyms: *Alsine*; *Alsinopsis*; *Minuartia* (in part)

Distribution: global: Africa; South America; circumboreal; pantemperate; **North America:** widespread

Diversity: global: 200 species; **North America:** 22 species

Indicators (USA): OBL: *Arenaria paludicola*

Habitat: freshwater; palustrine; **pH:** 5.8–8.3; **depth:** <1 m; **life-form(s):** emergent (herb)

Key morphology: stems (to 1 m) angled, trailing, rooting at nodes; leaves (to 3 cm) opposite, lanceolate, acute, slightly connate at base, with a single midvein; flowers solitary, axillary, on long peduncles (to 4 cm); petals (to 6 mm) white; stamens 10; capsules (to 4 mm) ovoid, with 15–20 shiny, compressed seeds (to 0.9 mm)

Life history: duration: perennial (rhizomes); **asexual reproduction:** shoot fragments; **pollination:** unknown; **sexual condition:** hermaphroditic; **fruit:** capsules (infrequent); **local dispersal:** seeds, shoot fragments; **long-distance dispersal:** seeds

Imperilment: (1) *Arenaria paludicola* [G1]; SX (WA); S1 (CA)

Ecology: general: *Arenaria* species grow in shady or open conditions at sea level to alpine (to 2600+ m) elevations and often in disturbed sites. An affinity for dry, sandy sites is evidenced by the common name of "sandwort." Their habitats include cliffs, dunes, roadsides, rock outcrops, and woodlands, with a few species occurring along moist stream margins. Only one North American species (<5%) is a wetland indicator, and it is designated as OBL. The genus includes summer annuals, winter annuals, and perennial species. The flowers are self-compatible and can vary from being strongly protandrous or protogynous and outcrossing, to cleistogamous and self-pollinating, sometimes even within a single species. Outcrossing is facilitated by insect pollinators (Insecta), which include bees (Hymenoptera: Andrenidae: *Andrena*; Halictidae), flies (Diptera: Anthomyiidae; Bombyliidae; Chloropidae; Ephydridae; Stratiomyidae; Syrphidae), and occasionally butterflies and moths (Lepidoptera), all of which feed on the pollen and/or nectar. The sweat bees (Andrenidae: *Andrena*) are particularly important pollinators in western North America. Most of the seeds are believed to be dispersed passively. The seeds of some species are dispersed by ants (Insecta: Hymenoptera: Formicidae) and possibly also by birds (Aves).

Arenaria paludicola **B.L. Rob.** is a perennial species that grows in coastal, freshwater cienega, dunes, marshes, meadows, swamps, thickets, and along the margins of lakes at elevations of up to 304 m. Despite their OBL status, the plants do not tolerate complete immersion for prolonged periods but can be found in a few centimeters of standing water. They can tolerate salinities of up to 3 ppt, but experience low survival rates at sites with levels from 4 to 5 ppt. The plants occur within a Mediterranean climate and occupy saturated, predominantly sandy, acidic soils (pH: 5.8–8.3) with a high organic content. The sites have open to partially shaded exposures (5%–80% openness) with some disturbance. Flowering and fruiting occur from May to June; otherwise, information on the reproductive ecology of this species is scarce. Each capsule produces 20–25 seeds, which germinate well under ambient greenhouse light and temperature conditions but not following wet vernalization treatments. The seeds remain viable in the seed bank for an undetermined length of time. The plants are able to root at their nodes, making them easy to propagate vegetatively. They are known to grow from the tussocks of sedges (*Carex*) in some cases. The species is threatened by habitat loss, lowered water tables, and the encroachment of woody species. **Reported associates:** *Athyrium, Berula erecta, Calamagrostis nutkaensis, Carex aurea, C. cusickii, Epilobium ciliatum, Hydrocotyle, Juncus effusus, Lonicera involucrata, Lupinus polyphyllus, Mimulus guttatus, Myrica californica, Oenanthe sarmentosa, Potentilla anserina, Ribes divaricatum, Rubus ursinus, Salix lasiolepis, Schoenoplectus pungens, Scirpus microcarpus, Sisyrinchium angustifolium, Sparganium eurycarpum, Typha domingensis, T. latfolia, Urtica dioica.*

Use by wildlife: *Arenaria paludicola* is probably of little wildlife value because of its extreme rarity.

Economic importance: food: none; **medicinal:** none; **cultivation:** *Arenaria paludicola* is not cultivated and is restricted from commerce under the provisions of the Endangered Species Act; **misc. products:** none; **weeds:** none; **nonindigenous species:** none

Systematics: Preliminary molecular data (*ndhF* sequences) indicated that *Arenaria* formed a paraphyletic association with *Moehringia* with both genera together comprising a distinct clade within subfamily Alsinoideae. More comprehensive surveys using combined molecular data resolved *Arenaria* as monophyletic (but for a few misplaced species of *Moehringia*) and a sister clade to *Moehringia*. A subsequent analysis of 132 taxa demonstrated the monophyly of *Arenaria* (s.s.) and *Eremogone*, which has been treated as a subgenus of the former. However, despite the intensive sampling of *Arenaria* in these phylogenetic analyses, *A. paludicola* has not yet been included, perhaps because of its extreme rarity. The basic chromosome number of *Arenaria* is uncertain ($x = 7, 8, 10, 11$); no counts have been obtained for *A. paludicola*.

Comments: *Arenaria paludicola* is extremely rare in North America, known only from two populations (10 and 85 individuals, respectively) in San Luis Obispo County, California (it is believed to be extirpated from Washington state). It also occurs along the Pacific coast of Mexico, extending southward to Guatemala.

References: Bonilla-Barbosa & Novelo Retana, 1995; Bontrager et al., 2014; Hartman et al., 2005; Nepokroeff et al., 2001, 2002; USFWS, 1998a; Wyatt, 1986.

2. *Honckenya*

Seaside sand plant

Etymology: after G. H. Honckeny (1724–1805)

Synonyms: *Adenarium*; *Ammonalia*; *Arenaria* (in part); *Halianthus*; *Honkenya* (orthographic variant); *Minuartia* (in part)

Distribution: global: Asia; Europe; North America; **North America:** northern
Diversity: global: 1 species; **North America:** 1 species
Indicators (USA): OBL; FACU: *Honckenya peploides*
Habitat: saline (coastal) palustrine; **pH:** 6–7; **depth:** <1 m; **life-form(s):** emergent (herb)
Key morphology: shoots (to 5 dm) fleshy, much-branched, freely rooted; leaves (to 5 cm) opposite, succulent, acute to acuminate; flowers solitary in upper axils; petals 5, small (to 3 mm), white; stamens 10, each with a basal nectary
Life history: duration: Perennial (tap root, vertical rhizomes); **asexual reproduction:** Stolons, rhizomes; **pollination:** Insect; **sexual condition:** Dioecious or hermaphroditic; **fruit:** Capsules (common); **local dispersal:** Seeds, rhizomes, stolons; **long-distance dispersal:** Capsules (sea currents)
Imperilment: (1) *Honckenya peploides* [G5]; SH (DE, MD); S1 (VA); S2 (MB); S3 (CT, PE)
Ecology: general: Monotypic (see next)

   Honckenya peploides **(L.) Ehrh.** inhabits beaches (coastal), blowouts, coppice dunes, crevices, dunes, flats, gravel bars, marshes, meadows, mudflats, outwash (glacial), salt marshes, sand flats, seepage slopes, shingle beaches, shores, silt bars, tundra, woodlands, and the margins of lakes and rivers at elevations of up to 600 m. This species is regarded as an OBL indicator only in the Alaskan region [facultative upland (FACU) otherwise]. The sites are brackish to saline (electrical conductivities up to 56 dS/m) and generally occur above the tidal limit. The plants often form low mounds (25–50 cm high) on beaches, which are called embryo dunes. The substrates are infertile (low in organic matter content) and include cobble, gravel, rocky gravel, sand, sandy gravel, sandy rock, silty stone, and stones. Flowering and fruiting occur from April to September. Sex expression in *H. peploides* is "subdioecious" with male, female, and hermaphrodite-flowered individuals. The female flowers have reduced perianths and staminodes, whereas the male flowers have a larger, white perianth and reduced gynoecium. The fruits mature (on female plants) from July onward, with most male clones being seed sterile. Occasionally, seeds develop on the first flowers produced by plants, which afterward function exclusively as males. Female flowers always produce more seeds than the hermaphrodites. The seeds have physiological dormancy and germinate after 122 days of cold stratification followed by a 15°C/5°C temperature regime. They germinate better in the light, under alternating temperatures, and in NaCl concentrations up to about 0.20 M. The capsules float in seawater and facilitate seed dispersal along shorelines. The seedlings are intolerant of sand burial and high salinities. Plant emergence commences in March and peaks in April. Seeds that germinated from open-pollinated female plants yielded 55% male offspring. Seeds collected from the early hermaphroditic phase of the male plants yielded female offspring in a 1:3 ratio, which indicates heterogametic sex determination in the male plants. The male plants initially allocate twice the resources to reproduction than females, but the pattern reverses throughout the season. Above-ground biomass is comparable for both the sexes, but allocation to below-ground structures varies, with females usually having higher values. Proportionally, female plants allocate more biomass to reproduction but grow less and more slowly than males, independent of nutrient levels or stress (e.g., under salt spray conditions). Most plants become dormant by November. The plants are highly clonal and spread effectively by runners. The branching stems develop adventitious roots as they are buried by the sand and provide effective protection against erosion. Despite their clonal growth, the plants retain unusually high levels of genetic variation within their populations. *Honckenya peploides* was one of the first angiosperm species to colonize the newly formed volcanic island of Surtsey and high levels of genetic variation also persist there. Carbon fixation occurs by means of a C$_3$ pathway. The plants become fairly rare at the southern limits of their distributional range. **Reported associates:** *Abronia latifolia, Alnus viridis, Alopecurus, Ammophila breviligulata, Angelica, Arctopoa eminens, Artemisia norvegica, Atriplex dioica, A. gmelinii, Barbarea, Bidens amplissima, Botrychium yaaxudakeit, Cakile edentula, C. maritima, Calystegia sepium, Carex glareosa, C. macrocephala, Castilleja, Chamerion angustifolium, Chenopodium berlandieri, Chrysanthemum arcticum, Cochlearia officinalis, Conioselinum, Deschampsia cespitosa, Elymus mollis, Empetrum, Equisetum arvense, Festuca baffinensis, Galium triflorum, Gaultheria, Gnaphalium obtusifolium, Lathyrus japonicus, Leymus arenarius, L. mollis, Ligusticum scothicum, Limonium carolinianum, Lonicera japonica, Lupinus nootkatensis, Lysimachia maritima, Mertensia maritima, Panicum virgatum, Papaver lapponicum, Picea, Populus trichocarpa, Potentilla anserina, P. fragiformis, P. villosa, Pseudotsuga menziesii, Puccinellia nutkaensis, P. phryganodes, P. tenella, Rosa nutkana, R. rugosa, Rumex, Salsola kali, Senecio pseudoarnica, Solidago sempervirens, Stellaria humifusa, Suaeda maritima.*

Use by wildlife: *Honckenya peploides* has been used as fodder for pigs (Mammalia: Suidae: *Sus scrofa*) and sheep (Mammalia: Bovidae: *Ovis aries*). The foliage provides nesting cover that is used by common and king eiders (Aves: Anatidae: *Somateria mollissima, S. spectabilis*). Mounds of the plants resemble female eiders and provide camouflage for them. The plants are hosts of rust Fungi (Basidiomycota: Pucciniaceae: *Uromyces accuminatus*).

Economic importance: food: Leaves and shoots of *H. peploides* (reputedly high in vitamins A and C) are eaten as a vegetable (similar to sauerkraut) or with dry fish by the Inupiat Eskimos. Chopped, cooked, and soured leaves of *H. peploides* are an ingredient in some Eskimo ice cream; **medicinal:** *Honckenya peploides* has been attributed with unspecified medicinal properties; **cultivation:** *Honckenya peploides* is cultivated as a salinity-tolerant ornamental; **misc. products:** none; **weeds:** none; **nonindigenous species:** none

Systematics: *Honckenya* has been placed within subfamily Alsinoideae, tribe Alsineae of Caryophyllaceae. However, phylogenetic studies indicate that it should be assigned to tribe Sclerantheae (Alsonoideae) instead. Although *Honckenya* has been synonymized with both *Arenaria* and *Minuartia*, preliminary molecular data (*matK, trnLC-F* sequences) indicated

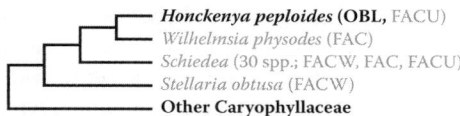

Honckenya peploides (**OBL,** FACU)
Wilhelmsia physodes (FAC)
Schiedea (30 spp.; FACW, FAC, FACU)
Stellaria obtusa (FACW)
Other Caryophyllaceae

FIGURE 3.13 The phylogenetic position of *Honckenya* as indicated by analysis of combined DNA sequence data for a large number of Caryophyllaceae taxa. North American wetland indicator categories are given in parentheses, with the taxa that contain OBL species highlighted in bold. (Modified after Greenberg, A.K. & Donoghue, M.J., *Taxon*, 60, 1637–1652, 2011.)

that it was more closely related to *Wilhelmsia* with both genera comprising a sister clade to the Hawaiian representatives (i.e., *Schiedea*) of subfamily Alsinoideae. Those relationships have been confirmed by subsequent, more comprehensive studies (e.g., Figure 3.13). *Honckenya peploides* has variable chromosome numbers reported of $2n = 48, 66, 68$, and 70. However, the correct number is $2n = 68$, which indicates a tetraploid derived from the basic number of $x = 17$.
Comments: *Honckenya peploides* occurs in northern coastal areas of North America and extends into Eurasia.
References: Anderson, 1939; Árnason et al., 2014; Aronson, 1989; Baillie, 2012; Bittrich, 1993; Brysting et al., 2001; Fridriksson & Johnsen, 1968; Gagné & Houle, 2002; Greenberg & Donoghue, 2011; Malling, 1957; Nepokroeff et al., 2001; Sánchez-Vilas & Retuerto, 2012; Sánchez-Vilas et al., 2010, 2012; Schamel, 1977; Wagner, 2005b; Welsh, 1974; Yun et al., 2010.

3. *Spergularia*

Sand-spurry, sandwort, sea-spurry; spergulaire
Etymology: probably from the Latin *spargere*, meaning "to scatter"
Synonyms: *Alsine*; *Arenaria* (in part); *Balardia*; *Buda*; *Delia*; *Lepigonum*; *Tissa*
Distribution: global: cosmopolitan; **North America:** widespread
Diversity: global: 60 species; **North America:** 12 species
Indicators (USA): OBL: *Spergularia canadensis, S. echinosperma, S. platensis, S. salina*; **FACW:** *S. canadensis, S. platensis, S. salina*; **FACU:** *S. echinosperma*
Habitat: freshwater, saline palustrine; **pH:** 4.1–7.0; **depth:** <1 m; **life-form(s):** emergent (herb)
Key morphology: stems (to 3.5 dm) few to numerous from base; leaves (to 4.5 cm) opposite, linear, often fleshy, blunt tipped; stipules deltoid; flowers in cymes; petals (to 4.0 mm) 5, white, pink, or rose colored; capsules (to 6.4 mm) ovoid, dehiscent by three valves; seeds (to 150+; to 1.4 mm) margins winged or wingless
Life history: duration: annual (fruit/seeds); **asexual reproduction:** none; **pollination:** insect, self; **sexual condition:** hermaphroditic; **fruit:** capsules (common); **local dispersal:** seeds (animals, wind); **long-distance dispersal:** seeds (animals, water)
Imperilment: (1) *Spergularia canadensis* [G5]; SX (NY); S1 (CT, RI, SK); S3 (ON); (2) *S. echinosperma* [GNR]; (3) *S. platensis* [GNR]; (4) *S. salina* [G5]; SH (OK); S1 (NC, ON); S2 (AB); S3 (QC)

Ecology: general: *Spergularia* includes annuals, biennials, and perennial species that often grow in alkaline or saline habitats characterized by sandy substrates (hence their common names). Many of the species occupy sites associated with wet and disturbed conditions. Ten of the North American species (83%) are wetland indicators, with four (33%) designated as OBL. The flowers are regarded as "promiscuous" in terms of their pollination biology, which does not rely on a specific vector. Some flowers remain open for long periods, produce abundant quantities of nectar, and are outcrossed by various insects (Insecta); others are cleistogamous and self-pollinating. The seeds are often winged, which facilitates their abiotic dispersal by water or wind. They are also transported externally or endozoically by birds (Aves), especially by ducks (Anatidae), and mammals (Mammalia). Several species produce seeds that are both winged and wingless, for which a satisfactory explanation has not been forthcoming.

Spergularia canadensis **(Pers.) G. Don** is an annual species of northern sites that include tidal or intertidal bays, beaches, crevices, deltas, dikes, ditches, estuaries, flats, gravel bars, hollows, lagoons, marshes, meadows, mudflats, pannes, roadsides, salt marshes (coastal), sand spits, seashores, shingle beaches, strands, tidal pools, and the margins of ponds, rivers, and streams at elevations of up to 122 m. The plants typically occur just below the high tide limit and endure summer flooding for up to 30% duration. The substrates (pH: 4.1–4.8) can be fresh, brackish, or saline and include clay, cobble, gravel, muck, mud, muddy pebbles, rock, sand, sandy gravel, shells, silt, silty clay, and silty sand. Sand typically is the major component (80%–81%), with lower silt (14%) and clay (5%–6%) fractions. Flowering and fruiting occur from June to October. The seeds are heteromorphic, being winged or wingless. They are dispersed primarily by wind and water. They germinate well (96%–100%) in salt concentrations up to 1500 mg NaCl/L; a small proportion (1%) remains viable in concentrations up to 20,000 mg NaCl/L. For germination, the seeds are air-dried and stored at 4°C in the dark and then placed under cold, wet stratification at 5°C for 20 days. Germination is induced (in 20 days) after incubation under a 12/12 h, 5°C/15°C light/temperature regime. The plants frequently colonize bare substrates and readily exploit areas where the vegetation has been disturbed by Canada geese (Aves: Anatidae: *Branta canadensis*). Carbon fixation occurs by means of a C_3 pathway. **Reported associates:** *Achillea millefolium, Agropyron repens, Alnus, Anthoxanthum nitens, Atriplex alaskensis, A. patula, Betula, Bolboschoenus maritimus, Carex lyngbyei, C. macrocephala, C. maritima, Castilleja ambigua, Cladophora gracilis, Cochlearia officinalis, Conioselinum chinense, Cordylanthus maritimus, Cotula coronopifolia, Cuscuta salina, Cytisus, Deschampsia cespitosa, Distichlis spicata, Eleocharis acicularis, Elymus glaucus, E. repens, E. trachycaulus, Enteromorpha intestinalis, Festuca rubra, Fucus, Grindelia hirsutula, G. stricta, Hippuris tetraphylla, Hordeum brachyantherum, H. jubatum, Isolepis cernua, Jaumea carnosa, Juncus balticus, J. leseurii, Leymus mollis, Lilaeposis occidentalis, Limonium californicum, L. carolinianum, Lysimachia maritima, Oenanthe sarmentosa, Phalaris arundinacea, Plantago major, P. maritima, Pluchea sericea, Poa nemoralis, Polygonum* *fowleri, Populus, Potentilla anserina, Puccinellia angustata, P. grandis, P. nutkaensis, P. nuttalliana, P. phryganodes, P. pumila, Ranunculus cymbalaria, Ruppia maritima, Sagittaria latifolia, Salicornia depressa, S. rubra, Salix, Saururus cernuus, Schedonorus arundinaceus, Schoenoplectus americanus, Solidago sempervirens, Sonchus arvensis, Spartina alterniflora, S. densiflora, S. patens, Spergularia macrotheca, Stellaria humifusa, Suaeda calceoliformis, S. maritima, Symphyotrichum praealtum, S. subspicatum, Trifolium wormskioldii, Triglochin maritimum, T. palustre, Vicia nigricans.*

***Spergularia echinosperma* Celak.** is a nonindigenous annual species that inhabits dunes, flats, prairies, roadsides, salt marshes (coastal), sand flats, shores, swales, and the margins of reservoirs and rivers at elevations of up to 1433 m. The exposures receive full sunlight. The substrates (North America) have been described as brackish clay and sand. Flowering begins in spring with fruiting observed in October (North America). The seeds are heteromorphic, being either winged or wingless. They are dispersed by the wind or by water. **Reported associates (North America):** *Atriplex cristata, Bromus catharticus, Chloris texensis, Crassula aquatica, Cynodon, Festuca, Gnaphalium, Gratiola flava, Houstonia rosea, Hymenoxys texana, Iva angustifolia, Juncus, Lechea sansabeana, Lepidium densiflorum, Nostoc, Phemeranthus parviflorus, Plantago aristata, Polygonum, Schoenolirion wrightii, Sporobolus pyramidatus, Thurovia triflora, Valerianella florifera, Vulpia octoflora, Willkommia texana.*

***Spergularia platensis* (Cambess.) Fenzl** is a nonindigenous annual species that occurs in depressions, ditches, flats, floodplains, meadows, mudflats, playas, and vernal pools at elevations of up to 470 m. The substrates are alkaline/saline and have been described (North America) as clay, gravel, and Solano soil series. Flowering (in North America) has been observed in April to August. The flowers produce small, monomorphic seeds. They are dispersed in the feces of rabbits (Mammalia: Leporidae) and averaged 98 germinating seedlings/1000 fecal pellets in one study. The seeds are also major contaminants in soils that are shipped along with container-grown ornamental plants and are potentially dispersed in that way. **Reported associates (North America):** *Anagallis minima, Arthrocnemum subterminale, Atriplex coronata, Clarkia purpurea, Cotula coronopifolia, Crassula aquatica, C. connata, Cressa, Deschampsia danthonioides, Eleocharis macrostachya, Frankenia salina, Hordeum depressum, H. intercedens, H. marinum, H. pusillum, Juncus bufonius, Lasthenia californica, L. conjugens, L. ferrisiae, Lepidium dictyotum, Lythrum hyssopifolia, Marsilea, Myosurus minimus, Navarretia hamata, Parapholis incurva, Plagiobothrys leptocladus, Plantago bigelovii, P. elongata, Pogogyne abramsii, Polypogon monspeliensis, Psilocarphus brevissimus, P. tenellus, Salicornia depressa, Vulpia myuros.*

***Spergularia salina* J. Presl & C. Presl** is an annual species that occupies freshwater, brackish or saline, tidal or nontidal sites in or on bayshores, beaches, berms, cienega, coasts, coulee bottoms, crevices, depressions, dikes, ditches, dunes, estuaries, flats (inland), floodplains, gravel bars, marshes (coastal),

meadows, mudflats, pans, playas, pools, potholes, prairies, salt flats, salt marshes, seeps, shores, slopes (to 20%), spits, springs, strands, swales, thickets, vernal pools, washes, and along the margins of canals, lagoons, lakes, ponds, reservoirs, rivers, streams, and sloughs at elevations of up to 2195 m. The plants normally occur below the high tide limit and can tolerate fairly high salinity levels (electrical conductivities up to 30.2 dS/m). Exposures range from full sun to shade. The substrates (often disturbed) are characterized as alkaline and include alluvium, clay, clay loam, glysol, gravel, loam, mud, Neihart quartzite, peaty sand, rock, sand, sandy clay, sandy loam, silt, silty clay, silty loam, silty mud, silty sand, stones, and Watsonville soil. The plants are in flower and fruit from April to November. The flowers are self-pollinating and produce heteromorphic seeds (larger, winged and smaller, wingless). The seeds are produced in prodigious quantities, averaging in some cases up to 23,108 seeds/m^2. They are reported both as lacking dormancy and also as persisting in a dormant state. At least in some localities, dormancy appears to be enforced by high salinities and low temperatures. Seeds that have been kept under cold stratification for 21 days germinated when exposed to alternating 15°C/5°C temperatures. Germination can occur at salinities up to 0.34 M NaCl but decreases as temperatures increase. The winged seeds disperse further in the wind and provide flotation for water dispersal; the wingless seeds are dispersed more effectively (over broader distances) in dense vegetation. The seeds are also eaten and dispersed by rabbits (Mammalia: Leporidae). The unwinged seeds occur more frequently in the seed bank, which commonly contains more than 1000 seeds/m^2 and can reach 471,135 seeds/m^2. The plants are known to colonize devegetated, hypersaline mudflats (where their seeds persist in the seed bank) along with *Atriplex patula* and *Salicornia borealis*. Stands of *S. marina* often follow in succession after the removal of *Spartina* colonies. The roots are colonized by arbuscular mycorrhizal Fungi. Carbon is fixed via a C$_3$ photosynthetic pathway. **Reported associates:** *Abronia maritima, Abutilon theophrasti, Acacia salicina, Agrostis palustris, Allenrolfea occidentalis, Amblyopappus pusillus, Ambrosia artemisiifolia, A. chamissonis, A. dumosa, Amsinckia menziesii, Artemisia californica, A. tridentata, Arthrocnemum subterminale, Asclepias verticillata, Atriplex argentea, A. canescens, A. confertifolia, A. glabriuscula, A. lentiformis, A. leucophylla, A. micrantha, A. patula, A. polycarpa, A. prostrata, A. semibaccata, A. watsonii, Avena fatua, Baccharis salicifolia, Bahiopsis parishii, Bassia hyssopifolia, Bebbia juncea, Beta, Bidens connatus, Bolboschoenus maritimus, B. robustus, Brassica nigra, Bromus carinatus, B. diandrus, B. hordeaceus, B. rubens, B. sterilis, Cakile maritima, Camissoniopsis cheiranthifolia, C. micrantha, Carex lyngbyei, Carnegiea gigantea, Carpobrotus edulis, Cenchrus clandestinus, Centaurea melitensis, Centaurium pulchellum, Centromadia, Chenopodium chenopodioides, C. glaucum, C. murale, C. rubrum, Chrysothamnus, Cichorium intybus, Cirsium vulgare, Claytonia perfoliata, Conium maculatum, Convolvulus arvensis, Conyza bonariensis, C. canadensis, Coreopsis gigantea, Cotula australis, C. coronopifolia, Cressa truxillensis, Croton setiger, Crypsis vaginiflora, Cryptantha* *pterocarya, Cylindropuntia acanthocarpa, Cynodon dactylon, Cyperus, Dactylis glomerata, Daucus carota, Deinandra fasciculata, D. kelloggii, Deschampsia cespitosa, D. danthonioides, Descurainia sophia, Distichlis littoralis, D. spicata, Echinochloa crus-galli, Eleocharis erythropoda, E. palustris, E. parvula, E. rostellata, Elymus repens, Encelia farinosa, Epilobium ciliatum, Ericameria laricifolia, Erodium, Euphorbia maculata, Extriplex californica, Fallopia scandens, Festuca rubra, Foeniculum vulgare, Forestiera, Frankenia salina, Gnaphalium palustre, Grindelia hirsutula, Helianthus annuus, Heliotropium curassavicum, Hesperochiron californicus, Heterotheca grandiflora, Hierochloe odorata, Hordeum intercedens, H. jubatum, H. murinum, Hornungia procumbens, Hyptis, Ipomoea lacunosa, Isocoma acradenia, I. menziesii, I. veneta, Iva axillaris, I. frutescens, Jaumea carnosa, Juncus balticus, J. bufonius, J. filiformis, J. gerardi, Juniperus monosperma, Kochia scoparia, Lactuca serriola, Lamarckia aurea, Larrea tridentata, Lasthenia california, L. glabrata, Lepidium dictyotum, L. draba, L. latifolium, L. latipes, L. perfoliatum, Leptochloa fusca, Limonium carolinianum, Lolium multiflorum, L. perenne, Lotus corniculatus, Lupinus truncatus, Lycium andersonii, L. californicum, Lysimachia maritima, Malacothrix glabrata, Malephora crocea, Malva parviflora, Marrubium vulgare, Marsilea, Medicago lupulina, M. polymorpha, Melilotus indicus, Mesembryanthemum crystallinum, M. nodiflorum, Monolepis nuttalliana, Morus alba, Muhlenbergia utilis, Myosurus, Oligomeris linifolia, Oncosiphon piluliferum, Opuntia engelmannii, Panicum capillare, P. dichotomiflorum, P. virgatum, Parapholis incurva, Parkinsonia microphylla, Pascopyrum smithii, Perityle emoryi, Persicaria amphibia, P. lapathifolia, P. maculosa, P. pensylvanica, Phacelia, Phalaris arundinacea, P. minor, Phragmites australis, Pinus edulis, Plagiobothrys leptocladus, P. parishii, Plantago coronopus, P. eriopoda, P. lanceolata, P. major, Pluchea sericea, Poa compressa, P. nemoralis, Polygonum aviculare, Polypogon monspeliensis, P. viridis, Populus fremontii, Portulaca oleracea, Potentilla anserina, P. norvegica, Prosopis velutina, Pseudognaphalium, Psilocarphus brevissimus, P. tenellus, Puccinellia distans, P. nutkaensis, P. nuttalliana, P. parishii, Rafinesquia neomexicana, Ranunculus cymbalaria, Rorippa calycina, Rosa nutkana, Rubus laciniatus, R. ulmifolius, Rumex crispus, R. maritimus, R. salicifolius, R. stenophyllus, Sagina decumbens, Salicornia bigelovii, S. depressa, S. rubra, Salix exigua, S. gooddingii, S. lasiolepis, Salsola soda, S. tragus, Sarcobatus vermiculatus, Sarcocornia perennis, Schedonorus arundinaceus, Schismus barbatus, Schoenoplectus americanus, S. pungens, Senecio vulgaris, Sesuvium verrucosum, Setaria faberi, Sisymbrium irio, Solanum douglasii, Solidago sempervirens, Sonchus arvensis, S. oleraceus, Spartina alterniflora, S. patens, Spergularia bocconi, S. macrotheca, S. media, S. villosa, Sporobolus airoides, Suaeda calceoliformis, S. depressa, S. maritima, S. nigra, S. taxifolia, Suckleya suckleyana, Symphyotrichum ciliatum, S. ericoides, S. frondosum, S. lanceolatum, S. pilosum, S. tenuifolium, Tamarix chinensis, T. ramosissima, Taraxacum officinale, Trifolium fucatum, T. repens, T. resupinatum, Triglochin maritimum, Triphysaria eriantha, Triteleia hyacinthina, Typha angustifolia, T.* *domingensis, T. latifolia, Typha ×glauca, Veronica anagallis-aquatica, V. arvensis, Vicia villosa, Vulpia myuros, V. octoflora, Xanthium strumarium, Ziziphus obtusifolia.*

Use by wildlife: *Spergularia canadensis* provides habitat for the cackling Canada goose (Aves: Anatidae: *Branta canadensis minima*). The plants are hosts to leafhoppers (Insecta: Hemiptera: Cicadellidae: *Macrosteles inundatus*). *Spergularia platensis* is eaten by rabbits (Mammalia: Leporidae).

Economic importance: food: The aquatic species of *Spergularia* are not known to be used as food; **medicinal:** *Spergularia salina* contains triterpenes and saponines and has been used as an expectorant; **cultivation:** *Spergularia salina* is available occasionally in cultivation; **misc. products:** *Spergularia canadensis* has been considered as a candidate for treating roadsides contaminated by runoff from deicing salts; **weeds:** Both *S. canadensis* and *S. salina* are aggressive colonizers and are considered to be potentially weedy; **nonindigenous species:** *Spergularia echinosperma* is an Old World species, which was introduced into the southern United States. *Spergularia platensis* is native to South America and has been introduced to California and Texas.

Systematics: Although recognized traditionally within tribe Polycarpeae, various molecular data place *Spergularia* closer phylogenetically to members of tribes Alsineae, Caryophylleae, Sileneae, and Sclerantheae, which has resulted in its assignment to a distinct tribe Spergulareae (Figure 3.12). Molecular data also resolve *Spergularia* as the sister group to a clade comprising *Spergula* and *Thylacospermum* (Figure 3.14). There has not yet been a comprehensive phylogenetic analysis of *Spergularia* (e.g., only *S. salina* has been sampled among the OBL taxa), so relationships among most of the species remain largely unresolved. However, preliminary results indicate that the genus is monophyletic. The base chromosome number for *Spergularia* is $x = 9$. *Spergularia echinosperma* has diploid ($2n = 18$) and tetraploid ($2n = 36$) populations; *S. canadensis* and *S. salina* ($2n = 36$) are both tetraploid. Chromosome numbers remain unreported for the other OBL species. *Spergularia salina* was long known under the name *S. marina*, to which much of the literature refers.

Comments: *Spergularia canadensis* is distributed along the northern portions of the east and west coasts. *S. salina* is widespread except for the central, interior portions of North

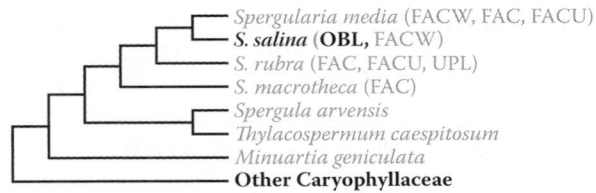

FIGURE 3.14 Phylogenetic position of *Spergularia* in Caryophyllaceae based on an analysis of combined DNA sequence data. *Spergularia salina* (the only OBL *Spergularia* species sampled) resolves within a clade containing three FACW congeners. The wetland indicator status is provided in parentheses, with taxa containing OBL species shown in bold. (Adapted from Greenberg, A.K. & Donoghue, M.J., *Taxon*, 60, 1637–1652, 2011.)

America and extends into Eurasia. The nonindigenous *S. echinosperma* (Alabama, Georgia, Louisiana, Texas, Wyoming) was introduced from Europe and *S. platensis* (California, Texas) from South America.

References: Aronson, 1989; Barbour et al., 2007; Bittrich, 1993; Campbell & Bradfield, 1989; Chang et al., 2001; Conn et al., 2008; Dawe et al., 2011; Dellafiore et al., 2007; Dodd & Coupland, 1966; Druva-Lusite & Ievinsh, 2010; Eallonardo, Jr. & Leopold, 2014; Frenkel et al., 1981; Gill et al., 1996; Grant, 1949; Greenberg & Donoghue, 2011; Hamilton & Langor, 1987; Knowlton, 1915; Mazer & Lowry, 2003; Morteau et al., 2009; Rabeler & Hartman, 2005b; Redbo-Torstensson & Telenius, 1995; Rossbach, 1940; Singhurst et al., 2014; Smissen et al., 2002; Sterk, 1969; Telenius & Torstensson, 1989; Ungar, 1984, 1988; Valkó et al., 2014; Vince & Snow, 1984; Zedler & Black, 1992.

4. *Stellaria*

Chickweed, starwort, stitchwort; stellaire

Etymology: from the Latin *stella* ("star"), in reference to the appearance of the flowers

Synonyms: *Alsine*; *Arenaria* (in part); *Fimbripetalum*; *Mesostemma*; *Sagina* (in part); *Tytthostemma*

Distribution: global: cosmopolitan; **North America:** widespread

Diversity: global: 120 species; **North America:** 30 species

Indicators (USA): OBL: *Stellaria alsine*, *S. borealis*, *S. rassifolia*, *S. humifusa*, *S. littoralis*, *S. longipes*; **FACW:** *S. borealis*, *S. crassifolia*, *S. longipes*; **FAC:** *S. alsine*; **FACU:** *S. longipes*

Habitat: freshwater, saline; palustrine; **pH:** 5.1–8.6; **depth:** <1 m; **life-form(s):** emergent (herb)

Key morphology: plants often tufted; stems (to 5 dm) weak, often fleshy, with swollen nodes; leaves (to 7 cm) opposite, simple, entire, often fleshy, sessile; flowers solitary or in few-flowered cymes; sepals 5 (to 7 mm), margins scarious; petals absent or 5 (to 8 mm), white, deeply two-cleft; styles 3; capsules (to 7 mm) globose, containing to 20+ seeds (to 1 mm)

Life history: duration: annual (fruit/seeds), perennial (rhizomes, stolons); **asexual reproduction:** bulbils, shoot fragments, rhizomes, stolons; **pollination:** insect; **sexual condition:** hermaphroditic, gynodioecious; **fruit:** capsules (common); **local dispersal:** seeds, bulbils, stolons; **long-distance dispersal:** seeds

Imperilment: (1) *Stellaria alsine* [G5]; SH (LA, WV); S1 (MD, QC, TN, VA, VT); S2 (DE, PE); S3 (NC, NJ); (2) *S. borealis* [G5]; S1 (NJ, PA, WV); S2 (MA, PE); S3 (WY); (3) *S. crassifolia* [G5]; SH (PE); S1 (MI, MT, NB, NS, UT); S2 (NF, WY); S3 (BC, QC); (4) *S. humifisa* [G5]; SH (OR); S1 (PE); S2 (MB, NS, ON, YT); S3 (NB, NF, BC); (5) *S. littoralis* [G3/G4]; S3 (CA); (6) *S. longipes* [G5]; S1 (AB, NB); S2 (MI, NF, NY); S3 (MN, SK)

Ecology: general: *Stellaria* is a large group containing annual, biennial, and perennial species that can be found in a variety of habitats including dry waste areas, dunes, moist woodlands, montane communities (to 4000 m elevation), rocky outcrops, talus slopes, tundra, and wetlands. Several species are characteristic weeds of agricultural pastureland and other disturbed sites. Nearly half of the North American species (57%) are categorized as wetland indicators, with six (20%) designated as OBL. The flowers are self-compatible, but breeding systems can vary from strict autogamy to allogamy. Some of the species are protandrous and are outcrossed by insects (Insecta: Diptera; Hymenoptera). In some cases, male sterile flowers occur along with the normally bisexual flowers, resulting in gynodioecious populations. The perennial species are often characterized by extensive vegetative reproduction. Some species shed physiologically dormant seeds, and the genus typically is well represented in seed banks. The seeds are not adhesive and lack specialized features to facilitate dispersal; they are generally categorized as barochorous (gravity dispersed). In a few species, they are transported by ants (Insecta: Hymenoptera: Formicidae). Some seeds are capable of floating for at least a short duration. Human-mediated dispersal is evidenced by the fairly large numbers of seeds that have been recovered from cakes of mud, which cling to automobiles. Viable seeds have also been recovered from the manure of horses (Mammalia: Equidae: *Equus*).

***Stellaria alsine* Grimm** is a freshwater (coldwater) annual or biennial species that inhabits brooksides, ditches, lawns, ledges, marshes, pools, riverbeds, roadsides, seeps, springs, and the margins of streams at elevations of up to 600 m. The substrates are acidic to neutral (pH: 5.1–7.3) and include cobble, gravel, peat, sand, and schist. Flowering and fruiting extend from May to October. Seed production is prolific. The small seeds are quite vagile and are dispersed abiotically by wind and water. Viable seeds have also been recovered from mud that clings to automobiles and from the fecal pellets of rabbits (Mammalia: Leporidae), which indicate other dispersal vectors. The stems are weak, prolonged, somewhat stoloniferous, and root at their nodes. **Reported associates:** *Acer macrophyllum*, *Asclepias exaltata*, *Bolboschoenus maritimus*, *Eleocharis palustris*, *Equisetum fluviatile*, *Fragaria virginiana*, *Ludwigia palustris*, *Lythrum salicaria*, *Myosotis scorpioides*, *Oenanthe sarmentosa*, *Penstemon smallii*, *Pseudotsuga menziesii*, *Ranunculus repens*, *Rubus bifrons*, *R. odoratus*, *Salix discolor*, *Scirpus microcarpus*, *Solidago uliginosa*, *Symphyotrichum eatonii*, *Thuja plicata*, *Tsuga heterophylla*, *Veronica officinalis*.

***Stellaria borealis* Bigelow** is a weak perennial species that inhabits brackish to freshwater beaches, berms, bogs, borrow pits, bottoms, depressions, dikes, ditches, draws, fens, flats, floodplains, gullies, gravel bars, heath, marshes, meadows, mires, muskeg, potholes, prairies (coastal), puddles, rivulets, roadsides, sand bars, seeps, slopes (to 20%), strand beaches, swamps, thickets, washes, woodlands, and the margins of brooks, lakes, ponds, salt marshes, sloughs, and streams at elevations of up to 3122 m. Exposures range from open areas with full sun to densely shaded sites. The substrates have been categorized as acidic (pH: 4.5–5.0) or calcareous and include alluvium, basalt, boulders, clay–gravel, cobble, decomposed granite, granite–diorite, gravel, gravelly clay loam, limestone breccia, loam, marl, mud, rocky till, sand, sandy gravel, sandy loam, sandy silt, silt, silty till, and talus. Flowering and fruiting extend from May to October. Disturbances apparently facilitate seed germination, as the plants frequently occur on soil heaps that have been freshly dug up by arctic foxes (Mammalia: Canidae: *Vulpes lagopus*). The plants also do well in open areas that have been grazed by gray red-backed voles (Mammalia: Rodentia: Cricetidae: *Myodes rufocanus*). Vegetative reproduction occurs by means of stoloniferous rhizomes. **Reported associates:** *Abies concolor, A. lasiocarpa, Acer glabrum, A. macrophyllum, Aconitum maximum, Agoseris aurantiaca, Agrostis scabra, A. stolonifera, Alnus rhombifolia, A. rubra, A. tenuifolia, A. viridis, Alopecurus aequalis, Ammophila arenaria, Anemone, Aquilegia formosa, Arabis, Arctagrostis latifolia, Arctostaphylos uva-ursi, Arnica cordifolia, A. ovata, Artemisia norvegica, A. tilesii, A. tridentata, Astragalus alpinus, Athyrium filix-femina, Beckmannia, Betula nana, B. papyrifera, Bistorta bistortoides, B. vivipara, Calamagrostis canadensis, C. purpurascens, Calocedrus decurrens, Carex aquatilis, C. bolanderi, C. lyngbyei, C. macrochaeta, C. microptera, C. norvegica, C. obnupta, C. pachystachya, C. rossii, C. scopulorum, C. stipata, Chamaecyparis, Chamerion angustifolium, C. latifolium, Chrysanthemum, Cicuta, Cinna latifolia, Circaea alpina, Cirsium scariosum, Claytonia sibirica, Comarum palustre, Cornus nuttallii, Corylus, Cupressus, Cystopteris fragilis, Dactylis, Deschampsia cespitosa, Diapensia lapponica, Digitalis purpurea, Draba stenoloba, Drepanocladus, Eleocharis palustris, Empetrum nigrum, Epilobium ciliatum, E. palustre, Equisetum arvense, Eurybia radulina, Festuca altaica, Floerkea proserpinacoides, Fragaria, Galium trifidum, G. triflorum, Gaultheria shallon, Geranium erianthum, Geum macrophyllum, G. peckii, Glyceria striata, Graphephorum wolfii, Grindelia, Gymnocarpium dryopteris, Heracleum sphondylium, Hieracium triste, Holcus lanatus, Juncus balticus, J. bufonius, J. castaneus, J. drummondii, J. ensifolius, Kalmia microphylla, K. procumbens, Leymus mollis, Ligusticum scoticum, Linnaea borealis, Lupinus nootkatensis, L. polyphyllus, Luzula parviflora, Lysichiton americanus, Marchantia, Meneyanthese trifoliata, Mertensia paniculata, Micranthes nelsoniana, Mimulus guttatus, M. lewisii, M. moschatus, Minuartia groenlandica, Montia chamissoi, M. fontana, Nasturtium officinale, Neottia convallarioides, Nuphar polysepala, Oplopanax horridus, Orthilia secunda, Osmorhiza berteroi, Packera, Parnassia palustris, Penstemon globosus, Petasites frigidus, Philadelphus lewisii, Phleum alpinum, Phlox idahonsis, Physocarpus malvaceus, Picea engelmannii, P. mariana, P. sitchensis, Pinus contorta, P. ponderosa, P. lambertiana, Plagiobothrys, Plantago major, P. maritima, Platanthera dilatata, P. hyperborea, Poa laxa, P. pratensis, Polemonium acutiflorum, Populus trichocarpa, Potentilla villosa, Potentilla ×diversifolia, Prunella vulgaris, Prunus emarginata, Pseudoroegneria spicata, Pseudotsuga menziesii, Pteridium aquilinum, Quercus, Ranunculus repens, Rhamnus alnifolia, Rhododendron columbianum, R. groenlandicum, Rhynchospora alba, Rhytidiadelphus triquetrus, Rubus arcticus, R. spectabilis, Salix alaxensis, S. eastwoodiae, S. hookeriana, S. planifolia, S. pulchra, S. reticulata, S. uva-ursi, Sambucus racemosa, Sanguisorba stipulata, Scrophularia, Senecio triangularis, Sibbaldiopsis tridentata, Solidago canadensis, Solidago lepida, Sphagnum, Spiraea douglasii, Stellaria crassifolia, S. crispa, Symphoricarpos albus, S. oreophilus, Symphyotrichum foliaceum, Tanacetum vulgare, Tellima grandiflora, Tephroseris palustris, Thuja plicata, Torreyochloa pallida, Trifolium longipes, Trollius albiflorus, Tsuga heterophylla, Typha latifolia, Urtica dioica, Vaccinium cespitosum, V. myrtillus, Valeriana sitchensis, Veratrum californicum, V. viride, Veronica americana, V. wormskjoldii, Viola langsdorffii.*

***Stellaria crassifolia* Ehrh.** is a perennial species that occurs in brackish or freshwater sites including alluvial flats, beach bluffs, bogs, cliffs, depressions, ditches, draws, fens, floodplains, gravel/mud/sand bars, heath tundra, hummocks, marshes, meadows, muskeg, prairies, scrub, slopes (10%), terraces, thickets, tundra flats, and the margins of channels, lakes, ponds, sloughs, springs, and streams at elevations of up to 3557 m. The plants have been observed to grow in up to 10 cm of standing water and occur often in shaded (e.g., 60%) sites. The substrates have been characterized as calcareous and include boulders, clay, cobbles, gravel, loam, logs, marl, muck, mud, peat, rock (Belt series), sand, sandy gravel, sandy rock, sandy silt, scree, silt, silty loam, silty mud, and silty organics. Flowering and fruiting occur from May to August. The small seeds are probably dispersed passively by wind and water. However, the plants are also ingested, and the seeds dispersed endozoically by bison (Mammalia: Bovidae: *Bison bonasus*). The plants reproduce vegetatively by rhizomes and often form mats around grasses (Poaceae) and sedges (Cyperaceae). The rhizomes lie just beneath the ground surface and can develop into clones up to a meter in diameter. They are shallow rooted (<15 cm) and possess numerous reduced leaves, whose petioles extend to 7.5 cm, approximately about every 15 cm along the rhizome. Sterile shoots produce fleshy terminal buds, which survive under the snow and function as propagules when dispersed by the spring runoff. **Reported associates:** *Abies lasiocarpa, Achillea alpina, Agrostis scabra, Alnus tenuifolia, Anthoxanthum nitens, Arctagrostis latifolia, Beckmannia syzigachne, Betula papyrifera, Bromus ciliatus, Calamagrostis canadensis, C. stricta, Caltha natans, Cardamine pratensis, Carex aquatilis, C. arcta, C. brunnescens, C. nebrascensis, C. saxatilis, C. simulata, C. utriculata, Catabrosa aquatica, Chrysanthemum arcticum, Chrysosplenium, Comarum palustre, Cystopteris fragilis, Dasiphora floribunda, Deschampsia cespitosa, Eleocharis palustris, Elymus trachycaulus, Empetrum, Epilobium clavatum, E. palustre, Equisetum arvense, E. fluviatile, Eriophorum angustifolium, Festuca, Galium trifidum, Glyceria grandis, Hippuris vulgaris, Hordeum brachyantherum, Juncus balticus, Koenigia islandica, Lonicera oblongifolia, Luzula confusa, Mentha, Menyanthes trifoliata, Mimulus guttatus, Myrica gale, Parnassia fimbriata, P. palustris, Pedicularis groenlandica, P. sudetica, Persicaria amphibia, Phippsia algida, Picea engelmannii, P. mariana, Pinus albicaulis, Potentilla norvegica, Puccinellia phryganodes, Ranunculus cymbalaria, R. flammula, R. gmelinii, R. hyperboreus, R. occidentalis, Rhodiola rhodantha, Rhododendron groenlandicum, Rumex maritimus, R. occidentalis, Salix alaxensis, S. hastata, S. pulchra, Saxifraga cernua, Schoenoplectus validus, Scutellaria galericulata, Senecio triangularis, Silene involucrata, Sinapis arvensis, Sphagnum, Stellaria borealis, S. longipes, Streptopus, Taraxacum officinale, Tephroseris palustris, Trichophorum* *pumilum, Triglochin maritimum, T. palustre, Trisetum spicatum, Utricularia intermedia, Vaccinium uliginosum, Veronica scutellata, Viola epipsila, V. renifolia, Wilhelmsia physodes.*

***Stellaria humifusa* Rottb.** is a salt-tolerant perennial of colder regions that occurs primarily in brackish, coastal, and tidal sites including beaches, crevices, depressions, dunes, estuaries, flats, floodplains, gravel/sand/silt bars, meadows, mudflats, salt marshes, salt meadows, seeps, shores, slopes, strand beach, swales, tundra, and the margins of bays, lagoons, lakes, ponds, rivers, and sloughs at elevations of up to 61 m. The substrates are described as brackish or saline and include cobble, cobbley gravel, gravel, mud, muddy gravel, muddy sand, peaty sand, sand, sandy gravel, sandy silt, silt, and silty sand. The plants are in flower and fruit from May to September. A seed bank develops but is not very dense (e.g., 13 seeds/m^2). It is not unusual to find that seeds are absent from the seed bank, even where the plants are well represented in the standing vegetation. Seeds stored for a year under simulated arctic conditions ($-2°C$ to $-14°C$) showed high germination (94%) after cold stratification for 35 days followed by incubation at 20°C. Seeds collected from the field also have germinated at 13°C under constant light conditions. The seedlings can reach average field densities of up to 24 seedlings/m^2. The plants are known to establish in exposed salt flats after initial colonization by *Puccinellia phryganodes, Carex subspathacea,* and *Ranunculus cymbalaria.* This is a cushion-forming species, which does not tolerate grazing well but can respond by producing prolific flowers and seeds once herbivores have been removed from sites. **Reported associates:** *Alopecurus, Arctagrostis latifolia, Arctopoa eminens, Arenaria, Artemisia norvegica, Atriplex gmelinii, Betula nana, Bolboschoenus maritimus, Calamagrostis canadensis, C. deschampsioides, C. lapponica, Cardamine pratensis, Carex glareosa, C. lachenalii, C. lyngbyei, C. ramenskii, C. subspathacea, C. ursina, Castilleja, Chrysanthemum arcticum, Cochlearia groenlandica, C. officinalis, Dupontia fisheri, Elymus, Empetrum nigrum, Equisetum arvense, Festuca rubra, Honckenya peploides, Juncus arcticus, Leymus mollis, Lysimachia maritima, Mertensia maritima, Micranthes nelsoniana, Montia fontana, Oxyria digyna, Petasites frigidus, Pinus pumila, Plantago maritima, Poa arctica, P. paucispicula, Polemonium acutiflorum, Potentilla anserina, Puccinellia nutkaensis, P. phryganodes, P. tenella, Ranunculus cymbalaria, R. pygmaeus, Rhodiola integrifolia, Salicornia maritima, Salix ovalifolia, Saxifraga bracteata, Schoenoplectus americanus, Spergularia canadensis, Stellaria umbellata, Triglochin maritimum, Trisetum spicatum, Vaccinium uliginosum, Wilhelmsia physodes.*

***Stellaria littoralis* Torr.** is a perennial halophyte known from bluffs, bogs, dunes, marshes, meadows, prairies, roadsides, seeps, slopes, swales, thickets, and the margins of lagoons along coastal California at elevations of up to 339 m. The substrates are described consistently as sand. Flowers and fruits are produced from March to July. Vegetative reproduction occurs by elongate rhizomes. The plants fix carbon by means of a C$_3$ pathway. Populations of this imperiled species are threatened by grazing, trampling, and invasive species. Additional life-history information for this species is scarce.

Reported associates: *Ammophila, Baccharis pilularis, Juncus lesueurii, Leymus mollis, Lupinus arboreus, Mimulus guttatus.*

***Stellaria longipes* Goldie** is a freshwater perennial species that inhabits alluvial fans, barrens, beaches, bogs (montane), caves, cliffs, depressions, ditches, draws, dunes, fellfields, fens, floodplains, gravel/sand bars, gravel pits, heath, knolls, lawns, meadows, mires, muskeg, prairies, roadsides, scrub, seeps, slopes (to 40%), sloughs, steppe, swales, thickets, tundra, tussocks, woodlands, and the margins of ponds, pools, rivers, and streams at elevations of up to 3850 m. Exposures range from full sunlight to partial shade. The substrates are described as acidic or calcareous and can consist of alluvium, argillite, basalt bedrock, boulders, clay, clay–humus, cobble, diorite, granite (pulverized), gravel, gravelly loam, limestone (Triassic), loam, logs, marble, mica-schist, mud, mudstone, organic silty sand, peat, quartzite, rhyolite, rock, rocky silt, sand, sandstone (weathered), sandy gravel, sandy loam, sandy silt, schist, serpentine, shale, shaley scree, silt (eolian), silty humus, silty loam, silty mud, silty pumice, slaty talus, stony sand, or talus. Flowering and fruiting extend from April to October. The plants are self-compatible but usually are protandrous and primarily outcrossed and pollinated by insects (see *Use by wildlife*). Outcrossed plants produce a higher percentage seed set. Although mainly hermaphroditic, some gynodioecious/gynomonoecious populations are known. On average, the hermaphroditic flowers produce more capsules (30 vs. 13) and slightly more seeds (21 vs. 17) than the female flowers; however, the mean seed weight of the female flowers is somewhat heavier (166 µg) than that of the hermaphroditic flowers (147 µg). In addition to passive abiotic vectors, the seeds can be transported endozoically within mammals (Mammalia) such as the arctic fox (Canidae: *Vulpes lagopus*), which can disperse them in its scat. The passed seeds exhibit higher germination rates (43%) than control seeds (22%). The seeds will germinate after 5 weeks of storage at $-5°C$, followed by a 3-day thaw at 0.5°C and a 4-day acclimatization period at 4°C. Germination proceeds at 18°C under a 24-h photoperiod over 11–12 weeks. A seed bank can develop but typically does not attain a high density (e.g., averaging only 2.6–5.0 seeds/m^2). The plants also occur at sites where they are not represented in the seed bank, because of their efficient vegetative reproduction. The plants are tufted and propagate vegetatively by means of rhizomes. The rhizomes grow close to the surface, which allows the ramets to emerge early in the spring before most of the other neighboring plants resume their growth. Phenotypic plasticity is extensive in this species and is attributed to complex interactions of multigene phytohormone families. **Reported associates:** *Abies grandis, A. lasiocarpa, A. magnifica, Achillea borealis, A. millefolium, Aconitum columbianum, A. delphiniifolium, Agrostis exarata, Agrostis scabra, Allium validum, Alnus rugosa, A. tenuifolia, A. viridis, Amelanchier alnifolia, Andromeda polifolia, Androsace septentrionalis, Anelsonia eurycarpa, Anemone multifida, A. narcissiflora, A. patens, Angelica lucida, Antennaria corymbosa, A. monocephala, A. rosea, Anthoxanthum monticola, Anticlea elegans, Aphragmus eschscholtzianus, Arctagrostis latifolia, Arctophila fulva, Arctous alpina, Arnica cordifolia, A. griscomii, A. lessingii, Artemisia alaskana, A. arbuscula, A. borealis, A. campestris, A. frigida, A. glomerata, A. ludoviciana,*

A. norvegica, A. tilesii, A. tridentata, Astragalus alpinus, A. johannis-howellii, A. miser, Betula glandulosa, B. nana, B. occidentalis, B. papyrifera, Bistorta bistortoides, B. vivipara, Boechera holboellii, Boykinia richardsonii, Bromus inermis, B. pumpellianus, Bupleurum americanum, Calamagrostis canadensis, C. lapponica, C. purpurascens, C. rubescens, Caltha leptosepala, Camassia, Campanula lasiocarpa, Cardamine bellidifolia, C. umbellata, Carex aquatilis, C. athrostachya, C. atrata, C. bigelowii, C. concinna, C. douglasii, C. elynoides, C. limosa, C. lyngbyei, C. macrochaeta, C. microchaeta, C. micropoda, C. nigricans, C. nova, C. obtusata, C. paysonis, C. pellita, C. petasata, C. petricosa, C. podocarpa, C. praegracilis, C. rostrata, C. rotundata, C. scirpoidea, C. scopulorum, C. simulata, C. supina, C. tenuiflora, C. utriculata, C. vernacula, C. vesicaria, Cassiope tetragona, Castilleja covilleana, C. cusickii, C. miniata, C. pallescens, Catabrosa aquatica, Ceanothus cordulatus, Cerastium arvense, C. beeringianum, C. fontanum, C. regelii, Ceratodon purpureus, Cercocarpus, Chaenactis douglasii, Chamerion angustifolium, C. latifolium, Chenopodium album, Cistanthe umbellata, Cladina, Cladonia pyxidata, Claytonia sarmentosa, Cochlearia groenlandica, Collinsia, Collomia debilis, Comarum palustre, Cornicularia cuvieri, Cornus canadensis, C. suecica, Crataegus douglasii, Crepis nana, C. tectorum, Cymopterus ibapensis, C. nivalis, Danthonia spicata, Dasiphora floribunda, Delphinium depauperatum, D. glaucum, Deschampsia brevifolia, D. cespitosa, D. danthonioides, Descurainia incana, D. sophioides, Diapensia lapponica, Draba stenoloba, Dryas alaskensis, D. crenulata, D. integrifolia, D. octopetala, Drymocallis, Dryopteris fragrans, Eleocharis palustris, E. quinqueflora, Elymus trachycaulus, Empetrum nigrum, Epilobium ciliatum, E. palustre, Equisetum arvense, E. fluviatile, E. palustre, E. scirpoides, Ericameria nauseosa, Erigeron compositus, E. glacialis, E. gracilis, E. grandiflorus, E. philadelphicus, E. rydbergii, Eriogonum vimineum, Eriophorum russeolum, E. vaginatum, Eriophyllum lanatum, Eritrichium nanum, Erysimum asperum, Eurybia sibirica, Euthamia, Festuca altaica, F. brachyphylla, F. idahoensis, F. ovina, F. saximontana, Fragaria virginiana, Galium boreale, G. triflorum, Gentiana affinis, G. algida, G. calycosa, Gentianopsis detonsa, Geranium, Geum rossii, G. triflorum, Glyceria elata, Gymnocarpium dryopteris, Hedysarum boreale, Heracleum sphondylium, Heuchera micrantha, Hierochloe alpina, Holcus lanatus, Hordeum brachyantherum, H. jubatum, Horkelia, Hulsea algida, Hydrurus foetidus, Hylocomium splendens, Hypericum formosum, H. scouleri, Ilex, Iris missouriensis, Juncus balticus, J. drummondii, J. mertensianus, J. nevadensis, J. triglumis, Juniperus communis, Kobresia myosuroides, Koeleria pyramidata, Koenigia islandica, Lathyrus japonicus, Lepidium, Leptarrhena pyrolifolia, Leucopoa kingii, Leymus innovatus, L. mollis, Ligusticum canbyi, L. grayi, L. scoticum, Linnaea borealis, Linum perenne, Lloydia serotina, Lomatium cous, Luetkea pectinata, Lupinus arcticus, L. argenteus, L. nootkatensis, Luzula arctica, L. arcuata, L. campestris, L. confusa, L. parviflora, L. spicata, L. wahlenbergii, Madia gracilis, Matricaria discoidea, Melilotus albus, Mertensia campanulata, M. paniculata, Micranthes ferruginea, M. nelsoniana, M. occidentalis, M. oregana, M. rhomboidea, Mimulus floribundus, M.

guttatus, M. primuloides, M. tilingii, Minuartia arctica, M. elegans, M. nuttallii, M. obtusiloba, M. rubella, Moehringia lateriflora, Monostroma, Montia chamissoi, M. linearis, Myosotis asiatica, Noccaea arctica, Oxyria digyna, Oxytropis arctica, O. campestris, O. maydelliana, O. nigrescens var. nigrescens, O. sericea, O. splendens, Packera pauciflora, P. paupercula, P. werneriifolia, Papaver alboroseum, P. lapponicum, P. macounii, Parnassia, Paxistima, Pedicularis capitata, P. contorta, P. groenlandica, P. lanata, Peltigera, Penstemon gormanii, P. humilis, P. rydbergi, Perideridia bolanderi, Petasites frigidus, Phacelia ramosissima, P. sericea, Phleum alpinum, P. pratense, Phlox idahonis, P. pulvinata, P. richardsonii, Phyllodoce glanduliflora, Physaria carinata, Picea engelmanii, P. glauca, P. mariana, Pinus albicaulis, P. contorta, Plantago eriopoda, Poa alpina, P. arctica, P. cusickii, P. glauca, P. leptocoma, P. palustris, P. paucispicula, P. pratensis, P. secunda, Podistera macounii, Polemonium caeruleum, P. occidentale, P. viscosum, Polystichum, Polytrichum piliferum, Populus tremuloides, P. trichocarpa, Potentilla anserina, P. gracilis, P. hyparctica, P. nivea, P. ovina, P. villosa, Potentilla ×diversifolia, Primula parryi, Prunella vulgaris, Prunus virginiana, Pseudoroegneria spicata, Pseudotsuga menziesii, Pyrola asarifolia, Pyrrocoma uniflora, Racomitrium, Ranunculus cooleyae, R. eschscholtzii, R. natans, R. nivalis, Rhododendron columbianum, R. tomentosum, Rhodiola integrifolia, R. rosea, Ribes inerme, R. oxyacanthoides, Rosa acicularis, Rubus arcticus, R. chamaemorus, R. idaeus, Rumex salicifolius, Salix alaxensis, S. arbusculoides, S. arctica, S. barclayi, S. barrattiana, S. bebbiana, S. exigua, S. glauca, S. lemmonii, S. melanopsis, S. myrtillifolia, S. nivalis, S. phlebophylla, S. planifolia, S. polaris, S. pulchra, S. reticulata, S. sitchensis, S. wolfii, Saxifraga bronchialis, S. hyperborea, S. tricuspidata, Scirpus microcarpus, Scrophularia lanceolata, Sedum lanceolatum, Senecio congestus, S. crassulus, S. fremontii, S. hydrophiloides, S. serra, S. sphaerocephalus, S. triangularis, Shepherdia canadensis, Sibbaldia procumbens, Sidalcea oregana, Silene acaulis, S. repens, S. uralensis, S. williamsii, Sisyrinchium angustifolium, S. idahoense, Smelowskia, Solidago lepida, S. multiradiata, Sphenosciadium capitellatum, Spiraea douglasii, Spiranthes romanzoffiana, Stellaria crassifolia, Stenotus acaulis, Stereocaulon, Stipa, Symphyotrichum foliaceum, Synthyris dissecta, Taraxacum officinale, T. phymatocarpum, Tephroseris lindstroemii, Thalictrum fendleri, Thermopsis montana, Tofieldia coccinea, Townsendia condensata, T. leptotes, Toxicoscordion venenosum, Trifolium haydenii, T. hybridum, T. longipes, T. wormskioldii, Tripleurospermum maritimum, Trisetum spicatum, Trollius laxus, Urtica, Vaccinium membranaceum, V. oxycoccos, V. uliginosum, V. vitis-idaea, Valeriana edulis, V. sitchensis, Vaucheria, Veratrum californicum, Veronica wormskioldii, Viburnum edule, Vicia americana, Viola glabella, Wilhelmsia physodes, Woodsia scopulina, Wyethia helianthoides.

Use by wildlife: Several smut Fungi (Basidiomycota: Ustilaginaceae) are hosted by Stellaria alsine (Microbotryum stellariae) and S. borealis (Microbotryum violaceum). Stellaria crassifolia provides nesting habitat for horned larks (Aves: Alaudidae: Eremophila alpestris) in alpine environments. The plants are grazed to a small extent by bison (Mammalia: Bovidae: *Bison bonasus*). *Stellaria humifusa* is eaten by lesser snow geese (Aves: Anatidae: *Chen caerulescens*). The roots of *S. humifusa* are colonized by mycorrhizal dark septate endophytic Fungi; the fungus *Septoria stellariae* (Ascomycota: Mycosphaerellaceae) occurs on its leaves. The flowers of *S. longipes* attract various insects (Insecta) including bumblebees (Hymenoptera: Apidae: *Bombus melanopygus*, *B. terricola*), flies (Diptera: Anthomyiidae: *Hylemya*; Muscidae: *Spilogona imitatrix*), and a variety of parasitoid wasps (Insecta: Hymenoptera: Braconidae' Chalcididae; Eulophidae: *Tetrastichus*; Ichneumonidae: *Atractodes*, *Cryptus arcticus*, *Mesoleptus*, *Stenomacrus*; Pteromalidae: *Tridymus*), which seek their nectar and function as pollinators. **Economic importance: food:** *Stellaria longipes* and other *Stellaria* species are reportedly edible and good sources of vitamin C and minerals. The young foliage is used as a salad green, potherb, or cooked vegetable, which is similar in taste to spinach. Chopped plants are made into soup and puree. Dried plants are brewed as a tea; **medicinal:** *Stellaria alsine*

has been used to treat acne, colds, snakebite, and trauma and also is said to relieve intestinal gas, promote lactation, purify blood, and reduce sweating; **cultivation:** none; **misc. products:** none; **weeds:** none of the OBL species is regarded as weedy; **nonindigenous species:** none.

Systematics: Molecular data (5′ *ndhF* sequences) resolved *Stellaria* as the sister genus to *Cerastium*. However, a more comprehensive phylogenetic study that incorporated multiple DNA loci and numerous *Stellaria* species indicated a somewhat more complex result. The preceding statement is not entirely accurate considering that neither *Cerastium* nor *Stellaria* is monophyletic as previously circumscribed; however, with the exception of a few wayward taxa, the two genera do resolve essentially as sister clades, which comprise the majority of species within tribe Alsineae. Five of the OBL indicators have been included in the more comprehensive analysis, which shows them to be scattered across the *Stellaria* clade rather than to represent a closely related group (Figure 3.15); this result indicates several independent origins of the OBL habit in the genus. *Stellaria crassifolia*,

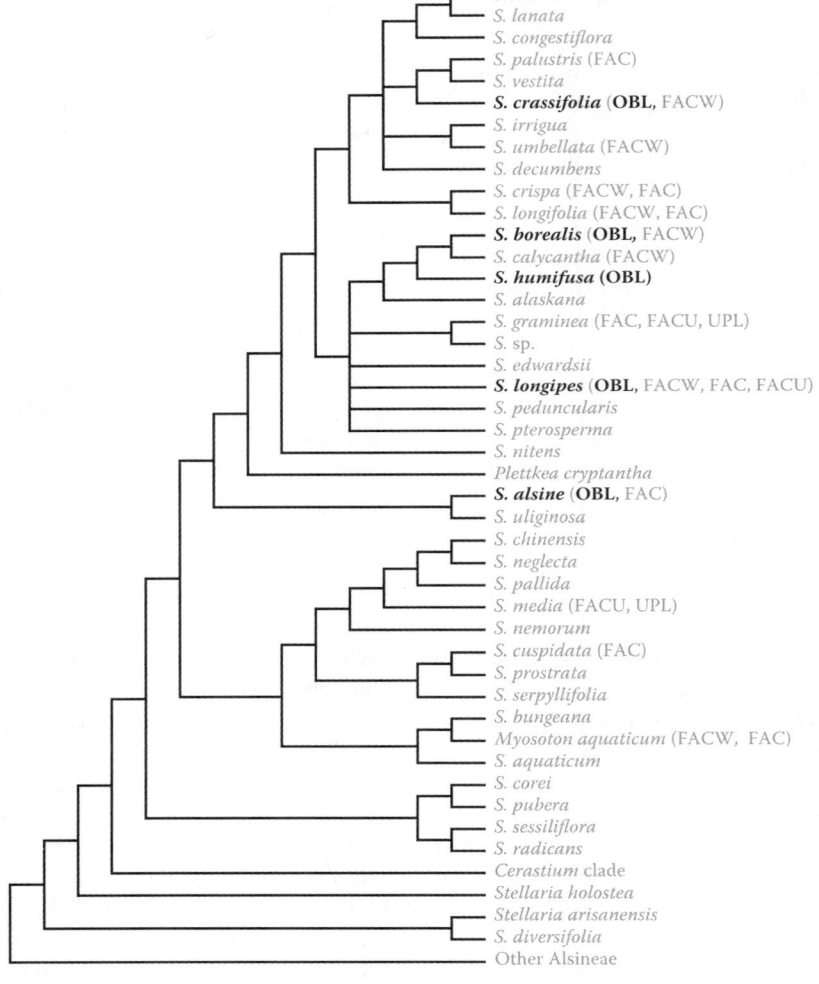

FIGURE 3.15 Phylogenetic relationships in *Stellaria* (as indicated by analysis of combined DNA sequence data). The genus is monophyletic if *Myosoton* and *Plettkea* are included and a few species (*S. arisanensis*, *S. diversifolia*, *S. holostea*) are excluded. The wetland indicator status of North American species is shown in parentheses. The OBL habit (taxa in bold) appears to have arisen sporadically throughout the genus (*S. littoralis* was not surveyed in the analysis). (Adapted from Greenberg, A.K. & Donoghue, M.J., *Taxon*, 60, 1637–1652, 2011.)

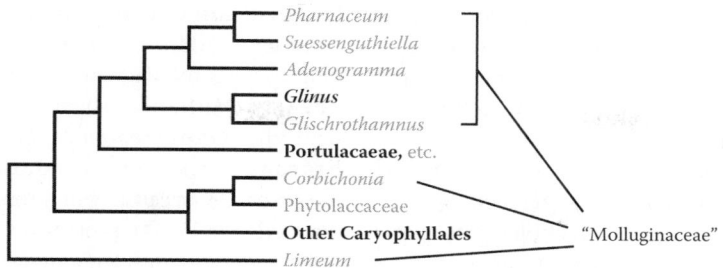

FIGURE 3.16 Phylogenetic relationships (based on *matK* sequence data) indicate a polyphyletic Molluginaceae, with a core of genera representing the family in a strict sense. Some of the outlying genera have been transferred to different families (e.g., *Limeum* to Limeaceae). Within this core, *Glinus* resolves as the sister genus to *Glischrothamnus* (see also Figure 3.17). North American taxa containing OBL indicators are highlighted in bold. (Adapted from Cuénoud, P. et al., *Amer. J. Bot.*, 89, 132–144, 2002.)

S. humifusa, and *S. longifolia* are diploid (2*n* = 26); *S. alsine* is also diploid but 2*n* = 24. *S. borealis* (subsp. *borealis* and subsp. *sitchana*) are tetraploid (2*n* = 52). *Stellaria longipes* is a highly polymorphic species and diverse cytologically with counts ranging from 2*n* = 51–107. The major ploidy levels of tetraploid (2*n* = 52), hexaploid (2*n* = 78), and octoploid (2*n* = 104) do not correlate with morphological features and show high interfertility. *Stellaria longipes* is an allopolyploid believed to have been derived from *S. longifolia* and *S. porsildii* (both 2*n* = 26). *Stellaria longipes* and *S. longifolia* produce viable hybrids when crossed artificially. The diploid *S. longifolia* has greater plasticity but less genetic variability than the polyploids of *S. longipes*. *Stellaria longipes* is also known to form natural hybrids with *S. borealis*. *Myosoton aquaticum* sometimes is transferred to *Stellaria* as *S. aquatica* (results of phylogenetic analyses are consistent with that result—see Figure 3.15); however, despite its specific epithet, this species is not categorized as an OBL indicator.

Comments: *Stellaria alsine* is native to North America (disjunct between eastern and western portions) but occurs also in Africa and Europe. *Stellaria borealis* (circumboreal), *S. crassifolia* (also Eurasia), *S. humifusa* (also Eurasia), and *S. longipes* (circumpolar) are native to northern North America and the western montane regions. *Stellaria littoralis* is narrowly restricted to coastal California.

References: Alsos et al., 2013; Bazely & Jefferies, 1986; Bittrich, 1993; Bledsoe et al., 1990; Bruun et al., 2005; Brysting et al., 2001; Chinnappa & Morton, 1984; Chinnappa et al., 2005; Cooper et al., 2004a; Dang & Chinnappa, 2007; Emery & Chinnappa, 1992; Freedman et al., 1982; Graae et al., 2004; Greenberg & Donoghue, 2011; Hambäck & Ekerholm, 1997; Handa et al., 2002; Holch et al., 1941; Jefferies & Rockwell, 2002; Kevan, 1973; Macdonald et al., 1987, 1988; Maillette et al., 2000; Morton, 2005; Nelson, 1912; Pakeman et al., 1999; Philipp, 1980; Purdy et al., 1994; Rabeler, 1993; Schmidt, 1989; Staniforth et al., 1998; Swales, 1979; Verbeek, 1967; Weaver & Adams, 1996; Wherry, 1920; White, 1998.

Family 4: Molluginaceae [9]

Molecular data (Cuénoud et al., 2002; Brockington et al., 2013) resolve Molluginaceae as polyphyletic by virtue of the errant placements of several genera (e.g., *Corbichonia*,

Hypertelis, *Limeum*). Although some authors have completely abandoned the family (e.g., Judd et al., 2016), other genera that have been recognized traditionally within Molluginaceae comprise a clade (e.g., Figure 3.16), which represents the family in the strict sense. However, because of the state of flux in the circumscription of Molluginaceae at this time, it seems futile to attempt any further description of the group until the taxonomy of the family has been settled more firmly. There is no major economic importance of the family regardless of its circumscription. In North America, OBL indicators are found only within a single genus:

1. *Glinus* L.

1. *Glinus*

Carpetweed, sweetjuice

Etymology: from the Greek *glinos* meaning "sweet juice"

Synonyms: *Damascisa*; *Nemallosis*; *Mollugo* (in part)

Distribution: global: Africa; Asia; Australia; Europe; North America*; **North America:** Southern

Diversity: global: 6 species; **North America:** 2 species

Indicators (USA): OBL; FACW: *Glinus lotoides*

Habitat: freshwater, saline; palustrine; **pH:** alkaline; **depth:** <1 m; **life-form(s):** emergent (herb)

Key morphology: stems (to 4 dm) decumbent, highly branched, densely stellate pubescent; leaves (to 2.4 cm) petioled (to 2.5 cm), in basal rosette or cauline and opposite to verticillate, obovate to oblong–spatulate, obtuse, densely stellate pubescent; flowers in sessile/subsessile groups; tepals (to 10 mm) 5, elliptic to oblong, yellowish; capsules (to 4.5 mm) ellipsoidal; seeds (to 0.6 mm) pappilate, up to 20 per locule

Life history: duration: annual (fruit/seeds); **asexual reproduction:** none; **pollination:** self; **sexual condition:** hermaphroditic; **fruit:** capsules (prolific); **local dispersal:** seed; **long-distance dispersal:** seed

Imperilment: (1) *Glinus lotoides* [G5]

Ecology: general: *Glinus* species are mainly tropical but do occur in temperate regions following introductions. The genus is not particularly helophytic, but the species often occur in dessicated sites formerly occupied by standing water. Both of the North American species share this affinity and are designated as wetland indicators, with just one as OBL.

The reproductive biology of the group has not been studied in any detail. At least some of the species are self-pollinating. Several species are myrmecochorous (dispersed by ants), and the dispersal of seeds in mud that adheres to automobiles has also been documented in some cases.

Glinus lotoides L. is a nonindigenous annual species that occurs in or on arroyos, depressions, drying beds (lakes, ponds, sloughs, washes), flats, mudflats, mudholes, pits, pools (ephemeral or vernal), roadsides, slopes, waste areas, woodlands, and along the receding margins of lakes, marshes, reservoirs, rivers, sloughs, strands, and swales at elevations of up to 1070 m. The plants are partially tolerant to salinity and occur in warm (often arid) but wet habitats. They are often found growing in full sunlight at sites devoid of other plant species in their vicinty. In some cases, they are mat forming. The reported substrates (North America) are alkaline and include alluvium, clay, clay mud, granite, gravel, humus, muck, mud, rocky sandstone, sand, sandy mud, sandy silt, serpentine, shale, silt, stones, and Willows silty clay. Flowering and fruiting (North America) occur from May to November. The seeds possess a strophiole and also a curved, sharp funiculus, which may facilitate their attachment to animal fur for dispersal. They have physiological dormancy but will germinate at room temperature. An extensive seed bank can develop. Because the foliage is unpalatable to livestock, the seeds can become a dominant feature of those seed banks associated with grazed areas. **Reported associates (North America):** *Amaranthus albus, Atriplex suberecta, Baccharis salicifolia, B. sarothroides, Bergia texana, Calibrachoa parviflora, Callitriche marginata, Chenopodium album, Conyza canadensis, Cotula coronopifolia, Croton setiger, Crypsis schoenoides, C. vaginiflora, Cynodon dactylon, Cyperus eragrostis, C. squarrosus, Echinodorus berteroi, Eleocharis macrostachya, Epilobium ciliatum, Eriogonum fasciculatum, Euphorbia, Glyceria declinata, Gnaphalium palustre, Heliotropium curassavicum, Hirschfeldia incana, Juncus, Lythrum hyssopifolia, Malosma laurina, Malvella leprosa, Marsilea, Nama stenocarpa, Nicotiana glauca, Persicaria lapthaifolia, Phyla nodiflora, Polygonum aviculare, Pseudognaphalium luteoalbum, Quercus berberidifolia, Rorippa curvisiliqua, Rumex, Salix gooddingii, Schoenoplectus acutus, S. californicus, Stuckenia pectinata, Tamarix ramosissima, Typha, Zannichellia palustris.*

Use by wildlife: *Glinus lotoides* few reported uses by North American wildlife. It is generally unpalatable and avoided by grazing animals in other regions. However, the plants are eaten by California ground squirrels (Mammalia: Rodentia: Sciuridae: *Otospermophilus beecheyi*).

Economic importance: food: *Glinus lotoides* is not eaten as a food probably because of its strong laxative properties. However, the seeds contin 20% protein, 47 total carbohydrates, along with calcium, folic acid, selenium, and vitamin E; **medicinal:** *Glinus lotoides* is a strong laxative and vermifuge, which is effective on humans and ruminants. The plants are used in Africa to treat diabetes and skin ailments and in India, as a remedy for abdominal disease, boils, and diarrhea. The foliage is known to contain alkaloids that are toxic to locusts (Insecta: Orthoptera) and to the snail (Mollusca: Gastropoda) vectors of Schistosomiasis. The extracts have been shown to surpress certain types of tumors. The seed extracts exhibit antioxidant activity and selectively inhibit the growth of cancer cells *in vitro*. However, despite these potentially beneficial properties, the plant extracts also contain hopane-type saponins, which cause DNA damage and may account for reports of genotoxicity associated with the crude plant extracts; **cultivation:** none; **misc. products:** none; **weeds:** *Glinus lotoides* is regarded as a weed when it occurs in cultivated sites; **nonindigenous species:** *Glinus lotoides* was introduced to North America from the Old World tropics.

Systematics: Molecular data (Figures 3.16 and 3.17) place *Glinus* within the Molluginaceae s.s. clade, which comprises at least nine genera, including *Mollugo* (the type of the family). The most comprehensive analyses to date resolve *Glinus* as the sister genus to a clade comprising *Glischrothamnus* (a monotypic Brazilian genus of terrestrial, dioecious shrubs) and *Mollugo* (Figure 3.17). A comprehensive phylogenetic study of *Glinus* species has not yet been carried out, and the circumscription of the genus (as well as its monophyly) deserves renewed scrutiny. The base chromosome number of Molluginaceae is $x = 9$. *Glinus lotoides* $(2n = 36)$ is a tetraploid.

Comments: *Glinus lotoides* occurs in the south-central and southwestern United States.

References: Abbassi et al., 2003; Belal et al., 1997; Berg, 1975; Brockington et al., 2013; Clifford, 1959; Demma et al., 2013; Evans & Holdenried, 1943; Kavimani et al., 1999; Mengesha & Youan, 2010; Nicol et al., 2007; Short, 2002; Thieret, 1966; Vincent, 2003.

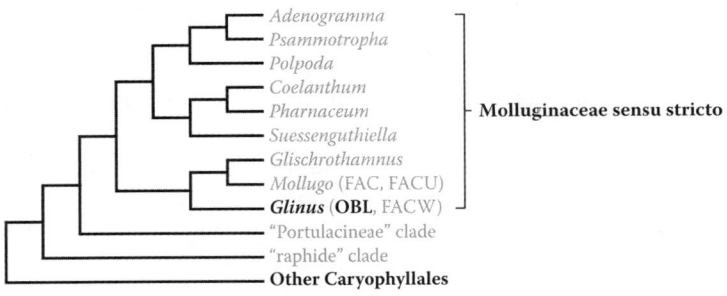

FIGURE 3.17 Phylogenetic placement of *Glinus* in the order Caryophyllales as indicated by analysis of combined DNA sequence data. The wetland indicator status for North American taxa is shown in parentheses. *Glinus* contains the only OBL indicators (highlighted in bold) within the Molluginaceae "sensu stricto" clade. (Adapted from Brockington, S. et al., *Amer. J. Bot.*, 100, 1757–1778, 2013.)

Family 5: Montiaceae [14]

Phylogenetic analyses of *matK* sequence data (Cuénoud et al., 2002) initially revealed that a long standing taxonomic concept of Portulacaceae was polyphyletic with respect to the relative positions indicated for a number of genera including *Portulaca Portulacaria*, and *Calyptrotheca*, whereas other surveyed genera of Portulacaceae comprised two clades, one of which has now been segregated as Montiaceae (Figure 3.18). Those results were confirmed subsequently by analyses of combined *matK*, *ndhF*, and *phyC* sequences (Nyffeler & Eggli, 2010; Ogburn & Edwards, 2015), which validated the transfer of *Claytonia*, *Lewisia*, and *Montia* (and other genera) to the resurrected family Montiaceae as has been followed here.

In this circumscription, Montiaceae comprise approximately 225 species assigned to 14 genera. They are distributed throughout the Americas, northern Eurasia, Australia, and New Zealand (Nyffeler & Eggli, 2010). Most of the species are stemless, succulent, annual or perennial, belalain-containing herbs with swollen roots and mucilage cells; some have CAM photosynthesis (Nyffeler & Eggli, 2010; Judd et al., 2016). Pollination typically is facilitated by nectar-seeking insects (Insecta). The fruits are capsules or utricles, which contain seeds having an embryo curved around the perisperm. Seed dispersal occurs mainly by means of wind or water (Judd et al., 2016).

Montiaceae contain few economically important plants. These are primarily garden ornamentals, which occur in the genera *Calandrinia*, *Claytonia*, *Phemeranthus*, and *Lewisia*, the latter of primary importance in this respect.

Claytonia sibirica was included among the OBL indicators in the 1996 list; however, it has since been recategorized as FACW, FAC and is excluded here. Although *Lewisia* contains some terrestrial species, a single major radiation of the aquatic habit in this family is indicated by the inclusion of both OBL North American genera within a single clade (Figure 3.18):

1. ***Lewisia*** Pursh
2. ***Montia*** L.

1. *Lewisia*
Bitterroot
Etymology: after Meriwether Lewis (1774–1809)
Synonyms: *Erocallis*; *Oreobroma*
Distribution: global: North America; **North America:** western
Diversity: global: 18 species; **North America:** 18 species
Indicators (USA): OBL: *Lewisia cantelovii*
Habitat: freshwater; palustrine; **pH:** unknown; **depth:** <1 m; **life-form(s):** emergent (herb)
Key morphology: plants succulent; leaves (to 8 cm) fleshy, spatulate, nearly sessile, in basal rosettes, margins fine to coarsely dentate; stems (to 4.5 dm) scapiform, bracteate; flowers numerous in open panicles; pedicelled (to 1 cm); sepals 2; petals (to 7 mm) 5–6, light pink, striped with dark pink veins; capsules (to 3 mm) with 1–3 shiny seeds (to 1.5 mm)
Life history: duration: perennial (caudex); **asexual reproduction:** none; **pollination:** insect; **sexual condition:** hermaphroditic; **fruit:** capsules (common); **local dispersal:** seeds (wind); **long-distance dispersal:** seeds (wind)
Imperilment: (1) *Lewisia cantelovii* [G3]; S3 (CA)
Ecology: general: *Lewisia* species are perennials that persist from a branching taproot. They occur in a variety of habitats including alpine meadows, rocky crevices, slopes, and woodlands. The substrates are often of a rocky or gravelly nature. This entirely North American genus is mainly terrestrial, with only three of the species (17%) listed as wetland indicators and just one as OBL. The flowers are highly

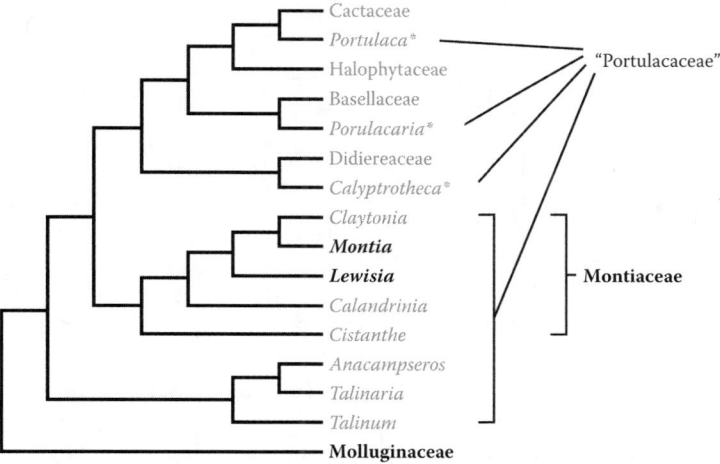

FIGURE 3.18 Phylogenetic relationships of Portulacaceae indicated by *matK* sequence data. Three genera assigned to Portulacaceae (asterisked) are distant phylogenetically from others, indicating the polyphyly of the family as formerly circumscribed. The remaining genera of Portulacaceae surveyed occurred within two distinct clades, with one corresponding to Montiaceae (as currently delimited) and the other now segregated as families Anacampserotaceae and Talinaceae. This result subsequently was confirmed by analyses of additional taxa by Ogburn and Edwards (2015). The North American genera containing OBL indicators (bold) occur within a single subclade in this otherwise principally terrestrial group. However, *Lewisia* is not entirely aquatic and also contains terrestrial species. (Adapted from Cuénoud, P. et al., *Amer. J. Bot.*, 89, 132–144, 2002.)

attractive to insects (Insecta) and are pollinated primarily by bees (e.g., Hymenoptera: Andrenidae: *Andrena*; Apidae: *Anthophora*; Halictidae: *Lasioglossum*) and to a lesser degree by butterflies (Lepidoptera). The seeds are physiologically dormant, but germination requirements vary among the species with some requiring optimal temperatures of only 1°C. The seeds of most of the species are unmodified for specialized dispersal but are very light and believed to be dispersed primarily by the wind. They have been collected from windborne debris at elevations of up to 3350 m. The seeds of one species at least have a fleshy caruncle, which might indicate their dispersal by ants (Insects: Hymenoptera: Formicidae). In some cases, higher seed densities are attained in the presence of "nurse plant" canopies, which help to trap and collect the propagules.

Lewisia cantelovii **J.T. Howell** inhabits cliffs, crevices, domes, ledges, ravines, river canyons, seeps, and slopes (40%–90%) at elevations of up to 1310 m. The plants are often at sites where cold air drainage occurs. They can tolerate full sunlight but grow best in dense shade, on steep, north-facing slopes of moss-covered rocks. The substrates include granite outcrops, krumholtz, serpentine, and shale. The nonserpentine substrates are acidic (pH: 4.5–5.9) and the serpentine more alkaline (pH: 6.4–7.5). Serpentine soils contain much higher values of Mg (to 1023 μg/g) than the nonserpentine (to 131 μg/g). Individuals grow considerably larger (more than twice the biomass) on nonserpentine substrates than those growing on serpentine soils. Flowering and fruiting occur from May to October. The physiologically dormant seeds require 14–21 days of cold stratification (4°C) to induce their germination, which occurs in 14–30 days after being placed under cool conditions; excessively warm temperatures will delay their germination. Genetic surveys using 22 isozyme loci showed that populations retain fairly high levels of heterozygosity (=0.208). The plants are not fire tolerant. Popualtions of the species are threatened by overcollection (for horticultural specimens) and road maintenance activities. **Reported associates:** *Acer macrophyllum, Aspidotis densa, Avena, Brickellia californica, Calocedrus decurrens, Dicranum howellii, Draperia systyla, Epilobium canum, E. minutum, Eriogonum ursinum, Gilia capitata, Grimmia trichophylla, Heuchera micrantha, Holodiscus microphyllus, Homalothecium pinnatifidum, Mielichhoferia mielichhoferiana, Penstemon deustus, Philadelphus lewisii, Pinus albicaulis, P. jeffreyi, P. ponderosa, Polystichum imbricans, P. munitum, Polytrichum juniperinum, Pseudotsuga menziesii, Quercus chrysolepis, Q. durata, Ribes nevadense, Scapania undulata, Sedum albomarginatum, S. spathulifolium, Selaginella hansenii, S. wallacei, Sericocarpus oregonensis, Streptanthus tortuosus, Symphyotrichum eatonii, Umbellularia californica.*

Use by wildlife: *Lewisia cantelovii* is a host of rust Fungi (Basidiomycota: Pucciniaceae: *Uromyces unitus*).

Economic importance: food: Other *Lewisia* species were eaten by Native North Americans, but there are no specific uses reported for *L. cantelovii*; **medicinal:** none; **cultivation:** *Lewisia cantelovii* is cultivated as an ornamental rock garden plant; **misc. products:** none; **weeds:** none; **nonindigenous species:** none.

Systematics: Some phylogenetic surveys (e.g., Figure 3.18) resolve *Lewisia* as the sister group to a clade consisting of *Claytonia* and *Montia* (the latter also including *Neopaxia*); however, more complete taxon sampling indicates *Lewisiopsis* to be the actual sister group of *Lewisia* (Figure 3.19).

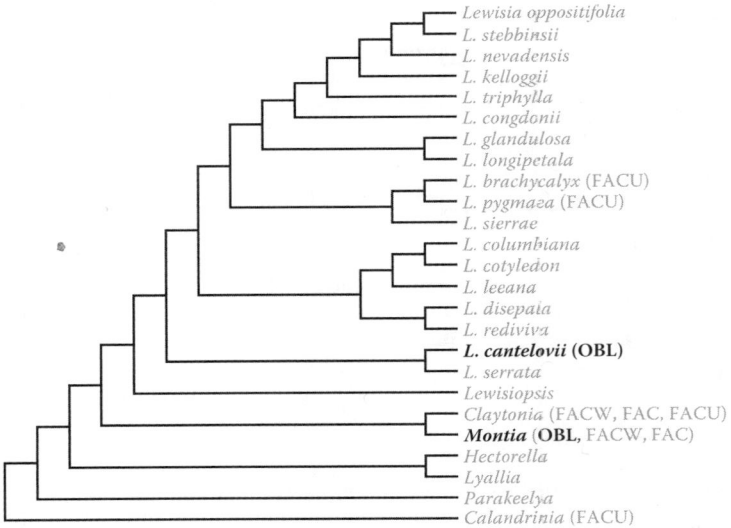

FIGURE 3.19 Phylogenetic analysis of Montiaceae using combined DNA sequence data indicates the relative positions of *Lewisia* and *Montia*, the two genera containing OBL indicators. The wetland indicator status for North American taxa is shown in parentheses (OBL in bold). These results indicate two separate origins of the OBL habit in Montiaceae. The OBL *L. cantelovii* resolves as the sister species of *L. serrata*, which some authors regard as a synonym. The intergeneric relationships indicated are consistent with previous results, which sampled fewer taxa (Applequist & Wallace, 2001; Cuénoud et al., 2002; Hershkovitz & Zimmer, 2000). (Adapted from Ogburn, R.M. & Edwards, E.J., *Mol. Phylogen. Evol.*, 92, 181–192, 2015.)

A comprehensive analysis of *Lewisia* (18 species) using combined DNA sequence data has demonstrated the monophyly of *Lewisia* and also a close relationship of *L. cantelovii* to *L. serrata* (Figure 3.19). Although *L. cantelovii* and *L. serrata* have been merged by some authors, isozyme data show these taxa to be distinct species with no evidence of gene flow. The base chromosome number of *Lewisia* is uncertain because of extensive variability ($x = 10-15$). *Lewisia cantelovii* ($2n = 28$) presumably is diploid.

Comments: *Lewisia cantelovii* is a narrow endemic restricted to the Klamath Ranges, High Cascade Range, and High Sierra Nevada region of California.

References: Bonde, 1969; Carroll, 1997; Davidson, 2000; Day & Wright, 1989; Foster et al., 1997; Hershkovitz & Hogan, 2003; Hershkovitz & Zimmer, 2000; O'Brien, 1980; Ogburn & Edwards, 2015; Richerson, 1997; Rogers et al., 1996; Wagstaff & Hennion, 2007; Woolhouse, 2012.

2. *Montia*

Miner's lettuce, water blinks

Etymology: after Giuseppe Monti (1682–1760)

Synonyms: *Claytonia* (in part); *Claytoniella*; *Crunocallis*; *Limnalsine*; *Maxia*; *Montiastrum*; *Mona*; *Naiocrene*; *Neopaxia*; *Paxia*

Distribution: global: Africa; Asia; Europe; North America; **North America:** widespread

Diversity: global: 15 species; **North America:** 8 species

Indicators (USA): OBL: *Montia chamissoi*, *M. fontana*; **FACW:** *M. fontana*

Habitat: freshwater; palustrine; **pH:** acidic; **depth:** <1 m; **life-form(s):** emergent (herb)

Key morphology: stems (to 3 dm) succulent, ascending, procumbent or prostrate; leaves (to 5 cm) opposite, simple, entire, petiole indistinct; racemes 3–9 flowered, pedicels recurved in fruit; flowers with two sepals (to 3 mm) and three to five unequal petals (to 3.5 mm), petals white or pink; capsules three-valved; seeds (to 1.5 mm) 3, black, muricate

Life history: duration: annual (fruit/seeds); perennial (rhizomes, stolons); **asexual reproduction:** bulbils, rhizomes, stolons; **pollination:** insect, self; **sexual condition:** hermaphroditic; **fruit:** capsules (common); **local dispersal:** rhizomes, stolons, bulblets; **long-distance dispersal:** seeds (animals)

Imperilment: (1) *Montia chamissoi* [G5]; SH (IA); S1 (MN, PA); S2 (BC, WY); (2) *M. fontana* [G5]; SH (NB, PE); S1 (MB, NS, UT, YT); S2 (ME, ON); S3 (BC, LB)

Ecology: general: *Montia* includes annuals, biennials, and perennial species that are commonly associated with wet or moist substrates. Seven of the North American species (88%) are wetland indicators, with two designated as OBL. The flowers are either pollinated by insects (Insecta) or self-pollinated. A substantial seed bank can develop in some species. The seeds are dispersed by a variety of vectors in addition to gravity, water, and wind. They can be dispersed in some species by harvester ants (Insecta: Hymenoptera: Formicidae: *Messor barbarus*) as they are dropped accidentally after being collected for food. They are also transported in viable state through the dung of grazing animals such as domestic sheep (Mammalia: Bovidae: *Ovis aries*).

***Montia chamissoi* (Ledeb. ex Spreng.) Greene** is a perennial that inhabits beaches, bogs, bottoms, canyons, channels, ditches, draws, fellfields, fens, flats, gravel bars, meadows, muskeg, sandstone ledges, seeps (riverbank), slopes (to 15%), springs, streambeds (dry), swales, swamps, thickets, vernal pools, and the margins of brooks, estuaries, lakes, ponds (including constructed trout ponds), rivers, and streams at elevations of up to 3659 m. The exposures range from open to fully shaded sites. The reported substrates include alluvium (mucky), basalt, clay loam, cobble, cobbly gravel, granite, gravel, gravelly mud, humus, loam (James Canyon), mud, muddy gravel, peat, pebbles, quartzite, rocks, rocky volcanic loam, rotting logs, sand, sandy gravel, sandy loam, scree, silt, stones, and talus. Flowering and fruiting extend from April to September. The plants reproduce vegetatively by means of slender rhizomes and by the production of bulblets that develop at the ends of spreading stolons. They grow erect or prostrate (often in tufts) and can form floating mats within the small eddies along stream channels. The flowers sometimes are replaced by vegetative bulbils, which can function as asexual propagules. Although designated as OBL, it is not unusual to find these plants growing along with species of drier habitats. **Reported associates:** *Abies concolor*, *A. lasiocarpa*, *Achillea millifolium*, *Agrostis exarata*, *A. heterophylla*, *Alnus rugosa*, *A. viridis*, *Alopecurus pratensis*, *Amelanchier utahensis*, *Androsace septentrionalis*, *Angelica*, *Anthriscus caucalis*, *Aquilegia chrysantha*, *A. formosa*, *A. saximontana*, *Arctostaphylos patula*, *Arnica chamissonis*, *Artemisia arbuscula*, *A. ludoviciana*, *A. tridentata*, *Astragalus kentrophyta*, *Athyrium filix-femina*, *Balsamorhiza sagittata*, *Barbarea orthoceras*, *Bistorta bistortoides*, *Bolandra oregana*, *Brickellia microphylla*, *Bromus inermis*, *Calamagrostis canadensis*, *Callitriche verna*, *Calocedrus decurrens*, *Camassia quamash*, *Campanula rotundifolia*, *Cardamine breweri*, *Carex abrupta*, *C. aquatilis*, *C. athrostachya*, *C. aurea*, *C. canescens*, *C. disperma*, *C. douglasii*, *C. ebenea*, *C. hoodii*, *C. jonesii*, *C. lenticularis*, *C. luzulina*, *C. lyngbyei*, *C. nebrascensis*, *C. nova*, *C. praegracilis*, *C. rostrata*, *C. stevenii*, *C. subfusca*, *C. utriculata*, *C. vesicaria*, *Castilleja applegatei*, *C. cusickii*, *C. lasiorhyncha*, *C. miniata*, *Cerastium arvense*, *C. fontanum*, *Chenopodium album*, *Claytonia sibirica*, *Collinsia parviflora*, *C. torreyi*, *Comarum palustre*, *Cornus*, *Cryptantha setosissima*, *C. simulans*, *Cystopteris fragilis*, *Danthonia intermedia*, *Dasiphora floribunda*, *Delphinium depauperatum*, *Deschampsia cespitosa*, *D. elongata*, *Descurainia*, *Draba*, *Drymocallis arizonica*, *D. glandulosa*, *D. lactea*, *Eleocharis acicularis*, *E. bella*, *E. palustris*, *E. quinqueflora*, *Epilobium ciliatum*, *E. clavatum*, *E. hornemannii*, *E. obcordatum*, *E. pallidum*, *Equisetum arvense*, *E. fluviatile*, *E. laevigatum*, *Ericameria nauseosa*, *Erigeron algidus*, *E. formosissimus*, *E. subtrinervis*, *Eriogonum racemosum*, *Erysimum capitatum*, *Festuca arizonica*, *F. rubra*, *Floerkea proserpinacoides*, *Fontinalis*, *Fragaria virginiana*, *Gaillardia aristata*, *Galium trifidum*,

Gayophytum diffusum, Geranium erianthum, G. richardsonii, Glyceria elata, G. striata, Gnaphalium palustre, Hackelia micrantha, Heracleum sphondylium, Hesperochiron, Hesperostipa comata, Heuchera micrantha, H. parvifolia, Hordeum brachyantherum, Horkelia fusca, H. rydbergii, Hosackia oblongifolia, Houstonia, Hymenoxys hoopesii, H. subintegra, Hypericum anagalloides, H. formosum, Iris missouriensis, Jamesia americana, Juncus balticus, J. bufonius, J. castaneus, J. ensifolius, J. macrandrus, J. nevadensis, J. orthophyllus, Juniperus occidentalis, J. osteosperma, Lactuca serriola, Larix, Lewisia kelloggii, L. nevadensis, L. pygmaea, L. triphylla, Leymus cinereus, L. triticoides, Ligusticum porteri, Lilium parryi, Limosella acaulis, Linum lewisii, Lithophragma glabrum, L. parviflorum, Lupinus argenteus, L. confertus, L. lepidus, L. pratensis, Luzula comosa, L. orestera, Maianthemum stellatum, Mentha arvensis, Menyanthes trifoliata, Mertensia lanceolata, Micranthes oregana, Microsteris gracilis, Mimulus androsaceus, M. breviflorus, M. breweri, M. floribundus, M. gemmiparus, M. glabratus, M. guttatus, M. moschatus, M. pilosus, M. primuloides, M. suksdorfii, Mitella, Monarda, Montia linearis, M. parvifolia, Muhlenbergia filiformis, M. richardsonis, Myosurus apetalus, Navarretia, Nemophila pedunculata, Oenothera flava, Olsynium douglasii, Oreochrysum parryi, Oreostemma alpigenum, Orobanche uniflora, Oxypolis occidentalis, Oxytropis deflexa, O. parryi, Packera neomexicana, P. plattensis, P. werneriifolia, Pedicularis groenlandica, Penstemon heterodoxus, Perideridia, Petasites, Phacelia exilis, P. heterophylla, Phalacroseris bolanderi, Phalaris arundinacea, Phleum alpinum, P. pratense, Phlox stansburyi, Phyllodoce breweri, Picea engelmannii, P. pungens, Pinus contorta, P. flexilis, P. jeffreyi, P. monophylla, P. ponderosa, Plagiobothrys hispidulus, P. scouleri, Platanthera dilatata, Poa bulbosa, P. pratensis, P. secunda, Polemonium eximium, Polygonum douglasii, P. polygaloides, Polypodium hesperium, Populus balsamifera, P. tremuloides, P. trichocarpa, Potentilla anserina, P. drummondii, P. gracilis, P. pseudosericea, P. wheeleri, Primula pauciflora, P. suffrutescens, P. tetrandra, Prunella vulgaris, Prunus, Pseudocymopterus montanus, Pseudoroegneria spicata, Pseudostellaria jamesiana, Pteridium aquilinum, Purshia tridentata, Quercus gambelii, Q. kelloggii, Ranunculus aquatilis, R. occidentalis, R. orthorhynchus, R. uncinatus, Rhodiola integrifolia, Ribes aureum, R. cereum, R. inerme, R. niveum, Rorippa curvisiliqua, Rosa californica, R. woodsii, Rudbeckia hirta, Rumex crispus, Sagina saginoides, Salix bebbiana, S. eastwoodiae, S. exigua, S. geyeriana, S. lasiandra, S. lasiolepis, S. lutea, S. planifolia, S. scouleriana, S. wolfii, Sanicula marilandica, Saxifraga mertensiana, Schoenoplectus, Senecio crassulus, S. hydrophilus, S. scorzonella, S. triangularis, Sidalcea malviflora, S. neomexicana, S. oregana, Silene invisa, Solidago elongata, Sphagnum warnstorfii, Sphenosciadium capitellatum, Stachys ajugoides, S. albens, S. rigida, Stellaria longipes, Swertia perennis, Symphoricarpos rotundifolius, Symphyotrichum spathulatum, Taraxacum californicum, T. officinale, Taraxia subacaulis, Thalictrum fendleri, Thelypodium laciniatum, Thermopsis divaricarpa, T.

montana, Toxicodendron diversilobum, Tragopogon dubius, Trifolium cyathiferum, T. eriocephalum, T. monanthum, T. wormskioldii, Triteleia hyacinthina, Typha, Urtica dioica, Vaccinium uliginosum, Veratrum californicum, Verbascum thapsus, Veronica americana, V. anagallis-aquatica, V. peregrina, V. serpyllifolia, Vicia, Viola canadensis, V. macloskeyi, V. palustris, V. sororia, Zigadenus.

***Montia fontana** L.* is an annual or biennial inhabitant of wet sites or fresh or brackish, flowing or still, shallow (to 30 cm) waters, which occur on beaches, bluffs, cliffs, crevices, depressions, ditches, domes, draws, dunes, exposed mud banks, fens, flats (tidal), gravel bars, gravel pits, hot springs, lagoons, lawns, ledges, marshes, meadows, mudflats, pools, prairies, roadsides, salt marshes, seeps, shores (tidal), slopes (to 50%), sloughs, springs, swales, thickets, tundra, vernal ponds/pools, and along the margins of brooks, lakes, ponds, and streams at elevations of up to 3352 m. The plants can grow as terrestrials when the temporary pools become dry. They will tolerate exposures from full sun to shade. The substrates have been characterized not only as acidic (e.g., pH: 4.5–5.0) but also as alkaline and include alluvium, basalt (rocky), breccia (volcanic), clay, clay loam, clayey silt, cobble, granite (decomposed), gravel, gravelly sand, humus, loamy clay, mud, sand, sandstone, sandy gravel, sandy loam, sandy silt, serpentine, silt (organic rich), and siltstone. Flowering and fruiting occur from March to August (the plants are day neutral with respect to their flowering requirements). The flowers are believed to be primarily autogamous. The seeds are shed at densities averaging 48.1/m^2 and are well represented in seed banks, which can attain mean densities of 1250 seeds/m^2 (upper 5 cm of soil). The small seeds are vagile and sometimes are dispersed by water, especially along the drift lines of watercourses. They are known to be carried (and occasionally dropped) by harvester ants (Insecta: Hymenoptera: Formicidae: *Messor barbarus*) and are dispersed in the dung of domestic sheep (Mammalia: Bovidae: *Ovis aries*), which graze the plants. The plants will grow among taller grasses and sedges or in open places where the stems then become prostrate and tufted. They have been described as a 'fugitive' species ecologically, where persistence at sites is facilitated by domination (here assisted by the extensive seed bank) during the early stages of vegetation development and gap dynamics. In this respect, they are greatly assisted by disturbances (e.g., fires), which remove thatch and enable the seeds to germinate more rapidly. The plants occur at significantly higher frequencies in burned relative to unburned sites. **Reported associates:** *Abies concolor, Achillea millefolium, Acmispon humistratus, A. strigosus, Agrostis scabra, A. stolonifera, Alchemilla, Allium tolmiei, Alnus viridis, Aphanes occidentalis, Arabidopsis thaliana, Arctophila fulva, Arctostaphylos columbiana, A. uva-ursi, Armeria maritima, Arnica chamissonis, Artemisia douglasiana, A. dracunculus, A. ludoviciana, A. tridentata, Athyrium filix-femina, Bellis perennis, Betula glandulosa, Brodiaea, Bromus tectorum, Calamagrostis canadensis, C. stricta, Calandrinia ciliata, Callitriche longipedunculata, C. marginata, Camassia leichtlinii, Capsella bursa-pastoris, Cardamine hirsuta, C. penduliflora, C. pensylvanica, C.*

pratensis, Carex aperta, C. aquatilis, C. echinata, C. jonesii, C. mackenziei, C. macrochaeta, C. microptera, C. rariflora, C. subspathacea, C. ursina, Castilleja tenuis, Cerastium fontanum, C. glomeratum, Chamerion angustifolium, Cicuta virosa, Cirsium parryi, Claytonia exigua, C. parviflora, C. perfoliata, C. rubra, C. sibirica, Cochlearia groenlandica, Collinsia parviflora, C. sparsiflora, Comarum palustre, Crassula aquatica, C. connata, C. tillaea, Cystopteris fragilis, Cytisus scoparius, Daucus pusillus, Deschampsia cespitosa, D. danthonioides, Dicentra formosa, Distichlis spicata, Downingia bella, D. cuspidata, Draba verna, Dupontia fisheri, Elatine californica, Eleocharis macrostachya, E. quinqueflora, Empetrum nigrum, Epilobium ciliatum, Equisetum arvense, E. fluviatile, Erigeron formosissimus, Eriogonum fasciculatum, E. wrightii, Eriophorum angustifolium, Erodium brachycarpum, E. cicutarium, Eryngium castrense, E. vaseyi, Eschscholzia californica, Festuca rubra, Fraxinus latifolia, F. velutina, Fremontodendron, Fritillaria camschatcensis, Galium aparine, G. trifidum, Gaultheria shallon, Geranium dissectum, G. richardsonii, Glyceria elata, Gratiola ebracteata, Hemizonia, Heracleum sphondylium, Hippuris vulgaris, Honckenya peploides, Hordeum brachyantherum, Horkelia rydbergii, Isoetes, Juncus balticus, J. bufonius, J. effusus, J. ensifolius, J. longistylis, J. nevadensis, Lasthenia californica, L. fremontii, L. glaberrima, Lepidium, Leymus mollis, Lilaea scilloides, Limnanthes douglasii, Lithophragma, Lomatium, Lupinus nootkatensis, Lysichiton americanus, Malus, Matricaria, Menyanthes trifoliata, Mertensia franciscana, Mimulus bicolor, M. glaucescens, M. guttatus, Montia dichotoma, M. howellii, M. linearis, Myosotis discolor, Myosurus minimus, Nassella pulchra, Nemophila heterophylla, Phippsia algida, Physocarpus, Picea pungens, Pilularia americana, Pinus coulteri, P. monophylla, P. ponderosa, Plagiobothrys collinus, P. stipitatus, Plantago elongata, P. lanceolata, Platanthera dilatata, Platanus racemosa, Plectritis congesta, Poa annua, P. arctica, P. bulbosa, P. compressa, P. palustris, P. secunda, Pogogyne ziziphoroides, Polypodium, Populus tremuloides, Primula, Pseudotsuga menziesii, Psilocarphus brevissimus, Puccinellia phryganodes, Pycnanthemum californicum, Quercus agrifolia, Q. chrysolepis, Q. douglasii, Q. garryana, Q. kelloggii, Ranunculus aquatilis, R. californicus, R. canus, R. flammula, R. hyperboreus, R. pallasii, R. testiculatus, Rhodiola rosea, Rhododendron, Rhus ovata, Ribes pinetorum, Rosa nutkana, Rubus chamaemorus, R. idaeus, R. spectabilis, Rumex hymenosepalus, Sagina procumbens, Salix barclayi, S. laevigata, S. lasiolepis, S. pulchra, Saxifraga radiata, Scirpus microcarpus, Sedella pumila, Sedum lanceolatum, Selaginella bigelovii, Senecio, Sidalcea neomexicana, Spergularia canadensis, S. media, Stellaria crassifolia, S. humifusa, S. media, Taraxacum officinale, Taraxia tanacetifolia, Tephroseris palustris, Toxicodendron rydbergii, Trifolium dubium, T. repens, T. variegatum, Trillium albidum, Trisetum spicatum, Uropappus lindleyi, Veratrum californicum, Verbascum thapsus, Veronica americana, V. anagallis-aquatica, V. arvensis, Vicia sativa, Vulpia, Woodwardia fimbriata.

Use by wildlife: The plants are eaten by some herbivores such as domestic sheep (Mammalia: Bovidae: *Ovis aries*).

Economic importance: food: The leaves of *Montia chamissoi* and *M. fontana* can be eaten raw in salads or cooked. A bitter taste may develop in plants when grown under warm, dry conditions; **medicinal:** none; **cultivation:** none; **misc. products:** none; **weeds:** *Montia fontana* is regarded as a weed in Labrador and other areas; **nonindigenous species:** none.

Systematics: Preliminary phylogenetic studies using nrITS sequence data indicated that *Montia* was monophyletic and included a clade containing *M. chamissoi, M. fontana*, and *M. parvifolia*. Subsequent analysis of all recognized *Montia* species (using combined DNA sequence data from five regions) confirmed the monophyly of the genus and also the close interrelationships among *M. chamissoi, M. fontana*, and *M. parvifolia* (Figure 3.20). Despite the morphological similarity observed between *M. fontana* (section *Montia*) and *M. howellii* (section *Maxia*) by numerical analyses, it is evident that *M. chamissoi* and *M. fontana* are more closely related. However, *M. chamissoi* is most similar morphologically to *M. calcicola*, which has not yet been included in phylogenetic analyses; the two species comprise section *Alsinastrum* which some authors have recognized as a distinct genus (*Crunocallis*). The basic chromosome number of *Montia* is variable ($x = 7, 8, 10, 11$). The widespread *M. chamissoi* is diploid ($2n = 22$). *Montia fontana* has diploid ($2n = 18, 20$) and tetraploid ($2n = 40$) cytotypes reported.

Comments: *Montia chamissoi* occurs throughout western North America and also is disjunct in the Delaware River watershed of Pennsylvania. *Montia fontana* is distributed worlwide and throughout northern North America.

References: Azcárate et al., 2005; Hedberg, 1969; McNeill, 1975; Miller, 2003b; Ogburn & Edwards, 2015; O'Quinn & Hufford, 2002; Pakeman & Small, 2005; Pakeman et al., 2002; Parachnowitsch & Elle, 2005; Rasran & Vogt, 2008; Tardío et al., 2011; York, 1997; Zutter, 2009.

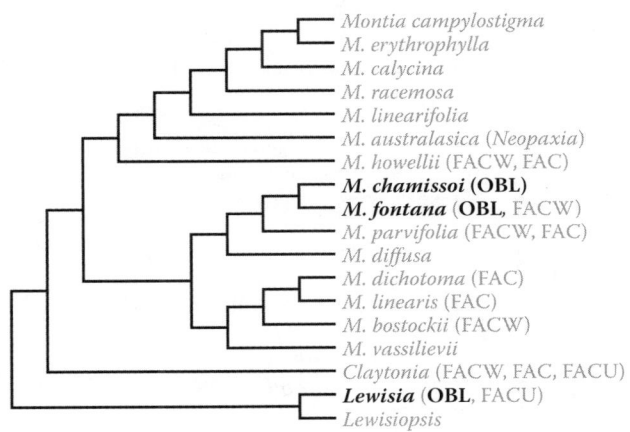

FIGURE 3.20 Cladogram showing relationships of *Claytonia, Montia*, and *Lewisia* based on combined DNA sequence data from five regions. The wetland indicator status is indicated in parentheses for any ranked North American taxon (OBL indicators are highlighted in bold). The OBL habit appears to have arisen only once in *Montia*. (Adapted from Ogburn, R.M. & Edwards, E.J., *Mol. Phylogen. Evol.*, 92, 181–192, 2015.)

ORDER 2: POLYGONALES [9]

A variety of molecular data (e.g., 18S, *atpB*, *matK*, *rbcL* sequences) resolve a clade that is positioned phylogenetically as the sister group to Caryophyllales (Figures 3.1 and 3.21). This clade, referred to by some authors as the "noncore Caryophyllales" or suborder Polygonineae (e.g., Judd et al., 2016), comprises the families that are included here in the order Polygonales. The group is characterized by secretory cells with plumbagin (a napthaquinone), an indumentum of glandular-headed hairs, basal placentation (rarely axile), and starchy endosperm (Judd et al., 2016).

The Polygonales contain about 2050 species. In North America, obligate aquatic species occur within the following three families (Frankeniaceae were included in the 1996 indicator list, but no longer have OBL indicators in the revised lists):

1. **Droseraceae** Salisb.
2. **Plumbaginaceae** Juss.
3. **Polygonaceae** Juss.

Those families containing OBL aquatic taxa are distributed in both major subclades of Polygonales (Figure 3.21). Although a concentration of aquatics appears to occur in families close to Polygonaceae, it is important to note that most Frankeniaceae and Plumbaginaceae, and many Polygonaceae are terrestrial. Perhaps there is a predisposition to aquatic life in these families, but the aquatic habit certainly has originated independently among them and is not ancestral as might be construed from the intergeneric relationships depicted in Figure 3.21. On the other hand, the entire family Droseraceae consists of species adapted to aquatic habitats.

Family 1: Droseraceae [3]

The family Droseraceae comprises about 109 species, all which characteristically possess adaptations to wet or aquatic habitat conditions. The species are carnivorous herbs, with sensitive, adaxially circinate leaves and pollen

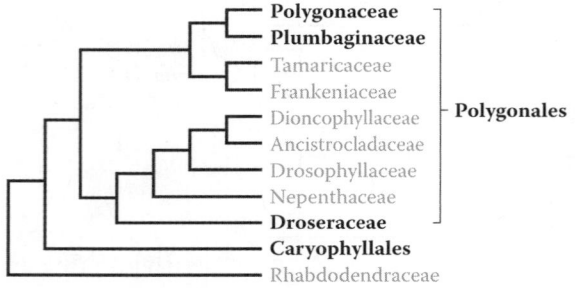

FIGURE 3.21 Phylogenetic relationships among the nine families of Polygonales as reconstructed from combined molecular data. An identical topology for the lower Polygonales clade was recovered by an analysis of combined nuclear and plastid DNA sequence data (Renner & Specht, 2011). The taxa that contain OBL indicators in North America are highlighted in bold. (Adapted from Cuénoud, P. et al., *Amer. J. Bot.*, 89, 132–144, 2002.)

released as tetrads (Judd et al., 2016). The family is quite unusual in that all of the species are carnivorous and possess various specialized contrivances for capturing minute animal prey. The carnivorous habit provides the plants with supplemental nitrogen and other nutrients to facilitate their survival in nutrient-poor habitats such as acidic bogs with organic (peat) substrates, where representatives of this family often thrive. To facilitate capture of insects (Insecta) and other invertebrate prey, the leaf blades are modified either as active "snap" traps or as passive traps that capture prey upon their surfaces, which are covered by sticky, mucilaginous, glandular hairs.

Ironically, the colorful, protandrous flowers are usually pollinated by insects (Insecta). The flowers are self-compatible and capable of self-pollination as well. The seeds are dispersed abiotically by water or wind. Asexual reproduction can occur by means of detached leaves, stem fragments, and the production of adventitious plantlets (Judd et al., 2016).

The family is cultivated widely as a source of curiosities and for novelty specimen plants: *Aldrovanda* (waterwheel plant) as an aquarium specimen, *Dionaea muscipula* (Venus flytrap) with over 20 cultivars, and *Drosera* (sundew) with dozens of cultivars.

All of the genera in the family (*Aldrovanda*, *Dionaea*, and *Drosera*) can be regarded as "aquatic" to some degree but only one North American genus (*Drosera*) includes OBL indicators. However, *Aldrovanda* is a submerged aquatic indigenous to the Old World; it has recently been introduced to North America (Lamont et al., 2013; Floyd et al., 2015) and is included here as an OBL indicator. *Dionaea* is endemic to the southeastern United States, but its wetland status is ranked for some reason only as facultative (FACW). *Dionaea* has also been included in the present treatment because the plants occur only in semipermanently wet sites and can tolerate even submerged conditions for short periods. Consequently, three North American genera are recognized here as containing OBL aquatics:

1. *Aldrovanda* L.
2. *Dionaea* Sol. ex J. Ellis
3. *Drosera* L.

Analyses of various DNA sequence data (e.g., Cuénoud et al., 2002; Renner & Specht, 2011) indicate that *Drosophyllum*, although formerly included in Droseraceae, is actually more distantly related. The intergeneric relationships in Droseraceae have been well established by a number of molecular phylogenetic studies. The most comprehensive treatments (based on various combinations of nuclear and plastid sequence data) indicate that the family is monophyletic with *Aldrovanda* and *Dionaea* forming a clade and *Drosera* resolving as the sister group of that clade (Figure 3.22).

1. *Aldrovanda*

Waterwheel plant

Etymology: after Ulisse Aldrovandi (1522–1605)

Synonyms: *Drosera* (in part)

FIGURE 3.22   Relationships among genera of Droseraceae as indicated by combined *rbcL* and 18S sequence data. The same result has been demonstrated using a combination of 18S, *atpB*, *matK*, and *rbcL* data (Cameron et al., 2002) and by analysis of other nuclear and cpDNA markers (Renner & Specht, 2011). Those North American genera containing OBL indicators are highlighted in bold; *not ranked in 2013 indicator list but unquestionably OBL; ** = categorized in this treatment as OBL, but as FACW in the 2013 indicator list. (Adapted from Rivadavia, F. et al., *Amer. J. Bot.*, 90, 123–130, 2003.)

Distribution: global: Africa; Asia; Australia; Europe; North America; **North America:** eastern United States
Diversity: global: 1 species; **North America:** 1 species
Indicators (USA): OBL: *Aldrovanda vesciculosa*
Habitat: freshwater; lacustrine; **pH:** 4.5–9.3; **depth:** <1 m; **life-form(s):** suspended (vittate)
Key morphology: stems (to 30 cm) green (Eurasia) or red (Australia), rootless, floating, branched irregularly, with specialized apical turions (to 7 mm); leaves (to 15 mm) in dense whorls (of 6–17), petiole (to 9 mm) cuneate, terminating in four to six bristles, blade (to 8 mm) reniform, hinged (closing rapidly as a trap), with medial trigger hairs; flowers solitary, axillary, peduncled (to 15 mm), emergent, or submersed (then cleistogamous); petals (to 5 mm) greenish-white; ovary superior, unilocular, with five styles; capsules (to 4 mm) with six to nine seeds (to 1.5 mm)
Life history: duration: perennial (turions); **asexual reproduction:** turions; **pollination:** insect, self; **sexual condition:** hermaphroditic; **fruit:** capsules (common); **local dispersal:** turions, seeds; **long-distance dispersal:** seeds (birds?)
Imperilment: (1) *Aldrovanda vesciculosa* [GNR]
Ecology: general: *Aldrovanda* is monotypic (see next).

Aldrovanda vesiculosa **L.** occurs in the still, shallow waters of lakes, ponds, pools, and river deltas at elevations of up to 263 m. It is not restricted to any particular depth because of its suspended habit, whereby it grows just beneath the water surface at depths from 10 to 50 cm. Its habitats are soft, acidic waters (optimal pH: 4.5–6.5), which can reach temperatures up to 30°C. North American populations occur in similar conditions (pH: 5.6–6.4; specific conductance: 25–61 μS/cm). The plants have tolerated pH values up to 9.3 under experimental conditions. The plants occur naturally in dystrophic sites but grow best experimentally under mesoeutrophic conditions. Water temperatures of at least 16°C and high light levels are required during the growing season. The plants grow quickly (up to 0.9 cm/day). Flowering occurs only when water temperatures are maintained above 25°C during late summer; otherwise, the plants reproduce vegetatively by turion formation. Relatively few seeds with low viability (<50%) are produced by the North American plants; however, one site yielded 98 fertile fruits. A persistent seed bank does not appear to form

in North American sites. Genetic studies using a variety of different markers have shown little variation to exist worldwide, with only two haplotypes detected, which essentially distinguish Eurasian populations from the others; overall polymorphism is low otherwise. The plants are believed to be dispersed over long distances by seeds or turions transported (externally) by birds (Aves), but they do not appear to survive passage through their digestive tract. Vegetative reproduction can occur by the branching of shoots or formation of turions (in temperate sites), which sink to the bottom where they overwinter. The plants can form dense colonies (averaging 150 individuals/m^2; reaching up to 1260 individuals/m^2) but compete poorly with filamentous algae. The leaf traps capture various aquatic invertebrates (primarily zooplankton), which contact the trigger hairs, causing them to close in less than 0.2 s. The captured prey provides supplemental nitrogen to the plants, which is not essential for survival but results in significantly enhanced growth rates. **Reported associates (North America):** *Carex alata, C. festucacea, C. lurida, Commelina virginica, Decodon verticillatus, Drosera intermedia, Dulichium arundinaceum, Eriocaulon decangulare, Iris virginica, Juncus tenuis, Lemna minor, Nuphar advena, Nymphaea odorata, Peltandra virginica, Sagittaria latifolia, Saururus cernuus, Scirpus cyperinus, Sparganium americanum, Sphagnum, Typha latifolia, Utricularia geminiscapa, U. gibba, U. inflata, U. macrorhiza, U. purpurea.*
Use by wildlife: unknown
Economic importance: food: unknown; **medicinal:** none; **cultivation:** *Aldrovanda vesiculosa* is cultivated infrequently as an aquarium plant; **misc. products:** none; **weeds:** *A. vesiculosa* is potentially a serious invasive weed in North America, despite the dissenting opinions of some authors (e.g., Lamont et al., 2013). The major impact likely would be the disruption of native bog plant communities; **nonindigenous species:** *Aldrovanda vesiculosa* was introduced intentionally to North America in a misguided attempt to provide refugia for the species, which is threatened with extirpation in the Old World. The plants were introduced to Virginia during the 1980s and to New Jersey and New York in 1999. They have since become naturalized and are thriving at numerous sites in all three states.
Systematics: Phylogenetic studies consistently resolve the monotypic *Aldrovanda* as the sister group of *Dionaea*, which also is monotypic (Figure 3.22). The basic chromosome number of *Aldrovanda* is $x = 8$; *A. vesiculosa* ($2n = 48$) is a hexaploid.
Comments: *Aldrovanda vesiculosa* is known currently in North America only from the United States (New Jersey, New York, and Virginia).
References: Adamec, 1995, 2000; Adamec & Kondo, 2002; Breckpot, 1997; Casper & Krausch, 1981; Cook, 1996a; Cross, 2012; Cross et al., 2015; Elansary et al., 2010; Floyd et al., 2015; Lamont et al., 2013; Maldonado San Martín et al., 2003; Mühlberg, 1982.

2. Dionaea

Venus flytrap; attrape-mouches
Etymology: after Dione, the mother of Aphrodite (Venus) in Greek mythology

Synonyms: none
Distribution: global: North America; **North America:** southeastern
Diversity: global: 1 species; **North America:** 1 species
Indicators (United States): OBL; FACW: *Dionaea muscipula*
Habitat: freshwater; palustrine; **pH:** 3.5–6.5; **depth:** <1 m; **life-form(s):** emergent (herb)
Key morphology: stem a bulblike rhizome, but elongating horizontally (to 6 cm); leaves (to 20 cm) arranged in rosettes (of up to seven), the blades valvate, clam-like, resembling miniature "bear-traps" and with long (to 8 mm), marginal, comb-like bristles, the valves closing together quickly following stimulation of internal trigger hairs, the petioles winged; flowering scapes (to 40 cm) with a terminal cyme of 1–15 flowers; petals (to 1.3 cm) five, white; capsules (to 6 mm) ovoid, with 20–30 shiny, black seeds (to 1 mm)
Life history: duration: perennial (rhizomes); **asexual reproduction:** adventitious plantlets, rhizomes; **pollination:** insect, self; **sexual condition:** hermaphroditic; **fruit:** capsules (common); **local dispersal:** rhizomes, seeds; **long-distance dispersal:** seeds
Imperilment: (1) *Dionaea muscipula* [G3]; S3 (NC, SC)
Ecology: general: *Dionaea* is monotypic (see next).

Dionaea muscipula **J. Ellis** occurs in bogs, borrow pits, depressions, ditches, pine savannas, pocosins, seeps, and sphagnum openings (often between hummocks) at elevations of up to 366 m. Although categorized as a FACW indicator, this species should be regarded as OBL because of its dependency on saturated soil conditions throughout most of the growing season. The substrates are acidic (pH: 3.5–6.5), infertile, and consist primarily of sand (St. John's series quartzite, 8% organic matter) or sphagnum peat. The plants require full sunlight but thrive where summer temperatures are below 26°C and winter temperatures fall below 4°C. Flowering and fruiting occur from May to June. Flowering initiates when the plants are at least 3 years old and have six or more traps. The flowers are cross-pollinated by insects (Insecta) and also are self-compatible and autogamous, perhaps to assure seed set if the pollinators become entrapped. When mature, seeds of *D. muscipula* lack dormancy and germinate readily. Optimum germination occurs at 26.7°C–29.5°C. Seed viability falls to 30% after 1 year of storage. The plants reproduce vegetatively from short rhizomes (growing in the upper 10 cm of soil), which bud off the stem, and in rare instances by adventitious plantlets, which replace flowers in some inflorescences. Individual plants produce no more than seven leaves and can live up to 25 years. Though small, the plants can reach high densities, forming "lawns" when the growth conditions are favorable. The populations rely on fire to suppress the competing vegetation. Following a fire, about 75% of the plant's nitrogen is derived from insects. Potential prey organisms are attracted to the plants by their release of terpenes, benzenoids, aliphatics, and other volatile organic compounds, and cues provided by ultraviolet (blue) fluorescence (366 nm), which is emitted at the trap sites. Documented prey items consist of various gastropods (Mollusca: Gastropoda) and arthropods

(Arthropoda) including centipedes (Diplopoda), millipedes (Diplopoda), mites and spiders (Arachnida: Acari; Araneae), springtails (Collembola), and various insects (Insecta: Blattoidea; Coleoptera; Diptera; Hemiptera; Hymenoptera: Formicidae; Lepidoptera; Orthoptera). The efficacy of prey capture is comparable in natural and introduced populations. Remarkably, the prey is digested using the same kind of digestive enzymes used by other plants to repel insects. Digestion of arthropods is also facilitated by the secretion of chitinase. Seedlings from which the prey was withheld exhibited substantially reduced growth rates. Once the graminoid layer grows up around plants, the insect-derived nitrogen level drops to about 45%, and the plants become more dependent on the soil nitrogen pools. Genomic studies have shown the transcriptsome of *Dionaea* to be highly represented in molecular functions related to antioxidant, catalytic, and electron carrier activities. Because of its relatively small native range and immense popularity, *D. muscipula* is threatened by over collecting. However, management activities that reduce the incidence of fire in habitats also threaten populations, which fare poorly as the taller vegetation establishes. It is illegal to collect *Dionaea* specimens from wild populations and the plants are regulated in international trade by the CITES convention. **Reported associates:** *Aristida stricta, Arundinaria gigantea, Calopogon pallidus, Carex cusickii, Ctenium aromaticum, Drosera capillaris, D. tracyi, Eriocaulon decangulare, Euphorbia inundata, Fimbristylis, Juncus effusus, Kalmia, Lachnanthes caroliniana, Lachnocaulon anceps, Lobelia, Lophiola aurea, Lycopodium carolinianum, Pinguicula, Pinus contorta, Polygala, Rhododendron groenlandicum, Rhynchospora alba, R. breviseta, R. chapmannii, Sarracenia alata, S. flava, S. leucophylla, S. minor, S. psittacina, S. purpurea, S. rubra, Sphagnum, Tephrosia, Vaccinium oxycoccus, Xyris ambigua.*
Use by wildlife: *Dionaea muscipula* is the host plant for larvae of the Venus flytrap cutworm moth (Insecta: Lepidoptera: Noctuidae: *Hemipachnobia subporphyrea*).
Economic importance: food: *Dionaea muscipula* has not been used as a food source, due perhaps to its rarity and unappetizing admixture of animal carcasses in the foliage. The plants also contain naphthoquinones, which can be highly toxic; **medicinal:** *Dionaea* contains plumbagin (a naphthoquinone) and other secondary metabolites, which reputedly have antibacterial, anticancer, and other medicinal properties. Extracts of the plant are sold commercially as the product "Carnivora" whose therapeutic value has not been substantiated; **cultivation:** *Dionaea muscipula* is a popular cultivated species because of its unusual carnivorous habit. Numerous cultivars are distributed including 'Akai Ryu' (an Award of Garden Merit plant), 'All Green,' 'B-52,' 'Big Mouth,' 'Bohemian Garnet,' 'Clayton's Red Sunset,' 'Cupped Trap,' 'Darwin,' 'Dente Traps,' 'Fused Tooth,' 'Green Dragon,' 'Holland,' 'Jaws,' 'Justin Davis,' 'Louchapates,' 'Mk1979,' 'Noodle Ladle,' 'Petite Dragon,' 'Pink Venus,' 'Red Burgundy,' 'Red Dragon,' 'Red Piranha,' 'Royal Red,' 'Sawtooth,' 'South West Giant' (an Award of Garden Merit plant), 'Spider,' 'Tiger Fangs,' 'Wacky

Traps,' and 'Whale'; **misc. products:** The Cherokee spat small pieces of chewed *Dionaea* plants on their fishing bait; **weeds:** none; **nonindigenous species:** *Dionaea muscipula* has become naturalized in Alabama, California, Florida, New Jersey, and Washington mainly as a result of intentional planting for experimentation and cultivated plant stocks.

Systematics: Some authors have placed the unusual *Dionaea* in a distinct family Dionaeaceae; however, results from several molecular phylogenetic studies indicate that *Aldrovanda*, *Dionaea*, and *Drosera* form a distinct clade (Figure 3.22), which is consistent with the circumscription of the family Droseraceae, once it is emended to exclude *Drosophyllum* (Figure 3.21). These studies also show that *Dionaea* is related more closely to the Old World genus *Aldrovanda* which shares its unusual "spring-trap" mechanism of prey capture. *Dionaea* is monotypic, containing only the species *D. muscipula*. The chromosomal base number for both *Dionaea* and *Aldrovanda* is $x = 8$; *D. muscipula* ($2n = 32$) is a tetraploid.

Comments: *Dionaea muscipula* is restricted to portions of North and South Carolina, with introduced populations in Alabama, California, Florida, New Jersey, and Washington.

References: Bailey, 2008; Cameron et al., 2002; Cheers, 1992; Devi et al., 2016; Evert, 1957; Gaascht et al., 2013; Hamilton & Kessinger, 1967; Hatcher & Hart, 2014; Hutchens, Jr. & Luken, 2015; Jensen et al., 2015; Kreuzwieser et al., 2014; Kurup et al., 2013; Luken, 2012; Mellichamp, 2015; Meyers-Rice, 2000; Ogihara et al., 2013; Paszota et al., 2014; Pietropaolo & Pietropaolo, 1993; Read, 1999; Schnell, 2002; Stokstad, 2016; Szpitter et al., 2014; Walker & Peet, 1984.

3. *Drosera*

Sundew; rossolis

Etymology: from the Greek *droseros* meaning "dewy"

Synonyms: *Dismophyla*; *Rorella*; *Rossolis*; *Sondera*

Distribution: global: cosmopolitan; **North America:** widespread

Diversity: global: 170 species; **North America:** 10 species

Indicators (USA): OBL: *Drosera aliciae*, *D. anglica*, *D. brevifolia*, *D. capensis*, *D. capillaris*, *D. filiformis*, *D. intermedia*, *D. linearis*, *D. rotundifolia*, *D. tracyi*

Habitat: freshwater; palustrine; **pH:** 3.2–8.0; **depth:** <1 m; **life-form(s):** emergent (herb)

Key morphology: stems short; leaves in rosettes, petioled, blades variable, ranging from nearly circular (to 1 cm) to narrowly linear (to 50 cm), clothed with red, sticky, stalked, glandular hairs, which entrap small animal prey; inflorescence an elongate cyme with flowers clustered near tip; petals (to 6 mm) 5, white to rose-red in color

Life history: duration: perennial (winter buds, rhizomes); **asexual reproduction:** adventitious plantlets, winter buds; **pollination:** insect, self; **sexual condition:** hermaphroditic; **fruit:** capsules (common); **local dispersal:** seeds, seedlings (water); **long-distance dispersal:** seeds (birds, water)

Imperilment: (1) *Drosera aliciae* [unranked]; (2) *D. anglica* [G5]; S1 (CO, ME, <u>NB</u>, WI); S2 (CA, YT); S3 (<u>AB</u>, <u>MB</u>, MI, MN, MT, <u>QC</u>, <u>SK</u>, WY); (3) *D. brevifolia* [G5]; S1 (KS, KY, OK); S2 (TN); S3 (NC, VA); (4) *D. capensis* [GNR]; (5) *D. capillaris* [G5]; SH (DE); S1 (MD, TN); S3 (VA); (6) *D. filiformis* [G4]; SX (DE); SH (CT, RI); S1 (FL, <u>NS</u>); S2 (NC); S3 (NY); (7) *D. intermedia* [G5]; S1 (ID, KY, OH, <u>PE</u>); S2 (IL, IN, LA, TN); S3 (FL, RI, VA); (8) *D. linearis* [G4]; S1 (<u>BC</u>, ME, <u>NB</u>, <u>NF</u>, <u>SK</u>, WI); S2 (<u>MB</u>, MT); S3 (<u>AB</u>, MN, <u>QC</u>); (9) *D. rotundifolia* [G5]; S1 (AL, GA, ID, <u>NB</u>, ND, TN); S2 (CO, DE, IL); S3 (IN, MD, MT, NC, OH, WV, <u>YT</u>); (10) *D. tracyi* [G3/G4]; SH (LA); S1 (GA)

Ecology: general: All of the North American *Drosera* species are OBL indicators. They include annual and perennial herbs that occur typically in acidic peatlands; however, several species are found in more alkaline habitats where at least weakly minerotrophic conditions exist. The species also often colonize wet, nutrient-poor, sandy substrates. All of the species are carnivorous and capture small invertebrate prey on their leaves, which are modified as passive, sticky, traps. The captured prey is digested to supplement nitrogen and other nutrients, which often occur in low concentrations in the nutrient-poor habitats occupied by these plants. Many of the species are fairly resistant to environmental perturbations and are known to effectively recolonize habitats that have been disturbed by disruptive activities such as peat mining. In general, sundews are shade intolerant, flood tolerant (up to several months) but are drought intolerant. They are not fire tolerant but quickly colonize suitable habitat in burned areas, where the fires have removed overstory and litter from the sites. From 5 to 20 flowers are produced by the different species. The flowers open during the morning (9:00–10:00 am) to early afternoon (closing by 2:00 pm), and only when adequate sunlight is available. Each flower remains open for about 5 h and few will reopen the successive day. The flowers are potentially cross-pollinated during the day by insects (Insecta: Diptera: Bombyliidae; Calliphoridae; Dolichopodidae; Muscidae; Syrphidae; Tachinidae; Hymenoptera: Apidae; Halictidae; Thysanoptera: Thripidae), but most of the species are also capable of self-pollination (apparently their predominant mode), which occurs at night after the flowers close. The flowers of many of the species are more adapted for autogamy (e.g., low pollen:ovule ratios), which may represent the principal breeding system. The seeds are dispersed by the water, and over longer distances by birds (Aves), which transport them in the mud that clings to their feet. The plants reproduce asexually by adventitious plantlets, which form along the stem. The northern species form turion-like winter buds having compact, specialized leaves; these remain attached to the dead roots. Sundews are threatened by habitat loss or alterations to drainage patterns, which affect the plants directly and also indirectly by reducing the abundance of their invertebrate prey species.

***Drosera aliciae* Raym.-Hamet** is a nonindigenous perennial that occurs in wet depressions at elevations of up to 160 m. The substrates include peat or sand. The flowers are self-pollinating. Reproduction in California occurs both sexually (by seed) and vegetatively. The plants can be propagated vegetatively by leaf cuttings and cuttings made from their thick roots. The following list of associates primarily identifies other carnivorous plants, which were also introduced intentionally to the same locality (see also *D. capensis* below). **Reported associates (North America)**: *Darlingtonia californica, Dionaea muscipula, Drosera binata, D. burmannii, D. capensis, D. capillaris, D. filiformis, D. intermedia, D. nitidula* × *D. occidentalis, D. slackii, Pinguicula lusitanica, Sarracenia flava, S. leucophylla, S. minor, S. purpurea, S. rubra, Sphagnum, Utricularia gibba, U. subulata.*

Drosera anglica **Huds.** occurs on wet ground or in shallow waters (to 10 cm) of bogs, channels, depressions, ditches, dunes, fens, flats, gravel pits, hanging bogs, lakeshores, meadows, moors, muskeg, patterned fens (in flarks), pools, ponds (drying), potholes, slopes (to 1%), and along the margins of lakes and ponds in colder portions of North America at elevations of up to 2628 m. Exposures typically receive full sunlight. The substrates are characterized as acidic or calcareous (pH: 4.5–7.2) and include cobble, gravel, logs, marl, muck (organic), mucky loam, mud, muddy peat, peat, peaty muck, sand, and sandy loam. The sites are nutrient poor [e.g., alkalinity: 4–16 mg/L; conductivity: 16–70 mmhos; hardness (CaCO$_3$): 10–25.5 mg/L; Ca: 2.0–6.2 mg/L; K: 0.8 mg/L; Mg: 0.5–2.4 mg/L]. Flowering occurs from June to September, with fruiting extending into October. Pollination occurs primarily by flies (Insecta: Diptera). Individual plants can produce up to 700 seeds on each cyme. Long-distance dispersal (e.g., to Hawaii) is attibuted to the carriage of seeds in mud attached to the feet of plovers (Aves: Charadriidae). The seeds require cold stratification (1°C for 12 weeks) in order to break dormancy and will achieve their highest germination in the light under a 25°C/15°C temperature regime. The plants overwinter by the formation of hibernaculae. They have not been found to be associated with arbuscular mycorrhizal Fungi. **Reported associates**: *Abies procera, Agrostis scabra, Amelanchier alnifolia, Andromeda polifolia, Antennaria corymbosa, Betula glandulosa, B. nana, Bistorta bistortoides, Calamagrostis canadensis, Calopogon tuberosus, Carex aquatilis, C. aurea, C. buxbaumii, C. canescens, C. chordorrhiza, C. cusickii, C. diandra, C. echinata, C. exilis, C. flava, C. interior, C. lapponica, C. lasiocarpa, C. lenticularis, C. limosa, C. livida, C. luzulina, C. magellanica, C. muricata, C. pauciflora, C. praeceptorum, C. prionophylla, C. rostrata, C. rotundata, C. sartwellii, C. scopulorum, C. utriculata, C. viridula, Chamaecyparis nootkatensis, Chamaedaphne calyculata, Cicuta bulbifera, C. douglasii, Cladium mariscoides, Comarum palustre, Crataegus douglasii, Cypripedium parviflorum, Dasiphora floribunda, Deschampsia cespitosa, Drepanocladus, Drosera intermedia, D. linearis, D. rotundifolia, Dulichium arundinaceum, Eleocharis compressa, E. palustris, E. rostellata, E. quinqueflora, E. tenuis, Epipactis gigantea, Equisetum, Eriophorum angustifolium, E. chamissonis, E. gracile, E. scheuchzeri, Gentiana douglasiana, Habenaria, Hypericum anagalloides, Juncus acuminatus, J. balticus, J. stygius, Juniperus communis, Kalmia microphylla, Larix laricina, Lobelia kalmii, Lonicera caerulea, Lycopodium inundatum, Lycopus uniflorus, Malaxis, Meesia triquetra, Mentha, Menyanthes trifoliata, Menziesia ferruginea, Mimulus moschatus, M. primuloides, Myrica*

gale, Nephrophyllidium crista-galli, Parnassia palustris, Pedicularis groenlandica, P. pennellii, Picea engelmannii, P. mariana, Platanthera, Pogonia ophioglossoides, Polytrichum, Primula mistassinica, Pyrola, Rhododendron albiflorum, R. groenlandicum, Rhynchospora alba, R. fusca, Sarracenia purpurea, Scheuchzeria palustris, Schoenoplectus acutus, S. pungens, Senecio triangularis, Scorpidium scorpioides, Sparganium natans, Sphagnum subsecundum, Spiraea douglasii, Spiranthes romanzoffiana, Thuja occidentalis, Triantha glutinosa, T. occidentalis, Trichophorum alpinum, T. cespitosum, Triglochin maritimum, T. palustre, Tsuga, Typha latifolia, Utricularia gibba, U. intermedia, U. minor, Vaccinium uliginosum, V. oxycoccos, Viola macloskeyi.

Drosera brevifolia **Pursh** is an annual or perennial species that occurs in barrens, bogs, depressions, ditches, dunes, flats, flatwoods, glades, hammocks, lawns (and cemeteries), pinelands, prairies, roadsides, savannas, seeps, slopes (to 5%), swales, swamps, woodlands, and along the margins of lakes, ponds, and streams at elevations of up to 300 m. Exposures usually receive full sunlight and less often are shaded. The substrates are described as clay, clay loam, Knox dolemite (karst), loam, muck, Ruston/Smithdale soils, sand, sandy clay, sandy loam, sandy peat, sandy rock, silty clay loam, silty loam, and Stilson soils (arenic plinthic paleudults). Flowering and fruiting extend from March to December. A persistent seed bank develops. The seeds reportedly germinate well when placed at 20°C under a 16-h photoperiod. The plants appear often in recently burned (or otherwise disturbed) sites. Experiments have shown that the plants do not attract prey items but simply capture them passively on their sticky leaf surfaces. **Reported associates**: *Acer rubrum, Agalinis fasciculata, Allium canadense, Ambrosia, Andropogon gerardii, A. mohrii, A. ternarius, A. virginicus, Aristida dichotoma, A. purpurascens, A. stricta, Aronia arbutifolia, Arundinaria gigantea, Asclepias hirtella, Axonopus fissifolius, Bacopa monnieri, Briza minor, Buchnera americana, Camassia angusta, Carex, Carphephorus odoratissimus, Centella asiatica, Cercis, Chaetopappa asteroides, Chaptalia tomentosa, Clethra alnifolia, Coleataenia anceps, Coreopsis basalis, C. tinctoria, Crataegus, Ctenium aromaticum, Cyperus lecontei, C. oxylepis, C. polystachyos, Cyrilla racemiflora, Danthonia sericea, Dichanthelium aciculare, D. acuminatum, D. dichotomum, D. ovale, D. scabriusculum, D. strigosum, Drosera capillaris, D. tracyi, Echinacea pallida, Eleocharis baldwinii, E. geniculata, E. flavescens, E. montevidensis, E. parvula, Erigeron tenuis, Eriocaulon decangulare, Eupatorium capillifolium, E. leucolepis, Fuirena longa, F. scirpoidea, Gaylussacia mosieri, G. tomentosa, Helenium flexuosum, Helianthus angustifolius, H. mollis, Houstonia pusilla, Hypericum brachyphyllum, Hypoxis hirsuta, Ilex glabra, I. vomitoria, Isoetes texana, Juncus brachycarpus, J. caesariensis, J. marginatus, Lachnocaulon minus, Liatris pycnostachya, Liquidambar styraciflua, Lobelia brevifolia, Ludwigia, Lycopodiella alopecuroides, Lyonia lucida, Magnolia, Muhlenbergia capillaris, Myrica cerifera, Nyssa sylvatica, Oldenlandia uniflora, Orbexilum simplex, Osmanthus, Packera tomentosa, Panicum hemitomon,*

Persea palustris, Physostegia digitalis, Pinguicula lutea, P. pumila, Pinus elliottii, P. palustris, P. serotina, P. taeda, Polygala nana, P. ramosa, P. rugelii, Pteridium aquilinum, Pycnanthemum tenuifolium, Quercus, Rhexia alifanus, R. mariana, Rhynchospora caduca, R. capitellata, R. chapmanii, R. divergens, R. gracilenta, R. microcarpa, R. nitens, R. plumosa, R. pusilla, R. recognita, Rubus, Rudbeckia grandiflora, R. hirta, R. maxima, Sarracenia alata, Schizachyrium scoparium, S. tenerum, Schoenolirion wrightii, Scleria lithosperma, S. pauciflora, S. triglomerata, Serenoa repens, Silphium laciniatum, Smilax glauca, Sorghastrum nutans, Sphagnum, Spiranthes vernalis, Sporobolus junceus, S. silveanus, Steinchisma hians, Syngonanthus flavidulus, Taxodium distichum, Tephrosia onobrychoides, Tridens strictus, Triodanis perfoliata, Utricularia subulata, Vaccinium myrsinites, V. stamineum, X. brevifolia, X. jupicai.

***Drosera capensis* L.** is a nonindigenous perennial species that occurs in wet depressions at elevations of up to 160 m. The plants will tolerate exposures of full to filtered sunlight. The substrates include peat or sand. The plants generally grow under phosphorous limited conditions, a deficit that is alleviated by carnivory, which results in improved photosynthetic performance. Flowering occurs during the spring. The flowers are believed to be largely self-pollinating, because of the entanglement of their anthers and stigmatic papillae as the petals wither and wrinkle. The plants can be propagated vegetatively by root cuttings (see also the comments for *D. aliciae* above). **Reported associates (North America):** *Darlingtonia californica, Dionaea muscipula, Drosera aliciae, D. binata, D. burmannii, D. capillaris, D. filiformis, D. intermedia, D. nitidula × D. occidentalis, D. slackii, Pinguicula lusitanica, Sarracenia flava, S. leucophylla, S. minor, S. purpurea, S. rubra, Sphagnum, Utricularia gibba, U. subulata.*

***Drosera capillaris* Poir.** is a perennial species that inhabits barrens, bogs, borrow pits, bottoms (pond), depressions, ditches, domes, dunes, flatwoods, floodplains, glades, marshes, pinelands, prairies, roadsides, savannas, seeps, swales, swamps, and the margins of baygalls, canals, lakes, pocosins, and ponds at elevations of up to 300 m. The substrates are described as acidic and are characterized as Basinger (spodic psammaquents), loam, loamy sand, muck, Nettles (alfic arenic haplaquods), peat, peaty sand, Placid (typic humaquepts), Pottsburg (grossarenic haplaquods), sand (Corolla–Duckston; Leon), sandy peat, Sellers (cumulic humaquepts), silt, silty clay, silty loam, and Wabasso (alfic haplaquods). Flowering and fruiting occur from April to November. A persistent seed bank develops. The seeds reportedly germinate well when placed at 20°C under a 16-h photoperiod. Although perennial, the plants do not form winter hibernaculae and lack a cormose base. Their seed bank enables them to colonize sites after fires, cutting, or other disturbances. In some sites, the plants are susceptible to soil disturbance and burial because of the activity of crayfish (Crustacea: Decapoda: Cambaridae: *Fallicambarus*). They are hosts of *Agalinis* (Orobanchaceae) species. **Reported associates:** *Acer rubrum, Agalinis flexicaulis, Aletris aurea, Alnus rugosa, Amphicarpum muhlenbergianum, Andropogon glaucopsis, A. glomeratus, Aristida,*

Arnoglossum ovatum, Aronia arbutifolia, Arundinaria tecta, Bacopa, Burmannia, Calamovilfa curtissii, Callicarpa americana, Calopogon multiflorus, Carex turgescens, Casuarina, Centella asiatica, Chrysopsis, Cliftonia monophylla, Coleataenia tenera, Coreopsis gladiata, Cyperus lecontei, Dichanthelium acuminatum, D. ensifolium, Doellingeria umbellata, Drosera brevifolia, D. intermedia, D. tracyi, Eleocharis tuberculosa, Eriocaulon decangulare, Eryngium aquaticum, Eupatorium capillifolium, E. rotundifolium, Fuirena breviseta, F. scirpoidea, Gaylussacia tomentosa, Helianthus angustifolius, Hypericum brachyphyllum, H. cistifolium, H. crux-andreae, H. galioides, Hypoxis, Hyptis alata, Ilex cassine, I. coriacea, I. glabra, I. vomitoria, Juncus effusus, J. marginatus, J. repens, J. trigonocarpus, Lachnanthes caroliniana, Lachnocaulon anceps, L. engleri, L. minus, Liatris acidota, L. pycnostachya, Linum medium, Lophiola aurea, Ludwigia linearis, Lycopodiella alopecuroides, L. appressa, Lyonia lucida, Magnolia virginiana, Malva, Marshallia graminifolia, Medicago minima, Mitreola sessilifolia, Myrica caroliniensis, M. cerifera, M. heterophylla, Nyssa, Osmundastrum cinnamomea, Panicum hemitomon, Persea palustris, Phyla nodiflora, Pinguicula pumila, Pinus elliottii, P. palustris, P. serotina, P. taeda, Pluchea baccharis, Pogonia ophioglossoides, Polygala lutea, P. nana, P. ramosa, P. rugelii, Proserpinaca, Pteridium aquilinum, Rhexia alifanus, R. lutea, R. mariana, Rhynchospora chapmanii, R. ciliaris, R. divergens, R. elliottii, R. filifolia, R. glomerata, R. inexpansa, R. plumosa, R. rariflora, Sabatia brevifolia, S. macrophylla, Sagittaria graminea, S. lancifolia, Sarracenia alata, S. flava, S. leucophylla, S. minor, S. psittacina, S. purpurea, S. rubra, Schizachyrium scoparium, S. tenerum, Scleria ciliata, S. georgiana, S. reticularis, Serenoa repens, Sisyrinchium atlanticum, Smilax laurifolia, Spartina patens, Sphagnum, Syngonanthus flavidulus, Taxodium distichum, Toxicodendron vernix, Triantha racemosa, Utricularia cornuta, U. gibba, U. juncea, U. simulans, U. subulata, Vaccinium myrsinites, Viburnum nudum, Viola primulifolia, Vitis, Xyris ambigua, X. caroliniana, X. elliottii, X. stricta.

***Drosera filiformis* Raf.** is a perennial species that inhabits barrens, beaches, bogs, borrow pits, depressions, ditches, dunes, fens, flatwoods, meadows, pocosins, roadcuts, roadsides, savannas, seeps, shores (exposed), swales, and along the margins of lakes, ponds, and swamps at elevations of up to 56 m. Exposures are characteristically in full sunlight. The substrates are acidic (pH: 3.8–4.5), contain low levels of Ca (0.4–1.4 mg/L) and Mg (0.8–2.0 mg/L), and include gravel, karst, peat, peaty sand, and sand. Prey items up to 10 mm in size can be trapped, but larger organisms are able to escape the sticky leaves. The capture of prey results in higher numbers of flowers and greater reproductive biomass than in the plants that remain unfed. Flowering and fruiting occur from May to October. The flowers are self-pollinating and have a peculiar behavior of opening consistently in an eastward direction. This species occurs in areas following burns or where other disturbances have removed the overstory vegetation to leave exposed peaty substrates. The plants are associated negatively with woody species but can reach large numbers in the ruts created by all-terrain vehicles. They overwinter by the formation of specialized hibernaculae. Vegetative propagation can be achieved by leaf cuttings.

Reported associates: *Acer rubrum, Agalinis purpurea, Alnus serrulata, Andromeda polifolia, Andropogon virginicus, Anemone americana, Aronia ×prunifolia, Calamagrostis canadensis, C. pickeringii, Calopogon tuberosus, Carex exilis, C. silicea, C. striata, Centella asiatica, Cephalanthus occidentalis, Chamaecyparis thyoides, Chamaedaphne calyculata, Cladina mitis, C. terrae-novae, Cladonia cervicornis, C. phylophora, Coreopsis rosea, Dichanthelium acuminatum, Dionaea muscipula, Drosera intermedia, D. rotundifolia, D. tracyi, Eleocharis melanocarpa, E. robbinsii, Eriophorum virginicum, Eupatorium leucolepis, Euphorbia inundata, Euthamia tenuifolia, Fuirena pumula, Galium obtusum, Gaultheria procumbens, Gaylussacia baccata, G. dumosa, Gratiola aurea, Hypericum lissophloeus, Hypoxis hirsuta, Ilex glabra, Iris prismatica, Juncus biflorus, J. militaris, Juniperus communis, Kalmia angustifolia, Lachnanthes caroliniana, Linum, Liquidambar styraciflua, Lycopodiella appressa, L. caroliniana, Lysimachia terrestris, Myrica gale, Oclemena nemoralis, Panicum hemitomon, Persicaria puritanorum, Platanthera blephariglottis, P. cristata, Pogonia ophioglossoides, Polygala cruciata, P. lutea, P. nuttallii, Polytrichum commune, Proserpinaca pectinata, Pyxidanthera, Rhexia mariana, Rhododendron viscosum, Rhynchospora alba, R. macrostachya, R. nitens, Rubus hispidus, Sabatia kennedyana, Sagittaria teres, Sarracenia alata, S. flava, S. leucophylla, S. minor, S. psittacina, S. purpurea, S. rubra, Schizaea pusilla, Scleria reticularis, Solidago uliginosa, Sphagnum cuspidatum, S. flavicomans, S. fuscum, S. magellanicum, S. palustre, S. pulchrum, S. rubellum, S. tenellum, Stachys hyssopifolia, Taxodium ascendens, Trichophorum cespitosum, Utricularia, Vaccinium angustifolium, V. corymbosum, V. macrocarpon, V. oxycoccos, Woodwardia virginica, Zygnema.*

***Drosera intermedia* Hayne** is a perennial species that grows in shallow standing water (<5 cm) or on saturated substrates associated with bays, bogs, depressions, ditches, dunes, fens, flatwoods, gravel pits, pinelands, roadsides, sandbars, sandpits, seeps, swales, swamps, and the margins of Carolina bays, lakes, ponds, pools, sloughs, and streams at elevations of up to 914 m. This species requires wetter conditions than many of its congeners and grows often on floating debris or in mats of floating vegetation (e.g., *Sphagnum*). The substrates are described as acidic and include clay, gravel, logs (floating), muck, peaty humus, sand, and stumps. Exposures range from full sun to partial shade. Flowering and fruiting extend from May to November. The flowers are self-pollinating. Individual plants produce more than 500 seeds on each cyme. A persistent seedbank develops at densities of about 500 seeds/m². The seeds remain viable for several years when stored under dry, refrigerated conditions. They germinate readily when sowed on peat. Occasionally, the flowers are replaced by vegetative plantlets, which are capable of rooting and establishing as propagules. The stems form winter hibernaculae (technically turions), which detach and can also function as propagules.

The supplementation of limiting nutrients is an obvious benefit of prey capture when the plants occur under nutrient-poor conditions. However, prey capture is also advantageous to the plants in more nutrient-rich habitats by reducing the effects of interspecific competition. Plant survival and recruitment are relatively low where extreme water table fluctuations occur compared to more hydrologically stable sites; however, the latter are characterized by higher turnover rates for both genets and ramets. The plants increase significantly in their percent cover following the removal of competing vegetation.

Reported associates: *Agalinis purpurea, Agrostis scabra, Alnus, Aristida longispica, Bartonia virginica, Betula populifolia, Bidens cernuus, B. tripartitus, Brasenia schreberi, Calamagrostis canadensis, Calopogon tuberosus, Carex lasiocarpa, Chamaecyparis thyoides, Chamaedaphne calyculata, Cladium mariscoides, Comptonia peregrina, Cyperus flavescens, Danthonia spicata, Dichanthelium acuminatum, D. ensifolium, Drosera anglica, D. capillaris, D. filiformis, D. rotundifolia, D. tracyi, Dulichium arundinaceum, Eleocharis acicularis, E. olivacea, E. tenuis, Eriocaulon aquaticum, E. compressum, E. decangulare, Eupatorium perfoliatum, Euthamia graminifolia, Fimbristylis autumnalis, Fuirena pumula, Glyceria canadensis, Hypericum boreale, H. canadense, H. majus, Ilex glabra, Juncus acuminatus, J. canadensis, J. effusus, J. greenei, J. militaris, J. pelocarpus, Kalmia, Lachnanthes caroliniana, Lachnocaulon anceps, L. digynum, Larix laricina, Leersia oryzoides, Linum striatum, Ludwigia palustris, Lycopodiella inundata, Lycopodium appressum, Lycopus uniflorus, Lysimachia terrestris, Magnolia, Mayaca fluviatilis, Menyanthes trifoliata, Muhlenbergia uniflora, Myrica gale, Myriophyllum tenellum, Nymphaea, Nyssa, Oenothera fruticosa, Osmunda regalis, Panicum verrucosum, Persicaria careyi, Platanthera blephariglottis, P. ciliaris, Pogonia ophioglossoides, Polygala cruciata, Pontederia cordata, Quercus, Rhexia virginica, Rhynchospora alba, R. capitellata, R. cephalantha, R. chalarocephala, R. fusca, R. knieskernii, R. macrostachya, R. scirpoides, Sagittaria engelmanniana, S. graminea, Salix, Sarracenia alata, S. purpurea, Schoenoplectus pungens, Scirpus cyperinus, Sparganium americanum, Spartina palustre, S. patens, Sphagnum compactum, S. portoricense, S. rubellum, Spiraea alba, S. tomentosa, Spiranthes cernua, Syngonanthus flavidulus, Toxicodendron vernix, Triadenum fraseri, T. virginicum, Utricularia cornuta, U. geminiscapa, U. juncea, U. resupinata, U. subulata, Vaccinium angustifolium, V. corymbosum, V. macrocarpon, Viola lanceolata, Xyris ambigua, X. baldwiniana, X. difformis, X. torta.*

***Drosera linearis* Goldie** is a perennial species that occurs in shallow waters (1–2 cm) of beaches, bogs, fens, marshes, pools, shores, and string bogs (on the flarks) at elevations of up to 400 m. The substrates range more toward alkaline conditions (pH: 5.6–8.0; typically 6.8–7.9) and typically are high in calcium (25–58.5 mg/L) and magnesium (10–18.8 mg/L). They include magnesium limestone, marl, marly sand, and logs and are comprised mainly of sand (75%–88%) with smaller proportions of silt (7%–13%) and clay (6%–12%). Ammonia nitrogen levels are low (2.5 ppm). Flowering and fruiting occur from June to September. The flowers have a low pollen:ovule ratio (18:7) and are primarily self-pollinating. Individual plants produce 5–10 flowers and up to 200 seeds on each cyme. The seeds germinate well in covered pots of wet peat following 6 weeks of cold (4°C) stratification. The seedlings float and help to disperse the species. The leaves die back in late autumn when the winter hibernaculae develop. These can detach and serve as propagules. Vegetative growth resumes the following May. The plants are poor competitors, even with mosses (Bryophyta), and typically occur at sites (e.g., hummocks) where no competitive vegetation is established. The linear leaves are efficient at capturing prey. When compared to *D. rotundifolia* (orbicular leaves), individuals of *D. linearis* trapped four times as many insects per leaf and 14.7 times as many insects per plant on average and overall was 2.3 times more efficient at prey capture than the former species. **Reported associates:** *Agalinis purpurea, Calopogon pulchellus, Campanula aparinoides, Campyllium stellatum, Carex chordorrhiza, C. exilis, C. interior, C. lasiocarpa, C. limosa, C. livida, C. sterilis, C. tetanica, Chamaedaphne calyculata, Comarum palustre, Drosera anglica, D. rotundifolia, Eleocharis rostellata, Equisetum fluviatile, Eriophorum viridicarinatum, Eupatorium perfoliatum, Galium labradoricum, Habenaria dilatata, H. leucophaea, Kalmia polifolia, Lobelia kalmii, Menyanthes trifoliata, Parnassia glauca, Phragmites australis, Picea, Pinguicula vulgaris, Platanthera hyperborea, Pogonia ophioglossoides, Rhynchospora alba, Scorpidium scorpioides, Solidago uliginosa, Sphagnum, Triantha glutinosa, Trichophorum cespitosum, Triglochin maritimum, Typha, Sarracenia purpurea, Schoenoplectus subterminalis, Utricularia cornuta, U. gibba, U. intermedia, U. macrorhiza, U. minor, Vaccinium oxycoccus.*

Drosera rotundifolia L. is a perennial species that occurs on moist substrates or in shallow (to 5 cm) standing waters of bogs, depressions, ditches, fens, floodplains, glades, gravel pits, meadows, mires, mudflats, muskeg, ponds, potholes, roadsides, sand pits, seeps, shorelines, swales, swamps, tundra, and the margins of lakes, ponds, and streams at elevations of up to 3200 m. Exposures can range from full sun to partial shade; however, plants grown in the shade have less viscid leaves and a reduced investment in carnivory. The substrates are acidic [pH: 4.0–7.3; e.g., total hardness (as $CaCO_3$): 10–11.3 mg/L; total alkalinity: 16–48 mg/L; Ca: 2.7 mg/L; Mg: 0.78 mg/L; conductivity: 70 mmhos] and can consist of gravel, humus, logs, marl, muck, peat, peaty gravel, peaty humus, sand, sandy loam, sandy peat, and silt. Flowering and fruiting occur from May to October. The plants rely relatively highly on sexual (vs. asexual) reproduction and seed production for survival. The low pollen:ovule ratio (9:0) indicates that reproduction is primarily autogamous. Individual plants can produce more than 450 seeds on each cyme. Air trapped in the seed coat enables the seeds to float for several days, which facilitates their dispersal by water. They are also dispersed by the wind. A semipersistent seed bank (<5 years) develops. The seeds are physiologically dormant and require cold stratification (1 month at 4°C) to induce germination. Successful germination has occurred at temperatures of

20°C–29°C and also under a variable (10°C/25°C; 12/12 h) temperature regime. Rates of 100% germination have been obtained at 20°C under a 8/16 h dark/light regime after the seeds have been scarified and placed on 1% agar + 250 mg/L gibberellic acid (GA3) germination medium. Prey capture provides supplemental nitrogen for the plants when growing under low nutrient conditions; however, the plants switch toward reliance on root-derived nitrogen as sediment nitrogen availability increases. The leaf glands produce class I chitinase, which facilitates the digestion of the prey. The plants are known to recolonize abandoned subarctic roads, which previously had been cut through peatlands. Several endophytic Fungi have been found to colonize the roots of European plants. **Reported associates:** *Abies grandis, A. lasiocarpa, Agrostis hyemalis, A. idahoensis, A. scabra, Alnus rugosa, A. tenuifolia, A. viridis, Anaphalis, Andromeda glaucophylla, A. polifolia, Anticlea elegans, Aronia melanocarpa, Asclepias incarnata, Aulacomnium palustre, Barbarea orthoceras, Bartonia virginica, Betula glandulosa, B. nana, Bistorta bistortoides, Blechnum spicant, Calamagrostis canadensis, Calla palustris, Caltha, Camassia leichtlinii, Carex angustata, C. aquatilis, C. atlantica, C. aurea, C. canescens, C. capillaris, C. capitata, C. chordorrhiza, C. comosa, C. crinita, C. cusickii, C. diandra, C. echinata, C. fissuricola, C. folliculata, C. heteroneura, C. hoodii, C. integra, C. interior, C. lasiocarpa, C. lemmonii, C. lenticularis, C. limosa, C. lurida, C. lyngbyei, C. magellanica, C. muricata, C. obnupta, C. pauciflora, C. pluriflora, C. rostrata, C. rotundata, C. scoparia, C. scopulorum, C. trisperma, C. utriculata, C. vesicaria, Castilleja, Cephalanthus occidentalis, Chamaedaphne calyculata, Chamerion angustifolium, Chara, Cicuta bulbifera, C. douglasii, Comarum palustre, Darlingtonia californica, Dasiphora floribunda, Deschampsia cespitosa, Drosera anglica, D. filiformis, D. intermedia, Drymocallis glandulosa, Dulichium arundinaceum, Eleocharis decumbens, E. palustris, E. quinqueflora, Epilobium ciliatum, E. palustre, Epipactis gigantea, Equisetum sylvaticum, E. variegatum, Erigeron glacialis, Eriophorum angustifolium, E. chamissonis, E. criniger, E. crinigerum, E. gracile, E. vaginatum, E. virginicum, E. viridicarinatum, Eutrochium maculatum, Galium trifidum, Gaultheria hispidula, Gentiana linearis, Gentianopsis simplex, Glyceria borealis, Gymnadeniopsis clavellata, Hastingsia alba, Helenium bigelovii, Hippuris vulgaris, Hosackia oblongifolia, Hymenophyllum tunbrigense, Hypericum anagalloides, Iris setosa, Juncus acuminatus, J. canadensis, J. effusus, J. subcaudatus, Juniperus communis, Kalmia microphylla, K. polifolia, Larix laricina, Lilium kelleyanum, L. michauxii, Ludwigia palustris, Lycopodium clavatum, Lycopodiella inundata, Lycopus uniflorus, Lysichiton americanus, Mentha, Menyanthes trifoliata, Micranthes oregana, Mimulus primuloides, Myrica gale, Narthecium californicum, Neottia smallii, Nephrophyllidium crista-galli, Onoclea sensibilis, Oreostemma alpigenum, Osmunda regalis, Oxypolis occidentalis, Parnassia palustris, Pedicularis attollens, P. labradorica, Penstemon confertus, Philonotis fontana, Picea engelmannii, P. glauca, P. mariana, P. sitchensis, Pinguicula villosa, Pinus contorta, P. monticola,*

Platanthera dilatata, P. sparsiflora, Pleurozium schreberi, Poa pratensis, Podagrostis humilis, Pogonia ophioglossoides, Polygala vulgaris, Polytrichum juniperinum, Potentilla gracilis, Primula jeffreyi, P. tetrandra, Pteridium aquilinum, Pyrola asarifolia, P. rotundifolia, Rhododendron columbianum, R. groenlandicum, R. tomentosum, Rhynchospora alba, R. capitellata, Rubus chamaemorus, R. hispidus, Sagittaria latifolia, Salix arbusculoides, S. eastwoodiae, S. fuscescens, S. niphoclada, Sanguisorba minor, S. stipulata, Sarracenia purpurea, Scheuchzeria palustris, Schoenoplectus acutus, Scirpus atrocinctus, S. cyperinus, S. diffusus, Senecio triangularis, Sisyrinchium elmeri, Solidago uliginosa, Sparganium eurycarpum, Sphagnum angustifolium, S. capillifolium, S. centrale, S. fuscum, S. papillosum, S. recurvum, S. russowii, S. teres, Sphenosciadium capitellatum, Spiraea douglasii, Spiranthes romanzoffiana, Stachys, Thelypteris palustris, Toxicodendron vernix, Trianta glutinosa, T. occidentalis, Trichophorum cespitosum, Trientalis arctica, Trifolium longipes, Tsuga heterophylla, Typha latifolia, Urticularia intermedia, U. minor, U. subulata, Vaccinium macrocarpon, V. oxycoccos, V. uliginosum, V. vitis-idaea, Veronica scutellata, Viola adunca, V. palustris.

***Drosera tracyi** Macfarl.* is a perennial species that grows in bogs, borrow pits, canals, ditches, flatwoods, meadows, pinelands, roadsides, savannas, seeps, and swales at elevations of up to 70 m. The substrates are acidic (e.g., pH: 4.5) and have been characterized as Alapaha (arenic plinthic paleaquults), Florala (plinthaquic paleudults), Fuquay (plinthic paleudults), Leefield–Stilson complex, muck, mucky clay, peat, Plummer (grossarenic paleaquults), sand, and sandy peat. Flowering occurs from April to June. Typically, a single flower opens each day at 8:00 am and closes by noon. While open, the flowers consistently face eastward in their orientation. The flowers are self-compatible, have fairly low pollen:ovule ratios (23:9), and do not produce nectar; yet they are pollinated primarily by bees (Insecta: Hymenoptera: Apidae: *Bombus*; Halictidae: *Agapostemon radiatus*). The plants often occur in sites that have been disturbed by fire or where the surface has been scraped. **Reported associates:** *Aletris lutea, Andropogon arctatus, Aristida stricta, Balduina uniflora, Bigelowia nudata, Calopogon barbatus, C. multiflorus, Carphephorus pseudoliatris, Chaptalia tomentosa, Cleistesiopsis bifaria, C. divaricata, Cliftonia monophylla, Coreopsis gladiata, Ctenium aromaticum, Dichanthelium acuminatum, Drosera brevifolia, D. capillaris, D. filiformis, D. intermedia, Eleocharis tuberculosa, Eriocaulon compressum, E. decangulare, Euphorbia inundata, Gymnadeniopsis nivea, Harperocallis flava, Helenium vernale, Helianthus heterophyllus, H. radula, Hypericum brachyphyllum, H. crux-andreae, H. fasciculatum, Ilex coriacea, I. glabra, I. myrtifolia, Lachnanthes caroliniana, Liatris spicata, Lilium catesbaei, Lophiola aurea, Ludwigia linearis, L. linifolia, L. pilosa, Lycopodium, Muhlenbergia expansa, Myrica heterophylla, M. inodora, Nyssa biflora, Oclemena reticulata, Oxypolis filiformis, Pinguicula lutea, Pinus elliottii, Pleea tenuifolia, Pogonia ophioglossoides, Polygala cruciata, P. cymosa, P. lutea, P. ramosa, Rhexia alifanus, R. lutea, Rhynchospora* *chapmanii, R. ciliaris, R. latifolia, R. oligantha, R. plumosa, Sabatia decandra, S. macrophylla, Sarracenia alata, S. flava, S. leucophylla, S. psittacina, Scleria baldwinii, S. triglomerata, Smilax laurifolia, Sphagnum, Stenanthium densum, Taxodium ascendens, Triantha racemosa, Utricularia juncea, Xyris ambigua, X. baldwiniana.*

Use by wildlife: The flowers of *D. anglica* attract various insects (Insecta: Diptera: Bombyliidae; Calliphoridae; Dolichopodidae; Muscidae; Syrphidae; Tachinidae; Hymenoptera: Halictidae; Thysanoptera: Thripidae), which represent different guilds than those trapped by the plants. *Drosera brevifolia, D. capillaris, D. filiformis, D. intermedia,* and *D. tracyi* are host plants for larval plume moths (Insecta: Lepidoptera: Pterophoridae: *Trichoptilus parvulus*). *Drosera capillaris* is a host plant of aphids (Insecta: Hemiptera: Aphididae: *Hyalomyzus jussiaeae*). *Drosera rotundifolia* is eaten by moose (Mammalia: Cervidae: *Alces alces*). *Drosera intermedia* and *D. rotundifolia* are an important food source for bog-dwelling ants (Insecta: Hymenoptera: Formicidae), which scavenge the insects that become trapped in their leaves.

Economic importance: food: *Drosera* species are not considered to be edible; **medicinal:** *Drosera aliciae* contains the naphthoquinone ramentaceone, which exhibits cytotoxic activity toward the leukemic U937 cell line. Extracts of *D. rotundifolia* and *D. intermedia* contain the naphthoquinones 7-methyljuglone and plumbagin, which reputedly confer antimicrobial, antispasmodic, antipertussive, antiasthmatic, demulcent, expectorant, and hypoglycemic properties. *Drosera rotundifolia* contains antibiotic substances that are effective against pneumococcal, staphylococcal, and streptococcal bacteria. It has commonly been used as a treatment for asthma, whooping cough, and other respiratory ailments. The Kwakiutl used *D. rotundifolia* to treat bunions, corns, and warts and also as an aphrodisiac. The Seminoles applied the glandular leaves of *D. capillaris* as an antibiotic and to relieve various sores; **cultivation:** Many sundews are available commercially as cultivars and novelty horticultural specimens. These include *D. aliciae* (an Award of Garden Merit plant), *D. capensis* ('Albino'; an Award of Garden merit plant), *D. anglica, D. capillaris, D. filiformis* ('California Sunset'), *D. intermedia* ('Carolina Giant'), and *D. rotundifolia*; **misc. products:** *Drosera capensis* is the source of a naturally secreted polysaccharide adhesive, which has high elasticity. The plumbagin obtained from *D. intermedia* effectively inhibits food-spoilage Fungi. *Drosera rotundifolia* is the source of a yellow dye used for coloring wool. The plants are also used to curdle milk for cheesemaking. The juice of several species will stain paper a purple color; **weeds:** none; **nonindigenous species:** The South African *D. aliciae* and *D. capensis* persist within a small area of California where they were introduced intentionally by horticulturists sometime before 1997.

Systematics: Molecular data resolve *Drosera* as the sister genus of *Aldrovanda* and *Dionaea* (Figure 3.22). A phylogenetic analysis of *Drosera* using *rbcL* data indicated that the genus is monophyletic (save for one misplaced species); however, section *Drosera* is not (Figure 3.23). The indigenous North American species are closely related and occur within a single

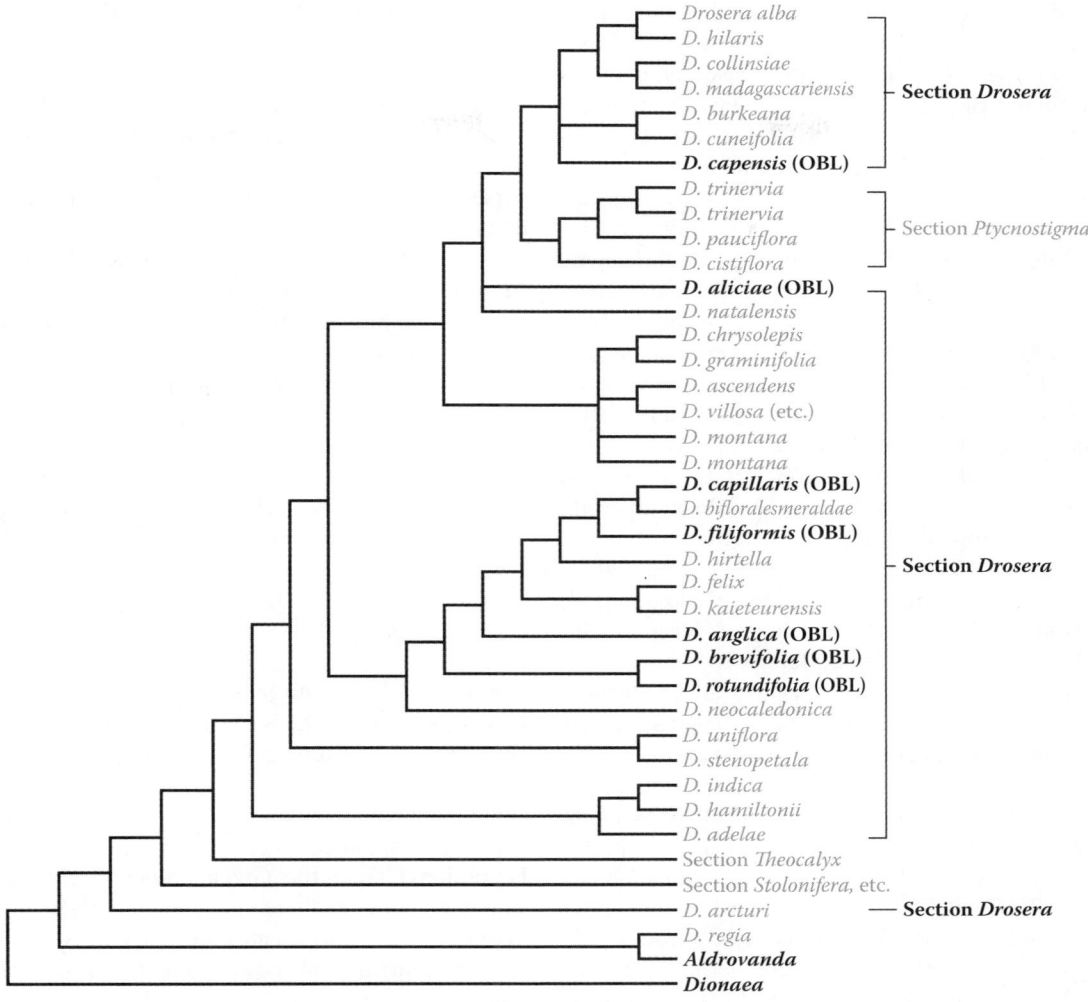

FIGURE 3.23 Phylogenetic relationships in *Drosera* as indicated by *rbcL* sequences. The indigenous North American *Drosera* species are fairly closely related and occur in one clade within section *Drosera*. Section *Drosera* is polyphyletic with respect to sections *Ptycnostigma* and *D. arcturi*. The clustering of *D. regia* with *Aldrovanda* is only weakly supported with *rbcL* data and is not maintained in combined data analyses (see Figure 3.22). Although all *Drosera* species should be regarded as aquatic, the OBL North American indicators are highlighted in bold. (Adapted from Rivadavia, F. et al., *Amer. J. Bot.*, 90, 123–130, 2003.)

clade (Figure 3.23). The phylogenetic relationships indicate that the OBL North American taxa occur broadly throughout the phylogeny and that they represent a diverse set of phyletic backgrounds within the North American group. A close relationship is supported between *D. rotundifolia* and *D. brevifolia* which resolve as a subclade (Figure 3.23). *Drosera capillaris* and *D. filiformis* occur within another subclade, which also contains the South American *D. esmeraldae*. *Drosera anglica* occurs in a position intermediate to these subclades. Together, the indigenous North American taxa associate in a clade with several other South American species (*D. felix*, *D. hirtella*, *D. kaieteurensis*). Sister to this North/South American group of species is *D. neocaledonica*, a New Caledonian species (Figure 3.23). *Drosera tracyi* sometimes is recognized as a variety of *D. filiformis* from which it is distinct morphologically. The taxa are sympatric (but do not co-occur at sites) and have been hybridized artificially; however, natural hybrids have not been found. *Drosera* is diverse chromosomally with base numbers

of $x = 6$, 7, 8, 10, 13, and 14 reported; however, the accepted base number is $x = 10$. *Drosera brevifolia*, *D. capillaris*, *D. filiformis*, *D. intermedia*, *D. linearis*, *D. rotundifolia*, and *D. tracyi* are diploid ($2n = 20$); *D. anglica* and *D. capensis* are tetraploid ($2n = 40$). *Drosera aliciae* has tetraploid and octoploid cytotypes reported ($2n = 40$, 80). Many *Drosera* hybrids have been reported. *D. rotundifolia* allegedly is involved in the parentage of several hybrids including *Drosera* ×*anglica* (*D. rotundifolia* × *D. linearis*), *Drosera* ×*belezeana* (*D. intermedia* × *D. rotundifolia*), *Drosera* ×*obovata* (a sterile but vegetatively vigorous F₁ hybrid between *Drosera anglica* × *D. rotundifolia*), and *Drosera* ×*woodii* (*D. linearis* × *D. rotundifolia*). Hybrids occur also between *D. rotundifolia* and *D. intermedia*. Other putative hybrids include *Drosera* ×*californica* (*D. filiformis* × *D. tracyi*), *Drosera* ×*hybrida* (*D. filiformis* × *D. intermedia*), and *Drosera* ×*linglica* (*D. anglica* × *D. linearis*).

Comments: *Drosera* species that occur in northern North America include *D. anglica* (also in Eurasia), *D. linearis*, and *D. rotundifolia* (also in Eurasia); *D. intermedia* occurs in eastern North America but extends also to Eurasia and South America. Several species occur along the coastal plain of the southeastern United States including *D. brevifolia* (extending to South America), *D. capillaris* (extending to South America), *D. filiformis*, and *D. tracyi*. The nonindigenous *D. aliciae* and *D. capensis* (both from South Africa) are currently known only from Mendocino County, California.

References: Baltensperger, 2004; Baskin et al., 2001; Brewer, 1999b; Budzianowski et al., 1993; Campbell & Bergeron, 2012; Carter et al., 1999; Cheers, 1992; Clark, 2003; Clark et al., 2008; Cohen et al., 2004; Freedman et al., 1992; Gibson, 1991a,b; Godefroid et al., 2010; Grevenstuk et al., 2012a,b; Grittinger, 1970; Harper, 1922; Hays, 2010; Hoyo & Tsuyuzaki, 2015; Kawiak et al., 2011; Keddy, 1989; Keddy & Reznicek, 1982; Kondo, 1976; Krafft & Handel, 1991; Landry & Cwynar, 2005; Lee, 1972; Lenaghan et al., 2011; LeResche & Davis, 1973; Linke, 1963; Matthews et al., 1990; Mellichamp, 2015; Millett et al., 2012; Mitchell, 2011; Murza & Davis, 2003; Murza et al., 2006; Pavlovič et al., 2014; Pietropaolo & Pietropaolo, 1993; Potts & Krupa, 2016; Quilliam & Jones, 2010; Rice, 2002; Renner & Specht, 2012; Ridder & D'hondt, 1992; Rivadavia et al., 2003; Robuck, 1985; Saunders, 1900; Schnell, 1982, 2002; Singhurst et al., 2011a,b, 2012; Sipple & Klockner, 1980; Stoetzel et al., 1999; Stromberg-Wilkins, 1984; Takahashi, 1988; Taylor & Raney, 2013; Thorén et al., 2003; Thum, 1986; Tsuyuzaki & Miyoshi, 2009; Wieder et al., 1984; Wilson, 1985, 1994, 1995.

Family 2: Plumbaginaceae [24]

Molecular data (e.g., *matK*, *rbcL* sequences; Lledó et al., 1998; Cuénoud et al., 2002) provide compelling evidence for the monophyly of Plumbaginaceae and its association as the sister clade to Polygonaceae (Figure 3.24). The two subfamilies (Plumbaginoideae and Staticoideae) also resolve as clades. The group is characterized mostly by perennial herbs and a few woody species (lianas, shrubs), which possess alternate leaves with epidermal glands that secrete mucilage or

calcareous salts. The flowers are bisexual, often heterostylous, and are pollinated by insects (Insecta: Diptera; Hymenoptera; Lepidoptera). The syncarpous, unilocular gynoecium contains a single pendulous ovule with basal placentation. The fruits are achenes, circumscissile or valvate capsules, nuts, or utricles containing a single-winged seed. The seeds are dispersed abiotically by gravity, water, or wind and occasionally by birds (Aves).

Horticulturally important genera include *Armeria*, *Ceratostigma*, *Goniolimon*, *Limonium*, *Plumbago*, and *Psylliostachys*, which provide showy ornamentals. No major medicinal plants occur here despite the derivation of the family name from the Latin *plumbum* (lead) from the erroneous belief that the plants provided a cure for lead poisoning (Morin, 2005).

Plumbaginaceae are cosmopolitan in distribution and contain many xerophytic and halophytic species. Only one genus in the family includes a single species regarded as an OBL indicator in North America:

1. ***Limonium*** Mill.

Among the genera surveyed phylogenetically (Figure 3.24), *Limonium* (worldwide) is closely related to *Goniolimon* (Asia) and *Afrolimon* (South Africa).

1. *Limonium*

Sea-lavender, statice

Etymology: from the Greek *leimon* meaning "marsh" or "meadow"

Synonyms: *Statice* (rejected name)

Distribution: global: cosmopolitan; **North America:** eastern, southern, western

Diversity: global: 300 species; **North America:** 8 species

Indicators (USA): OBL: *Limonium californicum*, *L. carolinianum*; **FACW:** *L. californicum*

Habitat: brackish, saline; palustrine; **pH:** 5–9; **depth:** <1 m; **life-form(s):** emergent (herb)

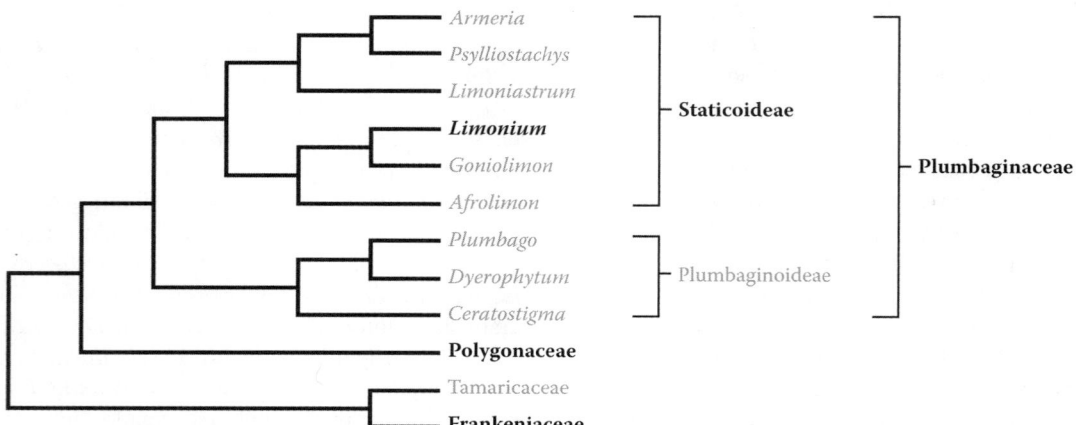

FIGURE 3.24 Relationships among nine Plumbaginaceae genera as indicated by analysis of *matK* sequence data. The family, as well as the subfamilies Plumbaginoideae and Staticoideae, resolve as clades in this analysis. The North American taxa that contain obligate indicators are shown in bold. (Adapted from Cuénoud, P. et al., *Amer. J. Bot.*, 89, 132–144, 2002.)

Key morphology: stems a short caudex; leaves in a basal rosette, petiolate (to 16.5 cm), blades (to 17 cm) fleshy, spatulate to obovoid, obtuse and slightly mucronate; scapes solid, leafless, bracteate, terminating in a panicle (to 6 dm) comprising numerous one-sided spikes; flowers five-parted, in one to three flowered "spikelets"; petals (to 9 mm) long-clawed, violet to lavender, fused basally to the stamens; fruit (to 5.5 mm) a utricle

Life history: duration: perennial (taproot, woody caudex); **asexual reproduction:** none; **pollination:** insect; **sexual condition:** hermaphroditic; **fruit:** utricles (common); **local dispersal:** seeds (water, wind); **long-distance dispersal:** seeds (birds)

Imperilment: (1) *Limonium californicum* [G4]; S1 (OR); S2 (NV); (2) *L. carolinianum* [G5]; S2 (NF); S3 (QC)

Ecology: general: *Limonium* species occupy a variety of brackish to saline habitats including cliffs, dunes, and waste areas, but most often occur in coastal wetland environments. Six North American species (75%) are wetland indicators, with two (25%) designated as OBL. The flowers are pollinated by insects (Insecta). Some of the species are self-incompatibile, but those in North America are self-compatible. Agamospermy has also been reported in some European species but in none from North America. *Limonium* seeds can withstand long periods in seawater and are dispersed by water, especially along coastal drift lines. It is not unusual for entire inflorescences to be dispersed along the beach strand. All of the species use a C_3 photosynthetic pathway for carbon fixation. *Limonium* species are somewhat threatened by collecting for flower arrangements, an activity, which would best be restricted to garden-grown, cultivated material.

Limonium californicum (**Boiss.**) **A. Heller** inhabits beaches, bluffs, coastal strands, depressions, ditches, dunes, mima mounds, mudflats, roadsides, salt flats, salt marshes, salt springs, sand spits, tidal lagoons, washes, and the margins of bays at elevations of up to 396 m. The plants occur in exposures of full sunlight. The substrates are alkaline (pH: 7.1–7.2) or saline (to at least 40 dS/m) and are low in organic matter (8.9%–10.4%). They consist of clay, clayey sand, humus, mud, sandy loam, and silty loam. The plants survive on substrates of high salinity by excreting salt crystals from their leaf surfaces. Flowering extends from April to December. The plants are effective at colonizing gaps caused by declining stands of *Sarcocornia perennis*, which have been parasitized by dodder (Convolvulaceae: *Cuscuta salina*). They are host plants for the rare hemiparasite *Chloropyron maritimus* (Orobanchaceae). **Reported associates:** *Ambrosia psilostachya, Amblyopappus pusillus, Anemopsis californica, Armeria maritima, Arthrocnemum subterminale, Atriplex patula, A. prostrata, A. semibaccata, A. watsonii, Baccharis, Batis maritima, Bromus diandrus, B. hordeaceus, Camissoniopsis cheiranthifolia, Chloropyron maritimum, Cotula coronopifolia, Cuscuta salina, Deschampsia, Distichilis littoralis, D. spicata, Euphorbia, Extriplex californica, Frankenia palmeri, F. salina, Grindelia humilis, G. stricta, Isocoma menziesii, Jaumea carnosa, Juncus acutus, Lilaeopsis occidentalis, Lolium multiflorum, Malva parviflora, Melilotus indicus, Parapholis incurva, Polypogon monspeliensis, Raphanus sativus, Rumex crispus, Salicornia bigelovii, S. depressa, Sarcocornia pacifica, S. perennis, Sesuvium verrucosum, Solanum xanti, Sonchus oleraceus, Spartina foliosa, Spergularia marina, S. media, Suaeda taxifolia, Symphyotrichum subulatum, Tamarix, Triglochin maritimum, Typha.*

Limonium carolinianum (**Walter**) **Britton** inhabits brackish to saline (to 47.1 dS/m) sites including beaches, coastal strands, ditches, dunes, estuaries, hammocks (mangrove), meadows, mudflats (intertidal), roadsides, salt flats, salt (tidal) marshes, salt meadows, salt pans, shores, sloughs, spits (dredge), swales, swamps (mangrove), thickets, and the margins of canals, lagoons, and lakes (dessicating) at elevations of up to 3 m. The substrates include clay, coral limestone, mud, muddy sand, muddy shells, sand, sandy gravel, sandy marl, sandy shells, and shell middens. The plants are tolerant of full sunlight exposures, fire, and periodic tidal inundation. Flowering extends from June to December. DNA fingerprinting studies indicate a mixed mating system. The seeds exhibit high plasticity for germination across different salinity gradients. They have physiological dormancy, which is broken after 30 days of cold stratification (3.5°C) and incubation under a 4-h photoperiod and a 20°C/15°C or 27°C/16°C day/night temperature regime. The seeds are short-lived (about 7 months) and do not survive beyond their first year of production. They are dispersed over a relatively small distance (many remaining attached to the inflorescence through winter). The seedlings reach their peak density at a distance of 20–30 cm from their parent plant and seldom exceed a distance of 80 cm. Genetic studies have also found population subdivision to occur at a spatial scale below 100 m. Some seeds have been found attached to the feet and plumage of waterfowl (Aves: Anatidae), which potentially serve as long-distance dispersal vectors. Populations can withstand substantial flower overharvesting (up to 90%); however, full recovery from such activites can take over 30 years once harvesting ceases. **Reported associates:** *Agalinis maritima, Agave, Amyris, Andropogon, Aristida, Atriplex cristata, A. patula, A. prostrata, Avicennia germinans, Baccharis halimifolia, Bacopa monnieri, Batis maritima, Blutaparon vermiculare, Bolboschoenus robustus, Borrichia frutescens, Bourreria succulenta, Bryum capillare, Bursera, Cakile geniculata, C. lanceolata, Calothrix crustacea, Calystegia sepium, Casuarina equisetifolia, Celtis, Chamaecrista fasciculata, Chenopodium berlandieri, Conocarpus erectus, Croton punctatus, Cyperus filicinus, C. lupulinus, Distichlis littoralis, D. spicata, Dysphania ambrosioides, Eleocharis, Eragrostis secundiflora, Eugenia, Eustoma exaltatum, Fimbristylis caroliniana, F. spadicea, Gouania, Guapira discolor, Heliotropium curassavicum, Heterotheca, Ipomoea pes-caprae, Iva frutescens, I. imbricata, Juncus gerardii, J. roemerianus, Juniperus virginiana, Kochia scoparia, Laguncularia racemosa, Limosella subulata, Lycium carolinianum, Lysimachia maritima, Metopium toxiferum, Myrica cerifera, M. pensylvanica, Oenothera drummondii, Panicum virgatum, Persea, Phragmites communis, Piscidia, Plantago maritima, Pluchea purpurascens,*

Potentilla anserina, Puccinellia maritima, Quercus virginiana, Rayjacksonia phyllocephala, Reynosia, Rhizoclonium riparium, Rhizophora mangle, Sabal palmetto, Sabatia dodecandra, S. stellaris, Salicornia bigelovii, S. depressa, Sarcocornia perennis, Schoenoplectus americanus, Serenoa repens, Sesuvium portulacastrum, Seutera angustifolia, Sideroxylon tenax, Solidago sempervirens, Sophora, Spartina alterniflora, S. patens, Spergularia canadensis, Sporobolus virginiana, Suaeda calceoliformis, S. linearis, Suriana maritima, Swietenia, Symphyotrichum subulatum, S. tenuifolium, Thrinax, Triglochin striata, Vigna, Waltheria indica, Yucca.

Use by wildlife: Some *Limonium* species (North American taxa not surveyed) contain phytoecdysteroids, which are known to deter phytophagous invertebrate predators. The epidermal cells of *L. carolinianum* contain lignin which may provide some resistance to herbivory. However, *L. carolinianum* is grazed by livestock (Mammalia: Bovidae) and is fed upon by aphids (Insecta: Hemiptera: Aphididae: *Staticobium staticis*) and moth larvae (Insecta: Lepidoptera: Tortricidae: *Gynnidomorpha romonana*), which in turn host wasp parasitoids (Insecta: Hymenoptera: Braconidae: Microgastrinae). It is regarded as a nectar plant for butterflies (Insecta: Lepidoptera) and is the major nectar source for the Maritime ringlet butterfly (Nymphalidae: *Coenonympha tullia nipisiquit*). The flowers also attract flower flies (Insecta: Diptera: Syrphidae: *Copestylum vittatum, Eristalinus aeneus, Lejops curvipes*).

Economic importance: food: *Limonium* species are generally not edible; **medicinal:** The Costanoan and Ohlone tribes used a tea made from *L. californicum*, which was reputed to purify and thin the blood. The Costanoan people inhaled the powdered plant to clear out congestion (by inducing sneezing) and used a decoction to treat internal injuries and as a remedy for urinary and venereal diseases. The Micmac people added ground roots of *L. carolinianum* to boiling water as a treatment for consumption and hemorrhage. Extracts of *L. californicum* are known to inhibit the production of verotoxin in hemorrhage-inducing strains of *Escherichia coli* (Bacteria: Enterobacteriaceae). The roots of *L. carolinianum* contain tannin and are highly astringent. Root decoctions are used as a remedy for diarrhea, dysentery, and mouth sores. Dried root powder is applied to ulcers and piles. *Limonium carolinianum* contains the naphthoquinone plumbagin, which is attributed with a number of medicinal properties; **cultivation:** *Limonium californicum* and *L. carolinianum* are cultivated as salt-tolerant ornamentals; **misc. products:** Inflorescences of *L. carolinianum* are used widely in dried flower arrangements and are harvested commercially in Nova Scotia for this purpose; **weeds:** none; **nonindigenous species:** none.

Systematics: A phylogenetic survey of the large genus *Limonium* has not yet been undertaken but will be necessary before interspecific relationships can be ascertained confidently. Neither of the OBL species has yet been included in a phylogenetic analysis, although a few sequences in GenBank indicate that such work may be in progress. Several species of *Limonium* (but none from North America) have recently been segregated as the genus *Myriolepis*. The characters used to differentiate *L. nashii* are unreliable, and it is usually placed in synonymy with *L. carolinianum*. White-flowered forms of the species have been recognized taxonomically as forma *albidiflorum*. The base chromosome number for Plumbaginaceae ranges from $x = 6–9$ and is $x = 8–9$ in *Limonium*. *Limonium californicum* ($2n = 18$) is presumably diploid and *L. carolinianum* ($2n = 36$) tetraploid.

Comments: *Limonium californicum* occurs in the western United States (California, Nevada, and Oregon) and extends southward into Mexico (Baja California). *L. carolinianum* occurs along the Atlantic and Gulf seacoasts of eastern North America and also extends southward into Mexico.

References: Aronson, 1989; Baker, 1953; Baltzer et al., 2002a,b; Boorman, 1968; Eiseman & Jensen, 2015; Garbary et al., 2008; Grieve, 1980; Hamilton, 1996, 1997; Lehman et al., 2009; Lledó et al., 2003; Luteyn, 1976; Maier, 2011; Rand, 2000; Sakagami et al., 2001; St. Omer, 2004; Sei & Porter, 2003; Whiting et al., 1998.

Family 3: Polygonaceae [43]

The Polygonaceae are a large family of 1100 annual or perennial herbaceous and woody species that occupy many diverse habitats (Freeman & Reveal, 2005; Judd et al., 2016). The family is characterized by the usual presence of tannins and oxalic acid, swollen nodes, and a modified stipular sheath (ocrea), which is diagnostic. The flowers are small, greenish, white, or pinkish red and contain a nectary disk. They are pollinated by insects (Insecta), notably by bees (Hymenoptera) and flies (Diptera). The fruits are lenticular or trigonously angled achenes or nutlets (sometimes winged), which are dispersed abiotically by water or wind (sometimes facilitated by a persistent calyx). The fruits are also consumed in quantity by various birds (Aves: Anatidae) (Martin & Uhler, 1951), which can disperse them endozoically over long distances.

Molecular phylogenetic analyses (e.g., Figures 3.21, 3.24, and 3.25) consistently implicate Plumbaginaceae as the sister group of Polygonaceae and also confirm the monophyly of both families (Chase et al., 1993; Cuénoud et al., 2002; Lamb-Frye & Kron, 2003). However, the systematic relationships within Polygonaceae have experienced a large upheaval over the past decade, as molecular data (e.g., Lamb-Frye & Kron, 2003; Kim & Donoghue, 2008a; Burke et al., 2010; Schuster et al., 2011) continued to reveal problems with the existing circumscriptions of many genera. The large genus *Polygonum* has been particularly problematic taxonomically, with many taxa (e.g., section *Persicaria*) now extracted from that genus and transferred elsewhere. Given their phyletic distance in molecular cladograms (e.g., Figure 3.25), it is hard to envision that *Persicaria* ever was recognized as a section of *Polygonum*. Some phylogenetic results showed other *Polygonum* species to be intermixed with several genera (*Atraphaxis, Fallopia, Muehlenbeckia, Polygonella*), which has led to further reevaluations of generic limits in the family. *Rumex* appears to be monophyletic from the species surveyed.

The family is of minor economic importance as the source of buckwheat (*Fagopyron*), rhubarb (*Rheum*),

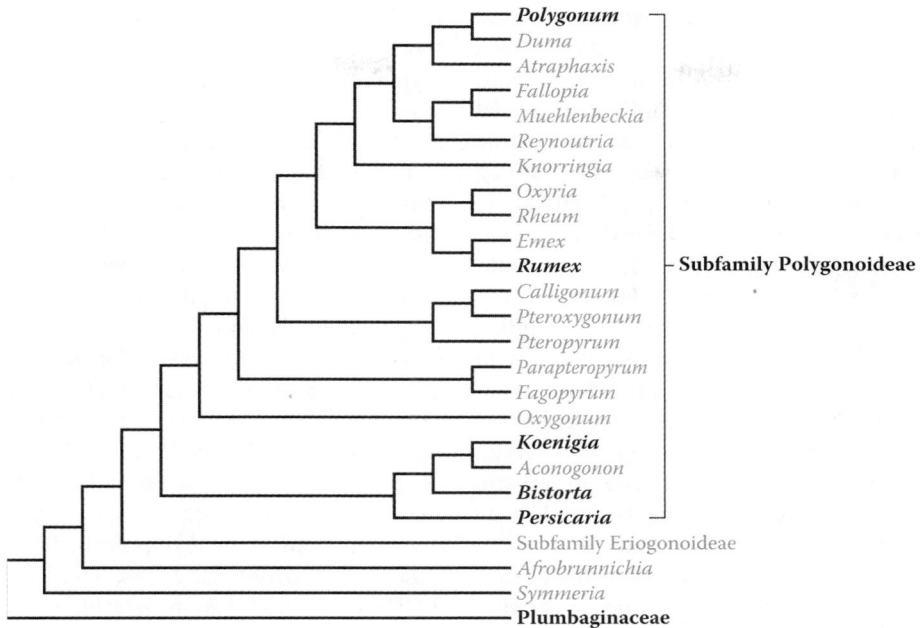

FIGURE 3.25 Intergeneric relationships in Polygonaceae as indicated by analysis of combined DNA sequence data. This and other studies have confirmed the monophyly of Polygonaceae, the monophyly of subfamilies Eriogonoideae and Polygonoideae (in exclusion of *Afrobrunnichia* and *Symmeria*), and the sister group relationship of the family to Plumbaginaceae. Taxa containing OBL North American indicators (highlighted in bold) are dispersed across Polygonoideae, indicating several independent origins of the habit in the subfamily. (Adapted from Schuster, T.M. et al., *Taxon*, 60, 1653–1666, 2011.)

and sorrel (*Rumex*). Ornamental cultivars have been derived from *Antigonon*, *Eriogonum*, *Fagopyron*, *Fallopia*, *Homalocladium*, *Muehlenbeckia*, *Persicaria*, *Polygonum*, *Rheum*, and *Rumex*.

Polygonaceae include a number of species adapted to wet or aquatic habitats of both fresh and saline conditions. Five genera contain OBL indicators in North America:

1. ***Bistorta*** (L.) Scopoli
2. ***Koenigia*** L.
3. ***Persicaria*** (L.) Mill.
4. ***Polygonum*** L.
5. ***Rumex*** L.

1. *Bistorta*

Bistort, smokeweed; bistorte

Etymology: from the Latin *bis tortus* (twice twisted), in reference to the rhizomes

Synonyms: *Persicaria* (in part); *Polygonum* (in part)

Distribution: global: Asia; Europe; North America; **North America:** northern

Diversity: global: 50 species; **North America:** 4 species

Indicators (USA): OBL; FACW: *Bistorta bistortoides*

Habitat: freshwater; palustrine; **pH:** alkaline; **depth:** <1 m; **life-form(s):** emergent (herb)

Key morphology: stems (to 75 cm) 1–3, rhizomes twisted; leaves basal (mostly) or cauline, petioled (to 5 cm), the blades (to 4.8 cm) elliptic to oblong–lanceolate/oblanceolate, margins entire; ocrea (to 32 mm) brown; inflorescence (to

50 mm) 1–2, cylindric or ovoid, peduncled (to 10 cm); flowers minute, pedicelled (to 11 mm), tepals (to 5 mm) 5, white or pale pink; stamens exserted; achenes (to 4.2 mm) brownish, shiny, smooth

Life history: duration: perennial (rhizomes); **asexual reproduction:** rhizomes; **pollination:** insect; **sexual condition:** hermaphroditic; **fruit:** achenes (common); **local dispersal:** seeds; **long-distance dispersal:** seeds

Imperilment: (1) *Bistorta bistortoides* [G5]; S3 (AB, BC)

Ecology: general: *Bistorta* is a genus of perennial herbs that inhabit heaths, meadows, shores, and woodlands. All four of the North American species (100%) are wetland indicators, with one designated as OBL. The flowers are small and in some species are self-compatible and capable of self-pollination; however, they are also protandrous and are normally outcrossed by generalist pollinators such as flies (Insecta: Diptera). Other species are self-incompatible and are pollinated by larger insects such as bees (Hymenoptera). In some species, the flowers are replaced by vegetative bulbils. Overall, there is a high rate of asexual reproduction (by bulbils or rhizomes); however, the seeds germinate well once they have been cold stratified. The seeds lack specialized adaptations for dispersal, which occurs not only probably by wind and water but also possibly by endozoic transport.

Bistorta bistortoides **(Pursh) Small** grows in alpine and subalpine bogs, cirques, crevices, depressions, ditches, draws, fell fields, fens, flats, heath, hummocks, marshes, meadows, mires, prairies, roadsides, seeps, slopes (to 90%; mainly south facing), springs, swales, swamps, tundra, woodlands, and along the margins of lakes, ponds, and streams at elevations of up to 3979 m (occasionally in coastal freshwater marshes below 20 m). Exposures range from full sunlight to partial shade. The substrates are usually characterized as alkaline and include clay, clay loam (granitic), cobbles (granitic), decomposed limestone, gneiss, granite, gravel, gravelly loam, gravelly sandy loam, gravelly silt, Grinell argyllite, gumbo soil, humus, loam (organic), krummholz, mud, peat, peridotite, peridotite–dunite, quartzite, rock, rocky clay, sand, sandstone, sandy clay, sandy loam (granitic), shale, silt (granitic), silty loam, and talus (basalt). Flowering occurs from May to September, with fruiting extending into October. Floral phenology is highly variable and does not appear to respond to a specific photoperiod, but sexual reproduction is observed most commonly between June and August. The flowers are self-incompatible and are pollinated by insects (see *Use by wildlife*). Nitrogen stored in the shoots and rhizomes accounts for 60% of the total N allocated during the growing season and facilitates accelerated seed production and maturation during the brief Arctic summers. The seeds germinate at 20°C following cold stratification or afterripening at 22°C. Numerous seedlings have been observed in meadow habitats. Vegetative reproduction occurs by rhizomes. **Reported associates:** *Abies amabilis, A. grandis, A. lasiocarpa, Achillea millefolium, Achnatherum nelsonii, Aconitum columbianum, Aconogonon phytolaccifolium, Agoseris glauca, Agrostis exarata, A. idahoensis, A. pallens, A. scabra, Allium fibrillum, A. schoenoprasum, A. validum, Alnus tenuifolia, Amelanchier, Androsace chamaejasme, A. septentrionalis, Anemone occidentalis, A. patens, Angelica arguta, Antennaria alpina, A. anaphaloides, A. corymbosa, A. lanata, A. media, A. microphylla, A. parvifolia, A. umbrinella, Anthoxanthum, Anticlea elegans, Arnica fulgens, A. gracilis, A. latifolia, A. longifolia, A. mollis, A. rydbergii, Artemisia cana, A. scopulorum, A. tridentata, Astragalus alpinus, Balsamorhiza sagittata, Barbarea orthoceras, Betula nana, B. occidentalis, Bistorta vivipara, Boechera microphylla, Bromus porteri, Calamagrostis canadensis, C. purpurascens, Calocedrus decurrens, Calochortus apiculatus, Caltha leptosepala, Camassia quamash, Campanula californica, C. parryi, C. rotundifolia, Carex aquatilis, C. atriformis, C. capitata, C. douglasii, C. ebenea, C. elynoides, C. hoodii, C. integra, C. luzulina, C. mariposana, C. microptera, C. nebrascensis, C. nigricans, C. nova, C. obnupta, C. pachystachya, C. paysonis, C. petasata, C. phaeocephala, C. podocarpa, C. raynoldsii, C. rostrata, C. rupestris, C. scopulorum, C. spectabilis, C. subnigricans, C. utriculata, C. vernacula, C. vesicaria, Cassiope mertensiana, Castilleja angustifolia, C. cusickii, C. miniata, C. parviflora, Catabrosa aquatica, Ceanothus, Cerastium arvense, C. beeringianum, Chamerion angustifolium, Chenopodium album, Cirsium drummondii, Cistanthe umbellata, Claytonia lanceolata, Collinsia parviflora, Crataegus, Crepis nana, C. runcinata, Danthonia intermedia, Darlingtonia californica, Dasiphora floribunda, Delphinium depauperatum, Deschampsia cespitosa, Draba, Drosera anglica, D. rotundifolia, Dryas octopetala,*

Eleocharis bernardina, E. palustris, Elymus elymoides, E. trachycaulus, Epilobium anagallidifolium, E. ciliatum, E. glaberrimum, E. oreganum, Equisetum, Eremogone aculeata, E. congesta, E. flavum, E. ovalifolium, Erigeron compositus, E. coulteri, E. glacialis, E. gracilis, E. formosissimus, E. nivalis, E. peregrinus, E. glacialis, E. simplex, E. ursinus, Eriogonum wrightii, Eritrichium nanum, Erythronium grandiflorum, Eucephalus ledophyllus, Festuca altaica, F. arizonica, F. calligera, F. idahoensis, F. ovina, F. thurberi, F. viridula, Fragaria vesca, Fritillaria pudica, Galium parishii, G. trifidum, Gentiana algida, G. calycosa, G. fremontii, Gentianopsis detonsa, G. holopetala, Geranium richardsonii, Geum rossii, G. triflorum, Hackelia, Hastingsia serpentinicola, Helenium bigelovii, Heracleum sphondylium, Hieracium caespitosum, H. triste, Holcus lanatus, Hymenoxys hoopesii, Hypericum anagalloides, H. formosum, Ipomopsis, Iris missouriensis, Ivesia gordonii, Juncus acuminatus, J. arcticus, J. balticus, J. drummondii, J. effusus, J. parryi, Juniperus horizontalis, J. occidentalis, Kalmia microphylla, K. polifolia, Kobresia myosuroides, Lathyrus nevadensis, Leucothoe davisiae, Lewisia pygmaea, L. triphylla, Ligularia dentata, Ligusticum canbyi, L. grayi, L. tenuifolium, Lilium parryi, L. philadelphicum, Lithophragma, Lithospermum, Lloydia serotina, Lomatium cous, L. dissectum, L. grayi, Luetkea pectinata, Lupinus arcticus, L. argenteus, L. depressus, L. caudatus, L. latifolius, L. polyphyllus, L. sericeus, L. wyethii, Luzula parviflora, L. spicata, Maianthemum stellatum, Melica, Mertensia campanulata, M. macdougalii, M. paniculata, Micranthes bryophora, M. occidentalis, M. odontoloma, M. oregana, M. rhomboidea, M. subapetala, Microsteris gracilis, Mimulus guttatus, M. lewisii, M. moschatus, M. primuloides, Minuartia obtusiloba, Mitella breweri, M. pentandra, Myosotis asiatica, Narthecium californicum, Noccaea fendleri, Oenanthe sarmentosa, Olsynium douglasii, Oreostemma alpigenum, Oreoxis humilis, Orthocarpus luteus, Osmorhiza berteroi, Oxalis decaphylla, Oxyria digyna, Parnassia californica, P. fimbriata, Paxistima myrsinites, Pedicularis attollens, P. bracteosa, P. contorta, P. groenlandica, P. racemosa, Penstemon attenuatus, P. confertus, P. globosus, P. hallii, P. procerus, P. wilcoxii, Perideridia parishii, Petasites, Phacelia lyallii, Phleum alpinum, P. pratense, Phlox austromontana, P. diffusa, P. kelseyi, P. pulvinata, Phyllodoce empetriformis, P. glanduliflora, Picea engelmannii, P. glauca, Pinus albicaulis, P. contorta, P. flexilis, P. monticola, P. ponderosa, Plantago major, Platanthera, Pleuropogon, Poa alpina, P. arida, P. cusickii, P. nemoralis, P. pratensis, P. secunda, Podagrostis humilis, Polemonium californicum, P. occidentale, P. viscosum, Polystichum munitum, Populus tremuloides, Potentilla anserina, P. drummondii, P. flabellifolia, P. glaucophylla, P. gracilis, P. hippiana, P. pulcherrima, Potentilla ×diversifolia, Primula fragrans, P. jeffreyi, P. parryi, P. pauciflora, P. tetrandra, Prunus, Pseudocymopterus montanus, Pseudoroegneria spicata, Pteridium aquilinum, Quercus garryana, Ranunculus acris, R. adoneus, R. cardiophyllus, R. eschscholtzii, R. gormanii, R. occidentalis, Rhodiola integrifolia, R. rosea, Rhododendron albiflorum, R. columbianum,

Ribes lacustre, Rosa, Rubus parviflorus, R. pedatus, R. ursinus, Rumex acetosella, R. occidentalis, R. paucifolius, Salix arctica, S. brachycarpa, S. eastwoodiae, S. geyeriana, S. glauca, S. nivalis, S. planifolia, S. wolfii, Sanguisorba officinalis, Saxifraga bronchialis, S. mertensia, S. subapetala, S. hydrophiloides, S. fremontii, S. integerrimus, S. scorzonella, S. triangularis, Shepherdia canadensis, Sibbaldia procumbens, Sidalcea malviflora, S. oregana, Sieversia pentapetala, Silene acaulis, S. scouleri, S. verecunda, Sisyrinchium bellum, S. idahoense, Smelowskia americana, Solidago multiradiata, Sorbus, Spergularia rubra, Sphagnum, Sphenosciadium capitellatum, Spiraea betulifolia, Spiranthes romanzoffiana, Stachys albens, Stellaria borealis, S. longipes, Stipa, Symphoricarpos oreophilus, Symphyotrichum foliaceum, Taraxacum officinale, Taraxia subacaulis, Thalictrum, Thermopsis, Torreyochloa erecta, Toxicoscordion venenosum, Trianthia glutinosa, Trichophorum, Trifolium eriocephalum, T. haydenii, T. longipes, T. parryi, T. repens, Trisetum spicatum, Trollius laxus, Vaccinium deliciosum, V. membranaceum, V. scoparium, V. uliginosum, V. vitis-idaea, Vahlodea atropurpurea, Valeriana edulis, V. occidentalis, V. sitchensis, Veratrum californicum, V. viride, Veronica americana, V. cusickii, V. scutellata, V. wormskjoldii, Vicia americana, Viola adunca, Wyethia helianthoides, Xerophyllum tenax.

Use by wildlife: The flowers and foliage of *P. bistortoides* are eaten by deer (Mammalia: Cervidae: *Odocoileus*), elk (Mammalia: Cervidae: *Cervus*), and sheep (Mammalia: Bovidae: *Ovis aries*). The roots are eaten by bears (Mammalia: Ursidae: *Ursus*) and rodents (Mammalia: Rodentia). The plants are a principal food of Olympic marmots (Mammalia: Sciuridae: *Marmota olympus*) and are a host for larval butterflies (Insecta: Lepidoptera: Nymphalidae: *Boloria napaea, B. titania*) and long-horned beetles (Insecta: Coleoptera: Cerambycidae: *Cosmosalia chrysocoma*). The flowers are visited (and likely pollinated) by bumblebees (Insecta: Hymenoptera: Apidae: *Bombus*) and butterflies (Insecta: Lepidoptera: Hesperiidae: *Pyrgus centaureae*; Lycaenidae: *Plebejus glandon, P. saepiolus*; Nymphalidae: *Boloria alaskensis halli, B. eunomia, B. frigga, B. improba harryi, B. titania, Chlosyne whitneyi damoetas, Erebia epipsodea*; Papilionidae: *Parnassius smintheus*; Pieridae: *Colias scudderii*).

Economic importance: food: The leaves of *B. bistortoides* can be eaten raw in salads or boiled or steamed and eaten like spinach or used in soups and stews. The roots are starchy and can be sliced and sautéed with onions or peeled and added to soup. They were used for soup and stew by the Blackfoot and boiled for flavor with meat by the Cherokee; **medicinal:** The Miwok prepared a poultice of the root to treat boils and sores; **cultivation:** *Bistorta bistortoides* is not cultivated; **misc. products:** none; **weeds:** none; **nonindigenous species:** none.

Systematics: Phylogenetic analysis of combined DNA sequence data for eight *Bistorta* species (but excluding *B. bistortoides*) resolves the genus as a clade, which is sister to a clade comprising *Aconogonon* and *Koenigia* (Figures 3.25 and 3.26). A more comprehensive sampling of *Bistorta* taxa in phylogenetic analyses will be necessary before the precise relationship of *B. bistortoides* can be determined. The basic chromosome number of *Bistorta* is $x = 11$ or 12. *Bistorta bistortoides* ($2n = 24$) is a diploid. No hybrids involving this species have been reported.

Comments: *Bistorta bistortoides* occurs throughout western North America, most often at high elevations (above 1500 m) in the Rocky Mountains.

References: Benoliel, 2011; Bills et al., 2015; Diggle et al., 2002; Elias & Dykeman, 1990; Fan et al., 2013; Forbis, 2009; Freeman & Hinds, 2005; Holway & Ward, 1965; Jaeger & Monson, 1992; O'Neill et al., 2008; Schuster et al., 2011; Scott, 2014.

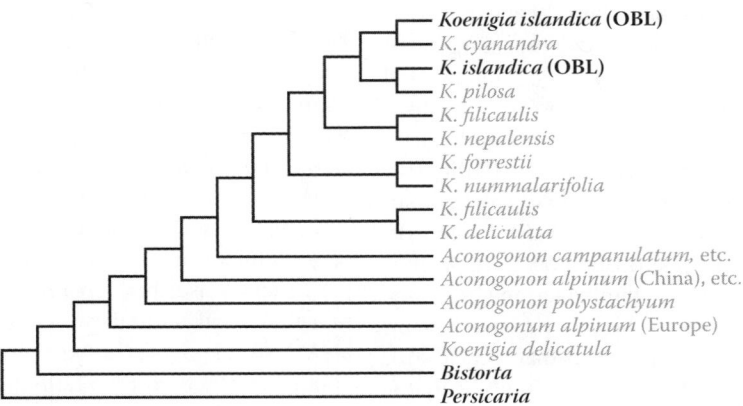

FIGURE 3.26 Phylogenetic relationships of *Koenigia* in Polygonaceae subfamily Polygonoideae derived by analysis of combined DNA sequence data. The anomalous placement of *K. delicatula* and resolution of *Aconogonon* as a paraphyletic grade have inspired some authors (e.g., Schuster et al., 2015) to merge *Aconogonon* and *Koenigia*. The placement of individuals sampled from different populations of *K. islandica* differs, perhaps because of the polyploid nature of that taxon (a similar discrepancy also appears for *K. alpinum*). Taxa containing OBL North American indicators are highlighted in bold. A different analysis (Fan et al., 2013) shows *K. islandica* to be more closely related to *K. fertilis*. (Adapted from Schuster, T.M. et al., *Taxon*, 64, 1188–1208, 2015.)

2. *Koenigia*

Purslane

Etymology: after Johan Gerhard Koenig (1728–1785)

Synonyms: *Macounastrum*; *Polygonum* (in part)

Distribution: global: Asia; Europe; New World; **North America:** northern

Diversity: global: 8 species; **North America:** 1 species

Indicators (USA): OBL: *Koenigia islandica*

Habitat: freshwater; palustrine; **pH:** acidic; **depth:** <1 m; **life-form(s):** emergent (herb)

Key morphology: stems (to 5 cm) tufted, ascending to prostrate, succulent, reddish; leaves (to 5 mm) alternate/suboppo-site, petiolate (to 3 mm), narrowly elliptic, succulent, reddish, minutely punctate; ocrea (to 2 mm) membranous; flowers minute, in small terminal clusters or upper leaf axils; sepals (to 1 mm) 3, greenish; stamens 3; petals absent; achenes (to 1.8 mm) two-lobed

Life history: duration: annual (fruit/seeds); **asexual reproduction:** none; **pollination:** self (cleistogamous); **sexual condition:** hermaphroditic; **fruit:** achenes (common); **local dispersal:** seeds; **long-distance dispersal:** seeds

Imperilment: (1) *Koenigia islandica* [G4]; SH (<u>ON</u>); S1 (<u>AB</u>, <u>MB</u>, UT, WY); S2 (CO, MT, <u>YT</u>) S3 (<u>BC</u>, <u>LB</u>, <u>QC</u>)

Ecology: general: *Koenigia* contains annual and perennial species that typically occur in moist environments of Arctic climates or at high montane elevations (to 5100 m). North America has a single species, which is designated as an OBL indicator. The life histories of the group can vary considerably, even with a single species. The flowers of several annual species studied are self-compatible, typically cleistogamous, and self-pollinating. Seed germination requirements also vary extensively. Their means of dispersal has not been elucidated but probably occurs abiotically by water or wind.

Koenigia islandica **L.** is an Arctic (or alpine) species that occurs in brackish or freshwater sites associated with alluvial fans, beach ridges, bogs, brooklets, depressions, dunes, fens, floodplains, gravel bars, marshes, meadows, meltwaters, mud-flats, pools (ephemeral), salt flats, sand bars, seeps, silt bars, slopes, strands (pools), swales, tundra, and the margins of lagoons, ponds, roadside pools, rivers, springs, streams, and lakes at elevations of up to 4049 m. The substrates are usually characterized as acidic and are described as bare earth, clay, cobbly gravel, frost/mud boils, granite, gravel (alluvial), limestone, loam, moss, muck, mucky silt, mud, rock, sand, sandy gravel, sandy rock, sandy silt, scree, shaley scree, silt, silty peat, and volcanic ash. The plants often occur in patches of moss that have become saturated with water and are plentiful in areas with prolonged snow cover. Their annual habit confines individuals to sunny, protected areas that receive still, warm air and minimal wind. The plants are small at maturity, which is beneficial by keeping their tissues close to the warm soil and by requiring only a minimal growth period. Arctic and alpine populations differ radically by their morphological, life history and phenological traits, indicating substantial genetic divergence (a result also indicated by some molecular phylogenetic studies). The flowering response is day neutral, with flowering and fruiting occurring from July to September. The flowers are self-compatible, cleistogamous, and self-pollinating. Any viable seeds will germinate within their first spring. Depending on their origin, the seeds either lack a light requirement for germination or have a strong light requirement for germination. Populations can also differ further in their seed germination requirements ranging from no pretreatment necessary, to cold stratification, or scarification being required. In some cases, unscarified seeds germinate optimally at 24°C under 10 h of daylight, and scarified seeds achieve 100% germination after 3 days at 21°C. Overall, the optimal seed germination temperature is 27°C. Large annual fluctuations in population numbers and survivorship because of adverse conditions indicate that the plants may be highly vulnerable to climatic fluctuations. **Reported associates:** *Arctagrostis latifolia, Arctophila fulva, Artemisia campestris, A. tilesii, Bistorta vivipara, Calamagrostis canadensis, Caltha leptosepala, Cardamine pratensis, Carex aquatilis, C. illota, C. lachenalii, C. media, C. microchaeta, C. scopulorum, C. subspathacea, Cassiope tetragona, Cerastium beeringia-num, Chrysosplenium tetrandrum, Cochlearia groenlandica, Dasiphora floribunda, Deschampsia brevifolia, D. cespitosa, Dryas, Eleocharis acicularis, Epilobium anagallidifolium, E. hornemannii, E. palustre, Equisetum arvense, Eriophorum angustifolium, Festuca altaica, F. baffinensis, F. rubra, Geum rossii, Hippuris vulgaris, Juncus biglumis, J. castaneus, J. mertensianus, Kobresia simpliciuscula, Mertensia paniculata, Micranthes nelsoniana, M. razshivinii, Minuartia rubella, Oxyria digyna, Parnassia fimbriata, Pedicularis groenland-ica, Petasites frigidus, Phippsia algida, Poa alpina, P. lep-tocoma, P. paucispicula, Podagrostis humilis, Polemonium caeruleum, Ranunculus flammula, R. hyperboreus, R. niva-lis, R. pygmaeus, Rhodiola rhodantha, R. rosea, Rorippa alpina, Rumex arcticus, Salix alaxensis, S. arctica, S. glauca, S. polaris, S. pulchra, S. stolonifera, Saxifraga cernua, S. hirculus, Silene uralensis, Stellaria crassifolia, S. longipes, S. umbellata, Trisetum spicatum, Wilhelmsia physodes.*

Use by wildlife: none reported

Economic importance: food: none; **medicinal:** none; **cultivation:** none; **misc. products:** none; **weeds:** none; **nonindigenous species:** none

Systematics: Several authors have merged *Koenigia* with *Polygonum*, a disposition that is not warranted by the outcome of several systematic investigations (e.g., Figure 3.25). To address discrepancies resulting from phylogenetic analyses of subfamily Polygonoideae (e.g., Figure 3.26), some authors have recircumscribed *Koenigia* to include *Aconogonon*. The morphological similarity of the two genera has been used as a justification to support their merger. However, debate continues whether to retain the genera as distinct, which could be achieved by transferring *K. delicatula* to a separate genus, and by accepting *Aconogonon* as a paraphyletic grade, which is resolved by some molecular analyses (e.g., Figure 3.26). *Koenigia* species (less *K. delicatula*) resolve as a well-supported clade in any case. According to phylogenetic results, individuals of *K. islandica* are related more closely to

K. cyanandra, *K. fertilis*, or *K. pilosa*. These discrepancies are possibly related to the polyploid nature of this species. The base chromosome number of *Koenigia* is *x* = 7. *Koenigia islandica* is uniformly tetraploid (2*n* = 28) throughout its widespread distributional range, but it has not been distinguished as an auto- versus allopolyploid.

Comments: *Koenigia islandica* is essentially an arctic plant, but it also occurs sporadically in high elevation alpine habitats of the Rocky Mountains. The species has a circumpolar distribution and also is disjunct in the Andes and Himalayas. **References:** Aiken et al., 2007; Costea & Tardif, 2003; Fan et al., 2013; Hedberg, 1997; Heide & Gauslaa, 1999; Löve & Sarkar, 1957; Qasair et al., 2003; Reynolds, 1984; Schuster et al., 2015; Wagner & Simons, 2008, 2009a,b.

3. *Persicaria*

Smartweed; renouée

Etymology: from the Latin *persica* ("peach") in reference to the resemblance of some leaves to those of the peach tree **Synonyms:** *Polygonum* (in part) **Distribution: global:** cosmopolitan; **North America:** widespread

Diversity: global: 100 species; **North America:** 24 species **Indicators (USA): OBL:** *Persicaria amphibia*, *P. arifolia*, *P. glabra*, *P. hirsuta*, *P. hydropiper*, *P. hydropiperoides*, *P. lapathifolia*, *P. meisneriana*, *P. minor*, *P. punctata*, *P. robustior*, *P. sagittata*, *P. setacea*; **FACW:** *P. hydropiper*, *P. lapathifolia*

Habitat: freshwater; lacustrine, palustrine; **pH:** 3.0–9.4; **depth:** 0–2 m; **life-form(s):** emergent (herb), floating-leaved **Key morphology:** stems (to 1.5 m) appearing jointed (sometimes square and prickly), nodes swollen; leaves (to 2.2 dm) alternate, the blades simple, entire, acute and lanceolate, ovate, sagittate or hastate; ocrea present; inflorescences axillary or terminal, their flowers in dense or loose, capitate to elongate, nodding or erect spikes; perianth green, pink, or white; fruit a trigonous or lenticular achene (to 3.2 mm) **Life history: duration:** annual (fruit/seeds); perennial (rhizomes, stolons); **asexual reproduction:** rhizomes, shoot fragments, stolons; **pollination:** insect, self; **sexual condition:** hermaphroditic; **fruit:** achenes (common); **local dispersal:** rhizomes, shoot fragments, stolons, achenes (water); **long-distance dispersal:** achenes (birds)

Imperilment: (1) *Persicaria amphibia* [G5]; S1 (DC, GA, MO, NC, PE); S2 (NB, NF, WV); S3 (BC, WY); (2) *P. arifolia* [G5]; S1 (IL, GA, MO, TN); S2 (NS, PE); S3 (NB, ON, QC); (3) *P. glabra* [G5] S1 (GA, MO, NJ); S3 (NC); (4) *P. hirsuta* [G3/G4]; S1 (NC); (5) *P. hydropiper* [GNR]; (6) *P. hydropiperoides* [G5]; S1 (AK, ND); S3 (IA, NB, QC, VT); (7) *P. lapathifolia* [G5]; S2 (YT); S3 (NC); (8) *P. meisneriana* [G5]; S1 (FL, GA); S2 (FL); (9) *P. minor* [GNR]; (10) *P. punctata* [G5]; S1 (SK); S2 (ND, PE); S3 (BC, NB, QC); (11) *P. robustior* [G4/G5]; SH (ME); S1 (NH, QC); S2 (OH, ON); S3 (NJ, NS); (12) *P. sagittata* [G5]; SH (ND); S1 (CO, KS); S3 (IL, MB); (13) *P. setacea* [G5]; SH (RI); S1 (IN, NY, OH); S2 (MA, PA)

Ecology: general: *Persicaria* comprises a widespread group of annual or perennial herbs that are encountered frequently in wet habitats. All of the North American species are wetland indicators, with 13 (54%) designated as OBL. The flowers of most species investigated are self-compatible, but self-incompatibility is indicated in some. At least one species is functionally gynodioecious. Many of the species appear to be primarily autogamous, with some bearing both cleistogamous and chasmogamous flowers. At least one species exhibits reciprocal herkogamy, which reduces the extent of self-pollination. Pollination can also occur by various insects (Insecta) including bees (Hymenoptera), butterflies (Lepidoptera), and flies (Diptera), which seek the floral nectar. The seeds are dispersed locally by gravity (especially those produced by the cleistogamous flowers) but can be distributed widely by waterfowl (Aves: Anatidae) and other birds, which transport them endozoically. Although the seeds naturally sink, they can also be shed within the dry perianths, which are capable of floatation (and water dispersal) for periods of up to 10 days, before becoming waterlogged and releasing the seeds.

***Persicaria amphibia* (L.) Delarbre** is an amphibious perennial species that inhabits beaches, bogs, bottomlands, cattle tanks, depressions, ditches, draws, flats, floodplains, gravel bars, gravel pits, kettles, lagoons, marshes, meadows, mudflats, outwash, oxbows, playas, ponds, pools (ephemeral, seasonal), potholes, prairies, quaking mats, roadsides, sand bars, seeps, shores, slopes (to 3%), sloughs (dry), swales, swamps, thickets, vernal pools, and the margins of lakes, ponds, reservoirs, and rivers at elevations of up to 4516 m. Exposures range from full sunlight to shaded conditions. The plants are tolerant to turbidity (secchi disc transparency: 2.5–7.0 m) and grow either as emergent herbs or as floating-leaved life-forms in waters that are shallower than 2 m. They can form extensive stands and spread rapidly when the water levels are lowered. The shoots are highly sensitive to frost. This species can be found under a broad range of habitat conditions (e.g., pH: 3.0–9.4; organic matter: 3%–31%) but occurs most often in waters that are still, warm (20°C–25°C) and of moderate hardness (e.g., 10–200 ppm $CaCO_3$; alkalinity: 45–66 mg/L; conductivity: 97 mmhos). The substrates include alluvium, boulders, clay, clay loam, gravel, humus, loam, muck, mud, muddy clay, ooze, organic loam, peat, Pilchuck soils, sand, sandy humus, sandy loam, sandy silt, and silt. Some stands occur on selenium-rich sites. Sexual reproduction is uncommon with some clones not setting seed during 28 years of observation. The clones can be male fertile or male sterile; however, low seed set occurs even on male-fertile plants, which indicates self-incompatibility. Flowering and fruiting extend from May to October. Observed insect (Insecta) pollinators include various Coleoptera, Diptera, Hymenoptera, and Lepidoptera. The seeds reach maturity within 3 weeks of fertilization. Few seeds are usually produced, but these are dispersed by water currents (the inflorescences can float for more than a week) or animals. The mature seeds sink in water. The seeds can retain viability after being frozen for a month. Seeds that have been stored in water at 1°C–3°C for 7 months exhibit moderate germination (>25%) but lose their viability when dried. Germination (higher in the light) occurs after 210 days of cold stratification at a 30°C/20°C temperature regime. Scarification greatly enhances seed germination. Plants form large clones that may extend for several kilometers along watercourse margins or cover hectares of marshland. The root system is effective at stabilizing the surrounding substrate. **Reported associates:** *Acalypha rhomboidea, Acer rubrum, A. saccharinum, Agrostis gigantea, A. scabra, A. stolonifera, Alisma subcordatum, A. triviale, Alnus rubra, A. viridis, Alopecurus aequalis, Ambrosia artemisiifolia, A. grayii, A. psilostachya, A. tomentosa, A. trifida, Ammannia coccinea, Amorpha fruticosa, Amphiscirpus nevadensis, Anaphalis margaritacea, Anemopsis californica, Apios americana, Apocynum cannabinum, Arctophila fulva, Artemisia ludoviciana, Asclepias incarnata, Azolla filicoides, Bacopa rotundifolia, Barbarea, Beckmannia syzigachne, Betula pumila, Bidens aristosus, B. cernuus, B. connatus, B. polylepis, Boehmeria cylindrica, Bolboschoenus fluviatilis, B. maritimus, Boltonia asteroides, Brasenia schreberi, Bromus ciliatus, Calamagrostis canadensis, Callitriche palustris, Calystegia sepium, Carex aquatilis, C. athrostachya, C. bebbii, C. buxbaumii, C. canescens, C. cusickii, C. flava, C. fuliginosa, C. haydenii, C. lacustris, C. lasiocarpa, C. lenticularis, C. lurida, C. muricata, C. nebrascensis, C. oligosperma, C. praegracilis, C. retrorsa, C. rostrata, C. simulata, C. stricta, C. utriculata, C. vesicaria, Cephalanthus occidentalis, Ceratophyllum demersum, C. echinatum, Chara, Chelone glabra, Chenopodium album, Cicuta bulbifera, C. douglasii, C. maculata, Cirsium vulgare, Comarum palustre, Cornus amomum, C. sericea, Crypsis schoenoides, Cyperus eragrostis, C. esculentus, C. strigosus, Decodon verticillatus, Deschampsia cespitosa, D. elongata, Distichlis spicata, Drosera anglica, D. rotundifolia, Dulichium arundinaceum, Dysphania botrys, Echinochloa crus-galli, Echinodorus berteroi, Eclipta prostrata, Elatine, Eleocharis acicularis, E. compressa, E. macrostachya, E. montana, E. obtusa, E. palustris, E. quinqueflora, Elodea canadensis, E. nuttallii, Epilobium ciliatum, E. palustre, E. strictum, Equisetum arvense, E. fluviatile, Erechtites hieracifolius, Eriophorum gracile, Eupatorium perfoliatum, Euphorbia dentata, Euthamia graminifolia, E. occidentalis, Fallopia scandens, Festuca altaica, Fontinalis, Forestiera acuminata, Fraxinus pennsylvanica, Galium tinctorium, G. trifidum, Gaultheria shallon, Geum, Glyceria borealis, G. elata, G. grandis, Glycyrrhiza lepidota, Gnaphalium exilifolium, Gratiola, Grindelia squarrosa, Helenium autumnale, Helianthus annuus, H. grosseserratus, Heteranthera limosa, Hibiscus laevis, H. lasiocarpos, Hippuris vulgaris, Hordeum, Iris missouriensis, I. pseudacorus, I. virginica, Iva annua, Juncus balticus, J. diffusisimus, J. howellii, J. nevadensis, J. nodosus, Justicia americana, Kalmia microphylla, Koeleria, Leersia lenticularis, L. oryzoides, Leitneria floridana, Lemna minor, L. valdiviana, Lepidium densiflorum, Leptochloa fascicularis, Limosella aquatica, Lindernia dubia, Lobelia cardinalis, L. siphilitica, Lotus corniculatus, Ludwigia palustris, L. peploides, L. polycarpa, Lupinus kingii, Luzula campestris, L. parviflora, Lycopus americanus, L. uniflorus, Lysimachia terrestris, L. thyrsiflora, Lythrum alatum, Marsilea vestita, Medicago sativa, Melilotus albus, Mentha arvensis, Menyanthes trifoliata, Mimulus guttatus, Mirabilis linearis, Muhlenbergia frondosa, Myosurus minimus, Myriophyllum sibiricum, M. spicatum, Nasturtium officinale, Navarretia intertexta, Nitella, Nuphar polysepala, N. variegata, Nymphaea odorata, Oenanthe sarmentosa, Onoclea sensibilis, Orthocarpus luteus, Oxypolis rigidior, Panicum dichotomiflorum, Pedicularis groenlandica, Peltandra virginica, Penstemon linarioides, Persicaria coccinea, P. hydropiperoides, P. lapathifolia, P. pensylvanica, P. punctata, P. sagittata, Phalaris arundinacea, Phleum pratense, Phragmites australis, Physostegia virginiana, Pinus contorta, Poa compressa, Potamogeton alpinus, P. amplifolius, P. berchtoldii, P. bicupulatus, P. diversifolius, P. epihydrus, P. friesii, P. gramineus, P. natans, P. pusillus, P. richardsonii, P. zosteriformis, Populus angustifolia, P. heterophylla, Porterella carnosula, Potentilla anserina, P. norvegica, Proserpinaca palustris, Pycnanthemum virginianum, Ranunculus aquatilis, R. flammula, R. macounii, R. occidentalis, R. subrigidus, R. trichophyllus, Rhododendron groenlandicum, Ribes triste, Ricciocarpus natans, Rorippa curvisiliqua, Rosa, Rubus pensilvanicus, Rumex britannica, R. crispus, R. mexicanus, R. triangulivalvis, Sagittaria cuneata, S. latifolia, Salix amygdaloides, S. candida, S. discolor, S. exigua, S. lasiandra, S. nigra, S. petiolaris, S. scouleriana, S. sitchensis, Sambucus nigra, Saururus cernuus, Schedonorus arundinaceus, Scheuchzeria palustris, Schoenoplectus acutus, S. hallii, S. pungens, S. tabernaemontani, Scirpus ancistrochaetus, S. cyperinus, Scutellaria epilobiifolia, S. galericulata, S. lateriflora, Sesuvium verrucosum, Setaria faberi, Sium suave, Solanum, Solidago altissima, S. velutina, Sonchus arvensis, Sparganium angustifolium, S. chlorocarpum, S. emersum, S. eurycarpum, Spartina gracilis, S. pectinata, Sphagnum, Spiraea alba, S. douglasii, Spiranthes romanzoffiana, Spirodela polyrhiza, Sporobolus cryptandrus, Stachys, Stellaria crassifolia, Stuckenia filiformis, S. pectinata, Symphoricarpos albus, Symphyotrichum pilosum, S. praealtum, S. subulatum, Taraxacum officinale, Taxodium distichum, Thelypteris palustris, Toxicodendron rydbergii, Triadenum virginicum, Trifolium, Typha angustifolia, T. domingensis, T. latifolia, Ulmus americana, Urtica dioica, Utricularia gibba, U. macrorhiza, Vaccinium oxycoccus, Verbascum thapsus, Verbena bracteata, Vernonia gigantea, Veronica americana, V. anagallis-aquatica, V. peregrina, V. scutellata, Viola, Xanthium strumarium, Zannichellia palustris, Zizania palustris.*

***Persicaria arifolia* (L.) Haraldson** is a weak annual species that grows in freshwater to brackish, tidal or nontidal sites including bogs, ditches, floodplains, marshes, meadows, ravines, seeps, swamps, woodlands, and the margins of ponds, rivers, and swamps at elevations of up to 600 m. The plants occur typically in shaded exposures and usually under flooded conditions (to at least 20 cm). The shoots and leaf midribs are covered by sharp prickles enabling the plants to develop a vinelike habit, which gains support by clinging to and growing upon other wetland vegetation. Relatively dense stands can form. The substrates are circumneutral (pH: 6.0–7.3) and most often are described as muck or mud. Flowering and fruiting occur from July to October. The method of pollination has not been reported but presumably is primarily autogamous. A transient seedbank develops, whereby the seeds germinate immediately after meeting their stratification requirements. The seeds will germinate after 150 days of cold stratification followed by incubation under a 25°C/15°C temperature regime. The seedlings are large, establish effectively, and can form dense stands (>4000 m^{-2}), which preempt other potentially competing vegetation. The experimental addition of nitrogen to sites affected percent cover negatively in this species (by providing a greater competitive edge to neighboring perennials), whereas the addition of phosphorous had a positive effect, indicating that the plants were phosphorous limited. Simultaneous addition of N and P had no effect on percent cover. **Reported associates:** *Acer rubrum, Acorus calamus, Aeschynomene virginica, Alnus serrulata, Alternanthera philoxeroides, Amaranthus cannabinus, Amphicarpa bracteata, Apios americana, Aronia melanocarpa, Asclepias incarnata, Athyrium filix-femina, A. thelypterioides, Bidens laevis, Bignonia capreolata, Boehmeria cylindrica, Bolboschoenus fluviatilis, Caltha palustris, Cardamine hirsuta, Carex atlantica, C. bromoides, C. comosa, C. lupulina, C. lurida, C. prasina, C. scoparia, C. stipata, Carpinus caroliniana, Chasmanthium latifolium, Chelone glabra, Cicuta maculata, Clematis crispa, Commelina virginica, Cyperus, Decumaria barbara, Dryopteris carthusiana, D. cristata, Eleocharis ambigens, E. obtusa, Equisetum arvense, Glyceria striata, Hibiscus moscheutos, Hydrocotyle bonariensis, Hymenocallis rotata, Impatiens capensis, Iris hexagona, Juncus effusus, J. marginatus, Justicia ovata, Larix laricina, Leersia oryzoides, Lobelia cardinalis, L. siphilitica, Lysimachia nummularia, Lythrum salicaria, Murdannia keisak, Myrica cerifera, Nuphar advena, Onoclea sensibilis, Osmunda regalis, Osmundastrum cinnamomeum, Peltandra virgininca, Persea palustris, Persicaria hydropiperoides, P. maculosa, P. punctata, P. sagittata, P. setacea, Phanopyrum gymnocarpon, Pilea fontana, P. pumila, Platanthera clavellata, Polystichum acrostichoides, Pontederia cordata, Ptilimnium costatum, Quercus michauxii, Ranunculus, Rumex verticillatus, Sabal minor, Saururus cernuus, Schoenoplectus tabernaemontani, Scirpus cyperinus, S. divaricatus, Sium suave, Solidago sempervirens, Symphyotrichum puniceum, Symplocarpus foetidus, Taxodium distichum, Toxicodendron radicans, Triadenum walteri, Typha angustifolia, T. latifolia, Veratrum viride, Viola, Zizania aquatica, Zizaniopsis miliacea.*

***Persicaria glabra* (Willd.) M. Gómez** is a coarse perennial of bogs, bottomlands, canals, depressions, ditches, domes, flatwoods, floodplains, marshes, mats (floating), meadows, mudflats, ponds, pools, roadsides, rock bars, sand bars, seeps, shores, sinks, solution holes, swamps, thickets, and the margins of lakes, ponds, and rivers at elevations of up to 300 m. The plants grow mainly along the southeastern coastal plain and can tolerate relatively deep water (to at least 1.5 m), fairly high currents, and fluctuating water levels. They occur in open to shaded exposures. The reported substrates include Copeland (typic argiaquolls), floating logs, limestone, marl, muck, sandy clay, sandy loam, sandy silt, Santee loam, and

silt. Flowering and fruiting extend from January to November. The pollination biology has not been studied. Seed production can be prolific (e.g., 167 seeds/m^2) with some studies estimating yields of more than 745 kg/ha. The seed germination requirements are unknown. Vegetative reproduction occurs by means of rhizomes. The plants stabilize banks by trapping sediments in their roots, which reduces the flushing of deposited organic matter. **Reported associates:** *Acer rubrum, Agalinis purpurea, Alternanthera philoxeroides, Azolla caroliniana, Bidens aristosus, B. bipinnatus, B. discoideus, B. pilosus, Carex atlantica, C. hyalinolepis, C. joorii, C. seorsa, Cephalanthus occidentalis, Cicuta maculata, Cladium jamaicense, Cuphea carthagenensis, Cyperus odoratus, Echinochloa muricata, Echinodorus cordifolius, Eichhornia crassipes, Eleocharis vivipara, Eleusine indica, Helenium autumnale, Hydrocotyle umbellata, Hydrolea quadrivalvis, Hymenachne amplexicaulis, Ipomoea coccinea, Juncus canadensis, Kyllinga brevifolia, Leersia virginica, Lemna, Limnobium spongia, Lobelia glandulosa, Ludwigia alternifolia, L. grandiflora, L. leptocarpa, L. palustris, L. peruviana, L. repens, Lycopus rubellus, Microstegium vimineum, Mikania scandens, Nuphar, Nyssa biflora, Oldenlandia corymbosa, Orontium aquaticum, Osmunda regalis, Oxycaryum cubense, Panicum dichotomiflorum, P. hemitomon, Paspalidium geminatum, Persicaria arifolia, P. punctata, Pistia stratiotes, Polygonum ramosissimum, Pontederia cordata, Ptilimnium capillaceum, Ranunculus sceleratus, Rumex verticillatus, Sacciolepis striata, Sagittaria lancifolia, Salvinia minima, Saururus cernuus, Scleria lacustris, Setaria parviflora, Taxodium distichum, Typha latifolia, Woodwardia virginica.*

***Persicaria hirsuta* (Walter) Small** is a perennial species that inhabits barrens, Carolina bays, depressions, ditches, flats, flatwoods, marshes, meadows, prairies, roadsides, savannahs, seeps, swales, and the margins of lakes, ponds, and sinkholes at elevations of up to 100 m. The plants occur primarily in open areas along the southeastern coastal plain of North America and often in shallow water (to 80 cm). The substrates have been described as Basinger (spodic psammaquents), loam, mucky sand, mud, peat, and sand. Flowering extends from June to November. Few details are available on the reproductive ecology of this species. A seed bank develops sporadically. The plants reproduce vegetatively by rhizomes and can form dense mats. **Reported associates:** *Acer rubrum, Bacopa monnieri, Carex longii, Centella asiatica, Cephalanthus occidentalis, Chasmanthium sessiliflorum, Cicuta mexicana, Cladium jamaicense, Coreopsis, Cyperus compressus, C. odoratus, C. retrorsus, C. surinamensis, C. virens, Digitaria serotina, Diodia, Echinochloa muricata, Eichhornia crassipes, Eleocharis baldwinii, E. equisetoides, E. vivipara, Eupatorium compositifolium, Fimbristylis caroliniana, Fuirena pumila, Hydrocotyle umbellata, Hypericum cistifolium, Juncus effusus, Leersia oryzoides, Lemna minor, Leptochloa, Ludwigia alternifolia, L. decurrens, L. leptocarpa, L. octovalvis, L. suffruticosa, Luziola fluitans, Magnolia virginiana, Nymphaea odorata, Nyssa, Oldenlandia uniflora, Panicum hemitomon, P. repens, P. verrucosum, Persicaria punctata, Pinus elliotii, P. palustris, Pontederia* *cordata, Rhexia mariana, Rhynchospora, Sacciolepis striata, Sagittaria graminea, S. latifolia, S. subulata, Salix nigra, Saururus cernuus, Schoenoplectus pungens, Sesbania macrocarpa, Setaria parviflora, Spartina bakeri, Sphagnum macrophyllum, Triadenum walteri, Typha domingensis, T. latifolia, Urochloa mutica, Utricularia olivacea, U. subulata, Xyris jupicai.*

***Persicaria hydropiper* (L.) Opiz** is a nonindigenous, annual species of tidal or nontidal, shallow water habitats (to 46 cm depth), which occur in bottomlands (alluvial), backwaters, beaches, bogs, depressions, ditches, fens, flats, flatwoods, floodplains, gravel bars, gravel pits, marshes (intertidal), meadows, mud flats, oxbows, pools (drying), prairies, roadsides, sand bars, savannas, seeps, shores, slopes (to 2%), sloughs, swales, swamps, thickets, vernal pools, washes (alluvial), waste ground, woodlands, and along the margins of lakes, ponds, reservoirs, rivers, and streams at elevations of up to 2052 m. Exposures can range from full sun to shade. The substrates have been described as acidic or alkaline (e.g., pH: 7.8) and include alluvium, ballast, clay, clay loam, cobbles, cobbly sand, crushed dolomite, gravel, Hepler silty loam, loess, muck, mud, peat, pebbles (coarse), rock, sand, sandstone, sandy clay, sandy gravel, sandy loam, sandy mud, silt, Tuscan mudflow, and Verdigris silty loam. Flowering and fruiting extend from May to November. Both chasmogamous and cleistogamous flowers are produced, with the latter being strictly autogamous but yielding larger fruits. The chasmogamous flowers are visited by flies (Insecta: Diptera: Syrphidae), which likely facilitate at least some outcrossing. Pollen production (averaging 1778 grains/flower) is not high, and the plants are believed to be primarily self-pollinating. The seeds germinate following 60–90 days of cold stratification (at 4°C) when placed under a 20°C/15°C temperature regime. However, their germination was found to be less dependant on cold, moist stratification than on fluctuating temperatures, with high germination achieved without stratification when the seeds were incubated under a 24°C/12°C temperature regime. Although the buried seeds of *P. hydropiper* are known to retain their viability for up to 50 years, the plants often do not emerge in seedbank studies. They have been recovered in fairly large numbers from sediment samples. Seed dispersal occurs locally by wind and water (especially along rivers) but over longer distances via endozoic transport by various animals. Aside from the many species of waterfowl (Aves: Anatidae) that eat the seeds (see *Use by wildlife*), studies in Europe have demonstrated that the seeds can also be dispersed endozoically by various mammals (Mammalia) including elk (Cervidae: *Cervus elaphus*), red foxes (Canidae: *Vulpes vulpes*), and wild boars (Suidae: *Sus scrofa*). **Reported associates (North America):** *Acalypha gracilens, A. rhomboidea, Acer negundo, A. saccharinum, Acorus calamus, Agalinis purpurea, Ageratina adenophora, Agrimonia parviflora, Agrostis exarata, A. gigantea, Alisma subcordatum, A. triviale, Alliaria petiolata, Allium tricoccum, Alnus rubra, A. serrulata, Alopecurus geniculatus, Amaranthus, Ambrosia artemisiifolia, A. trifida, Ammannia coccinea, Andropogon gerardii, Anemone virginiana, Apios americana, Arctium*

lappa, Arisaema triphyllum, Aronia ×prunifolia, Artemisia biennis, A. douglasiana, A. dracunculus, A. ludoviciana, Asclepias incarnata, A. speciosa, Betula nigra, Bidens cernuus, B. connatus, B. frondosus, B. trichospermus, B. tripartitus, B. vulgatus, Boehmeria cylindrica, Brasenia schreberi, Brassica nigra, Calamagrostis canadensis, Callitriche stagnalis, Carex flava, C. haydenii, C. hystricina, C. lacustris, C. lenticularis, C. pellita, C. rostrata, C. stricta, C. vulpinoidea, Carya ovata, Cephalanthus occidentalis, Ceratophyllum, Chenopodium, Cicuta bulbifera, Cirsium arvense, Comarum palustre, Coreopsis tinctoria, Cornus sericea, Cryptotaenia canadensis, Cyperus esculentus, C. strigosus, Datura, Daucus carota, Dichanthelium clandestinum, Doellingeria umbellata, Dysphania, Echinochloa crusgalli, E. muricata, Eclipta prostrata, Eleocharis obtusa, E. palustris, E. wolfii, Elymus elymoides, E. glaucus, E. repens, Epilobium brachycarpum, E. ciliatum, E. coloratum, E. densiflorum, Equisetum arvense, E. fluviatile, Erechtites hieracifolius, Eriogonum fasciculatum, Eupatorium perfoliatum, E. serotinum, Eutrochium maculatum, Fallopia scandens, Fimbristylis autumnalis, Fraxinus americana, F. pennsylvanica, Galium palustre, G. trifidum, Gaylussacia baccata, Gentiana andrewsii, Geranium maculatum, Glyceria borealis, G. elata, G. grandis, G. septentrionalis, Gratiola aurea, Helenium autumnale, Helianthus grosseserratus, Heracleum sphondylium, Hippuris, Hypericum adpressum, H. canadense, H. mutilum, Ilex verticillata, Impatiens capensis, I. pallida, Iris pseudacorus, I. virginica, Juncus acuminatus, J. articulatus, J. balticus, J. bolanderi, J. effusus, J. marginatus, J. supiniformis, J. tenuis, Juniperus scopulorum, Justicia americana, Kalmia latifolia, K. microphylla, Lactuca serriola, Laportea canadensis, Leersia oryzoides, Lemna, Lepidospartum squamatum, Lilaeopsis, Lindernia dubia, Lipocarpha micrantha, Lobelia siphilitica, Lotus corniculatus, Ludwigia polycarpa, Lycopus americanus, L. asper, L. uniflorus, Lysimachia lanceolata, L. nummularia, Lythrum alatum, L. salicaria, Maianthemum racemosum, Medicago lupulina, Melilotus albus, Mentha, Microstegium vimineum, Mimulus guttatus, M. ringens, Muhlenbergia asperifolia, M. frondosa, Myosotis laxa, M. scorpioides, Myriophyllum aquaticum, M. quitense, M. spicatum, M. ussuriense, Nasturtium officinale, Navarretia intertexta, Nuphar advena, N. polysepala, Nymphaea tuberosa, Nyssa biflora, Oenothera pilosella, O. villosa, Onoclea sensibilis, Oxalis corniculata, Oxypolis rigidior, Panicum capillare, P. virgatum, Parnassia asarifolia, Paspalum laeve, Peltandra virginica, Penthorum sedoides, Perideridia howellii, Persicaria amphibia, P. careyi, P. hydropiperoides, P. lapathifolia, P. maculosa, P. orientalis, P. pensylvanica, P. sagittata, Phalaris arundinacea, Phyla lanceolata, Pilea fontana, P. pumila, Plantago lanceolata, P. major, Poa annua, P. nemoralis, P. palustris, P. pratensis, Podophyllum peltatum, Polygonum polygaloides, P. monspeliensis, Populus balsamifera, P. tremuloides, Potamogeton amplifolius, Potentilla anserina, Proserpinaca palustris, Prunella vulgaris, Pseudotsuga menziesii, Quercus alba, Q. bicolor, Q. macrocarpa, Q. palustris, Q. rubra, Q. velutina, Ranunculus aquatilis, R. flabellaris, R. flammula, R.

*uncinatus, Rhamnus cathartica, Rhododendron groenlandi-
cum, Rhus copallinum, Rhynchospora alba, R. capitellata,
Rorippa aquatica, R. curvipes, R. sessiliflora, Rotala ramo-
sior, Rubus hispidus, R. laciniatus, R. setosus, R. ulmifolius,
Rudbeckia subtomentosa, Rumex crispus, R. obtusifolius,
Sagittaria latifolia, Salix discolor, S. exigua, S. lasiolepis,
S. scouleriana, S. sericea, S. sitchensis, Sambucus nigra,
Schoenoplectus acutus, S. heterochaetus, S. lacustris, S. pun-
gens, S. triqueter, Scirpus atrovirens, S. cyperinus, Silphium
perfoliatum, Sium suave, Solanum dulcamara, Solidago
canadensis, S. flexicaulis, S. gigantea, S. ulmifolia, S. velutina,
Sparganium androcladum, Spartina pectinata, Sphagnum,
Spiraea douglasii, S. tomentosa, Symphyotrichum firmum,
S. lanceolatum, S. lateriflorum, S. praealtum, S. puniceum,
Tanacetum, Teucrium canadense, Thalictrum dasycarpum,
Thelypteris palustris, Trifolium hybridum, T. pratense, T.
repens, Typha angustifolia, T. latifolia, Ulmus americana,
Urtica dioica, Utricularia gibba, Vaccinium macrocar-
pon, Veratrum californicum, Verbascum thapsus, Verbena
bonariensis, V. hastata, Vernonia fasciculata, V. missurica,
Veronica americana, V. anagallis-aquatica, Viburnum opu-
lus, V. rafinesqueanum, Viola lanceolata, V. primulifolia,
Xanthium strumarium, Xyris torta, Zizia aurea.*

***Persicaria hydropiperoides* (Michx.) Small** is a perennial
species, which inhabits shallow water sites (to 15 dm depth)
that are associated with backwaters, bogs, bottomlands, channels, depressions, ditches, dunes, flatwoods, floodplains,
marshes, meadows, mud flats, oxbows, ponds, pools, prairies,
roadsides, savannas, seeps, sinkholes, slopes (to 8%), sloughs,
swales, swamps, washes, woodlands, and the margins of
lakes, ponds, rivers, and streams at elevations of up to 1494 m.
The plants are common in full sunlight exposures and are
somewhat shade intolerant. The reported substrates range
broadly in acidity (pH: 4.5–8.8) and include clay, clay loam,
Corolla–Duckston sand, dolomite, granite, gravel, Harleston
fine sandy loam, muck, mud, peat, rock, sand, sandy clay
loam, sandy loam, silt, silty clay loam, silty loam, silty rock,
and Verdigris silt loam. Flowering and fruiting occur from
June to November. The reproductive ecology has not been
studied specifically, but the plants are visited by a variety of
bees (see *Use by wildlife*), which likely function as pollinators. This is one of few *Persicaria* species, which always
seems to have at least some open (chasmogamous) flowers
present during anthesis. However, because other flowers
apparently remain closed, a mixed mating system is indicated.
Local seed dispersal likely is facilitated by abiotic vectors
(water, wind). Seed dispersal over longer distances occurs via
endozoic transport by waterfowl (Aves: Anatidae), which
enthusiastically consume the propagules. A substantial seed
bank can develop, with seedling emergence being highest
under drawdown conditions. The seeds will germinate after
receiving 135 days of moist, cold stratification (at 4°C) and
then incubated under a 30°C/20°C temperature regime.
Germination percentages are higher in the light. Although the
plants often occur in monospecific stands, the list of their
associated species is impressive. **Reported associates:**
Acalypha gracilens, A. rhomboidea, Acer negundo, A.

*rubrum, A. saccharinum, Adiantum capillus-veneris,
Ageratina adenophora, Agrimonia parviflora, Agrostis
gigantea, Alisma subcordatum, A. triviale, Allium canadense,
Alnus rubra, Alopecurus geniculatus, Alternanthera philoxeroides, Ambrosia artemisiifolia, A. grayii, Ammannia coc-
cinea, A. robusta, Amorpha fruticosa, Ampelopsis arborea,
Amsonia rigida, Andropogon, Apocynum cannabinum,
Arisaema triphyllum, Aronia ×prunifolia, Asclepias incar-
nata, A. sullivantii, A. syriaca, Avena, Azolla caroliniana,
Baccharis halimifolia, B. salicifolia, Bacopa monnieri,
Betula nigra, Bidens cernuus, B. connatus, B. discoideus, B.
frondosus, B. mitis, B. polylepis, B. trichospermus, Boehmeria
cylindrica, Bolboschoenus fluviatilis, Brasenia schreberi,
Bromus, Calamagrostis canadensis, Callicarpa americana,
Callitriche heterophylla, Caltha palustris, Campsis radicans,
Cardamine bulbosa, Carex annectens, C. bicknellii, C. caroliniana, C. conjuncta, C. corrugata, C. crinita, C. cristatella,
C. crus-corvi, C. granularis, C. grayii, C. haydenii, C. hyalinolepis, C. lenticularis, C. longii, C. lupulina, C. muskingumensis, C. pellita, C. scoparia, C. shortiana, C. stricta, C.
tribuloides, C. typhina, C. vesicaria, C. vexans, Carya,
Centella asiatica, Centrosema virginianum, Cephalanthus
occidentalis, Ceratophyllum, Chasmanthium latifolium,
Cicuta bulbifera, C. maculata, Cinna arundinacea, Cladium
jamaicense, Coleataenia longifolia, Comarum palustre,
Commelina virginica, Cotoneaster lacteus, Cryptotaenia
canadense, Cuscuta gronovii, Cynodon dactylon, Cyperus
erythrorhizos, C. odoratus, C. squarrosus, C. strigosus, C.
virens, Daucus carota, Decodon verticillatus, Dichanthelium,
Diodella teres, Diodia virginiana, Diospyros virginiana,
Dulichium arundinaceum, Eclipta prostrata, Echinochloa
crus-galli, Elephantopus carolinianus, Eleocharis acicularis,
E. cellulosa, E. compressa, E. flavescens, E. montevidensis,
E. palustris, Elymus glaucus, E. repens, Epilobium ciliatum,
Epipactis gigantea, Equisetum arvense, E. fluviatile,
Eragrostis elliottii, E. hypnoides, Erechtites hieracifolius,
Erigeron vernus, Eriogonum fasciculatum, Erodium,
Eupatorium altissimum, E. perfoliatum, E. serotinum,
Euthamia graminifolia, Eutrochium fistulosum, Ficus carica,
Fimbristylis autumnalis, Fraxinus pennsylvanica, F. pro-
funda, Fuirena scirpoidea, Galium tinctorium, Gamochaeta
purpurea, Gaylussacia baccata, Gleditsia triacanthos,
Glinus lotoides, Glyceria borealis, G. grandis, G. septentrionalis, G. striata, Gratiola aurea, G. virginiana, Helenium flexuosum, Heuchera richardsonii, Hibiscus laevis, H.
moscheutos, Hippuris vulgaris, Hydrocotyle umbellata,
Hydrolea, Hypericum mutilum, Hypoxis, Hyptis alata, Ilex
verticillata, Impatiens capensis, Ipomoea lacunosa, Iris
pseudacorus, I. virginica, Iva annua, Juncus acuminatus, J.
articulatus, J. balticus, J. diffusissimus, J. effusus, J. roemerianus, J. torreyi, Justicia americana, J. ovata, Kalmia latifolia, K. microphylla, Lachnanthes caroliniana, Laportea
canadensis, Leersia hexandra, L. lenticularis, L. oryzoides,
L. virginica, Lemna minor, L. valdiviana, Lepidospartum
squamatum, Leptochloa panicoides, Lilaeopsis, Lindernia
dubia, Lipocarpha micrantha, Liquidambar styraciflua,
Lonicera japonica, Lotus corniculatus, Ludwigia*

alternifolia, L. erecta, L. glandulosa, L. palustris, L. peploides, L. polycarpa, Lupinus, Lycopus americanus, L. uniflorus, Lysimachia ciliata, L. radicans, Lythrum salicaria, Lysimachia lanceolata, L. terrestris, Lythrum, Malacothamnus densiflorus, Malus coronaria, Medicago minima, Melilotus albus, Mentha arvensis, Mikania scandens, Mimulus guttatus, M. ringens, Morus alba, Murdannia keisak, Myosotis laxa, Myrica cerifera, Myriophyllum aquaticum, M. quitense, Nuphar advena, N. polysepala, Nymphaea odorata, N. tuberosa, Nyssa biflora, Oenanthe, Oenothera pilosella, Onoclea sensibilis, Panicum hemitomon, P. virgatum, Parnassia asarifolia, Paspalum plicatulum, P. praecox, Peltandra virginica, Penthorum sedoides, Persicaria amphibia, P. hydropiper, P. lapathifolia, P. maculosa, P. pensylvanica, P. punctata, P. setacea, Phalaris arundinacea, Phlox glaberrima, Phragmites australis, Phyla lanceolata, Pinus echinata, P. elliottii, P. taeda, Plantago major, Platanus occidentalis, P. racemosa, Poa annua, P. nemoralis, P. pratensis, Polygonum polygaloides, P. ramosissimum, Polypremum procumbens, Populus deltoides, P. heterophylla, Potamogeton amplifolius, P. pusillus, Potentilla anserina, Proserpinaca palustris, Prunus serotina, Pteris vittata, Quercus macrocarpa, Q. michauxii, Q. nigra, Q. palustris, Ranunculus aquatilis, R. flabellaris, R. flammula, R. hispidus, R. pusillus, R. uncinatus, Rhododendron groenlandicum, Rhus copallinum, Rhynchospora alba, R. capitellata, R. fascicularis, Robinia pseudoacacia, Rorippa aquatica, R. palustris, Rubus hispidus, R. ulmifolius, R. ursinus, Rudbeckia triloba, Rumex crispus, R. obovatus, R. verticillatus, Sabal minor, Sacciolepis indica, Sagittaria graminea, Salix exigua, S. gooddingii, S. lasiolepis, S. nigra, S. scouleriana, Sambucus nigra, Saururus cernuus, Schoenoplectus acutus, S. heterochaetus, S. pungens, S. tabernaemontani, S. triqueter, Scirpus atrovirens, S. cyperinus, Scleria, Sclerolepis uniflora, Scutellaria lateriflora, Sida rhombifolia, Sium suave, Smilax, Solanum carolinense, S. dulcamara, Solidago canadensis, S. gigantea, S. velutina, Sonchus asper, Sparganium americanum, S. androcladum, S. eurycarpum, Spartina patens, S. pectinata, Sphagnum, Spiraea douglasii, S. tomentosa, Spirodela polyrhiza, Stachys hyssopifolia, Symphyotrichum lanceolatum, S. ontarionis, S. pilosum, S. praealtum, S. racemosum, Symplocarpus foetidus, Tamarix, Taxodium distichum, Thelypteris palustris, Torilis japonica, Toxicodendron pubescens, T. radicans, Tradescantia, Triadenum tubulosum, Triadica sebifera, Tripsacum, Typha angustifolia, T. domingensis, T. latifolia, Ulmus americana, Urtica dioica, Utricularia gibba, Vaccinium elliottii, Verbena bonariensis, V. hastata, Vernonia fasciculata, V. gigantea, Veronica anagallis-aquatica, Viburnum recognitum, Vigna luteola, Viola sororia, Vitis cinerea, V. girdiana, Wolffia brasiliensis, Xanthium strumarium, Xyris torta.

Persicaria lapathifolia (L.) Gray is an annual species that inhabits alluvial fans, backwaters, beaches, bogs, bottomlands, borrow pits, dikes, ditches, flats, floodplains, gravel bars, marshes, meadows, mudflats, orchards, oxbows, ponds (drying), ricefields, roadsides, sand bars, scrub, seeps, slopes, swamps, washes, waste ground, and the margins of canals,

lakes, ponds, reservoirs (drying), rivers, sloughs, and streams at elevations of up to 3100 m. The plants occur in exposures of full sunlight. This species is the most FAC among its North American congeners, being designated as OBL only in the Great Plains region (FACW elsewhere). It is not only quite suitably adapted to wetlands but also grows well in a variety of upland environments, including agricultural sites. Although the plants occur mainly in freshwater habitats, they are tolerant of mild salinity. The substrates typically are described as alkaline (pH: 6.6–7.2) and include as alluvium, ballast, basalt, clay, clay loam, cobble, diatomite, gravel, humusy silt, limerock, loam, muck, mud, peat, pebbles, rocky gravel, sand, sandy gravel, sandy loam, sandy loess, shells, silt, silty clay, silty loam, silty mud, stones, and travertine. The plants are often found on disturbed sediments such as those of low, dredge spoil dikes. Flowering and fruiting extend from May to October. The flowers are self-compatible, not dichogamous, have low pollen:ovule ratios (e.g., 200), and probably self-pollinated to some extent. They are outcrossed by insects with honeybees (Insecta: Hymenoptera: Apidae: *Apis mellifera*), observed as potential pollinators in one case. Abiotic factors (water, wind) facilitate local seed dispersal, with hydrochory being prevalent in riverine systems. The seeds are common in the gullets of migrating waterfowl (Aves: Anatidae), which probably serve as routine (endozoic) long-distance dispersal agents. A transient to long-term persistent seed bank develops, with seed densities of up to 642 m^{-2} documented. An average seedling emergence of up to 24.5 seedlings/400 cm^2 has been observed in some sites. Seeds that were stored for 2 months, then cold stratified in complete darkness at 3°C for 18 weeks, germinated well in the light under a 3.5°C/18°C temperature regime. Seeds stratified at 5°C for only 8 weeks germinated well under a 25°C/10°C, 8/16 h temperature/light regime. Seeds stratified for only 1 week germinated well under a 24°C/12°C temperature regime. Overall, temperature fluctuation is more important to the induction of germination than the duration of chilling. Light is also not a critical factor, with the seeds germinating equally well under dark conditions. However, reduced germination has been reported for seeds buried by 1 cm of soil. Seeds under the surface litter layer can retain over 62% germinability. Intentionally seeded stands of *P. lapathifolium* have been found to outcompete the highly invasive purple loosestrife (*Lythrum salicaria*). **Reported associates:** *Abies concolor, Acalypha phleoides, A. rhomboidea, Acer glabrum, A. negundo, A. saccharum, Achillea millefolium, Acmispon americanus, A. argophyllus, A. glabrus, A. grandiflorus, A. strigosus, Aconitum delphiniifolium, Aconogonon alpinum, Agalinis purpurea, Agoseris, Agrostis exarata, A. gigantea, A. stolonifera, Alnus oblongifolia, A. rhombifolia, A. rugosa, A. tenuifolia, Alopecurus aequalis, Amaranthus albus, A. blitoides, A. hybridus, A. powellii, A. retroflexus, A. tuberculatus, Ambrosia artemisiifolia, A. grayii, A. monogyra, A. psilostachya, A. trifida, Ammannia coccinea, A. robusta, Amorpha fruticosa, Anagallis arvensis, Andropogon gerardii, Anemone canadensis, Anthemis, Apium graveolens, Apocynum cannabinum, Arabidopsis lyrata, Artemisia annua, A. biennis,*

A. californica, A. douglasiana, A. ludoviciana, A. tilesii, A. tridentata, Arundo donax, Asclepias incarnata, A. lemmonii, A. speciosa, Astragalus lentiginosus, Atriplex canescens, A. lentiformis, A. micrantha, A. polycarpa, Avena barbata, Baccharis pilularis, B. salicifolia, B. salicina, B. sarothroides, Bacopa rotundifolia, Bassia hyssopifolia, Bebbia juncea, Beckmannia, Bergia texana, Berula erecta, Betula neoalaskana, Bidens cernuus, B. frondosus, B. connatus, B. tripartitus, B. vulgatus, B. pilosus, Boehmeria cylindrica, Boerhavia coccinea, Bolboschoenus fluviatilis, B. maritimus, Boltonia decurrens, Boschniakia rossica, Bouteloua aristidoides, B. barbata, B. curtipendula, B. hirsuta, Brassica nigra, Brickellia californica, Bromus arizonicus, B. ciliatus, B. diandrus, B. inermis, B. japonicus, B. secalinus, Calamagrostis canadensis, Calibrachoa parviflora, Callitriche hermaphroditica, Calocedrus decurrens, Calochortus splendens, Carduus nutans, C. pycnocephalus, Carex athrostachya, C. crawfordii, C. frankii, C. pellita, C. praegracilis, C. scoparia, C. simulata, C. stricta, C. vulpinoidea, Carpobrotus edulis, Castilleja applegatei, Ceanothus cordulatus, C. leucodermis, Centaurea diffusa, Cephalanthus occidentalis, Chamerion angustifolium, Chelone glabra, Chenopodium album, C. berlandieri, C. chenopodioides, C. fremontii, C. leptophyllum, C. pratericola, C. rubrum, Chilopsis linearis, Chloris, Chrysopsis villosa, Cichorium intybus, Cirsium vulgare, C. wheeleri, Commelina communis, C. erecta, Conium maculatum, Conoclinium coelestinum, Conyza canadensis, Cornus amomum, C. sericea, Croton californicus, C. texensis, Crypsis schoenoides, Cucurbita foetidissima, Cuscuta pentagona, Cynodon dactylon, Cyperus acuminatus, C. aristatus, C. eragrostis, C. esculentus, C. fendlerianus, C. involucratus, C. niger, C. odoratus, C. squarrosus, C. strigosus, Dactylis glomerata, Dasiphora floribunda, Datisca glomerata, Datura wrightii, Daucus, Deinandra fasciculata, D. floribunda, Descurainia sophioides, Desmanthus illinoensis, Dichanthelium oligosanthes, Dicoria canescens, Digitaria sanguinalis, Distichlis spicata, Drymocallis glandulosa, Dudleya variegata, Dysphania ambrosioides, D. botrys, D. pumilio, Echinochloa crus-galli, E. muricata, Echinodorus berteroi, E. rostrata, Eclipta prostrata, Elatine chilensis, Eleocharis acicularis, E. compressa, E. microcarpa, E. obtusa, E. palustris, E. parishii, E. wolfii, Elodea canadensis, Elymus elymoides, E. macrourus, E. virginicus, Encelia farinosa, Epilobium brachycarpum, E. canum, E. ciliatum, E. coloratum, E. palustre, Equisetum laevigatum, Eragrostis cilianensis, E. pectinacea, Ericameria nauseosa, Eriochloa acuminata, Eriodictyon trichocalyx, Eriogonum arborescens, E. fasciculatum, Erodium, Eryngium yuccifolium, Erysimum cheiranthoides, Eucalyptus camaldulensis, Eupatorium perfoliatum, E. serotinum, Euphorbia dentata, E. nutans, Euthamia occidentalis, Evolvulus arizonicus, Fallopia scandens, Ferocactus viridescens, Fimbristylis autumnalis, Fraxinus nigra, F. pennsylvanica, F. velutina, Galactia wrightii, Galium boreale, G. trifidum, Geocaulon lividum, Glandularia bipinnatifida, Glinus lotoides, Glyceria grandis, G. leptostachya, G. septentrionalis, Gnaphalium palustre, Grayia spinosa, Hedysarum alpinum, Helenium autumnale, Helianthus annuus, Heliotropium curassavicum, Hesperidanthus linearifolius, Hesperoyucca whipplei, Heteranthera limosa, Heteromeles arbutifolia, Heterotheca subaxillaris, Hibiscus laevis, Hirschfeldia incana, Holcus lanatus, Hordeum, Horkelia rydbergii, Hosackia alamosana, H. oblongifolia, Humulus japonicus, Hypericum adpressum, H. canadense, H. mutilum, H. perforatum, Impatiens capensis, Ipomoea hederacea, Ipomopsis thurberi, Iris pseudacorus, Isocoma menziesii, Iva annua, Juncus acuminatus, J. alpinoarticulatus, J. articulatus, J. balticus, J. bufonius, J. canadensis, J. diffusissimus, J. marginatus, J. repens, Juglans major, Juniperus californica, J. occidentalis, J. scopulorum, Justicia americana, Keckiella antirrhinoides, Kochia scoparia, Lactuca serriola, Leersia oryzoides, Lemna minor, L. valdiviana, Lepidium latifolium, Lepidospartum squamatum, Leptochloa fascicularis, L. fusca, Leucanthemum vulgare, Leymus condensatus, Limosella aquatica, Lindernia dubia, Lipocarpha micrantha, Lolium perenne, Lotus, Ludwigia hexapetala, L. peploides, Lycopus americanus, L. asper, Lysimachia nummularia, Lycium cooperi, Lythrum californicum, L. salicaria, Malosma laurina, Malva neglecta, Marrubium vulgare, Marsilea vestita, Matricaria discoidea, Melampodium sericeum, Melilotus albus, M. indicus, M. officinalis, Mentha arvensis, Mentzelia micrantha, Mertensia paniculata, Mimulus cardinalis, M. guttatus, M. pilosus, M. ringens, Mirabilis laevis, Mollugo verticillata, Monarda citriodora, Muhlenbergia asperifolia, M. rigens, Myriophyllum, Nasturtium officinale, Nepeta cataria, Nicotiana attenuata, N. quadrivalvis, Nymphaea tuberosa, Nuphar advena, N. polysepala, Oenothera curtiflora, O. elata, O. filiformis, O. suffrutescens, O. villosa, Oxalis stricta, Panicum capillare, P. dichotomiflorum, P. hirticaule, P. miliaceum, P. virgatum, Parthenocissus, Pascopyrum smithii, Paspalum dilatatum, P. distichum, P. laeve, Persicaria amphibia, P. careyi, P. coccinea, P. hydropiperoides, P. lapathifolia, P. maculosa, P. pensylvanica, Phacelia, Phalaris arundinacea, Phleum pratense, Phoenix canariensis, Phragmites australis, Physalis angulata, P. longifolia, Physostegia angustifolia, Picea glauca, P. mariana, Pinus contorta, P. jeffreyi, P. ponderosa, Plantago lanceolata, P. major, Platanus racemosa, P. wrightii, Pluchea odorata, P. sericea, Poa bulbosa, P. pratensis, Polygonum aviculare, P. ramosissimum, Polypogon monspeliensis, P. viridis, Populus angustifolia, P. deltoides, P. fremontii, Portulaca oleracea, Potamogeton natans, Potentilla anserina, P. recta, P. norvegica, Proboscidea parviflora, Prosopis pubescens, Prunus fasciculata, P. ilicifolia, P. virginiana, Pseudognaphalium canescens, P. luteoalbum, P. microcephalum, Pulicaria paludosa, Pycnanthemum virginianum, Quercus agrifolia, Q. berberidifolia, Q. corneliusmulleri, Q. emoryi, Ranunculus cymbalaria, R. flabellaris, R. macranthus, R. sceleratus, Raphanus, Rhamnus californica, R. crocea, Rhaponticum repens, Rhinanthus minor, Rhus integrifolia, R. ovata, Rhynchospora capitellata, Ribes aureum, R. roezlii, R. triste, Ricinus communis, Rorippa islandica, R. palustris, R. sinuata, Rosa acicularis, R. woodsii, Rotala ramosior, Rubus setosus, R. ulmifolius, Rudbeckia laciniata, R. subtomentosa, Rumex crispus, R. salicifolius, R.

triangulivalvis, R. violascens, Sagittaria latifolia, Salicornia depressa, Salix alaxensis, S. amygdaloides, S. arbusculoides, S. bonplandiana, S. discolor, S. exigua, S. exigua, S. gooddingii, S. interior, S. laevigata, S. lasiandra, S. lasiolepis, S. lucida, S. pseudomonticola, S. serissima, S. taxifolia, Salsola, Salvia apiana, S. leucophylla, S. mellifera, Sambucus nigra, Samolus vagans, Schinus molle, S. terebinthifolius, Schoenoplectus acutus, S. americanus, S. californicus, S. hallii, S. heterochaetus, S. pungens, S. tabernaemontani, Scirpus atrovirens, S. pendulus, Scutellaria lateriflora, Senegalia greggii, Sesbania, Sesuvium verrucosum, Setaria faberi, S. glauca, S. parviflora, Shepherdia canadensis, Silphium integrifolium, S. perfoliatum, Silybum, Sisymbrium irio, S. officinale, Sium suave, Solanum douglasii, S. elaeagnifolium, S. rostratum, Solidago canadensis, S. gigantea, Sonchus, Sorghum halepense, Spartina pectinata, Spergularia bocconi, Spermacoce glabra, Sphagnum fuscum, Spiraea douglasii, Spirodela polyrhiza, Sporobolus wrightii, Stuckenia pectinata, Swertia, Symphoricarpos, Symphyotrichum ascendens, S. lanceolatum, S. novae-angliae, S. ontarionis, S. pilosum, S. subulatum, Tamarix chinensis, T. ramosissima, Tephroseris palustris, Teucrium canadense, Thlaspi arvense, Torilis arvensis, Torreyochloa pallida, Toxicodendron diversilobum, T. radicans, Tragia nepetifolia, Tragopogon dubius, Trifolium ciliolatum, T. fragiferum, T. repens, Typha domingensis, T. angustifolia, Ulmus parvifolia, U. pumila, Urtica dioica, Vaccinium uliginosum, V. vitis-idaea, Verbascum thapsus, Verbena hastata, V. urticifolia, Verbesina alternifolia, V. rothrockii, Veronica anagallis-aquatica, Viburnum edule, Viola primulifolia, Vitis arizonica, V. girdiana, V. riparia, Washingtonia robusta, Xanthium strumarium, Xanthocephalum gymnospermoides, Yucca brevifolia, Zannichellia palustris.

***Persicaria meisneriana* (Cham. & Schltdl.) M. Gómez** is a weak annual species that occurs in shallow waters (0.25–1.5 m depth) of ditches, floodplains, marshes, roadsides, savannas, shores, swamps, and along the margins of lakes and rivers at elevations of up to at least 26 m (North America; to 1220 m elsewhere). These are prickly, scrambling plants, which are similar in habit to *P. arifolia* and *P. sagittata* by their reliance upon other vegetation for physical support. They are capable of forming dense mats. The substrates are described as muck (organic). Flowering has been observed in June. Otherwise, the reproductive ecology of these plants remains poorly documented. The Florida habitats of this species are threatened by the invasive spread of *Lygodium japonicum*. Because of the rarity of this species in North America, there exists little additional life-history information for the region. **Reported associates:** *Acer rubrum, Bidens laevis, Boehmeria cylindrica, Brasenia schreberi, Canna, Cephalanthus occidentalis, Conoclinium coelestinum, Decodon verticillatus, Eichhornia crassipes, Eleocharis, Hydrilla verticillata, Hydrocotyle ranunculoides, Juncus repens, Limnobium spongia, Liquidambar styraciflua, Ludwigia alternifolia, L. decurrens, L. leptocarpa, Luziola fluitans, Lycopus, Lygodium japonicum, Mayaca fluviatilis, Micranthemum umbrosum, Mikania scandens, Murdannia*

nudiflora, Nymphaea odorata, Nyssa biflora, Persicaria amphibia, P. hydropiper, P. hydropiperoides, P. lapathifolia, P. setacea, Pontederia cordata, Ricciocarpus, Rumex crispus, Sacciolepis striata, Sagittaria cuneata, S. latifolia, Salix nigra, Salvinia minima, Saururus cernuus, Schoenoplectus pungens, S. tabernaemontani, Scirpus atrovirens, Spartina pectinata, Spirodela polyrhiza, Taxodium distichum, Typha angustifolia, T. domingensis, T. latifolia.

***Persicaria minor* (Huds.) Opiz** is a nonindigenous annual species that inhabits nontidal or tidal bogs, deltas, landfills, meadows, mudflats, roadsides, shores, and the margins (often receding) of lakes, ponds, and streams at elevations of up to 100 m. North American records indicate open exposures and substrates consisting of gravel, mud, peat, rock, sand, or stones. The plants flower within 6–12 weeks of age. Flowering (North America) occurs from July to October. The flowers are self-compatible, not dichogamous, have high (91%–99%) pollen viability, and are primarily self-pollinating (pseudocleistogamous). One hundred percent germination is reported for seed that have been scarified mechanically and then incubated under a 25°C/10°C, 8/16 h temperature/light regime. Population genetic studies in Iran using ISSR markers have shown that populations contain high levels of genetic variation and are structured geographically. **Reported associates (North America):** *Eleocharis acicularis, E. palustris, Equisetum fluviatile, Hippuris, Iris pseudacorus, Juncus effusus, Limosella aquatica, Mentha, Myosotis scorpioides, Myriophyllum ussuriense, Navarretia squarrosa, Persicaria hydropiper, P. maculosa, Plantago major, Rorripa curvipes, Sagittaria latifolia, Salix, Schoenoplectus lacustris.*

***Persicaria punctata* (Elliott) Small** is an annual or perennial of shallow (to 0.6 m), fresh or brackish, tidal or nontidal waters associated with backwaters, bogs, bottomlands, channels, depressions, ditches, dunes, fens, floodplains, gravel bars, gullies, hammocks, levees, marshes, meadows, pools, prairies, roadsides, sand bars, savannas, seeps, shores (receding), slopes (to 15%), sloughs, swales, swamps, thickets, woodlands, and the margins of gravel pits, lagoons, lakes, ponds, reservoirs, rivers, and streams at elevations of up to 2276 m. It occurs in exposures of full sun to shade. The plants are broadly adapted to different substrates (pH: 6.0–8.6), which include Capay-Clear Lake soil, clay, gravel, gravelly alluvium, Harleston fine sandy loam, loam, loamy clay mud, muck, mucky sand, mud, Pescadero clay, Resota sand, rocks, rocky alluvium, sand, sandy alluvium, sandy gravel, sandy loam (Gaviota), sandy muck, sandy rock, sandy silt, shale, shale talus, silt, silty alluvium, and Taft silty loam. Flowering and fruiting extend from June to November. The flowers are pollinated by insects (Insecta). Observed floral visitors include several flies (Insecta: Diptera: Syrphidae; Tachinidae), which presumably serve as pollinators. The extent of self-pollination has not been determined. The seeds can remain afloat for a week or more, and even longer if their perianth remains attached. They are dispersed passively by the water and occur as prominent elements of river drift. Longer distance dispersal likely occurs by endozoic transport facilitated by waterfowl (Aves: Anatidae). A seed bank develops, with up to 145 seedlings/m² emerging on average in some seedbank studies. The seeds are physiologically dormant but will germinate well after 210 days of cold stratification followed by incubation under a 25°C/15°C temperature regime. They require at least 4 months of afterripening at 5°C or they will remain dormant until the second growing season subsequent to dispersal. Once afterripening requirements have been met, fair germination (39%) will occur at 5°C (under hypoxia) but will increase substantially (e.g., 79%) if incubated under a 25°C/15°C temperature regime. Sediment particle size appears to have no effect on the germination rate. The seedlings will emerge under flooded or nonflooded conditions but achieve higher densities under nonflooded conditions. They tend to occur primarily along banks, where their survival rates are high. The plants persist well in areas characterized by torrential flooding. The perennial forms reproduce vegetatively by means of rhizomes. **Reported associates:** *Abies balsamea, Acalypha rhomboidea, Acer rubrum, A. saccharinum, Acmispon americanus, Acorus americanus, Adenostoma fasciculatum, Adolphia californica, Aesculus glabra, Agalinis auriculata, Ageratina altissima, Agrimonia parviflora, Agrostis gigantea, Alisma subcordatum, Allophyllum divaricatum, Alnus rugosa, A. oblongifolia, A. viridis, Alopecurus aequalis, Alternanthera philoxeroides, Amaranthus tuberculatus, Ambrosia psilostachya, Ampelopsis arborea, Anagallis arvensis, Andropogon gerardii, Anemopsis californica, Apium graveolens, Arctostaphylos, Aristida purpurascens, Artemisia annua, A. biennis, A. californica, A. tridentata, Arundo donax, Asclepias incarnata, Asemeia grandiflora, Asimina triloba, Astragalus lentiginosus, Athyrium filixfemina, Atriplex patula, Azolla filiculoides, Baccharis pilularis, B. salicifolia, B. sarothroides, Bacopa caroliniana, B. monnieri, Berula erecta, Betula, Bidens cernuus, B. connatus, B. frondosus, B. laevis, B. polylepis, B. trichospermus, B. tripartitus, Boehmeria cylindrica, Bolboschoenus fluviatilis, B. maritimus, Bromus ciliatus, Calamagrostis canadensis, Calla palustris, Callitriche heterophylla, Calystegia malacophylla, C. sepium, Campanula americana, C. aparinoides, Carduus pycnocephalus, Carex aquatilis, C. atlantica, C. canescens, C. cristatella, C. exilis, C. hystericina, C. lasiocarpa, C. lurida, C. nudata, C. obnuta, C. praegracilis, C. stricta, C. tribuloides, C. trisperma, C. utriculata, C. vesicaria, C. vulpinoidea, Celtis, Centella asiatica, Centaurea solstitialis, Centrosema virginianum, Cephalanthus occidentalis, Chamaecyparis thyoides, Chelone glabra, Chenopodium desiccatum, C. pratericola, Cinna arundinacea, Cirsium discolor, C. vulgare, Coleataenia longifolia, Comarum palustre, Conioselinum chinense, Conyza canadensis, Coptis trifolia, Coreopsis tinctoria, C. tripteris, Cornus amomum, C. racemosa, C. sericea, Cotula coronopifolia, Croton setiger, Cryptotaenia canadensis, Cuscuta polygonorum, Cynodon dactylon, Cyperus acuminatus, C. diandrus, C. eragrostis, C. esculentus, C. flavescens, C. niger, C. strigosus, Dasiphora floribunda, Datisca glomerata, Desmodium paniculatum, Dichanthelium acuminatum, D. boscii, D. laxiflorum, Digitaria sanguinalis, Diodea virgininia, Distichlis spicata,* *Doellingeria umbellata, Dysphania pumilio, Echinochloa crus-galli, E. muricata, E. walteri, Eclipta prostrata, Elaeagnus angustifolia, Eleocharis acicularis, E. obtusa, E. ovata, E. palustris, E. tenuis, Elodea canadensis, Elymus trachycaulus, E. villosus, Epilobium ciliatum, E. coloratum, E. palustre, E. torreyi, Equisetum arvense, E. fluviatile, E. hyemale, Eragrostis frankii, E. hypnoides, E. pectinacea, E. spectabilis, Ericameria nauseosa, Erechtites hieracifolius, Eriodictyon californicum, Eriogonum fasciculatum, Eriophorum chamissonis, Eupatorium altissimum, E. perfoliatum, E. serotinum, Euphorbia nutans, Euthamia occidentalis, Eutrochium maculatum, E. purpureum, Fallopia scandens, Festuca, Fimbristylis, Fragaria virginiana, Frankenia grandifolia, Fraxinus americana, F. pennsylvanica, F. velutina, Galium labridoricum, G. tinctorium, G. triflorum, Gayophytum diffusum, Geum aleppicum, G. canadense, Glandularia bipinnatifida, Gleditsia triacanthos, Glyceria leptostachya, G. septentrionalis, Grindelia squarrosa, Helenium autumnale, Helianthus annuus, H. grosseserratus, Heliotropium curassavicum, Heterocodon rariflorus, Hibiscus laevis, Hippuris vulgaris, Hordeum jubatum, Humulus japonicus, Hydrocotyle umbellata, H. verticillata, Hypericum gentianoides, H. hypericoides, H. mutilum, H. sphaerocarpum, Impatiens capensis, Iris versicolor, I. virginica, Juglans major, Juncus articulatus, J. bufonius, J. effusus, J. mertensianus, J. nodosus, J. saximontanus, Juniperus occidentalis, Kalmia microphylla, Lactuca, Laportea canadensis, Larix laricina, Leersia hexandra, L. oryzoides, L. virginica, Lemna minor, Lepidium latifolium, Lepidospartum squamatum, Leptochloa fusca, Leucospora multifida, Lindernia dubia, Liparis loeselii, Lipocarpha micrantha, Liquidambar styraciflua, Lobelia cardinalis, L. siphilitica, Lotus corniculatus, Ludwigia leptocarpa, L. palustris, L. peploides, Luziola fluitans, Lycopus americanus, L. rubellus, L. uniflorus, L. virginicus, Lysichiton americanus, Lysimachia ciliata, L. nummularia, L. terrestris, L. thyrsiflora, Lythrum salicaria, Malosma laurina, Medicago lupulina, Melilotus albus, Menispermum canadense, Mentha arvensis, M. piperita, Menyanthes trifoliata, Mikania scandens, Mimulus alatus, M. floribundus, M. guttatus, M. moschatus, M. ringens, Muhlenbergia asperifolia, M. bushii, M. frondosa, M. mexicana, M. repens, M. schreberi, Myosotis laxa, M. scorpioides, Myrica californica, M. gale, Myriophyllum, Nasturtium officinale, Oenanthe sarmentosa, Oenothera elata, Onoclea sensibilis, Oryza sativa, Oxalis stricta, Panicum capillare, P. hemitomon, P. virgatum, Parthenocissus, Paspalum distichum, Pedicularis lanceolata, Peltandra virginica, Penthorum sedoides, Persicaria amphibia, P. arifolia, P. coccinea, P. hydropiper, P. hydropiperoides, P. lapathifolia, P. longiseta, P. maculosa, P. pensylvanica, P. sagittata, Phalaris arundinacea, Phragmites australis, Phryma leptostachya, Phyla lanceolata, P. nodiflora, Picea mariana, P. rubens, Pilea pumila, Pinus elliottii, P. glabra, P. jeffreyi, P. ponderosa, P. strobus, Plantago major, P. rugelii, Platanus occidentalis, P. racemosa, Poa palustris, Pogonia ophioglossoides, Polygonum aviculare, Polypogon monspeliensis, P. viridis, Populus deltoides,*

P. fremontii, P. trichocarpa, Potamogeton natans, Potentilla norvegica, Prosperpinaca palustris, Prunella vulgaris, Prunus virginiana, Pseudognaphalium luteoalbum, Pteridium aquilinum, Quercus agrifolia, Q. berberidifolia, Q. engelmannii, Q. macrocarpa, Q. palustris, Q. velutina, Ranunculus aquatilis, R. flabellaris, R. longirostris, Rhamnus frangula, Rhus copallinum, R. integrifolia, Rhynchospora glomerata, R. inundata, Ribes hirtellum, Rorripa curvipes, R. palustris, Rosa californica, R. multiflora, R. nutkana, R. palustris, R. woodsii, Rubus bifrons, R. ursinus, Rudbeckia laciniata, R. subtomentosa, Rumex acetosella, R. britannica, R. crispus, R. triangulivalvis, Sabal minor, Sacciolepis striata, Sagittaria brevirostra, S. graminea, S. latifolia, Salix amygdaloides, S. discolor, S. exigua, S. gooddingii, S. interior, S. laevigata, S. lasiandra, S. lasiolepis, S. nigra, Salvia mellifera, Sambucus nigra, Samolus parviflorus, Sanicula canadensis, S. odorata, Sapium sebiferum, Saponaria officinalis, Sarracenia purpurea, Schoenoplectus acutus, S. americanus, S. californicus, S. lacustris, S. pungens, S. tabernaemontani, Scirpus ancistrochaetus, S. atrovirens, S. cyperinus, S. microcarpus, Scleria, Scutellaria galericulata, S. lateriflora, Sesbania herbacea, Setaria faberi, S. glauca, S. magna, Silphium integrifolium, S. terebinthinaceum, Silybum marianum, Sium suave, Smilax bona-nox, S. tamnoides, Solanum ptychanthum, Solidago canadensis, S. gigantea, S. patula, Sorghastrum nutans, Sparganium americanum, S. angustifolium, S. eurycarpum, Spartina pectinata, Spergularia, Sphagnum, Spiraea alba, S. douglasii, Spiranthes cernua, Stachys pilosa, Staphylea trifolia, Stellaria longifolia, Stuckenia pectinata, Symphoricarpos albus, Symphyotrichum ericoides, S. lanceolatum, S. lateriflorum, S. novae-angliae, S. ontarionis, S. pilosum, S. puniceum, S. subulatum, Taraxacum officinale, Taxodium, Thalictrum dasycarpum, Thelypteris kunthii, T. palustris, Thuja occidentalis, Tilia americana, Toxicodendron diversilobum, Triadenum fraseri, T. virginicum, Tridens flavus, Trifolium fragiferum, Tsuga canadensis, Typha angustifolia, T. domingensis, T. latifolia, Ulmus americana, Urtica dioica, Vaccinium oxycoccus, Verbascum thapsus, Verbena hastata, V. urticifolia, Verbesina alternifolia, V. helianthoides, Veronica anagallis-aquatica, Viburnum lentago, V. prunifolium, Vitis girdiana, Xanthium strumarium, Xylococcus bicolor, Xyris jupicai, X. laxifolia, Yucca schidigera, Zannichellia palustris, Zizania aquatica.

***Persicaria robustior* (Small) E.P. Bicknell** is a coarse perennial species that grows in shallow (to 1.5 m), tidal or nontidal brooks, channels, depressions, fens, floodplains, marshes, pools, riverbeds (dry), sluiceways, swales, swamps, thickets, and along the margins or shores of lakes, ponds, rivers, and streams at elevations of up to 1500 m. Exposures include partial shade. The substrates are described as acidic, nutrient poor, and include cobble, muck, mud, gravel, and peat. Flowering occurs from July to October. Vegetative reproduction occurs by strong, forking rhizomes or by stolons. The stems root along their nodes and can extend along the water surface for up to 2 m, often leading to dense, clonal patches. Life-history information particular to this species is scarce

because of its historical treatment as a varietal synonym of *P. punctata*. The ecology of this species deserves renewed study.
Reported associates: *Apios americana, Bartonia paniculata, Eriophorum tenellum, Eutrochium, Juncus subcaudatus, Oclemena nemoralis, Persicaria hydropiperoides, P. punctata, Rosa palustris, Sisyrinchium angustifolium, Smilax rotundifolia, Thelypteris simulata, Woodwardia virginica.*

***Persicaria sagittata* (L.) H. Gross** is a weak annual species that occurs in shallow (to 0.3 m), freshwater or brackish, tidal or nontidal habitats including bogs, bottomlands, ditches, depressions, estuaries, draws, fens, flats, floodplains, marshes, meadows, pond bottoms, prairies, roadsides, seeps, slopes, sloughs, springs, swamps, thickets, woodlands, and the margins of clay pits, lakes, ponds, rivers, and streams at elevations of up to 3474 m. Exposures range from full sunlight to shade. The substrates span a broad range of acidity (pH: 5.0–8.4) and include clay, gravel, muck, mud, peat, sand, sandy peat, and sandy rock. The twining habit is similar to that of *P. arifolia*, with which it sometimes occurs. Flowering and fruiting extend from June to November. The reproductive ecology has not been studied in any detail. The achenes can become dominant elements of some seedbanks. The plants are not regarded as particularly vagile, and local seed dispersal presumably occurs abiotically via water or wind. The seeds are dispersed endozoically over longer distances by white-tailed deer (Mammalia: Cervidae: *Odocoileus virginianus*). The plants can grow into dense stands but have been observed to attain an average biomass of only about 3.4 g/m^2 year.
Reported associates: *Acalypha rhomboidea, Acer rubrum, A. saccharinum, Achillea millefolium, Acorus calamus, Actaea pachypoda, Agalinis maritima, Agrimonia parviflora, Agrostis alba, A. gigantea, Alnus rugosa, Alternanthera philoxeroides, Amaranthus australis, Ambrosia artemisiifolia, Andropogon glomeratus, Apios americana, Apocynum cannabinum, Asclepias incarnata, Aulacomnium palustre, Betula papyrifera, Bidens coronatus, B. laevis, B. vulgatus, Boehmeria cylindrica, Calamagrostis canadensis, Calla palustris, Campanula aparinoides, Carex crinita, C. debilis, C. echinata, C. lacustris, C. lurida, C. obnupta, C. stricta, C. trisperma, C. vulpinoidea, Carya cordiformis, Celastrus orbiculatus, Centaurea, Cephalanthus occidentalis, Cicuta bulbifera, C. maculata, Clintonia, Conoclinium coelestinum, Cuscuta gronovii, Cyperus polystachyos, C. strigosus, Daucus carota, Decodon verticillatus, Dichanthelium clandestinum, D. scoparium, Dulichium arundinaceum, Eleocharis rostellata, Epilobium coloratum, Erechtites hieraciifolius, Eupatorium perfoliatum, Eutrochium maculatum, Fallopia scandens, Fraxinus, Galium tinctorium, Glyceria canadensis, Helianthus grosseserratus, Hieracium scabrum, Hydrocotyle, Hypericum mutilum, Ilex verticillata, Impatiens capensis, Ipomoea sagittata, Iris versicolor, Juncus articulatus, J. effusus, J. marginatus, Kosteletzkya pentacarpos, Larix laricina, Leersia oryzoides, L. virginica, Lemna minor, Leptochloa fusca, Leptodictyum riparium, Lespedeza hirta, Leucanthemum vulgare, Lobelia siphilitica, Lonicera, Ludwigia alternifolia, L. leptocarpa, Lycopus americanus, L. uniflorus, Lythrum lineare, L. salicaria,*

Magnolia virginiana, Malus, Mimulus alatus, Muhlenbergia mexicana, Myrica cerifera, Onoclea sensibilis, Osmunda regalis, Panicum hemitomon, P. virgatum, Parthenium integrifolium, Parthenocissus quinquefolia, Paspalum vaginatum, Peltandra virginica, Penthorum sedoides, Persicaria amphibia, P. arifolia, P. hydropiper, P. longiseta, P. punctata, Phalaris arundinacea, Phragmites australis, Phytolacca americana, Pilea fontana, P. pumila, Platanthera ciliaris, Poa palustris, Pontederia cordata, Potentilla anserina, Prunus serotina, Pueraria montana, Quercus alba, Q. velutina, Rhamnus cathartica, Rhus copallinum, R. glabra, Rhynchospora, Robinia, Rosa multiflora, R. palustris, R. rugosa, Rubus, Rumex obtusifolius, Sacciolepis striata, Sagittaria lancifolia, S. latifolia, Salix cordata, S. discolor, S. eriocephala, S. exigua, S. nigra, S. petiolaris, Sambucus nigra, Saururus cernuus, Schoenoplectus, Scirpus cyperinus, Sium suave, Smilax, Solanum dulcamara, Solidago gigantea, S. sempervirens, Sparganium americanum, S. eurycarpum, Spartina pectinata, Sphagnum fimbriatum, Spiraea douglasii, Spiranthes cernua, Succisa pratensis, Symphyotrichum firmum, S. laeve, S. puniceum, Thelypteris palustris, Thuja occidentalis, Toxicodendron radicans, T. vernix, Triadenum fraseri, T. virginicum, Trifolium pratense, Typha latifolia, Verbena hastata, Verbesina alternifolia, Vaccinium angustifolium, Vigna luteola, Vitis, Xyris laxifolia, Zizania aquatica.

***Persicaria setacea* (Baldwin) Small** is a coarse perennial species that inhabits brackish or freshwater, tidal or nontidal sites in bottomlands, canals, depressions, ditches, draws, floodplains, hammocks, meadows, roadsides, sloughs, swamps, thickets, woodlands (alluvial), and the margins or shores of lakes, ponds, and streams, mainly along the eastern coastal plain and piedmont, at elevations of up to 300 m. The plants occur frequently where standing water persists, and the exposures are partially shaded. The substrates are circumneutral (pH: 5.9–8.2) and are usually described as sand. Flowering and fruiting occur from June to October. Otherwise, the reproductive ecology of this species is poorly documented. Constructed tidal wetlands have been found to contain seed banks averaging up to 220 seeds/m². Although specific germination requirements have not been described, a period of cold stratification is probably required. Seeds collected from the field in the spring (March to June) germinate well under greenhouse conditions. Vegetative reproduction occurs by means of rhizomes. **Reported associates:** *Acer rubrum, Amaranthus cannabinus, Ampelopsis arborea, Asclepias incarnata, Betula nigra, Bidens frondosus, B. laevis, Boehmeria cylindrica, Bolboschoenus robustus, Boltonia asteroides, Carex caroliniana, C. crus-corvi, C. grayi, C. hystericina, C. lupuliformis, C. lupulina, C. muskingumensis, Carya illinoinensis, C. laciniosa, Cephalanthus occidentalis, Cornus foemina, Crinum americanum, Cyperus virens, Diodia virginiana, Echinochloa walteri, Echinodorus cordifolius, Eleocharis intermedia, E. obtusa, E. quadrangulata, E. tuberculosa, Glyceria septentrionalis, Hibiscus moscheutos, Hydrolea uniflora, Hymenocallis occidentalis, Ilex verticillata, Impatiens capensis, Ipomoea sagittata, Iris virginica, Juncus diffusissimus, J. megacephalus, J. roemerianus,* *Justicia ovata, Leersia oryzoides, L. lenticularis, Limonium carolinianum, Liquidambar styraciflua, Lobelia cardinalis, Ludwigia alternifolia, L. sphaerocarpa, Lythrum lineare, Magnolia virginiana, Mikania scandens, Murdannia keisak, Nyssa, Osmunda regalis, Oxypolis rigidior, Panicum, Peltandra virginica, Persicaria arifolia, P. hydropiperoides, P. meisneriana, P. pensylvanica, P. punctata, P. sagittata, Pilea pumila, Platanthera flava, P. peramoena, Platanus occidentalis, Pluchea odorata, Pontederia cordata, Populus heterophylla, Proserpinaca pectinata, Ptilimnium capillaceum, P. costatum, Quercus bicolor, Q. lyrata, Q. michauxii, Q. palustris, Q. virginiana, Rhexia mariana, Rhynchospora corniculata, R. macrostachya, Rumex verticillatus, Sabatia calycina, Sagittaria lancifolia, S. latifolia, Saururus cernuus, Schoenoplectus americanus, S. tabernaemontani, Scirpus cyperinus, Sium suave, Sparganium americanum, Spartina cynosuroides, Symphyotrichum lanceolatum, Taxodium ascendens, T. distichum, Typha angustifolia, T. latifolia, Viburnum nudum, Zizania aquatica.*

Use by wildlife: The prickly stems of *P. arifolia* and *P. sagittata* deter predators and provide good wildlife cover. The leaves of *P. amphibia* produce shade and shelter for fish (Vertebrata: Osteichthyes) and structural habitat for various aquatic invertebrates. Leaves of various *Persicaria* species are eaten by raccoons (Mammalia: Procyonidae: *Procyon lotor*), but high amounts of calcium oxalate make them less palatable to livestock (Mammalia: Bovidae). Because of the oxalic acid content, consumption of large quantities of *Persicaria* foliage by livestock can disrupt their calcium metabolism and lead to phototoxicity. *Persicaria glabra* is eaten by beavers (Mammalia: Castoridae: *Castor canadensis*) and *P. hirsuta* by marsh rabbits (Mammalia: Leporidae: *Sylvilagus palustris paludicola*) and grasshoppers (Insecta: Orthoptera: Acrididae: *Paroxya atlantica*; Romaleidae: *Romalea microptera*). *Persicaria* achenes can contain up to 9.4% protein and rank among the most important foods eaten by waterfowl (Aves: Anatidae). The achenes of many species are consumed by waterfowl in large numbers: *P. amphibia* (eaten by 18 species), *P. glabra* (eaten by 3 species), *P. hydropiper* (eaten by 13 species), *P. hydropiperoides* (eaten by 15 species), *P. lapathifolia* (eaten by 17 species), *P. punctata* (eaten by 11 species), and *P. sagittata* (eaten by 14 species). The largest waterfowl consumers of *Persicaria* fruits are blue-winged teal (*Anas discors*), green-winged teal (*Anas crecca*), lesser scaups (*Aythya affinis*), mallards (*Anas platyrhynchos*), pintail ducks (*Anas acuta*), and ring-necked ducks (*Aythya collaris*). *Persicaria glabra* is especially prolific and can produce more than 700 kg of seeds per hectare. In one instance, 36,000 seeds of *P. hydropiper* were found within the stomach of a single mallard duck. Achenes of *P. amphibia*, *P. punctata*, and other smartweeds are also eaten by deer (Mammalia: Cervidae: *Odocoileus*), mice (Mammalia: Cricetidae), muskrats (Mammalia: Cricetidae: *Ondatra zibethicus*), and raccoons (Mammalia: Procyonidae: *Procyon lotor*). *Persicaria* species are important food plants for larval butterflies or "caterpillars" (Insecta: Lepidoptera). Several species are damaged by the cattail caterpillar (Noctuidae: *Simyra insularis*). Numerous other caterpillars are hosted by *P. amphibia*

(Arctiidae: *Estigmene acrea*; Lycaenidae: *Epidemia helloides*, *Hyllolycaena hyllus*; Pyralidae: *Nymphula nymphaeata*), *P. glabra* (Crambidae: *Ostrinia penitalis*, *Pleuroptya silicalis*, *Synclita obliteralis*; Gelechiidae: *Aristotelia absconditella*, *Chionodes discoocellella*; Tortricidae: *Argyrotaenia ivana*), *P. hydropiper* (Crambidae: *Ostrinia penitalis*; Geometridae: *Haematopis grataria*, *Orthonama centrostrigaria*, *Orthonama obstipata*, *Prochoerodes transversata*; Noctuidae: *Acronicta oblinita*, *Agrochola helvola*, *Hypena scabra*, *Melanchra picta*, *Papaipema nebris*, *Simyra henrici*; Papilionidae: *Battus philenor*), *P. hydropiperoides* (Crambidae: *Synclita obliteralis*; Gelechiidae: *Chionodes discoocellella*; Lycaenidae: *Epidemia helloides*; Noctuidae: *Simyra insularis*), *P. lapathifolia* (Lycaenidae: *Epidemia helloides*, *Strymon melinus*; Noctuidae: *Acronicta rumicis*, *Simyra henrici*), *P. punctata* (Arctiidae: *Apantesis nais*, *Diacrisia virginica*, *Estigmene acrea*; Coleophoridae: *Coleophora shaleriella*; Crambidae: *Ostrinia penitalis*, *Parapoynx obscuralis*, *Synclita obliteralis*; Erebidae: *Spilosoma virginica*; Gelechiidae: *Aristotelia absconditella*, *Chionodes discoocellella*, *Monochroa absconditella*; Geometridae: *Anacamptodes defectaria*; Lycaenidae: *Epidemia helloides*; Noctuidae: *Acronicta oblinita*, *Argyrogramma verruca*, *Neoerastria apicosa*, *Palthis asopialis*, *Spodoptera dolichos*, *S. eridania*; Tortricidae: *Argyrotaenia ivana*, *Choristoneura parallela*, *Platynota rostrana*, *Sparganothis sulfureana*), and *P. setacea* (Noctuidae: *Argyrogramma verruca*). The foliage of *P. amphibia* is also known to harbor populations of algae (Chlorophyta: Oocystaceae: *Chlorella*; Cladophoraceae: *Cladophora*; Zygnemataceae: *Spirogyra*), diatoms (Ochrophyta: Fragilariaceae: *Fragilaria*), and more than 100 individual invertebrates (Insecta: Diptera, Ephemeroptera, Trichoptera; Mollusca; Annelida: Oligochaeta) per gram of dry weight. In addition, *P. amphibia* is a host of leaf beetles (Insecta: Coleoptera: Chrysomelidae: *Galerucella nymphaeae*), and its floating leaves are eaten (and destroyed) by larval moths (Insecta: Lepidoptera: Pyralidae: *Elophila nymphaeata*). *Persicaria glabra* hosts beetles (Coleoptera: Chrysomelidae: *Disonycha conjugata*, *D. pensylvanica*; Curculionidae: *Bagous lunatoides*, *Lixus merula*, *L. punctinasus*, *Rhinoncus longulus*, *Tyloderma rufescens*) and flies (Diptera: Stratiomyidae: *Nothonomyia calopus*). *Persicaria hydropiperoides* is a host of beetles (Coleoptera: Cantharidae: *Chauliognathus pennsylvanicus*; Chrysomelidae: *Colaspis crinicornis*, *Disonycha limbicollis*; Coccinellidae: *Coccinella novemnotata*; Curculionidae: *Listronotus caudatus*). The plants are also associated with several bees (Insecta: Hymenoptera: Apidae: *Ceratina dallatorreana*; Colletidae: *Hylaeus concinnus*, *H. mesillae*; Halictidae: *Halictus tripartitus*, *Lasioglossum incompletum*). *Persicaria hirsuta* is eaten by grasshoppers (Orthoptera: Acrididae: *Paroxya atlantica*, *Romalea microptera*). The substance polygodial, which occurs in the foliage of *P. hydropiper*, is highly repellent to aphids (Insecta: Hemiptera: Aphididae). The foliage of *P. hydropiperoides* is damaged by beetles and weevils (Coleoptera: Chrysomelidae: *Galerucella nymphaea*; Curculionidae: *Tyloderma rufescens*, *T. subpubescens*). The plants also harbor several parasitic wasps (Insecta: Hymenoptera: Chrysididae: *Chrysis inaequidens*, *C. nitidula*, *Hedychrum violaceum*, *H. wiltii*, *Holopyga*; Mutillidae: *Sphaeropthalma*; Tiphiidae: *Myzinum quinquecinctum*, *Tiphia letalis*). Parasitic wasps have also been recovered from *P. lapathifolia* (Chrysididae: *Hedychrum wiltii*, Tiphiidae: *Myzinum maculatum*, *M. quinquecinctum*; Scoliidae: *Scolia bicincta*). Other insects hosted by *P. punctatum* include aphids and bugs (Hemiptera: Aphalaridae: *Aphalara persicaria*; Aphididae: *Hyalomyzus tissoti*), beetles (Coleoptera: Curculionidae: *Lixus merula*, *L. punctinasus*, *Rhinoncus longulus*, *Tyloderma rufescens*, *T. subpubescens*), flies (Diptera: Chloropidae: *Elachiptera willistoni*), grasshoppers (Orthoptera: Acrididae: *Paroxya atlantica*, *P. clavuliger*, *Romalea microptera*), and sawflies (Hymenoptera: Tenthredinidae: *Ametastegia articulata*). The foliage of *P. setacea* is eaten by leaf beetles (Coleoptera: Chrysomelidae: *Disonycha pensylvanica*).

Economic importance: food: The Lakota and Sioux tribes ate the young shoots of *P. amphibia* for food and as a relish. The plants have also been used as a substitute for sarsaparilla. The Cherokee ate the boiled and fried young shoots of *P. hydropiper* and used its ground foliage as a pepper-like spice. The plants contain approximately 7.5% protein, 1.9% fat, and 8% carbohydrate. *Persicaria* seeds sometimes are ground into a flour; **medicinal:** There have been many, long-standing traditional medicinal uses of *Persicaria* species. The Cree, Meskwaki, and Woodlands tribes made a poultice of fresh *P. amphibia* roots to treat mouth blisters. They (as did the Potawatomi) administered the powdered roots as a general remedy for various ailments. The Okanagan-Colville people took an infusion of the dried roots for chest colds. The Meskwaki found leaf and stem infusions of *P. amphibia* useful in treating diarrhea; the Ojibwa used similar infusions to treat stomach disorders. Infusions of *P. arifolia* are diuretic and have been used to treat various urinary disorders. Plant infusions of *P. glabra* have been used in Hawaii to "purify" the blood. *Persicaria hydropiper* has been used as an antiseptic, blistering agent, diaphoretic, diuretic, menstrual aid, and stimulant. The strong, spicy taste of the foliage is responsible for the common name of "water pepper" and is because of the presence of a sesquiterpene dialdehyde known as polygodial. The Cherokee used *P. hydropiper* as a treatment for urinary disorders, swelling and inflammation (as a poultice), and root infusions as an antidiarrheal agent. The Iroquois applied a poultice of *P. hydropiper* to the forehead to relieve headaches and used leaf decoctions to treat fever, chills, and indigestion. The Malecite people administered a dried leaf infusion of *P. hydropiper* to treat dropsy. Infusions of *P. lapathifolia* were taken by the Apache, Keres, Potawatomi, and Zuni as an emetic and as a treatment for fevers and stomach disorders. *Persicaria minor* is a commonly used medicinal plant in southeast Asia (where it is sometimes known as "kesum"). Aqueous leaf extracts have been shown to exhibit high antioxidant activity. A decoction made from the raw or cooked leaves is used to treat digestive disorders, and the oil is applied as a remedy for dandruff. Aqueous and ethanolic extracts exhibit significant antibacterial properties against several bacterial

(Bacteria) strains (Enterobacteriaceae: *Escherichia coli* ATCC 11229; Enterococcaceae: *Enterococcus faecalis* ATCC 29212; Staphylococcaceae: *Staphylococcus aureus* ATCC 6538). *Persicaria punctata* was employed by the Chippewa to treat stomach pain (leaf and flower decoction), by the Houma to alleviate swelling and pain in the joints (decoction), and by the Iroquois for psychological improvement. The foliage is spicy in tase but not as strong as in *P. hydropiper*; **cultivation:** *Persicaria amphibia* is cultivated as an ornamental water garden plant; *P. hydropiper* is grown occasionally as an ornamental wetland garden species; **misc. products:** *Persicaria hydropiper*, *P. lapathifolia*, and *P. punctata* exhibit tolerance to industrially polluted sites and potentially could be used for restoration projects in such areas. The roots of *P. amphibia* contain more than 20% tannin and are suitable for use in tanning. Members of the Thompson tribe used the flowers of *P. amphibia* as a bait when fishing. *Persicaria hydropiper* leaves were applied to the thumbs of Cherokee children to discourage thumb sucking. The Cherokee also used the plant as a fish poison. A yellow dye can be obtained from *P. amphibia* and *P. hydropiper*. Dilute aqueous extracts of *P. hydropiper* can be used to expel earthworms (Annelida: Oligochaeta) from their underground burrows (as a source of fishing bait); **weeds:** Dense stands of *Persicaria hydropiperoides* and *P. lapathifolia* can impede water flow in irrigation canals. These species, along with *P. amphibia*, *P. hydropiper*, *P. punctata* and *P. setacea*, can also occur as weeds in wet, cultivated sites. *Persicaria hydropiperoides* is a weed of Portuguese rice fields; **nonindigenous species:** *Persicaria hydropiper* was introduced to North America from Europe during the mid-19th century. *Persicaria lapathifolia* is naturalized from Eurasia. *Persicaria minor* was introduced from Europe. *Persicaria punctata* is presumably nonindigenous to North America, but its origin remains uncertain.

Systematics: The taxonomy of *Persicaria* has been particularly tumultuous, with numerous nomenclatural changes prompted by results of phylogenetic analyses. Although the group had long been treated as a section of *Polygonum* (section *Persicaria*), phylogenetic studies (e.g., Figure 3.25) have demonstrated conclusively that not only are the two genera distinct but also that they are not even closely related. Most molecular data resolve *Persicaria* as a monophyletic sister group to a clade consisting of *Bistorta*, *Aconogonon*, and *Koenigia* (e.g., Figure 3.25) in a position quite distant from *Polygonum*. Taxonomic issues within *Persicaria* are no better off. There is extensive disagreement between phylogenetic trees that have been derived from maternally inherited (i.e., cpDNA) DNA sequences compared to biparentally inherited (i.e., nuclear) DNA sequences. The incongruence primarily is a result of ancient hybridization and subsequent allopolyploidy, processes which have been involved in the origin of many of the species. In general, the ploidy level of a species generally corresponds to the number of different parental gene copies that are detected using low-copy nuclear markers. Autopolyploidy does not seem to occur. As one example, the tetraploid *P. punctata* shares nuclear DNA markers with *P. hydropiper* but cpDNA markers with *P. hirsuta* and

P. setacea. The complexity of the problem is well illustrated by *P. lapathifolia*, which has been implicated in at least six different allopolyploid speciation events. Continuing hybridization among the taxa complicates matters even further. With these factors in mind, a general overview of interspecific relationships is presented as the maternally derived (cpDNA) phylogenetic tree for the group (Figure 3.27). The literature should be consulted for further clarification. The basic chromosome number in *Persicaria* is $x = 10$, 11, or 12. Diploids include *P. hirsuta* ($2n = 20$), *P. hydropiper* ($2n = 20$), *P. lapathifolia* ($2n = 22$), and *P. setacea* ($2n = 20$). Polyploids include *P. amphibia* ($2n = 66$, 132), *P. glabra* ($2n = 40$, 60), *P. hydropiperoides* ($2n = 40$), *P. minor* ($2n = 40$), *P. punctata* ($2n = 44$), and *P. sagittata* ($2n = 40$). Counts remain unreported for *P. arifolia*, *P. meisneriana*, and *P. robustior*. Molecular data indicate that introgressive hybridization has occurred involving the indigenous *P. careyi* and *P. pensylvanica* with the introduced *P. lapathifolia*. Hybridization also has been reported between *P. lapathifolia* and *P. maculosa*. The recognition of several species (e.g., *P. densiflora*, *P. opelousana*) remains unsettled taxonomically. Experimental transplants indicate that *P. coccinea* should be merged with *P. amphibia* (where it often is treated as a variety); however, it would be desirable to confirm these results using molecular markers given the author's experience that the species are quite distinct, even with respect to the degree of phenotypic plasticity exhibited.

Comments: Species that are widespread throughout North America include *P. amphibia*, *P. hydropiper*, *P. hydropiperoides*, *P. lapathifolia*, and *P. punctata*; species of eastern North America include *P. arifolia* (also disjunct in Washington, United States), *P. robustior*, and *P. sagittata*; and *P. minor* occurs in northeastern North America. Species in the southeastern United States include *P. glabra*, *P. hirsuta*, and *P. setacea* (also disjunct in Washington, United States); *P. meisneriana* occurs throughout the southern United States.

References: Abubakar et al., 2015; Araki & Washitani, 2000; Askew & Wilcut, 2002; Baldwin, 2013; Baldwin et al., 2010; Bender et al., 2013; Bhowmik & Datta, 2013; Blair, 1936; Brock et al., 1994; Brown, 1987; Christapher et al., 2015; Collins et al., 2013; D'hondt et al., 2011; Dolan & Sharitz, 1984; Edelman, 2003; Efroymson et al., 2008; Eleuterius, 1972; Ensign et al., 2014; Fernald, 1921; Flowers et al., 1994; Furness & Upadhyaya, 2002; Goodson et al., 2003; Graham et al., 2012; Green et al., 2002; Gurnell et al., 2007a; Hagy & Kaminski, 2012; Harms & Grodowitz, 2009; Heppner & Habeck, 1976; Hernandez, 2013; Hickman, 1993; Hoagland, 2002; Hopfensperger et al., 2009; Kim & Donoghue, 2008a,b; Kim et al., 2003; Konuma & Terauchi, 2001; Landers et al., 1977; Leck, 2003; Leck & Brock, 2000; Leck & Simpson, 1993, 1995; Leck et al., 2008; McAtee, 1925; Miyajima, 1995; Mitchell, 1970; Miwa & Meinke, 2015; Mohlenbrock, 1959b; Neff et al., 2009; Pappers et al., 2002; Parker et al., 2007b; Partridge, 2001; Pasternack et al., 2000; Powell, 2000; Robbins, 1940; Schmidt et al., 2004; Seybold et al., 2002; Sharpe & Baldwin, 2009; Sheidai et al., 2016; Singleton, 1951; Squitier & Capinera, 2002; Staniforth & Cavers, 1976; Sultan & Bazzaz, 1993a,b,c; Tooker & Hanks, 2000; Tooker

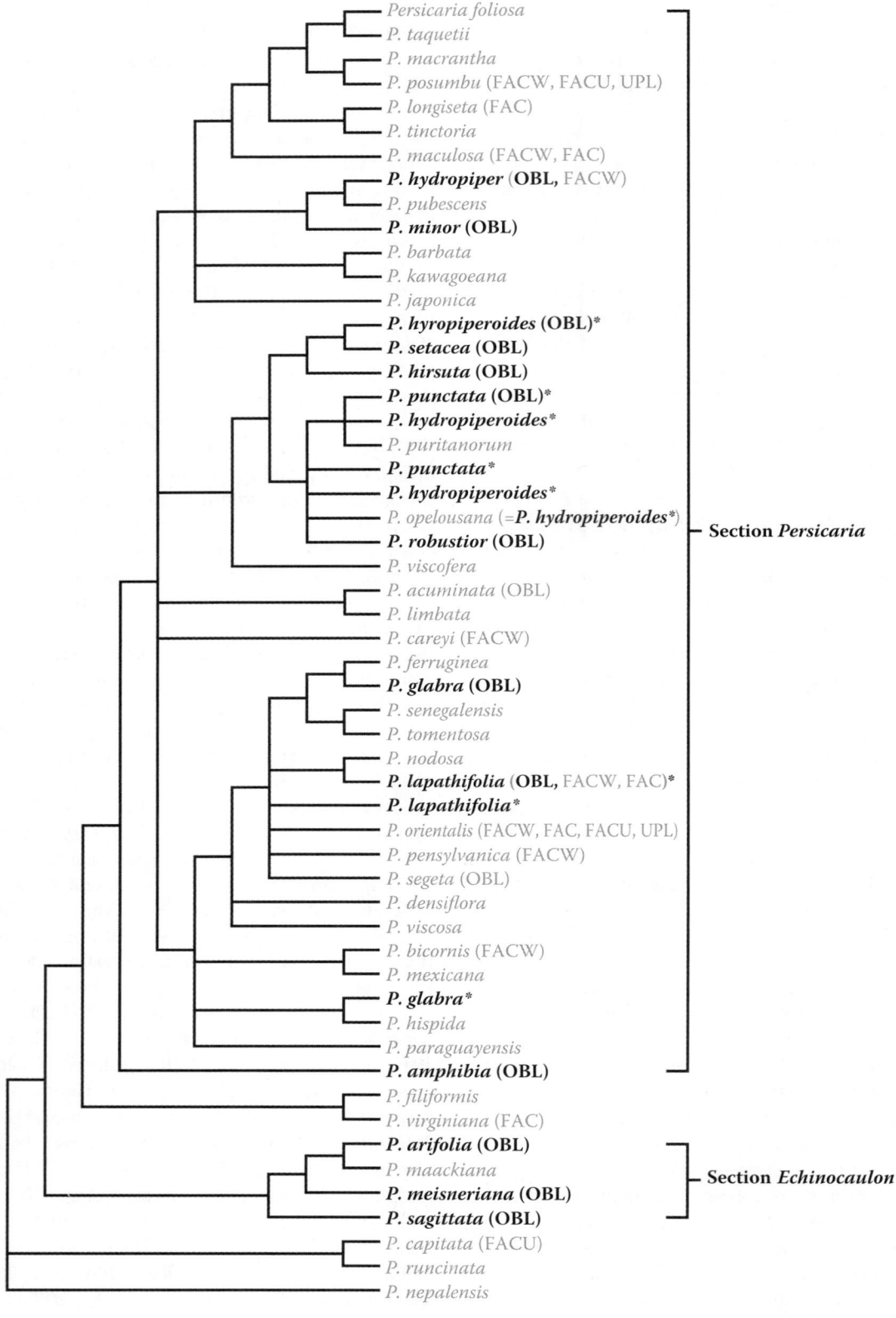

FIGURE 3.27 Maternally derived phylogeny for *Persicaria* species as indicated by analysis of combined cpDNA sequence data. The OBL indicators (highlighted in bold) are scattered throughout the group and within two different sections of the genus (sections *Echinocaulon* and *Persicaria*). Two species (*P. acuminata* and *P. segeta*) have OBL status but occur outside of the range covered by this book. Asterisks denote taxa, which resolve in more than one position in the cladogram, a likely consequence of hybridization events involving different maternal parents. The wetland indicator status is provided in parentheses for the North American species. (Adapted from Kim, S.-T. & Donoghue, M. J., *Amer. J. Bot.*, 95, 1122–1135, 2008b.)

et al., 2006; Van der Valk & Rosburg, 1997; Vimala et al., 2012; Visser & Sasser, 2009; Von Oheimb et al., 2005; Wetzel et al., 2001; Williams et al., 2008; Yasuda & Yamaguchi, 2005; Yoo & Park, 2001; Zomlefer et al., 2008.

4. *Polygonum*

Knotweed; renouée

Etymology: from the Greek *polys gony* meaning "many jointed"

Synonyms: *Bistorta* (in part); *Duravia*; *Persicaria* (in part); *Polygonella*; *Tracaulon*; *Tinaria*; *Pleuropterus*

Distribution: global: cosmopolitan; **North America:** widespread

Diversity: global: 76 species; **North America:** 44 species

Indicators (USA): OBL: *Polygonum argyrocoleon, P. marinense*; **FAC:** *P. argyrocoleon*

Habitat: freshwater, saline (coastal or inland); palustrine; **pH:** alkaline; **depth:** <1 m; **life-form(s):** emergent (herb)

Key morphology: stems (to 100 cm) green or reddish, appearing jointed, sometimes somewhat succulent, the nodes swollen, heterophyllous (stem leaves to 4× as long as branch leaves), prostrate to erect; leaves alternate, petioled (to 5 mm), the blades (to 35 mm) simple, entire, elliptic, lanceolate, linear–lanceolate, oblanceolate, or obovate, apex acute or rounded; ocrea (to 8 mm) cylindric or funnelform; flowers axillary or terminal, cymose (to six-flowered), pedicels (to 4 mm) enclosed within or exserted from the ocrea; perianth (to 4 mm) closed or partly open, of green or white overlapping tepals with pink, red, or white margins; stamens 7 or 8; achenes (to 5 mm) brown, ovate, shiny, two- to three-lobed, enclosed within or exserted from the perianth

Life history: duration: annual (fruit/seeds); **asexual reproduction:** none; **pollination:** insect, self; **sexual condition:** hermaphroditic; **fruit:** achenes (common); **local dispersal:** achenes (gravity); **long-distance dispersal:** achenes (animals)

Imperilment: (1) *Polygonum argyrocoleon* [GNR]; (2) *P. marinense* [G2]; S2 (CA)

Ecology: general: *Polygonum* species occupy a diverse range of habitats. They occur primarily as terrestrials on disturbed or dry, rocky sites and are also found in association with wetlands. Twenty of the North American species (45%) are wetland indicators, but only two species (<5%) are designated as OBL. The flowers are self-compatible or self-incompatible. They are either cleistogamous (and self-pollinated) or chasmogamous and pollinated by insects (Insecta). In one species, they are pollinated by ants (Hymenoptera: Formicidae: *Formica argentea*). The plants are heterocarpous, producing different seasonal fruit types. The autumn achenes are substantially (2–5×) larger than the summer achenes. Autumn achenes can germinate immediately at 20°C–25°C (lower temperatures will delay their germination until spring); summer achenes are dormant when shed. The seeds are dispersed locally by gravity and are dispersed over longer distances by granivorous birds (Aves), which transport them endozoically and by ungulates (e.g., Mammalia: Cervidae: *Capreolus capreolus*; Suidae: *Sus scrofa*), which transport them externally by adhesion to their fur. The seeds are also dispersed along

roadways by motor vehicles and are known to occur as contaminants in nursery stocks.

Polygonum argyrocoleon **Steud. ex Kunze** is a nonindigenous winter annual species that inhabits shallow (to 16 cm), fresh to brackish sites associated with bottomlands, channels, chapparal, depressions, ditches, draws, dunes, flats, floodplains, gardens, marshes, meadows, mudflats, playa lakes, riverbeds, roadsides, salt marshes, scrub, sinks, slopes, washes, waste ground, woodlands, and the margins of canals, lakes, and temporary ponds at elevations of up to 2988 m. The habitats are often disturbed and are characterized by full sunlight exposures. The plants often spread to drier, cultivated sites. The substrates (North America) are alkaline or saline and include clay, clay loam, granite, gravel, gravelly loam, marl, muck, mud, sand, sandy clay, sandy silt, silt, and silty clay. Flowering (North America) has been observed nearly throughout the year (December to October). The flowers are self-compatible, cleistogamous, and self-pollinating. The late-season achenes are unknown. Specific germination requirements are not known, but the seeds likely are not dormant. The seeds are eaten by birds (see *Use by wildlife*), which probably disperse them over considerable distances. Human-mediated dispersal occurs through the distribution of contaminated agricultural seed crops. The eclectic listing of associated species helps to explain the FAC designation, which is applied to some North American regions. **Reported associates (North America):** *Achnatherum hymenoides, Allenrolfea occidentalis, Almutaster pauciflorus, Amaranthus deflexus, Ambrosia psilostachya, A. salsola, Ammannia coccinea, Anagallis arvensis, Artemisia californica, Arundo donax, Atriplex confertifolia, A. polycarpa, A. prostrata, A. semibaccata, Baccharis salicifolia, B. sarothroides, Bahiopsis parishii, Bassia hyssopifolia, Bebbia juncea, Berula erecta, Brassica nigra, Bromus diandrus, B. hordeaceus, Calibrachoa parviflora, Carnegiea gigantea, Celtis reticulata, Centromadia pungens, Ceanothus crassifolius, Cephalanthus occidentalis, Chenopodium album, Convolvulus arvensis, Conyza bonariensis, C. canadensis, Cotula australis, Cressa truxillensis, Cryptantha angustifolia, Cuscuta, Cynodon dactylon, Cyperus eragrostis, C. involucratus, C. odoratus, Diplacus puniceus, Dipsacus sativus, Distichlis spicata, Dodonaea viscosa, Echinodorus berteroi, Eleocharis macrostachya, Encelia farinosa, Ephedra, Epilobium, Equisetum, Eragrostis cilianensis, Ericameria nauseosa, Erigeron, Eriogonum fasciculatum, Euphorbia albomarginata, E. glyptosperma, E. peplus, Foeniculum vulgare, Fouquieria splendens, Frankenia salina, Fraxinus velutina, Gutierrezia sarothrae, Heliotropium curassavicum, Heterotheca villosa, Hirschfeldia incana, Hymenopappus filifolius, Hymenoxys odorata, Isocoma menziesii, Juncus arcticus, Juniperus occidentalis, J. osteosperma, Larrea tridentata, Lepidium montanum, Leptochloa fusca, Linum aristatum, Lolium perenne, Lotus corniculatus, Ludwigia peploides, Lycium andersonii, L. pallidum, Lycopus, Lythrum californicum, Malosma laurina, Melilotus albus, M. indicus, Mentha, Mimulus guttatus, Muhlenbergia pungens, Nama stenocarpa, Navarretia fossalis, Oenothera pallida, Opuntia aurea, Panicum capillare,*

Paspalum distichum, Penstemon ambiguus, Persicaria amphibia, P. lapathifolia, P. maculosa, Phacelia crenulata, Phalaris, Plagiobothrys leptocladus, Platanus wrightii, Pluchea sericea, Polanisia dodecandra, Polygonum aviculare, Polypogon monspeliensis, Populus fremontii, Prosopis glandulosa, P. velutina, Pseudognaphalium luteoalbum, Quercus agrifolia, Q. gambelii, Ranunculus, Rhus integrifolia, Rorippa palustris, Rumex dentatus, Salicornia, Salix gooddingii, S. laevigata, S. lasiolepis, Sambucus, Sarcobatus vermiculatus, Schismus barbatus, Schoenoplectus acutus, S. americanus, S. californicus, Senegalia greggii, Sesbania herbacea, Simmondsia chinensis, Sisymbrium irio, Sonchus oleraceus, Sorghum halepense, Stemodia durantifolia, Stephanomeria exigua, Suaeda nigra, Symphyotrichum subulatum, Tamarix ramosissima, Taraxacum officinale, Tradescantia occidentalis, Triticum, Typha domingensis, T. latifolia, Urtica dioica, Vachellia constricta, Veronica anagallis-aquatica, V. peregrina, Xanthium spinosum, X. strumarium, Yucca baileyi.

***Polygonum marinense* Mert. & Raven** is a rare, restricted annual species that occurs in brackish tidal marshes, high salt marshes, mudflats, and swamps of coastal California at elevations of up to 20 m. The substrates have been characterized as brackish or saline mud. Flowering occurs from April to October. The flowers are chasmogamous but open only partially. Their reproductive ecology is unknown. Late-season achenes are produced but are uncommon. There is no information available regarding specific seed germination requirements or dispersal mechanisms. Some reports indicate that the species is spreading rapidly in the San Francisco Bay area. Additional study of this species is needed. **Reported associates:** *Anagallis arvensis, Atriplex prostrata, Brassica rapa, Bromus hordeaceus, Centaurea solstitialis, Chloropyron maritimum, C. molle, Cotula coronopifolia, Distichlis spicata, Eleocharis parvula, Frankenia, Grindelia hirsutula, Helminthotheca echioides, Jaumea carnosa, Limonium californicum, Lolium multiflorum, Lotus corniculatus, Polygonum aviculare, Polypogon monspeliensis, Raphanus sativus, Rumex crispus, Salicornia depressa, Sarcocornia pacifica, S. perennis, Spartina foliosa, Spergula arvensis, Spergularia macrotheca, S. rubra, Suaeda californica, Tetragonia tetragonioides, Triglochin maritimum, Vicia sativa, Vulpia myuros.*

Use by wildlife: The seeds of *P. argyrocoleon* are eaten (and relished) by mourning doves (Aves: Columbidae: *Zenaida macroura*). The plants are known to harbor soft rot plant pathogens (Bacteria: Enterobacteriaceae: *Erwinia*).

Economic importance: food: The parched seeds of *P. argyrocoleon* were ground and eaten by members of the Cocopa tribe; **medicinal:** no uses reported; **cultivation:** not cultivated; **misc. products:** none; **weeds:** *Polygonum argyrocoleon* is a serious weed in Arizona and California; **nonindigenous species:** The southwest Asian species *P. argyrocoleon* is nonindigenous to North America but has spread as a seed contaminant of alfalfa (Fabaceae: *Medicago*). It is believed to have been introduced to California from the Caspian Sea region around 1910.

Systematics: Previously, a much larger genus, *Polygonum* has been recircumscribed to accommodate the results of phylogenetic investigations, which revealed that the former genus concept was polyphyletic. The ensuing taxonomic changes included the transfer of many former *Polygonum* species to the genera *Bistorta, Fallopia,* and *Persicaria* as well as the incorporation of other genera (e.g., *Polygonella*) into *Polygonum.* Current phylogenetic results depict *Polygonum* as comprising three subclades, which are recognized as sections (sections *Duravia, Polygonum,* and *Pseudomollia*). Both *P. argyrocoleon* and *P. marinense* are assigned to section *Polygonum*; however, only the former species has been included in phylogenetic analyses at this time. Although phylogenetic studies of *Polygonum* continue, further sampling of taxa and genetic loci will be necessary to achieve a more complete picture of interspecific relationships in the group. Pertinent here is the taxonomic status of the presumably endemic *P. marinense,* which remains uncertain. Although regarded as a California endemic, there is some suspicion that North American occurrences actually may represent introduced plants of the European *P. robertii* (or related species). Further investigation in this regard is necessary. The basic chromosome number for the genus is $x = 10$. *Polygonum argyrocoleon* ($2n = 40$) is tetraploid; *P. marinense* ($2n = 60$) is hexaploid. There are no hybrids reported involving either species, at least in North America.

Comments: *Polygonum argyrocoleon* occurs across the southern United States; *P. marinense* is endemic to California.

References: Brenckle, 1941; Burr & Schroth, 1977; Costea et al., 2005b; Faber, 2004; Haase, 1972; Heinken & Raudnitschka, 2002; Hickman, 1974; Hodkinson & Thompson, 1997; Isely & Wright, 1951; Schuster et al., 2011; Simmonds, 1945; Whitcraft et al., 2011; Woo et al., 2004.

5. *Rumex*

Dock, sorrel, water dock; patience

Etymology: a name used by Pliny (of uncertain derivation)

Synonyms: *Acetosa; Acetosella*

Distribution: global: cosmopolitan **North America:** widespread

Diversity: global: 200 species; **North America:** 63 species

Indicators (USA): OBL: *Rumex britannica, R. dentatus, R. fueginus, R. lacustris, R. obovatus, R. occidentalis, R. tomentellus, R. verticillatus*; **FACW:** *R. dentatus, R. fueginus, R. occidentalis, R. verticillatus*; **FACU:** *R. dentatus*

Habitat: brackish (coastal); freshwater; palustrine, riverine; **pH:** 6.2–9.6; **depth:** <1 m; **life-form(s):** emergent (herb)

Key morphology: stems (to 2 m) erect, with swollen nodes; leaves (to 6 dm) alternate, lanceolate to orbiculate, petiolate; ocrea present; flowers greenish, stalked (to 1 cm), pendant, in verticillate whorls; petals absent; calyx (to 9 mm) six-parted, veiny, with a grain-like tubercle; achenes (to 4.5 mm) three-lobed, included in perianth

Life history: duration: perennial (rhizomes, taproots); annual (achenes); **asexual reproduction:** rhizomes; **pollination:** wind; **sexual condition:** hermaphroditic, dioecious; **fruit:** achenes (common); **local dispersal:** rhizomes, stem fragments; **long-distance dispersal:** seeds

Imperilment: (1) *Rumex britannica* [G5]; S2 (NF); S3 (AB, IA, IL); (2) *R. dentatus* [GNR]; (3) *R. fueginus* [G5]; SH (CT); S1 (NY); S3 (IL, QC, WY, YT); (4) *R. lacustris* [G5]; S1 (OR); (5) *R. obovatus* [G5]; (6) *R. occidentalis* [G5]; SH (VT); S1 (IA, MI, NB); S2 (NF); S3 (QC, WY); (7) *R. tomentellus* [GH]; SH (NM); SH (NM); (8) *R. verticillatus* [G5]; S1 (MA, NE); S2 (KS, WV); S3 (GA, OK, QC)

Ecology: general: *Rumex* is a large and diverse genus that occupies a wide range of habitats but often occurs on disturbed gravelly or rocky substrates in cultivated sites or waste areas. The plants grow from sea level to alpine elevations as annuals, biennials, or perennials. Many of the species can be found growing in or near wetlands or wet, disturbed sites (e.g., river margins) at least part of the time. Thirty-eight species (60%) are wetland indicators with eight (13%) designated as OBL. The flowers can be bisexual or unisexual, with the latter arranged in dioecious, synoecious, or occasionally polygamomonoecious sexual conditions. All *Rumex* species are wind pollinated. The flowers are generally self-compatible, but some species can also possess self-incompatible flowers. In some species, the flowers are protandrous and visited occasionally by bees (Insecta: Hymenoptera: Apidae: *Bombus*). Fruit production typically is prolific. Some species are heterocarpic and produce as many as four morphologically distinct fruit types, which are characterized by different dispersal characteristics. The seeds are dispersed abiotically not only by the water or wind but also by adhesion to animal fur, contamination of agricultural seed stocks, and other means of human-mediated transport. A variety of seed dormancy mechanisms exists.

Rumex britannica L. is a perennial species that occurs in shallow waters (to 5 dm) associated with bogs, carrs, channels, depressions, ditches, fens, gravel bars, hollows, marshes, meadows, oxbows, pools, prairie fens, prairies, roadsides, seeps, swamps, teardrop islands, thickets, and along the margins or shores of canals, lagoons, lakes, ponds, rivers, and streams at elevations of up to 1500 m. Exposures range from full sunlight to shade. The substrates are described as alluvium, gravel, marly peat, muck, peat, and sand. Flowering and fruiting occur from August to September. The seeds are dispersed primarily by water. They have a relatively high mean germination rate (70%) and can germinate (but to a lesser extent) even under low oxygen conditions. The highest seed germination (92%) has been obtained using a 20°C/30°C, 16/18 h temperature regime in the light. The seeds remain viable (up to 91% germination) for at least 4 years when stored in air. Seedlings can survive in water depths up to 2 cm but do best at depths of −6 to −4 cm (below soil surface). **Reported associates:** *Acer saccharinum, Agrimonia parviflora, Agrostis gigantea, Alisma triviale, Alnus rugosa, Amaranthus tuberculatus, Ambrosia, Amorpha fruticosa, Andromeda polifolia, Angelica atropurpurea, Apios americana, Apocynum cannabinum, Aronia melanocarpa, Asclepias incarnata, Aulacomnium palustre, Betula pumila, Bidens cernuus, B. frondosus, B. trichospermus, Boehmeria cylindrica, Bolboschoenus fluviatilis, Bromus ciliatus, B. secalinus, Calamagrostis canadensis, Calliergon stramineum, Caltha palustris, Calystegia sepium, Campanula aparinoides, Cardamine bulbosa, Carex atherodes, C. canescens, C. comosa, C. disperma, C. granularis, C. haydenii, C. hystericina, C. interior, C. intumescens, C. lacustris, C. lurida, C. pseudocyperus, C. rostrata, C. stipata, C. stricta, C. tenera, C. tenuiflora, C. trichocarpa, Chamaedaphne calyculata, Chelone glabra, Cicuta bulbifera, C. maculata, Cirsium arvense, C. muticum, Clintonia borealis, Conyza canadensis, Cornus amomum, C. drummondii, C. sericea, Cuscuta gronovii, Doellingeria umbellata, Dryopteris spinulosa, Echinocystis lobata, Eleocharis palustris, Epilobium ciliatum, E. coloratum, E. leptophyllum, E. strictum, Equisetum arvense, E. sylvaticum, Eragrostis spectabilis, Erigeron, Eupatorium perfoliatum, Eutrochium maculatum, Fallopia scandens, Galium boreale, G. obtusum, G. trifidum, G. triflorum, Geum rivale, Glyceria striata, Helenium autumnale, Helianthus grosseserratus, Hibiscus laevis, Hydrocotyle americana, Impatiens capensis, Iris versicolor, I. virginica, Juncus effusus, J. nodosus, J. torreyi, Kalmia polifolia, Laportea canadensis, Leersia oryzoides, Lemna minor, Lobelia siphilitica, Lonicera maackii, L. villosa, Lycopus americanus, L. rubellus, L. uniflorus, L. virginicus, Lysimachia terrestris, L. thyrsiflora, L. vulgaris, Lythrum salicaria, Mentha arvensis, Menyanthes trifoliata, Mimulus ringens, Muhlenbergia frondosa, M. mexicana, Myrica gale, Nasturium officinale, Nuphar variegata, Nymphaea tuberosa, Onoclea sensibilis, Ophioglossum pusillum, Osmunda claytoniana, Oxypolis rigidior, Panicum dichotomiflorum, Pedicularis lanceolata, Penthorum sedoides, Persicaria amphibia, P. coccinea, P. maculosa, P. punctata, P. sagittata, Phalaris arundinacea, Phragmites australis, Pilea fontana, Plantago, Poa palustris, Populus deltoides, Potentilla norvegica, Prunus virginiana, Pycnanthemum tenuifolium, P. virginianum, Quercus palustris, Rhamnus cathartica, R. frangula, Rhizomnium punctatum, Ribes americanum, Rosa palustris, Rubus pubescens, Sagittaria latifolia, Salix amygdaloides, S. bebbiana, S. candida, S. discolor, S. interior, S. nigra, S. petiolaris, Sambucus, Sarracenia purpurea, Schoenoplectus acutus, S. pungens, S. tabernaemontani, Scirpus atrovirens, S. cyperinus, Scutellaria galericulata, S. lateriflora, Senecio suaveolens, Sium suave, Solanum dulcamara, Solidago gigantea, S. patula, S. riddellii, Sparganium eurycarpum, S. glomeratum, Spartina pectinata, Sphagnum papillosum, S. recurvum, Spiraea tomentosa, Stachys palustris, Symphyotrichum firmum, S. lanceolatum, S. novaeangliae, S. praealtum, S. puniceum, Symplocarpus foetidus, Taraxacum officinale, Thelypteris palustris, Toxicodendron vernix, Triadenum fraseri, Tridens flavus, Trifolium, Typha angustifolia, T. latifolia, Urtica dioica, Vaccinium macrocarpon, V. oxycoccos, Verbena hastata, Veronica americana, Viburnum lentago, V. opulus, Viola cucullata, Zizania palustris.*

Rumex dentatus L. is a nonindigenous annual or biennial species that grows in alluvial fans, bottomlands, cultivated fields, ditches (irrigation), dunes, floodplains, pools, rice fields, riverbeds, saltmarshes, shores, sloughs, thickets, vernal pools, washes, waste areas, woodlands, and along the margins of channels, lakes, and rivers at elevations of up to 1275 m. The plants can tolerate exposures of full sunlight to shade.

This is the least "aquatic" species among its OBL congeners and is designated as a FACW or FACU indicator in most of its North American range. The plants occur on substrates that are categorized as alkaline or saline and are described as ballast, clay, gravel, muck, mud, rocky cobble, sand, sandy alluvium, silty loam, and Willows clay. Flowering and fruiting occur from March to June. The seeds do not break dormancy after being flooded. Cold stratification (or physical scarifiaction) and light are necessary for optimal germination to occur. Stratified seeds have germinated under 15°C, 18°C/8°C, and 21°C/10°C temperature regimes. Germination of seeds incubated at 25°C–35°C is promoted by their subsequent transfer to a 15°C temperature regime. The seeds are dispersed normally by abiotic vectors (water and wind). They have also been shown to survive digestion by several grazing animals (Mammalia) such as nilgai (Bovidae: *Bos tragocamelus*) and wild boars (Suidae: *Sus scrofa*), which can disperse the propagules endozoically. **Reported associates (North America):** *Abronia angustifolia, Acmispon glabrus, A. heermannii, Adenostoma fasciculatum, Ambrosia chamissonis, A. monogyra, A. psilostachya, Ammi visnaga, Artemisia californica, A. douglasiana, Atriplex lentiformis, A. polycarpa, A. semibaccata, Avena barbata, Baccharis pilularis, B. salicifolia, Baileya, Bebbia juncea, Beta, Brassica nigra, Bromus catharticus, B. diandrus, B. inermis, B. japonicus, Cakile maritima, Camissoniopsis cheiranthifolia, Capsella, Celtis reticulata, Centaurea melitensis, Chenopodium, Chilopsis linearis, Conium maculatum, Convolvulus simulans, Conyza canadensis, Cressa truxillensis, Cryptantha angustifolia, Cucurbita foetidissima, Cynodon dactylon, Cyperus eragrostis, C. involucratus, C. odoratus, Deinandra conjugens, D. fasciculata, Distichlis spicata, Echinochloa crus-galli, Elaeagnus angustifolia, Encelia farinosa, Ericameria nauseosa, Eriodictyon trichocalyx, Eriogonum fasciculatum, E. parvifolium, Eucalyptus microtheca, Frankenia salina, Fraxinus velutina, Glandularia, Glinus lotoides, Helianthus annuus, Heliotropium, Hesperoyucca whipplei, Hirschfeldia incana, Isocoma menziesii, Jaumea carnosa, Juglans major, Juncus acutus, J. torreyi, Juniperus californica, Lasthenia fremontii, Lepidium latipes, Leptochloa fusca, L. panicea, Lolium perenne, Ludwigia peploides, Lupinus, Lycium, Lythrum hyssopifolia, Malva parviflora, Marrubium vulgare, Melilotus albus, M. indicus, M. officinalis, Mimulus pilosus, Nasturtium officinale, Navarretia leucocephala, Nicotiana, Parkinsonia aculeata, P. florida, Pectocarya platycarpa, Perityle, Persicaria maculosa, Phalaris minor, Plagiobothrys stipitatus, Platystemon californicus, Pluchea sericea, Polygonum argyrocoleon, Polypogon monspeliensis, Populus deltoides, P. fremontii, P. racemosa, P. trichocarpa, Portulaca oleracea, Prosopis velutina, Prunus ilicifolia, Pseudognaphalium luteoalbum, Quercus agrifolia, Q. berberidifolia, Rhamnus crocea, Rhus ovata, Ricinus communis, Rubus bifrons, Rumex crispus, R. densiflorus, Salix exigua, S. gooddingii, S. laevigata, S. lasiolepis, S. lutea, Salsola kali, Salvia apiana, Sambucus, Schinus terebinthifolius, Schoenoplectus acutus, S. americanus, S. californicus, Senegalia greggii, Sesuvium portulacastrum, Sisymbrium irio, Solanum americanum,* *S. elaeagnifolium, Sonchus oleraceus, Sorghum halepense, Spergularia salina, Sporobolus airoides, Stuckenia pectinata, Suaeda, Tamarix aphylla, T. chinensis, T. ramosissima, Typha angustifolia, T. domingensis, Veronica anagallis-aquatica, Washingtonia, Xanthium strumarium.*

***Rumex fueginus* Phil.** is an annual (rarely biennial) species that inhabits backwaters, beaches, bogs, bottomlands, channels, cobblestone/gravel bars, ditches, draws, flats, floodplains, lake bottoms, marshes, meadows, mudflats, oxbows, prairies, ravines, roadsides, saltmarshes, salt pans, sand bars, scrub, seeps, sloughs, springs, strands, streambeds (dry), swales, swamps, thickets, woodlands, and the margins or drying shores of lagoons, lakes, ponds, potholes, reservoirs, and rivers at elevations of up to 2743 m. The plants can occur in shallow water (5 cm to 1 m deep), sometimes becoming completely submersed for short periods. The sites are often disturbed, with exposures ranging from full sunlight to partial shade. The substrates are described as alkaline or saline (pH: 8.3–9.6) and include alluvium, clay (calcareous), clay loam muck, clay muck, gravel, gravelly sand, loam, loamy clay, muck, mud, muddy clay silt, rocky alluvium, rocky organic muck, sand (granitic), sandy ash, sandy clay, sandy gravel, sandy loam, sandy mud, sandy silt, silt, and silty loam. Flowering occurs from June to September with fruiting extending into October. There is no reliable information available on the reproductive ecology or seed ecology of this species. **Reported associates:** *Agrostis, Alisma, Alnus, Alopecurus aequalis, Amaranthus albus, A. retroflexus, A. tuberculatus, Ambrosia artemisiifolia, A. psilostachya, Amorpha fruticosa, Amphiscirpus nevadensis, Artemisia californica, A. douglasiana, A. tridentata, A. vulgaris, Asclepias incarnata, Astragalus canadensis, Atriplex lentiformis, A. leucophylla, A. patula, A. prostrata, Baccharis pilularis, B. salicifolia, Bacopa, Bassia hyssopifolia, Berula erecta, Bidens cernuus, B. tripartitus, Bolboschoenus fluviatilis, B. maritimus, B. robustus, Calibrachoa parviflora, Callitriche palustris, Carex atherodes, C. praegracilis, C. rostrata, C. utriculata, Carpobrotus edulis, Cerastium beeringianum, Ceratophyllum demersum, Chenopodium album, C. chenopodioides, C. fremontii, C. glaucum, C. leptophyllum, C. rubrum, Cirsium arvense, Cleomella parviflora, Conyza canadensis, Crypsis schoenoides, C. vaginiflora, Cyperus bipartitus, C. erythrorhizos, C. odoratus, Deinandra floribunda, Distichlis spicata, Echinochloa crus-galli, Echinodorus berteroi, Elaeagnus angustifolia, Eleocharis acicularis, E. palustris, E. parishii, Elodea canadensis, Elymus elymoides, Epilobium ciliatum, E. coloratum, Eragrostis hypnoides, Erechtites hieraciifolius, Ericameria nauseosa, Erigeron, Eriogonum fasciculatum, Eucalyptus camaldulensis, Eupatorium serotinum, Euthamia occidentalis, Geum rivale, Glinus lotoides, Glyceria grandis, Gnaphalium palustre, G. uliginosum, Helenium, Helianthus petiolaris, Heliotropium curassavicum, Heteranthera dubia, Hippuris, Hirschfeldia incana, Hordeum jubatum, Hypericum majus, Isocoma menziesii, Jaumea carnosa, Juncus balticus, J. bufonius, J. tenuis, J. torreyi, Leersia oryzoides, Leptochloa fusca, L. panicea, Limosella acaulis, Lindernia dubia, Lipocarpha micrantha, Ludwigia palustris,*

Lycopus americanus, L. asper, Lythrum salicaria, Malvella leprosa, Melilotus albus, Mentha arvensis, Mimulus guttatus, Muhlenbergia asperifolia, Myoporum laetum, Myosurus minimus, Nuphar, Oenothera, Panicum capillare, Persicaria amphibia, P. coccinea, P. hydropiper, P. lapathifolia, Phoenix dactylifera, Phragmites australis, Plantago major, Polygonum aviculare, Polypogon monspeliensis, Pontederia cordata, Populus angustifolia, P. deltoides, P. fremontii, Potamogeton foliosus, P. natans, Potentilla anserina, P. rivalis, P. supina, Pseudognaphalium luteoalbum, P. stramineum, Puccinellia distans, P. nuttalliana, Quercus agrifolia, Ranunculus cymbalaria, R. sceleratus, Ricciocarpus, Rorippa calycina, R. curvipes, R. palustris, Rumex crispus, R. obtusifolius, R. stenophyllus, R. verticillatus, Sagittaria latifolia, Salicornia depressa, Salix amygdaloides, S. exigua, S. geyeriana, S. gooddingii, S. interior, S. laevigata, S. lasiolepis, Salsola kali, Sambucus nigra, Sarcobatus, Schoenoplectus acutus, S. americanus, S. californicus, S. pungens, S. tabernaemontani, Scolochloa festucacea, Scutellaria lateriflora, Solidago, Sonchus arvensis, Sparganium eurycarpum, Spartina pectinata, Sphagnum, Symphyotrichum ascendens, S. ciliatum, S. frondosum, Taraxia tanacetifolia, Toxicodendron diversilobum, Trifolium, Triglochin maritimum, T. palustre, Typha angustifolia, T. domingensis, T. latifolia, Urtica dioica, Verbascum thapsus, Veronica anagallis-aquatica, V. scutellata, Xanthium strumarium, Zannichellia palustris, Zizania palustris.

***Rumex lacustris* Greene** is a perennial species that grows in ditches, lake beds, marshes, meadows, mudflats, roadsides, seeps, sinks, thickets, vernal pools, and along the margins or drying shores of lakes, playas, ponds, pools, reservoirs, and streams at elevations of up to 2423 m. The plants occur in sites having open exposures and produce both an emergent (pubescent) and an aquatic (glabrous) form. The substrates are alkaline (sometimes slightly saline) and include adobe, clay, clay basalt, gravel, loam, peat, rock, sand, and volcanic ash. Flowering occurs from May to July, with fruiting extending into October. There is no information on the reproductive biology or seed ecology of this species; however, the seeds of the related *R. salicifolius* (under which it has been treated as a variety by some authors) require no stratification and germinate under a 32°C/22°C temperature regime. Vegetative reproduction occurs by means of thick rhizomes. Additional life-history information is lacking. **Reported associates:** *Acmispon americanus, Agoseris heterophylla, Ambrosia psilostachya, Artemisia ludoviciana, A. tridentata, Bromus tectorum, Carex, Centaurea melitensis, Collinsia sparsiflora, Conyza canadensis, Descurainia sophia, Eleocharis, Elymus canadensis, Epilobium, Erodium cicutarium, Gayophytum diffusum, Gnaphalium, Hordeum brachyantherum, Iva axillaris, Juncus dubius, J. mexicanus, Leymus mollis, Microsteris gracilis, Mimulus guttatus, Muhlenbergia richardsonis, M. rigens, Oenothera, Phalaris arundinacea, Polyctenium fremontii, Polygonum douglasii, Polypogon monspeliensis, Potentilla newberryi, Rumex, Salix exigua, Senecio serra, Solidago canadensis, Taraxia tanacetifolia, Xanthium strumarium.*

***Rumex obovatus* Danser** is a nonindigenous annual (biennial or even perennial in tropical latitudes) species that occurs sporadically in southern North America, where it is found in deltas, depressions, ditches, dunes, meadows, roadsides, sea shores, slopes (to 8%), swamps, and along the drying shores or margins of bayous, ponds, and rivers at elevations of up to 50 m. The plants are known mostly from coastal habitats in open to semishaded exposures and occur often in disturbed sites. The substrates (North America) are described as acidic and include Duckston sand, Harleston fine sandy loam, Leon sand, mucky clay loam, peaty sand, Plummer loamy sand, and sand. Flowering and fruiting (North America) occur from March to July. The seeds are known to be capable of endozoic dispersal by wild boars (Mammalia: Suidae: *Sus scrofa*). **Reported associates (North America):** *Allium canadense, Alternanthera, Amaranthus australis, Atriplex cristata, Bacopa monnieri, Calystegia sepium, Carex longii, C. stipata, Centella asiatica, Chamaecrista fasciculata, Chenopodium album, Cirsium horridulum, Colocasia esculenta, Cynodon dactylon, Diodia, Eleocharis montevidensis, E. obtusa, Eupatorium capillifolium, Euphorbia, Galium tinctorium, Hibiscus grandiflorus, Hydrocotyle bonariensis, H. ranunculoides, Juncus effusus, Lythrum lineare, Myrica cerifera, Nothoscordum bivalve, Nyssa biflora, Oenothera drummondii, Panicum hemitomon, P. repens, P. virgatum, Paspalum dilatatum, Persicaria hydropiperoides, Phyla nodiflora, Phytolacca americana, Pinus elliottii, Ranunculus sceleratus, Robinia pseudoacacia, Rubus trivialis, Rudbeckia, Rumex chrysocarpus, R. verticillatus, Sabal palmetto, Sambucus nigra, Saururus cernuus, Sesuvium portulacastrum, Sonchus asper, Sorghum, Spartina patens, Sphenopholis obtusata, Suaeda linearis, Tripsacum dactyloides, Vigna luteola.*

***Rumex occidentalis* S. Watson** is a perennial species that grows in tidal or nontidal bays, beaches, bogs, bottomlands, brooks, depressions, ditches, draws, fens, flats, floodplains, gravel bars, hummocks, marshes (estuarine, freshwater), meadows (tidal), outwash plains, oxbows, prairies, ravines, salt marshes, seeps, slopes (to 40%), springs, swales, swamps, thickets, tundra, washes, woodlands, and along the margins of brooks, canals, lagoons, lakes, ponds, reservoirs, rivers, sloughs, and streams at elevations of up to 3455 m. Site exposures range from full sun to partial shade. The plants can tolerate shallow standing waters (to 15 cm), low levels of salinity (coastal, brackish sites), and a wide range of substrates (e.g., pH: 3.5–6.7; specific conductance: 62–899 μS; Ca: 3–65 mg/L; ammonium nitrogen: 28–250 μg/L; total phosphorous: 221–1110 μg/L), which are described as clay, gravel, gravelly mud, loam, muck, mud, peat, peaty rock, quartz fragments, rock, rocky sand, sand, sandy gravel, sandy loam, sandy silt, serpentine, silt (alluvial), silty loam, and stones. Flowering and fruiting extend from March to September. The flowers are hermaphroditic. A single plant can produce upward of 23,000 seeds in a season. They are dispersed locally by water or wind and probably over longer distances by birds (Aves), which consume the seeds. The seeds require cold stratification (4°C for 60 days) for optimal germination (80%–95%) under a 16/8 h light/dark, 21°C/10°C temperature regime. Their germination is not influenced by potentially alleopathic leaf extracts of *Artemisia* species. The plants perennate from a vertical or oblique rootstock. **Reported associates:** *Abies lasiocarpa, Acer negundo, Achillea millefolium, Agastache urticifolia, Agrostis idahoensis, A. scabra, Allium acuminatum, Alnus rugosa, A. viridis, Alopecurus aequalis, Amsinckia, Angelica atropurpurea, A. lucida, Aquilegia chrysantha, Athyrium filix-femina, Atriplex confertifolia, A. dioica, Avena fatua, Barbarea, Betula, Bidens cernuus, Bistorta bistortoides, B. vivipara, Bouteloua dactyloides, Bromus diandrus, B. tectorum, Calamagrostis canadensis, C. stricta, Camassia leichtlinii, Campanula parryi, Capsella bursa-pastoris, Carex aquatilis, C. atherodes, C. athrostachya, C. bicolor, C. cusickii, C. diandra, C. disperma, C. douglasii, C. lasiocarpa, C. limosa, C. lyngbyei, C. macrochaeta, C. magellanica, C. membranacea, C. microptera, C. nebrascensis, C. obnupta, C. pachystachya, C. praegracilis, C. rostrata, C. simulata, C. spectabilis, C. stipata, C. utriculata, C. vecta, Castilleja miniata, C. unalaschcensis, Catabrosa aquatica, Cerastium arvense, Chenopodium, Cicuta douglasii, C. virosa, Cirsium, Clarkia amoena, Claytonia, Collomia linearis, Comarum palustre, Conium maculatum, Dactylis glomerata, Dasiphora floribunda, Delphinium ×occidentale, Deschampsia cespitosa, Descurainia, Dichanthelium oligosanthes, Distichlis spicata, Drepanocladus aduncus, Dulichium arundinaceum, Elaeagnus, Eleocharis palustris, Empetrum, Epilobium ciliatum, E. lactiflorum, E. palustre, Equisetum arvense, E. fluviatile, E. hyemale, E. pratense, E. variegatum, Erigeron flagellaris, Eriophorum chamissonis, Festuca arizonica, F. idahoensis, Forestiera, Fragaria virginiana, Fraxinus, Galium trifidum, G. triflorum, Gentianopsis holopetala, Geranium erianthum, Geum aleppicum, G. macrophyllum, Glyceria grandis, G. striata, Helianthella quinquenervis, Helodium blandowii, Heracleum sphondylium, Heuchera, Hippuris vulgaris, Holcus lanatus, Honckenya peploides, Hordeum brachyantherum, Hymenoxys cooperi, H. hoopesii, Hypericum perforatum, Hypnum pratense, Iris missouriensis, Juncus balticus, J. effusus, J. ensifolius, J. patens, Juniperus scopulorum, Larix, Lepidium, Leymus mollis, L. triticoides, Ligusticum scoticum, Lomatium nudicaule, Lonicera utahensis, Lupinus nootkatensis, L. polyphyllus, Lycopus uniflorus, Medicago lupulina, M. sativa, Linaria vulgaris, Melilotus officinalis, Mentha arvensis, Menyanthes trifoliata, Mertensia, Mimulus guttatus, Monolepis nuttalliana, Myosotis scorpioides, Myrica gale, Nephrophyllidium crista-galli, Osmorhiza occidentalis, Packera neomexicana, Parentucellia, Parnassia palustris, Perideridia, Petasites frigidus, Phalaris arundinacea, Phleum pratense, Physocarpus, Picea engelmannii, P. pungens, P. sitchensis, Pinus contorta, P. ponderosa, Platanthera dilatata, P. stricta, Poa annua, P. nervosa, P. pratensis, Polemonium occidentale, Polygonum aviculare, Populus angustifolia, P. deltoides, P. tremuloides, Potentilla anserina, P. biennis, P. hippiana, Prunella vulgaris, Pseudoroegneria spicata, Pseudotsuga menziesii, Pteridium aquilinum, Purshia tridentata, Quercus gambelii, Q. grisea, Ranunculus gmelinii,* *R. orthorhynchus, Rhinanthus minor, Rhus aromatica, Ribes inerme, Rosa, Rubus arcticus, R. ursinus, Rudbeckia occidentalis, Rumex salicifolius, Salicornia depressa, Salix barclayi, S. bebbiana, S. candida, S. exigua, S. gooddingii, S. lasiandra, S. monticola, S. planifolia, Salsola tragus, Sarcobatus vermiculatus, Schoenoplectus acutus, S. americanus, S. tabernaemontani, Senecio pseudoarnica, S. triangularis, Sidalcea neomexicana, S. oregana, Sisymbrium altissimum, Sisyrinchium angustifolium, Solanum dulcamara, Solidago canadensis, Sphagnum squarrosum, Sphenosciadium capitellatum, Spiranthes romanzoffiana, Stellaria longifolia, Symphoricarpos, Symphyotrichum eatonii, S. spathulatum, Taeniatherum caput-medusae, Tamarix, Taraxacum officinale, Thalictrum venulosum, Thermopsis montana, Tragopogon dubius, Trantha occidentalis, Trifolium repens, Triglochin maritimum, T. palustre, Typha angustifolia, T. latifolia, Urtica dioica, Vaccinium uliginosum, Veratrum californicum, Verbascum thapsus, Veronica americana, Vicia americana, Viola epipsila, Vitis, Vulpia myuros, Wyethia amplexicaulis, Xanthium strumarium, Zuloagaea bulbosa.*

***Rumex tomentellus* Rech. f.** (included as an OBL indicator in the 1996 list) is a perennial known only from the type locality that was a stream bank at an elevation of up to 2330 m. The species was declared extinct by the Center for Biological Diversity in 2001. **Reported associates:** none.

***Rumex verticillatus* L.** is a perennial species that occurs in bayous, beaches, bogs, bottomlands, canals, depressions, dikes, ditches, flats, flatwoods, floodplains, hammocks, marshes, meadows, mudflats, ponds (temporary), roadsides, salt marshes, sand bars, seeps, shores, slopes (to 8%), sloughs, springs, swamps, vernal pools, woodlands (alluvial), and along the margins of lakes, ponds, pools, reservoirs, rivers, and streams at elevations of up to 1098 m. The plants are usually found in brackish to fresh, shallow (to 1 m), standing water, in exposures ranging from full sun to dense shade. The substrates are described as calcareous (pH: 6.2–7.3) and include alluvium (sandy), clay, Harleston fine sandy loam, loam, loamy clay, loamy sand, marl, muck (alluvial), mud, Okeelanta (terric medisaprists), oyster shell fragments, peat (fibrous), sand (Corolla–Duckston; Leon), sandy gravel, sandy marl, sandy peat, sandy silt, and silty loam. Flowering and fruiting occur from February to July. The seeds are small (averaging 1.1 mg) and physiologically dormant when shed. They will germinate following 270 days of cold (4°C–5°C) stratification when incubated under fluctuating day/night temperatures. High seed germination (88%) has been obtained in the light using a 20°C/30°C, 16/18 h temperature regime. Seed viability decreases considerably (to 32%) after 3 years of dry air storage. Presumably, seed dispersal occurs primarily by water. The seedlings have a mean relative growth rate of 0.20 g/g day. Adult plants are common on the mounds built by muskrats (Mammalia: Cricetidae: *Ondatra zibethicus*). **Reported associates:** *Acer rubrum, A. saccharinum, Acorus calamus, Alisma subcordatum, Allium canadense, Alnus rugosa, Alternanthera philoxeroides, Amaranthus cannabinus, Ambrosia artemisiifolia, Anemone canadensis, Anethum graveolens, Apocynum cannibinum, Aquilegia,*

Arisaema dracontium, Aronia arbutifolia, Asclepias incarnata, Athyrium filix-femina, Azolla caroliniana, Baccharis halimifolia, Berula erecta, Betula nigra, Bidens cernuus, B. discoideus, Boehmeria cylindrica, Bolboschoenus fluviatilis, Boltonia asteroides, Butomus umbellatus, Calamagrostis canadensis, Callitriche heterophylla, Caltha palustris, Carex atlantica, C. comosa, C. conjuncta, C. crinita, C. crus-corvi, C. davisii, C. emoryi, C. granularis, C. grayi, C. grisea, C. hyalinolepis, C. joorii, C. leavenworthii, C. leptalea, C. lupuliformis, C. lupulina, C. muskingumensis, C. radiata, C. retrorsa, C. seorsa, C. stipata, C. tenera, C. tribuloides, C. tuckermanii, C. typhina, Carya cordiformis, C. glabra, C. illinoinensis, C. laciniosa, Carpinus, Celtis occidentalis, Cephalanthus occidentalis, Chaerophyllum tainturieri, Chamaecrista fasciculata, Chasmanthium latifolium, Cicuta bulbifera, C. maculata, Cinna arundinacea, Cirsium muticum, Coreopsis grandiflora, Crotalaria rotundifolia, Croton glandulosus, Cynodon, Doellingeria umbellata, Dryopteris cristata, D. spinulosa, Echinochloa walteri, Eleocharis montevidensis, E. obtusa, E. palustris, Elymus virginicus, Equisetum fluviatile, Eragrostis hypnoides, E. lugens, Forestiera acuminata, Fraxinus nigra, F. pennsylvanica, F. profunda, Galium obtusum, G. trifidum, Geum canadense, Glechoma hederacea, Glyceria grandis, G. septentrionalis, Helenium autumnale, Hibiscus moscheutos, Hottonia inflata, Hydrocotyle verticillata, Hydrolea quadrivalvis, Hypericum hypericoides, Ilex vomitoria, Impatiens capensis, Iodanthus pinnatifidus, Ipomoea pandurata, Iris hexagona, I. virginica, Juglans nigra, Juncus acuminatus, Juniperus virginiana, Kyllinga odorata, Laportea canadensis, Larix laricina, Leersia lenticularis, L. oryzoides, L. virginica, Lemna minor, Lepidium virginicum, Limnobium spongia, Limnodea arkansana, Lindera, Lindernia dubia, Liquidambar styraciflua, Ludwigia palustris, L. polycarpa, Lycopus virginicus, Lysimachia ciliata, L. nummularia, L. terrestris, Lythrum salicaria, Magnolia virginiana, Mimulus alatus, Nasturtium officinale, Nelumbo lutea, Nymphaea tuberosa, Onoclea sensibilis, Orontium aquaticum, Osmunda regalis, Osmundastrum cinnamomeum, Oxypolis rigidior, Panicum amarum, P. capillare, P. repens, P. virgatum, Pedicularis lanceolata, Peltandra virginica, Penthorum sedoides, Persea palustris, Persicaria arifolia, P. coccinea, P. densiflora, P. hydropiper, P. hydropiperoides, P. lapathifolia, P. pensylvanica, P. punctata, P. setacea, Phalaris arundinacea, Phlox divaricata, Physostegia virginiana, Pilea pumila, Pinus elliottii, Platanus occidentalis, Pluchea odorata, Polygala lutea, Pontederia cordata, Populus deltoides, P. heterophylla, Potamogeton obtusifolius, Potentilla norvegica, Proserpinaca pectinata, Ptilimnium ahlesii, P. capillaceum, Quercus bicolor, Q. palustris, Q. rubra, Q. virginiana, Ranunculus abortivus, R. flabellaris, R. hispidus, R. sceleratus, Rhapidophyllum hystrix, Rhododendron viscosum, Rorippa islandica, R. palustris, Rosa setigera, Rubus trivialis, Rudbeckia laciniata, Rumex crispus, R. fueginus, R. mexicanus, R. obovatus, R. pulcher, Sabal palmetto, Sagittaria lancifolia, S. latifolia, Salix amygdaloides, S. nigra, Saururus cernuus, Schoenoplectus heterochaetus, S. pungens, S. tabernaemontani, Scirpus cyperinus, Scrophularia marilandica, Sesbania punicea, Sium suave, Smilax auriculata, S. tamnoides, Solanum americanum, Solidago gigantea, S. sempervirens, Spartina cynosuroides, Sparganium androcladum, S. chlorocarpum, S. eurycarpum, S. glomeratum, Spirodela polyrhiza, Stachys tenuifolia, Stellaria, Symphyotrichum lanceolatum, S. lateriflorum, S. ontarionis, S. puniceum, Taxodium distichum, Teucrium canadense, Toxicodendron radicans, Tradescantia subaspera, Triadenum virginicum, Typha angustifolia, T. latifolia, Ulmus americana, Utricularia macrorhiza, Vernonia fasciculata, Veronica anagallis-aquatica, Viburnum recognitum, Vigna luteola, Viola cucullata, V. sororia, Vitis cinerea, V. riparia, Woodwardia virginica, Xanthium strumarium, Yucca aloifolia, Zanthoxylum, Zizania aquatica.

Use by wildlife: *Rumex* species seldom are eaten by livestock and can be toxic (because of oxalates) if consumed in quantity. *Rumex britannica* plants are eaten by voles (Mammalia: Rodentia: Muridae). Most *Rumex* fruits are of minor value as a waterfowl food but are of relatively higher importance in western North America. The achenes of *R. fueginus* are eaten by northern pintail ducks (Aves: Anatidae: *Anas acuta*) and those of *R. verticillatus* are an occasional food of the American wigeon (Aves: Anatidae: *Anas americana*). In some areas, *R. occidentalis* is a major food of Canada geese (Aves: Anatidae: *Branta canadensis*). Several *Rumex* species are used as host plants by larval moths and butterflies (Insecta: Lepidoptera). These include *R. fueginus* (Lycaenidae: *Lycaena hyllus*), *R. occidentalis* (Lycaenidae: *Chalceria rubida, Gaeides xanthoides, Lycaena helloides*; Noctuidae: *Luperina passer*); *R. paucifolius* (Lycaenidae: *Gaeides editha, Lycaena cuprea*), and *R. verticillatus* (Lycaenidae: *Hyllolycaena hyllus, Lycaena hyllus, L. phlaeas*; Lymantriidae: *Euproctis chrysorrhoea*; Noctuidae: *Mesapamea passer*). *Rumex occidentalis* also hosts aphids (Insecta: Hemiptera: Aphididae: *Pemphigus betae*), Fungi (Ascomycota: Erysiphaceae: *Phyllactinia corylea*; Mycosphaerellaceae: *Venturia rumicis*; Basidiomycota: Pucciniaceae: *Puccinia ornata*), and stink bugs (Insecta: Hemiptera: Pentatomidae). The chrysomelid beetle *Galerucella calmariensis* feeds to a limited degree on *R. verticillatus*. The plants also host leaf-spot Fungi (Ascomycota: Mycosphaerellaceae: *Septoria rumicis*).

Economic importance: food: Foliage of most *Rumex* species is rich in minerals and contains a high concentration of oxalic acid which imparts a sour, lemony taste. The plants should not be consumed in excess because of possible associated nutrient deficiencies (e.g., oxalic acid can bind with calcium). The foliage of *R. dentatus* can contain more than 15% protein, 9% fiber, and 50% carbohydrate. The stems and young leaves of *R. occidentalis* (raw or cooked) were eaten as a vegetable by the Apache, Bella Coola, Chiricahua, Hanaksiala, Heiltzuk, Kaigani Haida, Kitasoo, Klallam, Mescalero, Montana, Oweekeno, and Tanana tribes. The seeds were eaten by the Montana Indians. The Hanaksiala brewed a wine from *R. occidentalis* plants; **medicinal:** The Cree and Woodlands tribes applied a decoction of *R. britannica* plants to relieve painful joints. The Meskwaki consumed a root decoction of *R. britannica* as a poison antidote. The Potawatomi used the roots to treat various blood ailments. The Hocąk used the plants to make a tonic for treating consumption. *Rumex dentatus* contains anthraquinones, which are molluscicidal against several snail vectors (Mollusca: Gastropoda: Planorbidae: *Biomphalaria glabrata*, *Bulinus globosus*; Pomatiopsidae: *Oncomelania hupensis*), of schistosomiasis (Platyhelminthes: Schistosomatidae: *Schistosoma mansoni*, *S. haematobium*, *S. japonicum*). Alcoholic extracts from the plants are antioxidant and inhibitory to Bacteria (Enterobacteriaceae: *Klebsiella pneumoniae*; Pseudomonadaceae: *Pseudomonas aeruginosa*) and Fungi (Ascomycota: Candidaceae: *Candida albicans*). The leaves are diuretic and refrigerant. The roots are used as an astringent and for treating skin problems. The Bella Coola people of British Columbia used *R. occidentalis* in a general analgesic bath. The Bella Coola and Hanaksiala applied a poultice of the leaves and mashed roots to treat sores and wounds. *Rumex occidentalis* contains anthraquinones and was used as a laxative by the Haisla. The Kwakiutl people applied root extracts (which contain astringent tannins) as a medicinal wash and ate the roots as a remedy for stomach ache (excess consumption of *R. occidentalis* roots can cause gastric upset, nausea, and dermatitis). Extracts of *R. occidentalis* comprise the active element of Tyrostat-09, a tyrosinase inhibitor, which is used as a skin-lightening agent to treat melasma. The plants are regarded as hay-fever plants during the spring (June). The Choctaw people bathed in an infusion of *R. verticillatus* leaves to ward off smallpox. Extracts of the plants are highly antioxidative and are inhibitory to the growth of Bacteria (Enterobacteriaceae: *Escherichia coli*; Pseudomonadaceae: *Pseudomonas aeruginosa*; Staphylococcaceae: *Staphylococcus aureus*) and Fungi (Ascomycota: Candidaceae: *Candida albicans*). The plants have been used to prepare a treatment for jaundice. The pollen (April and November) is regarded as a hay-fever allergen; **cultivation:** none of the OBL *Rumex* species is cultivated; **misc. products:** A dark green, brown, or dark grey dye can be obtained from the roots of most *Rumex* species. The dried, powdered root of *R. occidentalis* is used for cleaning teeth; **weeds:** *Rumex obovatus* is listed as a federal noxious weed in the United States; **nonindigenous species:** The wide-ranging *R. occidentalis* is native to North America but sometimes is reported erroneously as being introduced. *Rumex britannica* is native to eastern North America but reportedly was introduced to some portions of western North America. Introduced species include *R. dentatus* (from southern Asia; originated by seeds in contaminated ballast), *R. kerneri* (from southeastern Europe), and *R. obovatus* (introduced in the 1930s with contaminated grain imported from Argentina, South America). *Rumex fueginus* has been introduced to Europe.

Systematics: *Rumex* comprises a clade related to the genera *Rheum*, *Oxyria*, and *Emex*. It is related more distantly to *Polygonum* (sensu lato) and *Fagopyrum* (Figure 3.25). A comprehensive interspecific phylogenetic study of the large genus *Rumex* has not yet been carried out but studies of several species groups have been conducted or appear to be underway. Few DNA sequences are yet avaiailable for the OBL species, and these have not yet been incorporated into a comprehensive phylogenetic study. From the existing studies, *Rumex* seems to be monophyletic; however, possibly requiring the merger of *Emex*, which nests within the genus in some analyses. *Rumex* is variable cytologically with a range of base numbers from $x = 7–10$ and many incidences of polyploidy. *R. britannica* and *R. lacustris* are diploid ($2n = 20$); *R. dentatus* and *R. fueginus* are tetraploid ($2n = 40$); *R. verticillatus* ($2n = 60$) is hexaploid; and *R. occidentalis* ($2n = 120$) is a dodecaploid. Counts are unavailable for *R. obovatus*. *R. obovatus* reportedly hybridizes with *R. crispus*. *Rumex britannica* was long known as *R. orbiculatus*, and some databases still list records separately under both names.

Comments: *Rumex britannica* occurs mainly in northern and northeastern North America and is disjunct in California. *R. fueginus* is widespread throughout North America; *R. occidentalis* occurs mainly in northern North America; *R. verticillatus* is widespread in eastern North America; *R. dentatus* occurs sporadically throughout the region (mainly in the central and western portions). More restricted distributions occur in *R. lacustris* (California, Nevada, and Oregon) and *R. obovatus* (Florida and Louisiana).

References: Al-Helal, 1996; Anderson, 1930; Barbour et al., 2007; Baskin & Baskin, 1998; Borchardt et al., 2008a; Brenckle, 1918; Brescacin, 2010; Brown & Marcus, 1998; Cooper, 1939a; Couvreur et al., 2004; Craven & Hunt, 1984b; Davidson, 1909; Dearness, 1926; Fisk et al., 2014; Fraser & Karnezis, 2005; Fraser & Madson, 2008; Fraser et al., 2014; Ghadiri & Niazi, 2005; Glaser et al., 1981; Greene, 1949; Hameed & Dastagir, 2009; Hickman, 1993; Hoffman & Hazlett, 1977; Humeera et al., 2013; Jones, 2006; Kangas & Hannan, 1985; Kindscher & Hurlburt, 1998; Lawrence, 1905; Liu et al., 1997; McPherson & McPherson, 2000; Middleton, 2000; Middleton & Mason, 1992; Miller, 1983; Mohlenbrock, 1959b; Mosyakin, 2005; Muller, 1979; Munz & Keck, 1973; Nicholson, 1995; Norton, 1981; Ouren, 1978; Penfound et al., 1930; Perry & Uhler, 1981; Rechinger, 1937; Reynolds et al., 1978; Sabancilar et al., 2011; Scott, 1978, 1986; Shipley & Parent, 1991; Simonot et al., 2002; Steinbauer & Grigsby, 1960; Stevens, 1932; Talavera et al., 2010, 2013; Thieret, 1969; Vallejo-Marín & Hiscock, 2016; Vasas et al., 2015; Vitt & Chee, 1990; Weakley & Nesom, 2004.

4 Core Eudicots: Dicotyledons IV
"Rosid" Tricolpates

FABID ROSIDS (EUROSIDS I)

A substantial amount of combined molecular data (Figure 4.1) now resolves a putatively monophyletic assemblage of 15 orders, known as the "rosid" or "superrosid" clade (APG, 2003; Judd et al., 2016), an allusion to the subclass Rosidae, wherein most of the orders were placed formerly. Most of the members share the presence of specialized mucilage cells in the flowers (Judd et al., 2016). These orders are further divided into four subclades: the "fabid rosids" (Eurosids I), "malvid rosids" (Eurosids II), Vitales, and Saxifragales (Figure 4.1).

Despite the large amount of data assembled for this group (e.g., Jansen et al., 2006; Moore et al., 2010), some phylogenetic relationships have not yet been settled with a high degree of certainty. Studies have not yet determined whether Saxifragales or Vitaceae are the sister group to the remaining orders in the clade and have been unable to resolve the interrelationships satisfactorily among Celastrales, Oxalidales, and Malpighiales (Figure 4.1) or their joint relationship to other members of the group. The order Saxifragales has been excluded from or included into the rosids proper, with some referring to the group containing Saxifragales as the "superrosids."

In North America, 12 of the rosid orders contain obligate indicators:

Fabid rosids (Eurosids I)

1. Celastrales
2. Cucurbitales
3. Fabales
4. Fagales
5. Malpighiales
6. Rosales
7. Saxifragales
8. Vitales

Malvid rosids (Eurosids II)

9. Brassicales
10. Malvales
11. Myrtales
12. Sapindales

All of these rosid orders are predominantly terrestrial. However, the wide dispersion of many hydrophytic taxa across the group indicates that the obligate wetland (OBL) habit has evolved repeatedly within the rosids.

FABID ROSIDS (EUROSIDS I)

ORDER 1: CELASTRALES [3]

The order Celastrales has been redefined considerably as an outcome of numerous molecular phylogenetic analyses (Judd et al., 2016; Soltis et al., 2005). The current circumscription (followed here) accepts the expanded concept of Celastraceae (to include Brexiaceae, Canotiaceae, Hippocrateaceae, Stackhousiaceae, and Plagiopteridaceae), which comprises a clade with Parnassiaceae that is sister to Lepidobotryaceae (Figures 4.2 and 4.3). These relationships are also supported by comparative floral anatomy and morphology (Matthews & Endress, 2005). So defined, Celastrales contain about 850 species (concentrated in a few large genera) and resolve phylogenetically within a clade together with Malpighiales and Oxalidales (Figure 4.1). In North America, obligate aquatic species are found in only one family:

1. **Parnassiaceae** Gray

Family 1: Parnassiaceae [2]

Parnassiaceae are a small family of about 70 species, with all but one occurring in *Parnassia* (the second genus *Lepuropetalon* is monotypic). In many earlier taxonomic treatments, *Parnassia* was included within Saxifragaceae, often as a subfamily (Parnassioideae). However, this disposition is inconsistent with the results of several molecular phylogenetic studies that clearly resolve it along with *Lepuropetalon* as a clade within Celastrales (Figure 4.3; Soltis & Soltis, 1997; Zhang & Simmons, 2006). Although some authors combine Parnassiaceae with Celastraceae, they appear to represent distinct groups, particularly with respect to their herbaceous habit (woody in Celastraceae).

The foliage of Parnassiaceae contains epidermal tannin sacs. The flowers possess five distinct and protruding staminodes, each with glandular tips, and are set upon a nectariferous base. Other characteristics of the group are summarized in the following treatment of *Parnassia*. Some species of *Parnassia* are grown as ornamental plants, especially because of their showy flowers and unusual staminodes (the perianth is reduced in the minute *Lepuropetalon*).

The family has a holarctic and temperate distribution centered mostly in central and southeast Asia.

In North America, obligate aquatic species are found only in one genus:

1. *Parnassia* L.

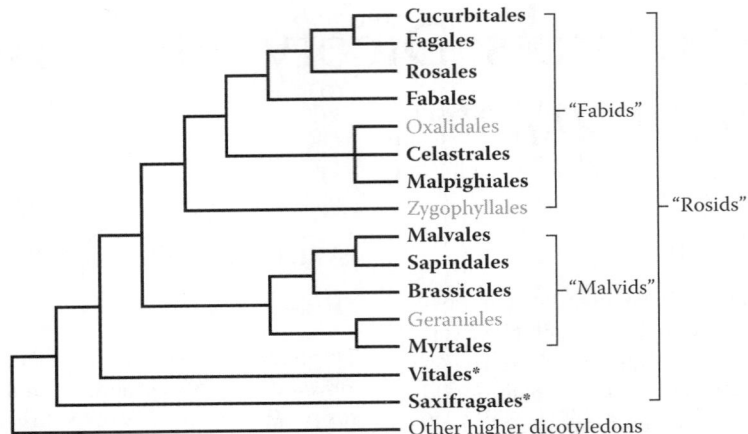

FIGURE 4.1 Phylogenetic relationships among "rosid" dicotyledon orders as indicated by analyses of combined DNA sequence data. Orders containing obligate aquatic North American taxa (indicated by bold type) are distributed widely throughout the group. (Adapted from Judd, W.S. et al., *Plant Systematics: A Phylogenetic Approach*, Sinauer Associates, Inc., Sunderland, MA, 2016.)

FIGURE 4.2 Phylogenetic relationships among families of Celastrales as indicated by analyses of DNA sequence data. Within the order, the occurrence of North America OBL indicators (indicated in bold) is confined to the family Parnassiaceae. (Adapted from Soltis, D.E. et al., *Phylogeny and Evolution of Angiosperms*, Sinauer Associates, Inc., Sunderland, MA, 2005.)

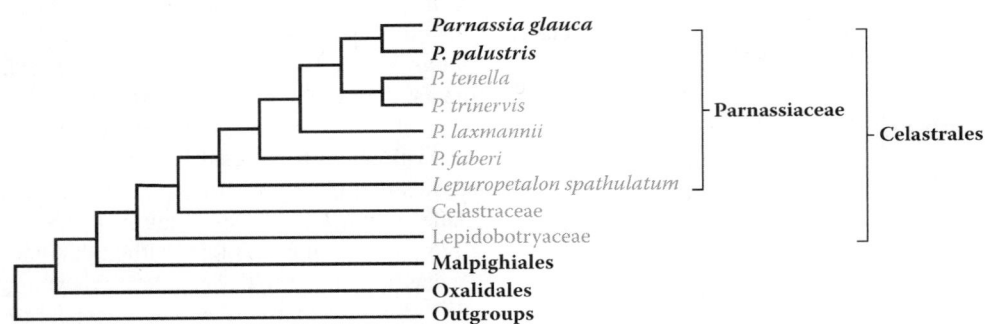

FIGURE 4.3 Phylogenetic relationships of Celastrales as indicated by analysis of combined DNA sequence data. Taxa containing OBL North American wetland indicators are indicated in bold. The two North American *Parnassia* species in the analysis associate as a clade (the remaining *Parnassia* species are Asian). These results resolve *Lepuropetalon* as the sister group to *Parnassia*, Celastraceae as the sister to Parnassiaceae, and Lepidobotryaceae as the sister to all other families of Celastrales. In this analysis, Malpighiales resolved as the sister group to Celastrales (see also Figure 4.1). (Adapted from Zhang, L.-B. & Simmons, M.P., *Syst. Bot.*, 31, 122–137, 2006.)

1. *Parnassia*

Grass-of-Parnassus; parnassie

Etymology: after a plant supposedly described by Dioscorides as growing on Mount Parnassus of Greece

Synonyms: none

Distribution: global: Asia; North America; **North America:** widespread

Diversity: global: 70 species; **North America:** 8 species

Indicators (USA): OBL: *Parnassia asarifolia*, *P. californica* [=*P. palustris*], *P. caroliniana*, *P. cirrata*, *P. fimbriata*, *P. glauca*, *P. grandifolia*, *P. kotzebuei*, *P. palustris*; **FACW:** *P. fimbriata*, *P. kotzebuei*, *P. palustris*

Habitat: brackish, freshwater; palustrine; **pH:** 5.0–8.0; **depth:** <1 m; **life-form(s):** emergent herb

Key morphology: stems reduced (rosulate); leaves (to 10 cm), long petioled (to 15 cm), cordate, ovate, or reniform; flowers showy (to 4 cm), solitary on 1–25 elongate scapes (to 6.0 dm), naked or subtended by a single leaf-like bract; petals (to 20 mm) 5, entire or fimbriate, white to pale yellow, with 3–15 prominent green or yellow veins; staminodia 5, alternating with stamens, each terminated by up to 24 glandular-tipped segments; capsules (to 15 mm) unilocular, with numerous, minute seeds

Life history: duration: perennial (rhizomes); **asexual reproduction:** rhizomes; **pollination:** insect; **sexual condition:** hermaphroditic; **fruit:** capsules (common); **local dispersal:** seeds (water, wind); **long-distance dispersal:** seeds (animals) **Imperilment:** (1) *Parnassia asarifolia* [G4]; S1 (AR, KY, MD, TX); S2 (AL, SC, WV); S3 (NC); (2) *P. caroliniana* [G3]; S2 (FL, NC, SC); (3) *P. cirrata* [G2]; S2 (CA); (4) *P. fimbriata* [G5]; S1 (WA); S3 (CA, NM, WY, <u>YT</u>); (5) *P. glauca* [G5]; SH (RI); S2 (<u>NF</u>, NH, PA, SD, <u>SK</u>); S3 (IA, ME, <u>NB</u>, <u>QC</u>); (6) *P. grandifolia* [G3]; S1 (AL, GA, KY, LA, OK, TX, WV); S2 (FL, MS, NC, SC, VA); S3 (AR, MO, TN); (7) *P. kotzebuei* [G5]; S1 (<u>SK</u>, WA); S2 (CO, ID, <u>NF</u>, <u>ON</u>, WY); S3 (<u>AB</u>, <u>LB</u>, MT); (8) *P. palustris* [G5]; SH (ND); S1 (<u>MB</u>, <u>PE</u>, SD, <u>SK</u>); S2 (<u>NS</u>); S3 (MT, <u>NF</u>, <u>QC</u>, WY)

Ecology: general: *Parnassia* species principally occur in wetlands and often in calcareous sites. All of the North American species are identified as OBL indicators in at least a portion of their range (but see comments on *P. californica*, which has been placed in synonymy with *P. palustris*). Generally, these plants appear to be highly dependent on insect pollinators to achieve maximum seed set. The pollinators are attracted to the flowers by their sharply contrasting petal veins and their glistening, glandular-tipped staminodia, which emit an attractive scent but actually produce the nectar at their base. Often categorized as "deceptive," the staminodes have been shown to increase floral visitation rates by 40%–46%. The flowers are generally described as protandrous. The seeds of many species are winged and are either dispersed abiotically by water or wind or transported in the mud that adheres to animals. Physiological seed dormancy is reported for most species. All North American species overwinter by means of a short rhizome. Some studies suggest that trampling by deer (Mammalia: Cervidae: *Odocoileus*) actually may create site conditions that are favorable to the persistence of *Parnassia* plants. Most species associate commonly with various sedges (Cyperaceae: *Carex*).

Parnassia asarifolia **Vent.** inhabits montane bogs, ravines, roadsides, seeps, swamps, and the margins of streams at elevations from 450 to 1920 m. Unlike its congeners, this species occupies organic substrates in acidic sites (pH: 5.0–6.0) and the plants will not persist long if transplanted to alkaline sites (e.g., pH: 7.8). They also occur in talus or rubble and are often found in partial shade. A proportion (up to 67%) of fresh, dry seed can germinate at 20°C (in light) over a 3-week period. However, higher germination is obtained after the seeds have been cold stratified (wet or dry) at 3°C and incubated at 16°C–29°C in the light (after 4 weeks) or dark (after 12 weeks). The most rapid germination occurs at 24°C–29°C following 8 weeks of cold stratification. New seedlings will grow optimally at 19°C. **Reported associates:** *Acer rubrum, Arethusa bulbosa, Aulacomnium palustre, Bazzania trilobata, Boykinia aconitifolia, Calamagrostis cainii, Calopogon tuberosus, Cardamine clematitis, Carex atlantica, C. collinsii, C. debilis, C. echinata, C. folliculata, C. gynandra, C.*

intumescens, C. leptalea, C. misera, C. ruthii, C. trisperma, Chelone cuthbertii, C. glabra, C. lyonii, Chrysosplenium americanum, Cicuta maculata, Dennstaedtia punctilobula, Diphylleia cymosa, Eriophorum virginicum, Gentiana linearis, Geum geniculatum, Glyceria nubigena, Helonias bullata, Houstonia serpyllifolia, Hypericum graveolens, Juncus effusus, J. gymnocarpus, J. subcaudatus, Krigia montana, Laportea canadensis, Lilium grayi, Lycopodium obscurum, Lysimachia terrestris, Medeola virginiana, Melanthium virginicum, Micranthes micranthidifolia, M. petiolaris, Mitchella repens, Nyssa sylvatica, Orontium aquaticum, Osmunda regalis, Osmundastrum cinnamomeum, Oxypolis rigidior, Packera aurea, Platanthera clavellata, P. grandiflora, P. integrilabia, Poa paludigena, Polytrichum commune, Rhizomnium appalachianum, Rhynchospora alba, R. capitellata, Sagittaria latifolia, Scirpus atrovirens, S. cyperinus, S. expansus, S. polyphyllus, Solidago glomerata, S. patula, Sphagnum affine, S. bartlettianum, S. fallax, S. palustre, S. recurvum, S. warnstorfii, Spiraea alba, S. tomentosa, Stellaria corei, Stenanthium gramineum, Symplocarpus foetidus, Thalictrum clavatum, Thelypteris noveboracensis, T. palustris, Tiarella cordifolia, Trautvetteria caroliniensis, Vaccinium macrocarpon, Veratrum viride, Viola cucullata, V. macloskeyi, V. primulifolia.

Parnassia californica **(A. Gray) Greene** [=*P. palustris*; see *Systematics*]

Parnassia caroliniana **Michx.** grows in bogs, depressions, flatwoods, prairies, savannas, and on seepage slopes. Characteristic habitats are open, sunny, and are underlain by limestone. Germination levels are higher (34%) when the seeds have been covered by a thin soil layer than when simply exposed (20%). Most seedlings emerge in February and March. Established plants are resistant to fire and show a high degree of resprouting (e.g., 82% of plants) following burns. Reports of this taxon beyond the southeastern United States are likely in error owing to misidentifications. **Reported associates:** *Agalinis, Aletris lutea, Allium, Amorpha georgiana, Aristida stricta, Balduina uniflora, Bartonia verna, Calopogon barbatus, Carex lutea, Carphephorus pseudoliatris, Chaptalia tomentosa, Ctenium aromaticum, Dichanthelium dichotomum, Drosera capillaris, Eriocaulon compressum, Eryngium aquaticum, Euphorbia inundata, Eurybia eryngiifolia, Gaylussacia dumosa, Helianthemum bicknellii, Hypericum, Justicia crassifolia, Lachnanthes caroliana, Lycopodiella alopecuroides, L. appressa, L. caroliniana, Lysimachia asperulifolia, Macbridea caroliniana, Myrica caroliniensis, M. heterophylla, Parnassia grandifolia, Pinguicula ionantha, P. lutea, P. pumila, Pinus palustris, P. serotina, Plantago sparsiflora, Pleea tenuifolia, Polygala lutea, P. polygama, Pteridium aquilinum, Pteroglossaspis ecristata, Rhexia alifanus, Rhynchospora globularis, R. oligantha, R. thornei, Rudbeckia, Sarracenia flava, Schwalbea americana, Scleria, Serenoa repens, Solidago pulchra, Sporobolus pinetorum, Syngonanthus flavidulus, Taxodium ascendens, Thalictrum cooleyi, Tofieldia glabra, Triantho racemosa, Utricularia subulata, Woodwardia virginica, Xyris ambigua, Zigadenus glaberrimus.*

Parnassia cirrata **Piper** grows in bogs, fens, marshes, meadows, seeps, slopes (to 25%), springs, washes, and along lake and stream margins at elevations from 778 to 2439 m. The site exposures range from sun to shade. The substrates have been described as calcareous, rocky, and serpentine. Flowering extends from July to October. Little other ecological information is available for this species. **Reported associates:** *Abies ×shastensis, Antennaria media, Caltha leptosepala, Carex, Chamaecyparis lawsoniana, Darlingtonia californica, Epilobium, Gentiana sceptrum, Gentianella amarella, Gentianopsis simplex, Hypericum anagalloides, Juncus, Lilium pardalinum, Luzula comosa, Malaxis brachypoda, Micranthes tolmiei, Mimulus cardinalis, Muhlenbergia andina, Narthecium californicum, Oreostemma alpigenum, Pinus contorta, P. lambertiana, P. ponderosa, Primula suffrutescens, Pseudotsuga menziesii, Quercus kelloggii, Ranunculus eschscholtzii, Rhododendron occidentale, Rubus leucodermis, R. ulmifolius, Rudbeckia californica, Sambucus nigra, Sibbaldia procumbens, Sphagnum, Tsuga mertensiana, Veratrum viride.*

Parnassia fimbriata **K.D. Koenig** occurs in montane bogs, bottomlands, ditches, floodplains, meadows, muskeg, rivulets, sand bars, seeps, slopes (to 55%), and along the banks and shores of ponds, lakes, and streams at mid to high elevations (especially subalpine zones) from 120 to 3424 m. The substrates generally are circumneutral (pH: 6.7–7.4) and are categorized as clay loam, gravel, humus, limestone breccia, loam, loomis, peat, rock, quartzite, sand, sandy loam, silt, silt loam (granitic), talus (granitic), and till. Exposures most often are described as shady. Flowering and fruiting extend from July to September. The seeds are physiologically dormant. Successful propagation has been reported after drying freshly collected seeds in paper bags and then sowing them outdoors on the surface for 5 months in full sunlight. Germination occurs when daytime temperatures reach 21°C. **Reported associates:** *Abies lasiocarpa, Achillea millefolium, Aconitum columbianum, Agrostis exarata, A. humilis, A. scabra, A. stolonifera, A. thurberiana, Allium schoenoprasum, Alnus sinuata, Alopecurus alpinus, Amerorchis rotundifolia, Androsace filiformis, Anemone parviflora, Angelica arguta, Antennaria media, A. parvifolia, Anticlea elegans, A. occidentalis, Aquilegia flavescens, Arnica chamissonis, A. diversifolia, A. latifolia, A. mollis, Aulacomnium palustre, Betula glandulosa, Bistorta bistortoides, B. vivipara, Boykinia major, Bromus carinatus, Bryum, Calamagrostis canadensis, Caltha leptosepala, Carex aquatilis, C. aurea, C. capillaris, C. disperma, C. flava, C. gynocrates, C. illota, C. interior, C. jonesii, C. laeviculmis, C. luzulina, C. macrochaeta, C. microptera, C. nigricans, C. phaeocephala, C. podocarpa, C. rostrata, C. scirpoidea, C. scopulorum, C. utriculata, C. vaginata, Cassiope mertensiana, Castilleja miniata, C. occidentalis, Catabrosa aquatica, Chamaecyparis nootkatensis, Cicuta douglasii, Cinna latifolia, Cirsium foliosum, C. hookerianum, Conioselinum scopulorum, Cornus sericea, Cypripedium calceolus, C. passerinum, Cystopteris montana, Dasiphora floribunda, Deschampsia cespitosa,*

Drosera rotundifolia, Eleocharis palustris, E. quinqueflora, Elymus glaucus, Empetrum nigrum, Epilobium ciliatum, E. halleanum, E. palustre, Equisetum arvense, E. scirpoides, Erigeron humilis, E. peregrinus, Eriophorum angustifolium, Erythronium grandiflorum, Fragaria vesca, Galium boreale, G. triflorum, Gentiana calycosa, G. sceptrum, Gentianella amarella, G. propinqua, Geranium richardsonii, Geum macrophyllum, Glyceria grandis, Hedysarum sulphurescens, Heracleum lanatum, Hieracium triste, Hypericum anagalloides, H. scouleri, Juncus balticus, J. conglomeratus, J. drummondii, J. mertensianus, Kalmia microphylla, K. polifolia, Kobresia myosuroides, Ligusticum filicinum, L. grayi, Linnaea borealis, Lonicera involucrata, Lophozia, Lupinus, Luzula campestris, L. hitchcockii, L. parviflora, Lysichiton americanus, Menyanthes trifoliata, Mertensia perplexa, Micranthes lyallii, M. odontoloma, M. subapetala, Mimulus guttatus, M. lewisii, M. primuloides, M. tilingii, Mitella pentandra, Moneses uniflora, Nephrophyllidium crista-galli, Oreostemma alpigenum, Orthilia secunda, Oxyria digyna, Packera streptanthifolia, Parnassia kotzebue, P. palustris, Pedicularis contorta, P. groenlandica, Phalaris arundinacea, Philonotis fontana, Phleum alpinum, Phyllodoce glanduliflora, Picea engelmannii, P. glauca, P. pungens, Pinus contorta, P. ponderosa, Platanthera dilatata, P. hyperborea, P. stricta, P. sparsiflora, Poa alpina, P. pratensis, Polemonium occidentale, Populus tremuloides, Potamogeton gramineus, Potentilla glandulosa, P. ×diversifolia, Primula pauciflora, P. jeffreyi, Prunella vulgaris, Pyrola uniflora, Pyrus fusca, Ranunculus acriformis, R. eschscholtzii, R. uncinatus, Rhododendron columbianum, R. groenlandicum, Ribes hudsonianum, Rumex crispus, Salix arctica, S. barclayi, S. commutata, S. drummondiana, S. eastwoodiae, S. geyeriana, S. monticola, S. nivalis, S. pedicellaris, S. planifolia, S. polaris, S. pseudomyrsinites, S. reticulata, S. vestita, S. wolfii, Sanguisorba officinalis, Sanicula graveolens, Senecio cymbalarioides, S. triangularis, Sibbaldia procumbens, Silene acaulis, Solidago multiradiata, Sphagnum, Spiraea douglasii, Spiranthes romanzoffiana, Stipa occidentalis, Streptopus amplexifolius, Sullivantia hapemanii, Swertia perennis, Symphyotrichum foliaceum, S. spathulatum, Taraxacum officinale, Thalictrum alpinum, T. occidentale, Triantha glutinosa, Trichophorum cespitosum, Trientalis europaea, Trifolium repens, Trollius laxus, Tsuga heterophylla, T. mertensiana, Vaccinium myrtillus, V. ovatum, V. oxycoccus, V. uliginosum, Valeriana dioica, V. sitchensis, Veratrum viride, Verbena verna, Veronica americana, V. wormskjoldii, Viola adunca, V. palustris.

Parnassia glauca **Raf.** is found in water (to 0.3 m) or on wet, open sites of bogs, ditches, fens, marshes, meadows, pools, prairies, prairie fens, seeps, shores, slopes, swamps, and along the margins of lakes, rivers, and streams at elevations of up to 300 m. The substrates are slightly acidic to (mostly) calcareous (pH: 6.8–8.2) and include gravel (limestone), marl, muck, peat, rock, and sand. Flowering and fruiting occur from July to October. The principal pollinators are insects (Insecta) and include bees (Hymenoptera: Apidae: *Apis mellifera, Bombus* *affinis*; Halictidae: *Lasioglossum zephyrus*; Andrenidae: *Andrena parnassiae*) and flies (Diptera: Bombyliidae: *Villa alternata*; Syrphidae: *Allograpta obliqua, Eristalis dimidiatus, E. tenax, Eupeodes americanus, Helophilus fasciatus, Melanostoma mellinum, Paragus bicolor, Sericomyia chrysotoxoides, Syritta pipiens, Toxomerus politus*; Tachinidae: *Eumea caesar, Myiopharus doryphorae*; Muscidae: *Musca domestica*; Calliphoridae: *Lucilia caesar*; Anthomyiidae: *Anthomyia, Delia platura*). The seeds have physiological dormancy and should germinate well after receiving 60 days of cold stratification. **Reported associates:** *Agalinis purpurea, A. tenuifolia, Alnus rugosa, A. viridis, Amphicarpa bracteata, Andropogon gerardii, Anemone cylindrica, Anticlea elegans, Apocynum sibiricum, Asclepias incarnata, Betula papyrifera, B. pumila, Bidens coronatus, Bromus kalmii, Cacalia tuberosa, Calamagrostis canadensis, C. stricta, Callitriche verna, Caltha palustris, Campanula aparinoides, C. rotundifolia, Campylium stellatum, Carex aquatilis, C. buxbaumii, C. crawei, C. cryptolepis, C. eburnea, C. exilis, C. flava, C. garberi, C. granularis, C. hassei, C. hystericina, C. interior, C. lacustris, C. languinosa, C. lasiocarpa, C. leptalea, C. livida, C. prairea, C. sterilis, C. stricta, C. tenuiflora, C. tetanica, C. viridula, Castilleja coccinea, Chamaelirium luteum, Chara, Chelone glabra, Cirsium muticum, Cladium mariscoides, Conocephalum conicum, Cornus alternifolia, C. amomum, C. racemosa, C. sericea, Cyperus, Cypripedium calceolus, C. reginae, Cystopteris bulbifera, Dasiphora floribunda, Deschampsia cespitosa, Dichanthelium acuminatum, Diervilla lonicera, Doellingeria umbellata, Drosera rotundifolia, Eleocharis acicularis, E. compressa, E. elliptica, E. rostellata, Equisetum arvense, E. fluviatile, E. variegatum, Eriophorum angustifolium, E. viridicarinatum, Eupatorium perfoliatum, Euthamia, Eutrochium maculatum, Floerkea proserpinacoides, Fragaria virginiana, Gentiana andrewsii, Gentianopsis crinata, G. procera, Geum aleppicum, G. rivale, Glyceria septentrionalis, G. striata, Helenium autumnale, Helianthus grosseserratus, Hieracium odorata, Hymenoxys herbacea, Hypericum kalmianum, Hypoxis hirsuta, Impatiens capensis, Iris versicolor, Juncus articulatus, J. brachycephalus, J. canadensis, J. nodosus, J. stygius, J. torreyi, Juniperus virginiana, Larix laricina, Liatris pycnostachya, Liparis loeselii, Lobelia kalmii, L. nuttallii, L. siphilitica, Lycopus americanus, L. asper, L. uniflorus, L. virginiana, Lysimachia quadriflora, Lythrum alatum, L. salicaria, Mentha arvensis, Micranthes pensylvanica, Mimulus glabratus, M. ringens, Muhlenbergia glomerata, M. richardsonis, M. uniflora, Myrica cerifera, M. gale, Onoclea sensibilis, Osmunda regalis, Oxalis, Oxypolis rigidior, Oxytropis campestris, Packera aurea, P. paupercula, Panicum flexile, P. virgatum, Parnassia palustris, Pedicularis lanceolata, Penthorum sedoides, Phlox glaberrima, P. maculata, Phragmites australis, Platanthera dilatata, Pogonia ophioglossoides, Polygala polygama, Porteranthus trifoliatus, Potentilla anserina, Primula mistassinica, Prunella vulgaris, Pycnanthemum virginianum, Ranunculus, Rhamnus alnifolia, Rhynchospora alba, R. capillacea, Rubus pubescens, Rudbeckia fulgida, R. hirta, Salix bebbii, S. candida, S. discolor, S. eriocephala, Sanguisorba*

canadensis, Sarracenia purpurea, Satureja arkansana, Schizachyrium scoparium, Schoenoplectus acutus, S. pungens, S. tabernaemontani, Scleria verticillata, Scutellaria lateriflora, Selaginella apoda, S. eclipes, Silphium integrifolium, S. terebinthinaceum, Smilax tamnoides, Solidago houghtonii, S. ohioensis, S. patula, S. ptarmicoides, S. purshii, S. riddellii, S. rigida, S. uliginosa, Sorghastrum nutans, Spartina pectinata, Spenopholis intermedia, Spiraea alba, Spiranthes cernua, S. lucida, S. romanzoffiana, Sullivantia renifolia, Symphyotrichum boreale, S. ericoides, S. firmum, S. laeve, S. lanceolatum, S. novae-angliae, S. puniceum, Symplocarpus foetidus, Thalictrum pubescens, Thelypteris palustris, Thuja occidentalis, Toxicodendron vernix, Triadenum fraseri, Triantha glutinosa, Trichophorum alpinum, T. cespitosum, Trifolium pratense, Triglochin maritimum, T. palustre, Typha angustifolia, T. latifolia, Utricularia cornuta, U. minor, Valeriana edulis, Verbena hastata, Veronicastrum, Viola nephrophylla, V. pratincola, Zannichellia palustris, Zizia aurea.

***Parnassia grandifolia* DC.** is found in or on bluffs, bogs, ledges, meadows, seeps, springs, wet cliffs, and along the margins of streams at elevations of up to 198 m. This is principally a species of limestone seepages. The habitats can be calcareous (e.g., pH: 7.8) but the plants also reportedly occur on serpentine deposits, which are more acidic (e.g., pH: 6.1). Typically, the substrates are deposits of organic matter, which overlay limestone gravel or various rocky outcrops (e.g., gabbro, gneiss, hornblende, or shale). Flowering and fruiting extend from August to October. There is little additional information on the reproductive ecology of this plant. **Reported associates:** *Acer rubrum, Adiantum capillus-veneris, A. pedatum, Agrostis perennans, Aletris farinosa, Alnus serrulata, Andropogon gerardii, Aquilegia canadensis, Aruncus dioicus, Asplenium resiliens, A. rhizophyllum, A. ruta-muraria, Calopogon tuberosus, Campylium stellatum, Cardamine bulbosa, Carex atlantica, C. buxbaumii, C. interior, C. leptalea, C. lurida, C. stricta, Castilleja coccinea, Chamaecyparis thyoides, Chelone glabra, Chenopodium simplex, Cirsium muticum, Cladium mariscoides, Cornus amomum, Cystopteris bulbifera, Dichanthelium dichotomum, Drosera rotundifolia, Eleocharis tenuis, Elymus virginicus, Eupatorium fistulosum, Fuirena squarrosa, Glyceria striata, Helenium autumnale, H. brevifolium, Helianthus angustifolius, Houstonia caerulea, H. serpyllifolia, Hypericum densiflorum, Impatiens capensis, Jamesianthus alabamensis, Juncus acuminatus, J. brachycephalus, J. coriaceus, J. diffusissimus, J. effusus, J. subcaudatus, Linum striatum, Lobelia cardinalis, L. puberula, Lyonia ligustrina, Lysimachia quadriflora, Marshallia trinervia, Mitreola petiolata, Muhlenbergia glomerata, Nasturtium officinale, Oenothera fruticosa, O. perennis, Osmunda regalis, Oxypolis rigidior, Packera aurea, Panicum virgatum, Parnassia caroliniana, Phlox glaberrima, Physocarpus opulifolius, Pilea pumila, Pinus palustris, Rhamnus alnifolia, Rhynchospora alba, R. capitellata, R. globularis, Rhytidium rugosum, Rudbeckia fulgida, R. laciniata, Salix caroliniana, S. humilis, Sanguisorba canadensis, Schizachyrium scoparium, Scirpus atrovirens, S. cyperinus,*

Selaginella apoda, Solidago caesia, S. flexicaulis, S. patula, S. uliginosa, Sphagnum bartlettianum, S. subsecundum, Spiraea alba, S. tomentosa, Symphyotrichum novi-belgii, S. pilosum, Thelypteris palustris, Trautvetteria caroliniensis, Triantha glutinosa, Viola cucullata, V. walteri, Xyris tennesseensis, X. torta.

***Parnassia kotzebuei* Cham. ex Spreng.** inhabits fresh to brackish sites associated with beaches, bluffs, cliffs, depressions, fellfields, flats, floodplains, gravel bars, gravel pits, gullies, heaths, hummocks, ledges, meadows, mudflats, muskeg, pools, roadsides, seeps, shores, slopes (to 45%), sloughs (dry), snowmelt channels, thickets, tundra, tussocks, and the margins of lakes, rivers, and streams at elevations of up to 3674 m. The plants frequently occur in snow patches at low arctic to subalpine elevations at latitudes as far north as 72°27′N. Site exposures range from open to shaded. The plants often are pioneers of wet gravel. The substrates have been described as alluvium, basalt gravel, boulders, calcareous rock, calcareous schist, clay, fibrous organic soil, granite, gravel, gravelly cobbley loam, limestone, limestone breccia, loam, marl, moss, mud, peat, quartzite, sand, sandstone, sandy clay, sandy rock, scree, shale, silt (0.01–0.1 m deep), and talus. Flowering and fruiting extend from June to August. Flowers are produced commonly and their seed maturation is high. **Reported associates:** *Abies lasiocarpa, Achillea millefolium, Aconitum, Agropyron, Allium bristylum, Alnus, Amerorchis rotundifolia, Anemone parviflora, Antennaria media, A. monocephala, Aquilegia flavescens, Arctagrostis latifolia, Arctostaphylos uva-ursi, Arnica lessingii, Artemisia arctica, A. norvegica, A. tilesii, Astragalus alpinus, A. umbellatus, Betula glandulosa, Bistorta bistortoides, B. vivipara, Braya glabella, Bromus inermis, Calamagrostis stricta, Caltha leptosepala, Cardamine bellidifolia, C. digitata, Carex aquatilis, C. aurea, C. capillaris, C. incurviformis, C. lenticularis, C. macrochaeta, C. membranacea, C. microchaeta, C. nigricans, C. nova, C. rupestris, C. saxatilis, C. scirpoidea, C. scopulorum, Cassiope mertensiana, C. tetragona, Castilleja applegatei, C. elegans, Cerastium beeringianum, Chamerion angustifolium, C. latifolium, Chrysanthemum bipinnatum, Claytonia megarhiza, Crepis nana, Cystopteris montana, Dasiphora floribunda, Deschampsia cespitosa, Draba, Dryas, Empetrum nigrum, Equisetum arvense, E. pratense, Erigeron humilis, E. peregrinus, E. purpuratus, Eriophorum angustifolium, E. scheuchzeri, Erysimum pallasii, Eurybia sibirica, Eutrema edwardsii, Festuca altaica, F. brachyphylla, F. rubra, Gentiana prostrata, Gentianella amarella, G. tenella, Geum rossii, Harrimanella stelleriana, Hedysarum alpinum, H. boreale, H. sulphurescens, Hierochloe alpina, H. odorata, Hylocomium, Juncus arcticus, J. castaneus, J. mertensianus, J. triglumis, Juniperus communis, Kobresia myosuroides, Leymus mollis, Linnaea borealis, Lomatogonium rotatum, Luetkea pectinata, Lupinus arcticus, Luzula confusa, Melandrium taylorae, Mertensia paniculata, Micranthes odontoloma, Minuartia macrocarpa, Myosotis asiatica, Oreostemma alpigenum, Oxyria digyna, Oxytropis, Papaver macounii, Parnassia fimbriata, P. parviflora, Pedicularis capitata, Petasites, Phleum alpinum, Phlox sibirica, Picea engelmannii, P. glauca, Platanthera dilatata, P. obtusata, Poa alpina, P. arctica, P. pratensis, Polemonium, Populus balsamifera, Primula egaliksensis, P. frigida, Puccinellia phryganodes, Racomitrium ericoides, Rhodiola rhodantha, R. rosea, Rhododendron tomentosum, Salix alaxensis, S. arbusculoides, S. arctica, S. barclayi, S. barrattiana, S. farriae, S. glauca, S. niphoclada, S. nivalis, S. planifolia, S. polaris, S. pulchra, S. reticulata, S. richardsonii, S. rotundifolia, S. stolonifera, S. tweedyi, Sanguisorba, Saussurea angustifolia, Saxifraga cernua, S. debilis, S. hirculus, S. oppositifolia, S. tricuspidata, Senecio congestus, S. crassulus, S. elmeri, S. lugens, Shepherdia canadensis, Sibbaldia procumbens, Silene acaulis, Solidago multiradiata, Stellaria, Swertia, Taraxacum ceratophorum, Thalictrum alpinum, Tofieldia pusilla, Trisetum spicatum, Vaccinium uliginosum, Valeriana capitata, V. sitchensis, Veronica wormskjoldii, Wilhelmsia physodes, Zygadenus elegans.*

***Parnassia palustris* L.** inhabits fresh to slightly saline beaches, bogs, bottomlands, brooklets, channels, deltas, ditches, fans, fens, floating mats, floodplains, gravel bars, gulches, gullies, heath, hot springs, hummocks, interdunal depressions, marshes, meadows, mires, moors, outwash plains, roadsides, rocky ledges, scrub, seeps, shores, silt bars, slopes (to 10%), springs, swamps, thickets, tundra, woodlands, and the margins of canals, lakes, ponds, rivers, sloughs, streams, and waterfalls at elevations of up to 3615 m. This species occurs northward to 69°42′N latitude. The plants will tolerate full sun but exposures typically are shady. The substrates range in specific conductivity from 14.39 to 31.24 dS/m and can be slightly acidic to neutral (e.g., 6.3–7.0), but more often are calcareous (pH: 7.0–7.9). They are described as alluvium, bentonite mud, Cedar mesa sandstone, clay, clay loam, cobble, cobbly humus, granite, gravel, gravelly rock, gravelly sand, limestone, loam, loess, marl, mud, peat, pebbly sand, sand, sandy gravel, sandy loam, scree, sedge peat, serpentine, silt, silty clay, tertiary claron colluvium, and till. The organic content can range from low to high. Flowering and fruiting occur from June to October. The flowers are not only self-compatible but also protandrous. Consequently, some populations are almost exclusively outcrossed (e.g., exhibiting a 95% reduction in seed set in flowers bagged to exclude pollinators), yet others show evidence of insect-mediated self-pollination. The main pollinators are flies (Insecta: Diptera: Anthomyiidae; Dolichopodidae; Empididae; Muscidae; Syrphidae; Tephritidae: *Euleia heraclei*). Population genetic studies indicate that a high genetic diversity exists within populations but low levels of gene flow and very restricted seed dispersal occur among populations. Fragmentation and loss of habitat are responsible for the severe decline of this species in some European populations. Seed germination is substantially increased with a period of dry after ripening but is highly suppressed by dark conditions. Successful seed germination has been achieved after 75 days of cold stratification (at 4°C) followed by incubation under 15°C/5°C, 25°C/5°C light and temperature regimes. Relatively low-salt concentrations (e.g., 0.06 M NaCl) can reduce seed germination rates by up to 90%. **Reported associates:** *Abies lasiocarpa, Achillea millefolium, Aconitum columbianum, A. delphiniifolium, Adiantum capillus-veneris, A. pedatum, Agrostis scabra, Allium schoenoprasum, Alnus rugosa, A. tenuifolia, A. viridis, Anemone richardsonii, Angelica arguta, A. pinnata, Antennaria microphylla, A. pulcherrima, Anticlea virescens, Apocynum, Aquilegia eximia, Arctostaphylos uva-ursi, Arnica lanceolata, Artemisia tridentata, Asclepias incarnata, Astragalus alpinus, A. eucosmus, Betula glandulosa, B. kenaica, B. nana, B. neoalaskana, B. occidentalis, B. pumila, Bistorta vivipara, Blechnum spicant, Bromopsis ciliata, Calamagrostis canadensis, C. scopulorum, C. stricta, Caltha leptosepala, Campanula parryi, C. rotundifolia, Carex aquatilis, C. athrostachya, C. aurea, C. concinna, C. crawei, C. deweyana, C. granularis, C. interior, C. lasiocarpa, C. lenticularis, C. limosa, C. nebrascensis, C. pluriflora, C. prairea, C. rariflora, C. rossii, C. rostrata, C. saxatilis, C. scirpoidea, C. simulata, C. stricta, C. utriculata, C. viridula, Castilleja caudata, C. miniata, C. minor, C. sulphurea, Centaurea maculosa, Chamaecyparis lawsoniana, Chamerion latifolium, Chelone glabra, Cirsium rydbergii, Clinopodium glabrum, Cornus amomum, C. sericea, Cypripedium californicum, Cystopteris fragilis, Darlingtonia californica, Dasiphora floribunda, Deschampsia cespitosa, Dichanthelium acuminatum, Doellingeria umbellata, Drosera rotundifolia, Dryas integrifolia, Eleocharis palustris, E. quinqueflora, E. tenuis, Empetrum nigrum, Epipactis gigantea, Equisetum arvense, E. fluviatile, E. scirpoides, E. sylvaticum, E. variegatum, Ericameria nauseosa, Erigeron coulteri, E. kachinensis, E. lonchophyllus, Eriogonum umbellatum, Eriophorum angustifolium, E. chamissonis, E. crinigerum, E. russeolum, E. viridicarinatum, Eupatorium perfoliatum, Eutrochium maculatum, Festuca rubra, Filipendula occidentalis, Fragaria virginiana, Frasera speciosa, Gentiana andrewsii, G. fremontii, Gentianella amarella, Gentianopsis detonsa, G. procera, Geranium erianthum, G. richardsonii, Geum aleppicum, G. macrophyllum, Glyceria elata, Halenia deflexa, Hastingsia alba, Hedysarum alpinum, Helenium autumnale, H. bigelovii, Helianthus grosseserratus, Heracleum sphondylium, Hieracium aurantiacum, H. florentum, Hippuris vulgaris, Hordeum brachyantherum, Iris missouriensis, Juncus balticus, J. ensifolius, J. vaseyi, Juniperus scopulorum, Kalmia microphylla, Koeleria macrantha, Larix, Lathyrus palustris, Lemna, Leymus mollis, Ligusticum, Lilium, Liparis loeselii, Lobelia kalmii, L. siphilitica, Lupinus arcticus, L. leucophyllus, Luzula multiflora, Lycopodium, Lythrum alatum, Matricaria discoidea, Menyanthes trifoliata, Micranthes odontoloma, M. pensylvanica, Muhlenbergia andina, M. filiformis, M. glomerata, M. richardsonis, Narthecium californicum, Oenothera elata, Onoclea sensibilis, Orthilia secunda, Oxypolis rigidior, Oxytenia acerosa, Oxytropis campestris, Packera pseudaurea, Parnassia fimbriata, P. glauca, Pedicularis capitata, P. groenlandica, P. lanceolata, P. verticillata, Petasites frigidus, Phlox maculata, Picea engelmannii, P. glauca, P. mariana, Pinguicula vulgaris, Pinus contorta, Plantago major,* *Platanthera dilatata, P. hyperborea, P. leucostachys, P. sparsiflora, P. stricta, P. zothecina, Poa alpina, P. pratensis, P. secunda, P. triflora, Polemonium acutiflorum, Populus balsamifera, P. tremuloides, P. trichocarpa, Potentilla anserina, P. gracilis, Primula egaliksensis, P. frigida, P. incana, P. mistassinica, P. tetrandra, Prunella vulgaris, Pseudotsuga menziesii, Pycnanthemum virginianum, Pyrola grandiflora, Ranunculus abortivus, R. cymbalaria, Rhamnus alnifolia, R. betulifolia, Rhinanthus minor, Rhodiola integrifolia, Rhododendron columbianum, R. occidentale, R. tomentosum, Rhynchospora capillacea, Ribes hudsonianum, Rosa acicularis, Rubus chamaemorus, R. parviflorus, Rudbeckia californica, R. hirta, Salix alaxensis, S. arbusculoides, S. arctica, S. barclayi, S. bebbiana, S. bebbii, S. boothii, S. brachycarpa, S. candida, S. discolor, S. drummondiana, S. farriae, S. geyeriana, S. glauca, S. monica, S. monochroma, S. niphoclada, S. planifolia, S. pseudomonticola, S. pseudomyrsinites, S. pulchra, S. richardsonii, S. serissima, S. wolfii, Sanguisorba microcephala, Schizachyrium scoparium, Schoenoplectus acutus, Scirpus microcarpus, Scutellaria galericulata, Shepherdia canadensis, Sisyrinchium montanum, S. pallidum, Solidago missouriensis, S. riddellii, Sorbus, Sphagnum, Sphenopholis obtusata, Sphenosciadium capitellatum, Spiraea alba, S. stevenii, Spiranthes cernua, S. romanzoffiana, Stellaria longipes, Swertia perennis, Symphyotrichum boreale, S. ericoides, S. puniceum, S. spathulatum, Taraxacum officinale, Thalictrum alpinum, T. sparsiflorum, Thelypteris palustris, Thuja occidentalis, Triantha glutinosa, Trichophorum cespitosum, Trifolium longipes, Triglochin maritimum, T. palustre, Trisetum spicatum, Vaccinium uliginosum, Valeriana capitata, V. edulis, Viola lanceolata, V. nephrophylla.*

Use by wildlife: *Parnassia fimbriata* is the host plant for a larval moth (Insecta: Lepidoptera: Yponomeutidae: *Zelleria parnassiae*) and a rust pathogen (Fungi: Basidiomycota: Pucciniaceae: *Puccinia parnassiae*). *Parnassia fimbriata* and *P. palustris* are both hosts for plant pathogenic chytrids (Fungi: Chytridiomycota: Synchytriaceae: *Synchytrium aureum*). *Parnassia glauca* is a habitat component of the bog turtle (Reptilia: Testudines: Emydidae: *Glyptemys muhlenbergii*), a rare dragonfly (Insecta: Odonata: Corduliidae: *Somatochlora incurvata*), and an extremely rare snail (Mollusca: Gastropoda: Succineidae: *Catinella exile*). Its flower-visiting insects (Insecta) include wasps (Hymenoptera: Ichneumonidae: *Lissonata clypeator, L. coriacinus*) and butterflies (Lepidoptera: Nymphalidae: *Vanessa virginiensis*; Pieridae: *Pieris rapae*), which collect the nectar.

Economic importance: food: *Parnassia* species are fairly tanniniferous and not edible as a result; **medicinal:** the Cheyenne made tea from the leaves of *P. fimbriata*, which was administered to young children with stomach ailments. The Gosiute applied a poultice of the plant as wash for treating venereal disease. It is sold in some modern markets in the form of a therapeutic, natural healing essence. *Parnassia palustris* is attributed with astringent, curative, diuretic, and sedative properties. It has been used as an ingredient in mouthwash to treat inflammation and as an eyewash; **cultivation:** despite the beauty of these plants, it is surprising that they are not often found in cultivation. *Parnassia palustris* appears most often in the cultivated plant trade and is sometimes distributed under the cultivar name 'Miniana.' Several endemic North American species (e.g., *P. californica*, *P. fimbriata*, *P. glauca*) are sold occasionally as native wildflowers; **misc. products:** *P. glauca* has been used in some habitat restoration projects; **weeds:** none; **nonindigenous species:** none.

Systematics: Molecular data firmly resolve *Parnassia* as the sister clade to the monotypic *Lepuropetalon*, and the close association of that larger clade (Parnassiaceae) to Celastraceae (Figure 4.3). Nine sections are recognized within this relatively large genus, for which a comprehensive phylogenetic survey of species relationships remains to be conducted. Preliminary studies (Figure 4.3) are consistent with the sectional divisions, by resolving the North American *P. glauca* and *P. palustris* (both section *Parnassia*) as a clade distinct from species in other sections. *Parnassia kotzebuei* reportedly is closely related to *P. palustris*. *Parnassia californica* has been placed in synonymy with *P. palustris*, and the data collected for the former taxon have been merged with the latter. Few other details of relationships among the North American species are known at this time, although their chromosome numbers provide some additional insight. The base chromosome number of *Parnassia* usually is reported as $x = 9$. However, the morphologically similar eastern North American species *P. asarifolia*, *P. caroliniana*, *P. glauca*, and *P. grandifolia* all are $2n = 32$, which indicates that they are tetraploids derived from a base number of $x = 8$. Because all other counts reported for *Parnassia* are based on $x = 9$, the four eastern North American taxa presumably comprise a distinct group of closely related species. *Parnassia fimbriata* ($2n = 36$) is tetraploid; *P. kotzebuei* ($2n = 18$) primarily is diploid, with tetraploid ($2n = 36$) cytotypes also reported. *Parnassia palustris* possesses several cytotypes ($2n = 18, 27, 36, 54$). In Europe, autoploid (tetraploid) populations (verified by tetrasomic inheritance) are formed repeatedly in *P. palustris* and are distributed more northerly than the diploids. The closely related genus *Lepuropetalon* has an anomalous chromosome number of $2n = 46$. *Parnassia parviflora* is often recognized as a variety of *P. palustris* (var. *parviflora*), which can create some uncertainty regarding literature accounts for these species. Confusion also exists in the earlier literature, where the name *P. caroliniana* was misapplied to some populations of *P. glauca*.

Comments: *Parnassia koteztbuei* occurs in northern portions of North America (it is disjunct in the southern Rockies) and in Asia. *Parnassia palustris* is circumpolar and occurs in northern and western North America. The remaining species are of more limited distribution: *P. glauca* (northeastern North America), *P. asarifolia*, *P. caroliniana*, and *P. grandifolia* (southeastern United States), *P. fimbriata* (western North America), *P. californica* (California, Nevada, Oregon), and *P. cirrata* (California and Mexico).

References: Aiken et al., 2007; Baskin & Baskin, 1998; Bonnin et al., 2002; Borgen & Hultgård, 2003; Evans et al., 2001a; Farmer, 1980; Gastony & Soltis, 1977; Glitzenstein et al., 1998; Howat, 2000; Jensen, 2004; Lenz, 2004; Lesica &

McCune, 2004; Mansberg & Wentworth, 1984; Matthews & Endress, 2005; Munz & Johnston, 1922; Pellerin et al., 2006; Phillips, 1982; Picking & Veneman, 2004; Rigg, 1940; Sandvik & Totland, 2003; Smiley, 1915; Spetzman, 1959; Wu et al., 2005a,b; Zhang & Simmons, 2006

ORDER 2: CUCURBITALES [7]

The Cucurbitales contain about 2250 species and together with Fagales, comprise a clade that is related phylogenetically to Fabales and Rosales (Figure 4.1). An amalgam of 10 molecular data sets [including chloroplast DNA (cpDNA), mitochondrial DNA (mtDNA), nuclear DNA loci] has provided strong evidence for the monophyly of the order (Figure 4.4). The order is almost entirely terrestrial, with only a few species well adapted to wet habitats. In North America, obligate aquatic species are restricted to one family:

    1. **Cucurbitaceae** Juss.

Family 1: Cucurbitaceae [4]

Molecular data (Zhang et al., 2006; Kocyan et al., 2007) show Cucurbitaceae to be monophyletic and placed centrally in Cucurbitales as the sister group of a clade comprising Begoniaceae, Datiscaceae, and Tetramelaceae (Figure 4.4). The family is readily recognized by its herbaceous species having spirally coiled and branched tendrils, which originate from the nodes (Judd et al., 2016). The flowers usually are unisexual (the plants dioecious or monoecious) and produce fruits containing distinctively flattened seeds. The flowers often are large and showy, and are pollinated by birds (Aves), insects (Insecta), or mammals (Mammalia). Animals are also the primary dispersal vectors, although the seeds of some species are dispersed ballistically (Judd et al., 2016).

The family is important as a source of edible vegetables, which include cantaloupe, cucumbers, and melons (*Cucumis*), gourds, pumpkins, and squash (*Cucurbita*), and watermelons (*Citrullus*). Cucurbitaceae mainly are tropical or subtropical, with few temperate representatives.

OBL indicators occur only within one North American genus:

    1. *Cucurbita* L.

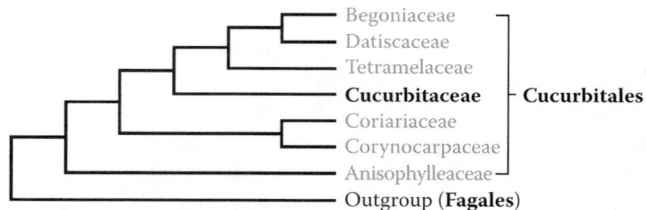

FIGURE 4.4 Phylogenetic relationships within Cucurbitales inferred from a combined analysis of cpDNA, mtDNA, and nuclear DNA sequences. Only one North American species in one family (Cucurbitaceae) has been designated as an obligate aquatic (taxa containing obligate indicators are shown in bold). (Adapted from Zhang, L.-B. et al., *Mol. Phyogen. Evol.*, 39, 305–322, 2006.)

1. *Cucurbita*

Gourd; courge

Etymology: from the Latin *cucurbita* meaning "gourd"

Synonyms: *Pepo*; *Tristemon*

Distribution: global: neotemperate; neotropical; **North America:** southern

Diversity: global: 14 species; **North America:** 8 species

Indicators (USA): OBL: *Cucurbita okeechobeensis*

Habitat: freshwater; palustrine; **pH:** unknown; **depth:** <1 m; **life-form(s):** emergent (herb; vine)

Key morphology: stems viny, twining, with unequally branched tendrils (probably modified shoots) that occur sub-opposite to the leaves; leaves (to 2 dm) alternate, three to seven palmately lobed; flowers solitary, showy, bell shaped, unisexual; hypanthium of male flowers densely pubescent; petals (to 7 cm) creamy white; ovary pubescent; fruits (to 9 cm) light green with 10 indistinct stripes; seeds (to 12 mm) flat, gray green

Life history: duration: annual (fruit/seeds); **asexual reproduction:** shoot fragments; **pollination:** insect; **sexual condition:** monoecious; **fruit:** berries (infrequent); **local dispersal:** fruits/seeds (water); **long-distance dispersal:** seeds (animals)

Imperilment: (1) *Cucurbita okeechobeensis* [G1]; S1 (FL)

Ecology: general: *Cucurbita* is almost entirely terrestrial, with species that occur most often in open, sandy sites. They are all herbaceous vines, which bear unisexual flowers that are pollinated primarily by insects (Insecta). There is only one OBL indicator in North America.

Cucurbita okeechobeensis **(Small) L.H.** Bailey grows in bottomlands, lowland forests swamps, thickets, and along the margins of canals and lakes at low elevations. The plants occur in areas of full sunlight, which are devoid of dense canopy growth. The substrates have been characterized as Torry muck. Flowering has been observed from March to November and fruiting from July to October. Pollination is believed to occur by various insects (Insecta) such as bees (Hymenoptera), flies (diptera), and beetles (Coleoptera). The viny plants must be supported physically by other species (often *Annona* branches). Some vegetative reproduction can occur during the growing season by the fragmentation of the stems, which will root at their nodes. The plants require fluctuating water levels. High waters promote their dispersal and reduce the occurrence of competing vegetation; whereas, low levels facilitate seed germination and seedling growth. This species is typically found rooting in the nesting sites of the American alligator (Reptilia: Alligatoridae: *Alligator mississippiensis*), which provide openings for seed germination to occur, while also affording reduced levels of competition. Germination occurs during the dry season and the seedlings cannot tolerate extended periods of inundation. Seed dispersal is impeded by dense stands of *Colocasia esculenta* (Araceae). These plants have been extirpated from all but a few sites along the shores of Lake Okeechobee and margins of the St. Johns River (Florida, USA). This rare species is threatened by the conversion of suitable swamp forest habitat to agricultural uses and by management practices, which maintain stable, high water levels. **Reported associates:** *Acer rubrum, Ampelaster carolinianus, Annona glabra, Arundo donax, Carya aquatica, Cephalanthus occidentalis, Colocasia esculenta, Fraxinus caroliniana, Gleditsia aquatica, Ipomoea alba, Melaleuca, Mikania scandens, Panicum hemitomon, Phragmites australis, Salix caroliniana, Sambucus nigra, Taxodium distichum, Thalia geniculata.*

Use by wildlife: *Cucurbita okeechobeensis* is a host plant for several larval Lepidoptera (Insecta) including saltmarsh caterpillars (Arctiidae: *Estigmene acrea*), melon worms (Pyralidae: *Diaphania hyalinata*), and the related pickle worms (Pyralidae: *Diaphania nitidalis*). The fruits attract beetles (Insecta: Coleoptera: Chrysomelidae) including southern corn rootworms (*Diabrotica undecimpunctata howardi*) and striped cucumber beetles (*Acalymma vittatum*). The seeds are eaten and dispersed by marsh rabbits (Mammalia: Leporidae: *Sylvilagus palustris*) and other rodents (Mammalia: Rodentia).

Economic importance: food: *Cucurbita okeechobeensis* contains high concentrations of bitter cucurbitacins (tetracyclic triterpenoids), which render the fruits inedible and toxic; however, the seeds reportedly are edible; **medicinal:** an infusion made from the fruits of *C. okeechobeensis* subsp. *martinezii* has been used as a remedy for stomach ache; the flesh of the fruit has been used as a burn ointment; **cultivation:** *C. okeechobeensis* is cultivated as a germplasm resource because it is resistant to many plant pathogens including bean yellow mosaic virus (Geminiviridae: *Begomovirus*), cucumber mosaic virus (Bromoviridae: *Cucumovirus*), squash mosaic virus (Secoviridae: *Comovirus*), tobacco ringspot virus (Secoviridae: *Nepovirus*), tomato ringspot virus (Secoviridae: *Nepovirus*), and powdery mildew (Fungi: Ascomycota: Erysiphales). The species reportedly is also susceptible to watermelon mosaic 2 potyvirus (Potyviridae: *Potyvirus*) and zucchini yellow mosaic potyvirus (Potyviridae: *Potyvirus*); **misc. products:** *C. okeechobeensis* has been considered as a commercial source of cucurbitans for use as "baits" to discourage crop infestations by rootworms and cucumber beetles. The fruit flesh has been used as a soap; **weeds:** none; **nonindigenous species:** none.

Systematics: Molecular phylogenetic analyses of Cucurbitaceae (*matK, rbcL, rpl20–rps12* intron, *trnL* intron/spacer sequence data) resolve two monophyletic groups, which coincide with the taxonomic subfamilies Cucurbitoideae and Nhandiroboideae. The former subfamily also includes a "CBC clade," which contains the monophyletic New World tribe Cucurbiteae, resolved with the respective sister groups Benincaseae and Coniandreae. Within Cucurbiteae, *Cucurbita* resides in a position closely related to *Peponopsis*. *Cucurbita okeechobeensis* originally was distinguished from the Mexican *C. martinezii* with which it apparently shares a close relationship. When allozyme analyses indicated that the two species differed only by a single detectable allele, they were reclassified as conspecific subspecies with *C. okeechobeensis* becoming *C. okeechobeensis* subsp. *okeechobeensis* and *C. martinezii* becoming *C. okeechobeensis* subsp. *martinezii*. Closely related to *C. okeechobeensis* is *C. lundelliana*, a species from the Yucatan region of Mexico. All three taxa

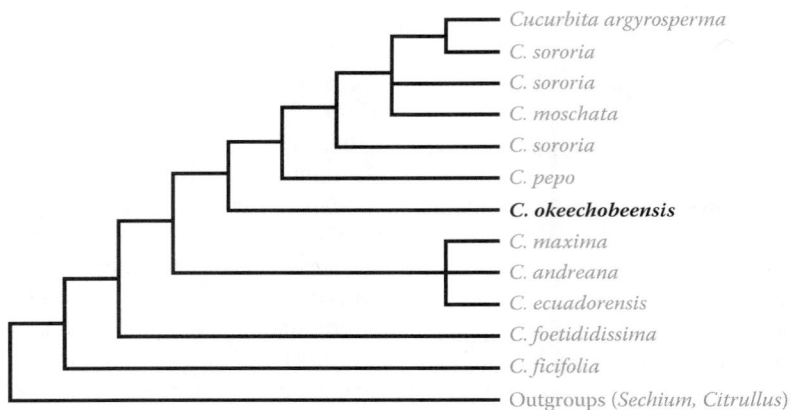

FIGURE 4.5 Phylogenetic relationships among *Cucurbita* species as indicated by analysis of mtDNA (*nad1*) sequence data. The phylogenetic position of *C. okeechobeensis*, the only species designated as an obligate aquatic (bold) in North America, is intermediate among species in the genus. A similar result was obtained by the analysis of 74 SSR loci, which also placed *C. okeechobeensis* centrally in *Cucurbita*, but as closely related to *C. lundelliana* (Gong et al., 2013). (Adapted from Sanjur, O.I. et al., *Proc. Natl. Acad. Sci. U.S.A*, 99, 535–540, 2002.)

are distinct morphologically and together form a group that is well-distinguished morphologically within *Cucurbita*.

Allozyme data yield a high genetic identity (*I*) between the subspecies of *C. okeechobeensis* (*I* = 0.91), with a reduced identity of the subspecies to *C. lundelliana* (*I* = 0.61–0.71). An analysis of 74 simple sequence repeat (SSR) loci also clearly showed *C. okeechobeensis* as being closely related to *C. lundelliana*. Phylogenetic analysis of mtDNA sequences confirmed that the group of species including *C. okeechobeensis* is allied most closely to *C. pepo*. (Figure 4.5). *Cucurbita okeechobeensis* is cross-compatible with both *C. lundelliana* and *C. pepo*. The base chromosome number of Cucurbitaceae is *x* = 7–14; the chromosome number for *C. okeechobeensis* has not been reported.

Comments: *Cucurbita okeechobeensis* is endemic to the Florida Everglades.

References: Andres & Nabhan, 1988; de Vaulx & Pitrat, 1979; Kocyan et al., 2007; Lira & Caballero, 2002; Robinson & Puchalski, 1980; Sanjur et al., 2002; USFWS, 1995a, 1999; Walters & Decker-Walters, 1993; Walters et al., 1992; Ward & Minno, 2002.

ORDER 3: FABALES [4]

Phylogenetically, the large order Fabales with nearly 19,000 species, is closely related to Rosales (Figure 4.1). Relationships among the four families of Fabales remain unsettled (Wojciechowski, 2003; Bello et al., 2009). Analyses of combined molecular data (18S rDNA, *rbcL*, *atpB* sequences) resolve Polygalaceae and Surianaceae as a clade, which is sister to Fabaceae (Soltis et al., 2000; Wojciechowski, 2003). However, other analyses using cpDNA (*rbcL*; *trnL* intron sequences) and a larger sampling of taxa resolve a clade comprising Fabaceae and Surianaceae, which is sister to Polygalaceae and subtended ultimately by Quillajaceae as the sister family to the remainder of the order (Forest et al., 2002). Combined *matK* and *rbcL* sequence data (Bello et al., 2009) favor yet a different topology, where Quillajaceae and

Surianaceae form a sister clade to Fabaceae, which is subtended by Polygalaceae (Figure 4.6). The Fabales principally are terrestrial but exhibit extreme diversity of habit. They are important economically for their many food plants and other products. Ellagic acid, a phenolic compound with anticarcinogenic properties, occurs throughout the group. Obligately aquatic North American species occur within the two largest families:

1. **Fabaceae** Lindl.
2. **Polygalaceae** Hoffmanns. & Link

Family 1: Fabaceae [4]

The Fabaceae with more than 18,000 species are the third largest angiosperm family (Judd et al., 2016). Analyses of much molecular and morphological data consistently depict the family as monophyletic (Wojciechowski, 2003; Bello et al., 2009; Judd et al., 2016). Many previous classifications distinguished three separate families within the group (i.e., Caesalpiniaceae, Fabaceae, Mimosaceae); however, subsequent phylogenetic analyses (e.g., Bruneau et al., 2001; Kajita et al., 2001) resolved only the latter two as clades. In current treatments, these same taxa often are treated as three subfamilies (Figure 4.7): Faboideae and Mimosoideae (each

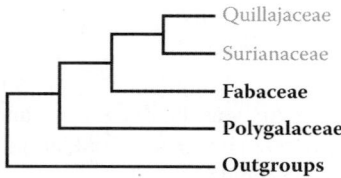

FIGURE 4.6 Phylogenetic relationships among the four families of Fabales as indicated by cpDNA data. Although other molecular data sets resolve different topologies, the two families containing aquatic representatives (shown in bold) do not resolve as a single clade, which indicates independent origins of the OBL habit in the order. (Adapted from Bello, M.A. et al., *Syst. Bot.*, 34, 102–114, 2009.)

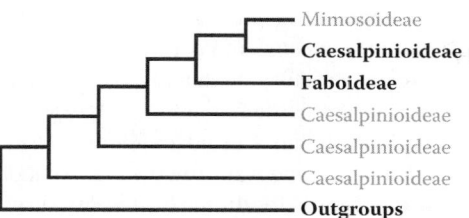

FIGURE 4.7 Phylogenetic relationships among subfamilies of Fabaceae as indicated by analysis of *rbcL* sequence data. Although Faboideae and Mimosoideae are monophyletic, subfamily Caesalpinioideae is highly paraphyletic and polyphyletic with respect to the taxa associated with Mimosoideae. Groups containing obligate aquatics in North America are indicated in bold. (Adapted from Kajita, T. et al., *Syst. Bot.*, 26, 515–536, 2001.)

monophyletic) and Caesalpinioideae (strongly paraphyletic, polyphyletic).

Phylogenetic analyses have demonstrated the lack of a direct relationship between the Old World and New World taxa formerly assigned to the large genus *Lotus* (Allan & Porter, 2000; Sokoloff et al., 2007; Degtjareva et al., 2006). As a consequence, the New World segregates have been assigned to four distinct genera: *Acmispon, Hosackia, Ottleya,* and *Syrmatium* (Brouillet, 2008). By following this convention, the two OBL indicators assigned previously to *Lotus,* have been included here in *Hosackia.*

Fabaceae contain high nitrogen metabolism plants, which generally possess nitrogen-fixing root nodules. Most of the species share a suite of morphological features including compound leaves with pulvini, perigynous flowers with a short hypanthium, corollas often with an enlarged "banner" petal, a single carpel, and legume fruits (Judd et al., 2016). Their distribution essentially is cosmopolitan.

There are numerous economically important species in Fabaceae (Judd et al., 2016). The edible plants include many familiar "legumes" such as beans (*Phaseolus*), lentils (*Lens*), peanuts (*Arachis*), peas (*Pisum*), and soybeans (*Glycine*). Several genera (*Medicago, Melilotus, Trifolium, Vicia*) are important as animal forage crops. The familiar blue indigo dye is derived from *Indigofera.* Many species are cultivated as garden or ornamental plants. The most notable of these occur in the genera *Acacia, Albizia, Anthyllis, Astragalus, Baptisia, Bauhinia, Caragana, Carmichaelia, Cassia, Cercis, Cytisus, Erythrina, Genista, Gleditsia, Glycine, Indigofera, Laburnum, Lathyrus, Lespedeza, Lotus, Lupinus, Medicago, Ononis, Oxytropis, Phaseolus, Pisum, Robinia, Senna, Sophora, Trifolium, Vicia,* and *Wisteria.*

Fabaceae principally are terrestrial with the few aquatics found mainly in subfamily Faboideae. In North America obligate aquatic species occur within eight genera. *Amorpha* (facultative wetland (FACW), facultative (FAC), facultative upland (FACU), upland (UPL)), a former OBL indicator, has been excluded:

1. *Aeschynomene* L.
2. *Astragalus* L.
3. *Gleditsia* L.
4. *Hoita* Rydb.
5. *Hosackia* Douglas ex Benth.
6. *Lathyrus* L.
7. *Trifolium* L.
8. *Vicia* L.

Further insight from phylogenetic studies will undoubtedly lead to revised classifications of Fabaceae. Following existing circumscriptions, *Gleditsia* would be assigned to subfamily Caesalpinioideae with the remaining aquatic genera assigned to subfamily Faboideae.

1. *Aeschynomene*

Jointvetch

Etymology: from the Greek *aeschyn mene* meaning "moon shy" (with reference to the sensitive leaves, which fold up at night or when touched)

Synonyms: *Bakerophyton; Climacorachis; Hedysarum; Herminiera; Hippocrepis* (in part); *Rueppellia; Secula*

Distribution: global: pantropical; **North America:** southern

Diversity: global: 150 species; **North America:** 10 species

Indicators (USA): OBL: *Aeschynomene pratensis, A. virginica;* **FACW:** *A. virginica*

Habitat: brackish, freshwater (tidal); palustrine; **pH:** 4.4–7.9; **depth:** <1 m; **life-form(s):** emergent (herb)

Key morphology: stems (to 2.4 m) herbaceous to suffrutescent, single or branching at tips, smooth to hairy; leaves (to 12 cm) alternate, even-pinnately compound with 20–56 gland-dotted leaflets (to 2 cm), petiolate (to 10 cm), slightly sensitive (folding upon touch); racemes axillary, with two to six flowers; flowers bilateral; corolla papilionaceous, petals (to 15 mm) yellow to orange (suffused with red); fruit (to 6 cm) stalked (to 25 mm), a four to nine segmented loment

Life history: duration: annual (fruit/seeds), perennial (persistent shoot bases); **asexual reproduction:** none; **pollination:** insect, self; **sexual condition:** hermaphroditic; **fruit:** legume (loment) (common); **local dispersal:** seeds (gravity); **long-distance dispersal:** seeds (water)

Imperilment: (1) *Aeschynomene pratensis* [G4]; S1 (FL); (2) *A. virginica* [G2]; SX (DE, PA); S1 (MD, NC, NJ); S2 (VA)

Ecology: general: Although only two species are designated as OBL indicators in North America, nearly half of all *Aeschynomene* species (and 40% of the North American species) are aquatic, or are at least adapted for FAC survival in wetland habitats. *Aeschynomene* includes annual and perennial species, which produce root nodules and are nitrogen fixing; several contain stem nodulating bacteria. Many have tactile-sensitive leaves, which fold up during the night, or when touched. The flowers are self-pollinated, or are outcrossed by insects such as bees (Insecta: Hymenoptera: Apidae), which can be quite specialized. The seeds of some species float well and are dispersed by water currents.

Aeschynomene pratensis **Small** is a perennial, which occurs in marshes, roadsides, sloughs, and along the margins of canals and pinelands throughout the Florida Everglades at low elevations. The extent of nitrogen fixation in *A. pratensis* is uncertain. Root nodules are present and the plants exhibit minor nodulation on submerged portions of the stem.

The leaves sometimes harbor low densities of heterologous rhizobia bacteria (i.e., those associated with other congeners). Little information exists on the life history of this species. **Reported associates:** *Bacopa caroliniana, Blechnum serrulatum, Cephalanthus occidentalis, Cladium jamaicense, Crinum americanum, Eleocharis cellulosa, Hydrolea corymbosa, Hymenocallis palmeri, Justicia angusta, Leersia hexandra, Nymphaea odorata, Panicum hemitomon, P. tenerum, Paspalidium geminatum, Peltandra virginica, Pontederia cordata, Utricularia purpurea.*

Aeschynomene virginica **(L.) Britton, Sterns & Poggenb.** is an annual, which inhabits the fresh to brackish (0.7–0.8 ppt salinity) margins of tidally influenced estuaries, lakes, rivers, and streams at low elevations. The plants grow in open, disturbed sites, which often occur in newly created habitats. The substrates can be fairly acidic (e.g., pH: 4.4) and include gravel, mud, peat, and sand. The plants usually grow along the margins of marshland, which is formed at the upper limit of tidal influence. Barren or sparsely vegetated areas, such as sites denuded by muskrat (Mammalia: Cricetidae: *Ondatra zibethicus*) activity, are necessary for establishment. Fluctuating water levels are essential for reducing competition by other species. The flowers are self-compatible and are highly self-pollinating; however, bumblebees (Insecta: Apidae: *Bombus*) have also been observed as pollinators. Individual plants can produce upward of 300 seeds, which fall mainly within a 1 m radius. The fruits break into small units that can remain afloat for more than 80 h and are capable of dispersal by water to distances of more than 2600 m. The water-transported seeds are very effective at establishing new populations. Germination rates are higher on wet than on waterlogged soils and often are high within deposits of flotsam. The seeds have physical dormancy and once scarified will germinate under a 30°C/15°C day/night temperature regime. The seedlings cannot establish on submerged soils. A substantial seed bank develops, which can persist for at least a season. Annual population sizes vary considerably, fluctuating in one instance from 50 to 2000 individuals over 3 years of observation. Population models indicate that the maintenance of open space for colonization via disturbance is more effective in promoting the long-term survival of populations than is enhanced seed dispersal. The plants are threatened by alteration or degradation of their habitats and by invasion of aggressive species such as *Phragmites australis* (Poaceae). **Reported associates:** *Bidens laevis, Bolboschoenus robustus, Chamaecrista fasciculata, Cicuta maculata, Cyperus strigosus, Echinochloa walteri, Juncus effusus, Leersia oryzoides, Lobelia cardinalis, Ludwigia palustris, Murdannia keisak, Panicum virgatum, Peltandra virginica, Persicaria arifolia, P. hydropiperoides, P. punctata, P. sagittatum, Pontederia cordata, Schoenoplectus americanus, S. tabernaemontani, Spartina cynosuroides, Typha angustifolia, Zizania aquatica.* **Use by wildlife:** Plants of *A. virginica* are associated with various insects (Insecta) including bumblebees and leaf-cutter bees (Hymenoptera: L Apidae: *Bombus*; Megachilidae) and least skipper butterflies (Lepidoptera: Hesperiidae: *Ancyloxypha numitor*). The seeds of *A. virginica* are eaten by several larvae (Insecta: Lepidoptera: Noctuidae) including tobacco budworms (*Heliothis virescens*) and corn earworms (*Helicoverpa zea*). *Aeschynomene* species (with a high sugar content and up to 20% protein) are regarded as good forage plants for birds (Aves), deer (Mammalia: Cervidae: *Odocoileus*), livestock (Mammalia: Bovidae), and waterfowl (Aves: Anatidae).

Economic importance: food: none; **medicinal:** none; **cultivation:** Seeds of *A. virginica* and other species are sold commercially for planting as wet area forage; **misc. products:** none; **weeds:** Although *A. virginica* has been reported as a weed, this unfortunate designation is owing to taxonomic confusion with *A. indica*, a distinct, nonindigenous species; the accounts of *A. virginica* as a weed of rice and other crops are erroneous; **nonindigenous species:** both of the obligate aquatic species are indigenous

Systematics: *Aeschynomene* is the type genus of tribe Aeschynomeneae in subfamily Faboideae of Fabaceae. Morphological cladistic studies have indicated a close relationship of *Aeschynomene* and *Ochopodium*, with the latter also treated as a section of the former. Molecular data [nuclear ribosomal internal transcribed spacer (nrITS), *matK*, *trnK* intron sequences] place *Aeschynomene* species within a monophyletic group referred to as the "*Dalbergia* clade," which occurs within a larger, monophyletic "dalbergioid clade." The constituent genera are diverse morphologically, but share the presence of aeschynomenoid root nodules. However, combined morphological and molecular data indicate that species in *Aeschynomene* section *Ochopodium* are distinct from other *Aeschynomene* and are closer to *Machaerium* and *Dalbergia*; the remaining *Aeschynomene* species (including *A. virginica*) are allied with a clade (section *Aeschynomene*) containing *Cyclocarpa, Soemmeringia,* and other genera, whose close relationships have also been indicated by morphological cladistic analyses (Figure 4.8). *Aeschynomene virginica* is morphologically similar to the introduced North America species *A. indica* and *A. rudis* and was once thought to be conspecific with the former. However, DNA sequence data (Figure 4.9) have clarified that *A. virginica* is an allopolyploid derived from the *A. rudis* and *A. scabra* lineages (Figure 4.9). Isozyme data clearly demonstrate the distinctness of all three species and experimental interspecific crosses among them are unsuccessful. Isozyme data also reveal extremely low levels of either intra- or interpopulational genetic variation in *A. virginica*. Taxonomically, the North American representatives of *A. pratensis* are recognized as variety *pratensis*; plants in the Caribbean, Central America, and South America are referred to as *A. pratensis* var. *caribaea*. Analyses of single-copy DNA sequences have shown that *A. pratensis* is an allopolyploid derivative of *A. sensitiva* and related but unknown lineage. The base chromosome number of *Aeschynomene* is $x = 10$. *Aeschynomene virginica* and *A. pratensis* are tetraploids ($2n = 40$) of allopolyploid origin.

Comments: *Aeschynomene virginica* is a rare, regional endemic with about 24 extant populations surviving in coastal areas of the eastern and southeastern United States. *Aeschynomene pratensis* var. *pratensis* is a rare taxon endemic to south Florida.

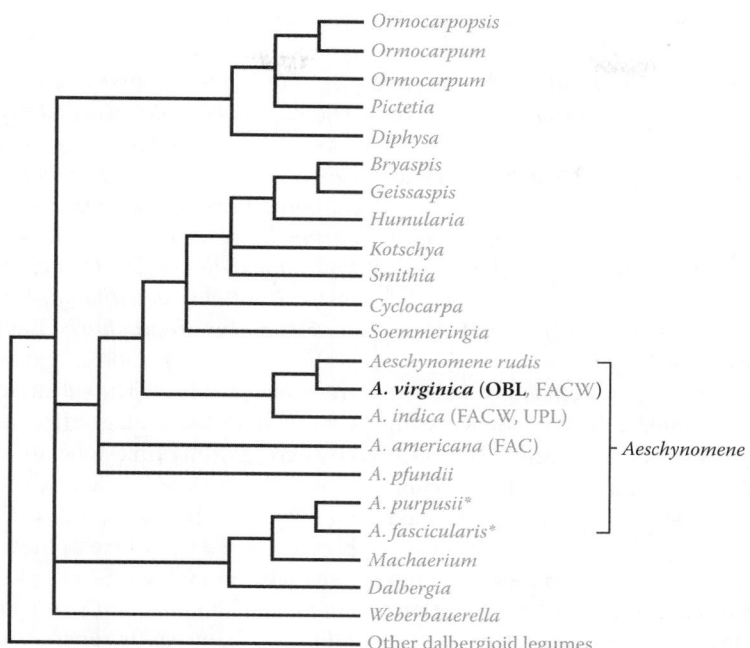

FIGURE 4.8 Phylogenetic relationships among "dalbergioid legumes" (Fabaceae) as indicated by cladistic analysis of combined morphological and molecular data. As formerly circumscribed, *Aeschynomene* is polyphyletic, with species in section *Ochopodium* (indicated by asterisks) distantly related to others in the genus. The North American obligate aquatic *A. virginica* (indicated in bold) is related closely to *A. rudis* and *A. indica* with which it has been confused taxonomically. Subsequent studies (see Figure 4.9) have shown that *A. virginica* is an allotetraploid, with *A. rudis* comprising one parental lineage. (Adapted from Lavin, M. et al., *Amer. J. Bot.*, 88, 503–533, 2001.)

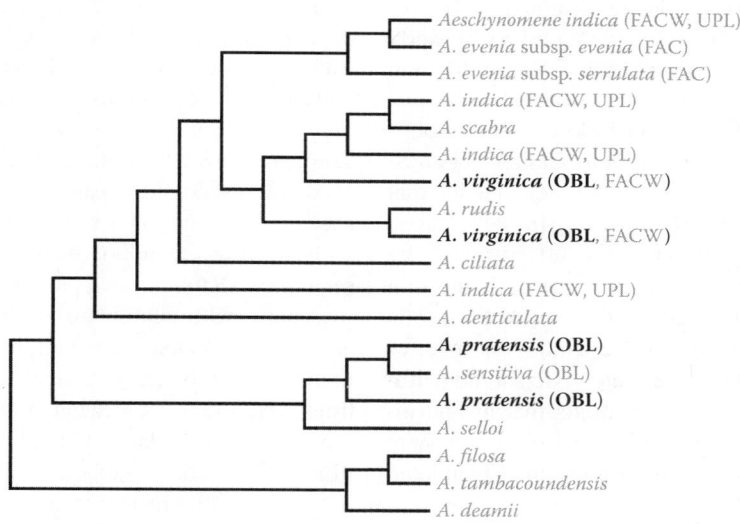

FIGURE 4.9 Phylogenetic relationships among *Aeschynomene* species based on analysis of single-copy nuclear DNA sequences. The two tetraploid OBL North American indicators (bold) are of allopolyploid origin with *A. virginica* derived from the *A. rudis* and *A. scabra* lineages, and *A. pratensis* derived from *A. sensitiva* and an unidentified lineage. Note: *A. sensitiva* is ranked as OBL, but is not North American as defined in this book. In any case, it is evident that the OBL habit has originated independently in *A. pratensis* and *A. virginica*. (Adapted from Arrighi, J.F. et al., *New Phytol.*, 201, 1457–1468, 2014.)

References: Adebayo, 1989; Arrighi et al., 2014; Bailey et al., 2006b; Baskin et al., 1998; Beyra-Matos & Lavin, 1999; Blanchard, 1992; Cardoso et al., 2013; Carleial et al., 2015; Carulli & Fairbrothers, 1988; Griffith, 2014; Griffith & Forseth, 2002, 2003, 2005, 2006; Isely, 1990; Lavin et al., 2001; McKellar et al., 1991; Ross et al., 2006; Rudd, 1955; USFWS, 1995b.

2. *Astragalus*

Milkvetch; astragale

Etymology: from the Greek *astragalos* meaning "anklebone" with reference to the seed shape

Synonyms: *Aragallus; Astracantha; Biserrula; Diplotheca; Hamosa; Homalobus; Jonesiella; Ophiocarpus; Orophaca; Phaca; Sewerzowia; Tragacantha*

Distribution: global: Northern Hemisphere; **North America:** widespread

Diversity: global: 3000 species; **North America:** 400 species

Indicators (USA): OBL: *Astragalus leptaleus, A. limnocharis, A. pycnostachyus*

Habitat: brackish (coastal), freshwater; palustrine; **pH:** alkaline; **depth:** <1 m; **life-form(s):** emergent (herb)

Key morphology: stems (to 9 dm) erect, sometimes hollow; leaves (2–15 cm) alternate, odd-pinnately compound, with 5–41 leaflets (to 30 mm); stipules present; racemes (to 6 cm) with 5–12 papilionaceous flowers; petals (to 15 mm) cream, greenish-white, lavender, or purple; fruit (to 26 mm) beaked (to 8 mm), inflated, strigose

Life history: duration: perennial (caudex, tap root); **asexual reproduction:** none; **pollination:** insect; **sexual condition:** hermaphroditic; **fruit:** legumes (common to infrequent); **local dispersal:** seeds; **long-distance dispersal:** seeds

Imperilment: (1) *Astragalus leptaleus* [G3/G4]; S1 (WY); S2 (CO); S3 (ID, MT); (2) *A. limnocharis* [G2]; S1/S2 (UT); (3) *A. pycnostachyus* [G2]; S1/S2 (CA)

Ecology: general: *Astragalus* is primarily a terrestrial genus, which is often found in dry sandy or gravelly sites. Twenty-one of the North American species (5%) are wetland indicators, but only three (<1%) are designated as OBL. *Astragalus lentiginosus* (FACU, UPL) was ranked formerly as an OBL indicator, but no longer retains that status and is excluded here. This diverse genus includes annuals and perennials, species that are self-compatible as well as obligately outcrossed, and flowers that are cleistogamous or chasmogamous. However, most of the several dozen species that have been studied are primarily outcrossed. In these cases, the flowers are pollinated by insects (Insecta), primarily by large bees (Hymenoptera: Anthophoridae; Apidae), and to a lesser degree by beetles (Coleoptera) and flies (Diptera). The seeds are dispersed by animals or by the wind. A number of *Astragalus* species effectively concentrate toxic soil constituents such as selenium and contain alkaloids that are harmful to livestock. Some of the species can also fix nitrogen; however, the extent of this process has not yet been evaluated adequately in the OBL aquatic species.

Astragalus leptaleus **A. Gray** occurs in bottomlands, gullies, hummocks, meadows, seeps, and along the margins of brooks, springs, and streams at elevations of up to 2978 m. Exposures can vary from full sun to shade. The substrates are alkaline and have been described as alluvium, loam, and sand. Flowering occurs from June to August with fruiting from July to September. Other life history information is lacking for this species. **Reported associates:** *Abies lasiocrpa, Antennaria anaphaloides, A. microphylla, Aquilegia formosa, Astragalus agrestis, A. alpinus, A. argophyllus, A. diversifolius, A. eucosmus, Betula glandulosa, Bistorta vivipara, Carex idahoa, Dasiphora floribunda, Deschampsia cespitosa, Gentiana fremontii, Gentianopsis thermalis, Hesperochiron pumilus, Hordeum brachyantherum, Iris missouriensis, Juncus balticus, Juniperus communis, Lysimachia maritima, Oxytropis deflexa, Packera debilis, Pedicularis groenlandica, P. racemosa, Phlox kelseyi, Pinus contorta, Poa pratensis, Populus, Potentilla anserina, Pseudotsuga menziesii, Pyrrocoma uniflora, Ranunculus cymbalaria, Salix boothii, S. brachycarpa, S. geyeriana, Shepherdia, Sisyrinchium idahoense, Spiranthes diluvialis, Thalictrum alpinum, Trifolium longipes, Triglochin maritimum, Zizia aptera.*

Astragalus limnocharis **Barneby** occurs along barrens, high elevation lakeshores, and slopes (to 70%) at elevations from 946 to 3143 m. The substrates are described as alluvium, Claron limestone, Flagstaff limestone, granular clay, gravelly clay, gravelly limestone, limestone shale, marl limestone, sandy clay, volcanic talus, white Wasatch limestone gravel, and Wasatch limestone talus. Flowering and fruiting have been observed from June to July. Additional life history information is lacking for this species. **Reported associates:** *Abies, Arctostaphylos patula, Castilleja revealii, Cercocarpus ledifolius, Cymopterus minimus, Dasiphora floribunda, Draba subalpina, Eriogonum panguicense, Phlox pulvinata, Picea, Pinus flexilis, P. longaeva, Silene petersonii.*

Astragalus pycnostachyus **A. Gray** occurs in fresh to brackish sites (salinity to 7.9 mmhos/cm) along the California coast including bluffs, cienega, dunes, flats, marshes, meadows, salt marshes (above tidal limit), seeps, slopes, and the margins of lagoons and rivers at elevations of up to 152 m. The plants are believed to persist primarily by colonizing disturbed or sparsely vegetated areas. Although the plants can grow under somewhat brackish conditions, they do not tolerate even brief periods of inundation by saltwater unless they are dormant. Site exposures are described as open. The substrates are characterized as gravel, loamy silt, sand, sandy clay, sandy hardpan, sandy loam, sandy rocky clay, and shells. Flowering occurs from mid-June to early September, with fruiting extending from June to October. The plants senesce in October. Each plant produces on average, 26 inflorescences, with each bearing about 37 flowers. The flowers are self-compatible and their percent of fruit maturation is high. The fruits are persistent, weakly dehiscent, and yield three seeds on average, but occasionally as many as seven. Open-pollinated plants generate slightly higher (but not significantly different) seed set than selfed plants, but the species is believed to be primarily autogamous. The flowers also are visited by insects (e.g., Hymenoptera: Apidae: *Bombus*), which might facilitate some level of cross-pollination. Seed production can be extremely high, with an estimated 3.7 million seeds produced by a group of only 29 plants studied during one growing season. The seeds possess a hard coat, making them resistant to water imbibition, and therefore germination. Scarification by physical abrasion is necessary to facilitate water imbibition. Presumably, a substantial, long-lived seed bank develops, facilitating the ready colonization of newly disturbed sites. This is a rare species, once thought to have gone extinct, but eventually rediscovered. **Reported associates:** *Achillea millefolium, Acmispon glabrus, Allium, Baccharis* *pilularis, B. salicifolia, Bromus madritensis, Carpobrotus edulis, Castilleja affinis, C. latifolia, Centaurea melitensis, Corethrogyne filaginifolia, Cortaderia selloana, Distichlis spicata, Dudleya farinosa, Ericameria ericoides, Eriogonum latifolium, Eriophyllum staechadifolium, Euthamia occidentalis, Frankenia salina, Grindelia, Heliotropium curassavicum, Heterotheca sessiliflora, Hirschfeldia incana, Jaumea carnosa, Limonium, Medicago, Melilotus albus, Myoporum laetum, Polypogon monspeliensis, Rhus ovata, Salicornia, Salix lasiolepis, Sarcocornia perennis.*

Use by wildlife: The foliage of some *Astragalus* species is toxic and can cause locoism or reproductive dysfunction in livestock that graze on it (hence the common name of "locoweed" often is applied to this genus). *Astragalus leptaleus* is grazed by cattle (Mammalia: Bovidae: *Bos taurus*). The foliage of *A. pycnostachyus* is grazed by milk snails (Mollusca: Gastropoda: Helicidae: *Otala lactea*) and by mammals (Mammalia), including rabbits (Leporidae) and small rodents (Rodentia) such as gophers (Geomyidae), ground squirrels (Sciuridae: *Spermophilus*), and meadow voles (Cricetidae: *Microtus*). *Astragalus pycnostchyus* is also fed on by aphids (Insecta: Hemiptera: Aphididae). Senescing leaves of *A. pycnostachyus* are susceptible to attack by a sooty fungus (Fungi: Ascomycota). The seeds of *A. pycnostachyus* often are infested with seed beetles (Insecta: Coleoptera: Bruchidae), which can seriously reduce plant fecundity.

Economic importance: food: The foliage is not edible and should be avoided because of potentially high concentrations of dangerous alkaloids and other toxic substances such as selenium; **medicinal:** none reported; **cultivation:** *Astragalus limnocharis* is listed as a species suitable for cultivation in rock gardens; **misc. products:** none reported; **weeds:** none; **nonindigenous species:** none.

Systematics: Phylogenetic relationships of *Astragalus* have been clarified to some degree by the application of molecular data; however, much work still needs to be carried out in this vastly speciose genus. Nuclear ITS sequence data indicate that *Astragalus* sensu stricto is monophyletic and related closely to a clade that includes *Biserrula, Colutea* (and other genera of the "Coluteoid clade"), *Oxytropis, Podlechiella,* and several "straggler" *Astragalus* species, which evidently are misplaced taxonomically (Figure 4.10). Molecular data also indicate that none of the traditionally defined subgenera is monophyletic. Studies incorporating various molecular data [nrITS and *trnL* sequences; cpDNA restriction fragment length polymorphisms (RFLPs)] resolve a clade of aneuploid, New World species, which has been labeled as the "Neo-Astragalus" group. *Astragalus leptaleus* (the only obligate aquatic species included in molecular studies) falls within this clade. *Astragalus limnocharis* is believed to be closely related to *A. jejunus,* a species of dry habitats; however, this relationship needs to be verified. The chromosomal base number of *Astragalus* is x = 8. North American and other New World "Neo-Astragalus" are aneuploids with chromosomal base numbers of x = 11–15. Chromosome counts for the OBL species have not been reported.

Comments: Although a widespread genus, the OBL taxa are restricted to western North America. *Astragalus limnocharis*

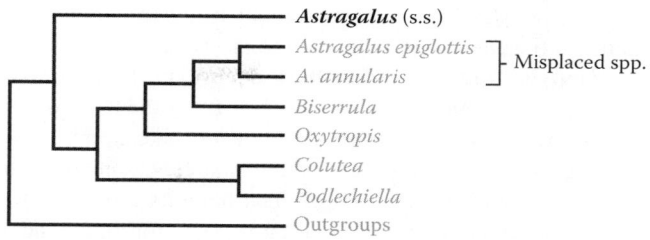

FIGURE 4.10 Phylogenetic relationships of *Astragalus* sensu stricto (s.s.) as indicated by nrITS sequence data. *Astragalus leptaleus,* the only OBL species (in bold) yet included in phylogenetic analyses of the genus, resolves within the upper clade of species. (Adapted from Wojciechowski, M.F. et al., *Syst. Bot.,* 24, 409–437, 1999; Osaloo, S.K. et al., *Plant Syst. Evol.,* 242, 1–32, 2003.)

(Utah) and *A. pycnostachyus* (California) are narrowly endemic. *Astragalus leptaleus* is somewhat more widespread in the central Rocky Mountain region of the United States.

References: Barneby, 1989; Green & Bohart, 1975; Jensen, 2007; Jones, 2001a; Molyneux & James, 1982; Moseley, 1991; Osaloo et al., 2003; Van der Pijl, 1969; Watrous & Cane, 2011; Wilken & Wardlaw, 2001; Wojciechowski et al., 1999, 2000; Ziemkiewicz & Cronin, 1981.

3. *Gleditsia*

Honeylocust, water locust

Etymology: after Johann Gottlieb Gleditsch (1714–1786)

Synonyms: *Garugandra*

Distribution: global: Africa; Asia; New World; **North America:** eastern, southeastern

Diversity: global: 13 species; **North America:** 2 species

Indicators (USA): OBL: *Gleditsia aquatica*

Habitat: freshwater; palustrine; **pH:** 5.0–8.0; **depth:** <1 m; **life-form(s):** emergent (tree)

Key morphology: deciduous trees (to 25 m), the branches and trunks armed with slender, mostly simple thorns (to 14 cm); leaves (to 20 cm) alternate, even-pinnate (or bipinnately) compound, with 9–18 pairs of leaflets (to 4 cm); spikes (to 9 cm) elongate; flowers hermaphrodite, pistillate, or staminate; petals (to 4 mm) 3–5, yellow-green; fruits (to 5 cm) flat, elliptic to ovoid, 1(-3)-seeded; seeds not surrounded by pulp

Life history: duration: perennial (buds); **asexual reproduction:** root, stump sprouts; **pollination:** insect; **sexual condition:** polygamomonoecious; **fruit:** legumes (common); **local dispersal:** seeds (water); **long-distance dispersal:** seeds (animals)

Imperilment: (1) *Gleditsia aquatica* [G5]; S1 (IN); S3 (KY)

Ecology: general: All *Gleditsisa* species are trees, which most often are associated with UPL rather than lowland habitats. Both North American species are wetland indicators, but only one is designated as OBL. The flowers can be male, female, or hermaphroditic on the same plant (i.e., polygamomonoecious). They are pollinated by insects (Insecta), which typically include small bees (Hymenoptera) or flies (Diptera). The seeds are dispersed primarily by birds (Aves) or other animals (Mammalia), but in some plants also by water.

They are physically dormant and require physical abrasion or chemical (i.e., acidic) scarification to promote germination.

***Gleditsia aquatica* Marshall** occurs in bottomlands, ditches, floodplains, hammocks, levees, roadsides, swales, swamps, and along the margins of bayous, lakes, ponds, rivers, salt marshes, sloughs, and streams at elevations of up to 150 m. It is hardy in zones 6–9. The habitats are subjected to periodic inundation, which often persists for an extended length of time. After prolonged inundation, the twigs will produce enlarged (hypertrophied) lenticels, which facilitate oxygen uptake. The plants are not shade tolerant but do tolerate atmospheric pollution. They form root nodules and fix nitrogen, which also becomes available to the surrounding vegetation. The substrates are described as alluvial clay, clay loam, loamy sand, muck, sand, sandy alluvium, and silt. The flower buds appear in May and open by June. The flowers are pollinated by small insects (Insecta). Fruits have been observed from July to November. The seeds have physical dormancy and require some form of scarification for germination (which is enhanced from passing through animal digestive tracts). Presoaking the seeds for 24 h in warm water causes them to swell, which then usually enables them to germinate at 20°C within 4 weeks. The seeds retain their viability under flooded conditions for up to 23 months. They are dispersed by water, birds (Aves), and other animals (Mammalia). **Reported associates:** *Acer rubrum, A. saccharinum, Betula nigra, Carya aquatica, C. ovata, Celtis laevigata, C. occidentalis, Cephalanthus occidentalis, Cornus foemina, Decodon verticillatus, Diospyros virginiana, Forestiera acuminata, Fraxinus caroliniana, F. pennsylvanica, F. profunda, Ilex verticillata, Itea virginica, Liquidambar styraciflua, Nyssa aquatica, N. sylvatica, Planera aquatica, Platanus occidentalis, Populus deltoides, P. heterophylla, Quercus laurifolia, Q. lyrata, Q. palustris, Q. texana, Salix nigra, Taxodium distichum, Ulmus americana.*

Use by wildlife: *Gleditsia aquatica* is often described as unimportant to wildlife; however, it is considered a masting species whose seeds are eaten by small mammals (Mammalia) and quail (Aves: Odontophoridae). These plants also provide valuable nesting cover and habitat for birds such as the eastern screech owl (Aves: Strigidae: *Megascops asio*). *Gleditsia* species are regarded as good nectar and pollen plants for honey bees (Insecta: Hymenoptera: Apidae: *Apis mellifera*), which also function as pollinators. Plants of *G. aquatica* can harbor "bracket" wood-rotting fungi (Basidiomycota: Gloeophyllaceae: *Veluticeps abietina*).

Economic importance: food: none; **medicinal:** none; **cultivation:** *Gleditsia aquatica* is rarely cultivated, probably owing to its wet habitat requirements; **misc. products:** The durable wood of *G. aquatica* is resistant to wood-decaying "honey" fungi (Basidiomycota: Physalacriaceae: *Armillaria* spp.) and is used for making fence posts. Owing to its extensive root system, the species sometimes is planted for bank stabilization. The wood has been used occasionally for cabinetry; **weeds:** none; **nonindigenous species:** *Gleditsia aquatica* reportedly is naturalized in New York.

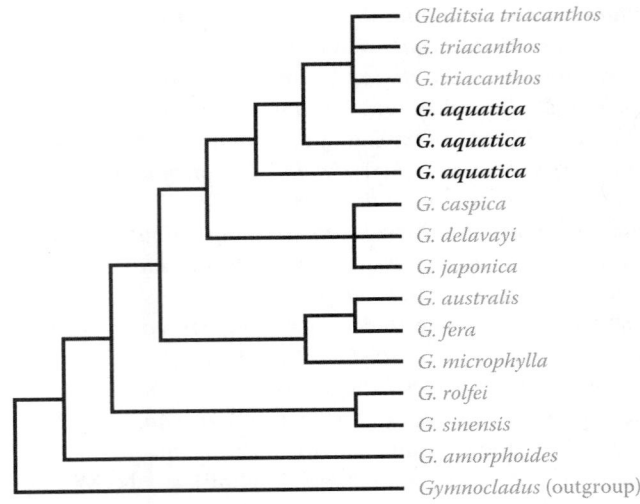

FIGURE 4.11 Phylogenetic relationships in *Gleditsia* as indicated by combined analysis of nrITS and cpDNA sequence data. Molecular data indicate that the aquatic *G. aquatica* (indicated in bold) and the more upland *G. triacanthos* are closely related, recently diverged species. (Adapted from Schnabel, A. et al., *Amer. J. Bot.*, 90, 310–320, 2003.)

Systematics: Traditionally, *Gleditsia* is placed within the tribe Caesalpinieae, subfamily Caesalpinioideae of Fabaceae (but see remarks on the paraphyly of Caesalpinioideae above). Molecular data (*matK, 3′-trnK,* and *trnL* intron sequences) indicate that *Gleditsia* is related most closely to *Gymnocladus,* with which it forms a clade (Figure 4.11). Phylogenetic relationships within *Gleditsia* have been elucidated quite completely based on cladistic analysis of nuclear and chloroplast DNA data (nrITS, *ndhF, rpl16* sequences) obtained for 11 of the 13 species (Figure 4.11). These studies indicate that *Gleditsia aquatica* and *G. triacanthos* are recently diverged sister species. The close relationship of these species is evidenced further by their ability to form interspecific hybrids that have been referred to taxonomically as *Gleditsia ×texana*. The chromosomal base number of *Gleditsia* is $x = 14$. Counts are unavailable for *G. aquatica*; however, the related *G. triacanthos* ($2n = 28$) is diploid.

Comments: *Gleditsia aquatica* occurs in the southeastern United States along the coastal plain and Mississippi embayment regions.

References: Bruneau et al., 2001, 2008; Elias, 1987; Fessel & Middleton, 2000; Middleton, 2000; Schnabel & Wendel, 1998; Schnabel et al., 2003.

4. Hoita

Leather root

Etymology: the Native American name for the plant

Synonyms: *Psoralea*

Distribution: global: North America; **North America:** western

Diversity: global: 3 species; **North America:** 3 species

Indicators (USA): OBL: *Hoita macrostachya, H. orbicularis*

Habitat: brackish (coastal), freshwater; palustrine; **pH:** 5.0–7.2; **depth:** <1 m; **life-form(s):** emergent (herb)

Key morphology: stems (to 30 dm) erect or creeping; foliage glandular, odorous; leaves (to 8 cm) alternate, trifoliate, petiolate (to 6 dm); stipules (to 1 cm) present; racemes (to 35 cm) spike-like, densely pubescent; petals (to 16 mm) purple or reddish purple; fruit (to 9 mm)

Life history: duration: perennial (creeping stems, woody rootstocks); **asexual reproduction:** shoot fragments; **pollination:** insect; **sexual condition:** hermaphroditic; **fruit:** legumes (common); **local dispersal:** creeping stems, seeds; **long-distance dispersal:** seeds

Imperilment: (1) *Hoita macrostachya* [G5]; (2) *H. orbicularis* [G4]

Ecology: general: *Hoita* is a small genus of only three species, of which two (OBL indicators) are adapted to wetland conditions. The third species grows in chaparral and woodlands. The flowers are pollinated by insects (Insecta). The plants reportedly produce root nodules and fix nitrogen. Otherwise, the genus is poorly known ecologically.

Hoita macrostachya (DC.) Rydb. grows in freshwater sites that occur in bottoms, chaparral, creek beds, ditches, flats, floodplains, gullies, marshes, meadows, ravines, riverbeds, roadsides, sandbars, seeps, slopes (to 5%), springs, swamps, thickets, washes, woodlands, and along the margins of ponds, rivers, sloughs, and streams at elevations of up to 1703 m. The substrates are described as clay, granite colluvium, gravel, gravelly loam, loamy sand, rock, rocky clay, rocky gravelly loam, rocky sand, rocky shale, sand, sandstone, sandy gravel, sandy loam, serpentine, silt, silty clay loam, and silty loam. Exposures range from open to shaded sites. Flowering occurs from May to August. The seeds have physical dormancy and require scarification to induce germination. Germination can be enhanced by soaking the seeds in warm water for 24 h prior to planting. **Reported associates:** *Acer macrophyllum, Acmispon glabrus, A. heermannii, Adenostoma fasciculatum, Agrostis stolonifera, Ailanthus altissima, Alnus rhombifolia, A. tenuifolia, Anemopsis californica, Arctostaphylos, Artemisia californica, A. douglasiana, A. dracunculus, Avena barbata, Baccharis pilularis, B. salicifolia, B. sarothroides, Bahiopsis laciniata, Barbarea orthoceras, Brickellia californica, Bromus rubens, B. tectorum, Calamagrostis ophitidis, Callitropsis sargentii, Calocedrus decurrens, Calochortus albus, Calystegia occidentalis, Carex spissa, Ceanothus crassifolius, C. jepsonii, C. spinosus, Cercis occidentalis, Cercocarpus betuloides, Cirsium vulgare, Clarkia rhomboidea, Clematis ligusticifolia, Conyza canadensis, Cornus, Cryptantha, Cyperus, Datisca glomerata, Datura quercifolia, D. wrightii, Diplacus aurantiacus, Dryopteris arguta, Echinochloa crus-galli, Encelia farinosa, Epilobium canum, E. ciliatum, Equisetum arvense, Eriastrum sapphirinum, Eriodictyon crassifolium, Eriogonum elongatum, E. fasciculatum, Foeniculum vulgare, Forestiera, Fraxinus dipetala, F. latifolia, Funastrum cynanchoides, Galium angustifolium, Gastridium ventricosum, Hazardia, Helianthus californicus, Hesperoyucca whipplei, Heteromeles arbutifolia, Hirschfeldia incana, Hosackia oblongifolia, Ibicella lutea,* *Juglans californica, Juncus balticus, J. effusus, J. xiphioides, Keckiella cordifolia, Lepidospartum squamatum, Lonicera subspicata, Lythrum hyssopifolia, Malacothamnus fasciculatus, Malosma laurina, Melilotus albus, Mimulus cardinalis, Muhlenbergia rigens, Nasturtium officinale, Oenothera elata, Paspalum distichum, Perideridia kelloggii, Persicaria, Phacelia imbricata, Physalis, Pinus jeffreyi, P. sabiniana, Platanus racemosa, Populus fremontii, Pteridium aquilinum, Quercus agrifolia, Q. berberidifolia, Q. dumosa, Q. kelloggii, Q. wislizeni, Rhamnus californica, R. ilicifolia, Ribes, Rosa californica, Rubus bifrons, R. ulmifolius, R. ursinus, Salix exigua, S. gooddingii, S. laevigata, S. lasiolepis, S. lucida, Salvia apiana, S. mellifera, Schoenoplectus acutus, S. californicus, Scirpus microcarpus, Scrophularia californica, Silene laciniata, Spartium junceum, Stachys rigida, Taraxacum officinale, Toxicodendron diversilobum, Typha domingensis, T. latifolia, Urtica dioica, Verbena lasiostachys, Veronica anagallis-aquatica, Viguiera purisimae, Vitis californica, V. girdiana, Woodwardia fimbriata, Xanthium strumarium.*

Hoita orbicularis (Lindl.) Rydb. occurs in canyon or creek bottoms, chaparral, cienaga, ditches, dunes, flats, hollows, marshes, meadows, roadsides, salt marshes, seeps, slopes (to 10%), springs, vales, woodlands, and along the margins of rivers, salt marshes, sloughs, and streams at elevations of up to 2134 m. The plants can grow in shallow water. They supposedly are shade intolerant, but have been reported from open to semishaded sites in habitats that receive 50–150 cm rainfall annually. The substrates have been characterized as boulders, clay, cobbles, decomposed granite, duff, gravel, James Canyon loam, loamy sand, rock, rocky sand, sand, sandy loam, serpentine, and silty loam. Flowering and fruiting extend from April to August. The seeds have physical dormancy and require scarification to induce their germination. Germination can be enhanced by soaking the seeds in warm water for 24 h prior to planting. The plants have a creeping habit and spread and propagate vegetatively by rhizomes. However, this species reportedly is difficult to transplant without incurring severe damage to its fragile root system. **Reported associates:** *Achillea millefolium, Adenostoma fasciculatum, Agrimonia gryposepala, Alnus rhombifolia, Ambrosia psilostachya, Anemopsis californica, Aquilegia formosa, Arctostaphylos, Artemesia douglasiana, A. dracunculus, Baccharis douglasii, B. salicifolia, Brodiaea orcuttii, Bromus, Calocedrus decurrens, Carex senta, Ceanothus crassifolius, C. palmeri, C. tomentosus, Cirsium vulgare, Clematis ligusticifolia, Cornus sericea, Cryptantha intermedia, Cyperus eragrostis, C. niger, Datisca glomerata, Dipsacus sativus, Eleocharis montevidensis, Epilobium ciliatum, Epipactis gigantea, Equisetum, Eriastrum densifolium, Eucalyptus, Festuca rubra, Gilia, Helianthus californicus, Helminthotheca echioides, Hosackia oblongifolia, Hypericum scouleri, Juncus balticus, J. effusus, J. macrophyllus, Lilium parryi, Lolium multiflorum, Lupinus latifolius, Mimulus cardinalis, M. guttatus, Muhlenbergia asperifolia, M. rigens, Nassella pulchra, Oenanthe sarmentosa, Oenothera elata, Perideridia parishii, Persicaria hydropiperoides, Phalaris arundinacea, Pinus ponderosa, P. radiata, Platanus racemosa, Poa*

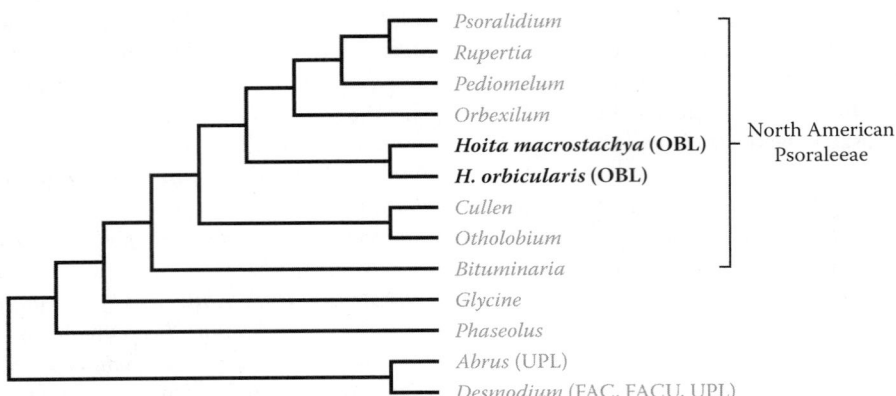

FIGURE 4.12 Phylogenetic relationships of *Hoita* as indicated by analysis of combined DNA sequence data. These results resolve the two OBL *Hoita* species (bold) as a clade situated centrally within the North American members of Fabaceae tribe Psoraleeae, which otherwise is devoid of wetland indicators (shown in parentheses). (Adapted from Egan, A.N. & Crandall, K.A., *BMC Biol.*, 6, 55, 2008.)

pratensis, Prunella vulgaris, Prunus, Pteridium aquilinum, Pycnanthemum californicum, Quercus agrifolia, Q. chrysolepis, Q. kelloggii, Rhamnus californica, Rhus ovata, Rosa californica, Rubus ursinus, Rumex salicifolius, Salix laevigata, S. lasiolepis, Schoenoplectus americanus, Scirpus microcarpus, Scutellaria californica, Sidalcea malviflora, Solidago confinis, Stachys albens, S. chamissonis, S. pycnantha, Symphoricarpos albus, Toxicodendron diversilobum, Typha latifolia, Umbellularia californica, Urtica dioica, Venegasia carpesioides, Vitis girdiana.

Use by wildlife: *Hoita macrostachya* is a host plant for larval moths (Insecta: Lepidoptera: Noctuidae: *Papaipema angelica, P. circumlucens*; Oecophoridae: *Agonopterix posticella, A. psoraliella*). Larvae of the orange sulfur butterfly (Insecta: Lepidoptera: Pieridae: *Colias eurytheme*) also feed on *Hoita* species. Furanocoumarins in the seeds of *H. macrostachya* (and possibly also *H. orbicularis*) provide some deterrent against herbivory.

Economic importance: food: The starchy roots of *Hoita macrostachya* can be eaten raw or cooked. Plants of *H. orbicularis* were eaten by the Luiseno people as a green vegetable. Its foliage has also been used to prepare tea; **medicinal:** *Hoita macrostachya* contains the furanocoumarins psoralen and angelicin, which can cause photosensitivity in humans. The Ohlone used tea made from the plants to treat fevers. The Luiseno tribe made a salve from the plants to treat ulcers and sores. The Costanoan people used decoctions of *H. orbicularis* to treat fevers and as a blood medicine; **cultivation:** *Hoita orbicularis* is sometimes grown as a rock garden and moist garden plant. It is also useful as a ground cover; **misc. products:** The Cahuilla and Luiseno people extracted a yellow dye (used to stain baskets) from the boiled roots of *H. macrostachya*. The Concow and Yokia tribes fashioned a sewing thread from stem fibers of *H. macrostachya*. The root fibers of *H. macrostachya* were used by the Mendocino Indians to make rope and hunting bags. The foliage of *Hoita orbicularis* produces a long-lasting, perfume-like scent; **weeds:** none; **nonindigenous species:** none.

Systematics: *Hoita* is assigned to tribe Psoraleeae of Fabaceae. Analyses of combined molecular data for two of the three species (including both OBL taxa) indicate that the genus resolves as a clade, which is situated centrally among the remaining North American members of the tribe (Figure 4.12). *Hoita* appears to represent a unique radiation of OBL species in tribe Psoraleeae, given that no wetland indicators have been designated in any of the other genera. Neither the base chromosome number nor counts for any of the *Hoita* species are known. Hybrids have not been reported in the genus.

Comments: In North America, *Hoita macrostachya* and *H. orbicularis* are restricted to California, but both extend into Baja California.

References: Calcagno et al., 2002; Egan & Crandall, 2008; Hickman, 1993; Mason, 1957.

5. Hosackia

Birdsfoot-trefoil, deervetch; lotier

Etymology: after David Hosack (1769–1835)

Synonyms: *Lotus* (in part)

Distribution: global: Central America, Mexico, North America; **North America:** western

Diversity: global: 11 species; **North America:** 9 species

Indicators (USA): OBL: *Hosackia alamosana, H. oblongifolia*

Habitat: freshwater; palustrine; **pH:** 5.5–7.5; **depth:** <1 m; **life-form(s):** emergent (herb)

Key morphology: stems erect or procumbent; leaves oddpinnately compound with 3–11 leaflets (to 3 cm); stipules (<1 cm) membranous; flowers papilionaceous, umbellate; petals (to 14 mm) white to yellow, purple veined; fruit (to 5 cm) subterete

Life history: duration: perennial (rhizomes); **asexual reproduction:** rhizomes; **pollination:** insect; **sexual condition:** hermaphroditic; **fruit:** legumes (common); **local dispersal:** seeds; **long-distance dispersal:** seeds

Imperilment: (1) *Hosackia alamosana* [G3]; S1 (AZ); (2) *H. oblongifolia* [G5]; S2 (CA)

Ecology: general: Formerly regarded as a subgenus of *Lotus* (a principally terrestrial genus), *Hosackia* recently has been maintained as distinct owing to the results of phylogenetic analyses. Unlike *Lotus*, this group has a much greater affinity for wetlands, with five of the North American species (56%) listed as wetland indicators, and two of them (22%) as OBL indicators. The genus occurs in grasslands, shrublands, and woodlands with temperate or Mediterranean climates. All of the species are perennials with dehiscent fruits. Little else is known regarding their specific life history information. The fairly large and showy flowers are pollinated by bees (Insecta: Hymenoptera: Megachilidae). Like *Lotus*, they are probably self-incompatible with seeds that are characterized by physical dormancy.

***Hosackia alamosana* Rose** inhabits canyon bottoms, seeps, stream banks, and the edges of washes at elevations from 1067 to 1700 m (to 2134 m in Mexico). The substrates have been described as rock and sand. Flowering and fruiting have been observed from April to July. The plants can become carpet-like under optimal conditions but can be dislodged extensively during periods of heavy flooding. Additional life history information is lacking for this species. **Reported associates:** *Amsonia grandiflora, Aquilegia chrysantha, Baccharis salicifolia, Eleocharis, Fraxinus velutina, Gomphocarpus fruticosus, Juglans major, Juncus, Mimulus guttatus, Platanus wrightii, Polypogon monspeliensis, Populus fremontii, Quercus arizonica, Q. emoryi, Salix bonplandiana, S. gooddingii, Scutellaria potosina, Toxicodendron radicans, Veronica anagallis-aquatica, Vitis arizonica.*

***Hosackia oblongifolia* Benth.** inhabits bottoms, depressions, ditches, draws, flats, floodplains, marshes, meadows, ravines, roadsides, seeps, slopes (to 25%), springs, thickets, washes, woodlands, and the margins of streams at elevations of up to 2896 m. The plants have been observed on occasion in shallow, flowing water. Exposures can vary from full sunlight to dense shade. The substrates include basalt loam, clay, clay loam, decomposed granite, granidiorite, granite, gravel, gravelly clay, gravelly clay loam, limestone loam, loam, loam porphyry, loamy sand, mud, olivine gabbro, periodotite, rock, sand, sandy alluvium, sandy loam, sandy silt, serpentine, and volcanics. Flowering and fruiting extend from March to October. The flowers are visited and pollinated by bees (e.g., Insecta: Hymenoptera: Megachilidae). **Reported associates:** *Abies concolor, Acer macrophyllum, Achillea millefolium, Adenostoma fasciculatum, Aesculus californica, Agrostis gigantea, Allium validum, Alnus rhombifolia, Amelanchier, Anaphalis margaritacea, Anemopsis californica, Apocynum cannabinum, Aquilegia formosa, Arabis eschscholtziana, Aralia, Arbutus, Arctostaphylos glandulosa, A. patula, A. viscida, Artemisia douglasiana, A. dracunculus, A. ludoviciana, A. tridentata, Avena barbata, Baccharis sergiloides, B. salicifolia, Berula erecta, Betula occidentalis, Brickellia californica, Brodiaea elegans, B. orcuttii, Bromus arenarius, B. diandrus, B. hordeaceus, B. tectorum, Calocedrus decurrens, Carduus pycnocephalus, Carex alma, C. athrostachya, C. aurea, C. fracta, C. hassei, C. jonesii, C. occidentalis, C. senta, C. subfusca, Castilleja miniata, C. minor, Ceanothus cordulatus, C. cuneatus, C. greggii, C. leucodermis, C. palmeri, Cercocarpus betuloides, C. ledifolius, Chamaecyparis, Cirsium vulgare, Clarkia, Collinsia tinctoria, Cynosurus echinatus, Cyperus eragrostis, Dactylis glomerata, Darlingtonia californica, Datisca glomerata, Deschampsia danthonioides, Dichanthelium acuminatum, Dieteria canescens, Diplacus aurantiacus, Distichlis spicata, Eleocharis parishii, E. quinqueflora, Epilobium ciliatum, E. glaberrimum, E. torreyi, Epipactis gigantea, Equisetum arvense, E. hyemale, E. laevigatum, Ericameria linearifolia, E. nauseosa, Eriodictyon, Eriogonum fasciculatum, E. spergulinum, Erodium cicutarium, Erysimum capitatum, Euthamia occidentalis, Festuca roemeri, F. rubra, Foeniculum vulgare, Fragaria, Fraxinus velutina, Fremontodendron californicum, Gayophytum, Gentiana fremontii, Gentianella amarella, Geranium richardsonii, Glyceria elata, Glycyrrhiza lepidota, Gutierrezia microcephala, Helenium bigelovii, Heracleum sphondylium, Hesperoyucca whipplei, Heterocodon rariflorus, Heteromeles arbutifolia, Hoita orbicularis, Holcus lanatus, Horkelia rydbergii, Hosackia crassifolia, Hypericum anagalloides, H. perforatum, Iris, Juncus balticus, J. dubius, J. macrandrus, J. orthophyllus, J. saximontanus, J. textilis, J. xiphioides, Keckiella breviflora, Lactuca serriola, Lathyrus latifolius, Lepidium virginicum, Lepidospartum squamatum, Leymus condensatus, Lilium parryi, Linanthus concinnus, Lithocarpus densiflorus, Lupinus latifolius, Lythrum californicum, Madia elegans, Maianthemum stellatum, Marrubium vulgare, Melilotus albus, Mentha arvensis, Mentzelia laevicaulis, Mimulus cardinalis, M. floribundus, M. gutattus, Muhlenbergia andina, M. richardsonis, M. rigens, Nasturium officinale, Oenanthe sarmentosa, Oreostemma, Penstemon grinnellii, Perideridia, Phacelia, Phlox, Physocarpus, Pinus attenuata, P. contorta, P. jeffreyi, P. lambertiana, P. monophylla, P. ponderosa, P. sabiniana, Plagiobothrys, Platanthera, Platanus racemosa, Poa annua, P. bulbosa, P. palustris, P. pratensis, Polygonum aviculare, Polypogon monspeliensis, Populus fremontii, P. trichocarpa, Primula jeffreyi, Prunella vulgaris, Pseudotsuga macrocarpa, P. menziesii, Pteridium aquilinum, Quercus agrifolia, Q. chrysolepis, Q. gambelii, Q. kelloggii, Ranunculus californicus, R. cymbalaria, Rhamnus californica, R. ilicifolia, Rhododendron columbianum, R. occidentale, Ribes cereum, R. nevadense, Rosa californica, R. woodsii, Rubus ulmifolius, Rumex crispus, Salix exigua, S. gooddingii, S. laevigata, S. lasiolepis, S. lemmonii, S. lutea, S. scouleriana, Sarcodes sanguinea, Schedonorus arundinaceus, Schoenoplectus acutus, Scirpus lineatus, Senecio flaccidus, S. triangularis, Sidalcea malviflora, S. oregana, Sisymbrium altissimum, Sisyrinchium bellum, Solidago confinis, Sonchus asper, Sphenosciadium capitellatum, Spiranthes romanzoffiana, Sporobolus flexuosus, Stachys albens, Symphyotrichum, Taraxacum officinale, Thalictrum polycarpum, Toxicodendron diversilobum, Tragopogon dubius, Trichostema oblongum, Trifolium wormskioldii, Turricula parryi, Typha domingensis, T. latifolia, Umbellularia californica, Urtica dioica, Veratrum californicum, Verbascum thapsus, Verbena lasiostachys, Veronica, Vitis girdiana, Vulpia myuros, Whipplea modesta, Xanthium strumarium.*

Use by wildlife: *Hosackia oblongifolia* is a larval host plant for several butterflies (Insecta: Lepidoptera: Lycaenidae: *Callophrys dumetorum, Lycaeides idas, Plebejus acmon, P. idas*; Papilionidae: *Papilio anna*). Its flowers are visited by leafcutting bees (Insecta: Hymenoptera: Megachilidae: *Anthidium atripes, A. illustre, A. mormonum, A. placitum*).

Economic importance: food: Because *Hosackia* species had long been placed in the genus *Lotus*, some brief clarification is appropriate. There is much confusion regarding the taxonomic identity of the plants consumed by the mythological Lotophagi ("lotus eaters") of Homer's Odyssey. Reputed to cause amnesia and indolence, the "lotus" plants mentioned in the epic probably refer to the jujube tree (*Zizyphus lotus*), a member of the family Rhamnaceae and not to *Celtis australis, Nelumbo nucifera,* or *Nymphaea lotus* as some have concluded. In any event, members of the botanical genus *Lotus* generally are not eaten by humans because all parts of some species contain cyanogenic glycosides, which can be lethal if consumed by humans (they are not as harmful to ruminants). The palatability of *Hosackia* species has not been determined specifically, but like *Lotus*, the plants should be regarded as inedible; **medicinal:** none reported (although the associated nomenclatural changes are known to cause taxonomic headaches); **cultivation:** *Hosackia oblongifolia* has been grown as an ornamental rock garden plant; **misc. products:** *Hosackia alamosana* is considered to be a good plant for controlling erosion; **weeds:** none; **nonindigenous species:** none.

Systematics: Long regarded as a subgenus of *Lotus*, *Hosackia* is now maintained as a distinct genus by virtue of molecular phylogenetic analyses, which show the two groups to be unrelated systematically (e.g., Figure 4.13). The *Lotus* clade is entirely Old World; whereas, *Hosackia* is closely allied to several other exclusively New World clades (also formerly within *Lotus*), which are now recognized as the genera *Acmispon, Ottleya,* and *Syrmatium*. *Ornithopus* (Old/New World) and *Dorycnopsis* (Old World) are also apparently closely related (Figure 4.13). Within *Hosackia, H. oblongifolia* resolves with a clade of FACW indicator species. It will be necessary to include *H. alamosana* in similar analyses to determine whether the OBL habit has arisen once or twice within the genus. Although the precise relationship of *H. alamosana* remains uncertain at this time, its general similarity to *H. oblongifolia* perhaps indicates their close relationship. The base chromosome number of *Hosackia* is $x = 7$, although few counts are available for the genus. The chromosome number remains unknown for both of the OBL indicator species.

Comments: Both of the OBL *Hosackia* species occur in western North America. *Hosackia alamosana* is endemic to southern Arizona and northern Mexico (Chihuahua, Durango, Sonora)), whereas, *H. oblongifolia* ranges more broadly from Oregon to Mexico.

References: Allan & Porter, 2000; Fernald & Kinsey, 1943; Gonzalez & Griswold, 2013; Gooding & Gooding, 1961; Grant, 1995; Kearney & Peebles, 1960; Lavin et al.,

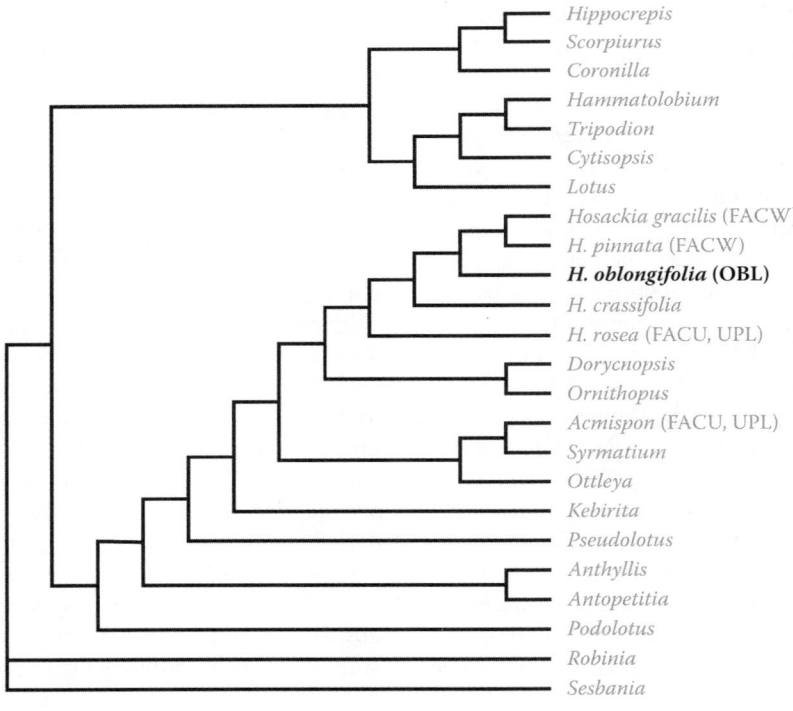

FIGURE 4.13 Phylogenetic relationships of *Hosackia* reconstructed from combined morphological and nrDNA (ITS) sequence data. Although formerly assigned to the Old World genus *Lotus* (upper clade), the molecular data resolve the New World segregates *Acmispon, Hosackia, Ottleya,* and *Syrmatium* quite distantly (lower clade) and reconcile their recognition as genera distinct from *Lotus*. Among those species surveyed, the OBL North American *H. oblongifolia* (in bold) is related closely to a clade containing FACW indicators (*H. alamosana* was not included in the analysis). (Adapted from Sokoloff, D.D. et al., *Int. J. Plant Sci.*, 168, 801–833, 2007.)

2003; Scott, 1986; Shapiro, 1990; Sokoloff et al., 2007; Wojciechowski et al., 2000.

6. *Lathyrus*

Marsh pea, marsh vetchling; gesse des marais
Etymology: from the Greek *lathyros* meaning "pea"
Synonyms: *Aphaca*; *Orobus*
Distribution: global: Asia; Europe; North America; **North America:** widespread
Diversity: global: 160 species; **North America:** 41 species
Indicators (USA): OBL: *Lathyrus jepsonii*, *L. palustris*; **FACW:** *L. palustris*
Habitat: brackish (coastal), freshwater; palustrine; **pH:** 7.2–7.5; **depth:** <1 m; **life-form(s):** emergent (herbaceous vine)
Key morphology: stems (to 3 m) climbing, angled, winged; leaves alternate, even-pinnate with 4–12 leaflets (to 7 cm), terminating in a coiled, branched or simple tendril; stipules sagittate; racemes axillary, 2–15 flowered, one sided; flowers papilionaceous, showy; petals (to 30 mm) blue, pink, purple, or rose; fruits (to 12 cm) flat
Life history: duration: perennial (rhizomes); **asexual reproduction:** rhizomes; **pollination:** insect; **sexual condition:** hermaphroditic; **fruit:** legumes (common); **local dispersal:** rhizomes, seeds; **long-distance dispersal:** seeds
Imperilment: (1) *Lathyrus jepsonii* [G5]; S2 (CA); (2) *L. palustris* [G5]; SH (DC, DE, WV); S1 (AB, GA, MD, PA, TN, VA); S2 (CA, KY, NC, NE, VT); S3 (BC, NJ)
Ecology: general: *Lathyrus* is primarily a genus of terrestrial plants, which inhabit diverse sites including beaches, chaparral, desert scrub, dunes, meadows, and woodlands. They frequently occur in disturbed sites. Some of the species extend along riparian habitats and into other marginally wet areas. Nine of the North American species (22%) are wetland indicators, with just two (5%) designated as OBL. The genus includes annuals and perennials and showy flowered species, which are either self-pollinating or are outcrossed by insects such as bees (Insecta: Hymenoptera: Apidae). With few exceptions, the annual species are self-pollinating while the perennials are outcrossed. The seeds are believed to have physical dormancy and exposure to fire may make them more permeable and facilitate water imbibition. The seeds can be highly toxic to humans and other animals. Most *Lathyrus* species probably fix nitrogen. Many species have a climbing or sprawling habit and rely on the presence of other species for physical support.

Lathyrus jepsonii **Greene** is a viny perennial, which occurs in freshwater, nontidal, or tidally influenced bottoms, chaparral, coastal deltas, ditches, flats, gullies, levees (outboard sides), marshes, meadows, open woodlands, ravines, rivulets, roadsides, salt marshes (middle and high zones), sandbars, sinks, slopes (to 20%), and along the margins of brooks, rivers, sloughs, and streams at elevations of up to 1982 m. Although the precise degree of salt tolerance has not been determined, this species prefers saturated soils, which occur above the zone of tidal influence. Exposures include full sun to partial shade. The substrates are described as alkaline and include adobe, alluvium, clay, cobble, decomposed granite,

loam, muck, mud, rock, rocky volcanics, sand, sandy loam, and serpentine. Flowering and fruiting occur from April to August. The flowers are pollinated by a variety of insects (Insecta) including bees (Hymenoptera: Apidae: *Bombus caliginosus*, *Eucera frater albopilosa*, *Xylocopa tabaniformis*; Andrenidae: *Andrena plana*) and beeflies (Diptera: Bombyliidae: *Bombylius major*). Seed production can vary from 4 to 440/m². The seeds are very large (averaging 6.99 mg) and do not float for any extended period of time, which limits their potential dispersal by water. The plants have a climbing habit and require other vegetation to support their stems in an upright position. They are threatened by losses of habitat to agricultural land and by hydrological alterations to existing habitats. **Reported associates:** *Abies*, *Alnus*, *Arctostaphylos patula*, *Baccharis douglasii*, *Brassica nigra*, *Bromus carinatus*, *Calocedrus decurrens*, *Calycanthus occidentalis*, *Carex barbarae*, *Ceanothus integerrimus*, *Chrysolepis sempervirens*, *Cirsium*, *Collinsia*, *Cordylanthus mollis*, *Cornus nuttallii*, *Cytisus*, *Distichlis spicata*, *Elymus glaucus*, *Emmenanthe penduliflora*, *Equisetum*, *Foeniculum vulgare*, *Fraxinus latifolia*, *Fritillaria*, *Galium aparine*, *Grindelia*, *Hibiscus lasiocarpus*, *Hydrocotyle verticillata*, *Hypericum perforatum*, *Jaumea carnosa*, *Lilaeopsis masonii*, *Limosella subulata*, *Linanthus*, *Lolium multiflorum*, *Lomatium*, *Lonicera interrupta*, *Marah oregana*, *Paspalum*, *Phragmites australis*, *Pinus jeffreyi*, *P. lambertiana*, *P. ponderosa*, *P. radiata*, *P. sabiniana*, *Plantago lanceolata*, *Populus fremontii*, *Prunus emarginata*, *Pseudotsuga menziesii*, *Pteridium aquilinum*, *Quercus agrifolia*, *Q. douglasii*, *Q. lobata*, *Raphanus raphanistrum*, *R. sativus*, *Rosa californica*, *Rubus discolor*, *R. ursinus*, *Sarcocornia perennis*, *Schoenoplectus acutus*, *S. californicus*, *Scrophularia*, *Symphoricarpos albus*, *S. mollis*, *Symphyotrichum lentum*, *Toxicodendron diversilobum*, *Triglochin*, *Typha*, *Vicia sativa*, *V. villosa*, *Vitis*.

Lathyrus palustris **L.** is a perennial, which inhabits shallow (e.g., 2 cm), fresh to brackish, coastal or inland, tidal or nontidal sites including barrens, beaches, bluffs, bogs, deltas, depressions, estuaries, fens, flats, floodplains, gravel bars, gulches, hummocks, marshes, meadows, muskeg, prairies, roadsides, saltmarsh, savannas, scrub, shores, slopes (to 12%), swales, swamps, thickets, woodlands, and the margins of lagoons, lakes, ponds, rivers, and streams at elevations of up to 1798 m. Exposures range from full sun to shade, but these plants typically prefer open sites (often wet prairies and fens) and have reappeared in marshes that were burned to manage shrub cover, after being absent for more than 20 years. The substrates are characterized as calcareous and include boulders, clay loam, gleyed silt, gravel, loam, marl, mud, muddy dolomite, pebbles, rock, sand, sandy loam, sandy silt, and shale. Flowering occurs from May to September, with fruit production from June to October. The flowers are pollinated and outcrossed by insects (Insecta). About 3.1 seeds are produced per plant on average, with an average seed mass of 15.6 mg. Recommended treatments for seed germination include cold, dry stratification (6–8 weeks), presoaking in warm water for 24 h, and physical scarification. High germination rates (71.2%) have been obtained following physical seed coat scarification (using sandpaper) and incubation under a 14/10 h, 25°C/20°C light/temperature regime. The plants vegetatively reproduce by slender, creeping rhizomes, which form root nodules and reportedly are nitrogen fixing. **Reported associates:** *Abies balsamea, Achillea millefolium, Agrimonia parviflora, Alnus, Alopecurus geniculatus, Ambrosia trifida, Andropogon gerardii, Anemone canadensis, Angelica lucida, Anthoxanthum monticola, A. nitens, A. odoratum, Apios americana, Apocynum cannabinum, Arbutus, Arnica, Arnoglossum plantagineum, Asclepias amplexicaulis, A. incarnata, A. sullivantii, A. syriaca, A. viridiflora, Athyrium filix-femina, Baptisia alba, Barbarea vulgaris, Betula, Bidens polylepis, Boehmeria cylindrica, Bolboschoenus maritimus, Bromus inermis, Calamagrostis canadensis, C. rubescens, Caltha palustris, Campanula aparinoides, C. rotundifolia, Carex buxbaumii, C. conoidea, C. granularis, C. haydenii, C. lacustris, C. longii, C. lyngbyei, C. macrochaeta, C. obnupta, C. pellita, C. pensylvanica, C. pluriflora, C. prairea, C. sartwelli, C. scoparia, C. sterilis, C. stricta, C. tetanica, C. trichocarpa, C. vulpinoidea, Castilleja unalaschcensis, Catalpa speciosa, Ceanothus integerrimus, Chelone glabra, Chenopodium rubrum, Cicuta douglasii, C. maculata, Cirsium discolor, C. muticum, Comarum palustre, Conioselinum, Coreopsis palmata, C. tripteris, Cornus amomum, C. racemosa, Cypripedium candidum, Dalea candida, Deschampsia cespitosa, Dichanthelium acuminatum, D. clandestinum, Distichlis, Eleocharis palustris, Elymus riparius, Equisetum arvense, Eragrostis spectabilis, Eriophorum angustifolium, Eupatorium altissimum, E. perfoliatum, Euthamia graminifolia, E. gymnospermoides, Eutrochium maculatum, Festuca rubra, Fragaria virginiana, Fraxinus nigra, Galium aparine, G. asprellum, G. obtusum, G. trifidum, Gentiana puberulenta, Geranium erianthum, G. maculatum, Geum laciniatum, Glyceria striata, Gratiola neglecta, Helianthemum canadense, Helianthus giganteus, H. grosseserratus, Heliopsis helianthoides, Heracleum sphondylium, Hierochloe odorata, Holcus lanatus, Hordeum brachyantherum, Iris virginica, Juncus alpinoarticulatus, J. dudleyi, J. effusus, J. greenei, Larix laricina, Lechea tenuifolia, Leymus mollis, Liatris pycnostachya, Ligusticum scoticum, Lilium michiganense, Lonicera tartarica, Ludwigia polycarpa, Lupinus, Lycopus americanus, Lysimachia quadriflora, L. terrestris, Malus fusca, M. sieboldii, Melilotus albus, Monarda fistulosa, M. punctata, Muhlenbergia glomerata, M. mexicana, M. richardsonis, Myrica gale, Oenanthe sarmentosa, Oenothera laciniata, Onoclea sensibilis, Opuntia humifusa, Oxypolis rigidior, Packera aurea, Parietaria pensylvanica, Parnassia palustris, Pastinaca sativa, Pedicularis lanceolata, Persicaria amphibia, Phalaris, Phlox glaberrima, P. maculata, P. pilosa, Physostegia virginiana, Picea sitchensis, Pinus contorta, Platanthera flava, Poa palustris, P. pratensis, Populus tremuloides, Potentilla anserina, Prenanthes aspera, Prunus serotina, Pseudotsuga menziesii, Pycnanthemum virginianum, Quercus garryana, Q. palustris, Q. velutina, Ranunculus cymbalaria, Ratibida pinnata, Rhamnus cathartica, R. frangula, Rhinanthus minor, Rorippa islandica, Rosa carolina, Rubus arcticus, R. hispidus,* *R. pensilvanicus, R. ursinus, Rumex altissimus, Salicornia, Salix barclayi, S. commutata, S. discolor, S. exigua, S. fuscescens, S. lasiandra, Sambucus nigra, Saponaria officinalis, Scirpus atrovirens, Silphium laciniatum, S. terebinthinaceum, Solanum dulcamara, Solidago altissima, S. canadensis, S. gigantea, S. nemoralis, S. riddellii, S. rigida, Spartina pectinata, Sphenopholis intermedia, Spiraea alba, Stachys palustris, S. pilosa, S. tenuifolia, Stellaria, Symphyotrichum drummondii, S. ericoides, S. novae-angliae, S. puniceum, Symplocarpus foetidus, Thalictrum dasycarpum, T. revolutum, Thelypteris palustris, Thuja plicata, Toxicodendron vernix, Tradescantia ohiensis, Typha latifolia, Vaccinium macrocarpon, V. vitis-idaea, Verbena hastata, Vernonia fasciculata, Veronicastrum virginicum, Zizia aurea.*

Use by wildlife: *Lathyrus jepsonii* is a host plant for several butterfly larvae (Insecta: Lepidoptera: Lycaenidae: *Everes amyntula, E. comyntas, Glaucopsyche lygdamus*; Pieridae: *Colias eurytheme*). *Lathyrus palustris* is generally attractive to butterflies (Insecta: Lepidoptera) and is also a larval host plant (Hesperiidae: *Epargyreus clarus*; Tortricidae: *Cydia nigricana*).

Economic importance: food: The Yokia tribe cooked and ate young green plants of *Lathyrus jepsonii* as a vegetable. The Chippewa and Ojibwa tribes cooked and ate the seeds (peas) as a vegetable. The entire legume of *L. palustris* is boiled (or the seeds eaten alone) in China. Although reportedly edible, extreme caution is advised. Excessive consumption of *Lathyrus* seeds (especially of *L. sativus*) can lead to debilitating leg paralysis or even fatality in humans (see next); **medicinal:** *Lathyrus* seeds contain β-aminopropionitrile, which inhibits the enzyme lysyl oxidase, leading to pain and paralysis (lathyrism) in humans and other animals. The Mendocino tribe applied a poultice made from boiled *L. jepsonii* plants

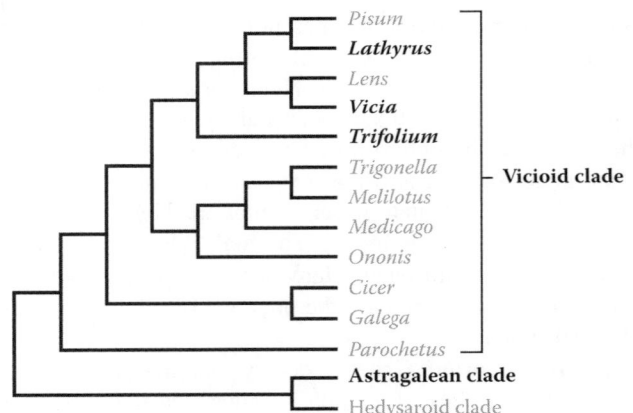

FIGURE 4.14 Phylogenetic relationships among the vicioid legumes based on analysis of *matK* data. *Lathyrus* and several other North American taxa containing OBL indicators (shown in bold) occupy a relatively derived position in the cladogram. However, each associated genus principally is terrestrial; thus, the aquatic habit most likely has arisen independently in *Lathyrus, Trifolium,* and *Vicia.* (Adapted from Wojciechowski, M.F., *Advances in Legume Systematics, Part 10, Higher Level Systematics,* Royal Botanic Gardens, Kew, UK, 2003; Wojciechowski, M.F. et al., *Advances in Legume Systematics,* Royal Botanic Gardens, Kew, UK, 2000.)

as a remedy for swollen joints; **cultivation:** *Lathyrus palustris* is sold commercially as an ornamental plant for bog and marsh gardens (the related *Lathyrus latifolius* and *L. odoratus* are the popular, cultivated sweet peas); **misc. products:** The Mendocino people cut plants of *L. jepsonii* as hay and fodder for cattle and horses. The Meskwaki tribe used the root of *L. palustris* as a lure to trap beaver and other wildlife. The Ojibwa fed plants of *L. palustris* to their ponies to fatten them; **weeds:** none; **nonindigenous species:** none.

Systematics: Phylogenetic studies using *matK* sequence data resolve *Lathyrus* within the "vicioid clade" of legumes as the sister genus of *Pisum* in close proximity to *Vicia* (Figure 4.14). The vicioids in turn are related to the astragalean and hedysaroid clades that together comprise a monophyletic group known as the "inverted repeat-lacking clade (IRLC)" legumes, wherein all of the genera lack one copy of the chloroplast inverted repeat region. *Lathyrus* is also among those legume genera, which lack both the *clpP* intron and the *rps12* intron. These results are consistent with a "supertree" analysis of 571 legume taxa, which is based on phylogenetic elements from a variety of molecular and nonmolecular data. Data from cpDNA restriction sites (*rpo*C, IR regions) indicate that section *Orobus* (which contains *Lathyus jepsonii* and *L. palustris*) resolves as a paraphyletic grade unless combined with section *Notolathyrus*. *Lathyrus jepsonii* and *L. palustris* are not sister species (Figure 4.15), but their closest relatives have not been adequately identified by existing cpDNA data analyses. Two varieties of *Lathyrus jepsonii* usually are distinguished: *L. jepsonii* var. *califonicus* (foliage pubescent), which occurs in higher elevation forests; and *L. jepsonii* var. *jepsonii* (foliage glabrous), which occurs in low elevation coastal estuaries and marshes. A study using finer-scale genetic markers might indicate whether these varieties are meaningful systematically. There have been many varietal names assigned to the wide-ranging *L. palustris*, but these are difficult to apply and deserve further scrutiny. The base chromosome number of *Lathyrus* is *x* = 7. *Lathyrus jepsonii* and *L. palustris* are diploids (2*n* = 14). *Lathyrus venosus* (2*n* = 28) is believed to have arisen through hybridization of *L. ochroleucus* and *L. palustris*, but this hypothesis requires verification. **Comments:** *Lathyrus jepsonii* is endemic to California. *Lathyrus palustris* is circumboreal and distributed widely in North America.

References: Asmussen & Liston, 1998; Buchmann et al., 2010; Diggory & Parker, 2011; Fiedler et al., 2007b; Jansen et al., 2008; Middleton, 2002b; Winter et al., 2008; Wojciechowski et al., 2000.

7. *Trifolium*

Clover; trèfle

Etymology: from the Latin *ter folium* meaning "three leaved"

Synonyms: *Amoria*; *Bobrovia*; *Chrysaspis*; *Lupinaster*; *Ursia*; *Xerosphaera*

Distribution: global: cosmopolitan; **North America:** widespread

Diversity: global: 240 species; **North America:** 100 species

Indicators (USA): OBL: *Trifolium amabile*, *T. bolanderi*

Habitat: brackish (coastal), freshwater; palustrine; **pH:** unknown; **depth:** <1 m; **life-form(s):** emergent (herb)

Key morphology: stems (to 40 cm) ascending to decumbent; leaves petiolate (to 6 cm), trifoliate, the leaflets (to 2.5 cm) somewhat serrate; stipules foliaceous, clasping; flowers numerous in pedunculate (to 20 cm) heads, papilionaceous; petals (to 12 mm) pink, lavender, purple, or rose; fruits (to 5 mm) oblong–elliptic

Life history: duration: perennial (persistent woody base, rhizomes); **asexual reproduction:** rhizomes; **pollination:** insect; **sexual condition:** hermaphroditic; **fruit:** legumes (common); **local dispersal:** rhizomes, seeds; **long-distance dispersal:** seeds

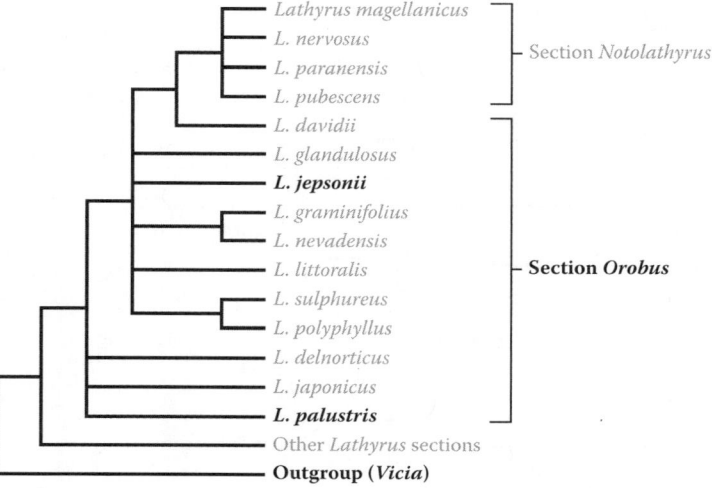

FIGURE 4.15 Evolutionary relationships among *Lathyrus* species as indicated by cpDNA restriction fragment data. These results reveal that section *Orobus* is paraphyletic unless merged with section *Notolathyrus*. Although the cladogram is not well resolved, the obligately aquatic North American indicators (shown in bold) do not form a single clade, indicating that the aquatic habit apparently has been derived independently in the two *Lathyrus* species. (Adapted from Asmussen, C.B. & Liston, A., *Amer. J. Bot.*, 85, 387–401, 1998.)

Imperilment: (1) *Trifolium amabile* [G4]; S1 (AZ); (2) *T. bolanderi* [G2/G3]; S2 (CA)

Ecology: general: Most *Trifolium* species are characteristic of terrestrial grasslands but many species do express morphological and physiological adaptations when grown under conditions of inundation. Over a third (38%) of the North American species are wetland indicators at some level. *Trifolium wormskioldii* was an OBL indicator in the 1996 list; however, it has since been reclassified as FACW, FAC, and is excluded here. The roots of *Trifolium* species are typically nodulated by symbiotic nitrogen-fixing bacteria (Bacteria: Rhizobiaceae: *Rhizobium leguminosarum*), although different species can be nodulated by different strains. *Trifolium* species are known to possess gametophytic self-incompatibility (GSI) and their flowers are usually outcrossed by insect pollinators, mainly bees (Insecta: Hymenoptera). The seeds of those *Trifolium* species studied have physical dormancy and require scarification in order to germinate (generally under a 20°C/15°C day/night temperature regime). The flower and seed ecology has not been studied for either of the OBL species, but their requirements are likely to be similar.

Trifolium amabile **Kunth** occurs in bottomlands, depressions, meadows, and near brooks, springs, and streams at elevations of up to 1890 m (to 3350 m in Mexico). The substrates are described as wet, sandy soil. The plants are highly sensitive to drought. Flowering (which can be profuse) has been observed (in North America) from August to October. Simple physical scarification of the seed coat (using sandpaper) does not effectively induce germination. However, scarification followed by treatment with 1.0 M thiourea (to overcome embryo dormancy) has resulted in germination when incubated at 25°C. The plants produce a deep taproot, from which stoloniferous stems (up to 48 cm in total) develop into a spreading habit. They have a relatively low annual growth rate compared to other clover species. Although this species is common in Mexico, Central America, and South America, its ecology in North America is poorly documented otherwise. **Reported associates:** *Juniperus deppeana, Pinus leiophylla.*

Trifolium bolanderi **A. Gray** is a perennial, which occurs in moist montane meadows of the central Sierra Nevada range at elevations from 2048 to 2287 m. The plants favor cool, wet sites, where snow cover remains until late spring. Most of the sites (~33% of total) are flat with southern exposures (31%). Approximately 42% of the occurrences are found in partially shaded exposures, with 35% in medium shade and 23% in

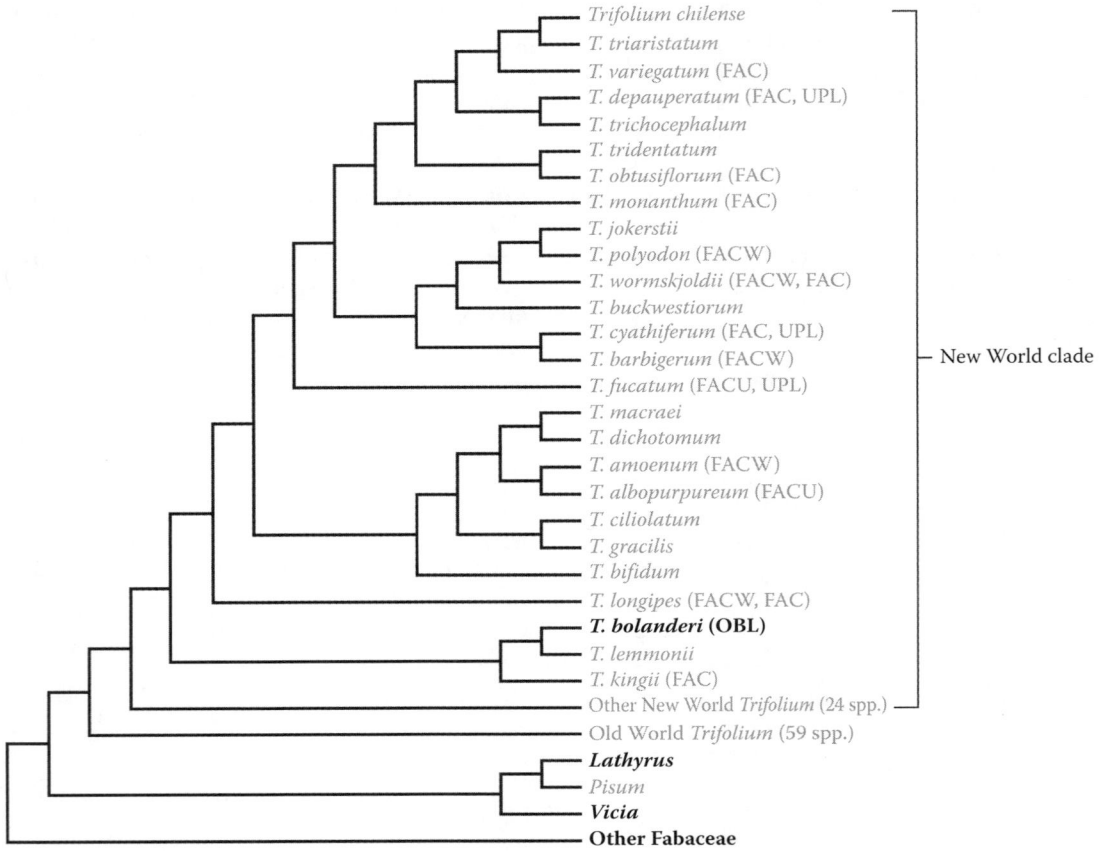

FIGURE 4.16 Phylogenetic relationships in *Trifolium* as indicated by molecular data. The obligate North American *T. bolanderi* resolves as the sister species of *T. lemmonii*, which is not a wetland indicator. *Trifolium amabile* was not surveyed. Many *Trifolium* species exhibit adaptations to aquatic conditions as evidenced by the broad distribution of taxa having wetland indicator status (shown in parentheses). Taxa containing OBL indicators are shown in bold. (Adapted from Wojciechowski, M.F. et al., *Advances in Legume Systematics,* Royal Botanic Gardens, Kew, UK, 2000.)

dense shade. The substrates are described as clay loam, granite, muck, and sandy loam with an average organic content (i.e., 15.3%). Typical substrate composition is 61.7% sand, 31.3% silt, and 7.0% clay. Flowering and fruiting occur from June to July. The plants develop a deeply rooted crown and can reproduce vegetatively by the formation of lateral offsets. They apparently assume a more prostrate growth form when growing in areas that are grazed by herbivores. Additional information on the ecology of this species is lacking. **Reported associates:** *Abies magnifica*, *Bistorta bistortoides*, *Carex nebrascensis*, *Deschampsia cespitosa*, *Eleocharis acicularis*, *E. quinqueflora*, *Galium brandegei*, *Hypericum anagalloides*, *Ivesia unguiculata*, *Juncus oxymeris*, *Mimulus primuloides*, *Muhlenbergia filiformis*, *Oreostemma alpigenum*, *Perideridia bolanderi*, *P. parishii*, *Phalacroseris bolanderi*, *Pinus contorta*, *Primula jeffreyi*, *Trifolium longipes*, *T. monanthum*, *Viola macloskeyi*.

Use by wildlife: *Trifolium amabile* plants contain 2.55% nitrogen, 18.9% protein, and 71.1% digestible matter, and probably are grazed by roaming herbivores. *Trifolium bolanderi* is grazed by livestock, which does not appear to threaten its survival.

Economic importance: food: *Trifolium amabile* is grown locally as a potherb in Mexico and Peru; **medicinal:** *Trifolium amabile* has some traditional use in Latin America for the treatment of eye diseases; **cultivation:** *Trifolium amabile* is grown as an ornamental garden plant in Latin America; **misc. products:** none; **weeds:** none; **nonindigenous species:** none.

Systematics: Molecular data (*matK* sequences) place *Trifolium* among the vicioid legumes as the sister group to a clade comprising *Lathyrus*, *Lens*, *Pisum*, and *Vicia* (Figures 4.14 and 4.16). Phylogenetic analyses of *Trifolium* based on a "super tree" constructed from different sources of molecular data (mainly ITS and *matK* sequences) resolve the New World species as a clade. Those phylogenetic associations (Figure 4.16) indicate a close relationship to exist between *T. bolanderi* and *T. lemmonii*, which is not a wetland indicator. The relationships of *T. amabile* (not included in that analysis) remain uncertain. Other wetland indicators are distributed broadly across the New World *Trifolium* clade. Isozyme data (16 loci) from 10 populations of *Trifolium bolanderi* indicate that it retains relatively high levels of genetic polymorphism (43.1% polymorphic loci) and moderate heterozygosity ($H_e = 0.17$). The chromosomal base number for *Trifolium* is $x = 8$. The chromosome numbers have not been determined for *T. amabile* or *T. bolanderi*.

Comments: In North America, *Trifolium amabile* occurs only in Arizona where it reaches its northernmost distributional limit, but its range extends southward through Mexico, Central America, and South America. *Trifolium bolanderi* is endemic to the Sierras of California.

References: Barneby, 1989; Denton, 2003; Dodd & Orr, 1995; Kuhnlein et al., 1982; Ratliff & Denton, 1993; Turner & Kuhnlein, 1982; Turner et al., 1983, 2003; Wernegreen et al., 1997; Wojciechowski et al., 2000; Zohary & Heller, 1984.

8. *Vicia*

Vetch; vesce

Etymology: from the Latin *vicia*, the classical name of the plant; probably derived from *vincire* ("to bind") in reference to the twining habit

Synonyms: *Bona*; *Ervum*; *Faba*

Distribution: global: Asia; Europe; North America; **North America:** widespread

Diversity: global: 130 species; **North America:** 30 species

Indicators (USA): OBL: *Vicia ocalensis*

Habitat: freshwater; palustrine; **pH:** unknown; **depth:** <1 m; **life-form(s):** emergent (viny herb)

Key morphology: stems (to 12 dm) angled and striate, reclining or twining on adjacent vegetation; herbage sparsely to densely hairy; leaves (to 10 cm) alternate, stipulate (stipules half-hastate), pinnately compound, with four to six leaflets (to 5 cm), terminating in an unbranched tendril; racemes with 12–18 papilionaceous flowers; petals (to 12 mm) bluish white, purple veined; fruit (to 4.5 cm) flat

Life history: duration: perennial (persistent stem bases, rhizomes); **asexual reproduction:** rhizomes; **pollination:** unknown; **sexual condition:** hermaphroditic; **fruit:** legumes (common); **local dispersal:** seeds; **long-distance dispersal:** seeds

Imperilment: (1) *Vicia ocalensis* [G1]; S1 (FL)

Ecology: general: Although *Vicia* mainly is a genus of terrestrial habitats, nine of the North American species (30%) are wetland indicators, with one designated as OBL. Different *Vicia* species can be selfed or cross-pollinated. The seeds of those *Vicia* species studied have physical dormancy and require scarification for germination.

Vicia ocalensis **R.K. Godfrey & Kral** grows in ditches, thickets, or along the marshy banks of brooks or streams at low elevations. The substrates have been characterized as sand or sandy peat. The sites are open and receive full sun. Flowering and fruiting extend from April to June. The plants produce up to 18 axillary racemes, each with up to 18 flowers. It has not been determined experimentally whether *V. ocalensis* is primarily selfed or cross-pollinated. However, the flowers reportedly are pollinated by larger bees (Insecta: Hymenoptera) and flies (Insecta: Diptera), which would indicate a prevalent outcrossing breeding system. Little information exists on the seed ecology of this rare species, but physical scarification of the seed coat is necessary to induce germination. These plants have a viny, scrambling habit, and use other plants for support; they can grow into a dense entangled mat. They are threatened to some degree by disturbance from off-the-road vehicles. Additional ecological information on this species is scarce. **Reported associates:** none.

Use by wildlife: *Vicia* species often are eaten by domestic livestock and by wildlife; however, the forage value of *V. ocalensis* (and its general use by wildlife) is unknown and probably is fairly low because of its rarity. The flowers of *V. ocalensis* are visited (and pollinated) by several bees (Insecta: Hymenoptera: Apidae: *Apis mellifera*, *Bombus californicus*, *B. occidentalis*, *B. vosnesenskii*) and flies (Insecta: Diptera: Syrphidae: *Toxomerus*).

Economic importance: food: The roots and seeds of several *Vicia* species are edible; however, the food value of *V.*

ocalensis remains unknown; **medicinal:** many *Vicia* species have been used medicinally; however, no medicinal uses of *V. ocalensis* have been reported; **cultivation:** none; **misc. products:** none; **weeds:** none; **nonindigenous species:** none.

Systematics: Phylogenetic studies (Figure 4.14) resolve *Vicia* within the "vicioid" legumes as a monophyletic sister genus to *Lens* in a clade adjacent to another comprising *Lathyrus* and *Pisum*. All four genera share a twining, viny habit. *Vicia ocalensis* has not yet been included in any phylogenetic analyses of the genus, thus its interspecific relationships remain uncertain. It is assigned taxonomically to section *Cracca*. Morphologically, it appears to be related most closely to *V. acutifolia* and *V. floridana*, two FACW indicators also from the southeastern United States. The chromosomal base number of *Vicia* is $x = 7$. The chromosome number of *V. ocalensis* ($2n = 14$) indicates that it is a diploid. Its karyotypic similarity to *V. acutifolia* and *V. floridana* (also section *Cracca*) supports a close relationship with those species. However, it differs from both by the presence of two (vs. one) satellite pairs of chromosomes.

Comments: *Vicia ocalensis* is endemic to the Ocala National Forest in north-central peninsular Florida.

References: Adams et al., 2010; Godfrey & Kral, 1958; Godfrey & Wooten, 1981; Veerasethakul & Lassetter, 1981; Zhang & Mosjidis, 1998.

Family 2: Polygalaceae [17]

Although molecular phylogenetic studies clearly place Polygalaceae within the Fabales, its precise relationship to other members of the order remains unresolved, the family being either the sister group to Fabaceae, the sister group to Surianaceae, or the sister to a clade containing Fabaceae and Surianaceae depending on the data analyzed (see discussion of Fabales above; Figure 4.6). Molecular studies on a subset of taxa from this large family of some 850 species (using plastid DNA sequences) indicate that it is monophyletic and subdivided phylogenetically into one clade consisting of the genus *Xanthophyllum* (assigned to tribe Xanthophylleae) and a second clade containing the remaining genera (Persson, 2001; Forest et al., 2007). Taxonomic issues within the family are also unsettled. Based on more recent phylogenetic studies of larger numbers of taxa, some authors (e.g., Abbott, 2011; Pastore & Abbott, 2012) have recommended the recognition of several segregate genera within the large genus *Polygala* (e.g., *Acanthocladus*, *Asemeia*, *Badiera*, *Hebecarpa*, *Heterosamara*, *Polygaloides*, and *Rhinotropus*), which roughly correspond with former sectional divisions in the genus. Currently, it is difficult to evaluate the taxonomic merit of such a scheme relative to one which would simply broaden the circumscription of *Polygala* itself. At least for the time being, the present treatment retains a more traditional taxonomic concept of the family, which places all of the North American taxa within two genera (*Monnina*, *Polygala*). In North America, OBL indicators occur only within the latter (one species, *P. grandiflora*, would be assigned to *Asemeia* in the alternate taxonomic scheme):

1. *Polygala* L.

Polygalaceae are widespread from temperate to tropical regions. The showy flowers produce distinctive (polycolporate) pollen and attract insects such as bees and wasps (Insecta: Hymenoptera) as pollinators. Some species are self-pollinating, and others even cleistogamous (Judd et al., 2016). The seeds can be dispersed by the wind or by animals. The latter group includes arillate seeds dispersed by ants (Insecta: Hymenoptera: Formicidae) and fleshy fruits dispersed by vertebrates.

The family is known for its ornamental horticultural specimens, which occur primarily in the genus *Polygala*.

1. *Polygala*

Milkwort; polygale

Etymology: from the Greek *polys gala* meaning "much milk" as plants were believed to stimulate milk secretion in grazing cattle

Synonyms: *Acanthocladus*; *Asemeia*; *Badiera*; *Galypola*; *Hebecarpa*; *Heterosamara*; *Phlebotaenia*; *Pilostaxis*; *Polygaloides*; *Rhinotropus*; *Trichlisperma*

Distribution: global: cosmopolitan; **North America:** widespread

Diversity: global: 500 species; **North America:** 65 species

Indicators (USA): OBL: *Polygala balduinii*, *P. brevifolia*, *P. chapmanii*, *P. cruciata*, *P. cymosa*, *P. grandiflora*, *P. ramosa*; **FACW:** *P. brevifolia*, *P. cruciata*, *P. ramosa*; **FACU:** *P. grandiflora*

Habitat: freshwater; palustrine; **pH:** uknown; **depth:** <1 m; **life-form(s):** emergent (herb)

Key morphology: stems (to 12 dm) erect; leaves (1–14 cm) simple and entire, elliptic to subulate, alternate, opposite or whorled; racemes (to 5 cm) many flowered; flowers bilateral; sepals 5, the two innermost wing-like and petaloid; petals (to 6 mm) 3, fused to stamen column, colored variously (creamy white, greenish white, lavender, purple, rose purple, yellow, white), the lowermost boat-like (keeled), sometimes fringed or lacerate apically; fruits (to 2 mm) flattened, often notched apically

Life history: duration: annual, biennial (fruit/seeds); **asexual reproduction:** none; **pollination:** insect or cleistogamous; **sexual condition:** hermaphroditic; **fruit:** capsules (common); **local dispersal:** seeds (ants); **long-distance dispersal:** seeds

Imperilment: (1) *Polygala balduinii* [G4]; S1 (AL, GA, TX); (2) *P. brevifolia* [G4/G5]; S1 (LA); S2 (NC); (3) *P. chapmanii* [G3/G5]; S1 (LA); (4) *P. cruciata* [G5]; SX (ON); SH (ME); S1 (IA, KY, MN, OK, PA, RI, WV); S2 (CT, DE, MD, OH); S3 (MI, TN, VA); (5) *P. cymosa* [G5]; SX (DE); S3 (NC); (6) *P. grandiflora* [G5]; S1 (NC); S2 (NC); S3 (LA); (7) *P. ramosa* [G5]; SX (DE, NJ); SH (VA)

Ecology: general: *Polygala* species are adapted to a wide diversity of habitats including moist or wet sites. Twenty-six of the North American species (40%) are wetland indicators, with seven (11%) designated as OBL. The species include annuals and perennials. *Polygala* flowers are ordinarily pollinated by bees (Insecta: Hymenoptera), beetles (Insecta: Coleoptera), and other insects; however, the reproductive ecology has not been studied in detail for any of the OBL species.

A few species (but none of the OBL species) are amphicarpous and produce subterranean cleistogamous flowers in addition to the normal (chasmogamous) aerial flowers. The seeds of most species are arillate and are dispersed by ants.

Polygala balduinii **Nutt.** is an annual or biennial inhabitant of bogs, borrow pits (dried), depressions, ditches, flats, flatwoods, lake bottoms (dry), glades, marshes, meadows, pine barrens, prairies, roadsides, savannas, swales, and the margins of canals and ponds at low elevations. This species is regarded as a calciphilic specialist. The substrates are alkaline or calcareous and are described as Basinger (spodic psammaquents), Estero (typic haplaquods), Felda (arenic ochraqualfs), Holopaw (grossarenic ochraqualfs), karst, limestone, loam, Malabar (grossarenic ochraquults), marl, Myakka (aeric haplaquods), peat, Pineda (arenic glossaqualfs), Plummer (grossarenic paleaquults), Riviera (arenic glossaqualfs), rocky, Rutledge (typic humaquepts), sand, sandy loam, and sandy peat. Flowering occurs year round (January to December). Otherwise, the species is poorly known ecologically. **Reported associates:** *Agalinis, Aletris, Andropogon, Annona, Aristida stricta, Asclepias, Centella asiatica, Cladium jamaicense, Clematis, Dichanthelium aciculare, Diospyros, Eriocaulon, Eryngium, Eupatorium, Ficus, Fuirena scirpoidea, Hyptis, Ilex glabra, Juncus, Liatris, Linum arenicola, Myrica cerifera, Panicum, Paspalum notatum, Persea, Pinus elliottii, P. palustris, Polygala ramosa, Pontederia cordata, Quercus chapmanii, Q. incana, Q. laevis, Rhexia, Rhynchospora, Sabal palmetto, Sabatia, Sagittaria, Serenoa repens, Spartina patens, Taxodium, Typha.*

Polygala brevifolia **Nutt.** is annual, which is found in barrens, bogs, depressions, dunes, swales and along the margins of pine savannas and pocosins at low, coastal elevations. The exposures are in full sun. The reported substrates include Corolla–Duckston sand and loam. Flowering has been observed from June to December. Few additional life-history details are known for this species. **Reported associates:** *Acer rubrum, Agalinis filifolia, Aletris aurea, A. farinosa, A. obovata, Alnus serrulata, Andropogon glomeratus, A. liebmannii, Aristida, Arundinaria gigantea, Baccharis halimifolia, Balduina uniflora, Baptisia tinctoria, Bartonia paniculata, Carex striata, Centella asiatica, Centrosema virginianum, Chamaecyparis thyoides, Croptilon divaricatum, Ctenium, Cynodon dactylon, Cyrilla racemiflora, Drosera capillaris, D. intermedia, D. rotundifolia, Dysphania ambrosioides, Echinochloa walteri, Eleocharis microcarpa, Eriocaulon compressum, Eryngium integrifolium, Eupatorium capillifolium, E. mohrii, E. serotinum, Euthamia graminifolia, Fuirena scirpoidea, Gamochaeta purpurea, Gaylussacia dumosa, G. frondosa, Ilex coriacea, I. glabra, I. opaca, Juncus effusus, J. pelocarpus, Kalmia angustifolia, Lachnocaulon anceps, Leucothoe axillaris, Lilium catesbaei, Liquidambar styraciflua, Liriodendron tulipifera, Ludwigia alata, L. alternifolia, Lycopodiella caroliniana, L. alopecuroides, Lyonia lucida, Magnolia virginiana, Mikania scandens, Mitreola sessilifolia, M. cerifera, M. heterophylla, Nyssa biflora, Oclemena nemoralis, Persea palustris, Persicaria hydropiperoides, Pinus elliottii, P. serotina, Platanthera blephariglottis, P.* *ciliaris, Pogonia ophioglossoides, Polygala incarnata, P. lutea, P. nana, P. polygama, Pseudognaphalium obtusifolium, Pteridium aquilinum, Pycnanthemum, Quercus, Rhexia alifanus, R. mariana, Rhododendron canescens, Rhynchospora fascicularis, Rubus, Sabatia stellaris, Sarracenia alata, S. purpurea, S. rubra, Schizachyrium scoparium, Schizaea pusilla, Scutellaria integrifolia, Smilax auriculata, S. glauca, S. laurifolia, Sonchus asper, Sorghastrum nutans, Spartina patens, Sphagnum, Spiranthes, Symphyotrichum dumosum, Symplocos tinctoria, Toxicodendron vernix, Triantha racemosa, Vaccinium elliottii, V. macrocarpon, Vigna luteola, Viola sagittata, Utricularia striata, Xyris chapmanii.*

Polygala chapmanii **Torr. & A. Gray** is an annual plant of bogs, flatwoods, prairies, roadsides, sandhills, savannas, seeps, and slopes (to 30°) at elevations of at least 31 m. Exposures include open to shaded sites. The substrates are acidic and have been described as Bladen (typic albaquults), Coxville soils, Dorovan–Johnston series soils, loamy sand, mucky clay, Pelham (arenic paleaquults), sand, sandy clay loam, and sandy loam. Flowering extends from May to August. The average root depth is 2.17 cm. Other ecological information is lacking for this species. **Reported associates:** *Acer rubrum, Andropogon, Aristida, Callicarpa americana, Eryngium yuccifolium, Panicum, Pinus elliottii, P. palustris, P. taeda, Quercus, Sarracenia, Sphagnum.*

Polygala cruciata **L.** is annual, which occurs in bogs, depressions, dessicated ponds, ditches, flatwoods, interdunal hollows, lake plains, meadows, moors, oak openings, pinelands, prairies, roadsides, sand pits, savannas, seeps, sloughs, springs, swales, swamps, woodlands and along the margins of ponds, rivers, and salt marshes at elevations of up to 205 m. The plants grow in areas of perennially standing water or in sites, which are inundated seasonally. They occur in exposures of full sunlight, are relatively shade intolerant, and thrive when the overstory vegetation is cleared away or burned. The substrates can be quite acidic (e.g., pH: 4.5–4.6) and tend to be organic (usually more than 60 cm muck). They include clay loam, gravel, peaty sand, sand, and Smithton soils. Flowering and fruiting extend from May to October. The flowers are mildly fragrant, but specific pollinators have not been reported. A persistent seed bank does not develop. The plants are not resistant to fire and recover poorly from burn-related disturbances. **Reported associates:** *Acalypha rhomboidea, Agalinis aphylla, A. purpurea, Agrimonia parviflora, Agrostis gigantea, A. hyemalis, Aletris farinosa, A. lutea, Andropogon gerardii, A. glomeratus, Anthaenantia rufa, Aristida palustris, Aronia arbutifolia, A. melanocarpa, A. ×prunifolia, Athyrium filix-femina, Aureolaria pedicularia, Balduina uniflora, Bartonia paniculata, B. virginica, Bidens polylepis, B. trichospermus, Bigelowia nudata, Bulbostylis capillaris, Burmannia capitata, Calamagrostis canadensis, Calamovilfa brevipilis, Calopogon barbatus, C. pallidus, C. tuberosus, Carex annectens, C. aquatilis, C. buxbaumii, C. cumulata, C. emoryi, C. glaucescens, C. haydenii, C. longii, C. scoparia, C. swanii, Carphephorus pseudoliatris, Carya tomentosa, Castilleja coccinea, Centella asiatica, Chamaecrista fasciculata, Chaptalia tomentosa, Cliftonia* *monophylla, Coleataenia longifolia, Comandra umbellata, Coreopsis gladiata, C. rosea, Ctenium aromaticum, Cyperus, Dichanthelium acuminatum, D. ensifolium, D. scoparium, Diospyros virginiana, Doellingeria umbellata, Drosera capillaris, D. filiformis, D. intermedia, D. tracyi, Dryopteris cristata, Eleocharis elliptica, E. melanocarpa, E. robbinsii, Epilobium leptophyllum, Eragrostis, Erigeron vernus, Eriocaulon compressum, E. decangulare, Eriophorum angustifolium, Eryngium integrifolium, Eupatorium linearifolium, E. serotinum, Euthamia graminifolia, Fimbristylis autumnalis, Fuirena scirpoidea, Galium tinctorium, G. trifidum, Gaylussacia mosieri, Gentiana andrewsii, Gymnadeniopsis nivea, Hartwrightia floridana, Helenium vernale, Helianthus angustifolius, H. giganteus, H. heterophyllus, Hypericum brachyphyllum, H. canadense, H. fasciculatum, H. gymnanthum, H. majus, H. mutilum, H. setosum, Hypoxis hirsuta, Ilex glabra, I. vomitoria, Juncus acuminatus, J. anthelatus, J. canadensis, J. dudleyi, J. effusus, J. greenei, J. marginatus, J. vaseyi, Lachnanthes caroliniana, Lachnocaulon anceps, Liatris spicata, Linum floridanum, L. striatum, Lobelia glandulosa, L. puberula, Lophiola aurea, Ludwigia alternifolia, Lycopodiella alopecuroides, Magnolia virginiana, Marshallia tenuifolia, Mitreola sessilifolia, Muhlenbergia expansa, Myrica cerifera, M. heterophylla, M. pensylvanica, Oclemena reticulata, Oenothera laciniata, Oldenlandia uniflora, Onoclea sensibilis, Osmunda claytoniana, O. regalis, Osmundastrum cinnamomeum, Oxypolis rigidior, Panicum wrightianum, Persea palustris, Persicaria hydropiperoides, Pinguicula lutea, P. planifolia, Pinus elliottii, P. palustris, P. taeda, Platanthera integra, P. lacera, Pluchea baccharis, Poa nemoralis, Pogonia ophioglossoides, Polygala cymosa, P. lutea, P. rugelii, P. sanguinea, Polytrichum, Potentilla simplex, Quercus falcata, Q. palustris, Q. phellos, Q. velutina, Rhexia alifanus, R. lutea, R. mariana, R. nuttallii, R. virginica, Rhus copallinum, Rhynchospora breviseta, R. capitellata, R. cephalantha, R. ciliaris, R. fascicularis, R. latifolia, R. macra, R. macrostachya, R. oligantha, R. rariflora, Rosa carolina, Rubus hispidus, R. pubescens, Sabatia campanulata, S. difformis, S. grandiflora, Saccharum giganteum, Sagittaria, Salix candida, S. humilis, S. myricoides, S. petiolaris, Sarracenia alata, S. flava, S. leucophylla, S. minor, S. purpurea, S. psittacina, Schizachyrium scoparium, Schoenoplectus pungens, Scleria muhlenbergii, S. pauciflora, S. reticularis, S. triglomerata, S. verticillata, Scutellaria galericulata, S. lateriflora, Serenoa repens, S. tenerum, Smilax laurifolia, Solidago nemoralis, S. speciosa, Sorghastrum nutans, Sphagnum, Spiraea tomentosa, Spiranthes cernua, Stachys hyssopifolia, Stenanthium densum, Symphyotrichum boreale, S. oolentangiense, Syngonanthus flavidulus, Thelypteris palustris, Tiedemannia filiformis, Toxicodendron vernix, Triadenum fraseri, Triantha racemosa, Typha, Utricularia cornuta, U. juncea, U. subulata, Vaccinium macrocarpon, Viburnum, Viola lanceolata, V. primulifolia, Xyris ambigua, X. baldwiniana, X. drummondii, X. elliotii, X. fimbriata, X. smalliana, X. torta, Zigadenus glaberrimus.*

***Polygala cymosa* Walter** is a biennial, which occurs in shallow (to 40 cm), standing waters of seasonally flooded bogs, borrow pits, depressions, ditches, flats, flatwoods, meadows, ponds, pools, prairies, roadsides, savannas, sloughs, and swales at elevations of up to 100 m. Typical exposures occur in full sun. The substrates are acidic and have been characterized as Bayboro soils, muck, peat, sand, sandy clay, and sandy peat. Flowering and fruiting extend from March to September. The flowers are pollinated by bees (e.g., (Insecta: Hymenoptera: Megachilidae). A seed bank develops and persists for an undetermined period of time. The seeds have germinated successfully after receiving cold stratification at 3.3°C for a period of 40 days followed by incubation at ambient greenhouse temperatures. **Reported associates:** *Agalinis linifolia, Aletris lutea, Andropogon arctatus, Aristida stricta, Balduina uniflora, Bartonia verna, Bigelowia nudata, Blechnum serrulatum, Carphephorus pseudoliatris, Chaptalia tomentosa, Cirsium lecontii, Cladium, Clethra alnifolia, Coleataenia tenera, Coreopsis gladiata, Ctenium aromaticum, Cyrilla racemiflora, Dichanthelium acuminatum, D. sphaerocarpon, Drosera capillaris, D. tracyi, Erigeron vernus, Eriocaulon compressum, E. decangulare, Eryngium prostratum, Euphorbia inundata, Fuirena scirpoides, Gymnadeniopsis nivea, Helenium pinnatifidum, H. vernale, Helianthus heterophyllus, H. radula, Hypericum brachyphyllum, H. fasciculatum, H. myrtifolium, Ilex coriacea, I. myrtifolia, Juncus elliottii, Lachnanthes caroliniana, Liatris spicata, Lilium catesbaei, Lobelia boykinii, L. floridana, Lophiola aurea, Ludwigia linearis, L. linifolia, Lycopodium, Lyonia lucida, Magnolia virginiana, Mitreola sessilifolia, Mnesithea rugosa, Muhlenbergia expansa, M. torreyana, Myrica heterophylla, Nyssa biflora, Panicum hemitomon, Pinguicula ionantha, P. lutea, P. planifolia, Pinus elliottii, P. palustris, P. serotina, Pleea tenuifolia, Pogonia ophioglossoides, Polygala cruciata, P. ramosa, P. rugellii, Rhexia alifanus, R. aristosa, R. cubensis, R. lutea, Rhynchospora chapmanii, R. ciliaris, R. latifolia, R. oligantha, R. perplexa, R. pleiantha, R. plumosa, R. tracyi, Rudbeckia graminifolia, R. mohrii, Sabatia bartramii, S. macrophylla, Saccharum giganteum, Sarracenia flava, S. psittacina, Scleria baldwinii, S. georgiana, S. triglomerata, Sphagnum, Stenanthium densum, Taxodium ascendens, Tiedemannia canbyi, T. filiformis, Tillandsia usneoides, Triantha racemosa, Utricularia subulata, Woodwardia virginica, Xyris ambigua, X. baldwiniana, X. elliottii.*

***Polygala grandiflora* Walter** is a biennial or perennial, which occurs in barrens, beaches, bogs, dunes, flatwoods, glades, hammocks, marshes, pinelands, prairies, roadsides, savannas, swales, swamps, and along the margins of canals and ponds at elevations of up to 116 m. The sites occur in open exposures that receive full sunlight. The substrates are described as alkaline or calcareous and include chalk, limestone, marl, sand, sandy loam, and shells. Flowering and fruiting have been observed nearly year round (from January to November). The flowers are pollinated by bees (e.g., Insecta: Hymenoptera: Megachilidae). The seeds possess an elaiosome and are readily collected (and dispersed) by ants (Insecta: Hymenoptera: Formicidae). Darker-colored seeds were found to be more viable (33% viable) than lighter seeds (8% viable). This species is highly fire adapted. It resprouts effectively after fires (from a deep tap root) and is often found in frequently (sometimes annually) burned areas. **Reported associates:** *Agalinis setacea, A. tenuifolia, Allium canadense, Amsonia ciliata, Andropogon gerardii, A. gyrans, A. glomeratus, A. virginicus, Anemia wrightii, Aristida lanosa, A. purpurascens, Asclepias viridiflora, Baccharis halimifolia, Blephilia ciliata, Buchnera longifolia, Cakile, Callirhoe alcaeoides, Carya tomentosa, Castanea pumila, Castilleja kraliana, Cenchrus, Chamaecrista deeringiana, Chenopodium, Chiococca alba, Cladium jamaicense, Clitoria, Cnidoscolus urens, Coleataenia tenera, Coreopsis grandiflora, Croton monanthogynus, Cyperus flavescens, Dalea cahaba, D. purpurea, Delphinium carolinianum, Dichanthelium, Eragrostis spectabilis, Erigeron strigosus, Eryngium yuccifolium, Eupatorium leptophyllum, Euphorbia corollata, Fimbristylis puberula, Funastrum clausum, Gaillardia aestivalis, Helenium, Hydrocotyle, Hypericum hypericoides, Hypoxis hirsuta, Hyptis alata, Ipomoea pes-caprae, Isoetes butleri, Juniperus virginiana, Kyllinga odorata, Leavenworthia exigua, L. uniflora, Liatris cylindracea, L. oligocephala, Linum carteri, L. sulcatum, Lithospermum canescens, Lobelia spicata, Marshallia mohrii, Mecardonia acuminata, Minuartia patula, Mirabilis albida, Monarda citriodora, Nothoscordum bivalve, Oenothera filipes, O. humifusa, Onosmodium decipiens, Oxalis priceae, Panicum flexile, P. virgatum, Paronychia virginica, Paspalum setaceum, Penstemon tenuiflorus, Persicaria punctata, Phyllanthopsis phyllanthoides, Pinus echinata, P. palustris, P. taeda, Polygala boykinii, Quercus incana, Q. stellata, Ratibida pinnata, Rhus copallinum, Rhynchospora colorata, R. globularis, R. glomerata, Rudbeckia triloba, Ruellia humilis, Sabal minor, Salvia azurea, Scaveola, Schizachyrium rhizomatum, S. scoparium, Schoenolirion croceum, Scirpus cyperinus, Scleria, Scutellaria parvula, Senna ligustrina, Silphium glutinosum, Solidago petiolaris, S. stricta, S. ulmifolia, Spartina, Sphenomeris clavata, Spigelia gentianoides, Spiranthes magnicamporum, Sporobolus compositus, S. junceus, Stenaria nigricans, Stylosanthes, Symphyotrichum concolor, Tephrosia, Tetragonotheca helianthoides, Thelypteris kunthii, Uniola, Viola walteri, Vitis rotundifolia, Xyris laxifolia, Yucca filamentosa.*

***Polygala ramosa* Elliott** is an annual, which inhabits barrens, bogs, depressions, ditches, flatwoods, hillside seeps, marshes, meadows, pinelands, pocosins, prairies, roadsides, savannas, and the margins of lakes, ponds, and streams at elevations of up to at least 33 m. The plants are often found in shallow (to 15 cm), standing waters. The sites are in open exposures, which receive full sunlight. The substrates are acidic and include clay, clay loam, loam, peaty sand, sand, sandy loam, sandy peat, silty loam, and silty sandy loam. Flowering occurs from February to October. The plants develop a seed bank of uncertain duration. They are well-adapted to fire and frequently occur in burned-over areas. **Reported associates:** *Acer rubrum, Aletris aurea, A. lutea, Amphicarpum muhlenbergianum, Andropogon glomeratus, Aristida stricta, Arnoglossum ovatum, Balduina uniflora, Bigelowia nudata, Calopogon, Carex, Centella asiatica, Chaptalia tomentosa, Cirsium lecontii, Coleataenia tenera, Ctenium aromaticum,*

Cynoctonum sessilifolium, Cyperus, Diospyros virginiana, Drosera capillaris, D. tracyi, Eriocaulon compressum, E. decangulare, Eryngium, Eupatorium recurvans, Fuirena breviseta, F. scirpoidea, Gaylussacia, Gratiola pilosa, Hartwrightia floridana, Helenium vernale, Helianthus angustifolius, H. heterophyllus, Hypericum fasciculatum, H. galioides, H. myrtifolium, Hyptis alata, Ilex glabra, I. vomitoria, Juncus, Lachnanthes caroliniana, Lachnocaulon anceps, Liatris acidota, L. pycnostachya, Lophiola aurea, Ludwigia linearis, Lycopodiella alopecuroides, L. appressa, Magnolia virginiana, Marshallia graminifolia, M. tenuifolia, Mitreola sessilifolia, Myrica cerifera, Panicum abscissum, Paspalum praecox, Persea palustris, Physostegia, Pinguicula ionantha, P. planifolia, P. pumila, Pinus elliottii, P. palustris, Pluchea, Polygala cymosa, P. lutea, P. nana, P. rugelii, Pterocaulon, Quercus laurifolia, Q. virginiana, Rhexia lutea, R. mariana, Rhynchospora cephalantha, R. chapmanii, R. ciliaris, R. compressa, R. divergens, R. elliottii, R. fascicularis, R. filifolia, R. gracilenta, R. plumosa, R. rariflora, Rubus, Rudbeckia graminifolia, Sabatia grandiflora, Sagittaria, Sarracenia alata, S. flava, S. minor, S. psittacina, Schizachyrium scoparium, S. tenerum, Scleria baldwinii, S. georgiana, S. hirtella, S. reticularis, Smilax, Sphagnum, Sphenostigma coelistina, Stenanthium densum, Symphyotrichum dumosum, Syngonanthus flavidulus, Taxodium distichum, Toxicodendron, Utricularia subulata, Vaccinium, Xyris ambigua, X. caroliniana, X. elliotii, X. stricta.

Use by wildlife: There are few reports of wildlife use for any of the OBL *Polygala* species, perhaps because of their relative rarity. *Polygala balduinii* and *P. ramosa* are reportedly of poor forage value. *Polygala cruciata* is host to leaf-spot fungi (Ascomycota: Mycosphaerellaceae: *Pseudocercospora grisea*). *Polygala cymosa* is a habitat component of Holbrook's chorus frog (Amphibia: Hylidae *Pseudacris ornata*). Its flowers are visited by leafcutting bees (Insecta: Hymenoptera: Megachilidae: *Megachile brevis*), which presumably function as pollinators. The plants are also hosts to boll weevils (Insecta: Coleoptera: Curculionidae: *Anthonomus grandis*). *Polygala grandiflora* is a component of nesting habitat for the Cape Sable seaside sparrow (Aves: Emberizidae: *Ammodramus maritimus mirabilis*). Its relatively lightweight seeds are readily collected by fire ants (Insecta: Hymenoptera: Formicidae: *Solenopsis invicta*), which seek the oil-bearing elaiosomes. The plants are also host to fungi (Ascomycota: Phyllachoraceae: *Phyllachora polygalae*). The flowers of *P. grandiflora* attract leafcutting bees (Insecta: Megachilidae: *Anthidiellum notatum*), which are assumed to function as pollinators.

Economic importance: food: *Polygala* plants are bitter, inedible, and often contain irritating saponins; **medicinal:** Although some *Polygala* species have been used in traditional medicine, there are no medical uses reported for any of the OBL species; **cultivation:** several *Polygala* species are cultivated, but rarely any of the aquatics. *Polygala cruciata* sometimes is grown as a wildflower; **misc. products:** none; **weeds:** none; **nonindigenous species:** none.

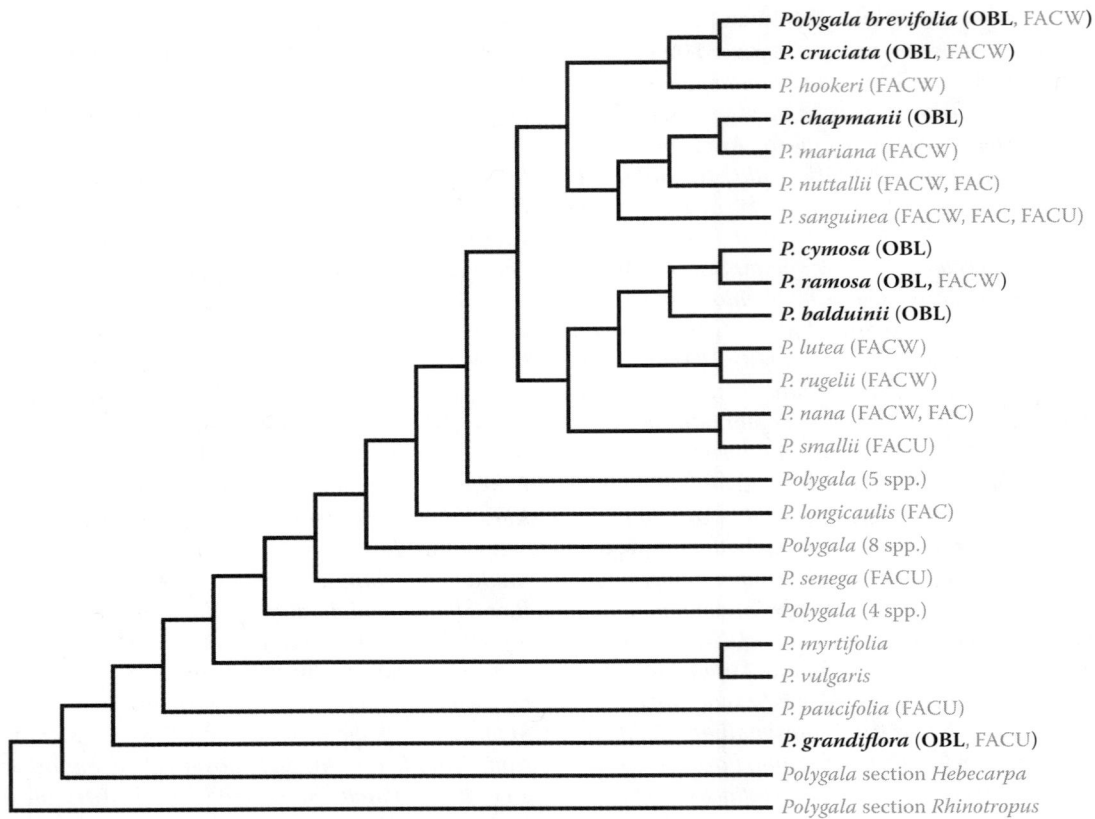

FIGURE 4.17 Interspecific relationships in *Polygala* (sensu lato), as indicated by analysis of nrITS sequence data. The North American wetland indicator status is given in parentheses, with OBL taxa shown in bold. Some authors have transferred *Polygala grandiflora* to the genus *Asemeia*. In any case, these results clearly indicate four distinct origins of the OBL taxa in the family. (Adapted from Abbott, J.R., Phylogeny of the Polygalaceae and a revision of *Badiera*, PhD dissertation, University of Florida, 2009.)

Systematics: Phylogenetic analysis of 25 *Polygala* species and related taxa (using *trnL-F* sequence data) indicate that the genus is not monophyletic as currently circumscribed. Subsequent analyses (which include all of the obligate species) also indicate that the genus is not monophyletic and have provided additional details of interspecific relationships as well. Sequence data from the nrITS region (Figure 4.17) resolve *P. cymosa* and *P. ramosa* as sister species, with *P. balduinii* as the sister group to that clade. *Polygala brevifolia* and *P. cruciata* also resolve as sister species. *Polygala chapmanii* is the sister species of *P. mariana*, a FACW indicator. The aforementioned groups are closely related, but represent three independent origins of the OBL habit, each apparently derived from different groups of FACW indicator species (Figure 4.17). A fourth OBL origin is indicated by *P. grandiflora*, which some authors have transferred to the genus *Asemeia*. Species boundaries in *Polygala* are often believed to be obscured by interactions of hybridization and polyploidy. Alleged hybrids include *Polygala balduinii* × *P. ramosa*, which resolve as closely related species. The base chromosome number for Polygalaceae is $x = 5$–11; for this group of *Polygala* it most likely is $x = 16$–18. Accordingly, *Polygala balduinii*, *P. cymosa*, and *P. ramosa* are tetraploids ($2n = 64$); anomalous counts of $n = 34$ have been reported for *P. ramosa*, which might be in error. *Polygala cruciata* ($2n = 36$) would

represent a diploid based on $x = 18$. Chromosome numbers have not been reported for the other aquatic species.

Comments: The obligately aquatic *Polygala* species generally are rare at least in portions of their ranges and often they occur with other imperiled plants. Most of the OBL species occur along the southeastern coastal plain (*P. balduinii*, *P. brevifolia*, *P. chapmanii*, *P. cymosa*, *P. grandiflora*, *P. ramosa*); *P. cruciata* is more widely distributed in eastern North America.

References: Abbott, 2009, 2011; Boyer & Carter, 2011; Brewer et al., 2011; Bridges & Orzell, 1989; Cohen et al., 2004; Cumberland & Kirkman, 2013; Deyrup et al., 2002; Entrup, 2015; Farlow & Seymour, 1888; Fell, 1957; Godfrey & Wooten, 1981; Hardee et al., 1999; Harper, 1937; Hinman & Brewer, 2007; Kirkman et al., 2000; Lockwood et al., 1999; Mitchell, 2011; Moyer & Bridges, 2015; Murphy & Boyd, 1999; Orzell & Bridges, 2006; Persson, 2001; Reid & Urbatsch, 2012; Sharma, 2012; Singhurst et al., 2012; Smith & Ward, 1976; Sorrie & Leonard, 1999.

ORDER 4: FAGALES [8]

The order Fagales consists of about 1100 species. Phylogenetic analysis of multiple data sets (e.g., 18s rDNA, *atpB*, *rbcL* sequences) provides strong support for the monophyly of the

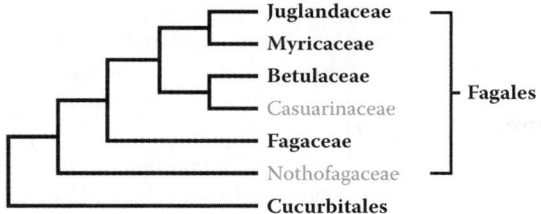

FIGURE 4.18 Phylogenetic relationships among six families of Fagales as inferred from analysis of multiple gene sequence data. Molecular data resolve Fagales as a clade that is sister to Cucurbitales. The families containing obligate aquatic plants in North America (shown in bold) occur throughout the order; however, all are primarily terrestrial. (Adapted from Soltis, D.E. et al., *Bot. J. Linn. Soc.*, 133, 381–461, 2000; Soltis, D.E. et al., *Amer. J. Bot.*, 98, 704–730, 2011.)

order as well as for relationships among the constituent families (Figure 4.18). The group is also well-defined morphologically, being represented by tanniferous trees or shrubs with glandular headed or stellate foliar hairs, and reduced, unisexual flowers (in catkins) that lack nectaries, have inferior ovaries that develop into indehiscent, single-seeded fruits, and pollen tubes that enter the chalazal region of the ovule. The association of Fagales with Cucurbitales is supported consistently, but only moderately by some data sets. Most species of Fagales are wind-pollinated plants of terrestrial forests.

In North America, OBL indicators are found within four principally terrestrial families:

1. **Betulaceae** Gray
2. **Fagaceae** Dumort
3. **Juglandaceae** DC. ex Perleb
4. **Myricaceae** Blume

Family 1: Betulaceae [6]

Betulaceae are a small family of six genera and about 160 species. The species are distinctive trees or shrubs (2-ranked, doubly serrate leaves, flowers in catkins), which are diverse ecologically, but characteristically inhabit early successional habitats, forests, or wetlands. Nitrogen fixation (by root nodulation) occurs in the genus *Alnus*. Phylogenetic studies of Betulaceae (morphology; ITS, *rbcL* sequences) resolve two monophyletic groups within the family (Figure 4.19), which are recognized taxonomically as distinct subfamilies: Betuloideae (with *Alnus*, *Betula*) and Coryloideae (with *Carpinus*, *Corylus*, *Ostrya*, *Ostryopsis*) (Chen et al., 1999). Some authors have elevated these subfamilies to familial level (as Betulaceae, Corylaceae).

The family is known for its production of edible filberts and hazelnuts (*Corylus*) and a variety of ornamental shrubs (*Alnus*, *Betula*, *Carpinus*, *Corylus*, *Ostrya*). Minor economic products include birch beer, which is flavored using the bark or sap of birch trees (*Betula*). One version of the "Black Cow" beverage employs a combination of chocolate ice cream and birch beer. The sap of birch is also concentrated to make a syrup noted for its spicy caramel taste. Birch bark was also

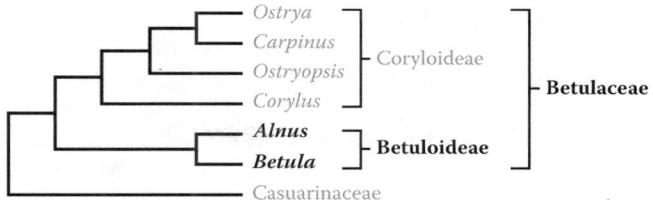

FIGURE 4.19 Phylogenetic relationships in Betulaceae as indicated by cladistic analyses of molecular and nonmolecular data. The few taxa containing OBL North American indicators (in bold) include only the genera *Alnus* and *Betula* (subfamily Betuloideae). (Modified from Chen, Z.-D. et al., *Amer. J. Bot.*, 86, 1168–1181, 1999.)

used by native American tribes to build sturdy, lightweight canoes.

The OBL indicator species are restricted to subfamily Betuloideae where they occur within the two component genera:

1. ***Alnus*** Mill.
2. ***Betula*** L.

1. *Alnus*

Alder; aulne, aune

Etymology: an ancient Latin name probably derived from the Celtic *al lan* meaning "near riverbanks"

Synonyms: *Alnaster*; *Betula* (in part); *Betula-alnus*; *Clethropsis*

Distribution: global: Africa; Asia; Europe; New World; **North America:** widespread

Diversity: global: 25 species; **North America:** 8 species

Indicators (USA): OBL: *Alnus maritima*, *A. serrulata*; **FACW:** *A. rugosa*, *A. serrulata*, *A. tenuifolia*; **FAC:** *A. tenuifolia*

Habitat: brackish (coastal), freshwater; palustrine; **pH:** 4.8–8.1; **depth:** <1 m; **life-form(s):** emergent (shrub, tree)

Key morphology: shrubs or trees (to 20 m); twigs with conspicuous lenticels, stalked buds, and triangular pith; leaves (to 9 cm) deciduous, stipulate (to 10 mm), alternate, simple, the margins serrate; flowers unisexual, in catkins or cones; fruits (to 1.6 cm) woody, cone-like, containing small, winged nutlets (samaras)

Life history: duration: perennial (buds); **asexual reproduction:** stolons, suckers; **pollination:** wind; **sexual condition:** monoecious; **fruit:** nutlets (common); **local dispersal:** seeds (wind); **long-distance dispersal:** seeds (water, wind)

Imperilment: (1) *Alnus maritima* [G3]; S1 (GA); S2 (OK); S3 (MD); (2) *A. rugosa* [G5]; S1 (IL); S2 (VA); S3 (IA, IN, NJ); (3) *A. serrulata* [G5]; SH (KS); S1 (QC); S2 (NB); S3 (IL, NS); (4) *A. tenuifolia* [G5]

Ecology: general: All eight North American *Alnus* species are wetland indicators, but only two currently are designated as OBL species. Because the previous (1996) indicator list included *A. rugosa* and *A. tenuifolia* (as subspecies of *A. incana*) among the OBL indicators, their treatments were completed prior to the release of the revised list and have been retained here. *Alnus* species all occur on nitrogen-poor soils and develop nitrogen-fixing root nodules in a symbiotic bacterial association (Bacteria: Actinobacteria: Frankiaceae: *Frankia alni*). As a consequence, alder communities are important ecologically because they fix large quantities of nitrogen annually. The plants contribute further to soil fertility by the deposition of their nitrogen-rich litter. Most alders are early successional species, which grow in at least seasonally wet habitats exposed to full sunlight. The flowers are wind pollinated. Alder seeds will release from detached cones after several weeks at ambient temperatures, or within a week if the cones are dried at 16°C–27°C. The winged fruits are dispersed in the air or on the water (they can remain afloat for a considerable duration of time). Vegetative reproduction occurs by suckering from their stoloniferous roots.

Alnus maritima (**Marshall**) **Muhl. ex Nutt.** occurs on floodplains, in swamps, and along the margins of estuaries, ponds, rivers, and streams at elevations of up to 250 m. The plants are typically found growing in fresh (<1 g/kg salinity), often flowing waters (to several cm deep) of circumneutral (pH: 6.1–8.1). Despite its specific epithet, *A. maritima* is only weakly tolerant of low-salt concentrations and cannot withstand salinities above 5 g/kg for prolonged periods. The habitats receive up to 40 dm of rainfall annually. The reported substrates include boulders, granite, limestone, mud, rock, and sand. The plants are moderately fast growing and live up to 40 years. The seeds reportedly germinate after receiving 6 weeks of cold stratification. **Reported associates:** *Acer negundo, A. rubrum, Amorpha fruticosa, Betula nigra, Cornus, Eleocharis equisetoides, Eriocaulon aquaticum, Juglans, Liquidambar styraciflua, Lobelia cardinalis, Nymphaea odorata, Panicum hemitomon, Parthenocissus, Platanus occidentalis, Salix caroliniana, S. nigra, Schoenoplectus etuberculatus, S. subterminalis, Taxodium distichum, Ulmus, Utricularia fibrosa, U. radiata.*

Alnus rugosa (**Du Roi**) **Spreng.** grows in or on beaches, bogs, depressions, ditches, dunes, fens, floodplains, gravel bars, marshes, meadows, roadsides, seeps, swales, swamps, thickets, woodlands, and along the margins of brooks, lakes, rivers, and streams at elevations of up to 736 m. *Alnus rugosa* commonly exists as an understory shrub, but the plants thrive in open sunlight. The plants occur mainly in open lowland areas where they often form dense "thickets." The substrates are acidic to somewhat alkaline (pH: 4.8–7.7) and include muck, peat, peaty sand, rock, and sandy loam. Flowering and fruiting extend from April to August. Pollination occurs by the wind. Newly shed seeds are physiologically dormant and require 60 days of stratification at 1°C–5°C followed by incubation at 24°C–27°C to induce germination. Those residing in the soil for a year or more often exhibit no dormancy. Seeds stored at 1°C–3°C have remained viable for up to 10 years. *Alnus rugosa* stands are known to fix over 43 kg nitrogen per hectare annually. New shoots will sprout quickly from persistent root crowns following fires. The plants are thought to form extensive clones by vegetative suckering but genetic data indicate that patches often contain numerous individuals (genets). **Reported associates:** *Abies balsamea, Acer rubrum, Alisma plantago-aquatica, Alnus viridis, Athyrium filix-femina, Betula nana, B. papyrifera, B. pumila, Bidens cernuus, B. frondosus, Bolboschoenus fluviatilis, Calamagrostis canadensis, Calla palustris, Callitriche verna, Caltha natans, C. palustris, Carex brunnescens, C. comosa, C. intumescens, C. lacustris, C. leptalea, C. paupercula, C. retrorsa, C. rostrata, C. stipata, C. tenera, C. trisperma, C. tuckermanii, Carpinus caroliniana, Chamaedaphne calyculata, Chelone glabra, Chrysosplenium americanum, Clintonia borealis, Comarum palustre, Coptis groenlandica, Cornus canadensis, C. sericea, Corylus cornuta, Doellingeria umbellata, Dryopteris cristata, D. spinulosa, D. goldiana, Eleocharis obtusa, Epilobium coloratum, Equisetum fluviatile, E. sylvaticum, Eutrochium maculatum, Fragaria, Fraxinus nigra, Galium tinctorium, Gaultheria hispidula, Gentianopsis crinita, Glyceria grandis, Gymnocarpium dryopteris, Ilex mucronata, I. verticillata, Impatiens capensis, Iris versicolor, Juncus effusus, Larix laricina, Lemna minor, Lycopus uniflorus, Lysimachia terrestris, Myrica gale, Nuphar variegata, Nymphaea tuberosa, Onoclea sensibilis, Osmunda claytoniana, Osmundastrum cinnamomeum, Persicaria sagittata, Petasites sagittatus, Phalaris arundinacea, Picea mariana, P. rubens, Pinus banksiana, Platanthera psycodes, Populus balsamifera, P. deltoides, P. grandidentata, Potamogeton, Potentilla norvegica, Quercus macrocarpa, Ranunculus gmelinii, R. pensylvanicus, Rhododendron canadense, R. groenlandicum, Rhus typhina, Rubus alleghaniensis, R. pubescens, R. strigosus, Rumex orbiculatus, Salix discolor, S. gracilis, S. lucida, S. planifolia, S. purpurea, S. sericea, S. serissima, Scirpus cyperinus, Scutellaria lateriflora, Sium suave, Solidago, Sparganium chlorocarpum, S. glomeratum, Sphagnum fallax, S. magellanicum, S. palustre, S. recurvum, Spiraea alba, S. tomentosa, Spirodela polyrhiza, Symphyotrichum puniceum, Thuja occidentalis, Tilia americana, Triadenum virginicum, Trientalis borealis, Typha latifolia, Ulmus americana, Utricularia intermedia, U. macrorhiza, Vaccinium angustifolium, Vernonia fasciculata, Viburnum nudum, V. trilobum.*

Alnus serrulata (**Aiton**) **Willd.** inhabits nontidal or tidal freshwater bogs, bottomlands, cobble bars, ditches, flats, flatwoods, floodplains, glades, gravel pits, lowlands, meadows, ravines, roadsides, savannahs, seeps, slopes, sloughs, swamps, thickets, woodlands, and the margins of bayous, lakes, ponds, rivers, sloughs, and streams at elevations of up to 1036 m. The plants occur within hardiness zones 5–8 where annual precipitation exceeds 81 cm. They can grow in shallow water and usually prefer full sun but will tolerate partial shade. The substrates are acidic to neutral (pH: 5.0–7.0) and include alluvium, Catahoula sandstone, clay, granite, gravel, loam, muck, mud, rock, sand, sandy clay, and sandy loam. Flowering and fruiting have been observed from April to December. The flowers are wind pollinated. The seeds are produced in quantity each year and require 4–6 months of cold stratification (2°C–4°C) to break dormancy. They do not maintain their viability when stored. They germinate best when air dried after collection and planted within 1 month of harvest. The plants can develop into clonal colonies by vegetative suckering. The stems are weak and easily damaged by wind and ice. **Reported associates:** *Acalypha rhomboidea, Acer negundo, A. rubrum, A. saccharinum, A. saccharum, Aesculus flava, Albizia, Alisma subcordatum, Alternanthera philoxeroides, Amphicarpaea bracteata, Andropogon, Antennaria plantaginifolia, Apios americana, Arundinaria gigantea, Athyrium angustum, A. filix-femina, Barbarea vulgaris, Betula alleghaniensis, B. lenta, B. nigra, Boehmeria cylindrica, Bulbostylis capillaris, Calycanthus floridus, Campsis radicans, Cardamine diphylla, Carex blanda, C. bromoides, C. debilis, C. laxiflora, C. prasina, C. projecta, C. scabrata, C. torta, Carya glabra, Carpinus caroliniana, Cephalanthus occidentalis, Chamaecrista nictitans, Cicuta maculata, Circaea canadensis, Clethra alnifolia, Coleataenia longifolia, Cornus amomum, C. florida, C. sericea, C. stricta, Corylus americana, Cyrilla racemiflora, Decodon verticillatus, Desmodium nudiflorum, D. rotundifolium, Diodia virginiana, Drosera rotundifolia, Elephantopus tomentosus, Elymus virginicus, Equisetum hyemale, Eragrostis hypnoides, Eryngium prostratum, Euphorbia, Eupatorium perfoliatum, E. serotinum, Eutrochium fistulosum, Fagus grandifolia, Fimbristylis autumnalis, Fraxinus americana, F. pennsylvanica, Geranium maculatum, Geum canadense, Hamamelis virginiana, Hexastylis shuttleworthii, Houstonia, Hydrangea arborescens, Hypericum densiflorum, H. hypericoides, H. mutilum, Ilex verticillata, Impatiens capensis, Ipomoea coccinea, Itea virginica, Juglans nigra, Juncus caesariensis, Kalmia latifolia, Leersia oryzoides, Leucothoe fontanesiana, Lindera benzoin, Linum striatum, Liquidambar styraciflua, Liriodendron tulipifera, Lonicera japonica, Luzula acuminata, Lycopus virgincus, Lyonia ligustrina, Lysimachia fraseri, Magnolia fraseri, M. tripetala, M. virginiana, Microstegium vimineum, Mikania, Mitella diphylla, Myrica cerifera, Nyssa sylvatica, Onoclea sensibilis, Ophioglossum vulgatum, Osmundastrum cinnamomeum, Oxydendrum arboreum, Panax quinquefolius, Panicum capillare, Parthenocissus quinquefolia, Paspalum, Perilla frutescens, Persicaria pensylvanica, P. posumbu, P. sagittata, P. virginiana, Phlox paniculata, Phyllanthus caroliniensis, Phytolacca americana, Pilea pumila, Pinus echinata, P. palustris, P. taeda, Platanus occidentalis, Polypremum procumbens, Polystichum acrostichoides, Populus deltoides, Pueraria, Quercus alba, Q. margaretta, Q. marilandica, Q. pagoda, Q. palustris, Q. rubra, Q. velutina, Ranunculus recurvatus, Rhexia virginica, Rhododendron maximum, Rhus copallina, Robinia pseudoacacia, Rorippa palustris, Rosa palustris, Rotala ramosior, Rubus argutus, Salix nigra, Sambucus nigra, Samolus valerandi, Sarracenia, Sassafras albidum, Scutellaria integrifolia, Sedum ternatum, Smilax hispida, S. laurifolia, Sphagnum, Symphyotrichum dumosum, S. lateriflorum, Taxodium distichum, Thelypteris noveboracensis, Tiarella cordifolia, Tilia americana, Toxicodendron radicans, T. vernix, Tsuga canadensis, Ulmus americana, Vaccinium, Valerianella radiata, Veronica, Viburnum nudum, V. prunifolium, Viola blanda, V. labradorica, Xanthium strumarium, Xanthorhiza simplicissima, Xyris tennesseensis.*

Alnus tenuifolia **Nutt.** occurs in or on bogs, bottoms, carrs, draws, fens, floodplains, gravel bars, gravel pits, hollows,

meadows, muskegs, ravines, roadsides, rock bars, scrub, seeps, slopes (to 30%), swamps, woodlands, and along the margins of lakes, rivers, springs, and streams at elevations of up to 3110 m. The plants occur typically where the water table remains within a meter of the surface. Exposures range from open sites to deep shade. The substrates are described as alluvial sand, basalt, basaltic cobble, clay, clay loam, cobble, granite–diorite, gravel, gravelly alluvium, gravelly silt, limestone, lithosols, loam, muck, Neihart quartzite, organic, rock, rocky clay loam, rocky sand, sand, sandy gravel, sandy loam, silty alluvium, silty loam, stones, and talus. Older sites often develop increasingly thicker upper layers of organic soil. Flowering and fruiting extend from March to November. The flowers are wind pollinated. This is an early successional species, which has a rapid growth rate and can colonize disturbed sites rapidly. It benefits from seasonal flood disturbances, which create sites suitable for seed germination and establishment. Dense, clone-like colonies may develop from the densely shed seeds. The plants are capable of producing more than 3300 seeds/m^2; however, up to 95% of the seeds may be inviable. The seeds lack dormancy, and under appropriate conditions, can germinate immediately after their dispersal. Germination is higher on mineral soils than on organic substrates. The plants are known to resprout from their root crown following fires. **Reported associates:** *Abies concolor, A. grandis, A. lasiocarpa, A. magnifica, Acer glabrum, A. negundo, Achillea millefolium, Aconitum, Actaea rubra, Agrostis scabra, A. stolonifera, Allium schoenoprasum, A. tolmiei, A. validum, Alnus rhombifolia, A. viridis, Amelanchier alnifolia, Amauriopsis dissecta, Apocynum cannabinum, Arctostaphylos uva-ursi, Arnica cordifolia, Artemisia carruthii, A. tridentata, Balsamorhiza sagittata, Betula fontinalis, B. nana, B. neoalaskana, B. occidentalis, B. papyrifera, Bouteloua gracilis, Bromus anomalus, B. inermis, Calamagrostis canadensis, C. stricta, Calocedrus decurrens, Caltha palustris, Campanula rotundifolia, Cardamine cordifolia, Carex aquatilis, C. disperma, C. festivella, C. foenea, C. interior, C. lanuginosa, C. leptalea, C. mertensii, C. microptera, C. nebrascensis, C. projecta, C. rostrata, C. stipata, Castilleja linariifolia, Ceanothus cordulatus, C. velutinus, Chamerion angustifolium, Chimaphila, Cinna latifolia, Clematis, Clintonia, Collinsia, Comarum palustre, Conyza canadensis, Coreopsis tinctoria, Cornus canadensis, C. sericea, Crataegus chrysocarpa, C. douglasii, Cypripedium calceolus, Cystopteris fragilis, Darlingtonia californica, Dianthus armeria, Dipsacus fullonum, Drosera, Dryas drummondii, Dryopteris cristata, D. expansa, Elaeagnus commutata, Eleocharis bella, E. palustris, Elymus elymoides, Epilobium brachycarpum, E. densiflorum, E. hirsutum, Equisetum arvense, E. laevigatum, E. pratense, E. variegatum, Ericameria, Euthamia occidentalis, Euphorbia chamaesula, Eutrochium maculatum, Festuca rubra, Fragaria ovalis, Galium boreale, Gayophytum decipiens, Geranium erianthum, Geum macrophyllum, Glyceria septentrionalis, G. striata, Gutierrezia sarothrae, Gymnocarpium dryopteris, Haplopappus parryi, Heracleum sphondylium, Heterotheca villosa, Heuchera, Holodiscus, Hypericum, Hylocomium splendens, Ilex verticillata, Impatiens capensis, Ipomopsis aggregata, Juglans nigra, Juncus balticus, J. bufonius, J. effusus, Juniperus occidentalis, J. scopulorum, Larix, Lepidium densiflorum, L.*

draba, Leymus cinereus, L. innovatus, Lilium philadelphicum, Linnaea borealis, Lomatium bicolor, Lonicera, Lupinus polyphyllus, Lysimachia ciliata, Madia glomerata, Maianthemum racemosum, M. stellatum, Matteuccia struthiopteris, Melilotus albus, Mentha arvensis, Menziesia ferruginea, Mertensia ciliata, Mimulus guttatus, M. lewisii, Mitella breweri, M. pentandra, Monarda, Moneses uniflora, Morus alba, Oenothera pubescens, Onoclea sensibilis, Opuntia polyacantha, Orthilia secunda, Osmorhiza depauperata, Oxypolis fendleri, Oxytropis lambertii, Parnassia palustris, Paxistima, Pedicularis groenlandica, Petasites frigidus, Phalaris, Philadelphus lewisii, Phleum pratense, Physocarpus malvaceus, Picea breweriana, P. engelmannii, P. glauca, P. mariana, P. pungens, P. sitchensis, Pinus contorta, P. echinata, P. edulis, P. jeffreyi, P. lambertiana, P. monophylla, P. monticola, P. ponderosa, P. strobus, Plantago, Platanthera dilatata, P. obtusata, Poa alpina, P. palustris, P. interior, P. pratensis, Populus angustifolia, P. balsamifera, P. deltoides, P. tremuloides, P. trichocarpa, Prosartes trachycarpa, Prunella, Prunus virginiana, Pseudotsuga menziesii, Ptilium, Purshia, Pycnanthemum californicum, Quercus gambelii, Q. garryana, Ranunculus, Rhamnus purshiana, Rhododendron groenlandicum, Rhytidiadelphus triquetrus, Ribes aureum, R. hudsonianum, R. inerme, R. lacustre, R. montigenum, R. nevadense, Rosa acicularis, R. woodsii, Rubus parviflorus, R. pedatus, R. strigosus, Rudbeckia laciniata, Rumex hymenosepalus, Salix amygdaloides, S. barclayi, S. bebbiana, S. boothii, S. drummondiana, S. exigua, S. fragilis, S. geyeriana, S. irrorata, S. jepsonii, S. lasiandra, S. lutea, S. prolixa, S. scouleriana, S. setchelliana, S. sitchensis, Sambucus nigra, Scirpus microcarpus, Senecio serra, S. triangularis, Shepherdia canadensis, Sidalcea candida, Silene, Solanum stoloniferum, Solidago missouriensis, Spiraea douglasii, S. stevenii, Streptopus amplexifolius, Symphoricarpos albus, S. occidentalis, S. oreophilus, Symphyotrichum lanceolatum, Symplocarpus foetidus, Tamarix, Taxus, Thalictrum fendleri, T. occidentale, T. pubescens, T. sparsiflorum, Thelypteris palustris, Thermopsis, Thuidium abietinum, Thuja plicata, Tiarella, Toxicodendron rydbergii, Trichophorum cespitosum, Trifolium repens, Triglochin palustre, Tsuga heterophylla, T. mertensiana, Typha latifolia, Ulmus pumila, Urtica dioica, Vaccinium scoparium, Valeriana edulis, V. occidentalis, Veratrum californicum, Viburnum edule, V. recognitum, Viola canadensis, Wyethia helianthoides.

Use by wildlife: *Alnus* is an extremely important wildlife genus. The plants provide cover for various mammals (Mammalia) such as elk (Cervidae: *Cervus elaphus*), moose (Cervidae: *Alces alces*), mule deer (Cervidae: *Odocoileus hemionus*), snowshoe hares (Leporidae: *Lepus americanus*; especially during winter), and white-tailed deer (Cervidae: *Odocoileus virginianus*). Beavers (Mammalia: Castoridae: *Castor canadensis*) eat the bark and use the twigs of *A. rugosa* and *A. tenuifolia* to construct their dams and lodges. The plants provide cover and shade for salmonid fish (Osteichthyes: Salmonidae). Alder thickets are also used as brooding, nesting, and resting sites by several birds (Aves) such as American redstarts (Parulidae: *Setophaga ruticilla*), MacGillivray's warbler (Parulidae: *Geothlypis tolmiei*), northern cardinals (Cardinalidae: *Cardinalis cardinalis*),

scarlet tanagers (Cardinalidae: *Piranga olivacea*), white-eyed vireos (Vireonidae: *Vireo griseus*), woodcocks (Scolopacidae: *Scolopax minor*), and as drumming sites by grouse (Phasianidae) and woodcocks (Scolopacidae: *Scolopax minor*). Numerous mammals graze or browse on the twigs and foliage of *A. maritima, A. rugosa, A. serrulata,* and *A. tenuifolia* including domestic cattle (Bovidae: *Bos taurus*), goats (Bovidae: *Capra hircus*), and sheep (Bovidae: *Ovis aries*) as well as beavers (Castoridae: *Castor canadensis*), cottontail rabbits (Leporidae: *Sylvilagus*), elk (Cervidae: *Cervus elaphus*), moose (Cervidae: *Alces alces*), muskrats (Cricetidae: *Ondatra zibethicus*), snowshoe hares (Leporidae: *Lepus americanus*), and whitetail deer (Cervidae: *Odocoileus virginianus*). The buds, cones, and seeds are eaten by several birds (Aves) including chickadees (Paridae: *Poecile*), goldfinches (Fringillidae: *Spinus tristis*), grouse (Phasianidae), pine siskins (Fringillidae: *Spinus pinus*), redpolls (Fringillidae: *Acanthis flammea*), warblers (Parulidae), woodcocks (Scolopacidae: *Scolopax minor*), and occasionally by ducks (Anatidae). The twigs also harbor numerous insects (see below), which provide food for downy woodpeckers (Picidae: *Picoides pubescens*), yellow-rumped warblers (Parulidae: *Setophaga coronata*), and other birds (Aves). Alder species are hosts to an impressive diversity of butterfly and moth larvae (Insecta: Lepidoptera). *Alnus maritima* is a host of false webworms (Arctiidae: *Hyphantria cunea*). *Alnus rugosa* hosts a vast array of Lepidopteran larvae (Apatelodidae: *Apatelodes torrefacta*; Arctiidae: *Hyphantria cunea, Lophocampa maculata*; Coleophoridae: *Coleophora serratella*; Drepanidae: *Drepana arcuata, D. bilineata*; Geometridae: *Anagoga occiduaria, Anavitrinella pampinaria, Antepione thisoaria, Biston betularia, Campaea perlata, Cyclophora pendulinaria, Ectropis crepuscularia, Ennomos magnaria, E. subsignaria, Eulithis xylina, Eupithecia ravocostaliata, Hydria undulata, Hydriomena furcata, H. renunciata, Hypagyrtis unipunctata, Iridopsis larvaria, Itame exauspicata, Lobophora nivigerata, Lycia rachelae, L. ursaria, Melanolophia canadaria, Nematocampa filamentaria, Nemoria mimosaria, Pero morrisonaria, Plagodis alcoolaria, P. phlogosaria, Plemyria georgii, Probole amicaria, Prochoerodes transversata, Protitame virginalis, Rheumaptera hastata, R. subhastata, Semiothisa aemulataria*; Gracillariidae: *Caloptilia alnivorella, C. pulchella, Phyllonorycter auronitens*; Lasiocampidae: *Malacosoma disstria, Phyllodesma americana*; Lycaenidae: *Feniseca tarquinius*; Lymantriidae: *Lymantria dispar, Orgyia antiqua, O. leucostigma*; Noctuidae: *Acronicta americana, A. dactylina, A. fragilis, A. grisea, A. impressa, A. innotata, A. leporina, A. oblinita, Amphipyra pyramidoides, Anaplectoides pressus, Cucullia intermedia, C. lucifuga, Enargia decolor, Eupsilia tristigmata, Eurois astricta, Homoglaea hircina, Lithophane amanda, L. thaxteri, Lomanaltes eductalis, Melanchra adjuncta, Orthosia hibisci, Palthis angulalis, Polia imbrifera, Raphia frater, Spiramater lutra, Zale minerea*; Notodontidae: *Clostera albosigma, Gluphisia septentrionis, Heterocampa biundata, Nadata gibbosa, Schizura concinna, S. unicornis*; Nymphalidae: *Basilarchia arthemis*; Oecophoridae: *Agonopterix argillacea, Bibarrambla allenella, Nites betulella, N. grotella*; Pantheidae: *Colocasia flavicornis*; Papilionidae: *Papilio canadensis, P.*

glaucus; Pyralidae: *Acrobasis rubrifasciella*, *Nealgedonia extricalis*, *Phlyctaenia coronata*; Saturniidae: *Eacles imperialis*, *Hemileuca nevadensis*, *Hyalophora cecropia*, *H. columbia*; Sphingidae: *Sphinx gordius*, *S. luscitiosa*; and Tortricidae: *Acleris braunana*, *A. caliginosana*, *A. cornana*, *A. fuscana*, *A. logiana*, *A. maccana*, *Amorbia humerosana*, *Archips cerasivorana*, *Choristoneura conflictana*, *C. rosaceana*, *Clepsis persicana*, *Epinotia rectiplicana*, *E. solandriana*, *Evora hemidesma*, *Gretchena semialba*, *Olethreutes appendiceum*, *O. submissana*, *Orthotaenia undulana*, Pandemis canadana, *Sparganothis reticulatana*). *Alnus serrulata* also is host to many larval Lepidoptera (Apatelodidae: *Apatelodes torrefacta*; Arctiidae: *Lophocampa caryae*; Bucculatricidae: *Bucculatrix loculples*; Geometridae: *Ennomos subsignaria*, *Hydriomena furcata*; Gracillariidae: *Phyllonorycter auronitens*; Lymantriidae: *Lymantria dispar*; Noctuidae: *Acronicta oblinita*, *Eurois astricta*; Pyralidae: *Acrobasis rubrifasciella*; Saturniidae: *Antheraea polyphemus*, *Eacles imperialis*, *Hyalophora cecropia*; and Sphingidae: *Dolba hyloeus*). Likewise, numerous Lepidoptera larvae are hosted by *A. tenuifolia* (Arctiidae: *Lophocampa maculata*; Drepanidae: *Pseudothyatira cymatophoroides*; Geometridae: *Anagoga occiduaria*, *Biston betularia*, *Campaea perlata*, *Cyclophora pendulinaria*, *Erannis tiliaria*, *Eupithecia misturata*, *Hydriomena furcata*, *Hydriomena renunciata*, *Iridopsis emasculata*, *Lycia ursaria*, *Plagodis phlogosaria*, *Plemyria georgii*, *Probole amicaria*, *Rheumaptera hastata*, *R. subhastata*, *Selenia alciphearia*, *Semiothisa ulsterata*, *Venusia cambrica*, *V. pearsalli*; Gracillariidae: *Caloptilia alnivorella*; Lasiocampidae: *Malacosoma californica*; Limacodidae: *Tortricidia testacea*; Nepticulidae: *Stigmella canadensis*; Noctuidae: *Acronicta grisea*, *A. hesperida*, *A. impleta*; Notodontidae: *Schizura ipomoeae*; Papilionidae: *Papilio rutulus*; Pyralidae: *Nealgedonia extricalis*; and Tortricidae: *Acleris braunana*, *A. caliginosana*, *Archips negundana*, *A. rosana*, *Epinotia solandriana*, *Pandemis limitata*, *Syndemis afflictana*). The white apple leafhopper (Insecta: Hemiptera: Cicadellidae: *Typhlocyba pomaria*) is associated with *Alnus serrulata*.

Economic importance: food: Alders generally are not edible, but their inner bark is regarded as an emergency food. Chewing on the buds of the plants (as practiced by some children) produces a brown-colored saliva; **medicinal:** Alders were used extensively by Native North Americans in an assortment of medicinal applications. The bark contains salicin, which is converted to salicylic acid (the functional constituent of aspirin) when ingested. *Alnus rugosa* can cause an allergic contact dermatitis in some people; however, a decoction of *A. serrulata* bark is said to relieve the symptoms of poison ivy. Infusions and decoctions derived from the inner bark and roots of *A. rugosa* were used by the Cherokee, Chippewa, Iroquois, and Menominee as a cathartic, cold remedy, emetic, salve for sore eyes, and to assist in labor. Decoctions made from the roots, twigs, and young plants were used by the Iroquois, Meskwaki, Mohegan, and Ojibwa as an analgesic or treatment for internal bleeding. Bark scraped from *A. rugosa* and *A. tenuifolia* was used by the Cree and Woodlands people as a laxative. Bark infusions or decoctions of *A. rugosa* were used as an eye ointment (Cherokee, Cree, Woodlands), a treatment for anemia

(Chippewa), to control diarrhea (Potawatomi), to induce sweating (Shuswap), as a remedy for mouth sores (Micmac), and to relieve urinary disorders (Iroquois). The Abnaki made a decoction of *A. rugosa* to treat skin irritations. The Algonquin and Quebec tribes prepared an infusion of the inner bark of *A. rugosa* as a laxative and mixed the root bark with molasses to prepare a toothache remedy. Bark decoctions of *Alnus tenuifolia* served as a soothing wash for sores (Shuswap) or sore eyes (Cree, Woodlands). The Gitksan people ate the catkins of *A. tenuifolia* as a diuretic or laxative. The Keres and Sanpoil used the powdered bark of *A. tenuifolia* to treat sores. The Bella Coola made a soothing poultice from the buds of the plant; **cultivation:** *Alnus maritima*, *A. rugosa*, and *A. serrulata* (e.g., 'Panbowl') are cultivated occasionally for use as wildlife plantings. *Alnus tenuifolia* is widely planted in wetland restoration projects; **misc. products:** The Algonquin, Chippewa, Cree, Ojibwa, Potawatomi, and Woodlands people obtained brown, orange, and yellow dyes (to color baskets, hides, moccasins, and quills) from infusions made from the inner bark of *A. rugosa*. A similar dye was derived from the bark of *A. tenuifolia* by the Apache, Blackfoot, Flathead, Isleta, Jemez, Keres, Klamath, Kutenai, Montana, Navajo, Nez Perce, Okanagan-Colville, Shuswap, Tewa, Western, White Mountain, and Zuni Indians. A yellow dye also was extracted by the Cree and Woodlands tribes as a decoction of *A. rugosa* catkins. They also caulked their canoes using a mixture of wood charcoal from *A. rugosa* and pitch. Wood was softened for bending by applying decoctions of the plant. The Menominee applied an infusion from the root bark of *A. rugosa* to horses as a wash for relieving saddle gall; the Potawatomi treated the same ailment using powdered bark from the plant as an astringent. The Iroquois made a hunting charm from decoctions of *A. rugosa*. Native North Americans sometimes used *A. tenuifolia* for firewood and obtained variously colored dyes from it including a brown dye (from the roots), a red dye (from powdered wood or the inner bark), and a yellow-green dye (from the leaves). The Montana Indians used its bark for tanning and the Thompson its wood for hunting bows. A hot drink made from the bark of *A. tenuifolia* was a Blackfoot treatment for scrofula. Burnt ashes obtained from *A. tenuifolia* were used by the Okanagan-Colville people to clean their teeth. *Alnus tenuifolia* is planted to restore and stabilize weak streambanks. The early settlers of western North America used *A. tenuifolia* to indicate the presence of water; **weeds:** none; **nonindigenous species:** *Alnus glutinosa* (FACW) has been introduced in North America but is not designated as an obligate indicator.

Systematics: *Alnus* plants should not be confused with alder buckthorn (*Rhamnus frangula*), a superficially similar but unrelated (and invasive) wetland shrub. Morphological and molecular data (*rbcL*, nrDNA 18S, nrITS sequences) confirm that *Alnus* comprises a monophyletic sister genus to *Betula*. The obligate aquatic North American species occur in subgenus *Alnus*, which molecular data (nrITS sequences) resolve as monophyletic or paraphyletic depending on the method of analysis used (Figure 4.20). The treatment of several North American shrub taxa as subspecies of the tree-like European *Alnus incana* seems to have met with overwhelming acceptance; however, this practice merits reconsideration and has not been followed here. Notably, preliminary molecular data (ITS sequences) indicate that the three widely recognized subspecies of *Alnus incana* (subsp. *incana*, subsp. *rugosa*, subsp. *tenuifolia*) do not form a unique clade, but associate with *A. hirsuta*, *A. inokumae*, and *A. rubra*, which many authors retain as distinct species. Moreover, one accession of *A. rugosa* resolves as the sister species to *A. rubra* (Figure 4.20), which is almost always regarded as a distinct species. Whether such results indicate that the inclusive circumscription of *Alnus incana* is incorrect, or is owing to a complicating factor (such as hybridization), has not been determined adequately. At least until more compelling evidence to the contrary is presented, the present treatment adopts a more conservative approach, which treats the taxa as distinct species, i.e., as *A. rugosa* and *A. tenuifolia*. These two taxa essentially are allopatric, with ranges that overlap only slightly in north-central Canada, where intermediate plants have been reported. Certainly the relationship among these taxa would benefit from further study. *Alnus maritima* is a distinctive species within the monophyletic section *Clethropsis*, where it is more closely related to Asian species (*A. formosana*, *A. nitida*) than to other North American species. Recently, it has been subdivided into three subspecies that correspond to its disjunct centers of distribution. *Alnus serrulata* is also distinct and resolves phylogenetically as the sister species of the Old World *A. japonica*. Natural

hybrids (known as *A. ×fallacina*) reportedly occur between *A. rugosa* and *A. serrulata*. Some suspected hybrids (from New England) were found to be apomictic and exhibited irregular meiosis but high seed set. The base chromosome number of *Alnus* is $x = 7$. All of the North American taxa (including *A. maritima*, *A. rugosa*, *A. serrulata*, *A. tenuifolia*) uniformly are tetraploid ($2n = 28$). Allozyme studies of *A. rugosa* in Canada revealed little genetic differentiation between populations ($G_{ST} = 0.052$), a reflection either of extensive gene flow or their recent establishment. Electrophoretic data have also indicated that dense stands of this species can comprise mixtures of genetic individuals and are not strictly clonal as was once believed.

Comments: *Alnus maritima* is rare and its occurrences are oddly disjunct between Oklahoma, Delaware, and Maryland. The pattern was believed to reflect intentional transplantation by early North American inhabitants; however, the morphological divergence observed among populations indicates that a much longer period of natural isolation has occurred. The distributions of the other aquatic species are fairly regional with *Alnus rugosa* mainly northeastern, *A. serrulata* eastern, and *A. tenuifolia* western in North America.

References: Chen & Li, 2004; Furlow, 1979, 1997; Graves & Gallagher, 2003; Huenneke, 1985; Huh, 1999; Hurd et al., 2001; Lahring, 2003; Navarro et al., 2003; Schrader & Graves,

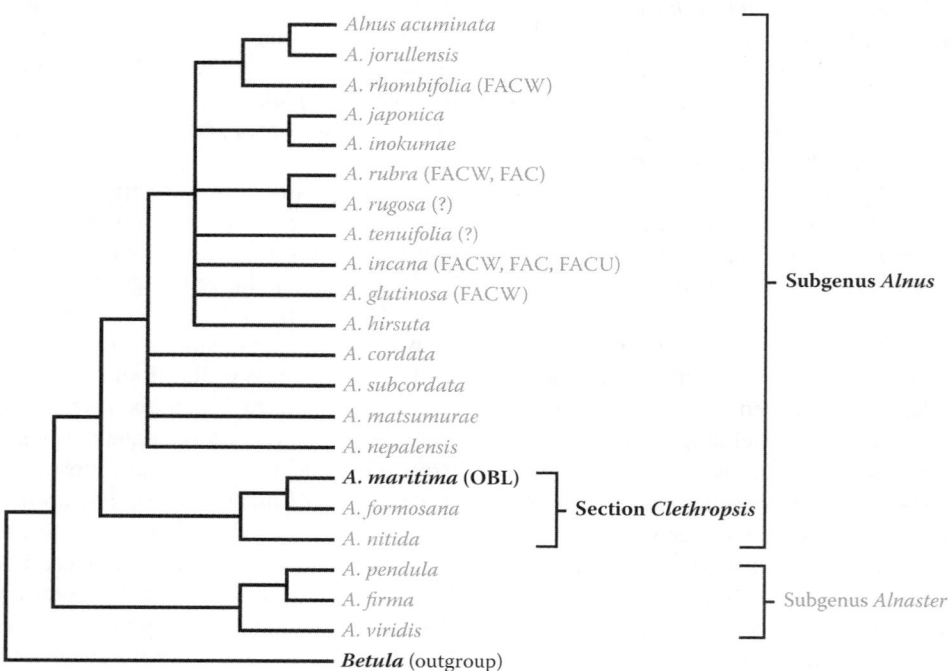

FIGURE 4.20 Phylogenetic relationships among 21 *Alnus* taxa as indicated by nrDNA (ITS) sequence data analyzed by maximum likelihood methods. The cladogram supports the monophyly of the two subgenera and section *Clethropsis*, but does not resolve the subspecies of *A. incana* within a single clade (maximum parsimony analyses resolve subgenus *Alnus* as paraphyletic). The prevailing (but questionable) tendency to treat *A. rugosa* and *A. tenuifolia* (both formerly OBL indicators) as subspecies of *A. incana* resulted in their loss of OBL status in the subsequent indicator list. Their taxonomic status as well as their wetland indicator status should be reconsidered. The OBL *A. maritima* is most closely related to Asian (i.e., *A. formosana*, *A. nitida*) rather than to other North American species. Although excluded in the earlier analysis, *A. serrulata* (OBL) resolved with a more distant group of Asian species as the sister to *A. japonica* in a similar analysis of nrITS data (Chen & Li, 2004). Thus, these results indicate that the OBL habit (bold) was derived independently in *A. maritima* and *A. serrulata*. (Adapted from Navarro, E. et al., *Plant Soil*, 254, 207–217, 2003.)

2002; 2004; Savard et al., 1993; Uchytal, 1989a; Van Deelen, 1991; Vandevender, 2008; Windell et al., 1986.

3. *Betula*

Birch; Bouleau
Etymology: from the Latin name for the plant
Synonyms: none
Distribution: global: Northern Hemisphere; **North America:** widespread
Diversity: global: 50 species; **North America:** 25 species
Indicators (USA): OBL: *Betula glandulosa*, *B. pumila*, *B. ×purpusii*, *B. ×sandbergii*; **FACW:** *B. nigra*; **FAC:** *Betula glandulosa*, *B. nana*
Habitat: freshwater; palustrine; **pH:** 2.0–8.2; **depth:** <1 m; **life-form(s):** emergent (shrub, tree)
Key morphology: deciduous shrubs (to 4 m) or trees (to 25 m), often with multiple trunks, the bark smooth or shredding and exfoliating in irregular sheets, winter buds sessile; leaves (to 8 cm) alternate, simple, 2-ranked, the margins crenate to serrate; male catkins (to 7.5 cm) pendulous; female cones (to 3 cm) leathery, erect; seeds winged.
Life history: duration: perennial (buds); **asexual reproduction:** stump sprouts; **pollination:** wind; **sexual condition:** monoecious; **fruit:** samaras (common); **local dispersal:** seeds (air); **long-distance dispersal:** seeds (air, water)
Imperilment: (1) *Betula glandulosa* [G5]; S1 (ME, NB, NS, NY); S2 (NH); S3 (NF, WY); (2) *B. nana* [G5]; S3 (MT); (3) *B. nigra* [G5]; S2 (KS, NH); S3 (IL, MA, NY); (4) *B. pumila* [G5]; S1 (IL, MA, NH, NS); S2 (CT, IA, ID, ME, NS, NJ, NY, OH, PE, YT); S3 (NB); (5) *B. ×purpusii* GNA; S1 (IL); (6) *B. ×sandbergii* GNA; SH (VT); S1 (IL)
Ecology: general: Birches (*Betula*) generally occur on sandy, well-drained soils. Although more than half of the North American species (56%) have status as wetland indicators at some level, only two are associated primarily with wetlands as OBL indicators. *Betula nana* (FAC) and *B. nigra* (FACW) were OBL indicators in the 1996 list but were reclassified in the revised 2013 list. Having been completed previously, the treatment for *B. nigra* (FACW) has been included here. Two spontaneous F$_1$ hybrids (*B. ×purpusii*, *B. ×sandbergii*) are currently designated as OBL indicators; but as hybrids, they are excluded from this treatment. All birches have unisexual flowers (in monoecious condition), which are wind pollinated. Despite their unisexual flowers, most birches appear to be only weakly self-compatible genetically. The winged seeds are dispersed abiotically by wind or by water. The seeds are dependent on sufficient levels of light for their germination.

Betula glandulosa Michx. occurs in Arctic and alpine barrens, bogs, borrow pits, bottoms, carrs, depressions, dunes, fellfields, fens, flarks, flats, floodplains, gravel bars, gullies, heath, hummocks, marshes, meadows, muskeg, outwash plains, roadsides, scrub, slopes (to 30%), swales, swamps, thickets, tundra, woodlands, and along the margins of lakes, ponds, rivers, and streams at elevations of up to 3703 m. The exposures are consistently described as open. The plants have high frost tolerance and are capable of withstanding extremely cold conditions, persisting (with adequate, insulative snowfall) where

average January low temperatures reach −27°C. Leaf growth commences directly after the winter snows have melted (e.g., by mid-June at mid-latitudes). The substrates are usually described as acidic but can also be somewhat alkaline (pH: 5.5–7.4, hardness: 10–21 mg/L; conductivity: 45–70 mmhos). They have been characterized as alluvium, Birch Creek schist, clay, clay loam, granite, granitic clay loam, gravel, humus, loam, mica-rich rock, muck, peat, peaty marl, quartzite, rock, scree, shale, silt, silty loam, and till. Flowering extends from May to August, with the fruits appearing from May to November. Each pistillate catkin can contain from 30 to 50 fruits. The flowers are self-incompatible, with a stronger effect as temperatures increase. Staminate catkins mature slightly earlier than the pistillate catkins. Juvenile plants begin to produce seeds after 2 years of age. Seed set in the southern portion of the range can be high (~70%), but far northern populations are pollen limited and reproduce virtually entirely by vegetative layering. The winged seeds are dispersed primarily by the wind at quantities from 350 to 13,000/m^2 depending on plant density. The seeds are physiologically dormant but have germinated well (>60%) after a period of cold stratification (several months at 4°C), when placed under a 10°C/5°C, 20/4 h temperature/light regime. Light is required for germination. The seeds are long lived and a persistent seed bank (229–2164 seeds/m^2) develops in the warmer portions of the range; however, a seed bank is nonexistent in the north of the range. The stems produce adventitious roots when overgrown by moss or other vegetation. Electrophoretic surveys of northern populations have indicated that a single genotype typically dominates a site owing to extensive clonal reproduction. The clones can reach up to 5 m in diameter. Adult plants can attain a total biomass of 5564 kg/ha, distributed as 53% (stems), 41% (roots), and 6% (leaves). The plants will resprout vigorously following fires. The roots are colonized by numerous mycorrhizal fungi. **Reported associates:** *Abies lasiocarpa, Allium schoenoprasum, Alnus tenuifolia, A. viridis, Anaphalis margaritacea, Andromeda polifolia, Angelica arguta, Antennaria corymbosa, A. pulcherrima, A. racemosa, Anticlea elegans, Arctagrostis latifolia, Arctostaphylos uva-ursi, Arnica longifolia, A. rydbergii, Artemisia arbuscula, A. tridentata, Betula nana, B. occidentalis, B. pumila, Bistorta bistortoides, Bromus marginatus, Calamagrotis canadensis, Caltha, Camassia quamash, Carex angustata, C. aquatilis, C. aurea, C. buxbaumii, C. dioica, C. disperma, C. interior, C. lasiocarpa, C. limosa, C. livida, C. luzulina, C. magellanica, C. muricata, C. nebrascensis, C. rostrata, C. scirpoidea, C. scopulorum, C. simulata, C. utriculata, C. vesicaria, Cassiope, Castilleja cusickii, C. miniata, Chamaedaphne calyculata, Cicuta douglasii, Cladonia, Comarum palustre, Cornus sericea, Dasiphora floribunda, Deschampsia cespitosa, Drepanocladus aduncus, Drosera anglica, Dryas integrifolia, D. punctata, Dryopteris, Dulichium arundinaceum, Eleocharis palustris, E. quinqueflora, E. tenuis, Empetrum nigrum, Epilobium ciliatum, Epipactis gigantea, Equisetum arvense, Eriophorum angustifolium, E. chamissonis, E. gracile, E. vaginatum, Evernia, Festuca altaica, Floerkea proserpinacoides, Galium trifidum, G. triflorum, Gentianopsis detonsa, Heracleum sphondylium, Hylocomium splendens, Hypericum formosum, Juncus* *balticus*, *J. ensifolius*, *Juniperus communis*, *Kalmia micro-phylla*, *Larix laricina*, *L. occidentalis*, *Ligusticum canbyi*, *L. tenuifolium*, *Lonicera caerulea*, *L. involucrata*, *L. utahensis*, *Lupinus*, *Luzula*, *Lysichiton americanus*, *Menyanthes trifo-liata*, *Menziesia ferruginea*, *Mimulus guttatus*, *Muhlenbergia richardsonis*, *Myrica gale*, *Oenanthe sarmentosa*, *Oxytropis campestris*, *Packera pseudaurea*, *Pedicularis groenland-ica*, *Phalaris arundinacea*, *Phleum alpinum*, *Phlox kelseyi*, *Phyllodoce*, *Picea engelmannii*, *P. glauca*, *P. mariana*, *Pinus albicaulis*, *P. contorta*, *P. flexilis*, *P. monticola*, *P. ponderosa*, *Platanthera obtusata*, *Pleurozium schreberi*, *Poa alpina*, *Polemonium occidentale*, *Populus tremuloides*, *P. trichocarpa*, *Potentilla ×diversifolia*, *Primula jeffreyi*, *Pseudotsuga menzie-sii*, *Puccinellia*, *Pyrola asarifolia*, *Racomitrium lanuginosum*, *Rhamnus alnifolia*, *Rhododendron albiflorum*, *R. columbia-num*, *R. groenlandicum*, *R. tomentosum*, *Ribes aureum*, *R. inerme*, *Rosa*, *Rubus chamaemorus*, *Salix alaxensis*, *S. boothii*, *S. brachycarpa*, *S. candida*, *S. drummondiana*, *S. eastwoodiae*, *S. exigua*, *S. geyeriana*, *S. glauca*, *S. lemmonii*, *S. myrtillifolia*, *S. pseudomonticola*, *S. pulchra*, *S. reticulata*, *S. richardsonii*, *S. sitchensis*, *S. wolfii*, *Saxifraga tricuspidata*, *Schoenoplectus acutus*, *Senecio hydrophiloides*, *S. serra*, *S. triangularis*, *Shepherdia canadensis*, *Solanum dulcamera*, *Sphagnum fus-cum*, *Spiraea betulifolia*, *S. douglasii*, *S. splendens*, *Stellaria longifolia*, *Swertia perennis*, *Symphoricarpos oreophilus*, *Symphyotrichum*, *Tephroseris integrifolia*, *Thalictrum alpi-num*, *T. occidentale*, *Thuja plicata*, *Triantha occidentalis*, *Trientalis europaea*, *Trifolium longipes*, *Triglochin maritimum*, *Trollius laxus*, *Tsuga heterophylla*, *Typha latifolia*, *Vaccinium cespitosum*, *V. membranaceum*, *V. ovalifolium*, *V. oxycoccos*, *V. scoparium*, *V. uliginosum*, *V. vitis-idaea*, *Veronica wormsk-joldii*, *Viburnum edule*, *Viola*, *Woodsia ilvensis*, *Xerophyllum tenax*, *Zizia aptera*.

***Betula nigra* L.** grows in or on bottomlands, cobble bars, depressions, ditches, dunes, floodplains, gravel bars, levees, meadows, prairies, ravines, roadsides, sand springs, savan-nahs, seeps, slopes, sloughs, swales, swamps, woodlands, and along the margins of canals, channels, lakes, ponds, rivers, and streams at elevations of up to 297 m. This pioneer species of new alluvial forests occurs in open habitats and is not shade tolerant. Typically, the habitats possess deep, rich, acidic soils, which are periodically inundated for part of the year; however, prolonged inundation (beyond 3 months) is detrimental. The optimal pH range of the sediments is from 4.0 to 6.5; however, the plants can also survive on extremely acidic soils (pH: 2–4) such as those associated with acid mining drainage. The substrates are char-acterized as clay, cobble, gravel, loam, mud, rock, sand, sandy alluvium, sandy clay, sandy loam, and silt. Flowering and fruit-ing occur from April to July, with the fruits being released by early summer. The seeds are produced in quantity, but are short lived and will germinate immediately after they have been shed. The seeds are large and germinate readily on moist alluvial sub-strates. Unstratified seeds exhibit about 35% germination. The seeds can also become physiologically dormant, then germinat-ing at 24°C–27°C following 30–60 days of cold stratification. Germination rates are substantially higher in the light than in the dark. The species often forms dense thickets and seedling

densities can exceed 6000/m². Seedlings are fast growing but require high levels of soil moisture. The adult plants normally live for up to 30–40 years. **Reported associates:** *Acalypha rhomboidea*, *Acer negundo*, *A. rubrum*, *A. saccharum*, *A. sac-charinum*, *Aesculus glabra*, *A. octandra*, *Ageratina altissima*, *Agrimonia parviflora*, *Agrostis perennans*, *Alliaria peti-olata*, *Alnus serrulata*, *Ambrosia trifida*, *Amorpha fruticosa*, *Ampelopsis arborea*, *Anemone canadensis*, *Angelica atropur-purea*, *Aronia ×prunifolia*, *Asarum canadense*, *Boehmeria cylindrica*, *Botrychium virginianum*, *Brunnichia ovata*, *Calamagrostis canadensis*, *Campsis radicans*, *Cardamine bul-bosa*, *Carex conjuncta*, *C. corrugata*, *C. crinita*, *C. debilis*, *C. granularis*, *C. grayi*, *C. grisea*, *C. hyalinolepis*, *C. haydenii*, *C. hystericina*, *C. interior*, *C. intumescens*, *C. lupulina*, *C. lurida*, *C. muskingumensis*, *C. pellita*, *C. pensylvanica*, *C. squar-rosa*, *C. radiata*, *C. tribuloides*, *C. typhina*, *Carpinus carolin-iana*, *Carya aquatica*, *C. cordiformis*, *C. ovata*, *C. scoparia*, *C. stipata*, *C. swanii*, *C. tomentosa*, *C. vulpinoidea*, *Celtis occidentalis*, *Cephalanthus occidentalis*, *Chasmanthium lati-folium*, *Cinna arundinacea*, *Circaea canadensis*, *Claytonia virginica*, *Cornus amomum*, *C. drummondii*, *Corylus ameri-cana*, *Dichanthelium clandestinum*, *Diospyros virginiana*, *Echinochloa muricata*, *Eleocharis obtusa*, *E. ovata*, *Elymus virginicus*, *Erigeron philadelphicus*, *Eupatorium perfoliatum*, *Euthamia graminifolia*, *Fagus grandifolia*, *Fallopia scandens*, *Festuca*, *Forestiera acuminata*, *Fragaria virginiana*, *Fraxinus pennsylvanica*, *Gaylussacia baccata*, *Geum canadense*, *G. laciniatum*, *Gleditsia aquatica*, *G. triacanthos*, *Glyceria stri-ata*, *Halesia carolina*, *Hypericum prolificum*, *Ilex verticillata*, *Impatiens capensis*, *Ipomoea hederacea*, *I. lacunosa*, *Iris virgi-nica*, *Juglans nigra*, *Juncus effusus*, *Krigia biflora*, *Leersia vir-ginica*, *Ligustrum japonicum*, *Lindera benzoin*, *Liquidambar styraciflua*, *Liriodendron tulipifera*, *Lobelia cardinalis*, *Lonicera maackii*, *L. ×minutiflora*, *Lycopus americanus*, *L. uniflorus*, *L. virginicus*, *Lysimachia lanceolata*, *L. nummu-laria*, *Morus alba*, *Nasturtium officinale*, *Nyssa sylvatica*, *Oenothera laciniata*, *Onoclea sensibilis*, *Osmanthus*, *Osmunda claytoniana*, *O. regalis*, *Ostrya virginiana*, *Parthenocissus quinquefolia*, *Penstemon pallidus*, *Persicaria hydropiperoi-des*, *Phalaris arundinacea*, *Phlox glaberrima*, *Pinus*, *Planera aquatica*, *Platanus occidentalis*, *Populus deltoides*, *P. grandi-dentata*, *P. heterophylla*, *Potentilla simplex*, *Prunus serotina*, *Quercus alba*, *Q. bicolor*, *Q. lyrata*, *Q. macrocarpa*, *Q. mich-auxii*, *Q. nigra*, *Q. palustris*, *Q. phellos*, *Q. rubra*, *Q. shumar-dii*, *Q. velutina*, *Ranunculus abortivus*, *R. hispidus*, *Rhamnus cathartica*, *R. frangula*, *Rhus aromatica*, *Rhynchospora capi-tellata*, *Ribes americanum*, *R. missouriense*, *Rosa carolina*, *R. multiflora*, *Rubus hispidus*, *Rumex acetosella*, *Salix discolor*, *S. exigua*, *S. nigra*, *S. rigida*, *Sanicula odorata*, *Sassafras albi-dum*, *Saururus cernuus*, *Schizachyrium scoparium*, *Scutellaria lateriflora*, *Setaria faberi*, *Sium suave*, *Smilax ecirrhata*, *S. hispida*, *Solidago canadensis*, *S. gigantea*, *Spartina pectinata*, *Symphyotrichum lanceolatum*, *S. ontarionis*, *S. racemosum*, *Symplocarpus foetidus*, *Taxodium distichum*, *Thalictrum revo-lutum*, *Thelypteris palustris*, *Tilia americana*, *Toxicodendron radicans*, *Trifolium repens*, *Typha latifolia*, *Ulmus alata*, *U. americana*, *Urtica dioica*, *Vaccinium*, *Verbesina alternifolia*,

Vernonia fasciculata, Viburnum recognitum, Viola lanceolata, V. sororia, Vitis riparia, Xyris torta, Zanthoxylum americanum, Zizia aurea.

Betula pumila L. inhabits bogs, carrs, ditches, fens, flats, gullies, hummocks, marshes, muskeg, seeps, slopes (to 5%), swamps, thickets, tundra, and the margins of lakes, ponds, and sloughs at elevations of up to 1240 m. Although regarded as indicators of calcareous, minerotrophic conditions (e.g., rich boreal fens), these plants can tolerate a broad range of acidity (pH: 5.8–8.2) and also occur in acidic communities such as *Sphagnum* bogs. The substrates (e.g., alkalinity: 45 mg/L; conductivity: 97 mmhos; hardness: 42.8 mg/L CaCO$_3$) have been described as Houghton muck, krummholz, loam, peat, sand, silt, and typic medisaprist. Flowering and fruiting occur from May to November. The leaves appear in mid-May and senesce in mid-October. The roots are colonized by dark septate endophytic fungi (Ascomycota: incertae sedis: *Leptodontidium orchidicola*). **Reported associates:** *Agrostis scabra, Alnus rugosa, Amelanchier alnifolia, Andromeda glaucophylla, A. polifolia, Andropogon scoparius, A. gerardii, Aralia nudicaulis, Arethusa bulbosa, Betula glandulosa, B. papyrifera, Botrychium, Bromus ciliatus, Calamagrostis canadensis, C. rubescens, Calla palustris, Campylium, Carex aquatilis, C. atherodes, C. aurea, C. bebbii, C. chordorrhiza, C. comosa, C. exilis, C. flava, C. interior, C. lacustris, C. lanuginosa, C. lasiocarpa, C. limosa, C. muricata, C. rostrata, C. scirpoidea, C. stricta, C. vesicaria, Chamaedaphne calyculata, Cirsium arvense, Cladium mariscoides, Comarum palustre, Conyza canadensis, Cornus amomum, C. sericea, Crataegus douglasii, Dasiphora floribunda, Dicranum, Drosera rotundifolia, Dulichium arundinaceum, Elaeagnus angustifolia, Eleocharis erythropoda, E. rostellata, Equisetum arvense, E. fluviatile, E. scirpoides, Eriogonum alpinum, Eriophorum alpinum, E. vaginatum, Eupatorium perfoliatum, Eutrochium maculatum, Fragaria virginiana, Geum rivale, Helenium autumnale, Helianthus grosseserratus, Hylocomium splendens, Ilex verticillata, Iris versicolor, Juncus bufonius, Juniperus horizontalis, Kobresia simpliciuscula, Larix laricina, Linnaea borealis, Liparis, Lobelia kalmii, Lonicera, Lupinus arcticus, Lysimachia quadriflora, Lythrum salicaria, Maianthemum stellatum, Mentha arvensis, Menyanthes trifoliata, Micranthes pensylvanica, Muhlenbergia glomerata, Myrica gale, Orthilia secunda, Parnassia glauca, Peltigera, Pentaphylloides floribunda, Petasites sagittatus, Phalaris arundinacea, Physocarpus opulifolius, Picea engelmannii, P. glauca, P. mariana, Pinguicula vulgaris, Pinus banksiana, P. contorta, P. strobus, Platanthera huronensis, P. leucophaea, Poa pratensis, Polemonium reptans, Populus tremuloides, Pseudotsuga menziesii, Ptilium crista-castrensis, Pycnanthemum virginianum, Pyrola asarifolia, Ranunculus gmelinii, Rhamnus alnifolia, R. frangula, Rhododendron canadense, R. groenlandicum, Rhynchospora alba, Rosa palustris, Rubus acaulis, R. idaeus, Rudbeckia hirta, Salix bebbiana, S. candida, S. myricoides, S. pedicellaris, S. serissima, Sanicula marilandica, Sarracenia purpurea, Schoenoplectus acutus, S. pungens, Shepherdia canadensis, Silphium terebinthinacuem, Solidago ohioensis, S. uliginosa, Sorghastrum nutans, Sphagnum, Spiraea douglasii, Spiranthes romanzoffiana, Symphoricarpos albus, Thelypteris palustris, Thuja occidentalis, Toxicodendron vernix, Trichophorum cespitosum, Triglochin maritimum, Typha latifolia, Ulmus americana, Utricularia intermedia, Vaccinium oxycoccos, V. vitis-idaea, Valeriana edulis, V. uliginosa, Viburnum acerifolium, Viola sororia, Zizia aurea.*

Use by wildlife: Birch seeds are eaten by rodents (Mammalia: Rodentia) and by various small birds (Aves) along with grouse (Phasianidae) and wild turkeys (Phasianidae: *Meleagris gallopavo*). The foliage and buds are browsed by mammals (Mammalia) including beavers (Castoridae: *Castor canadensis*), hares (Leporidae: *Lepus*), moose (Cervidae: *Alces alces*), porcupines (Erethizontidae: *Erethizon dorsatus*), and white-tailed deer (Cervidae: *Odocoileus virginianus*). The leaves and shoot tips of *B. glandulosa* are browsed by several mammals (Mammalia) including domestic cattle (Bovidae: *Bos taurus*), goats (Bovidae: *Capra hircus*), horses (Equidae: *Equus caballus*), sheep (Bovidae: *Ovis aries*), and by antelope (Antilocapridae: *Antilocapra americana*), elk (Cervidae: Cervus elaphus), hares (Leporidae: *Lepus*), moose (Cervidae: *Alces alces*), mule deer (Cervidae: *Odocoileus hemionus*), reindeer (Cervidae: *Rangifer tarandus*), and white-tailed deer (Cervidae: *Odocoileus virginianus*). The buds, catkins, and seeds are consumed by a number of birds (Aves) including chickadees (Paridae: *Poecile*), kinglets (Regulidae: *Regulus*), pine siskins (Fringillidae: *Spinus pinus*), redpolls (Fringillidae: *Acanthis*), ruffed grouse (Phasianidae: *Bonasa umbellus*), sharp-tailed grouse (Phasianidae: *Tympanuchus phasianellus*), spruce grouse (Phasianidae: *Falcipennis canadensis*), willow ptarmigans (Aves: Phasianidae: *Lagopus lagopus*), and various waterfowl (Anatidae). The plants are fed upon by aphids and other sapsucking bugs (Insecta: Hemiptera: Aphidae: *Euceraphis betulae, E. punctipennis*; Psyllidae: *Psylla betulaenanae*; Cicadellidae: *Colladonus youngi, Coulinus usnus, Deltocephalus lividellus, Oncopsis albicollis, Thamnotettix confinis*). The leaves are eaten or mined by larval butterflies and moths (Insecta: Lepidoptera: Eriocraniidae: *Eriocrania semipurpurella*: Geometridae: *Aethalura intertexta, Campaea perlata, Cyclophora pendulinaria, Eufidonia discospilata, Hydriomena furcata, Macaria notata, Operophtera bruceata, Pero morrisonaria, Plagodis pulveraria, Plemyria georgii, Rheumaptera hastata*; Lasiocampidae: *Malacosoma californicum*; *Phyllodesma americana*; Nymphalidae: *Nymphalis antiopa*; Tortricidae: *Acleris caryosphena, Epinotia solandriana, Eucosma indecorana*). The plants are also host to a variety of fungi (Ascomycota: incertae sedis: *Trimmatostroma betulinum*; Apiosporaceae: *Apiospora rosenvingei*; Dermateaceae: *Mollisia cinerea, Monostichella betularum*; Diatrypaceae: *Diatrypella decorata*; Euantennariaceae, *Antennatula arctica*; Herpotrichiellaceae: *Capronia apiculata*; Hyaloscyphaceae: *Dasyscyphus bicolor*; Mycosphaerellaceae: *Mycosphaerella harthensis, M. maculiformis*; Taphrinaceae: *Taphrina bacteriosperma, T. carnea, T. nana*; Valsaceae: *Gnomonia campylostyla, Ophiognomonia intermedia, Tympanis alnea*; Venturiaceae: *Atopospora betulina, Venturia ditricha*; Basidiomycota: Exidiaceae: *Exidia candida, E.*

repanda; Pucciniastraceae: *Melampsoridium betulinum*; Pluteaceae: *Pluteus cervinus*; Polyporaceae: *Polyporus varius*; Tricholomataceae: *Hygrocybe coccineocrenata*, *Mycena rubromarginata*). *Betula nigra* is used for roosting and nesting by cattle egrets (Aves: Ardeidae: *Bubulcus ibis*), but the plants can die within 2 years if guano deposition is excessive. The leaves and shoots are damaged by aphids (Insecta: Hemiptera: Aphidae), fungi (Ascomycota: Dermateaceae: *Monostichella betularum*), weevils (Insecta: Coleoptera: Curculionidae: *Cryptorhynchus lapathi*), and wood-boring sawflies (Insecta: Hymenoptera; Xiphydriidae: *Xiphydria decem*). Plants in the southern portion of their range are parasitized by mistletoe (Santalaceae: *Phoradendron serotinum*). *Betula nigra* is a host plant for numerous butterfly and moth caterpillars (Insecta: Lepidoptera: Arctiidae: *Hyphantria cunea*; Bucculatricidae: *Bucculatrix coronatella*; Erebidae: *Lymantria dispar* [gypsy moths]; Gelechiidae: *Pseudotelphusa betulella*; Geometridae: *Ennomos magnaria*, *Nemoria bistriaria*; Gracillariidae: *Parornix conspicuella*, *P. obliterella*; Lasiocampidae: *Malacosoma americana*; Limacodidae: *Phobetron pithecium*; Lymantriidae: *Lymantria dispar*, *Orgyia leucostigma*; Noctuidae: *Acronicta betulae*, *Morrisonia latex*, *Datana ministra*; Oecophoridae: *Nites betulella*; Pyralidae: *Acrobasis betulivorella*; Saturniidae: *Antheraea polyphemus*, *Hyalophora cecropia*). *Betula pumila* is browsed by white-tailed deer (Mammalia: Cervidae: *Odocoileus virginianus*), and to a lesser extent by moose (Cervidae: *Alces alces*) during the summer. The plants are damaged by aphids (Insecta: Hemiptera: Aphididae: *Calaphis manitobensis*, *Cepegillettea viridis*), weevils (Insecta: Coleoptera: Curculionidae: *Cryptorhynchus lapathi*), and are host to several moth larvae (Insecta: Lepidoptera: Lasiocampidae: *Malacosoma californica*; Saturniidae: *Hemileuca lucina*, *H. maia*, *H. nevadensis*; Tortricidae: *Adoxophyes orana* [summer fruit tortrix moth]), and fungi (Ascomycota: incertae sedis: *Bactrodesmium betulicola*; Erysiphaceae: *Microsphaera penicillata*; Taphrinaceae: *Taphrina carnea*; Venturiaceae: *Atopospora betulina*; Vibrisseaceae: *Phialocephala fortinii*). They attract adults of the frigga fritillary (Insecta: Lepidoptera: Nymphalidae: *Boloria frigga*).

Economic importance: food: Sap from *Betula nigra* can be drunk or concentrated into a sweetener or sweet syrup. Fermented sap from the plant is used to make birch beer and vinegar; **medicinal:** *Betula* pollen represents a significant component of wind-borne allergens. An oil infusion made from the leaf buds of *B. glandulosa* was used by the Kiluhikturmiut Inuinnait as a frostbite preventative and catalyst. A salve made from the boiled buds of *B. nigra* has been used to treat skin lesions. The plants also have been included as an ingredient in some homeopathic allergy medications. The Ojibwa administered an infusion from cones of *B. pumila* as a postpartum tonic. They also inhaled the smoke from burning *B. pumila* cones to alleviate congestion; **cultivation:** *Betula nigra* is highly resistant to birch borer (Insecta: Coleoptera: Buprestidae: *Agrilus anxius*) and often is planted as an ornamental. Cultivars of *B. nigra* include 'Black Star,' 'Bnmtf,' 'Cully,' 'Dickinson,' 'Heritage,' 'Little

King,' 'Northern Tribute,' 'Peter Collinson,' 'Shiloh Splash,' 'Studetec,' 'Summer Cascade,' and Wakehurst form. *Betula pumila* also is cultivated but to a lesser degree and includes the cultivar 'Leprechaun'; **misc. products:** The wood of *B. nigra* has been used to manufacture artificial limbs, baskets, inexpensive furniture, and tool handles. The tolerance of *B. nigra* to soil acidity has led to its use in strip mine reclamation projects. Its dense root system makes it a useful plant for erosion control. The Ojibwa used *B. pumila* twigs in basket making; **weeds:** none; **nonindigenous species:** none.

Systematics: Molecular phylogenetic studies (e.g., Figure 4.19) consistently corroborate the monophyly of *Betula* and its relationship as the sister genus to *Alnus*. However, relationships within the genus are highly unsettled and have been difficult to elucidate because of hybridization, introgression, polyploidy, and morphological convergences. Analyses of *Betula* species using *adh*, *matK* sequences show major incongruencies between the plastid (*matK*) and nuclear (*adh*) generated phylogenies. Furthermore, none of the analyses was compatible with recent classifications that arrange species into either subgenera or sections, i.e., they did not support those groups as being monophyletic. Analyses of diploid birch species using nuclear nitrate reductase sequences indicated that subgenera *Betula* and *Betulaster* were monophyletic, but did not resolve subgenera *Chamaebetula* and *Neurobetula* as clades. A similar result was obtained by analysis of amplified fragment length polymorphism (AFLP) data, which again indicated subgenera *Chamaebetula* and *Neurobetula* to be polyphyletic. It is equally difficult to evaluate the relationships of the OBL species. AFLP data resolve *B. pumila* (subgenus *Chamaebetula*) near *B. nana* (subgenus *Chamaebetula*) and *B. papyrifera* (subgenus *Betula*); whereas, nrITS data place it within an unresolved clade (polytomy) containing those as well as 11 other birch species. *Betula nigra* (subgenus *Neurobetula*) resolves near *B. nana* (subgenus *Chamaebetula*) using nitrate reductase sequence data, but is placed quite distant from it (and distinct from most other surveyed *Betula* species) in AFLP data analyses; nrITS sequence data resolve *B. nigra* and *B. alnoides* (remote from *B. nigra* in nitrate reductase trees) as sister species (and again remote from *B. nana*). *Betula glandulosa* (subgenus *Chamaebetula*) forms a subclade with *B. globispica* (subgenus *Betulenta*) in trees constructed from nrITS data. Such varied results instill little confidence in the existing taxonomy of this genus. Furthermore, *B. glandulosa* and *B. pumila* have been assigned most recently to section *Apterocaryon*, with *B. nigra* placed in section *Dahuricae*, neither section being defensible on phylogenetic grounds. It is apparent that further studies will be necessary before interspecific relationships in *Betula* (and its infrageneric taxonomy) can be clarified with any degree of certainty. Taxonomic difficulties also obfuscate the literature on North American birches, especially regarding the dwarf species. Authors have applied the name "*Betula nana*" both in a broad sense (as including the more boreal *B. glandulosa* and *B. michauxii*), or in a strict sense (as a subarctic species distinct from *B. glandulosa* and *B. michauxii*). Although most recent treatments (including the present) adopt the latter approach, the phylogenetic relationships among these taxa should be studied more intently with the objective of

distinguishing between these various hypotheses. Existing data support the distinctness of *B. glandulosa* and *B. nana*. Further complicating matters is *Betula pumila*, to which the names "*B. glandulosa* var. *glandulifera*" and "*B. glandulosa* var. *hallii*" have been applied. Accordingly, caution is advised when trying to ascertain the correct identity of species discussed in literature accounts. Despite this awareness, there is no assurance that this information has been parsed correctly in the present treatment as well, despite the honest attempt made. The base chromosome number of *Betula* is $x = 14$. *Betula glandulosa* and *B. nigra* are diploids ($2n = 28$), whereas, *B. pumila* ($2n = 56$) is a tetraploid. Pentaploid ($2n = 70$) hybrids (known as *B.* ×*purpusii*) occur between *B. alleghaniensis* and *B. pumila* where their ranges overlap. Similarly, hybrids known as *B.* ×*sandbergii* occur between *B. papyrifera* and *B. pumila* where the species form contact zones. Both hybrids are regarded as obligate aquatics. Because each of these three hybridizing species has been assigned to a different section of *Betula* taxonomically, the level of genetic differentiation between the sections as currently circumscribed does not appear to be very extensive. Hybrid swarms reportedly occur between *B. nana* and *B. glandulosa*, which have long been presumed to be closely related, but appear to be only remotely related by molecular phylogenetic analysis.
Comments: *Betula glandulosa* and *B. pumila* occur in northern North America, whereas *B. nigra* is southeastern in its distribution.
References: Addy et al., 2000; Alsos et al., 2003; Bret-Harte et al., 2001; Burns & Honkala, 1990; Caesar, 1916; Cannon, 2009; Clausen, 1966; Davis & Banack, 2012; Davis et al., 2005; de Groot et al., 1997; Furlow, 1997; Hagman, 1971; Halıcı et al., 2010; Hermanutz et al., 1989; Järvinen et al., 2004; Jumpponen & Trappe, 1998; Koevenig, 1976; Li et al., 2005, 2007; McClelland & Ungar, 1970; Mix, 1954; Nijland et al., 2014; Palmé, 2003; Pellerin et al., 2006; Portinga & Moen, 2015; Richards, 1969; Robinson & Bradley, 1968; Schenk et al., 2008; Spieles et al., 1999; Swengel & Swengel, 2010; Underwood, 1893; Walker et al., 1994; Weis & Hermanutz, 1993; Wolfe & Pittillo, 1977.

Family 2: Fagaceae [9]

Fagaceae contain about 900 species of tanniferous trees or shrubs, which bear their reproductive organs in catkins and have female flowers that are associated with a scaly cupule. The family contains some of the most familiar hardwood trees of North American temperate deciduous forests and is principally terrestrial, although a number of species can occur in lowland habitats. Phylogenetic analyses of morphological and molecular data indicate that Fagaceae are monophyletic and, in exclusion of Nothofagaceae, are sister to the remainder of families within the order (Figure 4.18).

Phylogenetic investigations using combined molecular data (nrITS, *matK* sequences) have provided some clarification of intergeneric relationships in Fagaceae (Manos et al., 2001a). The results indicate that the morphologically determined subfamilies proposed by some authors do not appear to represent clades (Figure 4.21); thus, the infrafamilial classification deserves further evaluation.

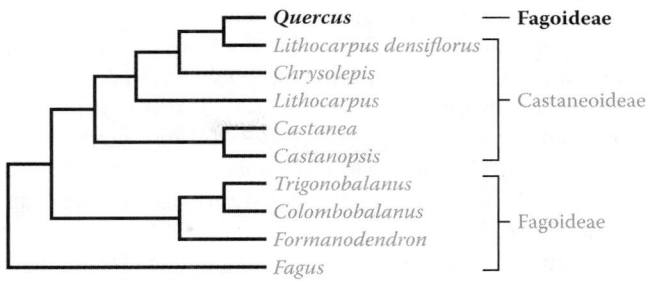

FIGURE 4.21 Phylogenetic relationships among genera of Fagaceae as indicated from combined *matK* and nrITS sequence data. *Quercus*, the only North American genus to contain obligate aquatic species (indicated in bold), occurs in a derived position in the family. The cladogram does not support subfamilial divisions as indicated by morphological studies, with Castaneoideae and Fagoideae shown to be para- or polyphyletic. (Adapted from Manos, P.S. et al., *Syst. Bot.*, 26, 585–602, 2001a.)

The flowers of many familiar species are reduced, unisexual, devoid of nectaries, and wind pollinated; however, those of some genera (*Castanea*, *Castanopsis*) are pollinated by insects (Insecta). The ubiquitous, cupule-enshrouded nuts are dispersed by birds (Aves) and mammals (Mammalia) (Judd et al, 2006).

The family is important as a source of edible chestnuts (*Castanea*) and beechnuts (*Fagus*). Natural cork is obtained from the outer bark of the cork oak (*Quercus suber*). Various oaks (*Quercus* spp.) provide a hard, durable wood, which has been used widely as firewood, lumber, hardwood flooring, and in the manufacture of barrels, cabinetry, and furniture (Nixon, 1997; Judd et al., 2016). Numerous ornamental woody plants are found in the genera *Castanea*, *Castanopsis*, *Chrysolepis*, *Fagus*, and *Quercus*.

Only one North American genus contains species designated as OBL indicators:

1. ***Quercus*** L.

In most analyses, *Quercus* resolves in a derived position relative to other genera of Fagaceae (Figure 4.21). Like the remainder of the family, the genus mainly is terrestrial.

1. *Quercus*
Oak; chêne
Etymology: the ancient Latin name for the oak or acorn
Synonyms: *Cyclobalanopsis*
Distribution: global: Africa, Asia, Europe, New World; **North America:** widespread
Diversity: global: 450 species; **North America:** 96 species
Indicators (USA): OBL: *Quercus lyrata*, *Q. texana*; **FACW:** *Q. bicolor*, *Q. texana*
Habitat: freshwater; palustrine, riverine; **pH:** 4.5–8.5; **depth:** <1 m; **life-form(s):** emergent (tree)
Key morphology: deciduous trees (to 30 m) having twigs with terminally clustered, five-angled buds; leaves (to 21 cm) alternate, simple, deeply or shallowly pinnately lobed, the lobes

rounded (blunt) or terminating in a bristle; staminate flowers in pendant catkins (to 18 cm); fruits (to 5 cm) developing into woody acorns, crowned by a cap of corky, overlapping scales. **Life history: duration:** perennial (buds); **asexual reproduction:** seedling and stump sprouts; **pollination:** wind; **sexual condition:** monoecious; **fruit:** nuts (acorns) (common); **local dispersal:** fruits (gravity); **long-distance dispersal:** fruits (animals, water)

Imperilment: (1) *Quercus bicolor* [G5]; S1 (KS, ME, SC); S2 (NC); S3 (IA, IL, QC); (2) *Q. lyrata* [G5]; S1 (NJ); S2 (DE); S3 (IL, IN); (3) *Q. texana* [G4/G5]; S1 (IL); S2 (MO, KY, OK); S3 (TN)

Ecology: general: Oaks (*Quercus*) typically are viewed as dry, UPL forest trees; yet 26 of the North American species (27%) have some status as wetland indicators and often appear as species associated with various wetland taxa. A few oak species are adapted to tolerate at least periodic, short-term inundation and have been designated as OBL indicators for their characteristic association with wet, low-lying habitats. *Quercus bicolor* formerly held status in the 1996 list as an OBL indicator. Although now categorized as FACW, the treatment (completed prior to the release of the revised list) has been included here. All oaks have small, unisexual flowers, which are wind pollinated. They all produce nuts for fruits, which are dispersed by animals, or occasionally by water. Oaks are important ecologically, and often represent the dominant overstory vegetation in temperate deciduous woodlands.

Quercus bicolor **Willd.** inhabits bottomlands, depressions, flatwoods, floodplains, knolls, marshes, prairies, savannahs, slopes, sloughs, swamps, thickets, woods, and the margins of lakes, ponds, sand ponds, and streams at elevations of up to 1000 m. The plants typically colonize flood-prone sites but not areas of permanent inundation (which they do not tolerate). They have an intermediate level of shade tolerance and can persist where the average annual temperatures range from 4°C to 16°C. The plants occur naturally on acidic soils (pH<5.9), but can also tolerate somewhat higher alkalinity (pH: 6.0–6.9), although leaf chlorosis may appear when growing on more alkaline soils. The substrates (principally entisols and inceptisols) include alluvium, clay, clay loam, Hosmer series, muck (organic), peat, sand, silt, and silty loam. These are fast-growing trees, which live for up to 350 years but do not flower until reaching 25–30 years of age. This is a mast fruiting species with major crops of acorns produced every 3–5 years. The seeds lack dormancy and germinate well (78%–98%) immediately after ripening. Levels of germination and seedling establishment are higher on better-drained (but still lowland) sites. **Reported associates:** *Acalypha rhomboidea, Acer negundo, A. rubrum, A. saccharinum, A. saccharum, Actaea pachypoda, Agrostis perennans, Alliaria petiolata, Allium canadense, A. tricoccum, Ambrosia trifida, Andropogon gerardii, Anemone quinquefolia, Apios americana, Arisaema dracontium, A. triphyllum, Asclepias incarnata, Berberis thunbergii, Betula nigra, Bromus inermis, Campsis radicans, Cardamine concatenata, Carex albursina, C. blanda, C. conjuncta, C. crus-corvi, C. davisii, C. gracilescens, C. grisea, C. grayi, C. jamesii, C. laxiculmis, C. leavenworthii, C. lupuliformis, C. lupulina, C. muskingumensis,* *C. pensylvanica, C. projecta, C. radiata, C. sparganioides, C. squarrosa, C. tribuloides, C. typhina, Carya illinoinensis, C. laciniosa, C. ovata, C. tomentosa, Celtis laevigata, C. occidentalis, Cephalanthus occidentalis, Chasmanthium latifolium, Circaea lutetiana, Claytonia virginica, Clethra alnifolia, Cornus amomum, C. drummondii, C. racemosa, Crataegus punctata, Dichanthelium leibergii, Diospyros virginiana, Doellingeria umbellata, Eleocharis palustris, Elymus villosus, Enemion biternatum, Euonymus alatus, Fraxinus americana, F. nigra, F. pennsylvanica, F. profunda, Galium aparine, Geranium maculatum, Gleditsia triacanthos, Glyceria striata, Hamamelis virginiana, Hibiscus laevis, Hydrophyllum virginianum, Ilex decidua, I. verticillata, Impatiens capensis, Iodanthus pinnatifidus, Iris virginica, Juglans nigra, Juncus acuminatus, Laportea canadensis, Larix laricina, Leersia virginica, Lemna, Ligustrum sinense, Liquidambar styraciflua, Liriodendron tulipifera, Lonicera japonica, Ludwigia palustris, Lysimachia ciliata, L. nummularia, Maianthemum racemosum, Micranthes pensylvanica, Nyssa aquatica, N. sylvatica, Osmorhiza longistylis, Ostrya virginiana, Oxypolis rigidior, Parthenocissus quinquefolia, Peltandra virginica, Platanus occidentalis, Poa pratensis, Polygonatum biflorum, Populus deltoides, P. heterophylla, Prenanthes crepidinea, Prunus serotina, P. virginiana, Quercus alba, Q. imbricaria, Q. lyrata, Q. macrocarpa, Q. marilandica, Q. muehlenbergii, Q. pagoda, Q. palustris, Q. rubra, Q. stellata, Q. velutina, Ranunculus abortivus, R. hispidus, Rhamnus cathartica, Rhus glabra, Rosa multiflora, R. setigera, Rubus alleghaniensis, Rudbeckia laciniata, Rumex verticillatus, Sagittaria latifolia, Salix nigra, Sambucus nigra, Sanicula, Sassafras albidum, Scutellaria lateriflora, Smilax bona-nox, S. lasioneura, Solidago flexicaulis, S. gigantea, S. patula, S. rugosa, Sorghastum nutans, Sparganium eurycarpum, Sphagnum, Spiraea tomentosa, Styrax americanus, Symphyotrichum lanceolatum, S. lateriflorum, S. ontarionis, Taxodium distichum, Tilia americana, Toxicodendron radicans, Tradescantia subaspera, Tridens flavus, Trillium grandiflorum, T. recurvatum, Ulmus americana, U. rubra, Urtica dioica, Vaccinium corymbosum, Viburnum opulus, V. recognitum, Viola pubescens, V. sororia, Vitis cinerea, V. palmata, V. riparia, V. vulpina, Zanthoxylum americanum.*

Quercus lyrata **Walter** grows in bayous, bottomlands, ditches, floodplains, hammocks, lagoons, ravines, roadsides, swamps, and along the margins of lakes, rivers, sloughs, and streams along the southeastern coastal plain at elevations of up to 200 m. It occurs where the summer temperatures average 28°C and the winter temperatures average 7°C. This species is relatively shade intolerant but is extremely flood tolerant and capable of surviving under inundated conditions (up to at least 0.6 m depth) for more than two growing seasons. Sites typically are inundated for 30%–40% of the growing season. Leaf out is delayed in the spring, which is thought to contribute to the high flood tolerance of this species. The plants are found mainly on poorly drained, acidic (pH<6) alluvial substrates (primarily alfisols or inceptisols), which include clay, marl, sand, sandy loam, sandy silt, silty clay, and silty loam. Individual trees can live up to 400 years, bearing seeds after 25 years of age, then producing seed masts every 3–4 years. Flowering occurs during spring. The seeds have epicotyl dormancy, i.e., the radicles are nondormant (emerging in fall); however, the shoots require cold stratification and remain dormant until the spring. Germination completes once the spring flood waters recede. The seeds germinate well in the shade, but the seedlings will die unless they become exposed to open conditions within 3 years. Soaking the acorns will delay germination but does not reduce their percent germination. The acorns contain spongy tissue, which makes them buoyant. They are dispersed readily along water courses and have been found several kilometers away from any potential source populations. This species resists competition by its high tolerance to flooding, which kills many potential competitors in early spring (and may account for the rather abbreviated list of associated species reported). The specific pattern of regeneration in stands is an outcome of compound interactions between available light and flooding frequency. Disturbances release the plants from suppression, allowing them to spread rapidly under such conditions. The species is only moderately fire tolerant. **Reported associates:** *Acer rubrum, A. saccharinum, Ageratina altissima, Arundinaria gigantea, Athyrium filix-femina, Callicarpa americana, Cardamine pensylvanica, Carex grisea, C. intumescens, C. joorii, Carya aquatica, Celtis laevigata, C. occidentalis, Cephalanthus occidentalis, Cornus drummondii, C. florida, C. stricta, Crataegus, Dichanthelium, Diodia virginiana, Diospyros virginiana, Forestiera acuminata, Fraxinus americana, F. caroliniana, F. pennsylvanica, F. profunda, Gleditsia aquatica, G. triacanthos, Gratiola virginica, Hackelia virginiana, Hydrocotyle verticillata, Hypericum crux-andreae, Justicia ovata, Lemna, Liquidambar styraciflua, Nyssa aquatica, N. biflora, Onoclea sensibilis, Paspalum urvillei, Persicaria virginiana, Pinus palustris, P. taeda, Planera aquatica, Platanus occidentalis, Populus deltoides, P. heterophylla, Quercus alba, Q. bicolor, Q. falcata, Q. imbricaria, Q. laurifolia, Q. macrocarpa, Q. michauxii, Q. nigra, Q. nuttallii, Q. pagoda, Q. palustris, Q. phellos, Q. rubra, Q. shumardii, Q. texana, Q. velutina, Rhynchospora corniculata, Sabal minor, Salix nigra, Sassafras albidum, Saururus cernuus, Smilax, Taxodium distichum, Triadica sebifera, Ulmus americana, U. crassifolia, U. rubra, Vitis, Viola.*

Quercus texana **Buckley** occurs in or on bottomlands, canyons, flats, floodplains, glades, and woodlands of the Mississippi embayment region at elevations of up to 389 m. The plants occur where the summer temperatures average 27°C and winter temperatures average 7°C–13°C. The habitats are not permanently flooded but normally retain 8–20 cm of standing water during the winter months. The plants generally are shade intolerant but the seedlings can survive under shaded conditions for 5–10 years. This species grows optimally on acidic soils (pH: 4.5–5.5) but can also persist on more alkaline soils (pH: 6.1–8.5) without exhibiting symptoms of chlorosis. The substrates have been described as clay, limestone, and sand. The trees begin to flower at about 20 years of age, producing masts of fruit every 3–4 years. Individual trees can produce from 6 to 35 kg of fruit annually. The seeds are physiologically dormant and require 60–90 days of cold stratification before germination is triggered. Germination (60%–90%) occurs at soil temperatures of 21°C–32°C. Periods of submergence for as long as 30 days do not

reduce germination. The fruits are dispersed by water and animals. Seedling establishment is high in populations. These are shallow rooting plants and the larger specimens frequently create canopy gaps as a result of windthrows. **Reported associates:** *Acer negundo, A. rubrum, A. saccharinum, Carya aquatica, Celtis laevigata, Cephalanthus occidentalis, Cornus drummondii, Crataegus, Forestiera acuminata, Fraxinus pennsylvanica, F. profunda, Gleditsia triacanthos, Juniperus, Liquidambar styraciflua, Magnolia virginiana, Nyssa aquatica, N. biflora, Planera aquatica, Quercus alba, Q. durandii, Q. fusiformis, Q. lyrata, Q. michauxii, Q. nigra, Q. palustris, Q. phellos, Sabal minor, Smilax laurifolia, Styrax texana, Taxodium distichum, Ulmus alatus, U. americana, U. crassifolia.*

Use by wildlife: *Quercus bicolor* is a host plant to the diverse fauna of butterfly and moth larvae (Insecta: Lepidoptera: Coleophoridae: *Coleophora atromarginata*; Gelechiidae: *Telphusa*; Gracillariidae: *Cameraria cincinnatiella, C. conglomeratella, C. hamadryadella, C. platanoidiella, Phyllonorycter aeriferella, P. albanotella, P. argentifimbriella, P. basistrigella, P. diaphanella, P. hagenii*; Lymantriidae: *Lymantria dispar*; Nepticulidae: *Stigmella flavipedella*; Saturniidae: *Anisota senatoria, Antheraea polyphemus*; Tischeriidae: *Tischeria castaneaeella*). *Quercus lyrata* is the larval host plant for several silk moths (Insecta: Lepidoptera: Saturniidae: *Anisota virginiensis, Antheraea pernyi, A. polyphemus*). *Quercus texana* also hosts larval Lepiodptera (Hesperiidae: *Erynnis horatius*; Lasiocampidae: *Gloveria sphingiformis*; Saturniidae: *Anisota virginiensis, Hemileuca grotei, H. maia*). *Quercus bicolor* is susceptible to fungi infections including wilt (Ascomycota: Pleosporaceae: *Alternaria*; Ceratocystidaceae: *Ceratocystis fagacearum*), canker (Ascomycota: Diaporthaceae: *Phomopsis*), and dieback (Ascomycota: Leptosphaeriaceae: *Coniothyrium*). The fruits of *Q. lyrata* and *Q. texana* are eaten (and often damaged severely) by acorn weevils (Insecta: Coleoptera: Curculionidae: *Curculio*). The wood of *Q. lyrata* and *Q. texana* is damaged by several boring insects (Insecta) including carpenterworms (Lepidoptera: Cossidae: *Prionoxystus robiniae*), red oak borers (Coleoptera: Cerambycidae: *Enaphalodes rufulus*), and white oak borers (Coleoptera: Cerambycidae: *Goes tigrinus*). Further damage to the wood of *Q. lyrata* is caused by spot-worm borers (Coleoptera: Buprestidae: *Agrilus acutipennis*) and to *Q. texana* by oak sapling borers (Coleoptera: Cerambycidae: *Goes tesselatus*), hardwood stump borers (Coleoptera: Cerambycidae: *Mallodon dasystomus*), and sap-feeding beetles (Coleoptera: Nitidulidae). Clearwing borers (Insecta: Lepidoptera: Sesiidae: *Paranthrene simulans*) can provide avenues of fungal infection in *Q. texana*. The foliage of *Q. lyrata* occasionally is attacked by leafminers (Insecta: Chrysomelidae: *Baliosus ruber*) and that of *Q. texana* by leafminers (*Baliosus nervosus*) as well as pinkstriped oakworms (Insecta: Lepidoptera: Saturniidae: *Anisota virginiensis*). Defoliation of *Q. texana* can also be a consequence of infestations by anthracnose (Ascomycota: Valsaceae: *Apiognomonia errabunda*) and leaf spot (Ascomycota: incertae sedis: *Tubakia dryina*) fungi. All three of the aquatic oaks are important wildlife plants because they are mast fruiting. The acorns of *Q. bicolor, Q. lyrata,* and *Q. texana* are eaten by numerous birds (Aves) including mallard ducks (Anatidae:

Anas platyrhynchos), wild turkeys (Phasianidae: *Meleagris gallopavo*), wood ducks (Anatidae: *Aix sponsa*), and woodpeckers (Picidae), and also by various mammals (Mammalia) such as hogs (Suidae), small rodents (Rodentia), squirrels (Sciuridae), and white-tailed deer (Cervidae: *Odocoileus virginianus*). The large fruits of *Q. lyrata* are used to a lesser degree by ducks (Aves: Anatidae) and fox squirrels (Mammalia; Sciuridae: *Sciurus niger*). *Quercus texana* is an important source of insects eaten by the golden-cheeked warbler (Aves: Parulidae: *Setophaga chrysoparia*).

Economic importance: food: The sweet acorns of *Q. bicolor* are edible and are used as a food by the Iroquois; **medicinal:** Iroquois medicine employed *Q. bicolor* for several remedies, including treatments for broken bones, cholera, and tuberculosis (compound bark decoctions), and for congestion (inhalation of smoke from burning leaves). *Quercus* pollen is a significant contributor to wind-borne allergens; **cultivation:** *Quercus bicolor* is planted as an ornamental; *Q. texana* and *Q. lyrata* are also cultivated but less frequently. The hybrid *Q. ×comptoniae* (see later) is cultivated occasionally. *Quercus lyrata* and *Q. texana* are planted as species for wildlife habitat enhancement; **misc. products:** The wood of *Quercus bicolor* ("white oak") is considered to be a quality material for use in cabinetry, furniture, and veneer. Wood from *Q. lyrata* is of poorer quality but is used sometimes as a rough lumber. *Quercus texana* is one source of lumber sold commercially as "red oak." The Chippewa cleaned rust from their hunting traps using a mixture containing boiled barks from *Q. bicolor*, hemlock (*Tsuga*), and maple (*Acer*). The Choctaw used the burned or boiled bark

of *Q. lyrata* as an ingredient to make red dye and paint; **weeds:** none; **nonindigenous species:** none.

Systematics: Traditionally, Fagaceae have been divided into two subfamilies (Castaneoideae, Fagoideae) with *Quercus* placed within the latter. However, molecular analyses (ITS, *matK* sequence data) indicate that *Quercus* instead is allied phylogenetically to the castaneoid genera in a position close to *Lithocarpus* and *Chrysolepis* (Figure 4.21). In North America, *Quercus* itself is represented by three sections containing the "white" oaks (section *Quercus*), the "black" or "red" oaks (section *Lobatae*), and the "golden" oaks (section *Protobalanus*). *Quercus bicolor* and *Q. lyrata* are placed in section *Quercus* with *Q. texana* placed in section *Lobatae*. Molecular (nrITS sequence) data indicate that the three sections form a clade and that each of the sections also represents a monophyletic group. However, even though up to 25 species have been included in cladistic analyses using DNA sequence data, no published analysis has yet included either of the obligate aquatic North American species. Discrepancies among studies reporting DNA sequence data have been attributed to the inadvertant analysis of paralogous alleles, a factor that must be considered in further studies of the genus. Some indications of relationships among the OBL (and FACW) species have been provided by allozyme studies. Allozyme data depict *Q. bicolor* as somewhat isolated and show a relatively close relationship to exist between *Q. lyrata* and *Q. stellata*. *Quercus texana* was excluded from that study; however, the entire section to which it belongs (section *Lobatae*) is distinct from the obligate species in section *Quercus* (Figure 4.22).

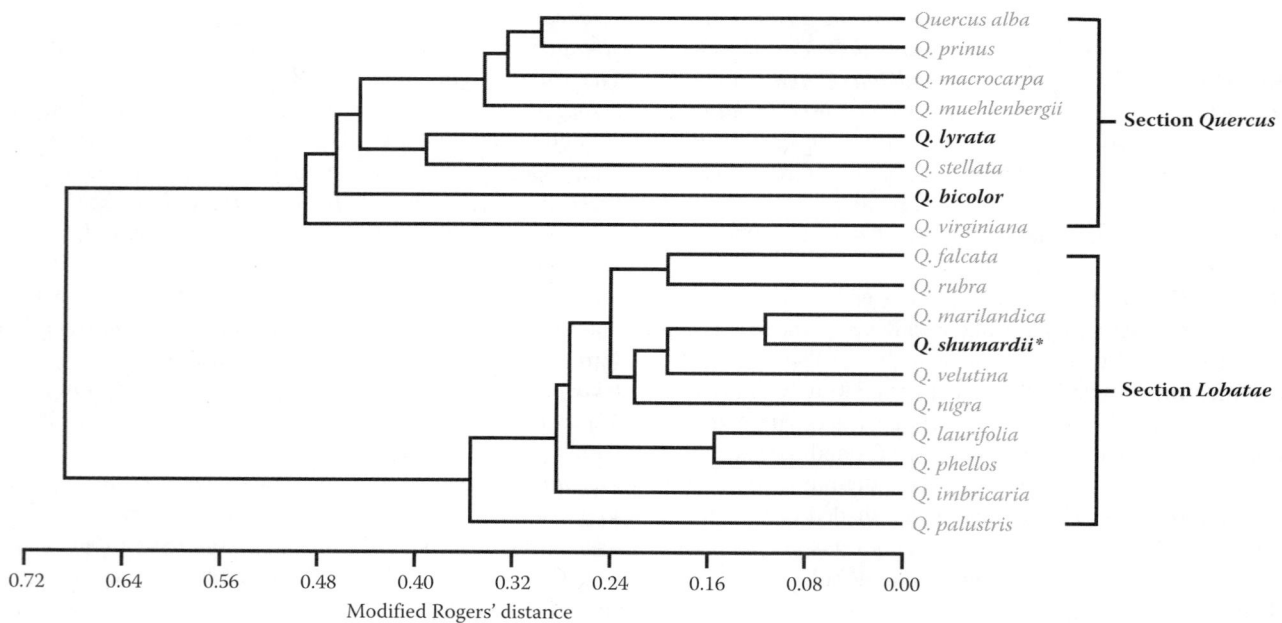

FIGURE 4.22 Phenogram showing overall genetic similarity among eastern North American oak species as indicated by allozyme data. Resolution of the two *Quercus* sections (section *Protobalanus* not surveyed) as distinct clusters supports their taxonomic delimitation, at least for the species included. Although this diagram is not a cladogram, the lack of clustering among the obligate North American taxa indicates that aquatic adaptations probably have arisen independently in these species (*Q. shumardii* [marked by an asterisk] is used as a proxy for the missing *Q. texana* because it is assigned to the same section and is believed to be closely related to that species). (Adapted from Guttman, S.I. & Weigt, L.A., *Can. J. Bot.*, 67, 339–351, 1989.)

Furthermore, *Q. shumardii* (believed to be closely related to *Q. texana*) is quite distant genetically from *Q. bicolor* and *Q. lyrata* (Figure 4.22). It is reasonable to conclude from these allozyme studies that there have been multiple origins of the obligate habit in the North American oaks. The base chromosome number of *Quercus* is $x = 12$. *Quercus bicolor* ($2n = 24$) is diploid; chromosome counts are not reported for *Q. lyrata* or *Q. texana*. *Quercus bicolor* and *Q. lyrata* both reportedly cross with *Q. virginiana*, producing hybrids known as *Q. ×nessiana* and *Q. ×comptoniae*, respectively. Hybrids between *Q. bicolor* and *Q. lyrata* are known as *Q. ×humidicola*.

Comments: *Quercus bicolor* occurs in east-central North America; *Q. lyrata* is southeastern; *Q. texana* is restricted to the southern Mississippi embayment region.

References: Burns & Honkala, 1990; Edgin et al., 2003; Guttman & Weigt, 1989; King & Antrobus, 2001; Larsen, 1963; Manos et al., 1999, 2001a; Mayol & Rossello, 2001; McCarthy & Evans, 2000; Nixon, 1997; Stein et al., 2003.

Family 3: Juglandaceae [8]

Juglandaceae contain about 60 species of aromatic, catkin-bearing, nut-producing, wind-pollinated trees (rarely shrubs). The flowers are reduced, unisexual (most often monoecious), and arranged in catkins. The foliage is typically pinnately compound and lacking stipules. Although the family was once viewed as being related to Anacardiaceae, phylogenetic analyses of morphological and molecular data have consistently resolved Juglandaceae as a monophyletic sister group to Myricaceae (Figure 4.18). The association of these families is evidenced by their synapomorphic aromatic glands, chains of cubic, crystal-containing wood cells, and single orthotropus ovule (Judd et al., 2016).

Cladistic analyses of morphological and molecular data have also provided reliable phylogenetic reconstructions of intergeneric relationships within Juglandaceae. These studies show the family to comprise two major clades, which are recognized as subfamilies (Engelhardioideae, Juglandoideae). Subfamily Juglandoideae is subdivided into two tribes (Juglandeae, Platycaryeae).

The fruits are dispersed by the wind (when winged), or by small animals (Mammalia: Rodentia) when drupe-like or nut-like (Judd et al., 2016).

The family includes several economically important plants, which produce edible hickory nuts and pecans (*Carya*) and walnuts (*Juglans*). It is also the source of hickory and walnut lumber, which is valued in cabinetry and furniture making. Numerous ornamental tree specimens are cultivated in *Carya*, *Juglans*, and *Pterocarya*.

The family is principally terrestrial and in North America, obligate aquatic species occur only within one genus:

1. *Carya* Nutt.

Carya resolves within the monophyletic tribe Juglandeae as the sister group of the remaining genera (Figure 4.23). Most *Carya* species (as well as the remainder of the family) are terrestrial.

1. *Carya*

Hickory; caryer, hicorier

Etymology: from the Greek *káryon* meaning "nut"

Synonyms: *Annamocarya*; *Hickoria*; *Hickorius*; *Juglans* (in part); *Rhamphocarya*

Distribution: global: Asia; New World; **North America:** eastern

Diversity: global: 25 species; **North America:** 13 species

Indicators (USA): OBL: *Carya aquatica*, *C. ×lecontei*

Habitat: freshwater; palustrine; **pH:** 4.5–7.0; **depth:** <1 m; **life-form(s):** emergent (tree)

Key morphology: deciduous trees (to 46 m), the bark exfoliating into long strips or plates; leaves (to 6 dm) alternate, odd-pinnately compound, petiolate (to 8 cm); staminate flowers in pendant catkins (to 21 cm); the fruit (to 3 cm) a compressed, obovate, thin-shelled nut

Life history: duration: perennial (buds); **asexual reproduction:** root, shoot sprouts; **pollination:** wind; **sexual condition:** monoecious; **fruit:** nuts (common); **local dispersal:** seeds (gravity); root/shoot sprouts; **long-distance dispersal:** fruit/seeds (animals, water)

Imperilment: (1) *Carya aquatica* [G5]; S1 (IL); S2 (KY, OK); S3 (VA)

Ecology: general: Like oaks, the hickories (*Carya*) usually grow in UPL habitats and are important hardwood trees of temperate deciduous forests. A few are found occasionally in temporarily wet sites but only the following species is associated obligately with periodically inundated habitats. *Carya ×lecontei* also is designated as an OBL indicator; however, as an F_1 hybrid it is excluded here. The fruits are nuts with a dehiscent outer husk, which are dispersed by small animals that feed on them.

***Carya aquatica* (F. Michx.) Elliott** inhabits bayous, bluffs, bottomlands, channels, ditches, flats, floodplains, hammocks, levees, marshes, mudflats, roadsides, sloughs, swamps, woodlands, and the margins of canals, rivers, and streams at elevations of up to 200 m. This is a slow-growing species, which occurs within warm, humid, flooded habitats, characterized by 1–1.5 m annual precipitation, average low winter temperatures of 2°C–16°C, and average high summer temperatures of 27°C. The adult plants have intermediate shade tolerance and will tolerate soil compaction. The substrates (mainly the vertic haplaquepts subgroup of inceptisols) are of acid to neutral pH and include alluvium, clay, clay loam, gumbo soil, loam, muck (organic), mucky clay, sand, sandy clay, sandy clay loam, sandy loam, silt, silty clay loam, and silty loam. The plants flower during the spring season. The flowers are wind pollinated. Self-pollination is possible (via geitonogamy), but much higher seed production is achieved through outcrossing. The plants bear seeds from about 20 years of age, with a peak output occurring in 40- to 75-year-old trees. Large seed crops are produced annually with individual trees capable of yielding up to 70 L of seeds in a single season. Like most hickories, *C. aquatica* has physiological seed dormancy and requires a 3- to 4-week period of cold stratification (at 4°C–5°C) as a prerequisite to germination (which can reach 80%)

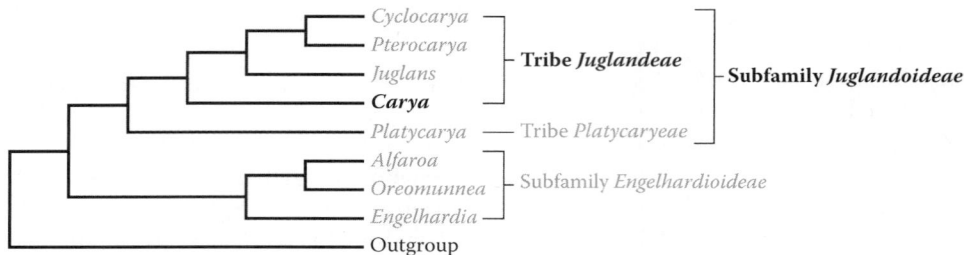

FIGURE 4.23 Phylogenetic relationships among genera of Juglandaceae as indicated by combined analysis of morphological and molecular (cpDNA, ITS sequences) data. Taxa containing OBL North American indicators are shown in bold. (Adapted from Manos, P.S. & Stone, D.E., *Ann. Missouri Bot. Gard.*, 88, 231–269, 2001.)

at 21°C–29°C. Hickory seeds are recalcitrant and quickly lose viability when their moisture content drops; however, they will survive for several months if stored on a cold, moist substrate. However, few seeds survive in natural habitats for more than one season. The seeds germinate on disturbed soil and the seedlings must be released from canopy cover and receive full sunlight in order to survive. The nuts are dispersed by water (especially during floods) and by animals. The plants can reproduce vegetatively from stump or severed root sprouts and can spread quickly through gaps. **Reported associates:** *Acer floridanum, A. rubrum, Aesculus avia, Ampelopsis arborea, Annona glabra, Aristolochia serpentaria, Arundinaria gigantea, Berchemia scandens, Boehmeria cylindrica, Brunnichia ovata, Callicarpa americana, Campsis radicans, Carex folliculata, Carya illinoensis, C. ovata, C. cordiformis, Celtis laevigata, C. occidentalis, Cephalanthus occidentalis, Ceratophyllum echinatum, Chasmanthium latifolium, C. laxum, Commelina virginica, Cornus drummondii, C. stricta, Crataegus, Dichanthelium, Diospyros virginiana, Forestiera acummata, Fraxinus caroliniana, F. pennsylvanica, Gleditsia aquatica, G. triacanthos, Hedyotis, Hydrocotyle verticillata, Hypericum crux-andreae, Ilex decidua, I. vomitoria, Itea virginica, Juncus, Leersia virginica, Liquidambar styraciflua, Mimosa strigillosa, Nyssa sylvatica, Panicum, Persea borbonia, Persicaria, Pinus taeda, Planera aquatica, Platanus occidentalis, Populus deltoides, Quercus falcata, Q. laurifolia, Q. lyrata, Q. nuttallii, Q. nigra, Q. palustris, Q. phellos, Q. stellata, Q. texana, Rhus copallinum, Sabal minor, S. palmetto, Salix caroliniana, Saururus cernuus, Smilax, Steinchisma hians, Styrax americana, Symplocos tinctoria, Taxodium distichum, Toxicodendron radicans, Trachelospermum difforme, Triadica sebifera, Ulmus americana, U. crassifolia, Vitis.*

Use by wildlife: Nuts of *C. aquatica* are eaten by many animals (Mammalia), including bears (Ursidae: *Ursus*), feral hogs (Suidae: *Sus scrofa*), foxes (Canidae: *Vulpes*), rabbits (Leporidae), raccoons (Procyonidae: *Procyon lotor*), small rodents (Rodentia), squirrels (Sciuridae: *Sciurus*), whitetail deer (Cervidae: *Odocoileus virginianus*), and by various birds (Aves) including bluejays (Corvidae: *Cyanocitta cristata*), crows (Corvidae: *Corvus*), grosbeaks (Cardinalidae), mallards (Anatidae: *Anas platyrhynchos*), pheasants (Phasianidae), quails (Odontophoridae), turkeys (Phasianidae:

Meleagris gallopavo), and wood ducks (Anatidae: *Aix sponsa*). The trees also attract yellow-bellied sapsuckers (Aves: Picidae: *Sphyrapicus varius*). *Carya aquatica* is a larval host plant for several moths (Insecta: Lepidoptera: Noctuidae: *Catocala agrippina, C. maestosa*). Wood of *C. aquatica* is excavated by hickory borers (Insecta: Coleoptera: Cerambycidae: *Goes pulcher*), and the leaves are eaten by tent caterpillars (Insecta: Lepidoptera: Lasiocampidae: *Malacosoma disstria*). The parasitic mistletoe (Santalaceae: *Phoradendron serotinum*) occurs commonly on plants in some areas.

Economic importance: food: The nuts of *C. aquatica* reportedly are edible raw or cooked but are not palatable to people owing to their bitterness; **medicinal:** the pollen of *C. aquatica* can represent a component of wind-borne allergens in some areas, but it is large and not transported very far from its source; **cultivation:** *Carya aquatica* is of minor importance as a cultivated ornamental or wildlife planting; **misc. products:** *Carya aquatica* produces a very hard wood but it yields an inferior, brittle, lumber, which often is marred by bacterially induced "shake." The wood is used locally for fenceposts and as charcoal and firewood for rendering "hickory smoked" flavoring to meats. Stands of *C. aquatica* provide important wildlife refuge and help to purify drainage waters; **weeds:** none; **nonindigenous species:** none.

Systematics: Phylogenetic analyses resolve *Carya* in a position between *Platycarya* and *Juglans* (Figure 4.23). Analysis of combined morphological and molecular data indicates that *Carya* is monophyletic, but only if the small genus *Annamocarya* (2 spp.) is included. No comprehensive molecular phylogenetic analysis of *Carya* species has been conducted, and a recent study of eight species unfortunately excluded *C. aquatica*. Traditionally, *Carya aquatica* is placed among the "pecan hickories" (section *Apocarya*), whose springwood contains bands of parenchyma tissue. However, analysis of seed oil fatty acids among North American hickory species indicated a high degree of similarity between *C. aquatica* and *C. myristiciformis* (section *Carya*). The chromosomal base number of *Carya* is $x = 16$. *Carya aquatica* ($2n = 32$) is diploid. The name *C. ×lecontei* is applied to hybrids involving *C. aquatica* and *C. illinoensis*. The average fatty acid composition of seed oils in *C. ×lecontei* is intermediate between *C. aquatica* and *C. illinoensis*. Hybrids known as *C. ×ludoviciana* occur reportedly between *C. aquatica* and *C. texana*, a tetraploid ($2n = 64$).

Comments: *Carya aquatica* occurs along the southeastern coastal plain and Mississippi embayment region.
References: Burns & Honkala, 1990; Elias, 1987; Manos & Stone, 2001; Stone, 1997; Stone et al., 1969.

Family 4: Myricaceae [2]

Myricaceae comprise 50 species of aromatic, resinous shrubs, and trees that occupy a wide range of habitats. The family is related phylogenetically as the sister group to Juglandaceae (Figure 4.18). The taxonomy of Myricaceae is in a state of flux as the result of phylogenetic analyses. As many as four genera (*Canacomyrica, Comptonia, Morella, Myrica*) have been recognized to subdivide this small family. In North America, only one genus contains obligate aquatics:

  1. ***Myrica*** L.

Preliminary molecular studies (Huguet et al., 2005) indicate that Myricaceae are monophyletic and resolve as three major subclades (Figure 4.24). Recently, these clades have been used to delimit three distinct genera, given that two species of *Myrica* (as traditionally defined) were found to resolve within a clade along with the monotypic *Comptonia*. In that interpretation, *Myrica* represents the clade containing the type species (*M. gale*) with the other clade defined as the genus *Morella* (Figure 4.24). Although this novel nomenclature has been adopted readily by a number of authors, it seems unnecessary. It would be much simpler to include the monotypic genus *Comptonia* within *Myrica* (as some authors already have done), which involves only a single nomenclatural change. The same fundamental clades could then be recognized as

subdivisions (e.g., subgenera) of *Myrica*. With a family of only 50 species, the incorporation of two additional genera to accommodate a total of only three species seems excessive. Consequently, this treatment considers *Myrica* in the broader (sensu lato) sense and does not recognize *Comptonia* or *Morella* at the genus level.

1. *Myrica*

Bayberry, sweet gale, wax-myrtle; bois-sent-bon, myrique
Etymology: from *myrike*, the Greek name for the plant, which is derived from *myrizei* (meaning "aromatic")
Synonyms: *Cerothamnus*; *Gale*; *Morella*
Distribution: global: cosmopolitan; **North America:** coastal, eastern, northern
Diversity: global: 50 species; **North America:** 8 species
Indicators (USA): OBL: *Myrica gale, M. inodora*
Habitat: freshwater; palustrine; **pH:** 3.8–6.7; **depth:** <1 m; **life-form(s):** emergent (shrub)
Key morphology: deciduous or evergreen shrubs or small trees (to 7 m), branchlets gland dotted; leaves (to 10.5 cm) alternate, simple, leathery, serrate distally; male inflorescences (to 2.2 cm) erect, cylindrical; female inflorescences (to 4.0 cm) ovoid; flowers inconspicuous, the females subtended by two or four bracteoles; fruit (to 8.0 mm) a globose, drupe-like nut, sometimes with a thin waxy coating or enclosed by spongy bracteoles
Life history: duration: perennial (buds); **asexual reproduction:** root suckers; **pollination:** wind; **sexual condition:** monoecious, dioecious; **fruit:** nutlets (common); **local dispersal:** fruits (water); **long-distance dispersal:** fruits (water)

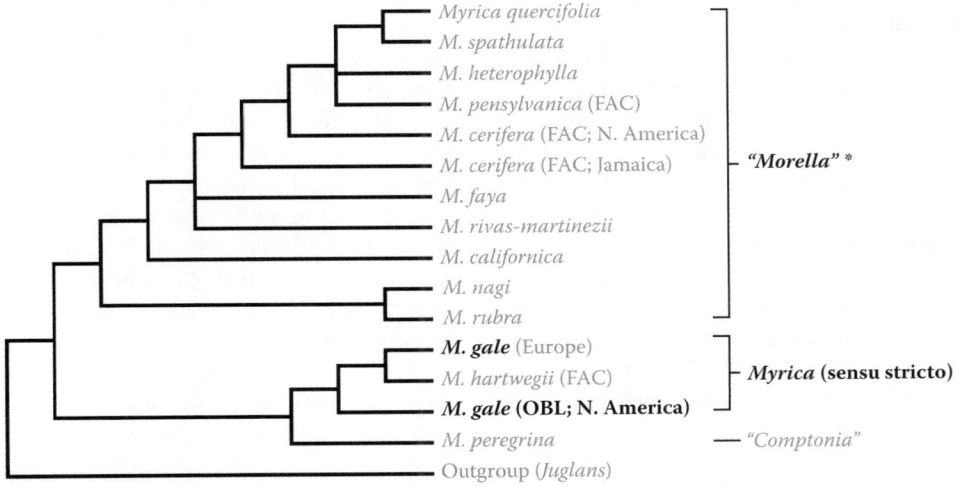

FIGURE 4.24 Phylogenetic tree for Myricaceae constructed from nrITS sequence data. Although many authors recognize the upper clade ("*Morella*") and the lower species ("*Comptonia*") as distinct genera, it seems more practical to retain all of the species in *Myrica*, which can be achieved simply by including the monotypic genus *Comptonia* (which formerly was placed in *Myrica*). The OBL North American *M. gale* (in bold) differs genetically from European accessions, which supports those taxonomic treatments that have distinguished varieties within this species. The "*Morella*" clade (asterisked) is considered here as a proxy for *M. inodora*, which was not surveyed but is assigned to the same section (Wilbur, 1994). Consequently, it is reasonable to conclude that there have been two independent origins of the OBL habit in the North American species. Analysis of combined DNA sequence data (Herbert et al., 2006) not only results in a similar general topology, but also resolves the genus *Canacomyrica* (not surveyed in above analysis) as the sister group to the remainder of Myricaceae. (Adapted from Huguet, V. et al., *Mol. Phylogen. Evol.*, 34, 557–568, 2005.)

Imperilment: (1) *Myrica gale* [G5]; S1 (NC, OR); S2 (PA); S3 (AB, NJ); (2) *M. inodora* [G4]; S1 (GA); S2 (LA)

Ecology: general: *Myrica* species occupy diverse habitats, which include dunes, forests, mountain slopes, scrub, and wetlands. All eight North American *Myrica* species have some degree of tolerance to wet habitats and are designated as wetland indicators; however, only two species occur consistently under such conditions and are regarded as obligate aquatics. The flowers of all species are reduced, unisexual (dioecious or monoecious), arranged in catkins, and wind pollinated. The fruits are dispersed by birds (Aves), or abiotically by water. The seeds can be nondormant or physiologically dormant, depending on the species.

Myrica gale L. occurs in or on barrens, beaches, bogs, carrs, ditches, dunes, fens, heaths, hummocks, marshes, meadows, muskeg, roadsides, scrub, sloughs, swales, swamps (coastal or inland), thickets, tundra, and along the margins of lakes, ponds, and streams at elevations of up to 944 m. The species often occurs in shallow water (e.g., 15 cm) in both tidal and nontidal sites, and can tolerate exposures ranging from full sunlight to dense shade. The substrates typically are shallow (50–80 cm), acidic (pH: 3.8–6.7) peats but also include gravel, muck, rock, sand, and silt. The plants survive well on acidic, nitrogen-poor soil because their roots develop nitrogen-fixing nodules (Bacteria: Frankiaceae: *Frankia*). Nitrogen fixation occurs optimally at low (pH: 5.4). Cluster roots also are produced, which enhance the localized mobilization of soil phosphorous. Flowering extends from March to August, and fruiting from April to November. *Myrica gale* is dioecious (populations tend to be male biased); however, individual plants can change sex expression, whereby monoecious or even hermaphroditic-flowered plants will occur occasionally. The fruits are subtended by small, spongy processes (bracteoles), which provide flotation when dispersed in water. Seeds that have floated on water at 5°C for several weeks achieve the highest germination rates. Their germination requires an extended exposure to light. Viability can be maintained for up to 6 years if seeds are refrigerated and stored dry. The plants spread vegetatively by the production of creeping, rhizome-like suckers and can develop into dense stands by this means. The leaves have relatively slow decomposition rates owing to the presence of various secondary compounds. This is one of the first shrubs to colonize the gradient from flark to strang in patterned wetlands (string bogs). The plants may occur on the margins of estuarine areas and salt marshes, but they do not grow directly on saline substrates. **Reported associates:** *Acer rubrum, Agrostis, Alnus glutinosa, A. rubra, A. rugosa, A. viridis, Andromeda glaucophylla, A. polifolia, Arctostaphylos alpina, Arctous rubra, Aulacomnium turgidum, Betula glandulosa, B. nana, B. papyrifera, B. populifolia, B. pumila, Bidens beckii, Calamagrostis canadensis, Calopogon tuberosus, Campylium stellatum, Carex aquatilis, C. canescens, C. chordorrhiza, C. exilis, C. lasiocarpa, C. limosa, C. lyngbyei, C. oligosperma, C. pluriflora, C.* *rostrata, C. saxatilis, C. stricta, C. utriculata, C. viridula, C. wiegandii, Cephalanthus occidentalis, Chamaedaphne calyculata, Cicuta, Cladium mariscoides, Clethra alnifolia, Comarum palustre, Dasiphora floribunda, Decodon verticillatus, Deschampsia cespitosa, Drosera anglica, Dulichium arundinaceum, Eleocharis palustris, Empetrum, Equisetum fluviatile, Eriophorum angustifolium, E. virginicum, Gentiana douglasiana, Geranium erianthum, Geum calthifolium, Ilex glabra, I. mucronata, I. verticillata, Iris, Juncus canadensis, J. stygius, Juniperus communis, Kalmia angustifolia, K. microphylla, Larix laricina, Lathyrus japonicus, Leptogium brebissonii, Lobelia kalmii, Lonicera villosa, Lycopus uniflorus, Lysimachia terrestris, Menyanthes trifoliata, Nephrophyllidium crista-galli, Oclomena nemoralis, Oenothera fruticosa, Parnassia palustris, Picea mariana, P. sitchensis, Pinus contorta, Platanthera clavellata, Poa, Pogonia ophioglossoides, Polemonium caeruleum, Rhinanthus minor, Rhododendron canadense, R. groenlandicum, Rhynchospora alba, Salix arctophila, S. barclayi, S. brachycarpa, S. candida, S. fuscescens, S. glauca, S. monticola, S. pedicellaris, S. planifolia, S. reticulata, Sanguisorba officinalis, Sarracenia purpurea, Scheuchzeria palustris, Schoenoplectus pungens, Shepherdia canadensis, Sibbaldia procumbens, Solidago nemoralis, Sparganium glomeratum, Sphagnum fimbriatum, S. girgensohnii, S. magellanicum, S. palustre, Spiraea alba, S. douglasii, S. tomentosa, Symphyotrichum boreale, Thalictrum sparsiflorum, Thuja plicata, Triadenum fraseri, T. virginicum, Trichophorum alpinum, T. cespitosum, Typha latifolia, Vaccinium corymbosum, V. macrocarpon, V. nudum, V. oxycoccos, V. uliginosum, Valeriana uliginosa.*

Myrica inodora W. **Bartram** grows in southern coastal habitats including bogs, flatwoods, marshes, savannahs, seeps, slopes, swamps, thickets, woodlands, and along the margins of ponds and streams at elevations below 10 m. The plants are shade tolerant and are assigned to hardiness zone 8b. The substrates are reported as sand. Flowering occurs in the late winter to early spring, with fruiting in mid-summer. Unlike most other Myrica species, *M. inodora* (as the name implies) is odorless. Further details on the ecology of this narrowly restricted species are lacking.

Reported associates: *Andropogon glomeratus, Baccharis halimifolia, Calamovilfa curtissii, Cephalozia lunulifolia, Cleistes divaricata, Cliftonia monophylla, Cyrilla racemiflora, Decumaria barbara, Drosera intermedia, D. traceyi, Gaylussacia mosieri, Harperocallis flava, Ilex coriacea, I. glabra, I. myrtifolia, Illicium floridanum, Lachnanthes caroliniana, Liquidambar styraciflua, Lophocolea martiana, Lyonia lucida, Magnolia virginiana, Myrica cerifera, M. heterophylla, Nyssa aquatica, Osmanthus americanus, Persea palustris, Pinus elliottii, P. palustris, P. serotina, Pleea tenuifolia, Rhexia alifanus, Rhododendron viscosum, Riccardia latifrons, Rosa bracteata, Sarracenia alata, S. psittacina, Scapania nemorosa, Smilax laurifolia, S. walteri, Sphagnum, Taxodium ascendens, T. distichum, Triadica sebifera, Viburnum.*

Use by wildlife: *Myrica gale* is browsed and eaten by several mammals (Mammalia) including goats (Bovidae: *Capra hircus*), hares (Leporidae), moose (Cervidae: *Alces alces*), and sheep (Bovidae: *Ovis aries*). It is a larval host plant for numerous butterflies and moths (Insecta: Lepidoptera: Arctiidae: *Spilosoma virginica, Utetheisa bella*; Bucculatricidae: *Bucculatrix paroptila*; Geometridae: *Biston betularia, Cingilia catenaria, Cleora projecta, Eupithecia miserulata, Itame sulphurea, Melanolophia canadaria, Nemoria rubrifrontaria, Pero zalissaria, Rheumaptera hastata, Tacparia atropunctata*; Gracillariidae: *Caloptilia asplenifoliatella, C. flavella*; Lymantriidae: *Lymantria dispar, Orgyia leucostigma*; Lyonetiidae: *Lyonetia ledi*; Noctuidae: *Acronicta impressa, A. oblinita, Autographa ampla, Catocala coelebs, Coenophila opacifrons, Eugraphe subrosea, Lithophane thaxteri, Melanchra assimilis, Mniotype ducta, Morrisonia confusa, Syngrapha epigaea, Xestia youngii*; Oecophoridae: *Agonopterix walsinghamella*; Pyralidae: *Acrobasis comptoniella, Ortholepis myricella*; Saturniidae: *Hemileuca nevadensis*; Sphingidae: *Sphinx gordius, S. poecila*; and Tortricidae: *Acleris bowmanana, A. fragariana, A. hastiana, A. kearfottana, A. minuta, Aphelia alleniana, Apotomis paludicolana, Archips myricana, Argyrotaenia repertana, A. velutinana, Olethreutes galevora, O. valdanum, Pandemis limitata, Sparganothis daphnana, Spilonota ocellana*). The herbage of *Myrica inodora* is eaten by black bears (Mammalia: Ursidae: *Ursus americanus*). The seeds are eaten by red-cockaded woodpeckers (Aves: Picidae: *Picoides borealis*).

Economic importance: food: People in Newfoundland use the leaves of *Myrica gale* for brewing into tea and the fruits to flavor roasted meats. The bark and branches of *M. gale* have been used in making "gruit" beer since the time of the Vikings; **medicinal:** The Bella Coola people prepared a decoction from the crushed branches of *M. gale* to use as a diuretic and to relieve symptoms of gonorrhea. *Myrica gale* contains dihydrochalcones which are antibacterial and fungicidal. Extracts of *M. gale* have been shown to suppress influenza A and a *Pseudomonas pyocyanea* bacteriophage. They have also been used to treat infections caused by *Herpes zoster* (Viruses: Herpesviridae). Pollen of *Myrica gale* is allergenic; **cultivation:** *Myrica gale* and *M. inodora* are cultivated as ornamental shrubs for attracting wildlife; **misc. products:** The Potawatomi preserved berries by placing them on *M. gale* plants. The Ojibwa extracted a brown dye from the boiled branch tips and a yellow dye from the boiled seeds of *M. gale*. The Cree and Woodlands tribes used the female catkins of *M. gale* as an ingredient in fishing lures. The Potawatomi burned *M. gale* plants to repel mosquitoes. Various oils produced by the plant have insecticidal properties to the extent that insect predation on other plants (e.g., *Lythrum salicaria*) growing in the presence of *M. gale* is reduced significantly. The bark, branches, and roots of the plant are used to tan leather. Boiling the female cones of *M. gale* in water produces a film of wax which can be used in candle making; **weeds:** none; **nonindigenous species:** none.

Systematics: Although affinities of the monotypic *Canacomyrica* (*C. monticola*) had long been disputed, molecular data have resolved the genus as the sister group to the remainder of Myricaceae. Aside from that distinction, molecular data also delimit subclades within Myricaceae (Figure 4.24), which some authors have equated taxonomically with distinct genera (i.e., *Comptonia, Morella,* and *Myrica*). Because phylogenetic results also support the more traditional recognition of a single genus (*Myrica*) to comprise these three taxa (perhaps recognizing the groups as subgenera), this approach has been followed here for several reasons. When there is no phylogenetic conflict, it is more prudent to adopt a taxonomic scheme that preserves nomenclature to facilitate access to the appropriate literature. Although several morphological features have been used to distinguish *Comptonia, Morella,* and *Myrica,* the extent of molecular divergence among these taxa is quite low compared to *Canacomyrica* or to other genus pairs in Fagales. Also, the additional segregate genera resulting from splitting the group (*Comptonia, Myrica*) each would accommodate only a small number of species (the majority simply moving from *Myrica* to *Morella*) in an already fairly small group. As a consequence that scheme would require a large nomenclatural disruption without any compelling taxonomic reason to do so. Essentially it comes down to a matter of choice. The chromosomal base number of *Myrica* is $x = 8$. *Myrica gale* ($2n = 48$ [Europe], 96 [N. America]) is highly polyploid and variable chromosomally (the chromosome number of *M. inodora* has not been reported). Several varieties of *M. gale* have been identified, which appear to represent relatively minor morphological variants. However, molecular data (Figure 4.24) indicate that the different ploidy levels of this species also are quite divergent genetically. It would be informative to study patterns of chromosomal, genetic, and morphological variation in more detail across the wide range of *M. gale* to evaluate the merit of recognizing infraspecific (or possibly distinct) taxa. *Myrica inodora* has not yet been included in molecular phylogenetic analyses; however, unlike *M. gale*, it is placed within the "*Morella*" clade (Figure 4.24). Consequently, the aquatic habit appears to have arisen independently in the two OBL North American species.

Comments: The distribution of *Myrica gale* occurs throughout the glaciated portions of North America and extends into Asia and Europe. *Myrica inodora* is restricted to the central gulf coast of the United States.

References: Baker & Parsons, 1997; Blanken & Rouse, 1996; Bornstein, 1997; Guerke, 1974; Hambäck et al., 2000; Herbert et al., 2006; Hess & James, 1998; Huguet et al., 2005; Lockwood et al., 2005; Maloney & Lamberti, 1995; Reese, 1976; Schwintzer & Ostrofsky, 1989; Simpson et al., 1996; Skene et al., 2000; Stratman & Pelton, 1999; Stuart & Stuart, 1998; Wilbur, 1994.

ORDER 5: MALPIGHIALES [36]

The Malpighiales are a diverse order containing more than 16,000 species. The monophyly of the order is clearly established by phylogenetic analysis of multiple molecular data sets (18s rDNA, *atpB*, *rbcL* sequences; Soltis et al., 2000); however, the group is very heterogeneous morphologically, without any clear nonmolecular synapomorphies (Judd et al., 2002, 2016). Many species possess leaves where a solitary vein extends into teeth that have a congested but deciduous apex (Judd et al., 2016). In the molecular analyses, Malpighiales resolve within a weakly supported clade together with Celastrales and Oxalidales (Figure 4.1). The clade is distinct from, but unresolved positionally between a small Zygophyllales clade and a large clade containing the eurosid orders Cucurbitales, Fabales, Fagales, and Rosales (Soltis et al., 2000). About 70 obligate indicator species occur interspersed among eight of the families having OBL North American representatives:

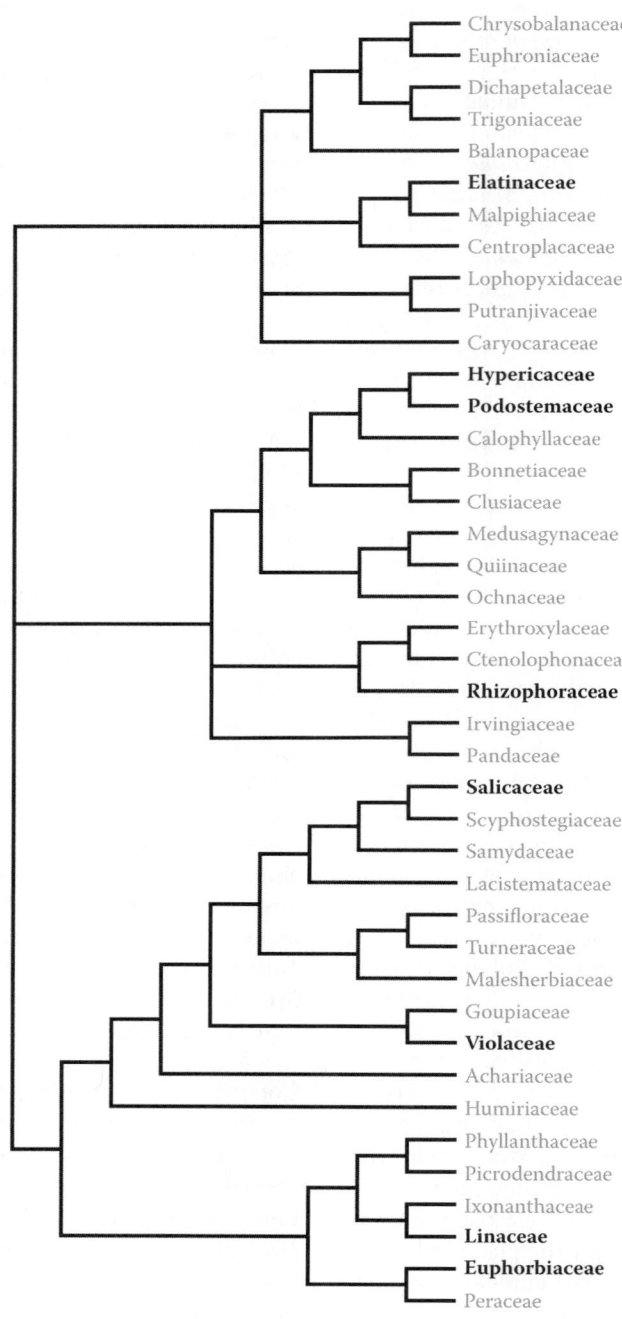

FIGURE 4.25 Phylogenetic relationships within the order Malpighiales as depicted by cladistic analyses of 82 plastid DNA gene sequences. The general pattern of relationships indicates that the North American families containing obligate aquatics (shown in bold) represent numerous independent origins of the OBL habit in the order. (Adapted from Xi, Z. et al., *Proc. Natl. Acad. Sci. U.S.A*, 109, 17519–17524, 2012.)

1. **Elatinaceae** Dumort.
2. **Euphorbiaceae** Juss.
3. **Hypericaceae** Juss.
4. **Linaceae** DC. ex Perleb
5. **Podostemaceae** Rich. ex C. Agardh
6. **Rhizophoraceae** Pers.
7. **Salicaceae** Mirb.
8. **Violaceae** Batsch

Molecular phylogenetic analyses have struggled to fully resolve relationships among the families of Malpighiales. Many molecular-based phylogenetic trees (Chase et al., 2002; Davis & Chase, 2004; Soltis et al., 2000; Tokuoka & Tobe, 2006; Wurdack & Davis, 2009) resolve a number of well-supported clades within the order, but depict the overall interrelationships among those clades as largely unresolved. A recent analysis of 82 plastid DNA loci has provided better resolution for much of the order (Figure 4.25), at least as hypothesized from combined cpDNA. Yet, the results are not particularly compelling, given numerous egregious topological discrepancies compared to studies using subsets of cpDNA and other molecular data.

Family 1: Elatinaceae [2]

The Elatinaceae (waterwort family) are a small but widely distributed group of 35–50 annual or short-lived perennial species, most of which are aquatic or at least are adapted to semi-aquatic conditions (Tucker, 2016a). The family includes herbaceous and shrubby species, with opposite or whorled leaves subtended by interpetiolar stipules. The flowers are small and inconspicuous, and sometimes cleistogamous. Waterworts (*Elatine*) are of minor commercial importance as aquarium plants. Both genera occasionally occur as ricefield weeds.

The precise relationship of the family has been debated, with some morphological data indicating a possible tie to Hypericaceae. However, recent molecular phylogenetic studies (Wurdack & Davis, 2009), which combine nuclear and plastid data sets, show that Elatinaceae (i.e., *Bergia* + *Elatine*) are monophyletic and also strongly support the placement of Elatinaceae within a larger, monophyletic clade as the sister group to Malpighiaceae (e.g., Figure 4.25). Both genera of Elatinaceae contain obligate aquatics in North America:

1. *Bergia* L.
2. *Elatine* L.

1. *Bergia*

Bergia
Etymology: after Peter Jonas Bergius (1730–1790)
Synonyms: *Elatine* (in part); *Merimea*
Distribution: global: New World; Africa; Asia; Australia; Europe*; **North America:** western
Diversity: global: 25 species; **North America:** 1 species
Indicators (USA): OBL: *Bergia texana*
Habitat: freshwater; palustrine; **pH:** 6.5–8.0; **depth:** <1 m;
life-form(s): emergent (herb)

Key morphology: stems (to 4.0 dm) ascending from a tap root, diffusely branched, reddish, glandular; leaves (to 3.0 cm) opposite (or appearing whorled), 4-ranked, petioled, glandular, the blade elliptic–oblong to oblong–oblanceolate, tapered to base, the margin serrulate; stipules (to 1.0 mm) present, scarious; flowers minute, axillary (1–3); sepals (to 4.0 mm) 5; petals (to 3.5 mm) 5, shorter than the sepals, white; stamens 5 or 10; styles 3–5; capsules (to 3 mm) globose; seeds glossy brown, obscurely reticulate
Life history: duration: annual, biennial (fruit/seeds); **asexual reproduction:** none; **pollination:** cleistogamous (self); **sexual condition:** hermaphroditic; **fruit:** capsules (common); **local dispersal:** seeds; **long-distance dispersal:** seeds
Imperilment: (1) *Bergia texana* [G5]; S1 (NE, UT); S2 (AR, KS, MO); S3 (OR)
Ecology: general: About half of the approximately 25 *Bergia* species are aquatics with the remainder found on wet ground. The biology of the genus is poorly known, especially with respect to pollination systems or its floral and seed ecology.

Bergia texana **(Hook.) Seub. ex Walp.** grows in channels, deltas, depressions, ditches, flats, floodplains, lakebeds, marshes, meadows, mud flats, playas, prairies, scrub, slough bottoms, strands, swamps, vernal pools, washes, and along the margins of lakes, ponds, reservoirs, and rivers at elevations of up to 1869 m. Exposures typically receive full sun. The substrates are somewhat alkaline or saline (pH: 6.5–8.0) and consist of adobe, clay, gravel, gravelly silt, limestone, mud, sand, shale, silt, silty clay, or silty sand. The plants are particularly common in muddy areas where standing water has receded. Flowering and fruiting occur from June to November. *Bergia texana* is self-compatible and adapted for autogamy. Many of the flowers are cleistogamous, which forces the anthers tightly against the stigma prior to anthesis, leading eventually to self-pollination. Insect-mediated cross-pollination of the chasmogamous flowers is possible but has not been reported. **Reported associates:** *Alisma plantago-aquatica, Amaranthus, Ambrosia pumilla, Ammannia coccinea, Atriplex coronata, A. suberecta, Avena barbata, Baccharis, Bacopa, Brodiaea filifolia, B. jolonensis, Calibrachoa parviflora, Carex barbarae, Centunculus minimus, Crassula aquatica, Cressa truxillensis, Croton setiger, Crypsis schoenoides, C. vaginiflora, Cyperus aristatus, C. eragrostis, Deinandra fasciculata, Deschampsia danthonioides, Downingia cuspidata, Echinochloa, Echinodorus berteroi, Eclipta prostrata, Elatine brachysperma, E. californica, Eleocharis acicularis, E. macrostachya, E. palustris, Epilobium campestre, E. densiflorum, E. pygmaeum, Eragrostis, Eryngium aristulatum, Euphorbia serpens, Glinus lotoides, Helenium, Helianthus annuus, Hirschfeldia incana, Hoffmannseggia, Hordeum jubatum, Hydrocotyle ranunculoides, Hymenoxys, Isoetes howellii, I. orcuttii, Iva axillaris, Juncus balticus, J. bufonius, J. xiphioides, Lachnagrostis filiformis, Lepidium latipes, L. nitidum, Leptochloa fusca, Leucospora multifida, Lilaea scilloides, Limosella acaulis, Lindernia dubia, Lolium perenne, Lotus corniculatus, Ludwigia, Lythrum hyssopifolia, Malvella leprosa, Marsilea vestita, Mimulus latidens, Mollugo verticillata, Myosurus minimus, Nama stenocarpa,*

Navarretia fossalis, N. hamata, Ophioglossum californicum, Orcuttia californica, O. pilosa, Panicum, Pascopyrum smithii, Paspalum distichum, Persicaria, Phalaris lemmonii, Phyla nodiflora, Physalis angulata, Pilularia americana, Plagiobothrys acanthocarpus, P. leptocladus, Plantago elongata, P. erecta, Pogogyne nudiuscula, Populus fremontii, Prosopis glandulosa, Psilocarphus brevissimus, P. tenellus, Rotala ramosior, Rumex persicarioides, Sagittaria sanfordii, Salix exigua, Schoenoplectus, Symphyotrichum subulatum, Trichocoronis wrightii, Tridens albescens, Verbena bracteata, Veronica peregrina, Xanthium strumarium.

Use by wildlife: The seeds of *Bergia texana* are eaten by various waterfowl (Aves: Anatidae).

Economic importance: food: none; **medicinal:** none; **cultivation:** none; **misc. products:** none; **weeds:** *Bergia capensis* is a weed of irrigation canals and ricefields in other parts of the world. Care should be taken to avoid the introduction of this potentially invasive species into North America; **nonindigenous species:** *Bergia capensis* is adventive in Portugal, Spain, and along the Pacific coast of South America but is not yet reported in North America.

Systematics: Combined molecular data (*ndhF, rbcL, PHYC* sequences) resolve *Bergia* and *Elatine* as sister genera; however, surveys of additional species in each genus will be necessary before the monophyly of each genus can be established conclusively. *Bergia texana* has been assigned to section *Dichasianthae*, which is characterized by axillary flowers that occur in dichasia. However, a systematic revision of *Bergia* is seriously needed, as the genus has not been evaluated taxonomically since the early 20th century. The base chromosome number of *Bergia* is $x = 6$. The chromosome number of *B. texana* remains unreported.

Comments: *Bergia texana* is the only species of the genus in North America and occurs throughout the western region.
References: Bauder & McMillan, 1998; Cook, 1996a; Tucker, 1986, 2016b.

2. *Elatine*

Waterwort; élatine
Etymology: from the Greek *elatinos* meaning "of the fir tree" alluding to the resemblance of some species to conifer seedlings
Synonyms: *Alsinastrum; Crypta; Peplis* (in part); *Potamopitys*
Distribution: global: cosmopolitan; **North America:** widespread
Diversity: global: 25 species; **North America:** 10 species
Indicators (USA): OBL: *Elatine ambigua, E. americana, E. brachysperma, E. californica, E. chilensis, E. heterandra, E. minima, E. ojibwayensis, E. rubella, E. triandra;* **FACW:** *E. brachysperma*
Habitat: freshwater; lacustrine, palustrine; **pH:** 4.8–8.6; **depth:** 0–2 m; **life-form(s):** emergent (herb), submersed (vittate)
Key morphology: stems (to 2 dm) upright when in water or procumbent when emergent, traversed by 5–11 air canals, rooting at the nodes; stipules present; leaves (to 12 mm) opposite, sessile or indistinctly petiolate, elliptic to orbiculate,

often notched apically; flowers sessile in upper leaf axils, globular, 2-, 3-, or 4-merous; sepals membranous, inconspicuous; petals membranous, pale greenish-white; capsules subglobose, the walls translucent; seeds reticulate, straight to curved
Life history: duration: annual (fruit/seeds); **asexual reproduction:** none; **pollination:** cleistogamous (self); **sexual condition:** hermaphroditic; **fruit:** capsules (common); **local dispersal:** seeds (water); **long-distance dispersal:** seeds (waterfowl)
Imperilment: (1) *Elatine ambigua* [GNR]; (2) *E. americana* [G4]; SH (DC, NH, VT); S1 (MA, MB, NY, RI, OK); S2 (DE, NB, NJ); S3 (MD, ON); (3) *E. brachysperma* [G5]; S1 (BC, GA, NE, OR); (4) *E. californica* [G5]; S1 (UT); (5) *E. chilensis* [GNR]; (6) *E. heterandra* [G3/G4]; (7) *E. minima* [G5]; S1 (MD, PE, VA, VT); S2 (DE, NF); S3 (NB, NJ); (8) *E. ojibwayensis* [G1/G2]; S1 (QC); (9) *E. rubella* [G5]; SH (ON); S2 (BC, WY); (10) *E. triandra* [G5]; SH (MO, OH); S1 (AB, CO, IA, KS, ND, UT, WI, WY); S2 (SK); S3 (QC)
Ecology: general: All *Elatine* species are aquatic or wetland plants with some species capable of growing amphibiously as both emergent and submersed life-forms. Many species occur in habitats where the standing waters have receded and few other competing species are established. In the water, the plants obtain much of their CO_2 from the layer directly over the sediments, where higher concentrations exist. These are annual plants, which recruit entirely by seed, and usually require clear water and minimal disturbance for good germination and seedling growth. The seeds are dispersed locally by water currents or to greater distances by waterfowl (Aves: Anatidae). Many species produce submersed, self-pollinating, cleistogamous flowers. Chasmogamous flowers, which appear on emergent plants, possess small nectaries but apparently do not attract insect visitors. Even here the anthers frequently make contact with the stigmas while the flowers open, with the stamens separating and breaking off as the fertilized capsules expand. The plants exhibit a broad range of phenotypic plasticity in vegetative characters, making it necessary to rely heavily on the more stable seed features for taxonomic purposes. Owing to much taxonomic confusion in the literature and electronic databases records, the proper assignment of information (especially for *Reported associates*) among *E. americana, E. rubella,* and *E. triandra* is doubtful in many cases and should be regarded appropriately. However, because the ecology and ranges of these species overlap substantially, the reported information applies to all three taxa in many cases.

Elatine ambigua **Wight** is a nonindigenous species, which occurs in canals, ditches, marshes, meadows, rice fields, streams, vernal pools, washes, and along the margins of ponds at elevations of up to 100 m. The plants can grow completely submersed to at least 1 m in depth. The substrates include clay and mud. Flowering occurs from June to August. The plants are not colonized by arbuscular mycorrhizal fungi or dark septate endophytes. There is little additional ecological information available for this species. **Reported associates (California):** *Bacopa eisenii, Cyperus difformis,*

Eleocharis, Lindernia dubia, Oryza sativa, Sagittaria monte-vidensis, Utricularia.

Elatine americana (Pursh) Arn. occurs in shallow (to 1 m) tidal or nontidal waters in or on estuaries, flats, lakes, meadows, mudflats, ponds, sloughs, swales, and along the margins of lakes, reservoirs, rivers, and streams at elevations of up to 900 m. The substrates are described as clayey peat, mud, muddy gravel, humus-rich sand, sand, and silt. Flowering extends throughout summer to early fall. The flowers are chasmogamous when emersed but cleistogamous when submersed. The plants appear to be primarily self-pollinated in either case. The seeds of *E. americana* have physiological dormancy but should germinate after receiving 210 days of cold stratification when placed under a 19°C/15°C temperature regime. **Reported associates:** *Aeschynomene virginica, Allium schoenoprasum, Alopecurus pratensis, Ammannia, Astragalus labradoricus, Bacopa innominata, Bidens eatoni, B. frondosus, B. hyperboreus, Callitriche palustris, C. stagnalis, C. verna, Carex viridula, Cephalanthus occidentalis, Ceratophyllum, Chara, Cornus amomum, Crassula aquatica, Cyperus bipartitus, Deschampsia cespitosa, Elatine minima, Eleocharis engelmannii, E. palustris, Epilobium ecomosum, Equisetum variegatum, Eriocaulon aquaticum, E. parkeri, Festuca, Gentiana victorinii, Gratiola neglecta, G. virginiana, Hypnum, Isoetes riparia, I. tuckermani, Juncus, Lemna, Limosella subulata, Lindernia dubia, Ludwigia palustris, Menyanthes trifoliata, Micranthemum micranthemoides, Myriophyllum, Najas, Nuphar, Peltandra virginica, Persicaria punctata, Pluchea odorata, Pontederia cordata, Potamogeton perfoliatus, Ruppia maritima, Sagittaria graminea, S. subulata, Schoenoplectus pungens, S. smithii, Spartina alterniflora, Spiranthes, Symphyotrichum puniceum, S. subulatum, Typha angustifolia, T. latifolia, Vallisneria americana, Viburnum lentago, Zizania aquatica.*

Elatine brachysperma A. Gray grows as an emergent or submersed plant in shallow (to 0.6 m), fresh to brackish waters of flats, depressions, gullies, lakes, marshes, meadows, mud flats, ponds, roadsides, shores, stock ponds, streams, swales, vernal pools, wallows, woodlands, and along the margins of lakes, ponds, and reservoirs at elevations of up to 3037 m. The plants typically are exposed to full sun at sites where the standing waters have receded, but can tolerate periods of inundation for up to 40 days in duration. The reported substrates include clay, cobble, mud, sand, and silt. Flowering and fruiting occur from May to October. The species is self-pollinating within its cleistogamous flowers. Previously shed seed germinates within 2–3 days when exposed to water, with natural germination rates observed as highest during February and March. **Reported associates:** *Agrostis viridis, Alopecurus saccatus, Ambrosia psilostachya, Ammannia coccinea, Anagallis minima, Artemisia californica, A. dracunculus, Asclepias fascicularis, Atriplex rosea, Baccharis pilularis, B. salicifolia, Brodiaea terrestris, Bromus hordeaceus, Callitriche heterophylla, C. longipedunculata, C. marginata, C. peploides, C. verna, Carex alma, C. pellita, Castilleja densiflora, Coreopsis tinctoria, Cotula coronopifolia, Crassula aquatica, C. drummondii, C. solieri, Crypsis schoenoides, Cuscuta*

pentagona, Cyperus squarrosus, Deinandra fasciculata, Deschampsia danthonioides, Distichlis spicata, Downingia laeta, Echinodorus berteroi, Elatine californica, Eleocharis acicularis, E. engelmannii, E. macrostachya, E. palustris, E. parishii, Elodea bifoliata, Epilobium ciliatum, E. pygmaeum, Eriogonum fasciculatum, Erodium botrys, E. brachycarpum, Eryngium armatum, E. vaseyi, Glyceria elata, Gnaphalium exilifolium, G. palustre, Hordeum brachyantherum, Horkelia truncata, Isoetes orcuttii, I. howellii, Juncus balticus, J. bufonius, J. dubius, J. phaeocephalus, Juniperus monosperma, Kochia californica, Lasthenia californica, L. conjugens, L. glaberrima, Lemna minuta, Lepidium virginicum, Leptochloa fusca, L. uninervia, Lilaea scilloides, Limosella acaulis, Linanthus dianthaflorus, Lindernia dubia, Lolium multiflorum, Lythrum hyssopifolia, L. portula, Madia glomerata, Malacothamnus fasciculatus, Malosma laurina, Malvella leprosa, Marsilea vestita, Mimulus floribundus, Mollugo verticillata, Myosurus minumus, Myriophyllum sibericum, Najas guadalupensis, N. marina, Navarretia hamata, N. prostrata, Orcuttia californica, Osmadenia tenella, Persicaria amphibia, P. lapathifolia, P. punctata, Phalaris lemmonii, Pilularia americana, Pinus edulis, Plagiobothrys bracteatus, P. chorisianus, P. hispidulus, P. scouleri, P. trachycarpus, P. undulatus, Poa pratensis, Pogogyne serpylloides, Polypogon monspeliensis, Potamogeton illinoensis, P. nodosus, Pseudognaphalium luteoalbum, Psilocarphus brevissimus, P. chilensis, Rhus integrifolia, Riccia nigrella, Rumex acetosella, R. crispus, R. maritimus, R. salicifolius, Salix gooddingii, S. laevigata, S. lasiolepis, Schoenoplectus acutus, S. californicus, Sedum smallii, Sisyrinchium bellum, Stachys ajugoides, Stuckleya pectinata, Suckleya suckleyana, Taraxacum officinale, Trifolium, Typha latifolia, Verbena bracteata, Verbesina encelioides.

Elatine californica A. Gray grows as submersed or emergent plants in shallow waters (0.5–1.0 m) of bottoms, depressions, dunes, embankments, marshes, meadows, mudflats, ponds, pools, rice fields, shores, swales, vernal pools, wallows, and along the margins of lakes, ponds, reservoirs, and streams at elevations of up to 2637 m. The substrates are described as alkaline and include clay, claypan, Corning gravelly loam, granite, mud (often drying), sand, sandstone, and vertisols. Flowering and fruiting occur from March to August. The plants root at the nodes and can form dense mats on shallow bottoms or when they grow on exposed sediments. **Reported associates:** *Acmispon americanus, Alisma, Alopecurus howellii, A. saccatus, Artemisia tridentata, Brodiaea filifolia, Callitriche longipedunculata, C. marginata, C. verna, Chara contraria, Crassula aquatica, Deinandra, Deschampsia danthonioides, Downingia bella, D. cuspidata, D. pulchella, Echinodorus cordifolius, Elatine brachysperma, E. chilensis, E. rubella, Eleocharis acicularis, E. macrostachya, E. palustris, Elodea, Eriogonum fasciculatum, Eryngium aristulatum, E. vaseyi, Festuca, Gleotrichia, Glyceria, Gnaphalium palustre, Helenium, Hordeum brachyantherum, H. marinum, Iris missouriensis, Isoetes howellii, I. orcuttii, Juncus bufonius, J. sphaerocarpus, J. uncialis, J. xiphioides, Lamarckia aurea, Lasthenia californica, L. fremontii, L. glabrata, Lilaea scilloides, Limosella acaulis, L. aquatica, Lomatium* *utriculatum, Lythrum hyssopifolium, Marsilia oligospora, M. vestita, Mimulus guttatus, M. tricolor, Monardella lanceolata, Montia fontana, Myosurus minimus, Myriophyllum, Navarretia breweri, N. intertexta, N. leucocephala, Orcuttia californica, Persicaria, Phalaris lemmonii, Pilularia americana, Plagiobothrys scouleri, P. stipitatus, P. trachycarpus, P. undulatus, Plantago bigelovii, Poa annua, Pogogyne douglasii, P. ziziphoroides, Potamogeton diversifolius, P. natans, P. pusillus, P. richardsonii, Psilocarphus brevissimus, Quercus agrifolia, Q. douglasii, Q. engelmannii, Ranunculus aquatilis, R. californicus, Rhamnus ilicifolia, Rhus aromatica, Rumex acetosella, Sagittaria, Salix, Schoenoplectus californicus, Sisyrinchium bellum, Stuckenia pectinata, Symphyotrichum frondosum, Taraxia tanacetifolia, Toxicodendron diversilobum, Trifolium variegatum, Typha, Utricularia, Veronica peregrina, Zannichellia palustris.*

***Elatine chilensis* Gay** is a nonindigenous species, which occurs as emergent or submersed forms in shallow waters of meadows, mudflats, ponds, stock ponds, vernal pools, and along the shores of ponds and reservoirs at elevations of up to 2690 m. Typical sites receive full sun and have muddy substrates. Flowering and fruiting extend from April to September. The plants probably are self-pollinating. Few additional details are known on the ecology of this introduced species. **Reported associates (North America):** *Elatine californica, Eleocharis acicularis, Epilobium, Juncus, Juniperus, Lilaeopsis, Limosella, Persicaria, Populus, Potamogeton, Rosa.*

***Elatine heterandra* H. Mason** grows emersed or in shallow (to 0.3 m) waters of ditches, lakes, marshes, meadows, mudflats, seeps, vernal pools, and along the margins of lakes, ponds, and reservoirs at elevations of up to 1523 m. The plants occur characteristically on exposed, drying bottoms or shores, where standing waters have receded. The substrates have been described as granite, gravel, Keefers loam, mud, sand, and stony Tuscan loam. Flowering and fruiting occur from April to July. Further details on the ecology of this species are scarce. **Reported associates:** *Anagallis minima, Callitriche heterophylla, Crassula aquatica, Cuscuta howelliana, Cyperus squarrosus, Deschampsia danthonioides, Downingia bella, Eleocharis macrostachya, Eryngium aristulatum, Juncus, Limosella acaulis, Lindernia dubia, Lythrum hyssopifolia, Marsilea vestita, Mimulus guttatus, Mollugo verticillata, Myriophyllum hippuroides, Orcuttia tenuis, Plagiobothrys, Potamogeton foliosus, P. diversifolius, Ranunculus muricatus, Rorippa curvisiliqua, Rotala ramosior, Sagittaria cuneata, Stachys ajugoides, Utricularia macrorhiza, Veronica peregrina, Zeltnera muehlenbergii.*

***Elatine minima* (Nutt.) Fisch. & C.A. Mey.** occurs as emersed or submersed growth forms in a wide range of habitats ranging from tidal to nontidal standing waters (to 2 m deep) of lakes and ponds to exposed flats, gravel pits, mudflats, and along the margins of lakes, rivers, and streams at elevations of up to at least 93 m. The plants occur in full sun on gravel, muck, mud, sand, sandy silt, sandy peat, silt, or stony substrates and in waters of broad acidity (pH: 4.6–8.6), low alkalinity (60 ppm CaCO₃), low color (5–60 Hazen units), and low conductivity (<150 μmhos/cm). Typical ion levels in the water

are: Al (1–16 μeq/L), Ca (20–43 μeq/L), Cl (107–138 μeq/L), Fe (1–8 μeq/L), HCO₃ (0–16 μeq/L), K (3–5 μeq/L), Mg (28–41 μeq/L), Na (109–130 μeq/L), NH₄ (0.7–2.1 μeq/L), and SO₄ (52–69). Flowering and fruiting occur from June to October. Chasmogamous flowers are produced by emergent plants; whereas, the flowers of submersed plants are cleistogamous. Both floral types are self-pollinating with their anthers coming into contact with the stigmas as the filaments elongate. The plants are highly sensitive to disturbance (e.g., conservatism C value of 9). In some lakes their colonization is facilitated (in August) by exploiting the nests abandoned (in June) by sunfish (Vertebrata: Osteichthyes: Centrarchidae), which clear vegetation annually during their construction, thereby creating open sites suitable for seedling establishment. Individuals reach substantially higher densities when growing in shallower (e.g., 1 m) water (500–2500 plants/m²) than when in deeper (e.g., 2 m) water (0.05–5 plants/m²). **Reported associates:** *Bidens beckii, Brasenia schreberi, Callitriche heterophylla, Cephalanthus occidentalis, Ceratophyllum demersum, C. echinatum, Chara, Cladium mariscoides, Cornus amomum, Crassula aquatica, Cyperus filicinus, Elatine americana, Eleocharis acicularis, E. palustris, E. parvula, Elodea canadensis, E. nuttallii, Equisetum fluviatile, Eriocaulon aquaticum, E. parkeri, Fontinalis novaeangliae, Glossostigma cleistanthum, Glyceria borealis, G. canadensis, Gratiola aurea, Heteranthera dubia, Isoetes acadiensis, I. ×eatoni, I. echinospora, I. lacustris, I. macrospora, I. riparia, I. tuckermanii, Juncus brevicaudatus, J. militaris, J. pelocarpus, Limosella aquatica, L. australis, Lindernia dubia, Littorella americana, Lobelia dortmanna, Ludwigia palustris, Marsilea quadrifolia, Micranthemum micranthemoides, Myriophyllum alterniflorum, M. farwellii, M. sibiricum, M. spicatum, M. tenellum, Najas flexilis, N. gracillima, Nitella, Nuphar variegata, Nymphaea odorata, Nymphoides cordata, Nyssa, Peltandra virginica, Persicaria amphibia, P. punctata, Pontederia cordata, Potamogeton alpinus, P. amplifolius, P. bicupulatus, P. confervoides, P. crispus, P. epihydrus, P. foliosus, P. gramineus, P. illinoensis, P. natans, P. oakesianus, P. perfoliatus, P. praelongus, P. pusillus, P. richardsonii, P. robbinsii, P. spirillus, P. vaseyi, P. zosteriformis, Ranunculus cymbalaria, R. longirostris, R. reptans, Sagittaria graminea, S. latifolia, S. subulata, Schoenoplectus subterminalis, Sium suave, Sparganium americanum, S. angustifolium, S. fluctuans, Sphagnum, Stuckenia pectinata, Subularia aquatica, Taxodium, Utricularia geminiscapa, U. intermedia, U. macrorhiza, U. minor, U. purpurea, U. radiata, U. resupinata, Vallisneria americana, Viburnum lentago.*

***Elatine ojibwayensis* Garneau.** No life history information is available at this time.

***Elatine rubella* Rydb.** grows in fresh (e.g., 160 μS/cm) to saline sites including cattle tanks, coulees, ditches, dunes, flats, lakes, marshes, mudflats, playas, prairies, rice fields, shores, sloughs, swales, swamps, vernal pools, and the margins of lakes, ponds, reservoirs, and streams at elevations of up to 3049 m. The plants grow both as amphibious emergents and submersed, shallow water forms (to 2 m depths). The substrates are described as alkaline (e.g., pH: 8.0) and include basalt, clay, gravelly loam, mud, sand, sandy loam, and vertisols. The stems root at the nodes and can develop into mats. Flowering and fruiting occur from April to July. **Reported associates:** *Alisma, Alopecurus carolinianus, A. pratensis, Ambrosia, Artemisia dracunculus, A. ludoviciana, Baccharis salicifolia, B. sergiloides, Bacopa rotundifolia, Callitriche verna, Carex occidentalis, C. rostrata, Cirsium neomexicanum, Conium maculatum, Coreopsis tinctoria, Crassula aquatica, C. longipes, Cyperus acuminatus, Datura wrightii, Distichlis, Echinochloa muricata, Elatine californica, Eleocharis acicularis, E. engelmannii, E. macrostachya, E. palustris, Elodea, Epilobium campestre, Eryngium aristulatum, Glyceria, Gnaphalium palustre, Gratiola neglecta, Heteranthera limosa, Hordeum, Iris missouriensis, Juniperus monosperma, Lasthenia glabrata, Limosella acaulis, L. aquatica, Lindernia dubia, Lobelia, Lythrum tribracteatum, Marsilea, Mimulus guttatus, M. pilosus, Mollugo verticillata, Muhlenbergia rigens, Myosurus minimus, Myriophyllum, Navarretia intertexta, Pascopyrum smithii, Phacelia distans, Pinus ponderosa, Persicaria amphibia, P. coccinea, P. pensylvanica, P. lapathifolia, Plagiobothrys hispidulus, P. scouleri, P. stipitatus, Poa fendleriana, Pogogyne douglasii, Populus fremontii, Potamogeton diversifolius, P. foliosus, P. nodosus, P. pusillus, Ranunculus aquatilis, R. flammula, Rorippa curvisiliqua, Rumex stenophyllus, Sagittaria calycina, Salix gooddingii, S. laevigata, Schoenoplectus californicus, S. mucronatus, Sparganium angustifolium, Spergularia, Stuckenia pectinata, Subularia aquatica, Typha angustifolia, Utricularia, Veronica americana, V. peregrina, Xanthium.*

***Elatine triandra* Schkuhr** is an amphibious species, which grows in tidal or nontidal, shallow waters (2–10 dm depths) of kettle ponds, lakes, mudflats, ponds, ricefields, slough bottoms, streams, swales, swamps, vernal pools, and along the margins of lakes, ponds, rivers, and streams at elevations of up to 2670 m. The habitats are circumneutral to somewhat alkaline (pH: 7.0–8.5). The substrates have been categorized as humus-rich sand and silt. The plants will form mats by rooting at the nodes of prostrate stems. Flowering and fruiting extend from summer to early fall. Self-pollination occurs in submersed, cleistogamous flowers, where the elongating stamens bring the anthers into contact with the stigmas. The pollen grains germinate *in situ* and the pollen tubes grow through the anther tissue and into the stigma. The seeds of *E. triandra* have been observed to persist in wetland seed banks (to depths over 3 m) even following fairly extensive periods of farming. Emergence of *E. triandra* seedlings can be prevented by the application of alfalfa (*Medicago sativa*) or Kava (*Piper methysticum*) pellets at the rate of 1–2 tons per hectare. *Elatine triandra* is known to survive in waters having a relatively high arsenic concentration but the mechanism of detoxification has not been determined. **Reported associates:** *Alopecurus pratensis, Ammannia, Bacopa innominata, Callitriche palustris, Cardamine pensylvanica, Ceratophyllum demersum, Chara, Cyperus, Egeria densa, Eleocharis engelmannii, E. palustris, Elodea nuttallii, Gratiola, Hydrodictyon, Justicia americana, Lemna, Limosella aquatica, Lindernia dubia, Mimulus*

guttatus, Myriophyllum aquaticum, M. sibiricum, M. spicatum, Najas flexilis, Nitella, Nuphar polysepala, Nymphaea odorata, Nymphoides peltata, Potamogeton crispus, P. foliosus, P. pusillus, P. richardsonii, Ranunculus aquatilis, Sagittaria latifolia, S. subulata, Stuckenia pectinata, Vallisneria americana.

Use by wildlife: The foliage and seeds of various *Elatine* species are eaten by several waterfowl (Aves: Anatidae), including black ducks (*Anas rubripes*), buffleheads (*Bucephala albeola*), canvasbacks (*Aythya valisineria*), and mallards (*Anas platyrhynchos*). The plants also provide habitat for zooplankton and young fish (Vertebrata: Osteichthyes). *Elatine americana* is a host plant for moth larvae (Insecta: Lepidoptera: Pyralidae: *Acentria nivea*).

Economic importance: food: none; **medicinal:** none; **cultivation:** Various *Elatine* species (including *E. americana* and *E. triandra*) are cultivated as aquarium plants; **misc. products:** *Elatine triandra* is considered to be a beneficial plant for stabilizing shorelines; **weeds:** *Elatine ambigua* can become a minor weed of rice fields as it has in California, Czech Republic and Slovakia, Italy, Russia, and southeast Asia. In Japan, strains of *Elatine triandra* are resistant to herbicides [acetolactate synthase (ALS) inhibitors] and have become serious rice-field pests since 1998; **nonindigenous species:** *Elatine ambigua*, a native of eastern and southern Asia, was introduced to California irrigation canals and rice fields sometime prior to the 1940s. Molecular data recently have verified its occurrence in Connecticut, Massachusetts, South Carolina, and Virginia, from among previously misidentified specimens. It also has been introduced to Australia, Czech Republic and Slovakia, Finland, Italy, Romania, and Russia. *Elatine chilensis* has been introduced to the western United States and Mexico from South America.

Systematics: Various molecular data resolve *Elatine* and *Bergia* as sister genera phylogenetically (see discussion for *Bergia* above), and as a sister clade to Malpighiaceae. Interestingly, the growth of pollen tubes through the anther tissue (as in *E. triandra*) is a rare phenomenon but also occurs in Malpighiaceae. Species delimitation in *Elatine* is difficult and has not been widely agreed upon with some authors preferring to merge taxa where others prefer to recognize several distinct species (e.g., *E. americana, E. brachysperma, E. triandra*). Because of their morphological simplicity and phenotypic plasticity, a practical taxonomy for *Elatine* has been difficult to elucidate. Previous taxonomic treatments have divided species into sections based on their number of leaves and stamens, but these classifications have been disputed recently on the basis of phylogenetic analyses. Combined molecular and morphological data resolve most interspecific relationships well (Figure 4.26), but indicate an allopolyploid origin for *E. americana* and *E. hexandra*. Patterns of morphological variation can be influenced by the reproductive system; for example, the seed morphology of the self-pollinating *E. brachysperma* is highly uniform within populations but quite variable between them. Because of extensive phenotypic plasticity, morphology can also vary considerably between submersed and emergent plants of the same species.

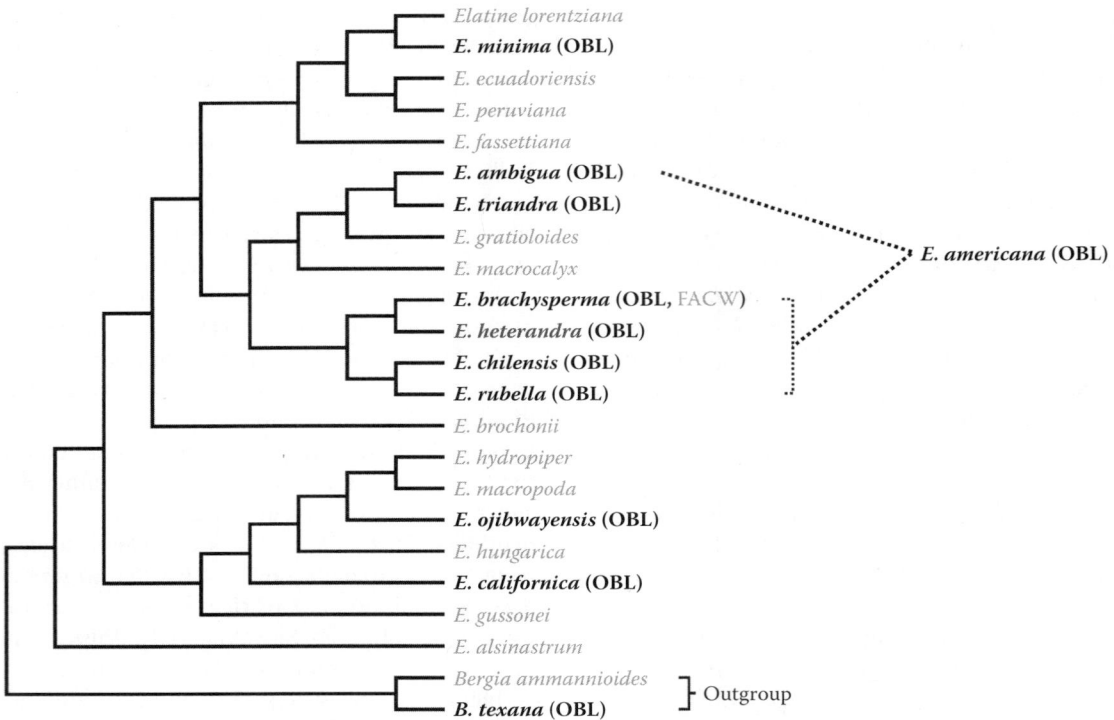

FIGURE 4.26 Phylogenetic relationships in *Elatine* as indicated by analysis of combined morphological and molecular data. Because all *Elatine* species are obligate aquatics, the scattered distribution of OBL North American indicators (in bold) is owing to their interspersal among non-North American taxa. *Elatine americana* is an allopolyploid derived from the lineages indicated by the dotted line. The Old World *E. hexandra* (not shown) is also of allopolyploid origin. (Adapted from Razifard, H., Systematics of *Elatine* L. (Elatinaceae), PhD dissertation, The University of Connecticut, 2016.)

The base chromosome number of *Elatine* is *x* = 9. *Elatine triandra* (2*n* = 40, 54) is a tetraploid or hexaploid, *E. ambigua* (2*n* = 54) is a hexaploid, and *E. americana* (2*n* = 70–72) is an octoploid. Chromosome numbers have not been determined for other North American *Elatine* species.

Comments: *Elatine triandra* is fairly widespread except for northwestern North America. *Elatine americana* is distributed in eastern and northern North America (but also is reported from Australia and New Zealand). *Elatine minima* is also northeastern. *Elatine brachysperma* is widespread except in the northeast, and also occurs in Argentina and Mexico. Western North American species include *E. californica* and *E. chilensis* (both widespread and extending to Mexico), *E. heterandra* (California, New Mexico, and Texas) and *E. rubella* (widespread). *Elatine ojibwayensis* is known only from a few localities in Ontario, Canada. The nonindigenous *E. ambigua* presently is restricted to California.

References: Barbour et al., 2005; Barrett & Seaman, 1980; Bliss & Zedler, 1998; Borman et al., 1997; Brouwer et al., 2002; Broyles, 1987; Carpenter & McCreary, 1985; Carpenter & Titus, 1984; Catling et al., 1986; Cook, 1973; Duncan, 1964; Fassett, 1939; Kai & Zhiwei, 2006; Keeley, 1999; Mason, 1955; Molnár et al., 2015; Nichols, 1999a,b; Razifard, 2016; Razifard et al., 2016; Rosman et al., 2016; Sheldon & Boylen, 1977; Swindale & Curtis, 1957; Tucker, 1986; Whitehouse, 1933; Wilcox & Meeker, 1981; Williams, 2006; Zheng et al., 2003.

Family 2: Euphorbiaceae [222]

Euphorbiaceae (spurges) are a large family of more than 6100 predominantly tropical species, which include only a few aquatics. More than half the species occur in *Croton* and *Euphorbia*, the two largest genera. The flowers are unisexual (in monoecious or dioecious arrangements), with bifid or multiple divided styles. In many species the small flowers aggregate into larger pseudanthia, which are known as cyathia or as "*Euphorbia*-type" inflorescences. There is a single ovule per locule, and the fruits usually are elastically dehiscent schizocarps (Judd et al., 2016). Many of the species are lactiferous, with some producing toxic alkaloids or cyanogenic glycosides (Judd et al., 2016).

The phylogenetic relationships of the family indicate that it should be included within Malpighiales rather than Malvales as some authors had concluded previously (Figure 4.25). Broader familial concepts of the family were shown to be polyphyletic, resulting in the current recognition of several former subfamilial segregates each at the rank of family (e.g., Phyllanthaceae, Picrodendraceae). A combined analysis of 82 plastid gene sequences (Xi et al., 2012) resolved Euphorbiaceae as the sister group of Peraceae, which is well supported and seems to be the most compelling result obtained thus far. Understandably, a comprehensive phylogenetic analysis of the speciose Euphorbiaceae has not yet been completed.

Euphorbiaceae have been divided into five subfamilies with *Euphorbia* and *Stillingia* (which contain the OBL species) both assigned to subfamily Euphorbioideae. The assignment of each genus to a different tribe (*Euphorbia*: tribe Euphorbieae; *Stillingia*: tribe Hippomaneae) indicates that their aquatic habit has arisen independently in the family; however, this conclusion should be substantiated once a more comprehensive phylogenetic analysis of intergeneric relationships in Euphorbiaceae is available. Tribe Euphorbieae includes the cyathium-bearing species (i.e., those taxa containing a cyathial inflorescence) and appears to be monophyletic on the basis of morphological phylogenetic analyses (Park & Backlund, 2002). The same analyses indicate that tribes Euphorbieae and Hippomaneae may comprise sister taxa phylogenetically. Molecular data (Dorsey et al., 2013) have provided compelling evidence to merge the former genus *Chamaesyce* within the larger genus *Euphorbia*, in which it is nested phylogenetically. Although one OBL species was attributed to *Chamaesyce* in the 1996 wetland indicator list, this taxon has since been transferred to *Euphorbia* (as *E. hooveri*), where it retains its OBL indicator status.

Euphorbiaceae are important economically as the source of natural rubber (*Hevea brasiliensis*) and ingredients used in the manufacture of tung oils (*Vernicii fordii*). Although the family (especially *Euphorbia*) contains a number of poisonous species, there are also several edible plants such as *Manihot esculenta* (cassava or manioc), which are rich in carbohydrates (starch). This plant is also the source of tapioca. Many species are grown as ornamentals, primarily in the genera *Acalypha*, *Codiaeum*, *Croton*, *Euphorbia* (which includes *Poinsettia*), *Jatropha*, and *Ricinis*. The latter genus is also notorious as the source of a highly toxic lectin known as ricin, which is produced by the seed endosperm of *Ricinus communis*. The pressed seeds of this species also yield castor oil, which is the primary source of ricinoleic acid.

Only two North American genera of Euphorbiaceae contain obligate aquatics:

1. ***Euphorbia*** L.
2. ***Stillingia*** Garden ex L.

1. Euphorbia

Spurge, sandmat

Etymology: after Euphorbus, the physician to king Juba II of Mauritania (1st century BC–AD).

Synonyms: *Chamaesyce*

Distribution: global: cosmopolitan; **North America:** widespread

Diversity: global: 2000 species; **North America:** 146 species

Indicators (USA): OBL: *Euphorbia hooveri*

Habitat: freshwater; lacustrine, palustrine; **pH:** broad range; **depth:** <1 m; **life-form(s):** emergent herb

Key morphology: Stems prostrate to decumbent, containing milky sap, arising from a tap root; leaves (to 5 mm) opposite, round, margins serrate; stipules fused, fringed; inflorescence a cyathium with a bell-shaped involucre (to 2 mm), containing white, three- to five-lobed appendages and petaloid, red to olive glands (<1 mm); flowers unisexual with 30–35 male flowers surrounding a single female flower; fruit (to 2 mm) spherical

Life history: duration: annual (fruit/seeds); **asexual reproduction:** none; **pollination:** insect; **sexual condition:** monoecious; **fruit:** schizocarps (common); **local dispersal:** ballistic; **long-distance dispersal:** birds

Imperilment: (1) *Euphorbia hooveri* [G2]; S2 (CA)

Ecology: general: *Euphorbia* is an extremely large and diverse genus of annuals and perennials, which mainly comprise tropical, terrestrial plants. The species contain an assortment of C_3, C_4, and crassulacean acid metabolism (CAM) photosynthetic mechanisms and colonize many types of habitats. Numerous members exhibit adaptations to hot, sunny conditions, with only 24 of the North American species (16%) designated as wetland indicators and just one species (<1%) ranked as OBL. Although the single OBL species does not grow as a submersed plant and cannot tolerate inundation, its life cycle is tied inextricably to deeper vernal aquatic habitats, which reduce the level of interspecific competition substantially and eventually (upon drying), provide open mud flat sites that are suitable for seed germination. The seeds are adapted well to prolonged inundation and germinate reliably once the ephemeral waters have receded. They are locally dispersed by ballistic ejection by the elastically dehiscent fruit walls, and to greater distances by animals, which consume them.

***Euphorbia hooveri* L.C. Wheeler** is a summer annual, which inhabits depressions, vernal pools (0.19–243 ha), and wallows, which occur on remnant alluvial fans and depositional stream terraces in the foothills of the Sierra Nevada at elevations of up to 131 m. The plants occupy acidic (iron silica cemented hardpan) or neutral to saline–alkaline (lime–silica cemented hardpan or claypan) soils ranging from clay to sandy loam (Anita, Laniger, Lewis, Madera, Meikle, Riz, Tuscan, Whitney, and Willows soil series). Competition from other species is reduced at these sites by the periodic, and prolonged cycles of inundation. Flowering and fruiting occur from May through October. Larger plants may produce several hundred seeds, which do not germinate until the water evaporates from the pools. The resulting plants can develop into mats extending from 5 to 100 cm on the mud flats but cannot grow in standing water. The seeds can remain dormant for at least 4 years. This species is threatened by habitat loss with only 30 known extant populations remaining. Some populations are also impacted by invasive species such as *Convolvulus arvensis* and *Xanthium strumarium*. **Reported associates:** *Alopecurus saccatus*, *Amaranthus albus*, *Anthoxanthum odoratum*, *Asclepias fascicularis*, *Convolvulus arvensis*, *Croton setiger*, *Crypsis schoenoides*, *Distichlis spicata*, *Downingia*, *Eleocharis macrostachya*, *Epilobium cleistogamum*, *Eryngium spinosepalum*, *E. vaseyi*, *Euphorbia ocellata*, *Frankenia salina*, *Gratiola heterosepala*, *Grindelia camporum*, *Hemizonia pungens*, *Hordeum depressum*, *H. geniculatum*, *Lilaea scilloides*, *Lippia nodiflora*, *Marsilea vestita*, *Navarretia leucocephala*, *Neostapfia colusana*, *Orcuttia pilosa*, *Plagiobothrys stipitatus*, *Polypogon monspeliensis*, *Proboscidea louisianica*, *Psilocarpus brevissimus*, *Sida hederacea*, *Trichostema lanceolatum*, *Tuctoria greenei*, *Xanthium strumarium*.

Use by wildlife: The flowers of *Euphorbia hooveri* are visited by various insects (Insecta), including bees and wasps (Hymenoptera), beetles (Coleoptera), butterflies and moths (Lepidoptera), and flies (Diptera). The seeds of *E. hooveri* are eaten by horned larks (Aves: Alaudidae: *Eremophila alpestris*), which presumably act as their dispersal agents.

Economic importance: food: *Euphorbia hooveri* is not edible. The milky juice of some species is at least partially toxic to mammalian herbivores and also can cause contact dermatitis in humans; **medicinal:** A number of North American *Euphorbia* species have been used medicinally but not *E. hooveri*, which likely is a consequence due at least in part to its rarity; **cultivation:** *E. hooveri* is not in cultivation; **misc. products:** none; **weeds:** none; **nonindigenous species:** none.

Systematics: Although the distinctness of *Chamaesyce* and *Euphorbia* has been debated for many years, the most recent phylogenetic analyses of molecular and morphological data indicate similarly that the genera should be merged (Figure 4.27). The former often is treated taxonomically as a subgenus of the latter (i.e., subgenus *Chamaesyce*), which is monophyletic and can be distinguished morphologically by the presence (unless lost secondarily) of petaloid appendages beneath the cyathial glands. Initially subdivided into three informal groups designated as the *Arcuta*, *Hypericifolia*, and *Peplis* subclades, subgenus *Chamaesyce* subsequently was revised to reflect formal sectional and subsectional taxa. As a result, *Euphorbia hooveri* has been assigned to subsection *Hypericifoliae* of section *Anisophyllum*, a group of 365 species. Molecular data also indicate that *E. hooveri* is likely to be of hybrid origin. Even though it is distinct morphologically, different nuclear alleles (for exon 9 of EMB2765) associate either with those most similar to *E. serpens* (FACW, FAC, FACU, UPL) or to *E. albomarginata*, which occur in closely related but distinct clades of subsection *Hypericifoliae*. The base chromosome number of *Euphorbia* is given as $x = 10$, but varies considerably among the different species. Although the chromosome number of *E. hooveri* has not been determined, counts for its putative ancestral species *E. albomarginata* ($2n = 24$, 36, 48) and *E. serpens* ($2n = 12$, 16, 22, 24) contain variable cytotypes.

Comments: *Euphorbia hooveri* is endemic to California.

References: Alexander & Schlising, 1998; Dorsey et al., 2013; Manson, 2003; Motley & Raz, 2004; Park & Elisens, 2000; Silveira, 2000a; Stone et al., 1988; Urbatsch et al., 1975; Webster et al., 1975; Yang & Berry, 2011; Yang et al., 2012.

2. *Stillingia*

Corkwood, toothleaf

Etymology: after Benjamin Stillingfleet (1702–1771)

Synonyms: none

Distribution: global: SE Asia; Madagascar; New World; **North America:** southern

Diversity: global: 30 species; **North America:** 7 species

Indicators (USA): OBL: *Stillingia aquatica*

Habitat: freshwater; palustrine; **pH:** alkaline; **depth:** <1 m; **life-form(s):** emergent (subshrub or shrub)

Key morphology: stem (to 15 dm) single, glabrous, with upper fascicles of branches; leaves alternate, crowded, petioled (to 4 mm), the blades (to 8 cm) lanceolate, margins minutely crenulate; glandular to filiform stipules present; inflorescence a terminal red, yellow, or green spike; flowers unisexual (the females occurring below the males), subtended by a bract, the female flowers solitary, the male flowers solitary or in small clusters; sepals 2–3; corolla absent; fruit (to 1 cm) a short ovate, septicidal capsule; seeds with a caruncle

Life history: duration: perennial (buds); **asexual reproduction:** none; **pollination:** insect; **sexual condition:** monoecious; **fruit:** capsules (common); **local dispersal:** gravity; **long-distance dispersal:** insects

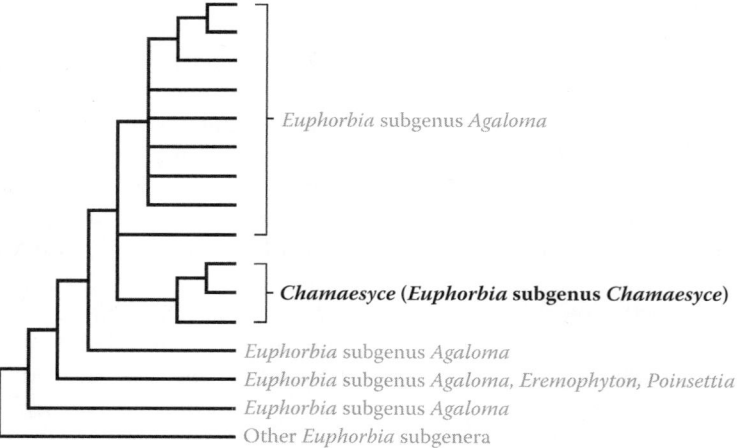

Euphorbia subgenus *Agaloma*

Chamaesyce (*Euphorbia* subgenus *Chamaesyce*)

Euphorbia subgenus *Agaloma*
Euphorbia subgenus *Agaloma, Eremophyton, Poinsettia*
Euphorbia subgenus *Agaloma*
Other *Euphorbia* subgenera

FIGURE 4.27 Cladogram derived from morphological data showing the phylogenetic placement of *Chamaesyce* (now treated as *Euphorbia* subgenus *Chamaesyce*) from within the former *Euphorbia* section *Agaloma*. Taxa containing obligate aquatics in North America are shown in bold. Although *Euphorbia* (*Chamaesyce*) *hooveri* was not included in the analysis, subsequent studies incorporating molecular data (Yang & Berry, 2011) substantiate the recognition of *Chamaesyce* as a subgenus within *Euphorbia*, and resolve *Euphorbia hooveri* within subsection *Hypericifoliae* of section *Anisophyllum* within the subgenus. (Adapted from Park, K.-R. & Elisens, W.J., *Int. J. Plant Sci.*, 161, 425–434, 2000.)

Imperilment: (1) *Stillingia aquatica* [G4/G5]; S1 (AL); S2 (SC); S3 (GA)

Ecology: general: Most *Stillingia* species grow in dry UPLs or under xerophytic conditions, so it is rather unusual to see an aquatic representative in the genus. However, although only one North American species is a wetland indicator, it is an OBL indicator. Few details exist on the reproductive ecology, but the presence of nectary glands indicates pollination by insects. One species reportedly is pollinated by ants (Insecta: Hymenoptera: Formicidae). The seeds are carunculate and are likely to be dispersed by insects.

Stillingia aquatica **Garden ex L.** occupies shallow (typically standing) waters in open, grassy habitats including barrens, barrier basins, canals, depressions, ditches, dunes, flatwoods, marshes, ponds, prairies, roadsides, savannahs, scrub, swales, swamps, and the margins of ponds at low elevations. The substrates have been described as calcareous and include muck, sand, Scranton series (Humaqueptic Psammaquents), and sandy peat. This species is a good indicator of natural wetlands that are characterized by low levels of disturbance. Flowering and fruiting occur from July to October. Further details on the reproductive biology and ecology of this species are poorly known and merit additional study. **Reported associates:** *Acrostichum danaeifolium, Amphicarpum muhlenbergianum, Annona glabra, Aristida affinis, A. palustris, A. patula, Asclepias longifolia, Calopogon, Cladium jamaicense, Cleistes, Clethra alnifolia, Cliftonia monophylla, Coreopsis nudata, Cyperus haspan, Cyrilla racemiflora, Dichanthelium acuminatum, D. erectifolium, D. sphaerocarpon, Drosera capillaris, Eleocharis baldwinii, E. vivipara, Eriocaulon compressum, E. decangulare, Fuirena scirpoidea, Gaylussacia dumosa, Hypericum brachyphyllum, H. denticulatum, H. fasciculatum, Ilex myrtifolia, Iris, Juncus scirpoides, Justicia ovata, Lachnanthes caroliniana, Leitneria floridana, Lobelia boykinii, L. floridana, Ludwigia, Lyonia lucida, Mnesithea rugosa, Muhlenbergia capillaris, Myrica heterophylla, Nyssa biflora, N. ursina, Oxypolis canbyi, O. greenmanii, Panicum hemitomon, P. tenerum, Pieris phillyreifolia, Pinus elliottii, P. palustris, P. serotina, Platanthera chapmanii, Pluchea baccharis, Polygala cymosa, Pontederia cordata, Rhexia aristosa, Rhynchospora decurrens, R. divergens, R. inundata, R. microcarpa, R. tracyi, Sabal palmetto, Sabatia decandra, S. quadrangula, S. stellaris, Sagittaria lancifolia, Scleria baldwinii, S. bellii, Syngonanthus flavidulus, Taxodium ascendens, T. distichum, Utricularia corniculata, Vaccinium myrsinites, Veratrum, Verbesina chapmanii, Xyris fimbriata, X. stricta.*

Use by wildlife: none reported

Economic importance: food: not edible; **medicinal:** The roots of some *Stillingia* species contain active compounds and have been used as various medicinals; however, no uses are reported specifically for *S. aquatica*; **cultivation:** *Stillingia aquatica* is planted occasionally as an ornamental shrub; **misc. products:** none; **weeds:** none; **nonindigenous species:** none

Systematics: The placement of *Stillingia* within tribe Hippomaneae of subfamily Euphorbioideae has been substantiated by several morphological phylogenetic analyses. With a few exceptions, the monophyly of tribe Hippomaneae has been confirmed by molecular analyses, which resolve two sister subclades (H1, H2) within the tribe (*Stillingia* falls within the H2 subclade). However, intergeneric relationships within Hippomaneae remain unsettled. Previous analyses have placed *Stillingia* either closest to *Falconeria* (Figure 4.28) or to *Excoecaria* and *Sebastiana* (*Falconeria* excluded); however, taxon sampling was sparse in these cases. In more densely sampled molecular phylogenetic analyses (but still excluding *Falconeria*), *Stillingia* resolved as paraphyletic, with different species associating also with *Adenopeltis, Sapium,* or *Spegazziniophytum*. All four genera have staminate flowers with two stamens and more or less sessile pistillate flowers. Arguably, *Stillingia* species can be distinguished by the presence of a woody, triangular fruit base known as the

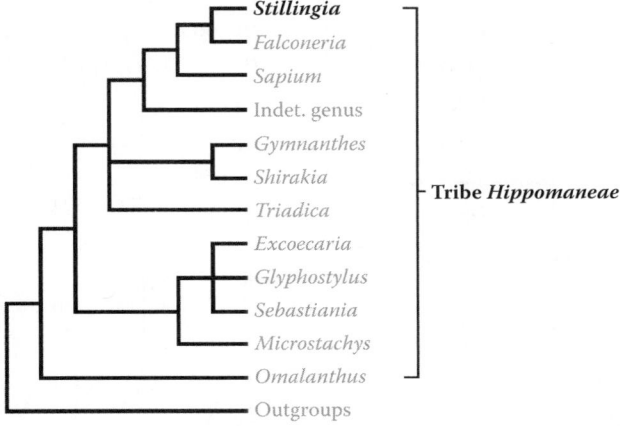

FIGURE 4.28 Intergeneric relationships within Euphorbiaceae subfamily Euphorbioideae (tribe Hippomaneae) reconstructed by phylogenetic analysis of morphological data. The position of *Stillingia* (shown in bold), which is the only genus of the tribe to contain obligate aquatic plants in North America, is relatively derived in the tribe (*Stillingia aquatica* was not included in the analysis). Subsequent molecular phylogenetic analyses (Wurdack et al., 2005) resolve *Stillingia* as paraphyletic. (Adapted from Esser, H.-J. et al., *Syst. Bot.*, 22, 617–628, 1997.)

carpidiophore (which also occurs in *Adenopeltis*), and a stami-nate calyx (lacking in *Adenopeltis*). Clearly, the circumscrip-tion of *Stillingia* should be reconsidered, which will require the sampling of additional taxa, including *S. aquatica*. Yet, phylogenetic studies consistently indicate that *Stillingia*, the only genus of tribe Hippomaneae to contain obligate aquat-ics, occupies a relatively derived position in Euphorbiaceae. The base chromosome number of *Stillingia* is $x = 6$; the chro-mosome number of *S. aquatica* has not been determined.

Comments: *Stillingia aquatica* is distributed narrowly in the southeastern United States.

References: Drewa et al., 2002; Esser et al., 1997; LeBlond et al., 2015; Park & Backlund, 2002; Urbatsch et al., 1975; Wurdack et al., 2005.

Family 3: Hypericaceae [9]

Hypericaceae comprise 540 species, which are mostly terres-trial and distributed primarily in temperate regions (Judd et al., 2016). The family exhibits a wide diversity of annual and peren-nial herbs, shrubs, and small trees. The species characteristi-cally have simple, opposite (or whorled) leaves marked by black or pellucid dots. The flowers are showy, with numerous, con-spicuous stamens having dorsifixed anthers, and small, punc-tate stigmas terminating distinct (or slightly fused) styles (Judd et al., 2016). The fruits include berries, capsules, and drupes.

The taxonomic circumscription of Hypericaceae has been problematic with the family being treated either as distinct, or as a subfamily of Clusiaceae (i.e., Clusioideae, Hypericoideae). Molecular analyses have clarified this question to some degree by indicating that Clusiaceae and Hypericaceae do not resolve as a clade. Analyses of *rbcL* sequence data (Gustafsson et al., 2002) placed Hypericaceae apart from Clusiaceae in a clade that includes the highly modified aquatic family Podostemaceae (Figure 4.29). However, the *rbcL* data provided essentially no support for the indicated monophyly of the larger group (i.e., Clusiaceae + Hypericaceae + Podos-temaceae). Other analyses using combined nuclear (*PHYC*) and chloroplast (*rbcL*, *ndhF*) sequence data (Figure 4.25) resolved Hypericaceae in a clade with Bonnetiaceae while placing Podostemaceae more distantly among members of Ochnaceae (which is remote from Podostemaceae by *rbcL* data alone). A fairly comprehensive analysis that included 82 cpDNA sequences (Xi et al., 2012; Figure 4.25) also resolved Hypericaceae and Podostemaceae as sister families in a clade distinct from both Clusiaceae and Ochnaceae. Although fur-ther analyses may be necessary to definitively resolve the issue of relationships within this group, thus far the phylo-genetic studies have consistently indicated the distinctness of Clusiaceae and Hypericaceae, which is the disposition fol-lowed here.

The large flowers are pollinated primarily by bees and wasps (Insecta: Hymenoptera), which seek their nutritious pollen. The small seeds are dispersed abiotically by water or wind (Judd et al., 2016). Numerous *Hypericum* species are cultivated as ornamentals. One species of "St. John's Wort" (*H. perforatum*) has become widely touted as a medicinal treatment for mild depression in humans, although its efficacy has not been supported adequately by clinical trials.

Among the North American representatives of Hypericaceae, obligate aquatics occur within two closely related genera (which are merged by some authors):

1. ***Hypericum*** L.
2. ***Triadenum*** Raf.

An analysis of nrITS sequence data (Nürk et al., 2013) con-cluded that *Triadenum* was nested among *Hypericum* spe-cies, which has inspired a number of authors to merge the genera. However, the species assigned to *Triadenum*, *Brathys*, and *Myriandra* (other segregate genera) resolve in a clade that is sister to remainder of *Hypericum* (with the exception of two Mediterranean species), which does not necessarily preclude their continued recognition as separate genera (see Figure 4.30), especially since the internal support for the branch resolving the placement of *Triadenum* was not well supported in that analysis. *Triadenum* is retained here, but the remainder of the OBL species, which resolve within the *Brathys* + *Myriandra* clade, have been retained in *Hypericum* provisionally, pending more refined circumscriptions for *Brathys* and *Myriandra*.

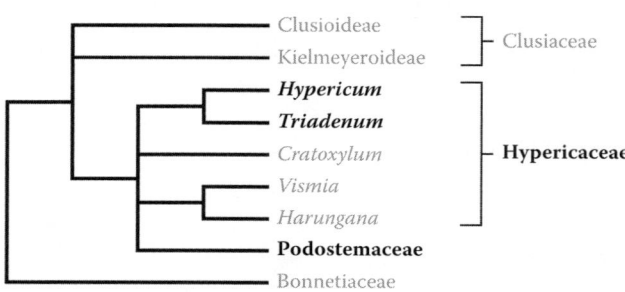

FIGURE 4.29 Phylogenetic relationships of Clusiaceae, Hypericaceae, and Podostemaceae as indicated by *rbcL* sequence data. The association of Hypericaceae and Podostemaceae is also obtained by analysis of combined cpDNA sequence data (Figure 4.25). Both gen-era of Hypericaceae that contain obligate aquatic species in North America (as indicated by bold type) resolve within a single clade with Podostemaceae, which are entirely aquatic. (Adapted from Gustafsson, M.H.G. et al., *Int. J. Plant Sci.*, 163, 1045–1054, 2002.)

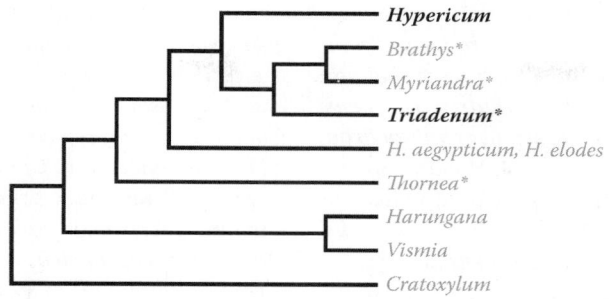

FIGURE 4.30 Phylogenetic relationships of *Hypericum* and several segregate genera (indicated by asterisks) condensed from analysis of nrITS sequence data from 206 species. Aside from two species (*Hypericum aegypticum, H. elodes*), the retention of the segregate genera as distinct from *Hypericum* is supported by these results. Taxa containing North American OBL indicators are indicated in bold. (Adapted from Nürk, N.M. et al., *Mol. Phylogen. Evol.*, 66, 1–16, 2013.)

1. *Hypericum*

St. John's-wort; millepertuis

Etymology: from the Greek *hyper eicon* meaning "above picture" with reference to the practice of hanging the plant over pictures as a charm

Synonyms: *Ascyrum; Lianthus; Sanidophyllum; Sarothra; Triadenia*

Distribution: global: cosmopolitan; **North America:** widespread

Diversity: global: 360 species; **North America:** 60 species

Indicators (USA): OBL: *Hypericum adpressum, H. anagalloides, H. boreale, H. chapmanii, H. edisonianum, H. ellipticum, H. galioides, H. gymnanthum, H. lissophloeus, H. nitidum, H. setosum; H. tetrapetalum;* **FACW:** *H. gymnanthum, H. setosum*

Habitat: freshwater; lacustrine, palustrine; **pH:** 3.6–7.0; **depth:** 0–2 m; **life-form(s):** emergent (herb, shrub)

Key morphology: stems herbaceous (to 1.0 m) or woody (to 4.0 m), limp (herbs) or spongy (shrubs) when growing in water; leaves (2.0–7.5 cm) sessile, broadly elliptic to needle-like, deciduous or evergreen, sometimes punctate dotted; flowers (to 3.0 cm) showy, solitary or in cymes; sepals (to 1.2 cm) 4–5, often deciduous; petals (to 1.8 cm) 4–5, yellow or orange; stamens numerous; styles (to 4.0 mm) 3–4 (or connivent as 1); capsules (to 8.0 mm) ovate to ellipsoid

Life history: duration: annual (fruit/seeds); perennial (buds, stolons); **asexual reproduction:** adventitious shoots, stolons; **pollination:** insect, self; **sexual condition:** hermaphroditic; **fruit:** capsules (common); **local dispersal:** seeds (wind, water), stolons; **long-distance dispersal:** seeds (wind, water)

Imperilment: (1) *Hypericum adpressum* [G3]; SX (PA); SH (CT, KY, NC, WV); S1 (AR, GA, IL, IN, MD, MI, MO, TN, VA); S2 (DE, MA, NJ, NY, RI, SC); (2) *H. anagalloides* [G4]; S1 (UT); S2 (AZ); (3) *H. boreale* [G5]; SH (WV); S1 (DE, IA); S2 (OH, VA); (4) *H. chapmanii* [G3]; S3 (FL); (5) *H. edisonianum* [G2]; S2 (FL); (6) *H. ellipticum* [G5]; SH (DE); S1 (PE, TN); S2 (NF, NJ, OH, RI); (7) *H. galioides* [G5]; (8) *H. gymnanthum* [G4]; SX (NY); S1 (IN, NJ, PA); S2 (DE, OH); S3 (MD, NC); (9) *H. lissophloeus* [G2]; S2 (FL); (10) *H. nitidum* [G4]; S1 (SC); S2 (AL); (11) *H. setosum* [G4/G5]; S1 (VA); (12) *H. tetrapetalum* [G5]

Ecology: general: *Hypericum* is highly diverse with species ranging from slight annual herbs to large evergreen shrubs. Although primarily a terrestrial genus, many species have fairly wide habitat tolerances and occur at times in wetlands. Forty of the North American species (67%) are wetland indicators, with 12 (20%) categorized as OBL. The species enumerated here show the greatest dependency on wet habitats for their continued survival, and several are able to survive and grow capably as truly submersed aquatic plants. The large, showy flowers of *Hypericum* are adapted for insect pollination by bees and wasps (Insecta: Hymenoptera); however, the extent of self-pollination has not been ascertained for many of the species. Few details are reported on the seed ecology of the OBL aquatic species. For most of the *Hypericum* species that have been studied, the seeds have physiological dormancy and require some period of cold stratification as a prerequisite to germination. In several species, the germination rates are enhanced when the seeds are immersed in water. Vegetative reproduction can occur in the perennial species by the formation of stolons or by adventitious shoots that arise from horizontally oriented roots.

Hypericum adpressum **W.P.C. Barton** is an erect perennial herb, which occurs mainly in ditches, ephemeral pools, or along the shores and shallow margins of ponds on the eastern coastal plain (but in disjunct sites further inland as well) at elevations of up to 173 m. The plants thrive in open sites on substrates described as gravel, peat, rocky gravel, and sand. Under stable hydrological conditions this species will persist as robust plants, but also will grow as diminutive plants along exposed shores, which experience fluctuating water levels. Flowering and fruiting extend from July to September. The plants survive prolonged periods of high water through a dormant seed bank. Vegetative reproduction occurs by stolons. This species survives in fewer than 50 populations and is threatened by recreational activities and competition by the invasive *Phragmites australis* (Poaceae). **Reported associates:** *Agalinis, Agrostis scabra, Aletris farinosa, Andropogon gerardii, Boltonia asteroides, Cladium mariscoides, Coleataenia longifolia, Coreopsis rosea, Cornus, Cyperus dentatus, Dichanthelium acuminatum, Drosera intermedia, Dulichium arundinaceum, Eleocharis engelmannii, E. microcarpa, E. obtusa, Eragrostis* *hirsuta, Eriocaulon septangulare, Eupatorium hyssopifolium, E. leucolepis, Euthamia graminifolia, Fragaria virginiana, Gratiola aurea, G. brevifolia, Hydrocotyle umbellata, Hypericum boreale, Juncus biflorus, J. militaris, J. repens, Liatris spicata, Lobelia dortmanna, Ludwigia sphaerocarpa, Lycopus amplectens, Lysimachia terrestris, Monarda punctata, Oldenlandia uniflora, Panicum longifolium, P. scabriusculum, Persicaria hydropiperoides, Phragmites australis, Polygala sanguinea, Pontederia cordata, Potentilla simplex, Pteridium aquilinum, Pycnanthemum setosum, Rhexia mariana, R. virginica, Rhynchospora capitellata, R. inundata, R. nitens, Sabatia, Sagittaria engelmanniana, S. teres, Scleria reticularis, Scirpus etuberculatus, Solidago tenuifolia, Spartina pectinata, Spiraea tomentosa, Strophostyles helvola, Toxicodendron vernix, Tradescantia ohiensis, Triadenum walteri, Viola lanceolata, V. sagittata, Xyris caroliniana, X. torta.*

Hypericum anagalloides Cham. & Schltdl. is a weak annual or perennial herb, which grows in or on beaches, bogs, bottoms, cienega, depressions, ditches, fens, flats, floodplains, gravel bars, marshes, meadows, prairies, roadsides, sand bars, seeps, slopes, springs, swales, swamps, thickets, washes, or along the margins of brooks, lakes, ponds, reservoirs, rivers, streams, and vernal pools at elevations of up to 3600 m. The stems float in shallow water and the plants also have been observed growing on floating logs. The sites are exposed to full sun or in light shade. The substrates are acidic (e.g., pH: 5.3–5.8) and include clay, clay loam, decomposed granite, gravel, Greenleaf manzanita serpentine, humus, loam, muck, mud, peat, peaty loam, rock, rocky loam, sand, sandy loam, silty clay loam, silt loam, slate, and Tuscan mudflow. Flowering and fruiting extend from May to November. The shoots are highly stoloniferous and are capable of covering the ground and developing into dense mats. **Reported associates:** *Abies concolor, A. lasiocarpa, A. magnifica, A. procera, Acer glabrum, Achillea millefolium, Achnatherum occidentale, Acmispon americanus, Aconitum, Agoseris glauca, Agrostis exarata, A. idahoensis, A. scabra, A. thurberiana, Aira caryophyllea, Alisma, Alnus rhombifolia, A. tenuifolia, A. viridis, Alopecurus, Ambrosia psilostachya, Angelica breweri, Antennaria corymbosa, A. rosea, Anthoxanthum odoratum, Anticlea occidentalis, Arctostaphylos nevadensis, A. patula, Arenaria, Arnica longifolia, A. mollis, Artemisia arbuscula, A. ludoviciana, A. tridentata, Asclepias cordifolia, Athyrium filix-femina, Aulacomnium palustre, Baccharis salicifolia, Barbarea orthoceras, Betula glandulosa, B. nana, Bidens frondosus, Bistorta bistortoides, Blechnum spicant, Botrychium multifidum, Boykinia major, B. occidentalis, Brasenia schreberi, Bromus marginatus, B. vulgaris, Calamagrostis breweri, C. canadensis, C. nutkaensis, Calandrinia, Callitriche, Calocedrus decurrens, Caltha leptosepala, Camassia quamash, Campanula rotundifolia, C. canescens, C. cusickii, C. densa, C. disperma, C. echinata, C. fracta, C. illota, C. integra, C. interior, C. jonesii, C. lasiocarpa, C. lenticularis, C. leptalea, C. limosa, C. livida, C. luzulina, C. microptera, C. nebrascensis, C. nigricans, C. occidentalis, C. pellita, C. praegracilis, C. schottii, C. scopulorum, C. senta, C. spectabilis, C. straminiformis, C. subfusca,*

C. unilateralis, C. utriculata, C. vesicaria, Cardamine oligosperma, Carex aquatilis, C. athrostachya, C. aurea, C. breweri, C. buxbaumii, Castilleja miniata, Ceanothus cordulatus, Chamerion angustifolium, Cheilanthes gracillima, Chrysolepis sempervirens, Cicuta douglasii, Cirsium vulgare, Claytonia lanceolata, Collinsia tinctoria, Comarum palustre, Coptis aspleniifolia, Cornus sericea, Cryptogramma acrostichoides, C. crispa, Cynosurus echinatus, Cyperus eragrostis, Danthonia californica, D. unispicata, Darlingtonia californica, Deschampsia cespitosa, D. elongata, Descurainia incisa, Dichanthelium acuminatum, Drepanocladus, Drosera anglica, D. rotundifolia, Drymocallis ashlandica, D. lactea, Dulichium arundinaceum, Eleocharis acicularis, E. quinqueflora, Elodea, Epilobium anagallidifolium, E. ciliatum, Epipactis, Equisetum arvense, E. fluviatile, E. hyemale, E. laevigatum, Ericameria nauseosa, Erigeron divergens, E. peregrinus, Eriogonum fasciculatum, Eriophorum chamissonis, Erodium cicutarium, Festuca rubra, Frasera speciosa, Galium trifidum, G. triflorum, Gaultheria humifusa, Gayophytum diffusum, Gentiana calycosa, G. fremontii, G. newberryi, G. sceptrum, Gentianella amarella, Gentianopsis simplex, Geranium richardsonii, Glyceria borealis, G. elata, Gratiola, Habenaria, Hackelia micrantha, Hastingsia alba, Helenium bigelovii, Heracleum sphondylium, Holcus lanatus, Horkelia rydbergii, H. truncata, Hosackia oblongifolia, Hypericum boreale, H. formosum, H. majus, H. scouleri, Isoetes occidentalis, Ivesia campestris, I. unguiculata, Juncus acuminatus, J. balticus, J. bufonius, J. dubius, J. ensifolius, J. hesperius, J. laccatus, J. macrandrus, J. macrophyllus, J. mertensianus, J. nevadensis, J. orthophyllus, J. parryi, J. phaeocephalus, J. supiniformis, J. textilis, J. xiphioides, Juniperus occidentalis, Kalmia microphylla, Keckiella breviflora, Kyhosia bolanderi, Lepidium, Leucanthemum vulgare, Leymus triticoides, Ligusticum canbyi, L. grayi, L. tenuifolium, Lilium kelleyanum, L. parryi, Liparis loeselii, Lonicera conjugialis, L. involucrata, Lotus corniculatus, Lupinus burkei, L. latifolius, L. polyphyllus, Luzula campestris, Lycopus americanus, L. uniflorus, Lysichiton americanus, Lythrum, Madia elegans, Maianthemum racemosum, M. stellatum, Marchantia, Medicago lupulina, Meesia triquetra, Melica, Mentha piperita, M. pulegium, Menyanthes trifoliata, Menziesia ferruginea, Mertensia ciliata, Micranthes ferruginea, M. odontoloma, M. oregana, Mimulus breweri, M. guttatus, M. moschatus, M. pilosus, M. primuloides, Montia chamissoi, Muhlenbergia andina, M. filiformis, M. richardsonis, M. rigens, Myosotis discolor, Myrica gale, Myriophyllum farwellii, M. hippuroides, M. verticillatum, Najas flexilis, Nasturtium officinale, Nuphar polysepala, Olsynium douglasii, Oreostemma alpigenum, Orobanche californica, Oxypolis occidentalis, Packera paupercula, P. pseudaurea, Parnassia palustris, Pedicularis attollens, P. groenlandica, P. racemosa, Penstemon heterodoxus, P. montanus, P. newberryi, P. procerus, Perideridia parishii, Philonotis fontana, Phleum alpinum, Phyllodoce empetriformis, Picea engelmannii, Pinus albicaulis, P. contorta, P. coulteri, P. jeffreyi, P. lambertiana, P. monticola, Plantago lanceolata, Platanthera dilatata, P. sparsiflora, Poa annua, P. pratensis, Polytrichum commune, Populus fremontii, P. tremuloides, Potamogeton berchtoldii, P. natans, P. epihydrus, Potentilla anserina, P. drummondii, P. flabellifolia, P. grayi, P. wheeleri, Primula jeffreyi, Prunella vulgaris, Pteridium aquilinum, Pterospora andromedea, Quercus agrifolia, Q. chrysolepis, Q. engelmannii, Ranunculus alismifolius, R. flammula, R. populago, R. repens, R. uncinatus, Rhododendron albiflorum, R. columbianum, R. groenlandicum, R. macrophyllum, R. occidentale, Rhynchospora alba, R. capitellata, Ribes cereum, R. nevadense, R. viscosissimum, Rosa woodsii, Rubus spectabilis, Rumex acetosella, Sagina saginoides, Salix drummondiana, S. exigua, S. geyeriana, S. gooddingii, S. lasiandra, S. lasiolepis, S. lemmonii, S. lutea, S. myrtillifolia, S. orestera, S. scouleriana, Sambucus racemosa, Sanguisorba officinalis, Scheuchzeria palustris, Schoenoplectus acutus, S. subterminalis, Scirpus microcarpus, Senecio integerrimus, S. triangularis, Sidalcea campestris, S. malviflora, Silene gallica, Sisyrinchium angustifolium, S. bellum, S. longipes, Solidago canadensis, S. confinis, S. velutina, Sparganium angustifolium, S. emersum, S. natans, Spergularia rubra, Sphagnum squarrosum, S. warnstorfii, Sphenosciadium capitellatum, Spiraea douglasii, Spiranthes porrifolia, S. romanzoffiana, Sporobolus cryptandrus, Stachys ajugoides, S. albens, Stellaria longifolia, S. longipes, Suksdorfia ranunculifolia, Symphoricarpos albus, Symphyotrichum spathulatum, Taraxacum officinale, Thalictrum fendleri, Thermopsis montana, Thuja plicata, Torreyochloa pallida, Toxicodendron diversilobum, Trianthia glutinosa, T. occidentalis, Trichophorum, Trifolium incarnatum, T. repens, T. willdenovii, T. wormskioldii, Trillium ovatum, Tsuga mertensiana, Typha latifolia, Urtica dioica, Utricularia intermedia, U. macrorhiza, U. minor, Vaccinium myrtillus, V. ovatum, V. oxycoccos, V. scoparium, V. uliginosum, Valeriana, Veratrum californicum, Verbascum thapsus, Veronica americana, V. serpyllifolia, Viburnum edule, Vicia americana, Viola adunca, V. macloskeyi, V. palustris, Xerophyllum tenax, Zeltnera venusta.

Hypericum boreale (Britton) E.P. Bicknell is a decumbent perennial herb, which occurs in shallow (to 60 cm depth) sites in bogs, cranberry fields, depressions, ditches, gravel pits, interdunal ponds, lakes, marshes, meadows, roadsides, ruts, sinkholes, swales, swamps, and the margins of lakes, ponds, and rivers at elevations of up to 180 m. Individuals also can grow as submersed plants, which become limp and elongate, resembling *Callitriche* in habit. Exposures range from full sun to partial shade. The substrates are acidic (pH: 5.0–6.5) and include clay, muck, peat, peaty sand, sand, and silty mud. The plants also can grow on floating sedge mats and rotting logs. Flowering and fruiting occur from July to October. The flowers produce many seeds, which germinate well to yield numerous seedlings when conditions permit. The seedbank can extend from 60 to 120 cm in depth and reach densities of 173 seeds/m². Seeds collected during May have germinated successfully after storage at 4°C followed by incubation under a 12/12 h, 20°C/5°C (with the latter increased to 10°C after 4 weeks) light and temperature regime. Shrub cover results in etiolation of the shoots and reduced or arrested sexual and vegetative reproduction. The plants increase significantly in

frequency following shrub removal treatments, which result in faster growth rates and higher seed masses. Fragments of the plants can be dispersed during periods of high water and can result in asexual reproduction. Quite different associates can be encountered when the plants grow as submersed vs. emergent life forms. **Reported associates:** *Agrostis capillaris, Alnus rubra, Ambrosia artemisiifolia, Andropogon virginicus, Anthoxanthum odoratum, Bidens cernuus, B. connatus, B. frondosus, Boehmeria cylindrica, Brasenia schreberi, Bulbostylis capillaris, Calamagrostis canadensis, Callitriche heterophylla, Carex howei, C. longii, C. obnupta, C. vulpinoidea, Ceratophyllum echinatum, Cyperus dentatus, C. diandrus, C. erythrorhizos, C. esculentus, C. strigosus, Cephalanthus occidentalis, Coleataenia longifolia, Deschampsia cespitosa, Dichanthelium acuminatum, Drosera intermedia, Dulichium arundinaceum, Echinochloa crus-galli, Echinodorus tenellus, Eleocharis acicularis, E. flavescens, E. obtusa, E. palustris, E. quadrangulata, E. tenuis, Epilobium ciliatum, Equisetum arvense, Erechtites hieracifolia, Euthamia graminifolia, Fimbristylis autumnalis, Fraxinus nigra, F. pennsylvanica, Galium trifidum, Glyceria borealis, Gratiola aurea, G. neglecta, Hypericum adpressum, H. anagalloides, H. canadense, H. majus, H. virginicum, Iris virginica, Isolepis cernua, Juncus acuminatus, J. articulatus, J. bufonius, J. canadensis, J. effusus, J. militaris, J. occidentalis, J. pelocarpus, J. supiniformis, Leersia oryzoides, Lemna minor, Lilaeopsis occidentalis, Lindernia dubia, Lipocarpha micrantha, Lobelia cardinalis, L. inflata, Lotus, Ludwigia alternifolia, L. palustris, Lycopodium appressum, Lycopus uniflorus, Lysimachia terrestris, Mentha arvensis, Myriophyllum humile, Najas canadensis, N. gracillima, Nuphar advena, N. variegata, Nymphaea odorata, Oenanthe, Onoclea sensibilis, Panicum dichotomiflorum, P. philadelphicum, P. verrucosum, Penthorum sedoides, Persicaria amphibia, P. hydropiperoides, P. maculosa, P. punctata, Pinus sylvestris, Poa annua, Pogonia ophioglossoides, Polygala cruciata, Potamogeton bicupulatus, P. diversifolius, Potentilla anserina, Proserpinaca palustris, Puccinellia fernaldii, Quercus palustris, Q. velutina, Ranunculus repens, Rhexia virginica, Rhynchospora alba, R. capitellata, R. macrostachya, R. scirpoides, Rorippa palustris, Rotala ramosior, Rubus spectabilis, Rumex acetosella, Sagittaria latifolia, Salix hookeriana, Scirpus cyperinus, S. torreyi, Schoenoplectus smithii, Scleria reticularis, Sium suave, Sisyrinchium californicum, Sphagnum rubellum, S. compactum, Sparganium americanum, S. androcladum, Spiraea douglasii, S. tomentosa, Symphyotrichum dumosum, Thelypteris palustris, Tillaea aquatica, Triadenum fraseri, Trifolium dubium, T. hybridum, T. wormskioldii, Triglochin maritimum, Utricularia radiata, U. subulata, Vaccinium macrocarpon, Veronica scutellata, Viola lanceolata.*

Hypericum chapmanii W.P. Adams is a large shrub, which occurs in shallow waters (to 60 cm) of borrow pits, depressions, flats, flatwoods, savannas, swamps, and along pond margins at low elevations to at least 2 m. Some sites are characterized by seasonal waters and can appear dry at times of the year. The habitats are fairly acidic (pH: 3.6–5.6). The flowering and fruiting period have not been characterized. The plants reach reproductive maturity at 10 years of age. Individuals produce relatively few flowers but the flowers produce numerous seeds. The seeds are dispersed by gravity or by birds (Aves) and develop into a persistent seed bank. Higher field germination rates have been observed immediately following disturbances that provide open space, with lower germination noted during periods free of disturbance. Adult plants are highly sensitive to fire and are killed completely by even low intensity blazes. Postfire recruitment occurs exclusively from seeds. Although seedlings can survive in UPL sites, they are destroyed during fire episodes. The lower fire frequency associated with lowland sites accounts for the predominantly wetland habit of this species. Although narrowly restricted in distribution (to a portion of the Florida "panhandle"), this species is long lived (>10 years) and can become abundant and dominant locally. **Reported associates:** *Aristida, Clethra alnifolia, Cliftonia monophylla, Cyrilla racemiflora, Dichanthelium, Drosera capillaris, D. tracyi, Eriocaulon compressum, E. lineari, Fraxinus caroliniana, Hypericum brachyphyllum, H. fasciculatum, H. galioides, H. nitidum, Ilex coriacea, I. decidua, I. glabra, I. myrtifolia, Leucothoe axillaris, Lophiola americana, Lycopodium alopecuroides, Lyonia lucida, Magnolia virginiana, Myrica cerifera, M. heterophylla, Nyssa biflora, Persea palustris, Pinguicula ionantha, P. planifolia, Pinus elliottii, Rhynchospora, Sarracenia flava, S. psittacina, Smilax, Styrax americanus, Taxodium ascendens, Utricularia, Xyris ambigua, X. caroliniana, X. serotina.*

***Hypericum edisonianum* (Small) W.P. Adams & N. Robson** is a small shrub, which grows in open depressions, flatwoods, prairies, seepage slopes, and along the margins of ephemeral (seasonal) lakes and ponds at low elevations. None of the sites is characterized by permanent standing water. The substrates are described as sandy. Flowering and fruit production extend throughout the year. However, few (1–3) flowers are produced individually by only a small proportion of plants (~30%) annually. Most of the flowers are produced by a small proportion of genets in the populations, with larger plants producing greater numbers of flowers. The perianths have alternating ultraviolet (UV) patterns that are visible to their insect pollinators (see *Use by wildlife*). The outer petal surfaces are UV absorbant, whereas the inner surfaces are UV reflective, which presumably enables the insects to distinguish between open and closed flowers. The flowers are self-incompatible and remain receptive for less than a day. Only about 58% of the flowers produce seed, averaging 62 seeds/capsule. Low seed production probably reflects populations that are structured with few genets, resulting in limited opportunities for cross compatibility. The seeds are dispersed passively by gravity or wind. Some dispersal might occur by ants (Insecta: Hymenoptera: Formicidae), which occasionally harvest them (but mostly consume them). The seed shadow develops in close proximity to the parental plants. Despite estimates that indicate the potential of plants to produce 30,422 seedlings/m², much lower observed levels (e.g., 277 seedlings/m²) indicate extensive mortality owing to desiccation, fire, inundation, pathogens, predation, and other factors. Fair germination (47%) has been achieved after 30 days for

seeds kept in moist petri dishes under ambient light and temperature conditions. However, transplanted seedlings exhibit poor survival. The plants can form dense, clonal thickets (more than 0.4 ha in extent) by the production of horizontal, adventitious rhizomes (to 210 cm long). Clonal growth is essential for the maintenance of local, natural stands, and is essential for recovery following periodic fires, which kill the aboveground portions. The plants exhibit a relatively continuous recovery following fire. Interconnectivity among ramets is about 56%. Periods of high water and interspecific competition both are linked to decreased plant density. A strong inverse relationship exists between plant density and accumulated aboveground biomass, which results from increased mortality during extended periods of high water levels or competition. This species is threatened by fire suppression, grazing, pasture improvement, and wetland drainage. **Reported associates:** *Andropogon brachystachyus, Calopogon multiflorus, Eupatorium recurvans, Gordonia lasianthus, Hartwrightia floridana, Hypericum cistifolium, H. fasciculatum, Ilex cassine, Lachnanthes caroliniana, Lilium catesbaei, Panicum abscissum, P. hemitomon, Persea palustris, Pinguicula caerulea, P. lutea, Pinus elliotii, Platanthera blephariglottis* var. *conspicua, Platanthera ciliaris, P. cristata, P. integra, Pluchea rosea, Pogonia ophioglossoides, Quercus geminata, Rhexia cubensis, Sabatia grandiflora, Sarracenia minor, Serenoa repens, Sphagnum.*

***Hypericum ellipticum* Hook.** is a stoloniferous perennial herb, which grows in bogs, meadows, shores, swamps, and along the margins of lakes, ponds, reservoirs, rivers, and streams at elevations of up to 519 m. In shallow waters the plants also occur as a submersed, sterile form having smaller, rounder leaves. The substrates are typically acidic gravel or sand with an organic content ranging from 0.2% to 4.6%. The highest biomass production has been observed at sites having an intermediate organic matter content (0.8%–2.7%). Flowering and fruiting occur from June to August. The seeds are small (averaging 0.03 mg) and do not appear to form a significant seed bank. A low germination rate (38%) was reported for seeds receiving 9 months of cold stratification (4°C), then incubated under a 20°C/30°C daily temperature cycle and 15/9 h light/dark cycle. However, 100% germination has been achieved for similarly stratified seeds placed under a 12/12 h, 25°C/15°C light/temperature cycle. The seedlings have a relative growth rate of 0.35 g/g d⁻¹. **Reported associates:** *Acer rubrum, Achillea millefolium, Acorus calamus, Agalinus paupercula, Agrostis hyemalis, A. scabra, Alnus rugosa, Betula lenta, Bolboschoenus fluviatilis, Cacalia suaveolens, Calamagrostis canadensis, Callitriche, Carex baileyi, C. crinita, C. debilis, C. hyalinolepis, C. interior, C. lasiocarpa, C. lurida, C. michauxiana, C. scoparia, C. stipata, C. stricta, C. tenera, C. tribuloides, C. utriculata, C. vulpinoidea, Chamaedaphne calyculata, Cirsium arvense, Cornus racemosa, Drosera rotundifolia, Dryopteris cristata, Eleocharis acicularis, E. obtusa, E. palustris, Elymus virginicus, Equisetum sylvaticum, Euthamia graminifolia, Fragaria virginiana, Galium asprellum, G. palustre, G. tinctorium, Gentiana linearis, Geranium maculatum, Glyceria canadensis, G. grandis, G. melicaria, G. striata, Hamamelis virginiana,*

Hydrocotyle americana, Hypericum mutilum, Impatiens capensis, Iris versicolor, Juncus acuminatus, J. effusus, Leersia oryzoides, Lindera benzoin, Lycopus americanus, Lyonia ligustrina, Lysimachia ciliata, L. terrestris, Mentha arvensis, Mimulus ringens, Myrica gale, Onoclea sensibilis, Ophioglossum vulgatum, Osmunda regalis, Osmundastrum cinnamomeum, Oxalis stricta, Panicum boreale, Persicaria sagittata, Picea, Platanthera flava, Potentilla norvegica, Prunella vulgaris, Rubus idaeus, Sabatia dodecandra, Sagittaria latifolia, Salix, Scirpus cyperinus, Scutellaria galericulata, S. lateriflora, Senecio pauperculus, Solidago rugosa, Sparganium americanum, Sphagnum, Spiraea alba, S. tomentosa, Thalictrum pubescens, Thelypteris palustris, Triadenum fraseri, Tsuga, Utricularia minor, Vaccinium corymbosum, Verbena hastata, Viola lanceolata.

Hypericum galioides Lam. is an evergreen shrub found in tidal or nontidal bogs, bottomlands, ditches, flats, flatwoods, floodplains, hammocks, marshes, meadows, prairies, roadsides, sandbars, savannah, seeps, sinkholes, sloughs, swales, swamps, woodlands, and along the margins of lakes, ponds, rivers, and streams across the eastern coastal plain at elevations of up to 92 m. The sites are seasonally ponded with shallow (e.g., 3–6 dm) water. The plants normally occur in full sun or in openings, but can tolerate partial shade. The substrates include Foreston/Woodington series, loamy sand, mud, peat, sand, sandy alluvium, sandy clay, sandy mud, sandy peat, sandy rock, sandy silt, silty clay loam, and silty sand. Flowering and fruiting extend from April to November. A persistent seedbank does not develop. **Reported associates:** *Acer rubrum, Aletris aurea, Alnus rugosa, A. serrulata, Andropogon glaucopsis, A. liebmannii, Anthaenantia rufa, Apios americana, Aristida palustris, A. purpurascens, Arnoglossum ovatum, Aronia arbutifolia, Arundinaria gigantea, Asclepias, Baccharis halimifolia, Betula nigra, Bigelowia nudata, Campsis radicans, Carex glaucescens, C. lutea, Carphephorus paniculatus, Carpinus caroliniana, Catalpa bignonioides, Centella asiatica, Cephalanthus occidentalis, Chaptalia tomentosa, Chrysopsis mariana, Clethra alnifolia, Cliftonia monophylla, Coleataenia tenera, Coreopsis gladiata, Crataegus aestivalis, Cyrilla racemiflora, Dichanthelium acuminatum, D. scabriusculum, D. scoparium, Diodia virginiana, Diospyros virginiana, Ditrysinia fruticosa, Drosera capillaris, Dulichium arundinaceum, Eleocharis melanocarpa, E. microcarpa, E. mutata, E. tuberculosa, Eragrostis refracta, Erigeron vernus, Eriocaulon compressum, E. decangulare, Eryngium, Eupatorium capillifolium, E. leucolepis, E. rotundifolium, Euthamia leptocephala, Fraxinus caroliniana, Fuirena breviseta, F. bushii, Gaylussacia mosieri, Gelsemium rankinii, Gentiana autumnalis, Gratiola brevifolia, Gymnadeniopsis nivea, Halesia, Harperella nodosa, Helenium drummondii, Helianthus angustifolius, Hymenocallis coronaria, Hypericum brachyphyllum, H. chapmanii, H. crux-andreae, H. fasciculatum, H. frondosum, H. hypericoides, H. nitidum, H. tenuifolium, Hyptis alata, Ilex coriacea, I. glabra, I. myrtifolia, I. opaca, I. verticillata, Itea virginica, Juncus validus, Justicia americana, Kalmia, Liatris acidota, L. pycnostachya, Ligustrum sinense, Liquidambar styraciflua, Liriodendron*

tulipifera, Lobelia flaccidifolia, Lophiola aurea, Ludwigia linearis, L. maritima, L. pilosa, Lycopodiella appressa, L. alopecuroides, Lyonia lucida, Lysimachia asperulifolia, Magnolia virginiana, Marshallia graminifolia, Mitreola sessilifolia, Myrica cerifera, M. heterophylla, Nyssa aquatica, N. biflora, Osmunda regalis, Osmundastrum cinnamomeum, Oxypolis canbyi, Panicum verrucosum, P. virgatum, Paspalum, Persea borbonia, P. palustris, Persicaria, Physostegia virginiana, Pieris phillyreifolia, Pinguicula pumila, Pinus elliottii, P. palustris, P. taeda, Platanthera cristata, Platanus occidentalis, Pluchea baccharis, P. foetida, Polygala cruciata, P. ramosa, Proserpinaca pectinata, Pycnanthemum flexuosum, Quercus durandii, Q. laevis, Q. laurifolia, Q. nigra, Rhexia alifanus, R. lutea, R. nashii, Rhododendron canescens, Rhynchospora divergens, R. elliottii, R. fascicularis, R. filifolia, R. gracilenta, R. inexpansa, R. latifolia, R. perplexa, R. plumosa, R. rariflora, Rosa palustris, Rubus, Sabal, Sabatia campanulata, Saccharum giganteum, Salix humilis, S. nigra, Sambucus nigra, Sarracenia alata, Saururus cernuus, Schizachyrium scoparium, Schoenoplectus pungens, Scleria georgiana, S. muehlenbergii, S. reticularis, Serenoa repens, Smilax laurifolia, S. rotundifolia, Sphagnum, Spiranthes laciniata, Stenanthium densum, Stylisma aquatica, Taxodium ascendens, T. distichum, Thalictrum cooleyi, Tillandsia usneoides, Tofieldia, Toxicodendron radicans, Triadenum walteri, Trifolium arvense, Ulmus rubra, Vaccinium arboreum, V. elliottii, Verbena brasiliensis, Vernonia angustifolia, Viburnum dentatum, V. nudum, Viola primulifolia, Vitis, Woodwardia areolata, Xyris ambigua, X. fimbriata, X. stricta, Zizaniopsis miliacea.

Hypericum gymnanthum Engelm. & A. Gray is a perennial herb, which occurs in borrow pits, depressions, ditches, dunes, flatwoods, glades, gullies, meadows, ponds, prairies, quarries, roadsides, seeps, shores, sinkholes, swales, swamps, woodlands, and along the margins of lakes, ponds, and streams across the eastern coastal plain at elevations of up to 506 m. Exposures typically are in full sunlight and the plants can succumb to shading owing to the successional growth of woody species. The substrates are described as acidic or calcareous and include clay loam, gravel, Iredell series, loam, loamy sand, mud, peat, sand, sandy clay, sandy loam, sandy peat, sandy silt, silty clay loam, silty loam, and silty sandy loam. Flowering and fruiting occur from May to August. A seedbank of undetermined duration is established. Specific seed germination requirements have not been established. **Reported associates:** *Acer rubrum, Agrostis hyemalis, Aletris aurea, Andropogon scoparius, Asclepias perennis, Baptisia tinctoria, Bartonia paniculata, Bidens polylepis, Callicarpa americana, Carex glaucescens, C. leptalea, C. scoparia, Carya, Cephalanthus occidentalis, Chamaecrista fasciculata, Chasmanthium, Coelorachis rugosa, Coreopsis lanceolata, C. major, Cornus foemina, Cyperus haspan, C. pseudovegetus, Dichanthelium acuminatum, Diodella teres, Diospyros virginiana, Drosera capillaris, Eleocharis engelmannii, E. tenuis, Elephantopus nudatus, Elymus canadensis, Erigeron, Eupatorium leucolepis, Eutrochium fistulosum, Fuirena squarrosa, Gratiola pilosa, Hedeoma* *hispida, Hibiscus moscheutos, Hydrolea ovata, Hypericum crux-andreae, H. denticulatum, Hyptis alata, Juncus, Lechea minor, Lespedeza capitata, Liatris spicata, Ligustrum sinense, Linum floridanum, Liquidambar styraciflua, Lobelia canbyi, Lolium perenne, Ludwigia alternifolia, L. hirtella, Lycopodiella alopecuroides, Lyonia ligustrina, Lysimachia lanceolata, Lythrum alatum, Malva, Melanthera nivea, Mikania scandens, Mimosa microphylla, Mitreola petiolata, Nyssa, Oenothera curtiflora, Packera anonyma, Panicum longifolium, P. verrucosum, P. virgatum, Paspalum, Persicaria, Phyla, Pinus elliottii, P. taeda, Plantago aristata, Platanthera cristata, Pluchea foetida, Polygala cruciata, P. mariana, Proserpinaca pectinata, Quercus, Rhexia mariana, Rhynchospora cephalantha, R. gracilenta, R. inexpansa, R. plumosa, R. torreyena, Rubus, Sabatia campanulata, S. dodecandra, Sarracenia, Scleria oligantha, S. reticularis, Silphium asteriscus, Smilax, Solanum carolinense, Sorghastrum nutans, Stachys aspera, Symphyotrichum dumosum, Taxodium distichum, Toxicodendron, Triadica sebifera, Ulmus americana, Vaccinium elliottii, V. tenellum, Verbena brasiliensis, V. urticifolia, Viola lanceolata, Vitis.*

***Hypericum lissophloeus* W.P. Adams** is a large shrub, which is narrowly restricted to the sandy margins of neutral to somewhat acidic bogs, depression ponds, or Karst sinkhole ponds at low elevations to at least 20 m. The plants occur in standing waters (to 1.8 m in depth) and extend up to the high water limit on exposed shorelines. Plants that grow in standing water can develop mangrove-like prop roots along their lower stem. Flowering and fruiting occur from June to October. The flowers are pollinated by insects (see *Use by wildlife*). Additional life history information for this species is scarce. **Reported associates:** *Amphicarpum muhlenbergianum, Andropogon virginicus, Axonopus, Bulbostylis barbata, B. ciliatifolia, Centella asiatica, Chrysopsis lanuginosa, Cliftonia monophylla, Drosera filiformis, Eleocharis robbinsii, Eragrostis refracta, Eriocaulon lineare, Eupatorium leptophyllum, Hypericum fasciculatum, H. reductum, Ilex myrtifolia, Lachnanthes carolinianam, Lachnocaulon anceps, Nymphoides, Panicum verrucosum, Paronychia chartacea, P. patula, Polypremum procumbens, Rhexia salicifolia, Rhynchospora globularis, R. pleiantha, Sagittaria isoetiformis, Scleria reticularis, Syngonanthus flavidulus, Utricularia cornuta, Vaccinium elliottii, Xyris isoetifolia, X. jupicai, X. longisepala, X. smalliana.*

***Hypericum nitidum* Lam.** is a shrub found in bogs, borrow pits, depressions, ditches, flatwoods, savannahs, thickets, and along the open margins of blackwater streams, borrow pits, lakes, and ponds at low elevations of up to 50 m. The substrates are described as sand or peat. Flowering and fruiting occur from May to November. A substantial seed bank develops, which is capable of attaining densities of up to 1167 seedlings/m². Dormant seeds are retained for a duration of at least 2 years. Seeds collected in June have germinated well after being placed at 4°C for 3 days, planted at a 2 cm depth, and then exposed to ambient greenhouse conditions. Areas that experience storm surges (salinity ~23 ppt) do not retain a seed bank. The plants occasionally form large masses when growing along the shores

of upper (freshwater) reaches of blackwater estuaries. They are sensitive to disturbances (e.g., fire, hurricanes), with this species being categorized as a strong predisturbance indicator. **Reported associates:** *Acer rubrum, Amphicarpum muhlenbergianum, Aristida stricta, Bartonia paniculata, Boltonia caroliniana, Burmannia biflora, Carex, Cephalanthus occidentalis, Cladium mariscoides, Cliftonia monophylla, Coreopsis rosea, Croton elliottii, Cuphea carthagensis, Cyrilla racemiflora, Diospyros virginiana, Drosera intermedia, Echinodorus parvulus, Eriocaulon, Eryngium, Fimbristylis vahlii, Helianthus, Hydrochloa caroliniensis, Hypericum adpressum, H. brachyphyllum, H. galioides, Ilex cassine, Iris tridentata, I. prismatica, I. hexagona, I. virginica, Juncus repens, Lachnanthes caroliniana, Litsea aestivalis, Lobelia boykinii, Ludwigia decurrens, L. sphaerocarpa, L. octovalis, L. alternifolia, L. suffruticosa, L. pilosa, L. spathulata, Myrica cerifera, Nyssa sylvatica, Oxypolis canbyi, Panicum hemitomon, P. verrucosum, P. dichotomiflorum, Paspalum, Physostegia, Pinus elliottii, P. palustris, Pluchea, Polygala nana, Proserpinaca pectinata, Pterocaulon, Ptilimnium nodosum, Rhexia aristosa, R. mariana, Rhynchospora inundata, R. tracyi, Rubus, Sabal, Sabatia bartramii, S. difformis, S. brevifolia, Sagittaria isoetiformis, Sarracenia alata, Scleria baldwinii, S. oligantha, Serenoa repens, Smilax, Spiranthes longilabris, S. laciniata, Stillingia aquatica, Symphyotrichum dumosum, Taxodium ascendens, Xyris.*

***Hypericum setosum* L.** is an annual or biennial herb, which grows in coastal plain bogs, ditches, flats, flatwoods, floodplains, meadows, roadsides, savannas, seepage slopes, swamps, and along the margins of rivers and streams at low elevations to at least 64 m. Exposures are open and receive full sun. The substrates are described as acidic and have been categorized as Alapaha (arenic plinthic paleaquults), loam, loamy sand, Pelham (arenic paleaquults), sand, sandy peat, and silty sand. Flowering and fruiting occur from January to October. The plants are observed often in frequently burned sites. Seeds collected from areas burned during the dormant season exhibited 13% (after 4 months) to 47% (after 1.5 years) germination when placed on moist filter paper at room temperature. The germination rate for these seeds was reduced (8%) when placed outdoors on moist sand under natural light and temperature conditions. **Reported associates:** *Acalypha, Acer rubrum, Alnus serrulata, Andropogon glomeratus, Anthaenantia villosa, Aristida lanosa, A. purpurascens, Arnoglossum, Bigelowia, Calamovilfa brevipilis, Carphephorus odoratissimus, Chelone cuthbertii, Ctenium aromaticum, Cuphea, Danthonia spicata, Eupatorium leucolepis, E. rotundifolium, Fimbrystylis perpusilla, Fuirena, Gaylussacia frondosa, Gymnopogon brevifolius, Harperocallis flava, Hottonia inflata, Hypericum cistifolium, H. fasciculatum, H. muticum, H. myrtifolium, Ilex vomitoria, Lachnanthes caroliniana, Liatris, Lilium, Liquidambar styraciflua, Liriodendron tulipifera, Litsea aestivalis, Lycopodiella alopecuroides, Magnolia virginiana, Myrica heterophylla, Nyssa biflora, N. sylvatica, Osmundastrum cinnamomea, Oxydendrum arboreum, Panicum virgatum, Photinia pyrifolia, Pinus elliottii, P. palustris, P. taeda,*

Platanthera, Pluchea, Pogonia ophioglossoides, Polygala cruciata, P. lutea, Pteridium aquilinum, Pycnanthemum virginianum, Quercus laevis, Q. marilandica, Rhexia, Rhynchospora inexpansa, Sabatia campanulata, Saccharum, Sarracenia flava, S. purpurea, S. ×catesbaei, Schizachyrium scoparium, Scleria georgiana, S. triglomerata, Smilax laurifolia, Solidago stricta, Sphagnum macrophyllum, Sporobolus, Symphyotrichum dumosum, Taxodium distichum, Tridens carolinianus, Vaccinium formosum, Viburnum rufidulum, Xylorhiza tortifolia, Xyris ambigua.

Hypericum tetrapetalum Lam. is a perennial shrub, which occurs in depressions, ditches, flatwoods, hammocks, marshes, meadows, prairies, roadsides, savannah, scrub, seeps, swamps, and along the margins of lakes, ponds, and streams at low elevations to at least 42 m. The site exposures are open with substrates that include loamy sand, sand, sandy loam, and sandy peat. Flowering and fruiting occur throughout the year (January to November). A seed bank develops with an average density of 9.5 seeds/m^2 observed in one study. Seeds retrieved from those cores germinated within 14 months after placing them under ambient greenhouse conditions (in January) with watering cycles that included two subsequent dry periods (21 July to 16 August; 11 February to 1 April). **Reported associates:** *Ampelopsis arborea, Andropogon brachystachyus, A. glomeratus, A. gyrans, A. virginicus, Aristida purpurascens, A. spiciformis, A. stricta, Arnoglossum ovatum, Asclepias pedicellata, Asimina reticulata, Axonopus furcatus, Balduina angustifolia, Bejaria racemosa, Bidens pilosus, Buchnera americana, Bulbostylis barbata, Callicarpa americana, Campsis radicans, Carex verrucosa, Carphephorus odoratissimus, Cephalanthus occidentalis, Chamaecrista fasciculata, Cirsium nuttallii, Coreopsis leavenworthii, Crotalaria pallida, C. rotundifolia, Croton glandulosus, Cyperus compressus, C. croceus, Dalea pinnata, Desmodium tortuosum, Dichanthelium acuminatum, D. ensifolium, D. portoricense, Drosera brevifolia, D. capillaris, Eleocharis baldwinii, E. robbinsii, E. vivipara, Elephantopus elatus, E. nudatus, Emilia fosbergii, Eragrostis secundiflora, Eriocaulon decangulare, Euphorbia cyathophora, Euthamia graminifolia, Fimbristylis autumnalis, F. puberula, Fuirena scirpoidea, Galactia elliottii, G. volubilis, Gaylussacia dumosa, G. frondosa, G. tomentosa, Gelsemium sempervirens, Gomphrena serrata, Gratiola hispida, Gymnopogon chapmanianus, Helianthemum carolinianum, H. corymbosum, Hypericum cistifolium, H. crux-andreae, H. fasciculatum, H. hypericoides, H. microsepalum, H. tenuifolium, Ilex cassine, I. decidua, I. glabra, I. opaca, I. vomitoria, Juncus effusus, J. marginatus, J. repens, Lachnanthes caroliniana, Lachnocaulon anceps, L. beyrichianum, Lechea minor, L. torreyi, Liatris tenuifolia, Licania, Liquidambar styraciflua, Lygodesmia aphylla, Lyonia ferruginea, L. fruticosa, L. lucida, Lythrum, Mitchella repens, Momordica charantia, Myrica cerifera, Oldenlandia uniflora, Osmunda regalis, Panicum hemitomon, Parthenocissus quinquefolia, Persea borbonia, Phytolacca americana, Piloblephis rigida, Pinus clausa, P. elliottii, P. palustris, Piriqueta cistoides, Pityopsis graminifolia, Pluchea baccharis, Polygala rugelii,* *P. setacea, Polygonella polygama, Polypremum procumbens, Pteridium aquilinum, Pterocaulon pycnostachyum, Quercus geminata, Q. laurifolia, Q. virginiana, Rhexia mariana, R. nuttallii, Rhus copallinum, Rhynchospora colorata, R. fascicularis, R. fernaldii, Rudbeckia, Sabal palmetto, Sabatia grandiflora, Sacciolepis indica, Schizachyrium stoloniferum, Senna occidentalis, Serenoa repens, Sericocarpus tortifolius, Sida rhombifolia, Sideroxylon reclinatum, Smilax auriculata, Stipulicida setacea, Symphyotrichum dumosum, Syngonanthus flavidulus, Taxodium distichum, Tillandsia ×floridana, T. recurvata, T. simulata, T. usneoides, Utricularia subulata, Vaccinium arboreum, V. corymbosum, V. darrowii, V. myrsinites, V. stamineum, Viburnum obovatum, Viola lanceolata, Vitis rotundifolia, V. shuttleworthii, Vulpia myuros, Woodwardia areolata, W. virginica, Xyris brevifolia, X. caroliniana, X. elliottii, X. flabelliformis.*

Use by wildlife: *Hypericum* seeds are eaten by ducks (Aves: Anatidae) and by ruffed grouse (Aves: Phasianidae: *Bonasa umbellus*). *Hypericum boreale* is a fungal host of rust (Basidiomycota: Pucciniaceae: *Aecidium hypericatum*). *Hypericum chapmanii* is a common element of breeding sites and larval habitat for the flatwoods salamander (Amphibia: Ambystomatidae: *Ambystoma cingulatum*); however, excessive plant growth can also reduce the amount of useful habitat. *Hypericum edisonianum* is the host plant for a leafminer moth (Insecta: Lepidoptera: Gelechiidae: *Coleotechnites nigra*) and webworms (Lepidoptera: Pyralidae: *Tallula watsoni*). It is also colonized by ants (Insecta: Hymenoptera: Formicidae: *Crematogaster clara*). The plants are hosts for a pathogenic fungus (Fungi: Ascomycota: Botryosphaeriaceae: *Sphaeropsis tumefaciens*), which normally infects citrus crops (Rutaceae). The flowers of *H. edisonianum* attract various bees (Insecta: Hymenoptera) including bumblebees and honeybees (Apidae: *Apis mellifera, Bombus impatiens, B. pensylvanicus*), leafcutting bees (Megachilidae: *Anthidiellum perplexum, Megachile brevis, M. mendica, M. petulans*), plasterer bees (Colletidae: *Hylaeus confluens, H. schwarzi*), and sweat bees (Halictidae: *Augochloropsis sumptuosa, Lasioglossum miniatulum, L. nymphale, L. surianae, L. tarponense*). Its other floral insect visitors include flies (Diptera: Syrphidae: *Volucella nigra, V. pusilla*), phasmids (Phasmida: Pseudophasmatidae: *Anisomorpha buprestoides*), wasps (Hymenoptera: Vespidae: *Vespa*), and weevils (Coleoptera: Curculionidae: *Odontocorynus larvatus, O. pulverulentus*). *Hypericum ellipticum* is a host of rust fungi (Basidiomycota: Pucciniaceae: *Aecidium hypericatum*). *Hypericum galioides* is a host of plant bugs (Insecta: Hemiptera: Miridae: *Parthenicus rufus*). Larvae of the lepidopteran *Peratophyga hyalinata* (Geometridae) are known to feed on plants of *H. galioides* grown in Japan. The flowers of *H. lissophloeus* are visited by flies (Insecta: Diptera: Syrphidae: *Toxomerus*) as well as several bees (Insecta: Hymenoptera) including bumblebees (Apidae: *Bombus auricomus, B. griseocollis, B. pensylvanicus, B. vagans*), leafcutting bees (Megachilidae: *Megachile rugifrons*), and sweat bees (Halictidae: *Agapostemon viridulus, Lasioglossum pilosum, L. versatum*). *Hypericum nitidum* is used occasionally as a nesting site by Cuba sandhill cranes

(Aves: Gruidae: *Grus canadensis nesiotes*). *Hypericum perforatum* (Klamath weed) causes livestock poisoning but it is not aquatic. Toxicity of the aquatic species has not been determined.

Economic importance: food: none; **medicinal:** Although some *Hypericum* species (e.g., *H. hypericoides*, *H. perforatum*) are widely used as herbal medicines and potentially as a source of antidepressant drugs, few medicinal uses are reported for any of the aquatic taxa. The leaves of *H. anagalloides* supposedly have been used medicinally. Shoots of *H. boreale* are made into a preparation for treating urinary disorders in parts of Asia. *Hypericum ellipticum* and other species contain hypericin, which can block melanin production and is known to possess high antiretroviral activity. The aerial portions of *H. ellipticum* contain acylphloroglucinol and several xanthones, all of which exhibit selective but moderate cytotoxicity against HCT-116 and Caco-2 human colon tumor cell lines. They also contain benzophenone glucoside, which inhibits proliferation of CNS tumor cell line (SF-268) and lipid peroxidation. The Iroquois used a decoction made from *H. ellipticum* to suppress menses. A chromanone derivative isolated from *H. lissophloeus* enhances current mediation by γ-aminobutyric acid type A (GABA$_A$) receptors, which are major inhibitory neurotransmitter receptors in mammalian central nervous systems; **cultivation:** *Hypericum anagalloides* is cultivated as an ornamental ground cover and rock garden plant. *Hypericum boreale* and *H. galioides* are marketed as ornamental water garden plants. One commercial cultivar of *H. galioides* is 'Brodie.' *Hypericum lissophloeus* has minor commercial importance as an ornamental (or for use in wetland restoration) and includes the cultivar 'Eglin.' *Hypericum nitidum* is sold occasionally as an ornamental shrub; **misc. products:** *Hypericum chapmanii* produces unusually high amounts (31.5%) of β-eudesmol, an uncommon volatile substance; **weeds:** *Hypericum boreale* is reported to be a weed along river banks and in cranberry (*Vaccinium*) bogs of British Columbia and the Pacific Northwest, and is also regarded as a weed in Wisconsin cranberry bogs; **nonindigenous species:** A native of eastern North America, *Hypericum boreale* was introduced to British Columbia sometime before 1961 and to Oregon more recently. Its introduction to western North America is attributable to plants and bedding soil, which originated in eastern North America for use in commercial cranberry culture. As many as 2177 seeds have been recovered from contaminated cranberry vines. Its spread in the western region may have been facilitated by the transfer of propagules by peat-harvesting equipment. The eastern North American *H. ellipticum* recently has been introduced to cranberry ponds in British Columbia and probably also originated through contaminated cranberry planting stocks brought from the east. *Hypericum ellipticum* and *H. mutilum* have been introduced recently to cranberry plantations in eastern Europe.

Systematics: An initial phylogenetic analysis of *rbcL* sequence data indicated strong support for a clade comprising *Hypericum* and *Triadenum* as sister genera (Figure 4.29), a result also recovered using low-copy nuclear markers. These results substantiated the presumed close relationship and also the distinctness of those genera from *Cratoxylum*, which some authors had allied to *Hypericum* because of anomalous chromosomal base numbers between the latter and *Triadenum*. However, a more comprehensive study of *Hypericum* accessions yielded yet a different result, placing *Hypericum* as the sister group to a clade containing species that had been assigned formerly to *Triadenum*, *Brachys*, and *Myriandra* (Figures 4.30 and 4.31). The clade comprising the latter three genera remained in a position remote from *Cratoxylum* in the analysis, but as a sister clade to all but two *Hypericum* species, a result prompting a number of taxonomists to treat the entire group as a single genus (*Hypericum*). That disposition seems extreme, given distinctness of this group and the lack of any major phylogenetic discordance (a statistically indistinguishable result in the resulting analysis) that would be incurred by retaining three separate genera, which already have been distinguished previously. Among the eight OBL taxa surveyed, none falls within the "core" *Hypericum* clade (Figure 4.31) but all are assigned among the following sections (which would fall outside of *Hypericum* if subdivided): *Brathys* (*H. setosum*); *Trigynobrathys* (*H. boreale*); and *Myriandra* (*H. adpressum*, *H. chapmanii*, *H. galioides*, *H. lissophloeus*, *H. nitidum*, *H. tetrapetalum*). If *Triadenum*, *Myriandra*, and *Brathys* are recognized as distinct genera, then all of the OBL species would need to be transferred out of *Hypericum*. Although four OBL *Hypericum* species are yet to be included in a phylogenetic analysis, the relationships depicted so far indicate multiple (at least four), independent origins of the habit (Figure 4.31). Some authors recognize *Hypericum boreale* at the subspecific level (*H. mutilum* L. subsp. *boreale* ([Britton)] J.M. Gillett), but the distinctness of these taxa deserves further consideration. This issue makes it difficult to interpret some literature accounts for "*H. mutilum*," which as a consequence, could apply to either taxon. *Hypericum* is variable cytologically and is characterized by a series of descending basic chromosome numbers ($x = 12, 10, 9, 8, 7$). *Hypericum setosum* ($2n = 12$), *H. boreale* ($2n = 16$), along with *H. adpressum*, *H. ellipticum*, *H. galioides*, *H. lissophloeus*, and *H. tetrapetalum* (all $2n = 18$), are diploid. Chromosome counts are unreported for the five remaining OBL species. *Hypericum boreale* reportedly hybridizes with *H. canadense* but this supposition has not been confirmed experimentally.

Comments: *Hypericum adpressum* occurs in the eastern and midwestern United States. *Hypericum anagalloides* occurs in western North America. *Hypericum chapmanii* and *H. lissophloeus* are endemic to the Florida panhandle, whereas, *H. edisonianum* is endemic to peninsular Florida (where it occurs only at 25 sites). *Hypericum tetrapetalum* is restricted to Florida and Georgia. *Hypericum boreale* is disjunct in eastern and northwestern North America (introduced to the latter) whereas, *H. ellipticum* is northeastern. *Hypericum galioides*, *H. nitidum*, and *H. setosum* occur in the southeastern United States while *H. gymnanthum* has a general southeastern North American distribution.

References: Abrahamson, 1984; Abrahamson & Vander Kloet, 2014; Adams et al., 2010; Aguilera et al., 2005; Allison, 2011; Anderson, 1991; Carr, 2007; Cohen et al., 2004;

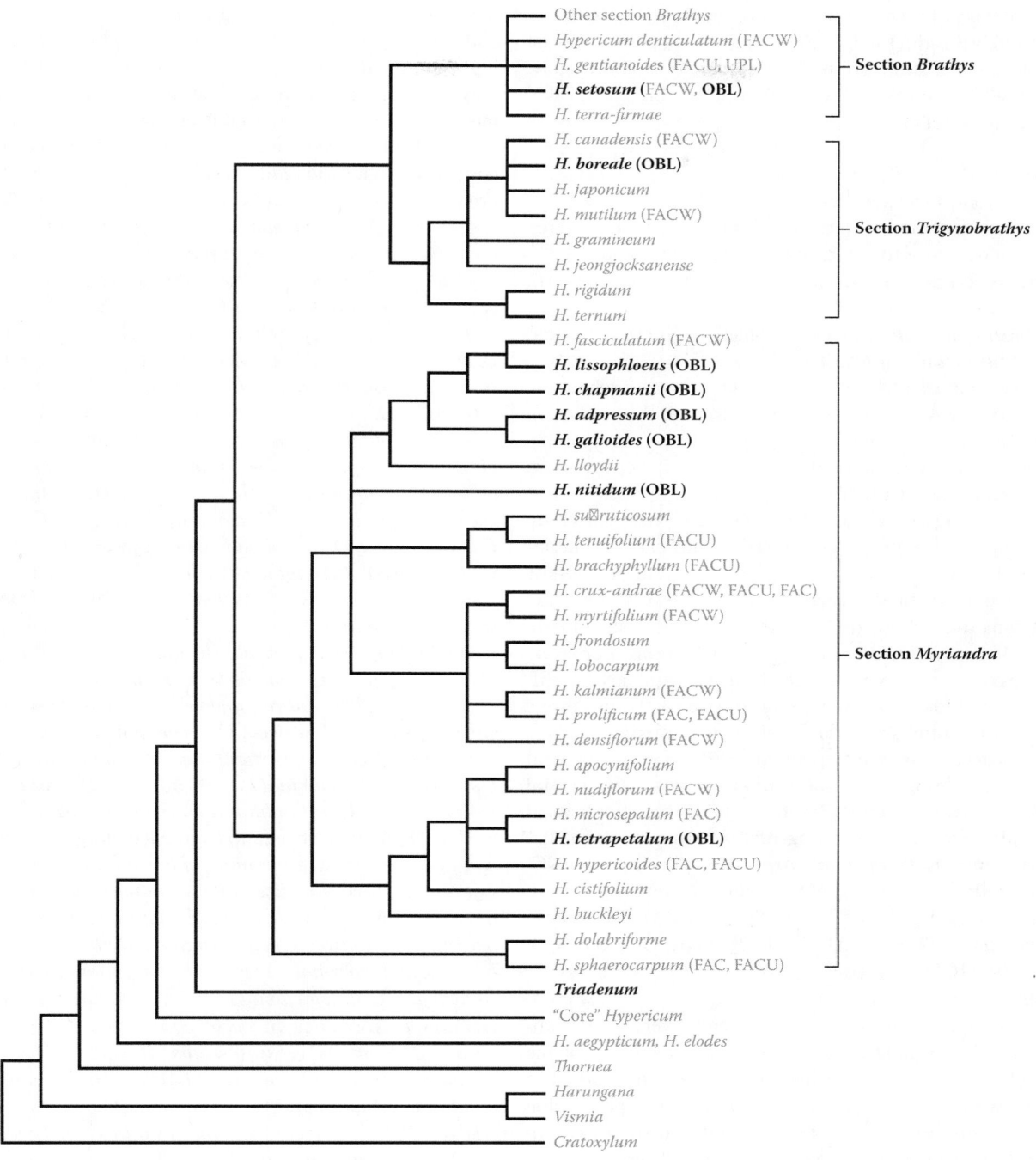

FIGURE 4.31 Phylogenetic relationships among *Hypericum* species, as indicated by analysis of nrITS sequence data, indicate at least four independent origins of OBL taxa (in bold) in the genus. Although some authors merge *Triadenum* and the taxa assigned to sections *Brathys*, *Myriandra*, and *Trigynobrathys* with the "core" *Hypericum* species, the merit of recognizing these groups as distinct from the latter deserves further consideration. However, such a disposition would require the taxonomic transfer of most (if not all) of the OBL *Hypericum* species to other genera (four species have not been included in phylogenetic analyses). (Modified from Nürk, N.M. et al., *Mol. Phylogen. Evol.*, 66, 1–16, 2013.)

Crandall & Platt, 2012; Crockett et al., 2008, 2016; Deka & Devi, 2015; Deyrup & Trager, 1986; Deyrup et al., 2002, 2004; Dillingham, 2005; Dressler et al., 1987; Dzhus, 2014; Eisner et al., 1973; Enser, 2001; Fruchter, 2005; Glitzenstein et al., 2001; Gronquist et al., 2001; Hanlon et al., 1998; Harper, 1914; Harper et al., 1998; Henry, 2007; Howard & Allain, 2012; Kalk, 2011; Kalmbacher et al., 2005; Keddy, 1989; Keddy & Reznicek, 1982; Keddy et al., 1998; Kral, 1966a; Landman & Menges, 1999; Le Page & Keddy, 1998; Manning et al., 2011; Martin, 2006; McMaster, 1994; Meseguer et al., 2014; Nürk et al., 2013; Parmelee & Savile, 1954; Petrunak et al., 2009; Prena, 2008; Robson & Adams, 1968; Schmidt,

2005; Shipley & Parent, 1991; Singhurst et al., 2012; Skroch & Dana, 1965; Smith, 1996; Van Alstine et al., 2001; Van de Kerckhove et al., 2002; Weiher et al., 1996; Wilson & Keddy, 1985, 1988; Wisheu & Keddy, 1991; Wright et al., 2003; Zika, 2003; Zomlefer et al., 2008.

2. *Triadenum*

Marsh St. John's-wort; millepertuis

Etymology: from the Greek *tri* and *adenos* meaning "three glands" in reference to the distinctive floral staminodes

Synonyms: *Elodes*; *Gardenia* (in part); *Hypericum* (in part); *Martia*

Distribution: global: Asia, North America; **North America:** eastern (disjunct in northwest)

Diversity: global: 10 species; **North America:** 4 species

Indicators (USA): OBL: *Triadenum fraseri*, *T. tubulosum*, *T. virginicum*, *T. walteri*

Habitat: freshwater; palustrine; **pH:** 3.0–6.5; **depth:** <1 m; **life-form(s):** emergent (herb)

Key morphology: stems (to 1 m) erect, glabrous; leaves (to 15 cm) opposite, simple, entire, usually with clear glandular dots on the lower surface (which darken when dried), sessile or short-petioled; inflorescences cymose; flowers complete, radial, 5-merous; petals (to 10 mm), pinkish or flesh colored; stamens 9, grouped in 3 fascicles of 3, each group alternating with one of 3 conspicuous orange glands (staminodes), the filaments united for ½ to ⅔ their length; styles 3, locules three; capsules (to 12 mm) septicidal, containing numerous seeds

Life history: duration: perennial (rhizomes); **asexual reproduction:** rhizomes; **pollination:** insect, self; **sexual condition:** hermaphroditic; **fruit:** capsules (common); **local dispersal:** rhizomes; seeds; **long-distance dispersal:** seeds

Imperilment: (1) *Triadenum fraseri* [G5]; S1 (DE, SK, NC, TN); S2 (NE, VA); S3 (IA, MB, NJ); (2) *T. tubulosum* [G4]; S1 (MD, MO, OK, WV); S2 (GA, NC, OH, VA); S3 (IN); (3) *T. virginicum* [G5]; SH (IL); S1 (AR, NB, OK, QC); S3 (NC); (4) *T. walteri* [G5]; S1 (NJ, WV); S2 (OH); S3 (IN)

Ecology: general: North American *Triadenum* entirely comprises OBL species, which occupy a broad diversity of freshwater and slightly brackish habitats. Asian members of the genus also have a strong affinity for wetland habitats. The fairly showy flowers are insect pollinated, but self-pollination can also occur. Tuberous shoots are produced from axillary cotyledonary buds in some species. In such cases, the original genet persists for only 1–2 years, with the axillary shoots proliferating subsequently as ramets. Additional details on the reproductive biology and seed ecology of *Triadenum* species remain poorly known.

Triadenum fraseri **(Spach) Gleason** grows in bogs, depressions, fens, marshes, meadows, prairie fens, prairies, river bars, shorelines, sloughs, swales, swamps, and along the margins of lakes, ponds, reservoirs, rivers, and streams at elevations of up to 907 m. The substrates are acidic (pH: 4.8–6.5) and include peat, sand, or silty humus. The plants grow best in open sites on substrates where there is a relatively high organic content and exhibit reduced biomass on sandy substrates or when growing under shrubby overstory. They occur occasionally on floating mats of vegetation. Flowering and fruiting occur from July to September. Although the highly nectariferous flowers remain closed in full sunlight, they open during the night when they are pollinated by crepuscular insects, which visit during the dawn and dusk hours. The seeds of *T. fraseri* have physiological dormancy but will germinate following 270 days of cold stratification when brought under a 30°C/20°C temperature regime. **Reported associates:** *Abies balsamea, Acalypha rhomboidea, Acer rubrum, Agrostis hyemalis, Alnus rugosa, Andromeda glaucophylla, Andropogon gerardii, Anticlea elegans, Apocynum cannabinum, Arethusa bulbosa, Aronia arbutifolia, A. melanocarpa, Asclepias incarnata, Aulacomnium palustre, Bartonia paniculata, B. virginica, Betula alleghaniensis, B. pumila, Bidens aristosus, B. cernuus, B. discoideus, B. frondosus, B. polylepis, B. trichospermus, Boehmeria cylindrica, Calamagrostis canadensis, C. stricta, Calliergon cordifolium, Campanula aparinoides, C. rotundifolia, Carex aquatilis, C. atlantica, C. barrattii, C. buxbaumii, C. canescens, C. chordorrhiza, C. comosa, C. diandra, C. echinata, C. exilis, C. flava, C. folliculata, C. gynandra, C. lacustris, C. lasiocarpa, C. leptalea, C. limosa, C. livida, C. oligosperma, C. pellita, C. pensylvanica, C. pseudocyperus, C. sterilis, C. stipata, C. stricta, C. trisperma, C. utriculata, Cephalanthus occidentalis, Chamaedaphne calyculata, Cicuta bulbifera, C. maculata, Cirsium muticum, Cladium mariscoides, Comandra umbellata, Comarum palustre, Cornus foemina, C. sericea, Cyprepedium acaule, Dasiphora floribunda, Decodon verticillatus, Deschampsia cespitosa, Dichanthelium acuminatum, D. boreale, Drosera anglica, D. intermedia, D. rotundifolia, Dryopteris cristata, D. intermedia, Dulichium arundinaceum, Eleocharis acicularis, E. elliptica, E. obtusa, Epilobium coloratum, E. leptophyllum, Equisetum fluviatile, Erechtites hieracifolius, Eriophorum angustifolium, E. vaginatum, E. virginicum, Eupatorium perfoliatum, Eutrochium maculatum, Euthamia graminifolia, E. gymnospermoides, Fragaria virginiana, Galium labradoricum, G. tinctorium, G. trifidum, Gentiana andrewsii, Gentianopsis crinita, Glyceria canadensis, G. grandis, G. striata, Helianthus grosseserratus, Hypericum boreale, H. canadense, H. densiflorum, H. ellipticum, H. kalmianum, Ilex montana, I. mucronata, I. verticillata, Impatiens capensis, Iris versicolor, I. virginica, Juncus balticus, J. brachycephalus, J. brevicaudatus, J. canadensis, J. effusus, J. greenei, J. marginatus, J. subcaudatus, J. supiniformis, J. torreyi, J. vaseyi, Kalmia latifolia, K. polifolia, Larix laricina, Lathyrus palustris, Leersia oryzoides, Lemna minor, Leptodictyum riparium, Liparis loeselii, Lobelia cardinalis, L. kalmii, L. spicata, Ludwigia alternifolia, L. polycarpa, Lycopodiella inundata, Lycopus americanus, L. uniflorus, Lysimachia terrestris, L. thyrsiflora, Lythrum alatum, L. salicaria, Maianthemum canadense, Menyanthes trifoliata, Monarda fistulosa, Muhlenbergia glomerata, Myrica gale, Nuphar variegata, Onoclea sensibilis, Ophioglossum pusillum, Osmunda regalis, Osmundastrum cinnamomeum, Oxalis montana, Oxypolis rigidior, Packera paupercula, Panicum virgatum, Pedicularis lanceolata, Penthorum sedoides, Persicaria* *amphibia, P. coccinea, P. hydropiperoides, P. punctata, P. sagittata, Picea mariana, P. rubens, Polygala cruciata, P. sanguinea, Polytrichum commune, P. pallidisetum, P. strictum, Populus tremuloides, Potentilla anserina, Proserpinaca palustris, Pycnanthemum virginianum, Rhamnus alnifolia, Rhododendron canadense, R. groenlandicum, R. maximum, Rhynchospora alba, R. inundata, Rosa carolina, R. woodsii, Rubus flagellaris, R. hispidus, Sagittaria latifolia, Salix pedicellaris, S. petiolaris, S. sericea, Sambucus nigra, Sarracenia purpurea, Scheuchzeria palustris, Schizachyrium scoparium, Schoenoplectus acutus, S. pungens, Scirpus atrovirens, S. cyperinus, S. microcarpus, Scleria verticillata, Scutellaria galericulata, Sisyrinchium albidum, Solidago gigantea, S. ohioensis, S. riddellii, S. rugosa, S. uliginosa, Sorghastrum nutans, Sparganium androcladum, Spartina pectinata, Sphagnum angustifolium, S. cuspidatum, S. fallax, S. fimbriatum, S. fuscum, S. henryense, S. magellanicum, S. palustre, S. papillosum, S. recurvum, Spiranthes cernua, Spirea alba, S. tomentosa, Sporobolus heterolepis, Symphyotrichum boreale, S. lanceolatum, S. subspicatum, Symplocarpus foetidus, Thalictrum dasycarpum, Thelypteris palustris, Thuja occidentalis, Toxicodendron radicans, T. vernix, Triadenum virginicum, Triantha glutinosa, Trichophorum cespitosum, Trifolium hybridum, Triglochin maritimum, Tsuga canadensis, Typha latifolia, Utricularia intermedia, Vaccinium macrocarpon, V. myrtilloides, V. oxycoccus, Verbena hastata, Viburnum nudum, V. recognitum, Viola cucullata, V. lanceolata, Vitis riparia.*

Triadenum tubulosum (Walter) Gleason occurs in bogs, bottomlands, channels, floodplains, marshes, mudflats, oxbows, ruts, seeps, shores, sloughs, swales, swamps, thickets, and along the vegetated margins of lakes, ponds, rivers, and streams at elevations of up to 195 m. The exposures most often are described as shaded, but the plants can also occur in canopy openings. The reported substrates include claypan, muck, mud, sandy alluvium, sandy loam, and silty sand. The plants can also grow on floating logs. Flowering and fruiting occur from August to October. *Triadenum tubulosum* distinctly lacks the abaxial glandular dots that occur on the leaves of other North American species. **Reported associates:** *Acer rubrum, Acmella oppositifolia, Boehmeria cylindrica, Carex hyalinolepis, C. louisiana, C. lupulina, C. oxylepis, Carpinus caroliniana, Carya aquatica, Cephalanthus occidentalis, Chasmanthium laxum, Commelina virginica, Diospyros virginiana, Dulichium arundinaceum, Elephantopus carolinianus, Fagus, Forestiera acuminata, Hibiscus laevis, Hypericum denticulatum, Illicium, Itea virginica, Justicia ovata, Leersia lenticularis, Liquidambar styraciflua, Ludwigia palustris, Lygodium japonicum, Lysimachia radicans, Mikania scandens, Nyssa aquatica, Onoclea sensibilis, Persicaria hydropiperoides, Planera aquatica, Quercus lyrata, Q. phellos, Ranunculus pusillus, Salix nigra, Scirpus atrovirens, Sideroxylon lycioides, Smilax, Spiranthes cernua, Styrax americanus, Taxodium distichum, Triadica sebifera, Vaccinium.*

Triadenum virginicum (L.) Raf. grows in shallow (e.g., 15–30 cm), fresh to brackish (salinity to 5.0 ppt) waters of bogs, canals, Carolina bays, channels, depressions, ditches,

fens, flats, floodplains, glades, gravel pits, marshes, meadows, savannahs, seeps, shores, sloughs, stumps, swales, swamps, and along the margins of borrow pits, lakes, ponds, and streams at elevations of up to 980 m. The plants can also occur on floating mats ("islands") of vegetation or on floating or submerged logs. Exposures range from full sun to partial shade. The substrates are acidic (pH: 3.0–6.6) and have been characterized as loamy muck, loamy sand, muck, mucky loam, mud, peat, sand, sandy loam, and sandy peat. The plants can tolerate moderate rates of sedimentation (0–8 cm/year). Flowering and fruiting extend from June to October. The flowers facilitate self-pollination by the proximity of the anthers and stigmas and their eventual contact upon the closure of petals. However, the relative proportions of selfing and outcrossing have not been determined for this species. The seeds are small (averaging 0.04 mg) but retain germinability after at least 130 days of submergence. The plants apparently tolerate fire as they have been collected at sites subjected to annual burning. Although sometimes appearing in brackish sites, the plants generally exhibit a marked reduction in cover when growing under high salt concentrations (>112 mg/L Na$^+$, >54 mg/L Cl$^-$). The roots are colonized (up to 64%) by arbuscular mycorrhizae and dark septate endophytic Fungi. Vegetative reproduction occurs by tuberous shoots, which are produced from cotyledonary buds, which can remain dormant during the first season of growth. **Reported associates:** *Acer rubrum, Agrostis gigantea, Alnus rugosa, Amelanchier canadensis, Andromeda polifolia, Andropogon glomeratus, Anthoxanthum odoratum, Arethusa bulbosa, Aronia ×prunifolia, Bacopa caroliniana, Bartonia paniculata, Bidens connatus, B. frondosus, B. trichospermus, Boehmeria cylindrica, Bolboschoenus maritimus, Bulbostylis capillaris, Burmannia, Calamagrostis canadensis, Calla palustris, Calopogon tuberosus, Celtis, Chamaecyparis thyoides, Chamaedaphne calyculata, Carex atlantica, C. canescens, C. crinita, C. echinata, C. emoryi, C. flava, C. folliculata, C. granularis, C. hormathodes, C. interior, C. intumescens, C. lacustris, C. lasiocarpa, C. lurida, C. magellanica, C. oligosperma, C. retroflexa, C. rostrata, C. scoparia, C. striata, C. stricta, C. trisperma, C. verrucosa, Centella asiatica, Cephalanthus occidentalis, Chelone glabra, Chenopodium rubrum, Chionanthus virginicus, Cladium jamaicense, C. mariscoides, Clethra alnifolia, Coleataenia longifolia, Comarum palustre, Coreopsis rosea, Cyperus pseudovegetus, C. strigosus, Cyrilla racemiflora, Decodon verticillatus, Dichanthelium clandestinum, D. dichotomum, D. spretum, Diospyros virginiana, Drosera filiformis, D. intermedia, D. rotundifolia, Dulichium arundinaceum, Eleocharis elongata, E. melanocarpa, E. obtusa, E. palustris, E. robbinsii, E. tenuis, E. vivipara, Epilobium leptophyllum, Equisetum arvense, E. hyemale, E. pratense, E. sylvaticum, E. variegatum, Erichtites, Eriocaulon aquaticum, E. decangulare, Eriophorum tenellum, E. virginicum, Eubotrys racemosa, Eupatorium perfoliatum, Eutrochium maculatum, Euthamia caroliniana, E. graminifolia, Fagus grandifolia, Fimbristylis autumnalis, Fragaria, Fuirena pumila, F. squarrosa, Galium palustre, G. tinctorium, Gaylussacia dumosa, Gentiana linearis, Habenaria repens, Helianthus angustifolius, Hieracium* *robinsonii, Hydrocotyle umbellata, Hypericum boreale, H. canadense, H. cistifolium, Ilex laevigata, I. verticillata, Impatiens capensis, Ipomoea cordatotriloba, Iris prismatica, I. versicolor, I. virginica, Itea virginica, Juncus balticus, J. canadensis, J. effusus, J. marginatus, J. pelocarpus, J. repens, Kalmia angustifolia, K. polifolia, Leersia hexandra, L. oryzoides, Lindernia dubia, Liquidambar styraciflua, Ludwigia sphaerocarpa, Lycopodiella alopecuroides, L. appressa, L. caroliniana, L. inundata, Lycopus rubellus, L. uniflorus, Lyonia ligustrina, Lysimachia terrestris, L. thyrsiflora, Lythrum hyssopifolia, L. salicaria, Magnolia virginiana, Mayaca fluviatilis, Mentha arvensis, Menyanthes trifoliata, Myrica caroliniensis, M. gale, Nuphar variegata, Nyssa sylvatica, Onoclea sensibilis, Osmunda regalis, Osmundastrum cinnamomeum, Oxypolis rigidior, Packera paupercula, Pancratium maritimum, Panicum hemitomon, P. repens, P. verrucosum, P. virgatum, Parnassia glauca, Peltandra sagittifolia, P. virginica, Persicaria punctata, P. sagittata, Phalaris arundinacea, Phanopyrum gymnocarpon, Phragmites australis, Phyla nodiflora, Picea mariana, Pinguicula primuliflora, Pinus rigida, Pogonia ophioglossoides, Polygala cruciata, P. lutea, Polytrichum, Pontederia cordata, Potentilla canadensis, Proserpinaca pectinata, Ptilimnium capillaceum, P. costatum, Quercus palustris, Q. phellos, Q. rubra, Rhexia virginica, Rhododendron canadense, R. groenlandicum, R. viscosum, Rhynchospora alba, R. capitellata, R. corniculata, R. glomerata, R. inexpansa, R. inundata, R. macrostachya, R. scirpoides, Rotala ramosior, Rubus hispidus, Saccharum giganteum, Sacciolepis striata, Sagittaria lancifolia, S. latifolia, Sarracenia minor, S. purpurea, S. rubra, Saururus cernuus, Schizaea pusilla, Scirpus ancistrochaetus, S. atrovirens, S. cyperinus, Schoenoplectus pungens, S. tabernaemontani, Scleria reticularis, Scutellaria galericulata, S. lateriflora, Selaginella apoda, Sisyrinchium montanum, Smilax rotundifolia, Solanum dulcamara, Solidago sempervirens, S. uliginosa, Sparganium americanum, Spartina patens, S. pectinata, Sphagnum palustre, Spirea alba, S. tomentosa, Spiranthes romanzoffiana, Stachys hyssopifolia, Symphyotrichum dumosum, S. novi-belgii, S. puniceum, Syngonanthus flavidulus, Taxodium ascendens, T. distichum, Thelypteris palustris, Toxicodendron radicans, T. vernix, Triadenum fraseri, Triantha racemosa, Typha angustifolia, T. latifolia, Utricularia cornuta, U. fibrosa, U. subulata, Vaccinium corymbosum, V. macrocarpon, V. oxycoccus, Viburnum nudum, Vigna luteola, Viola lanceolata, V. macloskeyi, Woodwardia areolata, W. virginica, Xyris caroliniana, X. fimbriata, X. laxifolia, X. torta.*

Triadenum walteri (J.F. Gmel.) Gleason inhabits shallow (e.g., 20–30 cm), tidal or nontidal waters in bogs, bottomlands, channels, depressions, flats, floodplains, hummocks, marshes, meadows, mudflats, ravines, sandbars, seeps, shores, sinkholes, slopes, sloughs, swales, swamps, and along the margins of borrow pits, lakes, ponds, rivers, and streams at elevations of up to 118 m. Although the plants often are found in standing water, persistent stands generally establish above the high water mark of seasonal waters. Where permanent waters occur, the plants can be found growing on elevated substrata

such as tussocks, cypress knees, rotting logs, or tree stumps. Exposures can range from full sunlight to fairly dense shade. The substrates are acidic (e.g., pH: 4.2) and include alluvium, clay, loam, loamy sand, muck, mud, peat, rock, sand, sandy loam, sandy silt, silt, silty loam, and silty clay loam. Flowering and fruiting extend from June to November. A persistent seed bank does not appear to develop. Vegetative reproduction occurs by the formation of rhizomes. **Reported associates:** *Acer negundo, A. rubrum, A. saccharinum, Agrimonia, Alisma subcordatum, Alnus rugosa, A. serrulata, Ammannia coccinea, Amorpha fruticosa, Amphicarpaea bracteata, Apios americana, Aronia arbutifolia, Arundinaria gigantea, Asclepias incarnata, A. perennis, Asplenium platyneuron, Athyrium filix-femina, Azolla caroliniana, Bacopa egensis, B. rotundifolia, Betula nigra, Bidens aristosus, B. cernuus, B. discoideus, Boehmeria cylindrica, Boltonia, Botrychium biternatum, Brunnichia ovata, Carex atlantica, C. bromoides, C. canescens, C. comosa, C. crinita, C. cruscorvi, C. decomposita, C. gigantea, C. grayi, C. intumescens, C. joorii, C. lupulina, C. seorsa, C. stricta, Carpinus caroliniana, Carya aquatica, Celtis laevigata, Cephalanthus occidentalis, Chamaecrista, Chelone glabra, Cicuta maculata, Cinna arundinacea, Cirsium muticum, Clethra alnifolia, Colocasia esculenta, Commelina virginica, Cornus amomum, Crataegus, Cuscuta compacta, C. gronovii, Cynanchum laeve, Cyperus erythrorhizos, C. frankii, C. iria, C. polystachyos, C. pumilus, C. squarrosa, C. typhina, Cypripedium reginae, Cyrilla racemiflora, Decodon verticillatus, Decumaria barbara, Dichanthelium, Digitaria, Diodella teres, Diospyros virginiana, Dulichium arundinaceum, Dysphania ambrosioides, Echinochloa crus-galli, Echinodorus cordifolius, Eclipta prostrata, Eichhornia crassipes, Eleocharis microcarpa, E. obtusa, Eubotrys racemosa, Eupatorium perfoliatum, Euphorbia purpurea, Eutrochium fistulosum, Fagus grandifolia, Fimbristylis littoralis, Forestiera acuminata, Fraxinus caroliniana, F. pennsylvanica, F. profunda, Galium obtusum, G. tinctorium, Gleditsia aquatica, G. triacanthos, Glyceria septentrionalis, G. striata, Gymnadeniopsis clavellata, Heteranthera reniformis, Hibiscus, Hottonia inflata, Hydrocotyle verticillata, Hydrolea quadrivalvis, H. uniflora, Hypericum mutilum, Ilex opaca, I. verticillata, Impatiens capensis, Ipomoea, Iris virginica, Itea virginica, Juncus canadensis, J. effusus, Lacnanthes caroliniana, Leersia oryzoides, L. virginica, Leptochloa fusca, Leucothoe axillaris, Ligustrum, Lilium, Limnobium spongia, Lindera benzoin, Lindernia dubia, Liquidambar styraciflua, Liriodendron tulipifera, Lobelia glandulosa, Lonicera japonica, Ludwigia alternifolia, L. decurrens, L. grandiflora, L. leptocarpa, L. palustris, L. peploides, L. pilosa, L. sphaerocarpa, Lycopus rubellus, L. virginicus, Macbridea caroliniana, Magnolia virginiana, Micranthemum umbrosum, Mikania scandens, Mimosa strigillosa, Mimulus alatus, M. ringens, Mitchella repens, Murdannia keisak, Myrica cerifera, Neottia bifolia, Nuphar advena, Nyssa aquatica, N. biflora, N. sylvatica, Osmunda regalis, Osmundastrum cinnamomeum, Oxypolis rigidior, Panicum dichotomiflorum, P. hemitomon, Paspalum notatum, Peltandra virginica, Penthorum sedoides, Persea* *palustris, Persicaria arifolia, P. hydropiperoides, P. maculosa, P. pensylvanica, P. punctata, P. sagittata, Pilea pumila, Pinus palustris, P. taeda, Planera aquatica, Platanthera flava, P. peramoena, Platanus occidentalis, Pluchea foetida, P. rosea, Poa annua, P. autumnalis, Pontederia cordata, Populus deltoides, P. heterophylla, Proserpinaca palustris, Quercus lyrata, Q. michauxii, Q. nigra, Q. palustris, Q. phellos, Rhododendron viscosum, Rhynchospora corniculata, Riccia, Ricciocarpos natans, Rosa palustris, Rudbeckia auriculata, Rumex verticillatus, Sacciolepis, Sagittaria graminea, S. lancifolia, S. latifolia, Salix nigra, Salvinia, Sambucus nigra, Saururus cernuus, Scirpus cyperinus, Sesbania herbacea, Sium suave, Smilax laurifolia, Solidago, Sparganium androcladum, Sphagnum, Sphenoclea zeylanica, Spiraea alba, Spiranthes, Spirodela polyrrhiza, Styrax americanus, Symphyotrichum novi-belgii, Taxodium ascendens, T. distichum, Thelypteris palustris, Torreyochloa pallida, Toxicodendron radicans, T. vernix, Typha latifolia, Ulmus americana, U. rubra, Utricularia gibba, Vaccinium corymbosum, Verbena brasiliensis, Viburnum dentatum, V. nudum, Viola villosa, Vitis rotundifolia, Woodwardia areolata, Xanthium strumarium, Xyris laxifolia, X. torta.*

Use by wildlife: *Triadenum virginicum* is a larval host plant for several caterpillars (Insecta: Lepidoptera: Noctuidae: *Nedra ramosula*; Oecophoridae: *Agonopterix hyperella*, *A. lythrella*; Tortricidae: *Ancylis maritima*) and spittlebugs (Insecta: Hemiptera: Aphrophoridae: *Lepyronia quadrangularis*). Fireworm eggs (Lepidoptera: Torticidae: *Choristoneura parallela*) are laid on its leaves; the egg masses are parasitized by wasps (Insecta: Hymenoptera: Trichogrammatidae: *Trichogramma minutum*). The lower stems and roots are fed upon by aphids (Insecta: Hemiptera: Aphididae: *Hyalomyzus pocosinus*). The stems can become infected by gall midges (Insecta: Diptera: Cecidomyiidae: *Neolasioptera triadenii*). The plants are also host to rust Fungi (Basidiomycota: Pucciniaceae: *Caeomurus hyperici*). *Triadenum walteri* is fed on by crayfish (Arthropoda: Crustacea: Cambaridae: *Procambarus acutus, P. spiculifer*).

Economic importance: food: Not reported as edible; **medicinal:** Leaf infusions of *Triadenum virginicum* were used by the Potawatomi as a drug to treat fevers. This species recently has seen a revived therapeutic use in nontraditional medicine for the treatment of depression-related disorders; **cultivation:** *Triadenum fraseri* is sold by some native plant nurseries and *T. virginicum* has also been encouraged for planting in native wildflower gardens; **misc. products:** none; **weeds:** *Triadenum virginicum* has been reported as a weed of cranberry marshes in Wisconsin; **nonindigenous species:** *Triadenum fraseri* was introduced to cranberry bogs in British Columbia sometime before 1913. It has also been introduced to cranberry bogs in Europe (Belarus).

Systematics: Previously, *Triadenum* either has been merged with *Hypericum* or is retained as a distinct genus regarded as more closely related to *Cratoxylum*. In the most comprehensive analysis to date, *Triadenum* resolves phylogenetically within a distinct sister clade to the core *Hypericum* species (Figure 4.30); however, with less than compelling internal support for its association with *Hypericum* species of sections *Brathys* and *Myriandra*. Until more compelling justification is presented, *Triadenum* has been retained here as a distinct genus. So defined, *Triadenum* species resolve as a strongly supported clade in phylogenetic analyses (Figure 4.32). *Triadenum* is distinct morphologically by its pink to flesh-colored petals (yellow in *Hypericum*), stamens grouped into three fascicles (ungrouped in *Hypericum*), presence of conspicuous staminodal glands (absent in *Hypericum*), and distinct chromosome number. The original base number of *Triadenum* ($x = 19$) is difficult to ascertain given its close relationship to *Hypericum* ($x = 7$–12). *Triadenum virginicum* and *T. fraseri* are diploids ($2n = 38$), whose "anomalous" counts were taken formerly as evidence of a more distant relationship to *Hypericum* and a possible derivation from *Cratoxylum*. However, molecular data have indicated otherwise (see discussion for *Hypericum* above; Figure 4.30). There are 6–10 species of *Triadenum* worldwide and further work is necessary to more concisely delimit taxa within the genus and to better evaluate their relationships. *Triadenum fraseri* was once thought to be closely related to *T. virginicum* (and even merged within that taxon as a variety or subspecies); however this conclusion is not supported phylogenetically where *T. fraseri* and *T. tubulosum* resolve as sister species (Figure 4.32). AFLP primers have been designed for *Hypericum*, which

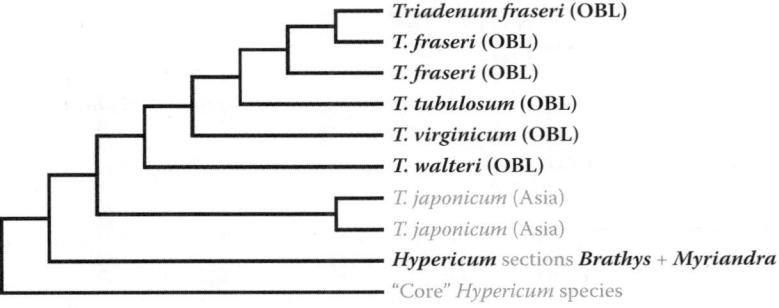

FIGURE 4.32 Interspecific relationships in *Triadenum* as indicated by results of phylogenetic analysis of nrITS sequence data. In that analysis, *Triadenum* resolved as a clade outside of the "core" *Hypericum* species, along with other taxa assigned formerly to several distinct genera. All of the North American taxa have been surveyed and appear to be distinct. The genus represents a single origin of OBL taxa (indicated in bold); the Asian *T. japonicum* is also a wetland plant. (Modified from Nürk, N.M. et al., *Mol. Phylogen. Evol.*, 66, 1–16, 2013.)

amplify markers in *H. adpressum* and *Triadenum walteri*. No natural *Triadenum* hybrids have been reported.

Comments: *Triadenum* is distributed disjunctly in eastern Asia and eastern North America. *Triadenum fraseri* extends broadly across the northern regions of North America; *T. virginicum* extends throughout eastern N. America; *T. tubulosum* and *T. walteri* are distributed mainly in the southeastern United States.

References: Arthur, 1898; Barnes, 1952; Belden, Jr. et al., 2003; Bender et al., 2013; Byers et al., 2007; Cherry & Gough, 2006; Doering, 1942; Dzhus, 2014; Ervin, 2005; Haines et al., 2004; Harper, 1920; Hellquist & Crow, 2003; Holm, 1925; Homoya & Hedge, 1982; Howard & Wells, 2009; Karlin & Lynn, 1988; Keddy et al., 2000; Kost et al., 2007; Lamont, 1988; McMaster, 2001; Mohlenbrock, 1959a; Nichols et al., 2013; Parker, 2005; Percifield et al., 2007; Picking & Veneman, 2004; Richburg et al., 2001; Robson & Adams, 1968; Sharp & Keddy, 1985; Shipley & Parent, 1991; Shull, 1914; Stuart & Polavarapu, 2000; Sundue, 2007; Tyndall et al., 1990; Voigt & Mohlenbrock, 1964; Weeks, 2009; Weishampel & Bedford, 2006; Westervelt et al., 2006; Wells, 1928; White & Simmons, 1988.

Family 4: Linaceae [13]

The Linaceae are a family of about 250 mostly terrestrial species, which are distributed throughout the world. Although Linaceae formerly were placed in Geraniales (or maintained as a distinct order), molecular data strongly support the placement of the family in Malpighiales. However, the sister group of Linaceae remains ambiguous. Combined analysis of 82 plastid gene sequences (Xi et al., 2012) strongly supports an association with Ixonanthaceae (in a clade also including Euphorbiaceae, Peraceae, Phyllanthaceae, and Picrodendraceae (Figure 4.25). Yet single plastid genes yield widely conflicting topologies; For example, *rbcL* data resolve its sister group as a clade comprising Balanopaceae, Chrysobalanaceae, Picrodendraceae, and Trigoniaceae; *matK* data resolve a clade of Clusiaceae and Hypericaceae as its sister group; combined *matK* and *rbcL* data resolve Phyllanthaceae and Picrodendraceae as the sister clade (all placing the family remote from Ixonanthaceae) (McDill & Simpson, 2011). The disparate placement of the family by these different data sets, which all originate from the same genome, do not lend much confidence in their ability to resolve higher level relationships in this case.

Results of molecular data analyses have also suggested the merger of Hugoniaceae (as subfamily Hugonioideae) with Linaceae (McDill et al., 2009); however, the placement of the group as a sister clade of the remaining Linaceae (McDill & Simpson, 2011; Schneider et al., 2016) could just as well argue for its continued maintenance as a distinct family. As formerly circumscribed, *Linum* is polyphyletic, unless members of several genera (*Cliococca, Herperolinon, Radiola, Sclerolinon*) are included (McDill et al., 2009; Figure 4.33). Phylogenetic studies have not yet included either of the OBL *Linum* species, so their precise relationship within the family remains to be clarified.

Linaceae include herbaceous plants and shrubs, as well as trees if Hugoniaceae are merged. The species are pollinated primarily by insects and contain both homostylous and heterostylous flowers. The seeds of some species are mucilaginous, which might facilitate their dispersal by animal vectors.

The family is known most prominently as the source of flax (*Linum usitatissimum*), which yields a versatile fiber as well as edible seeds, from which linseed oil (a preservative and ingredient used in the manufacture of linoleum, paint, putty, and varnish) is also derived. The showy flowers of numerous *Linum* species have led to their use as garden ornamentals. Lignans and α-linolenic acids derived from the plants are believed to have potential for the treatment of cardiovascular disease and cancer (McDill et al., 2009).

In North America, only one genus of Linaceae contains OBL indicators:

   1. ***Linum*** L.

1. *Linum*

Flax; lin

Etymology: from *linum* the Latin word for flax

Synonyms: *Adenolimon, Carthartolinum, Cliococca, Hesperolinon, Mesyniopsis, Radiola, Sclerolinon*

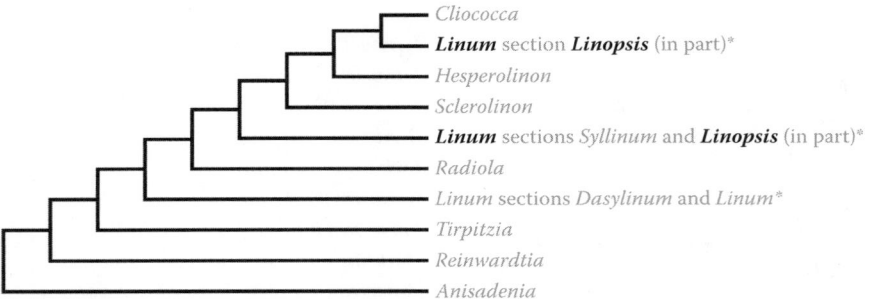

FIGURE 4.33 Phylogenetic relationships in *Linum* as indicated by analysis of *rbcL* sequence data. These data indicate that various *Linum* species (marked by an asterisk) do not comprise a single clade but are interspersed among other genera; thus the genus appears to be polyphyletic. None of the OBL *Linum* species has been included in phylogenetic analyses, but both have been assigned to section *Linopsis*, which resolves here as polyphyletic, thus precluding an assessment of their phylogenetic placement. (Adapted from McDill, J. et al., *Syst. Bot.*, 34, 386–405, 2009.)

Distribution: global: cosmopolitan; **North America:** widespread

Diversity: global: 180+ species; **North America:** 36 species

Indicators (USA): OBL: *Linum arenicola, L. westii*

Habitat: freshwater; palustrine; **pH:** alkaline; **depth:** <1 m; **life-form(s):** emergent (herb)

Key morphology: stems (to 7 dm) 2-several, striate to wing angled, glabrous; leaves (to 1.5 cm) opposite below, alternate above, linear subulate to narrowly elliptic, sometimes with reddish, glandular stipules; inflorescence a terminal cyme; flowers radial, 5-merous; petals (to 6 mm) pale to bright yellow; stamens 5, the filaments united basally; styles 5, distinct; capsules (to 3 mm) dehiscing into 10 single-seeded segments; seeds (to 1 mm) flat, with a yellowish, marginal cartilaginous band

Life history: duration: perennial (persistent base); **asexual reproduction:** none; **pollination:** insect, self; **sexual condition:** hermaphroditic; **fruit:** capsules (common); **local dispersal:** seeds; **long-distance dispersal:** seeds

Imperilment: (1) *Linum arenicola* [G1/G2]; S1 (FL); (2) *L. westii* [G2]; S2 (FL)

Ecology: general: Most of the 200 or so *Linum* species are terrestrial. Eight *Linum* species (22%) are North American wetland indicators (OBL, FACW, FAC, FACU); however, only two species (6%) are categorized as obligate aquatics (OBL). Both of the OBL species are rare and many details on their ecology, floral and seed biology remain poorly studied. *Linum* flowers are both self- and insect pollinated, but precise details of their reproductive biology are unknown for either of the OBL species. Heterostyly occurs in the genus but only within the Old World species. Many *Linum* species are self-compatible. All *Linum* species have C_3 photosynthesis.

***Linum arenicola* (Small) H.J.P. Winkl.** grows in clearings, ephemeral pools, hammocks, levees, marl prairies, pine rocklands, roadsides, and solution pits at low elevations. The plants often occur in disturbed sites with open to partially shaded exposures. The substrates are shallow calcareous soils consisting of gravel, limestone, oolite, or rock. The communities are maintained by periodic fires, which occur naturally every 2–15 years. Optimal habitats are young forests that support extensive graminoids, have a high level of pine (*Pinus*) regeneration, and possess large areas of exposed rock. Areas that remain unburned for several decades can retain good growths of the plants, but periodic fire is necessary to curtail the inevitable encroachment of large, woody species. Natural plant densities currently are quite low, reaching only 0.5 individuals/m². The plants are perennials, which often flower during their first year. Flowering occurs throughout the year, with the flowers opening during the afternoon hours. The seeds germinate readily and will establish fairly successfully in cultivation. This species is known only from about seven sites, where it is threatened by development, invasive plants, and modifications to the natural fire cycles. **Reported associates:** *Abildgaardia ovata, Acalypha chamaedrifolia, Alvaradoa amorphoides, Andropogon gracilis, Angadenia berteroi, Asemeia grandiflora, Ayenia euphrasiifolia, Bletia purpurea, Bouteloua gracilis, Buchnera americana, Byrsonima lucida, Cenchrus incertus, Centrosema virginianum, Chamaecrista lineata, C. nictitans, Cladium jamaicense, Coccothrinax argentata, Coccoloba uvifera, Conocarpus erectus, Crotalaria pumila, Desmanthus virgatus, Ditaxis argothamnoides, Eragrostis elliottii, Erithalis fruticosa, Ernodea littoralis, Euphorbia blodgettii, E. deltoidea, Evolvulus sericeus, Fimbristylis cymosa, F. spadicea, Flaveria linearis, Galactia parvifolia, G. volubilis, Linum carteri, Melanthera parvifolia, Metopium toxiferum, Morinda royoc, Mosiera longipes, Myrica cerifera, Paspalum caespitosum, P. setaceum, Phyla nodiflora, Pinus elliottii, Pisonia rotundata, Pithecellobium keyense, Polygala balduinii, Sabatia stellaris, Schizachyrium sanguineum, Serenoa repens, Sida ciliaris, Sideroxylon salicifolium, Sophora tomentosa, Spermacoce verticillata, Sporobolus pyramidalis, Stylosanthes hamata, Symphyotrichum adnatum, Thrinax morrissii, T. radiata, Tragia saxicola, Waltheria indica.*

***Linum westii* C.M. Rogers** occurs in shallow waters of bogs, depression ponds, ditches, flatwoods, and along the margins of swamps at low elevations. Exposures range from full sun to partial shade. The substrates are described as mud and undisturbed peaty soils. The flowers possibly are pollinated by beetles (see *Use by wildlife*). They open during the evening, between 6 and 7 p.m. (June to July). The plants are clump forming and are maintained by periodic episodes of fire. Additional life-history information is needed for this species. **Reported associates:** *Aristida palustris, Centella asiatica, Cliftonia monophylla, Cyrilla racemiflora, Dichanthelium sphaerocarpon, Eleocharis, Eriocaulon, Gratiola ramosa, Hypericum chapmanii, H. myrtifolium, Ilex myrtifolia, Lachnanthes caroliniana, Linum floridanum, Magnolia, Mitreola sessilifolia, Nyssa biflora, Panicum hemitomon, Pinguicula planifolia, Pinus elliottii, P. palustris, Pleea tenuifolia, Polygala cymosa, Proserpinaca, Rhexia lutea, Rhynchospora filifolia, R. latifolia, Sabatia grandiflora, Sarracenia minor, Scleria triglomerata, Serenoa repens, Sporobolus floridanus, Taxodium ascendens, T. distichum, Xyris elliottii.*

Use by wildlife: The flowers of *Linum westii* are visited by (and potentially pollinated by) soft-winged flower beetles (Insecta: Coleoptera: Melyridae: *Listrus*).

Economic importance: food: not edible; **medicinal:** No medicinal uses have been reported for either of the OBL aquatic species; however, other members of the genus produce a variety of secondary metabolites, which have medicinal applications; **cultivation:** The OBL species are not cultivated; **misc. products:** The related *Linum usitatissimum* is the source of commercial flax fiber; **weeds:** none; **nonindigenous species:** none.

Systematics: *Linum* has been shown to be monophyletic, once several former segregate genera (*Cliococca, Herperolinon, Radiola, Sclerolinon*) have been merged within it. Phylogenetic surveys of *Linum* have cast doubts on the taxonomic integrity of the proposed sections and series. The genus is subdivided into five sections (*Catharticum, Dasylinum, Linopsis, Linum, Syllinum*) with *L. arenicola* and *L. westii* both placed within section *Linopsis*, which is polyphyletic as originally circumscribed (Figure 4.33). Morphologically, *L. arenicola* is quite similar to *L. rupestre* and is closely related to it. Molecular data (Figure 4.34) resolve it in a clade together with *L. rupestre* and *L. flagellare*. *Linum arenicola* also shares a similar

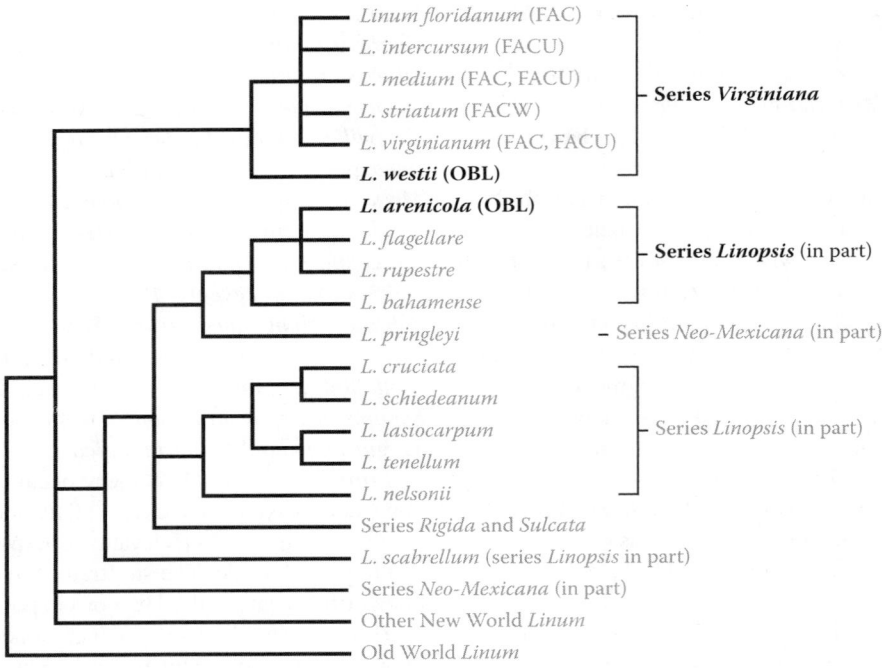

FIGURE 4.34 Phylogenetic relationships in *Linum* as reconstructed from chloroplast DNA sequence data. These results indicate the independent origin of the OBL North American species (in bold). Several series (*Linopsis*, *Neo-Mexicana*) are not monophyletic in this analysis. (Adapted from McDill, J., Molecular phylogenetic studies in the Linaceae and *Linum*, with implications for their systematics and historical biogeography, PhD dissertation, University of Texas, 2009.)

pollen morphology with *L. rupestre* and *L. bahamense*, the latter resolving as the sister group to the *L. arenicola*, *L. flagellare*, and *L. rupestre* clade (Figure 4.34). The relationships of *L. westii* are with species of section *Linopsis* series *Virginica*, where it resolves as the sister group to the remainder of the series (Figure 4.34). These results indicate that the OBL habit has arisen independently in the two North American taxa. The chromosomal base number of Linaceae (and presumably *Linum*) is *x* = 6 although modified numbers (e.g., *x* = 15, 18) occur commonly. *Linum arenicola* is 2*n* = 36; the chromosome number for *L. westii* is unreported.

Comments: *Linum arenicola* is restricted to pine rockland habitats of peninsular southern Florida and the keys. *Linum westii* is endemic to northeast Florida and the Florida Panhandle.

References: Adams et al., 2010; Bradley & Gann, 1999; Godfrey & Wooten, 1981; Hammer, 2004; Harris, 1968; Hodges & Bradley, 2006; Kearns & Inouye, 1994; Long & Lakela, 1971; McDill, 2009; McDill & Simpson, 2011; McDill et al., 2009; Osborne & Lewis, 1962; Rogers, 1963, 1972; Schneider et al., 2016; Xavier & Rogers, 1963.

Family 5: Podostemaceae [50]

Podostemaceae with approximately 270 species comprise the largest family of exclusively aquatic angiosperms (Philbrick & Novelo, 2004). All species occur in freshwater, riverine habitats that are characterized by rapid currents. Roughly 160 species in 20 genera occur in the New World, but mainly in the tropical regions. The family itself is one of the most enigmatic with respect to its phylogenetic affinities. Extreme morphological

divergence is mirrored by high levels of molecular divergence, making it difficult to reconstruct relationships to other angiosperms with a high degree of confidence. Analyses of various DNA sequence data sets have suggested possible associations of the family to such diverse groups as Hydrostachyaceae (Les et al., 1997) or various families within the order Malpighiales (Kita & Kato, 2001; Gustafsson et al., 2002; Davis & Chase, 2004; Xi et al., 2012; Figure 4.25). Because the phylogenetic placement of the family is so inconsistently resolved by different molecular data sets, it is difficult to conclude at this time any more specific details of relationships other than its apparent inclusion within the Malpighiales.

Morphologically, all Podostemaceae share a highly modified axial organ system, which is algal-like and difficult to interpret in the traditional sense as shoot, root, or leaves (Philbrick & Novelo, 2004). There are no economically important members in the family.

Both molecular (Kita & Kato, 2001) and morphological data (Philbrick & Novelo, 2004) support the recognition of three monophyletic subfamilies: Tristichoideae, Weddellinoideae, and Podostemoideae, of which only the latter is represented in North America, and there by a single genus.

1. *Podostemum* Michx.

1. *Podostemum*

Riverweed

Etymology: from the Greek *podos stemon* meaning "stamen foot" in reference to the floral andropodium

Synonyms: *Crenias, Devillea, Mniopsis*

Distribution: global: New World; **North America:** eastern
Diversity: global: 11 species; **North America:** 1 species
Indicators (USA): OBL: *Podostemum ceratophyllum*
Habitat: freshwater; riverine; **pH:** 6.6–8.2; **depth:** 0–1.5 m;
life-form(s): submersed (thalloid)
Key morphology: plants algal-like; roots (to 1.1 mm wide) photosynthetic and modified as holdfasts (haptera); stems (to 3 dm) procumbent; leaves (to 1.4 dm) distichous, petiolate, 1–13 times dichotomously divided or lobed; stipules (to 3.6 mm) present; flowers 1–12 per stem; tepals (to 2 mm) 3, stamens (to 2 mm) 2, arising from a common, stalk-like andropodium (to 3.3 mm); capsules (to 3.1 mm) 6-ribbed, pedicellate (to 9 mm); 0–42 seeds per capsule
Life history: duration: perennial (proliferation of roots); **asexual reproduction:** fragmentation of roots and stems; **pollination:** cleistogamy, self; **sexual condition:** hermaphroditic; **fruit:** capsules (prolific); **local dispersal:** seeds; vegetative growth; **long-distance dispersal:** seeds
Imperilment: (1) *Podostemum ceratophyllum* [G5]; S1 (DE, LA, NS, OH, VT); S2 (MA, ME, NB, NJ, NY, OK, ON, QC, RI); S3 (AR, CT, KY, MD)
Ecology: general: Members of Podostemaceae occupy a narrow ecological niche associated with extreme environmental conditions. All of the species are thalloid, obligately aquatic haptophytes, which grow attached to various stable, rocky substrates in association with a cyanobacterial biofilm, in clear, clean, lotic flowing waters (waterfalls and river rapids) whose levels exhibit seasonal periodicity. The species have a high light requirement. Few specifics are known about the general life history of many of the species and numerous aspects of their reproductive biology and seed ecology remain uncertain. The flowers of some species experience an initial xenogamous phase pollinated by bees (Insecta: Hymenoptera: Apidae), which then is followed by an autogamous phase.

Podostemum ceratophyllum Michx. is found most often in river rapids or riffles, although some populations also grow well along the bottom of deeper, rapidly flowing watercourses (15–128 cm/s) at elevations up to 1100 m. The plants thrive in hard, lotic waters where levels of dissolved inorganic carbon are high. The waters are circumneutral to alkaline (pH: 6.6–8.2) and shallow (<1 m depth; occasionally deeper in tidal sites). The substrates range from gravel and pebbles to cobble, rocks, and boulders (and occasionally logs), which can be acidic to circumneutral (e.g., granite, sandstone), or alkaline (e.g., limestone). Attachment to manmade structures (e.g., concrete) also occurs. Flowering and fruiting occur from June to October. Flowering is induced as the plants are exposed by falling water levels. Populations are cyclic, with periods of flowering, vegetative die back, seed dispersal, and establishment occurring annually during low-water conditions, and vegetative growth occurring mainly during high-water periods. *Podostemum ceratophyllum* primarily is inbreeding due largely to self-pollination resulting from preanthesis cleistogamy. Isozymes and noncoding cpDNA sequence data indicate that populations south of the glacial boundary have higher levels of genetic variation. The lower genetic diversity of northern populations probably reflects bottlenecks associated with northward post-Pleistocene migrations. The plants can produce more than 20 capsular fruits per cm^2, which mature in 2–3 weeks and produce upward of 4.5 million seeds/m^2. The seeds possess a mucilaginous coat, which binds them to their rocky substrates upon drying. The seeds have no dormancy and germinate readily at 23°C when emersed in water and exposed to high light condtions. The seeds retain nearly 100% germination when stored for 2 months at 9°C–12°C. Upon germination, the rapidly growing roots attach quickly to the substrate by means of haptera. The largest mature plants occur as sterile shoots, which grow in habitats where higher water levels are maintained rather constantly. Although typical vegetative propagules are lacking, the plants are able to disperse within a watercourse by means of detached root fragments, which reattach tenaciously to downstream substrates. *Podostemum ceratophyllum* often forms monocultures in its highly specialized habitat; however, a few other angiosperm, bryophyte, and algal species occasionally are found in association. **Reported associates:** *Callitriche, Cladophora, Elodea canadensis, Fontinalis novae-anglia, Justicia americana, Leersia, Lemanea australis, Lemna minor, Najas guadalupensis, Nitella, Orontium aquaticum, Potamogeton crispus, P. diversifolius, P. robbinsii, Saururus cernuus, Vaucheria.*
Use by wildlife: *Podostemum ceratophyllum* enhances habitat (up to an observed fourfold increase in surface area) for many riverine benthic macroinvertebrates including beetles (Insecta: Coleoptera: Elmidae: *Promoresia*), blackfly larvae (Insecta: Diptera: Simuliidae: *Simulium*), caddis fly larvae (Insecta: Trichoptera: Brachycentridae: *Brachycentrus*; Hydropsychidae: *Cheumatopsyche*, *Hydropsyche simulans*; Hydroptilidae: *Hydroptila*; Lepidostomatidae: *Lepidostoma*; Leptoceridae: *Ceraclea joannae*), dobsonflies (Insecta: Megaloptera: Corydalidae: *Corydalus*), flatworms (Protostomia: Platyhelminthes), mayfly larvae (Insecta: Ephemeroptera: Baetidae: *Baetis*; Ephemerellidae: *Serratella*; Isonychiidae: *Isonychia*), midge larvae (Insecta: Diptera: Chironomidae: Tanypodinae), snails (Mollusca: Gastropoda: Pleuroceridae: *Elimia*), stoneflies (Insecta: Plecoptera: Perlidae: *Paragnetina*, *Perlesta*; Pteronarcyidae: *Pteronarcys*), water mites (Arachnida: Acari: Trombidiformes: Hydrachnidiae), and worms (Annelida: Oligochaeta). The species is regarded as an important habitat feature for several fish (Chordata: Osteichthys: Perciformes: Percidae) including the banded darter (*Etheostoma zonale*), holiday darter (*Etheostoma brevirostrum*), and riverweed darter (*Etheostoma podostemone*); it is categorized as critical habitat for two endangered fish: the amber darter (*Percina antesella*) and Conosauga logperch (*Percina jenkinsi*). Adult tangerine darters (*Percina aurantiaca*) are known to feed extensively on invertebrates found on the foliage. The plants provide spawning habitat for spotfin chubs (Chordata: Osteichthys: Cypriniformes: Cyprinidae: *Erimonax monachus*) and frecklebelly madtoms (Chordata: Osteichthys: Siluriformes: Ictaluridae: *Noturus munitus*). The plants also harbor a diverse assemblage of epiphytic algae and diatoms (Chromista: Bacillariophyceae), which are consumed by various grazers including beavers (Mammalia: Rodentia: Castoridae: *Castor canadensis*), Canada geese

(Aves: Anatidae: *Branta canadensis*), crayfish (Crustacea: Decapoda: Cambaridae: *Procambarus spiculifer*), and turtles (Reptilia: Testudines).

Economic importance: food: none; **medicinal:** none; **cultivation:** none; **misc. products:** none; **weeds:** none; **nonindigenous species:** none.

Systematics: Phylogenetic studies have redefined *Podostemum* as an exclusively New World genus, with two previously assigned Old World species (*P. barberi*, *P. subulatum*) assigned to *Zeylanidium* by some authors but resolving as the sister group of *Willisia* in molecular analyses. Morphological data place *Podostemum* in a clade comprising all other New World genera except *Cipoia* and *Diamantina*; however resolution is insufficient to clarify its position among those genera more precisely. Phylogenetic analysis of *rbcL* data for a limited sample of six New World Podostemaceae resolved the Asian genus *Cladopus* as the sister group of *Podostemum*; however, *matK* data with increased taxon sampling resolved *Podostemum* among African/Malagasy clades of Old World genera, but not with compelling internal support. Additional sampling and analyses will be necessary before intrafamilial relationships of the genus can be clarified further; however, current studies point to an Old World affinity of the genus. Morphological data situate *P. ceratophyllum* in a basal position within *Podostemum*; however, several molecular data sets (e.g., Figure 4.35) indicate the species to be more derived and related to *P. comatum*, which associated closely in morphological cladistic analyses as well. The chromosome number is unreported for *Podostemum*. Counts for other Podostemaceae genera include $2n = 20, 26, 30$ & 34, with $x = 5$ being the probable base number.

Comments: *Podostemum ceratophyllum* occurs in eastern North America with disjunct populations in the Dominican Republic and Honduras.

References: Argentina, 2006; Argentina et al., 2010; Capers & Les, 2001; Connelly et al., 1999; Fehrmann et al., 2012; Hill & Webster, 1984; Hutchens et al., 2004; Khanduri et al., 2015; Kita & Kato, 2001; Kita et al., 2012; Les et al., 1997; Moline, 2001; Morse & Lenat, 2005; Parker et al., 2007a; Philbrick, 1984a, 1992; Philbrick & Novelo, 2004; Philbrick et al., 2006, 2015; Sobral-Leite et al., 2011; Tinsley, 2012.

Family 6: Rhizophoraceae [15]

The Rhizophoraceae (red mangroves) are a small but pantropical family of about 149 trees and shrub species, which inhabit sites ranging from moist montane forests to low tidal wetlands (Judd et al., 2016). The plants share opposite leaves with intrapetiolar stipules, vessel elements with vestured pits, and 4–5-merous bisexual flowers; an intrastaminal nectary disk usually is present on the ovary or hypanthium (Judd et al., 2016). Four of the genera (and about 17 species) are recognized as mangroves, i.e., a derived clade of woody plants, which form dense intertidal swamp communities along coastal and tidal shores. Although the flowers of most mangrove species are pollinated biotically by insects (e.g., Lepidoptera) or birds (Aves), those of *Rhizophora* are nectarless and wind pollinated (Judd et al., 2016).

Some analyses, even using multiple molecular data sets, do not provide much phylogenetic resolution within the Malpighiales; yet other studies using various combinations of molecular (nuclear and cpDNA) and morphological data support a close association of Rhizophoraceae with Ctenolophonaceae and Erythroxylaceae (Schwarzbach & Ricklefs, 2000; Xi et al, 2012; Figures 4.25 and 4.36). However, some caution is advised. Analyses derived from different molecular data produce tree topologies that are so inconsistent as to raise serious doubts that any firm conclusions can be made regarding higher level phylogenetics of the order at this time. As one example, a study analyzing three cpDNA loci along with 18S nrDNA (Tokuoka & Tobe, 2006) resolved Rhizophoraceae as the sister group to a clade comprising Clusiaceae, Elatinaceae, Hypericaceae, Ixonanthaceae, Malpighiaceae, Podostemaceae, and Ochnaceae; whereas, an analysis of 82 plastid gene sequences (Xi et al., 2012) scatters these families across the entire order.

Some *Rhizophora* species are a source of charcoal and tannins; otherwise the primary importance of this family is ecological, by providing protection and stabilization of shorelines and habitat for marine biota (Judd et al., 2016).

Combined morphological and molecular data (Schwarzbach & Ricklefs, 2000) resolve three clades within Rhizophoraceae (Figure 4.36), which have been designated as distinct subfamilies: *Gynotrocheae*, *Macarisieae*, *Rhizophoreae*. The four mangrove genera (*Bruguiera*,

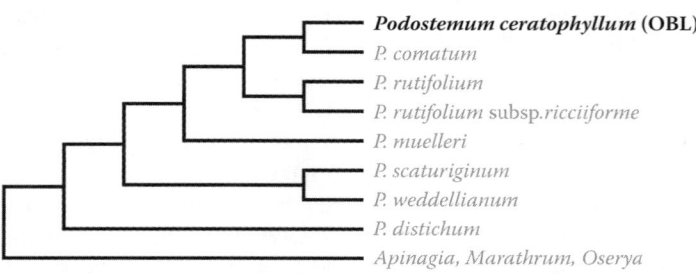

FIGURE 4.35 Interspecific phylogenetic relationships in *Podostemum* as indicated by analysis of nrITS sequence data *Podostemum ceratophyllum*, the only OBL North American species (shown in bold), occupies a relatively derived position in this cladogram. All members of the family are obligate aquatic plants. (Adapted from Moline, P., *Podostemum* and *Crenias* (Podostemaceae) – American riverweeds – infrageneric systematic relationships using molecular and morphological methods, MSc thesis, Institut für Systematische Botanik, Universität Zürich, 2001.)

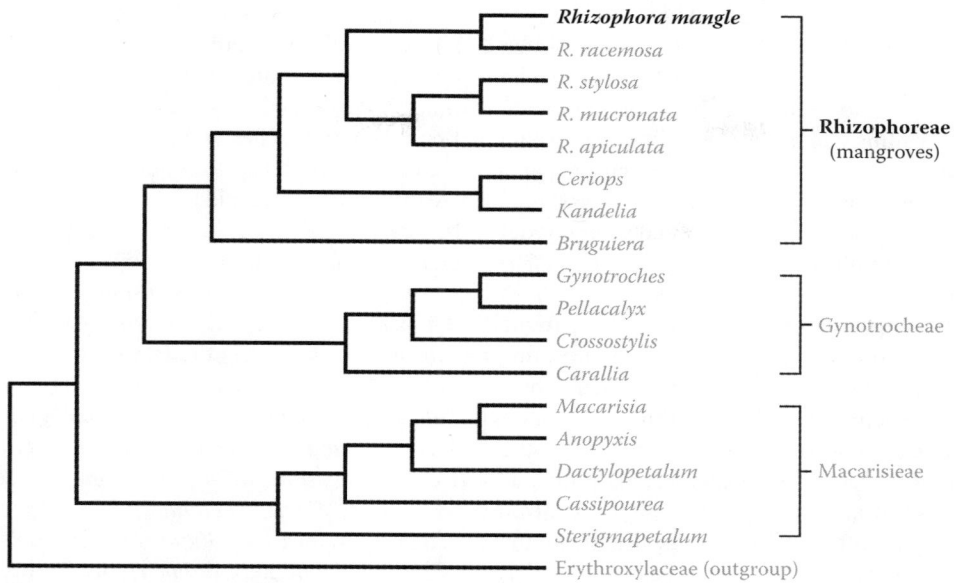

FIGURE 4.36 Phylogenetic relationships in Rhizophoraceae (and *Rhizophora*) as indicated by analysis of combined molecular and morphological data. The aquatic mangroves form the clade delimited as subfamily Rhizophoreae. The sole North American species *Rhizophora mangle* (shown in bold) occupies a relatively derived position in the family. (Adapted from Schwarzbach, A.E. & Ricklefs, R.E., *Amer. J. Bot.*, 87, 547–564, 2000.)

Ceriops, Kandelia, Rhizophora) are monophyletic and comprise subfamily *Rhizophoreae* (Figure 4.36). Only *Rhizophora* (red mangrove) occurs in North America where it is represented by a single species, which is an OBL indicator.

1. ***Rhizophora*** L.

1. *Rhizophora*
Red mangrove
Etymology: from the Greek *rhizos phoros* meaning "to carry roots" in reference to the prop roots of the stem
Synonyms: *Mangium*
Distribution: global: Africa; New World; **North America:** southeastern
Diversity: global: 6 species; **North America:** 1 species
Indicators (USA): OBL: *Rhizophora mangle*
Habitat: marine; palustrine, riverine; **pH:** 5.3–7.8; **depth:** <1 m; **life-form(s):** emergent shrub, emergent tree
Key morphology: stems (to 25 m) with a flat-topped crown, the bark gray externally, reddish internally, trunk and branches often with pendant, aerial prop roots; leaves (to 15 cm) opposite, simple, leathery, evergreen; inflorescence of 2–3 flowers (to 2.6 cm); sepals (to 10 mm) 4, yellow, acute tipped; petals (to 9 mm) 4, yellow white, hairy above; stamens 8; fruit (to 3.6 cm) baccate, conical, single seeded; seedlings (to 30 cm) viviparous, torpedo shaped
Life history: duration: perennial (buds, whole plants); **asexual reproduction:** none; **pollination:** self, wind; **sexual condition:** hermaphroditic; **fruit:** baccate with viviparous seedling (common); **local dispersal:** seedlings (animals, water); **long-distance dispersal:** seedlings (animals?, water)
Imperilment: (1) *Rhizophora mangle* [G5]; S3 (FL)

Ecology: general: All *Rhizophora* species are OBL plants, which develop into "mangrove" swamps along coastal areas. The stems produce woody "prop roots" that help to stabilize the plants against the severe wind and wave action associated with coastal storms and hurricanes, which can reduce plant coverage by an order of magnitude. The flowers are not only wind pollinated, but also attract bees (Insecta: Hymenoptera) that forage for the pollen, which is deposited on the petals before the flowers open. The fruits are fibrous, single-seeded berries described as "viviparous," because the seeds germinate while within the fruit. An embryo with tubular cotyledons develops in the fruit and can be retained on the maternal plant for up to 9 months. When mature, the hypocotyl/plumule of the embryo detaches, falls to the saline water, and floats on the surface, eventually producing adventitious roots. The young seedlings establish readily once they have rooted in the substrate.

Rhizophora mangle L. inhabits tidal, brackish and saline waters of beaches, canals, coastlines, ditches, dunes, estuaries, hammocks, lagoons, salt marshes, shoreline/tidal flats, sloughs, strands, swamps, thickets, and the margins of ponds, rivers, and streams at elevations near sea level (e.g., 1–4 m). The plants are highly frost sensitive and occur at latitudes below 30° in both the Northern and Southern Hemispheres. These are FAC halophytes, which not only tolerate salinities from 0 to 90 ppt and typically colonize anoxic sediments of the lowest intertidal zone, but also thrive above the limit of tidal inundation. The substrates are described as coral rock, marl, marl clay, muck, mucky sand, mud, Peckish (typic sulfaquents), rocky marl, sand, shells, and shell sand. The roots are equipped with ultrafilters, which exclude salt, allowing the plants to extract fresh water from their marine sediments. The aerial prop roots stabilize the plants in unconsolidated sediments and facilitate oxygen exchange to the roots by means of their lenticels and aerenchyma system. Extensive mangrove stands protect shorelines from erosion by reducing current velocity and by increasing sedimentation. Flowering and fruiting occur throughout the year. Young plants can flower as early as 1.5 years of age and primarily are autogamous. Genetic studies using phenotypic propagule pigmentation markers have documented a broad extent of outcrossing rates ranging from 0% to 41%, with 17% to 40% of sites characterized by complete selfing. The seeds are viviparous, i.e., germinating while still attached to the parental plant. Seed germination occurs at about 70 days after pollination, yet factors responsible for the breaking of their physiological dormancy are unknown. The resulting propagules (enlarged seedlings) can float for up to a year before rooting and are susceptible to excessive sediment burial. Propagule density declines markedly at distances greater than 2 km from the source plants and data from AFLP markers indicate that only low levels of dispersal occur between islands. Seedling establishment can occur in dense shade, but varies depending on the substrate, being anchored more strongly (up to 3.5×) on coarser substrates relative to sand. Intact seeds have also been recovered from the guts and feces of crocodilians (see *Use by wildlife*), which indicates the potential for at least local animal dispersal as well. The plants have C$_3$ photosynthesis, do not appear to grow clonally, and are not adapted to fire. The decaying leaves are involved in anaerobic nitrogen fixation. Populations are threatened naturally by hurricanes (which can devastate stands), by various pollutants (heavy metals, nutrients, oil spills, thermal pollution), and by deforestation practices related to agriculture, residential development, and timber harvesting. The plants have few angiosperm ecological associates owing to their specialized habitat niche. **Reported associates:** *Acrostichum aureum, A. danaeifolium, Atriplex, Avicennia germinans, Batis maritima, Blutaparon, Borrichia, Casuarina, Conocarpus erectus, Distichilis spicata, Euphorbia mesembrianthemifolia, Ipomoea, Juncus roemerianus, Laguncularia racemosa, Monanthochloe, Pinus elliottii, Pithecellobium, Quercus, Sapindus, Sarcocornia perennis, Schinus terebinthifolius, Serenoa repens, Sesuvium, Spartina alterniflora, Sporobolus, Suriana maritima, Thrinax morrisii.* **Use by wildlife:** Red mangrove is used by an incredible diversity of wildlife, which is summarized here only partially (see USFWS, 1999 for additional information). The foliage, viviparous propagules, and young seedlings of *R. mangle* provide food for crocodilians (Reptilia: Crocodilia: Alligatoridae: *Alligator mississippiensis*; Crocodylidae: *Crocodylus acutus*), swamp ghost crabs (Crustacea: Decapoda: Ocypodidae: *Ucides cordatus*), spotted mangrove crabs (Crustacea: Decapoda: Grapsidae: *Goniosis cruentata*), coffee bean snails (Mollusca: Gastropoda: Ellobiidae: *Malampus coffeus*), ladder horn snails (Gastropoda: Potamididae: *Cerithidea scalariformis*), manatees (Mammalia: Sirenia: Trichechidae: *Trichechus manatus*), and stem-boring beetles (Insecta: Coleoptera: Curculionidae: *Coccotrypes rhizophorae*). Red mangrove is severely impacted by wood-boring beetles (Insecta: Coleoptera: Cerambycidae: *Elaphidion mimeticum, Elaphidinoides*), which can girdle

and kill the stems. The hollow stems created by wood-boring beetles and moth larvae are used as shelter by numerous species of ants (Insecta: Hymenoptera: Formicidae), mites (Arthropoda: Arachnida: Acari), moths (Insecta: Lepidoptera), roaches and termites (Insecta: Blattodea), scorpions (Arthropoda: Arachnida: Scorpiones), and spiders (Arthropoda: Arachnida: Araneae). The leaves are eaten by several crabs (Crustacea: Decapoda) including blue land crabs (Gecarcinidae: *Cardisoma guanhumi*), mangrove tree crabs (Sesarmidae: *Aratus pisonii*), spotted mangrove crabs (Grapsidae: *Goniosis cruentata*), and also by many insects. *Rhizophora mangle* is the larval host plant for several caterpillars (Insecta: Lepidoptera: Cossidae: *Psychonoctua personalis*; Gelechiidae: *Compsolechia mangelivora*; Geometridae: *Oxydia cubana*; Heliozelidae: *Coptodisca rhizophorae*; Hesperiidae: *Phocides pigmalion*; Limacodidae: *Alarodia slossoniae, Euclea delphinii*; Megalopygidae: *Megalopyge krugii*; Notodontidae: *Gluphisia lintneri*; Psychidae: *Oiketicus abbotii*; Saturniidae: *Automeris io*; Tortricidae: *Ecdytolopha desotanum, E. insiticiana* and *E. punctidiscanum*). The plants also harbor serious pathogenic pests such as leaf spot (Fungi: Ascomycota: Phyllachoraceae: *Pterosporidium rhizophorae*) and lobate lac scale (Insecta: Hemiptera: Kerriidae: *Paratachardina lobata*). Also hosted are several pore fungi (Basidiomycota: Hymenochaetaceae: *Phellinus*; Meruliaceae: *Ceriporiopsis aneirina*; Polyporaceae: *Coriolopsis caperata, Trichaptum biforme*). The extensive prop root systems harbor algae (Chlorophyta: Caulerpaceae: *Caulerpa racemosa*; Rhodophyta: Rhodomelaceae: *Acanthophora spicifera*; Corallinaceae: *Lithophyllum*), barnacles (Crustacea: Sessilia: Balanidae: *Balanus amphitrite, B. eburneus, B. inexpectatus, Fistulobalanus suturaltus*; Chthamalidae: *Chthamalus fissus, Microeuraphia eastropacensis*), blue-green algae (Cyanobacteria: Scytonemataceae: *Scytonema polycystum*), cnidarians (Cnidaria: Anthozoa: Aiptasiidae: *Aiptasia pallida*), crabs (Crustacea: Decopoda), oysters (Mollusca: Ostreidae: *Crassostrea virginica*), shrimp (Crustacea: Decapoda), sponges (Porifera: Chalinidae: *Haliclona curacaoensis, H. implexiformis*; Coelosphaeridae: *Lissodendoryx*; Geodiidae: *Geodia papyracea*; Halichondriidae: *Hymeniacidon heliophila*; Mycalidae: *Mycale magnirhaphidifera*; Niphatidae: *Amphimedon viridis*; Scopalinidae: *Scopalina ruetzleri*; Tedaniidae: *Tedania ignis*), tunicates (Chordata: Tunicata: Ascidiacea: Ascidiiae: *Phallusia nigra*; Didemnidae: *Didemnum concyhliatum, Diplosoma glandulosum*; Perophoridae: *Perophora multiclathrata*), and juvenile fish (Vertebrata: Osteichthyes: Teleostei). The prop roots are damaged by feeding root weevils (Insecta: Coleoptera: Curculionidae: *Diaprepes*) and wood-boring isopods (Crustacea: Malacostraca: Isopoda: Limnoriidae: *Limnoria clarkae*; Sphaeromatidae: *Sphaeroma terebrans*). Red mangrove plants release carbon to root-inhabiting sponges (e.g., Porifera: Chalinidae: *Haliclona*; Tedaniidae: *Tedania*), which mutualistically transfer nitrogen to their roots and help to protect them from isopod feeding. Because of their extensive root systems, red mangroves are used as habitat by many fish (Vertebrata: Osteichthyes). These include adult flatfish (Pleuronectiformes: Achiridae: *Achirus lineatus*); herrings

(Clupeiformes) such as bay anchovy and finescale menhaden (Clupeidae: *Anchoa mitchilli, Brevoortia gunteri*); killifishes and livebearers (Cyprinodontiformes) including goldspotted killifish (Cyprinodontidae: *Floridichthys carpio*), Gulf and longnose killifish (Fundulidae: *Fundulus grandis, F. similis*), mangrove gambusia (Poeciliidae: *Gambusia rhizophorae*), mangrove rivulus (Aplocheilidae: *Rivulus marmoratus*) rainwater killifish (Fundulidae: *Lucania parva*), sailfin molly (Poeciliidae: *Poecilia latipinna*), and sheepshead minnows (Cyprinodontidae: *Cyprinodon variegatus*); perch-like fish (Perciformes) such as the Atlantic bumper (Carangidae: *Chloroscombrus chrysurus*), blackchin tilapia (Cichlidae: *Sarotherodon melanotheron*), clown and code goby (Gobiidae: *Microgobius gulosus, Gobiosoma robustum*), juvenile goliath grouper (Serranidae: *Epinephelus itajara*), mangrove snapper (Lutjanidae: *Lutjanus griseus*), Mayan cichlid (Cichlidae: *Cichlasoma urophthalma*), pinfish (Sparidae: *Lagodon rhomboides*), sheepshead (Sparidae: *Archosargus probatocephalus*), spot (Sciaenidae: *Leiostomus xanthurus*), and spotfin mojarra and striped mojarra (Gerreidae: *Eucinostomus argenteus, Eugerres plumieri*); and silversides (Atheriniformes: Atherinopsidae: *Menidia*). Decaying leaves of *R. mangle* release a proline-rich peptide that acts as an environmental cue for settling and metamorphosis in planula larvae of the upside down jellyfish (Cnidaria: Rhizostomeae: Cassiopeidae: *Cassiopea xamachana*). Pelicans (Aves: Pelecanidae: *Pelecanus*) and other seabirds use mangrove canopies for roosting. Bald eagles (Aves: Accipitridae: *Haliaeetus leucocephalus*) and black-whiskered vireos (Aves: Vireonidae: *Vireo altiloquus*) build their nests in the trees. The plants are also a known host of at least 34 species of xylophilous fungi (Basidiomycota).

Economic importance: food: The wood of *Rhizophora mangle* reputedly is edible and also can be fermented to make a light wine. The roots have been used as a famine food and the leaves to brew tea. The species is regarded as a honey plant; **medicinal:** The bark of *R. mangle* has been used to treat fever and leprosy, to control hemorrhage, and as an expectorant. The plants have been used as a blood purifier and to promote kidney health. A drink made from water, root pieces, and honey results in a reddish liquid that is drunk to treat liver problems and anemia. Similar preparations have been used as a dermatological aid, especially to treat excessive sun exposure. Extracts of *R. mangle* have been shown to possess antihyperglycemic activity and are used as an antidiabetic. They also have been shown to be gastroprotective through antioxidant and prostaglandin-dependent pathways; **cultivation:** Plants of *R. mangle* are sold as water clarifyers for saltwater ponds and aquariums and for native habitat restoration; **misc. products:** The Seminoles used the tanniniferous *R. mangle* (up to 30% tannins by dry weight) as a dye for buckskin. The boiled bark yields a furniture stain. The wood of *R. mangle* is resistent to decay fungi and is used for boat building, charcoal, fuel (burning even when green), railroad ties, timber, and planks. The ash can be used as a soap substitute. The dried fruits (i.e., seedlings) are smoked like cigars and the dried leaves can be smoked in pipes. *Rhizophora mangle* produces triterpenoids, which exhibit insecticidal activity. Bark

extracts from the plants are highly effective at preventing corrosion of concrete steel reinforcement; **weeds:** *Rhizophora mangle* is a serious pest in Hawaii where it infests fishponds and wetlands, and reduces habitat for the imperiled Hawaiian stilt (Aves: Recurvirostridae: *Himantopus mexicanus knudseni*); **nonindigenous species:** *Rhizophora mangle* was introduced to Hawaii more than a century ago.

Systematics: Several molecular studies demonstrate that *Rhizophora* is monophyletic and occurs in a relatively derived position within the monophyletic subfamily Rhizophoreae ("mangroves") of family Rhizophoraceae (Figure 4.36). Combined morphological and molecular data (Figure 4.36) implicate the African-Central/South American *R. racemosa* as a closely related sister species to *R. mangle*. This association is understandable given that the former had been recognized previously as a variety of *R. mangle*. Although *R. mangle* primarily is self-pollinating, interpopulational gene flow with some local differentiation has been indicated in a survey of southern Florida populations using RAPD markers. The chromosomal base number of Rhizophoraceae is $x = 8, 9$. *Rhizophora mangle* ($2n = 36$) presumably is tetraploid. Preliminary transcriptome analyses are available, which ultimately should shed light on the genetic basis of adaptive traits relative to the extreme environments colonized by these plants. Hybridization and backcrossing of *R. mangle* and *R. racemosa* have been documented using cpDNA sequences and microsatellite genetic markers. Population genetic studies using AFLP markers have documented low levels of gene flow between mainland and island populations, as well as some populations being characterized by low levels of genetic variation.

Comments: *Rhizophora mangle* occurs along the coastal regions of North Carolina to Florida, and extends into Mexico, South America, Puerto Rico, the Virgin Islands, and western Africa.

References: Alarcon-Aguilara et al., 1998; Albrecht et al., 2013; Allen & Krauss, 2006; Baltazar et al., 2009; Berenguer et al., 2006; Boizard & Mitchell, 2011; Brooks & Bell, 2005; Castelblanco-Martínez et al., 2009; Cerón-Souza et al., 2010; Dassanayake et al., 2010; Ellison et al., 1996; Fleck & Fitt, 1999; Fleck et al., 1999; Frias-Torres, 2006; Gilbert & Sousa, 2002; Gotto & Taylor, 1976; Judd et al., 2016; Klekowski et al., 1994; Lowenfeld & Klekowski, 1992; Mullin, 1995; Okeniyi et al., 2014; Platt et al., 2013; Porter-Utley, 1997; Proffitt et al., 2006; Schwarzbach & Ricklefs, 2000; Sengupta et al., 2005; Sousa et al., 2003; USFWS, 1999; Williams, 1999.

Family 7: Salicaceae [58]

The systematic delimitation of Salicaceae has changed markedly following numerous insights provided by molecular phylogenetic data. Once regarded as comprising only two genera (*Populus, Salix*) and roughly 485 species, the circumscription of the family now includes over 1200 species and many noncyanogenic genera that were assigned previously to Flacourtiaceae (Chase et al., 2002; Judd et al., 2016). Most members of the family are distinguished by the presence of characteristic "salicoid" teeth on the foliage. However, the relationships of

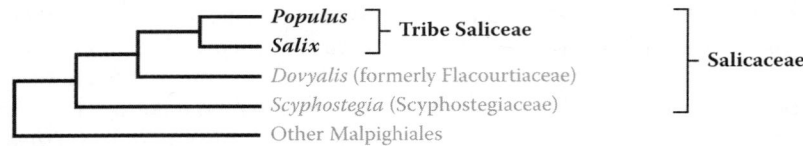

FIGURE 4.37 Molecular data (e.g., Azuma et al., 2000; Davis & Chase, 2004) consistently resolve *Populus* and *Salix* as a clade, which is recognized taxonomically as tribe Saliceae of Salicaceae. These taxa (shown in bold) are the only genera of the expanded Salicaceae (now containing 58 genera) to contain obligate aquatics in North America. (Redrawn from Davis, C.C. & Chase, M.W., *Amer. J. Bot.*, 91, 262–273, 2004.)

this broadly expanded Salicaceae remain largely unresolved among other families of Malpighiales. Combined cpDNA sequence data (Xi et al., 2012) place Scyphostegiaceae as the sister group of Salicaceae (Figure 4.25), a family that some authors merge with Salicaceae (Figure 4.37). The two "original" genera (*Populus* and *Salix*), with their woody habit and dioecious catkins of apetalous, unisexual flowers, form a well-supported clade (Figure 4.37), which is recognized as tribe Saliceae. Of the two included genera, *Populus* is relatively less specialized with multiple bud scales and a lack of floral nectaries in contrast to the single bud scale and nectariferous calyx found in *Salix*. Both are wind pollinated with *Salix* also being secondarily entomophilous.

The seeds of drupe, berry, or capsular-fruited taxa (the latter with arillate seeds) are dispersed by birds (Aves) or mammals (Mammalia); whereas, those of *Populus*, *Salix*, and several other genera release winged or plumed seeds, which are wind or water dispersed.

The family is a source of various ornamental trees and shrubs, which originate primarily from *Azara*, *Populus*, and *Salix*. It is of paramount importance medicinally, being the original source of salicylic acid (i.e., derived from *Salix*), which is used to manufacture acetylsalicylic acid, the major analgesic known widely as aspirin.

There are only two genera of Salicaceae that contain OBL indicators in North America.

1. ***Populus*** L.
2. ***Salix*** L.

1. *Populus*

Aspen, cottonwood, poplar; peuplier
Etymology: from the Latin *populus* meaning "poplar"
Synonyms: *Aigeiros*
Distribution: global: Africa; Asia; Europe; New World; **North America:** widespread
Diversity: global: 35 species; **North America:** 14 species
Indicators (USA): OBL: *Populus heterophylla*
Habitat: freshwater; palustrine, riverine; **pH:** 4.6–5.9; **depth:** <1 m; **life-form(s):** emergent tree
Key morphology: Stems (to 40 m) with apically pointed, gummy, ovate winter buds (to 1.5 cm); leaves (to 20 cm) deciduous, alternate, simple, with a cordate base and toothed margin, petiolate (to 8.8 cm); flowers in catkins, the males (to 6 cm) upright or drooping, the females (to 12 cm) pendant; capsules (to 1.3 cm) stalked (to 1.5 cm); seeds reddish, covered by a tuft of hairs (coma).

Life history: duration: perennial (buds, whole plants); **asexual reproduction:** root suckers, stump sprouts; **pollination:** wind; **sexual condition:** dioecious; **fruit:** capsules (prolific); **local dispersal:** seeds (wind, water); **long-distance dispersal:** seeds (wind, water)
Imperilment: (1) *Populus heterophylla* [G5]; SH (PA); S1 (MA, MI, <u>ON</u>, RI); S2 (AL, CT, NJ, NY); S3 (IL, KY, OH)
Ecology: general: Seven North American *Populus* species (50%) are wetland indicators (OBL, FACW, FACU, FAC), but only one is designated as OBL.

Populus heterophylla **L.** inhabits shallow waters (<1 m) in backwaters, baygalls, bottomlands, ditches, floodplain forests, hammocks, oxbows, ponds, ravines, roadsides, seeps, sloughs, swales, swamps, and the margins of lakes, rivers, and streams of the Atlantic coastal plain and Mississippi valley at elevations of at least 146 m. Although also reported as calcareous, the substrates generally are acidic (pH: 4.6–5.9) and have been described as Alfisols, alluvium, Entisols, Guthrie, heavy clay soils (24% to 65% clay in upper 0.3 m), humus, Inceptisols, loess, Maybid silt loam, muck, mud, sand, and silt. The plants are shade intolerant, shallow rooted, and thrive in deep, moist soils of shallow swamps and low-lying tidewater areas. Populations also occur in seasonally flooded depressions (vernal pools) perched on basaltic ("trap rock") ridges. This species is found in areas with an average annual temperature of 10°C–21°C (minimum −1°C to −29°C; 180–300 frost-free days), which receive 890–1500 mm rainfall. The dioecious plants are obligatory outcrossers. The proteranthous (appearing before the leaves) catkins of *P. heterophylla* generally appear by the time plants reach 10 years of age. Flowering occurs from March to May. Prolific seed production occurs in the spring (April to July) when habitats characteristically are flooded. The tufted, buoyant seeds are dispersed by the wind for upward of 100 m and are transported secondarily as they float on the surface of floodwaters. The seeds have no dormancy but retain viability for no more than a few weeks, resulting in the lack of a seed bank. Seedlings (which can reach densities of 209/ha) establish best on wet, open, mineral soils and require full sunlight and ample moisture for survival. The plants usually occur singly and rarely are found growing in patches. Although capable of clonal growth from root suckers, large clones do not develop in this species. The plants can be propagated *in vitro* by culture of the embryonic hypocotyls, from which adventitious shoots arise. Populations are threatened by native and nonnative species that compete for habitat and seedling establishment sites, and by hydrological alterations that disrupt the habitat available for adult plants (e.g., prolonged lowered water tables) or seedlings (e.g., prolonged high water tables). **Reported associates:** *Acer rubrum, A. negundo, A. saccharinum, Aesculus pavia, Alnus rugosa, A. serrulata, Amelanchier, Amorpha fruticosa, Aronia melanocarpa, Arundinaria gigantea, Asclepias perennis, Asimina triloba, Azolla caroliniana, Berchemia scandens, Betula lenta, B. nigra, B. populifolia, Bidens discoideus, Boehmeria cylindrica, Campsis radicans, Cardamine bulbosa, Carex amphibola, C. bromoides, C. conjuncta, C. corrugata, C. crinita, C. crus-corvi, C. granularis, C. grayi, C. hyalinolepis, C. intumescens, C. lurida, C. muskingumensis, C. socialis, C. squarrosa, C. tribuloides, C. typhina, C. vulpinoidea, Carpinus caroliniana, Carya aquatica, C. illinoinensis, C. tomentosa, Catalpa speciosa, Celtis laevigata, C. occidentalis, Cephalanthus occidentalis, Cercis canadensis, Cladrastis kentukea, Clethra alnifolia, Commelina virginica, Cornus amomum, Crataegus punctata, Cuscuta gronovii, Cyrilla racemiflora, Decodon verticillatus, Diospyros virginiana, Euonymus americanus, Eutrochium fistulosum, Fraxinus americana, F. caroliniana, F. pennsylvanica, F. profunda, Forestiera acuminata, Gleditsia aquatica, Glyceria septentrionalis, Gymnocladus dioicus, Ilex decidua, I. opaca, I. verticillata, Leersia lenticularis, Lindera benzoin, Liquidambar styraciflua, Liriodendron tulipifera, Lobelia cardinalis, Magnolia tripetala, Malus coronaria, Nyssa aquatica, N. biflora, N. ogeche, N. sylvatica, Onoclea sensibilis, Osmunda regalis, Osmundastrum cinnamomeum, Packera glabella, Peltandra virginica, Persea, Persicaria punctata, Phanopyrum gymnocarpon, Physostegia virginiana, Phyla lanceolata, Pilea pumila, Pinus palustris, Planera aquatica, Platanus occidentalis, Populus deltoides, Proserpinaca pectinata, Quercus alba, Q. bicolor, Q. falcata, Q. lyrata, Q. macrocarpa, Q. michauxii, Q. palustris, Q. rubra, Ranunculus abortivus, R. flabellaris, R. hispidus, Rhododendron viscosum, Rotala ramosior, Sagittaria, Salix amygdaloides, S. exigua, S. nigra, Sanicula canadensis, Saururus cernuus, Scutellaria lateriflora, Sium suave, Sphagnum, Symphyotrichum lanceolatum, S. ontarionis, Taxodium ascendens, T. distichum, Toxicodendron radicans, T. vernix, Tsuga canadensis, Ulmus americana, U. rubra, Vaccinium corymbosum, Viburnum dentatum, Viola sororia, Vitis palmata.*

Use by wildlife: *Populus heterophylla* is not regarded as an important wildlife food but it is eaten by birds (Aves), and various mammals (Mammalia) including beavers (Castoridae: *Castor canadensis*), deer (Cervidae: *Odocoileus*), rabbits (Leporidae), and squirrels (Sciuridae). The plants provide nesting sites for herons (Aves: Ardeidae). They are host to several larval Lepidoptera (Insecta) including Noctuidae (*Raphia frater*), Nymphalidae (*Basilarchia archippus, B. arthemis*), and Sphingidae (*Pachysphinx modesta*). This is also a food plant of the rare marbled underwing (Lepidoptera: Erebidae: *Catocala marmorata*). The plants are susceptible to damage by beetles (Insecta: Coleoptera), which commonly afflict other *Populus* species including bark beetles (Curculionidae: *Platypus compositus, Xyleborus affinis*), cottonwood borer beetles (Cerambycidae: *Plectrodera scalator*), cottonwood leaf beetles (Chrysomelidae: *Chrysomela scripta*), and poplar borer beetles (Cerambycidae: *Saperda calcarata*). They are also attacked by cottonwood

twig borer moths (Lepidoptera: Tortricidae: *Gypsonoma haimbachiana*). Also hosted are several grasshoppers (Insecta: Orthoptera: Acrididae: *Chloealtis conspersa, Paroxya hoosieri*). Disease causing Fungi include leaf rust (Basidiomycota: Melampsoraceae: *Melampsora abietis-canadensis, M. medusae*), powdery mildew (Ascomycota: Erysiphaceae: *Uncinula adunca*), and several canker-causing fungi (*Cytospora, Fusarium, Septoria*). The plants also host a wood-rotting fungus (Basidiomycota: Polyporaceae: *Lenzites elegans*). The roots are parasitized by various nematodes (Nematoda: Hoplolaimidae: *Helicotylenchus dihystera, Hoplolaimus galeatus, Scutellonema brachyurum*; Telotylenchidae: *Tylenchorhynchus claytoni*; Trichodoridae: *Trichodorus christiei*).

Economic importance: food: the bast (inner bark) of most poplars is edible, but is not highly palatable and used only as an emergency food; **medicinal:** no uses reported; **cultivation:** *Populus heterophylla* is cultivated infrequently for use in very wet sites; **misc. products:** The straight and tall *P. heterophylla* occasionally is harvested for lumber and veneer for use in boxes, crates, excelsior, and furniture parts. The pulpwood can produce a high-quality paper for books and magazines. The plants grow at densities that usually are too low to provide much erosion or flood control. The use for habitat restoration is limited by the difficulty in propagating saplings from twig cuttings in this species (but regeneration of plants from embryo culture has been successful); **weeds:** none; **nonindigenous species:** none

Systematics: Phylogenetic analyses of molecular data demonstrate that *Populus* is a monophyletic sister genus to *Salix*. Six sections are recognized in *Populus*: *Abaso, Aigeiros, Leucoides, Populus, Tacamahaca, Turanga*, although sections *Aigeiros* and *Tacamahaca* at least do not appear to be monophyletic. *Populus heterophylla* (along with the Asian *P. lasiocarpa* and *P. glauca*) is assigned to section *Leucoides*. Several phylogenetic studies of *Populus* using DNA sequences have been reported, but none has yet included any members of section *Leucoides*. Preliminary studies using AFLP markers have indicated a close relationship between sections *Leucoides* and *Tacamahaca*. Microsatellite markers have been developed that work broadly within *Populus* (including *P. heterophylla*); however, at this time, they have not yet been applied to the genetic study of *P. heterophylla*. Although hybridization is common in *Populus*, *P. heterophylla* reputedly hybridizes only occasionally with *P. deltoides*. The chromosome number of *P. heterophylla* has not been reported; however, *Populus* species ($x = 19$) are rather uniformly diploid ($2n = 38$), with polyploidy observed rarely in the genus.

Comments: *Populus heterophylla* is fairly widespread in eastern North America, but it is not particularly common anywhere throughout its range.

References: Beaver et al., 1980; Blackman, 1922; Blatchley, 1898; Britton, 1887; Burrill & Earle, 1887; Cervera et al., 2003; Eckenwalder, 1996; Grand et al., 2009; Hamzeh & Dayanandan, 2004; Hill, 1992; Homoya, 1983; Homoya & Rayner, 1987; Johnson, 1990a; McMaster, 2003; Peacock, 2000; Robertson et al., 1978; Ruehle, 1971; Searcy & Ascher, 2001; Terrel, 1972; Tuskan et al., 2004; Uphof, 1922.

2. *Salix*

Willow; saule

Etymology: from *salio* meaning "to leap" in reference to its rapid growth

Synonyms: *Chosenia*; *Toisusu*

Distribution: global: Asia; Europe; North America; **North America:** widespread

Diversity: global: 450+ species; **North America:** 113 species

Indicators (USA): OBL: *Salix arctophila*, *S. arizonica*, *S. athabascensis*, *S. barclayi*, *S. boothii*, *S. candida*, *S. caprea*, *S. caroliniana*, *S. commutata*, *S. farriae*, *S. geyeriana*, *S. interior*, *S. jepsonii*, *S. lutea*, *S. maccalliana*, *S. melanopsis*, *S. monticola*, *S. nigra*, *S. pedicellaris*, *S. petiolaris*, *S. planifolia*, *S. prolixa*, *S. purpurea*, *S. pyrifolia*, *S. sericea*, *S. serissima*, *S. viminalis*, *S. wolfii*; **FACW:** *Salix arctophila*, *S. arizonica*, *S. barclayi*, *S. bebbiana*, *S. boothii*, *S. caprea*, *S. caroliniana*, *S. drummondiana*; *S. eastwoodiae*, *S. gooddingii*, *S. interior*, *S. lasiandra*, *S. lasiolepis*, *S. lemmonii*, *S. lutea*, *S. maccalliana*, *S. melanopsis*, *S. myrtillifolia*, *S. nigra*, *S. petiolaris*, *S. planifolia*, *S. prolixa*, *S. purpurea*, *S. pyrifolia*, *S. viminalis*, *S. wolfii*; **FAC:** *S. arctica*, *S. barclayi*, *S. bebbiana*, *S. caprea*, *S. commutata*, *S. eriocephala*, *S. exigua*; **FACU:** *S. arctica*

Habitat: freshwater; palustrine, riverine; **pH:** 4.5–8.2; **depth:** <1 m; **life-form(s):** emergent tree, emergent shrub

Key morphology: Shrubs and trees (to 42 m) with slender branches; buds with a single, cap-like scale; leaves (to 18 cm) alternate, deciduous, stipulate (sometimes leaf-like), linear to lanceolate; inflorescence a catkin (to 10 cm); flowers reduced, apetalous, containing 1-several narrow nectaries (♀); stamens 1–8 (♂); fruit a 2-valved capsule (to 9 mm); seeds each with a coma of cottony, fine, white hairs

Life history: duration: perennial (buds, whole plants); **asexual reproduction:** layering (stolons), root sprouts, stem sprouts, detached twigs; **pollination:** insect, wind; **sexual condition:** dioecious; **fruit:** capsules (prolific); **local dispersal:** seeds (water, wind); **long-distance dispersal:** seeds (water, wind)

Imperilment: (1) *Salix arctica* [G5]; S2 (NF, SK); S3 (ON, QC); (2) *S. arctophila* [G5]; S1 (ME); S2 (NF, SK, YT); S3 (ON, QC); (3) *S. arizonica* [G2/G3]; S1 (CO, NM); S2 (AZ, UT); (4) *S. athabascensis* [G4/G5]; S2 (AK); S3 (AB, BC, MB, YT); (5) *S. barclayi* [G5]; S2 (WY); S3 (AB); (6) *S. bebbiana* [G5]; SH (MD); S1 (CA, NE); S2 (AZ); S3 (IA, IL, OH); (7) *S. boothii* [G5]; S2 (BC); S3 (AB); (8) *S. candida* [G5]; S1 (ME, NS, PA, PE, SD, WA); S2 (CO, ID, IL, NB, NJ, OH, WY, YT); S3 (AK, CT, IA, MA, MT, VT); (9) *S. caprea* [GNR]; (10) *S. caroliniana* [G5]; SH (DE); S1 (LA, NE, PA); S3 (IL, IN, MD, MS, OH); (11) *S. commutata* [G5]; S2 (AB); (12) *S. drummondiana* [G4/G5]; S3 (WY); (13) *S. eastwoodiae* [G5]; S2 (WY); S3 (MT); (14) *S. eriocephala* [G5]; S1 (FL, NH); S2 (IL, NF); (15) *S. exigua* [G5]; (16) *S. farriae* [G4]; S1 (ID); S2 (OR, WY, YT); S3 (BC, MT); (17) *S. geyeriana* [G5]; S3 (BC); (18) *S. gooddingii* [G5]; (19) *S. interior* [G5]; SH (DE, NH); S1 (ME, VA); S2 (MA, WY); S3 (AK); (20) *S. jepsonii* [G4]; (21) *S. lasiandra* [G5]; (22) *S. lasiolepis* [G5]; (23) *S. lemmonii* [G5]; S2 (WY); S3 (MT); (24) *S. lutea* [G4/G5]; (25) *S. maccalliana* [G5]; S1 (ND, WA, YT); S2 (QC); S3 (MN, ON); (26) *S. melanopsis* [G5]; S1 (UT); S3 (AB, BC, WY); (27) *S. monticola* [G4/G5]; S2 (WY); (28) *S. myrtillifolia* [G5]; S1 (CO, NB, WY); S3 (BC); (29) *S. nigra* [G5]; S3 (NB, QC); (30) *S. pedicellaris* [G5]; SH (RI); S1 (CT, NJ, OH, PA); S2 (IA, ID, MA, NF, NS, VT, YT); S3 (NB, ND); (31) *S. petiolaris* [G5]; S1 (MO, PE); S2 (BC, OH); S3 (IL, NS); (32) *S. planifolia* [G5]; SH (MI); S1 (AB, AK, ME, VT); S2 (NH, SK, UT, WI); S3 (BC, NF); (33) *S. prolixa* [G5]; S1 (AK); S2 (YT); S3 (AB); (34) *S. purpurea* [G5]; (35) *S. pyrifolia* [G5]; S2 (YT); S3 (BC, NF, NY); (36) *S. sericea* [G5]; SH (AR, WI); S1 (DE); S2 (GA, NS); S3 (IA, IL, QC); (37) *S. serissima* [G4]; S1 (CO, IL, NB, SD, WY); S2 (BC, IN, NF, NJ, PA); S3 (CT, MA, MT, OH, QC); (38) *S. viminalis* [GNR]; (39) *S. wolfii* [G5]; S2 (OR)

Ecology: general: Willows are pioneer species, which occur frequently in wetland habitats, especially along riverine banks and in wet, open meadows. Their affinity for wetlands is indicated by the 93 North American species (82%) recognized as wetland indicators, of which 28 (25%) currently are designated as OBL. These include eight OBL species newly designated since the release of the 1996 indicator list (*S. arizonica*, *S. barclayi*, *S. boothii*, *S. interior*, *S. maccalliana*, *S. purpurea*, *S. pyrifolia*, *S. viminalis*), but exclude 11 species (*S. arctica*, *S. bebbiana*, *S. drummondiana*, *S. eastwoodiae*, *S. eriocephala*, *S. exigua*, *S. gooddingii*, *S. lasiandra*, *S. lasiolepis*, *S. lemmonii*, *S. myrtillifolia*) that have since been reclassified. Because treatments for the latter group already had been prepared prior to the release of the 2013 indicator list, all 39 species (i.e., current and former OBL indicators) are treated here. Willows are fast growing, rapidly spreading trees and shrubs with highly branched, soil-binding roots. Most species have some form of vegetative reproduction and are able to regenerate quickly following disturbances by flood or fire. All of the species produce catkins with reduced flowers that are insect pollinated, or less frequently, wind pollinated. Willows produce prolific quantities of plumose seeds which, as in *Populus*, primarily are wind dispersed and secondarily water dispersed. The seeds of spring dispersing species typically lack dormancy and are extremely short lived; whereas, the fall dispersers (e.g., arctic species) produce dormant seeds, which germinate after a period of cold stratification. Many species are suspected to hybridize frequently, leading to much taxonomic confusion. Willows are so similarly adapted to disturbed riparian habitats that it is common to see several to many different species growing together under virtually identical environmental conditions.

Salix arctica **Pall.** is a dwarf, prostrate shrub (15–50 cm high) of cold, arctic habitats including beach ridges, bogs, cliffs, copses, fens, floodplains, heaths, hummocks, krummholz, meadows, outwash plains, ravines, salt flats, slopes, snow basins, snow beds, tundra, and the margins of pools and streams at elevations of up to 3754 m. Sites vary from open to shaded exposures. The plants can tolerate drier habitats at higher elevations, and can occur at higher elevations at lower latitudes. Their survival on permafrost is facilitated by the production of shallow root systems. Soil depths can be quite shallow (e.g.,

1–10 cm). The substrates are acidic to alkaline (or brackish) and have been described as frost-heaved clay, cobble, gneiss, gravel, gravelly loam, gypsum, humus, mud, rock, rocky loam, sand, sandstone, sandy gravel, scree, shale, silt, siltstone, and talus. Flowering proceeds from mid-June through August. The floral sex ratios are altered environmentally. Drier, less-fertile sites produce larger numbers of male plants and herbivores often feed preferentially on the male flowers. Seed production occurs in 28%–88% of the capsules. The seeds are physiologically dormant and require cold stratification for 30 days to induce germination, which occurs optimally at 25°C. An individual plant can live for up to 236 years. The plants can spread by layering; that is, producing long lateral branches that root at the nodes upon contact with the ground. They form a mutualistic association with ectomycorrhizal fungi. **Reported associates:** *Abies lasiocarpa, Achillea millefolium, Aconitum delphiniifolium, Agoseris aurantiaca, Alectoria ochroleuca, Alnus crispa, A. incana, A. viridis, Alopecurus magellanicus, Androsace chamaejasme, Anemone narcissiflora, A. parviflora, A. richardsonii, Antennaria, Anthoxanthum monticola, Arctagrostis latifolia, Arctostaphylos alpina, A. rubra, Arnica griscomii, A. latifolia, A. lessingii, Artemisia arctica, A. furcata, A. norvegica, A. tilesii, Astragalus eucosmus, Athyrium filix-femina, Aulacomnium turgidum, Barbarea orthoceras, Betula glandulosa, B. nana, B. neoalaskana, Bistorta bistortoides, B. vivipara, Calamagrostis canadensis, C. deschampsioides, C. purpurascens, C. stricta, Caltha leptosepala, C. palustris, Campanula lasiocarpa, Cardamine bellidifolia, Carex anthoxanthea, C. aquatilis, C. aurea, C. bigelowii, C. capillaris, C. chordorrhiza, C. engelmannii, C. fuliginosa, C. glareosa, C. lenticularis, C. lyngbyaei, C. macloviana, C. macrochaeta, C. maritima, C. membranacea, C. microcheata, C. nardina, C. paysonis, C. pluriflora, C. podocarpa, C. rariflora, C. rupestris, C. scirpoidea C. scopulorum, C. sitchensis, C. subnigricans, Cassiope mertensiana, C. tetragona, Castilleja miniata, C. unalaschcensis, Cerastium beeringianum, Chamerion latifolium, Chrysanthemum arcticum, Cladina stellaris, Claytonia sarmentosa, C. tuberosa, Cornus suecica, Dactylina arctica, Dasiphora floribunda, Delphinium glaucum, Deschampsia, Diapensia lapponica, Ditrichum flexicaule, Draba densifolia, Dryas alaskensis, D. drummondii, D. integrifolia, D. octopetala, Dupontia fisheri, Elymus violaceus, Empetrum nigrum, Epilobium anagallidifolium, E. latifolium, Equisetum arvense, E. scirpoides, Erigeron caespitosus, E. humilis, E. peregrinus, Eriophorum angustifolium, E. russeolum, E. scheuchzeri, E. triste, E. vaginatum, Erythronium grandiflorum, Festuca altaica, F. brachyphylla, F. rubra, F. viviparoidea, Gentiana calycosa, Gentianella amarella, Geranium erianthum, Geum rossii, Harrimanella stelleriana, Hedysarum alpinum, H. sulphurescens, Hippuris montana, Huperzia, Hylocomium splendens, Hypericum formosum, Juncus balticus, J. biglumis, J. drummondii, Kalmia microphylla, K. procumbens, Kobresia simpliciuscula, Lagotis glauca, Lecanora epibryon, Leptarrhena pyrolifolia, Leymus mollis, Lloydia serotina, Lomatogonium rotatum, Luetkea pectinata, Lupinus, Luzula confusa, L. multiflora, L. nivalis,*

L. piperi, L. spicata, Meesia triquetra, M. uliginosa, Mertensia paniculata, Micranthes ferruginea, M. lyallii, Minuartia obtusiloba, M. rossii, Myosotis asiatica, Nephrophyllidium crista-galli, Oplopanax horridus, Oxytropis maydelliana, O. nigrescens, Parnassia kotzebuei, P. palustris, Pedicularis lanata, P. oederi, Petasites frigidus, Phleum alpinum, Phlox pulvinata, P. richardsonii, Phyllodoce aleutica, P. empetriformis, P. glanduliflora, Pinus albicaulis, P. contorta, Platanthera dilatata, Poa alpina, P. arctica, P. glauca, Podistera macounii, Potentilla ×diversifolia, Pseudephebe pubescens, Puccinellia tenella, Racomitrium lanuginosum, Ranunculus cooleyae, R. nivalis, Rhododendron camtschaticum, R. columbianum, R. lapponicum, R. tomentosum, Rubus arcticus, Salix alaxensis, S. barclayi, S. barrattiana, S. bebbiana, S. commutata, S. glauca, S. lanata, S. nivalis, S. ovalifolia, S. phlebophylla, S. planifolia, S. polaris, S. pulchra, S. reticulata, S. richardsonii, S. rotundifolia, S. sitchensis, S. stolonifera, Sanguisorba, Saxifraga bronchialis, S. hirculus, S. oppositifolia, S. tricuspidata, Sedum divergens, Senecio cymbalarioides, S. resedifolius, S. triangularis, Sibbaldia, Silene acaulis, Smelowskia calycina, Solidago multiradiata, Sphagnum fuscum, S. papillosum, Stellaria longipes, Symphyotrichum foliaceum, Tephroseris yukonensis, Thalictrum alpinum, Tomentypnum nitens, Trichophorum cespitosum, Trisetum spicatum, Vaccinium cespitosum, V. scoparium, V. uliginosum, V. vitis-idaea, Valeriana capitata, V. sitchensis, Viola adunca.

***Salix arctophila* Cockerell ex A. Heller** is another dwarf, prostrate shrub (<15 cm high), which grows in arctic and alpine habitats including alluvial plains, barrens, copses, heaths, hummocks, meadows, sloughs, snow beds, terraces, thickets, tundra, and along the margins of ponds and streams at elevations of up to 1220 m. The substrates include conglomerate, granite, gravelly sand, sandstone, shale, and stones. Flowering extends from late May through July. The plants reproduce vegetatively by their long, trailing stems, which enable them to spread by layering. **Reported associates:** *Andromeda polifolia, Arctostaphylos, Betula glandulosa, B. nana, Bistorta vivipara, Campanula uniflora, Carex aquatilis, C. bigelowii, C. mackenzii, C. membranacea, C. rariflora, C. scirpoides, Cassiope tetragona, Cladina stellaris, C. rangiferina, Cochlearia officinalis, Dicranum elongatum, D. undulatum, Ditrichum flexicaule, Draba lactea, D. nivalis, Dryas crenulata, D. integrifolia, Empetrum nigrum, Eremonotus myriocarpus, Eriophorum angustifolium, Fissidens adianthoides, F. osmundioides, Limprichtia cossonii, Marsupella boeckii, Meesia uliginosa, Melandrium apetalum, Micranthes foliolosa, Orthothecium strictum, Oxyria digyna, Papaver radicatum, Parnassia kotzebuei, Petasites frigidus, Potentilla hyparctica, Rhododendron groenlandicum, R. lapponicum, R. tomentosum, Rubus chamaemorus, Salix alaxensis, S. arctica, S. cordifolia, S. glauca, S. herbacea, S. reticulata, S. richardsonii, Saxifraga aizoon, S. crenulata, S. oppositifolia, Stellaria humifusa, S. laeta, S. longipes, Taraxacum phymatocarpum, Tortella tortuosa, Trichophorum cespitosum, Vaccinium uliginosum, V. vitis-idaea.*

***Salix arizonica* Dorn** inhabits subalpine cienega, drainageways, marshes, meadows, and the margins of streams at elevations from 2600 to 3400 m. Exposures range from open sun to partial shade. The substrates are acidic to somewhat alkaine (pH: 5.2–7.7) and consist of complex mixtures of basaltic, epiclastic, and felsic formations or limestone. They include alluvium, basaltic gravel, loam, peat (to 44 cm depth), and rhyolitic gravel. Flowering extends from late May to June. The seeds are light and are dispersed by wind and water. Seedlings occur rarely in populations. Genetic analyses using RAPD markers, showed there to be low similarity (37.5%) between Arizona and Utah stands. The plants are believed to first establish along flowing streams, but attain their largest stature on substrates that are moist but not saturated. They can live to about 19 years of age. Although no rhizomes are produced, adventitious roots can develop on branches that become buried; so some vegetative reproduction by layering is possible. **Reported associates:** Abies concolor, A. lasiocarpa, Achillea millefolium, Achnatherum lettermanii, Aconitum columbianum, Agropyron, Agrostis scabra, A. stolonifera, Allium geyeri, A. macropetalum, Alnus, Angelica pinnata, Antennaria pulcherrima, Artemisia cana, Bistorta bistortoides, Bromus carinatus, Caltha leptosepala, Campanula parryi, Cardamine californica, C. cordifolia, Carex aquatilis, C. microptera, C. nebrascensis, C. rossii, C. utriculata, Castilleja linariifolia, C. miniata, Chenopodium, Cirsium, Conioselinum scopulorum, Conium maculatum, Dasiphora floribunda, Delphinium barbeyi, Deschampsia cespitosa, Epilobium ciliatum, E. saximontanum, Equisetum hyemale, Erigeron speciosus, Festuca ovina, Fragaria virginiana, Geranium richardsonii, Geum macrophyllum, Glyceria striata, Heracleum sphondylium, Hordeum brachyantherum, Hymenoxys hoopesii, Hypericum formosum, Iris missouriensis, Juncus balticus, J. halli, J. longistylis, Lathyrus, Lupinus argenteus, Luzula parviflora, Mertensia arizonica, Mimulus guttatus, M. primuloides, Oxypolis fendleri, Packera streptanthifolia, Parnassia palustris, Pedicularis groenlandica, Perideridia, Phleum alpinum, Picea engelmannii, P. pungens, Poa alpina, P. pratensis, Polemonium caeruleum, Populus angustifolia, P. tremuloides, Potentilla anserina, P. ×diversifolia, P. gracilis, P. pulcherrima, Primula pauciflora, P. tetrandra, Pseudotsuga menziesii, Pyrrocoma lanceolata, Ranunculus aquatilis, R. cardiophyllus, R. cymbalaria, R. flammula, R. macounii, Rhodiola rhodantha, Ribes cereum, R. inerme, R. leptanthum, R. montigenum, Rosa woodsii, Rubus idaeus, Salix bebbiana, S. boothii, S. brachycarpa, S. drummondiana, S. exigua, S. geyeriana, S. irrorata, S. monticola, S. planifolia, S. wolfii, Sambucus racemosa, Senecio bigelovii, S. triangularis, Sidalcea neomexicana, Swertia perennis, Symphyotrichum foliaceum, Taraxacum officinale, T. scopulorum, Thalictrum fendleri, Trifolium repens, Trisetum spicatum, Veronica americana, Veratrum californicum, Viola.

***Salix athabascensis* Raup** is a medium-sized shrub (to 1.5 m) that occurs mainly in the colder, montane portions of northern North America in shallow waters (e.g., 46 cm) of bogs, burns, fens, muskeg, patterned wetlands, roadsides, seeps, swamps, thickets, tundra, and along the margins of lakes, ponds, and rivers at elevations of up to 1800 m. The substrates often are alkaline and have been described as humus,

marl, peat, and sandy clay. Flowering extends from May to July. The catkins emerge simultaneously with the leaves. More detailed life-history information is needed for this species. **Reported associates:** Acer glabrum, A. macrophyllum, Alnus rubra, A. viridis, Andomeda, Arnica diversifolia, Aruncus dioicus, Betula glandulosa, Boykinia occidentalis, Calamagrostis stricta, Campylium stellatum, Cardamine angulata, Carex aquatilis, C. canescens, C. capillaris, C. rostrata, C. scirpoidea, C. subspathacea, Chamerion angustifolium, Chrysosplenium tetrandrum, Comarum palustre, Cornus sericea, Galium trifidum, Holodiscus discolor, Hylocomium splendens, Larix laricina, Leymus innovatus, Menyanthes trifoliata, Myrica gale, Picea glauca, P. mariana, Pinguicula vulgaris, Pinus contorta, Poa pratensis, Populus balsamifera, Potentilla anserina, Prunus virginiana, Ranunculus cymbalaria, Rhododendron groenlandicum, Ribes bracteosum, R. laxiflorum, R. sanguineum, Rosa acicularis, Rubus leucodermis, R. spectabilis, Salix alaxensis, S. brachycarpa, S. candida, S. pedicellaris, S. vestita, Sambucus racemosa, Shepherdia canadensis, Sphagnum, Spiraea douglasii, Stachys ciliata, S. mexicana, Stellaria crispa, Tofieldia pusilla, Tolmiea menziesii, Triantha glutinosa, Trichophorum cespitosum, Vaccinium, Veratrum californicum.

***Salix barclayi* Andersson** is a small shrub (to 5 m), which inhabits avalanche tracks, beaches, bogs, carrs, cirques, copses, depressions, draws, fens, flats, floodplains, fluvial fans, gravel bars, gullies, knolls, meadows, moraines, mudflats, muskeg, outwash plains, patterned peatlands, ravines, seeps, slopes, sloughs, snowbanks, swales, swamps, thickets, tundra, and the margins of lakes, ponds, rivers, and streams at elevations of up to 3354 m. The substrates can be acidic or alkaline (pH: 6.7–8.2), organic or mineral, and typically are fairly shallow (0.1–1.0 m deep). They are described as boulders, clay sand, conglomerate, gravel, gravelly loam, humus, loam, muck, mud, orthic gleysol, pumice, rock, sand, sandy loam, sandy silt, shale, silt, silt loam, stones, and talus. Flowering occurs from late May through early August. **Reported associates:** Abies lasiocarpa, Aconitum delphiniifolium, Allium schoenoprasum, Alnus rugosa, A. tenuifolia, A. viridis, Anaphalis margaritacea, Angelica arguta, Antennaria racemosa, Aquilegia formosa, Aralia nudicaulis, Arctagrostis, Arctostaphylos uva-ursi, Arctous rubra, Arnica lanceolata, Artemisia norvegica, A. tilesii, Athyrium filix-femina, Betula glandulosa, B. nana, B. neoalaskana, B. papyrifera, Bistorta vivipara, Brodoa oroarctica, Bryoria tenuis, Bryum weigelii, Calamagrostis canadensis, C. rubescens, Caltha leptosepala, Carex anthoxanthea, C. aquatilis, C. bigelowii, C. canescens, C. cusickii, C. disperma, C. flava, C. illota, C. jonesii, C. lenticularis, C. limosa, C. luzulina, C. macrochaeta, C. magellanica, C. mertensii, C. microchaeta, C. nigricans, C. pluriflora, C. rostrata, C. saxatilis, C. scopulorum, C. spectabilis, C. tenuiflora, C. utriculata, C. vesicaria, Chamaedaphne calyculata, Chamerion angustifolium, C. latifolium, Cicuta virosa, Cirsium, Comarum paluste, Cornus sericea, Cystopteris montana, Dasiphora floribunda, Delphinium glaucum, Deschampsia cespitosa, Drepanocladus, Dryas, Dryopteris expansa, Eleocharis *palustris, Empetrum nigrum, Equisetum arvense, E. fluviatile, E. sylvaticum, Erigeron peregrinus, Eriophorum angustifolium, E. chamissonis, E. vaginatum, Festuca altaica, F. rubra, Fritillaria camschatcensis, Gaultheria humifusa, Gentianopsis detonsa, Geranium erianthum, Geum calthifolium, G. rivale, Glyceria striata, Gymnocarpium dryopteris, Habenaria, Harrimanella stelleriana, Hedysarum alpinum, Heracleum sphondylium, Hylocomium splendens, Iris setosa, Juncus drummondii, Juniperus communis, Larix laricina, Leptarrhena pyrolifolia, Leymus mollis, Linnaea borealis, Lonicera caerulea, L. involucrata, L. utahensis, Luetkea pectinata, Lupinus nootkatensis, Luzula parviflora, Lysichiton americanus, Marchantia polymorpha, Melanelia hepatizon, Mertensia ciliata, M. paniculata, Micranthes lyallii, M. odontoloma, Mimulus lewisii, Mitella pentandra, Moneses uniflora, Montia fontana, Myrica gale, Nephrophyllidium crista-galli, Oxyria digyna, Parmelia saxatilis, Parnassia fimbriata, Pedicularis, Petasites frigidus, Phyllodoce aleutica, P. empetriformis, Picea engelmannii, P. glauca, P. mariana, P. sitchensis, Pinus contorta, Platanthera dilatata, Pleurozium schreberi, Poa alpina, P. pratensis, Populus balsamifera, P. tremuloides, P. trichocarpa, Primula pauciflora, Pseudotsuga menziesii, Pyrola asarifolia, Racomitrium, Ranunculus gmelinii, R. pensylvanicus, Rhamnus alnifolia, Rhizomnium punctatum, Rhodiola integrifolia, Rhododendron albiflorum, R. columbianum, R. groenlandicum, R. tomentosum, Ribes hudsonianum, R. lacustre, R. triste, Rosa acicularis, Rumex occidentalis, Salix alaxensis, S. arbusculoides, S. arctica, S. bebbiana, S. boothii, S. brachycarpa, S. cascadensis, S. commutata, S. drummondiana, S. eastwoodiae, S. exigua, S. farriae, S. geyeriana, S. glauca, S. interior, S. lucida, S. myrtillifolia, S. planifolia, S. pulchra, S. reticulata, S. richardsonii, S. scouleriana, S. stolonifera, S. tweedyi, S. wolfii, Sanguisorba canadensis, S. stipulata, Saxifraga hirculus, Schoenoplectus tabernaemontani, Scirpus microcarpus, Senecio triangularis, Shepherdia canadensis, Sibbaldia procumbens, Sorbus scopulina, S. sitchensis, Sphagnum, Spiraea douglasii, S. stevenii, Streptopus amplexifolius, Symphyotrichum foliaceum, Thalictrum occidentale, Trichophorum cespitosum, Trientalis europaea, Trollius albiflorus, T. laxus, Tuckermannopsis chlorophylla, Typha latifolia, Umbilicaria proboscidea, Vaccinium deliciosum, V. membranaceum, V. ovalifolium, V. scoparium, V. uliginosum, Valeriana dioica, V. sitchensis, Veratrum viride, Veronica wormskjoldii, Viburnum edule, Viola adunca.*

Salix bebbiana is a large shrub or tree (to 8 m), which occurs in alluvial flats, bogs, bottoms, canals, carrs, cienega, depressions, ditches, dunes, fens, flats, floodplains, gravel bars, meadows, muskeg, prairies, ravines, roadsides, scrub, seeps, slopes, sloughs, springs, swales, swamps, terraces, thickets, tundra, woodlands, and along the margins of lakes, ponds, rivers, and streams at elevations of up to 3232 m. This is a shade-intolerant pioneer species that moves rapidly into wet, disturbed sites where gaps occur in the vegetation. It grows on rich soils, which include alluvium, clay, clay loam, decomposed granite, gravel, gravelly loam, loam, muck, mucky alluvium, peat, rock, sand, sandy gravel, sandy loam,

sandy silty loam, silt, silty clay, silty loam, silty mud, stones, and talus. The plants thrive and often become the dominant willow species following fire, by their capacity to resprout rapidly from the stems as well as disperse plentiful seed into newly burned sites. These are fast growing but short-lived plants, which initiate flowering at a very young age (2-year old); however, maximum seed production occurs in trees that are 10–30 years of age. Flowering occurs from April to June. The flowers are pollinated by bees (Hymenoptera) and other insects. The capsule-bearing catkins are shed on the ground in July after the seeds have ripened. The seeds lack dormancy and remain viable for only a few days. Germination (90%–100%) occurs at 5°C–25°C and is highest on sites that recently have been burned. Vegetative reproduction occurs by means of root and basal stem sprouts. Layering will occur if the branches become buried. **Reported associates:** *Abies grandis, A. lasiocarpa, Acer negundo, A. saccharinum, Aconogonon alaskanum, Actaea rubra, Agrostis gigantea, A. scabra, Alnus rugosa, A. tenuifolia, A. viridis, Amelanchier alnifolia, Amphiscirpus nevadensis, Androsace septentrionalis, Angelica lucida, Antennaria corymbosa, A. racemosa, Aquilegia formosa, Arctostaphylos uva-ursi, Arctous rubra, Arisaema triphyllum, Arnica chamissonis, A. cordifolia, Artemisia campestris, A. frigida, A. ludoviciana, Asclepias incarnata, Aulacomnium palustre, Balsamorhiza sagittata, Beckmannia syzigachne, Berteroa incana, Betula glandulosa, B. kenaica, B. nana, B. neoalaskana, B. occidentalis, B. papyrifera, B. populifolia, Boechera holboellii, Bromus tectorum, Bryum pseudotriquetrum, Calamagrostis canadensis, C. rubescens, Calliergonella cuspidata, Campanula, Campylium stellatum, Carex aquatilis, C. aurea, C. buxbaumii, C. canescens, C. comosa, C. crinita, C. disperma, C. flava, C. interior, C. lasiocarpa, C. leptalea, C. limosa, C. lurida, C. mertensii, C. nebrascensis, C. praegracilis, C. rostrata, C. scoparia, C. simulata, C. stipata, C. stricta, C. utriculata, C. vesicaria, Ceanothus sanguineus, C. velutinus, Ceratodon purpureus, Chamerion angustifolium, Chrysanthemum leucanthemum, Circaea alpina, Cirsium arvense, Cladonia gracilis, C. rangiferina, Claytonia cordifolia, Clematis ligusticifolia, Clintonia uniflora, Cnidium cnidiifolium, Comarum palustre, Conioselinum chinese, Coptis occidentalis, Cornus amomum, C. canadensis, C. sericea, Corylus cornuta, Crataegus douglasii, Crepis runcinata, Cyperus bipartitus, Cypripedium parviflorum, Cystopteris fragilis, Dasiphora floribunda, Daucus carota, Decodon verticillatus, Dianthus repens, Distichlis spicata, Drepanocladus aduncus, Dryopteris fragrans, Dulichium arundinaceum, Elaeagnus angustifolia, Eleocharis palustris, E. rostellata, E. tenuis, Empetrum nigrum, Epilobium palustre, Equisetum arvense, E. fluviatile, E. hyemale, E. laevigatum, E. pratense, E. sylvaticum, E. variegatum, Erigeron annuus, Eriophorum gracile, Erythronium grandiflorum, Eupatorium perfoliatum, Euthamia graminifolia, Eutrochium maculatum, Festuca altaica, Fragaria virginiana, Fraxinus pennsylvanica, Galium boreale, G. trifidum, G. triflorum, Gentianella propinqua, Geocaulon lividum, Geranium bicknellii, Geum macrophyllum, Grimmia, Gutierrezia, Hedwigia*

ciliata, *Helodium blandowii*, *Heracleum*, *Hieracium aurantiacum*, *H. paniculatum*, *Holodiscus discolor*, *Hylocomium
splendens*, *Impatiens capensis*, *Iris missouriensis*, *Juncus
balticus*, *J. bufonius*, *Juniperus communis*, *Larix occidentalis*, *Lepidium*, *Leucanthemum vulgare*, *Leymus innovatus*, *L. mollis*, *Linnaea borealis*, *Lobelia kalmii*, *Lonicera
involucrata*, *L. utahensis*, *Lupinus arcticus*, *L. burkei*, *L.
nootkatensis*, *Lysimachia terrestris*, *Lythrum salicaria*,
Maianthemum stellatum, *M. trifolium*, *Marchantia polymorpha*, *Marsilea vestita*, *Medicago lupulina*, *Melilotus albus*,
Menyanthes trifoliata, *Mertensia paniculata*, *Mimulus lewisii*, *Muhlenbergia*, *Myrica gale*, *Nasturtium officinale*, *Nyssa
sylvatica*, *Onoclea sensibilis*, *Oplopanax*, *Orthilia secunda*,
Osmorhiza, *Osmunda regalis*, *Osmundastrum cinnamomeum*,
Oxytropis sericea, *Parthenocissus quinquefolia*, *Pedicularis
groenlandica*, *Peltigera aphthosa*, *Penstemon fruticosus*,
Petasites frigidus, *Phalaris arundinacea*, *Philadelphus lewi-
sii*, *Phleum pratense*, *Phlox kelseyi*, *Physocarpus capitatus*,
P. malvaceus, *Picea englemanii*, *P. glauca*, *P. mariana*, *Pinus
banksiana*, *P. contorta*, *P. flexilis*, *P. monticola*, *P. ponderosa*,
P. strobus, *Plagiomnium ellipticum*, *Plantago canescens*,
Pleurozium schreberi, *Poa glauca*, *P. nemoralis*, *P. palustris*,
P. pratensis, *Polemonium occidentale*, *Polygonatum biflo-
rum*, *Populus* ×*acuminata*, *P. angustifolia*, *P. balsamifera*, *P.
deltoides*, *P. tremuloides*, *P. trichocarpa*, *Potentilla anserina*,
P. rubricaulis, *Primula incana*, *P. jeffreyi*, *P. pauciflora*,
Prosartes trachycarpa, *Prunus virginiana*, *Pseudoroegneria
spicata*, *Pseudotsuga menziesii*, *Pyrola asarifolia*, *Quercus
rubra*, *Racomitrium lanuginosum*, *Ranunculus uncinatus*,
Rhamnus alnifolia, *R. cathartica*, *R. purshiana*, *Rhodiola
rhodantha*, *Rhododendron groenlandicum*, *R. tomentosum*,
Rhus typhina, *Ribes hudsonianum*, *R. lacustre*, *R. niveum*,
R. oxyacanthoides, *Rosa acicularis*, *R. gymnocarpa*, *R.
nutkana*, *R. woodsii*, *Rubus idaeus*, *R. parviflorus*, *R. pubescens*, *Rudbeckia laciniata*, *Sagittaria latifolia*, *Salix alaxensis*, *S. arbusculoides*, *S. barclayi*, *S. boothii*, *S. candida*,
S. discolor, *S. drummondiana*, *S. eriocephala*, *S. exigua*, *S.
geyeriana*, *S. glauca*, *S. lasiandra*, *S. lemmonii*, *S. lutea*,
S. nigra, *S. petiolaris*, *S. planifolia*, *S. pseudomonticola*, *S.
scouleriana*, *S. sericea*, *S. serissima*, *S. sitchensis*, *Sambucus
nigra*, *Saussurea angustifolia*, *Schedonorus arundinaceus*,
Schoenoplectus acutus, *S. tabernaemontani*, *Senecio triangularis*, *Shepherdia canadensis*, *Silene menziesii*, *S. nivea*,
Sisyrinchium angustifolium, *S. montanum*, *Solidago gigantea*, *Sparganium americanum*, *Spartina gracilis*, *Sphagnum
squarrosum*, *Sphenopholis*, *Spiraea alba*, *S. betulifolia*,
S. douglasii, *Stachys palustris*, *Streptopus amplexifolius*,
Sullivantia hapemanii, *Symphoricarpos albus*, *S. occidentalis*, *Symphyotrichum lanceolatum*, *S. praealtum*, *S.
puniceum*, *Taraxacum officinale*, *Thalictrum alpinum*, *T.
occidentale*, *Thelypteris palustris*, *Thuja plicata*, *Tiarella
trifoliata*, *Torreyochloa pallida*, *Toxicodendron rydbergii*, *Triadenum virginicum*, *Trifolium pratense*, *T. repens*,
Tsuga heterophylla, *T. mertensiana*, *Typha latifolia*, *Ulmus
americana*, *Vaccinium cespitosum*, *V. membranaceum*, *V.
scoparium*, *V. uliginosum*, *V. vitis-idaea*, *Valeriana edulis*, *Veronica americana*, *Viburnum edule*, *V. prunifolium*,

Viola canadensis, *V. glabella*, *V. macloskeyi*, *Vitis riparia*,
Xerophyllum tenax.

***Salix boothii* Dorn** is a shrub (to 6 m), which grows in bogs,
bottoms, carrs, cirques, depressions, ditches, fens, floodplains,
marshes, meadows, prairies, seeps, sloughs, springs, swamps,
terraces, thickets, and along the margins of lakes, ponds, and
streams at elevations (mainly subalpine) from 602 to 3597 m.
The plants occur on slopes of 3%–20% upon acidic to somewhat
alkaline substrates (pH: 5.5–7.5), which have been categorized
as alluvium, calcareous, clay, clay loam, cobble, gravel, gravelly clay, gravelly loam, gravelly sand, gravelly silt, limestone,
loam, muck, peat, quartzite, rock, rocky loam, sand, sandy clay
loam, sandy gravel, silt, silty gravel, silty loam, silty sand, and
travertine. The plants are pioneers of newly deposited alluvium
or recently disturbed sites. They are shade intolerant (thriving
in full sun) but well adapted to frost and flooding. Growth is
reduced where excessive acidity or alkalinity occur, or when the
root crown is inundated for a long duration. Flowering and fruiting extend from April through November, but can be arrested
in areas that have been overgrazed by herbivores. In such cases,
it can take up to 3 years to regain adequate sexual reproduction
once the grazing has been suspended. Maximum seed dispersal
occurs from July to August and is facilitated by wind. The seeds
are short lived and require disturbed, nutrient-rich sand or gravel
substrates for germination, which is inhibited where insufficient light and/or a continuous cover of tree litter exists. The
seeds germinate over a wide temperature gradient (5°C–25°C),
which somewhat compensates for their short life. No seed bank
is formed. Adult plants become senescent after 15–20 years.
The plants can regenerate vegetatively by sprouts from the root
crown or lower stem. They are fire tolerant, but more susceptible to rapid, hot fires, which result in more extensive sprouting.
Reported associates: *Abies grandis*, *A. lasiocarpa*, *Achillea
millefolium*, *Aconitum columbianum*, *Actaea rubra*, *Agastache
urticifolia*, *Agrostis gigantea*, *Allium validum*, *Alnus rugosa*,
A. viridis, *Amelanchier*, *Angelica arguta*, *Antennaria racemosa*, *Apocynum*, *Arctostaphylos uva-ursi*, *Arnica chamissonis*, *Artemisia cana*, *A. tridentata*, *Betula glandulosa*, *B.
nana*, *B. occidentalis*, *Bistorta bistortoides*, *Bromus carinatus*,
Calamagrostis canadensis, *C. rubescens*, *Caltha leptosepala*,
Carex angustata, *C. aquatilis*, *C. athrostachya*, *C. aurea*, *C.
buxbaumii*, *C. dioica*, *C. disperma*, *C. echinata*, *C. geyeri*,
C. interior, *C. jonesii*, *C. lasiocarpa*, *C. leptalea*, *C. luzulina*,
C. microptera, *C. nebrascensis*, *C. nigricans*, *C. pellita*, *C.
rostrata*, *C. saxatilis*, *C. scopulorum*, *C. simulata*, *C. stipata*, *C.
utriculata*, *C. vesicaria*, *Castilleja miniata*, *Chamerion angus-
tifolium*, *Chenopodium album*, *Comarum palustre*, *Cornus
sericea*, *Crataegus chrysocarpa*, *C. douglasii*, *C. rivularis*,
Cypripedium parviflorum, *Dasiphora floribunda*, *Delphinium*,
Deschampsia cespitosa, *Descurainia*, *Dimeresia howellii*,
Drymocallis, *Elaeagnus commutata*, *Eleocharis palustris*,
E. quinqueflora, *Elymus glaucus*, *Epilobium ciliatum*, *E. suffruticosum*, *Equisetum arvense*, *E. variegatum*, *Erigeron gla-
bellus*, *E. glacialis*, *Festuca idahoensis*, *Fragaria virginiana*,
Galium trifidum, *G. triflorum*, *Gaylussacia dumosa*, *Gentiana
calycosa*, *Geranium viscosissimum*, *Geum macrophyllum*,
G. rivale, *Glyceria striata*, *Glycyrrhiza lepidota*, *Heracleum*

sphondylium, Heterotheca villosa, Hydrurus foetidus, Hypericum formosum, H. scouleri, Ipomopsis congesta, Iris missouriensis, Juncus balticus, Juniperus, Kalmia microphylla, Ligusticum tenuifolium, Lonicera caerulea, L. involucrata, L. tatarica, Luzula campestris, Lysichiton, Maianthemum stellatum, Melilotus officinalis, Mentha arvensis, Micranthes oregana, Mimulus guttatus, M. primuloides, Mitella pentandra, Moehringia lateriflora, Monostroma, Muhlenbergia filiformis, Oxytropis sericea, Packera pseudaurea, P. subnuda, Parnassia fimbriata, Pedicularis groenlandica, Penstemon, Petasites frigidus, Phalaris arundinacea, Phleum alpinum, Phlox kelseyi, Physaria, Picea engelmannii, P. glauca, P. pungens, Pinus contorta, P. ponderosa, Platanthera dilatata, P. obtusata, Poa pratensis, Polemonium occidentale, P. viscosum, Populus angustifolia, P. balsamifera, P. tremuloides, P. trichocarpa, Potentilla ×diversifolia, P. flabellifolia, P. gracilis, Primula incana, P. jeffreyi, P. pauciflora, P. tetrandra, Prunus virginana, Pseudotsuga menziesii, Ranunculus inamoenus, Rhamnus alnifolia, Rhododendron columbianum, Ribes aureum, R. cereum, R. hudsonianum, R. inerme, R. lacustre, R. oxyacanthoides, Rosa gymnocarpa, R. woodsii, Rubus arcticus, R. parviflorus, Rudbeckia laciniata, R. occidentalis, Rumex salicifolius, Salix bebbiana, S. brachycarpa, S. candida, S. commutata, S. drummondiana, S. eastwoodiae, S. eriocephala, S. exigua, S. farriae, S. geyeriana, S. lasiandra, S. lasiolepis, S. lemmonii, S. lutea, S. melanopsis, S. planifolia, S. pseudomonticola, S. scouleriana, S. sitchensis, S. tweedyi, S. wolfii, Sanguisorba, Schoenoplectus pungens, Scirpus microcarpus, Senecio integerrimus, S. serra, S. triangularis, Sidalcea oregana, Solidago speciosa, Sphagnum, Spiraea betulifolia, Spiranthes diluvialis, S. romanzoffiana, Stephanomeria fluminea, Symphoricarpos albus, S. oreophilus, Symphyotrichum ascendens, S. foliaceum, S. spathulatum, Taraxacum officinale, Taxus brevifolia, Thermopsis montana, Triantha glutinosa, Trifolium longipes, Tsuga heterophylla, Typha latifolia, Urtica dioica, Vaccinium membranaceum, V. scoparium, V. uliginosum, Valeriana edulis, V. occidentalis, Vaucheria, Veratrum californicum, Vicia americana, Wyethia helianthoides.

Salix candida Flüggé ex Willd. is a low shrub (<1 m), which inhabits shallow waters (to 4 dm) in bogs, carrs, depressions, ditches, dunes, fens, floodplains, hummocks, marshes, meadows, mudflats, patterned peatlands, pools, prairies, roadsides, seeps, springs, swamps, and the margins of lakes, ponds, and streams at elevations of up to 2853 m. The plants require full sun and usually occur as scattered and solitary individuals. The substrates are calcareous (Ca^{2+} 43–94 mg/kg; pH: 6.2–8.7) or saline and include clay, cobble, gravelly sandy loam, loam, loamy alluvium, marl, marly peat, muck, mud, peat (e.g., fibric organic cryosol), and silty muck. The abundance of plants is reduced considerably on substrates associated with high salt concentrations (Na$^+$> 112 mg/L, Cl$^-$ >54 mg/L). Flowering occurs from April to August. The catkins appear simultaneously with, or slightly before the leaves. A lack of sufficient soil surface moisture can result in severe seedling mortality (up to 100%). No seed bank has been observed in studies of several fen habitats where the plants occur in the standing vegetation. **Reported associates:**

Agrostis gigantea, Alnus rugosa, A. viridis, Andromeda polifolia, Anemone parviflora, Antennaria pulcherrima, Anthoxanthum nitens, Anticlea elegans, Arctagrostis, Arctostaphylos uva-ursi, Arctous rubra, Asclepias incarnata, Astragalus diversifolius, Betula glandulosa, B. papyrifera, B. pumila, Bromus ciliatus, Calamogrostis canadensis, C. inexpansa, C. stricta, Calliergon trifarium, Caltha palustris, Campanula aparinoides, Campylium stellatum, Carex aquatilis, C. atherodes, C. aurea, C. bigelowii, C. buxbaumii, C. capillaris, C. comosa, C. disperma, C. flava, C. interior, C. laeviconica, C. lasiocarpa, C. leptalea, C. livida, C. lurida, C. microglochin, C. muricata, C. nebrascensis, C. praegracilis, C. prairea, C. rostrata, C. rotundata, C. sartwellii, C. scirpoidea, C. simulata, C. sterilis, C. stricta, C. utriculata, C. viridula, Cephalanthus occidentalis, Chamaedaphne calyculata, Chamerion angustifolium, Chelone glabra, Cicuta maculata, Cirsium muticum, Cladium mariscoides, Cornus amomum, C. foemina, C. racemosa, C. sericea, Cypripedium candidum, Dasiphora floribunda, Deschampsia cespitosa, Drosera rotundifolia, Dryas, Dryopteris spinulosa, Elaeagnus commutata, Eleocharis calva, E. palustris, E. quinqueflora, E. rostellata, E. tenuis, Elymus albicans, Epilobium leptophyllum, E. strictum, Equisetum arvense, E. fluviatile, E. palustre, E. variegatum, Eriophorum angustifolium, E. gracile, E. viridicarinatum, Eutrochium maculatum, Galium labradoricum, G. obtusum, G. trifidum, Gentiana procera, Gentianopsis detonsa, Geum macrophyllum, G. rivale, Glyceria striata, Hedysarum alpinum, Helianthus grosseserratus, H. rydbergii, Helodium blandowii, Hippuris montana, Hylocomium splendens, Hypnum pratense, Iris versicolor, Juncus alpinoarticulatus, J. balticus, J. torreyi, Juniperus communis, Kobresia myosuroides, K. simpliciuscula, Larix laricina, Lathyrus palustris, Leymus innovatus, Liatris ligulistylis, Lilium philadelphicum, Limprichtia revolvens, Lobelia kalmii, Lomatogonium rotatum, Lycopus americanus, L. uniflorus, Lysimachia quadriflora, L. terrestris, L. thyrsiflora, Lythrum salicaria, Mentha arvensis, Menyanthes trifoliata, Mertensia paniculata, Muhlenbergia frondosa, M. glomerata, M. richardsonis, Myrica gale, Oxypolis rigidior, Parnassia glauca, P. palustris, P. parviflora, Peltandra virginica, Persicaria amphibia, Petasites frigidus, Phalaris arundinacea, Phlox kelseyi, Phragmites australis, Physocarpus opulifolius, Picea glauca, P. mariana, Pinguicula vulgaris, Pinus contorta, Poa pratensis, Polemonium occidentale, Populus balsamifera, Potentilla norvegica, Primula alcalina, P. egaliksensis, P. incana, P. pauciflora, Prunus pensylvanica, Ptilagrostis porteri, Pycnanthemum virginianum, Pyrola asarifolia, Ranunculus septentrionalis, Rhamnus alnifolia, R. frangula, Rhododendron lapponicum, R. tomentosum, Rhynchospora capillacea, Ribes americanum, Rubus arcticus, Rumex britannica, R. occidentalis, Sagittaria, Salix alaxensis, S. amygdaloides, S. bebbiana, S. brachycarpa, S. geyeriana, S. glauca, S. interior, S. ligulifolia, S. maccalliana, S. monticola, S. myricoides, S. myrtillifolia, S. niphoclada, S. pedicellaris, S. petiolaris, S. planifolia, S. pseudomonticola, S. serissima, S. wolfii, Sarracenia purpurea, Schoenoplectus tabernaemontani, S. acutus, Scirpus atrovirens, S. cyperinus, S. hudsonianus, Scleria verticillata, Scorpidium scorpioides, S. turgescens, Scutellaria galericulata, S. lateriflora, Senecio cymbalarioides, S. streptanthifolius, Shepherdia canadensis, Sisyrinchium pallidum, Sium suave, Solidago canadensis, S. gigantea, S. graminifolia, S. riddellii, Sparganium, Spartina pectinata, Sphagnum, Spiraea alba, Stachys tenuifolia, Stellaria longifolia, Swertia perennis, Symphyotrichum boreale, S. foliaceum, Taraxacum officinale, Thalictrum alpinum, Thelypteris palustris, Thuja occidentalis, Tofieldia pusilla, Toxicodendron vernix, Triantha glutinosa, Trichophorum cespitosum, T. pumilum, Triglochin maritimum, T. palustre, Typha angustifolia, T. latifolia, Utricularia cornuta, U. ochroleuca, Vaccinium macrocarpon, V. uliginosum, Valeriana edulis, V. sitchensis, Viola nephrophylla, Vitis riparia, Zizia aptera.

Salix caprea L. is a nonindigenous shrub or small tree (to 15 m) that is not particularly invasive but occurs occasionally in marshes, meadows, ravines, roadsides, seeps, thickets, woodlands, and along the margins of rivers and streams at elevations of up to 4600 m. It is relatively short lived but can tolerate heavy shade and other fairly harsh growing conditions. It is found on substrates that include mud, rock, rocky loam, and sand. The pH in its native habitat ranges from 5.8 to 7.3. The flowers (produced from March to June) are both insect- and wind pollinated. Overall, they are somewhat better adapted for insect-pollination as a consequence of their greater pollen adhesion relative to more anemophilous willow species. Common insect pollinators are bees (Hymenoptera: Apidae: *Andrena, Apis mellifera, Bombus*), and moths (Lepidoptera), which are attracted to the flowers by olfactory and visual cues (e.g., yellow pollen masses). During daylight, bees are attracted by 1,4-dimethoxybenzene; whereas, at night, moths are attracted by increased emissions of lilac aldehyde. Up to 100% germination has been obtained for seeds placed on sterile Norstog media at 21°C, under a 12/12 h light regime. The roots are colonized by dark septate endophyte fungi. The plants do not survive well in disturbed sites like floodplains, dut to their inability to resprout vegetatively, a factor that also correlates with their difficult artificial propagation. **Reported associates (North America):** *Ailanthus altissima, Cornus, Prunus cerasifera, Rubus bifrons, Salix scouleriana, Typha latifolia.*

Salix caroliniana Michx. is a small, fast-growing tree (to 10 m), which occurs in fresh to brackish (oligohaline), shallow (to 1 m), standing waters of baygalls, borrow pits, bottomlands, canals, channels, depressions, ditches, dunes, flatwoods, floodplains, gravel bars, hammocks, hummocks, ledges, marshes, meadows, ponds, pools, prairies, ravines, roadsides, sandbars, seeps, shoals, sloughs, springs, streambeds, swales, swamps, thickets, trenches, washes, and along the margins of lakes, ponds, reservoirs, rivers, and streams at elevations of up to 740 m. This is a pioneer species that will tolerate partial shade but normally requires full sun. It will readily colonize disturbed habitats or unburned marshland. The substrates usually are calcareous but somewhat acidic (e.g., pH: 5.9) and include cobble, gravel, limestone, loam, loamy sand, rock, sand, sandstone, sandy loam, sandy muck, sandy silt, shale, and stony gravel. The flowers appear from December to May in the south, and from April to June in the northern extent of its range. Their nectar contains a glucose:fructose:sucrose ratio of 50:46:4, and they are insect-pollinated, primarily by bees (Hymenoptera: Apidae: Colletidae; Halictidae). Pollen has also been collected from stable flies (Insecta: Diptera: Muscidae: *Stomoxys calcitrans*), which may function as inadvertant pollinators. The seeds are dispersed during the summer by wind or by water and lack a dormancy requirement for germination. The highest germination and seedling survival occur on saturated (but drained), organic substrates and decrease on inundated sites or on dry, sandy substrates. Only green-colored seeds (vs. brown) are viable. Freshly collected seeds should be sown and watered immediately, as they begin to lose their viability within 5 days of being shed. They have germinated successfully under a 12/12 h 8°C/22°C light/temperature regime. The seeds will retain viability for several months when refigerated (at 4°C), which indicates that northern populations may exhibit greater longevity (to 10 days). However, seedling survival is low from refrigerated seeds. Although densities up to 76 seeds/m² have been reported from some seed bank studies (near surface), no persistent seed bank would be expected for the short-lived seeds. The plants tolerate flooding by the production of lenticels (for aeration) and adventitious roots (for anchorage). Stands of *S. caroliniana* are particularly important ecologically for riverine bank stabilization. **Reported associates:** *Acer rubrum, A. saccharinum, Aesculus glabra, Alnus serrulata, Alternanthera philoxeroides, Andropogon glomeratus, A. virginicus, Annona glabra, Apocynum sibiricum, Baccharis, Bacopa monnieri, Betula nigra, Boehmeria cylindrica, Cabomba caroliniana, Caloplaca lecanorae, Carex comosa, C. longii, Carpinus caroliniana, Carya aquatica, C. tomentosa, Cephalanthus occidentalis, Chara zeylanica, Chasmanthium, Chrysobalanus icaco, Cicuta maculata, Cladium jamaicense, Commelina diffusa, Cornus amomum, Crinum americanum, Cyperus odoratus, Dichanthelium clandestinum, Eichhornia crassipes, Eleocharis cellulosa, E. geniculata, Equisetum hyemale, Eriocaulon, Eupatorium capillifolium, Fraxinus caroliniana, F. pennsylvanica, F. profunda, Hydrocotyle ranunculoides, H. umbellata, Hypericum densiflorum, H. fasciculatum, Ilex cassine, I. verticillata, I. vomitoria, Itea virginica, Juncus marginatus, Juniperus virginiana, Justicia, Kosteletzkya pentacarpos, Leitneria floridana, Lemna, Liquidambar styraciflua, Lonicera japonica, Ludwigia peruviana, Magnolia virginiana, Megathyrsus maximus, Melaleuca quinquenervia, Metopium toxiferum, Mikania scandens, Morus alba, M. rubra, Muhlenbergia, Myrica cerifera, Myriophyllum pinnatum, Myrsine florida, Nuphar, Nymphaea odorata, Nyssa aquatica, N. sylvatica, Orontium aquaticum, Osmunda regalis, Panicum hemitomon, P. virgatum, Parietaria floridana, Peltandra virginica, Persea borbonia, P. palustris, Persicaria glabra, Pinus palustris, P. taeda, Pistia stratiotes, Planera aquatica, Plantago virginica, Platanus occidentalis, Pontederia cordata, Populus deltoides, Ptelea trifoliata, Quercus lyrata, Q. michauxii, Q. nigra, Q. phellos, Q. virginiana, Rhapidophyllum hystrix, Rhododendron, Rubus, Rumex hastatulus, R. verticillatus,*

Sabal palmetto, Sagittaria latifolia, Salix amygdaloides, S. nigra, Sambucus nigra, Saururus cernuus, Schoenoplectus tabernaemontani, Serenoa repens, Sphenoclea zeylanica, Symphoricarpos orbiculatus, Taxodium ascendens, T. distichum, Thalia geniculata, Thelypteris palustris, Toxicodendron radicans, Tradescantia ohiensis, Tripsacum dactyloides, Typha domingensis, T. latifolia, Ulmus alata, U. americana, Vaccinium, Viburnum prunifolium, Vitis rupestris, Woodwardia virginica, Xanthosoma sagittifolium, Zizania aquatica.

Salix commutata Bebb is a small shrub (to 3 m), which occurs at higher elevations (to 2789 m) or on alpine tundra in tidal or nontidal, shallow waters (to 40 cm deep) of bogs, carrs, coastal mudflats, fens, floodplains, glacial outwash plains, gravel benches, gulches, heath, meadows, moraines, muskeg, roadsides, scrub, seeps, slopes, swales, swamps, thickets, woodlands, and along the margins of lakes, rivers, and streams. This is a pioneer species, which tolerates fire and light shade. It occurs on substrates that often are acidic (pH: 5.0–7.0) and include clay loam, gravel, humus, loam, muck, pumice, rock, sand, sandy granitic loamy alluvium, sandy loam, silt, talus, till, and peat (>10 cm deep). Stands of *S. commutata* somewhat inhibit the establishment of other native plants and form dense thickets with other willow species, but create soil conditions that generally favor plant establishment. *Salix commutata* flowers after the leaves emerge, later than most other willows (May to August), and releases its seeds primarily from late June to July. The seeds have no dormancy and remain viable for only a few days. The roots are colonized by arbuscular and dark septate endophytic mycorrhizal fungi. **Reported associates:** *Abies lasiocarpa, A. ×shastensis, Achillea, Aconitum columbianum, Agrostis exarata, A. pallens, A. scabra, Allium, Alnus crispa, A. rugosa, A. sinuata, A. viridis, Anaphalis margaritacea, Angelica genuflexa, A. lucida, Antennaria argentea, Anthoxanthum nitens, Arabis lyrata, Arctostaphylos uva-ursi, Arctous rubra, Arnica latifolia, A. longifolia, Artemisia norvegica, Aruncus dioicus, Athyrium filix-femina, Aulacomnium palustre, Betula glandulosa, B. nana, Bidens cernuus, Bistorta bistortoides, B. vivipara, Calamagrostis canadensis, C. stricta, Caltha leptosepala, Cardamine cordifolia, Carex anthoxanthea, C. aperta, C. aquatilis, C. buxbaumii, C. cusickii, C. disperma, C. geyeri, C. jonesii, C. lasiocarpa, C. lenticularis, C. leptalea, C. limosa, C. luzulina, C. lyngbyei, C. macrochaeta, C. mertensii, C. microptera, C. nebrascensis, C. nigricans, C. obnupta, C. pluriflora, C. rostrata, C. scopulorum, C. sitchensis, C. spectabilis, C. tenuiflora, C. utriculata, C. viridula, Cassiope mertensiana, C. tetragona, Castilleja miniata, C. unalaschcensis, C. viridula, Chamerion angustifolium, C. latifolium, Cirsium arvense, Comarum palustre, Cornus sericea, Corydalis caseana, Dasiphora floribunda, Deschampsia cespitosa, Drosera rotundifolia, Dryas drummondii, Dryopteris assimilis, Eleocharis palustris, E. quinqueflora, Empetrum nigrum, Epilobium anagallidifolium, E. ciliatum, E. hornemannii, Equisetum arvense, E. fluviatile, E. palustre, E. variegatum, Eremogone, Erigeron acris, Eriophorum angustifolium, E. gracile, Erythronium grandiflorum,*

Festuca altaica, F. rubra, Galium trifidum, Gaultheria humifusa, Gentiana sceptrum, Geranium erianthum, Geum calthifolium, G. macrophyllum, Glyceria striata, Gymnocarpium dryopteris, Heracleum sphondylium, Hieracium albiflorum, Hordeum brachyantherum, Hypericum anagalloides, Hypochaeris radicata, Juncus balticus, J. mertensiana, Kalmia microphylla, Leymus arenarius, L. mollis, Ligusticum canbyi, L. grayi, L. verticillatum, Lonicera caerulea, L. involucrata, Luetkea pectinata, Lupinus latifolius, L. lepidus, L. nootkatensis, Luzula hitchcockii, Lycopus uniflorus, Lysichiton americanus, Menyanthes trifoliata, Menziesia ferruginea, Mertensia campanulata, M. ciliata, Micranthes ferruginea, M. integrifolia, M. lyallii, M. odontoloma, Mimulus guttatus, M. lewisii, M. primuloides, Muhlenbergia filiformis, Myrica gale, Nephrophyllidium crista-galli, Oenanthe sarmentosa, Oreostemma alpigenum, Packera streptanthifolia, Parnassia fimbriata, P. kotzebuei, Pedicularis attollens, P. groenlandicum, Penstemon cardwellii, Petasites frigidus, Phacelia, Phalaris arundinacea, Phyllodoce empetriformis, Picea engelmannii, Pinus contorta, Platanthera dilatata, Poa alpina, P. pratensis, Polemonium occidentale, Polytrichum, Populus balsamifera, P. trichocarpa, Potentilla flabellifolia, Primula jeffreyi, Pyrola asarifolia, Racomitrium canescens, Ranunculus alismifolius, Rhamnus alnifolia, Rhododendron albiflorum, R. columbianum, R. tomentosum, Ribes hudsonianum, R. lacustre, R. laxiflorum, R. montigenum, Rosa pisocarpa, Rubus arcticus, R. chamaemorus, R. laciniatus, R. spectabilis, R. ursinus, Salix alaxensis, S. arctica, S. barclayi, S. boothii, S. brachycarpa, S. drummondiana, S. fuscescens, S. geyeriana, S. glauca, S. hookeriana, S. lanata, S. lasiandra, S. lemmonii, S. monticola, S. myrtillifolia, S. planifolia, S. pulchra, S. reticulata, S. scouleriana, S. sitchensis, S. stolonifera, S. tweedyi, S. vestita, Sanguisorba officinalis, S. sitchensis, Scheuchzeria palustris, Schoenoplectus tabernaemontani, Scirpus microcarpus, Senecio congestus, S. crassulus, S. triangularis, Shepherdia canadensis, Sibbaldia, Sorbus sitchensis, Sparganium emersum, Sphagnum, Spiraea douglasii, S. splendens, Spiranthes romanzoffiana, Stachys rigida, Stereocaulon paschale, S. tomentosum, Streptopus amplexifolius, Symphoricarpos oreophilus, Thuja plicata, Triantha glutinosa, Trichophorum alpinum, Trientalis europaea, Trifolium longipes, Triglochin, Trisetum spicatum, Trollius laxus, Tsuga mertensiana, Typha latifolia, Vaccinium cespitosum, V. deliciosum, V. membranaceum, V. scoparium, V. uliginosum, Valeriana sitchensis, Veratrum californicum, V. viride, Veronica americana, V. serpyllifolia, V. wormskjoldii, Viola macloskeyi, V. palustris.

Salix drummondiana Barratt ex Hook. is a shrub (to 5 m), which occurs in or on bogs, bottomlands, brooks, carrs, depressions, ditches, fens, floodplains, forests, gravel bars, gravel pits, meadows, prairies, roadsides, seeps, slopes, swamps, thickets (frequently along very steep, rocky streams), vernal pools, and along the margins of lakes, ponds, rivers, and streams from 216 to 3413 m. Many occurrences are located in montane valleys or along subalpine watercourses on slopes of up to 35%. The substrates tend to be acidic (e.g., pH: 5.5; hardness: 10 mg/L; Ca^{2+}: 2.72 mg/L) and relatively deep (to 1+ m).

They have been described as clay, duff, gravel, gravelly clay, gravelly loam, humus, limestone, muck, peat, quartzite rock, rocky clay, rocky loam, sand, sandy alluvium, sandy granite, sandy loam, silt, silty gravel, silty sand, and talus. These are cold tolerant plants that occupy cooler sites such as shaded valleys and north-facing slopes. The flowers appear from late March to September, with fruit production occurring from mid-May through September. The seeds are dispersed by wind or by water. The plants are fast growing and can live for more than 30 years. The shoots can become seriously stunted by overgrazing but will recover rapidly upon its cessation. *Salix drummondiana* is easily propagated by stem cuttings which show a rooting rate of 98%–100%. The stems contain predeveloped primordia and will form roots along their length within 10 days of burial. **Reported associates:** *Abies grandis, A. lasiocarpa, Acer glabrum, Achillea lanulosa, A. millifolium, Actaea rubra, Agrostis stolonifera, Allium schoenoprasum, A. validum, Alnus rugosa, A. viridis, Amelanchier alnifolia, Angelica arguta, Antennaria corymbosa, A. racemosa, A. rosea, Anticlea elegans, Arctostaphylos rubra, A. uva-ursi, Arnica chamissonis, A. cordifolia, A. mollis, Artemisia cana, Astragalus alpinus, Balsamorhiza sagittata, Betula glandulosa, B. nana, B. occidentalis, Bistorta bistortoides, B. vivipara, Bromus, Calamagrostis canadensis, Cardamine cordifolia, Carex aperta, C. aquatilis, C. cusickii, C. dioica, C. festivella, C. flava, C. heteroneura, C. jonesii, C. lasiocarpa, C. limosa, C. microptera, C. nebrascensis, C. pachystachya, C. rostrata, C. saxatilis, C. scoparia, C. scopulorum, C. straminiformis, C. utriculata, Carum carvi, Castilleja miniata, C. septentrionalis, Ceanothus velutinus, Chenopodium album, Cladonia nivalis, Clintonia uniflora, Comarum palustre, Cornus canadensis, C. sericea, Crataegus douglasii, Crepis runcinata, Dasiphora floribunda, Delphinium glaucum, Deschampsia cespitosa, Dulichium arundinaceum, Eleocharis rostellata, Elymus glaucus, Empetrum nigrum, Epilobium ciliatum, E. suffruticosum, Equisetum arvense, E. fluviatile, E. pratense, E. scirpoides, E. variegatum, Erigeron glabellus, E. peregrinus, Eriophorum gracile, Fragaria ovalis, F. vesca, Galium triflorum, Gaultheria humifusa, Gentianella propinqua, Geranium richardsonii, Geum macrophyllum, G. rivale, Hedysarum alpinum, Heracleum sphondylium, Holodiscus discolor, Hydrurus, Hylocomium splendens, Juncus balticus, J. longistylus, Larix occidentalis, Leymus innovatus, Ligusticum verticillatum, Lonicera involucrata, Lysichiton americanus, Maianthemum stellatum, Melilotus officinalis, Mentha arvensis, Menyanthes trifoliata, Menziesia ferruginea, Mertensia ciliata, M. paniculata, Micranthes ferruginea, M. subapetala, Mimulus lewisii, M. primuloides, Monostroma, Orthilia secunda, Oxytropis sericeus, Packera subnuda, Paxistima myrsinites, Pedicularis groendlandica, Phleum alpinum, P. pratense, Phlox, Picea engelmanii, P. glauca, P. mariana, Pinus balfouriana, P. contorta, P. flexilis, P. jeffreyi, P. monticola, P. ponderosa, Plagiomnium ellipticum, Platanthera dilatata, P. hyperborea, Poa alpina, P. palustris, P. pratensis, P. reflexa, Polemonium caeruleum, P. occidentale, Populus angustifolia, P. balsamifera, P. tremuloides, P. trichocarpa, Potentilla ×diversifolia,*

P. pulcherrima, Primula pauciflora, Pseudotsuga menziesii, Pyrola asarifolia, Ranunculus orthorhyncus, Rhamnus, Rhodiola rhodantha, Rhododendron columbianum, R. groenlandicum, Ribes hudsonianum, R. inerme, R. oxyacanthoides, Rosa nutkana, Rumex crispus, Salix barclayi, S. barrattiana, S. bebbiana, S. boothii, S. brachycarpa, S. candida, S. commutata, S. eastwoodiae, S. eriocephala, S. exigua, S. farriae, S. geyeriana, S. glauca, S. lasiandra, S. lemmonii, S. lutea, S. melanopsis, S. monochroma, S. monticola, S. myrtillifolia, S. orestera, S. planifolia, S. pseudolapponica, S. pseudomonticola, S. scouleriana, S. sitchensis, S. vestita, S. wolfii, Sanguisorba canadensis, Schoenoplectus tabernaemontani, Scirpus microcarpus, Senecio serra, S. triangularis, Shepherdia canadensis, Sphagnum, Spiraea douglasii, Streptopus amplexifolius, Swertia perennis, Symphoricarpos albus, Symphyotrichum ascendens, S. foliaceum, S. lanceolatum, S. spathulatum, Taraxacum officinale, Thalictrum fendleri, T. venulosum, Thuja plicata, Torreyochloa pallida, Trifolium repens, Trisetum wolfii, Tsuga heterophylla, Typha latifolia, Urtica dioica, Vaccinium cespitosum, V. membranaceum, V. myrtillus, V. scoparium, Valeriana dioica, V. occidentalis, Vaucheria, Veratrum californicum, Veronica americana, Vicia americana, Viola palustris, Wyethia, Xerophyllum tenax.

***Salix eastwoodiae* Cockerell ex A.** Heller is a shrub (<4 m), which inhabits bogs, cirques, depressions, fens, flats, floodplains, ledges, marshes, meadows, moraines, prairies, scrub, seeps, shores, slopes, streams, swales, talus slopes, thickets, and the margins of lakes, ponds, rivers, and streams at elevations from 685 to 3505 m. Most of the plants occur within the subalpine or alpine zone in sites having up to a 30% slope. Exposures range from open to semishaded conditions. The substrates have been characterized as alluvium, Aquic Cumulic Cryoborolls, bouldery talus, clay, cobble, glacial till, granite (decomposed), limestone, loam, rock (basaltic), gravel, gravelly loam, Mollisol, peat, pumice, rocky gravelly loam, rocky silt, sandy loam, serpentine, silt, silty gravel, silty loam, silty rocky gravel, silty rocky loam, talus, typic cryaquoll, and volcanic. Flowering and fruiting have been observed from May to September; the catkins appear along with the leaves. Additional information is needed on the floral biology and seed ecology of this species. **Reported associates:** *Abies concolor, A. lasiocarpa, A. magnifica, Achnatherum occidentale, A. thurberianum, Agastache urticifolia, Agrostis idahoensis, A. pallens, Allium schoenoprasum, Alnus viridis, Angelica arguta, Aquilegia formosa, Arctostaphylos patula, Arnica chamissonis, A. mollis, Artemisia ludoviciana, A. tridentata, Balsamorhiza sagittata, Betula glandulosa, B. occidentalis, Bistorta bistortoides, Botrychium simplex, Bromus carinatus, Calamagrostis breweri, C. canadensis, Caltha leptosepala, Carex abrupta, C. aquatilis, C. atrata, C. capitata, C. echinata, C. integra, C. jonesii, C. lemmonii, C. lenticularis, C. mariposana, C. microptera, C. nebrascensis, C. neurophora, C. nigricans, C. norvegica, C. praegracilis, C. rostrata, C. simulata, C. scopulorum, C. subfusca, C. subnigricans, C. utriculata, C. vesicaria, Castilleja applegatei, C. covilleana, C. miniata, Ceanothus cordulatus,*

Chamerion angustifolium, Cornus sericea, Dasiphora floribunda, Delphinium bicolor, Deschampsia cespitosa, Descurainia californica, Drosera rotundifolia, Drymocallis glandulosa, D. lactea, Eleocharis acicularis, Elymus trachycaulus, Epilobium ciliatum, E. hornemannii, Equisetum arvense, Ericameria nauseosa, Erigeron coulteri, E. glacialis, E. peregrinus, Eriogonum umbellatum, Eriophorum crinigerum, Festuca rubra, Fragaria virginiana, Galium andrewsii, Gentiana algida, Gentianopsis holopetala, Geum macrophyllum, Helenium bigelovii, Heracleum sphondylium, Hosackia oblongifolia, Hymenoxys hoopesii, Hypericum formosum, Ivesia campestris, Jamesia americana, Juncus balticus, J. longistylis, J. mertensianus, J. parryi, Kalmia polifolia, Lagophylla glandulosa, Leptosiphon nuttallii, Lewisia pygmaea, Ligusticum grayi, Lilium kelleyanum, Lomatium grayi, Lonicera caerulea, L. conjugialis, L. involucrata, L. utahensis, Lupinus arbustus, L. caudatus, L. padrecrowleyi, L. polyphyllus, Luzula subcongesta, Maianthemum stellatum, Menziesia ferruginea, Micranthes integrifolia, M. oregana, M. subapetala, Mimulus floribundus, M. guttatus, M. lewisii, M. primuloides, Monardella glauca, Montia chamissoi, Oreostemma alpigenum, Oxypolis occidentalis, Parnassia fimbriata, Pedicularis attollens, P. cystopteridifolia, P. groenlandica, Penstemon azureus, P. heterodoxus, P. newberryi, P. rydbergi, Perideridia parishii, Phacelia hastata, Phleum alpinum, Phyllodoce empetriformis, Picea engelmannii, Pinus albicaulis, P. contorta, P. jeffreyi, P. monticola, Platanthera dilatata, P. sparsiflora, Poa cusickii, P. nervosa, P. palustris, P. pratensis, Podagrostis humilis, Populus tremuloides, Potentilla drummondii, P. gracilis, P. grayi, Primula jeffrey, P. tetrandra, Pteridium aquilinum, Ranunculus alismifolius, R. populago, Rhamnus alnifolia, Rhodiola rosea, Rhododendron albiflorum, R. columbianum, Ribes cereum, Rosa woodsii, Rumex paucifolius, Sagina saginoides, Salix barclayi, S. boothii, S. brachycarpa, S. drummondiana, S. farriae, S. geyeriana, S. glauca, S. jepsonii, S. lasiandra, S. lemmonii, S. melanopsis, S. nivalis, S. orestera, S. planifolia, S. sitchensis, S. tweedyi, S. wolfii, Sanguisorba stipulata, Senecio scorzonella, S. sphaerocephalum, S. triangularis, Sisyrinchium elmeri, Sorbus californica, Sphagnum, Sphenosciadium capitellatum, Spiraea douglasii, S. splendens, Spiranthes romanzoffiana, Swertia perennis, Symphoricarpos rotundifolius, Symphyotrichum foliaceum, Thalictrum fendleri, T. sparsiflorum, Torreyochloa erecta, Triantha occidentalis, Trichostema laxum, Trifolium longipes, Vaccinium cespitosum, V. membranaceum, V. uliginosum, Veratrum californicum, Veronica wormskjoldii, Viola adunca, V. macloskeyi, Xerophyllum tenax.

Salix eriocephala Michx. is a shrub or small tree (to 6 m), which occurs in or on beaches, bogs, borrow pits, bottomlands, depressions, ditches, dunes, embankments, fens (rich), floodplains, gravel pits, gullies, marshes, meadows, mudflats, prairies, roadsides, seeps, shores, sloughs, stream banks, swales, swamps, thickets, woodlands, and along the margins of ponds, rivers, and streams at elevations of up to 2104 m. The exposures range from open sites to partial shade. The substrates are calcareous and include alluvium, cobble, gravel,

gravelly silt, limestone, loam, muck, mucky sand, peat, rock, sand, sandy clay loam, and silt. This species often grows in monospecific stands (or with other willows) and is important in streambank stabilization. It is a fast growing, but relatively long-lived plant that is an indicator of depositional sediments having a high nitrogen content. The leaves contain high amounts of condensed tannins. The plants are known to survive well near sites that have been contaminated by iron cyanide. They also translocate soil-borne fluoride into their leaf tissue, but suffer from reduced leaf area and water use when exposed to concentrations exceeding 80 ppm. The flowers appear before the leaves in March to June with the fruits appearing in May to June. This is an outcrossing species, which is pollinated primarily by flies and bees (Insecta: Diptera; Hymenoptera). Isozyme analyses have demonstrated a positive association between heterozygosity and biomass production. Microsatellite markers have also been tested on this species, and yield an average of 2.95 alleles/locus. Microsatellite data indicate relatively high genetic diversity within populations, with few clonal individuals. However, some clonal reproduction occurs occasionally by stem fragmentation. Cuttings taken from the shoots are fast rooting and the wood is fairly durable. **Reported associates:** *Acer negundo, A. rubrum, A. saccharinum, A. saccharum, Acorus calamus, Agrimonia parviflora, Agrostis gigantea, Alnus rugosa, Alopecurus aequalis, Amorpha fruticosa, Andropogon gerardii, Apios americana, Apocynum cannabinum, Aronia melanocarpa, Asclepias incarnata, Bidens cernuus, Boehmeria cylindrica, Calamagrostis canadensis, Caltha palustris, Calystegia sepium, Campanula aparinoides, Cardamine bulbosa, C. douglassii, C. parviflora, Carex annectens, C. buxbaumii, C. frankii, C. granularis, C. hystericina, C. lacustris, C. lasiocarpa, C. lurida, C. molesta, C. nebrascensis, C. pellita, C. pensylvanica, C. sartwellii, C. scabrata, C. stipata, C. stricta, C. tribuloides, C. vulpinoidea, Cephalanthus occidentalis, Chelone glabra, Chrysosplenium americanum, Cicuta bulbifera, Cirsium, Cornus amomum, C. racemosa, C. sericea, Cryptotaenia canadensis, Doellingeria umbellata, Dryopteris carthusiana, D. cristata, D. spinulosa, Elaeagnus umbellata, Eleocharis compressa, E. erythropoda, E. obtusa, E. palustris, Equisetum arvense, Erigeron philadelphicus, Eupatorium perfoliatum, Euthamia graminifolia, Eutrochium maculatum, Fragaria virginiana, Fraxinus americana, F. pennsylvanica, Galium boreale, G. obtusum, Glyceria striata, Helianthus grosseserratus, Ilex verticillata, Impatiens capensis, Iris versicolor, I. virginica, Juncus acuminatus, J. brachycephalis, J. dudleyi, Laportea canadensis, Leersia oryzoides, Liriodendron tulipifera, Lobelia siphilitica, L. spicata, Lycopus americanus, Lysimachia ciliata, L. thyrsiflora, Lythrum alatum, L. salicaria, Muhlenbergia rigens, Nasturtium officinale, Nyssa sylvatica, Oenothera pilosella, Onoclea sensibilis, Ostrya virginiana, Panicum bulbosum, Peltandra virginica, Persicaria, Phalaris arundinacea, Pinus banksiana, Poa nemoralis, P. pratensis, Populus angustifolia, P. deltoides, P. tremuloides, Prunus americana, P. virginiana, Pycnanthemum virginianum, Quercus macrocarpa, Q. rubra, Ranunculus gmelinii, R.*

hispidus, Rhus typhina, Ribes americana, R. hirtellum, Rosa blanda, R. palustris, Rubus occidentalis, R. pubescens, Salix amygdaloides, S. bebbiana, S. discolor, S. exigua, S. gracilis, S. interior, S. lucida, S. lutea, S. myricoides, S. nigra, S. petiolaris, S. sericea, Sambucus nigra, Schizachyrium scoparium, Scirpus atrovirens, S. cyperinus, S. pendulus, Schoenoplectus acutus, S. pungens, S. tabernaemontani, Scutellaria galericulata, S. lateriflora, Shepherdia argentea, Silphium perfoliatum, S. terebinthinaceum, Solidago canadensis, S. riddellii, S. rigida, S. uliginosa, Sorghastrum nutans, Sparganium glomeratum, Spartina pectinata, Sphenopholis obtusata, Stachys palustris, Symphyotrichum lateriflorum, S. novae-angliae, S. puniceum, Symplocarpus foetidus, Thalictrum dasycarpum, T. pubescens, Thaspium trifoliatum, Thelypteris palustris, Tiarella cordifolia, Triadenum fraseri, Tussilago farfara, Typha angustifolia, T. latifolia, Ulmus americana, Urtica dioica, Vaccinium corymbosum, Verbena hastata, Veronica americana, Viburnum acerifolium, V. rafinesqueanum, Vitis aestivalis, V. riparia, Xanthium strumarium.

Salix exigua Nutt. is a shrub or small tree (to 8 m), which occurs in or on alluvial bars/benches, beaches, bottomlands, canals, carrs, depressions, dikes, ditches, draws, dunes, flats, floodplains, forests, gravel bars, marshes, meadows, mudflats, potholes, ravines, roadsides, sandbars, scrub, seeps, shores, slopes (to 40%), sloughs, springs, swales, thickets, washes, and along the margins of lakes, ponds, reservoirs, rivers, and streams at elevations of up to 2895 m. These plants characteristically form a zone along the water's edge in riparian habitats and are also found in shallow standing water (to 0.5 m). This species generally requires full sun and also occurs in partially shaded exposures. The sediments range in pH from 6 to 8 and have been categorized as alluvium, ashy sandy loam, boulders, clay, cobble (basaltic), granite, gravel, gravelly clay, humus, loam, loamy sand, muck, muddy loamy clay, pebbles, rock, rocky clay, sand, sandy clay, sandy gravel, sandy silt, schist, silt, silty gravel, silty loam, silty sand, stones, talus, travertine, and Tuscan mudflow. This is an aggressive, thicket-forming pioneer species that stabilizes sand bars and alluvial soils. The plants can be highly clonal (as evidenced by studies using AFLP markers) and will spread vegetatively by the production of underground root suckers. Pieces of stem that break from the plant can be transported by water currents, lodge into the sediment, root, and then develop into new plants. The insect-pollinated (Insecta: Hymenoptera) flowers appear during spring (March) along with the leaves and are produced through September. Fruits have been observed from April to October. The short-lived (<10 days) seeds of *S. exigua* lack dormancy and germinate (up to 83% within 96 h) at 22°C. They are wind dispersed and must land quickly on suitable moist, open sites in order to germinate. Unlike *S. eriocephala*, *S. exigua* shows no positive association between heterozygosity (determined from isozyme data) and biomass production. The plants can resprout from their roots following fire. Their prolific seed production can facilitate the rapid revegetation of burned sites. **Reported associates:** *Abies grandis, Abronia maritima, A. villosa, Acamptopappus sphaerocephalus, Acer glabrum, A. negundo,*

A. saccharinum, Achillea millefolium, Achnatherum hymenoides, Acmispon glabrus, A. grandiflorus, Ageratina adenophora, Agrostis scabra, A. stolonifera, Ailanthus altissima, Alisma, Allium geyeri, A. textile, A. tolmiei, Alnus rhombifolia, A. rugosa, A. tenuifolia, A. viridis, Ambrosia psilostachya, A. salsola, Amorpha fruitcosa, Amphiscirpus nevadensis, Amsinckia tessellata, Anagallis arvensis, Anemopsis californica, Antennaria microphylla, Anthriscus, Apocynum androsaemifolium, A. cannabinum, Arctostaphylos uva-ursi, Argemone, Aristida purpurea, Arnica chamissonis, Arrhenatherum elatius, Artemisia arbuscula, A. californica, A. cana, A. douglasiana, A. ludoviciana, A. nova, A. tilesii, A. tridentata, Arundo donax, Asclepias, Atriplex canescens, A. confertifolia, A. polycarpa, A. semibaccata, A. torreyi, Azolla filiculoides, Baccharis glutinosa, B. pilularis, B. salicifolia, B. sergiloides, Bahiopsis laciniata, B. parishii, Balsamorhiza sagittata, Berteroa incana, Betula occidentalis, Bouteloua aristidoides, Brassica nigra, Bromus diandrus, B. hordeaceus, B. inermis, B. pumpellianus, B. rubens, B. tectorum, Calamagrostis canadensis, Callitropsis forbsii, Camissoniopsis cheiranthifolia, Carex alma, C. densa, C. flava, C. lasiocarpa, C. nebrascensis, C. rostrata, C. saxatilis, C. spissa, C. utriculata, Ceanothus cuneatus, C. crassifolius, C. greggii, C. leucodermis, C. oliganthus, C. sanguineus, C. senta, C. velutinus, Celtis reticulata, Cenchrus longispinus, Centaurea, Cercocarpus betuloides, C. ledifolius, Chilopsis linearis, Chrysothamnus nauseosus, C. viscidiflorus, Cichorium, Cirsium arvense, C. vulgare, Clematis ligusticifolia, Coleogyne ramosissima, Conyza, Convolvulus arvensis, Conium maculatum, Coreopsis atkinsoniana, Cornus sericea, Crataegus douglasii, Crepis elegans, Croton californicus, Cryptantha confertiflora, Cyperus erythrorhizos, C. esculentus, C. involucratus, Dasiphora floribunda, Datisca glomerata, Delphinium parishii, Descurainia, Diplacus aurantiacus, Dipsacus fullonum, D. sativus, Distichlis spicata, Dryas drummondii, Dryopteris arguta, Dysphania botrys, Eclipta prostrata, Elaeagnus angustifolia, E. commutata, Eleocharis macrostachya, E. montevidensis, E. palustris, E. parishii, Elymus canadensis, E. elymoides, E. repens, Encelia californica, Ephedra trifurca, E. viridis, Epilobium canum, E. ciliatum, Epipactis gigantea, Equisetum arvense, E. hyemale, Eragrostis cilianensis, E. mexicana, Eremogone congesta, Ericameria cuneata, E. nauseosa, E. pinifolia, Erigeron glabellus, E. linearis, Eriodictyon crassifolium, E. trichocalyx, Eriogonum fasciculatum, E. saxatile, E. umbellatum, Eriophorum scheuchzeri, Eriophyllum confertiflorum, Erodium cicutarium, Eucalyptus camaldulensis, Eucrypta chrysanthemifolia, Euphorbia, Euthamia occidentalis, Festuca idahoensis, Foeniculum vulgare, Forestiera neomexicana, Frankenia salina, Fraxinus pennsylvanica, F. velutina, Galium aparine, Garrya flavescens, Gaura neomexicana, Geranium viscosissimum, Geum macrophyllum, Glyceria grandis, Glycyrrhiza lepidota, Gnaphalium palustre, Gratiola neglecta, Grayia spinosa, Hemerocallis, Heracleum, Hesperostipa comata, Hilaria rigida, Hordeum jubatum, Hylocomium splendens, Hypericum perfoliatum, Hyptis emoryi, Iris missouriensis, I. pseudacorus, I. setosa,

Isocoma acradenia, I. menziesii, Iva axillaris, I. hayesiana, Juglans californica, Juncus acutus, J. balticus, J. bufonius, J. dubius, J. xiphioides, Juniperus californica, J. occidentalis, J. osteosperma, Keckiella antirrhinoides, Larrea tridentata, Latuca serriola, Lathyrus japonicus, Layia glandulosa, Lemna minima, L. minuta, Lepidium latifolium, L. perfoliatum, Lepidospartum squamatum, Leptochloa fusca, Leptosiphon ciliatus, Lessingia filaginifolia, Leymus cinereus, L. innovatus, L. mollis, L. triticoides, Limosella acaulis, L. aquatilis, Lindernia dubia, Linnaea borealis, Loeflingia squarrosa, Lomatium insulare, Ludwigia hexapetala, L. peploides, Lupinus leucophyllus, L. polyphyllus, Lycium cooperi, Lyrocarpa coulteri, Lythrum californicum, L. salicaria, Maianthemum stellatum, Malacothamnus fasciculatus, Malacothrix saxatilis, Malosma laurina, Malva parviflora, Medicago lupulina, M. sativa, Melilotus albus, M. officinalis, Mentha arvensis, Mentzelia congesta, Mertensia paniculata, Mesembryanthemum crystallinum, Mimulus cardinalis, M. guttatus, M. pilosus, M. primuloides, Mirabilis tenuiloba, Mollugo cerviana, Monarda fistulosa, Monolepis nuttalliana, Morus alba, Muhlenbergia asperifolia, M. rigens, Nasturtium officinale, Nicotiana glauca, Oenothera elata, Opuntia chlorotica, Oxalis, Paeonia brownii, Panicum capillare, P. virgatum, Pascopyrum smithii, Penstemon eatonii, P. procerus, Persicaria amphibia, P. lapathifolia, Phacelia bicolor, P. ciliata, Phalaris arundinacea, Philadelphus lewisii, Phlox kelseyi, Phragmites australis, Phyla canescens, Physocarpus, Picea glauca, Pinus contorta, P. monophylla, P. ponderosa, P. sabiniana, Plantago lanceolata, P. major, Platanthera, Platanus racemosa, Pluchea sericea, Poa bulbosa, P. palustris, P. pratensis, P. secunda, Polypogon monspeliensis, Populus angustifolia, P. balsamifera, P. deltoides, P. fremontii, P. ×hinckleyana, P. tremuloides, P. trichocarpa, Potamogeton, Potentilla anserina, P. gracilis, Primula pauciflora, Prosopis glandulosa, P. pubescens, Prunus andersonii, P. pumila, P. virginiana, Pseudognaphalium beneolens, P. biolettii, P. canescens, P. leucocephalum, Pseudoroegneria spicata, Pseudotsuga menziesii, Pulicaria paludosa, Purshia tridentata, Quercus agrifolia, Q. cornelius-mulleri, Q. kelloggii, Q. lobata, Q. turbinella, Ranunculus acris, R. cymbalaria, R. pensylvanicus, Raphanus, Rhamnus californica, R. crocea, R. purshiana, Rhus aromatica, R. integrifolia, Ribes aureum, R. cereum, R. setosum, Ricinus communis, Robinia pseudoacacia, Rosa californica, R. pisocarpus, R. woodsii, Rubus bifrons, R. leucodermis, R. parviflorus, Rudbeckia laciniata, Rumex conglomeratus, R. crispus, R. paucifolius, Salix alaxensis, S. amygdaloides, S. barclayi, S. bebbiana, S. boothii, S. drummondii, S. eriocephala, S. geyeriana, S. goodingii, S. interior, S. laevigata, S. lasiandra, S. lasiolepis, S. lutea, S. melanopsis, S. pseudomyrsinites, S. ×sepulcralis, S. sessilifolia, Salvia apiana, S. columbariae, S. leucophylla, S. mellifera, Sambucus nigra, Schedonorus arundinaceus, Schoenoplectus acutus, S. californicus, S. lacustris, S. pungens, Scirpus microcarpus, Scrophularia villosa, Senegalia greggii, Sesuvium verrucosum, Shepherdia argentea, Sisymbrium altissimum, Sisyrinchium idahoense, Solidago canadensis, S. missouriensis, S. velutina, Sonchus asper, Sorghum halepense, Spartina gracilis, S. pectinata, Spergularia salina, Sphenosciadium capitellatum, Spiranthes diluvialis, Stachys albens, Stellaria, Stipa, Symphoricarpos occidentalis, S. rotundifolius, Tamarix chinensis, T. gallica, T. parviflora, T. pentandra, T. ramosissima, Taraxacum, Tetradymia axillaris, T. canescens, Thermopsis montana, Toxicodendron diversilobum, T. radicans, T. rydbergii, Trifolium fragiferum, Triglochin maritimum, Turricula parryi, Typha angustifolia, T. domingensis, T. latifolia, Ulmus pumila, Urtica dioica, Veratrum californicum, Verbascum thapsus, Veronica americana, V. anagallis-aquatica, V. peregrina, Viola adunca, Vitis arizonica, V. girdiana, Vulpia microstachys, V. octoflora, Wyethia helianthoides, Xanthium strumarium, Yucca brevifolia, Y. schidigera, Zizia aptera.

Salix farriae C.R. Ball is a small shrub (<1.5 m), which inhabits bogs, bottoms, brooks, carrs, draws, fens, flats, heath, hummocks, marshes, meadows, scrub, seeps, slopes (to 40%), swamps, thickets, tundra, woodlands, and the margins of lakes, ponds, rivers, and streams in montane and subalpine zones at elevations from 630 to 2927 m. It can tolerate exposures from full sun to partial shade. The substrates primarily are alkaline and have been described as alluvium, fluvaquentic cyroborolls, fluventic cyroborolls, gravel, loam, loamy sand, rocky loam, peat, sand, sandy alluvium, sandy loam, sandy loamy clay, silt, silty loam, and typic cryaquoll. Flowering occurs from May to August, with fruit production observed from June to September. There is little additional life-history information available for this species. **Reported associates:** *Abies lasiocarpa, Achillea millefolium, Allium validum, Alnus rugosa, Angelica, Antennaria racemosa, Anticlea elegans, Arctostaphylos rubra, Arctous alpina, Arnica chamissonis, A. cordifolia, Artemisia norvegica, Betula glandulosa, B. nana, Bistorta vivipara, Bromus ciliatus, Calamagrostis canadensis, Caltha leptosepala, Carex aperta, C. aquatilis, C. illota, C. interior, C. leptalea, C. limosa, C. livida, C. microglochin, C. praeceptorum, C. rossii, C. rostrata, C. saxatilis, C. scopulorum, C. subnigricans, C. utriculata, Cassiope tetragona, Castilleja miniata, Chamerion angustifolium, Cirsium scariosum, Crepis runcinata, Dasiphora floribunda, Delphinium ×occidentale, Deschampsia cespitosa, Dryas alaskensis, Eleocharis quinqueflora, Elymus glaucus, Equisetum arvense, E. scirpoides, E. variegatum, Erigeron coulteri, E. formosissimus, Eriogonum flavum, Festuca altaica, Fragaria virginiana, Gaultheria humifusa, Gentianella propinqua, Geranium richardsonii, Hedysarum alpinum, Heracleum sphondylium, Juncus balticus, Ligusticum tenuifolium, Linnaea borealis, Lonicera involucrata, Maianthemum stellatum, Mertensia paniculata, Minuartia arctica, Mitella pentandra, Moneses uniflora, Orthilia secunda, Oxytropis borealis, O. maydelliana, Packera subnuda, Parnassia fimbriata, Pedicularis groenlandica, Phleum alpinum, P. pratense, Picea engelmanii, P. glauca, Pinus contorta, Platanthera dilatata, Poa alpina, P. palustris, P. pratensis, Polemonium viscosum, Populus tremuloides, Potentilla flabellifolia, Primula tetrandra, Prunus, Ranunculus alismifolius, Rhododendron albiflorum, R. columbianum, R. lapponicum, Ribes lacustre, Rosa* *acicularis, Rudbeckia occidentalis, Salix arctica, S. barclayi, S. barrattiana, S. boothii, S. brachycarpa, S. commutata, S. drummondiana, S. eastwoodiae, S. geyeriana, S. myrtillifolia, S. planifolia, S. reticulata, S. tweedyi, S. vestita, S. wolfii, Shepherdia canadensis, Solidago canadensis, Sphagnum, Swertia perennis, Symphyotrichum foliaceum, S. spathulatum, Thalictrum occidentale, T. sparsiflorum, Tofieldia pusilla, Trifolium hybridum, Triglochin maritimum, Trollius laxus, Vaccinium cespitosum, V. scoparium, V. uliginosum.*

Salix geyeriana Andersson is a shrub (<6.5 m), which occurs in bottoms, carrs, chaparral, cienega, ditches, fens, flats, floodplains, gravel pits, gullies, meadows, seeps, slopes (to 10%), sloughs, springs, swamps, thickets, and along the margins of lakes, ponds, rivers, and streams (low velocity) at elevations of up to 3505 m. The plants require sites where the water table is less than a meter deep and can tolerate shallow standing waters (15–61 cm). Individuals generally are shade intolerant (but occasionally occur in semishade) and typically grow in widely spaced clumps. They are found typically on deep, fine textured, alluvial mineral soils with accumulated organic material (e.g., pH: 5.5; total hardness: 10 mg/L, Ca: 2.72 mg/L, Mg: 0.78 mg/L). However, the reported substrates are diverse and include adobe, ash, Claron limestone, clay, clayey sand, decomposed granite, gravel, gravelly clay, gravelly loam, humus gravel, lava rock, loam, muck, peat, rocky sandy loam, sand, sandy clay loam, silty clay, silty gravel, silty sand, pumice, and talus. Flowering and fruiting extend from April to September. The seeds lack dormancy and can germinate within 24 h if dispersed to suitable wet, open sites (on exposed mineral soils). Excessive leaf litter can inhibit germination and seedling establishment, which critically requires adequate soil moisture to succeed. The stems possess predeveloped root primordia and shoot cuttings will root along their length if they become buried. Vegetative reproduction occurs by root sprouts and from detached twigs that are dispersed by floodwaters and eventually root when deposited on alluvial soil. The plants will also resprout from their root crown following fires. However, reproduction is entirely by seed in many localities. They often are overgrazed but can recover in 5–6 years once grazing ceases. Grazing rates are correlated with population sizes of moose (Mammalia: Cervidae: *Alces alces*), a principal herbivore. **Reported associates:** *Abies lasiocarpa, Achillea lanulosa, A. millefolium, Achnatherum occidentalis, A. thurberianum, Agastache urticifolia, Agropyron caninum, Agrostis scabra, A. stolonifera, Alisma, Allium geyeri, A. tolmiei, A. validum, Alnus rugosa, A. viridis, Alopecurus aequalis, Amelanchier alnifolia, Angelica, Antennaria corymbosa, A. racemosa, Aquilegia formosa, Arctostaphylos patula, Arnica fulgens, Artemisia arbuscula, A. cana, A. tridentata, Astragalus alpinus, Balsamorhiza, Betula glandulosa, B. occidentalis, Bistorta bistortoides, Brasenia schreberi, Bromus ciliatus, Calamagrostis canadensis, Camassia quamash, Cardamine breweri, Carex angustata, C. aperta, C. aquatilis, C. athrostachya, C. aurea, C. buxbaumii, C. chordorrhiza, C. cusickii, C. disperma, C. echinata, C. eurycarpa, C. festivella, C. geyeri, C. hoodii, C. lasiocarpa, C. microptera, C.*

nebrascensis, C. obnupta, C. pachystachya, C. praegracilis, C. praticola, C. rostrata, C. scoparia, C. scopulorum, C. simulata, C. sitchensis, C. stricta, C. subfusca, C. utriculata, C. vesicaria, C. viridula, Carum carvi, Castilleja septentrionalis, Catabrosa aquatica, Centaurea maculosa, Cercocarpus, Chamerion angustifolium, Chenopodium album, Chrysolepis sempervirens, Cicuta douglasii, Cirsium arvense, C. scariosum, Claytonia lanceolata, Clematis, Comarum palustre, Cornus sericea, Crataegus chrysocarpa, C. douglasii, C. monogyna, Cytisus scoparius, Dasiphora floribunda, Deschampsia cespitosa, Drymocallis glandulosa, Dulichium arundinaceum, Eleocharis acicularis, E. palustris, E. quinqueflora, E. rostellata, Epilobium ciliatum, E. suffruticosum, Equisetum arvense, E. fluviatile, Ericameria nauseosa, Erigeron peregrinus, Eriodictyon californicum, Eriogonum flavum, E. latifolium, E. umbellatum, Erythronium grandiflorum, Euthamia, Festuca idahoensis, F. rubra, Floerkea proserpinacoides, Fontinalis antipyretica, Fragaria ovalis, F. vesca, F. virginiana, Galium boreale, G. trifidum, Gentianopsis detonsa, Geranium richardsonii, G. viscosissimum, Geum macrophyllum, Glyceria elata, G. grandis, G. maxima, G. striata, Heracleum sphondylium, Hordeum brachyantherum, Horkelia fusca, Hosackia oblongifolia, Hypericum anagalloides, Iris missouriensis, Juncus acuminatus, J. arcticus, J. balticus, J. effusus, J. ensifolius, J. longistylus, J. nevadensis, J. orthophyllus, Juniperus occidentalis, J. scopulorum, Kalmia microphylla, Koeleria macrantha, Lemna minor, Leptosiphon ciliatus, Leucothoe, Leymus cinereus, Ligusticum porteri, Lilium parvum, Linnaea borealis, Lomatium bicolor, L. cous, L. macrocarpum, Lonicera involucrata, L. utahensis, Lupinus argenteus, Luzula, Lycopus americanus, L. uniflorus, Lythrum salicaria, Maianthemum stellatum, Malus, Melilotus officinalis, Mentha arvensis, Menyanthes trifoliata, Mertensia ciliata, Micranthes odontoloma, M. oregana, Mimulus guttatus, M. lewisii, M. primuloides, Montia fontana, Myrica gale, Nasturtium officinale, Oenothera, Onopordum acanthium, Oxytropis sericeus, Paeonia brownii, Pedicularis groenlandicum, Phalaris arundinacea, Philadelphus lewisii, Phleum alpinum, P. pratense, Picea engelmannii, Pinus albicaulis, P. contorta, P. jeffreyi, P. monophylla, P. ponderosa, Plagiobothrys scouleri, Platanthera dilatata, P. stricta, Poa palustris, P. pratensis, P. reflexa, P. secunda, Polemonium caeruleum, P. occidentale, Populus angustifolia, P. balsamifera, P. fremontii, P. tremuloides, P. trichocarpa, Potentilla anserina, P. ×diversifolia, P. gracilis, P. pulcherrima, Primula pauciflora, Prunus virginiana, Pseudoroegneria spicata, Pseduotsuga menziesii, Ptilagrostis porteri, Pyrola asarifolia, Ranunculus sceleratus, Rhamnus alnifolia, Rhododendron albiflorum, R. columbianum, R. groenlandicum, Rhus aromatica, Rhynchospora alba, Ribes aureum, R. cereum, R. hudsonianum, R. inerme, R. lacustre, R. setosum, Rosa nutkana, R. woodsii, Rubus bifrons, R. sachalinensis, Rumex crispus, Salix amygdaloides, S. bebbiana, S. boothii, S. candida, S. drummondiana, S. eastwoodiae, S. eriocephala, S. exigua, S. hookeriana, S. lasiandra, S. lemmonii, S. ligulifolia, S. lutea, S. monticola, S. myrtillifolia, S. orestera, S. pedicellaris, S. planifolia, S. pseudomyrsinites,

S. scouleriana, S. sitchensis, S. wolfii, Schoenoplectus acutus, Scirpus microcarpus, Senecio hydrophilus, S. integerrimus, S. pauperculus, S. serra, S. sphaerocephalus, S. triangularis, Sidalcea oregana, Sium, Solidago canadensis, Sparganium eurycarpum, Sphagnum, Sphenosciadium capitellatum, Spiraea betulifolia, S. douglasii, Stellaria longipes, Streptopus amplexifolius, Symphoricarpos oreophilus, Symphyotrichum foliaceum, S. spathulatum, Taraxacum officinale, Taraxia subacaulis, Thermopsis montana, Thuja plicata, Trifolium hybridum, T. longipes, T. repens, Trisetum wolfii, Typha latifolia, Vaccinium membranaceum, Valeriana occidentalis, Veratrum californicum, Veronica americana, V. peregrina, Vicia americana, Viola canadensis, V. palustris, Wyethia amplexicaulis, W. helianthoides, W. mollis, Xerophyllum tenax, Zizia aptera.

***Salix gooddingii* C.R. Ball** is a fast growing, often dominant tree (to 30 m), which grows in alluvial fans, arroyos, chaparral, creek/river beds, depressions, ditches, flats, floodplains, forests, levees, marshes, meadows, mudflats, quarry pits, roadsides, sandbars, scrub, seeps, slopes, sloughs, springs, swales, thickets, washes, and along the margins of canals, lakes, ponds, reservoirs, rivers, and streams at elevations of up to 2439 m. The plants grow well at pH = 6–7 and also occur commonly under alkaline conditions. Their growth is inhibited at salinities of 1500 ppm. The substrates have been characterized as basalt, bouldery sandy granite, clay, cobble, granite, gravel, loam, rock, rocky clay, sand, sandstone, sandy clay, sandy gravel, sandy loam, schist, serpentine, silt, and silty clay alluvium. The plants require a long, hot growing season and abundant groundwater. They are shade intolerant but can form dense stands and will sprout from their base if they are cut or topped. This species is a phreatophyte, which produces numerous, small surface roots and deeper roots that can penetrate more than 2 m into the substrate. Flowering occurs from March to June. Seed production begins on trees as young as 2 years of age and peaks in trees that are 25–75 years old. The seeds are nondormant (germinating in 12–24 h) and retain their viability for only a few days. Moisture stress can reduce seed germination by more than 95%. Sealing moist seeds within a refrigerated (4°C) container can extend their viability for up to a month. Seed germination is optimal at 27°C and occurs best on moist, bare sites such as those created by fire and flood activity. The seedlings compete poorly with graminoids. The plants reproduce vegetatively by root crown sprouts. Populations are threatened by competition from the invasive *Tamarix ramosissima*. **Reported associates:** *Abutilon incanum, Acacia salicina, Acer negundo, Acmispon glabrus, Acourtia wrightii, Adenostoma fasciculatum, Agave deserti, A. utahensis, Alisma, Alnus oblongifolia, A. tenuifolia, Aloysia wrightii, Amaranthus, Ambrosia acanthicarpa, A. ambrosioides, A. deltoidea, A. dumosa, A. monogyra, A. psilostachya, Amelanchier utahensis, Amorpha fruticosa, Anemopsis californica, Anisacanthus thurberi, Anticlea elegans, Apium graveolens, Apocynum ×floribundum, Arctostaphylos glandulosa, A. pungens, Aristida purpurea, Artemisia californica, A. douglasiana, A. ludoviciana, Arundo donax, Atriplex canescens, A. torreyi, Avena barbata, A. fatua, Baccharis emoryi, B. glutinosa, B.*

salicifolia, B. sarothroides, B. sergiloides, Bahiopsis parishii, Bebbia juncea, Berberis haematocarpa, Boechera sparsiflora, Bothriochloa ischaemum, Bowlesia incana, Brassica nigra, B. tournefortii, Brickellia atractyloides, B. baccharidea, B. longifolia, Bromus diandrus, B. rubens, B. tectorum, Bulboschoenus maritimus, Calibrachoa parviflora, Calliandra eriophylla, Callitropsis forbsii, Capsella bursa-pastoris, Carex pellita, Carnegiea gigantea, Ceanothus crassifolius, C. tomentosus, Celtis reticulata, Cenchrus ciliaris, C. setaceus, Centaurea solstitialis, Cercocarpus montanus, Chamaebatia australis, Chara, Chilopsis linearis, Condalia lyciodes, Conium maculatum, Conyza canadensis, Coreopsis, Corethrogyne filaginifolia, Cornus sericea, Cortaderia, Crossosoma bigelovii, Cryptantha barbigera, C. muricata, Cylindropuntia ganderi, Cynodon dactylon, Cyperus eragrostis, C. involucratus, C. niger, Dasylirion, Datura wrightii, Deinandra fasciculata, Distichlis spicata, Dudleya edulis, Echinocereus fasciculatus, Echinochloa, Echinocystis lobata, Elaeagnus angustifolia, Eleocharis montevidensis, E. palustris, E. parishii, Elymus elymoides, Encelia farinosa, Ephedra aspera, Equisetum laevigatum, Eragrostis lehmanniana, Ericameria laricifolia, E. paniculata, E. pinifolia, Erigeron divergens, Eriogonum fasciculatum, Erodium, Eucalyptus microtheca, Eucrypta chrysanthemifolia, Eulobus californicus, Euphorbia abramsiana, E. eriantha, E. setiloba, Fallugia paradoxa, Ferocactus cylindraceus, Fouquieria splendens, Forestiera neomexicana, Frankenia, Fraxinus pennsylvanica, F. velutina, Galium aparine, Garrya, Gaura, Gilia sinuata, Glandularia bipinnatifida, Gutierrezia sarothrae, Helianthus annuus, Herniaria hirsuta, Hilaria belangeri, Hirschfeldia incana, Hordeum murinum, Hyemnoclea mongyra, Hyptis emoryi, Juglans major, Juncus articulatus, J. balticus, J. nodosus, Juniperus californica, J. deppeana, J. pinchotii, Keckiella antirrhinoides, Kochia scoparia, Krameria bicolor, Lactuca serriola, Larrea tridentata, Lepidium lasiocarpum, Lepidospartum squamatum, Leptochloa fusca, L. viscida, Leymus triticoides, Lolium perenne, Lonicera subspicata, Ludwigia peploides, Lupinus, Lycium andersonii, L. exsertum, L. fremontii, Lythrum californicum, Malosma laurina, Malus pumila, Malva parviflora, Marina parryi, Marrubium vulgare, Medicago lupulina, Melilotus, Mentha, Mimosa aculeaticarpa, Mimulus guttatus, Monolepis nuttalliana, Morus microphylla, Muhlenbergia rigens, Myosurus, Nama hispida, Nasturtium officinale, Navarretia, Nerium oleander, Nicotiana glauca, Oenothera primiveris, Opuntia basilaris, O. ficus-indica, O. littoralis, Oxytenia acerosa, Parietaria hespera, P. pensylvanica, Parkinsonia microphylla, Parthenocissus, Penstemon eatonii, P. pseudospectabilis, Perityle emoryi, P. gilensis, Persicaria lapathifolia, P. maculosa, Phacelia ciliata, P. distans, P. ramosissima, Phalaris minor, Phlox tenuifolia, Phragmites australis, Physaria purpurea, Pinus edulis, P. monophylla, Plantago patagonica, P. rhodosperma, Platanus racemosa, P. wrightii, Pluchea sericea, Polygonum argyrocoleon, P. aviculare, Polypogon monspeliensis, P. viridis, Populus fremontii, P. trichocarpa, Potamogeton nodosus, Prosopis glandulosa, P. juliflora, P. pubescens, P. velutina, Prunus fasciculata, Pseudognaphalium luteoalbum, Psilostrophe cooperi, Puccinellia parishii, Purshia, Quercus agrifolia, Q. berberidifolia, Q. chrysolepis, Q. engelmannii, Q. grisea, Q. turbinella, Q. wislizeni, Rhamnus californica, Rhus aromatica, R. integrifolia, R. microphylla, R. ovata, Ricinus communis, Robinia, Rosa woodsii, Rumex crispus, R. dentatus, Salicornia depressa, Salix amygdaloides, S. bonplandiana, S. exigua, S. hindsiana, S. irrorata, S. laevigata, S. lasiolepis, S. lucida, S. scouleriana, Salsola, Salvia apiana, S. mellifera, Sambucus nigra, Sapindus saponaria, Sarcobatus vermiculatus, Schedonorus arundinaceus, Schinus molle, Schismus barbatus, Schoenoplectus acutus, S. pungens, S. tabernaemontani, Senegalia greggii, Shepherdia argentea, Simmondsia chinensis, Sisymbrium irio, Solanum, Solidago, Sorghum halepense, Sphaeralcea ambigua, Sporobolus airoides, S. cryptandrus, S. wrightii, Stuckenia pectinata, Symphyotrichum subulatum, Tamarix chinensis, T. ramosissima, Taraxacum, Thysanocarpus curvipes, Toxicodendron diversilobum, Typha domingensis, T. latifolia, Ulmus pumila, Urtica holosericea, Vauquelinia californica, Verbena bracteata, Veronica anagallis-aquatica, V. peregrina, Vitis arizonica, Vulpia octoflora, Xanthium strumarium, Xylococcus bicolor, Ziziphus obtusifolia.

***Salix interior* Rowlee** inhabits bayous, beaches, bottoms, carrs, coulee bottoms, depressions, dikes, ditches, dunes, fens, flats, floodplains, gravel bars, gravel pits, levees, marshes, meadows, prairies, roadsides, sandbars, seeps, slopes, sloughs, spillways, swales, terraces, thickets, and the margins of lakes, ponds, reservoirs, rivers, and streams at elevations of up to 2621 m. The plants can occur in shallow standing water and often are found in disturbed sites. Exposures range from full sun to light shade. The substrates typically are calcareous (e.g., pH: 7.0–7.8) and have been categorized as alluvium, clay, gravel, gravelly loam, limestone, loam, muck, mucky silty loam over basalt, mud, peat, rock, sand, sandy clay, sandy loam, silt, silty mud, and stones. Flowering extends from April to August, with fruiting from June to August. The flowers are pollinated by insects, primarily bees (Hymenoptera: Andrenidae) and flies (Diptera: Syrphidae) (see *Use by wildlife*). The seeds germinate readily (to 100%) at temperatures from 5°C to 25°C. The shoots form numerous adventitious roots when inundated, which enables them to persist well in floodplain environments. Although numerous branches are produced each spring, the plants are known to "self-prune" by shedding branches later in the growing season. **Reported associates:** *Acer negundo, A. rubrum, A. saccharinum, Achillea alpina, Agrimonia parviflora, Agrostis gigantea, Alisma subcordatum, Alnus rugosa, Ambrosia artemisiifolia, A. trifida, Amelanchier interior, Ammannia robusta, Ammophila breviligulata, Amorpha fruticosa, Andropogon gerardii, A. glomeratus, Angelica, Apocynum cannabinum, Arabidopsis lyrata, Arabis hirsuta, Arctium minus, Artemisia campestris, A. tilesii, Asclepias incarnata, Betula nigra, B. papyrifera, Bidens cernuus, B. comosus, B. frondosus, B. trichospermus, Boehmeria cylindrica, Botrychium virginianum, Bromus inermis, B. tectorum, Bolboschoenus fluviatilis, Cakile edentula, Calamagrostis canadensis, Caltha palustris, Calystegia sepium, Campanula aparinoides, Campsis radicans, Capsella bursa-pastoris, Cardamine bulbosa, C. parviflora, Carex aquatilis, C. cristatella, C. emoryi, C. grayi, C. haydenii, C. lacustris, C. pellita, C. sartwellii, C. scoparia, C. stipata, C. stricta, C. vulpinoidea, Celastrus scandens, Cephalanthus occidentalis, Chelone glabra, Cicuta bulbifera, C. maculata, Circaea lutetiana, Cirsium discolor, C. vulgare, Comandra umbellata, Conoclinium coelestinum, Cornus amomum, C. racemosa, C. sericea, Crataegus mollis, Cyperus aristatus, C. erythrorhizos, Dactylis glomerata, Dalea purpurea, Daucus carota, Desmanthus illinoensis, Dipsacus laciniatus, Dryas drummondii, Elaeagnus angustifolia, Eleocharis acicularis, E. erythropoda, Elymus virginicus, Epilobium ×wisconsinense, Equisetum arvense, E. fluviatile, E. hyemale, E. laevigatum, Erigeron annuus, Eryngium yuccifolium, Eupatorium altissimum, E. perfoliatum, Euphorbia corollata, Eurybia sibirica, Euthamia graminifolia, Eutrochium maculatum, Fragaria virginiana, Fraxinus pennsylvanica, Galium aparine, G. obtusum, G. trifidum, Gentiana saponaria, Geum laciniatum, Glyceria striata, Glycyrrhiza lepidota, Hedysarum boreale, Helenium autumnale, Helianthus grosseserratus, Hordeum jubatum, Hypericum sphaerocarpum, Impatiens capensis, Imperata cylindrica, Iodanthus pinnatifidus, Ipomoea pandurata, Iris virginica, Juncus alpinoarticulatus, J. balticus, J. coriaceus, J. dudleyi, J. torreyi, Justicia americana, Lactuca, Laportea canadensis, Leymus arenarius, Lobelia cardinalis, Lonicera maackii, Lycopus americanus, L. uniflorus, L. virginicus, Lysimachia terrestris, L. thyrsiflora, Lythrum alatum, L. salicaria, Maclura pomifera, Mentha arvensis, Morus alba, Oenothera gaura, Onoclea sensibilis, Osmorhiza claytonii, Ostrya virginiana, Oxypolis rigidior, Panicum dichotomiflorum, P. virgatum, Parthenium integrifolium, Parthenocissus quinquefolia, Pastinaca sativa, Penstemon digitalis, P. pallidus, Persicaria amphibia, P. lapathifolia, P. pensylvanica, Phalaris arundinacea, Phlox glaberrima, Phragmites australis, Physostegia virginiana, Picea, Platanthera flava, Platanus occidentalis, Poa compressa, P. pratensis, Populus angustifolia, P. balsamifera, P. ×canescens, P. deltoides, P. tremuloides, Potentilla anserina, P. simplex, Proserpinaca palustris, Prunus americana, P. serotina, P. virginiana, Pycnanthemum virginianum, Ranunculus abortivus, R. sceleratus, Rhamnus cathartica, Rhus glabra, R. typhina, Ribes missouriense, Rorippa islandica, R. palustris, R. sessiliflora, Rosa carolina, Rubus hispidus, R. laciniatus, R. occidentalis, R. setosus, Rudbeckia laciniata, Rumex crispus, Sagittaria brevirostra, Salix alaxensis, S. amygdaloides, S. arbusculoides, S. candida, S. discolor, S. eriocephala, S. exigua, S. famelica, S. lasiandra, S. lutea, S. nigra, S. planifolia, S. pseudomonticola, S. sitchensis, Sambucus nigra, Sarcobatus, Sassafras albidum, Schedonorus arundinaceus, Schizachyrium scoparium, Schoenoplectus acutus, Scirpus atrovirens, S. cyperinus, Scolochloa festucacea, Scutellaria galericulata, S. lateriflora, Securigera varia, Senecio, Setaria faberi, Shepherdia canadensis, Silene antirrhina, Silphium perfoliatum, S. terebinthinaceum, Sium suave, Solanum dulcamara, Solidago altissima, S. canadensis, S. gigantea, S. ptarmicoides, Sorghastrum nutans, Sparganium eurycarpum,*

Spartina pectinata, Sphagnum, Sporobolus heterolepis, Symphoricarpos occidentalis, Symphyotrichum lanceolatum, S. novae-angliae, S. ontarionis, S. oolentangiense, S. pilosum, Symplocarpus foetidus, Taraxacum officinale, Thalictrum dasycarpum, Thelypteris palustris, Toxicodendron radicans, Tridens flavus, Typha angustifolia, T. latifolia, Ulmus americana, Verbena hastata, Vernonia fasciculata, Veronica peregrina, Veronicastrum virginicum, Viola lanceolata, Vitis riparia, Xanthium strumarium, Zanthoxylum americanum.

Salix jepsonii C.K. Schneid. is a small shrub (<3 m), which is found in ditches, meadows, seeps, thickets, and along the margins of lakes and streams at subalpine elevations from 1051 to 3520 m. The substrates include alluvium, boulders, granite, gravel, rock, and talus. Flowering occurs from June to July. This willow can become a dominant species of riverine scrub communities. More specific details on its life-history are lacking. **Reported associates:** *Abies concolor, A. magnifica, A. ×shastensis, Acer glabrum, Aconitum columbianum, Adenocaulon bicolor, Alnus tenuifolia, A. viridis, Amelanchier pallida, Aquilegia formosa, Arctostaphylos patula, Arnica lanceolata, Bistorta bistortoides, Boykinia major, Calocedrus decurrens, Camassia quamash, Campanula prenanthoides, Cardamine breweri, Carex aquatilis, C. fissuricola, C. integra, C. lemmonii, C. lenticularis, C. nebrascensis, C. scopulorum, C. simulata, C. utriculata, Castilleja miniata, Cornus nuttallii, C. sericea, Cryptogramma acrostichoides, Darlingtonia californica, Darmera peltata, Disporum hookeri, Elymus glaucus, Epilobium, Equisetum arvense, Eriogonum kennedyi, Euonymus occidentalis, Eupatorium occidentale, Festuca brachyphylla, Glyceria elata, Helenium bigelovii, Hippuris vulgaris, Hosackia oblongifolia, Hypericum anagalloides, Juncus chlorocephalus, J. howellii, J. xiphioides, Ligusticum californicum, Lilium kelleyanum, Lonicera conjugialis, Lorandersonia peirsonii, Lupinus latifolius, L. polyphyllus, Luzula comosa, Maianthemum racemosum, Micranthes nidifica, Oxyria digyna, Paxistima myrsinites, Pedicularis attollens, P. racemosa, Penstemon deustus, P. procerus, Pinus contorta, P. jeffreyi, P. lambertiana, P. monticola, Platanthera leucostachys, Poa pratensis, Populus trichocarpa, Primula jeffreyi, Prunus emarginata, Pteridium aquilinum, Quercus kelloggii, Rhamnus purshianus, Rhododendron columbianum, R. occidentale, Ribes nevadense, Rosa woodsii, Rubus parviflorus, Salix boothii, S. eastwoodiae, S. lemmonii, S. lasiandra, S. orestera, S. scouleriana, Sambucus racemosa, Scirpus diffusus, S. macrocarpus, Solidago canadensis, Spiraea densiflora, Stachys alba, Trientalis latifolia, Tsuga mertensiana, Vaccinium uliginosum, Veratrum californicum, Veronica americana, V. scutellata, V. serpyllifolia, Viola macloskeyi.*

Salix lasiandra Benth. is a shrub or tree (<10 m), which inhabits bottomlands, carrs, depressions, ditches, fens, flats, floodplains, gulches, gravel bars, gullies, marshes, meadows, roadsides, sandbars, seeps, shores, slopes (to 70%), sloughs, springs, swales, swamps, thickets, woodlands, and the margins of lakes, ponds, reservoirs, rivers, and streams at elevations of up to 3100 m. The substrates are usually categorized as alkaline and include alluvial clay, alluvial loam, clay loam, cobble, decomposing basaltic sand, gravel, gravelly

alluvium, gravelly clay, gravelly loam, loam, muck, peat, rock, rocky granite, sand, sandy gravel, sandy loam, sandy silt, sandy silty loam, silt, silty cobble, silty gravel, silty loam, silty sand, talus, and volcanic alluvium. The plants are fast growing but short lived, with an average age of 25 years. They are highly flood tolerant and somewhat shade tolerant. Flowering and fruiting occur from March to October, initiating in March in the southern portion of the range, and extending into May as one proceeds north. The seeds are short lived, lack dormancy, and germinate well (93%–100%) within 12–24 h at 5°C–25°C (best germination in high light). Prolific seed production facilitates the rapid colonization of disturbed sites such as burns, dredged areas, and mine spoils. Vegetative reproduction occurs from crown resprouts and by brittle, readily detached branches, which are dispersed by floodwaters. The branchlets root along their length (in about 10 days) when buried or when coming in contact with moist soil. The plants do not spread vegetatively because they lack the ability to produce root suckers. They rapidly lose vigor when they are overgrazed. **Reported associates:** *Abies grandis, A. lasiocarpa, Acer glabrum, A. macrophyllum, A. negundo, A. saccharum, Actaea rubra, Agastache urticifolia, Agrostis gigantea, Allium geyeri, Alnus rhombifolia, A. rugosa, A. rubra, A. tenuifolia, A. viridis, Amelanchier alnifolia, Amorpha californica, A. fruticosa, Angelica, Artemisia ludoviciana, A. tilesii, A. tridentata, Asarum caudatum, Baccharis viminea, Balsamorhiza sagittata, Betula occidentalis, B. papyrifera, Bidens amplissimus, Bromus tectorum, Calamagrostis canadensis, Calocedrus decurrens, Camassia leichtlinii, C. quamash, Carex angustata, C. aperta, C. athrostachya, C. disperma, C. geyeri, C. hoodii, C. jonesii, C. lasiocarpa, C. lenticularis, C. microptera, C. nebrascensis, C. pellita, C. rostrata, C. saxatilis, C. scopulorum, C. utriculata, Castilleja lasiorhyncha, Catalpa speciosa, Ceanothus sanguineus, C. velutinus, Celtis reticulata, Cirsium andersonii, C. vulgare, Clematis ligusticifolia, Comarum palustre, Coptis occidentalis, Cornus canadensis, C. sericea, Crategus douglasii, Cytisus scoparius, Dipsacus sylvestris, Elaeagnus angustifolia, E. commutata, Eleocharis macrostachya, E. obtusa, E. palustris, Elymus, Epilobium ciliatum, Equisetum arvense, E. fluviatile, E. variegatum, Festuca, Fraxinus latifolia, F. pennsylvanica, F. velutina, Galium, Gaultheria shallon, Geranium dissectum, Geum macrophyllum, Glyceria, Heliotropium curassavicum, Heracleum, Hippuris montana, Holcus lanatus, Horkelia fusca, Hypericum formosum, Iris missouriensis, I. pseudacorus, Juncus balticus, J. bufonius, J. effusus, J. ensifolius, J. tenuis, Juniperus scopulorum, Lomatium bicolor, L. grayi, Lupinus lepidus, L. polyphyllus, L. sericeus, Maianthemum stellatum, Medicago lupulina, Melilotus, Mentha, Mimulus lewisii, Monardella odoratissima, Morus alba, Nicotiana glauca, Oemleria cerasiformis, Oenanthe, Onopordum acanthium, Paxistima myrsinites, Phalaris arundinacea, Philadelphus lewisii, Physocarpus capitatus, Picea engelmannii, P. glauca, Pinus ponderosa, Plagiomnium ellipticum, Platanthera obtusata, Platanus racemosa, Poa palustris, P. pratensis, P. secunda, Polypodium glycyrrhiza, Polystichum, Polytrichum commune, Populus*

×acuminata, P. angustifolia, P. balsamifera, P. deltoides, P. tremuloides, P. trichocarpa, Potentilla anserina, P. gracilis, Prosopis, Prunus emarginata, P. virginiana, Pseudoroegneria spicata, Pseudotsuga menziesii, Pteridium aquilinum, Quercus kelloggii, Q. lobata, Ranunculus testiculatus, Rhus aromatica, Ribes aureum, R. hudsonianum, R. nevadense, R. oxyacanthoides, Robinia pseudoacacia, Rosa acicularis, R. gymnocarpa, R. nutkana, R. rubiginosa, R. woodsii, Rubus discolor, R. parviflorus, R. spectabilis, R. ursinus, Rudbeckia occidentalis, Rumex crispus, R. salicifolius, Salix alaxensis, S. alba, S. amygdaloides, S. barclayi, S. bebbiana, S. bebbii, S. boothii, S. candida, S. commutata, S. drummondiana, S. eriocephala, S. exigua, S. fragilis, S. geyeriana, S. glauca, S. hookeriana, S. lasiolepis, S. lemmonii, S. lutea, S. melanopsis, S. monticola, S. piperii, S. planifolia, S. pseudomonticola, S. scouleriana, S. sessifolia, S. sitchensis, Sambucus nigra, S. racemosa, Sanguisorba, Schoenoplectus acutus, S. pungens, Scirpus microcarpus, Senecio serra, S. triangularis, S. vulgaris, Sidalcea oregana, Solanum dulcamara, Solidago missouriensis, Sorbus californica, Spirea douglasii, S. splendens, Stellaria longipes, Symphoricarpos albus, S. occidentalis, S. oreophilus, Symphyotrichum ascendens, S. foliaceum, Taxus brevifolia, Thermopsis montana, T. rhombifolia, Thuja plicata, Toxicodendron diversilobum, T. radicans, Trientalis europaea, Trifolium repens, Triglochin maritimum, Tsuga heterophylla, Typha latifolia, Urtica dioica, Vaccinium uliginosum, Valeriana, Veratrum californicum, Verbascum thapus, Viburnum edule, Viola macloskeyi, Viscum album, Whipplea modesta, Wyethia angustifolia, W. helianthoides, Zizia aptera.

Salix lasiolepis Benth. is a shrub to small tree (<10 m), which inhabits bluffs, bogs, bottoms, draws, dunes, flats, floodplains, gravel bars, gulches, gullies, marshes, meadows, mudflats, ravines, roadsides, salt marshes, shores, slopes (to 70%), springs, terraces, thickets, washes, woodlands, and the margins of rivers and streams at elevations of up to 2758 m. The reported substrates are described as clay, clay loam, dolomite, Friant-Escondido sandy loam, gravel, gravelly sand, humus, loess, muck, rock, rocky clay, sand, sandy alluvium, sandy gravel, sandy loam, sandy rocky loam, schist, silt, and volcanic. The plants can tolerate wind but are very shade intolerant and typically occur in full sun. This is a fast growing but short-lived species, which can develop into clonal thickets or into areas of more widely spaced trees. The flowers, which appear from January to June, are pollinated by insects (Insecta). Insect pollination is highly effective and results in more than 80% seed set in natural populations. Wind pollination plays an exceptionally minor role and is estimated to contribute only 0.1% to the overall seed set. No evidence of apomixis has been found in this species. Fruits have been observed from February to September. The seeds remain viable only for a few days. The seedlings depend on abundant soil moisture and experience high mortality under drying conditions. Male plants have lower concentrations of herbivore-deterrent phenolglycosides than do the female plants. Vegetative reproduction occurs by stem sprouts. The plants are fire tolerant with 73%–83% of plants observed to resprout within 10–21 days following experimental burns. **Reported associates:** Abies concolor, Acacia cyclops, Acer glabrum, A. grandidentatum, A. macrophyllum, A. negundo, Achillea millefolium, Achnatherum lettermanii, A. parishii, Acmispon glabrus, A. rigidus, Actaea rubra, Adenostoma fasciculatum, A. sparsifolium, A. verbesina, Aesculus californica, Agropyron, Agrostis gigantea, Aira caryophyllea, Allium, Alnus rhombifolia, A. rubra, A. rugosa, A. tenuifolia, Alopecurus pratensis, Amauriopsis dissecta, Ambrosia psilostachya, Amorpha californica, Anagallis arvensis, Aquilegia formosa, Arbutus menziesii, Arctostaphylos patula, A. pringlei, Artemisia californica, A. douglasiana, A. ludoviciana, A. tridentata, Arundo donax, Asclepias fascicularis, Astragalus pycnostachyus, Avena, Baccharis douglasii, B. pilularis, B. salicifolia, B. sarothroides, Batis maritima, Berberis aquifolium, Betula occidentalis, Bidens leptocephalus, Brassica nigra, Brickellia atractyloides, B. desertorum, B. grandiflora, Bromus diandrus, B. rubens, B. tectorum, Calamagrostis nutkaensis, Callitropsis forbsii, Carex alma, C. athrostachya, C. barbarae, C. fracta, C. hassei, C. hirtissima, C. hoodii, C. jonesii, C. nebrascensis, C. pellita, C. schottii, C. subfusca, Carpobrotus edulis, Castilleja, Ceanothus crassifolius, C. greggi, C. lemmonii, C. leucodermis, C. sanguineus, Celtis reticulata, Cenchrus biflorus, Cercocarpus betuloides, C. ledifolius, Chilopsis linearis, Cirsium loncholepis, Clematis ligusticifolia, Coleogyne ramosissima, Commelina dianthifolia, Conium maculatum, Conyza canadensis, Cordylanthus rigidus, Cornus sericea, Cortaderia jubata, C. selloana, Crataegus douglasii, Cylindropuntia echinocarpa, C. ganderi, C. ramosissima, Cyperus eragrostis, Dactylis glomerata, Danthonia californica, Datisca glomerata, Delairea odorata, Deschampsia elongata, Descurainia incisa, Dichelostemma capitatum, Diplacus aurantiacus, Distichlis spicata, Drymocallis lactea, Dyssodia papposa, Eleocharis macrostachya, Elymus glaucus, E. repens, Encelia californica, Epilobium brachycarpum, E. ciliatum, Epipactus helelborine, Equisetum arvense, E. laevigatum, Ericameria cuneata, E. linearifolia, E. nauseosa, E. pinifolia, Erigeron neomexicanus, Eriodictyon crassifolium, E. trichocalyx, Eriogonum elongatum, E. fasciculatum, E. plumatella, E. wrightii, Eriophyllum confertiflorum, Eupatorium adenophorum, Festuca arundinacea, F. californica, Foeniculum vulgare, Frankenia palmeri, F. salina, Fraxinus dipetala, F. pennsylvanica, F. velutina, Fremontodendron californicum, Geranium caespitosum, G. dissectum, Hedera helix, Helenium, Hemerocallis, Hemizonia congesta, Heteromeles arbutifolia, Heuchera sanguinea, Hilaria rigida, Hirschfeldia incana, Holcus lanatus, Hordeum brachyantherum, Horkelia rydbergi, Hosackia oblongifolia, Isocoma menziesii, Juglans californica, Juncus acutus, J. balticus, J. bufonius, J. effusus, J. longistylis, J. macrophyllus, J. patens, J. phaeocephalus, J. saximontanus, J. xiphioides, Juglans major, Juniperus occidentalis, Keckiella antirrhinoides, Kickxia spuria, Larix occidentalis, Larrea, Lemna, Lepidium lasiocarpum, Lepidospartum squamatum, Limonium californicum, Lomatium, L. nudicaule, Lotus corniculatus, Leymus cinereus, L. condensatus, L. triticoides, Lupinus latifolius, Malosma laurina, Malva, Marah macrocarpa, Marrubium vulgare, Medicago polymorpha, Melica *imperfecta, Melilotus albus, Mimulus cardinalis, M. floribundus, M. guttatus, Mirabilis coccinea, M. laevis, Monardella sheltonii, Muhlenbergia richardsonis, M. rigens, Myosotis discolor, Myrica californica, Myriophyllum sibiricum, Nassella lepida, N. pulchra, Nicotiana glauca, N. obtusifolia, Oenanthe sarmentosa, Oenothera elata, Onopordum acanthium, Opuntia basilaris, Orobanche parishii, Pectocarya setosa, Pellaea mucronata, Penstemon eatonii, P. labrosus, P. rostriflorus, Perideridia lemmonii, Peritoma arborea, Persicaria amphibia, Phacelia hastata, P. ramosissima, Philadelphus lewisii, Physocarpus, Picris echioides, Pinus jeffreyi, P. lambertiana, P. monophylla, P. ponderosa, P. sabiniana, Piptatherum miliaceum, Plantago, Platanus racemosa, Poa annua, P. bulbosa, P. compressa, P. pratensis, Polypogon monspeliensis, Populus angustifolia, P. balsamifera, P. fremontii, P. nigra, P. tremuloides, P. trichocarpa, Potamogeton nodosus, Potentilla anserina, P. biennis, Prunus emarginata, P. fasciculata, P. ilicifolia, P. virginiana, Pseudotsuga menziesii, Purshia tridentata, Quercus agrifolia, Q. chrysolepis, Q. cornelius-mulleri, Q. kelloggii, Q. rugosa, Q. wislizeni, Rhamnus californica, R. ilicifolia, Rhus aromatica, R. integrifolia, R. ovata, Ribes aureum, R. nevadense, R. oxyacanthoides, R. speciosum, Ricinus communis, Robinia neomexicana, Rosa californica, R. woodsii, Rubus bifrons, R. discolor, R. laciniatus, R. ursinus, Rumex crispus, R. salicifolius, Salix bonplandiana, S. boothii, S. gooddingii, S. exigua, S. laevigata, S. lasiandra, S. lasiolepis, S. lemmonii, S. lutea, S. scouleriana, Salsola kali, Salvia mellifera, S. spathacea, Sambucus mexicana, S. nigra, Schoenoplectus acutus, S. californicus, Scirpus microcarpus, Scrophularia parviflora, Senecio serra, S. vulgaris, Sequoiadendron giganteum, Sidalcea oregana, Silybum marianum, Sisymbrium altissimum, Solanum americanum, Solidago canadensis, Sonchus oleraceus, Spartium junceum, Sphaeralcea ambigua, Stachys albens, Stanleya pinnata, Stephanomeria paniculata, Symphyotrichum foliaceum, S. subulatum, Tamarix, Tauschia arguta, Toxicodendron diversilobum, T. radicans, Typha domingensis, Umbellularia californica, Urtica dioica, U. holosericea, Verbesina longifolia, Veronica americana, Vinca major, Vitis girdiana, Woodsia phillipsii, Woodwardia fimbriata, Xantium strumarium, Xylorhiza tortifolia, Yucca brevifolia.*

Salix lemmonii Bebb is a shrub (<4 m), which occurs in bogs, bottoms, channels, depressions, ditches, flats, floodplains, forests, hot springs, meadows, ravines, roadsides, seeps, slopes (to 50%), springs, thickets, and along the margins of lakes, ponds, reservoirs, rivers, and streams at elevations from 244 to 3413 m. This is a moderately shade tolerant, riparian species found frequently along montane low-gradient floodplains and also in burned areas of subalpine pine forests, sometimes in shallow standing water (e.g., 8 cm deep). It occurs on acidic to alkaline (pH: 5.2–7.4) substrates, which have been characterized as granite, gravel, limestone, loam, loamy clay, muck, mud, rock, sand, sandy gravel, silt, silty clay loam, silty gravel, and silty sand, and where high water tables persist. The flowers emerge from March to August and are pollinated by bees (Insecta: Hymenoptera). The fruits are produced through September. The nondormant seeds are short lived (several

days) and germinate within 12–24 h of landing on a suitable site. They are produced in profusion. The plants reproduce vegetatively from stem sprouts and by detached twigs, which root upon contact with wet soil. Root suckers are not produced in this species. The plants are seriously impacted by grazing but can double their production of foliage when allowed to recover for 2–3 years. They will resprout from the stems and root crown within a few years following a fire. **Reported associates:** *Abies concolor, A. grandis, A. lasiocarpa, Acer glabrum, Achillea millifolium, Achnatherum occidentale, Aconitum columbianum, Allium tolmiei, A. validum, Alnus rhombifolia, A. rugosa, A. tenuifolia, Antennaria corymbosa, Arnica longifolia, Artemisia ludoviciana, A. tridentata, Athyrium filix-femina, Baccharis pilularis, Betula glandulosa, B. nana, Calamagrostis canadensis, C. rubescens, Camassia quamash, Carex aquatilis, C. lasiocarpa, C. limosa, C. microptera, C. nebrascensis, C. rostrata, C. scopulorum, C. sheldonii, C. simulata, C. utriculata, C. vesicaria, Cornus sericea, C. subfusca, C. utriculata, Crataegus douglasii, Dasiphora floribunda, Deschampsia cespitosa, Eleocharis quinqueflora, Epilobium ciliatum, Ericameria nauseosa, Eriophorum angustifolium, Erythronium grandiflorum, Festuca rubra, Galium trifidum, Gymnocarpium dryopteris, Heracleum sphondylium, Holodiscus discolor, Hordeum brachyantherum, Hypericum anagalloides, Iris missouriensis, Juncus balticus, J. ensifolius, J. nevadensis, Juniperus occidentalis, Linnaea borealis, Lomatium leptocarpum, Lonicera involucrata, L. utahensis, Lupinus burkei, L. latifolius, L. lepidus, Maianthemum stellatum, Menziesia ferruginea, Micranthes subapetala, Mimulus primuloides, Packera pseudaurea, Perideridia gairdneri, Phalaris arundinacea, Pinus contorta, P. jeffreyi, Poa pratensis, P. secunda, Populus tremuloides, P. trichocarpa, Potentilla gracilis, Prunus emarginata, P. virginiana, Pseudoroegneria spicata, Pseudotsuga menziesii, Purshia tridentata, Ranunculus orthorhynchus, R. sceleratus, Ribes aureum, R. cereum, R. hudsonianum, R. inerme, R. niveum, Rosa woodsii, Rudbeckia occidentalis, Salix bebbiana, S. boothii, S. drummondiana, S. eastwoodiae, S. exigua, S. fragilis, S. geyeriana, S. lasiandra, S. lasiolepis, S. lutea, S. melanopsis, S. myrtillifolia, S. orestera, S. planifolia, S. scouleriana, S. sitchensis, S. wolfii, Scirpus microcarpus, Senecio crassulus, S. serra, S. sphaerocephalus, Solidago canadensis, Sorbus scopulina, Sphagnum, Sphenosciadium capitellatum, Spiraea douglasii, Symphoricarpos oreophilus, Symphyotrichum spathulatum, Trifolium longipes, Typha latifolia, Urtica dioica, Vaccinium uliginosum, Valeriana occidentalis, Veratrum californicum, Veronica serpyllifolia, Wyethia.*

Salix lutea Nutt. is a small tree (<7 m), which inhabits alluvial terraces, ditches, draws, gravel bars, gulches, gullies, meadows, scrub, slopes (to 5%), sloughs, thickets, and the margins of canals, reservoirs, rivers, and streams at elevations from 229 to 3049 m. The substrates are described as basalt, cobble, decomposed granite, gravelly loam, pumice, rhyolite, rock, sand, sandy clay, sandy gravel, sandy silty loam, silt, silty sand, silty sandy gravel, and stoney basaltic clay loam. The plants thrive in full sun (no canopy), on wet, sandy soils with little leaf litter accumulation. Flowering and fruiting occur from April to September. The seeds are nondormant, short lived, and germinate within 24 h of their dispersal to a favorable site. Maximum seed germination (100%) is obtained using temperature regimes of 15°C/25°C and 15°C/30°C (under a 12/8 h period). Foliage production is greater when the plants grow on alluvial rather than on coarse-textured sediments. The plants reproduce vegetatively by stem sprouts (they do not produce root suckers) and by detached branchlets, which can produce stems and adventitious roots within 10 days of floating to moist sites suitable for establishment. The stems will resprout following a fire, although the plants will incur greater damage during prolonged burns. **Reported associates:** *Acer glabrum, A. negundo, A. saccharinum, Achillea millefolium, Achnatherum hymenoides, Acmispon grandiflorus, Adiantum, Agrostis exarata, A. gigantea, Alnus rhombifolia, Amorpha fruticosa, Aquilegia formosa, Artemisia cana, A. ludoviciana, A. rugosa, A. tridentata, Athyrium filix-femina, Barbarea orthoceras, Berteroa incana, Betula occidentalis, B. papyrifera, Bidens frondosus, Bromus carinatus, B. ciliatus, B. inermis, B. tectorum, Calamagrostis canadensis, C. inexpansa, Caragana arborescens, Cardaria draba, Carex alma, C. athrostachya, C. douglasii, C. fracta, C. hystericina, C. jonesii, C. lasiocarpa, C. lenticularis, C. nebrascensis, C. pellita, C. rostrata, C. senta, C. simulata, Catalpa speciosa, Cercocarpus ledifolius, Chamerion angustifolium, Chrysothamnus viscidiflorus, Cicuta maculata, Cirsium arvense, C. occidentale, C. vulgare, Clematis ligusticifolia, C. occidentalis, Collinsia callosa, Comarum palustre, Cornus sericea, Crataegus douglasii, Cryptantha confertiflora, Cyperus squarrosus, Dactylis glomerata, Danthonia, Descurainia sophia, Disporum, Drymocallis glandulosa, Echinocystis lobata, Elaeagnus angustifolia, E. commutata, Eleocharis parishii, E. quinqueflora, Epilobium ciliatum, E. glaberrimum, E. leptophyllum, Equisetum arvense, E. fluviatile, E. laevigatum, E. variegatum, Ericameria nauseosa, Eriogonum umbellatum, Euphorbia esula, Fragaria virginiana, Fraxinus pennsylvanica, Galium aparine, Gayophytum decipiens, Geranium richardsonii, G. viscosissimum, Geum macrophyllum, Glyceria striata, Glycyrrhiza lepidota, Gnaphalium palustre, Grayia spinosa, Heracleum sphondylium, Hesperostipa comata, Holcus lanatus, Hordeum brachyantherum, H. jubatum, Hosackia oblongifolia, Hypericum, Koeleria macrantha, Juglans regia, Juncus balticus, J. macrandrus, Juniperus scopulorum, Leersia oryzoides, Lepidium perfoliatum, Leymus cinereus, L. triticoides, Lilium parryi, Linanthus parryae, Lomatium mohavense, Lupinus excubitus, L. latifolius, L. polyphyllus, Lycopus americanus, L. asper, Maianthemum stellatum, Medicago lupulina, Mentha arvensis, Mimulus moschatus, M. primuloides, M. tilingii, Monardella odoratissima, Muhlenbergia asperifolia, M. mexicana, M. minutissima, M. richardsonis, Nama demissa, Nasturtium officinale, Nolina parryi, Opuntia basilaris, Orochaenactis thysanocarpha, Pascopyrum smithii, Persicaria lapathifolia, Phalaris arundinacea, Phyllodoce breweri, Pinus jeffreyi, P. monophylla, Platanthera dilatata, Poa palustris, P. pratensis, P. secunda, Polemonium occidentale, Populus ×acuminata,*

P. angustifolia, P. deltoides, P. tremuloides, P. trichocarpa, Potentilla anserina, P. gracilis, Primula, Prunus virginiana, Pseudotsuga menziesii, Purshia tridentata, Quercus chrysolepis, Rhamnus alnifolia, Ribes americanum, R. aureum, R. cereum, R. nevadense, R. oxyacanthoides, Robinia pseudoacacia, Rosa multiflora, R. woodsii, Rubus parviflorus, Rumex patientia, R. salicifolia, Salix amygdaloides, S. bebbiana, S. boothii, S. drummondiana, S. eriocephala, S. exigua, S. fragilis, S. geyeriana, S. lasiandra, S. lemmonii, Schoenoplectus pungens, Scirpus microcarpus, Scutellaria siphocampyloides, Senecio triangularis, Sidalcea malviflora, Silene menziesii, S. verecunda, Sisymbrium altissimum, Solanum dulcamara, Solidago missouriensis, S. spectabilis, Sphenosciadium capitellatum, Spiraea douglasii, S. splendens, Stachys albens, Stellaria, Symphoricarpos occidentalis, S. rotundifolius, Symphyotrichum ascendens, S. foliaceum, S. lanceolatum, S. spathulatum, Taraxacum officinale, Tetradymia canescens, Thalictrum, Thermopsis montana, Thlaspi arvense, Thysanocarpus laciniatus, Toxicodendron radicans, Tragopogon dubius, Trifolium repens, Typha latifolia, Ulmus americana, U. pumila, Urtica dioica, Valeriana, Veratrum californicum, Verbascum, Veronica americana.

Salix maccalliana **Rowlee** grows in bogs, bottomlands, carrs, depressions, fens, marshes, meadows, roadsides, string bogs (hummocks), thickets, and along the margins of lakes and sloughs at elevations of up to 1500 m. Exposures typically are full sunlight. The reported substrates include boulders, gleysols, marl, peat, regosols, sandy loam, sandy clay loam, silty clay loam, and silty loam. The flowers appear along with the leaves in early May and are produced through July. The anthers are red initially, but turn yellow prior to dehiscence. Other life history details on this species are lacking. **Reported associates:** *Achillea millefolium, Agoseris glauca, Agrostis hyemalis, A. scabra, Alnus, Anemone multifida, Arctostaphylos uva-ursi, Aulacomnium palustre, Betula glandulosa, Brachythecium, Bromus ciliatus, Calamagrostis canadensis, C. stricta, Carex aquatilis, C. buxbaumii, C. cusickii, C. flava, C. gynocrates, C. lasiocarpa, C. livida, C. praegracilis, C. prairea, C. rostrata, C. sartwellii, C. utriculata, Castilleja miniata, Cerastium arvense, Chamerion angustifolium, Deschampsia cespitosa, Drepanocladus aduncus, Eleocharis compressa, Elymus trachycaulus, Epilobium palustre, Equisetum pratense, Fragaria virginiana, Larix laricina, Lobelia kalmii, Lonicera involucrata, Lycopus uniflorus, Maianthemum stellatum, Petasites sagittatus, Picea mariana, Plagiomnium ellipticum, Poa palustris, P. pratensis, Populus balsamifera, P. tremuloides, Pyrola minor, Rhododendron groenlandicum, Ribes, Rosa acicularis, Rubus arcticus, R. sachalinensis, Rumex occidentalis, Salix arbusculoides, S. bebbiana, S. brachycarpa, S. candida, S. glauca, S. lucida, S. myrtillifolia, S. petiolaris, S. planifolia, S. prolixa, S. pseudomonticola, S. pyrifolia, S. serissima, Schoenoplectus acutus, Symphyotrichum boreale, S. ciliolatum, Tomenthypnum nitens, Typha latifolia, Valeriana dioica, Viola palustris.*

Salix melanopsis **Nutt.** is a shrub (<4 m), which occurs in or on beaches, bogs, bottoms, canals, carrs, channels, deltas, depressions, ditches, flats, floodplains, gravel bars, gullies, meadows, roadsides, sandbars, seeps, slopes (to 10%), sloughs, swamps, and along the margins of lakes, ponds, rivers, and streams at elevations of up to 3049 m. Exposures range from full sun to partial shade. The substrates have been characterized as alluvium, alluvial humus, boulders, clay loam, cobble, gravel, gravelly alluvium, gravelly loam, humus, limestone (Madison), muck, mud, peat, rocks, sand, sandy alluvium, sandy gravel, sandy loam, silt, and stones. They often are overlain by a layer of fertile loam. Flowering and fruiting extend from April to August. The plants are highly clonal by expansion of vigorous root sprouts, which enables them to quickly colonize sand bars and to form dense thickets in sunny, open areas. High-throughput DNA sequence data indicate that multiple inland regions served as source refugia for postglacial colonization events. **Reported associates:** *Abies lasiocarpa, Acer glabrum, A. negundo, Agrostis stolonifera, Alnus rugosa, A. viridis, Alopecurus aequalis, A. pratensis, Amaranthus blitoides, Amelanchier alnifolia, Antennaria, Apocynum androsaemifolium, Aquilegia formosa, Arnica longifolia, Artemisia douglasiana, A. tridentata, Arundo donax, Atriplex, Baccharis salicifolia, Betula occidentalis, Brassica nigra, Brickellia californica, Bromus, Calamagrostis canadensis, Carex aquatilis, C. densa, C. lenticularis, C. mertensii, C. nudata, C. praegracilis, C. rostrata, C. unilateralis, Centaurea melitensis, Cercocarpus ledifolius, Chamerion latifolium, Chrysothamnus viscidiflorus, Cirsium, Cornus sericea, Crepis elegans, Croton setiger, Cynodon dactylon, Dactylis glomerata, Dasiphora floribunda, Datura stramonium, Dysphania botrys, Elaeagnus commutata, Eleocharis palustris, Elymus, Equisetum arvense, E. variegatum, Eriophyllum, Fragaria virginiana, Geum macrophyllum, Heliotropium curassavicum, Heracleum sphondylium, Heterotheca oregona, Hirschfeldia incana, Holcus lanatus, Juncus, Juniperus communis, J. occidentalis, J. scopulorum, Lepidospartum squamatum, Leucanthemum vulgare, Linnaea borealis, Lonicera involucrata, Lupinus burkei, L. rivularis, Mahonia, Maianthemum stellatum, Mentzelia laevicaulis, Mimulus lewisii, Nasturtium officinale, Penstemon, Phalaris arundinacea, Picea engelmannii, P. pungens, Pinus contorta, P. flexilis, Plantago lanceolata, Poa palustris, P. pratensis, Polypogon monspeliensis, Populus angustifolia, P. balsamifera, P. fremontii, P. tremuloides, P. trichocarpa, Primula jeffreyi, Prunella vulgaris, Prunus virginiana, Pseudotsuga menziesii, Quercus chrysolepsis, Q. wislizenii, Rhododendron, Ribes inerme, R. nevadensis, R. oxyacanthoides, Rorippa, Rosa woodsii, Rubus bifrons, R. idaeus, Sairocarpus multiflorus, Salix bebbiana, S. boothii, S. caudata, S. drummondiana, S. eastwoodiae, S. eriocephala, S. exigua, S. farriae, S. geyeriana, S. irrorata, S. laevigata, S. lasiandra, S. lemmonii, S. lucida, S. lutea, S. melanopsis, S. monochroma, S. monticola, S. planifolia, S. prolixa, S. pseudocordata, S. pseudomonticola, S. purpurea, S. scouleriana, S. sessilifolia, S. sitchensis, Sambucus caerulea, Saponaria officinalis, Senecio flaccidus, S. serra, Shepherdia, Silene, Sorbus, Streptopus amplexifolius, Tanacetum vulgare, Taraxacum*

officinale, Tiarella trifoliata, Ulmus pumila, Vaccinium scoparium, Verbascum thapsus, Xanthium strumarium.

***Salix monticola* Bebb** is a shrub (to 6 m), which inhabits bogs, bottomlands, carrs, channels, cienega, ditches, fens, floodplains, gulches, heaths, marshes, meadows, riverbeds, slopes, springs, thickets, and the margins of canals, lakes, pools, reservoirs, rivers, and streams at elevations from 850 to 3561 m. The plants thrive on moderately alkaline, nutrient-rich alluvium, but the substrates can include adamellite, clay loam, gravel, loam, peat, quartz monzonite, rock, sand, sandy clay, sandy gravel, sandy loam, and silt. This is a shade-intolerant species, which sometimes occurs in shallow standing water, but cannot withstand prolonged flooding of its roots. It is also a fire-tolerant pioneer species, which tolerates high levels of iron and manganese and quickly colonizes sites that have been disturbed by fire, flooding, or logging. The flowers appear from April to July and are pollinated primarily by bees (Insecta: Hymenoptera). The seeds are short lived but germinate immediately when dispersed onto moist, mineral-rich substrates that are exposed to full sunlight. High seed germination (94%–100%) occurs across a broad range of temperatures (5°C–25°C) but is inhibited when the seeds are deposited in deep leaf litter. The plants are highly plastic morphologically and grow in dense, often clonal thickets owing to their ability to reproduce vegetatively from root sprouts and stem sprouts. Detached twigs also root and can be dispersed in the water, serving as efficient vegetative propagules. This species is important for streambank stabilization; however, establishment is difficult in areas that experience heavy grazing. **Reported associates:** *Abies concolor, A. lasiocarpa, Achillea lanulosa, A. millefolium, Achnatherum hymenoides, Agrostis gigantea, Alnus rugosa, A. tenuifolia, Amelanchier, Betula glandulosa, B. occidentalis, Breea arvense, Bromus inermis, Calamagrostis canadensis, Caltha leptosepala, Cardamine cordifolia, Carex aquatilis, C. festivella, C. interior, C. microptera, C. pachystachya, C. praticola, C. rostrata, C. scoparia, C. scopulorum, C. tahoensis, C. utriculata, Castilleja septentrionalis, Cirsium scariosum, Cladonia, Cornus sericea, Dasiphora floribunda, Deschampsia brevifolia, D. cespitosa, Distigia involucrata, Eleagnus angustifolia, Eleocharis quinquefolia, Equisetum arvense, E. pratense, Erigeron peregrinus, Eriophorum altaicum, E. angustifolium, Fragaria ovalis, Galium trifidum, Gentiana, Geranium richardsonii, Geum macrophyllum, Helianthella microcephala, Heracleum sphondylium, Hippochaete hyemalis, Iris, Juncus balticus, J. longistylis, Ligusticum, Limnorchis, Linaria vulgaris, Lonicera involucrata, Mentha arvensis, Mertensia ciliata, M. franciscana, Pedicularis bracteosa, Petradoria pumila, Phalaris arundinacea, Phleum pratense, Picea engelmannii, P. pungens, Pinus flexilis, P. ponderosa, Poa palustris, P. pratensis, P. reflexa, Polemonium occidentalis, Polytrichum, Populus angustifolia, P. tremuloides, Potentilla plattensis, P. pulcherrima, Prunus virginiana, Pseudotsuga menziesii, Pyrola asarifolia, Quercus, Ranunculus, Ribes aureum, R. inerme, Rosa woodsi, Rubus, Salix arizonica, S. bebbiana, S. boothii, S. brachycarpa, S. drummondiana, S. exigua, S.*

geyeriana, S. irrorata, S. lasiandra, S. ligulifolia, S. monticola, S. planifolia, S. scouleriana, S. wolfii, Sambucus racemosa, Senecio, Sibbaldia procumbens, Sphagnum, Swertia perennis, Tamarix chinensis, Taraxacum officinale, Trifolium repens, Trisetum wolfii, Urtica dioica, Vaccinium, Vicia americana.

Salix myrtillifolia Andersson is variable in habit, ranging from a low, profusely branched, often prostrate form (<60 cm) to a more erect, shrubby growth form (to 2.5 m). It occurs in bogs, depressions, draws, dunes, fens, floodplains, gravel bars, hummocks, knolls, lakeshores, meadows, mires, muskeg, prairies, roadsides, sandbars, scrub, siltbars, slopes (to 70%), string bogs, swales, swamps, thickets, tundra, woodlands, and along the margins of gravel pits, lakes, ponds, rivers, sloughs, and streams at elevations of up to 3049 m. The plants are shade intolerant, early successional colonizers of disturbed sites, and do not persist after a canopy begins to form. The substrates can be acidic or calcareous and have been described as alluvium, cobble, gravel, humus, limestone, loam, rock, sand, silt, terric fibrisol, and tufa. Flowering and fruiting occur from May to August with the seeds being produced in profusion. The seeds lack dormancy and will remain viable for only a week. They will germinate within 24 h of dispersal to moist, exposed mineral soils. High germination rates (97%–100%) have been observed within 1–3 days for seeds kept at temperatures between 5°C and 25°C. The plants reproduce vegetatively by their detached branches, which disperse in floodwaters and root when lodged in favorable sites. Prostrate plants reproduce vegetatively by layering, as their lower branches are overgrown by sphagnum moss. **Reported associates:** *Abies lasiocarpa, Achillea millefolium, Achnatherum thurberianum, Alnus rugosa, A. viridis, Anemone parviflora, A. richardsonii, Anticlea elegans, Arctagrostis latifolia, Arctostaphylos rubra, A. uva-ursi, Arctous, Astragalus, Athyrium filix-femina, Atremisia tridentata, Aulacomnium palustre, Betula glandulosa, B. nana, B. papyrifera, Bistorta vivipara, Calamagrostis canadensis, C. deschampsioides, C. rubescens, Calliergon trifarium, Carex aquatilis, C. aurea, C. bigelowii, C. buxbaumii, C. capillaris, C. membranacea, C. microglochin, C. rostrata, C. saxatilis, C. scirpoidea, C. utriculata, C. viridula, Chamerion angustifolium, Cinna latifolia, Comarum palustre, Cornus canadensis, Dasiphora floribunda, Delphinium glaucum, Drepanocladus, Dryas drummondii, D. integrifolia, Empetrum nigrum, Epilobium alpinum, Equisetum arvense, E. scirpoides, E. sylvaticum, E. variegatum, Erigeron, Eriophorum angustifolium, E. gracile, E. scheuchzeri, Festuca altaica, F. idahoensis, F. rubra, Galium boreale, Geocaulon lividum, Glyceria striata, Gymnocarpium jessoense, Hedysarum alpinum, Hylocomium splendens, Juncus arcticus, J. castaneus, J. triglumis, Kobresia myosuroides, K. simpliciuscula, Koeleria macrantha, Larix laricina, Lupinus argenteus, Maianthemum canadense, Menyanthes trifoliata, Mertensia paniculata, Mimulus guttatus, Myrica gale, Oxytropis, Packera pauciflora, Pedicularis sudetica, Peltigera aphthosa, Petasites frigidus, Picea glauca, P. mariana, Pinguicula vulgaris, Pinus contorta, Platanthera dilatata, P. obtusata, Pleurozium schreberi, Poa pratensis, Polemonium occidentale, Polygala, Populus balsamifera, P. tremuloides, P. trichocarpa, Potentilla anserina, Primula egaliksensis, Prunus virginiana, Ptilagrostis porteri, Ptilium crista-castrensis, Pyrola asarifolia, P. grandiflora, Rhododendron groenlandicum, R. tomentosum, Ribes, Rosa acicularis, Rubus arcticus, R. chamaemorus, Salix acutifolia, S. alaxensis, S. barclayi, S. brachycarpa, S. candida, S. commutata, S. fuscescens, S. geyeriana, S. glauca, S. lanata, S. planifolia, S. pseudomonticola, S. pulchra, S. reticulata, S. scouleriana, S. setchelliana, Scorpidium scorpioides, S. turgescens, Shepherdia canadensis, Sisyrinchium pallidum, Solidago multiradiata, Sphagnum fuscum, Spiraea douglasii, Streptopus amplexifolius, Symphoricarpos oreophilus, Taraxacum officinale, Thalictrum alpinum, Tomenthypnum nitens, Trichophorum pumilum, Triglochin maritimum, T. palustre, Vaccinium vitis-idaea, Viburnum edule.*

Salix nigra Marshall is a large tree (to 42 m) found on or in bayous, bogs, bottomlands, ditches, flats, floodplains, glades, gullies, levees, marshes, meadows, mudflats, prairies, ravines, roadsides, sand pits, seeps, sloughs, swales, swamps, thickets, washes, woodlands and along the margins of lakes, ponds, rivers, sloughs, and streams at elevations of up to 1400 m. It tolerates a variety of different substrates (pH: 4.5–7.0; salinity <0.5 ppt), where sufficient moisture (mainly stagnant water) is maintained during the growing season. The trees are flood and silt tolerant, but are highly susceptible to drought. Adult plants are highly freeze tolerant, and capable of withstanding temperatures to −60°C. The substrates include alluvium, Bodine–Brandon cherty silt loam, boulders, clay, clay loam, clay sand over dolomite, humus, limestone, loamy clay, mud, peat, sand, sandstone, sandy clay, sandy loam, silt, silty clay, silty sand, and till. This is a shade-intolerant (but tolerating some light shade), fast-growing, but short-lived pioneer tree that often is a codominant of floodplain communities. Its roots are shallow but extensive. The flowers open from February to July and are pollinated primarily by insects (e.g., Insecta: Hymenoptera: Apidae: *Apis mellifera, Bombus fervidus*). Seed production initiates when the plants reach about 10 years of age, and peaks in 25- to 75-year-old trees. The seeds ripen within 45–60 days (April to July) but lack dormancy and lose their viability quickly. They show optimal germination under a 30°C/20°C temperature regime. Natural germination occurs optimally on moist, open sites with full sunlight and no competing vegetation. The plants can reproduce vegetatively by root sprouts. Detached twigs can be dispersed in water and will root when they become stranded on a suitable substrate. Highly clonal stands of several hectares in extent can develop in this species. The plants will resprout from adventitious buds at their base following a fire, but are more susceptible to fire than many other willows. Populations are threatened by competition from the nonindigenous Chinese tallow tree (*Triadica sebifera*). **Reported associates:** *Acacia smallii, Acer negundo, A. rubrum, A. saccharinum, Ageratina altissima, Alisma subcordatum, Alliaria petiolata, Alnus serrulata, Ambrosia trifida, Anisostichus capreolata, Apios americana, Aralia spinosa, Arisaema dracontium, Aronia arbutifolia, Betula nigra, Bidens cernuus,*

B. discoideus, B. frondosus, B. polylepis, Boehmeria cylindrica, Botrychium virginianum, Briza minor, Calamagrostis canadensis, Cardamine douglassii, Carduus, Carex bebbii, C. conoidea, C. crinita, C. frankii, C. grayi, C. grisea, C. lacustris, C. pellita, C. vulpinoidea, Carpinus caroliniana, Celastrus scandens, Celtis laevigata, C. occidentalis, C. pallida, Cephalanthus occidentalis, Chelone glabra, Cicuta bulbifera, C. maculata, Cinna arundinacea, Claytonia virginica, Clethra alnifolia, Coleataenia longifolia, Cornus amomum, C. foemina, C. sericea, Crataegus mollis, Crotalaria sagittalis, Cyperus, Dichondra carolinensis, Diospyros virginiana, Echinocystis lobata, Eleocharis obtusa, E. tenuis, Elymus canadensis, E. virginicus, Epilobium, Equisetum arvense, E. hyemale, Erechtites hieracifolia, Erigeron annuus, Eupatorium serotinum, Floerkea proserpinacoides, Forestiera acuminata, Fragaria virginiana, Fraxinus pennsylvanica, F. profunda, Galium aparine, Gleditisa aquatica, G. triacanthos, Glyceria striata, Helenium autumnale, Hibiscus lasiocarpos, Ilex verticillata, Impatiens capensis, Isopyrum biternatum, Itea virginica, Juglans nigra, Juncus effusus, J. marginatus, J. torreyi, Juniperus virginiana, Laportea canadensis, Leersia oryzoides, L. virginica, Lindera benzoin, Lippia lanceolata, L. nodiflora, Liquidambar styraciflua, Lobelia cardinalis, Lonicera japonica, L. maackii, Ludwigia alterniflora, L. palustris, L. polycarpa, Lysimachia terrestris, Lythrum alatum, Maclura pomifera, Magnolia virginiana, Menispermum canadense, Micranthes pensylvanica, Mikania scandens, Morus alba, M. rubra, Myrica cerifera, Nyssa aquatica, N. sylvatica, Onoclea sensibilis, Osmunda regalis, Oxalis dillenii, O. stricta, Oxypolis rigidior, Panicum maximum, P. virgatum, Paspalum lividum, Penstemon pallidus, Persicaria coccinea, P. hydropiperoides, P. lapathifolia, P. pensylvanica, P. posumbu, Phalaris arundinacea, Phlox divaricata, P. paniculata, Phragmites australis, Physocarpus opufolius, Picea mariana, Pilea fontana, P. pumila, Pinus echinata, P. elliottii, P. palustris, P. taeda, Planera aquatica, Plantago aristata, Platanus occidentalis, Pluchea odorata, Poa pratensis, Polemonium reptans, Populus deltoides, P. heterophylla, Proserpinaca palustris, Prunus serotina, Pteridium aquilinum, Quercus alba, Q. bicolor, Q. coccinea, Q. lyrata, Q. macrocarpa, Q. michauxii, Q. nigra, Q. palustris, Q. phellos, Q. stellata, Ranunculus abortivus, R. sceleratus, R. septentrionalis, Rhamnus cathartica, R. frangula, Rhododendron maximum, Rhynchospora corniculata, Ribes americanum, R. missouriense, Rorippa islandica, Rosa multiflora, R. palustris, Rotala ramosior, Rubus pensilvanicus, R. setosus, Rudbeckia hirta, R. laciniata, Rumex altissimus, R. verticillatus, Sagittaria latifolia, Salix amygdaloides, S. caroliniana, S. exigua, S. ×glatfelteri, S. humilis, S. interior, S. myricoides, S. petiolaris, Sambucus nigra, Sanicula gregaria, Sassafras albidum, Saururus cernuus, Schizachyrium scoparium, Schoenoplectus pungens, S. tabernaemontani, Scirpus cyperinus, Scutellaria lateriflora, Sesbania drummondii, Silphium perfoliatum, Smilax, Solidago altissima, S. gigantea, Sparganium androcladum, Spiraea tomentosa, Spirodela polyrhiza, Symphyotrichum prenanthoides, Taraxacum officinale, Taxodium distichum,
Thelypteris palustris, Thuja occidentalis, Toxicodendron radicans, Triadenum walteri, Triadica sebifera, Tridens flavus, Tripsacum dactyloides, Typha angustifolia, T. domingensis, T. latifolia, Ulmus americana, Urtica dioica, U. procera, Vaccinium corymbosum, Verbesina alternifolia, Viburnum dentatum, V. lentago, V. recognitum, Viola lanceolata, V. sororia, Vitis riparia, Xanthium strumarium, Zanthoxylum americanum.

***Salix pedicellaris* Pursh** is a creeping shrub (to 2 m), which is found in bogs, carrs, ditches, fens, flats, glades, marshes, meadows, patterned fens, seeps, slopes (to 5%), swamps, and thickets at elevations of up to 1750 m. Most occurences are on organic (peat) soils (e.g., pH: 5.5–7.6; alkalinity: 45 mg/L; conductivity: 97 mmhos; hardness 10–42.8 mg/L $CaCO_3$). The flowers are produced from April to July, with fruit production extending into August. The shoots are stoloniferous and form vegetative colonies by layering. **Reported associates:** *Agrostis scabra, A. stolonifera, A. thurberiana, Alnus rugosa, Andromeda polifolia, Angelica atropurpurea, Aronia ×prunifolia, Aulacomnium palustre, Berula erecta, Betula glandulosa, B. pumila, Calamagrostis canadensis, Calliergonella cuspidata, Caltha leptosepala, Campanula aparinoides, Camylium stellatum, Cardamine bulbosa, Carex aquatilis, C. buxbaumii, C. canescens, C. chordorrhiza, C. comosa, C. cusickii, C. diandra, C. echinata, C. interior, C. lacustris, C. lasiocarpa, C. leptalea, C. limosa, C. livida, C. magellanica, C. neurophora, C. oligosperma, C. prairea, C. rostrata, C. saxatilis, C. trisperma, C. utriculata, C. viridula, Castilleja miniata, Cephalanthus occidentalis, Chamaedaphne calyculata, Cicuta douglasii, Cirsium muticum, Cladium mariscoides, Comarum palustre, Cornus amomum, Cypripedium, Dasiphora floribunda, Deschampsia cespitosa, Drepanocladus revolven, Drosera intermedia, D. rotundifolia, Dryopteris cristata, Dulichium arundinaceum, Eleocharis elliptica, E. quinqueflora, E. rostellata, E. palustris, Epilobium leptophyllum, E. palustre, E. strictum, Equisetum arvense, E. fluviatile, Eriophorum angustifolium, E. chamissonis, E. gracile, E. vaginatum, Eutrochium maculatum, Fraxinus nigra, Galium labradoricum, Gaultheria hispidula, Gentiana calycosa, Habenaria, Hamatocaulis vernicosus, Hylocomium splendens, Hypericum anagalloides, H. kalmianum, H. virginicum, Ilex verticillata, Iris versicolor, Juniper communis, J. horizontalis, Kalmia microphylla, K. polifolia, Larix laricina, Linnaea borealis, Liparis loeselii, Lobelia kalmii, Lonicera involucrata, L. oblongifolia, L. villosa, Lysichiton americanus, Lysimachia terrestris, L. thyrsiflora, Maianthemum trifolium, Menyanthes trifoliata, Micranthes oregana, Mimulus glabratus, Mitella nuda, Muhlenbergia glomerata, Myrica gale, Panicum lindheimeri, Parnassia fimbriata, Pedicularis groenlandica, P. lanceolata, Petasites frigidus, Phalaris arundinacea, Picea mariana, Pinus contorta, Platanthera dilatata, Pleurozium schreberi, Pogonia ophioglossoides, Polytrichum commune, Potentilla anserina, Primula jeffreyi, Proserpinaca palustris, Ptilium crista-castrensis, Pycnanthemum virginianum, Rhamnus alnifolia, Rhododendron groenlandicum, Rhynchospora alba, Ribes hirtellum, Rosa acicularis, Rubus acaulis, R.*

pubescens, Rumex orbiculatus, Sagittaria latifolia, Salix brachycarpa, S. candida, S. discolor, S. drummondiana, S. geyeriana, S. petiolaris, S. planifolia, S. pyrifolia, Sanguisorba officinalis, Sarracenia purpurea, Schoenoplectus acutus, S. tabernaemontani, Scirpus hudsonianus, S. longii, S. microcarpus, Scorpidium scorpioides, Scutellaria galericulata, Selaginella selaginoides, Senecio cymbalarioides, S. triangularis, Sium suave, Solidago uliginosa, Spartina pectinata, Sphagnum angustifolium, S. capillifolium, S. centrale, S. fallax, S. fuscum, S. girgensohnii, S. magellanicum, S. warnstorfii, Spiraea alba, S. douglasii, S. splendens, Spiranthes, Symphyotrichum boreale, S. spathulatum, Thelypteris palustris, Tomenthypnum nitens, Torreyochloa, Triadenum fraseri, Triantha glutinosa, Trichophorum cespitosum, Trientalis europaea, Trifolium longipes, Triglochin maritimum, Tsuga mertensiana, Typha latifolia, Utricularia intermedia, U. minor, Vaccinium macrocarpon, V. oxycoccos, V. uliginosum, Verbena hastata, Viola pallens, V. palustris, Xyris.

Salix petiolaris Sm. is a large shrub (to 7 m), which inhabits bogs, bottoms, carrs, channels, depressions, ditches, dunes, meadows, muskeg, prairies, ravines, roadsides, seeps, shores, swales, swamps, thickets, woods, and the margins of lakes, ponds, rivers, sloughs, and streams at elevations of up to 2637 m. The plants are highly shade intolerant. The substrates often are of histic (e.g., pH: 5.4) or mineral alluvial origin and include gravel, limestone, loamy sand, marl, muck, peat, peaty humus, sand, and sandy loam. The plants flower from April to June and are in fruit from July to September. The seeds lack dormancy and will germinate in several days following their burial. These plants are good colonizers of disturbed habitats by means of their highly effective seed dispersal. Vegetative reproduction can also occur by resprouting of stems; however, highly clonal patches do not develop as in some willow species. **Reported associates:** *Acer rubrum, Achillea, Agoseris, Alnus rugosa, Amorpha fruiticosa, Andropogon virginicus, Arnoglossum plantagineum, Aronia ×prunifolia, Asclepias incarnata, A. purpurascens, Betula glandulosa, B. papyrifera, B. populifolia, B. pumila, Bidens frondosus, Bromus ciliatus, Calamagrostis canadensis, C. stricta, Caltha palustris, Campanula aparinoides, Carex atherodes, C. buxbaumii, C. comosa, C. interior, C. lacustris, C. lasiocarpa, C. microptera, C. normalis, C. pellita, C. sartwellii, C. stricta, Chamaedaphne calyculata, Chelone glabra, Cladium mariscoides, Comptonia peregrina, Coreopsis tripteris, Cornus amomum, C. racemosa, C. sericea, Corylus americana, Crataegus rotundifolia, Dasiphora floribunda, Doellingeria umbellata, Elymus virginicus, Epilobium ciliatum, E. strictum, Eryngium yuccifolium, Euthamia graminifolia, Eutrochium maculatum, Fragaria, Fraxinus nigra, F. pennsylvanica, Galium boreale, G. trifidum, Gentiana quinquefolia, Glyceria canadensis, G. striata, Heliopsis helianthoides, Ilex verticillata, Iris virginica, Juncus canadensis, Krigia biflora, Larix laricina, Lycopus americanus, L. uniflorus, Lysimachia quadriflora, L. thyrsiflora, Lythrum salicaria, Micranthes pensylvanica, Muhlenbergia frondosa, M. glomerata, Myrica gale, Onoclea sensibilis, Osmunda regalis, Oxypolis rigidior, Persicaria amphibia, Petasites frigidus, Phalaris arundinacea, Physocarpus opulifolius, Picea glauca,*

P. mariana, Pilea fontana, Pinus banksiana, P. contorta, Populus balsamifera, P. deltoides, P. tremuloides, Potentilla anserina, P. norvegica, Prunus serotina, P. virginiana, Quercus macrocarpa, Q. palustris, Q. rubra, Rhamnus cathartica, R. frangula, Rhynchospora capitellata, Ribes americanum, R. triste, Rosa palustris, Rumex britannica, Salix amygdaloides, S. bebbiana, S. candida, S. discolor, S. eriocephala, S. exigua, S. glaucophylloides, S. interior, S. lucida, S. planifolia, S. pyrifolia, Scutellaria lateriflora, Scirpus atrovirens, S. cyperinus, S. longii, Sium suave, Solanum dulcamara, Spartina pectinata, Spiraea alba, S. tomentosa, Symphyotrichum lateriflorum, S. novae-angliae, Thelypteris palustris, Thuja occidentalis, Tilia americana, Toxicodendron vernix, Triadenum fraseri, Typha latifolia, Ulmus americana, Urtica dioica, Vaccinium angustifolium, Viburnum lentago, Viola lanceolata, Vitis labrusca, Zizia aurea.

Salix planifolia Pursh is a small shrub (<4 m), which inhabits alpine, arctic, boreal, and subalpine bogs, draws, fens, flats, gravel pits, heath, meadows, muskeg, seeps, slopes (to 40%), snow flush areas, swamps, thickets, tundra, and the margins of lakes, ponds, rivers, and streams at elevations of up to 4450 m. The plants are shade intolerant but more characteristic of stable rather than early successional communities. The substrates are acidic (pH: 5.0–6.9) and are characterized as alluvial loam, boulders (granite), clay, clay loam, granitic silty loam, gravelly loam, gravelly silt, igneous, krummholz, limestone (Claron), loam, mucky peat, peat, pumice alluvium, rock, rocky loam, sandy alluvium, sandy loam, scree, silty clay, silty loam, silty loamy peat, silty pumice, stones, till, and volcanic alluvium. Flowering and fruiting have been observed from April to August. The seeds are nondormant, short lived (for about 1 week), and germinate (95%–100%) within 24 h (optimally at 5°C–25°C) when dispersed to favorable sites (e.g., wet, mineral soils devoid of litter). Reproduction is primarily sexual with high seed output. Vegetative reproduction can occur from stem base/root crown sprouts or by detached branches, which are transported by water and root when coming in contact with soil. However, the plants do not reproduce by layering and consequently do not develop into large, clonal colonies. **Reported associates:** *Abies amabilis, A. concolor, A. lasiocarpa, Achillea lanulosa, A. millefolium, Achlys triphylla, Aconitum columbianum, Aira, Alectoria, Allium validum, Alnus rugosa, A. viridis, Alopecurus, Amelanchier, Angelica arguta, Antennaria corymbosa, A. lanata, Anthoxanthum nitens, Aquilegia formosa, Arctostaphylos uva-ursi, Arnica chamissonis, A. griscomii, A. latifolia, Artemisia norvegica, A. tridentata, Betula glandulosa, B. occidentalis, Bistorta bistortoides, B. vivipara, Calamagrostis breweri, C. canadensis, Calliergon giganteum, Caltha leptosepala, Campylium stellatum, Cardamine cordifolia, Carex aquatilis, C. athrostachya, C. atrata, C. aurea, C. buxbaumii, C. canescens, C. capillaris, C. chalciolepis, C. diandra, C. disperma, C. festivella, C. gynocrates, C. illota, C. interior, C. lacustris, C. lasiocarpa, C. limosa, C. livida, C. lyngbyei, C. microptera, C. nebrascensis, C. nigricans, C. pachystachya, C. paysonis, C. phaeocephala, C. praeceptorum, C. rostrata,*

C. saxatilis, C. scirpoidea, C. scoparia, C. scopulorum, C. siccata, C. simulata, C. subnigricans, C. utriculata, Carum carvi, Cassiope mertensiana, Castilleja occidentalis, C. septentrionalis, C. sulphurea, Ceanothus velutinus, Cetraria, Chamaedaphne calyculata, Chamerion angustifolium, C. latifolium, Cirsium eatonii, Cladina, Cladonia, Climacium dendroides, Comarum palustre, Cornus sericea, Crepis runcinataDanthonia intermedia, Dasiphora floribunda, Deschampsia cespitosa, Drepanocladus revolvens, Drosera anglica, Dryas integrifolia, D. octopetala, Eleocharis acicularis, E. quinqueflora, E. rostellata, Empetrum nigrum, Epilobium anagallidifolium, E. ciliatum, E. lactiflorum, Equisetum arvense, E. laevigatum, E. variegatum, Erigeron peregrinus, Eriophorum angustifolium, E. chamissonis, Festuca altaica, Fragaria ovalis, F. vesca, F. virginiana, Galium boreale, Gentiana calycosa, Gentianopsis detonsa, Geranium richardsonii, Geum macrophyllum, G. rivale, G. rossii, Glyceria striata, Heracleum sphondylium, Hylocomium splendens, Hymenoxys hoopesii, Iris missouriensis, Juncus balticus, J. drummondii, J. longistylis, Kalmia microphylla, K. polifolia, Kobresia myosuroides, Larix, Ligularia amplectens, L. pudica, Ligusticum canbyi, L. porteri, Lonicera caerulea, L. utahensis, Luetkea pectinata, Lupinus, Luzula, Lysichiton americanus, Maianthemum stellatum, M. trifolium, Menyanthes trifoliata, Mertensia alpina, M. ciliata, Micranthes integrifolia, M. odontoloma, M. subapetala, Mimulus lewisii, M. primuloides, Moehringia lateriflora, Muhlenbergia filiformis, Myrica gale, Nuphar polysepala, Oreostemma alpigenum, Packera subnuda, Parnassia fimbriata, Pedicularis bracteosa, P. groenlandicum, Phleum alpinum, P. pratense, Phyllodoce empetriformis, P. glanduliflora, Picea engelmannii, P. glauca, P. mariana, Pinus albicaulis, P. contorta, Poa alpina, P. leptocoma, P. palustris, P. pratensis, Plagiobothrys hispidulus, Pleurozium schreberi, Polemonium caeruleum, P. occidentale, Polytrichum, Populus angustifolia, P. tremuloides, Porterella carnosula, Potentilla ×diversifolia, P. pulcherrima, Primula jeffreyi, P. pauciflora, Ptilagrostis porteri, Pyrola asarifolia, Ranunculus alismaefolius, Rhamnus alnifolia, Rhodiola rhodantha, Rhododendron columbianum, R. tomentosum, Rhynchostegiella compacta, Ribes inerme, R. montigenum, R. oxyacanthoides, Rosa acicularis, R. woodsii, Rubus acaulis, R. chamaemorus, R. spectabilis, Sagittaria cuneata, Salix alaxensis, S. arbusculoides, S. barclayi, S. barrattiana, S. bebbiana, S. boothii, S. brachycarpa, S. candida, S. caudata, S. commutata, S. discolor, S. drummondiana, S. eastwoodiae, S. exigua, S. farriae, S. geyeriana, S. glauca, S. lasiandra, S. maccalliana, S. monticola, S. nivalis, S. orestera, S. pedicellaris, S. petiolaris, S. planifolia, S. polaris, S. pseudocordata, S. pseudomonticola, S. pulchra, S. reticulata, S. serissima, S. stolonifera, S. tweedyi, S. wolfii, Saxifraga tricuspidata, Senecio crassulus, S. crocatus, S. cymbalarioides, S. hydrophiloides, S. sphaerocephalum, S. triangularis, Sambucus, Sibbaldia, Sparganium glomeratum, Sphagnum, Sphenosciadium capitellatum, Spiraea, Stellaria longipes, Stereocaulon, Streptopus, Swertia perennis, Symphyotrichum foliaceum,

S. spathulatum, Taraxacum officinale, Thalictrum alpinum, T. sparsiflorum, Thermopsis montana, Thuja occidentalis, T. plicata, Trifolium parryi, T. repens, Triglochin maritimum, Trillium ovatum, Trisetum wolfii, Trollius laxus, Vaccinium scoparium, V. uliginosum, V. vitis-idaea, Valeriana occidentalis, Veratrum californicum, Veronica wormskjoldii, Vicia americana, Viola palustris.

***Salix prolixa* Andersson** is a shrub (<5 m), which grows in or on bogs, bottoms, carrs, dikes, ditches, draw, dunes, flats, floodplains, gravel bars, gulches, gullies, marshes, meadows, prairies, sandbars, seeps, slopes (to 30%), sloughs, swales, swamps, thickets, woodlands, and along the margins of canals, lakes, ponds, rivers, springs, and streams at elevations of up to 2530 m. Exposures vary from open to shaded sites. The substrates include alluvial sand, cobble, granite gravel, gravel, gravelly humus, gravelly loam, loam, muck, mucky sandy loam, rocky humus, sand, sandy gravel, sandy silt, silt, silty clay, silty sandy gravel, and stony alluvium. Flowering and fruiting occur from March to September. The seeds are short lived and retain viability for only a few days. The plants can resprout vegetatively from their stems (especially after fires) but do not develop into clonal colonies. **Reported associates:** *Acer glabrum, A. macrophyllum, A. negundo, A. saccharinum, Agropyron, Agrostis, Alnus rhombifolia, A. rubra, A. rugosa, A. tenuifolia, A. viridis, Amelanchier alnifolia, Apocynum androsaemifolium, Arnica latifolia, Artemisia campestris, A. ludoviciana, A. tridentata, Balsamorhiza sagittata, Betula occidentalis, Bromus inermis, Bryonia, Calamagrostis canadensis, Caltha leptosepala, Carex arcta, C. lenticularis, C. rostrata, C. scopulorum, C. utriculata, Chamerion angustifolium, Chrysothamnus, Cirsium arvense, Clematis ligusticifolia, Coptis occidentalis, Cornus canadensis, C. sericea, Crataegus douglasii, Cytisus scoparius, Dipsacus fullonum, Dryas drummondii, Elaeagnus commutata, Eleocharis macrostachya, Elymus, Epilobium palustre, Equisetum arvense, E. fluviatile, E. hyemale, E. pratense, E. variegatum, Eurybia sibirica, Euthamia, Fraxinus latifolia, Gaultheria humifusa, Geranium viscosissimum, Hemerocallis, Heracleum, Holcus mollis, Hypnum lindbergii, Juncus ensifolius, Juniperus scopulorum, Larix occidentalis, Lathyrus japonicus, Leymus mollis, Lomatium, Lysichiton americanus, Malus, Marsilea vestita, Melilotus albus, Mitella pentandra, Morus alba, Oenanthe, Petasites frigidus, Philadelphus lewisii, Picea engelmannii, Pinus contorta, P. ponderosa, Populus angustifolia, P. balsamifera, P. trichocarpa, Prunus virginiana, Pseudoroegneria spicata, Purshia tridentata, Rhamnus purshiana, Rhizomnium, Rhododendron columbianum, Ribes aureum, Rosa woodsii, Rubus bifrons, R. spectabilis, Rudbeckia laciniata, Salix alaxensis, S. amygdaloides, S. bebbiana, S. boothii, S. drummondiana, S. eriocephala, S. exigua, S. geyeriana, S. lasiandra, S. melanopsis, S. sitchensis, Scirpus microcarpus, Senecio serra, S. triangularis, Silene, Solidago canadensis, Sphagnum, Spiraea douglasii, Stephanomeria paniculata, Symphoricarpos albus, Tanacetum vulgare, Taraxacum officinale, Thermopsis montana, Toxicodendron radicans, Trifolium fragiferum, T. repens, Typha latifolia, Ulmas pumila, Vaccinium scoparium,*

Valeriana sitchensis, Veratrum californicum, Verbascum, Veronica americana, Vicia americana, Vulpia microstachys.

Salix purpurea L. is a nonindigenous species, which has spread to beaches, bogs, ditches, dunes, floodplains, fens, gravel pits, marshes, meadows, roadsides, seeps, shores, sloughs, springs, swamps, thickets, woodlands, and the margins of ponds, rivers, and streams at elevations of up to 1990 m. The plants typically are not shade tolerant. Their substrates are described as alkaline and include gravel, limestone, and sand. Flowering occurs from March to May. The shoots can withstand inundation at high-energy sites, and will resprout vegetatively from damaged trunks. Vegetative reproduction also reportedly occurs by rhizomes. The plants normally possess arbuscular mycorrhizae in their native range. **Reported associates (North America):** *Alopecurus aequalis, Betula nigra, Carex flacca, Cichorium intybus, Cornus sericea, Crataegus gaylussacia, Cyperus squarrosus, Daucus carota, Equisetum arvense, E. fluviatile, Erigeron annuus, Eriophorum viridicarinatum, Fraxinus latifolia, Glyceria striata, Justicia americana, Lonicera ×bella, L. maackii, Melilotus albus, M. officinalis, Mollugo verticillata, Phalaris arundinacea, Picea mariana, Plantago lanceolata, Poa compressa, P. nemoralis, Populus deltoides, Rhamnus cathartica, R. davurica, R. japonica, R. utilis, Rorippa, Rosa multiflora, Rubus occidentalis, Salix discolor, S. eriocephala, S. interior, S. nigra, S. sessilifolia, Scirpus microcarpus, Solidago altissima, Sphagnum, Symphyotrichum drummondii, S. pilosum, Trifolium pratense, Viburnum rafinesqueanum, Vitis riparia.*

Salix pyrifolia Andersson grows in bogs, depressions, ditches, dunes, fens, flats, hummocks, meadows, outwash plains, pools, roadsides, swales, swamps, woodlands, and along the margins of brooks, lakes, ponds, rivers, sloughs, and streams at elevations of up to 1600 m. Exposures range from open to partially shaded sites. The substrates have been characterized as acidic (e.g., pH: 5.5) or calcareous (but generally are more acidic) and include gravel, muck, peat, peaty sand, sand, and shale. Flowering occurs from May to July. The plants spread vegetatively by the production of rhizomatous branches. **Reported associates:** *Abies balsamea, A. lasiocarpa, Achillea millefolium, Alnus rugosa, Amelanchier, Andromeda, Betula glandulosa, B. papyrifera, B. pumila, Bromus ciliatus, Calamagrostis canadensis, Campanula aparinoides, Carex aquatilis, C. canescens, C. lacustris, C. lasiocarpa, C. magellanica, C. scoparia, C. stricta, C. trisperma, Chamaedaphne calyculata, Chamerion angustifolium, Comarum palustre, Comptonia peregrina, Cornus canadensis, Doellingeria umbellata, Dryas drummondii, Dryopteris cristata, Elymus glaucus, Epilobium leptophyllum, Equisetum arvense, E. fluviatile, E. sylvaticum, Erigeron annuus, Eriophorum, Eupatorium perfoliatum, Euthamia gymnospermoides, Eutrochium maculatum, Fragaria virginiana, Fraxinus nigra, Galium tinctorium, G. trifidum, Glyceria canadensis, Helianthus giganteus, Hylocomium splendens, Ilex mucronata, I. verticillata, Impatiens capensis, Juncus canadensis, J. effusus, Larix laricina, Leymus innovatus, Lycopus americanus, L. uniflorus, Lysimachia terrestris, L. thyrsiflora, Mertensia paniculata, Oenothera perennis, Onoclea sensibilis, Persicaria sagittata, Picea glauca, P. mariana, Pinus contorta,*

Poa palustris, Polytrichum strictum, Populus balsamifera, P. tremuloides, P. trichocarpa, Potentilla norvegica, Ptilium crista-castrensis, Rhododendron groenlandicum, Ribes, Rosa acicularis, Rubus idaeus, R. setosus, Rumex, Salix bebbiana, S. discolor, S. pedicellaris, S. petiolaris, S. rostrata, S. serissima, Scirpus cyperinus, Scutellaria galericulata, Solidago gigantea, S. uliginosa, Sphagnum capillifolium, S. centrale, S. cuspidatum, S. flexuosum, S. magellanicum, Spiraea alba, S. tomentosa, Symphyotrichum lanceolatum, S. puniceum, Thelypteris palustris, Thuja occidentalis, Toxicodendron vernix, Tradenum fraseri, Typha latifolia, Utricularia minor, Vaccinium cespitosum, V. ovalifolium, V. vitis-idaea, Viburnum edule, Viola primulifolia.

Salix sericea Marshall is a shrub to small tree (to 8 m), which occurs in bogs, bottomlands, ditches, dunes, fens, flats, floodplains, ledges, marshes, meadows, potholes, prairies, ravines, roadsides, seeps, shores, springs, swales, swamps, terraces, thickets, woodlands, and along the margins of lakes, ponds, pools, and streams at elevations of up to 1250 m. The plants will tolerate full sun to partial shade. The substrates are acidic (pH: 5.2–7.0) and include clay, gravel, humus, loam, mud, peat, peaty rock, sand, schist, serpentine, and silt. Flowering extends from March to June, with the catkins appearing before the leaves. The flowers are pollinated primarily by bees (Insecta: Hymenoptera: Andrenidae; Apidae). The plants reportedly possess both ecto- and arbuscular mycorrhizae, but some studies have failed to detect mycorrhizal root colonization. **Reported associates:** *Acer rubrum, Aesculus flava, A. glabra, Agrostis perennans, Ailanthus altissima, Alnus rugosa, A. serrulata, Anaphalis margaritacea, Apios americana, Apocynum androsaemifolium, Aronia arbutifolia, Asclepias, Betula alleghaniensis, B. lenta, Bidens cernuus, Boehmeria cylindrica, Calamagrostis pickeringii, Carex atlantica, C. debilis, C. folliculata, C. frankii, C. gynandra, C. lacustris, C. leptalea, C. lurida, C. scabrata, C. stricta, C. vulpinoidea, Cephalanthus occidentalis, Chamaecrista nictitans, Chamerion angustifolium, Collinsonia canadensis, Comptonia peregrina, Cornus amomum, C. sericea, Cypripedium reginae, Dalibarda repens, Decodon verticillatus, Desmodium glabellum, D. marilandicum, Dichanthelium clandestinum, Diervilla sessilifolia, Diphylleia cymosa, Eleocharis erythropoda, Epilobium strictum, Equisetum arvense, E. hyemale, Eriophorum virginicum, Eubotrys racemosa, Eupatorium perfoliatum, Filipendula rubra, Fragaria virginiana, Fraxinus pennsylvanica, Galium tinctorium, Gaultheria hispidula, Gentiana andewsii, Gleditsia triacanthos, Hamamelis virginiana, Houstonia caerulea, Hypericum mutilum, Juncus effusus, J. subcaudatus, Kalmia latifolia, Lactuca canadensis, Lechea mucronata, Leersia virginica, Leucothoe racemosa, Lilium grayi, Lindera benzoin, Liriodendron tulipifera, Lysimachia quadrifolia, Mentha arvensis, M. piperita, Micranthes micranthidifolia, M. pensylvanica, Microstegium vimineum, Mimulus alatus, Onoclea sensibilis, Orontium aquaticum, Osmunda regalis, Osmundastrum cinnamomeum, Panax quinquefolius, Parnassia asarifolia, Physocarpus opulifolius, Polemonium van-bruntiae, Populus tremuloides, Potentilla, Prunus americana, P. serotina, Pteridium aquilinum, Quercus rubra,*

Rhododendron canadense, R. catawbiense, R. maximum, Rhus copallinum, R. typhina, Rhynchospora alba, R. capitellata, Rosa gallica, R. palustris, Rubus flagellaris, Rumex verticillatus, Sagittaria latifolia, Salix discolor, S. myricoides, S. pentandra, Sambucus nigra, Sarracenia purpurea, Sassafras albidum, Scirpus cyperinus, S. expansus, Sium suave, Solidago patula, Spiraea, Symphoricarpos orbiculatus, Symphyotrichum lateriflorum, S. novae-angliae, Tilia americana, Toxicodendron radicans, T. vernix, Trifolium aureum, Typha angustifolia, Ulmus rubra, Vaccinium pallidum, Viburnum lentago, V. nudum, Viola blanda, Woodwardia virginica, Xanthorhiza simplicissima.

***Salix serissima* (L. H. Bailey) Fernald** is a shrub (to 5 m) of cold, fresh to brackish sites (sometimes in shallow water) including bogs, carrs, depressions, ditches, fens, flats, marshes, meadows, mires, pools, roadsides, shores, slopes, swamps, thickets, and the margins of lakes, ponds, rivers, and streams at elevations of up to 3000 m. The exposures range from open to partially shaded sites. The substrates most often are described as calcareous and include alluvium, clay, gravel, limestone, marl, peat (0.5–1.5 m deep), peaty sand, sand, and sandy loam. The flowers appear just as the leaves expand. Flowering occurs in late spring to summer (May to September) with fruit production observed in summer or fall (June to November). The catkins can persist on trees for more than a year, containing seeds that reportedly require cold stratification in order to break dormancy. Adult plants will succumb to fire, but burned areas can revegetate in several years from resprouting stems and by seedling establishment. This species does not form clonal stands or reproduce by stem fragmentation as do many other willows. **Reported associates:** *Abies balsamea, Achillea millefolium, Acorus americanus, Actaea rubra, Agoseris glauca, Alnus rugosa, Arctous rubra, Asclepias incarnata, Aulacomnium palustre, Betula glandulosa, B. nana, B. occidentalis, B. pumila, Bromus ciliatus, B. inermis, Calamagrostis canadensis, Calliergon cordifolium, C. giganteum, C. viridula, Caltha palustris, Campanula aparinoides, Campylium stellatum, Canadanthus modestus, Carex aquatilis, C. castanea, C. exilis, C. flava, C. garberi, C. interior, C. lacustris, C. leptalea, C. livida, C. nebrascensis, C. pauciflora, C. paupercula, C. pellita, C. retrorsa, C. rostrata, C. saximontana, C. scoparia, C. simulata, C. sterilis, C. stricta, C. utriculata, C. viridula, Castilleja cusickii, C. miniata, Chamerion angustifolium, Chara, Cirsium arvense, Climacium dendroides, Comarum palustre, Cornus sericea, Dasiphora floribunda, Drepanocladus, Dryopteris cristata, Eleocharis quinqueflora, Equisetum arvense, E. fluviatile, E. variegatum, Eriophorum angustifolium, Eutrochium maculatum, Fragaria virginiana, Fraxinus nigra, F. pennsylvanica, Galium boreale, G. labradoricum, G. palustre, G. trifidum, Gaillardia aristata, Gentiana affinis, Geum aleppicum, G. rivale, Hypnum lindbergii, Impatiens capensis, Iris versicolor, Juncus balticus, J. dudleyi, J. effusus, J. nodosus, Larix laricina, Lobelia kalmii, Lomatogonium rotatum, Lonicera involucrata, Lysimachia quadriflora, L. thyrsiflora, Maianthemum trifolium, Mentha arvensis, Menyanthes trifoliata, Mertensia paniculata, Mitella nuda, Muhlenbergia*

glomerata, Myrica gale, Onoclea sensibilis, Packera pseudaurea, Parnassia parviflora, Pedicularis groenlandica, P. labradorica, Petasites frigidus, Phleum pratense, Picea glauca, P. mariana, Platanthera stricta, Poa palustris, Populus balsamifera, P. tremuloides, Potentilla anserina, Ranunculus aquatilus, R. lapponicus, Rhamnus alnifolia, Rhododendron groenlandicum, Rhynchospora capillacea, Rosa palustris, Rubus arcticus, R. chamaemorus, Salix arbusculoides, S. bebbiana, S. brachycarpa, S. candida, S. commutata, S. discolor, S. eriocephala, S. geyeriana, S. interior, S. ligulifolia, S. maccalliana, S. monticola, S. myrtillifolia, S. pedicellaris, S. petiolaris, S. planifolia, S. pseudomonticola, Scirpus cyperinus, Senecio triangularis, Solidago canadensis, S. gigantea, S. patula, S. uliginosa, Sphagnum, Spiraea alba, Stellaria longifolia, Symphyotrichum boreale, S. ciliolatum, S. falcatum, S. lanceolatum, S. puniceum, Thalictrum, Thelypteris palustris, Thuja occidentalis, Triantha glutinosa, Trifolium pratense, Triglochin maritimum, Typha latifolia, Urtica dioica, Valeriana dioica, V. edulis, Viburnum lentago, Vicia americana, Viola cucullata, Zizia aptera.

***Salix viminalis* L.** is a nonindigenous shrub or small tree (to 6–9 m), which has been introduced to beaches, meadows, roadsides, wastelands, woodlands, and the margins of lakes, rivers, and streams at elevations of up to 300 m. Reported substrates include cobble, gravel, muck, and sand. Flowering has been observed from April to May. Successful seed germination (100%) has been achieved at 21°C under a 12/12 h light regime using sterile Norstog medium. There is a substantial body of ecological literature on this species but little is applicable to the nonindigenous North American populations. The species has been widely planted in experimental plantations in Quebec, but has shown to be highly susceptible to disease and insect attacks (see *Use by wildlife*). **Reported associates (North America):** *Acer macrophyllum, Alnus rubra, Lysichiton americanus, Malus fusca, Rubus bifrons, R. spectabilis.*

***Salix wolfii* Bebb** is a shrub (to 2 m), which grows in beaver ponds, bogs, bottoms, drainageways, fens, flats, floodplains, gravel bars, marshes, meadows, seeps, shores, slopes (to 30%), springs, tundra, and along the margins of lakes, rivers, and streams at elevations from 1067 to 3380 m. The plants typically occupy exposures receiving full sunlight. The substrates are described as alkaline (e.g., pH: 8.2) and include granitic loam, gravel, gravelly silt, humus, limestone, loam, marl, muck, peat, rock, rocky loam, rocky sandy loam, sandy clay loam, sandy granitic loam, siliceous metamorphic gravel, and silt. Flowering commences in May with fruiting extending from June to September. The seed ecology is unknown. **Reported associates:** *Abies lasiocarpa, Achillea lanulosa, Aconitum columbianum, Agoseris aurantiaca, Agropyron trachycaulon, Agrostis, Allium schoenoprasum, Alnus viridis, Angelica arguta, Antennaria corymbosa, A. pulcherrima, Aquilegia flavescens, Artemisia cana, Astragalus alpinus, Athyrium alpestre, Aulacomnium palustre, Betula glandulosa, B. nana, B. occidentalis, Bistorta bistortoides, Bromopsis inermis, Calamagrostis canadensis, C. rubescens, Calliergon cuspidatum, Caltha leprosepala, Cardamine cordifolia, Carex aquatilis, C. aurea, C. buxbaumii, C. canescens, C. capillaris, C.*

cusickii, C. disperma, C. douglasii, C. gynocrates, C. illota, C. interior, C. lasiocarpa, C. livida, C. microptera, C. nebrascensis, C. nigricans, C. norvegica, C. praegracilis, C. rostrata, C. scopulorum, C. simulata, C. utriculata, C. vesicaria, C. viridula, Carum carvi, Castilleja cusickii, C. septentrionalis, Climacium dendroides, Cornus sericea, Crataegus rivularis, Cynoglossum, Danthonia intermedia, Dasiphora floribunda, Deschampsia cespitosa, Dugaldia hoopesii, Eleocharis acicularis, E. palustris, E. quinqueflora, E. rostellata, Epilobium ciliatum, E. lactiflorum, E. palustre, Equisetum arvense, E. laevigatum, Erigeron coulteri, E. peregrinus, Eriophorum angustifolium, E. chamissonis, Eurybia integrifolia, Festuca brachyphylla, F. idahoensis, F. ovina, F. thurberi, Floerkea proserpinacoides, Fragaria vesca, F. virginiana, Gaultheria humifusa, Gentianopsis detonsa, Geranium richardsonii, Geum macrophyllum, Graphephorum wolfii, Helenium hoopesii, Juncus balticus, J. nevadensis, Kalmia microphylla, Koeleria macrantha, Ligusticum canbyi, L. porteri, L. tenuifolium, Linnaea borealis, Lonicera involucarta, Lupinus holosericeus, L. polyphyllus, Mertensia campanulata, M. ciliata, Micranthes integrifolia, M. subapetala, Mimulus guttatus, Mnium, Moneses uniflora, Muhlenbergia richardsonis, Packera subnuda, Parnassia fimbriata, Pedicularis groenlandica, Phleum alpinum, Phyllodoce empetriformis, P. glanduliflora, Picea engelmannii, P. glauca, Pinus albicaulis, P. contorta, Poa alpina, P. fendleriana, P. palustris, P. pratensis, P. secunda, Platanthera dilatata, P. obtusata, Pohlia nutans, Polemonium caeruleum, P. occidentale, Polytrichum commune, P. juniperum, Potamogeton gramineus, Potentilla ×diversifolia, P. gracilis, Primula pauciflora, Prunus virginiana, Pyrola asarifolia, Ranunculus alismaefolius, Rhamnus alnifolia, Rhodiola rhodantha, Rhododendron columbianum, R. groenlandicum, Ribes, Rosa woodsii, Salix arctica, S. barclayi, S. bebbiana, S. boothii, S. brachycarpa, S. candida, S. drummondiana, S. eastwoodiae, S. farriae, S. geyeriana, S. ligulifolia, S. monticola, S. myrtillifolia, S. orestera, S. planifolia, S. pseudomonticola, S. pseudomyrsinites, S. tweedyi, Sambucus nigra, Schoenoplectus acutus, S. tabernaemontani, Senecio crocatus, S. hydrophiloides, S. pauperculus, S. pseudaureus, S. sphaerocephalus, S. triangularis, Sisyrinchium idahoense, Solidago multiradiata, Sphagnum, Stellaria longipes, Stipa columbiana, Swertia perennis, Symphoricarpos oreophilus, Symphyotrichum foliaceum, Taraxacum officinale, Thalictrum alpinum, T. fendleri, T. sparsiflorum, Trifolium longipes, T. repens, Tsuga, Vaccinium scoparium, Valeriana edulis, V. occidentalis, Veronica wormskjoldii, Vicia americana, Viola adunca, V. palustris, Wyethia.

Use by wildlife: *Salix arctica* is a principal food of arctic hares (Mammalia: Leporidae: *Lepus arcticus*), collared lemmings (Mammalia: Rodentia: Cricetidae: *Dicrostonyx*), and muskox (Mammalia: Bocvidae: *Ovibos moschatus*) and is also eaten by ptarmigan (Aves: Phasianidae: *Lagopus*) and reindeer (Mammalia: Cervidae: *Rangifer tarandus*). The plants provide nesting habitat for greater snow geese (Aves: Anatidae: *Chen caerulescens*). They also host several larval butterflies (Insecta: Lepidoptera: Lymantriidae: *Gynaephora groenlandica, G. rossii*; Nymphalidae: *Boloria chariclea, Clossiana chariclea, C.*

frigga, C. improba harryi, C. titania; Pieridae: *Colias hecla, C. nastes*) and wasps (Hymenoptera: Tenthredinidae: *Pontania atrata*). *Salix arctophila* provides important nesting habitat for the Lapland longspur (Aves: Calcariidae: *Calcarius lapponicus*). It is a host plant for several caterpillars (Insecta: Lepidoptera: Lymantriidae: *Gynaephora rossii*; Nymphalidae: *Clossiana chariclea, C. titania*) and jumping plant lice (Insecta: Hemiptera: Pseudococcidae: *Atrococcus groenlandensis*; Psyllidae: *Cacopsylla groenlandica*). *Salix arizonica* is grazed by beavers (Mammalia: Rodentia: Castoridae: *Castor canadensis*) and by domestic and wild ungulates (Mammalia: Artiodactyla) including cattle (Bovidae: *Bos taurus*), mule deer (Cervidae: *Odocoileus hemionus*), Rocky Mountain elk (Cervidae: *Cervus elaphus nelsoni*), and pronghorns (Antilocapridae: *Antilocapra americana*). The roots are clipped by voles (Mammalia: Rodentia: *Microtus*). The plants host several insects (Insecta) including beetles (Coleoptera), grasshoppers (Orthoptera), and larval caterpillars of the mourning cloak butterfly (Lepidoptera: Nymphalidae: *Nymphalis antiopa*). The foliage is susceptible to infection by rust fungi (Basidiomycota: Melampsoraceae: *Melampsora epitea*). The herbage of *S. barclayi* is eaten by caribou (Mammalia: Cervidae: *Rangifer tarandus*) and collared pikas (Mammalia: Ochotonidae: *Ochotona collaris*). The twigs contain from 0.9% to 1.4% nitrogen, are 47%–54% digestible, and are important winter forage for moose (Mammalia: Cervidae: *Alces alces*). The plants are host to larval cottonwood dagger moths (Insecta: Lepidoptera: Noctuidae: *Acronicta lepusculina*), and wasps (Insecta: Hymenoptera: Pteromalidae: *Hyperimerus pusillus*). They also provide habitat for damselflies (Insecta: Odonata: Lestidae: *Lestes disjunctus, L. forcipatus*). Galls can be induced on the shoots by midges (Insecta: Diptera: Cecidomyiidae: *Rabdophaga*). *Salix bebbiana* is an important food (up to 15% of their winter diet) for moose (Mammalia: Cervidae: *Alces alces*) and snowshoe hares (Mammalia: Leporidae: *Lepus americanus*) in subarctic communities. The leaves and twigs can approach 18% in protein content. The foliage is highly palatable to livestock and can represent more than 10% of the total summer forage of grazing cattle (Mammalia: Bovidae: *Bos taurus*). Beavers (Mammalia: Rodentia: Castoridae: *Castor canadensis*), grouse (Aves: Phasianidae), hares (Mammalia: Leporidae), and white-tailed deer (Mammalia: Cervidae: *Odocoileus virginianus*) rely on the buds, inner bark, and young shoots as a source of food. The plants are larval hosts to many butterflies and moths (Insecta: Lepidoptera: Coleophoridae: *Coleophora atlantica*; Geometridae: *Campaea perlata, Eupithecia miserulata, Mesothea incertata*; Gracillariidae: *Phyllonorycter salicifoliella, P. scudderella*; Hesperiidae: *Erynnis icelus*; Lymantriidae: *Orgyia antiqua*; Noctuidae: *Acronicta dactylina, A. noctivaga, Autographa ampla, Eueretagrotis perattentu, Lithomoia solidaginis, Lithophane amanda, L. fagina, Melanchra assimilis, Scoliopteryx libatrix, Spiramater lutra, Trichordestra legitima, Xestia oblata, Zale minerea*; Notodontidae: *Clostera inclusa, Furcula cinerea, F. occidentalis*; Nymphalidae: *Basilarchia arthemis, Nymphalis antiopa*; Oecophoridae: *Agonopterix argillacea*; Saturniidae: *Antheraea polyphemus, Hemileuca nevadensi*; Sphingidae:*

Smerinthus cerisyi, S. jamaicensis; Tortricidae: *Archips argyrospila, A. mortuana, A. myricana, Gypsonoma salicicolana*). They also host beetles (Insecta: Coleoptera: Chrysomelidae: *Calligrapha suturella*; Scolytidae: *Trypophloeus striatulus*) and some plant bugs (Insecta: Heteroptera: Miridae: *Psallus aethiops, Salignus tahoensis*). The foliage is susceptible to infection by the willow rust (Basidiomycota: Melampsoraceae: *Melampsora epitea*) and is also known to host other Fungi (Ascomycota: Erysiphaceae: *Phyllactinia guttata, Uncinula adunca*; Valsaceae: *Valsa boreella*; Basidiomycota: Exidiaceae: *Exidia*; Fomitopsidaceae: *Daedalea confragosa*; Polyporaceae: *Fomes conchatus, F. igniarius, Polyporus varius, Trametes pubescens*). Larvae of the pine-cone gall midge (Insecta: Diptera: Cecidomyiidae: *Rabdophaga strobiloides*) stimulate the production of cone galls in *S. bebbiana*, in which they feed and develop. Black-capped and Carolina chickadees (Aves: Paridae: *Poecile atricapillus, P. carolinensis*) excavate the plants to construct their nesting cavities. Both *S. bebbiana* and *S. candida* provide important habitat for the northern bog lemming (Mammalia: Cricetidae: *Synaptomys borealis*). *Salix boothii* provides nesting and feeding cover for yellow warblers (Aves: Parulidae: *Setophaga petechia*) and a number of other birds. The twigs and foliage are browsed by ungulate mammals (Mammalia: Artiodactyla) including antelope (Antilocapridae: *Antilocapra americana*), deer (Cervidae: *Odocoileus hemionus, O. virginianus*), elk (Cervidae: *Cervus elaphus*), moose (Cervidae: *Alces alces*), domestic horses (Equidae: *Equus ferus*), and sheep (Bovidae: *Ovis* aries). The plants are larval hosts of tiger swallowtail butterflies (Insecta: Lepidoptera: Papilionidae: *Papilio glaucus*). They are also host to rust fungi (Basidiomycota: Melampsoraceae: *Melampsora*). *Salix candida* provides important cover and nesting habitat for birds (Aves) such as the northern yellowthroat (Parulidae: *Geothlypis trichas*), oldsquaw (Anatidae: *Clangula hyemalis*), and yellow rail (Rallidae: *Coturnicops noveboracensis*). The plants contain up to 1.95% nitrogen and are eaten by meadow voles (Mammalia: Rodentia: Cricetidae: *Microtus pennsylvanicus*), but contain toxins and are low in their dietary preference. They are also grazed by domestic cattle (Mammalia: Bovidae: *Bos taurus*). This is a host plant for several insects (Insecta) including beetles (Coleoptera: Chrysomelidae: *Lexiphanes saponatus*), bugs (Hemiptera: Miridae: *Agnocoris rubicundus, Atractotomus rubidus*), and larval caterpillars (Insecta: Lepidoptera: Geometridae: *Cingilia catenaria*; Nymphalidae: *Nymphalis antiopa*; Saturniidae: *Hemileuca lucina*). Although it is nonindigenous to North America, *Salix caprea* has become a larval host plant for numerous butterflies and moths (Insecta: Lepidoptera: Coleophoridae: *Coleophora albidella, C. atlantica*; Geometridae: *Hydriomena nubilofasciata*; Saturniidae: *Actias luna, Adeloneivaia subangulata, Automeris io, Eacles imperialis, Hemileuca maia, H. nevadensis, Hyalophora columbia, H. euryalus, Syssphinx molina, S. petersii*; Sesiidae: *Synanthedon culiciformis*; Yponomeutidae: *Argyresthia pygmaeella*). European plants of *S. caprea* ("goat willow") are known to carry the pathogen responsible for "sudden oak death" disease (Heterokontophyta: Oomycota: Pythiaceae: *Phytophthora ramorum*). The early spring twigs supposedly

are a favorite food of goats (Mammalia: Bovidae: *Capra hircus*). The plants are also eaten by moose (Mammalia: Cervidae: *Alces alces*), whose saliva has been found to stimulate the branching of browsed saplings. *Salix caroliniana* plants are eaten by deer (Mammalia: Cervidae: *Odocoileus*) and attract butterflies (Insecta: Lepidoptera) to their nectariferous flowers. They provide nesting substrate for a variety of birds (Aves) including the anhinga (Anhingidae: *Anhinga anhinga*), cattle egret (Ardeidae: *Bubulcus ibis*), little blue heron (Ardeidae: *Egretta caerulea*), roseate spoonbill (Threskiornithidae: *Platalea ajaja*), snail kite (Accipitridae: *Rostrhamus sociabilis plumbeus*), snowy egret (Ardeidae: *Egretta thula*), tricolored heron (Ardeidae: *Egretta tricolor*), white ibis (Threskiornithidae: *Eudocimus albus*), and wood stork (Ciconiidae: *Mycteria americana*). Birds and small mammals eat the flowers and fruits. The foliage is eaten by beetles (Insecta: Chrysomelidae: *Chrysomela scripta*) and the plants are hosts for numerous larval butterflies and moths (Insecta: Lepidoptera: Arctiidae: *Hyphantria cunea*; Lycaenidae: *Mitoura hesseli*; Lymantriidae: *Lymantria dispar* [gypsy moth], *Orgyia antiqua*; Noctuidae: *Homoglaea carbonaria, Simyra insularis*; Notodontidae: *Cerura multiscripta*; Nymphalidae: *Basilarchia archippus, Limenitis archippus, L. arthemis, Nymphalis antiopa, Polygonia comma*; Papilionidae: *Papilio glaucus*; Pyralidae: *Pyrausta*; Saturniidae: *Automeris io*) as well as lobate lac scale (Insecta: Hemiptera: Kerriidae: *Paratachardina lobata*). The flowers attract bees (Insecta: Hymenoptera: Apidae: *Apis mellifera*; Colletidae: *Hylaeus confluens*; Halictidae: *Augochlora pura mosieri, Lasioglossum tarponense, L. tegulare*) as pollinators. Several fungi (Basidiomycota: Cortinariaceae: *Cortinarius*; Hydnangiaceae: *Laccaria montana*) form an ectomycorrhizal association with the roots of *S. commutata*. The plants are susceptible to willow rust Fungi (Basidiomycota: Melampsoraceae: *Melampsora epitea*) and are associated with several "hymenomycete" fungi (Basidiomycota: Ganodermataceae: *Ganoderma lucidium*; Hydnangiaceae: *Laccaria proxima*; Hymenochaetaceae: *Fuscoporia gilva*; Lachnocladiaceae: *Asterostroma cervicolor*; Pleurotaceae: *Pleurotus ostreatus*; Polyporaceae: *Daedaleopsis confragosa, Lenzites elegans, Polyporus variabilis, Trichaptum biforme*; Strophariaceae: *Galerina autumnalis*). The species also harbors several insects (Insecta) including beetles (Coleoptera: Chrysomelidae), flies (Diptera: Cecidomyiidae: *Rabdophaga rigidae, R. rosaria*), and sawflies (Insecta: Hymenoptera: Tenthredinidae: *Phyllocolpa, Pontania*). The twigs of *S. commutata* are browsed by moose (Mammalia: Cervidae: *Alces alces*) and the foliage provides habitat for several rodents (Mammalia: Rodentia) including deer mice (Cricetidae: *Peromyscus maniculatus*), ground squirrels (Sciuridae: *Urocitellus columbianus*), and water voles (Cricetidae: *Microtus richardsoni*). The branches are drilled by red-breasted and yellow-bellied sapsuckers (Aves: Picidae: *Sphyrapicus ruber, S. varius*) for their sap, which contains from 13.4% to 16.2% sugars. The plants are a host of the willow rust (Fungi: Basidiomycota: Melampsoraceae: *Melampsora epitea*). *Salix drummondii* is palatable to sheep (Mammalia: Bovidae: *Ovis aries*) and cattle (Mammalia: Bovidae: *Bos taurus*) and is eaten in quantity by moose (Mammalia: Cervidae: *Alces alces*) during the winter months. The buds, catkins, leaves, and shoots are eaten by birds (Aves), especially ducks (Anatidae) and grouse (Phasianidae), as well as by deer (Mammalia: Cervidae: *Odocoileus*), elk (Cervidae: *Cervus elaphus*), and small mammals. Red-naped sapsuckers (Aves: Picidae: *Sphyrapicus nuchalis*) feed on the sap of *S. drummondiana* from wells drilled in their stems. The wells are also used by other birds such as hummingbirds (Trochilidae) and warblers (Phylloscopidae), and by chipmunks (Mammalia: Sciuridae: *Tamias*) and red squirrels (Sciuridae: *Tamiasciurus*) as a source of food. The plants are a larval host for butterflies (Insecta: Lepidoptera: Nymphalidae *Limenitis weidemeyerii*). They provide cover for moose (Mammalia: Cervidae: *Alces alces*) and nesting and foraging sites for various birds (Aves) including ducks (Anatidae), shore birds (Charadriiformes), sparrows (Emberizidae), vireos (Vireonidae: *Vireo*), and warblers (Parulidae). They comprise important habitat for the northern bog lemming (Mammalia: Rodentia: Cricetidae: *Synaptomys borealis*). The twigs are used as a building material by beavers (Mammalia: Castoridae: *Castor canadensis*). The overhanging branches of *S. drummondiana* provide shade for salmonid fish (Osteichthyes: Salmonidae). The pollen is an important food source for honeybees (Insecta: Hymenoptera: Apidae) in the spring. *Salix eastwoodii* is grazed on by elk (Cervidae: *Cervus elaphus*) and other ungulates (Mammalia: Artiodactyla). It is a host plant of leafhoppers (Insecta: Hemiptera: Psyllidae: *Cacopsylla curta*; *C. minor*) and leafminer flies (Insecta: Diptera: Agromyzidae: *Melanagromyza buccalis*). The twigs and foliage of *S. eriocephala* are eaten by herbivorous mammals (Mammalia) including rabbits (Leporidae: *Sylvilagus*), voles (Rodentia: Cricetidae: *Microtus pennsylvanicus*), and white-tailed deer (Cervidae: *Odocoileus virginianus*). It is a host plant for larval moths (Insecta: Lepidoptera: Noctuidae: *Autographa ampla*, *Catocala relicta*, *Melanchra assimilis*, *M. assimilis*, *Spiramater lutra*; Notodontidae: *Furcula modesta*). The plants are also susceptible to attack by rust fungi (Basidiomycota: Melampsoraceae: *Melampsora*), leaf beetles (Insecta: Coleoptera: Chrysomelidae: *Chrysomela knabi*; Scarabaeidae: *Popillia japonica*), and leaf-folding sawflies (Insecta: Hymenoptera: Tenthredinidae: *Phyllocolpa nigrita*). The seedlings can be damaged heavily by slugs (Mollusca: Gastropoda: Arionidae: *Arion subfuscus*). The plants also produce galls, which harbor willow gall midges (Diptera: Cecidomyiidae: *Mayetiola rigidae*, *Rhabdophaga strobiloides*). The flowers attract numerous insects (Insecta) including bees (Hymenoptera: Andrenidae: *Andrena algida*, *A. carlini*, *A. clarkella*, *A. cressonii*, *A. dunningi*, *A. frigida*, *A. hippotes*, *A. vicina*) and flies (Diptera: Syrphidae: *Dasysyrphus laticaudus*, *Eristalis dimidiata*, *Sphaerophoria contigua*, *Syrphus vitripennis*), which also function as pollinators. *Salix exigua* is eaten by several mammals (Mammalia) including beaver (Castoridae: *Castor canadensis*), elk (Cervidae: *Cervus elaphus*), moose (Cervidae: *Alces alces*), mule deer (Cervidae: *Odocoileus hemionus*), and white-tailed deer (Cervidae: *Odocoileus virginianus*); the leaves and bark are also palatable to domestic cattle (Bovidae: *Bos taurus*), horses (Equidae: *Equus ferus*), and sheep (Bovidae: *Ovis aries*), more so later in

the growing season. The plants host many butterfly and moth larvae (Insecta: Lepidoptera: Gelechiidae: *Aristotelia fungivorella*, *A. salicifungiella*); Gracillariidae: *Caloptilia stigmatella*, *Micrurapteryx salicifoliella*, *Phyllonorycter salicifoliella*; Lasiocampidae: *Malacosoma disstria*; Lycaenidae: *Satyrium acadicum*, *S. sylvinum*; Nymphalidae: *Basilarchia archippus*, *B. weidemeyerii*, *Nymphalis antiopa*; Papilionidae: *Papilio rutulus*; Saturniidae: *Antheraea polyphemus*, *Automeris io*, *Hemileuca eglanterina*, *H. maia*, *H. nevadensis*, *Hyalophora cecropia*, *H. columbia*, *H. euryalus*, *H. kasloensis*, *Rothschildia lebeau*, *Saturnia homogena*, *S. mendocino*; Sphingidae: *Hyles lineata*). *Salix exigua* is also attacked by leaf mining beetles (Insecta: Coleoptera: Chrysomelidae), sawflies (Insecta: Hymenoptera: Tenthredinidae: *Euura exiguae*), and a bud galling midge (Insecta: Diptera: Cecidomyiidae: *Rabdophaga*), which damages the terminal and lateral shoots. The plants are associated with several fungi (Ascomycota: Dermateaceae: *Marssonina kriegeriana*; Erysiphaceae: *Uncinula adunca*; Gnomoniaceae: *Discula brenckleana*; Basidiomycota: Melampsoraceae: *Melampsora epitea* [willow rust]; Pleurotaceae: *Pleurotus ostreatus*). Thickets of *S. exigua* provide cover for waterfowl (Aves: Anatidae), especially during the winter. Plants are a frequent nesting site for birds (Aves) including Forster's tern (Laridae: *Sterna forsteri*), yellow warblers (Parulidae: *Setophaga petechia*), and two endangered species: the Least Bell's Vireo (Vireonidae: *Vireo bellii pusillus*) and southwestern willow flycatcher (Tyrannidae: *Empidonax traillii extimus*). *Salix geyeriana* is more palatable to livestock than several other willows and can represent more than 11% of the summer browse eaten by cattle (Mammalia: Bovidae: *Bos taurus*). The stems contain roughly 6.8% protein. The plants are also eaten by other mammals (Mammalia) including beavers (Rodentia: Castoridae: *Castor canadensis*), elk (Cervidae: *Cervus elaphus*), and moose (Cervidae: *Alces alces*); the beavers also use it as a building material for their lodges. The foliage hosts the willow rust (Fungi: Basidiomycota: Melampsoraceae: *Melampsora epitea*). The widely spaced clumps provide good habitat for deer (Mammalia: Cervidae: *Odocoileus*) as well as foraging, nesting, and roosting habitat for assorted avifauna (Aves) such as blackbirds (Icteridae), ducks (Anatidae), the Merrill song sparrow (Emberizidae: *Melospiza melodia merrilli*), various shorebirds (Charadriiformes), the southwestern willow flycatcher (Tyrannidae: *Empidonax traillii extimus*), vireos (Vireonidae: *Vireo*), and warblers (Phylloscopidae). The stems are injured by feeding red-naped sapsuckers (Aves: Picidae: *Sphyrapicus nuchalis*), which create entrance wounds that facilitate infection by a potentially lethal pathogen (Fungi: Ascomycota: Sordariomycetidae: *Valsa sordida*). *Salix gooddingii* creates shade for fish (Vertebrata: Osteichthyes) and other animals and provides habitat for yellow-billed cuckoo (Aves: Cuculidae: *Coccyzus americanus*) and the endangered southwestern willow flycatcher (Aves: Tyrannidae: *Empidonax traillii extimus*). The leaves are eaten by deer (Mammalia: Cervidae) and the bark, buds, and catkins offer food for prairie fowl (Aves: Phasianidae: *Tympanuchus*) and small mammals (Mammalia: Rodentia). *Salix interior* is used as a nesting site by yellow warblers (Aves: Parulidae: *Setophaga petechia*) and provides habitat for clay-colored sparrows (Aves: Emberizidae: *Spizella pallida*). It is also important habitat for beavers (Mammalia: Castoridae: *Castor canadensis*) and a host plant of bagworm moths (Insecta: Lepidoptera: Psychidae: *Thyridopteryx ephemeraeformis*). The flowers are visited by long-horned beetles (Insecta: Coleoptera: Cerambycidae: *Analeptura lineola, Strangalepta abbreviata, S. famelica, Typocerus velutinus*), which feed on the pollen. They are pollinated by nectar-seeking insects (Insecta), which mainly include bees (Hymenoptera: Andrenidae: *Andrena carlini, A. clarkella, A. crataegi, A. cressonii, A. dunningi, A. frigida, A. hippotes, A. vicina*; Halictidae: *Agapostemon melliventris, A. nasutus*), and flies (Diptera: Syrphidae: *Dasysyrphus laticaudus, Eristalis dimidiata, Parasyrphus genualis, P. semiinterruptus, P. nigritarsus, Syrphus ribesii, S. torvus, S. vitripennis*). The foliage is eaten by leaf beetles (Insecta: Coleoptera: Chrysomelidae: *Neogalerucella calmariensis*), which have been introduced as biological control agents for the invasive purple loosestrife (*Lythrum salicaria*). Other leaf beetles (Chrysomelidae: *Plagiodera versicolora*) lay their eggs on the plants. The foliage is parasitized by rust fungi (Basidiomycota: Melampsoraceae: *Melampsora epitea*). *Salix jepsonii* is a larval host plant for mourning cloak butterflies (Insecta: Lepidoptera: Nymphalidae: *Nymphalis antiopa*). *Salix lasiolepis* is attacked by several sawflies (Insecta: Hymenoptera: Tenthredinidae) including foliage feeders (*Nematus*) and gall formers (*Euura lasiolepis, Pontania pacifica, Phyllocolpa excavata*). It is a larval host plant for many butterflies and moths (Insecta: Lepidoptera: Batrachedridae: *Batrachedra salicipomonella, Batrachedra striolata*; Coleophoridae: *Coleophora*; Gelechiidae: *Coleotechnites gallicola*; Geometridae: *Hydriomena quinquefasciata*; Gracillariidae: *Caloptilia palustriella, Marmara salictella, Micrurapteryx, Phyllonorycter apicinigrella*; Heliozelidae: *Coptodisca saliciella*; Lycaenidae: *Satyrium sylvinum*; Nepticulidae: *Stigmella*; Noctuidae: *Homoglaea dives*; Notodontidae: *Clostera apicalis*; Tortricidae: *Acleris hastiana, A. senescens, Epinotia columbia*; Nymphalidae: *Basilarchia lorquini*; Oecophoridae: *Agonopterix argillacea*; Papilionidae: *Papilio rutulus*; Saturniidae: *Hemileuca nevadensis*; Tortricidae: *Acleris hastiana, A. senescens, Archips argyrospila, Argyrotaenia franciscana, Choristoneura rosaceana, Epinotia columbia, E. keiferana, Pandemis pyrusana*; Yponomeutidae: *Argyresthia*). Observed insect pollinators include several flies (Diptera: Calliphoridae; Syrphidae; Tachinidae) as well as bees (Hymenoptera: Apidae: *Bombus*; Andrenidae: *Andrena striatifrons, A. frigida*; Halictidae: *Agapostemon, Dialictus perdificilis, D. pruiniformis, D. ruidosensis, Sphecodes*). Stands of *S. lasiolepis* provide cover for bighorn sheep (Mammalia: Bovidae: *Ovis canadensis*) as well as many birds (Aves) such as Bell's vireo (Virionidae: *Vireo bellii*), southwestern willow flycatcher (Tyrannidae: *Empidonax traillii*), summer tanager (Cardinalidae: *Piranga rubra*), yellow-breasted chat (Parulidae: *Icteria virens*), and yellow warbler (Parulidae: *Setophaga petechia*). This species forms part of the critical habitat for the endangered unarmored three-spine stickleback (Osteichthyes: Gasterosteidae: *Gasterosteus*

aculeatus). The twigs of *S. lasiolepis* and *S. lemmonii* are eaten by deer (Mammalia: Cervidae: *Odocoileus*) and elk (Mammalia: Cervidae: *Cervus*). *Salix lasiandra* is palatable but not widely eaten by cattle (Mammalia: Bovidae: *Bos taurus*). The twigs provide browse for mule deer (Mammalia: Cervidae: *Odocoileus hemionus*), elk (Mammalia: Cervidae: *Cervus*), and are a major food of beaver (Mammalia: Castoridae: *Castor canadensis*), especially during the winter months. The plants provide cover for various birds (Aves) such as the belted kingfisher (Alcedinidae: *Megaceryle alcyon*), great blue heron (Ardeidae: *Ardea herodias*), green heron (Ardeidae: *Butorides virescens*), osprey (Pandionidae: *Pandion haliaetus*), shorebirds (Charadriiformes), waterfowl (Anatidae), and other wildlife. They are used as nesting sites by green herons (Ardeidae: *Butorides virescens*) and the endangered least Bell's vireo (Vireonidae: *Vireo bellii pusillus*). The foliage is damaged by a number of insects (Insecta) including beetles (Chrysomelidae), which skeletonize the leaves, sawflies (Hymenoptera: Tenthredinidae), willow midge larvae (Diptera: Tephritidae: *Bactrocera*), and gypsy moths (Lepidoptera: Erebidae: *Lymantria dispar*). The flowers attract butterflies (Insecta: Lepidoptera) and the plants are a larval host for the western tiger swallowtail (Papilionidae: *Papilio rutulus*). They are also known to host many fungi (Ascomycota: Incertae sedis: *Myxofusicoccum salicis, Septogloeum salicis-fendlerianae*; Elsinoaceae: *Sphaceloma murrayae*; Erysiphaceae: *Uncinula adunca*; Rhytismataceae: *Marthamyces phacidioides, Rhytisma salicinum*; Valsaceae: *Valsa sordida*; Basidiomycota: Melampsoraceae: *Melampsora epitea*; Polyporaceae: *Bjerkandera adusta, Coriolopsis gallica, Fuscoporia gilva, Hapalopilus rutilans, Phellinus conchatus, P. ferreus, P. igniarius, Trametes pubescens, T. versicolor, Trichaptum biforme*; Stereaceae: *Aleurodiscus helveolus, Stereum hirsutum*). *Salix lasiandra* is also a host of *Xylella fastidiosa* (Bacteria: Xanthomonadaceae), an infectious blight of several economically important plants. *Salix lemmonii* is palatable to cattle (Mammalia: Bovidae: *Bos taurus*) and sheep (Mammalia: Bovidae: *Ovis aries*) and often is overgrazed by them. The plants provide cover for songbirds (Aves: Passeriformes) and shade for coldwater fish (Vertebrata: Osteichthyes) but can be damaged by various mammals (Mammalia) including beavers (Castoridae: *Castor canadensis*), mice (Rodentia: Muridae), muskrats (Cricetidae: *Ondatra zibethicus*), and voles (Cricetidae: *Microtus*). *Salix lemmonii* is also susceptible to damage by poplar and willow borers (Insecta: Coleoptera: Curculionidae: *Cryptorhynchus lapathi*). The plants are used as nesting sites by the willow flycatcher (Aves: Tyrannidae: *Empidonax traillii brewsteri*) and for cover by white-crowned sparrows (Aves: Emberizidae: *Zonotrichia leucophrys*). *Salix lutea* is a larval host for several butterflies (Insecta: Lepidoptera: Nymphalidae: *Limenitis lorquini, Nymphalis antiopa*; Pieridae: *Colias gigantea*), and beetles (Insecta: Chrysomelidae: *Chrysomela aeneicollis*). The twigs and foliage provide browse for elk (Mammalia: Cervidae: *Cervus*) and moose (Mammalia: Cervidae: *Alces alces*). Red-naped and red-breasted sapsuckers (Aves: Picidae: *Sphyrapicus nuchalis, S. ruber*) feed from holes drilled in the twigs. *Salix melanopsis* is preferred as a nesting site by the song sparrow (Aves: Emberizidae: *Melospiza melodia*). It is grazed by Rocky Mountain elk (Mammalia: Cervidae: *Cervus elaphus*) and is a host plant of leaf beetles (Insecta: Chrysomelidae: *Calligrapha multipunctata*) and weevils (Insecta: Coleoptera: Curculionidae: *Cryptorhynchus lapathi*). *Salix monticola* is browsed by beaver (Mammalia: Castoridae: *Castor canadensis*), deer (Mammalia: Cervidae: *Odocoileus*), elk (Mammalia: Cervidae: *Cervus*), moose (Mammalia: Cervidae: *Alces alces*), snowshoe hare (Mammalia: Leporidae: *Lepus americanus*), and other small mammals (Rodentia). It is also eaten by various songbirds (Aves: Passeriformes), ruffed grouse (Aves: Phasianidae: *Bonasa umbellus*), and ptarmigan (Aves: Phasianidae: *Lagopus*). The dense growth of plants furnishes cover for mammals (Mammalia) and assorted birds (Aves) including the cordilleran flycatcher (Tyrannidae: *Empidonax occidentalis*), house wren (Troglodytidae: *Troglodytes aedon*), Lincoln sparrow (Emberizidae: *Melospiza lincolnii*), song sparrow (Emberizidae: *Melospiza melodia*), and yellow warbler (Parulidae: *Setophaga petechia*). Dense thickets of plants provide shade for fish such as the Gila trout (Osteichthyes: Salmonidae: *Oncorhynchus gilae*) when growing along banks. Beavers (Mammalia: Castoridae: *Castor canadensis*) use the twigs to build dams and lodges. Honey bees (Insecta: Apidae: *Apis mellifera*) eat the pollen and nectar. The plants are palatable and are eaten readily by cattle (Mammalia: Bovidae: *Bos taurus*) and sheep (Mammalia: Bovidae: *Ovis aries*), especially in riparian areas. This species is a host of the willow rust (Fungi: Basidiomycota: Melampsoraceae: *Melampsora epitea*). *Salix myrtillifolia* foliage contains about 6.4% protein and is browsed heavily by moose (Mammalia: Cervidae: *Alces alces*) in some areas. It is a host for aphids (Insecta: Hemiptera: Aphididae) and willow rust (Fungi: Basidiomycota: Melampsoraceae: *Melampsora epitea*). *Salix nigra* is fairly palatable to livestock (Mammalia: Bovidae). The leaves and twigs are eaten by deer (Mammalia: Cervidae: *Odocoileus*), beavers (Mammalia: Castoridae: *Castor canadensis*), and rabbits (Mammalia: Leporidae: *Lepus*); the bark, buds, and twigs are eaten by snowshoe hares (Mammalia: Leporidae: *Lepus americanus*). It is not considered to be an important food of elk (Mammalia: Cervidae: *Cervus elaphus*), moose (Mammalia: Cervidae: *Alces alces*), or waterfowl (Aves: Anatidae). The sap is consumed by yellow-bellied sapsuckers (Aves: Picidae: *Sphyrapicus varius*) and the seed by various sparrows (Aves: Emberizidae). The plants offer suitable cover for many bird species (Aves), especially along watercourse margins. Bees (Hymenoptera), butterflies (Lepidoptera), and other insects (Insecta) feed on the nectar. The plants host numerous butterfly and moth larvae (Lepidoptera: Hesperiidae: *Erynnis icelus*; Lasiocampidae: *Malacosoma disstria*; Lycaenidae: *Satyrium acadicum, S. liparops*; Lymantriidae: *Lymantria dispar* [gypsy moth]; Noctuidae: *Catocala cara, C. parta*; Notodontidae: *Notodonta scitipennis*; Nymphalidae: *Basilarchia archippus, B. astyanax, Nymphalis antiopa, N. vau-album*; Papilionidae: *Papilio glaucus*; Saturniidae: *Automeris io, Hyalophora cecropia, H. columbia*). Other hosted Insecta include beetles (Coleoptera: Cerambycidae: *Leptura emarginata, Oberea ferruginea, Plectrodera scalator*;

Chrysomelidae: *Chrysomela scripta, Disonycha pluriligata, Plagiodera versicolora*), bugs (Hemiptera: Aphididae: *Fullawaya flocculosa, Pterocomma smithiae*; Diaspididae: *Chionaspis salicis-nigrae, Hemiberlesia lataniae, Melanaspis tenebricosa*; Miridae: *Ceratocapsus fuscinus*), and willow sawflies (Hymenoptera: Tenthredinidae: *Nematus ventralis*). Hosted pathogenic fungi include bleeding canker (Oomycota: Pythiaceae: *Phytophthora cactorum*), canker (Ascomycota: Valsaceae: *Sphaeria chrysosperma*), and leaf blight (Ascomycota: Incertae sedis: *Dendrophoma caespitosa*). The seedlings are afflicted by black canker (Ascomycota: Venturiaceae: *Pollaccia saliciperda*) and leaf rust (Basidiomycota: Melampsoraceae: *Melampsora epitea*). The branches are sometimes colonized by parasitic mistletoes (Santalaceae: *Phoradendron*). *Salix pedicellaris* provides habitat for the willow flycatcher (Aves: Tyrannidae: *Empidonax traillii*) and yellow warbler (Aves: Parulidae: *Setophaga petechia*). It is a host plant for several moth larvae (Insecta: Lepidoptera: Saturniidae: *Hemileuca lucina, H. nevadensis*) and leaf rust (Fungi: Basidiomycota: Melampsoraceae: *Melampsora epitea*). *Salix petiolaris* is browsed heavily by deer (Mammalia: Cervidae: *Odocoileus*) and muskrats (Mammalia: Cricetidae: *Ondatra zibethicus*) and provides nesting and feeding habitat for numerous birds (Aves) including the common yellowthroat (Parulidae: *Geothlypis trichas*), yellow warbler (Parulidae: *Setophaga petechia*), willow flycatcher (Tyrannidae: *Empidonax traillii*), and American woodcock (Scolopacidae: *Scolopax minor*). It is a host plant for many butterfly and moth larvae (Insecta: Lepidoptera: Geometridae: *Anagoga occiduaria*; Hepialidae: *Sthenopis thule*; Lycaenidae: *Satyrium acadicum*; Noctuidae: *Acronicta grisea, Anaplectoides pressus, Cerastis tenebrifera, Lacanobia radix, Morrisonia confusa, Polia nimbosa, Pyreferra citrombra, Scoliopteryx libatrix, Zale minerea*; Notodontidae: *Furcula occidentalis, Pheosia rimosa, Schizura unicornis*; Nymphalidae: *Chlosyne nycteis*; Saturniidae: *Hemileuca lucina, Hemileuca nevadensis*). *Salix planifolia* is palatable to cattle (Mammalia: Bovidae: *Bos taurus*) and can represent up to 5% of their summer browse. Beavers (Mammalia: Castoridae: *Castor canadensis*) are important for maintaining the high water tables vital for this species; however, overgrazing by beavers (or cattle) can result in the severe stunting of plants. This willow is an extremely important forage plant for moose (Mammalia: Cervidae: *Alces alces*), but is used less by elk (Mammalia: Cervidae: *Cervus elaphus*) and mule deer (Mammalia: Cervidae: *Odocoileus hemionus*). The plants provide excellent nesting and foraging habitat for various birds (Aves) including ducks (Anatidae), sandhill cranes (Gruidae: *Grus canadensis*), shorebirds (Charadriiformes), sparrows (Emberizidae), vireos (Vireonidae: *Vireo*), and yellow warblers (Parulidae: *Setophaga petechia*). The branches overhanging along streambanks provide cover and shade for salmonid fish (Osteichthyes: Salmonidae). The plants are hosts for larval sulfur butterflies (Insecta: Lepidoptera: Pieridae: *Colias scudderii*). The wildlife value of *S. prolixa* is unknown. *Salix pyrifolia* is a larval host plant of buckmoths (Insecta: Lepidoptera: Saturniidae: *Hemileuca*). It is also host of sac fungi (Ascomycota: Mycosphaerellaceae: *Ramularia rosea*). *Salix sericea* is a larval host plant for various butterflies and moths (Insecta: Lepidoptera: Gracillariidae: *Caloptilia, Phyllonorycter salicifoliella, Phyllocnistis*; Lycaenidae: *Satyrium acadicum*; Nymphalidae: *Basilarchia archippus, Nymphalis antiopa*; Sphingidae: *Smerinthus cerisyi*). The flowers are visited by bees (e.g., Insecta: Hymenoptera: Andrenidae: *Andrena salictaria*; Apidae: *Apis mellifera*), which function as primary pollinators. The small branches and shoots are eaten during winter by cottontail rabbits (Mammalia: Leporidae: *Sylvilagus*). Herbivorous insects (Insecta) include beetles (Coleoptera: Chrysomelidae: *Chrysomela knabi, Plagiodera versicolora*), gall-forming flies (Diptera: Cecidomyiidae: *Iteomyia salicifolius, Rhabdophaga rigidae, R. salicibrassicoides*), leaf-folding moths (Lepidoptera: Gracillariidae: *Caloptilia*), leaf-mining moths (Lepidoptera: Gracillariidae: *Phyllocnistis, Phyllonorycter salicifoliella*), mites (Arachnida: Acari: Eriophyidae: *Aculops tetanothrix*), and sawflies (Hymenoptera: Tenthredinidae: *Phyllocolpa eleanorae, P. nigrita, Pontania terminalis*). The flowers and fruits are eaten by birds (Aves). The foliage contains low amounts of condensed tannins but high concentrations of the phenolic glycoside salicortin (>10% dry leaf weight), which deters some herbivores (e.g., Insecta: Coleoptera: Chrysomelidae: *Calligrapha multipunctata*; Scarabaeidae: *Popillia japonica*; Mollusca: Gastropoda: Arionidae: *Arion subfuscus*) except those capable of metabolically processing the compounds (e.g., Insecta: Coleoptera: Chrysomelidae: *Chrysomela knabi, C. scripta*). The plants also host some sac fungi (Ascomycota: Dermateaceae: *Septogloeum salicinum*). *Salix serissima* is known to be a host of fungal leaf rust (*Melampsora ribesii-pupureae*). *Salix viminalis* is a larval food plant of silkmoths (Insecta: Lepidoptera: Saturniidae: *Hyalophora cecropia, H. euryalus*). It is also a host of various sac fungi (Ascomycota: Diatrypaceae: *Cryptosphaeria ligniota*; Nectriaceae: *Nectria cinnabarina*; Rhytismataceae: *Cryptomyces maximus*; Valsaceae: *Valsa salicina*). When grown in cultivated plantations in Quebec, *S. viminalis* clones exhibited high vulnerability to damge by insects such as the potato leafhopper (Insecta: Hemiptera: Cicadellidae: *Empoasca fabae*) and infections from up to 30 species of pathogenic fungi. *Salix wolfii* provides important nesting habitat for songbirds (Aves: Passeriformes) including the Lincoln's sparrow (Emberizidae: *Melospiza lincolnii*) and yellow warbler (Parulidae: *Setophaga petechia*). The foliage is palatable to livestock, but is less preferred than other willows. It is browsed by cattle (Mammalia: Bovidae: *Bos taurus*), especially in late summer, and by moose (Mammalia: Cervidae: *Alces alces*) during the early winter months.

Economic importance: food: The tender, early spring shoots (and young underground shoots) of *S. arctica* are edible once the outer bark has been removed. The leaves are a rich source of vitamin C, and a tea substitute (known to the Asian Yakuts as "chai-talak") can be brewed from them. *Salix arctophila* is also edible. Young twigs of *S. caroliniana* can also be brewed into tea. The inner bark and leaves of *S. commutata* are edible but are quite bitter and not highly palatable. The bark can be ground and added to flour for use in baking. The Navajo made a juice-like drink from the leaves of *S. exigua* and the

Kawaiisu produced a sticky, sweet candy-like substance from the plants. A sweet honey-dew obtained from cut branches of *S. gooddingii* was eaten by the Cocopa and Yuma tribes. The Mohave and Yuma people brewed tea from the young shoots and bark of this plant. The Yuma also ate the bark cooked or raw. The Pima ate the raw catkins of *S. gooddingii*; **medicinal:** Willows are extremely important medicinally. Their usage in traditional medicine for more than 2000 years led eventually to the discovery and isolation of the active ingredient salicin (i.e., salicylic acid; so named for its willow source) from willow bark in 1826. This discovery subsequently provided the basis for synthesis of modern aspirin (acetylsalicylic acid), which is a buffered form that reduces stomach irritation. Although not used medicinally, *S. arctophila* exhibits higher antioxidant potential than many of the willow species used as medicinal plants. In North America, the Cree and Okanagan-Colville tribes applied a poultice of the inner root bark and sap of *S. bebbiana* to treat broken bones and to heal severe cuts. The Okanagan-Colville people administered a decoction prepared from the branches to improve circulation in women recovering from childbirth. The inner bark of *S. candida* was used medicinally by the Meskwaki, Ojibwa, and South tribes to treat coughs, fainting, stomach ailments, and trembles. Extracts of *S. caprea* are effective antioxidants and prevent phorbol ester-induced tumor promotion. The inner bark of *S. caroliniana* has been used as an aspirin substitute and the bark and leaves as a malaria preventative, anti-inflammatory, and treatment for arthritis and rheumatism. The Houma and Seminole used various decoctions of the bark and roots as an emetic and to treat assorted aches, diarrhea, fever, and wounds. Although mainly insect pollinated, the pollen of *S. caroliniana* is highly allergenic once it becomes wind borne. The bark of *S. exigua* was used by the Montana tribe to treat fevers. The Navajo and Ramah used its leaves to make a ceremonial emetic. The Kashaya, Pomo, and Zuni used bark and leaf decoctions of *S. exigua* to treat coughs and sore throats. A decoction made from leaves of *S. gooddingii* was used by the Pima to relieve fevers. Foliar decoctions of *S. lasiandra* were used by the Pomo and Kashaya to treat coughs and colds and by the Navajo and Ramah as a ceremonial emetic. The Bella Coola administered the charred twigs to treat diarrhea and placed strips of bark in wounds to facilitate healing. The Okanagan-Colville tribes used a decoction of branch tips for a soak in which to relieve foot and leg cramps. Infusions made from the bark and leaves of *S. lasiolepis* were used by the Costanoan, Mendocino, and Mewuk tribes to treat ailments such as colds, diarrhea, fever, itching, and measles. The Montana people used the bark of *S. melanopsis* to treat fevers. The bark of *S. nigra* is effective at treating headaches, backpain, and osteoarthritis. It is used in cosmetics and is also a common ingredient of herbal remedies allegedly effective for treating a variety of ailments including ankylosing spondylitis, bursitis, fever, flu, mild diarrhea, menstrual pain, nocturnal enuresis, nonmalignant prostate disorders, poison ivy, sexual dysfunction, sores, and tendonitis. The bark and roots of *S. nigra* were widely employed by native Americans to treat coughs (Iroquois), diarrhea (Cherokee), fever (Cherokee, Houma, Koasati), headaches (Koasati), indigestion (Iroquois, Koasati), throat disorders (Cherokee, Iroquois), and as a hair tonic (Cherokee). Poultices made from the leaves or bark commonly were applied to bruises, sprains and broken bones (Cherokee, Micmac). The Ojibwa used the bark of *S. pedicellaris* to treat stomach disorders. *Salix sericea* was used by the Iroquois to treat mouth and throat infections. An extract of *S. wolfii* is sold as a natural remedy, but its efficacy has not been determined; **cultivation:** Several willows are widely cultivated for use in rock gardens including *S. arctica* var. *petraea*, *S. arctophila* 'Thunder Mountain,' *S. myrtillifolia*, and *S. pedicellaris*. Other cultivated or commercially available willows include (along with numerous hybrids): *S. bebbiana* ('Alba'), *S. candida*, *S. caprea* ('Black Stem,' 'Curlilocks,' 'Kilmarnock,' 'Pendula,' 'Select,' and 'Weeping Sally'), *S. caprea* var. *variegata*, *S. commutata*, *S. drummondiana* ('Curlew'), *S. eriocephala* ('American Mackay,' 'Kerksii,' 'Mawdesley,' and 'Russelliana'), *S. exigua*, *S. geyeriana*, *S. lemmonii* ('Palouse'), *S. lasiandra* ('Nehalem', and 'Roland'), *S. melanopsis* ('Multnomah'), *S. petiolaris*, *S. prolixa* ('Rivar'), *S. sericea*, and *S. viminalis* ('Aquatica Gigantea,' 'Black Osier,' 'Black Satin,' 'Brown Merriam,' 'Campbell,' 'Cane Osier,' 'Endeavour,' 'English Rod,' 'Gigantea,' 'Gigantea Korso,' 'Green Gotz,' 'Irish Rod,' 'Jorr,' 'Jorunn,' 'Mealy Top,' 'Mulattin,' 'Nora,' 'Reader's Red,' 'Red,' 'Regalis,' 'Riefenweide,' 'Romanin,' 'Roth Cheviot,' 'Stone Osier,' 'Stricta,' 'Suffolk Osier,' 'Superba,' 'Tora,' 'Tordis,' 'Utelescens,' and 'Yellow Osier'); **misc. products:** *Salix arctica* is used for basket weaving, clothing fibers, and as a fuel. Cultivars of *S. barclayi* have been used for the revegetation of riparian zones in the western United States. The decorative "diamond wood," which is carved into candlesticks, canes, furniture, and lamps, is derived from the trunks of *S. bebbiana* and results from a wood infection caused by *Valsa sordida* (Ascomycota: Valsaceae) and other fungi. The wood has also been used to construct baseball bats and for charcoal used in making gunpowder. The Cree and Woodlands tribes used the stems of *S. bebbiana* to trim their birch-bark baskets, to construct netting used for fishing, and to remove excessive pitch when sealing their canoes. The Okanagan-Colville people fashioned cord from the bark to make dresses, bags, rope, and a coarse sewing thread. The twigs were used by the Cree and Woodlands people as implements for roasting fish. The Cree and Woodlands tribes made bows, arrows, and toy whistles from the twigs of *S. bebbiana*. A hybrid of *S. caprea* (*S.* ×*smithiana*) was introduced to eastern Canada for use in basketry. The Seminoles fashioned game bats from the wood of *S. caroliniana*. *Salix bebbiana*, *S. boothii*, *S. caroliniana*, *S. drummondiana*, *S. eriocephala*, *S. exigua*, *S. geyeriana*, *S. gooddingii*, *S. lasiandra*, *S. lemmonii*, *S. lutea*, *S. monticola*, *S. nigra*, *S. planifolia*, and *S. sericea* are transplanted for use in stream stabilization projects. The wood of *S. eriocephala* is used for fence posts. The Ute people used the fibers of *S. eriocephala* for basket making and the Lakota fashioned the wood into canes and staffs. It has been evaluated as a plant for bioenergy-related biomass production in Canada. *Salix exigua* was widely used by native

North Americans and often is found at sites of former Indian habitation. Roots or twigs of *S. exigua* were made into baskets, mats, or rugs by the Costanoan, Flathead, Havasupai, Karok, Kashaya, Keres, Mandan, Paiute, Pomo, Tewa, Tolowa, Ukiah, and Western tribes. Fibers from the plant were commonly used as various building materials. *Salix exigua* produced a sturdy cordage and sewing fiber that was used by the Kashaya, Kawaiisu, Lakota, Montana, Okanagan-Colville, and Pomo people. Twigs were also fashioned into many miscellaneous products including bows and arrows, cooking tools, fishing weirs, sewing needles, and toy whistles. The Costanoan used the plants as kindling wood. Southwestern natives make tightly woven baskets from twigs of *S. gooddingii* to hold water. This plant was used by early pioneers as an indicator of water. The Cahuilla fashioned cradle boards from the wood of *S. gooddingii*. The Pima used the twigs to make baskets, bird cages, and hunting bows. Shoots of *S. lasiolepis* were used for basketry by the Costanoan, Diegueno, Kawaiisu, and Shoshoni tribes. The Mendocino people made a chewing tobacco substitute from the dried, powdered inner bark. The California and Mendocino people used the inner bark for rope fiber. The Round Valley Indians used the wood for fuel and thatch. Baskets fashioned from *S. lasiandra* fibers were made by the Gosiute, Kashaya, Panamint, Pomo, Shoshoni, and Ute tribes. Other miscellaneous products obtained from this plant include bows (Coast, Salish), building materials (Navajo, Ramah), fire drills (Cowlitz), fish weirs (Gosiute), looms (Navajo, Ramah), mallet heads (Haisla, Hanaksiala), roasting sticks (Nitinaht), string (Chehalis), and toy horses (Navajo, Ramah). Stems of *S. melanopsis* were woven into baskets by Dakota and Flathead people and fashioned into lodge poles by the Montana. Fibers from the plants were also used to make cordage (Montana) or woven into mats (Mandan). The wood of *S. nigra* is important commercially and is used to make boxes, cabinets, crates, doors, flooring, furniture, plywood, polo balls, toys, and as a source of paper pulp and fuel. It once was used extensively in the manufacture of artificial limbs and as a source of gunpowder charcoal. The Papago tribe used the twigs for basketry and weaving. *Salix viminalis* has also been used extensively in basket making; **weeds:** *Salix nigra* is now regarded as a weed of "National significance" in Australia where it was introduced for soil stabilization in 1962. Importation of *S. bebbiana* and other willows is prohibited in Australia where they are feared as potential weeds; **nonindigenous species:** *Salix caprea*, *S. purpurea*, and *S. viminalis* were introduced to North America from Eurasia as escapes from cultivation. *Salix nigra* was introduced to Utah where it is now common. More than 100 *Salix* taxa have been introduced to Australia. **Systematics:** Molecular data consistently resolve *Salix* as the sister genus to *Populus* (e.g., Figure 4.37). Analyses using combined nrDNA and cpDNA sequence data render *Salix* as a well-supported clade, and do not support the recognition of two monotypic segregate genera (*Chosenia*, *Toisusu*), which nest within *Salix*. The New World *Salix* species have been subdivided into four subgenera and 28 sections based upon the results of numerical cluster analysis (Table 4.1). However,

TABLE 4.1

Classification of Obligately Aquatic *Salix* Species in North America

I. *Salix* subgenus *Chamaetia* (8)

    a. Section *Diplodictyae*

        1. *S. arctica*

    b. Section *Myrtilloides*

        2. *S. athabascensis*

        3. *S. pedicellaris*

    c. Section *Myrtosalix*

        4. *S. arctophila*

II. *Salix* subgenus *Longifoliae* (1)

    a. Section *Longifoliae*

        5. *S. exigua*

        6. *S. interior*

        7. *S. melanopsis*

III. *Salix* subgenus *Salix* (4)

    a. Section *Humboldtianae*

        8. *S. caroliniana*

        9. *S. gooddingii*

        10. *S. nigra*

    b. Section *Maccallianae*

        11. *S. maccalliana*

    c. Section *Salicaster*

        12. *S. lasiandra*

        13. *S. serissima*

IV. *Salix* subgenus *Vetrix* (15)

    a. Section *Candidae*

        14. *S. candida*

    b. Section *Cinerella*

        15. *S. caprea*

    c. Section *Helix*

        16. *S. purpurea*

    d. Section *Cordatae*

        17. *S. eriocephala*

        18. *S. lutea*

        19. *S. prolixa*

    e. Section *Fulvae*

        20. *S. bebbiana*

    f. Section *Geyeriana*

        21. *S. geyeriana*

        22. *S. lemmonii*

        23. *S. petiolaris*

    g. Section *Griseae*

        24. *S. sericea*

    h. Section *Hastatae*

        25. *S. arizonica*

        26. *S. barclayi*

        27. *S. boothii*

        28. *S. commutata*

        29. *S. eastwoodiae*

        30. *S. farriae*

        31. *S. monticola*

        32. *S. myrtillifolia*

        33. *S. pyrifolia*

        34. *S. wolfii*

    i. Section *Mexicanae*

        35. *S. lasiolepis*

(*Continued*)

TABLE 4.1 (*Continued*)

Classification of Obligately Aquatic *Salix* Species in North America

  j. Section *Phylicifoliae*

    36. *S. drummondiana*

    37. *S. planifolia*

  k. Section *Sitchensis*

    38. *S. jepsonii*

  l. Section *Viminella*

    39. *S. viminalis*

Source: Argus, G. W., *Bot. Electron. News*, 227, 1, 1999.

Note: The total number of sections (including terrestrial species) is given in parentheses after each subgenus; species are numbered consecutively. Although this classification is widely followed, recent molecular data have challenged many aspects of the relationships depicted by it.

recent studies using molecular data have revealed a number of problems with that proposed classification. Molecular analyses provide no support for the recognition of subgenus *Chamaetia* (which consists mostly of plants with a dwarf shrubby habit), and indicate that only subgenera *Salix* and *Vetrix* are defensible phylogenetically, and even here only after extensive taxonomic rearrangements of species have been taken into account. Species assigned formerly to subgenus *Longifoliae* (section *Longifoliae*) are resolved either within subgenus *Vetrix* (nuclear ITS data) or subgenus *Salix* (chloroplast *matK* data), which suggests their possible hybrid origin. *Salix* always has been difficult taxonomically owing presumably to the reticulate effects of hybridization, which are assumed to pervade this genus. That assumption has been borne out by molecular phylogenetic studies, which show extensive discrepancies between trees constructed from maternally (e.g., cpDNA) vs. biparentally (e.g., nrITS) inherited DNA regions. Although several molecular phylogenetic studies of *Salix* have appeared recently, the taxon sampling of this large genus has been insufficient to provide much detailed insight into interspecific relationships. Plastid and nuclear data have resolved two distinct subclades within subgenus *Salix*, which distinguish the Old World from New World species. Within the New World subclade, *S. exigua* and *S. interior* consistently resolve as closely related, sister species, a result consistent with their merger (and recognition as subspecies) by some authors. Molecular data place *S. maccalliana* and *S. nigra* close to them as well. *Salix bebbiana*, *S. eriocephala*, *S. sericea*, and *S. wolfii* resolve within a clade containing species from subgenera *Chamaetia* and *Vetrix*. Otherwise, the remaining OBL *Salix* species have not been evaluated using molecular data and will require additional genetic studies in order to test the relationships that have been hypothesized largely on the basis of morphological similarities. Some of the presumed interspecific relationships based on morphological evaluations include the following: *S. arctica* to *S. brachycarpa*, *S. glauca* and *S. sphenophylla*; *S. arctophila* to *S. chamissonis*; *S. athabascensis* to *S. petiolaris*; *S. bebbiana* and *S. starkeana*; *S. eastwoodiae* and *S. commutata* (sometimes placed in synonymy); *S. farriae* and *S. hastata*; *S. gooddingii* and *S. nigra*; *S. geyeriana* and *S. lemmonii*; *S. lasiolepis* and *S. hookeriana*, *S. traceyi*; *S. lutea* and *S. eriocephala*; *S. myrtillifolia* with *S. arizonica*, *S.*

ballii, *S. boothii*, *S. pseudomyrsinites*; *S. planifolia* and *S. pulchra*. *Salix lutea* and *S. ligulifolia* have been merged by some authors with *S. eriocephala*. Hybrids are believed to occur commonly in *Salix*, and most of the polyploids are presumed to have arisen through allopolyploidy. Seasonal isolating mechanisms reduce the opportunities for hybridization among a number of sympatric species. Successful synthetic hybrids (e.g., *S. interior* × *S. bebbiana*, *S. eriocephala*, *S. petiolaris*; *S. eriocephala* × *S. exigua*, *S. petiolaris*) indicate that many willow species retain the ability to cross, but such hybrids often are short lived and sterile. However, several hybrids are vigorous vegetatively and are likely to persist as clones, which may be able to overcome sterility bottlenecks and eventually to introgress with parental populations. Yet, despite the widespread assumption of rampant hybridization in the genus, compelling genetic evidence often is lacking, and much additional work needs to be carried out to document willow hybrids. Reported natural hybrids (assessed mainly using morphological criteria) include: *S. arctica* × *S. arctophila*; *S. arctica* × *S. pedicellaris*; *S. arctophila* × *S. uva-ursi*; *S. arctophila* × *S. glauca* (in Greenland); *S. arizonica* × *S. boothii*; *S. athabascensis* × *S. pedicellaris*; *S. barclayi* × *S. arctica*; *S. barclayi* × *S. barrattiana*; *S. barclayi* × *S. brachycarpa*; *S. barclayi* × *S. cascadensis*; *S. barclayi* × *S. commutata*; *S. barclayi* × *S. farriae*; *S. barclayi* × *S. hastata*; *S. barclayi* × *S. hookeriana*; *S. barclayi* × *S. richardsonii*; *S. barclayi* × *S. stolonifera*; *S. caprea* × *S. viminalis* (*S.* ×*smithiana* Willd.); *S. candida* × *S. bebbiana*; *S. candida* × *S. brachycarpa*; *S. candida* × *S. calcicola*; *S. candida* × *S. eriocephala*; *S. candida* × *S. petiolaris*; *S. barclayi* × *S. commutata*; *S. drummondiana* × *S. alaxensis*; *S. eriocephala* × *S. sericea*; *S. eriocephala* × *S. petiolaris*; *S. exigua* × *S. interior*; *S. exigua* × *S. melanopsis*; *S. farriae* × *S. barclayi*; *S. interior* × *S. exigua*, *S. myrtillifolia* × *S. candida*; *S. nigra* × *S. alba*, *S. nigra* × *S. amygdaloides*, *S. nigra* × *S. bonplandiana*, *S. nigra* × *S. caroliniana*, *S. nigra* × *S. lucida*, *S. nigra* × *S. sericea*; *S. petiolaris* × *S. pedicellaris*; *S. petiolaris* × *S. candida*; *S. petiolaris* × *S. melanopsis*; and *S. pyrifolia* × *S. brachycarpa*. Combined morphological and molecular data have effectively documented introgressive hybridization involving *S. eriocephala* and *S. sericea* and similar studies would be highly useful for documenting other suspected willow hybrids. F_1 hybrids of these species exhibited heterosis during periods of adequate water availability, but showed suppressed growth during drought periods. Hybrid swarms attributed to introgression have also been reported for *S. atha-bascensis* × *S. pedicellaris* and for *S. arctica* × *S. stolonifera*. Isozyme analyses indicated that genetic variability in *Salix* often correlates with the extent of geographical distribution, showing moderate levels of genetic variation in *S. exigua* and *S. melanopsis*, both relatively widespread in western North America. *Salix melanopsis*, with numerous isolated populations, also shows a high degree of interpopulational genetic differentiation. As in its sister genus *Populus*, the base chromosome number of *Salix* is $x = 19$. Chromosome counts indicate both interspecific and intraspecific variation in ploidy levels including diploids ($2n = 38$: *S. arizonica, S. bebbiana, S. candida, S. caprea, S. commutata, S. eriocephala, S. exigua, S. geyeriana, S. gooddingii, S. interior, S. lutea, S. monticola, S. myrtillifolia, S. nigra, S. pedicellaris, S. petiolaris, S. purpurea, S. pyrifolia, S. viminalis*), triploids ($2n = 57$: *S. drummondiana, S. pedicellaris, S. planifolia*), tetraploids ($2n = 76$: *S. arctica, S. arctophila, S. athabascensis, S. barclayi, S. boothii, S. caprea, S. eastwoodiae, S. lasiolepis, S. lemmonii, S. lasiandra, S. pedicellaris, S. planifolia, S. serissima*), pentaploids ($2n = 95$: *S. athabascensis*), hexaploids ($2n = 114$: *S. arctica, S. athabascensis, S. monticola*), octaploids ($2n = 152$: *S. planifolia*), and higher level polyploids ($2n = 190$, ~228: *S. maccalliana*). Chromosome numbers remain unreported for *S. caroliniana, S. farriae, S. jepsonii, S. melanopsis, S. sericea*, and *S. wolfii*.

Comments: *Salix arctica* is distributed throughout northern portions of North America and Eurasia. *Salix arctophila* is found throughout the North American Arctic. *Salix athabascensis* occurs in western Canada and Alaska. *Salix bebbiana, S. maccalliana, S. petiolaris*, and *S. pyrifolia* are common and widespread throughout northern North America. The nonindigenous, Euarasian native *S. caprea* occurs sporadically in eastern portions of North America. *Salix caroliniana* is widely distributed throughout the southeastern United States. *Salix commutata, S. drummondiana* and *Salix lemmonii* are species of higher elevations in western North America. *Salix interior* is widespread throughout all but southwestern North America. *Salix eastwoodiae* and *Salix lasiolepis* occur in the western United States. *Salix eriocephala* and *S. sericea* are fairly common species of eastern North America. *Salix viminalis* occurs in northeastern North America. *Salix exigua, S. boothii, S. lasiandra, S. lutea, S. melanopsis, S. monticola, S. prolixa* and *S. wolfii* are species of western North America. *Salix arizonica* is restricted to Arizona, Colorado, New Mexico, and Utah. *Salix barclayi* and *S. farriae* are species of montane northwest-central North America. *Salix geyeriana* occurs throughout the Rocky Mountains. *Salix gooddingii* is distributed in the southwestern United States and Mexico. *Salix myrtillifolia* is a northern species of Alaska and western Canada. *Salix nigra* occurs mainly in the eastern United States. *Salix pedicellaris* and *S. serissima* are relatively uncommon species of boreal North America. *Salix planifolia* occurs throughout northern and western North America. *Salix purpurea* has been introduced to portions of eastern North America and the western United States.

References: Alison, 1975; Aravanopoulos, 2000; Aravanopoulos & Zsuffa, 1998; Argus, 1999, 2010; Argus et al., 2000; Azuma et al., 2000; Barrows & Gordh, 1974; Bart & Davenport, 2015; Baskin & Baskin, 1998; Bayley & Mewhort, 2004; Bergerud et al., 1984; Bergman, 2002; Boal & Andersen, 2005; Böcher et al, 2015; Boecklen et al., 1990; Breden & Wade, 1989; Broberg et al., 2001; Brookshire et al., 2002; Brambilla & Sutton, 1969; Brunsfeld & Anttila, 2004; Brunsfeld et al., 1991; Burke et al., 1994; Cannings & Simaika, 2005; Carstens et al., 2013; Castro-Morales et al., 2014; Cázares et al., 2005; Cha et al., 2009; Chambers, 1968; Chastant et al., 2015; Chen et al., 2010; Christy, 2004; Cobbaert et al., 2004; Cody et al., 2003; Collet, 2010; Cooke, 1997; Cooper, 1991, 1996; Crocker & Major, 1955; Crowe et al., 2004; Darby et al., 1996; David, 1994; Decker, 2006a,b; Densmore & Zasada, 1983; Deyrup et al., 2002; Dorn, 1970; Dötterl et al., 2014; Douhovnikoff & Dodd, 2003; Ellis & Everhart, 1885b; Esser, 1992a,b; Faubert et al., 2012a,b; Fellers et al., 2013; Fraser et al., 2007; Freeland, 1974; Fritz et al., 2001; Glawe, n.d.; Goodwin & Marion, 1979; Graham & Henry, 1933; Grand & Vernia, 2002; Hammond, 1943; Hardig et al., 2000, 2010; Henry, 1973; Heyes, 1979; Hoffmann & Taber, 1960; Hopkins, 1969; Hornbeck et al., 2003; Howard et al., 2006; Jarzen & Hogsette, 2008; Johnson & Steingraeber, 2003; Johnston, 2003; Jones, 1944; Jumpponen et al., 1998; Jürgens et al., 2014; Kaczynski et al., 2014; Karrenberg et al., 2002; Keigley et al., 2003; Kenaley et al., 2014; Krasny et al., 1988; Kufeld, 1973; Lack, 1982; Lauron-Moreau et al., 2013; Labrecque & Teodorescu, 2005; LeSage, 1984; Lewis, 1980; Lewis et al., 1928; Likar & Regvar, 2013; Lin et al., 2009; Lindsey, 1953; Lite & Stromberg, 2003; Long & Medina, 2007; Lutz, 1958; Markow, 2004; Maschinski, 2001; Maun, 1998; McClellan et al., 2003; McCormick & Gibble, 2014; Meiman et al., 2009; Merritt & Wohl, 2006; Moral & Jones, 2002; Mosseler, 1990; Mosseler & Papadopol, 1989; Munson, 1992; Orians & Floyd, 1997; Orians et al., 1997; Ostaff et al., 2015a,b; Ottenbreit & Staniforth, 1992; Percy et al., 2012; Pickwell & Smith, 1938; Porter, 1983; Prendusi et al., 1996; Price et al., 1989; Reed, 1993; Richburg et al., 2001; Roberts, 1984; Roche & Fritz, 1997; Rockwell & Stephens, 2014; Russell et al., 2003; Sacchi & Price, 1988, 1992; Salick & Pfeffer, 1999; Schender et al., 2014; Sealy, 1989; Seavey & Seavey, 2012; Simak, 1982; Smith, 1929; Smith et al., 1983, 2004; Sockman, 2008; Spaeth et al., 2002; Spencer, 1981; Stein et al., 1992; Stewart, 1953; Sultana & Saleem, 2004; Talbot et al., 2002; Tanner, 1958; Taylor et al., 1996; Tesky, 1992; Thompson et al., 2003; Titus, 1991; Todd, 1927; Trefry & Hik, 2009; Tsuyuzaki et al., 1997; Uchytil, 1989b,c,d,e, 1991a,b, 1992; Van Handel et al., 1972; Vannimwegen & Debinski, 2004; Vitt & Chee, 1990; Vroege & Stelleman, 1990; Walford et al., 2001; Walters et al., 2002; Warner & Hendrix, 1984; Weishampel & Bedford, 2006; Wetzel et al., 2001; Wiedenmann, 2005; Williams, 1990b; Wilson, 1968; Windell et al., 1986; Woodland, 1984; Woods & Cooper, 2005; Young & Clements, 2003.

Family 8: Violaceae [22]

The Violaceae are a widely distributed family of 950 species (Judd et al., 2016), of which only a few are adapted to aquatic habitats. Some molecular data have indicated a relationship of the family to a clade comprising Malesherbiaceae, Passifloraceae, and Turneraceae (Soltis et al., 2000), whereas other data weakly associate the family with Goupiaceae, Lacistemataceae, and Ctenolophonaceae (Chase et al., 2002), or resolve Violaceae and Goupiaceae as sister groups (Xi et al., 2012), in the most comprehensive evaluation of relationships (Figure 4.25).

Despite these uncertainties, all phylogenetic analyses thus far have indicated that Violaceae are monophyletic (e.g., Tokuoka, 2008; Wahlert et al., 2014). The family is united by a number of features including stamens, which form a ring around the gynoecium by their appressed margins; anthers with dorsal, glandular, or spur-like nectaries; introrse pollen dehiscence; and curved or hooked style, which is dilated or modified distally (Judd et al., 2016).

The group is noted for its showy flowers, which are all pollinated by insects (Insecta) such as bees and wasps (Hymenoptera), butterflies (Lepidoptera), and flies (Diptera). Most of the species are outcrossed, although some produce cleistogamous (selfed) flowers. The seeds are dispersed passively, by ballistic ejection from the capsules, or by ants (Insecta: Hymenoptera: Formicidae) (Judd et al., 2016).

Ornamental garden plants are found in *Hybanthus*, *Melicytus*, and *Viola*, with a large number of cultivars in the latter. The family also contains several medicinal species (*Anchietea salutaris*, *Corynostylis hybanthus*, and *Hybanthus ipecacuanha*) of minor use as emetics.

Viola is the largest genus of the family and the only one to contain OBL aquatics in North America.

1. ***Viola*** Tourn. ex L.

1. *Viola*

Violet; violette

Etymology: derived from the Greek *ion* meaning "violet" [color]

Synonyms: *Erpetion*; *Mnemion*

Distribution: global: Asia; Europe; New World; **North America:** widespread

Diversity: global: 500 species; **North America:** 78 species

Indicators (USA): OBL: *Viola cucullata, V. lanceolata, V. macloskeyi, V. novae-angliae, V. palustris*; **FACW:** *V. cucullata, V. langsdorfii, V. macloskeyi, V. palustris*; **FAC:** *V. novae-angliae*

Habitat: freshwater; palustrine; **pH:** 5.1–7.0; **depth:** <1 m; **life-form(s):** emergent herb

Key morphology: "stemless" rosettes of long-petioled (to 20 cm) leaves, the blades circular, cordate-ovate to reniform (to 10 cm) or lanceolate (to 15 cm); chasmogamous flowers (to 1.5 cm) solitary, pedunculate (to 21 cm), showy; corolla zygomorphic; petals 5, white, pale blue, or violet, the lowermost spurred; stalked, bud-like cleistogamous flowers present later in season; fruits (to 17 mm) capsular, explosively dehiscent

Life history: duration: perennial (rhizomes, stolons); **asexual reproduction:** rhizomes, stolons; **pollination:** insect or cleistogamous; **sexual condition:** hermaphroditic; **fruit:** capsules (common); **local dispersal:** seeds (ballistic to 2 m; ants); **long-distance dispersal:** seeds (birds)

Imperilment: (1) *Viola cucullata* [G4/G5]; S3 (MO); (2) *V. lanceolata* [G5]; S1 (PE, VT); S2 (CA, IA, MN, NE, NF, OR); S3 (NJ, OH, QC); (3) *V. langsdorfii* [G4]; S1 (CA); S2 (YT); (4) *V. macloskeyi* [G4]; S1 (DE, SK); S2 (AB, MO, WY); S3 (GA, IA, MT, NC, VA); (5) *V. novae-angliae* [G4]; S1 (MB, NY); S2 (ME, MI, NB); S3 (ON); (6) *V. palustris* [G5]; S1 (CA, ME, NH); S2 (NF); S3 (MT, QC, WY)

Ecology: general: Although a few violet species are categorized as OBL indicators, 43 of the North American species (55%) occur at least facultatively within wetlands, which is an indication of their fairly widespread tolerance of wet conditions. *Viola langsdorfii* (FACW) was recognized as OBL in the 1996 indicator list and it has been included here. The species are annual or perennial and commonly categorized either as "stemless" (their leaves arising from a rhizome) or "leafy stemmed" (their leaves borne on erect stems). Many violets produce cleistogamous flowers in addition to showy, chasmogamous ones. All of the obligate North American aquatics are "stemless" perennials, which produce both cleistogamous and chasmogamous flowers. The elaiosome-appendaged seeds of some species are dispersed locally by ants (Insecta: Hymenoptera: Formicidae). Long distance dispersal of seeds is facilitated by birds (Aves). The seed ecology is unknown for most of the OBL violets. Other investigated violet species mainly exhibit physiological dormancy.

***Viola cucullata* Aiton** grows in bogs, bottomlands, crevices, ditches, floodplains, glades, hummocks, meadows, prairies, seeps, slopes, springs, swales, swamps, thickets, woodlands, and along the margins of brooks and streams at elevations of up to 1638 m. The plants can tolerate partial to dense shade. The substrates are calcareous and described as alluvium, mafic, peat, rock, and sandy loam. Flowering occurs from January to June, with the fruits developing by June. A large proportion of the flowers are cleistogamous. The seeds are small with small elaiosomes. They occasionally are dispersed by ants (Insecta: Hymenoptera: Formicidae), but to much less of a degree than other violet species. A persistent seed bank develops, particularly in hummocks. The species can spread quickly by vegetative reproduction from underground stolons. **Reported associates:** *Acer negundo, A. rubrum, A. saccharinum, A. saccharum, Achillea millefolium, Actaea pachypoda, A. rubra, Adiantum pedatum, Alliaria petiolata, Ambrosia artemisiifolia, Amphicarpaea bracteata, Andropogon gerardii, Anemone canadensis, A. cylindrica, Angelica atropurpurea, Antennaria neglecta, Anthriscus sylvestris, Aplectrum hyemale, Arisaema triphyllum, Asclepias tuberosa, Athyrium thelypterioides, Betula alleghenien-sis, B. lenta, Boehmeria cylindrica, Boykinia aconitifolia, Brachyelytrium erectum, Caltha palustris, Cardamine angustata, C. bulbosa, C. clematitis, Carex albicans, C. bromoides, C. canescens, C. corrugata, C. crinita, C. crus-corvi, C. davisii, C. grisea, C. hystricina, C. lurida, C. laevivaginata,*

C. prasina, C. scabrata, C. squarrosa, C. stricta, Carya cordiformis, C. illinoinensis, C. laciniosa, C. ovata, Celastrus scandens, Chasmanthium latifolium, Chelone glabra, C. lyonii, Chrysosplenium americanum, Cicuta maculata, Circaea alpina, Cirsium discolor, Collinsia verna, Coptis trifolia, Corallorhiza wisteriana, Cornus alternifolia, C. amomum, Corylus americana, Cyperus filiculmis, Cypripedium parviflorum, Danthonia spicata, Desmodium canadense, Dicentra canadensis, D. cucullaria, Dichanthelium scabriusculum, Diphylleia cymosa, Dryoperis cartrhusiana, Elaeagnus umbellata, Elymus canadensis, Equisetum arvense, E. fluviatile, E. sylvaticum, Erigenia bulbosa, Erigeron strigosus, Euthamia graminifolia, Fagus grandifolia, Fragaria virginiana, Fraxinus americana, F. nigra, F. pennsylvanica, F. profunda, Galearis spectabilis, Galium obtusum, Gentianella quinquefolia, Geranium maculatum, Geum geniculatum, G. rivale, Glyceria striata, Goodyera pubescens, Hamamelis virginiana, Houstonia longifolia, H. serpyllifolia, Huperzia lucidula, Hydrocotyle americana, Impatiens capensis, Iris brevicaulis, I. versicolor, I. virginica, Juglans nigra, Juncus effusus, Laportea canadensis, Leersia oryzoides, Lespedeza capitata, Lilium grayi, Lindera benzoin, Liriodendron tulipifera, Lobelia cardinalis, Lycopus americanus, L. virginicus, Lysimachia quadriflora, L. terrestris, Maianthemum canadense, M. stellatum, Micranthes micranthidifolia, Mikania scandens, Monarda fistulosa, Nyssa biflora, Oenothera biennis, O. pilosella, Osmorhiza claytonii, Osmunda claytonia, Osmundastrum cinnamomeum, Oxalis stricta, Oxypolis rigidior, Packera aurea, Parnassia asarifolia, Phalaris arundinacea, Pinus strobus, Platanthera clavellata, P. grandiflora, Poa pratensis, Podophyllum peltatum, Polygonatum pubescens, Polystichum acrostichoides, Prunus serotina, P. virginiana, Pycnanthemum virginianum, Quercus alba, Q. bicolor, Q. rubra, Q. shumardii, Ranunculus septentrionalis, Rhododendron, Rhus typhina, Ribes missouriense, Rosa palustris, Rudbeckia hirta, Rumex orbiculatus, Salix bebbiana, S. humilis, S. nigra, S. pensylvanica, Sambucus nigra, Sanguinaria canadensis, Saururus cernuus, Schizachyrium scoparium, Solidago juncea, S. nemoralis, S. rugosa, S. uliginosa, Sorghastrum nutans, Sphagnum, Sporobolus cryptandrus, Staphylea trifolia, Symplocarpus foetidus, Stellaria corei, Symphyotrichum cordifolium, S. lanceolatum, S. oolentangiense, Thalictrum clavatum, Thelypteris novaboracensis, T. palustris, Tiarella cordifolia, Tilia americana, Toxicodendron radicans, Trautvetteria caroliniensis, Trientalis borealis, Trillium erectum, T. flexipes, T. recurvatum, T. sessile, Tsuga canadensis, Typha latifolia, Ulmus americana, U. rubra, Valeriana pauciflora, Viburnum acerifolium, V. lentago, V. prunifolium, Viola consparsa, V. macloskeyi, V. pubescens, Vitis riparia.

Viola lanceolata L. occurs in barrens, bogs, Carolina bays, depressions, ditches, fens, flatwoods, marshes, meadows, prairies, roadsides, sand prairies, savannas, swales, woodlands, and along the margins of kettle holes, lakes, ponds, and streams at elevations of up to 250 m. The plants grow in full sun to semishaded sites but typically inhabit open areas or clearings. They are often found in shallow, standing water.

The substrates primarily are acidic (e.g., pH: 5.1) and have been characterized as clay, loamy sand, muck, peat, sand, sandy loam, shale, and silty clay. The plants produce facultatively outcrossing chasmogamous (CH) flowers and obligately selfing cleistogamous (CL) flowers. The cleistogamous flowers appear from June to September and are correlated negatively with stolon production. Flowering and fruiting of the chasmogamous flowers have been observed from January to October (depending on latitude). Their primary pollinators are bees (see *Use by wildlife*). The largest plants can produce up to three fruits, each containing 16–68 seeds (44 on average). A seed bank is produced (e.g., 4 seeds/m^2), which can persist for several years and allows the plants to germinate immediately following a disturbance. This can be a mat-forming species as a result of its profusion of long, slender stolons, by which the plants overwinter. The roots are colonized by vesicular–arbuscular mycorrhizal fungi (Glomeromycota: Glomeraceae: *Funneliformis geosporum*; *Glomus fasciculatum*). **Reported associates:** *Acer rubrum, Agalinis maritima, A. purpurea, Agrimonia parviflora, Agrostis hyemalis, Aletris farinosa, Alisma subcordatum, Amphicarpum purshii, Andropogon gerardii, A. virginicus, Antennaria plantaginifolia, Anthoxanthum, Aronia melanocarpa, A. prunifolia, Baccharis halimifolia, Bartonia paniculata, B. virginica, Betula pumila, B. papyrifera, Bidens cernuus, B. frondosus, B. trichospermus, Briza minor, Bulbostylis capillaris, Calamagrostis canadensis, Calopogon pulchellus, Cardamine parviflora, C. pensylvanica, Carex albicans, C. buxbaumii, C. conoidea, C. folliculata, C. rostrata, C. scoparia, C. swanii, C. vesicaria, Castilleja coccinea, Centella asiatica, Cephalanthus occidentalis, Cerastium fontanum, Chaptalia tomentosa, Cicuta maculata, Cladium jamaicense, Coreopsis rosea, C. tripteris, Cornus amomum, Corylus americana, Crataegus marshallii, Cyperus dentatus, C. polystachyos, C. rivularis, C. squarrosus, C. strigosus, Cypripedium californicum, Cyrilla racemiflora, Darlingtonia californica, Daucus carota, Dichanthelium acuminatum, D. clandestinum, D. stigmosum, Drosera intermedia, D. rotundifolia, Dulichium arundinaceum, Elaeagnus umbellata, Eleocharis acicularis, E. intermedia, E. melanocarpa, E. microcarpa, E. obtusa, E. tenuis, Equisetum, Eragrostis spectabilis, Eryngium yuccifolium, Eupatorium perfoliatum, E. serotinum, Euphorbia corollata, Euthamia graminifolia, E. tenuifolia, Fimbristylis autumnalis, Fragaria virginiana, Fraxinus pennsylvanica, Fuirena squarrosa, Galium aparine, G. obtusum, Gaylussacia baccata, Gentiana saponaria, Gerardia paupercula, Glyceria acutiflora, G. borealis, G. striata, Gratiola aurea, Hedyotis boscii, Helianthus mollis, Hemicarpha micrantha, Hydrocotyle, Hydrolea quadrivalvis, Hypericum boreale, H. canadense, H. denticulatum, H. gentianoides, H. kalmianum, H. majus, H. mutilum, H. punctatum, Hypoxis hirsuta, Hyptis alata, Ilex coriacea, I. glabra, I. opaca, I. verticillata, I. vomitoria, Juncus acuminatus, J. alpinus, J. biflorus, J. brachycarpus, J. brevicaudatus, J. bufonius, J. canadensis, J. effusus, J. greenei, J. marginatus, J. militaris, J. pelocarpus, J. vaseyi, Juniperus virginiana, Lachnanthes caroliniana, Lechea maritima, Leersia hexandra, L. oryzoides, Lilium* canadense, Liparis loeselii, Lobelia inflata, L. kalmii, L. siphilitica, L. spicata, Lonicera ×bella, L. maackii, Ludwigia palustris, Lycopodiella alopecuroides, Lycopodium inundatum, Lycopus americanus, L. uniflorus, Lysimachia ciliata, L. terrestris, Magnolia grandiflora, M. virginiana, Maianthemum canadense, Malus sieboldii, Muhlenbergia torreyana, M. uniflora, Myrica cerifera, M. pensylvanica, Nyssa sylvatica, Onoclea sensibilis, Osmunda claytoniana, O. regalis, Osmundastrum cinnamomeum, Oxalis violacea, Packera plattensis, Panicum abscissum, P. columbianum, P. hemitomon, P. philadelphicum, P. virgatum, P. wrightianum, Parthenium integrifolium, Paspalum plicatulum, P. praecox, edicularis canadensis, Persea palustris, Persicaria hydropiperoides, P. maculosa, Phragmites australis, Pilea pumila, Pinus palustris, P. rigida, P. sylvestris, P. taeda, Platanthera flava, P. peramoena, Poa pratensis, Pogonia ophioglossoides, Polygala cruciata, P. sanguinea, Populus tremuloides, Potentilla anserina, P. simplex, Proserpinaca palustris, Prunus serotina, Pycnanthemum virginianum, Quercus alba, Q. palustris, Q. velutina, Q. virginiana, Ranunculus abortivus, R. rhomboidea, Rhamnus cathartica, R. frangula, Rhexia mariana, R. virginica, Rhododendron canescens, Rhynchospora alba, R. caduca, R. capitellata, R. glomerata, R. macrostachya, R. perplexa, Rosa multiflora, R. rugosa, Rotala ramosior, Rubus allegheniensis, R. flagellaris, R. hispidus, R. pubescens, Rumex acetosella, Sabal minor, Sabatia campanulata, Salix humilis, S. interior, Sarracenia alata, Schedonorus arundinaceus, Schoenoplectus hallii, S. smithii, Scirpus atrovirens, Scleria reticularis, S. triglomerata, Serenoa repens, Sisyrinchium atlanticum, Smilax laurifolia, Solidago canadensis, S. sempervirens, Sorghastrum nutans, Spartina pectinata, Sphagnum magellanicum, Spiraea alba, S. salicifolia, S. tomentosa, Spiranthes cernua, Stachys hyssopifolia, S. tenuiflora, Symphyotrichum lanceolatum, S. novi-belgii, Symplocos tinctoria, Taraxacum officinale, Taxodium distichum, Thelypteris palustris, Triadenum virginicum, Triadica sebifera, Trientalis borealis, Trifolium wormskioldii, Tripsacum, Typha latifolia, Utricularia cornuta, U. inflata, Vaccinium angustifolium, V. darrowi, V. macrocarpon, Verbena hastata, Veronica arvensis, Viburnum dentatum, Viola adunca, V. primulifolia, V. sagittata, Vitis labrusca, Woodwardia areolata, Xyris torta.

***Viola langsdorffii* Fisch. ex Ging.** occurs in bogs, crevices, depressions, dunes, flats, floodplains, heathlands, hummocks, lagoons, marshes, meadows, mires, muskeg, poor fens, salt marshes, seeps, shores, slopes (to 60°), snowbeds, thickets, tundra, and along the margins of ponds and streams at elevations of up to 1650 m. The plants are able to grow in shallow standing water (to 5–10 cm) and can tolerate exposures of full sun to partial shade. The substrates are acidic (pH: 5.8–6.2) and are described as basalt, boulders, cobble, granite, gravel, gravelly rock, humus over diorite cobble, limestone, loam, mud, organic (>10 cm), rock, sandy loam, silt, silty gravel, and volcanic. Flowering and fruiting occur from April to September. Other details on the reproductive ecology of this species are lacking. **Reported associates:** *Abies lasiocarpa, Achillea millefolium, Aconitum delphiniifolium,*

Alnus viridis, Andromeda polifolia, Anemone narcissiflora, Angelica lucida, Arctagrostis latifolia, Arctostaphylos uva-ursi, Arnica latifolia, A. unalaschcensis, Artemisia norvegica, Asplenium trichomanes-ramosum, Athyrium filix-femina, Betula glandulosa, Bistorta vivipara, Botrychium multifidum, Calamagrostis canadensis, C. nutkaensis, C. stricta, Caltha leptosepala, C. palustris, Camassia leichtlinii, Campanula lasiocarpa, Cardamine, Carex anthoxanthea, C. aquatilis, C. bigelowii, C. circinata, C. lyngbyei, C. macloviana, C. macrochaeta, C. obnupta, C. pauciflora, C. pluriflora, C. rotundata, C. saxatilis, Cassiope lycopodioides, Castilleja unalaschcensis, Chamerion angustifolium, C. latifolium, Cladina mitis, Cladonia cristatella, C. pyxidata, Claytonia sibirica, Conioselinum chinense, Coptis aspleniifolia, C. trifolia, Cornus canadensis, C. suecica, Cryptogramma stelleri, Cystopteris fragilis, Danthonia intermedia, Dasiphora floribunda, Deschampsia cespitosa, Dicranum scoparium, Drosera rotundifolia, Dryopteris expansa, Elliottia pyroliflora, Empetrum nigrum, Epilobium hornemannii, Equisetum arvense, Erigeron peregrinus, Eriophorum chamissonis, Festuca altaica, F. rubra, Fragaria chiloensis, Fritillaria camschatcensis, Galium boreale, Gentiana douglasiana, G. sceptrum, Geocaulon lividum, Geranium erianthum, Geum calthifolium, G. rossii, Harrimanella stelleriana, Hebeloma Helenium bolanderi, Heracleum sphondylium, Hierochloe odorata, Huperzia selago, Hylocomium splendens, Juncus arcticus, J. mertensianus, Kalmia microphylla, K. procumbens, Lathyrus torreyi, Listera cordata, Luetkea pectinata, Lupinus nootkatensis, Luzula parviflora, Lycopodium alpinum, L. annotinum, Lysichiton americanus, Maianthemum dilatatum, Menziesia ferruginea, Mertensia paniculata, Micranthes unalaschensis, Moehringia lateriflora, Nephrophyllidium crista-galli, Oxalis, Oxyria digyna, Parnassia, Petasites frigidus, Phalaris arundinacea, Phleum alpinum, Picea sitchensis, Pinus contorta, Plantago macrocarpa, Pleurozium schreberi, Poa macrocalyx, Polemonium caeruleum, Polystichum aleuticum, P. lonchitis, Polytrichastrum alpinum, Prenathes alata, Primula pauciflora, P. tschuktschorum, Pyrola asarifolia, Racomitrium, Ranunculus occidentalis, Rhodiola integrifolia, Rhododendron groenlandicum, R. tomentosum, Rhynchospora alba, Rhytidiadelphus triquetrus, Rubus arcticus, R. chamaemorus, R. pedatus, R. spectabilis, R. stellatus, Salix arctica, S. barclayi, S. barrattiana, S. sitchensis, S. rotundifolia, Sanguisorba officinalis, S. sitchensis, S. stipulata, Sanionia uncinata, Saussurea americana, Senecio triangularis, Sorbus sitchensis, Sphagnum aongstroemii, Spiraea stevenii, Streptopus amplexifolius, S. roseus, S. streptopoides, Swertia perennis, Tiarella trifoliata, Tofieldia coccinea, Trichophorum cespitosum, Trientalis arctica, T. europaea, Trisetum spicatum, Tsuga heterophylla, T. mertensiana, Vaccinium cespitosum, V. ovalifolium, V. oxycoccus, V. uliginosum, V. vitis-idaea, Vahlodea atropurpurea, Valeriana capitata, V. sitchensis, Veratrum viride, Veronica stelleri, V. wormskjoldii, Viola glabella.

***Viola macloskeyi* F.E. Lloyd** inhabits backwaters, bogs, bottomlands, depressions, ditches, fens, glades, gullies, hummocks, meadows, mudflats, prairies, roadsides, seeps, slopes (to 65%), swamps, woodlands, and the margins of lakes, ponds, and streams at elevations of up to 3606 m. The plants occur in exposures ranging from open to deeply shaded sites. The substrates (e.g., pH: 4.4–7.6; alkalinity 16–45 mg/L; conductivity 45–97 mmhos; hardness 11–110 mg/L CaCO$_3$) include alluvium, clay, gabbroic rock, granite, granodiorite, gravel, humus, loam, loamy alluvium, muck, peat, quartz monzonite, rocky loam, rocky serpentine, sandy loam, serpentine peridotite, and silty loam (also occasionally found growing on logs). Flowering and fruiting extend from April to August. The plants are highly stoloniferous, and their vegetative reproduction often results in dense patches. Additional information is needed on the reproductive ecology of this species. **Reported associates:** *Abies amabilis, A. balsamea, A. fraseri, A. grandis, A. lasiocarpa, A. magnifica, Acer rubrum, Achillea millefolium, Aconitum, Agrostis idahoensis, A. pallens, Allium validum, Alnus rhombifolia, A. rubra, A. rugosa, A. viridis, Amelanchier, Andromeda glaucophylla, Antennaria corymbosa, Anthoxanthum, Aquilegia formosa, Arctostaphylos patula, Arethusa bulbosa, Arisaema triphyllum, Arnica cordifolia, Artemisia dracunculus, A. tridentata, Athyrium alpestre, A. filix-femina, Barbarea, Betula alleghaniensis, B. glandulosa, B. papyrifera, B. pumila, Bistorta bistortoides, Botrychium, Boykinia aconitifolia, Calamagrostis canadensis, Calocedrus decurrens, Caltha leptosepala, C. palustris, Camassia, Cardamine breweri, C. clematitis, C. flexuosa, Carex amplifolia, C. aquatilis, C. arctata, C. bolanderi, C. brunnescens, C. canescens, C. capitata, C. chordorrhiza, C. cusickii, C. disperma, C. geyeri, C. heteroneura, C. illota, C. interior, C. jonesii, C. lasiocarpa, C. lemmonii, C. lenticularis, C. leptalea, C. leptonervia, C. limosa, C. luzulina, C. mertensii, C. muricata, C. nebrascensis, C. neurophora, C. nigricans, C. paysonis, C. rostrata, C. scopulorum, C. senta, C. stipata, C. trisperma, C. utriculata, C. vesicaria, Castilleja cusickii, C. miniata, Ceanothus, Cephalanthus occidentalis, Cerastium nutans, Chamaedaphne calyculata, Chelone glabra, C. lyonii, Chrysosplenium americanum, Cicuta maculata, Cinna arundinacea, Circaea alpina, Claytonia cordifolia, C. siberica, Clintonia borealis, C. uniflora, Comarum palustre, Coptis trifolia, Corallorhiza trifida, Cornus, Cryptantha simulans, Cyprepedium acaule, Danthonia unispicata, Darlingtonia californica, Delphinium nuttallianum, Deschampsia cespitosa, Diphylleia cymosa, Drosera anglica, D. rotundifolia, Drymocallis cuneifolia, D. glandulosa, Dryopteris cristata, Eleocharis bella, E. engelmannii, Epilobium anagallidifolium, E. ciliatum, E. glaberrimum, Equisetum arvense, E. hyemale, E. scirpoides, Erigeron coulteri, E. philadelphicus, Eriophorum chamissonis, E. crinigerum, E. gracile, E. virginicum, Erysimum capitatum, Erythronium grandiflorum, Floerkea proserpinacoides, Fragaria vesca, F. virginiana, Frasera speciosa, Fraxinus nigra, Galium trifidum, G. triflorum, Gaultheria humifusa, Gentiana affinis, Gentianella amarella, Geranium, Geum geniculatum, G. macrophyllum, Glyceria canadensis, G. elata, Gnaphalium palustre, Gymnocarpium dryopteris, Heracleum sphondylium, Hieracium albiflorum, H. triste, Hosackia oblongifolia, Houstonia serpyllifolia, Hydrocotyle americana, Hymenoxys hoopesii, Hypericum anagalloides, H. scouleri, Ilex verticillata, Impatiens capensis, Iris versicolor, Ivesia campestris, Juncus balticus, J. brevicaudatus, J. drummondii, J. effusus, J. ensifolius, J. mertensianus, J. nevadensis, Kalmia microphylla, K. polifolia, Laportea canadensis, Larix laricina, L. occidentalis, Lilium grayi, L. kelleyanum, L. parryi, Linnaea borealis, Lloydia serotina, Lupinus argenteus, L. burkei, L. latifolius, L. polyphyllus, Luzula comosa, L. glabrata, L. parviflora, Lycopodium annotinum, L. clavatum, L. obscurum, Lycopus uniflorus, Lysichiton americanus, Maianthemum canadense, M. racemosum, M. stellatum, Melampyrum lineare, Menyanthes trifoliata, Menziesia ferruginea, Mertensia ciliata, Micranthes micranthidifolia, M. odontoloma, M. oregana, Mimulus lewisii, M. moschatus, M. primuloides, M. tilingii, Mitella breweri, M. diphylla, M. nuda, M. pentandra, Moneses uniflora, Monotropa uniflora, Montia chamissoi, Muhlenbergia filiformis, M. richardsonis, Myosotis laxa, Myrica gale, Noccaea fendleri, Onoclea sensibilis, Oplopanax horridus, Oreostemma alpigenum, Orthilia secunda, Osmundastrum cinnamomeum, Oxalis montana, Oxypolis rigidior, Parnassia asarifolia, Paxistima myrsinites, Pedicularis attollens, P. bracteosa, P. groelandica, Perideridia parishii, Phleum alpinum, Phyllodoce breweri, P. empetriformis, Picea engelmannii, P. glauca, P. mariana, P. rubens, Pinus albicaulis, P. contorta, P. flexilis, P. jeffreyi, P. lambertiana, P. monticola, Plagiobothrys hispidulus, Platanthera dilatata, P. obtusata, P. sparsiflora, P. stricta, Poa pratensis, P. secunda, Polemonium pulcherrimum, Polytrichum, Populus deltoides, P. tremuloides, Potentilla drummondii, P. flabellifolia, P. gracilis, P. grayi, P. tridentata, Primula jeffreyi, P. tetrandra, Prunus serotina, Pseudotsuga menziesii, Pteridium aquilinum, Pyrola asarifolia, Quercus chrysolepis, Q. kelloggii, Q. vacciniifolia, Ranunculus alismifolius, R. gormanii, R. orthorhynchus, R. populago, Rhamnus frangula, R. purshiana, Rhododendron columbianum, R. groenlandicum, Rhynchospora capitellata, Ribes cereum, R. hudsonianum, R. nevadense, Rosa palustris, Rubus parviflorus, Rumex, Salix exigua, S. humilis, S. lasiolepis, S. lutea, S. orestera, S. planifolia, S. wolfii, Sambucus nigra, Sarracenia purpurea, Sassafras albidum, Scirpus diffusus, Senecio integerrimus, S. scorzonella, S. triangularis, Sidalcea oregana, Sisyrinchium bellum, Solidago glomerata, S. multiradiata, Sorbus californica, Sphagnum russowii, Sphenosciadium capitellatum, Spiraea douglasii, S. splendens, Spiranthes gracilis, S. romanzoffiana, Stachys rigida, Stellaria corei, S. longifolia, S. longipes, Streptopus amplexifolius, Symphyotrichum ciliolatum, S. novi-belgii, S. spathulatum, Symplocarpus foetidus, Synthyris reniformis, Taraxacum officinale, Thalictrum clavatum, T. dasycarpum, T. occidentale, Thelypteris palustris, Thuja occidentalis, T. plicata, Tiarella cordifolia, T. trifoliata, Tofieldia, Toxicodendron vernix, Trautvetteria caroliniensis, Triadenum fraseri, Triantha occidentalis, Trientalis borealis, T. europaea, Trifolium cyathiferum, T. longipes, T. monanthum, Triglochin maritimum, Trillium albidum, T. cernuum, T. ovatum, Trollius laxus, Triphysaria pusilla, Tsuga canadensis, T. heterophylla,*

Typha latifolia, Ulmus americana, Urtica dioica, Vaccinium corymbosum, V. macrocarpon, V. membranaceum, V. oxycoccus, V. scoparium, V. uliginosum, Vahlodea atropurpurea, Veratrum californicum, Veronica americana, V. serpyllifolia, V. wormskjoldii, Vicia tetrasperma, Viola adunca, V. blanda, V. cucullata, V. glabella, V. lanceolata, V. nephrophylla, V. orbiculata, V. pallens, Xerophyllum tenax.

***Viola novae-angliae* House** inhabits crevices, floodplains, ledges, meadows, prairies, shores, and the margins of lakes, rivers, and streams at lower elevations. The plants tend to occupy more mesic–xeric microsites within or adjacent to wet habitats and do not necessarily require that high levels of soil moisture are sustained. They often occur in disturbed sites, which are subjected to flooding or ice scour. The species is shade tolerant but its highest flowering frequency occurs in open, sunny sites. Flowering and fruiting occur in June to August. The plants produce both chasmogamous (outcrossed) and cleistogamous (selfing) flowers. The seeds are dispersed ballistically. The substrates are acidic to circumneutral and include gravel, rock, and sand. Stolons are not produced by this species. Only 106 confirmed extant populations of *V. novae-angliae* were located in a 1994 field survey. This species may be miscategorized as an OBL plant, given its apparent preference for xeric microsites and its common association with species of more mesic habitats. **Reported associates:** *Abies balsamea, Achillea millefolium, Allium schoenoprasum, Alnus rugosa, Andropogon gerardii, Antennaria, Aquilegia canadensis, Betula papyrifera, Calamograstis canadensis, Campanula rotundifolia, Carex conoidea, C. flava, C. hassei, C. lacustris, C. pensylvanica, Castilleja coccinea, Chrysanthemum leucanthemum, Comandra umbellata, Cornus racemosa, Corylus cornuta, Danthonia spicata, Dasiphora floribunda, Deschampsia cespitosa, Diervilla lonicera, Doellingeria umbellata, Eleocharis elliptica, Equisetum arvense, Erigeron hyssopifolia, Eurybia macrophylla, Fragaria virginiana, Fraxinus pennsylvanica, Galium boreale, G. triflorum, Geum, Hedyotis longifolia, Helianthus pauciflorus, Heuchera richardsonii, Hieracium, Juncus vaseyi, Lilium michiganense, Lobelia spicata, Lonicera dioica, L. morrowii, Luzula multiflora, Maianthemum canadense, Mentha arvensis, Muhlenbergia mexicana, Myrica asplenifolia, Oenothera perennis, Oxalis violacea, Pinus banksiana, P. resinosa, Picea glauca, Poa compressa, Polygala sanguinea, Primula mistassinica, Prunus pumila, Pteridium aquilinum, Ranunculus acris, Rubus, Salix gracilis, S. pyrifolia, Scirpus clintonii, Senecio pauperculus, Solidago hispida, S. houghtonii, Sporobolus heterolepis, Tanacetum vulgare, Taraxacum officinale, Triantha glutinosa, Trifolium pratense, Trisetum spicatum, Vaccinium angustifolium, Vicia americana, Viola adunca, V. conspersa, V. lanceolata.*

***Viola palustris* L.** grows primarily in boreal and montane habitats including arctic and alpine bogs, bottoms, brooks, carrs, depressions, dunes, fens, flats, gulches, hummocks, marshes, meadows, seeps, shrub swamps, slopes (to 25%), springs, thickets, washes, and along the margins and shores of lakes, ponds, potholes, rivers, springs, and streams at elevations of up to 3840 m. The substrates are usually described

as calcareous (but range in pH from 4.0 to 7.9) and include alluvium, clay, clay loam, decomposing granitic sand, granite, humus, limestone, loam, mud, peat, pumice, quartzite, sand, and volcanics. The plants are tolerant of fairly deep shade. Flowers and fruits are produced from April to September. Cleistogamous flowers are promoted under longer days, whereas short days induce more chasmogamous flowers. The seeds (averaging 0.67 mg) are conditionally physiologically dormant and require a 90-day-cold stratification period to induce germination (achieved under 20°C/15°C and 15°C/25°C temperature regimes). Germination requires light. Dry storage induces secondary dormancy, which lowers the germination rates of the seeds. The highest germination for secondarily dormant seeds has been obtained under a 15°C/25°C temperature regime. This species accumulates an extensive, persistent seed bank. The plants reproduce vegetatively by the production of rhizomes and stolons. **Reported associates:** *Abies amabilis, A. grandis, A. lasiocarpa, Acer, Achillea borealis, Aconitum columbianum, Actaea rubra, Agrostis capillaris, A. idahoensis, A. variabilis, Alnus rugosa, A. rubra, A. tenuifolia, A. viridis, Alopecurus geniculatus, Anaphalis margaritacea, Anemone parviflora, Antennaria umbrinella, Arnica cordifolia, Artemisia norvegica, Asclepias speciosa, Astragalus, Athyrium alpestre, A. filix-femina, Aulacomnium palustre, Berula erecta, Betula glandulifera, B. glandulosa, B. papyrifera, Bistorta bistortoides, Boechera lyallii, Brachythecium, Calamagrostis canadensis, C. nutkaensis, C. stricta, Caltha leptosepala, Campylium stellatum, Canadanthus modestus, Carex aquatilis, C. bigelowii, C. brunnescens, C. buxbaumii, C. canescens, C. capillaris, C. disperma, C. echinata, C. exilis, C. exsiccata, C. flava, C. garberi. C. hystericina, C. interior, C. kelloggii, C. lasiocarpa, C. lenticularis, C. limosa, C. livida, C. magellanica, C. membranacea, C. muricata, C. panicea, C. rostrata, C. scopulorum, C. utriculata, C. vesicaria, C. viridula, Cassiope, Castilleja miniata, C. septentrionalis, Circaea alpina, Claytonia cordifolia, Clintonia borealis, Comarum palustre, Coptis groenlandica, C. trifolia, Cornus canadensis, C. sericea, Cryptogramma acrostichoides, Cyperus bipartitus, Danthonia intermedia, Dasiphora floribunda, Daucus carota, Deschampsia cespitosa, Digitalis, Drosera rotundifolia, Dryopteris expansa, Eleocharis palustris, E. rostellata, Empetrum nigrum, Epilobium alpinum, Epipactis gigantea, Equisetum arvense, E. fluviatile, E. sylvaticum, E. variegatum, Erigeron eatonii, E. peregrinus, Eriophorum angustifolium, E. gracile, Euphrasia oakesii, Festuca altaica, Fragaria virginiana, Galium, Gentiana sceptrum, Geum macrophyllum, G. peckii, Glyceria borealis, G. maxima, Gymnocarpium dryopteris, Harrimanella hypnoides, Holodiscus, Houstonia caerulea, Hypericum anagalloides, H. formosum, Juncus ensifolius, Kalmia microphylla, Larix laricina, L. occidentalis, Leptarrhena pyrolifolia, Limprichtia revolvens, Linnaea borealis, Lobelia kalmii, Lonicera involucrata, Lupinus, Luzula parviflora, L. spicata, Lycopus uniflorus, Lysichiton americanus, Maianthemum canadense, Marchantia, Menyanthes trifoliata, Menziesia ferruginea, Mertensia ciliata, Micranthes oregana, Mimulus moschatus, Mitella nuda, M. pentandra, Moneses uniflora. Montia chamissoi, Myrica* *gale, Oenanthe, Oplopanax horridus, Orthilia secunda, Osmorhiza purpurea, Oxyria digyna, Parnassia fimbriata, P. parviflora, Paxistima myrsinites, Penstemon, Phegopteris connectilis, Phleum alpinum, Phyllodoce caerulea, P. empetriformis, Picea engelmannii, P. mariana, P. pungens, Pinguicula villosa, Pinus albicaulis, P. contorta, P. flexilis, P. monticola, P. ponderosa, Plagiomnium medium, Platanthera dilatata, P. sparsiflora, P. stricta, Pleurozium schreberi, Poa, Podagrostis thurberiana, Polystichum, Polytrichum, Populus balsamifera, P. trichocarpa, Potentilla erecta, Primula jeffreyi, Prosartes trachycarpa, Prunus emarginata, Pseudotsuga menziesii, Ptilium crista-castrensis, Pyrola chlorantha, P. minor, Ranunculus glaberrimus, R. grayi, Rhizomnium glabrescens, Rhododendron albiflorum, R. columbianum, R. groenlandicum, Rhus aromatica, Ribes lacustre, Rosa acicularis, R. woodsii, Rubus arcticus, R. pedatus, R. pubescens, Sagittaria latifolia, Salix bebbiana, S. commutata, S. geyeriana, S. maccalliana, S. planifloria, S. scouleriana, S. serissima, S. sitchensis, Sambucus, Sanguisorba officinalis, S. sitchensis, Schoenoplectus acutus, Scirpus cyperinus, S. microcarpus, Scorpiodium scorpioides, Senecio triangularis, Solidago macrophylla, Sorbus, Sparganium eurycarpum, Sphagnum angustifolium, S. centrale, S. subsecundum, S. teres, Sphenopholis obtusata, Spiraea douglasii, Stellaria obtusa, Streptopus amplexifolius, S. roseus, Taxus, Tellima grandiflora, Thalictrum alpinum, T. occidentale, Thuja plicata, Tiarella trifoliata, Tofieldia pusilla, Tomenthypnum falcifolium, Trichophorum alpinum, T. cespitosum, Tsuga heterophylla, T. mertensiana, Typha latifolia, Urtica dioica, Vaccinium cespitosum, V. membranaceum, V. ovalifolium, V. oxycoccos, V. uliginosum, Vahlodea atropurpurea, Valeriana sitchensis, Veratrum viride, Veronica americana, V. scutellata, Viburnum edule, Viola glabella, V. renifolia, V. sempervirens, Xerophyllum tenax.*

Use by wildlife: Violet seeds are eaten by many birds (Aves) and also by small mammals (Mammalia: Rodentia). The rhizomes are a favorite food of wild turkeys (Aves: Phasianidae: *Meleagris gallopavo*). The roots of *V. cucullata* are eaten by woodland voles (Mammalia: Cricetidae: *Microtus pinetorum*) and are also associated with Nematodes (Nematoda: Criconematidae: *Paratylenchus microdorus*). The foliage hosts numerous plant pathogens including leaf spots and rusts (Fungi: Ascomycota: Incertae sedis: *Ascochyta violae*; Botryosphaeriaceae: *Phyllosticta violae*; Dermateaceae: *Marssonina violae*; Elsinoaceae: *Sphaceloma violae*; Erysiphaceae: *Alphitomorpha fuliginea*; Glomerellaceae: *Colletotrichum violae-rotundifoliae*, *C. violae-tricoloris*, *Vermicularia violae*; Mycosphaerellaceae: *Cercospora violae*, *Passalora granuliformis*, *P. murina*, *Septoria violae*; Pleosporaceae: *Alternaria violae*; Basidiomycota: Pucciniaceae: *Puccinia ellisiana*, *P. violae*; Typhulaceae: *Sclerotium rolfsii*; Chytridiomycota: Synchytriaceae: *Synchytrium aureum*). The plants are a larval host of Diana fritillary butterflies (Insecta: Lepidoptera: Nymphalidae: *Speyeria diana*). *Viola lanceolata* is grazed by cattle (Mammalia: Bovidae: *Bos taurus*) and horses (Mammalia: Equidae: *Equus ferus*). It is a host plant of the silver-bordered

and regal fritillary butterflies (Insecta: Lepidoptera: Nymphalidae: *Boloria selene*; *Speyeria aphrodite*, *S. idalia*). The foliage is a host of leaf spot (Fungi: Ascomycota: Mycosphaerellaceae: *Septoria violae*). The flowers are visited by bees (Insecta: Hymenoptera: Andrenidae: *Andrena*; Halictidae: *Halictus poeyi*; Megachilidae: *Osmia*), which serve as their primary pollinators. They also attract skippers and pearl crescent butterflies (Insecta: Lepidoptera: Hesperiidae: *Hesperia*; Nymphalidae: *Phyciodes tharos*). A urediniomycete rust fungus (*Puccinia fergussonii*) is associated with *V. langsdorffii*. The flowers of *V. macloskeyi* attract brush-footed butterflies (Insecta: Lepidoptera: Nymphalidae: *Boloria bellona*, *B. titania*) and those of *V. palustris* attract great spangled and Mormon fritillaries (Insecta: Lepidoptera: Nymphalidae: *Speyeria cybele*, *S. mormonia*).

Economic importance: food: All violet (*Viola*) flowers or flower buds, and the leaves of many species, reportedly are edible; however, the plants should not be confused with the unrelated "African violets" (Gesneriaceae: *Saintpaulia*), which are *not* edible. Yellow-flowered species are said to be strongly laxative and should not be eaten in excess (none of the aquatic North American species has yellow flowers). Violet roots should also not be eaten as they can be emetic. Violet flowers are used often in salads or can be jellied and candied. The leaves are eaten as a salad green and contain vitamins A and C. A tea can be brewed from the leaves. Young leaves and flower buds of *V. cucullata* have been added to soup as a thickening agent. The young leaves and flowers of *V. langsdorffii* can be eaten raw or cooked, or dried and brewed into a tea. The flowers can be candied or added as a flavoring for vinegar; **medicinal:** The Cherokee administered a leaf poultice of *V. cucullata* to treat headaches and applied a poultice of crushed roots to soothe boils. They also used an infusion of the plant as a nose spray, tonic and to treat colds, coughs, and dysentery. *Viola cucullata* was a general medicinal plant used also by the Ute people. The leaves and flowers of *V. langsdorffii* are eaten as a preventative for bronchial disorders; **cultivation:** Violets such as "wild pansy" and "johnny-jump-up" (*V. tricolor*) represent some of the most popular and widely cultivated garden plants. Among the wetland species, *V. cucullata* (popular since the latter part of the 19th century) has been designated an "Award of Garden Merit" plant by the Royal Horticultural Society. It is also sold in trade as the cultivars 'Alba,' 'Freckles,' 'Rosea,' 'Striata Alba,' and 'Striata Emperor White.' *Viola langsdorfii* and *V. palustris* are sold occasionally for native wildflower gardens; **misc. products:** The Cherokee prepared corn for planting by soaking it in a root infusion of *V. cucullata* to ward off insects. The quick spreading habit of this species makes it useful as a ground cover, especially for sloping sites. *Viola cucullata* is the provincial flower of New Brunswick, Canada; **weeds:** *Viola lanceolata* is regarded as a weed of cranberry fields in western North America; **nonindigenous species:** All of the North American OBL violets are indigenous; however, the disjunct western distribution of *V. lanceolata* occurred most likely as an introduction from contaminated cranberry stock, sometime prior to 1936.

Systematics: Preliminary molecular analyses indicate that *Viola* is related most closely to *Hybanthus* and *Noisettia*. Phylogenetic analyses of nuclear ribosomal internal transcribed spacer (nrITS) sequence data for 43 *Viola* species indicate that the genus is monophyletic, but that several of the currently recognized sections (e.g., *Chamaemelanium*, *Viola*) are not (Figure 4.38). The same analyses indicate that *V. langsdorffii* is the sister species to the Hawaiian members of this genus. However, because other analyses have clarified that several of the species (e.g., *V. langsdorffii*, *V. palustris*) are allopolyploids with a reticulate ancestry, it is difficult to interpret their complex relationships by simple cladograms. As one example, the North American and European members of *V. palustris* are not monophyletic, but have arisen from different ancestral species groups. *Viola cucullata* is closely related to *V. affinis*, a FACW indicator. Some analyses resolve *V. macloskeyi* and *V. domingensis* (which have been merged by some authors) as sister species, while others resolve *V. macloskeyi* and *V. occidentalis* as a subclade. *Viola novae-angliae* is recognized by some authors as *V. sororia* var. *novae-angliae* (see later). Hybrids (including introgressive swarms) are reported widely in *Viola*. As many as 11 species reportedly hybridize with *V. cucullata* and more than four species with *V. lanceolata*, but documentation of these and other hybrids in the genus will require additional study incorporating genetic/molecular analyses. Among the better documented cases involving aquatic species are alleged hybrids (known as *V. ×primulifolia*) between *V. lanceolata* and *V. macloskeyi* and suspected instances of introgression between *V. cucullata* and *V. septentrionalis*, and among *V. incognita*, *V. macloskeyi*, and *V. renifolia*. *Viola novae-angliae* is postulated to be a hybrid derivative of *V. sagittata* and *V. sororia*. The chromosomal base number of *Viola* probably is *x* = 6 (although a few *x* = 3, 4 counts have been reported). Following this interpretation, *V. lanceolata* and *V. macloskeyi* (2*n* = 24) would be tetraploid and *V. palustris* (2*n* = 48) octoploid. *Viola cucullata* (2*n* = 54) is also polyploid (nonaploid). *Viola langsdorffii* is highly polyploid and variable cytologically (2*n* = 60, 100, 120). Counts for *V. novae-angliae* have not been reported, but *V. sororia* (with which it has been merged) is 2*n* = 54.

Comments: *Viola cucullata* is distributed throughout eastern North America, whereas *V. lanceolata* occurs disjunctly in eastern and western North America. *Viola langsdorfii* is restricted to western North America with an amphi-Beringian distribution extending into Asia. *Viola novae-angliae* is a relatively rare species restricted to northeastern North America. *Viola palustris* occurs in northern and western North America as well as in northern Eurasia. *Viola macloskeyi* is relatively widespread except for central and extreme southern North America.

References: Anderson et al., 1984; Ballard & Gawler, 1994; Ballard & Sytsma, 2000; Ballard et al., 1999; Bragazza & Gerdol, 2002; Braun, 1928; Craine & Orians, 2004; Culver &

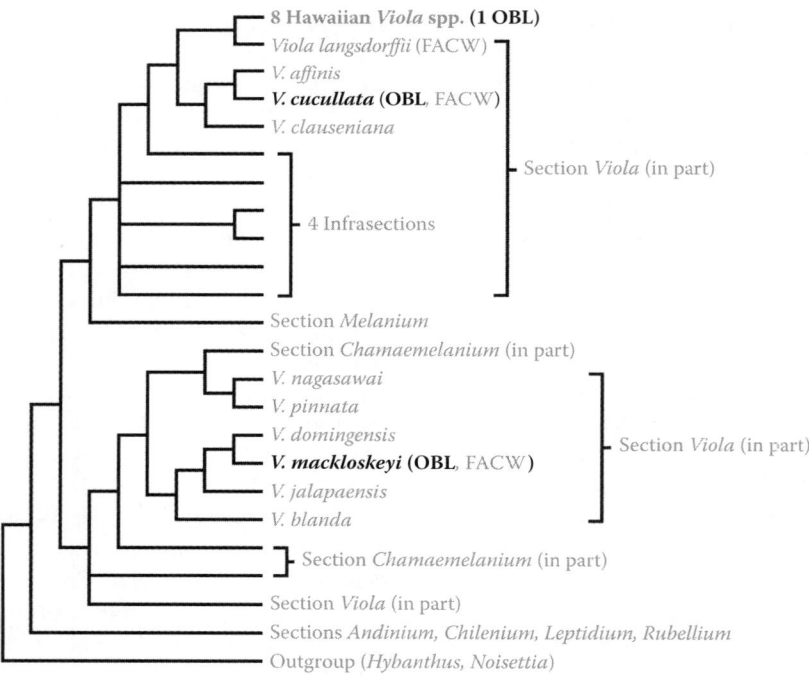

FIGURE 4.38 Phylogenetic relationships in *Viola* as indicated by nrITS DNA sequence data. The OBL North American species included (shown in bold) have arisen independently in the genus. The one obligately aquatic Hawaiian species (*V. kauaensis*—not shown) is nested within the Hawaiian clade, thereby indicating another independent origin of the aquatic habit in the genus. In this analysis, *V. langsdorffii* resolves as the sister species to the Hawaiian clade. The nrITS data indicate that several of the taxonomic sections currently recognized (e.g., *Chamaemelanium*, *Viola*) are polyphyletic. More recent analyses incorporating low-copy nuclear gene data (Marcussen et al., 2012) have indicated a more complex picture of relationships, particularly for the higher polyploids (e.g., *V. langsdorffii*, *V. palustris*), for which a reticulate allopolyploid ancestry is inferred. (Adapted from Ballard, Jr., H.E. & Sytsma, K.J., *Evolution*, 54, 1521–1532, 2000.)

Beattie, 1978; Erben, 1996; Greene, 1949; Hamilton, Jr., 1938; Jensen, 2004; Jeglum, 1971; Jones, 1937; Marcussen et al., 2012; Martin, 1887; Mohlenbrock, 1959b; Murrill, 1940a,b; Nieuwland, 1914; Peterson & Baldwin, 2004; Ranua & Weinig, 2010; Reznicek & Maycock, 1983; Rigg, 1937; Robertson, 1889b; Robuck, 1985; Russell, 1954, 1955; Schultz, 1946; Solbrig et al., 1988; Soltis et al., 2000; Sperduto, 1996; Swengel, 1997; Talbot et al., 1995; Wisheu & Keddy, 1991; Wu, 1962; Zartman & Pittillo, 1998; Zika, 2003.

ORDER 6: ROSALES [9]

Phylogenetically, members of Rosales occur among the rosid dicotyledons in a clade with Cucurbitales, Fagales, and Fabales being nearest the latter order (Figures 4.1 and 4.39). The order Rosales (Figure 4.39) with roughly 6300 species is well supported by various molecular data sets (Judd & Olmstead, 2004). However, potentially defining (synapomorphic) morphological characters for Rosales are few, among them are the presence of a hypanthium and lack (or reduction) of endosperm. Within the order occurs the well-defined "urticoid subclade," which is characterized by several molecular data sets, an intron in the mitochondrial *nad1* gene and a number of morphological characters such as urticoid leaf teeth, prominent prophyllar buds, reduced flowers, five or fewer stamens, bicarpellate, unilocular ovaries, and single apical/basal ovules (Sytsma et al., 2002; Judd & Olmstead, 2004).

Rosales are principally terrestrial. Family interrelationships in Rosales (Figure 4.39) indicate multiple origins of the obligate aquatic habit, with the families containing aquatics dispersed throughout the clade. In all, 13 genera in four families contain species categorized as obligate aquatics in North America:

1. **Rosaceae** Adans.
2. **Rhamnaceae** Juss.
3. **Ulmaceae** Mirb.
4. **Urticaceae** Juss.

Family 1: Rosaceae [85]

Rosaceae are a large family containing some 3000 species, which are mainly terrestrial. Despite their size, only two dozen species of Rosaceae are recognized among the obligate aquatics in North America. The family includes a diverse assemblage of herbs, trees, and shrubs, which lack alkaloids and possess showy, 5-merous flowers with a hypanthium, numerous stamens, and different fruit types (see later).

Several sources of data (molecular and morphological) support the monophyly of the family; however, delimitation of traditional subfamilies (Maloideae, Prunoideae, Rosoideae, Spiraeoideae), once based principally on fruit types (pomes, drupes, achenes, and follicles, respectively), has been more problematic. A number of studies incorporating molecular data have indicated that fruit-based subfamily limits in Rosaceae do not always effectively represent monophyletic groups (e.g., Morgan et al., 1994; Potter et al., 2002).

Although phylogenetic studies of Rosaceae are yet to resolve a consistent subfamilial arrangement of genera, certain details are emerging. Various molecular data have generated similar phylogenetic topologies where subfamily Rosoideae is basal to clades consisting of genera placed traditionally within subfamilies Spiraeaoideae, Prunoideae, and Maloideae that successively diverge from Rosoideae, respectively; interspersed among those clades are a number of misplaced genera once assigned taxonomically to subfamilies Spiraeoideae and Rosoideae (Figure 4.40). Most of the aquatics (occurring in seven genera) occur within subfamily Rosoideae (Figure 4.41), which appears to be a reasonably well-supported, natural clade (Eriksson et al., 1998, 2003).

Many species of economic importance occur within the family. Among these are various fruit trees such as *Malus* (apple, crabapple), *Prunus* (almond, apricot, cherry, peach, plum), and *Pyrus* (pear) as well as other widely cultivated fruits such as *Fragaria* (strawberry) and *Rubus* (blackberry, raspberry). The family is also rich in ornamental species in addition to many

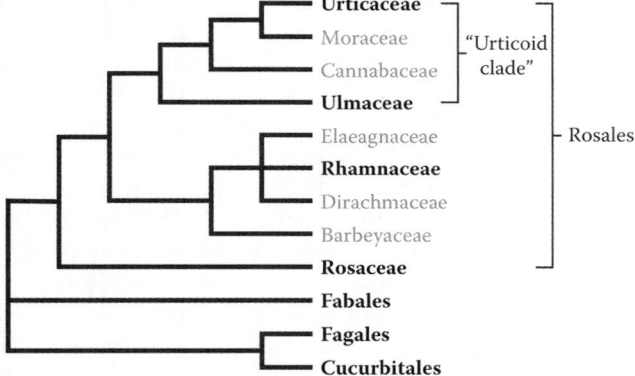

FIGURE 4.39 Phylogenetic relationships among families of Rosales as indicated by phylogenetic analysis of multiple DNA data sets. Taxa containing obligate aquatics (shown in bold) occur throughout the order. (Modified from Judd, W.S. et al., *Plant Systematics: A Phylogenetic Approach*, Sinauer Associates, Inc., Sunderland, MA, 2002; Sytsma, K.J. et al., *Amer. J. Bot.*, 89, 1531–1546, 2002; APG, *Bot. J. Linn. Soc.*, 141, 399–436, 2003.)

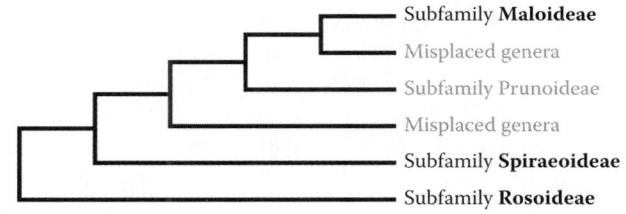

FIGURE 4.40 Emerging consensus of relationships among traditional subfamilies of Rosaceae. More recent analyses (Evans et al., 2002a; Potter et al., 2002) indicate that further refinements to this pattern of relationships and to the resulting classification undoubtedly will be necessary. Obligate aquatics occur within the subfamilies in bold, indicating multiple and widespread origins of the hydrophytic habit in the family. Seven of the genera containing obligate aquatics occur within subfamily Rosoideae, with an additional genus each in subfamilies Maloideae and Spiraeoideae. (Adapted from Morgan, D.R. et al., *Amer. J. Bot.*, 81, 890–903, 1994.)

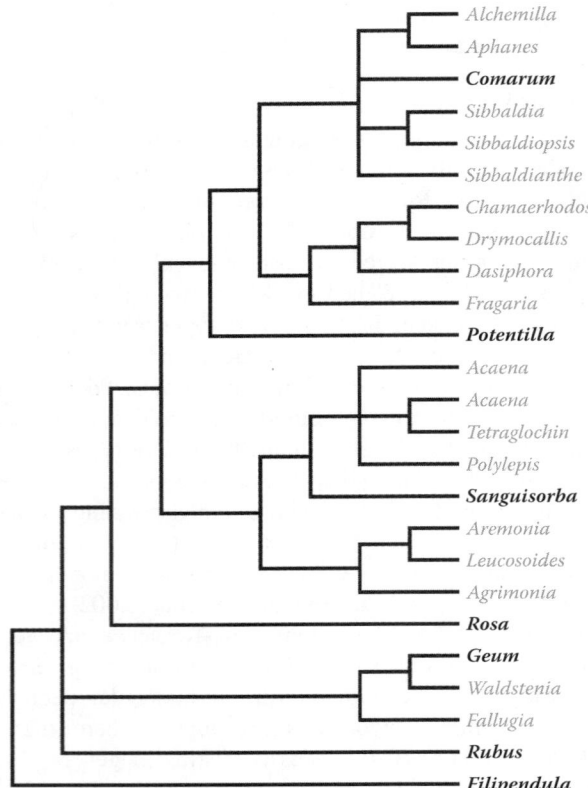

FIGURE 4.41 Intergeneric relationships in Rosaceae subfamily Rosoideae as indicated by phylogenetic analysis of combined data. Seven genera (in bold) contain species that are categorized as obligate aquatics in North America. Because all of these genera (except *Comarum*; 1–2 species) are principally terrestrial, it is reasonable to conclude that the aquatic habit has evolved independently in each case, in a pattern scattered throughout the subfamily. (Adapted from Eriksson, T. et al., *Int. J. Plant Sci.*, 164, 197–211, 2003.)

of the plants just mentioned. The most significant genus is *Rosa* (rose), which has provided thousands of cultivars to the commercial flower industry. Other important genera include *Cotoneaster*, *Crataegus*, *Pyracantha*, and *Spiraea*, with hundreds of cultivars planted as ornamental shrubs. The sugar alcohol sorbitol (a sugar substitute) was first extracted from the berry juice of mountain ash (*Sorbus*). Some species (e.g., *Rosa multiflora*: FACU, UPL) are noxious weeds, which can extend at least into the periphery of wetlands.

Obligate aquatics are found within nine North American genera of Rosaceae:

    1. *Comarum* L.
    2. *Crataegus* L.
    3. *Filipendula* Mill.
    4. *Geum* L.
    5. *Potentilla* L.
    6. *Rosa* L.
    7. *Rubus* L.
    8. *Sanguisorba* L.
    9. *Spiraea* L.

1. *Comarum*

Marsh cinquefoil, marshlocks; comaret
Etymology: from *komaros*, the Greek name for the unrelated *Arbutus unedo*, which has a superficially similar fruit
Synonyms: *Potentilla* (in part)
Distribution: global: Asia; Europe; North America; **North America:** northern
Diversity: global: 1 species; **North America:** 1 species
Indicators (USA): OBL: *Comarum palustre*
Habitat: brackish, freshwater; palustrine; **pH:** 5.2–8.9; **depth:** <1 m; **life-form(s):** emergent herb
Key morphology: stems (to 6 dm) decumbent; leaves (to 20 cm) pinnate, with 5–7 toothed leaflets (2–10 cm), petiole bases sheathing, stipules membranous; inflorescence few-flowered (<10), epicalyx present; flowers 5-merous, with a hypanthium (to 10 mm); sepals (to 15 mm) reddish-purple inside; petals (to 6 mm) maroon; stamens numerous (20–25); gynoecium apocarpous, carpels 40–60; achenes (to 1.5 mm) numerous, in aggregates
Life history: duration: perennial (rhizomes); **asexual reproduction:** rhizomes; **pollination:** insect; **sexual condition:** hermaphroditic; **fruit:** aggregates (achenes) (common); **local dispersal:** seeds; **long-distance dispersal:** seeds (water)
Imperilment: (1) *Comarum palustre* [G5]; S1 (CO, NJ, UT, WY); S2 (ND, OH); S3 (IA)
Ecology: general: monotypic (see next).

    Comarum palustre **L.** is rooted in the substrate but floats along the surface or emerges from shallow water in a diverse range of habitats including beaches, beaver ponds, bogs, bottoms, channels, depressions, dunes, fens, floodplains, ditches, kettle holes, marshes, meadows, mires, mudflats, muskeg, oxbows, prairies, sloughs, swales, swamps, thickets, tidal depressions, tundra, and the margins of lakes, ponds, and rivers at elevations of up to 3100 m. It is tolerant to a wide range of (pH: 5.2–8.9) and occurs on various substrates including calcareous muck, cobble, mud, gravel, limestone, loam, peat, sand, schist, silt clay, and occasionally within the interstices of rocks. It is sensitive to salinity, but at high latitudes, it is known to persist in coastal depressions and in intertidal salt marshes. The plants typically occur in full sun in habitats with low alkalinity (<130 mg/L CaCO$_3$) and low conductivity (<230 μmhos/cm). Flowering occurs from May to August. The flowers emit a fetid odor, which attracts flies for pollination. Insect pollinators reported in Old World populations include several bees (Hymenoptera), beetles (Coleoptera), and Lepidoptera. The plants are cold tolerant and can reportedly survive temperatures as low as −25°C. Plant vigor is stimulated by periodic mowing but not by drawdown conditions. A high foliar tannin and alkaloid content may be responsible for the low usage of this species as forage by grazing birds (Aves). *Comarum palustre* is not grazed by geese (Aves: Anatidae) in the Arctic, where many plants invade open mossy sites that result from foraging by lesser snow geese (*Chen caerulescens*). The seeds are dispersed by water along with the drift. They require no pretreatment for germination and can be sown in a cold frame during the fall or spring. Seeds sown below 5°C exhibit irregular germination after several months. Good germination is reported for seeds when brought to temperatures of 21°C–27°C. **Reported associates:** *Abies balsamea, A. lasiocarpa, Achillea alpina, Acorus calamus, Agrostis canina, A. scrabra, Alisma, Alnus rugosa, A. tenuifolia, Andromeda glaucophylla, Antennaria corymbosa, Apocynum cannabinum, Arctophila fulva, Aronia melanocarpa, Artemisia campestris, A. norvegica, Aulacomnium palustre, Barbarea orthoceras, Betula glandulifera, B. glandulosa, B. lutea, B. nana, B. papyrifera, B. pumila, Bidens cernuus, Brasenia schreberi, Bromus ciliatus, B. tectorum, Calopogon pulchellus, Calamagrostis canadensis, C. stricta, Calla palustris, Caltha palustris, Campanula aparinoides, Carex arcta, C. aquatilis, C. buxbaumii, C. canescens, C. comosa, C. cusickii, C. diandra, C. disperma, C. echinata, C. flava, C. interior, C. lacustris, C. lasiocarpa, C. leptalea, C. limosa, C. lyngbyei, C. oligosperma, C. pluriflora, C. rostrata, C. stricta, C. tenuiflora, C. trisperma, C. utriculata, C. vesicaria, Ceratophyllum, Chamaedaphne calyculata, Chamerion angustifolium, Chrysosplenium tetrandrum, Cicuta bulbifera, C. douglasii, C. virosa, Cladium mariscoides, Climacium dendroides, Coptis trifolia, Cornus canadensis, C. rugosa, C. sericea, Cypripedium parviflorum, C. passerinum, Dasiphora floribunda, Deschampsia, Drosera anglica, D. intermedia, D. rotundifolia, Dryopteris cristata, Dryas drummondii, Dulichium arundinaceum, Eleocharis palustris, Elodea canadensis, Elymus glaucus, E. trachycaulus, Epilobium ciliatum, E. leptocarpum, E. leptophyllum, E. palustre, Equisetum arvense, E. fluviatile, E. variegatum, Eriophorum angustifolium, E. chamissonis, E. gracile, E. virginianum, Eurybia sibirica, Eutrochium maculatum, E. purpureum, Fragaria virginiana, Fraxinus nigra, Galearis rotundifolia, Galium bifolium, G. boreale, G. tinctorium, G. trifidum, G. triflorum, Glyceria canadensis, G. grandis, G. striata, Hippuris montana, H. vulgaris, Ilex verticillata, Impatiens capensis, Iris versicolor, I. virginica, Juncus balticus, J. canadensis, J. filiformis, Kalmia polifolia, Larix laricina, Lemna minor, Leymus mollis, Linnaea borealis, Liparis loeselii, Lonicera involucrata, Lycopus americanus, L. uniflorus, Lysimachia terrestris, L. thyrsiflora, Lythrum salicaria, Maianthemum canadense, M. trifolium, Mentha arvensis, Menyanthes trifoliata, Mitella nuda, Moneses uniflora, Muhlenbergia glomerata, Myrica gale, Nuphar polysepala, N. variegata, Nymphaea, Parnassia palustris, Pedicularis parviflora, Persicaria amphibia, P. hydropiper, Petasites sagittatus, Phalaris arundinacea, Physocarpus capitatus, P. opulifolius, Picea engelmannii, P. mariana, Pinus pumila, P. strobus, P. resinosa, Platanthera dilatata, P. hyperborea, P. stricta, Poa alsode, P. palustris, Pogonia ophioglossoides, Populus trichocarpa, Potamogeton, Potentilla anserina, P. norvegica, Pseudoroegneria spicata, Pyrola asarifolia, P. chlorantha, Pyrus floribunda, Ranunculus gmelini, R. pensylvanicus, Rhamnus alnifolia, Rhinanthus minor, Rhododendron groenlandicum, Rhynchospora alba, Ribes hudsonianum, Rorippa palustris, Rubus arcticus, R. chamaemorus, R. pubescens, Rudbeckia hirta, Rumex altissimus, Sagittaria cristata, S. latifolia, Salix alaxensis, S. arctophila, S. barclayi, S. bebbiana, S. boothii, S. discolor, S. drummondiana, S. eriocephala, S. geyeriana, S. glauca, S. lasiandra, S. lucida, S. niphoclada, S. pedicellaris, S. petiolaris, S. planifolia, S. reticulata, Sarracenia purpurea, Schoenoplectus tabernaemontani, Scirpus atrovirens, S. cyperinus, Scutellaria galericulata, Sibbaldia procumbens, Sium suave, Solidago gigantea, S. patula, S. uliginosa, Sorbus decora, Sparganium chlorocarpum, S. eurycarpum, S. glomeratum, S. hyperboreum, S. natans, Sphagnum fuscum, Spiraea alba, S. betulifolia, S. tomentosa, Spiranthes romanzoffiana, Symphyotrichum boreale, S. eatonii, S. lanceolatum, S. spathulatum, Symplocarpus foetidus, Thalictrum sparsiflorum, Thelypteris palustris, Thuja occidentalis, T. plicata, Toxicodendron vernix, Triadenum fraseri, Trianthaglutinosa, Trichophorum pumilum, Trientalis borealis, T. europaea, Triglochin palustre, Typha angustifolia, T. latifolia, Usnea, Utricularia gibba, U. cornuta, U. intermedia, U. macrorhiza, Vaccinium macrocarpon, V. oxycoccus, V. uliginosum, V. vitisideae, Viburnum edule, Viola glabella, V. palustris, V. sagittata.*

Use by wildlife: Seeds of *Comarum palustre* are eaten by shore birds such as grouse (Aves: Phasianidae) and waterfowl (Aves: Anatidae). The leaves are grazed by hares and rabbits (Mammalia: Leporidae) and constitute a part of the seasonal diet for wild moose (Mammalia: Cervidae: *Alces alces*). The foliage comprises 18.8% crude protein and is a preferred food of caribou and reindeer (Mammalia: Cervidae: *Rangifer tarandus*). The plants serve as a host for butterfly larvae (Lepidoptera: Lycaenidae: *Epidemia dorcas*), plant bugs (Miridae: *Macrotylus josephinae*), and for several Ascomycota fungi (Dermateaceae: *Marssonina potentillae*; Erysiphaceae: *Sphaerotheca macularis*; Mycosphaerellaceae: *Mycosphaerella fragariae, Ramularia arvensis, Septoria purpurascens*; Valsaceae: *Gnomonia fragariae*; Insertae sedis: *Septogloeum potentillae*). In Eurasia, the plants reportedly are grazed by deer (Mammalia: Cervidae), sheep (Mammalia: Bovidae: *Ovis aries*), and beavers (Mammalia: Castoridae: *Castor fiber*).

Economic importance: **food:** The Eskimo brewed dried leaves of *Comarum palustre* as a tea; **medicinal:** The Chippewa used decoctions made from the astringent roots of *C. palustre* to treat dysentery. The plants were used by the Ojibwa to relieve abdominal cramps. The anti-inflammatory properties of the plants are attributed to the substance comaruman, which inhibits adhesion of human neutrophils to fibronectin; **cultivation:** *Comarum palustre* is cultivated as an ornamental water garden plant; **misc. products:** The flowers of *C. palustre* yield a red dye and its roots are a rich source of tannins; **weeds:** none; **nonindigenous species:** none

Systematics: Recent systematic studies support the recognition of *Comarum* (subfamily Rosoideae) as distinct from *Potentilla* (with which it long had been allied), as it is quite distant from other members of that genus in the analyses of molecular data (e.g., Figure 4.41). Although *Comarum* comprises only one species as treated here, the delimitation of this genus is complex, considering the inclusion of the morphologically similar *Farinopsis salesoviana* (as *Comarum salesoviana*) by some authors. Some studies have identified the closest allies of *Comarum* as *Alchemilla, Aphanes, Sibbaldia,*

Sibbaldiopsis, and *Sibbaldianthe*. Although more distantly related to *Fragaria* (strawberry), *Comarum* is able to hybridize with that genus, a development that led to the production of pink-flowered strawberry cultivars. Moreover, recent analyses of nuclear (nrITS) and plastid (*trnL-F*) sequence data have shown discrepancies between phylogenetic trees derived from the biparentally vs. maternally inherited markers. The nuclear data resolve *Comarum palustre* and *Farinopsis salesoviana* as closely related species; whereas the cpDNA markers resolve them in different clades, with *C. palustre* associating as the sister species of *Fragaria virginiana*. Those results suggest problems relating to hybrid material, where "chloroplast capture" by the maternal parent may be responsible for generating such disparate phylogenetic placements. Until the status of *Farinopsis salesoviana* is settled more definitively, a conservative approach has been followed here, which retains it as a distinct genus. The chromosomal base number of *Comarum* is x = 7. *Comarum palustre* is widespread geographically and variable cytologically. A variety of ploidy levels is represented among populations including (with aneuploid derivatives) tetraploids (2n = 28), pentaploids (2n = 35–36), hexaploids (2n = 42), octaploids (2n = 56), and nonaploids (2n = 62–64).

Comments: *Comarum palustre* is circumpolar in distribution. In North America it ranges across the United States from the mid-latitudes northward. Populations are threatened mainly along the southern extent of its range.

References: Aiken et al., 2007; Batt, 1997; Dieterich & Morton, 1990; Ertter & Reveal, 2014; Faghir et al., 2014; Jansson et al., 2005; Jensen & Meyer, 2001; Ngai & Jefferies, 2004; Nichols, 1999a; Popov et al., 2005.

2. *Crataegus*

Hawthorne, Mayhaw; cenellier, pomettes
Etymology: from the Greek *kratos* meaning "strength," in reference to the hardness of the wood.
Synonyms: *Anthomeles*; *Mespilus* (in part)
Distribution: global: Asia; Europe; New World; **North America:** widespread
Diversity: global: 230 species; **North America:** 169 species
Indicators (USA): OBL: *Crataegus aestivalis*, *C. brachyacantha*, *C. opaca*, *C. viridis*; **FACW:** *C. brachyacantha*, *C. viridis*; **FAC:** *C. viridis*
Habitat: freshwater; palustrine, riverine; **pH:** 4.2–7.0; **depth:** <1 m; **life-form(s):** emergent shrub, emergent tree
Key morphology: shrubs or trees (to 15 m); branches having straight or curved thorns (to 4 cm); leaves (to 6.5 cm) subcoriaceous, the margins entire to crenate/serrate, petiolate (0.5–25 mm); flowers showy, 5-merous, epigynous, few to numerous in corymbs or in few-flowered umbels; petals (to 1 cm) white or pinkish (sometimes fading to orange); stamens numerous (5–25); pomes (to 15 mm) black, blue, orange, or red, berry-like, with 1–5 nutlets
Life history: duration: perennial (buds); **asexual reproduction:** agamospermy, root suckers; **pollination:** insect; **sexual condition:** hermaphroditic; **fruit:** pomes (common); **local dispersal:** seeds (gravity, water); **long-distance dispersal:** fruit/seeds (birds)

Imperilment: (1) *Crataegus aestivalis* [G5]; S2 (NC); S3 (GA); (2) *C. brachyacantha* [G4]; SH (GA); S1 (MS); S2 (AR); (3) *C. opaca* [G5]; (4) *C. viridis* [G5]; S1 (WV); S2 (IN, KS); S3 (TX)
Ecology: general: *Crataegus* (hawthorns) are mainly terrestrial trees and shrubs that inhabit well-drained sites. Seventeen species (10%) are wetland indicators, but only four (2%) are designated as OBL in some portion of their range. Most of the species are insect pollinated and reproduce sexually. Pollen:ovule ratios are fairly high, indicating general adaptation for FAC outcrossing. However, many species also reproduce asexually by gametophytic apomixis. The reproductive ecology of the obligate aquatic species has not been investigated in much detail. Two of the aquatic species (*C. aestivalis*, *C. opaca*) are known as Mayhaws (May haws) because their fruit matures earlier in the season (spring, early summer) than most other species (late summer). The fleshy *Crataegus* fruits are commonly dispersed by birds (Aves). *Crataegus* seeds frequently have double dormancy, which requires acid scarification (e.g., 2 h in concentrated sulfuric acid) as a pregermination treatment followed by a period of cold stratification (to break embryo dormancy).

***Crataegus aestivalis* (Walter) Torr. & A. Gray** occurs locally in ponded or regularly flooded sites such as depressions, ditches, flatwoods, floodplains, limesinks, marshes, meadows, oxbows, ponds, pools, sinkholes, sloughs, swamps, and the margins of lakes, ponds, and rivers at elevations of up to 100 m. The substrates are acidic to slightly alkaline (optimal pH: 6.0–6.5) and include clay, loam, marl, and sand. The trees occur in areas of prolonged standing water. Adult trees can withstand full sun to partial shade, are moderately drought tolerant, and are very cold tolerant (down to −32°C). The flowers emit a fetid odor, which attracts midges (Insecta: Diptera: Chironomidae) for pollination. The seeds are physiologically dormant and must be stratified for germination. To promote germination, the seeds should be placed for 3 months at 15°C, then for 3 months at 4°C. Subsequent germination may take up to 18 months. Germination can be enhanced by scarification or by several days of fermentation (in the pulp) prior to stratification. It takes 5–8 years for a seedling to reach maturity and produce flowers, which develop from February to March. The trees are slow growing but will produce fruit for over 50 years, with a single tree capable of yielding more than 18 kg of fruit. The fruits ripen early in the season (April to July). The plants can be propagated by stem cuttings that have been placed in an intermittently misted or humid environment. Rooting of stem cuttings (up to 35%) can be achieved by the application of potassium salt of indole butyric acid (8000 ppm) combined with potassium salt of naphthalene acetic acid (2000 ppm). **Reported associates:** *Acer rubrum, Betula nigra, Campsis radicans, Carpinus caroliniana, Carya aquatica, Chasmanthium latifolium, Diospyros virginiana, Ditrysinia fruticosa, Fraxinus caroliniana, Ilex decidua, I. vomitoria, Juniperus virginiana, Lemna, Nyssa aquatica, N. biflora, Panicum, Pinus elliottii, Planera aquatica, Platanus occidentalis, Quercus laurifolia, Q. lyrata, Q. nigra, Q. virginiana, Sabal minor, S. palmetto, Salix nigra,*

Saururus cernuus, Serenoa repens, Sideroxylon reclinatum, Smilax bona-nox, S. rotundifolia, Solidago, Taxodium ascendens, T. distichum, Tillandsia usneoides, Toxicodendron radicans, Ulmus alata, U. crassifolia.

***Crataegus brachyacantha* Sarg. & Engelm.** grows where ponded waters occur in alluvial flats, bottomlands, depressions, ditches, flatwoods, floodplains, prairies, swamps, and along the margins of streams at elevations of up to 200 m. The sites typically receive full sun and in some cases, the soils may contain a fairly high salt content. Flowering occurs from March to May with fruiting in August to November. Additional life-history information for this species would be desirable. **Reported associates:** *Baccharis halimifolia, Carex louisianica, Carya texana, Chasmanthium laxum, C. sessiliflorum, Cissus trifoliata, Crataegus marshallii, C. opaca, C. spathulata, C. viridis, Dichanthelium boscii, Diospyros virginiana, Forestiera ligustrina, Gelsemium sempervirens, Gleditsia triacanthos, Ilex decidua, I. vomitoria, Nyssa biflora, Panicum rigidulum, Quercus marilandica, Q. phellos, Q. similis, Q. stellata, Ranunculus fascicularis, Sabal minor, Sphagnum lescurii, Ulmus alata, Vaccinium arboreum.*

***Crataegus opaca* Hook. & Arn.** is found in sites where water is ponded for much of the year including bayous, depressions, flatwoods, floodplains, swamps, and along the margins of rivers and streams at elevations of up to 100 m. The plants are long lived and thrive on acidic (pH: 4.2–4.7) substrates (e.g., sand, sandy loam), which are exposed to full sun or in partial shade. Container-grown plants can tolerate an irrigation solution of up to 2 dS m^{-1} salinity (having moderately high NaCl salinity of 10 meq l^{-1}), but solution salinities above 3 dS m^{-1} are detrimental to root function unless supplemental Ca^{2+} is provided. Flowering occurs from January to March with fruits produced from April to June. Seed germination (up to 93%) can be enhanced by the fermentation of fresh, undried fruit pulp for 4 or 8 days. Apomictic seeds are produced by agamospermy resulting from nucellar embryo production. The plants can be propagated by stem cuttings as for *C. aestivalis* (see earlier). **Reported associates:** *Ampelopsis arborea, Amsonia tabernaemontana, Arundinaria gigantea, Brunnichia ovata, Carex glaucescens, Cephalanthus occidentalis, Chasmanthium laxum, Crataegus brachyacantha, C. viridus, Dichanthelium, Diospyros virginiana, Eleocharis microcarpa, Forestiera acuminata, Fraxinus caroliniana, Ilex decidua, I. vomitoria, Juncus repens, Liquidambar styraciflua, Magnolia grandiflora, Nyssa biflora, Pinus glabra, P. taeda, Planera aquatica, Quercus laurifolia, Q. michauxii, Q. nigra, Q. pagoda, Q. phellos, Q. virginiana, Rhynchospora corniculata, R. mixta, Sabal minor, Smilax bona-nox, S. rotundifolia, Styrax americanus, Trachelospermum difforme, Triadica sebifera.*

***Crataegus viridis* L.** inhabits alluvial woodlands, bottomlands, depressions, flats, floodplains, levees, marshes, meadows, oxbows, prairies, ravines, roadsides, spillways, swales, swamps, thickets, woodlands, and the margins of lakes, ponds, rivers, and streams at elevations of up to 200 m. The plants can survive in standing water up to 20 cm deep, at least for short duration. The substrates are described as calcareous (often highly) and include Blackbelt clay, claypan loam,

Dorovan muck, Guthrie, loam, Montevallo gravelly loam, sandy clay loam, sandy gravel, sandy loam, shale, Sharkey clay, silt, silty clay loam, silty sandy loam, and Yazoo clay. The flowers occur from March to May and are visited by bees (Insecta: Hymenoptera). The fruits are often shed while still green (September to January) and persist poorly on the ground. They are buoyant and are dispersed over short distances by water and over longer distances by birds (e.g., Aves: Anatidae) and small mammals (Mammalia). The plants have a fairly large seed rain, averaging upwards of 31–34 seeds/m^2 year, which tend to fall close to the maternal plants. The seed rain for water dispersed seeds can be substantially higher (e.g., ×5.7) than for aerial dispersed seeds. There does not seem to be a persistent seed bank for this species. **Reported associates:** *Acer negundo, A. rubrum, A. saccharinum, Allium canadense, Ampelopsis arborea, Andropogon virginicus, Arisaema dracontium, Arundinaria tecta, Baccharis halimifolia, Berchemia scandens, Boehmeria cylindrica, Campsis radicans, Carex amphibola, C. bromoides, C. caroliniana, C. cherokeensis, C. frankii, C. lupulina, C. oxylepis, Carpinus caroliniana, Carya aquatica, C. cordiformis, C. illinoinensis, Celtis laevigata, C. occidentalis, Cephalanthus occidentalis, Cercis canadensis, Cinnamomum camphora, Cirsium muticum, Cornus asperifolia, C. drummondii, C. florida, C. foemina, Crataegus brachyacantha, C. crus-galli, C. marshallii, C. mollis, C. opaca, C. spathulata, Cryptotaenia canadensis, Cynosciadium digitatum, Danthonia, Dichanthelium, Dichondra repens, Dicliptera brachiata, Diospyros virginiana, Eleocharis macrostachya, Elymus, Eupatorium perfoliatum, Forestiera acuminata, Fraxinus pennsylvanica, F. profunda, Galium aparine, G. tinctorium, Geranium carolinianum, Geum canadense, Gleditsia triacanthos, Gymnocladus dioicus, Hydrocotyle verticillata, Hypericum mutilum, Ilex decidua, I. verticillata, Ipomoea hederacea, Iris hexagona, Itea virginica, Juglans nigra, Leersia oryzoides, Liquidambar styraciflua, Lonicera tatarica, Ludwigia palustris, Maclura pomifera, Magnolia virginiana, Mikania scandens, Mimosa strigillosa, Morus rubra, Myosotis macrosperma, Myrica cerifera, M. heterophylla, Nyssa aquatica, Oplismenus hirtellus, Packera glabella, Paspalum, Persicaria virginiana, Pinus elliottii, P. taeda, Planera aquatica, Platanus occidentalis, Poa annua, Populus deltoides, Quercus falcata, Q. laurifolia, Q. lyrata, Q. macrocarpa, Q. marilandica, Q. nigra, Q. pagoda, Q. phellos, Q. rubra, Q. shumardii, Q. stellata, Q. texana, Q. virginiana, Ranunculus hispidus, R. recurvatus, Rhus copallinum, Rubus flagellaris, Ruellia strepens, Sabal minor, Sagittaria papillosa, Salix nigra, Sambucus nigra, Samolus valerandi, Sapindus saponaria, Saururus cernuus, Sesbania drummondii, S. vesicaria, Sideroxylon lanuginosum, Smilax bona-nox, Sphenopholis pensylvanica, Sporobolus indicus, Stellaria cuspidata, Symphoricarpos orbiculatus, Taxodium distichum, Toxicodendron radicans, Triadica sebifera, Ulmus alata, U. americana, U. crassifolia, Urtica chamaedryoides, Vaccinium arboreum, Veronica peregrina, Viola, Vitis cinerea.*

Use by wildlife: Song birds (Aves), cedar wax wings (Bombycillidae: *Bombycilla cedrorum*), rodents (Rodentia), small mammals (Mammalia), and other wildlife eat the fruits of *Crataegus aestivalis* and other *Crataegus* species. The plants also provide good nesting cover for birds (Aves). White-tailed deer (Mammalia: Cervidae: *Odocoileus virginianus*) browse the leaves and twigs. The foliage, flowers, fruit, and wood of *C. aestivalis* are fed upon by a number of insects (Insecta) including apple maggot flies (Diptera: Tephritidae: *Rhagoletis pomonella*), plum curculio weevils (Coleoptera: Curculionidae: *Conotrachelus nenuphar*), roundheaded appletree borer beetles (Coleoptera: Cerambycidae: *Saperda candida*), whitefringed beetles (Coleoptera: Curculionidae: *Naupactus*), hawthorn lace bugs (Hemiptera: Tingidae: *Corythucha cydoniae*), flower thrips (Thysanoptera: Thripidae: *Frankliniella tritici*), leafminers (Hymenoptera: Tenthredinidae: *Profenusa canadensis*), mealybugs, and scale insects (Homoptera). It is a larval butterfly (Lepidoptera) host plant for hairstreaks (Lycaenidae) and red-spotted purple butterflies (Nymphalidae: *Basilarchia astyanax*). *Crataegus aestivalis* and *C. opaca* are hosts for bacterial fire blight (Bacteria: Enterobacteriaceae: *Erwinia amylovora*), quince rust (Fungi: Basidiomycota: Pucciniaceae: *Gymnosporangium clavipes*), and hawthorn leaf blight (Fungi: Ascomycota: Sclerotiniaceae: *Monilinia johnsonii*). *Crataegus brachyacantha* is fed on by apple maggot flies (Insecta: Diptera: Tephritidae: *Rhagoletis pomonella*) and also reportedly is susceptible to rust infections (Fungi: Basidiomycota: Pucciniaceae). *Crataegus brachyacantha* and *C. opaca* are categorized as being susceptible to gypsy moth (Insecta: Lepidoptera: Lymantriidae: *Lymantria dispar*) feeding. *Crataegus opaca* is used moderately by beavers (Mammalia: Castoridae: *Castor canadensis*). The fruits of *C. viridis* are an important source of food for wood ducks (Aves: Anatidae: *Aix sponsa*). The plants are parasitized by apple maggot flies (Diptera: Tephritidae: *Rhagoletis pomonella*), hawthorn lace bugs (Hemiptera: Tingidae: *Corythucha cydonia*), and several aphids (Insecta: Hemiptera: Aphididae: *Hyalomyzus eriobotryae*, *H. jussiaeae*, *H. sensoriatus*). The dead branches and flowers host a number of larval beetles (Insecta: Coleoptera) from the families Buprestidae (*Acmaeodera tubulus*, *Chrysobothris azurea*, *C. purpureovittata*, *C. sexsignata*) and Cerambycidae (*Anelaphus parallelus*, *Batyle suturalis*, *Euderces pini*, *E. reichei*, *Liopinus misellus*, *L. punctatus*, *Molorchus bimaculatus*, *Neoclytus acuminatus*, *Obrium maculatum*, *Psenocerus supernotatus*, *Tilloclytus geminatus*, *Xylotrechus convergens*). This is also a host plant for several cicada species (Hemiptera: Cicadellidae: *Erythridula sagittata*, *E. furcillata*, *E. repleta*).

Economic importance: food: The juicy fruits of *Crataegus aestivalis* and *C. opaca* (which are high in citric acid, fructose and glucose) have long been used for making jelly, preserves, syrup, and wine in the southern United States and their popularity has led to the recent development of several commercial orchards for their production. Blended juices from the fruits of *C. opaca* and grape (*Vitis*) have resulted in a flavorful beverage found to be suitable for the consumer fruit juice market; **medicinal:** Although not verified for any of the obligate aquatic species, the fruits and flowers of several *Crataegus* taxa are hypotensive when ingested and have been used as a

mild heart tonic. The juice of *C. opaca* contains anthocyanins and phenolics, which impart antioxidant activity. Other bioactive constituents include butyl hexanoate, hexanal, linalool, pentyl hexanoate, and methyl hexanoate; **cultivation:** *Crataegus aestivalis* and *C. opaca* are grown commercially and are among the few ornamental trees adapted for wet landscaping use. Because these species are sometimes merged, the specific assignment of cultivar names is problematic. Cultivars attributed to *C. aestivalis* include 'Crimson,' 'G-5,' 'Lindsey,' 'Lori,' 'Mason's Super Berry,' 'Saline,' 'Texas Star,' 'Turnage 57,' 'Turnage 88,' and 'Yellow Gem.' Cultivars attributed to *C. opaca* include 'Heavy Mason's Super Berry,' 'Mississippi Beauty,' 'No. 1 Big,' 'Super Berry,' 'Texas Super Berry,' and 'Yellow.' Cultivar names assigned to both species include: 'Big Red,' 'Big V,' 'Heavy,' 'Highway Super Berry,' 'Red & Yellow,' 'Super Spur,' and 'T.O. Super Berry.' The seed of *C. brachyacantha* is sold commercially. 'Winter King' is a popular cultivar of *C. viridis*; **misc. products:** *Crataegus aestivalis* is a recommended screening plant for buffer strips, wind breaks, and highway median strips. The thorny stems make *C. aestivalis* and *C. opaca* useful as hedges. Their wood is heavy and strong and has been used to fashion tool handles, mallets, and miscellaneous wooden items. A mayhaw festival is held each May in El Dorado, Arkansas; **weeds:** none; **nonindigenous species:** none

Systematics: *Crataegus* has a notorious reputation as being one of the more difficult angiosperm genera taxonomically. Pervasive polyploidy, a widespread penchant for hybridization, high morphological variability, and apomictic reproduction have combined to blur species boundaries and to create havoc with identification in the genus. Molecular data consistently indicate the phylogenetic placement of *Crataegus* within Rosaceae subfamily Maloideae as the sister genus to *Mespilus* (Figure 4.42). Intergeneric hybrids are known between *Crataegus* and both *Mespilus* and *Sorbus*, and some authors have merged *Mespilus* with *Crataegus*. There currently exists no comprehensive interspecific phylogeny for this large and difficult genus, and this work is sorely needed to help clarify the taxonomy. Preliminary molecular surveys of taxa consistently resolve *C. brachyacantha* as the sister to the remaining *Crataegus* species (unless *Mespilus* is merged). The early fruiting Mayhaws (including *C. aestivalis* and *C. opaca*) are segregated taxonomically within series *Aestivales*. Certainly *C. aestivalis* and *C. opaca* are closely related and have been merged by some authors as a single species. *Crataegus brachyacantha* is a fairly distinctive species with its shiny blue fruit and short, curved thorns. It is placed within series *Brevispinae*. *Crataegus aestivalis* and many *Crataegus* species hybridize freely. The chromosomal base number of *Crataegus* is $x = 17$. Chromosome numbers have been reported for *C. brachyacantha* ($2n = 34$, 51), *C. opaca* ($2n = 34$), and *C. viridis* ($2n = 34$, 51).

Comments: *Crataegus aestivalis* and *C. viridis* occur in the southeastern United States. *Crataegus brachyacantha* and *C. opaca* are distributed in the south central United States.

References: Baker, 1991; Battaglia et al., 2008; Campbell et al., 1991, 1995; Cha et al., 2011; Chabreck, 1958; Chapman &

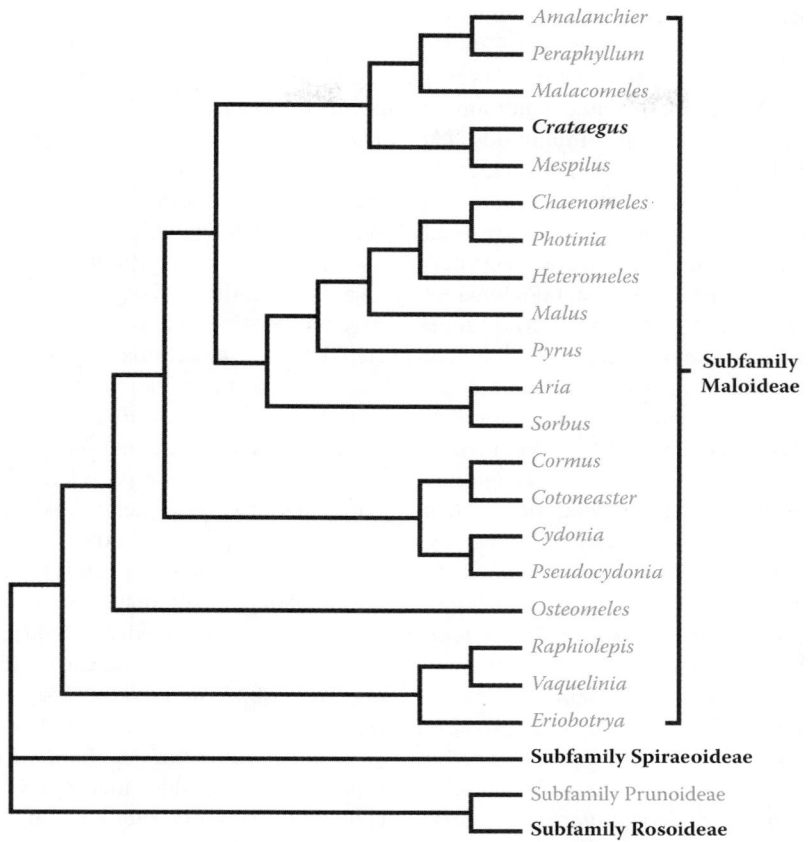

FIGURE 4.42 Phylogenetic relationships among genera of Rosaceae subfamily Maloideae as indicated by DNA sequence data. Taxa containing obligate aquatics are shown in bold. *Crataegus*, the only genus of Maloideae with obligate aquatics, is relatively derived in the subfamily. (Adapted from Campbell, C.S. et al., *Amer. J. Bot.*, 82, 903–918, 1995.)

Horvat, 1993; Correll & Correll, 1975; Denslow & Battaglia, 2002; Dickinson et al., 2000; Dirrigl, Jr. & Mohlenbrock, 2012; Dmitriev & Dietrich, 2009; Elias, 1987; Godfrey, 1988; Godfrey & Wooten, 1981; Graham & Kuehny, 2003; Horvat & Chapman, 2007; Kirkman et al., 2000; Krewer, 2000; Little, Jr., 1938; Lo et al., 2007, 2009; MacRae & Nelson, 2003; MacRae & Rice, 2007; MacRoberts et al., 2014; McCarter & Payne, 1993; McGilvrey, 1966; Middleton, 2000; Neyland & Meyer, 1997; Payne & Krewer, 1990; Payne et al., 1990; Phipps, 1988, 1998, 1999; Picchioni & Graham, 2001; Robertson et al., 1991; Scherm & Savelle, 2003; Skeate, 1987; Stoetzel et al., 1999; Szczepaniec & Raupp, 2007; Thieret, 1971; Trappey et al., 2007; White, 1983; White & Skojac, 2002.

3. *Filipendula*

Queen-of-the-prairie; reine des prés rose

Etymology: from the Latin *filum pendulus* meaning "hanging thread" for the thread-like attachment of tubers in some species

Synonyms: *Spiraea* (in part); *Thecanisia*; *Ulmaria*

Distribution: global: Asia; Europe; North America; **North America:** E/W disjunct

Diversity: global: 15 species; **North America:** 4 species

Indicators (USA): OBL; FACW: *Filipendula rubra*

Habitat: freshwater; palustrine; **pH:** 4.1–8.2; **depth:** <1 m; **life-form(s):** emergent herb

Key morphology: stems (to 25 dm) erect; stipules (to 16 mm) reniform, serrate; leaves (to 1 m) alternate, pinnately divided into 7–9 palmately incised leaflets (laterals to 1 dm; terminal to 2 dm), their margins toothed; inflorescence a panicle with many (200–1000) small (to 10 mm), 5-merous flowers; petals (to 4 mm) pink; stamens numerous (20–40); gynoecium apocarpous with 5–15 pistils; achenes (to 8 mm) reddish

Life history: duration: perennial (rhizomes); **asexual reproduction:** rhizomes; **pollination:** insect; **sexual condition:** hermaphroditic; **fruit:** aggregates (achenes) (infrequent); **local dispersal:** achenes; **long-distance dispersal:** achenes

Imperilment: (1) *Filipendula rubra* [G4]; SH (NJ); S1 (IA, IL, MD, NC, PA); S2 (MI, MO, VA); S3 (IN)

Ecology: general: *Filipendula* species occur often in wetlands or at least in damp sites, with three of the four North American species (75%) ranked as FACW indicators, and one species also as an OBL indicator in a part of its range. Our species at least are rhizomatous perennials with insect-pollinated flowers, which are self-incompatible. The seeds are physiologically dormant.

Filipendula rubra **(Hill) B.L. Robins.** occurs locally in ditches, fens, marshes, meadows, prairies, prairie fens, rocky outcrops, and seeps. Its habitats typically provide full sun to partial shade and receive alkaline groundwater (pH: 7.0–7.5) having a high calcium content (mean: 4049 ppm); however, the plants are also known to occur under acidic conditions in unusual high-elevation wet, rocky outcrop communities. The flowering period is short, lasting only about 3 weeks. The flowers are pollinated by various insects (see *Use by wildlife* later) and are self-incompatible (selfed plants yield only 0–2.8 seeds/1000). Because populations are highly clonal (see later), the scarcity of compatible genotypes often results in low seed set. As a result, seed production in natural populations ranges from 0.2% to 57% (mean: 24%). The seeds have physiological dormancy and require 9 months of cold stratification for germination. Seedling recruitment is believed to be limited by competition with grasses (Poaceae) and sedges (Cyperaceae).

Reported associates: *Agalinis purpurea, Agrostis gigantea, Allium cernuum, Alnus rugosa, Amelanchier, Andropogon gerardii, Apocynum cannabinum, Arnoglossum plantagineum, Asclepias purpurascens, A. syriaca, Berula erecta, Bouteloua gracilis, Calamintha arkansana, Caltha palustris, Campanula aparinoides, Carex aquatilis, C. buxbaumii, C. frankii, C. haydenii, C. hystericina, C. interior, C. prairea, C. leptalea, C. sterilis, C. stricta, C. trichocarpa, Chelone glabra, C. obliqua, Cirsium muticum, Cladium mariscoides, Coreopsis palmata, Cornus amomum, C. obliqua, C. sericea, Crataegus, Cypripedium candidum, C. reginae, Dasiphora floribunda, Desmodium canadense, Doellingeria umbellata, Drosera anglica, Echinacea purpurea, Eleocharis compressa, E. geniculata, Equisetum arvense, Erigeron strigosus, Eryngium yuccifolium, Eupatoriadelphus maculatus, Eupatorium perfoliatum, Euphorbia purpurea, Eurybia divaricata, Euthamia graminifolia, Eutrochium maculatum, Fragaria virginiana, Fraxinus nigra, Galium asprellum, G. triflorum, Gentianopsis crinita, G. virgata, Helianthus strumosus, Hemerocallis fulva, Hierochloe odorata, Holcus lanatus, Houstonia serpyllifolia, Huperzia porophila, Impatiens capensis, Iris versicolor, Juncus balticus, J. brachycephalus, J. dudleyi, J. nodosus, Lathyrus palustris, Liatris aspera, L. spicata, Lilium michiganense, Lobelia kalmii, L. siphilitica, Lonicera, Lycopus americanus, Lysimachia quadriflora, Lythrum alatum, Maianthemum canadense, Malus, Mentha arvensis, Menyanthes trifoliata, Micranthes petiolaris, Mitchella repens, Monarda fistulosa, Muhlenbergia glomerata, M. mexicana, M. richardsonis, Myrica pensylvanica, Oenothera biennis, Onoclea sensibilis, Osmunda claytonia, Oxalis, Oxypolis rigidior, Packera aurea, Paeonia hybrida, Panicum virgatum, Parnassia glauca, P. grandifolia, Pedicularis lanceolata, Penstemon digitalis, Phalris arundinacea, Philonotis muehlenbergii, Phlox maculata, P. paniculata, Photinia melanocarpa, Pilea fontana, Polemonium reptans, Populus tremuloides, Primula meadia, Pycnanthemum muticum, P. virginianum, Rhamnus lanceolata, Rosa blanda, Rubus, Rudbeckia hirta, Salix discolor, S. petiolaris, S. sericea, Sanguisorba canadensis, Schizachyrium scoparium, Schoenoplectus tabernaemontani, Scirpus atrovirens, S. pendulus, Scleria verticillata, Silphium integrifolium, S. terebinthinaceum, S. trifoliatum, Solidago canadensis, S. ohioensis, S. patula, S. uliginosa, Sorghastrum nutans, Spartina pectinata, Spiranthes lucida, Sporobolus heterolepis, Symphyotrichum firmum, S. laeve, S. novae-angliae, S. puniceum, Thalictrum dasycarpum, Thelypteris palustris, Toxicodendron radicans, T. vernix, Tradescantia ohioensis, Typha latifolia, Vaccinium corymbosum, Valeriana edulis, Verbena hastata, Veronica scutellata, Viburnum lentago, Viola sororia, Woodwardia virginica, Zizania aquatica.*

Use by wildlife: Plants of *Filipendula rubra* will attract various bees (Insecta: Hymenoptera), butterflies (Insecta: Lepidoptera), and birds (Aves). They are grazed on (and impacted negatively) by white-tailed deer (Mammalia: Cervidae: *Odocoileus virginianus*). The principal pollinators of *F. rubra* are insects and include sweat bees (Hymenoptera: Halictidae: *Augochlorella; Halictus ligatus; Lassioglossum*), honey bees (Hymenoptera: Apidae: *Apis mellifera; Bombus*), and flies (Diptera: Muscidae; Syrphidae). *Filipendula rubra* is susceptible to powdery mildew (Ascomycota: Erysiphaceae: *Microsphaera; Oidium*) and to Japanese beetles (Insecta: Coleoptera: Scarabaeidae: *Popillia japonica*). Adult leaf-eating beetles (Coleoptera: Chrysomelidae: *Galerucella calmariensis*), which are used for biological control of other species (e.g., *Lythrum salicaria*), also will feed moderately on *F. rubra*.

Economic importance: food: *Filipendula rubra* is not reported to be edible; **medicinal:** The Meskwaki used the root of *Filipendula rubra* as an important heart medicine. The roots contain tannins and have also been used to treat bleeding, diarrhea, and dysentery. Salicylic acid (cf. aspirin) has also been extracted from the flowers and roots and used as a remedy for arthritis, influenza, and fever. The leaf extracts are moderately inhibitory to some bacteria (e.g., Eubacteria: Staphylococcaceae: *Staphylococcus aureus*); **cultivation:** *Filipendula rubra* is cultivated frequently as a garden ornamental, especially for wetter sites such as water garden or pond margins. 'Albicans' is a white-flowered cultivar. The cultivar 'Venusta' (also known as 'Magnifica') has deep pink to red flowers and is a Royal Horticultural Society Award of Garden Merit plant; **misc. products:** Roots of *F. rubra* were an ingredient in a love potion used by the Meskwaki. *Filipendula rubra* is used as an ingredient in some dermatalogical products. The wild plants are said to have the scent of cucumbers; **weeds:** none; **nonindigenous species:** *Filipendula rubra* has spread northward and westward beyond its indigenous range because of its wide use as an ornamental garden plant.

Systematics: Molecular data resolve *Filipendula* as the basal genus of Rosaceae subfamily Rosoideae (Figure 4.41). *Filipendula* has not yet been studied phylogenetically and such work would help to define species limits and to clarify interspecific relationships in the genus. The latest revision by Schanzer (1994) subdivided the genus into four sections. In that treatment, *F. rubra* was placed within section *Albicoma* along with the Asian species *F. angustiloba* and *F. palmata* (which also grow in wet habitats). It is believed to represent a species transitional between this section and section *Schalameya*. Allozyme studies indicate that low levels of genetic variation exist in natural populations, which are highly clonal in structure. A survey of 25 populations detected only 1–15 genotypes (mean: 5.5), with the genotype diversity being a function of population size. Many populations apparently lack adequate numbers of s-alleles (self-incompatibility alleles) to sustain adequate levels of sexual reproduction; thus, attempts to re-establish populations of this species using vegetative (rhizomatous) stock should strive to maximize the number of genetic individuals by acquiring plants from different, widespread populations. The chromosomal base number of *Filipendula* is $x = 7$. Chromosome counts have not been reported for *F. rubra*.

Comments: *Filipendula rubra* occurs throughout eastern North America and is disjunct in Montana and Wyoming. Many northern (i.e., Canadian) populations may represent escapes from cultivation.

References: Aspinwall & Christian, 1992a,b; Baker & Baker, 1967; Belden et al., 2004; Borchardt et al., 2008b; Crête et al., 2001; Gordon, 1933; Ruch et al., 2008, 2009; Schanzer, 1994, 2014; Wiser et al., 1996.

4. *Geum*

Avens; benoîte

Etymology: from the Greek *geuo*, meaning "food," in reference to edibility of some species

Synonyms: *Acomastylis*; *Coluria*; *Erythrocoma*; *Neosieversia*; *Novosieversia*; *Oncostylus*; *Oreogeum*; *Orthurus*; *Parageum*; *Sieversia*

Distribution: global: Asia; Europe; North America; **North America:** widespread

Diversity: global: 45 species; **North America:** 16 species

Indicators (USA): OBL: *Geum peckii*, *G. rivale*; **FACW:** *G. rivale*

Habitat: freshwater; palustrine; **pH:** 4.1–7.7; **depth:** <1 m; **life-form(s):** emergent herb

Key morphology: stems (to 6 dm) erect; basal leaves (to 3 dm) lyrate-pinnate with up to 15 lateral leaflets (1 to 10 cm) and a larger terminal leaflet (to 10 cm) that is simple, 3-lobed or 3-parted; cauline leaves similar but reduced; petioles winged by the leaf-like stipules; cymes of 1–9 5-merous, nodding or erect flowers (to 3 cm); hypanthium (to 5 mm) campanulate, often purplish; petals (to 15 mm) ascending or spreading, yellow to pink with purple veins; stamens numerous; ovaries (to 4 mm) numerous, plumose in fruit; styles (to 10 mm) jointed or straight, plumose or glabrous

Life history: duration: perennial (rhizomes); **asexual reproduction:** rhizomes; **pollination:** insect; **sexual condition:** hermaphroditic; **fruit:** aggregates (achenes) (common); **local dispersal:** seeds (wind), rhizomes; **long-distance dispersal:** seeds (animals, wind)

Imperilment: (1) *Geum peckii* [G2]; S1 (<u>NS</u>); S2 (NH); (2) *G. rivale* [G5]; SX (IL); SH (ND); S1 (IN, <u>LB</u>, WV, WY); S2 (WA); S3 (<u>BC</u>, NJ, OH)

Ecology: general: *Geum* species principally are terrestrial, but often occur in moist woodland habitats. Twelve of the species (75%) are North American wetland indicators, but only two species (12.5%) have been designated as OBL aquatics. Sexual reproduction occurs frequently and all North American species

also reproduce vegetatively by rhizome production. Seed dispersal in most *Geum* species is facilitated by the jointed styles, which form hook-like processes and result in a bur-like aggregate of achenes that attaches easily to animal fur. However, jointed styles are absent in *G. peckii*. The styles of both aquatic species become plumose, which facilitates their dispersal by wind.

***Geum peckii* Pursh** occurs in bogs, depressions, ditches, seeps, slopes, snowbank meadows, and along stream margins of alpine and boreal coastal habitats at elevations of up to 1830 m. The cup-like flowers move to "track" the sun throughout the day, which accelerates fruit production and helps to attract pollinators. The flowers are self-compatible. Both autogamy and outcrossing can occur, with pollination accomplished primarily by flies (insecta: Diptera). The seeds require cold stratification for germination; however, observed germination rates can be low (<5%). The seeds will germinate at 13°C–18°C after 3 months of burial. Dried or refrigerated seed reportedly exhibits good germination when sowed. The fruits probably are wind dispersed, although only the ovary (not the style) becomes plumose. Although adapted for alpine conditions, *G. peckii* requires high-light exposure and warm microenvironments to attain maximum productivity. The species is believed to be particularly sensitive to habitat change relating to global warming. It is quite rare with the populations threatened by land drainage, shrub encroachment, and trampling by gulls (Aves: Laridae) in coastal populations. **Reported associates:** *Agrostis mertensii*, *Betula glandulosa*, *B. minor*, *Bistorta vivipara*, *Calamagrostis canadensis*, *C. pickeringii*, *Campanula rotundifolia*, *Cardamine bellidifolia*, *Carex bigelowii*, *C. scirpoidea*, *Castilleja septentrionalis*, *Chamaedaphne calyculata*, *Clintonia borealis*, *Deschampsia flexuosa*, *Diapensia lapponica*, *Draba incana*, *Empetrum eamesii*, *E. nigrum*, *Epilobium hornemanni*, *Hieracium robbinsonii*, *Hierochloe alpina*, *Houstonia caerulea*, *Juncus trifidus*, *Kalmia angustifolia*, *Loiseleuria procumbens*, *Luzula spicata*, *Phleum alpinum*, *Phyllodoce caerulea*, *Pinguicula vulgaris*, *Pinus rigida*, *Polytrichum strictum*, *Potentilla robbinsiana*, *Prenanthes boottii*, *Rhododendron lapponicum*, *Salix uva-ursi*, *Sarracenia purpurea*, *Saxifraga rivularis*, *Solidago cutleri*, *S. macrophylla*, *Sphagnum capillifolium*, *S. compactum*, *S. girgensohnii*, *S. russowii*, *Trichophorum cespitosum*, *Vaccinium cespitosum*, *V. uliginosum*, *Veratrum viride*.

***Geum rivale* L.** grows in bogs, ditches, fens, marshes, meadows, prairies, seepage swamps, and along lake, river, and stream margins. The plants occur on a variety of substrates ranging from organic peat to sand (pH: 4.7–7.5). Sites often are fertile and calcareous with open exposure or light shade. The self-compatible flowers are strongly protogynous. They are autogamous or outcrossed by bumblebees (Insecta: Hymenoptera: Apidae: *Bombus*). Fresh seeds will germinate (to 70%) without treatment and will retain a similar level of viability after a year in dry storage. Seed germination can occur at temperatures from 11°C to 27°C, but is optimal at 22°C. The germination rate increases for seeds that have been stored. The achenes are plumose throughout and are wind dispersed. **Reported associates:** *Abies balsamea*, *Acer pensylvanicum*, *A. rubrum*, *A. saccharum*, *Actaea rubra*, *Allium brevistylum*, *Alnus rugosa*,

Angelica arguta, Arisaema triphyllum, Athyrium felix-femina, Aulacomnium palustre, Betula alleghaniensis, B. glandulosa, B. papyrifera, B. pumila, Brachythecium rivulare, Bromus kalmii, Bryhnia novae-angliae, Bryum pseudotriquetrum, Calliergon cordifolium, C. cuspidata, C. giganteum, C. trifarium, Caltha palustris, Campylium stellatum, Canadanthus modestus, Cardamine bulbosa, C. pensylvanica, Carex aquatilis, C. aurea, C. bromoides, C. buxbaumii, C. castanea, C. flava, C. gynandra, C. hystericina, C. interior, C. lacustris, C. lasiocarpa, C. leptalea, C. livida, C. prairea, C. scabrata, C. sterilis, C. stricta, C. utriculata, Carpinus caroliniana, Chamaecyparis thyoides, Chamerion angustifolium, Chelone glabra, Chrysosplenium americanum, Cinna arundinacea, Circaea alpina, Climacium dendroides, Clintonia borealis, Comarum palustre, Coptis groenlandica, Corallorhiza striata, Cornus sericea, Cypripedium parviflorum, C. reginae, Dasiphora floribunda, Drepanocladus revolvens, Drosera rotundifolia, Dryopteris cristata, Eleocharis tenuis, Elymus glaucus, Equisetum arvense, E. fluviatile, E. hyemale, E. palustre, Eriophorum virginicum, E. viridicarinatum, Fragaria virginiana, Fraxinus nigra, Galium boreale, G. labradoricum, G. triflorum, Geranium richardsonii, Geum macrophyllum, Glyceria borealis, G. melicaria, G. striata, Hieracium odorata, Hydrocotyle americana, Ilex verticillata, Impatiens capensis, Iris versicolor, Juncus, Juniperus communis, Laportea canadensis, Larix laricina, Lindera benzoin, Linnaea borealis, Lobelia kalmii, Luzula campestris, Lycopus uniflorus, Maianthemum canadense, M. stellatum, Meesia triquetra, Micranthes pensylvanica, Mimulus ringens, Mitella nuda, Muhlenbergia glomerata, Onoclea sensibilis, Osmunda claytoniana, Osmundastrum cinnamomeum, Packera aurea, P. pseudaurea, P. schweinitziana, Paludella squarrosa, Parnassia glauca, Philonotis fontana, Picea mariana, P. rubens, Pilea pumila, Pinus banksiana, P. resinosa, P. strobus, Platanthera dilatata, P. grandiflora, P. hyperborea, P. psycodes, Poa paludigena, Polemonium occidentale, Populus tremuloides, Pyrola chlorantha, Quercus rubra, Ranunculus hispidus, Rhamnus alnifolia, Rhizomnium appalachianum, R. punctatum, Rhynchospora alba, Rhynchostegium serrulatum, Rhytidiadelphus triquetrus, Rosa acicularis, Rubus arcticus, Salix bebbiana, S. candida, S. discolor, S. lucida, S. pseudomonticola, Sarracenia purpurea, Scorpidium scorpioides, Shepherdia canadensis, Solidago patula, S. uliginosa, Sphagnum warnstorfii, Spiranthes romanzoffiana, Symphyotrichum ciliolatum, Symplocarpus foetidus, Thalictrum pubescens, Thuidium delicatulum, Thuja occidentalis, T. plicata, T. noveboracensis, T. palustris, Tiarella cordifolia, Tomentypnum nitens, Toxicodendron radicans, T. vernix, Trianthia glutinosa, Trichophorum alpinum, T. cespitosum, Trientalis borealis, Trollius laxus, Tsuga canadensis, Typha, Ulmus americana, Vaccinium corymbosum, Valeriana edulus, Veratrum viride, Viburnum lantanoides, Viola cucullata.

Use by wildlife: *Geum rivale* is a host plant for larval plume moths (Lepidoptera: Pterophoridae: *Geina*).

Economic importance: food: The dried or fresh roots of *Geum rivale* are fragrant and used as a seasoning or can be boiled in water (and added to milk) to make a chocolate-like drink. The plant also was used as a flavoring for ale; **medicinal:** The roots of *Geum rivale* were used by several Native North American tribes as a medicinal. The Algonquin and Tete-de-Boule made a decoction to control bleeding. The Iroquois, Malecite, and Micmac treated diarrhea using an infusion. The Iroquois administered an infusion to reduce fever and the Micmac consumed a decoction to relieve colds, coughs, and dysentery; **cultivation:** *Geum rivale* frequently is grown as an ornamental and includes the cultivars: 'Album,' 'Apricot,' 'Barbra Lawton,' 'Cream Drop,' 'Deep Rose,' 'Elfenbein,' 'Leonard's Double,' 'Leonard's Variety,' 'Lionel Cox,' 'Marmalade,' 'Oxford Marmalade,' 'Paso Doble,' 'Snowflake,' and 'Variegatum' as well as several hybrids; **misc. products:** The dried roots of *G. rivale* are said to repel moths. The cultivar 'Leonard's Variety' is used as a ground cover; **weeds:** none; **nonindigenous species:** none

Systematics: Molecular data resolve *Geum* (subfamily Rosoideae) in a clade together with *Fallugia* and *Waldstenia* as the sister genus to a clade containing the latter (Figure 4.41). The morphologically distinct *Geum peckii* (along with *G.*

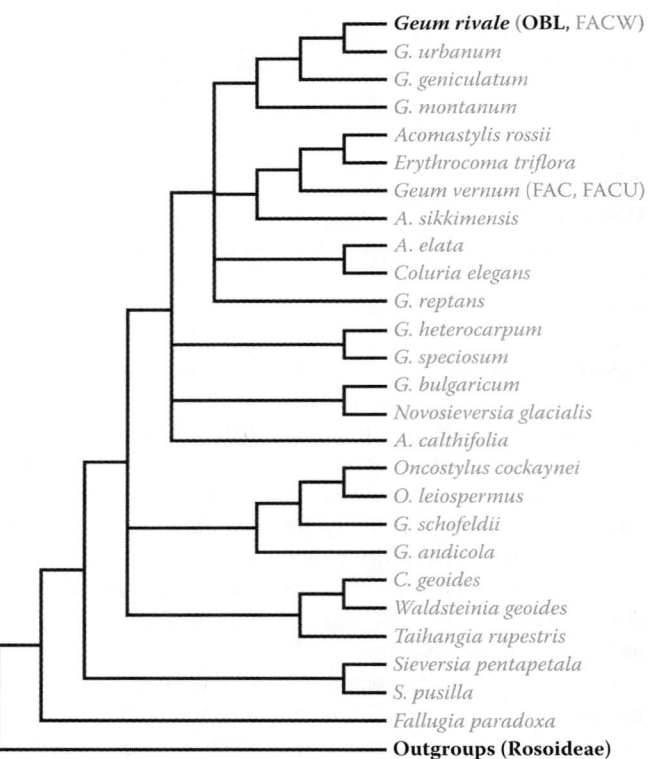

FIGURE 4.43 Phylogenetic relationships of *Geum* and allied genera as indicated by combined molecular data. These results indicate that *Geum* is not a natural clade as circumscribed, but includes various species currently assigned to other genera (*Acomastylis, Erythrocoma, Coluria, Novosieversia, Oncostylus*). Further refinements to the circumscription of *Geum* undoubtedly will be necessary as additional species are surveyed. The one obligate North American aquatic included in the study (*G. rivale*; shown in bold) occurs in a highly derived position in the genus. (Adapted from Smedmark, J.E.E. & Eriksson, T., *Syst. Bot.*, 27, 303–317, 2002.)

radiatum) often has been distinguished from *Geum* and even placed within a separate genus (e.g., *Acomastylis, Parageum, Sieversia*). Chromosomally, however, these species do not differ from other members of *Geum*. *Geum peckii* and *G. radiatum* are quite similar morphologically and undoubtedly are closely related. Some authors consider these taxa to be conspecific; however, evidence from DNA (RAPD) markers indicates that they are distinct genetically. Phylogenetic studies of 11 *Geum* species using combined molecular data have included *G. rivale* but not *G. peckii* (Figure 4.43). The relationships depicted show *G. rivale* to occur in a highly derived clade with *G. urbanum* (Europe; introduced to N. America) and *G. geniculatum* (USA). The close relationship of *G. rivale* and *G. urbanum* is evidenced further by their ability to hybridize in native European populations. In North America, hybrids reportedly occur with *G. aleppicum* (*G. ×aurantiacum*) and *G. macrophyllum* (*G. ×pervale, G. ×pulchrum*). The chromosomal base number of *Geum* is $x = 7$. *Geum peckii* and *G. rivale* are hexaploid ($2n = 42$). Microsatellite markers developed for *G. urbanum* are functional across the genus and would provide an excellent tool for studying the population genetics of *G. peckii* and *G. rivale*.

Comments: *Geum peckii* is an extremely rare endemic to eastern North America, known only to occur in 28 populations in northern New Hampshire (USA) and two populations in Nova Scotia (Canada). *Geum rivale* extends across much of northern North America, Europe, and Asia.

References: Arens et al., 2004; Brackley-Tolman, 2001; Graves & Taylor, 1988; Hadley & Bliss, 1964; Kimball & Weihrauch, 2000; Paterson & Snyder, 1999; Raynor, 1952; Rohrer, 2014; Smedmark & Eriksson, 2002.

5. *Potentilla*

Cinquefoil; potentille

Etymology: a diminutive of the Latin *potens* meaning "powerful" with reference to perceived medicinal properties

Synonyms: *Argentina; Duchesnea; Horkelia; Horkeliella; Ivesia; Sibbaldia* (in part); *Trichothalamus; Tylosperma*

Distribution: global: Asia; Europe; New World; **North America:** widespread

Diversity: global: 430 species; **North America:** 128 species

Indicators (USA): OBL: *Potentilla anserina, P. multijuga, P. newberryi, P. pickeringii, P. supina;* **FACW:** *P. anserina, P. plattensis, P. rivalis, P. supina;* **FAC:** *P. glandulosa, P. lycopodioides*

Habitat: brackish, freshwater; palustrine; **pH:** alkaline; **depth:** <1 m; **life-form(s):** emergent herb

Key morphology: stems absent (tufted), erect (to 90 cm) or decumbent (to 50 cm), sometimes with stolons (to 80 cm); leaves (to 75 cm) stipulate, alternate, pinnate, with up to 50 pairs of leaflets (to 50 mm); flowers 1 to numerous, 5-merous, with a hypanthium (to 15 mm); petals (to 20 mm) yellow, yellow-white, white/pinkish or cream; stamens (to 4.5 mm) 5–25; achenes (to 3 mm) numerous in aggregates, with or lacking a corky ridge or process.

Life history: duration: annual, biennial (fruit/seeds); perennial (rhizomes, stolons); **asexual reproduction:** rhizomes, stolons; **pollination:** insect; **sexual condition:** hermaphroditic; **fruit:** aggregates (achenes) (common); **local dispersal:** wind, water (seeds); **long-distance dispersal:** animals (seeds)

Imperilment: (1) *Potentilla anserina* [G5]; S1 (IA, NE); S2 (IN); S3 (PA, WY); (2) *P. glandulosa* [G5]; S1 (CA); S2 (UT); S3 (<u>AB</u>); (3) *P. lycopodioides* [G3]; (4) *P. multijuga* [GX]; SX (CA); (5) *P. newberryi* [G3]; SH (WA); S2 (CA); (6) *P. pickeringii* [G2]; S2 (CA); (7) *P. plattensis* [G4]; S1 (<u>AB</u>, MT, NE, UT); S2 (<u>MB</u>); S3 (WY); (8) *P. rivalis* [G5]; SH (<u>ON</u>); S1 (OK, UT); S2 (WY); S3 (<u>AB, BC</u>, IA); (9) *P. supina* [G5]; S1 (<u>BC</u>, ID, OH, NY, PA); S2 (<u>AB</u>, KS, <u>SK</u>, WY); S3 (<u>ON</u>)

Ecology: general: *Potentilla* mainly comprises terrestrial species. Less than a quarter (about 23%) of the North American species are wetland indicators and only five (about 4%) currently are categorized as obligate indicators in some portion of their range. A number of changes have resulted from the revised 2013 indicator list. These include the loss of OBL status for *P. glandulosa* (FAC; which also has been transferred to the genus *Drymocallis*), *P. lycopodioides* (FAC), *P. plattensis* (FACW), and *P. rivalis* (FACW). However, because the treatments for these species had been completed prior to the release of the 2013 list, they have been retained here for reference. The species are herbaceous annuals or perennials with showy (usually yellow-petaled), insect-pollinated flowers. Studies on reproductive biology indicate that most of the species are self-compatible, obligate outcrossers, which are nonapomictic. Seed dispersal occurs by wind or water (some seeds can float for more than 2 weeks) and also by the transport of seeds consumed by birds (Aves) or by large herbivores such as cattle (Mammalia: Bovidae: *Bos*) and horses (Mammalia: Equidae: *Equus*).

Potentilla anserina **L.** is a perennial, which occurs in a wide variety of freshwater or brackish habitats including beaches, depressions, ditches, dunes, fens, forests, lake shores, marshes, meadows, prairies, salt marshes, sea shores, wet woods, and along river margins at elevations of up to 3000 m. The sites typically are open with calcareous gravelly or sandy substrates. The plants will tolerate slight acidity and partial shade and will grow larger in the shade where they are less overheated. Flower production is proportional to the rosette number. *Potentilla anserina* is an obligate outcrosser. The pollen fertility varies depending on the ploidy; i.e., diploids are fully fertile but the hexaploids are pollen and seed sterile. The yellow petals possess honey guides and show a distinctive "bullseye" pattern under ultraviolet light, which helps to attract bees (Insecta: Hymenoptera: Halictidae: *Halictus, Lasioglossum imitatum*) and other insect pollinators. The flowers mature an average of 0.36 seeds/ovule. Typically, there are 1–10 seeds/flower (max = 50) and 10–100 seeds/plant. The achenes are eaten by birds (Aves) and by large herbivores (Mammalia), which disperse them in their excrement. The seeds (which have corky ridges) are also water dispersed. They persist variably in seed banks from <1 year to several years in duration. Successful germination has been achieved using a 20°C/30°C temperature regime for 14 days. Sexual reproduction does not reduce the extent of clonal growth in this species. Plants growing in sediments containing heavy metals (copper, nickel) exhibit highly reduced flowering and vegetative growth. Maintenance of clonal integrity among the ramets has been shown to enhance their survival when they become buried by sand. Seedlings exhibit a higher level of mortality than the daughter ramets and remain in a juvenile (nonreproductive) state for at least 4 years; whereas, ramets are capable of sexual and vegetative reproduction by the next successive season. The plants spread vegetatively by stolons, which can reach 80 cm in length. **Reported associates:** *Achillea millefolium, Agoseris glauca, Agropyron dasystachyum, A. repens, A. spicatum, Agrostis alba, A. gigantea, A. stolonifera, Alisma gramineum, Allium cernuum, Ambrosia artemisiifolia, A. trifida, Andropogon gerardii, Antennaria parvifolia, Arabis lyrata, Arenaria congesta, Armoracia rusticana, Artemisia caudata, A. frigida, Asclepias incarnata, Astragalus miser, A. purshii, Atriplex, Besseya wyomingensis, Betula papyrifera, Bidens cernuus, B. frondosus, Bistorta bistortoides, Boehmeria cylindrica, Brachyactis ciliata, Butomus umbellatus, Cakile edentula, Calamagrostis canadensis, Calamovilfa longifolia, Campanula rotundifolia, Campylium stellatum, Carex aurea, C. brevior, C. crawei, C. duriuscula, C. eburnea, C. filifolia, C. flava, C. lanuginosa, C. lasiocarpa, C. luzulina, C. nebrascensis, C. simulata, C. viridula, Castilleja coccinea, C. luteovirens, Centaurea maculosa, Centaurium pulchellum, Cerastium arvense, Chrysopsis villosa, Cicuta bulbifera, Cirsium arvense, C. pitcheri, C. scariosum, C. vulgare, Cladium mariscoides, Clinopodium arkansanum, Comarum palustre, Convolvulus sepium, Coriospermum, Cornus sericea, Crepis acuminata, Cuscuta, Cyperus polystachyos, C. rivularis, Dasiphora floribunda, Deschampsia cespitosa, Dichanthelium acuminatum, Distichlis spicata, Echinochloa crus-galli, E. walteri, Eleocharis parvula, Elymus canadensis, Epilobium watsonii, Equisetum arvense, E. variegatum, Erigeron acris, E. canadensis, E. humilis, Eupatorium perfoliatum, Euphorbia corallata, E. polygonifolia, Festuca idahoensis, Fragaria virginiana, Fritillaria pudica, Gaillardia aristata, Galium trifidum, Geum triflorum, Glaux maritima, Hackelia floribunda, Helinathus petiolaris, Helictotrichon hookeri, Heteranthera dubia, Heuchera parvifolia, Hieracium aurantiacum, Hordeum brachyantherum, Hymenoxys herbacea, Hypericum kalmianum, Ilex verticillata, Impatiens capensis, Iris missouriensis, I. virginica, Juncus articulatus, J. balticus, J. bufonius, J. dudleyi, J. gerardii, J. torreyi, Lactuca canadensis, Lathyrus maritimus, L. palustris, Lechea, Lobelia kalmii, Lomatium cous, Ludwigia polycarpa, Lupinus argenteus, Lychnis alba, Lycopus americanus, L. asper, Lysimachia nummularia, Maianthemum stellatum, Medicago lupulina, Melilotus alba, Mimulus ringens, Monolepis nuttalliana, Muhlenbergia asperifolia, M. filiformis, Oenothera biennis, Oxytropis campestris, O. deflexa, O. sericea, Panicum capillare, P. lindheimeri, P. virgatum, Persicaria amphibia, P. lapathifolia, P. maculosa, Phalaris arundinacea, Phleum pratense, Phlox hoodii, Poa annua, P. compressa, P. pratensis, P. secunda, Pontederia cordata, Populus balsamifera, P. deltoides, Potamogeton gramineus, Potentilla hippiana, P. rivalis, P. paradoxa, Primula conjugens, P. mistassinica, Puccinellia nuttalliana, Ranunculus scleratus, Rhamnus cathartica, Rhynchospora, Rorippa palustris, Rubus parviflorus, Rumex crispus, R. mexicanus, Salix glaucophylloides, Saxifraga rhomboidea, Schizachyrium scoparium, Schoenoplectus pungens, S. tabernaemontani, Scutellaria galericulata, Selaginella densa, Senecio pauciflorus, Setaria lutescens, Sisyrinchium, Solidago nemoralis, S. ohioensis, S. ptarmicoides, Sorghastrum nutans, Spartina patens, Spergularia marina, Sphagnum, Sphenopholis intermedia, Spiraea alba, Spirodela polyrhiza, Sporobolus, Symphyotrichum lanceolatum, S. laurentianum, S. spathulatum, S. subulatum, Taraxacum eriophorum, T. officianale, Thuja occidentalis, Toxicoscordion venenosum, Trifolium longipes, Triglochin maritimum, Typha latifolia, Ulmus americana, Utricularia macrorhiza, U. purpurea, Vernonia fasciculata, Veronica peregrina, Viola nephrophylla, Xanthium strumarium.*

***Potentilla (Drymocallis) glandulosa* Lindl.** is a perennial, which grows in a diverse spectrum of habitats including roadsides, slopes, meadows, seeps, woodlands, and the margins of streams at elevations of up to 3800 m. Although previously categorized as an obligate aquatic in part of its range, the plants also can grow in fairly dry sites such as sagebrush shrublands and subalpine forests, which reflects its current reclassification as a FAC indicator. The plants grow well in full sunlight to partial shade but generally are shade intolerant. Substrates include clay loam, granitic soils, or sandy loam, having an optimal soil depth of 25–51 cm. *Potentilla glandulosa* is an early successional species that is well adapted to disturbance from fire or grazing and often dominates in heavily grazed sites. The fruits are produced in abundance, but tend to fall near the parental plant. Large seeds banks (up to 840 viable seeds/m^2) can be produced. The seeds retain 75%–85% viability after 3–5 years and those in the upper soil layers exhibit a higher germination rate (24%) than those buried deeper (13%). Scarification (mechanical or by fire) greatly enhances seed germination. Maximum germination is achieved after a 2- to 3-week after ripening period in light at 25°C. Rhizomes are produced in some populations. Classic experiments on ecotypes have shown that single clones transplanted to different elevations (over a 0–3500 m range) show phenotypic plasticity but are best adapted to conditions that are characteristic of their altitude of origin. **Reported associates:** *Abies grandis, Acer glabrum, Achillea millefolium, Achnatherum nelsonii, A. richardsonii, Agastache urticifolia, Agoseris glauca, Agropyron dasystachyum, A. smithii, A. spicatum, Allium cernuum, A. textile, Alnus rugosa, Amelanchier alnifolia, Antennaria parvifolia, Anticlea elegans, Apocynum androsaemifolium, Aquilegia formosa, Arabis drummondii, Arctostaphylos uva-ursi, A. viscida, Arenaria congesta, Arnica cordifolia, A. mollis, A. sororia, Artemisia dracunculoides, A. ludoviciana, A. tridentata, Aspidotis densa, Astragalus miser, Balsamorhiza sagittata, Besseya wyomingensis, Bistorta bistortoides, B. viviparum, Brodiaea douglasii, Bromopsis inermis, Bromus brizaeformis, B. carinatus, B. tectorum, Calamagrostis rubescens, Camassia quamash, Campanula rotundifolia, Carex backii, C. canescens, C. elyniformis, C. geyeri, C. hoodii, C. illota, C. microptera,*

C. obtusata, C. petasata, C. praegracilis, C. raynoldsii, Castilleja linoides, C. longispica, C. lutescens, C. miniata, C. pulchella, Ceanothus integerrimus, C. sanguineus, C. velutinus, Cerastium arvense, Chamerion angustifolium, Chrysothamnus nauseosus, C. viscidiflorus, Cirsium foliosum, Clarkia pulchella, Claytonia perfoliata, Clematis hirsutissima, Collinsia tenella, Collomia grandiflora, C. linearis, Comandra lividum, C. umbellata, Cynoglossum officinale, Cystopteris fragilis, Danthonia intermedia, D. unispicata, Dasiphora floribunda, Delphinium bicolor, Descurainia pinnata, Elymus glaucus, Epilobium paniculatum, Erigeron speciosus, E. trifidus, Eriogonum caespitosum, E. chrysocephalum, E. heracleoides, E. umbellatum, Eucephalus elegans, Eurybia integrifolia, Festuca campestris, F. idahoensis, F. scabrella, Fragaria vesca, F. virginiana, Frasera speciosa, Galium aparine, G. bifolium, G. boreale, Geranium viscosissimum, Geum triflorum, Glycosma occidentalis, Hackelia micrantha, Helianthus uniflorus, Heuchera cylindrica, H. parvifolia, Holodiscus discolor, Hypericum perforatum, Iris missouriensis, Juncus balticus, Juniperus communis, Leymus cinereus, Lithophragma parviflorum, Lithospermum ruderale, Lomatium dissectum, Lupinus argenteus, L. sericeus, Mahonia repens, Maianthemum stellatum, Melica bulbosa, Mertensia perplexa, Mimulus nanus, Myosotis sylvatica, Nemophila breviflora, Orthocarpus hispidus, Oryza, Osmorhiza chilensis, Penstemon diphyllus, P. procerus, P. radicosus, P. rydbergii, P. watsonii, Phacelia hastata, Philadelphus lewisii, Phleum alpinum, P. pratense, Phlox longifolia, Physocarpus malvaceus, Pinus flexilis, P. ponderosa, Poa compressa, P. cusickii, P. fendleriana, P. nervosa, P. pratensis, P. secunda, Polemonium viscosum, Polygonum majus, Populus tremuloides, Potamogeton gramineus, Potentilla rivalis, Prunus emarginata, P. virginiana, Pseudotsuga menziesii, Pteridium aquilinum, Purshia tridentata, Ranunculus glaberrimus, Ribes aureum, R. inerme, R. nevadense, R. viscosissimum, Rosa gymnocarpa, R. woodsii, Rumex paucifolius, Salix brachystachys, Saxifraga bronchialis, Sedum lanceolatum, S. stenopetalum, Senecio integerrimus, S. pseudaureus, S. serra, Silene acaulis, S. oregana, Sitanion hystrix, Solidago missouriensis, S. occidentalis, Spiraea betulifolia, Stellaria jamesiana, Stipa comata, S. lettermanii, S. occidentalis, Symphoricarpos albus, S. oreophilus, Symphyotrichum laeve, S. spathulatum, Taraxacum erythrospermum, T. officinale, Tetradymia canescens, Thalictrum fendleri, T. occidentale, Tortula muralis, Toxicoscordion venenosum, Tragopogon dubius, Trifolium dubium, T. longipes, Trisetum spicatum, Valeriana occidentalis, V. sitchensis, Verbascum thapsus, Viola nuttallii, Vulpia microstachys.

***Potentilla lycopodioides* (A. Gray) Baill. ex J.T. Howell** is a perennial, which occurs in alpine meadows and rocky seeps at elevations from 2300 to 4000 m. Plants are often found on north- or east-facing slopes where snow patches persist. They produce a long taproot that extends 20–40 cm in the soil. Although formerly regarded as OBL in a portion of its range, this species grows in a diverse range of substrates (e.g., dolomite barrens, Barcroft granites, Poleta shales) and also under some fairly dry conditions. It is now regarded only as a FAC wetland indicator. When sown under shallow soil, the seeds will germinate within 1–3 months at 18°C–21°C. **Reported associates:** *Agoseris glauca, Antennaria media, A. rosea, Arabis, Calamagrostis purpurascens, Carex filifolia, C. helleri, C. pseudoscirpoidea, Cistanthe monosperma, Danthonia intermedia, Dasiphora floribunda, Deschampsia cespitosa, Erigeron linearis, Eriogonum incanum, E. ovalifolium, Elymus elymoides, Erigeron algidus, Eriophyllum, Juncus parryi, Lupinus lepidus, Minuartia nuttallii, Muhlenbergia richardsonis, Penstemon heterodoxus, Phlox condensata, P. pulvinata, Poa glauca, P. secunda, Podistera nevadensis, Potentilla diversifolia, P. drummondii, P. muirii, Raillardella argentea, Rhodiola rosea, Sedum stenopetalum, Selaginella watsonii, Silene sargentii, Solidago multiradiata, Trisetum spicatum.*

***Potentilla multijuga* Lehm.** occurred formerly in brackish marshes, meadows, and seeps along the southern coast of California (Ballona Marsh) at elevations from sea level to 10 m. It is now presumed to be extinct. The flowering period extended from April to July.

***Potentilla newberryi* A. Gray** is a prostrate biennial or short-lived perennial, which occurs in sites characterized by ephemeral water levels such as channels, ditches, mudflats, sand flats, slopes, and the receding shorelines of lakes, ponds, reservoirs, rivers, streams, vernal pools, and water holes, at elevations from 1300 to 2200 m. The habitats are often exposed to full sunlight and include alkaline (sometimes somewhat saline) substrates derived from basalt, clay, gravel, humus, mud, pumice, sand, or sandy loam. Flowering occurs from May to August. The petals differ from most *Potentilla* species by their creamy-white color. The plants grow from a rosette with a taproot and are devoid of rhizomes. By virtue of the harsh habitat conditions, the total plant cover of sites typically is low (<5%), frequently being less than 1%. **Reported associates:** *Amaranthus, Artemisia cana, A. tridentata, Bromus tectorum, Camissonia tanacetifolia, Carex douglasii, Downingia, Eleocharis, Epilobium, Gnaphalium, Heliotropium curassavicum, Hordeum jubatum, Iva axillaris, Juncus balticus, Lupinus argenteus, Muhlenbergia richardsonis, Myosurus minimus, Nama, Oenothera, Phacelia lutea, Plagiobothrys kingii, Polyctenium williamsiae, Persicaria, Primula, Psilocarphus brevissimus, Rorippa, Rumex, Salsola, Sarcobatus, Taraxia tanacetifolia, Veronica.*

***Potentilla pickeringii* (Torr. ex A. Gray) Greene** is a perennial, which grows in ephemeral drainages and on seasonally wet, grassy slopes, and swales in rocky meadows of coniferous woodlands at elevations from 800 to 1500 m. The substrates usually are serpentine clay. Flowering occurs during the summer. Unlike most *Potentilla* species, the flower petals are white to pinkish. There is little ecological information available for this rare species. **Reported associates:** none reported.

***Potentilla plattensis* Nutt.** is a perennial, which occurs in or on drains, grassy meadows, hummocks, plateaus, ridges, steppe, swamps, tundra, and along the margins of reservoirs and streams at elevations of up to 2900 m. Although sometimes found in drier steppe shrubland, the plants generally occupy wet sites. Nevertheless, they no longer are regarded as OBL indicators, but have been recategorized as FACW. The substrates are variable. Plants have been found on alkaline (e.g., pH: 7.7) mucky peat of fairly low conductivity (200 μmhos/cm) but also on rocky or stony sediments. **Reported associates:** *Achillea millefolium, Antennaria microphylla, Artemisia frigida, Astragalus terminalis, Beckmannia syzigachne, Bistorta bistortoides, Bromus inermis, Calamagrostis canadensis, Caltha leptosepala, Carex aquatilis, C. nebrascensis, C. praegracilis, C. utriculata, Cirsium arvense, C. scariosum, Dasiphora floribunda, Deschampsia cespitosa, Distichlis spicata, Gentiana affinis, G. fremontii, Geum triflorum, Glaux maritima, Glyceria striata, Halerpestes cymbalaria, Hordeum brachyantherum, H. jubatum, Iris missouriensis, Juncus arcticus, J. balticus, Maianthemum stellatum, Mentha arvensis, Mertensia ciliata, Muhlenbergia asperifolia, Pascopyrum smithii, Phleum alpinum, Phlox kelseyi, Poa pratensis, Potentilla anserina, Primula incana, P. pauciflora, Ribes inerme, Rosa woodsii, Rumex aquaticus, Salix brachycarpa, S. monticola, Sporobolus airoides, Stephanomeria spinosa, Taraxacum offcinale, Thermopsis montana, Trifolium repens.*

Potentilla rivalis **Nutt.** is an annual or biennial, which grows in or on channels, creek beds, depressions, ditches, meadows, mudflats, prairies, sand flats, seeps, and along the margins of lakes, ponds, reservoirs, rivers, sloughs, and streams at elevations of up to 2400 m. Although designated previously as OBL in a portion of its range, this species is now regarded as a FACW indicator. The plants often occur in disturbed sites on clay, gravel, sand, sandy loam, sandy silt, or talus. The flowering period extends from March through November. The flowers lack nectar and are facultatively outcrossed by small solitary bees (Insecta: Hymenoptera: Halictidae: *Halictus confusus, Lasioglossum*), which forage for the pollen. Small flies (Insecta: Diptera) may serve as occasional pollinators. If cross-pollination does not occur, then selfing is promoted by the movement of thrips (Insecta: Thysanoptera) among the reproductive organs, which can account for up to 50% of the seed set. **Reported associates:** *Achillea millefolium, Agoseris aurantiaca, Agrostis scabra, Alopecurus aequalis, Amaranthus, Antennaria lanulosa, A. luzuloides, Arnica latifolia, Artemisia candicans, A. tridentata, Astragalus canadensis, Bidens amplissimus, Brachyactis ciliata, Bromus, Calamagrostis canadensis, Caltha biflora, Carex atherodes, C. brevior, C. concinnoides, C. retrorsa, C. rossii, C. sychnocephala, C. utriculata, Centaurea, Chrysopsis, Chylismia scapoidea, Conyza canadensis, Crypsis alopecuroides, Danthonia intermedia, Eleocharis acicularis, E. palustris, Epilobium brachycarpum, E. ciliatum, E. glandulosum, Eriogonum heracleoides, E. hookeri, Euphorbia serpyllifolia, Fragaria vesca, Galium trifidum, Gentianella amarella, Geum triflorum, Glyceria grandis, G. maxima, Hippuris vulgaris, Hordeum jubatum, Juncus bufonius, J. conglomeratus, Luzula hitchcockii, Madia gracilis, Melilotus alba, Mentha arvensis, Micranthes nelsoniana, Mimulus, Mitella pentandra, Monolepis nuttalliana, Oenothera, Oxytropis deflexa, Panicum capillare, Penstemon procerus, Pera, Persicaria amphibia, P. hydropiperoides, Phleum alpinum, P. pratense,*

Picea engelmannii, Pinus contorta, P. ponderosa, Poa secunda, Polemonium pulcherrimum, Potentilla anserina, P. glandulosa, P. supina, Pseudoroegneria spicata, Puccinellia nuttalliana, Ranunculus flammula, Rhododendron columbianum, Rorippa islandica, R. palustris, Rumex maritimus, R. stenophyllus, Salix bebbiana, S. geyeriana, Salsola, Schoenoplectus acutus, Senecio integerrimus, S. macounii, S. triangularis, Sium suave, Sphenopholis obtusata, Stellaria media, Streptopus roseus, Symphyotrichum frondosum, Taraxacum officinale, Taraxia tanacetifolia, Thalictrum occidentale, Trifolium repens, Typha latifolia, Vaccinium scoparium, Valeriana sitchensis, Veratrum viride, Veronica americana, V. peregrina, Viola canadensis, V. glabella.

Potentilla supina **L.** is an annual or biennial plant that is found on sites characterized by water level fluctuations including beaches, bottomlands, depressions, ditches, flats, floodplains, hummocks, lake and pond shores, marshes, meadows, prairies, riverbanks, roadsides, sandplains, and seeps at elevations from 100 to 2000 m. Some sites can retain standing water (up to 1 m deep) for at least short durations. The plants occur typically in full sun on alkaline substrates that often are muddy, sandy, or gravelly; e.g., mudflats and sandbars. The flowering period extends from late spring into summer. A single plant is capable of producing up to 22,500 achenes, which possess a prominent corky ridge. *Potentilla paradoxa* is described as early-seral species characteristic of bare mineral soil. The size of populations can fluctuate considerably in successive years. **Reported associates:** *Alopecurus aequalis, Ammophilia breviligulata, Bidens comosus, B. frondosus, Bolboschoenus maritimus, Cakile edentula, Carex viridula, Chenopodium glaucum, Crassula aquatica, Crepis capillaris, Cyperus engelmanii, C. erythrorhizos, C. squarrosus, Elaeagnus, Eleocharis palustris, E. quadrangulata, Eragrostis hypnoides, Euphorbia polygonifolia, Gleditsia triacanthos, Grindelia squarrosa, Hibiscus moscheutos, Hordeum jubatum, Juncus alpinus, J. torreyi, Lactuca serriola, Leersia oryzoides, Leptochloa panicea, Lipocarpha micrantha, Lobelia siphilitica, Lythrum salicaria, Marsilea vestita, Myosurus minimus, Osmorhiza longistylis, Panicum capillare, Peltandra virginica, Persicaria amphibia, P. lapathifolia, Phalaris arundinacea, Plantago lanceolata, Poa annua, Polygonum aviculare P. polygaloides, Populus deltoides, Potentilla anserina, P. rivalis, Poteridium occidentale, Psilocarphus elatior, Ranunculus sceleratus, Rumex crispus, R. maritimus, Salix eriocephala, S. exigua, S. interior, S. nigra, Schoenoplectus acutus, S. smithii, Subularia aquatica, Symphyotrichum ciliatum, S. lanceolatum, Tamarix, Triplasis purpurea, Triticum aestivum, Typha latifolia, Verbena bracteata.*

Use by wildlife: The high tannin content of *Potentilla* species renders them as a relatively poor wildlife food. However, *P. glandulosa* is fed on by mule deer (Mammalia: Cervidae: *Odocoileus hemionus*) in the fall and it also provides a fair food resource for cattle (Mammalia: Bovidae: *Bos*), elk (Mammalia: Cervidae: *Cervus*), sheep (Mammalia: Bovidae: *Ovis*), small mammals (Mammalia), and UPL birds (Aves) during the summer. The new growth of *P. newberryi* is more nutritious than sagebrush and is also fed upon by mule deer. *Potentilla anserina* is a host for several fungi including Ascomycota (Dermateaceae: *Marssonina potentillae*, Dothideomycetidae: *Septoria purpurascens*) and Oomycota (Pythiaceae: *Phytophthora fragariae*). It is moderately fed upon by beetles (Insecta: Coleoptera: Chrysomelidae: *Galerucella calmariensis*, *G. pusilla*) that are used for biological control of *Lythrum salicaria*. Spiders (Arthropoda: Arachnida) have been observed to lay their eggs on foliage of *P. anserina*. *Potentilla glandulosa* is host to a number of fungi including Ascomycota (Dermateaceae: *Marssonina potentillae*; Dothideomycetidae: *Ramularia arvensis*; Erysiphaceae: *Sphaerotheca macularis*; Taphrinomycetidae: *Taphrina tormentillae*), Chytridiomycota (Synchytriaceae: *Synchytrium potentillae*), and Oomycota (Pythiaceae: *Phytophthora fragariae*). *Potentilla newberryi* is a host of the fungus *Ramularia arvensis* (Ascomycota: Dothideomycetidae). *Potentilla supina* and *P. rivalis* reportedly attract butterflies (Insecta: Lepidoptera). *Potentilla plattensis* is the host to a plant bug (Insecta: Hemiptera: Miridae: *Chlamydatus pallidicornis*). *Potentilla rivalis* is another host of the fungus *Marssonina potentillae* (Ascomycota: Dermateaceae).

Economic importance: food: The Montana, Okanagon, Shuswap, and Thompson tribes ate the young shoots and roots (raw or cooked) of *Potentilla anserina*, the latter supposedly tasting like parsnips or sweet potatoes; **medicinal:** The roots of *P. anserina* have been used to treat diarrhea (Blackfoot, Iroquois), as a poultice for sores (Blackfoot), as an emetic for stomach disorders (Blackfoot) and as an analgesic (Kwakiutl). The Iroquois prepared an infusion of its leaves to administer as a diuretic. A tea made from *P. anserina* is antispasmodic and was drunk to reduce cramps associated with childbirth. The Gosiute, Okanagon, and Thompson tribes used the root of *P. glandulosa* as a general medicinal tonic. The Gosiute applied a poultice made from the plant to relieve swelling. The Okanagon and Thompson tribes administered an infusion of the plant as a stimulant; **cultivation:** *Potentilla anserina* is cultivated for production of its edible roots (see earlier) and as a garden ornamental. Cultivars include 'Golden Treasure,' 'Ortie,' 'Shine,' and 'Variegata.' *Potentilla glandulosa* is sold occasionally as a native ornamental; **misc. products:** Sprigs of *Potentilla anserina* are placed in shoes to absorb sweat and to prevent blisters. Leaf infusions of the plants have been used to make a cleansing lotion due to their tannin content. The Blackfoot used the stolons of *P. anserina* as cordage. *Potentilla glandulosa* was one of the plants studied in the infamous experiments of phenotypic plasticity carried out by J. Clausen, D. D. Keck, and W. M. Hiesey; **weeds:** *Potentilla anserina* can become weedy, especially in the western United States. *Potentilla supina* is reported as a weed in South Korean landfills; **nonindigenous species:** *Potentilla rivalis* was introduced to northwest Europe sometime before the mid-1970s.

Systematics: *Potentilla* (subfamily Rosoideae) has recently undergone a number of different circumscriptions, mainly as a consequence of new insight provided by molecular systematic investigations. The actual tally of species is uncertain, but the number estimated earlier reflects the inclusion of *Argentina*,

Duchesnea, *Horkelia*, and *Ivesia* (whose species are merged) and the exclusion of *Comarum*, *Dasiphora*, *Drymocallis*, and *Sibbaldiopsis* (whose species are transferred out of the genus). In this circumscription, *Potentilla* is the sister genus to two clades, which contain the segregate genera as well as others (e.g., *Fragaria*) with which it has been associated historically (Figure 4.44). The relatively close relationship among these genera is further evidenced by a number of successful (but sterile) intergeneric hybrids (e.g., *Fragaria* × *Comarum*; *Fragaria* × *Dasiphora*; *Fragaria* × *Potentilla*). Interspecific relationships within *Potentilla* have been investigated for 98 species, which represent less than a quarter of this large genus. Three of the OBL North American species (*P. anserina*, *P. newberryi*, *P. supina*) have been included, which each indicate an independent origin of the habit (Figure 4.44). *Potentilla glandulosa* is now assigned to the genus *Drymocallis* (Figure 4.44), which is justified phylogenetically. *Potentilla anserina* and *P. egedii* (sometimes treated as a subsp. of the former) have been segregated as the genus *Argentina* (=*Potentilla* section *Gymnocarpae*, subsection *Leptostylae*), a result supported by preliminary molecular studies; however, more recent molecular analyses place *P. anserina* close to the Asian species *P. peduncularis* and *P. stenophylla*, or within a sister clade to that containing most of the *Potentilla* species, making it a matter of preference whether to merge or segregate these genera. Overall morphological similarities and cytogenetic evidence (successful growth of *P. anserina* pollen tubes in the styles of diverse *Potentilla* species) support the retention of this group within *Potentilla*. Seven varieties of *P. glandulosa* have been recognized, but these are poorly defined and are associated with a wide diversity of habitats. *Potentilla lycopodioides*, *P. newberryi*, and *P. pickeringii* were placed formerly in *Ivesia*, which recent molecular studies (Figure 4.44) show (at least for four species) to be included within *Potentilla*. Three subspecies have been recognized for *Potentilla lycopodioides* with subsp. *megalopetala* showing the highest affinity for wet sites. *Potentilla supina*, is conspecific perhaps, with *P. paradoxa*, and has also been treated as *P. supina* var. *paradoxa* (Nutt.) Wolf and as *P. supina* L. subsp. *paradoxa* (Nutt.) Soják. Three varieties of *Potentilla rivalis* have been recognized, but their taxonomic integrity is doubtful. The base chromosome number of *Potentilla* is $x = 7$. *Potentilla* (*Drymocallis*) *glandulosa* ($2n = 14$) is a diploid; whereas, *P. lycopodioides* and *P. pickeringii* are tetraploids ($2n = 28$). *Potentilla anserina* occurs as tetraploid, pentaploid, and hexaploid cytotypes ($2n = 28, 35, 42$); whereas, *P. rivalis* has diploid and decaploid cytotypes ($2n = 14, 70$). *Potentilla plattensis* ($2n = 70$) is a decaploid; *P. supina* ($2n = 28$) is tetraploid. Pentaploid ($2n = 35$) intergeneric hybrids involving *Fragaria chiloensis* and *P. glandulosa* have been synthesized, but these are sterile.

Comments: *Potentilla anserina* is circumboreal and is distributed throughout northern North America; whereas, the circumboreal *P. supina* occurs mainly in central regions of North America. *Potentilla rivalis* is disjunct between eastern North America (where it is uncommon), and the central and western regions, where it is relatively widespread. The

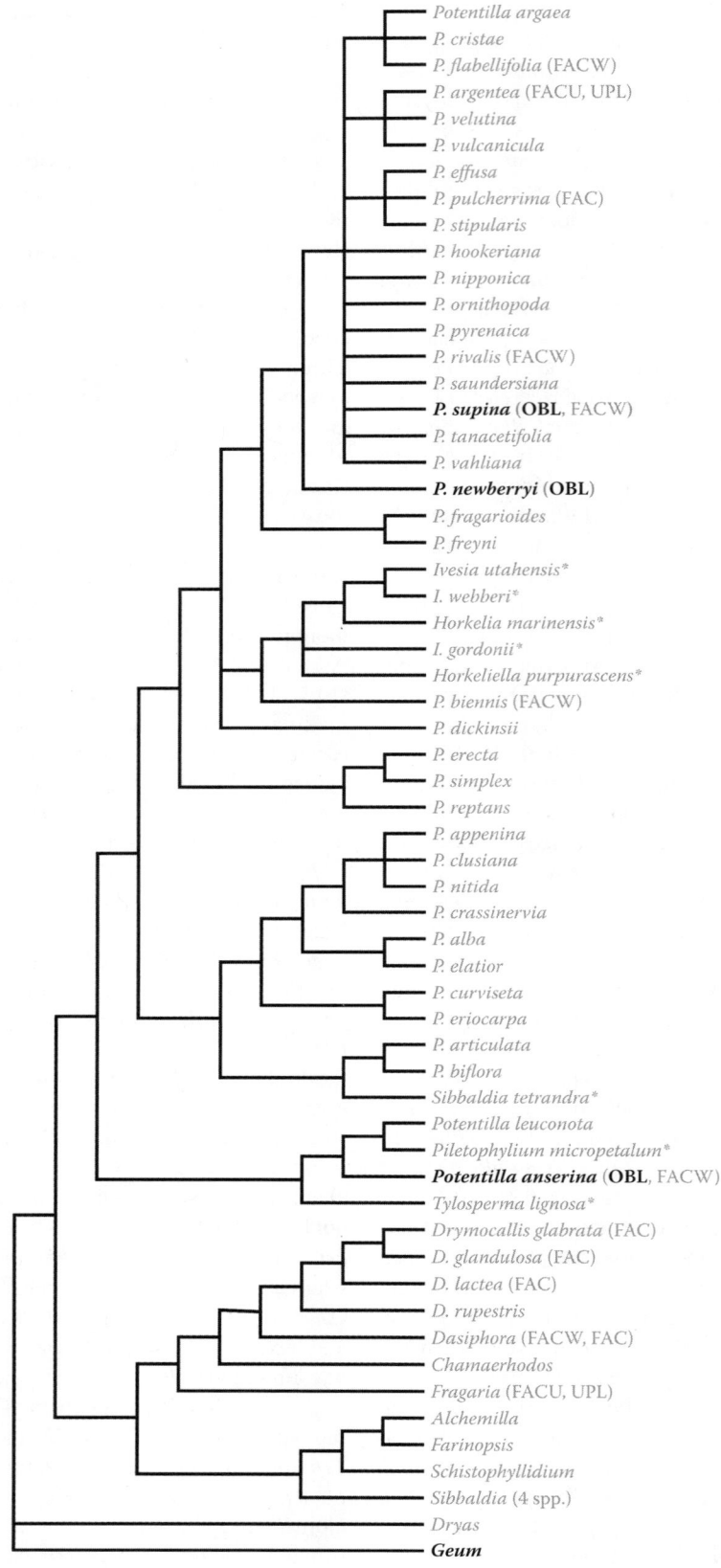

FIGURE 4.44 Phylogenetic relationships in *Potentilla* and allied genera as indicated by analysis of combined cpDNA sequence data. The OBL taxa (bold) have evolved several times within the genus. Asterisks denote misplaced groups, which should be transferred to *Potentilla*. (Adapted from Dobeš, C. & Paule, J., *Mol. Phylogen. Evol.*, 56, 156–175, 2010.)

distributions of several species show western geographical affinities. *Potentilla* (*Drymocallis*) *glandulosa* occurs throughout western North America, *P. plattensis* in west-central North America, and *P. newberryi* in the western United States. Several species have restricted distributions. *Potentilla lycopodioides* occurs only in California and Nevada and *P. pickeringii* is endemic to the Klamath ranges of northern California. *Potentilla multijuga* was endemic to the Ballona Marsh of California (Los Angeles Co.), where it has not been observed since 1890. It was proclaimed to be extinct by The Center for Biological Diversity in 2001.

References: Baker & Cruden, 1991; Clausen et al., 1940; Clausen & Hiesey, 1958; Dobeš & Paule, 2010; Eriksson, 1986, 1987, 1988; Eriksson et al., 1998, 2003; Ertter et al., 2015; Fernald & Kinsey, 1943; Goswami & Matfield, 1975; Holland & Morefield, 2002; Kim, 2002; Maillette, 1992; Rousi, 1965; Saikkonen et al., 1998; Schuh & Schwartz, 2005; Senanayake & Bringhurst, 1967; Stevens, 1932; Stevens et al., 1996; Tirmenstein, 1987a,b; Yu et al., 2001.

6. *Rosa*

Rose

Etymology: the ancient Latin name for the rose plant derived from the Greek *rhodon* meaning "red" in reference to the typical color of the petals.

Synonyms: *Hulthemia*; *Hulthemosa*

Distribution: global: Africa; Asia; Europe; North America; **North America:** widespread

Diversity: global: 140 species; **North America:** 33 species

Indicators (USA): OBL: *Rosa palustris*

Habitat: freshwater; palustrine; **pH:** 4.0–7.3; **depth:** <1 m; **life-form(s):** emergent shrub

Key morphology: stems (to 2.5 m) reddish, with hard, often decurved, infrastipular prickles (to 6 mm); leaves odd pinnate with 2–9 pairs of serrate leaflets (to 5 cm), petiolate (to 2 cm), the stipules (to 1.5 cm) fused to the petiole for ⅔ their length; inflorescence solitary to few flowered; flowers (to 8 cm) perigynous, 5-merous; hypanthium urceolate; sepals (to 2 cm) with foliaceous, dilated tips; petals (to 3 cm) pink; stamens numerous; achenes (to 3.5 mm) numerous, enclosed in the hypanthium to form an aggregate, reddish "hip" (to 2 cm) when mature

Life history: duration: perennial (buds, rhizomes); **asexual reproduction:** winter buds; **pollination:** insect; **sexual condition:** hermaphroditic; **fruit:** aggregates (achenes) (common); **local dispersal:** fruits (water); **long-distance dispersal:** fruits (birds)

Imperilment: (1) *Rosa palustris* [G5]; S1 (IA); S2 (NB); S3 (NS)

Ecology: general: Roses principally are plants of well-drained habitats. Eighteen North American species (55%) are designated wetland indicators but just one species is categorized as OBL. Roses are known for their showy, fragrant flowers, which are pollinated by bees (Hymenoptera) and other insects. The diploid species generally are self-incompatible; whereas, the polyploids are self-fertile. The aggregates of achenes remain within the fleshy floral receptacle to yield a fruit known as a "hip." The seeds can be dispersed in the air for several meters, or the hips may be eaten by birds and other wildlife, which can disperse the achenes by defecation. The seeds of some species float and are dispersed on the water surface.

***Rosa palustris* Marshall** occurs in tidal and nontidal freshwater habitats including bogs, bottomlands, depressions, ditches, flatwoods, floodplains, interdunal ponds, lake shores, marshes, meadows, prairies, roadsides, sand prairies, sloughs, swales, swamps, and along the margins of brooks, rivers, seeps, and streams at elevations of up to 700 m. It sometimes occurs in shallow standing water (to 10 cm) and occupies habitats that primarily are acidic (pH: 4.5–7.3) with open to slightly shaded exposures. The substrates typically have a high organic matter content and include loam, sand, sandy loam, sandy peat, and silt. The plants can also grow directly on exposed logs and cypress (*Taxodium*) knees. Although the plants are regarded as being salt-intolerant, they will withstand periodic influxes of salinity from 0.5 to 11.0 ppt. *Rosa palustris* can grow in full sun but is also shade tolerant. It will survive cold temperatures down to −36°C. The shoots are strongly rhizomatous (rooting to a depth of up to 46 cm), clonal, and produce suckers readily. *Rosa palustris* flowers later in the season (June to August) compared to all other North American *Rosa* species. The large, showy flowers are self-incompatible, pollinated mainly by bees (Hymenoptera), and are obligately outcrossed. Self-pollination results in 0% seed/fruit set, whereas, seed set is 74% (fruit set 80%) in outcrossed plants. Natural pollen fertility ranges from 64% to 95%. Specific information on seed germination is lacking, but may require a combination of acid scarification (due to the hard seed coat) and a period of cold stratification at 0°C–4°C. Seeds for propagation should be collected in the fall. The fruits will float for a prolonged period, which enables their dispersal by water.

Reported associates: *Acer negundo, A. rubrum, A. saccharinum, Acorus calamus, Agrimonia parviflora, Agrostis gigantea, Aletris farinosa, Alnus rugosa, A. serrulata, Amelanchier canadensis, Amorpha fruticosa, Andropogon gerardii, A. glomeratus, Apios americana, Apocynum cannabinum, Aronia arbutifolia, A. melanocarpa, Asclepias hirtella, A. incarnata, Baccharis halimifolia, Baptisia lactea, Betula, Bidens laevis, Boehmeria cylindrica, Bromus ciliatus, Bolboschoenus fluviatilis, Calamagrostis canadensis, Calopogon pulchellus, Cardamine bulbosa, Carex annectens, C. bushii, C. buxbaumii, C. conjuncta, C. conoidea, C. crinita, C. crus-corvi, C. davisii, C. exilis, C. festucacea, C. frankii, C. gravida, C. grisea, C. haydenii, C. hormathodes, C. intumescens, C. jamesii, C. lacustris, C. laevivaginata, C. lasiocarpa, C. leptalea, C. longii, C. lurida, C. muskingumensis, C. pellita, C. scoparia, C. shortiana, C. squarrosa, C. straminea, C. stricta, C. swanii, C. vulpinoidea, Carya illinoinensis, C. laciniosa, Celtis occidentalis, Cephalanthus occidentalis, Chamaecrista fasciculata, Chamaecyparis thyoides, Chamaedaphne calyculata, Chasmanthium latifolium, Cicuta maculata, Cinna arundinacea, Cladium jamaicense, C. mariscoides, Clematis pitcheri, C. virginiana, Clethra alnifolia, Comarum palustre, Convolvulus sepium, Cornus amomum, C. racemosa, C. sericea, Corylus americana, Cuscuta gronovii, Dasiphora floribunda, Daucus carota, Decodon verticillatus, Desmodium glabellum, D. paniculatum, Dichanthelium* *clandestinum, Doellingeria umbellata, Drosera intermedia, Dulichium arundinaceum, Elaeagnus umbellata, Eleocharis fallax, E. ovata, E. rostellata, E. tenuis, Eragrostis spectabilis, Erecthtites hieracifolia, Erigeron philadelphicus, E. strigosus, Eriocaulon decangulare, Eriophorum virginicum, Eryngium integrifolia, Eupatorium perfoliatum, Euphorbia corollata, Eutrochium maculatum, Festuca subverticillata, Fragaria virginiana, Fraxinus nigra, F. pennsylvanica, F. profunda, Galium circaezans, G. obtusum, G. tinctorium, Gaylussacia baccata, Gentiana andrewsii, Geum canadense, Glyceria canadensis, G. laxa, G. striata, Gratiola neglecta, Hamamelis, Helenium autumnale, Helianthus angustifolius, H. mollis, Hibiscus moscheutos, Holcus mollis, Hypericum ellipticum, H. mutilum, Ilex verticillata, Impatiens capensis, Iris brevicaulis, I. virginica, Juncus acuminatus, J. canadensis, J. dichotomus, J. effusus, J. marginatus, J. pelocarpus, J. scirpoides, J. subcaudatus, J. vaseyi, Kalmia angustifolia, Kosteletzkya pentacarpos, Larix laricina, Leersia oryzoides, L. virginica, Lemna minor, Lespedeza virginica, Leucothoe racemosa, Liatris pycnostachya, L. spicata, L. squarrulosa, Lilaeopsis carolinensis, Lindera benzoin, Linum striatum, Lobelia spicata, Lycopodiella appressa, Lycopus americanus, L. uniflorus, L. virginicus, Lyonia ligustrina, Lysimachia lanceolata, L. quadrifolia, L. terrestris, Lythrum alatum, L. salicaria, Malus ioensis, M. sieboldii, Mikania scandens, Myrica gale, M. cerifera, Napaea dioica, Nyssa aquatica, N. biflora, N. ogeche, Oenothera gaura, O. pilosella, Onoclea sensibilis, Osmunda regalis, Osmundastrum cinnamomeum, Oxypolis rigidor, Packera aurea, Panicum rigidulum, P. virgatum, Parthenocissus quinquefolia, Peltandra virginica, Penstemon hirsutus, Persicaria arifolia, P. coccinia, P. punctata, P. sagittata, P. virginiana, Phalaris arundinacea, Phlox glaberrima, P. pilosa, Phragmites australis, Picea mariana, Pilea pumila, Platanthera flava, Platanus occidentalis, Pluchea foetida, Poa palustris, P. pratensis, Pontederia cordata, Populus deltoides, P. tremuloides, Potentilla simplex, Prunella vulgaris, Prunus serotina, Ptilimnium capillaceum, Pycnanthemum verticillatum, P. virginianum, Quercus alba, Q. bicolor, Q. palustris, Q. rubra, Q. shumardii, Ranunculus septentrionalis, Rhamnus frangula, Rhododendron canadense, R. viscosum, Rhynchospora alba, R. capitellata, R. corniculata, Ribes americanum, Robinia pseudoacacia, Rubus flagellaris, R. hispidus, R. idaeus, R. pubescens, Ruellia humilis, Sagittaria latifolia, Salix bebbiana, S. discolor, S. eriocephala, S. fragilis, S. humilis, S. interior, S. nigra, S. petiolaris, Sambucus nigra, Sanguisorba canadensis, Sanicula canadensis, Sarracenia oreophila, Sassafras albidum, Saururus cernuus, Schoenoplectus americanus, S. pungens, Scirpus ancistrochaetus, S. atrovirens, S. cyperinus, S. pendulus, Scutellaria galericulata, Sium suave, Smilax rotundifolia, Solidago canadensis, S. gigantea, S. graminifolia, S. missouriensis, S. rugosa, Sorghastrum nutans, Spartina, Sphagnum palustre, Sphenopholis obtusata, Spiraea alba, S. tomentosa, Spiranthes cernua, S. romanzoffiana, Stachys tenuifolia, Symphyotrichum ericoides, S. lanceolatum, S. lateriflorum, S. novi-belgii, S. ontarionis, S. pilosum, S. puniceum, Symplocarpus foetidus, Taxodium distichum, Teucrium canadense, Thalictrum dasycarpum, T. polygamum, Thelypteris noveboracensis, T. palustris, Tradescantia ohiensis, Triadenum virginicum, Tridens flavus, Toxicodendron radicans, Typha angustifolia, T. latifolia, Ulmus americana, Utricularia cornuta, Vaccinium corymbosum, V. macrocarpon, Verbena hastata, Vernonia missurica, Veronicastrum virginicum, Viburnum dentatum, V. lentago, V. nudum, V. recognitum, Viola cucullata, Vitis riparia, Xyris torta, Zizania aquatica.*

Use by wildlife: Beavers (Mammalia: Castoridae: *Castor canadensis*) and deer (Mammalia: Cervidae) have been observed grazing on the foliage of *Rosa palustris*. Its fleshy fruits ("hips") are eaten and dispersed by birds (Aves) and small mammals (Mammalia). *Rosa palustris* is a pupal host of the rose hip fly (Insecta: Diptera: Trypetidae: *Rhagoletis basiola*) and is the larval host of a leaf-mining moth (Insecta: Lepidoptera: Tischeriidae: *Coptotriche admirabilis*). Gall wasps (Insecta: Hymenoptera: Cynipidae: *Rhodites bicolor*) produce stem galls on the plants. The plants also serve as a host for mites (Acari: Eriophyidae: *Phyllocoptes adalius, P. fructiphilus*), which have been implicated as vectors of rose rosette disease; however, they are resistant to that disease. The plants are also highly resistant to rose black spot, a disease caused by a fungus (Ascomycota: Dermateaceae: *Diplocarpon rosae*).

Economic importance: food: The petals of many rose species are edible and sometimes are candied or used in salads. The dried and ground "hips" of *R. palustris* are added to wheat flour for making a leavened bread; **medicinal:** The Cherokee made an infusion from the bark and roots of *R. palustris* as a treatment for worms and also administered a decoction prepared from the roots for treating dysentery; **cultivation:** *Rosa palustris* var. *scandens* is a double-flowered form sometimes found in cultivation. The cultivar 'Coriandre' reputedly is derived from a hybrid between *R. palustris* and *R. rugosa* (double-flowered). Native *R. palustris* plants are recommended as alternative ornamentals to the invasive (but terrestrial) *R. multiflora*; **misc. products:** none; **weeds:** none; **nonindigenous species:** none

Systematics: Phylogenetic analyses of molecular data (Figure 4.41) have indicated that *Rosa* occupies a relatively isolated position near the base of subfamily Rosoideae, where it occurs between *Rubus* and a clade containing *Geum* and its allies, and the remainder of the subfamily. Traditionally, the genus has been subdivided into four subgenera: *Hulthemia, Hesperhodos, Platyrhodon,* and *Rosa; R. palustris* is placed within section *Carolinae* of subgenus *Rosa*. However, systematic studies of *Rosa* using both plastid and nuclear sequence data indicate that the subgenera and many of the sections are not delimited naturally. Results from recent molecular studies suggest the placement of *R. palustris* within section *Cinnamomeae*, which merges species from the former section *Carolinae*. Molecular phylogenetic studies also show highly inconsistent interspecific relationships among the roses. Analysis of nrITS data resolves *R. palustris* as being most closely related to *R. bracteata* from China (a species of dry habitats that was introduced to North America). However, evidence from nuclear *GAPDH* sequence data indicates that the North American (and ecologically similar) diploid *R. nitida* is most closely related to *R. palustris*. Yet, combined cpDNA sequences (*atpB–rbcL*–IGS) place *R. palustris* distant phylogenetically from either *R. bracteata* or *R. nitida* in an isolated position nearer to *R. rugosa* and *R. inodora* in some cases, or in an unresolved cluster with *R. carolina*, *R. virginiana*, etc., in others. Although the association of *R. palustris* and *R. nitida* perhaps is one of the more palatable results to be obtained thus far, additional studies are necessary to explain the phylogenetic incongruency that exists among the various molecular data sets analyzed for the genus. *Rosa palustris* long has been presumed to be related closely to the tetraploid *R. virginiana*. *GAPDH* sequence data support the origin of the polyploid *R. virginiana* from members of the *R. foliolosa–nitida–palustris* group. *Rosa palustris* initially was confused taxonomically with *R. virginiana* and the older literature often is difficult to interpret due to this misconception. The chromosomal base number of *Rosa* is $x = 7$. *Rosa palustris* is a diploid ($2n = 14$) but has been mistaken for the tetraploid *R. virginiana*, especially in determining the parentage of some hybrids. The hybrid cultivar 'Coriandre' allegedly involves *Rosa rugosa* × *R. palustris*. Natural hybrids involving *R. palustris* include *R. blanda* × *R. palustris* (diploid), *R. carolina* × *R. palustris* (triploid), and *R. virginiana* × *R. palustris* (triploid; 90% pollen sterility). "*Rosa bracteata* ×" supposedly is a hybrid formed between *R. blanda* and the F_1 hybrids of *R. rugosa* × *R. palustris*.

Comments: *Rosa palustris* occurs throughout eastern North America.

References: Balduf, 1958; Bruneau et al., 2007; Erlanson, 1929, 1938; Hurst, 1928; Joly et al., 2006; Lewis et al., 2015; Nybom et al., 2005; Rydberg, 1920; Starr & Bruneau, 2002; Ueda & Akimoto, 2001; Wissemann & Ritz, 2005.

7. Rubus

Blackberry, raspberry; framboises, ronces

Etymology: an ancient name for brambles derived from the Latin root "*rub-*" meaning "red" with reference to the common color of the berries or stems

Synonyms: *Batidaea*; *Comarobatia*; *Dalibarda*

Distribution: global: cosmopolitan; **North America:** widespread

Diversity: global: 250–750 species; **North America:** 37–230 species

Indicators (USA): OBL: *Rubus hypolasius*, *R. plexus*, *R. spectatus* (but see *Ecology* later); **FACW:** *R. arcticus*, *R. setosus*; **FAC:** *R. arcticus*

Habitat: freshwater; palustrine; **pH:** 4.0–8.0; **depth:** <1 m; **life-form(s):** emergent herb, emergent shrub

Key morphology: stems (to 15 cm), armed with prickles and/or bristles or unarmed, erect to arching, with woody rhizomes; leaves stipulate (to 38 mm), 3–5-foliate, the leaflets (to 11.5 cm) toothed; inflorescences axillary or terminal, 3–20-flowered; flowers 5 or 8-merous, showy; sepals (to 13 mm) green, lanceolate; petals (to 2 cm) pink, red, rose, or white; stamens 30–40; pistils 20–40; fruit (to 1.5 cm) a berry-like aggregate of 5–40 black or red drupelets

Life history: duration: perennial (buds, rhizomes); **asexual reproduction:** rhizomes; **pollination:** insect; **sexual condition:** hermaphroditic; **fruit:** aggregates (drupelets) (common); **local dispersal:** animals (small mammals); **long-distance dispersal:** animals (birds)

Imperilment: (1) *Rubus arcticus* [G5]; S1 (CO, MI, WA); S2 (WY, YT); S3 (NF); (2) *R. hypolasius* [G1]; S1 (NJ); (3) *R. plexus* [GU, GHQ]; (4) *R. spectatus* [G5]

Ecology: general: Although *Rubus* principally is a terrestrial genus, 48 North American species are listed as wetland indicators, with three species designated as OBL. However, there are immense discrepancies regarding the taxonomic delimitation of species in this genus, due to a variety of factors involving apomixis, hybridization, and polyploidy. Interpretive difficulties materialize immediately upon consulting the recent *Flora of North America* treatment, where only 37 North American species are recognized, a number far less than that of the recognized wetland indicators! Further confusion arises from the proposed taxonomic synonymy. *Rubus lawrencei* (OBL in the 1996 indicator list) is now regarded as synonymous with *R. setosus* (FACW), which also includes *R. spectatus* (designated as OBL in the 2013 indicator list). To comply with this recommendation, the treatments for both *R. lawrencei* and *R. spectatus* (completed prior to the release of the 2013 indicator list) have been merged and presented here under the name *R. setosus*, despite its rank as a FACW indicator. It is more difficult to address the status of *R. plexus* (OBL in the 2013 indicator list), which is now regarded as a synonym of *R. flagellaris* (FACU, UPL); consequently, it has been excluded from this treatment. In addition, *R. hypolasius* (OBL in the 2013 list) is now considered to represent a hybrid (as *R.* ×*hypolasius*) between *R. flagellaris* (FACU, UPL) and *R. pensilvanicus* (FAC, FACU, UPL). Accordingly, this taxon has also been excluded. Last, *R. arcticus* (OBL, FAC in the 1996 indicator list) is designated as FACW, FAC in the revised 2013 list. Its treatment has been retained here, given its completion prior to the release of the 2013 indicator list. As a consequence of these proposed taxonomic modifications, it is doubtful whether any North American *Rubus* species should actually be regarded as an OBL indicator. *Rubus* typically is associated with a prickly "bramble" habit, but some species are low-growing, herbaceous plants. The flowers are showy and are pollinated by insects. Pseudogamous agamospermy occurs in many of the polyploid species. The seeds are dispersed endozoically by animal vectors, which feed upon the sweet berries.

Rubus arcticus **L.** is a herbaceous species, which ranges from wet to dry habitats and is found in bogs, depressions, fens, marshes, meadows, muskeg, stringbogs, swamps, thickets, and along the margins of rivers and streams in the boreal forest and tundra at elevations of up to 3000 m. The plants occur in full sun or partial shade in both acidic (pH: 4.0–5.5) and calcareous (pH: 6.4–8.0) sites with peat, gravel, or rich sandy loam substrates. The plants flower from May to August. The flowers produce low levels (0.15–0.37 µl) of nectar and are pollinated by honey bees and bumble bees (Insecta: Hymenoptera: Apidae: *Apis mellifera*, *Bombus*). The pollinators move between closely spaced plants, a behavior that favors seed production where different clones occur. The low-growing plants often are pollinator limited. The percentage of flowering (up to 39%) is higher in open marsh habitat than in forested areas (up to 27%). Colder temperatures (2°C–5°C) are detrimental to pollen viability; however, the plants are cold tolerant to at least −17.6°C. Their fruits produce about 17 seeds on average and contain about 20% sugar. The seeds require 1 month of cold stratification at 3°C for germination. Higher fruit production has been observed along lakeshores than in drier sites. The plants grow clonally from slender, woody rhizomes, which allow them to colonize areas rapidly following fires. Population size can vary considerably, ranging from as few as 25 to as many as 70,000 stems. **Reported associates:** *Abies lasiocarpa, Aconitum columbianum, Alnus rugosa, A. viridis, Andromeda polifolia, Angelica arguta, Aralia nudicaulis, Arenaria lateriflora, Astragalus alpinus, A. americanus, Athyrium filix-femina, Aulacomnium palustre, Betula papyrifera, B. pumila, Bistorta vivipara, Botrychium lanceolatum, Bromus vulgaris, Calamagrostis canadensis, C. stricta, Caltha palustris, Carex aquatilis, C. canescens, C. cephalantha, C. chordorrhiza, C. diandra, C. disperma, C. exilis, C. lacustris, C. lasiocarpa, C. leptalea, C. rostrata, C. trisperma, C. utriculata, C. vesicaria, Chamaedaphne calyculata, Chamerion angustifolium, Comarum palustre, Cornus canadensis, C. sericea, Cypripedium arietinum, Dasiphora floribunda, Drosera rotundifolia, Empetrum nigrum, Epilobium glandulosum, Equisetum arvense, E. fluviatile, E. scirpoides, Eriophorum angustifolium, E. polystachion, E. vaginatum, E. viridicarinatum, Fragaria virginiana, Galium bifolium, G. boreale, G. trifidum, G. triflorum, Geranium richardsonii, Geum macrophyllum, G. rivale, Glyceria, Iris versicolor, Kalmia polifolia, Larix laricina, Linnaea borealis, Lobelia kalmii, Lonicera involucrata, L. villosa, Luzula parviflora, Lysimachia thyrsiflora, Maianthemum canadense, M. stellatum, M. trifolium, Menyanthes trifoliata, Menziesia ferruginea, Mertensia ciliata, M. paniculata, Micranthes subapetala, Mitella nuda, Moneses uniflora, Myrica gale, Parnassia palustris, Pedicularis bracteosa, P. groenlandica, Petasites frigidus, P. sagittatus, Phleum alpinum, Phragmites australis, Picea engelmannii, P. glauca, P. mariana, P. sitchensis, Platanthera dilatata, Pleurozium schreberi, Populus balsamifera, Potentilla gracilis, Prenanthes sagittata, Pyrola asarifolia, P. minor, Ranunculus lapponicus, Rhamnus alnifolia, Rhododendron groenlandicum, Rhynchospora capillacea, Ribes triste, Rosa gymnocarpa, R. sayi, R. woodsii, Rubus idaeus, R. pubescens, Salix bebbiana, S. boothii, S. discolor, S. geyeriana, S. pedicellaris, S. planifolia, S. scouleriana, Sarracenia purpurea, Shepherdia canadensis, Solidago uliginosa, Sphagnum angustifolium, S. capillifolium, S. centrale, S. fallax, S. girgensohnii, S. magellanicum, Streptopus amplexifolius, Symphyotrichum boreale, Thalictrum sparsiflorum, Triandenum fraseri, Trichophorum alpinum, T. cespitosum, Trientalis borealis, Trifolium repens, Vaccinium oxycoccos, V. uliginosum, V. vitis-idaea, Valeriana dioca, V. sitchensis, Veronica wormskjoldii, Vicia americana, Viola nephrophylla, V. pallens.*

***Rubus setosus* Bigelow** is a shrubby species, which occurs in bogs, depressions, ditches, dunes, lakeshores, meadows, pond margins, prairies, roadsides, savannas, swamps, thickets, and woodlands at elevations of up to 1000 m. It occupies habitats on dry to wet sites with acidic substrates described as gravel, humus, igneous outcrops, peat, and sand. Exposures can range from full sun to partial shade. The flowering period extends from June to August. The common name "sphagnum blackberry" reflects the frequent association of this species with peat bogs; however, further details of its ecology are scarce in the literature. **Reported associates:** *Abies, Acer, Agrimonia parviflora, Alisma subcordatum, Andropogon gerardii, Betula, Bromus inermis, Carex buxbaumii, C. crawfordii, C. haydenii, C. scoparia, Cladonia, Coreopsis tripteris, Fagus, Fimbristylis autumnalis, Galium obtusum, Helianthus mollis, Hypericum canadense, Ilex verticillata, Juncus marginatus, Larix laricina, Lysimachia terrestris, Onoclea sensibilis, Panicum virgatum, Parietaria pensylvanica, Parthenium integrifolium, Persicaria careyi, P. hydropiperoides, Picea mariana, Pinus banksiana, Poa nemoralis, P. pratensis, Populus alba, P. balsamifera, P. tremuloides, Prunus serotina, Quercus palustris, Rhynchospora capitellata, Rumex acetosella, Schizachyrium scoparium, Solidago altissima, S. gigantea, S. juncea, Sorghastrum nutans, Spartina pectinata, Sphagnum, Spiraea tomentosa, Thelypteris palustris, Thuja occidentalis, Tilia americana, Ulmus americana, Vaccinium corymbosum, Viola primulifolia, V. sororia.*

Use by wildlife: The fruits of *Rubus arcticus* are eaten by a number of mammals (Mammalia) including bears (Ursidae: *Ursus*), chipmunks (Sciuridae: *Tamias*), hares (Leporidae: *Lepus*), mice (Cricetidae), raccoons (Procyonidae: *Procyon lotor*), squirrels (Sciuridae: *Sciurus*), and a variety of birds (Aves) including catbirds (Passeriformes: Mimidae), grosbeaks (Passeriformes: Cardinalidae), grouse (Galliformes: Phasianidae), orioles (Passeriformes: Icteridae), robins (Passeriformes: Turdidae), sparrows (Passeriformes: Emberizidae), and thrushes (Passeriformes: Turdidae). *Rubus arcticus* is a host plant for a larval tussock moth (Lepidoptera: Erebidae: *Gynaephora rossii*). The raspberry bud moth (Lepidoptera: Adelidae: *Lampronia corticella*) oviposits in the flowers of *R. arcticus*. The plants are susceptible to downy mildew (Oomycota: Peronosporaceae: *Peronospora rubi*). They have endotrophic mycorrhizae, which improves nutrient absorption by the roots. The pollen is collected by several bees (Hymenoptera: Apidae: *Bombus*; *Apis melifera*).

Economic importance: food: The berries of *Rubus arcticus* are richly flavored, high in vitamins C and A, and are eaten raw, or made into jam and jelly. They have also been used to flavor liquors. The berries were eaten as food, fruit, and in pies by various Native North Americans including the Eskimo, Inuktitut, Koyukon, Upper Tanana, Vuntut Gwitchin, and Woodlands Cree tribes. The flavor of the berries was regarded highly by Linnaeus. The flowers are also sweet and edible and the leaves can be brewed as a tea substitute; **medicinal:** The Shuswap made an antidiarrheal remedy from the leaves of *Rubus arcticus*; **cultivation:** There are several cultivars of *Rubus arcticus* grown as ornamentals or for their edible fruits including 'Anna,' 'Astra,' 'Aura,' 'Beata,' 'Kaansoo,' 'Linda,' 'Mesma,' 'Mespi,' 'Pima,' 'Sofia,' and 'Valentina'; **misc. products:** Fruits of *Rubus arcticus* yield a blue/purple dye; **weeds:** none; **nonindigenous species:** none

Systematics: Phylogenetic analysis of DNA sequences including about 40 *Rubus* species (subfam. Rosoideae) indicates that *Dalibarda* is either the sister to or is nested within the genus. Otherwise, *Rubus* is related closely to a clade consisting of *Geum, Waldstenia,* and *Fallugia* (Figure 4.41). Molecular data indicate that many subgenera within *Rubus* (e.g., *Cylactis*) probably are not monophyletic. The taxonomic issues in *Rubus* are myriad and provide much frustration. *Rubus arcticus* (subgenus *Cylactis*) is considered to be a widely ranging circumpolar species comprising several subspecies. Among these, *R. arcticus* subsp. *acaulis* (known also as *R. acaulis*) has the highest affinity for aquatic habitats. *Rubus arcticus* is believed to be related most closely to *Rubus pedatus.* Molecular phylogenetic analyses (that exclude *R. pedatus*) show *R. pubescens* as being most closely related to *R. arcticus* of the *Rubus* species surveyed (Figure 4.45). Several formerly recognized species (all placed within subgenus *Rubus*) doubtfully are distinct taxonomically, and further clarification of their status is needed (see earlier comments under *Ecology*). *Rubus hypolasius* is regarded as a hybrid (*R.* ×*hypolasius*) of *R. flagellaris* and *R. pensilvanicus.* Some authors merge *R.*

hypolasius with *R. enslenii*, which in turn has been merged with *R. flagellaris. Rubus lawrencei* has been considered to be related to *R. hispidus* but most recently has been merged with *R. setosus. Rubus spectatus* has been merged with *R. vermontanus*, which has been merged with *R. setosus.* The taxonomic status of these and other *Rubus* species remains uncertain; thus, their morphological circumscriptions are vague. As a consequence, it is extremely difficult to provide a comprehensible summary of those species that might be OBL indicators. Even if the taxonomic issues were settled with certainty, the confusion of these taxa in the literature provides little confidence in extracting accurate life-history information in many cases. Much of the taxonomic difficulty in *Rubus* has been attributed to widespread hybridization and polyploidy. Evidence from DNA sequence data shows that hybridization can occur even between relatively distantly related species in the genus. The close relationship of *R. arcticus* and *R. pubescens* indicated by molecular data is evidenced further by their ability to hybridize. The hybrids have been known as *R. propinquus* and as *R.* ×*paracaulis.* The chromosomal base number of *Rubus* is $x = 7$. Both diploid ($2n = 14$) and triploid ($2n = 21$) cytotypes are reported for *Rubus arcticus. Rubus setosus* has a broad range of reported counts ($2n = 14, 21, 28, 35$). Genotypic variation [as measured using amplified fragment length polymorphism (AFLP) markers] is high ($D = 0.72$ to 0.94) in European populations of *R. arcticus*, which indicates

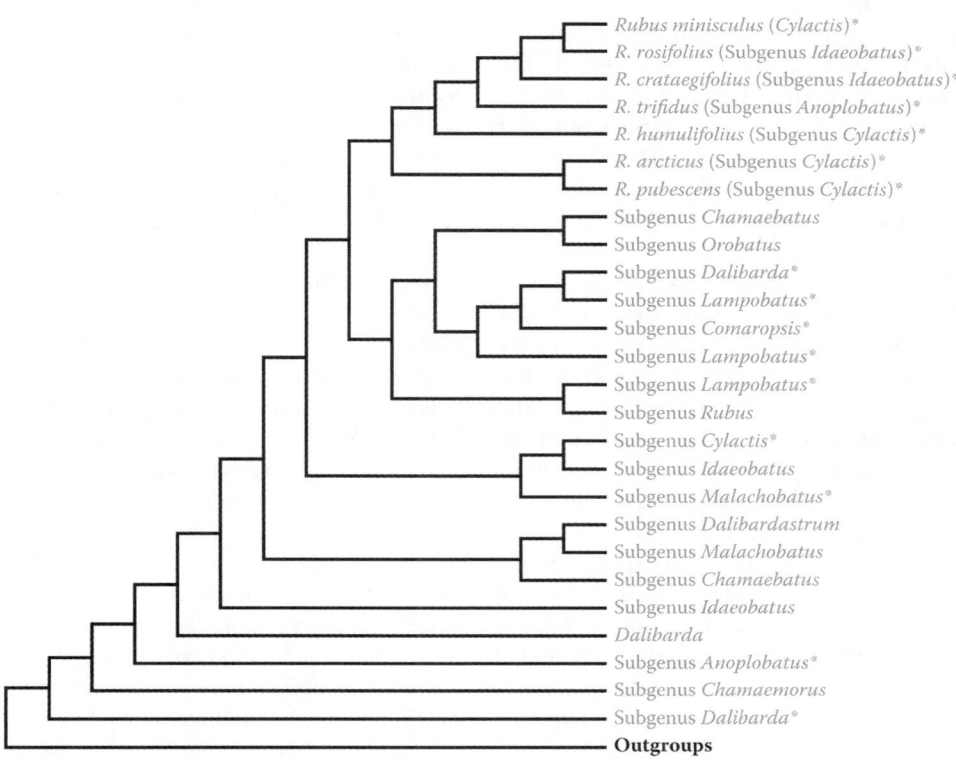

FIGURE 4.45 Phylogenetic relationships in Rubus (Rosaceae; subfamily Rosoideae) as indicated by analysis of DNA sequences. Asterisks indicate subgenera (subgenus) that do not resolve as monophyletic. (Adapted from Alice, L.A. & Campbell, C.S., *Amer. J. Bot.*, 86, 81–97, 1999.)

frequent sexual reproduction. Nevertheless, much genetic variation (e.g., 48%) is partitioned among populations, indicating low interpopulational gene flow and high population differentiation.

Comments: *Rubus arcticus* is circumpolar in distribution and occurs throughout northern North America. *Rubus setosus* occurs in northeastern North America.

References: Alice & Campbell, 1999; Alice et al., 2001, 2014; Fertig, 2000; Karp et al., 2004; Lahring, 2003; Lindqvist-Kreuze et al., 2003; Newmaster et al., 1997; Robuck, 1985; Ryynanen, 1973; Traveset et al., 2004; USDA, 2002.

8. *Sanguisorba*

Burnet; pimprenelle

Etymology: from the Latin *sanguis sorbere* meaning "to absorb blood" in reference to the alleged styptic properties of the plants

Synonyms: none

Distribution: global: Asia; Europe; North America; **North America:** widespread

Diversity: global: 15 species; **North America:** 4 species

Indicators (USA): OBL; FAC: *Sanguisorba menziesii*

Habitat: freshwater; palustrine; **pH:** unknown; **depth:** <1 m; **life-form(s):** emergent herb

Key morphology: leaves (to 2.5 dm), pinnate, with up to 15 coarsely serrate leaflets (to 5 cm); inflorescence (to 1 m) simple or branched, terminated by dense spikes (to 7 cm) of flowers; hypanthium 4-winged, pubescent; sepals (to 3 mm) 4, reddish-purple; petals absent; stamens 4, clavate, the filaments (to 6 mm) flattened; pistil of 1 carpel; achene included in hypanthium

Life history: duration: perennial (persistent rootstock, rhizomes); **asexual reproduction:** rhizomes; **pollination:** insect; **sexual condition:** hermaphroditic; **fruit:** achenes (common); **local dispersal:** achenes, rhizomes; **long-distance dispersal:** achenes

Imperilment: (1) *Sanguisorba menziesii* [G3]; S1 (WA); S2 (BC); S3 (AK)

Ecology: general: *Sanguisorba* species occur most commonly in forested habitats. Three of the four North American species (75%) are wetland indicators, but only one is considered to be OBL in a portion of its range. The flower clusters are showy and attract insect pollinators.

Sanguisorba menziesii **Rydb.** occurs in fresh or brackish waters of beaches, coastal bogs, fens, marshes, meadows, muskeg, seeps, slopes, and tidal marshes at elevations of up to 1200 m. The habitats typically occur on acidic or calcareous substrates such as limestone or peat, with exposures from full sun to partial shade. Flowering proceeds from May to July, with the fruiting period extending from June to August. The seeds are physiologically dormant and require a period of winter stratification (6 months at 4°C) to promote germination. Germination is higher in the light than in dark and occurs best at a constant temperature of 24°C–25°C. This species is a relatively minor component of communities, producing less than 2 kg/ha biomass. Otherwise, there is little known on the ecology of this rare species. **Reported associates:** *Agrostis, Alnus oregona, Anemone oregana, Arenaria paludicola, Arnica lanceolata, Blechnum spicant, Calamagrostis nutkaensis, Carex aquatilis, C. obnupta, C. stylosa, Coptis asplenifolia, Cornus canadensis, Deschampsia cespitosa, D. danthoides, Drosera rotundifolia, Dryas, Empetrum nigrum, Erigeron, Eriophorum, Euphrasia, Fritillaria, Gaultheria shallon, Gentiana, Hemitomes congestum, Hypopites monotropa, Juncus, Kalmia occidentalis, Ligusticum calderi, Lilium columbianum, Linnaea borealis, Luzula, Maianthemum dilatatum, Menyanthes trifoliata, Menziesia ferruginea, Myrica gale, Oxypolis occidentalis, Parnassia fimbriata, Picea sitchensis, Pinus contorta, Platanthera leucostachys, Potentilla egedii, Prunella vulgaris, Pteridium aquilinum, Pyrus diversifolia, Rhamnus purshiana, Rhododendron groenlandicum, Rhyncospora alba, Rosa acicularis, Rubus chamaemorus, Salix geyeriana, S. hookeriana, S. lemmonii, Senecio triangularis, Sphagnum balticum, S. mendocinum, S. squarrosum, S. warnstorfii, Spiraea douglasii, S. splendens, Trichophorum cespitosum, Trientalis latifolia, Veratrum viride, Veronica scutellata, Vaccinium cespitosum, V. ovalifolium, V. oxycoccos, V. parvifolium, V. uliginosum.*

Use by wildlife: The plants are browsed by deer (Mammalia: Cervidae: *Odocoileus*).

Economic importance: food: The leaves of *Sanguisorba menziesii* can be eaten when cooked, but are not particularly palatable; **medicinal:** The leaves and roots of *S. menziesii* have astringent properties; **cultivation:** *Sanguisorba menziesii* is a popular cultivated garden ornamental due to its attractive flowers. Cultivars include 'Burnet' and 'Dali Marble'; **misc. products:** none; **weeds:** none; **nonindigenous species:** *Sanguisorba minor* is introduced, but it is not often found in wet areas.

Systematics: *Sanguisorba* currently typifies tribe Sanguisorbeae (subtribe Sanguisorbinae) of subfamily Rosoideae. A complete phylogenetic survey of *Sanguisorba* has not yet been carried out; however, preliminary molecular studies of other Rosoideae (including 4–9 *Sanguisorba* species) indicated that the genus was polyphyletic as formerly circumscribed. In those studies (e.g., Figure 4.46), the North American taxa resolved as a sister clade to a group containing *Acaena, Cliffortia, Margyricarpus,* and *Polylepis,* whereas the Eurasian and Meditteranean species are associated with various Macaronesian and Meditteranean taxa (*Bencomia, Marcetella, Sarcopoterium*). Consequently, the latter *Sanguisorba* species have since been transferred to the genera *Poteridium* and *Poterium. Sanguisorba menziesii* is believed to be closely related to *S. officinalis* and *S. sitchensis,* but a better understanding of relationships awaits a more comprehensive survey of species, which would include *S. menziesii* and *S. sitchensis.* Some authors have suggested that *S. menziesii* arose as a hybrid between *S. officinalis* and *S. stipulata,* but this possibility requires further evaluation. Analysis of *Adh* sequence data indicates that *Sanguisorba* itself is likely to be of hybrid origin with *Poterium* being its paternal ancestor. The chromosomal base number of *Sanguisorba* is $x = 7$; however, counts for *S. menziesii* have not been reported.

Comments: *Sanguisorba menziesii* is restricted to northwestern North America (Alaska, British Columbia, Washington).

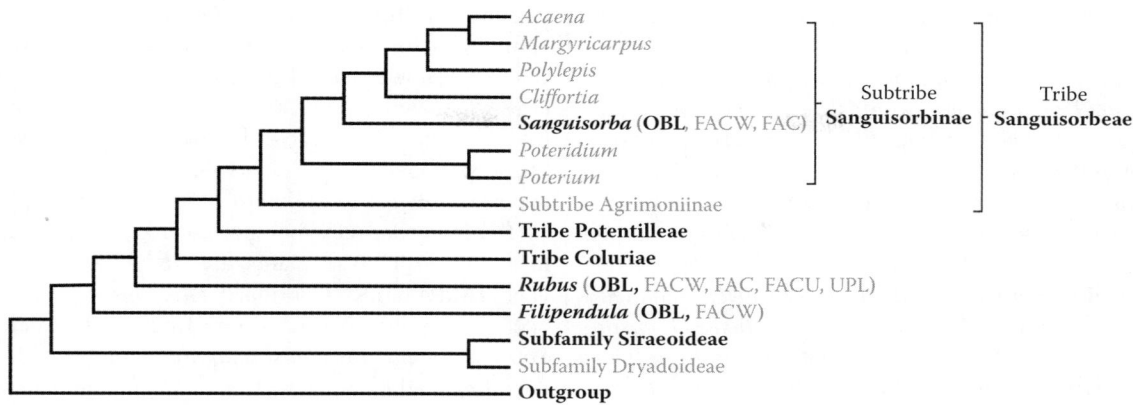

FIGURE 4.46 Phylogenetic position of *Sanguisorba* within Rosaceae subfamily Rosoideae. Taxa containing OBL North American indicators are in bold; other wetland indicators are listed in parentheses). (Adapted from Potter, D. et al., *Plant Syst. Evol.*, 266, 5–43, 2007a.)

References: Helfgott et al., 2000; Holloway & Matheke, 2003; Kerr, 2004; Mead, 1998; Potter et al., 2007a; Rigg, 1937; Weakley, 2015.

9. *Spiraea*

Hardhack, meadowsweet; spirée

Etymology: from the Greek *speira* meaning "wreath," in reference to the use of the plant for garlands

Synonyms: *Eleiosina*; *Pentactina*

Distribution: global: Asia; Europe; North America; **North America:** widespread

Diversity: global: 100–120 species; **North America:** 17 species

Indicators (USA): OBL: *Spiraea salicifolia*; **FACW:** *S. douglasii*, *S. salicifolia*; **UPL:** *S. douglasii*

Habitat: freshwater; palustrine; **pH:** 5.6–7.8; **depth:** <1 m; **life-form(s):** emergent shrub

Key morphology: stems (to 2 m) erect; leaves (to 12 cm) alternate, simple, petiolate (to 10 mm), the margins serrate; flowers (to 7 mm) 5-merous, perigynous, in dense, terminal panicles; petals (to 3.5 mm) pink or rose; stamens (to 7 mm) numerous (~30); follicles 5, each with 2-several small (<2 mm) seeds

Life history: duration: perennial (rhizomes); **asexual reproduction:** rhizomes; **pollination:** insect; **sexual condition:** hermaphroditic; **fruit:** follicles (common); **local dispersal:** rhizomes, seeds (wind); **long-distance dispersal:** seeds (animals)

Imperilment: (1) *Spiraea douglasii* [G5]; S2 (AK, MT); *S. salicifolia* [GNR]

Ecology: general: Although *Spiraea* is often encountered in wetlands, the genus typically has a wide tolerance for substrate conditions and the species most often are categorized as FACW indicators (FACW, FACU). Overall, nine of the North American *Spiraea* species (53%) are wetland indicators but only one is ranked as OBL in a portion of its range. Two species were designated as OBL in the 1996 indicator list. Since then, *S. douglasii* has been reclassified as a FACW, UPL indicator. However, its treatment has been retained here, given its completion prior to the release of the revised list. *Spiraea* species are pollinated by bees (Insecta: Hymenoptera). Dispersal

of their small seeds is facilitated by wind. The seeds remain viable for at least a year; however, they probably do not form a major component of regenerative seed banks. Most seeds do not require any pretreatment for germination, especially if they are sown fresh under lighted conditions.

Spiraea douglasii **Hooker** grows in varied habitats including bogs, marshes, mudflats, rock balds, shrub carrs, seeps, swamps, and along the margins of lakes, ponds, rivers, springs, and streams at elevations of up to 2500 m. The plants generally are shade intolerant and occur mainly on sandy loam, but also grow on clay loam, gravel, peat, and silty clay (pH: 5.6–7.8). *Spiraea douglasii* is described as a seral, pioneer species, which can become dominant in some situations. A rhizomatous habit facilitates the development of dense (sometimes impenetrable) clonal colonies. When buried, the stems will root adventitiously and can resprout quickly following incidences of fire or other disturbance. Rhizome extension has been observed even in the volcanic ash that was ejected by Mt. St. Helens, where 1–5 roots/cm stem were produced within a year after burial by the eruption. *Spiraea douglasii* often replaces willows (*Salix*) in sites that have been overgrazed. The seeds will germinate without pretreatment when they are fresh; however, seeds that have dried must be sown in the fall or first stratified for 1–3 months under cold, moist conditions to promote germination. The unusual "UPL" indicator status assigned to some regions (see earlier) reflects occurrences in nonnative portions of the species range. **Reported associates:** *Abies grandis, A. lasiocarpa, Abronia latifolia, Acer circinatum, A. glabrum, A. grandifolia, A. macrophyllum, Achillea millefolium, Aconitum columbianum, Actaea rubra, Adenocaulon bicolor, Agrostis exarata, A. scabra, A. stolonifera, Alnus rugosa, A. viridis, Amelanchier alnifolia, Angelica arguta, A. genuflexa, Arabis furcata, Arctostaphylos uva-ursi, Athyrium filix-femina, Berberis aquifolium, Betula papyrifera, Blechnum spicant, Calamagrostis canadensis, Canadanthus modestus, Carex aquatilis, C. arcta, C. aurea, C. brunnescens, C. diandra, C. disperma, C. eurycarpa, C. geyeri, C. lanuginosa, C. microptera, C. obnupta, C. phaeocephala, C. retrorsa, C. rostrata, C. stipata, C. utriculata, C. vesicaria, Castilleja miniata, Cerastium vulgatum,*

Chamerion angustifolium, Cicuta douglasii, Cinna latifolia, Circaea alpina, Cornus nuttallii, C. sericea, Crataegus douglasii, Cryptogramma acrostichoides, Danthonia, Delphinium pavonaceum, Deschampsia cespitosa, Drosera anglica, D. rotundifolia, Drymocallis arguta, Dryopteris spinulosa, Eleocharis palustris, Elymus glaucus, Equisetum arvense, E. fluviatile, E. palustre, E. praealtum, Eucephalus ledophyllus, Festuca occidentalis, F. ovina, F. viridula, Galium aparine, G. triflorum, Gaultheria shallon, Geum aleppicum, Glyceria elata, Heracleum lanatum, Hypericum anagalloides, Juncus balticus, J. drummondii, J. effusus, J. ensifolius, J. torreyi, Larix occidentalis, Lemna minor, Ligusticum canbyi, Linnaea borealis, Lonicera involucrata, Luzula hitchcockii, Lysichiton americanus, Maianthemum stellatum, Mentha canadensis, Menyanthes trifoliata, Micranthes ferruginea, Microseris alpestris, Mimulus guttatus, Montia cordifolia, Myrica californica, M. gale, Osmorhiza chilensis, Pedicularis bracteosa, Petasites sagittatus, Phalaris arundinacea, Phleum pratense, Phlox diffusa, Physocarpus malvaceus, Picea engelmannii, Pinus contorta, P. monticola, P. ponderosa, Platanthera dilatata, Poa pratensis, P. triflora, Polemonium occidentale, Populus balsamifera, P. tremuloides, Potentilla, Prunella vulgaris, Pseudotsuga menziesii, Pteridium aquilinum, Pyrola asarifolia, Ranunculus aquatilis, R. macounii, Rhamnus purshiana, Rhododendron columbianum, R. occidentale, Ribes lacustre, Rosa eglanteria, R. gymnocarpa, R. nutkana, R. woodsii, Rubus idaeus, R parviflorus, R. pubescens, R. spectabilis, Salix barclayi, S. bebbiana, S. boothii, S. brachystachys, S. commutata, S. drummondiana, S. farriae, S. geyeriana, S. hindsiana, S. hookeriana, S. sitchensis, Sambucus racemosa, Sanguisorba officinalis, Scirpus microcarpus, Senecio pseudaureus, S. triangularis, Solidago canadensis, Sorbus scopulina, S. sitchensis, Sphagnum, Stachys cooleyae, Stellaria longifolia, Symphoricarpos albus, Symphyotrichum spathulatum, Taraxacum officinale, Thalictrum occidentale, Thuja plicata, Trautvetteria caroliniensis, Trientalis arctica, Trifolium agrarium, T. pratense, T. repens, Trisetum cernuum, Typha latifolia, Urtica dioica, Vaccinium deliciosum, V. globulare, V. occidentale, Valeriana sitchensis, Veratrum californicum, V. viride, Veronica americana, V. cusickii, Viburnum edule, Viola glabella, V. orbiculata, Xerophyllum tenax.

***Spiraea salicifolia* L.** is an introduced species that occurs in floodplains, meadows, thickets, and along riverbanks at elevations of up to 300 m. The plants require an acidic soil and full sunlight. They are only weakly cold tolerant. Flowering occurs from June to August with fruits produced from June to September. One stem can yield about 115 tiny (0.00008 g) seeds on average. The shoots sucker readily, which enables the plants to develop into dense thickets. They will withstand cold temperatures down to −25°C. There are few additional details on the ecology of this species in the introduced North American portion of its range. **Reported associates:** Due to taxonomic confusion in the older literature, many of the reported associations actually apply to *S. alba* (see *Systematics* later); consequently, they have been omitted here.

Use by wildlife: *Spiraea douglasii* contributes important habitat structure and cover for wildlife in many communities. The dried flower spikes of *S. douglasii* are eaten by grouse (Aves: Phasianidae) and the plants provide a nesting site for marsh wrens (Aves: Troglodytidae: *Cistothorus palustris*). Shade provided by the plants during the hot periods of the dry season helps to maintain remnant pools used as habitat by the imperiled Oregon spotted frog (Amphibia: Ranidae: *Rana pretiosa*). The plants are grazed on by black-tailed deer (Mammalia: Cervidae: *Odocoileus hemionus*) and occasionally by livestock (more frequently in late summer). *Spiraea douglasii* is a host plant for a number of Lepidoptera including admiral butterflies (Nymphalidae: *Limenitis lorquini*), cutworms (Noctuidae: *Adelphagrotis indeterminata, Orthosia praeses, O. revicta, Spiramater lutra*), tussock moths (Arctiidae: *Lophocampa maculata*), and western sheep moths (Saturniidae: *Hemileuca eglanterina*). It is also a host for a diverse group of sac fungi (Ascomycota: Incertae sedis: *Camarosporium coronillae*; Botryosphaeriaceae: *Diplodia constricta*; Dothideomycetidae: *Dothidea sambuci*; Leotiomycetidae: *Cylindrosporium filipendulae, Godronia spiraeae, Podosphaera clandestina*; Nectriaceae: *Nectria cinnabarina*; Pleosporomycetidae: *Rhopalidium cercosporelloides, Stagonospora spiraeae*; Valsaceae: *Cryptodiaporthe macounii*). In Asia, *Spiraea salicifolia* is known to host a pathogenic phytoplasma known as "*Candidatus Phytoplasma ziziphi*" (Tenericutes: Mollicutes: Acholeplasmataceae). Because of taxonomic confusion between *S. alba* and *S. salicifolia* in the older North American literature (see *Systematics* later) some of the following accounts might actually apply to *S. alba* and should be regarded as tentative. *Spiraea salicifolia* is reportedly used occasionally as a nesting site for alder flycatchers (Aves: Tyrannidae: *Empidonax alnorum*). It has been cited as a host plant for various Lepidoptera (Geometridae: *Cingilia catenaria, Eupithecia strattonata, Probole amicaria*; Lycaenidae: *Celastrina ladon*; Noctuidae: *Actebia fennica, Diachrysia aereoides, Pseudohermonassa bicarnea*; Saturniidae: *Hemileuca lucina, H. maia, Hyalophora cecropia, Paonias excaecata*; Tortricidae: *Endopiza spiraeifoliana, Olethreutes albiciliana, O. permundana*). The flowers are host to many beetles (Insecta: Coleoptera) from a wide diversity of families including Bruchidae (*Bruchus calvus*), Carabidae (*Lebia atriventris*), Cerambycidae (*Analeptura lineola, Brachyleptura rubrica, B. vagans, Euderces picipes, Leptura bifloris, L. subhamata, Metacmaeops vittata, Pachyta monticola, Stictoleptura canadensis, Strangalia acuminata, S. famelica, S. luteicornis, Trigonarthris proxima, Typocerus velutinus*), Cleridae (*Trichodes nuttalli*), Coccinellidae (*Hippodamia parenthesis*), Chrysomelidae (*Diachus auratus, Nodonota puncticollis*), Cryptophagidae (*Antherophagus ochraceus*), Dermestidae (*Anthrenus museorum. A. scrophulariae, Cryptorhopalum haemorrhoidale*), Kateretidae (*Heterhelus abdominalis*), Mordellidae (*Mordella marginata, M. melaena, M. scutellaris, Mordellistena biplagiata, M. comata, M. limbalis*), Mycteridae (*Mycterus scaber*), Scarabaeidae (*Aphodius fimetarius, Macrodactylus subspinosus, Trichiotinus affinis, T. piger*), and Tenebrionidae

(*Isomira sericea*). Chinch bugs (Insecta: Hemiptera: Lygaeidae: *Kleidocerys resedae*) have also been reported on North American plants.

Economic importance: food: The young leaves of *Spiraea salicifolia* are rich in vitamin C and supposedly are edible when cooked; **medicinal:** The Mahuna used the roots of *Spiraea salicifolia* to treat coughs and colds. The Lummi administered the seeds of *S. douglasii* as a remedy for diarrhea. Seeds of *S. salicifolia* were used similarly by the Meskwaki. The Potawatomi used the bark of *S. salicifolia* as a general remedy; **cultivation:** *Spiraea douglasii* (and its hybrid *S.* ×*pachystachys*) and *S. salicifolia* are cultivated as ornamental shrubs. 'Bashaw' is a cultivar of *S. douglasii*. *Spiraea* ×*billiardii* 'Triumphans' is a cultivar selected from hybrids made between *S. douglasii* and *S. salicifolia*. *Spiraea salicifolia* is planted as a hedge within its native range; **misc. products:** The Thompson tribe used the branches of *Spiraea douglasii* for making brooms. The Bella Coola and Lummi tribes also used the branches as utensils when cooking and for smoking salmon. The peeled stems were used by the Quinault to string together clams for roasting. The Ojibwa used the root of *S. salicifolia* as a trapping medicine. *Spiraea douglasii* and *S. salicifolia* are effective at stabilizing lake and river margins; **weeds:** The cultivated *S. salicifolia* is nonindigenous, but rarely escapes and is not reported as invasive where introduced in North America; **nonindigenous species:** The western North American *Spiraea douglasii* has been introduced to Missouri where it is an adventive escape from cultivation. The Eurasian *S. salicifolia* was introduced to North America as an ornamental garden plant.

Systematics: *Spiraea* (Figure 4.40) is the type genus of subfamily Spiraeoideae, which molecular data indicate are not monophyletic as traditionally circumscribed. It is the only genus of the subfamily to contain obligate aquatics. Phylogenetic analyses of combined molecular data indicate that *Spiraea* itself is monophyletic and place it in a clade with *Kelseya* and *Petrophyton*, which together are included within a larger clade with *Aruncus, Holodiscus, Luetkea, Sibiraea,* and *Xerospiraea* (Figure 4.47). Those analyses have also

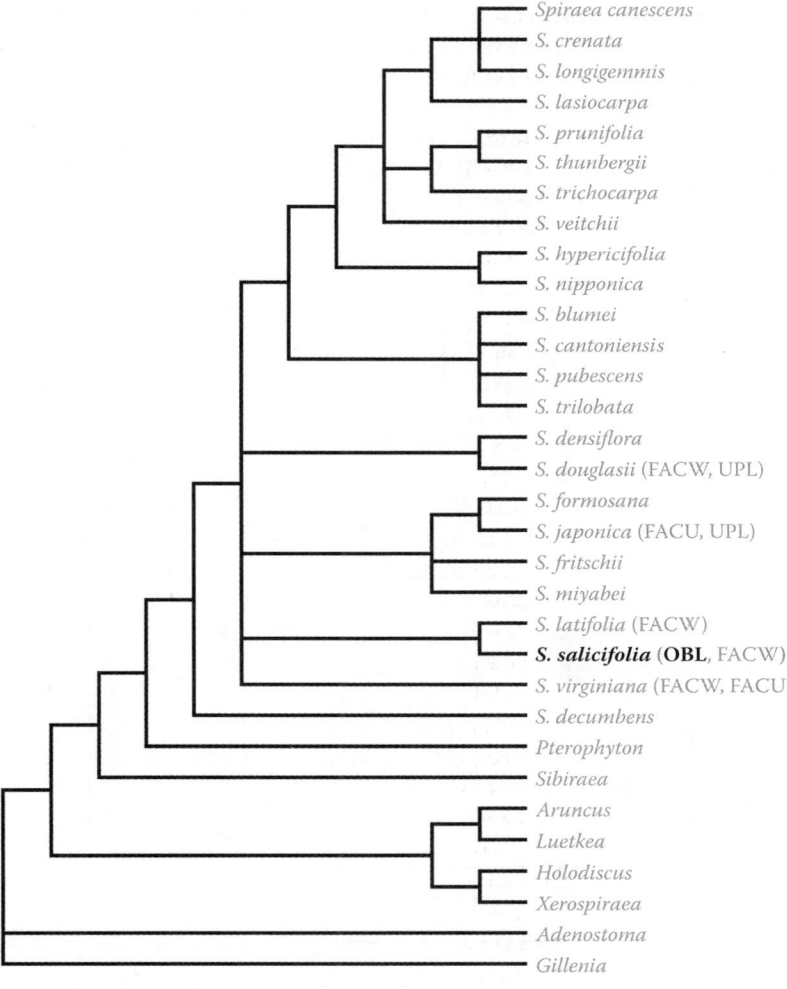

FIGURE 4.47 Interspecific relationships in *Spiraea* based on phylogenetic analysis of combined DNA sequence data. Only about a quarter of the species are represented and relationships remain weakly resolved among many of those included. *Spiraea salicifolia* (bold), the only North American OBL indicator in subfamily Spiraeoideae, exhibits a close relationship to *S. latifolia* (FACW). (Adapted from Potter, D. et al., *Plant Syst. Evol.*, 266, 105–118, 2007b.)

indicated that the sections defined previously within *Spiraea* are not monophyletic. A comprehensive interspecifc phylogenetic investigation of *Spiraea* has not yet been conducted, leaving many questions of relationships unsettled in this relatively large genus. However, studies that have included 24 species (about 20%–25% of the total) have contained both *S. douglasii* and *S. salicifolia*, which do not resolve as being particularly closely related (Figure 4.47). There is some degree of taxonomic misunderstanding regarding the North American occurrences of *S. salicifolia*, because this introduced species had been confused for some time with the native *S. alba*. Consequently, much of the older literature is unreliable when referring to "*S. salicifolia*." The base chromosome number of *Spiraea* is *x* = 9. *Spiraea douglasii* and *S. salicifolia* (both *2n* = 36) are tetraploids. *Spiraea salicifolia* reputedly hybridizes with *S. douglasii* and a number of other species. Care should be taken to avoid planting the nonindigenous *S. salicifolia* in proximity to any native populations of *S. douglasii* to avoid the possibility of genetic contamination. *Spiraea roseata* is another putative hybrid, which allegedly represents a cross between *S. densiflora* × *S. douglasii*.

Comments: Native populations of *Spiraea douglasii* occur in western North America but the species has also been introduced eastward. The nonindigenous *S. salicifolia* is scattered throughout portions of eastern North America.

References: Berger & Parmelee, 1952; Darris, 2002; Esser, 1995a; Fan & Wang, 2011; Li et al., 2010; Lovell, 1915; Potter et al., 2007b; Slater, 1952; Stevens, 1932; Still & Potter, 2005.

Family 2: Rhamnaceae [50]

Rhamnaceae are a fairly large family containing about 950 species of mostly trees and shrubs. Only a few species are found growing under wet conditions with most of the family exhibiting xeromorphic adaptations (Richardson et al., 2000). Within Rosales, the family is related closely to Barbeyaceae, Dirachmaceae, and Elaeagnaceae (Figure 4.39). Morphologically, Rhamnaceae can be distinguished by their concave, hooded petals that enclose the anthers in flower, stamens that are opposite of the petals and are fused to their base, and drupes (also berries and nuts). The fruits are dispersed by birds (Aves), mammals (Mammalia), or by the wind. Nitrogen fixation occurs within Rhamnaceae as well as in some Elaeagnaceae, Rosaceae, and Ulmaceae.

Rhamnaceae are of minor economic importance, but contain several genera (*Berchemia*, *Ceanothus*, *Colletia*, *Frangula*, *Rhamnus*, *Ziziphus*), which are cultivated as ornamental shrubs. In ancient times, the wood was made into charcoal for use in gunpowder. In China, jujube fruits (*Ziziphus jujuba*) are candied and eaten as snacks, or made into a juice.

Phylogenetic analyses of chloroplast DNA sequences from 42 genera (Richardson et al., 2000) have provided a fairly comprehensive assessment of intergeneric relationships in the family, which currently are depicted as 11 tribes (Figure 4.48). Molecular data support the placement of *Rhamnus* within tribe Rhamneae as the sister genus of *Frangula* (Figure 4.48), which contains several FACW species. Phylogenetically,

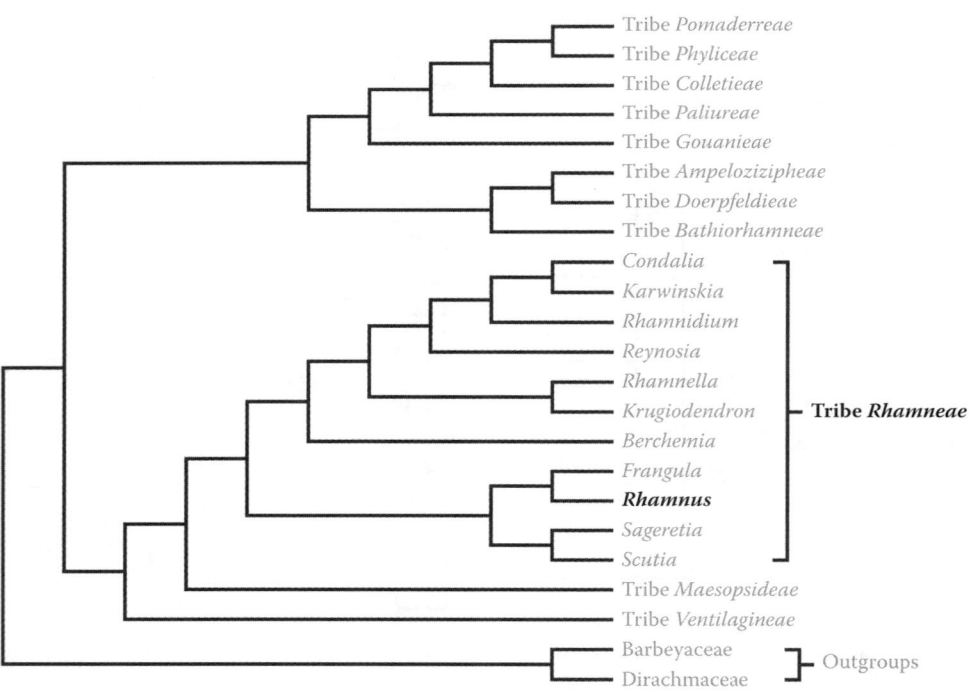

FIGURE 4.48   Relationships among tribes of Rhamnaceae, and genera of tribe Rhamneae as inferred from combined cpDNA sequence data. Only *Rhamnus* (in bold) contains obligate North American aquatics in this predominantly terrestrial family. The clade containing *Rhamnus* and *Frangula* is supported further by analyses of 20 species in those genera using combined nuclear and cpDNA sequences (Bolmgren & Oxelman, 2004). (Adapted from Richardson, J.E. et al., *Amer. J. Bot.*, 87, 1309–1324, 2000.)

Rhamnus and other members of *Rhamneae* occur within the basal clade of Rhamnaceae. Only one genus of Rhamnaceae contains obligate aquatics in North America:

1. *Rhamnus* L.

1. *Rhamnus*

Buckthorn; nerprun
Etymology: from the Greek *rhamnos* meaning "branch"
Synonyms: *Alaternus*; *Apetlorhamnus*; *Frangula* (in part); *Girtanneria*; *Oreoherzogia*; *Oreorhamnus*
Distribution: global: Africa; Asia; Europe; New World; **North America:** widespread
Diversity: global: 110 species; **North America:** 12 species
Indicators (USA): OBL; FACW: *Rhamnus alnifolia*
Habitat: freshwater; palustrine; **pH:** 5.5–8.2; **depth:** <1 m; **life-form(s):** emergent shrub
Key morphology: stems (to 1.5 dm) ascending, often gray; leaves (to 11 cm) alternate, simple, deciduous, petiolate (to 1.5 cm), the margins serrate, venation arcuate; flowers in axillary clusters (1–3), bisexual (rarely) or (mostly) unisexual, perigynous, 5-merous; sepals (to 1.5 mm) yellowish-green; petals absent; stamens (♂) 5; pistil (♀) 1, syncarpous, trilocular, each locule with one ovule; drupes (to 8 mm) globular, black, with 3 pyrenes
Life history: duration: perennial (buds); **asexual reproduction:** none; **pollination:** insect; **sexual condition:** dioecious (polygamous); **fruit:** drupes (common); **local dispersal:** drupes (mammals); **long-distance dispersal:** drupes (birds)
Imperilment: (1) *Rhamnus alnifolia* [G5]; SH (SD); S1 (IA, IL, TN, UT, VA, WV); S2 (WY); S3 (AB, CA, CT, IN, NS, PE)
Ecology: general: Only one species of *Rhamnus* is categorized as an obligate aquatic in North America. Two other species have status as FACW indicators (FACW, FAC, FACU, UPL) and the remaining species occupy terrestrial habitats. *Rhamnus* species are referred to as being "polygamous," a condition where the plants possess mixtures of bisexual and unisexual flowers. Reproduction occurs entirely by seed.

Rhamnus alnifolia **L'Hér.** occurs in bogs, ditches, fens, forests, marshes, meadows, prairies, seeps, stream bottoms, swamps, and along the margins of ponds and watercourses at elevations of up to 1800 m. Its habitats span a broad range of acidity (pH: 5.5–8.2) but often are alkaline with mineral-rich groundwater (e.g., 300 mg CaCO₃/L). The reported substrates include alluvial clay, basalt, belt series, loam, marl, sand, sandy loam, talus, and peat that is derived from sedges. This species often is considered to be an indicator of fen communities. The plants thrive in full sun but can tolerate partial shade. On average, each shoot produces nine leaves, which each live for 110 days. The plants usually are dioecious but the flowers are polygamous (both hermaphroditic and unisexual forms present) and are pollinated by bees and flies. The fruits are dispersed rapidly by birds (Aves) from August onwards. There are about 114 seeds in a gram. The seeds are physiologically dormant and require at least 60 days

of cold, moist stratification at 5°C to promote germination, which begins within three weeks afterwards. For seeds collected in late July, a 90-day period of stratification is optimal (13% germination after 5 weeks), with the rate falling to 5% after 120 days of stratification. Higher germination rates (48%) were observed following a 30-day period of cold stratification, for seeds collected in early September. A subsequent stratification period or scarification of seeds reduced their germination rates. Dry seeds kept in sealed containers have retained high viability after 2 years in storage at 5°C. The plants also can be propagated vegetatively by softwood stem cuttings either as untreated (75% rooting frequency), or (to 85% rooting frequency) when treated with indole-3-butyric acid (IBA) in talc (3–8 g/kg; 3000–8000 ppm). About 321 fine roots are produced annually, which live longest during the spring and shortest during the fall. **Reported associates:** *Abies balsamea*, *A. grandis*, *A. lasiocarpa*, *Acer glabrum*, *A. rubrum*, *A. saccharinum*, *Actaea rubra*, *Agrostis exarata*, *A. scabra*, *A. stolonifera*, *Alnus rugosa*, *A. sinuata*, *A. viridis*, *Alopecurus alpinus*, *Amelanchier alnifolia*, *Amerorchis rotundifolia*, *Angelica arguta*, *A. atropurpurea*, *Antennaria corymbosa*, *Aralia nudicaulis*, *Arctium minus*, *Arnica cordifolia*, *A. latifolia*, *A. longifolia*, *A. parryi*, *Aronia ×prunifolia*, *Athyrium filix-femina*, *Aulacomnium palustre*, *Betula glandulosa*, *B. nana*, *B. occidentalis*, *B. papyrifera*, *B. pumila*, *Bistorta vivipara*, *Bromus anomalus*, *B. carinatus*, *B. ciliatus*, *Bryum pseudotriquetrum*, *Calamagrostis canadensis*, *C. rubescens*, *Caltha palustris*, *C. stricta*, *Calla palustris*, *Caltha palustris*, *Carex aquatilis*, *C. arcta*, *C. aurea*, *C. bebbii*, *C. buxbaumii*, *C. canescens*, *C. capillaris*, *C. cusickii*, *C. disperma*, *C. flava*, *C. geyeri*, *C. gynocrates*, *C. hystricina*, *C. interior*, *C. lasiocarpa*, *C. lenticularis*, *C. leporinella*, *C. leptalea*, *C. luzulina*, *C. magellanica*, *C. norvegica*, *C. pachystachya*, *C. pellita*, *C. rostrata*, *C. scirpoidea*, *C. seorsa*, *C. spectabilis*, *C. sterilis*, *C. stipata*, *C. stricta*, *C. utriculata*, *Centaurea stoebe*, *Chamaedaphne calyculata*, *Chamerion angustifolium*, *Chelone glabra*, *Chrysosplenium americanum*, *Cinclidium stygium*, *Cinna latifolia*, *Cirsium arvense*, *C. muticum*, *C. vulgare*, *Claytonia cordifolia*, *Clintonia uniflora*, *Comarum palustre*, *Coptis groenlandica*, *Cornus alternifolia*, *C. amomum*, *C. canadensis*, *C. racemosa*, *C. sericea*, *Corylus cornuta*, *Crataegus columbiana*, *C. douglasii*, *C. gaylussacia*, *Cypripedium candidum*, *Dactylis glomerata*, *Danthonia intermedia*, *Dasiphora floribunda*, *Deschampsia cespitosa*, *Disporum hookeri*, *Drosera rotundifolia*, *Eleocharis compressa*, *E. erythropoda*, *E. pauciflora*, *E. rostellata*, *E. tenuis*, *Elymus glaucus*, *Equisetum arvense*, *E. fluviatile*, *E. pratense*, *E. scirpoides*, *E. sylvaticum*, *Erigeron lonchophyllus*, *Eriophorum chamissonis*, *E. vaginatum*, *E. viridicarinatum*, *Erythronium grandiflorum*, *Eurybia conspicua*, *Fallopia scandens*, *Festuca occidentalis*, *F. subulata*, *Fragaria vesca*, *F. virginiana*, *Frangula alnus*, *Fraxinus nigra*, *Galium asprellum*, *G. bifolium*, *G. boreale*, *G. labradoricum*, *G. triflorum*, *Gentiana andrewsii*, *Gentianopsis crinita*, *G. detonsa*, *Geum macrophyllum*, *Glyceria grandis*, *Gymnocarpium dryopteris*, *G. robertianum*, *Habenaria*, *Hackelia floribunda*, *Heracleum lanatum*, *H. maximum*,

Hieracium caespitosum, Hypericum kalmianum, Hypnum lindbergii, Ilex verticillata, Impatiens capensis, Juncus balticus, Juniperus communis, J. horizontalis, Larix laricina, Leucanthemum vulgare, Ligusticum tenuifolium, Linnaea borealis, Liparis loeselii, Lomatium dissectum, Lonicera caerulea, L. involucrata, L. oblongifolia, L. utahensis, L. villosa, Lycopus uniflorus, Lysichiton americanus, Mahonia repens, Maianthemum stellatum, M. trifolium, Malaxis paludosa, Marchantia polymorpha, Medicago falcata, Melica smithii, Mentha arvensis, M. canadensis, Menyanthes trifoliata, Mertensia paniculata, Micranthes pennsylvanica, Mimulus guttatus, Mitella breweri, M. nuda, Mnium, Moneses uniflora, Myrica gale, Nasturtium officinale, Onoclea sensibilis, Oryza, Osmorhiza chilensis, O. occidentalis, Osmundastrum cinnamomeum, Panicum, Parnassia glauca, Paxistima myrsinites, Pedicularis lanceolata, Penstemon rydbergii, Persicaria sagittata, Phalaris arundinacea, Phleum pratense, Phlox idahonis, Physocarpus malvaceus, P. opulifolifolius, Picea glauca, P. engelmannii, P. mariana, Pinus contorta, P. ponderosa, Plagiomnium ellipticum, P. medium, Platanthera dilatata, P. obtusata, Poa pratensis, P. triflora, Polemonium occidentale, P. reptans, P. van-bruntiae, Populus balsamifera, P. tremuloides, P. trichocarpa, Potamogeton gramineus, Prenanthes racemosa, Primula jeffreyi, Prunus virginiana, Pseudotsuga menziesii, Ptelidium pulcherrimum, Pyrola asarifolia, P. chlorantha, Pyrus melanocarpa, Quercus bicolor, Q. palustris, Ranunculus aquatilis, R. populago, Rhamnus purshiana, Rhododendron columbianum, R. groenlandicum, Rhynchospora capillacea, Ribes americanum, R. hirtellum, R. hudsonianum, R. inerme, R. lacustre, Rosa acicularis, R. gymnocarpa, R. nutkana, R. woodsii, Rubus arcticus, R. idaeus, R. parviflorus, R. pubescens, Rumex orbiculatus, Salix bebbiana, S. boothii, S. candida, S. commutata, S. discolor, S. drummondiana, S. farriae, S. geyeriana, S. humilis, S. lutea, S. maccalliana, S. monica, S. pedicellaris, S. petiolaris, S. pseudocordata, S. pseudomyrsinites, S. scouleriana, S. serissima, S. vestita, S. wolfii, Sambucus racemosa, Sarracenia purpurea, Saururus cernuus, Schoenoplectus acutus, S. tabernaemontani, Senecio crassulus, S. integerrimus, S. pseudoarnica, S. triangularis, Shepherdia canadensis, Silene vulgaris, Solidago canadensis, S. gigantea, S. ohioensis, S. riddellii, Sorbus scopulina, Sphagnum warnstorfii, Spiraea alba, S. betulifolia, S. densiflora, Spiranthes cernua, Stellaria calycantha, Stipa occidentalis, S. richardsonii, Symphoricarpos albus, S. occidentalis, Symphyotrichum boreale, S. foliaceum, S. spathulatum, S. puniceum, Symplocarpus foetidus, Tanacetum vulgare, Taraxacum officinale, Thalictrum dasycarpum, T. occidentale, T. venulosum, Thelypteris palustris, Thuidium recognitum, Thuja occidentalis, T. plicata, Tiarella trifoliata, Tomenthypnum falcifolium, Toxicodendron vernix, Trichophorum alpinum, T. cespitosum, Trientalis borealis, Trillium ovatum, Trisetum wolfii, Trollius laxus, Typha latifolia, Urtica dioica, Usnea, Utricularia intermedia, Vaccinium corymbosum, V. oxycoccus, V. scoparium, V. uliginosum, V. vitis-idaea, Valeriana uliginosa, Veratrum viride, Verbascum thapsus, Veronica americana, Viburnum lentago, Vicia americana, Viola glabella, V. nephrophylla, V. labradorica, V. pallens, Xerophyllum tenax, Zanthoxylum americanum.

Use by wildlife: *Rhamnus alnifolia* is a habitat component of several rare land snails (Gastropoda: Vertiginidae: *Vertigo hubrichti, V. occulta*). The fruits are high in lipids (26%) and are eaten by a variety of birds (Aves) and mammals (Mammalia) including black bears (Ursidae: *Ursus americanus*), grizzly bears (Ursidae: *Ursus arctos horribilis*), hares (Leporidae: *Lepus americanus*), martens (Mustelidae: *Martes americana caurina*), and white-tailed deer (Cervidae: *Odocoileus virginianus*). The latter also find the shoots to be quite palatable and can browse significant numbers of them. *Rhamnus alnifolia* also is a minor food source for mule deer (Cervidae: *Odocoileus hemionus hemionus*) and beavers (Rodentia: Castoridae: *Castor canadensis*). The foliage contains the anthraquinone emodin, which is an effective feeding deterrent against many phytophagous insects. Yet, it is known to be a host plant for several insects including Lepidoptera (Arctiidae: *Hyphantria cunea*; Lymantriidae: *Orgyia leucostigma*; Papilionidae: *Papilio eurymedon*; Saturniidae: *Hemileuca eglanterina*; Tortricidae: *Ancylis brauni*) and soybean aphids (Homoptera: Aphididae: *Aphis glycines*). It is an alternate host of crown rust (Basidiomycota: Pucciniaceae: *Puccinia coronata*) and other fungi (e.g., Ascomycota: Xylariaceae: *Sphaerographium niveum*).

Economic importance: food: Although readily consumed by wildlife, the fruits of *Rhamnus alnifolia* (and many other *Rhamnus* species) are mildly toxic to humans and should never be eaten; **medicinal:** The fruits and foliage of *R. alnifolia* contain glycosides that are purgative, and can result in diarrhea, nausea, and vomiting if ingested (The well known purgative "Cascara Sagrada" is derived from the related *R. purshianus*). The Iroquois administered infusions and applied a poultice from *R. alnifolia* as a poison antidote. An infusion of the bark was given to Iroquois children as a laxative, tonic, and wash. The Iroquois also used a root decoction to treat gonorrhea. A decoction of bark was used as a laxative by the Meskwaki and by the Potawatomi as a purgative; **cultivation:** *Rhamnus alnifolia* is not in cultivation; **misc. products:** none; **weeds:** Although some *Rhamnus* species are weedy, *R. alnifolius* has not been reported as a nuisance; **nonindigenous species:** *Rhamnus alnifolius* has not been reported as introduced beyond its native North American range.

Systematics: Molecular data resolve *Rhamnus* in a clade as the sister group to *Frangula* (Figure 4.48). A number of species have been assigned to each genus on occasion, and there is still some debate whether to merge the groups (as *Rhamnus*) or retain them as separate. Although *Rhamnus* and *Frangula* are distinct morphologically and phylogenetically, it remains a matter of taxonomic preference whether to recognize these clades as distinct genera or simply as monophyletic subgenera of *Rhamnus*. Current usage has favored the former, which is the disposition that is followed here. A recent molecular phylogenetic analysis of *Rhamnus* and *Frangula* did not, unfortunately, include *R. alnifolia*; thus its interspecific relationships remain uncertain. *Rhamnus alnifolia* is placed within section *Rhamnus*, which indicates some degree of relationship to species such as *R. cathartica* and to *R. lanceolata*. There are no reports of interspecific hybridization involving *R. alnifolia*. The base chromosome number of *Rhamnus* is $x = 12$. No chromosome number has been reported for *R. alnifolia*; however, counts published for 14 other *Rhamnus* species are uniformly diploid ($2n = 24$).

Comments: *Rhamnus alnifolia* occurs broadly throughout the central portion of North America.

References: Bolmgren & Oxelman, 2004; Christy & Meyer, 1991; Dietz, 1923; Izhaki, 2002; Gallant et al., 2004; Groves, 1947; Locky et al., 2005; Mace & Jonkel, 1986; Murie, 1961; Noyce & Coy, 1990; Pellerin et al., 2006; Richardson et al., 2000; Sharma & Graves, 2005; Smith et al., 2014; Stiles, 1980; Trial, Jr. & Dimond, 1979; Tryon & Easterly, 1975; Voegtlin et al., 2004; Wiese et al., 2012; Wilkins, 1957; Woodcock, 1925.

Family 3: Ulmaceae [6]

The Ulmaceae are a small family of about 40 tree species, which are characteristic of terrestrial habitats, and also occur occasionally in wetlands. The monophyly of the family is supported strongly by cpDNA restriction fragment length polymorphism (RFLP) data (Figure 4.49), as well as by their common chromosome number ($x = 14$), and similar pollen and seed coat morphology (Wiegrefe et al., 1998). Other common features include alternate, serrate, and two-ranked leaves with unequal bases, whose veins terminate in the marginal teeth. Recent systematic investigations (based mainly on molecular data) support the abolishment of the former subfamily Celtidoideae, given that this group is clearly more closely allied to Cannabaceae while relatively distant from Ulmaceae.

Within North America, obligate aquatics occur only within one genus:

2. *Planera* J. F. Gmel.

Phylogenetic analyses (Figure 4.49) indicate *Planera* to occur within a relatively derived clade within the family along with

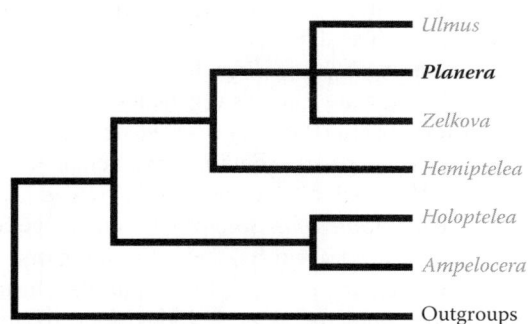

FIGURE 4.49 Phylogenetic relationships among the six genera of Ulmaceae as indicated by analysis of cpDNA RFLP data. The North American endemic *Planera* (in bold) is an obligate aquatic and resolves in a position that is relatively derived in the family. A similar topology was obtained by Ueda et al. (1997) using *rbcL* sequence data. (Adapted from Wiegrefe, S.J. et al., *Plant Syst. Evol.*, 210, 249–270, 1998.)

Ulmus and *Zelkova*. The close relationship of *Planera*, *Ulmus*, and *Zelkova* also is evidenced by their shared possession of stipitate ovaries and by their Pc-type sieve-element plastids, which occur nowhere else in the family. CpDNA RFLP data were unable to resolve more precise relationships among the three genera. Their next closest relative is *Hemiptelea* (Figure 4.49).

1. *Planera*

Planer-tree, water elm

Etymology: after Johann Jakob Planer (1743–1789)

Synonyms: none

Distribution: global: North America; **North America:** southeastern

Diversity: global: 1 species; **North America:** 1 species

Indicators (USA): OBL: *Planera aquatica*

Habitat: freshwater; palustrine; **pH:** 5.0–8.5; **depth:** 0–2 m; **life-form(s):** emergent tree

Key morphology: stems (to 12 m) with grayish bark peeling in long plates, the inner bark reddish-brown; leaves (to 8 cm) alternate, petiolate (to 6 mm), the bases oblique, the margins unevenly serrate; plants polygamous, the flowers 5-merous, bisexual (rarely) or (mostly) unisexual, with ♂ flowers in fascicles and ♀ flowers in groups of 1–3; corolla absent; stamens (♂) 5; drupes (to 8 mm) nut-like, leathery, compressed, ribbed with warty or knob-like processes

Life history: duration: perennial (buds); **asexual reproduction:** none; **pollination:** wind; **sexual condition:** monoecious; **fruit:** drupes (common); **local dispersal:** fruit, seeds (water); **long-distance dispersal:** fruit, seeds (birds)

Imperilment: (1) *Planera aquatica* [G5]; S1 (NC); S2 (IL, OK); S3 (KY)

Ecology: general: *Planera* is monotypic (see later).

Planera aquatica **J. F. Gmelin** grows in canals, ditches, floodplains, lakes, oxbows, sloughs, swamps, and along the margins of rivers and streams at elevations of up to 200 m. The plants occur on acidic to alkaline soil (pH: 5.0–8.5), but usually are found under more calcareous conditions (pH: 7.0–8.0). They often grow on sand or gravel bars. The trees are slow growing, have low drought tolerance, but are shade tolerant. They can withstand cold temperatures of up to −19°C. Individuals are polygamous (but typically monoecious), having a few perfect flowers interspersed along with the unisexual flowers. The plants flower in early spring (February to May), whereafter the seeds ripen within 4–6 weeks, thus providing an early food source for wildlife. The seeds are dispersed by the water, where they can reach densities up to 7.1/m². Seed germination requirements are not known but this species does not appear to form a persistent seed bank. *Planera* has no means of vegetative reproduction and does not root well from cuttings. *Planera* seedlings were observed to regenerate rapidly in a number of study plots surveyed following hurricane Andrew. Stem densities can reach 6–9.9/ha and a basal area of 0.06–0.07 m²/ha. **Reported associates:** *Acer negundo*, *A. rubrum*, *A. saccharinum*, *Alnus serrulata*, *Ampelopsis arborea*, *Arundinaria gigantea*, *Asimina triloba*, *Berchemia scandens*, *Betula nigra*, *Bidens*, *Bignonia* *capreolata*, *Boehmeria cylindrica*, *Botrychium*, *Brunnichia ovata*, *Campsis radicans*, *Carex bromoides*, *C. glaucescens*, *C. grayi*, *C. joorii*, *C. lupulina*, *Carya aquatica*, *C. cordiformis*, *C. illinoensis*, *Carpinus caroliniana*, *Celtis laevigata*, *Cephalanthus occidentalis*, *Commelina diffusa*, *C. virginica*, *Cornus drummondii*, *C. foemina*, *Crataegus aestivalis*, *C. opaca*, *C. spathulata*, *C. viridis*, *Cyrilla racemiflora*, *Dichanthelium commutatum*, *D. dichotomum*, *Diospyros virginiana*, *Eupatorium coelestinum*, *Forestiera acuminata*, *Fraxinus americana*, *F. caroliniana*, *F. pennsylvanica*, *F. profunda*, *Gleditsia aquatica*, *Heliotropium curassavicum*, *H. indicum*, *Hibiscus moscheutos*, *Hypericum hypericoides*, *Ilex cassine*, *I. coriacea*, *I. decidua*, *I. opaca*, *I. vomitoria*, *Itea virginica*, *Justicia ovata*, *Leersia lenticularis*, *Lemna*, *Leucothoe racemosa*, *Ligustrum sinense*, *Lindera benzoin*, *Liquidambar styraciflua*, *Lobelia cardinalis*, *Magnolia virginiana*, *Morus rubra*, *Myrica cerifera*, *Nyssa aquatica*, *N. biflora*, *N. ogeche*, *N. sylvatica*, *Onoclea sensibilis*, *Osmunda regalis*, *Panicum rigidulum*, *Parthenocissus quinquefolia*, *Paspalum laeve*, *Persea borbonia*, *P. palustris*, *Persicaria hydropiperoides*, *Phanopyrum gymnocarpon*, *Pilea pumila*, *Platanus occidentalis*, *Populus heterophylla*, *Porella*, *Quercus laurifolia*, *Q. lyrata*, *Q. nigra*, *Q. pagoda*, *Q. shumardii*, *Q. texana*, *Sabal minor*, *S. palmetto*, *Saccharum baldwinii*, *Salix caroliniana*, *S. nigra*, *Sambucus nigra*, *Saururus cernuus*, *Sebastiania fruticosa*, *Sericea lespedeza*, *Smilax tamnoides*, *S. walteri*, *Solidago leavenworthii*, *Styrax americanus*, *Symphyotrichum*, *Taxodium distichum*, *Toxicodendron radicans*, *Trachelospermum difforme*, *Ulmus alata*, *U. americana*, *U. rubra*, *Viburnum obovatum*, *Viola*, *Vitis palmata*, *V. rotundifolia*.

Use by wildlife: The foliage of *P. aquatica* has low palatability but is browsed moderately by white-tailed deer (Mammalia: Cervidae: *Odocoileus virginianus*). The seeds are eaten by many different kinds of wildlife including mice (Mammalia: Cricetidae: *Peromyscus*) and birds (Aves). They represent a large part of waterfowl (Aves: Anatidae) diets, especially mallards (*Anas platyrhynchos*) and wood ducks (*Aix sponsa*), which can consume up to 200–300 fruits when feeding. The trees are used as nesting sites by Prothonotary warblers (Aves: Parulidae: *Protonotaria citrea*) and for roosting by Acadian flycatchers (Aves: Tyrannidae: *Empidonax virescens*). *Planera* is a host plant of mistletoe (Santalaceae: *Phoradendron serotinum*), leafhoppers (Insecta: Hemiptera: Cicadellidae: *Erythridula planerae*), and several fungi (Ascomycota: Phlyctidaceae: *Phlyctis boliviensis*; Basidiomycota: Corticiaceae: *Dendrothele americana*).

Economic importance: food: none; **medicinal:** no uses reported; **cultivation:** *P. aquatica* has been recommended as an ornamental tree for use in wet areas; **misc. products:** The wood of *P. aquatica* has a specific gravity of 0.52 and an average moisture content of 75% (oven dry weight). It is fragrant and sometimes used in cabinetmaking, but generally is weak and brittle. The native Miami tribe used the wood and bark for making boxes and tables and for building canoes. The wood also has a low usage locally as a fuel and

occasionally is harvested for pulp; **weeds:** none; **nonindigenous species:** none

Systematics: *Planera* is related closely to *Ulmus* and *Zelkova* and somewhat more distantly to *Hemiptelea* (Figure 4.49). The genus is monotypic. The chromosome number of *P. aquatica* ($2n = 28$) indicates a base chromosome number of $x = 7$. There are no reported studies of the reproductive ecology or population genetics of *P. aquatica*, which are two areas that might provide useful information to assist with the conservation of this species.

Comments: *Planera aquatica* is endemic to the southeastern United States.

References: Barker, 1997; Dmitriev & Dietrich, 2009; Duberstein et al., 2014; Elias, 1987; Keeland & Gorham, 2009; Keeland et al., 2010; King, 2003; Lendemer & Harris, 2014; Loos, 2014; McAtee, 1939; Meitzen, 2009; Middleton, 1995a,b; Miles & Smith, 2009; Nakasone, 2006; Oginuma et al., 1990; Overlease & Overlease, 2011; Schneider & Sharitz, 1986; Ueda et al., 1997; Wiegrefe et al., 1998.

Family 4: Urticaceae [45]

Urticaceae (nettles) are a cosmopolitan family containing about 800 species with 21 of them distributed among eight genera in North America (Boufford, 1997). The group consists mainly of herbaceous plants (some shrubs, trees, vines), which occupy terrestrial habitats. The family is monophyletic with the species characteristically possessing elongate cystoliths, laticifers (yielding milky or clear sap), and stamens whose filaments are incurved and elastically reflexed in bud (Judd et al., 2008). Phylogenetic studies (Figure 4.39) resolve Moraceae as the sister family to Urticaceae. Some genera (e.g., *Cecropia*, *Pokilospermum*) have been segregated as a distinct family (Cecropiaceae); however, that interpretation is inconsistent with the results from molecular phylogenetic analyses (Figure 4.50). The flowers are small and wind pollinated, a mechanism that is assisted by the recoil of the stamens at dehiscence. The achenes are dispersed by birds (Aves), wind, or ballistically.

Urticaceae are important economically as a source of fibers such as ramie, which is derived from the bast of *Boehmeria nivea*. The fibers are extremely long and can have a tensile strength surpassing that of steel. *Girardinia*, *Laportea*, and *Urtica* are eaten as vegetables in some parts of the world. A few genera (*Boehmeria*, *Elatostema*, *Parietaria*, *Pilea*, *Soleirolia*, *Urtica*) contain ornamental cultivars, which include the popular "aluminum plant" (*Pilea cadierei*) and "baby's tears" (*Soleirolia soleirolii*). Some genera (e.g., *Laportea*, *Urtica*) are known for their stinging hairs, which when contacted can inject a painful mixture of histamine and acetylcholine into the skin. Several genera (*Laportea*, *Parietaria*, *Urtica*) include species having FACW indicator status in North America, and which occur commonly in wet woodlands and on floodplains. Only two genera contain obligate aquatics:

1. ***Boehmeria*** Jacquin
2. ***Pilea*** Lindley

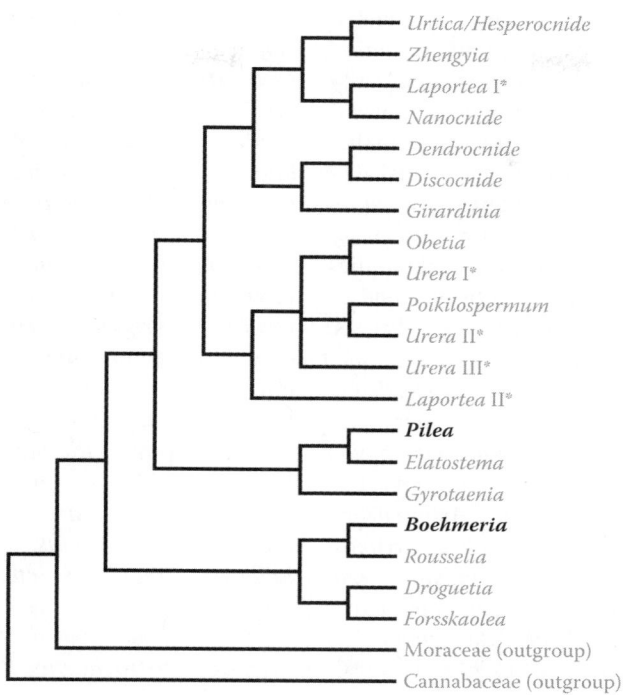

FIGURE 4.50 Phylogenetic relationships among 18 genera of Urticaceae as indicated by combined nrITS and cpDNA sequence data. Asterisks denote those genera that resolved as polyphyletic. This topology is similar to that reported earlier by Sytsma et al. (2002), except for the placement of *Pilea*, which was sister to *Laportea* in their analyses. The two North American genera containing OBL indicators (bold) indicate independent origins of the habit in these groups. (Adapted from Kim, C. et al., *Taxon*, 64, 65–78, 2015.)

Phylogenetic analyses (Figure 4.50) show that these genera are not closely related, an indication that their adaptations to aquatic conditions represent independent evolutionary events.

1. *Boehmeria*

False nettle; ortie de savane

Etymology: after Georg Rudolph Bohme (1723–1803)
Synonyms: *Splitgerbera*; *Urtica* (in part)
Distribution: global: Africa, Asia, Australia, New World; **North America:** eastern and southern
Diversity: global: 50 species; **North America:** 2 species
Indicators (USA): OBL; FACW: *Boehmeria cylindrica*
Habitat: freshwater; palustrine, riverine; **pH:** 5.2–8.6; **depth:** <1 m; **life-form(s):** emergent herb
Key morphology: stems (to 16 dm) erect; leaves (to 18 cm) petiolate (to 10 cm), opposite, simple, ovate-lanceolate, with three main palmate veins, margins coarsely toothed; stipules present; flowers minute (to 2 mm), unisexual, 4-merous, arranged in distinct (♂) or continuous (♀) spherical clusters on elongate axillary spikes (to 8 cm), which often bear reduced, terminal leaves; plants monoecious (spikes with mixed sexes or unisexual with staminate spikes occurring lower) or dioecious; perianth absent; achenes (to 1.6 mm) invested by the persistent calyx
Life history: duration: perennial (rhizomes); **asexual reproduction:** rhizomes; **pollination:** wind; **sexual condition:** dioecious/monoecious; **fruit:** achenes (common); **local dispersal:** seeds (wind); **long-distance dispersal:** seeds (water)

Imperilment: (1) *Boehmeria cylindrica* [G5]; S2 (<u>NB</u>); S3 (<u>QC</u>)
Ecology: general: With the exception of *B. cylindrica* (OBL, FACW), most *Boehmeria* species occur in tropical or subtropical climates, and primarily in relatively dry places, albeit occasionally in riparian sites. Some species are somewhat weedy and inhabit roadsides and waste areas. A number of them are shrubby or suffrutescent. Pollination is abiotic by wind and is associated with reduced perianths, explosive anther dehiscence, large, feathery stigmas, and high pollen/ovule ratios. Agamospermous reproduction has been found to occur in several triploid species.

Boehmeria cylindrica **(L.) Swartz** inhabits a fairly wide spectrum of sites including alluvial woods, bogs, ditches, floodplains, marshes, meadows, prairies, seeps, swamps, and the margins of ponds, rivers, and streams at elevations of up to 1800 m. It is a characteristic species of the riparian communities found in peripheral riverine wetlands. The plants tolerate acidic to alkaline substrates (pH: 5.2–8.6), but more often occur under moderately acidic conditions. The substrates often have a fairly high organic matter content and include gravel, muck, peat, rock, sand (e.g., sandbars), and silt. This species occupies shaded habitats and is not often found in open sites. The plants are in flower from summer through fall. The unisexual flowers (in monoecious or dioecious arrangement) are wind pollinated. Germination rates typically appear to be low (<30%). Germination is highest in freshwater (30%) and decreases with salinity (8% at 15 ppt); rates are also higher under humid (8%–30%) rather than inundated (5%–14%) conditions at all salinity levels from

0 to 15 ppt. Seeds collected from soil samples in early fall (October) reportedly will germinate following a brief period of cold stratification (4 days at 4°C). The achenes and corky seeds float and are dispersed prolifically by wind and water (on the surface). Observations of seeds in horse (Mammalia: Equidae: *Equus ferus*) dung indicate that some animal dispersal may occur as well. The extent of seed bank formation is unclear; however, *B. cylindrica* can reach a mean density of up to 7.2 seedlings/2750 cm^3 of soil in seedbank studies of cypress (*Taxodium*) swamps and has also been found as a minor seedbank component of secondary UPL forests.

Reported associates: *Acalypha rhomboidea, Acer negundo, A. rubrum, A. saccharinum, Agrimonia parviflora, Alisma triviale, Alliaria petiolata, Alnus rugosa, A. serrulata, Ampelopsis arborea, Amphicarpaea bracteata, Andropogon gerardii, Angelica atropurpurea, Arisaema dracontium, Arundinaria gigantea, Asclepias incarnata, Asimina triloba, Betula alleghaniensis, B. lutea, B. nigra, B. papyrifera, B. pumila, Bidens cernuus, B. vulgatus, Brunnichia ovata, Calamagrostis canadensis, Calla palustris, Caltha palustris, Campanula aparinoides, Campsis radicans, Cardamine bulbosa, C. pensylvanica, Carex amphibola, C. buxbaumii, C. comosa, C. crinita, C. cristatella, C. crus-corvi, C. gravida, C. grayi, C. hystericina, C. intumescens, C. lasiocarpa, C. limosa, C. lupuliformis, C. lupulina, C. lurida, C. muskingumensis, C. retrorsa, C. squarrosa, C. stricta, C. tribuloides, C. tuckermanii, C. typhina, Carya aquatica, C. cordiformis, C. ovata, Cassia hebecarpa, Celastrus orbiculatus, Celtis laevigata, C. occidentalis, Cephalanthus occidentalis, Chasmanthium laxum, Cicuta bulbifera, C. maculata, Cinna arundinacea, Circea alpina, Cirsium muticum, Cocculus carolinus, Comarum palustre, Coptis trifolia, Cornus foemina, C. sericea, Corylus americana, Crataegus viridis, Cryptotaenia canadensis, Cuscuta cephalanthi, C. gronovii, Cynosciadium digitatum, Cyperus aristatus, Dichanthelium acuminatum, Doellingeria umbellata, Dulichium arundinaceum, Echinocystis lobata, Eleocharis, Elymus virginicus, Epilobium leptophyllum, Equisetum arvense, Eragrostis pectinacea, Euonymus atropurpureus, Eupatorium perfoliatum, Eutrochium maculatum, Fimbristylis autumnalis, Fraxinus americana, F. nigra, F. pennsylvanica, Galium obtusum, G. trifidum, G. labradoricum, Geranium maculatum, Geum canadense, G. laciniatum, Glechoma hederacea, Gleditsia triacanthos, Glyceria canadensis, G. striata, Gratiola neglecta, Helenium autumnale, Hibiscus militaris, H. moscheutos, Hydrocotyle bonariensis, Hypericum majus, Ilex cassine, I. decidua, I. opaca, Impatiens capensis, Iris, Itea virginica, Juncus effusus, Koeleria macrantha, Laportea canadensis, Larix laricina, Lathyrus palustris, Leersia lenticularis, L. oryzoides, L. virginica, Lemna, Lilium michiganense, Lindera benzoin, Lipocarpha micrantha, Liquidambar styraciflua, Liriodendron tulipifera, Lobelia cardinalis, L. kalmii, L. siphilitica, Lonicera japonica, Ludwigia palustris, Lycopus americanus, L. uniflorus, L. virginicus, Lygodium microphyllum, Lysimachia ciliata, L. hybrida, L. nummularia, L. quadriflora, Lythrum salicaria, Magnolia virginiana, Matteuccia struthiopteris, Menispermum canadense, Mentha arvensis, Microstegium vimineum, Mikania scandens, Mimulus ringens, Mitella nuda, Muhlenbergia glomerata, Myosoton aquaticum, Nyssa biflora, Onoclea sensibilis, Osmunda regalis, Packera aurea, Panicum virgatum, Parthenocissus quinquefolia, Pedicularis lanceolata, Peltandra virginica, Persea palustris, Persicaria hydropiper, P. hydropiperoides, P. lapathifolia, P. punctata, P. sagittata, P. setacea, Phalaris arundinacea, Phlox divaricatus, Phyla lanceolata, Physalis longifolia, Physostegia virginiana, Pilea fontana, P. pumila, Pinus banksiana, P. resinosa, P. rigida, P. strobus, Plantago rugelii, Platanus occidentalis, Polygonatum, Populus deltoides, P. tremuloides, Prunus serotina, Quercus alba, Q. bicolor, Q. borealis, Q. ellipsoidalis, Q. lyrata, Q. macrocarpa, Q. palustris, Q. rubra, Q. texana, Ranunculus abortivus, R. ficaria, R. septentrionalis, Rhexia mariana, R. virginica, Ribes americanum, Rorippa sylvestris, Rubus, Rudbeckia laciniata, Rumex crispus, R. verticillatus, Sagittaria latifolia, Salix amygdaloides, S. candida, S. interior, Sambucus nigra, Saponaria officinalis, Saururus cernuus, Schizachyrium scoparium, Schoenoplectus acutus, S. pungens, S. tabernaemontani, Scirpus ancistrochaetus, S. cyperinus, Scutellaria galericulata, S. lateriflora, Sicyos angulatus, Smilax hispida, S. tamnoides, Solanum dulcamara, Solidago flexicaulis, S. gigantea, S. patula, S. riddellii, S. uliginosa, Sparganium eurycarpum, Spartina pectinata, Sphagnum, Spiraea, Sporoblus heterolepis, Stachys tenuifolia, Symphyotrichum boreale, S. drummondii, S. lateriflorum, S. puniceum, Symplocarpos foetidus, Taxodium distichum, Teucrium canadense, Thuja occidentalis, Thelypteris noveboracensis, T. palustris, Tilia americana, Toxicodendron radicans, T. vernix, Trachelospermum difforme, Triadenum fraseri, Typha angustifolia, T. latifolia, Ulmus americana, U. rubra, Urtica dioica, Verbena hastata, Viola cucullata, Vitis aestivalis, V. cinerea, V. palmata, V. riparia, V. rotundifolia, Wisteria frutescens, Woodwardia areolata, W. virginica, Zanthoxyllum americanum.*

Use by wildlife: Grazing of *B. cylindrica* by larger mammals probably occurs only at low levels, but the plants can provide substantial forage for white-tailed deer (Mammalia: Cervidae: *Odocoileus virginianus*). The seeds are a minor dietary component of shoveler ducks (Aves: Anatidae: *Anas clypeata*) and other waterfowl. *Boehmeria cylindrica* is a host plant for several Lepidoptera (Nymphalidae: *Nymphalis milberti, Polygonia comma, P. interrogationis, Vanessa atalanta*; Pyralidae: *Pilocrocis ramentalis, Pleuroptya fluctuosalis, P. silicalis*). It is also a host of a gall midge (Insecta: Diptera: Cecidomyiidae: *Neolasioptera boehmeriae*).

Economic importance: food: *Boehmeria cylindrica* has been suggested as an alternate to stinging nettle (*Urtica dioica*) for use as a leafy green vegetable; **medicinal:** Although there are no medical uses reported for *B. cylindrica*, the plants contain alkaloids (cryptopleurine and 3,4-dimethoxy-ω-(2′-piperidyl) acetophenone), which are known to be cytotoxic to certain carcinomas or to have antiviral and antimicrobial properties. The compounds are highly effective against certain yeasts (e.g., Ascomycota: Debaryomycetaceae: *Candida albicans*); **cultivation:** *Boemeria cylindrica* is not cultivated but sometimes is used in flower arrangements; **misc. products:** The durable fiber known as "ramie" is obtained from the related species *B. nivea*. Fibers of *B. cylindrica* reportedly have similar properties and were used as twine by various Native American tribes to make bowstrings and to affix spearpoints and arrowheads to their shafts; **weeds:** none; **nonindigenous species:** *Boemeria nivea* was introduced intentionally to the United States, but it is not a wetland plant.

Systematics: Molecular phylogenetic analyses have demonstrated that *Boehmeria* is not monophyletic as currently circumscribed. Furthermore, various studies have resolved *Boehmeria* as closely related to various genera including *Debregeasia*, *Parietaria*, or *Rousselia* (e.g., Figure 4.50) depending on the taxa and loci sampled. In some cases, *Boehmeria* appears to be only distantly related to *Parietaria* or *Rousselia*. More definitive work on intergeneric relationships in this large family is necessary. Elucidation of relationships between *B. cylindrica* and other members of *Boehmeria* awaits also a more comprehensive phylogenetic analysis of the genus than is currently available. The base chromosome number of *Boehmeria* is $x = 14$, which probably is derived ultimately from $x = 7$, as in the related *Parietaria*. *Boehmeria cylindrica* is a diploid ($2n = 28$). A better ecological understanding of *B. cylindrica* might be gained by conducting studies of its reproductive biology and population structure, especially given that some members of the genus are known to be agamospermous.

Comments: *Boehmeria cylindrica* occurs throughout eastern North America and is disjunct in South America.

References: Ailstock et al., 2001; Al Shamma et al., 1982; Anderson et al., 2001; Beauchamp et al., 2013; Boufford, 1997; Campbell & Gibson, 2001; Farnsworth et al., 1969; Garciá, 2013; Hanberry et al., 2014; Kim et al., 2015; Krmpotic et al., 1972; McAtee, 1939; Middleton & McKee, 2011; Musselman & Wiggins, 2013; Neff & Baldwin, 2005.

2. Pilea

Clearweed; petite ortie

Etymology: from the Latin *pileus* meaning "felt cap" in reference to the cap-like calyx

Synonyms: *Adicea*; *Neopilea*

Distribution: global: Africa; Asia; Europe; New World; **North America:** eastern

Diversity: global: 400 species; **North America:** 5 species

Indicators (USA): OBL: *Pilea fontana*, *P. herniarioides*; **FACW:** *P. fontana*

Habitat: freshwater; palustrine, riverine; **pH:** alkaline; **depth:** <1 m; **life-form(s):** emergent herb

Key morphology: stems (to 1 dm) erect or prostrate, opaque or translucent; leaves (to 10 cm) opposite, simple, petiolate (to 4.5 cm), the margins entire or coarsely dentate, the blade with three palmate veins arising from base; stipules (to 3 mm) deciduous; flowers minute (to 1 mm), unisexual, arranged in compound, axillary cymes either with the sexes mixed (androgynous), or unisexual with the ♂ cymes occurring lower on the plant; achenes (to 1.7 mm) black or light brown, partially enclosed by a tepal at maturity

Life history: duration: annual (fruit/seeds); **asexual reproduction:** none; **pollination:** wind; **sexual condition:** monoecious; **fruit:** achenes (common); **local dispersal:** achenes (ballistic); **long-distance dispersal:** achenes

Imperilment: (1) *Pilea fontana* [G5]; SH (VT); S2 (QC); S3 (IA, IL, MD, NC, NJ); 1. *P. herniarioides* [G5];

Ecology: general: Of the five *Pilea* species that occur in continental North America, all have status as wetland indicators with two listed as obligately aquatic. In addition, several other species have FAC or OBL status in the Caribbean region. The species are annuals or short-lived perennials. Like most Urticaceae, the flowers are wind pollinated. The seeds of *Pilea* species are ejected ballistically from the female flowers as the tensioned staminodes spring apart.

Pilea fontana (Lunell) **Rydb.** grows in bogs, ditches, fens, marshes, meadows, prairies, seeps, springs, swamps, and along the margins of lakes, rivers, and streams (sometimes in floating mats) at elevations of up to 300 m. Habitats usually are described as alkaline (e.g., limestone cobble, marly sand), but specific pH values for this species rarely are reported in the literature and one value (pH: 6.0–6.9) is slightly acidic. It is considered to be a good indicator of fens. The plants occupy a range of substrates from cobble and sand to rich muck or peat. This species grows in shade and seldom is found persisting in open sites. A substantial reduction in percent cover has been observed for this species in ungrazed meadows surveyed over a 10-year period. Germination of *Pilea* seeds in general reportedly is sporadic after 14–60 days at 18°C–24°C. **Reported associates:** *Abies*, *Acer negundo*, *A. rubrum*, *Acorus calamus*, *Agrimonia parviflora*, *Agrostis alba*, *Alisma triviale*, *Alnus serrulata*, *Amaranthus tuberculatus*, *Ambrosia trifida*, *Arctium minus*, *Asclepias incarnata*, *Berula erecta*, *Betula pumila*, *Bidens cernuus*, *B. coronatus*, *B. discoideus*, *B. frondosus*, *B. laevis*, *Boehermia cylindrica*, *Bromus ciliatus*, *Calamagrostis canadensis*, *Caltha palustris*, *Campanula aparinoides*, *Cardamine bulbosa*, *Carex aquatilis*, *C. buxbaumii*, *C. comosa*, *C. hystericina*, *C. lacustris*, *C. lasiocarpa*, *C. limosa*, *C. lurida*, *C. pellita*, *C. stricta*, *Cephalanthus occidentalis*, *Chelone glabra*, *Cicuta bulbifera*, *Cinna arundinacea*, *Cornus amomum*, *C. foemina*, *C. racemosa*, *C. sericea*, *Cryptotaenia canadensis*, *Cuscuta gronovii*, *Cyperus strigosus*, *Dasiphora floribunda*, *Decodon verticillatus*, *Decumaria barbara*, *Digitaria cognata*, *Doellingeria umbellata*, *Echinocystis lobata*, *Eleocharis tenuis*, *Epilobium glandulosum*, *E. leptophyllum*, *Equisetum arvense*, *E. hyemale*, *Eupatorium perfoliatum*, *Eutrochium maculatum*, *Fraxinus nigra*, *F. pennsylvanica*, *Galium boreale*, *G. labradoricum*, *G. tinctorium*, *G. trifidum*, *Glyceria striata*, *Helenium autumnale*, *Helianthus strumosus*, *Heliopsis helianthoides*, *Hypericum pyramidatum*, *Impatiens capensis*, *Iris viginica*, *Juncus torreyi*, *Lactuca canadensis*, *Larix laricina*, *Lathyrus palustris*, *Leersia oryzoides*, *Lemna minor*, *Lindera benzoin*, *Liquidambar styraciflua*, *Lobelia kalmii*, *L. syphilitica*, *Ludwigia alternifolia*, *Lycopus americanus*, *L. uniflorus*, *Lysimachia quadriflora*, *L. terrestris*, *L. thyrsiflora*, *Lythrum salicaria*, *Maianthemum canadense*, *Marchantia polymorpha*, *Mentha* *arvensis, Muhlenbergia glomerata, Myosoton aquaticum, Myrica cerifera, Panicum capillare, Pedicularis lanceolata, Penthorum sedoides, Persicaria coccinia, P. lapathifolia, P. punctata, P. sagittata, Phalaris arundinacea, Phragmites australis, Physostegia virginiana, Potentilla norvegica, Pycnanthemum virginianum, Ribes americanum, Rosa multiflora, Rudbeckia laciniata, Rumex orbiculatus, Sagittaria latifolia, Salix candida, S. interior, S. petiolaris, S. rigida, Sambucus, Saururus cernuus, Schoenoplectus acutus, S. pungens, S. tabernaemontani, Scutellaria galericulata, S. lateriflora, Sedum sexangulare, Sium suave, Solanum dulcamara, Solidago altissima, S. canadensis, S. gigantea, S. graminifolia, S. patula, Sparganium eurycarpum, Spartina pectinata, Stachys palustris, Sullivantia, Symphyotrichum boreale, S. lanceolatum, S. lateriflorum, S. puniceum, Thalictrum dasycarpum, Thelypteris palustris, Thuja occidentalis, Toxicodendron radicans, T. vernix, Triadenum fraseri, Typha angustifolia, T. latifolia, Ulmus americana, Urtica dioica, Verbena hastata, Woodwardia areolata, Zizania aquatica.*

Pilea herniarioides (Sw.) Lindl. inhabits hammocks and waste places at elevations of up to 300 m. In its native range, it can grow on wet rocks in montane habitats. Exposures include shaded sites. Flowering persists from fall to spring and intermittently throughout the summer. The OBL status of this species in North America seems unusual, given its common appearance as a weed in numerous UPL sites. **Reported associates:** none reported.

Use by wildlife: The leaves of *Pilea herniarioides* are eaten by feral goats (Mammalia: Bovidae: *Capra*).

Economic importance: food: some *Pilea* species are used as potherbs; **medicinal:** Although no medical uses are reported specifically for *P. fontana*, the juice from the stems of the closely related *P. pumila* (with which it can easily be confused) was used to relieve itching between the toes (Cherokee) and to relieve sinusitis (Iroquois). The plants have also been used as a general diuretic; **cultivation:** The OBL indicators are not cultivated; **misc. products:** none; **weeds:** *Pilea herniarioides* is a common weed of nurseries and parks in southern Florida; *P. microphylla*, although not an aquatic species, is a familiar greenhouse weed; **nonindigenous species:** Populations of *Pilea herniarioides* near Mobile, Alabama reportedly are escapes from cultivation.

Systematics: Results from molecular phylogenetic analyses that included roughly a quarter of the Urticaceae genera initially indicated that *Pilea* was related most closely to *Laportea*; however, a subsequent study that sampled a broader representation of taxa and genetic loci indicated that *Pilea* is the sister genus to *Elatostema*, and is not closely related to *Laportea* (Figure 4.50). Yet another study resolved *Lecanthus* as the sister genus of *Pilea*, placing *Elatostema* in a more distant position. From these inconsistent results, it is evident that more definitive evidence will be necessary before any firm conclusions regarding intergeneric relationships can be reached. Also, a comprehensive phylogenetic investigation of this rather large genus has not yet been undertaken but must be conducted before interspecific relationships and species

delimitations can be reasonably determined. Preliminary phylogenetic analyses using DNA sequence data have indicated a close relationship between *P. herniarioides* and *P. quercifolia*. Because of its similar morphology, *P. fontana* is considered to be closely related to (if not conspecific with) *P. pumila*. However, the distinctness of *P. fontana* and *P. pumila* has been debated for many years, with seed morphology providing the only fairly consistent means of separating the taxa. Although their distributions in North America are sympatric, the two species rarely are found growing at the same site. Clarification of the relationship between these two taxa would benefit greatly by the application of genetic data. The chromosomal base number of *Pilea* is uncertain with several numbers listed ($x = 8, 12, 13, 15, 18$). The chromosome number of *P. fontana* is not known, but counts of $2n = 24, 26$ are reported for the putatively related *P. pumila*. Because of their taxonomic uncertainty as distinct species and inconsistent characters used to delimit them in past treatments, it is likely that at least some of the information reported for *P. fontana* pertains instead to *P. pumila* and vice versa.

Comments: *Pilea fontana* occurs throughout eastern North America, extending somewhat into the central region; *P. herniarioides* occurs in Florida and Alabama, but extends into the West Indies and to South America.

References: Boufford, 1997; Kim et al., 2015; Lamont & Young, 2004; Meléndez-Ackerman et al., 2008; Middleton, 2002b; Mohr, 1901; Monro, 2006; Nekola, 2004; Steury & Davis, 2003.

ORDER 7: SAXIFRAGALES [15]

Phylogenetic relationships of and within Saxifragales have been studied intensively using a combination of morphological and molecular data. Yet, despite the large amount of molecular data directed toward its study, the overall placement of the Saxifragales among the "core eudicots" remains unsettled. Depending on the data set(s) analyzed, the order resolves as the sister group of all rosids (*rbcL*; *rbcL*+18S rDNA+*atpB*), as the sister to most core eudicots (*matK*; *rbcL*+18S rDNA+26S rDNA+*atpB*), or as the sister to Vitaceae (full chloroplast genomes); however, support for any of the resulting topologies remains weak (Soltis et al., 2005; Ruhfel et al., 2014). On the other hand, these studies have greatly clarified the circumscription of the order, which now is delimited to include a strongly supported clade of 13–15 families (Figure 4.51), of which several (e.g., Haloragaceae, Hamamelidaceae, Paeoniaceae, Peridiscaceae) had never been considered previously as being allied with the group (Davis & Chase, 2004; Soltis et al., 2005; Ruhfel et al., 2014).

Saxifragales are diverse ecologically with some families (e.g., Crassulaceae) being widely known for their xerophytic members and others (e.g., Haloragaceae) for their aquatics. The families that include obligate aquatic species in North America are:

1. **Crassulaceae** J. St.-Hil.
2. **Grossulariaceae** DC.

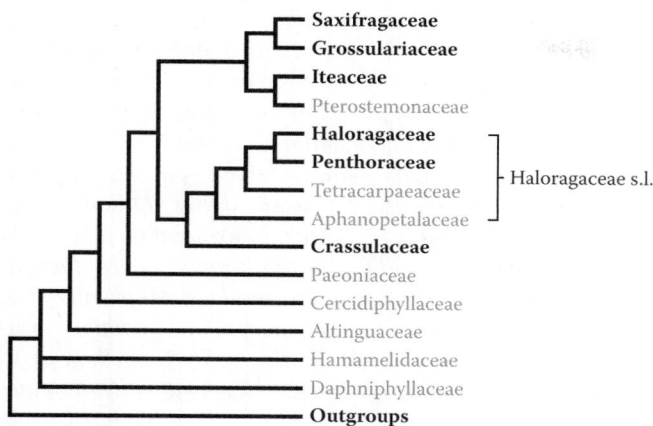

FIGURE 4.51 Phylogenetic relationships within Saxifragales as indicated by analyses of combined molecular data. Some authors prefer a broad "sensu lato" (s.l.) concept of Haloragaceae, which is indicated. Multiple molecular data sets indicate that Peridiscaceae (not included in the survey shown) also belong to the order (Davis & Chase, 2004). Families with obligate aquatic species in North America are indicated in bold. This analysis indicates that most of the families with aquatics are relatively derived in the order. (Adapted from Fishbein, M. et al., *Syst. Biol.*, 50, 817–847, 2001.)

3. **Haloragaceae** R. Br.
4. **Iteaceae** J. Agardh
5. **Penthoraceae** Rydb.
6. **Saxifragaceae** Juss.

Family 1: Crassulaceae [35]

Because nearly all of the estimated 1500 species of Crassulaceae are typically associated with arid habitats, it is an unusual family to discuss in the context of aquatic plants. Analysis of *matK* sequence data for 112 species shows the family to be monophyletic (Mort et al., 2001) and the representatives can be characterized as succulent plants with C_4 (CAM) photosynthesis and distinct to basally connate carpels that are subtended by scale-like nectaries. Two subfamilies currently are distinguished, the Crassuloideae (with a single stamen whorl and thin megasporangial wall) and Sedoideae (with two stamen whorls, thick megasporangial wall, and ridged seed coat). Each subfamily represents a well-supported clade (Mort et al., 2001). Aquatic plants occur only within the Crassuloideae, which contains one or two genera (*Crassula*, *Tillaea*). Although it has become common practice to merge these genera (as *Crassula*), results from cpDNA restriction-site analysis show them to be quite distinct and also to differ with respect to their fruit dehiscence and ovule number (Van Ham & Hart, 1998). However, an analysis of DNA sequence data from a much larger sampling of species from both genera does not maintain the distinctness of *Tillaea* (Figure 4.52), and the merger of these genera appears to be warranted (Mort et al., 2009). Only one genus of Crassulaceae contains obligate aquatics in North America:

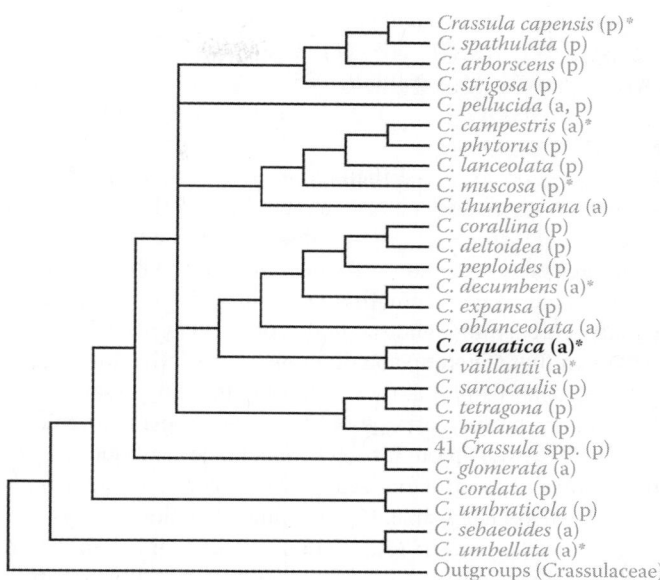

FIGURE 4.52 Phylogenetic relationships among *Crassula* species as indicated by analysis of combined nrDNA and cpDNA sequence data. Species marked by an asterisk (*) are those that have been placed in the genus *Tillaea*. Two major clades are resolved, one containing the annual (a) *C. glomerata* and 41 perennial (p) *Crassula* species, and the other containing a mixture of annuals and perennials along with those taxa also assigned to *Tillaea*. These results indicate multiple origins of the annual habit in *Crassula*, and do not resolve *Tillaea* as a clade; thus, the merger of *Tillaea* and *Crassula* is warranted. *Crassula aquatica* (bold), an obligate aquatic in North America (and Eurasia), resolves as closely related to the Eurasian/African *C. vaillantii*, also an aquatic annual [*C. connata* and *C. solierii* were not surveyed]. (Adapted from Mort, M.E. et al., *Plant Syst. Evol.*, 283, 211–217, 2009 and M. Mort, unpublished.)

1. ***Crassula*** L.

1. Crassula

Water pygmyweed

Etymology: from the Latin *crassus* meaning "thick" with reference to the succulent habit

Synonyms: *Bulliardia*; *Hydrophila*; *Tillaea*; *Tillaeastrum*

Distribution: global: cosmopolitan; **North America:** widespread

Diversity: global: 150 species; **North America:** 7 species

Indicators (USA): OBL: *Crassula aquatica*, *C. connata*, *C. solierii*; **FAC:** *C. connata*

Habitat: brackish (coastal), freshwater; palustrine; **pH:** 6.0–8.3; **depth:** <1 m; **life-form(s):** emergent herb, submersed (vittate)

Key morphology: stems succulent, branched, erect (to 10 cm) or mat-forming, rooting at lower nodes; leaves (to 7 mm) opposite, fleshy, entire, their bases connate/perfoliate; flowers minute (to 2 mm), solitary, axillary, sessile or pedicelled (to 8 mm), 4-merous; carpels 4, distinct, subtended by a basal scale; follicles (to 2 mm) 8–10-seeded, spreading in fruit

Life history: duration: annual (fruit/seeds); **asexual reproduction:** shoot fragments; **pollination:** self; **sexual condition:** hermaphroditic; **fruit:** follicles (common); **local dispersal:** seeds; **long-distance dispersal:** seeds, fragments (birds)

Imperilment: (1) *Crassula aquatica* [G5]; SX (PA); SH (CO, MD); S1 (CT, ID, <u>NF</u>, NH, NY, OK, <u>ON</u>, <u>PE</u>, UT, <u>YT</u>); S2 (AR, MA, ME, MN, <u>NB</u>, <u>NS</u>, VT, WA); S3 (AK, <u>BC</u>, <u>QC</u>); (2) *C. connata* [G5]; S1 (WA); S2 (<u>BC</u>); (3) *C. solierii* [G4]; S1 (OR, WY)

Ecology: general: It is difficult to think of *Crassula*, which is known for its majestic succulent species like the ornamental jade plant (*C. argentea*), as anything but xerophytic. However, a number of species are well adapted to wet habitats. These species tend to be diminutive annuals, which occur primarily along receding muddy shores or other vernal habitats. Although flies (Diptera) and other insects visit the flowers, the plants probably are mainly self-pollinating. Dispersal occurs by seeds or by the carriage of fruit-bearing plant fragments by birds (Aves). Although annual, vegetative reproduction can occur during the growing season by dispersal of stem fragments, which are capable of producing adventitious roots.

Crassula aquatica **(L.) Schönl.** is an annual found in brackish (intertidal) to freshwater habitats as an erect submergent in shallow water or as a prostrate, mat-forming emergent in borrow pits, depressions, ditches, estuaries, hot springs, marshes, mudflats, salt marshes, swales, tidal shores, tidal streams, vernal pools, and the margins of pools and rivers, at elevations of up to 2896 m. It is found on various substrates including basalt, clay, gravel, mud, rocks, sand, sandy mud, and silt. *Crassula aquatica* is adapted to the low CO_2 conditions that often prevail in its oligotrophic habitats. It is not a bicarbonate user but has a C_4 (CAM) photosynthetic pathway, which enhances carbon uptake by nocturnal CO_2 fixation. The plants can survive temperatures down to −40°C. Syrphid flies (Diptera: Syrphidae: *Somula decora*) have been observed visiting the flowers, but self-pollination is probably prevalent. The seeds of Crassulaceae are known to possess physiological dormancy, but germination requirements of *C. aquatica* have not been studied. **Reported associates:** *Alisma triviale, Alopecurus howellii, A. saccatus, Ambrosia pumilla, Anagallis minima, Anthemis, Atriplex coronata, Avena fatua, Bergia texana, Bidens cernuus, B. eatonii, B. hyperboreus, Blennosperma nanum, Bolboschoenus fluviatilis, B. maritimus, B. robustus, Brodiaea filifolia, B. jolonensis, B. orcuttii, Bromus hordeaceus, Callitriche hermaphroditica, C. longipedunculata, C. marginata, C. stagnalis, Cardamine longii, Carex athrostachya, C. utriculata, C. viridula, Centunculus minimus, Chara contraria, Cicendia quadrangularis, Conioselinum chinense, Cryptantha, Cyperus aristatus, C. bipartitus, C. filicinus, Damasonium californicum, Deschampsia cespitosa, D. danthonioides, Downingia bella, D. cuspidata, Echinodorus berteroi, Elatine americana, E. brachysperma, E. californica, E. minima, Eleocharis acicularis, E. flavescens, E. halophila, E. macrostachya, E. montevidensis, E. obtusa, E. palustris, E. parvula, E. quinqueflora, E. rostellata, Epilobium densiflorum, E. pygmaeum, Equisetum fluviatile, E. palustre, Eragrostis hypnoides, Eriocaulon parkeri, Erodium botrys, Eryngium aristulatum, E. vaseyi, Euphorbia nutans, Gnaphalium palustre, Gratiola neglecta, G. virginiana, Helenium, Heteranthera reniformis, Hordeum intercedens, Hydrocotyle verticillata, Hypericum majus, Iris pseudacorus, Isoetes howellii, I. maritima, I. orcuttii, I. riparia, Juncus alpinus, J. articulatus, J. bufonius, J. dubius, J. nevadensis, J. occidentalis, J. oxymeris, J. supiniformis, J. triformes, Koenigia islandica, Lasthenia fremontii, Lepidium latipes, L. nitidum, Lilaea scilloides, Lilaeopsis chinensis, Limnanthes gracilis, Limosella acaulis, L. aquatica, L. australis, Linanthus ciliatus, Lindernia dubia, Lolium multiflorum, Ludwigia palustris, Lythrum hyssopifolia, Malvella leprosa, Marsilea vestita, Micranthemum micranthemoides, Microsteris gracilis, Mimulus latidens, M. primuloides, Montia fontana, Muhlenbergia asperifolia, Myosotis scorpioides, Myosurus apetalus, M. minimus, Myriophyllum ussuriense, Nama stenocarpum, Navarretia fossalis, N. hamata, N. intertexta, N. leucocephala, N. prostrata, Ophioglossum californicum, Orcuttia californica, Oreostemma alpigenum, Orontium aquaticum, Oxypolis occidentalis, Panicum capillare, Parvisedum pumilum, Persicaria hydropiperoides, P. punctatum, Phalaris arundinacea, P. lemmonii, Phleum alpinum, Pilularia americana, Plagiobothrys acanthocarpus, P. leptocladus, P. scouleri, P. undulatus, Plantago elongata, P. erecta, Pleuropogon californicus, Pogogyne abramsii, P. nudiuscula, Polypogon monspeliensis, Pontederia cordata, Prunella, Psilocarphus brevissimus, P. tenellus, Ranunculus aquatilis, R. cymbalaria, R. flammula, Rorippa curvipes, R. curvisiliqua, Rotala ramosior, Sagittaria calycina, S. graminea, S. latifolia, S. montevidensis, S. subulata, Samolus valerandi, Schedonorus pratensis, Schoenoplectus pungens, S. smithii, S. triqueter, Spartina alterniflora, S. cynosuroides, Spergularia salina, Symphyotrichum subspicatum, Teucrium, Trifolium depaupertaum, Triglochin palustre, Verbena bracteata, Veronica peregrina, Vulpia bromoides, Zannichellia palustris.*

Crassula connata **(Ruiz & Pavón) Berger** is an annual, which inhabits alluvial fans, beaches, bluffs, chaparral, cliffs, coastal scrub, depressions, ditches, flats, floodplains, marshes, meadows, mudflats, prairies, ravines, riverbottoms, roadsides, seeps, slopes, strands, streambeds, swales, vernal pools, wallows, washes, woodlands, and the margins of lakes, sinkholes, and streams at elevations of up to 2700 m. This species has an extremely wide habitat range, which includes those listed previously as well as desert scrub, sand dunes, and dry mesas. The plants occur commonly in burned areas and in other open sites exposed to full sun but also can tolerate partial shade. The substrates typically are alkaline (but also to pH: 5.6) and have been described as adobe, basalt, clay, clay loam, cobbly clay, granite sand, gravel, gravelly loam, Hideaway soil series, lava, limestone, loam, loamy alluvial sand, pulverized rock, rocky clay, rocky loam, sand, sandstone, sandy clay, sandy clay loam, sandy gravel, sandy loam, serpentine, shale, silty loam, and volcanic. Even when growing in vernal pool environments, the plants occur at the dry end of the soil moisture spectrum. Because of its eclectic array of non–wetland (and arid)-associated species and frequent occurrences in dry habitats, its designation as an OBL species (in the Great Plains area) seems inappropriate. Sexual reproduction occurs from February to May. Otherwise, the reproductive ecology is poorly documented. The seeds are small and can dominate the seedbank in abundance (e.g., densities up to 2080 seeds/m²). Germination of daily watered seeds has been achieved at a daytime temperature of 24°C. Winter germination has also been observed to occur regardless of the watering regime.

Reported associates: *Abronia maritima, Acalypha californica, Achillea millefolium, Achyrachaena mollis, Acmispon argophyllus, A. glabrus, A. humistratus, A. prostratus, A. strigosus, A. wrangelianus, Adenostoma fasciculatum, A. sparsifolium, Aegilops triuncialis, Agave deserti, Ageratina adenophora, Aira caryophyllea, A. praecox, Allium vineale, Alopecurus saccatus, Ambrosia chamissonis, A. deltoidea, A. dumosa, A. psilostachya, A. salsola, Amorpha fruticosa, Amsinckia intermedia, A. menziesii, Anagallis arvensis, Andropogon glomeratus, Anthemis cotula, Aphanes occidentalis, Apiastrum angustifolium, Arbutus menziesii, Arctostaphylos catalinae, A. glandulosa, A. glauca, A. insularis, Artemisia californica, A. dracunculus, Asclepias subulata, Astragalus gambelianus, A. lentiginosus, A. tener, A. trichopodus, Atriplex fruticulosa, A. hymenelytra, A. lentiformis, A. leucophylla, A. polycarpa, A. semibaccata, Avena barbata, A. fatua, Baccharis pilularis, B. salicifolia, B. sarothroides, Bahiopsis laciniata, B. parishii, Bebbia juncea, Bothriochloa barbinodis, Bowlesia incana, Brassica nigra, B. tournefortii, Brickellia californica, Bromus diandrus, B. hordeaceus, B. rubens, Cakile maritima, Calandrinia ciliata, Calliandra eriophylla, Callitriche marginata, Calochortus plummerae, C. splendens, Calystegia macrostegia, Camassia leichlinii, Camissonia campestris, Camissoniopsis bistorta, C. cheiranthifolia, C. hirtella, C. intermedia, C. micrantha, C. pallida, Cardamine oligosperma, Carex praegracilis, Carnegiea gigantea, Carpobrotus chilensis, C. edulis, Castilleja exserta, Caulanthus heterophyllus, C. lasiophyllus, Ceanothus arboreus, C. crassifolius, C. cuneatus, C. greggii, C. leucodermis, C. megacarpus, C. verrucosus, Celtis pallida, Centaurea melitensis, C. solstitialis, Centromadia pungens, Cerastium glomeratum, C. semidecandrum, Cercocarpus betuloides, Chilopsis linearis, Chlorogalum pomeridianum, Cicendia quadrangularis, Cistanthe monandra, Claytonia parviflora, C. perfoliata, Cneoridium dumosum, Coreopsis gigantea, Corethrogyne filaginifolia, Crassula aquatica, C. tillaea, Cressa truxillensis, Crossosoma californicum, Croton setiger, Cryptantha intermedia, C. micrantha, C. micromeres, C. microstachys, C. muricata, C. pterocarya, Cylindropuntia acanthocarpa, C. californica, C. ganderi, C. leptocaulis, C. prolifera, Cyperus eragrostis, Dactylis glomerata, Daucus pusillus, Deinandra fasciculata, Deschampsia danthonioides, Descurainia pinnata, Dicentra chrysantha, Dichelostemma capitatum, Diplacus aridus, D. aurantiacus, D. parviflorus, Distichlis spicata, Dryopteris arguta, Dudleya arizonica, D. multicaulis, D. pulverulenta, Echinocereus engelmannii, Eleocharis acicularis, E. macrostachya, Emmenanthe penduliflora, Encelia californica, E. farinosa, E. virginensis, Ephedra torreyana, E. viridis, Epilobium brachycarpum, E. canum, Ericameria laricifolia, E. linearifolia, E. palmeri, E. pinifolia, Eriodictyon crassifolium, E. trichocalyx, Eriogonum fasciculatum, E. inflatum, E. parvifolium, E. thurberi, Eriophyllum confertiflorum, Erodium botrys, E.*

brachycarpum, E. cicutarium, E. moschatum, Eryngium vaseyi, Eschscholzia californica, Eucrypta chrysanthemifolia, Eulobus californicus, Ferocactus cylindraceus, Foeniculum vulgare, Fouquieria splendens, Frankenia salina, Fritillaria biflora, Funastrum cynanchoides, Galium angustifolium, G. aparine, G. nuttallii, G. stellatum, Gazania linearis, Gilia angelensis, G. capitata, G. modocensis, G. tricolor, Glebionis coronarium, Gnaphalium palustre, Grindelia hirsutula, G. integrifolia, Gutierrezia californica, Hazardia detonsa, Hedypnois cretica, Helianthemum scoparium, Helianthus annuus, Hemizonia congesta, Hesperoyucca whipplei, Heteromeles arbutifolia, Hilaria rigida, Hirschfeldia incana, Hordeum depressum, H. murinum, H. vulgare, Hyptis emoryi, Isocoma acradenia, I. menziesii, Jaumea carnosa, Juglans californica, Juncus acutus, J. balticus, J. bufonius, J. patens, Juniperus californica, Justicia californica, Keckiella antirrhinoides, Krameria bicolor, Lamarckia aurea, Larrea tridentata, Lasthenia californica, L. coronaria, L. fremontii, L. glabrata, L. platycarpha, Layia platyglossa, Lepidium latipes, L. nitidum, L. lasiocarpum, Lepidospartum squamatum, Leymus condensatus, Lithophragma affine, Loeflingia squarrosa, Logfia depressa, L. filaginoides, Lolium multiflorum, Lomatium utriculatum, Lupinus albifrons, L. arboreus, L. bicolor, L. concinnus, L. nanus, L. sparsiflorus, L. truncatus, Lycium andersonii, L. berlandieri, L. californicum, L. exsertum, Lyrocarpa coulteri, Lythrum hyssopifolia, Malacothrix foliosa, M. incana, Malephora crocea, Malosma laurina, Marah gilensis, M. macrocarpa, Marrubium vulgare, Medicago polymorpha, Melica imperfecta, Mentzelia, Mesembryanthemum crystallinum, M. nodiflorum, Micropus californicus, Microseris campestris, M. douglasii, Mimulus bigelovii, M. brevipes, M. douglasii, M. fremontii, Minuartia douglasii, Mirabilis laevis, M. tenuiloba, Monoptilon bellioides, Montia fontana, Mucronea californica, Myosurus minimus, Nassella cernua, N. lepida, N. pulchra, Navarretia tagetina, Neogaerrhinum strictum, Nuttallanthus canadensis, N. texanus, Oenothera californica, Ophioglossum californicum, Opuntia engelmannii, O. littoralis, O. phaeacantha, Orthocarpus, Parietaria floridana, Parkinsonia aculeata, P. florida, P. microphylla, Pectocarya linearis, P. setosa, Pellaea mucronata, Penstemon subulatus, Pentagramma triangularis, Peucephyllum schottii, Phacelia brachyloba, P. campanularia, P. cicutaria, P. distans, P. grandiflora, P. minor, Phalaris lemmonii, Phoradendron californicum, Physaria gordonii, Pilularia americana, Pinus muricata, Pityrogramma, Plagiobothrys acanthocarpus, P. bracteatus, P. canescens, P. collinus, P. leptocladus, P. stipitatus, Plantago bigelovii, P. coronopus, P. elongata, P. erecta, P. lanceolata, P. ovata, P. patagonica, Platanus racemosa, Pleurocoronis pluriseta, Poa annua, P. bulbosa, P. secunda, Polygonum bidwelliae, Porophyllum gracile, Primula clevelandii, Prosopis juliflora, P. velutina, Prunus fremontii, P. ilicifolia, Pseudognaphalium biolettii, P. californicum, Psilocarphus brevissimus, P. chilensis, P. oregonus, Psorothamnus schottii, Pterocarya stenoptera, Pterostegia drymarioides, Quercus ×acutidens, Q. agrifolia, Q. berberidifolia, Q. dumosa, Q. engelmannii, Q. john-tuckeri, Q. lobata, Q.

tomentella, Rafinesquia neomexicana, Raphanus sativus, Rhamnus crocea, R. ilicifolia, R. pirifolia, Rhus aromatica, R. integrifolia, R. ovata, Ribes aureum, R. quercetorum, R. speciosum, Riccia sorocarpa, Ricinus communis, Romneya trichocalyx, Rosa californica, Rostraria cristata, Rumex hymenosepalus, Sagina apetala, S. decumbens, S. maxima, Sairocarpus nuttallianus, Salix exigua, S. lasiolepis, Salvia apiana, S. leucophylla, S. mellifera, Sambucus nigra, Sanicula arguta, S. crassicaulis, Schismus barbatus, Scrophularia californica, Sedella pumila, Sedum album, S. lanceolatum, S. spathulifolium, Selaginella bigelovii, Senecio aphanactis, S. vulgaris, Senegalia greggii, Senna armata, Sidalcea malviflora, Silene gallica, Silybum marianum, Simmondsia chinensis, Sisyrinchium, Solanum douglasii, Sonchus oleraceus, Spergularia salina, Sphaeralcea ambigua, S. emoryi, Sporobolus airoides, Stellaria media, Stephanomeria pauciflora, Stylocline, Suaeda nigra, S. taxifolia, Symphoricarpos mollis, Toxicodendron diversilobum, Toxicoscordion fremontii, Trifolium albopurpureum, T. depauperatum, T. gracilentum, T. willdenovii, Triphysaria pusilla, Triteleia hyacinthina, Trixis californica, Typha, Uropappus lindleyi, Vaccinium ovatum, Vachellia constricta, Veronica peregrina, Vulpia bromoides, V. microstachys, V. myuros, V. octoflora, Xylococcus bicolor, Yucca schidigera, Ziziphus parryi.

Crassula solieri (Gay) **Meigen** occurs in flats, marshes, meadows, mudflats, playa, prairies, vernal pools, vernal streams, and along the shores of lakes, reservoirs, rivers, and streams at elevations of up to 2100 m. The substrates have been described as alkaline and include basalt, Capay silty clay, clay, and mud. Flowering occurs during the spring season. Additional ecological information for this species is scarce. **Reported associates:** Alopecurus saccatus, Atriplex coronata, Callitriche longipedunculata, C. verna, Castilleja densiflora, Crassula aquatica, Cressa, Deschampsia danthonioides, Downingia, Elatine brachysperma, E. chilensis, Eleocharis palustris, Erodium brachycarpum, Ernygium aristulatum, Frankenia salina, Hordeum intercedens, Isoetes orcuttii, Juncus bufonius, J. hemiendytus, Lasthenia californica, Lepidium dictyotum, Lilaea scilloides, Lythrum, Marsilea, Myosurus apetalus, M. minimus, Navarretia prostrata, Pilularia americana, Plagiobothrys stipitatus, Psilocarphus brevissimus, Ranunculus aquatilis, R. bonariensis, Sisyrinchium bellum, Veronica peregrina.

Use by wildlife: none reported.

Economic importance: food: not reported as edible; **medicinal:** none; **cultivation:** Crassula aquatica rarely is cultivated as an aquarium plant because of its annual habit. Crassula solierii is not found in cultivation; **misc. products:** none; **weeds:** Crassula helmsii, not yet known in North America, could become a serious aquatic weed in the region; **nonindigenous species:** none; however, the aquatic C. helmsii (indigenous to Australia and New Zealand) is marketed as an aquarium plant and should be watched for.

Systematics: Several molecular data sets (matK sequences, cpDNA restriction fragments), which represent up to 30 of the 35 recognized genera, strongly support a phylogenetic position of Crassula (including Tillaea) that is basal in Crassulaceae.

Past taxonomic treatments have advocated the recognition of Tillaea as being closely related to, but distinct from Crassula, to accommodate a number of reduced, mostly annual species with an affinity for wet habitats. However, an expanded phylogenetic analysis of approximately 45% of species in the genus using combined molecular data has indicated that the annual habit has evolved repeatedly within one major clade in the genus and also that species assigned to Tillaea are not particularly closely related and do not form a clade (Figure 4.52). Although a greater sampling of "Tillaea" species may provide additional insight, it is unlikely that increased taxon representation would alter these preliminary results. Consequently, the merger of Tillaea with Crassula has been followed in this treatment. Preliminary analyses (Figure 4.52) also show C. aquatica to be closely related to the Eurasian/African C. vaillantii, a similar aquatic annual. More definitive estimates of relationship await the inclusion of additional species, notably other North American taxa such as C. solierii. A number of varieties have been proposed for C. aquatica and C. connata; however, they are not well defined. The basic chromosome number of Crassula is $x = 7, 8$. Crassula aquatica ($2n = 42$) is a hexaploid (the related C. vaillantii [$2n = 28$] is tetraploid). Crassula connata ($2n = 16$) is a diploid. Counts are not available for C. solierii. There are no reported hybrids involving either C. aquatica or C. solierii. Studies of the population genetic structure of these aquatic annuals in North America would be beneficial, particularly because of their rare occurrences.

Comments: Crassula aquatica is widespread except for central portions of North America and also occurs throughout northern Eurasia; whereas, C. connata occurs in western North America and C. solierii is found in the western United States and extends southward into Baja California and Chile. Crassula aquatica and C. solierii are rare throughout most of their North American range.

References: Angoa-Román et al., 2005; Bauder, 2000; Bauder & McMillan, 1998; Boyd, 2012; Bywater & Wickens, 1984; Cody, 1954; Fassett, 1928; Ferren et al., 1998; Keeley, 1999; Keeley & Zedler, 1998; Mort et al., 2001; Murphy, 2009; Schlising & Sanders, 1982; Wainwright et al., 2012.

Family 2: Grossulariaceae [1]

Combined molecular data (Figure 4.51) resolve the small family Grossulariaceae (about 150 species) as the sister clade of Saxifragaceae, a group with which it often had been merged in older taxonomic treatments. The family consists of the single genus Ribes, which contains two subgenera (Grossularia, Ribes) known commonly as gooseberries and currants. Although these subgenera have been treated as distinct genera, recent molecular studies (Figure 4.53) indicate that subgenus Grossularia is monophyletic, but is nested within Ribes, with subgenus Ribes resolved as a paraphyletic grade (Messinger et al., 1999; Senters & Soltis, 2003; Schultheis & Donoghue, 2004). Grossulariaceae are characterized by their shrubby (sometimes prickly) habit, small flowers with scalelike petals, hypanthium and inferior ovary, and pulpy berries crowned by a persistent perianth. The flowers are insect

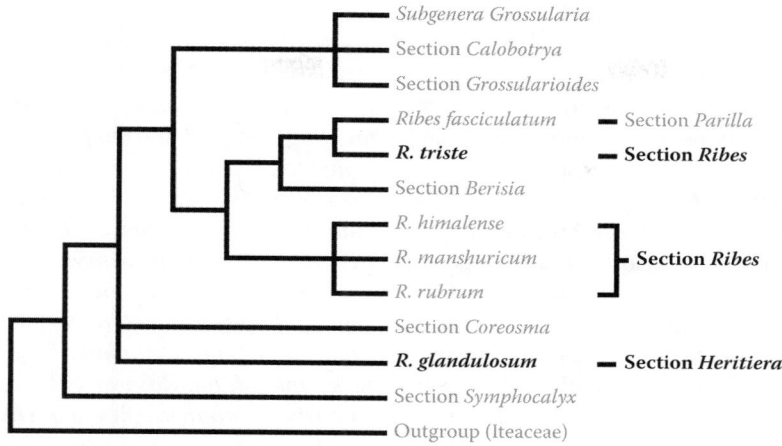

FIGURE 4.53 Phylogenetic relationships in *Ribes* as indicated by combined nuclear external transcribed spacer (ETS) and cpDNA (*psbA–trnH*) sequence data. Although more definitive conclusions regarding interspecific relationships will require the sampling of additional taxa, these results place the obligate aquatic *R. triste* (in bold) closest to *R. fasciculatum* of section *Parilla*. However, a study using nrITS sequence data (Senters & Soltis, 2003) placed *R. triste* as the sister species to *R. manshuricum*, which resolved remotely from *R. fasciculatum*. Such substantial inconsistencies must be clarified before any confident conclusions regarding the relationship of these species can be reached. *Ribes glandulosum* (section *Heritiera*) represents an independent origin of the OBL habit in this phylogeny. *Ribes hudsonianum* was not included in either study. (Modified from Schultheis, L.M. & Donoghue, M.J., *Syst. Bot.*, 29, 77–96, 2004.)

pollinated, and the fleshy fruits are dispersed by various animals, including many birds.

Ribes is of minor economic importance due to the tart, palatable berries of some species (commonly known as red or black currants), which are used in pastries and to make jelly and syrup. A number of species are grown horticulturally as ornamentals.

Ribes species occur mainly in open, UPL sites and have C_3 photosynthesis. However, a fairly large number can tolerate wetland conditions, at least to some degree. In North America, there are only a few obligate aquatics that occur within the single genus of this monotypic family:

1. ***Ribes*** L.

1. *Ribes*

Currant; Gadellier

Etymology: from the Persian *ribas* ("acid-tasting") in reference to the fruit

Synonyms: *Botrycarpum*; *Botryocarpium*; *Calobotrya*; *Cerophyllum*; *Chrysobotrya*; *Coreosma*; *Grossularia*; *Liebichia*; *Rebis*; *Ribesium*; *Rolsonia*

Distribution: global: Africa; Asia; Europe; New World; **North America:** widespread

Diversity: global: 150–200 species; **North America:** 55 species

Indicators (USA): OBL: *Ribes glandulosum*, *R. hudsonianum*, *R. triste*; **FACW:** *R. glandulosum*, *R. hudsonianum*; **FAC:** *R. hudsonianum*, *R. triste*; **FAC:** *R. glandulosum*, *R. hudsonianum*, *R. triste*

Habitat: freshwater; palustrine; **pH:** broad; **depth:** <1 m; **lifeform(s):** emergent shrub

Key morphology: stems erect (to 1 m) or reclining (the branches rooting freely), not spiny, the herbage sometimes with a fetid odor; leaves (to 15 cm) alternate, palmately lobed (3–7) and veined, petiolate (to 9.5 cm), the margins uni- or bicrenate to serrate; racemes erect (to 17 cm) or pendant (to 10 cm), bracts (to 3 mm) ovate to lanceolate or linear; flowers (6–50) epigynous, pedicelled (to 11 mm), 5-merous, hypanthium (to 1 mm) greenish, purple, or white; sepals pink or white and spreading (to 7 mm) or grayish-purple (to 3 mm), petals white (to 2 mm), pinkish (10 1.5 mm), or reddish-purple (to 1 mm), the nectary disc thin and pale green or prominent and reddish-purple or yellowish; berries dark red or black, waxy (to 12 mm) or red (to 10 mm).

Life history: duration: perennial (buds); **asexual reproduction:** rooting branches (layering); **pollination:** insect; **sexual condition:** hermaphroditic; **fruit:** berries (common); **local dispersal:** berries (mammals); **long-distance dispersal:** berries (birds)

Imperilment: (1) *Ribes glandulosum* [G5]; SH (OH); S1 (NJ); S2 (CT); S3 (MD, NC, VA, WV); (2) *R. hudsonianum* [G5]; S1 (IA); S2 (CA, UT, WY, QC); S3 (WI); (3) *R. triste* [G5]; S1 (CT, MT, NJ, WV); S2 (OH, PA); S3 (MA)

Ecology: general: Of the 55 or so *Ribes* species in North America, only three (5%) are categorized as OBL in some regions; the remainder principally includes terrestrial taxa of woodlands, meadows, grasslands, and slopes. Although *Ribes* species most often occur in UPL communities, their affinity for wetter sites is indicated by 21 of the North American species (38%) having some designation as wetland indicators, with many reported from swampy areas and along the margins of water courses. In most situations, the plants occur as nondominant understory shrubs. The flowers are either unisexual (then dioecious) or bisexual; all three OBL species are bisexual. The principal pollinators include bees (Insecta: Hymenoptera), butterflies (Insecta: Lepidoptera), flies (Insecta: Diptera), and hummingbirds (Aves: Trochilidae). The fleshy fruits are eaten and dispersed by a large variety of animals. Small mammals relish the seeds, but often destroy them in the process. Several *Ribes* species are self-incompatible (none of the obligate aquatics has been surveyed) and the reproductive biology and ecology of additional species deserve further study.

Ribes glandulosum **Grauer** inhabits beaches, bogs, cliffs, depressions, floodplains, heath, krummholz, meadows, mudflats, muskeg, ravines, riverbottoms, roadsides, seeps, shores, sloughs, swales, swamps, taiga, thickets, woodlands, and the margins of brooks, lakes, rivers, and streams at elevations of up to 2200 m. Occurrences have been reported in exposures ranging from open to densely shaded conditions. The substrates often are acidic (pH: 3.6–6.0) and have been described as basalt, clay, cobble, gravel, Lackawanna and Swartswood soils, limestone, quartzite talus, rocks, sand, sandstone sliderock, sandstone talus, sandy rock, sandy silt, silty till, and volcanic. The plants are in flower from May to June and their fruits mature from July to August. Although the reproductive ecology of this species has not been studied in any detail, the flowers presumably are pollinated (as in most *Ribes*) primarily by insects or hummingbirds (Aves: Triochilidae). The fruits ripen within 27.6 days on average. A seed bank reportedly exists, but no specific details on its duration are available. The fruits are dispersed locally by gravity and over longer distances by various frugivorous animals. Because the plants reproduce primarily by seed, they recover well following fires or clear-cut logging events and occasionally will dominate in wet, severely burned areas. Their frequency and percent cover declines markedly in sites that have not experienced fire in more than 26 years. Asexual reproduction can occur by the self-rooting of branches when they come in contact with the soil. The roots are non-mycorrhizal. **Reported associates:** *Abies balsamea, A. lasiocarpa, Acer pensylvanicum, A. rubrum, A. saccharum, A. spicatum, Aconitum reclinatum, Actaea podocarpa, Aesculus flava, Alnus rugosa, A. viridis, Amelanchier laevis, A. spicata, Aralia nudicaulis, Arctostaphylos uva-ursi, Aristolochia macrophylla, Avenella flexuosa, Betula alleghaniensis, B. lenta, B. neoalaskana, B. papyrifera, Calamagrostis canadensis, Calopogon tuberosus, Caltha palustris, Cardamine clematitis, Carex aestivalis, C. laxiflora, Cassiope, Chamedaphne calyculata, Chamerion angustifolium, Cladonia, Claytonia caroliniana, Clematis occidentalis, Clintonia borealis, Coptis trifolia, Cornus alternifolia, C. canadensis, C. rugosa, Corydalis sempervirens, Crataegus punctata, Dicranum polysetum, Diervilla lonicera, D. sessilifolia, Dryopteris campyloptera, D. expansa, D. intermedia, D. marginalis, Epigaea repens, Equisetum arvense, Erythronium americanum, Eurybia chlorolepis, E. macrophylla, Fagus grandifolia, Fallopia cilinodis, Fragaria virginiana, Fraxinus americana, F. nigra, Galium, Gaultheria hispidula, Gaylussacia baccata, Geranium erianthum, Grimmia, Gymnocarpium dryopteris, Halesia carolina, Huperzia lucidula, Hydrangea arborescens, Hylocomium splendens, Ilex montana, I. mucronata, Impatiens capensis, Juncus trifidus, Kalmia angustifolia, Larix laricina, Lathyrus, Linnaea borealis, Lonicera*

canadensis, L. involucrata, Maianthemum canadense, M. trifolium, Monotropa uniflora, Myrica pensylvanica, Oclemena acuminata, Oplopanax horridus, Oxalis montana, Parthenocissus quinquefolia, Phyllodoce, Picea engelmannii, P. glauca, P. mariana, P. rubens, Pinus banksiana, P. contorta, P. strobus, Piptatherum pungens, Poa pratensis, Pleurozium schreberi, Polypodium appalachianum, P. virginianum, Polytrichum commune, Populus balsamifera, P. tremuloides, Prunus pensylvanica, P. serotina, Pteridium aquilinum, Ptilidium ciliare, Pyrola chlorantha, Quercus rubra, Ranunculus, Rhododendron groenlandicum, Ribes cynosbati, R. hudsonianum, R. lacustre, R. oxyacanthoides, R. rotundifolium, Rosa acicularis, R. rugosa, Rubus allegheniensis, R. arcticus, R. canadensis, R. idaeus, R. parviflorus, R. pubescens, Salix scouleriana, Sambucus racemosa, Sarracenia purpurea, Sibbaldiopsis tridentata, Solidago canadensis, S. simplex, Sorbus americana, S. decora, Sphagnum angustifolium, Streptopus amplexifolius, S. lanceolatus, Thalictrum dasycarpum, Thuja occidentalis, Tilia americana, Trientalis borealis, Trillium erectum, T. undulatum, Tsuga canadensis, T. mertensiana, Umbilicaria, Vaccinium angustifolium, V. cespitosum, V. erythrocarpum, V. myrtilloides, V. oxycoccus, Viburnum edule, V. lantanoides, Woodsia ilvensis.

Ribes hudsonianum **Richardson** occurs in bogs, depressions, seeps, swamps, and along the margins of springs. The plants occur on acidic to calcareous substrates, usually in partial shade. They can survive low temperatures down to −20°C. The seeds have physiological dormancy and require 3–5 months cold stratification at 0°C–9°C for maximum germination (under heated greenhouse conditions). Unstratified seeds exhibited 57% germination compared to 85% when stratified. Seeds stored at 21°C retained 40% viability after 17 years. The glandular plants emit an odor that has been described as sweetish or even cat-like. **Reported associates:** *Abies balsamea, A. lasiocarpa, Acer rubrum, A. spicatum, Aconitum columbianum, Alnus rugosa, A. viridis, Amelanchier alnifolia, Astragalus miser, Athyrium filixfemina, Betula alleghaniensis, B. occidentalis, Calamagrostis canadensis, C. rubescens, C. stricta, Calypso bulbosa, Caltha palustris, Canadanthus modestus, Carex atrosquama, C. aurea, C. disperma, C. flava, C. gynocrates, C. interior, C. intumescens, C. leptalea, C. microptera, C. rossii, C. stipata, C. trisperma, C. utriculata, C. vaginata, Chamerion angustifolium, Chelone glabra, Chrysosplenium americanum, Cinna latifolia, Circaea alpina, Cirsium muticum, Clintonia borealis, Comarum palustre, Coptis trifolia, Cornus canadensis, C. sericea, Cypripedium calceolus, Diervilla lonicera, Dryopteris spinulosa, Equisetum arvense, E. fluviatile, E. scirpoides, E. sylvaticum, Eurybia macrophylla, Eutrochium maculatum, Fragaria virginiana, Fraxinus nigra, Galium triflorum, Gaultheria hispidula, Geocaulon lividum, Geum macrophyllum, Glyceria elata, G. grandis, Gymnocarpium, Habenaria, Halenia deflexa, Heracleum lanatum, Hylocomium splendens, Impatiens capensis, Juniperus communis, Larix laricina, Linnaea borealis, Listera cordata, Lonicera involucrata, L. oblongifolia, L. villosa, Luzula acuminata, Lycopodium,*

Maianthemum canadense, M. stellatum, M. trifolium, Mentha canadensis, Mertensia paniculata, M. perplexa, Micranthes odontoloma, M. pensylvanica, Mimulus guttatus, Mitella nuda, M. pentandra, Mnium, Moneses uniflora, Oplopanax horridus, Orthilia secunda, Persicaria amphibia, Petasites frigidus, Picea engelmannii, P. glauca, P. mariana, Pinus contorta, P. flexilis, P. strobus, Platanthera obtusata, Pleurozium schreberi, Poa pratensis, P. triflora, Populus tremuloides, Ptilium crista-castrensis, Pyrola asarifolia, P. chlorantha, Ranunuculus lapponicus, Rhamnus alnifolius, Rhododendron groenlandicum, Ribes glandulosum, R. hirtellum, R. lacustre, R. triste, Rosa, Rubus idaeus, R. parviflorus, R. pubescens, R. strigosus, Salix bebbiana, S. brachystachys, S. drummondiana, S. geyeriana, S. pseudomyrsinites, S. pseudocordata, Sambucus nigra, Scirpus microcarpus, Scutellaria galericulata, Senecio triangularis, Sparganium glomeratum, Sphagnum capillaceum, Stellaria calycantha, Streptopus amplexifolius, Symphoricarpos albus, S. oreophilus, Symphyotrichum puniceum, S. spathulatum, Thalictrum occidentale, Thuja occidentalis, Tsuga canadensis, Urtica dioica, Vaccinium oxycoccos, V. scoparium, V. vitis-idaea, Veratrum viride, Verbena verna, Veronica americana, Viola macloskeyi, V. renifolia.

***Ribes triste* Pall.** is found in bogs, floodplains, meadows, swamps, and along the margins of beaver ponds, creeks, springs, and streams at elvations of up to 1500 m. The substrates range from rich sandy loam to muck and can be slightly acidic (pH: 6.1–6.5) or calcareous. Ecological requirements are similar to *R. hudsonianum*, but broader, with the ability to occur also on well-drained sites; the two species are known to co-occur in wetter habitats. This is a straggly shrub whose lower branches produce adventitious roots along their length. **Reported associates:** *Abies balsamea, Acer rubrum, A. saccharum, A. spicatum, Allium tricoccum, Alnus rugosa, Anemone canadensis, Aralia nudicaulis, Betula lutea, B. papyrifera, Caltha palustris, Calypso bulbosa, Cardamine douglassii, Carex arctata, C. aurea, C. disperma, C. flava, C. interior, C. leptalea, C. leptonervia, C. stipata, C. trisperma, C. vaginata, Chelone glabra, Chrysosplenium americanum, Circaea alpina, Clintonia borealis, Coptis groenlandica, C. trifolia, Corallorhiza trifida, Cornus canadensis, Deschampsia cespitosa, Dryopteris cristata, Equisetum arvense, E. scirpoides, E. sylvaticum, Erythronium americanum, Eurybia macrophylla, Floerkea proserpinacoides, Fraxinus nigra, Gaultheria hispidula, Glyceria striata, Goodyeara repens, Gymnocarpium dryopteris, Habenaria, Larix laricina, Linnaea borealis, Listera cordata, Lonicera dioica, Maianthemum canadense, M. trifolium, Mertensia paniculata, Mitella nuda, Moneses uniflora, Orthilia secunda, Osmundastrum cinnamomeum, Oxalis montana, Panax trifolia, Picea glauca, P. mariana, Populus balsamifera, P. tremuloides, Pterites pensylvanica, Pyrola chlorantha, Quercus macrocarpa, Q. rubra, Rhamnus alnifolius, Ribes americanum, R. cynosbati, R. glandulosum, R. hudsonianum, R. hirtellum, R. lacustre, Rosa acicularis, Rubus pedatus, R. pubescens, Salix amygdaloides, S. discolor, Sarracenia purpurea, Sphagnum, Stellaria calycantha,* *Streptopus amplexifolius, S. roseus, Symplocarpus foetidus, Taxus canadensis, Thuja ocidentalis, Thelypteris phegopteris, Tholurna dissimilis, Tilia americana, Trientalis borealis, Trillium cernuum, T. grandiflorum, Ulmus americana, Veronica americana, Viola blanda.*

Use by wildlife: *Ribes* plants provide important food and cover for various wildlife species. *Ribes glandulosum* is eaten by black bears (Mammalia: Ursidae: *Ursus americanus*), moose (Mammalia: Cervidae: *Alces alces*), snowshoe hares (Mammalia: Leporidae: *Lepus americanus*), and white-tailed deer (Mammalia: Cervidae: *Odocoileus virginianus*). The plants are host to larval butterflies and moths (Insecta: Lepidoptera: Nymphalidae: *Polygonia gracilis*; Geometridae: *Xanthotype urticaria*) as well as many fungi including Ascomycota (Dermateaceae: *Drepanopeziza ribis*; Dothideomycetidae: *Dothidea ribesia*; Erysiphaceae: *Sphaerotheca mors-uvae*; Helotiaceae: *Godronia cassandrae, G. davidsonii*; Mycosphaerellaceae: *Mycosphaerella grossulariae*), Basidiomycota (Cronartiaceae: *Cronartium ribicola* [white pine blister rust]; Melampsoraceae: *Melampsora epitea*; Pucciniaceae: *Puccinia caricina, P. caricis*), and Oomycota (Peronosporaceae: *Plasmopara ribicola*). *Ribes hudsonianum* plants provide nesting habitat and cover in addition to food (berries) for an assortment of birds (Aves) including American robins (Turdidae: *Turdus migratorius*), black-billed magpies (Corvidae: *Pica hudsonia*), Bohemian waxwings (Bombycillidae: *Bombycilla garrulus*), hermit and varied thrushes (Turdidae: *Catharus guttatus, Ixoreus naevius*), pine grosbeaks (Fringillidae: *Pinicola enucleator*), ruffed grouse (Phasianidae: *Bonasa umbellus*), Steller's jays (Corvidae: *Cyanocitta stelleri*), and white-crowned sparrows (Emberizidae: *Zonotrichia leucophrys*). *Ribes hudsonianum* is browsed by moose (Mammalia: Cervidae: *Alces alces*) and its berries eaten by bears (Mammalia: Ursidae: *Ursus*) and other animals. It is also a host plant for several fungi including Ascomycota (Botryosphaeriaceae: *Phyllosticta ribesicida*; Erysiphaceae: *Phyllactinia corylea*; *Sphaerotheca mors-uvae*; Mycosphaerellaceae: *Mycosphaerella grossulariae*; *Ramularia*) and Basidiomycota (Cronartiaceae: *Cronartium ribicola* [white pine blister rust]; Pucciniaceae: *Puccinia caricina*). *Ribes triste* is known to be a host plant for Lepidoptera (Insecta) in several families (Geometridae: *Eupithecia miserulata*; Nymphalidae: *Polygonia gracilis*; Oecophoridae: *Agonopterix antennariella*). It is also the host for various fungi including Ascomycota (Dermateaceae: *Drepanopeziza ribis*), Basidiomycota (Cronartiaceae: *Cronartium ribicola* [white pine blister rust]; Pucciniaceae: *Puccinia caricina, P. ribis*), and Oomycota (Peronosporaceae: *Plasmopara ribicola*).

Economic importance: food: The fruits of *R. glandulosum* contain 78.5% water, 232.2 g/kg mean total sugars (19.1 g/kg glucose; 20.0 g/kg fructose; 5.99 g/kg sucrose), and an average citric acid content of 17.4 g/kg. Their juice has a mean total phenolic compound content of 408 g/kg and possesses antioxidant and free radical scavenging properties. The fruits were eaten by the Algonquin, Cree, Quebec, and Woodlands tribes. The Cree and Woodlands people brewed a bitter tea from the stems. Although the plants are malodorous, the fruits can be made into a tasteful jelly. The berries of *R. hudso-nianum* are bitter but edible. Both *R. hudsonianum* and *R. triste* were gathered as food or made into jam by the Cree, Gwich'in, Montana, Ojibwa, Secwepemc, Shuswap, Tanana, Thompson, Woodlands, and Yukon people. The Salish and Coast tribes boiled the berries and dried them into rectangular cakes. Berries of *R. triste* were eaten by the Alaska, Chippewa, Eskimo, Inupiat, Iroquois, Ojibwa, and Tanana tribes. They are acidic in flavor and their high pectin content makes them excellent for preparing jellies; **medicinal:** The Chippewa people administered a compound root decoction of *R. glandulosum* for treating back pain and "female weakness." A stem decoction was used by the Cree and Woodlands tribes to prevent blood clotting after childbirth. The Cree, Ojibwa, Tanana, Thompson, and Woodlands tribes ate the berries or administered decoctions from the bark, leaves, roots, and stems of *R. hudsonianum* as a general remedy for colds, sore throats, stomach problems, or other general ailments. The Thompson also made a medicine from the roots as a treatment for tuberculosis and placed sprigs of the plants in cribs as a sedative to quiet infants. Leaves, roots, and stems of *R. triste* were also used medicinally (as an eye wash or treatment for gynecological and urinary problems) by the Chippewa, Ojibwa, and Tanana tribes; **cultivation:** *R. glandulosum* has been distributed under the cultivar name 'Dart's Coverboy.' *Ribes hudsonianum* sometimes is cultivated for native gardens. None of the OBL species should be planted near white pine (Pinaceae: *Pinus stobus*) because they are intermediate hosts for blister rust (Fungi: Basidiomycota: Cronartiaceae: *Cronartium ribicola*); **misc. products:** *Ribes hudsonianum* has been recommended as a native plant to control streambank erosion; **weeds:** none; **nonindigenous species:** none.

Systematics: *Ribes* is the only genus "currantly" recognized in Grossulariaceae (see family treatment earlier). Consequently, the monophyly of the family and genus is well established. However, a standard infrageneric classification of *Ribes* has not been established and various treatments recognize quite different species groupings. No phylogenetic study yet has included all three OBL species in a single analysis. Data from cpDNA RFLP analysis indicated *Ribes hudsonianum* to be closely related to the Eurasian *R. nigrum* (both assigned to section *Coreosma*). Results of nrETS/cpDNA sequence analyses show *R. triste* (section *Ribes*) to be more closely related to *R. fasciculatum* (section *Parilla*) than to other members of section *Ribes* (Figure 4.53). However, nrITS sequence data are in major conflict by placing *R.* triste with *R. manshuricum* while distant from *R. fasciculatum*. Unfortunately, these anomalous results, as well as those from combined nrITS/ETS data (adjacent nuclear loci), which place *R. triste* with *R. aureum* (section *Symphocalyx*), inspire little confidence in the relationships depicted using these markers. *Ribes glandulosum* (section *Heritiera*) resolves as distinct from *R. triste* (Figure 4.53), which at least suggests the independent derivation of the OBL habit in these species. Additional clarification on the phylogenetic relationships among the OBL species is needed. The basic chromosome number of *Ribes* is $x = 8$. *Ribes hudsonianum* and *R. triste* ($2n = 16$) are diploid. Although

there appear to be only weak crossing barriers among many of the *Ribes* species, natural hybridization is rare in the genus, and no hybrids have been reported that involve any of the OBL species. However, little specific information is known regarding their pollination biology and reproductive ecology. Care should be taken not to confuse *R. glandulosum* Grauer with *R. glandulosum* Ruiz & Pavon, the latter being a synonym of *R. ruizii* Rehder and often appearing in the literature simply as "*Ribes glandulosum*."

Comments: *Ribes glandulosum* extends across northern North America and into the eastern United States. *Ribes hudsonianum* and *R. triste* occur throughout northern portions of North America, with the former extending further south in the western part of its range. *Ribes triste* also occurs in northeastern Asia.

References: Ahlgren, 1960; Anderson et al., 2005; Arasu, 1970; Atwood, 1941; Baskin & Baskin, 1998; Bell et al., 2011; Burn & Friele, 1989; De Grandpré et al., 1993; Dodds, 1960; Harris et al., 2014; Hébert et al., 2008; Humbert et al., 2007; Légaré et al., 2001; LeResche & Davis, 1973; Malloch & Malloch, 1982; McDonald & Andrews, 1981; Messinger et al., 1999; Mikulic-Petkovsek et al., 2012; Morin, 2009; Murray et al., 2005; Offord et al., 1944; Schultheis & Donoghue, 2004; Simon & Schwab, 2005; Sinnott, 1985; Uprety et al., 2015; Viereck, 2010; Wein & Freeman, 1995; Weigend et al., 2002; Zielinski, 1953.

Family 3: Haloragaceae [8]

Haloragaceae are recognized here as the sister group to Penthoraceae (Figure 4.51). Some authors prefer to include both families along with Tetracarpaeaceae and Aphanopetalaceae in a broader (i.e., *sensu lato*) concept of Haloragaceae. However, doing so would create a highly irregular assemblage of species while either circumscription is compatible with the results of molecular phylogenetic analyses. Molecular (DNA sequence) data have greatly helped to clarify the circumscription of Haloragaceae and also the delimitation of its genera (Moody & Les, 2002; Moody, 2004). The family is small (approximately 120 species) yet extremely diverse with species that range from diminutive terrestrial trees to submerged herbaceous aquatic plants. The highly variable features found in the family make it difficult to circumscribe morphologically. Generally, Haloragaceae have dichasial inflorescences and flowers whose hooded, keeled petals are folded conduplicately, stamens that are twice the number of petals or sepals, free styles, and an inferior ovary. However, even these features can be highly reduced. Pollination is typically by wind (anemophily), although the flowers of some species also attract insects.

Few members of this family are of major economic importance. Ornamental cultivars of *Haloragis* include 'Wanganui Bronze' and 'Wellington Bronze.' Several *Myriophyllum* species are grown as water garden or aquarium plants; however, some have escaped cultivation to become notorious aquatic and wetland weeds.

Four genera of this cosmopolitan family are entirely terrestrial (*Glischrocaryon, Gonocarpus, Haloragis, Trihaloragis*)

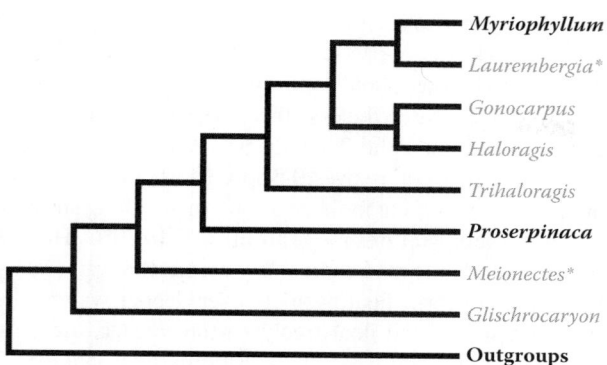

FIGURE 4.54 Intergeneric relationships in Haloragaceae as indicated by phylogenetic analysis of combined DNA sequence data. Genera that contain obligate aquatic species in North America are shown in bold. Genera marked by an asterisk are not North American and are also principally aquatic. These results indicate the aquatic habit to have evolved several times within the family. (Adapted from Moody, M.L., Systematics of the angiosperm family Haloragaceae R. Br. emphasizing the aquatic genus *Myriophyllum*: phylogeny, hybridization and character evolution, PhD dissertation, University of Connecticut, 2004; Moody, M.L. & Les, D.H., *Amer. J. Bot.*, 94, 2005–2025, 2007.)

and four contain aquatics (*Laurembergia*, *Meionectes*, *Myriophyllum*, *Proserpinaca*). Only the latter two genera are represented in North America with obligate aquatic species occurring in both:

1. *Myriophyllum* L.
2. *Proserpinaca* L.

DNA sequence analyses (Moody, 2004; Moody & Les, 2007) indicate *Myriophyllum* as a relatively derived group in the family, with *Proserpinaca* resolved in a more intermediate position among the other genera (Figure 4.54).

1. *Myriophyllum*

Water milfoil; myriophylle

Etymology: from the Greek *myrios phyllon* meaning "countless leaved" with respect to the pinnately dissected foliage
Synonyms: *Burshia*; *Enydria*; *Vinkia*
Distribution: global: cosmopolitan; **North America:** widespread
Diversity: global: 60 species; **North America:** 14 species
Indicators (USA): OBL: *Myriophyllum alterniflorum, M. aquaticum, M. farwellii, M. heterophyllum, M. hippuroides, M. humile, M. laxum, M. pinnatum, M. quitense, M. sibiricum, M. spicatum, M. tenellum, M. ussuriense, M. verticillatum*
Habitat: brackish, freshwater; lacustrine, palustrine, riverine; **pH:** 4.8–9.9; **depth:** 0–5 m; **life-form(s):** emergent herb, submersed (vittate)
Key morphology: stems variable, naked (to 2.5 cm) or leafy (to >7 m), submersed and flexuous or emergent and erect, rooting adventitiously, turions (to 2 cm) present or absent; leaves (to 5 cm) scattered or in whorls (of 3–6), pinnately compound (3–24 pairs of filiform segments); flowers reduced, bisexual

or unisexual (plants dioecious or monoecious), axillary or on emergent spikes (to 3.7 dm), foliose bracts (to 3 cm) present or absent; stamens 4 or 8; schizocarps (to 2.5 mm) splitting into four, 1-seeded mericarps

Life history: duration: perennial (dormant apices, rhizomes, turions); **asexual reproduction:** rhizomes, shoot fragments, turions (stem); **pollination:** cleistogamous (self), wind; **sexual condition:** hermaphroditic, dioecious, monoecious; **fruit:** schizocarps (common); **local dispersal:** seeds, vegetative propagules (water; waterfowl); **long-distance dispersal:** seeds (waterfowl)

Imperilment: (1) *Myriophyllum alterniflorum* [G5]; SH (RI); S1 (CT, MA, <u>SK</u>); S2 (<u>MB</u>, MI, NY, VT); S3 (NH); (2) *M. aquaticum* (GNR); (3) *M. farwellii* [G5]; SH (NH); S1 (AK, <u>MB</u>, MA, <u>NF</u>, PA); S2 (MI, <u>NB</u>, <u>NS</u>, NY, VT); S3 (<u>BC</u>, <u>QC</u>, WI); (4) *M. heterophyllum* [G5]; S1 (DE, IA, MD, <u>NB</u>, OH, PA); S2 (KS, NJ, <u>QC</u>); S3 (KY, NC, WV); (5) *M. hippuroides* [G5]; S3 (<u>BC</u>); (6) *M. humile* [G5]; S1 (<u>NB</u>, <u>QC</u>, VA); S2 (VT); S3 (<u>NS</u>); (7) *M. laxum* [G3]; S1 (MS, NC, VA); S2 (AL, GA, SC); S3 (FL); (8) *M. pinnatum* [G5]; SH (KY); S1 (<u>BC</u>, CT, IA, IN, NE, NJ, NY, RI, <u>SK</u>, TN, WV); S2 (DE, NC, ND, NM); S3 (MA, VA); (9) *M. quitense* [G4]; S1 (WY); S2 (<u>BC</u>); (10) *M. sibiricum* [G5]; S1 (CT, NJ, PA); S2 (KS, <u>LB</u>, <u>NB</u>, <u>NF</u>, OH, WV); S3 (<u>NS</u>, OR, WY); (11) *M. spicatum* (GNR); (12) *M. tenellum* [G5]; SH (MD); S1 (IN, <u>LB</u>, NC, NJ, VA); S2 (PA); S3 (CT, <u>NB</u>, <u>QC</u>); (13) *M. ussuriense* [G3]; S1 (OR); S3 (<u>BC</u>); (14) *M. verticillatum* [G5]; SH (NJ, <u>PE</u>, WV); S1 (CO, IA, NE, OH, PA, UT, VT, WY); S2 (IN, <u>NF</u>, <u>NS</u>, VT); S3 (AK, <u>AB</u>, <u>SK</u>, <u>QC</u>)

Ecology: general: *Myriophyllum* is entirely aquatic (all North American species are designated as OBL) and comprises species that are diverse both ecologically and morphologically. North American species can be separated ecologically into a "northern" group (e.g., *M. alterniflorum, M. farwellii, M. tenellum*) and "widespread" group (e.g., *M. heterophyllum, M. sibiricum, M. spicatum, M. verticillatum*). The former have a more boreal distribution and possess leaves with low mass, volume, and surface area, but with high surface area:volume ratios. These adaptations enable the plants to thrive in the oligotrophic, acidic to circumneutral, low calcium waters, and sandy, nutrient-poor substrates that commonly exist at higher latitudes. The latter are distributed more broadly (not restricted to northern latitudes) and have higher leaf mass, volume, and surface areas and relatively lower surface area:volume ratios. These species are better adapted for mesotrophic to eutrophic, alkaline, high-calcium waters, and nutrient-rich clay or silt substrates. The *Myriophyllum* species that occupy harder waters (e.g., *M. aquaticum, M. spicatum*) often are able to use bicarbonate ions directly as a carbon source. Those species common to waters of intermediate hardness (e.g., *M. verticillatum*) or soft waters lack this ability. Presumably, the species with flowers on emergent spikes are wind pollinated. The lower (♀) flowers are protogynous, maturing slightly before the upper (♂) flowers. However, flower production in some species (*M. farwellii, M. humile*) is axillary and occurs entirely underwater. A few species are dioecious. Facilitation of pollen transfer may also involve insects (aphids, bees, etc.), which have been observed on the emergent flowers of some species. Overall, the pollination biology of *Myriophyllum* remains poorly under-stood and is in need of critical study. Plants growing in cal-careous water often are coated by "marl" (calcium carbonate), which precipitates on the foliage as a consequence of photo-synthetic activity. Growth of plants in highly alkaline water can become limited by iron and manganese, which precipitate as hydroxides at high pH. Many species possess hydathodes on their leaves and stems, which facilitate nutrient uptake from the water column. Diminutive "mud-flat" forms can occur in several species (e.g., *M. alterniflorum, M. aquaticum, M. heterophyllum, M. humile, M. pinnatum, M. spicatum, M. verticillatum*) when the plants become stranded on exposed shorelines (often due to water-level fluctuations). Compared to other aquatic plants, most *Myriophyllum* species contain very high levels of polyphenolics, which are known to deter some herbivores.

Myriophyllum alterniflorum **DC** occurs in cool, clear waters at depths to 4.5 m in northern lakes and streams at elevations of up to 500 m. The plants can tolerate fairly swift currents and are found in oligotrophic to mesotrophic, cir-cumneutral waters (pH: 6.0–8.8; mean 7.4) with low alka-linity (<90 mg/L CaCO$_3$) and conductivity (<190 μS/cm). This species is intolerant to turbidity and generally occurs where conditions provide at least 14% of the surface illumi-nation. Substrates typically are nutrient-poor sands some-times mixed with muck. The plants are monoecious with their flowers borne in simple, emergent spikes. Seed pro-duction occurs but its relative role in plant persistence is not well understood. This species persists and reproduces vegetatively by means of rhizomes. The plants also disperse by stem fragments, which can remain afloat for more than 3 days. **Reported associates:** *Ceratophyllum demersum, Chara, Eleocharis acicularis, E. palustris, Elodea canaden-sis, Equisetum fluviatile, Eriocaulon aquaticum, Gratiola aurea, Hippuris vulgaris, Isoetes macrospora, Juncus pelocarpus, Littorella americana, Lobelia dortmanna, Megalodonta beckii, Myriophyllum sibiricum, M. tenel-lum, M. verticillatum, Najas flexilis, Nuphar rubrodisca, Nymphaea odorata, Potamogeton amplifolius, P. epihydrus, P. gramineus, P. natans, P. obtusifolius, P. praelongus, P. pusillus, P. richardsonii, P. robbinsii, P. spirillus, P. stric-tifolius, Ranunculus reptans, R. trichophyllus, Sagittaria cuneata, S. graminea, Schoenoplectus acutus, Sparganium angustifolium, Stuckenia pectinata, Utricularia, Vallisneria americana.*

Myriophyllum aquaticum **(Vell.) Verdc.** is a nonindig-enous species, which inhabits still or slow-flowing, fresh, intertidal, or nontidal waters of bayous, canals, ditches, dunes, lakes, marshes, sloughs, swamps, and the shores of lakes and ponds at elevations of up to 1500 m. It is found on various sub-strates (e.g., mud, peaty muck, sandy gravel, sandy loam, silty loam, silty clay loam) that tend to be high in organic matter. Occurrences typically are in open sites but exposures of par-tial shade are tolerated. Plants typically grow in alkaline (pH: 5.2–9.0; median: 7.5), hardwater (median calcium: 25.3 mg/L; median alkalinity [CaCO$_3$]: 24 mg/L; median specific conduc-tivity: 126 μS/cm), nutrient-rich conditions (median nitrogen:

840 μg/L). They can tolerate waters with fairly high turbidity (median Secchi depth: 0.8 m) because the submersed foliage generally is shed as photosynthesis occurs primarily by the emersed shoots. Nevertheless, the emergent portions of the plants require high light. This species cannot tolerate brack-ish water (i.e., salinities over 0.5 ppt). Most often it is found growing as an emergent in the shallow waters along shorelines where water temperatures range from 10°C to 29°C. However, its stems can also spread extensively across the water surface and are able to form floating mats over deeper water. These mats can dislodge and float freely on the surface. Stems and leaves are heavily cutinized and are highly hydrophobic. The flowers are axillary (but always emergent) and this is the only milfoil species in North America that is dioecious. Monoecious individuals (male flowers above) have been observed rarely, but not in North America. Only the female plants have been introduced outside of its native range (where male plants are rare). Fruits are not produced in North America where repro-duction is entirely asexual. Vegetative reproduction occurs by fragmentation/rooting of the stems. This species is capable of using bicarbonate as a carbon source. Plant growth is highest in shallow water (<0.5 m) and can be limited by nitrogen (early in the season) or phosphorous (late in the season). Adventitious roots facilitate the uptake of nutrients from the water col-umn. **Reported associates (North America):** *Alternanthera philoxeroides, Ammophila arenaria, Azolla caroliniana, Bidens, Brasenia schreberi, Cabomba caroliniana, Carex, Cephalanthus occidentalis, Ceratophyllum demersum, Egeria densa, Eichhornia crassipes, Elodea canadensis, Epilobium, Eupatorium, Hydrilla verticillata, Hydrocotyle ranunculoi-des, Hypericum, Iris pseudacorus, Isoetes, Juncus effusus, Leersia oryzoides, Lemna minor, L. minuta, Limnobium spongia, Ludwigia hexapetala, L. peploides, Myriophyllum heterophyllum, M. laxum, Nymphaea, Panicum hemitomon, Persicaria amphibia, P. hydropiperoides, Phalaris arundi-nacea, Pistia stratiotes, Pontederia, Potamogeton nodosus, Potentilla, Salix hookeriana, Salvinia, Sparganium emersum, Spirodela punctata, Stuckenia pectinata, Taxodium distichum, Typha latifolia, Utricularia macrorhiza, Wolffia columbiana.*

Myriophyllum farwellii **Morong** is an uncommon plant of shallow waters (but up to 5 m depth) of northern bog ponds, ditches, lakes, river backwaters, slow streams, and oxbows at elevations of up to 600 m. The habitats mainly are acidic (pH: 4.9–8.5; mean 6.4) with low alkalinity (<80 mg/L CaCO$_3$) and conductivity (<180 μS/cm). The bottom sediments can range from highly organic muck, mud, or peat to rocky, gravelly, and sandy substrates. The plants are completely submersed. The axillary flowers are usually perfect and produce fruits below the water surface. Their method of pollination is uncertain. Cleistogamous selfing has been hypothesized as one possible explanation (given the proximity of anthers and stigmas in bud), but the reproductive biology of this species requires study and further clarification. Vegetative reproduc-tion occurs by small turions that are produced at the shoot extremities. Other information on the ecology of this spe-cies is scarce. **Reported associates:** *Brasenia schreberi, Ceratophyllum demersum, C. echinatum, Elatine minima,*

Eleocharis acicularis, E. palustris, Elodea canadensis, Eriocaulon aquaticum, Heteranthera dubia, Isoetes, Juncus pelocarpus, Lemna triscula, Littorella americana, Lobelia dortmanna, Megalodonta beckii, Myriophyllum heterophyllum, M. sibiricum, M. spicatum, M. tenellum, Najas flexilis, Nitella, Nymphaea odorata, Nuphar variegata, Persicaria amphibia, Potamogeton amplifolius, P. confervoides, P. diversifolius, P. epihydrus, P. foliosus, P. friesii, P. gramineus, P. natans, P. oakesianus, P. pusillus, P. richardsonii, P. robbinsii, P. zosteriformis, Riccia fluitans, Sagittaria graminea, Schoenoplectus subterminalis, Sparganium angustifolium, S. chlorocarpum, S. fluctuans, Sphagnum, Stuckenia pectinata, Utricularia gemniscapa, U. gibba, U. intermedia, U. macrorhiza, U. purpurea, U. resupinata, Vallisneria americana, Zizania aquatica.

***Myriophyllum heterophyllum* Michx.** occurs in relatively clear waters (mean Secchi depth: 2.4 m) at depths to 5 m in ditches, lakes, marshes, ponds, rivers, sloughs, and streams at elevations of up to 600 m. This species grows under a broad range of ecological conditions that can vary substantially from one region to another [e.g., pH: 4.7–7.5; mean 6.0 (Florida); 6.5–7.8; mean: 6.9 (New England); 5.0–9.0; mean: 7.9 (Wisconsin); mean $CaCO_3$ alkalinity: 130 mg/L (Wisconsin); 18.0 mg/L (New England); 9.7 mg/L (Florida); mean conductivity: 220 μS/cm (Wisconsin); 112 μS/cm (Florida)]. It is often found on highly organic substrates (muck, silt). Emergent, flowering, leafy-bracted spikes are produced when the plants are growing in calm, shallow water. Little information exists on the pollination/reproductive biology of this species. The upper stems often are dilated by air-space tissue, which facilitates their flotation. Vegetative reproduction occurs by means of shoot fragmentation and rhizome production. Rhizomatous winter buds have also been reported. Growth is rapid when conditions are favorable, and plants frequently develop into dense, monospecific stands. Rhizosphere bacterial colonies promote nutrient (nitrogen) uptake by nitrification processes. Dislodged fragments deposited along exposed shorelines frequently develop into diminutive mudflat plantlets that are resistant to dessication. These plants can revert to the typical submersed form if inundated by rising water levels. There is some confusion in the literature as a result of hybridization involving this species (see *Systematics* section later). **Reported associates:** *Alisma subcordatum, Brasenia schreberi, Cabomba caroliniana, Callitriche heterophylla, Ceratophyllum demersum, C. echinatum, Echinodorus rostratus, Egeria densa, Eleocharis acicularis, Elodea canadensis, E. nuttallii, Eriocaulon aquaticum, Fontinalis, Glossostigma cleistanthum, Hydrilla verticillata, Hypericum boreale, Isoetes lacustris, Leersia hexandra, L. oryzoides, Lemna minor, Myriophyllum alterniflorum, M. aquaticum, M. humile, M. laxum, M. sibiricum, M. spicatum, M. verticillatum, Najas flexilis, N. guadalupensis, Nitella flexilis, Nuphar advena, N. variegata, Nymphaea odorata, Nymphoides cordata, Persicaria amphibia, P. coccinea, Pontederia cordata, Potamogeton amplifolius, P. bicupulatus, P. epihydrus, P. foliosus, P. gramineus, P. illinoensis, P. natans, P. pulcher, P. pusillus, P. richardsonii, P. robbinsii, P. zosteriformis,* *Ranunculus trichophyllus, Sagittaria latifolia, Sclerolepis uniflora, Stuckenia pectinata, Utricularia geminiscapa, U. gibba, U. intermedia, U. macrorhiza, U. purpurea, U. radiata, U. striata, Vallisneria americana.*

***Myriophyllum hippuroides* Nutt. ex Torr. & A. Gray** grows submersed in shallow waters of ditches, lakes, marshes, ponds, sloughs, streams, and vernal pools at elevations of up to 1800 m. The substrates usually are described as muck or mud. Habitat data are scarce, but plants have been cultured in circumneutral water (pH: 6.5–7.5) at temperatures from 18°C to 28°C. Flowers are produced on emergent spikes. Vegetative reproduction is by stem fragmentation and rhizome proliferation. Although there is not much information published on the ecology of this species, it appears to be similar to that of *M. heterophyllum* (see earlier), to which it is very closely related (see *Systematics* section below). **Reported associates:** *Brasenia schreberi, Callitriche stagnalis, Ceratophyllum demersum, Chara, Egeria densa, Eleocharis ovata, Elodea canadensis, E. nuttallii, Fontinalis antipyretica, Heteranthera dubia, Isoetes, Juncus planifolius, Lemna minor, L. trisulca, Ludwigia palustris, Myriophyllum sibiricum, M. spicatum, M. ussuriense, Najas flexilis, N. guadalupensis, Nitella, Nuphar polysepala, Nymphaea odorata, Oenanthe sarmentosa, Persicaria amphibia, P. minor, Phalaris arundinacea, Potamogeton amplifolius, P. epihydrus, P. foliosus, P. gramineus, P. illinoensis, P. natans, P. nodosus, P. praelongus, P. pusillus, P. richardsonii, P. zosteriformis, Ranunculus aquatilis, Sagittaria rigida, Sparganium emersum, Spirodela polyrhiza, Stuckenia pectinata, Utricularia inflata, U. macrorhiza, Vallisneria americana, Zannichellia palustris.*

***Myriophyllum humile* (Raf.) Morong** is found in shallow, softwater lakes, ponds, streams, and their shorelines at elevations of up to 700 m. It has a broad pH tolerance (pH: 4.9–8.6) but typically occupies acidic sites (e.g., pH: 6.3) with low to moderate alkalinity (to 110 mg/L $CaCO_3$) and conductivity (<200 μS/cm). The substrates vary from sand to mud and peat. This species is extremely variable and occurs as a diminutive mudflat form on exposed shorelines, as a completely submersed form, or as an intermediate form with submersed shoots and an emergent inflorescence. As in *M. farwellii* (see earlier), fruit production in *M. humile* is enigmatic. The flowers have been described variously as perfect or (less frequently) unisexual (the plants then monoecious). Regardless, they typically occur on submersed stems below the water surface where fruiting takes place. Pollination possibly occurs by cleistogamous selfing, but the floral biology of this species requires further investigation. Occurrences of *M. humile* are associated negatively with the presence of *M. spicatum* and indicate habitats not particularly suitable for colonization by that invasive species. **Reported associates:** *Brasenia schreberi, Callitriche, Ceratophyllum, Eclipta prostrata, Elatine minima, Eleocharis acicularis, E. obtusa, E. robbinsii, Elodea nuttallii, Eriocaulon aquaticum, Glossostigma cleistanthum, Gratiola aurea, Isoetes echinospora, Juncus militaris, J. pelocarpus, Lindernia dubia, Ludwigia palustris, Myriophyllum heterophyllum, Najas flexilis, N. gracillima, N. minor, Nymphaea odorata, Nuphar variegata, Potamogeton* *bicupulatus, P. epihydrus, P. oakesianus, P. pusillus, P. zosteriformis, Ranunculus flabellaris, Rhynchospora capitellata, Sagittaria graminea, Schoenoplectus subterminalis, Utricularia fibrosa, U. gibba, U. macrorhiza, U. purpurea, U. resupinata, Vallisneria americana.*

Myriophyllum laxum Shuttlw. ex Chapm. inhabits the shallow waters of backwaters, beaver ponds, blackwater streams, brooks, canals, depressions, ditches, floodplains, lakes, marshes, ponds, pools, sinkholes, sloughs, and springs at elevations of up to 150 m. The sites tend to be clear, cool (spring-fed), and oligotrophic with sandy substrates and flowing water. The flowers are produced on emergent spikes but details on the reproductive ecology are lacking. Vegetative reproduction presumably involves stem fragmentation and rhizome production. There is very little ecological information available for this species, despite the fact that it is imperiled (S1–S3) throughout its entire distributional range. Existing reports should be viewed cautiously, given that this species is likely to be misidentified frequently. **Reported associates:** *Chamaecyparis thyoides, Lilaeopsis carolinensis, Litsea aestivalis, Nuphar advena, N. ulvacea, Nymphaea odorata, Nymphoides aquatica, Panicum hemitomon, Potamogeton floridanus, Rhynchospora crinipes, Sagittaria subulata, Utricularia floridana, U. olivacea, U. purpurea, Websteria confervoides, Xyris smalliana.*

Myriophyllum pinnatum (Walter) Britton, Sterns & Poggenb. grows submersed in shallow waters or emergent along the margins of borrow pits, canals, ditches, lakes, marshes, ponds, pools, sloughs, streams, and swamps at elevations of up to 700 m. Its habitats tend to be fairly acidic (pH: 4.1–7.3). The plants usually grow on substrates that have a high organic matter content (e.g., muck, mud, peat) and can also be found on rocky loam or sand. They tolerate only low salinity (up to 0.4%) but can withstand water temperatures up to 25°C and require high light levels. Hermaphroditic (rare) or unisexual flowers are produced on emergent stems. They are axillary and subtended by leaf-like bracts or extend down to the axils of normal leaves. Fruiting is common. The plants reproduce vegetatively by stem fragmentation and rhizome proliferation. Mud-flat forms ranging from dwarf to fairly large plants occur along open shoreline sites. The list of associates indicates that this species grows more commonly among emergent wetland plants than do most of the other milfoils. Additional ecological information on this species is scarce. **Reported associates:** *Acer rubrum, Alisma subcordatum, Alternanthera philoxeroides, Amorpha fruticosa, Baccharis halimifolia, Bacopa caroliniana, B. monnieri, Brasenia schreberi, Bolboschoenus fluviatilis, Cabomba caroliniana, Cephalanthus occidentalis, Ceratophyllum demersum, Crinum americanum, Didiplis diandra, Egeria densa, Eleocharis elongata, E. equisetoides, E. olivacea, E. robbinsii, Elodea canadensis, E. nuttallii, Eriocaulon compressum, Gratiola virginiana, Heteranthera dubia, Hibiscus, Hottonia inflata, Hydrilla verticillata, Hydrocotyle umbellata, H. verticillata, Hymenocallis rotata, Hypericum boreale, Iris virginica, Iva frutescens, Juncus pelocarpus, Kosteletzkya pentacarpos, Limosella aquatica, Ludwigia palustris, L. polycarpa, Lysimachia terrestris,*

Marsilea mutica, Micranthemum umbrosum, Myosotis laxa, Myrica cerifera, Nasturtium officinale, Nelumbo lutea, Nitella flexilis, Nuphar orbiculata, Nymphaea odorata, Nymphoides aquaticum, Nyssa biflora, Osmunda regalis, Panicum anceps, P. virgatum, Penthorum sedoides, Persicaria hirsuta, P. hydropiperoides, P. punctata, Phragmites australis, Pluchea camphorata, P. foetida, Pontederia cordata, Potamogeton diversifolius, P. epihydrus, P. nodosus, Proserpinaca pectinata, Rhynchospora scirpoides, Rumex verticillatus, Sabal minor, Sabatia campanulata, Sacciolepis striata, Sagittaria lancifolia, Salix nigra, Samolus parviflorus, Schoenoplectus etuberculatus, Solidago sempervirens, Sparganium americanum, Sphagnum, Spirodela polyrhiza, Stuckenia pectinata, Taxodium distichum, Thalia geniculata, Utricularia gibba, U. macrorhiza, U. purpurea, Veronica scutellata, Wolffia columbiana, Xyris smalliana, Zannichella palustris, Zizaniopsis miliacea.

Myriophyllum quitense Kunth is a submersed inhabitant of cold, clear, oligotrophic waters (to depths of 2.5 m) of high elevation backwaters, beaches, lakes, rivers, streams, and swamps to elevations of 4800 m. The substrates are generally described as gravel, mud, sand, or silt. The pH range of this species is uncertain but appears to be primarily alkaline. The plants are known to thrive in clear, highly alkaline (pH: 10.0–10.4), calcareous waters, where their foliage can become coated by marl deposits. Unlike most other milfoils, this species sometimes produces multiple floral spikes and flowers later in the season (July to August). Typically, the spikes are emergent and monoecious (rarely with bisexual transitional flowers) with the male flowers occurring above the females. Flowers and seeds are produced rarely. Terrestrial forms are known (South America), which possess perfect flowers and can develop into extensive colonies along shores that become exposed by receding water levels. The shoots can withstand dessication for prolonged periods of time (at least for several weeks). Vegetative reproduction occurs by shoot fragmentation and by the spread of the highly branched rhizomes. Vegetative winter buds sometimes are produced, but plants will often overwinter as intact evergreen shoots. **Reported associates:** *Ceratophyllum demersum, Elodea, Hippuris vulgaris, Isoetes occidentalis, Lobelia dortmanna, Myriophyllum sibiricum, M. verticillatum, Nuphar polysepala, Potamogeton illinoensis, P. pusillus, P. richardsonii, Ranunculus reptans, Spaganium angustifolium, S. fluctuans, Stuckenia filiformis, Utricularia intermedia.*

Myriophyllum sibiricum Kom. occurs submersed (at depths to 4.2 m) in brackish or freshwater sites including beaver ponds, channels, ditches, gravel bars, gravel pits, lakes, mudflats, ponds, potholes, rivers, sloughs, and streams at elevations of up to 3300 m. Its habitats are primarily alkaline (pH: 5.5–9.8; mean: 7.8) and are typically characterized by high alkalinity (to 376 mg/L $CaCO_3$) and conductivity (to 600 μS/cm). The plants are not tolerant to turbidity and generally occur in clear water. The substrates are described as clay, gravel, marl, muck, mud, organic, sand, silt, and silty muck. Marl precipitates on the foliage in highly calcareous water. Vegetative reproduction occurs from shoot fragments (which quickly develop adventitious roots) and by dispersal of turions, which form in the fall and persist upon the decay of the attached stem. A characteristic vertical and unbranched growth habit allows for adequate light penetration to enable the survival of other submersed species. Although laboratory experiments indicate a superior preemptive competitive ability over *M. spicatum*, the ability of the latter to branch at the water surface eventually reduces light levels, causing the decline of the former and other submersed species. *Myriophyllum sibiricum* is distributed mainly north of the mean January 0°C isotherm. Axenic laboratory culture of plants has been successful using modified Andrew's culture medium (pH: 5.8). **Reported associates:** *Alisma gramineum, Alopecurus aequalis, Aphanizomenon flos-aquae, Beckmannia syzigachne, Bidens beckii, Bolboschoenus fluviatilis, Brasenia schreberi, Calla palustris, Callitriche hermaphroditica, C. verna, Caltha natans, Carex aquatilis, C. lenticularis, C. saxatilis, C. subspathacea, C. vesicaria, Ceratophyllum demersum, C. echinatum, Chara globularis, C. vulgaris, Dichelyma uncinatum, Drepanocladus, Egeria densa, Eleocharis acicularis, E. palustris, Elodea canadensis, Elymus repens, Enteromorpha, Equisetum fluviatile, E. variegatum, Eriocaulon aquaticum, Heteranthera dubia, Hippuris tetraphylla, H. vulgaris, Iris pseudacorus, Isoetes echinospora, I. lacustris, Juncus balticus, J. pelocarpus, Lemna minor, L. trisulca, Lobelia dortmanna, Menyanthes trifoliata, Myriophyllum alterniflorum, M. farwellii, M. heterophyllum, M. hippuroides, M. quitense, M. spicatum, M. tenellum, M. verticillatum, Najas flexilis, Nitella flexilis, Nuphar advena, N. polysepala, N. variegata, Nymphaea odorata, N. tuberosa, Persicaria amphibia, Potamogeton alpinus, P. amplifolius, P. berchtoldii, P. crispus, P. epihydrus, P. foliosus, P. friesii, P. gramineus, P. illinoensis, P. natans, P. nodosus, P. perfoliatus, P. pusillus, P. praelongus, P. richardsonii, P. robbinsii, P. spirillus, P. strictifolius, P. subsibiricus, P. vaseyi, P. zosteriformis, Ranunculus aquatilis, R. flammula, R. longirostris, R. trichophyllus, Rumex maritimus, Sagittaria cuneata, S. graminea, S. latifolia, Schoenoplectus acutus, S. subterminalis, Sium suave, Sparganium angustifolium, S. chlorocarpum, S. eurycarpum, S. minimum, Spirodela polyrhiza, Stuckenia filiformis, S. pectinata, S. vaginata, Tephroseris palustris, Typha latifolia, Utricularia cornuta, U. geminiscapa, U. intermedia, U. macrorhiza, U. minor, Vallisneria americana, Volvox, Wolffia, Zannichellia palustris.*

Myriophyllum spicatum L. is a submersed, nonindigenous plant, which grows at depths up to 8 m in brackish (≤20 ppt once established) or fresh waters in many habitats including bayous, canals, lakes, mudflats, oxbows, ponds, and along the margins of gravel pits, lakes, and rivers at elevations of up to 1500 m. An optimal depth of 2.0 m (maximum biomass production) has been determined in relatively clear, circumneutral lakes. This species can tolerate a wide range of environmental conditions (e.g., pH: 5.4–11.0); however, its habitats generally are alkaline (pH>7.0) with high alkalinity (to 300 mg/L $CaCO_3$) and conductivity (to 610 µS/cm). The plants usually occur in eutrophic systems and are tolerant to turbidity. An increase in frequency has been observed at sites that have undergone eutrophication. Most exposures occur in full sunlight. The substrates have been described as clay, gravel, mucky sand, sand, silt, and silty sand. *Myriophyllum spicatum* is a bicarbonate user with its maximum photosynthetic rate occurring at pH = 8.5–9.0. The plants are capable of withstanding very warm temperatures, theoretically even prolonged exposures up to 35°C. This species does not produce turions or rhizomes but overwinters by means of dormant apical meristems that persist at the base of the plant when the shoots have died back, or by persisting as intact plants throughout the winter. As the plants grow and elongate, primary nutrient uptake shifts from the roots (sediment) to the shoots (water). Shoot uptake is facilitated by production of water roots and by hydropoten, which occur at the bases and tips of the leaves and pinnae. Sediment nitrogen (e.g., <50 mg/kg) can be limiting. Higher values enable the shoots to reach the surface of shallow lakes, where they can spread across the surface to shade out native submersed species. The unisexual flowers are produced on emergent, monoecious spikes (the ♂ flowers above) and are protogynous; yet self-pollination (geitonogamy) occurs frequently. Seeds can be produced in quantity and are able to remain afloat for up to a day. They are dispersed by water or by waterfowl (Aves: Anatidae). The seeds either exhibit prolonged dormancy (which can be broken by scarification or stratification) or they can germinate without any after-ripening treatment. In one study, seeds collected in the fall, and sterilized using 0.625% sodium hypochlorite, germinated readily in sterile tap water (22°C–25°C). Other studies report the germination of fresh seeds at temperatures above 10°C. Cold stratification of seeds for 120 days is recommended for optimal germination when brought subsequently to a temperature of 20°C. Germination is enhanced by white light, but is inhibited by blue light. Coverage of seeds by 2 cm of sediment can reduce their germination by as much as 30%. Dried seeds can retain their viability for at least 7 years. Yet, recruitment from seedlings rarely has been observed in natural populations and deserves further study. In deep water, vegetative reproduction occurs by the layering of older stems, which arch over and root apically. Fragmentation facilitates dispersal along the shoreline, especially during periods of water level fluctuation. Fragments stranded on newly exposed substrates will root and develop into semiterrestrial mud-flat plantlets. These will develop into typical, elongate stems as normal water levels are restored. By a combination of layering and fragmentation, this species can colonize littoral habitats from both lakeward and shoreward directions, leading eventually to complete monospecific stands. Establishment of plants is thwarted where lake bottoms are covered by well-established populations of native species (especially dense *Chara* beds); consequently, establishment often occurs under disturbed conditions, or where habitats are unsuitable for sustaining healthy native species. The plants grow rapidly once they have been introduced to a lake. Field studies have shown that stands initiated from small (10 cm) fragments can reach carrying capacity in a lake by the end of the next year subsequent to their establishment. **Reported associates**

(North America): *Beckmannia syzigachne, Bidens beckii, B. cernuus, Cabomba caroliniana, Callitriche hermaphroditica, C. verna, Ceratophyllum demersum, Chara, Cyperus bipartitus, Eichhornia crassipes, Eleocharis palustris, Elodea canadensis, E. nuttallii, Equisetum fluviatile, Heteranthera dubia, Hippuris vulgaris, Hydrilla verticillata, Hydrocotyle ranunculoides, Juncus supiniformis, Lemna minor, L. minuta, L. obscura, L. trisulca, Limnobium spongia, Ludwigia, Lythrum salicaria, Myriophyllum heterophyllum, M. hippuroides, M. sibiricum, Najas flexilis, N. guadalupensis, N. marina, N. minor, Nelumbo lutea, Nitella tenuissima, Nuphar advena, N. polysepala, N. variegata, Nymphaea odorata, Peltandra virginica, Pontederia cordata, Potamogeton amplifolius, P. berchtoldii, P. crispus, P. foliosus, P. friesii, P. gramineus, P. illinoensis, P. natans, P. nodosus, P. perfoliatus, P. pulcher, P. pusillus, P. richardsonii, P. robbinsii, P. strictifolius, P. zosteriformis, Ranunculus aquatilis, R. flabellaris, R. flammula, R. longirostris, R. sceleratus, Rumex maritimus, Ruppia maritima, Sagittaria cuneata, S. latifolia, S. platyphylla, Salix exigua, Schoenoplectus acutus, Sium suave, Sparganium angustifolium, S. emersum, Spirodela polyrrhiza, Stuckenia pectinata, Trapa natans, Typha latifolia, Utricularia macrorhiza, U. minor, Vallisneria americana, Veronica anagallis-aquatica, Zannichellia palustris.*

Myriophyllum tenellum **Bigelow** is found submersed in fresh (or brackish) waters at depths to 3.8 m in beach pools and lakes, or in shallow water along the exposed shorelines of lakes and ponds at elevations of up to 900 m. The surrounding waters are oligotrophic, circumneutral (pH: 5.1–8.6), and are characterized by low alkalinity (<70 mg/L CaCO$_3$) and conductivity (<150 μS/cm). This species is an indicator of relatively pristine conditions and occurs where the water is fairly clear and provides at least 5.9% of the surface illumination. The sediments are gravel, mud, peat, or sand. The plants can tolerate some salinity (to 1200 mg/L chloride). They do not withstand any appreciable burial by sediments. *Myriophyllum tenellum* lacks leaves and usually occurs as a submersed, sterile form. In very shallow water (e.g., 5 cm depth), where the stems are able to extend above the surface, aerial flowering spikes (up to 30 cm) sometimes are produced. Here again the flowers are unisexual (♂ above, ♀ below), the latter sometimes reduced to one. More often, the plants reproduce vegetatively by vigorous rhizome proliferation, which can result in dense, turf-like mats of intertwined plants that are several meters in extent. In lakes subject to disturbance (e.g., power boating), huge mats may become dislodged and can appear on the surface as floating islands of plants. The roots oxidize the substrate, which enhances the availability of iron and phosphorous. Invasions of *M. spicatum* have resulted in serious declines of this species. **Reported associates:** *Agrostis scabra, Bidens beckii, B. cernuus, Brasenia schreberi, Carex, Ceratophyllum demersum, C. echinatum, Chara globularis, C. zeylanica, Cladium mariscoides, Drepanocladus fluitans, Drosera intermedia, Dulichium arundinaceum, Elatine minima, E. triandra, Eleocharis acicularis, E. palustris, E. robbinsii, E. smallii, Elodea canadensis, E. nuttallii, Equisetum fluviatile, Eriocaulon aquaticum,*

Glossostigma cleistanthum, Glyceria borealis, G. canadensis, Gratiola aurea, Heteranthera dubia, Hypericum boreale, H. canadense, H. majus, Isoetes echinospora, I. lacustris, I. macrospora, I. muricata, Juncus canadensis, J. effusus, J. pelocarpus, Lemna trisulca, Linum striatum, Littorella americana, Lobelia dortmanna, Ludwigia palustris, Lycopus uniflorus, Lysimachia terrestris, Myriophyllum alterniflorum, M. farwellii, M. sibiricum, M. spicatum, M. verticillatum, Najas flexilis, Nuphar variegata, Nymphaea odorata, Panicum acuminatum, Persicaria amphibia, P. careyi, Pontederia cordata, Potamogeton alpinus, P. amplifolius, P. epihydrus, P. gramineus, P. natans, P. obtusifolius, P. pusillus, P. praelongus, P. richardsonii, P. robbinsii, P. spirillus, Rancunculus flammula, R. reptans, R. trichophyllus, Rhexia virginica, Rhynchospora capitellata, Sagittaria cuneata, S. graminea, S. latifolia, Schoenoplectus acutus, S. pungens, S. smithii, S. subterminalis, Scutellaria galericulata, Sparganium angustifolium, S. chlorocarpum, S. eurycarpum, S. fluctuans, Stuckenia pectinata, Subularia aquatica, Typha latifolia, Utricularia cornuta, U. gibba, U. intermedia, U. macrorhiza, U. minor, U. purpurea, U. resupinata, Vallisneria americana, Viola lanceolata, Xyris difformis.

Myriophyllum ussuriense **(Regel) Maxim.** is a submersed plant that grows in freshwater or brackish, nontidal, or intertidal localities that are characterized by fluctuating water levels, including depressions, floodplains, and the margins of lakes, rivers, and sloughs at elevations of up to 600 m. The plants are tolerant to daily tidal inundation. The natural pH range of the plants has not been surveyed, but they grow well at slightly acidic values (pH: 6.0–7.0). The substrates include mud, sand, sandy mud, and silt. Like *M. aquaticum*, this species is primarily dioecious although some hermaphroditic flowers are occasionally observed and monoecious plants have been found in Japan. Female populations occur most commonly in North America. This species also produces a diminutive mud-flat form on exposed shores. Vegetative reproduction is by rhizomes or apically produced turions. The aggressive spread of the introduced *Schoenoplectus triqueter* represents a potentially serious threat to this species. **Reported associates:** *Alisma triviale, Bidens cernuus, Callitriche stagnalis, Ceratophyllum demersum, Crassula aquatica, Elatine rubella, Eleocharis acicularis, E. palustris, Elodea canadensis, Equisetum arvense, E. fluviatile, Gratiola ebracteata, G. neglecta, Hypericum boreale, Juncus bulbosus, J. nevadensis, J. oxymeris, Lilaeopsis occidentalis, Limosella aquatica, Lindernia dubia, Ludwigia palustris, Menyanthes trifoliata, Mimulus guttatus, Myosotis scorpioides, Myriophyllum hippuroides, M. spicatum, Oenanthe sarmentosa, Persicaria amphibia, P. hydropiperoides, P. minor, Potamogeton, Ranunculus flammula, R. reptans, Rumex obtusifolius, Sagittaria latifolia, Samolus floribundus, Schoenoplectus americanus, S. triqueter, Sium suave, Sparganium emersum, Subularia aquatica, Symphyotrichum subspicatum, Veronica beccabunga.*

Myriophyllum verticillatum **L.** grows submersed at depths to 4.0 m in fresh waters of beaver ponds, bog pools, ditches, lakes, mudflats, ponds, potholes, reservoirs, rivers, sloughs, and streams at elevations of up to 2700 m. This species has a broad pH tolerance (pH: 5.2–9.9; mean = 7.7) but occurs frequently in habitats with high alkalinity (to 260 mg/L CaCO₃) and conductivity (to 580 µS/cm). The substrates include marl, muck, peat, sand, and silt. The plants usually grow in calm water or in slow currents where the clarity is high enough to provide at least 10% of the surface illumination. This species is unable to use bicarbonate. The flowers (hermaphrodite or unisexual) are produced on emergent spikes in the axils of foliaceous (but reduced) bracts. Seed production occurs often but is of minor importance for year-to-year persistence. Vegetative reproduction occurs by means of rhizomes and stem fragments, or by stem turions, which are produced as the water temperatures fall below 15°C and day length decreases to 8–12 h. When attached, the turions are not dormant but require a long day length (16 h) to elongate. Abscised turions are dormant initially, but germinate after they have been exposed to cold water temperatures (0°C–4°C) for 17–60 days. After 30 days of low temperatures, the turions are able to germinate even in the dark conditions of lake bottoms, often while still under a cover of ice. The turions gain buoyancy as they germinate and elongate. As in *M. sibiricum*, the mature stems are not highly branched. Reduced, terrestrial forms are produced when shoot fragments become stranded on exposed shorelines. The plants reportedly can survive cold temperatures down to −15°C. They alter the redox potential (and nutrient availability) of sediments by releasing oxygen from their roots. **Reported associates:** *Alisma, Bidens beckii, Brasenia schreberi, Calla palustris, Caltha natans, Carex utriculata, Ceratophyllum demersum, Chara vulgaris, Damasonium californicum, Decodon verticillatus, Drepanocladus exannulatus, Eleocharis acicularis, E. palustris, Elodea canadensis, E. nuttallii, Epilobium ciliatum, Equisetum fluviatile, Gratiola aurea, Heteranthera dubia, Hippuris vulgaris, Hydrocharis morsus-ranae, Isoetes echinospora, I. macrospora, Juncus pelocarpus, Lemna gibba, L. minor, L. trisulca, Littorella americana, Lobelia dortmanna, Ludwigia palustris, Menyanthes trifoliata, Mimulus glabrata, Myriophyllum alterniflorum, M. heterophyllum, M. quitense, M. sibiricum, M. tenellum, Najas flexilis, Nasturtium officinale, Nuphar polysepala, N. variegata, Nymphaea leibergii, N. odorata, N. tetragona, N. tuberosa, Persicaria amphibia, Phragmites australis, Potamogeton amplifolius, P. crispus, P. epihydrus, P. friesii, P. gramineus, P. natans, P. obtusifolius, P. praelongus, P. pusillus, P. richardsonii, P. robbinsii, P. spirillus, P. strictifolius, P. zosteriformis, Ranunculus gmelinii, R. reptans, R. trichophyllus, Rorippa palustris, Ruppia cirrhosa, Sagittaria cuneata, S. graminea, Schoenoplectus acutus, S. subterminalis, S. tabernaemontani, Sparganium angustifolium, S. chlorocarpum, S. fluctuans, S. minimum, Spirodela polyrhiza, Stuckenia filiformis, S. pectinata, Typha latifolia, Utricularia gibba, U. intermedia, U. macrorhiza, Vahlodea atropurpurea, Vallisneria americana, Wolffia borealis, W. columbiana, Zannichellia palustris, Zizania.* **Use by wildlife:** The complex leaves and branching stems of *Myriophyllum* plants provide valuable underwater habitat and often are correlated positively with the presence of

crustaceans (amphipods, copepods), gastropods, insect larvae (e.g., Chironomidae, Ephemeroptera, Odonata, Trichoptera), and periphyton. Several species (e.g., *M. farwellii, M. heterophyllum, M. hippuroides*) are important as cover plants for fish and also for many invertebrates, which are eaten by fish. *Myriophyllum heterophyllum* hosts an abundance of mayfly larvae (Insecta: Ephemeroptera), which graze the algae that occur on the foliage. *Myriophyllum humile* provides habitat for rotifers (e.g., Rotifera: Notommatidae: *Cephalodella asarcia*) and other freshwater invertebrates. A copepod (Crustacea: Daphniidae: *Daphnia magna*) uses *M. sibiricum* for cover when fish are present. *Myriophyllum verticillatum* plants also harbor large numbers of rotifers and crustaceans. A flea beetle (Insecta: Coleoptera: Chrysomelidae: *Lysathia ludoviciana*) occasionally uses *M. aquaticum* as a larval host plant. Two moths (Insecta: Lepidoptera: Tortricidae: *Argyrotaenia ivana, Choristoneura parallela*) have been observed on this species, but their interactions remain unknown. The leaves of *M. aquaticum* are mined by caterpillars (Lepidoptera: Pyralidae: *Parapoynx allionealis*). *Myriophyllum sibiricum* is a host plant for two aquatic weevils (Insecta: Curculionidae: *Euhrychiopsis, Phytobius leuogaster*). *Myriophyllum spicatum* is a host plant for several Lepidoptera (Pyralidae: *Parapoynx allionealis, P. badiusalis, P. diminutalis*). Both *M. spicatum* and *M. sibiricum* are hosts for *Acentria nivea* (Lepidoptera: Pyralidae). Herbivory by *Acentria ephemerella* (Lepidoptera: Pyralidae) has led to the decline of *M. spicatum* populations in New York, but the high levels of polyphenolic compounds found in most milfoil species generally deter herbivory by these larvae. The North American "water milfoil" weevil (Insecta: Coleoptera: Curculionidae: *Euhrychiopsis lecontei*) feeds preferentially on *M. spicatum* (vs. *M. sibiricum*) and is being researched for biological control programs. The weevil also appears to be less selective for hybrids between these species. Decomposition of milfoil plants releases nitrogen and phosphorous, which promote the growth of microflora communities. Plants of *M. spicatum* provide food for many aquatic animals as they rot over winter. Decaying leaves of *M. verticillatum* are associated with nematodes (e.g., Nematoda: Plectidae: *Plectus cirratus*). A fungus (Chytridiomycota: Cladochytriaceae: *Cladochytrium replicatum*) has been found in decaying cells of this species. *Myriophyllum ussuriense* is a host for a leaf spot fungus (Blastocladiomycota: Physodermataceae: *Physoderma myriophylli*). *Myiophyllum heterophyllum* is a colonization substrate for algae (e.g., Chlorophyta: Aphanochaetaceae: *Aphanochaete confervicola*; Charophyta: Zygnemataceae: *Zygnema*) and other epiphytic periphyton. *Myriophyllum laxum* is among the food plants preferred by grass carp (Osteichthyes: Cyprinidae: *Ctenopharyngodon idella*) in Florida lakes. It is also an important cover species for the least killifish (Osteichthyes: Poeciliidae: *Heterandria formosa*). Grass carp feed variably on *M. verticillatum*. *Myriophyllum pinnatum* and *M. spicatum* are not a preferred food source of grass carp or its hybrids. *Myriophyllum tenellum* has been identified as an important component of habitat for the banded killifish (Osteichthyes: Fundulidae: *Fundulus diaphanus*).

Myriophyllum heterophyllum is important habitat for eared grebe (Aves: Podicipedidae: *Podiceps nigricollis*) and its seeds and shoots are eaten by waterfowl (Anatidae) and other birds (Aves). Seed production in most water milfoil species is quite low; however, the seeds (especially those of *M. exalbescens, M. heterophyllum, M. pinnatum, M. sibiricum, M. tenellum, M. verticillatum*) are relished by waterfowl when available. Twenty-one different duck species (Aves: Anatidae), mainly mallard (*Anas platyrhynchos*), lesser scaup (*Aythya affinis*), and green-winged and blue-winged teal (*Anas crecca, A. discors*), are known to consume them, sometimes in high numbers; e.g., 7760 mericarps observed in the stomach of a single ring-necked duck (*Aythya collaris*). The foliage and fruits of *M. farwellii* are also eaten occasionally by waterfowl. Reports from outside of North America indicate that *M. aquaticum* and *M. quitense* occasionally are eaten by cattle (Mammalia: Bovidae: *Bos*) and *M. aquaticum* is grazed by various waterfowl. Analyses of *M. sibericum* and *M. spicatum* show them to be fairly high in protein (20%–25%) and fiber (10%–13%), although crude protein content is lower in *M. aquaticum* (14.1%) and *M. heterophyllum* (8.5%–13.5%). Also, *M. aquaticum* contains over 10% tannins, which reduces the digestibility of proteins and decreases its palatability to many herbivores.

Economic importance: food: The shoots of *Myriophyllum aquaticum* are eaten as a vegetable in Java. In North America, the Tanana tribes froze the rhizomes of *M. sibiricum* as a reserve food to be eaten fried, raw, or roasted. *Myriophyllum verticillatum* is eaten as a pot herb in some countries; **medicinal:** The Iroquois obtained medicine (to improve circulation) and a strong emetic from infusions of *M. sibiricum*. The Menominee also used the plant in preparing several general medicines. The Iroquois administered a decoction made from *M. verticillatum* as a stimulant for lethargic children; **cultivation:** *Myriophyllum aquaticum* is very popular as a water garden plant because of its attractive, emergent growth habit. Many milfoil species (*M. heterophyllum, M. hippuroides, M. humile, M. pinnatum, M. spicatum, M. ussuriense, M. verticillatum*) appear for sale as aquarium plants. However, these names often are associated with plants that have been misidentified (e.g., *M. aquaticum* commonly appears in catalogues as "*M. verticillatum*") and it is difficult to tell which species are being distributed. 'Red stem' is a milfoil cultivar of uncertain affinity; **misc. products:** *Myriophyllum quitense* has been considered for use as a water quality biomonitor; **weeds:** *Myriophyllum aquaticum* is a serious weed, especially along shallow lake and river margins. It usurps native habitat and species, shades underwater sites, and impedes navigation and recreational activities. *Myriophyllum heterophyllum* (and its hybrid with *M. laxum*) can grow so dense as to impede navigation and recreational activities. Infestations have been linked to reduced property values for lakefront homes. Although native, *M. hippuroides* has been reported as a pest in California irrigation canals and *M. humile* as a weed in Massachusetts. *Myriophyllum laxum* sometimes has been reported as weedy; however, such accounts may be due to erroneous identifications with *M. heterophyllum* (with

which it hybridizes) given the rarity and specialized habitat of this species. *Myriophyllum spicatum* is ranked among the worst aquatic weeds in North America and is found in nearly every state and province in the region. Millions of dollars are spent annually in attempts to control or eradicate it. Hybrids between *M. spicatum* and the native *M. sibiricum* have been discovered relatively recently in the northern United States (extending from Michigan to Washington state) and are at least as invasive as their nonindigenous parent. Preliminary field surveys indicate that *M. spicatum* rarely coexists in lakes containing hybrid plants. There is also evidence that the weevils (Insecta: Coleoptera: Curculionidae: *Euhrychiopsis lecontei*) used for biocontrol of *M. spicatum* are less specific for the hybrid plants, creating the possibility that this method of management could lead to selection for hybrids having equal or greater invasive qualities. *Myriophyllum verticillatum* is a troublesome weed of canals in Ireland where it is native; **nonindigenous species:** In every known case, introductions of water milfoil species have resulted from careless disposal of cultivated aquarium, pond, or water garden specimens. *Myriophyllum aquaticum* was introduced to North America in the late 19th-century (Washington, DC area) and spread to the west coast by the 1940s. *Myriophyllum heterophyllum* is native to much of eastern North America, but it (along with hybrids) was introduced to the northeastern United States during the 1930s, presumably as aquarium plants. *Myriophyllum heterophyllum* also was introduced to Europe, where it was discovered in Britain in the 1940s and to Austria and Germany during the 1970s. *Myriophyllum spicatum* was introduced to North America in the 1940s (again in the Washington, DC area) and likely escaped cultivation from plantings in fish culture ponds.

Systematics: Phylogenetic analyses of combined DNA sequence data (Figure 4.54) resolve *Myriophyllum* as the sister genus to *Laurembergia*. A comprehensive survey of *Myriophyllum* using combined molecular data has identified two major clades in the genus, each representing relatively heterogeneous geographical affinities. The results (Figure 4.55) show North American species to occur in both clades and among several subclades within those groups. Seven North American species (*M. farwellii, M. heterophyllum, M. hippuroides, M. humile, M. laxum, M. pinnatum, M. tenellum*) comprise a monophyletic group of endemics whose closest relatives are Australian (Figure 4.55). *Myriophyllum ussuriense* (North America, Asia) originates from a different clade as the sister species to the Australian *M. crispatum*. *Myriophyllum sibiricum* occurs in yet another clade along with the nonindigenous *M. spicatum* (its sister species) and *M. alterniflorum*, whose closest relatives are also Australian. A fourth clade includes the primarily South American *M. quitense*, and *M. triphyllum* from Australia/New Zealand. The widespread *M. verticillatum* comprises a clade with the Asian *M. oguraense*, its sister species. The nonindigenous *M. aquaticum* associates with *M. mattagrossensis* in a South American clade. Phylogenetic analyses indicate that the diversity of *Myriophyllum* species in North America is a consequence of repeated dispersal to the region, mainly from Australia.

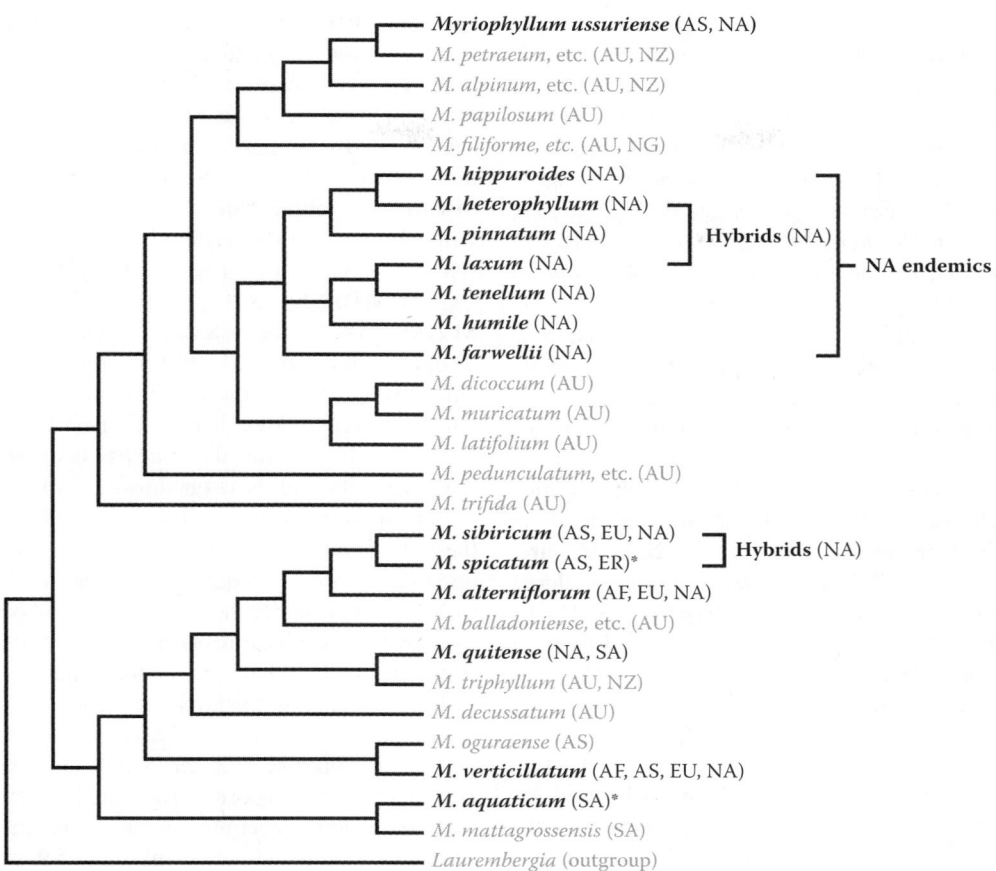

FIGURE 4.55 Relationships among *Myriophyllum* species as indicated by phylogenetic analyses of combined molecular data. All *Myriophyllum* species are obligate aquatics. Species occurring in North America are highlighted in bold (the clade of endemics is indicated). Species nonindigenous to North America are marked by "*." Parents of the two known North American hybrids are identified by brackets. These results indicate multiple origins of *Myriophyllum* species present in the North American flora, primarily from Australian progenitors. It is evident that long-distance dispersal of *Myriophyllum* species, probably by means of migrating waterfowl, has strongly influenced the distributional history of the genus (geographical regions are abbreviated as: AF = Africa, AS = Asia, AU = Australia, EU = Europe, NA = North America, NG = New Guinea, NZ = New Zealand; SA = South America). (Adapted from Moody, M.L., Systematics of the angiosperm family Haloragaceae R. Br. emphasizing the aquatic genus *Myriophyllum*: phylogeny, hybridization and character evolution, PhD dissertation, University of Connecticut, 2004.)

This pattern indicates a series of natural long-distance dispersal events (probably due to transport of propagules via migrating birds) from Australia through eastern Asia and into western North America. The taxonomic status of *M. sibiricum* remains somewhat unresolved. Although most recent taxonomic treatments have merged the former *M. exalbescens* (North America) with this species, molecular data indicate that some divergence has occurred between North American and European populations. Difficulties in distinguishing *M. sibiricum* from *M. spicatum* in North America have been exacerbated by bidirectional hybridization of these species in northern North America. Leaves of the native *M. sibiricum* characteristically have less than 12 pairs of segments, whereas, those of *M. spicatum* possess more than 13 segment pairs. The difference in pinnae number reliably distinguishes these species in the absence of hybrids. However, *M. sibiricum* × *M. spicatum* hybrids possess intermediate pinnae numbers and, when present, can be difficult to distinguish from either parent. The hybrids are highly invasive. Molecular data have also revealed the occurrence of unidirectional hybrids between *M. heterophyllum* (♀) and *M. laxum* (♂) in the eastern United States. Both *M. heterophyllum* and the hybrids (but not *M. laxum*) have been introduced to the New England region, where they exhibit invasive characteristics. Population genetic analyses could better clarify the reproductive biology of *Myriophyllum* species, but have been conducted in detail only for *M. alterniflorum*. Isozyme data indicate that sexual reproduction occurs frequently in *M. alterniflorum* and that inbreeding does not occur within populations, where fairly high levels of genetic variation are maintained. However, genetic subdivision between populations occurs as a consequence of wind pollination and the existence of numerous, moderately isolated sites. The basic chromosome number of *Myriophyllum* is $x = 7$. *Myriophyllum alterniflorum*, *M. tenellum*, and *M. ussuriense* are diploid ($2n = 14$), *M. verticillatum* ($2n = 28$) is tetraploid and *M. quitense*, *M. sibiricum*, and *M. spicatum* are hexaploid ($2n = 42$). Triploid cytotypes ($2n = 21$) have also been reported for *M. ussuriense* (North America) and *M. spicatum*.

Comments: The general geographical distributions of *Myriophyllum* species are indicated in Figure 4.55. In North America, the geographical affinities of species are northern (*M. farwellii*), northeastern (*M. alterniflorum, M. humile, M. tenellum*), eastern (*M. heterophyllum*), southeastern (*M. laxum, M. pinnatum*), western (*M. quitense*), and northwestern (*M. ussuriense*). *Myriophyllum hippuroides* is disjunct in west and central regions. *Myriophyllum aquaticum* is fairly widespread along both coasts, especially in the south. *Myriophyllum sibiricum, M. verticillatum,* and *M. spicatum* are widespread, the former two especially throughout the north.

References: Aiken, 1981; Aiken & Walz, 1979; Blotnick et al., 1980; Boylen et al., 1999; Ceska & Warrington, 1976; Ceska et al., 1986; Chen et al., 2014; Christy, 2004; Conard, 1935; Cook, 1988; Couch & Nelson, 1988; Creed & Sheldon, 1993, 1994; Crow & Hellquist, 1983; DiTomaso & Healy, 2003; Eyles, 1941; Flessa, 1994; Gerber, 1994; Gerber & Les, 1994, 1996; Halstead et al., 2003; Harris et al., 1992; Hoyer et al., 1996; Hutchinson, 1970, 1975; Jaynes & Carpenter, 1986; Johnson et al., 1997; Keddy & Reznicek, 1982; Knepper et al., 2002; Kunkel, 1984; Lauridsen & Lodge, 1996; Les & Gerber, 1991; Les & Mehrhoff, 1999; Little, 1979; Magdych, 1979; Martin & Uhler, 1951; Moody, 2004; Moody & Les, 2002, 2007; Nichols, 1999a; Nichols & Buchan, 1997; O'Neill, 1986; Orchard, 1975, 1980, 1981, 1986; Patten, 1956; Penfound & Hathaway, 1938; Philbrick & Les, 1996; Preston & Croft, 1997; Rho & Gunner, 1978; Ritter & Crow, 1998; Roshon et al., 1996; Schloesser, 1986; Scribailo & Alix, 2006; Sheldon & Jones, 2001; Sheridan, 1990; Skogerboe et al., 2003; Solarz & Newman, 1996; Stanley, 1970; Sytsma & Anderson, 1993; Ueno & Kadono, 2001; Valley & Newman, 1998; Weber, 1972; Weber & Noodén, 1974, 1976a,b.

2. *Proserpinaca*

Mermaidweed; proserpinie

Etymology: from *Proserpine* the Latin form of the Greek Persephone, the mythical daughter of Zeus and Demeter

Synonyms: none

Distribution: global: North America; **North America:** eastern

Diversity: global: 3 species; **North America:** 3 species

Indicators (USA): OBL: *Proserpinaca intermedia, P. palustris, P. pectinata*

Habitat: brackish; freshwater; palustrine; **pH:** 4.3–7.3; **depth:** 0–2 m; **life-form(s):** emergent herb, submersed (vittate)

Key morphology: stems (to 1 m) erect to decumbent, homophyllous or heterophyllous; leaves alternate, all pinnate (to 2.5 cm) with 4–9 segment pairs (to 7.5 mm), all pinnatifid (to 4 cm) with 7–12 segment pairs, or with emergent leaves (to 8.5 cm) simple and serrate to pinnately lobed, grading to pinnate submersed leaves (to 6 cm) with 8–14 segment pairs (to 3 cm); flowers axillary, single or in groups (2–5), 3-merous,

apetalous; fruit (to 6 mm) achene-like, a trigonal to rounded, 3-loculed, and 3-seeded nutlet

Life history: duration: perennial (rhizomes); **asexual reproduction:** rhizomes, shoot fragments; **pollination:** wind; **sexual condition:** hermaphroditic; **fruit:** achenes (common); **local dispersal:** rhizomes, stem fragments, fruits; **long-distance dispersal:** fruits

Imperilment: (1) *Proserpinaca intermedia* [G4]; S1 (NS); S2 (NC); S3 (NJ); (2) *P. palustris* [G5]; S1 (IA, VT); S2 (NB, QC); S3 (NJ, NS); (3) *P. pectinata* [G5]; SX (PA); SH (NH); S1 (ME, MI); S2 (NY); S3 (NS, VA)

Ecology: general: All *Proserpinaca* species are obligate aquatics, which occupy an extremely broad range of sites within various hydric habitats. They are excellent wetland indicators. In general, the species occur in acidic sites with low alkalinity and grow on substrates that tend to be high in organic matter. Pollination has not been investigated in any of the species but reportedly occurs by wind. There have been no studies on the reproductive ecology of any of the species to determine whether the plants are outcrossers or inbreeding. Likewise, specific elements of fruit dispersal are not known in any detail. The chambered nutlets may float for a short time and are eaten by waterbirds (Aves), which probably facilitate their dispersal over longer distances.

***Proserpinaca intermedia* Mack.** is a homophyllous species that has been reported from ditches, marshes, and stream margins at low elevations along the Atlantic Coastal Plain. The sites tend to be acidic (pH: 4.9–5.9) and the substrates sandy. The plants generally are not found in water that exceeds a few cm in depth. This plant, often regarded as a hybrid between the next two species (see *Systematics* section below), is reported from sites where both suspected parents co-occur; thus it is similar to them ecologically. Otherwise, information on the ecology of this species is rarely reported.

Reported associates: *Eleocharis microcarpa, Juncus effusus, J. repens, Ludwigia linearis, Mecardonia acuminata, Proserpinaca palustris, P. pectinata, Rhynchospora corniculata, Sphagnum cuspidatum, Xyris laxifolia.*

***Proserpinaca palustris* L.** is a heterophyllous species that occurs in fresh or brackish (tidal) waters or on wet ground of bogs, borrow pits, ditches, fens, floodplains, marshes, lakeshores, ponds, prairies, swales, swamps, and vernal pools at elevations of up to 700 m. It tolerates a fairly broad range of acidity (pH: 4.3–7.3; mean: 6.5) but typically occurs in low alkalinity sites (mean: 13.0 mg/L $CaCO_3$). The substrates include organic soils, marly or mineral clay, and sand. The plants can withstand low levels of salinity (0.5–5 ppt) but thrive in freshwater sites exposed to moderate or high light levels and temperatures ranging from 10°C to 28°C. *Proserpinaca palustris* can grow either as a completely submersed plant (with pinnately compound leaves and appearing much like a giant *Myriophyllum*), or as a heterophyllous emergent in shallow water or on exposed but hydric substrates. Heterophylly in *P. palustris* is determined by photoperiodic responses that are mediated by water temperatures. Lanceolate, serrate, aerial leaves develop under long-day conditions (16 h light); whereas, short days (10 h light) will induce dissected leaf formation.

Low temperatures (15°C) will prevent the development of undissected leaves. It is not unusual to see stems with dissected leaves below, entire leaves in between, and dissected leaves apically at the end of the growing season. Heterophylly enables the efficient uptake of nutrients from both the air and the water simultaneously. However, submergence of the aerial leaves reduces the photosynthetic compensation point in comparison to emergent conditions. Ecological studies also show that the heterophyllous plants have a significantly higher vegetative growth rate and greater flower/fruit production than do the homophyllous plants. *Proserpinaca palustris* obtains most of its nitrogen and phosphorous from its roots, but potassium is taken up primarily from the water and can be limiting to growth. **Reported associates:** *Acer rubrum, Agalinis, Alisma subcordatum, Allium cernuum, Alnus serrulata, Alopecurus aequalis, Amsonia hubrichtii, Andropogon virginicus, Asclepias incarnata, A. perennis, Azolla caroliniana, Bacopa caroliniana, B. monnieri, Berchemia scandens, Bidens discoideus, B. frondosus, Blechnum serrulatum, Boehmeria cylindrica, Boltonia asteroides, Calamagrostis canadensis, Calamintha glabella, Carex frankii, C. glaucescens, C. lupulina, C. oligosperma, C. sartwellii, C. stricta, Calystegia sepium, Centella asiatica, Cephalanthus occidentalis, Chelone glabra, Cicuta bulbifera, Cirsium, Cladium jamaicense, C. mariscoides, Cornus amomum, C. racemosa, Crinum americanum, Cyperus haspan, Decodon verticillatus, Dichanthelium dichotomum, D. sphaerocarpon, D. spretum, Dryopteris cristata, Dulichium arundinaceum, Echinodorus cordifolius, E. tenellus, Eleocharis cellulosa, E. elliptica, E. equisetoides, E. erythropoda, E. fallax, E. obtusa, E. palustris, E. quadrangulata, E. rostellata, Elymus virginicus, Equisetum arvense, Erianthus giganteus, Eriocaulon decangulare, Eryngium aquaticum, Eupatorium capillifolium, Forestiera acuminata, Fuirena squarrosa, Galium tinctorium, Gentiana andrewsii, G. crinita, Glyceria acutiflora, G. septentrionalis, Gratiola aurea, G. brevifolia, Heliotropium indicum, Hibiscus moscheutos, Hydrocotyle umbellata, H. verticillata, Hymenocallis caroliniana, Hypericum canadense, H. mutilum, H. virginicum, Ilex cassine, I. glabra, Impatiens capensis, Iris hexagona, I. versicolor, I. virginica, Isoetes, Itea, Juncus effusus, J. repens, J. roemerianus, Justicia ovata, Larix laricina, Leersia oryzoides, Leitneria floridana, Lemna minor, Leptochloa fusca, Limnobium spongia, Lindernia grandiflora, Lobelia cardinalis, L. kalmii, Ludwigia alata, L. alternifolia, L. palustris, L. peruviana, L. polycarpa, L. repens, Lycopus americanus, L. rubellus, L. uniflorus, L. virginicus, Lysimachia lanceolata, L. quadriflora, Lythrum salicaria, Magnolia grandiflora, Mentha arvensis, M. spicata, Micranthemum umbrosum, Mikania scandens, Myriophyllum verticillatum, Nelumbo lutea, Nuphar variegata, Nymphaea odorata, Nyssa aquatica, Onoclea sensibilis, Orontium aquaticum, Oxydendrum arboreum, Oxypolis rigidior, Panicum gymnocarpon, P. hemitomon, P. verrucosum, Paspalidium geminatum, Peltandra virginica, Penthorum sedoides, Persicaria amphibia, P. arifolia, P. coccinea, P. hydropiperoides, P. pensylvanica, P. punctata, P. sagittata, Phalaris arundinacea, Phragmites australis, Physostegia purpurea, Pluchea foetida, Pontederia cordata, Potamogeton bicupulatus, Ptilimnium nodosum, Pycnanthemum tenuifolium, Quercus, Ranunculus flabellaris, Rhexia virginica, Rhus copallina, Rhynchospora capillacea, R. capitellata, R. colorata, R. macrostachya, R. miliacea, Riccia fluitans, Rubus, Rumex orbiculatus, Sabatia dodecandra, Sagittaria graminea, S. lancifolia, S. latifolia, Salix interior, S. nigra, Samolus parviflorus, Sarcostemma clausum, Saururus cernuus, Schoenoplectus acutus, S. americanus, Scripus cyperinus, S. divaricatus, S. lineatus, Scleria verticillata, Sclerolepis uniflora, Sium suave, Smilax, Solanum dulcamara, Solidago ptarmicoiides, S. riddellii, S. rugosa, Sparganium androcladum, Sphagnum, Stachys hyssopifolia, Taxodium distichum, Thalia geniculata, Thelypteris palustris, Tillandsia bartramii, Torreyochloa pallida, Triadenum virginicum, Triantha glutinosa, Typha angustifolia, T. latifolia, Utricularia cornuta, U. gibba, U. macrorhiza, Verbena hastata, Vicia cracca, Viola pallens, Vitis rotundifolia, Woodwardia areolata, Xanthium strumarium, Xyris smalliana.*

***Proserpinaca pectinata* Lam.** is a homophyllous species of very shallow water or saturated substrates of bogs, canals, depressions, ditches, marshes, pools, savannas, swales, swamps, and the margins of lakes, ponds, and springs at elevations of up to 500 m. The sites are acidic (pH: 5.0–6.2) with substrates of clay/silty loam, gravel, mud, or sand. Additional details on the ecology of this species are wanting. **Reported associates:** *Acer rubrum, Ambrosia artemisiifolia, Amphicarpum muhlenbergianum, Amsonia tabernaemontana, Andropogon virginicus, Aristida palustris, Axonopus compressus, Brasenia schreberi, Buchnera, Carex bromoides, C. bullata, C. gigantea, C. glaucescens, C. striata, Centella asiatica, Cephalanthus occidentalis, Chamaecrista nictitans, Cladium jamaicense, Cliftonia, Coelorachis rugosa, Cyperus stenolepis, Cyrilla racemiflora, Dichanthelium longiligulatum, D. sabulorum, D. spretum, Diodia, Diospyros virginiana, Drosera, Dulichium arundinaceum, Eleocharis microcarpa, E. robbinsii, E. tuberculosa, Erianthus giganteus, Erigeron vernus, Eriocaulon aquaticum, E. compressum, E. decangulare, Eryngium baldwinii, Eupatorium capillifolium, Euthamia, Fimbristylis autumnalis, Fuirena breviseta, Gratiola aurea, Helenium flexuosum, Helianthus radula, Hibiscus moscheutos, Hypericum fasciculatum, H. hypericoides, Hyptis alata, Ilex cassine, I. glabra, I. myrtifolia, Juncus abortivus, J. effusus, J. megacephalus, J. militaris, J. pelocarpus, J. repens, J. trigonocarpus, Lachnanthes caroliniana, Leersia hexandra, L. lenticularis, Litsea aestivalis, Lobelia, Ludwigia linearis, L. octovalvis, L. pilosa, Lysimachia terrestris, Magnolia grandiflora, M. virginiana, Mecardonia acuminata, Mikania scandens, Mitreola, Myrica, Myriophyllum, Nuphar, Nymphaea odorata, Nymphoides cordata, Nyssa aquatica, N. biflora, Oldenlandia uniflora, Orontium aquaticum, Osmunda regalis, Oxypolis filiformis, Panicum hemitomon, P. hirstii, P. rigidulum, P. tenerum, P. verrucosum, Persea palustris, Persicaria hydropiperoides, Phyla nodiflora, Physostegia purpurea, Pinus elliottii, Pleopeltis polypodioides, Pluchea, Polygala cruciata, P. lutea, Pontederia* *cordata, Potamogeton, Proserpinaca intermedia, P. palustris, Ptilimnium nodosum, Pterocalon, Quercus laurifolia, Q. virginiana, Rhexia alifanus, R. aristosa, R. brevifolia, R. nashii, R. virginica, Rhus copallina, Rhynchospora cephalantha, R. chalarocephala, R. chapmanii, R. corniculata, R. harperi, R. inundata, R. mixta, R. perplexa, Rubus, Sabatia difformis, Saccharum baldwinii, Sarracenia alata, Schizachyrium scoparium, Schoenoplectus etuberculatus, S. subterminalis, Scirpus cyperinus, Scleria baldwinii, S. muehlenbergii, S. reticularis, Smilax walteri, Solidago, Stylisma aquatica, Taxodium distichum, Triadenum fraseri, T. virginicum, Triadica sebifera, Typha latifolia, Utricularia foliosa, U. juncea, U. purpurea, Vaccinium, Woodwardia virginica, Xyris fimbriata, X. laxifolia, X. smalliana.*

Use by wildlife: *Proserpinaca* fruits are eaten by spoonbills (Aves: Threskiornithidae: *Platalea ajaja*), swans (Aves: Anatidae: *Cygnus*), and 15 species of duck (Aves: Anatidae), notably black ducks (*Anas rubripes*), mallards (*Anas platyrhynchos*), teals (*Anas*), and wood ducks (*Aix sponsa*). Although mallard ducks have been observed to consume up to 220 fruits at one time, the genus is of relatively minor importance overall as a waterfowl food. The fruits sometimes are eaten by muskrats (Mammalia: Cricetidae: *Ondatra zibethicus*). *Proserpinaca* contains extremely high levels of polyphenolic compounds, which are known to deter feeding by *Acentria ephemerella* (Insecta: Lepidoptera: Pyralidae) and other aquatic herbivores. *Proserpinaca palustris* has been identified as a component of the habitat for the bluehead shiner (Osteichthyes: Cyprinidae: *Pteronotropis hubbsi*), an imperiled minnow.

Economic importance: food: none reported; **medicinal:** none reported; **cultivation:** *Proserpinaca palustris* and *P. pectinata* are cultivated as aquarium and water garden plants; **misc. products:** none; **weeds:** none; **nonindigenous species:** none

Systematics: Aside from *Myriophyllum, Proserpinaca* is the only other North American genus of Haloragaceae; however, the two genera are not closely related. Phylogenetic analysis (Figure 4.54) indicates that *Proserpinaca* is relatively basal in the family, occupying a position between the endemic Australian genera *Meionectes* (also aquatic) and *Trihaloragis* (terrestrial). This pattern imitates that found within *Myriophyllum* (see earlier), which has experienced repeated dispersal events from Australian source populations. Relationships within *Proserpinaca* are straightforward. Molecular phylogenetic studies that include *P. palustris* and *P. pectinata* verify that the genus is monophyletic, an issue that never was in doubt given their distinctive morphological characteristics. Considered here tentatively as a distinct species, *P. intermedia* is often regarded simply as an interspecific hybrid between *P. palustris* and *P. pectinata*. There has been much circumstantial evidence in support of this interpretation (e.g., *P. intermedia* is said to occur only where *P. palustris* and *P. pectinata* are sympatric), but genetic analyses unfortunately are lacking. Hybridization between *P. palustris* and *P. pectinata* would be expected given their abiotic (wind) pollination, inelaborate flowers, sympatric distribution, and

similar ecological preferences (though *P. palustris* occurs often in slightly more alkaline sites). *Proserpinaca palustris* also can occupy deeper water than *P. pectinata*, but both can occur in wet, exposed sites. The base chromosome number of *Proserpinaca* is $x = 7$. *Proserpinaca palustris* and *P. pectinata* both are diploid ($2n = 14$).

Comments: All three *Proserpinaca* species essentially are sympatric in eastern North America; however, *P. palustris* is more common to the north and *P. pectinata* to the south.

References: Barko & Smart, 1981; Catling, 1998; Davis, 1967; Dennis & Wofford, 1976; Fassett, 1953c; Kane & Albert, 1982, 1987; Keller, 2000; Pigliucci, 2004; Salvucci & Bowes, 1981; Schmidt & Millington, 1968; Wallenstein & Albert, 1963; Wells & Pigliucci, 2000.

Family 4: Iteaceae [2]

Following the relationships depicted by compelling molecular phylogenetic analyses (Figure. 4.51), Iteaceae (*Choristylis, Itea*) are treated here as distinct from both Grossulariaceae and Saxifragaceae, where they often have been included traditionally. Some treatments (Soltis et al., 2005) also merge the Mexican Pterostemonaceae (monotypic: *Pterostemon*) within the family. In any case, *Pterostemon* (with only two species) is the sister group to Iteaceae (Figure 4.51; Fishbein & Soltis, 2004). The close relationship of these families also is evidenced by their similar flavonoid chemistry, which differs from that of Saxifragaceae (Bohm et al., 1999). Iteaceae are a small family of approximately 17 species of cyanogenic trees and shrubs distributed primarily in Africa and Asia.

Itea species are grown as ornamental plants with about a dozen cultivars in the horticultural trade. The family is otherwise of little economic importance. It is represented in North America by a single genus and species, which is also regarded as an obligate aquatic.

1. *Itea* J. Agardh

1. *Itea*

Sweet-spires, Virginia willow
Etymology: from the Greek *itea* meaning "willow"
Synonyms: *Diconangia; Kurrimia; Reinia*
Distribution: global: Africa; Asia; North America; **North America:** southeastern
Diversity: global: 15 species; **North America:** 1 species
Indicators (USA): OBL; FACW: *Itea virginica*
Habitat: freshwater; palustrine; **pH:** 4.0–7.5; **depth:** <1 m; **life-form(s):** emergent shrub
Key morphology: stems (to 3.0 m) with arching branches and chambered pith; leaves (to 10 cm) alternate, simple, minutely serrate, short-petioled; racemes (to 2.0 dm) pendant, arching, with numerous, small flowers, their pedicels (to 3 mm) diverging at right angles to the axis; flowers 5-merous; petals (to 7 mm) white; capsules (to 10 mm) 2-celled, longitudinally 2-grooved, with several seeds
Life history: duration: perennial (buds); **asexual reproduction:** root suckers; **pollination:** insect; **sexual condition:**

hermaphroditic; **fruit:** capsules (common); **local dispersal:** root suckers, seeds; **long-distance dispersal:** seeds
Imperilment: (1) *Itea virginica* [G4]; S1 (IN, OK, PA)
Ecology: general: There is only one North American species (see next).

Itea virginica L. inhabits depressions, floodplains, pools, river and stream margins, seeps, and swamps at elevations of up to 463 m. It can tolerate a fairly broad range of (pH: 4.0–7.5) but does best when grown under moderately acidic conditions (pH: 5.5–7.0). The substrates include clay loam, muck, rock, sand, sandstone, sandy loam, and silty loam. The plants can grow in full sunlight to partial shade, but will flower more prolifically in sunny (>6 h) sites. The shrubs attain a very high importance value in the Okefenokee Swamp, where they are part of a successional stage of vegetation that colonizes the so-called "batteries" (floating or dislodged peat islands). Individual plants can tolerate some inundation but their occurrence is associated negatively with increasing flooding depth. Their long-term survival generally ranges from 5 to 25 years. The plants have a fairly long flowering period, which extends from May to July. The principal pollinators probably are insects such as bees (Hymenoptera) and butterflies (Lepidoptera). The mature seeds germinate readily without requiring any pretreatment. Propagation by seed is the most common method; however, cuttings (taking about 4 weeks to root) or root suckers can also be used. The suckering habit enables plants to form thickets, which helps to stabilize substrates. The plants are cold tolerant down to −31°C. They usually are deciduous but will retain their leaves as long as the temperatures remain above −7°C to −9°C. **Reported associates:** *Acer rubrum, Adiantum capillus-veneris, Alnus serrulata, Ampelopsis arborea, Andropogon glomeratus, A. virginicus, Aralia spinosa, Arundinaria gigantea, Berchemia scandens, Boehmeria cylindrica, Callicarpa americana, Carex crinita, C. debilis, C. glaucescens, C. intumescens, C. lonchocarpa, C. lurida, Carpinus caroliniana, Carya, Cephalanthus occidentalis, Chamaelirium luteum, Chasmanthium, Cladium jamaicense, Clethra alnifolia, Cliftonia monophylla, Cornus amomum, C. foemina, C. stricta, Crataegus viridis, Cyrilla racemiflora, Decodon verticillatus, Decumaria barbara, Dichanthelium scoparium, Dryopteris, Dulichium arundinaceum, Eichhornia crassipes, Eleocharis tortilis, Elymus virginicus, Erechtites hieracifolia, Fagus grandifolia, Forestiera acuminata, Fraxinus caroliniana, F. profunda, Galium, Gleditsia aquatica, Glyceria striata, Gordonia lasianthus, Habenaria repens, Hamamelis, Hydrocotyle verticillata, Hypericum densiflorum, H. fasciculatum, H. hypericoides, Ilex cassine, I. decidua, I. verticillata, Jamesianthus alabamensis, Juncus coriaceus, J. effusus, Lachnanthes caroliniana, Leersia lenticularis, L. virginica, Leucothoe racemosa, Linum striatum, Liquidambar styraciflua, Liriodendron tulipifera, Ludwigia alternifolia, L. decurrens, L. leptocarpa, L. palustris, Lycopus virginicus, Lyonia ligustrina, L. lucida, Magnolia virginiana, Microstegium vimineum, Mikania scandens, Mithcella repens, Murdannia keisak, Myrica cerifera, M. heterophylla,* *Nymphaea odorata, Nyssa aquatica, N. biflora, N. sylvatica, Onoclea sensibilis, Orontium aquaticum, Osmunda regalis, Oxydendrum arboreum, Packera glabella, Panicum hemitomon, P. rigidulum, Peltandra virginica, Persea palustris, Persicaria hydropiperoides, Phanopyrum gymnocarpon, Phytolacca americana, Pinus elliottii, P. taeda, Planera aquatica, Populus heterophylla, Quercus alba, Q. laurifolia, Q. lyrata, Q. palustris, Q. phellos, Q. velutina, Rhexia mariana, R. virginica, Rhododendron, Rhynchospora capitellata, R. chalarocephala, R. corniculata, R. macrostachya, Rosa palustris, Rubus argutus, Sabal palmetto, Sagittaria calycina, Salix nigra, Sambucus nigra, Saururus cernuus, Scirpus cyperinus, Scleria, Smilax laurifolia, S. rotundifolia, S. walteri, Sparganium americanum, Sphagnum, Taxodium ascendens, T. distichum, Tiarella, Tillandsia bartramii, T. usneoides, Toxicodendron radicans, Triadenum virginicum, T. walteri, Typha latifolia, Ulmus rubra, Utricularia gibba, Vaccinium, Viburnum nudum, V. obovatum, Vitis, Woodwardia areolata, W. virginica, Xyris laxifolia, X. smalliana, X. tennesseensis.*

Use by wildlife: The flowers of *Itea virginica* are known to attract butterflies (Insecta: Lepidoptera) and carpenter bees (Insecta: Hymenoptera: Anthophoridae: *Xylocopa virginica*). The seeds occasionally are eaten by birds (Aves). The foliage is grazed only moderately by herbivores and seldom by whitetail deer (Mammalia: Cervidae: *Odocoileus virginianus*). The plants provide cover for various birds (Aves), especially wood ducks (Anatidae: *Aix sponsa*) during molting. *Itea virginica* is a host of the strawberry rootworm (Insecta: Coleoptera: Chrysomelidae: *Paria fragariae*), which can severely damage the plants.

Economic importance: food: *Itea virginica* has no edible uses reported; **medicinal:** There are no medical uses reported for *I. virginica*; however, the leaves exhibit diurnally fluctuating concentrations of acyclic polyols (a type of sugar-free sweetener), with levels of allitol increasing in the light and those of allulose in the dark; **cultivation:** *Itea virginica* is widely cultivated as an ornamental either as the species or the following cultivars: 'Beppu,' 'Henry's Garnet' [an Award of Garden Merit plant], 'Sprich' ('Little Henry'), 'Long Spire,' 'Merlot,' 'Morton' 'Scarlet Beauty', 'Sarah Eve,' 'Saturnalia,' and 'Shirley's Compact.' It is a recommended planting alternative for invasive shrubs such as *Berberis vulgaris* (Berberidaceae) or *Euonymus alatus* (Celastraceae); **misc. products:** *Itea virginica* is recommended as a plant for stream restoration; **weeds:** none; **nonindigenous species:** none

Systematics: Molecular phylogenetic analyses have verified that *Itea* and *Choristylis* are sister genera and comprise a clade that is the sister group to *Pterostemon* (Figure 4.51); however, a comprehensive phylogenetic survey of *Itea* has not yet been carried out. The base chromosome number of *Itea* is $x = 11$. *Itea virginica* (and other species counted) is diploid ($2n = 22$). There is little published information regarding the reproductive ecology or population genetics of *Itea virginica*.

Comments: *Itea virginica* (the only species of the genus in North America) occurs mainly throughout the southeastern United States.

References: Bowden, 1940; Cypert, 1972; Dorr, 1981; Fishbein & Soltis, 2004; Fishbein et al., 2001; Hesselein & Boyd, Jr., 2003; Lewis & Smith, 1967; Monk & Brown, 1965; Schlesinger, 1978.

Family 5: Penthoraceae [1]

Penthoraceae are maintained here as distinct from Haloragaceae (Figure 4.51) despite the advocated merger of these taxa by some authors (see comments for Haloragaceae earlier). Penthoraceae are monotypic and include only *Penthorum*, a genus variously assigned to either Crassulaceae or Saxifragaceae in past taxonomic treatments. However, molecular data clearly resolve Penthoraceae as the sister group to Haloragaceae, which together comprise a strongly supported clade.

The family is not significant economically. The sole genus of the family contains obligate aquatics in North America:

1. *Penthorum* L.

1. *Penthorum*

Ditch stonecrop; faux-orpin

Etymology: from the Greek *pente horos* meaning "five marked" with reference to the 5-merous flowers

Synonyms: none

Distribution: global: Asia; North America; **North America:** E/W disjunct

Diversity: global: 2 species; **North America:** 1 species

Indicators (USA): OBL: *Penthorum sedoides*

Habitat: freshwater; palustrine; **pH:** 5.2–7.2; **depth:** <1 m; **life-form(s):** emergent herb

Key morphology: stems (to 1 m) erect; leaves (to 10 cm) alternate, lanceolate, serrate; inflorescence a terminal cluster of 2–6 spike-like, apically recurved cymes (to 8 cm), each bearing minute green/cream, star-like flowers on the upper side; sepals (to 2 mm) 5; petals absent; stamens 10; carpels (to 4 mm) 5, horned, fused basally; fruit (to 6 mm) a red/orange circumscissile capsule, dehiscent apically; seeds minute, numerous

Life history: duration: perennial (rhizomes, stolons); **asexual reproduction:** stolons; **pollination:** self; **sexual condition:** hermaphroditic; **fruit:** capsules (prolific); **local dispersal:** rhizomes, stolons, seeds; **long-distance dispersal:** seeds (wind, water)

Imperilment: (1) *Penthorum sedoides* [G5]; S1 (MB); S2 (NB); S3 (DE)

Ecology: global: There is only one North American species (see next).

Penthorum sedoides L. grows in tidal, intertidal, or nontidal ditches, floodplains, marshes, prairies, roadsides, sandbars, seeps, sloughs, springs, streambeds, swamps, vernal depressions, and along the margins of lakes, ponds, rivers, and streams at elevations of up to 700 m. The habitats range from acidic to slightly alkaline conditions (pH: 5.2–7.2) and include substrates of clay, muck, mud, sand, or silt, which generally contain a high amount of organic matter. The plants thrive in full sun to partial shade and their rootstocks will withstand freezing. The small flowers are produced from July

to October and reportedly are self-pollinated. *Penthorum sedoides* produces an enormous quantity of seeds (~320,000/plant) for a perennial plant. Initially, the seeds are conditionally dormant; however, once dormancy is broken they will germinate over a wide temperature range. Seed germination can be induced by 28 days of cold stratification followed by a 35°C/20°C temperature regime under a 14-h photoperiod. Successful germination (23%–52%) has also been reported for dried seeds (collected September to November), which have been stored at 20°C–25°C, then soaked in sterile water and mashed to remove the seed coats and placed in the light for at least 18 days. In the field, seed germination levels are highest in soils of intermediate organic content, nonflooded conditions, and high sediment levels of nitrate nitrogen. Germination of the related *P. chinense* (also a wetland plant) is optimal (65%) when its seeds are placed in distilled water at 4°C for 8 months and then incubated for 4–6 days in the light at 15°C. The mechanism of seed dispersal is unknown for *P. sedoides* but may be similar to that of the related *P. chinense*. In the later species, 40%–80% of the shed seeds will float (due to an oily coating) and are dispersed by water. Seeds of *P. sedoides* are known to persist in the substrate (for an undetermined period of time); but even where plants occur in high density, they often make up less than 1% of the total seed bank. *Penthorum sedoides* has a high relative growth rate, but competes poorly with many wetland species (e.g., *Lythrum salicaria*, *Phalaris arundinacea*, *Spartina pectinata*, *Typha ×glauca*), especially when growing on higher nutrient sites. Photosynthesis occurs via a C_3 pathway.

Reported associates: *Acer rubrum*, *Agrimonia parviflora*, *Agrostis alba*, *A. stolonifera*, *Alisma subcordatum*, *Alnus rugosa*, *Althernanthera philoxeroides*, *Amaranthus cannabinus*, *Ammannia coccinea*, *Ampelopsis arborea*, *Andropogon gerardii*, *Apocynum cannabinum*, *Asclepias incarnata*, *Berchemia scandens*, *Betula nigra*, *B. sandbergi*, *Bidens cernuus*, *B. laevis*, *Boehmeria cylindrica*, *Boltonia asteroides*, *Bulboschoenus fluviatilis*, *Calamagrostis canadensis*, *Campanula aparinoides*, *Campsis radicans*, *Carex crinita*, *C. frankii*, *C. grayi*, *C. gynandra*, *C. haydenii*, *C. intumescens*, *C. lacustris*, *C. leptalea*, *C. lupuliformis*, *C. lupulina*, *C. lurida*, *C. muskingumensis*, *C. stricta*, *C. tribuloides*, *C. tuckermanii*, *C. typhina*, *Cephalanthus occidentalis*, *Chelone glabra*, *Cicuta bulbifera*, *Coreopsis tinctoria*, *Cornus foemina*, *Cryptotaenia canadensis*, *Cyperus aristatus*, *C. diandrus*, *C. erythrorhizos*, *C. esculenta*, *C. odoratus*, *C. squarrosus*, *C. strigosus*, *Desmanthus illinoensis*, *Dichanthelium*, *Echinochloa crus-galli*, *E. muricata*, *Echinodorus tenellus*, *Eleocharis acicularis*, *E. erythropoda*, *E. ovata*, *E. palustris*, *Elymus virginicus*, *Epilobium ciliatum*, *Equisetum arvense*, *E. fluviatile*, *Eragrostis hypnoides*, *E. pectinacea*, *Eupatorium perfoliatum*, *Fimbristylis autumnalis*, *Galium tinctorium*, *G. trifidum*, *Gentianopsis*, *Geum virginianum*, *Glechoma hederacea*, *Glyceria*, *Gratiola neglecta*, *Hackelia virginiana*, *Hamamelis virginiana*, *Helenium autumnale*, *Hemicarpha micrantha*, *Hypericum boreale*, *H. canadensis*, *H. majus*, *H. prolificum*, *H. punctatum*, *Ilex opaca*, *Impatiens capensis*, *I. ecalcarata*, *Itea virginica*, *Juncus acuminatus*, *J. bufonius*, *J.*

diffusissimus, J. effusus, J. marginatus, J. nodosus, J. tenuis, J. torreyi, Justicia americana, Kyllinga gracillima, Leersia lenticularis, L. oryzoides, Lemna minor, Lilaeopsis, Limosella aquatica, Lindernia dubia, Liquidambar styraciflua, Lobelia siphilitica, Lonicera japonica, Ludwigia alternifolia, L. decurrens, L. palustris, L. peploides, L. polycarpa, Lycopus americanus, Lysimachia nummularia, L. quadriflora, L. terrestris, L. thyrsiflora, Lythrum alatum, L. salicaria, Magnolia virginiana, Mentha arvensis, M. pulegium, Mikania scandens, Mimulus alatus, M. ringens, Mitchella repens, Myriophyllum ussuriense, Nuphar advena, Nyssa biflora, Oenothera pilosella, Onoclea sensibilis, Osmunda regalis, Osmundastrum cinnamomea, Panicum capillare, P. dichotomiflorum, P. gymnocarpon, P. rigidulum, P. virgatum, Parietaria pennsylvanica, Paspalum, Persicaria arifolia, P. coccinia, P. hydropiper, P. hydropiperoides, P. lapathifolia, P. pensylvanica, P. punctata, P. sagittata, Phalaris arundinacea, Phragmites australis, Phyla lanceolata, P. nodiflora, Physostegia virginiana, Pilea pumila, Pinus taeda, Plantago, Platanthera flava, Prunella vulgaris, Ranunculus pensylvanicus, Rorippa curvipes, Rudbeckia hirta, Rumex orbiculatus, R. verticillatus, Sagittaria latifolia, Salix discolor, S. lucida, S. nigra, S. scouleriana, Scirpus cyperinus, S. microcarpus, Scoparia dulcis, Scutellaria lateriflora, Setaria, Sium suave, Smilax laurifolia, S. rotundifolia, Solidago patula, Sparganium emersum, Spartina pectinata, Sphagnum, Stachys tenuifolia, Symphyotrichum ontarionis, S. puniceum, Taxodium distichum, Thuja occidentalis, Thyrsanthella difformis, Toxicodendron radicans, Triadenum fraseri, Typha angustifolia, T. latifolia, Vaccinium macrocarpon, Verbena hastata, V. litoralis, V. rigida, V. urticifolia, Veronica anagallisaquatica, Veronicastrum virginicum, Viburnum nudum, Viola lanceolata, Vitis aestivalis, V. riparia, Woodwardia areolata, W. virginica, Xanthium strumarium.

Use by wildlife: The seeds of *Penthorum sedoides* are minute and do not offer much as a food resource for wildlife. However, the fruits are reportedly eaten by some waterfowl (Aves: Anatidae). The high tannin content of the foliage makes it resistant to grazing by some herbivores such as deer (Mammalia: Cervidae: *Odocoileus*). Muskrats (Mammalia: Cricetidae: *Ondatra zibethicus*) use the plants in building their mounds. *Penthorum sedoides* is a host of several fungi including (Ascomycota: Dermateaceae: *Hainesia lythri*; Mycosphaerellaceae: *Pseudocercospora sedoides*).

Economic importance: food: The leaves of *Penthorum sedoides* were eaten as a vegetable by the Cherokee. In China, the mucilaginous shoots and leaves of the related *P. chinense* are eaten; **medicinal:** Most medical applications of *Penthorum sedoides* are attributable to its high tannin content. The plants have been used as a laxative or to treat chronic pharyngeal and nasal disorders, diarrhea, hemorrhoids, and inflammations of the mucous membranes. They are also used in preparing an astringent wash. The Meskwaki people used the seeds as an ingredient in a cough remedy; **cultivation:** *Penthorum sedoides* occasionally is planted as an ornamental water garden plant and has also been used in aquariums (grown then as a submersed aquatic); **misc. products:** *Penthorum sedoides* is

a component of some wetland seed mixes; **weeds:** *Penthorum sedoides* often is reported as weedy in cultivated cranberry fields; **nonindigenous species:** *Penthorum sedoides* has been introduced to British Columbia, Oregon, and Washington through cranberry culture.

Systematics: Although its systematic position had been debated for more than a century, molecular data clearly place *Penthorum* as the closest extant relative of Haloragaceae (Figure 4.51). With only two species recognized worldwide, interspecific relationships in the genus are incontrovertible. The North American *P. sedoides* and Asian *P. chinense* have an allozymically determined genetic identity of 0.53% and exhibit 1.7% divergence in their nrITS sequences, which translates into an estimated divergence time of 6.0–12.6 million years for these intercontinentally disjunct species. It is not known if the species are capable of hybridization. The base chromosome number of *Penthorum* is $x = 8$. Both *P. sedoides* and *P. chinense* are diploid ($2n = 16$), although an anomalous count of $2n = 18$ has also been reported for *P. sedoides*. Little information exists on the pollination biology, reproductive ecology, or population genetic structure of either species.

Comments: *Penthorum sedoides* occurs throughout eastern North America and is disjunct in Oregon, Washington, and British Columbia due to introductions.

References: Baskin & Baskin, 1998; Baskin et al., 1989; Downton, 1975; Felter & Lloyd, 1898; Freeman, 2009; Haskins & Hayden, 1987; Ikeda & Itoh, 2001; Kangas & Hannan, 1985; Keddy et al., 2000; Kellogg et al., 2003; Leck, 2003; Lee et al., 1996; Lovell, 1912; Middleton, 2000; Mitchell, 1926; Shipley & Peters, 1990; Soltis & Bohm, 1982; Speichert & Speichert, 2004; Stevens, 1932.

Family 6: Saxifragaceae [30]

Phylogenetic analyses (Figure 4.51) resolve Saxifragaceae as the sister group to Grossulariaceae, which some authors treat as a subfamily of the former. The systematic relationships of Saxifragaceae have been clarified greatly by molecular phylogenetic analyses, which have provided evidence for the removal of a number of discordant genera (e.g., *Hydrangea, Lepuropetalon, Parnassia, Penthorum, Philadelphus*) and their placement elsewhere. In the circumscription followed here, Saxifragaceae are herbaceous, exstipulate plants having vessel elements with simple perforations and capsular or follicular fruits (Judd et al., 2002). This fairly small family comprises approximately 550 species (mainly in *Saxifraga*), and is dominated by terrestrial plants; however, aquatic species are scattered throughout.

Other phylogenetic analyses (Figure 4.56) have indicated that the family comprises two major clades, an arctic/alpine "*Saxifraga* sensu stricto" clade (*Saxifraga, Saxifragella*) and the temperate "*Heuchera*" clade (all remaining genera). The former clade has a uniform floral morphology (actinomorphic, 5 sepals, 5 petals, 10 stamens, 2 carpels); whereas, the latter varies with respect to all of its floral characters. Similar analyses have also demonstrated that *Mitella* and *Saxifraga* were polyphyletic as formerly circumscribed (Soltis et al., 2001a; Okuyama et al., 2008; Deng et al., 2015), which has

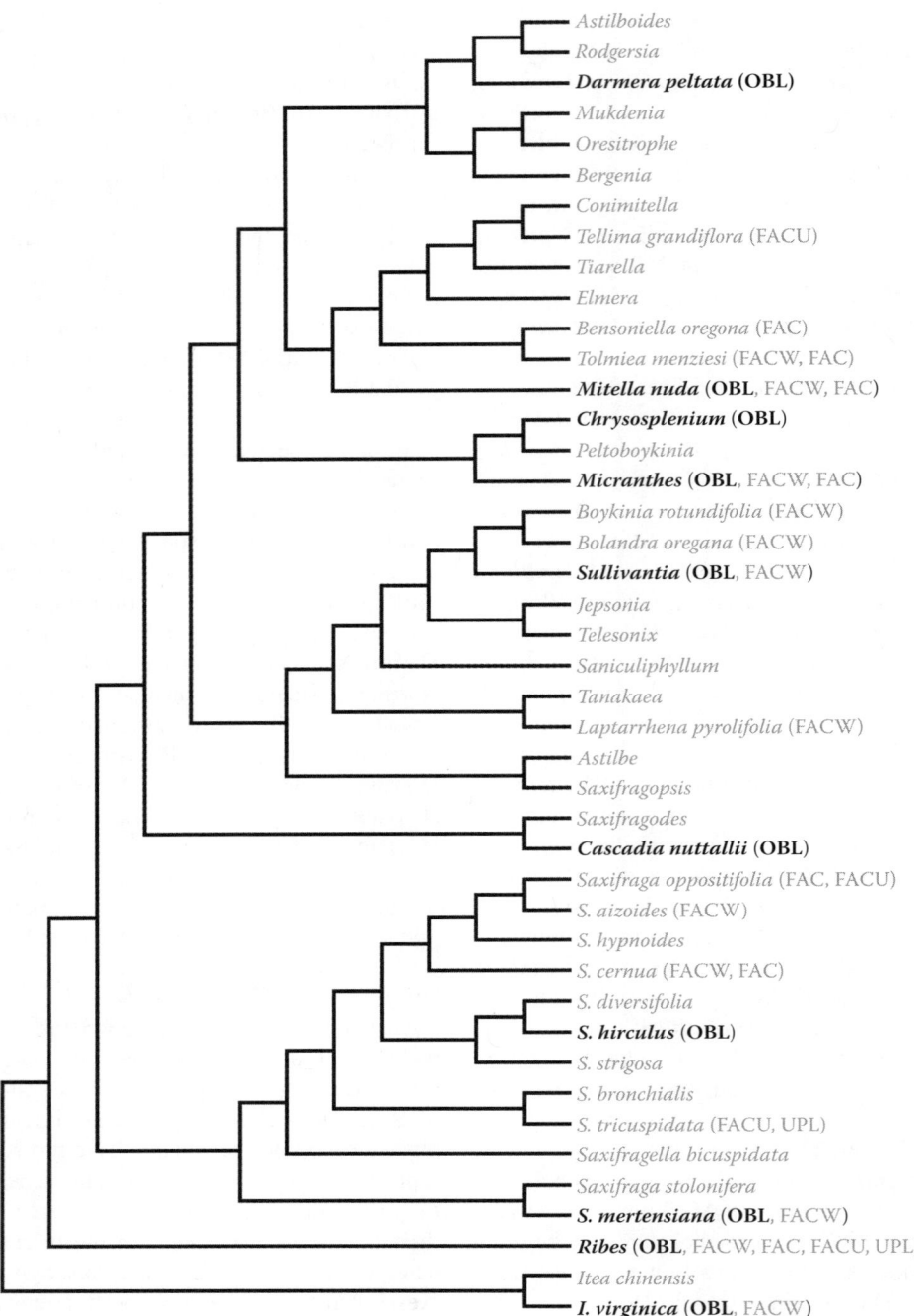

FIGURE 4.56 Phylogenetic relationships in Saxifragaceae as resolved using combined cpDNA data. This toplogy depicts genera with obligate aquatics in North America (bold type) as distributed throughout the family, with the habit evolving at least seven separate times (*Pectiantia* was not surveyed). The wetland indicator status of the North American species is shown in parentheses. (Adapted from Deng, J.-B. et al., *Mol. Phylogen. Evol.*, 83, 86–98, 2015.)

resulted in the recent transfer of several species into the genera *Cascadia*, *Micranthes*, and *Pectiantia*.

Saxifragaceae contain a number of economically important genera (e.g., *Astilbe*, *Boykinia*, *Darmera*, *Heuchera*, *Micranthes*, *Saxifraga*, *Tellima*, *Tiarella*, *Tolmiea*), which are grown as ornamental garden plants. The family is distributed primarily in the north temperate regions and in western South America. Two genera (*Hemieva* Raf. [as

Suksdorfia A. Gray]; *Tolmiea* Torr. & A. Gray) contained OBL aquatics in the 1996 indicator list, but have been recategorized as non-OBL in the 2013 list. Treatments for these genera have been retained here, as a consequence of their completion prior to the release of the later list. Eight genera currently contain obligate aquatics in North America. *Suksdorfia* and *Tolmiea* contained OBL species in the 1996 indicator list, but these have since been recategorized as

FACW, FAC indicators. However, the treatments for both genera have been retained given their completion prior to the release of the 2013 list:

1. *Cascadia* A.M. Johnson
2. *Chrysosplenium* L.
3. *Darmera* Voss
4. *Micranthes* Haw.
5. *Mitella* L.
6. *Pectiantia* (Greene) Rydb.
7. *Saxifraga* L.
8. *Suksdorfia* A. Gray [no longer OBL]
9. *Sullivantia* Torr. & A. Gray ex A. Gray
10. *Tolmiea* Torrey & A. Gray [no longer OBL]

Phylogenetic relationships among most genera of Saxifragaceae have been established fairly consistently as a result of analyses that incorporate combinations of several molecular data sets (Soltis et al., 2001a; Deng et al., 2015). These studies have indicated that the genera containing obligately aquatic species are dispersed throughout the family with at least seven separate origins of the habit (Figure 4.56).

1. Cascadia

Cascade saxifrage
Etymology: after North America's Cascade Mountains
Synonyms: *Saxifraga* (in part)
Distribution: global: North America; **North America:** western (Cascade Ranges)
Diversity: global: 1 species; **North America:** 1 species
Indicators (USA): OBL: *Cascadia nuttallii*
Habitat: freshwater; palustrine, riverine; **pH:** unknown; **depth:** <1 m; **life-form(s):** emergent herb
Key morphology: Stems (to 40 cm) trailing or suberect; leaves cauline, the blades (to 20 mm) ovate to lanceolate; petioled (to 6 mm); the margins entire, often 3-toothed apically; inflorescences solitary or 2–4-flowered panicle-like or racemose thyrses, pedicels filiform; flowers 5-merous, hypanthium (to 2 mm) green, adnate to proximal 1/3 of ovary, nectary disc present; petals (to 6 mm) white, elliptic or obovate; stamens 10, with flattened filaments; ovary partially inferior; capsules (to 5 mm) follicle-like, 2-beaked; seeds brown, with longitudinal rows of spines
Life history: duration: perennial (rhizomes, stolons); **asexual reproduction:** rhizomes, stolons; **pollination:** unknown; **sexual condition:** hermaphroditic; **fruit:** follicular capsules (common); **local dispersal:** seeds, rhizomes, stolons; **long-distance dispersal:** seeds
Imperilment: (1) *Cascadia nuttallii* [G4]; S1 (CA)
Ecology: general: *Cascadia* is monotypic (see next).
Cascadia nuttallii **(Small) A.M. Johnston** inhabits cliffs, crevices, ledges, mossy mats, ravines, seeps, spray zones of waterfalls, and coastal streambank springs in wet, shaded, rocky sites at elevations of up to 1200 m. Some authors incorrectly refer to the plants as annuals; however, they are perennial and produce fine rhizomes (<1 mm in diameter) or creeping stolons (from the lower leaf axils). The flowers are produced from April to May. Their pollination biology has not been described. Seeds that are planted during the winter (and barely covered by soil) reportedly will germinate within 1–3 months when temperatures reach 16°C–21°C. As in *Saxifraga*, local seed dispersal probably occurs by means of a wind-induced vibratory resonation of the capsules, a method effective only over very short distances. Additional ecological information on this rare species is scarce. **Reported associates:** *Alnus rubra, Castilleja hispida, Claytonia parviflora, Collinsia parviflora, Delphinium menziesii, Filipendula occidentalis, Mimulus alsinoides, M. guttatus, Primula austrofrigida, Romanzoffia californica, R. thompsonii, Rubus parviflorus, R. spectabilis, Saxifraga occidentalis.*
Use by wildlife: none reported
Economic importance: food: not reported as edible; **medicinal:** none; **cultivation:** not cultivated; **misc. products:** none; **weeds:** none; **nonindigenous species:** none
Systematics: The monotypic genus *Cascadia* resolves phylogenetically as the sister group of *Saxifragodes* (Figure 4.56), a southern South American genus, although the sole species (*C. nuttallii*) long had been placed in *Saxifraga*. The base chromosome number of *Cascadia* is *x* = 8, with *C. nuttallii* (2*n* = 16) a diploid. No hybrids involving *C. nuttallii* have been reported.
Comments: *Cascadia nuttallii* occurs along the west slopes of the Cascade Ranges in California, Oregon, and Washington.
References: Brouillet, 2009; Johnson, 1927; Parks & Elvander, 2012.

2. Chrysosplenium

Golden saxifrage, water carpet; cresson doré, dorine
Etymology: from the Greek *chrysos splen* meaning "golden spleen" with reference to the color of the plants and resemblance of their foliage to the spleen
Synonyms: none
Distribution: global: Asia; Europe; North America; South America; **North America:** northern, eastern, and western
Diversity: global: 57 species; **North America:** 6 species
Indicators (USA): OBL: *Chrysosplenium americanum, C. glechomifolium, C. iowense, C. tetrandrum*
Habitat: freshwater; palustrine, riverine; **pH:** 4.9–7.5; **depth:** <1 m; **life-form(s):** emergent herb submersed (vittate)
Key morphology: stems (to 25 cm) erect or creeping, somewhat succulent; leaves (to 17 mm) alternate or opposite, petioled (to 2.5 cm), cordate, reniform or rounded, crenate; flowers, small (to 5.0 mm), solitary or in cymose clusters of 2–6; calyx yellow or greenish, 4 or 5-merous; petals absent; stamens 4, 8, or 10, inserted on a conspicuous disk; capsule (to 6 mm) flattened, bilobed, unilocular, many seeded (25–50)
Life history: duration: perennial (rhizomes, stolons); **asexual reproduction:** rhizomes, shoot fragments, stolons; **pollination:** insect or self; **sexual condition:** hermaphroditic; **fruit:** capsules (common); **local dispersal:** seeds (rain), rhizomes, stolons; **long-distance dispersal:** seeds
Imperilment: (1) *Chrysosplenium americanum* [G5]; SH (AL); S1 (GA, SC); S2 (IN, KY); S3 (NC); (2) *C. glechomifolium* [G5]; (3) *C. iowense* [G3]; S1 (BC, MB, MN, SK); S2

(IA); S3 (<u>AB</u>); (4) *C. tetrandrum* [G5]; S1 (ID, <u>LB</u>); S2 (<u>MB</u>, WA); S3 (<u>AB</u>, MT, <u>ON</u>, QC)

Ecology: general: *Chrysosplenium* species commonly inhabit wet sites within shaded forest understory communities, which occur in cold, northern climates. Four of the North American species (67%) are categorized as OBL indicators. They often are associated with cold flowing waters and can grow submersed in shallow watercourses, or as emergents along their margins. Few details have been reported on their reproductive ecology. From the general information available, most flowers are insect pollinated, seed dispersal appears to be fairly local (occurring by a rain-induced splash cup method), and the seeds do not appear to require stratification for germination. Water splash has also been implicated in facilitating pollination.

Chrysosplenium americanum **Schweinitz ex Hooker** occurs either as a submersed plant in cold, fast/slow-flowing shallow waters, or as an emergent in bogs, brooks, depressions, ditches, marshes, seeps, springs, streams, swamps, or along the margins of streams at elevations of up to 1500 m. The substrates include loam, mud, sand, silt, and stones. The pH tolerance of this species is not known in detail, but the few sites reported range from being fairly acidic (pH: 5.5–5.8) to slightly alkaline (pH: 7.2). The sites typically are shaded. A relatively broad range of tolerance is indicated by the fairly diverse list of associated species. Flowering occurs from March to July. Fairly large seed banks (up to 1400 seeds m²) have been found for this species. **Reported associates:** *Abies balsamea, Acer rubrum, A. saccharum, A. spicatum, Allium tricoccum, Alnus, Arisaema triphyllum, Asplenium trichomanes, Betula alleghaniensis, B. lutea, B. papyrifera, Boehmeria cylindrica, Brachythecium rivulare, Bryhnia novae-angliae, Calliergon cordifolium, Caltha palustris, Cardamine bulbosa, C. clematitis, C. pensylvanica, Carex aurea, C. bromoides, C. flava, C. gynandra, C. interior, C. leptalea, C. prasina, C. scabrata, C. stipata, C. stricta, Chelone glabra, C. lyonii, Cinna arundinacea, C. latifolia, Circaea alpina, Clintonia borealis, Coptis groenlandica, Cornus canadensis, Diphylleia cymosa, Drosera rotundifolia, Dryopteris carthusiana, D. disjuncta, D. goldiana, D. intermedia, D. spinulosa, Equisetum arvense, E. scirpiodes, E. sylvaticum, Eupatorium rugosum, Eurybia schreberi, Fagus grandifolia, Floerkia proserpinacoides, Fraxinus nigra, Galium kamtschaticum, G. triflorum, Gaultheria hispidula, Geum rivale, Glyceria melicaria, G. striata, Hamamelis virginiana, Hydrocotyle americana, Hylocomnium splendens, Ilex, Impatiens capensis, I. pallida, Laportea canadensis, Larix laricina, Lindera benzoin, Linnaea borelis, Liriodendron tulipifera, Listera convallarioides, Lobelia siphilitica, Maianthemum trifolium, Mertensia paniculata, Micranthes micranthidifolia, M. pensylvanica, M. petiolaris, Mimulus glabratus, M. ringens, Mitella diphylla, M. nuda, Monarda didyma, Moneses uniflora, Nasturtium officinale, Oclemena acuminata, Onoclea sensibilis, Osmunda regalis, Osmundastrum cinnamomea, Oxalis montana, Packera aureus, P. schweinitziana, Phegopteris connectilis, Picea rubens, Pilea pumila, Pinus strobus, Platanthera dilatata, P. psycodes, Poa paludigena, Quercus alba, Q. bicolor,* *Q. borealis, Ranunculus recurvatus, Rhamnus alnifolius, Rhizomnium appalachianum, R. punctatum, Rhynchostegium serrulatum, Ribes lacustre, Rubus pubescens, Rudbeckia laciniata, Rumex orbiculatus, Scirpus microcarpus, Solidago flexicaulis, S. patula, Sphagnum girgensohnii, S. russowii, S. squarrosum, S. warnstorfii, Sphenopholis pensylvanica, Stellaria calycantha, Sullivantia, Symplocarpus foetidus, Taxus canadensis, Thalictrum clavatum, T. pubescens, Thelypteris noveboracensis, T. palustris, Thuidium delicatulum, Thuja occidentalis, Tiarella cordifolia, Tilia americana, Toxicodendron vernix, Trientalis borealis, Tsuga canadensis, Ulmus americana, Veratrum viride, Veronica americana, V. officinalis, Viola cucullata, V. macloskeyi.*

Chrysosplenium glechomifolium **Nutt.** is found either submersed or emergent in bogs, depressions, gullies, marshes, seeps, swales, swamps and along stream margins, at elevations up to 900 m. The substrates often are highly humic but underlain by sandstone bedrock. They include clay–humus, humus, and sandy loam. The habitats typically are characterized by relatively dense shade. Flowering occurs from February to May. **Reported associates:** *Acer macrophyllum, Adiantum pedatum, Alnus rubra, Athyrium filix-femina, Blechnum spicant, Bromus vulgaris, Cardamine angulata, Carex amplifolia, C. deweyana, Cirsium, Claytonia sibirica, Corydalis, Dentaria, Dryopteris carthusiana, Equisetum telmateia, Galium aparine, G. triflorum, Glyceria grandis, G. striata, Lysichiton americanus, Mimulus dentatus, M. guttatus, M. moschatus, Mitella caulescens, M. ovalis, Nasturtium officinale, Oenanthe sarmentosa, Oxalis, Picea sitchensis, Poa trivialis, Polystichum munitum, Pseudotsuga menziesii, Ribes bracteosum, Rubus spectabilis, Salix hookeriana, S. sitchensis, Sambucus racemosa, Scirpus microcarpus, Scoliopus hallii, Senecio, Sorbus dumosa, Stachys mexicana, Stellaria crispa, Stenanthium occidentale, Tellima grandiflora, Tiarella trifoliata, Tolmiea menziesii, Tsuga heterophylla, Typha latifolia, Urtica dioica, Veratrum, Viola glabella.*

Chrysosplenium iowense **Rydb.** grows on algific tallus slopes, in marshes, meadows, seeps, swamps, and along river and stream margins at elevations from 500 to 1500 m. Populations occur typically at arctic or boreal latitudes where the surface temperatures are maintained below 16°C. The habitats are often shaded. The pH range is not well documented, but some sites are fairly acidic (e.g., pH: 4.9). The plants are moderately self-incompatible (exhibiting highly reduced seed set when selfed experimentally) and require insects to achieve successful pollination in natural populations. Allozyme data indicate very low levels of genetic variation in populations, possibly due to a genetic bottleneck. The pollinators are unknown, but Collembola (Arthropoda: Hexapoda) have been observed with pollen attached. Maximum flower production occurs when substrate temperatures are between 11°C and 12°C. Seed dispersal is by means of a splash-cup method with a maximum range of about one meter from the maternal plant. The plants possess a rhizome and also reproduce vegetatively by stolons, which can result in the formation of large clonal patches. **Reported associates:** *Abies balsamea, Acer spicatum, Aconitum noveboracense, Adoxa moschatellina, Asarum, Betula lutea, Brachythecium acuminatum,*

Carex media, C. peckii, Circaea alpina, Climacium dendroides, Cornus canadensis, Cystopteris bulbifera, C. protrusa, Equisetum scirpoides, Gymnocarpium robertianum, Impatiens pallida, Linnaea borealis, Maianthemum canadense, Mertensia paniculata, Micranthes penslyvanica, Mitella nuda, Montia chamissoi, Poa paludigena, P. wolfii, Primula fassettii, P. mistassinica, Pyrola asarifolia, Rhamnus alnifolia, Rhododendron lapponicum, Rhytidiadelphus triquetris, Ribes hudsonianum, Rosa acicularis, Sambucus pubens, Streptopus roseus, Sullivantia renifolia, Taxus canadensis, Thuidium delcatulum, Trillium nivale, Viburnum opulus, V. trilobum, Viola renifolia.

***Chrysosplenium tetrandrum* Th. Fries** occurs in shallow water (up to 10 cm) along the margins of boreal lakes, ponds, rivers and streams, and in channels, crevices, depressions, ditches, dune ridges, fens, floodplains, gravel bars, gullies, heath, mossy hummocks, muskeg, puddles, scree slopes, scrub, seeps, slopes, sloughs, streams, thickets, and in low areas of tundra to elevations of 3300 m. The pH range is not known, but the sites range from acidic to slightly alkaline (e.g., pH: 7.3). The substrates include cobble, gravel, limestone, mud, peat, rock, quartzite, sand, sandy clay, schist, and silt. Habitat exposures usually are shaded. The flowers reportedly self-pollinate. The seeds germinate without stratification under a 15°C/5°C temperature regime. This is a nitrophilous species, which often grows in soils that have been enriched by animal manure. It spreads vegetatively by the production of slender stolons, which can attain a length of up to 15 cm.

Reported associates: *Aconitum delphiniifolium, Adoxa moschatellina, Alnus rugosa, A. viridis, Alopecurus alpinus, Arctagrostis latifolia, Artemisia norvegica, A. tilesii, Betula glandulosa, Brachythecium, Braya glabella, Calamagrostis canadensis, Caltha palustris, Cardamine pensylvanica, C. pratensis, Carex aquatilis, C. canescens, C. fuliginosa, C. lachenalii, C. microchaeta, C. utriculata, Cerastium jenisejense, Chamerion angustifolium, Circaea alpina, Claytonia sibirica, C. tuberosa, Comarum palustre, Cornus canadensis, Cystopteris fragilis, Dasiphora floribunda, Delphinium glaucum, Dryas integrifolia, D. octopetala, Dupontia fisheri, Epilobium anagallidifolium, E. hornemannii, Equisetum arvense, E. fluviatile, Erigeron elatus, Eriophorum angustifolium, Galium trifidum, Geum macrophyllum, Glyceria, Gymnocarpium dryopteris, Hippuris vulgaris, Juncus castaneus, Koenigia islandica, Larix occidentalis, Leymus mollis, Lonicera involucrata, Marchantia polymorpha, Mertensia paniculata, Micranthes hieraciifolia, M. nelsoniana, Mitella nuda, Mnium, Neottia borealis, Oncophorus wahlenbergii, Oxyria digyna, Parnassia kotzebuei, Petasites frigidus, Picea engelmannii, P. mariana, Pinus contorta, Platanthera obtusata, Poa alpina, P. arctica, P. palustris, P. stenantha, Polemonium caeruleum, Polytrichastrum alpinum, Polytrichum strictum, Populus tremuloides, Primula pumila, Pseudotsuga menziesii, Pyrola, Ranunculus flammula, Ribes lacustre, R. triste, Rubus arcticus, R. chamaemorus, R. pedatus, Salix alaxensis, S. arctica, S. barclayi, S. polaris, S. pulchra, S. reticulata, S. richardsonii, Sarmentypnum sarmentosum, Saxifraga cernua, S. hirculus, Senecio lugens,* *Sphagnum, Stellaria crispa, S. laeta, Streptopus amplexifolius, Thalictrum sparsiflorum, Tsuga heterophylla, Valeriana capitata, Viola selkirkii, Wilhelmsia physodes.*

Use by wildlife: There are few wildlife uses reported for any *Chrysosplenium* species. The unusual algific tallus slope habitat occupied by *C. iowense* is also shared by a number of imperiled snails (Mollusca: Gastropoda) including the Iowa Pleistocene snail (Discidae: *Discus macclintocki*), frigid ambersnail (Succineidae: *Catinella gelida*), Iowa and Minnesota Pleistocene ambersnails (Succineidae: *Novisuccinea*), Briarton Pleistocene vertigo (Pupillidae: *Vertigo briarensis*), Midwest Pleistocene vertigo (*V. hubrichti*), bluff vertigo (*V. meramecensis*), and occult vertigo (*V. occulta*). Cattle (Mammalia: Bovidae: *Bos*) reportedly graze on *C. tetrandrum* when it grows on hummocks; however other accounts state that the plants are poisonous to livestock.

Economic importance: food: The leaves of *Chrysosplenium iowense* and other *Chrsosplenium* species are eaten in salads (raw or cooked) and are described as having a "peppery" flavor (although some authors regard them as poisonous). Naturally, care must be taken to disinfect plants, especially if they have been collected from waters that might be contaminated. *Chrysosplenium tetrandrum* has been found in archeological Thule burial sites and meat caches; **medicinal:** Medicinal uses have not been reported for North American species. Some folk recipes include *C. iowense* in teas to treat urinary infections. *Chrysosplenium carnosum* and *C. nepalense* are used medicinally in Tibet. Compounds with cytotoxic (potentially antitumor) properties have been isolated from *C. flagelliferum* and *C. grayanum*. *Chrysosplenium tosaense* contains axillarin, chrysosplenol B, and chrysosplenol C, which exhibit antiviral activity, especially against rhinovirus; **cultivation:** Several *Chrysosplenium* species are cultivated as ornamentals, although only one North American species (*C. iowense*) is cultivated routinely; **misc. products:** none; **weeds:** none; **nonindigenous species:** none

Systematics: Several molecular data sets have consistently supported the monophyly of *Chrysosplenium* as well as its association with *Peltoboykinia* as sister genera (Figure 4.56). This clade occupies a relatively derived position in the family. *Chrysosplenium* is characterized by two clades (recognized as section *Alternifolia* and section *Oppositifolia*), which are distinguishable by their leaf arrangement (alternate or opposite) as the sectional names imply. North American species occur in both sections (Figure 4.57). *Chrysosplenium americanum* and *C. glechomifolium* resolve as sister species related to the European *C. oppositifolium* (section *Oppositifolia*), whereas *C. tetrandrum* falls within a clade that includes *C. iowense* and *C. japonicum* (section *Alternifolia*). These results indicate at least two distinct origins of the OBL habit in the genus. Some taxonomic issues require additional clarification. *Chrysosplenium iowense* has been known formerly as *C. alternifolium* var. *sibiricum* and as *C. alternifolium* var. *iowense*. *Chrysosplenium tetrandrum* has also been recognized as a subspecies of *C. alternifolium*. Molecular data (Figure 4.57) indicate that *C. iowense* and *C. tetrandrum* are very closely related, and their distinction as distinct species vs. infraspecific taxa remains debatable. The

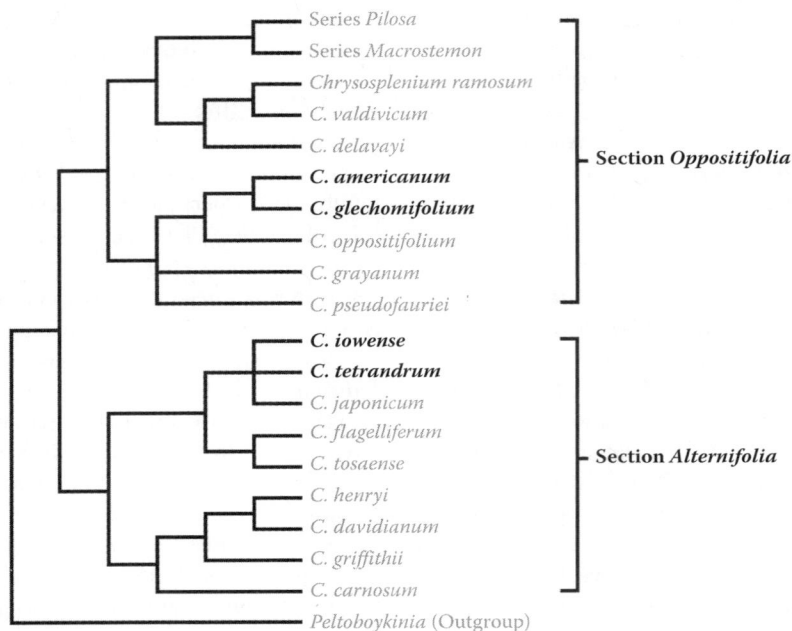

FIGURE 4.57 Relationships among *Chrysosplenium* species as indicated by phylogenetic analysis of *matK* sequence data. The four obligately aquatic North American species (in bold) occur within two discrete sections (clades) in the genus. (Adapted from Soltis, D.E. et al., *Amer. J. Bot.*, 88, 883–893, 2001b.)

base chromosome number for *Chrysosplenium* is difficult to determine because of heterogeneity among different species. Phylogenetic reconstructions indicate that it probably is $x = 11$, with $x = 12$ (as well as several other anomalous numbers) being derived repeatedly due to aneuploidy. Chromosome numbers also exhibit intraspecific variation. Multiple counts ($2n = 18$, 24, 36, 48, 66, c.120) have been attributed to *C. iowense*. *Chrysosplenium americanum* ($2n = 18$) and *C. glechomifolium* ($2n = 18$) presumably are diploid, but derived from a reduced base number of $x = 9$. An anomalous count of $2n = 24$ is also reported for *C. americanum*. *Chrysosplenium tetrandrum* reportedly has diploid and tetraploid cytotypes ($2n = 24$, 48). Hybridization has not been reported for any species in North America (or elsewhere). Further studies of the reproductive ecology and population structure of all species are encouraged, and for the rarer taxa in particular.

Comments: *Chrysosplenium iowense* and *C. tetrandrum* are imperiled throughout most of their northern North American range. Both species also occur in Europe. *Chrysosplenium americanum* is fairly widespread throughout eastern North America, whereas, *C. glechomifolium* is restricted to the western United States (California, Oregon, Washington).

References: Aiken et al., 2007; Arisawa et al., 1991, 1992, 1997; Baskin & Baskin, 1998; Packer, 1963; Perez, 2003; Savile, 1953, 1975; Schwartz, 1985; Smith, 1981b; Soltis et al., 2001a,b; Sperduto & Nichols, 2004; Weber, 1979; Wells & Elvander, 2009.

3. *Darmera*

Umbrella plant; Darméria

Etymology: after Karl Darmer (1843–1918)

Synonyms: *Leptarrhena* (in part); *Peltiphyllum*; *Saxifraga* (in part)

Distribution: global: North America; **North America:** western

Diversity: global: 1 species; **North America:** 1 species

Indicators (USA): OBL: *Darmera peltata*

Habitat: freshwater; riverine; **pH:** 6.0–7.5; **depth:** <1 m; **life-form(s):** emergent herb

Key morphology: rhizomes (to 1 m) fleshy (to 5 cm thick), stemless; leaves basal, the blades (to 1 m) peltate, funnel-like, roundish, margins shallow to deeply lobed, the lobes irregularly toothed, petioles (to 0.5 m) fleshy; inflorescence (to 15 dm) precocious, the naked stalks terminated by roundish (or flat-topped) cymes of numerous, 5-merous flowers; sepals (to 4 mm) reflexed; petals (to 7 mm) showy, white or pink; stamens 10; follicles (to 12 mm) two, reddish terminally

Life history: duration: perennial (rhizomes); **asexual reproduction:** rhizomes; **pollination:** insect; **sexual condition:** hermaphroditic; **fruit:** follicles (common); **local dispersal:** seeds, rhizomes; **long-distance dispersal:** seeds

Imperilment: (1) *Darmera peltata* [G4]

Ecology: general: *Darmera* is monotypic (see next).

Darmera peltata **(Torrey ex Bentham) Voss** occurs along lake, river or stream margins, streambanks, and in rock pools or wet depressions at elevations of up to 2286 m. It is mainly a species of cool riparian montane habitats where it occurs in swift-flowing streams in silt or on rocky substrates (e.g., Amphibole gneiss, olivine gabbro, or serpentine), frequently growing among boulders. The pH range has not been determined adequately, although sites tend to be more acidic (pH: 6.0–7.5). Plants can tolerate full sun to partial shade and are quite resistant to dessication when grown in cultivation. However, they do not tolerate prolonged inundation or high temperatures. The flowers (April to September) emerge early in the spring on leafless stalks and are protandrous (the leaves are produced later in the season). Pollination is by insects. The fresh seeds germinate irregularly (over 1–3 months) after being sown (uncovered on damp soil) in the light, at temperatures of 12°C–15°C. Germination has also been reported using a warm–cold–warm cycle, with the seeds sowed at 22°C for 6 weeks, then brought to 4°C for 6–8 weeks, followed by an incubation at 10°C–22°C. The mechanism of seed dispersal is not known. The plants root firmly, even in rocky substrates, and are an effective sediment stabilizer. Propagation is by seed or division of the stout rhizomes, which can reach lengths of 76–100 cm. The rhizomes spread rapidly and extensively to strongly anchor the plants.

Reported associates: *Abies concolor, Acmispon americanus, Aconitum, Adiantum pedatum, Agrostis exarata, A. gigantea, Allium, Alnus rhombifolia, A. rugosa, Aquilegia, Arbutus menziesii, Arctostaphylos, Aristolochia californica, Artemesia douglasiana, Asarum lemmonii, Athyrium filix-femina, Boykinia major, Brickellia californica, Briza, Calocedrus decurrens, Calycanthus occidentalis, Carex aquatilis, C. senta, C. vesicaria, Ceanothus integerrimus, Chlorogalum pomeridianum, Claytonia perfoliata, Clematis ligusticifolia, Cornus glabrata, C. sericea, C. sessilis, Deschampsia elongata, Dicentra formosa, Dryopteris arguta, Eleocharis acicularis, Epilobium, Epipactis gigantea, Equisetum arvense, Erigeron cervinus, E. philadelphicus, Galium trifidum, Geum macrophyllum, Geranium, Glyceria elata, Helenium bigelovii, Horkelia, Hosackia oblongifolia, Hypericum anagalloides, H. formosum, Juncus effusus, J. ensifolius, J. howellii, J. tenuis, J. xiphioides, Leersia oryzoides, Luina hypoleuca, Lupinus polyphyllus, Luzula, Mimulus cardinalis, M. guttatus, Montia parviflora, Myrica hartwegii, Panicum acuminatum, Perideridia parishii, Phacelia, Philadelphus lewisii, Pinus contorta, P. jeffreyi, P. ponderosa, Platanthera leucostachys, Polygonum, Populus tremuloides, Potentilla, Prunella vulgaris, Pteridium aquilinum, Ranunculus, Rhododendron occidentale, Ribes cereum, R. nevadense, Rubus leucodermis, R. parviflorus, R. ursinus, R. vitifolius, Rumex acetosella, Salix exigua, S. geyeriana, S. jepsonii, S. lasiolepis, S. lucida, Sambucus, Scirpus microcarpus, Senecio triangularis, Solidago canadensis, Spiraea densiflora, Spiranthes romanzoffiana, Stachys alba, Stellaria, Symphoricarpos rivularis, Symphyotrichum campestre, S. eatonii, Typha, Verbascum thapsus, Viola macloskeyi, Vitis californica, Woodwardia fimbriatum.*

Use by wildlife: none reported

Economic importance: food: The young petioles of *Darmera peltata* (known as káaf) were eaten raw as a green vegetable by the Karuk people. When peeled, the petioles can be eaten raw or cooked or added to soup and stew (although cooking causes a reduction in flavor). The Miwok added the pulverized root to acorn meal as a whitening agent; **medicinal:** The Karuk administered a root infusion of *D. peltata* to aid women during their pregnancy. Specimens of *D. peltata* from California produced no antileukemia activity when tested in rodent systems by the National Cancer Institute; **cultivation:** *Darmera peltata* is a popular cultivated ornamental and was given the Award of Garden Merit by the Royal Horticultural Society. A dwarf form is distributed as the cultivar 'Nana' (or var. *nana*) and the larger plants have been known as var. *maxima*; **misc. products:** *Darmera peltata* is planted as a ground cover or to stabilize banks. It has also been considered for use as an additive to ruminant feed; **weeds:** none; **nonindigenous species:** *Darmera peltata* possibly was introduced to Utah, but this issue requires clarification.

Systematics: Combined molecular data resolve *Darmera* either with *Astilboides* as a clade that is sister to *Rodgersia* (cpDNA) or as the sister genus to a clade comprising *Astilboides* and *Rodgersia* (nrDNA) (Figure 4.56). Although combined data favor the latter topology, the result is not strongly supported. Regardless, these three genera are closely related and associate with a clade comprising *Bergenia, Mukdenia,* and *Oresitrophe* to comprise what has been called the "*Darmera* group" (Figure 4.56). *Darmera* is monotypic, consisting only of *D. peltata*. The base chromosome number of *Darmera* is $x = 17$ as *D. peltata* ($2n = 34$) presumably is diploid. Studies are needed on the reproductive ecology and population structure of *Darmera*. Older literature has referred to this species as *Peltiphyllum peltatum* and as *Saxifraga peltata*.

Comments: *Darmera peltata* occurs infrequently and only in California, Oregon, and Utah, principally in the Cascade, Klamath, North Coast, and Sierra Nevada Ranges.

References: Deng et al., 2015; Soltis et al., 2001a; Speichert & Speichert, 2004; Tarbell, 2004; Wells & Elvander, 2009.

4. Micranthes

Pseudosaxifrage

Etymology: from the Greek *mikros anthos*, meaning "small flower"

Synonyms: *Saxifraga* (in part)

Distribution: global: Eurasia; New World; **North America:** widespread

Diversity: global: 68–93 species; **North America:** 45 species

Indicators (USA): OBL: *Micranthes ferruginea, M. foliolosa, M. lyallii, M. marshallii, M. micranthidifolia, M. oregana, M. pensylvanica;* **FACW:** *M. foliolosa, M. lyallii, M. marshallii, M. oregana, M. pensylvanica;* **FAC:** *M. ferruginea*

Habitat: freshwater; palustrine; **pH:** 5.2–8.0; **depth:** <1 m; **life-form(s):** emergent herb

Key morphology: Shoots herbaceous, rosulate, caudices bearing scaly rhizomes or stolons, occasionally producing bulbils; leaves in basal rosettes, estipulate, sessile or petiolate (to 15 cm), the blade (to 35 cm) often fleshy, weakly to strongly ciliate, lanceolate, spatulate, oblanceolate, obovate, or ovate, the margins entire, crenate, serrate, or distally toothed; flowering scapes (to 125 cm) leafless, inflorescence solitary or a 2–30-flowered thyrse (to 125 cm), the flowers 5-merous, radial or bilateral, sometimes replaced by bulbils; sepals reflexed; petals (to 8 mm) slightly to strongly clawed, white, cream, or rarely purple, with or without 1–2 basal, yellow or greenish spots; stamens 10, the filaments linear, club shaped, or petaloid; pistils 2 or more, distinct nearly to base or connate for more than ½ their length, ovary superior to half-inferior, two- to three-locular, styles 2–3; capsules follicle-like or valvate, two- to three-beaked, green, yellow, or purplish; seeds brown, with longitudinal ribs

Life history: duration: perennial (bulbils, rhizomes); **asexual reproduction:** bulbils, rhizomes; **pollination:** insect; **sexual condition:** hermaphroditic; **fruit:** capsules, follicles (common); **local dispersal:** bulbils, rhizomes; **long-distance dispersal:** seeds

Imperilment: (1) *Micranthes ferruginea* [G5]; S3 (<u>AB</u>); (2) *M. foliolosa* [G4]; S1 (CO, ME, <u>NF</u>); S2 (<u>LB</u>, <u>YT</u>); S3 (<u>QC</u>); (3) *M. lyallii* [G5]; (4) *M. marshallii* [G5]; S1 (CA); (5) *M. micranthidifolia* [G5]; S1 (KY); S2 (SC); S3 (GA, NC); (6) *M. oregana* [G4/G5]; (7) *M. pensylvanica* [G5]; SX (KY); S1 (DE, <u>MB</u>, NC, <u>ON</u>, RI, <u>SK</u>, TN); S2 (WV); S3 (IA, ME, NJ, VA)

Ecology: general: Twenty-six of the North American *Micranthes* species (58%) have some designation as wetland indicators, with seven (16%) categorized as OBL in at least a portion of their range. Many aspects of the natural history of this genus are poorly documented. The flowers are bisexual, but can be functionally androdioecious in some species. In the few taxa studied, the flowers are pollinated by insects. In many cases, the flowers are replaced by vegetative bulbils that facilitate survival under extreme (typically frigid) conditions, and can function on occasion as vegetative propagules. Several of the species produce rhizomes or stolons to accommodate vegetative reproduction. All of the species are perennial. No information is available regarding the seed ecology or dispersal mechanisms; however, seed production tends to be high when the plants undergo sexual reproduction.

***Micranthes ferruginea* (Graham) Brouillet & Gornall** occupies balds, beaches, bogs, cliffs, crags, crevices, duffs, flats, hanging gardens, heath, knolls, ledges, meadows, mossy boulders, runnels, seeps, slopes, snowbeds, streambeds, thickets, tundra, woodlands, and the margins of lakes, streams, and vernal pools at elevations of up to 2744 m. Exposures include shade or partial shade. The substrates usually are described as acidic, have a relatively low organic matter content (5.4–10.5 mg g⁻¹ dry weight), and include argillite, basalt, Belt series mudstone and quartzite, boulders, clay, cobble, gneiss, granite, gravel, limestone, loam sand, mica-schist, mud, organic matter, pumice, quartz diorite, Raft-Batholith metamorphics, rock, rocky loam, sand, sandy loam, scoria, scree, silt, silt loam, slate, talus, and volcanic ash. The plants flower from summer through early autumn. The reproductive ecology has not been described, but the plants typically produce a fairly large quantity of seeds. In the southern portion of their distribution, it is common to find plants with the flowers replaced by vegetative bulbils. The plants produce an extensive root system that can reach 10 cm in depth. Vegetative reproduction occurs by means of slender rhizomes. The plants are nonmycorrhizal and appear to be highly resistant to copper. The presence of *M. ferruginea* has been shown to correlate negatively with the occurrence of willow species (*Salix*), despite their reports as associated species (see later). The species has been described as a pioneer of newly deglaciated terrain and barren volcanic plains, where it establishes mainly in sites that trap the seeds and protect them from dessication. **Reported associates:** *Abies grandis, A. lasiocarpa, A. magnifica, A. ×shastensis, Acer circinatum, A. glabrum, Achillea millefolium, Aconogonon davisiae, A. phytolaccifolium, Adenocaulon bicolor, Adiantum pedatum, Agrostis pallens, Aira, Allium cernuum, A. crenulatum, Alnus viridis, Anaphalis margaritacea, Angelica dawsonii, Antennaria alpina, A. lanata, Anthoxanthum monticola, Aquilegia formosa, Arabis nuttallii, Arctostaphylos uva-ursi, Arnica cordifolia, A. latifolia, A. rydbergii, Artemisia norvegica, Aruncus dioicus, Athryium filix-femina, Bistorta bistortoides, Boechera divaricarpa, Boykinia major, Callitropsis nootkatensis, Calocedrus decurrens, Calochortus, Caltha leptosepala, Campanula wilkinsiana, Cardamine bellidifolia, Carex circinata, C. lachenalii, C. macrochaeta, C. microchaeta, C. nigricans, C. pachystachya, C. pyrenaica, C. rossii, C. rupestris, C. spectabilis, Cassiope mertensiana, C. tetragona, Castilleja hispida, C. miniata, C. parviflora, C. rhexiifolia, C. suksdorfii, Chamaecyparis, Chamerion angustifolium, C. latifolium, Cheilanthes gracillima, Cladonia, Claytonia cordifolia, C. sibirica, Clintonia uniflora, Cryptogramma acrostichoides, C. crispa, Cystopteris fragilis, Danthonia, Dasiphora floribunda, Deschampsia cespitosa, Dicentra formosa, Digitalis, Draba, Dryopteris, Dryptodon patens, Empetrum nigrum, Epilobium anagallidifolium, E. minutum, Eremogone capillaris, E. congesta, Erigeron glacialis, E. peregrinus, Eriogonum ovalifolium, Eriophorum angustifolium, Erythronium grandiflorum, E. montanum, Eucephalus ledophyllus, Festuca altaica, F. brachyphylla, F. viridula, Galium oreganum, Gaultheria shallon, Geum calthifolium, Glyceria, Harrimanella stelleriana, Heuchera cylindrica, H. glabra, Hieracium triste, Holodiscus discolor, Hypericum scouleri, Hypochaeris, Juncus drummondii, J. mertensianus, J. parryi, Juniperus communis, Kalmia procumbens, Larix lyallii, Leucanthemum vulgare, Lewisia columbiana, L. cotyledon, L. triphylla, Leymus mollis, Lomatium ambiguum, L. grayi, L. martindalei, Lonicera involucrata, Luetkea pectinata, Lupinus latifolius, Luzula arcuata, L. campestris, L. hitchcockii, L. parviflora, L. piperi, L. spicata, L. wahlenbergii, Maianthemum stellatum, Menziesia ferruginea, Mertensia paniculata, Micranthes nelsoniana, M. occidentalis, M. odontoloma, M. tolmiei, Mimulus guttatus, M. lewisii, M. tilingii, Minuartia rubella, Mitella breweri, Myosotis asiatica, Nardia scalaris, Neottia banksiana, Nothocalais, Oplopanax horridus, Oreostemma alpigenum, Orthocarpus imbricatus, Osmorhiza purpurea, Oxyria digyna, Parnassia fimbriata, Pedicularis bracteosa, Penstemon davidsonii, P. fruticosus, P. montanus, P. newberryi, P. procerus, Phleum alpinum, Phlox diffusa, Phyllodoce aleutica, P. empetriformis, Picea breweriana, P. engelmannii, Pinus albicaulis, P. contorta, P. monticola, Plectritis, Poa alpina, Podagrostis thurberiana, Polemonium pulcherrimum, Polystichum lonchitis, P. munitum, Potentilla flabellifolia, P. ×diversifolia, Primula jeffreyi, P. pauciflora, Pseudotsuga menziesii, Puccinellia rupestris, Pyrola secunda, Quercus garryana, Ranunculus cooleyae, R. eschscholtzii, Rhacomitrium, Rhodiola rosea,*

Rhododendron albiflorum, Ribes acerifolium, Romanzoffia sitchensis, Rosa gymnocarpa, Rubus lasiococcus, R. parviflorus, R. pedatus, R. spectabilis, Salix arctica, S. commutata, S. drummondiana, S. planifolia, S. polaris, S. sitchensis, S. stolonifera, Sambucus racemosa, Sanguisorba, Saussurea americana, Saxifraga bronchialis, S. mertensiana, S. tricuspidata, Sedum divergens, S. lanceolatum, S. obtusatum, S. stenopetalum, Selaginella wallacei, Senecio triangularis, Sibbaldia procumbens, Silene acaulis, Sorbus sitchensis, Spiraea splendens, Stellaria calycantha, S. crispa, Stereocaulon, Suksdorfia ranunculifolia, Synthyris platycarpa, Taxus brevifolia, Thuja plicata, Triantha glutinosa, Trichophorum cespitosum, Trisetum spicatum, Tsuga heterophylla, T. mertensiana, Turritis glabra, Vaccinium cespitosum, V. deliciosum, V. membranaceum, V. uliginosum, V. vitis-idaea, Vahlodea atropurpurea, Valeriana sitchensis, Veratrum californicum, V. viride, Veronica cusickii, V. wormskjoldii, Viola adunca, Xerophyllum tenax.

Micranthes foliolosa (R. Br.) Gornall grows in or on alluvial fans, beaches, bluffs, depressions, flats, floodplains, frost boils, hummocks, ledges, meadows, scree slopes, seepage fens, snowbeds, swales, tundra, and along the margins of boreal lakes, rivers, streams, and ponds at elevations of up to 1800 m. The substrates more often are described as acidic but include calcite, clay, granitic sandy loam, gravel, humus, limestone, mud, peat, rock, sand, and silty gravel with the plants often found growing among mosses. This species occurs mainly at Arctic latitudes as far as 83°03′N. Individuals contain morphological males or hermaphrodites (i.e., they are androdioecious). Most of the flowers on an inflorescence (except the terminal ones) are replaced by small, vegetative bulbils. More flowers are produced in warmer years. There is a very high incidence of meiotic abnormalities (see *Systematics* later) resulting in populations that are highly asexual. Plants propagate vegetatively by short, leafy horizontal stems (rhizomes) that terminate in leafy rosettes. They also proliferate vegetatively by the inflorescence bulbils (which detach from the plant and are capable of independent establishment) or by shoot fragmentation. Fewer, shorter, and more delicate stems are produced under harsher conditions. The efficient means of vegetative reproduction facilitates short-term (<20 years) tolerance to habitat disturbance. Field studies indicate that plants occur more commonly in sites that have been grazed by muskoxen (Mammalia: Bovidae: *Ovibos moschatus*). **Reported associates:** *Aconitum delphiniifolium, Alnus viridis, Alopecurus alpinus, Anemone narcissiflora, A. parviflora, Arctagrostis latifolia, Arctophila fulva, Arnica lanceolata, Artemisia norvegica, Athyrium alpestre, Aulacomnium palustre, Betula glandulosa, B. nana, Bistorta vivipara, Blepharostoma trichophyllum, Braya glabella, Bryum pseudotriquetrum, Calamagrostis canadensis, C. holmii, Campanula lasiocarpa, Cardamine bellidifolia, C. pratensis, Carex aquatilis, C. atrofusca, C. bicolor, C. bigelowii, C. lachenalii, C. macrochaeta, C. magellanica, C. membranacea, C. misandra, C. podocarpa, C. rariflora, C. rotundata, C. scirpoidea, C. subspathacea, C. tenuiflora, C. williamsii, Cassiope tetragona, Cerastium*

alpinum, C. beeringianum, Chrysosplenium tetrandrum, Cinclidium arcticum, Claytonia sarmentosa, C. scammaniana, Cochlearia officinalis, Comarum palustre, Dasiphora floribunda, Deschampsia brevifolia, Dicranum angustum, Ditrichum flexicaule, Draba corymbosa, D. lactea, D. micropetala, Dryas integrifolia, Dupontia fisheri, Empetrum nigrum, Epilobium anagallidifolium, E. hornemannii, E. lactiflorum, Equisetum arvense, Eriophorum angustifolium, E. callitrix, E. chamissonis, E. russeolum, E. scheuchzeri, E. triste, E. vaginatum, Festuca baffinensis, F. brachyphylla, Gnaphalium norvegicum, G. supinum, Hierochloe alpina, H. pauciflora, Huperzia miyoshiana, Juncus biglumis, J. castaneus, J. trifidus, J. triglumis, Lagotis glauca, Lewisia pygmaea, Lupinus nootkatensis, Luzula arctica, L. confusa, L. wahlenbergii, Melandrium apetalum, Micranthes nelsoniana, M. nivalis, M. tenuis, Montia bostockii, Oxyria digyna, Papaver hultenii, P. lapponicum, Pedicularis kanei, P. sudetica, Peltigera aphthosa, Petasites frigidus, Phegopteris connectilis, Phippsia algida, Phleum alpinum, Plueropogon sabinei, Poa arctica, P. malacantha, Pogonatum alpinum, Pohlia juniperinum, Polytrichum juniperinum, Potentilla hyparctica, Prenanthes alata, Primula pumila, Ranunculus eschscholtzii, R. glacialis, R. hyparctica, R. hyperboreus, R. nivalis, R. pygmaeus, R. sulphureus, Rhododendron tomentosum, Rubus chamaemorus, Sagina intermedia, S. nivalis, Salix arctica, S. argyrocarpa, S. chamissonis, S. herbacea, S. pulchra, S. reticulata, S. rotundifolia, Sanguisorba canadensis, S. stipulata, Saxifraga cernua, S. cespitosa, S. flagellaris, S. hieracifolia, S. hirculus, S. hyperborea, S. oppositifolia, S. paniculata, S. polaris, S. pulchra, S. rivularis, Scorpidium turgescens, Senecio atropurpureus, Sibbaldia procumbens, Solidago macrophylla, Sphagnum, Stellaria borealis, S. humifusa, S. laeta, S. longipes, S. umbellata, Streptopus amplexicaulis, Timmia austriaca, Tofieldia pusilla, Trichophorum cespitosum, Tritomaria quinquedentata, Vaccinium cespitosum, V. uliginosum, V. vitis-idaea, Vahlodea atropurpurea, Veronica wormskjoldii, Viola palustris, Wilhelmsia physodes.

Micranthes lyallii (Engl.) Small inhabits bogs, cirques, cliffs, crevices, fens, floodplains, gravel bars, gullies, hummocks, krummholz, marshes, meadows, ridges, rivulets, runnels, seeps, slopes, streambeds, and the margins of lakes, ponds, springs, and streams at elevations (often alpine) of up to 2805 m. The plants occur commonly on late snowbeds, which can persist until mid-August. The sites can range from full sun to shady exposures. The substrates frequently are described as acidic but site measurements indicate alkaline conditions (pH: 7.1–8.0). Analyzed substrates indicate a high sand (46%–51%) and silt (41%–46%) composition with low levels of organic matter (~4%), total nitrogen (0.1%–0.2%), and available phosphorous (5–8 ppm). Reported types include boulders, clay, cobble, granite talus, krummholz, limestone, limestone breccia, organic, peat, Raft–Batholith metamorphics, rock, rocky scree, rubble, sand, sandy gravel, shale, silt, silty alluvium, and volcanic ash. Specific soil types include alpine eutric brunisols, cumulic regosols, eutric cryochrepts, orthic regosols, and typic cryumbrepts. The plants are in flower from summer through early fall. Little information exists on the reproductive biology or seed ecology. The plants reproduce vegetatively by the formation of rhizomes and can develop into fairly extensive mats. **Reported associates:** *Abies lasiocarpa, Achillea millefolium, Aconitum columbianum, A. delphiniifolium, Actaea rubra, Agrostis scabra, Alopecurus magellanicus, Anemone narcissiflora, A. occidentalis, A. parviflora, Antennaria rosea, Aquilegia flavescens, Arnica mollis, Artemisia norvegica, A. tilesii, Bistorta bistortoides, B. vivipara, Bromus carinatus, Calamagrostis purpurascens, Caltha leptosepala, Camassia quamash, Cardamine umbellata, Carex anthoxanthea, C. bigelowii, C. lachenalii, C. macrochaeta, C. microchaeta, C. nigricans, C. podocarpa, Cassiope tetragona, Castilleja hyperborea, C. miniata, Cetraria nivalis, Chamerion latifolium, Chrysosplenium tetrandrum, Cirsium scariosum, Claytonia lanceolata, C. sarmentosa, Collinsia parviflora, Cystopteris fragilis, Dactylis glomerata, Delphinium depauperatum, Dryas integrifolia, Equisetum, Erigeron aureus, E. peregrinus, Eriophorum, Erythronium grandiflorum, Festuca altaica, Geranium erianthum, Gymnocarpium dryopteris, Harrimanella stelleriana, Heracleum, Heuchera glabra, Juncus drummondii, Juniperus, Larix lyalli, Luetkea pectinata, Lupinus nootkatensis, Luzula hitchcockii, Melica bulbosa, Mertensia ciliata, M. paniculata, Micranthes ferruginea, M. nelsoniana, M. occidentalis, M. odontoloma, M. tolmiei, Mimulus lewisii, Minuartia rubella, Moneses uniflora, Myosotis asiatica, Osmorhiza purpurea, Oxyria digyna, Packera cymbalaria, Parnassia fimbriata, Pedicularis contorta, P. verticillata, Penstemon fruticosus, Petasites frigidus, Philonotis, Phleum alpinum, Phyllodoce empetriformis, Phippsia algida, Picea engelmanii, P. glauca, Pinus, Platanthera dilatata, Poa alpina, Pohlia wahlenbergii, Polemonium caeruleum, P. pulcherrimum, Polystichum lonchitis, P. munitum, Polytrichum proliferum, Potamogeton gramineus, Prenanthes alata, Primula frigida, Pseudotsuga menziesii, Ranunculus eschscholtzii, R. glaberrimus, Rhodiola integrifolia, R. rosea, Rumex crispus, Salix arctica, S. barclayi, S. nivalis, S. polaris, S. reticulata, S. rotundifolia, Sanguisorba stipulata, Saxifraga oppositifolia, S. serpyllifolia, S. tricuspidata, Selaginella densa, Senecio fremontii, S. serra, S. triangularis, Sibbaldia procumbens, Silene acaulis, Solidago multiradiata, Taraxacum officinale, Telesonix heucheriformis, Thalictrum occidentale, Thuja plicata, Trifolium haydenii, T. pratense, T. repens, Trisetum spicatum, Trollius laxus, Tsuga heterophylla, T. mertensiana, Vaccinium uliginosum, Vahlodea atropurpurea, Valeriana sitchensis, Veratrum viride, Veronica cusickii, V. wormskjoldii, Viola, Woodsia oregana, Xerophyllum tenax.*

***Micranthes marshallii* (Greene) Small** grows in or on bluffs, cliffs, crevices, draws, meadows, overhangs, ridges, river bottoms, seeps, slopes, terraces, tundra, and along the margins of rivers and streams at elevations of up to 2133 m. The sites range from open exposures to deep shade. The substrates are described as gravel, humus, limestone, rock, sandstone, sandy loam, and talus. Flowering occurs during the spring months. Additional life-history information for this species is scarce. **Reported associates:** *Acer, Adiantum, Amelanchier, Calamagrostis rubescens, Camassia, Cerastium, Claytonia*

perfoliata, Daucus, Digitalis, Galium, Heuchera, Lomatium dissectum, Luzula parviflora, Micranthes integrifolia, M. odontoloma, Mimulus, Pentagramma triangularis, Pinus ponderosa, Pityrogramma, Polypodium calirhiza, Primula pauciflora, Pseudoroegneria spicata, Pseudotsuga, Quercus chrysolepis, Rubus, Salix, Saxifraga mertensiana, Sedum, Selaginella, Sonchus, Suksdorfia violacea, Vicia.

***Micranthes micranthidifolia* (Haw.) Small** inhabits montane brook margins, seeps, spray cliffs, and swamps at elevations of up to 2100 m. The substrates are acidic to circumneutral and consist of rock or gravel (often moss covered) that is underlain by granite, calcareous shale, or limestone. Sites generally are shady. There is essentially no information known regarding the reproductive ecology of this species or other aspects of its natural history. **Reported associates:** *Acer saccharum, Acontium reclinatum, Aesculus flava, Ageratina altissima, Agrostis perennans, Allium tricoccum, Alnus serrulata, Bazzania trilobata, Betula alleghaniensis, B. lenta, Boehmeria cylindrica, Boykinia aconitifolia, Brachyelytrum erectum, Brotherella recurvans, Caltha palustris, Calypogeia muelleriana, Cardamine clematitis, Carex bromoides, C. brunnescens, C. debilis, C. folliculata, C. gynandra, C. leptalea, C. plantaginea, C. prasina, C. scabrata, C. stricta, C. vulpinoidea, Chelone glabra, C. lyonii, Chrysosplenium americanum, Cicuta maculata, Cimicifuga racemosa, Cinna latifolia, Circaea alpina, Cladium mariscoides, Collinsonia canadensis, Cornus amomum, Cryptotaenia canadensis, Deparia acrostichoides, Diphylleia cymosa, Drosera rotundifolia, Dryopteris intermedia, Euonymus obovata, Eupatorium purpureum, E. rugosum, Euphorbia purpurea, Fagus grandifolia, Festuca subverticillata, Fragaria virginica, Galium mollugo, G. triflorum, Geum geniculatum, Glyceria, Helenium autumnale, Heuchera parviflora, Houstonia caerulea, H. serpyllifolia, Huperzia porophila, Hydrangea arborescens, Hydrocotyle americana, Hypericum mutilum, Hypnum curvifolium, H. pallescens, Impatiens capensis, I. pallida, Juncus effusus, J. subcaudatus, Laportea canadensis, Lilium grayi, Lindera benzoin, Micranthes caroliniana, M. pensylvanica, M. petiolaris, Microstegium vimineum, Mitella diphylla, Mitrula paludesa, Monarda didyma, Osmunda regalis, Osmundastrum cinnamomea, Oxypolis rigidior, Packera aurea, Parnassia asarifolia, Persicaria virginiana, Platanthera grandiflora, Poa paludigena, Polystichum acrostichoides, Ptilidium pulcherrimum, Ranunculus recurvatus, Rhododendron maximum, Rhynchospora capitellata, Rudbeckia lacinata, Salix sericea, Sambucus nigra, Saururus cernuus, Scapania undulata, Solidago curtisii, S. patula, Sphagnum, Stachys, Stellaria corei, Symphyotrichum novaeangliae, S. subulatum, Symplocarpus foetidus, Tetraphis pellucida, Thalictrum clavatum, T. dioicum, Tiarella cordifolia, Tilia americana, Trautvetteria caroliniensis, Ulmus americana, Uvularia sessilifolia, Veratrum viride, Veronica americana, Viola blanda, V. cucullata, V. macloskeyi, V. sororia, Xanthorhiza simplicissima.*

***Micranthes oregana* (T.J. Howell) Small** grows in bogs, marshes, meadows, prairies, and along stream margins at elevations of up to 2500 m. Substrates include basalt and clay.

The plants flower from early spring through summer. They persist and reproduce vegetatively by means of stout rhizomes. Otherwise, there is a paucity of information on the reproductive ecology and natural history of this species. **Reported associates:** *Abies amabilis, A. lasiocarpa, Achillea millefolia, Achlys triphylla, Aconitum columbianum, Agoseris glauca, Aira caryophyllea, Allium validum, Alnus rugosa, Angelica arguta, Antennaria parvifolia, Aquilegia formosa, Arabis drummondii, Athyrium filix-femina, Besseya rubra, Betula nana, Bistorta bistortoides, Boykinia major, Calamagrostis canadensis, Caltha leptosepala, Camassia quamash, Canadanthus modestus, Cardamine penduliflora, Carex aquatilis, C. brunnescens, C. canescens, C. capitata, C. cusickii, C. disperma, C. exsiccata, C. illota, C. jonesii, C. laeviculmis, C. lenticularis, C. luzulina, C. nigricans, C. pachystachya, C. scopulorum, C. spectabilis, C. utriculata, Cerastium arvense, Cornus sericea, Danthonia intermedia, Deschampsia cespitosa, Drosera rotundifolia, Eleocharis palustris, E. quinqueflora, Epilobium ciliatum, Equisetum arvense, Erigeron peregrinus, Eriophorum altaicum, E. gracile, Erythronium grandiflorum, Festuca idahoensis, Galium aparine, Glyceria grandis, G. striata, Hosackia oblongifolia, Hypericum anagalloides, Hypochaeris radicata, Juncus balticus, J. bufonius, J. xiphoides, Kalmia polifolia, Lemna minor, Ligusticum grayi, Lonicera caurina, L. involucrata, Lupinus argenteus, L. polyphyllus, Luzula comosa, Lysichiton americanus, Maianthemum stellatum, Mimulus guttatus, M. primuloides, Mitella pentandra, Montia, Muhlenbergia filiformis, Myosotis sylvatica, Oenanthe sarmentosa, Oreostemma alpigenum, Packera cymbalarioides, Parnassia fimbriata, Pedicularis cystopteridifolia, P. groenlandica, Penstemon procerus, Petasites frigidus, Phleum alpinum, Picea engelmannii, Platanthera dilatata, Poa alpina, P. cusickii, P. pratensis, P. trivialis, Potentilla diversifolia, P. flabellifolia, Primula jeffreyi, P. pauciflora, Ranunculus alismifolius, R. occidentalis, R. populago, Rhododendron columbianum, Ribes bracteosum, Rubus spectabilis, Salix commutata, S. geyeriana, S. hookeriana, S. myrtillifolia, S. orestera, S. sitchensis, Sanguisorba officinalis, Senecio crassulus, S. triangularis, Sidalcea campestris, S. virgata, Sorbus sitchensis, Spiraea douglasii, Stachys ajugoides, S. ciliata, Stellaria, Symphyotrichum spathulatum, Tiarella trifoliata, Tolmiea menziesii, Torreyochloa pallida, Triantha glutinosa, T. occidentalis, Trientalis europaea, Trifolium longipes, Triteleia hyacinthina, Tsuga heterophylla, Vaccinium occidentale, V. ovalifolium, Valeriana sitchensis, Viburnum edule, Viola adunca, V. glabella, V. macloskeyi.*

***Micranthes pensylvanica* (L.) Haw.** inhabits bogs, cliffs, depressions, ditches, marshes, meadows, prairies, seeps, swamps, and the margins of brooks, ponds, rivers, and streams at elevations of up to 1400 m. The diverse substrates vary from alkaline to acidic (pH: 5.2–7.7) and include calcareous shale, clay, decaying logs, granite boulders, moss hummocks, limestone or sandstone ledges, loam, muck, peat, and sand. The plants can tolerate full sun to partial shade. The flowers are pollinated by flies (Insecta: Diptera). For seed germination, it is recommended that they be sown on the surface at −4°C to 4°C for 12 weeks, and then brought to 20°C. **Reported associates:** *Abies balsamea, Acer rubrum, A. saccharinum, A. saccharum, A. spicatum, Achillea millefolium, Adiantum pedatum, Agrimonia parviflora, Allium canadense, Alnus rugosa, Amelanchier, Amerorchis rotundifolia, Ambrosia trifida, Amphicarpa bracteata, Andropogon gerardii, Angelica atropurpurea, Apios americana, Apocynum cannabinum, Aquilegia canadensis, Arabis lyrata, Aralia nudicaulis, Arisaema triphyllum, Asarum canadense, Athyrium felix-femina, Betula allegheniensis, B. papyrifera, B. nigra, B. pumila, B. ×sandbergi, Bromus ciliatus, Bryoxiphium norvegicum, Calamagrostis canadensis, Calla palustris, Caltha palustris, Campanula rotundifolia, Carex bebbii, C. bromoides, C. buxbaumii, C. gracillima, C. gynandra, C. interior, C. intumescens, C. lacustris, C. leptalea, C. prasina, C. retrorsa, C. stricta, C. trisperma, Castilleja coccinea, Chrysosplenium americanum, Cicuta maculata, Cinna latifolia, Circaea alpina, Clintonia borealis, Conocephalum conicum, Convolvulus sepium, Coptis groenlandica, C. trifolia, Corallorhiza trifida, Cornus canadensis, C. racemosa, C. sericea, Cryptotaenia canadensis, Cyperus, Cypripedium parviflorum, Cystopteris bulbifera, C. fragilis, Dicranella heteromalla, Diervilla lonicera, Dryopteris carthusiana, D. cristata, D. disjuncta, D. goldiana, D. marginalis, D. spinulosa, Equisetum fluviatile, E. sylvaticum, Erigeron annuus, Eutrochium maculatum, Fallopia cilinodis, Fragaria virginiana, Fraxinus nigra, Galium aparine, G. triflorum, Gaultheria, Gentianopsis, Geum aleppicum, G. canadense, G. rivale, Goodyera repens, Glyceria canadensis, G. striata, Hamamelis virginiana, Helianthus giganteus, H. grosseserratus, Heracleum lanatum, Heuchera richardsonii, Hydrocotyle americana, Hypoxis hirsuta, Ilex verticillata, Impatiens capensis, Juglans cinerea, Koeleria macrantha, Laportea canadensis, Larix laricina, Leonurus cardiaca, Liatris, Lilium michiganense, Lobelia syphilitica, Lycopus americanus, L. uniflorus, Lysimachia ciliata, L. thyrsiflora, Maianthemum canadense, M. racemosum, M. stellatum, Melanthium virginicum, Melilotus alba, Mentha arvensis, Menyanthes trifoliata, Micranthes micranthidifolia, Milium effusum, Mitella nuda, Mnium, Muhlenbergia mexicana, Nepeta cataria, Oenothera biennis, Onoclea sensibilis, Osmorhiza claytonii, O. longistylis, Osmunda claytoniana, O. regalis, Osmundastrum cinnamomeum, Ostrya, Oxalis montana, Oxypolis rigidior, Packera aurea, Persicaria arifolia, Phalaris arundinacea, Picea rubens, Pilea pumila, Pinus strobus, Platanthera hyperborea, Poa paludigena, Polemonium reptans, P. van-bruntiae, Polygonatum biflorum, Populus tremuloides, Potentilla simplex, Prenanthes alba, Primula meadia, Prunus serotina, Pycnanthemum virginianum, Quercus bicolor, Q. rubra, Ranunculus acris, R. septentrionalis, Rhododendron groenlandicum, Rhynchospora alba, Ribes americanum, R. missouriense, Rosa palustris, Rubus hispida, R. occidentalis, Salix bebbiana, S. candida, S. interior, Scheuchzeria palustris, Scutellaria galericulata, Smilax ecirrhata, Solidago gigantea, S. riddellii, S. sciaphila, Sorghastrum nutans, Spartina pectinata, Sphagnum magellanicum, S. russowii, S. squarrosum, Spiraea, Stellaria longifolia, Streptopus roseus, Sullivantia renifolia, Symphyotrichum lanceolatum, S. lateriflorum, Symplocarpus foetidus, Taraxacum officinale, Tetraphis pellucida, Thalictrum dasycarpum, Thaspium trifoliatum,*

Thelypteris palustris, Thuja occidentalis, Tilia americana, Toxicodendron vernix, Trientalis borealis, Trollius laxus, Tsuga canadensis, Typha, Ulmus americana, U. thomasii, Urtica dioica, Vaccinium corymbosum, V. macrocarpon, Verbena hastata, Veratrum viride, Veronica arvensis, Veronicastrum virginicum, Viburnum lentago, V. recognitum, V. trilobum, Viola pallens, Vitis riparia, Woodsia ilvensis, Zizia aurea.

Use by wildlife: *Micranthes micranthidifolia* is a habitat component of the rare Blue Ridge Mountain amphipod (Crustacea: Amphipoda: Crangonyctidae: *Stygobromus spinosus*). It is grazed on (and impacted negatively) by white-tailed deer (Mammalia: Cervidae: *Odocoileus virginiana*). Larvae of a rare caddisfly (Insecta: Trichoptera: Limnephilidae: *Desmona bethula*) have been observed feeding on *M. oregana*. *Micranthes marshallii* and *M. oregana* are host plants for rust fungi (Basidiomycota: Pucciniaceae: *Puccinia heucherae*).

Economic importance: food: The foliage of *Micranthes foliolosa* contains fairly high amounts of riboflavin, thiamine, niacin, and ascorbic acid. The leaves of *M. micranthidifolia* and *M. pensylvanica* (when gathered before the flowers appear) are edible raw in salads or cooked as greens, sometimes with wild leek (Alliaceae: *Allium tricoccum*). The crisp, hollow flower stalks of *M. pensylvanica* also are edible and reportedly are very tasty (when young), despite their somewhat unpleasant hairy texture; **medicinal:** Decoctions made from the roots and leaves of *M. ferruginea* have been used to treat urinary disorders. The Cherokee applied a root poultice made from *M. pensylvanica* to relieve sore muscles. The Iroquois administered infusions of its roots and leaves to treat dropsy, to purify the blood, and as a remedy for weak kidneys. The plants were also used as a general medicinal remedy by the Menominee; **cultivation:** *Micranthes micranthidifolia* and *M. pensylvanica* are cultivated, often as rock garden ornamentals; **misc. products:** *Micranthes oregana* has been planted as a "biofilter" for pollutant removal in stormwaters; **weeds:** none; **nonindigenous species:** none

Systematics: Although *Micranthes* species long had been recognized as a discrete section (section *Micranthes*) within the genus *Saxifraga*, phylogenetic analyses showed the two taxa to be quite distant (Figures 4.56 and 4.58), which has resulted in the recognition of *Micranthes* as a distinct genus. Some studies (Figure 4.58) resolve *Micranthes* close to the *Darmera* and *Heuchera* "groups" of Saxifragaceae; however, other studies place it as the sister group of *Chrysosplenium* (also OBL) *Peltoboykinia* (Figures 4.56 and 4.59), which seems to be more accurate. Recent phylogenetic studies recommend the recognition of eight sections within *Micranthes*, with the OBL species distributed as follows: section *Micranthes* (*M. marshallii, M. micranthidifolia, M. oregana, M. pensylvanica*), section *Rotundifoliatae* (*M. lyallii*), and section *Stellares* (*M. ferruginea, M. foliolosa*). The close relationship proposed between *M. lyallii* and *M. odontoloma* (Piper) A. Heller by their ability to hybridize is corroborated by phylogenetic analysis, which resolves these as sister species (Figure 4.59). *Micranthes foliolosa* often has been included among

an "aggregate" of species with *M. ferruginea, M. redofskyi* (Adams) Elven & D.F. Murray, and *M. stellaris* (L.) Galasso, Banfi & Soldano, along with the suggestion that it may represent a series of asexual polyploid races derived from these species. A relatively close association of these species is supported by phylogenetic analysis (Figure 4.59). Earlier phylogenetic analyses of a small subset of species resolved *M. foliolosa* as the sister species to *M. ferruginea*, and then in turn to *M. stellaris* (Figure 4.58); however, a larger survey of taxa indicates a more distant relationship among these three taxa (Figure 4.59). Additional sampling of these taxa and inclusion of nuclear (biparentally inherited) DNA markers will be necessary to further test hypotheses regarding the origin of *M. foliolosa*. *Micranthes ferruginea* is similar morphologically and in its flavonoid chemistry to *M. petiolaris* (Raf.) Bush of the eastern United States. Phylogenetic analysis (Figure 4.59) confirms their close relationship. *Micranthes ferruginea* is a variable species, but infraspecific subdivisions are not justified taxonomically. There is some disagreement whether to recognize *M. idahoensis* as distinct from or as a subspecies of *M. marshallii*. The basic chromosome number of *Micranthes* is regarded as $x = 8$, but is difficult to determine due to the extensive variation found among species. The following counts and cytotypes have been reported: *M. ferruginea* ($2n = 20$, 38 [the former cytotype is found mainly south of and the latter mainly north of the glacial boundary]); *M. foliolosa* ($2n = 40$, 48, 56 [common], 64, 66); *M. lyallii* ($2n = 56$, 58); *M. marshallii* ($2n = 20$); *M. micranthidifolia* ($2n = 22$); *M. oregana* ($2n = 38$, 72, 76); and *M. pensylvanica* ($2n = 56$, 84, 112). Chromosome number variation in *M. pensylvanica* has been attributed to autopolyploidy and intercytotypic hybridization.

Comments: *Micranthes foliolosa* is a circumpolar species that grows in far northern North America and in northern Europe. *Micranthes pensylvanica* is widespread in northeastern North America and *M. micranthidifolia* is limited to the eastern United States. *Micranthes ferruginea* and *M. oregana* occur in western North America, whereas *M. lyallii* is found in northwestern North America (also extending into Russia); *M. marshallii* is restricted to the western United States (California, Idaho, Montana, & Oregon).

References: Aiken et al., 2007; Alsos et al., 2003; Arft et al., 1999; Brochmann & Håpnes, 2001; Brouillet & Evander, 2009; Brouillet et al., 1998; Burns, 1942; Calder & Savile, 1960; Cázares et al., 2005; Christy, 2004; Cooper et al., 2004; Crête et al., 2001; Del Moral & Wood, 1993; Fernald & Kinsey, 1943; Forbes, 1996; Forbes et al., 2001; Guard, 1995; Gustafson, 1954; Henry, 1998; Hollingsworth et al., 1998; Hollister & Webber, 2000; Jackson, 1918; Jones & Del Moral, 2005; Jumpponen et al., 1998, 1999; Knapik et al., 1973; McGregor, 2008; Miller & Bohm, 1980; Molau, 1992; Molau & Prentice, 1992; Nikolin & Petrovskii, 1988; Ohtonen et al., 1999; Olesen & Warncke, 1989a,b,c; Oliver et al., 2006; Pannell, 2002; Peinado et al., 2005b; Quattrocchi, 2012; Randhawa & Beamish, 1970, 1972; Savile, 1975; Scotter, 1975; Shacklette, 1961; Soltis et al., 1996b; Teeri, 1976; Tkach et al., 2015; Walker & Everett, 1991.

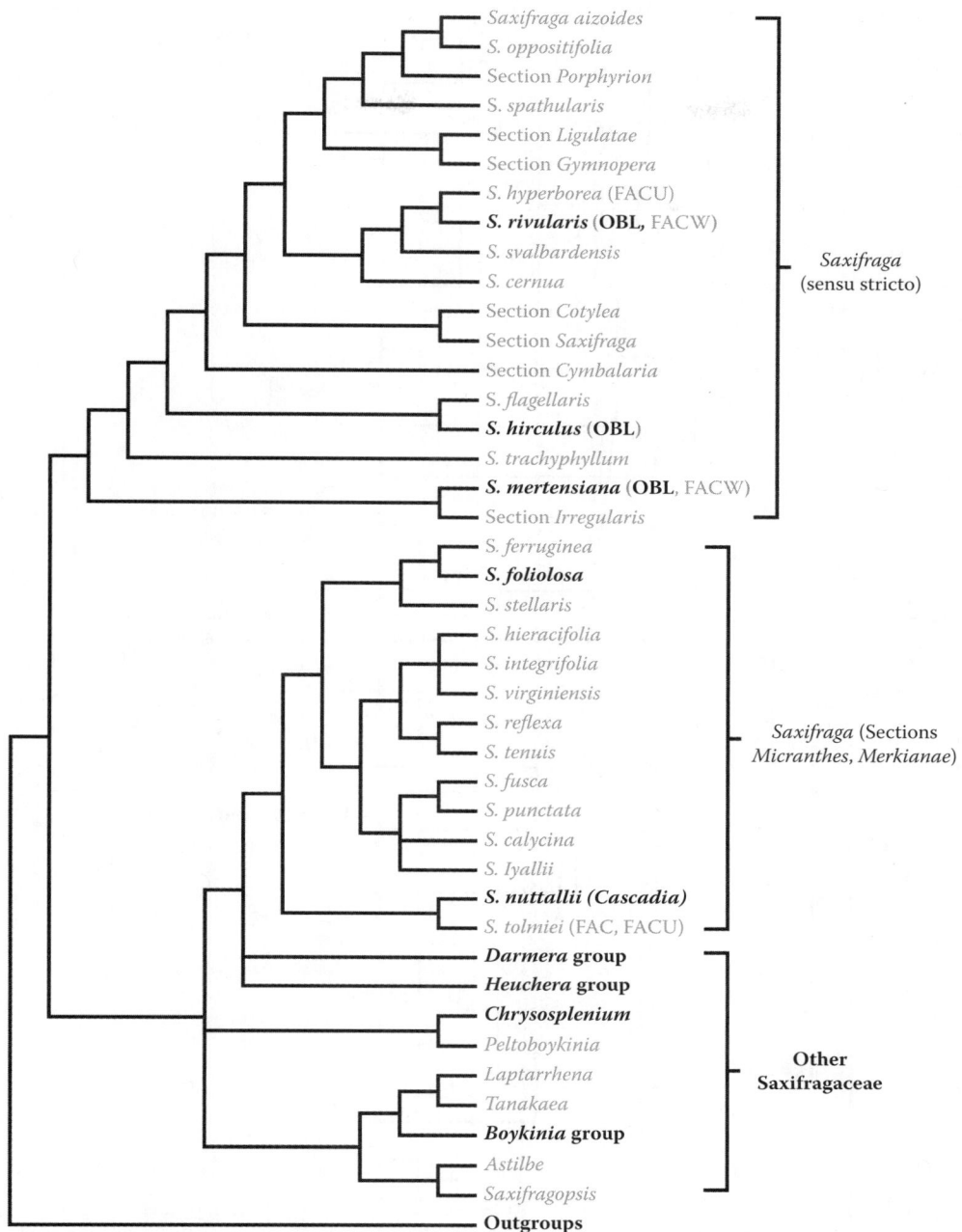

FIGURE 4.58 Phylogenetic relationships among *Saxifraga* species as indicated by analysis of *matK* sequence data. Cladistic analysis demonstrates that *Saxifraga* is polyphyletic and resolves as two distinct clades. It also indicates that the taxa containing obligate aquatics in North America (bold) occur scattered throughout the group and in each of the two main clades. Species assigned formerly to *Saxifraga* section *Micranthes* have since been placed within the distinct genus *Micranthes*. *Saxifraga nuttallii* has been transferred to the monotypic genus *Cascadia* as *C. nuttallii*. (Adapted from Soltis, D.E. et al., *Amer. J. Bot.*, 83, 371–382, 1996b.)

5. *Mitella*

Miterwort, bishop's cap; mitrelle

Etymology: from the diminutive of the Latin *mitra*, literally "small turban" for the resemblance of the fruit to a bishop's miter

Synonyms: *Drummondia*; *Mitellastra*; *Mitellopsis*; *Ozomelis*; *Pectiantia*

Distribution: global: Asia; North America; **North America:** eastern, northern, western

Diversity: global: 20 species; **North America:** 9 species

Indicators (USA): OBL: *Mitella nuda*; **FACW:** *M. nuda*, *M. pentandra*; **FAC:** *M. nuda*, *M. pentandra*

Habitat: freshwater; palustrine; **pH:** 4.1–8.1; **depth:** <1 m; **life-form(s):** emergent herb

Key morphology: shoots often stoloniferous; leaves (to 8.5 cm) basal from the rhizome, petioled (to 10 cm), cordate or rounded, crenate, palmately veined; flowering stalks (to 3 dm) naked or with a few reduced, sessile leaves, terminating in a

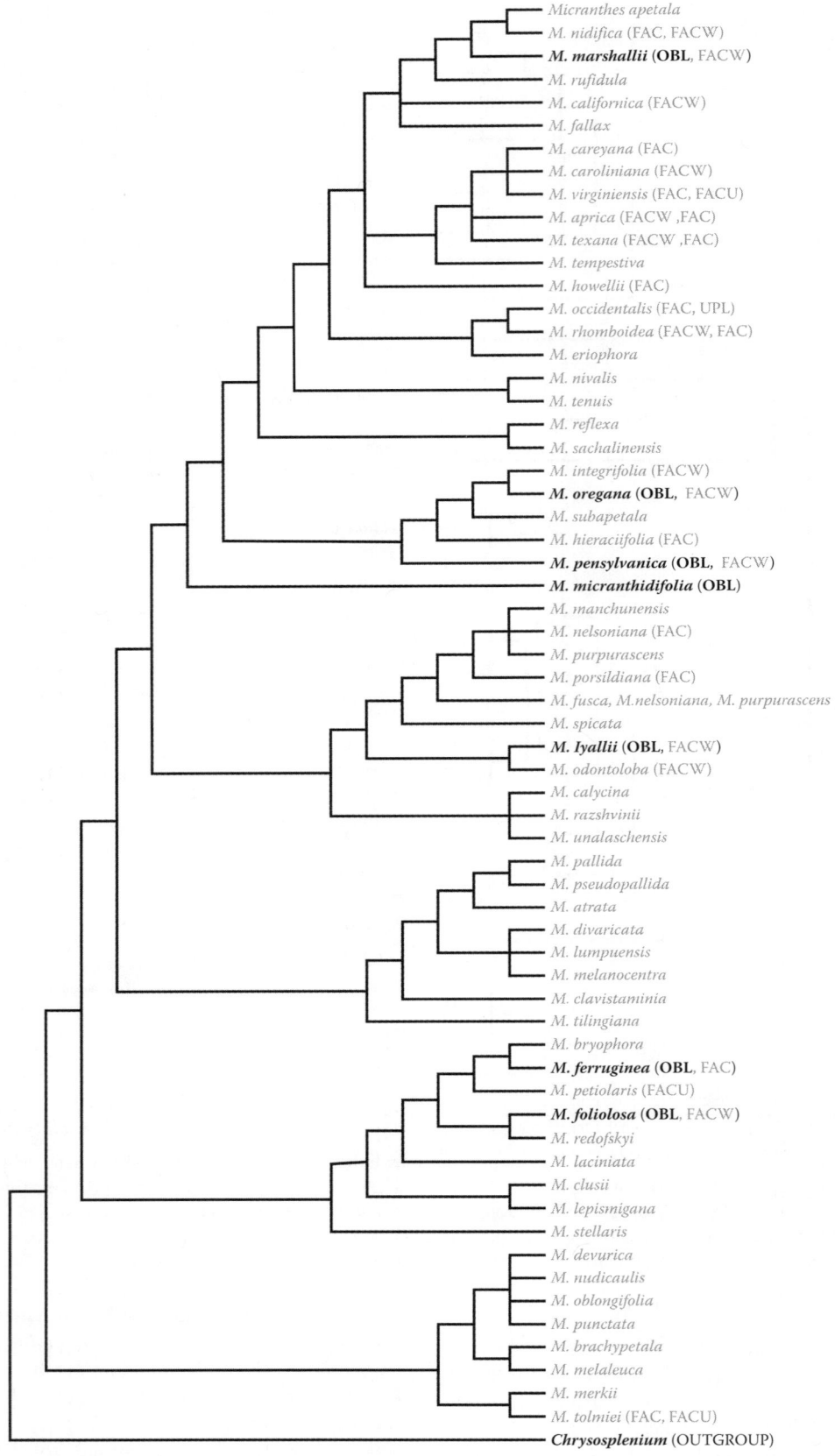

FIGURE 4.59 Phylogenetic relationships among 62 *Micranthes* species based on ML analysis of combined DNA sequence data. This tree indicates that the OBL habit might have arisen as many as seven times within the genus. (Adapted from Tkach, N. et al., *Bot. J. Linn. Soc.,* 178, 47–66, 2015.)

raceme (to 12.5 cm) of 2–25, pedicelled (to 8 mm), 5-merous flowers; hypanthium saucer-like, 5-lobed (to 3.5 mm); petals (to 5 mm) green, whitish or yellowish-green, pinnately divided into 5–9 laciniate segments; stamens 5 or 10; capsules (to 3 mm) mitre-like when dehisced; seeds few (2–5), blackish or reddish-brown, shiny or pitted

Life history: duration: perennial (rhizomes, stolons); **asexual reproduction:** rhizomes, stolons; **pollination:** insect; **sexual condition:** hermaphroditic; **fruit:** capsules (common); **local dispersal:** rhizomes, stolons, seeds (raindrops); **long-distance dispersal:** seeds

Imperilment: (1) *Mitella nuda* [G5]; S1 (IA, PA); S2 (AK, LB); S3 (CT, ND); (2) *M. pentandra* [G5]; S1 (NM); S2 (YT); S3 (AB, WY)

Ecology: general: *Mitella* species occur typically in the understory of boreal, subarctic, temperate, and montane forests over a broad range of soil moisture conditions. The 2013 list recognized only one *Mitella* species (*M. nuda*) as an OBL indicator. Although *M. pentandra* was originally recognized as OBL in a portion of its range, it has since been categorized as FACW, FAC. However, the treatment has been retained here because it was completed prior to the release of the 2013 indicator list. The species categorized as wetland indicators do not occur in areas of standing water, and often occur in sites other than wetlands (see wetland status earlier). Most species grow where there is shade. Although some authors have speculated that *Mitella* flowers are adapted for pollination by nocturnal moths or by wind; studies of Asian species indicate the principal pollinators to be fungus gnats (Insecta: Mycetophilidae), which perch on the dissected petals with their long, spiny legs. These are efficient pollinators, with only 1.3–2.6 gnat visits/inflorescence/day resulting in fruit sets exceeding 63%. Inadvertent pollination by nectar-feeding moths has been reported for one species (*M. stauropetala*), but not for any of the wetland indicators. Seed dispersal is highly local, occurring by a "splash cup" method. The seeds are categorized as physiologically dormant; however, one study found little difference in germination between refrigerated and unrefrigerated seeds for at least one species (*M. nuda*). All North American species are rhizomatous.

Mitella nuda L. grows in bogs, depressions, fens, floodplains, marshes, meadows, muskeg, ravines, seeps, slopes, swales, swamps, thickets, woodlands, and along stream margins elevations of up to 3460 m. The plants can occupy a fairly broad range of soil moisture (see indicator status earlier) but usually occur in cool and shady, forested sites. This species is described as "edge negative" with respect to its reduced occurrence near the lakeshore forest edge. Sites of occurrence primarily are acidic but the species is also reported from more alkaline sites (pH: 4.1–8.1). The substrates are described as clay, humus, loam, muck, sand, sandy loam, or organic (e.g., fibric fibosol, leaf mold, muck, mossy rocks, peat). Specific pollinators have not been identified, but likely are insects and possibly fungus gnats as found for several Asian *Mitella* species (see earlier). A rate of 50% germination is reported for seeds that have been stratified for 5 months (2 months cold moist stratification; 3 months warm moist stratification), then sowed when temperatures are 12°C–16°C daytime and 0°C–10°C at night.

However, one study found only slightly better germination for refrigerated seeds (63%; 64–98 days) compared to unrefrigerated seeds (42%; 48–139 days). *Mitella nuda* does not develop a large seed bank (up to 100 seeds/m²) and occurs mainly in areas undisturbed by fire. The plants spread vegetatively by means of their long, slender rhizomes and stolons. **Reported associates:** *Abies balsamea, A. lasiocarpa, Acer rubrum, A. saccharinum, A. saccharum, A. spicatum, Adiantum pedatum, Adoxa moschatellina, Alnus tenuifolia, A. viridis, Amerorchis rotundifolia, Andromeda polifolia, Anemone quinquifolia, Aralia nudicaulis, Arctostaphylos uva-ursi, Arisaema triphyllum, Arnica latifolia, Asarum canadense, Asplenium trichomanes-ramosum, Athyrium filix-femina, Aulacomnium palustre, Bazzania trilobata, Betula alleghaniensis, B. neoalaskana, B. occidentalis, B. papyrifera, B. pumila, Botrychium virginianum, Byrum, Calamagrostis canadensis, C. rubescens, Calliergon cordifolium, C. giganteum, Caltha palustris, Calypso bulbosa, Campylium stellatum, Cardamine douglassii, Carex aquatilis, C. arctata, C. aurea, C. bromoides, C. capillaris, C. concinna, C. disperma, C. eburnea, C. gracillima, C. interior, C. intumescens, C. leptalea, C. leptonervia, C. novae-angliae, C. peckii, C. pedunculata, C. rosea, C. rostrata, C. scabrata, C. stipata, C. stricta, C. trisperma, C. vaginata, C. vesicaria, Chamaedaphne calyculata, Chamerion angustifolium, Chrysosplenium americanum, C. iowense, C. tetrandrum, Cinna latifolia, Circaea alpina, Cirsium muticum, Claytonia, Clematis virginiana, Climaceum dendroides, Clintonia borealis, Coptis groenlandica, C. trifolia, Corallorhiza maculata, C. trifida, Cornus canadensis, C. sericea, Cryptogramma stelleri, Cyprepedium acaule, C. arietinum, C. calceolus, C. parviflorum, C. passerinum, C. reginae, Cystopteris bulbifera, C. fragilis, Deparia acrostichoides, Dicranum flagellare, D. polysetum, Drepanocladus uncinatus, Drosera rotundifolia, Dryopteris carthusiana, D. cristata, D. disjuncta, D. expansa, D. spinulosa, Epigaea repens, Equisetum arvense, E. fluviatile, E. hyemale, E. pratense, E. scirpoides, E. sylvaticum, Eriophorum, Erythronium americanum, Eurybia conspicua, E. macrophyllus, Fragaria virginiana, Fraxinus nigra, Galearis rotundifolia, Galium boreale, G. triflorum, Gaultheria hispidula, Geocaulon lividum, Geum macrophyllum, G. rivale, Glyceria striata, Goodyera oblongifolia, G. repens, Gymnocarpium dryopteris, Halenia deflexa, Huperzia lucidula, Hydrocotyle americana, Hylocomnium splendens, Hypericum ellipticum, Impatiens capensis, Iris lacustris, I. versicolor, Juniperus communis, Larix laricina, Lathyrus ochroleucus, Leymus innovatus, Linnaea borealis, Liparis loeselii, Listera borealis, L. cordata, Lonicera canadensis, L. involucrata, Lupinus, Lycopodium annotinum, L. obscurum, Maianthemum canadense, M. racemosum, M. stellatum, M. trifolium, Malaxis monophylla, M. unifolia, Menyanthes trifoliata, Mertensia paniculata, Micranthes pensylvanica, Milium effusum, Mimulus glabratus, Mitchella repens, Mitella caulescens, M. diphylla, Moneses uniflora, Muhlenbergia cuspidata, Nemopanthus mucronatus, Onoclea sensibilis, Oplopanax horridus, Orthilia secunda, Oryzopsis asperifolia, Osmorhiza chilensis, Osmunda regalis, Osmundastrum cinnamomeum, Ostrya virginiana, Oxalis acetosella, O. montana, Packera* *aurea, Panax trifolius, Pedicularis racemosa, Peltigera aph-thosa, P. canina, Petasites frigidus, Phegopteris connecti-lis, Picea engelmannii, P. glauca, P. mariana, Pilea pumila, Pinus contorta, P. ponderosa, P. strobus, Plagiomnium cus-pidatum, P. medium, Platanthera hyperborea, P. obtusata, P. orbiculata, Pleurozium schreberi, Poa, Polemonium van-bruntiae, Polygala pauciflora, Polygonatum pubes-cens, Populus balsamifera, P. grandidentata, P. tremuloides, Primula pauciflora, Prunus serotina, Pseudotsuga menzie-sii, Pteridium aquilinum, Ptilium crista-castrensis, Pyrola asarifolia, P. chlorantha, P. secunda, P. uniflora, Quercus macrocarpa, Rhamnus alnifolia, Rhizomnium punctatum, Rhododendron groenlandicum, Rhytidiadelphus triquetrus, Ribes americanum, R. cynosbati, R. hirtellum, R. lacustre, R. triste, Rosa acicularis, Rubus idaeus, R. pubescens, Salix candida, S. pedicellaris, S. planifolia, S. scouleriana, S. ves-tita, Sambucus racemosa, Scutellaria galericulata, Sedum stenopetalum, Shepherdia canadensis, Solidago patula, Sorbus americana, Sphagnum centrale, S. girgensohnii, S. magellanicum, S. nemoreum S. palustre, S. russowii, S. warn-storfii, Spiraea betulifolia, Streptopus amplexifolius, S. roseus, Symphyotrichum ciliolatum, S. foliaceum, S. puniceum, S. subspicatum, Symplocarpus foetidus, Taraxacum officinale, Taxus, Thalictrum dioicum, T. pubescens, Thuidium delicatu-lum, Thuja occidentalis, T. plicata, Tiarella cordifolia, Tilia americana, Toxicodendron vernix, Toxicoscordion panicu-latum, Trichocolea tomentella, Trientalis borealis, Trillium cernuum, T. grandiflorum, Trollius laxus, Tsuga canadensis, Ulmus americana, Vaccinium angustifolium, V. corymbosum, V. oxycoccus, V. scoparium, V. vitis-idaea, Valeriana uligi-nosa, Veratrum viride, Viburnum cassinoides, V. edule, Viola canadensis, V. renifolia.*

***Mitella pentandra* Hooker** inhabits bogs, fens, flood-plains, glades, meadows, muskeg, ravines, scree slopes, talus slopes, seeps, thickets, woodlands, and the margins of lakes, rivers, and streams at elevations of up to 3476 m. The sites usually are shady (but also can be sunny) with substrates that tend to be acidic (pH: 5.1–7.3) and nitrogen rich, but can include basalt, boulders, clay, granite, granitic loamy sand, gravel, limestone, loam, loamy sand, organics (moder, mull humus), quartz diorite, rubble, Sericite shist (Savoy Schist), and shale. The plants have also been reported to grow on fallen logs in bogs. A warm–cool–warm cycle of 18°C–22°C (2–4 weeks), −4°C to 4°C (4–6 weeks), and a final tempera-ture of 5°C–12°C is recommended to promote seed germina-tion. The plants spread vegetatively by slender rhizomes and have been known to persist (or emerge from the seed bank) following logging or fire disturbance. **Reported associ-ates:** *Abies concolor, A. grandis, A. lasiocarpa, Acer circi-natum, A. glabrum, A. macrophyllum, A. rubrum, Achillea millefolium, Achlys triphylla, Aconitum columbianum, A. delphiniifolium, Adenocaulon bicolor, Adiantum pedatum, Agrostis exarata, Allium textile, Alnus rhombifolia, A. rubra, A. rugosa, A. sinuata, A. tenuifolia, A. viridis, Amelanchier alnifolia, A. utahensis, Anemone canadensis, A. occidentalis, A. piperi, Angelica arguta, A. canbyi, A. lucida, Antennaria racemosa, Anthoxanthum odoratum, Aquilegia flavescens,*

Arctostaphylos uva-ursi, Arnica cordifolia, A. latifolia, A. mollis, Asarum caudatum, Athyrium, Balsamorhiza sagit-tata, Berberis nervosa, B. repens, Bistorta bistortoides, B. vivipara, Blechnum spicant, Boykinia major, Bromus cari-natus, Calamagrostis canadensis, C. rubescens, Caltha lep-tosepala, Canadanthus modestus, Cardamine constancei, C. cordifolia, Carex anthoxanthea, C. aquatilis, C. deweyana, C. disperma, C. geyeri, C. macrochaeta, C. nigricans, C. podo-carpa, C. prionophylla, C. rossii, C. scopulorum, Cassiope mertensiana, Castanopsis sclerophylla, Castilleja min-iata, C. parviflora, Ceanothus, Chamaecyparis, Chamerion angustifolium, Cinna latifolia, Circaea alpina, Cladonia, Claytonia cordifolia, C. sibirica, Clematis occidentalis, Clintonia uniflora, Conioselinum scopulorum, Coptis occi-dentalis, C. trifolia, Cornus canadensis, C. nuttallii, C. seri-cea, Corydalis caseana, Corylus cornu, Crepis, Cystopteris fragilis, Cytisus scoparius, Delphinium barbeyi, D. nut-tallianum, Deschampsia cespitosa, Dicentra, Dryopteris filix-mas, D. gymnocarpa, Empetrum nigrum, Epilobium alpinum, E. anagallidifolium, E. ciliatum, E. glaberrimum, E. hornemannii, E. latifolium, E. minutum, Equisetum arvense, E. fluviatile, E. telmateia, Erigeron peregrinus, Eriophorum, Erythronium grandiflorum, Eurybia conspicua, E. radulina, Festuca altaica, F. idahoensis, F. occidentalis, Fragaria vesca, F. virginiana, Galium bifolium, G. boreale, G. trifidum, G. triflorum, Gaultheria humifusa, G. shallon, Gentiana affinis, Geranium richardsonii, Geum macro-phyllum, Glyceria striata, Glycyrrhiza, Goodyera repens, Gymnocarpium dryopteris, Hedysarum boreale, Heracleum sphondylium, Hieracium albiflorum, Holodiscus discolor, Huperzia occidentalis, Hydrophyllum fendleri, Hypochaeris radicata, Ilex verticillatus, Juncus balticus, J. drummondii, Juniperus communis, J. scopulorum, Larix lyallii, L. occi-dentalis, Leptarrhena pyrolifolia, Ligusticum filicinum, L. grayi, L. porteri, Limnorchis dilatata, Linnaea borealis, Listera cordata, Lithophragma, Lonicera involucrata, L. uta-hensis, Lophozia, Lupinus latifolius, Luzula hitchcockii, L. parviflora, L. wahlenbergii, Lysichiton americanus, Mahonia nervosa, Maianthemum racemosum, M. stellatum, Medicago lupulina, Menziesia ferruginea, Mertensia ciliata, M. panicu-lata, Micranthes lyallii, M. nelsoniana, M. odontoloma, M. oregana, Mimulus guttatus, M. lewisii, Mitella breweri, M. stauropetala, Moehringia macrophylla, Moneses uniflora, Montia chamissoi, Noccaea fendleri, Oenanthe sarmentosa, Oplopanax horridus, Orobanche uniflora, Orthilia secunda, Osmorhiza chilensis, O. depauperata, O. purpurea, Oxalis oregana, Oxypolis fendleri, Parnassia fimbriata, P. palustris, Pedicularis bracteosa, P. groenlandica, P. ornithorhyncha, Petasites frigidus, Philadelphus lewisii, Phyllodoce empet-riformis, P. glanduliflora, Physocarpus malvaceus, Picea engelmannii, P. glauca, P. mariana, P. sitchensis, Pinus albicaulis, P. contorta, Platanthera dilatata, P. stricta, Poa leptocoma, P. pratensis, Polemonium pulcherrimum, Polystichum munitum, Populus balsamifera, P. tremuloides, Prosartes hookeri, P. trachycarpa, Prunus, Pseudoroegneria spicata, Pseudotsuga menziesii, P. taxifolia, Pteridium aqui-linum, Pyrola asarifolia, P. minor, P. uniflora, Ranunculus eschscholtzii, Rhododendron albiflorum, R. columbianum, R. groenlandicum, Ribes hudsonianum, R. inerme, R. lacustre, R. viscosissimum, R. wolfii, Rosa acicularis, R. nutkana, Rubus idaeus, R. parviflorus, R. pedatus, R. procerus, R. spectabilis, R. ursinus, Rumex acetosella, Salix arctica, S. barclayi, S. boothii, S. commutata, S. drummondiana, S. farriae, S. reticulata, Sambucus racemosa, Saussurea americana, Senecio sylvaticus, S. triangularis, Sphagnum, Spiraea betulifolia, Stellaria crassifolia, Stenanthella occidentalis, Streptopus amplexifolius, Symphoricarpos albus, Symphyotrichum foliaceum, S. lanceolatum, Taraxacum officinale, Tellima grandiflora, Thalictrum occidentale, Thuja plicata, Tiarella trifoliata, Tolmiea menziesii, Trautvetteria caroliniensis, Trifolium longipes, Trillium, Trollius laxus, Tsuga heterophylla, Urtica dioica, Vaccinium membranaceum, V. myrtillus, V. ovalifolium, V. scoparium, Vahlodea, Valeriana sitchensis, Vancouveria hexandra, Veratrum viride, Veronica wormskjoldii, Viburnum edule, Viola canadensis, V. glabella, Xerophyllum tenax.

Use by wildlife: Coast deer (Mammalia: Cervidae: *Odocoileus hemionus columbianus*) browse the leaves, stems, and flowers of *Mitella pentandra* in the spring, but it is rated low overall in its importance as a deer food. *Mitella pentandra* is a host of rust fungi (Basidiomycota: Pucciniaceae: *Puccinia heucherae*).

Economic importance: food: none; **medicinal:** The leaves of *Mitella nuda* are diuretic and were once used to treat inflammation of the bladder and kidneys. Crushed, cloth-wrapped leaves of *M. nuda* were placed in the ear by the Cree and Woodlands tribes to treat earaches. Roots of *M. pentandra* are astringent and have been used to treat diarrhea; **cultivation:** *Mitella pentandra* is cultivated as an ornamental garden plant; **misc. products:** none; **weeds:** none; **nonindigenous species:** none

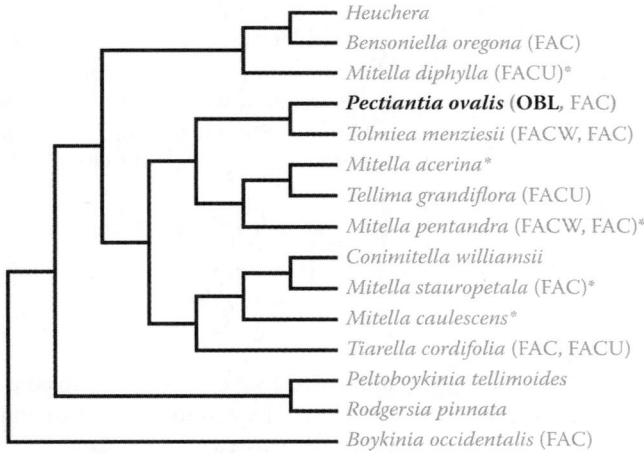

Heuchera
Bensoniella oregona (FAC)
Mitella diphylla (FACU)*
Pectiantia ovalis (**OBL**, FAC)
Tolmiea menziesii (FACW, FAC)
Mitella acerina*
Tellima grandiflora (FACU)
Mitella pentandra (FACW, FAC)*
Conimitella williamsii
Mitella stauropetala (FAC)*
Mitella caulescens*
Tiarella cordifolia (FAC, FACU)
Peltoboykinia tellimoides
Rodgersia pinnata
Boykinia occidentalis (FAC)

FIGURE 4.60 Phylogenetic relationships of *Mitella* and *Pectiantia* as resolved by analysis of combined nuclear gene sequence data. These results clearly indicate the polyphyly of *Mitella* as currently circumscribed (asterisked taxa), and also call to question the generic limits for many of the related genera. The wetland indicator status is provided in parentheses. (Adapted from Folk, R.A. & Freudenstein, J.V., *Amer. J. Bot.* 101, 1532–1550, 2014.)

Systematics: Although cytology, flavonoid chemistry, morphology, and molecular data have resolved *Mitella* within the "*Heuchera* group" (also *Heucherina* clade) of Saxifragaceae, its precise relationships within that assemblage have been difficult to ascertain because of complications with analyses of genetic data due to intergeneric hybridization (e.g., with *Conimitella* and *Tiarella*). Phylogenetic reconstructions based on various molecular data sets clearly indicate the polyphyletic nature of *Mitella*. Those studies have also presented a highly unstable picture of phylogeny, where the species resolve inconsistently, often among other genera (e.g., *Tellima, Tolmiea, Tiarella*), as different data or analyses are used. As a consequence, there is no current consensus regarding the circumscription of *Mitella* or *Pectiantia* (also other closely related genera in this group), and their disposition as followed here should be regarded as tentative. As a consequence, any discussion of interrelationships among these poorly defined genera would be fruitless at this time. A general overview of relationships is presented from a recent study, which nicely illustrates the complexities described earlier (Figure 4.60). Here are evident several conflicts with previous studies such as analysis of cpDNA restriction site data, which resolved *M. pentandra* and *M. caulescens* as sister species. Other studies incorporating different arrays of combined DNA sequence data resolve *M. nuda* in a clade with *M. diphylla*, which complicates matters even further by indicating a fairly distant relationship to *M. pentandra*. It is apparent that further phylogenetic studies of this group are imperative. As currently defined, the base chromosome number of *Mitella* is $x = 7$. *Mitella pentandra* is diploid ($2n = 14$), whereas *M. nuda* ($2n = 14, 28$) has diploid and tetraploid cytotypes. The chromosome morphology of both species is uniform. Molecular studies have confirmed that hybridization and introgression are relatively common in *Mitella*. Hybrids (known as *M. ×intermedia*) have been reported between *M. nuda* and *M. diphylla*, but should be verified by genetic analyses. There are no reports of hybrids involving *M. pentandra*. Despite the highly unusual pinnate petals of *Mitella*, there have been few studies to elucidate pollinators or other aspects of reproductive biology, which remain poorly understood for most North American species.

Comments: *Mitella nuda* is widespread throughout northern North America (and also in eastern Asia). *Mitella pentandra* occurs in western North America.

References: Cowan, 1945; Folk & Freudenstein, 2014; Hamilton & Peterson, 2003; Harmon & Franklin, 1995; Harper & MacDonald, 2001; Klinka et al., 1995; Leckie et al., 2000; Lee, 2004; Nichols, 1934; Okuyama et al., 2004, 2005, 2012; Raible, 2004; Savile, 1953; Soltis, 1988; Soltis et al., 1991; USDA, 2005.

6. *Pectiantia*

Bishop's cap

Etymology: from the Greek *pec anti* meaning "opposite comb," in reference to opposition of stamens and petals in some species

Synonyms: *Mitella* (in part)
Distribution: global: Asia; North America; **North America:** western
Diversity: global: 13 species; **North America:** 3 species
Indicators (USA): OBL; FACW: *Pectiantia ovalis*
Habitat: freshwater; palustrine; **pH:** acidic (e.g., pH: 5.0) **depth:** <1 m; **life-form(s):** emergent herb
Key morphology: flowering shoots (to 40 cm) erect, naked; leaves in a basal rosette, petioled (to 11.3 cm), blades (to 7 cm) cordate-ovate to cordate-oblong, the margins shallowly five- to nine-lobed, doubly crenate–dentate or dentate; flowers (to 60) pedicelled (to 2 mm), in racemes (1–6), 5-merous; hypanthium (to 3.5 mm) saucer-like; sepals (to 1.9 mm) recurved; petals (to 2 mm), greenish yellow, three- to nine-lobed, the lobes linear; stamens (to 0.7 mm) opposite the sepals; ovary inferior, the styles (to 0.3 mm) divergent; capsules 2-beaked; seeds (to 1.1 mm) pitted, reddish purple to black
Life history: duration: perennial (rhizomes); **asexual reproduction:** rhizomes; **pollination:** insect; **sexual condition:** hermaphroditic; **fruit:** capsules (common); **local dispersal:** rhizomes, seeds; **long-distance dispersal:** seeds
Imperilment: (1) *Pectiantia ovalis* [G5]
Ecology: general: Due to the uncertain circumscription of this genus (see *Systematics* later), a general ecological overview is not provided.

Pectiantia ovalis **(Greene) Rydb.** occurs in or on bogs, cliffs, depressions, fens, floodplains, mudflats, ravines, slopes, springs, woodlands (alluvial), and along the margins of rivers and streams at elevations of up to 1768 m. The sites are typically shady. The substrates are acidic (e.g., pH: 5.0) and usually consist of an organic layer (moder, mull humus) that may overlay boulders or rocks. Other reported substrates include clay loam, duff, loam, marine clay, muck, mud, quartz diorite, sand, sandstone, slates, and talus. This is an indicator species of nitrogen-rich soils and is characteristic of areas having a cool, mesothermal climate. Flowering occurs from March to July. Many aspects of the reproductive biology and seed ecology of this species remain unknown. Sites with a relatively low cover (<1.5%) of *P. ovalis* can develop poor to moderately developed seed banks (to 115 seeds m²). This species occurs more consistently in hydric sites than many of the related *Mitella* species. **Reported associates:** *Acer circinatum, A. macrophyllum, Achlys triphylla, Adenocaulon bicolor, Adiantum pedatum, Agrostis exarata, A. longiligula, Allium validum, Alnus rubra, Anemone deltoidea, Aquilegia formosa, Aralia californica, Asarum caudatum, Athyrium filix-femina, Berberis nervosa, Blechnum spicant, Boletus, Boykinia elata, B. occidentalis, Bromus vulgaris, Calamagrostis foliosus, Cardamine angulata, Carex deweyana, C. hendersonii, C. lenticularis, C. ormantha, C. viridula, Castanopsis sclerophylla, Ceanothus, Chamaecyparis, Chrysosplenium glechomifolium, Circaea alpina, Claytonia sibirica, Collomia heterophylla, Crepis, Darlingtonia californica, Dicentra formosa, Digitalis purpurea, Equisetum, Euonymus occidentalis, Festuca subulata, Galium aparine, G. triflorum, Gaultheria shallon, Gaylussacia, Glyceria elata, G. striata, Heracleum lanatum, Holcus lanatus,*

Hydrophyllum tenuipes, Juncus oreganus, Lactuca muralis, Luzula parviflora, Lysichiton americanus, Mahonia nervosa, Maianthemum dilatatum, Marah oreganus, Mimulus cardinalis, M. dentatus, M. guttatus, M. moschatus, Mitella caulescens, Montia parvifolia, Oenanthe sarmentosa, Oplopanax horridum, Oxalis oregana, O. trilliifolia, Petasites frigidus, P. palmatus, Picea sitchensis, Pinus contorta, Poa trivialis, Polystichum munitum, Prunella vulgaris, Pseudotsuga menziesii, Pteridium aquilinum, Ranunculus flammula, R. repens, R. uncinatus, Rhamnus purshiana, Rhododendron, Ribes bracteosum, Rubus spectabilis, Rubus spectabilis, Rumex obtusifolius, Sambucus racemosa, Scirpus microcarpus, Scoliopus hallii, Senecio triangularis, Stachys chamissonis, S. mexicana, Stellaria crispa, Streptopus amplexifolius, Taxus, Tellima grandiflora, Thuja plicata, Tiarella trifoliata, Tolmiea menziesii, Trientalis, Tsuga heterophylla, Urtica dioica, Usnea barbata, Vaccinium parvifolium, Viola glabella, Woodwardia fimbriata.

Use by wildlife: Coast deer (Mammalia: Cervidae: *Odocoileus hemionus columbianus*) browse the leaves, stems, and flowers of *Pectiantia ovalis*, but it is not highly rated as a deer food. *Pectiantia ovalis* is a host of rust fungi (Basidiomycota: Pucciniaceae: *Puccinia heucherae*).

Economic importance: food: not reported as edible; **medicinal:** No medical uses reported; **cultivation:** *Pectiantia ovalis* is cultivated as an ornamental garden plant; **misc. products:** none; **weeds:** none; **nonindigenous species:** none

Systematics: Phylogenetic studies using DNA sequence data have demonstrated that the once familiar genus *Mitella* is polyphyletic as previously defined (e.g., Figure 4.60). As a result, recent floristic treatments have begun to recognize a number of the smaller segregates as distinct genera (e.g., *Mitellastra, Ozomelis, Pectiantia*). However, the taxonomy of *Mitella* and related genera remains unsettled and inconsistencies among different studies have made it difficult to accept the proposed circumscriptions of *Pectiantia* and other offshoots of *Mitella* with confidence. No strong taxonomic preference is expressed here, but this disposition should be regarded as tenuous until the taxonomy of this group can be better clarified. The base chromosome number of *Pectiantia* is $x = 7$, given the count for *P. ovalis* ($2n = 14$), which indicates a diploid.

Comments: *Pectiantia ovalis* occurs in western North America.

7. Saxifraga

Saxifrage
Etymology: from the Latin *saxum fragere*, meaning "to break stone" with reference to the early perception that medicinal extracts of some species could dissolve urinary stones
Synonyms: *Antiphylla; Boecherarctica; Heterisia; Hirculus; Leptasea; Lobaria; Muscaria; Zahlbrucknera*
Distribution: global: Africa, Eurasia, New World; **North America:** widespread
Diversity: global: 390 species; **North America:** 25 species
Indicators (USA): OBL: *Saxifraga hirculus, S. mertensiana, S. rivularis;* **FACW:** *S. mertensiana, S. rivularis*

Habitat: freshwater; palustrine; **pH:** 5.4–7.4; **depth:** <1 m; **life-form(s):** emergent herb

Key morphology: stems a compressed caudex, forming foliar rosettes, rhizomatous, and/or stoloniferous, sometimes bearing bulbils; leaves sessile or petiolate (to 5 cm), the blades (to 10 cm) linear to spatulate, reniform, or rounded, the margins serrate or entire; flowering stem (to 7 cm) naked or with reduced leaves, terminating in a cyme or thyrse of 1–30+, 5-merous flowers, which frequently are replaced by bulbils; stamens 10; petals (to 18 mm) cream, pinkish, white, or yellow; with or without proximal orange spots; ovary superior to half-inferior; capsules 2 or 3-beaked; seeds brown, tuberculate to papillate

Life history: duration: perennial (bulbils, rhizomes, stolons); **asexual reproduction:** bulbils, fragments, rhizomes, stolons; **pollination:** insect, self; **sexual condition:** hermaphroditic; **fruit:** capsules (common); **local dispersal:** seeds (gravity, meltwater, wind); **long-distance dispersal:** seeds (animals)

Imperilment: (1) *Saxifraga hirculus* [G5]; S1 (MT, UT); S2 (BC, MB); S3 (QC); (2) *S. mertensiana* [G5]; S3 (AB, MT); (3) *S. rivularis* [G5]; S1 (MB, NH, ON); S2 (AK, NF); S3 (QC, WA, WY)

Ecology: general: Saxifrages are commonly regarded as "rock garden" plants but they often show affinities for wet habitats. In North America, 17 species (68%) are designated as wetland indicators, with 12% of them regarded as OBL in at least some portion of their range. The species are notorious for their high level of chromosomal variability (see *Systematics* later) and also their propensity for hybridization. However, much remains to be learned about the pollination biology and reproductive ecology of many species. The outcrossed species are pollinated by insects, usually flies. In all *Saxifraga*, local dispersal occurs by means of a wind-induced vibratory resonation of the capsules, a method effective only over very short distances. Mechanisms for their long distance dispersal are yet to be elucidated.

Saxifraga hirculus **L.** grows in Arctic to boreal bogs, fens, tundra meadows, and along brook, lake, and river margins, extending northward to 82°31′N. The plants usually occur in alkaline (e.g., pH: 7.4) sites on clay, sand, or on mossy organic substrates. The flowers are strongly protandrous (for a duration of 9 days) but are somewhat self-compatible; however, higher seed set occurs among ramets than within a ramet. The petals possess UV reflective honey guides at their base. The flowers contain small amounts of nectar and in Europe are pollinated by various insects (Insecta) including beetles (Coleoptera), flies (Diptera: Syrphidae [mainly in the short-term]; Mycetophilidae), Lepidoptera (Zygaenidae), and a fungus gnat. One European study found 57 Diptera species from 16 families visiting the flowers. The varied flight behavior of these diverse pollinators facilitates gene flow between and among populations. Insect floral visitors in Canadian Arctic populations include parasitic wasps (Hymenoptera), springtails (Collembola), and flies (Muscidae: *Spilogona*). Pollinator activity is minimal on overcast and rainy days. Seed production is moderate overall, with one study reporting a seed set of 30% and pollen inviability of 11%. The seeds have physiological dormancy, but details on their germination requirements have not been specified. There are no special adaptations to disperse the seeds, which drop near the parent plant (to an average distance of 13 cm) when they are jarred from the capsules by rain or wind. Longer dispersal distances may be facilitated by deer (Mammalia: Cervidae) or other herbivores. Seedling recruitment depends on the availability of bare ground for establishment. The plants reproduce vegetatively by bulbils (replacing flowers in the inflorescence or forming in the leaf axils), by fragmentation of the rhizomes (the ends develop into new plants), or by short stolons. Taller plants (over 15 cm) indicate more favorable environments. Plants are more common in sites that have been grazed by muskoxen (Mammalia: Bovidae: *Ovibos moschatus*). Haplotype analysis has indicated that European populations retain much lower genetic diversity (18% polymorphism) than North American populations (71% polymorphism). The highest genetic diversity occurs in areas that were unglaciated during the last ice age and which probably served as refugia. European and North American populations do not share haplotypes; however, they are not structured geographically between those continents, a pattern indicating extensive postglacial dispersal. **Reported associates:** *Alectoria nigricans, Alopecurus alpinus, Anastrophyllum minutum, Andromeda polifolia, Arctagrostis latifolia, Arctophila fulva, Arctostaphylos rubra, Aulacomnium turgidum, Betula glandulosa, B. nana, Bistorta officinalis, B. vivipara, Calamagrostis holmii, Campylium stellatum, Cardamine bellidifolia, C. pratensis, Carex aquatilis, C. atrofusca, C. bigelowii, C. lugens, C. membranacea, C. misandra, C. rupestris, C. stans, C. subspathacea, Cassiope tetragona, Catoscopium nigritum, Cerastium beeringianum, Cetraria aculeata, C. islandica, Chrysosplenium tetrandrum, Cladina rangiferina, Cladonia pyxidata, Cochlearia officinalis, Dactylina arctica, Dicranum, Distichium capillaceum, Ditrichum flexicaule, Draba lactea, D. micropetala, D. nivalis, Dryas integrifolia, Dupontia fisheri, Empetrum nigrum, Encalypta, Epilobium latifolium, Equisetum arvense, E. variegatum, Eriophorum angustifolium, E. russeolum, E. scheuchzeri, E. triste, E. vaginatum, Flavocetraria, Hierochloe alpina, H. pauciflora, Hylocomium splendens, Hypnum, Juncus biglumis, Kobresia myosuroides, Luzula arctica, L. confusa, L. nivalis, Masonhalea richardsonii, Meersia uliginosa, Melandrium apetalum, Micranthes foliolosa, M. hieracifolia, M. nelsoniana, Minuartia arctica, Novosieversia glacialis, Ochrolechia frigida, Oncophorus wahlenbergii, Orthothecium chryserum, Oxytropus maydelliana, Papaver hultenii, Parrya nudicaulis, Pedicularis capitata, P. kanei, P. sudetica, Petasites frigidus, Plueropogon sabinei, Poa arctica, Pohlia, Polytrichum, Potentilla hyparctica, P. uniflora, Puccinellia vahliana, P. wrightii, Racomitrium lanuginosum, Ranunculus hyparctica, R. nivalis, R. pygmaeus, Rhododendron tomentosum, Rubus chamaemorus, Salix arctica, S. phleobophylla, S. pulchra. S. reticulata, S. rotundifolia, Saxifraga cespitosa, S. cernua, S. oppositifolia, S. phlebophylla, S. polaris, S. reticulata, S. rotundifolia, Sanionia uncinata, Scorpidium turgescens, Senecio atropurpureus, Silene acaulis, Sphaerophorus globosus, Sphagnum, Stellaria humifusa, S. laeta, S. longipes, Tetralophozia* setiformis, *Thamnolia vermicularis*, *Tomenthypnum nitens*, *Vaccinium vitis-idaea*, *Warnstorfia sarmentosa*.

Saxifraga mertensiana Bongard inhabits cliffs, crevices, ledges, meadows, runnels, seeps, slopes, snowfields, waterfall sprays, and the margins of brooks, lakes, rivers, and streams at elevations of up to 3284 m. Site exposures typically vary from partial to dense shade. The substrates have been described as argellite (red and green), basalt, belt series quartzite, decomposing marble, friable loam duff, gneiss, granite, humus, igneous diorite, limestone, loam, organic, rocky loam, sand, sandstone, sandy loam, scree, serpentine, silt, and talus. Flowering occurs from February to August. Little has been reported on the reproductive biology or seed ecology of this species. The seeds are dispersed downslope by meltwaters, snow, or wind. Some of the flowers in the inflorescence (except in California plants) can be replaced by bulbils (also produced in leaf axils), which can detach and vegetatively propagate asexually derived plantlets. Vegetative reproduction by means of short rhizomes can also occur. **Reported associates:** *Abies lasiocarpa*, *Acer glabrum*, *A. macrophyllum*, *Adiantum pedatum*, *Amelanchier alnifolia*, *Antennaria monocephala*, *Anticlea occidentalis*, *Aquilegia formosa*, *Arabis nuttallii*, *Arbutus*, *Arnica latifolia*, *A. mollis*, *Aruncus dioicus*, *Aspidotis densa*, *Betula*, *Bistorta bistortoides*, *B. vivipara*, *Blechnum*, *Brachythecium*, *Brassica rapa*, *Bryum*, *Calamagrostis rubescens*, *Callitropsis nootkatensis*, *Calochortus apiculatus*, *Carex geyeri*, *Castilleja miniata*, *Ceanothus*, *Cerastium beeringianum*, *Chamaecyparis lawsoniana*, *Claytonia parviflora*, *Cornus nuttallii*, *Corydalis*, *Cryptogramma stelleri*, *Cystopteris fragilis*, *Dactylis glomerata*, *Delphinium glareosum*, *Deschampsia cespitosa*, *Dicentra formosa*, *Draba*, *Dryas hookeriana*, *Epilobium anagallidifolium*, *Erigeron compositus*, *E. peregrinus*, *Erythronium grandiflorum*, *Eucephalus ledophyllus*, *Festuca altaica*, *F. viridula*, *Fragaria vesca*, *Galium*, *Garrya*, *Gaultheria shallon*, *Gentiana calycosa*, *Goodyera oblongifolia*, *Harrimanella*, *Heuchera cylindrica*, *H. glabra*, *Holodiscus discolor*, *Juniperus communis*, *Larix lyallii*, *Lasthenia*, *Linnaea borealis*, *Lithocarpus*, *Lomatium*, *Luetkea pectinata*, *Luina*, *Lupinus latifolius*, *Mertensia longiflora*, *M. paniculata*, *Micranthes ferruginea*, *M. integrifolia*, *M. lyallii*, *M. marshallii*, *M. nelsoniana*, *M. occidentalis*, *M. odontoloma*, *M. rufidula*, *Mimulus guttatus*, *Minuartia obtusiloba*, *M. rubella*, *Moneses uniflora*, *Montia parvifolia*, *Nemophila*, *Oplopanax horridus*, *Osmorhiza berteroi*, *Oxyria digyna*, *Pedicularis bracteosa*, *P. racemosa*, *Penstemon fruticosus*, *Pentagramma triangularis*, *Philonotis fontana*, *Phlox diffusa*, *Phyllodoce glanduliflora*, *Picea engelmannii*, *Pinus albicaulis*, *P. contorta*, *Plectritis*, *Poa leibergii*, *Polemonium pulcherrimum*, *Polypodium hesperium*, *Polystichum lonchitis*, *Potentilla ×diversifolia*, *P. nivea*, *Primula pauciflora*, *Pseudotsuga menziesii*, *Quercus*, *Ranunculus eschscholtzii*, *Rhodiola rosea*, *Rhododendron albiflorum*, *Ribes*, *Romanzoffia sitchensis*, *Rubus parviflorus*, *Salix stolonifera*, *Sanguisorba*, *Saxifraga bronchialis*, *S. cespitosa*, *Sedum leibergii*, *S. stenopetalum*, *Selaginella*, *Sibbaldia procumbens*, *Sorbus scopulina*, *Tellima grandiflora*, *Thuja plicata*, *Tonestus lyallii*, *Toxicodendron*, *Trifolium*, *Tsuga heterophylla*, *Umbellularia californica*, *Vaccinium membranaceum*, *V. parvifolium*, *V. scoparium*, *Valeriana*, *Vicia*, *Xerophyllum tenax*.

Saxifraga rivularis L. occurs on alpine tundra, cliffs, crevices, limestone and granite ledges, slopes, streambanks, seeps, and talus near snowbanks at elevations of up to 4500 m. The habitats usually are shady, with acidic (e.g., pH: 5.4) substrates that include clay, gravel, rock, and silt. The flowers of *S. rivularis* are self-compatible, have a low pollen/ovule ratio, and are highly autogamous. However, the plants rarely set seed but instead reproduce asexually by the proliferation of vegetative bulbils (formed in the leaf axils) or by fragmentation. Both rhizomes and stolons (horizontally disposed to 6 cm) can develop. Plants in the high Arctic require nearly continuous illumination (23+ h/day) to fully produce flowers, and even short dark periods can prevent a flowering response altogether. When flowers are produced, the seeds mature within 30 days. The high level of vegetative reproduction enables plants to withstand disturbance, at least in the short term (<20 years). Inbreeding and high levels of asexual reproduction are consistent with the results of population genetic analyses, which disclosed extremely low levels of genetic variation as detected by allozymes and various DNA markers. The seeds have physiological dormancy. Successful germination within 12 weeks has been achieved by placing seeds at −5°C for 5 weeks, thawing them at 0.5°C for 3 days, acclimatizing them at 4°C for 4 days, and then germinating them at 18°C under continuous light conditions. Where they do occur, seed banks can yield more than 60 seedlings/m^2. **Reported associates:** *Alnus viridis*, *Alopecurus alpinus*, *Amphidium lapponicum*, *Arctagrostis latifolia*, *Arnica lanceolata*, *Besseya alpina*, *Bistorta vivipara*, *Brachythecium*, *Braya*, *Bryum pseudotriquetrum*, *Bucklandiella sudetica*, *Calamagrostis canadensis*, *Cardamine bellidifolia*, *C. nymanii*, *Carex atrofusca*, *C. bicolor*, *C. scirpoidea*, *Cerastium alpinum*, *C. beeringianum*, *Cirriphyllum cirrosum*, *Cochlearia officinalis*, *Cryptogramma stelleri*, *Deschampsia cespitosa*, *Desmatodon heimii*, *D. leucostoma*, *Ditrichum flexicaule*, *Draba corymbosa*, *D. fladnizensis*, *Drepanocladus uncinatus*, *Epilobium anagallidifolium*, *E. arcticum*, *E. hornemannii*, *Equisetum arvense*, *E. variegatum*, *Eutrema edwardsii*, *Festuca brachyphylla*, *Grimmia elatior*, *Gymnomitrium concinnatum*, *G. corallioides*, *Hulteniella integrifolia*, *Juncus biglumis*, *J. trifidus*, *Koenigia islandica*, *Leptobryum pyriforme*, *Luzula arctica*, *L. confusa*, *Marchantia polymorpha*, *Micranthes foliolosa*, *M. nivalis*, *M. occidentalis*, *M. tenuis*, *Minuartia rossii*, *M. stricta*, *Orthothecium speciosum*, *Orthotrichum rupestre*, *Oxyria digyna*, *Papaver radicatum*, *Peltigera aphthosa*, *Philonotis fontana*, *Phippsia algida*, *Phleum alpinum*, *Plagiomnium medium*, *Poa arctica*, *P. glauca*, *Pogonatum alpinum*, *Pohlia cruda*, *P. nutans*, *Polytrichastrum alpinum*, *Polytrichum juniperinum*, *Potentilla hyparctica*, *Puccinellia angustata*, *P. langeana*, *Ranunculus hyperboreus*, *R. pygmaeus*, *Sagina nivalis*, *S. saginoides*, *Salix arctica*, *Saxifraga cernua*, *S. mertensiana*, *S. oppositifolia*, *S. paniculata*, *Silene uralensis*, *Splachnum vasculosum*, *Stellaria edwardsii*, *S.*

longipes, S. umbellata, Tortula ruralis, Trichophorum cespitosum, Trisetum spicatum, Vaccinium uliginosum.

Use by wildlife: *Saxifraga hirculus* is eaten by domestic sheep (Mammalia: Bovidae: *Ovis aries*). Its foliage hosts several fungi (Ascomycota: Mycosphaerellaceae: *Mycosphaerella allicina, Pseudocercosporella saxifragae*; Polystomellaceae: *Dothidella sphaerelloides*; Basidiomycota: Melampsoraceae: *Melampsora hirculi*). *Saxifraga mertensiana* grows in the habitat of Van Dyke's salamander (Amphibia: Plethodontidae: *Plethodon vandykei*). The plants also host various fungi (Ascomycota: Dermateaceae: *Marssonina saxifragae*; Mycosphaerellaceae: *Ramularia saxifragae*; Basidiomycota: Pucciniaceae: *Puccinia aspera, P. heterisiae, P. heucherae, P. pazschkei*). Fungi hosted by *S. rivularis* include Ascomycota (Mycosphaerellaceae: *Mycosphaerella allicina, Pseudocercosporella saxifragae*; Sclerotiniaceae: *Botrytis cinerea*), Basidiomycota (Melampsoraceae: *Melampsora epitea*; Pucciniaceae: *Puccinia heucherae, P. saxifragae*), and Chytridiomycota: Synchytriaceae: *Synchytrium rubrocinctum*).

Economic importance: food: The raw or cooked leaves of *S. mertensiana* are reportedly edible, especially when collected prior to flowering; **medicinal:** no medicinal uses reported; **cultivation:** *Saxifraga mertensiana* and *S. rivularis* are cultivated as an ornamentals, often in rock gardens; **misc. products:** none; **weeds:** none; **nonindigenous species:** none

Systematics: Formerly containing a larger number of North American species, *Saxifraga* was found to be polyphyletic as a result of phylogenetic studies using DNA sequence data. The majority of species were retained in one clade ("Saxifragoids"), which was distinct from a clade containing the remaining genera ("Heucheroids"). A number of former *Saxifraga* species resolved within the Heucheroid clade, which necessitated their transfer to other genera such as *Cascadia, Micranthes* (Figure 4.56), and *Pectiantia*. The former clade is related closest to the *Heuchera* group of Saxifragaceae; the latter is sister to a clade containing that group together with the remainder of genera in the family (Figure 4.58). Molecular data indicate that *S. hirculus* is related to the terrestrial *S. flagellaris*. Although *S. rivularis* has been regarded as most closely related to *S. cernua* (FACW, FACU), phylogenetic analyses resolve it as the sister species to *S. hyperborea* (FACU), with *S. cernua* being somewhat more distantly related (Figure 4.58). Nevertheless, *S. rivularis* does cross successfully with *S. cernua*, forming a hybrid known as *S. ×opdalensis*. *Saxifraga mertensiana* has been placed in a distinct section (*Heterisia*), where it resolves as the sister group to section *Irregulares* (Figure 4.58). *Saxifraga rivularis* has an unusual Atlantic/Beringian disjunction, which AFLP data indicate has occurred at least two times during its history. There are two clearly differentiated subspecies, *S. rivularis* subsp. *rivularis* and *S. rivularis* subsp. *arctolitoralis*. The basic chromosome number of *Saxifraga* is difficult to determine due to the extensive variation found among species. Several base numbers have been suggested including: $x = 6, 8, 11, 13,$ and 14. The following chromosome counts and cytotypes have been reported: *Saxifraga hirculus* ($2n = 16$ [common], 24, 28, 32 [common]); *S. mertensiana* ($2n = 36,$ ca. 48,

50); and *S. rivularis* ($2n = 26, 43, 47, 48, 50, 52$ [common], 56, 85 and 95).

Comments: *Saxifraga hirculus* and *S. rivularis* are circumpolar species that grow in far northern North America. *Saxifraga rivularis* extends southward into western North America. *Saxifraga hirculus* is disjunct in the Colorado Rockies and *S. rivularis* is disjunct in Colorado and Utah. *Saxifraga mertensiana* occurs in western and northwestern North America.

References: Aiken et al., 2007; Alsos et al., 2003; Arft et al., 1999; Brochmann & Håpnes, 2001; Brouillet & Evander, 2009b; Brouillet et al., 1998; Burns, 1942; Christy, 2004; Cooper et al., 2004; Crête et al., 2001; Fernald & Kinsey, 1943; Forbes, 1996; Forbes et al., 2001; Guard, 1995; Gustafson, 1954; Henry, 1998; Hollingsworth et al., 1998; Hollister & Webber, 2000; Inouye et al., 2015; Jacques, 1973; Jorgensen et al., 2000; McIntyre et al., 2006; Molau, 1992; Molau & Prentice, 1992; Olesen & Warncke, 1989a,b,c; Oliver et al., 2006; Pannell, 2002; Peinado et al., 2005b; Savile, 1975; Soltis et al., 1996b; Teeri, 1976; Walker & Everett, 1991; Westergaard et al., 2010.

8. *Suksdorfia*

Mock brookfoam

Etymology: after Wilhelm Nikolaus Suksdorf (1850–1932)

Synonyms: *Boykinia* (in part); *Hemieva*

Distribution: global: North America; **North America:** western

Diversity: global: 1–2 species; **North America:** 1–2 species

Indicators (USA): FACW; FAC: *Suksdorfia ranunculifolia*

Habitat: freshwater; palustrine; **pH:** acidic; **depth:** <1 m; **life-form(s):** emergent herb

Key morphology: stems condensed at base, forming a foliar rosette, vegetative bulblets numerous at base; basal leaves (to 4 cm) fleshy, cordate or reniform, deeply 3-lobed, the margins coarsely crenate, petioles (to 11 cm) with broad, sheathing bases; flowering stems (to 35 cm) with 4–9 reduced leaves, terminating in a flat-topped inflorescence with numerous, 5-merous flowers; petals (to 4 mm) spreading, white or purple tinged at base; capsules (to 4 mm) with brown, warty seeds (to 0.6 mm)

Life history: duration: perennial (bulbils, rhizomes); **asexual reproduction:** bulbils, rhizomes; **pollination:** insect; **sexual condition:** hermaphroditic; **fruit:** capsules (common); **local dispersal:** bulbils, seeds (wind); **long-distance dispersal:** seeds

Imperilment: (1) *Suksdorfia ranunculifolia* [G5]; S2 (AB); S3 (BC)

Ecology: general: Both species of *Suksdorfia* occupy similar moist sites along rocky seeps and stream margins. One species formerly was given OBL status because of the high level of inundation associated with its habitats in the spring. However, the revised 2013 indicator list no longer regards either species as OBL. The treatment of *S. ranunculifolia* has been retained here because it was completed prior to the release of the 2013 indicator list. Few studies provide details regarding the reproductive biology or ecology of either species.

Suksdorfia ranunculifolia (**Hook.**) **Engl.** is found in or on alluvial flats, cliffs, crevices, hanging gardens, meadows, runnels, seeps, slopes, swales, and along waterfalls, often at montane to subalpine elevations (to 3232 m). Full sun to semi-shaded exposures are tolerated. The sites are rocky (often moss covered) and are characterized by alternating wet (spring) and dry (summer) cycles. Substrates are acidic and include granite, gravel, loamy sand, sand, scree, and talus. Little else is known about the ecology of this species and mechanisms of pollination, germination, etc., have not been elucidated. Local dispersal occurs by a "censer" mechanism, where the seeds are strewn from the capsules as a result of wind agitation. **Reported associates:** *Abies lasiocarpa, Achillea mille-folium, Aconogonon phytolaccifolium, Agastache, Agoseris heterophylla, Agrostis scabra, Allium cernuum, A. schoeno-prasum, Alnus, Amelanchier alnifolia, Antennaria lanata, A. luzuloides, Anticlea elegans A. occidentalis, Aquilegia formosa, Arctostaphylos, Arenaria capillaris, Aspidotis densa, Botrychium simplex, Bromus suksdorfii, Bryum, Calamagrostis rubescens, Camassia quamash, Carex podo-carpa, C. scirpoidea, C. spectabilis, Cassiope, Castilleja elmeri, C. hispida, C. miniata, C. parviflora, Ceanothus, Cerastium arvense, Ceratodon purpureus, Cheilanthes gracil-lima, Cirsium hookerianum, Cladonia pocillum, C. pyxi-data, Clarkia amoena, Collinsia parviflora, Cryptogramma acrostichoides, C. crispa, C. stelleri, Cymopterus glaucus, Cystopteris fragilis, Dactylis glomerata, Danthonia interme-dia, Dasiphora floribunda, Delphinium burkei, Deschampsia cespitosa, D. elongata, Dicentra uniflora, Dichanthelium, Ditrichium flexicaule, Draba, Drymocallis glandulosa, Erigeron glacialis, Eriogonum flavum, Eriophyllum lanatum, Erythronium grandiflorum, Festuca idahoensis, F. scabrella, F. viridula, Geum canadense, Heracleum, Hesperochiron pumilus, Holodiscus discolor, Hypericum anagalloides, H. formosum, Juncus bufonius, Juniperus communis, Kalmia microphylla, Koeleria macrantha, Lewisia triphylla, Lithophragma parviflorum, Lomatium ambiguum, L. triterna-tum, L. utriculatum, Mertensia ciliata, Micranthes bryophora, M. ferruginea, M. idahoensis, M. integrifolia, M. odontoloma,* *Microsteris gracilis, Mimulus breweri, M. floribundus, M. guttatus, M. lewisii, Minuartia nuttallii, Montia parvi-flora, Orobanche uniflora, Parnassia fimbriata, Peltigera, Penstemon fruticosus, P. montanus, Philonotis fontana, Phlox diffusa, Phyllodoce empetriformis, Physocarpus, Picea engel-mannii, Pinguicula vulgaris, Pinus albicaulis, P. contorta, P. ponderosa, Piptatherum exiguum, Plectritis congesta, Poa gracillima, P. secunda, Polygonum austiniae, Polytrichum juniperinum, Populus tremuloides, Potentilla diversifo-lia, Preissia quadrata, Primula hendersonii, P. pauciflora, Prunella vulgaris, Pseudoroegneria spicata, Pseudotsuga menziesii, Rhodiola rosea, Ribes cereum, Romanzoffia sitch-ensis, Rubus, Saxifraga cespitosa, S. occidentalis, Sedum lanceolatum, S. stenopetalum, Selaginella standleyi, Senecio cymbalarioides, S. triangularis, Sorbus, Spiraea splendens, Suksdorfia violacea, Symphyotrichum foliaceum, Thuja pli-cata, Tortella fragilis, T. tortuosa, Toxicoscordion venenosum, Trientha glutinosa, Trichophorum, Trifolium, Trisetum spi-catum, Tsuga heterophylla, Vaccinium scoparium, Woodsia alpina, Xerophyllum tenax.*

Use by wildlife: none reported.

Economic importance: food: none; **medicinal:** none; **cultivation:** not cultivated; **misc. products:** none; **weeds:** none; **nonindigenous species:** none

Systematics: *Suksdorfia* has been placed in the genus *Boykinia* to which it is closely related (Figure 4.61). However, the position of *Suksdorfia* is strongly supported in opposing relationships by nuclear (with *S. violacea* and *Bolandra*) vs. chloroplast DNA (with *Boykinia*) data, indicating the possibility of ancient hybridization and chloroplast capture (from *Boykinia*) by *S. ranunculifolia* (Figure 4.61). There are also nomenclatural issues involving the genus name *Hemieva*. Although published prior to *Suksdorfia*, the latter name has been conserved nomenclaturally, which makes it the appropriate generic name if both *S. ranunculifolia* and *S. violacea* are placed within the same genus. However, if placed in different genera (as some authors have done), the genus name *Hemieva* would apply to *S. ranunculifolia* [i.e., *Hemieva ranunculi-folia*]. The base chromosome number of *Suksdorfia* is *x* = 7.

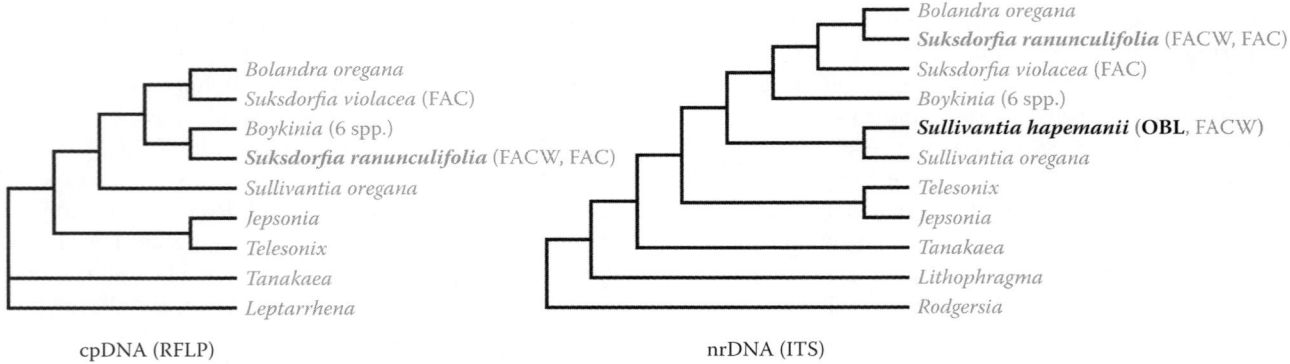

FIGURE 4.61 Different relationships indicated for *Suksdorfia ranunculifolia* by analyses of cpDNA (left) and nrDNA (right) data. Morphologically, *S. ranunculifolia* is closest to *S. violacea* and members of *Bolandra*. Its association with *Boykinia* by cpDNA data may reflect an ancient hybridization and "chloroplast capture." Obligate aquatics are indicated in dark bold. (Adapted from Soltis, D.E. et al., *Syst. Bot.*, 21, 169–185, 1996a.)

Suksdorfia ranunculifolia ($2n = 14$) is diploid and shares a specific karyotype with *Bolandra*, *Boykinia*, and *Sullivantia*, a group resolved as a clade by molecular data.

Comments: *Suksdorfia ranunculifolia* occurs in western North America where the genus is endemic.

References: Klinkenberg, 2004; Savile, 1975; Soltis, 1987, 1988; Soltis et al., 1996a,b; Wells & Elvander, 2009.

9. *Sullivantia*

Coolwort

Etymology: after William Starling Sullivant (1803–1873)

Synonyms: *Boykinia* (in part); *Heuchera* (in part); *Saxifraga* (in part); *Therofon*

Distribution: global: North America; **North America:** midwestern, western

Diversity: global: 3 species; **North America:** 3 species

Indicators (USA): OBL; FACW: *Sullivantia hapemanii*

Habitat: freshwater; palustrine; **pH:** alkaline; **depth:** <1 m; **life-form(s):** emergent herb

Key morphology: stems (to 60 cm) erect; basal leaves (to 11 cm) reniform, palmately incised into 5–13 crenate to dentate lobes, petioled; cauline leaves progressively reduced apically; inflorescence a panicle of numerous, 5-merous flowers; petals (to 3.1 mm) white; stamens 5; ovary inferior; capsule (to 8 mm) containing numerous, winged (rarely wingless) seeds (to 1.7 mm)

Life history: duration: perennial (persistent stem base); **asexual reproduction:** none; **pollination:** insect, self; **sexual condition:** hermaphroditic; **fruit:** capsules (common); **local dispersal:** seeds (water, wind); **long-distance dispersal:** seeds

Imperilment: (1) *Sullivantia hapemanii* [G3]; S2 (ID, MT); S3 (CO, WY)

Ecology: general: All three *Sullivantia* species inhabit moist sites along small watercourses or seeps. However, persistent conditions of saturation are required only for the following species, which is designated as an obligate aquatic in a portion of its range.

Sullivantia hapemanii **(J.M. Coult. & Fisher) J.M. Coult.** grows in shallow water or in exposed sites on wet substrates of bottoms, cliffs, draws, grottos, hanging gardens, seeps, slopes, springs, stream margins, and waterfall splash zones at elevations of up to 3200 m. The sites are cool and densely to partially shaded with calcareous substrates that include boulders, clay, gravel, limestone, loam, quartzite, shale, and travertine. Substrates are often mossy, which assures that the rooting zone remains saturated through most of the year. The plants are short-lived perennials and lack any means of vegetative reproduction. Flowering occurs from June to August, and its frequency is higher when the plants are not water stressed. The flowers are somewhat protandrous but are self-compatible and highly self-pollinating. Insect pollinators include flies (Diptera) and small bees (Hymenoptera) but are seldom observed on flowers. The winged seeds are dispersed by water or by wind. There is no obvious mechanism for very long dispersal. Coupled with their often small size, high rates of selfing, lack of vegetative propagules, and rather limited seed dispersal, the populations exhibit extremely low levels of genetic variation as indicated by allozyme data. More detailed studies of seed dispersal, germination, and seedling establishment need to be conducted. At least some occurrences are threatened by displacement from weedy species such as *Arctium minus*, *Cirsium arvense*, *Festuca arundinacea*, *Phalaris arundinacea*, and *Solanum dulcamara*. **Reported associates:** *Acer glabrum*, *A. negundo*, *Aconitum columbianum*, *Actaea rubra*, *Adiantum capillus-veneris*, *A. pedatum*, *Adoxa moschatellina*, *Agrostis stolonifera*, *Allium acuminatum*, *Anticlea elegans*, *Apocynum androsaemifolium*, *Aquilegia barnebyi*, *Arenaria rubella*, *Arnica cordifolia*, *A. lonchophylla*, *Asplenium trichomanes-ramosum*, *Betula occidentalis*, *Boykinia heucheriformis*, *Bromus inermis*, *Calamagrostis canadensis*, *Campanula rotundifolia*, *Cardamine cordifolia*, *C. oligosperma*, *Carex aurea*, *C. diandra*, *C. hassei*, *C. utriculata*, *Catabrosa aquatica*, *Cirsium rydbergii*, *Clematis ligusticifolia*, *Cornus sericea*, *Cryptogramma stelleri*, *Cypripedium calceolus*, *Cystopteris fragilis*, *Epilobium ciliatum*, *Epipactis gigantea*, *Equisetum arvense*, *E. hyemale*, *E. laevigatum*, *Erigeron acris*, *E. allocotus*, *E. compositus*, *E. kachinensis*, *E. subtrinervis*, *Festuca saximontana*, *Fragaria virginiana*, *Galium boreale*, *G. triflorum*, *Glyceria striata*, *Hackelia davisii*, *Heuchera parvifolia*, *Juncus balticus*, *Leucopoa kingii*, *Limnorchis ensifolia*, *Lupinus*, *Lysimachia ciliata*, *Maianthemum stellatum*, *Marchantia*, *Mertensia ciliata*, *Micranthes odontoloma*, *Mimulus eastwoodiae*, *M. glabratus*, *M. guttatus*, *Musineon*, *Parnassia fimbriata*, *Pellaea breweri*, *Petrophyton*, *Physocarpus monogynous*, *Picea engelmannii*, *Pinus ponderosa*, *Piptatherum micranthum*, *Platanthera hyperborea*, *Poa interior*, *Populus angustifolia*, *P. balsamifera*, *Pseudotsuga menziesii*, *Ribes lacustre*, *R. oxyacanthoides*, *Rosa woodsii*, *Rubus*, *Rumex*, *Salix bebbiana*, *S. exigua*, *Sambucus*, *Saxifraga occidentalis*, *Senecio pseudoaureus*, *S. streptanthifolius*, *Silene menziesii*, *Solidago canadensis*, *Sorbus*, *Spiraea betulifolia*, *Symphyotrichum foliaceum*, *S. molle*, *S. molle* × *S. spathulatum*, *Taraxacum officinale*, *Telesonix heucheriformis*, *Urtica*, *Veronica*, *Viola canadensis*, *V. nephrophylla*.

Use by wildlife: *Sullivantia hapemanii* reportedly is not eaten by insects, livestock, or by wildlife. The plants are host to leaf rust fungi (species unidentified).

Economic importance: food: none; **medicinal:** none; **cultivation:** not cultivated; **misc. products:** none; **weeds:** none; **nonindigenous species:** none

Systematics: Molecular data clearly have established that *Sullivantia* occurs within the "*Boykinia* group" of Saxifragaceae as the sister group to a clade consisting of *Bolandra*, *Boykinia*, and *Suksdorfia* (Figure 4.61). *Sullivantia hapemanii* has also been regarded as a variety of *S. oregana*, to which it is closely related (Figure 4.61). Two varieties of *S. hapemanii* are recognized (var. *hapemanii*, var. *purpusii*), distinguished mainly by the shape of the ovary and fruit. Experimental hybrids between *S. hapemanii* and *S. oregana* exhibited pollen fertility of 50%–86%, but only 31%–43% between either species and *S. sullivantii*. The base chromosome number for *Sullivantia* is $x = 7$. *Sullivantia hapemanii* ($2n = 14$) is diploid.

Comments: *Sullivantia hapemanii* is restricted to Colorado, Idaho, Montana, and Wyoming.
References: Heidel, 2004; Savile, 1975; Soltis, 1982, 1991, 2009; Soltis et al., 1996a.

10. *Tolmiea*

Piggyback plant
Etymology: after William Fraser Tolmie (1812–1886)
Synonyms: *Tierella* (in part)
Distribution: global: North America; **North America:** western
Diversity: global: 2 species; **North America:** 2 species
Indicators (USA): FACW; FAC: *Tolmiea menziesii*
Habitat: freshwater; palustrine; **pH:** 4.5–7.5; **depth:** <1 m; **life-form(s):** emergent herb
Key morphology: rhizomatous, producing a foliar rosette and an erect stem (to 8 dm); basal leaves (to 10 cm) cordate, palmately 5–7 lobed, long-petioled (to 30 cm), adventitious plantlets sometimes developing at junction of blade and petiole, cauline leaves progressively reduced upward, stipules (to 8 mm) leaf-like; racemes (to 30 cm) with numerous bilateral flowers; petals (to 12 mm) 4, filamentous, recurved, brown-purple; stamens 3, unequal; ovary superior; capsule (to 14 mm) with numerous (up to 155) spiny seeds
Life history: duration: perennial (rhizomes); **asexual reproduction:** adventitious plantlets, rhizomes; **pollination:** insect; **sexual condition:** hermaphroditic; **fruit:** capsules (common); **local dispersal:** seeds (animals); **long-distance dispersal:** seeds (animals)
Imperilment: (1) *Tolmiea menziesii* [G5]
Ecology: general: Although regarded previously as OBL in a portion of its range, the 2013 indicator list recategorized *Tolmiea menziesii* as FACW, FAC. Regardless, the treatment has been retained here because it was completed prior to the issuance of the revised indicator list. *Tolmiea diplomenziesii* is not a wetland indicator but does occur in moist woodlands and also along streams.

Tolmiea menziesii **(Pursh) Torrey & A. Gray** inhabits bottoms, cliffs, fens, floodplains, gravel bars, hillsides, meadows, rainforest, ravines, roadsides, woodlands, and the margins of brooks, lakes, rivers, and streams at elevations of up to 1800 m. The substrates tend to be acidic (pH: 4.5–7.5) and include alluvial silt, clay, humus, loam, muck, mud, Newburg soils, pumice, and sandy loam. Individuals have also been reported to grow on rotting logs. The sites are characterized by full shade to partial sun. The plants will tolerate low temperatures to −17°C. The flowers are strongly protandrous (3–4 days), self-incompatible, produce a dilute nectar, and are pollinated principally by *Gnoriste megarrhina*, a fungus gnat (Diptera: Mycetophilidae). Insect-mediated pollination is essential for seed set. *Tolmiea* can produce a sizeable, sometimes dominant seed bank. A seed density of 8.7/m² and buried seed bank of 40.5/m² have been observed in old-growth forests where plants occurred at a frequency of 4.5%. Gravel bars (where plant densities were slightly higher at 8.7/m²) yielded up to 612 seeds/m². Seeds stored at 0°C–5°C for 2 months germinate under 16 h light at 15°C–21°C. The seeds are bristly

and dispersed by birds or mammals. The plants persist vegetatively by a well-developed rhizome. The common name (piggyback plant) refers to the ability to produce adventitious plantlets on the leaves (at the petiole junction). Sites must be moist for detached plantlets to develop successfully. Plants can also be propagated artificially by leaf cuttings. **Reported associates:** *Abies grandis, Acer circinatum, A. macrophyllum, Achillea millefolium, Achlys, Adiantum pedatum, Alliaria petiolata, Alnus rubra, Aquilegia formosa, Aralia californica, Arbutus menziesii, Asarum, Athyrium filix-femina, Berberis nervosa, Blechnum spicant, Boykinia intermedia, Bromus, Cardamine angulata, Carex aquatilis, C. canescens, C. deweyana, C. microptera, C. nigricans, C. obnupta, Cassiope mertensiana, Castanopsis sclerophylla, Ceanothus, Chamaecyparis, Chamerion angustifolium, Chrysosplenium glechomifolium, Circaea alpina, Claytonia sibirica, Cornus sericea, Corydalis scouleri, Corylus cornuta, Delphinium glaucum, Dicentra formosa, Digitalis purpurea, Disporum hookeri, Equisetum hyemale, Eurhynchium oreganum, Festuca subulata, Galium trifidum, G. triflorum, Geranium robertianum, Glycyrrhiza, Gymnocarpium dryopteris, Hedera helix, Heracleum lanatum, Heuchera micrantha, Holodiscus discolor, Hydrophyllum tenuipes, Juncus parryi, Lactuca muralis, Lapsana communis, Lithocarpus densiflorus, Lonicera involucrata, Luetkea pectinata, Luzula campestris, Lysichiton americanus, Maianthemum dilatatum, M. stellatum, Marah oreganus, Melica subulata, Mimulus dentatus, M. guttatus, M. lewisii, Mitella caulescens, M. ovalis, Montia parvifolia, Nemophila parviflora, Oemleria cerasiformis, Oenanthe sarmentosa, Oplopanax horridum, Osmorhiza berteroi, O. chilensis, Oxalis oregona, Petasites frigidus, Phyllodoce empetriformis, Picea sitchensis, Plantago major, Platanthera stricta, Pleuropogon refractus, Poa trivialis, Polypodium glycyrrhiza, Polystichum munitum, Populus balsamifera, P. trichocarpa, Prunella vulgaris, Pteridium aquilinum, Quercus chrysolepis, Ranunculus repens, R. uncinatus, Rhododendron, Ribes bracteosum, Rubus parviflorus, R. spectabilis, R. ursinus, Rumex acetosella, Sambucus racemosa, Saxifraga, Senecio triangularis, Sphagnum, Stachys chamissonis, S. cooleyae, S. mexicana, S. rigida, Stellaria crispa, S. media, Streptopus amplexifolius, Symphoricarpos, Synthyris reniformis, Taxus brevifolia, Tellima grandiflora, Thalictrum occidentale, Thuja plicata, Tiarella trifoliata, T. unifoliata, Trillium ovatum, Trisetum cernuum, Tsuga heterophylla, Umbellularia californica, Urtica dioica, Vaccinium ovalifolium, V. parvifolium, Valeriana sitchensis, Vancouveria hexandra, Viola glabella.*
Use by wildlife: *Tolmiea menziesii* is the host plant for a moth larva (Lepidoptera: Prodoxidae: *Greya punctiferella*) and for the fungi *Puccinia heucherae* (Basidiomycota: Pucciniaceae) and *Sphaerotheca mors-uvae* (Ascomycota: Erysiphaceae). The flowers are visited by various insects including blowflies (Diptera: Calliphoridae: *Melastoma mellinum*), hoverflies (Diptera: Syrphidae: *Parasyrphus insolitus, P. macularis, Platycheirus obscurus, Syrphus*), and bumblebees (Hymenoptera: Apidae: *Bombus caliginosus, B. flavifrons, B. sitkensis*), which forage for the pollen

but rarely serve as pollinators. The plants are not palatable to native (*Ariolimax columbianus*) or introduced (*Arion ater*) slugs (Gastropoda: Arionidae). *Tolmiea* is a habitat feature of the long-toed salamander (Amphibia: Ambystomatidae: *Ambystoma macrodactylum*).

Economic importance: food: The early spring shoots of *Tolmiea menziesii* are bitter but were eaten raw by the Makah people. However, the plants are best avoided as a food because they can produce a minor skin irritation in some people who contact the foliage; **medicinal:** The Cowlitz applied a poultice of the fresh leaves to treat boils; **cultivation:** *Tolmiea menziesii* is grown as an ornamental plant (both indoors and outdoors) and includes the cultivars 'Cool Gold' and 'Taffs Gold'; **misc. products:** *Tolmiea menziesii* often is planted as a ground cover. It has also been used as an experimental system for studying autopolyploidy; **weeds:** none; **nonindigenous species:** none

Systematics: Phylogenetic analysis resolves *Tolmiea* as the sister genus to a clade comprising *Elmera*, *Tellima*, and *Heuchera* (Figure 4.56). A relatively close relationship between *Tolmiea* and *Tellima* is also supported by nrDNA RFLP data, which indicate that natural intergeneric hybridization occurs between these genera. The base chromosome number of *Tolmiea* is *x* = 7. *Tolmiea* was originally regarded as monotypic, with *Tolmiea menziesii* comprising both diploid (2*n* = 14) and tetraploid (2*n* = 28) cytotypes, the latter a result of autopolyploidy. However, the diploid and tetraploid cytotypes are clearly distinguishable by cpDNA RFLP patterns, an indication of their divergence subsequent to the onset of autoploidy (the chloroplast genome is inherited maternally as in most plants). The cytotypes are now recognized as distinct species, with *T. menziesii* regarded as an autotetraploid derivative of *T. diplomenziesii*. Allozyme data indicate higher levels of heterozygosity in the tetraploid (i.e., *T. menziesii*) populations.

Comments: *Tolmiea menziesii* occurs in coastal western North America.

References: Aller, 1956; Cates & Orians, 1975; Doyle et al., 1985; Goldblatt et al., 2004; Harmon & Franklin, 1995; Kellman, 1974; Pabst & Spies, 1998; Savile, 1975; Soltis, 1984; Soltis & Rieseberg, 1986; Soltis et al., 1989, 1990, 2009; Yarbrough, 1936.

Order 8: Vitales [1–2]

Family 1: Vitaceae Juss. [14]

Some authors merge *Leea* (Leeaceae) with Vitaceae, which is a matter of taxonomic preference given that molecular data (Ingrouille et al., 2002; Rossetto et al., 2002; Soejima & Wen, 2006) consistently resolve Leeaceae and Vitaceae as sister clades, thus reconciling either their merger or their maintenance as distinct families (Figure 4.62). These families are also linked by several morphological synapomorphies including "pearl" glands and raphides crystals (Judd et al., 2002; Soejima & Wen, 2006). The phylogenetic placement of Vitales remained inscrutable, even after analyses of multiple molecular data sets. Analyses of different DNA sequences

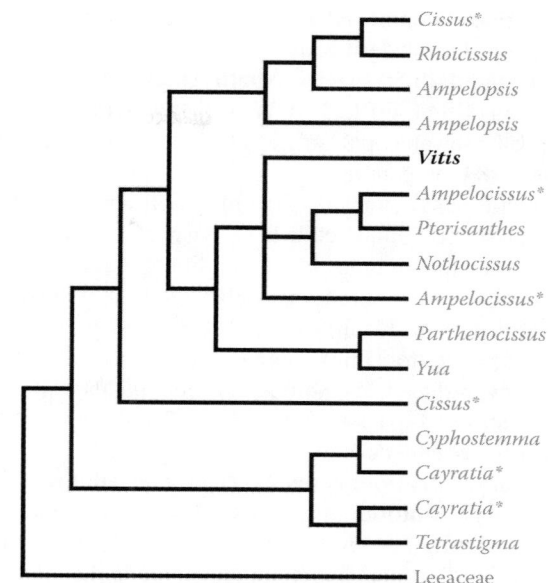

FIGURE 4.62 Phylogenetic relationships in Vitaceae as indicated by combined cpDNA sequence data. Genera marked by an asterisk (*Ampelocissus*, *Cayratia*, *Cissus*) are not monophyletic. *Vitis* (bold) is the only genus of Vitaceae in North America with obligate aquatic plants. (Adapted from Soejima, A. & Wen, J., *Amer. J. Bot.*, 93, 278–287, 2006.)

placed the order variously as the sister group to Asteridae or Caryophyllales (*rbcL*), Saxifragales (*atpB*), Dilleniaceae (*matK*), or to all rosids (combined *atpB*, *rbcL*, 18S nrDNA) (Jansen et al., 2006; Soejima & Wen, 2006; Soltis et al., 2005). Multiple (3–4) gene analyses indicated the position of the order as the sister group to all rosids (Figure 4.1), a result that was not well-supported and accepted only tentatively (Soltis et al., 2005). However, more recent phylogenetic analyses incorporating complete chloroplast genome sequences (Jansen et al., 2006) have strongly corroborated the placement of Vitaceae as the sister group to all rosids, the disposition which has been accepted here.

In addition to the various molecular data sets mentioned earlier, the monophyly of Vitaceae is indicated by their liana habit, leaf-opposed tendrils, seeds with an adaxial cord-like raphe, 3-lobed endosperm, and fleshy fruits (Judd et al., 2002). The flowers are reduced and pollinated by various insects including Hymenoptera (bees and wasps) Coleoptera (beetles), and Diptera (flies). The fruits are eaten and dispersed endozoically by animals.

Vitaceae are a common element of terrestrial forests and can be fairly common in wetter areas such as floodplain swamps. Only one genus contains species designated as obligate aquatic plants in North America:

  1. *Vitis* L.

1. *Vitis*

Grape; raisin, vigne
Etymology: from the Latin *vitis* meaning "grape-vine"
Synonyms: none

Distribution: global: Africa; Asia; Europe; New World; **North America:** widespread

Diversity: global: 50 species; **North America:** 18 species

Indicators (USA): OBL; FACW: *Vitis palmata*

Habitat: freshwater; palustrine; **pH:** alkaline; **depth:** <1 m; **life-form(s):** emergent liana

Key morphology: stems (to 20 m) high-climbing, the young shoots herbaceous and reddish or purplish, with broad (to 5 mm) nodal diaphragms in longitudinal section, the older stems woody with shredding bark; tendrils reddish, leaf opposed; leaves (to 12 cm) palmately (3–5) lobed, petioled (to 10 cm), stipulate (to 3 mm); panicles (to 18 cm) leaf opposed, with many reduced, 5–merous flowers; petals minute (to 1 mm), valvate, fugacious; berries (to 8 mm) round, black, glossy, with 1–4 seeds

Life history: duration: perennial (buds); **asexual reproduction:** none; **pollination:** insect; **sexual condition:** polygamo-dioecious; **fruit:** berries (common); **local dispersal:** fruits (water); **long-distance dispersal:** fruits, seeds (birds)

Imperilment: (1) *Vitis palmata* [G4]; SH (GA); S2 (IN)

Ecology: general: Grapes are common forest lianas that often occur in wet areas. Thirteen of the 20 North American *Vitis* species (65%) are designated by some type of wetland indicator status. However, even though most grape species are associated with permanent water, a few can withstand prolonged conditions of standing water where the sediment oxygen levels are low. Only one North American species can tolerate such prolonged, hydric conditions. Grapes are polygamous, having plants with bisexual or unisexual flowers. Pollination is not well studied in this group except for a few economically important species where insect (Coleoptera, Diptera, Hymenoptera) and self-pollination are reported.

Vitis palmata **Vahl** grows in ditches, floodplains, swamps, and along the margins of ponds, rivers, sloughs, and streams. The substrates are described as calcareous and include clay, limestone, rocks, and sand. The flowers are bisexual or unisexual and the plants often are dioecious. Flowering occurs relatively late in the season (mid-late June) with respect to other grape species. There are no accounts of pollination or reproductive ecology published for *V. palmata*. The seeds are physiologically dormant and their germination is promoted by 6 months of cold stratification; however, it may be delayed for more than a year afterwards. As presumed for most grape species, the juicy fruits are likely to be dispersed by animals, particularly birds (Aves). Dispersal by water over short distances is also likely.

Reported associates: *Acer barbatum, A. rubrum, Berchemia scandens, Brunnichia ovata, Cardiospermum halicacabum, Carex debilis, C. vulpinoidea, Carpinus caroliniana, Carya aquatica, Celtis, Celtis laevigata, Cephalanthus occidentalis, Cornus foemina, Crataegus opaca, C. viridis, Diospyros virginiana, Forestiera acuminata, Fraxinus americana, F. pennsylvanica, F. profunda, Gleditsia aquatica, Ilex decidua, I. opaca, Impatiens capensis, Itea virginica, Liquidambar styraciflua, Lonicera japonica, Ludwigia alternifolia, Lycopus virginicus, Lygodium japonicum, Microstegium vimineum, Mimosa strigilosa, Myosotis macrosperma, Quercus laurifolia, Q. lyrata,* *Q. nigra, Q. phellos, Pinus taeda, Planera aquatica, Platanus occidentalis, Populus heterophylla, Styrax americanus, Taxodium distichum, Tradescantia hirsutiflora, Triadenum walteri, Ulmus americana.*

Use by wildlife: The fruits of *Vitis palmata* reportedly are eaten by many wildlife species including gamebirds, songbirds, and mammals. Generally, grape vines provide shelter and nesting sites for birds (Aves) and their bark sometimes is used as a nesting material. *Vitis palmata* is a component of habitat for the Henslow's sparrow (Aves: Emberizidae: *Ammodramus henslowii*).

Economic importance: food: Grapes are important economically for winemaking (*Vitis vinifera*) and their edible fruits (fresh or dried as raisins). The mature fruits of *V. palmata* are sweet and edible and often are used to make wild grape jelly. The young tendrils are also edible either raw or cooked. The young leaves are wrapped around foods as a flavor enhancer when baking; **medicinal:** none reported; **cultivation:** *Vitis palmata* is cultivated on occasion; **misc. products:** A yellow dye can be extracted from the leaves of *V. palmata*. Because of its ability to tolerate hydric conditions, *V. palmata* has been considered as a candidate for breeding resistance to root rot (Chromista: Oomycota: Pythiaceae: *Phytophthora*) in other grape species; **weeds:** Some states regard all grapes as potentially noxious weeds; **nonindigenous species:** none occurring in aquatic habitats.

Systematics: *Vitis* is characterized morphologically by polygamodioecy, calyptrate petals, and 5-merous flowers. Although limited analyses using only *rbcL* data weakly suggest the paraphyly of the genus, analyses of combined cpDNA sequence data strongly indicate that *Vitis* is monophyletic and resolves within a larger clade containing *Ampelocissus, Pterisanthes,* and *Nothocissus* (Figures 4.62 and 4.63). A phylogenetic analysis using combined *trnL* intron, *trnL*-F IGS, and *trnK* intron sequence data recovered four major *Vitis* clades, individually having geographical affinities to either Europe, Asia, or North America (Figure 4.63). One of the two North American clades (represented by *V. rotundifolia*) corresponds to the small subgenus *Muscadinia*, which resolved as a sister group to the remaining species (all assigned to subgenus *Vitis*). Most *Vitis* species (including *V. palmata*) occur within subgenus *Vitis*, which is characterized by vines having shredding bark, inconspicuous lenticels, forked tendrils, and a pith interrupted by nodal diaphragms. Interestingly, the molecular phylogeny (Figure 4.63) placed *V. palmata* within the clade of principally Asian species, closest to *V. betulifolia* (China) and *V. coignetiae* (Japan, Korea, and Russia), neither being associated with wetland habitats. However, because there was some question regarding the taxonomic integrity of the *V. palmata* accession used in that study, it is possible the subsequent analyses might place this species differently. The subgenera also differ by their base chromosome number ($x = 19$ in subgenus *Vitis*; $x = 20$ in subgenus *Muscadinia*). Although *V. palmata* has no chromosome number reported, those available for other species in subgenus *Vitis* are uniformly diploid ($2n = 38$). Although all hybrids between the subgenera of *Vitis*

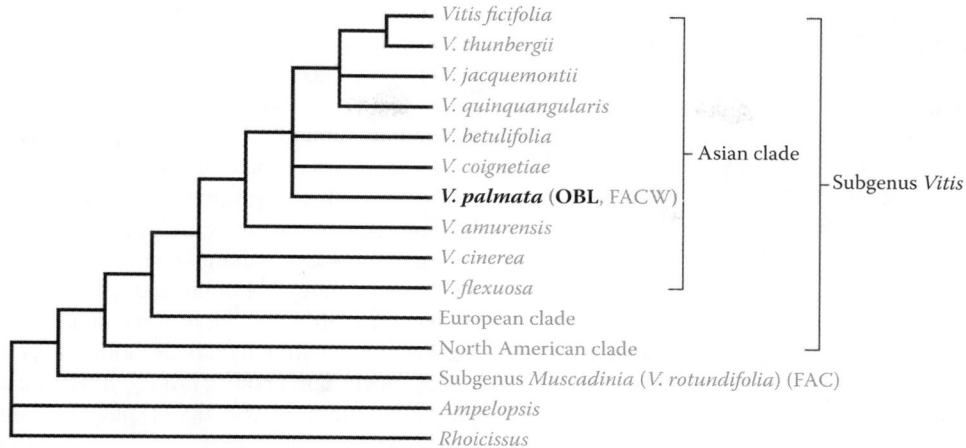

FIGURE 4.63 Phylogenetic relationships in *Vitis* as indicated by analysis of combined cpDNA sequence data. The OBL *V. palmata* (bold) resolves within a clade comprising mainly Asian species of subgenus *Vitis*. (Adapted from Tröndle, D. et al., *Amer. J. Bot.*, 97, 1168–1178, 2010.)

are inviable, *V. palmata* also reportedly does not hybridize with cultivated varieties of *V. vinifera*, which are in subgenus *Vitis*. A number of microsatellite markers developed for *V. vinifera* show excellent cross reactivity with North American species from both subgenera, thereby opening up the possibility of conducting much-needed population genetic studies of *V. palmata*. **Comments:** *Vitis palmata* occurs in the Mississippi embayment region and also is oddly disjunct in the northeastern United States.
References: Arroyo-García et al., 2002; Di Gaspero et al., 2000; Goto-Yamamoto et al., 2006; Hardie & O'Brien, 1988; Ingrouille et al., 2002; Soejima & Wen, 2006; Tröndle et al., 2010.

MALVID ROSIDS (EUROSIDS II)

ORDER 9: BRASSICALES [19]

The Brassicales contain about 4400 species and represent one of the more advanced rosid orders (Figure 4.1). The group is known for the widespread occurrence of species that produce sulfur-containing mustard-oil glycosides (glucosinolates), which are compounds that deter pathogens and herbivores. Glucosinolates and their associated myrosin cells are regarded as synapomorphic traits for the order (Judd et al., 2016). In North America, OBL indicators occur within three families:

1. **Bataceae** Mart. ex Perleb
2. **Brassicaceae** Burnett
3. **Limnanthaceae** R. Br.

Phylogenetic relationships among these families (Figure 4.64) indicate several separate origins of the OBL habit in the Brassicales. Most of the aquatic taxa occur in the Brassicaceae, which occupy a relatively derived position in the cladogram of the order.

Family 1: Bataceae [1]

Bataceae (occasionally appearing as Batidaceae) are a small family of succulent, suffruticose, maritime plants,

which contains a single genus (*Batis*) with only two species. Although sometimes described as xerophytes, Bataceae are coastal halophytes, which typically inhabit salt marshes. The family is related closely to Salvadoraceae (Figure 4.64), a small family of dry habitats, and somewhat more distantly to Koeberliniaceae, another family of desert and semidesert

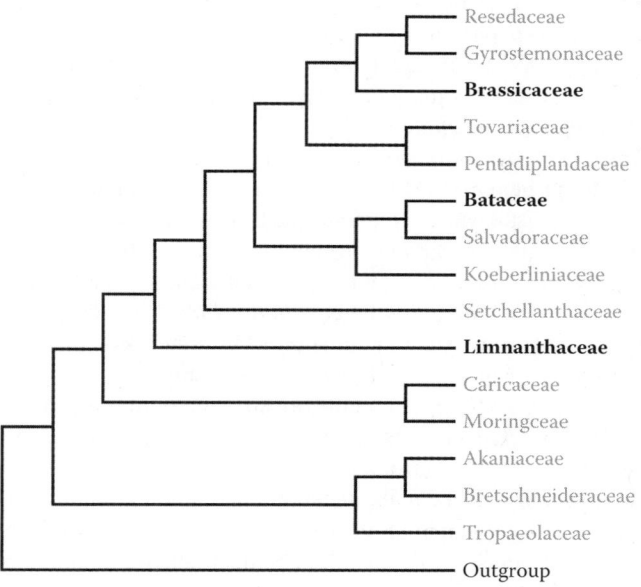

FIGURE 4.64 Phylogenetic relationships among families of Brassicales as indicated by combined *rbcL* and 18S rDNA sequences. The topology agrees well with an analysis of fewer families (but including *atpB* data) except for the positions of Setchellanthaceae and Limnanthaceae, which are reversed (Soltis et al., 2000). Additional studies (e.g., Hall, 2008) have provided evidence for maintaining Capparaceae and Cleomaceae as distinct families allied with Brassicaceae (see Figure 4.65). Families containing OBL indicators in North America (shown in bold) are separated widely in the order. (Adapted from Rodman, J.E. et al., *Amer. J. Bot.*, 85, 997–1006, 1998.)

habitats. The features of the family are essentially those of the genus, which contains one North American OBL indicator:

1. *Batis* P. Browne

With only two species, *Batis* arguably is monophyletic; however, *Batis argillicola* has not yet been included in a phylogenetic analysis for confirmation.

1. *Batis*
Saltwort
Etymology: from an ancient Greek name of a seashore plant
Synonyms: none
Distribution: global: Australia; New Guinea; New World; **North America:** southern
Diversity: global: 2 species; **North America:** 1 species
Indicators (USA): OBL: *Batis maritima*
Habitat: saline (coastal); palustrine; **pH:** 6.0–9.0; **depth:** <1 m; **life-form(s):** emergent (subshrub)
Key morphology: shoots (to 1.5 m) fleshy, 4-angled, arising from a woody base, branches ascending; leaves (to 3 cm) opposite, sessile, fleshy, linear-oblanceolate, simple, entire, bases with small, downward points, the tips pointed to mucronate; inflorescences axillary, 4-angled, fleshy, cone-like spikes; flowers reduced, unisexual, enclosed by bracts; fruit berry-like pyrenes, from coherent pistils
Life history: duration: perennial (woody base/stems); **asexual reproduction:** stems (layering); **pollination:** wind; **sexual condition:** dioecious; monoecious; **fruit:** multiple (pyrenes) (common); **local dispersal:** fruits; **long-distance dispersal:** fruits (water)
Imperilment: *Batis maritima* [G5]
Ecology: general: Both *Batis* species are halophytes, which inhabit coastal warm temperate and tropical salt marshes. They are perennial, evergreen, shrubby perennials. The flowers are highly reduced, lack a perianth, and are wind pollinated. *Batis maritima* is dioecious; whereas, *B. argillicola* is monoecious. The fruits are buoyant and dispersed by water. The plants fix carbon by means of a C_3 photosynthetic pathway.

Batis maritima L. occurs in coastal, tidal sites, which include beaches, ditches, dunes, mangrove swamps, marshes, prairies, roadsides, saline flats, salt marshes, salt pans, swales, and the margins of canals at elevations near sea level (to 10 m). The plants are shade intolerant and occur in open exposures on wet, brackish, saline, or hypersaline, coarse mineral soils (up to 59.0 dS/m), which are inundated periodically by salt water. The substrates have been described as Handsboro association, muck, sand, and shells. Optimal growth occurs at a salinity of 200 mM NaCl, but the plants can survive levels up to 1000 mM NaCl. Flowering and fruiting occur from June to September. The plants are dioecious and the flowers are pollinated abiotically by the wind. Rapid germination of 2-month-old seeds has been achieved by placing them in an equal mixture of sand and watering. The seeds are dispersed by water. The compound fruit structures (infructescences) can float in salt water for up to 2 months before releasing the individual pyrene fruits, which also float and remain viable

for up to three additional months. This species often forms large monocultures comprising 80%–90% of the vegetational cover. It is known to produce mats more than 90 m in diameter. The plants are resistant to disturbance and quickly can recolonize habitats that have been damaged by hurricanes. They tend to dominate the saline coastal landscape following the aftermath of mangrove community destruction by tropical storms. *Batis maritima* also thrives in areas of natural wrack deposition where other species do poorly. However, it is sensitive to pollution and plants will die if the leaves, stems, or their substrate become coated with oil. Restoration projects have found that the addition of kelp compost and tidal creek networks positively influence establishment and increase the survivorship of plants. **Reported associates:** *Achyranthes rotundata, Amaranthus australis, Atriplex pentandra, Avicennia germinans, Blutaparon vermiculare, Bolboschoenus robustus, Borrichia frutescens, Chamaesyce vaginulatum, Conocarpus erectus, Cuscuta salina, Cyperus odoratus, C. virens, Distichlis spicata, Fimbristylis castanea, F. ferruginea, Frankenia salina, Iva frutescens, Jaumea carnosa, Juncus roemerianus, Laguncularia racemosa, Limonium carolinianum, Lycium carolinianum, L. tweedianum, Monanthochloe littoralis, Muhlenbergia capillaris, Oenothera humifusa, Paspalum vaginatum, Persicaria punctata, Portulaca rubricaulis, Rhachicallis americana, Rhizophora mangle, Rumex obovatus, Salicornia bigelovii, Sarcocornia perennis, Sesuvium maritima, S. portulacastrum, Setaria parviflora, Seutera angustifolia, Solidago sempervirens, Spartina alterniflora, S. cynosuroides, S. foliosa, S. patens, S. spartinae, Sporobolus virginicus, S. wrightii, Suaeda linearis, Symphyotrichum tenuifolium, Typha domingensis.*

Use by wildlife: *Batis maritima* is eaten by white-tailed deer (Mammalia: *Odocoileus virginianus*). It is a host plant for larval butterflies (Insecta: Lepidoptera) including the great southern white butterfly (Pieridae: *Ascia monuste*) and eastern pygmy blue (Lycaenidae: *Brephidium exilis*). It is also a nectar food plant for the eastern pygmy blue butterfly. The plants provide habitat for salt marsh and floodwater mosquito larvae (Insecta: Diptera: Culicidae: *Aedes sollicitans*, *A. taeniorhynchus*) and also for several birds (Aves) including horned larks (Alaudidae: *Eremophila alpestris*), whooping cranes (Gruidae: *Grus americana*), willets (Scolopacidae: *Tringa semipalmata*), and Wilson's plover (Charadriidae: *Charadrius wilsonia*). Salt flats that are dominated by *B. maritima* are used as feeding grounds by long-billed curlews and whimbrels (Aves: Scolopacidae: *Numenius americanus*; *N. phaeopus*). *Batis maritima* has also been identified as an element of critical habitat for the endangered rice rat (Mammalia: Rodentia: Cricetidae: *Oryzomys palustris*) and the lower keys rabbit (Mammalia: Leporidae: *Sylvilagus palustris hefneri*). Wild rabbits (Mammalia: Leporidae) readily eat the salty shoot tips of the plants.

Economic importance: food: The fresh young shoots and leaves of *Batis maritima* can be eaten as a vegetable, either raw or boiled (to reduce saltiness); however, the plants can be toxic if consumed in large amounts. The plants are also used as a source of salt. The seeds are also edible and can contain up to 25% oil, which is rich in essential fatty acids; **medicinal:** Foliage of *B. maritima* has been used to make a medicinal tea. The plants have been used to treat skin disorders (eczema, psoriasis), and as a remedy for gout, rheumatism, and scurvy; **cultivation:** *Batis maritima* is cultivated as a specimen for saline, sunny sites; **misc. products:** Ashes obtained from *B. maritima* plants have been used to manufacture glass and soap. *Batis maritima* plants have been used in ecological studies of metal halide emissions; **weeds:** *Batis maritima* is invasive in Hawaiian estuaries and saltmarshes where it can form expansive colonies; **nonindigenous species:** *Batis maritima* was introduced to Hawaii sometime before 1859.

Systematics: With only two, similarly distinctive species, *Batis* presumably is monophyletic, with *B. maritima* and *B. argillicola* (Australia and New Guinea) comprising sister species. Only *B. maritima* has been included in phylogenetic studies thus far. The base chromosome number of *Batis* is $x = 11$. *Batis maritima* is a diploid ($2n = 22$). Other reported counts of $2n = 18$ are likely erroneous.

Comments: *Batis maritima* occurs along the coasts of California and the southeastern United States.

References: Aronson, 1989; Atia et al., 2010; Cronquist, 1981; Davis et al., 2005c; Debez et al., 2010; De Craene, 2005; Goldblatt, 1976b; Johnson, 1935; Kunza & Pennings, 2008; Liogier, 1985; McKee et al., 2004; O'Brien & Zedler, 2006; Pennings & Richards, 1998; Shay, 1990; Stutzenbaker, 1999.

Family 2: Brassicaceae [321]

Brassicaceae (=Cruciferae) are a large (~3780 species), diverse, and monophyletic family (Al-Shehbaz, 2010; Judd et al., 2016). They essentially are a cosmopolitan group characterized by a herbaceous habit and the presence of a replum in the fruit (Judd et al., 2016). As strictly delimited the family contains 97 genera with 744 species in North America (Al-Shehbaz, 2010). Typically these plants possess 4-merous flowers with six (tetradynamous) stamens, and a gynoecium of two carpels connected by the replum, a false septum derived from the placental tissue. The fruits are capsular siliques (or silicles when not elongate). The flowers are pollinated by a diverse array of insects (Insecta). They generally are protogynous and can also be self-pollinating, especially in the weedier taxa (Judd et al., 2016). The seeds are dispersed abiotically, by wind or water.

Molecular data initially resolved the members of Capparaceae as a paraphyletic grade, which resulted in the recommended merger of the group within Brassicaceae (Hall et al., 2002). However, subsequent studies (Hall, 2008) were able to resolve three distinct clades, which corresponded with Brassicaceae, Capparaceae, and Cleomaceae (Figure 4.65). In that sense, Cleomaceae would represent the sister group of Brassicaceae (although some authors would still prefer to treat the latter as a subfamily of the former). Molecular data (*rbcL*, 18S nrDNA sequences) indicate that Brassicaceae (sensu lato) are related closely to a clade comprising Resedaceae and Gyrostemonaceae (Figure 4.64).

Tribal delimitation in Brassicaceae remains problematic and was not even attempted in a fairly comprehensive treatment of the North American taxa (Rollins, 1993). Although

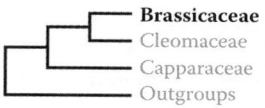

FIGURE 4.65 Relationships of Brassicaceae, Capparaceae, and Cleomaceae as indicated by analysis of *matK* and *ndhF* sequence data. All three groups resolve as well-supported clades, with Cleomaceae as the sister group of Brassicaceae. Taxa containing OBL indicators are shown in bold. (Adapted from Hall, J.C., *Botany*, 86, 682–696, 2008.)

more intensive study will be necessary to address the issue adequately, preliminary findings (e.g., Koch et al., 2001; Mitchell & Heenan, 2000) have indicated that a number of tribes (e.g., *Arabideae*, *Lepidieae*, *Sisymbrieae*) were polyphyletic as formerly circumscribed. Ongoing phylogenetic studies continue to clarify the tribal limits, with the most recent estimate recognizing 49 tribes in the family (Al-Shehbaz, 2012).

Many Brassicaceae are important economically, particularly because of their acrid taste. The family is the source of horseradish (*Armoracia*), mustard (*Brassica*), radish (*Raphanus*), and a rich variety of cultivars derived from *Brassica oleracea*, which include broccoli, brussels sprouts, cabbage, cauliflower, kale, and kohlrabi. The wasabi mustard (popular at sushi bars) is derived from *Eutrema*. Numerous cultivated ornamentals occur as well, particularly in the genera *Aethionema*, *Alyssum*, *Arabis*, *Aubrieta*, *Brassica*, *Cardamine*, *Draba*, *Erysimum*, *Iberis*, *Matthiola*, and *Raphanus*. Many weedy species are also found here, especially in the genera *Alliaria*, *Capsella*, and *Lepidium*. Because of their rapid developmental cycles, several plants in the family are used in molecular genetic research including *Arabidopsis thaliana* (regarded as a model system for studying plant development), and various "fast plants" (*Brassica campestris*, *B. rapa*) with extremely short life cycles.

Iodanthus and *Neobeckia* were among the OBL indicators in the 1996 list. The former no longer contains any OBL species, and the latter has been merged with *Rorippa*. *Rorippa coloradensis* (an OBL indicator in the 1996 list) has since been placed in synonymy with the cultivated, nonindigenous white mustard *Sinapis alba*. Perhaps due to this transfer (but surely in error), the revised (2013) indicator list retains *S. alba* as an OBL indicator (OBL, FAC); however, this widely cultivated crop plant can hardly be regarded as a wetland species and consequently has not been included in the present treatment. With these modifications, there are eight North American genera recognized as containing obligate aquatic taxa:

1. ***Barbarea*** W. T. Aiton
2. ***Cardamine*** L.
3. ***Eutrema*** R. Br.
4. ***Leavenworthia*** Torr.
5. ***Lepidium*** L.
6. ***Nasturtium*** W. T. Aiton
7. ***Rorippa*** Scop.
8. ***Subularia*** L.

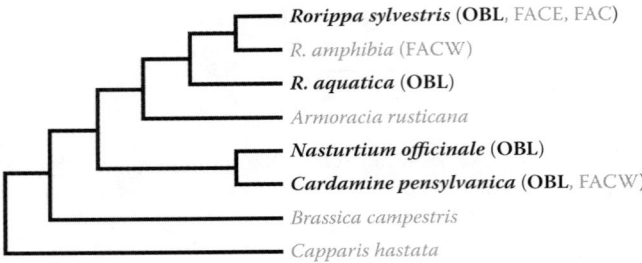

FIGURE 4.66 Phylogenetic relationships among seven genera of Brassicaceae as indicated by *rbcL* sequence data. The North American indicator status is given in parentheses (OBL indicators are in bold). (Adapted from Les, D.H., *Aquat. Bot.*, 49, 149–165, 1994.)

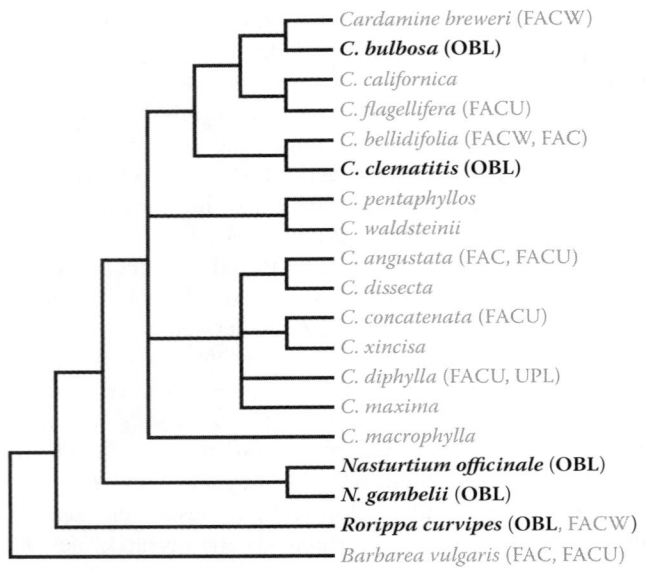

FIGURE 4.67 Phylogenetic relationships of *Cardamine* from combined analysis of *trnL* and *ndhF* sequences. Wetland indicator designations are given in parentheses for the North American species with the OBL taxa shown in bold. A subsequent study, which evaluated combined DNA sequence data for 39 *Cardamine* species (Ali et al., 2012) yielded a similar result where the OBL indicators were concentrated toward the more derived portions of the cladogram. (Adapted from Sweeney, P.W. & Price, R.A., *Syst. Bot.*, 25, 468–478, 2000.)

Intergeneric relationships within this large family have not yet been resolved reliably, and even some genera appear to be highly polyphyletic (e.g., O'Kane, Jr. & Al-Shehbaz, 2003; Hall, 2008). Eventually, more comprehensive investigations incorporating molecular data should greatly facilitate the taxonomic delimitation of tribes and genera in Brassicaceae.

A preliminary phylogenetic study of five aquatic mustard genera using *rbcL* sequence data (Figure 4.66) indicated the distinctness of several genera once merged taxonomically (Les, 1994). The molecular data resolved *Nasturtium* (watercress) as closely related to *Cardamine*, a result that recommended it should not be combined with *Rorippa* (a convention adopted in many recent treatments). Subsequent studies using

ITS and *trnL* and *ndhF* sequences (Sweeney & Price, 2000; Yang et al., 1999; Nakayama et al., 2014) verified the distinction of *Rorippa* and *Nasturtium* and the sister-group relationship of *Nasturtium* and *Cardamine* (Figures 4.66 and 4.67). Morphometric analyses (Khalik et al., 2002) also indicated that *Nasturtium* and *Rorippa* are distinct morphologically.

1. *Barbarea*

Winter-cress; Barbarée, cresson d'hiver
Etymology: after St. Barbara (4th century AD)
Synonyms: *Campe*
Distribution: global: Asia; Europe; North America; **North America:** northern
Diversity: global: 22 species; **North America:** 4 species
Indicators (USA): OBL; FACW: *Barbarea orthoceras*
Habitat: freshwater; palustrine; **pH:** unknown; **depth:** <1 m; **life-form(s):** emergent (herb)
Key morphology: stems (to 6 dm) angled, purplish, freely branched; basal leaves (to 12 cm) long-petioled, oblong/elliptic to lyrate-pinnatifid (or pinnate); cauline leaves similar but reduced above; racemes simple or compound; flowers pedicelled (to 3 mm); petals (to 5 mm) 4, yellow, cruciform; siliques (to 5 cm) somewhat 4-angled
Life history: duration: biennial/perennial (taproot, caudex); **asexual reproduction:** none; **pollination:** insect; **sexual condition:** hermaphroditic; **fruit:** siliques (common); **local dispersal:** seeds; **long-distance dispersal:** seeds
Imperilment: (1) *Barbarea orthoceras* [G5]; SH (ME, NH); S1 (NB, NI); S2 (AB); S3 (WY)
Ecology: general: *Barbarea* is a temperate genus of biennial or perennial plants, which occupy habitats including rocky alpine sites, disturbed or waste areas, moist forests, and riparian wetlands. Two of the North American species (50%) are wetland indicators, but only one is designated as OBL. The flowers can be self-incompatible and pollinated mainly by insects (e.g., Insecta: Hymenoptera), or self-compatible and autogamous. Seed production in large populations can be prolific, estimated to reach several billion seeds/hectare in some cases. Seeds can be dispersed by animals (which consume them), by water (due to their buoyancy), and by adhesion of their seed coat, which becomes mucilaginous when wet in some species. The seeds can remain dormant in the soil for up to 20 years in some species. Seed dormancy can be induced by drying.

***Barbarea orthoceras* Ledeb.** inhabits fresh to brackish sites in beaches, bogs, bottoms, chaparral, cienega, cliffs, depressions, ditches, draws, dunes, fellfields, flats, floodplains, gravel bars, gravel pits, ledges, marshes, meadows, mud flats, outwash channels, prairies, roadsides, sand bars, scrub, seeps, sinkholes, slopes (to 18%), sloughs, springs, swales, swamps, thickets, tundra, woodlands, and along the margins of lakes, ponds, rivers, rivulets, and streams at elevations of up to 3809 m. The exposures range from open to shaded sites. The substrates are alkaline and have been described as alluvium, clay, clay loam, cobble, decomposed granite, gravel, gravelly loam, gravelly sand, humus, karst, loam, mud, organic gravelly muck, Pilchuck soils, rock, rocky sand, sand, sandy gravel, sandy loamy silt, scree, serpentine, shale, shaley rock, silt, stony clay, and typic cryaquent. Flowering occurs from March to August, with the fruiting period extending into September. The flowers produce a relatively low pollen:ovule ratio (1100:1) but are outcrossed obligately by insects (see *Use by wildlife*). Each flower produces about 24 ovules on average. The seeds of *B. orthoceras* are dormant but will germinate at 20°C after stratification in the dark at 3°C for 2–4 weeks. **Reported associates:** *Abies concolor, A. grandis, Acer glabrum, A. macrophyllum, A. negundo, Achillea millefolium, Actaea rubra, Adenostoma fasciculatum, Agoseris glauca, Agropyron macrourum, Agrostis gigantea, Allium, A. rhombifolia, Alnus rugosa, A. tenuifolia, Alopecurus aequalis, Amelanchier utahensis, Androsace filiformis, Angelica genuflexa, A. lucida, Antennaria racemosa, Aquilegia formosa, Arabidopsis lyrata, Arabis hirsuta, Arbutus menziesii, Arctagrostis, Arctostaphylos obispoensis, A. viscida, Arenaria physodes, Arnica longifolia, Artemisia arbuscula, A. douglasiana, A. ludoviciana, A. packardiae, A. tilesii, A. tridentata, A. tripartita, Astralagus alpinus, Athyrium filixfemina, Baccharis salicifolia, Berberis, Bistorta bistortoides, Brassica, Brickellia, Bromus arenarius, B. ciliatus, B. inermis, B. marginatus, B. tectorum, Calamagrostis canadensis, Calocedrus decurrens, Calycanthus occidentalis, Camassia quamash, Cardamine breweri, C. nuttallii, C. oligosperma, C. pensylvanica, Carduus pycnocephalus, Carex arcta, C. atherodes, C. disperma, C. geyeri, C. gmelini, C. macrochaeta, C. nebrascensis, C. praegracilis, C. saxatilis, C. sprengelii, C. utriculata, Castanopsis sclerophylla, Castilleja foliolosa, C. lasiorhyncha, C. pallida, Ceanothus crassifolius, C. leucodermis, C. palmeri, Centaurea melitensis, C. stoebe, Cerastium, Cercocarpus betuloides, C. ledifolius, Chamaebatia foliolosa, Chamaedaphne calyculata, Chamerion angustifolium, C. latifolium, Chorizanthe membranacea, Cirsium arvense, C. scariosum, C. vulgare, Cladonia rangiferina, Claytonia exigua, C. perfoliata, C. sibirica, Clematis ligusticifolia, Collinsia torreyi, Conium maculatum, Cornus sericea, C. sessilis, Corylus, Cynosurus echinatus, Cytisus, Dactylis glomerata, Dasiphora floribunda, Delphinium hesperium, D. menziesii, Deschampsia cespitosa, D. danthonioides, Descurainia californica, Distichlis, Draba, Drepanocladus, Drymocallis glandulosa, Eleocharis bernardina, E. montevidensis, E. palustris, E. parishii, Elymus elymoides, E. glaucus, E. repens, Epilobium ciliatum, E. densiflorum, Equisetum arvense, E. palustre, E. variegatum, Eremogone congesta, Ericameria discoidea, E. nauseosa, Eriogonum nudum, E. strictum, Eriophorum vaginatum, Eriophyllum lanatum, Festuca idahoensis, F. rubra, F. saximontana, Foeniculum vulgare, Forestiera, Fragaria virginiana, Fraxinus latifolia, Fritillaria micrantha, Geranium carolinianum, G. richardsonii, Geum calthifolium, G. macrophyllum, G. triflorum, Glyceris grandis, G. striata, Grindelia hirsutula, Hedysarum alpinum, H. boreale, Heracleum sphondylium, Heteromeles arbutifolia, Heuchera parviflora, Holcus lanatus, Holodiscus discolor, Horkelia rydbergii, Hypericum anagalloides, H. perforatum, Iris missouriensis, I. setosa, Juncus balticus,* *J. bufonius, J. ensifolius, J. nevadensis, J. parryi, J. textilis, Juniperus occidentalis, Lactuca serriola, Larix, Leptosiphon bicolor, Leymus condensatus, L. mollis, Ligusticum canbyi, Lilium parryi, Limnanthes alba, Lithophragma parviflorum, Lomatium, Lupinus argenteus, L. bicolor, Luzula comosa, Lysimachia, Lythrum, Maianthemum stellatum, Malus, Marrubium vulgare, Matricaria, Mertensia franciscana, M. paniculata, Micranthes lyallii, Mimulus guttatus, M. pilosus, M. tilingii, Monardella glauca, M. odoratissima, Montia chamissoi, M. fontana, M. linearis, Muhlenbergia rigens, Myosotis laxa, Nasturtium officinale, Nemophila heterophylla, Nephrophyllidium crista-galli, Oreostemma alpigenum, Osmorhiza berteroi, O. occidentalis, Paeonia brownii, Pedicularis, Penstemon ellipticus, Perideridia bolanderi, P. parishii, Phacelia hastata, Phalaris, Phleum alpinum, P. pratense, Phlox dispersa, Picea mariana, Pickeringia montana, Pinus albicaulis, P. balfouriana, P. contorta, P. jeffreyi, P. lambertiana, P. ponderosa, P. sabiniana, Plantago lanceolata, Platanthera hyperborea, Platanus racemosa, Plectritis congesta, Poa annua, P. palustris, P. pratensis, P. secunda, Polygonum aviculare, P. douglasii, Populus angustifolia, P. tremuloides, Potentilla anserina, P. gracilis, P. ×diversifolia, Primula fragrans, P. hendersonii, Prunus virginiana, Pseudotsuga menziesii, Pteridium aquilinum, Purshia, Pyrola picta, Quercus agrifolia, Q. chrysolepis, Q. douglasii, Q. garryana, Q. kelloggii, Q. wislizeni, Ranunculus cardiophyllus, R. sceleratus, R. uncinatus, Rhamnus purshiana, Rhododendron, Rhus aromatica, Ribes aureum, R. inerme, R. malvaceum, R. montigenum, R. pinetorum, Rorippa curvipes, R. palustris, Rosa californica, R. nutkana, R. woodsii, Rubus spectabilis, R. ulmifolius, Rumex crispus, Sagina maxima, S. saginoides, Salicornia, Salix alaxensis, S. arbusculoides, S. bebbiana, S. exigua, S. geyeriana, S. lasiandra, S. lasiolepis, S. pulchra, S. scouleriana, Salvia apiana, Sambucus nigra, Schedonorus arundinaceus, Scirpus microcarpus, Sedum lanceolatum, Senecio crassulus, S. flaccidus, S. hydrophiloides, S. integerrimus, S. triangularis, Sherardia arvensis, Sidalcea oregana, S. sparsifolia, Sisyrinchium bellum, Sonchus asper, Sphenosciadium capitellatum, Stachys albens, Stellaria longipes, S. media, Stipa, Streptopus amplexifolius, Symphoricarpos albus, Symphyotrichum foliaceum, S. spathulatum, Taraxacum ceratophorum, T. officinale, Thalictrum fendleri, Thermopsis californica, T. montana, Toxicodendron diversilobum, Tragopogon dubius, Trifolium longipes, T. monanthum, T. wormskioldii, Triglochin, Triteleia hyacinthina, Typha, Umbellularia californica, Urtica dioica, Vaccinium membranaceum, V. uliginosum, Valeriana sitchensis, Veratrum californicum, Verbascum thapsus, Verbena, Veronica, Vicia americana, Viola adunca, V. canadensis, V. sororia, Wyethia helianthoides, Zeltnera venusta.*

Use by wildlife: *Barbarea orthoceras* is a host plant for several larval butterflies and moths (Insecta: Lepidoptera: Pieridae: *Anthocharis sara, Euchloe ausonides, Pieris napi, P. rapae*; Yponomeutidae: *Plutella porrectella*). The flowers are visited by various adult insects (Insecta) including bees (Hymenoptera: Andrenidae: *Andrena nigrocaerulea*), beetles

(Coleoptera), flies (Diptera), and numerous moths and butterflies (Lepidoptera: Hesperiidae: *Amblyscirtes phylace, Erynnis persius, E. telemachus, Hesperia juba, Pyrgus centaureae, Thorybes pylades*; Lycaenidae: *Callophrys dumetorum, C. eryphon, C. polios, Celastrina humulus, C. lucia, Glaucopsyche lygdamus oro, G. piasus daunia, Plebejus glandon rustica, P. melissa, P. saepiolus*; Nymphalidae: *Aglais milberti, Chlosyne gorgone, Coenonympha tullia ochracea, Erebia .epipsodea, Phyciodes cocyta selenis, P. pallida, P. pulchella camillus, Polygonia faunus, P. gracilis zephyrus, P. satyrus, Speyeria coronis, Vanessa cardui*; Pieridae: *Colias scudderii, Euchloe ausonides coloradensis, E. olympia, Pieris rapae*), which serve as pollinators.

Economic importance: food: Native Alaskans and others have eaten the leaves of *B. orthceras* as a vegetable either cooked or raw in salads. Boiling the leaves reduces their bitter properties. The seeds have a 27% oil content (16% erucic acid); **medicinal:** none; **cultivation:** *Barbarea orthoceras* is recommended as a butterfly nectar plant for wet areas; **misc. products:** none; **weeds:** *Barbarea orthoceras* sometimes is regarded as weedy, but some reports may be due to confusion with the similar *B. vulgaris*; **nonindigenous species:** *Barbarea verna* and *B. vulgaris* are nonindigenous but neither is categorized as an OBL indicator in North America

Systematics: Analysis of DNA sequence data from up to eight *Barbarea* species indicates that the genus is monophyletic. Phylogenetic studies using plastid (*matK*) and nuclear (*Chs*) sequence data indicate that *Barbarea* is related closely to *Rorippa*; whereas, ITS1 sequences resolve the group as sister to a clade containing *Armoracia, Cardamine*, and *Rorippa*. A weak association of *Barbarea* and *Planodes* has been retrieved using combined data in a "supermatrix" analysis. The exclusion of *B. orthoceras* from phylogenetic analyses of *Barbarea* has precluded a definitive evaluation of its interspecific relationships. However, analyses of several species using AFLP and SSR markers indicate a close relationship between *B. orthoceras* and *B. stricta*. The chromosomal base number of *Barbarea* is $x = 8$. *Barbarea orthoceras* ($2n = 16$) is a diploid.

Comments: The native distribution of *Barbarea orthoceras* extends from central and eastern Asia to northern and western North America.

References: Al-Shehbaz, 2010a; Couvreur et al., 2010; Filiz et al., 2014; Goering et al., 1965; Heller, 1953; Laberge & Ribble, 1975; MacDonald & Cavers, 1991; Mulligan, 1964; Nakano & Washitani, 2003; Preston, 1986; Rollins, 1993; Scott, 2014; Toneatto et al., 2012.

2. *Cardamine*

Bitter cress; cresson

Etymology: from the Greek *kardia damao* meaning "to subdue the heart"

Synonyms: *Arabis* (in part); *Dentaria*; *Dracamine*

Distribution: global: cosmopolitan; **North America:** widespread

Diversity: global: 200 species; **North America:** 39 species

Indicators (USA): OBL: *Cardamine bulbosa, C. clematitis, C. cordifolia, C. douglassii, C. flexuosa, C. longii, C.*

micranthera, C. penduliflora, C. pensylvanica, C. rotundifolia; **FACW:** *C. cordifolia, C. douglassii, C. pensylvanica*; **FAC:** *C. flexuosa*; **FACU:** *C. flexuosa*

Habitat: freshwater, brackish (coastal); palustrine; **pH:** circumneutral; **depth:** <1 m; **life-form(s):** emergent (herb)

Key morphology: stems (to 10 dm) decumbent, erect or trailing, herbaceous; leaves (to 1 dm) cauline or in basal rosettes, simple, lobed, trifoliate or pinnate (to 15 leaflets); flowers in racemes, pedicelled (to 3 cm); petals (to 16 mm) 4 or absent, clawed, cruciform, pink, purple, rose-purple, or white; fruits siliques (to 4 cm)

Life history: duration: annual, biennial, perennial (rhizomes, stolons); **asexual reproduction:** adventitious plantlets, rhizomes, shoot fragments, stolons; **pollination:** insect, self; **sexual condition:** hermaphroditic; **fruit:** siliques (common); **local dispersal:** seeds, rhizomes, stolons; **long-distance dispersal:** seeds (water), vegetative propagules (birds)

Imperilment: (1) *Cardamine bulbosa* [G5]; S1 (KS, <u>MB</u>, ND, NH, VT); S2 (NE, <u>QC</u>); S3 (NC); (2) *C. clematitis* [G2]; SH (AL); S1 (VA); S2 (NC, TN); (3) *C. cordifolia* [G5]; S2 (WY); (4) *C. douglassii* [G5]; S1 (AR, MA); S2 (AL, CT, NC, NJ); S3 (IA, MD, NY, VA); (5) *C. flexuosa* [G]; S1 (); S2 (); S3 (); (6) *C. longii* [G3]; SH (CT, NH, NJ); S1 (DE, MA, MD, NC, RI); S2 (ME, NY); S3 (VA); (7) *C. micranthera* [G1]; S1 (NC, VA); (8) *C. penduliflora* [G4]; (9) *C. pensylvanica* [G5]; S1 (NE); S2 (CO, <u>LB</u>); S3 (WY); (10) *C. rotundifolia* [G4]; S1 (DE, NC, NJ, NY); S2 (TN); S3 (KY, MD)

Ecology: general: The general affinity of *Cardamine* species for aquatic habitats is evidenced by the designation of a quarter of the North American species as obligate aquatics. Thirty-one of the North American species (79%) are wetland indicators, with ten (26%) designated as OBL. Four species were included among the OBL indicators in the 1996 list, but have since been recategorized. These include (with current indicator status): *C. breweri* (FACW), *C. flagellifera* (FACU), *C. gemmata* [=*C. nuttallii*] (FACW, FAC), and *C. occidentalis* (FACW). Two former OBL indicators (*C. pratensis* and *C. regeliana*) have been excluded from the current list; all six species have been excluded from the present treatment. *Cardamine* includes annuals, biennials, or perennials, the latter producing tubers, rhizomes, or stolons. Most of the species flower early in the spring. The flowers can be self-compatible or self-incompatible. They are self-pollinated or cross-pollinated by various insects (Insecta) including bees (Hymenoptera), butterflies and moths (Lepidoptera), and flies (Diptera). The seeds are released from the dehiscing capsules, sometimes with enough ballistic force to dislodge or even kill predatory caterpillars (Insecta: Lepidoptera) in their proximity. Some of the seeds can retain viability for at least 30 days when cryopreserved in liquid nitrogen. The seeds of some species can float for several days, which facilitates their water dispersal. Vegetative bulbils or plantlets form at the bases of detached leaflets in a few species, which promotes their vegetative reproduction and dispersal.

***Cardamine bulbosa* (Schreb. ex Muhl.) Britton, Sterns & Poggenb.** is a perennial, which occurs in bottomlands, creek bottoms, depressions, ditches, fens, floodplains, hummocks, marshes, meadows, pinelands, prairies, roadsides, savannas, seeps, slopes, springs, swales, swamps, woodlands, and along the margins of rivers, sloughs, and streams at elevations of up to 900 m. The plants can occur in shallow, fresh (rarely brackish, tidal) standing water and often occupy shaded sites in habitats with closed canopies. They also grow in sites that receive full sunlight. The substrates are circumneutral or calcareous and are commonly underlain by limestone or mafic rocks. They are described as clay, clay loam, Glendon formation, muck, mud, peaty muck, sand, sandy loam, silt, and silty loam. The plants are in flower and fruit from March to June. Several insect pollinators have been observed (see *Use by wildlife*), but the plants are also self-fertile. Grazing mammalian herbivores can reduce flower and seed production by more than 50%. This species does persist in the seed bank but for an undetermined length of time. The plants perennate from tuberous rhizomes. **Reported associates:** *Acer negundo, A. rubrum, A. saccharinum, A. saccharum, Actaea, Ageratina altissima, Agrimonia parviflora, Agrostis alba, Alisma subcordatum, Alliaria petiolata, Allium canadense, Alnus, Amorpha fruticosa, Anemone canadensis, Angelica atropurpurea, Apios americana, Arisaema dracontium, A. triphyllum, Asarum canadense, Asimina triloba, Barbarea vulgaris, Bidens, Boehmeria cylindrica, Bolboschoenus fluviatilis, Botrychium virginianum, Calamagrostis canadensis, Caltha palustris, Camassia scilloides, Campsis radicans, Cardamine concatenata, C. diphylla, C. douglassii, C. pensylvanica, C. pratensis, Carex bromoides, C. buxbaumii, C. conjuncta, C. corrugata, C. crinita, C. crus-corvi, C. diandra, C. emoryi, C. festucacea, C. grayi, C. haydenii, C. hyalinolepis, C. hystericina, C. hitchcockiana, C. interior, C. jamesii, C. lacustris, C. laxiculmis, C. laxiflora, C. lupulina, C. muskingumensis, C. nebrascensis, C. pellita, C. radiata, C. sartwellii, C. stipata, C. stipitata, C. stricta, C. trichocarpa, Carpinus caroliniana, Carya cordiformis, C. glabra, C. tomentosa, Celastrus scandens, Celtis occidentalis, Cephalanthus occidentalis, Chaerophyllum procumbens, Cicuta bulbifera, Cinna arundinacea, Cirsium horridulum, Claytonia caroliniana, C. virginica, Clematis socialis, Collinsia verna, Coreopsis lanceolata, Cornus amomum, C. racemosa, C. sericea, Corylus americana, Cryptotaenia canadensis, Cuscuta pentagona, Diarrhena obovata, Dichanthelium scoparium, Dioscorea, Eleocharis erythropoda, Elymus virginicus, Epilobium leptophyllum, Equisetum arvense, Erigeron tenuis, Erythronium albidum, Eupatorium perfoliatum, E. rotundifolium, Euphorbia spathulata, Euthamia tenuifolia, Eutrochium maculatum, Fagus grandifolia, Floerkea proserpinacoides, Fraxinus americana, F. nigra, F. pensylvanica, F. profunda, F. quadrangulata, Galium aparine, G. tinctorium, G. triflorum, Gaylussacia baccata, Gentianopsis virgata, Geranium maculatum, Geum canadense, G. vernum, Glechoma hederacea, Gleditsia triacanthos, Glyceria declinata, G. striata, Hedyotis crassifolia, Helianthus grosseserratus, Houstonia micrantha, Hydrophyllum appendiculatum, H. virginianum, Hypericum punctatum, Ilex opaca, I. verticillata, Illicium floridanum, Impatiens capensis, Iodanthus pinnatifidus, Iris virginica, Juglans nigra, Juncus coriaceus, Laportea canadensis, Lathyrus palustris, Ligustrum sinense, Lilium michiganense, Lindera benzoin, Liriodendron tulipifera,* *Lobelia kalmii, Lonicera japonica, L. maackii, Lycopus americanus, Lysimachia nummularia, Maclura pomifera, Magnolia grandiflora, M. macrophylla, Maianthemum racemosum, Malus ioensis, Menispermum canadense, Mentha arvensis, Micranthes pensylvanica, Myosoton aquaticum, Onoclea sensibilis, Osmorhiza longistylis, Osmunda regalis, Ostrya, Packera aurea, P. glabella, Panicum, Parnassia glauca, Parthenocissus quinquefolia, Pedicularis lanceolata, Phalaris arundinacea, Pilea pumila, Pinus taeda, Plantago virginica, Platanthera psycodes, Platanus occidentalis, Poa pratensis, P. sylvestris, Polemonium reptans, Polygonatum biflorum, Polystichum, Populus tremuloides, Prenanthes alba, Prunella vulgaris, Prunus serotina, P. virginiana, Pycnanthemum virginianum, Quercus alba, Q. bicolor, Q. falcata, Q. lyrata, Q. macrocarpa, Q. michauxii, Q. nigra, Q. palustris, Q. phellos, Q. rubra, Q. texana, Q. velutina, Ranunculus abortivus, R. flabellaris, R. hispidus, R. parviflorus, R. recurvatus, R. sceleratus, Rhamnus alnifolia, R. cathartica, Rhus copallinum, R. glabra, Ribes americanum, R. missouriense, Robinia pseudoacacia, Rorippa palustris, Rosa multiflora, Rubus allegheniensis, R. pensilvanicus, R. pubescens, Rudbeckia laciniata, Sabal minor, Sagittaria latifolia, Salix amygdaloides, S. eriocephala, S. exigua, Sambucus nigra, Sanicula canadensis, S. odorata, Saururus cernuus, Schoenoplectus pungens, Scirpus atrovirens, Scutellaria lateriflora, Silene nivea, Silphium perfoliatum, Sium suave, Smilax, Solidago canadensis, S. flexicaulis, S. gigantea, S. patula, Sparganium eurycarpum, Stellaria media, Symphyotrichum lanceolatum, S. ontarionis, S. praealtum, S. puniceum, Symplocarpos foetidus, Taraxacum officinale, Taxodium distichum, Tilia americana, Toxicodendron radicans, Trichostema brachiatum, Tridens strictus, Trifolium, Trillium cuneatum, T. erectum, T. grandiflorum, T. recurvatum, Typha latifolia, Ulmus americana, U. rubra, Urtica dioica, Uvularia grandiflora, Vaccinium, Valeriana edulis, Veratrum viride, Verbesina alternifolia, Viburnum lentago, V. prunifolium, V. rafinesquianum, Viola pubescens, V. sororia, V. striata, Vitis, Zanthoxylum americanum, Zephyranthes atamasco, Zizia aurea.*

***Cardamine clematitis* Shuttlw. ex A. Gray** is restricted to seeps, slopes, springs, and the margins of streams in mountainous portions of the southeastern United States at elevations above 1200 m [1219–1646 m]. The sites are characterized by shaded exposures. The plants are in flower from May to June. They perennate by means of vertical rhizomes. Additional life-history information is required for this species. **Reported associates:** *Acer rubrum, Acontium reclinatum, Alnus serrulata, Carex folliculata, C. gynandra, C. leptalea, C. stricta, Chelone, Chrysosplenium americanum, Cicuta maculata, Cladium mariscoides, Diphylleia cymosa, Drosera rotundifolia, Fagus grandifolia, Helenium autumnale, Hydrocotyle americana, Hypericum, Impatiens capensis, Juncus subcaudatus, Kalmia latifolia, Lilium grayi, Micranthes micranthidifolia, Monarda didyma, Osmundastrum cinnamomeum, Persicaria, Picea rubens, Rhododendron maximum, Rhynchospora capitellata, Rudbekia lacinata, Solidago patula, Symphyotrichum novae-angliae, Thalictrum clavatum, T. dioicum, Veratrum, Veronica americana, Viburnum nudum, Viola.*

Cardamine cordifolia **A. Gray** is a perennial, which inhabits bottoms, depressions, fens, flats, gullies, meadows, ravines, rivulets, seeps, slopes (to 20%), springs, swamps, thickets, tundra, woodlands, and the margins of brooks, lakes, ponds, rills, rivers, and streams throughout the Rocky Mountains at elevations of up to 4024 m. The plants occur nearly exclusively in shaded sites, often under the cover of willows (*Salix*), or near the edge of spruce (*Picea*) forests. They sometimes grow in shallow, standing, or flowing water. The substrates are alkaline (pH: 7.6–7.7) and include alluvium, basalt, clay, colluvium, granite, granodiorite, gravel, humus, Krumholz, limestone, loam, muck, peat, quartzite, rock, rocky gravel, rocky loam, loamy cryumbrept, Mancos shale, sand, sandy alluvium, sandy clay, sandy gravel, sandy loam, shale, silt, stones, and talus. Flowering and fruiting extend from April to August. The flowers are pollinated primarily by Lepidoptera (See *Use by wildlife*). Vegetative reproduction occurs by the production of slender rhizomes. The plants suffer higher levels of herbivory when growing in sunny, drier sites where they are more stressed. **Reported associates:** *Abies amabilis, A. concolor, A. grandis, A. lasiocarpa, A. procera, Acer circinatum, A. negundo, Achillea millefolium, Aconitum columbianum, Aconogonon phytolaccifolium, Agrostis, Allium validum, Alnus rubra, A. rugosa, A. tenuifolia, Arnica parryi, Athyrium, Betula glandulosa, B. occidentalis, Bistorta bistortoides, Bromus tectorum, Calamagrostis canadensis, Calocedrus decurrens, Caltha biflora, C. leptosepala, Carex aquatilis, C. aurea, C. leporinella, C. microptera, C. paysonis, C. scopulorum, Castilleja, Cercocarpus ledifolius, Chaenactis, Cicuta douglasii, C. maculata, Clintonia, Corydalis caseana, Dasiphora floribunda, Delphinium, Deschampsia cespitosa, Eleocharis, Elymus glaucus, Epilobium, Equisetum arvense, Erigeron speciosus, Erythronium grandiflorum, Galium trifidum, Geranium richardsonii, G. viscosissimum, Geum macrophyllum, Glyceria elata, G. striata, Habenaria, Heracleum sphondylium, Heuchera, Hydrophyllum fendleri, Juncus, Juniperus communis, J. osteosperma, Larix occidentalis, Ligusticum canbyi, L. porteri, Lonicera involucrata, Lupinus, Luzula piperi, Maianthemum, Marchantia, Medicago lupulina, Mertensia arizonica, Micranthes odontoloma, M. tolmiei, Microseris, Mimulus guttatus, M. lewisii, Minuartia, Mitella caulescens, Osmorhiza berteroi, O. depauperata, Oxypolis fendleri, Pedicularis bracteosa, P. groenlandica, Penstemon, Phacelia sericea, Philadelphus lewisii, Phlox gladiformis, Picea engelmannii, P. pungens, Pinus aristata, P. contorta, P. lambertiana, P. ponderosa, Plantago major, Poa pratensis, Polemonium occidentale, Populus angustifolia, P. tremuloides, Potentilla flabellifolia, Primula jeffreyi, P. pauciflora, Pseudoroegneria spicata, Pseudostellaria jamesiana, Pseudotsuga menziesii, Pyrola, Ranunculus alismaefolius, R. inamoenus, R. testiculatus, Rhamnus alnifolia, Rhodiola rhodantha, R. rosea, Ribes hudsonianum, R. lacustre, R. roezlii, R. viscosissimum, Rorippa teres, Rosa, Rubus parviflorus, Rudbeckia laciniata, Salix arizonica, S. commutata, S. drummondiana, S. irrorata, S. lemmonii, S. planifolia, Scirpus microcarpus, Senecio triangularis, Sequoiadendron giganteum, Sibbaldia procumbens, Solidago, Stellaria umbellata,* *Streptopus amplexicaulis, S. amplexifolius, Symphoricarpos albus, S. oreophilus, Symphyotrichum foliaceum, Thuja plicata, Trifolium, Trollius laxus, Tsuga heterophylla, Vaccinium membranaceum, Valeriana, Veratrum californicum, Veronica americana, Viola adunca, V. epipsila, V. sororia, Wyethia.*

Cardamine douglassii **Britton** is a perennial, which inhabits bluffs, bottomlands, floodplains, sandbars, savannas, seeps, slopes, springs, swamps, washes, woodlands, and the margins of rivers and streams, at elevations of up to 400 m. The plants occur in shaded to semiopen exposures. The substrates are calcareous, nutrient-rich, and described as alluvium, gravel, loam, rock, sand, and sandy loam. Flowering and fruiting occur from March to May. The flowers are pollinated by bees (see *Use by wildlife*). The plants reproduce vegetatively from fleshy rhizomes. As the following list of associated species illustrates, this species is far more common in mesic sites than in wetlands throughout its range, making it difficult to rationalize its OBL indicator status. **Reported associates:** *Acer negundo, A. saccharum, Aesculus flava, Alliaria petiolata, Allium canadense, A. tricoccum, Amphicarpa bracteata, Anemone acutiloba, Arisaema trivphyllum, Arundinaria gigantea, Asarum canadense, Asimina triloba, Caltha palustris, Campanula americana, Cardamine bulbosa, C. concatenata, Carex albursina, C. lacustris, Carya ovata, Celtis occidentalis, Claytonia virginica, Collinsia verna, Cornus racemosa, Crataegus, Cryptotaenia canadensis, Cystopteris protrusa, Dicentra cucullaria, Enemion biternatum, Erigenia bulbosa, Erythronium albidum, E. americanum, Euonymus obovatus, Fagus grandifolia, Floerkea proserpinacoides, Fraxinus americana, Galium aparine, G. obtusum, Geranium maculatum, Geum canadense, G. vernum, Hamamelis virginiana, Hemerocallis fulva, Heracleum sphondylium, Hydrophyllum canadense, H. virginianum, Hystrix patula, Impatiens capensis, Juglans nigra, Laportea canadensis, Lindera benzoin, Lithospermum latifolium, Lysimachia ciliata, Maianthemum racemosum, Mertensia virginica, Osmorhiza claytoni, O. longistylis, Ostrya virginiana, Parthenocissus quinquefolia, Perideridia americana, Persicaria virginiana, Phlox divaricata, Pilea pumila, Pinus strobus, Poa secunda, Podophyllum peltatum, Polygonatum biflorum, Polymnia canadensis, Polystichum acrostichoides, Populus deltoides, Prenanthes crepidinea, Prunus serotina, Quercus alba, Q. macrocarpa, Q. muehlenbergii, Q. rubra, Q. shumardi, Ranunculus abortivus, R. hispidus, Rhus typhina, Rudbeckia laciniata, Salix, Sanguinaria canadensis, Sanicula odorata, Smilax lasioneura, Solidago flexicaulis, Staphylea trifolia, Taraxacum officinale, Thalictrum dioicum, Tilia americana, Toxicodendron radicans, Trillium flexipes, T. grandiflorum, T. recurvatum, Ulmus americana, Valerianella chenopodiifolia, Verbascum, Viola papilionacea, V. pubescens, V. sororia, Zanthoxylum americanum.*

Cardamine flexuosa **With.** is an introduced annual or biennial, which occurs sporadically in or on banks, bottoms, ditches, floodplains, gardens, lawns, marshes, meadows, roadsides, slopes, swamps, thickets, woodlands, and the margins of streams at elevations of up to 700 m. This species occurs in fens in its native range but is frequent in disturbed sites within North America. Exposures can range from full sun to shade, or partial shade. The substrates reported in North America are various and include bark mulch, gravelly sand, humus, leaf litter, muck, mud, rock, sandy loam, and stoney mud. Flowering and fruiting can occur from April to December. The flowers are self-fertile, allowing for self-pollination. Summer seed dormancy is weak and plants that germinate in early summer will reproduce by the autumn. However, plants that germinate in late summer will behave as winter annuals. Submergence of seeds during the summer will prevent their germination and result in a winter annual habit. **Reported associates (North America):** *Acer circinatum, A. macrophyllum, Agrostis capillaris, Alnus rubra, Arctium, Athyrium filix-femina, Blechnum spicant, Calystegia, Cardamine hirsuta, C. oligosperma, Carex deweyana, Cirsium arvense, Claytonia sibirica, Coleanthus subtilis, Conium maculatum, Crassula aquatica, Cynodon, Dicentra formosa, Dichondra, Draba verna, Elatine rubella, Epilobium ciliatum, Equisetum arvense, E. telmateia, Galium aparine, Geranium robertianum, Geum macrophyllum, Juncus, Lamium amplexicaule, L. purpureum, Lysichiton americanus, Myriophyllum ussuriense, Nemophila parviflora, Phalaris arundinacea, Planodes virginica, Poa annua, P. pratensis, Polystichum munitum, Populus, Prosartes hookeri, Prunus emarginata, Pseudotsuga menziesii, Ranunculus repens, R. uncinatus, Rubus bifrons, R. spectabilis, Sagina procumbens, Senecio vulgaris, Stellaria media, Stenotaphrum secundatum, Tellima grandiflora, Thuja plicata, Tolmiea menziesii, Trifolium pratense, Typha, Urtica dioica, Veronica americana, Viola.*

Cardamine longii **Fernald** is a perennial, which occurs in crevices (inundated at high tide), estuaries, marshes (tidal), mud flats, shores, streams, swamps, and along the margins of lakes and rivers (tidal) at elevations of up to 10 m along the Atlantic coast. The plants will grow in shallow water, usually in shaded sites. The substrates are slightly acidic (pH: 6.2–6.7) with low salinity (1.5 ppt at high tide, undetectable at low tide) and include cobble, muck, sandy muck, and rock. The plants are reproductive from June to September. The apetalous flowers, lack of observed floral visitors, and high seed production (97% mean seed set) indicate that *C. longii* is self-pollinating. When stored for 1–2 months, the seeds will germinate at 25°C without prior stratification. Wet-stored seeds achieve up to 87% germination, whereas, the germination of dried seed is markedly lower (8%–15%). Although sometimes described as annual, the plants produce thin, cylindrical rhizomes. **Reported associates:** *Bidens eatoni, B. hyperborea, Eriocaulon parkeri, Lilaeopsis chinensis, Nyssa, Sagittaria montevidensis, S. subulata, Taxodium.*

Cardamine micranthera **Rollins** is a rare perennial, which grows in crevices, sandbars, seeps, slopes (to 60%), woodlands and along the margins of small to medium-sized streams at low elevations in the Dan River drainage of the North Carolina and Virginia piedmont. The plants occur in fully to partially shaded sites. The substrates have been characterized as gravel, and Rion, Pacolet, or Wateree soil series. Flowering occurs from April to May. The flowers often are visited by ants (Insecta: Hymenoptera: Formicidae), but appear to be mainly self-pollinated. Suitable habitats are threatened by invasive species

such as *Lonicera japonica*, conversion to pasture land, livestock trampling, sedimentation, and by dams and other hydrological perturbations associated with beavers (Mammalia: Castoridae: *Castor canadensis*). **Reported associates:** *Acer rubrum, Betula, Cardamine rotundifolia, Carex laevivaginata, C. prasina, Carpinus caroliniana, Cicuta maculata, Fagus grandifolia, Houstonia caerulea, Impatiens capensis, Juncus effusus, Kalmia latifolia, Lindera benzoin, Liriodendron tulipifera, Micranthes micranthidifolia, Pinus strobus, Rhododendron nudiflorum, Viburnum prunifolium, Xanthorhiza simplicissima.*

Cardamine penduliflora **O.E. Schulz** is a narrowly restricted perennial (western Oregon), which occurs in shallow water (to 6 cm deep) or wet ground of channels, ditches, flats, marshes, meadows, ponds, pools, prairies, roadsides, streams, swales, and swamps at elevations of up to 975 m. Exposures can range from full sunlight to light shade. The substrates consist of clay or clay loam. Flowering and fruiting occur from March to July. The plants spread vegetatively by tuberous rhizomes. More detailed life-history information is needed for this species. **Reported associates:** *Agrostis capillaris, A. exarata, Alopecurus pratensis, Beckmannia syzigachne, Briza minor, Brodiaea coronaria, Callitriche, Camassia quamash, Carex densa, C. pellita, C. unilateralis, Centaurium erythraea, Claytonia sibirica, Danthonia californica, Deschampsia cespitosa, Eleocharis acicularis, E. palustris, Epilobium ciliatum, Equisetum, Eryngium petiolatum, Fraxinus latifolia, Galium, Geum macrophyllum, Grindelia integrifolia, Holcus lanatus, Hosackia pinnata, Hypericum perforatum, Juncus balticus, J. effusus, J. nevadensis, J. patens, Lomatium bradshawii, Madia glomerata, Mentha piperita, M. pulegium, Microsteris gracilis, Montia fontana, M. linearis, Oenanthe, Phalaris arundinacea, Pyrus communis, Quercus garryana, Ranunculus alismifolius, R. occidentalis, R. orthorhynchus, R. uncinatus, Rosa nutkana, R. rubiginosa, Schedonorus arundinaceus, Spiraea douglasii, Torreyochloa pallida, Trillium ovatum, Triteleia hyacinthina, Veronica scutellata.*

Cardamine pensylvanica **Muhl. ex Willd.** is a wide-ranging winter annual or biennial, which is found in or on beaches, bluffs, bogs, bottomlands, depressions, ditches, draws, flats, flatwoods, floodplains, lawns, ledges, marshes, meadows, mud bars, pools, ravines, ricefields, roadsides, sand bars, savannas, seeps, slopes (to 40%), springs, streams, swales, swamps, thickets, waste grounds, woodlands, and along margins of arroyos, bayous, borrow pits, brooks, canals, lakes, ponds, rivers, sloughs, and streams at elevations of up to 3200 m. The plants often occur in shallow (to 8 cm), flowing, or standing water in sites with exposures ranging from full sunlight to shade. The substrates are calcareous and have been characterized as alluvium, Ashlar sandy loam, chert, clay, Fausse, gravel, humus, limestone, loam, mollisol, muck, mud, mudstone, peaty loam, Pinedale glacial alluvium, rock, rocky loam, rotting logs, sand, sandstone, sandy clay, sandy humus, sandy loam, Sharkey, silt, silty clay, silty loam, silty muck over sand, and volcanic. Flowering and fruiting can extend from January to October. The flowers are self-fertile and capable of self-pollination and are also cross-pollinated by insects (see *Use by wildlife*). Some seeds will germinate readily (within 2 weeks) without any specific treatment, when placed in a moist environment. Nevertheless, seed dormancy occurs at some level in natural populations because a seed bank does develop. Fecundity is reduced in plants that are grown at high densities. It takes about 2 months for new seedlings to reach reproductive maturity. **Reported associates:** *Abies grandis, A. lasiocarpa, Acer glabrum, A. macrophyllum, A. rubrum, A. saccharinum, A. saccharum, Adenocaulon bicolor, Adiantum pedatum, Agrimonia parviflora, Allium canadense, Alnus rubra, A. rugosa, A. viridis, Alopecurus geniculatus, Alternanthera, Angelica arguta, Antennaria corymbosa, A. plantaginifolia, Aphanes occidentalis, Aralia nudicaulis, Arisaema triphyllum, Asarum, Asplenium platyneuron, Athyrium felix-femina, Bararea orthoceras, B. vulgaris, Berberis thunbergii, Bidens, Bistora bistortoides, Calamagrostis canadensis, Calandrinia ciliata, Calla palustris, Callitriche heterophylla, C. terrestris, Caltha palustris, Cardamine angulata, C. bulbosa, C. oligosperma, Carex albicans, C. buxbaumii, C. canescens, C. disperma, C. glaucodea, C. jamesii, C. lacustris, C. lenticularis, C. limosa, C. lupulina, C. microptera, C. nigricans, C. obnupta, C. rostrata, C. saximontana, C. scabrata, C. spectabilis, C. stipata, C. stricta, C. trichocarpa, C. utriculata, Carya ovata, C. tomentosa, Catabrosa aquatica, Celtis laevigata, C. occidentalis, C. reticulata, Cerastium fontanum, Chamaecyparis, Chara, Chrysosplenium americanum, Cinna latifolia, Circaea alpina, Cirsium, Claytonia exigua, C. perfoliata, C. sibirica, C. virginica, Clematis ligusticifolia, Clintonia, Collinsia, Comandra umbellata, Comarum palustre, Coptis trifolia, Cornus alternifolia, C. amomum, C. florida, C. sericea, Corydalis aurea, Cynodon dactylon, Danthonia spicata, Dennstaedtia punctilobula, Deschampsia cespitosa, Dichanthelium clandestinum, Digitaria, Draba nemorosa, Dryopteris intermedia, Eleocharis elliptica, Epilobium palustre, Equisetum arvense, E. fluviatile, Erigeron philadelphicus, Erodium cicutarium, Euonymus, Eutrochium maculatum, Festuca, Fontinalis antipyretica, Fragaria, Fraxinus latifolia, F. pennsylvanica, Galium aparine, G. tinctorium, G. trifidum, Gaultheria procumbens, Gaylussacia baccata, Gelsemium sempervirens, Geranium, Geum macrophyllum, G. vernum, Glyceria melicaria, G. striata, Glycyrrhiza, Hamamelis, Heracleum sphondylium, Hippuris vulgaris, Holodiscus, Huperzia lucidula, Hydrocotyle americana, Hypericum dentatum, Hypoxis hirsuta, Ilex verticillata, Impatiens capensis, I. noli-tangere, Iris virginica, Juglans nigra, Juncus balticus, J. effusus, J. ensifolius, J. filiformis, J. hesperius, J. supiniformis, Juniperus virginiana, Kalmia latifolia, Lamium, Laportea canadensis, Larix, Leucolepis acanthoneuron, Ligustrum sinense, Lilaeopsis, Lindera benzoin, Lindernia dubia, Linnaea borealis, Lotus corniculatus, Luzula multiflora, Lysichiton americanus, Lysimachia nummularia, L. quadrifolia, Maianthemum racemosum, Medeola virginiana, Medicago polymorpha, Mimulus guttatus, M. moschatus, Moneses uniflora, Montia fontana, M. linearis, Morus alba, Mycelis muralis, Myosotis laxa, Oenanthe, Onoclea sensibilis, Oplopanax horridus, Oryza sativa, Osmunda claytoniana, Osmundastrum cinnamomeum, Ostrya virginiana, Oxalis, Packera aurea, Parthenocissus quinquefolia, Paspalum, Paxistima myrsinites, Persicaria maculosa, Petasites frigidus, Phalaris arduninacea, Philadelphus, Physocarpus capitatus, P. malvaceus, Phytolacca americana, Picea engelmannii, P. mariana, P. sitchensis, Pinus monticola, P. ponderosa, P. sabiniana, P. strobus, P. taeda, Plantago elongata, P. major, Platanthera dilatata, Poa annua, P. pratensis, Podophyllum peltatum, Polystichum munitum, Populus balsamifera, P. grandidentata, P. tremuloides, P. trichocarpa, Potentilla simplex, Primula jeffreyi, Prunus pensylvanica, P. serotina, Pseudotsuga menziesii, Quercus alba, Q. chrysolepis, Q. falcata, Q. garryana, Q. georgiana, Q. kelloggii, Q. lyrata, Q. michauxii, Q. palustris, Q. phellos, Q. rubra, Q. velutina, Ranunculus abortivus, R. aquatilis, R. flammula, R. hispidus, R. parviflorus, R. repens, Rhamnus cathartica, Rhus copallinum, R. glabra, R. typhina, Ribes curvatum, R. triste, Ricciocarpus natans, Rorippa curvisiliqua, R. palustris, R. sessiliflora, Rosa carolina, R. nootkana, Rubus occidentalis, R. parviflorus, R. spectabilis, Rudbeckia laciniata, Salix discolor, S. scouleriana, Salvinia, Sambucus nigra, Sassafras albidum, Saururus cernuus, Schizachyrium, Scirpus cyperinus, Scutellaria lateriflora, Senecio triangularis, S. vulgaris, Sisyrinchium albidum, Smilax lasioneura, Solanum dulcamara, Solidago canadensis, S. juliae, S. rugosa, Spergularia media, Sphagnum, Sporobolus, Stellaria crispa, Streptopus amplexifolius, Symphoricarpos orbiculatus, Symphyotrichum prenanthoides, Symplocarpus foetidus, Thalictrum pubescens, T. thalictroides, Thelypteris palustris, Thuja plicata, Tilia americana, Toxicodendron radicans, Tradescantia virginiana, Tridens strictus, Trifolium dubium, T. repens, Trillium ovatum, Tsuga canadensis, T. heterophylla, Typha latifolia, Ulmus pumila, U. rubra, Urtica dioica, Vaccinium angustifolium, V. cespitosum, V. myrtillus, V. scoparium, Veronica americana, V. peregrina, V. serpyllifolia, Viburnum lantanoides, Viola canadensis, V. cucullata, V. macloskeyi, V. sororia, Vitis rotundifolia, Yucca flaccida.*

***Cardamine rotundifolia* Michx.** is a perennial, which grows in bottoms, seeps, streams, swamps, woodlands, and along the margins of brooks and streams at elevations of up to 762 m. The plants often occur in cold, shallow water. They can tolerate light to heavy shade. The substrates include clay, loam, rock, and sand. Flowering occurs from April to June. The flowers are self-fertile, allowing for self-pollination. They are also pollinated by various insects (see *Use by wildlife*). The seeds are physiologically dormant. They reportedly will germinate within 1–3 weeks when incubated at 15°C, presumably after having undergone a period of cold stratification. The plants are stoloniferous and mat-forming; shoots branching from the upper nodes can root and proliferate vegetatively. **Reported associates:** *Acer rubrum, A. saccharum, Betula alleghaniensis, Carex bromoides, C. gynandra, Cephalanthus occidentalis, Chelone, Chrysosplenium americanum, Circaea alpina, Dryopteris cristata, Fraxinus americana, F. nigra, Glyceria melicaria, Impatiens capensis, Juglans cinerea, Laportea, Micranthes pensylvanica, Quercus rubra, Ranunculus hispidus, Salix nigra, Solidago* *patula, Tiarella cordifolia, Tilia americana, Tsuga canadensis, Veratrum viride, Viola cucullata, V. sororia*.

Use by wildlife: *Cardamine bulbosa* is grazed heavily by white-tailed deer (Mammalia: Cervidae: *Odocoileus virginianus*). It is a host plant for moth larvae (Insecta: Lepidoptera: Pieridae: *Paramidea midea*). The flowers are visited and pollinated by various insects (Insecta) including bees (Hymenoptera: Andrenidae: *Andrena*; Apidae: *Apis mellifera, Ceratina*; Halictidae: *Lasioglossum*), flies (Diptera: Syrphidae: *Orthonevra pictipennis, Sphaerophoria contiqua, Toxomerus marginatus*), and butterflies (Lepidoptera). Plants of *C. cordifolia* are insect hosts to herbivorous beetles (Coleoptera: Chrysomelidae: *Phaedon oviformis, Phyllotreta ramosa, P. polita*), bugs (Hemiptera: Psyllidae: *Aphalara confusa, A. nubifera*), leafmining flies (Diptera: Drosophilidae: *Scaptomyza nigrita*), moths (Lepidoptera: Pieridae: *Pieris napi*; Plutellidae: *Plutella xylostella*), and weevils (Coleoptera: Curculionidae: *Ceutorhynchus americanus, C. moznettei, C. pusio, C. subpubescens*). They provide habitat for the rare coffee pot snowfly (Insecta: Plecoptera: Capniidae: *Capnia nelsoni*). The flowers are visited and pollinated by butterflies and moths (Insecta: Lepidoptera) including several fritillaries (Nymphalidae: *Boloria eunomia caelestis, B. frigga sagata, B. titania helena, Speyeria zerene*), large marbles (Pieridae: *Euchloe ausonides coloradensis*), and whites (Pieridae: *Pieris marginalis mcdunnoughii*). The flowers of *C. douglassii* are visited and pollinated by several bees (Insecta: Hymenoptera: Andrenidae: *Andrena*; Apidae: *Apis mellifera, Ceratina*). *Cardamine flexuosa* is eaten by tortoises (Reptilia: Testudines: Testudinidae) in its native range. *Cardamine pensylvanica* is a host for the West Virginia white moth (Insecta: Lepidoptera: Pieridae: *Pieris virginiensis*). Flowers of *C. pensylvanica* are visited by various insects (Insecta) including bees (Hymenoptera), butterflies and moths (Lepidoptera), and flies (Diptera), which can function as pollinators. *Cardamine rotundifolia* is a common habitat element of the bog turtle (Reptilia: Testudines: Emydidae: *Glyptemys muhlenbergii*). Its flowers are visited by various insects (Insecta) including bees (Hymenoptera), butterflies and moths (Lepidoptera), and flies (Diptera), which are potential pollinators.

Economic importance: food: The young leaves of *Cardamine bulbosa* are used in salads and have a piquant flavor described as similar to cabbage or horseradish. The grated roots mixed with vinegar can be used as a horseradish substitute. The flavor of *C. cordifolia* reportedly is stronger than that of watercress. *Cardamine flexuosa, C. pensylvanica*, and *C. rotundifolia* are also used in salads or as greens (cooked or raw) and are similar in taste to watercress; **medicinal:** The Iroquois used the roots of *C. bulbosa* as a poison and the mashed roots of *C. douglassii* as a poison antidote. *Cardamine pensylvanica* has been used to relieve gas and as a digestive aid; **cultivation:** *Cardamine bulbifera* is planted as a water garden ornamental. *Cardamine rotundifolia* is grown as a submersed, coldwater aquarium plant; **misc. products:** roots of *C. douglassii* were used as a magic charm by the Iroquois; **weeds:** *Cardamine pensylvanica* is a widespread and persistent weed of greenhouses and plant nurseries;

nonindigenous species: *Cardamine flexuosa* was introduced to North America (including Hawaii) from Eurasia.

Systematics: Phylogenetic studies incorporating molecular data have resulted in a number of changes to the taxonomy of *Cardamine* in recent years. A notable modification is the merger of *Dentaria* and *Cardamine*, which is well justified. Most studies indicate that *Cardamine* (including *Dentaria*) is monophyletic and related most closely to *Nasturtium* (e.g., Figure 4.67). It is no surprise that some species (e.g., *C. pensylvanica*) strongly resemble *Nasturtium* (watercress) and grow in similar habitats, which sometimes complicates their identification. A fairly comprehensive phylogenetic study of interspecific relationships has been carried out using nrITS sequence data, which represent seven of the 10 OBL indicators. Those results indicate that most of the OBL taxa are fairly closely related, with the exception of the nonindigenous (Old World) *C. flexuosa*, which understandably resolves in a distantly related clade (Figure 4.68). Altogether, there appear to be at least several origins of the OBL habit in *Cardamine* (Figure 4.68). Many species are variable morphologically and several varieties have been recognized for *C. cordifolia*. *Cardamine* (x = 7, 8) is notoriously variable cytologically and contains many polyploids. Extensive intraspecific

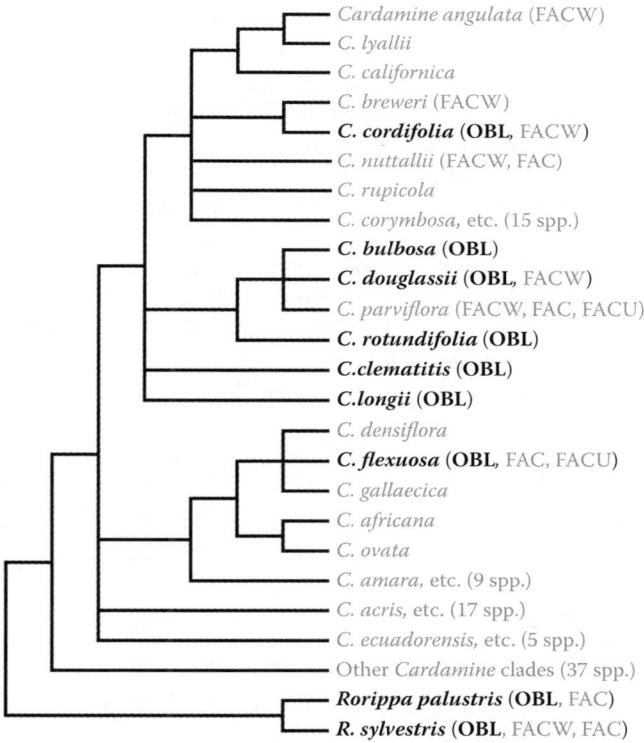

FIGURE 4.68 Interspecific relationships in *Cardamine* as indicated by analysis of nrITS sequence data. The OBL North American species (bold) occur in two closely related groups (*C. bulbosa, C. cordifolia, C. clematitis, C. douglassii, C. longii, C. rotundifolia*) and in a distantly related clade (*C. flexuosa*—nonindigenous, Old World), which indicate several origins of the habit in the genus. The wetland indicator status is provided in parentheses. (Abridged from Carlsen, T. et al., *Ann. Missouri Bot. Gard.*, 96, 215–236, 2009.)

chromosomal variation occurs in *C. bulbosa* ($2n$ = 16, 56, 64, 80, 96, 112) and in *C. douglassii* ($2n$ = 60, 64, 96, 144). Although *C. cordifolia* is $2n$ = 24, genomic studies have shown that it is not a triploid, but a diploidized tetraploid, wherein the ancestral tetraploid number ($2n$ = 32) has been reduced through four terminal chromosomal translocations. *Cardamine flexuosa* and the widespread *C. pensylvanica* (both $2n$ = 32) are tetraploids. *Cardamine bulbosa* hybridizes spontaneously when coming into contact with the morphologically similar *C. douglassii*, but the two species normally are isolated ecologically. The $2n$ = 96 chromosomal races of *C. douglassii* exhibit reduced intrapopulational seed set. *Cardamine longii* is similar to *C. pensylvanica* and sometimes is regarded as an estuarine form of that species, but its lack of petals and other differences are distinctive. Chloroplast DNA data (Figure 4.67) indicate that the *Cardamine* species designated as OBL indicators occur in a relatively derived position within *Cardamine*, indicating that the habit is derived independently from the OBL *Nasturtium* species. Comparable results have been obtained in several other analyses.

Comments: *Cardamine* is distributed throughout North America. *Cardamine pensylvanica* is widespread. *Cardamine cordifolia* occurs in the west; *C. bulbosa*, *C. douglassii*, *C. longii*, and *C. rotundifolia* (central Appalachians) in the east; and *C. clematitis* in the southeast. Once thought to be extinct, *Cardamine micranthera* is a rare endemic restricted to 13 known populations surviving in North Carolina and Virginia. *Cardamine penduliflora* is narrowly endemic in western Oregon. *Cardamine flexuosa* has been introduced throughout North America, but mainly in the eastern region.

References: Aiken et al., 2007; Ali et al., 2012; Al-Shehbaz et al., 2010a; Andersson et al., 2000; Beauchamp et al., 2013; Carlsen et al., 2009; Crone & Taylor, 1996; Fernald & Kinsey, 1943; Frankland & Nelson, 1999; Franzke et al., 1998; Garbary & Taylor, 2007; Gordon, 2013; Hart & Eshbaugh, 1976; Heinold et al., 2013; Koch et al., 2001; Leck & Leck, 2005; Louda & Rodman, 1996; Mandáková et al., 2016; Marhold et al., 2004; Molofsky et al., 2014; Muenscher, 1955; Murdock & Weakley, 1991; Newmaster et al., 1997; Padgett et al., 2004; Rollins, 1993; Scott, 2014; Sweeney & Price, 2000; Tooker et al., 2006; Williams, 2015; Wilson et al., 1993; Yano, 1997; Yatsu et al., 2003.

3. *Eutrema*

Alpine fen mustard

Etymology: from the Greek *eutrema* meaning "complete hole" (i.e., the perforated septum of the siliques)

Synonyms: *Glaribraya*; *Neomartinella*; *Platycraspedum*; *Taphrospermum*; *Thellungiella*; *Wasabia*

Distribution: global: Northern Hemisphere; **North America:** northern

Diversity: global: 26 species; **North America:** 2 species

Indicators (USA): OBL; FAC: *Eutrema edwardsii*

Habitat: freshwater; palustrine; **pH:** alkaline; **depth:** <1 m; **life-form(s):** emergent (herb)

Key morphology: stems (to 8 cm) decumbent to erect; basal leaves (to 1 cm) simple, ovate to elliptic, petiolate (to 2.5 cm);

cauline leaves (to 1.5 cm) fleshy, crowded, narrowly oblong, sessile, midrib well-defined; flowers in crowded racemes (to 3 cm); sepals (to 2 mm) 4, purplish; petals (to 3.5 mm) 4, white, clawed; siliques (to 8 mm) with perforate to nearly absent septa

Life history: duration: perennial (taproot); **asexual reproduction:** none; **pollination:** unknown; **sexual condition:** hermaphroditic; **fruit:** siliques (common); **local dispersal:** seeds; **long-distance dispersal:** seeds

Imperilment: (1) *Eutrema edwardsii* [G4]; S1 (CO, LB, MB); S2 (BC); S3 (QC)

Ecology: general: The circumscription of *Eutrema* has been altered substantially as a result of phylogenetic studies (see *Systematics*). As currently defined, the genus includes annual, biennial, or perennial herbs, which grow from fusiform roots, rhizomes, or a caudex. Many of the species have an affinity for moist sites. There are two species recognized in North America, but only one is a wetland indicator (OBL). The other North American species was assigned formerly to *Thellungiella*, a group known for their high tolerances to salinity. It is more typical of alkaline, saline (but drier) habitats. The flowers of all *Eutrema* species possess nectary glands, which indicate their potential pollination by insects (Insecta). Some of the seeds produce an adhesive, mucilaginous coat when wet, which likely aids in their dispersal by facilitating attachment to animal vectors. One species produces vegetative bulbils in its leaf axils.

Eutrema edwardsii **R. Br.** is a perennial, which inhabits alpine or arctic barrens, bluffs, bogs, depressions, draws, fens, floodplains, gullies, heath, hummocks, marshes, meadows, mires, ridges, scrub, seeps, silt/gravel bars, slopes (to 5%), snow-melt areas, solifluction slopes, swales, tundra, tussocks, and the margins of lakes, ponds, rivers, rivulets, and streams at elevations of up to 3875 m. Most of the habitats represent oligotrophic, rheotrophic, sites. The plants are able to grow from peat mats that obtain water from persistent snowfields and develop on small, flat-to-gently sloping ledges in leeward cirques. They also occur in deep organic soils formed along areas of clear, flowing water from snow melt. Occurrences are rarely found more than 0.5 m away from flowing meltwaters. Alpine habitats can be particularly harsh with growing seasons of 0–70 days, summer temperatures reaching only 16°C, and continuously cool, windy conditions. The substrates have been characterized as acidic or calcareous, and are described as sandstone, clay, cobbly schist, granite, gravel, humus, limestone, loam, marble, mica-schist, mudstone, peat, sand, serpentine, silt, talus, and wacke. Flowering proceeds from June to August, with fruiting from July to September. Additional information on the reproductive or seed ecology are unknown. The shoots perennate from a fleshy taproot. Populations are threatened by mining activities, recreational land uses, and by any activity that would alter the hydrology of the habitat. **Reported associates:** *Aconitum delphiniifolium, Alnus, Alopecurus magellanicus, Anemone parviflora, Anthoxanthum monticola, Arctagrostis latifolia, Artemisia globularia, A. tilesii, Berylsimpsonia vanillosma, Betula, Bistorta bistortoides, B. vivipara, Boykinia richardsonii, Braya glabella, Caltha leptosepala, Cardamine pratensis, C.*

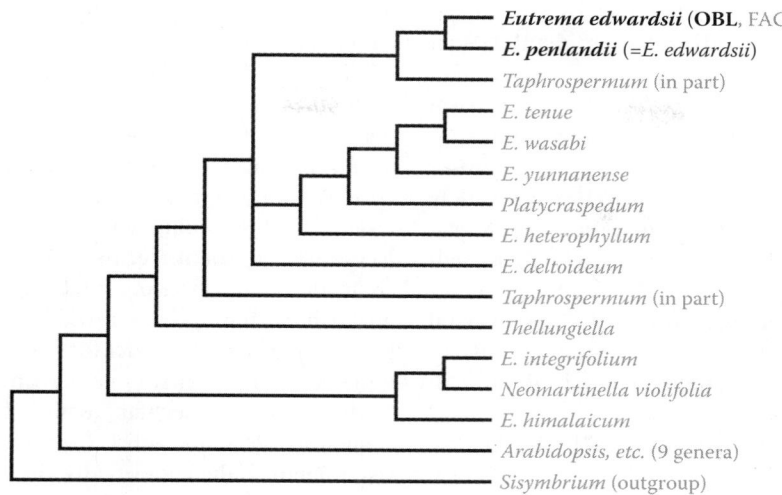

FIGURE 4.69   Phylogenetic relationships of *Eutrema* as indicated by analysis of nrITS sequence data. These results reveal the polyphyletic nature of *Eutrema*, which subsequently has been recircumscribed to include members of *Neomartinella*, *Platycraspedum*, *Taphrospermum*, and *Thellungiella*. *Eutrema penlandii* also has been merged with *E. edwardsii*, due to a lack of any significant morphological or genetic differences. The OBL indicators are shown in bold. (Adapted from Warwick, S.I. et al., *Canad. J. Bot.*, 84, 269–281, 2006.)

purpurea, Carex aquatilis, C. bigelowii, C. chordorrhiza, C. fuliginosa, C. microchaeta, C. podocarpa, Cassiope tetragona, Cerastium regelii, Chamerion latifolium, Corydalis pauciflora, Deschampsia brevifolia, D. cespitosa, Draba, Dryas alaskensis, D. integrifolia, D. octopetala, Dupontia fisheri, Epilobium arcticum, Equisetum variegatum, Eriophorum angustifolium, E. russeolum, E. scheuchzeri, E. vaginatum, Eritrichium nanum, Festuca altaica, F. baffinensis, F. brachyphylla, F. rubra, Geum rossii, Juncus biglumis, J. castaneus, J. triglumis, Micranthes foliolosa, M. lyallii, Minuartia macrocarpa, Oxyria digyna, Papaver radicatum, Pedicularis groenlandica, P. lanata, P. langsdorffii, P. sudetica, Petasites, Phippsia algida, Picea mariana, Pleuropogon sabinei, Poa arctica, P. glauca, Polemonium caeruleum, Populus balsamifera, Potentilla rubricaulis, P. villosa, Puccinellia arctica, Ranunculus nivalis, R. sulphureus, Salix arctica, S. fuscescens, S. lanata, S. myrtillifolia, S. ovalifolia, S. phlebophylla, S. polaris, S. pulchra, S. reticulata, S. stolonifera, Saxifraga cernua, S. hirculus, Selaginella sibirica, Silene acaulis, S. uralensis, Stellaria longipes, Tephroseris integrifolia, Vaccinium uliginosum, V. vitis-idaea.

Use by wildlife: *Eutrema edwardsii* is a host for plant pathogenic fungi (Ascomycota: Mycosphaerellaceae: *Mycosphaerella confinis*). The seeds are cached by Holarctic ground squirrels (Mammalia: Rodentia: Sciuridae: *Urocitellus parryii*).

Economic importance: food: *Eutrema wasabi* is the source of wasabi horseradish, a popular condiment of Japanese sushi bars; however, no culinary uses have been reported for *E. edwardsii*; **medicinal:** No medicinal properties are reported for *Eutrema edwardsii*; however the related *E. wasabi* contains 6-methylsulfinylhexyl isothiocyanates, which have anticarcinogenic properties; **cultivation:** *Eutrema edwardsii* is recommended as a rock garden plant, but never should be collected from the wild; **misc. products:** none; **weeds:** none; **nonindigenous species:** none

Systematics: Phylogenetic analysis employing nrDNA sequence data resolves a distinct clade, where various *Eutrema* species are dispersed among the genera *Neomartinella*, *Platycraspedum*, *Taphrospermum*, and *Thellungiella* (Figure 4.69). As a consequence, all of the latter genera have been merged with *Eutrema*, which then becomes monophyletic. The closest relatives of *Eutrema* include *Arabidopsis* and allied genera. *Eutrema edwardsii* formerly was treated as *E. penlandii*, which was considered to be very closely related. Although some morphometric analyses indicate that the two taxa are distinct, a broader examination of material indicates that they should be regarded as a single morphologically variable species, which has been done here. The base chromosome number of *Eutrema* is $x = 7$. *Eutrema edwardsii* is variable cytologically and includes tetraploids, hexaploids, and octaploids ($2n = 28, 42, 56$).

Comments: *Eutrema edwardsii* occurs across northern North America and is also disjunct in the Colorado Rocky Mountains. Its distribution extends worldwide in a circumpolar fashion.

References: Al-Shehbaz, 2010b; Al-Shehbaz & Warwick, 2005; Fayette & Bruederle, 2001; Fuke et al., 1998; Glenn & Woo, 1997; Heenan et al., 2002; Holm & Holm, 1996; Marr et al., 2012; McKendrick, 1987; O'Kane, Jr. & Al-Shehbaz, 2003; Ostenfeld, 1925; Rollins, 1993; Savile, 1964; Smith, 1993; Talbot et al., 2001; Warwick et al., 2006; Zazula et al., 2011.

4. *Leavenworthia*

Glade cress

Etymology: after Melines Conklin Leavenworth (1796–1862)
Synonyms: none
Distribution: global: North America; **North America:** south-central
Diversity: global: 8 species; **North America:** 8 species

Indicators (USA): OBL: *Leavenworthia torulosa*
Habitat: freshwater; palustrine; **pH:** alkaline; **depth:** <1 m;
life-form(s): emergent (herb)
Key morphology: stems reduced; leaves (to 9 cm) in basal
rosettes, lyrately pinnatifid at maturity, the lobes entire to
broadly dentate; early flowers borne on scapes (to 1 dm) or
later on decumbent lateral branches (to 2 dm) bearing loosely
flowered racemes; petals (to 1 cm) 4, emarginate, spatulate,
white to lavender (occasionally yellow); siliques (to 3 cm) pedicelled (to 7 cm), strongly torulose
Life history: duration: winter annual (fruit/seeds); **asexual**
reproduction: none; **pollination:** self; **sexual condition:**
hermaphroditic; **fruit:** siliques (common); **local dispersal:**
seeds; **long-distance dispersal:** seeds (water)
Imperilment: (1) *Leavenworthia torulosa* [G4]; SX (AL); S2
(KY)
Ecology: general: Almost all of the *Leavenworthia* species
occur in moist sites such as glades, seeps, and stream margins. Four of the species (50%) are wetland indicators, but
only one is designated as OBL. All of the species are winter
annuals, which include both self-compatible (primarily selfing) and self-incompatible (primarily outcrossed) plants. The
outcrossed species are pollinated primarily by bees (Insecta:
Hymenoptera). The self-compatible species have been derived
independently from an ancestral sporophytic incompatibility
system. The seeds are dormant and require a period of warm
stratification to induce germination.

Leavenworthia torulosa **A. Gray** occurs in cedar glades,
depressions, meadows, prairies, roadsides, and seeps at elevations of up to 218 m. The sites typical represent seasonal pools
with open exposures that receive full sunlight. The substrates
are thin layers (1–5 cm) of soil (e.g., clay loam) that are underlain by limestone rock. Flowering occurs from February to
April (peaking in late March to April), with fruiting extending to May. The flowers are self-compatible, highly self-
pollinating (rate of autogamy = 0.66), and strictly inbreeding.
Allozyme analyses of a population containing 31 individuals detected only one allele and a dearth of polymorphic loci
(0%). The seeds are dispersed in May, but remain dormant
until September to October. Seed dormancy in *L. torulosa* is
broken by warm stratification and optimum germination is
achieved at 15°C in the light. Natural germination commences
in September, reaching its maximum level near the end of that
month. Germination can continue until early October. The
buoyant seeds can be dispersed by water during floods. The
plants overwinter as rosettes. They are poor competitors and
do not fare well in interactions with larger plants. **Reported**
associates: *Allium cernuum, Andropogon gerardii, Aristida
purpurea, Baptisia australis, Cardamine hirsuta, Cerastium
brachypodum, C. nutans, Chamaesyce maculata, Cyperus
squarrosus, Daucus carota, Delphinium carolinianum, Draba
verna, Eleocharis compressa, Erigeron strigosus, Erodium
cicutarium, Euphorbia corollata, Forestiera, Geranium
carolinianum, G. dissectum, Helenium amarum, Isoetes,
Juniperus virginiana, Lactuca serriola, Lamium amplexicaule, Leavenworthia exigua, L. stylosa, L. uniflora, Lepidium,
Lespedeza capitata, Minuartia patula, Nostoc, Nothoscordum*

*bivalve, Oenothera filipes, Ophioglossum engelmanni,
Opuntia humifusa, Phacelia dubia, Physaria, Plantago virginica, Portulaca oleracea, Potentilla, Ranunculus sardous, Rhus
aromatica, Rumex crispus, Schizachyrium scoparium, Sedum
pulchellum, Silene antirrhina, Sorghastrum nutans, Stellaria
media, Trifolium, Ulmus alata, Valerianella radiata, V. umbilicata, Veronica arvensis, Viola bicolor, Yucca.*
Use by wildlife: unknown
Economic importance: food: none; **medicinal:** none; **cul-**
tivation: none; **misc. products:** The annual habit and variety of breeding systems in *Leavenworthia* have made this
group an important model for research on reproductive system evolution; **weeds:** none; **nonindigenous species:** none
Systematics: Molecular data (*LD*, nrITS, *psbJ–petA*
sequences) clearly establish *Leavenworthia* as a monophyletic
sister genus to the ecologically similar *Selenia* (Figure 4.70).
All *Leavenworthia* species are diploids but represent two
distinct chromosomal series ($x = 11$; $x = 15$), which associate
with different clades (Figure 4.70). *Leavenworthia torulosa*
($2n = 30$) is highly inbreeding, and morphologically and chromosomally similar to the sympatric but self-incompatible,
outcrossing *L. stylosa* ($2n = 30$), which occupies similar
habitats. Sequences from *PgiC* and the previously mentioned
loci consistently resolve *L. torulosa* as the sister species of
L. stylosa, from which it undoubtedly has been derived.
Molecular data also indicate that *Leavenworthia aurea* and
L. texana are hybrids derived from the same (but maternally
reciprocal) parental taxa: *L. torulosa* and some member of the
$x = 11$ clade (with most evidence implicating *L. alabamica*).
The maternal parent of *L. texana* is within the $x = 11$ clade;
whereas, *L. aurea* originated at least twice, with *L. torulosa*
and some member of the $x = 11$ clade comprising its maternal
parents.
Comments: *Leavenworthia* is endemic to North America;
L. torulosa is narrowly endemic to central Kentucky and
Tennessee (extirpated from Alabama).
References: Al-Shehbaz & Beck, 2010; Anderson & Busch,
2006; Baskin & Baskin, 1971a,b, 1977b; Beck et al., 2006;

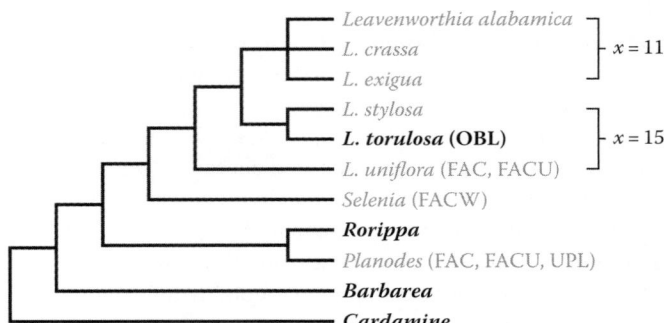

FIGURE 4.70 Phylogenetic relationships of *Leavenworthia* as
indicated by combined DNA sequence data. Taxa containing OBL
North American indicators (designations in parentheses) are shown
in bold. These results indicate that the OBL habit arose independently in *Leavenworthia*. Chromosomal series in *Leavenworthia*
($x = 11, 15$) are delimited by brackets. (Adapted from Urban, L. &
Bailey, C.D., *Syst. Bot.*, 38, 723–736, 2013.)

Busch & Urban, 2011; Charlesworth & Yang, 1998; Filatov & Charlesworth, 1999; Koch et al., 2000; Lloyd, 1965; Rollins, 1993; Urban & Bailey, 2013.

5. *Lepidium*

Pepper-grass

Etymology: diminutive form of the Greek *lepidos* meaning "scale" in reference to the fruit

Synonyms: *Carara*; *Cardaria*; *Coronopus*; *Nasturtium* (in part); *Neolepia*; *Physolepidion*; *Senebiera*; *Sprengeria*; *Stroganowia*; *Winklera*

Distribution: global: cosmopolitan; **North America:** widespread

Diversity: global: 220 species; **North America:** 42 species

Indicators (USA): OBL: *Lepidium latipes*, *L. oxycarpum*; **FACW:** *L. latipes*

Habitat: freshwater, saline; palustrine; **pH:** alkaline; **depth:** <1 m; **life-form(s):** emergent (herb)

Key morphology: stems (to 25 cm) procumbent to erect, densely pubescent, hirsute, or glabrous; basal leaves (to 1 dm) linear, pinnatifid or pinnate with cleft or laciniate lobes; cauline leaves, linear, entire; racemes many flowered; pedicels (to 3.5 mm) strongly flattened; siliques (to 7 mm) elliptical, flattened, their apex notched or winged

Life history: duration: annual (fruit/seeds); **asexual reproduction:** none; **pollination:** self; **sexual condition:** hermaphroditic; **fruit:** siliques (common); **local dispersal:** seeds; **long-distance dispersal:** seeds (adhere to animals)

Imperilment: (1) *Lepidium latipes* [G4]; S2 (CA); (2) *L. oxycarpum* [G4]; SX (BC); S1 (WA)

Ecology: general: *Lepidium* primarily is a terrestrial genus with only a few species adapted to aquatic conditions. Their habitats include deserts, forests, montane sites, prairies, and waste grounds. They occur often in disturbed sites and in abandoned agricultural fields. Thirteen North American species (31%) are wetland indicators, but only two (5%) are OBL. *Lepidium dictyotum* (FACW, FAC) was designated as OBL in the 1996 indicator list but is excluded here. The genus includes annuals, biennials, and perennials. *Lepidium* species predominantly are autogamous. The seeds of most *Lepidium* species become mucilaginous when wet, which enables them to adhere readily to animal fur, notably to the wool of sheep (Mammalia: Bovidae: *Ovis aries*).

Lepidium latipes **Hook.** is an annual, which inhabits bluffs, flats, grasslands, meadows, playas, roadsides, slopes (to 40%), swales, woodlands, and the margins of reservoirs, salt marshes, and vernal pools at elevations of up to 914 m. The site exposures are open. The substrates are alkaline and consist of adobe, Capay silty clay, clay, Clearlake clay loam, decomposed schist, gravel, Monterey shale, Olivenhain cobbly loam, mud, Pescadero clay, sand, serpentine, silty clay, and Willows silty clay. Flowering occurs from March to June. The plants reportedly are autogamous and inbreeding. **Reported associates:** *Achyrachaena mollis*, *Acmispon wrangelianus*, *Alopecurus saccatus*, *Anagallis arvensis*, *Artemisia californica*, *Astragalus tener*, *A. trichopodus*, *Atriplex argentea*, *A. coronata*, *A. parishii*, *A. semibaccata*, *A. serenana*, *Avena barbata*, *A.* *fatua*, *Bassia hyssopifolia*, *Blennosperma nanum*, *Brodiaea coronaria*, *Bromus hordeaceus*, *B. rubens*, *Calandrinia ciliata*, *Capsella bursa-pastoris*, *Centromadia fitchii*, *C. pungens*, *Cicendia quadrangularis*, *Coreopsis gigantea*, *Cotula australis*, *Crassula connata*, *Cressa truxillensis*, *Crypsis schoenoides*, *Cuscuta*, *Deinandra fasciculata*, *Deschampsia danthonioides*, *Dichelostemma capitatum*, *Distichlis spicata*, *Downingia*, *Dudleya blochmaniae*, *Eleocharis macrostachya*, *Erodium cicutarium*, *Eryngium aristulatum*, *E. castrense*, *E. pendletonensis*, *Euphorbia crenulata*, *Extriplex joaquinana*, *Frankenia grandifolia*, *F. salina*, *Hemizonia congesta*, *Hesperevax caulescens*, *H. sparsiflora*, *Hordeum depressum*, *H. intercedens*, *H. marinum*, *H. murinum*, *Juncus bufonius*, *Lasthenia californica*, *L. fremontii*, *L. glaberrima*, *L. glabrata*, *L. gracilis*, *L. platycarpha*, *Layia chrysanthemoides*, *L. platyglossa*, *Lepidium dictyotum*, *L. nitidum*, *L. strictum*, *Logfia filaginoides*, *Lolium multiflorum*, *Lupinus bicolor*, *L. succulentus*, *Lycium californicum*, *Lythrum hyssopifolia*, *Malacothrix saxatilis*, *Marrubium vulgare*, *Marsilea*, *Matricaria discoidea*, *Medicago polymorpha*, *Mesembryanthemum nodiflorum*, *Microseris douglasii*, *M. elegans*, *Myosurus minimus*, *Nassella pulchra*, *Orcuttia californica*, *Oxalis pes-caprae*, *Phalaris lemmonii*, *P. minor*, *P. paradoxa*, *Plagiobothrys acanthocarpus*, *P. collinus*, *P. fulvus*, *P. leptocladus*, *P. stipitatus*, *Plantago bigelovii*, *P. coronopus*, *P. elongata*, *P. erecta*, *P. virginica*, *Poa annua*, *Pogogyne douglasii*, *P. ziziphoroides*, *Primula clevelandii*, *Psilocarphus brevissimus*, *P. oregonus*, *Raphanus sativus*, *Rumex*, *Salicornia*, *Sanicula arguta*, *Schismus arabicus*, *Sisymbrium irio*, *Sonchus asper*, *Spergularia macrotheca*, *S. marina*, *Stebbinsoseris heterocarpa*, *Suaeda moquinii*, *Taeniatherum caput-medusae*, *Trifolium fucatum*, *T. willdenovii*, *Triphysaria eriantha*, *Veronica peregrina*, *Vulpia bromoides*, *Xanthium strumarium*.

Lepidium oxycarpum **Torr. & A. Gray** is an annual, which grows in or on bluffs, ditches, flats, forests, levees, meadows, mud flats, roadsides, scalds, sinks, swales, terraces, and along the margins of Pacific coastal salt marshes, streams, and vernal pools at elevations of up to 747 m. The exposures receive full sunlight. The substrates are alkaline saline and comprise adobe, alluvium, clay, clay loam, heavy loam, Lethent series, Pescadero series, sand, and Solano loam. Flowering and fruiting occur from February to May. The plants presumably are autogamous. The habitat of *L. oxycarpum* possibly is threatened by invasions of *Spartina* (Poaceae). **Reported associates:** *Astragalus tener*, *Bromus rubens*, *Callitriche hermaphroditica*, *Carex*, *Cotula coronopifolia*, *Deschampsia danthonioides*, *Distichlis*, *Frankenia*, *Galium aparine*, *Hordeum marinum*, *Juncus bufonius*, *Lasthenia conjugens*, *L. platycarpha*, *Lepidium dictyotum*, *L. nitidum*, *Lolium multiflorum*, *Lupinus arboreus*, *Lythrum hyssopifolia*, *Medicago polymorpha*, *Myosurus minimus*, *Nassella cernua*, *Pinus sabiniana*, *Plagiobothrys stipitatus*, *P. trachycarpus*, *Plantago elongata*, *P. heterophylla*, *Poa annua*, *Quercus agrifolia*, *Q. douglasii*, *Q. lobata*, *Salicornia depressa*, *Senecio vulgaris*, *Sonchus oleraceus*, *Spergularia marina*, *Zannichellia palustris*.

Use by wildlife: *Lepidium latipes* is a host of the beet leafhopper (Insecta: Hemiptera: Cicadellidae: *Circulifer tenellus*),

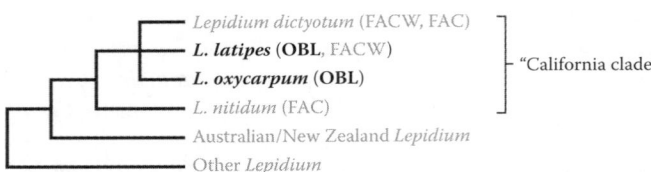

FIGURE 4.71 Phylogenetic relationships of obligate North American indicator species of *Lepidium* as indicated by nrDNA and cpDNA sequence data. The OBL species (bold) are closely related to each other and to *L. dictyotum*, a FACW indicator (indicator ranks are shown in parentheses). A close relationship between *L. dictyotum* and *L. nitidum* is supported by intron sequence data from the nuclear pistillata (PI) gene (Lee et al., 2002). Some species within the "California clade" actually have a broader distribution, extending along western coastal North America from Washington to Mexico. (After Mummenhoff, K. et al., *Amer. J. Bot.*, 88, 2051–2063, 2001; Mummenhoff, K. et al., *Amer. J. Bot.*, 91, 254–261, 2004; Mummenhoff, K. et al., *J. Exp. Bot.*, 60, 1503–1513, 2009.)

which is a vector of the bacterial agent that causes citrus stubborn disease.

Economic importance: food: none; **medicinal:** none; **cultivation:** none; **misc. products:** none; **weeds:** *Lepidium latipes* is a weed of cultivated wheat (*Triticum aestivum*); **nonindigenous species:** *Lepidium oxycarpum* was introduced to Vancouver Island (Canada) in the late 19th century, but did not persist there.

Systematics: Molecular data indicate that *Lepidium* is monophyletic, but only when combined with members of *Cardaria*, *Coronopus*, *Stroganowia*, *Stubendorffia*, and *Winklera*. The genus occupies a phylogenetic position intermediate between a clade containing *Arabidopsis* and its associated genera, and one containing *Barbarea*, *Cardamine*, *Rorippa*, and their associated genera. A "supermatrix" analysis of combined DNA sequence data specifically resolved *Smelowskia* as the sister genus of *Lepidium*, but only with weak support. *Lepidium latipes* and *L. oxycarpum* are very closely related (also to *L. dictyotum*) within a single clade (sometimes called the "California clade"), which comprises western North American species whose nearest relatives occur in Australia and/or New Zealand (Figure 4.71). Although better resolution within the "California clade" is necessary, it is conceivable that the obligate habit might have arisen only once in *Lepidium*. Taxonomic distinctiveness between the alleged varieties of *L. latipes* (i.e., var. *heckardii*; var. *latipes*) is maintained to some degree by autogamy, but recent treatments have abandoned their recognition. The chromosomal base number for *Lepidium* is *x* = 8. Chromosome numbers have not been reported for either of the OBL species.

Comments: In North America, *Lepidium latipes* and *L. oxycarpum* are restricted to California (with *L. latipes* extending into Mexico).

References: Al-Shehbaz, 1986; Al-Shehbaz & Gaskin, 2010; Bailey et al., 2006a; Cook, 1967; Lee et al., 2002; McLaughlin, 1974; Mummenhoff et al., 2001, 2004, 2009; Rollins, 1993.

6. *Nasturtium*

Watercress; cresson d'eau, cresson de fontaine

Etymology: from the Latin *nasus tortus*, meaning "twisted nose" with reference to the pungent properties

Synonyms: *Baeumerta*; *Erysimum* (in part); *Rorippa* (in part); *Sisymbrium* (in part)

Distribution: global: Africa; Asia; Europe; New World*; **North America:** widespread

Diversity: global: 5 species; **North America:** 4 species

Indicators (USA): OBL: *Nasturtium floridanum*, *N. gambelii*, *N. microphyllum*, *N. officinale*

Habitat: freshwater; riverine; **pH:** 6.0–8.0; **depth:** <1 m; **life-form(s):** emergent (herb), floating leaved

Key morphology: stems (to 15 dm) succulent, procumbent to prostrate, rooting at the nodes; leaves pinnate, with 3–13 leaflets (to 2.5 cm), petioles auriculate at base; racemes loose, elongate; petals (to 8 mm) 4, white; siliques (to 3 cm), curved upward; seeds uni- or biseriate, the surface areolate

Life history: duration: perennial (stolons, whole plants); **asexual reproduction:** shoot fragments, stolons; **pollination:** insect, self; **sexual condition:** hermaphroditic; **fruit:** siliques (common); **local dispersal:** vegetative fragments, seeds (water); **long-distance dispersal:** seeds (birds)

Imperilment: (1) *Nasturtium floridanum* [G3/G4]; S3 (FL); (2) *N. gambelii* [G1]; S1 (CA); (3) *N. microphyllum* [GNR]; (4) *N. officinale* [GNR]; S1 (AK)

Ecology: general: As treated here, the North American *Nasturtium* comprises four, closely related species, which occupy similar habitats. All of the species have an affinity for flowing waters and each is regarded here as an OBL indicator, including *N. floridanum*, which inexplicably was excluded from the 2013 indicator list. The flowers are either self-pollinated, or are outcrossed by insect (Insecta) vectors. The stems root freely along their nodes, which facilitates the dispersal and establishment of vegetative fragments along watercourse habitats. The seeds float and are dispersed by water, or are transported over longer distances in the mud that adheres to the feet of waterfowl (Aves: Anatidae).

***Nasturtium floridanum* (Al-Shehbaz & Rollins) Al-Shehbaz & R.A.** Price grows in ditches, roadsides, springs, streams, swamps, and along the margins of hummocks and streams at elevations of up to 50 m. The plants occur often in shallow (to 60 cm) flowing waters and on substrates that have been described as loam, muck, and mud. Flowering extends from February to October. No other life-history information is currently available for this species. **Reported associates:** *Cephalanthus occidentalis*, *Cicuta maculata*, *Lobelia cardinalis*, *Mikania scandens*.

***Nasturtium gambelii* (S. Watson) O.E. Schulz** inhabits freshwater cienega, marshes, swamps, and the margins of lakes and streams at elevations of up to 305 m along the California coast. The plants occur occasionally in slightly brackish water are found on acidic substrates (humus, mud, peat, or sand) with a high organic content. The flowers are self-pollinated. Individual plants can produce from 25 to 200 fruits and are capable of yielding upward of 800 seeds. Seeds collected from natural populations germinate well under ambient greenhouse light and temperature conditions, but few will germinate under wet/vernalization treatments. The plants perennate from a horizontal rootstock. **Reported associates:** *Arenaria paludicola, Athyrium filix-femina, Berula erecta, Carex cusickii, Epilobium ciliatum, Juncus, Lonicera involucrata, Mimulus guttatus, Myrica californica, Oenanthe sarmentosa, Oenothera, Rubus ursinus, Salix lasiolepis, Schoenoplectus californicus, S. pungens, Sparganium eurycarpum, Typha domingensis, T. latifolia, Urtica dioica.*

Nasturtium microphyllum **Boenn. ex Rchb.** occurs in ditches, flats, marshes, meadows, ponds, pools, seeps, springs, spring-fed streams, swales, thickets, and along the margins of brooks, lakes, rivers, and streams at elevations of up to 1657 m. It is more tolerant to acid conditions (not confined to calcareous substrates), lower temperatures (high frost tolerance), and tends to be more northerly distributed than *N. officinale* (see next). The plants can inhabit shallow waters (to 20 cm) and are found on circumneutral substrates (e.g., pH: 6.8–7.6), which include marl, muck, rock, rubble, and sand. Flowering extends from January to August. The flowers are self-compatible. Spontaneous autogamy and seed set will occur when pollinators are excluded. Seed production is high, averaging 29 seeds per fruit and 20+ fruits per inflorescence. The seeds are not dormant when shed and can germinate (92%–100%) immediately (within 1 week). They will retain reasonable germination rates after 5 years of dry storage. A large seed bank can develop, with emergence densities observed between 35 and 395 seedlings/m². The seeds do not germinate in the dark or under flooded conditions. Recently shed seeds float (for at least 12 h and longer if the water surface is not disturbed) and can germinate without further treatment. They are dispersed readily by humans (e.g., by adhering to footwear) as well as by birds (Aves) and water. **Reported associates:** *Alisma, Alnus rugosa, Anemone canadensis, Carex atlantica, C. leptalea, C. stricta, Chara, Cicuta maculata, Cornus, Decodon verticillatus, Epilobium, Equisetum, Hippuris vulgaris, Juncus, Lemna minor, Mentha piperita, Mimulus, Myosotis scorpioides, Phalaris arundinacea, Populus deltoides, Rosa palustris, Scutellaria, Spirodela polyrrhiza, Typha, Utricularia, Veronica, Zizaniopsis miliacea.*

Nasturtium officinale **W.T. Aiton** occupies cold (10°C–20°C), clear, still or flowing, shallow (e.g., 5–100 cm) waters of bogs, bottoms, brooks, canals, channels, cienega, ditches, draws, dunes, fens, flats, floodplains, gullies, marshes, mudflats, rivers, sandbars, seeps, springs, streams, swales, swamps, washes, and the margins of lakes, ponds, and rivers at elevations of up to 3565 m. The plants can withstand temperatures of −15°C, are not shade tolerant, and mainly inhabit alkaline waters (e.g., pH: 7.2–8.2) of moderate hardness (e.g., 50–200 ppm CaCO₃). This species may persist throughout winter in some northern localities but is more tolerant to warmer conditions than *N. microphylla*. The site exposures typically are open and receive full sunlight, but occasionally include shaded conditions. Phenotypic plasticity enables the plants to acclimate to low light by increasing their leaf area and canopy surface area; however, their biomass generally increases proportionally with light levels. The substrates include alluvium, caliche, clay, clay loam, cobble, decomposed granite, gravel, Esquatzel silty loam, gravelly shale, humus, limestone, loam, ooze, rock, sand, sandy alluvium, sandy clay loam, sandy humus, sandy silty loam, shale

talus, silt, silty loam, and travertine. Flowering and fruiting can occur from February to October. The flowers are self-compatible. Some level of cross-pollination occurs by various insects (see *Use by wildlife*), but the flowers mainly are self-pollinating and set seed spontaneously if pollinators are excluded. They often remain closed during rains, with self-pollination occurring within. Seed production is high, averaging 26 seeds per fruit and 15+ fruits per inflorescence. The seeds retain over 76% germination after 5 years of dry storage or storage in water at 1°C–3°C. Optimal seed germination occurs under a 19°C/15°C temperature regime. Recently shed seeds fall close to their source and can also float for at least 12 h, and for longer periods if the water surface is not excessively disturbed. They can germinate without further treatment, with rates of 67% reported within 1 week. The seedlings are unable to establish under closed vegetation. Vegetative reproduction (rhizomes, shoot fragments) is well developed in this species. The plants are known to perennate intact throughout the winter, even in some northern localities. The S1 imperilment category of *N. officinale* in Alaska is unusual, given that it is regarded as introduced in the state. **Reported associates:** *Abies concolor, A. lasiocarpa, Abronia maritima, Acalypha phleoides, Acer negundo, A. rubrum, Achillea millefolium, Adiantum capillus-veneris, Aesculus glabra, Ageratina adenophora, Agrostis exarata, A. gigantea, A. stolonifera, Alisma subcordatum, Alnus oblongifolia, A. rhombifolia, A. rubra, Alopecurus arundinacea, Ambrosia dumosa, A. monogyra, A. psilostachya, Amelanchier utahensis, Amorpha fruticosa, Androsace filiformis, Anemone canadensis, Anemopsis californica, Antirrhinum, Apium graveolens, Apocynum, Aquilegia chrysantha, A. formosa, Artemisia tridentata, Arundo donax, Asclepias fascicularis, A. lemmonii, Athyrium filix-femina, Baccharis salicifolia, Bensoniella oregona, Berula erecta, Betula occidentalis, Bidens laevis, Bolboschoenus maritimus, Bouteloua curtipendula, B. hirsuta, Brachythecium rivulare, Brickellia longifolia, Bromus arizonicus, B. diandrus, B. hordeaceus, B. inermis, B. japonicus, B. rigidus, B. rubens, B. tectorum, Cakile maritima, Calamagrostis canadensis, Callitriche heterophylla, Calocedrus decurrens, Caltha palustris, Calycanthus occidentalis, Camissoniopsis cheiranthifolia, Capsella bursa-pastoris, Carduus nutans, Carex aquatilis, C. aurea, C. bushii, C. emoryi, C. hystericina, C. interior, C. laeviconica, C. lurida, C. molesta, C. nebrascensis, C. pellita, C. praegracilis, C. rostrata, C. scoparia, C. scopulorum, C. senta, C. stipata, C. utriculata, C. vesicaria, C. viridula, Catabrosa aquatica, Celtis reticulata, Cerastium, Ceratophyllum demersum, Cercis occidentalis, Chara, Chenopodium fremontii, Chysosplenium americanum, Cicuta douglasii, C. maculata, Cirsium arizonicum, C. rydbergii, C. vinaceum, C. vulgare, C. wheeleri, Claytonia perfoliata, Clematis ligusticifolia, Comarum palustre, Commelina erecta, Conioselinum scopulorum, Conium maculatum, Corallorhiza maculata, Coreopsis tinctoria, Cornus florida, Crataegus douglasii, Cyperus eragrostis, C. fendlerianus, C. involucratus, C. niger, Dasiphora floribunda, Deschampsia cespitosa, Dieteria canescens, Draba corrugata, Dulichium arundinaceum, Echinochloa, Elaeagnus angustifolia, Eleocharis erythropoda, E. montevidensis, E. palustris, E. parishii, Encelia farinosa,*

Epilobium ciliatum, Epipactis gigantea, Equisetum arvense, E. hyemale, E. laevigatum, Ericameria nauseosa, Erigeron rybii, Eriogonum wrightii, Eupatorium perfoliatum, Evolvulus arizonicus, Fontinalis neomexicana, Fraxinus anomala, F. velutina, Galactia wrightii, Galearis spectabilis, Galium parishii, G. trifidum, G. triflorum, Geranium richardsonii, Glandularia bipinnatifida, Glyceria elata, G. grandis, G. striata, Grindelia squarrosa, Helenium bigelovii, H. puberulum, Helianthus annuus, Hesperidanthus linearifolius, Hesperis matronalis, Hippuris vulgaris, Hosackia alamosana, Hulsea vestita, Hydrocotyl verticillata, Hymenoxys hoopesii, Impatiens capensis, Ipomopsis thurberi, Isocoma menziesii, Juglans major, Juncus acuminatus, J. articulatus, J. balticus, J. brachycephalus, J. bufonius, J. dubius, J. effusus, J. ensifolius, J. longistylis, J. marginatus, J. nodosus, J. torreyi, J. xiphioides, Juniperus communis, J. deppeana, J. occidentalis, J. virginiana, Justicia americana, Laportea canadensis, Leersia oryzoides, Lemna minor, L. trisulca, Lepidium latifolium, Lindera benzoin, Lindernia dubia, Liriodendron tulipifera, Lithocarpus densiflorus, Lobelia cardinalis, Lotus corniculatus, Ludwigia alternifolia, L. palustris, Lupinus padre-crowleyi, Lycopus americanus, Lysichiton americanus, Lythrum californicum, L. salicaria, Maianthemum stellatum, Malosma laurina, Malva, Marrubium vulgare, Medicago polymorpha, Melampodium sericeum, Melilotus albus, M. officinalis, Mentha arvensis, Menyanthes trifoliata, Mimulus cardinalis, M. glabratus, M. guttatus, Mirabilis multiflora, Monarda citriodora, Morus, Muhlenbergia asperifolia, M. rigens, Myosotis scorpioides, Myrica, Myriophyllum, Napaea dioica, Nitella, Nymphaea odorata, Oenanthe sarmentosa, Oenothera elata, Onoclea sensibilis, Oxalis stricta, Parkinsonia microphylla, Parthenocissus vitacea, Paspalum dilatatum, P. distichum, Pedicularis groenlandica, Persicaria amphibia, P. hydropiperoides, P. lapathifolia, P. punctata, P. sagittata, Phacelia ramosissima, Phalaris arundinacea, Phleum, Phlox kelseyi, Picea pungens, Pilea fontana, Pinus contorta, P. jeffreyi, P. ponderosa, Piptatherum miliaceum, Plantago major, Platanthera dilatata, P. hyperborea, Platanus racemosa, P. wrightii, Poa, Podostemum ceratophyllum, Polemonium occidentale, Polypogon monspeliensis, P. viridis, Populus angustifolia, P. fremontii, P. trichocarpa, Potamogeton alpinus, P. berchtoldii, P. natans, Potentilla anserina, P. pulcherrima, Prosopis, Prunella vulgaris, Pseudognaphalium canescens, P. luteoalbum, Pseudoroegneria spicata, Pseudotsuga menziesii, Pteridium aquilinum, Puccinellia distans, Purshia tridentata, Pyrola asarifolia, Quercus agrifolia, Q. emoryi, Q. gambelii, Q. grisea, Q. turbinella, Q. wislizeni, Ranunculus longirostris, R. macranthus, R. sceleratus, R. trichophyllus, Rhamnus betulifolia, Rhus aromatica, Ribes americanum, R. cereum, R. nevadense, Robinia neomexicana, Rorippa aquatica, R. sylvestris, Rosa californica, R. woodsii, Rubus leucodermis, R. ursinus, Rudbeckia laciniata, Rumex acetosella, R. crispus, R. verticillatus, Sagittaria cuneata, Salix amygdaloides, S. bebbiana, S. bonplandiana, S. exigua, S. goodingii, S. laevigata, S. lasiandra, S. lasiolepis, S. taxifolia, Salvia apiana, S. lyrata, Sambucus nigra, Samolus vagans, S. valerandi, Sanicula bipinnata, Saururus cernuus, Schoenoplectus acutus, S. californicus, S. pungens, S. tabernaemontani, Scirpus pallidus, Scutellaria lateriflora, Senecio integerrimus, S. vulgaris, Senegalia greggii, Sidalcea malviflora, Silene verecunda, Sisyrinchium idahoense, Solanum dulcamara, Solidago canadensis, Sonchus oleraceus, Sparganium americanum, S. chlorocarpum, S. emersum, S. eurycarpum, Spergularia bocconi, Sphagnum, Sphenosciadium capitellatum, Stachys albens, Stuckenia pectinata, Symphoricarpos oreophilus, Symphyotrichum subulatum, Symplocarpus foetidus, Taraxacum officinale, Thalictrum alpinum, T. fendleri, Thelypteris palustris, Thlaspi arvense, Thuja plicata, Tiarella cordifolia, Torreyochloa pallida, Toxicodendron diversilobum, T. radicans, T. rydbergii, Tragia nepetifolia, Trichophorum cespitosum, Trifolium dubium, T. fragiferum, T. repens, T. wormskioldii, Triglochin palustre, Tsuga, Typha angustifolia, T. domingensis, T. latifolia, Ulmus americana, Umbellularia californica, Urtica dioica, Utricularia macrorhiza, Vaccinium membranaceum, Valerianella radiata, Verbascum thapsus, Verbena hastata, V. macdougalii, Verbesina rothrockii, Veronica americana, V. anagallis-aquatica, V. beccabunga, V. catenata, Viola canadensis, Vitis arizonica, Woodwardia fimbriata, Xanthium strumarium, Xanthocephalum gymnospermoides, Zizia aptera.

Use by wildlife: The pollen of *Nasturtium gambelii* is eaten by thrips (Insecta: Thysanoptera). The foliage of *Nasturtium officinale* is grazed by deer (Mammalia: Cervidae: *Odocoileus*), muskrats (Mammalia: Cricetidae: *Ondatra zibethicus*), and ducks (Aves: Anatidae). *Nasturtium officinale* provides habitat for many fish (Osteichthyes; a number of them rare) including the Arkansas darter (Percidae: *Etheostoma cragini*), vermilion darter (*Etheostoma chermocki*), watercress darter (*Etheostoma nuchale*), beautiful shiner (Cyprinidae: *Cyprinella formosa*), brook trout (Salmonidae: *Salvelinus fontinalis*), fathead minnow (Cyprinidae: *Pimephales promelas*), Hiko White River springfish (Goodeidae: *Crenichthys baileyi* subsp. *grandis*), Topeka shiner (Cyprinidae: *Notropis topeka*), and least chub (*Iotichthys phlegethontis*). *Nasturtium officinale* is also identified as habitat for the bog turtle (Reptilia: Emydidae: *Glyptemys muhlenbergii*), cricket frog (Amphibia: Hylidae: *Acris crepitans*), Columbia spotted frog (Amphibia: Ranidae: *Rana pretiosa*), Gila spring snail (Mollusca: Hydrobiidae: *Pyrgulopsis gilae*), Kanab ambersnail (Mollusca: Succineidae: *Oxyloma haydeni kanabensis*), Page Springsnail (Mollusca: Hydrobiidae: *Pyrgulopsis morrisoni*), and for many invertebrate trout foods. *Nasturtium officinale* is a host plant for several butterfly larvae (Insecta: Lepidoptera: Pieridae: *Pieris napi, P. rapae*). The flowers are visited by assorted insects (Insecta) including bees (Hymenoptera: Apidae), beetles (Coleoptera), and flies (Diptera: Conopidae; Empididae; Syrphidae; Tachinidae). A foliar glucosinolate–myrosinase system deters herbivory by amphipods (Crustacea: Amphipoda), caddisflies (Insecta: Trichoptera), and snails (Mollusca: Gastropoda); however, the foliage is eaten readily by the introduced apple-snail (Mollusca: Ampullariidae: *Pomacea canaliculata*) and the giant ramshorn snail (Ampullariidae: *Marisa cornuarietis*). *Nasturtium officinale* is susceptible to the crook-root plasmodiophore (Rhizaria: Plasmodiophoraceae: *Spongospora subterranea* f. sp. *nasturti*) and watercress yellow spot virus (WYSV).

Economic importance: food: Watercress species (*Nasturtium microphyllum*, *N. officinale*) and their hybrid (*N. ×sterile*) are important economically as salad crops worldwide. The seeds can be ground and used as a mustard substitute or sprouted and used in salads. However, any plants gathered from the wild should be disinfected before eating. Watercress gathered near geothermal sources has been found to contain high levels of arsenic. Contaminated plants are also known to spread parasites such as vectors of the fasciolosis liver fluke. The herbage of *N. officinale* is rich in calcium and iron and also contains copper, iodine, manganese, phosphorus, and vitamins A, B, C, E, and G. The foliage consists approximately of 3.6% protein, 9.4% crude fiber, 1.1% fat, and contains high amounts of β-carotene (209.6 mg 100⁻¹g). The leaves (cooked or raw) have been eaten as a relish, added to salads, made into sauces, or used as a vegetable by various native North American tribes including the Algonquin, Cahuilla, Cherokee, Diegueno, Gosiute, Havasupai, Iroquois, Karok, Kawaiisu, Luiseno, Mendocino, Okanagan-Colville, Saanich, and Tubatulabal; **medicinal:** Extracts of *Nasturtium microphyllum* have exhibited antimicrobial activity against various strains of bacteria and fungi in laboratory experiments. As the specific epithet "*officinale*" (i.e., "official"; used pharmacologically) indicates, *N. officinale* has a lengthy medicinal history; however, its alleged curative properties remain under investigation (the literature in this area is extensive). Watercress foliage contains various glucosinolates and nitriles. The former are responsible for its pungent flavor and therapeutic effects. The extracts reportedly have antibacterial, anticarcinogenic (lung and oral cancers), and expectorant properties. The plants contain the glucosinolate gluconasturtiin, which on hydrolysis releases phenylethyl isothiocyanate (PEITC), a known anticarcinogenic agent. Watercress shoot extracts show high free-radical scavenging activity. Watercress plants have long been eaten as a preventative for scurvy (caused by vitamin C deficiency). The seeds have been boiled in vinegar and then applied to treat goiters. The Costanoans used a cold infusion of *N. officinale* leaves to relieve fever and decoctions of the plant to treat kidney ailments. Both the Costanoan and Mahuna tribes used *N. officinale* to treat liver ailments. The Okanagan-Colville tribes applied a poultice of *N. officinale* plants to the forehead to relieve headaches and dizziness; **cultivation:** *Nasturtium officinale* is cultivated as an ornamental water garden and as a coldwater aquarium plant. A variegated variety is known as *N. officinale* var. *variegata*. *Nasturtium officinale* and *N. ×sterile* are grown for their edible sprouts and foliage. Watercress has been in commercial production since at least 1750 (in Europe). However, prolonged vegetative propagation of watercress for commercial production has resulted in severe fungal and viral infections that must be overcome by production from seed. Commercial watercress populations have low levels of genetic variation as determined by random amplified polymorphic DNA (RAPD) markers; **misc. products:** The juice of *N. officinale* is a nicotine solvent and also the source of a purple dye; **weeds:** *Nasturtium officinale* is designated as an invasive weed in many parts of North America. It is likely that *N. microphylla* and *N. ×sterile* are also invasive, but are not reported because of taxonomic oversight, i.e., they are often not recognized as distinct from *N. officinale*; **nonindigenous species:** *Nasturtium microphyllum* and *N. officinale* were introduced to North America from Europe as food plants, probably in the early to mid-19th century. Aside from their natural dispersal agents, the seeds of the former are known to have been introduced to the Arctic by adhering to the footwear of human travelers.

Systematics: Although it has become common practice to merge *Nasturtium* with *Rorippa*, a number of phylogenetic studies indicate that the genera are not closely related (e.g., Figure 4.66). Rather, phylogenetic evidence implicates *Cardamine* (including *Dentaria*) as the sister genus to *Nasturtium* (Figure 4.67). Three of the species (*N. gambelii*, *N. microphyllum*, *N. officinale*) have been included in several phylogenetic analyses, but not simultaneously. Nevertheless, both *N. gambelii* and *N. microphyllum* resolve in a clade with *N. officinale* in individual studies, which indicates that the genus is monophyletic. The similarity of *Nasturtium* to some *Cardamine* species (e.g., *C. pensylvanica*) is striking at times and can lead to difficult identifications. The chromosomal base number of *Nasturtium* is $x = 8$. *Nasturtium floridanum* and *N. officinale* (both $2n = 32$) are regarded (genetically) as diploids and *N. microphyllum* ($2n = 64$) as an allotetraploid; *N. gambelii* has not been counted. *Nasturtium ×sterile* ($2n = 48$) is the name given to pollen and seed-sterile allotriploid hybrids, which arise from crosses involving *N. microphyllum* (female parent) and *N. officinale* (pollen parent). The parental species and their hybrid can be identified readily by their different DNA contents using flow cytometry. When in fruit, *N. officinale* (biseriate seeds) is readily distinguished from *N. gambelii*, *N. floridanum*, and *N. microphyllum* (uniseriate seeds). However, the distinctions among the latter three species are not as clearly marked. Further genetic studies that focus on the distinctness of these species from one another (and *N. ×sterile*) should be undertaken. Identification of *Nasturtium* species in North America has been problematic because of the inconsistent recognition of *N. microphyllum* in the flora before the 1950s. Consequently, it is sometimes difficult to ascertain whether a specific published report refers correctly or erroneously to *N. officinale* or to another watercress taxon.

Comments: *Nasturtium floridanum* is endemic to Florida. *Nasturtium gambelii* occurs in California, but extends into Mexico and Central America. *Nasturtium microphyllum* is distributed primarily in eastern North America and is also known from some western localities. *Nasturtium officinale* is widespread throughout North America.

References: Al-Shehbaz, 2010b; Al-Shehbaz & Price, 1998; Bleeker, 2004; Bryson et al., 1994; Caires et al., 1992; Duncan et al., 2010; Franzke et al., 1998; Furlong & Pill, 1980; Going et al., 2008; Green, 1962; Hall et al., 2001; Hecht et al., 1999; Howard & Lyon, 1952a,b; Imtiaz et al., 2012; Les & Mehrhoff, 1999; Morozowska et al., 2015; Newman et al., 1992; Pieroni et al., 2002; Preston & Croft, 1997; Rollins, 1993; Rondelaud et al., 2000; Runkel & Roosa, 1999; Shad et al., 2013; Sheridan et al., 2001; Tenorio & Drezner, 2006; USFWS, 1998a; Ware et al., 2012; Welsh, 1974.

7. *Rorippa*

Lakecress, yellowcress; rorippe

Etymology: Latin form of *Rorippen*, the old Saxon name

Synonyms: *Armoracia* (in part); *Brachiolobos*; *Cochlearia* (in part); *Kardamoglyphos*; *Myagrum* (in part); *Nasturtium* (in part); *Neobeckia*; *Radicula*; *Sisymbrium* (in part); *Tetrapoma*

Distribution: global: cosmopolitan; **North America:** widespread

Diversity: global: 85 species; **North America:** 20 species

Indicators (USA): OBL: *Rorippa aquatica, R. calycina, R. columbiae, R. curvipes, R. curvisiliqua, R. palustris, R. sessiliflora, R. sphaerocarpa, R. subumbellata, R. sylvestris, R. teres*; **FACW:** *R. calycina, R. curvipes, R. curvisiliqua, R. sphaerocarpa, R. sylvestris, R. teres*; **FAC:** *R. palustris, R. sphaerocarpa, R. sylvestris*

Habitat: freshwater; palustrine; **pH:** 5.5–7.5; **depth:** <1 m; **life-form(s):** emergent (herb); submersed (rosulate, vittate)

Key morphology: stems (to 14 dm), weak, rosulate, or vittate, often rhizomatous; leaves (to 15 cm) sometimes heterophyllous (then the blades fugacious), in basal rosettes or cauline and alternate; the blades simple, pinnatifid, pinnate, or pinnately decompound into capillary segments (lower, submersed foliage), the margins entire, dentate, laciniate, serrate, or pectinate; racemes (to 6 dm) terminal or lateral; flowers pedicellate; pedicels (to 13 mm) usually recurved; petals (to 1 cm) 4 (or absent), clawed, pale to sulfur yellow or white; siliques (to 2 cm) obovoid or ellipsoid to elongate-cylindrical, straight to falcate, replum present or absent; seeds numerous (10–200 per silique)

Life history: duration: annual (fruit/seeds); perennial (rhizomes); **asexual reproduction:** leaf, root, and shoot fragments, creeping roots; **pollination:** insect, self; **sexual condition:** hermaphroditic; **fruit:** siliques (common); **local dispersal:** fragments (water), fruits/seeds (water, wind); **long-distance dispersal:** seeds (waterfowl)

Imperilment: (1) *Rorippa aquatica* [G4]; SH (IA, ME, NJ, VA); S1 (AL, GA, IN, KS, KY, MD, MS, OK, QC, TX, VT, WI); S2 (MI, MO, NY, OH, TN); S3 (IL, ON); (2) *R. calycina* [G3]; SH (ND); S1 (BC, NE, MT); S2 (WY); (3) *R. columbiae* [G3]; S1 (CA, WA); S3 (OR); (4) *R. curvipes* [G5]; SH (KS); S1 (KS); S2 (SK, WY); S3 (WY); (5) *R. curvisiliqua* [G5]; S1 (AK); S2 (WY); (6) *R. palustris* [G5]; S1 (KS); S2 (IL, KS); S3 (NJ, PE, WY); (7) *R. sessiliflora* [G5]; SH (DC); S1 (NC, VA, WV); S3 (MN); (8) *R. sphaerocarpa* [G5]; S1 (WY); S2 (UT); (9) *R. subumbellata* [G1]; S1 (CA, NV); (10) *R. sylvestris* [G5]; (11) *R. teres* [G5]; S1 (NC, OK)

Ecology: general: Virtually all *Rorippa* taxa have an affinity for moist habitats and they often can be found growing under fairly wet conditions. All of the North American species (100%) are wetland indicators at some level with over half of the species (55%) categorized as OBL aquatics. The genus includes annuals, biennials, and perennials. Most are pioneer species, which are among the first to colonize wet, denuded areas. They are generally shade intolerant and poor competitors. The flowers can be self-compatible or self-incompatible. Accordingly, they are primarily self-pollinated or are outcrossed by insects (Insecta), including bees and wasps (Hymenoptera), beetles (Coleoptera), and flies (Diptera). The seeds germinate readily in warm, wet, open sites. Vegetative reproduction is common in the perennials. Both the seeds and vegetative propagules are dispersed by water or waterfowl (Aves: Anatidae).

***Rorippa aquatica* (Eaton) E.J. Palmer & Steyerm.** is a perennial occurring across a wide spectrum of freshwater habitats ranging from exposed substrates to deep waters (to 2 m) associated with ditches, flood plains, oxbows, ponds, streams (slow-flowing), springs, swamps, and lakes at elevations of up to 300 m. The most vigorous growth has been observed under colder (18°C–25°C), oligotrophic conditions; however, hydrological conditions can vary from clear and slow moving to turbid and stagnant waters. Exposures also vary from open conditions to shaded sites (e.g., under *Taxodium/Nyssa* canopies). The substrates (pH: 5.5–7.5) most often are described as alkaline and include clay, gravel, marl, muck, mud, sand, and silt. These plants are heterophyllous, with increasingly dissected foliage produced with respect to their degree of inundation. The homeobox gene KNOX1, which promotes compound leaf development under low light conditions, is expressed only in the submersed, dissected leaves. The common name "lake" cress is somewhat of a misnomer as these plants perhaps are more commonly found in riverine systems. A key habitat feature appears to be areas of natural disturbance from fluctuating water levels such as those found along floodplains and the margins of larger lacustrine systems. Reproduction (both sexual and asexual) is most effective on shallowly inundated substrates. However, only the populations south of the glacial limit are diploid, fertile, and capable of seed production (those north of the boundary are sterile triploids and rely entirely on vegetative reproduction). Preliminary studies indicate that sporophytic self-incompatibility (SSI) is maintained in the fertile diploids, thus complicating seed set in clonal populations, or in those with low numbers of self-incompatibility alleles (s-alleles). Crosses between genetically distinct diploids can result in substantial fruit set (to 85%). Both diploid and triploid populations harbor considerable genetic variation as detected using random amplified polymorphic DNA (RAPDS) markers; however much of this variation (at least in the triploids) is attributable to somatic mutation and does not necessarily indicate relative s-allele diversity, which is necessary to maintain successful sexual reproduction. When grown under common greenhouse conditions, the triploid plants flower rarely and irregularly; whereas, diploids flower commonly and regularly (from April to August). The flowers remain open for up to 6 days. In nature, the diploids are most likely pollinated by small insects (Insecta). The germination of seeds arising from fertile plants is low (13%–25%; at 12 h light/dark, 65% humidity, 18°C/24°C temperature regime) and those cultured on moist filter paper often succumb to fungal infections. The extent of vegetative reproduction is remarkable. The leaves are shed from plants at the slightest perturbation (e.g., wave splash, wind), and when detached, will float and eventually produce adventitious plantlets from the petiole or any severed portion. Nearly any fragment of lake cress herbage (or roots) will produce adventitious plantlets, even pieces that only are a few millimeters in length. In contiguous waters (e.g., the Great Lakes), dispersal of the vegetative fragments can occur across distances of more than 50 km. However, vegetative propagules are inefficient in dispersal between isolated water bodies (smaller lakes and river systems), which is accomplished most effectively by the transport of seeds. This species has suffered a loss of about 68% of its historical populations in the United States. Most of the extant populations range in size from 50 to 100 individuals, although some contain several thousand individuals. The populations are threatened by habitat loss and by invasions of nonindigenous species such as *Butomus umbellatus*, *Lythrum salicaria*, *Myriophyllum spicatum*, and *Trapa natans*. **Reported associates:** *Alisma triviale*, *Butomus umbellatus*, *Cardamine*, *Carex crus-corvi*, *Cephalanthus occidentalis*, *Ceratophyllum demersum*, *C. echinatum*, *Eleocharis acicularis*, *E. palustris*, *Glyceria borealis*, *Heteranthera dubia*, *Hippuris vulgaris*, *Juncus canadensis*, *Landoltia punctata*, *Leersia oryzoides*, *Lemna minor*, *Ludwigia palustris*, *Lysimachia nummularia*, *Lythrum salicaria*, *Megalodonta beckii*, *Myriophyllum sibiricum*, *M. spicatum*, *Nasturtium officinale*, *Nuphar variegata*, *Nymphaea odorata*, *Nyssa sylvatica*, *Persicaria*, *Potamogeton natans*, *Proserpinaca palustris*, *Ranunculus longirostris*, *R. reptans*, *Rorippa*, *Sagittaria latifolia*, *Samolus floribundus*, *Scirpus acutus*, *S. tabernaemontani*, *Sium suave*, *Sparganium*, *Spirodela polyrhiza*, *Taxodium distichum*, *Trapa natans*, *Typha*, *Wolffia*, *Zizania aquatica*.

Rorippa calycina (**Engelm.**) **Rydb.** is a perennial occurring near the high water mark along beaches, dunes, flats, mudflats, playas, swales, washes, and the shores of lakes, ponds, reservoirs, rivers, and streams at elevations of up to 2079 m. The habitats are seasonally flooded, with open exposures (typically <25% cover), an annual rainfall of 20–30 cm, and average annual temperatures ranging from a minimum of −15.5°C (January) to a maximum of 31.1°C (July). Formerly wet but dessicated sites and the margins of artificial reservoirs provide particularly suitable habitat for the species, perhaps because of the fluctuating water levels. The substrates are level (slope <5%) and consist of loosely textured, alkaline, or nonalkaline clay, mud, muddy clay, gravel, sand, sandy gravel, and shaley sand. Flowering occurs mainly from May to August (to October in favorable years) with fruiting extending from June to September. Prolonged periods of high water can delay or even postpone flowering. The plants are self-compatible, but their reproductive biology has not been studied in detail. The flowers are visited by small bees (Insecta: Hymenoptera), which presumably effect outcrossing. They produce an average of 20 seeds per fruit. Seeds collected from dried, 4-year-old specimens have germinated within 2 days without receiving any type of treatment. The seeds are dispersed by waterfowl (Aves: Anatidae), apparently over fairly long distances. Vegetative reproduction occurs by means of the elongating rhizomes and spreading roots. **Reported associates:** *Achnatherum hymenoides*, *Ambrosia tomentosa*, *Artemisia*

cana, *A. frigida*, *Astragalus*, *Atriplex micrantha*, *Bidens cernuus*, *Carex douglasii*, *Chenopodium berlandieri*, *C. glaucum*, *C. rubrum*, *Cirsium arvense*, *Dieteria canescens*, *Eleocharis*, *Ericameria nauseosa*, *Gnaphalium*, *Grindelia squarrosa*, *Halogeton glomeratus*, *Hordeum jubatum*, *Iris*, *Iva axillaris*, *Juncus balticus*, *J. compressus*, *Lepidium latifolium*, *Lupinus*, *Monolepis nuttalliana*, *Musineon*, *Pascopyrum smithii*, *Physaria*, *Plagiobothrys*, *Poa secunda*, *Polygonum aviculare*, *Potentilla anserina*, *P. norvegica*, *P. rivalis*, *P. supina*, *Rumex crispus*, *R. maritimus*, *R. salicifolius*, *Salix exigua*, *Salsola australis*, *Spergularia rubra*, *Suckleya suckleyana*, *Tamarix chinensis*, *Verbena*, *Veronica peregrina*, *Xanthium strumarium*.

Rorippa columbiae (**S. Watson**) **Howell** is a perennial occurring in ditches, ephemeral mountain streams, flats, gravel bars, meadows, playas, roadsides, sand bars, seasonally flooded lake beds, swamps, vernal pools, and along the margins of lakes, ponds, rivers, and streams at elevations of up to 1737 m. The exposures must remain open with little competition. Prolonged periods of inundation are detrimental to the growth of this species. These plants require habitats that are wet (but not inundated) throughout their growing season and are characterized by water level fluctuations. They are common in the lower vegetated riparian zone and are well adapted to sites where periodic flooding (especially where slow, late seasonal drawdowns occur) and unstable substrates contribute to reduced competition. The substrates are described as alkaline and include alluvial cobble, andesite boulders, clay, clay loam, cobble, gravel, gravelly cobble, rock, sand, sandy loam, sandy pumice, sandy silt, and silt. Flowering occurs from June to August. The plants spread locally and can form large clones by their creeping roots, which produce adventitious, rooting stems. **Reported associates:** *Agrostis*, *Amaranthus*, *Arabidopsis*, *Arnica chamissonis*, *Cerastium*, *Deschampsia danthonioides*, *Downingia*, *Draba*, *Eriogonum ovalifolium*, *Eryngium alismifolium*, *Euphorbia*, *Hordeum jubatum*, *Lupinus lepidus*, *Mentha pulegium*, *Oenothera*, *Polygonum polygaloides*, *Portulaca*, *Potentilla newberryi*, *Rumex*, *Salix*, *Taraxia tanacetifolia*, *Veronica*, *Xanthium*.

Rorippa curvipes **Greene** is an annual or short-lived perennial, which inhabits bogs, bottomlands, channels, depressions, ditches, draws, drying lake beds, flats, gravel bars, gullies, hummocks, meadows, mud flats, roadsides, savannas, seeps, slopes, strands, stream beds, thickets, tundra, vernal pools, and the margins of lakes, ponds, reservoirs, sloughs, and streams at elevations of up to 4016 m. Site exposures can range from full sun to dense shade. The substrates are characterized as alkaline, calcareous, or saline and include alluvium, basaltic sand, clay, cobbles, gravel, gravelly loam, Laketown dolomite, loamy mud, muck, mud, quartzite, rocks, rocky volcanics, sand, sandy gravel, sandy loam, sandy silt, silt, silty gravel, stones, and Swan Peak quartzite. Flowering and fruiting occur from April to November. The reproductive ecology of this species is poorly documented. Adventitious basal rosettes sometimes develop from the roots, which can result in a perennial growth habit. **Reported associates:** *Abies* concolor, A. lasiocarpa, Achillea millefolium, Acmispon utahensis, Agrostis scabra, Alisma triviale, Allium schoenoprasum, Alnus rugosa, A. tenuifolia, Alopecurus aequalis, Antennaria, Apera interrupta, Apocynum, Aquilegia coerulea, Arnica mollis, Artemisia biennis, A. carruthii, A. dracunculus, A. ludoviciana, A. tridentata, Azolla, Bidens cernuus, B. frondosus, Calamagrostis canadensis, Callitriche palustris, Calycanthus occidentalis, Carex alma, C. aquatilis, C. athrostachya, C. canescens, C. filifolia, C. haydeniana, C. lenticularis, C. nebrascensis, C. pellita, C. rostrata, C. scopulorum, C. subnigricans, C. utriculata, C. vesicaria, Castilleja exserta, C. miniata, Cerastium, Chara, Chenopodium chenopodioides, C. glaucum, C. rubrum, Chorizanthe membranacea, Cirsium arvense, Conyza canadensis, Coreopsis tinctoria, Cornus, Crataegus, Dasiphora floribunda, Deschampsia cespitosa, Descurainia, Dieteria canescens, Echinochloa crus-galli, Eleocharis acicularis, E. palustris, E. parvula, Epilobium ciliatum, Equisetum arvense, E. fluviatile, Eragrostis minor, Eremogone fendleri, Ericameria nauseosa, Erigeron coulteri, E. divergens, E. flagellaris, Eriogonum racemosum, Floerkea proserpinacoides, Galium trifidum, Gayophytum diffusum, Gentianopsis holopetala, Geum macrophyllum, Glyceria borealis, G. elata, Gnaphalium exilifolium, G. palustre, G. uliginosum, Helenium autumnale, Heracleum sphondylium, Heterotheca villosa, Hippuris vulgaris, Holcus lanatus, Hordeum jubatum, Hymenoxys subintegra, Hypericum boreale, Juncus balticus, J. bufonius, J. dudleyi, J. parryi, J. supiniformis, J. tenuis, Juniperus occidentalis, J. scopulorum, Koenigia islandica, Lactuca serriola, Lappula occidentalis, Leerzia oryzoides, Leptochloa, Limosella aquatica, Lonicera, Ludwigia, Lupinus argenteus, Luzula, Maianthemum stellatum, Marsilea vestita, Mazus pumilus, Melilotus, Mentha arvensis, Mertensia ciliata, Micranthes subapetala, Mimulus primuloides, M. tilingii, Mitella pentandra, Montia, Muhlenbergia mexicana, M. montana, M. tricholepis, Nuphar polysepala, Oenothera flava, Orthilia secunda, Orthocarpus luteus, Panicum capillare, Parnassia fimbriata, Penstemon barbatus, Perideridia parishii, Persicaria amphibia, P. coccinea, P. lapathifolia, P. maculosa, Phleum, Picea engelmannii, P. pungens, Pinus contorta, P. ponderosa, Plagiobothrys hispidulus, Plantago tweedyi, Platanthera hyperborea, Poa nervosa, P. pratensis, Polygonum aviculare, P. douglasii, P. polygaloides, Polypogon, Populus angustifolia, P. tremuloides, Porterella carnosula, Potamogeton amplifolius, P. illinoensis, P. pusillus, Potentilla anserina, P. hippiana, P. norvegica, Primula fragrans, Prunus, Pseudotsuga menziesii, Pteridium, Purshia, Quercus wislizeni, Ranunculus flammula, R. macounii, Rhodiola rhodantha, Ribes montigenum, Rorippa sinuata, R. sphaerocarpa, R. sylvestris, R. tenerrima, Rosa woodsii, Rumex crispus, R. fueginus, R. maritimus, R. salicifolius, R. triangulivalvis, Sagittaria cuneata, Salix bebbiana, S. eastwoodiae, S. exigua, S. geyeriana, S. glauca, S. irrorata, S. monticola, S. planifolia, Sambucus nigra, Scirpus microcarpus, Senecio flaccidus, S. triangularis, Solidago, Sparganium angustifolium, S. emersum, Spergula, Spergularia, Swertia perennis, Symphyotrichum lanceolatum, Taraxacum officinale, Taraxia breviflora, T. tanacetifolia,

Thalictrum fendleri, Thermopsis, Thinopyrum ponticum, Torreyochloa pallida, Trisetum spicatum, Typha latifolia, Umbellularia californica, Urtica dioica, Utricularia macrorhiza, Vaccinium, Valeriana acutiloba, V. occidentalis, Verbascum thapsus, Verbena bracteata, V. lasiostachys, Veronica americana, V. anagallis-aquatica, V. peregrina, Xanthium strumarium.

***Rorippa curvisiliqua* (Hook.) Bessey ex Britton** is an annual or biennial (in coastal areas) occurring in or on beaches, bogs, bottomlands, depressions, ditches, ephemeral ponds, floodplains, gravel bars, marshes, meadows, mud flats, prairies, roadsides, seeps, slopes (to 10%), swales, thickets, vernal pools, washes, and the margins or shores of lakes, ponds, pools, reservoirs, sloughs, and streams at elevations of up to 3505 m. Sites typically are reported as having open exposures that receive full sun. The substrates usually are described as alkaline (e.g., pH: 7.3) and include alluvium, clay, clay loam, decomposed granite, granite, gravel, gravelly loam, humus, loam, mucky clay loam, mud, muddy clay, peat, quartzite, rock, sand, sandy loam, sandy loess, sandy rock, silt, and silty loam. Flowering and fruiting occur from April to October. Additional life history information is scarce for this species. **Reported associates:** *Abies concolor, A. lasiocarpa, Acer circinatum, A. negundo, Achillea, Acmispon glabrus, Ageratina adenophora, Agoseris glauca, Agrostis scabra, Alisma, Allium geyeri, A. schoenoprasum, Alnus rhombifolia, A. rubra, A. rugosa, Alopecurus aequalis, Anagallis minima, Anthemis cotula, Anthoxanthum nitens, A. odoratum, Arabis, Artemisia arbuscula, A. californica, A. tridentata, Baccharis salicifolia, Balsamorhiza sagittata, Beckmannia, Betula occidentalis, Bistorta bistortoides, Calamagrostis canadensis, Callitriche heterophylla, Calocedrus decurrens, Camassia quamash, Campanula rotundifolia, Capsella bursa-pastoris, Cardamine oligosperma, Carex athrostachya, C. densa, C. hoodii, C. lenticularis, C. microptera, C. nebrascensis, C. pachystachya, C. rostrata, C. scopulorum, C. unilateralis, C. utriculata, C. vesicaria, Castilleja attenuata, Centromadia fitchii, Cerastium, Chenopodium, Chrysopogon zizanioides, Circaea alpina, Cirsium vulgare, Collinsia, Cornus sericea, Crataegus douglasii, Crucianella angustifolia, Crypsis schoenoides, Cyperus esculentus, Dactylis glomerata, Dasiphora floribunda, Datura wrightii, Delphinium nuttallianum, Deschampsia cespitosa, D. danthonioides, Digitaria sanguinalis, Draba stenoloba, Echinochloa crus-galli, Eleocharis acicularis, E. engelmannii, E. macrostachya, E. obtusa, E. palustris, Encelia farinosa, Epilobium campestre, E. ciliatum, E. cleistogamum, E. glaberrimum, Equisetum, Eragrostis, Eriochloa acuminata, Eriogonum, Euthamia, Ficus carica, Fontinalis neomexicana, Fragaria, Fraxinus latifolia, Galium tricornutum, G. trifidum, Gamochaeta argyrinea, Gastridium phleoides, G. ventricosum, Gentianopsis detonsa, Geranium, Glyceria, Gnaphalium palustre, Hackelia, Heliotropium curassavicum, Hordeum brachyantherum, Iris, Isoetes bolanderi, I. howellii, Juncus acuminatus, J. balticus, J. bufonius, J. ensifolius, Kickxia elatine, Koeleria macrantha, Lasthenia glaberrima, Lathyrus angulatus, Lemna minor, Lepidium latifolium, L. virginicum,*

Lewisia pygmaea, Leymus cinereus, Ligusticum canbyi, Lilaeopsis, Lindernia dubia, Lithophragma, Lolium multiflorum, Lomatium grayi, Lupinus caudatus, L. lepidus, Lythrum hyssopifolia, Madia sativa, Malosma laurina, Melica subulata, Mentha arvensis, Menziesia ferruginea, Micranthes subapetala, Mimulus guttatus, M. lewisii, M. pilosus, M. primuloides, Mollugo verticillata, Monardella villosa, Montia chamissoi, M. linearis, Morus alba, Myosotis laxa, Nasturtium officinale, Oxyria digyna, Panicum capillare, Penstemon laetus, Perideridia, Persicaria amphibia, P. lapathifolia, P. maculosa, Phalaris arundinacea, Phleum alpinum, Picea engelmannii, Pinus albicaulis, P. contorta, P. flexilis, P. ponderosa, Plagiobothrys bracteatus, P. figuratus, P. hispidulus, P. scouleri, P. stipitatus, Plantago coronopus, P. major, P. virginica, Platanus racemosa, Poa annua, P. pratensis, Polygonum aviculare, P. polygaloides, Polypogon monspeliensis, Populus fremontii, P. tremuloides, P. trichocarpa, Potamogeton alpinus, P. berchtoldii, Potentilla gracilis, Pseudognaphalium stramineum, Pseudostellaria, Pseudotsuga menziesii, Psilocarphus oregonus, Quercus garryana, Q. kelloggii, Q. lobata, Ranunculus aquatilis, R. flammula, R. orthorhynchus, R. pusillus, R. repens, R. sceleratus, Ribes, Ricciocarpus natans, Robinia pseudoacacia, Rorippa palustris, R. subumbellata, R. sylvestris, Rosa nutkana, Rubus ursinus, Rumex conglomeratus, R. crispus, R. paucifolius, Salix exigua, S. geyeriana, S. gooddingii, S. laevigata, S. lasiandra, S. lasiolepis, S. planifolia, S. wolfii, Salvia mellifera, Sambucus nigra, Schedonorus arundinaceus, Senecio hydrophiloides, S. triangularis, Setaria parviflora, Sium suave, Sorghum halapense, Sparganium angustifolium, S. emersum, Spergularia rubra, Sphagnum, Spiraea douglasii, Symphoricarpos, Symphyotrichum spathulatum, Taraxacum officinale, Taraxia tanacetifolia, Thalictrum, Thermopsis, Thlaspi arvense, Torreyochloa pallida, Trichostema lanceolatum, Trifolium cernuum, T. dubium, T. pratense, T. repens, Triodanis perfoliata, Triteleia hyacinthina, Typha latifolia, Urtica dioica, Vaccinium membranaceum, V. scoparium, Veratrum californicum, V. viride, Verbascum thapsus, Veronica anagallis-aquatica, V. peregrina, Viola adunca, Vitis girdiana, Wyethia amplexicaulis.

***Rorippa palustris* (L.) Besser** is an annual, biennial, or short-lived perennial, which inhabits beaches, bogs, borrow pits, bottomlands, cienega, depressions, ditches, draws, estuaries, fens, flats, floating hummocks, floodplains, marshes, meadows, mudflats, muskeg, playas, pools, potholes, ravines, roadsides, sand bars, seeps, slopes (to 20%), sloughs, springs, streambeds, swales, swamps, thickets, tundra, woodlands, and along the margins of canals, lakes, reservoirs, rivers, sloughs, and streams at elevations of up to 3580 m. The plants are reported frequently from agricultural fields (active and abandoned) and other disturbed sites. They occur often in standing, shallow waters (to 50 cm). Exposures usually are in full sun, but can include partial shade. The substrates have been described as alkaline or calcareous (but the full pH range is undetermined) and include alluvium, boulders, clay, clay loam, cobble, decomposed granite, granitic alluvium, gravel, gravelly clay, gravelly sand, humus, limestone, loamy clay, mucky clay, muck, mud, muddy peat, peat, rock, rocky sand, sand, sandy gravel, sandy humus, sandy loam, sandy muck, silty gravel, gumbo soil, silty loam, silty mud, silty sand, stony silt, and Wasatch alluvium. Flowering and fruiting extend from April to October. The flowers are self-fertile and highly autogamous and are cross-pollinated occasionally by bees and flies (Insecta: Hymenoptera; Diptera). The seeds will germinate within the same season that they are produced, often reaching fairly high germination rates (to 90%). Germination is highest when incubated under a 30°C/15°C temperature regime. The viability of stored seeds is low with 2-year-old seeds exhibiting reduced viability and seeds over 5 years old generally being inviable; thus, a long-term seed bank is not likely to develop. The activity of digging beetles (Insecta: Coleoptera) can bury the seeds to depths of 5–10 cm within 6 months. The plants do not reproduce vegetatively.

Reported associates: *Abies concolor, A. grandis, A. glabrum, Acalypha rhomboidea, Acer negundo, A. saccharinum, Achillea millefolium, Acmispon glabrus, Acorus, Adenostoma fasciculatum, Agoseris heterophylla, Agropyron repens, Agrostis alba, A. gigantea, A. stolonifera, Alisma, Allium geyeri, Alnus rubra, A. tenuifolia, A. viridis, Alopecurus aequalis, A. carolinianus, A. geniculatus, Amaranthus arenicola, A. blitoides, A. californicus, A. graecizans, A. tuberculatus, Ambrosia artemisiifolia, A. psilostachya, Ammannia coccinea, A. robusta, Androsace septentrionalis, Angelica atropurpurea, Anthemis cotula, Apocynum cannabinum, Aquilegia, Arabidopsis lyrata, Arabis hirsuta, Arctagrostis, Arctostaphylos uva-ursi, Artemisia annua, A. biennis, A. californica, A. douglasiana, A. tridentata, Asclepias amplexicaulis, A. incarnata, Atriplex gmelinii, Avena barbata, A. sativa, Baccharis halimifolia, B. salicifolia, B. sarothroides, Betula glandulosa, Bidens cernuus, B. laevis, Bistorta bistortoides, Bolboschoenus fluviatilis, Bromus diandrus, Calamagrostis canadensis, C. stricta, Caltha natans, Calystegia, Cardamine parviflora, C. pensylvanica, C. pratensis, Carex aquatilis, C. atherodes, C. cristatella, C. emoryi, C. festucacea, C. foenea, C. granularis, C. grayi, C. lacustris, C. lasiocarpa, C. lenticularis, C. lyngbyei, C. molesta, C. nebrascensis, C. normalis, C. obnupta, C. pellita, C. retrorsa, C. rostrata, C. sartwellii, C. scoparia, C. scopulorum, C. stipata, C. sychnocephala, C. tribuloides, C. utriculata, C. vesicaria, C. vulpinoidea, Castilleja miniata, Ceanothus tomentosus, Centaurium tenuiflorum, Cephalanthus occidentalis, Chamerion angustifolium, C. latifolium, Chasmanthium latifolium, Chenopodium album, C. rubrum, Cicuta bulbifera, Cinna arundinacea, Cirsium, Conyza canadensis, Coreopsis tinctoria, Cornus canadensis, C. sericea, Crataegus crus-galli, Crypsis, Cynodon dactylon, Cyperus eragrostis, C. odoratus, C. squarrosus, C. strigosus, Dactylis glomerata, Danthonia unispicata, Deschampsia cespitosa, Descurainia, Dichanthelium oligosanthes, Digitaria ciliaris, D. sanguinalis, Diospyros virginiana, Distichlis spicata, Dracocephalum parviflorum, Drepanocladus, Dryas crenulata, Echinochloa crus-galli, E. muricata, Echinodorus berteroi, Eclipta prostrata, Elaeagnus angustifolia, Eleocharis erythropoda, E. obtusa, E. palustris, E. quinqueflora, Elodea canadensis, Elymus trachycaulus,*

Epilobium ciliatum, E. hornemannii, Equisetum arvense, E. fluviatile, E. hyemale, Eragrostis cilianensis, E. hypnoides, Erechtites hieracifolia, Eriodictyon californicum, Eriogonum fasciculatum, Eucalyptus camaldulensis, Eutrochium maculatum, Fallopia scandens, Fraxinus pennsylvanica, Galium trifidum, Gamochaeta argyrinea, Gaultheria shallon, Gentianella propinqua, Geranium carolinianum, Geum laciniatum, G. macrophyllum, Glinus lotoides, Glyceria grandis, G. striata, Glycyrrhiza, Gratiola neglecta, Gutierrezia californica, Helenium microcephalum, Helianthus annuus, Heliotropium curassavicum, H. indicum, Herniaria hirsuta, Heterotheca grandiflora, Hippuris montana, H. vulgaris, Hirschfeldia incana, Honckenya peploides, Hordeum arizonicum, H. brachyantherum, H. jubatum, Humulus japonicus, Hydrocotyle, Impatiens capensis, Iodanthus pinnatifidus, Ipomoea pandurata, Iris pseudacorus, Isocoma menziesii, Iva, Juncus alpinoarticulatus, J. alpinus, J. arcticus, J. balticus, J. bufonius, J. effusus, J. nodosus, J. supiniformis, J. tenuis, J. torreyi, Juniperus scopulorum, Kickxia elatine, Koenigia islandica, Laportea canadensis, Larix occidentalis, Leersia lenticularis, L. oryzoides, L. virginica, Lemna minor, Lepidium virginicum, Leptochloa fusca, Leymus mollis, Lindernia dubia, Lonicera interrupta, Ludwigia peploides, Lycopus americanus, L. uniflorus, Lysichiton americanus, Lysimachia nummularia, L. thyrsiflora, Lythrum salicaria, Marrubium vulgare, Marsilea vestita, Medicago lupulina, Melilotus indicus, M. officinalis, Mentha arvensis, Mikania scandens, Mimulus guttatus, M. ringens, Myoporum laetum, Myosotis scorpioides, Myriophyllum sibiricum, Nasturtium officinale, Nepeta cataria, Nyssa, Onoclea sensibilis, Oreostemma alpigenum, Osmorhiza berteroi, Oxalis stricta, Panicum dichotomiflorum, P. virgatum, Paspalum dissectum, P. setaceum, Paxistima myrsinites, Pennisetum glaucum, Penthorum sedoides, Persicaria bicornis, P. coccinea, P. hydropiper, P. hydropiperoides, P. lapathifolia, P. maculosa, P. pensylvanica, P. posumbu, P. punctata, Phalaris arundinacea, Phragmites australis, Physostegia virginiana, Picea engelmannii, P. glauca, P. mariana, P. sitchensis, Pilea pumila, Pinus contorta, P. edulis, P. monticola, P. ponderosa, P. strobiformis, Plagiobothrys scouleri, Plantago lanceolata, P. major, Platanus occidentalis, P. racemosa, Poa nemoralis, P. palustris, P. pratensis, P. wheeleri, Polygonum aviculare, Polypogon monspeliensis, Populus balsamifera, P. deltoides, P. fremontii, P. tremuloides, Portulaca oleracea, Potamogeton gramineus, Potentilla anserina, P. norvegica, P. rivalis, P. supina, Pseudognaphalium luteoalbum, P. stramineum, Pseudoroegneria spicata, Pseudostellaria, Pseudotsuga menziesii, Quercus ×acutidens, Q. agrifolia, Q. gambelii, Q. macrocarpa, Ranunculus acris, R. cymbalaria, R. flammula, R. gmelinii, R. hyperboreus, R. pensylvanicus, R. pusillus, R. sceleratus, Rhododendron tomentosum, Rhus integrifolia, Ricciocarpus natans, Rorippa curvisiliqua, R. sylvestris, Rosa acicularis, Rotala ramosior, Rubus ulmifolius, Rudbeckia laciniata, Rumex acetosella, R. altissimus, R. arcticus, R. crispus, R. maritimus, R. pulcher, R. salicifolius, Sagittaria cuneata, S. latifolia, Salix alaxensis, S.

amygdaloides, S. arbusculoides, S. bebbiana, S. exigua, S. gooddingii, S. hastata, S. interior, S. laevigata, S. lasiolepis, S. niphoclada, S. pseudomonticola, S. sitchensis, Salvia mellifera, Sambucus nigra, Schedonorus arundinaceus, Schoenoplectus acutus, S. californicus, S. pungens, S. tabernaemontani, Scirpus atrovirens, Scleranthus annuus, Scolochloa festucacea, Scutellaria galericulata, S. lateriflora, Senecio congestus, S. hydrophilus, Setaria faberi, Sibbaldia procumbens, Sium suave, Solanum dulcamara, S. ptychanthum, Solidago canadensis, S. gigantea, S. graminifolia, Sonchus arvensis, S. asper, Sorghum halepense, Sparganium eurycarpum, Spergularia salina, Sphagnum, Sphenopholis obtusata, Spiraea douglasii, Stellaria crassifolia, S. graminea, Symphoricarpos, Symphyotrichum ascendens, S. eatonii, S. ontarionis, S. patens, S. spathulatum, S. subulatum, Tamarix chinensis, T. ramosissima, Taraxacum officinale, Taxodium, Thalictrum dasycarpum, T. sparsiflorum, Thlaspi arvense, Toxicodendron diversilobum, T. radicans, Trichophorum pumilum, Triodanis perfoliata, Typha angustifolia, T. domingensis, T. latifolia, Uropappus lindleyi, Urtica dioica, Vaccinium macrocarpon, Vahlodea atropurpurea, Valeriana, Verbena bonariensis, Veronica americana, V. anagallis-aquatica, V. catenata, V. peregrina, V. serpyllifolia, Vitis riparia, Vulpia octoflora, Wyethia amplexicaulis, Xanthium strumarium, Xylococcus bicolor, Zeltnera muehlenbergii.

***Rorippa sessiliflora* (Nutt.) Hitchc.** is an annual or biennial occurring in bottomlands, channels, depressions, dikes, ditches, floodplains, gravel bars, marshes, meadows, mud flats, roadsides, stagnant pools, swales, swamps, and along the margins of bayous, lakes, pools, reservoirs, rivers, sloughs, and streams at elevations of up to 255 m. The plants occur in open sites with exposures of full sunlight to partial shade and can tolerate inundation in shallow waters (to 20 cm deep). The reported substrates include alluvium, clay, gravel, mud, sandy clay, sand, sandy loam, silt, and silty sand. Flowering and fruiting occur from March to October. The flowers are self-compatible and primarily autogamous but are cross-pollinated occasionally by bees, wasps, and flies (Insecta: Hymenoptera: Halictidae; Vespidae; Diptera: Syrphidae). This species can persist for several years as fruits buried (at 15–30 cm depth) in the seed bank, where densities can reach upward of 800 seeds/m^2. However, the seeds will germinate only when exposed on mud or sand substrates. Long-range dispersal occurs by water. The fruits are buoyant and can remain afloat for more than 60 days. The roots are colonized to a slight degree by dark septate endophyte fungi. **Reported associates:** *Acer negundo, A. saccharinum, Alisma, Alliaria petiolata, Amaranthus powellii, A. tuberculatus, Ammannia coccinea, A. robusta, Ampelopsis cordata, Apocynum cannabinum, Artemisia annua, Asclepias amplexicaulis, A. incarnata, Asimina triloba, Betula nigra, Bidens cernuus, B. frondosus, B. tripartitus, Campsis radicans, Cardamine bulbosa, Carex caroliniana, C. cristatella, C. crus-corvi, C. davisii, C. emoryi, C. festucacea, C. granularis, C. grayi, C. grisea, C. jamesii, C. lupulina, C. muskingumensis, C. shortiana, C. squarrosa, C. typhina, Cephalanthus* occidentalis, Chaerophyllum procumbens, Chasmanthium latifolium, Cicuta maculata, Claytonia virginica, Coleataenia longifolia, Conium maculatum, Conoclinium coelestinum, Convolvulus arvensis, Cornus, Cryptotaenia canadensis, Cyperus acuminatus, C. aristatus, C. erythrorhizos, C. ferruginescens, C. odoratus, C. squarrosus, Desmanthus illinoensis, Dichanthelium oligosanthes, Diospyros virginiana, Echinochloa crus-galli, Eclipta prostrata, Eleocharis obtusa, Eleusine indica, Ellisia nyctelea, Eragrostis hypnoides, E. pectinacea, Erigeron philadelphicus, Floerkea proserpinacoides, Forestiera acuminata, Fraxinus pennsylvanica, Galium aparine, Glyceria striata, Gratiola neglecta, Hordeum jubatum, Humulus japonicus, Impatiens capensis, Iodanthus pinnatifidus, Ipomoea lacunosa, I. pandurata, Iris virginica, Juglans nigra, Juncus, Laportea canadensis, Leersia oryzoides, Lepidium virginicum, Leptochloa panicea, Leucospora multifida, Lindernia dubia, Lipocarpha micrantha, Lysimachia ciliata, Matricaria chamomilla, Medicago sativa, Melilotus officinalis, Mertensia virginica, Mimulus, Mollugo verticillata, Morus alba, Muhlenbergia sobolifera, Myosoton aquaticum, Packera glabella, Panicum capillare, P. dichotomiflorum, P. virgatum, Persicaria hydropiperoides, P. lapathifolia, P. maculosa, P. pensylvanica, Phalaris arundinacea, Planera aquatica, Platanus occidentalis, Poa sylvestris, Polygonum ramosissimum, Populus deltoides, Potentilla supina, Quercus bicolor, Q. macrocarpa, Q. palustris, Ranunculus hispidus, R. sceleratus, Rorippa palustris, R. sylvestris, R. teres, Rotala ramosior, Rudbeckia laciniata, Rumex altissimus, R. verticillatus, Salix amygdaloides, S. exigua, S. interior, S. nigra, Sanicula odorata, Saururus cernuus, Schoenoplectus hallii, Scirpus cyperinus, Scutellaria lateriflora, Senecio, Sicyos angulatus, Sium suave, Solanum nigrum, Sphenopholis intermedia, Symphyotrichum lanceolatum, S. ontarionis, Taraxacum officinale, Tilia americana, Toxicodendron radicans, Triodanis perfoliata, Ulmus americana, Urtica dioica, Veronica peregrina, Viola sororia, Vitis cinerea, V. riparia, Vulpia octoflora, Xanthium strumarium.

***Rorippa sphaerocarpa* (A. Gray) Britton** is an annual or biennial, which grows in or on depressions, flats, floodplains, gravel bars, levees, marshes, meadows, mud flats, pools, seeps, slopes, swamps, thickets, and along the margins (often drying) of canals, lakes, ponds, reservoirs, rivers, and streams at elevations of up to 3202 m. The substrates are categorized as cinders, clay, granite, gravel, Kaibab limestone, muck, mud, rhyolite, sand, sandy cobble, sandy loam, silt, and silty clay loam. Flowering occurs from May to August. The ecology of this species is poorly documented otherwise. **Reported associates:** Abies concolor, Acer glabrum, Achillea millefolium, Acmispon utahensis, Agrostis, Alisma triviale, Alnus rugosa, Alopecurus aequalis, A. geniculatus, Ambrosia tomentosa, Amorpha californica, Apocynum, Arctostaphylos patula, A. pringlei, Artemisia biennis, A. frigida, A. ludoviciana, Asclepias, Betula, Bouteloua curtipendula, Brickellia californica, Bromus ciliatus, B. inermis, B. tectorum, Callitriche, Capsella bursa-pastoris, Carex abrupta, C. athrostachya, C. bolanderi, C. brevior, C. heteroneura, C. sheldonii, C. subfusca, C. utriculata, Ceanothus cordulatus, Cercocarpus,

Chara, Chenopodium album, C. rubrum, Chrysolepis sempervirens, Cicuta douglasii, Cirsium arvense, C. vulgare, Conyza canadensis, Cordylanthus rigidus, Coreopsis, Cornus, Cyperus, Dactylis glomerata, Deschampsia cespitosa, D. elongata, Echinochloa, Elatine, Eleocharis acicularis, E. macrostachya, E. palustris, Elodea canadensis, Elymus elymoides, E. glaucus, Epilobium ciliatum, Equisetum, Eragrostis pectinacea, Ericameria nauseosa, Erigeron divergens, E. flagellaris, Eriophyllum confertiflorum, Euphorbia, Festuca, Forestiera pubescens, Fraxinus velutina, Galium trifidum, Geranium richardsonii, Glyceria borealis, G. striata, Gnaphalium exilifolium, Grindelia, Heracleum, Heterotheca villosa, Holcus lanatus, Holodiscus microphyllus, Hordeum jubatum, Horkelia rydbergii, Iris missouriensis, Juncus balticus, J. bufonius, J. effusus, J. ensifolius, J. interior, J. laccatus, J. longistylis, J. nevadensis, J. saximontanus, Juniperus communis, J. monosperma, J. occidentalis, J. osteosperma, J. scopulorum, Lactuca serriola, Lappula occidentalis, Limosella acaulis, Lobelia, Lupinus argenteus, Lysimachia, Maianthemum stellatum, Malva neglecta, Marsilea, Medicago lupulina, Melilotus officinalis, Mentha, Microsteris, Mimulus guttatus, Mirabilis oxybaphoides, Montia chamissoi, Muhlenbergia richardsonis, M. rigens, M. wrightii, Myosurus, Myriophyllum sibiricum, Navarretia, Panicum, Penstemon barbatus, P. caesius, P. labrosus, Pericome caudata, Persicaria amphibia, P. lapathifolia, P. pensylvanica, Phacelia, Phalaris arundinacea, Phleum pratense, Picea engelmannii, Pinus contorta, P. flexilis, P. jeffreyi, P. ponderosa, Plagiobothrys scouleri, Plantago major, Poa annua, P. compressa, P. pratensis, Polygonum aviculare, P. douglasii, Polypogon monspeliensis, Populus angustifolia, P. tremuloides, Portulaca, Potamogeton gramineus, P. nodosus, Potentilla biennis, P. hippiana, P. norvegica, Prunella vulgaris, Prunus, Pseudostellaria, Pseudotsuga menziesii, Pteridium, Quercus gambelii, Q. kelloggii, Ranunculus macounii, Rhus aromatica, Ribes cereum, Rorippa curvipes, Rosa californica, R. woodsii, Rudbeckia laciniata, Rumex crispus, R. fueginus, Sagina saginoides, Salix bebbiana, S. exigua, S. gooddingii, S. lasiolepis, Salvia pachyphylla, Sarcobatus vermiculatus, Schoenoplectus acutus, Scirpus microcarpus, Setaria viridis, Sidalcea, Sisyrinchium, Sparganium angustifolium, Spartium junceum, Sporobolus cryptandrus, Stachys albens, Stellaria umbellata, Stipa, Symphoricarpos mollis, Taraxacum officinale, Thalictrum fendleri, Thermopsis, Trifolium longipes, T. wormskioldii, Typha domingensis, T. latifolia, Urtica dioica, Veratrum californicum, Verbascum thapsus, Verbena, Veronica americana, V. peregrina, Viola sororia, Vitis, Xanthium.

***Rorippa subumbellata* Rollins** is a rare perennial, which is restricted to the beaches and shores of Lake Tahoe at elevations from 1800 to 2000 m. The exposures are open (cover <3%) and receive full sunlight. The sites are characterized by fluctuating water levels and substrates that comprise coarse or shifting sand. The plants do not tolerate permanent inundation, but exhibit their best growth on wet sand where the water table remains 5 cm below the surface. The habitats experience summer temperatures below 18.3°C and winter temperatures above

−3.3°C, with 70–120 frost-free days per year. Flowering occurs continuously from May to October. The flowers are visited by bees and small flies (Insecta: Hymenoptera; Diptera), but the breeding system of this species has not yet been elucidated. Seed germination is high, approaching 100% under laboratory conditions. Temperature and light are important factors for germination, which can reach nearly 80% at a 24°C/10°C temperature regime under constant illumination. Higher germination rates (to 88%) have been achieved by treating the seeds with cytokinin [0.5 mg/L 6-benzylaminopurine (BAP)] to break dormancy. Cold stratification (5°C) beyond a week decreases germination rates. Seeds that are stored dry for up to 4 years maintain high germinability and are less sensitive to low temperature effects. The seeds are buoyant (up to 75% remaining afloat after 1 month), but their germination rates decline in water. Some seeds will germinate while floating. Vegetative reproduction occurs from horizontal rhizomes. Isozyme studies (23 loci surveyed) indicate that the populations are depauperate genetically (proportion of polymorphic loci = 13%; mean alleles/locus = 1.13; observed heterozygosity = 0.0003). Genetic analyses using microsatellite markers have also shown that there are fairly widespread subpopulational differentiations among the occurrences around the Lake Tahoe shoreline. The plants are threatened by human disturbance and conditions that promote prolonged high water levels.

Reported associates: *Achillea millefolium, Agrostis scabra, Alnus tenuifolia, Arnica chamissonis, Bromus hordeaceus, Carex douglasii, Chrysothamnus viscidiflorus, Eleocharis, Juncus balticus, J. nevadensis, Lepidium virginicum, Lupinus lepidus, Mentha spicata, Mimulus primuloides, Myriophyllum spicatum, Phacelia hastata, Poa pratensis, Rorippa curvisiliqua, Rumex acetosella, R. crispus, Salix exigua, Sulcaria badia, Symphyotrichum spathulatum, Texosporium sanctijacobi, Verbascum thapsus.*

Rorippa sylvestris (**L.**) **Besser** is a nonindigenous perennial, which inhabits beaches, borrow pits, bottomlands, ditches, flats, floodplains, gardens, gravel bars, gravel pits, lawns, marshes, meadows, mudflats, puddles, roadsides, scrub, sloughs, swamps, thickets, washes, waste places, and the margins of lakes, ponds, pools, rivers, and streams at elevations of up to 2438 m. The plants occur frequently in disturbed or agricultural sites. They can tolerate prolonged periods of complete submergence (at least 20 days) without noticeable negative effects. Site exposures can vary from full sun to shade. The reported substrates include clay, cobble, Coconino sandstone, gravel, loam, mud, organic muck, rock, rocky limestone, sand, sandy clay, sandy gravel, sandy loam, sandy mud, and stones. Flowering and fruiting occur from May to September. The plants are strongly self-incompatible (SI) and are highly outcrossed by nonspecialist pollinators. Plants in the field often exhibit low seed set due to an insufficient pool of compatible genotypes to overcome their SI reproductive system. The introduced North American plants rarely produce seed and their potential pollinators have not been elucidated. Vegetative reproduction occurs by means of stem fragments or by fragments originating from the root runners, upon which are formed numerous adventitious

shoots. The plants have a sprawling habit and often are mat forming. **Reported associates (North America):** *Abies, Abutilon theophrasti, Acalypha rhomboidea, Acer negundo, A. saccharinum, Aesculus glabra, Alliaria petiolata, Allium canadense, Alopecurus aequalis, Alternanthera, Amaranthus tuberculatus, Ammannia coccinea, Anemone canadensis, Apocynum cannabinum, Arenaria serpyllifolia, Artemisia biennis, A. rigida, Asclepias incarnata, Astragalus humistratus, Barbarea vulgaris, Bidens cernuus, Boehmeria cylindrica, Bouteloua simplex, Caltha palustris, Calystegia sepium, Campanula americana, Cardamine oligosperma, Carex athrostachya, C. blanda, C. grayi, C. siccata, Celtis occidentalis, Chenopodium album, Cicuta bulbifera, Cirsium vulgare, Convolvulus arvensis, Conyza canadensis, Coreopsis, Cornus sericea, Cynodon dactylon, Cyperus acuminatus, C. esculentus, C. squarrosus, Dracocephalum parviflorum, Elaeagnus angustifolia, Eleocharis acicularis, Elymus repens, Eragrostis hypnoides, Erigeron, Euonymus alatus, Eupatorium serotinum, Forestiera acuminata, Fraxinus pennsylvanica, Geum, Glechoma hederacea, Glyceria grandis, Hackelia floribunda, Helianthus, Hibiscus laevis, Hordeum jubatum, Hydrocotyle, Hydrophyllum virginianum, Hymenoxys subintegra, Impatiens capensis, Iva annua, I. axillaris, Justicia, Lactuca serriola, Larix, Leersia oryzoides, L. virginica, Leymus cinereus, Lipocarpha micrantha, Lupinus kingii, Lycopus americanus, Lysimachia nummularia, Matteuccia struthiopteris, Medicago lupulina, Melilotus officinalis, Mertensia, Myosotis scorpioides, Myosoton aquaticum, Napaea dioica, Nasturtium officinale, Oxalis stricta, Panicum capillare, P. dichotomiflorum, P. repens, Pascopyrum smithii, Paspalum dilatatum, P. notatum, Penstemon linarioides, Penthorum sedoides, Persicaria amphibia, P. pensylvanica, Phalaris arundinacea, Phyla lanceolata, Physostegia virginiana, Pinus contorta, P. ponderosa, Plantago major, P. rugelii, Poa compressa, P. palustris, P. pratensis, Polygonum douglasii, Populus angustifolia, P. deltoides, Portulaca oleracea, Potamogeton, Potentilla anserina, Pseudoroegneria spicata, Quercus bicolor, Ranunculus hispidus, R. sceleratus, Rhamnus cathartica, Ribes americanum, Rorippa curvipes, R. curvisiliqua, R. palustris, R. sessiliflora, Rudbeckia laciniata, Rumex altissimus, R. crispus, R. maritimus, R. verticillatus, Sagittaria montevidensis, Salix exigua, S. interior, S. nigra, Setaria faberi, Sherardia arvensis, Sida spinosa, Sium suave, Sparganium angustifolium, S. eurycarpum, Spermacoce glabra, Stenotaphrum secundatum, Taraxacum officinale, Toxicodendron radicans, Trifolium, Triteleia grandiflora, Typha, Ulmus americana, Urtica dioica, Verbascum thapsus, Verbena bracteata, V. urticifolia, Veronica arvensis, V. peregrina, Vitis riparia, Xanthium strumarium.*

Rorippa teres (**Michx.**) **Stuckey** is an annual or biennial, which is found in clay pans, depressions, ditches, flatwoods, floodplains, gardens, glades, hammocks, lawns, marshes, meadows, prairies, roadsides, sand bars, savannas, swales, swamps, waste areas, and along the margins of canals, lakes, ponds, streams, and thickets at elevations of up to 600 m. The plants often occur in disturbed sites and will tolerate exposure of full sunlight to shade. The substrates are described as Dania (lithic medisaprists), gravel, humus, leaf mold, loam, marl, muck, mud, peat, sand, and sandy peat. Flowering extends from December to May. Other specific details on the ecology of this species are scarce. **Reported associates:** *Acer, Agrostis hyemalis, Amaranthus viridis, Argemone, Betula nigra, Callitriche terrestris, Celtis occidentalis, Chaerophyllum, Cladium jamaicense, Conoclinium coelestinum, Corydalis, Cynodon dactylon, Cyperus acuminatus, Digitaria ciliaris, Diospyros virginiana, Ditrysinia fruticosa, Echinochloa crus-galli, Eleocharis palustris, Eleusine indica, Eragrostis hypnoides, Eriochloa michauxii, Eryngium prostratum, Eupatorium capillifolium, Euphorbia, Euploca procumbens, Eustachys petraea, Ficus, Forestiera acuminata, Fraxinus caroliniana, Fuirena pumila, Geranium, Gnaphalium, Helenium autumnale, Houstonia procumbens, Hydrocotyle, Ilex decidua, Juniperus virginiana, Lepidium, Linaria, Lindernia dubia, Lobelia feayana, Ludwigia, Marsilea vestita, Melaleuca quinquenervia, Melilotus albus, Mollugo verticillata, Muhlenbergia capillaris, Nama jamaicensis, N. stenocarpa, Parietaria, Pentodon pentandrus, Phyllanthus urinaria, Pinus palustris, Planera aquatica, Pluchea, Polypremum procumbens, Portulaca, Rorippa sessiliflora, Rudbeckia mollis, Sabal palmetto, Salix, Samolus valerandi, Schoenoplectus saximontanus, Senna, Serenoa repens, Solanum, Sonchus, Stachys crenata, Taxodium distichum, Triodanis, Ulmus alata, U. americana, U. crassifolia, Youngia japonica.*

Use by wildlife: *Rorippa aquatica* provides habitat for invertebrates and fish (Vertebrata: Teleostei). It is a host plant for moth larvae (Insecta: Lepidoptera: Geometridae: *Xanthorhoe designata*). The flowers of *R. calycina* and *R. columbiae* attract small beetles (Insecta: Coleoptera), which might function as pollinators. *Rorippa curvisiliqua* is a host plant for moth larvae (Insecta: Lepidoptera: Pieridae: *Pontia protodice*). The flowers of *R. subumbellata* are visited by bees (Insecta: Hymenoptera: Andrenidae; Megachilidae; Tiphiidae) and flies (Insecta: Diptera: Dolichopodidae; Muscidae; Syriphidae). *Rorippa teres* is a host to parasitic wasps (Insecta: Hymenoptera).

Economic importance: food: The foliage and sprouts of *Rorippa palustris* can be eaten when cooked or in salads as a suitable substitute for watercress (*Nasturtium*). The plants contain vitamin C and were eaten to prevent scurvy. They are added to fish soup as a condiment by the Inuktitut Eskimos. *Rorippa curvisiliqua* was eaten by the Paiute people; **medicinal:** The Navajo used an infusion of *R. curvipes* as a postpartum tonic. Iroquois mothers used a decoction of *Rorippa sylvestris* to treat fevers; **cultivation:** *Rorippa aquatica* is grown as an aquarium plant for its interesting, dissected foliage; **misc. products:** The heterophyllous *R. aquatica* has been used as a genetic model for studying leaf development. The Ramah tribe made a ceremonial eye wash from *Rorippa palustris*. The Navajo and Ramah tribes used *Rorippa teres* as a food for sheep (Mammalia: Bovidae: *Ovis aries*); **weeds:** Prescott (1980) stated that *R. aquatica* often is weedy and crowds out more desirable plants; however, that assessment

is a gross misconception (perhaps based on confusion with another plant) given the rare, declining status of the species. *Rorippa columbiae* has been reported as a weed of alfalfa (*Medicago sativa*) fields. *Rorippa sessiliflora* is a common garden weed. *Rorippa sylvestris* and *R. teres* are weeds of agricultural fields and gardens; **nonindigenous species:** *Rorippa calycina* reportedly was introduced to the Canadian Northwest Territories; however, that claim is unsubstantiated and most likely erroneous. *Rorippa sylvestris* was introduced from Europe before 1818, probably in shipping ballast. It has been transported throughout North America as seed in contaminated nursery stock. It arrived to California via contaminated *Delphinium* material that originated from Holland.

Systematics: Preliminary molecular phylogenetic studies (Figure 4.66) indicated that *Rorippa* was most closely related to the monotypic *Neobeckia* [=*Rorippa aquatica*] and somewhat more distantly to *Armoracia* and *Barbarea*. Although a close relationship of *Neobeckia* and *Rorippa* was evident, the former was retained as distinct because of several distinctive morphological features (e.g., white petals, incomplete fruit replum). However, a more comprehensive phylogenetic study of *Rorippa* taxa (using cpDNA sequence data) showed subsequently that *Neobeckia* was nested within *Rorippa* (at least by plastid DNA data), which has led its merger with *Rorippa*. Other genera (e.g., *Armoracia, Nasturtium*) have also had a history of taxonomic interchanges with *Rorippa*, although a number of phylogenetic analyses have maintained their distinctness from *Rorippa* and implicate *Sisymbrella* as its sister genus. One species name (*R. coloradensis*) has been transferred to *Sinapis* (as a synonym of *S. alba*). *Rorippa* appears to be monophyletic as long as *Nasturtium* is excluded and *Neobeckia* is included. The North American species usually are divided into two sections, section *Rorippa* (mainly rosette-forming annuals) and section *Sinuatae* (vittate perennials). Of the OBL aquatic species, *Rorippa calycina, R. columbiae,* and *R. subumbellata* are regarded as closely related and are placed together with *R. coloradensis* in the latter section; whereas, the remaining species are placed within the former section. Several phylogenetic hypotheses have been presented for *Rorippa* using cpDNA sequence data analysis. Similar results have been shown in a study of 25 species (*trnL* intron, *trnL/F* data) and 15 species (*trnL* intron, *trnG–trnM, psbC–trnS* data), with a few species (e.g., *R. amphibia, R. sylvestris*) resolving in different portions of a phylogeny presumably as a consequence of hybridization. Both studies show a close relationship among *R. amphibia, R. palustris,* and *R. sylvestris.* Only the latter study included *R. aquatica*, which resolved most closely either with the North American *R. sessiliflora* or the European *R. pyrenaica* depending on the method of analysis (neighbor joining vs. maximum likelihood). Overall, the molecular data indicate that sectional limits within *Rorippa* are defined poorly (especially section *Rorippa*) and warrant a reconsideration. *Rorippa curvisiliqua* is extremely variable with up to seven varieties recognized in some treatments. Although it was once believed to be related to *R. curvipes* and *R. sinuata*, molecular data place *R. curvisiliqua* distant from either species but related closely to the South American

R. philippiana (Figure 4.72). Formerly regarded as synonymous with the Old World *R. islandica*, *R. palustris* latterly was considered to be distinct and related to *R. curvipes*. Molecular data have confirmed the distinctness of *R. islandica* and *R. palustris* but indicate the relationship of *R. curvipes* to be with *R. islandica* instead of *R. palustris* from which it is remote (e.g., Figure 4.72). *Rorippa alpina* sometimes is regarded as distinct, otherwise as a variety of *R. curvipes*. Several subspecies and varieties of the variable *R. palustris* have been recognized and require further evaluation. More comprehensive phylogenetic studies are needed to test other hypotheses such as the presumed close relationship between *R. columbiae* and *R. subumbellata* or the proposed relationship of *R. teres* to *R. mexicana*, or *R. portoricensis*; although once assumed to have a close association, *R. teres* is only remotely related to *R. sessiliflora*. It will also be necessary to include data from nuclear DNA markers in these investigations before any definitive conclusions can be reached. It is possible that some taxa (e.g., *R. aquatica*) may be of allopolyploid origin, whereby the trees from cpDNA would indicate only the maternal parent (see also remarks under cytology later). With hybridization fairly common in *Rorippa*, further analyses in this direction are necessary. The chromosomal base number of *Rorippa* is $x = 8$. *Rorippa calycina*, *R. curvipes*, *R. curvisiliqua*, and *R. sessiliflora* are diploid ($2n = 16$). *Rorippa palustris* is tetraploid ($2n = 32$). *Rorippa sylvestris* is variable chromosomally with tetraploid ($2n = 32$), pentaploid ($2n = 40$), and hexaploid ($2n = 48$) counts reported. *Rorippa*

aquatica presents a special case. Both diploid ($2n = 16$) and triploid ($2n = 24$) cytotypes have been recorded for this species, which correspond to fertile and sterile material, respectively. However, recent studies have shown that some material (thought to have been triploid) has an anomalous count of $2n = 30$ (which differs from the base series in all other *Rorippa*). Because the differing counts appear to be accurate, it might be that some unusual cytotypes are generated as the result of somatic chromosomal mutations. A possible allopolyploid origin of *R. aquatica* should also be investigated. Because putatively diploid material has not been analyzed phylogenetically; it could be that the morphologically dissimilar *Neobeckia* actually is a distinct sister genus, with the triploid plants representing allopolyploid hybrids involving a *Rorippa* species. Several instances of hybridization have been described from Europe that involve species found in North America. Bidirectional introgressive hybridization has been documented to occur between *Rorippa sylvestris* and *R. amphibia*. The intermediate hybrids (known taxonomically as *R. ×anceps*) do not persist. Unidirectional introgression from *R. palustris* to *R. amphibia* has been documented using nuclear and plastid molecular markers.

Comments: *Rorippa aquatica* is endemic to eastern North America. *Rorippa calycina* is a rare North American endemic with few, scattered localities in North Dakota, Montana, Wyoming, and the Canadian Northwest Territories. *Rorippa columbiae*, *R. curvipes*, and *R. curvisiliqua* are distributed across western North America. The North American distribution of *R. sessiliflora* is principally eastern, that of *R. sphaerocarpa* southwestern, and that of *R. teres* southeastern. *Rorippa palustris* and *R. sylvestris* are widespread in North America, the latter being nonindigenous. *Rorippa subumbellata* is a rare endemic of the Lake Tahoe region.

References: Akman et al., 2012; Al-Shehbaz, 2010d; Bharathan et al., 2002; Bleeker & Hurka, 2001; Bleeker et al., 2002; Borman et al., 1997; Christy, 2013; Couvreur et al., 2010; Crone & Gehring, 1998; Evans, 1979; Fertig & Welp, 1998; Gabel & Les, 2001; Gilreath & Gilreath, 1983; Ingolia et al., 2008; Jarolímová, 1998, 2005; Jonsell, 1968; Judziewicz & Nekola, 1997; Koch et al., 2001; Lake et al., 2000; La Rue, 1943; Les, 1994; Les et al., 1995; Marie-Victorin, 1930; Mulligan & Porsild, 1966; Nakayama et al., 2014; Parrish & Bazzaz, 1979; Pavlik et al., 2002; Pence et al., 2006; Prescott, 1980; Raynal & Bazzaz, 1973; Rollins, 1993; Stanton & Pavlik, 2009; Stevens et al., 2010; Strother et al., 2012; Stuckey, 1966a,b; 1972; Sweeney & Price, 2000; Tooker & Hanks, 2000; Van der Valk, 2013; Whyte & Cain, 1981.

8. *Subularia*

Awl-wort; subulaire

Etymology: from the Latin *subula* meaning "awl"

Synonyms: none

Distribution: global: Africa; Asia; Europe; North America; **North America:** northern

Diversity: global: 2 species; **North America:** 1 species

Indicators (USA): OBL: *Subularia aquatica*

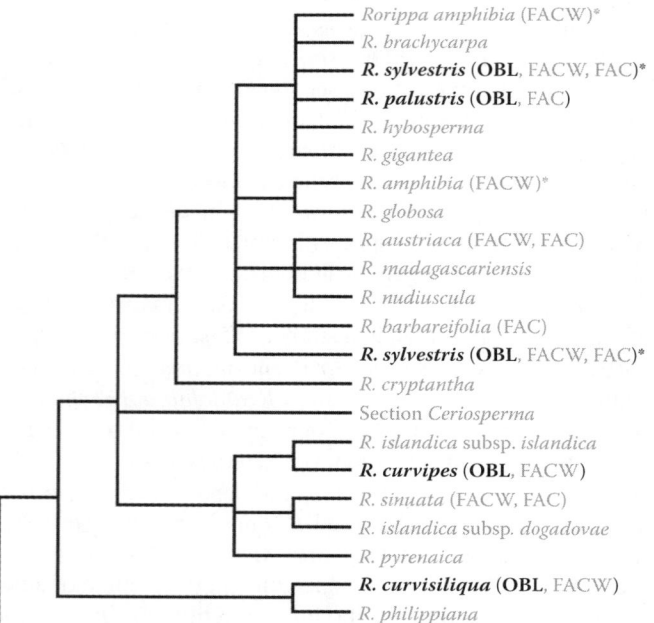

FIGURE 4.72 Phylogenetic relationships in *Rorippa* as indicated by analysis of cpDNA sequence data. Taxa designated as OBL North American indicators are shown in bold (indicator status in parentheses). Multiple sequences for two species (designated by asterisks) are attributed as the result of introgressive hybridization. (Adapted from Bleeker, W. et al., *Plant Biol.*, 4, 104–111, 2002.)

Habitat: freshwater; lacustrine, palustrine; **pH:** 5.3–7.1; **depth:** <1.5 m; **life-form(s):** emergent (herb), submersed (rosulate)

Key morphology: plants rosulate; leaves (to 5 cm) 10–20 in a rosette, subulate, terete to flattened dorsiventrally, with numerous, irregular lacunae, apices acute; racemes (to 15 cm) with 2–12 flowers; petals (to 1.5 mm) white; siliques (to 5 mm) subglobose, inflated

Life history: duration: annual (fruit/seeds); **asexual reproduction:** none; **pollination:** cleistogamous (self); **sexual condition:** hermaphroditic; **fruit:** siliques (common); **local dispersal:** fruit/seeds (water); **long-distance dispersal:** fruit/seeds (water)

Imperilment: (1) *Subularia aquatica* [G5]; SH (VT); S1 (CO, MI); S2 (MB, MN, WY); S3 (ME, LB, NB, QC, ON, SK, YT)

Ecology: general: *Subularia* comprises just two species, with one North American representative. Both species are annuals of small stature, which have a strong affinity for wet habitats. The African *S. monticola* inhabits wet, high elevation sites typically as a marginal plant and rarely grows under submersed conditions. In contrast, the North American (and Eurasian) *S. aquatica* grows primarily as a submersed aquatic. The habitats of each species usually are associated with clear, acidic, oligotrophic waters. Both chasmogamous and cleistogamous flowers can be produced, the latter autogamous and produced by submersed plants.

Subularia aquatica **L.** grows either completely submersed in the littoral zone of nutrient-poor water bodies or as an emergent along wet shorelines. Most plants are found submersed because stranded shoreline plants are dislodged easily from the substrate. The habitats include beaches, bogs, bottoms, depressions, channels, flats, lakes, marshes, meadows, mires, mudflats, ponds, pools, salt marshes, spits, tundra, and the shallow margins or shores of lakes, ponds, pools, rivers, streams, and tarns at elevations of up to 3181 m. The substrates are categorized as gravel, gravelly silt, muck, mud, muddy gravel, muddy sand, rock, rocky organic muck, and sandy silt. The plants generally grow in shallow water depths of 15–45 cm (but up to 1.5 m) in the erosional or neutral shore zone. The waters (tidal or nontidal) typically are acidic (pH: 5.3–7.1), oligotrophic, with low conductivity (60–82 μS), low alkalinity, and high clarity (e.g., secchi depth >1.7 m). Flowering occurs from July to October. The plants will produce autogamous, cleistogamous flowers when submersed and also self-pollinating chasmogamous flowers, which develop when emergent or stranded. The seeds are dispersed abiotically (in water) along with the floating, inflated fruits. Little information on the seed ecology is available, but the seeds are described as easy to germinate and the species is believed to persist from a well-developed seed bank. The roots have been found to contain a small number of associated fungi. The submersed ecological associates typically represent other rosulate species, which are indicators of oligotrophic, soft-water lakes. **Reported associates:** *Callitriche anceps, C. palustris, Carex lenticularis, C. limosa, C. rostrata, Crassula aquatica, Elatine americana, E. triandra, Eleocharis acicularis, E. palustris, Elodea canadensis, Equisetum fluviatile, Eriocaulon aquaticum, Eriophorum angustifolium, Gnaphalium, Gratiola aurea, Hippuris vulgaris, Isoetes bolanderi, I. occidentalis, Juncus militaris, J. pelocarpus, Koenigia islandica, Littorella uniflora, Lobelia dortmanna, Myriophyllum tenellum, Nuphar polysepala, Persicaria amphibia, P. pensylvanica, Potamogeton berchtoldii, P. epihydrus, P. gramineus, P. perfoliatus, P. praelongus, P. robbinsii, Ranunculus flammula, R. trichophyllus, Sagittaria cuneata, S. graminea, Sparganium angustifolium, Vaccinium uliginosum, Wahlenbergia.*

Use by wildlife: Asian populations of *S. aquatica* are eaten by amphipods (Crustacea: Amphipoda: Micruropodidae: *Gmelinoides fasciatus*).

Economic importance: food: none; **medicinal:** none; **cultivation:** *Subularia aquatica* has minor value as an unusual, ornamental aquarium specimen; **misc. products:** none; **weeds:** none; **nonindigenous species:** none

Systematics: Phylogenetic relationships of *Subularia* in the Brassicaceae remain to be elucidated. The genus had been assigned to tribe Lepidieae, which presumably indicated a more distant relationship to other aquatic mustards such as *Cardamine, Nasturtium,* and *Rorippa,* (tribe Arabideae). The distinctive *Subularia,* with only two species, arguably is monophyletic; however, both species have not yet been included simultaneously in a phylogenetic analysis to confirm that assumption. Moreover, the highly anomalous positions of *Subularia monticola* near *Idahoa* (tribe Aphragmeae) in some phylogenetic analyses [nuclear ribosomal internal transcribed spacer (nrITS) sequence data], but far remote from the latter and associated instead with *Tauschenia* (tribe Isatideae) in other analyses (combined cpDNA, mtDNA, and nuclear DNA data), have not helped to settle questions regarding relationships of the genus. Currently, *Subularia* remains unassigned to a tribe with some authors suggesting the placement within its own tribe (i.e., Subularieae). *Subularia aquatica* comprises two varieties with *S. aquatica* var. *aquatica* distributed in northern Eurasia, and North American plants assigned to *S. aquatica* var. *americana*. The North American plants differ by their persistent sepals and more obovate fruits. The base chromosome number of *Subularia* is uncertain (perhaps $x = 7$) and the chromosome numbers need to be verified by additional counts. Numbers reported for *S. aquatica* include $2n = 28$, 30 (North America) and $2n = 36$ (Europe). *Subularia monticola* reportedly is $2n = 28$. In any case, the species appear to be polyploids (presumably tetraploid).

Comments: *Subularia aquatica* is distributed in northern North America or at higher elevations in the western Rocky Mountains. The species is relatively rare throughout its range.

References: Al-Shehbaz, 2010e, 2012; Barkov & Kurashov, 2011; Cook, 1996; Couvreur et al., 2010; Glück, 1924; Kohout et al., 2012; Lacoul & Freedman, 2006a; Mulligan & Calder, 1964; Nelson & Harmon, 1993; Pfauth & Sytsma, 2005; Rand, 1899; Rollins, 1993; Voss, 1985; Warwick et al., 2010.

Family 3: Limnanthaceae [2]

Limnanthaceae are a small family of annual, herbaceous plants comprising only two genera and 5–8 species. All of the species exhibit some degree of affinity for wet habitats.

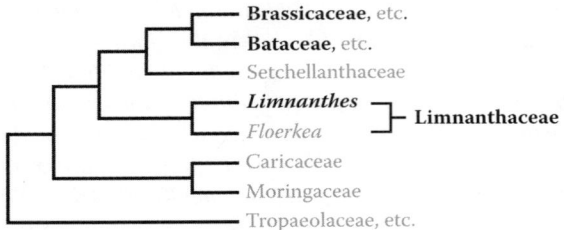

FIGURE 4.73 The phylogenetic position of Limnanthaceae as indicated by molecular cladistic analyses. Some data (e.g., Soltis et al., 2000) show the positions of Limnanthaceae and Setchellanthaceae to be reversed. Taxa containing OBL North America indicators are shown in bold. A general adaptation to wet habitats appears to have arisen once in Limnanthaceae given that all of the species have some affinity for such conditions. (Adapted from Rodman, J.E. et al., *Amer. J. Bot.* 85, 997–1006, 1998.)

Previous to the advent of molecular studies, the placement of Limnanthaceae was uncertain, and it often was assumed to be related to Geraniaceae. However, molecular data have clarified that the family occupies an intermediate phylogenetic position as a clade within Brassicales (Figures 4.64 and 4.73; Judd et al, 2016). A number of molecular cladistic analyses have also corroborated the monophyly of both genera, confirming an association that has long been upheld by traditional taxonomic treatments. However, comprehensive phylogenetic studies (Meyers et al., 2010) also indicate that several *Limnanthes* species represent fewer polymorphic taxa, with as few as four actual species in the genus.

The flowers are bisexual or gynodioecious and include both autogamous and outcrossing breeding systems. The fruits are schizocarps.

The group is of little economic importance, with a few cultivars distributed in the genus *Limnanthes*. *Floerkea* sometimes is eaten as a salad herb. The flower of *Floerkea proserpinacoides* serves as the logo for the *Flora of North America* project.

The family is endemic to temperate North America. Two North American genera contained OBL indicators in the 1996 list; however, the 2013 list excluded *Floerkea*, leaving only one OBL genus:

1. ***Limnanthes*** R. Br.

1. *Limnanthes*

Meadowfoam; limnanthe
Etymology: from the Greek *limne anthos* meaning "marsh flower"
Synonyms: *Floerkea* (in part)
Distribution: global: North America; **North America:** western
Diversity: global: 7 species; **North America:** 7 species
Indicators (USA): OBL: *Limnanthes bakeri, L. douglasii, L. floccosa, L. macounii, L. montana, L. vinculans*
Habitat: freshwater; palustrine; **pH:** 6.4–7.2; **depth:** <1 m; **life-form(s):** emergent (herb)

Key morphology: stems (to 50 cm) erect, foliage glabrous to villous; leaves (to 25 cm) alternate, pinnate/bipinnate (to 13 leaflets); flowers solitary, axillary; petals (to 18 mm) 5, white, yellow, or white with a yellow base, sometimes aging to pink; stamens 10; carpels 5, fused basally (appearing distinct); style gynobasic; nutlets 1-seeded, smooth to tuberculate
Life history: duration: winter annual (fruit/seeds); **asexual reproduction:** none; **pollination:** insect, self; **sexual condition:** gynodioecious, hermaphroditic; **fruit:** schizocarps (common); **local dispersal:** seeds (water); **long-distance dispersal:** seeds
Imperilment: (1) *Limnanthes bakeri* [G1]; S1 (CA); (2) *L. douglasii* [G4]; S1, S2 (CA); (3) *L. floccosa* [G4]; S1, S3 (CA); S1, S2 (OR); (4) *L. macounii* [G2]; S2 (BC); (5) *L. montana* [G3]; (6) *L. vinculans* [G1]; S1 (CA)
Ecology: general: *Limnanthes* species are a distinctive element of freshwater vernal pools and other western North American habitats that are characterized by highly ephemeral water. All of the species are either obligate or FACW indicators; however, recent taxonomic changes have necessitated some modifications. *Limnanthes pumila* (OBL in the 2013 list) has been merged with *L. floccosa* and is not recognized here as a separate species. On the contrary, *L. macounii* (excluded from the 2013 list) is treated here as an OBL species because of its vernal pool habitats. These short-lived winter annuals survive in open sites, which lack any significant cover by shrubs or grasses. The onset of flowering occurs in the spring. Specialist bees (Insecta: Hymenoptera) often are necessary to achieve adequate pollination. Studies using fluorescent markers indicate that most of the pollen is dispersed within 5 m of *Limnanthes* plants, but occasionally can be transported up to 80 m. The fruits are deeply lobed because of the gynobasic style and develop into nutlet-like (single-seeded) mericarps. The nutlet surfaces range from smooth to highly tuberculate (sometimes polymorphic within species); the most tuberculate surfaces provide the highest surface area and will float for the longest period of time due to the air that becomes trapped within the surface irregularities. Seed dispersal in most species is of very limited range (mainly within pools), although most of the nutlets can remain afloat for more than 8 h. Consequently, dispersal toward the periphery of pools occurs in wetter years; whereas, dispersal is concentrated more centrally during the drier years. The seeds of some species are known to have secondary dormancy and most species develop seed banks, which facilitate their persistence at sites through periods of unfavorable conditions. Seed germination usually occurs during late autumn. All *Limnanthes* seeds contain phytoecdysteroids that are believed to provide some protection against herbivory. Most of the species are threatened by habitat destruction, which eliminates the ecologically essential ephemeral water conditions necessary for their persistence and survival.

***Limnanthes bakeri* J.T. Howell** inhabits ditches, meadows, roadsides, swales, vernal pools, and the margins of pools at elevations of up to 502 m. Site exposures usually are characterized by full sun but occasionally by shade. The substrates are described as clay or mud. Flowering occurs in April to May. The flowers are highly self-compatible with stamens and styles that come in close contact within the funnel-like corolla, which results in a high level of self-pollination (outcrossing rates <50%). The flowers are spontaneously autogamous in the absence of pollinators. Although the plants tend to be distributed narrowly, they are clumped and often abundant at the sites where they do occur. **Reported associates:** *Camassia quamash, Carex, Juncus, Lolium perenne, Plagiobothrys, Pleuropogon, Ranunculus orthorhynchus.*

Limnanthes douglasii **R. Br.** occurs in depressions, ditches, gravel bars, meadows, prairies, roadsides, seeps, sinks, streams (ephemeral), swales, vernal pools, and along the margins of streams at elevations of up to 1829 m. The exposures are open sites that receive full sunlight. The substrates are described as alkali and include adobe, Capay–Clear Lake association, clay, Corning gravelly loam, gravel, loamy clay, muck, mud, Pescadero clay, rock, sandy loam, schist, stony Tuscan loam over hardpan, Tuscan mudflow, Tuscan series, and volcanics. Flowering and fruiting occur from March to July. *Limnanthes douglasii* was one of the species whose reproductive biology was studied by Charles Darwin. The species includes hermaphroditic and gynodioecious (individuals with bisexual and female flowers) populations, the latter resulting from nucleo-cytoplasmic male sterility. The flowers normally are protandrous and only weakly self-compatible, with outcrossing rates approaching 80%–95%; however, gynodioecious hermaphrodites are characterized by higher selfing rates than are the strict hermaphrodites. Gynodioecious populations also exhibit greater inbreeding depression and lower heterozygosity. The amino acid composition of female and hermaphroditic flowers is essentially identical. Pollination is facilitated by bees (Hymenoptera: Andrenidae: *Andrena pulverea, Panurginus occidentalis*; Apidae: *Apis mellifera*), and other insects (Insecta). The bees sometimes will visit other species during the morning, but then forage for nectar on *L. douglasii* during the afternoon. Seed set can be reduced substantially by heterospecific pollen deposition when congeners occur close by. The plants are especially pollen limited during the spring due to pollinator competition with co-occurring *Lasthenia* species. The nutlets can float (some for upward of 100 h) and are dispersed readily by water. The populations sometimes are suppressed by invasive growths of *Taeniatherum caput-medusae*. Once established, the plants are vulnerable to livestock grazing and trampling. However, there is also evidence that grazing can greatly increase the numbers of these plants by removing the competing vegetation. **Reported associates:** *Agoseris grandiflora, Allocarya stipitata, Alopecurus howellii, A. saccatus, Amsinckia, Anthoxanthum odoratum, Avena, Blennosperma nanum, Boisduvalia glabella, Brassica rapa, Briza, Brodiaea elegans, B. hyancinthina, Bromus hordeaceus, B. madritensis, Callitriche marginata, C. trochlearis, Camassia, Capsella bursa-pastoris, Carex densa, C. unilateralis, Cerastium glomeratum, Cotula coronopifolia, Cryptantha, Cynosurus cristatus, Cyperus eragrostis, Deschampsia danthonoides, Downingia bicornuta, D. concolor, D. cuspidata, Eleocharis acicularis, E. macrostachya,*

E. quinqueflora, Epilobium minutum, Erodium, Eryngium aristulatum, E. vaseyi, Eschscholzia californica, Evax caulescens, Geranium, Gratiola ebracteata, Hirschfeldia incana, Hordeum marinum, Howellia aquatilis, Isoetes howellii, I. orcuttii, Juncus bufonius, J. effusus, J. uncialis, J. xiphioides, Lasthenia chrysostoma, L. fremontii, Layia fremontii, L. platyglossa, Lepidium didymum, Lilaea scilloides, Limnanthes alba, L. floccosa, L. vinculans, Linanthus, Lolium multiforum, Lupinus bicolor, Lythrum hyssopifolia, Marsilea, Medicago hispida, Mimulus guttatus, M. tricolor, Minuartia douglasii, Myosurus, Nasturtium officinale, Navarretia intertexta, N. leucocephala, Nemophila menziesii, Noccaea fendleri, Oenanthe sarmentosa, Orthocarpus erianthus, Plagiobothrys bracteatus, P. stipitatus, Platystemon californicus, Pleuropogon californicus, Poa annua, P. secunda, Pogogyne ziziphoroides, Polypogon maritimus, P. monspeliensis, Potentilla anserina, Psilocarphus brevissimus, Quercus kelloggii, Ranunculus aquatilus, R. bonariensis, R. californicus, Romanzoffia, Rumex crispus, Spergula arvensis, Stellaria littoralis, Taeniatherum caput-medusae, Trifolium depauperatum, T. variegatum, T. wormskioldii, Triphysaria, Triteleia laxa, Veronica persica, Viola, Zeltnera muehlenbergii.

Limnanthes floccosa **Howell** inhabits balds, depressions, flats, gullies, meadows, pools, prairies, roadsides, ruts, savannas, slopes (to 8%), swales (ephemeral), woodlands, vernal pools, and the margins of ephemeral streams at elevations of up to 2039 m. The populations typically develop in open, treeless areas with exposures of full sunlight. The substrates are characterized as adobe, Agate–Winlow complex, clay, clay loam, cobbles, congloritic lahar, Corning gravelly loam, gravel, gravelly clay loam, Laguna formation, Redding–Igo complex, rock, rocky clay, sand, sandstone, sandy loam, stones, stony volcanics, Tuscan–Anita complex, Tuscan flows, and Tuscan stony clay loam. The soils typically are underlain by cemented and indurated hardpan. Flowering occurs from March to May. The flowers are self-compatible, may or may not be protandrous (depending on the subspecies), and are well adapted for autogamy. The evolution of autogamy in this species is associated with increased levels of genetic recombination. Allozyme data indicate that depending on their breeding system, some populations can retain a substantial level of genetic variability and also that many populations are relatively monomorphic. Allozymes and microsatellite markers also indicate that local populations can harbor unique allelic diversity. Levels of spontaneous self-pollination range from 69% to 100% in insect-excluded plants; however, outcrossing levels in field populations can approach 50%. The pollinators are primarily bees (Insecta: Hymenoptera). Dispersal is facilitated by buoyancy of the nutlets, which can remain afloat for more than 100 h. Germination of seeds that have been stored over the summer can be achieved at 10°C with the addition of water; however roughly 40% of seeds can remain dormant after 7 weeks. The plants are threatened by invasive grasses (Poaceae), whose dense cover reduces their percent seed germination. **Reported associates:** *Allium amplectans,*

Alopecurus saccatus, Arctostaphylos viscida, Aschrachena mollis, Blennosperma nanum, Brodiaea californica, Bromus hordeaceus, Ceanothus cuneatus, Cerastium viscosum, Chamaesyce hooveri, Chrysopsis, Collinsia, Cynosurus echinatus, Drymocallis glandulosa, Erodium botrys, Eryngium alismifolium, E. vaseyi, Geranium dissectum, Juncus uncialis, Lamium amplexicaule, Lasthenia californica, L. chrysostoma, L. fremontii, L. gracilis, Layia fremontii, Lepidium nitidum, Limnanthes alba, L. douglasii, Logfia filaginoides, Lolium multiflorum, Lomatium cookii, Lupinus bicolor, L. micranthus, Myosurus minimus, Navarretia heteranda, N. leucocephala, Nemophila, Orcuttia pilosa, Orthocarpus erianthus, Parvisedum pumilum, Plagiobothrys austiniae, P. bracteatus, P. glyptocarpus, P. stipitatus, Plantago, Plectritis congesta, Poa bulbosa, P. secunda, P. tenerrima, Pogogyne zizyphoroides, Polygonum polygaloides, Quercus douglasii, Ranunculus canus, Sedum spathulifolium, Sidalcea calycosa, S. hartwegii, Taeniatherum caput-medusae, Taraxacum officinale, Toxicoscordion fremontii, Trifolium amplectens, T. depauperatum, T. variegatum, T. willdenovii, Triphysaria eriantha, Tuctoria greenei.

Limnanthes macounii Trel. occurs in depressions (seasonal), flats, pools, seeps, and along the margins of vernal pools at elevations of up to 200 m. The plants occur near the coast (West), sometimes within 20 m of the high tide limit. The substrates (15–30 cm deep) are acidic and often have a high nutrient content. They have been described as mud or rocky. Flowers occur from March to May. The plants are exclusively autogamous. Additional life history information is scarce for this species, perhaps due to its uncertain taxonomic status (see *Systematics*). Its populations are susceptible to overgrowth by the invasive *Trifolium subterraneum* (Fabaceae). **Reported associates:** *Acmispon americanus, Armeria maritima, Bellis perennis, Castilleja ambigua, Festuca rubra, Fragaria chiloensis, Geranium molle, Grindelia integrifolia, Holcus lanatus, Hosackia gracilis, Hypochaeris radicata, Isoetes nuttallii, Mimulus guttatus, Montia fontana, Myosurus minimus, Orthocarpus bracteosus, Plagiobothrys scouleri, Plantago bigelovii, P. elongata, P. lanceolata, Primula pauciflora, Prunella vulgaris, Trifolium depauperatum, T. subterraneum, Vulpia bromoides.*

Limnanthes montana Jeps. grows in gullies, hollows, ledges, meadows, ravines, receding pools, roadside cuts, roadsides, seeps, slopes, swales, and along the margins of brooks, lakes, and streams at elevations of up to 1676 m. Site exposures can vary from full sun to shade. The substrates often are characterized as granitic and include boulders, clay, mud flow breccia volcanics, phyllite, and sand. Flowering occurs from February to June. The flowers are self-compatible but reportedly outcrossed. Nevertheless, data from simple sequence repeat (SSR) markers indicate low levels of intraspecific genetic variation. Other life history information for this species is unavailable. **Reported associates:** *Aesculus, Arctostaphylos viscida, Calocedrus decurrens, Cercocarpus betuloides, Chamaebatia foliolosa, Ericameria cuneata, Eriogonum saxatile, Lewisia congdonii, Mimulus guttatus, Pinus ponderosa, P. sabiniana, Primula hendersonii, Quercus chrysolepis, Q. douglasii, Q.*

wislizeni, Ranunculus hystriculus, Trifolium monanthum, T. willdenovii, Woodwardia fimbriata.

Limnanthes vinculans Ornduff inhabits ditches, marshes, meadows, swales, and vernal pools at elevations of up to 305 m. The sites represent open exposures in full sunlight. The substrates include alluvium and mud. Flowering occurs from April to June, about 5 months after sufficient rainfall has triggered seed germination and the pools fill. The plants are predominantly outcrossed by specialist bees (Insecta: Hymenoptera: Andrenidae: *Andrena pulverea*) but populations can also exhibit moderate to high levels of inbreeding. Microsatellite markers revealed that populations contain relatively high levels of genetic variation, but suffer from heterozygote deficiency attributable to bottlenecks, drift, or past inbreeding. The tuberculate seeds are believed to be dispersed by water or by animals. The seeds retain their viability after 13 years in storage and are believed to persist for up to 20 years. A diverse, long-lived seed bank develops. Each year only a portion of the seed bank germinates, but the precise environmental factors that initiate germination have not been determined. Natural populations can consist of as few as 100 or as many as 100,000+ individuals (>20,000 in some created pools). They are vulnerable to grazing and trampling by livestock. **Reported associates:** *Alopecurus pratensis, Blennosperma bakeri, Carex, Downingia concolor, D. humilis, Eleocharis acicularis, Juncus, Lasthenia burkei, L. glaberrima, Layia chrysanthemoides, Limnanthes douglasii, Navarraetia plieantha, Perideridia gairdneri, Pleuropogon californicus, P. davyi, Ranunculus lobbii.*

Use by wildlife: Several *Limnanthes* species (e.g., *L. bakeri, L. floccosa, L. vinculans*) are grazed by cattle (Mammalia: Bovidae: *Bos taurus*); however, *L. macounii* is unpalatable to grazing rodents (Vertebrata: Rodentia). The slightly fragrant flowers of *Limnanthes* attract hoverflies (Insecta: Diptera: Syrphidae) and bees (Insecta: Hymenoptera). Specialist bees (e.g., Andrenidae: *Andrena limnanthis, A. pulverea, Panurginus occidentalis*) forage on the flowers and pollinate several outcrossing species (*Limnanthes douglasii, L. gracilis, L. montana, L. vinculans*). More generalist species observed as floral visitors include honeybees (*Apis mellifera*) and solitary bees (Andrenidae: *Andrena caerulea, A. cuneilabris, A. layiae, A. prunorum; Calliopsis*). The flowers of *L. montana* attract sweat bees (Halictidae: *Lasioglossum tuolumnense*). Populations of *L. douglasii* that have been cultivated in the Netherlands are colonized by powdery mildew fungi (Ascomycota: Erysiphaceae: *Golovinomyces orontii*). Several species (*Limnanthes douglasii, L. floccosa, L. montana*) are susceptible to fruit flies (Insecta: Diptera: Drosophilidae: *Scaptomyza*), which destroy the plants by boring into their developing buds.

Economic importance: food: none; **medicinal:** none; **cultivation:** *Limnanthes douglasii* (known as poached egg flower) is cultivated widely and received the Royal Horticultural Society's Award of Garden Merit in 1993. *Limnanthes douglasii* 'Sulfurea' is an entirely yellow-petaled cultivar. *Limnanthes* 'Floral' is a cultivar derived from an *L. alba* × *L. floccosa* hybrid. *Limnanthes montana* and

L. vinculans are also reportedly in cultivation; **misc. products:** All *Limnanthes* seeds contain a unique oil (to 24%–30%), which is characterized by long-chain fatty acids that render it stable when exposed to heat or pressure. These qualities make the oil a potential substitute for sperm whale oil, which is used as an industrial lubricant. *Limnanthes* oil can be converted into a liquid wax having properties similar to jojoba oil. It is also used in various cosmetics such as hair conditioner, lip balm, and shampoo. Several OBL species have been considered for use as commercial oil plants; however, nonshattering cultivars of the FAC species *L. alba* (FACW) are grown most widely for oil production; **weeds:** *Limnanthes macounii* has been observed in fallow cabbage fields, but the species ordinarily is rare and not regarded as weedy; **nonindigenous species:** *Limnanthes douglasii* naturalizes occasionally in England where it has been introduced as an ornamental. Increased cultivation of *Limnanthes* species for oil production eventually may result in further extensions of their native distributions.

Systematics: *Limnanthes* and *Floerkea* (Figure 4.73) are very similar morphologically and their merger (as in some taxonomic treatments) is supported by numerical analyses. Phylogenetic analyses incorporating combined DNA sequence data (Figure 4.74) resolve each genus as distinct sister clades, considering that the latter is monotypic. *Limnanthes macounii* is the species most closely resembling *Floerkea*, but it is not its closest relative (Figure 4.74). Numerical analyses and molecular phylogenetic analyses each support the subdivision of *Limnanthes* into two groups recognized taxonomically as section *Inflexae* (petals inflexed after anthesis) and section *Limnanthes* (petals reflexed after anthesis). *Limnanthes alba*, *L. floccosa*, and *L. montana* are placed within the former section and the remaining OBL species in the latter. *Limnanthes gracilis* (formerly OBL) is now recognized as a subspecies of *L. alba*. *Limnanthes floccosa* presumably is an autogamous derivative of the outcrossing *L. alba*, which is also assigned to section *Inflexae*. Several infraspecific taxa have been recognized including subspecies or varieties of *L. douglasii*, *L. floccosa*, and *L. gracilis*. A comprehensive

molecular phylogeny for *Limnanthes* (Figure 4.74) provides a reasonable framework for evaluating the taxonomy of this genus, but several issues remain unsettled. Subspecies recognized for *L. alba* and *L. floccosa* resolve within their respective species clades, whereas those of *L. douglasii* resolve along with accessions of *L. macounii* and *L. vinculans*, which has inspired some authors to contemplate their merger under a single species name. However, strong isolating barriers exist among some of these taxa, which indicates that the lack of phylogenetic which suggests that the lack of phylogenetic resolution as distinct clades indicated by molecular sequence data might be due to their recent divergence. Such issues will require additional evaluation before appropriate taxonomic decisions can be made. The base chromosome number of *Limnanthes* is $x = 5$. All *Limnanthes* species are diploids ($2n = 10$). Isozyme surveys have indicated that different breeding systems exist among *Limnanthes* species as indicated by the proportion of polymorphic loci (P), mean number of alleles/locus (A), and proportion of total variation residing among populations (G_{ST}). *Limnanthes douglasii* is highly outcrossing ($P = 0.84$; $A = 3.05$; $G_{ST} = 0.167$); mixed mating systems characterize *L. bakeri* ($P = 0.37$; $A = 1.47$; $G_{ST} = 0.39$), *L. gracilis* ($P = 0.69$; $A = 2.11$), and *L. vinculans* ($P = 0.41$; $A = 1.71$; $G_{ST} = 0.08$); whereas, *L. floccosa* exhibits a pattern associated with inbreeding ($P = 0.23$–0.29; $A = 1.43$–1.44; $G_{ST} = 0.96$). Populations of *L. floccosa* are extremely depauperate genetically and are highly differentiated from one another. Levels of genetic variation are much higher in *L. alba*, the presumed progenitor of *L. floccosa*. Genetic data from SSR markers also indicate very low levels of intraspecific variation in *L. bakeri*. The extent of natural hybridization in *Limnanthes* has not been studied sufficiently. Several synthetic interspecific hybrids have been made (*L. alba* × *L. floccosa*; *L. bakeri* × *L. macounii*; *L. bakeri* × *L. vinculans*; *L. macounii* × *L. vinculans*), all involving species from within the same sections. However, *L. alba* and *L. montana* (both section *Inflexae*) are not interfertile. Moreover, two subspecies of *L. floccosa* (subspecies *floccosa*, subspecies *grandiflora*) yield sterile hybrids, which indicates that they are effectively isolated reproductively.

Comments: All *Limnanthes* species are distributed along the western North American coast from southern British Columbia to California. Most of the species are restricted either to California (*L. bakeri*, *L. montana*, *L. vinculans*), or to California and Oregon (*L. douglasii*, *L. floccosa*); *L. macounii* occurs only in British Columbia.

References: Arroyo, 1973, 1975; Arry, 1998; Baker & Baker, 1977; Braun et al., 2013; Brown & Jain, 1979; Collin & Shykoff, 2003; Dole & Sun, 1992; Donnelly et al., 2008; Elam, 1998; Fairbarns, 2004; Free, 1963; Gibbs, 2009; Hauptli et al., 1978; Jain, 1978; Karron, 1987, 1991; Kesseli & Jain, 1984, 1985; Kishore, 2002; Lam, 1998; Meyers et al., 2012; Ornduff, 1969c; Ornduff & Crovello, 1968; Panasahatham, 2000; Ritland & Jain, 1984; Runquist, 2012, 2013; Sarker et al., 1997; Sloop et al., 2011, 2012b; Smith, 1991; Thomsen et al., 1993; Thorp & LeBerge, 2005; Thorp & Leong, 1998; Witmer et al., 2009.

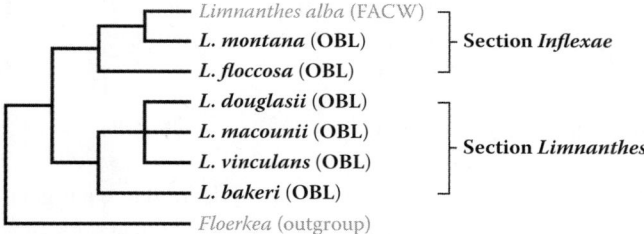

FIGURE 4.74 Interspecific relationships in *Limnanthes* as hypothesized by analysis of combined nuclear and plastid DNA sequence data. The wetland indicator status is provided in parentheses (OBL indicators are highlighted in bold). These data support the long-standing sectional divisions recognized in the genus and indicate that the OBL habit arose only once in the group. (Adapted from Meyers, S.C. et al., *Syst. Bot.*, 35, 552–558, 2010.)

ORDER 10: MALVALES [10]

Results from a number of molecular phylogenetic studies (Alverson et al., 1998, 1999; Bayer et al., 1999; Soltis et al., 2000) have clarified the circumscription of Malvales and convincingly demonstrate its monophyly. The order is also delimited by a suite of morphological characters that include stratified phloem, cuneate rays, mucilage canals, "malvoid" leaf teeth, and cyclopropenoid fatty acids (Judd et al., 2002). Malvales are related most closely to Sapindales and Brassicales, which together with Malvales, resolve as a clade (sometimes identified as "Malvids" or "Eurosids II") that is well supported by various molecular data (Figure 4.1).

Relationships within Malvales have proved to be quite difficult to elucidate, despite a considerable amount of molecular data available for the group (e.g., Alverson et al., 1998, 1999; Bayer et al., 1999; Pfeil et al., 2002, 2004). Resolution among the families within the order is mixed, with some strongly supported clades intermixed among a number of equivocal associations. One well-supported clade includes Malvaceae, the only family within the order to contain OBL North American indicators. However, the precise circumscription of Malvaceae has become increasingly difficult to attain, because molecular data have indicated a complex pattern of relationships among several related taxa within this clade, which were recognized previously as the distinct families Bombacaceae, Sterculiaceae, and Tiliaceae. Notably, these three families appear to be polyphyletic; whereas, the Malvaceae as traditionally defined (sensu stricto) are monophyletic and associate as a sister group to some (but not all) Bombacaceae (Figure 4.75). Given

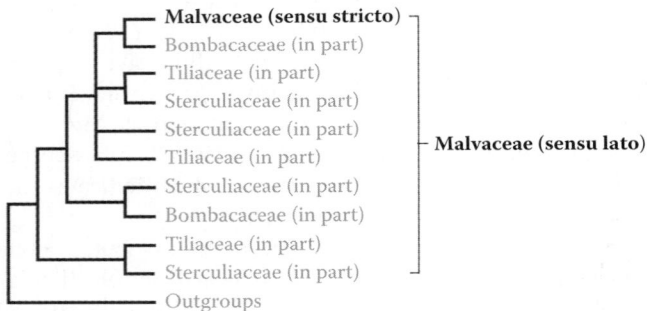

FIGURE 4.75 The taxonomic concept of the traditional Malvaceae (sensu stricto) has been redefined to comply with the results of molecular phylogenetic studies, which support the adoption of a broader (sensu lato) concept of the family. This convention eliminates the need to reconcile the taxonomic limits of Bombaceae, Sterculiaceae, and Tiliaceae, which are polyphyletic as previously circumscribed. All obligate North American aquatics occur within a derived clade (shown in bold) that corresponds with the traditionally delimited Malvaceae. However, many contemporary authors (e.g., Judd et al., 2016) have adopted a sensu lato concept of Malvaceae, in which Bombaceae, Sterculiaceae, and Tiliacae are redefined (along with a number of other taxa) as subfamilies of Malvaceae. (Redrawn from Alverson, W.S. et al., *Amer. J. Bot.*, 86, 1474–1486, 1999.)

these results, many contemporary authors have adopted a much broader (sensu lato) circumscription of Malvaceae to include all members of Bombacaceae, Sterculiaceae, and Tiliaceae as well (Figure 4.75). By this interpretation (which has been followed here), the aquatic North American taxa would occur within a relatively derived clade within the family.

Family 1: Malvaceae [243]

The Malvaceae (mallow family), as recently redefined to include members of Bombacaceae, Sterculiaceae, and Tiliaceae (Figure 4.75), are a cosmopolitan assemblage of about 4225 principally terrestrial species (Judd et al., 2016). The family includes herbaceous species as well as various trees, shrubs, and lianas. Generally, the species have palmately lobed or compound leaves, with malvoid teeth (if serrate), stipules, and mucilage canals. The flowers are arranged in repeating units that include three bracts, where one is sterile while the others subtend solitary flowers or cymes; the sepals are valvate and possess glandular nectaries; an epicalyx often is present (Judd et al., 2016).

The flowers typically are showy and are outcrossed by various nectar-seeking insects including ants, bees, and wasps (Hymenoptera), flies (Diptera), moths (Lepidoptera), as well as by birds (Aves) and bats (Mammalia: Chiroptera). In those species with capsular fruits, the seeds mainly are dispersed abiotically by water or wind, with some possessing specialized hairs or wings. The species that produce follicles or fleshy fruits are dispersed similarly and also by birds or mammals (Judd et al., 2016).

The family is important economically as the source of chocolate (*Theobroma cacao*), cola (*Erythroxylum coca*), and cotton (*Gossypium* spp.) and also for its many showy-flowered ornamentals, notably within the genera *Abutilon*, *Alcea*, *Gossypium*, *Hibiscus*, *Lavatera*, *Sidalcea*, *Sphaeralcea*, and *Tilia*. The common name "marsh mallow" has been applied to various wetland mallow species within *Althaea*, *Hibiscus*, and *Kosteletzkya*. Edible "marshmallows" are so named because they were made originally from the mucilage extracted from the roots of *Althaea officinalis*, a FACW plant (FACW, FAC) introduced to North America from Europe.

Four North American genera contain obligate indicators:

1. *Hibiscus* L.
2. *Kosteletzkya* C. Presl
3. *Pavonia* Cav.
4. *Sidalcea* A. Gray

Phylogenetic analyses (Figure 4.76) indicate that each genus probably represents an independent origin of the aquatic habit in the family, at least with respect to the North American species. Additional sampling of genera (e.g., *Kosteletzkya*, *Sidalcea*) in broader phylogenetic surveys of Malvaceae would help to clarify this issue. Preliminary molecular studies have indicated that *Kosteletzkya* and *Pavonia* might not be distinct from *Hibiscus*.

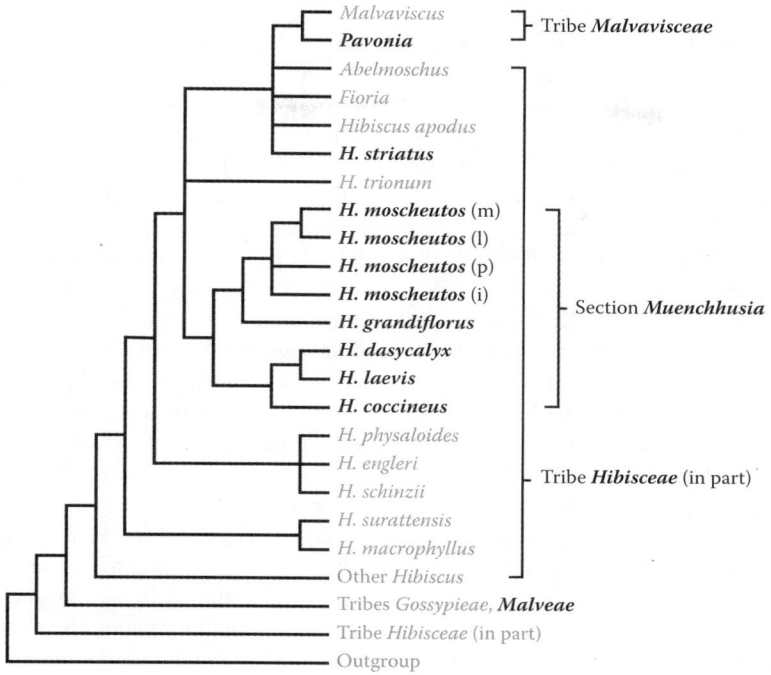

FIGURE 4.76 Phylogenetic relationships within Malvaceae as indicated by combined DNA sequence data. Entries in bold type indicate North American taxa that contain obligate aquatics. These results indicate that tribe Hibisceae is polyphyletic and that *Hibiscus* is paraphyletic. All but one of the OBL *Hibiscus* species (*H. striatus*) occurs within a clade that is recognized taxonomically as section *Muenchhusia*. Epithets of *Hibiscus moscheutos* subspecies are abbreviated as: (i) = *incanus*, (l) = *lasiocarpos*, (m) = *moscheutos*, (p) = *palustris*. *Kosteletzkya* (not surveyed by Small) is placed in tribe Hibisceae and is regarded as a segregate of *Hibiscus*. It possibly is related closely to *Fioria* (a segregate that nests within *Hibiscus*) which shares its winged, capsular fruit. *Pavonia* is another segregate of *Hibiscus* that nests within the latter. The one obligately aquatic North American species (*P. paludicola*) was not surveyed by Small, but presumably occurs within the genus as indicated. *Sidalcea* (also not surveyed by Small) is a member of tribe Malveae, which other studies have shown to be monophyletic. These relationships indicate at least three separate origins of obligate aquatic species within Malvaceae. (Adapted from Small, R.L., *Syst. Bot.*, 29, 385–392, 2004.)

1. *Hibiscus*

Rose mallow; ketmie

Etymology: the ancient Greek and Latin name for the marshmallow plant

Synonyms: *Abelmoschus*; *Bombycidendron*; *Bombycodendron*; *Brockmania*; *Papuodendron*; *Pariti*; *Wilhelminia*

Distribution: global: Africa; Asia; Europe; New World; **North America:** widespread

Diversity: global: 300 species; **North America:** 25 species

Indicators (USA): OBL: *Hibiscus coccineus, H. dasycalyx, H. grandiflorus, H. laevis, H. moscheutos, H. striatus*

Habitat: brackish, freshwater; palustrine, riverine; **pH:** 4.0–8.0; **depth:** <1 m; **life-form(s):** emergent herb

Key morphology: Multistemmed, shrub-like herbs (to 3 m); leaves (to 25 cm) petiolate (to 15 cm), alternate, palmately veined, frequently palmately lobed or divided; flowers very showy, axillary, long stalked (to 6 cm), with a 7–15 parted involucral epicalyx; petals (to 14 cm) 5, cream, white, pink, or deep red, often with a darker (red to purple) base; stamens numerous, fused into a monodelphous tube surrounding the apically 5-parted style; capsules (to 10 cm) striate.

Life history: duration: perennial (persistent caudex); **asexual reproduction:** none; **pollination:** insect or self; **sexual condition:** hermaphroditic; **fruit:** capsules (common); **local dispersal:** seeds (water); **long-distance dispersal:** seeds (birds, water)

Imperilment: (1) *Hibiscus coccineus* [G4]; S1 (AL); S2 (MS); (2) *H. dasycalyx* [G1]; S1 (TX); (3) *H. grandiflorus* [G4]; S2 (GA); (4) *H. laevis* [G5]; SX (ON); SH (MI); S2 (NC, WV); S3 (MD); (5) *H. moscheutos* [G5]; S1 (IN, NH); S2 (CA, KS, KY, RI); S3 (MI, ON); *H. striatus* [G5];

Ecology: general: Fourteen of the North American *Hibiscus* species (56%) occur at least facultatively in wetlands, with six species (24%) recognized as obligate indicators. All six of the latter are ranked exclusively as OBL throughout their range. One additional OBL species (*Hibiscus trilobus* Aubl.) occurs in the Caribbean (not included here). The OBL *Hibiscus* species occur commonly along floodplains, or in similar sites where standing water persists for much of the season. These species grow rather large and attain a shrub-like habit, but are herbaceous perennials, which die back to a caudex each year. All possess showy flowers that primarily are insect pollinated, although autogamy and geitonogamous selfing occurs in some species. The floating seeds are dispersed locally by water and probably over greater distances by waterfowl (Aves: Anatidae) and other birds that feed on them.

Hibiscus coccineus **Walter** is a herbaceous perennial, which grows in fresh to brackish waters in depressions, ditches, floodplains, marshes, sloughs, swamps, tidal wetlands, and along the margins of ponds, rivers, and streams at elevations of at least 284 m. The plants tolerate full sun to partial shade. The reported substrates are high in organic matter and include loam, sand, and sandy clay. Flowering and fruiting have been observed from February to October, and perhaps occur throughout the year. White-petaled forms of this normally scarlet-petaled species occur sporadically. These phenotypes are determined by a single diallelic locus with the complete dominance of red petal color over white. The flowers are pollinated and outcrossed by butterflies (Insecta: Lepidoptera), but they are self-compatible, which also allows for FAC autogamy. Good germination has been achieved by placing seeds at a constant temperature of 21°C. Although the plants can grow to be quite large (to 3 m), they die back each year to the persistent caudex. There is no mechanism for vegetative reproduction or clonal proliferation. Populations are threatened by the invasive spread of nonindigenous climbing fern (*Lygodium microphyllum*). **Reported associates:** *Acer rubrum, Amaranthus, Carex comosa, C. stipata, Carya aquatica, Celtis laevigata, Cephalanthus occidentalis, Cicuta maculata, Colocasia esculentum, Cornus foemina, Echinochloa walteri, Eichhornia crassipes, Fraxinus caroliniana, Hydrocotyle verticillata, Hymenocallis rotata, Iris hexagona, Lactuca floridana, Lemna valdiviana, Liquidambar styraciflua, Lobelia cardinalis, Magnolia virginiana, Nyssa, Osmunda regalis, Panicum hemitomon, Persicaria punctata, Pistia stratiotes, Pluchea odorata, Pontederia cordata, Sabal palmetto, Solidago, Taxodium distichum, Ulmus americana, Zizania aquatica.*

Hibiscus dasycalyx **S.F. Blake & Shiller** is a shrubby, herbaceous perennial, which occurs in bogs, floodplains, marshes, oxbows, ponds, sloughs, and along the margins of canals, rivers, and streams. Typical habitats exist as sunny openings in sites that are subject to annual flooding, and retain standing water until late in the season. Flowering and fruiting occur from May to October, but can be delayed if high water levels persist well into the spring season. The flowers usually have creamy-white petals, but pink forms are encountered occasionally. Although the plants are self-compatible, they are only facultatively autogamous and primarily are outcrossed and pollinated by bees (Insecta: Hymenoptera: Apidae: *Ptilothrix bombiformis*). The seeds remain buoyant in the water for several hours, facilitating their dispersal. High seed germination rates (70%–84%) have been achieved for greenhouse-grown seed that has been scarified with sulfuric acid (98% H_2SO_4) for 25 min and then incubated at 22°C. High germination (54%–95%) has also been observed following hot water scarification of seeds for 5 min (but poor rates if treated for longer periods). The plants have no means of vegetative spread, but can be propagated easily by cuttings. The habitats are threatened by grazing, herbicide application, hydrological alteration, and mowing. The genetic integrity of this species is threatened by natural hybridization with the closely related *H. laevis*. Small-scale

reintroduction programs for this species have been successful. Despite the rarity of this species, there is little other ecological information published in the literature. **Reported associates:** *Baccharis halimifolia, Boehmeria cylindrica, Brunnichia ovata, Carex lupulina, Carya aquatica, C. illinoinensis, Celtis laevigata, Cephalanthus occidentalis, Chasmanthium laxum, Coleataenia longifolia, Croton capitatus, Diodia virginiana, Eichhornia crassipes, Eupatorium serotinum, Heliotropium indicum, Hibiscus laevis, H. moscheutos, Hydrocotyle ranunculoides, Hydrolea ovata, Iva angustifolia, Juncus effusus, Liquidambar styraciflua, Ludwigia leptocarpa, Mikana scandens, Nuphar, Persicaria hydropiperoides, Phanopyrum gymnocarpon, Pluchea foetida, Poncirus trifoliata, Pontederia cordata, Quercus lyrata, Rhynchospora corniculata, Salix nigra, Scirpus cyperinus, Sesbania drummondii, S. herbacea, Thalia dealbata, Thyrsanthella difformis.*

Hibiscus grandiflorus **Michx.** is a large (to 3 m), shrublike, herbaceous perennial, which grows in brackish to freshwater habitats including bayous, canals, ditches, glades, marshes, sloughs, swales, swamps, and along the margins of lakes, ponds, and rivers at elevations of at least 372 m. The habitats are characterized by an excess of more than 200 days of above ground-level inundation per year, and the plants often occur in shallow standing water. The substrates are described as neutral to strongly acidic and include Corolla–Duckson sands, Plummer loamy sand, and St. Lucie sand. Flowering and fruiting occur from July to October. The petals are pink (rarely white) with reddish or purple bases and contain cyanidin, kaempferol, and quercetin. Salt stress (NaCl concentration >0.8%) will significantly inhibit seed germination and reduce seedling survival. **Reported associates:** *Acer rubrum, Allium canadense, Alternanthera philoxeroides, Amaranthus australis, Ampelopsis arborea, Annona glabra, Baccharis halimifolia, Bacopa monnieri, Boehmeria cylindrica, Calystegia sepium, Canna, Cephalanthus occidentalis, Chenopodium album, Cicuta mexicana, Cirsium horridulum, Cladium jamaicense, Colocasia esculenta, Cynodon dactylon, Cyperus haspan, C. odoratus, Echinochloa walteri, Eclipta prostrata, Eleocharis baldwinii, E. flavescens, E. montevidensis, Eupatorium capillifolium, E. serotinum, Galium tinctorium, Gleditsia aquatica, Hydrocotyle ranunculoides, H. umbellata, Ilex cassine, Kosteletzkya pentacarpos, Lemna obscura, Limnobium spongia, Leersia hexandra, Lythrum alatum, Ludwigia octovalvis, Melaleuca quinquenervia, Micranthemum umbrosum, Mikania scandens, Myrica cerifera, Nelumbo lutea, Nothoscordum bivalve, Nyssa biflora, Osmundastrum cinnamomeum, Panicum hemitomon, P. repens, P. virgatum, Paspalidium geminatum, Persicaria hydropiperoides, P. punctata, Phragmites australis, Phyla nodiflora, Phytolacca americana, Pinus elliottii, Pluchea rosea, Pontederia cordata, Rubus trivialis, Rumex obovatus, R. verticillatus, Sagittaria lancifolia, Salix caroliniana, Sambucus nigra, Schoenoplectus californicus, S. pungens, S. tabernaemontani, Scirpus cubensis, Sesbania vesicaria, Sesuvium portulacastrum, Smilax, Sonchus asper, Spartina bakeri, S. patens, Sphenopholis obtusata,*

Strophostyles helvola, Symphyotrichum elliotii, Taxodium ascendens, T. distichum, Teucrium canadense, Thalia geniculata, Toxicodendron radicans, Typha domingensis, T. latifolia, Urochloa mutica, Vigna luteola, Vitis rotundifolia.

Hibiscus laevis All. is a shrub-like, perennial herb, which occurs in shallow, brackish or freshwater batture, bayous, borrow pits, bottoms, depressions, dikes, ditches, floodplains, marshes, meadows, mudflats, prairies, roadsides, sandbars, sloughs, swamps, or along the margins of lakes, ponds, rivers, and streams at elevations of up to 191 m. The plants tolerate only partial shade and typically establish in open sites with standing water, on substrates of clay loam, gravel, loam, peat, sand, sandy clay, sandy loam, silt, silty sand, or mud (often black, fluvial clay; pH: 6–8). Flowering and fruiting have been observed from June to October. The white or pink-petaled flowers (with a purple center) are pollinated and outcrossed by generalist bees (Insecta: Hymenoptera: Apidae) including bumble bees (*Bombus*), mallow bees (*Melitoma taurea*), and solitary bees (*Ptilothrix bombiformis*). Self-compatibility also allows for FAC autogamy (mainly in more northern populations), where stylar recurvature eventually brings pollen in contact with the stigmas in flowers that remain unpollinated. Each flower can produce up to 75 seeds. The plants can grow in clumps (from seed) but have no means of clonal reproduction. **Reported associates:** *Acalypha rhomboidea, Acer negundo, A. saccharinum, Acorus americanus, Agalinis, Alisma subcordatum, Alnus serrulata, Alternanthera philoxeroides, Amaranthus tuberculatus, Ambrosia trifida, Ammannia, Amorpha fruticosa, Apocynum cannabinum, A. sibiricum, Asclepias incarnata, Baccharis, Betula nigra, Bidens cernuus, Boehmeria cylindrica, Bolboschoenus fluviatilis, Boltonia latisquama, Brunnichia ovata, Cardamine rhomboidea, Carex muskingumensis, Cephalanthus occidentalis, Cicuta maculata, Cornus amomum, Cyperus strigosus, C. virens, Echinochloa crus-galli, E. muricata, Eclipta prostrata, Eleocharis tenuis, Eragrostis pectinacea, Eupatorium perfoliatum, Fraxinus pennsylvanica, Helenium, Heliotropium, Heteranthera limosa, Hibiscus dasycalyx, H. moscheutos, Humulus japonicus, Ilex decidua, Impatiens capensis, Iris virginica, Iva annua, Juncus acuminatus, J. torreyi, Kosteletzkya pentacarpos, Laportea canadensis, Lathyrus hirsutus, Leersia lenticularis, L. oryzoides, Lindernia dubia, Liquidambar styraciflua, Lobelia cardinalis, Ludwigia peploides, Lycopus americanuus, Lysimachia ciliata, Mentha arvensis, Mimulus ringens, Morus alba, Muhlenbergia bushii, Nelumbo lutea, Nyssa aquatica, Panicum virgatum, Persicaria amphibia, P. hydropiperoides, P. lapathifolia, P. maculosa, P. pensylvanica, P. punctata, Phalaris arnudinacea, Phragmites australis, Phyla lanceolata, Physostegia virginiana, Pilea pumila, Planera aquatica, Platanus occidentalis, Pluchea camphorata, Populus deltoides, Potentilla norvegica, Quercus bicolor, Q. lyrata, Q. palustris, Q. texana, Rhynchospora caduca, Rubus, Rudbeckia laciniata, Rumex britannica, R. verticillatus, Sagittaria brevirostra, S. latifolia, Salix nigra, Schoenoplectus acutus, S. pungens, Scirpus cyperinus, Scutellaria galericulata, Salix interior, S. nigra, Saururus cernuus, Sium suave, Solidago canadensis,*

S. gigantea, Sparganium eurycarpum, Stachys hispida, Symphyotrichum lanceolatum, S. ontarionis, Taxodium distichum, Toxicodendron radicans, Typha latifolia, Ulmus americana, Verbena hastata, Vernonia gigantea, Vitis riparia, Zizaniopsis milacea.

Hibiscus moscheutos L. grows in shallow, tidal or nontidal, fresh, brackish, or saline waters (salinity <18 ppt) of bottomlands, channels, ditches, floodplains, levees, marshes, meadows, oxbows, prairies, ravines, roadsides, seeps, sloughs, swales, swamps, and along the margins of lakes, ponds, reservoirs, rivers, and streams at elevations of up to 1158 m. The plants thrive in full sun (or in slight shade) on rich substrates (pH: 6.2–7.7) including muck, mud, peaty silica sand, sand, silty clay, and silty sand. Flowering and fruiting extend from May to October. Petal color varies from pink to white and the flowers generally open for a single day. Pollinators include generalist bumble bees (Insecta: Hymenoptera: Apidae: *Bombus pennsylvanicus*) and specialist bees (Insecta: Hymenoptera: Apidae: *Ptilothrix bombiformis*), which outcross the plants. Autogamy is prevented by herkogamy (spatial separation of sexes), but geitonogamous transfer of pollen among flowers on a single individual can result in significant levels of inbreeding (up to 36%). Inbred progeny exhibit lower growth rates than outcrossed progeny. The fruits mature within 3–4 weeks, with each flower producing up to 130 seeds. The buoyant seeds are hydrochorous, capable of dispersing for more than 1200 m along tidal streams. Dispersal along watercourses leads to the establishment of genetically well-mixed metapopulations. Seedling mortality is high unless favorable sites (e.g., openings created by disturbances) are available. The seeds are physiologically dormant and remain viable for up to 4 years. They will germinate within 2–4 weeks when incubated at 21°C–27°C. To enhance their germination, seeds should be soaked overnight in warm water before planting. The highest germination rates and most vigorous seedling growth have been observed from young, outcrossed seeds. The plants are long-lived perennials but contain no means of vegetative reproduction and disperse entirely by their seeds. **Reported associates:** *Acmella oppositifolia, Acorus calamus, Alisma subcordatum, Alnus serrulata, Amaranthus tuberculatus, Ambrosia artemisiifolia, Ammannia, Andropogon glomeratus, Apios americana, Apocynum sibiricum, Asclepias incarnata, A. lanceolata, Atriplex patula, Baccharis halimifolia, Betula nigra, Bidens connatus, B. frondosus, B. polylepis, Bolboschoenus maritimus, B. robustus, Boltonia asteroides, Calystegia sepium, Carex cristatella, C. lupulina, C. pellita, C. stricta, C. tribuloides, C. vulpinoidea, Carya, Centella asiatica, Cephalanthus occidentalis, Chamaecrista fasciculata, Chenopodium rubrum, Cicuta maculata, Cinna arundinacea, Cirsium arvense, C. discolor, Cladium jamaicense, Clethra alnifolia, Coleataenia longifolia, Cornus amomum, Cuscuta gronovii, Cyperus haspan, Dichanthelium oligosanthes, D. sphaerocarpon, Distichlis spicata, Drosera intermedia, Echinochloa muricata, Eclipta prostrata, Eleocharis erythropoda, E. parvula, Eryngium yuccifolium, Eupatorium coelestinum, E. perfoliatum, E. serotinum, Eutrochium dubium, E. maculatum, Epilobium coloratum, Euthamia leptocephala,*

Festuca, Fimbristylis castanea, Forestiera acuminata, Galactia volubilis, Galium tinctorium, Gratiola aurea, Helenium autumnale, Helianthus angustifolius, Heteranthera limosa, Hibiscus dasycalyx, H. laevis, Hypericum canadense, H. mutilum, Impatiens capensis, Ipomoea sagittata, Iva annua, I. frutescens, Juncus bufonius, J. effusus, J. marginatus, J. roemerianus, J. scirpoides, J. tenuis, J. torreyi, Kosteletzkya pentacarpos, Lactuca canadensis, Leersia oryzoides, Lilaeopsis chinensis, Lindernia dubia, Lobelia cardinalis, Ludwigia alternifolia, L. linearis, L. palustris, L. peploides, L. polycarpa, L. uruguayensis, Lycopus americanus, L. amplectens, Lythrum alatum, L. salicaria, Mentha arvensis, Mikania scandens, Mimulus ringens, Neptunia pubescens, Nuphar, Nyssa, Panicum dichotomiflorum, P. hemitomon, P. virgatum, Paspalum plicatulum, Peltandra virginica, Penthorum sedoides, Persicaria arifolia, P. hydropiperoides, P. lapathifolia, P. pensylvanica, P. punctata, P. sagittata, Phragmites australis, Phyla lanceolata, Platanthera peramoena, Platanus occidentalis, Pluchea odorata, Poa nemoralis, Polygonum ramosissimum, Pontederia cordata, Populus deltoides, Proserpinaca palustris, Psilocarya nitens, Ptilimnium capillaceum, Quercus lyrata, Rhexia mariana, R. virginica, Rhynchospora caduca, R. globularis, R. macrostachya, Rosa palustris, Rubus trivialis, Rudbeckia laciniata, Rumex britannica, R. maritimus, Sagittaria latifolia, Salix, Saururus cernuus, Schoenoplectus acutus, S. americanus, S. pungens, S. tabernaemontani, Scirpus atrovirens, S. cyperinus, Scutellaria lateriflora, S. nervosa, Sium suave, Solanum dulcamara, Sorghastrum, Spartina cynosuroides, S. patens, S. pectinata, Symphyotrichum lanceolatum, S. ontarionis, S. racemosum, S. subulatum, Taxodium distichum, Teucrium canadense, Toxicodendron radicans, Triadenum virginicum, Tripsacum dactyloides, Typha angustifolia, T. latifolia, Urtica procera, Verbena brasiliensis, V. hastata, Xyris difformis, Zizania aquatica, Zizaniopsis miliacea.

Hibiscus striatus Cav. is a large (to 3 m) herbaceous perennial, which grows along canals and in marshes at elevations of at least 122 m. The petals are light purple with a small red center and contain cyanidin, kaempferol, and quercetin. The plants reportedly can withstand temperatures to −5°C. Otherwise, this species is poorly known ecologically.

Reported associates: none.

Use by wildlife: Many of the native *Hibiscus* species (*H. laevis*, etc.) are eaten by deer (Mammalia: Cervidae: *Odocoileus*) and are browsed by livestock (Mammalia: Bovidae). The plants generally provide cover for birds (Aves), frogs (Amphibia), snakes (Reptilia), and small mammals (Mammalia). The foliage and seeds of *Hibiscus dasycalyx* are eaten by sawflies (Insecta: Argidae: *Atomacera decepta*). The flowers of *Hibiscus coccineus*, *H. laevis*, and *H. moscheutos* attract butterflies (Insecta: Lepidoptera). Those of *H. coccineus*, *H. grandiflorus*, *H. laevis*, and *H. moscheutos* attract hummingbirds (Aves: Trochilidae). *Hibiscus coccineus* is afflicted by stalk borers (Insecta: Lepidoptera) and by grasshoppers (Insecta: Orthoptera), which feed on the foliage. Seeds of *H. laevis* and *H. moscheutos* reportedly are eaten by ducks (Aves: Anatidae) and northern bobwhite (Aves: Odontophoridae:

Colinus virginianus). Various long-tongued bees (Insecta: Hymenoptera: Apidae: *Bombus pensylvanica, Melitoma taurea, Ptilothrix bombiformis*) feed on the nectar or collect pollen from the flowers of *H. laevis*. *Hibiscus laevis* is a host plant for several larval Lepidoptera (Gelechiidae: *Chionodes hibiscella*; Noctuidae: *Helicoverpa zea*; Tortricidae: *Crocidosema plebejana*). Caterpillars of the gray hairstreak butterfly (Lepidoptera: Lycaenidae: *Strymon melinus*) feed on the flower buds and seeds of *H. laevis*. The foliage is consumed by caterpillars (Insecta: Lepidoptera) of the checkered skipper (Hesperiidae: *Pyrgus communis*), the pearly wood nymph moth (Noctuidae: *Eudryas unio*), and painted lady butterfly (Nymphalidae: *Vanessa cardui*). Japanese beetles (Insecta: Coleoptera: Scarabaeidae: *Popillia japonica*) eat the flowers and foliage of *H. laevis*. *Hibiscus moscheutos* is a host plant for larval Lepidoptera (Gelechiidae: *Chionodes hibiscella*; Noctuidae: *Acontia delecta, Bagisara brouana, B. rectifascia, Helicoverpa zea*; Saturniidae: *Automeris io*). Plants of *H. moscheutos* host aphids (Insecta: Hemiptera: Aphididae), "cucumber" beetles (Insecta: Coleoptera: Chrysomelidae), various fungi (leaf spots, blights, rusts, and canker), and whiteflies (Insecta: Hemiptera: Aleyrodidae). Japanese beetles (Insecta: Coleoptera: Scarabaeidae: *Popillia japonica*) are known to cause severe foliage damage. Seed beetle larvae (Insecta: Coleoptera: Bruchidae: *Althaeus hibisci*) and a weevil (Insecta: Coleoptera: Curculionidae: *Conotrachelus fissunguis*) are seed predators of *H. moscheutos*; adults of *Althaeus hibisci* also feed on the pollen. A nonsocial bee (Insecta: Hymenoptera: Apidae: *Ptilothrix bombiformis*) is a specialized pollinator of various *Hibiscus* species.

Economic importance: food: The flowers and foliage of *Hibiscus* species reportedly are edible, but their mucilaginous texture compromises their palatability; **medicinal:** The Shinnecock tribe used an infusion of dried *Hibiscus moscheutos* stalks to treat bladder inflammations. The mucilaginous roots and foliage of many species have emollient and soothing properties; **cultivation:** *Hibiscus coccineus* is a popular ornamental plant and is used as a parent of hybrids that produce some of the showiest flowers known among the cold-tolerant perennial cultivars. Some cultivars include 'Alba' (a white-flowered form), 'Davis Creek,' 'Great Red Hibiscus,' 'Red Flyer,' 'Texas Star,' and 'Scarlet Mallow.' Although very rare in nature, *H. dasycalyx* is propagated by cuttings and is grown as an ornamental. The giant-flowered *Hibiscus* 'Moy Grande' is a hybrid between *H. moscheutos* 'Southern Belle' and *H. grandiflorus*. *Hibiscus laevis* is sold (occasionally as *H. militaris*) as an ornamental native wildflower or as the cultivar 'White Blush.' *Hibiscus moscheutos* has been grown in American gardens since the 18th century. The numerous cultivars include 'Anne Arundel,' 'Blue River II,' 'Crimson Wonder,' 'Disco Belle Mix,' 'Disco Belle Pink,' 'Flare,' 'Frisbee,' 'Lady Baltimore,' 'Lord Baltimore,' 'Luna Blush,' 'Luna Red,' 'Mallow Marvels,' 'Old Yella,' 'Pink Clouds,' 'Pink Giant,' 'Poinsettia,' 'Rose,' 'Rouge,' 'Southern Belle,' 'Super Glow,' 'Super Rose,' 'Sweet Caroline,' and 'Turn of the Century.' Compact "vintage" cultivars include 'Bordeaux,' 'Carafe,' 'Chablis,' 'Grenache,' 'Pinot Grigio,' 'Pinot Noir' and 'Splash'; **misc. products:** none; **weeds:** none; **nonindigenous species:** none

Systematics: *Hibiscus* is assigned to tribe Hibisceae, which molecular data resolve as polyphyletic. It is difficult to evaluate the closest relatives of the genus given that *Hibiscus* itself does not resolve as a distinct clade unless members of other Hibisceae (i.e., *Abelmoschus, Alyogyne* [in part], *Fioria, Macrostelia*), as well as *Malvaviscus* and *Pavonia* (tribe Malvavisceae; Figure 4.76), are included. The closest relatives of this heterogeneous clade include other *Hibisceae* (*Alyogyne* [in part], *Howittia, Lagunaria*) and members of tribes *Gossypieae* and *Malveae. Hibiscus* and related genera possess a duplicated nuclear *rpb2* gene, which has proven to be useful for corroborating the results obtained from phylogenetic analyses of cpDNA sequences. Relationships of the OBL *Hibiscus* species have been worked out in detail by phylogenetic analysis of combined molecular data [*rpL16, ndhF*, nrITS, granule-bound starch synthase (GBSSI) sequences]. Five of the species resolve as a clade, which has been recognized taxonomically as section *Muenchhusia* (Figure 4.76). Within this section are two clades, one including *H. coccineus*, *H. laevis*, and *H. dasycalyx* and the other with *H. grandiflorus* and *H. moscheutos* (Figure 4.76). *Hibiscus laevis* was known widely in the older literature under the synonym *H. militaris*. Current taxonomic treatments recognize *H. moscheutos* with four subspecies (Figure 4.76), each treated formerly as distinct species (*H. incanus, H. lasiocarpos, H. moscheutos, H. palustris*). The lack of molecular divergence observed among these taxa (which resolve as a clade) supports morphological evidence to warrant their continued recognition at subspecific rank. The accepted assignment of *H. striatus* to section *Striati* is consistent with results from phylogenetic analyses, which place it in a position isolated from section *Muenchhusia* (Figure 4.76). Thus, molecular data indicate two separate origins of the obligate aquatic habit in North American *Hibiscus*. Hybridization is relatively common in *Hibiscus* and has been exploited in the development of many cultivars. Among the obligate aquatics, *Hibiscus dasycalyx* is fully interfertile with both *H. laevis* (identified as its sister species in phylogenetic analyses) and with *H. moscheutos*, yielding fully fertile hybrids when crossed artificially with either species. Allozyme markers indicate that *H. dasycalyx* may have arisen through introgressive hybridization with *H. laevis*. Allozyme data have also documented natural interspecific hybrids between *H. dasycalyx* and the more distantly related *H. moscheutos*, an indication that weak isolating barriers exist among species in section *Muenchhusia. Hibiscus laevis* and *H. moscheutos* are also interfertile, but hybridization in sympatric populations is reduced (to 7%–8%) by differential pollen competition, which acts as an isolating mechanism. Crossing studies indicate the close relationship of *H. coccineus* and *H. laevis*, a result consistent with molecular phylogenetic analyses, which resolve these as closely related species (Figure 4.76). Hybrid *Hibiscus* cultivars include 'Red Flyer' (*H. coccineus* × *H. grandiflorus*) and 'Moy Grande' (*H. moscheutos* × *H. grandiflorus*). Numerous interspecific hybrid cultivars have been produced involving *H. coccineus* × *H. moscheutos* crosses. All members of *Hibiscus*

section *Muenchhusia* reportedly are diploid (2*n* = 38); however, an anomalous septaploid count (2*n* = 133) has also been reported for *H. coccineus. Hibiscus striatus* differs with a count of 2*n* = 52.

Comments: Several of the aquatic *Hibiscus* species occur in the southern United States including *Hibiscus coccineus* (southeastern region) and *H. grandiflorus* (throughout). *Hibiscus dasycalyx* and *H. striatus* are restricted to Texas; the former is endemic there, but the latter extends into Mexico and South America. *Hibiscus dasycalyx* is extremely rare, known from only a few extant populations. *Hibiscus laevis* occurs in central–eastern North America; whereas, *H. moscheutos* is found across North America in areas south of approximately 42°N latitude.

References: Gettys, 2012; Grace et al., 2000; Hubbard & Judd, 2013; Jin-Gui et al., 2012; Kennedy & Strong, 2012; Klips, 1995, 1999; Klips & Snow, 1997; Kudoh & Whigham, 1997, 1998, 2001; Liu & Spira, 2001; Oldham, 1983; Pfeil et al., 2002, 2004; Poole et al., 2007; Sakhanokho, 2009; Shimamura et al., 2005; Snow & Spira, 1993, 1996; Snow et al., 1996, 2000; Spencer & Bousquin, 2014; Spira et al., 1992, 1996; Von Kesseler, 1932; Wise & Menzel, 1971; Yan et al., 2006.

2. *Kosteletzkya*

Seashore mallow

Etymology: after Vincenz Franz Kosteletzky (1801–1887)

Synonyms: *Hibiscus* (in part)

Distribution: global: Africa; Asia; Europe; New World; **North America:** eastern coastal plain

Diversity: global: 17 species; **North America:** 2 species

Indicators (USA): OBL: *Kosteletzkya pentacarpos*

Habitat: brackish, freshwater, saline; palustrine; **pH:** 6.0–8.7; **depth:** <1 m; **life-form(s):** emergent herb

Key morphology: shrub-like herb with multiple, stellate-pubescent stems (to 2.5 m), branched above; leaves (to 15 cm) alternate, petiolate (to 10 cm), stellate-pubescent, lowermost often hastate or sagittate; flowers (to 8 cm) axillary, with 7–10 parted epicalyx; petals (to 4.5 cm) 5, pink (rarely white) with a yellow base; stamens numerous, fused into a yellow, monodelphous tube (to 2.5 cm) that surrounds the apically 5-parted style; capsules (1.2 cm broad) hirsute, strongly 5-angled, 5-parted, each segment single seeded

Life history: duration: perennial (persistent rootstock); **asexual reproduction:** none; **pollination:** insect; self; **sexual condition:** hermaphroditic; **fruit:** capsules (common); **local dispersal:** seeds (water); **long-distance dispersal:** seeds (animals, water)

Imperilment: (1) *Kosteletzkya pentacarpos* [G5]; SX (NY); S3 (NJ)

Ecology: general: Most *Kosteletzkya* species are terrestrial plants with C_3 photosynthesis, but many show an affinity for habitats that are wet at least periodically. These habitats can range from brackish to fresh water sites and include ditches, marshes, springs, and seeps. Of the two North American species, one is an OBL indicator and the other is ranked as FAC. The short-day flowers are self-compatible, but generally are pollinated either by insects (Insecta) or by hummingbirds (Aves: Trochilidae). Most remain open for only 1 day. The fruits exhibit a number of structural adaptations indicating that they are dispersed passively over long distances by animals.

Kosteletzkya pentacarpos (L.) **Ledeb.** grows along irregularly flooded coastal areas in or on brackish or tidal freshwater beaches, ditches, dunes, flats, glades, levees, marshes, meadows, mudflats, roadsides, saltmarshes, sloughs, swamps, and along the margins of canals, ponds, rivers, and streams at low elevations of at least 4 m. The plants occur on moderately organic substrates described as muck, Plummer loamy sand, sand, shell, and silt. They often occur in shallow standing water and will tolerate full sun to partially shaded exposures. The adult plants can tolerate salinities from 0 to 10 ppt, but the seeds and seedlings are more sensitive to higher salinity levels. The pink flowers (which contain quercetin, kaempferol, and cyanidin) are produced from May to October and remain open for a single day, withering by early afternoon. Although normally outcrossed by insects (Insecta) or hummingbirds (Aves: Trochilidae), more than 4% of flowers in a population can undergo autogamous selfing because the styles eventually will recurve to come in contact with the stamens (recurvature ceases once foreign pollen has been deposited). A comparable level of seed set in selfed and outcrossed flowers indicates that this species is well adapted to selfing. High nutrient availability results in greater root biomass and higher fruit production. A maximum of five seeds is produced by each flower. Although quite large (to 4 mm), the mature seeds develop an air space that makes them buoyant and facilitates their dispersal on the surface of water. Observed natural germination rates range from 15% to 96% but are highest at 0% salinity and decline proportionally to less than 4% at 20 ppt salinity. The seeds are physically dormant and require scarification or dry storage/low temperatures (e.g., 4 years at 5°C) to trigger water imbibition and germination. Optimal germination occurs at 28°C–30°C under either light or dark conditions. A germination rate of 95% has been attained for seeds placed under a 14/10 h, 30°C/20°C light and temperature regime. Scarified seeds are more susceptible to germination inhibition as salinity levels increase. No significant persistent seedbank develops in this species. Instead, a transient seed bank develops in the shallow soil layers, which is depleted after the first year of deposition. The seedlings have high survivorship, are shade tolerant, and respond positively (by higher growth rates) to vegetative canopy cover, which is believed to increase water availability, thereby reducing salinity stress. Some references describe *K. pentacarpos* as rhizomatous, but the persistent base does not elongate and the plants do not form clonal colonies. The average lifespan of plants is 11 years. At least some populations are considered to be endangered by habitat destruction. **Reported associates:** *Acer, Agalinis maritima, Alternanthera philoxeroides, Amaranthus australis, A. cannabinus, Ambrosia artemisiifolia, Andropogon, Annona glabra, Aristida stricta, Atriplex patula, Baccharis angustifolia, B. halimifolia, Bacopa monnieri, B. rotundifolia, Boehmeria cylindrica, Bolboschoenus robustus, Borrichia frutescens,* *Calystegia sepium, Campsis radicans, Carex crus-corvi, Cephalanthus occidentalis, Cicuta maculata, Cinna arundinacea, Cladium jamaicense, Colocasia esculenta, Conyza bonariensis, Crotalaria, Cuscuta indecora, Cyperus filicinus, C. involucratus, C. odoratus, C. retrorsus, C. surinamensis, Distichlis spicata, Echinochloa walteri, Eleocharis fallax, E. palustris, E. parvula, E. rostellata, Eupatorium capillifolium, Euthamia graminifolia, Fimbristylis castanea, Fraxinus, Heliotropium curassavicum, Hibiscus laevis, H. moscheutos, Hydrocotyle verticillata, Ilex glabra, Ipomoea sagittata, Iva frutescens, Juncus roemerianus, Juniperus virginiana, Leersia oryzoides, Leptochloa fascicularis, Limonium, Ludwigia grandiflora, L. peruviana, Lythrum lineare, Mikania scandens, Myrica cerifera, Panicum dichotomiflorum, P. hemitomon, P. virgatum, Paspalum vaginatum, Peltandra virginica, Persicaria punctata, Phragmites australis, Pluchea camphorata, P. odorata, P. rosea, Ptilimnium capillaceum, Quercus virginiana, Rumex verticillatus, Rosa palustris, Sabal palmetto, Sabatia stellaris, Sagittaria falcata, S. lancifolia, S. latifolia, Salix caroliniana, S. interior, Sambucus nigra, Samolus valerandi, Schinus teribinthifolius, Schoenoplectus americanus, S. tabernaemontani, Scirpus cyperinus, Sesbania exaltata, S. herbacea, S. macrocarpa, S. punicea, Sesuvium portulacastrum, Setaria glauca, S. magna, Solidago sempervirens, Spartina alterniflora, S. bakeri, S. cynosuroides, S. patens, S. spartinae, Symphiotrichum novi-belgii, S. subulatum, Taxodium, Thelypteris palustris, Typha angustifolia, Typha ×glauca, Vigna luteola, V. repens, Zizaniopsis miliacea.*

Use by wildlife: The flowers of *Kosteletzkya pentacarpos* attract butterflies (Insecta: Lepidoptera) and hummingbirds (Aves: Trochilidae). The plants provide general cover for various wetland animals. **Economic importance: food:** Although the flowers and fruits of many Malvaceae are edible, the pubescent herbage of this species renders it unpalatable. However, oil from the seeds of *K. pentacarpos* is edible and mucilage from the seeds has been considered for industrial use in the making of candy and gum. *Kosteletzkya pentacarpos* has been introduced to China as an agricultural grain crop for planting in saline areas; **medicinal:** No uses have been reported for *K. pentacarpos*, but the mucilage may provide a soothing effect as in other mallow species; **cultivation:** *Kosteletzkya pentacarpos* is sold as an ornamental garden plant and is becoming more popular in recent years. 'Immaculate' is a white-petaled cultivar. Plants have been regenerated successfully using tissue culture; **misc. products:** The seeds of *K. pentacarpos* contain 20%–22% oil (unsaturated fatty acids) and 10%–32% crude protein, which makes them suitable for use as animal feed. The plants have also been considered for industrial generation of biodiesel fuel and ethanol; **weeds:** none; **nonindigenous species:** *Kosteletzkya pentacarpos* has been introduced to China and other parts of Asia for agricultural purposes; however, it is also native to Eurasia.

Systematics: *Kosteletzkya* originally was segregated from *Hibiscus* mainly because of its fruit morphology, which resembles that of *Fioria*. However, recent molecular

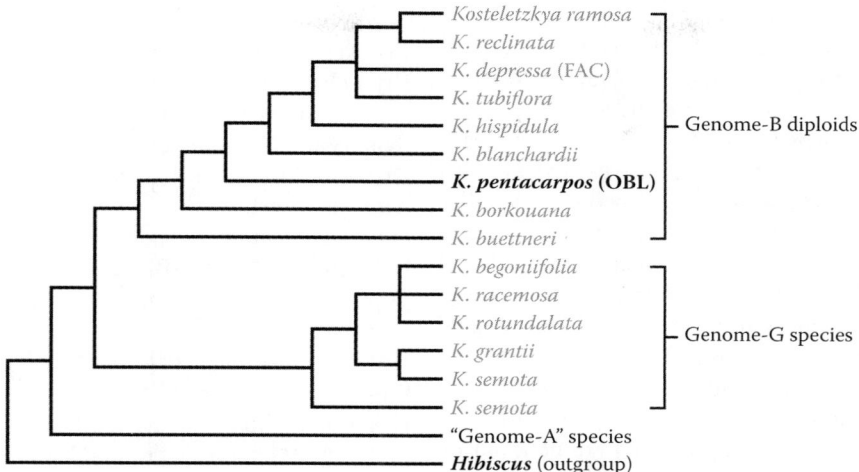

FIGURE 4.77 Phylogenetic relationships among *Kosteletzkya* species as indicated by analysis of nuclear DNA sequence data. Taxa containing OBL indicators are shown in bold. The New World clade containing *K. pentacarpos* is a group of "genome-B" diploid, which apparently is derived from African species (exemplified here by *K. buettneri*). The genome-A and genome-G species each comprises one diploid species with the remainder being of allopolyploid origin. (Adapted from Neubig, K.M. et al., *Bot. J. Linn. Soc.*, 179, 421–435, 2015.)

phylogenetic studies indicate, as they have for several other segregates (including *Fioria*), that the genus may not be distinct from *Hibiscus*. Once eight anomalous Malagasy species are excluded, the remaining taxa comprise a well-defined *Kosteletzkya* clade (Figure 4.77), although one that is possibly nested within *Hibiscus*. The genus is believed to have originated in Africa, with crossing studies showing that all of the diploid New World species share a genome (designated as "B") with the African *Kosteletzkya buettneri* Gürke. *Kosteletzkya pentacarpos* was long known as "*Kosteletzkya virginica*," a name which has since been synonymized. Evaluations of subspecific groups have also found no reason to maintain the latter taxon as distinct at any rank. Natural hybrids involving *K. pentacarpos* have not been reported; however, experimental crosses have shown successful crossability (2%–66% fruit set; 15%–98% pollen stainability) among all of the diploid (B-genome) New World taxa. The base chromosome number for *Kosteletzkya* is $x = 19$. *Kosteletzkya pentacarpos* ($2n = 38$) presumably is diploid.

Comments: *Kosteletzkya pentacarpos* occurs mainly in warmer areas along the eastern coastal plain of the United States, and extends into Eurasia.

References: Alexander et al., 2012; Blanchard, Jr., 1974, 2008, 2012, 2013; Blits & Gallagher, 1990a,b; Cook et al., 1989; Halchak et al., 2011; Islam et al., 1982; Koopman & Baum, 2005; Moser et al., 2013; Neubig et al., 2015; Pino & de Roa, 2007; Poljakoff-Mayber et al., 1992, 1994; Ruan et al., 2004, 2010.

3. *Pavonia*

Mangrove mallow, swampbush
Etymology: after José Antonio Pavon (1754–1844)
Synonyms: *Althaea* (in part); *Asterochlaena*; *Blanchetiastrum*; *Cancellaria*; *Goethea*; *Lass*; *Lopimia*; *Malache*; *Pseudopavonia*; *Pteropavonia*; *Thorntonia*; *Triplochlamys*

Distribution: global: Africa; Asia; neotropics; **North America:** southern
Diversity: global: 250 species; **North America:** 4 species
Indicators (USA): OBL: *Pavonia paludicola*
Habitat: freshwater, brackish, marine (coastal); palustrine; **pH:** unknown; **depth:** <1 m; **life-form(s):** emergent shrub
Key morphology: stems (to 4 m) stellate-pubescent; leaves (to 18 cm) alternate, broadly ovate, palmately veined, bases cordate, petiolate (to 8 cm); flowers stalked (to 4 cm), 3–12 in loose leafless racemes, with a 6–10 parted involucral epicalyx; petals (to 2 cm) 5, greenish-yellow; stamens numerous, fused into a monodelphous tube that surrounds the apically 5-parted style; schizocarps (to 1.3 cm) bowl shaped, 5-parted, each segment (mericarp) keeled and 3-pointed apically
Life history: duration: perennial (buds); **asexual reproduction:** none; **pollination:** bird, insect; **sexual condition:** hermaphroditic; **fruit:** schizocarps (common); **local dispersal:** mericarps; **long-distance dispersal:** mericarps
Imperilment: (1) *Pavonia paludicola* [G5]; S1 (FL)
Ecology: general: *Pavonia* is a genus comprised mainly of terrestrial, subtropical, or tropical plants. All four North American species are wetland indicators, but only one is ranked as OBL within the range treated here (the remaining species are Caribbean and are ranked FACW or FAC). Although most of the species presumably are pollinated by generalist insects (Insecta), the flowers of a few investigated species (including green-flowered species) are pollinated by birds (Aves), are genetically self-incompatible, and are incapable of autogamy. The seeds of some *Pavonia* species have physical dormancy but germinate well in the light at room temperature. Little additional ecological information exists for this group, especially for the North American species.

***Pavonia paludicola* D.H. Nicols. ex Fryxell** occurs in coastal mangrove forests, brackish to freshwater tidal marshes, saltmarshes, and coastal swamps at low elevations of at least 2 m. There is little information available on the reproductive ecology of this species or its mechanism of seed (i.e., mericarp) dispersal. Because *P. paludicola* shares a greenish colored corolla with several other species including *P. bahamensis*, which is pollinated by Bananaquits (Aves: Coerebidae: *Coereba flaveola*) and Bahama woodstars (Aves: Trochilidae: *Calliphlox evelynae*), it was suspected as being bird pollinated, possibly by hummingbirds. Also, the flowers have a deep corolla (averaging 16.8 mm) and produce a large volume of nectar (averaging 29.6 mL; 16.6%), which are features associated with bird pollination. Subsequent work substantiated at least for Caribbean populations, that the flowers are visited by bee hummingbirds (Aves: Trochilidae: *Mellisuga helenae*) and by Cuban emeralds (Aves: Trochilidae: *Chlorostilbon ricordii*). The mericarps are indehiscent and are believed to have prolonged dormancy, which increases their potential dispersal distance. Other details on the ecology of this species are scarce. **Reported associates:** *Conocarpus erectus, Dalbergia ecastophyllum, Funastrum clausum, Laguncularia racemosa, Rhabdadenia biflora, Rhizophora mangle*.

Use by wildlife: Hummingbirds (Aves: Trochilidae: *Chlorostilbon ricordii, Mellisuga helenae*) are highly attracted to the flowers of *Pavonia paludicola* and probably serve as their primary pollinators. The plants are hosts of the lobate lac scale (Insecta: Hemiptera: Kerriidae: *Paratachardina lobata*).

Economic importance: food: none; **medicinal:** none; **cultivation:** *Pavonia paludicola* reportedly is cultivated; **misc. products:** none; **weeds:** none; **nonindigenous species:** *Pavonia hastata* has been introduced to Florida and Georgia, but it is not an OBL indicator.

Systematics: *Pavonia*, regarded as the largest genus of Malvaceae, is assigned to tribe Malvavisceae, which cpDNA data resolve as a clade (with *Malvaviscus*) that is embedded within the genus *Hibiscus*. Analysis of sequences from the duplicated nuclear *rpb2* gene indicates a close association of *Pavonia* with *Hibiscus trionum*. Because most *Pavonia* species (including *P. paludicola*) have not yet been included in phylogenetic analyses, further indications of their relationships will require the addition of taxa in molecular studies. Although a phylogenetic analysis of the large genus *Pavonia* has not been carried out, *P. paludicola* is believed to be closely related to *P. bahamensis, P. rhizophorae*, and *P. troyana*, which similarly possess green or yellowish flowers and occur in comparable mangrove habitats. These four species comprise section *Malache* of *Pavonia* subgenus *Malache*. *Pavonia paludicola* appears in much of the literature under the synonyms *P. spicata* (an illegitimate name) or as *P. racemosa*. The chromosomal base number of *Pavonia* is x = 7. *Pavonia paludicola* has not been counted but reports for other species include numbers ranging from 2n = 28, 42, 56, to 112.

Comments: *Pavonia paludicola* extends into Central America and the West Indies but in North America occurs only in Florida where it is listed as endangered.

References: Dalsgaard et al., 2012; Fryxell, 1999; Godfrey & Wooten, 1981; Howard et al., 2006; McDade & Weeks, 2004; Pfeil et al., 2002, 2004; Rathcke, 2000.

4. *Sidalcea*

Checker mallow, checkerbloom

Etymology: a combination of the Greek words *sida* and *alkea*, both meaning "mallow"

Synonyms: *Hesperalcea, Malvastrum* (in part), *Sida* (in part)

Distribution: global: Mexico; North America; **North America:** western

Diversity: global: 25 species; **North America:** 25 species

Indicators (USA): OBL: *Sidalcea calycosa, S. hirsuta, S. pedata, S. ranunculacea, S. reptans, S. stipularis;* **FACW:** *S. oregana*

Habitat: freshwater; palustrine; **pH:** unknown; **depth:** <1 m; **life-form(s):** emergent herb

Key morphology: stems (to 1.5 m) erect, or decumbent at base; stipules (to 2 cm) present, leaves palmately veined, long-petioled (to 3 dm), dimorphic, the lower blades (to 10 cm) crenate, the upper blades deeply palmately lobed (to 5 cm); flowers short-pedicelled (to 3 mm), epicalyx absent; petals (to 2.5 cm) 5, purple, rose, pink, or white; stamens numerous, fused into a column, the filaments occurring in two distinct (an inner and outer) series; fruit capsular, but dehiscing into 5–10 single-seeded segments (to 4.5 mm)

Life history: duration: annual (fruit/seeds); perennial (rhizomes, persistent taproots); **asexual reproduction:** rhizomes; **pollination:** insect; **sexual condition:** hermaphroditic; gynodioecious; **fruit:** schizocarps (common); **local dispersal:** seeds (gravity); **long-distance dispersal:** seeds (animals)

Imperilment: (1) *Sidalcea calycosa* [G5]; S2 (CA); (2) *S. hirsuta* [G3/G4]; (3) *S. oregana* [G5]; S1 (BC, CA, MT); S2 (WY); (4) *S. pedata* [G1]; S1 (CA); (5) *S. ranunculacea* [G3]; (6) *S. reptans* [G4]; (7) *S. stipularis* [G1]; S1 (CA)

Ecology: general: Fifteen *Sidalcea* species (60%) occur at least facultatively within wetlands, with six species (24%) ranked as OBL indicators. *Sidalcea oregana* (now FACW) was ranked originally as OBL in part of its range, and has been included here because its treatment was completed prior to the release of the revised list. The flowers usually are bisexual and protandrous (sometimes functionally unisexual); occasionally the plants are gynodioecious. There is little information on the pollination biology or seed ecology for most of the obligate aquatic species. The seeds of at least the montane species are believed to possess physical dormancy.

***Sidalcea calycosa* M.E. Jones** is an annual or perennial plant, which is found in freshwater to somewhat brackish habitats including bogs, chaparral, depressions, flats, marshes, meadows, roadsides, seeps, swales, vernal pools, wallows, woodlands, and the margins of streams at elevations of up to 1310 m. This species prefers full sun but is partially shade tolerant. The substrates are described as basalt, clay, granitic sand, metamorphic, muck, rock, sandy loam, serpentine, Tuscan mudflow, and Tuscan stony clay loam. The plants occur as pale purple to white-flowered vernal pool annuals (recognized as subspecies *calycosa*), where they bloom along the pool margins from March to June, or as weak perennials (recognized as subspecies *rhizomata*), which flower from May to July in low, coastal marshes (elevations <30 m), where they spread vegetatively from a long rhizome. The plants are gynodioecious, with individuals bearing bisexual or pistillate flowers. This species apparently is adapted well to fire, with the plants observed to produce a significantly higher number of fruits in the first season following a burn. **Reported associates:** *Aira caryophyllea, Alisma, Alopecurus aequalis, Arbutus menziesii, Arctostaphylos, Avena barbata, Blennosperma nanum, Briza minor, Brodiaea appendiculata, B. minor, Bromus diandrus, B. hordeaceus, Callitriche marginata, Calycadenia oppositifolia, Campanula californica, Castilleja attenuata, C. campestris, Ceanothus cuneatus, Centaurium erythraea, Cerastium glomeratum, Chamaebatia foliolosa, Clarkia purpurea, Cyperus strigosus, Dichelostemma congestum, D. multiflorum, Epilobium densiflorum, E. torreyi, Erodium botrys, Eryngium vaseyi, Eschscholzia lobbii, Gilia tricolor, Heterocodon rariflorus, Heteromeles arbutifolia, Holocarpha virgata, Hordeum marinum, Hypericum anagalloides, H. perforatum, Hypochaeris glabra, Isoetes, Juncus bufonius, J. capitatus, J. effusus, Lasthenia californica, L. fremontii, Layia fremontii, Limnanthes alba, L. douglasii, L. floccosa, Logfia gallica, Lolium multiflorum, Lupinus bicolor, Lythrum hyssopifolia, Mimulus glaucescens, M. guttatus, Montia fontana, Navarretia intertexta, N. tagetina, Oenanthe sarmentosa, Pinus sabiniana, Plagiobothrys fulvus, Polystichum munitum, Quercus douglasii, Ranunculus, Sanicula bipinnatifida, Scirpus microcarpus, Taeniatherum caput-medusae, Taraxacum officinale, Thysanocarpus radians, Trifolium barbigerum, T. depauperatum, T. dubium, T. hirtum, T. microcephalum, T. variegatum, Triphysaria eriantha, Triteleia hyacinthina, Vulpia bromoides, Zeltnera abramsii, Z. muehlenbergii, Z. venusta.*

Sidalcea hirsuta **A. Gray** is an annual, which grows in chaparral, depressions, ditches, flats, meadows, playa, ravines, roadsides, savannas, slopes, swales, vernal pools, woodlands, and along the margins of ponds and streams, at elevations of up to 975 m. The substrates have been characterized as alluvium, Anita clay loam, clay, Corning (CgB) soil series, Corning gravelly loam, gravel, gravelly clay, Hideaway soil series, mud, Riz silt loam, rocky clay loam, serpentine, and stony Tuscan loam. Flowering occurs from April to June. The flowers are bisexual with either white or rose-colored petals and presumably are insect-pollinated. **Reported associates:** *Achyrachaena mollis, Adenostoma, Aesculus californica, Alopecurus saccatus, Arctostaphylos, Brodiaea matsonii, Bromus, Castilleja rubicundula, Ceanothus cuneatus, Chamaesyce hooveri, Coronilla, Cotula, Deschampsia danthonioides, Dichelostemma volubile, Downingia bicornuta, Epilobium torreyi, Eryngium castrense, Holozonia filipes, Hordeum marinum, Isoetes, Juncus, Lagophylla glandulosa, Lasthenia fremontii, Layia fremontii, Lessingia, Limnanthes alba, Lolium multiflorum, Lupinus bicolor, L. microcarpus, Lythrum hyssopifolia, Marsilea vestita, Mimulus guttatus, M. pulchellus, M. tricolor, Navarretia, Odontostemon hartwegii, Orcuttia pilosa, O. tenuis, Paronchia ahartii, Pinus ponderosa, Plagiobothrys greenei, P. stipitatus, Plantago lanceolata, Polypogon maritimus, Psilocarphus brevissimus, Quercus douglasii, Ranunculus arvensis, Rhamnus californica, Rubus laciniatus, Taraxacum officinale, Triteleia hyacinthina, Tuctoria greenei, Zeltnera venusta.*

Sidalcea oregana **(Nutt. ex Torr. & A. Gray) A. Gray** is a perennial, which occurs in freshwater to brackish sites including "biscuits," bogs, bottoms, cienega, ditches, draws, fens, flats, floodplains, levees, marshes, meadows, prairies, roadsides, seeps, slopes, sloughs, springs, swales, swamps, thickets, tidal flats, and the margins of arroyos, lakes, rivers, and streams at elevations of up to 2887 m. Site exposures range from full sun to partial shade. The substrates can be acidic or alkaline and often have a high organic content. They are described as adobe, alluvium, basalt, clay, clay humus, clay loam, gravel, gravelly sand, humus, loam, olivine gabbro, peat, peridotite, pumice, quartzite, rhyolitic rock, rocky basalt, rocky clay loam, rocky limestone, rocky loam, sand, sandy clay loam, sandy loam, serpentine, silicic ash-flow tuffs, silt, silty clay loam, silty loam, stones, and volcanic loam. Flowering occurs from June to August. The plants are gynodioecious with bisexual or pistillate flowers having rose-pink petals. The hermaphroditic flowers are self-compatible (but protandrous), have larger petals, produce 50% more nectar sugar, and have higher pollinator visitation rates than the female flowers. However, the female flowers are longer-lived, allocate more resources to seed production, and produce seeds with higher germination rates. Outcrossed plants yield seeds that produce plants that are 30%–50% larger and with 40% more flowers than those derived from inbred plants. Pollinators include several bees (Insecta: Hymenoptera: see *Use by wildlife*), beeflies (Insecta: Diptera: Bombyliidae), and skippers (Insecta: Lepidoptera: Hesperiidae). The seeds will germinate within 2 weeks if they first are scarified and then placed in moist peat–moss at 9°C. The seeds are dispersed close to the maternal plants, which often results in clumped distributions. The plants regenerate from vegetative rosettes, which are produced by overwintering woody taproots. They are long-lived, but their vegetative spread is minimal. Suitable habitat for this species presumably is maintained by periodic episodes of fire. Populations of this species are threatened by fire suppression, grazing, habitat fragmentation, and invasive species. **Reported associates:** *Abies concolor, A. grandis, A. lasiocarpa, A. ×shastensis, Achillea millefolium, Achnatherum richardsonii, Acmispon americanus, Aconitum, Adenocaulon bicolor, Adenostoma, Adiantum pedatum, Agastache urticifolia, Agoseris glauca, Agropyron, Agrostis gigantea, Aira caryophyllea, Allium acuminatum, A. tolmiei, Alnus rhombifolia, A. tenuifolia, A. viridis, Alopecurus pratensis, Amelanchier alnifolia, Anemopsis californica, Angelica lucida, Antennaria corymbosa, Anthoxanthum, Apera interrupta, Aquilegia formosa, Arbutus, Arctostaphylos patula, Arenaria, Arnica chamissonis, Artemisia arbuscula, A. cana, A. ludoviciana, A. packardiae, A. rigida, A. tridentata, A. tripartita, Asclepias, Astragalus canadensis, A. lentiginosus, Balsamorhiza hookeri, B. sagittata, Barbarea orthoceros, Betula occidentalis, Bistorta bistortoides, Blepharipappus scaber, Briza maxima, Brodiaea californica, B. elegans, Bromus hordeaceus, B. inermis, B. marginatus, B. sterilis, B. tectorum, B. vulgaris, Calamagrostis canadensis, C.*

rubescens, Calocedrus decurrens, Calochortus longebarbatus, C. luteus, Camassia leichtlinii, C. quamash, Campanula parryi, Carex angustata, C. athrostachya, C. douglasii, C. hoodii, C. jonesii, C. lasiocarpa, C. lyngbyei, C. microptera, C. nebrascensis, C. pellita, C. petasata, C. praegracilis, C. preslii, C. rostrata, C. scopulorum, C. subfusca, C. tumulicola, Castanopsis sclerophylla, Castilleja cusickii, C. densiflora, C. miniata, C. tenuis, C. unalaschcensis, Ceanothus cordulatus, C. integerrimus, C. prostratus, C. velutinus, Cenchrus biflorus, Centaurea ×moncktonii, Cercocarpus ledifolius, Chaenactis douglasii, Chamaecyparis, Chamerion angustifolium, Chenopodium album, Chrysothamnus viscidiflorus, Cirsium arvense, C. vulgare, Cistanthe umbellata, Clarkia amoena, C. arcuata, C. unguiculata, Claytonia parviflora, Collomia grandiflora, C. linearis, Cornus nuttallii, Corylus, Crataegus douglasii, Crepis runcinata, Cryptantha simulans, Cynoglossum officinale, Cytisus scoparius, Dactylis glomerata, Danthonia californica, Darlingtonia californica, Daucus carota, Delphinium depauperatum, D. hesperium, D. viridescens, Deschampsia cespitosa, D. danthonioides, D. elongata, Dicentra, Dichelostemma congestum, Digitalis, Diplacus aurantiacus, Dipsacus, Distichlis, Drymocallis arguta, D. glandulosa, Eleocharis bella, E. macrostachya, Elymus elymoides, E. glaucus, Epilobium ciliatum, E. densiflorum, E. watsonii, Equisetum laevigatum, Ericameria nauseosa, E. teretifolia, Erigeron decumbens, E. linearis, E. pumilus, Eriodictyon, Eriogonum umbellatum, Eriophorum crinigerum, Eriophyllum lanatum, Erodium, Eschscholzia californica, Eurybia integrifolia, Festuca campestris, F. idahoensis, F. rubra, Fragaria vesca, Fraxinus, Fritillaria camschatcensis, Galium aparine, G. boreale, Gaultheria shallon, Gayophytum decipiens, G. diffusum, Geranium erianthum, G. oreganum, G. richardsonii, G. viscosissimum, Geum, Glyceria elata, G. grandis, G. striata, Gnaphalium palustre, Helenium bigelovii, Helianthella uniflora, Helianthus, Heliotropium, Hemizonella minima, Heracleum maximum, H. sphondylium, Heuchera micrantha, Hieracium scouleri, Holcus, Holodiscus discolor, Hordeum brachyantherum, Horkelia fusca, Hosackia pinnata, Hypericum anagalloides, H. formosum, H. scouleri, Hypochaeris, Ipomopsis aggregata, Iris missouriensis, I. tenax, Juncus acuminatus, J. balticus, J. effusus, J. ensifolius, J. nevadensis, J. orthophyllus, J. oxymeris, J. tenuis, Juniperus occidentalis, J. scopulorum, Koeleria nitida, Lactuca serriola, Lasthenia californica, Lathyrus palustris, L. pauciflorus, Leptosiphon ciliatus, L. parviflorus, Leucanthemum vulgare, Leymus cinereus, L. triticoides, Lomatium dissectum, L. nudicaule, Lonicera involucrata, Lupinus latifolius, L. lepidus, L. nootkatensis, L. polyphyllus, L. sericeus, Luzula parviflora, Lythrum salicaria, Madia elegans, M. glomerata, Maianthemum dilatatum, M. stellatum, Medicago lupulina, Melica, Melilotus, Mentha arvensis, M. pulegium, Menyanthes trifoliata, Mertensia ciliata, Micropus californicus, Microsteris gracilis, Mimulus guttatus, M. moschatus, M. primuloides, Montia chamissoi, M. linearis, Muhlenbergia asperifolia, Myosotis, Navarretia leptalea, Oenanthe, Paxistima myrsinites, Pedicularis racemosa, Penstemon

confertus, P. globosus, P. heterodoxus, P. pratensis, P. rydbergii, P. speciosus, P. wilcoxii, Perideridia bolanderi, P. erythrorhiza, P. gairdneri, P. lemmonii, P. parishii, Phacelia heterophylla, P. linearis, Phalaris arundinacea, Philadelphus lewisii, Phleum alpinum, P. pratense, Phlox idahonis, Physocarpus malvaceus, Pickeringia montana, Pinus contorta, P. jeffreyi, P. lambertiana, P. monticola, P. ponderosa, Plagiobothrys hispidulus, Plantago erecta, P. major, Platanthera dilatata, Platystemon californicus, Poa bulbosa, P. pratensis, P. secunda, P. wheeleri, Polemonium foliosissimum, Polygonum douglasii, Polystichum munitum, Populus angustifolia, P. tremuloides, P. trichocarpa, Potentilla anserina, P. gracilis, P. recta, Primula tetrandra, Prosartes hookeri, Prunella vulgaris, Prunus cerasifera, P. emarginata, Pseudoroegneria spicata, Pseudostellaria, Pseudotsuga menziesii, Pteridium aquilinum, Purshia tridentata, Pycnanthemum, Pyrus, Quercus chrysolepis, Q. douglasii, Q. durata, Q. garryana, Q. kelloggii, Q. vacciniifolia, Ranunculus californicus, R. orthorhynchus, Rhacomitrium canescens, Rhinanthus minor, Rhododendron occidentale, Ribes aureum, R. cereum, R. inerme, R. roezlii, Rosa nutkana, R. woodsii, Rubus discolor, R. laciniatus, R. parviflorus, R. spectabilis, R. ursinus, Rudbeckia occidentalis, Rumex crispus, Salix boothii, S. hookeriana, S. laevigata, S. lasiandra, S. lemmonii, S. scouleriana, Salvia apiana, Sanguisorba, Sanicula, Sarcobatus vermiculatus, Schedonorus arundinaceus, Schoenoplectus acutus, S. americanus, Scirpus microcarpus, Scutellaria angustifolia, S. galericulata, Sedum oreganum, S. stenopetalum, Senecio crassulus, S. integerrimus, S. serra, S. triangularis, Sherardia arvensis, Sidalcea hendersonii, Silene montana, Sisyrinchium angustifolium, S. bellum, S. idahoense, Solidago elongata, Sonchus arvensis, Sorbus scopulina, Sphenosciadium capitellatum, Spiraea douglasii, Spiranthes, Stachys pycnantha, S. rigida, Stellaria longipes, Stipa, Symphoricarpos albus, S. oreophilus, Symphyotrichum foliaceum, S. spathulatum, S. subspicatum, Taraxacum officinale, Thalictrum fendleri, T. sparsiflorum, Thuja plicata, Toxicodendron diversilobum, Toxicoscordion, Tragopogon, Tridens flavus, Trifolium cyathiferum, T. dubium, T. hybridum, T. longipes, T. wormskioldii, Trillium ovatum, Triteleia crocea, T. grandiflora, T. hyacinthina, T. ixioides, T. laxa, T. pratense, Umbellularia californica, Urtica dioica, Valeriana, Veratrum californicum, V. viride, Veronica americana, Vicia hirsuta, Viola macloskeyi, Wyethia amplexicaulis, W. mollis, Xylococcus bicolor.

Sidalcea pedata A. Gray is a rare perennial species of wet, montane habitats including lake shores, meadows, roadsides, swales, woodlands, and the margins of streams at elevations from 1523 to 2438 m. The substrates are alkaline or saline and comprise gravel, quartzite, sand, or silty loam, which typically is underlain by impermeable clay lenses. The flowers open from May to August and have pink to magenta petals. The plants are gynodioecious and are pollinated by insects, which visit the bisexual flowers preferentially to the pistillate flowers. The seeds are dispersed in close proximity to the parental plants. Mechanisms for longer range dispersal are unknown. Seedling production can range from 38% to 120% of the standing number of adult plants in a population. Seedling survivorship (indicated by a 3-year study of 66 plants) is high (97%). The plants can live for three to 5 years, persisting by means of a fleshy taproot. *Sidalcea pedata* is extremely rare with only 17 extant locales and fewer than 8.1 hectares of habitat remaining. Multiple factors identified as threats to populations include competition from invasive weeds, construction activities, damage by recreational vehicles, development, grazing by animals, and habitat destruction. **Reported associates:** *Abies concolor, Achillea millefolium, Agropyron desertorum, A. pubescens, Artemisia ludoviciana, A. nova, A. tridentata, Bromus, Carex, Castilleja cinerea, C. lasiorhyncha, C. tenuis, Descurainia, Distichlis spicata, Eleocharis, Eriogonum nudum, Gutierrezia sarothrae, Horkelia bolanderi, Iris missouriensis, Ivesia argyrocoma, Juncus balticus, Juniperus, Leymus triticoides, Linanthus killipii, Mimulus exiguus, M. purpureus, Pinus contorta, P. jeffreyi, Poa atropurpurca, P. pratensis, P. secunda, Perideridia parishii, Potentilla glandulosa, P. gracilis, Pyrrocoma uniflora, Quercus kelloggii, Salix lutea, Senecio bernardinus, Symphyotrichum spathulatum, Taraxacum californicum, Thelypodium stenopetalum, Veratrum californicum.*

***Sidalcea ranunculacea* Greene** is a perennial, which grows in wet, grassy sites including crevices, flats, marshes, meadows, roadsides, slopes, switchbacks, and along riverbanks or streams at elevations from 954 to 3047 m. Exposures range from sun to partial shade. The substrates are described as granite, loam, sand, sandy loam, and silty mud. Flowering occurs from June to August. Individuals are gynodioecious, bearing bisexual or pistillate flowers with magenta–pink petals (more deeply colored in the latter). The plants persist by means of a deep tap root. Clonal growth and vegetative reproduction occur by production of short, slender rhizomes. More detailed ecological information is not available for this species. **Reported associates:** *Abies, Carex, Castilleja, Ceanothus cordulatus, Delphinium, Elymus, Juncus, Leptosiphon ciliatus, Mertensia, Mimulus whitneyi, Oxypolis occidentalis, Pinus contorta, P. jeffreyi, Poa, Populus tremuloides, Primula, Pteridium aquilinum, Ranunculus californicus, Salix, Senecio triangularis, Sequoiadendron giganteum, Stipa, Veratrum californicum, Viola purpurea.*

***Sidalcea reptans* Greene** is a perennial, which occurs in depressions, moist meadows, and on dried streambeds with other principally herbaceous vegetation at elevations from 1127 to 2498 m. Exposures range from full sun to mixed shade. The substrates are described as igneous, loam, and mud. Flowering extends from June to August. The petals are deep pink to lavender, turning darker with age. The plants can spread and propagate clonally by the production of long rhizomes. **Reported associates:** *Abies magnifica, Achillea millefolium, Agastache urticifolia, Agrostis exarata, A. oregonensis, A. scabra, Bistorta bistortoides, Camassia quamash, Carex abrupta, C. angustata, C. feta, C. lanuginosa, C. leptopoda, C. macloviana, Deschampsia cespitosa, Eleocharis engelmannii, Elymus glaucus, Epilobium ciliatum, Galium bifolium, G. trifidum, G. triflorum, Geranium incisum, Glyceria elata, Helenium bigelovii, Heracleum maximum, Hosackia*

oblongifolia, Hypericum anagalloides, H. galioides, Iris hartwegii, Juncus chlorocephalus, J. macrandrus, Kelloggia galioides, Lilium parvum, Lupinus latifolius, Mimulus guttatus, M. primuloides, Oxypolis occidentalis, Perideridia parishii, Phalacroseris bolanderi, Phleum pratense, Pinus contorta, P. jeffreyi, P. ponderosa, Platanthera leucostachys, Populus tremuloides, Potentilla gracilis, Pteridium aquilinum, Ribes roezlii, Rudbeckia californica, Rumex salicifolius, Senecio triangularis, Stachys albens, Stellaria crispa, S. longipes, Toxicoscordion, Trifolium wormskioldii, Veratrum californicum, Viola macloskeyi, Wyethia angustifolia.

***Sidalcea stipularis* J.T. Howell & G.H.** True is a rare perennial herb, which grows in freshwater montane marshes at elevations of 731–732 m. The pink flowers open from June to August. Vegetative reproduction occurs by means of slender rhizomes. Little other ecological information exists for this extremely rare species, which is known only from a single locality. There the plants are threatened by grazing, nonnative plant species, deteriorating water quality, road construction, and vehicular destruction. **Reported associates:** *Juncus, Pinus ponderosa, Rhynchospora capitellata, Rubus ulmifolius, Typha latifolia.*

Use by wildlife: *Sidalcea oregana* is grazed by sheep (Mammalia: Bovidae: *Ovis aries*). It is also a host for several fungi (Ascomycota: Mycosphaerellaceae: *Ramularia sidalceae*; Basidiomycota: Pucciniaceae: *Puccinia interveniens, P. sherardiana*) and is the host of a paratype of *Neoramularia oregana* (Fungi: Ascomycota). The flowers of *S. oregana* are visited by a number of bees (Insecta: Hymenoptera: Apidae: *Bombus bifarius, B. flavifrons, B. insularis, B. mixtus, B. vandykei, Diadasia nigrifrons*; Megachilidae: *Hoplitis albifrons, Osmia densa, O. malina*), which serve as pollinators. The fruits are eaten (and damaged) by weevils (Insecta: Coleoptera: Curculionidae: *Macrorhoptus sidalceae*). *Sidalcea oregana* is a host plant of several butterfly (Insecta: Lepidoptera) larvae including the common checkered skipper (Hesperiidae: *Pyrgus communis*) and the West Coast lady (Nymphalidae: *Vanessa annabella*). *Sidalcea pedata* is eaten by burros (Mammalia: Equidae: *Equus asinus*) and horses (Mammalia: Equidae: *Equus ferus*).

Economic importance: food: The Luiseno people used *Sidalcea malviflora* as vegetable greens. The Yana tribe added mashed leaves of the plant as a flavoring for black manzanita (*Arctostaphylos*) berries; **medicinal:** A cold infusion of *S. neomexicana* was used by the Ramah Navajo as a drug to treat internal injuries. Extracts of *S. oregana* are sold as a botanical for "natural healing"; **cultivation:** *Sidalcea oregana* (also as *S. oregana* subspecies *spicata*) is sold as an ornamental and includes the cultivars 'Brilliant' and 'Little Princess.' *Sidalcea pedata* is used as a garden ornamental. *Sidalcea reptans* appears in cultivation occasionally; **misc. products:** none; **weeds:** none; **nonindigenous species:** none

Systematics: *Sidalcea* is assigned to tribe Malveae of Malvaceae and traditionally was regarded as being closely related to *Callirhoë, Eremalche, Napaea*, and *Sphaeralcea*. Molecular phylogenetic studies (e.g., Figure 4.78) resolve *Eremalche* as the sister group of *Sidalcea*, but only with a fairly weak level of internal support. Relationships of all six OBL indicator species have been examined by phylogenetic analyses (Figure 4.78). Incongruent resolution of *S. ranunculacea* and *S. reptans* as indicated by nuclear vs. maternal (cpDNA) sequence data has indicated issues of hybridization that have not yet been resolved, and both species have been excluded from summary phylogenetic trees (e.g., Figure 4.78). However, the molecular data have indicated that those similar species do appear to be closely related in any case. *Sidalcea stipularis* is the sister to *S. hickmanii*, which is a species of chaparral and coniferous forest. *Sidalcea pedata* resolves within a clade containing *S. malviflora* (dry forest and scrub) and *S. neomexicana* (FACW), which inhabits alkaline springs and marshes. *Sidalcea oregana* (FACW) is weakly supported as the sister species of *S. nelsoniana* (FAC) in a clade that includes *S. hendersonii* (FACW, UPL) and a number of other FACW indicators. *Sidalcea calycosa* and *S. hirsuta* (both annuals) resolve as closely related sister species (Figure 4.78). Comparative studies of nuclear ribosomal DNA (nrDNA) sequences indicate that molecular evolutionary rates of the annual *Sidalcea* species are significantly higher than in the perennials. Experimental hybridizations have shown that high interfertility exists among many *Sidalcea* species, even for those species having different chromosome numbers.

Hybridization in several *Sidalcea* species is evidenced by DNA polymorphisms, which occur in sequences derived from biparentally inherited loci (ITS, ETS regions of 18S–26S nrDNA). Phylogenetic analysis of cloned DNA sequences has indicated possible past hybridization in the ancestry of *S. reptans*, *S. malviflora*, and species within the "*glaucescens*" clade (Figure 4.78). Hybridization between *S. oregana* × *S. asprella* has also been indicated by molecular data. The basic chromosome number of *Sidalcea* is $x = 10$. *Sidalcea pedata*, *S. ranunculacea*, and *S. reptans* ($2n = 20$) are diploid; whereas, *S. oregana* exists at different ploidy levels ($2n = 20, 40, 60$). Polyploidy is common among the more derived species of *Sidalcea* where it has originated multiple times. Counts have not been reported for *S. calycosa*, *S. hirsuta*, or *S. stipularis*.

Comments: *Sidalcea calycosa*, *S. hirsuta*, *S. pedata*, *S. ranunculacea*, *S. reptans*, and *S. stipularis* are restricted to California. *Sidalcea oregana* is distributed more broadly across western North America.

References: Andreasen, 2012; Andreasen & Baldwin, 2001, 2003a,b; Ashman, 1992a,b, 1994; Ashman & Stanton, 1991; Austin & Leary, 2008; Buck-Diaz et al., 2012; Christy, 2013; Hill, 2012; Hunter, 1986; Jones, 2001b; Kruckeberg, 1957; Preston, 2011; USFWS, 1998b; Zimmerman & Reichard, 2005.

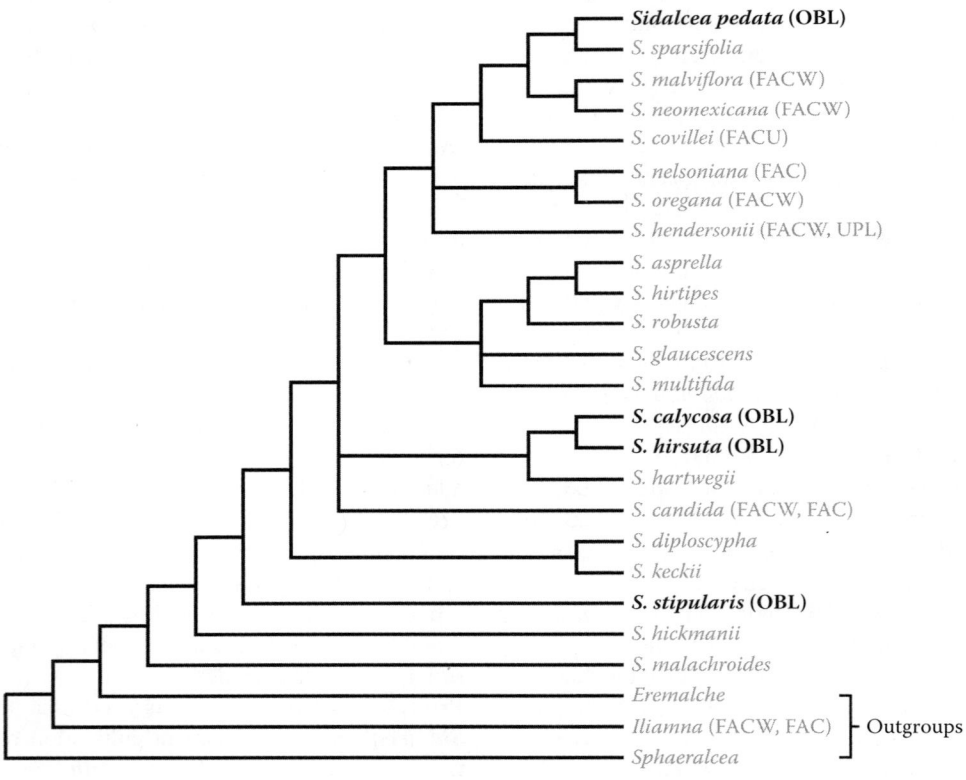

FIGURE 4.78 Phylogenetic relationships within *Sidalcea* as resolved by combined DNA sequence data. These relationships indicate that the OBL North American aquatics (shown in bold) are distributed widely throughout the genus where they have evolved at least three separate times (the status for other wetland indicators is indicated in parentheses). The OBL *Sidalcea ranunculacea* and *S. reptans* were included in the study and display a close relationship; however, they were excluded from the earlier analysis due to incongruencies attributed to hybridization. (Redrawn from Andreasen, K., *Int. J. Plant Sci.*, 173, 532–548, 2012.)

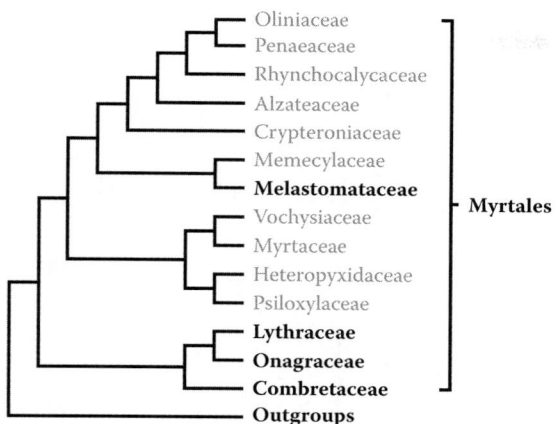

FIGURE 4.79 Phylogenetic relationships of Myrtales families. The families that contain OBL North American wetland indicators are highlighted in bold. (Adapted from Conti, E. et al., *Evolution*, 56, 1931–1931, 2002; Schönenberger, J. & Conti, E., *Amer. J. Bot.*, 90, 293–309, 2003.)

ORDER 11: MYRTALES [14]

The order Myrtales, with about 9000 species, is a well-defined clade as evidenced by a combination of morphological characters (bicollateral bundles, vestured pits) and the results from several molecular phylogenetic studies (e.g., Conti et al., 1997; Schönenberger & Conti, 2003). Myrtales unquestionably ally with the rosids; however, the precise affinity of the clade to either of the larger groups of rosid families (i.e., eurosids I, eurosids II) remains unresolved and the order generally is placed in a position intermediate to those groups.

The majority of Myrtales comprises terrestrial species. Although only four North American families contain obligate aquatics, several families include species that occur at least facultatively in or near wetlands. *Melaleuca quinquenervia* (Cav.) Blake (Myrtaceae) is a nonindigenous species introduced to the southern United States, which often occurs in swamps; however, its wetland status (FAC) reflects the broader habitat tolerance of this plant. *Conocarpus erectus* L. (Combretaceae) occurs near the upper edge of coastal mangrove swamps but is categorized as FACW species in North America. In North America, OBL indicators occur in four families:

1. **Combretaceae** R. Br.
2. **Lythraceae** J. St.-Hil.
3. **Melastomataceae** Juss.
4. **Onagraceae** Adans.

Phylogenetic studies (Figure 4.79) indicate that the obligate aquatic habit has evolved at least twice within Myrtales. The highest proportion of aquatic species occurs within the Combretaceae/Onagraceae/Lythraceae clade.

Family 1: Combretaceae [20]

Combretaceae include approximately 600 species of woody plants, which represent a diverse assortment of shrubs, trees, and vines (Judd et al., 2008). Phylogenetically, the family resolves as the sister group to a clade containing Lythraceae and Onagraceae (Figure 4.79). Phylogenetic analyses of morphological and molecular data support the monophyly of the family, which is divided into two subfamilies: Combretoideae and Strephonematoideae (Maurin et al., 2010). However, these analyses have also demonstrated that the two largest genera (*Combretum* and *Terminalia*) are not monophyletic (Maurin et al., 2010) (Figure 4.80). The group shares a unilocular, inferior ovary with apical placentation, few ovules attached to a long, pendulous funiculus, and drupes containing a single, large, more or less flattened seed with a fibrous outer coat (Judd et al., 2008).

The highly nectariferous flowers are pollinated by a variety of animals including birds (Aves), insects (Insecta), and mammals (Mammalia). Most species (except those of the dioecious *Conocarpus*) are bisexual, protogynous, and outcrossed; some genera contain species having both bisexual and staminate flowers and are androdioecious (*Laguncularia*) or andromonoecious (*Terminalia*) (Judd et al., 2008). The drupes are dispersed by animals (then fleshy), by water (then spongy), or by the wind (then winged).

Terminalia catappa (known as tropical almond) produces acidic but edible fruits. A number of genera (e.g., *Combretum*, *Quisqualis*, *Terminalia*) are grown as ornamentals.

There are four North American genera (*Bucida*, *Conocarpus*, *Laguncularia*, and *Terminalia*), of which only one is categorized as an OBL indicator:

1. ***Laguncularia*** C.F. Gaertn.

1. *Laguncularia*

White mangrove

Etymology: from the Latin *laguncula* (diminutive of *lagoena*) meaning "small flask," probably in reference to the small, flask-shaped flowers.

Synonyms: *Conocarpus* (in part); *Rhizaeris*, *Schousboea*

Distribution: global: Africa, New World; **North America:** southern

Diversity: global: 1 species; **North America:** 1 species

Indicators (USA): OBL; FACW: *Laguncularia racemosa*

Habitat: brackish, saline (coastal); palustrine; **pH:** 4.5–8.5; **depth:** m; **life-form(s):** emergent shrub or tree

Key morphology: shrub or tree (to 20 m), leaves (to 18 cm) opposite, simple, evergreen, leathery, margins entire, petioles short (to 13 mm), apically bearing 2–4 sugar-secreting glands (extra-floral nectaries); inflorescence terminal (panicle) or axillary (spike); flowers bisexual, small (to 5 mm), campanulate, 5-merous; petals greenish white; stamens 10; fruit (to 2 cm) a single-seeded, elliptical, flattened, leathery drupe, with 2 spongy wings

Life history: duration: perennial (buds); **asexual reproduction:** layering by rooting lower branches; **pollination:** insect or self; **sexual condition:** hermaphroditic; **fruit:** drupes (common); **local dispersal:** fruits (water); **long-distance dispersal:** fruits (water)

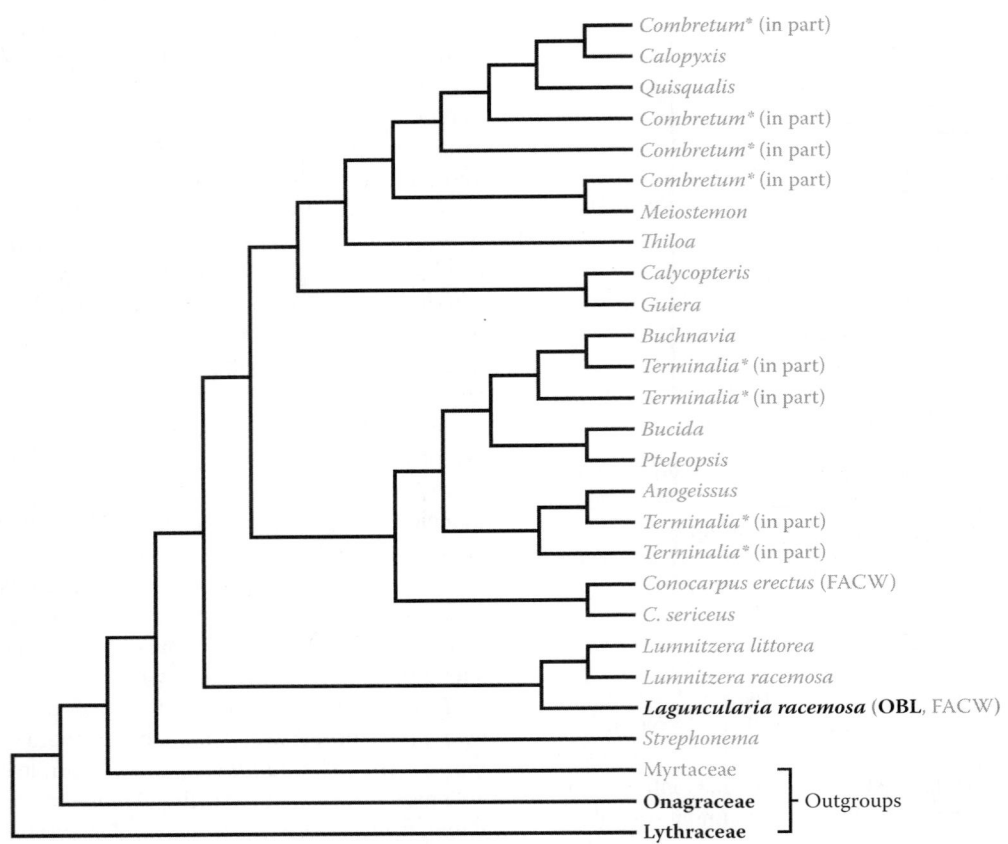

FIGURE 4.80 Phylogenetic relationships in Combretaceae as indicated by analyses of combined cpDNA and nrITS sequence data. The two largest genera (*Combretum* and *Terminalia*) are polyphyletic (asterisks). *Laguncularia*, the only North American OBL indicator (bold), resolves in a clade with *Lumnitzera* near the base of the family. (Adapted from Maurin, O. et al., *Bot. J. Linn. Soc.*, 162, 453–467, 2010.)

Imperilment: (1) *Laguncularia racemosa* [G5]
Ecology: general: *Laguncularia* is monotypic (see later).

***Laguncularia racemosa* (L.) Gaertn. f.** occurs in or on beaches, tidal coves, creeks, depressions, dunes, estuaries, lagoons, marshes, ponds, roadsides, salt marsh, scrub, shore flats, spoil islands, strands, swamps, thickets, and along the margins of canals, hammocks, rivers, and streams at elevations near sea level (<5 m). This species is categorized as a "mangrove," a term referring generally to woody plants that inhabit coastal wetlands. The plants are tolerant of salinity (at least 40 dS/m) but grow mainly in the higher marsh. They usually are found on the landward edge of mangrove communities, well above the high tide line, although they extend occasionally into more regularly flooded sites. Depending on conditions, salinities of up to 105 ppt can be tolerated. Adaptations to inundation and salinity occur in the leaves, which contain salt excretory glands on the blades, and by the production of pneumatophores and pneumathodes, which facilitate aeration of the root system. The plants also accumulate mannitol and proline to maintain an osmotic balance. Most of the plants occur in areas where the coldest winter temperatures remain above 15.5°C. They are shade tolerant and thrive when exposed to only intermittent periods of light. The substrates are acidic to alkaline (pH: 4.5–8.5) and include clay, coral limestone, marl, marl clay, oolitic limestone, peat, sand, shell mounds, and silt. Individuals possess nectar-producing flowers that either are bisexual or entirely staminate. About 60% of those populations surveyed are androdioecious, and contain from 1% to 68% male flowered plants. The staminate flowers open for only 1 day, whereas the bisexual flowers remain open for 2 days. Flowering can proceed year round, but peaks from May to July. Ordinarily, the flowers are outcrossed by a variety of insects (see *Use by wildlife*) including bees and wasps (Hymenoptera), butterflies (Lepidoptera), and flies (Diptera), which are attracted by the floral nectar. When growing with *Avicennia germinans*, pollinator competition reduces the visitation rate to *Laguncularia*. The hermaphroditic flowers are self-compatible and capable of autogamous self-pollination when pollinators are limited. Young plants (<2 years of age) are capable of fruit production. The fleshy fruits (formed from July to October) average 0.41 g and mature within 3 months from flowering. They are buoyant (possessing spongy wings) and are dispersed by water. The seeds can germinate while the fruits are still attached to the tree (then described as viviparous), or when detached and while floating. Their germination requires no pretreatment and occurs within 5–10 days of dispersal. The fruits remain viable for about 35 days. New seedlings will grow to 60–90 cm within a year. This is a ruderal species that readily colonizes disturbed sites, where pure stands can develop. The plants are tolerant to chromium (Cr^{3+}) levels up to at least 0.5 ppm. Ants (Hymenoptera: Formicidae) and flies (Diptera) are attracted to the plants by extrafloral nectaries, which occur on the petiole near the leaf base. The greatest nectar secretion occurs in more apically disposed leaves. Vegetative reproduction is insignificant but can occur by layering of rooting lower branches. Vegetative regrowth (coppicing) has also been reported from stump sprouts. The plants can be propagated readily by cuttings, most effectively if they are allowed first to root before their removal from the parental plant. Burrowing by fiddler crabs (Crustacea: Decapoda: Ocypodidae) has been shown to increase plant height (by 27%), trunk diameter (by 25%), and leaf production (by 15%). **Reported associates:** *Acrostichum aureum, A. danaeifolium, Ammannia latifolia, Avicennia germinans, Baccharis halimifolia, Bacopa monnieri, Batis maritima, Blutaparon vermiculare, Bolboschoenus robustus, Borrichia arborescens, B. frutescens, Bursera, Cladium jamaicense, Conocarpus erectus, Croton glandulosus, Cyperus strigosus, C. surinamensis, Dalbergia, Distichlis littoralis, Dysphania ambrosioides, Eleocharis cellulosa, Encyclia tampensis, Eupatorium serotinum, Euphorbia mesembrianthemifolia, Eustoma exaltatum, Ficus aurea, Fimbristylis thermalis, Gouania, Hydrocotyle bonariensis, Ipomoea violacea, Iva frutescens, Jacquinia keyensis, Kosteletzkya pentacarpos, Lycium carolinianum, Melothria pendula, Metopium toxiferum, Persea palustris, Phlebodium aureum, Pleopeltis polypodioides, Rhabdadenia biflora, Rhizophora mangle, Richardia brasiliensis, Sabal palmetto, Salicornia depressa, Sarcocornia perennis, Schinus, Schoenoplectus americanus, Sesuvium portulacastrum, Sida acuta, S. rhombifolia, Sideroxylon celastrinum, Spartina alterniflora, Spermacoce verticillata, Sporobolus virginicus, Suaeda linearis, Swietenia, Taxodium ascendens, Thespesia populnea, Tillandsia balbisiana, T. fasciculata, T. flexuosa, T. paucifolia, T. ×smalliana, T. utriculata, Tribulus cistoides, Vanilla barbellata, Vitex trifolia.*

Use by wildlife: *Laguncularia racemosa* is a host plant for several larval moths (Lepidoptera: Gelechiidae: *Anacampsis lagunculariella*; Nolidae: *Nola apera, N. lagunculariae*; Pyralidae: *Sarasota plumigerella*). The flowers are visited by various insects (many of them pollinators) including bees (Hymenoptera: Apidae: *Bombus, Xylocopa virginica*; Halictidae: *Agapostemon, Augochlora*; Megachilidae: *Megachile alleni*), beetles (Coleoptera: Buprestidae), butterflies and moths (Lepidoptera: Arctiidae: *Empyreuma*; Hesperiidae: *Phocides pigmalion*; Lycaenidae; Nymphalidae: *Heliconius charitonius, Junonia evarete*; Pyralidae), flies (Diptera: Bombyliidae: *Villa*; Calliphoridae; Syrphidae: *Baccha, Palpada albifrons*), and wasps (Hymenoptera: Scoliidae: *Scolia nobilitata*; Sphecidae: *Sphex jamaicensis*; Tiphiidae: *Myzinum*; Vespidae: *Euodynerus*). The plants also host a number of pathogenic fungi including causative agents for leaf necrosis (Ascomycota: Glomerellaceae: *Colletotrichum*), leaf spot (Ascomycota: Mycosphaerellaceae: *Cercospora; Septoria*), and seedling root rot (Oomycota: Pythiaceae: *Pythium*). A much broader fungal flora is reported from tropical regions.

Economic importance: food: The fruits of *Laguncularia racemosa* are eaten in parts of Africa; **medicinal:** The leaves of *L. racemosa* contain endophytic fungi, which produce secondary metabolites having antimicrobial activity. The plants have been used as an astringent and tonic and to prepare a remedy for dysentery. The bark extract reportedly has antitumor activity due to its tannin content, and is used to treat fever, scurvy, and oral ulcers. Ethyl acetate and butanolic extracts significantly inhibit human thrombin activity and plasma coagulation; **cultivation:** not cultivated; **misc. products:** the wood of *Laguncularia racemosa* is closely grained, hard, heavy (specific gravity: 0.6–0.8), strong, and can retain a high polish, but it is not durable (lasting only 2–3 years when exposed to the weather). It has been used to make charcoal, fence posts, poles, and tool handles, and also as a source of fuel. The bark and leaves are rich in tannin (12%–24%) and are the source of a brown dye. Bark extracts have been used to preserve fishing nets. The bark is also a source of gum that contains various sugars (arabinose, galactose, and rhamnose), and has been mixed with agar to make an inexpensive culture medium for fungi. The plants have been used to produce honey; **weeds:** The plants often are ruderal and can be weedy; **nonindigenous species:** none

Systematics: *Laguncularia racemosa* resolves phylogenetically as the sister genus to *Lumnitzera*, which represents another group of mangroves (Figure 4.80). Although this genus is considered to be monotypic, a broad genetic survey of populations should be conducted given their extensive geographical range and major continental disjunctions, especially between Old and New World localities. Populations that have been studied were found to contain low levels of genetic variation but are thought to exhibit heritable phenotypic differences that are induced by epigenetic responses to environmental variation. The chromosome number of *Laguncularia* has not been reported. There are no accounts of hybridization involving this genus.

Comments: *Laguncularia racemosa* occurs along southern coastal North America (Florida and Texas) and extends into Mexico, Central and South America as well as western Africa. **References:** Allen, 2002; Aronson, 1989; Dressler et al., 1987; Elster & Perdomo, 1999; Francini & Rovati, 2011; Geissler et al., 2002; Judd et al, 2008; Landry, 2013; Landry & Rathcke, 2007; Landry et al., 2009; Lira-Medeiros et al., 2010; Maurin et al., 2010; McKee, 1995; McMillan, 1975; Medina et al., 2007; Polania, 1990; Rabinowitz, 1978; Rocha et al., 2009; Rodrigues et al., 2015; Silva et al., 2011; Tomlinson, 1986; Wilcox et al., 2009.

Family 2: Lythraceae [30]

Lythraceae with approximately 600 species include a fairly sizeable number of aquatic plants. Several phylogenetic studies (Graham et al., 1993, 2005, 2011; Huang & Shi, 2002; Shi et al., 2000) have shown Lythraceae to be monophyletic once Punicaceae, Sonneratiaceae, and Trapaceae are merged within an expanded circumscription of the family (Figure 4.81). Molecular phylogenetic analyses have also convincingly supported the merger of *Ammannia*, *Hionanthera*, and *Nesaea* into a single genus under the former name (Graham et al., 2011). Two OBL species, once included in *Peplis*, are now placed in *Didiplis* and *Lythrum*.

Representatives of Lythraceae typically possess perigynous, heterostylous flowers with crumpled petals, distinct hypanthium, and a multilayered outer seed coat. A number of species exhibit heterostyly. The flowers often are showy and their pollination occurs commonly by insects. Although morphological characteristics have been used to divide Lythraceae into four subfamilies: Sonneratoideae (*Sonneratia*), Duabangoideae (*Duabanga*), Punicoideae (*Punica*), and Lythroideae (remaining genera), molecular data do not support these groups and the subfamilial classification of the family needs to be reconsidered.

Lythraceae include a fair number of economically important plants. Many cultivated ornamentals are derived from *Cuphea*, *Lagerstroemia*, *Lythrum*, and *Punica*. The latter

is also the source of the edible pomegranate (*P. granatum*). Leaves of *Lawsonia inermis* are the commercial source of henna dye. The decorative "crapemyrtles" that ornament the streets of Myrtle Beach, South Carolina are not members of the myrtle family (Myrtaceae), but specimens of *Lagerstroemia indica*. Several genera (e.g., *Lythrum*, *Rotala*) contain invasive species.

Numerous species also show an ecological affinity for wet habitats. Currently, six North American genera contain obligate indicators. *Cuphea micrantha* is designated as OBL in the Caribbean (Puerto Rico), but is excluded because it occurs outside of the geographical region covered here. However, *Cuphea aspera* (FACW; OBL in the 1996 list) has been included here because its treatment was completed prior to the issuance of the revised indicator list:

1. ***Ammannia*** L.
2. ***Cuphea*** P. Br. [FACW]
3. ***Decodon*** J. F. Gmel.
4. ***Didiplis*** Raf.

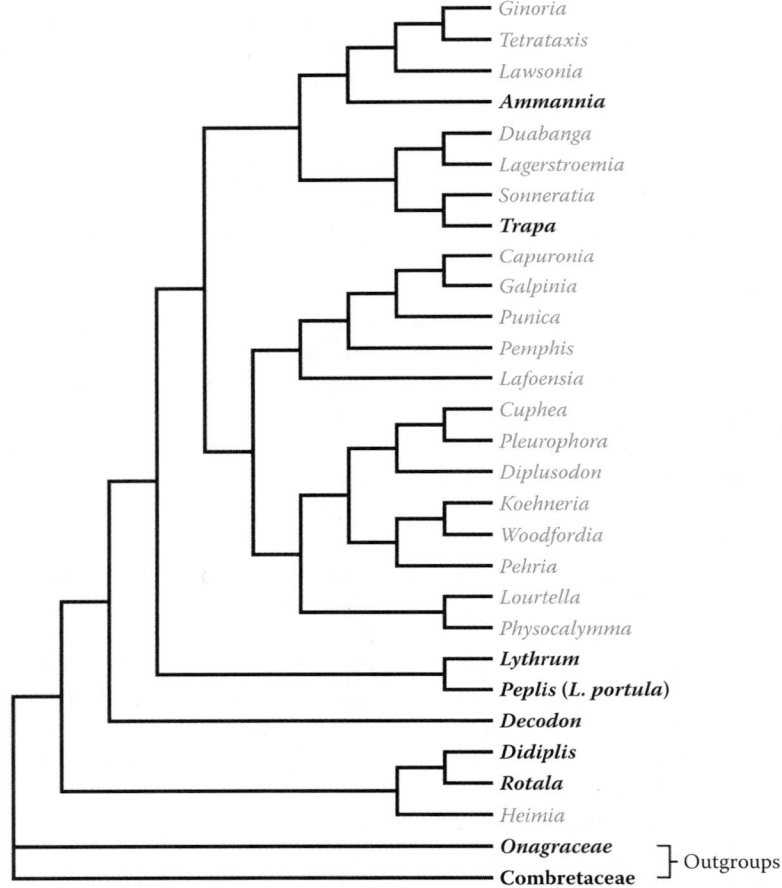

FIGURE 4.81 Phylogenetic relationships in Lythraceae as reconstructed from nrITS sequence data. This result clearly resolves *Punica*, *Sonneratia*, and *Trapa* (once assigned to three separate families) within the Lythraceae clade. Genera that contain obligate aquatics in North America are indicated in bold (the OBL species of *Cuphea* occurs outside of the range of this book). Aquatic plants are widespread in Lythraceae with at least six separate origins (including *Cuphea*) indicated here at the generic level. The indicator status of other North American genera that contain wetland species is indicated in parentheses. *Peplis portula* (the species surveyed in the study) has since been transferred to *Lythrum*. (Redrawn from Graham, S.A. et al., *Bot. J. Linn. Soc.*, 166, 1–19, 2011.)

5. *Lythrum* L.
6. *Rotala* L.
7. *Trapa* L.

1. *Ammannia*

Redstem

Etymology: after Paul Amman (1634–1691)

Synonyms: *Amannia; Ammanella; Cornelia; Hionanthera*

Distribution: global: Africa; Australia; Eurasia; New World; **North America:** widespread

Diversity: global: 25 species; **North America:** 4 species

Indicators (USA): OBL: *Ammannia auriculata, A. coccinea, A. latifolia, A. robusta*

Habitat: brackish, freshwater; palustrine; **pH:** 6.0–7.3; **depth:** <1 m; **life-form(s):** emergent herb

Key morphology: stems (to 10 dm) angled, glabrous, prostrate to erect; leaves (to 11 cm) opposite, 4-ranked, sessile, linear to oblanceolate, the bases tapering or auriculate, clasping; flowers axillary (1–10); perianth 4-merous, radial, fugacious, hypanthium present; petals (to 8 mm), lavender, pink, or rose-purple; stamens 4; capsules (to 6 mm) spherical, many seeded

Life history: duration: annual (fruit/seeds); **asexual reproduction:** none; **pollination:** insect, cleistogamous or self; **sexual condition:** hermaphroditic; **fruit:** capsules (common); **local dispersal:** seeds (water); **long-distance dispersal:** seeds (birds)

Imperilment: (1) *Ammannia auriculata* [G5]; S1 (AZ, NE); (2) *A. coccinea* [G5]; SH (MT); S1 (DC, DE); S2 (NC, PA, WV); S3 (GA, VA); (3) *A. latifolia* [G5]; SH (DC); S1 (NC, NJ); S2 (MD); S3 (GA); (4) *A. robusta* [G5]; S1 (BC, ON, WA, WI, WY)

Ecology: general: Most *Ammannia* species inhabit wet, open habitats at elevations extending to 1500 m. About 10 of the 25 *Ammannia* species typically occur in truly aquatic habitats with all four North American species categorized as obligate aquatics. A wide extent of substrates ranging from nearly pure sand to clay are colonized. The flowers are self-compatible and most species are self-pollinating (autogamous), with only occasional outcrossing. These are annual plants that produce numerous seeds (averaging 250/capsule) but have no means of clonal reproduction. The seeds possess an aerenchymatous float and surface hairs that evaginate when wet, both features that confer buoyancy. The floating seeds are transported by the water and can occur as a contaminant of rice seed, leading to weedy infestations in cultivated rice fields. The seeds can remain viable (>5%) for many (at least 27) years although their germination is reduced considerably (<50%) after 1 year. However, 12-year-old seeds still exhibit vigorous germination. Germination occurs in 6–14 days if the seeds are kept at 28°C with 100% humidity (in sealed bags). Field conditions (full sunlight, 20°C) result in germination within 10 days through 10 weeks. The plants require ample light and a lack of competition in order to establish. These conditions often are features of fluctuating water habitats, with which these species commonly are associated. *Ammannia* species are threatened by invasive plants such as *Lythrum salicaria*

and by hydrological alterations that alter the periodical water regime of their habitats.

Ammannia auriculata **Willd.** occurs in ditches, flats, floodplains, marshes, meadows, mudflats, playas, prairies, rice fields, river bottoms, swamps, and along the margins of ponds and rivers at elevations of up to 1350 m. The plants grow on exposed sediments or in shallow standing water. The substrates include alluvial silt, mud, sand, sandy muck, and silty sand. The simultaneously receptive flowers self-pollinate at the onset of anthesis. This is a late germinating species under drawdown conditions but the seeds will germinate continuously under flooded conditions. A large seed bank can be formed, with densities averaging 147–206 seeds/m². The roots are colonized by arbuscular mycorrhizal fungi and dark septate endophytes. **Reported associates:** *Ammannia coccinea, A. robusta, Bacopa repens, Bidens, Cephalanthus occidentalis, Cyperus erythrorhizos, Echinodorus, Gratiola virginiana, Juncus marginatus, Leptochloa panicoides, Leucospora multifida, Lindernia dubia, Ludwigia palustris, Neeragrostis reptans, Nyssa aquatica, Oryza sativa, Penthorum sedoides, Persicaria, Rotala ramosior, Taxodium distichum, Veronica peregrina, Xanthium strumarium.*

Ammannia coccinea **Rottb.** is a summer annual, which occurs in fresh to brackish (to 3.5 ppt salinity) shallow water (to 0.5 m deep) in bison wallows, channels, ditches, flats, gravel bars, lagoons, marshes, meadows, playas, rice fields, river bottoms, sandbars, seeps, sloughs, streams, washes, woodlands, or along the drying margins of lakes, ponds, reservoirs, rivers, sloughs, tidal marshes, and vernal pools at elevations of up to 910 m. The plants prefer high light but are fairly shade tolerant and will recover quickly when light conditions are restored. The substrates include clay, clay loam, gravel, muck, mud, sand, sandy silt, silt, silty clay (Capay), and Tuscan mudflow. The nectariferous flowers either are self-pollinated (at the onset of anthesis) or are outcrossed by insects such as small bees (Hymenoptera) and butterflies (Lepidoptera: Hesperiidae). Approximately 65%–100% of the newly produced seeds remain dormant over the winter and can persist in seed banks. The plants have represented as much as 24% of the seedlings emerging from seedbank studies. The seeds have physiological dormancy and require at least 100 days of cold stratification followed by day/night temperature regimes of 30°C/15°C or 35°C/20°C in the light (which is essential) for optimal (83%–93%) germination. Flooding increases germination rates at cooler daytime temperatures. The seeds lose their dormancy as the mudflats dry out during the summer. In competition with rice (*Oryza sativa*), more resources are allocated to shoot than to root growth (presumably as a response to maximize light exposure), resulting in taller plants, which can topple over in late season. Competition with rice can also result in up to a 97% decrease in dry seed weight. **Reported associates:** *Ageratina adenophora, Alisma Ambrosia psilostachya, Ammannia auriculata, A. latifolia, A. robusta, Atriplex coronata, Azolla filiculoides, Baccharis salicifolia, B. sarothroides, Bergia texana, Bidens cernuus, B. laevis, B. tripartitus, Bromus japonicus, Callitriche heterophylla, Chamaesyce humistrata, Conyza canadensis,*

Crassula aquatica, Croton setiger, Crypsis schoenoides, C. vaginiflora, Cyperus acuminatus, C. aristatus, C. difformis, C. diffusus, C. eragrostis, C. odoratus, C. setigerus, C. squarrosus, Cynodon dactylon, Distichlis spicata, Echinochloa colona, E. crus-galli, E. muricata, Echinodorus berteroi, Eclipta prostrata, Eleocharis acicularis, E. coloradoensis, E. engelmannii, E. macrostachya, E. montevidensis, E. obtusa, E. palustris, E. quadrangulata, E. rostellata, Epilobium ciliatum, Euphorbia nutans, Fimbristylis vahlii, Fuirena simplex, Gastridium ventricosum, Hedeoma hispida, Heteranthera limosa, Hydrocotyle ranunculoides, Juncus bufonius, J. dubius, J. torreyi, Lactuca saligna, Leersia oryzoides, Leptochloa fusca, Limosella acaulis, Lindernia dubia, Ludwigia peploides, Lythrum hyssopifolia, Marsillea vestita, Mentha, Mimulus guttatus, Muhlenbergia asperifolia, Najas flexilis, Nama stenocarpa, Oenothera elata, Persicaria hydropiperoides, P. lapathifolia, P. punctata, Phalaris caroliniana, Phyla nodiflora, Phyllanthus caroliniensis, Pilularia americana, Pluchea odorata, Poa arida, Polypogon viridis, Populus fremontii, Potentilla paradoxa, Pseudognaphalium luteoalbum, Ranunculus cymbalaria, R. sceleratus, Rotala indica, R. ramosior, Rumex altissimus, R. conglomeratus, R. crispus, R. maritimus, Sagittaria sanfordii, Salix gooddingii, S. laevigata, S. lasiolepis, Schoenoplectus acutus, S. californicus, S. hallii, S. mucronatus, S. pungens, S. saximontanus, S. tabernaemontani, Setaria parviflora, S. pumila, Stuckenia pectinata, Symphyotrichum subulatum, Tamarix, Typha domingensis, T. latifolia, Veronica anagallis-aquatica, V. peregrina, Xanthium strumarium.

***Ammannia latifolia* L.** occupies open, fresh to brackish (to 10.0 ppt salinity) sites in or on depressions, ditches, flatwoods, floodplains, hammocks, marshes, prairies, roadsides, salt flats, salt marsh, scrub, sloughs, spoil banks, swales, tidal marshes, wet prairies, and along the margins of canals, drying shores, and swamps at elevations of up to 28 m. The substrates have been described as marl, peat, sand, sandy muck, and Vero (Alfic Haplaquods). This species principally is an inbreeder that can produce chasmogamous or cleistogamous flowers, but is self-pollinating in either case. Both petalous and apetalous populations occur. The flowers yield relatively large numbers of small seeds that germinate quickly and produce numerous, small seedlings. **Reported associates:** *Acrostichum danaeifolium, Ammannia coccinea, Asclepias perennis, Bacopa monnieri, Batis maritima, Blutaparon, Boehmeria cylindrica, Carex lupulina, C. typhina, Commelina virginica, Conocarpus erectus, Cyperus compressus, C. odoratus, Distichlis spicata, Eleocharis fallax, Fimbristylis autumnalis, Genipa americana, Hydrocotyle verticillata, Hydrolea uniflora, Hypoxis hirsuta, Juncus roemerianus, Justicia ovata, Krugiodendron ferreum, Leersia lenticularis, Leptochloa fascicularis, Lindernia dubia, Lycium carolinianum, Metopium toxiferum, Mikania scandens, Onoclea sensibilis, Panicum, Persicaria punctata, P. setacea, Phanopyrum gymnocarpon, Phyla nodiflora, Proserpinaca palustris, Sabal palmetto, Sagittaria lancifolia, Salix, Saururus cernuus, Schoenoplectus pungens, S. tabernaemontani, Serenoa repens, Sesuvium, Spartina alterniflora, S. patens, Spiranthes*

cernua, Suaeda, Symphyotrichum subulatum, Taxodium, Uniola, Viola affinis.

***Ammannia robusta* Heer & Regel** grows in shallow water (to 0.5 m deep) in or on benches, depressions, ditches, ephemeral ponds, flats, floodplains, meadows, prairies, rice fields, riparian mud flats, sandbars, vernal pools, washes, and along the shores and margins of lakes, ponds, rivers, and streams at elevations of up to 1707 m. The plants occur in full sun, primarily where fluctuating water levels alternately produce flooding and drying cycles. The substrates usually are reported as alkaline and include adobe, alluvium, clay, clay loam muck, cobble, dolomite, gravel, loamy clay, loam muck, muck, mud, peat, sand, and silt. The nectariferous, simultaneously receptive flowers self-pollinate at the onset of anthesis and are visited by small bees (Hymenoptera) and butterflies (Lepidoptera: Hesperiidae). **Reported associates:** *Acer negundo, A. saccharinum, Alisma subcordatum, Amaranthus tuberculatus, Ambrosia artemisiifolia, A. trifida, Ammannia auriculata, A. coccinea, Amorpha, Apocynum sibiricum, Asclepias incarnata, Atriplex coronata, A. rosea, Baccharis salicifolia, Bacopa rotundifolia, Betula nigra, Bidens cernuus, B. comosus, B. connatus, B. frondosus, B. polylepis, Bolboschoenus fluviatilis, B. maritimus, Brassica nigra, Bromus japonicus, Campsis radicans, Carex athrostachya, C. caroliniana, C. cristatella, C. grayii, C. haydenii, C. muskingumensis, C. stipata, C. typhina, Cephalanthus occidentalis, Chasmanthium latifolium, Chenopodium bushianum, Coleataenia longifolia, Conobea multifida, Conoclinium coelestinum, Convolvulus simulans, Conyza canadensis, Coreopsis, Crypsis alopecuroides, C. schoenoides, Cyperus acuminatus, C. aristatus, C. bipartitus, C. difformis, C. ferruginescens, C. odoratus, C. squarrosus, C. strigosus, Deinandra conjugens, Desmanthus illinoensis, Distichlis spicata, Echinochloa crus-galli, E. muricata, E. walteri, Echinodorus berteroi, Eclipta prostrata, Eleocharis acicularis, E. erythropoda, E. intermedia, E. macrostachya, E. obtusa, E. ovata, E. palustris, Equisetum, Eragrostis frankii, E. hypnoides, Eupatorium perfoliatum, Euphorbia supina, Forestiera acuminata, Fraxinus pennsylvanica, Galium obtusum, Gratiola neglecta, Helenium, Helianthus tuberosus, Helminthotheca echioides, Hordeum jubatum, Humbertacalia, Ipomoea lacunosa, Iris virginica, Iva axillaris, Juncus acutus, J. articulatus, J. bufonius, J. dudleyi, J. torreyi, Laportea canadensis, Leersia oryzoides, Lemna minor, L. turionifera, Leptochloa fusca, L. panicoides, Leucospora multifida, Lilaeopsis occidentalis, Limosella acaulis, Lindernia dubia, Lipocarpha aristulata, Lobelia siphilitica, Ludwigia peploides, Lycopus americanus, Lysimachia ciliata, L. quadriflora, Lythrum alatum, L. hyssopifolium, L. tribracteatum, Melilotus albus, Mimulus ringens, Muhlenbergia frondosa, Nymphaea odorata, Panicum capillare, P. dichotomiflorum, P. virgatum, Parthenocissus vitacea, Pascopyrum smithii, Peltandra virginica, Persicaria amphibia, P. hydropiperoides, P. lapathifolia, P. maculosa, P. pensylvanica, P. punctata, Phalaris arundinacea, Phyla lanceolata, Physalis acutifolia, Platanus occidentalis, Poa pratensis, Polygonum ramosissimum, Populus deltoides, Potamogeton crispus, Proserpinaca, Quercus macrocarpa,*

Q. palustris, Ranunculus, Rorippa columbiae, R. curvisili-qua, R. palustris, Rotala ramosior, Rumex crispus, Sagittaria brevirostra, S. cuneata, S. latifolia, Salix amygdaloides, S. exigua, S. gooddingii, S. lucida, S. nigra, Schoenoplectus acutus, S. pungens, S. tabernaemontani, Scirpus cyperinus, Sesuvium verrucosum, Setaria faberi, Sium suave, Sonchus oleraceus, Spartina pectinata, Suaeda, Symphyotrichum ontarionis, S. pilosum, S. spathulatum, S. subulatum Tamarix, Trichocoronis wrightii, Trichostema lanceolatum, Typha latifolia, Verbena hastata, Verbesina alternifolia, Vernonia fasciculata, Veronica anagallis-aquatica, Vitis cinerea, V. riparia, Xanthium strumarium.

Use by wildlife: The foliage of *Ammannia coccinea* report-edly is palatable and is eaten by Canada geese (Aves: Anatidae: *Branta canadensis*), nutria (Mammalia: Myocastoridae: *Myocastor coypus*), and muskrats (Mammalia: Cricetidae: *Ondatra zibethicus*). In the fall and winter the floating seeds are consumed in quantity by ducks (Aves: Anatidae) such as green-winged teal (*Anas crecca*), mallard (*Anas platyrhyn-chos*), and northern pintail (*Anas acuta*). The floral nectar and pollen of *A. coccinea* attracts various bees (Hymenoptera: Andrenidae: *Calliopsis andreniformis*; Anthophoridae: *Ceratina dupla*; Halictidae: *Augochlorella aurata, Lasioglossum imitatus, L. pilosus, L. versatus*), butterflies (Lepidoptera: Hesperiidae (various); Lycaenidae: *Everes comyntas*; Pieridae: *Eurema lisa, Pontia protodice*), and flies (Diptera: Syrphidae: *Syritta pipiens, Toxomerus marginatus*; Tachinidae: *Acroglossa hesperidarum*).

Economic importance: food: The Mohave and Yuma tribes collected the seeds of *Ammannia coccinea* as food. The ground seeds of several species have been used in baking. The seeds of *A. auriculata* contain 15% oil, 16% protein, and are rich in linoleic acid; **medicinal:** Extracts of the Asian *Ammannia baccifera* are effective at treating urinary stones, but none of the North American species has any widespread medicinal use. *Ammannia auriculata* contains oleananes and polyphenols, which have high antioxidant activity (comparable to ascorbic acid) and exhibit mild cyto-toxicity against HaCaT cells; **cultivation:** *Ammannia lati-folia* is cultivated as an aquarium plant as it can withstand prolonged periods of submergence; **misc. products:** none; **weeds:** *Ammannia coccinea* is a rice field weed in the west-ern United States. Some strains of cultivated rice (*Oryza sativa*) have strong allelopathic effects on *A. coccinea*. *Ammannia auriculata* is a weed of rice fields in Sardinia and has also been introduced to South America. *Ammannia coccinea* has been introduced to Afghanistan, Hawaii, Italy, Japan, Korea, Philippines, South Pacific, and Spain through rice culture, and to Panama as a contaminant of hay. The South American distribution of *Ammannia latifolia* also probably originated through human introduction. *Ammannia robusta* has been introduced to Hawaii, Philippines, South America, and the South Pacific as a contaminant of rice. *Ammannia coccinea* populations introduced to California rice fields have been known to acquire resistance to ben-sulfuron methyl herbicides within as little as 4 years; **non-indigenous species:** *Ammannia auriculata* is regarded as nonindigenous to Australia and has also been introduced to Italy and Japan; *Ammannia auriculata* and *A. coccinea* are naturalized in the Hawaiian Islands. *Ammannia latifolia* occurs from the United States to South America, but was introduced to northern shipyards in the 19th century, prob-ably as a consequence of ballast disposal.

Systematics: Prior phylogenetic analyses of morphologi-cal data placed *Ammannia* closest to *Nesaea*, which was nearly identical morphologically and palynologically. That result was corroborated by molecular data, which resolved *Ammannia* and *Nesaea* (together with *Hionanthera*) as a strongly supported clade that in turn formed the sister group to a clade comprising *Ginoria*, *Lawsonia* (distant from *Ammannia* and *Nesaea* in the morphological analyses), and *Tetrataxis* (Figure 4.81). *Nesaea* also included species that are categorized as aquatic and wetland plants; whereas, the monotypic *Lawsonia* (*L. inermis*; cultivated henna) is terres-trial, occurs in dry deciduous forests, and can tolerate even arid conditions. Consequently, the current taxonomic disposi-tion is to merge *Hionanthera* and *Nesaea* with *Ammannia*, which has been followed here. Literature reports for indi-vidual *Ammannia* species should be viewed with some cau-tion. *Ammannia coccinea* often appears in the older literature under the synonym *A. teres* Raf.; whereas, *A. robusta* often was misreported as *A. coccinea* in publications prior to 1979. *Ammannia auriculata* has also been confused with *A. coc-cinea*, especially in California, where only the latter has been authenticated as resident. *Ammannia* is subdivided into two subgenera: *Cryptotheca* (including only *A. microcarpa*) and subgenus *Ammannia*, which contains the remainder of spe-cies including those following. *Ammannia latifolia* is placed in section *Ammannia* which is characterized by flowers hav-ing short, included styles. *Ammannia auriculata*, *A. coc-cinea*, and *A. robusta* are placed within section *Eustylia*, which is characterized by flowers having long, exserted styles. However, because no phylogenetic analysis of *Ammannia* has yet been carried out, neither the existing classification nor these implied relationships are tenable. A large amount of cytological data has enabled the elucidation of $x = 8$ as the chromosomal base number for Lythraceae. However, counts recorded within *Ammannia* are too diverse to ascertain a base number for this genus with any certainty. *Ammannia coccinea* ($2n = 66$) is believed to be an amphidiploid hybrid derivative of the tetraploids *A. auriculata* ($2n = 30, 32$) and *A. robusta* ($2n = 34$); however, molecular genetic data in support of this hypothesis would be desirable. *Ammannia latifolia* is a hexa-ploid ($2n = 48$).

Comments: *Ammannia auriculata* extends from the cen-tral and western United States to South America as well as Africa. *Ammannia coccinea* ranges from much of the United States through Central and South America. *Ammannia lati-folia* extends from the southeastern United States to South America. *Ammannia robusta* occurs from southern Canada to Mexico.

References: Baskin et al., 2002; Caton et al., 1997; Collins et al., 2013; Gibson et al., 2001; Gohar et al., 2012; Gordon, 1999; Graham, 1979, 1985; Graham & Cavalcanti, 2001;

Graham & Gandhi, 2013; Haukos & Smith, 2001; Middleton, 2003; Nawwar et al., 2014; Prasad et al., 1994; Stevens et al., 2010.

2. *Cuphea*
Waxweed

Etymology: from the Greek *cyphos* meaning "gibbous," with reference to the calyx
Synonyms: *Cuphaea*; *Melanium*; *Melvilla*; *Parsonsia*
Distribution: global: New World; **North America:** eastern, southern
Diversity: global: 260 species; **North America:** 6 species
Indicators (USA): FACW: *Cuphea aspera*
Habitat: freshwater; palustrine; **pH:** unknown; **depth:** <1 m; **life-form(s):** emergent herb
Key morphology: stems (to 4 dm) pubescent and sticky above; leaves (to 2.5 cm) whorled or opposite, sessile, simple, lanceolate to linear elliptic, 1-nerved, short pubescent; flowers in racemes, pedicellate (to 15 mm); perianth bilateral, a tubular hypanthium (to 9 mm) present; calyx gibbous above, sepals 6, green, alternating with bristle-tipped appendages; petals 6, lavender to white, the upper 2 largest; stamens 11; capsules with few (typically 3) seeds
Life history: duration: perennial (rhizomes, tuberous roots); **asexual reproduction:** rhizomes; **pollination:** insect; **sexual condition:** hermaphroditic; **fruit:** capsules (common); **local dispersal:** rhizomes, seeds; **long-distance dispersal:** seeds
Imperilment: (1) *Cuphea aspera* [G2]; S2 (FL)
Ecology: general: Only a few species in this large genus are aquatic or wetland plants. Of the species in our range, just two occur facultatively in wetlands (FACW), including *C. aspera*, which had been ranked formerly as an OBL indicator. Despite its size, little information exists on the floral biology or seed ecology of this genus. Flowers of at least some *Cuphea* species are self-compatible, but members of section *Euandra* (the section to which *C. aspera* is assigned) have exserted stamens and primarily are outcrossed. Pollinators in the genus include bees (Insecta: Hymenoptera) or occasionally hummingbirds (Aves: Trochilidae).

Cuphea aspera Chapm. occurs in seasonally wet depressions, flatwoods, savannahs, seepage slopes, wet prairies, and along moist roadsides or the margins of shrub bogs. The substrates have been described as loam, loamy sand, organic sand, or sandy peat. The plants are shade intolerant but are unable to thrive in areas that have been managed by fire suppression. The plants are also threatened by habitat alterations due to herbicides and mowing, as well as by encroaching development. Little other ecological information has been reported for this extremely rare species and more detailed studies should be undertaken. **Reported associates:** *Andropogon*, *Aristida*, *Asclepias viridula*, *Boltonia diffusa*, *Cliftonia monophylla*, *Cyrilla racemiflora*, *Eurybia spinulosa*, *Hypericum*, *Ilex*, *Justicia crassifolia*, *Lachnanthes caroliniana*, *Macbridea alba*, *Nyssa*, *Pinus elliottii*, *Platanthera cristata*, *Pteridium*, *Rhexia*, *Rhynchospora oligantha*, *Sarracenia*, *Scleria*, *Scutellaria floridana*, *Serenoa repens*, *Vaccinium*, *Verbesina chapmanii*, *Xyris*.

Use by wildlife: none reported
Economic importance: food: none; **medicinal:** none; **cultivation:** There are many cultivated species of *Cuphea*; however, *C. aspera* is not among them. Seeds of *C. aspera* have been deposited for conservation purposes in the National Seed Storage Laboratory; **misc. products:** none; **weeds:** none; **nonindigenous species:** none
Systematics: Phylogenetic analysis of combined molecular data resolve *Cuphea* and *Pleurophora* as sister genera in a clade with *Diplusodon* (Figure 4.81). *Cuphea* represents an independent origin of the aquatic habit (the Puerto Rican *C. micrantha* is OBL) within this clade of three genera, which otherwise lacks obligate aquatic species. The classification of *Cuphaea*, last revised comprehensively over a century ago, is in need of revision. *Cuphea aspera* has been assigned to subgenus *Bracteolatae*, section *Euandra* (the largest of 13 sections currently recognized), subsection *Oidemation*. About 10% of *Cuphea* species (but not *C. aspera*) have been surveyed in preliminary molecular phylogenetic studies that have indicated section *Euandra* and its subsections to be polyphyletic. Broader sampling of taxa will be necessary to obtain a better idea of relationships for *C. aspera*. *Cuphea* is diverse chromosomally, and speciation in the genus is believed to be facilitated by isolating barriers that arise from ploidy level differences. Whether the basic chromosome number of *Cuphea* is $x = 8$, 11, or 12 has been difficult to determine and the resolution of this question awaits a more comprehensive phylogenetic study of the genus. *Cuphea aspera* ($2n = 48$) either is tetraploid or hexaploid, but ranks among the highest chromosome numbers reported in the genus. The species also differs from most other members of subsection *Oidemation* whose chromosome number are based on $x = 9$.
Comments: *Cuphea aspera* is a rare species of about 20 extant populations, which are endemic to the Florida panhandle region.
References: Graham, 1989; Luker et al., 1999.

3. *Decodon*
Swamp loosestrife

Etymology: from the Greek *deka* meaning "ten" and *odous* meaning "tooth" in reference to the 10-toothed calyx
Synonyms: *Lythrum* (in part); *Nesaea* (in part)
Distribution: global: North America; **North America:** eastern
Diversity: global: 1 species; **North America:** 1 species
Indicators (USA): OBL: *Decodon verticillatus*
Habitat: freshwater; lacustrine, palustrine; **pH:** 4.6–9.9; **depth:** 0–2 m; **life-form(s):** emergent shrub
Key morphology: stems (to 2.5 m) arching, 4–6 angled, woody basally, rooting distally, spongy when inundated; leaves (to 20 cm) opposite or whorled, lanceolate; flowers in axillary verticels, tristylous; hypanthium terminating in 5–7 bristle-tipped calyx segments alternating with horn-like projections; petals (to 1.2 cm) 5, magenta; stamens 10, in two different lengths within a flower; capsules (to 5 mm) roundish
Life history: duration: perennial (buds, rhizomes); **asexual reproduction:** shoot layering; **pollination:** insect; **sexual**

condition: hermaphroditic; heterostylous; **fruit:** capsules (common); **local dispersal:** layering, seeds; **long-distance dispersal:** seeds

Imperilment: (1) *Decodon verticillatus* [G1]; S1 (IA, MO, NB, PE, WV); S2 (MS); S3 (KY, MN, NS, QC, TN)

Ecology: general: *Decodon* is monotypic (see later).

Decodon verticillatus **(L.) Elliott** occurs in a wide range of habitats including bogs, floodplains, flowages, levees, marshes, meadows, oxbows, ravines, seeps, sloughs, swamps, and the margins of lakes, ponds, rivers, and streams at elevations of up to 241 m. Although occurring mainly in freshwater, populations can tolerate slightly brackish conditions. The plants usually occur in shallow water (to ~1 m) but the shoots can extend over waters exceeding 3 m in depth. The substrates range from acidic to calcareous types, which include gravel, muck, mud, peat, sand, and silt of broad alkalinity (0–340 ppm total $CaCO_3$), conductivity (20–850 μmhos/cm), and (pH: 4.6–9.9). The plants are fairly shade tolerant and can establish under wetland forest canopies. *Decodon* is highly clonal and spreads vegetatively by layering of its arching shoot tips, which form adventitious roots when coming in contact with water. Shoot buoyancy is facilitated by a well-formed spongy aerenchyma tissue that forms around the periphery of any parts of the stem that become immersed. The shoots are severed from the parental plant during winter senescence. The detached vegetative shoots can disperse, but usually remain within a few meters of the parental plant from which they originate. The plants often grow so dense as to form a monospecific zone that encircles shallow margins of lakes and ponds; however, they also grow in association with a large number of other aquatics (see later). *Decodon* is tristylous but populations at the northern portions of its range frequently are monomorphic for style length. The plants are self-compatible (but possesses "cryptic" self-incompatibility favoring outcrossed pollen); however, they are not autogamous. Geitonogamy accounts for roughly 30% of progeny and typically results in inbreeding depression. The monomorphic populations have severely limited sexual reproduction and are depauperate genetically compared to tristylous populations, which are distributed in more southern areas. The flowers are pollinated by insects such as bumble bees (Hymenoptera: Apidae: *Bombus*), honey bees (Hymenoptera: Apidae: *Apis mellifera*), and butterflies (Lepidoptera). The seeds of *D. verticillatus* have physiological dormancy. Germination can be moderate after 90 days of stratification, but is optimal following 18 months of cold storage. **Reported associates:** *Acer rubrum*, *A. saccharinum*, *Alnus rugosa*, *Amorpha fruticosa*, *Andromeda glaucophylla*, *Apios americana*, *Bacopa caroliniana*, *Berula erecta*, *Boehmeria cylindrica*, *Bolboschoenus fluviatilis*, *Brasenia schreberi*, *Calla palustris*, *Calopogon pulchellus*, *C. tuberosus*, *Campanula aparinoides*, *Carex aquatilis*, *C. buxbaumii*, *C. comosa*, *C. interior*, *C. lasiocarpa*, *C. livida*, *C. pellita*, *C. prairea*, *C. sartwellii*, *C. stricta*, *Calamagrostis canadensis*, *Celtis occidentalis*, *Cephalanthus occidentalis*, *Ceratophyllum echinatum*, *Chamaecyparis thyoides*, *Chamaedaphne calyculata*, *Chara*, *Chelone glabra*, *Cicuta bulbifera*, *C. maculata*, *Cinna arundinacea*, *Cladium jamaicense*, *Colocasia esculenta*, *Comarum*

palustre, *Cuscuta gronovii*, *Cyperus erythrorhizos*, *Drosera intermedia*, *Dulichium arundinaceum*, *Eichhornia crassipes*, *Eleocharis baldwinii*, *E. elliptica*, *E. equisetoides*, *E. erythropoda*, *E. olivacea*, *E. palustris*, *E. quadrangulata*, *E. robbinsii*, *E. tuberculosa*, *Elodea nuttallii*, *Epilobium strictum*, *Eriocaulon aquaticum*, *Eriophorum angustifolium*, *E. virginicum*, *Eupatorium perfoliatum*, *E. serotinum*, *Eutrochium maculatum*, *Fraxinus nigra*, *F. pennsylvanica*, *Galium trifidum*, *Glyceria grandis*, *Hackelia virginiana*, *Hydrocotyle umbellata*, *Hypericum adpressum*, *Ilex verticillata*, *Impatiens capensis*, *Juncus militaris*, *Larix laricina*, *Leersia oryzoides*, *L. virginica*, *Lemna minor*, *Limnobium spongia*, *Lobelia dortmanna*, *L. siphilitica*, *Ludwigia peruviana*, *L. sphaerocarpa*, *Lysimachia terrestris*, *L. thyrsiflora*, *Lythrum salicaria*, *Mayaca fluviatilis*, *Menyanthes trifoliata*, *Mnium*, *Muhlenbergia glomerata*, *Myriophyllum verticillatum*, *Najas guadalupensis*, *N. marina*, *Nasturtium officinale*, *Nuphar advena*, *Nymphaea odorata*, *Nymphoides cordata*, *Nyssa aquatica*, *N. biflora*, *Onoclea sensibilis*, *Osmunda regalis*, *Panicum hemitomon*, *Paspalum urvillei*, *Peltandra virginica*, *Penthorum sedoides*, *Persicaria arifolia*, *P. punctata*, *P. virginiana*, *Phalaris arundinacea*, *Phragmites australis*, *Picea mariana*, *Pilea fontana*, *P. pumila*, *Pinus rigida*, *Platanus occidentalis*, *Pogonia ophioglossoides*, *Pontederia cordata*, *Populus heterophylla*, *Potamogeton amplifolius*, *P. crispus*, *P. nodosus*, *Quercus rubra*, *Rhexia mariana*, *Rhododendron groenlandicum*, *R. viscosum*, *Rhynchospora chalarocephala*, *R. inundata*, *Ribes americanum*, *Robinia pseudoacacia*, *Rosa palustris*, *Rubus*, *Sabal minor*, *Sacciolepis*, *Sagittaria lancifolia*, *S. latifolia*, *S. platyphylla*, *Salix candida*, *Sarracenia purpurea*, *Sassafras albidum*, *Saururus cernuus*, *Schoenoplectus acutus*, *S. deltarum*, *S. etuberculatus*, *S. subterminalis*, *S. tabernaemontani*, *Scirpus cyperinus*, *Solidago uliginosa*, *Spaganium americanum*, *S. eurycarpum*, *Sphagnum*, *Spiraea tomentosa*, *Stachys tenuifolia*, *Symphyotrichum boreale*, *S. puniceum*, *Symplocarpus foetidus*, *Taxodium ascendens*, *T. distichum*, *Thelypteris palustris*, *Toxicodendron radicans*, *Triadenum fraseri*, *T. virginicum*, *Typha angustifolia*, *T. latifolia*, *Urtica dioica*, *Utricularia intermedia*, *U. gibba*, *U. purpurea*, *U. macrorhiza*, *Vaccinium corymbosum*, *V. macrocarpon*, *V. oxycoccos*, *Viburnum lentago*, *Vigna luteola*, *Woodwardia virginica*, *Xyris congdoni*, *X. elliottii*, *Zizania aquatica*.

Use by wildlife: *Decodon* plants provide cover and food for muskrats (Mammalia: Cricetidae: *Ondatra zibethicus*). The seeds are eaten by various waterfowl (Aves: Anatidae) including black ducks (*Anas rubripes*), blue-winged teal (*Anas discors*), green-winged teal (*Anas carolinensis*), mallards (*Anas platyrhynchos*), and wood ducks (*Aix sponsa*). The least bittern (Aves: Ardeidae: *Ixobrychus exilis*) and Virginia rail (Aves: Rallidae: *Rallus limicola*) use *Decodon* plants as nesting sites. *Decodon* generally attracts bees (Hymenoptera), butterflies (Lepidoptera), and other insects as a nectar source. It is the larval host plant for several butterflies (Lepidoptera: Noctuidae: *Acronicta oblinita*, *Eudryas unio*, *Papaipema cataphracta*, *P. sulphurata*) and moths (Lepidoptera: Saturniidae: *Hyalophora cecropia*; Sphingidae:

Darapsa versicolor). Notably, *Papaipema sulphurata* (the decodon stem borer) is an extremely rare moth found only in southeastern Massachusetts. As a consequence of its fairly close relationship to *Lythrum* (Figure 4.81), *Decodon* plants are somewhat susceptible to phytophagous beetles (Insecta: Coleoptera: Chrysomelidae: *Galerucella*) that have been introduced to control the invasive *Lythrum salicaria*. *Decodon* plants are also a host of leaf spot fungi (Ascomycota: Mycosphaerellaceae: *Pseudocercospora*).

Economic importance: food: none; **medicinal:** *Decodon verticillatus* is rich in alkaloids including decine (a hydrodiuretic) and the cardiovascularly active cryogenine and vertine (hypotensives); **cultivation:** *Decodon verticillatus* is cultivated as an ornamental pond plant; **misc. products:** none; **weeds:** none; **nonindigenous species:** none

Systematics: Molecular data resolve the monotypic *Decodon verticillatus* as the sister group to the clade comprising the majority of Lythraceae sans a basal group that includes *Didiplis*, *Heimia*, and *Rotala*. It is placed near to a clade comprising *Lythrum* and *Peplis*, which also contains obligate aquatics (Figure 4.81). However, *Decodon* differs from those genera, which possess a derived chromosome number (based on x = 5) and share a 54 base pair deletion in the ITS-1 region of nrDNA. *Decodon verticillatus* (2n = 16) is a diploid based on x = 8, the presumed ancestral base number of Lythraceae. This species has provided an excellent study model for tristylous reproductive systems. It is not known to hybridize with other Lythraceae.

Comments: *Decodon* is distributed widely throughout eastern North America.

References: Dorken & Eckert, 2001; Eckert, 2002; Eckert & Allen, 1997; Eckert & Barrett, 1992, 1993a,b, 1994a,b,c, 1995; Eckert et al., 1999; Ferris et al., 1966a,b; Ghisalberti et al., 1998; Hoyer et al., 1996; Lempe et al., 2001; Winstead & King, 2006.

4. *Didiplis*

Water purslane, water hedge
Etymology: from the Greek *dis diplos* meaning "twice paired," referring either to the stamens or to the strongly decussate foliage
Synonyms: *Peplis* (in part)
Distribution: global: North America; **North America:** eastern
Diversity: global: 1 species; **North America:** 1 species
Indicators (USA): OBL: *Didiplis diandra*
Habitat: freshwater; lacustrine, palustrine; **pH:** 5.8–7.2; **depth:** <1 m; **life-form(s):** submersed (vittate), emergent herb
Key morphology: shoots (to 4 dm) weak, green but often reddish at tips; leaves heterophyllous, submersed (to 22 mm) or emersed (to 8.4 mm), opposite, decussate, sessile, linear (submersed) to subelliptic (emersed); flowers solitary, axillary, small (<3 mm), greenish; sepals (to 3 mm) 4; petals absent; stamens 2–4; fruit (to 2 mm) globose, indehiscent, 2-celled; seeds numerous
Life history: duration: annual (fruit/seeds); **asexual reproduction:** shoot fragments; **pollination:** unknown; **sexual**

condition: hermaphroditic; **fruit:** capsules (common); **local dispersal:** seeds; **long-distance dispersal:** seeds
Imperilment: (1) *Didiplis diandra* [G5]; SH (MN); S1 (KY, NC, OK, TN, TX, VA, WI); S2 (AR, IN, KS, LA); S3 (IA)
Ecology: general: *Didiplis* is monotypic (see later).

***Didiplis diandra* (Nutt. ex DC.) Alph. Wood** grows either as a submersed plant in shallow waters (to 0.5 m) of ponds and temporary pools, or as stranded plants exposed in depressions, on mudflats, in marshes, or along the margins of ditches, lakes, ponds, and streams. The plants thrive in abundant light but can tolerate slight shade. The substrates have been described as silt loam. Good growth occurs in water temperatures from 18°C to 28°C with optimal growth occurring (in aquaria) at 22°C–26°C. The plants remain sterile when growing under submersed conditions and produce flowers only when the shoots are emergent. Information on the reproductive ecology of this species is minimal. The seeds reportedly float initially, and do not germinate until they eventually sink. Although the plants are annuals, they can be propagated successfully from detached shoot fragments. **Reported associates:** *Alisma subcordatum*, *Bidens cernuus*, *B. frondosus*, *Carex obnupta*, *Ceratophyllum demersum*, *Cyperus erythrorhizos*, *Digitaria ischaemum*, *Echinochloa crus-galli*, *Eleocharis engelmannii*, *E. palustris*, *Elodea canadensis*, *Eragrostis hypnoides*, *Glyceria*, *Gratiola neglecta*, *Leersia oryzoides*, *Lemna minor*, *L. trisulca*, *Lindernia dubia*, *Ludwigia palustris*, *Myriophyllum pinnatum*, *Nasturtium officinale*, *Nuphar*, *Persicaria hydropiperoides*, *Phyla lanceolata*, *Proserpinaca palustris*, *Ranunculus flabellaris*, *R. longirostris*, *Ricciocarpus natans*, *Saururus cernuus*, *Scirpus cyperinus*, *Sparganium angustifolium*, *S. eurycarpum*, *Spirodela polyrhiza*, *Wolffia columbiana*.

Use by wildlife: none reported
Economic importance: food: none; **medicinal:** none; **cultivation:** *Didiplis diandra* is highly prized and widely distributed as an aquarium specimen due to its attractive foliage, but is somewhat difficult to grow and requires high light in culture; **misc. products:** none; **weeds:** none; **nonindigenous species:** none
Systematics: Many authors regard the monotypic *Didiplis diandra* as related so closely to *Lythrum* and *Peplis* that it often is merged with the latter. All three genera are assigned to subfamily Lythroideae, tribe Lythreae, subtribe Lythrinae. However, phylogenetic analyses of molecular data (*atpB–rbcL* intergenic spacer, nrITS sequences) indicate that *Didiplis* is not related either to *Peplis* or *Lythrum*, but occurs in a distant clade (with *Heimia*) as the sister genus to *Rotala* (also assigned to subtribe Lythrinae). This result is consistent with the pollen morphology of *Didiplis*, which is distinct from that of either *Lythrum* or *Peplis*. Prior morphological cladistic analyses had placed *Didiplis* within a clade of otherwise unresolved genera including *Rotala*, *Hionanthera*, and *Peplis* (but quite distant from *Lythrum*), thus being somewhat consistent with the results indicated by molecular data. Details on the reproductive biology and genetic structure of *D. diandra* populations are scarce in the literature. *Didiplis* (2n = 32) is

believed to represent a paleopolyploid based on the presumed basic number of $x = 8$ for Lythraceae.

Comments: *Didiplis diandra* is rare throughout its range which includes the eastern United States (except for the northeast) and disjunct stations in Utah.

References: Graham et al., 1987, 2011; Horn, 2011; Morris et al., 2005a; Tobe et al., 1986.

5. *Lythrum*

Loosestrife; salicaire

Etymology: from the Greek *lythron* meaning "blood" and referring either to the petal color or to the medicinal use of the plant as a styptic

Synonyms: *Peplis* (in part); *Salicaria*

Distribution: global: cosmopolitan; **North America:** widespread

Diversity: global: 35 species; **North America:** 12 species

Indicators (USA): OBL: *Lythrum alatum*, *L. californicum*, *L. curtissii*, *L. flagellare*, *L. hyssopifolia*, *L. lineare*, *L. portula*, *L. salicaria*, *L. tribracteatum*; **FACW:** *L. alatum*, *L. salicaria*

Habitat: brackish (coastal), freshwater; palustrine; **pH:** 4.0–9.1; **depth:** <1 m; **life-form(s):** emergent herb

Key morphology: stems (to 15 dm) decumbent, prostrate or erect, terete or angular; leaves sessile, alternate (above), opposite or whorled (below), the blades (1–14 cm) linear, lanceolate, elliptic or oblong; flowers pedicellate (to 2 cm), 1–2 in leaf axils or in elongate spikes, monomorphic, distylous or tristylous; hypanthium (to 7 mm) with 4–6 sepals; petals (1–14 mm) 4–6, crumpled, white to purple; stamens 4, 6, or 12; capsules (to 4 mm) with 2-valved dehiscence

Life history: duration: annual (fruit/seeds); perennial (persistent shoot bases); **asexual reproduction:** creeping rootstocks, shoot fragments; **pollination:** insect; **sexual condition:** heterostylous, homostylous; **fruit:** capsules (common); **local dispersal:** seeds (water, wind); **long-distance dispersal:** seeds (waterbirds)

Imperilment: (1) *Lythrum alatum* [G5]; SH (VT); S1 (CT, DC, KY, MA, MD, NC, PA, WY); S2 (NJ, VA, WV); S3 (ON); (2) *L. californicum* [G4]; S2 (UT); (3) *L. curtissii* [G1]; S1 (FL, GA); (4) *L. flagellare* [G2]; S2 (FL); (5) *L. hyssopifolia* [G5] (6) *L. lineare* [G5]; S1 (NY); S2 (GA); S3 (NC, NJ); (7) *L. portula* [GNR]; (8) *L. salicaria* [G5]; (9) *L. tribracteatum* [GNR].

Ecology: general: *Lythrum* species commonly are associated with wet habitats. About six species worldwide extend into deeper aquatic habitats, with most of the other species found in wetlands. Nine of the North American species (75%) are categorized as obligate aquatic plants. The flowers are insect pollinated and either homostylous (annuals) or heterostylous (di- or tristylous) (perennials). The monomorphic annuals mostly are self-pollinated. In distylous species, the long-styled floral morphs are called "pins" and the short-styled morphs "thrums." Individual plants produce only one type of morph. In *Lythrum*, self-incompatibility tends to be more strongly developed in the pins which typically outnumber thrums in natural populations (although exact ratios of morphs in populations can change considerably throughout the season). Charles Darwin studied the floral polymorphisms of *Lythrum* in detail and was the first to elucidate that heterostylous reproduction was a means of enhancing outcrossing. The seeds typically float (at least for a short time) and are dispersed locally by water currents. The seeds also adhere to mud on the feet of waterbirds, which can facilitate their long-distance dispersal.

Lythrum alatum **Pursh** inhabits wet open sites including ditches, fens, glades, meadows, prairies, seeps, swales, washes, and the margins of lakes, ponds, and streams at elevations of up to 1838 m. The substrates are acidic (pH: 5.7–6.7) and include clay, gravel, muck, peat, sand, sandy loam, and silty loam. Visitation rates of bumble bees and honey bees (Insecta: Hymenoptera: Apidae) to the distylous, entomophilous flowers of *L. alatum* are reduced in the presence of nonindigenous species, which compete for pollinators. However, butterflies (Insecta: Lepidoptera) are the most common pollinators of this species. Furthermore, mixtures of conspecific and foreign congeneric pollen can reduce seed set by nearly 30%. *Lythrum alatum* forms persistent seed banks. The seeds are physiologically dormant but they can be germinated by sowing them at temperatures below 5°C (−4°C to +4°C) for 12–13 weeks, followed by incubation at 20°C (in light). The slightly winged seeds presumably are dispersed by the wind and/or water. **Reported associates:** *Acer negundo, A. saccharinum, Agalinis purpurea, A. tenuifolia, Agrimonia striata, Agrostis gigantea, A. hyemalis, A. stolonifera, Aletris farinosa, Allium cernuum, Ambrosia trifida, Ammannia robusta, Amorpha fruticosa, Andropogon gerardii, A. scoparius, A. virginicus, Angelica atropurpurea, Apios americana, Apocynum cannabinum, A. sibiricum, Asclepias hirtella, A. incarnata, A. speciosa, A. sullivantii, A. syriaca, A. verticillata, Baccharis, Baptisia lactea, Betula sandbergi, Bidens frondosus, B. vulgatus, Boltonia asteroides, Cacalia plantaginea, C. tuberosa, Calamagrostis canadensis, Calystegia sepium, Campanula aparinoides, Campsis radicans, Carex annectens, C. aquatilis, C. atherodes, C. aurea, C. bebbii, C. bicknellii, C. brevior, C. bushii, C. buxbaumii, C. conoidea, C. crawei, C. cristatella, C. cryptolepis, C. festucacea, C. frankii, C. granularis, C. haydenii, C. interior, C. lacustris, C. longii, C. lurida, C. meadii, C. missouriensis, C. molesta, C. pellita, C. prairea, C. sartwellii, C. scoparia, C. stipata, C. stricta, C. tribuloides, C. trichocarpa, C. vulpinoidea, Castilleja coccinea, Cephalanthus occidentalis, Chelone glabra, Chenopodium, Cicuta maculata, Cirsium arvense, C. muticum, Cladium mariscoides, Coreopsis tripteris, Cornus amomum, Cynoctonum mitreola, Cynodon dactylon, Cyperus lupulinus, C. odoratus, C. squarrosus, C. strigosus, Dasiphora floribunda, Daubentonia, Daucus carota, Deschampsia cespitosa, Desmanthus illinoensis, Dichanthelium acuminatum, D. clandestinum, Echinochloa, Elaeagnus umbellata, Eleocharis compressa, E. elliptica, E. erythropoda, E. ovata, E. palustris, E. rostellata, E. tenuis, E. wolfii, Elymus canadensis, Epilobium leptophyllum, Equisetum fluviatile, E. hyemale, Eragrostis, Erigeron philadelphicus, E. strigosus, Erucastrum gallicum, Eryngium yuccifolium, Eupatorium*

perfoliatum, E. serotinum, Euphorbia corollata, Euthamia graminifolia, Eutrochium maculatum, Filipendula rubra, Fimbristylis decipiens, F. puberula, Fragaria virginiana, Galium boreale, G. obtusum, G. trifidum, Gentiana andrewsii, Gentianella crinita, Gentianopsis, Geum aleppicum, G. canadense, G. laciniatum, Glyceria grandis, G. striata, Glycerrhiza, Gratiola neglecta, Helenium autumnale, Helianthus grosseserratus, H. mollis, Hesperostipa spartea, Heterotheca camporum, Hierochloe odorata, Holcus lanatus, Hordeum jubatum, Hypericum ellipticum, H. kalmianum, H. majus, H. punctatum, H. sphaerocarpum, Hypoxis hirsuta, Iris shrevei, I. versicolor, Iva frutescens, Juncus acuminatus, J. balticus, J. brachycarpus, J. dudleyi, J. greenei, J. interior, J. nodosus, J. secundus, J. tenuis, J. torreyi, Larix laricina, Lathyrus palustris, Leersia oryzoides, Lepidium densiflorum, Leucospora, Liatris aspera, L. pycnostachya, L. spicata, Lindernia dubia, Lobelia kalmii, L. spicata, Ludwigia palustris, L. peploides, L. polycarpa, Lychnis alba, Lycopus americanus, L. uniflorus, Lysimachia hybrida, L. lanceolata, L. quadriflora, L. terrestris, Lythrum salicaria, Maianthemum stellatum, Malus ioensis, Melilotus albus, Mentha arvensis, Micranthes pensylvanica, Mimulus ringens, Monarda fistulosa, M. punctata, Muhlenbergia richardsonis, Oenothera pilosella, Onoclea sensibilis, Onosmodium hispidissium, Osmunda regalis, Oxalis dillenii, Oxypolis rigidior, Packera plattensis, Panicum dichotomiflorum, P. repens, P. virgatum, Parnassia glauca, Paspalum plicatulum, P. urvillei, Pastinaca sativa, Pedicularis lanceolata, Peltandra virginica, Penstemon digitalis, Penthorum sedoides, Persicaria amphibia, P. coccinea, P. hydropiper, P. hydropiperoides, P. lapathifolia, Phalaris arundinacea, Phlox glaberrima, P. maculata, Physalis heterophylla, Physostegia virginiana, Platanthera aquilonis, P. flava, P. lacera, P. leucophaea, Poa palustris, P. pratensis, Polygala polygama, Polygonum ramosissimum, Polyprenum, Populus tremuloides, Potentilla anserina, P. norvegica, P. simplex, Prenanthes racemosa, Proserpinaca palustris, Pycnanthemum tenuifolium, P. virginianum, Ratibida pinnata, Rhamnus cathartica, R. frangula, Rhynchospora, Rorippa palustris, Rosa carolina, R. multiflora, R. palustris, Rubus hispidus, R. setosus, Rudbeckia fulgida, R. hirta, R. subtomentosa, Rumex altissimus, R. crispus, Sagittaria latifolia, Salix amygdaloides, S. discolor, S. exigua, S. interior, Sambucus nigra, Sanicula canadensis, Schizachyrium scoparium, Schedonorus arundinaceus, Schoenoplectus acutus, S. pungens, S. tabernaemontani, Scirpus atrovirens, S. lineatus, S. pallidus, S. pendulus, Scleria triglomerata, S. verticillata, Scutellaria galericulata, Senecio pauperculus, Sesbania exaltata, Setaria faberi, S. parviflora, Silphium integrifolium, S. laciniatum, S. perfoliatum, S. terebinthinaceum, Sium suave, Solanum carolinense, S. dulcamara, Solidago canadensis, S. gigantea, S. ohioensis, S. ptarmicoides, S. riddellii, S. rigida, Sorghastrum nutans, Spartina patens, S. pectinata, Spiraea alba, Spiranthes diluvialis, Sporobolus heterolepis, Stachys palustris, S. pilosa, S. tenuifolia, Stellaria longifolia, Symphyotrichum boreale, S. dumosum, S. ericoides, S. lanceolatum, S. novae-angliae, S. puniceum, Thalictrum dasycarpum, Thelypteris palustris,
Toxicodendron radicans, Tradescantia ohiensis, Tragopogon porrifolius, Typha angustifolia, T. latifolia, Valeriana edulis, Verbena hastata, V. stricta, Vernonia fasciculata, V. missurica, Veronicastrum virginicum, Viburnum lentago, Vicia americana, Vigna luteola, Viola lanceolata, Vitis riparia, V. vulpina, Xanthium.

***Lythrum californicum* Torr. & A. Gray** is a glabrous perennial, which occurs on or in acequia, arroyos, bottoms, chaparral, cienega, cliffs, ditches, flats, floodplains, marshes, meadows, sandbars, scrub, seeps, shores, sloughs, springs, streambeds, washes, and along the margins of ponds, rivers, streams, and vernal pools at elevations of up to 2517 m. The plants can grow with their bases inundated by shallow standing water to at least 15 cm deep. The substrates are alkaline and have been categorized as adobe, alluvium, basalt, boulders, clay, cobble, granite, gravel, hardpan, Hobog–Tidwell soil, limestone, loamy Cumulic Haplustoll, mud, rock, rocky loam, sand, sandy loam, sandy rock, silt, and volcanics. The plants are found principally in freshwater and also tolerate fairly saline and tidal habitats. Exposures typically receive full sun, but the plants also occur in shade. The insect-pollinated flowers are distylous with purple petals. Honey bees (Hymenoptera: Apidae: *Apis mellifera*) are the most common pollinator and preferentially collect the smaller, usually yellow pollen produced by pin morphs (while avoiding the thrum pollen, which typically is green and larger in diameter). Other pollinators include cabbage butterflies (Lepidoptera: Pieridae: *Pieris rapae*), skippers (Lepidoptera: Hesperiidae), and flies (Diptera: Syrphidae). Considerable pollen flow occurs within the morphs of this species. Crossing thrum morphs results in low seed set (thus enabling selfing in thrums); whereas, pins do not produce seed when crossed. Vegetative reproduction occurs by means of a creeping, woody rootstock. The plants are susceptible to feeding damage by adult beetles (Coleoptera: Chrysomelidae: *Galerucella calmariensis*), which have been used to control *L. salicaria*, an invasive congener. The inclusion of xerophytic species (e.g., cacti) in the following list of associates appears to reflect more the seasonal transitional nature of some habitats (e.g., washes), than a FACW (rather than OBL) association. Nevertheless, it might be prudent to reconsider the indicator status of this species. **Reported associates:** *Abutilon, Acalypha phleoides, Acer negundo, Achillea, Acourtia wrightii, Adenostoma fasciculatum, A. sparsifolium, Adiantum capillus-veneris, Agave lecheguilla, Agrostis exarata, A. stolonifera, Alhagi maurorum, Almutaster pauciflorus, Alnus oblongifolia, Amaranthus palmeri, Ambrosia ambrosioides, A. monogyra, A. psilostachya, Amorpha fruticosa, Andropogon glomeratus, Anemopsis californica, Apocynum, Aquilegia chrysantha, Arctostaphylos glauca, A. pungens, Arida carnosa, Artemisia douglasiana, A. dracunculus, Asclepias lemmonii, Atriplex canescens, A. triangularis, Avena barbata, Baccharis emoryi, B. salicifolia, B. sarothroides, B. sergiloides, Bebbia juncea, Berberis haematocarpa, Berula erecta, Boerhavia wrightii, Bothriochloa barbinodis, Bouteloua curtipendula, B. hirsuta, Brickellia coulteri, Brodiaea orcuttii, Bromus arizonicus, B. tectorum, Callitropsis arizonica, C.*

forbsii, Carex hystericina, C. praegracilis, C. spissa, Celtis laevigata, C. reticulata, Cenchrus ciliaris, Centaurium calycosum, C. namophilum, Cephalanthus occidentalis, Chara, Cheilanthes wrightii, Chenopodium, Chilopsis linearis, Chylismia walkeri, Cicuta maculata, Cirsium neomexicanum, C. wheeleri, Clematis, Commelina erecta, Convolvulus arvensis, Conyza canadensis, Cordylanthus tecopensis, Croton texensis, Cucurbita foetidissima, Cuscuta salina, Cylindropuntia acanthocarpa, Cynodon dactylon, Cyperus fendlerianus, C. involucratus, C. niger, C. odoratus, C. parishii, Cupressus, Dasylirion wheeleri, Datura wrightii, Deinandra fasciculata, Deschampsia cespitosa, Desmodium, Dieteria asteroides, Distichlis spicata, Echinocereus triglochidiatus, Echinochloa crus-galli, Eleocharis macrostachya, E. palustris, E. parishii, E. parvula, E. rostellata, Elymus elymoides, Encelia farinosa, Ephedra, Epilobium ciliatum, Equisetum laevigatum, Eragrostis intermedia, Ericameria nauseosa, Eriogonum fasciculatum, Erythrina flabelliformis, Euphorbia antisyphilitica, E. revoluta, Eustoma exaltatum, Evolvulus arizonicus, Fallugia paradoxa, Ferocactus cylindraceus, Fimbristylis thermalis, Forestiera pubescens, Frankenia salina, Fraxinus velutina, Galactia wrightii, Garrya wrightii, Glandularia bipinnatifida, Grindelia fraxinopratensis, Guaiacum angustifolium, Gutierrezia microcephala, G. sarothrae, Gymnosperma glutinosum, Hedeoma, Helenium bigelovii, Helianthus annuus, H. nuttallii, Heliotropium curassavicum, Hesperidanthus linearifolius, Heteromeles arbutifolia, Hirschfeldia incana, Hosackia alamosana, H. oblongifolia, Hypericum formosus, Ipomoea cristulata, Ipomopsis thurberi, Isolepis carinata, I. cernua, Iva axillaris, Jatropha dioica, Jaumea carnosa, Juglans major, Juncus acuminatus, J. balticus, J. interior, J. nevadensis, J. saximontanus, J. tenuis, J. torreyi, Juniperus coahuilensis, J. deppeana, Keckiella antirrhinoides, K. cordifolia, Koeberlinia, Lactuca serriola, Leersia oryzoides, Lemna minor, Leptochloa dubia, Leucophyllum candidum, Leymus triticoides, Lilaeopsis masonii, L. occidentalis, Limonium californicum, Lobelia cardinalis, Lolium perenne, Ludwigia hexapetala, L. peploides, Lysimachia maritima, Mammillaria lasiacantha, M. pottsii, Melampodium sericeum, Melilotus, Mentzelia leucophylla, M. multiflora, Mimosa aculeaticarpa, M. dysocarpa, Mimulus cardinalis, M. guttatus, Monarda citriodora, Monardella stoneana, Muhlenbergia richardsonis, M. rigens, M. rigida, Nasturtium officinale, Nicotiana obtusifolia, Nitrophila occidentalis, Nolina microcarpa, Oenothera curtiflora, O. elata, O. rosea, O. sarmentosa, O. suffrutescens, Oxalis stricta, Panicum hirticaule, Parkinsonia aculeata, Paspalum dilatatum, P. distichum, Penstemon pseudospectabilis, Perityle emoryi, Persicaria lapathifolia, Peucephyllum schottii, Phragmites australis, Pinus cembroides, P. edulis, Plantago lanceolata, Platanus racemosa, P. wrightii, Pleurocoronis pluriseta, Pluchea odorata, P. sericea, Polygala, Polypogon monspeliensis, P. viridis, Populus fremontii, Proboscidea parviflora, Prosopis glandulosa, P. juliflora, P. pubescens, P. velutina, Prunus ilicifolia, Pseudognaphalium canescens, Pulicaria paludosa, Quercus agrifolia, Q. emoryi, Q. grisea, Q. oblongifolia, Q.

rugosa, Q. turbinella, Ranunculus cymbalaria, R. macranthus, Rhamnus californica, R. crocea, R. ilicifolia, Rhus aromatica, R. glabra, R. microphylla, R. virens, Ribes aureum, R. malvaceum, R. speciosum, Rosa californica, Rubus ursinus, Salix bonplandiana, S. exigua, S. gooddingii, S. laevigata, S. lasiolepis, S. taxifolia, Sambucus nigra, Samolus vagans, Sapindus saponaria, Sarcocornia perennis, Schoenoplectus acutus, S. americanus, S. pungens, Schoenus nigricans, Senegalia greggii, Senna lindheimeriana, Setaria leucopila, Sium suave, Solanum americanum, S. douglasii, S. elaeagnifolium, S. nigrum, S. rostratum, Solidago confinis, Spartina gracilis, Sphaeralcea, Spiranthes infernalis, Sporobolus airoides, S. contractus, S. wrightii, Stanleya pinnata, Suaeda nigra, Symphyotrichum laeve, Tamarix chinensis, T. ramosissima, Thelypodium integrifolium, Tiquilia, Toxicodendron rydbergii, Tragia nepetifolia, Triglochin concinnum, T. maritimum, T. striatum, Trixis californica, Typha domingensis, T. latifolia, Verbascum thapsus, Verbesina rothrockii, Veronica americana, V. anagallis-aquatica, Viguiera dentata, V. stenoloba, Vitis arizonica, V. girdiana, Xanthium strumarium, Xanthocephalum gymnospermoides, Yucca brevifolia, Y. treculeana, Y. ×schottii, Zeltnera calycosa, Ziziphus obtusifolia.

***Lythrum curtissii* Fernald** is a rare perennial, which occurs on wet substrates or in shallow waters of boggy pinelands, depressions, ditches, flats, floodplains, seepage slopes, swamps, thickets, and occasionally in openings maintained along powerline, railway or highway rights-of-way at low elevations. The sites have open to semishaded exposures with calcareous substrates that include loam, loamy sand, muck, rock, sand, and sandy peat. The flowers are distylous with deep purple petals. Both pins and thrums are capable of intramorph crossing that results in a low level of self-pollination. Selfed seed set is higher for thrums than for pins. The seeds are slightly winged and presumably are wind and/or water dispersed. The plants are somewhat woody at their base. Agricultural drainage has resulted in extensive habitat loss for *L. curtissii*, which also is damaged by cattle (Mammalia: Bovidae: *Bos*) and insects (Insecta). **Reported associates:** *Acer, Aristida stricta, Arnoglossum, Baccharis, Bumelia thornei, Calopogon pallidus, C. tuberosus, Carex chapmanii, C. gigantea, C. lupuliformis, Chamaecyparis thyoides, Chasmanthium, Cicuta maculata, Commelina virginica, Coreopsis integrifolia, Cornus, Cyrilla racemiflora, Drosera filiformis, D. intermedia, Fraxinus, Hartwrightia floridana, Hypericum galioides, Iris hexagona, Itea virginica, Juncus coriaceus, Lilium iridollae, Liquidambar styraciflua, Liriodendron tulipifera, Ludwigia, Lygodium japonicum, Lythrum alatum, Magnolia virginiana, Nyssa biflora, Panicum nudicaule, Parnassia grandiflora, Peltandra virginica, Physotegia veroniciformis, Platanthera integra, Pogonia ophioglossoides, Sabal palmetto, Sarracenia flava, S. leucophylla, S. rubra, Scirpus, Taxodium, Woodwardia, Xyris drummondii, X. scabrifolia.*

***Lythrum flagellare* Shuttlew. ex Chapm.** is a prostrate or creeping, freshwater perennial of low, open habitats including depressions, ditches, flats, floodplains, hammocks, marshes, prairies, roadsides, sandbars, sloughs, swamps, thickets, and the margins of canals, ponds, rivers, and streams at low elevations. Most exposures are reported as full sun. The substrates include sand, sandy muck, and sandy marl. The flowers (presumably distylous) have lavender to purple petals. The plants often appear on freshly mown sites; otherwise, there is little ecological information available for this rare species. **Reported associates:** *Acer rubrum, Andropogon, Axonopus affinis, Bacopa monnieri, Brachiaria mutica, Carex alata, Cassia nictitans, Centella asiatica, Cladium jamaicense, Coreopsis leavenworthii, Cynodon dactylon, Cyperus, Digitaria serotina, Eclipta prostrata, Eleocharis baldwinii, Erigeron quercifolius, Eryngium prostratum, Fimbristylis cymosa, Gnaphalium, Iris, Leersia, Lindernia dubia, L. grandiflora, Lobelia feayana, Ludwigia repens, Oxalis, Panicum hemitomon, P. repens, Paspalum caespitosum, P. notatum, P. setaceum, Persicaria punctata, Phyla nodiflora, Pluchea odorata, Pontederia cordata, Rhynchospora, Sacciolepis, Salix, Sisyrinchium rosulatum, Stenotaphrum secundatum, Tripsacum dactyloides, Vicia, Vigna luteola.*

***Lythrum hyssopifolia* L.** is a nonindigenous, glabrous, ascending to prostrate annual (rarely biennial). In its native range, it frequently inhabits calcareous (pH: 7.2–7.9) clay or peat soils as well as saline saltmarsh substrates to a lesser degree. In North America it inhabits freshwater or saline substrates associated with depressions, ditches, flats, gullies, marshes, meadows, mudflats, receding shores of ponds and vernal pools, roadsides, saltmarshes, seeps, swales, washes, and the margins (or dried beds) of lagoons, rivers, sloughs, and streams at elevations of up to 1600 m. There the substrates are described as alluvium, boulders, clay, clay loam, cobbley clay, Friant rocky sandy loam, granitic, gravel, gravelly sand, mud, Pilchuck soil, sand, sandy clay, sandy rock, and volcanic. The plants are long-day flowering. The flowers are homostylous, slightly protogynous, with pink to lavender petals. Pollinators include various insects such as bees (Hymenoptera), butterflies and moths (Lepidoptera), and flies (Diptera: Syrphidae). Self-pollination occurs regularly in any flowers that are not pollinated by insects. The plants produce numerous, nondormant seeds (3200 on average), which germinate rapidly. The seeds float when dry but sink when they become wet. They are found in the mud attached to the feet of waterfowl (Aves: Anatidae), which probably disperse them over longer distances. The seeds germinate immediately (as long as daily temperatures exceed 16°C), or can remain dormant in dry mud for as long as 14 years. Much higher germination occurs in the light than under dark conditions. When decumbent or prostrate, the shoots can root at the nodes and may perennate if water levels are maintained throughout the year. The roots are thin but richly branched and nonmycorrhizal. *Lythrum hyssopifolia* is described as highly competitive during early successional stages but is unable to maintain long-term competitive ability. The plants reportedly can grow in areas of mine waste where high metal concentrations are present in the soil. **Reported associates (North American range):** *Acacia salicina, Acmispon americanus, A. glabrus, Adenostoma fasciculatum, Agrostis stolonifera, Alisma, Allocarya stipitata, Alopecurus howellii, A.*

saccatus, Amaranthus californicus, Ambrosia artemisiifolia, Ammannia robusta, Anagallis arvensis, A. minima, Anthemis, Arctostaphylos, Artemisia californica, A. douglasiana, Arthrocnemum subterminale, Atriplex patula, Baccharis pilularis, B. salicifolia, B. sarothroides, Baileya, Bidens cernuus, B. tripartitus, Blennosperma nanum, Boisduvalia glabella, Bolboschoenus robustus, Brassica nigra, Briza, Brodiaea hyancinthina, Callitriche marginata, Calochortus splendens, Carduus pycnocephalus, Carex praegracilis, C. sychnocephala, Cirsium arvense, Convolvulus simulans, Conyza canadensis, Cotula coronopifolia, Crassula aquatica, Cressa truxillensis, Cynodon dactylon, Cyperus eragrostis, C. esculentus, C. squarrosus, Daucus carota, Deinandra conjugens, Deschampsia danthonoides, Didiplis, Diplacus aurantiacus, Downingia, Dudleya variegata, Echinochloa, Eclipta prostrata, Elatine brachysperma, Eleocharis acicularis, E. bella, E. erythropoda, E. macrostachya, E. montevidensis, E. obtusa, E. pachycarpa, E. parvula, Elymus repens, Epilobium brachycarpum, E. densiflorum, E. pygmaeum, Eriogonum fasciculatum, Eriophyllum confertiflorum, Eryngium aristulatum, E. castrense, E. vaseyi, Eucalyptus microtheca, Evax caulescens, Frankenia salina, Fraxinus, Geranium corecore, G. dissectum, Glandularia gooddingii, Gnaphalium palustre, Gratiola ebracteata, Heliotropium curassavicum, Heteromeles arbutifolia, Hirschfeldia incana, Holcus lanatus, Hordeum brachyantherum, H. californicum, H. marinum, Hypericum, Iris versicolor, Isoetes orcuttii, Isolepis cernua, Iva hayesiana, Juncus acutus, J. balticus, J. bufonius, J. dubius, J. effusus, J. torreyi, J. uncialis, J. xiphioides, Lasthenia, Leontodon saxatilis, Leptochloa fusca, L. uninervia, Lilaea scilloides, Limnanthes douglasii, Limonium, Limosella aquatica, Lindernia dubia, Lolium mulitiflorum, L. perenne, Ludwigia palustris, Lycopus, Lythrum salicaria, L. tribracteatum, Malosma laurina, Marsilea vestita, Melia azedarach, Melilotus albus, Mentha pulegium, Mimulus cardinalis, M. guttatus, Monardella lanceolata, Myosotis laxa, Myosurus minimus, Nassella pulchra, Navarretia intertexta, N. prostrata, Nerium oleander, Onoclea sensibilis, Orcuttia, Panicum capillare, Parkinsonia aculeata, Pasapalum dilatatum, P. distichum, Persicaria hydropiper, P. hydropiperoides, P. lapathifolia, P. maculosa, Phalaris arundinacea, P. caroliniana, P. lemmonii, Phragmites australis, Pilularia americana, Pinus sabiniana, Plagiobothrys trachycarpus, P. undulatus, Plantago lanceolata, P. major, Platanus racemosa, Pleuropogon californicus, Pluchea sericea, Poa nemoralis, Pogogyne abramsii, Polypogon monspeliensis, Populus fremontii, Prosopis, Pseudognaphalium luteoalbum, Psilocarphus brevissimus, Puccinellia distans, Quercus agrifolia, Q. douglasii, Q. engelmannii, Q. garryana, Ranunculus aquatilis, Rhamnus crocea, R. ilicifolia, Rhus integrifolia, Ribes, Rorippa islandica, Rotala ramosior, Rubus bifrons, R. laciniatus, Rumex crispus, R. salicifolius, Sagina decumbens, Sagittaria calycina, S. latifolia, Salicornia, Salix gooddingii, S. laevigata, S. lasiolepsis, Salvia apiana, Sarcocornia pacifica, Schedonorus arundinaceus, Schinus terebinthifolius, Schoenoplectus americanus, S. californicus, S. pungens, S. tabernaemontani, Senna artemisioides, Sida leprosa,

Sonchus oleraceus, Sorghum halepense, Spergularia marina, Spiraea douglasii, Symphyotrichum lanceolatum, Thelypteris palustris, Toxicodendron, Trifolium fucatum, T. hybridum, Typha angustifolia, T. domingensis, T. latifolia, Ulmus parvifolia, Urtica holoericea, Veronica anagallis-aquatica, V. peregrina, Vicia, Vitis girdiana, Xanthium strumarium.

***Lythrum lineare* L.** is a perennial, which occurs in tidal, brackish or saline (occasionally nearly freshwater) beaches, ditches, dunes, flats, hammocks (openings), marshes, meadows, pools, roadsides, saltmarsh, swales, and along the margins of ponds, rivers, sloughs, and streams at elevations of up to 10 m. Exposures generally receive full sunlight. The plants most frequently inhabit the upper intertidal zone but can be found in shallow water. The substrates are alkaline (reported pH: 8.3–8.6) and can be of mineral or organic composition. They include clay loam, Duckson soil, Handsboro mucky silt loam, rocks, sand, and sandy shell. The flowers are distylous with pale pink to white petals. In one studied population a pin:thrum ratio of 5.7:1 was observed. There the average pollen production of pins (14,421 grains) was just over twice that of thrums (7099 grains). The stigmas contained up to 34% foreign pollen loads, with conspecific pollen averaging 2.6 times greater loads on pins than on thrums. Of the conspecific pollen, pin stigmas received only 9% thrum pollen, but thrum stigmas received a much higher proportion (42%) of pin pollen. About 92% of the pollen produced in the population was pin pollen, but only 78% of combined thrum and pin pollen loads was pin pollen, indicating a higher pollinator pollen removal from pins. Data from this population indicated a lower than expected level of disassortative (intermorph) mating. A seed bank representing 1.5%–3.2% of the total pool of species can develop, even where the plants are represented in less than 1% of the standing vegetation. Seeds (from seed bank cores) have germinated successfully (within 16 days) after receiving 2 days of dark storage at 4°C, then being planted in sand and kept under a 30°C/25°C, 14/10-h day/night temperature and light regime. The plants form colonies of few individuals, which tend to reappear in the same locality in subsequent years. Vegetative reproduction occurs by means of creeping, basal rhizomatous offshoots. Individuals are tolerant to burning and flooding, but will decline with annual burning, mowing, or extensive livestock grazing. **Reported associates:** *Agalinis maritima, Amaranthus australis, Ambrosia artemisiifolia, Ampelopsis arborea, Atriplex patula, A. pentandra, Baccharis halimifolia, Bacopa monnieri, B. rotundifolia, Boltonia diffusa, Borrichia frutescens, Carex glaucescens, C. hormathodes, C. longii, Centella asiatica, Chamaecrista fasciculata, Chasmanthium latifolium, Cladium jamaicense, C. mariscoides, Crotalaria spectabilis, Cyperus esculentus, C. filicinus, C. odoratus, C. polystachyos, C. virens, Diodella teres, Diodia virginiana, Distichlis spicata, Echinochloa walteri, Eleocharis cellulosa, E. fallax, E. geniculata, E. montevidensis, E. palustris, E. rostellata, Eragrostis elliottii, Fimbristylis castanea, Fuirena scirpoidea, Galium tinctorium, Geocarpon minimum, Helianthus angustifolius, H. grosseserratus, Heliotropium curassivicum, Hibiscus moscheutos, Hydrocotyle bonariensis, Iris*
brevicaulis, *I. prismatica, Iva angustifolia, I. frutescens, Juncus gerardii, J. roemerianus, Kosteletzkya pentacarpos, Lachnanthes caroliniana, Limonium carolinianum, Ludwigia alata, Medicago minima, Morus rubra, Myrica cerifera, Oenothera biennis, O. humifusa, Panicum amarum, P. repens, P. virgatum, Paspalum dissectum, P. monostachyum, Phyla nodiflora, Plantago maritima, Pluchea camphorata, P. foetida, P. purpurascens, Polygonum aviculare, P. glaucum, Proserpinaca pectinata, Pycnanthemum virginianum, Quercus virginiana, Rhynchospora corniculata, Rumex obovatus, Sabal palmetto, Sabatia stellaris, Sacciolepis striata, Sagittaria lancifolia, Salicornia bigelovii, S. depressa, Samolus valerandi, Sarcocornia perennis, Schizachyrium maritimum, Schoenoplectus pungens, Sesuvium portulacastrum, Smilax bona-nox, Solidago sempervirens, Spartina cynosuroides, S. patens, S. pectinata, Strophostyles helvola, S. leiosperma, Suaeda linearis, Symphyotrichum ericoides, S. novae-angliae, S. subulatum, S. tenuifolium, Thalictrum dasycarpum, Toxicodendron radicans, Tradescantia occidentalis, Tridens strictus, Typha angustifolia, Ulmus alata, Viburnum nudum, Vicia ludoviciana, Vigna luteola.*

***Lythrum portula* (L.) D.A. Webb** is a nonindigenous creeping annual, which occurs in submerged or on the exposed muddy substrates of deltas, depressions, drying ponds, marshes, meadows, mudflats, roadside ruts, sandbars, seeps, shorelines, strands, vernal pools, and receding lake, pond, river, or reservoir margins at elevations of up to 2200 m. The plants can grow in water up to 1 m deep, but usually are found in tidal habitats or on substrates characterized by alternating flooding and drying cycles. The species is a calcifuge (reported pH: 6.4–7.2), which occurs on mineral-poor nonalkaline sediments described as clay, gravel, humus, lava, mud, mulch, Redding/Corning soil series, rock, sand, silt, silt clay muck, and stone. It also does not tolerate extreme acidity and rarely is found growing on peat. The reproductive ecology is similar to that of *L. hyssopifolia* (see earlier). The flowers are homostylous, have white to rose-pink petals and primarily self-pollinate. The fruits occur much less frequently on plants growing in the water than on those stranded on mud. The seeds germinate quickly (within 6 days) or can remain dormant for many years. The small seeds have no obvious dispersal mechanism but have high vagility. The shoots are prostrate (rarely erect) and root at the nodes. Fragmented plants may disperse vegetatively during the growing season. Although annual, some perennation occurs occasionally from the rooted stems. **Reported associates (North American range):** *Acer circinatum, Agrostis exarata, Alisma, Alnus rubra, Alopecurus pratensis, Amaranthus, Artemisia tridentata, Beckmannia syzigachne, Bidens frondosus, Callitriche stagnalis, Carex canescens, C. obnupta, Cirsium vulgare, Crassula aquatica, Cyperus squarrosus, Deschampsia cespitosa, Dipsacus fullonum, Downingia bicornuta, Elatine brachysperma, Eleocharis acicularis, E. ovata, E. palustris, Epilobium ciliatum, E. densiflorum, Eragrostis hypnoides, Eryngium castrense, Fraxinus, Gaultheria shallon, Glyceria, Gnaphalium palustre, G. uliginosum, Gratiola ebracteata, G. neglecta, Holcus lanatus, Isoetes nuttallii,*

Juncus acuminatus, J. articulatus, J. bufonius, J. effusus, J. patens, Lactuca seriola, Leontodon taraxacoides, Lindernia dubia, Ludwigia palustris, Lythrum salicaria, Marsilea vestita, Matricaria discoidea, Mazus pumilus, Mentha pulegium, Navarretia myersii, Panicum capillare, Persicaria hydropiperoides, Phalaris arundinacea, Plagiobothrys scouleri, P. stipitatus, Populus trichocarpa, Potentilla anserina, Pseudotsuga menziesii, Psilocarphus, Ranunculus aquatilis, Rorippa curvipes, R. curvisiliqua, Rubus armeniacus, R. laciniatus, Sagina, Salix sitchensis, Sisyrinchium californicum, Sparganium, Spergularia rubra, Spiraea douglasii, Trifolium repens, T. wormskioldii, Triglochin maritimum, Typha latifolia, Vaccinium corymbosum, V. macrocarpon, Veronica peregrina, V. scutellata.

***Lythrum salicaria* L.** is a nonindigenous, pubescent perennial, which occurs in fresh or brackish, intertidal or nontidal shallow water (to 2 dm), or on exposed substrates of bogs, bottoms, canals, ditches, fens, floodplains, levees, marshes, meadows, outwash, pools, prairies, prairie fens, roadsides, swales, swamps, vernal ponds, and along the margins of brooks, lakes, ponds, rivers, and streams at elevations of up to 1866 m. The plants thrive in full sun but can tolerate as much as 50% shade. This species grows frequently on sand but also on substrates described as basalt rimrock, boulders, clay, cobble, gravel, histosols, loam, marl, muck, mucky peat, mucky sand, mud, peat, Pilchuck soils, rock, sandy clay loam, silt, and silt clay. The substrates span a broad range of acidity (pH: 4.3–8.0). This is the only North American *Lythrum* species with tristylous flowers, which are distinguished by long-, mid-, and short-styled morphs. Many introduced North American populations show a higher incidence of missing morphs and uneven morph ratios (normally maintained by frequency-dependent selection) than populations in the native range of the species, a consequence attributed to random genetic drift operating in the founding populations. The flowers are large with purple to rose-purple petals and are pollinated by bumblebees, carpenter bees, honeybees (Insecta: Hymenoptera: Apidae), leaf-cutter bees (Hymenoptera: Megachilidae), and butterflies (Lepidoptera: Nymphalidae: *Cercyonis pegala*; Pieridae: *Colias philodice, Pieris rapae*). The presence of *L. salicaria* has been linked to pollinator competition with the native *Lythrum alatum*. Seed production is usually prolific with single plants producing from 100,000 to 2,700,000 seeds each growing season. Dense patches of *L. salicaria* can contain more than 400,000 seeds/m² in the upper 5 cm of soil, resulting in persistent seed banks. New seedlings can achieve seed production within the first growing season. Under optimal conditions flowering plants can occur within 8–10 weeks of seed germination. Crowded plants produce fewer seeds than more widely spaced individuals. Seeds disperse unassisted (i.e., via gravity and wind) to about 10 m from the parent plant. Further dispersal of the buoyant seeds and seedlings is achieved by water or by transport in fur, waterfowl plumage, or in mud attached to various vectors (including abiotic transport by vehicles). Dispersal by birds (Aves) through ingestion and subsequent defecation of viable seeds is also probable. The seeds germinate over an extremely wide range of (pH: 4.0–9.1). Light, wet conditions, and soil temperatures above 15°C–20°C are optimal for natural germination. Seeds can germinate underwater and their germination is reduced by only 20% after 2 years of storage under continuously waterlogged conditions. The seeds also germinate after 3 years of dried, refrigerated storage. Burial below sediment depths of 2 cm inhibits germination. Bare, open mudflats (e.g., drawdown conditions) provide optimal sites for seedling establishment (up to 20,000/m²); however, up to 50 seedlings/m² can establish in vegetated sites. Seedlings of *L. salicaria* are known to thrive in pots receiving acidic water (pH: 4.0) and to survive and grow under even more extreme acidity (pH: 2.5). The plants perennate from a persistent tap root, which can develop into a root mass reaching 0.5 m in extent. Buds on the lower shoot can also sprout resulting in successive shrub-like growth each year (producing up to 130 stems). The woody rootstock can also branch but vegetative reproduction by this means is limited to events that cause breakage of the rootstock (e.g., root feeding by carp) followed by subsequent dispersal of the pieces. The plants are not rhizomatous as many references report. Single individuals can live for up to 22 years. *Lythrum salicaria* is extremely aggressive and often develop into dense, monospecific stands that can cover more than a thousand hectares. It is a pioneer species that readily colonizes gaps by recruitment from massive existing seed banks. Although extensive, the fewer than expected list of associates likely reflects the tendency of this species to occur in monocultures. **Reported associates (North American range):** *Acalypha rhomboidea, Acer rubrum, A. saccharinum, Achillea millefolium, Agrostis gigantea, A. stolonifera, Alisma subcordatum, A. triviale, Allium canadense, A. cernuum, Alnus rubra, A. rugosa, Ambrosia acanthicarpa, A. artemisiifolia, A. trifida, Ammannia robusta, Amorpha, Amphicarpaea bracteata, Anagallis minima, Andropogon gerardii, Angelica atropurpurea, Apios americana, Apocynum cannabinum, Aronia arbutifolia, Artemisia ludoviciana, A. tridentata, Asclepias incarnata, A. syriaca, Barbarea vulgaris, Bidens cernuus, B. frondosus, Bromus inermis, Calamagrostis canadensis, Calla palustris, Calystegia sepium, Campanula aparinoides, Carex aquatilis, C. bebbii, C. comosa, C. crinita, C. granularis, C. hystericina, C. lurida, C. lyngbyei, C. nebrascensis, C. rostrata, C. stricta, C. vesicaria, Castilleja minor, Celastrus orbiculatus, Centaurea, Cephalanthus occidentalis, Chamerion angustifolium, Chelone glabra, Chenopodium album, C. rubrum, Cicuta bulbifera, Cirsium arvense, Clematis virginiana, Comarum palustre, Convolvulus arvensis, Conyza canadensis, Cornus sericea, Cotula, Cycloloma atriplicifolium, Cyperus bipartitus, C. squarrosus, C. strigosus, Daucus carota, Decodon verticillatus, Diervilla lonicera, Digitaria sanguinalis, Dipsacus laciniatus, Echinochloa crus-galli, E. muricata, E. walteri, Echinocystis lobata, Elaeagnus umbellata, Eleocharis acicularis, E. erythropoda, E. palustris, Elymus virginicus, Epilobium brachycarpum, E. ciliatum, Equisetum arvense, E. hyemale, Eragrostis hypnoides, Erechtites hieracifolius, Erigeron annuus, E. philadelphicus, Eupatorium perfoliatum, E. serotinum, Euthamia* *occidentalis, Eutrochium maculatum, E. purpureum, Fraxinus latifolia, F. pennsylvanica, Galium asprellum, G. obtusum, G. trifidum, Geum aleppicum, G. canadense, Glyceria canadensis, G. striata, Helianthus annuus, H. grosseserratus, Hibiscus moscheutos, Hippuris vulgaris, Holcus lanatus, Hypericum pyramidatum, Impatiens capensis, Iris pseudacorus, I. virginica, Juncus articulatus, J. balticus, J. brachycephalus, J. brevicaudatus, J. effusus, J. ensifolius, Juniperus deppeana, Lactuca, Larix laricina, Leersia oryzoides, Lemna minor, Lepidium latifolium, Leptochloa fusca, Leucanthemum vulgare, Lilaeopsis, Limosella aquatica, Lindernia dubia, Lobelia kalmii, L. spicata, Lonicera, Lotus corniculatus, Ludwigia palustris, L. peploides, Lycopus americanus, L. rubellus, L. uniflorus, Lysimachia quadriflora, L. terrestris, Lythrum alatum, Malus, Mentha arvensis, Mimulus glabratus, M. guttatus, M. ringens, Monarda, Muhlenbergia asperifolia, M. mexicana, Myrica gale, Napaea dioica, Nasturtium officinale, Nuphar polysepala, Oenanthe sarmentosa, O. villosa, Ostrya virginiana, Parthenocissus quinquefolia, Paspalum dilatatum, Peltandra virginica, Peritoma serrulata, Persicaria amphibia, P. hydropiper, P. hydropiperoides, P. lapathifolia, P. maculosa, P. punctata, P. sagittata, Phalaris arundinacea, Phleum pratense, Phragmites australis, Phytolacca americana, Picea mariana, Pinus ponderosa, Plantago lanceolata, P. rugelii, Poa nemoralis, P. pratensis, Polypogon monspeliensis, Populus balsamifera, P. fremontii, Potamogeton crispus, Potentilla, Proserpinaca palustris, Prunus serotina, Pseudoroegneria spicata, Pycnanthemum virginianum, Ranunculus bulbosus, R. pensylvanicus, R. sceleratus, Rhamnus frangula, Rhododendron viscosum, Rhus glabra, Robinia, Rorippa palustris, Rosa multiflora, R. pisocarpa, R. rugosa, Rubus bifrons, R. laciniatus, Rudbeckia hirta, Rumex crispus, R. salicifolius, Sagittaria latifolia, Salix amygdaloides, S. columbiana, S. eriocephala, S. exigua, S. interior, S. lasiandra, S. lutea, S. nigra, Sambucus nigra, Sanguisorba, Saururus cernuus, Schedonorus arundinaceus, S. pratensis, Schoenoplectus acutus, S. americanus, S. pungens, S. tabernaemontani, S. triqueter, Scirpus atrovirens, S. cyperinus, Scutellaria galericulata, Securigera varia, Silphium terebinthinaceum, Sisyrinchium, Smilax, Solanum dulcamara, Solidago canadensis, S. gigantea, S. graminifolia, Sparganium emersum, S. eurycarpum, Spartina pectinata, Spermacoce glabra, Spiraea alba, S. douglasii, S. latifolia, Stachys tenuifolia, Symphoricarpos, Symphyotrichum firmum, S. puniceum, Thelypteris palustris, Toxicodendron radicans, Triadenum fraseri, Trifolium repens, Typha angustifolia, T. latifolia, T. ×glauca, Urtica dioica, Utricularia intermedia, Verbascum blattaria, V. thapsus, Verbena hastata, V. urticifolia, Veronica scutellata, Veronicastrum virginicum, Vicia cracca, Vitis riparia, Xanthium strumarium, Zizania aquatica.*

***Lythrum tribracteatum* Salzm. ex Spreng.** is a nonindigenous prostrate annual, which inhabits ditches, marshes, meadows, mudflats, vernal pools, washes, and the margins of receding lakes, ponds, reservoirs, sloughs, or streams at elevations of up to 1485 m. The plants can occur in shallow water or on exposed substrates, which are alkaline or saline,

and include alluvium, clay, gravelly loam, mucky clay, mud, Riz soil series, sand, San Joaquin soil series, silt, vertisols, and Willows clay soil series. Exposures range from full to partial sun. The homostylous flowers have lavender petals. In its native European habitats, *L. tribracteatum* is described as "weakly competitive." **Reported associates (North American range):** *Alisma triviale, Alopecurus saccatus, Ambrosia psilostachya, Atriplex argentea, Bassia hyssopifolia, Bolboschoenus maritimus, Bromus japonicus, Callitriche verna, Centaurea solstitialis, Convolvulus arvensis, Cressa truxillensis, Crypsis vaginiflora, Damasonium californicum, Deschampsia danthonioides, Distichlis spicata, Downingia insignis, Elatine rubella, Eleocharis acicularis, E. macrostachya, E. palustris, E. parvula, Epilobium pygmaeum, Erodium botrys, E. cicutarium, E. moschatum, Eryngium aristulatum, E. vaseyi, Euphorbia hooveri, Extriplex joaquinana, Frankenia salina, Gnaphalium, Gratiola neglecta, Grindelia camporum, G. hirsutula, Hemizonia pungens, Holocarpha macradenia, Hordeum jubatum, Juncus bufonius, Lactuca serriola, Lasthenia fremontii, L. glabrata, Leymus triticoides, Limosella acaulis, Lythrum hyssopifolium, Malvella leprosa, Melilotus indicus, Myosurus sessilis, Navarretia leucocephala, Orcuttia pilosa, Pascopyrum smithii, Phalaris arundinacea, Phyla nodiflora, Plagiobothrys acanthocarpus, P. hispidulus, P. leptocladus, P. stipitatus, P. undulatus, Plantago elongata, Poa pratensis, Pogogyne douglasii, Polygonum aviculare, Polypogon monspeliensis, Populus fremontii, Psilocarphus brevissimus, P. oregonus, Puccinellia simplex, Rorippa, Rumex crispus, Salix exigua, Schismus arabicus, Schoenoplectus acutus, Spergularia bocconi, Symphyotrichum subulatum, Typha, Veronica americana, V. anagallis-aquatica, V. peregrina, Xanthium strumarium.*

Use by wildlife: Several Lepidoptera occur generally on *Lythrum* (Insecta: Noctuidae: *Eudryas unio*; Pyralidae: *Conogethes, Cryptoblabes*; Tortricidae: *Aterpia approximana*). Many bees (Insecta: Hymenoptera) have been observed collecting pollen or seeking nectar from *L. alatum* flowers. These include long-tongued bees (Anthophoridae: *Ceratina dupla, Epeolus bifasciatus, Florilegus condigna, Melissodes bimaculata, M. coloradensis, M. communis, M. comptoides, M. tepaneca, Svastra atripes, S. obliqua, Synhalonia speciosa, Triepeolus concavus, T. lunatus concolor*; Apidae: *Apis mellifera, Bombus griseocallis, B. impatiens*; Megachilidae: *Coelioxys modesta, C. octodentata, Hoplitis pilosifrons, Megachile brevis, M. campanulae, M. inimica sayi, M. mendica, M. petulans*) and short-tongued bees (Halictidae: *Agapostemon sericea, A. virescens, Augochlorella striata, Halictus confusus, H. ligatus, Lasioglossum versatus*; Andrenidae: *Calliopsis andreniformis*). Several wasps (Hymenoptera: Sphecidae: *Ammophila nigricans, A. procera, Prionyx atrata*) have also been observed on the flowers of *L. alatum*. Flower flies (Insecta: Diptera: Syrphidae: *Eristalis stipator, Helophilus latifrons, Tropidia mamillata*) will feed on the pollen but usually are nonpollinating. Beeflies (Diptera: Bombyliidae: *Exoprosopa fasciata, E. fascipennis, E. meigenii, Systoechus vulgaris*), tachina flies (Diptera: Tachinidae: *Archytas analis*), and other insects feed on nectar from the flowers. The flowers of *L. alatum* selectively attract a few butterflies (Lepidoptera). Among these are brush-footed butterflies (Nymphalidae: *Chlosyne nycteis*, *Vanessa cardui*), gossamer-winged butterflies (Lycaenidae: *Lycaena hyllus*), and pierids (Pieridae: *Colias philodice*, *Eurema lisa*, *Pieris rapae*, *Pontia protodice*). Skippers (Lepidoptera: Hesperiidae: *Ancyloxypha numitor*, *Pholisora catullus*, *Polites peckius*, *P. themistocles*) are also commonly observed on the flowers. *Lythrum alatum* is the larval host plant of a concealer moth (Lepidoptera: Oecophoridae: *Agonopterix lythrella*), and also of a grasshopper (Orthoptera: Acrididae: *Schistocerca emarginata*). The foliage of *Lythrum curtissii* and *L. flagellare* can be damaged severely by domestic cattle (Mammalia: Bovidae: *Bos taurus*) and various insects. In the European portion of its range, *L. hyssopifolia* is damaged by aphids (Insecta: Hemiptera: Aphididae: *Myzus lythri*), caterpillars (Lepidoptera), nymphs of plant-feeding bugs (Hemiptera: Miridae: *Lygus*), thrips (Insecta: Thysanoptera), and a smut fungus (Fungi: Basidiomycota: Doassansiaceae: *Heterodoassansia punctiformis*). Australian plants are infected by the cucumber mosaic virus (Group IV: Bromoviridae: *Cucumovirus*). The species reportedly is toxic to lambs (Mammalia: Bovidae: *Ovis aries*) and can be fatal to them if grazed in quantity. The seeds of *L. lineare* are eaten by ducks (Aves: Anatidae), while the foliage and seeds are consumed by small mammals (Mammalia). With the exception of red-winged blackbirds (Aves: Icteridae: *Agelaius phoeniceus*), which use stands of *L. salicaria* (purple loosestrife) as nesting sites, the presence of the plant generally is considered to be detrimental to most other wildlife. It supposedly eliminates nesting habitat for many native birds (Aves) such as American bittern (Ardeidae: *Botaurus lentiginosus*), black terns (Laridae: *Chlidonias niger*), least bittern (Ardeidae: *Ixobrychus exilis*), marsh wrens (Troglodytidae: *Cistothorus palustris*), sora (Rallidae: *Porzana carolina*), and Virginia rail (Rallidae: *Rallus limicola*). However, other studies have found that American coot (Rallidae: *Fulica americana*), American goldfinch (Fringillidae: *Spinus tristis*), black-crowned night heron (Ardeinae: *Nycticorax nycticorax*), gray catbird (Mimidae: *Dumetella carolinensis*), pied-billed grebe (Podicipedidae: *Podilymbus podiceps*), and swamp sparrow (Emberizidae: *Melospiza georgiana*) will also nest in *L. salicaria* stands, sometimes preferentially over other vegetation such as cattails (*Typha*). Hummingbirds (Aves: Trochilidae: *Archilochus colubris*) occasionally visit the flowers for their nectar. The seeds are eaten by various birds (Aves) including waterfowl (Anatidae), pigeons (Columbidae), red-winged blackbirds (Icteridae: *Agelaius phoeniceus*), and ring-necked pheasants (Phasianidae: *Phasianus colchicus*). The shoots are grazed to some extent by mammals (Mammalia) such as muskrats (Cricetidae: *Ondatra zibethicus*), rabbits (Leporidae: *Lepus*), and white-tailed deer (Cervidae: *Odocoileus virginianus*). However, dense stands provide poor habitat for muskrats and waterfowl overall and also usurp breeding sites used by other species such as the endangered bog turtle (Reptilia: Emydidae: *Glyptemys muhlenbergii*). Livestock do not graze it. Carp (Vertebrata: Osteichthyes: Cyprinidae: *Cyprinus carpio*)

are known to eat the roots of *L. salicaria*. Although there are numerous accounts of entire wetland plant communities that were converted into monocultures of purple loosestrife following its invasion, some studies indicate that its presence is associated with no change or even with an increase in plant species diversity in some instances. The long-term effect of purple loosestrife on native plant communities still requires further study for clarification. In its nonindigenous North American range *L. salicaria* has become a larval host plant for several Lepidoptera (Pyralidae: *Cryptoblabes gnidiella*; Saturniidae: *Actias luna*, *Antheraea polyphemus*, *Automeris io*, *Callosamia angulifera*, *C. promethea*, *Citheronia regalis*, *Eacles imperialis*, *Hemileuca nevadensis*, *Hyalophora cecropia*, *H. columbia*, *Rothschildia lebeau*).

Economic importance: food: The cooked or raw leaves of *Lythrum portula* reportedly are edible; **medicinal:** The Cherokee administered an infusion of *L. alatum* for kidney ailments. The Kawaiisu used *L. californicum* as a general medicinal. The Iroquois prepared a compound decoction from *L. salicaria* plants to treat fevers. *Lythrum salicaria* was used by early North American settlers as a medicinal herb for treating diarrhea, dysentery, sores and ulcers, and to stop bleeding wounds; **cultivation:** *Lythrum alatum* is cultivated commonly and hybridizes with *L. salicaria*. *Lythrum salicaria* is widely cultivated under its species name, as *L. virgatum* (see later) and also as many cultivars, including: 'Blush,' 'Brightness,' 'Croftway,' 'Dropmore Purple,' 'Dropmore scarlet,' 'Feuerkerze,' 'Florarose,' 'Happy,' 'Lady Sackville,' 'Morden Pink,' 'Prichard's Variety,' 'Red Gem,' 'Red Wings,' 'Robert,' 'Robin,' 'The Rocket,' 'Rose,' 'Rosencaule,' 'Rose Queen,' 'Rosy Gem,' 'Stichflamme' and 'Swirl,' 'The Beacon,' 'The Bride,' Ulverscroft form, and 'Zigeunerblut.' Some confusion exists regarding the identity of cultivar names that have been attributed to *Lythrum virgatum*, a name now regarded as synonymous with *L. salicaria*. Contrary to many reports, no "sterile hybrid" cultivar of *L. salicaria* has yet been developed that is safe to grow without fear of seed production. Even the cultivar 'Morden Pink,' which originated from a male-sterile plant of *L. salicaria*, is itself not male sterile, has high pollen fertility, and successfully hybridizes with the native North American *L. alatum*. Misconceptions of perceived "sterility" arise when only one morph of this tristylous species is grown in a garden. Because intramorph selfing rates are quite low, such plants generally will be barren. However, any vector providing pollen from a compatible source invariably will induce high seed set; **misc. products:** The Kawaiisu incorporated *Lythrum californicum* in a hair wash. *Lythrum salicaria* is regarded as an important plant for keeping bees (Hymenoptera: Apidae) and honey production because it is among the few species that remain in flower for prolonged periods during the summer months; **weeds:** *Lythrum hyssopifolia* has been reported as a weed of greenhouse pots. *Lythrum salicaria* is considered to be one of the most notorious of invasive wetland plants and it is listed as a noxious weed in the United States and Canada. The plants are known to displace native and rare species (leading to reduced growth for more than 44 native wetland plants) and eliminate open wetland habitat by forming vast, virtual monocultures. In the United States, wetland losses attributable to *L. salicaria* are estimated at nearly 200,000 hectares annually. Control of plants has been attempted using fire, hand pulling, herbicides, and mowing; none of these methods has proven to be very effective beyond short-term abatement. Several insects have been introduced throughout North America, intended for long-term biological control of *L. salicaria*. These include root-mining weevils (Coleoptera: Curculionidae: *Hylobius transversovittatus*), which destroy the root stock, flower-feeding weevils (Coleoptera: Curculionidae: *Nanophyes marmoratus*, *N. brevis*), which severely reduce seed production, and leaf-eating beetles (Coleoptera: Chrysomelidae: *Galerucella calmariensis*, *G. pusilla*), which damage the foliage, as well as a gall midge (Diptera: Cecidomyiidae: *Bayeriola salicariae*), which attacks the flower buds. Plantings of Japanese millet (*Echinochloa frumentacea*) have had limited success in outcompeting *L. salicaria*. *Lythrum alatum*, *L. californicum*, *L. hyssopifolia*, and *L. lineare* are variously susceptible to grazing by the same beetles (e.g., Coleoptera: Chrysomelidae: *Galerucella calmariensis*) that have been introduced intentionally as a potential means of controlling *L. salicaria*; **nonindigenous species:** *Lythrum hyssopifolia*, presumably native to southern Europe, was introduced to North America during the 19th century. It is a rice field weed in Egypt and may have been introduced to some parts of North America through rice culture. *Lythrum portula* is a European species introduced to the Americas and New Zealand. *Lythrum salicaria* originally was introduced nearly two centuries ago as a consequence of contaminated shipping ballast, and subsequently through repeated escapes from cultivation. *Lythrum tribracteatum* is native to southern Europe. Its means of introduction to North America is uncertain.

Systematics: Molecular phylogenetic studies have improved the circumscription of *Lythrum* by advocating the inclusion

of one species (*L. portula*) assigned previously to the genus *Peplis* (Figures 4.81 and 4.82), which now results in a monophyletic genus. Additional systematic studies of *Lythrum* indicate that the native North American species are monophyletic and resolve as the sister group to a Eurasian clade (Figure 4.82). The indigenous North American species of *Lythrum* are assigned to section *Euhyssopifolia*, subsection *Pythagorea*, which uniformly contains distylous tetraploids. Distyly in the native North American *Lythrum* is believed to have originated from tristylous ancestors even though pollen flow appears to be more efficient between morphs of the tristylous species. The indigenous North American species appear to be quite closely related, at least as evidenced by their low degree of molecular differentiation among the loci surveyed. *Lythrum virgatum* is regarded as conspecific with *L. salicaria* by some authors. Similarly, *L. anceps* has been treated variously as a distinct species, or as a variety of *L. salicaria* (*L. salicaria* var. *anceps*). Results from preliminary molecular analyses (Figure 4.82) provide some support for the continued recognition of *L. anceps* and *L. virgatum* as distinct from *L. salicaria*. Though the parental species are fairly distantly related, experimental and natural hybrids have been reported, which involve the native *L. alatum* and nonindigenous *L. salicaria*/*L. virgatum*. The potential for introgression is high among these taxa due to their high degree of intercrossability. Consequently, it is possible that introgression with *L. salicaria* could threaten the integrity of the native *L. alatum* gene pool and that hybridization could also yield offspring with enhanced vigor (heterosis) for traits already associated with vigor and invasiveness. *Lythrum* has a chromosomal base number of $x = 5$, which has been derived secondarily from the original $x = 8$ base number for the family. The annual *L. portula* and *L. tribracteatum* ($2n = 10$) are diploid. The perennial native North American species *L. alatum*, *L. californicum*, *L. curtissii*, *L. flagellare*, and *L. lineare* (all

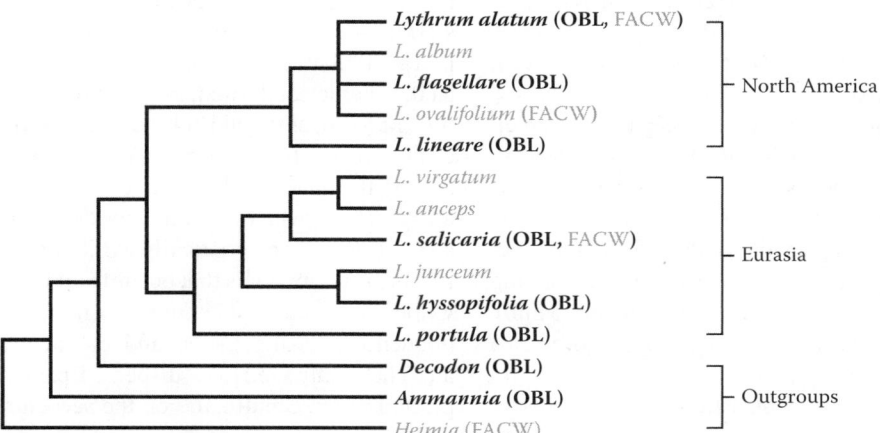

FIGURE 4.82 Phylogenetic relationships in *Lythrum* as indicated by analysis of combined DNA sequence data. A low level of molecular divergence characterizes several of the North American species. Many of the species are wetland indicators (shown in parentheses) including a wide distribution of obligate indicators (shown in bold). Two of the non-North American taxa (*L. virgatum* and *L. junceum*) inhabit wetlands. (Modified from Morris, J.A., A molecular phylogeny of the Lythraceae and inference of the evolution of heterostyly, PhD dissertation, Kent State University, 2007.)

$2n = 20$) are tetraploid. *Lythrum hyssopifolia* ($2n = 20$, 30) and *L. salicaria* ($2n = 30$, 50, 60) are high polyploids, which exhibit cytological variability.

Comments: Among the native species, *Lythrum alatum* (eastern North America), *L. californicum* (southwestern US and northern Mexico), and *L. lineare* (southern and eastern US coastal plain) have relatively widespread distributions compared to *L. curtisii* (restricted to Florida and Georgia) and *L. flagellare* (endemic to Florida). Current distributions of the nonindigenous *Lythrum* species reflect quite different patterns. *Lythrum tribractearum* occurs only in the western United States, *L. portula* in extreme western North America, *L. hyssopifolia* is disjunct between northeastern and western North America, and *L. salicaria* is widespread throughout much of North America.

References: Anderson & Ascher, 1993; Baldwin et al., 1996; Brown & Mitchell, 2001; Brown et al., 2002; Callaghan, 1998; Darwin, 1877; DiTomaso & Healy, 2003; Duever, 1984; Eckert et al., 1996; Gaudet & Keddy, 1988; Graham et al., 1987; Hager & Vinebrooke, 2004; Johnson & Rothfels, 2001; Lindgren & Clay, 1993; Morris, 2007; Morris et al., 2005b; Munger, 2002; Nagel & Griffin, 2001; Ornduff, 1978; Ottenbreit & Staniforth, 1994; Patrick et al., 1995; Preston & Croft, 1997; Rawinski, 1982; Stebbins & Daly, 1961; Strefeler et al., 1996a,b; Thompson et al., 1987; Welling & Becker, 1990.

6. Rotala

Toothcup

Etymology: from the Latin *rota* meaning "wheel" in reference to the whorled leaves of some species

Synonyms: *Ameletia*; *Ammannia* (in part); *Boykiana*; *Ditheca*; *Hoshiarpuria*; *Hydrolythrum*; *Hypobrichia*; *Nexilis*; *Nimmonia*; *Peplis* (in part); *Quartinia*; *Rhyacophila*; *Sellowia*; *Suffrenia*; *Tritheca*; *Winterlia*

Distribution: global: Africa; Asia; Europe*; New World; **North America:** widespread

Diversity: global: 44 species; **North America:** 3 species

Indicators (USA): OBL: *Rotala indica*, *R. ramosior*, *R. rotundifolia*

Habitat: freshwater; palustrine; **pH:** 5.5–9.0; **depth:** <1 m; **life-form(s):** emergent herb

Key morphology: stems (to 5.5 dm) creeping to erect, branched or simple, weakly 4- to 6-angled; leaves (to 4.5 cm) decussate (sometimes whorled), sessile to subsessile, lanceolate, obovate, orbicular or spatulate; flowers sessile, solitary in axils of leaf-like bracts, appearing axillary or in a distinct raceme, paired foliaceous bracteoles (to 6 mm) subtending each 4-merous flower; petals absent or present (to 3 mm), pinkish, rose or white; capsules (to 4.5 mm) globose, 2–4 valved, many seeded

Life history: duration: annual (fruit/seeds), perennial (buds); **asexual reproduction:** creeping stems; **pollination:** insect; self; **sexual condition:** hermaphroditic; **fruit:** capsules (common); **local dispersal:** seeds (wind); **long-distance dispersal:** seeds (birds)

Imperilment: (1) *Rotala indica* [GNR]; (2) *R. ramosior* [G5]; SH (NH); S1 (AZ, BC, CO, CT, MA, MT, ON, RI, WA); S2 (MN, NY, OR); S3 (IA, MI, NE, NJ, PA, WV); (3) *R. rotundifolia* [GNR]

Ecology: general: *Rotala* consists entirely of plants that inhabit aquatic or wetland habitats. Many species (especially the annuals) occupy sites that are relatively free from competition such as the temporarily exposed shores of drying ponds. However, such sites must also be inundated for part of the year for the plants to complete their life cycles. Within the genus occur large flowered, heterostylous, self-incompatible, entomophilous, outcrossed perennials and smaller flowered, homostylous, self-compatible, self-pollinating, inbreeding annuals. Two species found in North America represent the latter category, which overall, comprises the only group of species to achieve relatively widespread distributions. Many species are rice field weeds, with their minute seeds appearing as frequent contaminants of rice seed. The seeds are sticky when wet and can adhere to birds (Aves) in their plumage or attached to their feet in mud. The annuals have no means of vegetative reproduction.

Rotala indica (**Willd.**) **Koehne** is a nonindigenous annual, which grows submersed in shallow water (to 15 cm) or as an emergent on exposed substrates (often mud) in ditches, levees, marshes, rice fields, and wet prairies at elevations of up to 50 m (to 1000 m in its native range). The plants are self-compatible with homostylous, self-pollinating flowers. Information on the ecological impact of this species in North America is scarce. **Reported associates (North American range):** *Ammannia coccinea*, *Chara*, *Eleocharis atropurpurea*, *Heteranthera limosa*, *Oryza sativa*, *Rotala ramosior*, *Sphenoclea zeylanica*.

Rotala ramosior (**L.**) **Koehne** is an indigenous annual, which inhabits open sites in arroyo bottoms, bogs, bottomlands, depressions, ditches, flats, floodplains, lakeshores, marshes, meadows, mudflats, pools, prairies, puddles, ricefields, roadsides, sand pits, savannahs, seeps, silt flats, sloughs, swales, swamps, shores of receding lakes, ponds, and reservoirs, and along the margins of rivers and streams at elevations of up to 1590 m. The substrates typically are circumneutral (pH: 6.3–7.6) and have been described as alluvial mud, buckshot soil, clay, clay loam, cobble, gravel, Iredell soil, muck, mud, organic sand, sand, sandy clay, sandy gravel, sandy silt, silt, silt loam, silty mud, silty sand, and silty–sandy loam. The plants are relatively intolerant of competition and typically occupy the ephemeral zone between the annual high and low water marks, which is relatively devoid of other vegetation. Populations characteristically exhibit wide fluctuations in the number of individuals that occur from year to year. The homostylous, self-compatible flowers primarily are self-pollinated, frequently cleistogamous, and characterized by high seed set. The small seeds are dispersed passively by wind, gravity, and water. Small hairs on the seed surface facilitate their transport in the plumage of waterbirds (Aves: Anatidae) or in mud that is attached to their feet. Approximately 65%–100% of the seeds remains dormant following their Autumn production. The seeds have physiological dormancy and require 21–56 days of cold stratification followed by a 35°C/20°C day/night temperature regime for optimal (up to 98%)

germination. A day/night temperature regime of 20°C/10°C is optimal for breaking dormancy. Seed germination requires high daytime temperatures and high light levels (burial to even a few mm typically will inhibit germination). The seeds persist in seed banks. Although germination rates can fall by 50% within 1 year, some seeds retain their viability for upward of 10 years. The seeds will germinate when they are completely submersed. The plants are fire tolerant (they will reappear following burning of an area), but are not necessarily fire dependent. This species is widespread but relatively rare throughout its North American range and is threatened by invasive species (e.g., *Lythrum salicaria*), perturbations that interrupt the cyclic hydrological regime (alternating wet/dry conditions) of habitats, and sedimentation, which can reduce seed germination. **Reported associates:** *Acalypha gracilens, Acer negundo, A. saccharinum, Agalinis purpurea, A. tenuifolia, Agrostis capillaris, A. hymenalis, A. stolonifera, Alisma subcordatum, A. triviale, Alternanthera philoxeroides, Ambrosia trifida, Ammannia coccinea, A. robusta, Aristolochia tomentosa, Aronia ×prunifolia, Artemisia ludoviciana, Asclepias hirtella, Azolla caroliniana, Bacopa, Bergia, Bidens cernuus, B. connatus, B. frondosus, B. vulgatus, Boehmeria cylindrica, Boltonia asteroides, Brasenia schreberi, Bulboschoenus fluviatilis, Bulbostylis capillaris, Calamagrostis canadensis, Campsis radicans, Carex granularis, C. hyalinolepis, C. lacustris, C. pellita, C. scoparia, C. tribuloides, Cephalanthus occidentalis, Chamaesyce maculata, Cladium mariscoides, Commelina diffusa, Coreopsis tinctoria, Cyperus acuminatus, C. albomarginatus, C. aristatus, C. bipartitus, C. cumulata, C. dentatus, C. diandrus, C. difformis, C. erythrorhizos, C. esculentus, C. flavicomus, C. inflexus, C. odoratus, C. pseudovegetatus, C. squarrosus, C. strigosus, Danthonia californica, Datura stramonium, Deschampsia danthonioides, Dichanthelium, Drosera intermedia, Dulichium arundinaceum, Echinochloa crus-gallii, E. muricata, Echinodorus parvulus, Eclipta prostrata, Eleocharis acicularis, E. engelmannii, E. intermedia, E. melanocarpa, E. obtusa, E. ovata, E. pachycarpa, E. palustris, E. smallii, Epilobium coloratum, Eragrostis hypnoides, E. pectinacea, Erechtites hieracifolia, Eriocaulon aquaticum, Eriophorum virginicum, Eupatorium capillifolium, Euphorbia hirta, Euthamia graminifolia, Festuca subverticillata, Fimbristylis autumnalis, F. miliacea, F. perpusilla, F. vahlii, Forestiera acuminata, Fraxinus pennsylvanica, Galium obtusum, Gaylussacia baccata, Glyceria striata, Gnaphalium palustre, Gratiola aurea, G. neglecta, Helenium autumnale, H. thurberi, Heteranthera dubia, Hieracium caespitosum, Hoita macrostachya, Humbertacalia, Hypericum adpressum, H. boreale, H. canadense, H. majus, H. mutilum, Hypochaeris radicata, Ilex verticillata, Impatiens capensis, Ipomoea costellata, Iris virginica, Juncus acuminatus, J. balticus, J. brachycephalus, J. bufonius, J. effusus, J. marginatus, J. orthophyllus, J. pelocarpus, J. repens, J. tenuis, Justicia americana, Kyllinga pumila, Lasthenia glaberrima, Leersia lenticularis, L. oryzoides, L. virginica, Lemna minor, L. trisulca, Leptochloa panicoides, Leucospora, Lilaeopsis occidentalis, Limosella acaulis, Lindernia dubia, Lipocarpha*

aristulata, L. micrantha, Lobelia cardinalis, Lophotocarpus calycina, Ludwigia alternifolia, L. decurrens, L. palustris, L. polycarpa, L. repens, Lycopus americanus, L. uniflorus, L. virginicus, Lysimachia lanceolata, L. terrestris, Lythrum salicaria, Marsilea vestita, Mecardonia procumbens, Melissa officinalis, Mentha arvensis, M. pulegium, Micranthemum umbrosum, Mikania scandens, Mimulus ringens, Mollugo verticillata, Myosotis, Nepeta cataria, Nyssa sylvatica, Oenothera, Oldenlandia boscii, O. uniflora, Panicum agrostoides, P. capillare, P. dichotomiflorum, P. philadelphicum, P. rigidulum, Paspalum fluitans, Penthorum sedoides, Persicaria careyi, P. coccinea, P. hydropiper, P. hydropiperoides, P. lapathifolia, P. pensylvanica, P. punctata, Phalaris arundinacea, Phyla lanceolata, Physostegia virginiana, Pilea pumila, Pinus ponderosa, Plantago lanceolata, Polygala cruciata, P. sanguinea, Pontederia cordata, Populus heterophylla, Portulaca oleracea, Potamogeton diversifolius, P. natans, Potentilla recta, Poteridium annuum, Proserpinaca palustris, Prunella vulgaris, Psilocarya scirpoides, Quercus palustris, Ranunculus flabellaris, R. longirostris, Rhexia mariana, R. virginica, Rhynchospora capitellata, R. macrostachya, R. scirpoides, Rorippa columbiae, R. islandica, R. palustris, Rotala indica, Rubus hispidus, Rumex altissimus, Sagittaria graminea, S. latifolia, Salix nigra, S. petiolaris, S. sitchensis, Schoenoplectus hallii, S. pungens, S. saximontanus, Scirpus atrovirens, S. cyperinus, Scleria reticularis, Scrophularia marilandica, Scutellaria lateriflora, Sida spinosa, Sium suave, Solanum lycopersicum, Solidago tenuifolia, Spiraea tomentosa, Spiranthes diluvialis, Stachys hyssopifolia, Symphyotrichum lanceolatum, S. lateriflorum, Teucrium canadense, Thelypteris palustris, Toxicodendron radicans, Trifolium subterraneum, Typha latifolia, Verbena hastata, Veronica peregrina, Viola lanceolata, V. sororia, Wolffia columbiana, Xanthium strumarium, Xyris torta.

***Rotala rotundifolia* (Buch.-Ham. ex Roxb.) Koehne** is a nonindigenous, creeping perennial, which grows in canals, meadows, ponds, or along the margins of ponds and reservoirs at elevations of up to 2650 m (in its native range). The plants can occur as emergent, floating, or submersed life forms and have a medium to high light requirement. They grow optimally at temperatures from 24°C to 28°C and can tolerate broad extremes of water (pH: 5.5–9.0) and hardness (150–200 ppm). The plants remain in flower from March to August. The flowers are homostylous but are self-incompatible and insect pollinated. They produce viable seed in North America, which germinate readily. The plants can become somewhat woody at the base and may spread somewhat by their creeping stems. Detached shoot fragments will root readily when lodged in the sediment and are capable of vigorous vegetative growth. Adventitious shoots have been cultured from leaf explants on MS [Murashige & Skoog] medium (with 0.20 mg/L GA_3 [Gibberellic acid]; 1.00 mg/L BA [6-benzylaminopurine]; pH: 6–9), as a means of vegetative propagation. **Reported associates (North American range):** none reported.

Use by wildlife: No native wildlife uses are reported for North American *Rotala*. *Rotala ramosior* is susceptible to feeding damage by the beetles (Insecta: Coleoptera: Chrysomelidae: *Galerucella calmariensis*) used to control *Lythrum salicaria*. The herbivorous grass carp (Osteichthyes: Cyprinidae: *Ctenopharyngodon idella*) is known to feed on *Rotala indica*.

Economic importance: food: Shoots of *Rotala indica* are eaten in China as a famine food; **medicinal:** Ethanol extracts of *R. rotundifolia* contain flavonols, which reportedly are antioxidant and also inhibit human hepatitis B virus (HBV) activity; **cultivation:** The annual *Rotala indica* is grown (with difficulty) as an aquarium plant, although plants sold under this name include specimens of the perennial *R. rotundifolia*. *Rotala rotundifolia* is popular as an aquarium plant and is grown as a water garden ornamental; **misc. products:** none; **weeds:** *Rotala* species (including *R. indica* and *R. ramosior*) are frequent weeds of rice fields. In 1998, a strain of *R. indica* that is resistant to ALS inhibiting herbicides evolved in Japanese rice fields that had been exposed to persistent herbicide use; **nonindigenous species:** *Rotala indica* (native to eastern Asia) has been introduced through rice culture to Africa, Europe, and the United States. The New World *R. ramosior* has been introduced to rice fields in Italy, Java, Philippines, and Sulawesi. The Asian *R. rotundifolia* recently (1996) was introduced to North America, presumably as an escape from cultivation.

Systematics: Molecular data resolve *Rotala* as the sister genus to *Didiplis diandra* (formerly a species of *Peplis*), in a clade sister to *Heimia* (Figure 4.81). Results from morphological cladistic analyses also place *Rotala* close to *Didiplis*. Although further sampling of genera in Lythraceae might further clarify the intergeneric position of *Rotala*, the *Rotala/Didiplis* clade is well supported by combined molecular data. A comprehensive phylogenetic study of *Rotala* species has not yet been carried out. The existing classification, which groups species by highly artificial characters (e.g., leaf arrangement), is inappropriate and in need of revision. Because most species are highly similar morphologically, it will be difficult to achieve a meaningful classification without first elucidating interspecific relationships with a fair degree of certainty. The high degree of inbreeding in *Rotala* deters hybridization, even between morphologically distinctive races of *R. ramosior*, which occur sympatrically in central California. The chromosomal base number of *Rotala* is $x = 8$. *Rotala ramosior* is diploid ($2n = 16$) in the southern portion of its range (Mexico and South America) but tetraploid ($2n = 32$) in North America; *R. indica* ($2n = 32$) is also a tetraploid. The chromosome number reported for *R. rotundifolia* ($2n = 36$) indicates an anomalous tetraploid count.

Comments: *Rotala indica* currently is known only from rice fields in the United States (California and Louisiana). *Rotala ramosior* occurs from Canada to South America. *Rotala rotundifolia* currently is known only from the southeastern United States (Alabama and Florida).

References: Baskin et al., 2002; Blancaver et al., 2002; Cook, 1979; Graham, 1992; Jacono & Vandiver, 2007; Karatas et al., 2014; Kasselmann, 2003; Marhold, 2012; Mattrick, 2001; Reese & Haynes, 2002; Soerjani et al., 1987; Zhang et al., 2011.

7. Trapa

Water chestnut; châtaigne d'eau, macre negeante
Etymology: contracted from the Latin *calcitrappa* (*calcis trappa*) meaning "heel snare" in reference to the resemblance of the fruit to the caltrop, a weapon of medieval warfare
Synonyms: none
Distribution: global: Africa; Eurasia; North America*; **North America:** northeastern
Diversity: global: 5 species; **North America:** 1 species
Indicators (USA): OBL: *Trapa natans*
Habitat: freshwater; lacustrine, riverine; **pH:** 6.7–8.2; **depth:** 0–5 m; **life-form(s):** floating-leaved, free-floating
Key morphology: stems (to 40 dm) flexuous, submersed, bearing opposite nodal pairs of pinnate, leaf-like (green, photosynthetic) adventitious roots (to 15 cm), heterophyllous; submersed leaves (to 5 cm) opposite (below) or alternate (above), linear, fugacious; floating leaves alternate, forming a terminal rosette (to 50 cm), the petioles (to 21 cm) expanded into an inflated spongy region near the blade, the blades (to 5 cm) rhomboid, the margins dentate; flowers in axils of floating leaves, 4-merous, epigynous; petals (to 8 mm) 4, white, caducous; fruit (to 3 cm) drupe/nut-like, with four horn-like projections (derived from the accrescent sepals), single seeded
Life history: duration: annual (fruit/seeds); **asexual reproduction:** shoot fragments, stolons; **pollination:** self; **sexual condition:** hermaphroditic; **fruit:** drupes/nuts; **local dispersal:** fruits; **long-distance dispersal:** shoot fragments, possibly fruits
Imperilment: (1) *Trapa natans* [GNR]
Ecology: general: Only one species in North America (see next).

Trapa natans L. occurs on the surfaces of water bodies that include canals, lakes, ponds, mud flats, and margins of slow-moving rivers and streams at elevations of up to at least 145 m. The waters are fresh (or slightly brackish in some estuaries), calm, shallow (usually <2 m deep), circumneutral (pH: 6.7–8.2; mean: 7.1), nutrient rich, and characterized by relatively low alkalinity (12–128 ppm $CaCO_3$). The substrates typically are of highly organic muck and also include Gneissic boulders, mud, and sand. The plants grow from a submersed shoot system that terminates into a floating rosette of leaves, which can detach and become free floating. The flowers are aerial, self-compatible and apomictic (at least in the Asian var. *japonica*) with high seed set. Additional studies of *T. natans* in its adventive range are needed to clarify its means of reproduction and dispersal in nonindigenous localities. Individual plants are capable of producing up to 300 seeds/year with seed densities reaching 180/m². The seeds can germinate in complete anoxia and will ripen within 1 month. They have physiological dormancy and can remain dormant anywhere from 4 months to 5–12 years. Seed germination has been achieved after a 210-day exposure to a 19°C/15°C day/night temperature regime and also after 120 days at a constant 18°C. A 100-fold variation in size has been observed among seeds that retain germination capability. Following germination, the empty husks can be seen floating on the water surface where they often are mistaken for viable propagules. The seeds will lose their viability if dessicated. Drying for 2 weeks at room temperature will kill them. Although production of a few but large seeds is unusual for an annual plant, reproductive allocation to seeds in *Trapa* (upward of 36% of the total biomass) is comparable. The large seeds result in rapid, early seedling growth, which expedites the spread of plants across the water surface, eventually resulting in the pre-emption of other aquatic vegetation. The heavy nuts (up to 6 g) sink quickly once they have been shed and ordinarily do not disperse far from the parent plants. Although there are reports of *Trapa* fruits becoming attached to waterfowl (Aves: Anatidae) plumage, they doubtfully are dispersed in this fashion; these reports most likely refer to the empty husks given that the much heavier, viable fruits would not remain attached to a bird's plumage for long. When *Trapa* plants grow at low densities they will produce more than ten times the number of vegetative shoots than those growing at high densities (which can attain up to 50 rosettes and 1 kg/m^2 of dry matter). Vegetative shoot fragments and severed rosettes are dispersed by water currents, waterfowl, and humans (e.g., attached to boat trailers), but do not survive beyond their first growing season. The waters below dense *Trapa* stands often are hypoxic (<2.5 mg/L O$_2$) or even anoxic, thus providing unsuitable habitat for fish and other aquatic life. The plants can accumulate high levels of metals when growing in polluted waters. **Reported associates (North American range):** *Butomus umbellatus, Ceratophyllum demersum, Chara, Elodea canadensis, E. nuttallii, Heteranthera dubia, Hydrocharis morsus-ranae, Lemna minor, Lythrum salicaria, Myriophyllum spicatum, Najas flexilis, N. guadalupensis, N. minor, Potamogeton berchtoldii, P. crispus, P. ogdenii, Rorippa aquatica, Schoenoplectus tabernaemontani, Spirodela polyrhiza, Stuckenia pectinata, Typha latifolia, Vallisneria americana, Wolffia, Zannichellia palustris.*

Use by wildlife: *Trapa natans* reportedly is of no value to waterfowl and shades out aquatic plant species that are more useful to wildlife. However, the plants do provide substrate for a number of aquatic invertebrates and may actually increase the numbers of organisms that are eaten by fish. In one study, 58 invertebrate species were associated with *Trapa* plants including various annelid worms (Annelida: Oligochaeta), crustaceans (Crustacea: Malacostraca: Amphipoda; Branchiopoda: Cladocera), hydras (Cnidaria: Hydridae), insects (Insecta: Chironomidae, Coleoptera, Ephemeroptera), molluscs (Mollusca: Bivalvia), and nematodes (Nematoda). Yet, such invertebrates often are not readily available to fish, given the low oxygen environments that are associated with dense growths of *Trapa*, which fish will avoid. Several insects (Insecta) including adult water lily leaf beetles (Coleoptera: Chrysomelidae: *Galerucella nymphaeae*), aphids (Hemiptera: Aphididae), and some caterpillars (Lepidoptera) will feed on the foliage of *Trapa natans*. The fruits of *T. natans* are eaten by muskrats (Mammalia: Cricetidae: *Ondatra zibethicus*) and squirrels (Mammalia: Sciuridae: *Sciurus*).

Economic importance: food: Asians and Europeans have candied, roasted, and eaten *Trapa* seeds or have ground them into flour for centuries. *Trapa* husks discarded during the 15th century AD have been recovered from the Vasca Ducale rubbish pit of the ducal palace of Ferrara in northern Italy. The seeds are a key ingredient in some Italian risotto. In North America, the nuts are sold only occasionally in specialty shops as a canned food. However, the "waterchestnuts" typically eaten in Asian restaurants are unrelated and are obtained from the corms of *Eleocharis dulcis* (Cyperaceae), a monocotyledon; **medicinal:** Compounds isolated from *Trapa bispinosa* (sometimes regarded as a variety of *T. natans*) possess strong antibacterial and antioxidant properties. In India, *T. bispinosa* is made into a dish called "Halwa" which is used to treat women who are suffering from leucorrhoea. Liniments made from the fruits have been used to treat elephantiasis, rheumatism, sores, and sunburn. The related *Trapa japonica* contains up to 240 ppm of Germanium-132, a substance with reputed antitumor and antiviral properties; **cultivation:** *Trapa natans* is cultivated as an ornamental water garden plant, but its importation for such purposes is prohibited in Canada and much of the United States and should be discouraged; **misc. products:** *Trapa* extracts are antioxidant and have been used in some cosmetics. The empty husks can be strung together to make ornamental necklaces; **weeds:** *Trapa natans* is a serious pest, which forms extensive surface mats along the margins of lakes and rivers. The decline of several rare native aquatic plants (e.g., *Potamogeton ogdenii, Rorippa aquatica*) has been attributed at least in part to displacement of habitat by dense *Trapa* growths. Infestations often lead to hypoxic water conditions that reduce habitat use by fish and other aquatic organisms. Because of its surface growth habit, invasive outbreaks also can eliminate the use of waters from other recreational purposes. Furthermore, the hard, spiny nuts are hazardous and will cause lacerations (even penetrating some footwear) if they are stepped upon in the water; **nonindigenous species:** All *Trapa* species and varieties are nonindigenous to the New World. *Trapa natans* (native to Eurasia) was introduced to eastern Massachusetts as an ornamental plant sometime before 1879.

Systematics: Traditionally *Trapa* has been placed either within Onagraceae or segregated as the closely related but distinct family Trapaceae. However, combined molecular data clearly resolve the genus within Lythraceae as the sister genus to *Sonneratia* (Figure 4.81). Although the relationship of *Trapa* to Lythraceae is understandable (given its traditional association with Onagraceae), its specific association with *Sonneratia* is somewhat surprising given that the latter (a mangrove shrub having flowers with numerous carpels, stamens, and seeds), bears no obvious resemblance to the water chestnut. Both genera do share determinate inflorescences and wet stigmas. *Trapa* itself is highly unsettled taxonomically with species estimates ranging from 1 to 20. Authors often recognize a broad taxonomic concept of *T. natans* that includes several varieties (e.g., *T. natans* var. *natans, T. natans* var. *japonica, T. natans* var. *pumila*), sometimes including *T. bispinosa* (which appears to be distinctive) among the varieties. Some taxa (often differentiated by their fruit morphology) may represent ancient selected cultivars. Present taxonomic treatments recognize about five distinct species worldwide as well as several varieties of *T. natans*. Allozyme data indicate that the *T. japonica* "group" may be of a hybrid origin involving the *T. incisa* and *T. natans* groups. The genus desperately is in need of a comprehensive systematic investigation to more accurately determine species limits and to clarify interspecific relationships. The chromosomal base number of *Trapa* ($2n = 36, 48, 96$) is $x = 9$ or 12; that of the related *Sonneratia* presumably is $x = 8$ or 12. *Trapa natans* contains tetraploid ($2n = 48$) and octaploid ($2n = 96$) cytotypes.

Comments: *Trapa natans* occurs in eastern North America from Quebec to Virginia.

References: Bosi et al., 2009; Caraco & Cole, 2002; Cozza et al., 1994; Groth et al., 1996; Haber, 1999; Hellquist & Straub, 2001; Judd et al., 2016; Kadono & Schneider, 1986; Kim et al., 1997, 2010; Kunii, 1988; Les & Mehrhoff, 1999; Martin et al., 1957; Menegus et al., 1992; Methé et al., 1993; Rahman et al., 2001; Strayer et al., 2003; Takano & Kadono, 2005.

Family 3: Melastomataceae [188]

The large Melastomataceae (5100 species) are distributed almost exclusively in tropical world regions, with exception of the endemic North American *Rhexia* (Renner, 2004; Judd et al., 2016). The family has been subdivided into several subfamilies and into as many as 13 tribes. The sister group of Melastomataceae are Memecyclaceae, which some authors treat as a subfamily of the former. *Rhexia* is placed within tribe Rhexieae, which is allied closely with tribe Melastomeae (Clausing & Renner, 2001; Renner, 2004; Figure 4.83).

Melastomataceae are recognized by several vegetative and reproductive features including their characteristic opposite, simple leaves in which the primary veins ascend in a subparallel fashion from the base with the tertiary veins oriented perpendicular to the midrib. The stamens usually possess geniculate (elbow-shaped) filaments which curve to displace the anthers to one side of the flower. Pollination is effected mainly by bees (Insecta: Hymenoptera), which gather pollen from the typically nectarless flowers.

Phylogenetic analyses of Melastomataceae (Clausing & Renner, 2001; Renner et al., 2004; Fritsch et al., 2004) indicate

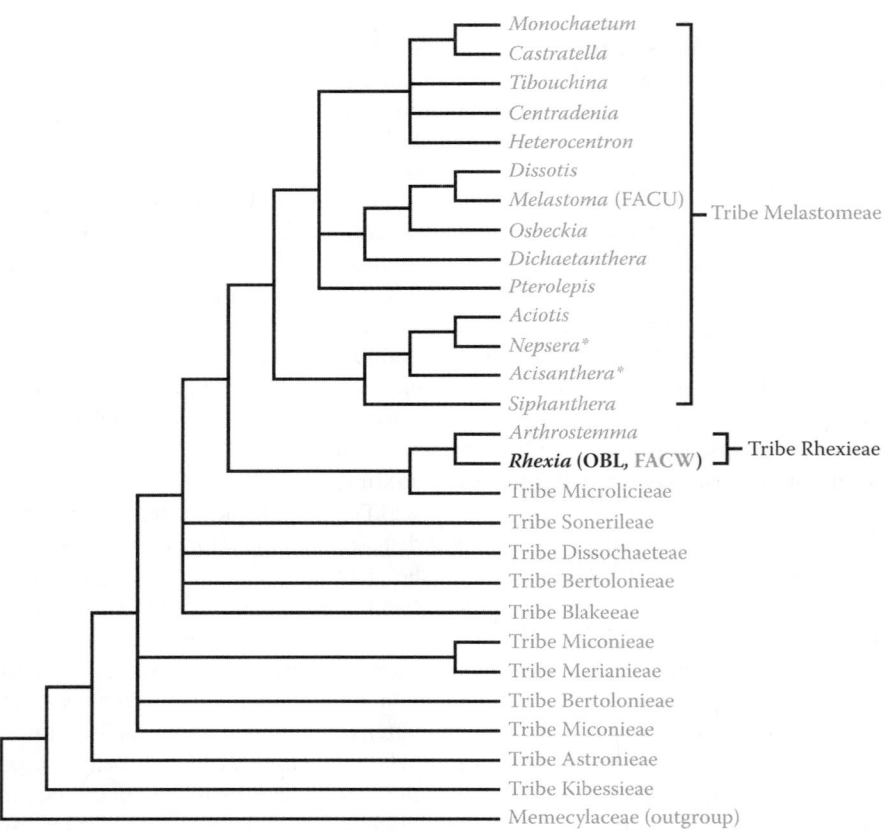

FIGURE 4.83 Phylogeny of Melastomataceae based on parsimony analysis of cpDNA (*rpl*16) sequence data. Several tribes (Bertolonieae, Miconieae) do not appear to be monophyletic. *Rhexia* (in bold), the only genus with obligately aquatic species in North America, occurs within a relatively derived clade. Several origins of hydrophytes in Melastomataceae are indicated by the placement of the Neotropical *Acisanthera* and *Nepsera* (asterisked), which also contain aquatics (but are excluded from the present work). The remainder of this large family is terrestrial. (Modified from Fritsch, P.W. et al., *Amer. J. Bot.*, 91, 1105–1114, 2004.)

the monophyly of the family and many of the tribes. In some analyses (Clausing & Renner, 2001) tribes Rhexieae and Melastomeae are sister groups and comprise a clade defined by apomorphic cochleate (spiral-shaped) seeds. Other data sets (Fritsch et al., 2004) include tribe Microlicieae within the group. In any case, this clade (which includes the obligately aquatic North American species) occupies a relatively derived position in the phylogeny of Melastomataceae (Figure 4.83). The Neotropical genera *Acisanthera* and *Nepsera* (tribe Melastomeae) also contain aquatics (Cook, 1996). Otherwise, the family essentially is terrestrial.

Obligate aquatic species occur in only one North American genus:

1. *Rhexia* L.

1. *Rhexia*

Deergrass, meadow beauty

Etymology: From the Greek *rhexis* meaning "rupture" in reference to the plant's purported ability to cure such ailments

Synonyms: *Alifanus*

Distribution: global: North America; **North America:** E/W disjunct

Diversity: global: 11–13 species; **North America:** 11–13 species

Indicators (USA): OBL: *Rhexia aristosa*, *R. mariana*, *R. nashii*, *R. parviflora*, *R. petiolata*, *R. salicifolia*, *R. virginica*; **FACW:** *R. mariana*, *R. nashii*, *R. petiolata*, *R. virginica*

Habitat: freshwater; palustrine; **pH:** 4.5–6.7; **depth:** <1 m; **life-form(s):** emergent herb

Key morphology: Stems erect (to 15 dm) hirsute, 4-angled; leaves (to 7 cm) opposite (decussate), simple, sessile or subsessile, with three principal veins; flower number variable; hypanthium (to 12 mm) urceolate; petals (to 2.5 cm) 4, white, pink, lavender or purple; stamens 8; filaments with an appendage/protuberance at juncture of anther; anthers unilocular, poricidal; ovary inferior; seeds spiral shaped (snail-like).

Life history: duration: perennial (woody caudex, rhizomes, tubers); **asexual reproduction:** rhizomes, tubers; **pollination:** insect; **sexual condition:** hermaphroditic; **fruit:** capsules (common); **local dispersal:** rhizomes, seeds, tubers; **long-distance dispersal:** seeds

Imperilment: (1) *Rhexia aristosa* [G3/4]; S1 (AL, DE, NJ); S2 (GA); S3 (NC, SC); (2) *R. mariana* [G5]; SX (NY); S1 (IN, KS, MA, MI, NJ, PA, WV); S3 (IN, KY, NC, VA); (3) *R. nashii* [G5]; S3 (LA); (4) *R. parviflora* [G2]; SH (GA); S1 (AL); S2 (FL); (5) *R. petiolata* [G5]; S1 (VA); (6) *R. salicifolia* [G2]; S1 (AL, GA); S2 (FL); (7) *R. virginica* [G5]; S1 (IA, VT); S3 (MI, NS, OH, WI)

Ecology: general: *Rhexia* species typically show an affinity for wet, acidic habitats. Eleven species are categorized at least as FACW indicators with seven of those (54%–64%) designated as OBL indicators. The flowers of *Rhexia* species lack nectar but are pollinated mainly by bees (Insecta: Hymenoptera), which elicit pollen from the poricidal anthers by means of a vibratory ("buzz") mechanism. All of the *Rhexia* species but two (*R. lutea*, *R. petiolata*) are self-incompatible. Vegetative reproduction occurs from the local spread of rhizomes (which can form extensive clones in some species), or by the dispersal of fleshy tubers, which sometimes occur on the rhizomes or at the ends of the roots.

***Rhexia aristosa* Britton** is distributed locally in barrens, bays, bogs, depressions, ditches, flats, flatwoods, limesink ponds, meadows, pond beds, savannahs, and vernal ponds at low elevations. Reported substrates include clay, sand, and sandy peat. The flowers (produced from June to September) presumably are pollinated by bees (Insecta: Hymenoptera) but specific pollinators have not been reported. The long-term persistence of this rare species depends on the maintenance of a large, viable seed bank. Water level fluctuations (with periods of drawdown) are necessary for promoting seed germination. Periodic burns and flooding cycles appear to be important in reducing competition from woody species. The plants can form clones by extension of their tuber-bearing roots. **Reported associates:** *Andropogon capillipes*, *A. glaucopsis*, *Aristida palustris*, *Bacopa*, *Bartonia verna*, *Centella asiatica*, *Coleataenia longifolia*, *Dichanthelium erectifolium*, *D. wrightianum*, *Eleocharis elongata*, *E. equisetoides*, *E. melanocarpa*, *E. robbinsii*, *Eriocaulon compressum*, *Eupatorium leptophyllum*, *Juncus abortivus*, *Lachanthes caroliniana*, *Lachnocaulon beyrichianum*, *L. minus*, *Lobelia boykinii*, *Ludwigia linearis*, *Lycopodiella alopecuroides*, *Oxypolis canbyi*, *Panicum hemitomon*, *P. rigidulum*, *P. tenerum*, *P. verrucosum*, *Pinus*, *Polygala cymosa*, *Quercus*, *Rhexia cubensis*, *Rhynchospora careyana*, *R. inundata*, *Rudbeckia mohrii*, *Sagittaria*, *Taxodium ascendens*, *Xyris smalliana*.

***Rhexia mariana* L.** occurs in open to partially shaded sites in beaver ponds, bogs, bottoms, depressions, ditches, draws, flatwoods, floodplains, glades, meadows, pocosins, prairies, roadsides, savannahs, seeps, shores, swales, swamps, and along the margins of borrow pits, channels, lakes, ponds, pools, rivers, and streams at elevations of up to 600 m. The substrates are acidic (pH: 5.1–6.7) and have been described as claypan, clay sand, clay/sand/silt loam, Corolla–Duckston sand, muck, sand, sandy clay, sandy clay loam, sandy loam, sandy peat, sandy silt, and silty clay loam. Generally the sediments are nutrient poor, and fairly high in organic matter. The flowers open from May to October and are buzz pollinated (and normally outcrossed) by bumblebees (Insecta: Hymenoptera: Apidae: *Bombus*). *Rhexia mariana* is also known to be agamospermous. The plants often form highly clonal patches by their prolific, stoloniferous rhizomes (which lack tubers). The seeds can retain viability for more than 32 months and are reported to reach densities of 900 seeds/m² in lake sediments. The seeds require 56 days of cold stratification to break their physiological dormancy. Optimal seed germination occurs in sunlight under a 35°C/20°C temperature regime. The rhizomes can also remain viable in the soil for several years until conditions are favorable for them to sprout. **Reported associates:** *Acer rubrum*, *Aletris*, *Allium*, *Andropogon*, *Aristida*, *Asclepias connivens*, *A. variegata*, *Baptisia simplicifolia*, *Betula*, *Bidens polylepis*, *Boehmeria*, *Callicarpa americana*, *Calopogon pallidus*, *Carex*, *Centella* *asiatica, Cephalanthus occidentalis, Chamaechrista, Chamaecyparis thyoides, Cladium jamaicense, Clethra alnifolia, Clitoria mariana, Commelina, Conoclinium, Coreopsis rosea, Cyperus lecontei, Cyrilla racemiflora, Dichanthelium, Diodia teres, Diospyros virginiana, Drosera capillaris, D. intermedia, Eleocharis melanocarpa, Elephantopus elatus, Eriocaulon, Eryngium yuccifolium, Eupatorium capillifolium, E. coelestinum, E. leucolepis, E. perfoliatum, Eustachys glauca, Euthamia leptocephala, E. tenuifolia, Fraxinus, Gleditsia, Gordonia lasianthus, Helianthus radula, Hibiscus, Hydrolea, Hypericum adpressum, H. canadense, H. cistifolium, Ilex glabra, I. vomitoria, Impatiens capensis, Juncus biflorus, J. canadensis, J. marginatus, J. militaris, Lachnanthes caroliniana, Lachnocaulon engleri, Lactuca canadensis, Liatris chapmanii, L. spicata, Licania michauxii, Linum medium, Liquidambar styraciflua, Lonicera, Lophiola aurea, Ludwigia sphaerocarpa, Lycopodiella, Lyonia mariana, Malva, Mecardonia acuminata, Mikania, Myrica cerifera, Nolina atopocarpa, Panicum anceps, P. hemitomum, P. scoparium, Paspalum urvillei, Persicaria, Phyla nodiflora, Physostegia, Pinus elliottii, P. palustris, P. taeda, Piriqueta caroliniana, Pluchea, Polygala, Polypremum procumbens, Pseudognaphalium helleri, Pteridium aquilinum, Pterocaulon, Quercus bicolor, Q. geminata, Q. laevis, Q. laurifolia, Q. marilandica, Q. nigra, Q. phellos, Q. pumila, Q. virginiana, Rhexia alifanus, R. nashii, R. petiolata, R. virginica, Rhynchospora capitellata, R. divergens, R. inundata, R. nitens, Rosa bracteata, Rubus, Sabatia brachiata, S. kennedyana, Sacciolepis, Sagittaria lancifolia, Sarracenia, Schizachyrium tenerum, Schoenoplectus etuberculatus, Scleria, Serenoa repens, Sesbania vesicaria, Sisyrinchium atlanticum, Smilax auriculata, Solanum carolinense, Sorghum, Spartina patens, Sphagnum, Spiraea tomentosa, Triantha racemosa, Utricularia cornuta, Vaccinium corymbosum, Veronica, Viola lanceolata, Vitis, Woodwardia virginica, Xyris elliottii.*

***Rhexia nashii* Small** inhabits bogs, bottoms, Carolina bays, depressions, ditches, flats, flatwoods, floodplains, lake shores, meadows, pocosins, pond margins, powerline right of ways, ricefields, roadsides, savannahs, seepage bogs, seeps, sink ponds, slopes, stream margins, swales, and swamps at elevations up to at least 38 m. The substrates typically are acidic and include Bayou sandy loam, sand, sandy loam, sandy peat, muck, mucky peat sand, and peat. The plants can form large clones by the spread of their tuberiferous rhizomes. **Reported associates:** *Acalypha rhomboidea, Acer, Amphicarpum muhlenbergianum, A. purshii, Andropogon glomeratus, Aristida beyrichiana, Axonopus furcatus, Bacopa, Bulbostylis, Carex complanata, Chamaecrista fasciculata, Cirsium nuttallii, Cladium, Ctenium aromaticum, Cuphea carthagenensis, Cyperus compressus, C. globulosus, C. haspan, C. polystachyos, C. retrorsus, C. surinamensis, Cyrilla arida, C. racemiflora, Desmodium triflorum, Dichanthelium chamaelonche, D. dichotomum, D. scabriusculum, Digitaria serotina, Eleocharis, Elephantopus elatus, Eleusine indica, Erechtites hieracifolia, Erigeron vernus, Eupatorium capillifolium, Fimbristylis autumnalis,*

F. dichotoma, F. schoenoides, Fuirena breviseta, Hypericum brachyphyllum, Hyptis alata, Ilex cassine, I. glabra, Impatiens capensis, Juncus biflorus, J. scirpoides, Kummerowia striata, Kyllingia brevifolia, Liquidambar styraciflua, Ludwigia microcarpa, L. octavalvis, L. pilosa, Magnolia virginiana, Mitreola sessilifolia, Muhlenbergia capillaris, Murdannia nudiflora, Myrica caroliniensis, M. cerifera, M. heterophylla, Nyssa sylvatica, Osmundastrum cinnamomeum, Oxalis corniculata, Oxypolis filiformis, Panicum abscissum, Parnassia caroliniana, Paspalum notatum, P. setaceum, P. urvellii, Passiflora incarnata, Pinus echinata, P. elliottii, Piriqueta caroliniana, Platanthera blephariglottis, Polypremum procumbens, Pteridium aquilinum, Pycnanthemum tenuifolium, Quercus geminata, Rhexia alifanus, R. mariana, R. petiolata, Rhynchospora capitellata, R. fascicularis, R. globularis, R. glomerata, R. gracilenta, R. microcephala, R. rariflora, R. recognita, R. virginica, Sabal etonia, Sabatia grandiflora, Sacciolepis indica, Sarracenia alata, Schizachyrium scoparium, Scleria minor, S. reticularis, S. triglomerata, Scoparia dulcis, Sesbania vesicaria, Solanum viarum, Taxodium distichum, Thelypteris palustris, Vaccinium formosum, V. fuscatum, Woodwardia virginica, Xyris caroliniana, X. fimbriata, X. laxifolia.

***Rhexia parviflora* Chapm.** is distributed locally in open or lightly shaded areas of bogs, depressions, pond margins, seepage slopes, and swales at low elevations. The substrates are sands that have a high peat content. The flowers open from June to August and are distinctive by their bright white petals. These plants require habitats where periodic fires remove competing vegetation. Vegetative reproduction occurs by means of slender rhizomes, which run shallowly under the soil surface. The shallow rhizomes make these plants particularly vulnerable to damage by recreational vehicles. **Reported associates:** *Burmannia biflora, Carex glaucescens, Clethra alnifolia, Erigeron vernus, Eriocaulon compressum, Euphorbia inundata, Hypericum, Ilex myrtifolia, Lophiola americana, Lycopodium alopecuriodes, Nyssa biflora, Panicum, Polygala cymosa, Proserpinaca pectinata, Rhynchospora corniculata, Sabatia bartramii, Smilax walteri, Sphagnum, Taxodium ascendens, Xyris.*

***Rhexia petiolata* Walter** occurs in open to partially shaded sites in bogs, bottoms, creeks, depressions, ditches, flats, flatwoods, interdunal swales, pocosins, pond and pool margins, prairies, roadsides, savannahs, scrub, seeps, swamps, and along the margins of streams at elevations of up to 152 m. The substrates are acidic and include Atmore series soil, clay, sand, sandy loam, sandy peat, and sandy peat loam. The petals are quite ephemeral, often falling when the flower is simply touched. Unlike most members of the genus, the flowers are self-compatible. Field studies have demonstrated that the representation of *R. petiolata* in the seed bank can be poor, even when it is a common component among the standing vegetation. This species lacks root tubers and is not reported to be clonal. **Reported associates:** *Acer rubrum, Aletris aurea, Andropogon arctatus, Aristida beyrichiana, A. palustris, A. purpurascens, Arundinaria gigantea, Balduina uniflora, Calopogon tuberosus, Carphephorus pseudoliatris,*

Chamaesyce hirta, Clethra alnifolia, Commelina erecta, Ctenium aromaticum, Cyrilla racemiflora, Dichanthelium dichotomum, Drosera capillaris, D. filiformis, Eriocaulon decangulare, Eupatorium rotundifolium, Eurybia eryngiifolia, Hypericum brachyphyllum, H. stans, Hyptis mutabilis, Ilex coriacea, I. glabra, Kalmia hirsuta, Lachnanthes caroliniana, Lachnocaulon digynum, Liatris spicata, Lobelia, Lophiola americana, Ludwigia alternifolia, L. hirtella, Lycopodiella alopecuroides, Magnolia virginiana, Marshallia tenuifolia, Myrica caroliniensis, Oldenlandia uniflora, Osmundastrum cinnamomea, Oxypolis filiformis, Panicum abscissum, P. nudicaule, Pinus elliottii, P. serotina, Platanthera integra, Pleea tenuifolia, Polygala cruciata, P. ramosa, Portulaca amilis, Rhexia alifanus, R. mariana, R. nashii, R. nuttallii, Rhynchospora latifolia, R. macra, R. oligantha, Sabatia macrophylla, Sarracenia flava, S. leucophylla, S. psittacina, Serenoa repens, Sporobolus pinetorum, Stillingia aquatica, Symphyotrichum dumosum, Triantha racemosa, Toxicodendron vernix, Xyris ambigua, Zigadenus.

***Rhexia salicifolia* Kral & Bostick** grows locally in open sites of depressions, exposed lake bottoms, interdunal swales, flatwoods, and along the margins of lakes, ponds, and sinkholes at low elevations. The substrates are described as sand or sandy peat. Changes in water levels can impact the plants negatively when growing in sinkhole sites. Even small disturbances like ruts created by recreational vehicles can alter the hydrology enough to reduce the occurrences of this rare plant. The plants can reproduce vegetatively by means of their elongating tuber bearing roots. **Reported associates:** *Amphicarpum muhlenbergianum, Bulbostylis barbata, B. ciliatifolia, Chrysopsis lanuginosa, Drosera intermedia, Eleocharis melanocarpa, Eriocaulon lineare, Fuirena pumila, Hypericum fasciculatum, H. lissophloeus, H. reductum, Juncus repens, Lachnanthes caroliniana, Lachnocaulon anceps, L. minus, Ludwigia suffruticosa, Panicum hemitomon, P. tenerum, Paronychia chartacea, P. patula, Polypremum procumbens, Rhynchospora globularis, R. nitens, R. pleiantha, Sacciolepis striata, Sagittaria isoetiformis, Sarracenia, Scleria reticularis, Syngonanthus flavidulus, Xyris jupicai, X. longisepala.*

***Rhexia virginica* L.** grows in bogs, borrow pits, bottomlands, Carolina bays, depressions, ditches, draws, fens, flats, flatwoods, floodplains, glades, hummocks, marshes, meadows, oxbows, pocosins, pools, potholes, prairies, roadsides, savannahs, seeps, shores, stream bottoms, swales, thickets, and along the drying margins of lakes, ponds, rivers, sloughs, and streams at elevations of up to 250 m. The plants occur in full sun to partial shade in shallow standing water (to 10 cm) or on exposed acidic (e.g., pH: 4.5) substrates, which can include cobble, gravel, mud, peat, peaty sand, rock, rocky cobble, sand, sandy clay, and sandy peat. The flowers (open from June to October) are highly self-compatible but remain fertile for only a single day. The loss of floral receptivity is marked by a change in color of the stamen filament, which turns from bright yellow to red on the second day of anthesis. The plants are outcrossed and are pollinated by bumblebees (see *Use by wildlife*), which remove the pollen by means of a "buzz"

mechanism. Self-pollination potentially can occur by geitonogamy (rarely) or by intrafloral selfing as the bees forage. The bees are relatively inefficient as pollinators with a fairly low level of seedset (52%–56%) observed in natural populations. Greater pollinator visits (due to facilitation) have been observed on plants growing in the presence of *Spiraea tomentosa*; however, high densities of *S. tomentosa* will reduce pollinator visits to *R. virginica* because of pollinator competition. The seeds require several months of cold, moist stratification in order to break dormancy. Subsequent germination will occur within 2–3 weeks if incubated at 21°C. Vegetative reproduction can occur by the tuberous roots (formed in spring), which survive only for one season. Shading by other plants (e.g., by *Juncus effusus*) will result in reduced biomass. **Reported associates:** *Acer rubrum, Achillea millefolium, Agalinis purpurea, Agrimonia parviflora, Agropyron repens, Agrostis hyemalis, Aletris aurea, A. farinosa, Allium, Alnus rugosa, A. serrulata, Ambrosia artemisiifolia, Andropogon gerardii, A. virginicus, Antennaria neglecta, Apios americana, Aristida stricta, A. virgata, Arnoglossum ovatum, Aronia arbutifolia, A. melanocarpa, A. prunifolia, Ascelpias incarnata, A. lanceolata, Baptisia alba, B. lactea, Bartonia paniculata, B. virginica, Bidens cernuus, B. coronatus, Bromus inermis, Bulbostylis capillaris, Calamagrostis canadensis, Calopogon pulchellus, C. tuberosus, Campanula aparinoides, Carex annectens, C. atlantica, C. bebbii, C. bicknelli, C. cryptolepis, C. cumulata, C. echinata, C. folliculata, C. intumescens, C. livida, C. longii, C. lurida, C. oligosperma, C. pellita, C. scoparia, C. stricta, C. swanii, C. vesicaria, Celastrus rotundifolia, Centaurea, Cephalanthus occidentalis, Chamaecrista fasciculata, Chamaecyparis thyoides, Clethra alnifolia, Coelorachis rugosa, Coleataenia longifolia, Coreopsis nudata, C. rosea, Cornus, Corylus americana, Croton michauxii, Cuphea carthagenensis, Cyperus bipartitus, C. strigosus, Danthonia spicata, Desmodium sessilifolium, Dichanthelium acuminatum, D. clandestinum, D. erectifolium, D. ovale, D. scoparium, D. wrightianum, Drosera capillaris, D. rotundifolia, Dulichium arundinaceum, Eleocharis acicularis, E. engelmannii, E. melanocarpa, E. obtusa, E. palustris, E. tenuis, E. tuberculosa, Erechtites hieracifolia, Erianthus giganteus, Eriocaulon compressum, E. decangulare, Eriophorum virginicum, Eupatorium perfoliatum, E. serotinum, Euphorbia corollata, Euthamia graminifolia, E. gymnospermoides, Fagus, Fimbristylis autumnalis, Fragaria virginiana, Fuirena pumila, F. scirpoidea, F. squarrosa, Gaylussacia baccata, Gentiana saponaria, Helianthus mollis, Hemicarpha micrantha, Hibiscus, Hydrolea, Hypericum adpressum, H. canadense, H. cistifolium, H. gentianoides, H. hypericoides, H. majus, H. mutilum, H. myrtifolium, H. setosum, Ilex glabra, I. myrtifolia, I. verticillata, Iris virginica, Juncus acuminatus, J. alpinus, J. anthelatus, J. brevicaudatus, J. canadensis, J. effusus J. greenei, J. marginatus, J. militaris, J. polycephalus, Leersia oryzoides, Liatris pychnostachya, Lobelia cardinalis, L. floridana, L. kalmii, Lonicera, Ludwigia pilosa, Lycopus americanus, L. uniflorus, Lyonia ligustrina, Lysmachia lanceolata, Lythrum alatum, Magnolia virginana, Malus, Mikania scandens,*

Mimulus alatus, Muhlenbergia mexicana, M. uniflora, Myrica, Nepeta cataria, Nyssa, Oxypolis filiformis, Panicum chamelonche, P. hemitomon, P. verrucosum, P. virgatum, Parthenocissus quinquefolia, Paspalum laeve, Persea palustris, Persicaria careyi, P. coccinea, P. hydropiperoides, P. pensylvanica, P. sagittata, Phlox glaberrima, Phragmites australis, Phytolacca americana, Pinguicula planifolia, Pinus banksiana, P. taeda, Plantanthera lavera, Poa pratensis, Polygala cruciata, P. cymosa, P. sanguinea, Polytrichum, Pontederia cordata, Populus tremuloides, Potentilla simplex, Prunus serotina, Pueraria montana, Pycnanthemum virginianum, Quercus pagoda, Rhamnus frangula, Rhexia cubensis, R. mariana, R. parviflora, Rhus glabra, Rhynchospora capitellata, R. careyana, R. corniculata, R. fascicularis, R. filifolia, R. fusca, R. glomerata, R. macrostachya, Robinia, Rosa multiflora, R. rugosa, Rotala ramosior, Rubus allegheniensis, R. flagellaris, R. hispidus, Rudbeckia hirta, Sabatia bartramii, S. campanulata, S. kennedyana, Sacciolepis, Salix humilis, Sambucus, Sarracenia flava, S. psittacina, S. purpurea, Sassafras albidum, Schizachryium scoparium, Schoenoplectus smithii, Scirpus cyperinus, Scleria baldwinii, S. triglomerata, Serenoa palmetto, Setaria geniculata, S. viridula, Smilax, Solidago chapmanii, S. gigantea, S. graminifolia, S. missouriensis, S. remota, Sorghastrum nutans, Spartina pectinata, Sphagnum, Spiraea alba, S. tomentosa, Spiranthes, Stachys tenuiflora, Symphyotrichum dumosum, Taxodium ascendens, Thelypteris palustris, Toxicodendron, Triadenum fraseri, Vaccinium angustifolium, V. corymbosum, Veronicastrum virginicum, Viola lanceolata, V. primulifolia, Woodwardia areolata, W. virginica, Xyris caroliniana, X. fimbriata, X. torta.

Use by wildlife: Caterpillars (Insecta: Lepidoptera) are hosted by *Rhexia petiolata* (Geometridae: *Cymatophora approximaria*) and *R. virginica* (Geometridae: *Eupithecia miserulata*; Noctuidae: *Heliothis virescens*. Bumblebees (Insecta: Hymenoptera: Apidae: *Bombus bimaculatus, B. impatiens*) forage for pollen on and serve as pollinators of *Rhexia virginica*. In the southeastern United States, *Rhexia* species reportedly make up an important portion of the diet for grazing steers (Mammalia: Bovidae: *Bos taurus*).

Economic importance: food: The Montagnais made a sour drink from the leaves and stems of *Rhexia virginica*. Leaves of several species (e.g., *R. virginica*) are sweetly tart in taste and have been eaten in salads; however, the leaves of all *Rhexia* species tested (including *R. mariana* and *R. virginica*) show accumulation of aluminum. The chopped roots of *R. virginica* sometimes are added to salads; **medicinal:** Brewed leaves and stems of *R. virginica* were used by the Micmac and Montagnais tribes as a throat cleanser; **cultivation:** *Rhexia mariana, R. nashii,* and *R. virginica* occasionally are grown as ornamentals; **misc. products:** A newsletter entitled "The *Rhexia*" is published by the Paynes Prairie Chapter of the Florida Native Plant Society; **weeds:** none; **nonindigenous species:** none

Systematics: Molecular data (Figure 4.83) place *Rhexia* as the sister group of the Central American genus *Arthrostemma*, which shares with it a strongly costate–tuberculate seed coat.

Opinions vary whether *Rhexia* should be classified within a separate tribe (Rhexieae) or among the members of tribe Melastomeae. Although some results (e.g., Figure 4.83) would favor acceptance of the former disposition, other analyses have resolved the genus as being nested within *Melastomeae*. Discrepancies in the number of species (11 vs. 13) occur because some authors treat *R. interior* and *R. ventricosa* as distinct species; whereas, others regard both taxa as varieties of *R. mariana*. A recent taxonomic treatment subdivided the genus into four putatively monophyletic sections: section *Rhexia*, section *Cymborhexia* [*Rhexia alifanus*], section *Brevianthera*, and section *Luteorhexia* [*Rhexia lutea*]. Interspecific relationships in *Rhexia* (including 11 species) have been studied using morphological data as well as cpDNA and nrITS sequences. These data have yielded well resolved but discordant phylogenies, a result that supports earlier presumptions of widespread hybridization in the history of the genus. The molecular data provide consistent evidence of relationships for six species and also indicate the derivation of *R. nashii* through hybridization of the *R. mariana* and *R. virginica* lineages (Figure 4.84). However, the placements of several other species resolve inconsistently among the different data sets analyzed. In one case, *R. salicifolia* is interpreted as being of hybrid origin, originating from parental species within the *R. aristosa* and *R. cubensis* lineages. The placement of *R. parviflora* remains unresolved, due to DNA sequencing artifacts and other technical complications. The base chromosome number of *Rhexia* is $x = 11$. Diploids ($2n = 22$) include *R. alifanus, R. aristosa, R. nuttallii, R. parviflora, R. petiolata,* and *R. salicifolia*; *R. lutea* ($2n = 44$) is tetraploid. Mixed cytotypes are reported for *R. cubensis* and *R. mariana* ($2n = 22, 44, 66$), *R. virginica* ($2n = 22, 44$), and *R. nashii* ($2n = 44, 66$). Natural hybrids have been observed and synthetic hybrids (generally infertile) have been produced between *R. mariana* and *R. virginica*. Other reported hybrids include *R. nashii* × *R. mariana, R. nashii* × *R. virginica* and *Rhexia* ×*brevibracteata*, which is a natural hybrid of *R. aristosa* × *R. virginica* from the coastal plain of New Jersey.

Comments: *Rhexia mariana* occurs throughout the eastern United States; whereas, *R. virginica* is distributed throughout eastern North America and is disjunct in British Columbia.

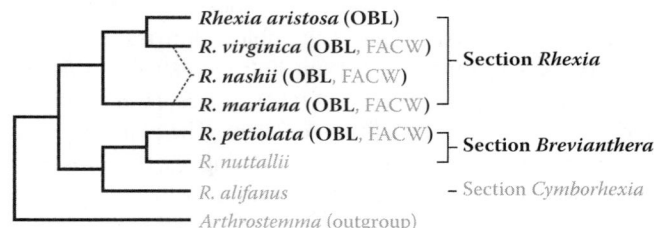

FIGURE 4.84 Interspecific relationships in *Rhexia*, as inferred by phylogenetic analyses of morphological and molecular data. If accurate, these relationships would indicate that OBL habit (bold) might have arisen twice within the genus. The dotted lines designate *R. nashii* as being of hybrid origin involving the *R. mariana* and *R. virginica* lineages. (Adapted from Ionta, G.M. et al., *Int. J. Plant Sci.*, 168, 1055–1066, 2007.)

Rhexia aristosa, *R. nashii* and *R. petiolata* occur along the coastal plain of the southeastern United States. More restricted species include *R. parviflora* (AL, FL, GA) and *R. salicifolia* (AL, FL). Suitable habitats for rare species like *R. aristosa*, *R. mariana* (New England), *R. parviflora*, and *R. salicifolia* are threatened by recreational uses of ponds and lakes, residential development, timber cutting, and by activities that alter natural hydrological processes.

References: Baskin et al., 1999a; Chiari, 2005; Clausing & Renner, 2001; Cohen et al., 2004; Craine, 2002; Godfrey & Wooten, 1981; Ionta et al., 2004, 2007; Jansen et al., 2002; Kalmbacher et al., 1984; Keddy & Reznicek, 1982; Kral & Bostick, 1969; LaRosa et al., 2004; Larson & Barrett, 1999; Nesom, 2012; Posluszny et al., 1984; Renner, 1989; Sharp & Keddy, 1985; Snyder, 1996; Sutter & Boyer, 1994; Thunhorst, 1995; Wurdack & Kral, 1982.

Family 4: Onagraceae [22]

Onagraceae (657 species) are among the more distinctive and easily recognized dicotyledon families by virtue of their consistent 4-merous floral morphology, inferior ovary, and (except for *Ludwigia*) the presence of an elongate, deciduous hypanthium. The nectariferous flowers are pollinated biotically by insects (Insecta) and birds (Aves) or are self-pollinated. In the capsular-fruited species the seeds are dispersed by wind or water and by birds for those with berries.

Cladistic analyses of various molecular and morphological data sets indicate the monophyly of Onagraceae and the basal placement of *Ludwigia* as the sister group to the remainder of the genera (Hoch et al., 1993; Levin et al., 2003, 2004; Wagner et al., 2007; Figure 4.85). Six tribes currently are recognized, which reflect the major generic diversifications in the family (Figure 4.85; Wagner et al., 2007). Some relatively recent modifications to the classification of Onagraceae include the merger of *Boisduvalia* with *Epilobium* but the distinction of *Chamerion* from *Epilobium*, dispositions that are both followed here. Phylogenetic studies by Hoggard et al. (2004) suggested the merger of the monotypic *Stenosiphon linifolius* and *Gaura*; however, both genera (as well as *Calylophus*) now have been transferred to *Oenothera* (Wagner et al., 2007).

Onagraceae primarily are terrestrial from an ecological standpoint with the species occurring in dry to wet sites ranging from artic regions to deserts. Only two genera (*Epilobium*, *Ludwigia*) are considered to include true aquatics on a worldwide basis (Cook, 1996a); however, three genera (and just over 40 species) contain OBL indicators in North America. The former genus *Gaura* included one OBL species (*G. neomexicana*) in the 1996 indicator list; however that species was transferred to *Oenothera* (as *Oenothera coloradensis* subspecious *neomexicana*) and has also been recategorized as a nonobligate (FACW, FAC) indicator. The genera with OBL North American indicators are:

1. *Epilobium* L.
2. *Ludwigia* L.
3. *Oenothera* L.

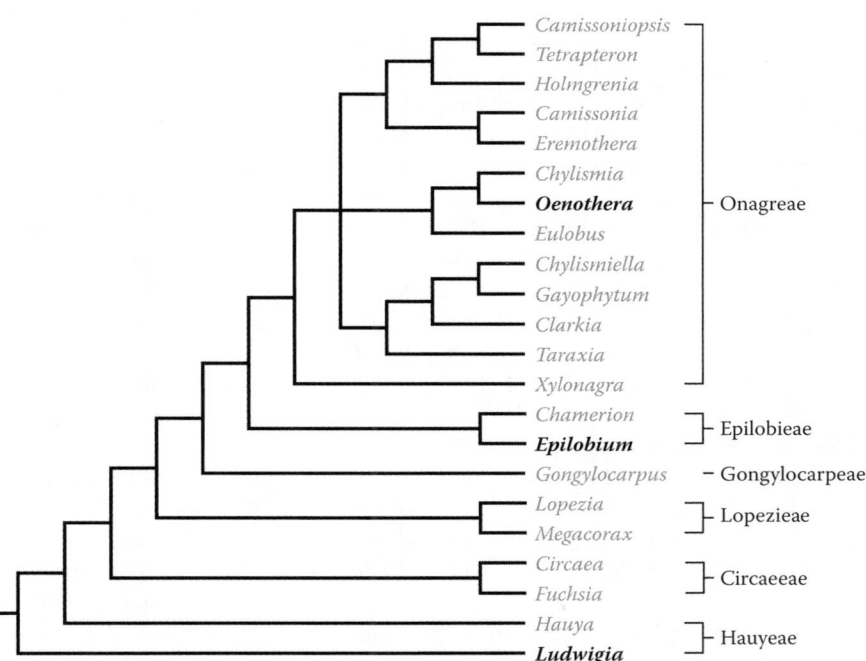

FIGURE 4.85 Phylogenetic relationships among the 22 recognized genera of Onagraceae as deduced from analyses of combined nrITS, *trnL–trnF*, and *rps*16 sequence data. Tribal designations are indicated on the right. The genera including OBL indicators in North America are indicated by bold type. *Ludwigia*, with a relatively large proportion of aquatic and wetland species, is basal in the phylogeny. Overall, the OBL aquatic species have originated independently at various levels of evolutionary specialization in Onagraceae. (After Levin, R.A. et al., *Amer. J. Bot.*, 90, 107–115, 2003; Levin, R.A. et al., *Syst. Bot.*, 29, 147–164, 2004; Wagner, W.L. et al., *Syst. Bot. Monogr.*, 83, 1–240, 2007.)

Phylogenetic analyses indicate separate origins of the OBL habitat for each genus at various levels of specialization (from basal to highly derived positions) within the family (Figure 4.85).

1. *Epilobium*

Willow-herb; épilobe

Etymology: from the Greek *epi lobon* meaning "upon a pod" with reference to the attachment of the hypanthium at the end of the capsule

Synonyms: *Boisduvalia*; *Chamaenerion*; *Chamaenerium*; *Cordylophorum*; *Cratericarpium*; *Crossostigma*; *Pyrogennema*; *Zauschneria*

Distribution: global: cosmopolitan; **North America:** widespread

Diversity: global: 165 species; **North America:** 40 species

Indicators (USA): OBL: *Epilobium campestre*, *E. cleistogamum*, *E. coloratum*, *E. leptophyllum*, *E. oreganum*, *E. oregonense*, *E. palustre*, *E. parviflorum*, *E. strictum*; **FACW:** *E. campestre*, *E. ciliatum*, *E. coloratum*, *E. densiflorum*, *E. glaberrimum*, *E. leptophyllum*, *E. oregonense*; **FAC:** *E. ciliatum*

Habitat: freshwater; palustrine; **pH:** 3.8–8.3; **depth:** <1 m; **life-form(s):** emergent herb

Key morphology: stems decumbent, prostrate or erect (to 2 m), often with basal rosettes; herbage glabrous to strigose; cauline leaves (to 15 cm) opposite, rounded, lanceolate or ovate; flowers in racemes, pedicellate (to 65 mm), 4-merous, epigynous, with a hypanthium (to 3 mm); petals (to 15 mm) white or pink to rose-purple; stamens 8; capsules (to 10 cm) many seeded.

Life history: duration: annual (fruit/seeds); perennial (bulblets, stolons); **asexual reproduction:** bulblets, stolons; **pollination:** insect or cleistogamous; **sexual condition:** hermaphroditic; **fruit:** capsules (common); **local dispersal:** seeds (water, wind); **long-distance dispersal:** seeds (animals, wind)

Imperilment: (1) *Epilobium campestre* [G5]; S1 (BC, ND, UT); S2 (AB, SK, WY); (2) *E. ciliatum* [G5]; SX (IN); SH (NY); S1 (DE, KY, MD, TN); S2 (BC, NC, NH, WV); S3 (NJ, WY); (3) *E. cleistogamum* [G3]; (4) *E. coloratum* [G5]; S2 (NB, NS); S3 (NC); (5) *E. densiflorum* [G5]; SH (MT); S1 (BC, UT); (6) *E. glaberrimum* [G5]; S1 (AB, UT); S2 (BC, WY); (7) *E. leptophyllum* [G5]; S1 (MO, TN); S2 (MD, NJ, VA, WY); S3 (AB, BC, NC, NF, WV); (8) *E. oreganum* [G2]; S1 (OR); S2 (CA); (9) *E. oregonense* [G5]; S1 (AZ, UT); S2 (BC, WY); (10) *E. palustre* [G5]; SH (CT, VT); S1 (CA, PA, RI); S2 (WY); S3 (AB, ID, MI, MT, WI); (11) *E. parviflorum* [GNR]; (12) *E. strictum* [G5]; S1 (IL, MD); S2 (NB, NJ, OH, WI); S3 (NS, PA, PE)

Ecology: general: *Epilobium* primarily is characteristic of mesic habitats, and often occurs in boreal or montane north temperate regions. Nearly a quarter of the North American species of *Epilobium* are designated as obligate indicators in at least a part of their range. However, even the more "aquatic" species normally are associated more with wetlands than with deeper water habitats. In the 1996 indicator list, the categorization of several OBL species as FAC indicators in portions of their ranges implied a fair tolerance to semiterrestrial conditions. Three of those species (*E. ciliatum*, *E. densiflorum*, *E. glaberrimum*) no longer retained OBL status in the 2013 list. Nevertheless, they are included here because their treatments had been completed prior to the release of the updated list. The nonindigenous *E. parviflorum* has been added as an OBL indicator since 1996. Even though most *Epilobium* species are pollinated by bees (Insecta: Hymenoptera), they primarily are self-compatible and autogamous. Other pollinators include butterflies (Lepidoptera), flies (Diptera), and hummingbirds (Aves: Trochilidae). The seeds of many species possess a coma of epidermal hairs, which assists in their dispersal by wind. The genus includes annuals and perennials.

Epilobium campestre (**Jepson**) **Hoch & W. L. Wagner** is an annual species restricted to sites of ephemeral waters that are associated with channels, depressions, ditches, flats, marshes, meadows, mud flats, playas, sloughs, vernal pools, washes, and the drying margins of ponds, reservoirs, and streams at elevations of up to 3181 m. The plants occur where they are exposed to full sunlight and occasionally are found in shallow water. The substrates are alkaline and are described as Alamo adobe clay, Capay clay, Capay silty clay loam, clay, clay loam, clay muck, gravel, hardpan, lava, loam, loamy clay, mud, Pescadero clay, rock, sand, sandy loam, serpentine, silt, and silty loam. The self-compatible flowers occur from May to September, are usually cleistogamous, and primarily are self-pollinating. The capsules remain closed until they are mature, a factor that could limit seed dispersal. **Reported associates:** *Alisma triviale*, *Allium nevii*, *Alopecurus aequalis*, *A. arundinaceus*, *Amaranthus albus*, *Ambrosia pumila*, *Anthemis*, *Artemisia cana*, *Atriplex fruticulosa*, *A. pacifica*, *Beckmannia syzigachne*, *Bergia texana*, *Bidens*, *Brodiaea jolonensis*, *B. orcuttii*, *Calandrinia maritima*, *Callitriche longipedunculata*, *Carex atherodes*, *Castilleja densiflora*, *Centromadia fitchii*, *Centunculus minimus*, *Chorizanthe procumbens*, *Cotula coronopifolia*, *Crassula aquatica*, *Cressa truxillensis*, *Croton setiger*, *Crypsis vaginiflora*, *Cyperus bipartitus*, *Deinandra kelloggii*, *Deschampsia cespitosa*, *D. danthonoides*, *Distichlis spicata*, *Downingia bella*, *D. cuspidata*, *D. insignis*, *Echinodorus berteroi*, *Elatine brachysperma*, *E. californica*, *Eleocharis acicularis*, *E. macrostachya*, *E. palustris*, *Epilobium cleistogamum*, *Eryngium aristulatum*, *E. vaseyi*, *Gnaphalium palustre*, *Heliotropium curassavicum*, *Hirschfeldia incana*, *Hordeum intercedens*, *H. jubatum*, *H. marinum*, *H. pusillum*, *Isoetes howellii*, *I. orcuttii*, *Juncus articulatus*, *J. bufonius*, *J. tenuis*, *Lasthenia californica*, *L. ferrisiae*, *L. glaberrima*, *L. glabrata*, *L. minor*, *Lepidium latipes*, *L. nitidum*, *Liliaea scilloides*, *Limosella acaulis*, *Linanthus dianthaflorus*, *Lolium multiflorum*, *Ludwigia palustris*, *Lupinus*, *Lythrum hyssopifolia*, *Madia glomerata*, *M. sativa*, *Malvella leprosa*, *Marah fabacea*, *Marsilea vestita*, *Mimulus latidens*, *Myosurus minimus*, *Nama stenocarpum*, *Navarretia fossalis*, *N. hamata*, *N. prostrata*, *N. sinistra*, *Neostapfia colusana*, *Ophioglossum californicum*, *Orcuttia californica*, *Pascopyrum smithii*,

Phalaris arundinacea, P. lemmonii, Pilularia americana, Pinus sabiniana, Plagiobothrys acanthocarpus, P. bracteatus, P. hispidulus, P. leptoclada, P. scouleri, P. stipitatus, P. undulatus, Plantago elongata, P. erecta, P. major, Pogogyne douglasii, P. nudiuscula, P. serpylloides, Polygonum aviculare, P. polygaloides, Polypogon monspeliensis, Potentilla anserina, Prunus ilicifolia, Psilocarphus brevissimus, P. tenellus, Pyrrocoma carthamoides, Quercus, Ribes quercetorum, Rotala ramosior, Rumex crispus, R. triangulivalvis, Sairocarpus cornutus, Salvia columbariae, Spergularia marina, S. villosa, Toxicodendron diversilobum, Triphysaria eriantha, Tuctoria, Typha, Verbena bracteata, V. menthifolia, Veronica anagallis-aquatica, V. peregrina, Zeltnera muehlenbergii.

***Epilobium ciliatum* Raf.** is a short-lived perennial found in a diverse array of habitats including bogs, ditches, fens, marshes, pools, seeps, sloughs, swamps, wet meadows, and along the shallow margins of brooks, rivers, and lakes at elevations of up to 4100 m. It is common on coarse-textured substrates (gravel, sand) and also occurs in muck and peat. The plants (formerly OBL in a portion of their range) sometimes are found growing on rocks or tree stumps and thrive in disturbed sites. The habitats span a wide range of pH and vary from fully open to deeply shaded sites. Sexual reproduction principally is autogamous with the pollen being released while the self-compatible flowers (sometimes cleistogamous) remain in bud. Visits by potential pollinators (Insecta: Hymenoptera: Apidae: *Bombus*; Halictidae) are infrequent, but larger-flowered plants of subspecies *glandulosum* reportedly are cross-pollinated. The plants are adapted to selfing and exhibit low levels of inbreeding depression. The populations are uniform genetically as indicated by a lack of detectable electrophoretic variation at surveyed allozyme loci. The seeds germinate without stratification within 2 weeks after they have been placed on wet filter paper and are kept under a 15°C/10°C day/night temperature regime; they also will germinate well after several months of storage. The seeds possess a coma and are wind dispersed. The plants perennate by persistent rosettes, which sometimes produce fleshy bulblets. **Reported associates:** *Abies balsamea, Acer rubrum, A. saccharum, A. spicatum, Achillea sibirica, Agropyron caninum, Agrostis exarata, A. stolonifera, Alisma plantago-aquatica, Allium schoenoprasum, Alnus rugosa, Ambrosia artemisiifolia, Amelanchier alnifolia, Andropogon scoparius, Angelica arguta, Apium graveolens, Arnica mollis, Artemisia biennis, A. douglasiana, Arundo donax, Asclepias verticillata, Baccharis pilularis, B. salicifolia, Barbarea vulgaris, Berula erecta, Betula lutea, B. occidentalis, B. papyrifera, Brachythecium, Bromus ciliatus, Bulboschoenus fluviatilis, B. maritimus, Calamagrostis canadensis, Caltha palustris, Campanula aparinoides, C. rotundifolia, Carex aquatilis, C. atherodes, C. aurea, C. bebbii, C. flava, C. hystericina, C. interior, C. lacustris, C. leptalea, C. nebraskensis, C. rostrata, C. stipata, C. stricta, C. utriculata, C. vesicaria, C. vulpinoidea, Castilleja miniata, C. rhexiifolia, Chamerion angustifolium, Chelone glabra, Chenopodium album, Chrysosplenium americanum, Cichorium intybus, Cicuta* *bulbifera, C. maculata, Circaea alpina, Cirsium arvense, C. muticum, C. vulgare, Clintonia borealis, Comarum palustre, Coptis groenlandica, Cornus sericea, Cotula coronopifolia, Cratoneuron commutatus, Cryptotaenia canadensis, Cuscuta gronovii, Cynodon dactylon, Cyperus eragrostis, Deschampsia cespitosa, Dryopteris cristata, Dulichium arundinaceaum, Echinochloa crus-galli, Eleocharis macrostachya, E. palustris, E. pauciflora, Elymus virginicus, Equisetum fluviatile, Epilobium anagallidifolium, E. coloratum, E. glaberrimum, Equisetum arvense, E. fluviatile, E. scirpoides, E. sylvaticum, Erigeron peregrinus, Eupatorium perfoliatum, Fallopia convolvulus, Galium obtusum, G. trifidum, G. triflorum, Gentianopsis crinita, Geum canadense, G. macrophyllum, Gnaphalium uliginosum, Glyceria striata, Habenaria, Helenium autumnale, H. puberulum, Heracleum lanatum, Heterotheca depressa, Hirschfeldia incana, Hordeum jubatum, Huperzia lucidula, Hypericum majus, Impatiens capensis, Iris versicolor, Juncus balticus, J. bufonius, J. torreyi, J. xiphioides, Laportea canadensis, Larix laricina, Lathyrus latifolius, Leersia oryzoides, Lemna minor, Leptochloa uninvervia, Ligusticum tenuifolium, Ludwigia polycarpa, Lycopodium obscurum, Lycopus americanus, L. uniflorus, Lysimachia nummularia, Lythrum salicaria, Maianthemum trifolium, Marchantia polymorpha, Melilotus alba, Mentha arvensis, Mertensia paniculata, Mimulus glabratus, M. lewisii, Mitella diphylla, M. nuda, Muhlenbergia mexicana, Nasturtium officinale, Nicotiana glauca, Onoclea sensibilis, Paspalum dilatatum, Pedicularis groenlandica, Persicaria amphibia, P. hydropiperoides, P. lapathifolia, P. punctata, Philonotis fontana, Phleum pratense, Physostegia virginiana, Picea glabra, P. mariana, Picris echioides, Pilea fontana, Pinus resinosa, P. strobus, Plantago lanceolata, P. rugelii, Pluchea odorata, Poa compressa, P. palustris, Polypogon monspeliensis, Populus angustifolia, P. tremuloides, Prunus virginiana, Quercus alba, Ranunculus sceleratus, Rhamnus alnifolius, Ribes lacustre, Rorippa palustris, Rubus idaeus, Rudbeckia laciniata, Rumex crispus, R. orbiculatus, R. salicifolius, Sagittaria cuneata, S. latifolia, Salix bebbiana, S. boothii, S. discolor, S. exigua, S. geyeriana, S. interior, S. lasiolepis, S. lutea, S. sitchensis, Schoenoplectus acutus, S. tabernaemontani, Scirpus pendulus, Scutellaria galericulata, Senecio hydrophilus, S. integerrimus, S. triangularis, Sium suave, Solidago canadensis, Sparganium emersum, Sphagnum, Spiraea alba, S. tomentosa, Stachys albens, Stellaria calycantha, Stephanomeria fluminea, Sullivantia renifolia, Symphyotrichum cordifolium, S. firmum, S. foliaceum, S. lanceolatum, S. pilosum, S. puniceum, S. spathulatum, S. subulatum, Taraxacum officinale, Teucrium canadense, Thelypteris palustris, Thuja occidentalis, Tilia americana, Tradescantia ohiensis, Trifolium hybridum, T. pratense, Tsuga canadensis, Typha angustifolia, T. domingensis, T. ×glauca, T. latifolia, Valeriana sitchensis, Veronica americana, V. anagallis-aquatica, Viburnum, Viola canadensis, Xanthium strumarium.*

***Epilobium cleistogamum* (Curran) Hoch & Raven** is a narrowly distributed annual, which inhabits channel beds, depressions, dry rivulets, floodplains, meadows, mud flats, playas, roadsides, stream courses, swales, and vernal pools at elevations of up to 1676 m. The substrates are alkaline and described as adobe, Alamo adobe clay, alluvium, Capay silty clay, clay, gravelly loam, mudflow, silt, and volcanic. The cleistogamous, self-compatible flowers are autogamous and produce glabrous seeds, which probably are dispersed only locally. **Reported associates:** *Achyrachaena mollis, Alopecurus saccatus, Amaranthus albus, Anthemis cotula, Avena fatua, Bromus hordeaceus, Crassula aquatica, Croton setiger, Crypsis schoenoides, Cuscuta howelliana, Deschampsia danthonioides, Downingia bicornuta, D. insignis, D. ornatissima, Eleocharis macrostachya, E. palustris, Epilobium campestre, Eryngium aristulatum, E. castrense, E. spinosepalum, E. vaseyi, Euphorbia hooveri, Gnapahalium palustre, Hemizonia congesta, Hesperevax caulescens, Hordeum marinum, Juncus bufonius, Lasthenia fremontii, L. glaberrima, L. glabrata, Layia chrysanthemoides, Lepidium latipes, Lessingia, Lolium multiflorum, Lupinus bicolor, Lythrum hyssopifolia, Medicago polymorpha, Mimulus tricolor, Myosurus minimus, Navarretia cotulifolia, N. leucocephala, Orcuttia inaequalis, O. pilosa, Pilularia americana, Plagiobothrys acanthocarpus, P. bracteatus, P. stipitatus, Pleuropogon californicus, Pogogyne douglasii, P. ziziphoroides, Polypogon monspeliensis, Psilocarphus brevissimus, P. chilensis, P. oregonus, Rumex crispus, Trifolium depauperatum, T. willdenovii, Triphysaria eriantha, Veronica peregrina.*

***Epilobium coloratum* Biehler** is a widespread perennial, which occurs in bogs, brooks, clay pits, ditches, dune ponds, fens, marshes, meadows, prairies, ravines, roadsides, seepage slopes, sloughs, streambeds, swales, swamps, and along the margins of lakes, ponds, rivers, and streams at elevations of up to 835 m. The plants have broad habitat tolerances (similar to *E. ciliatum*), often thriving in disturbed sites, and occur in full sun to shaded exposures. The substrates (pH: 5.5–8.3) can consist of gravel, marl, muck, peat, sand, sandy loam, or silty organics. The flowers are self-compatible but are pollinated by bees (Insecta: Hymenoptera). The plants produce numerous seeds, which possess a coma and are wind dispersed; however, their documented germination from fecal droppings of deer (Mammalia: Cervidae: *Odocoileus*) indicates that viable seeds also can be transported endozooically by grazing animals. Burial by sedimentation does not impact seed germination negatively. Vegetative persistence occurs by the formation of leafy, rosette-like offshoots in the fall. The plants frequently are found on mud flats that have become exposed by receding waters. **Reported associates:** *Abutilon theophrasti, Acalypha rhomboidea, Acer negundo, A. nigrum, A. rubrum, A. saccharinum, A. saccharum, Aconitum noveboracense, Ageratina altissima, Agrimonia gryposepala, A. parviflora, Agrostis hyemalis, Alliaria petiolata, Allium cernuum, Alnus rugosa, Amaranthus cannabinus, Ambrosia artemisiifolia, A. trifida, Ampelopsis cordata, Amphicarpa bracteata, Andropogon ternarius, Angelica atropurpurea, Apios americana, Arctium minus, Arisaema dracontium, A. triphyllum, Aronia melanocarpa, Asclepias incarnata, Bartonia virginica, Betula lutea, B. nigra, B. papyrifera, B. pumila, Bidens cernuus, B. frondosus, B. polylepis, B. vulgatus, Boehmeria cylindrica, Boltonia asteroides, Bouteloua curtipendula, Bromus ciliatus, Calamagrostis canadensis, Callitriche verna, Caltha palustris, Campanula rotundifolia, C. uliginosa, Cardamine bulbosa, C. pensylvanica, Carex aquatilis, C. atlantica, C. conjuncta, C. corrugata, C. crinita, C. cristatella, C. crus-corvi, C. davisii, C. diandra, C. festucacea, C. frankii, C. grisea, C. haydenii, C. hyalinolepis, C. hystericina, C. lacustris, C. laevivaginata, C. longii, C. lupulina, C. lurida, C. muskingumensis, C. nebrascensis, C. projecta, C. scoparia, C. shortiana, C. squarrosa, C. stipata, C. stricta, C. tenera, C. tribuloides, C. vulpinoidea, Carya cordiformis, C. illinoinensis, C. laciniosa, C. ovata, Cephalanthus occidentalis, Chara, Chasmanthium latifolium, Chelone glabra, C. obliqua, Chenopodium album, Cicuta bulbifera, C. maculata, Cinna arundinacea, Circaea lutetiana, Cirsium arvense, C. pumilum, C. vulgare, Clematis pitcheri, C. virginiana, Conocephalum, Conyza canadensis, Coreopsis tripteris, Cornus alternifolia, C. amomum, C. racemosa, C. sericea, Crataegus, Cuscuta cuspidata, C. gronovii, Cyperus diandrus, C. erythrorhizos, C. strigosus, Cypripedium acaule, Dasiphora floribunda, Decodon verticillatus, Dichanthelium clandestinum, D. dichotomum, Doellingeria umbellata, Drosera rotundifolia, Dryopteris cristata, D. fragrans, D. thelypteris, Echinochloa crus-galli, E. muricata, Echinocystis lobata, Eleocharis acicularis, E. compressa, E. erythropoda, E. obtusa, E. palustris, E. pauciflora, Epilobium ciliatum, E. leptophyllum, Equisetum hyemale, Eragrostis spectabilis, Erechtites hieracifolia, Erigeron annuus, E. philadelphicus, Eupatorium perfoliatum, E. serotinum, Euthamia graminifolia, Eutrochium maculatum, Fallopia scandens, Fragaria virginiana, Frangula alnus, Fraxinus americana, F. nigra, F. pennsylvanica, F. profunda, Galium brevipes, G. lanceolatum, G. tinctorium, G. trifidum, Gentiana andrewsii, Gentianopsis virgata, Geum aleppicum, G. canadense, Glyceria grandis, G. striata, Helenium autumnale, Helianthus grosseserratus, H. mollis, Hieracium canadense, Hypericum mutilum, H. perforatum, H. punctatum, H. virginicum, Impatiens capensis, Iris brevicaulis, I. virginica, Juglans nigra, Juncus acuminatus, J. balticus, J. brachycarpus, J. brevicaudatus, J. canadensis, J. effusus, J. marginatus, J. torreyi, Juniperus, Lactuca canadensis, Laportea canadensis, Larix laricina, Lathyrus palustris, Leersia oryzoides, L. virginica, Lemna, Leucospora multifida, Liatris pychnostachya, Lindera benzoin, Lobelia cardinalis, L. inflata, L. kalmii, L. siphilitica, Lonicera, Ludwigia alternifolia, L. palustris, L. polycarpa, Lycopus americanus, L. asper, L. rubellus, L. uniflorus, Lysimachia nummularia, L. quadriflora, L. thyrsiflora, Lythrum alatum, L. salicaria, Maianthemum stellatum, Malvastrum hispidum, Mentha arvensis, Micranthes pensylvanica, Microstegium vimineum, Mimulus alatus, M. glabratus, M. ringens, Monarda fistulosa, Morus alba, Muhlenbergia frondosa, M. glomerata, M. mexicana, M. schreberi, M. sylvatica, Myosotis laxa, Nasturtium officinale, Nymphaea odorata, Onoclea sensibilis, Osmunda regalis, Osmundastrum cinnamomeum, Oxypolis rigidior, Packera aurea, Panicum dichotomiflorum, Parthenocissus* *inserta, P. quinquefolia, Pedicularis lanceolata, Peltandra virginica, Penthorum sedoides, Persicaria amphibia, P. coccinea, P. hydropiperoides, P. lapathifolia, P. pensylvanica, P. punctata, P. sagittata, P. virginiana, Phalaris arundinacea, Phragmites australis, Phryma leptostachya, Physostegia virginiana, Phytolacca americana, Pilea fontana, P. pumila, Pinus strobus, Polygonum ramosissimum, Pontederia cordata, Populus deltoides, P. tremuloides, Potentilla norvegica, Proserpinaca palustris, Prunella vulgaris, Prunus serotina, Pycnanthemum virginianum, Quercus alba, Q. bicolor, Q. borealis, Q. macrocarpa, Q. palustris, Q. rubra, Q. shumardii, Q. velutina, Ranunculus hispidus, R. pensylvanicus, R. sceleratus, Rhus copallinum, Rhynchospora, Ribes americanum, R. missouriense, Rosa arkansana, R. palustris, Rubus hispidus, R. pensilvanicus, R. pubescens, Rudbekia laciniata, Rumex martimus, R. obtusifolius, R. orbiculatus, Sagittaria latifolia, S. rigida, Salix amygdaloides, S. bebbiana, S. discolor, S. exigua, S. interior, S. nigra, S. rigida, Sambucus nigra, Sarracenia purpurea, Saururus cernuus, Schoenoplectus hallii, S. pungens, Scirpus atrovirens, S. cyperinus, S. lineatus, S. pedicellatus, Scleria verticillata, Scutellaria galericulata, S. lateriflora, Setaria faberi, Sicyos angulatus, Silphium perfoliatum, Sium suave, Smilax ecirrhata, Solanum carolinense, S. dulcamara, Solidago canadensis, S. flexicaulis, S. gigantea, S. patula, S. riddellii, S. ulmifolia, Sparganium americanum, S. chlorocarpum, S. eurycarpum, Spartina pectinata, Sphagnum, Spiraea alba, S. tomentosa, Stachys hispida, S. pilosa, Stellaria longifolia, Stipa spartea, Symphyotrichum boreale, S. dumosum, S. firmum, S. lanceolatum, S. lateriflorum, S. novae-angliae, S. ontarionis, S. pilosum, S. prenanthoides, S. puniceum, S. shortii, Symplocarpus foetidus, Taxus canadensis, Thalictrum dasycarpum, T. dioicum, T. revolutum, Thelypteris palustris, Thuja occidentalis, Tilia americana, Toxicodendron radicans, T. vernix, Triadenum fraseri, Tridens flavus, Typha angustifolia, T. latifolia, Ulmus americana, U. rubra, Urtica dioica, Vaccinium macrocarpon, Verbena hastata, V. urticifolia, Verbesina alternifolia, Vernonia fasciculata, V. missurica, V. noveboracensis, Veronica officinalis, Viburnum lentago, V. recognitum, Viola cucullata, V. sororia, Vitis riparia, Xyris torta, Zannichellia palustris, Zanthoxylum americanum.*

***Epilobium densiflorum* (Lindl.) Hoch & Raven** is an annual of canyon botytoms, channels, chaparral, depressions, flats, floodplains, forest openings, gravel bars, marshes, meadows, mudflats, outwashes, roadsides, sandbars, seeps, slopes, stream beds, washes, and the margins of ponds, reservoirs, rivers, sloughs, streams, and vernal pools, especially in cismontane and montane communities, at elevations of up to 2591 m. Exposures can range from full sun to shade. The substrates have been described as alluvium, basalt, clay, clay loam, clay sand, granite, gravel, gravelly alluvium, gravelly loam, Julian schist, loamy sand, mud, rhyolitic soil, rock, rocky loam, sand, sandy clay loam, sandy loam, serpentine, silt, silty loam, and Tuscan mudflow. The plants occur commonly in areas of standing water. The flowers and fruits are produced from May to October. The flowers are self-compatible and probably autogamous. The seeds lack a coma and probably are dispersed fairly locally. **Reported associates:** *Acer negundo, Achillea millefolium, Acmispon americanus, Adenostoma fasciculatum, A. sparsifolium, Aesculus californica, Agnorhiza ovata, Agrostis stolonifera, Aira, Allium amplectens, A. tolmiei, Alnus rhombifolia, A. rubra, Alopecurus aequalis, Amaranthus retroflexus, Ambrosia psilostachya, Amelanchier, Amorpha californica, Anagallis arvensis, Anthemis cotula, Anthoxanthum odoratum, Apocynum cannabinum, Aquilegia formosa, Arctostaphylos glandulosa, A. glauca, A. pringlei, A. pungens, A. viscida, Artemisia arbuscula, A. californica, A. douglasiana, A. dracunculus, A. ludoviciana, A. rigida, A. tridentata, A. tripartita, Asclepias fascicularis, Astragalus, Athyrium filix-femina, Baccharis salicifolia, Bidens, Blennosperma nanum, Brassica, Briza minor, Brodiaea minor, B. terrestris, Bromus arvensis, B. diandrus, B. hordeaceus, B. inermis, B. tectorum, Callitriche bolanderi, C. marginata, Calocedrus decurrens, Calochortus nitidus, C. palmeri, Calystegia occidentalis, Camassia quamash, Camissonia tanacetifolia, Carex athrostachya, C. nebrascensis, C. praegracilis, Castilleja foliolosa, C. lasiorhyncha, C. minor, C. tenuis, Ceanothus crassifolius, C. leucodermis, Centaurea, Centromadia fitchii, Cephalanthus occidentalis, Cercocarpus betuloides, C. ledifolius, C. montanus, Chamaebatia foliolosa, Cicendia quadrangularis, Cirsium arvense, Clematis ligusticifolia, Cornus sericea, Crataegus douglasii, Cressa truxillensis, Croton setiger, Crypsis schoenoides, Cryptantha torreyana, Cynosurus echinatus, Cyperus eragrostis, C. strigosus, Danthonia californica, Datisca glomerata, Deschampsia danthonioides, Descurainia, Dianthus armeria, Dieteria canescens, Dipsacus fullonum, Drymocallis glandulosa, Echinochloa crus-galli, Elatine, Eleocharis macrostachya, Elymus glaucus, Epilobium brachycarpum, E. ciliatum, E. glaberrimum, E. hornemannii, E. torreyi, Equisetum laevigatum, Eriastrum densifolium, Ericameria nauseosa, Eriogonum fasciculatum, E. umbellatum, E. wrightii, Eriophyllum confertiflorum, Eryngium castrense, E. vaseyi, Euphorbia albomarginata, Euthamia occidentalis, Forestiera pubescens, Fragaria, Fraxinus velutina, Fremontodendron californicum, Gaillardia pulchella, Gayophytum diffusum, Geranium carolinianum, Glyceria, Gnaphalium palustre, Grindelia hirsutula, Heliomeris multiflora, Heliotropium europaeum, Heteromeles arbutifolia, Heuchera micrantha, Holocarpha virgata, Holozonia filipes, Hordeum murinum, Horkelia rydbergii, H. yadonii, Hosackia oblongifolia, Hypericum anagalloides, Hypochaeris radicata, Juncus balticus, J. bufonius, J. dubius, J. ensifolius, J. kelloggii, J. oxymeris, J. patens, J. phaeocephalus, J. uncialis, J. xiphioides, Juniperus occidentalis, Lactuca serriola, Lagophylla glandulosa, Lasthenia fremontii, Lepidium, Leptosiphon ciliatus, L. liniflorus, Linum lewisii, Lithocarpus densiflorus, Lolium, Lomatium utriculatum, Lonicera, Lupinus bicolor, L. latifolius, Luzula, Lycopus americanus, Lythrum californicum, L. hyssopifolia, Madia elegans, M. glomerata, Malvella leprosa, Medicago sativa, Melilotus albus, M. indicus, Mentha arvensis, M. spicata, Mimulus cardinalis, M. floribundus, M. guttatus, M. palmeri, M. parishii, M. pilosus, Montia parvifolia, Muhlenbergia rigens, Myosurus ×clavicaulis, Nasturtium officinale, Navarretia intertexta,*

N. leptalea, N. prostrata, Oenothera elata, Oxalis oregana, Paspalum distichum, Penstemon heterophyllus, Perideridia gairdneri, P. parishii, Persicaria amphibia, P. coccinea, P. lapathifolia, P. maculosa, Phacelia mohavensis, Phalaris arundinacea, P. paradoxa, Philadelphus lewisii, Phleum pratense, Pinus contorta, P. coulteri, P. jeffreyi, P. ponderosa, P. sabiniana, Piptatherum miliaceum, Plagiobothrys bracteatus, P. fulvus, P. trachycarpus, P. undulatus, Plantago major, Platanus racemosa, Plectritis congesta, Poa bulbosa, P. pratensis, Pogogyne clareana, Polanisia dodecandra, Polygonum polygaloides, Polypogon maritimus, P. monspeliensis, Populus fremontii, P. trichocarpa, Potentilla, Poteridium occidentale, Prunella vulgaris, Psilocarphus brevissimus, P. elatior, P. oregonus, Pseudoroegneria spicata, Pseudotsuga, Purshia tridentata, Quercus agrifolia, Q. berberidifolia, Q. chrysolepis, Q. garryana, Q. kelloggii, Q. wislizeni, Q. ×acutidens, Ranunculus cymbalaria, R. occidentalis, Ribes aureum, Rosa californica, R. pisocarpa, R. woodsii, Rubus bifrons, R. laciniatus, Rumex crispus, R. salicifolius, R. venosus, Salix exigua, S. gooddingii, S. laevigata, S. lasiolepis, S. lucida, Salsola tragus, Saltugilia splendens, Salvia apiana, Schedonorus arundinaceus, Schoenoplectus acutus, Scutellaria, Sedella pumila, Sidalcea campestris, S. pedata, Silene antirrhina, Sisymbrium altissimum, Sonchus asper, Spiraea betulifolia, Spiranthes romanzoffiana, Stachys albens, Symphoricarpos rotundifolius, Symphyotrichum, Tamarix, Thalictrum fendleri, Thinopyrum intermedium, Torilis arvensis, Toxicodendron diversilobum, Toxicoscordion venenosum, Trichostema micranthum, Trifolium cyathiferum, T. depauperatum, T. variegatum, Triteleia hyacinthina, Turricula parryi, Typha latifolia, Urtica dioica, Veratrum californicum, Verbascum blattaria, Verbena bracteata, Veronica anagallis-aquatica, V. peregrina, Vicia sativa, Vitis californica, Vulpia, Zeltnera venusta.

Epilobium glaberrimum Barbey is a perennial species of cienega, cliff benches, ditches, escarpments, estuaries, fens, flats, gravel bars, gullies, meadows, peatlands, prairies, roadsides, seeps, slopes, springs, stream bottoms, thickets, vernal pools, woodlands, and the margins of lakes and streams at elevations of up to 3399 m. The plants are common in open montane communities and also occur in shaded sites. The substrates (pH: 4.5–8.2), are characterized as alluvium, clay, gravel, Idaho batholith granite, limestone, loam, muck, mud, peat, rock, sand, sandy loam, scree, silty loam, silty rocky loam, slate, talus, ultramafic, and volcanic. Flowering and fruiting extend from June to September. The flowers are self-compatible and probably autogamous. Individuals often occur in clumps as a result of their short stolons or soboles. They are capable of accumulating large amounts of arsenic when growing on metal-contaminated sites. The seeds possess a coma and are dispersed by wind and water. Their germination requirements have not been determined. **Reported associates:** *Abies concolor, A. grandis, A. lasiocarpa, A. ×shastensis, Acer macrophyllum, Achillea millefolium, Achnatherum lettermanii, A. parishii, Aconitum columbianum, Aconogonon phytolaccifolium, Adenocaulon bicolor, Adiantum pedatum, Agoseris glauca, Agrostis exarata, A.*

scabra, A. stolonifera, Allium, Alnus rhombifolia, A. tenuifolia, Alopecurus aequalis, Amorpha californica, Anaphalis margaritacea, Anemone occidentalis, Aquilegia formosa, Arctostaphylos patula, A. pringlei, Arnica gracilis, Artemisia douglasiana, A. ludoviciana, A. tridentata, Astragalus alpinus, Athyrium filix-femina, Barbarea orthoceras, Betula occidentalis, B. glandulosa, B. nana, Bistorta bistortoides, Boechera davidsonii, Bolandra oregana, Botrychium ascendens, B. multifidum, Boykina, Brickellia californica, Bromus ciliatus, B. inermis, Calamagrostis canadensis, C. rubescens, Callitropsis nootkatensis, Calocedrus decurrens, Caltha leptosepala, Carex abrupta, C. alma, C. aquatilis, C. aurea, C. canescens, C. crawfordii, C. diandra, C. fracta, C. hassei, C. interior, C. jonesii, C. lasiocarpa, C. lemmonii, C. limosa, C. lyngbyei, C. magellanica, C. muricata, C. nebrascensis, C. obnupta, C. praeceptorum, C. rostrata, C. schottii, C. scopulorum, C. senta, C. simulata, C. subfusca, C. utriculata, C. vesicaria, Cassiope mertensiana, Castilleja applegatei, C. miniata, Ceanothus cordulatus, C. leucodermis, Cerastium, Cercocarpus ledifolius, Chrysolepis sempervirens, Cicuta douglasii, Cinna latifolia, Circaea alpina, Cirsium vulgare, Claytonia cordifolia, Collomia linearis, Conyza canadensis, Cordylanthus rigidus, Cornus nuttallii, Corydalis caseana, Cryptantha affinis, Dactylis glomerata, Danthonia intermedia, Dasiphora floribunda, Datisca glomerata, Delphinium polycladon, Deschampsia cespitosa, D. danthonioides, D. elongata, Distichlis, Drosera anglica, D. rotundifolia, Drymocallis ashlandica, D. glandulosa, D. lactea, Dulichium arundinaceum, Eleocharis macrostachya, E. palustris, E. quinqueflora, Elymus elymoides, E. glaucus, Epilobium ciliatum, E. densiflorum, E. hornemannii, E. obcordatum, E. suffruticosum, Epipactis gigantea, Equisetum arvense, Eremogone congesta, Ericameria cuneata, E. nauseosa, Erigeron acris, E. eatonii, Eriogonum wrightii, Eriophorum angustifolium, E. chamissonis, Erysimum capitatum, Eucephalus ledophyllus, Festuca rubra, Fremontodendron californicum, Galium parishii, G. trifidum, Gayophytum diffusum, Gentiana calycosa, Geranium richardsonii, Glyceria elata, G. striata, Hastingsia alba, Helenium bigelovii, Heracleum sphondylium, Hesperoyucca whipplei, Hieracium albiflorum, Holcus lanatus, Holodiscus microphyllus, Horkelia rydbergii, Hosackia oblongifolia, Hypericum anagalloides, H. formosum, Juncus balticus, J. bufonius, J. drummondii, J. effusus, J. ensifolius, J. longistylis, J. macrandrus, J. macrophyllus, J. mertensianus, J. saximontanus, Juniperus occidentalis, Kalmia microphylla, Koeleria macrantha, Larix occidentalis, Leymus triticoides, Lilium parryi, Lomatium, Lupinus polyphyllus, L. sericeus, Luzula comosa, L. subcongesta, Madia elegans, Maianthemum racemosum, M. stellatum, Mentha arvensis, Mentzelia laevicaulis, Menyanthes trifoliata, Mertensia paniculata, Micranthes odontoloma, Mimulus breweri, M. cardinalis, M. floribundus, M. guttatus, M. moschatus, M. pilosus, M. primuloides, M. tilingii, Minuartia, Mitella pentandra, Muhlenbergia filiformis, M. richardsonis, M. rigens, Oplopanax horridus, Orthocarpus imbricatus, Oxypolis occidentalis, Packera pseudaurea, Paxistima myrsinites, Pedicularis groenlandica,

Penstemon newberryi, P. procerus, P. rostriflorus, Persicaria amphibia, P. maculosa, Phacelia hastata, Philadelphus lewisii, Phleum alpinum, P. pratense, Phyllodoce empetriformis, Physocarpus, Picea breweriana, P. engelmannii, Pinus albicaulis, P. contorta, P. jeffreyi, P. lambertiana, P. ponderosa, Platanthera dilatata, Poa annua, P. compressa, P. palustris, P. pratensis, P. secunda, Polemonium occidentale, Polypogon monspeliensis, Polystichum munitum, Populus trichocarpa, Potentilla anserina, P. biennis, P. drummondii, P. flabellifolia, P. wheeleri, Primula fragrans, P. jeffreyi, Prunella vulgaris, Pseudognaphalium thermale, Pseudotsuga menziesii, Pteridium aquilinum, Quercus chrysolepis, Q. kelloggii, Ranunculus alismifolius, R. populago, Rhododendron columbianum, R. occidentale, Ribes cereum, R. nevadense, R. roezlii, Ricciocarpos natans, Rorippa curvisiliqua, Rosa californica, R. woodsii, Rubus idaeus, R. parviflorus, R. sachalinensis, Rumex acetosella, R. crispus, Sagina saginoides, Salix drummondiana, S. exigua, S. lasiandra, S. lasiolepis, S. monticola, S. myrtillifolia, S. scouleriana, Sarcobatus, Schedonorus arundinaceus, Scirpus microcarpus, Secale cereale, Sedum, Senecio integerrimus, S. triangularis, Sidalcea celata, S. malviflora, Silene menziesii, Sisyrinchium bellum, S. idahoense, Solidago velutina, Sorbus, Spartium junceum, Spergularia rubra, Sphagnum, Sphenosciadium capitellatum, Spiraea douglasii, S. splendens, Stachys ajugoides, S. albens, S. rigida, Stellaria longifolia, S. media, Suksdorfia ranunculifolia, Symphyotrichum spathulatum, Thalictrum fendleri, T. occidentale, Thlaspi arvense, Triantha glutinosa, Trifolium longipes, T. monanthum, T. repens, T. wormskioldii, Tsuga heterophylla, T. mertensiana, Typha domingensis, T. latifolia, Urtica dioica, Vaccinium scoparium, V. uliginosum, Valeriana, Veratrum californicum, V. viride, Veronica serpyllifolia, Vicia americana, Viola glabella.

***Epilobium leptophyllum* Raf.** is a perennial found in shallow waters of diverse habitats including bogs, depressions, ditches, draws, fens, floating mats of vegetation, floodplains, gullies, hollows, hummocks, ledges, marshes, meadows, oxbows, prairies, river beds, seeps, shores, sloughs, springs, swamps, thickets, and along the margins of lakes, ponds, and streams at elevations of up to 3390 m. The habitats are acidic, circumneutral (e.g., pH: 6.6), or alkaline/saline with substrates that include clay, clay loam, clay mud, gravel, loamy sand, muck, peat, sand, schist, and even dead logs. The plants usually colonize open areas but can tolerate partial shade. The flowers are self-compatible and appear from June to September. The seeds have a coma and are wind dispersed. Overwintering is facilitated by the production of turions at the stem base and by thread-like stolons, which are tipped with vegetative bulblets. **Reported associates:** *Acalypha rhomboidea, Acer rubrum, Achillea millefolium, Acorus calamus, Agalinis tenuifolia, Agrimonia gryposepala, Agrostis gigantea, A. scabra, A. stolonifera, Aletris, Alnus rugosa, A. serrulata, Amorpha fruticosa, Andromeda glaucophylla, Apios americana, Apocynum androsaemifolium, Arnoglossum plantagineum, Asclepias hirtella, A. incarnata, Bartonia virginica, Betula glandulosa, B. lutea, B. nana, B. papyrifera, B. pumila, B. sandbergi, Bidens aristosus, B. cernuus, B.* *connatus, B. trichospermus, Boehmeria cylindrica, Bromus ciliatus, Calamagrostis canadensis, Calla palustris, Calopogon pulchellus, Caltha palustris, Campanula aparinoides, C. uliginosa, Cardamine bulbosa, Carex aquatilis, C. aurea, C. buxbaumii, C. comosa, C. conoidea, C. crawei, C. cusickii, C. echinata, C. flava, C. granularis, C. haydenii, C. hystericina, C. interior, C. lacustris, C. lanuginosa, C. lasiocarpa, C. leptalea, C. nebrascensis, C. paupercula, C. prairea, C. projecta, C. rostrata, C. sartwellii, C. scoparia, C. scopulorum, C. simulata, C. stipata, C. stricta, C. tetanica, C. utriculata, Chamaedaphne calyculata, Chamerion angustifolium, Chelone glabra, Chrysosplenium americanum, Cicuta bulbifera, C. maculata, Cirsium arvense, C. canadense, C. muticum, C. vulgare, Cladium mariscoides, Comandra umbellata, Comarum palustre, Comptonia peregrina, Conyza canadensis, Cornus amomum, C. racemosa, C. sericea, Corylus cornuta, Cuscuta gronovii, Cyperus strigosus, Cypripedium candidum, Dasiphora floribunda, Deschampsia cespitosa, Desmodium canadense, Distichlis spicata, Drypoteris cristata, D. spinulosa, Echinocystis lobata, Eleocharis elliptica, E. erythropoda, E. obtusa, E. palustris, E. quinqueflora, E. tenuis, Epilobium ciliatum, E. coloratum, E. leptophyllum, E. strictum, Epipactis gigantea, Equisetum arvense, E. fluviatile, E. scirpoides, E. sylvaticum, Erigeron annuus, Eriophorum angustifolium, Eupatorium perfoliatum, Euthamia graminifolia, Eutrochium maculatum, Fimbristylis puberula, Fragaria virginiana, Frangula alnus, Galium asprellum, G. brevipes, G. labradoricum, G. tinctorium, G. trifidum, G. triflorum, Gentiana andrewsii, Gentianopsis procera, Geum rivale, Glyceria canadensis, G. grandis, Glycyrrhiza, Gymnocarpium dryopteris, Habenaria, Helenium autumnale, Helianthus giganteus, H. grosseserratus, Hypericum canadensis, H. kalmianum, H. majus, H. sphaerocarpum, Impatiens capensis, Iris versicolor, I. virginica, Juncus balticus, J. brevicaudatus, J. canadensis, J. effusus, J. marginatus, J. nodosus, J. torreyi, Kalmia polifolia, Laportea canadensis, Larix laricina, Lathyrus palustris, Leersia oryzoides, Lemna, Liatris pychnostachya, Liparis loeselii, Lobelia kalmii, L. siphilitica, Ludwigia palustris, L. polycarpa, Lycopus americanus, L. asper, L. uniflorus, L. virginicus, Lysimachia ciliata, L. maritima, L. quadriflora, L. terrestris, L. thyrsiflora, Lythrum alatum, L. salicaria, Maianthemum canadense, M. trifolium, Mentha arvensis, Menyanthes trifoliata, Menziesia ferruginea, Mertensia paniculata, Mimulus glabratus, M. ringens, Mitella nuda, Muhlenbergia glomerata, Myrica gale, Mysotis scorpioides, Nemopanthus mucronata, Oenothera perennis, Onoclea sensibilis, Oplopanax horridus, Osmunda regalis, Oxypolis rigidior, Panicum boreale, P. virgatum, Parnassia glauca, P. palustris, Parthenocissus inserta, Pedicularis canadensis, P. groenlandica, P. lanceolata, Penthorum sedoides, Peritoma multicaulis, Persicaria amphibia, P. coccinea, P. lapathifolia, P. punctata, P. sagittata, Phalaris arundinacea, Picea engelmannii, P. mariana, Pilea fontana, Pinus strobus, Platanthera lacera, Poa palustris, P. pratensis, Polygala sanguinea, Polytrichum, Populus tremuloides, Potentilla anserina, P. norvegica, P. simplex, Prunus serotina,*

Pycnanthemum virginianum, Rhamnus alnifolia, R. frangula, Rhododendron groenlandicum, Rhynchospora capillacea, Ribes americanum, R. lacustre, Rosa palustris, Rubus flagellaris, R. hispidus, R. idaeus, R. odoratus, R. pubescens, R. setosus, Rumex britannica, Sagittaria latifolia, Salix amygdaloides, S. candida, S. discolor, S. eriocephala, S. exigua, S. humilis, S. pedicellaris, S. petiolaris, S. planifolia, S. pyrifolia, S. serissima, Sarracenia purpurea, Schoenoplectus acutus, S. pungens, S. tabernaemontani, Scirpus cyperinus, Scutellaria galericulata, S. lateriflora, Senecio suaveolens, Silphium terebinthenaceum, Sium suave, Solidago canadensis, S. gigantea, S. gymnospermoides, S. ridellii, S. uliginosa, Sorghastrum nutans, Sparganium eurycarpum, Spartina gracilis, S. pectinata, Sphagnum, Sphenopholis intermedia, Spiraea alba, S. tomentosa, Spiranthes romanzoffiana, Stachys palustris, S. pilosa, Stellaria borealis, S. palustris, Symphyotrichum boreale, S. firmum, S. lanceolatum, S. novae-angliae, S. puniceum, Thalictrum dasycarpum, Thuja occidentalis, Thelypteris palustris, Toxicodendron vernix, Triadenum fraseri, Triglochin maritimum, Tsuga canadensis, Typha angustifolia, T. domingensis, T. latifolia, T. ×glauca, Utricularia intermedia, Vaccinium angustifolium, V. macrocarpon, V. oxycoccus, Verbena hastata, Veronica americana, V. peregrina, Viburnum dentatum, V. lentago, V. nudum, Viola cucullata, V. pratincola.

***Epilobium oreganum* Greene** is an infrequent perennial of bogs, ditches, fens, meadows, roadsides, and the margins of ponds and slow streams at elevations of up to 2682 m. The plants occur mainly on ultramafic serpentine substrates and also on gravel or peat loam. The flowers are self-compatible and are produced from June to July. Vegetative perennation occurs by means of persistent shoots. Little other information is available on the ecology of this rare species. **Reported associates:** *Abies concolor, Artemisia ludoviciana, Bistorta bistortoides, Carex leptalea, C. praticola, C. serratodens, C. viridula, Castelleja miniata, Chamaecyparis lawsoniana, Darlingtonia californica, Drosera anglica, Gentiana setigera, Hastingsia atropurpurea, H. bracteosa, Helenium bolanderi, Juncus, Lathyrus palustris, Lewisia oppositifolia, Lilium pardilinum, Lupinus latifolius, Microseris howellii, Mimulus primuloides, Montia howellii, Narthecium californicum, Packera hesperia, Parnassia palustris, Persicaria, Pinguicula vulgaris, Pinus jeffreyi, P. murrayana, Populus, Rhododendron occidentale, Ribes cereum, Sagittaria sanfordii, Salix lasiolepis, Sanguisorba officinalis, Senecio triangularis, Sisyrinchium idahoense, Spergularia rubra, Sphagnum, Typha, Veratrum californicum, Viola primulifolia.*

***Epilobium oregonense* Hausskn.** is a low, matted, or sprawling perennial of montane bogs, depressions, fens, flats, ledges, marshes, meadows, mudflats, rock bars, seepage slopes, stream bottoms, and margins of lakes, ponds, rivers, and streams at elevations of up to 3506 m. The plants occur in the shade or in sun on decomposed granite, humus, loam, mud, pumice, sand, sandy loam, scree, silt, or on peaty, waterlogged substrates. The flowers are self-compatible and are evident from July to August. The seeds possess a coma and are wind dispersed. Vegetative reproduction occurs by

extension of the thread-like stolons. **Reported associates:** *Abies concolor, A. grandis, A. magnifica, Achillea millefolium, Acmispon nevadensis, Agoseris heterophylla, Agrostis exarata, A. idahoensis, A. lepida, A. pallens, A. scabra, A. variabilis, Alnus tenuifolia, Aquilegia formosa, Arabis rectissima, Arctostaphylos patula, Arnica chamissonis, Barbarea orthoceras, Betula glandulosa, Bistorta bistortoides, Boechera rectissima, Calocedrus decurrens, Carex alma, C. aquatilis, C. aurea, C. buxbaumii, C. canescens, C. capitata, C. chordorrhiza, C. diandra, C. echinata, C. fissuricola, C. heteroneura, C. integra, C. interior, C. jonesii, C. lemmonii, C. lenticularis, C. magellanica, C. microptera, C. multicostata, C. paupercula, C. scopulorum, C. simulata, C. subnigricans, C. utriculata, Castilleja lemmonii, C. miniata, Ceanothus cordulatus, C. leucodermis, Cicuta douglasii, Cirsium scariosum, Comarum palustre, Darlingtonia californica, Deschampsia cespitosa, D. elongata, Drepanocladus aduncus, Drosera anglica, D. rotundifolia, Drymocallis glandulosa, D. lactea, Eleocharis acicularis, E. bella, E. quinqueflora, Ericameria, Erigeron coulteri, Eriogonum spergulinum, Eriophorum angustifolium, E. crinigerum, E. gracile, Festuca rubra, Galium trifidum, Gayophytum diffusum, Gentianopsis simplex, Geum macrophyllum, Glyceria striata, Gnaphalium palustre, Helenium bigelovii, Hordeum brachyantherum, Horkelia rydbergii, Hymenoxys hoopesii, Hypericum anagalloides, H. formosum, Iris missouriensis, Ivesia campestris, Juncus balticus, J. bufonius, J. effusus, J. ensifolius, J. macrandrus, J. orthophyllus, Juniperus occidentalis, Kalmia, Larix, Lepidium virginicum, Limosella acaulis, Lupinus latifolius, Madia glomerata, Mertensia, Micranthes oregana, Mimulus breweri, M. guttatus, M. moschatus, M. primuloides, Monardella linoides, Montia chamissoi, Muhlenbergia filiformis, M. richardsonis, Navarretia divaricata, Oreostemma alpigenum, Packera indecora, P. pseudaurea, Pedicularis attollens, Penstemon parvulus, Perideridia parishii, Phleum alpinum, P. pratense, Picea engelmannii, Pinus albicaulis, P. contorta, P. jeffreyi, Plagiobothrys hispidulus, Platanthera dilatata, P. sparsiflora, Poa bolanderi, P. pratensis, Polemonium occidentale, Potentilla anserina, P. drummondii, P. gracilis, P. grayi, Primula jeffreyi, P. tetrandra, Pseudotsuga menziesii, Ptilagrostis kingii, Quercus kelloggii, Ranunculus alismifolius, R. orthorhynchus, Ribes cereum, Rosa californica, Rumex salicifolius, Sagina saginoides, Salix eastwoodiae, S. lasiolepis, S. lutea, S. myrtillifolia, S. orestera, S. planifolia, Scirpus diffusus, S. microcarpus, Sedum, Senecio scorzonella, S. triangularis, Sidalcea malviflora, S. oregana, Sisyrinchium bellum, Solidago multiradiata, Sphagnum fuscum, Sphenosciadium capitellatum, Spiranthes romanzoffiana, Stachys albens, Stellaria longifolia, S. longipes, Symphoricarpos rotundifolius, Symphyotrichum foliaceum, S. frondosum, S. spathulatum, Taraxacum californicum, T. officinale, Triantha, Trichophorum, Trifolium longipes, T. monanthum, T. wormskioldii, Tsuga mertensiana, Urtica dioica, Vaccinium cespitosum, V. uliginosum, Veratrum californicum, Veronica americana, Viola adunca, V. macloskeyi.*

***Epilobium palustre* L.** is a common, wide-ranging perennial, which inhabits bogs, borrow pits, bottomlands, draws, fens, floodplains, hummocks, meadows, pools, seeps, springs, string bogs, swales, swamps, thickets, and the margins of lakes, ponds, and streams at elevations of up to 3390 m. These plants occur in open habitats (or in partial shade) on acidic substrates (pH: 4.9–6.3; pH extremes: 3.8–7.5), which can consist of clay, clay loam, clay mud, gravel, loam, loamy clay, muck, rock, sand, sandy loam, sandy rocky granite, or highly organic peat (sometimes layered over sandy loam). The self-compatible flowers are insect pollinated. The seeds possess a long coma of hairs and are wind dispersed. The seeds are known to germinate (up to 97%) immediately after collection, at temperatures ranging from 6°C to 34°C. However, extensive seed banks can be formed, with viable seed sometimes present in areas where the species does not occur in the standing vegetation. Conversely, the species has been recorded in areas where it does not occur in the seed bank. Vegetative reproduction is facilitated by thread-like stolons, which can produce fleshy bulblets. **Reported associates:** *Abies balsamea, Acer rubrum, A. spicatum, Agrostis alaskana, A. stolonifera, Alnus rugosa, Andromeda polifolia, Aneura pinguis, Angelica arguta, Antennaria microphylla, Anticlea elegans, Arctophila fulva, Arethusa bulbosa, Asclepias speciosa, Aulacomnium palustre, Avenella flexuosa, Betula nana, B. papyrifera, B. pumila, Breea arvense, Bryum pseudotriquetrum, Calamagrostis canadensis, C. stricta, Calliergon giganteum, Calopogon tuberosus, Caltha palustris, Campanula aparinoides, Campylium stellatum, Cardamine pratensis, Carex aquatilis, C. buxbaumii, C. capillaris, C. chordorrhiza, C. diandra, C. disperma, C. gynocrates, C. interior, C. lacustris, C. lasiocarpa, C. leptalea, C. limosa, C. livida, C. lyngbyei, C. magellanica, C. nebraskensis, C. oligosperma, C. pauciflora, C. rostrata, C. scopulorum, C. simulata, C. stans, C. trisperma, C. utriculata, C. vaginata, Chamaedaphne calyculata, Chamerion angustifolium, Cicuta douglasii, Cinna latifolia, Circaea alpina, Cirsium arvense, Clintonia borealis, Comarum palustre, Coptis trifolia, Cornus canadensis, C. sericea, Crepis runcinata, Dasiphora floribunda, Deschampsia cespitosa, Dipsacus fullonum, Doellingeria umbellata, Drepanocladus revolvens, Drosera anglica, D. rotundifolia, Dryopteris cristata, Eleocharis palustris, Empetrum nigrum, Epilobium leptophyllum, E. strictum, Equisetum arvense, E. fluviatile, E. sylvaticum, E. variegatum, Eriophorum angustifolium, E. virginicum, E. viridicarinatum, Fraxinus nigra, Galium labradoricum, G. trifidum, G. triflorum, Gaultheria hispidula, Geum macrophyllum, Glyceria canadensis, Glycyrrhiza, Gymnadeniopsis clavellata, Habenaria lacera, Helianthus, Hylocomium splendens, Juncus balticus, Juniperus horizontalis, Larix laricina, Lemna minor, Leucanthemum vulgare, Linnaea borealis, Lonicera villosa, Luzula acuminata, Lycopus uniflorus, Maianthemum stellatum, M. trifolium, Malaxis brachypoda, Meesia triquetra, Mentha arvensis, Menyanthes trifoliata, Mitella nuda, Moneses uniflora, Monotropa uniflora, Muhlenbergia asperfolia, M. filiformis, M. glomerata, Neottia borealis, Onoclea sensibilis, Osmundastrum cinnamomeum, Paludella squarrosa, Panicum virgatum, Parnassia fimbriata, P. palustris, Pedicularis groenlandica, P. parviflora,* *Petasites sagittatus, Phalaris arundinacea, Picea glauca, P. mariana, Pinus contorta, P. resinosa, P. strobus, Platanthera dilatata, P. huronensis, P. hyperborea, P. lacera, Pleurozium schreberi, Poa pratensis, Pogonia ophioglossoides, Populus tremuloides, Primula pauciflora, Pseudotsuga menziesii, Rhamnus alnifolia, Rhododendron groenlandicum, R. tomentosum, Rhynchospora alba, R. fusca, Ribes hudsonianum, Rubus acaulis, R. idaeus, R. pubescens, Salix barclayi, S. boothii, S. brachycarpa, S. candida, S. commutata, S. geyeriana, S. pedicellaris, S. planifolia, S. pyrifolia, S. serissima, S. wolfii, Sarracenia purpurea, Scheuchzeria palustris, Schoenoplectus pungens, Scirpus cespitosus, S. hudsonianus, Scorpidium scorpioides, Selaginella selaginoides, Senecio congestus, Solidago uliginosa, Sphagnum angustifolium, S. warnstorfii, Spiranthes romanzoffiana, Symphyotrichum, Thuja occidentalis, Tomenthypnum nitens, Triadenum fraseri, Triantha glutinosa, Trientalis borealis, Triglochin maritimum, T. palustre, Typha angustifolia, T. latifolia, Utricularia intermedia, U. minor, Vaccinium myrtilloides, V. oxycoccus, Verbena, Viola nephrophylla.*

***Epilobium parviflorum* Schreb.** is a nonindigenous perennial, which grows in ditches, fens, landfills, marshes, meadows, roadsides, shores, stream banks, swamps, thickets, and waste sites at elevations of up to 311 m (but up to 2605 m in its native range). The substrates are acidic or calcareous (pH: 5.6–7.5) and include cinders, gravel, gravelly sand, humus, muck, rubble (strongly correlated), sand, and shale sandstone. Flowering has been observed from June to September, with anthesis occurring when the average daily temperatures range from 17.6°C to 22.8°C. Rain will limit anthesis. The flowers are self-compatible, produce nectar, and present pollen from 9:00 am through 4:00 pm, but mainly during the morning (peaking from 11:00 am to noon). The flowers normally are self-pollinating, but various bees (Insecta: Hymenoptera: Apidae: *Apis mellifera*; *Bombus* spp.) have also been observed as pollinators. In one study, the rate of cross pollination (using pathogen sensitivity as a marker) was determined to be between 2.4% and 8.3%. The seeds (averaging 0.11 mg) are dispersed (in October) primarily by the wind, and fall at a velocity of 15.3 cm/s. The seeds float and can be dispersed on the water surface. They are believed to be dispersed over greater distances by waterfowl (Aves: Anatidae). The seeds require light but no after-ripening stratification for germination. Seeds have germinated well (95%–100%) when incubated in the light at 10°C–25°C. A persistent seedbank can develop with reported densities ranging from 2 to 333 seeds/m². Viable seeds have been recovered from riverbeds, where the species does not occur in the standing bank vegetation. The plants are categorized as hemicryptophytes, with their overwintering buds located just below the soil surface. **Reported associates (North America):** *Acer negundo, Agrostis gigantea, A. stolonifera, Ambrosia artemisiifolia, Ammannia robusta, Apocynum cannabinum, Barbarea vulgaris, Betula alleghaniensis, B. pumila, Bidens amplissimus, B. frondosus, Boehmeria cylindrica, Calystegia sepium, Carex hystericina, C. lacustris, Cirsium arvense, Cornus racemosa, Cyperus bipartitus, C. erythrorhizos, C. odoratus, Eleocharis* *acicularis, E. intermedia, E. obtusa, E. rostellata, Epilobium ciliatum, E. coloratum, E. hirsutum, Eupatorium perfoliatum, Eutrochium maculatum, Fraxinus nigra, Glyceria striata, Impatiens capensis, Juncus effusus, Lindernia dubia, Lycopus americanus, Lythrum salicaria, Madia glomerata, Medicago lupulina, Mentha arvensis, Oenothera biennis, Osmunda, Oxypolis rigidior, Penthorum sedoides, Persicaria punctata, Phalaris arundinacea, Pilea pumila, Pluchea odorata, Poa nemoralis, Prunella vulgaris, Rhamnus cathartica, Rubus pubescens, Salix bebbii, S. eriocephala, S. interior, Solanum dulcamara, Solidago canadensis, S. gigantea, S. patula, Stachys, Symphyotrichum lanceolatum, Thuja occidentalis, Typha latifolia, Valeriana sitchensis.*

***Epilobium strictum* Muhl. ex Spreng.** is a perennial, which inhabits bogs, fens, marshes, meadows, prairie fens, springs, swamps, and the margins of brooks, lakes, and ponds at elevations of up to 360 m. The plants occur in open to partially shaded sites on substrates described as humus and muck. The flowers are self-compatible and the plants probably are primarily autogamous. The seeds have a coma and are wind dispersed. Vegetative reproduction occurs by means of filiform stolons. **Reported associates:** *Acer rubrum, Agrostis gigantea, Alnus rugosa, Andromeda, Arethusa bulbosa, Asclepias incarnata, Bidens coronatus, Boehmeria cylindrica, Calamagrostis canadensis, Calopogon, Campanula aparinoides, Carex aquatilis, C. buxbaumii, C. comosa, C. lacustris, C. lasiocarpa, C. limosa, C. livida, C. sartwellii, C. tenuiflora, Cicuta bulbifera, Comarum palsutre, Cornus sericea, Cypripedium reginae, Dasiphora floribunda, Decodon verticillatus, Deschampsia flexuosa, Doellingeria umbellata, Drosera intermedia, D. linearis, D. rotundifolia, Dulichium arundinaceum, Eleocharis erythropoda, E. quinqueflora, Epilobium leptophyllum, E. palustre, Equisetum fluviatile, Eupatorium perfoliatum, Eutrochium maculatum, Fraxinus nigra, Galium labradoricum, G. trifidum, Gentianopsis virgata, Glyceria striata, Ilex verticillata, Iris versicolor, Larix laricina, Leersia oryzoides, Lemna minor, Liparis loeselii, Lobelia kalmii, Lycopus americanus, L. uniflorus, Lysimachia quadriflora, L. terrestris, Mentha arvensis, Menyanthes trifoliata, Muhlenbergia glomerata, Osmundastrum cinnamomea, Persicaria amphibia, Phalaris arundinacea, Phragmites australis, Pilea fontana, Pinus strobus, Platanthera dilatata, P. lacera, Pogonia ophioglossoides, Rhamnus frangula, Rhododendron groenlandicum, Rhynchospora alba, R. fusca, Rubus pubescens, Rumex orbiculatus, Sagittaria latifolia, Salix candida, S. discolor, S. interior, S. pedicellaris, Sarracenia purpurea, Scheuchzeria palustris, Schoenoplectus tabernaemontani, Scleria verticillata, Scutellaria galericulata, S. lateriflora, Solidago gigantea, S. patula, S. riddellii, Sparganium chlorocarpum, S. eurycarpum, Spartina pectinata, Sphagnum, Stellaria calycantha, Symphyotrichum boreale, S. lanceolatum, Thelypteris palustris, Thuja occidentalis, Toxicodendron vernix, Triadenum fraseri, Trichophorum alpinum, Triglochin maritimum, Typha angustifolia, T. latifolia, Utricularia cornuta, U. intermedia, U. macrorhiza, U. minor, Vaccinium, Viola pallens.*

Use by wildlife: *Epilobium cilatum* is a host plant of various fungi including Ascomycota (Erysiphaceae: *Erysiphe polygoni, Podosphaera macularis*; Mycosphaerellaceae: *Phaeoramularia punctiformis*) and Basidiomycota (Doassansiaceae: *Doassansia epilobii*; Pucciniastraceae: *Pucciniastrum epilobii*). It is a larval host plant for several Lepidoptera (Momphidae: *Mompha franclemonti, M. powelli*; Noctuidae: *Eudryas brevipennis*; Sphingidae: *Hyles lineata*). The seeds of *E. cleistogama* are eaten by northern pintail ducks (Aves: Anatidae: *Anas acuta*). The plants are a component of habitat for the California tiger salamander (Amphibia: Ambystomatidae: *Ambystoma californiense*), and the California red-legged frog (Amphibia: Ranidae: *Rana draytonii*). *Epilobium coloratum* is grazed by white-tailed deer (Mammalia: Cervidae: *Odocoileus virginianus*). The foliage is eaten by various moth caterpillars (Insecta: Lepidoptera) including the many-lined carpet (Geometridae: *Anticlea multiferata*), pearly wood nymph (Noctuidae; *Eudryas unio*), scythridid moth (Scythrididae: *Scythris magnabella*) and white-lined sphinx (Sphingidae: *Hyles lineata*). It is a larval host plant for several other Lepidoptera (Geometridae: *Ecliptopera silaceata*; Sphingidae: *Amphion floridensis, Hyles gallii*). The plants are hosts for several fungi (Ascomycota: Erysiphaceae: *Podosphaera macularis*; Basidiomycota: Pucciniastraceae: *Puccinia epilobii-tetragoni, Pucciniastrum epilobii*). *Epilobium glaberrimum* and *E. palustre* are also hosts of *Pucciniastrum epilobii* (Basidiomycota: Pucciniastraceae). *Epilobium leptophyllum* is the larval host plant of several Lepidoptera (Scythrididae: *Scythris inspersella, S. noricella*). *Epilobium strictum* hosts *Pucciniastrum epilobii* (Basidiomycota: Pucciniastraceae). The flowers of *E. parviflorum* are visited by bumblebees (Hymenoptera: Apidae: Bombus Apidae: *Bombus balteatus, B. fernaldae, B. flavifrons, B. frigidus, B. insularis, B. jonellus, B. melanopygus, B. mixtus, B. moderatus, B. perplexus, B. sylvicola*) and honey bees (Apidae: *Apis mellifera*).

Economic importance: food: The seeds of *Epilobium densiflorum* were eaten and made into bread and gruel by the Mendocino, Miwok, and Pomo tribes. The leaves and young shoots of *E. palustre* reportedly are edible when cooked; **medicinal:** *Epilobium ciliatum* was used by the Hopi as an analgesic to treat leg pains. The Kayenta and Navajo applied a root poultice to alleviate muscle cramps. The Potawatomi administered a root infusion to treat diarrhea. Dilute alcoholic tinctures of *E. palustre* have been used to treat chronic diarrhea and dysentery. This species is sold as a homeopathic remedy in Canada. *Epilobium parviflorum* is a source of Oenothein B, a 5-α-reductase-inhibitor, which is used to treat prostate disorders. The aqueous acetone extracts exhibit high antioxidant activity (EC 50 = 1.71 ± 0.05 μg/mL); the aqueous and ethanolic extracts are inhibitory to Bacteria (Enterobacteriaceae: *Escherichia coli*). Cold infusions have been used in veterinary medicine to treat various diseases in pigs (Mammalia: Suidae: *Sus scrofa*); **cultivation:** None of the obligate aquatic species currently is propagated commercially; **misc. products:**

Epilobium ciliatum has potential for use in the restoration of peat-mined areas; **weeds:** Some strains of *Epilobium ciliatum* are resistant to atrazine herbicides and powdery mildews (Fungi: Erysiphales); **nonindigenous species:** *Epilobium ciliatum* was introduced to the Alaskan arctic. It has been introduced to Australia and Europe (to the United Kingdom in 1891) and is regarded as weedy in Tasmania. The FACW species *Epilobium hirsutum* was introduced to North America and often is found in wetlands. *Epilobium parviflorum* was introduced to North America from Eurasia. It was documented among the nonindigenous plants germinating from ballast disposal sites in New York harbor as early as 1879.

Systematics: As currently circumscribed, *Epilobium* represents a monophyletic sister group to the genus *Chamerion*, a clade formerly included within *Epilobium* as section *Chamaenerion*. Together the genera comprise tribe Epilobieae, which is situated phylogenetically between the tribes Onagreae and Gongylocarpeae (Figure 4.85). *Epilobium* is subdivided into eight sections (Figure 4.86), with obligate aquatic species occurring only in section *Epilobiopsis* (both species) and section *Epilobium* (6/168 species). Thus, at least two origins of OBL aquatic species are indicated in *Epilobium*; however, with only slightly more than 3% of species surveyed for section *Epilobium*, (where most of the obligate aquatics occur), this estimate must be regarded as preliminary. The two annual, autogamous species of section *Epilobiopsis* both are aquatic and comprise a sister group to a larger clade that includes section *Boisduvalia* (formerly a segregate genus) with several FAC indicators. The other aquatic species (section *Epilobium*) are more distantly related to these (Figure 4.86). *Epilobium ciliatum*

(section *Epilobium*) includes *E. glandulosum*, which often is treated as a subspecies or variety. The base chromosome number of *Epilobium* is $x = 18$. *Epilobium ciliatum, E. coloratum, E. glaberrimum, E. leptophyllum, E. oreganum, E. oregonense, E. palustre, E. parviflorum,* and *E. strictum* all are $2n = 36$; *E. densiflorum* is $2n = 20$; *E. campestre* and *E. cleistogamum* are $2n = 30$. *Epilobium ×treleaseanum* [OBL] is a hybrid derived from *E. luteum × E. ciliatum* subspecies *glandulosum*, two non-OBL taxa. *Epilobium parviflorum* hybridizes with *E. ciliatum* and with *E. hirsutum* (*E. ×subhirsutum*).

Comments: *Epilobium ciliatum* is widespread throughout much of North America except for the southeastern United States; *E. leptophyllum* is widespread but is absent from extreme northern or southern portions of North America. *Epilobium coloratum* is widespread in eastern North America; *E. strictum* occurs throughout northeastern North America but is disjunct in Idaho. *Epilobium palustre* is circumboreal and widespread through central and northern North America. *Epilobium densiflorum, E. glaberrimum, E. oregonense* and *E. pygmaeum* all occur in western North America. Geographically restricted species include *E. oreganum* (California and Oregon) and *E. cleistogamum* (California). The nonindigenous *Epilobium parviflorum* occurs in northeastern North America and western Canada.

References: Aiken et al., 2007; Akerreta et al., 2010; Bartleman et al., 2001; Bauder & McMillan, 1998; Baum et al., 1994; Bossuyt et al., 2007; Brown, 1879; Buck-Diaz et al., 2012; Castelli et al., 2000; Clay et al., 1991; Cobbaert et al., 2004; Gardner et al., 2009; Godefroid et al., 2007; Gurnell et al., 2007a; Hevesi et al., 2009; Keating et al., 1982; Lam, 1998; Lesuisse et al., 1996; Levin et al., 2003,

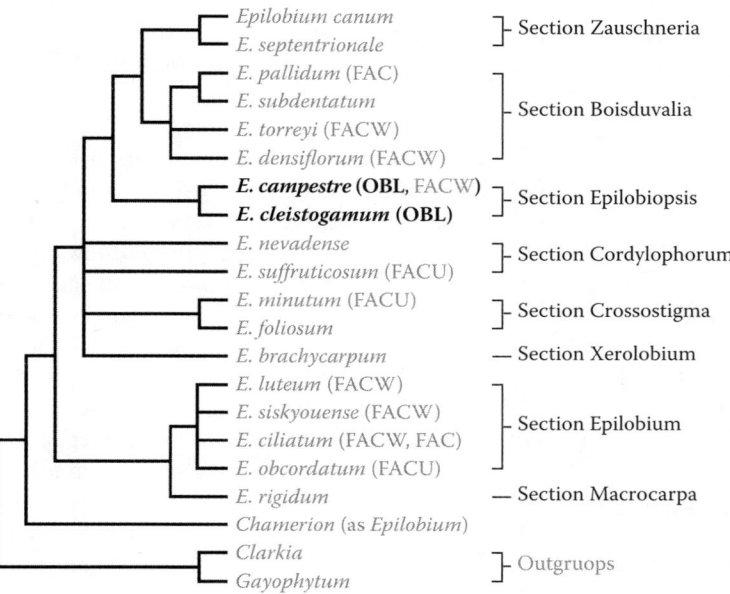

FIGURE 4.86 Phylogenetic analysis of *Epilobium* as indicated by nrDNA (ITS) data. Although this study included only two of the North American species recognized as obligate indicators (indicated in bold), six OBL species (not sampled) are assigned to section *Epilobium*, which indicates at least two independent origins of the OBL habit in the genus. The status of other wetland indicator status is given in parentheses. (Adapted from Baum, D.A. et al., *Syst. Bot.*, 19, 363–388, 1994.)

2004; Myers et al., 2004; Parker et al., 1995; Percival, 1955; Purcell, 1976; Schulz et al., 1991; Seavey & Raven, 1977a,b; Steenkamp et al., 2006; Stuckey, 1970; Tóth et al., 2009.

2. *Ludwigia*

Primrose willow, water primrose; ludwigie

Etymology: after Christian Gottleib Ludwig (1709–1773)

Synonyms: *Corynostigma*; *Cubospermum*; *Danthia*; *Dantia*; *Diplandra*; *Fissendocarpa*; *Isnardia*; *Jussiaea*; *Ludwigiantha*; *Nemotopyxis*; *Oocarpon*; *Prieuria*; *Quadricosta*

Distribution: global: cosmopolitan; **North America:** widespread

Diversity: global: 82 species; **North America:** 32 species

Indicators (USA): OBL: *Ludwigia alata*, *L. alternifolia*, *L. arcuata*, *L. bonariensis*, *L. brevipes*, *L. curtissii*, *L. decurrens*, *L. erecta*, *L. glandulosa*, *L. grandiflora*, *L. hexapetala*, *L. hirtella*, *L. lanceolata*, *L. leptocarpa*, *L. linearis*, *L. linifolia*, *L. longifolia*, *L. microcarpa*, *L. octovalvis*, *L. palustris*, *L. peploides*, *L. peruviana*, *L. pilosa*, *L. polycarpa*, *L. ravenii*, *L. repens*, *L. spathulata*, *L. sphaerocarpa*, *L. suffruticosa*, *L. virgata*; **FACW:** *L. alternifolia*, *L. hirtella*

Habitat: brackish (coastal); freshwater; lacustrine, palustrine, riverine; **pH:** 4.7–10.0; **depth:** 0–3 m; **life-form(s):** emergent herb, emergent shrub, floating leaved

Key morphology: Stems (1–30 dm) erect or procumbent and rooting at the nodes; leaves (2–15 cm) alternate or opposite, simple, entire, variable (e.g., elliptic, lanceolate, linear, obovate, round, etc.); stipules deciduous; inflorescence a bracteate spike; flowers radial, epigynous, hypanthium absent; sepals (1–19 mm) 4–5, persistent; petals (1–30 mm) 4 or 5, yellow or white, sometimes absent; stamens 4 or 10; capsules (2–45 mm) angular, pedicellate (0.5–90 mm).

Life history: duration: annual (fruit/seeds); perennial (caudex, rhizomes, stolons); **asexual reproduction:** creeping stems, rhizomes, stolons, shoot fragments; **pollination:** insect; **sexual condition:** hermaphroditic; **fruit:** capsules (common); **local dispersal:** seeds (water); **long-distance dispersal:** seeds (waterbirds)

Imperilment: (1) *Ludwigia alata* [G3/G5]; S1 (LA, VA); S2 (NC); S3 (GA); (2) *L. alternifolia* [G5]; S1 (NE, ON); S3 (IA, MI); (3) *L. arcuata* [G4/G5]; S1 (AL); (4) *L. bonariensis* [GNR]; S1 (AR); (5) *L. brevipes* [G2/G3]; SX (NJ); SH (GA); S1 (AR, NC, SC); S2 (VA); (6) *L. curtissii* [G3/G4]; (7) *L. decurrens* [G5]; SH (DC); S1 (PA); S2 (IN, MD); (8) *L. erecta* [G5]; (9) *L. glandulosa* [G5]; S1 (KS, MD); S2 (IN); S3 (VA); (10) *L. grandiflora* [GU]; SH (DC); S3 (NC); (11) *L. hexapetala* [GU]; S2 (NC); (12) *L. hirtella* [G5]; SX (DC); S1 (DE, KY, MD, NJ, OK, VA); (13) *L. lanceolata* [G3]; S1 (NC, SC); S3 (GA); (14) *L. leptocarpa* [G5]; S2 (MO, WV); S3 (NC, VA); (15) *L. linearis* [G5]; S1 (NJ, OK); S3 (DE); (16) *L. linifolia* [G4]; S2 (NC, SC); (17) *L. longifolia* [GNR]; (18) *L. microcarpa* [G5]; S1 (AR, LA); S2 (MO); S3 (NC, TN); (19) *L. octovalvis* [G5]; S1 (AR); S2 (GA, NC); (20) *L. palustris* [G5]; S1 (AZ); S2 (NM); S3 (IA, BC, QC); (21) *L. peploides* [G5]; SH (IA); S1 (NC); S3 (GA, WV); (22) *L. peruviana* [G5]; (23) *L. pilosa* [G5]; S1 (AR, VA); (24) *L. polycarpa* [G4]; SH (CT); S1 (ID, KS, MA, PA, VT, WV); S2 (ON); S3 (IA); (25) *L. ravenii* [G1/G2]; SH (FL); S1 (NC, VA); (26) *L. repens* [G5]; SH (KS); S2 (NC); S3 (GA); (27) *L. spathulata* [G2]; S1 (AL, FL); S2 (GA, SC); (28) *L. sphaerocarpa* [G5]; SX (PA); S1 (CT, IN, MA, MI, NC, RI, TN); S2 (LA, NY); S3 (VA); (29) *L. suffruticosa* [G5]; S2 (NC); (30) *L. virgata* [G5]; SH (VA); S3 (NC)

Ecology: general: Most *Ludwigia* species are aquatic or at least FACW inhabitants (only a few species in the United States are not categorized as obligate aquatics). The genus includes annual and perennial herbs as well as some subshrubs and one tree. The flowers of most species are showy, nectariferous, and attract various insects (Insecta). These include ants (Hymenoptera: Formicidae), beetles (Coleoptera), bumblebees and honeybees (Hymenoptera: Apidae), moths (Lepidoptera) and wasps (Hymenoptera: Vespidae), which potentially function as pollinators. Bees (Hymenoptera), butterflies (Lepidoptera), and flowerflies (Diptera: Syrphidae) are the most important pollinators. All North American species of *Ludwigia* are self-compatible with about eight self-incompatible species restricted to South America. Although often outcrossed, the flowers will self-pollinate within a few hours if they remain unpollinated, as their stamens gradually curve inward, bringing their anthers in contact with the stigma. All annual *Ludwigia* species are self-pollinating. Although some plants may grow as submerged aquatics, all flowering occurs above the water on emergent shoots. Vegetative propagation and overwintering by means of stolons is common, as are adventitiously rooted stems, which creep along muddy substrates or float along the water surface. Stolons enable some plants to produce vegetative colonies early in the season before other species become established. Vegetative dispersal by detached, floating stems occurs in some species. Hypodermal tissue that remains attached to dispersed seeds is believed to facilitate their flotation and dispersal by water. Two species not currently recognized as OBL indicators are included here because of taxonomic disagreement (*L. hexapetala*; see *Systematics*) or inexplicable exclusion (*L. longifolia*).

***Ludwigia alata* Ell.** is a perennial inhabitant of shallow water (to 31 cm) or exposed substrates in freshwater or brackish (0.5–5 ppt salinity), tidal or non-tidal sites including canals, depressions, ditches, flats, flatwoods, interdunal swales, marshes, meadows, pools, prairies, roadsides, seeps, shores, swamps, and the margins of lakes, rivers, and streams at elevations of up to at least 14 m. The plants are found in sites that receive full sunlight or partial shade. The substrates are calcareous or acidic (pH: 5.6–6.5) and include limestone, Meadowbrook (grossarenic ochraqualfs), mud, peaty soil, Rutledge (typic humaquepts), and sand. When submerged, the stems produce spongy aerenchyma tissue. The self-compatible flowers lack petals but probably are insect-pollinated as well as autogamous. Although the anthers reflex toward the stigmas during anthesis, herkogamy (spatial separation of the sexes) precludes self-pollination unless mediated by an insect pollinator. The pollen is shed as single grains. Vegetative reproduction is facilitated by stolons, which are formed at the stem bases during the autumn. **Reported associates**: *Acrostichum danaeifolium*, *Aletris lutea*, *Amaranthus australis*, *Ambrosia* *artemisiifolia, Ammania latifolia, Ampelopsis arborea, Amphicarpum muhlenbergianum, Andropogon brachystachyus, A. glomeratus, A. gyrans, Annona glabra, Ardisia solanacea, Aristida beyrichiana, A. palustris, Asclepias connivens, A. pedicellata, Asimina reticulata, Axonopus furcatus, Baccharis halimifolia, Bacopa caroliniana, B. monnieri, Bartonia virginica, Bigelowia nudata, Blechnum serrulatum, Boehmeria cylindrica, Calopogon barbatus, C. pallidus, Carphephorus carnosus, C. paniculatus, Centella asiatica, Cephalanthus occidentalis, Chaptalia tomentosa, Chrysobalanus icaco, Cirsium horridulum, Cladium jamaicense, C. mariscoides, Coreopsis floridana, Crinum americanum, Ctenium aromaticum, Cynoctonum mitreola, Cyperus haspan, C. ligularis, C. surinamensis, Dichanthelium aciculare, D. ensifolium, D. leucothrix, D. portoricense, D. scabriusculum, D. strigosum, Diodia virginiana, Drosera capillaris, Eleocharis baldwinii, E. cellulosa, E. fallax, E. geniculata, E. rostellata, E. vivipara, Erigeron quercifolius, E. vernus, Eriocaulon decangulare, Eryngium aquaticum, E. yuccifolium, Eupatorium capillifolium, E. coelastinum, E. pilosum, E. recurvans, E. rotundifolium, Euphorbia inundata, Eustachys glauca, Euthamia tenuifolia, Fuirena breviseta, F. scirpoidea, Galium obtusum, Gratiola pilosa, Hartwrightia floridana, Hydrocotyle umbellata, H. verticillata, Hypericum cistifolium, H. fasciculatum, H. myrtifolium, H. tetrapetalum, Hypoxis juncea, Hyptis alata, Ilex cassine, I. glabra, Imperata brasiliensis, Ipomoea alba, I. sagittata, Juncus marginatus, J. megacephalus, Justicia angusta, Kosteletzkya pentacarpos, Lachnanthes caroliniana, Lachnocaulon anceps, Liatris spicata, Lilium catesbaei, Lobelia glandulosa, L. paludosa, Ludwigia brevipes, L. curtissii, L. maritima, L. microcarpa, L. octovalvis, L. repens, Lycopodiella alopecuroides, Lythrum lineare, Magnolia virginiana, Marshallia tenuifolia, Melanthera angustifolia, Metopium toxiferum, Mikania scandens, Muhlenbergia filipes, Myrica cerifera, Myrsine floridanum, Najas, Oclemena reticulata, Oldenlandia uniflora, Oxypolis filiformis, Panicum abscissum, P. caerulescens, P. dichotomiflorum, P. hemitomon, P. polycaulon, P. rigidulum, Paspalum laeve, P. praecox, Passiflora suberosa, Peltandra virginica, Persea borbonia, Persicaria punctata, Phragmites australis, Phyla nodiflora, Phyllanthus amarus, Physostegia purpurea, Pinguicula caerulea, Platanthera blephariglottis, Pluchea foetida, P. odorata, P. rosea, Pogonia ophioglossoides, Polygala cruciata, P. lutea, P. ramosa, P. rugelii, P. setacea, Pontederia cordata, Proserpinaca palustris, P. pectinata, Pterocaulon pycnostachyum, Quercus minima, Rhexia cubensis, R. mariana, R. nashii, R. nuttallii, Rhynchosia minima, Rhynchospora breviseta, R. cephalantha, R. chapmanii, R. ciliaris, R. colorata, R. decurrens, R. fascicularis, R. inundata, R. latifolia, R. rariflora, R. wrightiana, Sabatia difformis, S. dodecandra, S. grandiflora, Sacciolepis indica, Sagittaria graminea, Salix caroliniana, Sarracenia minor, Schizachyrium stoloniferum, Schrankia microphylla, Scleria reticularis, Serenoa repens, Setaria geniculata, Sisyrinchium angustifolium, S. miamiense, Solidago fistulosa, S. leavenworthii, S. sempervirens, Spartina patens, Spermacoce prostrata, S. tetraquetra, Sphagnum, Spilanthes americana,* Spiranthes vernalis, Sporobolus domingensis, Stillingia sylvatica, Symphyotrichum carolinianum, S. dumosum, S. tenuifolium, Syngonanthus flavidulus, Taxodium distichum, Teucrium canadense, Thelypteris kunthii, Tillandsia fasciculata, Trismeria trifoliata, Typha dominguensis, Utricularia biflora, U. subulata, Verbena bonariensis, Vicia acutifolia, Vigna luteola, Viola lanceolata, Woodwardia virginica, Xyris ambigua, X. elliottii, X. fimbriata, X. platylepis, X. smalliana, Zeuxine strateumatica.*

***Ludwigia alternifolia* L.** is a perennial with shoots that arise from a tuberous-rooted caudex. The plants occur in shallow water (e.g., 0.6 m) or on exposed sites that include barrens, bogs, borrow pits, bottoms, creekbeds, depressions, ditches, floodplains, glades, gullies, marshes, meadows, mudflats, oxbows, prairies, roadsides, seeps, sloughs, swales, swamps, thickets, woodlands, and the margins of lakes, ponds, rivers, and streams at elevations of up to 311 m. The habitats are fully open to partially shaded sites on substrates described as clay, clay loam, Okaw silt loam, rock, Ruston sandy loam, sand, sandy clay, sandy clay loam, sandy silt, and silty clay loam. The sediments typically have a high organic matter content and represent a broad range of acidity (pH: 4.7–10.0). The periods of flowering and fruiting extend from June to September. The seeds are dispersed from a terminal pore of the mature capsule and require a 12-week period of cold stratification at 4°C followed by a 35°C/20°C incubation temperature regime for optimal germination. The plants can be common in agricultural sites and in waste areas, persisting even under some fairly dry conditions. **Reported associates:** *Acalypha rhomboidea, Acer rubrum, A. saccharinum, Achillea millefolium, Agalinis purpurea, Agrimonia parviflora, Agrostis alba, A. gigantea, Alnus serrulata, Andropogon gerardii, A. glomeratus, A. virginicus, Apios americana, Apocynum cannabinum, Aristida oligantha, Asclepias hirtella, A. incarnata, A. perennis, A. rubra, A. sullivantii, A. syriaca, A. verticillata, A. viridis, Baptisia alba, B. lactea, B. sphaerocarpa, Betula nigra, Bidens aristosus, B. coronatus, B. frondosus, Boehmeria cylindrica, Boltonia asteroides, B. diffusa, Bulbostylis capillaris, Burmannia capitata, Calamagrostis canadensis, C. stricta, Calamovilfa arcuata, Callicarapa americana, Calopogon tuberosus, Campanula americana, C. aparinoides, Cardamine bulbosa, Carex bushii, C. caroliniana, C. conjuncta, C. corrugata, C. cristatella, C. granularis, C. grayi, C. grisea, C. gynandra, C. haydenii, C. hyalinolepis, C. lacustris, C. lupulina, C. lurida, C. muskingumensis, C. radiata, C. scoparia, C. squarrosa, C. stipata, C. tribuloides, C. vulpinoidea, Catalpa speciosa, Celosia argentea, Cephalanthus occidentalis, Cerastium fontanum, Chamaecrista fasciculata, C. nictitans, Chasmanthium latifolium, Cicuta bulbifera, C. maculata, Cirsium discolor, Coleataenia anceps, Commelina virginica, Conoclinium coelestinum, Coreopsis lanceolata, C. palmata, C. tripteris, Cornus amomum, C. racemosa, Croton linearis, Cuphea carthagenensis, Cyperus compressus, C. croceus, C. iria, C. strigosus, Dasiphora floribunda, Daucus carota, Desmanthus illinoensis, Desmodium ciliare, D. glabellum, D. paniculatum, D. sessilifolium, Dichanthelium acuminatum, D. laxiflorum, D. spretum,*

Diodia virginiana, Diospyros virginiana, Drosera, Drymaria cordata, Drypoteris thelypteris, Dulichium arundinaceum, Echinochloa colona, Elaeagnus umbellata, Eleocharis erythropoda, E. obtusa, E. palustris, E. tenuis, E. wolfii, Epilobium ciliatum, E. leptophyllum, Erechtites hieracifolia, Eriocaulon decangulare, Eryngium yuccifolium, Eupatorium perfoliatum, E. serotinum, Euphorbia corollata, Euthamia graminifolia, E. gymnospermoides, Eutrochium dubium, E. maculatum, Fagus grandifolia, Fimbristylis, Fraxinus pennsylvanica, Galactia volubilis, Galium obtusum, G. tinctorium, Gelsemium sempervirens, Gillenia stipulata, Glyceria striata, Gnaphalium obtusifolium, Hackelia virginiana, Hamamelis virginiana, Helianthus grosseserratus, H. mollis, Heuchera richardsonii, Hibiscus moscheutos, Hieracium gronovii, Hypericum adpressum, H. drummondii, H. galioides, H. majus, H. mutilum, H. prolificum, H. punctatum, Impatiens capensis, Ipomoea cordatotriloba, Iris virginica, Isoetes melanopoda, Juncus acuminatus, J. brachycarpus, J. greenei, J. gymnocarpus, J. interior, J. marginatus, J. repens, J. torreyi, Juniperus virginiana, Justicia ovata, Kalmia latifolia, Lathyrus palustris, Leersia hexandra, L. lenticularis, L. oryzoides, L. virginica, Lespedeza capitata, Liatris pycnostachya, L. spicata, Lilium michiganense, Lindera benzoin, Lindernia dubia, Liquidambar styraciflua, Lobelia cardinalis, Ludwigia decurrens, L. palustris, L. polycarpa, Lycopus americanus, L. uniflorus, L. virginicus, Lygodium japonicum, Lyonia ligustrina, Lysimachia lanceolata, L. quadriflora, Lythrum alatum, L. salicaria, Magnolia grandiflora, M. virginiana, Marshallia, Mayaca fluviatilis, Mecardonia acuminata, Mentha arvensis, Mikania scandens, Mimulus ringens, Monarda fistulosa, Myrica asplenifolia, Oenothera pilosella, Onoclea sensibilis, Orbexilum pedunculatum, Osmunda regalis, Osmundastrum cinnamomeum, Oxydendrum arboreum, Oxypolis rigidior, Packera tomentosa, Panicum anceps, P. clandestinum, P. dichotomiflorum, P. rigidulum, P. virgatum, Paronychia fastigiata, Paspalum laeve, Penstemon calycosus, Penthorum sedoides, Persicaria amphibia, P. careyi, P. coccinea, P. hydropiperoides, P. pensylvanica, P. virginiana, Phanopyrum gymnocarpon, Phlox glaberrima, Phragmites australis, Physostegia virginiana, Pilea pumila, Pinus echinata, P. elliottii, P. taeda, Platanus occidentalis, Pluchea foetida, Poa compressa, Pogonia ophioglossoides, Polygala nuttallii, P. sanguinea, Polypremum procumbens, Populus tremuloides, Potentilla norvegica, P. simplex, Proserpinaca palustris, Prunus serotina, Pycnanthemum tenuifolium, P. virginianum, Pyrrhopappus carolinianus, Quercus bicolor, Ranunculus hispidus, Rhamnus frangula, Rhexia mariana, R. virginica, Rhus copallina, Rhynchospora capitellata, R. macrostachya, Rubus pensilvanicus, R. setosus, Rudbeckia laciniata, Rumex altissimus, R. crispus, Sabatia angularis, Saccharum giganteum, Salix discolor, S. exigua, S. humilis, Sambucus nigra, Sarracenia alata, Sassafras albidum, Saururus cernuus, Schenodorus arundinaceus, Schizachyrium scoparium, Scirpus atrovirens, S. cyperinus, S. georgianus, Scleria reticularis, S. triglomerata, Sclerolepis uniflora, Scutellaria lateriflora, S. nervosa, Sida rhombifolia, Silphium integrifolium, S. laciniatum, Sium suave, Smilax, Solanum carolinense, Solidago altissima, S. canadensis, S. gigantea, S. gymnospermoides, S. missouriensis, S. nemoralis, S. patula, S. speciosa, S. tenuifolia, Sorghastrum nutans, Spartina pectinata, Sphagnum, Spiraea alba, S. tomentosa, Sporobolus compositus, Stachys pilosa, Symphyotrichum dumosum, S. lanceolatum, S. novae-angliae, S. ontarionis, Taxodium distichum, Thelypteris noveboracensis, T. palustris, Toxicodendron radicans, Trachelospermum difforme, Tradescantia ohiensis, Triadenum fraseri, Triadica sebifera, Tridens flavus, Tripsacum dactyloides, Typha domingensis, Ulmus alata, U. americana, Utricularia cornuta, Vaccinium arboreum, V. fuscatum, Verbena hastata, V. scabra, Vernonia missurica, V. noveboracensis, Veronicastrum virginicum, Viola cucullata, V. primulifolia, V. sagittata, V. sororia, Vitis riparia, Vulpia octoflora, Xyris smalliana, Zinnia violacea, Zizia aurea.

Ludwigia arcuata Walter is a perennial, which grows variably either submerged and erect in water depths to 1 m, or as a prostrate emergent that roots at the nodes and creeps along exposed substrates. Habitats include creekbeds, ditches, flats, gator holes, marshes, meadows, ricefields, sandbars, and shallow shorelines along lakes, pools, and streams at low elevations. The plants occasionally form mats on the surface of shallow lakes. Some shade is tolerated. The substrates can include muck, peat, sand, and silt. The sites are characterized by relatively low values of alkalinity (mean = 6.3 mg/L as $CaCO_3$), conductivity (mean = 93 μS/cm), and pH (mean = 6.3; range = 5.8–7.5). The flowering period extends into August. The plants are self-compatible but are outcrossed if pollinators visit the nectariferous flowers when they are newly opened. Otherwise, eventual selfing occurs as the anthers atop the incurving stamens make contact with the stigma, except in some populations where the stigmas are elevated enough to prevent self-pollination. The seeds are dispersed by gravity from the irregularly dehiscing capsules. The plants can be propagated from seed or vegetatively by stem cuttings. Heterophylly of the submerged (narrow) and emergent (broad) foliage is regulated hormonally. Treatment with ethylene gas induces the submerged morphology in terrestrial leaves; whereas, treatment with abscissic acid induces leaves with terrestrial morphology on submergent shoots. The plants require high light levels and an optimal temperature between 24°C and 26°C when grown in aquariums. In autumn, the lowermost leaves are shed and the shoot fragments become more difficult to propagate vegetatively. **Reported associates:** *Baccharis halimifolia, Bacopa caroliniana, Brasenia schreberi, Cabomba caroliniana, Cephalanthus occidentalis, Cyperus odorata, C. lanceolatus, C. polystachyos, C. surinamensis, Diodia virginiata, Eleocharis baldwinii, Eupatorium capillifolium, Fuirena scirpoidea, Hydrocotyle umbellata, Juncus effusus, Lemna, Ludwigia octovalvis, Myrica cerifera, Najas guadalupensis, Nelumbo lutea, Nymphaea odorata, Nymphoides aquatica, Panicum hemitomon, P. repens, Persicaria punctata, Pontederia cordata, Sacciolepis striata, Sagittaria filiformis, S. graminea, Taxodium ascendens, T. distichum, Utricularia floridana, Xyris.*

Ludwigia bonariensis (**M. Micheli**) **Hara** is a nonindigenous perennial, which inhabits shallow waters in ditches, floodplains, marshes, ponds, riverbanks, and waste ground at low elevations to at least 7 m. Unlike most *Ludwigia* species, the flowers are self-incompatible. Otherwise, there is little ecological information available for this species. **Reported associates (North America):** *Acer rubrum, Alnus serrulata, Amaranthus cannabinus, Amorpha fructicosa, Apios americana, Arundinaria gigantea, Baccharis halimifolia, Bidens laevis, B. polylepis, Boehmeria cylindrica, Boltonia asteroides, Campsis radicans, Carex annectens, C. comosa, C. glaucescens, C. hyalinolepis, Cephalanthus occidentalis, Cicuta maculata, Cinna arundinacea, Clematis crispa, Clethra alnifolia, Commelina virginica, Cornus foemina, Cuphea carthagenensis, Cuscuta campestris, Cyrilla racemiflora, Decodon verticillatus, Decumaria barbara, Elymus virginicus, Eryngium yuccifolium, Eubotrys racemosa, Ilex opaca, Impatiens capensis, Fraxinus caroliniana, F. pennsylvanica, F. profunda, Galium obtusum, Habenaria repens, Juncus roemerianus, Liquidambar styraciflua, Lyonia lucida, Magnolia virginiana, Mikania scandens, Murdannia keisak, Myrica cerifera, Nyssa sylvatica, Onoclea sensibilis, Osmunda regalis, Osmundastrum cinnamomeum, Oxydendrum arboreum, Packera glabella, Parthenocissus quinquefolia, Peltandra virginica, Persea palustris, Persicaria arifolia, P. hydropiper, P. setacea, Phoradendron serotinum, Pinus serotina, Pontederia cordata, Quercus lyrata, Rosa palustris, Rubus, Rumex verticillatus, Sabatia calycina, Sagittaria lancifolia, Saururus cernuus, Sium suave, Smilax bona-nox, S. laurifolia, Taxodium distichum, Toxicodendron radicans, Typha angustifolia, T. latifolia, Ulmus americana, Viburnum dentatum, Viola pedata, Vitis rotundifolia, Woodwardia areolata, Zizania aquatica, Zizaniopsis miliacea.*

Ludwigia brevipes (**Long**) **Eames** is a mat-forming perennial with prostrate stems, which root at the nodes. Its habitats include gravel pits, interdunal pools, marshes, and the borders of lakes, ponds, and rivers at low elevations (to at least 11 m). The substrates are acidic (pH: 5.3–6.8) and include peat or sand. The plants require bright light conditions and occur mainly in habitats where alternating inundated/drying conditions exist. The self-compatible, nectariferous flowers are outcrossed if they are visited by pollinators when newly opened; otherwise selfing eventually occurs. The seeds are dispersed by gravity from the irregularly dehiscing capsules. A seed bank of undetermined persistence is formed. After receiving an adequate period of cold stratification, the seeds can be germinated at 21°C–26°C. **Reported associates:** *Eleocharis vivipara, Ludwigia alata, L. palustris.*

Ludwigia curtissii **Chapman** is a perennial, which produces short, leafy offshoots from its base during the autumn. It occurs in depressions, ditches, flats, flatwoods, pinelands, ponds, prairies, savannas, solution pits, and along the margins of canals at low elevations. The plants often are observed in areas that have been burned. The substrates are described as calcareous and include lime rock sand and sandy peat. The self-compatible flowers produce nectar but lack petals (which sometimes persist as vestiges) and are facultatively autogamous.

The pollen is shed as single grains. Stolons do not develop, but decumbent or ascending new shoots emerge from the lower stem nodes and eventually become erect. **Reported associates:** *Aeschynomene pratensis, Andropogon glomeratus, Annona glabra, Asclepias lanceolata, Bacopa caroliniana, Bidens mitis, Blechnum serrulatum, Boehmeria cylindrica, Bursera simaruba, Celtis laevigata, Cephalanthus occidentalis, Chrysobalanus icaco, Chrysophyllum oliviforme, Cirsium nuttallii, Cissus verticillata, Cladium jamaicense, Coccoloba diversifolia, Coleataenia longifolia, Crinum americanum, Cyperus haspan, C. polystachyos, Dalbergia ecastaphyllum, Dichanthelium dichotomum, Diodia virginiana, Eleocharis cellulosa, Erechtites hieraciifolius, Eugenia axillaris, Ficus aurea, Fuirena breviseta, Funastrum clausum, Helenium pinnatifidum, Hibiscus grandiflorus, Hypericum cistifolium, Hyptis alata, Ipomoea sagittata, Iva microcephala, Juncus roemerianus, Justicia angusta, Leersia hexandra, Lobelia glandulosa, Ludwigia alata, L. microcarpa, L. repens, Magnolia virginiana, Mecardonia acuminata, Mikania scandens, Mitreola petiolata, M. sessilifolia, Muhlenbergia capillaris, Myrica cerifera, Myrsine floridana, Nymphaea odorata, Nymphoides aquatica, Osmunda regalis, Oxypolis filiformis, Panicum hemitomon, Parthenocissus quinquefolia, Paspalidium geminatum, Peltandra virginica, Persea borbonia, Phyla nodiflora, Pluchea baccharis, Pontederia cordata, Proserpinaca pectinata, Pteridium aquilinum, Rhynchospora colorata, R. divergens, R. globularis, R. inundata, R. microcarpa, R. perplexa, R. tracyi, Sabal palmetto, Saccharum giganteum, Sacciolepis indica, Sagittaria lancifolia, Salix caroliniana, Saururus cernuus, Setaria parviflora, Simarouba glauca, Thelypteris interrupta, Typha domingensis, Utricularia foliosa, U. gibba, U. purpurea, Xyris jupicai.*

Ludwigia decurrens (**DC.**) **Walter** is an annual or perennial, which occurs in shallow waters (to 3 dm) of borrow pits, bottomlands, channels, depressions, ditches, flats, floodplains, gravel bars, levees, marshes, meadows, mudflats, pools, puddles, ravines, rice fields, roadsides, sandbars, seeps, sloughs, swales, swamps, and along the margins of bayous, lakes, ponds, reservoirs, rivers, and streams at elevations of up to 513 m. Exposures can range from full sun to semi-shade. The sites can remain inundated for several months out of the year. The substrates (pH: 5.1–7.8) can comprise alluvium, clay, clay loam, gravel, loam, mud, rock, sand, sandy clay loam, sandy loam, silt, silty loam, or silty sand and often are high in organic matter. Flowering and fruiting have been observed from May to November. The flowers are self-compatible and autogamous. The seeds require an 8-week period of cold stratification at 4°C followed by incubation under a 35°C/20°C temperature regime to promote optimal germination. The plants occur often in disturbed sites. **Reported associates:** *Acalypha rhomboidea, Acer rubrum, Alisma, Alopecurus carolinianus, Alternanthera philoxeroides, Amaranthus tuberculatus, Ambrosia, Ammannia coccinea, Ampelopsis arborea, Amsonia hubrichtii, A. illustris, Apios americana, Asclepias incarnata, Bacopa rotundifolia, Bidens coronatus, B. frondosus, B. polylepis, Boehmeria cylindrica, Boltonia diffusa, Callicarpa americana, Carex gynandra, C.*

lupulina, Carya, Cephalanthus occidentalis, Chasmanthium latifolium, Commelina diffusa, C. erecta, Conyza canadensis, Cornus amomum, Crateagus, Cynodon dactylon, Cyperus acuminatus, C. erythrorhizos, C. esculentus, C. iria, C. strigosus, Dalea leporina, Dichanthelium, Digitaria sanguinalis, Diodia teres, D. virginiana, Echinochloa crus-galli, E. muricata, Echinodorus berteroi, E. cordifolius, Eclipta prostrata, Eleocharis macrostachya, E. obtusa, E. ovata, Elymus virginicus, Eragrostis hypnoides, Erechtites, Eupatorium capllifolium, E. fistulosum, Fimbristylis autumnalis, F. littoralis, F. perpusilla, Fraxinus pennsylvanica, Glottidium vesicarium, Gratiola aurea, G. neglecta, Habenaria repens, Hamamelis vernalis, Helenium flexuosum, Helianthus, Heteranthera limosa, H. multiflora, H. reniformis, Hibiscus laevis, H. moscheutos, Hydrocotyle umbellata, Hydrolea, Hypericum mutilum, H. prolificum, Ilex, Ipomoea lacunosa, Iva annua, Jacquemontia tamnifolia, Juncus canadensis, J. effusus, J. repens, Kyllinga pumila, Leersia, Lemna minor, Leptochloa fusca, L. panicoides, Lespedeza, Leucospora multifida, Lindernia dubia, Lipocarpha micrantha, Liquidambar styraciflua, Ludwigia alternifolia, L. leptocarpa, L. linearis, L. palustris, L. pepoides, L. pilosa, L. sphaerocarpa, Lycopus americanus, Lygodium japonicum, Micranthemum umbrosum, Mikania scandens, Mimulus alatus, Mollugo verticillata, Nelumbo lutea, Nyssa sylvatica, Packera glabella, Panicum anceps, P. capillare, P. dichotomiflorum, P. verrucosum, P. virgatum, Paspalum urvilei, Penthorum sedoides, Perilla frutescens, Persicaria bicornis, P. glabra, P. lapathifolia, P. maculosa, P. pensylvanica, P. punctata, Phalaris arundinacea, Phanopyrum gymnocarpon, Phyla lanceolata, Pinus taeda, Pluchea camphorata, Portulaca oleracea, Quercus falcata, Ranunculus sceleratus, Rhexia virginica, Rhus copallina, Rhynchospora corniculata, R. macrostachya, Rorippa palustris, Rotala ramosior, Rubus, Rumex crispus, Sabal minor, Sabatia kennedyana, Saccharum giganteum, Sacciolepis, Sagittaria brevirostra, S. latifolia, S. montevidensis, Salix nigra, Saururus cernuus, Scirpus cyperinus, Scleria, Sida spinosa, Smilax, Solidago, Sorghum halepense, Spirodela polyrhiza, Strophostyles helvula, Stylosanthes albiflora, Symphyotrichum lanceolatum, Taxodium distichum, Toxicodendron pubescens, Triadenum walteri, Triadica sebifera, Typha latifolia, Ulmus alata, Vaccinium arboreum, Vernonia lettermannii, Veronica peregrina, Vitis, Xanthium strumarium, Xyris.

***Ludwigia erecta* (L.) H. Hara** is an annual or short-lived perennial, which grows in bogs, canals, depressions, ditches, flood plains, marshes, prairies, and along the margins of lakes, ponds, and rivers at elevations of up to 362 m. Exposures generally are reported as full sun. The substrates have been described as cobble, gravel, mud, sand, and sandy peat. Flowering and fruiting have been observed from August to November. The flowers are self-compatible and autogamous. The seeds are dispersed by floodwaters. Six-month-old seeds germinate earlier and have uniformly higher germination rates (87%–95%) on clay, loam or sand, than do freshly shed seeds (16%–44%). **Reported associates**: *Amaranthus palmeri, Ammannia coccinea, Baccharis salicifolia, B. sarothroides, Bidens mitis, Bolboschoenus maritimus, Brasenia schreberi, Caperonia palustris, Cenchrus ciliaris, Chenopodium ambrosioides, Cynodon dactylon, Cyperus croceus, C. difformis, C. erythrorhizos, C. involucratus, C. pygmaeus, C. odoratus, C. polystachyos, Echinochloa colona, Eclipta prostrata, Eleocharis baldwinii, E. elongata, E. geniculata, E. interstincta, Eustoma exaltatum, Fuirena scirpoidea, Habenaria repens, Hydrocotyle umbellata, Hypericum fasciculatum, Itea virginica, Leptochloa fusca, L. viscida, Ludwigia lanceolata, L. octovalvis, L. suffruticosa, Nymphaea ampla, N. odorata, Nuphar advena, Nymphoides aquatica, Phytolacca americana, Pluchea foetida, P. longifolia, P. odorata, P. sericea, Populus fremontii, Potamogeton nodosus, Prosopis, Rhynchospora microcephala, Salix gooddingii, Salvinia minima, Sida cordifolia, Spartina bakeri, Spirodela polyrhiza, Symphyotrichum subulatum, Tamarix, Triadenum virginicum, Typha, Utricularia foliosa, Xanthium strumarium, Youngia japonica.*

***Ludwigia glandulosa* Walter** is a perennial inhabitant of bottoms, channels, depressions, ditches, flats, floodplains, limesinks, marshes, pools, prairies, ravines, roadsides, ruts, savannahs, shores, slopes, sloughs, swales, swamps, woodlands, and the margins of ponds, rivers, sloughs, and streams at elevations of up to 101 m. The sites are characterized by shallow, circumneutral waters (pH: 6.0–7.4) and substrates described as alluvium, clay, claypan, cobble, gravel, limestone, marl, mud, peat, Ruston sandy loam, sand, or silt. The plants typically occur in open habitats such as exposed shorelines, but commonly are found in partial shade and sometimes in dense shade. Flowering and fruiting extend from June to August. The flowers produce nectar, but are apetalous and probably are mainly self-pollinated. The pollen is shed as tetrads. The seeds are dispersed from irregularly dehiscing capsules. A seed bank is formed, but the seeds will not germinate under flooded conditions. The plants spread vegetatively by elongate, leafy stolons, which are produced in the fall. Submersed portions of the shoots will develop extensive, spongy aerenchyma tissue. Cultivated specimens are slow growing and require high light levels and temperatures below 25°C to thrive. **Reported associates**: *Acer rubrum, Alternanthera philoxeroides, Amaranthus, Arundinaria gigantea, Betula, Bidens laevis, Boehmeria cylindrica, Carex, Centella asiatica, Chamaecrista fasciculata, C. nictitans, Colocasia esculenta, Commelina virginica, Coreopsis major, Cynodon dactylon, Cyperus erythrorhizos, C. haspan, C. pseudovegetus, C. virens, Digitaria sanguinalis, Eclipta prostrata, Eichhornia crassipes, Eleocharis microformis, E. quadrangulata, Eragrostis, Eupatorium, Fimbristylis vahlii, Fraxinus pennsylvanica, Fuirena squarrosa, Hyptis alata, Ipomoea, Jacquemontia tamnifolia, Juncus, Leersia hexandra, L. oryzoides, Lindernia dubia, Ludwigia alternifolia, L. linearis, Lycopus rubellus, Marsilea, Mecardonia acuminata, Mikania scandens, Mitreola petiolata, Mollugo verticillata, Myriophyllum pinnatum, Nyssa aquatica, Oenothera curtiflora, Osmunda regalis, Panicum dichotomiflorum, P. hians, P. virgatum, Persicaria punctata, Phyla nodiflora, Platanus, Pluchea, Polypremum procumbens,*

Polystichum acrostichoides, Portulaca oleracea, Rhexia mariana, Rhynchospora, Rumex chrysocarpus, Sagittaria longiloba, S. papillosa, Salix nigra, Saururus cernuus, Senna obtusifolia, Sesbania vesicaria, Setaria, Stenanthium densum, Strophostyles helvula, Symphyotrichum subulatum, Taxodium distichum, Typha, Ulmus americana, Vaccinium tenellum, Verbena brasiliensis, Viola blanda, Xanthium strumarium, Xyris jupicai.

***Ludwigia grandiflora* (Michx.) Greuter & Burdet** is a nonindigenous perennial, which occurs in fresh to brackish (0–3.5 ppt salinity), circumneutral (pH: 6.1–7.3), waters (to 2.4 m deep), of beaches, bogs, canals, channels, depressions, ditches, flats, lakes, levees, marshes, meadows, ponds, pools, prairies, rice fields, sand bars, shores, sloughs, swamps, and the margins of lakes, rivers and streams at elevations of up to 102 m. Typically it is found on organic substrates and on chewacla series, muck, sand, sandy clay, or wehadkee series. The plants readily colonize disturbed areas and thrive in wet habitats that undergo periodic drying. They sometimes will form dense, floating mats, which can extend upward of 23 m across the water surface. The species is tolerant to burning, flooding, grazing, and mowing unless subjected to these conditions for prolonged periods of time. Flowering has been observed from April to September. The pollen is released in monads. The seeds are not freely dispersed, but remain imbedded within the woody exocarp of the capsules. **Reported associates (North America):** *Alternanthera philoxeroides, Artemisia douglasiana, Arundo donax, Baccharis salicifolia, Bacopa monnieri, Boehmeria cylindrica, Cabomba caroliniana, Carex intumescens, C. joorii, Carya aquatica, Chara, Cladium jamaicense, Diodia virginiana, Eclipta prostrata, Eichhornia crassipes, Elodea canadensis, Gratiola virginiana, Hydrilla verticillata, Juncus acutus, Justicia ovata, Leersia lenticularis, Lemna minuta, Leptochloa uninervia, Limnobium spongia, Lobelia cardinalis, Ludwigia palustris, Lythrum, Mentha, Mikania scandens, Myriophyllum aquaticum, Nasturtium officinale, Panicum hemitomon, Persicaria lapathifolia, P. punctata, Pilea pumila, Pistia stratiotes, Platanus racemosa, Polypogon monspeliensis, Pontederia cordata, Populus fremontii, Quercus agrifolia, Q. lyrata, Ricinus communis, Sagittaria falcata, S. lancifolia, Salix nigra, Saururus cernuus, Schoenoplectus tabernaemontani, Symphyotrichum lateriflorum, Toxicodendron diversilobum, Typha domingensis, Wolffia brasiliensis, W. columbiana, Zizaniopsis miliacea.*

***Ludwigia hexapetala* (Hook. & Arn.) Zardini, H.Y. Gu & P.H. Raven** is a nonindigenous emergent (at depths to 1 m) or floating-stemmed perennial, which inhabits bottoms, canals, flats, lagoons, marshes, mudflats, pools, sloughs, thickets, and the margins of lakes, ponds, rivers, and streams at elevations of up to 405 m. The habitats are exposed to full sunlight. The substrates often are high in organic matter and include gravel, mud, sand, silt, and silty loam. The sprawling stems root adventitiously at the nodes and can form dense mats along the margins of ditches, lakes, ponds, and streams, often extending out into the water. Dense stands have been observed to alter associated communities of diatoms (Protista:

Chromalveolata: Stramenopila) toward the presence of more shade-tolerant species. This species will tolerate anaerobic as well as seasonally dry conditions. Because the plants readily absorb nutrients, habitats that are rich in nitrogen and phosphorous will result in accelerated growth and higher productivity. In addition to normal roots that absorb substrate nutrients, the plants produce specialized roots capable of extracting nutrients directly from the water. Flowering has been observed throughout the summer until November. The seeds remain embedded within the capsule wall and are dispersed with the fruit. Seed germination occurs at 20°C after a 6-week period of cold stratification at 4°C. Seedlings are not commonly observed. Vegetative reproduction occurs by shoot fragments, which are dispersed by water and waterfowl. **Reported associates (North America):** *Baccharis pilularis, B. salicifolia, Bolboschoenus fluviatilis, Cabomba caroliniana, Callitriche stagnalis, Carex nudata, Ceratophyllum demersum, Egeria densa, Eichhornia crassipes, Eleocharis macrostachya, Elodea canadensis, Epilobium ciliatum, Hibiscus moscheutos, Iris pseudacorus, Juncus, Lemna minor, Ludwigia brevipes, L. peploides, Lythrum hyssopifolia, L. salicaria, Myriophyllum aquaticum, Najas, Nasturtium officinale, Nelumbo lutea, Nuphar polysepala, Paspalum distichum, Persicaria amphibia, P. hydropiperoides, Phalaris arundinacia, Platanus racemosa, Potamogeton, Pulicaria paludosa, Quercus agrifolia, Sagittaria latifolia, Salix exigua, S. laevigata, S. lasiolepis, S. sessilifolia, Schoenoplectus acutus, S. americanus, S. californicus, Sparganium eurycarpum, Tamarix, Typha domingensis, T. latifolia, Veronica americana, V. anagallis-aquatica, Xanthium strumarium.*

***Ludwigia hirtella* Raf.** is a perennial species, which inhabits open, sunny sites in bayous, bogs, borrow pits, depressions, ditches, flats, flatwoods, floodplains, meadows, prairies, roadsides, savannahs, swales, swamps, and the margins of ponds, rivers, and streams at elevations of up to 149 m. The substrates are acidic and include clay, peat, and sand. Flowering and fruiting have been observed from April to December. The pollen is shed as tetrads. The seeds are freely dispersed through a terminal pore of the mature capsules. The plants reproduce vegetatively by the production of tuberous roots. **Reported associates:** *Alnus serrulata, Amphicarpum purshii, Andropogon glomeratus, A. perangustatus, Asclepias longifolia, A. rubra, Calamagrostis coarctata, Callicarpa americana, Carex, Coreopsis linifolia, Ctenium aromaticum, Dichanthelium dichotomum, D. scoparium, Drosera brevifolia, D. filiformis, Eleocharis tortilis, Ericaulon decangulare, Erigeron, Eriophorum virginicum, Eryngium integrifolium, Eupatorium pilosum, E. rotundifolium, Fimbristylis puberula, Fuirena squarrosa, Gymnopogon brevifolius, Hypericum adpressum, H. brachyphyllum, H. canadense, H. crux-andreae, H. setosum, Ilex glabra, Juncus elliottii, Lachnanthes caroliniana, Lobelia, Lophiola americanum, Ludwigia alternifolia, Lycopodiella appressa, Magnolia virginiana, Marshallia tenuifolia, Mitreola sessilifolia, Monarda fistulosa, Myrica inodora, Nyssa aquatica, Oenothera linifolia, Oxypolis filiformis, Panicum rigidulum, Parthenocissus quinquefolia, Pinus palustris, P. taeda,*

Platanthera blephariglottis, Pluchea baccharis, Polygala cruciata, P. lutea, Pteridium aquilinum, Quercus velutina, Rhexia lutea, R. petiolata, Rhynchospora chapmanii, R. knieskernii, R. oligantha, Rudbeckia grandiflora, R. maxima, Sabatia macrophylla, Sarracenia flava, S. purpurea, Schizachyrium scoparium, S. tenerum, Sphagnum, Taxodium distichum, Toxicodendron vernix, Utricularia, Vaccinium corymbosum, V. fuscatum, V. formosum, Viburnum nudum, Xyris caroliniana.

***Ludwigia lanceolata* Elliott** is a perennial, which occurs in acidic (pH: 6.0), freshwater to brackish "batteries," depressions, ditches, hammocks, "houses," marshes, and swamps at low elevations. The plants are tolerant to some shade. The flowers are self-compatible and facultatively autogamous. Few additional details are known regarding the ecology of this rare species. **Reported associates:** *Acer rubrum, Agalinis purpurea, Amphicarpum muhlenbergianum, Andropogon glomeratus, A. gyrans, A. virginicus, Aristida palustris, Arnoglossum ovatum, Aronia arbutifolia, Axonopus fissifolius, Bartonia verna, Bidens mitis, Carex glaucescens, Cephalanthus occidentalis, Coleataenia longifolia, Cyperus croceus, C. erythrorhizos, C. haspan, Cyrilla racemiflora, Decodon verticillatus, Dichanthelium acuminatum, D. hirstii, D. sphaerocarpon, Diospyros virginiana, Dulichium arundinaceum, Eleocharis baldwinii, E. elongata, Eragrostis hypnoides, Eriocaulon compressum, Eubotrys racemosa, Eupatorium recurvans, Fimbristylis autumnalis, Gratiola ramosa, Habenaria repens, Helenium pinnatifidum, Hydrocotyle umbellata, Hymenocallis palmeri, Hypericum crux-andreae, H. fasciculatum, H. harperi, H. myrtifolium, H. tetrapetalum, H. virginicum, Ilex glabra, I. myrtifolia, Iris, Itea virginica, Juncus repens, Justicia ovata, Lachnanthes caroliniana, Lobelia boykinii, Ludwigia erecta, L. linearis, L. pilosa, L. sphaerocarpa, Lycopodium, Lyonia fruticosa, L. lucida, Melaleuca quinquenervia, Mnesithea rugosa, Myrica cerifera, Nuphar advena, Nymphaea odorata, Nymphoides aquatica, Nyssa biflora, Oclemena reticulata, Orontium aquaticum, Osmundastrum cinnamomeum, Panicum abscissum, P. hemitomon, Paspalum dissectum, P. repens, Peltandra virginica, Pinguicula caerulea, P. lutea, Pinus elliottii, Pluchea baccharis, Polygala cruciata, Pontederia cordata, Proserpinaca pectinata, Rhexia nuttallii, Rhynchospora careyana, R. cephalantha, R. chapmanii, R. fascicularis, R. filifolia, R. harperi, R. inundata, R. microcephala, R. tracyi, Sabatia decandra, Saccharum giganteum, Sacciolepis striata, Sagittaria australis, S. graminea, Scleria reticularis, Sclerolepis uniflora, Serenoa repens, Smilax laurifolia, S. walteri, Solidago fistulosa, Sphagnum, Stillingia aquatica, Symphyotrichum subulatum, Syngonanthus flavidulus, Taxodium distichum, Utricularia, Vaccinium corymbosum, V. myrsinites, Vitis rotundifolia, Woodwardia virginica, Xyris ambigua, X. brevifolia, X. elliottii, X. fimbriata, X. floridana, X. jupicai, X. smalliana.*

***Ludwigia leptocarpa* (Nutt.) H. Hara** is an annual, which grows in shallow (to 1 m), fresh to brackish, tidal or nontidal waters of bayous, borrow pits, dikes, ditches, flats, floodplains, hummocks, marshes, meadows, mudflats, outwash,

pools, rice fields, sandbars, seeps, sloughs, swamps, and along the margins of lakes, ponds, reservoirs, and rivers at elevations of up to 75 m. Exposures are reported as full sun. The substrates (pH: 5.4–7.6) include alluvium, gravel, muck, mud, peat, sand, and rotting logs; they sometimes are oligohaline. The plants can grow upon floating mats of logs or vegetation. Flowering and fruiting have been observed from July through December. The flowers are self-compatible and selfing can occur, but the plants primarily are outcrossed. The seeds are dispersed from the capsules by gravity and can develop into a seedbank. The seeds possess a horseshoe-shaped endocarp that may confer flotation and facilitate their dispersal by water. Freshly collected seeds will germinate when incubated in moist conditions at 22°C under a 16/8 h light regime. Larger seeds have higher germination rates and earlier emergence times than the smaller ones. Although there are reports to the contrary, seedlings have been observed emerging from wetland seed banks. **Reported associates:** *Acer rubrum, Aeschynomene indica, Alnus serrulata, Anemone virginiana, Antigonon leptopus, Bacopa, Betula nigra, Bidens cernuus, B. coronatus, B. frondosus, B. laevis, Brachiaria ramosa, Carya aquatica, Catalpa bignonioides, Cephalanthus occidentalis, Chenopodium ambrosioides, Cladium jamaicense, Commelina virginica, Conyza canadensis, Cornus amomum, Cyperus erythrorhizos, Cyrilla racemiflora, Digitaria, Diodia virginiana, Diospyros virginiana, Dulichium arundinaceum, Echinochloa, Eclipta prostrata, Eichhornia crassipes, Eleocharis obtusa, Elephantopus carolinianus, Eragrostis hypnoides, Eupatorium capillifolium, E. coelestinum, Fraxinus profunda, Fuirena scirpoidea, F. simplex, Geranium carolinianum, Helenium autumnale, Hibiscus, Hydrocotyle, Hypericum densiflorum, H. mutilum, Indogifera spicata, Ipomoea trichocarpa, Itea virginica, Juncus canadensis, J. effusus, Leersia hexandra, Leptochloa, Leucospora multifida, Limonium carolinianum, Liquidambar styraciflua, Ludwigia decurrens, L. linearis, L. octovalvis, L. peploides, L. peruviana, L. pilosa, L. sphaerocarpa, Lycopus, Melaleuca quinquenervia, Mikania scandens, Mimosa, Mollugo verticillata, Murdannia keisak, Myrica cerifera, Nyssa aquatica, N. biflora, Orontium aquaticum, Oxalis stricta, Oxycaryum cubense, Oxypolis rigidior, Panicum hemitomon, P. maximum, P. repens, P. verrucosum, Paspalum vaginatum, Persicaria, Phlox carolina, Phyla, Phytolacca americana, Planera aquatica, Platanus occidentalis, Pluchea odorata, Pontederia cordata, Rhexia mariana, R. virginica, Rhynchospora colorata, R. macrostachya, R. traceyi, Rudbeckia laciniata, R. subtomentosa, Rumex pulcher, Saccharum giganteum, Sacciolepis striata, Sagittaria falcata, S. lancifolia, S. latifolia, Salix caroliniana, S. nigra, Salvia riparia, Sarcostemma clausum, Schoenoplectus californicus, S. tabernaemontani, Scirpus cyperinus, Sesbania, Solanum americanum, Sparganium americanum, Spartina bakeri, S. patens, Spermacoce surgens, Sphenoclea zeylandica, Strophostyles, Symphyotrichum, Taxodium distichum, Tripsacum dactyloides, Typha, Ulmus rubra, Verbena, Vernonia gigantea, Vigna luteola, Zizania aquatica, Zizaniopsis miliacea.*

Ludwigia linearis **Walter** is a perennial found in shallow waters of barrens, bogs, borrow pits, canals, depressions, ditches, flats, flatwoods, floodplains, marshes, pocosins, ponds, pools, prairies, ruts, savannahs, seeps, sloughs, swales, swamps, and along the margins of lakes and streams at elevations of up to 620 m. The plants normally occur in habitats that receive full sun. The substrates (pH: 4.7–6.6) include Atmore series, clay loam, loam, sand, sandy clay loam, sandy loam, and silt. Flowering has been observed from July to November. The flowers produce nectar and have showy petals but they are facultatively autogamous. The pollen is shed as tetrads. **Reported associates:** *Acer rubrum, Andropogon glomeratus, A. virginicus, Calamagrostis coarctata, Carex bullata, C. gigantea, C. glaucescens, C. joorii, C. striata, Carya, Centella asiatica, Cephalanthus occidentalis, Chamaecyparis thyoides, Cornus florida, Cynodon dactylon, Cyperus haspan, C. pseudovegetus, C. virens, Dichanthelium dichotomum, Diodia virginica, Drosera tracyi, Eleocharis microcarpa, E. quadrangulata, E. tuberculosa, Erechtites hieraciifolia, Eupatorium capllifolium, Euthamia graminifolia, Fagus grandifolia, Fimbristylis, Fuirena breviseta, F. squarrosa, Hedyotis, Helenium amarum, Helianthus angustifolius, Hypericum crux-andreae, Ilex, Jacquemontia tamnifolia, Juncus acuminatus, J. debilis, J. dichotomus, J. effusus, J. marginatus, J. repens, J. scirpoides, J. validus, Leersia hexandra, Lindernia dubia, Liquidambar styraciflua, Lobelia canbyi, L. glandulosa, Ludwigia glandulosa, L. linifolia, L. microcarpa, L. octovalvis, L. pilosa, L. sphaerocarpa, Magnolia virginiana, Mecardonia acuminata, Mikania scandens, Nyssa biflora, Panicum hemitomon, P. hians, P. rigidulum, P. verrucosum, P. virgatum, Paspalum notatum, P. urvillei, Pinus elliottii, P. taeda, Polygonella gracilis, Polypremum procumbens, Pontederia cordata, Proserpinaca palustris, P. intermedia, P. pectinata, Quercus, Rhexia mariana, R. virginiana, Rhus copallina, Rhynchospora caduca, R. capitellata, R. corniculata, R. glomerata, R. inexpansa, R. inundata, R. microcarpa, R. perplexa, Rubus, Sabal minor, Saccharum baldwinii, S. giganteum, Sagittaria graminea, S. latifolia, S. longiloba, S. papillosa, Sarracenia alata, Scirpus cyperinus, Sesbania, Smilax laurifolia, S. rotundifolia, Spartina patens, Sphagnum, Symphyotrichum subulatum, Taxodium distichum, Thelypteris palustris, Torreyochloa pallida, Toxicodendron, Triadica sebifera, Utricularia cornuta, Vaccinium, Xyris jupicai, X. laxifolia, Xyris torta.*

Ludwigia linifolia **Poir.** is a perennial, which grows as an emergent or in shallow (e.g., 0.5 m) brackish to fresh, still or flowing waters of bogs, borrow pits, depressions, ditches, flatwoods, meadows, ponds, savannahs, swamps, and along the margins of ponds at low elevations of at least 17 m. Reported soil types include Grady and Samsula–Myakka variants, sand, sandy peat, and silt loam. Flowering and fruiting have been reported from July to August. The self-compatible flowers possess petals and produce nectar but are facultatively autogamous. The pollen is shed as tetrads. Vegetative reproduction is facilitated by stolons that are produced from the base of plants during the Autumn. **Reported associates:** *Andropogon virginicus, Asclepias pedicellata, Axonopus furcatus, Bartonia verna, Calopogon tuberosus, Carex glaucescens, Dichanthelium ensifolium, D. erectifolium, D. strigosum, Drosera capillaris, D. intermedia, Eupatorium capillifolium, E. compositifolium, Euthamia, Hypericum chapmanii, H. denticulatum, Ilex glabra, Lachnanthes caroliniana, Ludwigia linearis, L. microcarpa, L. pilosa, Lycopodiella caroliniana, Nymphaea, Nyssa, Oldenlandia uniflora, Panicum abscissum P. hemitomon, P. tenerum, P. verrucosum, Physostegia purpurea, Pinus elliottii, Pluchea camphorata, P. rosea, Polygala cymosa, Rhexia mariana, R. virginica, Rhynchospora baldwinii, R. ciliaris, R. fascicularis, R. filifolia, R. tracyi, R. wrightiana, Sarracenia minor, Scirpus cyperinus, Scleria georgiana, Spiranthes longilabris, Taxodium ascendens, Viola lanceolata, V. primulifolia, Woodwardia virginica, Xyris jupicai.*

Ludwigia longifolia **(DC.) H. Hara** is a nonindigenous, tropical and subtropical annual of swamps and marshes that was introduced recently to central Florida. Although annual, the plants can grow to shrub stature (3 m), often becoming partially woody at the base, and can form dense thickets within a growing season. The flowers are self-compatible but primarily outcrossed. The North American ecology of this recent, nonindigenous species is virtually unknown at this time and deserves serious study. **Reported associates (North America):** none.

Ludwigia microcarpa **Michx.** is a diminutive perennial, which grows in shallow (e.g., 5 cm), brackish or freshwater water in sites that include bogs, depressions, ditches, fens, flats, flatwoods, marshes, meadows, mudflats, pocosins, prairies, roadsides, ruts, savannahs, seeps, strands, swales, swamps, and the margins of ponds and streams, at low elevations of up to 244 m. Exposures usually receive full sun. The substrates typically are alkaline and include bayou sandy loam, clay, Iredell, muck, mud, Newham, peat, sand, sandy alluvium, sandy clay, sandy peat, and silt. Flowering and fruiting have been observed from July to September. The self-compatible flowers are apetalous, produce little nectar, and are facultatively autogamous. The pollen is shed as single grains. At least some level of outcrossing is indicated by the ability of this species to form natural hybrids (see *Systematics* later). Vegetative reproduction occurs by stolons, which are produced during the Autumn. **Reported associates:** *Acacia pinetorum, Acrostichum danaeifolium, Aletris bracteata, Ambrosia artemisiifolia, Ammania latifolia, Ampelaster carolinianus, Ampelopsis arborea, Andropogon glomeratus, Annona glabra, Ardisia solanacea, Aristida purpurascens, Asclepias lanceolata, Baccharis halimifolia, Bacopa caroliniana, B. monnieri, Bidens aristosus, Boehmeria cylindrica, Borreria terminalis, Bumelia reclinata, B. salicifolia, Calopogon tuberosus, Caperonia casteneifolia, Centella asiatica, Cephalanthus occidentalis, Chrysobalanus icaco, Cirsium horridulum, Cladium jamaicense, Crinum americanum, Croton linearis, Cynoctonum mitreola, Cyperus esculentus, C. ligularis, C. surinamensis, Cyrilla racemiflora, Dalea foliosa, Diodia virginiana, Diospyros virginiana,*

Drosera tracyi, Dyschoriste angusta, Echinochloa paludigena, Eleocharis geniculata, E. microcarpa, E. tuberculosa, Elytraria caroliniensis, Erigeron quercifolius, Eupatorium capillifolium, E. coelastinum, E. mikaniodes, Eustachys glauca, Evolvulus sericeus, Ficus aurea, Fimbristylis littoralis, Fuirena breviseta, Galium obtusum, Helianthus angustifolius, Heliotropium polyphyllum, Hibiscus moscheutos, Hydrocotyle umbellata, H. verticillata, Hymenocallis palmeri, Hypericum crux-andreae, Hyptis alata, Ilex cassine, I, glabra, Imperata brasiliensis, Ipomoea sagittata, Iva microcephala, Juncus effusus, J. marginatus, J. megacephalus, Justicia angusta, Kosteletzkya pentacarpos, Lachnanthes caroliniana, Liatris spicata, Linum medium, Lobelia glandulosa, Ludwigia alata, L. curtissii, L. ebracteata, L. linearis, L. linifolia, L. octovalvis, L. pilosa, L. repens, Lythrum lineare, Macroptilium lathyroides, Mecardonia acuminata, Melanthera angustifolia, M. nivea, Metopium toxiferum, Mikania scandens, Mitreola petiolata, Muhlenbergia filipes, Myrica cerifera, Myrsine floridana, Nymphoides aquatica, Oenothera curtiflora, O. simulans, Oxypolis filiformis, Panicum caerulescens, P. dichotomiflorum, P. hemitomon, P. polycaulon, P. tenerum, Paspalum monostachyum, P. urvillei, Passiflora suberosa, Persea borbonia, Persicaria hydropiperoides, Phyla nodiflora, Phyllanthus amarus, P. caroliniensis, Pinus elliottii, Piriqueta caroliniana, Pityopsis graminifolia, Pluchea baccharis, P. odorata, Polygala balduinii, P. grandiflora, Polypremum procumbens, Proserpinaca intermedia, P. palustris, P. pectinata, Ptilimnium capillaceum, Randia aculeata, Rhynchosia minima, Rhynchospora colorata, R. divergens, R. inundata, R. microcarpa, R. perplexa, R. thornei, Rudbeckia triloba, Ruellia caroliniana, Sabatia grandiflora, S. stellaris, Saccharum giganteum, Sagittaria graminifolia, S. lancifolia, Salix caroliniana, Samolus ebracteata, Sarcostemma clausum, Sarracenia alata, Schinus terebinthifolius, Schizachyrium maritimum, S. rhizomatum, Scoparia dulcis, Serenoa repens, Setaria geniculata, Sida elliottii, S. rhombifolia, Silphium asteriscus, Sisyrinchium miamiense, Smilax laurifolia, Solidago leavenworthii, S. sempervirens, S. stricta, Spartina bakeri, Spermacoce assurgens, S. prostrata, S. tetraquetra, Spilanthes americana, Sporobolus domingensis, Stenandrium dulce, Stillingia sylvatica, Symphyotrichum dumosum, S. tenuifoliun, Teucrium canadense, Thelypteris kunthii, Tillandsia fasciculata, T. recurvata, T. setacea, Toxicodendron radicans, Trismeria trifoliata, Utricularia biflora, U. inflata, Verbena bonariensis, V. scabra, Vernonia blodgettii, Vicia acutifolia, Vigna luteola, Xyris laxifolia, Zeuxine strateumatica.

***Ludwigia octovalvis* (Jacq.) P.H. Raven** is a herbaceous annual or perennial subshrub with stems that are woody at the base. The plants inhabit a broad variety of conditions including fresh to brackish sites in borrow pits, bottomlands, canals, depressions, ditches, dredge spoil islands, flatwoods, levees, marshes, prairies, rice fields, roadsides, ruts, savannahs, scrub, shores, stream terraces, swales, swamps, and along the margins of lakes, ponds, rivers, and streams at low elevations of up to 158 m (to 1500 m in Mexico). Exposures range from full sun to light shade. Water conditions can

vary, with the plants often occupying slightly elevated sites in wet areas, which otherwise dry out during the growing season. The habitats typically are alkaline (mean pH = 7.2; pH range = 5.6–8.8; mean alkalinity = 31.2 mg/L as $CaCO_3$) and eutrophic (mean total P = 56 µg/L; mean total N = 980 µg/L). The substrates are of mineral derivation and include alluvium, clay, marl, mud, sand, sandy loam, and sandy peat. Flowering and fruiting have been observed from April to October. The flowers are self-compatible. The seeds are produced in quantity, germinate readily under moist conditions, and are known to be a contaminant of some potting soil mixes. Germination rates are higher (88.6%) for 6-month-old seeds than for recently shed seeds (74.2%). The older seeds germinate more quickly (3.4–4.4 days) than recently shed seeds (3.6–6.8 days). The plants will readily invade disturbed sites, especially along shorelines. They are tolerant to burning, grazing, and infrequent mowing, but do succumb to prolonged flooding conditions. The annual or perennial habit is determined by the severity of frost encountered where the plants occur. **Reported associates:** *Acacia pinetorum, Acrostichum danaeifolium, Alternanthera philoxeroides, Ambrosia artemisiifolia, Ammania latifolia, Ampelaster carolinianus, Ampelopsis arborea, Andropogon glomeratus, Annona glabra, Ardisia solanacea, Baccharis glomeruliflora, B. halimifolia, Bacopa monnieri, Bidens pilosus, Boehmeria cylindrica, Bumelia celastrina, B. salicifolia, Caperonia casteneifolia, Centella asiatica, Chamaesyce hyssopifolia, Chrysobalanus icaco, Cirsium horridulum, Cladium jamaicense, Commelina diffusa, Crinum americanum, Ctenium, Cynoctonum mitreola, Cynodon, Cyperus globulosus, C. ligularis, C. surinamensis, Desmanthus, Diodia virginiana, Echinochloa walteri, Eichhornia crassipes, Eleocharis geniculata, E. montevidensis, Erichtites hieracifolia, Erigeron quercifolius, Eupatorium capillifolium, E. coelestinum, E. mikaniodes, Eustachys glauca, Fimbristylis vahlii, Fuirena scirpoidea, F. simplex, Galium obtusum, Guettarda elliptica, Hydrocotyle umbellata, H. verticillata, Hypericum hypericoides, Hyptis alata, Ilex cassine, Imperata brasiliensis, Ipomoea sagittata, Jacquemontia tamnifolia, Juncus megacephalus, Justicia americana, J. angusta, Kosteletzkya pentacarpos, Liquidambar styraciflua, Lobelia cardinalis, L. glandulosa, Ludwigia alata, L. curtissii, L. erecta, L. leptocarpa, L. linearis, L. microcarpa, L. palustris, L. peploides, L. repens, Lythrum lineare, Macroptilium lathyroides, Magnolia virginiana, Melanthera angustifolia, Melothria pendula, Metopium toxiferum, Mikania scandens, Morinda royoc, Muhlenbergia filipes, Myrica cerifera, Myrsine floridana, Nymphaea odorata, Panicum caerulescens, P. dichotomiflorum, P. hemitomon, P. polycaulon, P. repens, Paspalum urveilli, Passiflora suberosa, Persea borbonia, Persicaria hydropiperoides, P. punctata, Phyla nodiflora, Phyllanthus amarus, P. caroliniensis, Pinus elliottii, Pistia stratiotes, Pluchea odorata, P. rosea, Pontederia cordata, Potamogeton illinoensis, Proserpinaca palustris, Psidium longipes, Psychotria sulzneri, Ptilimnium, Rhynchosia minima, Rhynchospora colorata, R. corniculata, R. inundata, R. microcarpa, R. nivea, Sabatia grandiflora, Saccharum giganteum, Sagittaria lancifolia, Salvinia* *rotundifolia, Salix caroliniana, S. nigra, Sarcostemma clausum, Schinus terebinthifolius, Scoparia dulcis, Sesbania exaltata, Setaria geniculata, Sida acuta, Sisyrinchium miamiense, Solidago leavenworthii, S. sempervirens, Spermacoce assurgens, S. prostrata, S. tetraquetra, Spilanthes americana, Sporobolus domingensis, Symphyotrichum dumosum, S. tenuifolium, Tetrazygia bicolor, Teucrium canadense, Thelypteris kunthii, Tillandsia fasciculata, Toxicodendron radicans, Triadica sebifera, Trismeria trifoliata, Utricularia biflora, U. purpurea, Verbena bonariensis, V. scabra, Vicia acutifolia, Vigna luteola, Vitis aestivalis, V. munsoniana, Zeuxine strateumatica.*

***Ludwigia palustris* (L.) Elliott** is a perennial, which is found in brackish or freshwater, tidal or nontidal habitats including beaches, bogs, borrow pits, canals, channels, depressions, ditches, draws, flats, flatwoods, floodplains, gravel bars, marshes, meadows, mudflats, oxbows, prairies, rice fields, river bottoms, roadsides, ruts, sandbars, seeps, sloughs, swamps, and the margins of flowages, lakes, ponds, rivers, and streams at elevations of up to 2042 m. The plants are quite versatile and will grow in full sun to partial shade, either fully submerged in shallow water (to ~1 m) or creeping along exposed, diverse substrates (pH: 5.0–8.6) that include alluvium, clay loam, cobble, floating logs, gravel, Iredell soil, loamy sand, muck, mud, peat, sand, sandy clay, sandy loam, silt, silty loam, and Tuscan mudflow. The dense, prostrate shoots often form mats both in exposed sites and in shallow waters. In deeper sites, the shoots ascend or float in the water. Flowering and fruiting extend from May to October. The apetalous, nectariferous flowers are self-compatible, but are outcrossed if pollinators visit them when they have newly opened. Selfing eventually occurs in unpollinated flowers. The seeds are dispersed by water and by birds (Aves), either while remaining within detached, ripened capsules, or by scattering from attached dehiscing capsules. The seeds can germinate in situ through the thin pericarps. Germination has been achieved within 6 weeks using freshly collected seeds (stored at 4°C), which have been placed under a 12-h photoperiod and exposed to a 20°C/5°C temperature regime (night temperatures increased to 10°C after 4 weeks). Extensive seed banks have been documented in beaver ponds and other sites. The plants reproduce vegetatively by the fragmentation of their creeping shoots, which root at the nodes. Although perennial, these plants can adopt an annual-like habit by producing abundant seeds rapidly when growing in severe (northern) climates. **Reported associates:** *Acer rubrum, A. saccharinum, Acorus americanus, Agalinis tenuifolia, Agrostis gigantea, Alisma subcordatum, A. triviale, Alnus rugosa, A. serrulata, Alternanthera philoxeroides, Amaranthus cannabinus, A. tuberculatus, Ammannia coccinea, Anagallis minima, Anemone canadensis, Anemopsis californica, Angelica atropurpurea, Apios americana, Arisaema dracontium, A. triphyllum, Asclepias incarnata, Athyrium filix-femina, Bacopa monnieri, Betula nigra, Bidens cernuus, B. eatonii, B. frondosus, Boehmeria cylindrica, Bolboschoenus fluviatilis, Boltonia asteroides, Briza minor, Bromus, Bulbostylis capillaris, Butomus umbellatus,*

Cabomba caroliniana, Cacalia, Calamagrostis canadensis, Callicarpa americana, Callitriche hermaphroditica, Caltha palustris, Calystegia sepium, Campanula aparinoides, Cardamine pensylvanica, Carex aperta, C. aquatilis, C. atlantica, C. bebbii, C. bromoides, C. comosa, C. cristatella, C. emoryi, C. gynandra, C. haydenii, C. hysterica, C. intumescens, C. joorii, C. lacustris, C. laeviculmis, C. leptalea, C. longii, C. lupuliformis, C. lupulina, C. lurida, C. muskingumensis, C. nebrascensis, C. obnupta, C. pellita, C. stipata, C. stricta, C. tenera, C. vesicaria, Carya ovata, Centella asiatica, Cephalanthus occidentalis, Ceratophyllum demersum, Chasmanthium laxum, Chelone glabra, Chenopodium glaucum, Cicuta bulbifera, C. maculata, Cinna arundinacea, Cirsium altissimum, Cladium jamaicense, Colocasia esculenta, Commelina virginica, Conobea multifida, Conyza canadensis, Cornus alternifolia, C. amomum, C. sericea, Crinum, Cynodon dactylon, Cyperus acuminatus, C. aristatus, C. bipartitus, C. esculentus, C. ferruginescens, C. inflexus, Dasiphora floribunda, Daucus carota, Deschampsia cespitosa, Dichanthelium acuminatum, D. clandestinum, Didiplis diandra, Digitaria sanguinalis, Diodia virginiana, Diospyros, Doellingeria umbellata, Drosera rotundifolia, Echinochloa crus-galli, E. muricata, E. walteri, Eclipta prostrata, Egeria densa, Eleocharis acicularis, E. engelmannii, E. erythropoda, E. geniculata, E. intermedia, E. montevidensis, E. obtusa, E. ovata, E. pachycarpa, E. palustris, E. smallii, E. tortilis, Elodea canadensis, Epilobium coloratum, E. leptophyllum, E. minutum, Equisetum fluviatile, Eragrostis hypnoides, E. pectinacea, E. pilosa, Eriocaulon parkeri, Eryngium prostratum, Erysimum cheiranthoides, Eupatorium perfoliatum, Euphorbia nutans, Eutrochium maculatum, Fimbristylis autumnalis, F. perpusilla, Forestiera acuminata, Fraxinus americana, F. pennsylvanica, Fuirena simplex, Galium obtusum, G. tinctorium, G. triflorum, Gleditsia aquatica, Glyceria borealis, G. striata, Gratiola aurea, G. neglecta, G. virginiana, Helenium autumnale, H. flexuosum, Helianthus grosseserratus, Heliomeris hispida, Hemicarpha micrantha, Hibiscus moscheutos, Hordeum pusillum, Hydrilla verticillata, Hydrocotyle umbellata, H. verticillata, Hypericum anagalloides, H. majus, H. mutilum, H. sphaerocarpum, Hyptis alata, Ilex coriacea, Impatiens capensis, Iris pseudacorus, I. virginica, Isoetes echinospora, I. riparia, Juglans nigra, Juncus acuminatus, J. alpinus, J. articulatus, J. brevicaudatus, J. bufonius, J. canadensis, J. effusus, J. gymnocarpus, J. marginatus, J. nodosus, J. repens, J. tenuis, J. torreyi, Justicia americana, J. ovata, Kalmia latifolia, Krigia, Kyllinga pumila, Laportea canadensis, Larix laricina, Lathyrus palustris, Leersia lenticularis, L. oryzoides, L. virginica, Lemna minor, Leucospora multifida, Leucothoe axillaris, Lilaeopsis, Lilium michiganense, Limosella acaulis, Lindera benzoin, Lindernia dubia, Lipocarpha micrantha, Lobelia cardinalis, Lotus corniculatus, Ludwigia alternifolia, L. brevipes, L. decurrens, L. grandiflora, L. octovalvis, L. polycarpa, Lycopus americanus, L. asper, L. rubellus, L. virginicus, Lyonia ligustrina, L. lucida, Lysimachia ciliata, L. quadriflora, L. vulgaris, Lythrum alatum, L. salicaria, Macbridea caroliniana, *Magnolia virginiana, Maianthemum stellatum, Marsilea, Mazus pumilus, Melilotus albus, Mentha arvensis, M. pulegium, Micranthemum umbrosum, Mikania scandens, Mimulus glabratus, M. ringens, Mitchella repens, Modiola caroliniana, Muhlenbergia glomerata, Myosotis laxa, Myrica cerifera, M. gale, M. heterophylla, Myriophyllum aquaticum, M. hippuroides, Nuphar polysepala, Nymphaea odorata, Nyssa aquatica, N. biflora, N. sylvatica, Oenothera pilosella, Onoclea sensibilis, Ophioglossum pusillum, Osmunda regalis, Osmundastrum cinnamomeum, Oxypolis rigidior, Panicum capillare, P. dichotomiflorum, P. hemitomon, P. virgatum, Parnassia glauca, Paspalum plicatulum, P. praecox, P. urvillei, Peltandra virginica, Penthorum sedoides, Persea, Persicaria amphibia, P. arifolia, P. careyi, P. hydropiper, P. hydropiperoides, P. maculosa, P. pensylvanica, P. punctata, P. sagittata, Phalaris arundinacea, Phragmites australis, Phyla lanceolata, Phystostegia virginiana, Pilea pumila, Planera aquatica, Pluchea odorata, Poa palustris, P. pratensis, Pontederia cordata, Populus angustifolia, P. fremontii, Potamogeton alpinus, P. illinoensis, P. pusillus, Potentilla anserina, Proserpinaca palustris, Prunella vulgaris, Prunus virginiana, Ptilimnium capillaceum, Pycnanthemum virginianum, Pyrrhopappus, Quercus alba, Q. bicolor, Q. lyrata, Q. macrocarpa, Q. rubra, Ranunculus flabellaris, R. gmelinii, R. pensylvanicus, R. recurvatus, R. sceleratus, R. trichophyllus, Rhamnus cathartica, Rhexia virginica, Rhynchospora colorata, R. miliacea, R. nivea, Ribes americanum, Rorippa curvisiliqua, Rosa multiflora, R. setigera, Rotala ramsior, Rubus, Rudbeckia laciniata, Rumex verticillatus, Sabal minor, Sabatia kennedyana, Sagittaria cuneata, S. fasciculata, S. lancifolia, S. latifolia, Salix candida, S. discolor, S. exigua, S. gooddingii, S. interior, S. petiolaris, S. serissima, Sambucus nigra, Samolus valerandi, Saururus cernuus, Schoenoplectus acutus, S. hallii, S. pungens, S. smithii, S. tabernaemontani, Scirpus atrovirens, S. cyperinus, S. georgianus, S. lineatus, S. pendulus, Scutellaria lateriflora, Sida spinosa, Silene nivea, Sisyrinchium, Sium suave, Smilax bona-nox, S. ecirrhata, S. laurifolia, Solanum dulcamara, Solidago altissima, S. canadensis, S. flexicaulis, S. gigantea, S. patula, S. rugosa, S. uliginosa, S. ulmifolia, Sparganium americanum, S. angustifolium, S. eurycarpum, Spartina pectinata, Spergularia bocconi, Sphagnum, Spiraea alba, S. douglasii, S. tomentosa, Spiranthes cernua, Stachys pilosa, Stuckenia vaginata, Symphyotrichum lanceolatum, S. lateriflorum, S. ontarionis, S. pilosum, S. puniceum, S. spathulatum, Tamarix chinensis, Taxodium distichum, Thalictrum dasycarpum, T. dioicum, T. pubescens, Thelypteris noveboracensis, T. palustris, Tilia americana, Torreyochloa pallida, Toxicodendron pubescens, T. radicans, T. vernix, Triadenum fraseri, T. tubulosum, T. walteri, Triadica sebifera, Trifolium wormskioldii, Tripsacum dactyloides, Trollius laxus, Typha angustifolia, T. domingensis, T. latifolia, Ulmus americana, Vaccinium fuscatum, Verbena hastata, Vernonia noveboracensis, Veronica anagallis-aquatica, V. catenata, V. scutellata, Veronicastrum virginicum, Viburnum dentatum, V. lentago, V. nudum, Viola cucullata, V. lanceolata, Vitis riparia, Vulpia octoflora, Woodwardia areolata, Xanthium strumarium, Xyris iridifolia, X. jupicai, Zannichellia palustris, Zizania aquatica.*

***Ludwigia peploides* (Kunth) P. H. Raven** grows either as an erect or sprawling emergent, or with its stems and leaves floating on the surface (extending to 7 m) in still or slow moving, shallow waters (e.g., 30–50 cm), in or on alluvial fans, channels, depressions, ditches, flats, floodplains, levees, marshes, pools, river bottoms, sandbars, scrub, seeps, sloughs, swales, swamps, thickets, vernal pools, and along the margins of canals, lakes, ponds, rivers, and streams at elevations of up to 2103 m. Exposures typically are characterized by full sun, but some occurrences are in the shade. The plants occur in fresh or somewhat brackish waters (tidal or nontidal) on substrates described as alluvium, Capay–Clear Lake association, clay, clay loam, gravel, gravelly loam, loamy clay, muck, mud, muddy loam, Pescadero clay, sand, sandy alluvium, sandy gravelly loam, sandy mud, silt, silty loam, silty mud, and vertisol. Reproduction can occur by seed production, creeping stems, or by vegetative stem fragmentation; however, sexual reproduction is diminished when the plants are growing in the floating phase. Flowering and fruiting occur from April to October. The flowers are self-compatible, but are insect-pollinated and mainly outcrossed. The seeds are not shed individually but are imbedded in the capsule wall and are dispersed along with the fruit. The roots are dimorphic anatomically (the aerial ones floating and spongy), and maximize gas exchange under aquatic conditions. These plants are known to spread rapidly along canals. Dense surface mats have been associated with high mosquito (Insecta: Diptera; Culicidae) populations (and the incidence of West Nile virus) in some areas because they deny insectivorous fish access to the larvae. **Reported associates:** *Acacia, Acer saccharinum, Alisma lanceolatum, A. subcordatum, A. triviale, Alternanthera philoxeroides, Amaranthus rudis, Ambrosia monogyra, A. psilostachya, Ammannia coccinea, A. robusta, Amorpha fruticosa, Artemisia californica, A. dracunculus, Arundo donax, Asclepias incarnata, Atriplex canescens, A. lentiformis, Azolla, Baccharis pilularis, B. salicifolia, B. sarothroides, Bacopa monnieri, Bebbia juncea, Bergia texana, Berula erecta, Bidens frondosus, B. laevis, Boehmeria cylindrica, Boerhavia coccinea, Bolboschoenus maritimus, Brickellia floribunda, Bromus rubens, Calibrachoa parviflora, Callitriche heterophylla, Carex frankii, C. lupulina, C. senta, Carya, Centaurea solstitialis, Cephalanthus occidentalis, Chaetopappa asteroides, Chilopsis linearis, Claytonia, Conium maculatum, Conyza canadensis, Cotula coronopifolia, Cressa truxillensis, Cynodon dactylon, Cyperus difformis, C. eragrostis, C. erythrorhizos, C. haspan, C. involucratus, C. odoratus, C. pygmaeus, C. squarrosus, C. strigosus, C. virens, Echinochloa colona, Echinodorus berteroi, E. cordifolius, Eclipta prostrata, Egeria densa, Eichhornia crassipes, Eleocharis acicularis, E. geniculata, E. obtusa, E. ovata, Encelia farinosa, Equisetum, Ericameria nauseosa, Eriogonum fasciculatum, Eryngium aristulatum, E. hookeri, Eucalyptus, Euthamia occidentalis, Fimbristylis, Fraxinus velutina, Gnaphalium palustre, Gratiola, Habenaria repens, Helianthus, Helminthotheca echioides, Heteranthera* *dubia, Hibiscus laevis, Hordeum, Hydrocotyle umbellata, Hydrolea quadrivalvis, Iris virginica, Juncus effusus, J. torreyi, Lactuca serriola, Larrea tridentata, Lasthenia glabrata, Leersia oryzoides, Lemna minor, Lepidospartum squamatum, Leptochloa fusca, L. viscida, Lindernia dubia, Liquidambar styraciflua, Ludwigia decurrens, L. hexapetala, L. leptocarpa, L. octovalvis, L. palustris, L. repens, Lycopus, Lythrum californicum, L. hyssopifolia, L. portula, Malacothamnus fasciculatus, Malosma laurina, Marsilea vestita, Melia azedarach, Melilotus albus, M. officinalis, Mentha arvensis, M. piperita, M. pulegium, Mikania scandens, Mimosa strigillosa, Mimulus guttatus, M. ringens, Morus alba, Myrica cerifera, Myriophyllum aquaticum, M. spicatum, Nasturtium officinale, Nelumbo lutea, Nicotiana glauca, Oncosiphon piluliferum, Oryza sativa, Oxycaryum cubense, Panicum hemitomon, Paspalum dilatatum, P. distichum, P. repens, Peltandra virginica, Persicaria amphibia, P. coccinea, P. densiflora, P. hydropiperoides, P. lapathifolia, P. maculosa, P. punctata, Phalaris arundinacea, Phyla lanceolata, Pilularia americana, Pinus taeda, Plagiobothrys stipitatus, Plantago lanceolata, P. major, Platanus racemosa, P. wrightii, Pluchea odorata, P. sericea, Pogogyne douglasii, Polygonum argyrocoleon, Polypogon monspeliensis, Populus fremontii, Potamogeton diversifolius, P. natans, P. nodosus, Proserpinaca, Prosopis velutina, Quercus agrifolia, Rhus integrifolia, Rhynchospora colorata, R. corniculata, Ribes aureum, Ricinus communis, Rorippa palustris, Rumex dentatus, Ruppia maritima, Saccharum giganteum, Sacciolepis striata, Sagittaria lancifolia, S. latifolia, S. platyphylla, S. sanfordii, Salix amygdaloides, S. exigua, S. gooddingii, S. laevigata, S. lasiolepis, S. lucida, Salvinia minima, Saururus cernuus, Schoenoplectus acutus, S. americanus, S. californicus, S. pungens, S. smithii, S. tabernaemontani, Scirpus cyperinus, Searsia lancea, Sesbania, Sidalcea hartwegii, S. hirsuta, Sphaeralcea, Symphyotrichum subulatum, Tamarix chinensis, T. ramosissima, Taxodium distichum, Toxicodendron diversilobum, Triadenum walteri, Trichocoronis wrightii, Trifolium, Typha angustifolia, T. domingensis, T. latifolia, Veronica anagallis-aquatica, V. catenata, Vitis riparia, Vulpia myuros, Washingtonia filifera, Xanthium strumarium, Zizaniopsis miliacea.*

***Ludwigia peruviana* (L.) H. Hara** is a nonindigenous, perennial subshrub, which grows in the shallow waters of canals, depressions, ditches, meadows, scrub, sloughs, swales, swamps, and along the margins of ponds, rivers, and streams at low elevations of at least 49 m. The habitats comprise freshwater to brackish, tidal or nontidal sites, often with sandy substrates, and also on mud. The herbaceous foliage dies back to the woody stem bases each season, but when grown in extremely cold conditions, the plant may die back completely and behave as an annual. Flowering and fruiting occur year round (observed from January to December). The flowers can self-pollinate, but primarily are outcrossed when they first open. The seeds are released from the capsules at maturity, or can remain viable within old capsules throughout the winter. Individual plants can produce up to 450,000 seeds/m². Most of the seeds germinate readily, but

a portion (~20%) may remain dormant for over 10 years and can form seed banks of up to 65,000 seeds/m² in the soil. Seed germination occurs on floating vegetation and on exposed mudflats. The seeds are dispersed by water and by waterfowl (Aves: Anatidae). Dense stands of plants can shade out smaller plants. *Ludwigia peruviana* has been found to sequester nutrients more effectively than most of the common wetland species studied. **Reported associates (North America):** *Acer negundo, A. rubrum, Acrostichum daniaefolium, Amaranthus australis, Ampelopsis arborea, Andropogon virginicus, Annona glabra, Ardisia crenata, Bacopa caroliniana, Blechnum serrulatum, Boehmeria cylindrica, Celtis laevigata, Cephalanthus occidentalis, Chrysobalanus icaco, Cladium jamaicense, Colocasia esculenta, Conoclinium coelestinum, Crinum americanum, Cynanchum, Cyperus haspan, Dichanthelium dichotomum, Diodia virginiana, Distichlis spicata, Echinochloa crus-galli, Eleocharis cellulosa, E. elongata, Erechtites hieraciifolia, Eriocaulon, Eupatorium capillifolium, Ficus aurea, Gnaphalium obtusifolium, Hydrocotyle umbellata, Hymenocallis, Juncus, Justicia ovata, Kosteletzkya pentacarpos, Lachnanthes caroliana, Leersia hexandra, Liquidambar styraciflua, Ludwigia octovalvis, L. repens, Melaleuca quinquenervia, Metopium, Mikania scandens, Murdannia nudiflora, Myrica cerifera, Nephrolepis, Nymphaea odorata, Nymphoides aquatica, Osmunda regalis, Panicum hemitomon, P. repens, Paspalidium geminatum, Peltandra virginica, Pennisetum purpureum, Persicaria hirsuta, P. hydropiperoides, P. setacea, Phalaris angusta, Phragmites australis, Pluchea rosea, Pontederia cordata, Proserpinaca palustris, Ranunculus, Rhynchospora inundata, R. microcarpa, R. tracyi, Rubus argutus, Rumex crispus, R. pulcher, Sagittaria lancifolia, Salix caroliniana, Sarcostemma clausum, Saururus cernuus, Schoenoplectus tabernaemontani, Scleria triglomerata, Smilax, Solidago levenworthii, Spartina bakeri, Taxodium distichum, Thelypteris, Tillandisa fasciculata, T. setacea, T. usneoides, Toxicodendron radicans, Typha domingensis, Veronica peregrina, Vitis aestivalis, Woodwardia areolata, W. virginica, Xyris.*

***Ludwigia pilosa* Walter** is a perennial, which inhabits barrens, bogs, borrow pits, canals, depressions, ditches, flats, flatwoods, marshes, meadows, pools, roadsides, savannahs, sinks, stream bottoms, swamps, woodlands, and the margins of lakes, pocosins, ponds, rivers, and streams at low elevations of up to 290 m. The plants occur in full sunlight on acidic (pH: 4.8–7.7; mean = 6.8) substrates described as Bayboro soils, Bayou sandy loam, clay loam, muck, mud, peaty sand, rubble, sand, and sandy loam. Flowering and fruiting have been observed from June to October. The self-compatible flowers are apetalous but produce nectar and attract numerous insects such as ants (Hymenoptera: Formicidae), bumblebees and honeybees (Hymenoptera: Apidae), moths (Lepidoptera), and wasps (Hymenoptera: Vespidae), which probably serve to outcross them. They are facultatively autogamous. Although the anthers reflex toward the stigmas during anthesis, herkogamy prevents self-pollination unless facilitated by insect vectors. The pollen is shed as tetrads. Vegetative reproduction occurs by layering of the stems, which arch and root at the tips during the autumn. **Reported associates:** *Acer rubrum, Amsonia tabernaemontana, Andropogon, Aristida palustris, Arnoglossum ovatum, Asclepias lanceolata, Bacopa caroliniana, Bidens coronatus, B. frondosus, Cacalia, Carex decomposita, C. glaucescens, C. verrucosa, Coelorachis rugosa, Coreopsis nudata, Croton capitatus, Cyrilla racemiflora, Dichanthelium erectifolium, D. scoparium, D. wrightianum, Diodella teres, Dulichium arundinaceum, Eleocharis equisetoides, E. quadrangulata, E. tuberculosa, Eriocaulon compressum, E. decangulare, Diodia virginiana, Gratiola brevifolia, Hypericum brachyphyllum, H. mutilum, H. myrtifolium, Hyptis alata, Ilex myrtifolia, Juncus canadensis, J. polycephalus, Lobelia floridana, Ludwigia decurrens, L. glandulosa, L. leptocarpa, L. linearis, L. linifolia, L. microcarpa, L. pilosa, L. sphaerocarpa, Magnolia virginiana, Mitreola sessilifolia, Myrica cerifera, Nyssa biflora, Oxypolis canbyi, O. filiformis, Panicum hemitomon, P. rigidulum, P. tenerum, P. verrucosum, P. virgatum, Paspalum, Pinguicula planifolia, Pinus echinata, P. elliottii, P. palustris, P. taeda, Polygala cymosa, Proserpinaca pectinata, Quercus, Rhexia nashii, R. virginica, Rhynchospora careyana, R. cephalantha, R. corniculata, R. filifolia, R. glomerata, R. macrostachya, R. mixta, R. perplexa, R. scirpoides, Rubus, Sabatia bartramii, Saccharum giganteum, Sacciolepis striata, Sagittaria latifolia, Sarracenia alata, Scirpus cyperinus, Scleria baldwinii, Sida, Smilax, Solidago, Stylisma aquatica, Taxodium ascendens, T. distichum, Woodwardia virginica, Xyris fimbriata, X. laxifolia.*

***Ludwigia polycarpa* Short & Peter** is a perennial, which grows either submerged in shallow water, or as an emergent along drying or exposed shores of creekbeds, ditches, floodplains, lagoons, lakes, marshes, meadows, oxbows, ponds, prairies, reservoirs, rivers, sloughs, and swamps at elevations of up to 229 m. The substrates are described as dolomite, muck, peat, and sand. Flowering and fruiting occur from July to October. The flowers lack petals and reportedly are self-pollinated, and are also probably outcrossed given that short-tongued bees (Hymenoptera: Apidae) and other insects visit the flowers to collect their nectar and pollen. The pollen is shed as tetrads. The plants reproduce vegetatively by producing leafy stolons (to 15 cm) from their base during the autumn. The seeds are dispersed from the capsules and will form seed banks. **Reported associates:** *Abutilon theophrasti, Acalypha gracilens, A. rhomboidea, Acer rubrum, A. saccharinum, Agalinis tenuifolia, Alisma gramineum, A. subcordatum, A. triviale, Allium cernuum, Alnus rugosa, Ambrosia artemisiifolia, Ammannia coccinea, Anthoxanthum nitens, Apocynum cannabinum, A. sibiricum, Asclepias incarnata, A. syriaca, A. verticillata, Azolla mexicana, Betula nigra, Bidens aristosus, B. cernuus, B. coronatus, B. discoideus, B. frondosus, B. polylepis, Boehmeria cylindrica, Boltonia asteroides, Butomus umbellatus, Calamagrostis canadensis, Campanula aparinoides, Carex buxbaumii, C. cristatella, C. festucacea, C. grayi, C. haydenii, C. lacustris, C. lupuliformis, C. lupulina, C. muskingumensis, C. pellita, C. tribuloides, C. vulpinoidea, Celastrus, Cephalanthus occidentalis,* *Cicuta bulbifera, C. maculata, Cinna arundinacea, Cirsium arvense, Clinopodium glabrum, Coleataenia longifolia, Coreopsis tripteris, Cornus amomum, C. racemosa, Cuscuta gronovii, Cyperus erythrorhizos, C. esculentus, C. strigosus, Danthonia, Desmanthus illinoensis, Dulichium arundinaceum, Echinochloa crus-galli, E. muricata, Eclipta prostrata, Eleocharis compressa, E. engelmannii, E. intermedia, E. obtusa, E. palustris, Elymus riparius, Epilobium coloratum, E. glandulosum, E. leptophyllum, Eragrostis hypnoides, Erechtites hieracifolia, Eupatorium perfoliatum, E. serotinum, Euthamia gymnospermoides, Eutrochium maculatum, Fallopia japonica, Fimbristylis autumnalis, F. vahlii, Fraxinus nigra, F. pennsylvanica, Galium obtusum, G. tinctorium, Glyceria grandis, Gratiola neglecta, Heteranthera dubia, Hibiscus moscheutos, Hypericum canadense, H. mutilum, H. sphaerocarpum, Impatiens pallida, Iris virginica, Juncus effusus, J. marginatus, J. nodosus, J. torreyi, Lactuca canadensis, Larix laricina, Lathyrus palustris, Leersia oryzoides, L. virginica, Lemna minor, Lindernia dubia, Lipocarpha micrantha, Ludwigia alternifolia, L. palustris, Lycopus americanus, L. asper, L. uniflorus, Lysimachia quadriflora, L. terrestris, L. thyrsiflora, Lythrum alatum, L. salicaria, Malvastrum hispidum, Matteuccia struthiopteris, Mentha arvensis, Mimulus ringens, Muhlenbergia uniflora, Oenothera pilosella, Onoclea sensibilis, Oxypolis rigidior, Panicum dichotomiflorum, Pedicularis lanceolata, Penthorum sedoides, Persicaria amphibia, P. coccinea, P. hydropiper, P. hydropiperoides, P. lapathifolia, P. pensylvanica, P. punctata, P. sagittata, Phalaris arundinacea, Phragmites australis, Phyla lanceolata, Platanthera peramoena, Poa palustris, Pontederia cordata, Populus deltoides, Potentilla anserina, P. norvegica, Proserpinaca palustris, Quercus bicolor, Q. palustris, Ranunculus flabellaris, R. pensylvanicus, Rosa carolina, R. multiflora, Rotala ramosior, Rumex verticillatus, Sagittaria latifolia, Salix discolor, S. exigua, S. interior, S. nigra, Sambucus nigra, Schedonorus arundinaceus, Schoenoplectus acutus, S. hallii, S. tabernaemontani, Scirpus cyperinus, S. pedicellatus, Scutellaria galericulata, S. lateriflora, S. nervosa, Setaria faberi, S. lutescens, Silphium integrifolium, Sium suave, Solanum dulcamara, Solidago gigantea, S. graminifolia, S. riddellii, Sorghastrum nutans, Sparganium androcladum, S. eurycarpum, Spartina pectinata, Sphagnum, Spiraea alba, Stachys pilosa, Symphoricarpos, Symphyotrichum lanceolatum, S. ontarionis, S. racemosum, Thelypteris palustris, Toxicodendron radicans, Triadenum fraseri, T. virginicum, Trichostema brachiatum, Typha latifolia, Ulmus, Urtica dioica, Vaccinium, Verbena hastata, Vernonia fasciculata, Veronica anagallis-aquatica, Veronicastrum virginicum, Viburnum lentago, V. recognitum, Viola lanceolata, Xanthium strumarium, Xyris torta.*

***Ludwigia ravenii* C.I. Peng** is a perennial, which occurs on wet peat or sand, or along the margins of bogs, ditches, ponds, and swamps at low elevations. Flowering and fruiting occur in September. The self-compatible flowers are apetalous and reproduction is facultatively autogamous. The seeds are dispersed freely from the capsules. The plants reproduce vegetatively by producing short, slender stolons at their base during late autumn. Additional life history information is needed for this species. **Reported associates:** *Cyperus, Eriocaulon, Eupatorium, Hypericum, Liatris, Ludwigia hirtella, L. linearis, Panicum, Paspalum, Polygonum, Rhexia, Sphagnum subsecundatum, Xyris.*

Ludwigia repens **J. R. Forst.** is a perennial, which grows almost entirely submerged (with only the leafy apex of the floating stems evident on the water surface) in shallow (to 1.8 m), still or flowing, clear or "black" waters of lakes and streams, or as a prostrate plant that roots at the nodes and creeps over the substrate in canals, channels, ditches, ephemeral pools, hammocks, lagoons, mudflats, swamps, and along the margins of lakes, ponds, reservoirs, and streams at elevations of up to 1372 m. The plants can form dense mats on the surfaces of shallow lakes. They occur in softwater (mean alkalinity = 17 mg/L as $CaCO_3$), and slightly acidic habitats (pH: 4.7–8.3; mean = 6.3) in full sun to partial shade. Substrates include mud, sand, sandy clay, sandy peat, and silt. Flowering and fruiting can occur year long. The plants are self-compatible, but are outcrossed if pollinators visit the nectariferous flowers when newly opened. Selfing eventually occurs in unpollinated flowers. The seeds are dispersed from the irregularly dehiscing capsules and can represent minor components of seed banks. They probably are dispersed by ducks and other waterfowl (Aves: Anatidae). **Reported associates:** *Acer rubrum, Alternanthera philoxeroides, Ambrosia artemisiifolia, Ammania latifolia, Ampelopsis arborea, Andropogon glomeratus, A. virginicus, Annona glabra, Apios americana, Arundo donax, Baccharis halimifolia, B. salicifolia, Bacopa caroliniana, B. monnieri, Blechnum serrulatum, Boehmeria cylindrica, Brachiaria mutica, Callicarpa americana, Carex alata, C. lupulina, Celtis laevigata, Centella asiatica, Cephalanthus occidentalis, Chara, Cladium jamaicense, Cornus foemina, Cupressus, Cynoctonum mitreola, Cynodon, Cyperus ligularis, C. surinamensis, Diodia virginiana, Eleocharis baldwinii, E. cellulosa, E. geniculata, Erigeron quercifolius, Eupatorium coelastinum, E. leptophyllum, Eustachys glauca, Ficus aurea, Fraxinus caroliniana, F. velutina, Fuirena squarrosa, Galium obtusum, Gleditsia aquatica, Hydrocotyle umbellata, Hypoxis curtissii, Hyptis alata, Ilex cassine, Imperata brasiliensis, Ipomoea sagittata, Juncus effusus, J. megacephalus, J. polycephalus, Justicia ovata, Kosteletzkya pentacarpos, Leersia hexandra, Leptochloa fusca, Liquidambar styraciflua, Ludwigia alata, L. microcarpa, L. octovalvis, L. peploides, L. suffruticosa, Lythrum hyssopifolium, L. lineare, Magnolia virginiana, Melanthera angustifolia, Mikania scandens, Muhlenbergia filipes, Myrica cerifera, Myriophyllum heterophyllum, Myrsine floridana, Nephrolepis, Nuphar, Oxycaryum cubense, Panicum caerulescens, P. dichotomiflorum, P. gymnocarpon, P. hemitomon, P. polycaulon, P. repens, P. rigidulum, P. virgatum, Parthenocissus quinquefolia, Persea borbonia, P. palustris, Persicaria punctata, Phyla nodiflora, Pistia stratiotes, Pleopeltis polypodioides, Pluchea odorata, P. rosea, Pontederia cordata, Populus, Proserpinaca palustris, Prosopis, Quercus laurifolia, Q.*

nigra, Rhynchosia minima, Rhynchospora colorata, R. inundata, R. microcarpa, R. tracyi, Roystonea elata, Sabal palmetto, Sabatia grandiflora, Saccharum giganteum, Sagittaria graminea, S. lancifolia, Salix caroliniana, S, exigua, S. lasiolepis, S. nigra, Schoenoplectus acutus, S. tabernaemontani, Setaria geniculata, Sisyrinchium miamiense, Solidago sempervirens, Spartina bakeri, Spermacoce prostrata, S. tetraquetra, Spilanthes americana, Sporobolus domingensis, Symphyotrichum dumosum, S. tenuifolium, Taxodium distichum, Thelypteris kunthii, Tillandsia, Toxicodendron radicans, Trismeria trifoliata, Typha domingensis, T. latifolia, Ulmus americana, Urtica dioica, Utricularia foliosa, U. gibba, Verbena bonariensis, Vicia acutifolia, Vigna luteola, Xyris elliottii, Youngia japonica, Zeuxine strateumatica.

Ludwigia spathulata **Torr. & A. Gray** is a prostrate, creeping perennial, which occurs during periods of low water on the exposed substrates of bogs, Carolina bays, ditches, drying ponds, granite quarry pools, low savannahs, river bottoms, sinks, swales, swamp margins, and along the margins of lakes or ponds. The substrates are sandy. The plants are self-compatible, but are outcrossed if pollinators visit the nectariferous flowers when they have recently opened; selfing eventually occurs in unpollinated flowers. An extensive seed bank can develop. The sprawling stems root at the nodes and can intertwine among plants, resulting in the formation of dense, extensive mats. **Reported associates:** *Ammannia coccinea, Andropogon virginicus, Axonopus furcatus, Azolla caroliniana, Bacopa caroliniana, Bidens tripartitus, Boehmeria cylindrica, Cardamine hirsuta, Croton elliotti, Dichanthelium acuminatum, Eclipta prostrata, Eleocharis minima, Elephantopus carolinianus, Elodea canadensis, Erechtites hieraciifolia, Eupatorium capillifolium, E. leucolepis, Gamochaeta falcata, Geranium carolinianum, Gratiola brevifolia, G. ramosa, Hydrocotyle umbellata, Hypericum harperi, H. mutilum, Ilex myrtifolia, Iva microcephala, Justicia ovata, Leersia hexandra, Lindernia dubia, Ludwigia alternifolia, L. decurrens, L. glandulosa, L. leptocarpa, L. palustris, Lycopus rubellus, Micranthemum umbrosum, Mitreola petiolata, Mollugo verticillata, Murdannia keisak, Nyssa, Oldenlandia uniflora, Panicum hemitomon, Persicaria hydropiper, P. punctata, Pluchea camphorata, Polypremum procumbens, Rhexia virginica, Rhynchospora globularis, Rotala ramosior, Sagittaria, Samolus valerandi, Solidago, Sphenoclea zeylandica, Typha latifolia, Viola.*

Ludwigia sphaerocarpa **Elliott** is a perennial species, which grows in bogs, Carolina bays, depressions, ditches, dunes, marshes, meadows, roadsides, savannahs, seeps, sloughs, streams, swales, swamps, and along the margins of lakes and ponds at elevations of up to 620 m. The plants tolerate full sun to deep shade and typically occur in shallow (e.g., 15 cm) water. Flowering and fruiting occur from May to December. The substrates are described as acidic (e.g., pH: 6.7) and include mucky sand, muck, mud, peat, sand, sandy loam, sandy peat, and silt. The self-compatible flowers are apetalous but produce nectar and appear to be pollinated by wasps (Hymenoptera: Vespidae) and other insects. The pollen is shed as tetrads. If unpollinated, the flowers can become autogamous; however, the extent of autogamy is limited by herkogamy, which then requires insect vectors to facilitate self-pollination. Dispersal of fruits and seeds presumably occurs by water. Many capsules remain undehisced and attached to the maternal plants well into the subsequent spring season when high water levels prevail. The fruits possess a spongy mesocarp which facilitates their flotation. Seed production is high. The seeds are neutrally buoyant and float when they are shed; a seed bank is formed. The seeds have germinated successfully (within 6 weeks) after 3 months of storage at 7°C followed by incubation at greenhouse temperatures kept below 35°C. The stems develop extensive aerenchyma tissue in response to inundation. The plants reproduce vegetatively by the production of leafy stolons during the autumn. Vegetative reproduction facilitates early emergence and growth in habitats (e.g., fluctuating water levels) that typically are dominated by annuals. Denser stands will develop in more sheltered sites. **Reported associates:** *Acer rubrum, Achnatherum lemmonii, Alnus tenuifolia, Alternanthera philoxeroides, Ammannia coccinea, Amphicarpum muehlenbergianum, Andropogon, Aristida palustris, A. purpurascens, Artemisia tridentata, Bacopa caroliniana, B. monnieri, Bidens aristosus, B. coronatus, B. discoideus, B. frondosus, Boltonia asteroides, Bouteloua gracilis, Brasenia schreberi, Cabomba caroliniana, Calamagrostis canadensis, Carex glaucescens, C. joorii, C. striata, Centella asiatica, Cephalanthus occidentalis, Cicuta bulbifera, Cladium mariscoides, Clethra alnifolia, Coreopsis nudata, Crinum americanum, Ctenium aromaticum, Cyperus erythrorhizos, C. esculentus, C. iria, C. strigosus, Decodon verticillatus, Dichanthelium acuminatum, D. hirstii, Diodia virginiana, Diospyros virginiana, Drosera, Dulichium arundinaceum, Echinochloa crus-galli, Eclipta prostata, Eleocharis elongata, E. equisetoides, E. flavescens, E. microcarpa, E. obtusa, E. olivacea, E. tuberculosa, Eragrostis hypnoides, Eriocaulon compressum, Eriogonum sphaerocephalum, Eubotrys racemosa, Eupatorium capillifolium, Euphorbia humistrata, Euthamia tenuifolia, Fimbristylis autumnalis, F. littoralis, F. perpusilla, Fuirena scirpoidea, F. squarrosa, Galium tinctorium, Glyceria obtusa, Gratiola aurea, G. virginiana, Habenaria repens, Hibiscus moscheutos, Hottonia inflata, Hydrocotyle ranunculoides, Hypericum hypericoides, H. mutilum, Ipomoea, Juncus canadensis, J. militaris, J. pelocarpus, J. repens, Kochia americana, Lachnanthes caroliniana, Leersia virginica, Lemna, Limnobium spongia, Lindernia dubia, Liquidambar styraciflua, Lobelia spicata, Lophiola aurea, Ludwigia decurrens, L. grandiflora, L. leptocarpa, L. linearis, L. maritima, L. palustris, L. peploides, L. pilosa, Lycopus uniflorus, Lythrum salicaria, Megalodonta beckii, Nelumbo lutea, Nitella gracilis, Nymphaea mexicana, N. odorata, N. tuberosa, Nymphoides aquatica, Nyssa biflora, N. sylvatica, Oldenlandia uniflora, Opuntia polyacantha, Oxypolis filiformis, Panicum hemitomon, P. rigidulum, P. verrucosum, P. virgatum, Pascopyrum smithii, Paspalum setaceum, Persicaria amphibia, P. coccinea, P. hydropiperoides, P. setacea, Pluchea, Poa, Polygala cymosa, Pontederia cordata, Potamogeton pusillus, Proserpinaca*

palustris, P. pectinata, Ptilimnium capillaceum, Quercus laurifolia, Q. phellos, Rhexia virginica, Rhynchospora cephalantha, R. chalarocephala, R. corniculata, R. filifolia, R. latifolia, R. macrostachya, R. scirpoides, Riccia, Rumex verticillatus, Sabatia bartramii, S. kennedyana, Saccharum baldwinii, S. giganteum, Sacciolepis striata, Sagittaria lancifolia, S. latifolia, Salix nigra, Sarracenia flava, S. purpurea, Schoenoplectus acutus, Scirpus cyperinus, Scleria baldwinii, S. reticularis, Sclerolepis uniflora, Smilax rotundifolia, Solidago, Sphenoclea zeylanica, Spirodela polyrrhiza, Symphyotrichum, Taxodium ascendens, Torreyochloa pallida, Triadenum virginicum, Typha, Utricularia, Vaccinium corymbosum, Viola lanceolata, Woodwardia virginica, Xyris serotina.

***Ludwigia suffruticosa* Walter** is a perennial, which colonizes depressions, ditches, marshes, meadows, pools, roadsides, savannahs, swales, woodlands, and the margins of lakes, rivers, ponds, and sinkholes at low elevations of at least 20 m. The plants occur in wet, exposed or shaded sites or in shallow, fresh to brackish water. The substrates can be alkaline or quite acidic (e.g., pH: 4.9), fairly high in organic matter, and include alluvium, loam, and sand. Flowering has been observed from May to September. The self-compatible flowers lack petals but attract wasps (Insecta: Hymenoptera: Vespidae) and are primarily outcrossed. They are only facultatively autogamous because of herkogamy, which prevents self-pollination in the absence of mediating insect vectors. The pollen is shed as single grains. The highly clonal plants are rhizomatous (to a lesser extent) but commonly produce leafy stolons during the autumn. **Reported associates:** *Acer rubrum, Amphicarpum muhlenbergianum, Andropogon brachystachyus, A. capillipes, A. glaucopsis, A. glomeratus, A. virginicus, Asclepias pedicellata, Axonopus furcatus, Bartonia virginica, Bulbostylis ciliatifolia, B. stenophylla, Burmannia capitata, Calopogon tuberosus, Centella asiatica, Cephalanthus occidentalis, Ceratopteris pteridoides, Chrysoma pauciflosculosa, Cyperus erythrorhizos, C. filicinus, C. haspan, C. odoratus, C. oxylepis, C. polystachyos, C. surinamensis, Dichanthelium ensifolium, D. erectifolium, Drosera capillaris, Eleocharis albida, E. baldwinii, E. flavescens, E. geniculata, E. melanocarpa, E. tricostata, E. vivipara, Eriocaulon compressum, E. decangulare, Eupatorium leptophyllum, E. mohrii, Euphorbia, Euthamia graminifolia, Fimbristylis autumnalis, F. puberula, Fuirena scirpoidea, Gratiola pilosa, G. ramosa, Hartwrightia floridana, Hypericum cistifolium, H. fasciculatum, H. lissophloeus, H. mutilum, H. myrtifolium, H. reductum, Ilex cassine, I. glabra, Juncus repens, Lachnanthes caroliniana, Lachnocaulon minus, Leptochloa fusca, Lindernia dubia, Lobelia glandulosa, Ludwigia alata, L. lanceolata, L. linifolia, L. repens, Lupinus westianus, Lycopodiella alopecuroides, L. caroliniana, Lyonia ligustrina, L. lucida, Marshallia tenuifolia, Myrica heterophylla, Nymphaea odorata, Nyssa sylvatica, Oldenlandia uniflora, Osmundastrum cinnamomeum, Panicum abscissum, P. hemitomon, P. virgatum, Paspalum setaceum, Persicaria hirsuta, P. hydropiperoides, Pinguicula caerulea, Pinus palustris, Pluchea foetida, P. rosea, Pogonia ophioglossoides,*

Polygala cruciata, P. cymosa, P. lutea, Polygonella gracilis, Proserpinaca pectinata, Ptilimnium capillaceum, Quercus, Rhexia chapmanii, R. *cubensis,* R. *mariana,* R. *nuttallii,* R. *salicifolia, Rhynchospora cephalantha,* R. *ciliaris,* R. *corniculata,* R. *fascicularis,* R. *filifolia,* R. *inundata,* R. *microcarpa,* R. *nitens, Rubus argutus, Sabatia difformis,* S. *grandiflora, Sacciolepis striata, Sagittaria isoetiformis,* S. *lancifolia, Salix caroliniana, Sarracenia minor, Scleria baldwinii,* S. *georgiana,* S. *reticularis,* S. *triglomerata, Spartina bakeri, Syngonanthus flavidulus, Utricularia juncea, U. subulata, Vaccinium arboreum, Woodwardia virginica, Xyris ambigua, X. difformis, X. elliottii, X. fimbriata, X. longisepala, X. platylepis.*

Ludwigia virgata Michx. is a perennial, which grows in fresh to brackish sites in bogs, Carolina bays (dried), depressions, ditches, flatwoods, floodplains, roadsides, savannahs, shores, swamps, and along the margins of borrow pits, lakes, and ponds at low elevations of up to at least 46 m. The plants can also occur on fairly dry sites, at locations where the water has withdrawn. Exposures are in full sun. The substrates are described as marly sand or sand. Flowering and fruiting have been observed from June to September. Although self-compatible, the flowers are outcrossed. The pollen is shed in tetrads. The seeds are dispersed from a terminal pore of the mature capsules. The plants spread vegetatively by a rhizomatous rootstock, which produces tuberous roots either singly or in fascicles. **Reported associates:** *Alnus serrulata, Aneilema keisak, Aristida stricta, Boehmeria cylindrica, Cephalanthus occidentalis, Coreopsis leavenworthii, Crataegus aestivalis, Ctenium aromaticum, Dichanthelium dichotomum, Diospyros virginiana, Drosera capillaris, D. tracyi, Fuirena scirpoidea, Helenium brevifolium, Hibiscus aculeatus, Hypericum harperi, Hyptis alata, Ilex cassine, I. coriacea, I. glabra, Impatiens capensis, Juncus marginatus, Lachnanthes caroliniana, Leersia hexandra, Linum floridanum, Lophiola aurea, Lycopodiella alopecuroides, Magnolia virginiana, Myrica caroliniensis, M. cerifera, Nyssa biflora, N. sylvatica, Panicum hemitomon, Pinus elliottii, P. palustris, P. serotina, P. taeda, Piriqueta caroliniana, Pleea tenuifolia, Proserpinaca pectinata, Rhexia alifanus, Rhynchospora ciliaris, R. gracilenta, Rubus trivialis, Sabatia campanulata, Sagittaria isoetiformis, S. latifolia, Salix, Sarracenia flava, S. leucophylla, S. leucophylla × S. purpurea, S. rosea, Solidago canadensis, S. sempervirens, Spartina patens, Sporobolus pinetorum, Taxodium ascendens, Utricularia inflata, U. subulata, Xyris ambigua, X. difformis.*

Use by wildlife: The seeds of many *Ludwigia* species (including *L. decurrens, L. grandiflora, L. hexapetala, L. leptocarpa, L. octovalvis, L. palustris, L. peploides*) are eaten by a variety of birds (Aves) including ducks (Anatidae), ground birds, waterbirds, and other wildfowl. A substantial portion of the diet of migrating blue-winged teal (Aves: Anatidae: *Anas discors*) consists of *Ludwigia repens* seeds. *Ludwigia repens* is reportedly grazed by manatees (Mammalia: Trichechidae: *Trichechus manatus latirostris*). *Ludwigia octovalvis* provides nesting habitat for various songbirds (Aves: Passeriformes: Passeri) and extensive stands are preferred resting areas

for migrating black-crowned night herons (Aves: Ardeidae: *Nycticorax nycticorax*). Dragonflies (Insecta: Odonata) commonly are seen perching on its foliage. *Ludwigia hexapetala* reportedly is eaten by muskrats (Mammalia: Cricetidae: *Ondatra zibethicus*). *Ludwigia octovalvis* is known to provide a good source of fodder for cows (Mammalia: Bovidae: *Bos*) whereas, *L. bonariensis* and *L. peploides* are eaten by white-tailed deer (Mammalia: Cervidae: *Odocoileus virginianus*). The foliage of *L. peploides* provides basking cover for the giant garter snake (Reptilia: Colubridae: *Thamnophis gigas*). *Ludwigia bonariensis* is a host plant for aphids (Insecta: Hempitera: Aphididae: *Hyalomyzus jussiaeae*). The flowers of *L. polycarpa* are visited by various insects (Insecta) including bees (Hymenoptera: Halictidae: *Augochlorella aurata, Lasioglossum cressonii, L. pectoralis, L. versatus, L. zephyrus*), flies (Diptera: Muscidae: *Limnophora narona*; Sarcophagidae: *Sarcophaga sinuata*; Syrphidae: *Toxomerus marginatus, T. politus*; Tachinidae: *Acroglossa hesperidarum, Archytas analis, Belvosia unifasciata*), and wasps (Hymenoptera: Philanthidae: *Cerceris compacta*; Pompilidae: *Anoplius atrox, A. lepidus*; Sapygidae: *Sapyga interrupta*; Vespidae: *Polistes fuscata*). *Ludwigia polycarpa* is a host plant for parasitic Hymenoptera. Adult and larval leaf beetles (Insecta: Coleoptera: Chrysomelidae: *Lysathia ludoviciana*) feed extensively on the foliage of *L. grandiflora* and have been considered as a potential biological control agent for that species. Adults and larvae of the leaf beetle *Altica litigata* (Coleoptera; Chrysomelidae) feed on the leaves of *Ludwigia octovalvis* and *L. palustris*. *Ludwigia alternifolia* is a host plant of seed beetles (Coleoptera: Chrysomelidae: Bruchinae: *Acanthoscelides alboscutellatus*). Adults and larvae of the primrose leaf weevil (Coleoptera: Curculionidae: *Perigaster cretura*) feed on the foliage of *Ludwigia alternifolia, L. octovalvis, L. peploides*, and *L. repens*. Slight infestations of the pink hibiscus mealybug (Insecta: Homoptera: Pseudococcidae: *Maconellicoccus hirsutus*) have been observed on *Ludwigia octovalvis* in southern Florida. *Ludwigia* species are hosts of several hawkmoths (Insecta: Lepidoptera: Sphingidae) including *Eumorpha fasciatus* (on *Ludwigia decurrens, L. erecta, L. leptocarpa, L. octovalvis, L. peruviana*) and *Eumorpha vitis* (on *Ludwigia decurrens* and *L. erecta*). *Ludwigia palustris* is a host of a fungus (Ascomycota: Mycosphaerellaceae: *Melanops ludwigiae*). In Costa Rica, *Ludwigia erecta* is a host plant of an aphid (Insecta: Hemiptera: Aphididae: *Aphis gossypii*), which is a viral vector of melons (Cucurbitaceae: *Cucumis melo*). *Ludwigia peruviana* is known to attract various butterflies (Insecta: Lepidoptera). A moth (Lepidoptera: Crambidae: *Desmia ploralis*) has been observed feeding on the foliage of *L. peruviana* in Florida. *Ludwigia peruviana* supports Hymenoptera parasitoids (Aphelinidae) of the silverleaf whitefly (Insecta: Homoptera: Aleyrodidae: *Bemisia argentifolii*), a pest of numerous wild and cultivated plants. It is a host plant for several Homoptera including a leafhopper (Cicadellidae: *Agallia constricta*) and the sunflower spittlebug (Cercopidae: *Clastoptera xanthocephalai*), as well as a three-cornered alfalfa hopper (Hemiptera: Membracidae: *Spissistilus festinus*). The foliage of *L. suffruticosa* is eaten by herbivorous grasshoppers (Insecta: Orthoptera: Acrididae *Gymnoscirtetes pusillus, Leptysma marginicollis, Paroxya atlantica, P. clavuliger, Romalea microptera, Stenacris vitreipennis*) and weevils (Coleoptera: Curculionidae: *Tyloderma punctatum*).

Economic importance: food: *Ludwigia erecta* is eaten in Africa as a vegetable of "minor importance." The seeds of *L. pilosa* contain more than 25% oil, with lesser amounts occurring in *L. longifolia* (13.9%) and *L. peruviana* (10.1%); **medicinal:** Preparations made from *Ludwigia bonariensis* have been used to treat skin diseases. In Hawaii, infusions from the plants are taken as a blood purifier, and the seeds or root pulp are used to alleviate the discomfort of small abrasions. N-butanol extracts of *L. decurrens* exhibit effective antibacterial activity. A tea made from the boiled leaves of *L. erecta* is used in Guyana to treat weak heart rate. The boiled liquid is mixed with various flours into a porridge used to treat yeast infections (candidiasis). Several oleanane-type triterpenes: (23Z)-coumaroylhederagenin, (23E)-coumaroylhederagenin, and (3Z)-coumaroylhederagenin, which were isolated from the foliage of *L. octovalvis*, are cytotoxic against human oral epidermoid carcinoma KB and colorectal carcinoma H tumor cell lines. Studies have demonstrated that extracts of *L. octovalvis* are a highly effective antibiotic against *Streptococcus mutans* (Bacteria: Streptococcaceae). The plants are used in Africa as an antibiotic, anti-inflammatory agent, and vermifuge. *Ludwigia palustris* has been used to treat asthma, chronic coughs, and pulmonary tuberculosis. *Ludwigia peruviana* is used as a medicinal plant in Columbia (South America). High antibacterial activity has been reported for ethyl acetate and n-butanol extracts of *L. suffruticosa*. The Florida Seminoles prepared a decoction of *Ludwigia virgata* roots as a treatment to relieve itching and as a remedy for "snake sickness"; **cultivation:** *Ludwigia* is an important genus of ornamental aquatics. *Ludwigia arcuata, L. brevipes, L. glandulosa, L. grandiflora, L. hexapetala, L. longifolia, L. palustris, L. repens*, and occasionally *L. peploides*, are grown as ornamental aquarium or water garden plants; **misc. products:** *Ludwigia octovalvis* has been used to facilitate nutrient removal in constructed waste water treatment wetlands in China. *Ludwigia peploides* has been efficient at removing nutrients from sewage and storm water in mesocosm experiments. When grown in drainage ditches, *L. peploides* readily takes up pesticides, thereby reducing their concentration in agricultural runoff. *Ludwigia sphaerocarpa* has been planted in wetland restoration projects; **weeds:** *Ludwigia hexapetala* and *L. peploides* are reported as a nuisance in western North American canals and drainage ditches where heavy growth reduces water flow. *Ludwigia peruviana* is invasive wherever it has been introduced. *Ludwigia palustris* often is reported as weedy. Dense shoreline growths of *L. octovalvis* interfere with navigational and recreational lake uses in Florida. *Ludwigia grandiflora* often is perceived as weedy in the southeastern United States. *Ludwigia erecta* was introduced to Africa and is categorized as a prohibited weed species in Australia. *Ludwigia longifolia* recently has become an invasive weed of wetlands near Sydney, Australia. It is commonly

reported as weedy even within its native range (e.g., Brazil). *Ludwigia palustris* and *L. peruviana* are other invasive weeds in Australia. *Ludwigia hexapetala* is weedy in France, Spain, and in other parts of Europe; **nonindigenous species:** *Ludwigia octovalvis* and *L. palustris* have been introduced to Hawaii; *Ludwigia bonariensis, L. longifolia*, and *L. peruviana* are nonindigenous introductions to the southeastern United States. *Ludwigia bonariensis* is believed to have been introduced to some sites as a result of contaminated seed used in construction projects. In one case, viable seeds were found in bags of peat sold commercially near Atlanta, Georgia in 1976. *Ludwigia grandiflora* and/or *L. hexapetala* (formerly combined as *L. uruguayensis*) often are regarded as nonindigenous, but both species are native to North America. However, *L. hexapetala* has been introduced to Washington state and to other portions of the northern United States as an escape from cultivation. It was introduced to France in 1823 and has since become an invasive weed there. *Ludwigia peploides* is native to southeastern North America, but was introduced to California. It is invasive in Australia, France, Spain, and New Zealand. *Ludwigia peruviana* was introduced to North America and near Sydney, Australia. *Ludwigia palustris* is nonindigenous to Australia, Eurasia, New Zealand, and South Africa. *Ludwigia repens* has become naturalized in Eurasia.

Systematics: *Ludwigia* is the sister genus of the remainder of Onagraceae (Figure 4.85) and is distinguished from the rest of the family by the 5-merous flowers of some species, persistent sepals, and several other anatomorphological features. No comprehensive phylogenetic survey of the genus has yet been completed; however this objective presently is being pursued by Hoch, Raven, and colleagues (personal communication). A sound phylogeny of *Ludwigia* is necessary to assess the existing classification and to further clarify the evolution of reproductive systems in the genus. According to the current classification, *Ludwigia* is divided into 23 sections, 13 of them monotypic. Sections *Microcarpium* and *Myrtocarpus* are the largest with 14 and 20 species respectively. The 28 obligate North American species occur within the following eight sections: **Isnardia** (*Ludwigia arcuata, L. brevipes, L. ×lacustris, L. palustris, L. repens, L. spathulata*); **Ludwigia** (*L. alternifolia, L. hirtella, L. virgata*); **Macrocarpon** (*L. bonariensis, L. octovalvis*); **Microcarpium** (*L. alata, L. curtisssii, L. glandulosa, L. lanceolata, L. linearis, L. linifolia, L. microcarpa, L. pilosa, L. polycarpa, L. ravenii, L. sphaerocarpa, L. suffruticosa*); **Myrtocarpus** (*L. peruviana*); **Oligospermum** (*L. grandiflora, L. hexapetala, L. peploides*); **Pterocaulon** (*L. decurrens, L. erecta, L. longifolia*) and **Seminuda** (*L. leptocarpa*). Although most *Ludwigia* species are self-compatible, self-incompatibility has been found in seven species of sections *Macrocarpon* and *Myrtocarpus*, all being of South American distribution. The base chromosome number of *Ludwigia* is $x = 8$. Aneuploidy is absent in the genus, but polyploidy is extensive. *Ludwigia alternifolia, L. decurrens, L. erecta, L. linearis, L. linifolia, L. longifolia, L. microcarpa, L. palustris, L. peploides*, and *L. virgata* are diploid ($2n = 16$); *Ludwigia arcuata, L. glandulosa, L. lanceolata, L. leptocarpa, L. octovalvis, L. pilosa, L. polycarpa, L. ravenii, L.*

spathulata, *L. sphaerocarpa*, and *L. suffruticosa* are tetraploid (2n = 32); *Ludwigia alata*, *L. brevipes*, *L. grandiflora*, and *L. repens* are hexaploid (2n = 48); *Ludwigia hexapetala* is decaploid (2n = 80). *Ludwigia peruviana* is highly polyploid and variable cytologically (2n = 64, 80, 96, 128). Triploid cytotypes (2n = 24) are reported for *L. leptocarpa* and *L. octovalvis*, but not for North American populations. The taxon formerly known as *Ludwigia uruguayensis* has been split into two related but distinct species: the hexaploid *L. grandiflora* and decaploid *L. hexapetala*. Octoploid (2n = 64) hybrids between these species have been identified from southern Brazil. Molecular data indicate that *L. spathulata* is an autopolyploid. There are few barriers to hybridization in *Ludwigia*. Generally, the diploid and the tetraploid species all are interfertile within their respective ploidy levels, but natural *Ludwigia* hybrids occur both within and among various ploidy levels. Diploid hybridization occurs between *L. palustris* and *L. microcarpa*. Hybrids at the tetraploid level include *L. brevipes* × *L. repens*; *L. sphaerocarpa* × *L. pilosa*, *L. polycarpa*, or *L. ravenii*; and *L. arcuata* × *L. pilosa*. Vigorous natural hybrids occur between *L. repens* (4n) and *L. arcuata* (6n) and resemble *L. brevipes*. *Ludwigia palustris* (2n) hybridizes with *L. glandulosa*, *L. polycarpa*, *L. repens*, and *L. spathulata* (all tetraploid). *Ludwigia repens* (6n) hybridizes with *L. simpsonii* (6n) and *L. curtissii* (8n). *Ludwigia* ×*lacustris* (4n) is a sterile interspecific hybrid between *L. brevipes* (6n) and *L. palustris* (2n), which is restricted in distribution to southern New England where it grows erect or prostrate on gravelly or silted, sandy bottoms in acidic waters up to 2.5 m in depth, or in shallow water along the receding, muddy shores of lakes and streams. Two populations of this sterile hybrid have persisted for more than 75 years by vegetative reproduction, which occurs by means of creeping rootstocks and vegetative fragmentation. Crossing studies indicate that the tetraploid *Ludwigia arcuata* shares two diploid genomes with the hexaploid *L. brevipes*, which may have arisen as a hybrid between the former and the diploid *L. palustris*. Other reported hybrids include *L. alata* (6n) × *L. pilosa* (4n), and *L. alata* (6n) × *L. suffruticosa* (4n). It is difficult to provide a simple phylogenetic summary for *Ludwigia* due to extensive hybridization and polyploidy in the histories of many of the species. One estimation (Figure 4.87), which includes 12 of the OBL indicator species, indicates several origins of the OBL habit in the genus. DNA sequence data indicate that *L. brevipes* is possibly a polyploid derivative of hybrid origin involving *L. palustris* and *L. arcuata*. The chloroplast DNA of *L. arcuata* (an allopolyploid) appears to have been contributed to *L. repens* and *L. brevipes* (hexaploids). Sequence data also implicate *L. palustris* (diploid) as the paternal genome donor of *L. spathulata* (tetraploid).

Comments: *Ludwigia palustris* is widespread in North America except for the Rocky Mountain region. *Ludwigia peploides* is widespread through the central and southern portions of the USA. *Ludwigia alternifolia* occurs in eastern North America. *Ludwigia polycarpa* occurs in eastern North America but is disjunct in Idaho. *Ludwigia sphaerocarpa* occurs in the eastern United States. *Ludwigia leptocarpa*

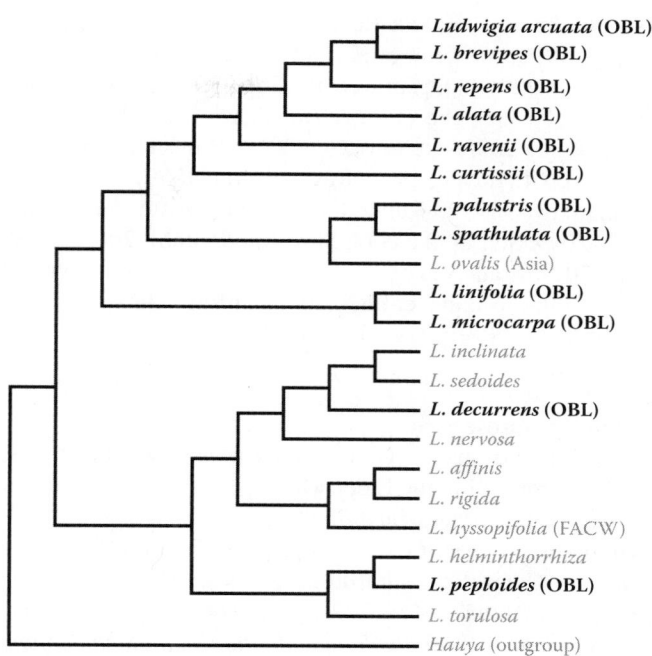

FIGURE 4.87 Phylogenetic relationships among *Ludwigia* species as estimated by analysis of nrITS sequence data. The OBL North American wetland indicators are shown in bold. Because most species in the lower clade lack aquatic adaptations, the OBL habit has arisen at least 2–3 times within the genus. The relationships depicted here should be considered carefully because many of the species are allopolyploids of hybrid origin (see *Systematics*). (Adapted from Bedoya, A.M. & Madriñán, S., *Aquat. Bot.*, 120, 352–362, 2015.)

ranges through the eastern and southern United States. *Ludwigia alata*, *L. arcuata*, *L. bonariensis* (extending to South America), *L. brevipes*, *L. decurrens*, *L. glandulosa*, *L. hexapetala*, *L. hirtella*, *L. lanceolata*, *L. linearis*, *L. linifolia*, *L. microcarpa*, *L. octovalvis*, *L. peruviana*, *L. pilosa*, *L. ravenii*, *L. spathulata*, *L. suffruticosa*, *L. virgata* all are distributed in the southeastern United States. *Ludwigia grandiflora* is found in both the southeastern and western United States. *Ludwigia repens* occurs in the southern United States. *Ludwigia curtissii*, *L. erecta*, and *L. longifolia* are restricted to Florida (USA). Several species have been impacted by the spread of invasive plants such as *Lythrum salicaria* and *Phragmites australis* in their habitats.

References: Aliyu et al., 2008; Beauchamp et al., 2013; Bedoya & Madriñán, 2015; Berkowitz et al., 2014; Bostick, 1977; Brown et al., 2011; Carr, 2007; Carter et al., 2009; Chang et al., 2004; Chen et al., 1989; Childers et al., 2003a; Choi et al., 2012; Cooper et al., 2004; Cypert, 1972; Deaver et al., 2005; DeBerry & Perry, 2007; Dolan, 1984; Dolan & Sharitz, 1984; Eames, 1933; Ellmore, 1981; Estes & Thorp, 1974; Fleckenstein, 2007; Green et al., 2002; Grubben & Denton, 2004; Harms & Grodowitz, 2009; Hill, 2006; Hoch et al., 1993; Howard & Wells, 2009; Howard et al., 2011; Hoyer et al., 1996; Hung et al., 2009; Hutton, 2010; Jacobs et al., 1993; Jing et al., 2002; Kasselmann, 2003; Kirkman & Sharitz, 1994; Kirkman et al., 2000; Krauss,

1987; Lake et al., 2000; Landman et al., 2007; Le Page & Keddy, 1998; McAvoy et al., 2015; McGregor et al., 1996; Middleton, 2009; Opsahl et al., 2010; Ott, 1991; Oyedeji et al., 2011; Oziegbe et al., 2010; Peng, 1984, 1988, 1989; Peng et al., 2005; Quattrocchi, 2012; Ramstetter & Mott-White, 2001; Raven, 1979; Raven & Tai, 1979; Reid, 2001; Rejmánková, 1992; Rundell & Woods, 2001; Sah et al., 2012, 2013; Self et al., 1974; Shrestha & Stahl, 2008; Smith et al., 2002; Stansly et al., 1997; Stoetzel et al., 1999; Tyndall et al., 1990; Wagner et al., 2007; Zardini & Raven, 1992; Zardini et al., 1991; Zygadlo et al., 1994.

3. Oenothera

Evening primrose

Etymology: from the Greek *oinos thera* meaning "wine catcher" from Pliny the Elder who reported that infusions of the root could sedate wild animals

Synonyms: *Anogra*; *Baumannia*; *Blennoderma*; *Calylophus*; *Gaura*; *Gaurella*; *Gauropsis*; *Hartmannia*; *Kneiffia*; *Lavauxia*; *Megapterium*; *Onagra*; *Onosuris*; *Onosurus*; *Pachylophus*; *Peniophyllum*; *Raimannia*; *Stenosiphon*; *Usoricum*; *Xylopleurum*

Distribution: global: New World; **North America:** widespread

Diversity: global: 145 species; **North America:** 84 species

Indicators (USA): OBL: *Oenothera longissima*, *O. organensis*

Habitat: freshwater; palustrine, riverine; **pH:** alkaline; **depth:** <1 m; **life-form(s):** emergent herb

Key morphology: stems (to 30 dm) erect to sprawling, arising from a taproot; leaves in basal rosettes (to 40 cm) or cauline (to 22 cm); flowers in spikes; hypanthium elongate (to 19 cm), greenish but fading to orange-red after anthesis; sepals (to 55 mm) 4, sometimes with tips (to 6 mm) free in bud; petals (to 65 mm) 4, yellow, often fading to reddish orange (sometimes drying lavender); stamens (to 4 cm) 8; styles (to 25 cm) terminating in 4 linear lobes (to 9 mm), subtended by a peltate indusium; fruit a capsule (to 55 mm); seeds numerous (to 160).

Life history: duration: biennial (taproot), perennial (adventitious shoots); **asexual reproduction:** none; **pollination:** insect; **sexual condition:** hermaphroditic; **fruit:** capsules (common); **local dispersal:** seeds; **long-distance dispersal:** seeds

Imperilment: (1) *Oenothera longissima* [G4]; S1 (CA, CO); (2) *O. organensis* [G2]; S2 (NM)

Ecology: general: *Oenothera* is a diverse genus of annuals, biennials, or perennials, which principally comprises terrestrial plants. Disturbed sites are commonly inhabited and species can occur at altitudes from sea level to 5000 m. There are 21 North American wetland indicators (25%); however, only two species (2%) have status as obligate indicators and these occupy distinctive sites such as desert riparian and scrub habitats. The flowers can be self-compatible or self-incompatible, usually last for less than 1 day, and are pollinated by a variety of insects (Insecta) including bees (Hymenoptera), butterflies and moths (Lepidoptera), and flies (Diptera). Those species with flowers that open at night or during the evening are pollinated by hawkmoths (Lepidoptera: Sphingidae) or wasps (Hymenoptera: Vespidae). Some species are cleistogamous or otherwise autogamous.

***Oenothera longissima* Rydb.** is a suffrutescent biennial or short-lived perennial, which is found in seasonally wet sites that are sporadically associated with alcoves, beaches, canyons, chaparral, cliff bases, creekbeds, desert washes, ditches, gulches, hanging gardens, marshes, meadows, roadsides, seeps, slopes, springs, and the margins of rivers and streams at elevations of up to 2881 m. Occurrences generally are in open exposures and less frequently in light shade. The substrates can be cryptogamic or consist of gravel, loam, Navajo sandstone, rock, sand, sandy loam, or shale. They often are highly alkaline or are associated with limestone deposits. Flowering and fruiting occurs from July to November. The plants are self-compatible but are insect pollinated, bivalent-forming outcrossers. The seeds contain 13.5% oil by weight and retain 92%–96% germination rates when stored for up to 14 years. A germination rate of 95% has been obtained for seeds kept at 21°C under a 12/12-h photoperiod. Collection sites often are disturbed and several collections have been made from recently burned areas. Although localities frequently are described as riparian, the predominantly mesic and xeric associates hardly typify wetland taxa, thus making the OBL status of this species unusual. **Reported associates:** *Abronia elliptica, A. fragans, Acer negundo, Achnatherum hymenoides, Acourtia wrightii, Adiantum capillus-veneris, Agave utahensis, Agrostis scabra, Aliciella hutchinsifolia, Allionia incarnata, Amaranthus, Amauriopsis dissecta, Ambrosia acanthicarpa, Amelanchier utahensis, Amsonia tomentosa, Andropogon glomeratus, Aquilegia micrantha, Arctostaphylos patula, Artemisia bigelovii, A. campestris, A. dracunculus, A. ludoviciana, A. tridentata, Asclepias asperula, A. latifolia, Astragalus mollissimus, Atriplex canescens, Baccharis salicifolia, B. salicina, B. sergiloides, Berberis fremontii, Bothriochloa barbinodis, B. laguroides, Bouteloua curtipendula, B. eriopoda, B. gracilis, Brickellia longifolia, B. microphylla, Bromus rubens, B. tectorum, Bursera, Carex nebrascensis, C. specuicola, Castilleja linariifolia, Celtis reticulata, Cercis occidentalis, Cercocarpus ledifolius, Chrysothamnus nauseosus, Cirsium arizonicum, C. rydbergii, C. wheeleri, Clematis ligusticifolia, Coleogyne ramosissima, Conyza canadensis, Cryptantha confertiflora, C. fendleri, Dalea candida, D. flavescens, Datura wrightii, Distichlis, Dysphania graveolens, Echinocactus polycephalus, Echinocereus triglochidiatus, Elaeagnus angustifolia, Elymus canadensis, Encelia virginensis, Ephedra, E. viridis, Epipactis gigantea, Equisetum arvense, E. hyemale Eragrostis, Ericameria nauseosa, Erigeron divergens, Eriogonum alatum, E. corymbosum, E. inflatum, E. microthecum, Euphorbia brachycera, Euthamia occidentalis, Fallugia paradoxa, Fendlera rupicola, Ferocactus cylindraceus, Frasera speciosa, Fraxinus anomala, F. velutina, Funastrum cynanchoides, Gaillardia, Galium, Gilia aggregata, Gutierrezia microcephala, G. sarothrae, Herrickia glauca, H. wasatchensis, Hesperostipa neomexicana, Heterotheca villosa, Hordeum jubatum, Hymenopappus* *filifolius, Hymenoxys, Imperata brevifolia, Ipomopsis longiflora, I. roseata, Isocoma acradenia, Juncus balticus, J. xiphioides, Juniperus communis, J. osteosperma, Lactuca serriola, Lepidium, Lobelia cardinalis, Lorandersonia salicina, Machaeranthera canescens, Melilotus albus, M. officinalis, Mentzelia puberula, Mimulus cardinalis, M. eastwoodiae, Mirabilis multiflora, Muhlenbergia andina, M. rigens, Nolina microcarpa, Oenothera cespitosa, O. pallida, Opuntia engelmannii, O. phaeacantha, O. polyacantha, Oryzopsis hymenoides, Packera multilobata, Paxistima myrsinites, Penstemon ambiguus, P. eatonii, P. rostriflorus, P. strictus, Perityle congesta, Petradoria pumila, Petrophytum caespitosum, Phacelia, Phlox, Phragmites australis, Physaria newberryi, Picea, Pinus edulis, P. monophylla, P. ponderosa, Platanthera zothecina, Pluchea sericea, Poa fendleriana, Polypogon monspeliensis, Populus fremontii, P. tremuloides, Prosopis glandulosa, P. velutina, Prunus, Pseudognaphalium, Psilostrophe sparsiflora, Psorothamnus fremontii, Purshia mexicana, Quercus chrysolepis, Q. gambelii, Q. turbinella, Q. ×undulata, Rhamnus betulifolia, Rhus aromatica, Robinia neomexicana, Rosa woodsii, Salix bonplandiana, S. exigua, S. gooddingii, S. lutea, S. scouleriana, Schizachyrium scoparium, Sclerocactus parviflorus, Senegalia greggii, Shepherdia rotundifolia, Solidago sparsiflora, S. velutina, Sphaeralcea grossulariaefolia, S. leptophylla, Sporobolus cryptandrus, S. flexuosus, Stanleya pinnata, Stephanomeria pauciflora, S. tenuifolia, Tamarix chinensis, Tetraneuris argentea, Thelypodium integrifolium, T. wrightii, Thymophylla pentachaeta, Typha domingensis, Verbena bracteata, Veronica anagallis-aquatica, Vitis arizonica, Xanthium strumarium, Xylorhiza tortifolia, Yucca angustissima, Y. baccata, Y. elata, Zeltnera calycosa.*

***Oenothera organensis* Munz** is a perennial herb with spreading or prostrate stems, which is restricted to canyon streambeds, hillside seeps, and springs in montane scrub and woodland communities at elevations from 1480 to 2378 m where surface water occurs during at least part of the growing season. The substrates consist of coarse granitic sands and gravelly loam. Flowering occurs from June to September. The flowers open from dusk through early morning and receive ample pollen for full seed set by insects, which pollinate the plants during a 2-h period each night. The flowers are self-incompatible (gametophytic) and are obligately outcrossed by hawkmoths (Insecta: Lepidoptera: Sphingidae). The exceptionally long (to 25 cm) floral styles are suited to the four hawkmoth pollinators (see *Use by wildlife* later). Studies have demonstrated that heavy pollen loads in this species will produce higher seed quality than do lighter loads. The seeds can be dispersed through their consumption and subsequent defecation by deer (Mammalia: Cervidae: *Odocoileus virginianus*). The plants are able to tolerate seasonal sediment deposition that results from autumnal rains. Individuals are said to be more frequent during El Niño years. **Reported associates:** *Aquilegia chrysantha, Asplenium resiliens, Astrolepis sinuata, Bommeria hispida, Carex mariposana, C. scoparia, Cheilanthes eatonii, C. lindheimeri, C. tomentosa, Eragrostis imtermedia, Lesquerella, Mimulus guttatus, Notholaena*

standleyi, Panicum bulbosum, Parietaria pensylvanica, Pellaea atropupurea, P. wrightiana, Phanerophlebia auriculata, Sphaeralcea.

Use by wildlife: *Oenothera organensis* is palatable to livestock (Mammalia: Bovidae), which feed on the shoots. Four hawkmoths (Insecta: Lepidoptera: Sphingidae) are associated with *O. organensis* as pollinators: *Hyles lineata, Manduca quinquemaculata, M. rustica* and *Sphinx chersis*.

Economic importance: food: Edibility of *O. longissima* or *O. organensis* has not been reported specifically; however, the roots of many *Oenothera* species are edible, and the shoots sometimes are used in salads; **medicinal:** *Oenothera longissima* contains fatty acids and α and γ tocopheroles, which are antioxidants; **cultivation:** *Oenothera organensis* occasionally is sold as an ornamental garden plant; **misc. products:** none; **weeds:** none; **nonindigenous species:** none

Systematics: The large genus *Oenothera* is subdivided into 18 sections. Both *O. longissima* and *O. organensis* are assigned to section *Oenothera*, but within different subsections. *Oenothera longissima* is placed within subsection *Oenothera*. It is believed to be related to (and possibly derived from) *O. elata*, which shares with it a similar genome (AA) and plastome (I). Hybrids between *O. elata* and *O. longissima* have not yet been verified conclusively, but are indicated by highly variable floral tubes in suspected hybrid plants, which originated from seed obtained where the two species grew sympatrically. *Oenothera organensis* is placed in subsection *Emersonia* along with three other species (*O. macrosceles, O. maysillesii, O. stubbei*), which are interfertile and readily form fertile hybrids. Morphologically, it is the most distinctive species in the subsection, a result that is supported by preliminary molecular phylogenetic studies of the genus (Figure 4.88). The base number of *Oenothera* is $x = 7$. *Oenothera longissima* and *O. organensis* both are diploid ($2n = 14$). Reciprocal translocations occur commonly in natural populations of both species. Allozyme surveys have indicated that populations of *Oenothera organensis* harbor low levels of genetic variability (monomorphic at 14/15 loci). However, polymorphism at the self-incompatibilty (s-allele) locus is maintained at a fairly high level as a consequence of frequency dependent selection. As a result, populations of 5000 plants can possess up to 45 different self-incompatibility (s) alleles. Studies using transgenic plants have shown that the degree of ovule fertilization does not always correlate with successful seed maturation in *O. organensis*.

Comments: *Oenothera longissima* occurs in the southwestern United States; whereas, *O. organensis* is endemic to the Organ Mountains of New Mexico (USA) where a total of approximately 2000–5000 individuals exist.

References: DeBruin, 1996; Dietrich et al., 1985; Evans et al., 2005; Flanagan et al., 1997; Havens & Delph, 1996; Levin et al., 1979; Raven et al., 1979; Velasco & Goffman, 1999; Wagner, 2005a; Wagner et al., 2007.

ORDER 12: SAPINDALES [11]

Sapindales are close phylogenetically to Malvales and encompass nearly 6000 species of woody plants (Figure 4.89). The

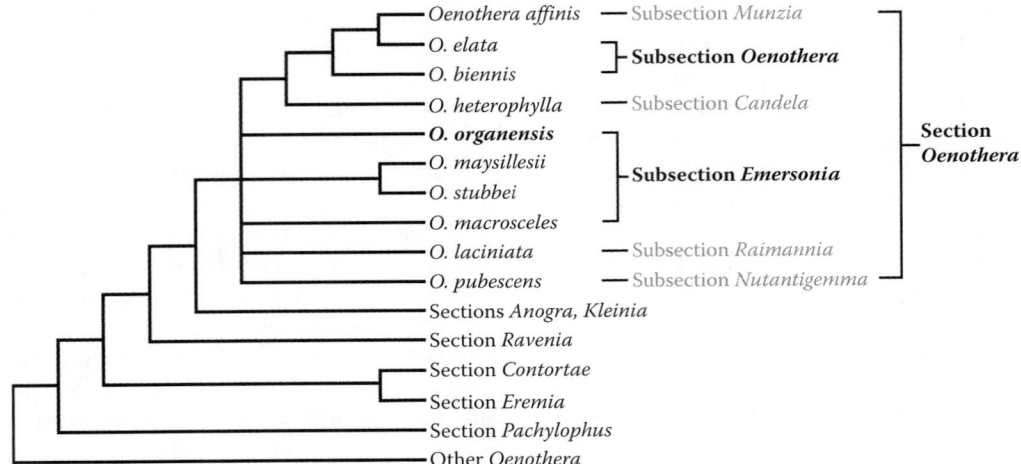

FIGURE 4.88 Relationships in *Oenothera* as evidenced by combined nrITS and *trnL–trnF* DNA sequence data. Taxa including North American obligate aquatics are indicated in bold (note: *Oenothera longissima* was not surveyed in that study; however, it is placed taxonomically within subsection *Oenothera*, where it is considered to be closely related to *O. elata*). These phylogenetic relationships resolve the aquatic *Oenothera* species in a relatively derived section (*Oenothera*) of the genus. If the presumed relationship of *O. longissima* is correct, then this cladogram also indicates that the hydrophytic habit has arisen independently in the two OBL North American *Oenothera* species. (Adapted from Wagner, W.L. et al., *Syst. Bot. Monogr.*, 83, 1–240, 2007.)

monophyly of Sapindales is indicated by distinctive features such as compound leaves and flowers with a nectary disk (Judd et al., 2002), and by results from phylogenetic analysis of *rbcL* sequence data (Gadek et al., 1996). More controversial is the delimitation of families within the order, notably recent proposals to revise the circumscription of Sapindaceae to include well-known families such as Aceraceae and Hippocastanaceae. Combined *matK* and *rbcL* DNA sequence data resolve a strongly supported clade corresponding to Aceraceae (*Acer*, *Dipteronia*), which resolves as the sister group of a well-supported clade corresponding to Hippocastanaceae (*Aesculus*, *Billia*, *Handeliodendron*); together, these clades are sister to the clade corresponding to Sapindaceae in a more traditional sense (Harrington et al., 2005). Consequently, one could include Aceraceae and Hippocastanaceae within

Sapindaceae (e.g., as subfamilies), or retain them as separate families as either disposition is compatible with results from phylogenetic analyses (Figures 4.89 and 4.90). Because the literature has long referred to these families as distinct, they have been maintained separately in this treatment.

Most members of Sapindales are terrestrial and those containing aquatics are dispersed throughout the order (Figure 4.89). Three families contain obligate aquatic species in North America:

1. **Aceraceae** Juss.
2. **Anacardiaceae** Lindl.
3. **Simaroubaceae** DC.

Family 1: Aceraceae [2]

Aceraceae are a small family of woody plants totaling about 150 species of predominantly terrestrial temperate forest trees and shrubs. Characteristics of Aceraceae include opposite leaves, nonappendaged petals, papillose stamens arising from a nectar disk, and bi-ovulate carpels (Judd et al., 2002). *Dipteronia* is an Asian genus of two species, which grow in UPL deciduous forests. The remaining species are in *Acer*, with most inhabiting UPL sites, but some found fairly often in swamps. The family is an important source of a durable, fine-grained wood that is used for furniture and cabinetry. Maple sugar candy and maple syrup are refined from the sap. Pollination is facilitated either by insects or by the wind. The seeds are dispersed within broadly winged mericarps (samaras), which can be carried for considerable distances by the wind.

Obligate aquatics occur within a single North American genus of Aceraceae:

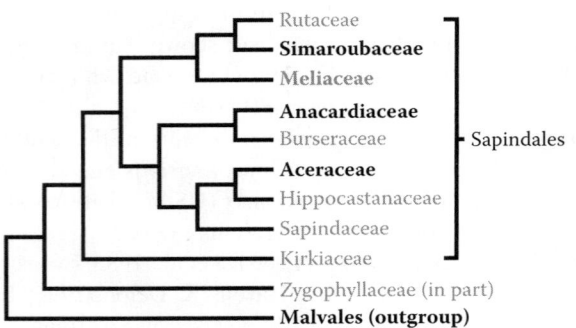

FIGURE 4.89 Phylogenetic relationships among families of Sapindales as indicated by analysis of *rbcL* sequence data. Relationships among the families containing aquatic species (bold) indicate that adaptations to hydric habitats have been acquired independently throughout the order. (Redrawn from Gadek, P.A. et al., *Amer. J. Bot.*, 83, 802–811, 1996.)

1. *Acer* L.

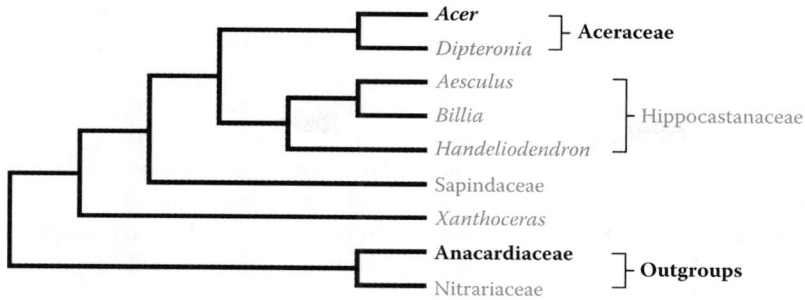

FIGURE 4.90 Relationships of Sapindaceae and segregate families as indicated by combined cpDNA sequence data. The small family Aceraceae is distinct from Sapindaceae, despite the merger of these families by some authors. Taxa containing OBL indicators in North America are shown in bold. (Adapted from Harrington, J. et al., *Syst. Bot.*, 30, 366–382, 2005.)

1. *Acer*

Maple; érable

Etymology: from the Greek *a keras* meaning "without a horn" in reference to the spurless flowers

Synonyms: *Argentacer*; *Negundo*; *Rufacer*; *Rulac*; *Saccharodendron*

Distribution: global: Asia; Europe; North America; **North America:** widespread

Diversity: global: 125 species; **North America:** 21 species

Indicators (USA): FAC: *Acer rubrum* (formerly OBL)

Habitat: freshwater; palustrine, riverine; **pH:** 4.4–7.5; **depth:** 0–2 m; **life-form(s):** emergent tree

Key morphology: trunks (to 28 m straight), the winter buds with dark red scales; leaves (to 15 cm) deciduous, opposite, palmately 3–5-lobed, the margins toothed; flowers small, bisexual or unisexual, in stalkless clusters; calyx (to 1 mm) 5-parted; petals (to 1.5 mm) 5, inconspicuous; stamens (when present) 5–8; fruit a v- or u-shaped samara (to 2.6 cm)

Life history: duration: perennial (buds); **asexual reproduction:** stump sprouts; **pollination:** wind, insect; **sexual condition:** hermaphroditic, dioecious, monoecious; **fruit:** samara (mericarps) (common); **local dispersal:** mericarps (wind); **long-distance dispersal:** mericarps (wind)

Imperilment: (1) *Acer rubrum* [G5]; S1 (IA); (2) *A. rubrum* var. *drummondii* [G5]; S3 (NJ); (3) *A. rubrum* var. *trilobum* [G5]; S3 (WV)

Ecology: general: The ecological tolerances of many maples are quite broad, which allows them to survive across an extreme range of habitats from well-drained UPL sites to wetlands. Maples often are encountered along floodplains and as a component of riverine swamps. Twelve maple species are listed as wetland indicators in North America, but none is designated as obligate in the revised (2013) indicator list. However, because the treatment for *Acer rubrum* (FAC) was completed while still included among the obligate indicators in the 1996 list, it is retained here.

Acer rubrum **L.** is a very common species of both terrestrial and aquatic habitats. At the more hydric extent of its range it occurs in bogs, depressions, floodplains, seeps, swamps, and along lake, river, and stream margins at elevations of up to 1400 m. The plants exhibit a very wide pH tolerance, but appear more commonly on somewhat acidic soils (pH<7.0). Substrates can vary extremely (from granitic to limestone), and range from inorganic (rock, gravel, sand), to organic (peat, loam, rich organic soil) composition. It is no more unusual to find plants growing on sandbars than it is to see them on pure peat. Plants are shade tolerant and occur where the minimum annual temperature is above −40°C. They grow rapidly and can live up to 150 years. The plants are polygamous and their gender can change from one year to another. The bisexual or unisexual flowers emerge early in the spring (before most other tree species) and are precocious with respect to leaf emergence. They mainly are wind pollinated and are visited by bees and butterflies (see *Use by wildlife* later). The unisexual flowers are borne both in monoecious and dioecious arrangements. Young saplings can produce seeds at 4 years of age. A mature tree commonly produces 12,000–91,000 seeds each year, which require no pretreatment and are able to germinate immediately after ripening (81%–95% germination). The seeds can be germinated by first soaking them, then sowing for 9 weeks at 4°C, then incubating them at 21°C. Little light is required for germination; however, seeds produced in deep shade can delay their germination for 1 year. The winged fruits are extremely light (although heavier in northern latitudes) and are dispersed considerable distances by the wind. *Acer rubrum* often exploits disturbed sites, replacing species afflicted by disease, fire, or windfall. The plants will sprout vigorously from dormant stump buds, but are difficult to propagate by stem cuttings. Benzoic acid (found in black cherry leaves) is allelopathic to *A. rubrum*. The prior designation of *A. rubrum* as an obligate aquatic species was problematic because it is one of the most common trees in eastern North America and only certain varieties (see *Systematics* later) normally are associated with wet habitats. Genetic studies have shown that differential adaptation (e.g., degree of lenticel proliferation, adventitious root formation, recovery from inundation) occurs among sites having different water regimes. The species occurs more frequently in swamps in the southern portion of its range. Because a list of associated species would include nearly every plant that grows in eastern woodlands, the following list specifies primarily those species that occur as associates of *A. rubrum* in lowland (wetland) habitats while excluding many species that coexist in more well-drained sites. **Reported associates (lowland sites):** *Abies balsamea, Acer saccharinum, A. spicatum, Alnus rugosa, A. serrulata, A. viridis, Ampelopsis arborea, Andropogon glomeratus, A. virginicus, Apios americana, Aralia nudicaulis, Aristida purpurascens, Aronia melano-carpa, Arundinaria gigantea, Asimina triloba, Athyrium filix-femina, Atrichum, Bazzania trilobata, Berchemia scandens, Betula alleghaniensis, B. glandulifera, B. nigra, B. papyrifera, B. populifolia, Bidens frondosus, Boehmeria cylindrica, Brunnichia ovata, Bryum pseudocapillare, Calamagrostis canadensis, Caltha palustris, Campsis radicans, Carex amphibola, C. atlantica, C. baileyi, C. bromoides, C. brunnescens, C. crinita, C. echinata, C. folliculata, C. gigantea, C. glaucescens, C. gynandra, C. gynocrates, C. intumescens, C. joorii, C. laevivaginata, C. leptalea, C. lonchocarpa, C. lupuliformis, C. lupulina, C. lurida, C. magellanica, C. rostrata, C. stricta, C. tribuloides, C. trisperma, Carpinus caroliniana, Carya aquatica, C. illinoinensis, C. myristiciformis, Celtis laevigata, Cephalanthus occidentalis, Cercis canadensis, Chasmanthium laxum, C. sessiliflorum, Chelone glabra, Cinna latifolia, Cladium mariscoides, Clethra acuminata, Clintonia borealis, Coptis groenlandica, C. trifolia, Cornus canadensis, C. drummondii, C. foemina, C. sericea, Corylus americana, Crataegus phaenopyrum, C. viridis, Cyrilla racemiflora, Decumaria barbara, Dennstaedtia punctilobula, Dichanthelium clandestinum, D. scoparium, Dicranum, Diospyros virginiana, Doellingeria umbellata, Drosera rotundifolia, Dryopteris cristata, Dulichium arundinaceum, Eleocharis microcarpa, E. tortilis, E. tuberculosa, Eriophorum tenellum, E. virginicum, Euonymus americana, Eupatorium dubium, E. pilosum, Festuca rubra, Fraxinus caroliniana, F. nigra, F. pennsylvanica, Gaylussacia baccata, G. dumosa, G. frondosa, G. ursina, Gleditsia triacanthos, Glyceria canadensis, G. melicaria, G. obtusa, G. striata, Halesia tetraptera, Hamamelis virginiana, Hypericum mutilum, Hypnum imponens, Hypoxis curtissii, Ilex cassine, I. coriacea, I. decidua, I. glabra, I. opaca, I. verticillata, Impatiens capensis, Iris hexagona, I. tridentata, Itea virginica, Juncus coriaceus, J. effusus, J. gymnocarpus, Juniperus virginiana, Justicia ovata, Kalmia angustifolia, K. latifolia, Laportea, Larix laricina, Leersia lenticularis, Leucothoe axillaris, L. fontanesiana, L. racemosa, Lindera benzoin, Linnaea borealis, Liquidambar styraciflua, Liriodendron tulipifera, Lobelia cardinalis, L. siphilitica, Ludwigia decurrens, L. leptocarpa, L. palustris, Lycopus uniflorus, L. virginicus, Lyonia ligustrina, L. lucida, Lysimachia terrestris, Magnolia virginiana, Maianthemum trifolium, Microstegium vimineum, Mnium, Muhlenbergia torreyana, Myrica caroliniensis, M. cerifera, M. pensylvanica, Nemopanthus mucronatus, Nyssa biflora, N. ogeche, N. sylvatica, Oclemena nemoralis, Onoclea sensibilis, Orontium aquaticum, Osmunda claytoniana, O. regalis, Osmundastrum cinnamomeum, Ostrya virginiana, Oxalis monata, Oxydendrum arboreum, Panax trifolia, Panicum hemitomon, Parthenocissus quinquefolia, Peltandra virginica, Persea palustris, Persicaria hydropiperoides, P. sagittatum, P. setaceum, Phalaris arundinacea, Phanopyrum gymnocarpon, Photinia pyrifolia, Picea glauca, P. mariana, P. rubens, Pilea pumila, Pinus elliottii, P. palustris, P. serotina, P. strobus, P. taeda, Planera aquatica, Platanthera integrilabia, Polytrichum juniperinum, P. strictum, Pontederia cordata, Populus balsamifera, P. grandidentata, P. tremuloides, Prunus serotina, Pteridium aquilinum, Pycnanthemum virginianum, Quercus alba, Q. laurifolia, Q. lyrata, Q. marilandica, Q. michauxii, Q. nigra, Q. phellos, Q. rubra, Q. shumardii, Q. texana, Q. virginiana, Rhexia alifanus, R. petiolata, R. virginica, Rhododendron arborescens, R. groenlandicum, R. maximum, R. viscosum, Rhynchospora capitellata, R. corniculata, R. globularis, R. macrostachya, R. miliacea, Rosa multiflora, Rubus alleghaniensis, R. argutus, R. canadensis, R. hispidus, R. idaeus, R. pubescens, Rumex obtusifolius, Sabal minor, S. palmetto, Saccharum, Sagittaria calycina, Salix nigra, S. sericea, Sarracenia purpurea, S. rubra, Saururus cernuus, Scirpus cyperinus, Scleria oligantha, Scutellaria nervosa, Sebastiania fruticosa, Serenoa repens, Smilax bona-nox, S. glauca, S. laurifolia, S. rotundifolia, S. tamnoides, Solidago patula, Sparganium americanum, Sphagnum capillifolium, S. girgensohnii, S. magellanicum, S. palustre, S. papillosum, S. recurvum, S. wulfianum, Spiraea alba, S. tomentosa, Streptopus roseus, Styrax americanus, Symphyotrichum puniceum, Symplocarpus foetidus, Symplocos tinctoria, Taxodium ascendens, T. distichum, Thalictrum dioicum, Thelypteris kunthii, T. noveboracensis, T. palustris, Thuidium delicatulum, Thuja occidentalis, Tillandsia bartramii, T. usneoides, Toxicodendron radicans, T. vernix, Triadenum walteri, Triadica sebifera, Trientalis borealis, Trillium, Tsuga canadensis, Typha latifolia, Ulmus americana, U. crassifolia, U. rubra, Urtica dioica, Vaccinium angustifolium, V. corymbosum, V. elliottii, V. erythrocarpum, V. virgatum, Vernonia noveboracensis, Viburnum dentatum, V. nudum, Viola cucullata, Vitis aestivalis, V. labrusca, V. rotundifolia, Woodwardia areolata, W. virginica, Xanthorhiza simplicissima, Xyris ambigua, X. caroliniana, X. laxifolia.*

Use by wildlife: The foliage and/or seeds of *Acer rubrum* are eaten by elk (Mammalia: Cervidae: *Cervus elaphus*), beavers (Mammalia: Castoridae: *Castor canadensis*), chipmunks (Mammalia: Sciuridae: *Tamias striatus*), downy woodpeckers (Aves: Picidae: *Picoides pubescens*), gray squirrels (Mammalia: Sciuridae: *Sciurus carolinensis*), muskrats (Mammalia: Cricetidae: *Ondatra zibethicus*), snowshoe hares (Mammalia: Leporidae: *Lepus americanus*), white-footed mice (Mammalia: Cricetidae: *Peromyscus leucopus*), white-tailed deer (Mammalia: Cervidae: *Odocoileus virginianus*), and white-throated sparrows (Aves: Emberizidae: *Zonotrichia albicollis*). The trunks are excavated by insect-seeking birds such as the yellow-bellied sapsucker (Aves: Picidae: *Sphyrapicus varius*), which can severely damage the trees. The resulting cavities are used as nesting sites by bluebirds (Aves: Turdidae: *Sialia sialis*), brown bats (Mammalia: Vespertilionidae: *Eptesicus fuscus*), Carolina chickadees (Aves: Paridae: *Parus carolinensis*), gray squirrels (Mammalia: Sciuridae: *Sciurus carolinensis*), pileated woodpeckers (Aves: Picidae: *Dryocopus pileatus*), rat snakes (Reptilia: Colubridae: *Elaphe obsoleta*), screech owls (Aves: Strigidae: *Megascops asio*), and wood ducks (Aves: Anatidae: *Aix sponsa*). The trees provide general shelter for many different animals such as beavers (Mammalia: Castoridae: *Castor canadensis*), white-tailed deer (Mammalia: Cervidae: *Odocoileus virginianus*), harvestman spiders (Arthropoda: Arachnida: Phalangiidae: *Phalangium opilio*), skunks (Mammalia: Mephitidae: *Mephitis mephitis*), and spring peepers (Amphibia: Hylidae: *Pseudacris crucifer*). Their branches are nesting sites for blackbirds (Aves: Icteridae: *Agelaius phoeniceus*), mockingbirds (Aves: Mimidae: *Mimus polyglottos*) and other species. The trees host an enormous diversity of insects. The nectar and pollen are gathered by bumblebees (Hymenoptera: Apidae: *Bombus fervidus*) and honeybees (Hymenoptera: Apidae: *Apis mellifera*) and the nectar by tiger swallowtails (Lepidoptera: Papilionidae: *Papilio glaucus*) and mourning cloaks (Lepidoptera: Nymphalidae: *Nymphalis antiopa*). A vast assemblage of lepidoptera is hosted by the plants (Arctiidae: *Halysidota tessellaris*, *Hyphantria cunea*, *Lophocampa caryae*, *Spilosoma virginica*; Coleophoridae: *Coleophora alniella*; Geometridae: *Alsophila pometaria*, *Anacamptodes ephyraria*, *Anavitrinella pampinaria*, *Besma endropiaria*, *Biston betularia*, *Campaea perlata*, *Cingilia catenaria*, *Ectropis crepuscularia*, *Ennomos magnaria*, *E. subsignaria* [elm spanworm], *Erannis tiliaria* [linden looper], *Eutrapela clemataria*, *Hypagyrtis unipunctata*, *Itame pustularia* [red maple spanworm], *Lambdina fiscellaria*, *Melanolophia canadaria*, *M. signataria*, *Metanema determinata*, *Nematocampa filamentaria*, *Operophtera brumata*, *Paleacrita vernata*, *Phigalia titea*, *Plagodis alcoolaria*, *P. serinaria*, *Probole alienaria*, *P. amicaria*, *Prochoerodes transversata*, *Protoboarmia porcelaria*, *Selenia kentaria*, *Semiothisa aemulataria*, *S. ulsterata*, *Tetracis cachexiata*, *Xanthotype sospeta*; Gracillariidae: *Caloptilia aceriella*, *C. bimaculatella*, *C. speciosella*, *C. umbratella*, *Cameraria aceriella*, *C. saccharella*, *Phyllonorycter trinotella*; Incurvariidae: *Paraclemensia acerifoliella*; Lasiocampidae: *Malacosoma americana*; Limacodidae: *Lithacodes fasciola*; Lymantriidae: *Dasychira plagiata*, *Lymantria dispar* [gypsy moth], *Orgyia antiqua*, *O. definita*, *O. leucostigma*, *Megalopyge crispata*; Noctuidae: *Acronicta americana*, *A. retardata*, *Amphipyra pyramidoides*, *Catocala cerogama*, *Crocigrapha normani*, *Eupsilia sidus*, *E. tristigmata*, *Hypena baltimoralis*, *Lithophane bethunei*, *L. grotei*, *L. innominata*, *L. laticinerea*, *L. petulca*, *Morrisonia confusa*, *M. latex*, *Orthosia revicta*, *Parallelia bistriaris*, *Phlogophora periculosa*, *Zale galbanata*, *Z. minerea*; Notodontidae: *Datana ministra*, *Heterocampa biundata*, *H. guttivitta*, *Nadata gibbosa*, *Schizura ipomoeae*, *Symmerista leucitys*; Oecophoridae: *Antaeotricha leucillana*, *Dafa formosella*, *Machimia tentoriferella*, *Nites betulella*; Pantheidae: *Colocasia flavicornis*; Psychidae: *Thyridopteryx ephemeraeformis*; Pyralidae: *Herpetogramma pertextalis*, *Oreana unicolorella*, *Pococera asperatella*; Saturniidae:

Actias luna, *Anisota oslari*, *Antheraea polyphemus* [polyphemus moth], *Automeris io*, *Dryocampa rubicunda*, *Eacles imperialis*, *Hemileuca nevadensis*, *Hyalophora cecropia*; Sesiidae: *Synanthedon acerni* [maple callus borer], *S. acerrubri*; and Tortricidae: *Acleris chalybeana*, *Archips argyrospila*, *A. rosana*, *Catastega aceriella*, *Choristoneura fractivittana*, *C. parallela*, *C. rosaceana*, *Episimus tyrius*, *Olethreutes appendiceum*, *Orthotaenia undulana*, *Pandemis lamprosana*, *Proteoteras aesculana*, *P. moffatiana*, *P. willingana*, *Sparganothis acerivorana*, *S. pettitana*). The trees host several Coleoptera including Hercules beetles (Scarabaeidae: *Dynastes tityus*), maple borers (Cerambycidae: *Xylotrechus aceris*), and Columbian timber beetles (Scolytidae: *Corthylus columbianus*). The plants are attacked by scale insects (Hemiptera) such as cottony maple scale (Coccidae: *Pulvinaria vitis*), maple leaf scale (Coccidae: *Pulvinaria acericola*), and oystershell scale (Diaspididae: *Lepidosaphes ulmi*). The leaves are eaten by katydids (Orthoptera: Tettigoniidae: *Pterophylla camellifolia*) and often show the reddish tuberculate growths caused by the maple gall mite (Arachnida: Acari: Eriophyidae: *Vasates quadripedes*). *Acer rubrum* is a host to several rotting fungi (Basidiomycota: Hymenochaetaceae: *Inonotus glomeratus*, *Phellinus igniarius*; Physalacriaceae: *Armillaria mellea*, Schizoporaceae: *Oxyporus populinus*). Cankers are formed on the trees by various fungal Ascomycota species (Diatrypaceae: *Eutypella*; Nectriaceae: *Nectria*; Sarcosomataceae: *Strumella*; Stictidaceae: *Schizoxylon*; Xylariaceae: *Hypoxylon*).

Economic importance: food: A fairly good quality sugar can be extracted from the sap of *Acer rubrum*, which was used by the Abnaki and Algonquin tribes as a food sweetener and syrup. French Canadians make a type of molasses from the sap. The Iroquois prepared bread from the dried, pulverized, and sifted bark. The seeds (with wings removed) can be boiled and eaten; **medicinal:** An infusion made from *Acer rubrum* bark the was administered by the Cherokee to treat cramps, dysentery, hives, and measles, by the Seminoles to relieve various pains and hemorrhoids, and by the Koasati to clean gunshot wounds. Bark extracts were used as a soothing eye wash by the Cherokee, Iroquois, Ojibwa, and Potawatomi; **cultivation:** Its brilliant red fall foliage makes *A. rubrum* a prized ornamental tree. The many cultivars include: 'Armstrong,' 'Autumn Flame,' 'Autumn Radiance,' 'Autumn Blaze,' 'Bowhall,' 'Brandywine,' 'Burgundy Belle,' 'Candy Ice,' 'Columnar,' 'Elstead,' 'Fairview Flame,' 'Franksred,' 'Gerling,' 'Northwood,' 'October Brilliance,' 'October Glory,' 'Red Sunset,' 'Scarlet Sentinel,' 'Scanlon,' 'Schlesingeri,' 'Shade King,' 'Somerset,' 'Sun Valley,' 'Tilford,' and 'V. J. Drake'; **misc. products:** Wood from *A. rubrum* was used for arrowheads (Seminole), baskets (Cherokee, Malecite, Micmac), bowls (Iroquois), carvings (Cherokee), ox yokes (Seminole), and spoons (Seminole). The wood is used for making furniture and as lumber. Wood from very old trees sometimes produces an undulate grain known as "curly maple" which is used as a decorative material for gun stocks. The Potawatomi cleaned their hunting traps in a solution of boiled bark. The maple leaf is a common motif observed in Ojibwa beadwork. Freshness of apples and various root crops is said to be extended by packing them in *A. rubrum* leaves. Boiling the inner bark produces a purple solution that can be mixed with lead sulfate to yield a black dye or ink. A maple leaf appears on the national flag of Canada; **weeds:** none; **nonindigenous species:** *Acer rubrum* has been introduced to many countries as an ornamental tree. It was first brought to Europe in 1656.

Systematics: Molecular data demonstrate that *Acer* and *Dipteronia* unite as a well-supported clade (Aceraceae) whose sister group is the family Hippocastanaceae (Figure 4.90). Several investigations of phylogenetic relationships within *Acer* (for 28 and 34 species) have been conducted using nuclear (nrITS) DNA sequence data. Results from those analyses indicate that some sections (sections *Glabra*, *Parviflora*) do not represent natural groups. Data from cpDNA sequences (*trn*L intron, 3′-*trn*L exon, *trn*L/F intergenic spacer) from 57 *Acer* taxa show further evidence for the artificiality of some sections but unfortunately did not include representatives of section *Rubra* (which contains *A. rubrum*). The nrITS data resolve section *Rubra* as being most closely related to section *Hyptiocarpa* (Figure 4.91a). *Acer rubrum* is similar morphologically to *A. saccharinum*

(also section *Rubra*) and the species are closely related as evidenced by results from analysis of nrITS sequence data from 34 *Acer* species (Figure 4.91b) as well as cpDNA restriction site analysis of 22 maple species. Additional species from section *Rubra* should be surveyed to further clarify relationships. Both natural and synthetic hybrids (e.g., *Acer ×freemanii*) are reported between *A. rubrum* and *A. saccharinum* (FAC, FACW), which occurs commonly along river margins. *Acer rubrum* hybridizes successfully with the Asian *A. pycnanthum*. The base chromosome number of *Acer* is $x = 13$. *Acer rubrum* ($2n = 26$) is a diploid; whereas, *A. saccharinum* ($2n = 52$) is tetraploid and *A. pycnanthum* ($2n = 78$) reportedly is hexaploid. Three varieties of *A. rubrum* have been recognized based on relatively minor differences in their leaf morphology. The typical variety (var. *rubrum*) is less prevalent in wetlands than either var. *drummondii* or var. *trilobum*. Analyses to determine whether these varieties (or plants growing under different soil moisture regimes) are distinguishable genetically would be helpful. Preliminary studies have indicated that local adaptation to habitat conditions does occur in this species.

Comments: *Acer rubrum* (vars. *rubrum*, *trilobum*) is distributed widely throughout eastern North America. Variety *drummondii* grows mainly in the southeastern United States.

References: Ackerly & Donoghue, 1998; Burns & Honkala, 1990; Elias, 1987; Farmer & Cunningham, 1981; Hasebe et al., 1998; Nixon & Ely, 1969; Peroni, 1994; Primack & McCall, 1986; Shimizu & Uchida, 1993; Suh et al., 2000; Whitlow & Anella, 1999.

Family 2: Anacardiaceae [70]

Anacardiaceae comprise a fairly large, mainly pantropical family of approximately 600 species of trees, shrubs and woody vines. The species (usually dioecious) are characterized by small, nectariferous, unisexual flowers that are insect pollinated. DNA sequence data (Figure 4.89) resolve the family as the sister group to Burseraceae, which shares with it the presence of resin canals and biflavones (Judd et al., 2002). Anacardiaceae appear to be monophyletic as indicated by DNA sequence data and drupes with a reduced ovule number. Two major clades subdivide the family (Pell & Urbatsch, 2001), the larger of these including *Rhus* and allied genera. Combined nrITS and cpDNA sequence data (Miller et al., 2001; Yi et al., 2004a), demonstrate that *Rhus* sensu stricto (35 species) is monophyletic, but is distinct from several genera (*Actinocheita*, *Cotinus*, *Malosma*, *Searsia*, *Toxicodendron*) that have been merged with it (Figure 4.92). The distinctness of *Rhus* from *Cotinus* and *Toxicodendron* is supported by cladistic analysis of anatomical and morphological data (Aguilar-Ortigoza et al., 2004). Although primarily a family of UPL species, several genera (*Metopium, Rhus, Schinus*) occur facultatively in North American wetlands.

Anacardiaceae are important economically for cashews, mango fruits, and pistachios. The drupes are dispersed by birds, bats, and other mammals. One genus includes obligate aquatics in North America:

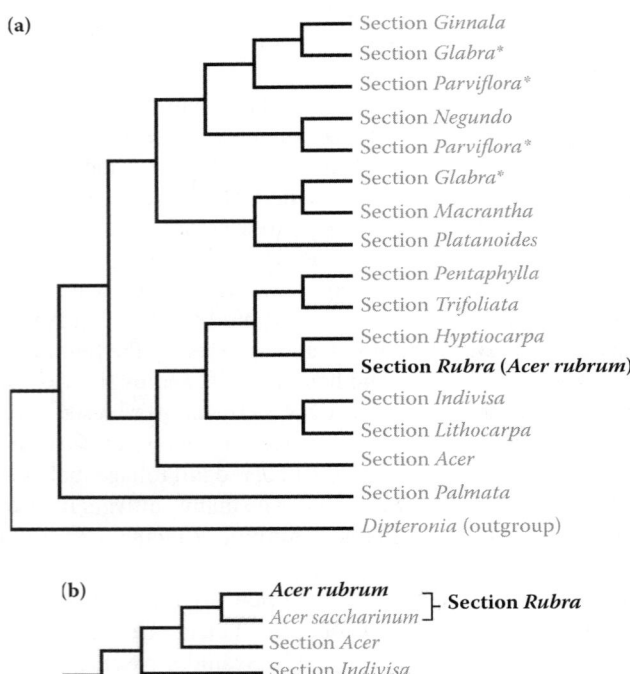

(a)

Section *Ginnala*
Section *Glabra**
Section *Parviflora**
Section *Negundo*
Section *Parviflora**
Section *Glabra**
Section *Macrantha*
Section *Platanoides*
Section *Pentaphylla*
Section *Trifoliata*
Section *Hyptiocarpa*
Section *Rubra* (*Acer rubrum*)
Section *Indivisa*
Section *Lithocarpa*
Section *Acer*
Section *Palmata*
Dipteronia (outgroup)

(b)

Acer rubrum
Acer saccharinum — **Section *Rubra***
Section *Acer*
Section *Indivisa*
Section *Pentaphylla*
Other *Acer*

FIGURE 4.91   (a) Phylogenetic relationships within *Acer* as indicated by nrITS sequence data for 28 species. (b) Relationship of *A. rubrum* as indicated by analysis of nrITS sequence data for 34 *Acer* species. Sections marked by an asterisk are not monophyletic. *Acer rubrum* (bold) is the only North American species previously designated as an obligate aquatic (now FAC). ((a) Adapted from Suh, Y. et al., *J. Plant Res.*, 113, 193–202, 2000; (b) Adapted from Ackerly, D.D. & Donoghue, M.J., *Amer. Nat.*, 152, 767–791, 1998.)

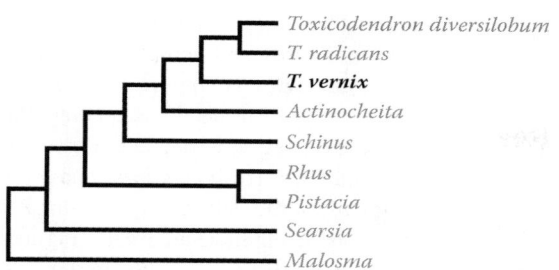

FIGURE 4.92 Phylogenetic relationships among a subset of genera from Anacardiaceae as indicated by combined DNA sequence data. These results are consistent with phylogenetic analyses of morphological data that also show the distinctness of *Toxicodendron* from *Rhus* (with which it has been merged in the past). *Toxicodendron vernix* (in bold) is the only member of the family categorized as an obligate aquatic plant in North America. (Adapted from Yi, T. et al., *Mol. Phylogen. Evol.*, 33, 861–879, 2004a.)

1. *Toxicodendron* Mill.

1. *Toxicodendron*

Poison sumac; sumac à vernis

Etymology: from the Greek *toxikon dendron* meaning "poison tree"

Synonyms: *Philostemon*; *Pocophorum*; *Rhus* (in part); *Rhus-Toxicodendron*; *Vernix*

Distribution: global: Asia; New World; **North America:** widespread

Diversity: global: 40 species; **North America:** 5 species

Indicators (USA): OBL: *Toxicodendron vernix*

Habitat: freshwater; palustrine; **pH:** 5.1–8.2; **depth:** <1 m; **life-form(s):** emergent shrub

Key morphology: stems (to 7 m) slender, branchlets drooping; leaves (to 35 cm) alternate, pinnate with 7–13 leaflets (to 6.5 cm); flowers minute, unisexual, yellowish, in elongate panicles arising from leaf bases; drupes (to 1.3 cm) white, waxy, berry-like, borne in branched clusters

Life history: duration: perennial (buds); **asexual reproduction:** stem suckers; **pollination:** insect; **sexual condition:** dioecious; **fruit:** drupes (common); **local dispersal:** drupes; **long-distance dispersal:** drupes (birds)

Imperilment: (1) *Toxicodendron vernix* [G5]; S1 (KY, NS); S2 (QC, WV); S3

Ecology: general: Several *Toxicodendron* species are found occasionally in North American wetlands but the genus mainly occurs in UPL habitats. One species (*T. vernix*) maintains status as an obligate aquatic throughout its distributional range and is an excellent indicator of wetland conditions. The resinous sap of all species contains toxic phenolic substances that are severely allergenic to most people.

Toxicodendron vernix **(L.) Kuntze** grows in bogs, depressions, ditches, fens, marshes, meadows, prairies, seeps, swamps, and along lake, pond, and stream margins. The plants can occupy either acidic or calcareous sites on substrates ranging from marly muck to sand, sandy loam, sandy peat, or peat. The species is shade tolerant and fast growing but short lived. Pollination is by bees (Insecta: Hymenoptera), which must transport pollen between the male and female plants. Further details on the reproductive ecology of *T. vernix* are scarce. The fruits ripen in the fall, but often persist throughout the winter. The drupes of *Toxicodendron* have physical dormancy and require scarification of the endocarp to promote seed germination. Low germination (3%) has been obtained from unscarified fruits (dry, endocarp removed) that were stored for several months at 4°C, then sowed (natural light) under a 26°C–29°C/18°C–20°C temperature regime; germination occurred after several months. Presoaking the seeds in hot water (80°C–90°C) for 24 h prior to sowing may help to leach out germination inhibitors. Seed dispersal occurs by their ingestion and subsequent defecation by various animals, notably bobwhite (Aves: Odontophoridae: *Colinus virginianus*) and ring-necked pheasant (Aves: Phasianidae: *Phasianus colchicus*). Passage of the fruits through the digestive tract of frugivores enhances seed germination. The plants have no means of vegetative spread, but the stems can produce suckers. **Reported associates:** *Acer rubrum, A. saccharinum, A. saccharum, Alnus rugosa, Aristida stricta, Aronia arbutifolia, A. melanocarpa, Arundinaria gigantea, Betula alleghaniensis, B. lutea, B. papyrifera, B. pumila, Botrychium simplex, Calla palustris, Carex echinata, C. lasiocarpa, C. leptalea, C. oligosperma, C. prairea, C. scoparia, C. seorsa, C. stipata, C. stricta, C. tenuiflora, Carya cordiformis, Castilleja coccinea, Cephalanthus occidentalis, Chamaedaphne calyculata, Chionanthus virginicus, Cladium mariscoides, Clethra alnifolia, Cornus amomum, C. foemina, C. sericea, Cypripedium acaule, C. reginae, Dasiphora floribunda, Drosera rotundifolia, Eriophorum tenellum, E. virginicum, Filipendula rubra, Fraxinus nigra, Gaylussacia baccata, G. frondosa, Huperzia lucidula, Hypericum kalmianum, H. virginicum, Ilex coriacea, I. glabra, I. verticillata, Juncus, Juniperus communis, Kalmia hirsuta, Larix laricina, Lindera benzoin, Liriodendron tulipifera, Listera cordata, Lyonia lucida, Lysimachia terrestris, Lythrum alatum, Magnolia virginiana, Micranthes pensylvanica, Myrica cerifera, M. heterophylla, Nemopanthus mucronatus, Nyssa aquatica, N. ogeche, Oenothera perennis, Onoclea sensibilis, Osmunda regalis, Osmundastrum cinnamomeum, Parnassia glauca, Pedicularis canadensis, Peltandra virginica, Persea borbonea, P. palustris, Physocarpus opulifolius, Picea mariana, Platanthera clavellata, P. integrilabia, P. obtusata, Poa paludigena, Potentilla simplex, Pteridium aquilinum, Pycnanthemum virginianum, Quercus laurifolia, Q. macrocarpa, Rhamnus alnifolia, R. lanceolata, Rhododendron canescens, R. maxima, R. viscosum, Ribes americanum, Rubus hispidus, Salix candida, S. discolor, S. sericea, S. serissima, Sarracenia purpurea, Serenoa repens, Solidago graminifolia, S. ohioensis, S. riddellii, Sorghastrum nutans, Sphagnum, Spiraea alba, S. tomentosa, Symplocarpus foetidus, Thelypteris palustris, Thuja occidentalis, Tilia americana, Trichophorum alpinum, Trientalis borealis, Trollius laxus, Ulmus americana, Vaccinium corymbosum, V. elliottii, V. macrocarpon, V. myrsinites, Viburnum lentago, V. nudum, Viola sagittata, Woodwardia virginica, Zizania aquatica.*

Use by wildlife: The fruits of *Toxicodendron vernix* are eaten by bobwhite (Aves: Odontophoridae: *Colinus virginianus*), ruffed grouse (Aves: Phasianidae: *Bonasa umbellus*), wild turkey (Aves: Phasianidae: *Meleagris gallopavo*), ring-necked pheasant (Aves: Phasianidae: *Phasianus colchicus*), assorted songbirds (Aves), and cottontail rabbits (Mammalia: Leporidae: *Sylvilagus*). They are known to be eaten and dispersed by various woodpeckers (Aves: Picidae) including the downy woodpecker (*Picoides pubescens*), hairy woodpecker (*P. villosus*), Nuttall's woodpecker (*P. nuttallii*), yellow-bellied sapsucker (*Sphyrapicus varius*), red-shafted flicker (*Colaptes auratus*), and yellow-shafted flicker (*C. auratus*). It is a host plant for several Lepidoptera (Arctiidae: *Lophocampa maculata*; Noctuidae: *Eutelia pulcherrimus*; Tortricidae: *Episimus argutanus*). It is a host for the coral-spot fungus (Ascomycota: Nectriaceae: *Nectria cinnabarina*). Some have suggested that the occurrence of catechol (see later) in the leaves of *Toxicodendron* may deter feeding by hemipteran insects.

Economic importance: food: *Toxicodendron vernix* is poisonous and never should be eaten (or even touched). Ingestion of any *Toxicodendron* species (even as an ingredient of herbal remedies) can produce serious (even fatal) gastroenteritis; **medicinal:** The resin of *T. vernix* contains the catechol urushiol (3-*n*-pentadecylcatechol), a toxic phenolic compound that can causes severe allergic contact dermatitis. Although the Cherokee regarded it as poisonous, they administered *T. vernix* as a treatment for ague, asthma, fever, gonorrhea, and ulcerated bladder. The inhalation of smoke from burning *Toxicodendron* plants can be life threatening; **cultivation:** *Toxicodendron vernix* is not cultivated because of its toxic properties; **misc. products:** A brown dye or mordant has been obtained from the tannin-rich leaves of *T. vernix*. Oil extracted from the seeds has been used to make candles, but their pungent smoke may be toxic if inhaled. The sap can be made into an indelible black ink or varnish. Sap from the related *Toxicodendron vernicifluum* oxidizes to a black color when exposed to air and gives oriental lacquerware its characteristic color. Any product derived from this plant should be regarded as potentially toxic; **weeds:** The related (but FAC) species *Toxicodendron radicans* (FAC, FACU) often occurs in wetlands where it can develop into dense, weedy patches, particularly along floodplains; **nonindigenous species:** none

Systematics: Although it has long been merged with the genus *Rhus*, *Toxicodendron* is distinct and actually is related more closely to *Actonicheita* and *Schinus* (Figure 4.92). A fair portion of the genus has been monographed, yet a comprehensive phylogenetic survey of the species is needed. Three sections are recognized as *Simplicifolia*, *Toxicodendron*, and *Venenata*, with *T. vernix* assigned to the latter. Of the three North American species surveyed using molecular data (Figure 4.92), *T. vernix* (section *Venenata*) is distinct from the two species that occur in section *Toxicodendron*. The base chromosome number of *Toxicodendron* is $x = 15$. *Toxicodendron vernix* ($2n = 30$) is a diploid. *Toxicodendron vernix* is not often common where it occurs and studies of its population genetics and reproductive ecology would provide a better understanding of genetic diversity in this dioecious species.

Comments: *Toxicodendron vernix* occurs throughout eastern North America but is not a particularly common species

References: Aguilar-Ortigoza et al., 2003; Elias, 1987; Gillis, 1971; Krefting & Roe, 1949; Kujawski, 2001; McAtee, 1947; Yi et al., 2004a.

Family 3: Simaroubaceae [21]

Results from molecular phylogenetic studies (e.g., Fernando et al., 1995; Gadek et al., 1996) have greatly helped to redefine the rather small family Simaroubaceae (100 species) as a natural, monophyletic group (Judd et al., 2002). One necessary modification has been to include the previously segregated Leitneriaceae within the family (Figure 4.93), a recommendation supported by results of serological (Petersen & Fairbrothers, 1983), phytochemical (Readel et al., 2003), and embryological (Tobe, 2011) analyses. As currently circumscribed, Simaroubaceae are mainly tropical to subtropical trees and shrubs containing quassinoid triterpenoids and having drupes or samara fruits (with one ovule/locule), carpels united by their styles, and unisexual flowers that are pollinated by birds (Aves), insects, or the wind. The family is closely related to Rutaceae (Figure 4.93). The family predominantly is terrestrial aside from *Ailanthus* (FACU, UPL), which occasionally is found in North American wetlands, and one genus that is designated as an obligate aquatic:

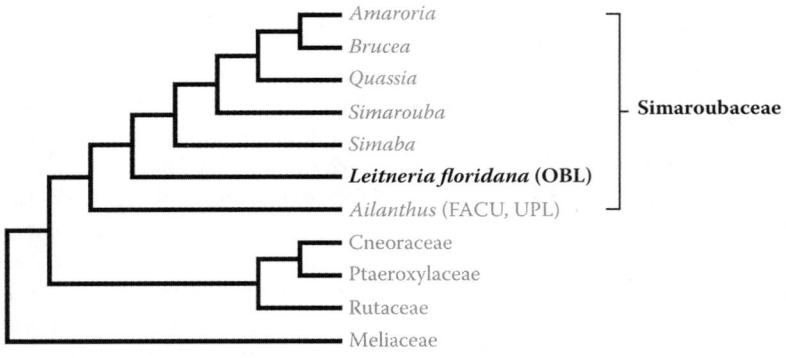

FIGURE 4.93 Intergeneric relationships among seven genera of Simaroubaceae as indicated by analysis of *rbcL* sequence data. Once segregated in the monotypic family Leitneriaceae, *Leitneria* (bold) resolves within the clade corresponding to Simaroubaceae. It is the only genus in the family to contain obligate aquatic plants in North America. (Adapted from Gadek, P.A. et al., *Amer. J. Bot.*, 83, 802–811, 1996.)

1. *Leitneria* Chapm.

1. *Leitneria*

Corkwood

Etymology: after Edward Frederick Leitner (1812–1838)

Synonyms: *Myrica* (in part)

Distribution: global: North America; **North America:** southern

Diversity: global: 1–2 species; **North America:** 1–2 species

Indicators (USA): OBL: *Leitneria floridana*

Habitat: brackish, freshwater; palustrine; **pH:** acidic; **depth:** <1 m; **life-form(s):** emergent shrub, emergent tree

Key morphology: stems (to 8 m) usually unbranched, lenticels numerous; leaves (to 20 cm) simple, alternate, 5-ranked, petiolate (to 7 cm), lanceolate; flowers unisexual, arranged in catkins (precocious), the ♂ drooping, the ♀ erect; perianth absent, stamens 3–15 (♂) or perianth present, ovary (to 2 mm) pubescent (♀); drupes (to 2.5 cm) elliptic, green (fresh) to brown (ripe), with a single seed

Life history: duration: perennial (buds); **asexual reproduction:** adventitious root buds, stem suckers; **pollination:** wind; **sexual condition:** dioecious (rarely bisexual); **fruit:** drupes (common); **local dispersal:** spreading shoots from roots; **long-distance dispersal:** drupes

Imperilment: (1) *Leitneria floridana* [G3]; S1 (GA, TX); S2 (MO); S3 (AR, FL)

Ecology: general: *Leitneria* is monotypic (but see *Systematics*).

Leitneria floridana Chapm. grows in shallow (e.g., 15–100 cm), brackish or fresh waters in depressions, ditches, estuaries, hammocks, marshes, roadsides, sand ponds, swamps, thickets, and along the margins of rivers and streams at elevations of up to 100 m. Exposures range from full to partial sunlight. The substrates have been described as alluvium, clay, clay over limestone, marl, sand, sandy clay, or silt. This species tolerates prolonged inundation and drastic water level fluctuations, but requires wet ground for much of the year. Technically the plants are polygamodioecious (producing some bisexual flowers), but chiefly they are dioecious. The catkins are produced from the previous year's growth and appear prior to the leaves. The flowers are wind pollinated. Little is known regarding the reproductive ecology of this species. The seed viability of *Leitneria* often is low (0%–7%) and natural germination of freshly sown, ripe drupes can be quite low (≤5%). Higher germination (21%–32%) has been achieved by leaching stored drupes (originally collected in Florida) with water and treating them (or excised seeds) with GA_3 (750 mg/L^{-1} for 24 h); similarly treated seeds originating from Missouri did not surpass ≤5% germination. A rate of 48% was obtained for unripened drupes originating from Texas, after scarification with H_2SO_4 and treatment with 1000 mg/L^{-1} GA_3. A germination rate of 29% was observed for seeds that were excised from unripe drupes (collected in Arkansas and Missouri) and then treated for 24 h with 750/1000 mg/L^{-1} GA_3. Thus, it appears that the best germination rate is obtained using unripened drupes. Others have observed that germination increases after 3 months of cold stratification. The plants can form dense patches by clonal growth resulting from a proliferation of stems from the adventitious buds that form along the spreading roots. **Reported associates:** *Acer rubrum, A. saccharinum, Agrimonia gryposepala, Asclepias incarnata, Baccharis halimifolia, Berchemia scandens, Brunnichia ovata, Carex lupuliformis, C. lupulina, Carpinus caroliniana, Carya aquatica, C. illinoensis, Celtis laevigata, Cephalanthus occidentalis, Cirsium vulgare, Cladium jamaicense, Cuphea carthagenensis, Cyperus, Diodia virginiana, Diospyros virginiana, Epidendrum conopseum, Erigeron annuus, Eryngium aquaticum, Eubotrys racemosa, Fraximus pensylvanica, F. tomentosa, Gleditsia aquatica, Ilex decidua, Impatiens capensis, Iva frutescens, Juncus effusus, J. repens, Juniperus virginiana, Lindera melissifolia, Liquidambar styraciflua, Lythrum lanceolatum, Mecardonia acuminata, Mimulus ringens, Myrica cerifera, Nymphaea odorata, Nyssa aquatica, N. biflora, Oxalis, Panicum gymnocarpon, P. hemitomon, Parthenocissus quinquefolia, Persea palustris, Phalaris arundinacea, Plantago lanceolata, Pluchea, Pontederia cordata, Populus heterophylla, Proserpinaca palustris, Prunella vulgaris, Quercus hemisphaerica, Q. laurifolia, Q. lyrata, Q. nigra, Q. nuttallii, Q. palustris, Q. phellos, Q. virginiana, Rhapidophyllum hystrix, Rhynchospora careyana, Rosa palustris, Sabal minor, S. palmetto, Salix nigra, Saururus cernuus, Sesbania vesicaria, Smilax rotundifolia, Spartina bakeri, Styrax americana, Taxodium ascendens, Teucrium canadense, Toxicodendron radicans, Vitis*.

Use by wildlife: The wildlife value of *Leitneria floridana* is not well documented. It is a host of the gypsy moth (Lepidoptera: Lymantriidae: *Lymantria dispar*).

Economic importance: food: none; **medicinal:** *Leitneria floridana* contains 1-methoxycanthinone (a powerful anti-HIV agent), and quassimarin, similakalactone, and 5-methoxycanthinone, which are effective at suppressing growth of human tumors; **cultivation:** *Leitneria floridana* is sold occasionally as a native ornamental plant; **misc. products:** *Leitneria floridana* has the lightest known wood of any North American tree (0.2 kg/dm^3) with a specific gravity (0.21) even less than that of cork (0.24). It is used to make fishing floats and bottle stoppers; **weeds:** none; **nonindigenous species:** none

Systematics: Once regarded as an isolated genus of dubious affinity, molecular phylogenetic studies confirm the association of *Leitneria* within Simaroubaceae. Preliminary surveys of that family indicate a relatively basal position of the genus near *Ailanthus* and *Simaba* (Figure 4.93); however, additional genera need to be surveyed. *Leitneria* usually is regarded as monotypic; however, recent studies have argued for the recognition of a second species (*L. pilosa*) on the basis of leaf morphology, ISSR, and nrITS sequence data analyses. A chromosomal base number of $x = 16$ is indicated by the count reported for *Leitneria floridana* ($2n = 32$).

Comments: *Leitneria floridana* occurs sporadically across the southern United States where it has been rare since its original discovery. Some populations are threatened by lumbering activity and recreational vehicle traffic.

References: Channell & Wood, 1962; Day, 1975; Godfrey & Clewell, 1965; Schrader & Graves, 2011; Sharma & Graves, 2004a,b; Tobe, 2011; Xu et al., 2000.

5 Core Eudicots: Dicotyledons V
"Asterid" Tricolpates

The "asterids" (Figure 5.1) comprise 10 orders (APG, 2003), which resolve as a clade that is well supported by both molecular and nonmolecular characters (Soltis et al., 2005). This group corresponds much to the old Sympetalae (Monopetalae), a taxon uniting many families that were characterized by species with fused petals. Molecular data have been very helpful in refining the delimitation of the group, which is now circumscribed to include various families placed formerly in several disparate groups (Dilleniidae, Hamamelidae, Rosidae). Nearly a third of all angiosperm species occurs within the asterids, which include some 80,000 species in 114 families (Soltis et al., 2005). Asterid species commonly produce iridoids, tropane alkaloids, and caffeic acid and possess flowers with an equal number of petals and stamens and unitegmic ovules with cellular endosperm (Judd et al., 2002; Soltis et al., 2005).

Although the asterids are dominated by terrestrial plants, North America contains more than 550 species that are recognized as obligate aquatics within the group. The aquatic taxa have evolved repeatedly within the clade and are dispersed broadly across the group (Figure 5.1).

Phylogenetic interrelationships have been elucidated fairly well among most asterid orders (Figure 5.1). Those nine orders with obligate aquatic representatives in North America are arranged here in the following sequence:

Basal Asterids	Core Asterids	
	Euasterids I	Euasterids II
1. Cornales	3. Gentianales	6. Apiales
2. Ericales	4. Lamiales	7. Aquifoliales
	5. Solanales	8. Asterales
		9. Dipsacales

BASAL ASTERIDS

ORDER 1: CORNALES [10]

Molecular data from a number of studies have clearly established the monophyly of Cornales, an order containing about 660 species in 40 genera (Soltis et al., 2005). The delimitation of families in the order remains unsettled. Although Alangiaceae, Davidiaceae, and Mastixiaceae have been merged with Cornaceae by many authors, current phylogenetic interpretations (Xiang et al., 2011) favor their individual recognition (Figure 5.2). Generally, these are plants with flowers having a reduced calyx, inferior ovary, an epigynous nectary

disk, and drupes (Judd et al., 2002). The aquatic Old World family Hydrostachyaceae has been placed both in and out of the order by molecular data (Fan & Xiang, 2003; Hufford, 1997; Les et al., 1997); however, the most comprehensive analyses (Xiang et al., 2011) support its placement here. In North America, there are two families within the order that contain obligate aquatic species:

1. **Hydrangeaceae** Dumort.
2. **Nyssaceae** Juss. ex Dumort.

The occurrence of these families within different subclades of Cornales indicates that North American aquatic plants evolved independently in each family (Figure 5.2). Species in the dogwood genus *Cornus* (Cornaceae) often are encountered in wetlands; however, their broad ecological amplitude probably is responsible for their lack of obligate wetland (OBL) status. Some species can tolerate such extremes of habitat as sand dunes and wetlands (Les & Miller, 1979).

Family 1: Hydrangeaceae [17]

Approximately 250 species occur within this relatively small family that often had been recognized as a subfamily of Saxifragaceae. Phylogenetic analyses indicate that Hydrangeaceae are monophyletic, but are distant from Saxifragaceae, occurring instead in Cornales as the sister group to Loasaceae (Judd et al., 2002; Soltis et al., 2005; Figures 5.2 and 5.3).

Many Hydrangeaceae are woody plants and most species are terrestrial. The flowers are nectariferous, protogynous, and are self-pollinated or insect pollinated by beetles (Coleoptera), bees and wasps (Hymenoptera), butterflies and moths (Lepidoptera), or flies (Diptera) (Judd et al., 2002). They are showy or are reduced in clusters that are surrounded by showy, sterile florets. The seeds often are winged and wind dispersed.

Several genera (e.g., *Cardiandra*, *Carpenteria*, *Decumaria*, *Deinanthe*, *Deutzia*, *Hydrangea*, *Kirengeshoma*, *Philadelphus*, and *Schizophragma*) are cultivated as garden ornamentals. Only one North American genus of Hydrangeaceae contains obligate aquatic plants:

1. *Decumaria* L.

1. *Decumaria*

Climbing hydrangea, wood-vamp
Etymology: from the Latin *decimarius*, meaning "tenth" with reference to the often 10-merous flowers

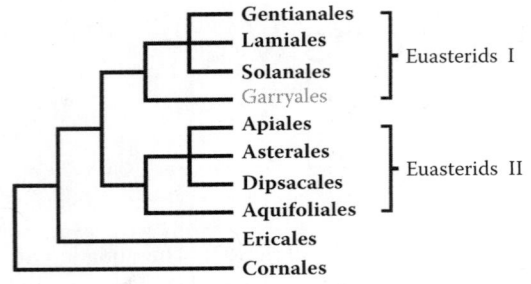

FIGURE 5.1 Relationships among the asterid orders as indicated by results of numerous phylogenetic studies. In North America, OBL indicators (bold) occur in all orders but Garryales. (Adapted from Judd, F.W. et al., *J. Coastal Res.*, 18, 751–759, 2002; Soltis, D.E. et al., *Phylogeny and Evolution of Angiosperms*, Sinauer Associates, Inc., Sunderland, MA, 2005.)

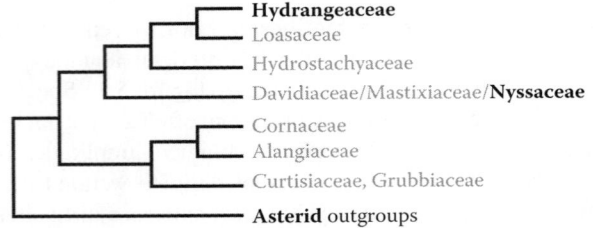

FIGURE 5.2 Phylogenetic relationships among families of the order Cornales as indicated by analysis of six cpDNA data sets. In North America, obligate aquatic plants originated independently in two families within the order (bold). The placement of the aquatic Hydrostachyaceae (Old World) in this order is unusual, and its position shifts when different molecular data are analyzed. (From Xiang, Q.-Y. et al., *Mol. Phylogen. Evol.*, 59, 123–138, 2011.)

Synonyms: none

Distribution: global: Asia; North America; **North America:** southeastern

Diversity: global: 2 species; **North America:** 1 species

Indicators (USA): OBL; FACW: *Decumaria barbara*

Habitat: freshwater; palustrine; **pH:** broad; **depth:** <1 m; **life-form(s):** emergent liana

Key morphology: stems (to 18 m) woody, climbing by fine, adventitious aerial roots (and eventually covered by them); leaves (to 12 cm) opposite, simple, elliptic, shiny above, petioled (to 4 cm); cymes (to 10 cm) compound, terminal on recent growth; flowers epigynous; sepals minute, 7–10; petals (to 6 mm), 7–10, white; stamens 20–30; capsules (to 6 mm) 7–10-locular, 10–15-ribbed; seeds (to 2 mm) numerous

Life history: duration: perennial (buds); **asexual reproduction:** stolons; **pollination:** insect; **sexual condition:** hermaphroditic; **fruit:** capsules (common); **local dispersal:** seeds (gravity); **long-distance dispersal:** seeds (wind)

Imperilment: (1) *Decumaria barbara* [G5]; S1 (DE)

Ecology: general: Only two species of *Decumaria* are known. The Asian *D. sinensis* is not aquatic but inhabits mountain slopes and rocky crevasses. Despite its occasional ranking as a FACW species, the North American *D. barbara* persists only where soils are wet or saturated for most of the season.

Decumaria barbara L. climbs upon vegetation that occurs in floodplains, ravines, seeps, swamps, and along brook, river, and stream margins. The vines attach by aerial roots that penetrate the bark, but rarely cause damage to the supporting plant. A broad diversity of habitats is tolerated, ranging from acidic to calcareous clay to sandy loam. The specific pH range has not been determined for this species. Plants can withstand full sun but thrive in shade and are hardy down to −10°C. Presumably, the flowers are insect pollinated; however, information on their

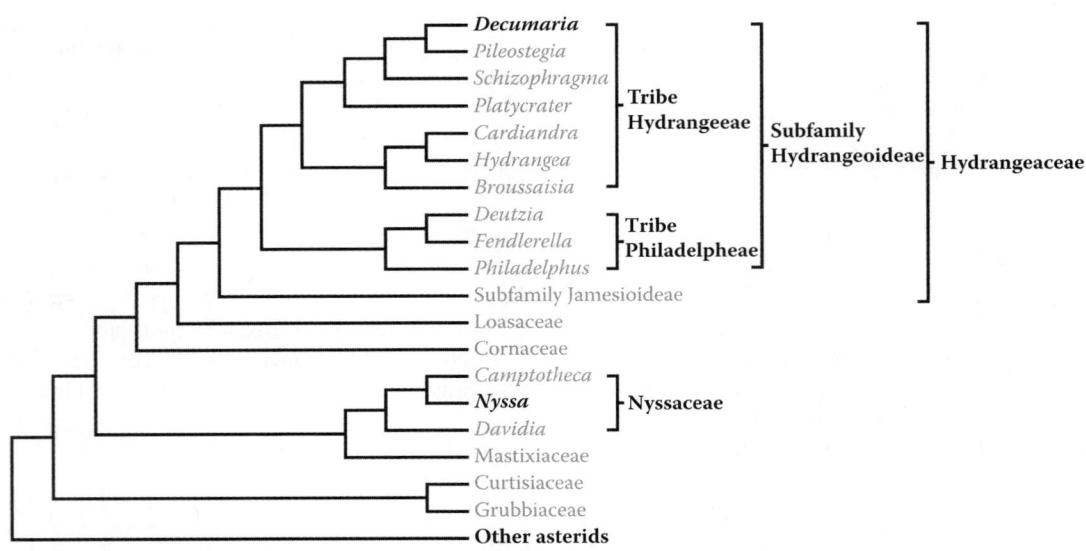

FIGURE 5.3 Relationships among genera of Cornales as indicated by combined DNA sequence data. *Decumaria* (bold), the only genus of Hydrangeaceae with obligate aquatic plants in North America, occurs in a relatively derived position in the family. (Adapted from Fan, C. & Xiang, Q.-Y., *Amer. J. Bot.*, 90, 1357–1372, 2003.)

reproductive ecology is wanting. The seeds are winged and dispersed by wind. Requirements for seed germination have not been determined. Stolon-like branches with reduced leaves are sometimes produced, and creep over rocks and around the bases of nearby trees. The importance of these branches as a potential means of vegetative reproduction is unknown. **Reported associates:** *Acer barbatum, A. rubrum, Adiantum capillus-veneris, Aesculus ×neglecta, A. pavia, Agrimonia pubescens, Alnus serrulata, Alternanthera philoxeroides, Amorpha fruticosa, Ampelopsis arborea, Apios americana, Apocynum, Aquilegia canadensis, Aralia spinosa, Ardisia escallonioides, Arisaema dracontium, A. triphyllum, Arundinaria gigantea, Asimina parviflora, Asplenium bradleyi, A. montanum, A. platyneuron, Athyrium filix-femina, Baccharis halimifolia, Berchemia scandens, Betula nigra, Bignonia capreolata, Boehmeria cylindrica, Brachyelytrum erectum, Callicarpa americana, Calycanthus floridus, Campsis radicans, Carex bromoides, C. debilis, C. intumescens, C. leptalea, C. picta, Carpinus caroliniana, Carya glabra, C. ovalis, Celtis laevigata, Cephalanthus occidentalis, Chasmanthium laxum, C. sessiliflorum, Chiococca alba, Clethra alnifolia, Cocculus carolinus, Commelina virginica, Conocephalum conicum, Coreopsis major, Cornus florida, C. foemina, Crataegus viridis, Cyrilla racemiflora, Desmodium nudiflorum, Dichanthelium dichotomum, Dioscorea villosa, Diospyros virginiana, Elephantopus tomentosus, Euonymus americanus, Fagus grandifolia, Fraxinus americana, F. caroliniana, F. pennsylvanica, F. profunda, Gaylussacia frondosa, Gelsemium sempervirens, Gleditsia aquatica, Hamelia patens, Heuchera americana, H. parviflora, Hexastylis arifolia, Hydrangea quercifolia, Hydrocotyle umbellata, H. verticillata, Hypericum hypericoides, Hypoxis hirsuta, Ilex cassine, I. coriacea, I. decidua, I. glabra, I. opaca, Illicium parviflorum, Iris verna, Itea virginica, Juglans nigra, Juncus effusus, Juniperus virginiana, Kalmia latifolia, Leersia, Leucothoe axillaris, L. racemosa, Lindera benzoin, Liquidambar styraciflua, Liriodendron tulipifera, Lonicera japonica, L. sempervirens, Lycopus rubellus, Lyonia ferruginea, L. ligustrina, L. lucida, Lysimachia, Macbridea caroliniana, Magnolia macrophylla, M. virginiana, Medeola virginiana, Menispermum canadense, Mikania scandens, Mitchella repens, Myrica cerifera, Myrsine floridana, Nyssa aquatica, N. biflora, N. ogeche, N. sylvatica, Onoclea sensibilis, Osmunda regalis, Osmundastrum cinnamomeum, Ostrya virginiana, Oxydendrum arboreum, Parthenocissus quinquefolia, Passiflora lutea, Persea palustris, Persicaria setacea, Pilea pumila, Pinus taeda, Planera aquatica, Poa autumnalis, Polystichum acrostichoides, Pontederia cordata, Populus heterophylla, Ptelea trifoliata, Pteridium aquilinum, Quercus alba, Q. falcata, Q. laurifolia, Q. muehlenbergii, Q. nigra, Q. phellos, Q. rubra, Q. shumardii, Q. stellata, Rhamnus caroliniana, Rhapidophyllum hystrix, Rhododendron canescens, R. nudiflorum, R. viscosum, Rhynchospora capitellata, Rosa carolina, Sabal minor, S. palmetto, Salix nigra, Sambucus nigra, Samolus valerandi, Sanicula canadensis, Saururus cernuus, Serenoa repens, Silene rotundifolia, Silphium compositum, Smilax bona-nox, S. ecirrata, S. glauca, S. laurifolia, S. rotundifolia, S. tamnoides, S. walteri, Smilicina racemosa, Solidago caesia, Sphagnum lescurii, Styrax grandifolius, Symphyotrichum lateriflorum, Symplocos tinctoria, Taxodium ascendens, T. distichum, Thalictrum thalictroides, Thelypteris kunthii, T. noveboracensis, T. palustris, T. pilosa, Tilia americana, Toxicodendron radicans, T. vernix, Triadenum virginicum, Tsuga canadensis, Ulmus americana, U. crassifolia, U. rubra, Uvularia perfoliata, Vaccinium corymbosum, V. elliottii, V. fuscatum, V. pallidum, V. stamineum, Viburnum acerifolium, V. nudum, V. obovatum, Vitis cinerea, V. rotundifolia, Woodwardia areolata, W. virginica.*

Use by wildlife: The sweetly fragrant flowers of *Decumaria barbara* are attractive to butterflies (Insecta: Lepidoptera). Its foliage provides cover and nesting sites for birds (Aves) and small mammals (Mammalia). The plants comprise an element of habitat for the Swainson's warbler (Aves: Parulidae: *Limnothlypis swainsonii*). *Decumaria* is categorized as a staple browse plant for deer (Mammalia: Cervidae: *Odocoileus*). The leaves are host to a leafspot fungus (Ascomycota: Mycosphaerellaceae: *Pseudocercospora decumariae*).

Economic importance: food: none; **medicinal:** none; **cultivation:** *Decumaria barbara* is widely grown as an ornamental vine. Cultivars include 'Barbara Ann,' 'Barber Creek,' 'Chattanooga,' 'Chattooga,' 'Chauga,' 'Margaret,' 'Savannah River,' and 'Vicki'; **misc. products:** *Decumaria barbara* is planted as a groundcover; **weeds:** none; **nonindigenous species:** none

Systematics: *Decumaria* is assigned to tribe Hydrangeeae, subfamily Hydrangeoideae of Hydrangeaceae. Phylogenetic analysis of morphological data clearly indicates that *Decumaria* is monophyletic and with only two species, interspecific relationships are unambiguous. Within Hydrangeaceae, morphological and molecular data (Figure 5.3) place *Decumaria* in a relatively derived position as the sister genus to *Pileostegia*, a small Asian genus of forest species. These two genera, with *Schizophragma*, comprise a clade of plants having a climbing habit. The base chromosome number of *Decumaria* is unknown as counts have not yet been made for either species.

Comments: *Decumaria barbara* occurs in the southeastern United States where it is relatively common.

References: Godfrey & Wooten, 1981; Golden, 1979; Golley et al., 1965; Hufford, 1995, 1997; Light et al., 2002; Tracy & Earle, 1899; Wells, 1928.

Family 2: Nyssaceae [3]

Although Nyssaceae often are merged with Cornaceae, phylogenetic relationships indicated by analyses of different molecular data sets do not support such a disposition reliably. Nyssaceae are monophyletic and consistently group within a clade that includes Davidiaceae (sometimes merged with Nyssaceae) and Mastixiaceae (Figures 5.2 and 5.3). However, the placement of this clade is depicted inconsistently when different molecular data are analyzed (e.g., Figure 5.2 vs. Figure 5.3). Because recognition of Nyssaceae as a separate family is compatible with all recent analyses, the group is maintained here as distinct. Regardless, Cornaceae and Nyssaceae are relatively closely related as evidenced by their shared possession of seeds with germination valves, transseptal ovule bundles, ellagitannins, and a base chromosome number of $x = 11$ (Soltis et al., 2005).

Nyssaceae as recognized here are a small family containing about a dozen species distributed among three genera of which two (*Camptotheca* [2 species]; *Davidia* [monotypic]) are Asian in distribution. *Davidia* sometimes is given family status. All of the species are woody plants, which exhibit a strong affinity for sites with ample soil moisture. *Camptotheca* is the source of the anticancer drug Camptothecin. Only one North American genus contains obligate aquatic species:

1. *Nyssa* L.

1. *Nyssa*

Black gum; water tupelo

Etymology: after *Nysa*, a water nymph of Greek mythology
Synonyms: *Agathisanthes*; *Ceratostachys*; *Daphniphyllopsis*; *Streblina*; *Tupelo*
Distribution: global: Asia; North America; **North America:** eastern
Diversity: global: 10 species; **North America:** 5 species
Indicators (USA): OBL: *Nyssa aquatica*, *N. biflora*, *N. ogeche*; **FACW:** *N. biflora*
Habitat: freshwater; palustrine; **pH:** 2.5–7.0; **depth:** 0–4 m; **life-form(s):** emergent shrub, emergent tree
Key morphology: swollen-based trees (to 40 m) or multi-trunked shrubs (to 18 m) with diaphragm pith; leaves (to 30 cm) alternate, simple, elliptical to ovoid, petioled (to 6 cm), margins entire or with a few large, widely spaced teeth; flowers minute, greenish, unisexual (mainly) or bisexual, numerous in axillary, stalked, spherical heads or racemes (♂), or 1–3 flowers (♀, ♂/♀) terminating a stalked (to 4 cm) cluster; drupes (to 4.0 cm) black, blue, purple, or red, with a ridged or winged endocarp.
Life history: duration: perennial (buds); **asexual reproduction:** stump sprouts; **pollination:** insect or wind; **sexual condition:** polygamodioecious; **fruit:** drupes (common); **local dispersal:** seeds (birds, gravity, water); **long-distance dispersal:** seeds (birds)
Imperilment: (1) *Nyssa aquatica* [G5]; S3 (KY); (2) *N. biflora* [G5]; SH (MO); S2 (AR); (3) *N. ogeche* [G4]
Ecology: general: *Nyssa* species commonly inhabit wet sites and often dominate the vegetation of wooded swamps in the southeastern United States. *Nyssa sylvatica* [facultative (FAC)] and *N. ursina* (no wetland status) can occur in wet habitats but to a lesser degree than the three species ranked as OBL. It is difficult to study the reproductive ecology of large trees, and there is some question regarding the relative importance of wind versus insect pollination in at least some of the species. The ecology of the three OBL species is similar and they can co-occur at some sites and in general share similar associates. Yet, each has subtle ecological differences that are compared later.

Nyssa aquatica L. is a tree that inhabits bayous, depressions, flats, floodplains, lake margins, oxbows, sloughs, and swamps at relatively low elevations. The habitats are

characterized by nearly continuous flooding and once established, the plants can tolerate prolonged water depths up to 4 m. Sites are acidic (pH: 2.5–8.0; mean: 4.9) with substrates having a high clay content (mean: 23%), high cation exchange capacity (10.4–17.9) and substantial organic matter content (1%–29%). Plants are shade intolerant and withstand temperatures down to −29°C. Although principally dioecious, they bear hermaphroditic flowers occasionally. Field studies indicate that male trees allocate more resources, grow to a larger size, and occupy shallower sites than female trees. The flowers are wind pollinated, and possibly also bee (Hymenoptera) pollinated. The larger size of male trees facilitates pollen dispersal by wind. Seed production generally initiates when plants approach 30 years of age (although seed are produced occasionally in much younger plants) and occurs each year. The seeds are relatively heavy and are dispersed locally by gravity; however, they are also buoyant (remaining afloat for up to 3 months) and are dispersed much more effectively by water, often to distances of 1800 m. Even longer dispersal distances are achieved by bird (Aves) vectors. The seeds can retain their viability after 14 months of inundation but are relatively short lived and do not form a seed bank. They will germinate without pretreatment, or can be stratified (at 2°C–4°C) and stored for up to 30 months to delay germination. Productivity (biomass) of established trees can be substantially higher in sites that are nearly permanently flooded than those where flooding is cyclic; however, periods of drawdown are necessary for seed germination and seedling establishment. The seeds will not germinate if they are inundated and the highest germination occurs when they are shallowly buried (to 3 cm) in sites receiving full sunlight. Seedling establishment is much higher in natural swamps than in disturbed sites. The seedlings often establish near cypress "knees" or logs, which prevent them from washing away during flood events. High seedling mortality can occur if flooding occurs during their early growth phase. Good seedling growth has been observed on soils having a relatively high (e.g., 22%) organic matter content. The seedlings do not tolerate fly ash and exhibit up to 75% reduction in biomass where fly-ash concentrations reach 5%–10%. The roots are highly flood tolerant and accumulate ethanol when grown under inundated conditions. *Nyssa aquatica* is a dominant tree of many southeastern seasonally flooded swamps. Adult trees often occur in dense stands but site thinning does not appear to enhance their growth. Survival of seedlings is also not influenced by canopy thinning even though associated herbaceous vegetation may triple in biomass. *Nyssa aquatica* readily produces stump sprouts, which facilitates the vegetative regeneration of damaged or injured plants. **Reported associates:** *Acer negundo*, *A. rubrum*, *Alisma subcordatum*, *Alnus serrulata*, *Alternanthera philoxeroides*, *Ampelaster carolinianus*, *Ampelopsis arborea*, *Andropogon virginicus*, *Angelica*, *Apios americana*, *Arisaema triphyllum*, *Aronia arbutifolia*, *Arundinaria tecta*, *Asclepias perennis*, *Asimina triloba*, *Athyrium filix-femina*, *Bacopa caroliniana*, *B. monnieri*, *Berchemia scandens*, *Betula nigra*, *Bidens discoideus*, *B. frondosus*, *Bignonia capreolata*, *Boehmeria cylindrica*, *Boltonia apalachicolensis*, *Botrychium*, *Brunnichia ovata*,

Campsis radicans, Cardamine, Carex alata, C. albolutescens, C. amphibola, C. bebbii, C. bromoides, C. comosa, C. crinita, C. crus-corvi, C. gigantea, C. glaucescens, C. glaucodea, C. grayi, C. intumescens, C. joorii, C. louisianica, C. lupulina, C. lurida, C. pseudocyperus, C. rosea, C. scoparia, C. seorsa, C. typhina, Carpinus caroliniana, Carya aquatica, C. cordiformis, Celtis laevigata, Centella asiatica, Cephalanthus occidentalis, Chamaecyparis thyoides, Chasmanthium latifolium, C. laxum, Cicuta maculata, Clematis crispa, Clethra alnifolia, Climacium americanum, Coleataenia anceps, Commelina virginica, Cornus foemina, Crataegus, Cuscuta compacta, Cyperus pseudovegetus, Cyrilla racemiflora, Decumaria barbara, Dichanthelium commutatum, D. dichotomum, Dicliptera brachiata, Dioscorea, Diospyros virginiana, Dulichium arundinaceum, Echinochloa walteri, Echinodorus cordifolius, Elymus virginicus, Epidendrum magnoliae, Erechtites hieracifolius, Eryngium baldwinii, Eubotrys racemosa, Eupatorium capillifolium, E. compositifolium, Eutrochium purpureum, Forestiera acuminata, F. ligustrina, Fraxinus caroliniana, F. pennsylvanica, F. profunda, Galium obtusum, G. tinctorium, Gelsemium sempervirens, Gleditsia aquatica, Glyceria striata, Gratiola virginiana, Gymnadeniopsis clavellata, Habenaria repens, Hibiscus moscheutos, Hydrocotyle ranunculoides, H. umbellata, H. verticillata, Hymenocallis crassifolia, Hymenocallis pygmaea, H. densiflorum, H. hypericoides, Hypochaeris radicata, Hypoxis curtissii, H. hirsuta, Ilex amelanchier, I. cassine, I. decidua, I. opaca, I. verticillata, Impatiens capensis, Iris virginica, Itea virginica, Juncus coriaceus, J. effusus, J. repens, J. tenuis, Justicia ovata, Leersia lenticularis, L. virginica, Lemna gibba, Ligustrum sinense, Liquidambar styraciflua, Lobelia cardinalis, L. inflata, Lonicera japonica, Ludwigia palustris, L. repens, Lycopus rubellus, Lygodium japonicum, Lyonia ligustrina, L. lucida, Magnolia virginiana, Menispermum canadense, Micranthemum umbrosum, Microstegium vimineum, Mikania cordifolia, M. scandens, Mimulus alatus, Mitchella repens, Morus rubra, Murdannia keisak, Myrica cerifera, Myriophyllum heterophyllum, M. pinnatum, Nymphoides aquatica, N. cordata, Nyssa aquatica, N. biflora, N. ogeche, Onoclea sensibilis, Osmunda regalis, Osmundastrum cinnamomeum, Packera glabella, Panicum hemitomon, P. rigidulum, Parthenocissus quinquefolia, Peltandra virginica, Penthorum sedoides, Persea borbonia, P. palustris, Persicaria arifolia, P. glabra, P. hydropiperoides, P. maculosa, P. punctata, P. sagittata, P. setacea, P. virginiana, Phanopyrum gymnocarpon, Phoradendron serotinum, Phytolacca americana, Pilea pumila, Pinus taeda, Planera aquatica, Platanus ocidentalis, Pleopeltis polypodioides, Pluchea camphorata, Pontederia cordata, Populus deltoides, P. heterophylla, Proserpinaca palustris, P. pectinata, Prunus serotina, Pseudognaphalium, Quercus falcata, Q. laurifolia, Q. lyrata, Q. michauxii, Q. nigra, Q. phellos, Rhododendron viscosum, Rhynchospora corniculata, R. miliacea, Riccia fluitans, Rosa palustris, Rubus, Rudbeckia laciniata, Rumex, Sabal minor, Sabatia, Saccharum baldwinii, S. giganteum, Sacciolepis striata, Sagittaria falcata, S. graminea, S. lancifolia, S. latifolia, Salix caroliniana, Sambucus nigra, Saururus cernuus, Scirpus cyperinus, Scutellaria lateriflora, Sebastiania fruticosa, Senecio glabellus, Smilax auriculata, S. bona-nox, S. laurifolia, S. rotundifolia, S. smallii, S. tamnoides, S. walteri, Solidago rugosa, Sparganium americanum, Spirodela polyrrhiza, Styrax americanus, Symphyotrichum dumosum, S. lateriflorum, Taxodium ascendens, T. distichum, Thelypteris palustris, Thyrsanthella difformis, Tillandsia usneoides, Toxicodendron radicans, Triadenum tubulosum, T. walteri, Triadica sebifera, Ulmus alata, U. americana, U. rubra, Utricularia purpurea, Vaccinium corymbosum, V. elliottii, V. formosum, V. fuscatum, V. pallidum, Viburnum dentatum, V. nudum, V. rufidulum, Viola palmata, V. primulifolia, V. pubescens, V. tripartita, Vitis aestivalis, V. rotundifolia, Wisteria frutescens, Wolffiella gladiata, Woodwardia areolata, W. virginica, Zizaniopsis miliacea.

***Nyssa biflora* Walter** is a tree that grows in bayous, bays, canals, coves, estuaries, ponds, sloughs, swamps, lake margins, river bottoms, and margins of rivers at low elevations. Although it grows in sites that are inundated for much of the growing season, it does not occur in deeper areas but thrives where water is shallow and slowly moving. Decreased growth occurs under conditions of prolonged deep inundation (unlike *N. aquatica*), stagnant water, or periodic drying cycles. Sites are acidic (pH: 4.5–5.7) and substrates include organic muck, heavy clay, and wet sand (entisols, inceptisols, ultisols). Plants are shade intolerant, can withstand temperatures down to −28°C, and are highly susceptible to fire damage if sites dry out enough to allow the enveloping peat layers to burn. They are quickly injured by salt spray. Plants essentially are dioecious, but bisexual flowers may also occur. The nectariferous flowers are pollinated primarily by bees (Hymenoptera) and secondarily by wind. Seed production is prolific, averaging nearly a million seeds per hectare annually. Seed viability typically exceeds 60%. Unlike *N. aquatica*, the seeds do not float and their dispersal is achieved primarily by birds, which disseminate more than 50% of the annual seed crop. Viability of seeds that pass through the digestive tract of birds is reduced somewhat to around 44%. Germination requires no stratification or pretreatment but will not occur while seeds are submerged in flooded soils. This strategy assures that seeds germinate only when suitable noninundated conditions are available for seedling establishment. However, seeds normally overwinter. In moist but drained soil, the highest and most rapid germination occurs at temperatures above 21°C (as high as 78% germination within 1 week at 33°C). Germination can be enhanced somewhat by removal of the seed coats. Extensive seedling mortality will occur if inundation persists during their early growth or in waters high in sulfate. The plants regenerate quickly from stump sprouts, which grow rapidly and are capable of producing seeds within 2 years. Genetic studies have identified seedling genotypes that are adapted differently to blackwater rivers, headwater swamps and ponds. **Reported associates:** *Acer barbatum, A. negundo, A. rubrum, A. saccharinum, Aesculus pavia, Akebia quinata, Alisma subcordatum, Alnus serrulata, Alternanthera philoxeroides, Amelanchier canadensis, Ampelopsis arborea, Amphicarpum muhlenbergianum, Andropogon glaucopsis,*

A. glomeratus, A. gyrans, A. virginicus, Apios americana, Apteria aphylla, Aralia spinosa, Arisaema triphyllum, Aristida palustris, Aristolochia macrophylla, Aronia arbutifolia, A. melanocarpa, Arundinaria gigantea, A. tecta, Asclepias perennis, Asimina triloba, Asplenium platyneuron, Athyrium filix-femina, Baccharis halimifolia, Bacopa caroliniana, Bartonia paniculata, Berchemia scandens, Betula nigra, Bidens frondosus, Bignonia capreolata, Boehmeria cylindrica, Botrychium biternatum, B. dissectum, Brachyelytrum erectum, Briza minor, Brunnichia ovata, Callicarpa americana, Campsis radicans, Canna flaccida, Cardamine, Carex abscondita, C. alata, C. albolutescens, C. aquatilis, C. atlantica, C. bromoides, C. comosa, C. crebriflora, C. crinita, C. crus-corvi, C. debilis, C. digitalis, C. festucacea, C. folliculata, C. frankii, C. gigantea, C. glabra, C. glaucescens, C. grayi, C. grisea, C. gynandra, C. intumescens, C. joorii, C. leptalea, C. lonchocarpa, C. louisianica, C. lupulina, C. lupuliformis, C. lurida, C. lutea, C. meadii, C. muehlenbergii, C. oxylepis, C. radiata, C. retroflexa, C. rosea, C. seorsa, C. stipata, C. striata, C. stricta, C. styloflexa, C. tribuloides, C. typhina, C. venusta, C. verrucosa, Carpinus caroliniana, Carya alba, C. aquatica, C. cordiformis, C. ovata, Celtis laevigata, Centella asiatica, Cephalanthus occidentalis, Cerastium glomeratum, Cercis canadensis, Chamaecyparis thyoides, Chamaedaphne calyculata, Chasmanthium latifolium, C. laxum, C. nitidum, Chelone glabra, Chimaphila maculata, Chionanthus virginicus, Cicuta maculata, Cladium jamaicense, Clematis virginiana, Clethra acuminata, C. alnifolia, Cliftonia monophylla, Climacium americanum, Coleataenia anceps, Commelina virginica, Conoclinium coelestinum, Conyza canadensis, Coreopsis gladiata, Cornus florida, C. foemina, Crataegus crus-galli, C. marshallii, C. opaca, C. viridis. Crinum americanum, Cynodon dactylon, Cyperus, Cyrilla racemiflora, Decodon verticillatus, Decumaria barbara, Desmodium glutinosum, Dichanthelium acuminatum, D. boscii, D. clandestinum, D. commutatum, D. dichotomum, D. erectifolium, D. laxiflorum, D. strigosum, Dicliptera brachiata, Diodia virginiana, Dioscorea floridana, D. villosa, Diospyros virginiana, Ditrysinia fruticosa, Doellingeria umbellata, Drosera brevifolia, D. rotundifolia, Dryopteris ludoviciana, Dulichium arundinaceum, Echinodorus cordifolius, Elaeagnus pungens, Eleocharis equisetoides, E. quadrangulata, E. robbinsii, E. tortilis, E. tricostata, E. tuberculosa, E. vivipara, Elephantopus carolinianus, E. elatus, E. tomentosus, Eleusine indica, Elymus hystrix, Encyclica tampensis, Epidendrum magnoliae, Erechtites hieracifolius, Eriocaulon compressum, E. decangulare, Eryngium baldwinii, Erythrina herbacea, Eubotrys racemosa, Euonymus americanus, Eupatorium capillifolium, E. compositifolium, E. leucolepis, E. mohrii, E. perfoliatum, E. pilosum, E. rotundifolium, E. semiserratum, E. serotinum, Euthamia graminifolia, Eutrochium dubium, E. fistulosum, E. purpureum, Fagus grandifolia, Festuca subverticillata, Forestiera acuminata, Frangula caroliniana, Fraxinus americana, F. caroliniana, F. pennsylvanica, F. profunda, Galium bermudense, G. obtusum, G. pilosum, G. tinctorium, G. triflorum, Gaylussacia frondosa, G.

mosieri, Gelsemium rankinii, G. sempervirens, Gentiana saponaria, Geum canadense, Gleditsia, Glyceria canadensis, G. septentrionalis, G. striata, Gonolobus suberosus, Gordonia lasianthus, Gratiola ramosa, G. virginiana, Gymnadeniopsis clavellata, Habenaria repens, Halesia diptera, Hamamelis virginiana, Helenium autumnale, Heliopsis helianthoides, Hexastylis arifolia, H. shuttleworthii, Hydrocotyle bonariensis, H. umbellata, H. verticillata, Hymenocallis crassifolia, H. occidentalis, H. pygmaea, Hypericum canadense, H. chapmanii, H. cistifolium, H. fasciculatum, H. galioides, H. gymnanthum, H. hypericoides, H. mutilum, H. myrtifolium, Hypoxis curtissii, H. hirsuta, Ilex ambigua, I. amelanchier, I. cassine, I. coriacea, I. decidua, I. glabra, I. laevigata, I. myrtifolia, I. opaca, I. verticillata, I. vomitoria, Illicium floridanum, Impatiens capensis, Iris hexagona, I. prismatica, I. tridentata, I. versicolor, I. virginica, Itea virginica, Juncus coriaceus, J. effusus, J. pelocarpus, J. polycephalus, J. repens, J. subcaudatus, Juniperus virginiana, Justicia ovata, Kalmia angustifolia, K. latifolia, Lachnanthes caroliniana, Lactuca, Lechea racemulosa, Leersia hexandra, L. lenticularis, L. virginica, Leucothoe axillaris, L. racemosa, Ligustrum canadense, L. lucidum, L. sinense, Lindera benzoin, L. melissifolia, Linum sulcatum, Liquidambar styraciflua, Liriodendron tulipifera, Litsea aestivalis, Lobelia cardinalis, L. georgiana, L. puberula, Lonicera japonica, L. sempervirens, Ludwigia alternifolia, L. leptocarpa, L. palustris, L. pilosa, L. sphaerocarpa, Lycopodiella alopecuroides, Lycopus rubellus, L. virginicus, Lyonia lucida, L. mariana, Lysimachia quadrifolia, Magnolia grandiflora, M. macrophylla, M. tripetala, M. virginiana, Malaxis spicata, M. unifolia, Malus, Matelea carolinensis, Melanthera nivea, Menispermum canadense, Micranthemum umbrosum, Microstegium vimineum, Mikania scandens, Mitchella repens, Mnesithea rugosa, Morus rubra, Muhlenbergia, Murdannia keisak, Myrica caroliniensis, M. cerifera, M. inodora, Myriophyllum, Nymphaea odorata, Nyssa aquatica, N. ogeche, Oenothera, Oldenlandia uniflora, Onoclea sensibilis, Oplismenus hirtellus, Orontium aquaticum, Osmanthus americanus, Osmunda regalis, Osmundastrum cinnamomeum, Ostrya virginiana, Oxalis stricta, Oxydendrum arboreum, Oxypolis rigidior, Packera glabella, Pallavicinia lyellii, Panicum hemitomon, P. rigidulum, P. verrucosum, Parthenocissus quinquefolia, Passiflora lutea, Peltandra sagittifolia, P. virginica, Penthorum sedoides, Persea borbonia, P. palustris, Persicaria glabra, P. hirsuta, P. hydropiperoides, P. maculosa, P. punctata, P. sagittata, P. setacea, P. virginiana, Phanopyrum gymnocarpon, Phegopteris hexagonoptera, Phlox glaberrima, Phoradendron serotinum, Photinia pyrifolia, Phytolacca americana, Pieris phillyreifolia, Pilea pumila, Pinus glabra, P. caribaea, P. echinata, P. elliottii, P. palustris, P. serotina, P. taeda, Planera aquatica, Platanthera cristata, P. flava, P. integrilabia, Pleopeltis polypodioides, Pluchea baccharis, P. camphorata, Poa autumnalis, Polygala cymosa, Polypodium virginianum, Polystichum acrostichoides, Pontederia cordata, Ponthieva racemosa, Populus heterophylla, Prenanthes serpentaria, Proserpinaca palustris, P. pectinata, Prunus caroliniana, P. serotina, Pteridium aquilinum, Pycnanthemum setosum, P. tenuifolium, Quercus alba, Q. falcata, Q. hemisphaerica, Q. laurifolia, Q. lyrata, Q. michauxii, Q. nigra, Q. pagoda, Q. phellos, Q. velutina, Q. virginiana, Rhapidophyllum hystrix, Rhexia aristosa, R. nashii, R. virginica, Rhododendron alabamense, R. periclymenoides, R. viscosum, Rhus copallinum, Rhynchospora caduca, R. careyana, R. cephalantha, R. chalarocephala, R. corniculata, R. elliottii, R. fascicularis, R. filifolia, R. gracilenta, R. harperi, R. inundata, R. latifolia, R. microcarpa, R. microcephala, R. miliacea, R. perplexa, R. pleiantha, R. pusilla, R. tracyi, Riccia fluitans, Rosa palustris, Rubus argutus, R. cuneifolius, R. flagellaris, R. hispidus, R. trivialis, Ruellia, Rumex verticillatus, Sabal minor, S. palmetto, Sabatia brachiata, S. difformis, Saccharum baldwinii, S. giganteum, Sagittaria fasciculata, S. lancifolia, S. latifolia, Salix caroliniana, Sambucus nigra, Samolus valerandi, Sanicula canadensis, Sarracenia, Sassafras albidum, Saururus cernuus, Schoenoplectus subterminalis, Scirpus cyperinus, Scleria georgiana, S. oligantha, S. reticularis, S. triglomerata, Sclerolepis uniflora, Scutellaria elliptica, S. integrifolia, S. lateriflora, Sebastiania fruticosa, Selaginella apoda, Serenoa repens, Sideroxylon lycioides, S. reclinatum, Sisyrinchium, Sium suave, Smilax auriculata, S. bona-nox, S. glauca, S. latissimifolia, S. laurifolia, S. rotundifolia, S. smallii, S. tamnoides, S. walteri, Solidago caesia, S. patula, S. rugosa, S. stricta, Sorghastrum nutans, Sphagnum macrophyllum, S. portoricense, Spiranthes odorata, Stellaria media, Styrax americanus, Symphyotrichum dumosum, S. lateriflorum, S. novi-belgii, S. pilosum, Symplocos tinctoria, Taxodium ascendens, T. distichum, Thaspium barbinode, Thelypteris kunthii, T. palustris, T. noveboracensis, Thyrsanthella difformis, Tilia americana, Tillandsia bartramii, T. recurvata, T. usneoides, Toxicodendron radicans, T. vernix, Triadenum tubulosum, T. virginicum, T. walteri, Triadica sebifera, Typha latifolia, Ulmus alata, U. americana, U. rubra, Usnea, Utricularia gibba, U. inflata, U. juncea, U. purpurea, Uvularia sessilifolia, Vaccinium arboreum, V. corymbosum, V. elliottii, V. formosum, V. fuscatum, V. pallidum, V. soraria, V. stamineum, V. tenellum, V. virgatum, Vernonia missurica, Viburnum dentatum, V. nudum, V. recognitum, V. rufidulum, Viola primulifolia, Vitis aestivalis, V. cinerea, V. labrusca, V. rotundifolia, Wisteria frutescens, Woodwardia areolata, W. virginica, Xanthorhiza simplicissima, Xyris ambigua, X. fimbriata, X. stricta, Yeatesia viridiflora, Zenobia pulverulenta, Zephyranthes simpsonii, Zizaniopsis miliacea.

Nyssa ogeche **W. Bartram ex Marshall** is a shrub or tree, which occurs in depressions, floodplains, marshes, swamps, and along lake, pond, river, and stream margins at low elevations. Sites are acidic (pH: 4.5–7.0) and typically remain inundated for long periods of time by slightly flowing water. Substrates predominantly are alluvial soils (mainly inceptisols). The plants are shade intolerant and can withstand temperatures down to −22°C. They appear to be relatively short lived with fruit production occurring in as little as 3 years. The flowers (entirely unisexual or mixed with hermaphrodites) are nectariferous and pollinated by bees. The fresh

fruits (or seeds) are quite large and sink; however, dried fruits retain some buoyancy and can be dispersed locally by water. Longer dispersal distances are the result of fruit consumption and dissemination by birds and small mammals. Seedling mortality occurs when the surface dries out and their growth slows in densely grassy sites. The plants can regenerate extensively from stump and root sprouts. **Reported associates:** *Acer rubrum, Alnus serrulata, Ampelopsis arborea, Asclepias perennis, Berchemia scandens, Betula nigra, Bignonia capreolata, Boehmeria cylindrica, Boltonia apalachicolensis, Campsis radicans, Carex glaucescens, C. striata, C. verrucosa, Centella asiatica, Cephalanthus occidentalis, Chamaecyparis thyoides, Chasmanthium laxum, Cladium jamaicense, Clethra, Cliftonia, Cornus foemina, C. stricta, Crataegus phaenopyrum, Crinum americanum, Cyperus, Cyrilla racemiflora, Decumaria barbara, Diospyros virginiana, Echinochloa, Eleocharis, Eriocaulon compressum, Forestiera acuminata, Fraxinus caroliniana, F. pennsylvanica, Fuirena, Gelsemium rankinii, Gleditsia aquatica, Habenaria repens, Hedyotis boscii, Hypoxis hirsuta, Ilex cassine, I. coriacea, I. myrtifolia, I. opaca, I. vomitoria, Itea virginica, Juncus effusus, Juniperus virginiana, Justicia ovata, Leersia lenticularis, Leucothoe racemosa, Lipocarpha, Liquidambar styraciflua, Lygodium japonicum, Lyonia ligustrina, L. lucida, Magnolia virginiana, Mitchella repens, Morus rubra, Myrica cerifera, Nyssa aquatica, N. biflora, Osmunda regalis, Osmundastrum cinnamomeum, Packera glabella, Panicum hemitomon, Persea borbonia, P. palustris, Persicaria hydropiperoides, Physocarpus opulifolius, Pinus taeda, Planera aquatica, Pleopeltis polypodioides, Pluchea camphorata, Pontederia cordata, Populus heterophylla, Quercus laurifolia, Q. lyrata, Q. nigra, Q. phellos, Rhexia mariana, Rosa palustris, Sabal minor, S. palmetto, Salix nigra, Sebastiania fruticosa, Serenoa repens, Smilax laurifolia, S. rotundifolia, S. walteri, Sorbus, Sphagnum, Taxodium ascendens, T. distichum, Tillandsia usneoides, Toxicodendron radicans, T. vernix, Triadenum walteri, Vaccinium corymbosum, Viburnum nudum, Woodwardia areolata, W. virginica, Xyris.*

Use by wildlife: The fleshy drupes of all *Nyssa* species are eaten by birds (Aves), which facilitate their dispersal. Fruits of *N. aquatica* are eaten by mockingbirds (Mimidae: *Mimus polyglottos*), robins (Turdidae: *Turdus migratorius*), starlings (Sturnidae: *Sturnus vulgaris*), thrashers (Mimidae: *Toxostoma*), thrushes (Turdidae: *Hylocichla mustelina*), and wild turkeys (Phasianidae: *Meleagris gallopavo*). The fruits of *N. biflora* are consumed extensively by robins and are also eaten by bluebirds (Turdidae: *Sialia sialis*), finches (Fringillidae: *Carduelis tristis, Carpodacus purpureus*), and wood ducks (Anatidae: *Aix sponsa*). The trees are highly susceptible to injury by sapsuckers (Picidae: *Sphyrapicus varius*), which forage on them for food. Critical recent video, suggesting the rediscovery of the presumably extinct ivory-billed woodpecker (Picidae: *Campephilus principalis*), showed the bird perched on a *N. aquatica* trunk. Fruits of *N. biflora* are also eaten by bears (Mammalia: Ursidae: *Ursus*) and by small mammals such as foxes (Canidae: *Vulpes*), opossums (Didelphidae: *Didelphis*), and squirrels (Sciuridae: *Sciurus*). Fruits of *N. aquatica* are eaten by woodchucks (Mammalia: Sciuridae: *Marmota monax*) and those of *N. ogeche* by squirrels. The foliage and twigs of *N. aquatica* and *N. biflora* are browsed by white-tailed deer (Mammalia: Cervidae: *Odocoileus virginianus*). Rafinesque's big-eared bat (Mammalia: Vespertilionidae: *Corynorhinus rafinesquii*) and southeastern bat (Vespertilionidae: *Myotis austroriparius*) roost in trees of *N. aquatica*. Seedlings of *N. aquatica* show a fairly high incidence of general herbivory. Both *N. aquatica* and *N. biflora* are preferred hosts of the forest tent caterpillar (Lepidoptera: Lasiocampidae: *Malacosoma disstria*), which causes serious defoliation. *Nyssa aquatica* is the host of a rust (Basidiomycota: Chaconiaceae: *Aplopsora nyssae*), a wood decay fungus (Basidiomycota: Hymenochaetaceae: *Phellinus gilvus*), and a sac fungus (Ascomycota: Mycosphaerellaceae: *Mycosphaerella nyssaecola*), which can cause minor defoliation. *Nyssa biflora* is the host to a wilt fungus (Ascomycota: Hypocreaceae: *Fusarium solani*), which causes "Tupelo lesion" of the stem. It is also host to a variety of woodrotting fungi (Basidiomycota: Fomitopsidaceae: *Daedalea ambigua*; Hericiaceae: *Hericium erinaceus*; Pleurotaceae: *Pleurotus ostreatus*; Polyporaceae: *Fomes, Lentinus tigrinus, Polyporus*), which cause heartrot. The fallen logs of *N. aquatica* provide habitat for a large number and diversity of aquatic invertebrates.

Economic importance: food: The fruit of some *Nyssa* species (e.g., *N. sylvatica*) is edible and acidic in flavor; however, fruits of *N. aquatica* have a poor flavor and those of *N. biflora* are bitter. The fruit of *N. ogeche* (known as "Ogeechee lime") is tangy and is used to make preserves and beverages; **medicinal:** Various medicinal uses (treatments for diarrhea, worms, etc.) have been reported for *N. sylvatica*, but it is difficult to distinguish those that apply specifically to *N. biflora*, which was regarded taxonomically as a variety of *N. sylvatica* in those reports. Such applications usually involve the bark or roots; **cultivation:** *Nyssa aquatica, N. biflora,* and *N. ogeche* all are cultivated occasionally; **misc. products:** The wood of *N. aquatica* is clear and defect free and is used for veneer and box lumber. Flowers of *N. biflora* and *N. ogeche* are a nectar source for commercial honey producing beekeepers and tupelo honey can be an important product locally. The Choctaw tribe combined the burned bark of *N. aquatica* with ash derived from red oak (Fagaceae: *Quercus rubra*) and water to make a red dye. Some 19th-century dispensatories recommended the use of *N. aquatica* roots "for surgical tents"; however, the meaning of this application is unclear and possibly intended as "stents" rather than "tents"; **weeds:** none; **nonindigenous species:** none

Systematics: Phylogenetic analysis of molecular data resolves *Nyssa* and *Camptotheca* as a clade, which is the sister group to *Davidia* (Figure 5.3). The monophyly of *Nyssa* is supported by several morphological characters including polygamodioecy, an axillary inflorescence, a reduced number of female flowers, and petals that are imbricate in bud. The taxonomy of *Nyssa* species remains unsettled. *Nyssa ursina* has been merged with *N. biflora*, which in turn often is regarded as a variety of *N.*

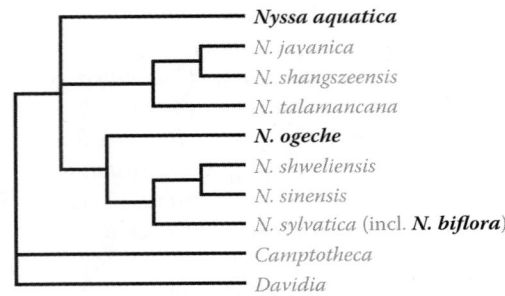

FIGURE 5.4 Phylogenetic relationships in *Nyssa* as indicated by analysis of morphological characters. The placement of *N. aquatica* is unresolved by morphological data. *Nyssa biflora* was not distinguished from *N. sylvatica* in this analysis. The OBL North American species (bold) do not associate as a distinct clade, but are intermixed with Asian species, possibly indicating several independent origins. (Adapted from Wen, J. & Stuessy, T.F., *Syst. Bot.*, 18, 68–79, 1993.)

sylvatica. *Nyssa biflora* and *N. sylvatica* reportedly hybridize; however, they are retained here as distinct species pending the outcome of more definitive studies. A narrow-leaved variety of *Nyssa ogeche* (var. *acuminata*) has also been recognized. Morphological cladistic analysis (Figure 5.4) indicates a relatively isolated position for *N. aquatica* in the genus. *Nyssa ogeche* is placed in a clade with *N. biflora* (as *N. sylvatica*), together with several Asian species. *Nyssa* would benefit from additional phylogenetic analyses using molecular data to not only clarify species relationships, but also to assist with the delimitation of taxa, as this issue remains unsettled. Thus far only *N. aquatica* and *N. ogeche* have been included in molecular phylogenetic analyses, which verify only that they are monophyletic and distinct. The base chromosome number of *Nyssa* is $x = 22$ but has been established only from a small number of counts. Counts have not been reported for any of the OBL North American species.

Comments: *Nyssa aquatica, N. biflora,* and *N. ogeche* occur in the southeastern United States, with the latter restricted to Florida, Georgia, and South Carolina.

References: Burke et al., 2003; Conner et al., 1981; DeBell & Naylor, 1972; Fitzpatrick et al., 2005; Gooding & Langford, 2004; He et al., 2004; Hook & Brown, 1973; Huenneke & Sharitz, 1986; Johnson, 1990b; Karlson et al., 2004; Kennedy, Jr., 1983; Kossuth & Scheer, 1990; Mains, 1921; McLeod & Ciravolo, 1997; McLeod et al., 2001; Middleton, 2003; Outcalt, 1990; Schneider & Sharitz, 1988; Shea et al., 1993; Stiles, 1980; Wen & Stuessy, 1993.

ORDER 2: ERICALES [24]

Although there are few evident morphological synapomorphies for Ericales (possibly "theoid" leaf teeth and vessels with scalariform perforations), the monophyly of this large order of nearly 9500 species is well supported by phylogenetic analyses of DNA data from chloroplast, mitochondrial, and nuclear genes comprising 18S, *atp1, atpB, matR, ndhF,* and *rbcL* sequences (Anderberg et al., 2002; Judd et al., 2002, 2008; Soltis et al., 2005). The order varies from most asterids (having unitegmic ovules and an equal number of petals and stamens) by its inclusion of some families with bitegmic ovules or stamens that are twice the number of petals or even numerous.

Several important aquatic groups occur here such as the Ericaceae and Sarraceniaceae, which are prominent elements of bog vegetation in North America. Also found in Ericales are several genera such as *Clethra* L. (FACW, FAC) and *Impatiens L.* (FACW, FAC, facultative upland (FAC) UPL), which include some fairly common wetland species, but have not been designated as OBL indicators. Perhaps the most surprising exclusion from this group is *Impatiens capensis* Meerb. (FACW), which seldom is found growing anywhere but in wetlands.

Eight families within the order contain OBL-ranked species in North America:

1. **Cyrillaceae** Endlr.
2. **Ericaceae** Juss.
3. **Myrsinaceae** R. Br.

4. **Polemoniaceae** Juss.
5. **Primulaceae** Vent.
6. **Sarraceniaceae** Dumort.
7. **Styracaceae** Dumort.
8. **Theophrastaceae** Link

Phylogenetic analyses of combined molecular data (Figure 5.5) indicate that these families are fairly dispersed throughout the order, with a minor concentration indicated by the clade of "primuloid" families (Myrsinaceae, Primulaceae, and Theophrastaceae). Although as many as 11 molecular data sets have now been incorporated (Schönenberger et al., 2005), resolution along the "backbone" of the cladogram still remains poor.

Family 1: Cyrillaceae [2]

Formerly circumscribed as comprising three genera (*Cliftonia, Cyrilla, Purdiaea*), recent analyses of Cyrillaceae using combined DNA sequence data (*atpB, ndhF, rbcL*) indicate that *Purdiaea* belongs instead to the closely related Clethraceae (Anderberg & Zhang, 2002). This disposition leaves only *Cliftonia* (monotypic) and *Cyrilla* (two species) to represent this small family, which DNA sequence data resolve as the sister group to Ericaceae (Figures 5.5 and 5.6).

One North American genus contains obligate aquatic plants:

1. *Cliftonia* (Lam.) Britton ex Sarg.

1. *Cliftonia*
Black titi, buckwheat tree
Etymology: after William Clifton (ca. 1750–1803)
Synonyms: *Mylocaryum*; *Ptelea* (in part)
Distribution: global: North America; **North America:** southeastern
Diversity: global: 1 species; **North America:** 1 species
Indicators (USA): OBL: *Cliftonia monophylla*
Habitat: freshwater; palustrine; **pH:** 4.6–5.7; **depth:** <1 m; **life-form(s):** emergent shrub, emergent tree
Key morphology: stems (to 8 m) woody, multiple; leaves (to 10 cm) evergreen, gland-dotted, leathery; 1-several racemes (to 9 cm) terminating prior season's growth; flowers 5-merous; petals (to 8 mm) white or pinkish, stamens 10; drupes (to 8 mm) 2–5-winged

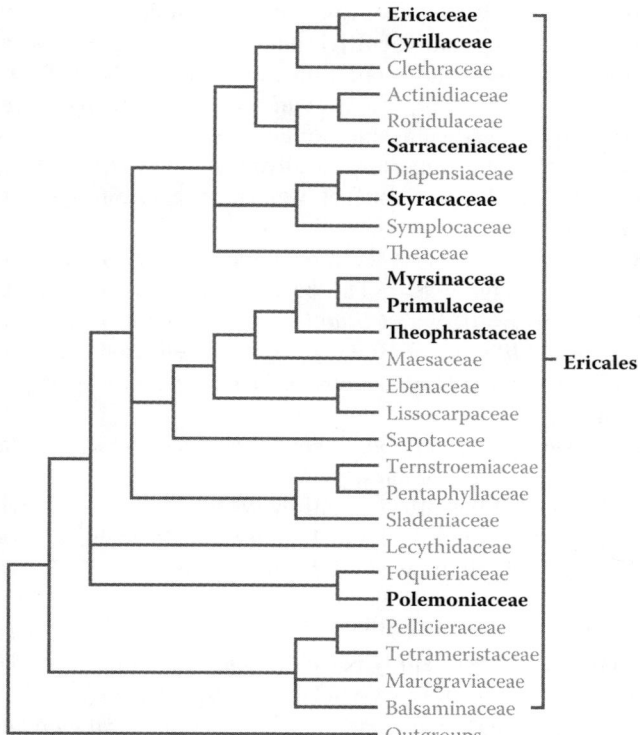

FIGURE 5.5 Phylogeny of Ericales families as indicated by combined analysis of 11 sets of DNA sequence data. A current circumscription of the order recognizes 23 families by the merger of Lissocarpaceae (with Ebenaceae), Pellicieraceae (with Tetrameristaceae), and Sladeniaceae and Ternstroemeriaceae (with Pentaphyllaceae). The eight families that contain obligate aquatic plants in North America (bold) are scattered throughout the clado-gram, indicating at least five independent origins of the aquatic habit in the order. (Adapted from Anderberg, A.A. et al., *Amer. J. Bot.*, 89, 677–687, 2002; Schönenberger, J. et al., *Int. J. Plant Sci.*, 166, 265–288, 2005.)

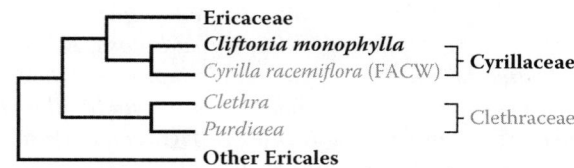

FIGURE 5.6 Relationships of Cyrillaceae as indicated by phylogenetic analyses of combined cpDNA sequence data. The OBL habit (indicated by bold type) appears to have arisen independently in *Cliftonia*. (Adapted from Anderberg, A.A. & Zhang, X., *Org. Divers. Evol.*, 2, 127–137, 2002.)

Life history: duration: perennial (buds); **asexual reproduction:** root sprouts; **pollination:** insect; **sexual condition:** hermaphroditic; **fruit:** drupes (common); **local dispersal:** root sprouts; drupes (gravity, wind); **long-distance dispersal:** drupes (birds, water)

Imperilment: (1) *Cliftonia monophylla* [G4]; S1 (LA)

Ecology: general: *Cliftonia* is monotypic.

Cliftonia monophylla **(Lam.) Britton ex Sarg.** grows in bogs, depressions, ditches, prairies, seeps, swamps, and along rivers and blackwater streams at elevations of up to 100 m. Its habitats are characterized by acidic peat or by sandy substrates that contain organic matter (loam). Sites often are shady although the plants reportedly can tolerate full sunlight. The fragrant flowers are pollinated mainly by nectar-foraging bees (Hymenoptera). The winged drupes are adapted for wind dispersal and also attach to the feet of birds (Aves) when wet. They probably float for at least a short time, especially when dried. Germination requirements for the seeds and other details on the reproductive ecology have not been reported. The plants can reproduce vegetatively by root sprouts, which leads to the formation of large, clonal thickets. Periodic episodes of fire help to maintain a balanced habitat and also to facilitate regeneration. When fire is suppressed, the species can spread so densely into savannahs and seepage bogs so as to exclude grasses and other herbaceous plants. Although *Cliftonia monophylla* is an important plant of coastal southern wetlands, few studies have focused on the ecology of this species in particular. **Reported associates:** *Acer rubrum, Aletris lutea, Alnus serrulata, Andropogon, Aristida beyrichiana, A. palustris, Balduina atropurpurea, B. uniflora, Bignonia capreolata, Buchnera, Carex lonchocarpa, C. striata, C. turgescens, Cephalanthus occidentalis, Chamaecyparis thyoides, Chaptalia tomentosa, Chionanthus virginicus, Cladium jamaicense, Cleistes bifaria, Clethra alnifolia, Ctenium aromaticum, Cyrilla racemiflora, Dichanthelium, Diodia, Drosera capillaris, D. tracyi, Epidendrum conopseum, Eriocaulon compressum, E. decangulare, E. nigrobracteatum, Fimbristylis, Fraxinus caroliniana, Galax urceolata, Gaylussacia mosieri, Gelsemium rankinii, Hamamelis virginiana, Harperocallis flava, Hymenocallis henryae, Hypericum brachyphyllum, H. chapmanii, H. fasciculatum, H. galioides, Ilex cassine, I. coriacea, I. myrtifolia, I. opaca, Illicium floridanum, Iris, Kalmia latifolia, Lachnanthes caroliana, Lachnocaulon, Leucothoe racemosa, Liriodendron tulipifera, Lobelia, Lophiola aurea, Ludwigia, Lycopodiella alopecuroides, Lyonia ferruginea, L. lucida, Magnolia virginiana, Mitchella repens, Mitreola, Myrica cerifera, M. heterophylla, Nyssa biflora, N. ogeche, N. sylvatica, N. ursina, Orontium aquaticum, Osmunda regalis, Osmundastrum cinnamomeum, Oxypolis filiformis, Pallavicinia, Panicum nudicaule, P. rigidulum, Peltandra sagittifolia, Persea borbonia, P. palustris, Photinia pyrifolia, Phyla, Pieris phillyreifolia, Pinguicula planifolia, Pinus elliottii, P. serotina, Pleea tenuifolia, Pleopeltis polypodioides, Pogonia ophioglossoides, Polygala cruciata, P. lutea, Quercus nigra, Rhexia alifanus, R. brevifolia, R. lutea, Rhododendron viscosum, Rhynchospora chapmanii, R.* *corniculata, R. latifolia, R. macra, R. microcephala, R. oligantha, R. stenophylla, Sabatia bartramii, S. macrophylla, Salix caroliniana, Sarracenia flava, S. psittacina, S. rubra, Saururus cernuus, Scleria baldwinii, Serenoa repens, Smilax laurifolia, Solidago sempervirens, Sphagnum, Stillingia aquatica, Styrax americanus, Symphyotrichum chapmanii, Syngonanthus flavidulus, Taxodium ascendens, Tillandsia usneoides, Utricularia subulata, Vaccinium corymbosum, Viburnum nudum, Vitis rotundifolia, Woodwardia virginica, Xyris stricta.*

Use by wildlife: *Cliftonia monophylla* is a general forage plant for wildlife and occasionally is browsed by deer (Mammalia: Cervidae: *Odocoileus*). It provides habitat for several birds (Aves) including the great-crested flycatcher (Tyrannidae: *Myiarchus crinitus*), yellow-billed cuckoo (Cuculidae: *Coccyzus americanus*), prothonotary warbler (Parulidae: *Protonotaria citrea*), and the endangered Mississippi sandhill crane (Gruidae: *Grus canadensis pulla*). It is also associated with habitat for the Florida bog frog (Amphibia: Ranidae: *Rana okaloosae*). The plants are host to a sac fungus (Ascomycota: Asterinaceae: *Lembosia cliftoniae*).

Economic importance: food: not reported as edible; **medicinal:** *Cliftonia* does not contain alkaloids and has not been used medicinally; **cultivation:** *Cliftonia* is grown occasionally as an ornamental. Cultivars include 'Berry Pink,' 'Chipola Pink,' 'Pink,' and 'Van Cleve'; **misc. products:** *Cliftonia* is regarded as a minor nectar plant for beekeepers, but can be quite important locally. The wood is brittle but burns well and is used for fuel; **weeds:** none; **nonindigenous species:** none

Systematics: *Cliftonia* is monotypic and related closely to *Cyrilla* (Figure 5.6). The base chromosome number of *Cliftonia* is $x = 10$, with *Cliftonia monophylla* ($2n = 20$) being a diploid. *Cliftonia monophylla* would benefit from field and genetic analyses of populations to elucidate patterns of dispersal and gene flow.

Comments: *Cliftonia monophylla* is endemic to coastal southeastern United States.

References: Elias, 1987; Godfrey & Wooten, 1981; Lewis et al., 1962; Raffauf, 1996; Sanford, 1988; Ståhl, 2004a; Tracy & Earle, 1896; Wade & Mengak, 2010.

Family 2: Ericaceae [130]

Ericaceae, a large family of nearly 3000 species, are a conspicuous element of European heathlands, comprise a distinctive vegetational zone ("ericaceous belt") of tropical montane habitats, and are prevalent vegetational components of North American bog (peatland) communities. Ecologically, they are a reliable indicator of open, acidic habitats. The species often are associated with mycorrhizae (Luteyn, 2004).

From extensive analyses of morphological and molecular data (Kron et al., 2002a), the family currently is circumscribed to comprise eight subfamilies, which include members of several former families (Empetraceae, Epacridaceae, Monotropaceae, Pyrolaceae, Vacciniaceae). Only two of the subfamilies (Ericoideae, Vaccinioideae) are represented by obligate aquatic plants in North America (Figure 5.7).

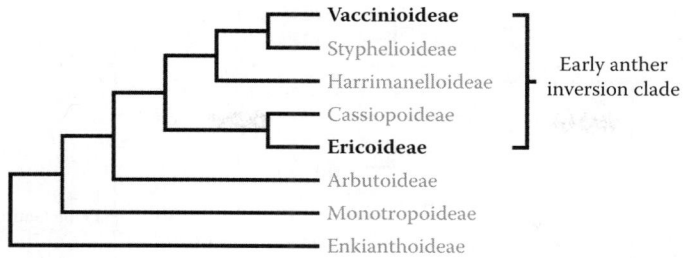

FIGURE 5.7 Subfamilial divisions of Ericaceae as indicated by phylogenetic analyses of DNA sequence data. In North America, obligate aquatics (in bold) occur in two isolated subfamilies. (Adapted from Kron, K.A., *Bot. Rev.*, 68, 335–423, 2002a.)

Flowers of Ericaceae often are showy and insect pollinated (temperate species) although some species are bird or wind pollinated. In most species, the flowers are pendulous with their petals fused into campanulate, cylindrical, or urceolate corollas. The stamens often possess a pair of projecting spurs at the point of anther attachment (Judd et al., 2002). Inverted anthers define a large clade within the family (Figure 5.7). The seeds are dispersed by wind (capsules) or by animals (berries).

Several genera of Ericaceae are important economically as ornamentals (e.g., *Rhododendron*) or for their edible berries (*Vaccinium*). Oil of wintergreen (methyl salicylate) was originally obtained from *Gaultheria procumbens* (Judd et al., 2002). All members of Ericaceae should be regarded as poisonous as all plant parts (including the nectar) contain grayanotoxins (andromedotoxin, deacetylandromedol, deacetylanhydroandromedol), which bind to cell membranes and cause their depolarization. Even the honey made by bees (Hymenoptera) that feed on *Rhododendron* nectar can be poisonous to humans.

Obligate aquatic plants occur in nine North American genera:

1. *Agarista* D. Don ex G. Don
2. *Andromeda* L.
3. *Chamaedaphne* Moench
4. *Gaylussacia* Kunth
5. *Kalmia* L.
6. *Leucothoe* D. Don
7. *Rhododendron* L.
8. *Vaccinium* L.
9. *Zenobia* D. Don

Most of the OBL genera occur within the relatively derived subfamily Vaccinioideae with *Kalmia* and *Rhododendron* placed within Ericoideae (Figure 5.7).

1. *Agarista*

Dog-hobble, hobblebush, pipestem

Etymology: after *Agarista*, the daughter of Cleisthenes, the Greek tyrant of Sicyon

Synonyms: *Agauria*; *Amechania*; *Andromeda* (in part); *Leucothoe* (in part)

Distribution: global: Africa; New World; **North America:** southeastern

Diversity: global: 32 species; **North America:** 1 species

Indicators (USA): OBL; FACW: *Agarista populifolia*

Habitat: freshwater; palustrine; **pH:** acidic; **depth:** <1 m; **life-form(s):** emergent shrub

Key morphology: stems (to 4 m) multiple, ridged longitudinally, pith chambered; leaves (to 10 cm) alternate, evergreen, short petioled (to 10 mm), the margins entire with a narrow cartilaginous band; racemes subsessile, axillary; flowers 5-merous, stalked (to 10 mm), fragrant; corolla (to 8 mm) urceolate, white; stamens 10, the filaments S-shaped, the anthers opening by terminal pores; capsules (to 6 mm) subglobose; seeds numerous

Life history: duration: perennial (buds); **asexual reproduction:** root sprouts; **pollination:** insect; **sexual condition:** hermaphroditic; **fruit:** capsules (common); **local dispersal:** suckers; seeds (wind); **long-distance dispersal:** seeds

Imperilment: (1) *Agarista populifolia* [G4]; SH (GA); S1 (SC)

Ecology: general: *Agarista* primarily is a genus of Neotropical species, which inhabit a variety of sites ranging from dry, open thickets to wet forests. North America has only one (endemic) species of the genus and it occurs consistently in wet sites.

Agarista populifolia (**Lam.**) **Judd** inhabits depressions, floodplains, seeps, swamps, and margins of springs or smaller watercourses at elevations of up to 50 m. Although the pH range has not been determined for this species, the plants occur mainly in sites that are described as acidic and in areas characterized by light to dense shade. However, they can also grow on marly substrates and in full sun. Substrates typically are sandy loam or peat. The flowers are pollinated by bees (Hymenoptera) but little specific information exists on the reproductive ecology of this species. The plants are fairly drought resistant and can develop into thickets by means of spreading root suckers, which render stability to their substrates. **Reported associates:** *Acer rubrum, Carex, Carpinus caroliniana, Cephalanthus occidentalis, Chamaecyparis thyoides, Chasmanthium laxum, C. nitidum, Cornus foemina, Cyrilla racemiflora, Dichanthelium, Dioscorea floridana, Diospyros virginiana, Dryopteris ludoviciana, Euonymus americana, Fraxinus caroliniana, Gaylussacia tomentosa, Gelsemium sempervirens, Gordonia lasianthus, Ilex cassine, I. coriacea, I. glabra, Illicium parviflorum, Iris virginica, Itea virginica, Leucothoe axillaris, Liquidambar styraciflua, Lyonia lucida, L. ligustrina, Magnolia virginiana, Mitchella repens, Myrica cerifera, Nyssa biflora, N.*

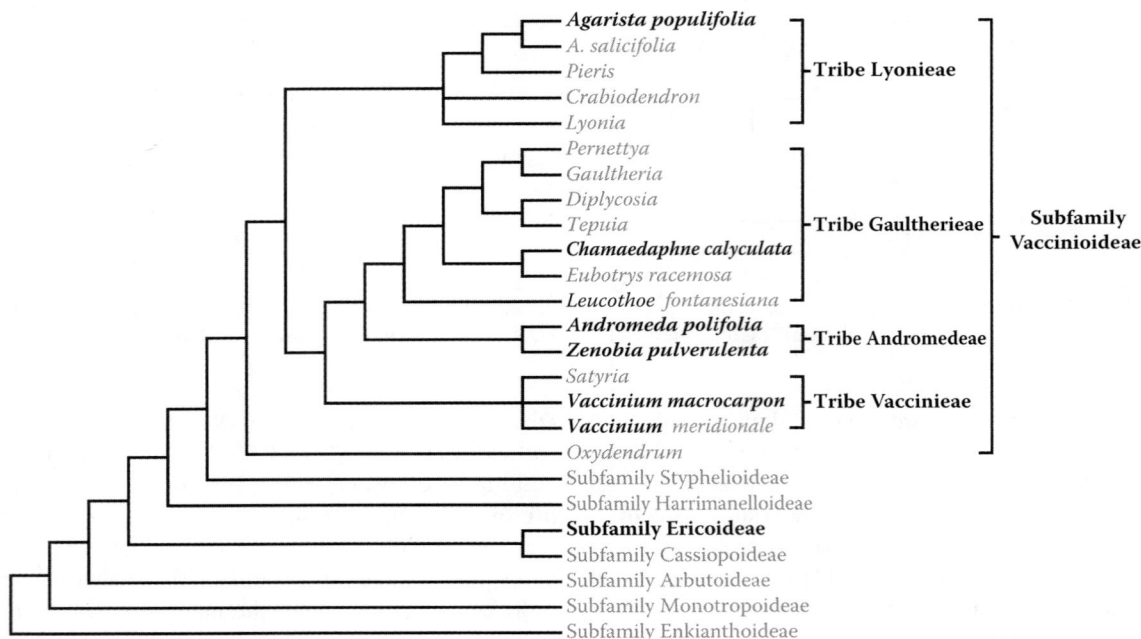

FIGURE 5.8   Relationships within Ericaceae subfamily Vaccinioideae as indicated by phylogenetic analysis of combined morphological and molecular data. Taxa with OBL North American species (indicated in bold) are concentrated in this relatively derived subfamily, but indicate at least several independent origins. (Adapted from Kron, K.A., *Amer. J. Bot.*, 89, 327–336, 2002b.)

sylvatica, Osmunda regalis, Osmundastrum cinnamomeum, Peltandra, Persea palustris, Phlebodium aureum, Pieris phillyreifolia, Pinus serotina, Pleopeltis polypodioides, Pteridium aquilinum, Rhododendron viscosum, Quercus chapmanii, Q. geminata, Q. laurifolia, Q. michauxii, Q. nigra, Q. virginiana, Rhapidophyllum hystrix, Rhododendron viscosum, Rhynchospora megalocarpa, R. microcarpa, R. miliacea, Rubus, Sabal minor, S. palmetto, Saururus cernuus, Serenoa repens, Taxodium ascendens, Toxicodendron radicans, Ulmus americana, Vaccinium corymbosum, V. elliottii, V. fuscatum, Vitis aestivalis, V. rotundifolia, Vittaria lineata, Woodwardia areolata, W. virginica.

Use by wildlife: In cultivation, *Agarista populifolia* reportedly is resistant to fungal infection and experiences a low incidence of herbivory, possibly due to its toxicity (see later). The plants are host to moth larvae (Lepidoptera: Tortricidae: *Zomaria interruptolineana*). The fragrant, nectariferous flowers attract bees (Hymenoptera) and butterflies (Lepidoptera) and the dense foliage provides cover for various wildlife species.

Economic importance: food: *Agarista populifolia* is highly poisonous and potentially fatal if ingested. The leaves contain andromedotoxin, which can induce vomiting, convulsions, and paralysis; **medicinal:** none; **cultivation:** *Agarista populifolia* is cultivated as an ornamental. Dwarf cultivars include 'Leprechaun' and 'Taylor's Treasure'; **misc. products:** *Agarista populifolia* is planted to control streambank erosion; **weeds:** none; **nonindigenous species:** none

Systematics: Systematic studies incorporating morphological and molecular data resolve *Agarista* as a monophyletic sister clade to *Pieris* in tribe Lyonieae, subfamily Vaccinioideae

of Ericaceae (Figure 5.8). Although some taxonomists have placed *Agarista populifolia* within *Andromeda* or *Leucothoe*, phylogenetic analyses clearly indicate that it is quite distantly related to either genus (Figure 5.8). A comprehensive interspecific phylogenetic analysis of *Agarista* has not yet been carried out. The basic chromosome number of *Agarista* (and tribe Lyonieae) is $x = 12$. *Agarista populifolia* ($2n = 24$) is diploid.

Comments: *Agarista populifolia* is endemic to the coastal plain of the southeastern United States.

References: Godfrey & Wooten, 1981; Judd, 1984; Kron & Judd, 1997; Kron et al., 2002; Middleton & Wilcox, 1990.

2. *Andromeda*

Bog-rosemary; andromède

Etymology: for *Andromeda*, the daughter of Cepheus and Cassiopeia and wife of Perseus in Greek mythology

Synonyms: none

Distribution: global: Asia; Europe; North America; **North America:** northern

Diversity: global: 1 species; **North America:** 1 species

Indicators (USA): OBL; FACW: *Andromeda polifolia*

Habitat: freshwater; palustrine; **pH:** 2.8–7.5; **depth:** <1 m; **life-form(s):** emergent shrub

Key morphology: stems (to 40 cm) few branched; leaves (to 4.0 cm) alternate, evergreen, leathery, the margins revolute, the undersides with a white to silvery indumentum, the apex often apiculate; racemes umbellate, terminal, 2–12-flowered; flowers 5-merous, pendant; corolla (to 8 mm) urceolate, terminating in 5 teeth, pink to white; stamens 10, filaments hairy and swollen above base, anthers opening by terminal pores; capsules (to 5 mm) subglobose, with up to 45 seeds

Life history: duration: perennial (buds, rhizomes); **asexual reproduction:** rhizomes; **pollination:** insect or self; **sexual condition:** hermaphroditic; **fruit:** capsules (common); **local dispersal:** rhizomes; seeds; **long-distance dispersal:** seeds (water, wind)

Imperilment: (1) *Andromeda polifolia* [G5]; SX (OH); S1 (ID, NJ, RI, WV); S2 (CT, IN, SK); S3 (PA, PE, QC)

Ecology: general: *Andromeda* is monotypic.

Andromeda polifolia **L.** grows in bogs, patterned peatlands, seeps, swamps, along lake and pond margins or occasionally on sand dunes at elevations of up to 1500 m. Its habitats are acidic (usually pH: 4.0–5.0) to neutral sites and occur in colder climates where soil fertility is low. Substrates are peat or sand. The plants can tolerate only light amounts of shade. A small number of flowers (up to 30) are produced late in the season (June to July). They are self-compatible and have a low pollen:ovule ratio (<350). The pollen is contained within the flowers and is shed onto the receptive stigma often while in bud. Self-pollination is prevalent and is facilitated by the movement of the flowers by wind. However, seed set in selfed flowers is substantially lower (2%–15%) than in open-pollinated or cross-pollinated flowers (26%–43%). Pollinators mainly are bees (Hymenoptera) and possibly some butterflies (Lepidoptera). Average seed set in natural populations approaches 30%. The small seeds (about 20 per capsule) are fairly short lived (19% germination after 4 years) and are water or wind dispersed. They do not form a seed bank. The seeds have physiological dormancy that is induced by prolonged cold stratification. Fresh seeds and those stratified for 30 days germinate well (65%–70%) at 25°C, but longer stratification (2–3 months) decreases their germination. Germination is enhanced by light and reduced at colder temperatures (e.g., 15°C). Moist storage of seeds (in wet sand at 4°C) accelerates their germination. The highest germination rate (72%) has been obtained at 30°C and full light after 1 month of dry storage at −20°C. Storage beyond this time (e.g., 2 months) will significantly reduce germination (to 4%). Burial to as little as a 5-mm depth can also substantially reduce germination. The plants are mycorrhizal, probably in association with an Ascomycota fungus (Helotiaceae: *Hymenoscyphus ericae*). The evergreen leaves live for up to 4 years. Seedlings are rare in natural populations where most reproduction is vegetative from proliferation of the long woody rhizomes, which represent about 75% of the total plant biomass. One study observed a higher incidence of *A. polifolia* at sites where deer (Mammalia: Cervidae: *Odocoileus virginianus*) were present compared to those where deer were absent. It would be informative to conduct population genetic studies of this species to determine the extent of genetic variability that resides within and among the populations. **Reported associates:** *Abies balsamea, Alnus rugosa, Amelanchier, Arctostaphylos, Arethusa bulbosa, Aulacomnium palustre, Betula papyrifera, B. populifolia, B. pumila, B. ×sandbergii, Calamagrostis canadensis, Calla palustris, Calliergon cordifolium, Calopogon pulchellus, Caltha palustris, Campanula aparinoides, Campylium stellatum, Carex buxbaumii, C. canescens, C. chordorrhiza, C. comosa, C. disperma, C. exilis, C. interior, C. lacustris, C. lasiocarpa, C. leptalea, C. limosa, C. livida, C. magellanica, C. oligosperma, C. pauciflora, C. paupercula, C. rostrata, C. stipata, C. stricta, C. tenuiflora, C. trisperma, Chamaedaphne calyculata, Cicuta bulbifera, Cladonia, Clintonia borealis, Comarum palustre, Coptis trifolia, Cornus canadensis, C. racemosa, C. rugosa, C. sericea, Corylus cornuta, Cypripedium acaule, Dasiphora floribunda, Dicranum polysetum, D. scoparium, Drosera intermedia, D. rotundifolia, Dryas integrifolia, Dryopteris cristata, Dulichium arundinaceum, Eleocharis compressa, Empetrum nigrum, Epilobium leptocarpum, E. palustre, Equisetum fluviatile, E. hyemale, E. pratense, Eriophorum chamissonis, E. gracile, E. vaginatum, E. viridicarinatum, Eupatorium perfoliatum, Fraxinus nigra, Galium labradoricum, Gaultheria hispidula, G. procumbens, Gaylussacia baccata, G. dumosa, Geocaulon lividum, Hudsonia, Iris versicolor, Juncus canadensis, Juniperus horizontalis, Kalmia polifolia, Larix laricina, Linnaea borealis, Listera australis, Lonicera hirsuta, L. oblongifolia, Lycopus asper, L. uniflorus, Lysimachia terrestris, L. thyrsiflora, Maianthemum canadense, M. trifolium, Menyanthes trifoliata, Mitella nuda, Monotropa uniflora, Muhlenbergia glomerata, M. uniflora, Myrica gale, Nemopanthus mucronata, Oclemena nemoralis, Orthilia secunda, Osmunda regalis, Osmundastrum cinnamomeum, Parnassia palustris, Picea mariana, Pinus resinosa, P. rigida, P. strobus, P. sylvestris, Platanthera dilatata, P. hyperborea, P. obtusata, P. orbiculata, Pleurozium schreberi, Pogonia ophioglossoides, Polytrichum commune, P. strictum, Pyrola asarifolia, Pyrus floribunda, Rhamnus alnifolia, Rhododendron canadense, R. groenlandicum, R. lapponicum, Rhynchospora alba, R. fusca, Ribes triste, Rubus arcticus, R. chamaemorus, R. hispidus, R. idaeus, R. parviflorus, R. pubescens, Sagittaria latifolia, Salix discolor, S. eriocephala, S. pedicellaris, S. planifolia, S. pyrifolia, S. reticulata, Sarracenia purpurea, Scheuchzeria palustris, Schoenoplectus tabernaemontani, Scutellaria galericulata, Selaginella rupestris, Senecio, Solidago uliginosa, Sphagnum angustifolium, balticum, S. capillifolium, S. contortum, S. fallax, S. flavicomans, S. flexuosum, S. fuscum, S. jensenii, S. magellanicum, S. majus, S. nitidum, S. palustre, S. papillosum, S. recurvum, S. rubellum, S. subsecundum, S. warnstorfii, S. wulfianum, Spiranthes cernua, S. romanzoffiana, Symphyotrichum puniceum, Symplocarpus foetidus, Thuja occidentalis, Thelypteris palustris, Toxicodendron vernix, Triadenum fraseri, Triantha glutinosa, Trichophorum cespitosum, Trientalis borealis, Typha angustifolia, T. latifolia, Usnea, Utricularia gibba, U. intermedia, Vaccinium angustifolium, V. macrocarpon, V. myrtilloides, V. oxycoccos, Valeriana uliginosa, V. vitis-idaea, Viola blanda, Warnstorfia exannulata, Xyris.*

Use by wildlife: *Andromeda polifolia* provides habitat and nesting cover for various birds (Aves) including the dunlin (Scolopacidae: *Calidris alpina*), golden plover (Charadriidae: *Pluvialis dominica*), Hudsonian curlew (Scolopacidae: *Numenius phaeopus*), Hudsonian godwit (Scolopacidae: *Limosa haemastica*), Lapland and Smith's longspur (Emberizidae: *Calcarius lapponicus*, *C. pictus*), least sandpiper (Scolopacidae: *Calidris minutilla*), savannah sparrow (Emberizidae: *Passerculus sandwichensis*), stilt sandpiper (Scolopacidae: *Calidris himantopus*), and yellow-bellied flycatcher (Tyrannidae: *Empidonax flaviventris*). The flowers of *A. polifolia* are foraged by a number of bees (Hymenoptera: Andrenidae: *Andrena*; Apidae: *Apis mellifera*, *Bombus affinis*, *B. fervidus*, *B. impatiens*, *B. pascuorum*, *B. ternarius*, *B. terricola*, *B. vagans*, *Psithyrus*), butterflies (Nymphalidae: *Polygonia interrogationis*), and flies (Bombyliidae: *Bombylius*; Syrphidae: *Eristalis*). The plants are hosts to aphids (Homoptera) and several larval Lepidoptera (Geometridae: *Cingilia catenaria*; Noctuidae: *Catocala andromedae*; Sphingidae: *Paonias astylus*, *Sphinx lugens*; Tortricidae: *Acleris hastiana*, *Zomaria andromedana*). They also host several fungi (Ascomycota: Rhytismataceae: *Rhytisma andromedae*; Basidiomycota: Exobasidiaceae: *Exobasidium vaccinii*) and the introduced andromeda lace bug (Insecta: Hemiptera: *Stephanitis takeyai*).

Economic importance: food: *Andromeda polifolia* should never be eaten because it contains grayanoside, a highly toxic andromedotoxin, which can be fatal if ingested. It is rather startling that the Ojibwa and Tanana made a tea from the fresh or dried leaves. It is likely that the tea was cold brewed because the toxin is released from the foliage when boiled in water; **medicinal:** *Andromeda polifolia* was used by the Mahuna tribe as a drug to reduce inflammation; **cultivation:** *Andromeda polifolia* is a popular garden ornamental and includes the cultivars 'Alba,' 'Blue Ice,' 'Compacta,' 'Compacta Alba,' 'Grandiflora,' 'Hayachine,' 'Kirigamine,' 'Macrophylla,' 'Major,' 'Minima,' 'Nana,' 'Nikko,' 'Red Winter,' and 'Shibutsu.' 'Compacta,' 'Compacta Alba,' and 'Macrophylla' are Royal Horticultural Society "Award of Garden Merit" plants; **misc. products:** The leaves and twigs of *A. polifolia* are a source of tannin; **weeds:** none; **nonindigenous species:** none

Systematics: Molecular phylogenetic analyses using combined DNA sequence data provide strong support for the association of *Andromeda* and *Zenobia* as sister genera (Figure 5.8); however, analysis of morphological data resolves these genera as being fairly closely related but not comprising a clade. Together, the two genera comprise tribe Andromedeae of subfamily Vaccinioideae in Ericaceae (Figure 5.8). Because both genera are monotypic, their systematic relationships are unambiguous. Both are regarded as obligate aquatics. Two species had been recognized previously in *Andromeda*; however, these are treated as varieties of *A. polifolia* (var. *glaucophylla*; var. *polifolia*) in recent taxonomic treatments. The base chromosome number of *Andromeda* is $x = 12$. *Andromeda polifolia* ($2n = 48$) is a tetraploid.

Comments: *Andromeda polifolia* is circumboreal in distribution and occurs in cooler northern portions of North America **References:** Campbell & Rochefort, 2003; Campbell et al., 2003; Hardikar, 1922; Jacquemart, 1998; Jehl, Jr., 1973;

Pellerin et al., 2006; Reader, 1975, 1977; Robuck, 1985; Small, 1976; Walkinshaw & Henry, 1957.

3. *Chamaedaphne*

Leatherleaf; petit-daphné
Etymology: from the Greek *chamai* meaning "ground" and the mythological *Daphne*, a nymph who was turned into a laurel tree by Gaia, hence "ground laurel"
Synonyms: *Andromeda* (in part); *Cassandra*
Distribution: global: Asia; Europe; North America; **North America:** northern
Diversity: global: 1 species; **North America:** 1 species
Indicators (USA): OBL; FACW: *Chamaedaphne calyculata*
Habitat: freshwater; palustrine; **pH:** 3.8–6.0; **depth:** <1 m; **life-form(s):** emergent shrub
Key morphology: stems (to 1.5 m) much branched; leaves (to 5 cm) alternate, evergreen, coriaceous, ascending, oblong to lanceolate, orangeish, and scaly beneath; racemes (to 10.5 cm) 1–15-flowered, one sided; in upper leaf axils; flowers (to 8 mm) 5-merous, pendulous, short-stalked (to 3 mm), subtended by reduced leaves (to 2 cm); corolla (to 6 mm) urceolate, white; stamens 10; capsules (to 4 mm) turbinate, many-seeded; seeds (to 1 mm) wedge shaped
Life history: duration: perennial (buds, rhizomes); **asexual reproduction:** rhizomes; **pollination:** insect or cleistogamous; **sexual condition:** hermaphroditic; **fruit:** capsules (common); **local dispersal:** rhizomes; seeds (wind); **long-distance dispersal:** seeds (water)
Imperilment: (1) *Chamaedaphne calyculata* [G5]; SH (DE); S1 (MD); S2 (IL); S3 (BC, NC, OH)
Ecology: general: *Chamaedaphne* is monotypic.

***Chamaedaphne calyculata* (L.) Moench** grows in bogs, ditches, swamps, vernal ponds, and along the margins of creeks, lakes, rivers, and streams and occasionally on sand dunes at elevations of up to 1600 m. Substrates typically are organic peat (often "quaking" mats), but also include sand, sandy loam, or fine loamy clay. *Chamaedaphne* is an indicator of wet, open, nutrient-poor sites with acidic (pH: 3.8–6.0; usually <5.0), organic substrates and is perhaps the most characteristic (and often dominant) plant of peat (*Sphagnum*) bogs (see associated *Sphagnum* species later). It is the first shrub species to colonize newly established *Sphagnum* sites. The plants can also occur on sandy substrates such as lakeshores. They do not tolerate shade but can withstand cold temperatures down to −33°C. The flowers are also extremely resistant to the cold, experiencing little mortality until exposed to temperatures of −10°C (14% mortality). This adaptability enables the plants to exploit a relatively early period of anthesis (see later). The flower buds are formed during the previous season. The flowers are self-compatible with high seed set occurring in experimental pollinations regardless of whether flowers are selfed (60%–100%) or outcrossed (70%–100%). In natural populations, self-pollination can occur while the flowers are in bud (when anthers and stigma are in close proximity) or before the style fully elongates beyond the corolla mouth. Otherwise, the flowers are cross-pollinated mainly by bees (Hymenoptera) and other foraging insects. Seed set is considerably higher in open-pollinated flowers (30%–100%) than in bagged flowers (0%–30%), perhaps because the flowers require some type of physical perturbation (e.g., shaking by wind) to release enough pollen to achieve selfing. Anthesis during cold, inclement weather deters pollinators, resulting in low pollen loads and poor seed set. Flowering occurs early in the season (May), which may reduce pollinator competition with other commonly co-occurring Ericaceae (e.g., *Andromeda*, *Kalmia*, *Rhododendron*, and *Vaccinium*), which all flower later (June to July). The seeds are physiologically dormant and require light and cold stratification for optimal germination. Germination will not occur in the dark, regardless of treatment. Unstratified seeds do not germinate under short day lengths and exhibit poor germination (10%–21%) under long day lengths at 15°C/20°C, respectively. Germination in stratified seeds improves to 41% at 15°C and to 65% at 20°C. Maximum germination is achieved when stratified seeds are subjected to alternating temperature regimes of 2.0°C–23°C (84%) or 6.0°C–29°C (100%). Germination occurs naturally on organic *Sphagnum* or sedge mats. No seed bank is formed. About 60 wedge-like seeds are produced by each capsule and are dispersed locally by the wind and over longer distances by the water. Productivity of *Chamaedaphne* is relatively consistent among sites, but varies with climate and substrate. Roots of the plants are associated with mycorrhizal fungi (Ascomycota: Myxotrichaceae: *Oidiodendron maius*; Vibrisseaceae: *Phialocephala fortinii*). The evergreen leaves normally persist for two seasons; however, their premature loss does not significantly reduce new shoot growth. *Chamaedaphne* reproduces vegetatively by means of woody rhizomes. For organs of a cold climate species, the rhizomes are surprisingly adaptable to warm temperatures (likely as an adaptation to fire) with maximum shoot production observed when grown at 50°C and significant mortality not occurring until temperatures reach 55°C. Vegetative reproduction can be efficient, resulting in dense thickets of plants under optimal conditions (up to 200 plants/m²). These dense patches are broken up by ice, which stimulates further spreading. The plants are highly resistant to fire because the rhizomes survive deeply (average of 32 cm depth) in saturated substrates and the shoots are insulated by *Sphagnum* plants and other organic debris. Resprouting occurs from the root crown and rhizome. *Chamaedaphne* is a rapid and effective recolonizer of windfalls and other disturbed sites. It would be informative to conduct population genetic studies to evaluate the significance of sexual reproduction in this species. **Reported associates:** *Abies balsamea, Acer rubrum, Agrostis scabra, Alnus rugosa, A. viridis, Amelanchier canadensis, Andromeda polifolia, Aneura pinguis, Arctostaphylos uva-ursi, Arethusa bulbosa, Aronia melanocarpa, Aulacomnium palustre, Betula papyrifera, B. populifolia, B. pumila, B. ×sandbergii, Bromus ciliatus, Calamagrostis canadensis, Calopogon tuberosus, Calla palustris, Calliergon giganteum, C. stramineum, Caltha palustris, Campanula aparinoides, Campylium, Carex alata, C. aquatilis, C. brunnescens, C. buxbaumii, C. canescens, C. chordorrhiza, C. diandra, C. disperma, C. echinata, C. exilis,*

C. gynocrates, C. interior, C. lacustris, C. lasiocarpa, C. leptalea, C. limosa, C. livida, C. magellanica, C. oligosperma, C. pauciflora, C. rostrata, C. stricta, C. tenuiflora, C. trisperma, C. utriculata, C. vesicaria, Cladina arbuscula, C. rangiferina, Cladium mariscoides, Climacium dendroides, Clintonia borealis, Comarum palustre, Coptis trifolia, Cornus canadensis, C. racemosa, C. rugosa, C. sericea, Corylus cornuta, Cypripedium acaule, C. reginae, Dasiphora floribunda, Dicranum polysetum, D. undulatum, Doellingeria umbellata, Drepanocladus aduncus, Drosera intermedia, D. rotundifolia, Dryopteris carthusiana, D. cristata, Dulichium arundinaceum, Empetrum nigrum, Epilobium coloratum, E. leptocarpum, E. leptophyllum, Equisetum fluviatile, E. hyemale, E. pratense, Eriophorum alpinum, E. angustifolium, E. chamissonis, E. tenellum, E. vaginatum, E. virginicum, E. viridicarinatum, Fissidens adianthoides, Fragaria, Fraxinus nigra, Galium labradoricum, Gaultheria hispidula, G. procumbens, Gaylussacia baccata, G. dumosa, Geocaulon lividum, Glyceria striata, Grimmia, Gymnocarpium dryopteris, Hudsonia, Hypnum imponens, H. lindbergii, Ilex mucronata, I. verticillata, Iris versicolor, Juncus canadensis, J. pelocarpus, Juniperus horizontalis, Kalmia angustifolia, K. polifolia, Larix laricina, Lilium philadelphicum, Limprichtia revolvens, Linnaea borealis, Listera australis, Lonicera canadensis, L. oblongifolia, L. villosa, Lycopodium annotinum, Lycopus uniflorus, L. virginicus, Lysimachia terrestris, L, thyrsiflora, Lythrum salicaria, Maianthemum canadense, M. trifolium, Malaxis unifolia, Melampyrum lineare, Menyanthes trifoliata, Mitella nuda, Mnim, Muhlenbergia uniflora, Myrica gale, Oclemena nemoralis, Oligoneuron album, Orthilia secunda, Osmundastrum cinnamomeum, Pellia, Persicaria amphibia, P. sagittata, Phragmites australis, Picea mariana, Pinguicula vulgaris, Pinus strobus, P. sylvestris, Platanthera hyperborea, P. psycodes, Pleurozium schreberi, Pogonia ophioglossoides, Polytrichum commune, P. juniperinum, P. longisetum, P. strictum, Ptilium crista-castrensis, Pyrola chlorantha, Rhamnus alnifolia, Rhododendron canadense, R. groenlandicum, R. viscosum, Rhynchospora alba, R. fusca, R. inundata, Rhytidiadelphus triquetrus, Ribes glandulosum, R. triste, Riccardia multifida, Rosa acicularis, R. carolina, R. virginiana, Rubus arcticus, R. hispidus, R. idaeus, R. pubescens, Rumex orbiculatus, Salix candida, S. discolor, S. pedicellaris, S. petiolaris, S. planifolia, S. pyrifolia, S. serissima, Sarracenia purpurea, Scheuchzeria palustris, Scirpus cyperinus, Scutellaria galericulata, Sibbaldiopsis tridentata, Sisyrinchium angustifolium, Solidago uliginosa, Sorbus decora, Sparganium, Sphagnum angustifolium, S. capillifolium, S. centrale, S. cuspidatum, S. fallax, S. fimbriatum, S. flavicomans, S. fuscum, S. girgensohnii, S. magellanicum, S. majus, S. nitidum, S. palustre, S. papillosum, S. recurvum, S. rubellum, S. russowii, S. squarrosum, S. subsecundum, S. teres, S. warnstorfii, S. wulfianum, Spiraea alba, S. tomentosa, Spiranthes romanzoffiana, Symphyotrichum puniceum, Symplocarpus foetidus, Thelypteris palustris, Thuja occidentalis, Tofieldia pusilla, Tomentypnum, Triadenum fraseri, Triantha glutinosa, Trichophorum cespitosum, Trientalis borealis, Trisetum spicatum, Typha latifolia, Usnea, Utricularia cornuta, U. gibba, U. interme-dia, Vaccinium angustifolium, V. corymbosum, V. macrocar-pon, V. myrtilloides, V. oxycoccos, V. uliginosum, Viburnum nudum, Viola renifolia, V. sagittata, Warnstorfia exannulata, Xyris.

Use by wildlife: *Chamaedaphne calyculata* provides browse (7%–8% protein), cover, and nesting sites for various wildlife species. The foliage is eaten by caribou (Mammalia: Cervidae: *Rangifer tarandus*), moose (Mammalia: Cervidae: *Alces alces*), sharp-tailed grouse (Aves: Phasianidae: *Tympanuchus phasianellus*), and white-tailed deer (Mammalia: Cervidae: *Odocoileus virginianus*). Plants provide essential cover for ruffed grouse (Aves: Phasianidae: *Bonasa umbellus*) and nesting sites for Mallard ducks (Aves: Anatidae: *Anas plat-yrhynchos*). The flowers of *C. calyculata* attract many for-aging insects including bees (Hymenoptera: Andrenidae: *Andrena bradleyi, A. carlini, A. cressonii, A. mandibularis, A. regularis, A. sigmundi, A. vicina*; Apidae: *Apis mellifera, Bombus affinis, B. bimaculatus, B. borealis, B. fervidus, B. griseocollis, B. impatiens, B. perplexus, B. rufocinctus, B. sandersoni, B. ternarius, B. terricola, B. vagans, Nomada bella, N. subrutila, Psithyrus*; Colletidae: *Colletes inaequalis*; Halictidae: *Augochlorella, Dialictus pilosus, Lasioglossum*; Melittidae: *Macropis ciliata*), butterflies (Lycaenidae: *Lycaena, Lycaenopsis argiolus*), and flies (Diptera: Bombyliidae: *Bombylius*; Muscidae: *Spilogona*; Syrphidae: *Eristalis, Helophilus fasciatus, Melanostoma, Sphaerophoria fatima, Volucella*). The plants are host to numerous butterfly and moth larvae (Lepidoptera: Coleophoridae: *Coleophora ledi*; Geometridae: *Cingilia catenaria, Ematurga amitaria, Metarranthis obfirmaria*; Lycaenidae: *Incisalia augusti-nus*; Noctuidae: *Coenophila opacifrons, Eugraphe subro-sea, Lithophane thaxteri, Melanchra assimilis, Syngrapha microgamma, Xestia dilucida, X. youngii*; Saturniidae: *Hemileuca nevadensis*; Tortricidae: *Acleris maculidor-sana, A. oxycoccana, Archips myricana, Choristoneura obsoletana, Olethreutes nanana, Sparganothis daphnana*). *Chamaedaphne* is also host to a spittlebug (Insecta: Hemiptera: Cercopidae: *Clastoptera saintcyri*) and a rust fungus (Fungi: Basidiomycota: Coleosporiaceae: *Chrysomyxa cassandrae*).

Economic importance: food: Like several other bog eri-coids, the leaves of *Chamaedaphne* contain poisonous andromedotoxin and should never be eaten or used in any type of food. Boiling the leaves in water facilitates release of the toxin. However, like *Andromeda* (earlier), the Ojibwa used the fresh and dried leaves of *Chamaedaphne* to make tea, perhaps by cold brewing so as to minimize the release of the toxins; **medicinal:** The Potawatomi used the leaves of *Chamaedaphne* as a poultice to reduce inflammation and as an infusion for fevers; **cultivation:** *Chamaedaphne calycu-lata* is cultivated as an ornamental bog plant. Cultivars include 'Angustifolia' and 'Nana'; **misc. products:** none; **weeds:** none; **nonindigenous species:** none

Systematics: Relationships ascertained by phylogenetic anal-ysis of molecular and morphological data (Figure 5.8) place the monotypic *Chamaedaphne* within tribe Gaultherieae, subfamily Vaccinioideae as the sister genus to *Eubotrys* (two

species: *Eubotrys racemosa*, FACW; *E. recurva*, FACU). Thus, evolution of the obligate habit in *Chamaedaphne* appears to represent an independent event in the family. The base chromosome number of *Chamaedaphne* is $x = 11$ with *C. calyculata* ($2n = 22$) being diploid. It would be interesting to examine *Chamaedaphne* genetically across its broad intercon-tinental distributional range, to determine whether any signifi-cant divergence has occurred within this widespread species.

Comments: *Chamaedaphne calyculata* is circumboreal and occurs throughout the cooler regions of northern North America.

References: Campbell et al., 2003; Crane, 2001; Densmore, 1997; Flinn & Pringle, 1983; Jumpponen & Trappe, 1998; Kron et al., 1999, 2002a; Lovell & Lovell, 1935; Pavek, 1993; Reader, 1975, 1977, 1987, 1979, 1982, 1983; Sigler & Gibas, 2005; Small, 1976; Thompson & Mohd-Saleh, 1995.

4. *Gaylussacia*

Huckleberry; airelle

Etymology: after Louis Joseph Gay-Lussac (1778–1850)

Synonyms: *Buxella; Decachaena; Decamerium; Lasiococcus; Lussacia; Vaccinium* (in part)

Distribution: global: North America; South America; **North America:** eastern

Diversity: global: 50 species; **North America:** 10 species

Indicators (USA): OBL: *Gaylussacia bigeloviana; G. orocola*

Habitat: freshwater; palustrine; **pH:** 3.8–6.7; **depth:** <1 m; **life-form(s):** emergent shrub

Key morphology: Shrubs (to 10 dm), the stems erect, branches ascending or spreading; leaves alternate, deciduous, the blades (to 4 cm) subcoriaceous, obovate, ovate, oblong or oblanceo-late, the margins entire, petioles (to 1.5 mm) short; racemes 3–8-flowered, axillary or terminal, erect or arching, bracts (to 10 mm) leaf-like; flowers pedicelled (to 4 mm), 5-merous, bisexual, petals connate, corollas (to 7.5 mm) pink, reddish, or white, campanulate, triangularly lobed (to 1.7 mm); anthers included, dehiscent by terminal pores; ovary inferior, stipi-tate, 5–10-carpellate; drupes (to 8 mm) black, juicy, globose or ovoid, each with 10 ellipsoid pyrenes (to 2 mm)

Life history: duration: perennial (buds, rhizomes); **asex-ual reproduction:** rhizomes; **pollination:** insect; **sexual condition:** hermaphroditic; **fruit:** drupes (common); **local dispersal:** drupes (animals, gravity, water); **long-distance dispersal:** drupes (animals)

Imperilment: (1) *Gaylussacia bigeloviana* [currently unranked]; (2) *G. orocola* [currently unranked]

Ecology: general: *Gaylussacia* is a genus of woody shrubs with its greatest species diversity in South America. The species inhabit a broad ecological range, which includes dry savannas, mesic woodlands, and xeric sandhills, as well as bogs, swamps, and other wetlands. Eight North American *Gaylussacia* species (80%) have some status as wetland indi-cators, but only two (20%) are ranked as OBL. Like most Ericaceae, the genus is common in acidic environments. The flowers produce nectar that is high in fructose and glucose and are known to be pollinated by bees (Hymenoptera: Apidae:

Bombus; Halictidae), hummingbirds (Aves: Trochilidae), and wasps (Hymenoptera: Verspidae). They are marginally protandrous but self-compatible and capable of self-pollination (although higher seed set occurs when outcrossed). The seeds of some North American species germinate well (80%–90%) following 2–3 months of cold stratification, whereas others exhibit higher germination rates for unstratified seed. The drupes are eaten by various animals and the successful germination of seeds from animal excrement indicates that endozoic transport contributes to their long-distance dissemination. The drupes are also likely to be dispersed at least short distances by water, but their floatation capability has not been evaluated. Seed banks can range from nonexistent to large, but do not necessarily contribute substantially to plant recruitment. Some species possess a large number of adventitious buds, which facilitate rapid resprouting following fires.

Gaylussacia bigeloviana (**Fern.**) **Sorrie & Weakley** is a resident of barrens, bogs, fens (acidic), heaths, pocosins, seepage swamps, and margins of boggy ponds at elevations of up to 500 m. Substrates are described as acidic (pH: 3.8–6.7) peat. The plants are in flower from June to early summer, reaching their peak in July, with fruits produced from July through September. Other life history information for this species is scarce, with at least some of it likely attributed to *Gaylussacia dumosa*, with which *G. bigeloviana* was long treated as synonymous (see *Systematics* later). Consequently, a detailed life history study focusing directly on this species is needed. **Reported associates:** *Acer rubrum, Alnus serrulata, Andromeda polifolia, Arethusa bulbosa, Aronia arbutifolia, A. melanocarpa, Bartonia paniculata, Betula cordifolia, B. michauxii, Calamagrostis pickeringii, Calopogon tuberosus, Carex atlantica, C. barrattii, C. bullata, C. canescens, C. collinsii, C. diandra, C. echinata, C. exilis, C. folliculata, C. gynandra, C. lasiocarpa, C. leptalea, C. limosa, C. oligosperma, C. pauciflora, C. magellanica, C. michauxiana, C. sterilis, C. striata, C. trisperma, C. utriculata, Chamaecyparis thyoides, Chamaedaphne calyculata, Cladina mitis, C. rangiferina, Cladium mariscoides, Cladonia squamosa, Cladopodiella fluitans, Clethra alnifolia, Corema conradii, Cyrilla racemiflora, Drosera anglica, D. intermedia, D. rotundifolia, Dulichium arundinaceum, Empetrum nigrum, Eriophorum angustifolium, E. callitrix, E. ×medium, E. tenellum, E. vaginatum, E. virginicum, Gaylussacia baccata, Gelsemium sempervirens, Geocaulon lividum, Gymnadeniopsis clavellata, Helonias bullata, Hypericum densiflorum, Ilex coriacea, I. glabra, I. mucronata, I. verticillata, Isoetes caroliniana, Juncus caesariensis, J. effusus, Juniperus communis, Kalmia angustifolia, K. latifolia, K. polifolia, Larix laricina, Liriodendron tulipifera, Lycopodiella appressa, Lyonia ligustrina, L. lucida, Magnolia virginiana, Maianthemum trifolium, Malaxis unifolia, Melampyrum lineare, Menyanthes trifoliata, Menziesia pilosa, Moneses uniflora, Mylia anomala, Myrica gale, M. pensylvanica, Narthecium americanum, Nyssa biflora, N. sylvatica, Oclemena nemoralis, Osmunda regalis, Osmundastrum cinnamomeum, Packera aurea, Parnassia asarifolia, Persea borbonia Picea mariana, Pinus rigida, P. serotina, P. strobus, Platanthera blephariglottis,*

P. dilatata, Pogonia ophioglossoides, Polytrichum juniperinum, P. strictum, Rhododendron arborescens, R. canadense, R. groenlandicum, R. maximum, R. viscosum, Rhynchospora alba, Rosa palustris, Rubus chamaemorus, Sagittaria fasciculata, Salix sericea, Sarracenia jonesii, S. purpurea, Scheuchzeria palustris, Schizaea pusilla, Scirpus cyperinus, S. longii, Sisyrinchium atlanticum, Smilax bona-nox, S. laurifoia, S. rotundifolia, Solidago patula, S. uliginosa, Sphagnum affine, S. bartlettianum, S. capillifolium, S. flavicomans, S. fuscum, S. imbricatum, S. magellanicum, S. palustre, S. papillosum, S. pulchrum, S. recurvum, S. rubellum, Symplocarpus foetidus, Taxodium distichum, Thelypteris palustris, Triadenum virginicum, Tsuga canadensis, Trichophorum alpinum, T. cespitosum, Utricularia cornuta, Vaccinium corymbosum, V. macrocarpon, V. oxycoccus, Viburnum nudum, Woodwardia areolata, W. virginica, Xerophyllum asphodeloides, Xyris torta, Zenobia pulverulenta.

Gaylussacia orocola **Small** inhabits montane seepage bogs at elevations from 500 to 1800 m. The substrates are derived from peat. Flowering occurs from late spring to early summer. As in the preceding, *G. orocola* had long been treated as a synonym of *G. dumosa*, making it difficult to identify from the literature any information that might apply specifically to the former. Especially given its rarity, a comprehensive study of the life history and ecology of *G. orocola* should be undertaken. **Reported associates:** *Myrica gale, Sarracenia jonesii, S. purpurea.*

Use by wildlife: The drupes of *Gaylussacia bigeloviana* are eaten by foxes (Mammalia: Canidae: *Vulpes vulpes*), quail (Aves: Phasianidae), ruffed grouse (Aves: Phasianidae: *Bonasa umbellus*), squirrels (Mammalia: Sciuridae), and turkeys (Aves: Phasianidae: *Meleagris gallopavo*). *Monilinia gaylussaciae* (Fungi: Ascomycota: Sclerotiniaceae) is a known pathogen of huckleberries but has not been reported to occur specifically on either *Gaylussacia bigeloviana* or *G. orocola*.

Economic importance: food: Although the fruits of some *Gaylussacia* species (e.g., *G. baccata*) are palatable and eaten on occasion, those of *G. bigeloviana* and *G. orocola* are described as "insipid" despite their juicy texture; **medicinal:** no uses reported; **cultivation:** *Gaylussacia bigeloviana* and *G. orocola* are not in cultivation; **misc. products:** none reported; **weeds:** *Gaylussacia bigeloviana* and *G. orocola* are not weedy; **nonindigenous species:** none.

Systematics: *Gaylussacia* has long been regarded as a close relative of *Vaccinium*; however, recent molecular phylogenetic analyses resolve both genera as polyphyletic and indicate the need for a major taxonomic reassessment of each group. Moreover, many authors do not recognize either *G. bigeloviana* or *G. orocola* as distinct species but treat both taxa as synonymous with *G. dumosa* (Andrews) Torr. & A. Gray. Morphological data distinguish the narrowly endemic *G. orocola* from *G. dumosa* phylogenetically, but resolve the former as the sister species of two South American taxa (*G. virgata* Mart. ex Meissn and *G. pseudogaultheria* Cham. & Schlecht.), a less than reassuring result. Because the relationships of neither *G. bigeloviana* nor *G. orocola* have yet been evaluated using molecular data, it is currently difficult to reconcile their taxonomic status with any degree of certainty. The base chromosome number of *Gaylussacia* is *x* = 12. *Gaylussacia bigeloviana* (2*n* = 24) is a diploid. Chromosome counts have not been reported for *G. orocola*. No hybrids involving either *G. bigeloviana* or *G. orocola* have been reported.

Comments: *Gaylussacia bigeloviana* occurs in northeastern North America, whereas *G. orocola* is endemic to North Carolina.

References: Almquist & Calhoun, 2003; Anderson & Davis, 1998; Angelo & Boufford, 2011; Batra, 1988; Dehgan et al., 1989; Fernald, 1911, 1921; Floyd, 2002; Freitas et al., 2006; Graham & Rebuck, 1958; Hayden & Hownsell, 2012; Knowlton, 1915; Kron et al., 2002a,b 2002; Kutcher et al., 2004; Matlack et al., 1993; Nichols, 1934; Rathcke, 1988; Sorrie, 2014; Sorrie & Weakley, 2007; Sorrie et al., 2009; Sperduto, 2011; Vander Kloet et al., 2012; Weakley & Schafale, 1994; Young & Weldy, 2004.

5. *Kalmia*

Alpine laurel, bog-laurel; laurier des marais
Etymology: after Pehr Kalm (1715–1779)
Synonyms: *Chamaedaphne* (in part); *Dendrium*; *Kalmiella*; *Leiophyllum*; *Loiseleuria*
Distribution: global: North America; **North America:** widespread
Diversity: global: 10 species; **North America:** 8 species
Indicators (USA): OBL: *Kalmia microphylla*, *K. polifolia*
Habitat: freshwater; palustrine; **pH:** 4.2–7.3; **depth:** <1 m; **life-form(s):** emergent shrub
Key morphology: stems (to 6 dm) erect or spreading, the branches 2-edged; leaves (to 35 mm) opposite, evergreen, coriaceous, subsessile, the margins revolute, the lower surface whitish; inflorescence terminal, racemose to umbellate, with 5-merous, showy flowers; corolla (to 20 mm) rotate, with 10 pockets, pink to crimson or white; stamens 10, the anthers dehiscent by apical pores; capsules (to 6 mm) globose, with numerous (to 200), winged seeds.
Life history: duration: perennial (buds, rhizomes); **asexual reproduction:** layering, rhizomes; **pollination:** insect; **sexual condition:** hermaphroditic; **fruit:** capsules (common); **local dispersal:** seeds (wind); **long-distance dispersal:** seed (animals, water)
Imperilment: (1) *Kalmia microphylla* [G5]; S1 (MB); S2 (ON); S3 (WY); (2) *K. polifolia* [G5]; S1 (CT, MT, RI); S3 (AB)
Ecology: general: All *Kalmia* species have some tolerance for wet habitats and occur often in lowlands; however, most are designated as FACW plants and tend to grow more frequently in sites where saturated conditions do not persist. As for many Ericaceae, their habitats primarily are acidic (usually pH: 4.0–5.5). The aquatic species occur in open sites at colder (northern) latitudes or at higher (alpine) elevations. *Kalmia* flowers are highly specialized for insect pollination. While in bud, the stamens are contained within ten pouches situated around the corolla margin. As the elastic filaments elongate within the pockets, they acquire tension, which increases

as the flower opens. The weight or movement of a pollinator on an open flower releases the filaments from the pockets, causing the anthers to spring forward, dusting the insect with pollen. The pollen (released in tetrads) can be discharged at distances for up to 15 cm in this fashion. Common pollinators include bumblebees (Hymenoptera: Apidae: *Bombus ternarius*) and mining bees (Hymenoptera: Andrenidae: *Andrena vicina*), but not honeybees (Hymenoptera: Apidae: *Apis*). The species are self-incompatible to partially self-compatible, but normally are outcrossed, with selfed seedlings exhibiting inbreeding depression. Most seed dispersal is abiotic, but some animal dispersal also occurs. The roots are associated with endophytic and ectophytic mycorrhizae. All *Kalmia* species reproduce vegetatively by layering. Dense rhizome networks characterize some species.

***Kalmia microphylla* (Hook.) A. Heller** grows in open alpine bogs, meadows, swamps, and rocky crevices at higher elevations from 2000 to 3500 m. The flowers produce up to 200 seeds per capsule. The seeds are nondormant and germinate without stratification at 22°C (21°C–24°C). The seeds are winged and are dispersed by wind or water. The plants spread vegetatively by layering and often form dense mats. Propagation by small, untreated cuttings (5 cm) is effective with roots developing within 3 weeks. **Reported associates:** *Abies amabilis*, *Abronia latifolia*, *Agrostis thurberiana*, *Alnus*, *Angelica arguta*, *Antennaria alpina*, *A. lanulosa*, *Bistorta bistortoides*, *Caltha leptosepala*, *Carex aquatilis*, *C. canescens*, *C. deweyana*, *C. illota*, *C. lenticularis*, *C. limosa*, *C. luzulina*, *C. nigricans*, *C. rossii*, *C. rupestris*, *C. scopulorum*, *C. spectabilis*, *C. tumulicola*, *C. utriculata*, *Cassiope mertensiana*, *Castilleja miniata*, *C. parviflora*, *Cicuta douglasii*, *Danthonia intermedia*, *Deschampsia atropurpurea*, *Drosera rotundifolia*, *Eleocharis pauciflora*, *Epilobium alpinum*, *E. minutum*, *E. watsonii*, *Equisetum arvense*, *Erigeron peregrinus*, *Eriophorum chamissonis*, *Gaultheria ovatifolia*, *Gentiana calycosa*, *Gnaphalium chilense*, *Hieracium gracile*, *Hypericum anagalloides*, *Juncus balticus*, *J. drummondii*, *J. mertensianus*, *J. parryi*, *Larix lyallii*, *Leucothoe davisiae*, *Ligusticum grayi*, *L. tenuifolium*, *Lonicera involucrata*, *Luetkea pectinata*, *Lupinus latifolius*, *L. lepidus*, *Luzula hitchcockii*, *Maianthemum stellatum*, *Micranthes ferruginea*, *Mimulus primuloides*, *Oreostemma alpigenum*, *Packera cymbalaria*, *P. pseudaurea*, *Pedicularis ornithorhyncha*, *Phleum alpinum*, *Phyllodoce empetriformis*, *P. glanduliflora*, *Picea engelmannii*, *Pinus contorta*, *Platanthera dilatata*, *Potentilla drummondii*, *P. flabellifolia*, *Primula jeffreyi*, *Rhododendron neoglandulosum*, *Salix farriae*, *Senecio fremontii*, *S. triangularis*, *Solidago canadensis*, *Spiraea densiflora*, *S. douglasii*, *Spiranthes romanzoffiana*, *Symphyotrichum foliaceum*, *S. spathulatum*, *Tephroseris lindstroemii*, *Triantha glutinosa*, *Trientalis arctica*, *Tsuga mertensiana*, *Vaccinium deliciosum*, *V. occidentale*, *V. scoparium*, *Veratrum viride*, *Veronica wormskjoldii*, *Viola macloskeyi*, *V. pallens*, *Xerophyllum tenax*.

***Kalmia polifolia* Wangenh.** inhabits ditches, bogs, lake or pond margins, meadows, poor fens, and swamps at elevations of up to 2000 m. Habitats are characterized by full sun, acidic (pH: 4.2–6.9) conditions and substrates that usually are organic (including quaking mats) but occasionally are rocky or sandy. Plants can withstand temperatures as low as −33°C. Their flowers produce about 150–200 seeds per capsule depending on pollinator availability and site fertility. The seeds are non-dormant or become physiologically dormant and germinate at 21°C–24°C after 71–112 days of cold stratification (normal outdoor winter conditions). The seeds possess large wings and are dispersed by the wind. They also remain afloat (by air bubbles) for more than 72 h. Seeds kept at 5°C have retained up to 59% viability after 15 years of storage. When growing near open water, the plants can form dense mats that extend over the water surface by layering. Small cuttings (5 cm) propagate readily and develop roots within 3 weeks with no special treatment. **Reported associates:** *Abies balsamea, A. concolor, Alnus rugosa, Andromeda polifolia, Antennaria media, A. rosea, Arethusa bulbosa, Argentina anserina, Aulacomnium palustre, Betula papyrifera, B. pumila, Bistorta bistortoides, Botrychium simplex, Calamagrostis breweri, C. canadensis, Calla palustris, Calopogon pulchellus, Carex aquatilis, C. buxbaumii, C. canescens, C. chordorrhiza, C. echinata, C. exserta, C. lacustris, C. lasiocarpa, C. limosa, C. magellanica, C. oligosperma, C. pauciflora, C. rossii, C. rostrata, C. scopulorum, C. stricta, C. subnigricans, C. trisperma, C. utriculata, Castilleja lemmonii, Chamaedaphne calyculata, Cladina portentosa, Cladonia, Clintonia borealis, Comarum palustre, Coptis trifolia, Cornus canadensis, Cypripedium acaule, Danthonia intermedia, D. unispicata, Deschampsia cespitosa, Dicranum polysetum, Drosera intermedia, D. rotundifolia, Dulichium arundinaceum, Eleocharis quinqueflora, Epilobium, Equisetum fluviatile, Eriophorum tenellum, E. vaginatum, E. virginicum, Fragaria virginiana, Gaultheria hispidula, G. procumbens, Gentiana newberryi, Hypericum anagalloides, Iris versicolor, Ivesia lycopodioides, Juncus filiformis, J. nevadensis, J. parryi, J. stygius, Larix laricina, Ligusticum grayi, Linnaea borealis, Lonicera caerulea, Lupinus lepidus, Luzula congesta, Lycopus uniflorus, Lysimachia terrestris, Maianthemum trifolium, Menyanthes trifoliata, Mimulus primuloides, Monotropa uniflora, Muhlenbergia filiformis, Myrica gale, Nemopanthus mucronata, Oreostemma alpigenum, Osmundastrum cinnamomeum, Pedicularis groenlandica, Perideridia parishii, Phyllodoce breweri, Picea mariana, Pinus contorta, P. resinosa, P. strobus, Platanthera clavellata, Pleurozium schreberi, Poa keckii, P. secunda, Pogonia ophioglossoides, Polytrichum strictum, Potentilla grayi, Ptilagrostis kingii, Rhododendron groenlandicum, Rhynchospora alba, R. capillacea, Ribes montigenum, Rubus arcticus, R. pubescens, Salix arctica, S. eastwoodiae, S. orestera, S. pedicellaris, S. pyrifolia, Sarracenia purpurea, Scheuchzeria palustris, Solidago uliginosa, Sphagnum capillifolium, S. fuscum, S. magellanicum, S. recurvum, S. subsecundum, Symphyotrichum boreale, Triadenum fraseri, Trichophorum alpinum, T. clementis, Triglochin maritimum, Trisetum spicatum, Vaccinium cespitosum, V. macrocarpon, V. oxycoccos, V. uliginosum, Viola, Warnstorfia exannulata, Xyris montana.*

Use by wildlife: *Kalmia microphylla* is poisonous to birds (Aves) and mammals (Mammalia) and is known to cause death in grazing livestock such as cattle (Bovidae: *Bos*), goats (Bovidae: *Capra*), horses (Equidae: *Equus*), and especially sheep (Bovidae: *Ovis*). *Kalmia polifolia* is reportedly unpalatable to wildlife, but is browsed by deer (Mammalia: Cervidae) in some areas. Rabbits (Mammalia: Leporidae) readily graze the plants and may be less susceptible to the toxins. The short, evergreen stature of *K. microphylla* provides cover for various small mammals. The flowers of *Kalmia* species are visited by bumble bees (Hymenoptera: Apidae: *Bombus*), mining bees (Hymenoptera: Andrenidae: *Andrena*), butterflies (Lepidoptera), hawkmoths (Lepidoptera: Sphingidae), and beetles (Coleoptera). *Kalmia polifolia* is a host for larval Lepidoptera (Noctuidae: *Acronicta lanceolaria, Apharetra dentata, Xylotype arcadia*) as well as several plant pathogenic fungi (Ascomycota: Erysiphaceae: *Microsphaera penicillata*; Mycosphaerellaceae: *Septoria angustifolia*; Phyllachoraceae: *Dothidella kalmiae*; Basidiomycota: Exobasidiaceae: *Exobasidium vaccinii*).

Economic importance: food: All species of *Kalmia* are highly poisonous and contain potentially lethal grayanotoxins (andromedotoxins) and the glycoside arbutin. Yet, the Hanaksiala people made tea from the leaves of *K. microphylla*, a dangerous practice given the toxicity of the plants. Honey derived from the nectar of *K. polifolia* is also poisonous; **medicinal:** The Kwakwaka'wakw tribe used a decoction made from the leaves of *K. microphylla* to induce vomiting and for spitting blood. The Kwakiutl used *K. polifolia* similarly and also to promote the healing of wounds. The Tlingits prepared infusions of the plant to treat skin disorders; **cultivation:** *Kalmia microphylla* is cultivated as a garden ornamental under the species name or as var. *occidentalis* (=*K. occidentalis*) or forma *alba*. One cultivar is 'Mount Shasta.' *Kalmia polifolia* is also cultivated as the species, var. *compacta*, forma *leucantha* (white-flowered) or cultivar 'Nana.' 'Rocky top' is an F_1 hybrid of *K. polifolia* × *K. microphylla*; **misc. products:** Some North American tribes reportedly made a suicidal poison from *K. microphylla* and other species. A yellow-brown dye can be extracted from the leaves of this plant; **weeds:** none; **nonindigenous species:** *Kalmia polifolia* has become naturalized in Scotland.

Systematics: Phylogenetic analyses of DNA sequence data provide compelling arguments to merge with *Kalmia* two monotypic genera (*Leiophyllum, Loiseleuria*), which, in several analyses, form a subclade nested within the genus (Figure 5.9). Although this merger has been followed here, it is worth noting that relationships among *Kalmia* species remain inadequately resolved, and it is conceivable that additional data might place *Leiophyllum* and *Loiseleuria* as a sister clade to *Kalmia*. Both genera lack the distinctive staminal pouches of other kalmias, which are regarded as a synapomorphy for the genus. They also differ by their reduced number of carpels (2–3) compared to five in other *Kalmia* species. Regardless, molecular data (DNA sequences and an informative insertion in the *matK* gene) provide strong support for the monophyly of a clade consisting of *Kalmia, Leiophyllum*, and *Loiseleuria*. This clade nests within the "phyllodocoid" members of subfamily Ericoideae of Ericaceae, in a position between *Bejara*

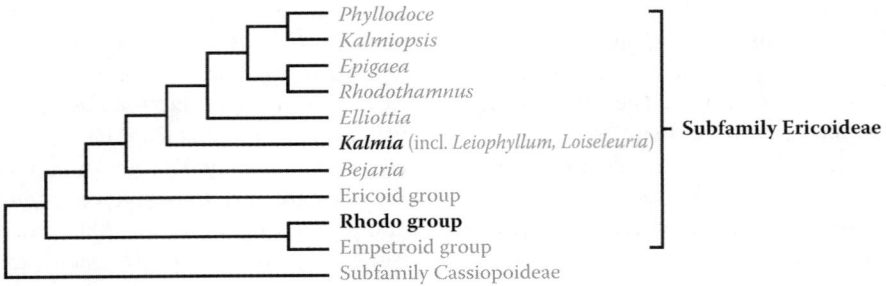

FIGURE 5.9 Relationships in Ericaceae subfamily Ericoideae as indicated by analysis of *matK* sequence data. These relationships indicate that there have been several independent origins of taxa containing OBL designated species in North America (bold) within the subfamily. (Adapted from Kron, K.A., *Amer. J. Bot.*, 84, 973–980, 1997.)

and *Elliottia* (Figure 5.9). Although genetic barriers to hybridization distinguish most *Kalmia* species, some taxonomic limits have been disputed. Notably, *K. occidentalis* has been treated as a subspecies or variety of both *K. microphylla* and *K. polifolia* and *K. microphylla* has been treated as a variety of *K. polifolia*. The status of *K. occidentalis* is unclear. DNA sequence analyses resolve this taxon as distinct from both *K. microphylla* and *K. polifolia*, but place it closer to the latter. *Kalmia polifolia* and *K. microphylla* are interfertile and hybridize readily under experimental manipulations. However, their hybrids are triploid (see later) and are generally sterile with highly reduced pollen fertility (<15%). The two taxa differ morphologically (mainly by the smaller stature of the latter), flower at different times (when grown in a common garden), occupy different elevational habitats, possess different chromosome numbers, and exhibit different crossing relationships with other *Kalmia* species. Thus, it seems appropriate to regard these as distinct species. Crossing relationships of *K.*

occidentalis have not been reported but could help to clarify the status of this taxon. Together, the obligate aquatic taxa form a clade (Figure 5.10), which is strongly isolated from other species genetically. Natural interspecific hybrids have not been reported in *Kalmia* and are unlikely to occur given the intrinsic genetic barriers that exist between the species. The base chromosome number of *Kalmia* is $x = 12$. *Kalmia polifolia* ($2n = 48$) is tetraploid; whereas, *K. microphylla* (and other North American species) is diploid ($2n = 24$).

Comments: *Kalmia microphylla* occurs in northern and western North America; *K. polifolia* is widespread throughout the northern half of the continent.

References: Baskin & Baskin, 1998; Campbell et al., 2003; Clawson, 1933; Jaynes, 1968, 1969, 1988; Knight & Walter, 2001; Kron & King, 1996; Lahring, 2003; Levri, 2000; Liu et al., 2009; Pellerin et al., 2006.

6. *Leucothoe*

Doghobble; Sierra-laurel

Etymology: after Leucothoë, the sister of Clytia and daughter of King Orchamus in Greek mythology

Synonyms: *Cassiphone*; *Eubtryoides* (in part); *Oreocallis*

Distribution: global: Asia; North America; **North America:** E/W disjunct

Diversity: global: 5 species; **North America:** 3 species

Indicators (USA): OBL; FACW: *Leucothoe davisiae*

Habitat: freshwater; palustrine; **pH:** 4.5–6.5; **depth:** <1 m; **life-form(s):** emergent shrub

Key morphology: stems (to 20 dm) erect; leaves (to 7 cm) alternate, coriaceous, evergreen; racemes (to 10 cm) terminal, with numerous, pendulous, 5-merous flowers; sepals (to 3.3 mm) five, distinct; corolla (to 6 mm) white, urceolate; stamens 10, dehiscent by terminal pores; capsules (to 6 mm) globose, 5-lobed; seeds winged

Life history: duration: perennial (buds); **asexual reproduction:** none; **pollination:** insect; **sexual condition:** hermaphroditic; **fruit:** capsules (common); **local dispersal:** seeds (wind); **long-distance dispersal:** seeds (water, wind)

Imperilment: (1) *Leucothoe davisiae* [G4]; S2 (OR)

Ecology: general: *Leucothoe* species are more typical of dry slopes and forests than of wet areas. A few species occur facultatively in low, wet woods. *Leucothoe davisiae* was ranked

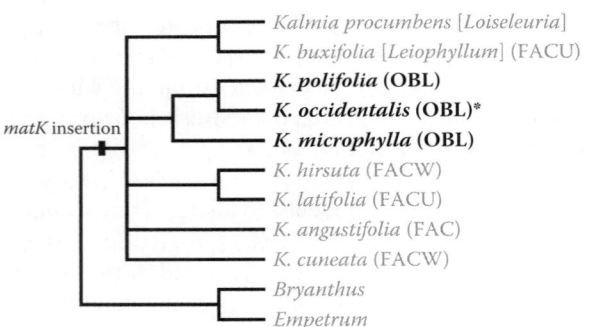

FIGURE 5.10 Phylogenetic relationships in *Kalmia* consistently indicated by DNA sequence analysis. Although overall resolution of species relationships remains poor in *Kalmia*, the OBL species (in bold) associate within a clade. **Kalmia occidentalis* is without wetland designation because it is treated as a synonym of *K. microphylla*; an OBL ranking has been assigned here in accord with that disposition. Given the poorly resolved nature of *Kalmia*, the inclusion of *Loiseleuria* and *Leiophyllum* within the genus deserves further consideration (see text), especially in light of their distinctive morphology. However, they certainly are no more distant than a sister clade to *Kalmia*. (Modified from Kron, K.A. & King, J.M., *Syst. Bot.*, 21, 17–29, 1996.)

as an obligate aquatic in the 1996 wetland indicator list, but has since been reclassified as FACW. It has been retained here because the treatment already was completed prior to the release of the 2013 list.

***Leucothoe davisiae* Torr. ex A. Gray** inhabits montane bogs and seeps at elevations of up to 2600 m. The habitats are acidic (pH: 4.5–6.5) with substrates representing various mixtures of clay, sand, and loam. Exposure ranges from partial sun to full shade. The plants are ectomycorrhizal and will tolerate serpentine soils. Information on the biology and ecology of this species is scarce. It is uncertain whether the plants possess any mechanism of vegetative reproduction. The flowers are fragrant and presumably insect pollinated; however, studies on the reproductive ecology of this species should be undertaken to elucidate the breeding system. No reports on seed germination are available. The seeds are winged and presumably dispersed by wind and also by water; however, their ability to float has not been confirmed. **Reported associates:** *Abies concolor, A. magnifica, Acer circinatum, Achlys triphylla, Aconitum columbianum, Adenocaulon bicolor, Allium validum, Alnus rubra, A. tenuifolia, Amelanchier alnifolia, Anemone deltoidea, A. quinquefolia, Aquilegia formosa, Arctostaphylos mewukka, A. nevadensis, A. patula, A. viscida, Asarum caudatum, Athyrium filix-femina, Berberis aquifolium, B. nervosa, Boykinia elata, B. major, Calocedrus decurrens, Caltha leptosepala, Carex hendersonii, Ceanothus velutinus, Chamaecyparis lawsoniana, Chamaescyce serpyllifolia, Chamerion angustifolium, Chimaphila menziesii, C. umbellata, Chrysolepis sempervirens, Circaea alpina, Clintonia uniflora, Corallorhiza maculata, Cornus canadensis, C. nuttallii, C. sericea, Darlingtonia californica, Disporum hookeri, Drosera, Gaultheria humifusa, G. ovatifolia, G. shallon, Glyceria elata, Goodyera oblongifolia, Hosackia crassifolia, Juncus, Kalmia microphylla, Lilium pardalinum, L. washingtonianum, Linnaea borealis, Listera convallarioides, Lonicera conjugialis, Lysichiton americanus, Maianthemum racemosum, M. stellatum, Mimulus cardinalis, M. moschatus, Mitella breweri, M. ovalis, M. pentandra, Orobanche, Orthilia secunda, Pedicularis semibarbata, Phyllodoce breweri, Physocarpus capitatus, Pinus contorta, P. monticola, Platanthera, Polystichum munitum, Pseudotsuga menziesii, Pteridium aquilinum, Pterospora andromedea, Pyrola picta, P. secunda, Rhamnus alnifolia, Quercus sadleriana, Rhododendron macrophyllum, R. neoglandulosum, R. occidentale, Ribes bracteosum, R. lacustre, R. sanguineum, Rosa gymnocarpa, Rubus parviflorus, R. ursinus, Salix ligulifolia, S. lucida, S. orestera, Sambucus nigra, S. racemosa, Sarcodes sanguinea, Scoliopus hallii, Senecio triangularis, Solidago, Sorbus californica, Spiraea douglasii, Streptopus amplexifolius, Symphoricarpos albus, Taxus brevifolia, Tellima grandiflora, Thalictrum fendleri, Tiarella trifoliata, Trientalis latifolia, Trillium ovatum, T. rivale, Tsuga heterophylla, T. mertensiana, Vaccinium arbuscula, V. membranaceum, V. parvifolium, V. scoparium, V. uliginosum, Veratrum viride, Viola glabella, Xerophyllum tenax.*

Use by wildlife: The leaves of *Leucothoe davisiae* contain grayanotoxin but sometimes are grazed by livestock, which can be poisoned. Sheep (Bovidae: *Ovis*) are particularly susceptible and can be killed by ingesting as little as 30 g of leaves. The plants are considered to be potentially susceptible to infection by water molds (Fungi: Oomycota: Pythiaceae: *Phytophthora ramorum*).

Economic importance: food: All parts of *Leucothoe davisiae* contain highly poisonous grayanotoxin and should never be eaten; **medicinal:** none; **cultivation:** *Leucothoe davisiae* is cultivated as a shade tolerant ornamental garden plant; **misc. products:** none; **weeds:** none; **nonindigenous species:** none

Systematics: Morphological cladistic analysis indicates that *Leucothoe* as traditionally circumscribed is polyphyletic, unless redefined to exclude *L. racemosa* and *L. recurva* (*Eubotrys*) and *L. grayana* (*Eubotryoides*). The remaining species comprise a clade that contains *Leucothoe davisiae* along with *L. axillaris, L. fontanesiana, L. griffithiana,* and *L. keiskei*. This clade (*Leucothoe* sensu stricto) represents a basal element of tribe Gaultherieae, subfamily Vaccinioideae. Phylogenetic analyses using molecular data are incomplete, but provide corroborative evidence for the transfer of *L. racemosa* to *Eubotrys* and the basal position of *Leucothoe* sensu stricto in tribe Gaultherieae (Figure 5.8). The base chromosome number of *Leucothoe* is $x = 11$. Counts have not been reported for *L. davisiae*.

Comments: *Leucothoe davisiae* is relatively uncommon and is restricted to higher elevation habitats in California and Oregon.

References: Judd & Waselkov, 2006; Kaye et al., 2003; Kron et al., 2002a,b; Largent et al., 1980; Lawrence & Kaye, 2003; Mason, 1957.

7. Rhododendron

Labrador/trapper's tea, swamp azalea; bois de savane

Etymology: from the Greek *rhodon dendron* meaning "rose tree"

Synonyms: *Azalea; Azaleastrum; Biltia; Chamaecistus; Chamaeledon; Hochenwartia; Hymenanthes; Ledum; Loiseleuria; Menziesia; Osmothamnus; Plinthocroma; Rhododendros; Rhodora; Rhodothamnus; Stemotis; Therorhodion; Tsusiophyllum; Tsutsiusi; Vireya*

Distribution: global: Asia; Europe; North America; **North America:** northern

Diversity: global: 1000 species; **North America:** 30 species

Indicators (USA): OBL; FACW: *Rhododendron columbianum, R. groenlandicum, R. viscosum*; **FAC:** *R. groenlandicum*

Habitat: freshwater; palustrine; **pH:** 2.9–7.7; **depth:** <1 m; **life-form(s):** emergent shrub

Key morphology: stems erect, short (to 1.5 m) or tall (to 5.0 m), with fragrant foliage; leaves (to 7 cm) alternate, deciduous or evergreen, subsessile, the undersides glabrous or softly pubescent to densely wooly (white turning brown), the margins flat or involute; corymbs/umbels terminal, compressed, bearing pedicellate (to 2 cm), 5-merous flowers, which arise from scaly buds; petals distinct (to 8 mm) or fused basally (to 5.5 cm), cream to white or pink to red; stamens 5–10, anthers dehiscent by terminal pores; capsules (<6 mm or up to 2 cm) elongate, many seeded

Life history: duration: perennial (buds, rhizomes, stolons); **asexual reproduction:** layering, rhizomes, root sprouts; **pollination:** insect, self; **sexual condition:** hermaphroditic; **fruit:** capsules (common); **local dispersal:** seeds (wind); **long-distance dispersal:** seeds (water, wind)

Imperilment: (1) *Rhododendron columbianum* [G5]; S2 (WY); S3 (AB); (2) *R. groenlandicum* [G5]; SX (SD); S1 (NJ, OH); S2 (CT); S3 (PA); (3) *R. viscosum* [G5]; SH (VT); S1 (ME, WV); S3 (NH)

Ecology: general: Rhododendrons mainly are terrestrial plants with few species being common in wet habitats. About 18 of the 30 North American species (60%) are given some wetland designation, but mostly as FAC or upland (UPL) ranks. Three species are designated as OBL, with two of these known formerly in the genus *Ledum* (see *Systematics* later).

Rhododendron columbianum (**Piper**) **Harmaja** inhabits bogs, mashes, meadows, and swamps, often occurring in the montane to subalpine zone, at elevations of up to 3600 m. The plants generally occur at the edges of bogs and swamps on acidic substrates (pH: 3.9–7.0) and can tolerate serpentine or ash soils. They are shade tolerant and can withstand temperatures down to −33°C. There is no specific information regarding the reproductive ecology of this species and additional information is needed. Presumably, the plants are insect pollinated and the winged seeds are dispersed primarily by the wind. The roots typically are colonized (55%–66%) by ericoid mycorrhizae and to a lesser degree by ectomycorrhizae. Experiments have shown that high soil fertility results in a significant increase in total biomass, foliar nitrogen, and root concentrations of condensed tannins and phenolics across a pH gradient of 3.5–6.5; however, foliar concentrations of condensed tannins and phenolics decline with increasing fertility irrespective of pH. **Reported associates:** *Abies amabilis, A. grandis, A. lasiocarpa, Acer glabrum, Achillea millefolium, Aconitum columbianum, Actaea rubra, Adenocaulon bicolor, Agrostis exarata, A. idahoensis, A. scabra, A. thurberiana, Alnus rubra, A. sinuata, Amelanchier alnifolia, Amerorchis rotundifolia, Anaphalis margaritacea, Anemone piperi, Angelica arguta, Antennaria flavescens, A. lanulosa, A. racemosa, Aquilegia formosa, Arctostaphylos uva-ursi, Arnica cordifolia, A. latifolia, A. mollis, Athyrium filix-femina, Berberis aquifolium, Bistorta bistortoides, Blechnum spicant, Boykinia major, Bromopsis vulgaris, Calamagrostis canadensis, C. rubescens, C. stricta, Calochortus nitidus, Caltha leptosepala, Camassia quamash, Campanula scabrella, Canadanthus modestus, Carex aquatilis, C. aurea, C. brunnescens, C. canescens, C. concinnoides, C. cusickii, C. disperma, C. echinata, C. geyeri, C. illota, C. leptalea, C. mariposana, C. microptera, C. obnupta, C. pachystachya, C. paupercula, C. prionophylla, C. pyrenaica, C. rossii, C. scopulorum, C. utriculata, C. vesicaria, Cassiope mertensiana, Castilleja rhexiifolia, Catabrosa aquatica, Chamaecyparis lawsoniana, Chamerion angustifolium, Chimaphila umbellata, Cicuta douglasii, Cinna latifolia, Claytonia lanceolata, Clematis occidentalis, Clintonia uniflora, Comarum palustre, Conium maculatum, Coptis occidentalis, Cornus canadensis, C. sericea, Cupressus goveniana, Cypripedium passerinum, Darlingtonia californica, Deschampsia cespitosa, Disporum hookeri, D. trachycarpum, Drosera rotundifolia, Elymus glaucus, Empetrum nigrum, Epilobium ciliatum, E. glaberrimum, E. watsonii, Equisetum arvense, E. scirpoides, Erigeron peregrinus, Eriophorum chamissonis, Erythronium grandiflorum, Eurybia conspicua, Festuca occidentalis, Fragaria vesca, F. virginiana, Frangula purshiana, Fritillaria pudica, Galium triflorum, Gaultheria humifusa, G. shallon, Gentiana calycosa, G. propinqua, Geranium richardsonii, Geum macrophyllum, Glyceria borealis, Goodyera oblongifolia, Hackelia diffusa, Heracleum lanatum, Hieracium albiflorum, H. gracile, Juncus balticus, J. conglomeratus, J. drummondii, J. parryi, Juniperus communis, Kalmia microphylla, Larix lyallii, L. occidentalis, Lewisia pygmaea, Ligusticum canbyi, L. tenuifolium, Linnaea borealis, Listera cordata, Lomatium utriculatum, Lonicera caerulea, L. involucrata, L. utahensis, Luetkea pectinata, Luina nardosmia, Lupinus argenteus, L. breweri, L. latifolius, L. polyphyllus, L. wyethii, Luzula glabrata, L. hitchcockii, L. parviflora, Lycopodium annotinum, Lysichiton americanus, Maianthemum stellatum, Malus fusca, Melampyrum lineare, Menziesia ferruginea, Mertensia paniculata, M. perplexa, Micranthes nelsoniana, M. odontoloma, Microseris troximoides, Mitella breweri, M. nuda, M. pentandra, M. stauropetala, Mitellastra caulescens, Montia cordifolia, Muhlenbergia filiformis, Myrica californica, M. gale, Oenanthe sarmentosa, Oreostemma alpigenum, Oryzopsis exigua, Osmorhiza chilensis, O. purpurea, Parnassia fimbriata, Paxistima myrsinites, Pedicularis bracteosa, P. contorta, P. groenlandica, Penstemon procerus, Pera, Phleum alpinum, Phyllodoce empetriformis, Picea engelmannii, P. sitchensis, Pinus albicaulis, P. contorta, P. jeffreyi, P. monticola, Platanthera dilatata, P. hyperborea, P. obtusata, P. stricta, Poa pratensis, Polemonium pulcherrimum, Potamogeton gramineus, Potentilla anserina, P. diversifolia, P. drummondii, P. flabellifolia, P. rivalis, Primula jeffreyi, P. latiloba, Prunella vulgaris, Prunus virginiana, Pseudotsuga menziesii, Pyrola asarifolia, P. minor, P. uniflora, Ranunculus uncinatus, Rhamnus purshiana, Rhododendron albiflorum, R. macrophyllum, R. occidentale, Ribes hudsonianum, R. lacustre, R. montigenum, R. viscidulum, Rosa gymnocarpa, Rubus parviflorus, R. pedatus, R. ursinus, Salix bebbiana, S. brachystachys, S. commutata, S. drummondiana, S. farriae, S. glaucops, S. hookeriana, S. pseudomyrsinites, Sambucus cerulea, Sanguisorba officinalis, S. sitchensis, Sedum stenopetalum, Senecio cymbalaria, S. pauciflorus, S. pseudaureus, S. triangularis, Shepherdia canadensis, Sisyrinchium californicum, Solidago canadensis, Sorbus scopulina, Sphagnum angustifolium, S. palustre, S. henryense, Spiraea betulifolia, S. densiflora, S. douglasii, S. splendens, Spiranthes romanzoffiana, Stenanthella occidentalis, Stipa occidentalis, Streptopus amplexifolius, Symphoricarpos albus, S. oreophilus, Symphyotrichum foliaceum, S. lanceolatum, S. spathulatum, Taraxacum officinale, Thalictrum occidentale, Thermopsis montana, Thuja plicata, Tiarella trifoliata, Trautvetteria caroliniensis, Triantha glutinosa, Trifolium longipes, Trillium ovatum, Trisetum* *cernuum, T. spicatum, Trollius laxus, Tsuga heterophylla, Vaccinium cespitosum, V. deliciosum, V. globulare, V. myrtillus, V. ovatum, V. oxycoccos, V. parvifolium, V. scoparium, V. uliginosum, Valeriana dioica, V. sitchensis, Veratrum californicum, V. viride, Verbascum thapsus, Verbena verna, Veronica americana, V. wormskjoldii, Viola aduncoides, V. glabella, V. macloskeyi, V. orbiculata, Xerophyllum tenax.*

Rhododendron groenlandicum **(Oeder) Kron & Judd** grows in bogs, depressions, ditches, interdunal swales, marshes, swamps, along the margins of lakes, ponds, and streams, and on cliffs or ledges at elevations of up to 2800 m. Although this species is regarded as an indicator of cold, sphagnous bogs, it has a broader ecological amplitude. Occurrences range from lowland sites to alpine elevations, often where permafrost exists. This species occurs commonly in open peat bogs, on acidic (pH: 2.9–7.7; typically <4.0) organic substrates (often quaking peat mats); however, also on clay, humus, muck or even gravel, rock, sand, and scree. Some habitats are characterized by deep shade. Perhaps the most unusual occurrences are on sandstone cliffs and ledges, which often are described as quite dry. It would be informative to conduct a detailed genetic study of this species in its different habitats to determine whether any ecological divergence may have occurred. The flowers are self-compatible and have been described both as weakly protogynous and slightly protandrous, depending on the study. They are fragrant and produce a small amount of nectar. The filaments elongate to become bowed under tension while in bud. The anthers dehisce in bud, liberating pollen from the pore and onto the stigma, thereby enabling self-pollination. When the flowers open, the stamens spring outward to contact pollinators, which include flies (Diptera) and bees (Hymenoptera). Despite the ability to self-pollinate, field studies show that a much higher seed set occurs in open-pollinated flowers (20%–100%) compared to those where pollinators were excluded (0%–50%). Mechanical perturbation of the flowers (e.g., by wind) may facilitate self-pollination as hand-pollinated flowers yielded a 70% seed set, which was independent of the pollen source (i.e., from the same or different plant). The capsules dehisce upwards from their base, perhaps as an adaptation to their often pendant habit when mature. Each produces an average of 133 seeds, which are winged, light (<0.05 mg), and are dispersed primarily by the wind. About 20% of the seeds can float, which may allow for some water dispersal. Germination requires light and acidic (e.g., pH: 5.5) conditions but is reduced under far-red light. A seed bank is not produced. Unstratified seeds generally germinate within 30 days. Germination of 1-month-old seeds is about 58% with that of 13-month-old seed reduced to only 20%. However, germination of cold-stratified seed (30 days) increased to 78%–98% at temperatures from 9°C to 26°C. Dried seed can be stored at 4°C for up to 3 years. Seedling establishment is optimal where fire or other disturbance has produced a discontinuous cover of mosses. Seedlings grow slowly at an initial rate of about 1 mm per month. Vegetative regeneration by sprouting is an important means of surviving disturbance and for persistence in this species. The rhizomes (at 15–50 cm depth) are protected from shallow burns. Reproduction by layering enables plants to develop into clones up to 10 m² in area. Plants are relatively resistant to fire and recover rapidly by sprouting. The plants readily colonize disturbed areas such as clear-cuts. The genetic population structure has not been investigated in this species.

Reported associates: *Abies balsamea, Acer rubrum, A. saccharum, A. spicatum, Actaea rubra, Agalinis paupercula, Agrostis scabra, Alnus rugosa, Amalanchier, Andromeda polifolia, Anemone quinquefolia, Aneura pinguis, Aralia hispida, A. nudicaulis, Arethusa bulbosa, Athyrium filix-femina, Aulacomnium palustre, Bartonia virginica, Betula alleghaniensis, B. papyrifera, B. pumila, B. ×sandbergii, Bidens amplissima, Botrychium virginianum, Bryoxiphium norvegicum, Calamagrostis canadensis, Calla palustris, Calliergon stramineum, Calopogon pulchellus, Caltha palustris, Campanula aparinoides, Cardamine diphylla, Carex brunnescens, C. buxbaumii, C. canescens, C. chordorrhiza, C. diandra, C. disperma, C. echinata, C. interior, C. intumescens, C. lacustris C. lasiocarpa, C. leptalea, C. limosa, C. magellanica, C. oligosperma, C. pauciflora, C. pedunculata, C. rostrata, C. stricta, C. tenuiflora, C. trisperma, C. vaginata, Chamaecyparis nootkatensis, Chamaedaphne calyculata, Cinna latifolia, Circaea alpina, Cladina rangiferina, Cladonia alpestris, Clintonia borealis, Comarum palustre, Conocephalum conicum, Coptis trifolia, Cornus canadensis, C. racemosa, C. rugosa, C. sericea, Corylus cornuta, Cypripedium acaule, C. reginae, Danthonia spicata, Dicranella heteromalla, Dicranum polysetum, Diervilla lonicera, Drepanocladus aduncus, Drosera intermedia, D. rotundifolia, Dryopteris carthusiana, D. cristata, Dulichium arundinaceum, Empetrum nigrum, Epilobium leptocarpum, E. palustre, Equisetum fluviatile, E. pratense, Eriophorum angustifolium, E. tenellum, E. vaginatum, E. virginicum, E. viridicarinatum, Eurybia macrophylla, Eutrochium maculatum, Fragaria virginiana, Galium labradoricum, G. obtusum, G. trifidum, G. triflorum, Gaultheria hispidula, G. procumbens, Glyceria canadensis, Goodyera repens, Halenia deflexa, Heuchera richardsonii, Huperzia lucidula, H. porophila, Hylocomium splendens, Hypnum lindbergii, Ilex mucronata, Impatiens capensis, Iris versicolor, Kalmia angustifolia, K. polifolia, Larix laricina, Leucobryum glaucum, Limprichtia revolvens, Linnaea borealis, Lonicera canadensis, L. oblongifolia, L. villosa, Lycopodiella inundata, Lycopodium clavatum, Lycopus americanus, L. uniflorus, Lysimachia thyrsiflora, Maianthemum canadense, M. trifolium, Mentha arvensis, Menyanthes trifoliata, Mertensia paniculata, Micranthes pensylvanica, Mitella nuda, Monotropa uniflora, Muhlenbergia uniflora, Myrica gale, Orthilia secunda, Osmunda claytoniana, Osmundastrum cinnamomeum, Oxalis montana, Persicaria amphibia, P. sagittata, Phegopteris connectilis, Photinia melanocarpa, Picea glauca, P. mariana, P. sitchensis, Pinus banksiana, P. contorta, P. strobus, Plagiomnium cuspidatum, Platanthera hyperborea, P. obtusata, P. orbiculata, Pleurozium schreberi, Poa, Pogonia ophioglossoides, Polypodium virginianum, Polytrichum commune, P. juniperinum, P. strictum, Prunus pensylvanica, Ptilidium ciliare, Ptilium crista-castrensis, Pyrola chlorantha, Quercus* *rubra, Rhamnus alnifolia, Rhizomnium magnifolium, R. pseudopunctatum, Rhododendron canadense, R. viscosum, Rhynchospora alba, Rhytidiadelphus, Ribes hirtellum, R. hudsonianum, R. triste, Riccardia multifida, Rubus arcticus, R. chamaemorus, R. idaeus, R. pubescens, Rumex orbiculatus, Salix discolor, S. eriocephala, S. pedicellaris, S. planifolia, S. pyrifolia, Sarracenia purpurea, Scheuchzeria palustris, Scutellaria galericulata, S. lateriflora, Shepherdia canadensis, Solidago hispida, S. uliginosa, Sorbus decora, Sphagnum capillifolium, S. fuscum, S. girgensohnii, S. magellanicum, S. recurvum, S. russowii, S. teres, S. warnstorfii, S. wulfianum, Spiraea tomentosa, Stellaria, Streptopus lanceolatus, Sullivantia sullivantii, Symphyotrichum puniceum, Symplocarpus foetidus, Tetraphis pellucida, Thelypteris palustris, Thuja occidentalis, T. plicata, Trientalis borealis, Tsuga heterophylla, T. mertensiana, Ulmus americana, Usnea, Vaccinium angustifolium, V. corymbosum, V. macrocarpon, V. myrtilloides, V. oxycoccos, V. uliginosum, V. vitis-idaea, Viola lanceolata, V. renifolia, V. sagittata, Xyris torta.*

***Rhododendron viscosum* (L.) Torr.** occurs in bogs, depressions, ditches, floodplains, seeps, swamps, and along pond, river, and stream margins at elevations of up to 1500 m. Substrates are acidic (pH: 3.1–6.0) and range from sphagnum peat to loam or sand. The plants tolerate partial shade and withstand low temperatures to −32°C. The large, fragrant flowers remain open for 3–7 days, with their stamens and style extending beyond the petals. They are self-compatible, but do not self-pollinate spontaneously. In hand-pollinated flowers, higher seed set occurs from cross-pollination (34%) than from self-pollination (14%). Natural pollination is achieved by honey and bumble bees (Hymenoptera: Apidae: *Apis mellifera, Bombus*) and possibly by hawkmoths (Lepidoptera: Sphingidae) and hummingbirds (Aves: Trochilidae), which feed on the nectar. Plants may be slightly pollinator limited, but field results were inconclusive. The corolla tube is covered by sticky glands that may function to deter nectar thieves. The scale-like seeds are dispersed by the wind. Seeds collected during the fall will germinate within 2–4 weeks at 18°C–21°C. Moss reportedly provides the best germination substrate. Other details regarding germination are not known. The plants can become strongly rhizomatous or stoloniferous, enabling them to develop into thickets. **Reported associates:** *Acer rubrum, Agarista populifolia, Alnus serrulata, Amelanchier canadensis, Andropogon virginicus, Ardisia escallonioides, Aronia arbutifolia, A. melanocarpa, Asimina parviflora, Athyrium filix-femina, Bidens bipinnata, Calamagrostis coarctata, Callicarpa americana, Callisia repens, Carex grayi, C. leptalea, Carya glabra, Celtis laevigata, Chamaecyparis thyoides, Chasmanthium laxum, Chelone cuthbertii, Chimaphila maculata, Chiococca alba, Citrus aurantium, Clethra alnifolia, Collinsonia tuberosa, Cornus florida, Cyperus tetragonus, Cyrilla racemiflora, Decumaria barbara, Dichanthelium boscii, Diodia virginianum, Dioscorea quaterna, Diospyros virgininiana, Epidendrum conopseum, Eupatorium fistulosum, Gaylussacia tomentosa, Gordonia lasianthus, Habeneria odontopetala, Hamelia patens, Houstonia purpurea, Ilex cassine, I. glabra, I. opaca, I. verticillata, Illicium parviflorum, Iresine diffusa, Juncus gymnocarpus, Kalmia hirsuta, Leucothoe racemosa, Ligusticum canadense, Liquidambar styraciflua, Liriodendron tulipifera, Listera cordata, Lobelia puberula, Lyonia ferruginea, L. ligustrina, L. lucida, L. mariana, Magnolia grandiflora, M. virginiana, Medeola virginiana, Mitchella repens, Morus rubra, Myrica cerifera, M. gale, Myrsine floridana, Nyssa biflora, N. sylvatica, Osmunda regalis, Osmundastrum cinnamomeum, Ostrya virginiana, Oxydendrum arboreum, Panicum joori, Parnassia asarifolia, Parthenocissus quinquefolia, Persea palustris, Pharus lappulaceus, Photinia pyrifolia, Pinus elliottii, P. rigida, P. serotina, Platanthera clavellata, P. integrilabia, Polypodium polypodioides, Populus heterophylla, Prunus serotina, Psychotria nervosa, Quercus alba, Q. laurifolia, Q. nigra, Q. phellos, Q. virginiana, Rhamnus caroliniana, Rhapidophyllum hystrix, Rhododendron maximum, Rosa palustris, Rubus cuneifolias, R. hispidus, Sabal palmetto, Sassafras albidum, Scleria triglomerata, Serenoa repens, Smilax glauca, S. rotundifolia, Solidago arguta, S. caesia, S. leavenworthii, Sphagnum palustre, S. recurvum, Symphyotrichum lateriflorum, Syngonanthus flavidulus, Taxodium ascendens, Thelypteris noveboracensis, T. palustris, Thuidium delicatulum, Tillandsia, Toxicodendron radicans, T. vernix, Triadenum virginicum, Trientalis borealis, Uvularia sessilifolia, Vaccinium corymbosum, Viburnum cassinoides, V. nudum, V. recognitum, Viola ×primulifolia, Vitis rotundifolia, Woodwardia areolata, W. virginica, Zanthoxylum clava-herculis.*

Use by wildlife: The flowers of *Rhododendron groenlandicum* are visited by a variety of bees (Hymenoptera: Andrenidae: *Andrena alleghaniensis, A. carlini, A. carolina, A. hippotes, A. mandibularis, A. regularis, A. vicina*; Apidae: *Apis mellifera, Bombus affinis, B. borealis, B. fervidus, B. griseocollis, B. impatiens, B. rufocinctus, B. sandersoni, B. ternarius, B. terricola, B. vagans*; Colletidae: *Colletes inaequalis*; Megachilidae: *Osmia atriventris*) and flies (Diptera: Calliphoridae: *Lucilia*; Conopidae: *Zonidon*; Muscidae: *Orthellia cornicina*; Stratiomyidae: *Odontomyla*; Syrphidae: *Cheilosia leucoparea, Eristalis, E. Dimidiata, Helophilus fasciatus, H. laetus, H. latifrons, Pyrophaena rosarum, Sericomyia ternsversa, Sphaerophoria, Volucella*), which seek nectar and/or pollen and function as pollinators. Occasional floral visitors include beetles (Coleoptera: Elateridae: *Agriotes stabilis*), and butterflies and moths (Insecta: Lepidoptera: Nymphalidae: *Vanessa atalanta*; Sphingidae: *Hemaris thysbe*). The plants are hosts for several larval Lepidoptera (Coleophoridae: *Coleophora ledi*; Lycaenidae: *Incisalia augustinus*; Noctuidae: *Lithomoia solidaginis, Melanchra assimilis, Syngrapha microgamma, S. montana*; Tortricidae: *Archips argyrospila, Epinotia septemberana*). They are also host to several Fungi including various Ascomycota (Amphisphaeriaceae: *Seimatosporium arbuti, S. ledi*; Elsinoaceae: *Elsinoë ledi*; Gnomoniaceae: *Diplodina fructigena*; Incertae sedis: *Ascochyta ledi*; Rhytismataceae: *Lophodermium sphaerioides*; Venturiaceae: *Antennularia variisetosa*) and Basidiomycota (Coleosporiaceae: *Chrysomyxa nagodhii*,

C. reticulata, C. woroninii; Exobasidiaceae: *Exobasidium japonicum*). *Chrysomyxa reticulata* is capable of infecting cultivated rhododendrons. Overall, *R. groenlandicum* foliage is not very palatable (it contains potentially toxic alkaloids, terpenoids, etc.) and seldom is grazed by wildlife. One component of the foliage is the sesquiterpene germacrone, which is a feeding deterrent to snowshoe hares (but see next). However, its protein content can exceed 9% and it is eaten by ptarmigans (Aves: Phasianidae: *Lagopus*), snowshoe hares (Mammalia: Leporidae: *Lepus americanus*), and is grazed incidentally by caribou (Mammalia: Cervidae: *Rangifer tarandus*). It sometimes is used as a winter food by deer (Mammalia: Cervidae: *Odocoileus virginiana*) and mountain goats (Mammalia: Bovidae: *Oreamnos americanus*) but is not eaten by moose (Mammalia: Cervidae: *Alces alces*). The plants provide cover, feeding, and/or nesting habitat for various frogs (Amphibia: Ranidae), heather voles (Mammalia: Muridae: *Phenacomys ungava*), grouse (Aves: Phasianidae: *Bonasa umbellus, Falcipennis canadensis, Tympanuchus phasianellus*), and waterfowl (e.g., Aves: Anatidae: *Anas platyrhynchos*). *Rhododendron columbianum* is a food plant of the mountain beaver (Mammalia: Aplodontiidae: *Aplodontia rufa*). It is a larval host for a leaf-mining caterpillar (Lepidoptera: Gracillariidae: *Phyllonorycter ledella*) and also the host of several fungi including a number of Ascomycota (Amphisphaeriaceae: *Seimatosporium arbuti*; Elsinoaceae: *Elsinoë ledi*; Erysiphaceae: *Microsphaera penicillata, Mycosphaerella punctiformis*; Gnomoniaceae: *Diplodina fructigena*; Niessliaceae: *Niesslia exilis*; Rhytismataceae: *Coccomyces coronatus, Rhytisma*) and rusts (Basidiomycota: Coleosporiaceae: *Chrysomyxa neoglandulosi*; Exobasidiaceae: *Exobasidium vaccinii*). *Chrysomyxa neoglandulosi* can spread to cultivated rhododendrons. The plants are an important component of habitat for the greater white-fronted goose (Aves: Anatidae: *Anser albifrons*) and the lotis blue butterfly (Lepidoptera: Lycaenidae: *Plebejus argyrognomon*). *Rhododendron viscosum* is a larval host plant for several Lepidoptera (Lymantriidae: *Lymantria dispar*; Noctuidae: *Acronicta tritona*; Sphingidae: *Darapsa choerilus, D. versicolor*). It is also a host of the azalea plant bug (Insecta: Heteroptera: Miridae: *Rhinocapsus vanduzeei*) and several leaf and flower gall fungi (Basidiomycota: Exobasidiaceae: *Exobasidium decolorans, E. discoideum E. vaccinii*). The plants are fed on only lightly by Japanese beetles (Insecta: Coleoptera: Scarabaeidae: *Popillia japonica*). They provide important breeding habitat for yellow warblers (Aves: Parulidae: *Setophaga petechia*) and their flowers attract nectar-seeking butterflies (Lepidoptera) and ruby-throated hummingbirds (Aves: Trochilidae: *Archilochus colubris*). The plants may provide some browse for deer (Mammalia: Cervidae: *Odocoileus*) during the winter. **Economic importance: food:** The common name of "Labrador tea" for *Rhododendron groenlandicum* indicates its widespread use as a beverage by numerous native North American tribes including the Alaska, Algonquin, Anticosti, Arctic, Bella Coola, Chippewa, Coast, Cree, Eskimo, Gitksan, Haisla, Hanaksiala, Hesquiat, Kitasoo, Kwakiutl, Makah, Malecite, Micmac, Nitinaht, Ojibwa, Okanagan-Colville, Oweekeno, Potawatomi, Quebec, Saanich, Salish, Shuswap, Southern, Thompson, and Woodlands people. *Rhododendron columbianum* was used similarly by the Tolowa and Yurok tribes for beverages. The Nez Perce use it to make a drink called *písqu* or mountain tea. Although both species possess a small amount of andromedotoxin and other potentially poisonous substances, the toxic effects are believed to be lessened if leaves are brewed without boiling. In any case, the plants should be avoided as food altogether or used only in sparing amounts; **medicinal:** *Rhododendrom groenlandicum* has been used extensively as a medicinal plant because of its variety of secondary compounds including andromedotoxin, monoterpenes (limonene, sabinene), sesquiterpenes, (α-selinene, and β-selinene, germacrone), phytosterols (campesterol, β-sitosterol [which lowers plasma low-density lipoprotein concentrations], taraxasterol), and pentacyclic triterpenoids (α-amyrin, β-amyrin). Taraxasterol and β-sitosterol reportedly exhibit anti-tumor properties. The plants were used by many native North American tribes as a general medicinal tonic (Anticosti, Makah, Micmac, Montagnais, Nitinaht, Nootka, Potawatomi, Salish), as a diuretic (Cree, Gitksan, Micmac, Woodlands), and emetic (Cree), to treat asthma (Micmac), colds (Abnaki, Algonquin, Haisla, Hanaksiala, Kitasoo, Micmac, Oweekeno), jaundice (Montagnais), pneumonia and whooping cough (Cree, Woodlands), tuberculosis (Haisla, Hanaksiala), and to relieve fever (Montagnais), headaches (Algonquin), kidney ailments (Makah, Malecite, Micmac, Okanagan-Colville), nasal inflammation (Abnaki), rheumatism (Cree, Hudson Bay, Quinault), scurvy (Micmac), sore throats (Oweekeno), and stomach aches (Bella Coola). The leaves were regarded as having narcotic properties by the Kwakiutl. The plants were used as an appetite stimulant by the Haisla, Hanaksiala, and Nitinaht people. *Rhododendron groenlandicum* had many dermatological uses including the treatment of burns and scalds (Chippewa, Cree, Woodlands), chafed skin, rashes, sores and wounds (Chippewa, Cree, Hudson Bay, Woodlands), and poison ivy rash (Shuswap); **cultivation:** *Rhododendron groenlandicum* and the cultivar 'Compactum' are distributed as ornamental garden shrubs along with the hybrid *R. ×columbianum*. *Rhododendron viscosum* (also as var. *montanum,* forma *rhodanthum*) is cultivated extensively as an ornamental and received the award of garden merit from the Royal Horticultural Society. Cultivars include 'Antilope,' 'Arpege,' 'Carat,' 'Cassley,' 'Creek Side,' 'Daviesii' (*R. molle* × *R. viscosum;* award of garden merit), 'Diorama,' 'Grey Leaf,' 'Helena Pratt' ('Flaming June' × *R. viscosum*), 'Lemon Drop,' 'Moidart' (*R. viscosum* × 'Rosella' or 'Sylphide'), 'Pink Mist,' 'Pink Rocket,' 'Rosata' (award of garden merit), 'Roseum,' 'Soir de Paris,' 'Solway' ([*R. viscosum* × 'Sylphides'] × 'Sugared Almond'), 'Torridon' (*R. viscosum* × 'Rosella' or 'Sylphide'), and *R. viscosum* × *R. arborescens*. It was the first North American *Rhododendron* to be grown in England (in 1680); **misc. products:** The leaves of *R. groenlandicum* were used by the Iroquois and Potawatomi tribes to make a brown dye. They are placed in corn cribs to deter feeding by mice or rats and are also used as a moth repellant. Because the plants concentrate gold, they have been used by mining companies as an indicator of gold deposits. They have also been used to revegetate sites disturbed by mining and other activities. A mosquito repellent is made by mixing the leaves with alcohol and glycerin. Plantings of *R. viscosum* are recommended for controlling streambank erosion; **weeds:** none; **nonindigenous species:** *Rhododendron groenlandicum* is naturalized locally in Germany and Great Britain.

Systematics: Molecular studies using plastid (*matK, trnK*) and nuclear [nuclear ribosomal internal transcribed spacer (nrITS); RPB2–1] DNA sequences confirm the placement of *Ledum* within the genus *Rhododendron* as had been determined previously by morphological cladistic analysis. Although *Ledum* still is recognized as a distinct genus in many treatments, the evidence to include it within *Rhododendron* (as has been done here) is overwhelming. Species of the former *Ledum* are recognized more recently as a subsection of *Rhododendron*; however, no single phylogenetic survey has included all former *Ledum* species simultaneously to determine whether they associate as a clade. Preliminary phylogenetic studies place some former *Ledum* species with members of section *Rhododendron*, but in a position that is fairly distant from members of section *Pentanthera*, which includes *R. viscosum*. Although this result would indicate at least two separate origins of an OBL habit in the genus, none of the North American OBL species has yet been included in these phylogenetic surveys for confirmation. Further studies of the large genus *Rhododendron*, which include additional taxa, will be necessary before more precise relationships can be determined for the OBL rhododendrons. *Rhododendron groenlandicum* and *R. columbianum* presumably are closely related as indicated by their ability to hybridize. *Rhododendron ×columbianum* (also OBL) is the name given to the natural hybrid between these species. Section *Pentanthera*, which includes *R. viscosum*, is known to have a propensity for hybridization. The many hybrids associated with cultivars of *R. viscosum* would attest to this generalization. The base chromosome number of *Rhododendron* is x = 13. *Rhododendron groenlandicum* and *R. viscosum* are diploid (2n = 26).

Comments: *Rhododendron groenlandicum* ranges throughout boreal North America, *R. columbianum* is distributed in western North America, and *R. viscosum* occurs along the eastern coastal plain of the United States.

References: Belleau & Collin, 1993; Calmes & Zasada, 1982; Christy, 2004; Crane, 2001; Gao et al., 2002; Goetsch et al., 2005; Gucker, 2006; Halsted, 1893; Harmaja, 2002; Karlin, 1978; Karlin & Bliss, 1983; Kawamura, 2004; Kraus et al., 2004; Kron, 1997; Kron & Judd, 1990; Kron & King, 1996; Kurashige et al., 2001; Lahring, 2003; Li, 1957; Lovell & Lovell, 1936; Mitra, 1999; Morrison, 1929; Ovesna et al., 2004; Pojar, 1974; Rathcke, 1988; Reader, 1975, 1977; Reichardt et al., 1990; Sax, 1930; Schultz et al., 2001; Small, 1976; Small & Catling, 2000; Stewart et al., 2002; Stuckey & Gould, 2000; Wurzburger & Bledsoe, 2001.

8. Vaccinium

Blueberry, cranberry; atocas, bleuet, canneberges, myrtille
Etymology: probably derived from *baccinium*, a form of the Latin *baca* meaning "berry."

Synonyms: *Acosta; Batodendron; Cavinium; Cyanococcus; Epigynium; Herpothamnus; Hornemannia; Hugeria; Metagonia; Neojunghuhnia; Oxycoca; Oxycoccus; Picrococcus; Polycodium; Rhodococcum; Rigiolepis; Schollera; Symphysia; Vitis-idaea*

Distribution: global: Eurasia; New World; **North America:** widespread

Diversity: global: 500 species; **North America:** 25 species

Indicators (USA): OBL: *Vaccinium caesariense, V. formosum, V. macrocarpon, V. oxycoccos, V. ×marianum*; **FAC:** *V. formosum, V. ×marianum*

Habitat: freshwater; palustrine; **pH:** 2.5–6.8; **depth:** <1 m; **life-form(s):** emergent shrub

Key morphology: stems (to 4 m) erect and clumped or wiry, trailing, decumbent (to 1.5 m), and rooting at the nodes; leaves alternate, deciduous (to 8 cm) or evergreen (to 15 mm), subsessile, glaucous or whitish below; racemes axillary and condensed or terminal and elongate; flowers nodding, 4–5-merous, epigynous; corollas (to 12 mm) urceolate, 5-lobed and white or strongly reflexed (petals to 10 mm), 4-parted and rose-pink; stamens 8–10; berries blue to black (to 12 mm) or red (to 2 cm), many seeded

Life history: duration: perennial (buds, rhizomes); **asexual reproduction:** rhizomes, suckers; **pollination:** insect, self; **sexual condition:** hermaphroditic; **fruit:** berries (common); **local dispersal:** fruit (birds, mammals, water); **long-distance dispersal:** fruit (birds)

Imperilment: (1) *Vaccinium caesariense* [G4]; (2) *V. formosum* [G5]; S3 (NJ); (3) *V. macrocarpon* [G4]; S1 (IL); S2 (NC, TN, VA, WV); S3 (DE, IN, MD); (4) *V. oxycoccos* [G5]; S1 (IL); S2 (ID, IN, MD, NJ, OH, RI, WV)

Ecology: general: Generally, the genus *Vaccinium* comprises acidiphilous, shade tolerant, manganese accumulating species. The roots are colonized by endophytic "ericoid" mycorrhizae (Ascomycota: Helotiaceae: *Pezizella ericae*), which enhance survival in nutrient-poor habitats by facilitating the uptake of nitrogen and phosphorous. The plants mainly are outcrossed by bees (Hymenoptera) but occasionally are self-compatible, which can allow for low levels of self-fertilization. However, high levels of inbreeding depression often have been observed in the latter instances. Population genetic studies of most *Vaccinium* species generally have indicated panmictic pollination resulting in Hardy–Weinberg equilibrium, even in species capable of extensive clonal growth. Ripe seeds will germinate well if stratified at 4°C for 1–10 weeks and those of some species can retain their viability for more than 15 years. The fruits are dispersed by a variety of animals that feed upon them, notably by birds (Aves). It is difficult to attribute ecological information accurately to the first two species (*V. caesariense, V. formosum*) because they have been merged taxonomically with *V. corymbosum* (FACW) by many taxonomists (see *Systematics*). The allotetraploid *V. ×marianum* (derived from *V. caesariense* and *V. fuscatum*) has seldom been recognized as a distinct species and has not been treated here in any detail.

***Vaccinium caesariense* Mack.** inhabits bogs, swamps, and thickets at lower elevations along the eastern coastal plain of the United States. Substrates typically are acidic peat (pH: 2.7–6.6). The plants are shade intolerant. Details on the reproductive ecology are not reported specifically; however, they are likely similar to *V. corymbosum*, which ranges from self-compatible to incompatible but is insect pollinated and outcrossed primarily by bees (Hymenoptera). The plants are able to resprout from suckers and can proliferate into clones that are several meters in diameter. Seed germination and dispersal are as described earlier (see *Ecology*: general). **Reported associates:** *Chamaecyparis thyoides, Chamaedaphne calyculata, Gaylussacia baccata, Kalmia angustifolia, Nemopanthus mucronata, Osmundastrum cinnamomeum, Photinia melanocarpa, Rhododendron viscosum, Vaccinium attrococcum* (additional associates probably are similar to those of the following species).

Vaccinium formosum **Andr.** grows in bogs, depressions, interdunal swales, marshes, seeps, and swamps at low elevations along the southeastern coastal plain. Substrates are acidic (pH: 2.7–6.6) peat or sandy loam. The plants tolerate only partial shade and can endure cold temperatures down to −26°C. The flowers are pollinated by bumblebees (Hymenoptera: Apidae: *Bombus*) and honey bees (Hymenoptera: Andrenidae: *Andrena*). The seeds can be planted immediately after ripening; however, higher germination rates can be achieved after 3 months of cold stratification (4°C). The seeds are dispersed by various birds (Aves) and mammals (Mammalia). The plants are clonal by means of resprouting suckers. **Reported associates:** *Acer rubrum, Agalinis setacea, Agarista populifolia, Alnus serrulata, Andropogon capillipes, Arisaema triphyllum, Arundinaria gigantea, Boehmeria cylindrica, Carex crinita, C. glaucescens, C. joorii, C. striata, Carpinus caroliniana, Cephalanthus occidentalis, Chamaecyparis thyoides, Clethra alnifolia, Cyrilla racemiflora, Dichanthelium dichotomum, Eleocharis, Fagus grandifolia, Gaylussacia frondosa, G. mosieri, G. tomentosa, Gordonia lasianthus, Hypericum, Ilex amelanchier, I. coriacea, I. glabra, I. opaca, I. verticillata, Illicium floridanum, Itea virginica, Juncus biflorus, J. dichotomus, J. scirpoides, Kalmia latifolia, Leucothoe racemosa, Liquidambar styraciflua, Liriodendron tulipifera, Lyonia ligustrina, L. lucida, L. mariana, Magnolia virginiana, Mitchella repens, Myrica cerifera, M. heterophylla, Nyssa biflora, Osmunda regalis, Osmundastrum cinnamomeum, Oxydendrum arboreum, Panicum amarum, P. chamaelonche, P. hemitomon, Pinus echinata, P. palustris, P. serotina, P. taeda, P. virginiana, Quercus alba, Q. lyrata, Q. michauxii, Q. pagoda, Q. prinus, Q. virginiana, Rhexia mariana, Rhododendron viscosum, Rhynchospora, Rosa palustris, Rubus cuneifolius, Sabal, Salix caroliniana, S. nigra, Scirpus cyperinus, Scleria pauciflora, Smilax laurifolia, S. walteri, Sphagnum cuspidatum, Spiraea alba, Symplocos tinctoria, Taxodium ascendens, Thelypteris palustris, Tillandsia usneoides, Toxicodendron radicans, Utricularia gibba, Vaccinium corymbosum, V. elliottii, V. fuscatum, V. myrsinites, Viburnum nudum, V. rufidulum, Woodwardia areolata, W. virginica, Zenobia pulverulenta.*

Vaccinium macrocarpon **Aiton** occurs in bogs, ditches, lakeshores, marshes, meadows, mires, sand pits, shores, swales, swamps, and along the margins of bog ponds at elevations of up to 1400 m. The plants grow in open sites with acidic (pH: 4.0–6.8) substrates, typically on organic peat (often on quaking sedge mats) but also can occur on sand. The flowers are produced biennially, with the buds formed during the previous year. They are protandrous and sensitive to frost. From 2 to 7 flowers are produced, with higher fruit set in the lower flowers than in the upper ones. The plants are self-compatible but are predominantly outcrossed. Tested cultivars have exhibited highly reduced seed set in self-pollinated flowers; however other studies have shown fairly high tolerance to inbreeding. Effectively, vibrations caused by insects ("buzz pollination") are necessary to release pollen from the poricidal anthers. The flowers are pollinated mainly by bumble bees (Hymenoptera: Apideae: *Bombus fervidus, B. impatiens, B. ternaries, B. vegans*), leaf-cutting bees (Hymenoptera: Megachilidae: *Megachile addenda*), and sweat bees (Hymenoptera: Halictidae: *Dialictus*), but many other bees also visit the plants. Daily foraging by a single female leaf-cutting bee is estimated to yield up to 1440 berries. Plants do not appear to be pollinator limited in natural populations. A minute amount of seed set has been observed on plants where pollinators were excluded, which has been attributed to wind pollination or to geitonogamy (staminate and pistillate organs do not mature simultaneously within a single flower, but may be receptive among flowers of a single plant). Fresh seeds will germinate but at a very low rate. Maximum germination is achieved after 3 months of cold (0–4°C) stratification. Dispersal is biotic, being achieved by endozooic transport of the berries by various frugivores. The plants are highly clonal by proliferation of the rhizomes (which root at the nodes and can reach lengths of 2 m), and populations are characterized by extremely low levels of genetic variation. Some clones are estimated to be more than 350 years old. **Reported associates:** *Acer rubrum, Agalinis paupercula, Alnus rugosa, Andromeda polifolia, Andropogon, Arethusa bulbosa, Aronia melanocarpa, Aulacomnium palustre, Bartonia virginica, Betula papyrifera, B. populifolia, B. pumila, Calamagrostis canadensis, Calopogon tuberosus, Carex aquatilis, C. atlantica, C. bullata, C. buxbaumii, C. canescens, C. chordorrhiza, C. cryptolepis, C. echinata, C. exilis, C. flava, C. folliculata, C. lacustris, C. lasiocarpa, C. leptalea, C. limosa, C. livida, C. lurida, C. magellanica, C. michauxiana, C. oligosperma, C. prairea, C. rostrata, C. striata, C. stricta, C. trisperma, C. utriculata, C. vesicaria, Chamaecyparis thyoides, Chamaedaphne calyculata, Cladina rangiferina, Cladium mariscoides, Clethra alnifolia, Comarum palustre, Cuscuta compacta, C. gronovii, Cyperus, Cyrilla racemiflora, Decodon verticillatus, Dichanthelium acuminatum, Drosera filiformis, D. intermedia, D. rotundifolia, Dryopteris cristata, Dulichium arundinaceum, Eleocharis elliptica, Empetrum nigrum, Eriophorum angustifolium, E. virginicum, Euthamia tenuifolia, Fimbristylis, Gaylussacia baccata, G. dumosa, Houstonia serpyllifolia, Hypericum virginicum, Ilex glabra, I. mucronata, Iris versicolor, Juncus balticus, J. biflorus, J.*

canadensis, J. greenei, J. pelocarpus, J. scirpoides, J. stygius, Kalmia angustifolia, K. polifolia, Larix laricina, Linum striatum, Lycopodiella inundata, Lycopus uniflorus, Lysimachia asperulifolia, L. terrestris, Maianthemum trifolium, Menyanthes trifoliata, Muhlenbergia glomerata, M. uniflora, Myrcia paganii, Myrica gale, M. pensylvanica, Narthecium americanum, Nyssa sylvatica, Oclemena nemoralis, Oenothera perennis, Osmunda regalis, Osmundastrum cinnamomeum, Oxypolis rigidior, Packera aurea, Panicum rigidulum, P. virgatum, Parnassia asarifolia, Peltandra virginica, Photinia melanocarpa, Phragmites australis, Picea mariana, Pinus banksiana, P. rigida, P. strobus, P. taeda, Platanthera clavellata, P. dilatata, Pogonia ophioglossoides, Polygala brevifolia, P. cruciata, Polytrichum commune, Prunus virginiana, Quercus palustris, Rhododendron viscosum, Rhynchospora alba, R. capillacea, R. capitellata, R. fusca, Rubus hispidus, R. setosus, Sagittaria, Sarracenia purpurea, Schoenoplectus pungens, Scheuchzeria palustris, Schizachyrium scoparium, Schoenoplectus tabernaemontani, Scirpus atrovirens, S. cyperinus, S. expansus, S. longii, S. polyphyllus, Smilax rotundifolia, Solidago nemoralis, S. patula, S. uliginosa, Sphagnum affine, S. angustifolium, S. bartlettianum, S. centrale, S. compactum, S. cuspidatum, S. fallax, S. fimbriatum, S. flavicomans, S. flexuosum, S. fuscum, S. lescurii, S. magellanicum, S. palustre, S. papillosum, S. pulchrum, S. recurvum, S. rubellum, S. subsecundum, S. subtile, S. torreyanum, S. warnstorfii, Spiraea alba, S. tomentosa, Thelypteris palustris, Toxicodendron radicans, Triadenum fraseri, T. virginicum, Trichophorum alpinum, T. cespitosum, Triglochin maritimum, Typha, Utricularia cornuta, U. intermedia, U. subulata, Vaccinium angustifolium, V. corymbosum, V. myrtilloides, V. oxycoccos, V. vitis-idaea, Viola lanceolata, V. sagittata, Xyris difformis, X. montana, X. torta, Zenobia pulverulenta.

Vaccinium oxycoccos L. grows in bogs, ditches, fens, lakeshores, marshes, muskeg, seeps, sloughs, swales, swamps, thickets, tundra, and along the boggy margins of ponds, rivers, and streams at elevations of up to 1500 m. Substrates are acidic (pH: 2.5–5.1) organic (*Sphagnum*) peat, quaking mats (frequently), or sand. Plants are shade intolerant, and often are among the first plant species to recolonize recently burned bogs. The flowers are protandrous and last from 5 to 18 days. They are self-compatible and capable of self-pollination by wind-induced shaking, which often occurs when the rate of insect visitation is low. In some studies, bagged or self-pollinated flowers exhibited reduced seed set (2%) compared to either cross-pollinated (6%) or open pollinated (10%) flowers. Seed production begins in 6–7-year-old plants, with an average of less than 500 seeds/m^2 produced. The berries that remain over winter are dispersed by water, birds, and mammals. The seeds are poorly represented in the seed bank, probably due to their low longevity in the substrate, with a 50% reduction in germination after 2 years and 0% germination after 3 years. Some fresh seeds can germinate; however 30 days of cold stratification can increase germination rates to 88% compared to 4% for unstratified seeds. Germination of stratified seeds is optimized in light or shade at 14 h of day length and

a 20°C/15°C temperature regime. High and rapid germination has also been observed for dried seeds that have been stored for 6 months. Germination can be facilitated by soaking seeds in 10% Na$_2$CO$_3$ for 24 h prior to sowing. Seedling development is slow. Clonal reproduction by rhizomes (which can reach 23 cm) appears to be more important for persistence and regeneration than sexual reproduction in this species. The root system remains close to the surface and plants compete for nutrients with surrounding *Sphagnum* species. The leaves typically live for 2 years. **Reported associates:** *Abies balsamea, Acer spicatum, Actaea rubra, Alnus rugosa, Andromeda polifolia, Arethusa bulbosa, Aronia melanocarpa, Aulacomnium palustre, Betula alleghaniensis, B. papyrifera, B. populifolia, B. pumila, Calamagrostis canadensis, Calla palustris, Calopogon tuberosus, Campylium stellatum, Cardamine diphylla, Carex aquatilis, C. buxbaumii, C. canescens, C. chordorrhiza, C. comosa, C. cusickii, C. echinata, C. exilis, C. flava, C. interior, C. lacustris, C. lasiocarpa, C. leptalea, C. limosa, C. livida, C. magellanica, C. michauxiana, C. nebrascensis, C. oligosperma, C. pauciflora, C. pedunculata, C. pellita, C. prairea, C. rostrata, C. trisperma, C. utriculata, C. vaginata, Chamaecyparis thyoides, Chamaedaphne calyculata, Cladina arbuscula, C. mitis, C. rangiferina, C. terraenovae, Cladium mariscoides, Cladonia crispata, C. uncialis, Clintonia borealis, Comarum palustre, Cornus canadensis, Cypripedium acaule, Darlingtonia californica, Dasiphora floribunda, Decodon verticillatus, Dicranum polysetum, D. scoparium, Doellingeria umbellata, Drosera intermedia, D. rotundifolia, Dulchium arundinaceum, Eleocharis elliptica, E. rostellata, Empetrum nigrum, Equisetum fluviatile, Eriophorum angustifolium, E. chamissonis, E. tenellum, E. vaginatum, E. virginicum, E. viridicarinatum, Euthamia graminifolia, Gaultheria hispidula, Gaylussacia baccata, G. dumosa, Hypericum kalmianum, Ilex mucronata, Iris versicolor, Juncus canadensis, J. filiformis, J. pelocarpus, Kalmia angustifolia, K. polifolia, Larix laricina, Limprichtia revolvens, Linnaea borealis, Lobelia kalmii, Lonicera involucrata, L. villosa, Lycopus uniflorus, Lyonia ligustrina, Lysichiton americanus, Lysimachia terrestris, Maianthemum trifolium, Menyanthes trifolia, Mertensia paniculata, Muhlenbergia glomerata, Myrica gale, Oclemena nemoralis, Osmunda regalis, Osmundastrum cinnamomeum, Parnassia glauca, Photinia floribunda, P. melanocarpa, Picea glauca, P. mariana, P. sitchensis, Pinus contorta, P. rigida, P. strobus, Pleurozium schreberi, Pogonia ophioglossoides, Polytrichum commune, P. strictum, Rhamnus alnifolia, Rhododendron canadense, R. columbianum, R. groenlandicum, R. viscosum, Rhynchospora alba, R. fusca, Rubus arcticus, R. chamaemorus, R. parviflorus, R. pubescens, Sagittaria latifolia, Salix discolor, S. pedicellaris, S. petiolaris, S. pyrifolia, Sanguisorba menziesii, S. officinalis, Sarracenia purpurea, Scheuchzeria palustris, Schoenoplectus acutus, S. tabernaemontani, Scorpidium scorpioides, Solidago nemoralis, S. uliginosa, Sphagnum affine, S. angustifolium, S. bartlettianum, S. capillifolium, S. cuspidatum, S. fallax, S. flavicomans, S. fuscum, S. girgensohnii, S. magellanicum, S. pacificum, S. palustre, S. papillosum, S. pulchrum, S. recurvum,*

S. rubellum, S. torreyanum, Spiraea alba, S. douglasii, S. tomentosa, Symphyotrichum boreale, Thuja occidentalis, Thelypteris palustris, Tomentypnum nitens, Toxicodendron vernix, Triadenum fraseri, T. virginicum, Triantha glutinosa, Trichophorum alpinum, T. cespitosum, Trientalis borealis, T. europaea, Tsuga canadensis, T. heterophylla, Typha latifolia, Utricularia cornuta, U. gibba, Vaccinium angustifolium, V. corymbosum, V. macrocarpon, V. myrtilloides, V. uliginosum, Viola, Woodwardia virginica, Xyris.

***Vaccinium ×marianum* S. Watson** is an allotetraploid hybrid, which seldom has been recognized as a distinct species. It is included here because of its OBL wetland ranking; however, virtually no meaningful ecological or life history information is available for this taxon in the literature.

Use by wildlife: Blueberry fruits (*Vaccinium caesariense, V. formosum*) are eaten (and dispersed) by many birds (Aves) including catbirds and thrashers (Mimidae), grouse (Tetraonidae), thrushes (Musicapidae), and towhees (Emberizidae). Fruits of *V. formosum* (and probably also *V. caesariense*) are eaten by American robins (Turdidae: *Turdus migratorius*), bobwhite quail (Odontophoridae: *Colinus virginianus*), brown thrashers (Mimidae: *Toxostoma rufum*), eastern bluebirds (Turdidae: *Sialia sialis*), gray catbirds (Mimidae: *Dumetella carolinensis*), northern cardinals (Cardinalidae: *Cardinalis cardinalis*), northern mockingbirds (Mimidae: *Mimus polyglottos*), rufous-sided towhees (Emberizidae: *Pipilo erythrophthalmus*), scarlet tanagers (Thraupidae: *Piranga olivacea*), scrub jays (Corvidae: *Aphelocoma coerulescens*), and wild turkeys (Phasianidae: *Meleagris gallopavo*). The berries are also eaten by a number of mammals (Mammalia) including black bears (Ursidae: *Ursus americanus*), chipmunks (Sciuridae: *Tamias striatus*), eastern cottontail rabbits (Leporidae: *Sylvilagus floridanus*), fox squirrels (Sciuridae: *Sciurus niger*), red foxes (Canidae: *Vulpes vulpes*), striped skunks (Mephitidae: *Mephitis mephitis*), white-footed mice (Muridae: *Peromyscus leucopus*), and white-tailed deer (Cervidae: *Odocoileus virginianus*). *Vaccinium formosum* is susceptible to the blueberry red ringspot virus (Caulimoviridae: *Caulimovirus*), an actinobacterium (Nocardiaceae: *Nocardia vaccinii*), and several fungi including gray-mold rot (Ascomycota: Leotiomycetidae: *Botrytis cinerea*), powdery mildew (Ascomycota: Erysiphomycetidae: *Microsphaera vaccinii*), phytophthora blight (Oomycota: Pythiaceae: *Phytophthora*), root and stem rot (Basidiomycota: Ceratobasidiaceae: *Rhizoctonia*), and rusts (Basidiomycota). It is also a host plant for various insects (Insecta) including the cranberry fruitworm (Lepidoptera: Pyralidae: *Acrobasis vaccinii*), fire ant (Formicidae: *Solenopsis geminata*), plum curculio weevil (Coleoptera: Curculionidae: *Conotrachelus nenuphar*), southern yellowjacket (Hymenoptera: Vespidae: *Vespula squamosa*), and yellownecked caterpillar (Lepidoptera: Notodontidae: *Datana ministra*). Cranberries (*V. macrocarpon, V. oxycoccos*) are eaten (and dispersed) by bears (*Ursus*), foxes (*Vulpes*), and many birds. The fruits are also eaten by crickets and grasshoppers (Insecta: Orthoptera), and mice (*Peromyscus*). Although many references indicate that the berries provide food for wildlife, it is unfortunate that examples of specific frugivores (e.g., bird species) seldom are provided in the literature. *Vaccinium macrocarpon* is a host to a plethora of lepidopteran larvae (Gelechiidae: *Dichomeris vacciniella*; Geometridae: *Anavitrinella pampinaria, Cingilia catenaria, Ematurga amitaria, Eupithecia miserulata, Eutrapela clemataria, Hesperumia sulphuraria, Macaria sulphurea, Phigalia titea*; Heliozelidae: *Coptodisca negligens*; Lycaenidae: *Epidemia epixanthe, Lycaena epixanthe*; Lymantriidae: *Lymantria dispar*; Megalopygidae: *Megalopyge crispata*; Nepticulidae: *Stigmella corylifoliella*; Noctuidae: *Acronicta tritona, Agrotis ipsilon, Anagrapha falcifera, Chaetaglaea sericea, Epiglaea apiata, Euxoa detersa, Hemipachnobia monochromataea, Hyppa xylinoides, Melanchra picta, Phalaenostola metonalis, Pseudaletia unipuncta, Spodoptera eridania, S. frugiperda, Xestia c-nigrum, Xylena nupera*; Pyralidae: *Acrobasis vaccinii* [cranberry fruitworm], *Chrysoteuchia topiaria* [cranberry girdler]; Saturniidae: *Hemileuca lucina, H. maia, Hemileuca nevadensis*; Sphingidae: *Sphinx drupiferarum, S. gordius*; Tortricidae: *Acleris minuta, Grapholita conversana, Orthotaenia undulana, Rhopobota finitimana, R. naevana* [blackheaded fireworm], *R. unipunctana, Sparganothis sulfureana*). The plants are eaten and damaged by black vine weevils (Coleoptera: Curculionidae: *Otiorhynchus sulcatus*) and larvae of the cranberry tipworm fly (Diptera: *Dasineura oxycoccana*). Some nematodes (e.g., Nematoda: Criconematidae: *Hemicycliophora*) are often abundant on cranberry plants but do not harm them. In other cases they can cause diseases such as lesion (Pratylenchidae: *Pratylenchus*), ring (Criconematidae: *Criconemoides*), sheath (Criconematidae: *Hemicycliophora*), spiral (Hoplolaimidae: *Helicotylenchus*), stubby root (Trichodoridae: *Trichodorus*), and stunt (Telotylenchidae: *Tylenchorhynchus*). A wealth of fungi has been reported from *V. macrocarpon* including Ascomycota (Amphisphaeriaceae: *Discosia atrocreas, Pestalotia vaccinii*; Arthoniaceae: *Naevia oxycocci*; Ascodichaenaceae: *Pseudophacidium callunae*; Botryosphaeriaceae: *Diplodia vaccinii, Guignardia vaccinii, Phyllosticta putrefaciens*; Davidiellaceae: *Cladosporium herbarum, C. oxycocci*; Dermateaceae: *Gloeosporium minus*; Erysiphaceae: *Microsphaera penicillata*; Glomerellaceae: *Glomerella cingulata*; Gnomoniaceae: *Gnomonia setacea*; Gymnoascaceae: *Talaromyces trachyspermus*; Helotiaceae: *Sporonema oxycocci*; Hyponectriaceae: *Physalospora vaccinii*; Incertae sedis: *Dematium, Strasseria oxycocci*; Mycosphaerellaceae: *Mycosphaerella nigro-maculans*; Nectriaceae: *Fusarium*; Phacidiaceae: *Allantophomopsis cytisporea, Phacidium vaccinii*; Phaeosphaeriaceae: *Leptosphaeria coniothyrium*; Pleosporaceae: *Alternaria, Epicoccum, Stemphylium*; Rhytismataceae: *Lophodermium hypophyllum, L. oxycocci*; Sclerotiniaceae: *Botrytis cinerea, Sclerotinia oxycocci*; Thelebolaceae: *Discohainesia oenotherae*; Valsaceae: *Diaporthe vaccinii, Valsa delicatula*; Venturiaceae: *Gibbera conferta*; Xylariaceae: *Sordaria destruens*) and Basidiomycota (Exobasidiaceae: *Exobasidium vaccinii*; Pucciniastraceae: *Naohidemyces vaccinii*). The plants also host many phytopathogenic Fungi including

Ascomycota such as berry speckle/fruit rot (Botryosphaeriaceae: *Guignardia vaccinii, Phyllosticta elongata*), berry speckle/leaf spot (Venturiaceae: *Gibbera myrtilli*), bitter rot (Glomerellaceae: *Colletotrichum acutatum, C. gloeosporioides, Glomerella cingulata*), black rot (Incertae sedis: *Strasseria geniculata*; Phacidiaceae: *Allantophomopsis cytisporea, A. lycopodina, Phacidium lunatum*), black spot (Mycosphaerellaceae: *Mycosphaerella nigromaculans, Ramularia nigromaculans*), blotch rot (Hyponectriaceae: *Physalospora vaccinii*), cottonball (Sclerotiniaceae: *Monilinia oxycocci*), early fruit rot (Botryosphaeriaceae: *Phyllosticta vaccinii*), end rot (Botryosphaeriaceae: *Fusicoccum putrefaciens*; Helotiaceae: *Godronia cassandrae*), fruit rot (Amphisphaeriaceae: *Pestalotia vaccinii*; Dermateaceae: *Gloeosporium minus*; Incertae sedis: *Synchronoblastia crypta*; Trichocomaceae: *Penicillium*), leaf spot (Davidiellaceae: *Stenella oxycocci*), leaf spot/leaf drop (Amphisphaeriaceae: *Pestalotia vaccinii*; Botryosphaeriaceae: *Botryosphaeria vaccinii*; Dermateaceae: *Eupropolella oxycocci*; Venturiaceae: *Pyrenobotrys compacta*), ripe/white rot (Incertae sedis: *Coleophoma empetri*), twig blight (Rhytismataceae: *Lophodermium hypophyllum, L. oxycocci*), upright dieback (Diaporthaceae: *Phomopsis vaccinii*; Incertae sedis: *Synchronoblastia crypta* Valsaceae: *Diaporthe vaccinii*), viscid rot (Diaporthaceae: *Phomopsis vaccinii*), and yellow rot (Sclerotiniaceae: *Botrytis*). Basidiomycota pathogenic to *V. macrocarpon* include fairy ring (Strophariaceae: *Psilocybe agrariella*), powdery mildew (Erysiphaceae: *Microsphaera vaccinii*), red leaf spot (Exobasidiaceae: *Exobasidium rostrupii*), red shoot (Exobasidiaceae: *Exobasidium perenne*), rose bloom (Exobasidiaceae: *Exobasidium oxycocci*), and rust (Pucciniastraceae: *Naohidemyces vaccinii*). Other pathogens include cranberry gall (Chytridiomycota: Synchytriaceae: *Synchytrium vaccinii*) and root rot (Oomycota: Pythiaceae: *Phytophthora cinnamomi*). False blossom disease in *V. macrocarpon* is caused by a bacterium (Bacteria: Anaeroplasmataceae: *Phytoplasma*). The plants also host a few parasitic angiosperms (Cuscutaceae: *Cuscuta compacta, C. gronovii*). *Vaccinium oxycoccos* is a larval host plant for a number of Lepidoptera (Lycaenidae: *Epidemia dorcas, E. epixanthe, Vacciniina optilete*; Noctuidae: *Actebia fennica*; Nymphalidae: *Clossiana eunomia*; Pyralidae: *Acrobasis vaccinii*; Tortricidae: *Rhopobota finitimana*). It also hosts several fungi including Ascomycota (Helotiaceae: *Godronia cassandrae, Sporonema oxycocci*; Hyponectriaceae: *Physalospora vaccinii*; Mycosphaerellaceae: *Mycosphaerella nigromaculans*; Phacidiaceae: *Allantophomopsis cytisporea*; Rhytismataceae: *Lophodermium hypophyllum*; Sclerotiniaceae: *Monilinia oxycocci*; Valsaceae: *Diaporthe vaccinii*; Venturiaceae: *Gibbera conferta, G. vaccinii*) and Basidiomycota (Exobasidiaceae: *Exobasidium vaccinii*; Pucciniastraceae: *Naohidemyces vaccinii*).

Economic importance: food: Both blueberries and cranberries are well known for their edible fruits, which are eaten raw or made into beverages, jams, jellies, pies, and sauces. The former are sweet tasting when ripe and the latter quite acidic and sour. Edible fruits are obtained from most segregates of the "highbush" blueberry complex (*V. corymbosum* sensu lato; see *Systematics*). The blue fruits of *V. caesariense* are relatively small (to 7 mm) but of good flavor. The black fruits of *V.* ×*marianum* are larger (to 10 mm) but reputedly of inferior flavor. The blue berries of *V. formosum* are larger (to 12 mm) and of superior flavor. Cranberries are a source of benzoic acid, pectin, polyphenol antioxidants, and vitamin C. Traditionally, they are believed to have been a food item served by the Native Americans during the first Thanksgiving feast. Berries of *V. macrocarpon* were eaten by the Algonquin, Anticosti, Chippewa, Iroquois, Ojibwa, Quebec, and Tete-de-Boule tribes. The Pequods made them into a relish to flavor venison. Fruits of *V. oxycoccos* were consumed as fruit or used in various food preparations by the Alaska, Algonquin, Anticosti, Cree, Eskimo, Haisla, Hanaksiala, Hesquiat, Inupiat, Iroquois, Kitasoo, Klallam, Makah, Menominee, Nitinaht, Ojibwa, Oweekeno, Potawatomi, Quebec, Quinault, Salish, Tanana, Tete-de-Boule, Thompson, and Woodlands tribes. The leaves were used by the Clallam people to brew tea; however, they contain potentially toxic glycosides (arbutin) and should not be ingested. The "cranberry scare of 1959" was prompted by reports that some crops in Oregon and Washington had been contaminated with the herbicide aminotriazole, a carcinogen; **medicinal:** The Montagnais people treated pleurisy using an infusion of *V. macrocarpon* branches as a medicine. The berry juice is a common treatment for urinary infections because it inhibits the adhesion of *Escherichia coli* (Eubacteria: Enterobacteriaceae) and other bacteria to the bladder's epithelial cells. It may also help to prevent infection by *Helicobacter pylori* (Bacteria: Helicobacteraceae), which causes gastrointestinal ulcers and dental plaque. The fruits have been eaten to prevent scurvy and as a diuretic. *Vaccinium oxycoccos* was used as a general medicinal plant by the Mohegans and as a treatment for nausea by the Ojibwa. Tea made from its leaves is astringent and has been used to treat diabetes and to control diarrhea; **cultivation:** Most presently cultivated highbush blueberries (known as *V.* ×*covilleanum*) resemble *V. formosum* and originated from hybridization between selections of *V. formosum* with *V. corymbosum* and *V. angustifolium*. 'Rubel' is a cultivar of *V. formosum*. North American cranberry cultivation began in Massachusetts around 1816, but did not become successful until after 1850. *Vaccinium macrocarpon* is the source of all commercially cultivated cranberry varieties, which include 'Ben Lear,' 'Bennett Jumbo,' 'Centennial,' 'Centerville,' 'CN,' 'Early Black,' 'Early Red,' 'Franklin,' 'Hamilton,' 'Howes,' 'McFarlin,' 'Metallic Bells,' 'Olson's Honkers,' 'Pennant,' 'Pilgrim,' 'Red Star,' 'Searles,' 'Searles Jumbo,' 'Stevens,' 'Thunderlake,' and 'Wilcox.' *Vaccinium oxycoccos* sometimes is cultivated as the species or as the variety *rubrum*; **misc. products:** Cranberry juice is used to make a purplish dye for linen and paper. Berries of *V. oxycoccos* were used by the Hesquiat people as forage for geese. Some Native American tribes applied a cranberry poultice to draw out the venom in wounds inflicted by poisoned arrows; **weeds:** none; **nonindigenous species:** *Vaccinium macrocarpon* was introduced to Britain accidentally in shipping ballast sometime before 1869. It was introduced from Massachusetts to western North America during the 19th century.

Systematics: Molecular data place *Vaccinium* within subfamily Vaccinioideae, tribe Vaccinieae of Ericaceae (Figures 5.7 and 5.8), which is characterized by species with an inferior ovary. Traditionally, *Vaccinium* was subdivided into the subgenera *Oxycoccus* (cranberries), which included red-fruited plants with trailing shoots and reflexed petals, and *Vaccinium*, which contained the remaining species having upright woody stems and campanulate/urceolate corollas. Most contemporary authors have abandoned this subgeneric classification and treat the major groups as sections, e.g., section *Oxycoccus* (cranberries) and section *Cyanococcus* (blueberries). Phylogenetic analyses of combined molecular data provide no support for the two traditionally recognized subgenera, and also indicate that *Vaccinium* is not even monophyletic, with a number of species associating among other genera of tribe Vaccinieae. Most *Vaccinium* species surveyed to date resolve within clades that also include species of *Agapetes, Costera, Gaylussacia, Notopora,* and *Orthaea* (Figure 5.11).

A nomenclatural revision of this entire group is necessary. The taxonomy of blueberries (section *Cyanococcus*), which consists of morphologically similar and interfertile diploids (e.g., *Vaccinium caesariense, V. elliotii*), tetraploids (*V. corymbosum, V. formosum*), and hexaploids (*V. ashei*), also remains unsettled. This group has been regarded as comprising a single polyploid compilospecies, which many authors now recognize under the name *V. corymbosum*. However, the present treatment has maintained *V. caesariense* and *V. formosum* (formerly known as *V. australe*) as distinct species to partition the information reported for these taxa, which differs ecologically from *V. corymbosum* (FACW) by its pattern of fruit maturation and wetland affinity. Because of taxonomic discrepancies in the literature, it is difficult to ascertain correctly the information that is specific to *V. caesariense* or *V. formosum*. At worst, all of the information reported here for *V. caesariense* or *V. formosum* simply may correspond to *V. corymbosum*. The basic chromosome number in *Vaccinium* is $x = 12$. *Vaccinium caesariense* ($2n = 24$) and *V. macrocarpon* ($2n = 24$) are diploid; whereas, *V. formosum* and

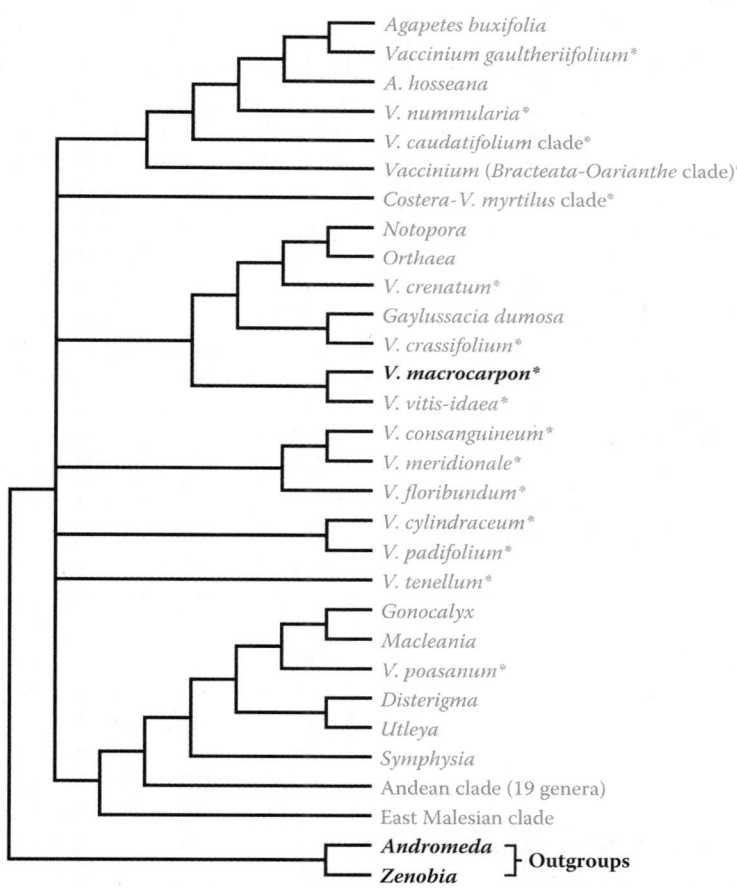

FIGURE 5.11 Phylogenetic relationships within tribe Vaccinieae (subgenus *Vaccinioideae*) of Ericaceae as indicated by analyses of combined molecular data representing 28 of the 30–35 genera. Taxa containing OBL North American species are indicated in bold. These results indicate that species of *Vaccinium* (asterisked) are not monophyletic, but associate with several other genera (*Agapetes, Costera,* etc.). Furthermore, the cranberries (represented by *V. macrocarpon*) are imbedded among other *Vaccinium* species, and support their recognition as a section (section *Oxycoccus*) rather than as a distinct subgenus. The blueberries (section *Cyanococcus*; represented by *V. tenellum*) do not associate closely with the cranberries, indicating that the OBL habit has evolved independently in each group. Additional sampling of *Vaccinium* species will be necessary before more precise relationships can be ascertained in the genus. (Adapted from Kron, K.A., *Amer. J. Bot.,* 89, 327–336, 2002b.)

V. ×*marianum* (2*n* = 48) are tetraploid. *Vaccinium oxycoccus* is a polyploid complex consisting of diploids (2*n* = 24), tetraploids (2*n* = 48), hexaploids (2*n* = 72), and many aneuploid derivatives (2*n* = 36, 42, 44, 46, 50, 52, 58, 64, 68, 70). Allozyme data indicate that the tetraploids are autopolyploids. Few sterility barriers exist among *Vaccinium* species of the same ploidy level and even different ploidy levels often are interfertile. *Vaccinium caesariense* hybridizes freely with *V. fuscatum* giving rise to the allotetraploid *V.* ×*marianum* (OBL, FAC). It also hybridizes with *V. pallidum*. *Vaccinium formosum* is believed to be an autotetraploid (2*n* = 48) derivative of *V. caesariense* and hybridizes with *V. lamarckii*. *Vaccinium macrocarpon* and diploid cytotypes of *V. oxycoccos* are cross-compatible and yield offspring with high seed set. Allozyme data indicate that diploid blueberries are primarily outcrossed and characterized by relatively high levels of genetic variation, although somewhat less than the levels found in plants with comparable breeding systems. Whether this reduction in diversity reflects a recent diversification of the group, or the consequence of limited self-pollination (which can occur in some individuals), has not been determined. Microsatellite markers have been developed that show good cross-reactivity among various blueberry and cranberry species and are available to facilitate further genetic studies of natural *Vaccinium* populations.

Comments: *Vaccinium caesariense* occurs in the eastern United States and *V. formosum* in the southeastern United States. *Vaccinium macrocarpon* occurs in northeastern North America and is disjunct (where introduced) in western North America. *Vaccinium oxycoccos* is circumboreal.

References: Boches et al., 2005; Brown & McNeil, 2006; Bruederle et al., 1991; Burger et al., 2000; Butkus & Pliszka, 1993; Camp, 1945; Cane et al., 1996; Eastwood, 1856; Eck & Childers, 1966; Foo et al., 2000; Hill & Vander Kloet, 2005; Hokanson & Hancock, 2000; Jacquemart, 1997; Krebs & Hancock, 1991; Kron et al., 2002a,b; Lahring, 2003; Loose et al., 2005; Mahy et al., 2000; Robuck, 1985; Sarracino & Vorsa, 1991; Sobota, 1984; Uchytil, 1993; Uttal, 1987; Vander Kloet, 1980, 1992, 2009; Vander Kloet & Austin-Smith, 1986; Vorsa, 1997; White, 1885.

9. *Zenobia*

Honeycup

Etymology: after Zenobia, queen of Palmyra (AD 3rd century)
Synonyms: *Andromeda* (in part)
Distribution: global: North America; **North America:** southeastern
Diversity: global: 1 species; **North America:** 1 species
Indicators (USA): OBL; FACW: *Zenobia pulverulenta*
Habitat: freshwater; palustrine; **pH:** 3.6–6.5; **depth:** <1 m; **life-form(s):** emergent shrub
Key morphology: stems (to 3 m) with a white bloom on the branchlets; leaves (to 8 cm) alternate, leathery, deciduous; flowers stalked (to 3 cm), pendant, 1–8 in showy, umbellate, axillary clusters disposed along leafless terminal shoots; corolla (to 12 mm) white, broadly campanulate, 5-lobed;

stamens 10, the filaments dilated basally, the anthers dehiscent by pores; capsules (to 7 mm broad) 5-lobed, keeled; seeds (to 1 mm) angular
Life history: duration: perennial (buds, rhizomes); **asexual reproduction:** rhizomes; **pollination:** insect; **sexual condition:** hermaphroditic; **fruit:** capsules (common); **local dispersal:** rhizomes, seeds; **long-distance dispersal:** seeds
Imperilment: (1) *Zenobia pulverulenta* [G4]; S1 (GA, VA)
Ecology: general: *Zenobia* is monotypic.

Zenobia pulverulenta **(W. Bartram ex Willd.)** Pollard inhabits bogs, Carolina bays, depressions, pocosins, savannas, and the margins of lakes and swamps at elevations of up to 100 m. The substrates are acidic sands, clay–loam, and peat. Plants commonly occur in ombrotrophic sites that receive full sun to partial shade. The flowers contain a small amount of nectar and emit a scent that has been described as anise-like or citric. They are weakly self-compatible, somewhat protandrous, and partially self-pollinating. The principal pollinators are bumblebees (Hymenoptera: Apidae: *Bombus*). Natural pollen fertility ranges from 87% to 99%. Fruit set where pollinators were excluded experimentally was considerable (9%–15%), but much lower than for open-pollinated plants (28%–62%). Little information exists on the seed ecology. The seeds reportedly germinate well on a peat moss substrate at 21°C. Vegetative reproduction occurs by rhizomes, which can proliferate to form clonal thickets. The plants rely on fire to replenish nutrients and are early colonizers of recently burned sites where they exploit the temporarily elevated nutrient levels. The roots are associated with ericoid mycorrhizae, which facilitate nutrient uptake in poor sites. **Reported associates:** *Acer rubrum, Andropogon glomeratus, A. virginicus, Arundinaria gigantea, Carex striata, Chamaedaphne calyculata, Clethra alnifolia, Cyrilla racemiflora, Gaylussacia dumosa, G. frondosa, Gordonia lasianthus, Ilex amelanchier, I. coriacea, I. glabra, Itea virginica, Kalmia angustifolia, K. carolina, K. cuneata, Liquidambar styraciflua, Lyonia lucida, L. mariana, Lysimachia asperulifolia, Magnolia virginiana, Myrica, Nyssa sylvatica, Osmundastrum cinnamomeum, Oxypolis canbyi, Panicum hemitomon, P. verrucosum, Peltandra virginica, Persea palustris, Photinia pyrifolia, Pinus serotina, Polygala brevifolia, Rhynchospora alba, R. fascicularis, R. wrightiana, Sarracenia flava, S. purpurea, S. rubra, Schizachyrium scoparium, Smilax laurifolia, Sphagnum, Taxodium ascendens, Utricularia subulata, Vaccinium crassifolium, V. formosum, V. fuscatum, V. macrocarpon, V. tenellum, Woodwardia virginica, Xyris.*
Use by wildlife: *Zenobia* plants provide cover to various amphibians, birds (Aves) such as common yellowthroat (Parulidae: *Geothlypis trichas*), mammals (Mammalia), and reptiles (Reptilia). The nectariferous flowers of *Zenobia pulverulenta* attract more than 60 species of bees (Hymneoptera), beetles (Coleoptera), butterflies (Lepidoptera: Hesperiidae: *Euphyes bimacula*), flies (Diptera), and Thysanoptera. The most important insect visitors, which serve as pollinators, are bumblebees and honeybees (Apidae: *Apis mellifera, Bombus bimaculatus, B. griseocollis, B. impatiens*), carpenter bees (Anthophoridae: *Xylocopa micans, X. virginica*),

leafcutter bees (Megachilidae: *Megachile georgica*), mining bees (Andrenidae: *Andrena hilaris*), plasterer bees (Colletidae: *Colletes productus, C. thoracicus*), short-tongued bees (Melittidae: *Mellita mellitoides*), and soldier beetles (Coleoptera: Cantharidae: *Chauliognathus marginatus*). The floral ovaries are eaten by larval flies (Diptera: Cecidomyiidae: *Dasyneura*) and moths (Lepidoptera: Noctuidae: *Acronicta*; Olethreutidae: *Epinotia*). The plants are a potential host of a fungus (Oomycota: *Phytophthora ramorum*), which is the causative agent of sudden oak death.

Economic importance: food: No reported edible uses; **medicinal:** There are no medicinal uses reported for *Zenobia*. However, the leaves contain 0.2%–1.0% polyprenols, which are physiologically active and sometimes used as an alternative to antibiotics; **cultivation:** *Zenobia pulverulenta* is a difficult plant to grow, but is prized as an ornamental for its blue-colored foliage and attractive flowers. Cultivars include 'Blue Sky,' 'Christoph's Blue,' 'Misty Blue,' 'Oree,' 'Raspberry Ripple,' 'Viridis,' 'Winter Star,' 'Woodlander's Blue,' and forma *nitida*; **misc. products:** none; **weeds:** none; **nonindigenous species:** none

Systematics: Molecular data resolve the monotypic *Zenobia pulverulenta* within tribe Andromedeae of subfamily Vaccinioideae of Ericaceae, as the sister species to *Andromeda polifolia* (Figure 5.8). The base chromosome number of *Zenobia* is $x = 11$. Estimated counts of $2n = 66$ reported for *Z. pulverulenta* indicate that it is a polyploid.

Comments: *Zenobia pulverulenta* is a coastal plain endemic of the southeastern United States

References: Dorr, 1981, 2009; Kron et al., 1999; Middleton & Wilcox, 1990; Ranjan et al., 2001; Simms, 1985, 1987.

Family 3: Myrsinaceae [40]

Phylogenetic analyses of various data sets (Anderberg et al., 1998; Källersjö et al., 2000; Martins et al., 2003; Hao et al., 2004; Manns & Anderberg, 2005) consistently indicate that several former genera of Primulaceae (including *Anagallis, Lysimachia, Trientalis*) are nested phylogenetically within Myrsinaceae and should be transferred to that family as has been followed here. Prior to these taxonomic modifications, Primulaceae essentially comprised herbaceous species with showy, bisexual flowers and capsular fruits; whereas, Myrsinaceae mainly comprised dioecious or monoecious woody plants with evergreen leaves, small flowers, and fleshy fruits (berries or drupes). The current circumscription of Myrsinaceae (now with more than 1000 species) presents a much more heterogeneous assemblage of habit, flower, and fruit types and it is premature to provide any meaningful overview of family characteristics until a final circumscription for the family is settled upon. The relatively uncommon feature of free-central placentation occurs here as it does also in the Primulaceae.

The circumscriptions of several Myrsinaceae genera also remain in a state of flux as a consequence of conflicting results from phylogenetic analyses; thus, their interrelationships have not been settled satisfactorily. It is particularly difficult to clarify the circumscriptions of *Lysimachia* or *Anagallis*, which both resolve in part as core clades, but also have several species that fall outside of those clades. Yet, all results point to their relatively close relationship (Figure 5.12) and some authors essentially have transferred most of the *Anagallis* species to *Lysimachia*. The two genera have been kept separate in the present account pending the outcome of more definitive systematic investigations. Another noteworthy change has been the transfer of the monotypic genus *Glaux* (*G. maritima*) to *Lysimachia* (Banfi et al., 2005), which is supported by several, independent analyses (Hao et al., 2004; Manns & Anderberg, 2005) and seems appropriate.

The wetland indicator status of several species has also changed since the publication of the 1996 list. These include the reclassification of *Anagallis minima* (regarded by some as *Lysimachia minima*) from (OBL, FACW, FACU) to (FACW) status, *Lysimachia nummularia* from (OBL, FACW) to (FACW) status, and *Trientalis europaea* from (OBL, FAC) to (FAC, FACU) status. The treatments of *Anagallis minima, Lysimachia nummularia*, and *Trientalis europaea* are included because they were completed prior to the release of the 2013 list.

Numerous ornamental plants are found in Myrsinaceae, primarily in the genera *Anagallis, Ardisia, Cyclamen, Lysimachia*, and *Myrsine*. Included in the former genus is *Anagallis arvensis*, the infamous "scarlet pimpernel" of the literature. *Ardisia japonica* is a herb used in traditional Chinese medicine. Although only one genus of Myrsinaceae (*Lysimachia*) presently contains obligate aquatic plants in North America, *Anagallis* and *Trientalis* are also treated here as explained earlier:

1. ***Anagallis*** L.
2. ***Lysimachia*** L.
3. ***Trientalis*** L.

1. *Anagallis*

Chaffweed, pimpernel; centenille, mouron

Etymology: from the Greek *an agallomai* meaning "without pride"

Synonyms: *Centunculus*; *Micropyxis*

Distribution: global: Africa; Asia; Europe; New World; **North America:** widespread

Diversity: global: 31 species; **North America:** 4 species

Indicators (USA): FACW: *Anagallis minima*

Habitat: freshwater, saline (coastal); palustrine; **pH:** unknown; **depth:** <1 m; **life-form(s):** emergent herb

Key morphology: stems (to 10 cm) decumbent, semisucculent, rooting at the nodes; leaves (to 5 mm) alternate, sessile; flowers 4–5-merous, minute, solitary, axillary, subsessile; perianth segments united basally; calyx (to 3 mm) linear; corolla (to 1.5 mm) translucent, pinkish, adherent to fruit when withered; stamens 4–5; capsule (to 2 mm) globose, circumscissile, many seeded

Life history: duration: annual (fruit/seeds); **asexual reproduction:** none; **pollination:** self; **sexual condition:** hermaphroditic; **fruit:** capsules (common); **local dispersal:** seeds (water, wind); **long-distance dispersal:** seeds (water, wind)

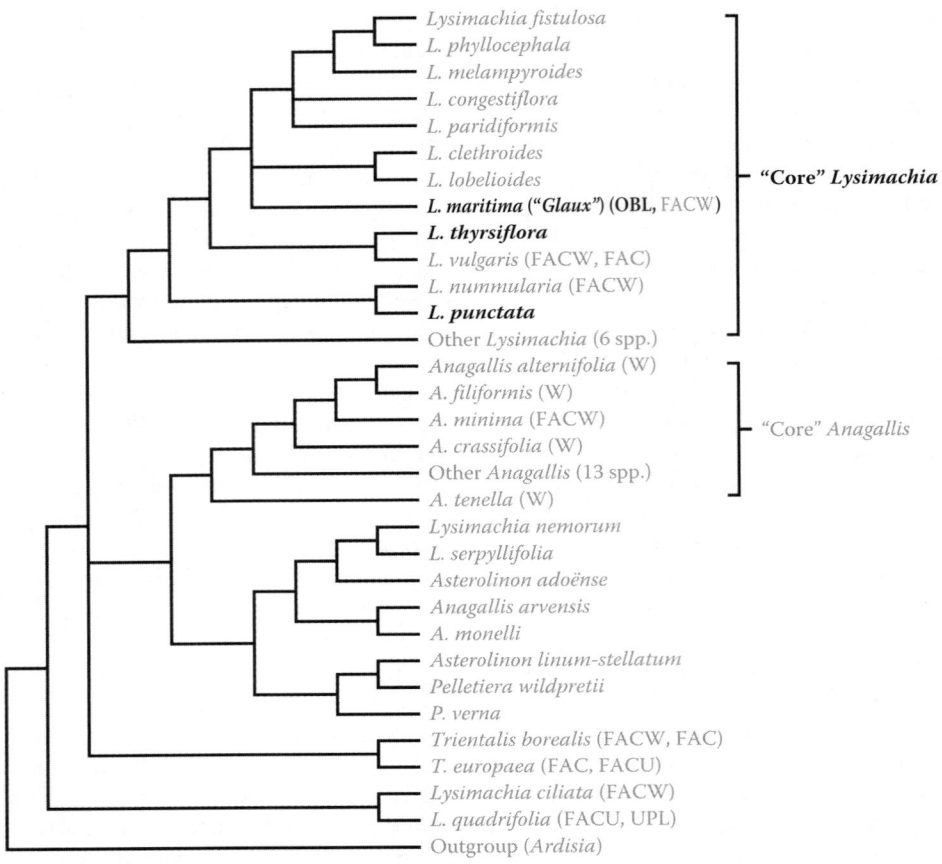

FIGURE 5.12 Phylogenetic relationships among several genera of Myrsinaceae as indicated by analyses of combined ITS, *trn*L-F, and *Ndh*F DNA sequence data. Several genera (*Anagallis, Asterolinon, Lysimachia*) do not resolve as monophyletic, although core clades are indicated for *Anagallis* and *Lysimachia*. North American species with OBL designations (bold) occur in different portions of the cladogram, indicating several independent origins of the habit in the family. *Anagallis minima* (previously ranked as OBL) associates with wetland plants from regions outside of North America (marked by a "W"). *Lysimachia maritima* formerly was segregated in the genus *Glaux* (indicated), which is not warranted by these results. (Adapted from Manns, U. & Anderberg, A.A., *Int. J. Plant Sci.*, 166, 1019–1028, 2005.)

Imperilment: (1) *Anagallis minima* [G5]; SH (DC, DE, KY, OH, VA, WV); S1 (<u>AB</u>, CO, NE, <u>NS</u>, WY); S2 (<u>BC</u>, KS, <u>SK</u>, MT); S3 (GA)

Ecology: general: *Anagallis* is a genus of low-growing annuals, often with an affinity for wet habitats. Most species would be ranked as FACW plants including *A. minima*, which was designated formerly as an obligate aquatic in a portion of its North American range.

Anagallis minima (L.) E.H.L. Krause grows in depressions, ditches, floodplains, mudflats, prairies, salt marshes, seeps, vernal pools, and along lake and pond margins at elevations of up to 950 m. Substrates are sand or mud and usually are low in fertility. The plants are not shade tolerant but do tolerate some degree of salinity. They commonly inhabit sites with fluctuating water levels and occur as an edge species in zones of intermediate inundation (35–65 days). It is not unusual to see the plants inhabiting drier sites, as its dynamic wetland indicator status history would suggest. The flowers are homomorphic, self-compatible, and self-pollinating. About 30 seeds are produced per flower and up to 1000 per plant. The seeds require partial light for germination, which typically achieves high rates

and occurs naturally in the spring. The seeds form a seed bank and can persist where the species no longer exists in the standing vegetation. Populations are known to fluctuate dramatically between years in some sites, making it difficult to obtain accurate census data. The plants reportedly are impacted positively by mowing or grazing, which reduces competing species. More detailed studies of habitat characteristics (e.g., pH, salinity tolerances) and seed biology (dispersal, germination) are needed for this species. **Reported associates:** *Achillea millefolium, Agrostis gigantea, Anemone caroliniana, Aristida dichotoma, Bigelowia nuttallii, Brodiaea orcuttii, Bromus hordeaceus, Callirhoe papaver, Callitriche marginata, Camassia quamash, Carex athrostachya, C. duriuscula, Centaurium muehlenbergii, Coreopsis atkinsoniana, C. tinctoria, Crassula aquatica, Cynosurus echinatus, Cytisus scoparius, Dactylis glomerata, Deschampsia cespitosa, D. danthonioides, Distichlis spicata, Downingia cuspidata, Drosera brevifolia, Eleocharis macrostachys, E. ovata, Eryngium aristulatum, Fritillaria camschatcensis, Geocarpon minimum, Gnaphalium, Gratiola neglecta, Grindelia integrifolia, Hedeoma hispida, Hemizonia fasciculata, Heterocodon rariflorum, Holcus lanatus,*

Hordeum brachyantherum, H. pusillum, Houstonia micrantha, H. pusilla, H. rosea, Hypochaeris glabra, Isoetes orcuttii, Isolepis carinata, Juncus balticus, J. bufonius, J. confusus, J. gerardii, J. kelloggii, J. nevadensis, Krigia occidentalis, Lepuropetalon spathulatum, Limosella acaulis, Lythrum hyssopifolium, Medicago, Mimosa strigillosa, Minuartia muscorum, Neptunia lutea, Nothoscordum bivalve, Oenothera linifolia, Ophioglossum crotalophoroides, Phleum pratense, Plantago, Pogogyne abramsii, Polypremum procumbens, Potentilla hippiana, P. pacifica, Psilocarphus brevissimus, P. elatior, Sagina, Samolus ebracteatus, S. valerandi, Sanicula crassicaulis, Sarcocornia perennis, Schizachyrium scoparium, Schoenolirion wrightii, Schoenoplectus hallii, Spartina pectinata, Spergularia echinosperma, Sporobolus coromandelianus, Stenotaphrum secundatum, Talinum parviflorum, Tradescantia occidentalis, Triglochin maritimum, Sapium glandulosum, Vulpia myuros.

Use by wildlife: none reported

Economic importance: food: none reported; **medicinal:** none reported; **cultivation:** not cultivated; **misc. products:** none; **weeds:** none in North America; **nonindigenous species:** *Anagallis minima* was naturalized in New Zealand sometime between 1940 and 1970. It has also been introduced to Western Australia.

Systematics: Although some taxonomic accounts place *Anagallis* in Primulaceae, a number of recent phylogenetic studies using molecular data consistently resolve the genus among members of Myrsinaceae. The same studies also indicate that *Anagallis* is not monophyletic as formerly circumscribed and would require that some representatives of other genera (*Asterolinon, Lysimachia, Pelletiera*) be included to distinguish a clade (Figure 5.12). *Anagallis minima* was long placed in the former genus *Centunculus* (as *C. minimus*); however, there is no phylogenetic justification for segregating this species from *Anagallis*, as it is nested among the core members of that genus (Figure 5.12). However, some authors have merged the genus with *Lysimachia. Anagallis minima* is distantly related to the only other native North American species, *A. pumila*, which occurs within a clade of 13 principally African species (Figure 5.12). Instead, cladograms constructed from DNA sequence data (compiled for ⅔ of all *Anagallis* species) place *Anagallis minima* as the sister species to *A. alternifolia* and *A. filiformis*, two South American wetland plants (Figure 5.12). The closest relative to this clade is *A. crassifolia*, another wetland plant from Africa. These results indicate that *A. minima* originated from within an existing group of species already adapted to aquatic conditions. The base chromosome number in *Anagallis* is *x* = 10–11. *Anagallis minima* (2*n* = 22) is a diploid. No hybrids have been reported that involve *A. minima*, perhaps because of its self-pollinating nature. There are no population genetic studies of this species reported in the literature.

Comments: *Anagallis minima* has nearly a cosmopolitan distribution and is widespread in North America

References: Bauder, 2000; Godfrey & Wooten, 1981; Looney & Gibson, 1995; Manns & Anderberg, 2005.

2. *Lysimachia*

Creeping Jenny, loosestrife, sea milkwort; herbe aux écus, lysimaque

Etymology: after King Lysimachos (ca. 355–281 BC) of Thrace, who reportedly subdued a leopard while brandishing these plants

Synonyms: *Apochoris; Glaux; Naumburgia; Nummularia; Steironema*

Distribution: global: cosmopolitan; **North America:** widespread

Diversity: global: 160 species; **North America:** 20 species

Indicators (USA): OBL: *Lysimachia asperulifolia, L. hybrida, L. loomisii, L. maritima, L. punctata, L. quadriflora, L. radicans, L. terrestris, L. thyrsiflora;* **FACW:** *L. maritima, L. nummularia, L. quadriflora, L. radicans*

Habitat: brackish (coastal), freshwater; palustrine, riverine; **pH:** 4.0–7.6; **depth:** <1 m; **life-form(s):** emergent herb

Key morphology: stems round or 4-angled, succulent in one species, either creeping (to 1.5 m) and rooting along nodes, or erect (to 1 m); elongate axillary vegetative bulbils produced in one species; leaves cauline (sometimes persisting in a basal rosette), roundish (to 3 cm) or lanceolate/ovate (to 15 cm), opposite or in whorls of 3–7, sessile or petioled, the foliage usually brown dotted or streaked (best seen when held to the light); flowers showy (to 3 cm), solitary to whorled and axillary, or in axillary, long peduncled (to 4 cm), capitate racemes, or in elongate (to 3 dm) terminal racemes, subsessile or pedicelled (to 4 cm); petals five (absent in one species where the calyx is petaloid), linear (to 5 mm) or broad (to 12.5 mm), yellow, sometimes with a red blotch basally; capsules (to 6.5 mm) with up to 45 seeds

Life history: duration: perennial (bulbils, rhizomes); **asexual reproduction:** bulbils, stolons, rhizomes; **pollination:** insect; **sexual condition:** hermaphroditic; **fruit:** capsules (common); **local dispersal:** rhizomes, bulbils/buds (water), seeds (water); **long-distance dispersal:** seeds (animals)

Imperilment: (1) *Lysimachia asperulifolia* [G3]; S1 (SC); S3 (NC); (2) *L. hybrida* [G5]; SH (KS); S1 (NB, NC, NY, ON, PA, SC, VT, WV); S2 (AB, AZ, DE, MD, MI, QC, VA); S3 (KY, MB, NJ, NM); (3) *L. loomisii* [G3]; SH (GA); S3 (NC); (4) *L. maritima* [G5]; SX (NJ); SH (MD, RI); S1 (MN, NE); S2 (YT); S3 (NH, ON, WY); (5) *L. nummularia* [GNR]; (6) *L. punctata* [GNR]; (7) *L. quadriflora* [G5]; SX (PA); SH (DC); S1 (NY, OK, SD, TN, VA, WV); S2 (MB); S3 (AL, GA, KY); (8) *L. radicans* [G4]; SH (KY, VA); S1 (IL); (9) *L. terrestris* [G5]; S1 (GA, KY, LB, TN); S3 (MB, NC); (10) *L. thyrsiflora* [G5]; SH (CO); S1 (MD, MO, NF, UT, WV, WY); S2 (NH, YT); S3 (AB, IA, NB, NJ, NS)

Ecology: general: *Lysimachia* often is encountered in aquatic habitats, with nearly half of the North American species are categorized as OBL in at least a portion of their range and virtually all of the North American species designated as wetland indicators at some level. The flowers of most species are showy and attract insects, which function as their pollinators. Their stamens are wrapped within the petals until the flowers open fully, a feature that deters self-pollination.

Both self-compatibility and self-incompatibility have been reported among the different species. Pollination in many *Lysimachia* species (but few of those in North America) is associated intimately with melittid bees (Hymenoptera: Melittidae: *Macropis*), which collect not only pollen, but also the oils produced by the specialized floral elaiophores. Nearly all of the North American species can reproduce vegetatively by rhizomes, and some also by means of shoot-borne bulbils.

Lysimachia asperulifolia **Poir.** occurs in bogs, depressions, pocosins, savannas, seeps, and swamps on the southeastern coastal plain at elevations of up to 300 m. The substrates typically are acidic, seasonally inundated sands, layers of organic peat that overlay sand, or deep peat. The species is shade intolerant. Plants are self-incompatible and produce few seeds in natural populations as a consequence of interactions between habitat fragmentation, pollinator limitation, self-incompatibility, and clonal growth. Flowers are produced early in the summer. The principal pollinators are sweat bees (Halictidae: *Augochlorella striata, Lasioglossum coreopsis, L. rohweri*). Each stem yields an average of three capsules, which dehisce late in the season (October). The seeds have no specialized adaptations for dispersal and most fall close to the maternal plant. Low germination (10%) has been achieved after 30 days of moist stratification at 4°C at ambient greenhouse temperatures; however, other studies report germination rates of up to 85%. This species typically occurs in the open ecotones between pine savannahs and pocosins. The plants reproduce vegetatively by extension and fragmentation of their fire-resistant rhizomes. Sites that are characterized by periodic fire disturbance are necessary to maintain persistence of the species by removal of larger-growing shrubs, which compete for resources. **Reported associates:** *Andropogon glomeratus, Aristida stricta, Calamovilfa brevipilis, Carex striata, Chamaedaphne calyculata, Cyrilla racemiflora, Dionaea muscipula, Eriophorum virginicum, Ilex glabra, Lyonia lucida, Peltandra sagittifolia, P. virginica, Rhexia alifanus, R. lutea, Rhynchospora alba, Smilax laurifolia, Sporobolus teretifolius, Utricularia subulata, Woodwardia virginica, Zenobia pulverulenta.*

Lysimachia hybrida **Michx.** grows in depressions, ditches, fens, floodplains, hammocks, marshes, meadows, prairies, sloughs, swamps, vernal pools, and along lake, pond, river, and stream margins at elevations of up to 2300 m. Habitats include freshwater intertidal and slightly brackish sites, which range in exposure from full sun to shade. The substrates are sand, gravel, or mud. Flowering extends from late summer to fall. Flowers are insect pollinated and the plants are self-incompatible. Seeds that have been cold stratified reportedly germinate at "warm" temperatures. The plants reproduce vegetatively by means of stout rhizomes. Additional studies on the reproductive biology and seed ecology are necessary to clarify the life history of this species. **Reported associates:** *Acer rubrum, A. saccharinum, Agalinis purpurea, Agrostis alba, A. hyemalis, Alisma subcordatum, Arisaema dracontium, Baptisia alba, Betula nigra, Bidens discoideus, B. frondosus, Boehmeria cylindrica, Campanula uliginosa, Campsis radicans, Cardamine bulbosa, Carex lupuliformis, C. lupulina, C.*

muskingumensis, C. parryana, C. squarrosa, C. stricta, C. vulpinoidea, Cephalanthus occidentalis, Circaea quadrisulcata, Cirsium discolor, C. flodmannii, Crotalaria, Cyperus diandrus, C. strigosus, Dioscorea villosa, Drosera rotundifolia, Echinochloa walteri, Echinocystis lobata, Eleocharis engelmanii, E. smallii, Eragrostis hypnoides, Erechtites hieraciifolia, Eryngium yuccifolium, Eupatorium perfoliatum, Eurybia radula, Fraxinus pennsylvanica, F. profunda, Galium asperellum, Glyceria striata, Helenium flexuosum, Helianthemum, Heliopsis helianthoides, Hypericum majus, H. sphaerocarpum, Hypoxis hirsuta, Ilex verticillata, Impatiens capensis, Ionactis linariifolius, Iris missouriensis, I. virginica, Ivesia multifoliata, Juncus greenei, J. longistylis, Justicia americana, Lechea, Leersia oryzoides, Lindera benzoin, Lindernia dubia, Lobelia cardinalis, Lycopus americanus, L. uniflorus, Lysimachia ciliata, Lythrum alatum, Marshallia grandiflora, Melilotus alba, M. officinalis, Menispermum canadense, Monarda didyma, Muhlenbergia richardsonis, Najas guadalupensis, Nyssa biflora, Onoclea sensibilis, Osmunda regalis, Oxalis stricta, Panicum villosissimum, P. virgatum, Peltandra virginica, Persicaria amphibia, P. arifolia, P. posumbu, P. hydropiper, P. hydropiperoides, P. punctata, P. sagittata, P. virginiana, Phalaris arundinacea, Phyla lanceolata, Poa palustris, Populus deltoides, Potentilla arguta, P. simplex, Proserpinaca palustris, Prunella vulgaris, Quercus palustris, Ranunculus septentrionalis, Ratibida pinnata, Rhododendron viscosum, Robinia neomexicana, Rorippa palustris, Rosa setigera, Rubus allegheniensis, Rumex altissimus, Salix interior, S. nigra, S. rigida, Sambucus nigra, Sanguisorba canadensis, Saururus cernuus, Schoenoplectus heterochaetus, S. torreyi, Scirpus atrovirens, Scutellaria galericulata, S. lateriflora, Sium suave, Smilax hispida, S. rotundifolia, Solidago bicolor, Sparganium androcladum, Spartina pectinata, Spiraea alba, Stenanthium leimanthoides, Strophostyles helvula, Thalisctrum fendleri, Thermopsis, Toxicodendron radicans, Triantha glutinosa, Verbena hastata, Viburnum nudum, V. recognitum, Zizia aptera.

Lysimachia loomisii **Torr.** is a rare inhabitant of bogs, depressions, pocosins, roadsides, and savannas along the southeastern coastal plain at elevations of up to 200 m. Biological and ecological information on this species is scarce. The plants flower in early summer and reproduce vegetatively by means of slender rhizomes. **Reported associates:** *Amphicarpum purshii, Andropogon glaucopsis, Anthaenantia rufa, Arnoglossum ovatum, Balduina uniflora, Bigelowia nudata, Calamovilfa brevipilis, Carphephorus odoratissimus, C. odoratissimus, Cirsium horridulum, Coreopsis falcata, Ctenium aromaticum, Dionaea muscipula, Drosera brevifolia, Helenium pinnatifidum, H. vernale, Helianthus heterophyllus, Lilium catesbaei, Muhlenbergia expansa, Paspalum praecox, Pinguicula caerulea, P. pumila, Pinus palustris, P. serotina, Platanthera integra, P. nivea, Pleea tenuifolia, Polygala brevifolia, P. hookeri, Pterocaulon pycnostachyum, Pyxidanthera barbulata, Rhynchospora chapmanii, R. latifolia, Sarracenia minor.*

Lysimachia maritima **(L.) Galasso, Banfi & Soldano** grows in tidal or inland sites including beaches and shores, depressions, estuaries, marshes/salt marshes, meadows, pannes, and seeps at elevations of up to 2300 m. Substrates are circumneutral (e.g., pH: 6.0–7.2) clay, clay/loam, sand, or sand overlain by peat. The plants sometimes are found to grow within rocky crevices. This is a shade intolerant, but salt-tolerant hydrohalophyte, which occurs on alkaline, brackish, or saline sites and can tolerate salinities up to 52.0 dS m^{-1}. The plants thrive at NaCl concentrations of 0.9%. The excretion of salt through specialized glands facilitates survival under saline conditions. Maximum flowering occurs from spring to summer in low salinity sites. The flowers are anomalous in the genus by lacking a corolla and by having whitish petaloid sepals. They are self-compatible and generally self-pollinated although insect pollination also may occur. Each flower produces an average of 10 seeds, which can reach densities of 175–23,000 seeds per m^2 (5400/m^2 on average). The seeds are physiologically dormant and retain viability in salinities as high as 0.09 M NaCl. Their germination can be triggered by fluctuating temperatures that result from flooding, which is an environmental cue that correlates with the damage of standing vegetation, which ordinarily competes with the plants for resources. Good germination has been reported under a 25°C/20°C temperature regime for cold-stratified seeds. The roots are associated with vesicular arbuscular mycorrhizal fungi. The plants are highly clonal and spread rapidly in response to environmental variation by means of slender rhizomes and winter buds. This species is described as a "pseudoannual," where the individual ramets live for only one season, but produce overwintering vegetative propagules, which germinate into new (but detached) ramets in the spring. The plants form a persistent propagule bank as a result of the vegetative winter buds and seeds. The seeds are dispersed locally by water and over longer distances from endozooic transport by animals such as geese (Aves: Anatidae: *Branta*) and hares (Mammalia: Leporidae: *Lepus*), which ingest them. *Lysimachia maritima* is efficient at taking up nitrate by having enhanced nitrate reductase activity, and it competes well among other salt marsh species when nitrate levels are low. The seedlings are effective at colonizing sites that are disturbed by ice scour. In cold climates, sexual reproduction is more common at sites where salinity is low; whereas, vegetative reproduction is prevalent where higher salinity levels occur. The rhizome internodes decrease in length relative to increased levels of competition, which results in the formation of denser clumps of plants. **Reported associates:** *Agalinis maritima, Agropyron, Almutaster pauciflorus, Amaranthus pumilus, Ammophila arenaria, Argentina anserina, A. egedii, Artemisia campestris, Atriplex patula, A. prostrata, Blysmus rufus, Cakile edentula, Caltha leptosepala, Carex aquatilis, C. mackenziei, C. nebrascensis, C. praegracilis, C. ramenskii, C. silicea, Chamaesyce polygonifolia, Chenopodium rubrum, C. salinum, Cladium mariscoides, Crepis runcinata, Cuscuta salina, Deschampsia cespitosa, Distichlis spicata, D. stricta, Eleocharis fallax, E. palustris, E. rostellata, Elymus repens, E. trachycaulus, Erigeron lonchophyllus, Festuca rubra, Frankenia salina, Galium tinctorium, Gentianopsis detonsa, Glyceria striata, Grindelia integrifolia, G. oregana,* *Honckenya peploides, Hordeum jubatum, Impatiens capensis, Iva axillaris, I. frutescens, Jaumea carnosa, Juncus arcticus, J. effusus, J. gerardii, Lathyrus japonicus, Ligusticum scothicum, Lilaeopsis chinensis, Limonium californicum, L. nashii, L. vulgare, Lythrum lineare, Maianthemum stellatum, Mertensia maritima, Monolepis nuttalliana, Muhlenbergia asperifolia, M. richardsonis, Oenothera humifusa, Pascopyrum smithii, Persicaria maculosa, Phleum alpinum, Pityopsis falcata, Plantago eriopoda, P. maritima, Poa eminens, Polygonum glaucum, Primula pauciflora, Puccinellia nuttalliana, P. tenella, Ranunculus cymbalaria, Rumex maritimus, Salicornia rubra, Salix candida, Salsola kali, Samolus valerandi, Sarcobatus vermiculatus, Sarcocornia pacifica, Schoenoplectus maritimus, S. pungens, Sesuvium maritimum, Solidago sempervirens, Spartina alterniflora, S. patens, Spergularia salina, Sporobolus airoides, Suaeda calceoliformis, S. linearis, S. maritima, Symphyotrichum ericoides, S. laurentianum, Triglochin gaspense, T. maritimum, T. palustre, Typha angustifolia.*

***Lysimachia nummularia* L.** grows in shallow water (<50 cm) or in depressions, ditches, fens, floodplains, oxbows, prairies, seeps, swamps, and along lake, pond, river, and stream margins at elevations of up to 1700 m. It is most common along the floodplains and marginal riverine wetlands but is broadly adapted to a wide range of conditions. Substrates include clay, loam, mud, peat, or sand of fairly broad acidity (pH: 4.0–7.2), but often represent the higher end of this range (pH >7.0). This species can grow in open sites or in shade and survives cold temperatures down to −25°C. The plants usually are sterile, but can produce showy flowers that mainly attract bees (Hymenoptera) when growing in exposed sites. Sexual reproduction in this species is enigmatic. Years ago Charles Darwin observed that *L. nummularia* almost always is barren of seed and reproduces virtually exclusively by vegetative reproduction. The plants reportedly possess gametophytic self-incompatibility, which could explain the absence of seeds in clonal populations where compatible mating types would not occur. The introduction of this nonindigenous species to North America would exacerbate the situation by limiting the pool of compatible genotypes available for reproduction. However, there have also been some reports of self-pollination, which obviously need to be confirmed. The plants have been characterized as apomicitc, but evidence of apomixis (other than vegetative) is wanting. Various chromosomal races exist (see *Systematics* later), but their association with fertility has not been determined. Some North American authors remark that the plants are dispersed by seeds. However, such reports typically are made without any direct evidence of, or specific examples of seed production. Yet, others indicate that seed production has not been observed in any of the nonindigenous North American populations, perhaps due to a scarcity of appropriate pollinators. The reproductive biology of this species must be better elucidated. Regardless, dispersal of plants in North America has been extremely effective, likely by the transport of vegetative fragments. The creeping stems root along the nodes and establish easily if they become dislodged and are transported to new, favorable sites. Transport of stem fragments by water is likely the primary means of dispersal along river corridors where ample habitat (i.e., floodplain) exists. *Lysimachia nummularia* is moderately competitive with other members of the North American wetland flora, with a relative competitive performance ranking of 58% when compared to 44 (mostly native) species. Plants inoculated with arbuscular mycorrhizal Fungi (Glomeromycota: Archaeosporaceae: *Archaeospora trappei, Simiglomus hoi*) exhibited enhanced growth. Occurrences often are associated with fairly high sediment nitrogen levels. **Reported associates:** *Abutilon theophrasti, Acalypha rhomboidea, Acer rubrum, A. negundo, A. saccharinum, Aegopodium podagraria, Agalinis tenuifolia, Agrimonia parviflora, Alisma, Alliaria petiolata, Alnus serrulata, Amaranthus cruentus, A. rudis, A. tuberculatus, Ambrosia artemissiifolia, A. trifida, Apocynum cannabinum, Argentina anserina, Arisaema triphyllum, Asclepias incarnata, Athyrium filix-femina, Betula nigra, Bidens cernuus, B. frondosus, B. laevis, B. vulgatus, Boehmeria cylindrica, Bolboschoenus fluviatilis, Callitriche terrestris, Cardamine pensylvanica, Carex amphibola, C. cristatella, C. granularis, C. hystericina, C. lupulina, C. muskingumensis, C. pellita, C. retrorsa, C. tetanica, C. typhina, C. vulpinoidea, Carpinus caroliniana, Carya cordiformis, Celtis occidentalis, Cephalanthus occidentalis, Cerastium vulgatum, Cichorium intybus, Cinna arundinacea, Circaea lutetiana, Cirsium arvense, C. discolor, Convolvulus sepium, Conyza canadensis, Crataegus punctata, Cryptotaenia canadensis, Cyperus erythrorhizos, C. strigosus, Daucus carota, Desmanthus illinoensis, Echinochloa crusgalli, Eleocharis obtusa, Elymus virginicus, Epilobium glandulosum, Eragrostis, Erigeron philadelphicus, Euonymus atropurpureus, Eupatorium perfoliatum, E. rugosum, E. serotinum, Fallopia convolvulus, F. japonica, Festuca arundinacea, Fraxinus pennsylvanica, Galium obtusum, G. trifolium, Geum canadense, G. laciniatum, Glechoma hederacea, Glyceria striata, Helenium autumnale, Hemerocallis fulva, Hesperis matronalis, Hypericum boreale, H. punctatum, Impatiens pallida, Iris virginica, Juglans cinerea, Juncus canadensis, J. dudleyi, J. effusus, Laportea canadensis, Lapsana communis, Leersia oryzoides, L. virginica, Lemna, Lespedeza intermedia, Lindera benzoin, Lobelia cardinalis, Lonicera morrowii, Lycopus americanus, L. virginicus, Lysimachia ciliata, L. lanceolata, L. quadriflora, L. terrestris, L. thyrsiflora, Lythrum salicaria, Maianthemum canadense, Matteuccia struthiopteris, Medicago lupulina, Menispermum canadense, Mentha arvensis, Mimulus ringens, Myosoton aquaticum, Oenothera biennis, Onoclea sensibilis, Osmundastrum cinnamomeum, Oxalis europaea, Panicum capillare, P. rigidulum, P. virgatum, Parthenocissus quinquefolia, Penthorum sedoides, Persicaria lapathifolia, P. maculosa, P. pensylvanica, P. punctata, P. setacea, P. virginiana, Phalaris arundinacea, Phlox divaricata, P. stolonifera, Phyla lanceolata, Physalis longifolia, Physostegia virginiana, Pilea pumila, Platanus occidentalis, Poa pratensis, Populus deltoides, Quercus bicolor, Q. rubra, Ranunculus abortivus, R. septentrionalis, Rhamnus cathartica, Rosa multiflora, Rudbeckia hirta, R. laciniata, Rumus crispus, R. obtusifolius,* *R. verticillatus, Salix alba, S. interior, S. nigra, Sambucus nigra, Schoenoplectus pungens, S. tabernaemontani, Scirpus atrovirens, Scutellaria lateriflora, Sedum sarmentosum, Setaria viridis, Sicyos angulatus, Silphium perfoliatum, Smilax rotundifolia, S. tamnoides, Solanum carolinense, S. dulcamara, S. nigrum, Solidago gigantea, S. rugosa, Spartina pectinata, Spirea alba, Stachys tenuifolia, Symphyotrichum firmum, S. lanceolatum, S. lateriflorum, Teucrium canadense, Thalictrum dasycarpum, T. pubescens, Thelypteris noveboracensis, Toxicodendron radicans, Trifolium pratense, T. repens, Ulmus americana, Urtica dioica, Verbena hastata, Veronia fasciculata, V. missurica, Veronica peregrina, Viola cucullata, Xanthium strumarium, Zanthoxyllum, Zizia aurea.*

***Lysimachia punctata* L.** inhabits ditches, meadows, roadsides, shores, and the margins of lakes and streams at elevations of up to 600 m. The preferred substrates are clay or loam of moderate acidity (pH: 5.5–7.0). The plants tolerate full sun to partial shade, but grow better in the latter. Flowering occurs during the summer. The flowers are insect pollinated and the seeds reportedly germinate optimally at 18°C–20°C either immediately or after 2–4 weeks of cold stratification at 4°C. Although this is a popular garden plant, there is surprisingly little information available on its ecology either in its native European range or in any of its introduced localities. The plants can be highly invasive in garden settings, but apparently do not escape often into natural communities. **Reported associates:** *Vinca minor.*

***Lysimachia quadriflora* Sims** occurs in bogs, depressions, ditches, fens, marshes, meadows, prairies, roadsides, seeps, sloughs, springs, swamps, and along pond and stream margins at elevations of up to 600 m. Substrates are calcareous peat (sedges), loam, marl, sand (sometimes gravel or rock), and usually alkaline (pH: 5.7–7.6). This species is regarded as a reliable indicator of alkaline fens and prairies. The plants are self-incompatible and their showy flowers (produced during summer) are pollinated by oil-collecting bees (Hymenoptera: Melittidae: *Macropis steironematis*). The seeds require 9–12 weeks of cold stratification at 4°C followed by 20°C for optimal germination. Vegetative reproduction occurs by rhizomes. Although this species is characteristic of prairies, there is no specific information reported on the fire resistance associated with either the rhizomes or the seeds. **Reported associates:** *Acer saccharum, Adiantum pedatum, Agalinis paupercula, A. purpurea, A. tenuifolia, Agrimonia gryposepala, Agrostis gigantea, Allium cernuum, Alnus serrulata, Andropogon gerardii, Anemone canadensis, Angelica atropurpurea, Anticlea elegans, Aquilegia canadensis, Arnoglossum plantagineum, Asclepias hirtella, A. incarnata, A. sullivantii, Asplenium rhizophyllum, A. ruta-muraria, A. resiliens, Betula papyrifera, B. ×sandbergii, Boehmeria cylindrica, Bromus ciliatus, Cacalia tuberosa, Calamagrostis canadensis, C. stricta, Caltha palustris, Campanula aparinoides, Cardamine bulbosa, Carex annectens, C. aquatilis, C. atherodes, C. aurea, C. buxbaumii, C. complanata, C. conoidea, C. crawei, C. eburnea, C. emoryi, C. garberi, C. granularis, C. haydenii, C. hystericina, C. interior, C. lasiocarpa, C. leptalea, C. limosa, C. livida, C. lurida, C. pellita,*

C. prairea, C. sartwellii, C. scirpoidea, C. scoparia, C. sterilis, C. stricta, C. suberecta, C. tetanica, C. trichocarpa, C. utriculata, C. viridula, C. vulpinoidea, Carpinus caroliniana, Castilleja coccinea, Cephalanthus occidentalis, Chara, Chelone glabra, Chenopodium simplex, Cicuta bulbifera, C. maculata, Cirsium discolor, C. muticum, Cladium mariscoides, Clinopodium arkansanum, C. glabellum, Comandra umbellata, Cornus amomum, C. sericea, Cystopteris bulbifera, Dasiphora floribunda, Deschampsia cespitosa, Dichanthelium sabulorum, D. scoparium, Doellingeria umbellata, Drepanocladus aduncus, Drosera rotundifolia, Dulichium arundinaceum, Eleocharis compressa, E. elliptica, E. erythropoda, E. intermedia, E. pauciflora, E. quinqueflora, E. rostellata, E. tenuis, Elymus canadensis, Epilobium coloratum, E. leptophyllum, Equisetum arvense, E. fluviatile, Erigeron, Eriophorum angustifolium, Eryngium yuccifolium, Eupatorium perfoliatum, Euthamia graminifolia, Eutrochium maculatum, Fagus grandifolia, Filipendula rubra, Frangula alnus, Galium boreale, G. labradoricum, G. trifidum, Gentianopsis virgata, Geum aleppicum, G. laciniatum, Glyceria grandis, G. striata, Helenium autumnale, Helianthus grosseserratus, H. mollis, H. ×verticillatus, Hierochloe hirta, H. odorata, Holcus lanatus, Hordeum jubatum, Hydrangea arborescens, Hydrocotyle ranunculoides, Hypericum kalmianum, Hypnum lindbergii, Hypoxis hirsuta, Impatiens capensis, Iris versicolor, Juncus anthelatus, J. arcticus, J. balticus, J. biflorus, J. brachycarpus, J. brachycephalus, J. dudleyi, J. filipendulus, J. nodosus, J. scirpoides, J. tenuis, J. torreyi, Koeleria macrantha, Larix laricina, Lathyrus palustris, Liatris pycnostachya, L. spicata, Liparis loeselii, Lobelia kalmii, L. siphilitica, Ludwigia microcarpa, Lycopus americanus, L. uniflorus, Lysimachia nummularia, L. thyrsiflora, Lythrum alatum, Maianthemum stellatum, Marshallia mohrii, Mecardonia acuminata, Mentha arvensis, Menyanthes trifoliata, Mitreola petiolata, Muhlenbergia glomerata, Nasturtium, Onoclea sensibilis, Osmunda regalis, Oxypolis rigidior, Packera aurea, P. paupercula, Panicum anceps, P. flexile, P. microcarpon, P. virgatum, Parnassia glauca, P. grandifolia, Pedicularis lanceolata, Persicaria amphibia, Philonotis muehlenbergii, Phlox glaberrima, P. maculata, P. pilosa, Physocarpus opulifolius, Physostegia virginiana, Pilea fontana, P. pumila, Platanthera peramoena, Poa compressa, Pogonia ophioglossoides, Prenanthes racemosa, Primula meadia, P. mistassinica, Ptilimnium costatum, Pycnanthemum tenuifolium, P. virginianum, Ranunculus sceleratus, Rhynchospora capillacea, R. capitellata, R. caduca, R. thornei, Rosa arkansana, Rubus pubescens, Rudbeckia fulgida, R. hirta, Sabatia angularis, Salix bebbiana, S. candida, S. caroliniana, S. discolor, S. humilis, S. serissima, Sarracenia purpurea, Schizachyrium scoparium, Schoenolirion croceum, Schoenoplectus acutus, S. pungens, S. tabernaemontani, Scirpus atrovirens, S. pendulus, Scleria verticillata, Scolochloa festucacea, Scutellaria galericulata, S. integrifolia, S. lateriflora, Selaginella apoda, S. eclipes, Silphium integrifolium, S. laciniatum, S. terebinthinaceum, Sium suave, Solidago caesia, S. canadensis, S. flexicaulis, S. gigantea, S. ohioensis, S. ptarmicoides, S. riddellii, S. rigida, *S. uliginosa, Sorghastrum nutans, Spartina pectinata, Sphenopholis intermedia, Spiranthes lucida, Stachys palustris, S. tenuifolia, Staphylea trifolia, Symphyotrichum boreale, S. firmum, S. laeve, S. lanceolatum, S. lateriflorum, S. novae-angliae, S. puniceum, Thalictrum dasycarpum, Thelypteris palustris, Toxicodendron radicans, T. vernix, Triadenum fraseri, Triantha glutinosa, Triglochin palustre, Typha latifolia, Valeriana ciliata, Vernonia fasciculata, Veronica anagallis-aquatica, V. scutellata, Veronicastrum virginicum, Viola cucullata, V. nephrophylla, Zizia aptera, Z. aurea.*

***Lysimachia radicans* Hook**. inhabits depressions, floodplains, marshes, seeps, sloughs, swamps, streambanks, and river margins at elevations of up to 200 m. The habitats often are heavily shaded. Substrates reportedly are acidic and mainly sandy. The flowers appear in summer and apparently are insect pollinated, but little information on the floral biology or seed ecology exists for this relatively rare species. The plants reproduce vegetatively by rhizomes and by trailing stems that root at their nodes. A thorough ecological study of this species is needed. **Reported associates:** *Acer rubrum, Arundinaria gigantea, Asclepias, Azolla mexicana, Boehmeria cylindrica, Carex crus-corvi, C. decomposita, C. hyalinolepis, C. intumescens, C. louisianica, Carpinus caroliniana, Chasmanthium latifolium, C. laxum, Cinna arundinacea, Cyperus, Diospyros virginiana, Glyceria arkansana, Gratiola virginiana, Hydrolea uniflora, Hygrophila lacustris, Hypericum, Ilex decidua, Iris fulva, Itea virginica, Juncus, Justicia ovata, Lemna, Limnobium spongia, Lindera melissifolia, Liquidambar styraciflua, Ludwigia glandulosa, Lycopus rubellus, L. virginicus, Melothria pendula, Nyssa, Onoclea sensibilis, Osmunda regalis, Persicaria virginiana, Phanopyrum gymnocarpon, Pilea pumila, Planera aquatica, Populus deltoides, Quercus lyrata, Q. palustris, Q. phellos, Rhynchospora, Sabatia calycina, Saccharum giganteum, Salix nigra, Saururus cernuus, Scirpus cyperinus, Solidago gigantea, Spirodela polyrhiza, Symphyotrichum lateriflorum, Taxodium distichum, Thelypteris palustris, Triadenum walteri, Vernonia gigantea, Wolffia, Woodwardia areolata, W. virginica.*

***Lysimachia terrestris* (L.) Britton, Sterns & Poggenb.** grows in shallow water (15 cm) or upon the exposed substrates of beach pools, bogs, ditches, fens, floodplains, marshes, prairies, sloughs, swamps, and along lake, pond, river, and stream margins at elevations of up to 1000 m. The habitats typically are open and unshaded. Substrates are acidic (pH: 4.4–6.6) to circumneutral and include gravel, muck, peat, and sand. The showy flowers are produced during summer and attract various insects; however, there is little specific information on either the floral or seed ecology of this species. In one study the seeds represented only a minor component of the seed bank. Vegetative reproduction occurs by stolons and also by specialized bulbils that are formed (often prolifically) in the upper leaf axils. The bulbils detach from the plants during the fall and are dispersed readily by water. However, aspects of their dormancy, longevity, and persistence in the environment have not been determined. **Reported associates:** *Abies balsamea,*

Acer rubrum, A. saccharinum, Acorus calamus, Agalinis paupercula, Agrostis hyemalis, Alisma subcordatum, Alnus rugosa, A. serrulata, Andromeda polifolia, Andropogon, Asclepias incarnata, Aulacomnium palustre, Bartonia virginica, Betula populifolia, Bidens connatus, Brasenia schreberi, Calamagrostis canadensis, Calla palustris, Calopogon tuberosus, Caltha palustris, Calystegia sepium, Campanula aparinoides, Carex atlantica, C. bullata, C. canescens, C. crinita, C. echinata, C. folliculata, C. haydenii, C. lacustris, C. lasiocarpa, C. lenticularis, C. leptalea, C. lurida, C. oligosperma, C. rostrata, C. stricta, C. utriculata, C. wiegandii, Chamaedaphne calyculata, Cicuta bulbifera, Cinna arundinacea, Cirsium discolor, Cladium mariscoides, Comarum palustre, Cornus sericea, Cyperus dentatus, C. rivularis, Dasiphora floribunda, Decodon verticillatus, Doellingeria umbellata, Drosera intermedia, Dryopteris cristata, Dulichium arundinaceum, Echinodorus tenellus, Eleocharis acicularis, E. olivacea, E. palustris, E. quadrangulata, E. robbinsii, E. rostellata, Elymus riparius, E. virginicus, Epilobium leptocarpum, Equisetum fluviatile, Erigeron canadensis, Eriocaulon aquaticum, Eriophorum chamissonis, E. vaginatum, E. virginicum, Eryngium yuccifolium, Eutrochium maculatum, Fimbristylis autumnalis, Fraxinus nigra, F. pennsylvanica, Galium palustre, G. tinctorium, Glyceria canadensis, Gratiola aurea, Houstonia serpyllifolia, Hydrocotyle umbellata, Hypericum kalmianum, H. mutilum, H. sphaerocarpum, Ilex mucronata, I. verticillata, Impatiens capensis, Iris versicolor, Juncus acuminatus, J. canadensis, J. filiformis, J. militaris, J. pelocarpus, J. scirpoides, Kalmia latifolia, Lachnanthes caroliana, Lactuca scariola, Larix laricina, Leersia oryzoides, Liatris, Liparis loeselii, Lobelia dortmanna, Ludwigia alternifolia, L. polycarpa, Lycopodiella inundata, Lycopus americanus, L. uniflorus, L. virginicus, Lythrum salicaria, Lyonia ligustrina, Lysimachia ciliata, L. nummularia, L. thyrsiflora, Maianthemum canadense, Matteuccia struthiopteris, Mentha arvensis, Menyanthes trifoliata, Mimulus ringens, Muhlenbergia glomerata, M. uniflora, Myrica gale, Myriophyllum, Nymphaea odorata, Nymphoides cordata, Onoclea sensibilis, Osmunda claytoniana, O. regalis, Osmundastrum cinnamomeum, Oxypolis rigidior, Packera aurea, Panicum rigidulum, Parnassia asarifolia, Persicaria coccinea, P. punctata, P. sagittata, Phegopteris connectilis, Photinia melanocarpa, Phragmites australis, Phyla lanceolata, Physostegia virginiana, Picea mariana, Pinus strobus, Platanthera psychodes, Polygala sanguinea, Polytrichum commune, P. longisetum, Pontederia cordata, Potentilla arguta, P. norvegica, Ranunculus, Ratibida pinnata, Rhododendron maximum, Rhynchospora alba, R. capitellata, Rosa palustris, Rubus flagellaris, R. idaeus, Sagittaria latifolia, S. teres, Salix discolor, S. lucida, S. myricoides, S. nigra, S. sericea, Sarracenia purpurea, Saururus cernuus, Scheuchzeria palustris, Schoenoplectus acutus, S. pungens, Scirpus atrovirens, S. cyperinus, S. expansus, S. longii, S. polyphyllus, Scutellaria galericulata, Sium suave, Solidago patula, S. uliginosa, Sparganium angustifolius, S. eurycarpum, Spartina pectinata, Sphagnum affine, S. bartlettianum, S. cuspidatum, S. flexuosum, S.

fimbriatum, Sp. henryense, S. magellanicum, S. palustre, S. recurvum, Spiraea alba, S. tomentosa, Thalictrum pubescens, Thelypteris palustris, T. simulata, Thuja occidentalis, Triadenum fraseri, T. virginicum, Typha angustifolia, T. latifolia, Ulmus americana, Utricularia cornuta, U. intermedia, Vaccinium macrocarpon, V. oxycoccos, Verbena hastata, Viola lanceolata, Woodwardia virginica, Xyris torta.

***Lysimachia thyrsiflora* L.** occurs in shallow water (15 cm) or on exposed substrates of bogs, ditches, marshes, meadows, prairies, swamps, and along lake and river margins at elevations of up to 2000 m. Substrates include clay, marl, muck, and sand and usually are acidic (pH: 4.0–6.3). Plants occur in open sites and bloom in mid-summer when many other wetland species are not in flower. They are insect pollinated; however, the secretory glands of the flowers are highly reduced and this is one *Lysimachia* species that does not rely on oil-collecting bees (Hymenoptera: Melittidae: *Macropis*) for pollination. The flowers are protogynous and are either outcrossed or self-pollinated. More information on the floral biology of this species is needed. The seeds are dispersed by water and can float for up to 22 days. The plants can yield up to 155 seeds per m^2 but only a transient seed bank is produced. Conditions for seed germination are not known. The stems are weak and often are supported by the surrounding vegetation. Vegetative reproduction occurs by rhizomes, which tightly bind the soil by their tillering roots. **Reported associates:** *Acer rubrum, A. saccharinum, Alisma subcordatum, Alnus rugosa, Amaranthus tuberculatus, Amphicarpaea bracteata, Andromeda polifolia, Apocynum cannabinum, Aralia nudicaulis, Arethusa bulbosa, Arisaema, Asclepias incarnata, Azolla mexicana, Barbarea vulgaris, Betula lutea, B. pumila, B. sandbergi, Boehmeria cylindrica, Bolboschoenus fluviatilis, Calamagrostis canadensis, C. stricta, Calla palustris, Caltha palustris, Campanula aparinoides, Cardamine bulbosa, C. pratensis, Carex aquatilis, C. canescens, C. comosa, C. haydenii, C. intumescens, C. lacustris, C. lasiocarpa, C. leptalea, C. lupulina, C. paupercula, C. pedunculata, C. pellita, C. pseudocyperus, C. retrorsa, C. sartwellii, C. stipata, C. stricta, C. trisperma, C. vesicaria, Chamaedaphne calyculata, Chelone glabra, Cicuta bulbifera, Clintonia borealis, Comarum palustre, Coptis trifolia, Cornus canadensis, C. sericea, Cyperus, Cypripedium parviflorum, Dasiphora floribunda, Decodon verticillatus, Doellingeria umbellata, Drosera rotundifolia, Dulichium arundinaceum, Eleocharis erythropoda, E. obtusa, Epilobium glandulosum, Equisetum arvense, E. fluviatile, Eragrostis, Eriophorum, Eupatorium perfoliatum, Euthamia graminifolia, Eutrochium maculatum, Fraxinus nigra, Galium labradoricum, G. tinctorum, G. trifidum, Gaultheria hispidula, Gentianopsis, Glyceria canadensis, G. grandis, G. striata, Hypericum ascyron, Hypericum kalmianum, Ilex mucronata, I. verticillata, Impatiens capensis, Iris versicolor, I. virginica, Kalmia polifolia, Larix laricina, Lemna minor, Liatris pychnostachya, Lilium michiganense, Lindera benzoin, Lobelia kalmii, Ludwigia polycarpa, Lycopus rubellus, L. uniflorus, L. virginicus, Lysimachia nummularia, L. quadriflora, Lythrum alatum, Maianthemum canadense, M. trifolium, Mentha* *arvensis, Menyanthes trifoliata, Micranthes pensylvanica, Mimulus ringens, Mitella nuda, Myrica gale, Onoclea sensibilis, Osmunda regalis, Osmundastrum cinnamomeum, Oxypolis rigidior, Panicum virgatum, Parthenocissus quinquefolia, Pedicularis lanceolata, Penthorum sedoides, Persicaria amphibia, P. coccinea, P. punctata, P. sagittata, Phalaris arundinacea, Phlox glaberrima, Photinia melanocarpa, Physocarpus opulifolius, Picea mariana, Pinus strobus, Pogonia ophioglossoides, Rhododendron groenlandicum, Riccia fluitans, Rosa palustris, Rubus pubescens, Rumex orbiculatus, Sagittaria latifolia, Salix candida, S. eriocephala, S. myricoides, S. petiolaris, S. lucida, Sanicula odorata, Sarracenia purpurea, Schizachyrium scoparium, Schoenoplectus acutus, S. tabernaemontani, Scirpus microcarpus, S. pedicellatus, Scutellaria galericulata, S. lateriflora, Silphium integrifolium, S. terebinthinaceum, Sium suave, Solanum dulcamara, S. nigrum, Solidago ohioensis, S. ptarmicoides, S. ridellii, S. rigida, Sorghastrum nutans, Sparganium eurycarpum, Spartina pectinata, Sphagnum, Spiraea alba, Stellaria longifolia, Symphyotrichum ericoides, S. novae-angliae, S. puniceum, Symplocarpus foetidus, Thalictrum pubescens, Thuja occidentalis, Thelypteris palustris, Toxicodendron vernix, Triadenum fraseri, Trientalis borealis, Typha angustifolia, T. latifolia, Ulmus americana, Vaccinium macrocarpon, V. oxycoccus, Valeriana edulis, Verbena hastata, Viola cuculata, Vitis riparia.*

Use by wildlife: Plants of *Lysimachia asperulfolia* are associated with the Saint Francis' satyr butterfly (Nymphalidae: *Neonympha mitchellii*). *Lysimachia hybrida* is susceptible to aphids (Insecta: Homoptera) and is a nectar plant of the federally (USA) endangered Karner blue butterfly (Lepidoptera: Lycaenidae: *Plebejus melissa samuelis*). *Lysimachia maritima* is an important feature of the breeding habitat for the Wilson's phalarope (Aves: Scolopacidae: *Phalaropus tricolor*) and is a host plant for several fungi (Ascomycota: Pleosporaceae: *Pleospora herbarum*; Basidiomycota: Pucciniaceae: *Puccinia glaucis, P. distichlidis*). *Lysimachia nummularia* is the larval host plant for some Lepidoptera (Noctuidae: *Catocala consors*; Tortricidae: *Epiblema discretivana*). Its foliage contains saponins and tannins, which may deter herbivory. However, it is grazed infrequently by rabbits (Mammalia: Leporidae) and groundhogs (Mammalia: Sciuridae: *Marmota monax*) and its leaves are eaten by the woodland jumping mouse (Mammalia: Dipodidae: *Napaeozapus insignis*). Pollen from the flowers of *L. quadriflora* is collected by short-tongued bees (Hymenoptera: Halictidae: *Lasioglossum versatus*) and the pollen and oils by mellitid bees (Hymenoptera: Melittidae: *Macropsis steironematis*). *Lysimachia terrestris* is a larval host plant for the loosestrife borer (Lepidoptera: Noctuidae: *Papaipema lysimachiae*) and fruitworm (Lepidoptera: Tortricidae: *Sparganothis sulphureana*). The flowers attract adults of the pink-edged sulfur butterfly (Lepidoptera: Pieridae: *Colias interior*). The plants are also host to a fungus (Ascomycota: Venturiaceae: *Fusicladium lysimachiae*).

Economic importance: food: The fleshy roots of *Lysimachia maritima* were eaten by the Coast, Kwakiutl, Salish, and Southern tribes. The young stems and leaves have also been pickled and eaten. A tea has been made from the leaves and flowers of *L. nummularia*; **medicinal:** The Iroquois included *L. thyrsiflora* in a compound decoction that was used as a wash or applied as a poultice to cease lactation. The Kwakiutl boiled and ate the roots of *L. maritima* as a sedative and to induce sleep. A tea from the plant was used by some North American tribes to induce lactation. An infusion of *L. nummularia* has been used to treat diarrhea, internal bleeding, and wounds. The plants are mildly astringent, diuretic, and have been used to prevent scurvy. The cytotoxic benzoquinone embelin has been isolated from the underground parts of *L. punctata*. *Lysimachia thyrsiflora* has been made into a sedative tonic and the juice has been used to treat skin irritations; **cultivation:** *Lysimachia hybrida* is grown as a garden ornamental with 'Snow Candle' as one cultivar. *Lysimachia nummularia* is a popular garden plant with its creeping habit and large flowers. The cultivar 'Aurea' has earned the Royal Horticultural Society's Award of Garden Merit. Other cultivars include 'Alexander,' 'Creeping Jenny,' and 'Fire Cracker.' This species occasionally is grown as a submersed aquarium plant, sometimes under the name of '*Lloydiella*.' *Lysimachia punctata* is a very popular garden ornamental. The cultivars include 'Alexander' (variegated), 'Gaulthier Brousse,' 'Walgoldalex,' 'Golden Glory,' 'Ivy Maclean,' 'Senior,' 'Snow Lady,' and 'Sunspot.' *Lysimachia terrestris* and *L. thyrsiflora* are cultivated as marginal plants for water gardens; **misc. products:** *Lysimachia nummularia* and *L. radicans* can be planted as a ground cover for wet sites. Plants of *L. punctata* are burned on barbecues to drive away flies and gnats. The juice of *L. thyrsiflora* has been used to bleach hair; **weeds:** *Lysimachia* is known commonly as "loosestrife," but it should not be confused with its unrelated namesake the "purple loosestrife" (Lythraceae: *Lythrum salicaria*). Several lysimachias are also weedy. *Lysimachia nummularia* is a weed of fields, gardens, and lawns where soil moisture is adequate and it often is regarded as invasive. *Lysimachia punctata* sometimes can become weedy in gardens. *Lysimachia terrestris* is a weed of cranberry fields and of natural communities in British Columbia, Oregon, and Washington; **nonindigenous species:** *Lysimachia nummularia* was introduced from Europe to eastern North America in the 19th century (around 1882) as an ornamental and continues to be a popular garden plant. The European *L. punctata* was also introduced to North America as an escape from cultivation and reached Illinois between 1922 and 1955. It has also been introduced to the British Isles, Poland, and other parts of Europe where it was not originally native. *Lysimachia terrestris* was introduced to the Pacific Northwest through cranberry culture and *L. thyrsiflora* to the same region probably as a garden escape. *Lysimachia terrestris* has also been introduced to the British Isles.

Systematics: Phylogenetic analyses (e.g., Figure 5.12) consistently place *Lysimachia*, once classified among Primulaceae, within the family Myrsinaceae. Combined DNA (nrITS, *trn*L-F) sequence data indicate that the core *Lysimachia* species probably are monophyletic, but only if *Glaux maritima* is included (Figure 5.13). A more comprehensive analysis of

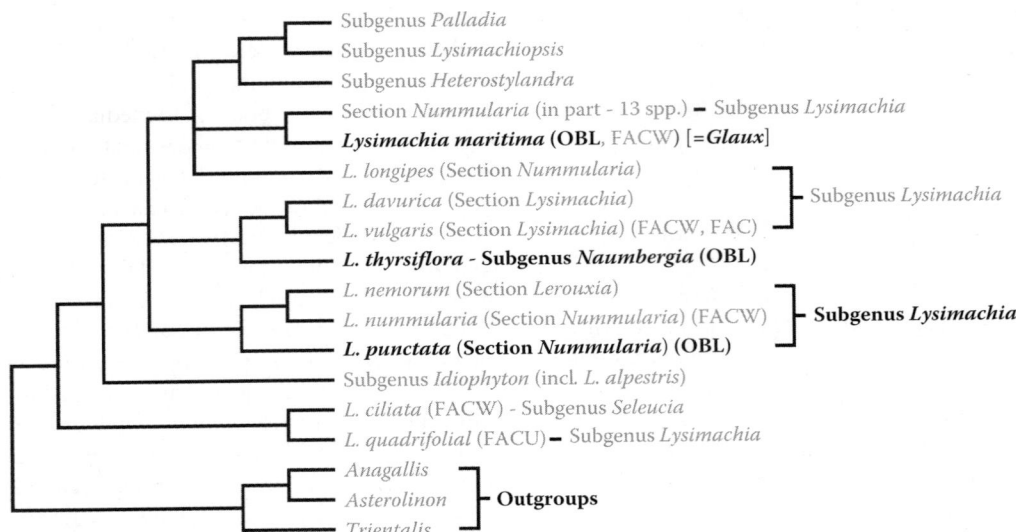

FIGURE 5.13 Phylogenetic relationships in *Lysimachia* as indicated by analysis of combined *trn*L-F and nrITS sequence data. The polyphyly of subgenus *Lysimachia* and section *Nummularia* is clearly indicated by these results. These relationships indicate that *Glaux* should be merged with *Lysimachia*, wherein it is rather deeply nested. Several independent origins of obligate aquatic North American species (indicated in bold) are likely to have occurred within the family. (Adapted from Hao, G. et al., *Mol. Phylogen. Evol.*, 31, 323–339, 2004.)

taxa based on *ndhF* sequence data also supports the inclusion of *Glaux* within *Lysimachia*, but the lack of resolution along the "backbone" of the tree does not allow for a definitive conclusion regarding the monophyly of *Lysimachia* with respect to *Anagallis*, *Trientalis*, and several other genera. Traditionally, *Lysimachia* has been divided into 5–8 subgenera. Most of the native North American species have been placed into four subgenera: *Lysimachia* (section *Verticillatae*; 6 spp.), *Naumburgia* (1 sp.), *Palladia* (section *Lubinia*; 1 sp.), and *Seleucia* (7 spp.). Although several of the subgenera appear to be natural, phylogenetic analyses of morphological and molecular data thoroughly demonstrate that subg. *Lysimachia* is polyphyletic and artificial, representing species in as many as five different lineages. Subgenera *Naumburgia* and *Seleucia* each associate with elements of subgenus *Lysimachia*. *Lysimachia maritima* (formerly the monotypic genus *Glaux*) resolves among various species assigned to section *Nummularia*, which itself is polyphyletic (Figure 5.13). *Lysimachia thyrsiflora* (subg. *Naumburgia*) is closely related to *L. davurica* and *L. vulgaris* (section *Lysimachia*) (Figure 5.13). Two species of section *Nummularia* that were introduced from the Old World (*L. nummularia*, *L. punctata*) are closely related and are sister to *L. nemorum* (section *Lerouxia*) (Figure 5.13). Relationships among the remaining *Lysimachia* species found in North America await studies that include a broader sampling of taxa. Using morphological criteria, some authors have suggested that *L. hybrida* is closely related to (and possibly conspecific with) *L. lanceolata*. The base chromosome number of *Lysimachia* is *x* = 15 or 17. *Lysimachia hybrida*, *L. quadriflora*, and *L. radicans* (2*n* = 34) are diploid (*x* = 17). *Lysimachia maritima* (2*n* = 30), *L. nummularia* (2*n* = 30, 32, 36, 43–45), and *L. punctata* (2*n* = 30) are diploid (*x* = 15), with triploid and aneuploid cytotypes also reported in *L. nummularia*. *Lysimachia asperulifolia* (2*n* = 42),

L. loomisii (2*n* = 42), *L. thyrsiflora* (2*n* = 40, 42, 54), and *L. terrestris* (2*n* = 84) are anomalous polyploids. There is a propensity for hybridization in the genus, especially among species within a particular ploidy level. Fertile hybrids have been synthesized by crossing *L. hybrida* with *L. cilata*, *L. graminea*, *L. lanceolata*, *L. quadriflora*, *L. radicans*, and *L. tonsa*; likewise, *L. quadriflora* has been crossed successfully with the same species. *Lysimachia ×producta* allegedly is a natural hybrid between *L. terrestris* and *L. quadrifolia*. *Lysimachia terrestris* also reportedly hybridizes with *L. thyrsiflora*. Hybridization has also been reported between *L. loomisii* and *L. quadrifolia*. Hybridization in natural populations appears to be prevented primarily by differences in floral phenology and by habitat isolation. Population genetic studies have been carried out for several Old World species, but are wanting for those of the New World.

Comments: *Lysimachia hybrida* is widespread in North America. *Lysimachia thyrsiflora* is circumpolar and widespread in northern North America; *L. maritima* is also circumpolar and widespread in northern and western North America. *Lysimachia quadriflora* occurs in eastern North America and *L. radicans* in the southeastern United States. The introduced *L. nummularia* and *L. punctata* are disjunct in eastern and western North America. Two species are restricted in the southeastern United States: *L. asperulifolia* (NC, SC) and *L. loomisii* (GA, NC, SC).

References: Anderberg et al., 2007; Andersson et al., 2000; Aronson, 1989; Arthur, 1910; Brotherson & Barnes, 1984; Chang et al., 2005; Choesin & Boerner, 2002; Cholewa, 2009a; Coffey & Jones, Jr., 1980; Cooke, 1997; Dahlgren, 1922; Darwin, 1892; Dodd & Coupland, 1966; Ewanchuk & Bertness, 2003; Fernald & Kinsey, 1943; Franklin, 2001; Frantz, 1995; Gaudet & Keddy, 1995; Grandin, 1971; Hao et al., 2004; Hughes & Cass, 1997; Jerling, 1988a,b;

Keddy & Reznicek, 1982; Lahring, 2003; Lindsey, 1953; Maki & Galatowitsch, 2004; Michez & Patiny, 2005; Milton, 1939; Muenscher, 1955; Nekola, 2004; Petty & Petty, 2005; Podolak et al., 2005; Ray, 1956; Rozema & Riphagen, 1977; Rozema et al., 1986; Salisbury, 1978; Schubert & Braun, 2005; Simpson et al., 1983; Speichert & Speichert, 2004; Stewart et al., 1973; Timoney, 2001; Van Strien et al., 1989; Vince & Snow, 1984; Voss, 1954; Whitaker, Jr., 1963; Wolters & Bakker, 2002; Zika, 2003.

3. Trientalis

Starflower; trientale

Etymology: from the Latin *trientalis* meaning "a third of a foot," alluding to the height of the plant

Synonyms: none

Distribution: global: Asia; Europe; North America; **North America:** E/W disjunct

Diversity: global: 2 species; **North America:** 2 species

Indicators (USA): FAC; FACU: *Trientalis europaea*

Habitat: freshwater; palustrine; **pH:** 3.7–6.7; **depth:** <1 m; **life-form(s):** emergent herb

Key morphology: stems (to 35 cm) slender, arising from tubers (to 1 cm); a few alternate and reduced (to 4 cm) leaves along the stem, the uppermost larger (to 8 cm) and whorled (4–6), short petioled (1–4 mm); flowers (to 1.6 cm) stellate, arising from the terminal leaf whorl in a cluster of 1–3; petals (to 7 mm) 5–9, white to pinkish, fused basally; capsules (to 2 mm) globose, with 1–18 seeds

Life history: duration: perennial (tubers); **asexual reproduction:** rhizomes, tubers; **pollination:** insect or self; **sexual condition:** hermaphroditic; **fruit:** capsules (infrequent); **local dispersal:** tubers; **long-distance dispersal:** seeds

Imperilment: (1) *Trientalis europaea* [G5]; S1 (CA, SK); S3 (AB, ID)

Ecology: general: *Trientalis* species all occur facultatively in wet areas; however, none of the North American species attains status as an obligate aquatic. *Trientalis europaea* was classified as an OBL in a portion of its range in the 1996 wetland indicator list. Although no longer ranked as OBL, it has been included here because the treatment was completed prior to the release of the 2013 indicator list.

Trientalis europaea **L.** inhabits bogs, forests, meadows, muskeg, pingos (ice mounds), streambanks, swamps, and tundra, typically at elevations below 10 m, but occasionally as high as 1500 m. This species is a calcifuge and requires acidic (pH: 3.7–6.7), infertile (nitrogen-poor), sites with a humic or peat substrate. The plants usually occur in partial shade in cool to arctic or alpine sites. The flowers are weakly protogynous and produce dimorphic (differently sized) pollen. They are self-compatible and either are insect pollinated and outcrossed or self-pollinated. The petal number (usually 7), varies with respect to the habitat conditions. Pollinators have been observed rarely on the flowers and represent a rather diverse, unspecialized insect fauna (Diptera, Coleoptera). Seed set in experimentally selfed and outcrossed flowers is similar (26%). Plants frequently are devoid of seed. In one study, plants produced two capsules, each with only eight seeds on average. Seeds are retained within the capsule by a reticulate membrane where they can persist until the following spring. Fresh seeds can germinate immediately if contacting moist substrates before becoming dormant. Germination is slow and progressive. Rates observed for unstratified seeds (at 20°C in diffuse light or darkness) are fair (42%) and increase moderately (56%) for seeds that are cold stratified for 18 weeks. After 14 months of observation, germination rates ranged from 75% to 87% and averaged 78%. Seeds remain viable for 1–5 years and can persist in the seed bank. Overall, seedling recruitment is impacted by low seed output and limited microsite availability. Vegetative reproduction occurs by means of tuber-forming rhizomes. When nutrient conditions are low, rhizomes elongate and produce few, larger tubers; at higher nutrient levels, rhizomes are short and branched, and produce numerous, small tubers. The shoots are short lived and the tubers separate from the decaying rhizomes in late summer. During the subsequent growing season, they produce 1–5 stolon-like rhizomes that extend 2–5 cm below the soil surface, each tip eventually forming a new tuber. Seedling establishment is almost entirely dependent on disturbance and rarely occurs in undisturbed environments. This life-history is described as "pseudoannual" because most plants reestablished annually from the tubers rather than by seed. The plants often appear following burns. Plants achieve their maximum productivity at temperatures of 11°C–18°C and moderate light levels. Although plants readily form abundant arbuscular mycorrhizal structures when inoculated, their growth rate does not appear to be influenced by their presence. A population genetic study would be helpful in evaluating the clonal nature of this species. **Reported associates:** *Abies amabilis, Acer circinatum, Aconitum delphiniifolium, Agrostis aequivalis, A. exarata, A. oregonensis, A. scabra, A. thurberiana, Alnus viridis, Andromeda polifolia, Anemone oregana, Angelica lucida, Arctagrostis latifolia, Argentina anserina, Artemisia, Athyrium filix-femina, Aulacomnium palustre, Bazzania trilobata, Betula papyrifera, Bistorta officinalis, Blechnum spicant, Boykinia intermedia, B. occidentalis, Calamagrostis canadensis, Caltha leptosepala, Calypogeja sphagnicola, C. nutkaensis, C. stricta, Camassia, Carex aquatilis, C. canescens, C. cusickii, C. echinata, C. interior, C. leptalea, C. livida, C. luzulina, C. obnupta, C. pauciflora, C. pluriflora, C. utriculata, Chamaecyparis nootkatensis, Chamerion angustifolium, Chimaphila umbellata, Cladina portentosa, C. rangiferina, Cladonia, Comarum palustre, Coptis aspleniifolia, C. laciniata, C. trifolia, Cornus unalaschkensis, C. suecica, Danthonia spicata, Deschampsia cespitosa, Dicranum majus, Drosera anglica, D. rotundifolia, Dryopteris expansa, Empetrum nigrum, Equisetum arvense, E. sylvaticum, Eriophorum angustifolium, E. chamissonis, E. gracile, Frangula purshiana, Galium boreale, G. trifidum, Gaultheria shallon, Gentiana douglasiana, G. sceptrum, Geranium, Geum calthifolium, Goodyera oblongifolia, Gymnocarpium dryopteris, Hylocomium splendens, Hypericum anagalloides, Hypnum callichroum, Iris setosa, Juncus acuminatus, J. balticus, J. ensifolius, J. nevadensis, J. supiniformis, Juniperus communis, Kalmia microphylla, Linnaea borealis, Listera cordata,*

Luzula, Lycopodium alpinum, L. annotinum, L. clavatum, Lycopus uniflorus, Lysichiton americanus, Mahonia nervosa, Maianthemum dilatatum, Malus fusca, Menyanthes trifoliata, Menziesia ferruginea, Mertensia paniculata, Micranthes nelsoniana, Moneses uniflora, Mylia anomala, Myrica californica, M. gale, Nephrophyllidium crista-galli, Oenanthe sarmentosa, Packera cymbalarioides, Panicum occidentale, Parnassia fimbriata, P. palustris, Pedicularis groenlandica, Pellia neesiana, Petasites frigidus, P. hyperboreus, Pinus contorta, P. monticola, Plagiochila porelloides, Plagiothecium undulatum, Plantago macrocarpa, Plantanthera dilatata, Pleurozium schreberi, Poa arctica, Polemonium acutiflorum, Polystichum munitum, Polytrichum, Primula jeffreyi, Prunella vulgaris, Pseudotsuga menziesii, Pteridium aquilinum, Ptilium crista-castrensis, Pyrola asarifolia, P. secunda, Racomitrium lanuginosum, Ranunculus flammula, R. lapponicus, Rhodiola integrifolia, R. rosea, Rhododendron macrophyllum, R. groenlandicum, R. neoglandulosum, Rhynchospora alba, Rhytidiadelphus loreus, Rhizomnium glabrescens, Romanzoffia, Rubus articus, R. chamaemorus, R. nivalis, R. pedatus, R. spectabilis, R. ursinus, Rumex, Salix bebbiana, S. commutata, S. monticola, Sambucus racemosa, Sanguisorba menziesii, S. officinalis, S. stipulata, Saxifraga bronchialis, Scapania undulata, Scheuchzeria palustris, Senecio triangularis, Silene acaulis, Siphula ceratites, Solidago multiradiata, Sphagnum austinii, S. capillifolium, S. compactum, S. fuscum, S. lindbergii, S. magellanicum, S. pacificum, S. palustre, S. papillosum, S. recurvum, S. rubellum, S. tenellum, Spiraea douglasii, S. stevenii, Spiranthes romanzoffiana, Stellaria crassifolia, S. crispa, Streptopus amplexifolius, S. lanceolatus, S. streptopoides, Swertia perennis, Taxus brevifolia, Thalictrum alpinum, T. occidentale, T. sparsiflorum, Thelypteris phegopteris, Thuja plicata, Tiarella trifoliata, T. unifoliata, Tofieldia coccinea, T. pusilla, Tolmiea menziesii, Triantha glutinosa, Trichophorum cespitosum, Trisetum canescens, Tsuga heterophylla, T. mertensiana, Typha latifolia, Vaccinium alaskaense, V. cespitosum, V. ovalifolium, V. ovatum, V. oxycoccos, V. parvifolium, V. uliginosum, V. vitis-ideae, Valeriana capitata, V. sitchensis, Veratrum viride, Veronica scutellata, Viola epipsila, V. glabella, V. langsdorfi, V. palustris, V. sempervirens, Whipplea modesta, Woodsia ilvensis, Xerophyllum tenax.

Use by wildlife: *Trientalis europaea* is a host for several fungi (Ascomycota: Mycosphaerellaceae: *Septoria increscens*; Basidiomycota: Pucciniaceae: *Puccinia caricina*). Details on herbivory of North American populations are scarce; however, the plants are eaten by slugs, scale insects (Hemiptera) and small rodents (mice, voles) in their European range.

Economic importance: food: The tubers of *Trientalis europaea* may have been used as food (as they were in the related *T. latifolia*) by some native North American tribes; **medicinal:** The roots of *Trientalis europaea* have emetic properties and the plants were made into a soothing ointment for treating wounds. An infusion was used to treat blood poisoning and eczema; **cultivation:** *Trientalis europaea* and the variety *rosea* are cultivated occasionally; **misc. products:** none; **weeds:** none; **nonindigenous species:** none

Systematics: Molecular phylogenetic analyses resolve *Trientalis* within the Myrsinaceae near *Anagallis* and *Asterolinon* (Figures 5.12 and 5.13). Although the genus contains few taxa, a comprehensive phylogenetic study of the two or three recognized species and their subspecies has not yet been carried out to confirm the taxonomic disposition used at present; however, studies including *T. borealis* and *T. europaea* indicate strong support for the monophyly of the genus but do not resolve whether the group should be merged with *Lysimachia*. *Trientalis latifolia* is recognized either as a distinct species or as a subspecies of *T. borealis*. Similarly, *T. arctica* sometimes is treated as distinct, but more often as a subspecies of *Trientalis europaea*. It is this taxon (as *T. europaea* subspecies *arctica*) that was once given OBL status in North America; whereas, the typical subspecies was ranked as FAC. The base chromosome number of *Trientalis* is reported as $x = 35$ or 43. *Trientalis europaea* exhibits numerous aneuploid/polyploid races that include $2n = 84, 90, 100, 110, 112, 128, 130, 160, 170$. There are no hybrids reported in the genus.

Comments: *Trientalis europaea* is nearly circumboreal but occurs only in western North America.

References: Anderberg et al., 2007; Anderson & Beare, 1983; Anderson & Loucks, 1973; Asada et al., 2003; Christy, 2004; Cooke, 1997; Gashwiler, 1970; Mead, 1998, 2000; Piqueras, 1999; Robuck, 1985; Ruotsalainen & Aikio, 2004; Sigafoos, 1951; Taylor et al., 2002; Tikhodeev & Tikhodeeva, 2001.

Family 4: Polemoniaceae [26]

Although Polemoniaceae were classified formerly within the Solanales, a substantial amount of molecular data consistently resolves the family within the order Ericales as the sister group to Foquieriaceae (Figure 5.5). This medium-sized family of about 400 species consists primarily of herbaceous plants (although woody species comprise subfamily Cobaeoideae), which possess showy tubular flowers with a nectary disk and three carpels. The flowers are pollinated by a wide range of vectors including various bats (Mammalia: Chiroptera), birds (Aves), and insects (Insecta). The seeds often become mucilaginous when wet, which enables them to adhere to animal dispersal agents. The small seeds are also dispersed by water or wind (Judd et al., 2002).

A number of species are grown as ornamental horticultural specimens, especially in the genera *Cobaea, Gilia, Ipomopsis, Linanthus, Phlox,* and *Polemonium*.

The greatest diversity of species occurs within temperate western North America. Most species are terrestrial and obligate aquatics occur only within one North American genus (*Navarretia*). One species of *Polemonium* was ranked as OBL in the 1996 indicator list but has since been recategorized as FACW. The treatment has been retained because it was completed prior to the release of the revised 2013 list:

1. ***Navarretia*** Ruiz & Pav.
2. ***Polemonium*** L.

1. *Navarretia*

Pincushionplant

Etymology: after Francisco Fernandez Navarrete (ca. 1680–1742)

Synonyms: *Aegochloa*

Distribution: global: North America; South America; **North America:** western

Diversity: global: 32 species; **North America:** 31 species

Indicators (USA): OBL: *Navarretia cotulifolia, N. fossalis, N. heterandra, N. leucocephala, N. myersii, N. prostrata*

Habitat: freshwater; palustrine; **pH:** 5.8–8.3; **depth:** <1 m; **life-form(s):** emergent herb

Key morphology: stems (to 30 cm) erect, prostrate or spreading, the herbage pubescent, often proliferating radially and terminating in flower clusters; leaves alternate, awliform (juvenile) or pinnate to bipinnate when mature (to 9 cm); bracts (to 1.7 cm) foliar, spiny, pinnate or bipinnate; flowers 4–5-merous, subsessile, in dense heads (2–50 flowers); calyx (to 9 mm) spiny tipped; corolla (to 21 mm) funnelform, blue, cream, white (sometimes with purple spots), violet, or yellow; capsules (to 2.5 mm) dehiscent or indehiscent, seeds 1-many

Life history: duration: annual (fruit/seeds); **asexual reproduction:** none; **pollination:** insect, self; **sexual condition:** hermaphroditic; **fruit:** capsules (common); **local dispersal:** seeds (water); **long-distance dispersal:** seeds (animals)

Imperilment: (1) *Navarretia cotulifolia* [G3]; S3 (CA); (2) *N. fossalis* [G2]; S2 (CA); (3) *N. heterandra* [G3]; S1 (OR); S3 (CA); (4) *N. leucocephala* [G4]; S1 (CA); S3 (<u>AB</u>, <u>SK</u>); (5) *N. myersii* [G1]; S1 (CA); (6) *N. prostrata* [G2]; S2 (CA)

Ecology: general: A little more than half of the *Navarretia* species (19) are assigned at some level as wetland indicators, with six species (~19%) designated as OBL. Most of the species occupy open sites on shallow clay or rocky soils and a few are serpentine endemics. The aquatic species inhabit vernal pools or similar sites characterized by ephemeral water conditions. All are annuals and rely exclusively on seed production for their reestablishment. The seeds usually become gelatinous when wet, which facilitates their adherence to animals as a potential means of long-range dispersal. The seeds of most species germinate within 3 days following a two week period of stratification at 2°C–4°C in distilled water and low light, followed by a 16/8 h 21°C/15°C light and temperature regime with a minimum relative humidity of 45%. The vernal pool specialists produce more pollen per flower, have higher pollen:ovule ratios, and possess broader corollas than their counterparts. The plants grow slowly while submerged, until their leaf tips come into contact with the air as they reach the surface.

Navarretia cotulifolia **(Benth.) Hook. & Arn.** occurs in depressions, grasslands, meadows, plains, roadsides, sloughs, and vernal pools at elevations of up to 426 m. Substrates are described as alkaline, heavy adobe soils. Flowering occurs from May to June. The capsules are upwardly dehiscent and contain a single seed. Detailed ecological information virtually is non-existent for this species. **Reported associates:** none.

Navarretia fossalis **Moran** grows in depressions, ditches, floodplains, mudflats, swales, and vernal pools at elevations of up to 1300 m. The plants occur on saline/alkali ground or on thin soils underlain by acidic rock (pH: 5.8–8.2). Substrates include clay, clay loam, sandy clay, sandy loam, or silty clay, which either are weakly cemented or hardpan (iron-silica cemented). Flowering occurs from April to June. The floral features indicate that the plants are adapted for self-pollination. The capsules are indehiscent and contain 5–25 seeds, which are released as they take up water (e.g., following heavy rains) and expand to rupture the fruit wall. The seeds become mucilaginous when wet. The plants deteriorate rapidly following seed set, making it difficult to inventory populations. Also, population size is proportional to rainfall and dwindles in drought years. An isoetid growth form (which may facilitate carbon uptake) is induced in juvenile plants when they are grown under submerged conditions. This species is not narrowly specialized to vernal pools and often is found in other wet habitats. However, it is very rare in California with fewer than 30 known populations concentrated in only three main sites (1998 data). The populations are threatened by habitat loss, urbanization, agriculture, and damage from off-road vehicles. **Reported associates:** *Alopecurus saccatus, Ambrosia pumila, Anagallis minima, Anthemis cotula, Atriplex argentea, A. coronata, Avena barbata, A. fatua, Bassia hyssopifolia, Bergia texana, Boisduvalia densiflora, Branchinecta sandiegonensis, Brodiaea filifolia, B. jolonensis, B. orcuttii, Callitriche marginata, Crassula aquatica, Cressa truxillensis, Crypsis schoenoides, Deschampsia danthoniodes, Downingia cuspidata, Echinodorus berteroi, Elatine brachysperma, E. californica, Eleocharis acicularis, E. macrostachya, Epilobium pygmaeum, Erodium cicutarium, Eryngium aristulatum, Frankenia grandiflora, F. salina, Helianthus annuus, Hemizonia fasciculata, H. parryi, H. pungens, Hirschfeldia incana, Hordeum depressum, H. intercedens, Isoetes howellii, I. orcuttii, Juncus bufonius, Lasthenia californica, L. glabrata, Lepidium dictyotum, L. latipes, L. nitidum, Lilaea scilloides, Lolium perenne, Lythrum hyssopifolia, Malvella leprosa, Marsilea vestita, Mimulus latidens, Montia fontana, Myosurus minimus, Nama stenocarpum, Nassella pulchra, Navarretia hamata, N. prostrata, Ophioglossum californicum, Orcuttia californica, Phalaris lemmonii, Phalaris minor, P. paradoxa, Pilularia americana, Plagiobothrys acanthocarpus, P. leptocladus, Plantago elongata, P. erecta, Pogogyne abramsii, P. nudiuscula, Polygonum argyrocoleon, P. aviculare, Psilocarphus brevissimus, P. tenellus, Rotala ramosior, Rumex crispus, R. martimus, R. persicariodes, Sida leprosa, Suaeda moquinii, Trichostema austromontanum, Trifolium depauperatum, Verbena bracteata, Veronica peregrina.*

Navarretia heterandra **H. Mason** inhabits depressions, drying flats, grasslands, slopes, swales, or vernal pools at elevations of up to 1081 m. The plants occur on thin but heavy soils having a high clay content and include Anita clay loam, Corning gravelly loam, gravelly mud, Olcott soil series, Peters clay, silty clay loam, stony Tuscan loam, Tuscan mudflow, Tuscan stony clay loam, and vertisols. Flowers are produced from May to June. The fruits dehisce from below and produce only 1–2 seeds. Otherwise, the reproductive biology of this species is poorly known. Occurrences range from 20 to 10,000 plants and average about 2,000 individuals at a given site. The best habitats exist at sites where little grass cover occurs. **Reported associates:** *Achyrachaena mollis, Acmispon humistratus, A. rubriflorus, A. wrangelianus, Aegilops, Ancistrocarphus filagineus, Athysanus pusillus, Avena, Brodiaea elegans, Bromus hordeaceus, Calochortus luteus, Ceanothus cuneatus, Centromadia fitchii, Cirsium, Clarkia affinis, Deschampsia danthonioides, Dichelostemma capitatum, Distichlis, Erodium, Eryngium castrense, Galium, Gastridium, Hesperevax caulescens, Hesperolinon, Hordeum marinum, Lasthenia, Layia fremontii, Logfia filaginoides, Madia exigua, Medicago polymorpha, Micropus californicus, Monardella douglasii, Navarretia eriocephala, N. intertexta, N. leucocephala, N. nigelliformis, N. pubescens, N. tagetina, Odontostomum hartwegii, Plagiobothrys scriptus, Plantago erecta, Pogogyne zizyphoroides, Psilocarphus tenellus, Sherardia arvensis, Sidalcea diploscypha, Taeniatherum caput-medusae, Tetrapteron graciliflorum, Trichostema lanceolatum, Trifolium ciliolatum, T. depauperatum, T. hirtum, Vulpia, Zeltnera muehlenbergii.*

***Navarretia leucocephala* Benth**. inhabits channels, depressions, ditches, meadows, playas, seeps, swales, vernal pools, and lake margins at elevations of up to 2194 m. Substrates are acidic or alkaline (pH: 6.1–8.3) and include heavy clay (adobe), iron-silica hardpan, Riz soil series, sandy loam, silty loam, Tuscan (TuB) soil series, vertisols, and volcanic ash. The floral features indicate a primarily outcrossing breeding system. The flowers produce abundant nectar and attract honey bees (Hymenoptera: Apidae) and butterflies (Lepidoptera), which function as pollinators. The flower clusters frequently are also found to contain flowers of the parasitic vernal pool dodder (*Cuscuta howelliana*). The capsules are indehiscent and contain 1–8 seeds, which are released as they imbibe water (e.g., following heavy rains) and expand to rupture the fruit wall. Germination occurs under water. An isoetid growth form (which may facilitate carbon uptake) is induced in juvenile plants when grown under submerged conditions. In some instances, microdepressions created by the hooves of grazing animals may facilitate establishment of plants. This species attains its highest ecological importance in pools (delimited by the high water level maintained during periods of rain), with a much reduced occurrence outside of the pool zone. The plants deteriorate rapidly (3–4 weeks) after the ponded water recedes. Populations of *N. leucocephala* have been impacted by drainage, grazing, and rooting activities of feral pigs. **Reported associates:** *Achyrachaena mollis, Acmispon wrangelianus, Agrostis hendersonii, Airya caryophyllea, Allium geyeri, A. lemmonii, Alopecurus saccatus, Alyssum, Aristida oligantha, Artemesia cana, Avena barbata, Blennosperma nanum, Briza minor, Brodiaea minor, Bromus hordeaceus, B. mollis, Callitriche, Castilleja attenuata, C. campestris, Camissonia tanacetifolia, Chamaesyce hooveri, Cicendio quadrangularis, Clarkia purpurea, Cuscuta howelliana, Cyperus squarrosus, Deschampsia danthonioides, Distichlis spicata, Downingia bella, D. bicornuta, D. concolor, D. cuspidata, D. elegans, D. ornatissima, D. pulchella,*

D. pusilla, D. yina, Elatine heterandra, Eleocharis macrostachya, E. palustris, Elymus elymoides, Epilobium torreyi, Eremocarpus setigerus, Erodium botrys, Eryngium aristulatum, E. constancei, E. petiolatum, E. vaseyi, Erysimum repandum, Gratiola ebracteata, G. heterosepala, Hemizonia congesta, H. fitchii, Hesperevax caulescens, Hordeum marinum, Hypochaeris glabra, Isoetes howellii, I. nuttallii, Iva axillaris, Ivesia sericoleuca, Juncus balticus, J. bufonius, J. capitatus, J. hemiendytus, J. leiospermus, J. uncialis, Lasthenia burkei, L. californica, L. conjugens, L. fremontii, L. glaberrima, L. glabrata, Layia fremontii, Legenere limosa, Lepidium nitidum, Leymus triticoides, Lilaea scilloides, Limnanthes alba, L. floccosa, Limosella aquatica, Lindernia dubia, Lolium multiflorum, Lupinus bicolor, Lythrum hyssopifolium, Mimulus, Minuartia, Mollugo verticillata, Muhlenbergia richardsonis, Myosurus apetalus, M. clavicaulis, M. minimus, Nassella pulchra, Navarretia heterandra, H. myersii, N. nigelliformis, N. tagetina, Neostapfia colusana, Oenothera tanacetifolia, Orcuttia inaequalis, O. pilosa, O. tenuis, O. viscida, Orthocarpus campestris, O. erianthus, Parvisedum leiocarpum, Phalaris lemmonii, Pilularia americana, Pinus jeffreyi, Plagiobothrys austiniae, P. greenei, P. leptocladus, P. stipitatus, P. undulata, Pleuropogon californicus, Poa bulbosa, Pogogyne zizyphoroides, Polyctenium fremontii, Polygonum polygaloides, Polypogon maritimus, P. monspeliensis, Psilocarphus brevissimus, P. chilensis, Puccinellia rupestris, Rorippa curvisiliqua, Rotala ramosior, Rumex venosus, Sanguisorba, Stachys ajugoides, Taeniatherum caput-medusae, Trichostema lanceolatum, Trifolium depauperatum, T. hirtum, T. microcephalum, T. willdenovii, Triphysaria eriantha, Triteleia hyacinthina, Tuctoria greenei, Veronica arvensis, V. peregrina, Vulpia bromoides, V. myuros.

***Navarretia myersii* Allen & A.G. Day** occurs in swales and vernal pools at elevations of up to 96 m. Soils are acidic and described as clay loam, Corning series, hardpan, or Redding series. The reproductive biology of this species has not been studied but presumably is similar to the closely related *N. leucocephala*. The plants flower in April to May. The floral features indicate adaptations for outcrossing. The capsules are indehiscent and contain 2–4 seeds, which are released as they take up water (e.g., following heavy rains) and expand to rupture the fruit wall. The wet seeds are adhesive. Juvenile plants of *N. myersii* consistently have an isoetid growth form, which may facilitate carbon uptake. Years that experience seasonally high rainfalls will yield plants with larger leaves and more flowers than in dry years. **Reported associates:** *Castilleja campestris, Downingia bicornuta, D. concolor, Eleocharis macrostachya, Eryngium aristulatum, E. castrense, E. vaseyi, Gratiola ebracteata, G. heterosepala, Isoetes howelii, Juncus bufonius, J. leiospermus, Lasthenia, Legenere limosa, Lilaea scilloides, Lythrum, Marsilea vestida, Mimulus tricolor, Navarretia leucocephala, Orcuttia viscida, Plagiobothrys stipitatus, Pogogyne douglasii, P. ziziphoroides, Polypogon maritimus, Psilocarphus brevissimus, Trichostema lanceolatum.*

***Navarretia prostrata* (A. Gray) Greene** occurs in depressions, floodplains, swales, and vernal pools at elevations of up to 701 m. Substrates range from acidic to alkaline (pH: 4.5–8.2) and include adobe, clay, and gravel over clay. Floral features of the plants indicate a primarily outcrossing breeding system. The capsules are indehiscent and contain 5–25 seeds, which are released as they take up water (e.g., following heavy rains) and expand to rupture the fruit wall. As in other species, the seeds become mucilaginous when wet. Additional information on the reproductive biology and seed ecology of this species is needed. **Reported associates:** *Allium haematochiton, Anagallis minima, Alopecurus saccatus, Blennosperma nanum, Brodiaea filifolia, B. orcuttii, B. terrestris, Callitriche heterophylla, C. longipedunculata, C. marginata, Calochortus albus, C. luteus, Chlorogalum purpureum, Cotula coronopifolia, Crassula aquatica, Cressa truxillensis, Crypsis schoenoides, C. vaginiflora, Deschampsia danthonioides, Downingia bella, D. bicornuta, D. cuspidata, D. ornatissima, D. pulchella, Elatine brachysperma, E. californica, E. chilensis, Eleocharis acicularis, E. macrostachya, E. montevidensis, Epilobium densiflorum, E. pygmaeum, Eryngium aristulatum, Euphorbia spathulata, Fritillaria biflora, Hesperolinon micranthum, Holocarpha virgata, Hordeum intercedens, H. marinum, Isoetes howellii, I. orcuttii, Juncus bufonius, Lasthenia glaberrima, Lepidium nitidum, Limnanthes douglasii, L. gracilis, Lolium perenne, Lomatium utriculatum, Lythrum hyssopifolia, Malvella leprosa, Marsilea vestita, Mimulus guttatus, M. tricolor, Monardella hypoleuca, Montia fontana, Myosurus minimus, Navarretia hamata, N. intertexta, N. prostrata, Orcuttia californica, Penstemon heterophyllus, Persicaria lapathifolia, Phalaris caroliniana, Pilularia americana, Plagiobothrys acanthocarpus, P. fulvus, P. humistratus, P. leptocladus, P. undulatus, Plantago elongata, Psilocarphus brevissimus, P. tenellus, Puccinellia simplex, Ranunculus aquatilis, R. occidentalis, Stebbinsoseris heterocarpa, Verbena bracteata, Veronica peregrina, Vulpia bromoides.*

Use by wildlife: Flowers of *Navarretia leucocephala* attract the painted lady butterfly (Insecta: Lepidoptera: Nymphalidae: *Vanessa cardui*) and common honey bee (Insecta: Hymenoptera: Apidae: *Apis mellifera*).

Economic importance: food: The Miwok tribe used the dried, pulverized seeds of *Navarretia* for food; **medicinal:** The Miwok boiled plants of *Navarretia cotulifolia* and applied the decoction to alleviate swelling; **cultivation:** not cultivated; **misc. products:** none; **weeds:** none; **nonindigenous species:** none

Systematics: Phylogenetic analyses place *Navarretia* as the sister genus to *Collomia* within subfamily Polemonioideae of Polemoniaceae (Figure 5.14). Four sections have been recognized in *Navarretia*, but only section *Mitracarpium* (including *N. cotulifolia* and *N. heterandra*) and section *Navarretia* (including *N. fossalis, N. leucocephala, N. myersii,* and *N. prostrata*) appear to be monophyletic (Figure 5.15). The currently recognized subspecies of *N. leucocephala* (*N. l. bakeri, N. l. leucocephala, N. l. minima, N. l. pauciflora, N. l. plientha*) all have been recognized as distinct species in older treatments. Neither DNA nor morphological data

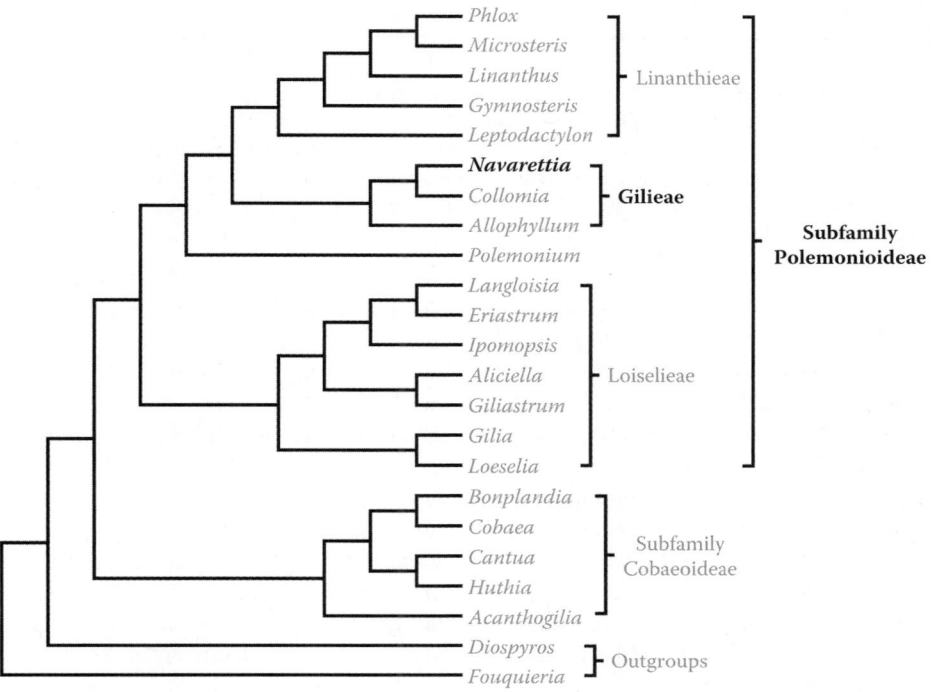

FIGURE 5.14 Relationships among genera of Polemoniaceae as indicated by phylogenetic analysis of *ndhF* sequence data. The results differ in some details but convey the general details of relationships as indicated by other molecular data sets, perhaps with exception of the placement of *Gilia*, a result of different species sampling (e.g., Steele & Vilgalys, 1994; Johnson et al., 1996). *Navarretia*, the only genus of the family to contain OBL species in North America (bold), occurs in a relatively derived position in the family. (Adapted from Prather, L. A. et al., *Amer. J. Bot.,* 87, 1300–1308, 2000.)

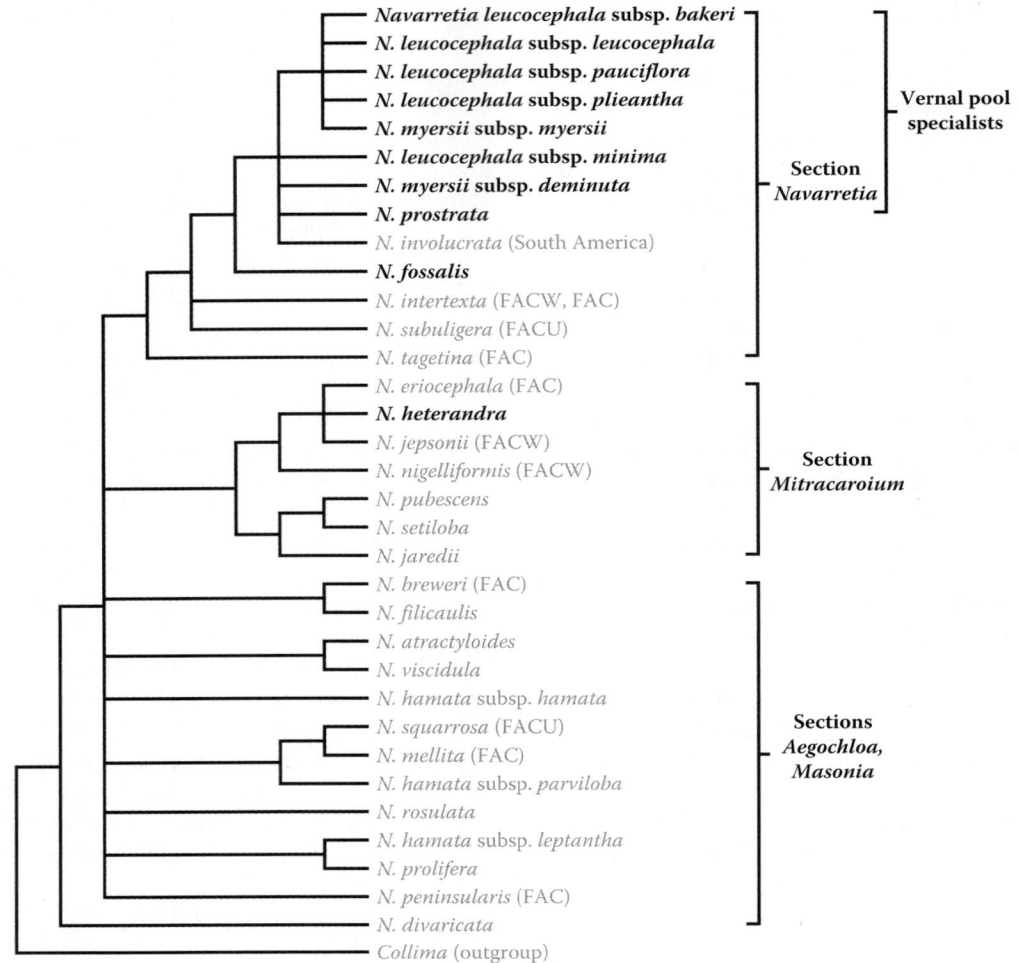

FIGURE 5.15 Interspecific relationships in *Navarretia* as indicated by phylogenetic analysis of nrITS sequence data. These results indicate that there were at least two separate origins of OBL North American species (indicated in bold). The South American *N. involucrata* is similar to *N. fossalis* ecologically. Analysis of morphological data in the same study (not shown) placed *N. cotulifolia* at the base of section *Mitracarpium* (not with *N. heterandra*), perhaps indicating a third origin of the OBL habit; however, this possibility should be tested further using molecular data. The vernal pool specialists are highly derived in the genus while FACW species occur throughout the group. (Adapted from Spencer, S.C. & Porter, J.M., *Syst. Bot.*, 22, 649–668, 1997.)

clearly distinguish the infraspecific taxa of *N. leucocephala* or *N. myersii* as monophyletic and studies using more sensitive genetic markers are needed to better resolve questions of their overall relationships. The true vernal pool specialists occur as highly derived species within section *Navarretia* (Figure 5.15); whereas, the two OBL species of section *Mitracarpa* are more FAC vernal pool inhabitants. Phylogenetic relationships (Figure 5.15) indicate that there have been two or three independent origins of OBL species in North America; whereas, FACW species have evolved repeatedly throughout the genus. A population genetic survey of *N. leucocephala* populations using random amplified polymorphic DNA (RAPD) markers indicated moderate levels of genetic variation (proportion of polymorphic loci 10%–55%; mean gene diversity 0.05–0.30), which are values consistent with the floral characteristics that indicate a predominantly outcrossing breeding system. Populations in close proximity (<250 m) are similar genetically; however, widely spaced populations exhibit significant genetic differentiation.

Population genetic studies should be carried out in detail for other *Navarretia* species, as many details regarding their reproductive biology remain unknown. The base chromosome number of *Navarretia* is $x = 9$. *Navarretia cotulifolia*, *N. fossalis*, and *N. leucocephala* are diploid ($2n = 18$) but counts are unknown for the remaining OBL species. Hybrids reportedly occur between *N. leucocephala* subsp. *pauciflora* and *N. leucocephala* subsp. *plieantha* or *N. leucocephala* subsp. *bakeri*.

Comments: *Navarretia leucocephala* occurs in western North America and *N. heterandra* in California and Oregon. *Navarretia cotulifolia*, *N. myersii*, and *N. prostrata* are restricted to California. *Navarretia fossalis* occurs in California and extends into Mexico. *Navarretia leucocephala* subsp. *pauciflora* and *N. leucocephala* subsp. *plieantha* are listed as endangered species in the United States.

References: Barry, 1998; Bauder & McMillan, 1998; Boose et al., 2005; Clausnitzer & Huddleston, 2002; Clausnitzer et al., 2003; Keeley & Zedler, 1998; Mason, 1957; Rogers,

1997; Schlising & Sanders, 1982; Spencer & Porter, 1997; Spencer & Rieseberg, 1998; Wacker & Kelly, 2004.

2. *Polemonium*
Jacob's ladder; polémoine
Etymology: probably after the ancient city of Polemonium, which was built by King Polemon II on the coast of Pontus
Synonyms: *Polemoniella*
Distribution: global: Asia; Europe; North America; South America; **North America:** widespread
Diversity: global: 28 species; **North America:** 20 species
Indicators (USA): FACW: *Polemonium occidentale*
Habitat: freshwater; palustrine; **pH:** circumneutral; **depth:** <1 m; **life-form(s):** emergent herb
Key morphology: stems (to 10 dm) solitary, unbranched; leaves (to 16 cm) alternate, pinnately compound with up to 13 pairs of leaflets (to 4 cm), the terminal three often fused; flowers showy, 5-merous, clustered in paniculate cymes; corolla (to 2 cm) rotate campanulate, blue; filaments pubescent basally; capsule 3-chambered
Life history: duration: perennial (rhizomes); **asexual reproduction:** rhizomes; **pollination:** insect; **sexual condition:** hermaphroditic; **fruit:** capsules (common); **local dispersal:** seeds; **long-distance dispersal:** seeds
Imperilment: (1) *Polemonium occidentale* [G5]; S1 (MN, WI); S2 (BC, WY)
Ecology: general: *Polemonium* species mainly are terrestrial, with over a third of the North American species having some type of wetland indicator status (FACW, FAC, FACU). One species (*P. occidentale*) was designated as an obligate aquatic in the 1996 list, but has since been categorized as FACW in the 2013 list. Most of the species are protandrous and are pollinated by bees (Hymenoptera) or flies (Diptera). The perennial species mainly are outcrossed while the annual species predominantly are self-pollinating.

Polemonium occidentale **Greene** grows in bogs, marshes, meadows, swamps, and along streams at elevations of up to 3300 m. In the west, the plants often extend into alpine or subarctic regions. They are capable of withstanding cold temperatures up to −40°C, but also do well where summer temperatures are warm. The few reports of soil chemistry indicate medium nutrient levels and a circumneutral pH; however, specific data are lacking. Some plants occur on silty muck or organic peat (alkaline in reaction due to groundwater upwellings); whereas, others thrive on rocky substrates. The plants tolerate open sun to partial shade and spread into gaps created by natural disturbances. The flowers are visited by bees (Hymenoptera: Apidae: *Apis mellifera*), but the breeding system of this plant remains unknown. Isozyme data indicate a complete lack of detectable genetic variation among the disjunct populations at the eastern extent of the distributional range. There is no information on the seed ecology of this species; however, seeds of related species germinate slowly (1–3 months) when sown at 20°C, presumably following a period of cold stratification. **Reported associates:** *Achillea millefolium, Aconitum columbianum, Agrostis exarata, A. stolonifera, Allium validum, Alnus rugosa, Alopecurus alpinus, A.*

pratense, Angelica arguta, Arabis glabra, Arethusa bulbosa, Arnica chamissonis, Artemisia ludoviciana, Betula glandulosa, Betula pumila, Bistorta bistortoides, Calamagrostis canadensis, Calla palustris, Caltha leptosepala, C. palustris, Calypso bulbosa, Canadanthus modestus, Cardamine pratensis, Carex angustata, C. aquatilis, C. aurea, C. cusickii, C. gynocrates, C. interior, C. jonesii, C. lacustris, C. lanuginosa, C. microptera, C. nebrascensis, C. rossii, C. rostrata, C. scopulorum, C. simulata, C. stipata, C. tenuiflora, C. utriculata, Clintonia borealis, Comarum palustre, Coptis groenlandica, Cornus canadensis, Cypripedium arietinum, C. parviflorum, C. reginae, Dasiphora floribunda, Delphinium ×occidentale, Deschampsia cespitosa, Dryopteris goldiana, Epilobium anagallidifolium, E. watsonii, Equisetum arvense, E. fluviatile, E. hyemale, Festuca occidentalis, Fragaria virginiana, Galium bifolium, G. triflorum, Geum macrophyllum, G. rivale, Glyceria striata, Gymnocarpium robertianum, Heracleum maximum, Hydrophyllum tenuipes, Juncus balticus, J. ensifolius, Laportea canadensis, Larix laricina, Ligusticum canbyi, Listera auriculata, Lonicera caerulea, Lupinus latifolius, Maianthemum racemosum, M. trifolium, Malaxis brachypoda, M. paludosa, Matteuccia struthiopteris, Mertensia ciliata, Micranthes pensylvanica, Mimulus guttatus, Mitella nuda, Muhlenbergia filiformis, Myosotis laxa, Pedicularis groenlandica, Petasites sagittatus, Picea mariana, Plantanthera clavellata, P. dilatata, Poa pratensis, P. triflora, Potentilla flabellifolia, P. gracilis, Primula jeffreyi, Pyrola asarifolia, Ranunculus orthorhynchus, Rhamnus alnifolia, Rhododendron groenlandicum, Ribes cereum, R. inerme, Salix bebbiana, S. commutata, S. geyeriana, S. lasiolepis, S. wolfii, Scirpus microcarpus, Senecio cymbalarioides, S. pseudaureus, S. triangularis, Solidago canadensis, Sphagnum, Spiraea densiflora, S. douglasii, Stellaria longifolia, Swertia perennis, Symphyotrichum foliaceum, S. spathulatum, Taxus canadensis, Thuja occidentalis, Torreyochloa pallida, Triglochin maritimum, Valeriana sitchensis, Veratrum californicum, Veronica americana, V. wormskjoldii, Viola pallens, V. palustris.

Use by wildlife: *Polemonium occidentale* is a host for fungi (Basidiomycota: Pucciniaceae: *Puccinia polemonii*; Chytridiomycota: Synchytriaceae: *Synchytrium polemonii*). The plants comprise a minor component of habitat associated with the willow flycatcher (Aves: Tyrannidae: *Empidonax traillii*).

Economic importance: food: *Polemonium occidentale* is not edible, and the leaves reportedly emit a rather unappetizing, skunk-like odor when crushed; **medicinal:** no uses reported; **cultivation:** *Polemonium occidentale* occasionally is cultivated as a native wildflower for gardens; **misc. products:** none; **weeds:** none; **nonindigenous species:** none
Systematics: Phylogenetic analyses resolve *Polemonium* in an isolated but central position within subfamily *Polemoniodeae* of Polemoniaceae (Figure 5.14). Investigations using nrITS data indicate that *Polemonium* is monophyletic, but the currently defined sections are not. Two allopatric subspecies are quite different ecologically, with *P. occidentale* subsp. *occidentale* typically occurring on rocky soils in open, often

alpine meadows; whereas, *P. occidentale* subsp. *lacustre* inhabits organic substrates of shaded, low-elevational cedar and spruce swamps. Some authors place the species in synonymy with *P. acutiflorum*, which molecular data indicate to be very closely related. The different accessions of *P. occidentale* subsp. *occidentale* from California and those of many other species, as well as the different subspecies of *P. caeruleum* do not associate as clades, indicating the need for additional systematic studies (using more sensitive genetic markers) aimed at better clarifying relationships among the closely related *P. chinense, P. occidentale, P. acutiflorum,* and *P. caeruleum* (Figure 5.16). It would also be desirable to compare these phylogenetic results with those based on a maternally inherited [e.g., chloroplast DNA (cpDNA)] marker, to test for other possible explanations (e.g., hybridization, paralogy) for these discrepancies. The base chromosome number of *Polemonium* is $x = 9$. A count has not been reported for *P. occidentale*; however, the related *P. acutiflorum* and *P. caeruleum* both are diploid ($2n = 18$). Isozyme surveys indicate that populations of subspecies *lacustre* are monomorphic. A genetic comparison of subspecies *occidentale* and *lacustre* has not been carried out but would be informative.

Comments: *Polemonium occidentale* occurs in western North America but is disjunct in Minnesota and Wisconsin.

References: Chadde, 1998; Cole, 1998, 2003; Dwire et al., 2006; Klinkenberg, 2004; Lahring, 2003; Mason, 1957; Schmidt, 2003a; Timme & De Geofroy, 2001; Timme & Patterson, 2001.

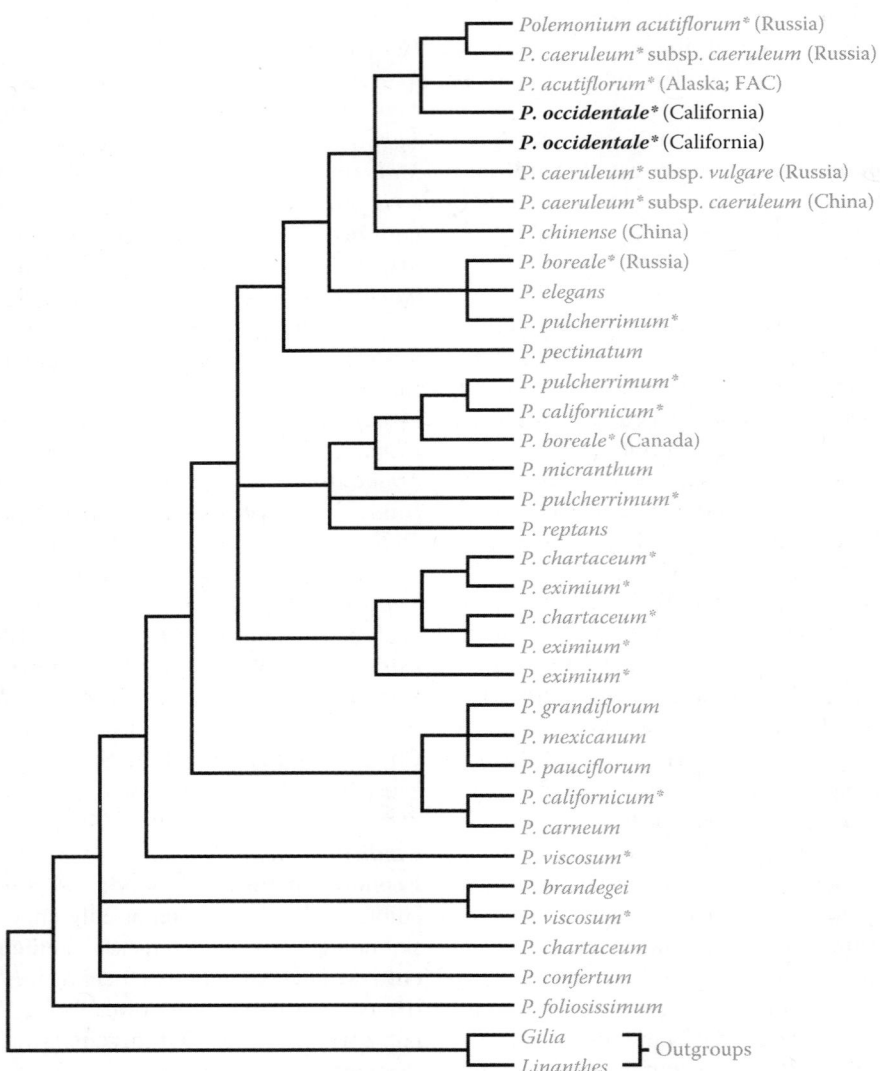

FIGURE 5.16 Interspecific phylogenetic relationships in *Polemonium* as indicated by analysis of nrITS sequence data. Different California accessions of the North American *Polemonium occidentale* subsp. *occidentale* (FACW; formerly OBL), and those of many other taxa, associate distantly in this analysis (some quite disparately), indicating that further systematic studies are needed to better clarify relationships in the group. Asterisks indicate species with multiple accessions that do not associate as clades. These unusual results might indicate confounding effects of paralogous copies of the ITS region in this group, and/or hybridization. (Adapted from Timme, R.E. & Patterson, R., The molecular phylogeny of *Polemonium* (Polemoniaceae), Thesis, San Francisco State University, San Francisco, CA, 2001.)

Family 5: Primulaceae [8]

Results from molecular phylogenetic analyses of morphological data and cpDNA sequences (e.g., Figure 5.5; Anderberg et al., 1998, 2000; Källersjö et al., 2000; Mast et al., 2001) have inspired a modified concept of Primulaceae to be followed here, which excludes various segregate genera (e.g., *Anagallis, Lysimachia, Samolus, Trientalis*) that have been transferred to Myrsinaceae and Theophrastaceae, families that together share the relatively uncommon trait of free-central placentation. The resulting circumscription of Primulaceae (which includes less than half of the original genera) delimits a group of roughly 600 species of herbaceous plants having a scapose inflorescence arising from a basal rosette of leaves, campanulate corollas with imbricate aestivation and isodiametric epidermal cells, and capsular fruits; however, without uniquely defining synapomorphies (Mast et al., 2001; Judd et al., 2002). The family occurs predominantly in cold and temperate regions of the Northern Hemisphere.

Analysis of nrITS data results from (Martins et al., 2003) does not entirely corroborate the results obtained using cpDNA data (Figure 5.17). Although nrITS sequence analyses reveal many similar associations (e.g., a clade placing tribe "Lysimachieae" within Myrsinaceae; a clade including *Hottonia, Omphalogramma,* and *Soldanella*; the placement of *Cortusa, Dionysia,* and *Dodecatheon* within *Primula*), they do not clearly separate Primulaceae from Mysinaceae (i.e., the latter are embedded within the former) as is the case with cpDNA data. However, the degree of internal branch support for the major clades provided by nrITS data is much lower than that obtained by analysis of cpDNA data. The present study accepts the transfer of all *Dodecatheon* species to *Primula* (Mast et al., 2006; Mast & Reveal, 2007), a result

that is consistent with phylogenetic analyses of seven molecular data sets (Mast et al., 2004).

The showy, insect-pollinated flowers of Primulaceae make it an important family economically as a source of ornamental plants, which abound in *Androsace, Cortusa, Dionysia, Primula* (incl. *Dodecatheon*), and *Soldanella*. Darwin (1862, 1877) used *Primula* to illustrate dimorphic heterostyly in plants. Other interesting pollination systems (e.g., "buzz" pollination) also occur within the family. The seeds are small and are dispersed by the wind or by water (Judd et al., 2002). One species of *Androsace* (*A. filiformis*) was ranked as (OBL, FACW) in the 1996 indicator list, but was reclassified as (FACW) in the 2013 list. The treatment for *Androsace* has been retained because it was completed prior to the release of the 2013 list. Overall, the family primarily is terrestrial, with just two genera containing obligate aquatic species in North America:

1. ***Androsace*** L. (FACW; formerly OBL, FACW)
2. ***Hottonia*** L.
3. ***Primula*** L.

Phylogenetic analyses (Figure 5.17) indicate that the OBL habit has evolved several times throughout the family, including multiple origins within the large genus *Primula*.

1. *Androsace*

Fairy candelabra, rock jasmine
Etymology: from the Greek *andros sakos* meaning "man's shield"
Synonyms: *Douglasia; Pomatosace; Vitaliana*
Distribution: global: Asia; Europe; North America; **North America:** western

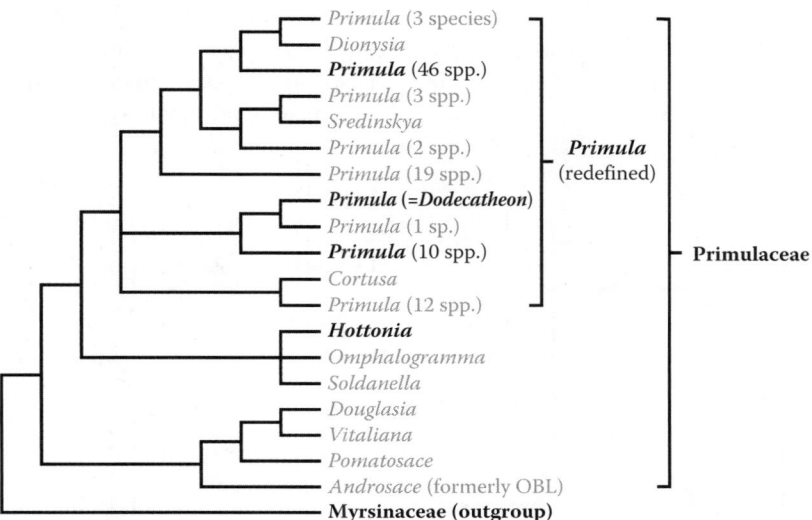

FIGURE 5.17 Relationships among genera of Primulaceae as indicated by phylogenetic analysis of cpDNA data. According to these results, the groups that contain OBL species in North America (bold) have evolved several times throughout the family. Note that the number of species indicated for taxa in bold does not reflect the number of OBL species, only that at least one OBL species occurs among the total number indicated. Species originally in *Dodecatheon* (and several other genera) have been transferred to *Primula* as molecular systematic studies have helped to redefine the genus with a circumscription that is more meaningful phylogenetically. (Adapted from Mast, A.R. et al., *Int. J. Plant Sci.*, 162, 1381–1400, 2001.)

Diversity: global: 120 species; **North America:** 6 species
Indicators (USA): FACW: *Androsace filiformis*
Habitat: freshwater; palustrine; **pH:** unknown; **depth:** <1 m;
life-form(s): emergent herb
Key morphology: rosettes (to 8 cm) solitary; leaves (to 3 cm) basal, petioled (to 3 cm), margin denticulate; scapes (to 20 cm) 3 to many, terminating in many-flowered (to 10) umbels; flowers 5-merous, pedicelled (to 7 cm); calyx (to 2 mm) persistent; corolla (to 3 mm) salverform, white; anthers subsessile; capsules (to 2 mm) spherical, 5-parted, enclosed by calyx
Life history: duration: biennial (fruit/seeds); **asexual reproduction:** none; **pollination:** insect; **sexual condition:** hermaphroditic; **fruit:** capsules (common); **local dispersal:** seeds; **long-distance dispersal:** seeds
Imperilment: (1) *Androsace filiformis* [G4]; S1 (CA, UT); S3 (WY)
Ecology: general: *Androsace* mostly is a genus of terrestrial plants. Of the four North American species having assigned some level of wetland indicator status, three are ranked between FAC and UPL and only one is designated as FACW. The genus contains annuals and perennials and a number of alpine plants having a "cushion" morphology.

Androsace filiformis **Retz**. is an annual or short-lived perennial, which grows in ditches, meadows, and along margins of lakes, rivers, and streams at elevations of up to 3000 m. Although this species often occurs at high elevations, it also grows much lower as well. Plants normally are annual, but can become short-lived perennials by developing a thickened root crown. There is no specific information on the reproductive biology or seed ecology of this species. Presumably, the flowers are insect pollinated; however, the annual members of the genus are associated with reduced corollas, which presumably indicate a tendency toward self-pollination. *Androsace filiformis* is only weakly mycotrophic. **Reported associates:** *Achillea millefolium, Agoseris glauca, Agropyron spicatum, Allium cernuum, Antennaria parvifolia, Arabis nuttallii, Arnica cordifolia, Artemisia frigida, Balsamorhiza sagittata, Besseya rubra, Bromopsis inermis, B. vulgaris,* *Calamagrostis rubescens, Carex, Castilleja luteovirens, Cerastium arvense, Cirsium foliosum, Claytonia lanceolata, Collinsia tenella, Comandra umbellata, Danthonia intermedia, Delphinium bicolor, Elymus canadensis, Erigeron trifidus, Erythronium grandiflorum, Festuca idahoensis, Fragaria vesca, Galium boreale, Geranium viscosissimum, Geum triflorum, Hydrophyllum capitatum, Iris missouriensis, Juniperus horizontalis, Linum perenne, Lithophragma tenellum, Lomatium cous, Lupinus argenteus, Maianthemum stellatum, Mertensia lanceolata, Muhlenbergia filiformis, Myosotis alpestris, Oryza, Oxytropis sericea, Penstemon eriantherus, Phlox hoodii, Picea engelmannii, Poa glauca, Potamogeton gramineus, Potentilla diversifolia, Primula conjugens, Pseudotsuga menziesii, Rosa woodsii, Rudbeckia occidentalis, Sedum lanceolatum, Senecio canus, S. crassulus, Shepherdia canadensis, Stipa richardsonii, Symphoricarpos oreophilus, Taraxacum officinale, Thalictrum occidentale, Trifolium longipes, Viola nuttallii, Wyethia helianthoides.*
Use by wildlife: none reported
Economic importance: food: not edible; **medicinal:** *Androsace filiformis* has been used to treat fevers in Korea; **cultivation:** *Androsace filiformis* is not among the many cultivated species of *Androsace*; **misc. products:** none; **weeds:** *Androsace filiformis* is reported to be a weed in southern Finland and Japan; **nonindigenous species:** *Androsace filiformis* was introduced in Japan
Systematics: Analyses of combined cpDNA and nrITS molecular data indicate that *Androsace* is not monophyletic as defined formerly (Figure 5.18) and must be modified to include several smaller genera: *Douglasia* (8 spp.), *Pomatosace* (1 sp.), and *Vitaliana* (1 sp.) in order to represent a phylogenetically meaningful clade. *Androsace filiformis* occurs as a relatively isolated species that is related to *A. axillaris, A. erecta* and *A. henryi* from southeastern China and *A. maxima* from central Asia (Figure 5.18). Thus, the North American plants probably represent an eastward range extension of the species. None of these related species shares an affinity for wet habitats despite the greater tendency towards such conditions by *A. filiformis*

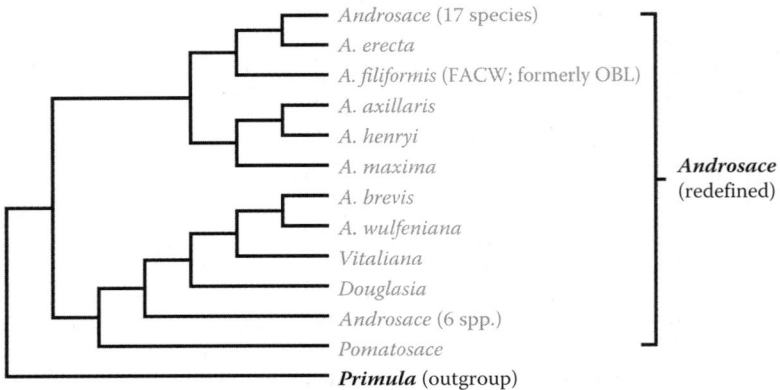

FIGURE 5.18 Phylogenetic relationships in *Androsace* as resolved using combined *trn*L-F and nrITS sequence data. These results indicate that several small genera (*Douglasia, Pomatosace, Vitaliana*) are embedded within the genus, which should be redefined to merge them within it. The FACW North American *A. filiformis* occurs centrally in the genus as an isolated species. Other *Androsace* species mainly occupy dry habitats. (Adapted from Wang, Y.-J. et al., *Acta Phytotax. Sin.,* 42, 481–499, 2004a.)

(FACW). The base chromosome number of *Androsace* is $x = 10$. *Androsace filiformis* ($2n = 18, 20$) is diploid, with two known chromosomal races. No hybrids have been reported for this species.

Comments: *Androsace filiformis* occurs in western North America and in northern Eurasia.

References: Kemppainen & Lampinen, 1987; Martins et al., 2003; Mukhin & Betekhtina, 2006; Robbins, 1944; Wang et al., 2004a.

2. *Hottonia*

Featherfoil, water-violet

Etymology: after Peter Hotton (1648–1709)

Synonyms: none

Distribution: global: Europe; North America; **North America:** eastern

Diversity: global: 2 species; **North America:** 1 species

Indicators (USA): OBL: *Hottonia inflata*

Habitat: freshwater; lacustrine, palustrine; **pH:** 4.6–7.3; **depth:** 0–3 m; **life-form(s):** emergent herb, free-floating, submersed (vittate)

Key morphology: stems (to 1.5 m) flexuous, the internodes compressed apically; leaves (to 18 cm) alternate, pinnate (to 30+ segments), submersed; inflorescence a rosette of up to 15 inflated, floating, hollow stalks (to 3 dm), each bearing several constricted, widely spaced verticels of 3–10 flowers; flowers inconspicuous (to 2.5 mm broad), pedicelled (to 15 mm), usually 5-merous (four petals or stamens occur occasionally); perianth parts subequal (to 5 mm), corolla white; capsules subglobose, 5-parted, with up to 170 seeds

Life history: duration: annual (fruit/seeds); **asexual reproduction:** none; **pollination:** self; **sexual condition:** hermaphroditic; **fruit:** capsules (common); **local dispersal:** seeds (water), fruiting plants (water); **long-distance dispersal:** seeds (waterfowl)

Imperilment: (1) *Hottonia inflata* [G4]; SX (PA); SH (OH); S1 (GA, MD, ME, MS, NH, NJ, RI, WV); S2 (AL, DE, IL, IN, MO, NC, NY, TN); S3 (CT, MA, TX, VA)

Ecology: general: There are only two *Hottonia* species worldwide and both are obligate aquatics that characteristically occupy still, shallow, acidic waters with somewhat organic substrates. The European *H. palustris* differs from *H. inflata* by its perennial habit and showy, heteromorphic (distylous), self-incompatible, outcrossed flowers.

Hottonia inflata **Elliott** inhabits waters from 0.1 to 2.7 m in depth, occurring in canals, ditches, lakes, oxbows, ponds, pools, swamps, and streams (occasionally in relatively fast currents). Habitats primarily are acidic (pH: 4.6–5.8 [sediments]; 5.7–7.3 [water]) with high aqueous phosphate levels, but with sediments low in nitrogen and phosphorous and containing a high percentage of organic matter. The plants cannot tolerate salinity above 0.5%. These are winter annuals that flower in early spring, senesce in early summer, germinate (seeds) in late summer, and remain vegetative throughout the winter. Green plants often can be seen below the ice in mid-winter, their upper portions sometimes becoming frozen within. Dense snow cover on the ice reduces light and negatively impacts the

plants as a consequence. At anthesis, the floral stalks develop as broadly inflated structures that act as floats to keep the flowers above the water surface. The flowers are homostylous, self-compatible, and are self-pollinated; however, reports of cleistogamy are erroneous. There are no known pollinators. Data from 15 isozyme loci also indicated a primarily autogamous breeding system by showing a complete lack of genetic variation (genetic diversity = 0), no heterozygosity (observed heterozygosity = 0), and complete genetic identity (genetic identity = 1) among the populations surveyed. Plants senesce and die completely by mid-July, leaving only seeds for regeneration. About ⅔ of the seeds will remain afloat for at least 3 weeks, with the remainder sinking near the parental plant. The seeds are enclosed within a mucilaginous matrix, which imparts adhesiveness, and enables them to attach to plant fragments or to other potential dispersal agents. They will adhere to their own glandular leaves, peduncles, and perianths and are also dispersed along with the floating rosettes of flower stalks if they become detached from the stems. About 95% of fresh seeds are conditionally dormant, but lose dormancy (>90% germination) by mid-summer if exposed to light and favorable temperature cycles (e.g., 30°C/15°C, 35°C/20°C). Field germination rates range from about 30% to 50%, with the highest rates occurring after 90 days. Germination occurs only on exposed sediments; thus, sites with fluctuating water levels are more likely to support persisting populations. The seeds remain viable for at least 4 years. A seed bank is absolutely necessary to maintain the species for any prolonged length of time. Population census numbers and seed production can vary dramatically from year to year, depending on the local precipitation patterns. Plants can disappear from a site completely in some years (remaining viable only in the seed bank), but can be abundant the next, making surveys difficult. More than 26 million seeds can be produced annually in larger populations; however, seed production in a given year is not related to the number of individuals produced in the following season, which is a function of favorable sites and conditions available for seed germination and seedling establishment. The populations are threatened primarily by habitat destruction and loss. **Reported associates:** *Acer rubrum, Alopecurus aequalis, Azolla, Bidens discoideus, B. frondosus, Boehmeria cylindrica, Cabomba caroliniana, Carex comosa, C. decomposita, C. tuckermanii, Cephalanthus occidentalis, Ceratophyllum demersum, C. echinatum, Cyperus strigosus, Dulichium arundinaceum, Elodea nuttallii, Eragrostis hypnoides, Fimbristylis autumnalis, F. perpusilla, Galium tinctorium, Glyceria acutiflora, G. arkansana, G. canadensis, G. septentrionalis, Heteranthera dubia, Hibiscus moscheutos, Hydrocotyle, Isoetes engelmannii, Itea virginica, Juncus pelocarpus, Leersia lenticularis, Lemna minor, Limnobium spongia, Ludwigia peploides, L. sphaerocarpa, Lythum salicaria, Nelumbo lutea, Nyssa aquatica, N. biflora, Oldenlandia uniflora, Orontium aquaticum, Persicaria amphibia, P. hydropiperoides, Phragmites australis, Planera aquatica, Populus heterophylla, Potamogeton diversifolius, Ranunculus flabellaris, R. longirostris, Rorippa aquatica, Sagittaria rigida, Saururus cernuus, Schoenoplectus tabernaemontani, Sium* *suave, Spirodela, Taxodium distichum, Torreyochloa pallida, Triadenum walteri, Typha latifolia, Utricularia, Vaccinium corymbosum, Viola lanceolata, Wolffia brasiliensis, Zizaniopsis miliacea.*

Use by wildlife: *Hottonia inflata* provides cover for amphibians, fish, and reptiles. The herbage (and seeds) of *H. inflata* are eaten by wood ducks (Aves: Anatidae: *Aix sponsa*). The plants also provide food directly (or indirectly by the associated fauna) to other birds (Aves) including coots (Rallidae: *Fulica americana*), gallinules (Rallidae: *Porphyrio martinica*), wading birds and various: "puddle ducks." The seeds are delicate and become inviable once they are digested by birds. Long-distance dispersal occurs mainly by the attachment of seeds to the feathers of wood ducks and mallards (Aves: Anatidae: *Anas platyrhynchos*). Beavers (Mammalia: Castoridae: *Castor canadensis*) often create ponds that are suitable habitats for *H. inflata*. Local dispersal by means of seeds lodged in their fur is also reported. The plants are a component of habitat associated with Blanchard's cricket frog (Amphibia: Hylidae: *Acris blanchardi*).

Economic importance: food: none; **medicinal:** none; **cultivation:** Although quite unusual and ornamental, *H. inflata* rarely is cultivated because of its winter annual habit. The related *H. palustris* is cultivated as an aquarium or water garden specimen, sometimes mistakenly as *H. inflata*; **misc. products:** none; **weeds:** *Hottonia inflata* occasionally is reported as weedy, which is puzzling (and likely due to misidentification) given that these winter annuals are completely absent during the summer; **nonindigenous species:** none

Systematics: Phylogenetic analysis of cpDNA sequence data demonstrates that *Hottonia* is monophyletic and occurs with *Bryocarpum, Omphalogramma,* and *Soldanella,* as a sister clade to *Primula* (Figure 5.19). The base chromosome number of *Hottonia* probably is *x* = 10, with *H. palustris* (2*n* = 20) being diploid. Chromosome counts are unavailable for *H. inflata*.

Comments: *Hottonia inflata* occurs sporadically in eastern North America and is regarded as imperiled nearly throughout the region.

References: Baker, 1959; Baskin et al., 1996; Copley, 1999; Kasselmann, 2003; Mast et al., 2006; NatureServe, 2003; Philbrick & Les, 1996; Stutzenbaker, 1999; Whitley et al., 1990.

3. *Primula*

Primrose, shootingstar; primevère

Etymology: a diminutive of the Latin *primus* meaning "first" in reference to the early-flowering habit of these plants

Synonyms: *Aretia; Cortusa; Dionysia; Dodecatheon; Exinia; Sredinskaya*

Distribution: global: Africa; Asia; Europe; New World; **North America:** widespread

Diversity: global: 475 species; **North America:** 39 species

Indicators (USA): OBL: *Primula alcalina, P. cuneifolia, P. egaliksensis, P. fragrans, P. incana;* **FACW:** *P. egaliksensis, P. incana, P. tetrandra*

Habitat: freshwater; palustrine; **pH:** 7.0–9.6; **depth:** <1 m; **life-form(s):** emergent herb

Key morphology: caudex compressed, rosulate; rosette leaves (to 60 cm) sometimes farinose; umbels with 1–15 flowers (4 or 5-merous), pedicelled (to 9 cm), terminating a leafless scape (to 60 cm); corolla (to 22 mm) tubular basally, the limb acute and strongly reflexed or emarginate and salverform; petals lavender, magenta, pink, rose, or white, their base often white or yellow and forming an "eye"; capsules (to 18 mm) ellipsoid or ovoid, many seeded (up to 200)

Life history: duration: perennial (caudex); **asexual reproduction:** none; **pollination:** insect, self; **sexual condition:** homostylous or heterostylous; **fruit:** capsules (common); **local dispersal:** seeds; **long-distance dispersal:** seeds

Imperilment: (1) *Primula alcalina* [G2]; S2 (ID, MT); (2) *P. cuneifolia* [G5]; S2 (BC); (3) *P. egaliksensis* [G4]; S1 (WY); S2 (AB, CO, YT); S3 (BC, LB, NF, QC); (4) *P. fragrans* [G3]; S1 (UT); (5) *P. incana* [G4]; S1 (ID, ND, ON, UT); S2 (MT, WY); S3 (BC); (6) *P. tetrandra* [G3]; S2 (UT)

Ecology: general: Some degree of wetland tolerance is indicated for 24 North American *Primula* species (62%), with

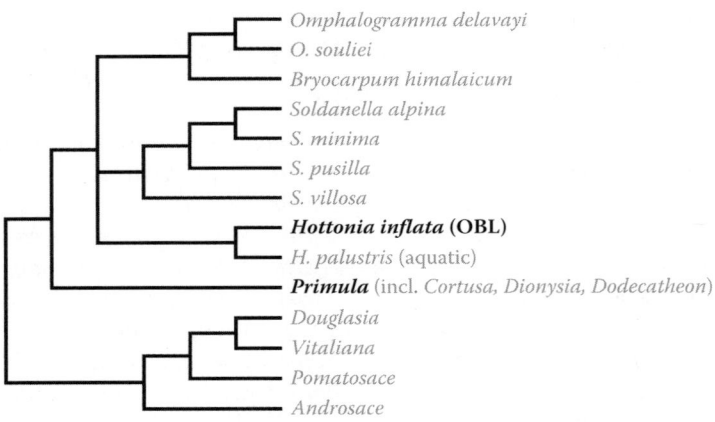

FIGURE 5.19 Phylogenetic relationships of *Hottonia* as indicated by analysis of cpDNA sequence data. Taxa with OBL North American representatives are shown in bold. The European *Hottonia palustris* is aquatic. (Adapted from Mast, A.R. et al., *New Phytol.*, 171, 605–616, 2006.)

five species (13%) achieving OBL status in all or a portion of their range. One species (*P. tetrandra*) was ranked as FACW, OBL in the 1996 indictor list, but later was reclassified as FACW. It has been retained here because the treatment was completed prior to the release of the 2013 list. Generally, the flowers are adapted for insect pollination, which in some species occurs by means of buzz pollination. As Charles Darwin elucidated, many species are heterostylous (distylous), having reciprocal floral morphs ("pins" with long styles; "thrums" with short styles) in populations. The heterostylous species are genetically self-incompatible and require pollen transfers between pin and thrum (either direction) to achieve subsequent fertilization. Homostylous species are self-compatible and are capable of selfing in the absence of pollinators. Of the few species examined for their seed ecology, some are nondormant and others have physiological dormancy requiring a period of cold stratification for germination. Most species (and all of the obligate aquatics) are perennials. Perennation is by dormant buds from the short lived, persistent caudex. Habitats characteristically are alkaline.

Primula alcalina **Cholewa & Douglass M. Hend**. grows in montane fens, meadows, seeps and on stream bars at elevations from 1900 to 2200 m. Soils are alkaline (pH: 8.9–9.6), fine textured and derived from carbonate alluvial outwash. Habitats are open and occur in areas of upwelling from spring-fed creeks. The plants often grow on hummocks where the graminoid cover is low. Plants growing on stream bars flower more regularly and produce taller scapes than plants growing on hummocks. The flowers are self-incompatible and heterostylous, with both pin and thrum morphs occurring in proximity. Pollination is diurnal and occurs mainly by flies (Diptera: Syrphidae). Seeds are physiologically dormant and require stratification for germination, which can be enhanced by the application of gibberellic acid (GA3). Although a rare species globally, the plants can be abundant locally, in some cases reaching numbers of 10,000+ individuals. Grazing during the flowering/fruiting period negatively impacts plants; whereas, grazing afterwards creates open areas suitable for seedling establishment. Beavers (Mammalia: Castoridae), which dam associated streams, help to maintain the permanently saturated conditions necessary for this plant's survival. **Reported associates:** *Agropyron trachycaulum, Anticlea elegans, Braya humilis, Carex idahoa, C. nebrascensis, C. praegracilis, C. scirpoidea, Crepis runcinata, Dasiphora floribunda, Deschampsia cespitosa, Hutchinsia procumbens, Juncus balticus, Kobresia simpliciuscula, Lomatogonium rotatum, Muhlenbergia richarsonis, Packera debilis, Phlox kelseyi, Poa pratensis, Potentilla ×diversifolia, Primula cusickii, P. incana, P. pauciflora, Salix, Senecio debilis, Taraxacum eriophorum, Thalictrum alpinum, Thelypodium sagittatum, Triglochin maritimum, Valeriana edulis.*

Primula cuneifolia **Ledeb**. inhabits artic or alpine bogs, depressions, heathlands, hummocks, meadows, mires, scree slopes, seeps, snowbeds, tundra, and the margins of ponds, rivers, and streams at subalpine or alpine elevations of up to 2100 m. Its substrates are described as alkaline (pH: 7.0–7.5) or acidic cobble, granite, gravel, greenstone, humus, loam,

loamy quartz, rock, sandy rock, shale, silty loam, talus, and ultra-mafic rock. The plants are short-lived perennials that tolerate partial to fairly deep shade. This species exhibits unusual floral variation with both self-compatible homostylous (long homostyles) and self-incompatible heterostylous morphs observed in populations. Flowering occurs rapidly (within 8 days) following snow melt and plants are more likely than other associated species to set fruit in sites where the growing season is shortened due to late snow melt. The seeds are physiologically dormant. Germination of untreated seeds is low (<20%) across a range of temperatures, but increases (~50%) under a 15°C/5°C temperature regime after 3–6 months of cold stratification. Warmer temperatures cause germination rates to decline and germination will not occur in the dark. This species is highly resistant to freezing with crowns that are capable of surviving temperatures down to −30°C and below. It is believed to process starch reserves by means of a specific amylase that is activated by extremely low temperatures. The plants have been regenerated successfully from tissue culture of leaf explants. Genetic analyses have found that the cpDNA haplotypes from the Alaskan Aleutian region match those from the Kuril Islands north of Japan. **Reported associates:** *Agrostis hiemalis, Alnus, Androsace chamaejasme, Anemone narcissiflora, Antennaria alpina, A. monocephala, Anthoxanthum monticola, Arabis lyrata, Arnica lessingii, A. unalaschcensis, Artemisia arctica, A. tilesii, Barbilophozia barbata, B. lycopodioides, Bistorta vivipara, Botrychium, Brachythecium salebrosum, Campanula lasiocarpa, Cardamine bellidifolia, C. oligosperma, Carex bigelowii, C. macrochaeta, C. nigricans, C. scirpoidea, Cassiope lycopodioides, C. tetragona, Castilleja elegans, Cerastium arcticum, C. fontanum, Cetraria islandica, Chamerion latifolium, Chrysosplenium wrightii, Cladina, Cladonia bellidiflora, C. gracilis, C. mitis, C. ustulata, Dactylorhiza viridis, Diapensia lapponica, Dicranella palustris, Draba nivalis, Dryas octopetala, Empetrum nigrum, Equisetum arvense, Erigeron peregrinus, Euphrasia mollis, Festuca altaica, F. brachyphylla, Gentiana frigida, Gentianella amarella, Geum calthifolium, G. rossii, Harrimanella stelleriana, Hierochloe alpina, Hippuris montana, Huperzia selago, Juncus triglumis, Lagotis glauca, Leontodon, Leptarrhena pyrolifolia, Lloydia serotina, Loiseleuria procumbens, Luetkea pectinata, Lupinus nootkatensis, Luzula arcuata, L. parviflora, L. spicata, L. wahlenbergii, Macounastrum islandicum, Micranthes ferruginea, Minuartia arctica, Myosotis asiatica, Oxytropis campestris, O. podocarpa, Papaver nudicaule, Pedicularis verticillata, Petasites frigidus, Phleum alpinum, Phyllodoce aleutica, P. glanduliflora, Pleurozium schreberi, Poa arctica, P. glauca, P. leptocoma, Polemonium pulcherrimum, Polytrichum piliferum, Potentilla alaskana, P. nana, P. villosa, Primula cuneifolia, P. frigida, Pseudephebe pubescens, Ranunculus flammula, Rhizocarpon geminatum, Rhodiola rosea, Rhododendron camtschaticum, Sagina saginoides, Salix stolonifera, Sanguisorba stipulata, Saxifraga bronchialis, S. cespitosa, S. flagellaris, S. rivularis, S. serpyllifolia, Senecio cymbalaria, S. lugens, Sibbaldia procumbens, Silene acaulis, Solorina crocea, Spiraea stevenii,*

Stereocaulon vesuvianum, Thalictrum alpinum, Thamnolia vermicularis, Tofieldia coccinea, Trisetum spicatum, Vaccinium uliginosum.

Primula egaliksensis Wormsk. occupies fens, flats, floodplains, fluvial terraces, gravel bars, heath, hummocks, marshes, meadows, seeps, shrub swamps, swales, thickets, tundra, and the margins of brooks, lakes, rivers, and streams, usually at higher elevations (>2000–3000 m). The substrates consistently are calcareous (pH: 7.4–7.7), occasionally brackish, and have been characterized as clay, gravel, limestone, marl, mica-rich, muck, mud, organic, rubble, sand, sandy clay, schist, scree, silt, silty clay, silty shale, and ultra-mafic. The flowers are homostylous and self-compatible; however, little else is known about the reproductive biology or seed ecology of this species. **Reported associates:** *Achillea borealis, Actaea rubra, Agrostis exarata, A. gigantea, A. scabra, Allium geyeri, Alnus crispa, Andromeda polifolia, Androsace chamaejasme, Anemone multifida, A. parviflora, A. richardsonii, Antennaria, Aquilegia coerulea, Arceuthobium americanum, Argentina anserina, Betula glandulosa, B. nana, Bistorta bistortoides, B. vivipara, Calamagrostis canadensis, C. stricta, Calliergon trifarium, Campylium stellatum, Carex aquatilis, C. aurea, C. bigelowii, C. capillaris, C. gynocrates, C. krausei, C. limosa, C. livida, C. lyngbyaei, C. membranacea, C. microptera, C. microglochin, C. parryana, C. rariflora, C. scirpoidea, C. simulata, C. utriculata, Chondrophylla aquatica, Chrysanthemum arcticum, Comarum palustre, Corallorhiza trifida, Cornus canadensis, Crepis runcinata, Cypripedium passerinum, Dasiphora floribunda, Deschampsia cespitosa, Distichum capillare, Drepanocladus aduncus, D. revolvens, Drosera, Dryas integrifolia, D. octopetala, Eleocharis acicularis, E. quinqueflora, Elymus trachycaulus, Empetrum nigrum, Epilobium adenocaulon, E. hornemannii, E. lactiflorum, E. palustre, Equisetum arvense, E. fluviatile, E. pratense, E. variegatum, Eriogonum umbellatum, Eriophorum angustifolium, E. gracile, Eurybia sibirica, Festuca altaica, F. brachyphylla, F. rubra, Fragaria vesca, F. virginiana, Galium boreale, G. trifidum, Geum macrophyllum, Goodyera repens, Hedysarum alpinum, Heuchera cylindrica, Hippuris vulgaris, Huperzia selago, Iris setosa, Juncus albescens, J. alpinoarticulatus, J. arcticus, J. bufonius, J. longistylus, Kobresia myosuroides, K. simpliciuscula, Lathyrus palustris, Leptosiphon septentrionalis, Limnorchis hyperborea, Lonicera involucrata, Mitella pentandra, M. stauropetala, Mniobryum albicans, Myrica gale, Packera pauciflora, Parnassia palustris, P. parviflora, Pedicularis groenlandica, Penstemon fruticosus, P. procerus, Persicaria amphibia, Philonotis fontana, Picea engelmannii × P. glauca, P. mariana, P. sitchensis, Pinguicula vulgaris, Pinus contorta, Plantago maritima, Platanthera aquilonis, P. dilatata, Poa glaucifolia, P. pratensis, Polemonium caeruleum, Polygonum achoreum, P. aviculare, P. douglasii, Populus tremuloides, P. trichocarpa, Potentilla gracilis, P. egedii, P. hippiana, Rhinanthus minor, Rhizomnium nudum, Rhododendron, Rosa acicularis, R. woodsii, Rubus idaeus, R. parviflorus, Rumex crispus, Stuckenia pectinata, Primula cusickii, Ptilagrostis porteri, Puccinellia nutkaenis, P. pumila,*

Ranunculus cymbalaria, R. sceleratus, Rubus arcticus, Salix alaxensis, S. arctica, S. barclayi, S. bebbiana, S. boothii, S. brachycarpa, S. candida, S. commutata, S. exigua, S. geyeriana, S. myrtillifolia, S. planifolia, S. pseudomonticola, S. reticulata, S. richardsonii, S. wolfii, Saxifraga aizoides, S. cespitosa, Scorpidium turgescens, S. scorpoides, Selaginella densa, Sibbaldia procumbens, Sisyrinchium pallidum, Sorbus scopulina, Sphagnum, Spiraea betulifolia, Stellaria calycantha, S. longipes, Swertia perennis, Symphyotrichum boreale, Thalictrum alpinum, T. occidentale, Tofieldia coccinea, Trichophorum cespitosum, T. pumilum, Triglochin maritimum, T. palustre, Typha latifolia, Urtica dioica, Utricularia ochroleuca, Vaccinium uliginosum, Valeriana edulis, Verbascum thapsus, Veronica americana, Viburnum edule, Viola adunca, V. macloskeyi, Warnstorfia exannulata, Zygadenus elegans.

Primula fragrans A.R. Mast & Reveal grows in montane marshes, meadows, seeps, springs, and along the margins of springs and streams at elevations from 2300 to 3600 m. The substrates consist of mud or pumice and sites vary from sunny to being partially shaded. There is little additional ecological information published on this species. **Reported associates:** *Abies lasiocarpa, Achillea millefolium, Aconitum columbianum, Allium validum, Angelica kingii, Antennaria soliceps, Aquilegia formosa, Botrychium crenulatum, Carex aurea, C. subfusca, Castilleja martinii, Chamerion angustifolium, Cirsium clokeyi, Cystopteris fragilis, Deschampsia cespitosa, Draba jaegeri, D. paucifructa, Eleocharis, Equisetum arvense, Gentianopsis holopetala, Geranium, Heuchera rubescens, Iris, Ivesia cryptocaulis, Juncus mertensianus, Juniperus scopulorum, Lilium, Maianthemum stellatum, Micranthes odontoloma, Mimulus primuloides, Platanthera sparsiflora, Populus tremuloides, Pyrola minor, Ribes montigenum, Rosa woodsii, Salix lasiolepis, Silene clokeyi, Sisyrinchium idahoense, Sphaeromeria compacta, Sphenosciadium capitellatum, Stellaria longipes, Synthyris ranunculina, Telesonix jamesii, Trifolium monanthum, Trisetum spicatum, Valeriana acutiloba, Veratrum californicum.*

Primula incana M.E. Jones inhabits fresh to saline boreal or montane bogs, ditches, fens, flats, floodplains, lakeshores, marshes, meadows, mudflats, muskeg, prairies, sandbars, seeps, sloughs, springs, stream margins, swales, and swamps at elevations of up to 2600 m. The plants colonize alkaline (pH: 7.0–8.2) calcareous clay, muck, muddy gravel, humus, sand, or silt on sunny to semi-shaded sites that often are disturbed and sometimes are quite saline (>10 dS/m). This is an early successional species and an indicator of calcareous groundwater. The flowers are elevated on long stalks, which orient them above the surrounding vegetation (typically grasses). However, they are homostylous, self-compatible, and autogamous with a low pollen:ovule ratio (245:1). The seeds reportedly germinate at 13°C–16°C after 3–6 months of cold stratification. The foliage is covered by a mealy whitish powder of uncertain function. **Reported associates:** *Achillea millefolium, Agoseris glauca, Agropyron smithii, A. trachycaulum, Agrostis scabra, A. stolonifera, Allium geyeri, Alopecurus borealis, Amelanchier alnifolia, Antennaria microphylla, A.*

parvifolia, A. rosea, Arabis, Argentina anserina, Artemisia frigida, Astragalus leptaleus, Betula occidentalis, Bistorta bistortoites, Brachythecium nelsonii, Calamagrostis canadensis, C. inexpansa, C. stricta, Calliergon giganteum, Carex aquatilis, C. atherodes, C. aurea, C. buxbaumii, C. capillaris, C. crawei, C. lanuginosa, C. muricata, C. nebrascensis, C. obtusata, C. parryana, C. praegracilis, C. pratericola, C. rostrata, C. sartwellii, C. scirpoidea, C. simulata, C. stenophylla, C. subspathacea, C. utriculata, C. viridula, Cirsium arvense, C. coloradense, C. flodmanii, Clementsia rhodantha, Conioselinum scopulorum, Cornus sericea, Crepis runcinata, C. tectorum, Dasiphora floribunda, Deschampsia cespitosa, Distichlis stricta, Drepanocladus revolvens, Eleocharis palustris, E. quinqueflora, Elymus trachycaulus, Epilobium leptophyllum, E. palustre, Equisetum arvense, E. variegatum, Erigeron lonchophyllus, E. speciosus, Eriophorum viridicarinatum, Festuca hallii, F. idahoensis, F. scabrella, F. trachyphylla, Gentiana fremontii, Gentianella amarella, Gentianopsis crinita, Glyceria striata, Grindelia squarrosa, Hierochloe hirta, H. odorata, Hordeum jubatum, Iris missouriensis, Juncus alpinoarticulatus, J. arcticus, J. balticus, J. longifolius, J. nodosus, Kobresia myosuroides, K. simpliciuscula, Larix, Lilium philadelphicum, Lobelia kalmii, Lomatogonium rotatum, Lycopus americanus, Lysimachia maritima, Mentha arvensis, Muhlenbergia glomerata, M. richardsonis, Parnassia palustris, Pedicularis crenulata, P. groenlandica, Persicaria amphibia, Petasites sagittata, Phlox kelseyi, Plantago eriopoda, Platanthera hyperborea, P. obtusata, Poa compressa, P. juncifolia, P. pratensis, Polemonium foliosissimum, Populus balsamifera, Potentilla anserina, P. subjuga, Primula alcalina, P. cusickii, P. nutans, P. pauciflora, Psychrophila leptosepala, Ptilagrostis porteri, Puccinellia nuttalliana, P. phryganodes, Pyrola asarifolia, Ranunculus, Rhizomnium pseudopunctatum, Rosa acicularis, R. woodsii, Rubus pubescens, Rumex, Salicornia, Salix alaxensis, S. bebbiana, S. boothii, S. brachycarpa, S. candida, S. eriocephala, S. geyeriana, S. glauca, S. maccalliana, S. myrtillifolia, S. novae-angliae, S. pedicellaris, S. petiolaris, S. planifolia, S. pseudomonticola, S. wolfii, Schoenoplectus pungens, Senecio pauperculus, Sisyrinchium idahoense, S. montanum, S. pallidum, Solidago canadensis, Sonchus uliginosus, Spergularia marina, Spiranthes diluvialis, Stellaria longipes, Stipa curtiseta, Suaeda calceoliformis, Symphoricarpos occidentalis, Symphyotrichum boreale, S. ciliatum, S. ericoides, S. spathulatum, Taraxacum officinale, Thalictrum alpinum, T. venulosum, Thermopsis rhombifolia, Triglochin maritimum, T. palustre, Valeriana edulis, Viola nephrophylla, Warnstorfia exannulata, Zizia aperta.

Primula tetrandra (Suksd. ex Greene) A.R. Mast & Reveal occurs in montane marshes, meadows, seeps, springs, and along the margins of lakes, springs, and streams at elevations from 1900 to 3500 m. Habitats often are shady with inorganic or organic substrates that include basalt, humus, mud, peat, sand, sandy gravel, and silty pumice. Information regarding the floral and seed ecology of this species is scarce. **Reported associates:** *Abies amabilis, A. concolor, A. lasiocarpa, Achillea millefolium, Achlys triphylla, Aconitum*

columbianum, Agoseris aurantiaca, Agrostis exarata, A. humilis, A. oregonensis, A. thurberiana, Allium validum, Alnus viridis, Anemone deltoides, A. oregana, Antennaria alpina, A. umbrinella, Aquilegia formosa, Arnica longifolia, A. mollis, Artemisia tridentata, Betula occidentalis, Bistorta bistortoides, Blechnum spicant, Bromus, Calamagrostis breweri, C. canadensis, Caltha leptosepala, Camassia quamash, Canadanthus modestus, Cardamine breweri, Carex abrupta, C. aquatilis, C. athrostachya, C. atrata, C. brunnescens, C. canescens, C. echinata, C. fracta, C. gymnoclada, C. hoodii, C. illota, C. jonesii, C. laeviculmis, C. lenticularis, C. leptalea, C. luzulifolia, C. luzulina, C. microptera, C. nebrascensis, C. nigricans, C. rostrata, C. scopulorum, C. spectabilis, C. straminiformis, C. utriculata, C. vernacula, C. vesicaria, Castilleja miniata, C. nana, C. suksdorfii, Chamerion angustifolium, Cirsium, Claytonia lanceolata, C. sibirica, Clintonia uniflora, Collinsia parviflora, Corallorrhiza maculata, Cornus sericea, Danthonia spicata, Deschampsia cespitosa, D. elongata, Draba stenoloba, Drosera rotundifolia, Eleocharis palustris, E. pauciflora, E. quinqueflora, Epilobium alpinum, E. brachycarpum, E. brevistylum, E. ciliatum, E. hornemannii, E. oregonense, Equisetum arvense, E. telmateia, Erigeron coulteri, E. peregrinus, Eriophorum crinigerum, E. gracile, Erysimum, Erythronium grandiflorum, Festuca idahoensis, Gaultheria ovatifolia, Gentiana calycosa, G. newberryi, G. prostrata, G. sceptrum, Gentianopsis holopetala, G. simplex, Geranium richardsonii, Geum macrophyllum, Glyceria elata, Helenium hoopesii, Hosackia pinnata, Hypericum anagalloides, Ivesia shockleyi, Juncus balticus, J. chlorocephalus, J. ensifolius, J. howellii, J. mertensianus, J. orthophyllus, Kalmia microphylla, Larix occidentalis, Leptarrhena pyrolifolia, Lewisia, Linnaea borealis, Lonicera involucrata, Lupinus latifolius, L. polyphyllus, Luzula campestris, L. multiflora, Ligusticum grayi, Maianthemum stellatum, Melica, Menziesia ferruginea, Micranthes aprica, M. bryophora, M. ferruginea, M. odontoloma, M. oregana, Mimulus floribundus, M. guttatus, M. primuloides, M. tilingii, Mitella breweri, M. pentandra, Montia linearis, Muhlenbergia filiformis, Oreostemma alpigenum, Orthilia secunda, Packera cymbalarioides, Parnassia fimbriata, Pedicularis groenlandica, Penstemon procerus, Phalacroseris bolanderi, Phleum alpinum, Phyllodoce breweri, P. empetriformis, Picea engelmannii, Pinus contorta, P. monticola, P. ponderosa, Platanthera dilatata, P. leucostachys, P. sparsiflora, P. stricta, Poa cusickii, P. glauca, P. secunda, Polemonium pulcherrimum, Polygonum minimum, Populus tremuloides, Potentilla breweri, P. drummondii, P. flabellifolia, Primula hendersonii, P. jeffreyi, Pteridium aquilinum, Pterospora andromedea, Quercus gambelii, Ranunculus alismifolius, R. occidentalis, R. orthorhynchus, R. populago, R. uncinatus, Rhododendron macrophyllum, Ribes lacustre, Rosa woodsii, Rubus lasiococcus, Salix anglorum, S. boothii, S. commutata, S. geyeriana, S. lemmonii, S. myrtillifolia, S. reticulata, S. sitchensis, Sanguisorba officinalis, Scirpus congdonii, S. microcarpus, Senecio hydrophilus, S. triangularis, Sibbaldia procumbens, Sphagnum, Sphenosciadium capitellatum, Spiraea densiflora, Spiranthes romanzoffiana, Stellaria crispa, S. longifolia,

Streptopus roseus, Symphyotrichum foliaceum, S. spathulatum, Swertia perennis, Taraxacum, Thuja plicata, Tofiledia glutinosa, Trautvetteria caroliniensis, Trichophorum cespitosum, Trientalis europaea, Trifolium cyathiferum, T. longipes, T. montanum, Trisetum spicatum, Tsuga heterophylla, T. mertensiana, Vaccinium cespitosum, V. deliciosum, V. membranaceum, V. ovalifolium, V. scoparium, V. uliginosum, Valeriana sitchensis, Veratrum californicum, V. viride, Veronica americana, V. wormskjoldii, Viburnum edule, Viola macloskeyi, V. orbiculata, V. palustris, Wyethia.

Use by wildlife: *Primula alcalina* is grazed by cattle and *P. incana* by deer. *Primula tetrandra* is a host for larval rustic blue and Sierra Nevada blue butterflies (Lepidoptera: Lycaenidae: *Agriades rusticus, Plebejus podarce*) and the fruits are eaten by larval caddis flies (Trichoptera: Limnephilidae: *Desmona bethula*). The plants are also host to a rust fungus (Basidiomycota: Pucciniaceae: *Puccinia ortonii*).

Economic importance: food: The leaves and roots of *Primula tetrandra* reportedly are edible when boiled or roasted; **medicinal:** Although primulas have a long traditional use as folk medicines, few accounts exist for any of the species treated here. The essence of *Primula tetrandra* is said to cure bad memories; **cultivation:** *Primula cuneifolia,*

P. fragrans, P. incana, and *P. tetrandra* are cultivated as ornamental garden plants; **misc. products:** none; **weeds:** none; **nonindigenous species:** none

Systematics: Phylogenetic analysis of cpDNA sequence data (Figure 5.20) clearly shows that species of the former genus *Dodecatheon* are embedded within *Primula*, where they now are recognized as section *Dodecatheon*. The same data also demonstrate the monophyly of section *Aleuritia* (with minor modifications), which includes *P. alcalina, P. egaliksensis* (assigned previously to section *Armerina*), and *P. incana* (Figure 5.20). The three taxa are relatively closely related, but not sister species (Figure 5.20). The OBL *Primula fragrans* (=*D. redolens*), is related closely to the FACW *P. tetrandra* (=*Dodecatheon alpinum*), which was ranked as (OBL) in the 1996 indicator list (Figure 5.20). *Primula cuneifolia* (section *Cuneifolia*) occupies a relatively isolated position in the genus (Figure 5.20). These results indicate that the OBL habit has arisen independently five times within North American *Primula*. The base chromosome number of *Primula* varies from $x = 8$–12. *Primula alcalina* ($2n = 18$) and *P. cuneifolia* ($2n = 22$) are diploid; whereas, *P. incana* ($2n = 18, 54, 72$) comprises diploid, hexaploid, and octaploid cytotypes. *Primula egaliksensis* ($2n = 36, 40$) and *P. tetrandra* ($2n = 44$)

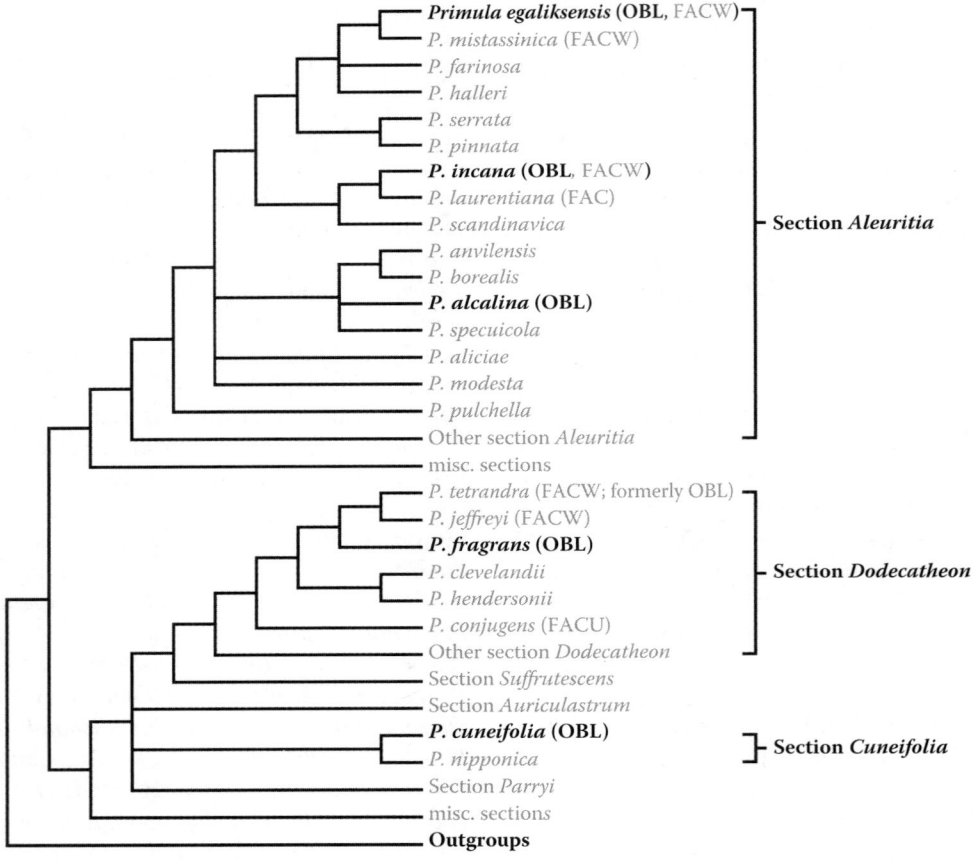

FIGURE 5.20 Phylogenetic relationships in *Primula* as evidenced by analysis of cpDNA data. The distribution of the five OBL North American species (bold) indicate that the aquatic habit has arisen as many as five independent times within the genus. Wetland status designations are provided for other North American species where applicable. (Adapted from Mast, A.R. et al., *New Phytol.*, 171, 605–616, 2006; condensed from 207 taxa.)

are tetraploid (the latter arguably diploid genetically). Several studies using genetic markers indicate that the tetraploid *P. egaliksensis* may be an allopolyploid derivative of the diploid *P. mistassinica*. *Primula egaliksensis* and *P. incana* reportedly hybridize regularly, making it difficult to differentiate the species from their hybrids. Hybridization between *Primula fragrans* and *P. tetrandra* has also been reported.

Comments: *Primula egaliksensis* (also known from Greenland) and *P. incana* are widespread throughout Canada and extend southward into the Rocky Mountains of the United States. *Primula fragrans* and *P. tetrandra* are limited to the western United States. Narrowly restricted species include *P. alcalina* (Idaho, Montana) and *P. cuneifolia* (Alaska, British Columbia); however, the latter also extends into eastern Asia.

References: Cholewa & Henderson, 1984a; Christy, 2004; Cook, 1986; Cruden, 1977; Darwin, 1862, 1877; Elzinga, 1997; Fitts, 1995; Fujii & Senni, 2006; Griggs, 1936; Guggisberg et al., 2006; Heusser et al., 1954; Johnson & Steingraeber, 2003; Kudo, 1992; Lahring, 2003; Lesica, 2003; Macdonald, 2003; Mast & Reveal, 2007; Mast et al., 2004; Mellmann-Brown, 2004; Muir & Moseley, 1994; Purdy et al., 2005; Reveal, 2006; Richards, 1986; Sakai & Otsuka, 1970; Shibata & Nishida, 1993; Shimada et al., 1997; Shimono & Kudo, 2005; Takahashi, 2001; Trift et al., 2002.

Family 6: Sarraceniaceae [3]

Although once thought to be related to Droseraceae and Nepenthaceae, analyses of multiple morphological and DNA sequence data sets have confirmed the monophyly of the small family Sarraceniaceae (with 15 species) as the sister group to a clade consisting of Actinidiaceae and Roridulaceae in the Ericales (Figure 5.5; Bayer et al., 1996). Sarraceniaceae are entirely carnivorous by means of their tubular, ascidiform (pouch-like) leaves ("pitchers"), which entrap small animals. The captured prey items are digested by resident micro/macroorganisms to provide supplemental nitrogen and phosphorous for the plants, which characteristically inhabit nutrient-poor sites. Similar leaves and a carnivorous habit occur elsewhere only in Cephalotaceae and Nepenthaceae; however, these families are quite distantly related to Sarraceniaceae (Albert et al., 1992). The closely related Roridulaceae contain carnivorous species that trap insects by means of viscid "flypaper"-like leaves rather than pitchers; Actinidiaceae are not carnivorous.

Relationships among the three genera of Sarraceniaceae have been established consistently by analyses of various molecular data (Albert et al., 1992; Bayer et al., 1996), which indicate *Heliamphora* and *Sarracenia* to be sister genera with *Darlingtonia* as their closest relative (Figure 5.21). The species are prized by collectors and widely cultivated because of their unusual morphology and carnivorous habit.

Ecologically, Sarraceniaceae are plants of acidic habitats and occur commonly on peat in bogs or on sandy substrates. The showy flowers are pollinated by insects, which collect the pollen and seek the nectar. The flowers produce copious amounts of nectar from specialized glands and also on the surface of the pitchers, which serves to attract prey into

FIGURE 5.21 Evolutionary relationships in Sarraceniaceae as indicated by phylogenetic analysis of molecular data. North American taxa containing obligate aquatics are indicated in bold. *Heliamphora* is aquatic; however it occurs only in South America. These results indicate a single origin of the OBL habit in the family. (Adapted from Bayer, R.J. et al., *Syst. Bot.*, 21, 121–134, 1996.)

the traps. Seed dispersal occurs by wind or water. All North American species are obligate aquatic plants, which occur in two genera:

1. ***Darlingtonia*** Torr.
2. ***Sarracenia*** L.

1. *Darlingtonia*
California pitcherplant, cobra-lily
Etymology: after William Darlington (1782–1863)
Synonyms: *Chrysamphora*
Distribution: global: North America; **North America:** western
Diversity: global: 1 species; **North America:** 1 species
Indicators (USA): OBL: *Darlingtonia californica*
Habitat: freshwater; palustrine; **pH:** 4.4–9.5; **depth:** <1 m; **life-form(s):** emergent herb
Key morphology: rhizomes and stolons producing clusters of 5–15 leaves; leaves (to 1 m) nectariferous, tubular, serpentiform, hooded, with a pair of broad, "fang-like" appendages (to 7 cm) attached above a lateral opening to the fluid-filled cavity; flowers pendent, solitary, 5-merous; sepals (to 5 cm) yellow-green; petals (to 3 cm) maroon or occasionally yellow-green; stamens (to 1 cm) 13–15; capsules (to 4 cm) turbinate, 5-celled, many-seeded
Life history: duration: perennial (rhizomes); **asexual reproduction:** rhizomes, stolons; **pollination:** insect; **sexual condition:** hermaphroditic; **fruit:** capsules (common); **local dispersal:** rhizomes, stolons; **long-distance dispersal:** seeds (animals, water)
Imperilment: (1) *Darlingtonia californica* [G3]; S3 (CA, OR)
Ecology: general: *Darlingtonia* is monotypic.

Darlingtonia californica **Torr.** grows in Pacific coastal bogs, depressions, fens, meadows, seeps, swamps, and along creeks at elevations of up to 2600 m. Substrates range from acidic to alkaline (pH: 4.4–9.5) peat, peridotite, sand, or often (but not always) ultramafic (serpentine) gravels containing high levels of magnesium but low levels of calcium. The plants typically occupy open sites and can become shaded out by the overgrowth of taller vegetation. The pendulous *Darlingtonia* flowers mature early and are held high over the leaves to reduce competition for insects that may serve as pollinators. They are protandrous and self-compatible; however, their unusual bell-shaped (apically dilated) ovary shields the stigma from self-pollen and also directs insects away from the stigma once they have encountered the anthers, thereby promoting outcrossing. Insects are rarely seen visiting the flowers (even under prolonged surveillance) of *Darlingtonia* but field studies indicate that bees (Hymenoptera: Andrenidae; Apidae), spiders (Arachnida: Arachnidae), and wasps (Hymenoptera: Vespidae) are at least potential pollinators. Pollinator-deprived flowers set virtually no seed, but viable seed can be produced by manipulated self- or cross-pollination. The capsules average over 1100 seeds. Fresh seed can be stored at 4°C if first dried at room temperature for 1 week. Cold stratification (4°C) for 1–2 months enhances germination, which occurs optimally in high light under damp conditions at 21°C–29°C. Seedlings grow ideally at 21°C. The plants can be propagated by separation of stolons or by rhizome cuttings, which will produce new growth in 3–6 weeks if kept under damp, high light conditions at 21°C. The seeds possess bristly projections that facilitate their dispersal by wind or animals and also enhance their flotation. *Darlingtonia* plants are tolerant of serpentine but accumulate only low amounts of nickel and zinc in their foliage. They are at least moderately fire-tolerant and resprout if habitats are not burned severely (small fires assist the plants by removing competing vegetation). The plants tolerate air temperatures from −11°C to 38°C, but require soil temperatures of 16°C–20°C for optimal growth. They must experience a 4–6 month dormant period (at 2°C–10°C) to persist and thrive. *Darlingtonia* habitats typically contain low levels of nitrogen and phosphorous in the substrate; however, most of the nitrogen is taken up directly by the plants from the soil. The carnivorous habit may provide critical nitrogen and/or phosphorous when conditions are extremely limiting for these nutrients. Approximately 2% of the organisms that visit the pitchers are captured. They are enticed by the coloration and copious nectar production of the tubular leaves. The leaves are covered by hoods that contain fenestrations, i.e., translucent areas that admit light and appear as open "windows" to flying insects. A pair of "fang-like" appendages extends in front of the leaf opening and may provide a landing surface for insects. Once inside the pitchers, insects encounter downward pointing hairs, which prevent their withdrawal and eventually direct them into the fluid contents. The plants implement a community of bacteria, insects, mites, and protozoa that reside within the pitcher fluid to decompose the captured prey, thereby liberating nutrients (mainly nitrogen and phosphorous) that are readily absorbed by the leaves. Populations are thought to be highly clonal; however, genetic studies are needed to effectively evaluate their genetic structure. **Reported associates:** *Achlys triphylla, Agrostis, Allium falcifolium, Allotropa virgata, Amelanchier utahensis, Angelica genuflexa, Arctostaphylos, Argentina egedii, Athyrium filix-femina, Blechnum spicant, Boykinia elata, Calamagrostis nutkaensis, C. stricta, Calocedrus decurrens, Camassia leichtlinii, Carex aquatilis, C. aurea, C. cusickii, C. leptalea, C. obnupta, Castilleja miniata, Chamaecyparis lawsoniana, Cicuta douglasii, Comarum palustre, Cornus sericea, Cypripedium californicum, Deschampsia cespitosa, Drosera rotundifolia, Epilobium ciliatum, Eriophorum chamissonis, E. crinigerum, Frangula purshiana, Galium aparine, G. boreale, G. trifidum, Gaultheria shallon, Gentiana setigera, Goodyera oblongifolia, Hastingsia alba, Helenium bigelovii, Horkelia sericata, Hosackia oblongifolia, Hydrocotyle ranunculoides, Hypericum anagalloides, Juncus effusus, J. ensifolius, J. xiphoides, Kalmia, Lathyrus, Leucothoe davisiae, Lilium bolanderi, L. pardalinum, L. washingtonianum, Linnaea borealis, Lonicera involucrata, Lotus corniculatus, Lycopus americanus, Lysichiton americanus, Lysimachia terrestris, Menyanthes trifoliata, Mimulus moschatus, M. primuloides, Mitella breweri, Myrica californica, Narthecium californicum, Oenanthe sarmentosa, Oreostemma alpigenum, Orobanche, Parnassia palustris, Picea sitchensis, Pinguicula macroceras, P. vulgaris, Pinus contorta, P. jeffreyi, P. lambertiana, P. monticola, P. ponderosa, Platanthera dilatata, P. sparsiflora, Polypodium glycyrrhiza, Potentilla breweri, Primula jeffreyi, Pseudotsuga, Quercus garryana, Q. vacciniifolia, Rhamnus californica, Rhododendron columbianum, R. occidentale, Rhychospora alba, Rubus ursinus, Rudbeckia californica, Salix hookeriana, Sanguisorba microcephala, Schoenoplectus acutus, Sisyrinchium, Sphagnum fuscum, S. palustre, S. papillosum, S. quinquefarium, Spiraea douglasii, Symphyotrichum spathulatum, Tiarella unifoliata, Triantha glutinosa, Trientalis europaea, T. latifolia, Trillium ovatum, T. rivale, Tsuga heterophylla, Vaccinium macrocarpon, V. oxycoccos, V. uliginosum, Veratrum viride, Veronica americana, Viola lanceolata, V. primulifolia.*

Use by wildlife: *Darlingtonia californica* is not known to be grazed. It is host to numerous arthropods including aphids (Homoptera: Aphididae: *Macrosiphum*), flies (Diptera: Chironomidae: *Metriocnemus edwardsi*; Chloropidae: *Caviceps darlingtoniae, Nartshukiella*; Phoridae: *Megaselia orestes*; Sciaridae: *Corynoptera*), mites (Astigmata: Histiostomatidae: *Sarraceniopus darlingtoniae*; Prostigmata: Eriophyidae: *Leipothrix*), moths (Lepidoptera: Pyralidae: *Udea*), spiders (*Eperigone trilobata*), and wasps (Hymenoptera: Braconidae: *Aspilota; Orgilus strigosus*). The plants also host a sac fungus (Ascomycota: Mycosphaerellaceae: *Septoria darlingtoniae*).

Economic importance: food: The leaves are not edible and emit a putrid odor when cut; **medicinal:** none; **cultivation:** *Darlingtonia californica* is widely cultivated as a botanical curiosity. It is distributed as the cultivars 'Othello' (an anthocyanin-lacking variety) and 'Siskiyou Dragon' and has been given the Award of Garden Merit by the Royal Horticultural Society; **misc. products:** The Yurok tribe used plants of *Darlingtonia californica* to make an insecticide; **weeds:** none; **nonindigenous species:** none

Systematics: Molecular data resolve *Darlingtonia* as the sister genus to a clade consisting of *Heliamphora* and *Sarracenia*, the two other genera of Sarraceniaceae (Figure 5.21). *Darlingtonia* is monotypic and narrowly restricted, which obviates the need for further elucidation of its relationships. All three genera of Sarraceniaceae consist entirely of aquatic species, thus indicating a single origin of the OBL habit for the family. Some *Darlingtonia* populations lack anthocyanins in the petals, which are greenish-yellow instead of the usual maroon coloration. The base chromosome number of *Darlingtonia* is $x = 15$, with *D. californica* ($2n = 30$) a diploid. It is not known to hybridize with other members of Sarraceniaceae. However, population genetic studies would be useful for characterizing the breeding system, dispersal range, and quantifying the level of genetic variation that resides within and among populations.

Comments: *Darlingtonia californica* is endemic to California, Oregon, and Washington.

References: Christy, 2004; Crane, 1990; Elder, 1994, 1997; Ellison & Farnsworth, 2005; Fashing, 1994; Fashing & O'Connor, 1984; Freeman, 1994, 1996; Kondo, 1969; Lloyd, 1976; Mellichamp, 2009; Meyers-Rice, 1998; Nielsen, 1990; Nyoka & Ferguson, 1999; Pietropaolo & Pietropaolo, 1993; Reeves et al., 1983; Schnell, 2002; Slack, 1981; Torrey, 1853.

2. Sarracenia

Pitcherplant; sarracénie

Etymology: after Michel Sarrazin (1659–1734)

Synonyms: *Sarazinia*; *Sarrazina; Sarrazinia*

Distribution: global: North America; **North America:** eastern, northern, western

Diversity: global: 11 species; **North America:** 11 species

Indicators (USA): OBL: *Sarracenia alabamensis, S. alata, S. flava, S. jonesii S. leucophylla, S. minor, S. oreophila, S. psittacina, S. purpurea, S. rosea S. rubra*

Habitat: freshwater; palustrine; **pH:** 3.3–8.9; **depth:** <1 m; **life-form(s):** emergent herb

Key morphology: stems (to 30 cm) rhizomatous; leaves in rosettes or clusters, phyllodial (to 40 cm) and/or tubular (to 120 cm), the latter modified into snake-like or horn-like structures consisting of a hollow, fluid-filled pitcher-like body covered by an apical hood that surmounts a terminal orifice; the remains of insects and other invertebrates often are found within the leaf cavities; flowers nodding, 5-merous, showy, solitary, terminating a long naked scape (to 80 cm) with three bracts (to 2 cm) that subtend the flower; sepals (to 6 cm) petaloid, maroon, purple, yellow, or yellow-green; petals (to 8.5 cm) maroon, pink, purple, red, rose, yellow, greenish-yellow, or whitish-yellow, drooping from the margins of the expanded stylar disk; stamens numerous, in 10–17 fascicles of 2–8; style expanded into an umbrella-like disk (to 8.5 cm), greenish, whitish, yellow, or yellow-green, 5-lobed, the stigmatic surfaces beneath each lobe; capsules (to 2 cm) globose or ovoid, many seeded; seeds winged laterally

Life history: duration: perennial (rhizomes); **asexual reproduction:** rhizomes; **pollination:** insect; **sexual condition:** hermaphroditic; **fruit:** capsules (common); **local dispersal:** rhizomes, seeds (gravity, water); **long-distance dispersal:** seeds (water, wind)

Imperilment: (1) *Sarracenia alabamensis* [G4]; S1 [AL]; (2) *S. alata* [G4]; (3) *S. flava* [G5]; S1 (VA); S3 (GA, NC); (4) *S. jonesii* [G4]; S1 (NC, SC); (5) *S. leucophylla* [G3]; S1 (GA); S2 (MS); S3 (AL, FL); (6) *S. minor* [G4]; S1 (NC); (7) *S. oreophila* [G2]; SX (TN); S1 (GA, NC); S2 (AL); (8) *S. psittacina* [G4]; S2 (GA); S3 (LA); (9) *S. purpurea* [G5]; SH (LA); S1 (GA, IL, VA); S2 (DE, MD, NC, OH); S3 (AB, AL,

IN, NC, NY, SC); (10) *S. rosea* [G5]; SX (GA); S1 (MS); S3 (FL); (11) *S. rubra* [G4]; S1 (MS); S2 (FL, GA); S3 (AL, NC, SC, FL)

Ecology: general: All *Sarracenia* species are categorized as OBL aquatics. The species are FAC carnivores, which can derive nutrients from animal prey that becomes captured within the fluid of their highly modified, tubuler, pitcher-like leaves. These leaves have significantly lower "construction costs" (i.e., the amount of carbon used to produce a gram of tissue) than leaves of non-carnivorous species, but require substantially longer photosynthetic periods to meet those costs. Prey is attracted to the openings of the pitcher traps by nectar and by scents derived from dozens of volatile substances, which are thought to mimic floral odors. Some species have window-like fenestrations in the pitchers, which were thought to induce entry into the traps as well as deceive prey regarding exit locations; however, experimental evidence indicates that they do neither, but function in the initial long distance attraction of insects to the plants. The species produce a variety of enzymes to assist with the decomposition of captured prey including acid phosphatase, amylase, esterase, and protease. Carnivory provides the plants with supplemental nutrients (e.g., nitrogen), which facilitate their survival in nutrient-poor sites (e.g., acidic peat or sand substrates) where the species typically occur. The flowers are self-compatible, but are designed to promote outcrossing by their pendent habit, broadened style with upturned stigmatic lobes, and petals that cover the stamens. Typically, the pollinators enter the flower from between two of the pendant petals, and proceed to deposit pollen on one of the v-shaped stigmas, while collecting any pollen that has dropped to the interior of the cup-like style. Because they exit by a different route, the deposition of self-pollen is unlikely. Pollination occurs mainly by bumblebees (Hymenoptera: Apidae: *Bombus*) and to a lesser degree by honey bees (Hymenoptera: Apidae: *Apis mellifera*). Natural interspecific hybridization is reduced by differences in seasonal flowering phenology, which often characterize sympatric species. The capsules dehisce apically in all species except *S. leucophylla*, in which the dehiscence is basal. Germination rates and pretreatment requirements are relative to seed size, with the larger-seeded species having the fastest germination rates but requiring longer periods of stratification. The seed surface is waxy and hydrophobic, which facilitates dispersal by water. The species are resistant to fire, which removes competing vegetation and promotes open conditions that are conducive to flowering, seed germination, and seedling establishment. All species reproduce vegetatively by means of rhizomes, which can live up to 35 years and are capable of division by fragmentation. Most of the species require a winter dormant period (induced below 10°C) to maintain their vitality.

Sarracenia alabamensis **Case & R.B. Case** occurs in bogs, canebreaks, depressions, meadows, seeps, slopes, swales, thickets, and along the margins of streams and swamps at elevations of up to 200 m. Most of the remaining extant sites are seepage bogs. The plants tolerate shade but grow best in full sunlight. Their substrates are acidic (pH: 3.9–4.7) and are described as gravel, muck, peat, sand, or sandy gravel. The substrates typically are high in aluminum (67–1300 ppm), low in nitrogen (mean = 0.18%), and are well drained with a high sand content (46%–91%; mean = 70%). Other substrate characteristics include (mean values): Ca (100 ppm), carbon (3.4%), Co (0.11 ppm), Fe (150 ppm), K (28 ppm), Mg (23 ppm), Mn (5 ppm), Mo (0.33 ppm), organic matter (6.7%), P (5 ppm), and Pb (2.8 ppm). Flowering extends from April to May, and can even occur in fairly dense shade, which reduces the vegetative growth. However, successful sexual reproduction (as evidenced by seedling occurrence) is low and has been observed only in about half of the extant populations. The "pitchers" (leaves) are dimorphic seasonally, with narrow, curved spring forms and broad, straight summer forms. Vegetative reproduction occurs by rhizomes, which can result in the development of clonal populations. Although frequent fires are believed necessary to maintain populations by reducing competition from woody species, attempts to manage populations of this rare species by the removal of potentially competing woody plants were unsuccessful, with no positive response observed afterwards. The populations are threatened primarily by habitat destruction due to various construction activities and by plowing. **Reported associates:** *Acer rubrum, Aletris aurea, A. farinosa, A. obovata, Alnus serrulata, Arundinaria gigantea, A. tecta, Balduina uniflora, Calopogon tuberosus, Carex, Cyrilla racemiflora, Drosera capillaris, Eleocharis microcarpa, Epigaea repens, Eriocaulon compressum, Eryngium integrifolium, Hexastylis speciosa, Ilex coriacea, I. glabra, I. opaca, Juncus effusus, Lachnocaulon anceps, Lilium catesbaei, Liquidambar styraciflua, Liriodendron tulipifera, Ludwigia alternifolia, Lycopodiella alopecuroides, L. carolinianum, Lyonia lucida, Magnolia virginiana, Mitreola sessilifolia, Myrica cerifera, M. heterophylla, Osmundastrum cinnamomeum, Pinguicula primuliflora, Pinus palustris, Platanthera ciliaris, Pogonia ophioglossoides, Polygala brevifolia, P. lutea, P. nana, Pteridium aquilinum, Pycnanthemum, Quercus, Rhexia alifanus, R. mariana, Rhododendron canescens, Rhynchospora, Rubus, Scutellaria integrifolia, Smilax auriculata, S. laurifolia, Sphagnum, Spiranthes, Stenanthium densum, Symplocos tinctoria, Toxicodendron vernix, Triantha racemosa, Typha latifolia, Utricularia subulata, Vaccinium elliottii, Xyris.*

***Sarracenia alata* (Alph. Wood) Alph. Wood** inhabits bogs, depressions, ditches, flatwoods, marshes, meadows, prairies, roadsides, savannas, seeps, slopes, and the margins of ponds and streams at elevations of up to 50 m. Slow water currents may characterize some localities. The sites are open to partially shaded, with acidic (pH: 3.8–5.8), substrates that include clay, Escambia sandy loam, muck, peat, and sand. The plants occur commonly in heavy sandy clay substrates on hillsides where seepages exist. Flowers are produced from March to April. The floral ecology of this species has not been described in any detail, but presumably is similar to that of the other species. Optimal seed germination occurs at 20°C after 2 weeks of cold, moist stratification at 4°C. Mass insect captures have been observed with the pitchers of approximately 7700 plants in a small (0.4 ha) bog found

to contain roughly two million "lovebugs" (Diptera: *Plecia nearctica*). However, the prey can reflect a diverse sample of the local fauna including ants (Hymenoptera: Formicidae), bees and wasps (Hymenoptera), beetles (Coleoptera), bugs (Hemiptera), flies (Diptera), moths and butterflies (Lepidoptera), nematodes (Nematoda), spiders (Araneae), and even larger specimens such as millipedes (Myriopoda: Diplopoda) and juvenile lizards (Reptilia: Polychrotidae: *Anolis*). *Sarracenia alata* plants respond plastically to light, reducing their nutrient requirements when shaded, and exploiting open periods of ample light (e.g., postfires) by carnivory. Periodic fires enhance the habitat conditions. After burning, plants showed an increase in leaf number compared to unburned control plots. Also, seedling recruitment is related exponentially to the number of floral scapes produced in the year before a fire and is greater in burned sites than in unburned sites. **Reported associates:** *Aletris, Andropogon gyrans, A. mohrii, Aristida palustris, A. stricta, Burmannia capitata, Calopogon tuberosus, Carex exilis, Carphephorus pseudoliatris, Chaptalia tomentosa, Cliftonia monophylla, Ctenium aromaticum, Cynodon, Cyrilla racemiflora, Doellingeria umbellata, Drosera brevifolia, D. capillaris, D. tracyi, Eriocaulon compressum, E. decangulare, E. lineare, E. texense, Eryngium cuneifolium, Helianthus heterophyllus, Hypericum, Ilex glabra, Lachnanthes caroliniana, Lachnocaulon digynum, Liatris, Lilium catesbaei, Lophiola, Ludwigia, Lycopodiella, Magnolia virginiana, Marshallia, Muhlenbergia capillaris, M. expansa, Myrica, Nyssa biflora, Orontium aquaticum, Panicum nudicaule, Persea borbonia, Pinguicula lutea, P. planifolia, P. primuliflora, Pinus elliottii, P. palustris, Platanthera integra, Pogonia ophioglossoides, Polygala cruciata, P. cymosa, P. hookeri, P. incarnata, P. lutea, P. nana, P. racemosa, P. ramosa, Rhexia alifanus, R. lutea, Rhynchospora chapmanii, R. gracilenta, R. macra, R. oligantha, R. rariflora, R. stenophylla, Rudbeckia scabrifolia, Sabatia macrophylla, Sarracenia jonesii, S. leucophylla, S. psittacina, Saururus cernuus, Schoenolirion croceum, Scleria muehlenbergii, Serenoa repens, Smilax laurifolia, Solidago patula, Sphagnum, Stokesia laevis, Toxicodendron vernix, Triantha racemosa, Utriculata cornuta, U. subulata, Xyris chapmanii, Zigadenus glaberrimus.*

***Sarracenia flava* L.** grows in baygalls, bogs, Carolina bays (peat based), depressions, flatwoods, meadows, savannas, seeps, slopes, swamps, thickets, and along pond and stream margins at elevations of up to 300 m. The habitats are acidic (pH: 4.0–5.5), and often have a peat (*Sphagnum*) substrate. The plants can tolerate shallow standing water but also are common in zones where the water has receded. The leaves do not persist over winter. The flowers are produced from March to April and are pollinated primarily by bumblebees (Hymenoptera: Apidae: *Bombus*) and secondarily by honey bees (Hymenoptera: Apidae: *Apis mellifera*). Although the plants are self-compatible, self-pollination can result in significant inbreeding depression (e.g., reduced seed mass and number; lower germination, growth, and survivorship) in some populations. In contrast, heterosis (enhanced vigor) has been observed as the consequence of crosses made using plants from different populations. Seed germination is optimized at 20°C–25°C following 4 weeks of cold, moist stratification at 4°C. Fires help to maintain the habitats by removing competitive shrubby vegetation. The pitchers produce coniine, an alkaloid that is capable of paralyzing insects. **Reported associates:** *Acer rubrum, Agalinis filicaulis, A. linifolia, Amianthium muscitoxicum, Andropogon glomeratus, A. gyrans, A. mohrii, A. virginicus, Aristida beyrichiana, A. palustris, Asclepias longifolia, Balduina uniflora, Bigelowia nudata, Calopogon pallidus, C. tuberosus, Carex exilis, C. striata, C. striatula, Carphephorus paniculatus, Castanea, Centella erecta, Chamaedaphne calyculata, Chaptalia tomentosa, Cleistes divaricata, Clethra alnifolia, Cliftonia monophylla, Coreopsis floridana, C. gladiata, C. linifolia, Ctenium aromaticum, Cyrilla racemiflora, Danthonia epilis, Dichanthelium acuminatum, D. dichotomum, D. longiligulatum, D. meridionale, D. scabriusculum, D. spretum, Drosera capillaris, D. intermedia, D. rotundifolia, D. tracyi, Eleocharis tuberculosa, Erigeron vernus, Eriocaulon compressum, E. decangulare, Eryngium integrifolium, Eupatorium leucolepis, E. mohrii, E. pilosum, E. resinosum, E. rotundifolium, Eurybia chapmanii, E. paludosa, Fimbristylis spadicea, Fuirena squarrosa, Gaylussacia dumosa, G. frondosa, Gentiana catesbaei, Gymnadeniopsis nivea, Helenium vernale, Helianthus angustifolius, H. heterophyllus, Hypericum brachyphyllum, H. chapmanii, H. cruxandreae, H. fasciculatum, H. hypericoides, Ilex glabra, I. myrtifolia, Ionactis linariifolia, Juncus canadensis, J. scirpoides, J. trigonocarpus, Lachnanthes caroliana, Lachnocaulon anceps, Liatris spicata, Lobelia glandulosa, L. paludosa, Lophiola americana, L. aurea, Lycopodiella alopecuroides, L. prostrata, Lyonia ligustrina, L. lucida, Magnolia virginiana, Marshallia graminifolia, Muhlenbergia capillaris, M. expansa, Myrica cerifera, M. heterophylla, Nyssa biflora, Osmundastrum cinnamomeum, Oxypolis filiformis, Panicum chamaelonche, Panicum hemitomon, P. virgatum, Parnassia caroliniana, Persea palustris, Photinia pyrifolia, Pinguicula caerulea, Pinus elliottii, P. palustris, P. serotina, Pleea tenuifolia, Pogonia ophioglossoides, Polygala cruciata, P. lutea, Rhexia alifanus, R. lutea, R. mariana, R. nashii, R. petiolata, Rhynchospora baldwinii, R. cephalantha, R. chalarocephala, R. inundata, R. latifolia, R. macra, R. oligantha, R. plumosa, R. rariflora, R. stenophylla, Rubus, Rudbeckia graminifolia, Sabatia difformis, S. macrophylla, Sarracenia leucophylla, S. minor, S. psittacina, S. purpurea, S. rubra, Schizachyrium scoparium, Scleria triglomerata, Silphium simpsonii, Sisyrinchium capillare, S. mucronatum, Smilax laurifolia, Solidago patula, S. stricta, Sphagnum, Sporobolus pinetorum, Stenanthium densum, Symphyotrichum dumosum, Syngonanthus flavidulus, Taxodium ascendens, Tofieldia glabra, Toxicodendron vernix, Trianthra racemosa, Utricularia juncea, U. subulata, Vaccinium formosum, V. fuscatum, Viola ×primulifolia, Woodwardia virginica, Xyris ambigua, X. baldwiniana, X. caroliniana, X. chapmanii, X. fimbriata, X. platylepis, Zenobia pulverulenta, Zigadenus glaberrimus.*

***Sarracenia jonesii* Wherry** grows in montane blanket bogs, meadows, seepage bogs, and seeps at elevations from 300 to 600 m. The plants have also been found in rock depressions and in soil pockets in the rock substrate adjacent to waterfalls. The substrates often are granitic. Flowering occurs in May. Genetic analyses (using allozymes) indicated reduced outcrossing rates and low levels of genetic variation in populations compared to the more widely distributed *S. purpurea*. Optimum germination (at 20°C; 14 h daylight) has been observed within 2 weeks for seeds subjected to a 4–6 week period of cold (4°C), moist stratification. Seedling establishment is low and was observed in only three of the 10 known populations, which were monitored over a 2-year period. Reproduction primarily is vegetative and occurs by long-lived rhizomes, which can develop into clones up to 0.5 m across. Some plants lack anthocyanins, which results in yellowish-green foliage and flowers. **Reported associates:** *Acer rubrum, Alnus serrulata, Aronia arbutifolia, Bignonia capreolata, Calamagrostis canadensis, Carex collinsii, C. echinata, C. folliculata, C. gynandra, C. leptalea, Chamaedaphne calyculata, Croton michauxii, Danthonia sericea, Dichanthelium dichotomum, D. ensifolium, D. sphaerocarpon, Drosera rotundifolia, Dulichium arundinaceum, Eriophorum virginicum, Eupatorium pilosum, E. rotundifolium, Fuirena breviseta, Gaylussacia bigeloviana, Hypericum densiflorum, H. hypericoides, Ilex verticillata, Isoetes caroliniana, Juncus caesariensis, J. effusus, Kalmia angustifolia, K. latifolia, Linum floridanum, Liriodendron tulipifera, Lobelia nuttallii, Ludwigia alternifolia, Lyonia ligustrina, Menziesia pilosa, Myrica gale, Nyssa sylvatica, Osmunda regalis, Osmundastrum cinnamomeum, Oxypolis rigidior, Packera aurea, Parnassia asarifolia, Pinus rigida, P. strobus, P. virginiana, Rhododendron arborescens, R. maximum, R. viscosum, Rhynchospora gracilenta, Rosa palustris, Rubus Sagittaria fasciculata, Salix sericea, Sarracenia purpurea, Scirpus cyperinus, Scleria triglomerata, Selaginella apoda, Smilax laurifolia, Solidago patula, Sphagnum affine, S. bartlettianum, S. palustre, S. recurvum, Thelypteris palustris, Tsuga canadensis, Vaccinium formosum, Viburnum nudum, Viola primulifolia, Vitis rotundifolia, Woodwardia areolata, W. virginica, Xyris platylepis.*

***Sarracenia leucophylla* Raf.** is found in baygall ecotones, bogs, depressions, ditches, flatwoods, prairies, savannas, seeps, swamps, thickets, and along the margins of branch bays and streams at elevations of up to 90 m. Habitats typically are open (including light gaps) and are characterized by relatively constant soil moisture levels. The substrates are acidic (pH: 4.1–5.1) sands or sandy peat. The flowers are produced from March to April and are pollinated and outcrossed mainly by bumblebees (Hymenoptera: Apidae: *Bombus*). Good seed germination has been reported at 20°C–25°C following 2–4 weeks of cold, moist stratification at 4°C. The seeds are dispersed by gravity or by water. The traps are extremely effective at retaining even large-bodied insects and preventing their escape. Surveys using allozymes indicate that populations exhibit extremely high levels of genetic diversity (highest among four *Sarracenia* species examined), especially for such a narrowly restricted species. However, heterozygote deficiencies in several populations indicate that some level of inbreeding may be occurring. Populations rely on periodic episodes of fire to remove competing woody vegetation and often show striking recoveries when overgrown areas are burned. **Reported associates:** *Acer rubrum, Agalinis, Aletris lutea, Andropogon capillipes, A. glomeratus, Aristida beyrichiana, A. stricta, Arnoglossum, Bartonia verna, Bigelowia nudata, Cacalia, Carphephorus pseudoliatris, Centella asiatica, Chamaecyperus thyoides, Chaptalia tomentosa, Cliftonia monophylla, Coreopsis gladiata, Ctenium aromaticum, Cyrilla racemiflora, Dichanthelium, Drosera capillaris, D. brevifolia, D. rotundifolia, D. tracyi, Dulichium arundinaceum, Eleocharis, Erigeron vernus, Eriocaulon compressum, E. decangulare, E. texense, Eupatorium rotundifolium, Fuirena, Gaylussacia mosieri, Helenium vernale, Helianthus heterophyllus, Hypericum brachyphyllum, H. chapmanii, H. fasciculatum, H. galioides, Ilex coriacea, I. glabra, I. myrtifolia, I. opaca, Lachnanthes caroliniana, Lachnocaulon anceps, Liquidambar styraciflua, Lophiola americana, Lycopodiella alopecuroides, L. appressa, L. caroliniana, Lyonia lucida, Magnolia virginiana, Mayaca fluviatilis, Mimosa strigillosa, Muhlenbergia, Myrica cerifera, Myriophyllum, Nymphaea, Nymphoides indica, Nyssa biflora, N. sylvatica, Orontium aquaticum, Osmundastrum cinnamomeum, Panicum, Persea palustris, Photinia pyrifolia, Pinguicula lutea, Pinus elliottii, P. glabra, P. palustris, Pleea tenuifolia, Polygala lutea, P. ramosa, Proserpinaca pectinata, Pteridium aquilinum, Quercus geminata, Rhexia alifanus, R. lutea, Rhynchospora chapmanii, R. oligantha, Rudbeckia auriculata, Sagittaria latifolia, Sarracenia alata, S. flava, S. psittacina, S. purpurea, S. rubra, Scleria, Smilax laurifolia, S. pumila, Sphagnum, Sporobolus floridanus, Stenanthium densum, Syngonanthus flavidulus, Taxodium ascendens, T. distichum, Thelypteris, Utricularia lutea, U. purpurea, Vaccinium fuscatum, Viburnum nudum, Vitis rotundifolia, Woodwardia areolata, W. virginiana, Xyris ambigua, X. brevifolia, X. caroliniana.*

Sarracenia minor **Walter** occurs in bogs, depressions, ditches, flatwoods, prairies, roadsides, savannas, seeps, and swamps at elevations of up to 90 m. The substrates are acidic (pH: 4.5–5.5) and are described as peat (sometimes quaking), peaty sand, sand, sandy peat, and spodic psammaquents. The plants typically grow in open areas but can also tolerate partial shade. They are fairly broadly adapted and will survive in standing water but often are found in relatively drier areas than other *Sarracenia* species. Unlike most *Sarracenia*, the pitchers and flowers expand simultaneously. Flowers are produced from March to May and are pollinated by bees (Hymenoptera). Seed germination is reported at 20°C–25°C following 4 weeks of cold, moist stratification at 4°C. In natural populations, germination and seedling establishment require bare, open ground. The plants require sites that experience periodic fire to remove competing woody vegetation, release nutrients, and provide open space necessary for seed germination. They survive burns by means of their resistant rhizomes; however, intense fires that burn into the peat layer will kill the plants. Increased plant density has been observed in areas that were burned annually for nearly 30 years. The trap leaves persist over the winter. The primary prey consists of ants (Inecta:

Hymenoptera: Formicidae), which are attracted to the pitchers by arrays of nectar droplets that form along their margins. **Reported associates:** *Agalinis linifolia, Aletris, Amianthium muscitoxicum, Andropogon glomeratus, Aristida purpurascens, Arundinaria gigantea, Bejaria racemosa, Bigelowia nudata, Burmannia, Calopogon tuberosus, Carex striata, Ctenium aromaticum, Cyrilla racemiflora, Dichanthelium scabriusculum, Drosera capillaris, Dulichium arundinaceum, Erigeron vernus, Eriocaulon compressum, E. decangulare, Eupatorium leucolepis, E. rotundifolium, Eurybia paludosa, Fimbristylis spadicea, Gaylussacia frondosa, G. tomentosa, Gymnadeniopsis nivea, Helenium vernale, Helianthus radula, Hypericum brachyphyllum, H. fasciculatum, Ilex coriacea, I. glabra, Ionactis linariifolia, Kalmia hirsuta, Lachnanthes caroliana, Lachnocaulon anceps, Leucothoe racemosa, Liatris spicata, Lilium catesbaei, Liquidambar styraciflua, Ludwigia lanceolata, Lycopodiella alopecuroides, Lyonia ligustrina, L. lucida, L. mariana, Marshallia graminifolia, Muhlenbergia expansa, Nyssa biflora, Orontium aquaticum, Panicum abscissum, P. hemitomon, Peltandra sagittifolia, Persea palustris, Pinguicula caerulea, Pinus elliottii, P. serotina, Polygala cruciata, P. lutea, Polygonum, Rhexia alifanus, Rhynchospora fascicularis, R. latifolia, Rubus cuneifolius, Sabatia difformis, Saccharum giganteum, Sarracenia flava, S. psittacina, Schizachyrium scoparium, Scleria reticularis, Serenoa repens, Sisyrinchium mucronatum, Smilax laurifolia, Solidago stricta, Sphagnum, Sporobolus pinetorum, Styrax americanus, Syngonanthus flavidulus, Taxodium ascendens, Triadenum virginicum, Trianthu racemosa, Utricularia inflata, Vaccinium, Woodwardia virginica, Xyris caroliniana, X. fimbriata.*

Sarracenia oreophila **Wherry** grows in depressions, seepage bogs, seeps, swamps, thickets, and along river and stream margins at elevations of up to 535 m. Its habitats are described as acidic (specific pH values not reported) with sandy substrates or sands mixed with heavy clay. The plants require open sites with full sunlight to thrive but occasionally are relictual in more shaded areas. Even if provided with ample moisture, the plants enter dormancy much earlier than other *Sarracenia* species (mid-July), which enables them to better endure dessicating habitat conditions. As the habitats dry, the pitchers die back and are replaced by phyllodial leaves. Dormant plants can withstand temperatures down to –7°C for short periods of time (~1 week). Flowering extends from April to June. Bumblebees (Insecta: Hymenoptera: Apidae: *Bombus*) are the major pollinators. Allozyme analysis indicates that *S. oreophila* predominantly is outcrossed, and that it possesses extremely low levels of genetic diversity. Genetic variation is lowest in the more isolated or smaller populations. Fire is an essential component of population persistence and is necessary to remove competition and to provide open areas necessary for seed germination. The seeds reportedly will not germinate unless they land on exposed, moist, mineral soils. Seed germination is high (to 90%) in natural populations and occurs within a few months after they are shed; however, germination is enhanced by cold, moist stratification (4 weeks at 4°C) and a germination temperature of 20°C–25°C. Stored seeds can survive for several years. The plants are relatively slow growing compared to other *Sarracenia* species. Reproduction primarily is asexual and occurs mainly by the vegetative growth of the long-lived rhizomes. **Reported associates:** *Acer rubrum, Alnus serrulata, Arundinaria gigantea, Aureolaria flava, Chamaecyparis thyoides, Cornus florida, Drosera intermedia, Eriophorum virginicum, Eryngium integrifolium, Eupatorium fistulosum, E. perfoliatum, E. pilosum, E. rotundifolium, Helianthus angustifolius, Hexastylis shuttleworthii, Hieracium gronovii, Ilex verticillata, Isotria verticillata, Juncus caesariensis, Kalmia latifolia, Lespedeza violacea, Liquidambar styraciflua, Liriodendron tulipifera, Lobelia spicata, Ludwigia alternifolia, Lyonia ligustrina, Lysimachia graminea, Nyssa sylvatica, Osmundastrum cinnamomeum, Oxydendrum arboretum, Photinia pyrifolia, Pinus echinata, P. palustris, P. taeda, Quercus coccinea, Q. falcata, Q. stellata, Q. velutina, Rhexia mariana, R. virginica, Rhododendron arborescens, R. canescens, Rhus copallinum, Rhynchospora rariflora, Rosa palustris, Rubus argutus, Sagittaria latifolia, S. secundifolia, Sanguisorba canadensis, Sassafras albidum, Smilax glauca, S. rotundifolia, Solidago odora, S. speciosa, Thelypteris palustris, Vaccinium arboreum, V. corymbosum, Vitis rotundifolia.*

Sarracenia psittacina **Michx.** inhabits bays, bogs, depressions, ditches, flatwoods, pocosins, prairies, roadsides, savannas, seepage slopes, seeps, sloughs, and swamps at elevations of up to 60 m. The substrates are acidic (pH: 3.8–5.8) and consist of peat, sand, or sandy peat. This species occurs typically in wetter areas and unlike most of its congeners, can tolerate brief periods of complete inundation. Flowering occurs from March to June and extends later than most of its congeners. Pollination presumably occurs by bumblebees (Hymenoptera: Apidae: *Bombus*), although the reproductive ecology of this species has not been studied in detail. Seeds germinate well after 2 weeks of cold, moist stratification at 4°C. The decumbent trap leaves overwinter and are modified with hooded tops that open to the tubular base (filled with inward-directed hairs) by means of a "collar," which prevents prey from escaping after entering. This "lobster-pot" type of trap is particularly effective at capturing aquatic prey when the traps are inundated. The plants alter their rhizome morphology in response to deepening substrate layers by sending up aerial "stolonoids," which ensures that new flowers and leaves will arise above the substrate. **Reported associates:** *Agalinis pinetorum, Aletris obovata, Andropogon glomeratus, A. gyrans, A. mohrii, Aristida beyrichiana, A. palustris, A. stricta, Balduina uniflora, Bigelowia nudata, Calopogon, Carex exilis, Carphephorus pseudoliatris, Chaptalia tomentosa, Cliftonia monophylla, Coreopsis linifolia, Ctenium aromaticum, Cyrilla racemiflora, Dichanthelium, Doellingeria umbellata, Drosera brevifolia, D. capillaris, D. tracyi, Eriocaulon compressum, E. decangulare, E. nigrobracteatum, E. texense, Fuirena breviseta, Helianthus heterophyllus, Hypericum brachyphyllum, H. chapmanii, H. fasciculatum, Ilex, Lachnanthes caroliniana, Lachnocaulon digynum, Lilium catesbaei, Linum, Lobelia paludosa, Lophiola americana, L. aurea, Lycopodiella alopecuroides, Lyonia lucida,*
Muhlenbergia expansa, Myrica heterophylla, Orontium aquaticum, Osmundastrum cinnamomeum, Oxypolis filiformis, Panicum nudicaule, Pinguicula lutea, P. planifolia, P. primuliflora, Pinus elliottii, Platanthera integra, Pleea tenuifolia, Polygala lutea, Rhexia alifanus, R. lutea, Rhynchospora chapmanii, R. corniculata, R. latifolia, R. macra, R. oligantha, R. stenophylla, Rudbeckia graminifolia, Sabatia macrophylla, Sarracenia alata, S. flava, S. leucophylla, S. purpurea, S. rubra, Scleria baldwinii, S. muehlenbergii, S. reticularis, Sphagnum, Sporobolus floridanus, Stokesia laevis, Syngonanthus flavidulus, Taxodium ascendens, Triantha racemosa, Utricularia juncea, U. subulata, Vaccinium, Woodwardia virginiana, Xyris ambigua, X. difformis, Zigadenus glaberrimus.

Sarracenia purpurea **L.** inhabits bays, bogs, depressions, ditches, fens, flatwoods, interdunal swales, lakeshores, pocosin ecotones, raised peatlands, savannas, seepage bogs, seeps, swamps, and the margins of lakes and ponds at elevations of up to 1000 m. The plants are widely adapted to grow in acidic to calcareous (pH: 3.3–8.9) conditions, on substrates that typically comprise peat (often quaking), sand, or sandy peat. Habitats range from open to shaded sites. The flowers are produced from March to July and are pollinated by large-bodied bees (Hymenoptera: Apidae: *Bombus*) and flies (Diptera). Approximately 500–1500 small seeds are produced by each capsule. The seeds are dormant when dispersed and require 4–6 weeks of cold, moist stratification at 4°C under lighted conditions for good germination at 28°C. Measured seed dispersal distances are low, with nearly 80% falling within 5 cm of the parental plant. The hydrophobic seeds may be dispersed further by water, but probably not to any significant distance given the isolated nature of many common habitats (i.e., bogs). Natural seedling establishment is also low with an estimated 95% seedling mortality rate. The seeds do not persist in the seed bank. The mechanism for long distance dispersal remains a mystery; however, effective dispersal to numerous isolated habitats throughout formerly glaciated areas has occurred over a relatively short time. Allozyme and inter-simple sequence repeats (ISSR) data indicate moderate levels of genetic variation and high interpopulational differentiation (due much to divergence associated with infraspecific taxa) relative to other *Sarracenia* species surveyed. Within-population levels of genetic variability are low. These genetic data indicate that dispersal has not effectively tempered population divergence across the range of the species. Once established at a favorable site, the plants can spread rapidly. A single plant introduced to one isolated bog achieved a population of 150,000 individuals within only 70 years. Where introduced to Sweden, one population has grown to more than 1000 individuals after 35 years. Plants introduced to one Irish bog spread to cover more than 32 ha within 20 years. Individuals are long lived with a lifespan of 30–50 years. The leaves are evergreen and can freeze solidly (with no permanent damage) during the winters in northern localities. This is the only *Sarracenia* species whose pitchers are not covered by the hood, which admits rainwater and promotes the development of bacterial communities that can facilitate digestion of prey; however, the plants also produce intrinsic digestive enzymes. The trap fluid also contains nitrogen-fixing bacteria (Azotobacteriaceae). The pitchers are inefficient at trapping prey, with only about 1% of the visiting organisms ultimately getting captured; however, plants often are not nutrient limited and can store the nutrients obtained during one season for use in subsequent years. The newer leaves (2–5 weeks of age) are more effective at prey capture than are the older leaves. Ants (Hymenoptera: Formicidae) are the most common potential prey (more than 26 species), but they are captured at extremely low rates (<0.5%). Ants typically comprise the majority of the prey in terms of both number and biomass, along with flies (Diptera), slugs and snails (Gastropoda), beetles (Coleoptera) and representatives from other insect orders (Collembola, Hemiptera, Homoptera, Lepidoptera, Neuroptera, Orthoptera, Plecoptera, Protura, Trichoptera) along with miscellaneous other invertebrates (Acarina, Araneae, Chilopoda, Diplopoda, Nematoda, Oligochaeta). The prey is attracted to the traps by their nectar, which contains amino acids. **Reported associates:** *Acer rubrum, Agalinis purpurea, Amelanchier, Andromeda glaucophylla, Andropogon glomeratus, Arethusa bulbosa, Aulacomnium palustre, Bartonia paniculata, Betula populifolia, B. pubescens, B. pumila, Calla palustris, Calopogon tuberosus, Campanula aparinoides, Campylium stellatum, Carex echinata, C. interior, C. flava, C. limosa, C. livida, C. magellanica, C. oligosperma, C. pauciflora, C. rostrata, C. stricta, C. trisperma, C. wiegandii, Chamaecyparis thyoides, Chamaedaphne calyculata, Cirsium, Cladium mariscoides, Cleistes divaricata, Comarum palustre, Coptis trifolia, Cornus amomum, Cypripedium acaule, Dasiphora floribunda, Decodon verticillatus, Drosera anglica, D. capillaris, D. intermedia, D. rotundifolia, D. tracyi, Dulichium arundinaceum, Eleocharis robbinsii, E. rostellata, Epilobium coloratum, Equisetum pratense, Eriocaulon decangulare, E. compressum, Eriophorum angustifolium, E. vaginatum, E. virginicum, Eupatorium perfoliatum, Eutrochium maculatum, Galium labradoricum, Gaultheria procumbens, Gaylussacia baccata, Glyceria canadensis, Ilex mucronata, Iris versicolor, Juncus canadensis, J. stygius, Kalmia angustifolia, K. polifolia, Larix laricina, Lobelia kalmii, Lophiola americana, Lycopus uniflorus, Lysichiton americanus, Lysimachia quadriflora, Maianthemum canadense, M. trifolium, Menyanthes trifoliata, Myrica gale, Nuphar advena, Nymphaea odorata, Osmunda regalis, Parnassia glauca, Peltandra virginica, Photinia melanocarpa, Phragmites australis, Picea mariana, P. pungens, Pinus strobus, P. taeda, Pogonia ophioglossoides, Pteridium aquilinum, Pycnanthemum virginianum, Rhexia lutea, Rhododendron canadense, R. groenlandicum, R. viscosum, Rhynchospora alba, R. capillacea, Sarracenia flava, S. leucophylla, S. psittacina, S. rubra, Scheuchzeria palustris, Schizachyrium scoparium, Schoenoplectus acutus, Scleria verticillata, Solidago spathulata, S. uliginosa, Sphagnum capillaceum, S. magellanicum, S. pylaesii, S. recurvum, Spiranthes cernua, Stenanthium densum, Symphyotrichum boreale, Syngonanthus flavidulus, Thuja occidentalis, Thelypteris palustris, Thuidium delicatulum, Thelypteris*

palustris, Toxicodendron vernix, Triadenum fraseri, Triantha glutinosa, Trichophorum cespitosum, Trientalis borealis, Triglochin maritimum, Typha angustifolia, Utricularia cornuta, U. intermedia, Vaccinium corymbosum, V. macrocarpon, V. oxycoccus, Viola lanceolata, V. primulifolia, Xyris ambigua, Zenobia pulverulenta, Zephyranthes atamasco.

***Sarracenia rosea* Naczi, F.W. Case, & R.B. Case** inhabits ditches, flatwoods, interdunal swales, savannas, seepage bogs, stream terraces, swales, swamps, thickets, and the margins of streams at elevations of up to 100 m. Exposures range from open sites, to lightly shaded conditions, to dense shade. Flowering has been documented to occur from March 14 to April 20, reaching its peak period in the last third of March. Although the reproductive biology has not been described in detail for this species, selfed plants are known to produce large numbers of seed having a germination rate higher than 95%, with the seedlings growing into normal adult plants. The seeds begin to germinate after 2 weeks (at 20°C; 14 h light) if initially subjected to 4–5 weeks of cold stratification (4°C). **Reported associates:** A complete list of associated species has not been attempted due to widespread confusion in the literature regarding the taxonomic distinction of *S. rosea* and *S. purpurea*. However, the following congeners have been documented to co-occur: *Sarracenia alabamensis, S. alata, S. flava, S. leucophylla, S. psitticina, S. rubra*.

***Sarracenia rubra* Walter** grows in bogs, ditches, flats, flatwoods, floodplains, hillsides, meadows, pocosins, prairies, savannas, seepage slopes, seeps, streamheads, streams, swamps, and along the margins of lakes and streams at elevations of up to 200 m. It is found occasionally on granite ledges by waterfalls. The habitats are acidic (pH: 4.3–6.5) with substrates of clay, peat (sometimes quaking), or sand. The plants thrive in full light, but can also tolerate partial shade. Flowering occurs from April to May. The flowers emit a sweet odor and are outcrossed by insect pollinators; however, high seed set has also been reported from the self-pollination of cultivated specimens. A small amount of inbreeding has been indicated by heterozygote deficiencies (detected by allozymes) in some populations. The seeds require 4 weeks of cold, moist stratification at 4°C for optimal germination. Allozyme surveys reveal that relatively high levels of genetic variation are maintained in this species, except for some of the very small populations. The plants occur typically within shrub-dominated subclimax communities, which require periodic disturbances (e.g., fire, hydrological fluctuations) in order to maintain open areas conducive to their long-term survival. The persistent rhizome is resistant to moderate fires; however, fire suppression leads to litter accumulation, thus to fires of greater intensity, which can damage the plants. Surprisingly, one study indicated that manual clipping and removal of neighboring vegetation had no detectable positive influence on the growth of the species. However, disturbances such as fire have other effects on the habitat (e.g., nutrient release) that were not duplicated in that study. Up to 20 trap leaves are produced annually. Ants (Hymenoptera: Formicidae) represent the majority of prey that is captured in the traps. Larger insects such as wasps (Hymenoptera) reportedly can escape the traps by chewing through them. In late summer and fall, the pitchers are replaced by phyllodial leaves. Genetic analyses of green anthocyanin-devoid mutants indicate that the plant's pigmentation is determined by two alleles at one locus, with red being dominant to green. **Reported associates:** *Acer rubrum, Agalinis, Alnus, Andropogon, Aristida beyrichianum, Arundinaria gigantea, Bartonia verna, Bigelowia nudata, Calopogon, Carex lonchocarpa, C. striata, Carphephorous pseudoliatrus, Chamaecyparis thyoides, Chamaedaphne calyculata, Chaptalia tomentosa, Cleistes, Clethera alnifolia, Cliftonia monophylla, Coreopsis gladiata, Ctenium aromaticum, Cyprepedium acaule, Cyrilla racemiflora, Dichanthelium, Drosera capillaris, D. intermedia, D. rotundifolia, D. tracyi, Erigeron vernus, Eriocaulon compressum, E. decangulare, Eupatorium rotundifolium, Gelsemium sempervirens, Gordonia lasianthus, Helenium brevifolium, H. vernale, Helianthus heterophyllus, Hypericum, Ilex coriacea, I. glabra, I. myrtifolia, Liquidambar styraciflua, Lophiola americana, Lycopodiella alopecuroides, Lyonia lucida, Lysimachia asperulifolia, Magnolia grandiflora, M. virginiana, Mitchella repens, Muhlenbergia capillaris, Myrica cerifera, M. heterophylla, Nyssa biflora, Orontium aquaticum, Osmunda regalis, Osmundastrum cinnamomeum, Peltandra virginica, Persea borbonia, P. palustris, Photinia pyrifolia, Pinguicula lutea, P. primuliflora, Pinus elliottii, P. palustris, P. serotina, Platanthera ciliaris, Pleea tenuifolia, Pogonia ophioglossoides, Polygala lutea, P. ramosa, Proserpinaca pectinata, Pteridium aquilinum, Quercus geminata, Rhexia alifanus, R. lutea, Rhynchospera chapmanii, R. plumosa, Sarracenia flava, S. leucophylla, S. psittacina, S. purpurea, Scleria, Serenoa repens, Smilax laurifolia, Sphagnum, Syngonanthus flavidulus, Taxodium ascendens, Toxicodendron vernix, Utricularia subulata, Vaccinium atrococcum, V. formosum, V. fuscatum, V. sempervirens, Viburnum nudum, Vitis rotundifolia, Woodwardia areolata, Xyris ambigua, X. difformis, Zenobia pulverulenta.*

Use by wildlife: *Sarracenia* species generally are not grazed by herbivores and appear to be unpalatable to livestock. The taller species are exploited by crab spiders (Arachnida: Thomisidae) and green tree frogs (Amphibia: Hylidae: *Hyla cinerea*), which commandeer some of the prey attracted to the plants. Despite their carnivorous habit, *Sarracenia*s host a large number of invertebrates. *Sarracenia alata, S. flava, S. leucophylla, S. minor, S. psittacina,* and *S. rubra* all are larval hosts of the plant mining moth (Lepidoptera: Noctuidae: *Exyra semicrocea*). Other larval Lepidoptera are hosted by *S. alata* (Psychidae: *Basicladus tracyi*), *S. flava* (Noctuidae: *Exyra fax, E. nigrocaput, E. ridingsii, Tarachidia semiflava*), *S. minor* (Noctuidae: *Tarachidia semiflava*), and *S. purpurea* (Noctuidae: *Exyra. fax, E. nigrocaput, E. ridingsii E. rolandiana, Papaipema appassionata*; Tortricidae: *Endothenia daeckeana, E. hebesana*). *Sarracenia purpurea* also provides habitat for rotifers (Rotifera: Habrotrochidae: *Habrotrocha rosa*) within the pitcher fluid. Various flesh flies (Diptera: Sarcophagidae) are found in association with *S. alata* (*Fletcherimyia abdita*), *S. flava* (*Fletcherimyia abdita, F. jonesi, F. rileyi, Sarcophaga sarraceniae*), *S. leucophylla*

(*Fletcherimyia abdita, F. celarata, F. rileyi, Sarcophaga georgiana, S. sarraceniae*), *S. minor* (*Fletcherimyia abdita, F. fletcheri, F. jonesi, F. rileyi, Sarcophaga sarraceniae*), *S. oreophila* (*Fletcherimyia oreophilae*), *S. purpurea* (*Fletcherimyia fletcheri, F. folkertsi, Sarcophaga sarraceniae*), *S. rosea* (*Fletcherimyia folkertsi*), and *S. rubra* (*Fletcherimyia abdita, F. fletcheri, F. papei*). A nonindigenous humpbacked fly (Diptera: Phoridae: *Dohrniphora cornuta*) has been found in the pitchers of *S. flava*. Larvae of a mosquito (Diptera: Culicidae: *Wyeomyia smithii*) are resistant to digestive enzymes and rely on the trap fluid of *S. purpurea* in which they develop. The related *Wyeomyia haynei* inhabits the pitchers of *S. alata* and *S. leucophylla*. A dipteran midge (Chironomidae: *Metriocnemus knabi*) also lives within the pitcher fluid of *S. purpurea* and releases a substance that suppresses Diptera other than *Wyeomyia*. *Sarracenia purpurea* plants are also host to aphids (Insecta: Hemiptera: Aphididae: *Macrosiphum jeanae*), several fungi (Ascomycota: Leptosphaeriaceae: *Leptosphaeria scapophila*; Massarinaceae: *Helminthosporium sarraceniae*; Pezizaceae: *Peziza*; Trichosphaeriaceae: *Brachysporium sarraceniae*), mites (Arachnida: Acari: Histiostomatidae: *Sarraceniopus gibsoni*), and a rare copepod (Crustacea: Copepoda: Cyclopidae: *Paracyclops canadensis*). Several species contain fungal endophytes including *S. minor* (Ascomycota: Glomerellaceae: *Colletotrichum gloeosporioides*), *S. oreophila* (Ascomycota: Glomerellaceae: *Colletotrichum gloeosporioides*), *S. psitticina* (Ascomycota: Glomerellaceae: *Colletotrichum gloeosporioides*; Trichocomaceae: *Penicillium*), and *S. purpurea* (Ascomycota: Dermateaceae: *Cryptosporiopsis actinidiae*; Diaporthaceae: *Phomopsis*; Glomerellaceae: *Colletotrichum acutatum*; Phaeosphaeriaceae: *Coniothyrium/Paraconiothyrium*). A wasp (Hymenoptera: Sphecidae: *Isodontia philadelphicus*) builds its nests in pitchers of *S. flava*. Leafcutting bees (Hymenoptera: Megachilidae) occasionally remove pieces of *Sarracenia* petals to use as nesting material. Several other flies (Diptera: Chloropidae: *Aphanotrigonum*; Sciaridae: *Bradysia macfarlanei*) are also found in association with various pitcher plant species. The traps of *S. minor* contain numerous Actinobacteria (Bacillaceae: *Bacillus cereus, B. thuringiensis*; Flavobacteriaceae: *Chryseobacterium*; Streptococcaceae: *Lactococcus lactis*; Micrococcaceae: *Micrococcus luteus*; Nocardiaceae: *Rhodococcus equi*) and Proteobacteria (Alcaligenaceae: *Achromobacter xylosoxidans*; Enterobacteriaceae: *Pantoea, Serratia marcescens*). Bacteria identified from traps of *S. purpurea* include *Bacteroidetes* (Chitinophagaceae; Porphyromonadaceae), *Chryseobacterium* (Flavobacteriaceae), *Clostridium algodixylanolyticum, C. saccharolyticum* (Lachnospiraceae), *C. butyricum, C. nitrophenolicum, C. chromoreductans, C. puniceum* (Clostridiaceae), *Cohnella, Emticicia ginsengisolt* (Cytophagaceae), *Escherichia coli* (Enterobacteriaceae), *Fluvicola taffensis* (Cryomorphaceae), *Paenibacillus borealis* (Paenibacillaceae), *Pedobacter heparinus, Pelosinus* (Vellionellaceae), and *Sphingobacterium* (Sphingobacteriaceae). *Sarracenia flava* attracts various ants as prey (Hymenoptera: Formicidae: *Crematogaster ashmeadi,*

Camponotus decipiens, Odontomachus brunneus), which ironically, also defend the plants against some herbivores.

Economic importance: food: not edible; **medicinal:** Root and leaf extracts of *Sarracenia flava* contain lupeol and the pentacyclic triterpene α-amyrin and exhibit antitumor activity against lymphocytic leukemia. They have been used to treat constipation, indigestion, and ailments of the kidneys and liver. Root extracts also contain betulin and display in vitro antitumor activity against human epidermoid carcinoma of the nasopharynx. Root extracts from *S. minor* have also been used to treat gastrointestinal ailments. *Sarracenia purpurea* was used extensively by native North Americans as a medicinal plant. A leaf decoction or infusion was used by several tribes (Algonquin, Cree, Ojibwa, Potawatomi, Quebec, Woodlands) for various gynecological applications. Root decoctions were used as a diuretic (Algonquin, Tete-de-Boule) for back pain (Cree, Woodlands), and to treat coughs (Iroquois), internal bleeding (Micmac, Penobscot), liver ailments (Iroquois), pneumonia (Iroquois), smallpox (Micmac), sore throats (Micmac), tuberculosis (Malecite), urinary disorders (Algonquin, Micmac, Penobscot, Quebec, Tete-de-Boule), and venereal disease (Cree, Woodlands). A leaf infusion was prepared by the Iroquois for fevers, by the Micmac for tuberculosis, and by the Montagnais to treat smallpox. The Montagnais called the plant "toad legging" and made a tea from the boiled leaves, which was applied as a wash to treat sores and childhood rashes. The efficacy of *S. purpurea* rhizome extracts to treat severe pain has been attributed to their ammonium chloride and ammonium sulfate content. The roots and rhizomes of *Sarracenia* species contain elevated levels of cholesterol and β-sitosterol, presumably as a consequence of their carnivorous habit; **cultivation:** *Sarracenias* are extremely popular plants, are hybridized easily, and are cultivated widely. Cultivars of *S. alata* include 'Black Tube,' 'Citronelle,' 'Copper Lid,' 'Nicolson,' 'Orange Sunset' and 'Red Lid.' *Sarracenia flava* received an Award of Garden Merit [AGM] by the Royal Horticultural Society. Its cultivars include 'Adrian Slack,' 'Burgundy,' 'Judith Hindle,' 'Prince George County,' *S. flava* var. *atropurpurea*, *S. flava* var. *cuprea*, *S. flava* var. *maxima*, *S. flava* var. *ornata*, *S. flava* var. *rubricorpora* and *S. flava* var. *rugelii*. *Sarracenia leucophylla* is also an AGM plant. Some cultivars are 'Adrian Slack,' 'Bug Scoop,' 'Dana's Delight,' 'Dixie Lace,' 'Hurricane Creek White,' 'Judith Hindle,' 'Juthatip Soper,' 'Ladies in Waiting,' 'Scarlet Belle,' 'Schnell's Ghost,' 'Tarnok,' 'Titan' and 'Yellow Flower.' 'Fat Chance' is a cultivar of *S. jonesii*. 'Lady Bug' and 'Okee Giant' are cultivars of *S. minor*. Cultivars attributed to *S. psittacina* include 'Dixie Lace,' 'Doodle Bug,' 'Hummer's Hammerhead,' 'Ladies in Waiting,' 'Lady Bug,' 'Scarlet Belle' and *S. psittacina* forma *heterophylla*. Cultivars of *S. purpurea* include 'Cobra Nest,' 'Dixie Lace,' 'Judith Hindle,' 'Lady Bug,' *S. purpurea* subsp. *purpurea* forma *heterophylla*, *S. purpurea* subsp. *venosa* and *S. purpurea* subsp. *venosa* var. *burkii*. *Sarracenia rubra* includes the cultivars 'Cobra Nest,' 'Dixie Lace,' 'Doodle Bug,' 'Golden Red Jubilee,' 'Hummer's Hammerhead,' 'John's Autumnal

Splendor,' 'Ladies in Waiting,' 'Long Lid,' 'Red Bug,' *S. rubra* subsp. *alabamensis*, *S. rubra* subsp. *gulfensis*, *S. rubra* subsp. *gulfensis* forma *heterophylla*, *S. rubra* subsp. *jonesii*, *S. rubra* subsp. *jonesii* forma *heterophylla* and *S. rubra* subsp. *wherryi*. In addition to many of the named cultivars, cultivated *Sarracenia* hybrids include *S.* ×*ahlesii* (=*S. alata* × *S. rubra*), *S.* ×*ahlesii* × (*S. rubra* × 'Red Lid'), *S. alata* × *S. flava*, *S. alata* × *S. flava* var. *maxima*, *S. alata* × *S.* ×*willisii*, *S. alata* 'Red Lid' × *S. flava*, *S.* ×*areolata* (=*S. leucophylla* × *S. alata*), *S.* ×*areolata* × (*S.* ×*areolata* × 'Red Throat'), *S.* ×*catesbyi* (=*S. flava* × *S. purpurea*; an AGM plant), *S.* ×*catesbyi* × (*S. flava* var. *maxima* × *S. purpurea* subsp. *venosa*), *S.* ×*chelsonii* (=*S. purpurea* × *S. rubra*; an AGM plant), *S.* ×*courtii* (=*S. psittacina* × *S. leucophylla*), *S.* ×*excellens* (=*S. leucophylla* × *S. minor*; an AGM plant), *S.* ×*excellens* 'Judy' (=*S. minor* × *S.* ×*excellens*), *S.* ×*exornata* (=*S. alata* × *S. purpurea*), *S.* ×*formosa* (=*S. minor* × *S. psittacina*), *S.* ×*gilpinii* (=*S. psittacina* × *S. rubra*), *S.* ×*harperi* (=*S. flava* × *S. minor*), *S. leucophylla* × *S. oreophila*, *S.* ×*miniata* (=*S. minor* × *S. alata*), *S. minor* × *S. oreophila*, *S.* ×*mitchelliana* (=*S. leucophylla* × *S. purpurea*; an AGM plant), *S.* ×*moorei* (=*S. leucophylla* × *S.* ×*catesbyi* × *S. flava*), *S.* ×*moorei* × (*S. leucophylla* × *S.* ×*moorei*), *S. oreophila* × *S. purpurea*, *S. oreophila* × *S. purpurea* subsp. *venosa*, *S.* ×*popei* (=*S. flava* × *S. rubra*), *S.* ×*popei* × (*S. flava* var. *maxima* × *S. rubra* subsp. *jonesii*), *S.* ×*readii* (=*S. leucophylla* × *S. rubra*), *S.* ×*readii* × (*S. leucophylla* × *S.* ×*readii*), 'Red Lid' × *S. flava*, *S.* ×*rehderi* (=*S. minor* × *S. rubra*), *S.* ×*swaniana* (=*S. minor* × *S. purpurea*), and *S.* ×*wrigleyana* (=*S. psittacina* × *S. leucophylla*; an AGM plant); **misc. products:** The leaves of *Sarracenia purpurea* were used as toys (kettles) by children of the Chippewa, Cree, and Woodlands tribes. The Iroquois made a love potion from the powdered plants. The Potawatomi would use the leaves as drinking cups while away from camp. *Sarracenia purpurea* has been identified as an effective organism for use in recombinant protein production; **weeds:** *Sarracenia purpurea* reportedly is invasive in Ireland and Switzerland; **nonindigenous species:** *Sarracenia purpurea* was introduced to California (in 1903) and to Washington state. It has also been introduced to England (around 1900), Ireland (late 19th century), Japan, Scotland, Switzerland (late 19th century), and southern Sweden (in 1948). *Sarracenia flava* and *S. leucophylla* were also introduced to Washington.

Systematics: Molecular data resolve *Sarracenia* as a sister clade of the South American genus *Heliamphora* (Figures 5.21 and 5.22). The species exhibit very low divergence at the molecular (DNA sequence) level, an indication that despite their distinct morphology, they are closely related and of relatively recent divergence (Figure 5.22). This conclusion is consistent with the fact that virtually all of the species are intercompatible and hybridize readily (see "cultivation" earlier). Consequently, there remains considerable disagreement on the delimitation of species in this genus and the species concepts applied in many cases have been based primarily on phenetic differences, even where considerable overlap in traits occurs between putative species. A number of infraspecific

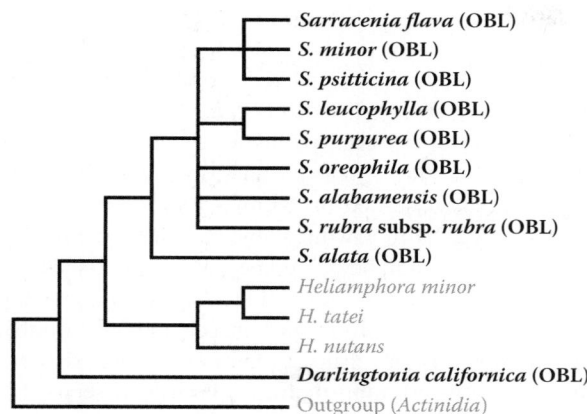

FIGURE 5.22 Estimated relationships among species of *Sarracenia* as indicated by phylogenetic analysis of nrITS sequence data. The close relationship of species has made it difficult to examine interspecific relationships due to their lack of genetic divergence. Because of the poor level of internal support for any clade resolved in *Sarracenia*, this tree is at best only an approximation of what might be the true interspecific relationships. All of the *Sarracenia* species (bold) are designated as OBL indicators in North America. *Darlingtonia* (bold) and *Heliamphora* (South America) also comprise obligate aquatics. (From Bayer, R.J. et al., *Syst. Bot.*, 21, 121–134, 1996.)

taxa have been recognized, with many of these representing relatively minor color variants or hybrids of questionable status. The morphology of two subspecies of the widely distributed *S. purpurea* (subsp. *purpurea*, subsp. *venosa*) correlates with environmental conditions (precipitation and temperature); however, allozyme analysis indicates that they are fairly distinct genetically. The disjunct taxon known as *S. purpurea* subsp. *venosa* var. *burkii* differs not only morphologically but also genetically from the other infraspecific taxa and many authors have recognized it as the distinct species *S. rosea*. This convention has been followed here to be consistent with the 2013 indicator list. *Sarracenia rubra* has also been subdivided to distinguish several localized, disjunct subspecies from the wider ranging subsp. *rubra*, i.e., subsp. *alabamensis* (Alabama), subsp. *gulfensis* (Florida), subsp. *jonesii* (Carolinas), and subsp. *wherryi* (Alabama, Mississippi). In at least one comparison using allozyme data (*S. rubra* subspecies *rubra* vs. *S. rubra* subsp. *alabamensis*), the populations of each taxon grouped in distinct clusters despite a fairly high intersubspecific genetic identity (*I* = 0.90). Analysis of nrDNA data does not resolve these subspecies as distinct; however, *S. rubra* subsp. *alabamensis* and *S. rubra* subsp. *jonesii* do exhibit a different pollen structure than *S. rubra* subsp. *rubra* and exhibit a comparable level of genetic differentiation at microsatellite loci as found in interspecific comparisons of uncontested *Sarracenia* species. Thus, they are recognized here as distinct species, at least to be consistent with the 2013 wetland indicator list. Numerous microsatellite markers have now been developed for 10 of the species, which should facilitate future genetic surveys of populations. Microsatellite data analysis indicates that widespread *Sarracenia* species

generally possess higher levels of genetic variation than the imperiled taxa. The base chromosome number of *Sarracenia* is $x = 13$. All species are diploid ($2n = 26$). Although synthetic hybrids have been produced between many *Sarracenia* species combinations (see "cultivation" earlier), natural hybridization is minimized by divergent flowering times and allopatric distributions. Yet, at least 22 natural instances of natural hybridization have been reported and hybridization has been shown to result in significant interspecific gene flow when populations occur in sympatry.

Comments: Most *Sarracenia* species occur in the southeastern United States (*S. flava*, *S. leucophylla*, *S. minor*, *S. oreophila*, *S. psittacina*, *S. rubra*) or in the southern United States (*S. alata*); whereas, *S. purpurea* is distributed widely throughout eastern and northern North America, with disjunct occurrences in California and Washington. *Sarracenia oreophila* (U.S. endangered) is very rare with only about 35 extant populations remaining. Other endangered species include *S. alabamensis* and *S. jonesii*. All of the species are threatened by overcollecting and most by fire suppression.

References: Almborn, 1983; Barker & Williamson, 1988; Brewer, 2003, 2005; Burr, 1979; Carter et al., 2006; Case, Jr. & Case, 1974; Cheers, 1992; Dahlem & Naczi, 2006; Dress et al., 1997; Ellison, 2001; Ellison & Gotelli, 2001; Ellison & Parker, 2002; Ellison et al., 2002, 2004, 2014; Evans et al., 2002b; Fish & Hall, 1978; Folkerts, 1989; Furches et al., 2013; Gaddy, 1982; Gibson, 1991a; Glenn & Bodri, 2012; Godt & Hamrick, 1996b, 1998, 1999; Gotsch & Ellison, 1998; Hamilton, IV et al., 2000; Heard, 1998; Juniper et al., 1989; Jürgens et al., 2009; Karagatzides & Ellison, 2009; Karberg & Gale, 2006; Krieger & Kourtev, 2012; Lawton & Gravatt, 2006; MacMillan, 1891; MacRoberts & MacRoberts, 2004; Mandossian, 1965; McDaniel, 1971b; Mellichamp & Case, 2009; Miles & Kokpol, 1976; Miles et al., 1974; Mody et al., 1976; Moon et al., 2008, 2010; Murphy & Boyd, 1999; Naczi et al., 1999; Neal & Norquist, 1992; Newell & Nastase, 1998; Neyland & Merchant, 2006; Oswald et al., 2011; Parisod et al., 2005; Pietropaolo & Pietropaolo, 1993; Prankevicius & Cameron, 1991; Rogers et al., 2010; Rosa et al., 2009; Schaefer & Ruxton, 2014; Schnell, 1977, 1983, 1993, 2002; Schwaegerle, 1983; Sheridan & Karowe, 2000; Sheridan & Mills, 1998; Siragusa et al., 2007; Tantaquidgeon, 1932; Thomas, 2002; Trzcinski et al., 2003; Wakefield et al., 2005; Walti, 1945; Wang et al., 2004b; Weakley & Schafale, 1994; Wherry, 1929; Whitman et al., 2005; Wichman et al., 2007.

Family 7: Styracaceae [11]

This small family of about 160 species of resinous trees and shrubs is distributed in warm temperate to tropical regions of Asia, the Mediterranean, and New World. Phylogenetic analyses of morphological and molecular data (Figures 5.5 and 5.23) demonstrate the monophyly of the family and place it in a clade with Diapensiaceae; Symplocaceae are their sister group. All three families clearly are embedded within the order Ericales. In Styracaceae the highest diversity occurs within the genus *Styrax* (about 130 species) and none of the

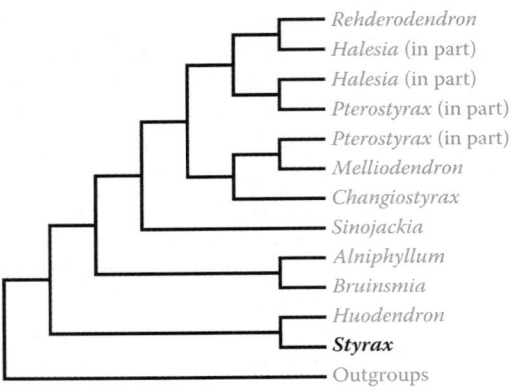

FIGURE 5.23 Relationships among 10 genera of Styracaceae as indicated by combined analysis of *trnL* and nrITS sequence data. *Styrax* (in bold) is the only genus to include OBL representatives in North America. These and other data clearly support the monophyly of *Styrax* and its sister relationship to *Huodendron*. Several other genera (*Halesia*, *Pterostyrax*) do not appear to be monophyletic. (Adapted from Fritsch, P.W. et al., *Mol. Phylogen. Evol.*, 19, 387–408, 2001.)

other genera contains more than five species. The flowers typically are showy, hermaphroditic and insect pollinated. The group is known for several ornamental trees and shrubs, notably in the genera *Halesia*, *Pterostyrax*, and *Styrax*. The family is also a source of gum benzoin (a balsamic resin derived from *Styrax benzoin*, *S. serrulatus*, *S. subpaniculatus*, or *S. tonkinensis*), and benzoin resin (obtained from *Styrax officinalis*), a resin used in the manufacture of stacte (a spice used in making incense) and perfume. *Styrax* resin, which was used in the discovery of polystyrene, originates not here, but from the family Hamamelidaceae (*Liquidambar*).

Ecologically, most of the species inhabit forested sites that range from dense, shaded conditions, to ravines or rain forest. Relatively few species occur on moist sites, where they tend to occupy wet areas adjacent to streams. *Halesia* includes some FACW species but only one genus contains obligate aquatics in North America:

1. ***Styrax*** L.

Phylogenetic studies of Styracaceae (Figure 5.23) confirm that *Styrax* is monophyletic and resolve it as a sister clade to *Huodendron* (4 species), a terrestrial Asian genus of dense forest species.

1. *Styrax*

Snowbell, storax
Etymology: from the Greek *sturax*, a pointed bronze cap attached to the base of a lance
Synonyms: *Adnaria*; *Anthostyrax*; *Benzoin*; *Cyrta*; *Darlingtonia*; *Epigenia*; *Foveolaria*; *Plagiospermum*; *Strigilia*
Distribution: global: Asia; Europe; New World; **North America:** eastern, southern, western
Diversity: global: 130 species; **North America:** 5 species
Indicators (USA): OBL; **FACW:** *Styrax americanus*

Habitat: freshwater; palustrine; **pH:** 4.5–6.5; **depth:** <1 m; **life-form(s):** emergent shrub
Key morphology: stems (to 6 m) woody, multiple, with smooth, gray bark; leaves (to 13 cm) deciduous, alternate, simple, elliptic or obovate, petioled (to 5 mm), the margins toothed; racemes (to 12 cm) axillary, 1–5-flowered; flowers showy, 5-merous, stalked (to 8 mm), often drooping; petals (to 15 mm) fused basally, white, downy, recurved; stamens 10; drupes (to 1 cm) subglobose, stellate-pubescent, 1-seeded
Life history: duration: perennial (buds); **asexual reproduction:** stem suckers; **pollination:** insect; **sexual condition:** hermaphroditic; **fruit:** drupes (common); **local dispersal:** drupes/seeds (water); **long-distance dispersal:** seeds (birds, waterfowl)
Imperilment: (1) *Styrax americanus* [G5]; S1 (OK); S2 (IL); S3 (IN, KY, NC, VA)
Ecology: general: *Styrax* species extend across a broad range of forest habitats ranging from semiarid regions to rainforests. Many of them are associated with riparian communities. Three of the native North American species are at least FACW plants, but only one has been designated as an obligate aquatic. The Puerto Rican *S. portoricensis* (not included here) is also designated as an OBL species. *Styrax* species are trees and shrubs with showy, insect-pollinated flowers, and capsular or drupe fruits. Information regarding the reproductive ecology of the genus exists for only a few species.

***Styrax americanus* Lam.** inhabits bogs, canals, depressions, ditches, floodplains, hammocks, oxbows, slopes, swamps, and the margins of lakes, ponds, and streams at elevations of up to 300 m. Habitats have acidic substrates (pH: 4.5–6.5), which include rich loam, peat, sand, or silt that is high in organic matter. The plants thrive in sun and are also very tolerant to shade. These are subcanopy understory plants of swamps that experience annual inundation or extended periods of standing water. Flowering occurs from March to June and fruiting from June to October. The flowers are pollinated mainly by bumblebees (Hymenoptera Apidae: *Bombus*) and honeybees (Hymenoptera Apidae: *Apis mellifera*). Other pollinators may include butterflies and moths (Lepidoptera), flies (Diptera), and other Hymenoptera such as wasps (Vespidae). Fresh seeds can germinate within three weeks; however, dormant seeds reportedly require alternating periods of warm, moist stratification and cool, moist stratification in order to germinate. Seeds supposedly do not germinate well when exposed to warm temperatures. Germination also can be achieved by stratifying seeds at 5°C for 2–3 months and then sowing them in a greenhouse during winter (January to February). Reports of seeds attached to the feet of waterfowl indicate that occasional dispersal may occur by this means. Seeds are also eaten by birds, which may transport them endozoically. As in several other species of wet habitats, dispersal of the seed-containing drupes by water also probably occurs. There is not a well-developed system of vegetative reproduction; however, stems will produce suckers after being cut down. **Reported associates:** *Acer rubrum*, *A. saccharinum*, *Aesculus sylvatica*, *Alnus serrulata*, *Ampelopsis arborea*, *Andropogon glomeratus*, *Aronia* *arbutifolia, Arundinaria gigantea, Asimina triloba, Athyrium filix-femina, Bartonia virginica, Berchemia scandens, Bignonia capreolata, Boehmeria cylindrica, Brunnichia cirrhosa, Callicarpa americana, Campsis radicans, Carex debilis, C. folliculata, C. gynandra, C. intumescens, C. joorii, C. leptalea, Carpinus caroliniana, Carya aquatica, C. cordiformis, C. glabra, Cephalanthus occidentalis, Chamaecyparis thyoides, Chasmanthium laxum, Clethra alnifolia, Cornus foemina, C. florida, Crataegus marshallii, C. opaca, C. viridis, Cyrilla racemiflora, Decumaria barbara, Dichanthelium dichotomum, Diospyros virginiana, Doellingeria umbellata, Eleocharis tuberculosa, Elephantopus carolinianus, Erechtites hieracifolia, Eupatorium fistulosum, E. pilosum, E. rotundifolium, Fagus grandifolia, Forestiera acuminata, Frangula caroliniana, Fraxinus caroliniana, F. pennsylvanica, Gelsemium rankinii, Gentiana saponaria, Halesia diptera, Hamamelis virginiana, Heliopsis helianthoides, Hypericum galioides, H. hypericoides, Ilex coriacea, I. decidua, I. glabra, I. opaca, I. verticillata, I. vomitoria, Itea virginica, Leucothoe axillaris, L. racemosa, Liquidambar styraciflua, Liriodendron tulipifera, Lobelia amoena, Lycopus, Lyonia lucida, Lysimachia quadrifolia, Magnolia grandiflora, M. virginiana, Malaxis unifolia, Mitchella repens, Myrica cerifera, M. heterophylla, Nyssa biflora, N. sylvatica, Onoclea sensibilis, Osmanthus americanus, Osmunda regalis, Osmundastrum cinnamomeum, Ostrya virginiana, Oxydendrum arboreum, Pallavicinia lyellii, Panicum hemitomon, Persea borbonia, Phlox glaberrima, Photinia melanocarpa, Photinia pyrifolia, Pinus elliottii, P. glabra, P. palustris, P. taeda, Planera aquatica, Platanthera clavellata, P. integrilabia, Populus heterophylla, Prunus serotina, Pteridium aquilinum, Quercus alba, Q. laurifolia, Q. lyrata, Q. marilandica, Q. michauxii, Q. nigra, Q. pagoda, Q. phellos, Q. virginiana, Rhododendron alabamense, R. canescens, R. oblongifolium, R. viscosum, Rhynchospora gracilenta, Rubus argutus, Saccharum baldwinii, Salix nigra, Sambucus nigra, Sapium sebiferum, Sarracenia, Saururus cernuus, Scirpus cyperinus, Sebastiania fruticosa, Smilax glauca, S. laurifolia, S. rotundifolia, Solidago patula, S. rugosa,* *Sphagnum, Sporobolus floridanus, Symphyotrichum lateriflorum, Taxodium ascendens, T. distichum, Thelypteris noveboracensis, Toxicodendron radicans, T. vernix, Triadenum walteri, Ulmus alata, U. americana, U. rubra, Uvularia sessilifolia, Vaccinium corymbosum, V. elliottii, V. fuscatum, Viburnum dentatum, V. nudum, Viola ×primulifolia, Vitis rotundifolia, Woodwardia areolata, W. virginica, Xanthorhiza simplicissima.*

Use by wildlife: The flowers of *Styrax americanus* attract butterflies (Insecta: Lepidoptera). Its leaves are eaten by larvae of the eastern tiger swallowtail butterfly (Papilionidae: *Papilio glaucus*) and by the promethea moth (Saturniidae: *Callosamia promethea*).

Economic importance: food: not edible; **medicinal:** A number of *Styrax* species produce resins, some which have various medicinal applications; however there are no medicinal uses reported for *S. americanus*; **cultivation:** *Styrax americanus* is cultivated as an ornamental shrub, which usually is distributed under the common name of American snowbell; **misc. products:** none; **weeds:** none; **nonindigenous species:** none

Systematics: Various morphological and molecular data sets (e.g., Figure 5.23) resolve *Huodendron* and *Styrax* as a clade, which occurs in a basal position within Styracaceae. All indications are that *Styrax* is monophyletic and it can be characterized by several morphological synapomorphies including a stamen tube attached high on the corolla, the presence of placental obturators, bitegmic ovules, and an indurate seed coat. Phylogenetic analyses of about a third of the species (incorporating morphological and molecular data) are congruent with a classification that subdivides the genus as section *Styrax* (series *Cyrta, Styrax*) and section *Valvatae* (series *Benzoin, Valvatae*) with these taxa resolved reasonably well as clades (Figure 5.24). Corolla aestivation is imbricate (or subvalvate) in section *Styrax* and valvate in section *Valvatae*. *Styrax americanus* is included within section *Styrax* series *Cyrta*; analyses of combined morphological and molecular data (Figure 5.24) resolve this species in a clade along with and closest to *S. glabrescens* (Central America & Mexico; essentially terrestrial but common along waterways) and

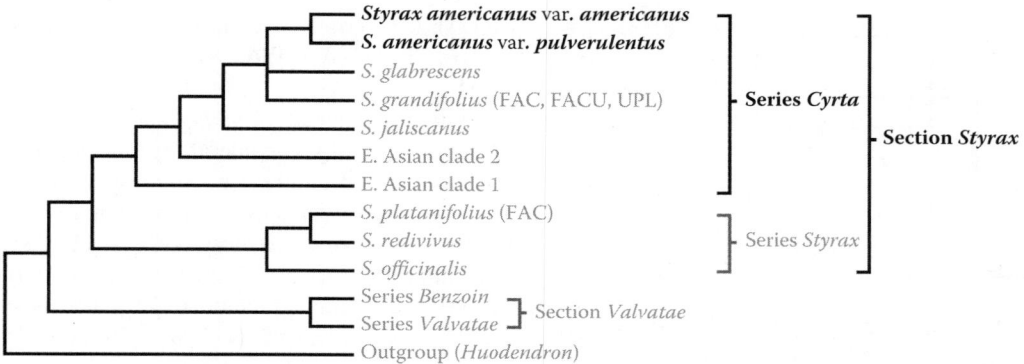

FIGURE 5.24 Interspecific relationships in *Styrax* as indicated by phylogenetic analysis of combined morphological and molecular data. The OBL habit of the North American *S. americanus* (in bold) appears to be uniquely derived in the genus. Wetland indicator designations are provided for other North American species, where appropriate. Most authors do not distinguish the varieties of *S. americanus* and they exhibited only minor genetic differences in this analysis. (Adapted from Fritsch, P.W. et al., *Mol. Phylogen. Evol.*, 19, 387–408, 2001.)

S. grandifolius (FAC, FACU, UPL). Consequently, the OBL habit of *S. americanus* appears to be uniquely derived. Some authors distinguish two varieties of *S. americanus*: *S. americanus* var. *americanus* and *S. americanus* var. *pulverulentus*, which are differentiated primarily by slight differences in leaf shape, and vestiture. Minor genetic differences have been detected between accessions of these varieties used in phylogenetic analyses, but most authors do not treat them as distinct taxa. The base chromosome number for *Styrax* is $x = 8$. The chromosome number of *S. americanus* ($2n = 16$) indicates that it is a diploid. Additional information is needed on the reproductive biology, seed ecology, and population genetics of *S. americanus*.

Comments: *Styrax americanus* occurs in the southeastern United States.

References: Fritsch, 1997, 2001, 2009; Gonsoulin, 1974; Huang et al., 2003; Tobe et al., 1998.

Family 8: Theophrastaceae [8]

Theophrastaceae are a small family, which contains about 110 species of mainly terrestrial Neotropical trees and shrubs. Recently, the circumscription of this family has been altered to include the herbaceous genus *Samolus* (formerly Primulaceae) as the result of various morphological and molecular phylogenetic analyses (see Källersjö & Ståhl, 2003), which resolve *Samolus* as the sister group to the remainder of Theophrastaceae (Figure 5.25). Because *Samolus* differs rather substantially from Theophrastaceae morphologically (although both groups possess staminodial flowers and similar placenta and ovule development), some authors (e.g., Caris & Smets, 2004; Ståhl, 2004b) prefer to recognize Samolaceae as a distinct family, which is also compatible phylogenetically. Here *Samolus* is retained within Theophrastaceae in a sensu lato concept of the family. Accordingly, Theophrastaceae are the sister family to a clade comprising Myrsinaceae and Primulaceae; Maesaceae are, in turn, their closest sister group (Figure 5.5). All of the genera are pollinated by insects (Insecta: Diptera; Hymenoptera), which forage for their floral nectar and pollen. There are no

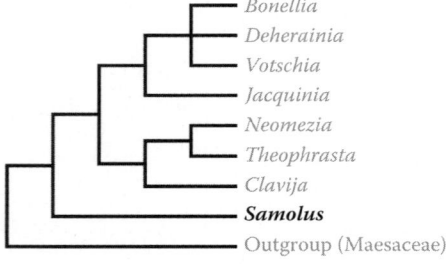

FIGURE 5.25 Phylogenetic relationships among the eight genera of Theophrastaceae (sensu lato) as indicated by analyses of combined morphological and DNA sequence data. Only *Samolus* (in bold) contains OBL species in North America. Some authors prefer to recognize *Samolus* within a separate family (Samolaceae). (Adapted from Källersjö, M. & Ståhl, B., *Int. J. Plant Sci.*, 164, 579–591, 2003.)

plants of major economic importance in the family. One species (*Deherainia smaragdina*) has dubiously earned the common name of "dog mess flower" due to its unpleasant floral odor (use your imagination).

Most Theophrastaceae (excluding *Samolus*) occupy habitats ranging from dry deciduous forests to rainforests. The species typically occupy lowland sites, but some can reach elevations up to 2000 m. A few species of *Clivija* can tolerate temporarily inundated conditions (Ståhl, 2004c). Only one genus of the family contains obligate aquatics in North America:

1. ***Samolus*** L.

1. *Samolus*

Brookweed, water pimpernel; samole

Etymology: from the Celtic *san* (beneficial) and *mos* (pig) in reference to the plant's use for treating diseases of swine

Synonyms: *Samodia*; *Sheffieldia*

Distribution: global: cosmopolitan; **North America:** widespread

Diversity: global: 10–12 species; **North America:** 3 species

Indicators (USA): OBL: *Samolus ebracteatus*, *S. vagans*, *S. valerandi*; **FACW:** *S. ebracteatus*

Habitat: brackish (coastal), freshwater; palustrine; **pH:** 6.0–8.0; **depth:** <1 m; **life-form(s):** emergent herb

Key morphology: stems (to 6 dm) erect or procumbent, branched or simple; leaves (to 15 cm) cauline and in basal rosettes, alternate, simple, somewhat succulent, sessile or with decurrent, winged petiolate bases; racemes (to 2 dm) arising from terminal leaf axils, bearing 5-merous, pedicellate (to 5 cm) flowers (to 1 cm); petals (to 9 mm) white to pink, fused basally; stamens included, staminodia present (alternate with stamens) or absent; ovary partly inferior; capsules (to 4 mm) globose, many seeded

Life history: duration: perennial (persistent rosettes); **asexual reproduction:** adventitious plantlets, rhizomes; **pollination:** insect, self; **sexual condition:** hermaphroditic; **fruit:** capsules (common); **local dispersal:** seeds (gravity, water); **long-distance dispersal:** seeds (animals)

Imperilment: (1) *Samolus ebracteatus* [G4]; SH (KS); S1 (LA); (2) *S. vagans* [G2]; S2 (AZ); (3) *S. valerandi* [G5]; S1 (NB, PE, QC, VT, WA); S2 (ME, NS, NH, PA, UT, WV); S3 (ME)

Ecology: general: *Samolus* species typically inhabit sites characterized by ephemeral inundation, such as the margins of lakes, ponds, and rivers. Some species tolerate salinity and can occur in brackish sites or in salt marshes. All of the North American species are designated as obligate aquatics. Much of the life history of this genus remains poorly studied. North American plants are short-lived perennials with basal leaf rosettes that persist over winter. Studies on the reproductive biology of *Samolus* essentially are nonexistent. The flowers appear to be adapted for insect pollination; however, at least some species reportedly self-pollinate.

***Samolus ebracteatus* Kunth.** occurs in brackish or freshwater arroyos, canals, cienega, ditches, dune swales, glades, marshes (often tidal), prairies, salt marshes, savannas, scrub, seeps, slopes, strands, swamps, along rivers, springs, and streams, and on shell mounds at elevations of up to 1656 m. Exposures in sun or shade are tolerated. The substrates are alkaline (specific pH values not reported) and have been characterized as clay, clay loam, gypsum, Miami oolite, rock (limestone), sand, and sandy gypsum. Flowering occurs from spring to fall. Otherwise, the reproductive ecology of this species is poorly known and ecological investigations are also scarce. One study found that the plants did not concentrate arsenic (As) when growing on high As sites, but that they contained As concentrations of 8–13 µg/g in their shoot and root tissues. Seedling emergence was observed to be higher in plots that were devoid of fallen pine crowns, which commonly serve as fuel for fires. **Reported associates:** *Acacia, Acrostichum danaeifolium, Aletris bracteata, Allenrolfea occidentalis, Ammannia latifolia, Andropogon glomeratus, A. virginicus, Aristida purpurascens, Asclepias perennis, Baccharis salicifolia, B. salicina, Bacopa monnieri, Blechum pyramidatum, Borreria terminalis, Bolboschoenus maritimus, Brickellia californica, Bumelia reclinata, Centella erecta, Cladium jamaicense, Croton linearis, Cynanchum, Distichlis spicata, Dyschoriste angusta, Eleocharis rostellata, Elytraria caroliniensis, Eupatorium serotinum, Evolvulus sericeus, Fallugia paradoxa, Fimbristylis caroliniana, F. castanea, Flaveria chlorifolia, Fraxinus velutina, Helianthus paradoxus, Hydrocotyle bonariensis, Juncus, Limonium limbatum, Linum medium, Lobelia glandulosa, Ludwigia ebracteata, Lysiloma latisiliquum, Melanthera angustifolia, Mikania scandens, Muhlenbergia asperifolia, M. filipes, M. porteri, Panicum virgatum, Phyla nodiflora, Pityopsis graminifolia, Pluchea foetida, P. odorata, P. rosea, Polygala balduinii, P. grandiflora, Polypogon monspeliensis, Populus deltoides, Pseudoclappia arenaria, Ptilimnium capillaceum, Quercus, Rhynchospora colorata, R. divergens, R. microcarpa, Russelia equisetiformis, Sabatia calycina, S. grandiflora, S. stellaris, Salix gooddingii, Saururus cernuus, Schoenoplectus americanus, Serenoa repens, Spartina patens, Spermacoce, Sporobolus airoides, Stenandrium dulce, Stillingia sylvatica, Tamarix ramosissima, Thelypteris palustris, Typha domingensis.*

***Samolus vagans* Greene** is a narrow endemic that occurs on hillsides, in marshes, meadows, seeps, and along brooks, springs, and streams at elevations from 900 to 2770 m. Occurrences can be found in open or in shaded sites. The substrates consist of limestone, mud, schistose clay, or sand. Flowering extends from spring to fall. The stem is procumbent and stolon-like. Otherwise, there is little additional information available on the ecology of this rare species. **Reported associates:** *Aquilegia, Baccharis, Carex, Cupressus, Cyperus, Dactylis, Eleocharis, Fraxinus, Geranium, Hydrocotyle ranunculoides, Juncus bufonius, J. saximontanus, Juniperus deppeana, Lobelia, Mimulus guttatus, Myosurus minimus, Nasturtium officinale, Oenothera rosea, Opuntia, Pinus cembroides, P. engelmannii, P. leiophylla, P. ponderosa, Piptochaetium fimbriatum, Platanus wrightii, Polypogon viridis, Prosopis velutina, Pseudotsuga menziesii, Pteridium,*

Quercus arizonica, Q. emoryi, Q. hypoleucoides, Rhus aromatica, Robinia, Salix gooddingii, Yucca madrensis.

***Samolus valerandi* L.** grows in brackish or freshwater sites including deltas, depressions, ditches, floodplains, glades, marshes, meadows, prairies, roadsides, savannas, seeps, swamps, and along the margins of brooks, lakes, ponds, rivers (often tidal), seeps, and streams at elevations of up to 1700 m. Its substrates are infertile and alkaline (pH: 6.0–8.0) and include calcareous (limestone) cliff faces, cobble, gypsum, muck, mud, peat, rock, sand, sandy clay, and silt. The plants even have been observed to grow in seepy cracks of old limestone block walls. They thrive in fully open sites and will tolerate partial shade. The species is highly salt tolerant, capable of withstanding NaCl concentrations of 2%–4%. The plants accumulate salt at low concentrations and grow well at fairly high salt concentrations. This is a long-day plant, which remains in flower from spring to fall. The flowers reportedly are self-compatible and are self-pollinated. The seeds require 75 days of cold stratification followed by a temperature regime of 15°C/5°C or 25°C/5°C (in light) for optimal germination. Maximum germination occurs earlier at 20°C (9 days) than at 10°C (13 days). This species forms a persistent seedbank that can consist of several hundred seeds per m². Natural germination tends to occur in a synchronous burst rather than over an extended time period. The seeds tolerate up to 0.25 M NaCl before their germination rate is reduced significantly. Carriage of the small seeds by adhesion to waterfowl feet is believed to be the principal means of their long-distance dispersal. Seeds have also been recovered from the dung of grazing horses and sheep, indicating some potential for endozoic transport; however, the seeds do not survive digestion by birds, possibly because their faceted surface renders them more susceptible to the grinding actions of the gizzard. The range of water dispersal is limited because dried seeds sink within a minute. Limited vegetative reproduction by rhizome division is possible and adventitious plantlets have been observed to develop from the peduncles. The plants exhibit high radial oxygen loss (ROL), which effectively aerates their otherwise anoxic substrate. The rate of ROL increases when the plants are shaded. Occurrences often are in proximity to other species that efficiently oxygenate the substrate. **Reported associates:** *Adiantum capillus-veneris, Agalinis maritima, Agrostis stolonifera, Alternanthera philoxeroides, Amaranthus cannabinus, Amorpha fruticosa, Ampelopsis arborea, Apium leptophyllum, Asclepias perennis, A. tuberosus, Atriplex patula, A. prostrata, Baccharis halimifolia, B. salicina, Bacopa monnieri, Berchemia scandens, Boehmeria cylindrica, Bolboschoenus maritimus, B. robustus, Calystegia sepium, Cardamine parviflora, C. pensylvanica, Carex longii, C. lyngbyei, C. paleacea, Carya, Cenchrus incertus, Centella erecta, Cercis canadensis, Chasmanthium laxum, Cladium mariscoides, Crassula aquatica, Cynodon dactylon, Cyperus retrorsus, Daucus, Deschampsia cespitosa, Dichanthelium acuminatum, Distichlis spicata, Eclipta prostrata, Eleocharis fallax, E. ovata, E. palustris, E. parvula, E. rostellata, E. uniglumis, Equisetum hyemale, Eryngium baldwinii, Eupatorium capillifolium, Fimbristylis castanea, Galium obtusum, G.*

tinctorium, Geranium carolinianum, Hydrocotyle umbellata, H. verticillata, Iris hexagona, Itea virginica, Juncus canadensis, J. effusus, J. gerardii, J. torreyi, Juniperus, Justicia lanceolata, Kosteletzkya pentacarpos, Leersia virginica, Lilaeopsis chinensis, Limonium carolinianum, Limosella australis, Lindernia grandiflora, Liquidambar styraciflua, Lobelia cardinalis, L. cliffortiana, Ludwigia palustris, Lycopus americanus, L. virginicus, Lythrum lineare, Medicago lupulina, Micromeria brownei, Mikania scandens, Myosotis, Myrica cerifera, M. inodora, Oenothera fruticosa, Onoclea sensibilis, Orontium aquaticum, Osmunda regalis, Panicum commutatum, P. virgatum, Paspalum dilatatum, Peltandra virginica, Penstemon, Persicaria pensylvanica, Phalaris arundinacea, Phyla lanceolata, Phragmites australis, Physostegia purpurea, Pluchea camphorata, P. odorata, P. rosea, Polypogon monspeliensis, Populus deltoides, Primula frenchii, Proserpinaca palustris, Ptilimnium capillaceum, Quercus laurifolia, Q. nigra, Ranunculus sceleratus, Rorippa aquatica, Rosa palustris, Rumex, Sagittaria, Salicornia, Salix caroliniana, S. nigra, Salvia penstemonoides, Sanguisorba canadensis, Schoenoplectus pungens, Scirpus divaricatus, S. lineatus, Scutellaria parvula, Senecio glabellus, Seutera angustifolium, Solidago sempervirens, Spartina alterniflora, S. patens, S. pectinata, Symphyotrichum novi-belgii, S. subspicatum, Taxodium distichum, Thelypteris ovata, T. palustris, Toxicodendron radicans, Trifolium repens, Typha angustifolia, Utricularia geminiscapa, Veronica, Vicia acutifolia, V. minutiflora, Woodwardia areolata, W. virginica.

Use by wildlife: *Samolus valerandi* is grazed by horses (Mammalia: Equidae: *Equus ferus*) and by sheep (Mammalia: Bovidae: *Ovis aries*).

Economic importance: food: The young leaves of *Samolus valerandi* are bitter in taste, but supposedly are edible either when raw or cooked; **medicinal:** *Samolus ebracteatus* is used as a medicinal plant in Mexico. *Samolus valerandi* is rich in vitamin C and was once used to treat scurvy; **cultivation:** *Samolus valerandi* is cultivated as a submersed aquarium plant but requires high light, a fertile substrate, hard water, and temperatures from 15°C to 26°C; **misc. products:** none; **weeds:** none; **nonindigenous species:** none

Systematics: Although *Samolus* clearly belongs to the Ericales, its precise placement therein has not been established unequivocally. Phylogenetic analysis of morphological data places the genus near or within Primulaceae, arguably on the basis of convergences. Combined analyses of DNA sequence data from three chloroplast regions provides strong support for the placement of *Samolus* as the sister group to Theophrastaceae, which is the result accepted here. However, because of its distinct morphology, some authors prefer to retain the genus within its own family (Samolaceae). Most contemporary authors treat *Samolus alyssoides* and *S. cuneatus* as subspecies of *S. ebracteatus*. *Samolus valerandi* is known in older North American literature as *S. floribundus* and as *S. parviflorus* (=*S. valerandi* subsp. *parviflorus*). However, molecular phylogenetic and morphological analyses of seven *Samolus* species (Figure 5.26), failed to distinguish either *S. parviflorus* or *S. vagans* from *S. valerandi* and

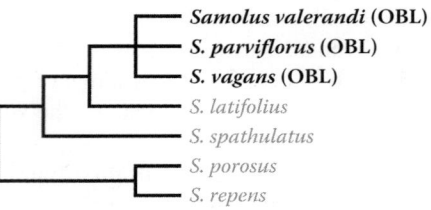

FIGURE 5.26 Interspecific relationships in *Samolus* as indicated by combined analysis of *trn*S–*trn*G and nrITS sequence data. Obligate North American taxa are indicated in bold. (From Jones, K. et al., *Plant Syst. Evol.*, 298, 1523–1531, 2012.)

indicated that they should not be maintained taxonomically as distinct species, but possibly as infraspecific taxa. Based on those results, this treatment applies the name *S. valerandi* rather than *S. parviflorus* to the North American plants (the 2013 wetland indicator list recognizes both taxa and assigns OBL status to both). Although *S. vagans* is also likely to be subsumed eventually as an infraspecific taxon of *S. valerandi*, it has been retained here as a species because it was included in the 2013 indicator list, and because the treatment already was completed prior to the publication of the 2012 phylogenetic analysis. The base chromosome number of *Samolus* is x = 12–13. *Samolus valerandi* (2n = 24, 26) is diploid. Otherwise, polyploidy (based on x = 13) appears to be rather common in the genus.

Comments: *Samolus ebracteatus* occurs in the southern United States as well as in Mexico and the West Indies. *Samouls valerandi* virtually is cosmopolitan. In North America it is widespread in all but the central and extreme northern portions and extends southward through Mexico to South America. *Samolus vagans* is a rare species locally endemic to southeastern Arizona.

References: Adam, 1990; Baldev & Lang, 1965; Baskin & Baskin, 1998; Bekker et al., 1999; Bossuyt & Hermy, 2004; Callaway & King, 1996; Chapman, 1936; Cholewa, 2009a, 2009b; Correll & Correll, 1975; Cosyns, 2004; DeVlaming & Proctor, 1968; Flores-Tavizón et al., 2003; Gibbs & Talavera, 2001; Hoagland & Buthod, 2005; Jones et al., 2012; Kasselmann, 2003; Koptur et al., 2000; Looney & Gibson, 1995; Salisbury, 1970; Schat & Scholten, 1986; Ståhl, 2004b; Van Bodegom et al., 2005; Zomlefer & Giannasi, 2005.

DICOTYLEDONS II ("ASTERID" TRICOPLATES)—CORE ASTERIDS: EUASTERIDS I ("LAMIIDS")

The "core" asterids comprise a clade of eight orders subdivided to represent two fundamental subclades designated as "euasterid I and euasterid II" (Figure 5.1). The monophyly of this clade is supported by several studies that incorporate molecular data (Olmstead et al., 1993; Soltis et al., 2000; Albach et al., 2001; Bremer et al., 2002; Judd et al., 2002; Soltis et al., 2005). Morphologically, the core asterids possess flowers with epipetalous stamens equal in number to the corolla lobes, and a syncarpous, bicarpellate gynoecium (Judd et al., 2002; Soltis et al., 2005). To better associate the subclades with their constituent families, the names "lamiids" and "campanulids" have been proposed for euasterid I and euasterid II respectively (Soltis et al., 2005). These groups are not easily delimited morphologically. However, the lamiids generally possess opposite, entire leaves, hypogynous flowers, filaments fused to the corolla tube and capsules; whereas, campanulids typically differ by their alternate, serrate leaves, epigynous flowers, free filaments and indehiscent fruits (Bremer et al., 2001; Soltis et al., 2005).

Relationships among orders within the euasterids I (lamiids) remain unclear; however, Garryales (which lacks aquatics) appears to be the sister clade to the remaining groups (Figure 5.1).

Order 3: Gentianales [5]

Gentianales represent a large order of more than 14,000 species, which are characterized by vestured pits, stipules, and glandular hairs on the stipules or petioles (Judd et al., 2002). The monophyly of the order has been corroborated by analyses of *atpB*, *rbcL*, *matK*, *ndhF*, and 18s DNA sequences (Judd et al., 2002).

Three families of Gentianales have species designated as obligate aquatics in North America:

1. **Apocynaceae** Adans.
2. **Gentianaceae** Juss.
3. **Rubiaceae** Juss.

Phylogenetic studies indicate that these families are dispersed throughout the order (Figure 5.27). Some FACW species occur in Gelsemiaceae. Several notable medicinal and ornamental plants occur in this order as well as *Coffea*, the economically important source of coffee.

Family 1: Apocynaceae [355]

Apocynaceae are treated here to include the former Asclepiadaceae as a result of molecular and nonmolecular data analyses (Judd et al., 1994; Endress & Bruyns, 2000; Endress, 2001). As currently defined, the family includes approximately 3700 species that are distributed worldwide, mainly in tropical and subtropical regions. The species possess laticifers

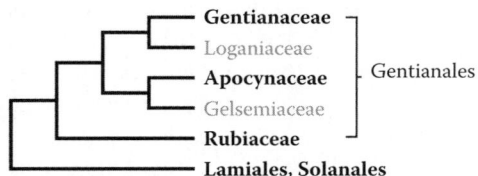

FIGURE 5.27 Phylogenetic relationships in Gentianales as deduced by various studies incorporating morphological and molecular data. Taxa that contain obligate aquatics in North America are shown in bold. (Adapted from Bremer, K. et al., *Plant Syst. Evol.*, 229, 137–169, 2001; Judd, F.W. et al., *J. Coastal Res.*, 18, 751–759, 2002; Soltis, D.E. et al., *Phylogeny and Evolution of Angiosperms*, Sinauer Associates, Inc., Sunderland, MA, 2005.)

with milky sap, and flowers with united styles/stigmas but free ovaries. The apical portion of the style is modified into a head-like, secretory structure (Judd et al., 2002). The flowers are adapted for a wide range of insect pollinators including bees (Hymenoptera), butterflies and moths (Lepidoptera), and flies (Diptera) (Judd et al., 2002). Some species have stamens that are modified to produce pollen in adherent masses (pollinia). The seeds are dispersed mainly by the wind; however, bird (Aves) and monkey (Mammalia: Primates) dispersal also occurs in some species (Judd et al., 2002).

Most species of Apocynaceae contain toxic alkaloids. Some taxa are useful medicinally, notably the Madagascar periwinkle (*Catharanthus*), which provides a treatment for leukemia. Several drugs for treating cardiac ailments and hypertension have also originated in this family. Many species in the genera *Adenium* (desert rose), *Allamanda* (golden trumpet), *Amsonia*, *Carissa* (Natal plum), *Mandevilla*, *Nerium* (oleander), *Plumeria* (pagoda tree), *Thevetia* (yellow oleander), and *Vinca* (periwinkle) are grown as ornamentals.

In addition to the modified family circumscription of Apocynaceae, a considerable amount of nomenclatural revision has also occurred within this family as a consequence of molecular phylogenetic studies. Several North American species formerly assigned to *Cynanchum* and *Sarcostemma* have been reassigned to *Seutera* and *Funastrum* respectively (Fishbein & Stevens, 2005; Liede & Täuber, 2000), and these transfers have been accepted here. *Funastrum clausum* (= *Sarcostemma clausum*) and *Rhabdadenia biflora* are designated as OBL in the Caribbean region, but only as FACW or FACU in the United States; thus they are excluded from the present treatment. *Amsonia tabernaemontana* originally was categorized as OBL, FACW in the 1996 indicator list, but has since been reclassified as FACW, FAC in the 2013 list. It has been retained here since the treatment was completed prior to the release of the 2013 list. *Seutera angustifolia* (see Fishbein & Stevens, 2005) is catalogued in the 2013 indicator list under the synonym *Funastrum angustifolium*; some authors recognize the plants as the synonym *Cynanchum angustifolium*.

The family is known mainly for its terrestrial species; however, obligate aquatics are found in three genera that occur in North America:

1. *Amsonia* Walter [FACW, FAC; previously OBL, FACW]
2. *Asclepias* L.
3. *Pentalinon* Voigt.
4. *Seutera* Rchb.

A representative phylogenetic survey of this large family (152 taxa representing all major tribes) has been carried out using data from *trn*L intron and *trn*L-F spacer regions (Potgieter & Albert, 2001). Results of that study indicate that only three of the five currently recognized subfamilies (Asclepioideae, Periplocoideae, and Secamonoideae) are monophyletic; whereas, subfamilies Apocynoideae and Rauvolfioideae are paraphyletic; the obligately aquatic genera occur within two of the subfamilies (Figure 5.28).

Subfamily **Asclepiadoideae** (*Asclepias, Seutera,* etc.)
Subfamily Secamonoideae
Subfamily Apocynoideae (in part)
Subfamily Periplocoideae
Subfamily **Apocynoideae** (*Pentalinon,* etc.)
Subfamily Rauvolfioideae (*Amsonia,* etc.)

FIGURE 5.28 Hypothetical relationships among subfamilies of Apocynaceae. In the full analysis, species assigned to subfamily Rauvolfioideae resolve as a large, paraphyletic grade (see Figure 5.29). Subfamily Periplocoideae is a clade embedded within subfamily Apocynoideae; however, its precise position was not determined in the study and will require the analysis of additional taxa for further clarification. The subfamilies with obligate aquatic genera in North America are shown in bold along with the placement of the genera. Despite the need for further study, it is apparent from this analysis that aquatic genera have evolved several times in the family. (Adapted from Potgieter, K. & Albert, V.A., *Ann. Missouri Bot. Gard.,* 88, 523–549, 2001.)

1. *Amsonia*

Blue dogbane, bluestar

Etymology: after John Amson (ca. 1698–1764)

Synonyms: *Anonymus, Ansonia, Tabernaemontana* (in part)

Distribution: global: Asia; North America; **North America:** southern

Diversity: global: 22 species; **North America:** 16 species

Indicators (USA): FACW; FAC: *Amsonia tabernaemontana* [also OBL in 1996 indicator list]

Habitat: freshwater; palustrine; **pH:** 7.0–7.8; **depth:** <1 m; **life-form(s):** emergent herb

Key morphology: stems (to 10 dm) ascending or spreading; leaves (to 5 cm) alternate, ovate-elliptic, entire, petioled (to 5 mm); cymes corymbose, dense, with many (to 50+) pedicelled (to 5 mm) flowers; corolla (to 8 mm) salverform, 5-lobed (to 6 mm), blue to whitish; style with a truncate stigmatic cap; fruit a pair of cylindrical follicles (to 10 cm) with 5–15 seeds (to 11 mm)

Life history: duration: perennial (woody roots); **asexual reproduction:** none; **pollination:** insect; **sexual condition:** hermaphroditic; **fruit:** follicles (common); **local dispersal:** seeds; **long-distance dispersal:** seeds

Imperilment: (1) *Amsonia tabernaemontana* [G5]; S1 (KS); S3 (TN, VA)

Ecology: general: *Amsonia* is not a characteristic genus of wet areas and several species grow in deserts. Although it was ranked formerly as OBL in a portion of its range, *A. tabernaemontana* has since been reclassified as FACW, FAC, likely because of its occasional occurrences in dry waste areas (see associated taxa later). Yet, this species often tolerates habitat conditions similar to those of OBL wetland species, which is not well reflected by its non-OBL status. The flowers are entomophilous and the seeds usually have physiological dormancy.

Amsonia tabernaemontana **Walter** inhabits canals, ditches, flatwoods, floodplains, prairies, riverbanks, sinks, sloughs and swamps at elevations of up to 350 m. The plants often grow in standing water and occur in full sun to partial shade on calcareous (often limestone) substrates including clay, gravel, sand, or moderately organic soil. They are drought tolerant but will not persist under prolonged dry soil conditions. Like most Apocynaceae, the flowers are adapted for pollination by insects, which mainly are bees

(Hymenoptera) and butterflies (Lepidoptera); however, few specific details exist on the floral biology of this species. Seed germination has been described as difficult. Germination within 3–4 weeks has been achieved following a 3-month period of moist stratification (at room temperature), succeeded by cold stratification (4°C) for 6–8 weeks and a final germination temperature of 21°C. The seeds can also be sown outdoors on the surface of compost when temperatures range from 13°C to 15°C. Experimental treatments of fresh and 1-year-old seeds with 10 mM GA3 (gibberellic acid) induced the highest percent germination and the fastest germination rate. The plants require 3 years to reach maturity from seed and can live up to 10 years. The mechanism of seed dispersal is unknown. Individuals perennate by means of a woody rootstalk, which makes propagation by division difficult. *Amsonia tabernaemontana* can grow in a broad range of communities as evidenced by its diverse ecological associates. **Reported associates:** *Acer barbatum, A. rubrum, Acmella oppositifolia, Agalinis, Agrostis hyemalis, Alopecurus geniculatus, Ambrosia bidentata, A. psilostachya, Amelanchier arborea, Amsonia rigida, Andropogon gerardii, A. glomeratus, A. virginicus, Anemone canadensis, A. virginiana, Apocynum androsaemifolium, A. cannabinum, Arabis virginica, Arisaema dracontium, A. triphyllum, Aristida palustris, Arnoglossum ovatum, Asarum canadense, Asclepias lanceolata, A. purpurascens, A. quadrifolia, A. tuberosa, A. viridis, Asplenium platyneuron, Aureolaria flava, A. grandiflora, Bacopa caroliniana, Baptisia bracteata, Bidens aristosa, B. comosa, B. coronata, Bignonia capreolata, Brunnichia ovata, Calamagrostis canadensis, Calystegia sepium, Capsella bursa-pastoris, Cardamine bulbosa, C. cancatenata, Carex glaucescens, C. intumescens, C. pensylvanica, C. verrucosa, Carya myristiciformis, C. ovata, Celtis laevigata, C. occidentalis, Centella erecta, Cephalanthus occidentalis, Cerastium arvense, Cercis canadensis, Chaerophyllum procumbens, Chamaecrista fasciculata, Chamaesyce humistrata, Chasmanthium latifolium, C. laxum, Cheilanthes, Claytonia virginica, Clematis crispa, Climacium, Cocculus carolinus, Commelina, Corallorhiza odontorhiza, C. wisteriana, Coreopsis tinctoria, Crataegus opaca, Cuphea viscosissima, Cynodon, Cynosciadium digitatum, Cyperus cephalanthus, Dalea, Danthonia sericea, D. spicata, Delphinium tricorne, Descurainia pinnata, Dichanthelium commutatum,*

D. scabriusculum, Diospyros virginiana, Echinacea purpurea, Eleocharis engelmannii, E. equisetoides, E. palustris, E. quadrangulata, E. wolfii, Elephantopus carolinianus, Elymus, Eragrostis, Erigeron philadelphicus, E. pulchellus, Eriocaulon compressum, Eriochloa punctata, Eryngium yuccifolium, Eupatorium serotinum, Euphorbia corollata, E. heterophylla, Euthamia, Fimbristylis littoralis, Forestiera ligustrina, Fraxinus caroliniana, F. pennsylvanica, Fuirena bushii, Galium tinctorium, Gaura coccinea, G. lindheimeri, Geranium carolinianum, G. maculatum, Gleditsia aquatica, G. triacanthos, Goodyera pubescens, Gratiola brevifolia, Hedyotis nigricans, Helianthus divericatus, H. mollis, H. occidentalis, H. petiolaris, H. salicifolius, Heuchera, Hibiscus moscheutos, Houstonia longifolia, Hydrolea ovata, Hydrophyllum virginianum, Hymenocallis liriosome, Hypericum hypericoides, H. prolificum, H. punctatum, Hyptis alata, Ilex amelanchier, I. coriacea, I. decidua, Ipomoea sagittata, Iva angustifolia, I. annua, Juncus effusus, J. polycephalus, J. roemerianus, Juniperus virginiana, Leersia hexandra, Liatris, Linum virginianum, Liquidambar styraciflua, Lolium pratense, Lonicera japonica, Ludwigia linearis, L. pilosa, L. sphaerocarpa, Lycopus rubellus, L. virginicus, Lysimachia ciliata, Lythrum alatum, L. lineare, Magnolia virginiana, Maianthemum racemosum, Mertensia virginica, Mitchella repens, Mnesithea rugosa, Monarda citriodora, M. fistulosa, M. punctata, Myosurus minimus, Myrica cerifera, Neptunia lutea, Nyssa biflora, Obolaria virginica, Oenothera drummondii, Osmorhiza claytoni, O. longistylis, Osmundastrum cinnamomeum, Ostrya virginiana, Oxalis corniculata, Oxypolis filiformis, Panicum hemitomon, P. rigidulum, P. tenerum, P. virgatum, Paronychia argyrocoma, Parthenocissus quinquefolia, Paspalum floridanum, P. plicatulum, Penstemon cobaea, P. hirsutus, Persicaria hydropiperoides, Philadelphus hirsutus, Phlox divaricata, Physostegia intermedia, Pinus taeda, P. virginiana, Pluchea, Podophyllum peltatum, Polygala leptocaulis, Polypodium polypodioides, Potentilla norvegica, Proserpinaca palustris, P. pectinata, Pteridium aquilinum, Pycnanthemum flexuosum, P. verticillatum, Ratibida columnifera, R. peduncularis, R. pinnata, Quercus laurifolia, Q. lyrata, Q. muhlenbergii, Q. nigra, Q. pagoda, Q. phellos, Q. prinus, Q. stellata, Ranunculus abortivus, R. hispidus, R. pusillus, R. recurvatus, R. sardous, R. septentrionalis, Rhexia mariana, R. virginica, Rhus aromatica, R. copallinum, Rhynchospora cephalantha, R. colorata, R. corniculata, R. elliottii, R. globularis, R. macrostachya, R. microcarpa, R. mixta, R. nitens, R. perplexa, Rorippa palustris, Rosa, Rubus trivialis, Rudbeckia texana, R. triloba, Ruellia humilis, Rumex acetosella, Sabal minor, Sabatia angularis, Sagittaria graminea, Salix nigra, Sanguinaria canadensis, Schizachyrium scoparium, S. tenerum, Scleria baldwinii, Scutellaria canescens, S. elliptica, Sedum ternatum, Selaginella rupestris, Silene regia, Silphium laciniatum, Smilax bona-nox, Solidago sempervirens, Sorghastrum nutans, Spartina patens, S. pectinata, Spiranthes laciniata, Sporobolus, Stachys tenuifolia, Stellaria media, Stylisma aquatica, Styrax americana, Symphyotrichum patens, S. turbinellum, Symplocos tinctoria, *Tephrosia virginiana, Thalictrum thalictroides, Thaspium barbinode, Thelypteris hispidula, Toxicodendron radicans, Tradescantia hirsutiflora, T. occidentalis, Trepocarpus aethusae, Tridens strictus, Trillium cuneatum, T. recurvatum, Triodanis, Triphora trianthophora, Tripsacum dactyloides, Ulmus alata, U. americana, U. crassifolia, Uvularia perfoliata, Valerianella radiata, Vernonia altissima, V. crinita, V. gigantea, V. missurica, Veronica officinalis, V. peregrina, Viola cucullata, Xyris fimbriata, X. laxifolia, Yucca.*

Use by wildlife: The alkaloid content of *Amsonia tabernaemontana* deters its use as a food source by wildlife. The foliage is highly resistant to herbivory by white-tailed deer (Mammalia: Cervidae: *Odocoileus virginianus*) and testing has shown that *A. tabernaemontana* is also insusceptible to carnation mottle virus (CarMV). The plants do host a lepidopteran larva (Sphingidae: *Hemaris diffinis*) and the roots are associated with an arbuscular mycorrhiza. The flowers are visited and pollinated by mourning cloak butterflies (Lepidoptera: Nymphalidae: *Nymphalis antiopa*) and various Hymenoptera (Apidae) including bumblebees (*Bombus pennsylvanicus*) and carpenter bees (*Xylocopa virginica*).

Economic importance: food: *Amsonia tabernaemontana* is not edible and no part of the plant should be ingested due to its toxic alkaloid content; **medicinal:** Many alkaloids have been extracted from the foliage, roots and seeds of *A. tabernaemontana*. The seeds contain tabersonine, an indole alkaloid precursor to vincamine and apovincamine, which are vasodilators that have been used as cerebrodilatory medicines. Rutin (a flavonoid) and vincadifformine (a hypotensive alkaloid) are obtained from the shoots and leaves; **cultivation:** *Amsonia tabernaemontana* is the only member of the genus widely in cultivation. Cultivars attributed to *A. tabernaemontana* include 'Blue Ice,' 'Blue Star,' 'Budakalaszi,' 'Montana,' and 'Short Stack'; **misc. products:** none; **weeds:** none; **nonindigenous species:** *Amsonia tabernaemontana* is naturalized in the northeastern United States as an escape from cultivation.

Systematics: *Amsonia* is assigned taxonomically to tribe Plumerieae, subfamily Rauvolfioideae of Apocynaceae; however, neither the tribe nor subfamily appears to be monophyletic (Figure 5.29). The genus once was believed to be related most closely to *Haplophyton*; however, results from molecular phylogenetic analysis place *Amsonia* quite distant from *Haplophyton* in a position close to *Thevetia* (Figure 5.29). Three subgenera (*Amsonia, Articularia, Sphinctosiphon*) have been recognized, with *A. tabernaemontana* assigned to subgenus *Amsonia*; however, the naturalness of these groups has not been tested explicitly and a thorough phylogenetic study of the entire genus is needed. Preliminary morphometric analyses indicate that several *Amsonia* species from the southeastern United States are not distinct morphologically from *A. tabernaemontana* and probably represent extreme variants of that species. These observations are supported by a preliminary phylogenetic study of *Amsonia* where DNA sequence data indicated that several species in the southeastern United States are conspecific with *A. tabernaemontana*, and that the species itself does not appear to be monophyletic. At least one hybrid (*A. tabernaemontana* × *A. ludoviciana*) has been reported, but it has not been well studied.

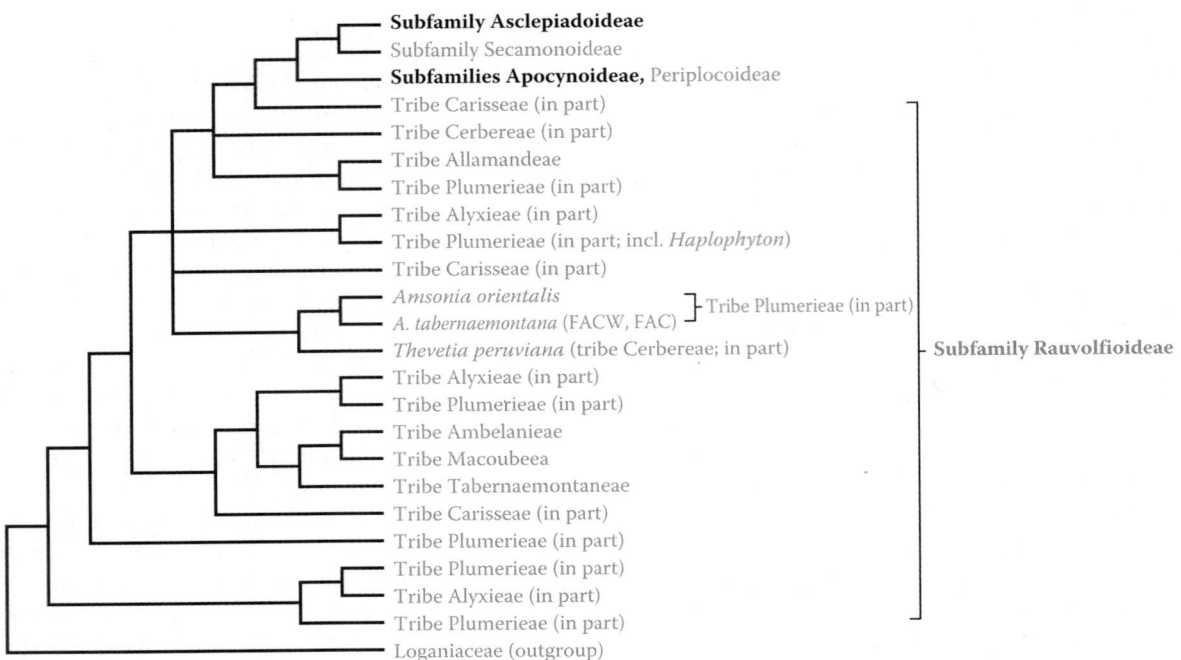

Subfamily Asclepiadoideae
Subfamily Secamonoideae
Subfamilies Apocynoideae, Periplocoideae
Tribe Carisseae (in part)
Tribe Cerbereae (in part)
Tribe Allamandeae
Tribe Plumerieae (in part)
Tribe Alyxieae (in part)
Tribe Plumerieae (in part; incl. *Haplophyton*)
Tribe Carisseae (in part)
Amsonia orientalis
A. tabernaemontana (FACW, FAC)] Tribe Plumerieae (in part)
Thevetia peruviana (tribe Cerbereae; in part)   **Subfamily Rauvolfioideae**
Tribe Alyxieae (in part)
Tribe Plumerieae (in part)
Tribe Ambelanieae
Tribe Macoubeea
Tribe Tabernaemontaneae
Tribe Carisseae (in part)
Tribe Plumerieae (in part)
Tribe Plumerieae (in part)
Tribe Alyxieae (in part)
Tribe Plumerieae (in part)
Loganiaceae (outgroup)

FIGURE 5.29 Relationships within Apocynaceae subfamily Rauvolfioideae as indicated by phylogenetic analyses of cpDNA sequence data. The subfamily resolves as a paraphyletic grade and only a few of the tribes recognized within it are monophyletic, indicating that several refinements to the classification of the family are necessary. Taxa with obligate aquatic representatives in North America are indicated in bold. *Amsonia tabernaemontana* currently is assigned to tribe Plumerieae, which is widely polyphyletic in the analysis shown. (Adapted from Potgieter, K. & Albert, V.A., *Ann. Missouri Bot. Gard.*, 88, 523–549, 2001.)

With a base chromosome number of $x = 11$ postulated for all Apocynaceae, *A. tabernaemontana* ($2n = 22$) presumably is diploid. There are no population genetic studies reported for *A. tabernaemontana*; however, microsatellite primers have been developed for other *Amsonia* species and should be screened for cross-reactivity with this species as a potential source of genetic markers to facilitate such studies.

Comments: *Amsonia tabernaemontana* occurs throughout the southeastern United States.

References: Darke, 2005; Doffitt & Fishbein, 2005; Ghisalberti et al., 1998; Kock, 1987; Máthé, Jr. et al., 1983; Potgieter & Albert, 2001; Pringle, 2004; Scocco et al., 1998; Topinka et al., 2004; Turner, 1934, 1936; Uphof, 1922; Van der Laan & Arends, 1985; Williams, 2005; Woodson, Jr., 1928.

2. *Asclepias*

Milkweed, swamp milkweed; asclépiade

Etymology: after *Asklepios*, a deity in Greek mythology symbolic of the healing arts

Synonyms: *Acerates*; *Anantherix*; *Anthanotis*; *Asclepiodella*; *Asclepiodora*; *Biventraria*; *Gomphocarpus* (in part); *Oligoron*; *Oxypteryx*; *Podostemma*; *Podostigma*; *Polyotus*; *Schizonotus*; *Solanoa*

Distribution: global: Africa; New World; **North America:** widespread

Diversity: global: 400 species; **North America:** 74 species

Indicators (USA): OBL: *Asclepias connivens*, *A. incarnata*, *A. lanceolata*, *A. perennis*, *A. rubra*; **FACW:** *A. incarnata*, *A. longifolia*

Habitat: brackish (coastal), freshwater; palustrine; **pH:** 4.2–8.0; **depth:** <1 m; **life-form(s):** emergent herb

Key morphology: herbage lactiferous; stems (9–15 dm) erect, simple or branched; leaves (7–25 cm) opposite, simple, entire, petioled (to 1 cm); cymes with umbellate clusters of radial, 5-merous flowers; corolla lobed (3–15 mm) rotate, orange, pink, red, white, or yellow; stamens and stigma united as a fleshy gynostegium (to 2 mm); corona present, comprising five nectariferous hoods (1.5–9 mm), which subtend the anthers; pollen in paired masses (pollinia) connected by yoke-like "translator arms"; carpels two, superior, fused apically at the stigma; fruit (to 15 cm) a pair of follicles (usually maturing only one); seeds (7–15 mm) with or without a fine coma of hairs (2–4.5 cm)

Life history: duration: perennial (persistent caudex); **asexual reproduction:** stem layering; **pollination:** insect; **sexual condition:** hermaphroditic; **fruit:** follicles (common); **local dispersal:** seeds (water, wind); **long-distance dispersal:** seeds (water, wind)

Imperilment: (1) *Asclepias connivens* [G4]; SH (MS); S3 (GA); (2) *A. incarnata* [G5]; SH (DE); S1 (AZ, MT, <u>PE</u>); S2 (AR, ID, LA, <u>NS</u>, WY); S3 (GA, <u>NS</u>, VA); (3) *A. lanceolata* [G5]; S1 (DE); S2 (NJ); S3 (GA, VA); (4) *A. longifolia* [G4]; SX (DE); S1 (VA); S2 (NC); (5) *A. perennis* [G5]; (6) *A. rubra* [G4]; SX (DC, NY, PA); SH (GA); S1 (AL, DE, MD); S2 (LA, NJ, VA); S3 (NC)

Ecology: general: Most *Asclepias* species are terrestrial; however, 22 of the North American species (30%) are assigned some level of wetland indicator status, with five regarded as obligately aquatic. *Asclepias longifolia* (FAC, UPL), which was classified as OBL, FACW in the 1996 indicator list, has also been included because its treatment was completed prior to the release of the 2013 list. Most of the species are self-incompatible and all are pollinated by insects, which tend to be larger types that are capable of extricating and transferring the bulky pollinia. The pollinators often lose their legs or become entrapped in the intricately designed floral structures. Although most *Asclepias* species are outcrossed, they possess anomalously low pollen:ovule ratios, which is believed to be a consequence of their aggregation of pollen into pollinia. The seeds typically are wind dispersed, facilitated by means of a coma, which also enhances their flotation. Seed size in some species has been shown to vary as a function of defoliation, light, and nutrients. The species are quite widely adapted to different sediment types. The milky latex found in most parts of the plants contains alkaloids, caoutchouc (natural rubber), and cardenolides. Some species are poisonous.

Asclepias connivens **Baldw.** grows in bogs, depressions, flatwoods, marshes, pocosins, prairies, savannas, seeps, and swamps of the coastal plain at elevations of at least to 12 m. Soils include muck, peat, sand, and sandy loam. The flowers are large and very fragrant. There is little information reported on the reproductive ecology of this species. The seeds have a coma present and presumably are wind dispersed. Information on germination or seed ecology is not available. **Reported associates:** *Aletris aurea, A. lutea, Aristida beyrichiana, A. stricta, Asclepias cinera, A. rubra, A. viridiflora, A. viridula, Balduina uniflora, Bigelowia nudata, Buchnera, Calopogon pallidus, Carex turgescens, Carphephorus pseudoliatris, Clethra alnifolia, Cliftonia monophylla, Coreopsis, Ctenium aromaticum, Drosera capillaris, D. tracyi, Eriocaulon compressum, E. decangulare, E. texense, Eryngium integrifolium, Eupatorium mohrii, Euphorbia telephioides, Eurybia eryngiifolia, Gaylussacia mosieri, Gymnadeniopsis nivea, Hypericum myrtifolium, Ilex coriacea, I. glabra, Juncus, Lachnanthes caroliniana, Liatris, Lilium catesbaei, Lophiola aurea, Ludwigia, Lycopodiella, Lyonia lucida, Magnolia virginiana, Marshallia tenuifolia, Muhlenbergia expansa, Myrica heterophylla, M. inodora, Nolina atopocarpa, Paspalum notatum, Physostegia godfreyi, Pinus elliottii, P. palustris, Pityopsis oligantha, Platanthera integra, Pleea tenuifolia, Polygala cruciata, P. lutea, P. ramosa, P. ruglii, Quercus, Rhexia alifanus, R. lutea, Rhynchospora oligantha, R. plumosa, Rudbeckia graminifolia, Sabatia macrophylla, Sarracenia flava, S. leucophylla, S. minor, S. psittacina, S. purpurea, Scleria baldwinii, Serenoa repens, Smilax laurifolia, Spiranthes, Sporobolus floridanus, Tofieldia, Verbesina chapmanii, Xyris ambigua, X. scabrifolia.*

Asclepias incarnata **L.** occurs in fresh and brackish, tidal or nontidal waters of bogs, depressions, ditches, estuaries, fens, floodplains, marshes, meadows, prairies, sloughs, swales, swamps, and along the margins of lakes, rivers, and streams at elevations of up to 2744 m. The plants can tolerate exposures from full sun to shade. They occupy extremely diverse sites and occur on a wide range of substrates, which often are alkaline but span a broad acidity gradient (pH: 5.0–8.0). These include gravel, humus, mud, peat, rock, sand, sandy clay or loam, and tree stumps. This species can occur in up to 1 m of standing water, but does not tolerate deep inundation for prolonged periods. Typically, it produces multiple stems with many umbels, which each produce few fruits. The flowering period is relatively short but plants can bloom within their first year of growth. *Asclepias incarnata* has a mixed breeding system comprising both self-compatible and self-incompatible plants. One study obtained a comparable level of successful self-pollination (29%) with plants from seven populations as compared to success rates for crosses made within or between populations (30%–56%, respectively). A diverse fauna of Hymenoptera and Lepidoptera (see *Use by wildlife* later) serve as the principal pollinators, depending on the locality and associated flora. Pollinator competition with *A. verticillata* can occur when the two species grow in proximity. Allozyme data indicate that natural populations are largely outcrossed; however the success of self-pollination increases in flowers that already have been outcrossed. The plants exhibit variation in seed set as a consequence of differential partitioning of maternal resources to developing, genetically distinct ovules. Each follicle typically produces from 300 to 400 mg of seeds. The seeds germinate within 1–3 months at 18°C after 2–3 weeks of cold stratification (4°C). The seeds survive prolonged inundation and can germinate throughout the summer if they become exposed on the mud by receding waters. The coma on the seeds provides for their dispersal by wind but also facilitates dispersal in water by increasing their buoyancy. Adult plants are extremely sensitive to damage from ozone but can withstand cold temperatures down to −39°C. The plants usually live from 5 to 10 years and are somewhat fire resistant. **Reported associates:** *Acer rubrum, A. saccharinum, Agrostis alba, Alisma, Alnus rugosa, A. serrulata, Amaranthus cannabinus, Amblystegium serpens, Ammannia coccinea, Amphicarpaea bracteata, Andropogon gerardii, Anemone canadensis, Angelica atropurpurea, Apios americana, Apocynum sibiricum, Arisaema triphyllum, Arnoglossum plantagineum, Asclepias hirtella, A. sullivantii, A. syriaca, Baccharis halimifolia, Bacopa monnieri, Baptisia alba, Betula glandulifera, B. lutea, B. papyrifera, Bidens aureus, B. cernuus, B. connatus, B. coronatus, B. frondosus, B. vulgatus, Boehmeria cylindrica, Bolboschoenus fluviatilis, B. robustus, Boltonia asteroides, Butomus umbellatus, Calamagrostis canadensis, Calla palustris, Calopogon pulchellus, Caltha palustris, Campanula aparinoides, Campylium stellatum, Carex annectens, C. aquatilis, C. atlantica, C. blanda, C. canescens, C. comosa, C. cryptolepis, C. emoryi, C. grayi, C. hormathodes, C. hystericina, C. interior, C. intumescens, C. lacustris, C. lasiocarpa, C. lurida, C. muskingumensis, C. pellita, C. prairea, C. sartwellii, C. scoparia, C. stipata, C. stricta, C. typhina, C. utriculata, C. vesicaria, C. viridula, C. vulpinoidea, Cephalanthus occidentalis, Chara, Chelone glabra, Cicuta bulbifera, C. maculata, Cladium jamaicense, Clematis virginiana, Clethra alnifolia, Comarum palustre, Cornus amomum, C. foemina, C. racemosa, C. sericea, Corylus americana, Crataegus, Cuphea viscosissima, Cyperus filicinus, C. haspan, C. odoratus, C. rivularis, C.*

strigosus, Decodon verticillatus, Dichanthelium acuminatum, Dioscorea villosa, Dipsacus laciniatus, Doellingeria umbellata, Dulichium arundinaceum, Echinochloa crusgalli, E. walteri, Eleocharis acicularis, E. erythropoda, E. fallax, E. macrostachya, E. palustris, E. parvula, E. quinqueflora, Elymus virginicus, Epilobium glandulosum, Equisetum arvense, Eragrostis frankii, E. hypnoides, Eriophorum virginicum, Eryngium yuccifolium, Eupatorium dubium, E. maculatum. E. perfoliatum, Fraxinus americana, F. caroliniana, F. nigra, F. pennsylvanica, Galium boreale, G. obtusum, G. tinctorium, Geum aleppicum, Glyceria canadensis, G. grandis, G. striata, Gratiola aurea, Hamamelis, Hasteola suaveolens, Helenium autumnale, Helianthus giganteus, H. grosseserratus, Hibiscus laevis, H. moscheutos, Hydrocotyle verticillata, Hypericum virginicum, Hypnum lindbergii, Hypoxis hirsuta, Ilex verticillata, Impatiens capensis, Iris versicolor, I. virginica, Juncus canadensis, J. effusus, J. interior, J. nodous, J. tenuis, J. torreyi, Koeleria macrantha, Kosteletzkya pentacarpos, Laportea canadensis, Larix laricina, Leersia lenticularis, L. oryzoides, Liatris pycnostachya, Lilium michiganense, Lindera benzoin, Lobelia cardinalis, L. elongata, L. kalmii, L. siphilitica, Ludwigia palustris, Lycopus americanus, L. uniflorus, Lysimachia nummularia, L. quadriflora, L. quadrifolia, L. terrestris, Lythrum alatum, L. lineare, L. salicaria, Matteuccia struthiopteris, Melilotus alba, Mentha arvensis, M. spicata, Mikania scandens, Mimulus ringens, Monarda fistulosa, Muhlenbergia mexicana, Myrica cerifera, M. gale, Onoclea sensibilis, Osmunda regalis, Osmundastrum cinnamomeum, Panicum capillare, P. hemitomon, P. rigidulum, P. virgatum, Parnassia glauca, Parthenocissus quinquefolia, Paspalum laeve, Pedicularis lanceolata, Peltandra virginica, Persicaria amphibia, P. arifolia, P. hydropiper, P. hydropiperoides, P. lapathifolia, P. longiseta, P. maculosa, P. pensylvanica, P. punctata, P. sagittata, Phalaris arundinacea, Phragmites australis, Phyla lanceolata, Physostegia virginiana, Picea mariana, Pilea fontana, P. pumila, Platanthera clavellata, P. psycodes, Pluchea odorata, P. purpurascens, Pontederia cordata, Populus balsamifera, P. deltoides, Ptilimnium capillaceum, Pycanthemum virginianum, Quercus, Ranunculus macranthus, Rhododendron viscosum, Rhynchospora macrostachya, Rorippa sylvestris, Rosa arkansana, R. palustris, Rudbeckia hirta, R. laciniata, Rumex altissimus, R. crispus, R. orbiculatus, R. verticillatus, Sacciolepis striata, Sagittaria falcata, S. lancifolia, S. latifolia, S. rigida, Salix bebbiana, S. candida, S. discolor, S. eriocephala, S. fragilis, S. gooddingii, S. interior, S. petiolaris, Salix ×bebbii, Sambucus nigra, Schedonorus arundinaceus, Schoenoplectus acutus, S. americanus, S. tabernaemontani, Scirpus atrovirens, S. cyperinus, Scrophularia marilandica, Scutellaria galericulata, S. lateriflora, Silphium laciniatum, S. perfoliatum, S. terebinthinaceum, Sium suave, Solanum dulcamara, Solidago canadensis, S. graminifolia, Sparganium americanum, S. eurycarpum, Spartina alterniflora, S. cynosuroides, S. pectinata, Sphenopholis intermedia, Spiraea alba, Stachys aspera, Symphyotrichum lanceolatum, S. novae-angliae, S. novi-belgii, S. puniceum, S. racemosum, S. subulatum, Symplocarpus foetidus, Teucrium

canadense, Thalictrum dasycarpum, Thelypteris palustris, Toxicodendron radicans, T. vernix, Triadenum fraseri, T. walteri, Typha angustifolia, T. latifolia, Ulmus americana, Vaccinium corymbosum, Verbena hastata, Vernonia baldwinii, V. fasciculata, Viburnum dentatum, Vicia villosa, Vigna luteola, Viola nephrophylla, Vitis riparia, Xanthium strumarium, Xyris torta, Zanthoxylum americanum, Zizania aquatica, Zizia aptera, Z. aurea.

***Asclepias lanceolata* Walter** inhabits brackish or freshwater bogs, depressions, ditches, flatwoods, floodplains, glades, margins of ponds and rivers, marshes, meadows, prairies, savannas, and swamps at elevations of up to 1951 m (from stations in Mexico). The plants thrive in full sun but tolerate partial shade. Substrates tend to be acidic but span a wide pH range (pH: 4.2–7.5). They include muck, sand, sandy loam, and sandy peat. The flowers are pollinated by various butterflies, bees, and wasps (see *Use by wildlife* later). Populations located amidst larger islands of trees (which harbor larger numbers of insects) are more effectively pollinated. The seeds are wind dispersed, facilitated by a coma of long hairs. Germination reportedly can be achieved using the method described earlier for *A. incarnata*. The optimal hydroperiod (length of inundation) is approximately 195 days, but plants can occur at sites having hydroperiods as short as 39 days. Adult plants can withstand cold temperatures down to −25°C. Further studies on the seed ecology of this species would provide useful ecological information. **Reported associates:** *Acer rubrum, Acmella oppositifolia, Agalinis fasciculata, A. purpurea, Agrostis hyemalis, Aletris aurea, A. farinosa, A. lutea, Amsonia tabernaemontana, Andropogon glomeratus, A. gyrans, A. mohrii, A. virginicus, Aristida palustris, A. purpurascens, Arnoglossum ovatum, Arundinaria gigantea, Asclepias longifolia, A. viridis, Bacopa caroliniana, Bigelowia nudata, Boehmeria cylindrica, Boltonia asteroides, Calopogon barbatus, C. pallidus, C. tuberosus, Carex canescens, C. hyalinolepis, C. lutea, C. striata, C. verrucosa, C. vulpinoidea, Carphephorus paniculatus, Centella asiatica, C. erecta, Cephalanthus occidentalis, Chaptalia tomentosa, Chasmanthium, Ciclospermum, Cladium jamaicense, Cleistes, Clethra alnifolia, Coreopsis falcata, C. linifolia, C. nudata, C. tinctoria, Ctenium aromaticum, Cyperus strigosus, Cyrilla racemiflora, Dichanthelium acuminatum, D. dichotomum, D. erectifolium, D. longiligulatum, D. scabriusculum, D. scoparium, D. strigosum, D. wrightianum, Drosera capillaris, D. tracyi, Eleocharis fallax, E. quadrangulata, E. rostellata, E. tuberculosa, Erigeron vernus, Eriocaulon compressum, E. decangulare, Eriochloa punctata, Eryngium aquaticum, E. integrifolium, E. yuccifolium, Eupatorium leucolepis, E. mohrii, E. purpureum, E. serotinum, Euthamia leptocephala, Fimbristylis littoralis, F. puberula, Galium, Gaura lindheimeri, Gaylussacia frondosa, Helianthus angustifolius, H. heterophyllus, Hibiscus aculeatus, H. moscheutos, Hypericum crux-andreae, H. myrtifolium, Hydrolea ovata, Hymenocallis liriosme, Hypoxis juncea, Hyptis alata, Ilex glabra, I. myrtifolia, Ipomoea sagittata, Iris tridentata, I. virginica, Iva angustifolia, Juncus polycephalus, J. roemerianus, Juniperus virginiana, Kyllinga pumila, Lachnanthes* caroliana, *Leersia hexandra, L. oryzoides, Liatris spicata, Lilium catesbaei, Linum striatum, Liquidambar styraciflua, Liriodendron tulipifera, Lobelia floridana, L. glandulosa, Lophiola americana, Ludwigia hirtella, L. pilosa, L. sphaerocarpa, L. virgata, Lycopodiella alopecuroides, Lycopus rubellus, Lyonia ligustrina, Lysimachia loomisii, Lythrum lineare, Magnolia virginiana, Marshallia graminifolia, Melanthium virginicum, Mentha spicata, Mnesithea rugosa, Muhlenbergia capillaris, Myrica caroliniensis, M. cerifera, Neptunia lutea, Nyssa biflora, Oenothera drummondii, Osmunda regalis, Osmundastrum cinnamomeum, Oxypolis filiformis, O. rigidior, Panicum anceps, P. hemitomon, P. rigidulum, P. virgatum, Parnassia asarifolia, Paspalum floridanum, P. plicatulum, P. praecox, Peltandra virginica, Persicaria punctata, Photinia pyrifolia, Physostegia leptophylla, Pinguicula planifolia, P. pumila, Pinus elliottii, P. palustris, P. serotina, Pityopsis graminifolia, Platanthera ciliaris, P. cristata, Pleea tenuifolia, Pluchea camphorata, Pogonia ophioglossoides, Polygala brevifolia, P. cruciata, P. cymosa, P. leptocaulis, P. lutea, P. ramosa, Pontederia cordata, Proserpinaca pectinata, Pycnanthemum flexuosum, Quercus laurifolia, Q. nigra, Q. phellos, Ratibida peduncularis, Rhexia alifanus, R. lutea, R. mariana, R. petiolata, R. virginica, Rhynchospora breviseta, R. careyana, R. cephalantha, R. chapmanii, R. colorata, R. corniculata, R. elliottii, R. fascicularis, R. filifolia, R. globularis, R. glomerata, R. gracilenta, R. latifolia, R. macrostachya, R. microcarpa, R. nitens, R. oligantha, R. plumosa, R. pusilla, R. rariflora, Rubus trivialis, Sabal palmetto, Sabatia angularis, S. bartramii, S. difformis, S. dodecandra, Saccharum giganteum. Sagittaria, Sarracenia alata, S. ×catesbaei, S. flava, Saururus cernuus, Schizachyrium scoparium, Schoenoplectus pungens, Scirpus cyperinus, Scleria baldwinii, S. georgiana, S. pauciflora, Scutellaria integrifolia, Sesbania vesicaria, Silphium asteriscus, Sisyrinchium angustifolium, Smilax glauca, S. laurifolia, Solidago sempervirens, S. stricta, Sorghastrum nutans, Spartina patens, Sphagnum, Spiranthes praecox, Symphyotrichum dumosum, Taxodium ascendens, Thelypteris palustris, Toxicodendron radicans, Tradescantia hirsutiflora, Triadenum walteri, Triantha racemosa, Tridens ambiguus, T. strictus, Typha angustifolia, T. latifolia, Vaccinium fuscatum, V. tenellum, Vernonia gigantea, V. noveboracensis, Viola lanceolata, V. ×primulifolia Woodwardia areolata, W. virginica, Xyris ambigua, X. caroliniana, X. fimbriata, X. smalliana, Zizania aquatica.*

***Asclepias longifolia* Michx.** is found in bogs, depressions, ditches, flatwoods, floodplains, glades, meadows, prairies, savannas, and swamps of the southeastern coastal plain at elevations of at least 29 m. The habitats mainly provide open exposures. Substrates range from acidic (e.g., 4.5–5.1) to calcareous (no specific pH range reported) and include clay, loam, sandstone outcrops, sandy loam, and sandy peat. There is little information reported on the reproductive or seed ecology of this species. The floral mechanism is modified such that it does not entrap insects by their legs as do other milkweeds; here, the hoods function primarily as nectaries. Bumblebees (Hymenoptera: Apidae: *Bombus*) are the most commonly observed pollinators, the floral pollinia

being removed by hairs on the underside of the bees. The seeds possess a coma and presumably are wind dispersed. Their germination requirements have not been reported. Adult plants can suffer substantial insect damage. The plants perennate from a tuberous rootstock. **Reported associates:** *Acer rubrum, Agalinis aphylla, Aletris aurea, A. farinosa, Amelanchier obovalis, Amphicarpum purshii, Andropogon virginicus, Aristida palustris, A. stricta, Arnoglossum ovatum, Asclepias lanceolata, A. michauxii, Balduina uniflora, Bigelowia nudata, Calopogon pallidus, C. tuberosus, Carex striata, Carphephorus paniculatus, C. tomentosus, Centella erecta, Chaptalia tomentosa, Cleistes, Coreopsis linifolia, Ctenium aromaticum, Cyrilla racemiflora, Dichanthelium acuminatum, D. dichotomum, D. longiligulatum, D. scabriusculum, D. spretum, Dionaea muscipula, Drosera capillaris, Eleocharis tuberculosa, Eriocaulon decangulare, E. texense, Eryngium integrifolium, Eupatorium leucolepis, E. mohrii, Eurybia paludosa, Fimbristylis puberula, Gaylussacia dumosa, G. frondosa, Helenium drummondii, Hypericum adpressum, H. galioides, Ilex glabra, I. myrtifolia, Iris tridentata, Juncus elliottii, Lachnanthes caroliana, Lachnocaulon anceps, Liatris spicata, Lobelia flaccidifolia, L. nuttallii, Ludwigia hirtella, L. virgata, Lycopodiella alopecuroides, Lyonia ligustrina, L. lucida, Magnolia virginiana, Marshallia caespitosa, M. graminifolia, Mitreola sessilifolia, Muhlenbergia capillaris, M. expansa, Myrica caroliniensis, M. cerifera, Nyssa, Osmundastrum cinnamomeum, Panicum rigidulum, P. virgatum, Parnassia caroliniana, Paspalum praecox, Persea palustris, Photinia pyrifolia, Pinus elliottii, P. palustris, P. taeda, Platanthera nivea, Pleea tenuifolia, Polygala crenata, P. lutea, P. nana, P. ramosa, Prunus serotina, Pteridium aquilinum, Quercus, Rhexia alifanus, R. lutea, R. petiolata, Rhynchospora baldwinii, R. cephalantha, R. elliottii, R. gracilenta, R. latifolia, R. microcarpa, R. plumosa, R. pusilla, Rubus, Rudbeckia, Sabatia campanulata, S. difformis, S. macrophylla, Sarracenia alata, S. flava, S. purpurea, Schoenolirion croceum, Scleria georgiana, S. pauciflora, Smilax glauca, S. laurifolia, Sisyrinchium capillare, Sporobolus pinetorum, Stenanthium densum, Symphyotrichum dumosum, Toxicodendron, Tragia, Triantha racemosa, Vaccinium fuscatum, Viola lanceolata, V. ×primulifolia, Woodwardia virginica, Xyris caroliniana, Zigadenus glaberrimus.*

***Asclepias perennis* Walter** occurs in bayous, depressions, ditches, drying ponds, floodplains, hammocks, margins of rivers and streams, marshes, meadows, prairies, seeps, sinkholes, sloughs, and swamps of the southeastern coastal plain at elevations of up to 120 m. Most sites of occurrence are characterized by full sun. The substrates span a broad range of pH (5.0–7.5) and often comprise fine silty clay, but range from organic muck to sand. *Asclepias perennis* is completely self-incompatible and natural populations have a rate of fruit set about 1%. Approximately 32 seeds are produced per fruit on average. The flowers are visited by various Diptera, Hymenoptera, and Lepidoptera, but little is known regarding their specific pollinators other than they are generalists. The species is well adapted for water dispersal with downward-oriented fruits that release the seeds with expanded seed coats but no coma as in other milkweeds. The seeds are buoyant and a large number (35%) can remain afloat for more than 6 months, comprising what has been described as a "floating seed bank." Viability remains high (~85%) for inundated seeds. High germination (100%) has been achieved for ripened seeds without any pretreatment, when placed for one week at 23°C (16 h light). Untreated seeds kept in dry storage also retain nearly 100% germinability. Genetic variation in *A. perennis* is lower than the level observed for many other milkweed species, but represents a pattern similar to that observed in other outcrossing, herbaceous perennials. Electrophoretic data from 16 loci indicated that 41% were polymorphic, averaging 1.6 alleles/locus, with a mean heterozygosity of 0.06. Most of the observed variation is partitioned within populations and gene flow among populations appears to be high. The plants possess a syndrome of defense characters that include a low C:N ratio, low specific leaf area, and high cardenolide content. However, specimens occasionally show substantial insect damage. Field observations indicate that unlike most other milkweeds, plants of *A. perennis* are capable of vegetative reproduction by stem layering. **Reported associates:** *Acer negundo, A. rubrum, A. saccharum, Agalinis tenuifolia, Agrostis stolonifera, Ampelopsis arborea, Apios americana, Arundinaria gigantea, Axonopus furcatus, Bacopa caroliniana, Berchemia scandens, Bidens comosa, B. connata, B. discoidea, B. vulgata, Bignonia capreolata, Boehmeria cylindrica, Brunnichia ovata, Campsis radicans, Cardamine pennsylvanica, Carex crus-corvi, C. debilis, C. glaucescens, C. hyalinolepis, C. intumescens, C. joorii, C. lonchocarpa, C. lupulina, C. stipata, C. vulpinoidea, Carpinus caroliniana, Celtis laevigata, Cephalanthus occidentalis, Chasmanthium latifolium, C. laxum, Circaea lutetiana, Cirsium, Commelina diffusa, C. virginica, Cornus amomum, C. foemina, Crataegus aestivalis, Croton capitatus, Cyperus erythrorhizos, C. ferruginescens, C. squarrosus, Dichanthelium dichotomum, Dyschoriste humistrata, Echinochloa muricata, Epilobium coloratum, Eragrostis hypnoides, Erechtites hieracifolia, Eryngium, Eupatorium coelestinum, Fraxinus caroliniana, F. pennsylvanica, F. profunda, Galium obtusum, Gelsemium rankinii, Gleditsia, Glyceria septentrionalis, G. striata, Helenium autumnale, Heliotropium indicum, Hibiscus moscheutos, Hydrocotyle verticillata, Hydrolea affinis, H. corymbosa, Hymenocallis duvalensis, Hypericum crux-andreae, H. galioides, H. perforatum, Hypoxis leptocarpa, Hyptis alata, Ilex cassine, I. decidua, I. opaca, I. verticillata, Ipomoea lacunosa, Itea virginica, Juncus, Justicia ovata, Leersia lenticularis, L. oryzoides, L. virginica, Lindernia dubia, Lippia nodiflora, Liquidambar styraciflua, Liriodendron tulipifera, Lobelia cardinalis, L. flaccidifolia, Ludwigia alternifolia, Lycopus americanus, L. rubellus, Lysimachia ciliata, Magnolia grandiflora, M. virginiana, Mentha spicata, Mikania scandens, Mimulus alatus, Mitchella repens, Mitreola petiolata, Muhlenbergia frondosa, M. schreberi, Myosurus minimus, Nyssa aquatica, N. biflora, N. ogeche, N. sylvatica, Onoclea sensibilis, Panicum gymnocarpon, P. rigidulum, Paspalum fluitans, P. pubiflorum, Peltandra virginica, Persicaria hydropiper, P. hydropiperoides, P. sagittata, Phytolacca americana, Pilea pumila, Planera aquatica, Platanthera flava, Platanus occidentalis, Pluchea camphorata, Polygala grandiflora, Polypodium polypodioides, Populus heterophylla, Proserpinaca palustris, P. pectinata, Ptilimnium costatum, Quercus alba, Q. laurifolia, Q. lyrata, Q. michauxii, Q. nigra, Q. nuttallii, Q. palustris, Q. prinus, Q. virginiana, Rhynchospora inundata, R. miliacea, Rudbeckia laciniata, Rumex crispus, R. verticillatus, Sabal minor, S. palmetto, Sabatia stellaris, Sagittaria latifolia, Salix nigra, Samolus ebracteatus, S. valerandi, Saururus cernuus, Scutellaria lateriflora, S. nervosa, Sebastiania fruticosa, Senecio glabellus, Serenoa repens, Silphium astericus, Sium suave, Smilax bona-nox, S. rotundifolia, Spiranthes cernua, S. odorata, Stachys tenuifolia, Symphyotrichum lanceolatum, S. lateriflorum, Taxodium distichum, Toxicodendron radicans, Trachelospermum difforme, Triadenum walteri, Triadica sebifera, Ulmus alata, U. americana, U. rubra, Vernonia, Viola affinis, V. esculenta, V. ×primulifolia, Vitis aestivalis, V. rotundifolia, Wisteria frutescens, Xanthium strumarium.*

***Asclepias rubra* L.** inhabits barrens, bogs, depressions, flatwoods, marshes, meadows, pocosins, prairies, roadsides, savannas, seeps, swales, and swamps at elevations of at least 68 m. The plants occur on sand or sandy loam substrates in sites that receive full sunlight. Little ecological information exists for this fairly rare species and no detailed studies have been made of its reproductive or seed ecology. As in similar milkweed species the flowers probably are pollinated by generalist Hymenoptera and/or Lepidoptera. The seeds possess a long (4 cm) coma, which presumably facilitates their dispersal by wind. **Reported associates:** *Acer rubrum, Agrostis hyemalis, Aletris aurea, A. lutea, Alisma triviale, Alnus serrulata, Andropogon virginicus, Arethusa bulbosa, Aristida beyrichiana, A. virgata, Asclepias connivens, Axonopus fissifolius, Balduina uniflora, Bigelowia nudata, Burmannia capitata, Calopogon tuberosus, Carex barrattii, C. buxbaumii, C. livida, C. lurida, C. striata, C. turgescens, C. venusta, Carphephorus pseudoliatris, Chamaecyparis thyoides, Coreopsis, Desmodium strictum, Dichanthelium dichotomum, D. scoparium, Doellingeria umbellata, Drosera intermedia, D. rotundifolia, D. tracyi, Eleocharis acicularis, E. melanocarpa, E. tortilis, E. tuberculosa, Erianthus contortus, Eriocaulon compressum, E. decangulare, E. texense, Eriophorum virginicum, Eryngium integrifolium, Eupatorium perfoliatum, E. pilosum, E. rotundifolium, Fuirena squarrosa, Helianthus angustifolius, Hypericum mutilum, Ilex laevigata, Iris versicolor, I. virginica, Juncus caesariensis, J. dichotomus, J. diffusissimus, J. effusus, J. nodatus, Linum intercursum, Ludwigia alternifolia, Lycopodiella appressa, Magnolia virginiana, Mayaca fluviatilis, Muhlenbergia expansa, Myrica cerifera, Nyssa biflora, N. sylvatica, Osmunda regalis, Osmundastrum cinnamomeum, Panicum anceps, P. brachyanthum, P. virgatum, Paspalum laeve, Pinus palustris, P. taeda, Platanthera blephariglottis, P. cristata, P. integra, Pogonia ophioglossoides, Polygala cruciata, P. lutea, Pteridium aquilinum, Rhexia lutea, R. mariana, R. virginica, Rhynchospora alba,*

R. cephalantha, R. glomerata, R. gracilenta, R. oligantha, R. pallida, R. plumosa, Rudbeckia scabrifolia, Sabatia macrophylla, Saccharum giganteum, Sagittaria latifolia, Sarracenia alata, S. leucophylla, S. psittacina, Scleria baldwinii, S. minor, S. reticularis, S. triglomerata, Solidago latissimifolia, Sparganium americanum, Sphagnum, Spiranthes odorata, Steinchisma hians, Symphyotrichum dumosum, Toxicodendron radicans, T. vernix, Triadenum virginicum, Utricularia cornuta, U. subulata, Viburnum nudum, Xyris difformis, X. fimbriata, X. scabrifolia, X. torta.

Use by wildlife: *Asclepias* species contain many alkaloids and resinoids that deter feeding by insects and mammals. Ingestion of *A. incarnata* by sheep (Mammalia: Bovidae: *Ovis aries*) reportedly can be fatal. The flowers of most species are attractive to butterflies (Insecta: Lepidoptera) and hummingbirds (Aves: Trochilidae). Some ducks (Aves: Anatidae: *Anas acuta, A. americana,* and *A. platyrhynchos*) occasionally eat the foliage. *Asclepias incarnata* is an important habitat element for the alder flycatcher (Aves: Tyrannidae: *Empidonax trailli*), which uses the stems in building its nest. The roots sometimes are eaten by muskrats (Mammalia: Muridae: *Ondatra zibethicus*) and the plants are a significant habitat component for the woodland jumping mouse (Mammalia: Dipodidae: *Napaeozapus insignis*). *Asclepias incarnata* is host to several larval Lepidoptera (Arctiidae: *Cycnia tenera*; Noctuidae: *Trichordestra legitima*; Nymphalidae: *Danaus plexippus*). More than seventy species of insects have been observed to forage on the flowers. Pollinators observed for *A. incarnata* are diverse and include Coleoptera (Cantharidae: *Chauliognathus pennsylvanicus*; Scarabaeidae: *Cotinus nitida, Popillia japonica*; Cerambycidae: *Tetraopes melanurus*), Diptera (Mydidae: *Mydas clavatus*; Syrphidae: *Platychelrus*), Hemiptera (Reduviidae: *Apiomerus crassipes*), Hymenoptera (Andrenidae: *Andrena*; Anthophoridae: *Melissodes bimaculata, Xylocopa virginica*; Apidae: *Apis mellifera, Bombus affinis, B. auricomus, B. bimaculatus, B. griseocollis, B. impatiens, B. nevadensis, B. pennsylvanicus, B. vagans*; Argidae: *Arge*; Pompilidae: *Anoplius, Pepsis elegans*; Scoliidae: *Scolia dubia*; Sphecidae: *Cerceris clypeata, C. fumipennis, Sphecius speciosus, Sphex ichneumoneus, S. nudus, S. pensylvanicus, Tachytes crassus*; Tiphiidae: *Myzinum carolinianum*; Vespidae: *Dolichovespula maculata, Polistes fuscatus, Vespula*), and Lepidoptera (Ctenuchidae: *Cisseps fulvicollis*; Hesperiidae: *Thorybes bathyllus*; Nymphalidae: *Danaus plexippus, Speyeria aphrodite*; Pieridae: *Pieris rapae*; Sphingidae: *Hemeris diffinis*; Yponomeutidae: *Atteva punctella*; Papilionidae: *Papilio glaucus, P. polyxenes*). Herbivores of *A. incarnata* include milkweed bugs (Hemiptera: Lygaeidae: *Lygaeus kalmii, Oncopeltus fasciatus*), yellow milkweed aphids (Hemiptera: Aphididae: *Aphis nerii*), swamp milkweed leaf beetles (Coleoptera: Chrysomelidae: *Labidomera clivicollis*), milkweed longhorn beetles (Coleoptera: Cerambycidae: *Tetraopes femoratus, T. tetrophthalmus*), milkweed stem weevils (Coleoptera: Curculionidae: *Rhyssomatus lineaticollis*), milkweed tiger moths (Lepidoptera: Arctiidae: *Cycnia tenera, Euchaetes egle*), monarch butterflies (Lepidoptera: Nymphalidae: *Danaus plexippus*), and leafminers (Diptera).

The plants also host several fungi (Ascomycota: *Cercospora clavata, C. incarnata*). More than 50 insect species (17 of them pollinators) have been observed as floral visitors of *A. lanceolata*. The plants are a host for larval Lepidoptera (Nymphalidae: *Danaus gilippus, D. plexippus*). Adult queen butterflies (Nymphalidae: *Danaus gilippus*) and swallowtail butterflies (Papilionidae: *Papilio cresphontes*) commonly act as long-range pollinators of this species along with the barred yellow butterfly (Pieridae: *Eurema daira*), Phaon crescent (Nymphalidae: *Phyciodes phaon*), and several Hymenoptera including bumblebees (Apidae: *Bombus pennsylvanicus*), carpenter bees (Apidae: *Xylocopa micans*), and paper wasps (Vespidae: *Polistes*). A longhorned beetle (Coleoptera: Cerambycidae: *Typocerus zebra*) occurs on the plants but is not a pollinator. *Asclepias longifolia* is also a host of monarch butterfly larvae (Nymphalidae: *Danaus plexippus*). The flowers are visited by bumblebees (Apidae: *Bombus*), digger wasps (Hymenoptera: Sphecidae: *Bembix nubillipennis*), and scarab beetles (Coleoptera: Scarabaeidae: *Trichiotinus piger*), which serve as pollinators, and various Lepidoptera, which forage for nectar but are not pollinators. *Asclepias rubra* is another host of the larval monarch butterflies (Nymphalidae: *Danaus plexippus*).

Economic importance: food: *Asclepias* species should never be eaten (especially if raw) due to their poisonous cardiac glycoside content. Despite this danger, the plants occasionally have been used as foods. The unopened flowers of *A. incarnata* have been boiled down to make sweet syrup. When cooked, they are said to have a pea-like flavor. The Menominee tribe added the dried flower heads to cornmeal mush and soup. The young shoots (or older stem tips) and young follicles have been eaten as vegetables; **medicinal:** Shoots of *Asclepias incarnata* contain numerous steroidal pregnane glycosides, which are substances supposedly possessing appetite suppressing activity. The Iroquois administered a decoction of the plant as a diuretic and as a treatment for back problems. They also used a cold root decoction to heal the navels of newly born children. The Meskwaki (and others) relied on a root infusion of the plant as a potent vermifuge and administered the root as a cathartic, diuretic, emetic and to relieve intestinal gas. This species causes dermatitis in some people. The Cherokee prepared several medicinal treatments from *A. perennis* including remedies for kidney and urinary disorders, laxatives, treatment for venereal disease and (root infusion) as an analgesic for back pain. They also rubbed the plants on warts to facilitate their removal. The Cherokee also administered the plant to treat mastitis (milk sickness) in cattle; **cultivation:** *Asclepias incarnata* is an attractive marginal plant for water gardens and is distributed under the cultivar names 'Alba,' 'Ice Ballet,' 'Iceberg,' 'Soulmate,' and 'White Superior.' *Asclepias lanceolata* and *A. rubra* are also cultivated; **misc. products:** The stem fibers of *A. incarnata* provide a substitute for flax or hemp, which was used by several native North American tribes including Ozark Bluff Dwellers (to construct baby cradles), Ohio late Woodland Period people (to make trapping nets), Iroquois (cord used for tooth extraction), and Chippewa (for twine). The seed coma is water repellent and useful as a stuffing (especially for nautical life vests). The follicles produce various oils and wax. *Asclepias incarnata* is also planted for wetland restoration projects. The Cherokee used the stem fibers of *A. perennis* to fabricate bowstrings; **weeds:** none among the OBL species; **nonindigenous species:** *Asclepias incarnata* has been introduced to Europe as an escape from cultivation.

Systematics: Phylogenetic analyses of *Asclepias* using DNA sequence data indicate a clear division between the African taxa (which include some species assigned to *Asclepias*) and New World taxa, which each form distinct clades. Thus, the New World *Asclepias* clade associates most closely with various African genera including *Gomphocarpus* and *Pachycarpus* (e.g., Figure 5.30). *Asclepias* nomenclaturally typifies Apocynaceae subfamily Asclepiadoideae and tribe Asclepiadeae, where it is placed taxonomically. Full plastome data support the position of Asclepiadoideae as the sister group to subfamily Secamonoideae, which together comprise the "milkweeds" clade. A mid-20th-century classification of North American *Asclepias* recognized nine subgenera with subgenus *Asclepias* divided into eight series. In that system *Asclepias longifolia* was placed in subgenus *Acerates* and *A. connivens* as the sole representative of subgenus *Anantheryix*. The other OBL species are assigned to subgenus *Asclepias*, either in series *Incarnatae* (*A. incarnata*, *A. perennis*) or series *Tuberosae* (*A. lanceolata*, *A. rubra*). Series *Incarnatae* was regarded as primitive in the genus, a result that is supported by preliminary phylogenetic studies of the genus using both nrITS and noncoding cpDNA (*rpl*16 intron; *trn*C–*rpo*B intergenic spacer) sequence data. The DNA data place series *Incarnatae* as a basal clade sister to a clade containing the remainder of species (Figure 5.31). Consistently poor resolution of relationships reported in molecular studies probably reflects the recent diversification

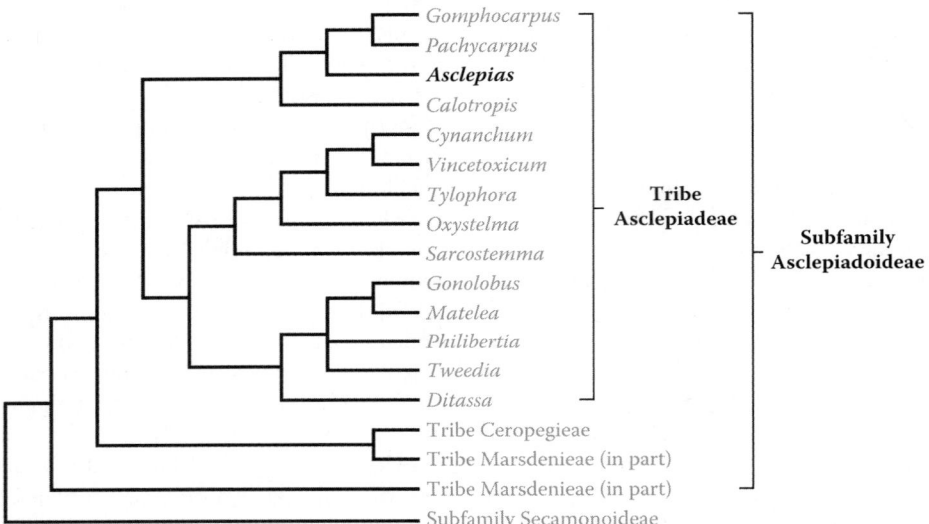

FIGURE 5.30 Intergeneric relationships within Apocynaceae subfamily Asclepiadoideae based on analysis of DNA sequence data. Preliminary molecular analyses that focus on *Asclepias* indicate that all North and South American species occur within one clade (here represented by *Asclepias*) and are sister to an African clade consisting of several genera and a few species currently assigned (apparently in error) to *Asclepias* (Fishbein et al., 2006). *Asclepias* (bold) was the only genus surveyed with OBL representatives in North America. The sister group relationship of subfamilies Asclepiadoideae and Secamonoideae is further supported by various data including analysis of complete plastomes (Livshultz, 2010; Straub et al., 2014). (Adapted from Potgieter, K. & Albert, V.A., *Ann. Missouri Bot. Gard.*, 88, 523–549, 2001.)

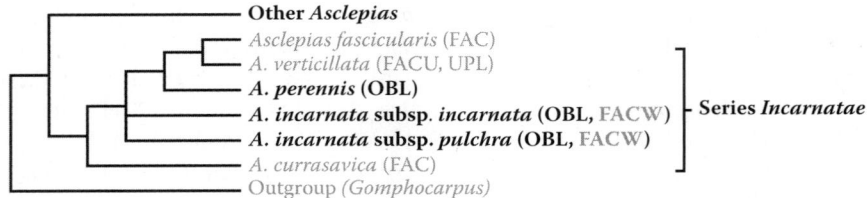

FIGURE 5.31 Relationships among milkweed (*Asclepias*) species as indicated by analysis of noncoding cpDNA sequences. In this analysis, series *Incarnatae* is monophyletic and resolves as a basal lineage, in agreement with prior phylogenetic interpretations. Only two OBL species were included in the analysis; however, it appears that the habit has arisen independently, at least with respect to the OBL species assigned to different series. OBL taxa are in bold. (Adapted from Agrawal, A.A. & Fishbein, M., *Ecology*, 87, S132–S149, 2006.)

of this group. Other preliminary studies of the genus using molecular data from several other loci show poor resolution. The preliminary results indicate a close relationship between *A. incarnata* and *A. perennis* as would be implied by their classification in series *Incarnatae* (Figure 5.31). Further phylogenetic studies of this large group would be informative and should provide an effective means of evaluating the classification currently in use. *Asclepias perennis* and *A. texana* (a species of UPL habitats) generally are regarded as sister species; however their genetic identity (as determined using allozyme data) is fairly reduced (*I* = 0.79). Yet, the two species are completely interfertile, with a level of fruit set (26%) as high as that obtained for intraspecific crosses. Fruit set is highest when *A. perennis* is used as the female parent. Natural interspecific hybrids have not been reported that involve any of the OBL *Asclepias* species. Interspecific hybridization generally is rare in *Asclepias* and among the OBL species has been achieved experimentally only between *A. perennis* and *A. texana* (both in series *Incarnatae*). These species possess virtually identical reproductive morphologies and are isolated ecologically by their habitat differences. Controlled crossing attempts between *A. perennis* and *A. curassavica* (also series *Incarnatae*) have failed, as have hybridization attempts between *A. incarnata* with *A. perennis*, *A. syriaca* (series *Syriacae*), *A. texana,* or *A. verticillata* (series *Incarnatae*). Hybrids can occur between subspecies of *A. incarnata* (subsp. *incarnata* vs. subsp. *pulchra*) but exhibit reduced pollen fertility. The base chromosome number of *Asclepias* is *x* = 11. All North American *Asclepias* species examined so far are diploid with a chromosome number of 2*n* = 22 (verified in *A. incarnata*).

Comments: *Asclepias incarnata* is widespread throughout North America except for extreme western and northwestern portions. *Asclepias connivens, A. lanceolata, A. longifolia, A. perennis,* and *A. rubra* all occur in the southeastern United States; *A. longifolia* extends into Mexico.

References: Agrawal & Fishbein, 2006; Artz & Waddington, 2006; Belden, Jr. et al., 2004; Bridges & Orzell, 1989; Dellinger, 1936; Edwards & Wyatt, 1994; Edwards et al., 1994; Felter & Lloyd, 1898; Fernald & Kinsey, 1943; Fishbein et al., 2006; Green, 1925; Hamerstrom, Jr. & Blake, 1939a; Harper, 1900; Ivey & Wyatt, 1999; Ivey et al., 1999, 2003; Kephart, 1981, 1983; Lieneman, 1929; Light et al., 1993; Lipow & Wyatt, 1999, 2000; Lynch et al., 2001; Mayfield, 1972; Mohlenbrock, 1959a; Pavek, 1992; Potgieter & Albert, 2001; Price & Wilson, 1979; Robertson, 1887; Straub et al., 2014; USDA, NRCS, 2004; Walkinshaw, 1966; Warashina & Noro, 2000; Wilbur, 1976; Woodson, Jr., 1954; Wyatt & Broyles, 1992; Wyatt et al., 1998, 2000.

3. *Pentalinon*

Wild alalmanda, hammock viperstail

Etymology: from the Greek *pente linon* meaning "five cord" in reference to the 5-merous flowers and twining habit of the plants

Synonyms: *Angadenia* (in part); *Apocynum* (in part); *Chariomma*; *Dipladenia* (in part); *Echites* (in part); *Haemadictyon* (in part); *Laseguea* (in part); *Neriandra*; *Rhabdadenia* (in part); *Urechites*; *Vinca* (in part)

Distribution: global: New World; **North America:** southeastern

Diversity: global: 2 species; **North America:** 1 species

Indicators (USA): OBL: *Pentalinon luteum*

Habitat: brackish (coastal), freshwater; palustrine; **pH:** broad; **depth:** <1 m; **life-form(s):** emergent vine (woody)

Key morphology: stems (to 15 m) twining, woody, with milky sap; leaves (to 9 cm) opposite, entire, evergreen, petioled (to 1.2 cm); cymes lateral, axillary, few flowered; flowers (to 5 cm) 5-merous, stalked (to 1.5 cm); corolla salverform, yellow to cream; follicles (to 20 cm) incurved, woody

Life history: duration: perennial (buds); **asexual reproduction:** none; **pollination:** insect; **sexual condition:** hermaphroditic; **fruit:** follicles (common); **local dispersal:** seeds (wind); **long-distance dispersal:** seeds (wind)

Imperilment: (1) *Pentalinon luteum* [G5]

Ecology: general: Only two species occur in this genus (one in North America) and both inhabit coastal areas. Information on the ecology of either species is extremely limited and additional study is needed. Both species are vines that rely on other associates (often mangroves) for structural support.

Pentalinon luteum **(L.) B.F. Hansen & Wunderlin** inhabits coastal flats, hammocks, and swamps at elevations of up to 915 m. Habitats receive full sun or partial shade. The substrates typically are alkaline (e.g., limestone) and include clay, loam, and sand; however plants also tolerate acidic soils (specific pH values not available). The plants are moderately salt tolerant and often are found on the margins of mangrove swamps or in saline flats. They are very drought tolerant. Little information exists on the reproductive or seed ecology of this species. There are few details reported on specific pollinators or seed germination requirements. The flowers are said to open throughout the year and are visited by hummingbirds (Aves: Trochilidae) and butterflies (Lepidoptera); however, the latter probably do not participate in pollination. The seeds possess a coma and presumably are wind dispersed. The plants can be propagated successfully from cuttings. **Reported associates:** *Agave decipiens, Alternanthera flavescens, Andropogon glomeratus, Avicennia germinans, Baccharis halimifolia, Bidens pilosa, Blutaparon vermiculare, Casuarina, Chamaesyce blodgettii, Cirsium horridulum, Conocarpus erectus, Croton glandulosus, Cynanchum bahamense, Dalbergia ecastaphyllum, Distichlis spicata, Eugenia foetida, Euphorbia heterophylla, Fimbristylis castanea, Flaveria floridana, F. linearis, Galactia volubilis, Heliotropium curassavicum, Indigofera trita, Ipomoea, Jacquemontia pentantha, Laguncularia racemosa, Muhlenbergia filipes, Neptunia, Pithecellobium unguis-cati, Randia aculeata, Rhizophora mangle, Rhynchosia minima, Schinus terebinthifolius, Schizachyrium sanguineum, Spartina, Sporobolus domingensis, S. virginicus, Symphyotrichum tenuifolium, Trema, Trichostema.*

Use by wildlife: *Pentalinon luteum* is poisonous to cattle (Mammalia: Bovidae: *Bos taurus*) but is eaten by feral goats (Mammalia: Bovidae: *Capra aegagrus hircus*). The flowers

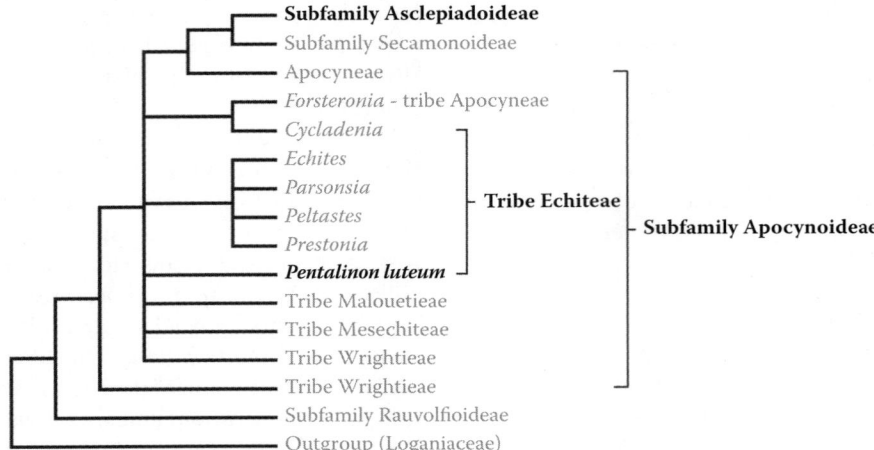

FIGURE 5.32 Phylogenetic position of *Pentalinon luteum* in Apocynaceae as indicated by analysis of DNA sequence data. *Pentalinon luteum* occurs in an isolated position with respect to other surveyed members of tribe Echiteae, which does not appear to be monophyletic as currently circumscribed. Taxa with OBL representatives in North America are indicated in bold. (Adapted from Potgieter, K. & Albert, V.A., *Ann. Missouri Bot. Gard.*, 88, 523–549, 2001.)

attract butterflies (Insecta: Lepidoptera) and streamertail hummingbirds (Aves: Trochilidae: *Trochilus polytmus*). The leaves are eaten by larvae of the oleander moth (Insecta: Lepidoptera: Arctiidae: *Syntomeida epilais*).
Economic importance: food: *Pentalinon luteum* should never be eaten because the plants reportedly contain poisonous cardiac glucosides (urechitin, urechitoxin), and a toxic alkaloid (urechitine); **medicinal:** The steroidal glycosides of *Pentalinon luteum* possess cardiac properties similar to those of digitalis. Extracts of the plant have been administered in some parts of the West Indies as a treatment for fevers. The milky sap can cause severe dermatitis in some individuals; **cultivation:** *Pentalinon luteum* is grown as an ornamental groundcover or espalier. It is sold under the cultivar names 'Variegata' and 'Yellow Allamanda'; **misc. products:** none; **weeds:** none; **nonindigenous species:** none
Systematics: *Pentalinon luteum* was long known in the genus *Urechites*, but was transferred to the current genus as a matter of nomenclatural priority. *Pentalinon* is assigned to subfamily Apocynoideae, tribe Echiteae. From the subset of taxa included in phylogenetic analyses, tribe Echiteaea is poorly resolved, does not appear to be monophyletic and *Pentalinon* is not closely related to other surveyed genera in the tribe (Figure 5.32). Additional sampling of taxa is needed to clarify relationships in this tribe. The base chromosome number of the genus probably is $x = 6$, with *P. luteum* ($2n = 12$) a diploid.
Comments: In North America, *Pentalinon luteum* occurs only in Florida but extends into the West Indies.
References: Burrows, 1996; Fritsch, 1971; Gilman, 1999; Godfrey & Wooten, 1981; Hansen & Wunderlin, 1986; Meléndez-Ackerman et al., 2008; Potgieter & Albert, 2001; Ross & Ruiz, 1998; Woodson, Jr., 1936.

4. *Seutera*
Seaside milkvine, swallow-wort
Etymology: after Bartholomäus Seuter (1678–1754)

Synonyms: *Amphistelma*; *Ceropegia*; *Cynanchum* (in part); *Funastrum* (in part); *Lyonia* (in part); *Metastelma* (in part); *Pattalias*; *Vincetoxicum* (in part)
Distribution: global: North America; Mexico (Baja); **North America:** southeastern
Diversity: global: 2 species; **North America:** 1 species
Indicators (USA): OBL; FACW: *Seutera angustifolia*
Habitat: brackish, saline; palustrine; **pH:** 3.4 to >7.0; **depth:** <1 m; **life-form(s):** emergent vine (herb)
Key morphology: stems (to 1.5 m) slender, twining, herbaceous; leaves (to 9 cm) opposite, linear, entire, sessile; the herbage with milky latex; flowers (to 6 mm) 5-merous, pediceled (to 5 mm), arising from axillary, umbellate, stalked (to 8 cm), clusters; corolla (to 7 mm) campanulate, greenish-white, hoods surpassing the gynostegium; follicles (to 7 cm) lanceolate
Life history: duration: perennial (rhizomes); **asexual reproduction:** rhizomes; **pollination:** insect; **sexual condition:** hermaphroditic; **fruit:** follicles (common); **local dispersal:** rhizomes; **long-distance dispersal:** seeds
Imperilment: (1) *Seutera angustifolia* [G5]; S3 (GA, NC)
Ecology: general: Both *Seutera* species are slender, herbaceous, lactiferous vines that occur in salt marshes and along beaches. The North American species usually occurs in wetlands but occasionally has been reported growing among terrestrial plants in rocky sites. The herbage emits a pungent odor when crushed.

Seutera angustifolia **(Pers.) Fishbein & W.D. Stevens** inhabits low elevation coastal, brackish to saline beach depressions, canal margins, dunes, flatwoods, hammocks, interdunal swales, salt marshes, shores, tidal marshes, and swamps, often those situated on barrier islands. The substrates range from acidic (e.g., pH: 3.4–5.8) to alkaline (specific pH values not reported) loamy sand, marl, sand, or shell middens. The plants occur in full sun, often just above the tidal limit. They are vines and must be supported by other vegetation (frequently found growing on *Juncus roemerianus*). Nothing is known regarding the reproductive or seed ecology of this species. The plants spread vegetatively by rhizomes. Other ecological information for this species is scarce and should be gathered, given that it is relatively uncommon. **Reported associates:** *Acmella oppositifolia, Aesculus pavia, Agalinis maritima, Ammophila brevigulata, Andropogon, Aristida, Baccharis angustifolia, B. glomeruliflora, B. halimifolia, Batis maritima, Bolboschoenus robustus, Borrichia frutescens, Bursera simaruba, Cakile edentula, Callicarpa americana, Capraria biflora, Celtis laevigata, Cenchrus echinatus, C. tribuloides, Centella asiatica, Chamaesyce polygonifolia, Chenopodium ambrosioides, Cladium jamaicense, Commelina erecta, Conyza canadensis, Croton punctatus, Cyperus, Dichanthelium aciculare, Diodia teres, Distichlis spicata, Ditrichum pallidum, Eleocharis rostellata, Eragrostis, Erechtites hieraciifolia, Euphorbia, Fimbristylis caroliniana, F. castanea, Forestiera segregata, Gamochaeta purpurea, Heterotheca subaxillaris, Hydrocotyle bonariensis, Ilex vomitoria, Ipomoea sagittata, Iva frutescens, I. imbricata, Juncus roemerianus, J. scirpoides, Juniperus virginiana, Kostletzkya virginica, Limonium carolinianum, Lycium carolinianum, Lythrum lineare, Mikania scandens, Muhlenbergia capillaris, Myrica cerifera, Oenothera humifusa, Opuntia humifusa, O. pusilla, Oxalis, Panicum amarum, P. repens, P. virgatum, Phragmites australis, Phyla nodiflora, Physalis angulata, Pinus elliottii, P. taeda, Pluchea odorata, Polygala, Prunus caroliniana, Quercus virginiana, Rhus, Rhynchospora colorata, Sabal palmetto, Sabatia stellaris, Sagittaria platyphylla, Salsola kali, Samolus valerandi, Sarcocornia perennis, Schizachyrium maritimum, Schoenoplectus pungens, Serenoa repens, Setaria parviflora, Solidago sempervirens, S. stricta, Sophora tomentosa, Spartina alterniflora, S. bakeri, S. cynosuroides, S. patens, S. spartinae, Sporobolus virginicus, Stenotaphrum secundatum, Strophostyles helvula, Suaeda linearis, Symphyotrichum subulatum, S. tenuifolium, Tilia heterophylla, Toxicodendron radicans, Triglochin striata, Triplasis purpurea, Uniola paniculata, Vigna, Yucca aloifolia, Y. gloriosa, Zanthoxylum clava-herculis.*

Use by wildlife: *Seutera angustifolia* is a nectar plant used by butterflies (Lepidoptera) and many other insects and by migrating ruby-throated hummingbirds (Aves: Trochilidae: *Archilochus colubris*). It is an important host plant for several larval Lepidoptera (Arctiidae: *Eucereon carolina, E. confine*; Nymphalidae: *Anetia briarea, Danaus gilippus, D. plexippus*; Pyralidae: *Glyphodes floridalis*; Sphingidae: *Erinnyis obscura*). Numerous ants (Hymenoptera: Formicidae) and aphids (Hemiptera: Aphididae) have been observed on the foliage.

Economic importance: food: not edible and probably toxic; **medicinal:** no uses reported; **cultivation:** not cultivated; **misc. products:** none; **weeds:** none; **nonindigenous species:** none

Systematics: *Seutera angustifolia* recently has been segregated from the large genus *Cynanchum* (=*C. angustifolium, C. palustre*), which has been described as a taxonomic "dustbin" containing remnants of numerous taxa that are difficult to classify. Molecular data indicate that *Cynanchum* essentially is an Old World taxon and that New World species assigned to that genus are not closely related and warrant reclassification. As currently circumscribed, *Seutera* (subfamily Asclepiadoideae; tribe Aslepiadeae; subtribe Oxypetalinae) associates with *Funastrum* and is included in that genus by some authors. Preliminary molecular data, although not definitive, indicate that *Seutera* might be nested within *Funastrum* (Figure 5.33); however, *S. palmeri* has not yet been included in a phylogenetic analysis. *Seutera* differs from *Funastrum* so extensively in its vegetative and floral characters that recognition as a distinct genus seems to be a reasonable disposition pending the outcome of additional investigations. The

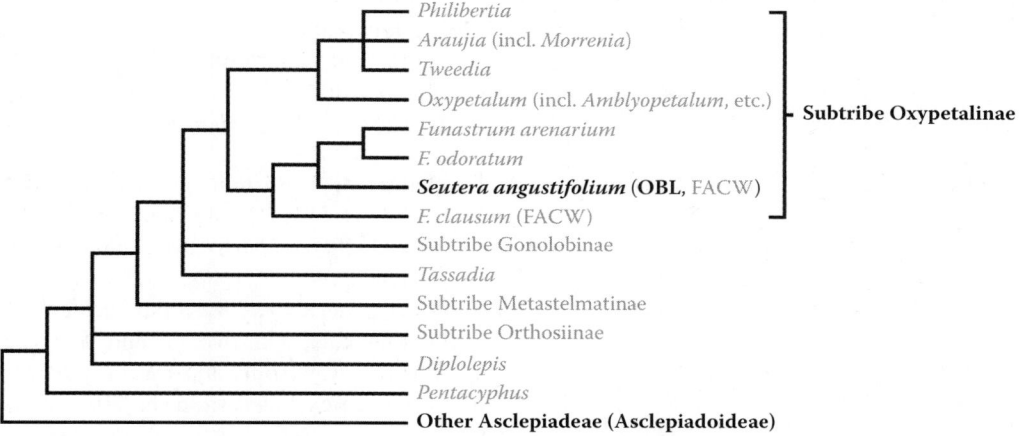

FIGURE 5.33 Relationships within tribe Asclepiadeae (subfamily Asclepiadoideae) of Apocynaceae as indicated by phylogenetic analysis of noncoding cpDNA data. Taxa with OBL North American representatives are shown in bold. The taxonomic limits of *Funastrum* and *Seutera* will require further refinement by the inclusion of additional species in such analyses. (Adapted from Liede-Schumann, S. et al., *Syst. Bot.*, 30, 184–195, 2005.)

chromosome number has not been determined for any species of *Seutera*. Further taxonomic study of this small genus is highly recommended.

Comments: *Seutera angustifolia* occurs along the coastal plain of the southeastern United States.

References: Fishbein & Stevens, 2005; Hosier & Eaton, 1980; Krings, 2005; Liede & Täuber, 2002; Liede-Schumann et al., 2005; Stallins, 2005; Whitaker et al., 2004.

Family 2: Gentianaceae [91]

The gentian family (Gentianaceae) represents a diverse group of nearly 1700 species of herbs, shrubs, trees, and mycoparasites, with many noted for their showy, ornamental flowers. The family mainly is temperate in its distribution, but extends into subtropical and tropical (montane) regions. The plants characteristically have opposite, simple leaves, four or five-merous nectariferous flowers, and deeply intruded, two-lobed placentas (Judd et al., 2002). The petals occasionally are ornamented with conspicuously fringed margins. The flowers tend to be protogynous and are outcrossed by bees (Hymenoptera) and butterflies (Lepidoptera). The seeds are quite small and are dispersed abiotically (water, wind) or biotically by birds (Aves), bats (Mammalia: Chiroptera), and other mammals. Unlike Apocynaceae, the herbage of gentians lacks latex.

Phylogenetic studies using molecular and morphological data (Downie & Palmer, 1992; Olmstead et al., 1992; Chase et al., 1993; Olmstead et al., 1993; Lundberg & Bremer, 2003) indicate that the family essentially is monophyletic once several aquatic genera (e.g., *Menyanthes*, *Nymphoides*, *Villarsia*) are removed (these have been transferred to Menyanthaceae). *Zeltnera exaltata* (FACW) was initially classified as OBL, FACW in the 1996 indicator list. Its treatment has been retained here nevertheless, because it was completed prior to the release of the 2013 list. Currently there are six tribes delimited within Gentianaceae as a result of the ongoing synthesis of existing phylogenetic information (Struwe & Albert, 2002). Two moderately specialized tribes: Chironieae (*Eustoma*, *Sabatia*) and Gentianeae (*Bartonia*, *Gentiana*, *Gentianella*, *Gentianopsis*, *Lomatogonium*) contain genera with OBL aquatic species in North America (Figure 5.34):

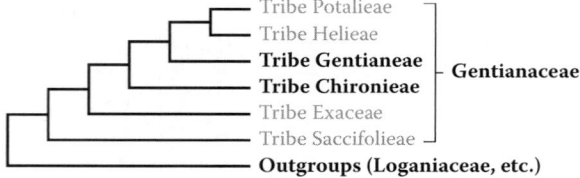

FIGURE 5.34 Phylogenetic relationships among tribes of Gentianaceae based on DNA sequence (*matK*, *trnL* intron) data. Independent origins of the aquatic habit are indicated by the occurrence of OBL North American species within two, predominantly terrestrial tribes (taxa containing OBL species are indicated in bold). (Adapted from Struwe, L. & Albert, V.A., eds., *Gentianaceae: Systematics and Natural History,* Cambridge University Press, Cambridge, UK, 2002.)

1. ***Bartonia*** Muhl. ex Willd.
2. ***Eustoma*** Salisb. ex G. Don
3. ***Gentiana*** L.
4. ***Gentianella*** Moench
5. ***Gentianopsis*** Ma
6. ***Lomatogonium*** A. Braun
7. ***Sabatia*** Adans.
8. ***Zeltnera*** G. Mans. [previously OBL]

Molecular phylogenetic studies of tribe Gentianeae (Chassot et al., 2001; Kadereit & von Hagen, 2003) do not support the monophyly of *Gentianella*. However, additional surveys will be necessary to clarify the correct taxonomic circumscription of *Gentianella*, which resolves as nested within some species of *Swertia* (also polyphyletic). However, *Gentianella* is retained here as distinct pending any contrary taxonomic recommendations that might arise subsequently from such studies.

1. *Bartonia*
Screwstem; Bartonie
Etymology: after Benjamin Smith Barton (1766–1815)
Synonyms: *Centaurella*
Distribution: global: North America; **North America:** eastern and southern
Diversity: global: 4 species; **North America:** 4 species
Indicators (USA): OBL: *Bartonia paniculata*, *B. texana*, *B. verna*
Habitat: freshwater; palustrine; **pH:** 4.5–5.9; **depth:** <1 m; **life-form(s):** emergent herb
Key morphology: stems (to 40 cm) wiry, erect, scrambling or twining, often purplish; leaves reduced to alternate (or opposite) scales (to 3 mm); inflorescence a terminal panicle or raceme (to 25 cm); flowers (to 5 mm) 4-merous, pedicelled (to 25 mm); calyx cleft nearly to base; corolla (2.5–10 mm) campanulate, white, cream-white, greenish-white, pink, purple, or yellow; stamens arising from sinuses between petal lobes; capsules (to 7 mm) 2-valved, with many (to 1500) small (<0.3 mm) seeds
Life history: duration: annual, biennial (fruit/seeds); **asexual reproduction:** none; **pollination:** insect; **sexual condition:** hermaphroditic; **fruit:** capsules (common); **local dispersal:** seeds (water, wind); **long-distance dispersal:** seeds (water, wind)
Imperilment: (1) *Bartonia paniculata* [G5]; S1 (DC, IL, ME, MO, NY, OH, OK, <u>ON</u>, WI); S2 (DE, MI, <u>NB</u>, NC); S3 (GA, IN, MD, PA, VA); (2) *B. texana* [G2]; S2 (TX); (3) *B. verna* [G5]; S1 (TX, VA); S2 (NC); S3 (GA, LA)
Ecology: general: All *Bartonia* species are associated with wetland habitats, with two currently designated as OBL and a third as FACW. *Bartonia texana* (OBL in 1996 indicator list) has been excluded from the 2013 list because of recent taxonomic studies that recommended its transfer to *B. paniculata* as a subspecies. Given that this portion of the treatment was completed prior to the publication of the revised taxonomic work, *B. texana* has been retained here "unofficially" as an OBL species, keeping in mind that the information

provided can simply be merged with that of *B. paniculata* (also OBL) if one prefers not to retain its species rank. These are inconspicuous plants that are difficult to find unless they are in flower; consequently, there has been limited ecological study of the group. The plants reportedly are saprophytic and obtain much of their nutrients by endomycorrhizal fungal associations; however, their nutritional ecology needs to be confirmed experimentally. Their sexual reproduction is orchid-like in producing numerous tiny seeds that are presumed to be dispersed abiotically. Virtually no other information exists regarding pollination, floral or seed ecology for any of the species.

***Bartonia paniculata* (Michx.) Muhl.** inhabits bogs, Carolina bays, depressions, ditches, flats, floodplains, marshes, meadows, pocosins, savannas, seeps, sloughs, springs, streamheads, stringbogs, swamps, and the margins of ponds, rivers, and streams at elevations to at least 44 m. The plants can tolerate shallow water and some shade but thrive in more open areas, often occurring along the dessicating margins of intermittent pools, and sometimes in former strip mined areas. The substrates are acidic (pH: 4.5–5.9) and nutrient-poor gravel, kaolin, peat, and sand. **Reported associates:** *Acer rubrum, Agalinis neoscotica, A. purpurea, Agrostis perennans, Aletris farinosa, Alnus, Andropogon glomeratus, Asclepias incarnata, Athyrium, Bartonia virginica, Bidens frondosus, Burmannia biflora, Calamagrostis pickeringii, Calopogon tuberosus, Campylopodiella stenocarpa, Carex atlantica, C. barrattii, C. collinsii, C. crinita, C. exilis, C. folliculata, C. incomperta, C. livida, C. lonchocarpa, C. lurida, C. oligosperma, Chamaecyparis thyoides, Chamaedaphne calyculata, Chelone glabra, Cladium mariscoides, Clethra alnifolia, Conium maculatum, Cyperus flavescens, C. polystachyos, C. strigosus, Danthonia epilis, Dichanthelium dichotomum, D. scoparium, D. sphaerocarpon, Drosera intermedia, D. rotundifolia, Eleocharis acicularis, E. olivacea, E. tenuis, Eriocaulon aquaticum, E. decangulare, Eriophorum virginicum, Eupatorium pilosum, Euthamia remota, Fagus, Fimbristylis autumnalis, Fuirena squarrosa, Gymnadeniopsis clavellata, Helenium flexuosum, Helonias bullata, Hypericum, Ilex glabra, I. opaca, Impatiens capensis, Itea virginica, Juncus acuminatus, J. canadensis, J. diffusissimus, J. effusus, J. interior, J. longii, J. pelocarpus, J. scirpoides, Lachnanthes caroliana, Larix laricina, Leersia virginica, Liatris pilosa, Lilium superbum, Linum medium, Lobelia canbyi, L. nuttallii, Lophiola aurea, Lycopodiella alopecuroides, L. inundata, L. subappressa, Lythrum lineare, Magnolia virginiana, Menyanthes trifoliata, Muhlenbergia uniflora, Nymphaea odorata, Nyssa biflora, N. sylvatica, Oclemena nemoralis, Oldenlandia uniflora, Onoclea sensibilis, Osmunda regalis, Osmundastrum cinnamomeum, Oxalis stricta, Oxypolis rigidior, Panicum flexile, P. verrucosum, Parnassia asarifolia, Persicaria arifolia, P. maculosa, P. punctata, P. sagittifolia, Photinia floribunda, P. melanocarpa, P. pyrifolia, Platanthera blephariglottis, P. ciliaris, Pogonia ophioglossoides, Polygala cruciata, P. mariana, P. sanguinea, Polytrichum, Potentilla simplex, Pycnanthemum tenuifolium, Rhexia petiolata, R. virginica, Rhododendron* *viscosum, Rhynchospora alba, R. capitellata, R. cephalantha, R. chalarocephala, R. gracilenta, R. oligantha, Rubus flagellaris, Sabatia angularis, S. difformis, Sagittaria latifolia, Sarracenia purpurea, Scheuchzeria palustris, Schizachyrium scoparium, Schoenoplectus purshianus, S. subterminalis, Scirpus polyphyllus, Scleria muhlenbergii, S. pauciflora, S. reticularis, Selaginella apoda, Senecio pauperculus, Sericocarpus linifolius, Solidago rugosa, S. uliginosa, S. speciosa, Sorghastrum nutans, Sparganium americanum, Sphagnum intricatum, Sphenopholis pensylvanica, Spiraea tomentosa, Spiranthes cernua, S. lacera, Symphyotrichum novi-belgii, S. patens, Symplocarpus foetidus, Symplocos tinctoria, Taxodium distichum, Thelypteris palustris, Thuidium recognitum, Triadenum virginicum, Triantha racemosa, Trisetum pensylvanicum, Utricularia cornuta, U. juncea, U. subulata, Vaccinium macrocarpon, Viburnum nudum, Viola lanceolata, Woodwardia areolata, W. virginica, Xyris montana, X. torta.*

***Bartonia texana* Correll** occurs in baygalls, bogs, savannas, seepage bogs, seeps, springs, and along stream margins. The substrates are described as acidic, usually organic, but include loam, peat, and sand. Some plants have also been observed growing in clumps of mosses at the bases of trees, and on roots or logs. The plants typically occur in open areas of forested sites, which often represent places that are burned frequently. The plants are in flower from September to November. Sparse information exists regarding the floral or seed ecology of this rare species, or other aspects of its ecology. **Reported associates:** *Acer rubrum, Alnus serrulata, Amsonia glaberrima, Apteria aphylla, Bartonia verna, Burmannia biflora, Calopogon barbatus, C. pulchellus, Carex glaucescens, Cyrilla racemiflora, Gentiana saponaria, Gymnadeniopsis nivea, Ilex coriacea, Itea virginica, Magnolia virginiana, Nyssa aquatica, N. biflora, N. sylvatica, Onoclea sensibilis, Osmunda regalis, Osmundastrum cinnamomeum, Pinus, Platanthera ciliaris, Pogonia ophioglossoides, Quercus, Sabatia campanulata, Sarracenia alata, Sphagnum, Woodwardia areolata.*

***Bartonia verna* (Michx.) Raf. ex Barton** occurs in bogs, ditches, flatwoods, prairies, savannas, seeps, swales, and along the margins of ponds at low elevations. The sites are acidic (specific pH values not reported) with substrates consisting primarily of sand, but sometimes mixed with silt. Plants occasionally are reported from recently burned or otherwise disturbed areas. Other ecological information is scarce. **Reported associates:** *Agalinis, Andropogon, Aristida beyrichiana, Balduina uniflora, Bartonia texana, Bidens mitis, Bigelowia nudata, Calopogon barbatus, C. pallidus, Carex, Carphephorus pseudoliatris, Centella erecta, Chaptalia tomentosa, Coleataenia longifolia, Coreopsis gladiata, Ctenium aromaticum, Dichanthelium erectifolium, D. wrightianum, Drosera brevifolia, D. capillaris, D. rotundifolia, Erigeron vernus, Eriocaulon compressum, Euphorbia inundata, Eurybia eryngiifolia, Gymnadeniopsis integra, Helenium vernale, Helianthus heterophyllus, Hypericum brachyphyllum, H. edisonianum, H. fasciculatum, H. tetrapetalum, Hypoxis juncea, Ilex glabra, I. myrtifolia, Juncus,*

Justicia crassifolia, Lachnanthes caroliana, Lachnocaulon minus, Lilium catesbaei, Liquidambar styraciflua, Listera australis, Lycopodiella alopecuroides, L. caroliniana, Magnolia virginiana, Ophioglossum crotalophoroides, Panicum rigidulum, P. verrucosum, Parnassia carolinianum, Photinia pyrifolia, Piloblephis rigida, Pinguicula lutea, P. planifolia, P. pumila, Pinus elliottii, P. palustris, Pleea tenuifolia, Polygala cymosa, P. lutea, P. polygama, P. ramosa, Pteridium aquilinum, Quercus geminata, Rhexia alifanus, R. cubensis, R. lutea, Rhynchospera chapmanii, R. filifolia, R. oligantha, Sabatia bartramii, Sarracenia flava, S. leucophylla, S. minor, S. psittacina, S. rubra, Scleria, Serenoa repens, Smilax laurifolia, Sphagnum, Spiranthes brevilabris, Symphyotrichum walteri, Syngonanthus flavidulus, Taxodium ascendens, Utricularia juncea, U. subulata, Vaccinium corymbosum, V. myrsinites, Viburnum nudum, Xyris ambigua, X. elliotti.

Use by wildlife: none reported

Economic importance: food: none reported; **medicinal:** no known medicinal uses; **cultivation:** not cultivated; **misc.**

products: The journal of the Philadelphia Botanical Club is named *Bartonia*; however, it commemorates William P. C. Barton, who is not the namesake of the genus; **weeds:** none; **nonindigenous species:** none

Systematics: This genus is not to be confused with *Bartonia* Pursh ex Sims, which is a synonym of *Mentzelia* (Loasaceae). Molecular data have better clarified the phylogenetic placement of *Bartonia* in Gentianaceae and confirm its inclusion within tribe Gentianeae, (Figures 5.35 and 5.36). Although, phylogenetic analysis of morphological data resolves *Bartonia* as the sister group to *Obolaria*, molecular data place it much closer to *Swertia* (Figure 5.36). All four *Bartonia* species have been included in molecular phylogenetic analyses, which have confirmed the monophyly of the genus and have better elucidated the interspecific relationships. The analysis of combined nrITS and cpDNA sequence data place *B. texana* among the subspecies of *B. paniculata*, indicating that it might better be treated taxonomically as a subspecies of the latter (Figure 5.36). The base chromosome number of tribe Gentianeae is *x* = 13. *Bartonia verna*

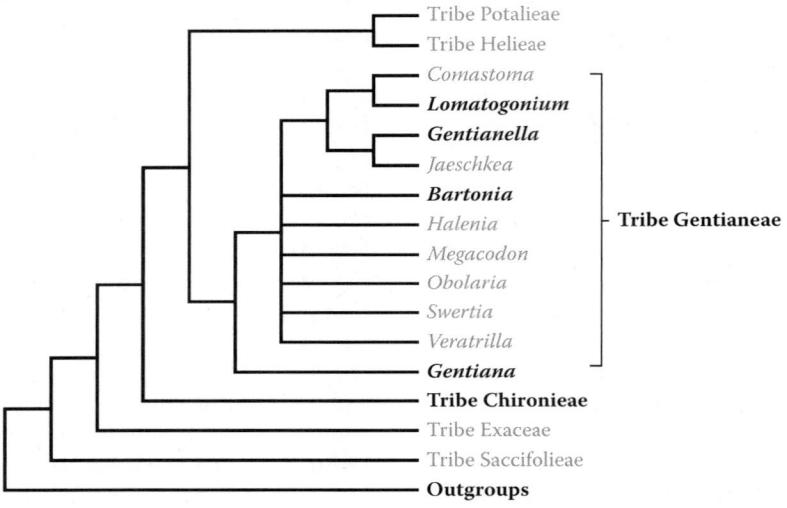

FIGURE 5.35 Phylogenetic relationships in Gentianaceae tribe Gentianeae as indicated by combined cpDNA sequence data. Although resolution is poor within the tribe, these results indicate that the OBL habit of taxa in North America (shown in bold) has arisen independently at least several times in the tribe. Additional sampling of taxa and loci is needed as other surveys have shown that several genera (e.g., *Gentianella*, *Swertia*) are not monophyletic. (Adapted from Struwe, L. et al., *Gentianaceae: Systematics and Natural History*. Cambridge University Press, Cambridge, UK, 2002.)

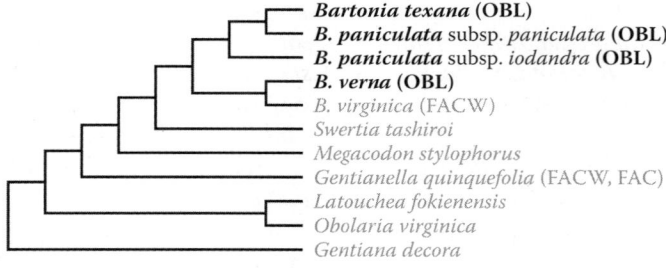

FIGURE 5.36 Interspecific relationships in *Bartonia* as indicated by analysis of combined nrITS and cpDNA sequence data. These results indicate that *Obolaria* is not closely related to *Bartonia* as previously thought and also suggest that *B. texana* might be better recognized taxonomically as a subspecies of *B. paniculata*. OBL North American taxa are indicated in bold. (Adapted from Mathews, K.G. et al., *Syst. Bot.* 34,162–172, 2009.)

($2n = 22$) is a diploid; whereas, *B. paniculata* ($2n = 52$) is a tetraploid. Chromosome numbers have not been reported for other *Bartonia* species.

Comments: *Bartonia paniculata* is widespread in eastern North America and *B. verna* occurs along the coastal plain of the southeastern United States. *Bartonia texana* is endemic to a small region of eastern Texas and western Louisiana.

References: Chassot et al., 2001; Garber, 1877; Higman & Penskar, 1996; Hill, 2003; Kadereit & von Hagen, 2003; Mathews et al., 2009; Nixon & Ward, 1981; Poole et al., 2007; Struwe et al., 2002.

2. Eustoma

Prairie gentian, catchfly gentian

Etymology: after the Greek *eu stoma* meaning "beautiful mouth" with reference to the colorful corolla

Synonyms: *Arenbergia*; *Bilamista*; *Dupratzia*; *Urananthus*

Distribution: global: Mexico; North America; **North America:** southern

Diversity: global: 2 species; **North America:** 2 species

Indicators (USA): OBL; FACW: *Eustoma exaltatum*

Habitat: brackish (coastal), freshwater; palustrine; **pH:** 6.6–8.5; **depth:** <1 m; **life-form(s):** emergent herb

Key morphology: stems (to 1 m) erect, branched, herbage glaucous; leaves (to 9 cm) broad (to 3 cm), opposite, sessile, clasping; flowers showy (to 4.5 cm), solitary or in open cymes, long-pedicelled (to 14 cm); corolla funnel-shaped, blue, lavender, or white, with a tube (to 1 cm) and 5 spreading lobes (to 25 mm); stamens 5–6, fused to throat of tube; style (to 5 mm) with 2-lipped stigma; capsule (to 2 cm) 2-valved, many seeded

Life history: duration: annual (fruit/seeds); **asexual reproduction:** none; **pollination:** insect; **sexual condition:** hermaphroditic; **fruit:** capsules (common); **local dispersal:** seeds; **long-distance dispersal:** seeds

Imperilment: (1) *Eustoma exaltatum* [G4]; S1 (MS, NV); S3 (NM)

Ecology: general: Neither *Eustoma* species is particularly characteristic of wet habitats, and even the sole OBL species (*E. exaltatum*), is ranked as FACW in some regions. The other species (*E. russellianum*) occurs commonly in dry grasslands. There is only minimal ecological information available for this genus. The flowers are pollinated by bees (Insecta: Hymenoptera). Seed germination has been reported to occur within 3 weeks when they are sowed on the surface in the light at 20°C–25°C.

Eustoma exaltatum **(L.) Salisb. ex G. Don** inhabits fresh or brackish beach depressions, cienega, ditches, floodplains, interdunal ponds, marshes, meadows, prairies, sand flats, seeps, shores, slopes, springs, stream beds, swamps, washes, and margins of lakes, ponds, rivers, and streams at elevations of up to 1767 m. The substrates are calcareous (pH: 6.6–8.5), often saline, and consist of clay, cobble, gravel, gypsum, limy gypsum, rock, sand, sandy clay loam, sandy gypsum, sandy loam, sandy rock, silt, or silty clay. Typical sites are open with full sunlight and often represent disturbed areas (e.g., levees, roadsides, spoil), where the plants can develop into monocultures. This species occurs to the upper reaches or edges of salt marshes. The plants will tolerate fire and minor drought but do not compete well with grasses and other species, especially C_4 taxa. Growth is optimal under moist soil conditions, but prolonged inundation may be detrimental. The seeds reportedly can be difficult to germinate. Although annual, the plants reportedly also can be propagated from cuttings. Detailed studies on the reproductive ecology and population genetics of this species are needed. **Reported associates:** *Acacia, Allenrolfea occidentalis, Amaranthus palmeri, Amblyolepis setigera, Andropogon glomeratus, Anemopsis californica, Aristida, Atriplex, Avicennia germinans, Baccharis neglecta, B. salicifolia, B. salicina, B. sarothroides, Bacopa, Batis maritima, Bebbia juncea, Bolboschoenus maritimus, Borrichia arborescens, B. frutescens, Carex praegracilis, Cenchrus ciliaris, Chloracantha spinosa, Chloris petraea, Chrysothamnus, Conocarpus erectus, Conyza, Crotalaria, Croton punctatus, Cynodon dactylon, Cyperus involucratus, C. onerosus, C. pseudovegetatus, Distichlis spicata, Elaeagnus angustifolia, Eleocharis rostellata, Encelia farinosa, Equisetum hyemale, Erigeron procumbens, Eryngium sparganophyllum, Fimbristylis castanea, Flaveria floridana, Fraxinus velutina, Helianthus paradoxus, Heliotropium currassavicum, Hibiscus moscheutos, Ipomoea imperati, I. sagittata, Juncus, Kosteletzkya virginica, Laguncularia racemosa, Lepidospartum, Limonium carolinianum, L. limbatum, L. nashii, Lythrum, Melilotus albus, Metopium toxiferum, Monanthochloe littoralis, Muhlenbergia asperifolia, Oenothera, Opuntia, Panicum amarum, P. capillare, P. virgatum, Paspalum urvillei, P. vaginatum, Pennisetum ciliare, Persea palustris, Phragmites australis, Phyla, Pinus elliottii, Platanus wrightii, Pluchea camphorata, P. odorata, P. sericea, Polypogon monspeliensis, Populus fremontii, Prosopis glandulosa, Pseudoclappia arenaria, Rayjacksonia phyllocephala, Rhizophora mangle, Rhynchospora colorata, Sabal palmetto, Salicornia bigelovii, Salix exigua, S. gooddingii, Samolus ebracteatus, Sarcocornia perennis, Schoenoplectus americanus, S. pungens, S. tabernaemontani, Sesbania drummondi, Sesuvium maritimum, S. portulacastrum, Sisyrinchium demissum, Solidago sempervirens, Spartina alterniflora, S. patens, S. spartinae, Spilanthes americana, Sporobolus airoides, S. pyramidatus, S. virginicus, Stillingia, Strophostyles, Suaeda linearis, Symphyotrichum subulatum, S. tenuifolium, Tamarix aphylla, T. ramosissima, Taxodium ascendens, Thelesperma megapotamicum, Tillandsia, Triglochin concinna, Typha angustifolia, T. latifolia, Uniola paniculata, Vigna, Viguiera, Washingtonia, Xanthium strumarium.*

Use by wildlife: *Eustoma exaltatum* is grazed by livestock, which also trample the plants. The plants are host to larval Lepidoptera (Tortricidae: *Cochylis caulocatax*). The flowers attract butterflies (Lepidoptera).

Economic importance: food: inedible; **medicinal:** no medicinal uses reported; **cultivation:** Most of the cultivated *Eustoma* material is assigned to *E. russellianum* (=*E. grandiflorum*); however, taxonomic distinctions with *E. exaltatum* are difficult and both species probably have been distributed as cultivated material; **misc. products:** A *Eustoma* having

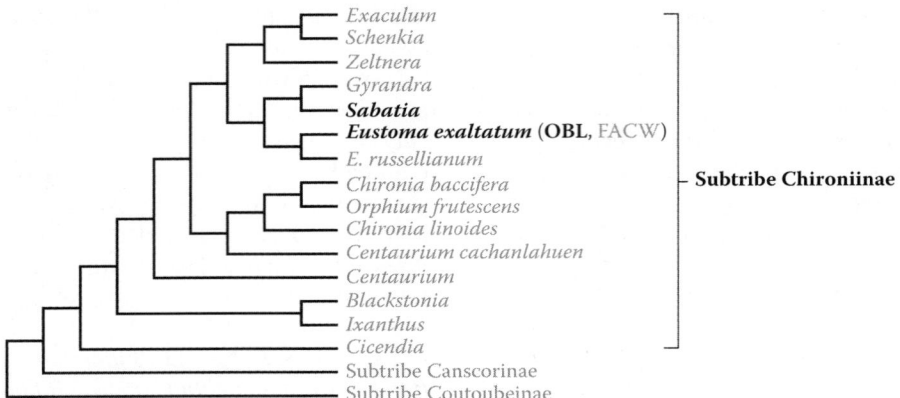

FIGURE 5.37 Phylogenetic relationships in subtribe Chironiinae (tribe Chironieae) of Gentianaceae based on analysis of combined nuclear and cpDNA sequence data. These results indicate that taxa with OBL North American representatives (shown in bold) have probably evolved several times within the subtribe. *Eustoma exaltatum* is closely related to *E. russellianum*, and some authors have merged the species. (Adapted from Mansion, G. & Struwe, L., *Mol. Phylogen. Evol.*, 32, 951–977, 2004.)

a deformed pistil has been patented (United States Patent 6878866) for breeding purposes; **weeds:** none; **nonindigenous species:** none

Systematics: Phylogenetic analysis of combined nuclear and cpDNA sequences (Figure 5.37) indicates that *Eustoma* is monophyletic, and is closely related to a clade containing *Gyrandra* and *Sabatia* within subtribe Chironiinae (tribe Chironieae) of Gentianaceae. Authors disagree on whether the genus should be considered as monotypic, or subdivided to contain two species. *Eustoma russellianum* sometimes has been recognized as a subspecies of *E. exaltatum* (e.g., as *E. exaltatum* subsp. *russellianum*). There is some merit to this disposition given that both taxa are broadly sympatric, interfertile, and produce fertile progeny (although natural hybrids rarely have been documented). However, the taxa exhibit various cytological, ecological, and morphological differences, and it seems prudent to retain them as distinct species until subsequent study might prove otherwise. The base chromosome number of *Eustoma* is $x = 18$. *Eustoma exaltatum* ($2n = 36$; 72) is diploid or tetraploid; whereas, *E. grandiflorum* ($2n = 36$) is diploid.

Comments: *Eustoma exaltatum* occurs across the southern United States (extending into Mexico, Central America, and the West Indies), mainly in coastal areas.

References: Correll & Correll, 1975; Godfrey & Wooten, 1981; Mansion & Struwe, 2004; Mason, 1957; Monson & Jaeger, 1991; Stutzenbaker, 1999; Turner, 2014; Turner & Turner, 2003.

3. Gentiana

Gentian; gentiane

Etymology: after Genthios (180–68 BC), last king of Illyria

Synonyms: *Asterias*; *Calathiana*; *Chiophila*; *Chondrophylla*; *Ciminalis*; *Coilantha*; *Cruciata*; *Dasystephana*; *Diploma*; *Ericala*; *Ericoila*; *Eudoxia*; *Eurythalia*; *Favargera*; *Gentianodes*; *Hippion*; *Holubogentia*; *Kuepferella*; *Kurramiana*; *Lexipyretum*; *Mehraea*; *Metagentiana*;

Pneumonanthe; *Qaisera*; *Ricoila*; *Selatium*; *Tretorhiza*; *Ulostoma*; *Varasia*; *Xolemia*

Distribution: global: Africa; Australia; Eurasia; New World; **North America:** widespread

Diversity: global: 360 species; **North America:** 28 species

Indicators (USA): OBL: *Gentiana catesbaei*, *G. douglasiana*, *G. fremontii*, *G. linearis*, *G. rubricaulis*, *G. sceptrum*, *G. setigera*; **FACW:** *G. douglasiana*, *G. fremontii*, *G. linearis*, *G. prostrata*

Habitat: freshwater; palustrine; **pH:** 3.7–7.7; **depth:** <1 m; **life-form(s):** emergent herb

Key morphology: stems decumbent/prostrate (10–45 cm) or erect (to 90 cm); leaves in a basal rosette or cauline and opposite, reduced (to 15 mm) or expanded (to 90 mm), narrow (4–6 mm) to broad (to 30 mm); flowers showy, open (campanulate, cup-like, flaring, funnelform) or nearly closed, 4–5-merous, terminal (or in upper axils), solitary or few (2–10); petals (14–50 mm) white, blue, or purple, "pleated," the lobes alternating with sinus "plaits" having fimbriate, lacerate, serrate, triangular, or truncate apices; style nearly lacking, stigma 2-branched; seeds winged or wingless

Life history: duration: annual/biennial (fruit/seeds); perennial (caudex, fleshy roots); **asexual reproduction:** none; **pollination:** insect; **sexual condition:** hermaphroditic; **fruit:** capsules (common); **local dispersal:** seeds (mammals, wind); **long-distance dispersal:** seeds (mammals, wind)

Imperilment: (1) *Gentiana catesbaei* [G5]; SX (PA); SH (NJ); S1 (AL); S3 (NC); (2) *G. douglasiana* [G4]; S2 (WA); (3) *G. fremontii* [G4]; S2 (<u>AB</u>, AZ, CA, <u>SK</u>); (4) *G. linearis* [G4]; SH (NJ); S1 (TN); S2 (MA, MI, <u>NB</u>, WV); S3 (MD, NY, VT); (5) *G. prostrata* [G4]; S2 (CA, OR, WY); S3 (<u>AB</u>, <u>BC</u>); (6) *G. rubricaulis* [G4]; SH (ME); S1 (<u>NB</u>); S2 (<u>MB</u>); (7) *G. sceptrum* [G4]; (8) *G. setigera* [G3]; S1 (CA); S2 (OR)

Ecology: general: Many North American gentians have an affinity for wet habitats with 21 of the species (75%) assigned to some wetland indicator category. Although most are rated only as FACW plants, seven North American species (25%)

are ranked as OBL. *Gentiana prostrata* was ranked as OBL, FACW in the 1996 indicator list but was reclassified as FACW in the 2013 list. It has been included here because the treatment was completed prior to the release of the updated list. For being as showy and popular a group of plants, it is unusual that a good deal of basic ecological information still is lacking for many of the gentian species. Details on pollination are surprisingly minimal, other than specifying that the process is carried out mainly by bumblebees (Hymenoptera: Apidae: *Bombus*). Generally, the intrinsic genetic isolating mechanisms of gentians are poorly developed, which often enables them to hybridize. Habitat and phenological isolating mechanisms are thought to be essential factors for maintaining distinctions among the species. For those *Gentiana* species studied, most have been shown to possess physiological seed dormancy, which requires a period of cold stratification for germination; however, the seed ecology is not well documented for any of the OBL species. Seed dispersal occurs mostly by the wind and is facilitated by breezes that agitate the capsules, and also by winged seeds, which occur in some of the species.

Gentiana catesbaei **Walter** is a perennial, which grows in ditches, depressions, flatwoods, pocosins, roadsides, seeps, swales, or along the margins of lakes, ponds, and swamps at low elevations. The plants occur mainly along the coastal plain in habitats exposed to full sun with acidic (pH: 5.6–6.5) substrates consisting of silty sand or sand. Overwintering is accommodated by the persistent, thickened roots. The winged seeds presumably are wind dispersed. **Reported associates:** *Acer rubrum, Andropogon glomeratus, Arundinaria tecta, Asclepias lanceolata, Carex glaucescens, C. striata, Centella erecta, Coreopsis falcata, Cyrilla racemiflora, Desmodium paniculatum, Dichanthelium acuminatum, D. consanguineum, D. ensifolium, D. scabriusculum, D. tenue, D. wrightianum, Eleocharis tuberculosa, Erigeron vernus, Eriocaulon decangulare, Eupatorium mohrii, E. pilosum, Gratiola pilosa, Hypericum crux-andrae, H. galioides, Iris tridentata, Lachnocaulon anceps, Leucothoe racemosa, Lilium catesbaei, Ludwigia hirtella, Lycopodiella alopecuroides, Magnolia virginiana, Myrica heterophylla, Osmunda regalis, Osmundastrum cinnamomeum, Panicum verrucosum, Photinia pyrifolia, Pinus taeda, Quercus phellos, Rhexia nashii, R. virginica, Rhynchospora elliottii, R. gracilenta, R. rariflora, R. torreyana, Saccharum baldwinii, Sarracenia flava, S. minor, Scleria ciliata, Smilax laurifolia, S. pseudochina, Trantha racemosa, Viburnum nudum, Viola primulifolia, Woodwardia areolata, Xyris ambigua, X. platylepis.*

Gentiana douglasiana **Bong.** is an annual, which inhabits Pacific coastal bogs, fens, meadows, muskeg, slopes, and the margins of streams at elevations of up to 1280 m. The plants are shade intolerant and occur characteristically on acidic (pH: 3.7–6.4) peat or on other nitrogen-poor soils, in sites that are flooded seasonally. The seeds are fusiform but unwinged, and their means of dispersal has not been determined. **Reported associates:** *Abies, Agrostis aequivalvis, Alnus viridis, Andromeda polifolia, Anemone oregana, Apargidium boreale, Artemisia tilesii, Aster, Bazzania trilobata, Blechnum*

spicant, Caltha biflora, C. palustris, Calypogeia sphagnicola, Campylopus japonicus, Carex aquatilis, C. interior, C. lenticularis, C. limosa, C. livida, C. luzulina, C. nigricans, C. obnupta, C. pauciflora, C. pluriflora, C. rostrata, C. utriculata, Cassiope mertensiana, Chamaecyparis, Cladina portentosa, Cladonia cristatella, C. pyxidata, Clintonia uniflora, Coptis aspleniifolia, C. trifolia, Cornus canadensis, C. suecica, Deschampsia cespitosa, Dichanthelium acuminatum, Drosera anglica, D. rotundifolia, Eleocharis pauciflora, Empetrum nigrum, Epilobium ciliatum, Equisetum arvense, Erigeron peregrinus, Eriophorum angustifolium, E. chamissonis, Fritillaria camschatcensis, Gaultheria shallon, Gentiana sceptrum, Geocaulon lividum, Geum calthifolium, Habenaria, Hypericum anagalloides, Hylocomium splendens, Iris setosa, Juncus, Juniperus communis, Kalmia microphylla, K. polifolia, K. procumbens, Listera cordata, Lomatogonium rotatum, Lupinus nootkatensis, Lycopodium annotinum, Lysichiton americanus, Maianthemum dilatatum, Menziesia ferruginea, Moneses uniflora, Mylia anomala, Myrica gale, Nephrophyllidium crista-galli, Orthilia secunda, Pedicularis ornithorhyncha, P. palustris, Picea sitchensis, Pinguicula vulgaris, Pinus contorta, Piperia unalascensis, Plagiothecium undulatum, Platanthera dilatata, P. hyperborea, Pleurozium schreberi, Podagrostis aequivalvis, Prenanthes alata, Primula jeffreyi, P. pauciflora, Pteridium aquilinum, Ptilium crista-castrensis, Racomitrium lanuginosum, Rhododendron groenlandicum, Rhynchospora alba, Rhytidiadelphus loreus, Rubus arcticus, R. chamaemorus, R. pedatus, Salix barclayi, Sanguisorba officinalis, S. stipulata, Selaginella selaginoides, Sphagnum austinii, S. capillifolium, S. compactum, S. fuscum, S. magellanicum, S. nemoreum, S. pacificum, S. papillosum, S. recurvum, S. rubellum, S. tenellum, Spiraea douglasii, Streptopus amplexifolius, S. roseus, S. streptopoides, Swertia perennis, Thuja plicata, Tiarella trifoliata, Triantha glutinosa, Trichophorum cespitosum, Trientalis europaea, Tsuga heterophylla, T. mertensiana, Vaccinium cespitosum, V. ovatum, V. oxycoccos, V. uliginosum, V. vitisidaea, Valeriana sitchensis, Veratrum viride, Viola langsdorfii, V. palustris.

Gentiana fremontii **Torr.** is an annual (or biennial) that inhabits montane fens, meadows, seeps, slopes, springs, and the margins of streams at elevations of up to 3810 m. The plants grow in sunny sites (often on hummocks) on substrates that are calcareous (pH: 6.4–7.7) and often have a high organic content (37%–60%). They are described as Greenhorn limestone, loam, peat, rock, rocky loam, silt, or silt loam. The capsules are sessile and mature from June to July; the seeds are dispersed by mid-August. These plants are quite small and often are decumbent, which makes them difficult to find when not in flower. The petals are white, but specific pollinators have not been reported. The seeds are unwinged and their means of dispersal has not been clarified. **Reported associates:** *Achillea millefolium, Agrostis stolonifera, Almutaster pauciflorus, Alopecurus borealis, Antennaria pulcherrima, Anticlea elegans, Argentina anserina, Artemisia frigida, Astragalus agrestis, A. leptaleus, Betula nana, B. occidentalis, Bistorta bistortoides, Calamagrostis canadensis,*

C. stricta, Calliergon trifarium, Campylium stellatum, Campanula parryi, Carex aquatilis, C. aurea, C. capillaris, C. crawei, C. echinata, C. lanuginosa, C. parryana, C. praeceptorum, C. scirpoidea, C. simulata, C. utriculata, C. viridula, Cirsium scariosum, C. tioganum, Crepis runcinata, Dasiphora floribunda, Deschampsia cespitosa, Distichum capillare, Drepanocladus aduncus, D. revolvens, Drymocallis glandulosa, Eleocharis bernardina, E. palustris, E. quinqueflora, Elymus trachycaulus, Epilobium lactiflorum, Erigeron lonchophyllus, Eriophorum angustifolium, Galium boreale, Gentianella amarella, Gentianopsis thermalis, Geranium richardsonii, Heracleum sphondylium, Hierochloe hirta, Hordeum brachyantherum, H. jubatum, Hymenoxys richardsonii, Iris missouriensis, Juncus balticus, J. longistylis, J. parryi, Kobresia myosuroides, K. simpliciuscula, Koeleria macrantha, Lilium parryi, Lycopus americanus, Mniobryum albicans, Muhlenbergia asperifolia, M. richardsonis, Packera debilis, P. paupercula, Parnassia palustris, Pedicularis crenulata, P. groenlandica, Philonotis fontana, Plantago eriopoda, Platanthera hyperborea, Poa pratensis, P. secunda, Potentilla plattensis, P. subjuga, Primula egaliksensis, P. incana, P. pauciflora, Ranunculus cymbalaria, Salix brachycarpa, S. candida, S. myrtillifolia, S. planifolia, S. pseudomonticola, Scorpidium turgescens, S. scorpioides, Sisyrinchium bellum, S. montanum, S. pallidum, Symphyotrichum spathulatum, Taraxacum officinale, Thalictrum alpinum, Triglochin maritimum, T. palustre, Valeriana edulis, Veratrum californicum, Viola adunca, Warnstorfia exannulata.

Gentiana linearis Froel. is a perennial inhabitant of higher elevation or boreal bogs, meadows, outcrops, shores, seeps, springs, and swamps from 1000 to 2000 m. The plants will not withstand prolonged inundation. Their habitats are characterized by open to semishaded exposures and often represent sites that are disturbed by natural fires. The substrates are acidic (pH: 4.0–6.7) and are derived from gneiss, granite, or silica. They typically are high in boron, copper, iron and manganese (but low in other cations), and high in organic content (~30%). Flowering occurs late in the season (August to September) and characteristically involves a very low abundance of plants. The flowers are "closed" and self-compatible but require pollination, which is carried out by bumblebees (Hymenoptera: Apidae: *Bombus*). The winged seeds are suited for wind dispersal. **Reported associates:** *Abies fraseri, Ageratina altissima, Agrostis hyemalis, A. perennans, Andromeda polifolia, Athyrium filix-femina, Bartonia virginica, Betula alleghaniensis, Calamagrostis cainii, C. canadensis, C. cinnoides, C. stricta, Carex baileyi, C. brunnescens, C. canescens, C. crinita, C. debilis, C. folliculata, C. interior, C. laevivaginata, C. michauxiana, C. misera, C. oligosperma, C. polymorpha, C. ruthii, C. scoparia, C. stricta, C. trisperma, C. utriculata, Chamaedaphne calyculata, Coptis trifolia, Danthonia compressa, Dennstaedtia punctilobula, Diervilla sessilifolia, Doellingeria umbellata, Drosera rotundifolia, Dulichium arundinaceum, Eleocharis tenuis, Epilobium leptophyllum, Eriophorum virginicum, Euthamia graminifolia, Galium asprellum, G. palustre, Gaultheria*

hispidula, Gentiana alba, Gentianella quinquefolia, Glyceria borealis, G. canadensis, G. nubigena, Houstonia caerulea, H. serpyllifolia, Huperzia appalachiana, Hypericum canadense, H. densiflorum, H. ellipticum, H. graveolens, H. punctatum, Ilex verticillata, Iris versicolor, Juncus acuminatus, J. brevicaudatus, J. effusus, J. filiformis, J. subcaudatus, J. tenuis, Kalmia angustifolia, K. polifolia, Krigia montana, Larix laricina, Leiophyllum buxifolium, Lonicera caerulea, Lycopus uniflorus, L. virginicus, Lygodium palmatum, Lyonia ligustrina, Lysimachia terrestris, Menyanthes trifoliata, Menziesia pilosa, Micranthes petiolaris, Myrica gale, Nemopanthus mucronata, Nyssa sylvatica, Oclemena acuminata, Osmundastrum cinnamomeum, Parnassia asarifolia, P. palustris, Photinia melanocarpa, P. pyrifolia, Picea rubens, Pinus rigida, Platanthera blephariglottis, Polygala sanguinea, Polytrichum commune, P. ohioense, Prunus pensylvanica, Quercus ilicifolia, Rhododendron canadense, R. carolinianum, R. catawbiense, R. groenlandicum, Rhynchospora alba, R. capillacea, Rubus canadensis, R. hispidus, Rugelia nudicaulis, Scapania nemorosa, Scirpus atrocinctus, S. cyperinus, Scutellaria galericulata, S. lateriflora, Senecio pauperculus, Sibbaldiopsis tridentata, Solidago glomerata, S. uliginosa, Sorbus americana, Sparganium americanum, Sphagnum capillifolium, S. compactum, S. fallax, S. girgensohnii, S. imbricatum, S. lindbergii, S. magellanicum, S. palustre, S. pylaesii, S. recurvum, S. tenellum, Spiraea alba, S. tomentosa, Stenanthium leimanthoides, Triadenum fraseri, T. virginicum, Trichophorum cespitosum, Utricularia cornuta, Vaccinium cespitosum, V. corymbosum, V. erythrocarpum, V. macrocarpon, V. myrtilloides, V. oxyccocus, V. uliginosum, Veratrum viride, Viburnum nudum, Viola lanceolata, V. palustris.

Gentiana prostrata Haenke is a biennial species found in western alpine or montane carrs, fens, hummocks, slopes, lake shores, meadows, terraces, tundra, and along the margins of streams at elevations from 2000 to 3963 m. The plants usually occur near or above the tree line, especially in areas of snowmelt. The substrates are circumneutral (pH: 6.2–7.6) and consist of clay loam, gravel, humus, limestone scree, loam, rocky basalt, sand, talus, or organic silt atop layers of gravel. The plants are self-compatible, generally self-pollinating and attain high seed set (up to 80%) in the absence of pollinators, which often are scarce in their alpine habitats. The flowers usually are open, but reportedly will close rather rapidly if they are touched or become shaded. They mature from 80 to 100% of their fruits (3–4/plant) annually, achieving an average seed set of around 80% (150–400 seeds/fruit). Natural seed viability is close to 100%. The seeds are wingless or winged along one edge and their germination requirements have not been elucidated. The dispersal mechanisms are undetermined, but may involve arctic ground squirrels (Mammalia: Sciuridae: *Urocitellus parryii*), which are known to cache the capsules or marmots (Mammalia: Sciuridae: *Marmota*), which feed on the flowers (see *Use by wildlife* later). Flower predation by mammals and insects can result in a nearly 10% loss of flowers in some seasons. About 36% of the plant biomass is allocated to reproduction. **Reported associates:** *Anemone parviflora, Antennaria lanata, A. media, Anticlea elegans,*

Aquilegia jonesii, Arctagrostis latifolia, Arenaria lanuginosa, Arnica alpina, Artemisia scopulorum, Astragalus alpinus, Aulacomnium acuminatum, A. palustre, Betula glandulosa, Bistorta bistortoides, B. vivipara, Campanula uniflora, Carex albonigra, C. bella, C. bigelowii, C. capillaris, C. capitata, C. chalciolepis, C. ebenea, C. elynoides, C. haydeniana, C. livida, C. norvegica, C. rupestris, C. scirpoidea, C. scopulorum, C. utriculata, Castilleja haydenii, C. occidentalis, Chamerion angustifolium, C. latifolium, Crepis runcinata, Dasiphora floribunda, Deschampsia brevifolia, D. cespitosa, Dicranum spadiceum, Distichium capillaceum, Ditrichum flexicaule, Dryas integrifolia, D. octopetala, Equisetum arvense, E. variegatum, Erigeron simplex, Eriophorum angustifolium, Euphrasia artica, E. disjuncta, Eurybia sibirica, Festuca altaica, F. brachyphylla, F. idahoensis, F. rubra, Gentiana algida, Gentianella propinqua, G. tenella, Geum rossii, Hedysarum alpinum, H. sulphurescens, Hylocomium splendens, Juncus arcticus, J. triglumis, Kobresia myosuroides, Ligusticum tenuifolium, Lloydia serotina, Luzula spicata, Mertensia, Minuartia obtusiloba, Moehringia macrophylla, Oxyria digyna, Parnassia fimbriata, P. palustris, Pedicularis sudetica, Petasites frigidus, Phlox, Poa alpina, P. nemoralis, Potentilla diversifolia, P. nivea, Primula pauciflora, Pseudocymopterus montanus, Ptilidium ciliare, Rhodiola rhodantha, R. rosea, Rhytidium rugosum, Salix alaxensis, S. arbusculoides, S. arctica, S. brachycarpa, S. glauca, S. nivalis, S. planifolia, S. pulchra, S. reticulata, S. richardsonii, S. rotundifolia, Selaginella densa, Sibbaldia procumbens, Silene acaulis, Smelowskia calycina, Solidago multiradiata, Thalictrum alpinum, Tofieldia pusilla, Tomentypnum nitens, Trifolium parryi, Trisetum spicatum, Utricularia minor, Vaccinium uliginosum, Viola sororia.

***Gentiana rubricaulis* Schwein.** is a boreal perennial, which occurs in depressions, ditches, fens, meadows, prairies, swamps, on lakeshores, or along the margins of rivers and streams. The plants do not tolerate prolonged inundation. They grow on alkaline (often highly calcareous) to moderately acidic substrates (e.g., pH: 6.9) including peaty muck or sand, but not on granitic ground. The habitats are open to semishaded and often are disturbed or drying sites. The flowers are "closed," sessile, and self-compatible, but require pollination by bumblebees (Hymenoptera: Apidae: *Bombus*) for seed set to occur. The seeds are winged and probably are wind dispersed. This species typically is found growing with grasses and sedges, but few specific taxa have been identified as associates. **Reported associates:** *Achillea millefolium, Aclepias incarnata, Agalinis paupercula, Alnus rugosa, Andropogon, Bidens cernuus, B. connatus, Calamagrostis canadensis, Carex, Chelone glabra, Cirsium arvense, Comarum palustre, Cornus racemosa, Eupatorium, Fragaria virginiana, Gnaphalium, Impatiens capensis, Juncus alpinus, J. balticus, Lathyrus, Liatris, Lycopus americanua, Lysimachia terrestris, Lythrum, Menyanthes trifoliata, Muhlenbergia glomerata, Onoclea sensibilis, Ophioglossum vulgatum, Parnassia glauca, Pedicularis lanceolata, Populus balsamifera, P. tremuloides, Salix rigida, Scirpus, Solidago gigantea, S. nemoralis, S. patula, S. uliginosa, Spiraea tomentosa, Spiranthes*

romanzoffiana, Symphyotrichum boreale, Triadenum, Viola nephrophylla.

***Gentiana sceptrum* Griseb.** is a perennial found in coastal montane bogs, fens, lagoons, lakeshores, meadows and at elevations of up to 1200 m. Substrates are acidic (pH: 5.5–6.5) peat, sand or loam with medium levels of nitrogen. The plants are shade intolerant. Seeds are winged at the ends and may be wind dispersed. Successful germination has been achieved by placing seeds at 15°C–25°C for 4 weeks, then moving to 4°C for 5 weeks and finally bringing to 5°C–13°C for germination. Germination at 21°C has been reported following treatment with gibberellic acid (GA). Plants overwinter by means of a persistent caudex. **Reported associates:** *Agrostis thurberiana, Alnus rubra, Andromeda polifolia, Anemone oregana, Angelica genuflexa, Aster, Aulacomnium palustre, Betula nana, Blechnum spicant, Caltha leptosepala, Camassia quamash, Campanula californica, Carex aquatilis, C. cusickii, C. livida, C. luzulina, C. obnupta, C. rostrata, C. utriculata, Cephaloziella, Chamaecyparis, Chrysanthemum, Cicuta, Comarum palustre, Coptis aspleniifolia, Cornus canadensis, C. unalaschkensis, Juncus balticus, J. ensifolius, Darlingtonia californica, Deschampsia cespitosa, Dichanthelium acuminatum, Drosera rotundifolia, Dulichium arundinaceum, Eleocharis quinqueflora, Empetrum nigrum, Epilobium angustifloium, Equisetum arvense, E. fluviatile, E. hyemale, Eriophorum angustifolium, E. chamissonis, E. gracile, Fraxinus latifolia, Gaultheria humifusa, G. shallon, Gentiana calycosa, G. douglasiana, Geum calthifolium, Hydrocotyle ranunculoides, Hylocomium splendens, Hypericum anagalloides, Juncus balticus, J. effusus, J. xiphoides, Kalmia microphylla, K. polifolia, Ligusticum apiifolium, Linnaea borealis, Loiseleuria procumbens, Lonicera involucrata, Luzula, Lycopodium clavatum, Lycopus uniflorus, Lysichiton americanus, Maianthemum dilatatum, Malus fusca, Menyanthes trifoliata, Menziesia ferruginea, Myrica californica, M. gale, Nephrophyllidium crista-galli, Oenanthe sarmentosa, Packera cymbalarioides, P. pseudaurea, Parnassia fimbriata, Pedicularis groenlandica, Phalaris arundinacea, Pinguicula vulgaris, Pinus contorta, Platanthera dilatata, P. leucostachys, Polytrichum juniperinum, Potentilla anserina, Primula jeffreyi, Pteridium aquilinum, Rhamnus purshiana, Rhododendron columbianum, R. groenlandicum, R. macrophyllum, R. palustre, Rosa nutkana, Rubus chamaemorus, R. laciniatus, R. pedatus, R. ulmifolius, R. ursinus, Salix commutata, Sanguisorba officinalis, Scutellaria, Senecio triangularis, Solidago canadensis, Sorbus sitchensis, Sphagnum mendocinum, S. papillosum, S. recurvum, Spiraea douglasii, Spiranthes romanzoffiana, Thuja plicata, Tofieldia, Triantha glutinosa, Trichophorum cespitosum, Trientalis europaea, Tsuga heterophylla, T. mertensiana, Vaccinium cespitosum, V. ovalifolium, V. ovatum, V. oxycoccos, V. parvifolium, V. uliginosum, V. vitis-idaea, Veratrum.*

***Gentiana setigera* A. Gray** is a perennial, which grows in montane bogs, meadows, and seeps at elevations near 1000 m. The plants are restricted to nutrient-poor, ultra-mafic (serpentine) substrates and occur in sites that are open to semishaded. The seeds have been germinated successfully at

10°C–13°C after sowing in the winter (under light soil cover) for 1–3 months. Detailed ecological studies of this species (e.g., population genetics and reproductive ecology) are seriously needed, especially given its rarity and restricted habitat type. **Reported associates:** *Arctostaphylos, Calamagrostis nutkaensis, Calocedrus decurrens, Caltha leptosepala, Carex halliana, C. rostrata, Castilleja miniata, Chamaecyparis lawsoniana, Cypripedium californicum, Darlingtonia californica, Deschampsia cespitosa, Drosera rotundifolia, Epilobium oreganum, Glyceria elata, Hastingsia alba, H. atropurpurea, H. bracteosa, Helenium bigelovii, Lilium wigginsii, Lupinus polyphyllus, Oreostemma alpigenum, Pedicularis attolens, Pinguicula macroceras, Pinus contorta, P. jeffreyi, P. monticola, P. ponderosa, Primula tetrandra, Quercus garryana, Ranunculus gormanii, Rhamnus californica, Rudbeckia californica, R. glaucescens, Sanguisorba officinalis, Sidalcea malviflora, Scirpus congdonii, Symphyotrichum spathulatum, Triantha glutinosa, Vaccinium oxycoccos, Veratrum viride, Viola macloskeyi, V. primulifolia.*

Use by wildlife: The flowers of *Gentiana prostrata* are eaten by marmots (Mammalia: Sciuridae: *Marmota flaviventris*) and moths (Lepidoptera: Noctuidae) and the capsules are cached by arctic ground squirrels (Mammalia: Sciuridae: *Urocitellus parryii*). However, studies have found that *G. sceptrum* is not eaten by mule deer (Mammalia: Cervidae: *Odocoileus hemionus*). Various Fungi are hosted by *Gentiana douglasiana* (Basidiomycota: Bolbitiaceae: *Hebeloma*), *G. linearis* (Ascomycota: Mycosphaerellaceae: *Cercospora gentianae*), and *G. sceptrum* (Ascomycota: Botryosphaeriaceae *Phyllosticta*; Nectriaceae: *Fusarium*; Basidiomycota: Pucciniaceae: *Puccinia gentianae*; Chytridiomycota: *Synchytrium* Synchytriaceae).

Economic importance: food: Children of the Hanaksiala tribe would suck the sweet nectar from flowers of *Gentiana douglasiana* as a treat; **medicinal:** The roots of the European gentian (*G. lutea*) have long been used as a medicinal agent and several North American species exhibit similar properties. *Gentiana catesbaei* was used by Confederate soldiers during the U.S. Civil War as a field remedy. The dried roots or their alcoholic extracts were used in the southern United States to treat various stomach disorders. Crude extracts of *G. linearis* exhibit germicidal activity against *Bacillus subtilis* (Bacteria: Bacillaceae). The roots contain several monoterpenes (iridoids and secoiridoids), which impart a bitter taste and are attributed with bile-stimulating, cardiovascular, gastric, and liver-detoxifying activities as well as having insect attractant and repellant properties; **cultivation:** *Gentiana linearis, G. prostrata, G. sceptrum,* and *G. setigera* are cultivated as ornamentals; **misc. products:** none; **weeds:** none; **nonindigenous species:** none

Systematics: *Gentiana* resolves as the sister group to a clade consisting of three polyphyletic genera (*Crawfurdia, Metagentiana,* and *Triptospermum*) whose classification requires extensive revision (Figure 5.38). *Gentiana* itself is monophyletic as currently circumscribed and has been subdivided into as many as 15 sections. Of the OBL North American species, *G. douglasiana* and *G. fremontii* are placed in section *Chondrophyllae, G. prostrata* (reclassified as FACW) in section *Dolichocarpa* (merged with section *Chondrophyllae* by some authors) and *G. catesbaei, G. linearis, G. rubricaulis, G. sceptrum,* and *G. setigera* are assigned to section *Pneumonanthe.* Relatively few species (especially from North America) have been investigated phylogenetically in this fairly large genus and specific relationships among the OBL North American species remain relatively unresolved, although the sections containing the OBL species appear to be closely related (Figure 5.39). Relationships have been hypothesized for some OBL species. *Gentiana catesbaei* is regarded as fairly primitive in section *Pneumonanthe* and presumably is most closely related to *G. saponaria* (FACW). The close relationship of these species is indicated by their similar morphology and ability to hybridize. Hybrids have also been reported between *G. catesbaei* and *G. villosa. Gentiana rubricaulis* formerly was regarded as a subspecies of the morphologically similar *G. linearis.* These two species can be crossed experimentally and natural hybrids have been reported. It has been hypothesized (but not tested experimentally) that *G. rubricaulis* may be of hybrid origin and possibly

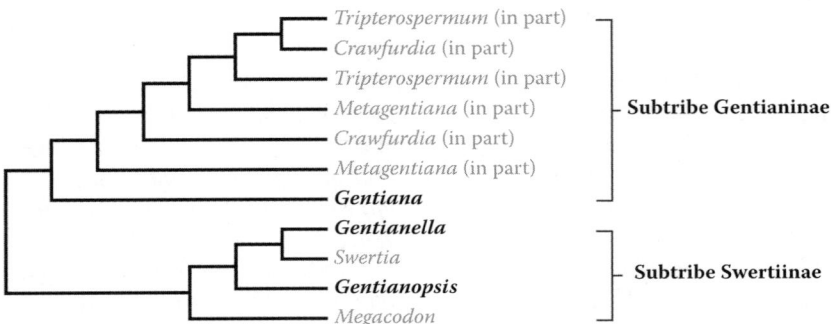

FIGURE 5.38 Phylogenetic relationships within tribe Gentianeae (Gentianaceae) as indicated by analysis of nrITS sequence data. The position of genera containing OBL species in North America (bold) indicates the independent origin of this habit in the tribe. *Gentiana* is the sister group to a clade containing several polyphyletic genera (*Crawfurdia, Metagentiana, Triptospermum*), which require considerable taxonomic revision. (Adapted from Chen, S. et al., *Ann. Bot.*, 96, 413–424, 2005.)

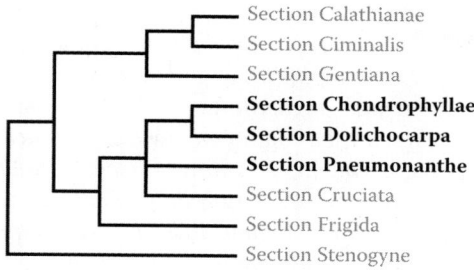

Section Calathianae
Section Ciminalis
Section Gentiana
Section Chondrophyllae
Section Dolichocarpa
Section Pneumonanthe
Section Cruciata
Section Frigida
Section Stenogyne

FIGURE 5.39 Phylogenetic relationships among sections of *Gentiana* as indicated by analysis of nrITS sequence data. Although none of the OBL North American species was included in the analyses, they are assigned to the three sections indicated in bold, which indicates that phylogenetically, they are relatively closely related. (Adapted from Yuan, Y.-M. et al., *Amer. J. Bot.*, 83, 641–652, 1996.)

descended from *G. linearis* and *G. alba*. *Gentiana* ×*grandilacustris* is a hybrid attributed to *G. andrewsii* × *G. rubricaulis*. An Asian segregate once included under *G. fremontii* is now treated as a separate species (*G. aquatica*). The two species arguably are closely related. Cross-compatibility is not in itself a reliable indicator of relationship because gentian species generally lack strong genetic isolating mechanisms and often are interfertile. Species distinctions are maintained by habitat isolation (e.g., acidic vs. alkaline substrates). Several base chromosome numbers occur in *Gentiana* including $x = 9, 10, 12, 13, 15$, and 23. The base number of *Gentiana* section *Chondrophyllae* is $x = 10$; however, the OBL North American species in this group have not yet been counted. Most of *Gentiana* section *Pneumonanthe* including *G. douglasiana*, *G. linearis*, *G. rubicaulis*, and *G. sceptrum* ($2n = 26$) are diploids based on an $x = 13$ series. *Gentiana prostrata* ($2n = 36$) is a tetraploid.

Comments: *Gentiana catesbaei* occurs in the southeastern United States; whereas, *G. linearis* and *G. rubricaulis* are distributed in northeastern North America. *Gentiana fremontii*, *G. prostrata* (also throughout Eurasia), and *G. sceptrum* occur in western North America, with *G. douglasiana* in northwestern North America. *Gentiana setigera* is endemic to California and Oregon.

References: Adams et al., 1920; Anderson, 2009; Asada et al., 2003; Bamberg & Major, 1968; Bergeron et al., 1996, 1997; Chen et al., 2005; Christy, 2004; Cooper, 1942; Costelloe, 1988; Cowan, 1945; Frahm, 1993; Gillis et al., 2005; Hanson, 1951; Hasegawa, 2000; Jankovsky-Jones, 1999; Johnson & Steingraeber, 2003; Klinka et al., 1995; Latham et al., 1996; Lieneman, 1929; Mead, 2002; Pringle, 1967, 1968; Rigg, 1925, 1937; Robuck, 1985; Rodriguez et al., 1998; Spira & Pollak, 1986; Walker et al., 1994; Wheeler et al., 1983; Wieder et al., 1984; Wiser et al., 1996; Wood & Bache, 1854; Yuan et al., 1996, 1998.

4. Gentianella

Dwarf gentian, felwort; gentiane
Etymology: a diminutive form of *Gentiana*

Synonyms: *Aliopsis*; *Aloitis*; *Amarella*, *Arctogentia*; *Chionogentias*; *Ericala*, *Gentiana* (in part); *Parajaeschkea*; *Pitygentias*; *Selatium*
Distribution: global: Africa; Australia; Eurasia; New World; **North America:** widespread
Diversity: global: 250 species; **North America:** 9 species
Indicators (USA): OBL: *Gentianella amarella*, *G. wrightii*; **FACW:** *G. amarella*
Habitat: freshwater; palustrine; **pH:** 6.2–7.8; **depth:** <1 m; **life-form(s):** emergent herb
Key morphology: stems (to 75 cm) erect, simple or branched, rectangular; leaves both in basal rosettes (to 4 cm; elliptic to spatulate, petioled) and cauline (to 6 cm), opposite, lanceolate, oblanceolate or ovate, sessile and partly clasping; cymes axillary or terminal; flowers showy, few to numerous, sessile or stalked (to 5 cm), 4–5-merous; corolla funnelform or salverform; petals (to 2.5 cm) blue, pink, white, or yellow, the inner base of lobes with a fringe of hairs; stigmas sessile, pistil sessile to short stalked; capsules (to 2 cm) dehiscing by terminal valves; seeds (to 1 mm) smooth, slightly flattened
Life history: duration: biennial (fruit/seeds); **asexual reproduction:** none; **pollination:** insect, self; **sexual condition:** hermaphroditic; **fruit:** capsules (common); **local dispersal:** seeds (wind); **long-distance dispersal:** seeds (animals)
Imperilment: (1) *Gentianella amarella* [G5]; SH (VT); S1 (ME, <u>NS</u>); S2 (<u>LB</u>, <u>NF</u>); S3 (MN, <u>NB</u>, NV); (2) *G. wrightii* [G5]
Ecology: general: *Gentianella* includes annuals, biennials, and perennials, which mostly are associated with UPL habitats. Four of the North American species are wetland indicators with two ranked as OBL. This is mainly a temperate genus that is especially common in alpine habitats.

***Gentianella amarella* (L.) Börner** is an annual or biennial, which inhabits beach or dune depressions, bogs, draws, flats, meadows, prairies, scrub, seeps, strands, swales, tundra, and the margins of lakes, rivers, and streams at elevations of up to 3500 m. The habitats (most commonly meadows) receive full sun or partial shade. The substrates are nutrient poor, circumneutral to alkaline (pH: 6.2–7.8), often saline, and can comprise various combinations of clay, cobble, gravel, humus, loam, rock, sand, silt, and talus. Flowering occurs from July to October. The plants are self-compatible and spontaneous selfing is as effective as open pollination (by insects) for attaining high seed set. Specific pollinators have not been identified for North American plants, but presumably are mainly bumblebees (Insecta: Hymenoptera). There is an average of up to 3.5 fruits/plant and 50 seeds/fruit. As biennials, all plants will flower during their second year, but those that are small at the time of flowering will have poor seed set compared to the larger plants. Minor grazing has been shown to stimulate (increase) fruit production. The seeds of *G. amarella* are physiologically dormant and form a persistent seed bank. Natural scarification (e.g., rotting of the seed coat) and a cold treatment are required to break dormancy. Germination has been achieved after placing seeds at 4°C for 12 weeks followed by 10 weeks at 20°C. Physical abrasion of the seed coat, followed by application of 10 mg/L gibberellic acid, will expedite germination. In natural habitats, germination occurs only at sites where the surrounding turf is from 1 to 3 cm in height. It is inhibited by taller grass cover. Demographic studies indicate that the plants experience roughly 0.5% mortality per day throughout the year. This species is characteristic of disturbed areas (e.g., beaches) and appears after short fire intervals, a factor that probably is related to an enhanced rate of seed germination following cover removal. The plants are associated with vesicular–arbuscular (VA) mychorrhizae. The seeds are dispersed locally (to about 1 m) as the capsules are shaken by the wind but reach greater distances as a consequence of animals, which ingest the capsules. The seeds probably are not dispersed to any extent by water, as they have been shown to remain afloat for less than 30 min. This species exhibits extensive stochasticity from year to year, tending to occur sporadically at different sites within an area of suitable habitat. Although common in wet sites, a fairly broad tolerance to soil moisture conditions is indicated by the list of associated species, many of them being only FACW plants.

Reported associates: *Achillea borealis, A. millefolium, Achnatherum nelsonii, Aconitum delphiniifolium, Agropyron dasystachyum, A. trachycaulum, Agrostis scabra, A. stolonifera, Allium cernuum, Alnus viridis, Amelanchier alnifolia, Androsace chamaejasme, Anemone narcissiflora, A. parviflora, A. patens, Angelica lucida, Antennaria corymbosa, A. microphylla, A. media, A. neglecta, A. parvifolia, A. pulcherrima, A. racemosa, A. rosea, Anticlea elegans, Apocynum cannabinum, Aquilegia coerulea, A. flavescens, Arctostaphylos uva-ursi, Arenaria congesta, A. hookeri, A. obtusiloba, Arnica cordifolia, A. lanceolata, A. lonchophylla, A. sororia, Artemisia cana, A. frigida, A. ludoviciana, A. tilesii, Astragalus alpinus, A. americanus, A. miser, A. vexilliflexus, Balsamorhiza sagittata, Berberis repens, Besseya wyomingensis, Betula glandulosa, B. occidentalis, Bistorta bistortoides, B. vivipara, Botrychium lunaria, B. tunux, B. yaaxudakeit, Bromus carinatus, B. ciliatum, B. inermis, B. vulgaris, Bupleurum americanum, Calamagrostis canadensis, C. purpurascens, C. rubescens, Caltha leptosepala, Calypso bulbosa, Campanula rotundifolia, Cardamine pratensis, Carex atratiformis, C. deweyana, C. disperma, C. elynoides, C. geyeri, C. hassei, C. macrochaeta, C. mertensii, C. nardina, C. nigricans, C. obtusata, C. rupestris, C. scirpoidea, C. simulata, C. subnigricans, Cassiope mertensiana, Castilleja miniata, C. septentrionalis, C. unalaschcensis, Cerastium arvense, Ceratodon purpureus, Chamerion angustifolium, C. latifolium, Chimaphila umbellata, Cicuta maculata, Cirsium arvense, Coeloglossum viride, Collomia linearis, Corallorhiza trifida, Cornus sericea, Crepis runcinata, C. tectorum, Dactylis glomerata, Danthonia intermedia, D. parryi, Dasiphora floribunda, Deschampsia cespitosa, Desmodium canadense, Draba graminea, Elaeagnus commutata, Eleocharis palustris, Elymus repens, E. trachycaulus, Empetrum nigrum, Epilobium anagallidifolium, Equisetum arvense, E. scirpoides, E. sylvaticum, E. variegatum, Eremogone fendleri, Erigeron caespitosus, E. hyssopifolia, E. lonchophyllus, E. peregrinus, Eriophorum angustifolium, Eritrichium nanum, Eurybia conspicua, Fallopia convolvulus, Festuca altaica, F. idahoensis, F. ovina, F. rubra, F. scabrella, Fragaria chiloensis, F. vesca, F. virginiana, Frasera fastigiata, Galium boreale, Gayophytum diffusum, Gentianella propinqua, Geranium erianthum, G. richardsonii, Geum aleppicum, G. macrophyllum, G. triflorum, Goodyera repens, Gutierrezia sarothrae, Halenia deflexa, Hedysarum alpinum, H. sulphurescens, Helianthus pauciflorus, Heracleum lanatum, H. maximum, Hesperostipa comata, Hieracium robinsonii, H. triste, H. umbellatum, Honckenya peploides, Hordeum jubatum, Iris versicolor, Juncus arcticus, J. drummondii, J. hallii, J. nevadensis, J. trifidus, J. vaseyi, Juniperus communis, Kalmia procumbens, Kobresia myosuroides, Koeleria macrantha, Lathyrus japonicus, L. ochroleucus, Lesquerella, Leucopoa kingii, Lewisia sierrae, Leymus innovatus, L. mollis, Ligusticum grayi, L. scoticum, Lilium columbianum, Linnaea borealis, Lomatium ambiguum, Lomatogonium rotatum, Lonicera utahensis, Lupinus lepidus, L. nootkatensis, Luzula parviflora, Madia minima, Maianthemum canadense, Melilotus, Mimulus primuloides, M. tilingii, Minuartia nuttallii, Moehringia lateriflora, Moneses uniflora, Muhlenbergia richardsonis, Nabalus racemosus, Oreostemma alpigenum, Oryzopsis pungens, Oxytropis campestris, Packera werneriifolia, Pedicularis attollens, P. furbishiae, P. groenlandica, P. racemosa, Penstemon confertus, P. heterodoxus, Phalaris arundinacea, Phleum alpinum, Phlox caespitosa, P. diffusa, P. multiflora, Picea engelmannii, Pinus contorta, Plantago major, Platanthera dilatata, P. obtusata, Poa alpina, P. glauca, P. palustris, P. pratensis, P. rupicola, Potentilla anserina, P. ×diversifolia, P. drummondii, P. flabelliformis, P. norvegica, P. ovina, Prenanthes racemosa, Primula alcalina, P. laurentiana, Prunella vulgaris, Prunus pumila, Pseudotsuga menziesii, Pyrola asarifolia, P. secunda, P. virens, Racomitrium canescens, Ranunculus abortivus, R. macounii, R. occidentalis, R. uncinatus, Rhinanthus borealis, R. cristagalli, R. minor, Rhodiola integrifolia, R. rhodantha, Rhytidiadelphus squarrosus, Rhytidium rugosum, Rosa blanda, Rubus parviflorus, R. stellatus, Rudbeckia hirta, R. occidentalis, Salix arctica, S. bebbiana, S. brachycarpa, S. candida, S. drummondiana, S. glauca, S. lucida, S. myrtillifolia, S. planifolia, S. scouleriana, Sanguisorba, Sanicula marilandica, Sedum lanceolatum, Selaginella densa, S. selaginoides, S. watsonii, Senecio canus, S. hydrophiloides, S. indecorus, Sibbaldia procumbens, Silene vulgaris, Solidago canadensis, S. juncea, S. lepida, S. missouriensis, S. multiradiata, S. simplex, Spartina pectinata, Spiraea alba, Spiranthes diluvialis, S. romanzoffiana, Stellaria longipes, Stenanthium occidentale, Stenotus acaulis, Stipa curtiseta, Symphoricarpos albus, Symphyotrichum anticostense, S. ciliolatum, S. falcatum, S. novi-belgii, Tanacetum bipinnatum, T. vulgare, Taraxacum officinale, Thalictrum alpinum, T. fendleri, T. venulosum, Thermopsis rhombifolia, Tofieldia, Toxicodendron* *rydbergii, Trautvetteria caroliniensis, Trichophorum pumilum, Trifolium hybridum, T. monanthum, T. pratense, Trisetum melicoides, T. spicatum, Vaccinium cespitosum, V. membranaceum, V. scoparium, Vahlodea atropurpurea, Valeriana dioica, V. edulis, V. sitchensis, Veronica wormskjoldii, Vicia americana, Viola canadensis, Zizia aptera.*

Gentianella wrightii (A. Gray) Holub is an annual, which grows in meadows and by springs at elevations from 2100 to 2400 m. Flowering occurs from August to November. There exists little life history information for this species, which many authors merge with *Gentianella amarella* (see *Systematics* later). **Reported associates:** none.

Use by wildlife: *Gentianella amarella* occasionally is grazed by livestock, rabbits (Mammalia: Leporidae: *Sylvilagus*), and deer (Mammalia: Cervidae: *Odocoileus*).

Economic importance: food: not edible; **medicinal:** *Gentianella amarella* is found as an ingredient in some herbal remedies that are alleged to alleviate despondency. The root has similar bitter properties as medicinal gentian (*Gentiana lutea*) as indicated by its specific epithet (*amarus* is the Latin word for "bitter"); **cultivation:** *Gentianella amarella* is cultivated; **misc. products:** none; **weeds:** none; **nonindigenous species:** none

Systematics: Although older literature did not distinguish *Gentianella* from *Gentiana* (tribe Gentianeae; subtribe Gentianinae), the former differs fundamentally from the latter by a lack of pleats (plicae) between the petals and no nectary disk, with the nectaries being inserted on the petals instead. Phylogenetic studies using DNA sequence data resolve *Gentianella* within subtribe Swertiinae as more closely related to *Swertia* and *Gentianopsis* than to *Gentiana* (Figure 5.38). This entire genus requires an extensive taxonomic reorganization and further phylogenetic study, as it is polyphyletic with the present circumscription of taxa. Within *Gentianella*, *G. amarella* is assigned to section *Gentianella*. The species of this section are characterized by very low levels of genetic differentiation, which indicates that they have diverged relatively recently. However, this factor has made it difficult to elucidate precise interspecific relationships using a molecular phylogenetic approach because of poor resolution from the genetic markers evaluated (Figure 5.40). Morphologically, *G. amarella* is similar to the European *G. uliginosa* (including *G. anglica*). Five subspecies of *G. amarella* have been recognized in North America (and Mexico) and there has been considerable variation observed among European populations. A fair amount of genetic differentiation has been detected among European *Gentianella* populations using amplified fragment length polymorphism (AFLP) markers. Molecular data indicate that species migrated from Asia into North America and eventually into South America, where the large extent of alpine habitat has resulted in the highest species diversity in their genus. Although the population genetics of *G. amarella* have been studied to some degree, most of that work has focused on European populations and the North American segregates (especially the different subspecies) warrant a detailed, directed investigation. Several chromosomal base numbers occur in the genus ($x = 5, 9, 11, 13$). *Gentiana amarella* ($x = 9$) has different cytotypes associated with subsp. *acuta* ($2n = 18$; diploid) and subsp. *amarella* ($2n = 36$; tetraploid). Chromosome counts are unavailable for *G. wrightii*. *Gentiana amarella* reportedly hybridizes extensively in Europe; however, hybrids have not been reported in North America.

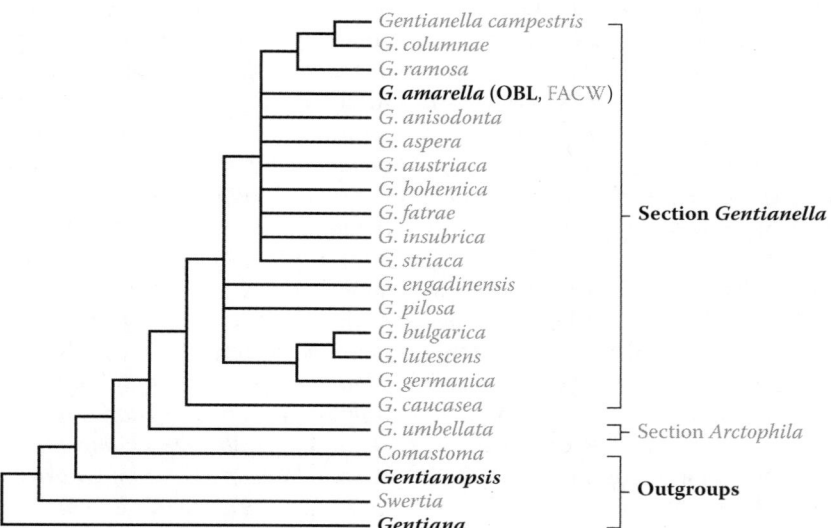

FIGURE 5.40 Phylogenetic relationships in *Gentianella* as indicated by analysis of combined molecular data. The recent diversification and correspondingly low level of genetic divergence that characterizes section *Gentianella* makes it difficult to resolve interspecific relationships. Taxa including OBL North American species are indicated in bold. (Adapted from Jang, C.-G. et al., *Bot. J. Linn. Soc.*, 148, 175–187, 2005.)

Comments: *Gentianella amarella* occurs in northern and western North America but extends into Mexico and northern Eurasia; *G. wrightii* occurs in Arizona and New Mexico but extends into Mexico.

References: Allen et al., 2002; Andersson et al., 2000; Fischer & Matthies, 1997; Gillett, 1957; Greimler et al., 2004; Holcroft Weerstra, 2001; Huhta et al., 2003; Jang et al., 2005; Jones & Fertig, 1999a,b; Kelly, 1984, 1989a,b; La Roi & Hnatiuk, 1980; Mason, 1998; McVeigh et al., 2005; Peinado et al., 2005a; Schoennagel et al., 2004; Stensvold et al., 2002; von Hagen & Kadereit, 2001, 2002; Winfield et al., 2003.

5. *Gentianopsis*

Fringed gentian; gentiane frangée

Etymology: a Latin form meaning "resembling the gentian"

Synonyms: *Crossopetalum*; *Gentiana* (in part)

Distribution: global: Asia; Europe; North America; **North America:** northern

Diversity: global: 24 species; **North America:** 9 species

Indicators (USA): OBL: *Gentianopsis crinita, G. holopetala, G. thermalis, G. virgata;* **FACW:** *G. crinita, G. detonsa, G. simplex, G. thermalis, G. virgata*

Habitat: freshwater; palustrine; **pH:** 6.2–8.2; **depth:** <1 m; **life-form(s):** emergent herb

Key morphology: stems (to 20–100 cm) erect, branched or simple; basal leaves (to 2–5 cm) in a rosette, cauline leaves (to 5 cm) opposite or whorled; inflorescence solitary or in terminal/axillary clusters (to 176 flowers), flowers showy, stalked (to 12 cm), 4-merous (rarely 3-merous); corolla (to 7.5 cm) campanulate to funnelform, tips of lobes lacerate to fimbriate, blue to purple (rarely white), sinuses lacking plaits; ovary sessile or stalked (to 12 mm); capsules (to 60 mm) apically valvate; seeds (to 1 mm) papillose, punctate, reticulate, or striate

Life history: duration: annual/biennial (fruit/seeds); perennial (fleshy roots); **asexual reproduction:** none; **pollination:** insect; **sexual condition:** hermaphroditic; **fruit:** capsules (common); **local dispersal:** seeds (water, wind); **long-distance dispersal:** seeds (waterbirds)

Imperilment: (1) *Gentianopsis crinita* [G5]; SX (DE, WV); S1 (GA, MD, NC, ND, QC, VA); S2 (NH, RI); S3 (AB, IA, ME, OH, VT); (2) *G. detonsa* G3]; SH (ON); S1 (AB, AL); S3 (YT); (3) *G. holopetala* [G4]; (4) *G. simplex* [G5]; S1 (MT, WY); (5) *G. thermalis* [G4]; S3 (WY); (6) *G. virgata* [G5]; SX (PA); S1 (NY); S2 (SD); S3 (IA, OH, WI)

Ecology: general: Most of the North American *Gentianopsis* species (78%) are either obligate (4 species) or FACW plants, which are characteristic of alkaline, calcareous meadows, especially at higher elevations. *Gentianopsis detonsa* and *G. simplex* were ranked in the 1996 indicator list as OBL, FACW; however, they were reclassified in the 2013 list as FACW. Having been completed prior to 2013, the treatments for these species have been retained here. *Gentianopsis* flowers have conspicuously fringed margins on their petals, which accounts for the common name. They are nectariferous and most are pollinated by bumblebees (Hymenoptera: Apidae: *Bombus*); however, the extent of self-pollination has not been investigated in many species. Usually, the flowers require full

sunlight in order to remain open. The seeds of some species are covered with fine hairs; however, none of them floats well in water. Dispersal is likely achieved by wind or by the adhesion of the seeds to mud that is attached to waterfowl feet. The seeds develop physiological dormancy and often can be difficult to germinate. Several species occur in naturally disturbed habitats; however, most will not survive at sites where a significant overstory canopy develops. The duration of species is not well documented with some described both as annual and perennial.

Gentianopsis crinita (**Froel.) Ma** is an annual or biennial found in ditches, dune depressions, fens, marshes, meadows, prairies, seeps, swamps, and margins of lakes and streams at elevations <500 m. Substrates are circumneutral (pH: 6.5–7.5), low nutrient, calcareous or serpentine (i.e., ultramafic) clay, loam, mud, or sand. The plants thrive in full sun but can tolerate partial shade; however, flowers will open only in full sunlight (closing each night and also on cloudy days). A single plant can produce up to 176 flowers, which are pollinated mainly by bumblebee (Hymenoptera: Apidae: *Bombus*) queens. The nectar contains about 12% fructose and glucose. Fresh seeds will germinate (~50%) at 7°C–16°C. When exposed to cold temperatures (3°C), the seeds acquire physiological dormancy and then require a moist, cold treatment (12 weeks at 3°C) for germination (at 16°C–24°C). Higher temperatures (24°C–29°C) result in reduced germination (~50%). The small papillose seeds are dispersed by the wind. Self-seeding does not occur well unless sites are disturbed to provide moist, bare soil conditions. The plants are adapted to fire. Early seedling growth is optimal at a soil temperature of 19°C. Plants rely on a symbiotic VA mycorrhizal Fungi association to regulate dormancy, overwintering, and reemergence; their transplant success has been increased by adding isolates of *Claroideoglomus etunicatum* (Glomeromycota: Glomeraceae), a VA fungus, to the transplant medium. Studies have not been conducted on the population genetics of this species. Although also described as biennial, tagged plants have shown this species to be a strict annual in some areas. **Reported associates:** *Agalinis acuta, A. purpurea, A. tenuifolia, Aletris farinosa, Allium cernuum, Andropogon gerardi, Antennaria, Asclepias verticillata, Berula erecta, Betula pumila, Bidens cernuus, B. coronatus, Bromus ciliatus, Calamagrostis canadensis, C. inexpansa, Campanula aparinoides, Carex eburnea, C. lanuginosa, C. lasiocarpa, C. muhlenbergii, C. pensylvanica, C. sartwelii, C. scirpoidea, C. viridula, Castilleja coccinea, Cenchrus, Cirsium discolor, C. muticum, Cladina, Cladium marsicoides, Cladonia, Clinopodium glabellum, Coreopsis major, Danthonia spicata, Dasiphora floribunda, Daucus carota, Deschampsia cespitosa, Dichanthelium acuminatum, D. lanuginosum, D. sphaerocarpon, Eleocharis elliptica, Elymus trachycaulus, Equisetum hyemale, Eriophorum angustifolium, Eupatorium perfoliatum, Euphorbia purpurea, Eutrochium maculatum, Filipendula rubra, Fragaria virginiana, Frasera caroliniensis, Fraxinus, Galium boreale, G. labradoricum, Gentiana andrewsii, Gentianopsis virgata, Gnaphalium uliginosum,*

Helianthemum bicknellii, H. propinquum, Helianthus, Hexastylis arifolia, Hierochloe odorata, Hypericum kalmianum, Ionactis linariifolius, Iris virginica, Juncus balticus, J. brachycephalus, J. canadensis, J. torreyi, Juniperus virginiana, Kalmia latifolia, Koeleria macrantha, Lechea, Liatris aspera, L. pilosa, L. spicata, Lilium michiganense, Linum sulcatum, Liparis loeselii, Liriodendron tulipifera, Lithospermum canescens, Lobelia kalmii, L. siphilitica, Lupinus perennis, Lycopus americanus, Lysimachia quadriflora, Lythrum alatum, Menyanthes trifoliata, Mimulus glabratus, Minuartia stricta, Muhlenbergia glomerata, Oxypolis rigidior, Panicum flexile, P. virgatum, Parnassia glauca, P. grandifolia, Pedicularis canadensis, Persicaria sagittata, Phlox subulata, Platanthera hyperborea, P. leucophaea, P. psycodes, Poa saltuensis, Polygala paucifolia, P. polygama, P. senega, Polygonum tenue, Potentilla simplex, Prunella vulgaris, Pteridium aquilinum, Pycnanthemum virginianum, Pyrola americana, Rhynochospora alba, R. capillacea, Rubus parviflorus, Sabatia angularis, Salix humilis, S. myricoides, S. pedicellaris, Sarracenia purpurea, Schizachyrium scoparium, Schoenoplectus acutus, S. pungens, Scleria verticillata, Senecio plattensis, Sibbaldiopsis tridentata, Sisyrinchium campestre, Solidago canadensis, S. graminifolia, S. juncea, S. nemoralis, S. ohioensis, S. patula, S. ptarmicoides, S. riddellii, S. rugosa, S. uliginosa, Sorghastrum nutans, Spartina, Spiraea tomentosa, Spiranthes cernua, S. lucida, S. romanzoffiana, Sporobolus heterolepis, Symphyotrichum depauperatum, S. lanceolatum, S. lateriflorum, S. novae-angliae, S. pilosum, S. puniceum, S. racemosum, Symplocarpus foetidus, Talinum teretifolium, Thalictrum macrostylum, Thelypteris palustris, Toxicodendron vernix, Triadenum fraseri, Triantha glutinosa, Typha angustifolia, Vaccinium corymbosum, V. stamineum, Vernonia gigantea, V. missurica.

Gentianopsis detonsa (Rottb.) Ma is an annual or biennial, which inhabits alpine and subalpine ditches, meadows, stream, lake, and river margins at elevations of up to 3615 m. Its substrates are alkaline (pH: 6.2–8.2) and include clay, gravel, peat, sand, and silt. The plants frequently occur in disturbed sites such as areas that are recovering from fire. There is little specific information on the reproductive or seed ecology of this species. As for several other members of the genus, the flowers open only in full sunlight. Presumably they are pollinated by bumblebees (Hymenoptera: Apidae: *Bombus*) and produce wind-dispersed seeds. Optimal July/January temperatures are reported as 11.0°C/−3.5°C respectively. **Reported associates:** *Achillea millefolium, Amelanchier alnifolia, Angelica arguta, A. dawsonii, Anticlea elegans, Arctostaphylos uva-ursi, Arenaria congesta, Arnica chamissonis, Betula glandulosa, Bistorta vivipara, Botrychium virginianum, Brachythecium rutabulum, Bromus carinatus, Calamagrostis canadensis, Carex aquatilis, C. buxbaumii, C. gynocrates, C. lasiocarpa, C. limosa, C. livida, C. muricata, C. pellita, C. rostrata, Castilleja rhexiifolia, Chamerion angustifolium, Cirsium arvense, Comarum palustre, Cornus canadensis, C. sericea, Danthonia intermedia, Deschampsia*

cespitosa, Dichanthelium languinosum, Dulichium arundinacea, Eleocharis quinqueflora, Elymus canadensis, E. glaucus, E. trachycaulus, Equisetum arvense, E. variegatum, Eriophorum angustifolium, E. gracile, Erysimum cheiranthoides, Eucephalus engelmannii, Eurybia conspicua, Festuca idahoensis, F. subulata, Fragaria virginiana, Galium boreale, G. trifidum, Gentianella propinqua, Geum aleppicum, Heracleum maximum, Iliamna rivularis, Juncus longistylis, Lathyrus ochroleucus, Leymus innovatus, Lonicera involucrata, Lycopodium inundatum, Maianthemum stellatum, Mimulus guttatus, Osmorhiza occidentalis, Packera subnuda, Pedicularis bracteosa, P. groenlandica, Phleum alpinum, P. pratense, Picea engelmannii, Pinus contorta, Plagiomnium medium, Platanthera, Poa palustris, Populus tremuloides, Potentilla diversiflolia, Rhamnus alnifolia, Rosa gymnocarpa, Rubus pubescens, Salix planifolia, S. scouleriana, Sanicula marilandica, Scheuchzeria palustris, Senecio cymbalarioides, S. flettii, S. hydrophilus, Shepherdia canadensis, Sisyrinchium idahoense, Solidago gigantea, S. multiradiata, Sphagnum, Spiraea betulifolia, Swertia perennis, Symphoricarpos albus, S. occidentalis, Symphyotrichum foliaceum, S. laeve, Taraxacum officinale, Thalictrum occidentale, Trifolium repens, Urtica dioica, Valeriana edulis, Veratrum viride, Vicia americana, Viola canadensis.

Gentianopsis holopetala (A. Gray) Iltis is an annual or perennial, which occurs in alpine and subalpine meadows and along lake or stream margins at elevations from 1800 to 4000 m. Substrates are described as sandy silt loam. Details on the reproductive ecology of this species are scarce. The seeds are small and papillose, and presumably are dispersed by wind or water. Germination at 10°C–13°C has been reported after 1–3 months for seeds sowed in winter under a thin cover of soil. The plants mainly are annuals, but reportedly can propagate from root sprouts or rooting stems as well. **Reported associates:** *Achillea lanulosa, Aconitum columbianum, Agrostis humilis, Allium validum, Antennaria umbrinella, Bistorta bistortoides, Botrychium simplex, Calamagrostis breweri, C. canadensis, Carex aquatilis, C. aurea, C. capitata, C. fissuricola, C. hassei, C. luzulifolia, C. praeceptorium, C. scopulorum, C. senta, C. subnigricans, Castilleja lemmonii, C. nana, Cirsium tioganum, Deschampsia cespitosa, Eleocharis quinqueflora, Gentiana newberryi, Ivesia campestris, Juncus arcticus, J. mertensianus, J. orthophyllus, Kalmia polifolia, Kobresia myosuroides, Lewisia, Lupinus lyallii, Mimulus primuloides, Muhlenbergia filiformis, Oreostemma alpigenum, Oxyria digyna, Pedicularis attollens, Penstemon heterodoxus, Perideridia parishii, Phleum alpinum, Pinus albicaulis, P. contorta, Polygonum douglasii, Potentilla diversifolia, P. drummondii, Primula fragrans, P. tetrandra, Ptilagrostis kingii, Ranunculus alismifolius, Salix orestera, Senecio scorzonella, S. triangularis, Sibbaldia procumbens, Stellaria longipes, Thalictrum alpinum, Trichophorum clementis, T. pumilum, Trifloium monanthum, Trisetum spicatum, Tsuga, Vaccinium cespitosum, Veronica wormskjoldii.*

Gentianopsis simplex (**A. Gray**) **Iltis** is an annual found in montane bogs, fens, marshes, meadows, seeps, springs (cold and thermal), and along the margins of ponds and streams at elevations from 1200 to 3400 m. The plants often occur on coarse substrates including boulders and cobble as well as on sand and sandy silt loam. This species is regarded as a reliable indicator of the *Pinus contorta* forest zone. Field observations found that neither bumblebees (Hymenoptera: Apidae: *Bombus*) nor other typical pollinating insects visited the flowers (as they did with a nearby *Gentiana* species), but that the flowers contained thrips (Insecta: Thysanoptera). The possibility of self-pollination should be investigated in this species. Other ecological information is unavailable. **Reported associates:** *Aconitum columbianum, Agrostis scabra, Alnus rugosa, Boykinia major, Carex breweri, C. rostrata, Castilleja pilosa, Centaurea, Cornus sericea, Deschampsia, Epilobium, Eucephalus engelmannii, Geum macrophyllum, Glyceria, Helenium, Horkelia, Hosackia oblongifolia, Hypericum anagalloides, H. scouleri, Juncus xiphioides, Lupinus latifolius, Madia bolanderi, Mimulus, Myrica hartwegii, Panicum acuminatum, Picea engelmannii, Pinus contorta, P. jeffreyi, P. monticola, Pterospora andromedea, Ranunculus alismifolius, R. cymbalaria, Ribes inerme, Salix, Saxifraga, Scirpus microcarpus, Senecio triangularis, Solidago canadensis, S. simplex, Spiranthes romanzoffiana, Symphyotrichum, Trifolium wormskjoldii, Trisetum spicatum, Veronica americana, V. serpyllifolia, Viola adunca.*

Gentianopsis thermalis (**Kuntze**) **Iltis** is an annual, which inhabits subalpine bogs, depressions, ditches, flats, hummocks, meadows, roadsides, seeps, slopes, swamps, thickets, tundra, and the margins of lakes, rivers, and streams at elevations from 1475 to 3597 m. The plants usually occur in open sites on peat and on substrates described as alluvium, clay, clay loam, colluvial loamy cryboralf, granite, gravel, loam, loamy sand, loomis soil, rock, silt, silty loam, or stones. The specific epithet ("*thermalis*") reflects their common occurrence in geyser basins near hot springs. The plants grow in other disturbed sites such as earthflows. The flowers are pollinated by bumblebees (Hymenoptera: Apidae) mainly *Bombus appositus* or *B. balteatus* and to a lesser extent by *B. flavifrons* or *B. occidentalis*. Other details on the ecology of this species are sparse. **Reported associates:** *Abies lasiocarpa, Achillea millefolium, Aconitum columbianum, Agrostis humilis, A. stolonifera, Antennaria corymbosa, Aquilegia caerulea, Arnica cordifolia, A. mollis, Betula nana, Bistorta vivipara, Bromus ciliatus, Calamagrostis canadensis, Caltha leptosepala, Cardamine cordifolia, Carex aquatilis, C. capillaris, C. disperma, C. interior, C. jonesii, C. nebrascensis, C. neurophora, C. praegracilis, C. simulata, C. utriculata, Castilleja sulphurea, Chamerion angustifolium, Cicuta maculata, Cirsium, Conioselinum, Corydalis caseana, Dasiphora floribunda, Delphinium barbeyi, D. nelsoni, Deschampsia cespitosa, Eleocharis quinqueflora, Elymus trachycaulus, Epilobium ciliatum, E. saximontanum, Equisetum arvense, Erigeron coulteri, E. lonchophyllus, E. peregrinus, Eucephalus engelmannii, Festuca arizonica, F.*

ovina, F. thurberi, Fragaria virginiana, Frasera speciosa, Galium triflorum, Gaultheria humifusa, Gentiana calycosa, Geranium richardsonii, Glyceria striata, Haplopappus croceus, H. parryi, Helenium hoopesii, Helianthella quinquenervis, Heracleum lanatum, Hydrophyllum capitatum, H. fendleri, Hymenoxys grandiflora, Ipomopsis aggregata, Iris missouriensis, Juncus arcticus, J. balticus, J. drummondii, J. tracyi, Kobresia myosuroides, K. simpliciuscula, Lathyrus leucanthus, Ligusticum porteri, L. tenuifolium, Listera cordata, Lomatogonium rotatum, Lonicera involucrata, Luzula parviflora, Lysichiton americanus, Menyanthes trifoliata, Mertensia ciliata, Micranthes odontoloma, Mitella pentandra, Muhlenbergia richardsonis, Orthilia secunda, Oxypolis fendleri, Packera streptanthifolia, Parnassia parviflora, Pedicularis groenlandica, Phacelia leucophylla, Phleum alpinum, P. pratense, Picea engelmannii, Platanthera dilatata, P. sparsiflora, Poa reflexa, Potentilla, Primula cusickii, P. egaliksensis, P. incana, Psilochenia runcinata, Ptilagrostis porteri, Rhodiola rhodantha, Salix bebbiana, S. candida, S. exigua, S. monticola, S. planifolia, S. serissima, S. wolfii, Senecio bigelovii, S. crassulus, S. fremontii, S. serra, S. triangularis, Sidalcea candida, Sisyrinchium pallidum, Solidago spathulata, Spiranthes romanzoffiana, Streptopus amplexifolius, Swertia perennis, Symphyotrichum foliaceum, S. spathulatum, Taraxacum officinale, Thalictrum alpinum, Trifolium hybridum, T. repens, Triglochin maritimum, T. palustre, Trisetum spicatum, Trollius laxus, Vaccinium myrtillus, V. scoparium, Veratrum californicum, Veronica wormskjoldii, Vicia americana, Viguiera multiflora, Viola, Wyethia amplexicaulis.

Gentianopsis virgata (**Raf.**) **Holub** is an annual or biennial, which grows on beaches, in depressions, bogs, ditches, fens, meadows, marshes, prairies, seeps, shores of lakes and rivers, and along stream margins. The plants tolerate full sun but very little shade. The substrates are alkaline (pH: 7.0–8.0) and comprise clay, cobble, gravel, marl, muck, peat, rock, or sand. They often occur in areas with fluctuating water levels such as tidal zones (e.g., St. Lawrence River shores) or Great Lakes shorelines. Although regarded as freshwater inhabitants, the plants can tolerate slightly brackish conditions in the short term. The flowers behave much like those of *G. crinita*. They open only in full sunlight and close in darkness or when submerged by tides. The flowers produce nectar to attract various insects, but are pollinated principally by bumblebees (e.g., Hymenoptera: Apidae: *Bombus terricola*), which can become trapped in flowers when they close at night. The seeds are papillose and are dispersed locally by the water; however, they sink once they become thoroughly wet and probably are transported over greater distances in mud that adheres to the feet of waterbirds. Successful germination has been obtained by rinsing fresh seeds with 70% ethyl alcohol for 3 min, followed by a distilled water rinse, and then placing them on moist filter paper in a petri dish for 14 days at ambient temperatures. Then they are refrigerated (4°C) for 30 days prior to germination at ambient temperatures. Newly emerging rosettes are intolerant of competition with other plants. Studies on the population biology and genetics of this species would be useful, given that many populations have indicated declines in recent years. **Reported associates:** *Agalinis purpurea, A. tenuifolia, Agrostis stolonifera, Allium cernuum, Amorpha fruticosa, Andropogon gerardii, Anemone canadensis, Anticlea elegans, Apocynum cannabinum, Aquilegia canadensis, Arctostaphylos uva-ursi, Arnoglossum plantagineum, Asclepias incarnata, A. sullivantii, Betula pumila, B. sandbergi, Bromus ciliatus, Calamagrostis canadensis, C. inexpansa, Calopogon tuberosus, Caltha palustris, Campanula aparinoides, C. rotundifolia, Cardamine bulbosa, Carex aquatilis, C. capillaris, C. crawei, C. garberi, C. granularis, C. haydenii, C. hystericina, C. interior, C. lanuginosa, C. leptalea, C. prairea, C. sartwellii, C. sterilis, C. stipata, C. stricta, C. tetanica, C. viridula, Castilleja sessiliflora, Chamaedaphne calyculata, Chara, Chelone glabra, Cicuta maculata, Cirsium muticum, Cladium mariscoides, Convolvulus sepium, Coreopsis palmata, Cornus sericea, Corylus, Cuscuta glomerata, Cyperus rivularis, Cypripedium candidum, C. parviflorum, Dasiphora floribunda, Doellingeria umbellata, Drosera rotundifolia, Eleocharis compressa, E. elliptica, E. erythropoda, E. rostellata, Elymus canadensis, Empetrum nigrum, Eriophorum angustifolium, E. leptophyllum, Epipactis helleborine, Equisetum arvense, E. pratense, E. scirpoides, Erigeron hyssopifolius, Eupatorium perfoliatum, Euthamia graminifolia, Eutrochium maculatum, Filipendula rubra, Frangula alnus, Galium labradoricum, Gentianopis crinita, Glyceria striata, Helenium autumnale, Helianthus grosseserratus, Hierchloe odorata, Hypericum kalmianum, Hypoxis hirsuta, Impatiens capensis, Iris, Juncus alpinoarticulatus, J. balticus, J. brachycephalus, Juniperus horizontalis, Kalmia polifolia, Larix laricina, Lathyrus palustris, Liatris ligulistylis, L. pycnostachya, L. spicata, Lilium philadelphicum, Linum, Liparis loeselii, Lithospermum incisum, Lobelia cardinalis, L. kalmii, L. nuttalli, L. siphilitica, Lycopus americanus, L. asper, L. rubellus, L. uniflorus, Lysimachia quadriflora, Lythrum salicaria, Melitotus, Maianthemum stellatum, Menyanthes trifoliata, Muhlenbergia glomerata, M. richardsonis, M. schreberi, Oxypolis rigidior, Parnassia glauca, Parthenocissus inserta, Pedicularis lanceolata, Phalaris arundinacea, Phlox glaberrima, Phragmites australis, Picea mariana, Pinguicula vulgaris, Platanthera hyperborea, Poa pratensis, Polygala senega, Potentilla anserina, Primula mistassinica, Pycnanthemum virginianum, Rhynchospora alba, R. capillacea, Ribes americanum, Rumex orbiculatus, Sabatia, Salix bebbiana, S. candida, S. discolor, S. exigua, Schoenoplectus acutus, S. pungens, S. tabernaemontani, Scleria verticillata, Scutellaria galericulata, Selaginella apoda, S. selaginoides, Senecio pseudaureus, Shepherdia, Silphium perfoliatum, S. terebinthinaceum, Solidago canadensis, S. gigantea, S. ohioensis, S. patula, S. riddelli, S. uliginosa, Sorghastrum nutans, Sparganium eurycarpum, Spartina pectinata, Spiranthes cernua, Stachys palustris, Symphyotrichum boreale, S. lateriflorum, S. novae-angliae, Thalictrum dasycarpum, Thelypteris palustris, Thuja occidentalis, Toxicodendron rydbergii, Trianthe glutinosa, Trichophorum alpinum, T. cespitosum, Trifolium pratense, Triglochin maritimum, T. palustre, Typha angustifolia, T. latifolia, Utricularia intermedia, U. minor, Vaccinium oxycoccos, Valeriana ciliata, V. edulis, Verbena hastata, Viola nephrophylla, V. pratincola, Zizia aurea.*

Use by wildlife: Flowers of *Gentianopsis crinita* are visited (and pollinated) by bumblebees (Insecta: Hymenoptera: Apidae: *Bombus fervidus, B. impatiens, B. perplexus, B. vagans*), which forage for nectar and pollen. *Bombus terricola* has been observed as a pollinator of *G. virgata*. Flowers of *G. thermalis* are visited (and pollinated) by *Bombus appositus, B. balteatus, B. flavifrons*, or *B. occidentalis*.

Economic importance: food: not edible; **medicinal:** *Gentianopsis crinita* was used by the Deleware and Rappahannock tribes as a medicine to purify the blood, and by the Deleware to relieve gastric ailments; **cultivation:** *Gentianopsis crinita* and *G. virgata* are cultivated as ornamental wild flowers; *G. holopetala* is grown as a rock garden alpine; **misc. products:** *Gentiana crinita* is planted for restoration of sand mined areas. *Gentianopsis detonsa* has been included in seed mixes for wetland restoration projects; **weeds:** *Gentianopsis* has been described as "weedy" in some wet meadows; **nonindigenous species:** none

Systematics: Most *Gentianopsis* species (tribe Gentianeae; subtribe Swertiinae) have at one time or another been placed in both *Gentiana* and *Gentianella* (as subgenus *Eublephis*); however, recent phylogenetic studies persuasively demonstrate that the genus should be maintained as distinct from either of those (Figure 5.40), especially *Gentiana*, which resolves within a different subtribe of the family (Figure 5.38). However, molecular phylogenetic studies have included only a few species of *Gentianopsis*; thus, interrelationships remain uncertain within the genus and its monophyly is yet to be demonstrated conclusively. Even the closest sister genus has not been determined with certainty, although *Obolaria* and *Pterygocalyx* have been implicated in studies using different representative species. Several *Gentianopsis* species appear to be closely related on the basis of their morphological similarity. *Gentianopsis detonsa* and *G. thermalis* have been synonymized by some authors. *Gentianopsis virgata* (*G. procera* of some authors) and *Gentianopsis crinita* arguably are closely related, as each is difficult to distinguish morphologically, and both are suspected to hybridize with one another. They sometimes are treated as subspecies. A close relationship has been suggested between *G. simplex* and the European *G. ciliata* on the basis of their similar seeds, which are tailed at each end. With only 24 species worldwide, a comprehensive phylogenetic study of the genus is feasible, and potentially would provide a much better perspective of relationships and species limits. The chromosomal base number of *Gentianopsis* is $x = 13$. *Gentianopsis crinita, G. detonsa, G. holopetala* and *G. virgata* (all $2n = 78$) are hexaploid.

Comments: *Gentiana crinita* and *G. virgata* occur in northern and eastern North America, *G. detonsa* (circumboreal) is distributed across northern North America and *G. holopetala*, *G. simplex* and *G. thermalis* (circumboreal) occur in the western United States.

References: Benedict, 1983; Chassot et al., 2001; Chen et al., 2005; Cooper & Sanderson, 1997; COSEWIC, 2004a; Costello, 1944; Costelloe, 1988; Farmer, 1978; Gillett, 1957; Hagen & Kadereit, 2002; Janssens & Johnson, 2001; Lammers, 2003; Lerner, 1997; Mellmann-Brown, 2004; Pyke, 1982; Saetersdal & Birks, 1997.

6. *Lomatogonium*

Marsh felwort

Etymology: from the Greek *lomato gono* meaning "fringed reproductive organ" in reference to the decurrent edge formed along the ovary margin by the stigma

Synonyms: *Pleurogyne*; *Pleurogynella*; *Swertia* (in part)

Distribution: global: Asia; Europe; North America; **North America:** northern

Diversity: global: 18 species; **North America:** 1 species

Indicators (USA): OBL: *Lomatogonium rotatum*

Habitat: freshwater, saline; palustrine; **pH:** 7.5–8.5; **depth:** <1 m; **life-form(s):** emergent herb

Key morphology: stems (to 40 cm) angular, erect, branched; leaves (to 4.5 cm) opposite, sessile, linear to lanceolate, somewhat fleshy; panicles racemose; flowers 5-merous, pedicelled (to 8 cm); sepals (to 2.7 cm) linear, alternating with and extending beyond the petals; corolla rotate; petals (to 2.5 cm) acute, pale blue, blue-lined, each with a pair of fringed nectaries at their base; stigmas triangular, decurrent to base of ovary; capsules (to 2.5 cm) ellipsoid; seeds (to 0.75 mm) smooth, numerous (to 100)

Life history: duration: annual (fruit/seeds); **asexual reproduction:** none; **pollination:** insect; **sexual condition:** hermaphroditic; **fruit:** capsules (common); **local dispersal:** seeds; **long-distance dispersal:** seeds

Imperilment: (1) *Lomatogonium rotatum* [G5]; S1 (ID, ME, MT, NB, UT, YT); S2 (AB, BC, SK, MB, WY); S3 (QC)

Ecology: general: *Lomatogonium* is found principally in China where the species grow in grasslands, on hillsides, in meadows, along streams and in alpine areas. Over a third of the species inhabit wet lakesides and streamsides. Only one wide-ranging species occurs in North America, where it is designated as OBL. Generally, the ecology of this genus is poorly known. The flowers are believed to be pollinated by insects, but few pollinators have been documented. The mechanisms of seed dispersal also remain largely unstudied.

Lomatogonium rotatum **(L.) Fr. ex Fernald** occurs in beaches, depressions, ditches, dunes, fens, floodplains, heaths, hummocks, lagoons, marshes, meadows, mires, mudflats, mudflows, muskeg, river bars (gavel, sand, or silt), roadsides, salt marshes, salt pans, seashores, swamps, and along the margins of lakes, rivers, and streams at elevations of up to 3048 m. The substrates typically are alkaline (pH: 7.5–8.5) and contain high levels of calcium (Ca^{2+}:

30–120 mg kg^{-1}) and a high amount of organic matter. The sediments include clay, coarse inorganic cobble, gravel, Leadville limestone, loamy silt, mica, Minturn-Belden formation, peat, sand, sandy clay loam, silt, silty sand, and schist. Some sites (e.g., edges of salt marshes) can be fairly saline, or at least exposed to temporarily saline conditions (e.g., by tides). The plants occur sometimes on talus and mudslides and inhabit sites from sea level (e.g., coastal maritime areas with fog banks) to subalpine elevations. They often grow in localities kept open by natural disturbances such as ice, salt spray, waves, and wind. Populations typically are small although the seeds can be locally abundant. The plants are influenced minimally by grazing birds (Aves). There is virtually no information regarding pollination or seed ecology for this species. Genetic studies are wanting and should be carried out to evaluate the breeding system of these plants. **Reported associates:** *Achillea millefolium*, *Agalinis neoscotica*, *Agrostis stolonifera*, *Andromeda polifolia*, *Anemone parviflora*, *Anthoxanthum hirtum*, *Astragalus diversifolius*, *A. leptaleus*, *Betula glandulosa*, *B. nana*, *Bistorta vivipara*, *Calamagrostis stricta*, *Campanula parryi*, *C. rotundifolia*, *Carex aquatilis*, *C. canescens*, *C. capillaris*, *C. lasiocarpa*, *C. magellanica*, *C. nebrascensis*, *C. pauciflora*, *C. pluriflora*, *C. recta*, *C. rostrata*, *C. rotundata*, *C. scirpoidea*, *C. simulata*, *C. utriculata*, *C. viridula*, *Castilleja elegans*, *C. sulphurea*, *Chrysanthemum arcticum*, *Comarum palustre*, *Conioselinum scopulorum*, *Dasiphora floribunda*, *Deschampsia cespitosa*, *D. flexuosa*, *Dryas*, *Elymus trachycaulus*, *Empetrum nigrum*, *Epilobium palustre*, *Equisetum fluviatile*, *E. palustre*, *E. variegatum*, *Eriophorum angustifolium*, *E. brachyantherum*, *E. gracile*, *Euphrasia canadensis*, *E. randii*, *Festuca arizonica*, *F. rubra*, *Fritillaria camschatcensis*, *Gentiana douglasiana*, *Gentianella amarella*, *Gentianopsis thermalis*, *Geum macrophyllum*, *Goodyera repens*, *Heracleum sphondylium*, *Honckenya peploides*, *Hordeum jubatum*, *Hylocomium splendens*, *Iris hookeri*, *I. setosa*, *Juncus alpinoarticulatus*, *J. arcticus*, *J. balticus*, *J. filiformis*, *Juniperus horizontalis*, *Kobresia myosuroides*, *K. simpliciuscula*, *Lathyrus japonicus*, *Ligusticum porteri*, *Luzula*, *Maianthemum stellatum*, *Mertensia maritima*, *Montia fontana*, *Muhlenbergia richardsonis*, *Oclemena nemoralis*, *Oxytropis*, *Parnassia kotzebuei*, *P. montana*, *P. palustris*, *P. parviflora*, *Pedicularis groenlandica*, *Picea glauca*, *P. mariana*, *Plantago juncoides*, *P. maritima*, *Platanthera huronensis*, *Poa arctica*, *P. pratensis*, *Polemonium pulcherrimum*, *Populus deltoides*, *P. tremuloides*, *Prenanthes trifoliata*, *Primula alcalina*, *P. cusickii*, *P. egaliksensis*, *P. incana*, *P. laurentiana*, *Psilochenia runcinata*, *Ptilagrostis porteri*, *Ptilium crista-castrensis*, *Puccinellia*, *Ranunculus pensylvanicus*, *Rhodiola rosea*, *Rhododendron groenlandicum*, *Rosa acicularis*, *Rubus chamaemorus*, *Sagina nodosa*, *Salix arctica*, *S. bebbiana*, *S. brachycarpa*, *S. candida*, *S. exigua*, *S. geyeriana*, *S. interior*, *S. ligulifolia*, *S. monticola*, *S. planifolia*, *S. pseudomyrsinites*, *S. serissima*, *S. wolfii*, *Senecio congestus*, *Shepherdia canadensis*, *Sisyrinchium idahoense*, *Solidago*

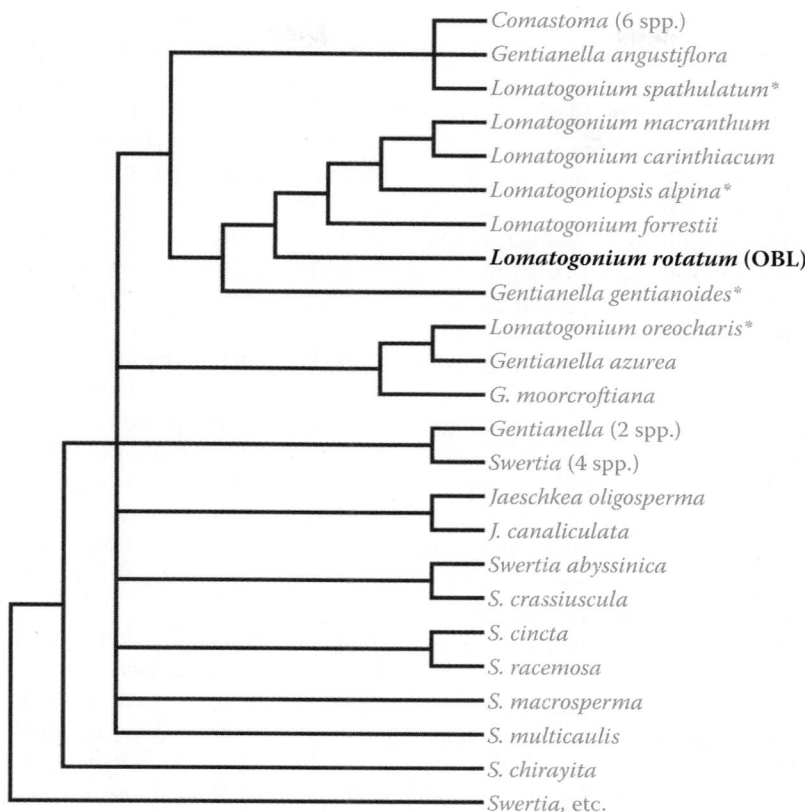

FIGURE 5.41 Relationships of *Lomatogonium* species as indicated by phylogenetic analysis of nrITS sequence data. These results indicate taxonomic problems with the circumscriptions for several genera of Gentianaceae, which are highly polyphyletic in such analyses. Several surveyed species of *Lomatogonium* resolve within a clade; however, others resolve with members of different genera and one genus (*Lomatogoniopsis*) is embedded within the group (anomalies are marked by asterisks). The OBL *Lomatogonium rotatum* (in bold) resolves near the base of this clade. Additional sampling of taxa and molecular loci will be necessary to resolve the limits of these problematic genera, but it is likely that several generic circumscriptions will change considerably as additional data are collected and analyzed. (Adapted from von Hagen, K.B. & Kadereit, J.W., *Syst. Bot.*, 27, 548–572, 2002.)

bicolor, S. sempervirens, Spiranthes romanzoffiana, Suaeda calceoliformis, Symphyotrichum boreale, S. novi-belgii, S. spathulatum, Tephroseris palustris, Thalictrum alpinum, Thermopsis montana, Triglochin maritimum, T. palustre, Trisetum spicatum, Typha latifolia, Vaccinium uliginosum, V. vitis-idea, Viburnum edule, Viola glabella, V. septentrionalis.

Use by wildlife: *Lomatogonium rotatum* plants are grazed minimally by snow geese (Aves: Anatidae: *Chen caerulescens*). The flowers are visited by bumblebees (Hymenoptera: Apidae: *Bombus*), several species of butterflies (Lepidoptera) and flies (Diptera: Syrphidae).

Economic importance: food: not edible; **medicinal:** *Lomatogonium rotatum* is used in Tibetan folk medicine. The plants contain c-glycosyl flavones, secoiridoids, xanthones and about 2.9% organic acids; **cultivation:** *Lomatogonium rotatum* is cultivated; **misc. products:** none; **weeds:** none; **nonindigenous species:** none

Systematics: *Lomatogonium* is assigned to subtribe Swertiinae of tribe Gentianeae (Gentianaceae). Several phylogenetic studies using molecular data have demonstrated that the genus does associate with other members of this subtribe; however, it is polyphyletic, with some species resolving with *Comastoma, Gentianella* (those species with two petal nectaries) and *Lomatogoniopsis* (Figure 5.41). The taxonomy of *Lomatogonium* and related genera will remain unsettled until additional taxa and molecular loci are surveyed and analyzed. The base chromosome number of *Lomatogonium* is uncertain, but some authors propose that it is $x = 8$. *Lomatogonium rotatum* ($2n = 10, 16$) presumably is diploid, but contains different cytotypes. Two varieties have been recognized, with var. *rotatum* (widespread in Eurasia and North America), and var. *floribundum* (restricted to Asia). Further genetic studies should be conducted across the entire range of this widespread, boreal taxon to better ascertain the extent of geographical diversification.

Comments: *Lomatogonium rotatum* (circumboreal) is distributed across northern North America with southward extensions in the Rocky Mountains.

References: Aiken et al., 2007; Chang et al., 2001; Cooper & Sanderson, 1997; Handa et al., 2002; Jensen & Schripsema, 2002.

7. Sabatia

Marsh-pink, rose-gentian; sabatie

Etymology: after Liberato Sabbati (1714–1779)

Synonyms: *Chironia* (in part); *Lapithea*; *Pleienta*; *Sabbatia*

Distribution: global: New World; **North America:** eastern

Diversity: global: 20 species; **North America:** 18 species

Indicators (USA): OBL: *Sabatia calycina*, *S. decandra*, *S. difformis*, *S. dodecandra*, *S. gentianoides*, *S. kennedyana*, *S. stellaris*; **FACW:** *S. gentianoides*, *S. stellaris*

Habitat: saline (coastal), freshwater; palustrine; **pH:** 5.4–8.6; **depth:** <1 m; **life-form(s):** emergent herb

Key morphology: stems (to 105 cm) erect, branching, partially hollow; basal rosettes present or absent; cauline leaves (to 10 cm) opposite, entire, sessile or clasping, sometimes fleshy; inflorescence a subcapitate to loosely spreading cyme, the bracts scale-like or leaf-like; flowers sessile or long pedicelled (to 15 cm); corolla rotate, 5- to 14-merous; petal lobes (to 3 cm) white or pink to deep rose, a yellow blotch (with or without a reddish margin) present or absent at the base; stamens 5–12, twisting or circinnately coiled after pollen release; capsules (to 14 mm) globose to cylindrical, many-seeded (to 2400)

Life history: duration: annual (fruit/seeds); perennial (rhizomes, stolons); **asexual reproduction:** rhizomes, stolons; **pollination:** insect, self; **sexual condition:** hermaphroditic; **fruit:** capsules (common); **local dispersal:** rhizomes, stolons, seeds (air, water); **long-distance dispersal:** seeds

Imperilment: (1) *Sabatia calycina* [G5]; S3 (NC, VA); (2) *S. decandra* [G4, G5]; S1 (SC); S2 (MS); (3) *S. difformis* [G4]; S1 (DE, MD, VA); (4) *S. dodecandra* [G4]; SX (NY); SH (CT); S1 (DE); S2 (LA, NJ); S3 (AL, MD, NC, VA); (5) *S. gentianoides* [G4]; S1 (AR); S3 (NC); (6) *S. kennedyana* [G3]; S1 (NC, <u>NS</u>, RI, SC); S3 (MA); (7) *S. stellaris* [G5]; SX (PA); S1 (CT, MA, RI); S2 (NY); S3 (NC)

Ecology: general: Most *Sabatia* species occur in wetlands, either facultatively or obligately. Several species are halophytes. The genus includes annual and perennial species. All *Sabatia* are self-compatible; however, protandry promotes outcrossing in most species. All pollen is removed within 24 h either by insects or wind, and the anthers often drop from the flowers several days before the stigmas are receptive. Pollinators are mainly bumblebees (Hymenoptera: Apidae: *Bombus*) in the larger flowers or sweat bees (Hymenoptera: Halictidae) in the smaller flowers. The few existing studies on seed germination have indicated physiological dormancy, with a period of cold stratification necessary to achieve optimal germination. Seeds of at least some species are buoyant and probably are dispersed locally by water. Long-distance transport of the small seeds by adherence to waterfowl (e.g., in mud on their feet) is also likely to occur.

Sabatia calycina (**Lam.**) **A. Heller** is a perennial, which inhabits depressions, flatwoods, floodplains, marshes, swamps, and the margins of ponds, rivers, and streams. Its substrates are acidic (pH: 5.6–6.5) and consist of muck or silt with relatively high levels of nutrients. The plants can tolerate full shade and also grow well in saline conditions (they are classified as halophytes). They often appear on shores during drawdown conditions. These plants are short-lived perennials that are capable of flowering in their first year. The flowers are not protandrous and pollen is transferred mechanically to the receptive stigmas for autogamy to occur. Exclusion of pollinators does not influence seed set, which can achieve 1000 seeds per capsule within 3–4 months after selfing occurs. There is no specific information on the seed ecology of this species. The plants spread vegetatively by means of slender to thick rhizomes (to 10 cm) and can be propagated from stem cuttings. **Reported associates:** *Acer rubrum*, *Ampelaster carolinianus*, *Ampelopsis arborea*, *Amsonia rigida*, *Aristolochia serpentaria*, *Arundinaria gigantea*, *Asclepias perennis*, *Baccharis glomeruliflora*, *Bacopa monnieri*, *Berchemia scandens*, *Bidens frondosus*, *B. mitis*, *Boehmeria cylindrica*, *Brunnichia cirrhosa*, *B. ovata*, *Campsis radicans*, *Carex amphibola*, *C. cherokeensis*, *C. decomposita*, *C. granularis*, *C. stipata*, *Carpinus caroliniana*, *Carya glabra*, *Celtis laevigata*, *Cephalanthus occidentalis*, *Chasmanthium latifolium*, *C. laxum*, *C. nitidum*, *C. sessiliflorum*, *Clematis crispa*, *Commelina diffusa*, *Conoclinium coelestinum*, *Cornus foemina*, *Crataegus opaca*, *C. viridis*, *Crinum americanum*, *Cyperus erythrorhizos*, *C. flavicomus*, *C. polystachyos*, *Decumaria barbara*, *Desmodium*, *Dichanthelium dichotomum*, *Dichondra carolinensis*, *Digitaria sanguinalis*, *Diodia virginiana*, *Diospyros virginiana*, *Echinochloa muricata*, *Echinodorus cordifolius*, *E. radicans*, *Eleocharis obtusa*, *Elephantopus carolinianus*, *Eragrostis hypnoides*, *Erechtites hieraciifolia*, *Euonymus americanus*, *Eupatorium altissimum*, *Fimbristylis autumnalis*, *F. perpusilla*, *Fraxinus caroliniana*, *F. pennsylvanica*, *Gratiola aurea*, *G. virginiana*, *Helenium flexuosum*, *Hibiscus*, *Hydrocotyle verticillata*, *Hygrophila lacustris*, *Hymenocallis rotata*, *Hypericum hypericoides*, *H. mutilum*, *Hypoxis curtissii*, *H. leptocarpa*, *Hyptis alata*, *Ilex decidua*, *I. opaca*, *Itea virginica*, *Juncus repens*, *Juniperus virginiana*, *Justicia ovata*, *Kosteletzkya pentacarpos*, *Leitneria floridana*, *Lindernia dubia*, *Lipocarpha micrantha*, *L. maculata*, *Liquidambar styraciflua*, *Ludwigia decurrens*, *L. palustris*, *L. repens*, *Lycopus rubellus*, *L. virginicus*, *Lysimachia radicans*, *Magnolia grandiflora*, *M. virginiana*, *Malaxis floridana*, *Micranthemum umbrosum*, *Mikania scandens*, *Mitreola petiolata*, *Muhlenbergia schreberi*, *Myrica cerifera*, *Nyssa biflora*, *N. sylvatica*, *Oldenlandia boscii*, *Onoclea sensibilis*, *Oplismenus setarius*, *Osmunda regalis*, *Panicum dichotomiflorum*, *P. rigidulum*, *P. verrucosum*, *Parthenocissus quinquefolia*, *Paspalum dissectum*, *P. fluitans*, *Persea palustris*, *Persicaria hydropiperoides*, *P. pensylvanica*, *Phanopyrum gymnocarpon*, *Phyla nodiflora*, *Pinus glabra*, *P. taeda*, *Pluchea camphorata*, *Polypodium polypodioides*, *Ptilimnium capillaceum*, *Pyrrhopappus*, *Quercus laurifolia*, *Q. michauxii*, *Q. nigra*, *Q. pagoda*, *Q. phellos*, *Rhynchospora corniculata*, *Rotala ramosior*, *Rubus trivialis*, *Ruellia caroliniensis*, *Sabal minor*, *Sabatia kennedyana*, *S. stellaris*, *Saccharum giganteum*, *Salix nigra*, *Sambucus nigra*, *Samolus parviflorus*, *Sanicula*, *Saururus cernuus*, *Scirpus cyperinus*, *Senecio glabellus*, *Smilax bona-nox*, *S. glauca*,

S. rotundifolia, S. tamnoides, Solidago gigantea, Spiranthes, Styrax americanus, Taxodium distichum, Thelypteris palustris, Toxicodendron radicans, Triadenum walteri, Ulmus alata, U. americana, Verbena, Vernonia gigantea, Viburnum dentatum, V. obovatum, Viola esculenta, Vitis aestivalis, V. rotundifolia, Woodwardia areolata.

***Sabatia decandra* (Walter) R.M. Harper** is a biennial, which occurs in barrens, bogs, cypress bays, cypress domes, depressions, ditches, flats, flatwoods, meadows, prairies, roadsides, savannas, seeps, swales, swamps, and along the margins of ponds and rivers at elevations of up to 17 m. The plants thrive in high light exposures, often occurring in standing water (to 8 dm) within seasonally to semipermanently flooded sites. Their substrates are described as calcareous and include alfic haplaquods, arenic glossaqualfs, muck, Myakka (aeric haplaquods), Pamlico-Dorovan (terric medisaprists), Pomona (ultic haplaquods), Rains (typic paleaquults), sand, and sandy peat. Flowers are produced from May to November. Other details on the life history of this species are unavailable. **Reported associates:** *Acer rubrum, Agalinis linifolia, Aletris lutea, Andropogon arctatus, Amphicarpum muhlenbergianum, Aristida purpurascens, A. stricta, Arnoglossum ovatum, Asclepias lanceolata, Asplenium platyneuron, Balduina uniflora, Bartonia verna, Bigelowia nudata, Boehmeria cylindrica, Calamovilfa curtissii, Carex glaucescens, Carphephorus pseudoliatris, Cassytha filiformis, Centella asiatica, Cephalanthus occidentalis, Diospyros virginiana, Chaptalia tomentosa, Cirsium horridulum, Cladium jamaicense, C. mariscoides, Clethra alnifolia, Cliftonia monophylla, Coleataenia tenera, Coreopsis gladiata, C. nudata, Ctenium aromaticum, Cuphea carthagenensis, Cyperus, Cyrilla racemiflora, Dichanthelium acuminatum, D. scoparium, D. sphaerocarpon, Drosera capillaris, D. tracyi, Eleocharis tuberculosa, Epidendrum magnoliae, Erigeron vernus, Eriocaulon compressum, E. decangulare, Euphorbia inundata, Fimbristylis, Fuirena scirpoidea, Gymnadeniopsis nivea, Helenium vernale, Helianthus heterophyllus, H. radula, Hymenocallis palmeri, Hypericum brachyphyllum, H. chapmanii, H. fasciculatum, H. myrtifolium, Ilex coriacea, I. glabra, I. myrtifolia, Iris hexagona, I. prismatica, I. tridentata, I. virginica, Juncus polycephalus, J. repens, Lachnanthes caroliniana, Leersia, Liatris spicata, Lilium catesbaei, Liquidambar styraciflua, Litsea aestivalis, Lobelia floridana, Lophiola aurea, Ludwigia alternifolia, L. decurrens, L. linifolia, L. octovalvis, L. pilosa, L. simpsonii, L. sphaerocarpa, L. suffruticosa, L. virgata, Luziola fluitans, Lycopodiella alopecuroides, Lyonia lucida, Magnolia virginiana, Mnesithea rugosa, Muhlenbergia expansa, Myrica heterophylla, Nyssa biflora, Osmanthus americanus, Oxypolis filiformis, O. greenmanii, Panicum dichotomiflorum, P. hemitomon, P. verrucosum, P. virgatum, Paspalum monostachyum, Persea palustris, Pieris phillyreifolia, Pinguicula lutea, P. planifolia, Pinus elliottii, P. palustris, Pleea tenuifolia, Pogonia ophioglossoides, Polygala cruciata, P. cymosa, P. ramosa, Proserpinaca pectinata, Rhexia alifanus, R. lutea, R. mariana, R. virginica, Rhynchospora careyana, R.*

cephalantha, R. chapmanii, R. ciliaris, R. colorata, R. corniculata, R. divergens, R. filifolia, R. harperi, R. inundata, R. latifolia, R. macra, R. microcarpa, R. oligantha, R. plumosa, R. stenophylla, R. tracyi, Rudbeckia mohrii, Sabatia brevifolia, S. difformis, S. macrophylla, Sarracenia flava, S. leucophylla, S. psittacina, Schizachyrium rhizomatum, Scleria baldwinii, S. triglomerata, Smilax laurifolia, S. walteri, Solidago stricta, Stillingia aquatica, Styrax americanus, Symphyotrichum dumosum, S. tenuifolium, Taxodium ascendens, Thelypteris kunthii, Tillandsia bartramii, Triantha racemosa, Woodwardia virginica, Xyris ambigua, X. baldwiniana, X. serotina, X. stricta.

***Sabatia difformis* (L.) Druce** is a perennial, which occurs in barrens, bogs, depressions, ditches, dune swales, flatwoods, floodplains, marshes, meadows, prairies, savannas, on shores, and along the margins of rivers and streams. The substrates typically are oligotrophic gravel and sand. Self-pollination in these self-compatible flowers is restricted by the curvature of stamens away from the stigmas after the pollen has been shed, and by protandry, with the stigmas remaining unreceptive until 2 days after pollen is shed. The outermost flowers of the inflorescence mature last and frequently are staminate. The white petals turn to a saffron or orange color upon drying. The seeds develop (about 165 per capsule) within 3 months of pollination. The plants reproduce vegetatively from a branching, twisted rhizome. The aerial stems die back in the fall and small rosettes develop on the rhizome, which eventually will expand with cauline leaves during the growing season. The mechanism of seed dispersal is unknown; however, plants can colonize new, open areas quite rapidly. **Reported associates:** *Acer rubrum, Agalinis purpurea, Agrostis perennans, A. scabra, Amphicarpum purshii, Andropogon glomeratus, Arundinaria tecta, Asclepias lanceolata, A. rubra, Bartonia paniculata, Boltonia asteroides, Calamagrostis pickeringii, Calamovilfa brevipilis, Calopogon tuberosus, Carex exilis, C. glaucescens, C. livida, C. striata, Centella erecta, Chamaecyparis thyoides, Cladium mariscoides, Coreopsis falcata, C. rosea, Ctenium aromaticum, Cuscuta compacta, Cyperus dentatus, Cyrilla racemiflora, Danthonia epilis, Dichanthelium acuminatum, D. dichotomum, D. ensifolium, D. hirstii, D. leucothrix, D. scabriusculum, D. spretum, D. wrightianum, Drosera filiformis, D. intermedia, Erigeron vernus, Eriocaulon aquaticum, E. compressum, E. decangulare, Eriophorum virginicum, Eryngium aquaticum, Eupatorium leptophyllum, E. leucolepis, E. mohrii, Euthamia caroliniana, Fimbristylis autumnalis, Gaylussacia dumosa, Gentiana autumnalis, G. catesbaei, Gratiola aurea, G. brevifolia, Helianthus radula, Hypericum adpressum, H. canadense, H. crux-andreae, H. denticulatum, H. galioides, H. mutilum, Ilex glabra, Iris tridentata, Iva microcephala, Juncus caesariensis, J. canadensis, J. militaris, J. pelocarpus, Kalmia angustifolia, Lachnanthes caroliana, Lachnocaulon anceps, Leucothoe racemosa, Lilium catesbaei, Liquidambar styraciflua, Lobelia canbyi, L. nuttallii, Lophiola aurea, Ludwigia hirtella, Lycopodiella alopecuroides, L. caroliniana, Magnolia virginiana, Mnesithea rugosa, Muhlenbergia torreyana, M. uniflora, Myrica heterophylla, M. pensylvanica,*

Narthecium americanum, Nyssa biflora, Oclemena nemoralis, Ophioglossum vulgatum, Osmundastrum cinnamomeum, Panicum hemitomon, P. hirstii, P. verrucosum, Pinus rigida, P. taeda, Platanthera cristata, Pogonia ophioglossoides, Polygala cruciata, P. cymosa, Proserpinaca pectinata, Rhexia aristosa, R. nashii, R. virginica, Rhynchospora alba, R. capitellata, R. cephalantha, R. chalarocephala, R. chapmanii, R. elliottii, R. filifolia, R. fusca, R. gracilenta, R. nitens, R. oligantha, R. rariflora, Sabatia kennedyana, Saccharum baldwinii, Sagittaria teres, Sarracenia flava, S. minor, S. purpurea, Schizachyrium scoparium, Schizaea pusilla, Scleria ciliata, S. reticularis, Sclerolepis uniflora, Smilax laurifolia, S. pseudochina, Solidago fistulosa, Sphagnum bartlettianum, S. magellanicum, S. portoricense, S. pulchrum, Spiranthes vernalis, Symphyotrichum novi-belgii, Taxodium ascendens, Triadenum virginicum, Triantha racemosa, Utricularia cornuta, U. juncea, U. subulata, Vaccinium macrocarpon, Viburnum nudum, Viola lanceolata, Xyris ambigua, X. difformis, X. platylepis, X. smalliana.

***Sabatia dodecandra* (L.) Britton, Sterns, & Poggenb.** is a perennial, which grows in bogs, depressions, ditches, flatwoods, marshes, meadows, roadsides, savannas, seeps, slopes, swamps, and along the margins of ponds, rivers, and streams at elevations of up to 37 m. The sites range from oligohaline (e.g., salinity = 0.5–5 ppt) to saline or tidal freshwater conditions, with the plants occurring often in shallow standing water. The habitats usually are open, with somewhat acidic (e.g., pH: 6.6) or calcareous substrates characterized as limestone, loam, marl, peat, sand, sandy loam, silt, and Surrency (umbric paleaquults). The self-compatible flowers are protandrous (var. *dodecandra*), with the stigmas not becoming receptive until a day after pollen is shed. In *S. dodecandra* var. *foliosa*, the anthers dehisce 3 days prior to stigma receptivity; however, pollen is not shed completely until after the stigmas are receptive. The seeds take from 2 to 4 months to develop and can number up to 500–2400 per capsule, depending on the variety. Some germination will occur without any pretreatment, but the seeds eventually will acquire physiological dormancy. Seeds that have been refrigerated (~4°C) for 70–100 days exhibit more than twice the percent germination, and also germinate more rapidly (within 25–49 days), than unrefrigerated seeds (48–69 days). Vegetative reproduction is facilitated by a branching rhizome, which also can produce stolons (to 10 cm) that bear small rosettes of leaves. **Reported associates:** *Acer rubrum, Asclepias lanceolata, Atriplex prostrata, Axonopus furcatus, Bacopa caroliniana, Borrichia frutescens, Buchnera floridana, Carex debilis, C. lonchocarpa, Centella erecta, Chamaecyparis henryae, Cladium mariscoides, Clethra alnifolia, Crataegus, Cyperus haspan, C. virens, Cyrilla racemiflora, Dichanthelium, Diodia, Distichlis spicata, Eleocharis fallax, E. rostellata, Eriocaulon decangulare, Eryngium aquaticum, E. yuccifolium, Eupatorium mohrii, Festuca rubra, Hibiscus moscheutos, Hyptis alata, Ipomoea sagittata, Juncus polycephalus, J. roemerianus, Juniperus virginiana, Lechea minor, Leersia hexandra, Limonium carolinianum, Ludwigia alata, L. pilosa, L. sphaerocarpa, Lyonia lucida, Lythrum alatum, L. lineare,* *Melanthera nivea, Mikania scandens, Mitreola petiolata, Myrica cerifera, Myriophyllum, Nyssa biflora, Oenothera filipes, Panicum virgatum, Persicaria, Phyla nodiflora, Polygala cymosa, Pontederia cordata, Proserpinaca palustris, Rhexia alifanus, R. petiolata, R. virginica, Rhynchospora crinipes, R. fascicularis, R. glomerata, Sabatia stellaris, Sagittaria lancifolia, Salix nigra, Schoenoplectus pungens, Scleria oligantha, Setaria geniculata, Silphium asteriscus, Smilax walteri, Spartina patens, S. spartinae, Symphyotrichum tenuifolium, Taxodium ascendens, Teucrium canadense, Toxicodendron radicans, Verbena urticifolia, Vigna luteola.*

***Sabatia gentianoides* Elliott** is an annual, which inhabits barrens, bogs, ditches, flatwoods, roadsides, savannas, and seepage slopes at elevations of up to 67 m. Exposures often are in full sun. Substrates typically are acidic and have been described as clay, Pamlico (terric medisaprists), sand, silt, and rock. The flowers of this species are protandrous with the stigma receptivity occurring 2 days after the pollen has been shed. The seeds (up to 1200 per capsule) develop after 2–3 months following pollination. Unlike most other *Sabatia* species, the plants arise from a basal rosette of leaves. There is little additional ecological information reported for this species. **Reported associates:** *Aletris aurea, A. farinosa, Aristida stricta, Asclepias, Calopogon tuberosus, Carex alata, Coreopsis linifolia, Cyrilla racemiflora, Drosera brevifolia, Eriocaulon decangulare, Eryngium integrifolium, Helianthus angustifolius, Hydrocotyle verticillata, Ilex, Liatris pycnostachya, Magnolia virginiana, Myrica, Nyssa biflora, Pinguicula pumila, Pinus elliottii, P. palustris, Platanthera nivea, Polygala cruciata, P. nana, P. ramosa, Rhexia lutea, R. petiolata, Rhynchospora, Sarracenia alata, Utricularia, Xyris.*

***Sabatia kennedyana* Fernald** is a perennial found in depressions, pond shores, and along the margins of rivers. The plants can withstand shallow inundation but their mortality increases as the water levels deepen. The habitats tend to be seasonally flooded sites, with gently sloping, oligotrophic cobble, gravel, peat, or sand substrates, which often contain low amounts of organic matter. Localities typically are rich in annual species and lack established shrubs or aggressive species. The plants typically produce two to three flowers, which are pollinated by sweat bees (Hymenoptera: Halictidae: *Augochloropsis metallica*), honey bees (Hymenoptera: Apidae: *Apis mellifera*), banded longhorn beetles (Coleoptera: Cerambycidae), and shining leaf chafer beetles (Coleoptera: Scarabaeidae: Rutelinae). The flowers are self-compatible but are slightly protandrous, with anther dehiscence occurring 3 days prior to stigma receptivity; however, some pollen is not shed completely until after the stigmas are receptive. Seed set (up to 1300 per capsule) typically ranges from 70% to 85% in natural populations. The seeds take about 3 months to develop. About 20% germination rate has been obtained for seeds incubated under a 20°C/30°C temperature regime after receiving 9 months of cold stratification (at 4°C) in wet sand. Optimal seedling germination and establishment occurs in wet years (when shores are exposed by receding water); however, optimal growth occurs during dry years. Overall, the seedlings exhibit relatively low growth rates compared to other wetland species and the plants are poor competitors with other vegetation. These are short-term perennials (living 3–4 years) that usually flower in their third year and die soon afterwards. A seed bank (observed to contain more than 150 seeds/m²) must be maintained at sites for populations to persist. Vegetative reproduction occurs by means of slender rhizomes (to 12 cm), which fragment easily. Population genetic studies using random amplified polymorphic DNA (RAPD) markers have shown there to be high genetic (haplotype) diversity within populations, low differentiation among populations, and high levels of migration. Habitats for this species are threatened by damage from off-the-road recreational vehicles. **Reported associates:** *Agalinis purpurea, Agrostis perennans, A. scabra, Artemisia campestris, Bartonia paniculata, Bidens discoideus, B. frondosus, Calamagrostis canadensis, Carex lenticularis, C. striata, Cladium mariscoides, Coreopsis rosea, Cyperus dentatus, C. strigosus, Dichanthelium rigidulum, Drosera filiformis, D. intermedia, Echinodorus tenellus, Eleocharis acicularis, E. melanocarpa, E. tenuis, E. tricostata, Eragrostis hypnoides, Eriocaulon aquaticum, Eupatorium dubium, E. leucolepis, Euthamia caroliniana, E. galetorum, E. tenuifolia, Fimbristylis autumnalis, F. perpusilla, Fuirena pumila, Gnaphalium uliginosum, Gratiola aurea, Helenium flexuosum, Hydrocotyle umbellata, Hymenocallis pygmaea, Hypericum adpressum, H. boreale, H. canadense, H. mutilum, Isoetes microvela, Juncus canadensis, J. filiformis, J. marginatus, J. militaris, J. pelocarpus, J. repens, Lachnanthes caroliniana, Lindernia dubia, Lipocarpha micrantha, Lobelia dortmanna, Ludwigia sphaerocarpa, Lycopus uniflorus, Lysimachia terrestris, Micranthemum umbrosum, Muhlenbergia uniflora, Nuphar sagittifolia, Oclemena nemoralis, Oldenlandia uniflora, Panicum rigidulum, P. verrucosum, P. virgatum, P. wrightianum, Persicaria amphibia, P. hydropiperoides, P. puritanorum, Platanthera flava, Pogonia ophioglossoides, Pontederia cordata, Ranunculus reptans, Rhexia virginica, Rhynchospora capillacea, R. capitellata, R. fusca, R. inundata, R. macrostachya, R. nitens, R. scirpoides, Sabatia calycina, S. campanulata, S. difformis, Sagittaria teres, Scirpus longii, Scleria reticularis, Solidago galetorum, Spartina pectinata, Stachys hyssopifolia, Symphyotrichum tradescantii, Triadenum virginicum, Utricularia subulata, Vaccinium macrocarpon, Viola lanceolata, Woodwardia areolata, Xyris caroliniana, X. difformis.*

Sabatia stellaris **Pursh** is an annual, which inhabits beaches, depressions, dunes, flats, hammocks, lakeshores, marshes, meadows, prairies, salt marshes, swales, and the margins of ponds and rivers at low elevations. The sites (typically salt marshes) often are brackish or saline (e.g., 20 ppt salinity), calcareous (pH: 5.4–8.6), with open exposures, and marl, sand, or shell substrates. The flowers are self-compatible but protandrous with all of the pollen being shed a full day before the stigmas become receptive. The seeds (about 600 per capsule) develop after 2 months and are physiologically dormant. A germination rate of 30% has been reported for seeds that have been cold stratified for 3 months at 4°C.

Seedlings are highly sensitive to competition with other species and must occur in sites with favorable growing conditions to survive. A seedbank is formed, which can contain up to 22 seeds/m². Habitats for this species are threatened by damage from off-the-road recreational vehicles. **Reported associates:** *Agalinis maritima, Aletris bracteata, Ammophila breviligulata, Amsonia rigida, Andropogon glomeratus, A. gyrans, A. virginicus, Aristida, Asclepias perennis, Atriplex patula, A. prostrata, Baccharis angustifolia, B. halimifolia, Bacopa monnieri, Bassia hirsuta, Batis maritima, Bletia purpurea, Borrichia frutescens, Cakile edentula, Carex alata, C. silicea, Cassytha filiformis, Cenchrus spinifex, C. tribuloides, Cephalanthus occidentalis, Ceratodon, Chamaecrista fasciculata, Chamaesyce bombensis, C. polygonifolia, Chloris, Cladium jamaicense, Conyza canadensis, Croton punctatus, Cynanchum angustifolium, C. palustre, Cyperus esculentus, C. filicinus, C. globulosus, C. haspan, C. polystachyos, C. strigosus, Dichanthelium aciculare, D. sabulorum, D. sphaerocarpon, Distichlis spicata, Drosera intermedia, Eclipta prostrata, Eleocharis baldwinii, E. olivacea, E. parvula, E. rostellata, Eleusine indica, Eragrostis oxylepis, E. pilosa, E. refracta, E. secundiflora, Erechtites hieraciifolia, Eupatorium capillifolium, E. mikanioides, Eustachys petraea, Eustoma, Euthamia tenuifolia, Festuca rubra, Fimbristylis autumnalis, F. caroliniana, F. castanea, F. spadicea, Fuirena scirpoidea, Gaillardia pulchella, Galactia canescens, Helianthus argophyllus, Heterotheca subaxillaris, Hibiscus moscheutos, Hydrocotyle bonariensis, H. umbellata, H. verticillata, Hypericum gentianoides, Ilex vomitoria, Ipomoea imperati, Iva frutescens, Juncus bufonius, J. canadensis, J. marginatus, J. roemerianus, J. scirpoides, Juniperus virginiana, Kosteletzkya pentacarpos, Lachnanthes caroliniana, Lathyrus japonicus, Leersia hexandra, Lilaeopsis carolinensis, Limonium carolinianum, Ludwigia octovalvis, L. repens, Lyonia mariana, Lythrum californicum, L. lineare, Mimosa, Muhlenbergia capillaris, M. filipes, Myrica cerifera, M. pensylvanica, Nostoc, Nyssa, Oenothera biennis, O. humifusa, Ophioglossum vulgatum, Opuntia humifusa, O. stricta, Panicum amarum, P. columbianum, P. hemitomon, P. virgatum, Parthenocissus quinquefolia, Paspalum distichum, P. monostachyum, Persicaria hydropiperoides, P. punctata, Phragmites australis, Phyla nodiflora, Pluchea foetida, P. odorata, P. rosea, Polygala grandiflora, Polygonum glaucum, Pontederia cordata, Pseudognaphalium stramineum, Ptilimnium capillaceum, Ranunculus, Rhexia mariana, Rhynchosia americana, Rhynchospora colorata, R. floridensis, R. inundata, R. tracyi, Rosa rugosa, Rubus trivialis, Rumex hastatulus, Sabal palmetto, Sabatia calycina, S. dodecandra, Sagittaria lancifolia, Salicornia bigelovii, S. depressa, Sarracenia, Saururus cernuus, Schizachyrium gracile, S. maritimum, S. scoparium, Schoenoplectus americanus, S. pungens, Schoenus nigricans, Serenoa repens, Sesuvium maritimum, S. portulacastrum, Setaria geniculata, S. magna, S. parviflora, Smilax bona-nox, Solanum pseudogracile, Solidago sempervirens, Spartina alterniflora, S. bakeri, S. cynosuroides, S. patens, Spergularia salina, Sphagnum, Spiranthes vernalis, Sporobolus virginicus, Strophostyles* *helvula, S. umbellata. Suaeda linearis, Symphyotrichum novi-belgii, S. subulatum, S. tenuifolium, Taxodium, Teucrium canadense, Toxicodendron radicans, Triglochin striata, Typha, Uniola paniculata, Utricularia subulata, Vaccinium corymbosum, Vigna, Vitis rotundifolia, Xyris.*

Use by wildlife: Some *Sabatia* species are hosts for larval caterpillars (Lepidoptera: Noctuidae: *Argyrostrotis anilis*). The flowers are visited by nonpollinating flies (Diptera) and by crab spiders (Arachnida: Thomisidae: *Misumenoides formosipes, Xyticus*). The pollen is eaten by several Coleoptera including blister beetles (Meloidae: *Epicauta lemniscata*), ladybird beetles (Coccinellidae: *Anisosticta*), and scarab beetles (Scarabaeidae: *Phyllophaga rugosa, Popillia japonica*). *Sabatia kennedyana* provides habitat for the comet darner dragonfly (Odonata: Aeshnidae: *Anax longipes*).

Economic importance: food: not edible; **medicinal:** *Sabatia decandra* was used by the Florida Seminoles as a bitter tonic for treating indigestion and as a remedy for "sun sickness". Infusions made from the roots of several other species reportedly provide remedies for a number of ailments; **cultivation:** *Sabatia kennedyana* and *S. stellaris* are cultivated; **misc. products:** *Sabatia stellaris* is recommended as a salt-tolerant plant for streambank restoration; **weeds:** *Sabatia difformis* has been described as having "weedy" growth in some instances; **nonindigenous species:** none

Systematics: The accepted name *Sabatia decandra* has been used here in lieu of *S. bartramii*, which appears in the 2013 indicator list. Molecular phylogenetic studies (Figure 5.37) resolve *Sabatia* as a monophyletic genus within tribe Chironieae subtribe Chironiinae of Gentianaceae. These studies indicate *Gyrandra* as the sister genus to *Sabatia*, with a close relationship of both genera to *Eustoma. Sabatia* is divided taxonomically into two sections (*Pseudochironia, Sabatia*) with the latter subdivided into five subsections. The OBL species are dispersed among both sections and three of the subsections, which is an indication of their multiple origins. Although a comprehensive phylogenetic analysis has not been conducted for *Sabatia*, some indications of interspecific relationships can be obtained from the results of crossing studies (evaluated by pollen fertility of F$_1$ hybrids). Artificial crosses between *S. difformis* and *S. macrophylla* (both assigned to section *Sabatia* subsection *Difformes*) yield viable seeds but attempts to raise hybrid plants have been unsuccessful. Most species within subsection *Campanulatae* can be crossed successfully; however, crosses made with species from other sections are unsuccessful or result in extremely low fertility. *Sabatia stellaris* crosses most successfully with *S. campanulata. Sabatia dodecandra* (subsection *Dodecandrae*) crosses most successfully with *S. angularis* (subsection *Angulares*); whereas, crosses attempted with other members of subsection *Dodecandrae* result in low fertility (e.g., *S. calycina*) or are unsuccessful (e.g., *S. kennedyana*). However, *S. angularis* also crosses quite successfully with *S. campanulata* (subsection *Campanulatae*). *Sabatia calycina* may be misplaced in subsection *Dodecandrae* as it cannot be crossed successfully with other members of the subsection, yet does cross with several species in subsection *Campanulatae*. Although some

authors recognize *S. kennedyana* as a variety of *S. dodecandra*, artificial hybrids between these species can be produced easily, but are sterile and their anthers are abortive. Two varieties of *S. dodecandra* are recognized, each differing morphologically and by chromosome number. The base chromosome number of *Sabatia* is difficult to determine, but is hypothesized as $x = 19$. Diploids include *S. kennedyana* ($2n = 40$), *S. dodecandra* var. *foliosa* ($2n = 38$), *S. difformis* ($2n = 36$), *S. stellaris* ($2n = 36$), *S. dodecandra* var. *dodecandra* ($2n = 34$), and *S. gentianoides* ($2n = 28$). *Sabatia calycina* ($2n = 64$) presumably is a tetraploid. Hybrids between *S. campanulata* and *S. kennedyana* are reported from New England. Otherwise, natural interspecific hybrids are rare in *Sabatia* and most species appear to be isolated fairly effectively by their different chromosome numbers. There is a trend that those species having lower chromosome numbers also tend to exhibit more of an annual life-history.

Comments: *Sabatia calycina, S. decandra, S. dodecandra*, and *S. gentianoides* occur in the southeastern United States; *S. difformis* and *S. stellaris* in the eastern United States; and *Sabatia kennedyana* sporadically along eastern coastal North America.

References: Aronson & Whitehead, 1989; Austin, 2004b; Avis et al., 1997; Bell & Lester, 1978; Enser, 2004; Franks, 2003; Hammer, 2002; Higman, 1972; Hill & Keddy, 1992; Hosier & Eaton, 1980; Kanouse, 2003; Keddy, 1984, 1985; Keddy et al., 1994; Light et al., 1993; Mansion & Struwe, 2004; Mitchell, 2011; Nelson, 1986; Nichols, 1934; Norcini, 2002; Orrell Elliston, 2006; Perry, 1971; Rosenfeld, 2004; Shipley & Parent, 1991; Silberhorn, 1982; Stalter & Lamont, 1990; Stuckey & Gould, 2000; Wilbur, 1955; Wisheu & Keddy, 1991; Zaremba, 2004.

8. *Zeltnera*

Centaury

Etymology: after Louis Zeltner (1938–) and Nicole Zeltner (1934–)

Synonyms: *Centaurion; Centaurium* (in part); *Cicendia; Erythraea*

Distribution: global: North America; South America; **North America:** western

Diversity: global: 25 species; **North America:** 16 species

Indicators (USA): FACW: *Zeltnera exaltata* (previously OBL)

Habitat: saline (inland); palustrine; **pH:** 6.3–9.0; **depth:** <1 m; **life-form(s):** emergent herb

Key morphology: stems (to 35 cm) erect, 4-angled, somewhat winged, unbranched; leaves (to 3 cm) opposite, entire, linear to lanceolate; cymes with numerous, long-pedicelled (to 50 mm), 3- to 4-merous flowers; corolla tubular, rotate; corolla (to 14 mm), the lobes (to 7 mm) pink, rose, or white; anthers coiling after dehiscence; stigmas fan shaped; capsules (to 2 cm) cylindrical

Life history: duration: annual (fruit/seeds); **asexual reproduction:** none; **pollination:** insect; **sexual condition:** hermaphroditic; **fruit:** capsules (common); **local dispersal:** seed; **long-distance dispersal:** seed

Imperilment: (1) *Zeltnera exaltata* [G5]; SH (MT); S1 (<u>BC</u>, CO, NE)

Ecology: general: Species of *Zeltnera* occur frequently in wet and unstable sites that are characterized by weak competition. They occupy habitats ranging from roadsides, fields, pastures, and open forests to stream banks. Nine of the species (36%) are FACW indicators, but none currently is designated as OBL. The following species was ranked as OBL, FACW in the 1996 indicator list, but now is ranked as FACW. It has been retained here because the treatment was prepared prior to the release of the 2013 list.

Zeltnera exaltata **(Griseb.) G. Mans.** grows in channels, depressions, flats, marshes, meadows, sloughs, springs, and along pond and stream margins at elevations of up to 1950 m. Its habitats occur mainly in areas surrounded by desert and have highly alkaline (pH: 6.3–9.0) and highly saline (NaCl, $MgSO_4$), often sandy substrates. Some localities are associated with hot springs but the plants can tolerate temperatures down to 11°C and require a minimum of 120 frost-free days for maturation. The flowers are pollinated by bees (Hymenoptera) and the anthers coil after the pollen is shed. The seeds reportedly do not require cold stratification for germination. More specific information on the reproductive ecology and population biology of this species is needed. **Reported associates:** *Agrostis gigantea, Arnica discoidea, Astragalus argophyllus, A. lentiginosus, Atriplex patula, A. truncata, Berula erecta, Bromus tectorum, Calochortus excavatus, C. palmeri, C. striatus, Carex nebrascensis, C. praegracilis, Castilleja minor, C. stenantha, Chrysothamnus albidus, Cirsium occidentale, Cressa truxillensis, Delphinium nudicaule, Distichlis spicata, Eleocharis palustris, E. rostellata, Elymus caninus, Ericameria nauseosa, Eriogonum argophyllum, Fimbristylis thermalis, Grindelia fraxino-pratensis, Helianthus annuus, Heliotropium curassavicum, Heterocodon rariflorum, Hordeum jubatum, Horkelia bolanderi, Ivesia kingii, Juncus balticus, J. mexicana, J. torreyi, Juniperus californica, Lactuca sativa, Leptodactylon californicum, Lythrum californicum, Mentha ×piperita, Mimulus guttatus, Monardella macrantha, Nasturtium officinale, Navarretia setiloba, Perideridia parishii, Plagiobothrys parishii, Poa pratensis, Polypogon monspeliensis, Populus fremontii, P. trichocarpa, Potentilla anserina, Psoralea orbicularis, Puccinellia nuttalliana, Quercus agrifolia, Q. chrysolepis, Q. wislizenii, Ranunculus cymbalaria, Rhus trilobata, Salicornia europaea, Salix exigua, Schoenoplectus americanus, S. robustus, Senecio canus, Spartina gracilis, Sporobolus airoides, Streptanthus cordatus, Tamarix ramosissima, Toxicoscordion venenosum, Triglochin maritimum, Typha domingensis.*

Use by wildlife: *Zeltnera exaltata* reportedly is grazed by livestock.

Economic importance: food: not edible; **medicinal:** The Miwok tribe prepared "Canchalagua," a decoction made from the stems and leaves of *Zeltnera exaltata*, to use as an analgesic for toothaches, stomach aches, and as a remedy for tuberculosis; **cultivation:** not cultivated; **misc. products:** none; **weeds:** none; **nonindigenous species:** none

Systematics: Although the species now assigned to *Zeltnera* were long recognized in the genus *Centaurium*, molecular phylogenetic analyses (Figure 5.37) have left little question that the latter should be subdivided to reflect more natural, monophyletic segregate genera. The closest sister group to *Zeltnera* is a clade comprising *Exaculum* and *Schenkia* (Figure 5.42). *Zeltnera* essentially is subdivided into three subclades having affinities to California, Texas, and Mexico (Figure 5.42), with *Z. exaltata* occurring in the "Californian" group. Some authors have placed *Zeltnera exaltata* in synonymy with *Z. namophila* var. *nevadensis* ($2n = 34$); however, a comprehensive phylogenetic analysis of the genus has indicated that *Z. exaltata, Z. namophila,* and *Z. nevadensis* are all distinct from one another (Figure 5.42). These analyses further indicate the sister to *Z. exaltata* ($2n = 40$) to be a morphologically similar taxon (not currently assigned to a particular species) that differs by its basic chromosome number ($2n = 74$ [80]). Thus, it appears that plants identified as *Z. exaltata* have included specimens from both ploidy groups. Plants possessing the larger chromosome number have been suspected to be of allopolyploid origin; however, both maternal and nuclear DNA markers resolve individuals with different chromosome complements within a single clade, which could reflect homogenization of the nuclear (nrITS) markers. In any case, it would be informative to ascertain the level of fertility that exists among these groups and also to compare their ecological tolerances. Estimates from molecular data indicate that *Zeltnera* diversified roughly six million years ago during a period of climatic aridification. The base chromosome number of *Zeltnera* is $x = 17$. *Zeltnera exaltata* includes tetraploid ($2n = 40$) and octoploid ($2n = 74, 80$) populations.

Comments: *Zeltnera exaltata* occurs in montane western North America.

References: Bolen, 1964; Broome, 1978; Correll & Corell, 1975; Mansion & Struwe, 2004; Mansion & Zeltner, 2004; Mason, 1957; Reveal et al., 1973; St. John & Courtney, 1924.

Family 3: Rubiaceae [650]

The Rubiaceae (madder) family rank among the largest of angiosperm families and contain nearly 13,000 species of herbs, lianas, shrubs, and trees. Most species in the family are woody and are distributed in tropical and subtropical regions. Typical members of the family have opposite (or whorled), simple, entire leaves, interpetiolar stipules, and 4- to 5-merous flowers with an inferior ovary. A nectary disk usually is present above the ovary and a large proportion of the species is heterostylous (Judd et al., 2002). Protandry is common in the homostylous species. Pollinators include bats (Mammalia: Chiroptera), bees (Insecta: Hymenoptera), birds (Aves), butterflies and moths (Insecta: Lepidoptera), and flies (Insecta: Diptera). The seeds are dispersed by birds (for those species with fleshy fruits), by wind, or by their attachment to animals.

Phylogenetically, Rubiaceae occupy a basal position within Gentianales (Figure 5.27). Analysis of DNA sequence data (Bremer et al., 1995; Rova et al., 2002) support the monophyly of Rubiaceae and the recognition of three subfamilies: Cinchonoideae (including *Cephalanthus, Strumpfia*), Ixoroideae (*Pinckneya*), and Rubioideae (including *Diodia, Galium, Oldenlandia, Pentodon*), with the former two

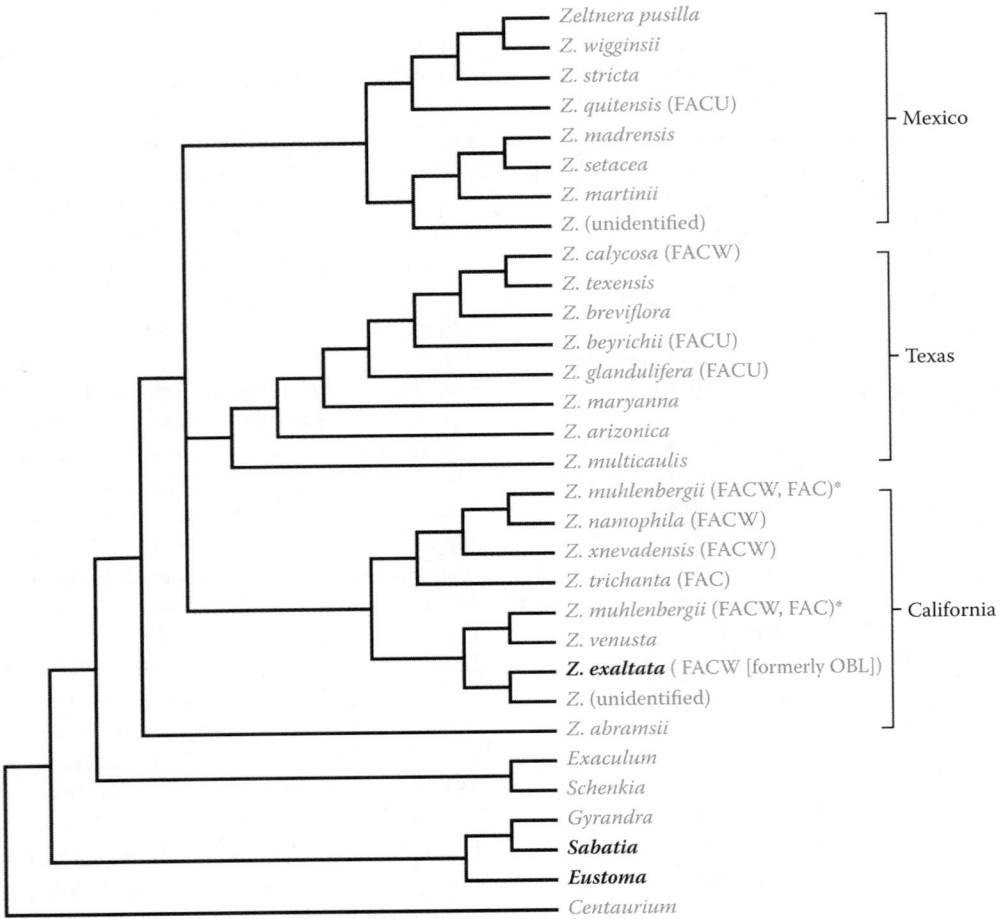

FIGURE 5.42 Phylogenetic relationships in *Zeltnera* as indicated by analysis of combined DNA sequence data. The formerly OBL North American *Z. exaltata* (bold) occurs with other species of "Californian" affinity, in a clade where several FACW species are concentrated. Anomalous positions of *Z. muhlenbergii* (asterisked) may be a consequence of hybridization. (Adapted from Mansion, G. & Zeltner, L., *Amer. J. Bot.*, 91, 2069–2086, 2004.)

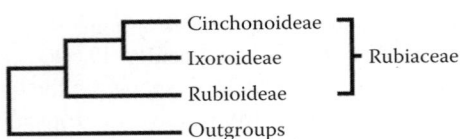

FIGURE 5.43 Phylogenetic relationships among subfamilies of Rubiaceae as indicated by analysis of DNA sequence data. OBL representatives occur within all of the groups shown. (Adapted from Bremer, B. et al., *Syst. Biol.*, 48, 413–435, 1999; Rova, J.H.E. et al., *Amer. J. Bot.*, 89, 145–159, 2002.)

subfamilies resolving as a sister group (Figure 5.43). Most members of subfamily Rubioideae (13 tribes) also share a deletion of the -488 plastid encoded polymerase (PEP) promoter of the *atpB* gene (Manen, 2000). Tribal associations within these clades remain somewhat unsettled, due in large part to the inability to sample such a large group thoroughly. However, fairly intensive studies have been made of all three subfamilies using various combinations of molecular and nonmolecular data (e.g., Andreasen & Bremer, 2000; Bremer & Manen, 2000;

Andersson & Antonelli, 2005). Members of tribe Rubieae also have an atypically altered region (usually highly conserved) in the −255 PEP promoter of the *atpB* gene (Manen, 2000).

Important ornamental genera include *Asperula*, *Bouvardia*, *Coprosma*, *Galium*, *Gardenia*, *Houstonia*, *Ixora*, *Luculia*, and *Pentas*. The plants contain a variety of alkaloids, which are important economically as caffeine (*Coffea*), quinine (*Cinchona*), and ipecac syrup (*Carapichea ipecacuanha*). Coffee is obtained from the roasted seeds of *Coffea* (*C. arabica*, *C. canephora*, and *C. liberica*) and ranks second only to petroleum among the world's most highly traded commodities.

Although the family is terrestrial for the most part, there are seven genera that contain OBL aquatics in North America:

1. ***Cephalanthus*** L.
2. ***Diodia*** L.
3. ***Galium*** L.
4. ***Oldenlandia*** L.
5. ***Pentodon*** Hochst.
6. ***Pinckneya*** Michx.
7. ***Strumpfia*** Jacq.

1. *Cephalanthus*

Buttonbush; bois bouton

Etymology: from the Greek *kephala anthos* meaning "head flower"

Synonyms: *Acrodryon*; *Axolus*; *Eresimus*; *Silamnus*

Distribution: global: Africa; Asia; North America; **North America:** eastern, southern

Diversity: global: 15 species; **North America:** 2 species

Indicators (USA): OBL: *Cephalanthus occidentalis, C. salicifolius*

Habitat: freshwater; palustrine; **pH:** 5.0–8.5; **depth:** 0–2 m; **life-form(s):** emergent shrub

Key morphology: stems (to 3 m) woody, multiple, often dilated at base, lenticels prominent; leaves (to 19 cm) opposite or whorled (in 3s or 4s), shiny, entire, ovate to lanceolate, petiolate (to 3 cm), stipulate (to 4 mm), venation arcuate; inflorescence in terminal clusters or axils, consisting of stalked (to 1 dm), dense, globose heads (to 3 cm) of 4-merous flowers; corolla (to 8 mm) white; styles (to 2 cm) strongly exserted; each fruit splitting into several 1-seeded, angular, obconic nutlets (to 8 mm); seeds brown with a white aril

Life history: duration: perennial (buds); **asexual reproduction:** shoot fragments; **pollination:** insect; **sexual condition:** hermaphroditic; **fruit:** nutlets (common); **local dispersal:** seeds (gravity/wind; water); **long-distance dispersal:** seeds (birds)

Imperilment: (1) *Cephalanthus occidentalis* [G5]; S1 (NB); S2 (NS); S3 (NE); (2) *C. salicifolius* [G5]; S1 (TX)

Ecology: general: The ecology of most *Cephalanthus* species is not well documented; however, both of the North American species are OBL aquatics and are always associated with wetland habitats.

Cephalanthus occidentalis L. inhabits canals, depressions, ditches, floodplains, marshes, oxbows, prairies, seeps, sloughs, swales, swamps, and occurs along lake, pond, river, and stream margins at elevations of up to 1000 m. The species is broadly adapted ecologically and occurs across a wide range of acidity (pH: 5.0–8.5) in substrates that include clay, muck, sand, or silt. The plants can withstand low temperatures down to −36°C but are fairly intolerant of salinity. Individuals can survive prolonged periods (3–15 months) of inundation (typically occurring in 0.3–1.2 m of water) but also grow well on exposed, drying sites. They tolerate a spectrum of exposures ranging from full sun to dense shade. The flowers emit an extremely strong floral scent, which attracts pollinating honeybees (Hymenoptera: Apidae: *Apis*), bumblebees (Hymenoptera: Apidae: e.g., *Bombus fervidus, B. ternarius, B. terricola*), and butterflies (Lepidoptera). This scent contains high levels of monoterpenes including the terpene alcohol linalool, which has been linked to butterfly pollination syndromes. There also is a rich production of nectar, which contains sucrose (53%), glucose (26%), and fructose (21%). The flowers are protandrous, with the stigmas not becoming receptive until the second day after the flowers open. Although pollen is presented secondarily on the stigmatic surface, it is removed prior to stigma receptivity, or the selfing ability of unremoved pollen is limited due to self-incompatibility. One study demonstrated that cross-pollination resulted in 92% seed set as compared to 0%–8% for selfed flowers. The nutlets float and are dispersed locally by water (up to 225 m) or by gravity/wind. They also are eaten by various birds (Aves), which likely disperse them across fairly long distances. The seeds reportedly have no dormancy and will germinate without pretreatment. This factor may explain why some preliminary studies found buttonbush to be absent from seed banks. The plants grow quickly and often develop into dense thickets; however, they typically do not reproduce vegetatively and reproduction and establishment occurs almost entirely by seed, which often dominate the propagule bank in wetland systems where the plants occur. They are relatively short-lived shrubs. Occasional dispersal may be achieved vegetatively by stem fragments. This species is a pioneer of flooded swamps and can even establish on rotting logs or stumps. Buttonbush thrives at sites where fire is withheld, but also will resprout within a few months following a fire. Nutrients released by fires are conducive to growth of the plants. The main (leader) shoots often die back, resulting in a scraggly appearance. Twigs have prominent lenticels, which facilitate gas exchange during winter after the leaves have been shed. The plants can be propagated from shoot cuttings, which should be rooted in moist sand. The "buttonbush swamp" is a relatively distinct community type, in which the composition of associated taxa may vary considerably due to the wide ecological tolerances of the species. **Reported associates:** *Acalypha, Acer grandidentatum, A. negundo, A. rubrum, A. saccharinum, Acorus calamus, Acrostichum, Agalinis fasciculata, Alisma, Allophyllum gilioides, Alnus serrulata, Alternanthera philoxeroides, Ambrosia trifida, Amorpha fruticosa, A. nitens, Ampelopsis arborea, Amsonia hubrichtii, Andropogon gerardii, Anemone virginiana, Annona glabra, Apios americana, Arisaema dracontium, Artemisia douglasiana, Arundinaria gigantea, Asclepias incarnata, A. tuberosa, Asimina triloba, Azolla caroliniana, Baccharis neglecta, Bacopa caroliniana, Baptisia alba, Berchemia scandens, Betula nana, B. nigra, Bidens aristosus, B. comosus, B. discoideus, B. frondosus, B. tripartitus, Bignonia capreolata, Blechnum serrulatum, Boehmeria cylindrica, Bolboschoenus fluviatilis, Boltonia asteroides, B. diffusa, Botrychium dissectum, Bromus diandrus, Brunnichia ovata, Calamagrostis canadensis, Callicarpa americana, Campsis radicans, Capsella bursa-pastoris, Cardamine pensylvanica, Carex alata, C. aquatilis, C. bullata, C. comosa, C. crinita, C. decomposita, C. frankii, C. glaucescens, C. grayi, C. joorii, C. lacustris, C. lupuliformis, C. lupulina, C. muskingumensis, C. oligosperma, C. striata, C. stricta, C. tribuloides, C. trisperma, C. tuckermanii, C. typhina, C. vesicaria, C. vulpinoidea, Carpinus caroliniana, Carya aquatica, C. illinoinensis, Catalpa bignonioides, Celtis laevigata, C. occidentalis, Centaurea cyanus, Centella erecta, Ceratophyllum echinatum, Chamaecyparis thyoides, Chamaedaphne calyculata, Chasmanthium laxum, C. sessiliflorum, Chenopodium album, Chionanthus virginicus, Cicuta bulbifera, C. maculata, Cinna arundinacea, Cladium jamaicense, C. mariscoides, Clarkia unguiculata, Claytonia perfoliata, Clethra* *alnifolia, Cliftonia monophylla, Climacium americanum, Collinsia heterophylla, Colocasia esculenta, Commelina virginica, Cornus amomum, C. foemina, C. obliqua, C. racemosa, C. sericea, Crataegus rufula, C. viridis, Crinum americanum, Crotalaria sagittalis, Croton capitatus, Cryptantha, Cuscuta compacta, Cyperus odoratus, C. polystachyos, Cyrilla racemiflora, Decodon verticillatus, Desmodium adscendens, Dichanthelium clandestinum, D. dichotomum, D. scoparium, D. sphaerocarpon, D. spretum, Diodia teres, Dioscorea villosa, Diospyros virginiana, Drepanocladus aduncus, Dulichium arundinaceum, Echinochloa crus-galli, E. muricata, Echinodorus, Eleocharis cellulosa, E. melanocarpa, E. microcarpa, E. quadrangulata, E. quinqueflora, Elephantopus carolinianus, Elymus virginicus, Epilobium coloratum, Eriocaulon, Eriophorum virginicum, Erodium cicutarium, Eryngium yuccifolium, Eupatorium coelestinum, E. perfoliatum, E. rotundifolium, Fagus grandifolia, Ficus aurea, Fimbristylis autumnalis, Fissidens fontanus, F. hallianus, Forestiera acuminata, Fraxinus caroliniana, F. latifolia, F. nigra, F. pennsylvanica, F. profunda, Fuirena pumila, F. scirpoidea, F. squarrosa, Galium obtusum, Gaylussacia baccata, Geum, Gilia capitata, Gleditsia aquatica, G. triacanthos, Glottidium vesicarium, Glyceria acutiflora, G. canadensis, G. obtusa, G. septentrionalis, G. striata, Gordonia lasianthus, Gratiola brevifolia, Habenaria repens, Halesia diptera, Hamamelis vernalis, Helenium autumnale, Helianthus angustifolius, H. grosseserratus, H. mollis, Heterotheca villosa, Hibiscus coccineus, H. laevis, H. moscheutos, Hordeum murinum, Hottonia inflata, Hydrocotyle umbellata, Hydrolea ovata, Hypericum densiflorum, H. hypericoides, H. mutilum, H. prolificum, Hypoxis curtissii, Ilex cassine, I. coriacea, I. decidua, I. mucronata, I. opaca, I. verticillata, Impatiens capensis, Iris hexagona, I. virginica, Itea virginica, Juglans microcarpa, Juncus acuminatus, J. brachycarpus, J. canadensis, J. effusus, J. marginatus, J. repens, Kosteletzkya pentacarpos, Lactuca serriola, Laportea canadensis, Larix laricina, Leersia lenticularis, L. oryzoides, Leitneria floridana, Lemna minor, Leptopus phyllanthoides, Lessingia leptoclada, Leucothoe racemosa, Liatris tenuifolia, Lilium superbum, Limnobium spongia, Lindera benzoin, Lindernia dubia, Liquidambar styraciflua, Lobelia cardinalis, Ludwigia alternifolia, L. decurrens, L. leptocarpa, L. palustris, L. peploides, L. peruviana, L. pilosa, L. repens, Lycopodiella appressa, Lycopus rubellus, L. uniflorus, Lysimachia lanceolata, L. nummularia, Lyonia ligustrina, Lythrum, Magnolia virginiana, Matteuccia struthiopteris, Menispermum, Menyanthes trifoliata, Mikania scandens, Mimulus, Mitchella repens, Myrica cerifera, M. heterophylla, Myrsine floridana, Nelumbo lutea, Nephrolepis, Nuphar advena, N. variegata, Nyssa aquatica, N. biflora, N. ogeche, N. sylvatica, Onoclea sensibilis, Orbexilum pedunculatum, Orontium aquaticum, Osmunda regalis, Osmundastrum cinnamomeum, Oxypolis rigidior, Packera tomentosa, Panicum anceps, P. hemitomon, P. rigidulum, P. verrucosum, P. virgatum, Parietaria, Parthenocissus quinquefolia, Paspalum setaceum, P. urvillei, Peltandra sagittifolia, P. virginica, Penthorum sedoides, Persea borbonia, P. palustris, Persicaria amphibia, P. arifolia, P. coccinea, P. densiflora, P. hydropiperoides, P. lapathifolia, P. maculosa, P. pensylvanica, P. punctata, P. sagittata, P. virginiana, Phacelia hydrophylloides, Phalaris arundinacea, Phanopyrum gymnocarpon, Phlox carolina, Photinia melanocarpa, P. pyrifolia, Phragmites australis, Physocarpus opulifolius, Physostegia virginiana, Phytolacca americana, Pilea pumila, Pinus elliotii, P. ponderosa, P. rigida, P. serotina, P. taeda, Pistia stratiotes, Planera aquatica, Platanthera ciliaris, Platanus occidentalis, Pleopeltis polypodioides, Pluchea odorata, Polygala cruciata, Polytrichum, Pontederia cordata, Populus deltoides, P. heterophylla, Porella, Potamogeton natans, Proserpinaca palustris, Prunus serotina, Psychotria nervosa, Pycnanthemum tenuifolium, Quercus bicolor, Q. chrysolepis, Q. kelloggii, Q. laurifolia, Q. lyrata, Q. michauxii, Q. nigra, Q. pagoda, Q. palustris, Q. phellos, Q. rubra, Q. texana, Q. velutina, Ranunculus flabellaris, Rhexia mariana, R. virginica, Rhododendron canescens, R. viscosum, Rhynchospora capitellata, R. colorata, R. corniculata, R. inundata, R. macrostachya, R. miliacea, R. scirpoides, Ribes cynosbati, Rorippa, Rosa palustris, Rotala ramosior, Roystonea elata, Rubus allegheniensis, R. discolor, Rudbeckia hirta, R. laciniata, R. subtomentosa, Rumex altissimus, R. crispus, R. verticillatus, Ruppia maritima, Sabal minor, S. palmetto, Saccharum baldwinii, Sagittaria brevirostra, S. falcata, S. lancifolia, S. latifolia, Salix caroliniana, S. exigua, S. interior, S. laevigata, S. lasiolepis, S. lucida, S. nigra, Salvinia minima, Sambucus nigra, Samolus, Saponaria officinalis, Saururus cernuus, Schizachyrium scoparium, Schoenoplectus acutus, S. californicus, S. tabernaemontani, S. torreyi, Scirpus ancistrochaetus, S. cyperinus, Scleria reticularis, Sclerolepis uniflora, Scutellaria lateriflora, Sebastiania fruticosa, Senna hebecarpa, Serenoa repens, Silphium astericus, Sium suave, Smilax glauca, S. laurifolia, S. rotundifolia, S. walteri, Solanum dulcamara, Solidago remota, S. rugosa, Sorghastrum nutans, Sparganium androcladum, Spartina patens, S. pectinata, Sphagnum cuspidatum, S. fallax, S. lescurii, S. palustre, S. pylaesii, S. recurvum, Spiraea alba, Spiranthes cernua, Spirodela, Stachys hyssopifolia, S. tenuifolia, Stellaria media, Strophostyles helvula, Styrax americanus, S. grandifolius, Symphyotrichum dumosum, S. novi-belgii, S. ontarionis, Taxodium ascendens, T. distichum, Teucrium canadense, Thalictrum pubescens, Thelypteris kunthii, T. palustris, Thuja occidentalis, Tillandsia usneoides, Torilis arvensis, Torreyochloa pallida, Toxicodendron radicans, T. vernix, Triadenum virginicum, T. walteri, Trifolium variegatum, Tripsacum dactyloides, Typha domingensis, T. latifolia, Ulmus americana, U. rubra, Urtica dioica, Utricularia gibba, Vaccinium corymbosum, V. elliottii, V. formosum, V. fuscatum, V. macrocarpon, V. myrtilloides, V. oxycoccos, Vernonia gigantea, Veronicastrum virginicum, Viburnum dentatum, V. nudum, Viola lanceolata, V. sagittata, Vitis californica, V. riparia, V. rotundifolia, Warnstorfia fluitans, Wolffia brasiliensis, Woodwardia areolata, W. virginica, Xanthium, Xanthorhiza simplicissima, Xyris, Zizaniopsis miliacea.*

***Cephalanthus salicifolius* Humb. & Bonpl.** inhabits canals, depressions, ditches, ponds, resacas (dried river beds), springs, and swamps at elevations of at least 563 m (to 1403 m in Mexico). The plants tolerate full sun to partial shade. The flowers are a nectar source for adult butterflies (Lepidoptera), which probably assist in their pollination. Other ecological information on this species in North America is scarce because of its extremely limited distribution. The modest list of associated taxa that follows represents those species that have been reported from the Mexican portion of its range, but also occur in the North American portion of its range. Additional ecological studies that focus specifically on the North American populations would be helpful. **Reported associates:** *Arundo donax, Baccharis salicifolia, B. salicina, Carya illinoinensis, Chilopsis linearis, Cynanchum racemosum, Fraxinus berlandieriana, Juglans microcarpa, Platanus occidentalis, Pluchea odorata, Prosopis glandulosa, Quercus fusiformis.*

Use by wildlife: *Cephalanthus* is extremely important ecologically and provides habitat for countless wetland organisms. The woody stems are gnawed by beaver (Mammalia: Rodentia: Castoridae: *Castor canadensis*) and muskrat (Mammalia: Rodentia: Muridae: *Ondatra zibethicus*) and are browsed by larger herbivores (Mammalia: Cervidae) such as pronghorn antelope (*Antilocapra americana*) and white-tailed deer (*Odocoileus virginianus*). However, the foliage contains toxic glucosides and is regarded as potentially poisonous to livestock. Plants of *C. occidentalis* are used occasionally by bison (Mammalia: Bovidae: *Bison bison*) for head rubbing and "horning" behavior. Seeds of *C. occidentalis* are eaten by a variety of UPL game birds (Aves) such as the California quail (Odontophoridae: *Callipepla californica*), and numerous waterfowl including 17 different species of ducks (Aves: Anatidae), especially mallards (*Anas platyrhynchos*) and wood ducks (*Aix sponsa*). Waterfowl usage is highest in the lower Mississippi region of the United States. Alder flycatchers (Tyrannidae: *Empidonax traillii*), cattle egrets (Ardeidae: *Bubulcus ibis*), little blue herons (Ardeidae: *Egretta caerulea*), and wood ducks are among a number of birds that use the plants as nesting sites. The nectariferous flowers of *C. occidentalis* are visited by long-tongued bees (Hymenoptera) including various Apidae (*Apis mellifera, Bombus auricomus, B. bimaculatus, B. fraternus, B. griseocallis, B. impatiens, B. pensylvanica, Psithyrus variabilis*), Anthophoridae (*Ceratina dupla, Ptilothrix bombiformis, Triepeolus concavus, T. lunatus, Florilegus condigna, Melissodes bimaculata, M. communis, M. tepaneca, Peponapis pruinosa, Svastra obliqua*), and Megachilidae (*Megachile brevis, M. inimica, M. mendica, M. parallela, M. petulans*), and also by an assortment of short-tongued bees including a number of Halictidae (*Agapostemon sericea, A. texanus, A. virescens, Halictus ligatus, H. rubicunda*), and Colletidae (*Hylaeus affinis*). They also attract wasps (Hymenoptera: Sphecidae: *Ammophila nigricans*; Tiphiidae: *Myzinum quinqecincta*; Pompilidae: *Entypus fulvicornis*; Vespidae: *Euodynerus annulatus*), flies (Diptera: Syrphidae: *Eristalis stipator, E. tenax, Sphaerophoria contiqua, Syritta pipiens, Volucella bombylans*; Conopidae: *Physocephala texana, P. tibialis, Physoconops brachyrhynchus*; Muscidae: *Musca domestica*), and a diverse assortment of Lepidoptera from many families including Arctiidae (*Utetheisa bella*), Ctenuchidae (*Cisseps fulvicollis*), Hesperiidae (*Anatrytone logan, Atalopedes campestris, Epargyreus clarus, Erynnis juvenalis, Poanes zabulon, Polites peckius, P. themistocles, Thorybes bathyllus*), Nymphalidae (*Cercyonis pegala, Danaus plexippus, Limenitis archippus, Phyciodes tharos, Speyeria cybele, Vanessa atalanta, V. cardui, V. virginiensis*), Lycaenidae (*Celastrina argiolus, Everes comyntas, Lycaena hyllus, Strymon melinus*), Papilionidae (*Battus philenor, Papilio polyxenes, P. troilus*), and Pieridae (*Colias cesonia, C. philodice, Pieris rapae, Pontia protodice*). Other flower visitors include beetles (Coleoptera: Buprestidae: *Taphrocerus gracilis*; Coccinellidae: *Hippodamia convergens*; Scarabaeidae: *Trichiotinus piger*) and plant bugs (Hemiptera: Lygaeidae: *Oncopeltus fasciatus*). *Cephalanthus salicifolius* also is regarded as an excellent nectar source for various adult butterflies. *Cephalanthus occidentalis* is a host plant for many larval lepidoptera (Arctiidae: *Hyphantria cunea*; Coleophoridae: *Mompha solomoni*; Cosmopterigidae: *Mompha cephalonthiella*; Lymantriidae: *Orgyia leucostigma*; Noctuidae: *Acronicta oblinita, Catocala connubialis, Eudryas grata, Harrisimemna trisignata, Ledaea perditalis, Melanomma auricinctaria*; Oecophoridae: *Machimia tentoriferella*; Saturniidae: *Callosamia promethea, Citheronia regalis, Hyalophora cecropia, H. euryalus, Rothschildia lebeau*; Sphingidae: *Aellopos titan, Darapsa versicolor*; and Tortricidae: *Aethes cephalanthana*). It also is a host to a weevil (Coleoptera: Curculionidae: *Plocetes ulmi*) and a fungus (Basidiomycota: *Aecidium cephalanthi*). Several species of arboreal ants (Insecta: Hymenoptera: Formicidae: *Pseudomyrmex brunneus, P. pallidus*) build their nests by excavating the soft pith of the stems.

Economic importance: food: *Cephalanthus occidentalis* is not edible and can cause gastrointestinal poisoning. The leaves contain cephalanthin, which destroys blood corpuscles and induces convulsions, paralysis, and vomiting if ingested; **medicinal:** The root and stem bark of *Cephalanthus occidentalis* contains several triterpenoid saponins, which are bioactive and potentially hemolytic. The stem bark has been used to treat fevers and the root is the source of a bitter glycoside, which was used for treating coughs and as a diuretic. The plants were used medicinally by several Native American tribes. Eye ailments were treated using a poultice of warmed roots applied to the head (Chickasaw) or by administering a bark decoction (Choctaw). Diarrhea from dysentery or other ailments was controlled using a bark decoction (Choctaw, Seminole). The inner bark was used as an emetic (Meskwaki). Bark from the stem and root was chewed to alleviate toothaches and fevers (Choctaw). The Kiowa administered a root decoction to stop bleeding. A decoction (made from leaves) was taken by the Koasati for rheumatism or (from roots) for enlarged muscles. The Seminoles relied on bark or fruit decoctions as a general medicine to treat constipation, fevers, headaches, menstrual pain, nausea, stomach aches, and urinary disorders; **cultivation:** *Cephalanthus* is grown as a pond or water garden ornamental. 'Keystone' is a cultivated variety of *C. occidentalis*; *C. salicifolius* also is cultivated;

misc. products: *Cephalanthus occidentalis* is planted to control erosion along shorelines. Some Native American tribes made their arrow shafts from the twigs of *C. occidentalis*. The Comanche fashioned game sticks from the wood; **weeds:** none; **nonindigenous species:** none

Systematics: Molecular data resolve *Cephalanthus* as a monophyletic subtribe (Cephalanthinae) within tribe Naucleeae of subfamily Cinchonoideae (Rubiaceae). Depending on the data examined, the Cephalanthinae clade is either basal or near basal (Figures 5.44 and 5.45) within tribe Naucleeae. Only three *Cephalanthus* species have been included simultaneously in any phylogenetic analysis, which nevertheless, show the genus to be monophyletic or a closely allied but unresolved grade; however, *C. occidentalis* always associates strongly as the sister species to *C. salicifolius* (e.g., Figure 5.45). This is not an unanticipated result given that some authors have considered *C. salicifolius* to be a variety (var. *salicifolius*) of *C. occidentalis*. Other varieties of *C. occidentalis* have been recognized to accommodate plants with leaves that are narrower (var. *californicus*) or pubescent (var. *pubescens*). No natural hybrids have been reported in North America, but this is understandable given that the two North American species are essentially allopatric. The basic chromosome number of *Cephalanthus* is *x* = 11. *Cephalanthus occidentalis* (*2n* = 44) is a tetraploid.

Comments: *Cephalanthus occidentalis* is common across the southern and eastern portions of North America; however, *C. salicifolius* occurs only in extreme southern Texas (Cameron and Hidalgo Counties) but extends southward into central Mexico.

References: Andersson et al., 2002; Battaglia et al., 2002; Bhatkar & Whitcomb, 1975; Bremer et al., 1999; Coppedge & Shaw, 1997; Correll & Correll, 1975; Felter & Lloyd, 1898; Hauser, 1964; Heinrich, 1975a; Imbert & Richards, 1993; Kiehn, 1995; Martin & Uhler, 1951; McAtee, 1939; McCarron et al., 1998; Middleton, 2000; Razafimandimbison & Bremer, 2002; Razafimandimbison et al., 2005; Rouse, 1941; Snyder, 1991; Spahn & Sherry, 1999; Stutzenbaker, 1999; Tooker et al., 2002; Tyrrell, 1987; Van Handel et al., 1972; Villarreal et al., 2006; Wagner et al., 2004; Woodcock, 1925; Zhang et al., 2005.

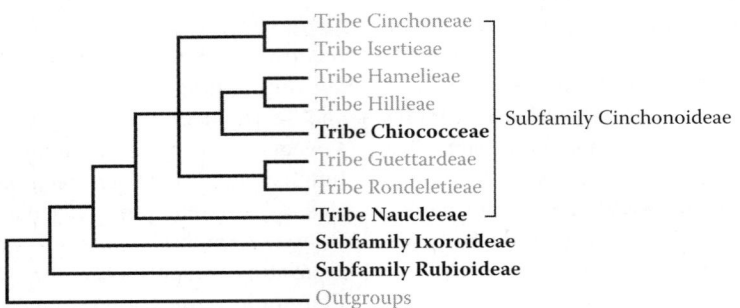

FIGURE 5.44 Relationships among tribes of Rubiaceae subfamily Cinchonoideae as indicated by analysis of DNA sequence data surveyed from five loci. Taxa continuing OBL North American aquatics (in bold) occur within two distinct tribes, namely Chiococceae (*Strumpfia*) and Naucleae (*Cephalanthus*), indicating independent origins of the OBL habit within the subfamily. (Adapted from Andersson, L. & Antonelli, A., *Taxon*, 54, 17–28, 2005.)

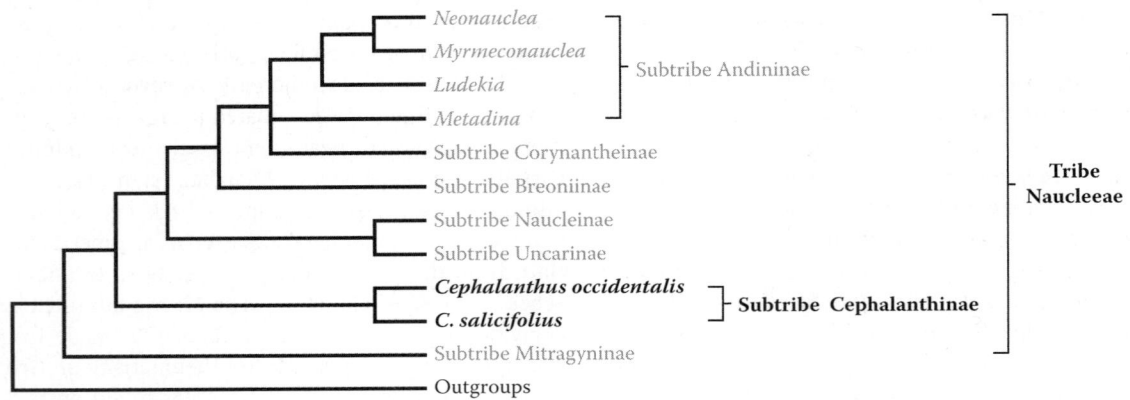

FIGURE 5.45 Subtribal relationships in Rubiaceae tribe Naucleeae as indicated by analysis of combined ETS/ITS sequence data. The OBL genus *Cephalanthus* (bold) resolves as a monophyletic subtribe (Cephalanthinae) near the base of the tribe. Analyses using combined molecular and morphological data (Razafimandimbison & Bremer, 2002) indicate similar subtribal limits, but associate them somewhat differently, i.e., placing Cephalanthinae at the base of tribe Naucleeae. (Adapted from Razafimandimbison, S.G. et al., *Mol. Phylogen. Evol.*, 34, 334–354, 2005.)

2. *Diodia*

Buttonweed

Etymology: from the Greek *diodos* meaning "pathway," in reference to the common roadside occurrence of plants

Synonyms: *Dasycephala*; *Decapenta*; *Diodella*; *Dioneiodon*; *Ebelia*; *Hexasepalum*; *Triodon*

Distribution: global: New World; **North America:** southeastern

Diversity: global: 5 species; **North America:** 1–5 species

Indicators (USA): OBL; **FACW:** *Diodia virginiana*

Habitat: freshwater; palustrine; **pH:** probably circumneutral; **depth:** <1 m; **life-form(s):** emergent herb

Key morphology: stems (to 8 dm) prostrate or spreading, ridged longitudinally; leaves (to 10 cm) opposite, sessile, margins finely serrulate, stipules bristle-like, fused with adjacent leaf bases; flowers aerial, 1–3 in axils (sometimes cleistogamous from underground shoots), epigynous, salverform; sepals (to 6 mm) 2, persisting in fruit; corolla 4-merous, the tube (to 1 cm) slender, lobes of the limb (to 6 mm) with finely ciliate margins, white, sometimes streaked with pink; stamens 4, exserted; ovary inferior; fruit (to 9 mm) globose, capsule-like, but sometimes breaking into two leathery, indehiscent segments, each 3-ribbed on back, and containing a single seed

Life history: duration: perennial (persistent roots); **asexual reproduction:** root/shoot fragments; **pollination:** insect or cleistogamous; **sexual condition:** hermaphroditic; **fruit:** schizocarps (common); **local dispersal:** seeds (water); vegetative fragments; **long-distance dispersal:** seeds (birds, waterfowl)

Imperilment: (1) *Diodia virginiana* [G5]; S1 (KS, NJ); S2 (IN)

Ecology: general: All *Diodia* species exhibit similar habitat affinities, which include wet sites along river or stream banks, or areas of moist sand in tropical to warm temperate regions. There is little detailed ecological information available for any of the species, especially regarding their reproductive biology.

Diodia virginiana **L.** grows in depressions, ditches, dunes, flatwoods, floodplains, marshes, meadows, prairies, savannas, seeps, sloughs, swamps, and along lake, pond, and river margins at elevations of up to 261 m. The range of acidity has not been determined for this species; however, many of the sites tend to be described as slightly acidic (e.g., pH: 6.7) with some occurrences reportedly on "calcareous" substrates. The substrates often are sandy but also include clay, clay-loam, muck, and silty clay. This species is particularly common in disturbed sites that are seasonally wet, which include drawdown zones, mudflats, and sandbars. Plants occur in full sun to partial shade. They are tolerant of fire, grazing, and mowing and occur occasionally in oligohaline marshes; however, they cannot withstand salinities over 0.5 ppt or prolonged periods of flooding. In the southern portion of their range, the plants remain in flower throughout most of the year. There are no detailed accounts of the reproductive ecology of *D. virginiana*. The pollination mechanism of aerial flowers (presumably insect mediated) has not been studied; however, self-pollinating, chasmogamous flowers develop from adventitious buds on the underground roots and hypocotyl. Vegetative reproduction occurs by root and shoot fragments.

Seedlings form adventitious buds on the roots within 4 weeks of germination. These buds can develop into aerial shoots especially if they become detached and encounter wet soil conditions. Root and stem regeneration ceases if plants are exposed to temperatures of −10°C. Seeds of *D. virginiana* exhibit optimal germination at 25°C–30°C (pH: 6.0) under 12 h of light alternating with 15–25 or 20°C–30°C in darkness. Natural germination is reduced under flooded or unaerated conditions but is inhibited allelopathically by a wilt fungus (*Cephalosporium diospyri*). The fruits are buoyant, dispersed by water and the seeds are well-represented in many marsh seedbanks. Detailed studies on the breeding system and population genetics of this species would provide useful information, given that it is considered to be quite rare in some areas but extremely weedy in others. **Reported associates:** *Acalypha virginica, Acer rubrum, A. saccharinum, Agalinis purpurea, Albizia julibrissin, Alnus serrulata, Alternanthera philoxeroides, Ambrosia artemisiifolia, Ammannia coccinea, Amorpha fruticosa, A. ouachitensis, Ampelopsis arborea, Amsonia hubrichtii, A. illustris, Andropogon gerardii, A. virginicus, Anthoxanthum odoratum, Apios americana, Apocynum cannabinum, Arundinaria gigantea, Asclepias purpurascens, A. verticillata, Athyrium filix-femina, Axonopus furcatus, Baccharis halimifolia, Bacopa caroliniana, B. monnieri, Baptisia sphaerocarpa, Berchemia scandens, Betula nigra, Bidens bipinnatifidum, B. coronatus, B. frondosus, B. polylepis, Blechnum serrulatum, Boehmeria cylindrica, Boltonia diffusa, Borrichia, Calamovilfa arcuata, Camassia scilloides, Campsis radicans, Carex atlantica, C. caroliniana, C. complanata, C. crinita, C. crus-corvi, C. debilis, C. glaucescens, C. grayi, C. intumescens, C. joorii, C. longii, C. lupuliformis, C. lupulina, C. lurida, C. muskingumensis, C. scoparia, C. vulpinoidea, Carpinus caroliniana, Carya aquatica, C. illinoensis, C. laciniosa, Catalpa bignonioides, Ceanothus herbaceus, Celtis laevigata, Cephalanthus occidentalis, Chasmanthium latifolium, Cheilanthes lanosa, C. tomentosa, Chelone glabra, Cicuta maculata, Cladium jamaicense, Clematis virginiana, Clitoria mariana, Colocasia esculenta, Commelina diffusa, C. erecta, C. virginica, Cooperia drummondii, Coreopsis major, C. tripteris, Cornus amomum, C. foemina, C. obliqua, Crataegus viridis, Croton willdenowii, Cucumis, Cynanchum laeve, Cynodon dactylon, Cyperus erythrorhizos, C. odoratus, C. polystachyos, C. pseudovegetus, C. strigosus, Cyrilla racemiflora, Dichanthelium acuminatum, D. dichotomum, D. scoparium, Diervilla rivularis, Digitaria ciliaris, D. sanguinalis, Diodia teres, Diospyros virginiana, Dracopis amplexicaulis, Drosera, Duchesnea indica, Dulichium arundinaceum, Echinochloa muricata, E. walteri, Echinodorus cordifolius, Eclipta prostrata, Eleocharis fallax, E. obtusa, E. palustris, E. parvula, E. tenuis, Elephantopus carolinianus, E. elatus, Elymus trachycaulus, Eragrostis hypnoides, E. refracta, Erechtites hieraciifolia, Erigeron strigosus, Eryngium yuccifolium, Eupatorium capillifolium, E. coelestinum, E. fistulosum, E. hyssopifolium, E. maculatum, E. perfoliatum, E. serotinum, Euphorbia corollata, Euthamia caroliniana,*

Fimbristylis autumnalis, F. perpusilla, Forestiera acuminata, Fraxinus pennsylvanica, F. profunda, Galactia regularis, Galium obtusum, G. tinctorium, Gleditsia triacanthos, Glyceria septentrionalis, G. striata, Gratiola aurea, G. virginiana, Hamamelis vernalis, Helenium amarum, H. flexuosum, Heterotheca pilosa, Hibiscus moscheutos, Holcus lanatus, Houstonia purpurea, Hydrochloa carolinensis, Hydrocotyle umbellata, H. verticillata, Hydrolea affinis, H. uniflora, Hymenocallis caroliniana, Hypericum cistifolium, H. gentianoides, H. gymnanthum, H. hypericoides, H. lobocarpum, H. mutilum, H. prolificum, H. punctatum, Hypoxis curtissii, H. hirsuta, Hyptis alata, Ilex cassine, I. decidua, I. vomitoria, Impatiens capensis, Ipomoea pes-caprae, I. sagittata, Iris virginica, Itea virginica, Juglans nigra, Juncus acuminatus, J. biflorus, J. canadensis, J. coriaceus, J. effusus, J. megacephalus, J. repens, J. roemerianus, J. scirpoides, J. tenuis, Juniperus virginiana, Justicia americana, J. ovata, Kalmia latifolia, Kosteletzkya pentacarpos, Kummerowia striata, Lachnanthes caroliana, Leersia hexandra, L. lenticularis, L. oryzoides, L. virginica, Leptochloa fascicularis, Lespedeza cuneata, Leucothoe racemosa, Liatris microcephala, L. squarrosa, Ligustrum sinense, Lindernia dubia, Linum floridanum, Lipocarpha micrantha, Liquidambar styraciflua, Liriodendron tulipifera, Lobelia cardinalis, L. inflata, L. nuttallii, L. puberula, Lolium pratense, Lonicera japonica, Ludwigia alternifolia, L. decurrens, L. linearis, L. palustris, L. repens, L. sphaerocarpa, L. uruguayensis, Luziola fluitans, Lycopus americanus, Lythrum lineare, Manfreda virginica, Marshallia obovata, Mazus pumilus, Mecardonia acuminata, Melaleuca quinquenervia, Micranthemum umbrosum, Microstegium vimineum, Mikania scandens, Mimosa strigillosa, Mimulus ringens, Mitreola petiolata, Modiola caroliniana, Murdannia keisak, Myrica cerifera, Nuphar, Nuttallanthus canadensis, Nyssa biflora, N. sylvatica, Oldenlandia boscii, Onoclea sensibilis, Osmunda regalis, Oxalis europaea, Oxypolis rigidior, Panicum anceps, P. dichotomiflorum, P. ensifolium, P. hemitomon, P. rigidulum, P. verrucosum, P. virgatum, Parthenium integrifolium, Paspalum dilatatum, P. fluitans, P. laeve, P. notatum, P. urvillei, P. vaginatum, Passiflora incarnata, Peltandra virginica, Perilla frutescens, Persicaria hydropiperoides, P. pensylvanica, P. posumbu, P. punctata, P. sagittata, P. setacea, Phalaris arundinacea, Phyla lanceolata, P. nodiflora, Phyllanthus caroliniensis, P. urinaria, Physostegia intermedia, P. leptophylla, Phytolacca americana, Pilea pumila, Pinus elliottii, P. taeda, P. virginiana, Piptochaetium avenaceum, Planera aquatica, Plantago, Platanthera peramoena, Platanus occidentalis, Pluchea camphorata, P. odorata, P. purpurascens, Polygala mariana, Polypremum procumbens, Pontederia cordata, Populus heterophylla, Prunella vulgaris, Pseudognaphalium canescens, Ptilimnium costatum, Pycnanthemum tenuifolium, Pyrrhopappus caroliniana, Quercus bicolor, Q. laurifolia, Q. lyrata, Q. michauxii, Q. nigra, Q. palustris, Q. phellos, Q. prinus, Q. stellata, Q. virginiana, Ranunculus bulbosus, Rhexia mariana, R. virginica, Rhododendron arborescens, Rhynchospora colorata, R. corniculata, R. glomerata, R. inexpansa, R. inundata, R. macrostachya, Rosa multiflora, R. palustris, Rotala ramosior, Rubus argutus, Rudbeckia hirta, R. laciniata, Ruellia humilis, Rumex acetosella, R. conglomeratus, Sabal minor, S. palmetto, Sabatia calycina, S. kennedyana, Saccharum baldwinii, S. giganteum, Sacciolepis striata, Sagittaria lancifolia, S. latifolia, Salix caroliniana, Salvia azurea, Sambucus nigra, Sarracenia, Saururus cernuus, Schizachyrium scoparium, Schoenoplectus americanus, S. tabernaemontani, Scirpus cyperinus, S. expansus, S. polyphyllus, Scleria oligantha, Scutellaria integrifolia, Selaginella rupestris, Senecio glabellus, Sesbania exaltata, Seteria geniculata, S. magna, S. parviflora, Sida rhombifolia, Silphium laciniatum, Sisyrinchium angustifolium, S. rosulatum, Sium suave, Smilax, Solanum caroliniense, Solidago canadensis, S. rugosa, S. sempervirens, Sorghum halepense, Sparganium americanum, Spartina patens, Spermacoce glabra, Spiranthes lacera, S. vernalis, Sporobolus compositus, Steinchisma hians, Stenotaphrum secundatum, Stylosanthes biflora, Symphyotrichum dumosum, S. lateriflorum, S. subulatum, S. tenuifolium, Talinum teretifolium, Taxodium ascendens, T. distichum, Toxicodendron radicans, Trachelospermum difforme, Tradescantia ohiensis, Tragia urticifolia, Trepocarpus aethusae, Triadenum walteri, Triadica sebifera, Tridens flavus, Trifolium pratense, Tripsacum dactyloides, Typha angustifolia, T. latifolia, Ulmus alata, U. americana, Vaccinium elliottii, Valerianella radiata, Verbena scabra, Verbesina officinalis, Vernonia lettermannii, V. noveboracensis, Viburnum cassinoides, Vigna luteola, Viola primulifolia, Vitis labrusca, V. rotundifolia, Woodwardia areolata, W. virginica, Xanthorhiza simplicissima, Xyris.

Use by wildlife: Seeds of *Diodia virginiana* are eaten by many birds (Aves) including bobwhite quail (Odontophoridae: *Colinus virginianus*), mourning doves (Columbidae: *Zenaida macroura*), ducks (Anatidae), geese (Anatidae: *Branta canadensis*), and pheasants (Phasianidae: *Phasianus colchicus*), and probably also by small rodents (Mammalia: Rodentia). The plants constitute a relatively small portion of the diets of deer (Mammalia: Cervidae: *Odocoileus virginianus*) and also are used as forage by the gopher tortoise (Reptilia: Testudinidae: *Gopherus polyphemus*). *Diodia virginiana* is a host to various organisms including bugs (Hemiptera: Miridae: *Polymerus basalis*), fungi (Ascomycota: *Cercospora diodiae-virginianae*), and a semiendoparasitic root-feeding nematode (Tylenchida: Heteroderidae: *Verutus volvingentis*). The diodia vein chlorosis virus (DVCV) is transmitted by the bandedwinged whitefly (Insecta: Hemiptera: Aleyrodidae: *Trialeurodes abutilonea*) and causes chlorosis in the foliage of *D. virginiana*. The plants also are susceptible to beet yellows virus (Closteroviridae: *Closterovirus*).

Economic importance: food: not edible; **medicinal:** none; **cultivation:** *Diodia virginiana* is cultivated as an ornamental but is not widely distributed; **misc. products:** *Diodia virginiana* is recommended as a native groundcover in some areas; **weeds:** *Diodia virginiana* is a persistent weed of lawns and cultivated turfgrass in the southeastern United States, where it is difficult to control, and may require extensive herbicide applications; **nonindigenous species:** *Diodia virginiana* has been introduced to Japan and Taiwan (Hsinchu). It also is reported from Redding, California.

Systematics: Several phylogenetic analyses have confirmed the placement of *Diodia* within tribe Spermacoceae of subfamily Rubioideae in Rubiaceae (Figure 5.46). Although some analyses have identified *Ernodea* as the sister group of *Diodia*, the limited number of genera in the tribe that has been analyzed simultaneously (about 50%), makes it difficult to elucidate intergeneric relationships within tribe Spermacoceae. Furthermore, preliminary molecular phylogenetic analyses have indicated that the circumscriptions of *Diodia*, *Hemidiodia*, *Spermacoce*, and other related Rubioideae are weakly defined and require additional investigation. Most problematical is that analysis of nrITS and *rps16* intron sequences have indicated that *Diodia*, as traditionally defined, is highly polyphyletic. In any event, *D. virginiana* will remain in the genus *Diodia*, given that

it is the type species. *Diodia* has recently been redefined to include only five New World species, with some species transferred to *Borreria* (often merged with *Spermacoce*) or *Galianthe* (e.g., *D. brasiliensis*) and some (e.g., *D. teres*) whose affinities remain uncertain. Under this revised circumscription of *Diodia*, *D. virginiana* is the only species to occur in North America. Given these complications, an accurate appraisal of phylogenetic relationships for *Diodia* will not be possible until the circumscription of the genus has been solidified, and additional representatives of tribe Spermacoceae (e.g., *Borreria*, *Galianthe*, *Ernodea*, additional species of *Diodia*) have been analyzed simultaneously along with other genera in the tribe. The basic chromosome number for *Diodia* is $x = 14$. *Diodia virginiana* ($2n = 28$) is diploid.

Comments: *Diodia virginiana* occurs in the central and southeastern United States.

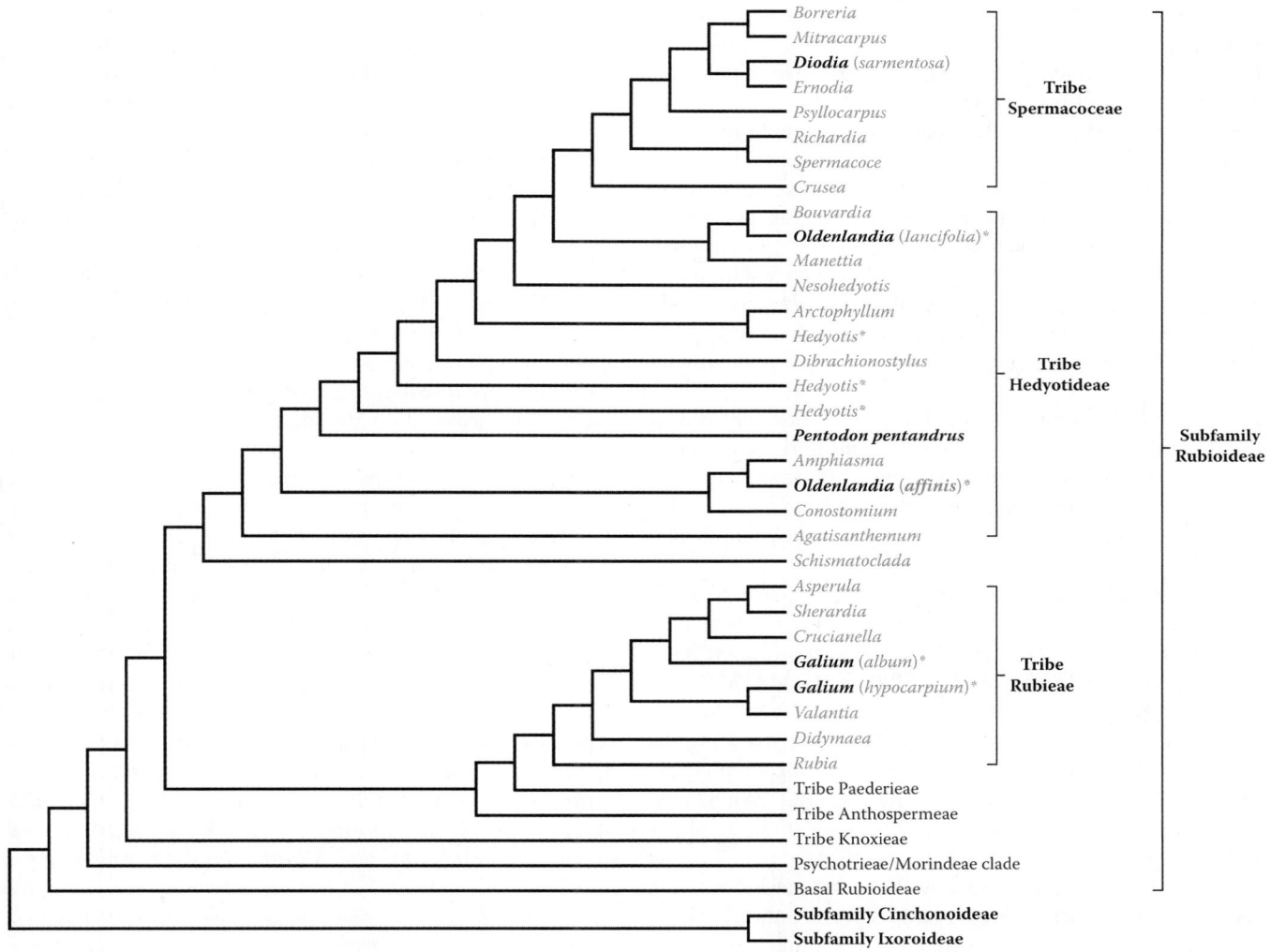

FIGURE 5.46 Phylogenetic relationships within Rubiaceae subfamily Rubioideae as indicated by analysis of *rps16* intron sequence data. The distribution of taxa having OBL North American representatives indicates multiple origins of the habit in the subfamily. Several genera (asterisked) are not monophyletic in this analysis and further sampling of species will be necessary to better clarify the generic limits. The survey of the polyphyletic *Galium* and *Oldenlandia* did not include OBL North American species; therefore, it is difficult to determine into which group these species would fall. The representative of *Diodia* also is not a North American species. (Adapted from Andersson, L. & Rova, J.H., *Plant Syst. Evol.*, 214, 161–186, 1999.)

References: Bacigalupo & Cabral, 1999; Baird & Dickens, 1991; Baird et al., 1992; Baldwin & Mendelssohn, 1998a; Bremer & Manen, 2000; Dessein, 2003; Esser, 1983; Hsieh & Chaw, 1987; Larsen et al., 1991; Lewis, 1962a; Mohlenbrock, 1959b; Negrón-Ortiz & Watson, 2003; Snodgrass et al., 1984a; Thulin & Bremer, 2004; Van Alstine et al., 2001.

3. *Galium*

Bedstraw, cleavers; gaillet

Etymology: from the Greek *gala*, meaning "milk," for the use of plants to curdle milk in cheese making

Synonyms: *Aparinanthus*; *Aparine*; *Aparinella*; *Aspera*; *Bataprine*; *Cruciata*; *Gallion*; *Gallium*; *Microphysa*; *Relbunium*; *Trichogalium*

Distribution: global: cosmopolitan; **North America:** widespread

Diversity: global: 400 species; **North America:** 82 species

Indicators (USA): OBL: *Galium asprellum*, *G. brevipes*, *G. labradoricum*, *G. palustre*, *G. tinctorium*, *G. trifidum*; **FACW:** *G. obtusum*, *G. tinctorium*, *G. trifidum*

Habitat: brackish (coastal), freshwater; palustrine; **pH:** 3.9–8.6; **depth:** <1 m; **life-form(s):** emergent herb

Key morphology: stems (to 20 dm) weak, simple to highly branched, square in cross-section, the angles often bristly or retrorsely barbed; leaves [including the leaf-like stipules] (to 3 cm) whorled (4–6 per node), sessile, margins smooth, or rough; flowers 3- to 4-merous, stalked (to 3 cm), in terminal, axillary inflorescences ranging from solitary, to few-flowered, to leafy, paniculate cymes; corolla rotate, the lobes (to 3.5 mm) white, greenish, or pinkish, ovary inferior; schizocarp breaking into two, globose, single-seeded, smooth, mericarps (to 5 mm)

Life history: duration: perennial (rhizomes); **asexual reproduction:** rhizomes, shoot fragments; **pollination:** insect or self; **sexual condition:** hermaphroditic; **fruit:** schizocarps (common); **local dispersal:** mericarps (gravity, water), rhizomes; **long-distance dispersal:** fragments, mericarps (water)

Imperilment: (1) *Galium asprellum* [G5]; S1 (TN); S2 (DE, IA, NC); S3 (NF); (2) *G. brevipes* [G4]; SH (ME, NY, VT); S1 (NB, WI); S2 (ON); S3 (QC); (3) *G. labradoricum* [G5]; S1 (CT, IA, OH, PA, PE, NJ, VT); S2 (BC, IL, MA, ME, NB, NS); S3 (AB, ND); (4) *G. obtusum* [G5]; SX (NY); SH (NH); S1 (NS, VT); S2 (NB); S3 (NC, NJ); (5) *G. palustre* [G5]; SH (WV); S1 (OH, TN, WI); S3 (NJ); (6) *G. tinctorium* [G5]; S3 (NF); (7) *G. trifidum* [G5]; S1 (NB, NJ); S2 (BC, PA); S3 (IA, VT, WY)

Ecology: general: Nineteen of the North American *Galium* species (23%) are designated as wetland indicators, with six species (7%) ranked as OBL in at least a portion of their range. *Galium obtusum* (FACW) formerly was ranked as OBL, FACW in the 1996 indicator list, but is retained here since the treatment was completed prior to the release of the 2013 list. A few species are annuals, but all of the OBL species are rhizomatous perennials. The pollination biology has not been studied thoroughly in this genus. The flowers are visited by bees (Hymenoptera: Andrenidae: *Andrena*; Halictidae), flies

(Diptera: Anthomyiidae: *Hylemya*; Syrphidae: *Mallota*), and wasps (Hymenoptera: Sphecidae: *Gorytini*; Pteromalidae), which may function as pollinators; however, several species are largely self-pollinating. The pollen often contains a fairly large proportion of abnormal grains (dwarf, giant, or inviable), which is attributed to environmental conditions. The fruits have been described as nutlets; however, they actually are schizocarps comprising two globose, indehiscent carpels, which separate as mericarps. The fruits of some species are bristly and are dispersed as they adhere to the fur of passing animals. However, the fruits of all OBL species are uniformly smooth, buoyant, and probably are dispersed mainly by water. The fruits of some species are known to retain their viability after passing through the digestive tract of herbivorous birds and mammals, which can disperse them in their feces. Fruit germination often is enhanced by shallow (2–10 mm) burial. Most species are rather weak, scrambling herbs, which cling by their bristly or barbed herbage to adjacent vegetation for support. Many of the species respond positively to grazing, which tends to reduce competing overstory vegetation.

***Galium asprellum* Michx.** occurs in bogs, ditches, fens, floodplains, marshes, meadows, mudflats, oxbows, prairies, seeps, springs, swales, swamps, and along margins of lakes, rivers, and streams at elevations of up to 1500 m. The plants are extremely adaptable and occur across a wide range of acidity (pH: 5.5–7.7) on substrates derived from limestone or sandstone, which include organic peat (e.g., *Sphagnum* mats), clay, gravel, loam, sand, silt, and mafic rock. The plants are tolerant to inundation and occur in sites with exposures ranging from full sun to substantial shade. Little is known about the reproductive ecology of this species. Presumably, the flowers are insect-pollinated. The means of dispersal has not been determined for the mericarps, which are devoid of hairs or other appendages. The grooved endosperm of the seeds contains an air space, which may confer buoyancy to facilitate their dispersal by water. The seeds are fairly well-represented in riparian forest seed banks. Their germination has been achieved at 22°C–30°C following a 14-week period of cold stratification at 4°C. There is some indication that the plants grow better in grazed areas than in ungrazed sites. The rhizomatous plants bear hooked prickles on the margins of their weak stems, which facilitate their ability to recline and to climb onto other vegetation, or assist in the dispersal of vegetative fragments. **Reported associates:** *Abies balsamea*, *A. fraseri*, *Acer negundo*, *A. rubrum*, *A. saccharum*, *A. spicatum*, *Actaea alba*, *Adiantum pedatum*, *Agalinis tenuifolia*, *Ageratina altissima*, *Agrostis alba*, *A. gigantea*, *A. perennans*, *Ajuga*, *Alisma subcordatum*, *Alliaria petiolata*, *Alnus rugosa*, *Amphicarpa bracteata*, *Anemone quinquefolia*, *Angelica atropurpurea*, *Apios americana*, *Arisaema triphyllum*, *Aronia melanocarpa*, *Artemisia serrata*, *Asarum canadense*, *Asclepias incarnata*, *Athyrium filix-femina*, *Berberis thunbergii*, *Berula*, *Betula alleghaniensis*, *B. lutea*, *B. papyrifera*, *B. pumila*, *B. ×sandbergii*, *Bidens connatus*, *Boehmeria cylindrica*, *Bolboschoenus fluviatilis*, *Botrychium virginianum*, *Brachyelytrum erectum*, *Bromus ciliatus*, *Bryoxiphium norvegicum*, *Calamagrostis canadensis*, *Caltha palustris*,

Cardamine diphylla, Carex aquatilis, C. atlantica, C. baileyi, C. debilis, C. echinata, C. festucacea, C. gynandra, C. incomperta, C. lacustris, C. leptalea, C. lurida, C. peckii, C. rosea, C. rostrata, C. stipata, C. stricta, C. trichocarpa, C. trisperma, Carpinus caroliniana, Carya cordiformis, Castilleja septentrionalis, Celastrus orbiculatus, Celtis occidentalis, Chamaedaphne calyculata, Chelone glabra, Chimaphila umbellata, Chrysosplenium americanum, Cicuta maculata, Cinna latifolia, Cirsium muticum, Claytonia caroliniana, Clematis virginiana, Clintonia borealis, Conioselinum chinense, Coptis trifolia, Cornus amomum, C. canadensis, C. racemosa, C. rugosa, C. sericea, Crataegus, Cryptotaenia canadensis, Cypripedium calceolus, Deparia acrostichoides, Dicentra canadensis, Dichanthelium clandestinum, D. ensifolium, Diervilla lonicera, Doellingeria umbellata, Dryopteris carthusiana, D. cristata, D. intermedia, Echinochloa crusgalli, Elaeagnus umbellata, Epigaea repens, Equisetum arvense, E. palustre, Erigeron annuus, Eriocaulon decangulare, Eriophorum virginicum, Eupatorium perfoliatum, Eutrochium maculatum, Floerkea proserpinacoides, Fraxinus americana, F. nigra, F. pennsylvanica, Galium aparine, G. tinctorium, G. trifidum, G. triflorum, Gentiana, Gentianopsis, Geum canadense, G. laciniatum, G. rivale, Glechoma hederacea, Glyceria canadensis, G. melicaria, G. striata, Goodyera oblongifolia, Gymnocarpium dryopteris, Hamamelis virginiana, Hasteola suaveolens, Helianthus giganteus, H. grosseserratus, Heracleum maximum, Houstonia serpyllifolia, Huperzia lucidula, H. porophila, Hydrocotyle americana, Hypericum pyramidatum, Ilex collina, I. verticillata, Impatiens capensis, Iris versicolor, I. virginica, Juglans cinerea, Juncus effusus, Juniperus communis, Kalmia angustifolia, K. latifolia, Laportea canadensis, Larix laricina, Leersia oryzoides, L. virginica, Lemna, Lilium michiganense, Lindera benzoin, Liriodendron tulipifera, Listera auriculata, Lobelia inflata, L. kalmii, L. siphilitica, Lonicera tatarica, Lycopodium digitatum, Lycopus americanus, L. uniflorus, Lyonia ligustrina, Lysimachia ciliata, L. quadriflora, Lythrum alatum, L. salicaria, Maianthemum canadense, M. trifolium, Medicago lupulina, Micranthes pensylvanica, Milium effusum, Mimulus glabratus, Mitchella repens, Mitella nuda, Nasturtium officinale, Oclemena acuminata, Onoclea sensibilis, Osmorhiza depauperata, Osmunda claytoniana, O. regalis, Osmundastrum cinnamomeum, Oxalis montana, O. stricta, Oxypolis rigidior, Packera aurea, Parnassia glauca, Parthenocissus quinquefolia, Pedicularis lanceolata, Peltandra virginica, Persicaria punctata, P. sagittata, P. virginica, Phalaris arundinacea, Phegopteris connectilis, Photinia melanocarpa, P. pyrifolia, Phragmites australis, Phryma leptostachya, Physostegia virginiana, Picea mariana, P. rubens, Pilea pumila, Pinus banksiana, P. strobus, Platanthera dilatata, P. psycodes, Platanus occidentalis, Poa palustris, Podophyllum peltatum, Polemonium vanbruntiae, Populus grandidentata, P. tremuloides, Potentilla simplex, Prunus serotina, P. virginiana, Pteridium aquilinum, Pycnanthemum virginianum, Quercus alba, Q. macrocarpa, Q. rubra, Ranunculus acris, R. recurvatus, Rhamnus alnifolia, Rhizomnium appalachianum,

Rhododendron canadense, R. catawbiense, R. groenlandicum, R. maximum, Rhus typhina, Rhynchospora capitellata, R. glomerata, Ribes americanum, R. hirtellum, Riccia, Rosa multiflora, R. palustris, Rubus hispidus, R. idaeus, R. occidentalis, R. pubescens, Rudbeckia laciniata, Rumex orbiculatus, R. verticillatus, Sagittaria, Salix candida, S. humilis, S. pyrifolia, S. sericea, Sambucus nigra, Scirpus atrovirens, S. cyperinus, S. expansus, S. polyphyllus, Scutellaria galericulata, S. lateriflora, Silphium perfoliatum, Smilax tamnoides, Solidago canadensis, S. gigantea, S. macrophylla, S. patula, Spartina pectinata, Sphagnum fuscum, S. recurvum, Spiraea alba, S. tomentosa, Sullivantia sullivantii, Symphyotrichum lateriflorum, S. puniceum, Symplocarpus foetidus, Taxus canadensis, Thalictrum dasycarpum, T. pubescens, Thuja occidentalis, Thelypteris noveboracensis, T. palustris, Tiarella cordifolia, Tilia americana, Toxicodendron radicans, T. vernix, Trifolium stoloniferum, Trillium erectum, T. grandiflorum, T. undulatum, Tsuga canadensis, Typha latifolia, Ulmus americana, U. rubra, Urtica dioica, Vaccinium, Verbena hastata, Verbesina alternifolia, Viburnum dentatum, V. nudum, V. trilobum, Viola blanda, V. cucullata, Vitis labrusca, Zanthoxylum americanum.

***Galium brevipes* Fernald & Wiegand** inhabits bogs, depressions, ditches, dune ponds, fens, marshes, meadows, prairies, swales, swamps, and the margins or shores of lakes and streams at elevations of up to 320 m. The plants often are found in shallow standing water but also occur on mudflats and at sites that dry out during the summer months. They generally occur in open areas but can tolerate some shade. Their substrates typically are described as calcareous loam, marl, or sand but occasionally as slightly acidic. Specific pH data are not available. There is a paucity of information on the ecology of this rare species and studies on its pollination and seed ecology are needed. The mericarps are smooth and probably are dispersed locally by water. These are fairly delicate, mat-forming plants, which reproduce vegetatively by prostrate or underground rhizomes. Their retrorsely barbed stems enable them to adhere to other vegetation for support. **Reported associates:** *Alnus rugosa, Aster, Betula pumila, Carex, Comarum palustre, Eleocharis intermedia, Galium labradoricum, G. trifidum, Hypericum boreale, Iris versicolor, Juncus balticus, Mentha arvensis, Myrica gale, Picea mariana, Salix bebbiana, S. lucida, S. pedicellaris, S. pyrifolia, Sphagnum, Thuja occidentalis, Veronica americana.*

***Galium labradoricum* (Wiegand) Wiegand** grows in bogs, carrs, depressions, ditches, fens, marshes, prairies, seeps, and along the margins of rivers. The habitats are open and alkaline (pH: 6.8–8.0) with substrates consisting of gravel, marl, muck, or peat. This species flowers relatively early in the season and the flowers have high pollen fertility (90%). Additional information on the reproductive ecology is not available. The smooth mericarps probably are dispersed locally by water; their germination requirements are unknown. These weak and slender-stemmed plants climb upon the surrounding vegetation by means of their somewhat scabrose stems; however, when neighboring vegetation is scarce, a mat-like form develops, which has fewer inflorescences and more lateral branches. The plants propagate vegetatively by prostrate or underground rhizomes. They occur much more frequently in fens that are grazed by deer (Mammalia: Cervidae: *Odocoileus virginianus*) than in ungrazed sites. **Reported associates:** *Acorus calamus, Agrostis scabra, Andromeda polifolia, Anemone canadensis, Aralia nudicaulis, Asclepias incarnata, Berula erecta, Betula papyrifera, B. pumila, Bromus ciliatus, Calamagrostis canadensis, C. stricta, Calla palustris, Calliergonella cuspidata, Caltha palustris, Campanula aparinoides, Campylium polygamum, C. stellatum, Cardamine pratensis, Carex aquatilis, C. arctata, C. atherodes, C. buxbaumii, C. chordorrhiza, C. comosa, C. diandra, C. exilis, C. flava, C. interior, C. lacustris, C. lasiocarpa, C. leptalea, C. limosa, C. livida, C. magellanica, C. oligosperma, C. pauciflora, C. pellita, C. prairea, C. rostrata, C. sartwellii, C. sterilis, C. stricta, C. tetanica, C. trisperma, C. utriculata, Chamaedaphne calyculata, Chamerion angustifolium, Chelone glabra, Cicuta bulbifera, C. maculata, Cirsium muticum, Cladium mariscoides, Clintonia borealis, Comarum palustre, Coptis trifolia, Cornus amomum, C. canadensis, C. racemosa, C. sericea, Cuscuta gronovii, Cypripedium reginae, Dasiphora floribunda, Decodon verticillatus, Dichanthelium acuminatum, Diphasiastrum sitchense, Doellingeria umbellata, Drepanocladus revolvens, Drosera rotundifolia, Dryopteris cristata, Dulichium arundinaceum, Eleocharis elliptica, E. palustris, E. rostellata, Epilobium, leptophyllum, E. strictum, Equisetum arvense, E. fluviatile, E. palustre, E. sylvaticum, Eriophorum angustifolium, E. vaginatum, E. viridicarinatum, Eupatorium perfoliatum, Euthamia graminifolia, Eutrochium maculatum, Fissidens adianthoides, Fragaria virginiana, Frangula alnus, Galium brevipes, G. tinctorium, G. trifidum, G. triflorum, Geum rivale, Glyceria grandis, G. striata, Gaultheria hispidula, Helianthus grosseserratus, Impatiens capensis, Iris versicolor, I. virginica, Juncus canadensis, J. stygius, Kalmia polifolia, Larix laricina, Lathyrus palustris, Leptodictyum riparium, Linnaea borealis, Lobelia kalmii, Lonicera oblongifolia, L. villosa, Lycopus americanus, L. uniflorus, Lysimachia quadriflora, L. thyrsiflora, Lythrum alatum, L. salicaria, Maianthemum trifolium, Mentha arvensis, Menyanthes trifoliata, Mimulus ringens, Mitella nuda, Muhlenbergia glomerata, M. mexicana, Myrica gale, Orthilia secunda, Osmunda regalis, Pedicularis lanceolata, Persicaria amphibia, P. punctata, P. sagittatum, Photinia melanocarpa, Phragmites australis, Physocarpus opulifolius, Picea mariana, Pilea fontana, Pinus banksiana, Polemonium reptans, Populus tremuloides, Pycnanthemum virginianum, Ranunculus lapponicus, Rhamnus alnifolia, Rhododendron groenlandicum, Rhynchospora alba, R. capillacea, Ribes americanum, R. hirtellum, Rubus acaulis, R. arcticus, R. pubescens, Rudbeckia hirta, Rumex orbiculatus, Sagittaria latifolia, Salix bebbiana, S. candida, S. discolor, S. lucida, S. pedicellaris, S. petiolaris, S. pyrifolia, S. rigida, Sarracenia purpurea, Scheuchzeria palustris, Schoenoplectus acutus, S. subterminalis, Scorpidium scorpioides, Scutellaria galericulata, Selaginella apoda, Sium suave, Solanum dulcamara, Solidago gigantea, S. ohioensis, S. patula, S. uliginosa,* *Sphagnum, Spiraea alba, S. tomentosa, Stellaria palustris, Symphyotrichum boreale, S. puniceum, Thalictrum dasycarpum, Thelypteris palustris, Thuja occidentalis, Tomentypnum nitens, Toxicodendron vernix, Triadenum fraseri, T. virginicum, Triantha glutinosa, Trichophorum alpinum, T. cespitosum, Trientalis borealis, Triglochin maritimum, Typha angustifolia, T. latifolia, Utricularia intermedia, U. macrorhiza, Vaccinium angustifolium, V. myrtilloides, V. oxycoccos, V. uliginosum, Valeriana edulis, Viola cucullata, Zizia aurea.*

***Galium obtusum* Bigelow** is found in ditches, dunes, fens, floodplains, marshes, prairies, seeps, swales, swamps, and along the margins of lakes and streams. The plants have a broad environmental tolerance to acidity (pH: 4.7–8.1) and also occur occasionally in slightly brackish soils. The substrates include loam, muck, peat, and sand. This species is regarded as a high light specialist. The small flowers are not believed to attract outcrossing insects. Pollen fertility is quite low (25%–50%) and the pollination biology and breeding system remain to be determined. Field studies indicate that a seed bank is formed. Seeds collected in the southern portion of the range germinate without pretreatment at 21.0°C–35.7°C. Vegetative reproduction occurs by the production and spread of the capillary rhizomes. Other details on the ecology of this species are sparse. **Reported associates:** *Acalypha virginica, Acer negundo, A. rubrum, A. saccharinum, Achillea millefolium, Acorus calamus, Agalinis purpurea, Agrimonia parviflora, Agrostis hyemalis, A. perennans, Alisma subcordatum, Allium canadense, A. cernuum, Alnus serrulata, Ambrosia trifida, Amorpha fruticosa, Amphicarpa bracteata, Andropogon gerardii, Anemone canadensis, Angelica atropurpurea, Antennaria neglecta, Anticlea elegans, Apios americana, Apocynum cannabinum, Arundinaria gigantea, Asclepias incarnata, A. verticillata, Baccharis halimifolia, Berchemia scandens, Berula erecta, Betula lenta, B. lutea, B. nigra, Bidens aristosa, B. connatus, B. frondosus, Bignonia capreolata, Boehmeria cylindrica, Calamagrostis canadensis, C. stricta, Calystegia sepium, Campanula aparinoides, Campsis radicans, Carex annectens, C. aquatilis, C. bebbii, C. brevior, C. bushii, C. buxbaumii, C. canescens, C. caroliniana, C. comosa, C. complanata, C. conoidea, C. crawei, C. cristatella, C. debilis, C. festucacea, C. glaucodea, C. gynandra, C. incomperta, C. interior, C. lacustris, C. longii, C. lupulina, C. lurida, C. molesta, C. normalis, C. pellita, C. rosea, C. rostrata, C. sartwellii, C. scoparia, C. squarrosa, C. stricta, C. stipata, C. tenera, C. tetanica, C. utriculata, C. vesicaria, C. vulpinoidea, Carpinus caroliniana, Centaurium pulchellum, Cephalanthus occidentalis, Chasmanthium laxum, Chelone glabra, Cicuta maculata, Cirsium altissimum, C. arvense, C. discolor, C. muticum, Comandra umbellatum, Comarum palustre, Cornus amomum, C. racemosa, C. sericea, Crataegus marshallii, Cuscuta glomerata, Cypripedium candidum, Dasiphora floribunda, Daucus carota, Dichanthelium acuminatum, D. clandestinum, D. dichotomum, D. scoparium, Diospyros virginiana, Dulichium arundinaceum, Echinocystis lobata, Eleocharis acicularis, E. compressa, E. erythropoda, E. tenuis, E. verrucosa, E. wolfii, Elymus canadensis, Epilobium strictum,*

Equisetum arvense, E. hyemale, Erechtites hieracifolia, Erigeron philadelphicus, E. strigosus, Eryngium yuccifolium, Eupatorium perfoliatum, Euphorbia corollata, Euthamia gymnospermoides, Eutrochium maculatum, Festuca paradoxa, Fragaria virginiana, Fraxinus nigra, F. pennsylvanica, Galium concinnum, G. trifidum, Gentiana andrewsii, Gentianopsis virgata, Geum aleppicum, Glyceria canadensis, G. striata, Glycyrrhiza lepidota, Helenium autumnale, Helianthus grosseserratus, H. mollis, Hibiscus moscheutos, Hierochloe odorata, Holcus lanatus, Hydrocotyle umbellata, H. verticillata, Hydrolea affinis, Hypericum kalmianum, H. mutilum, H. punctatum, H. sphaerocarpum, Hypoxis hirsuta, Ilex decidua, I. opaca, Impatiens capensis, Iris virginica, Juncus alpinus, J. brachycarpus, J. bufonius, J. canadensis, J. dudleyi, J. gymnocarpus, J. nodosus, J. tenuis, J. torreyi, Juniperus virginiana, Kalmia latifolia, Laportea canadensis, Lathyrus palustris, Leersia oryzoides, L. virginica, Liatris aspera, L. ligulistylis, L. pycnostachya, Leonurus cardiaca, Lilium michiganense, Lindera benzoin, Liparis loeselii, Liquidambar styraciflua, Lithospermum canescens, Lobelia cardinalis, L. kalmii, L. siphilitica, Ludwigia alternifolia, L. palustris, Lycopus americanus, L. asper, L. uniflorus, L. virginicus, Lyonia ligustrina, Lysimachia ciliata, L. lanceolata, L. quadriflora, L. terrestris, L. thyrsiflora, Lythrum alatum, L. salicaria, Magnolia tripetala, Maianthemum stellatum, Mikania scandens, Mimulus alatus, M. ringens, Monarda fistulosa, Muhlenbergia mexicana, Myosotis virginica, Myrica cerifera, M. pensylvanica, Nyssa aquatica, N. biflora, N. sylvatica, Onoclea sensibilis, Orbexilum pedunculatum, Osmunda regalis, Osmundastrum cinnamomeum, Oxalis stricta, Oxypolis rigidior, Packera aurea, P. paupercula, P. pseudaurea, Panicum rigidulum, P. virgatum, Parnassia glauca, Parthenocissus vitacea, Pedicularis lanceolata, Peltandra virginica, Penstemon digitalis, Persicaria arifolia, P. pensylvanica, P. punctata, P. sagittata, Phalaris arundinacea, Phlox glaberrima, P. pilosa, Phragmites australis, Physostegia angustifolia, P. virginiana, Pilea pumila, Pinus taeda, Platanthera flava, P. leucophaea, Pluchea odorata, Poa compressa, P. palustris, P. pratensis, Podophyllum peltatum, Polygala sanguinea, Populus deltoides, P. tremuloides, Potentilla anserina, P. arguta, P. simplex, Prunus americana, P. virginiana, Ptilimnium capillaceum, Pycnanthemum tenuifolium, P. virginianum, Quercus bicolor, Q. nigra, Q. phellos, Q. rubra, Q. velutina, Ranunculus cymbalaria, R. laxicaulis, R. pensylvanicus, R. sceleratus, Ratibida pinnata, Rhexia virginica, Rhus glabra, Rhynchospora capillacea, R. capitellata, Ribes americanum, Rosa arkansana, R. carolina, R. palustris, Rubus hispidus, Rudbeckia hirta, R. laciniata, Rumex mexicanus, Salix eriocephala, S. interior, S. nigra, S. petiolaris, S. rigida, Samolus parviflorus, Sanicula canadensis, Satureja arkansana, Saururus cernuus, Schizachyrium scoparium, Schoenoplectus pungens, S. tabernaemontani, Scirpus cyperinus, S. georgianus, S. pendulus, Scutellaria galericulata, S. lateriflora, Setaria geniculata, Silphium integrifolium, S. laciniatum, S. terebinthinaceum, Sisyrinchium angustifolium, S. campestre, Sium suave, Smilax bona-nox, S. rotundifolia, S. smallii, Solanum dulcamara, Solidago altissima, S. canadensis, S. gigantea, S. graminifolia, S. ohiensis, S. patula, S. ptarmicoides, S. riddellii, S. rigida, Sorghastrum nutans, Spartina pectinata, Sphenopholis obtusata, Spiraea alba, Spiranthes cernua, Sporobolus heterolepis, Stachys palustris, S. tenuifolia, Symphyotrichum ericoides, S. lanceolatum, S. novae-angliae, S. novi-belgii, S. pilosum, S. puniceum, Taxodium distichum, Thalictrum dasycarpum, T. polygamum, T. pubescens, Thaspium barbinode, Thelypteris noveboracensis, T. palustris, Toxicodendron radicans, Trachelospermum difforme, Tradescantia ohiensis, Triadenum virginicum, T. walteri, Tridens strictus, Triglochin maritimum, Triosteum aurantiacum, Typha latifolia, Ulmus americana, U. rubra, Utricularia macrorhiza, Vaccinium formosum, V. fuscatum, V. macrocarpon, Valeriana edulis, Vernonia baldwinii, V. fasciculata, V. noveboracensis, Veronica serpyllifolia, Veronicastrum virginicum, Viburnum dentatum, V. recognitum, Viola affinis, V. cucullata, V. lanceolata, V. nephrophylla, V. sagittata, Vitis riparia, V. rotundifolia, Woodwardia areolata, Zanthoxylum americanum, Zizia aurea.

***Galium palustre* L.** inhabits marshes, meadows, swamps, and the margins of lakes, ponds, and rivers. Although the plants often are associated with a lower pH (e.g., 5.0–6.0), they actually occur across a wide range of acidity (pH: 3.9–7.8) and can grow in fairly organic substrates (e.g., 30% organic matter). They usually occur in open areas but are tolerant to shade. The flowers are self-compatible, protandrous, and are either self-pollinating or are cross-pollinated by insects. In Europe, the flowers are visited by a parasitic hymenopteran (Pteromalidae); however, those insects appear to feed on the nectar and are not pollinators. The buoyant mericarps and vegetative fragments are dispersed by water and often are found floating in river drift, especially following floods. They germinate readily when favorable conditions are encountered. Seedlings also have been observed germinating from cattle (Mammalia: Bovidae: *Bos taurus*) dung, which implicates herbivorous mammals as potential dispersal agents. In Europe, viable seeds (germinating after a cold treatment) have been recovered from beneath cargos of pulpwood being transported aboard ships, indicating one probable means of human introduction to nonindigenous regions. The plants produce a substantial seed bank (up to 400 seeds/m²), which can vary from transient, to short-term persistent or persistent for up to 20 years or more. Seed germination reportedly can occur between 6°C and 34°C and natural seedling density can exceed 500/m². The ecology of this species has been studied extensively in Europe, but not to any great extent in North America. One study found the plants to occur more commonly in areas that are devoid of grazing deer (Mammalia: Cervidae). A study of plants in the Hudson Bay region indicated that they do not appear to be affected negatively (or positively) by grazing snow geese (Aves: Anatidae: *Chen caerulescens*). Some plants have been observed growing on mine wastes, which indicates at least partial tolerance to heavy metals. **Reported associates:** *Acer rubrum, Acorus americanus, A. calamus, Ageratina altissima, Agrostis alba, Alisma subcordatum, A. triviale, Alnus rugosa, Amorpha* *fruticosa, Anaphalis margaritacea, Anemone canadensis, Apocynum cannabinum, Arctium minus, Arisaema triphyllum, Asplenium trichomanes, Athyrium filix-femina, Bidens cernuus, B. connatus, Boehmeria cylindrica, Brassica nigra, Bromus inermis, Calamagrostis canadensis, Calliergon cordifolium, C. giganteum, Cardamine, Carex aquatilis, C. bebbii, C. crinita, C. cristatella, C. diandra, C. folliculata, C. frankii, C. hystericina, C. interior, C. intumescens, C. lasiocarpa, C. pedunculata, C. projecta, C. retrorsa, C. rostrata, C. shortiana, C. stipata, C. vulpinoidea, Cephalanthus occidentalis, Chaiturus marrubiastrum, Chamaecyparis thyoides, Chamaedaphne calyculata, Chelone glabra, Cirsium arvense, Cladium mariscoides, Clethra alnifolia, Climacium dendroides, Coptis trifolia, Cornus amomum, C. canadensis, Daucus carota, Decodon verticillatus, Desmodium, Dianthus armeria, Dichanthelium clandestinum, Drosera intermedia, D. rotundifolia, Dryopteris carthusiana, Dulichium arundinaceum, Echinochloa, Eleocharis acicularis, E. obtusa, E. ovata, E. smalliana, Epilobium ciliatum, E. coloratum, E. glandulosum, Equisetum arvense, Erigeron annuus, Eriophorum virginicum, Eupatorium perfoliatum, Euthamia tenuifolia, Eutrochium maculatum, Festuca elatior, Fraxinus nigra, Galium aparine, Gaultheria hispidula, Geum aleppicum, G. canadense, Glyceria acutiflora, G. canadensis, G. grandis, G. pallida, G. striata, Frangula alnus, Gratiola aurea, Hylocomium splendens, Hypericum boreale, H. majus, H. mutilum, H. perforatum, Ilex verticillata, Impatiens capensis, Iris versicolor, Juncus acutiflorus, J. brevicaudatus, J. canadensis, J. effusus, J. tenuis, Kalmia angustifolia, Leersia oryzoides, Lemna minor, Lindera benzoin, Lobelia spicata, Ludwigia palustris, Lycopus americanus, L. uniflorus, L. virginicus, Lysimachia ciliata, L. terrestris, L. thyrsiflora, Lythrum salicaria, Maianthemum canadense, Matteuccia struthiopteris, Melilotus alba, Mentha arvensis, M. spicata, Mimulus ringens, Mitella nuda, Mnium punctatum, Myosotis, Myrica gale, Nemopanthus mucronata, Nyssa sylvatica, Onoclea sensibilis, Osmunda regalis, Osmundastrum cinnamomeum, Peltandra virginica, Penthorum sedoides, Persicaria amphibia, P. coccinea, P. hydropiper, P. hydropiperoides, P. lapathifolia, P. sagittata, Phalaris arundinacea, Phleum pratense, Pilea pumila, Plantago major, Platanthera blephariglottis, Poa palustris, Populus deltoides, Potentilla norvegica, Prunus virginiana, Ranunculus acris, R. pensylvanicus, R. sceleratus, Rhamnus alnifolia, Rhododendron groenlandicum, Rhytidiadelphus triquetrus, Ribes americanum, Rorippa palustris, Rosa multiflora, R. palustris, Rubus allegheniensis, R. idaeus, R. pubescens, Rudbeckia hirta, Rumex crispus, Sagittaria brevirostra, S. latifolia, Salix amygdaloides, S. bebbiana, S. exigua, S. nigra, Sambucus nigra, Sarracenia purpurea, Schoenoplectus acutus, S. pungens, S. tabernaemontani, Scirpus atrovirens, S. cyperinus, Scutellaria lateriflora, Sisyrinchium montanum, Solanum dulcamara, Solidago altissima, S. gigantea, S. graminifolia, S. rugosa, Sparganium eurycarpum, Sphagnum, Spiraea alba, S. tomentosa, Spirodela polyrhiza, Stachys hispida, Symphyotrichum lateriflorum, S. puniceum, Thalictrum polygamum, T. pubescens, Thelypteris palustris, T. simulata, Thuidium delicatulum, Thuja occidentalis, Toxicodendron radicans, T. vernix, Tragopogon pratensis, Triadenum virginicum, Trifolium campestre, T. hybridum, T. pratense, T. repens, Tussilago farfara, Typha angustifolia, T. latifolia, T. ×glauca, Urtica dioica, Utricularia cornuta, Vaccinium macrocarpon, Verbena hastata, Viburnum dentatum, Viola lanceolata, V. palustris, Vitis riparia, Woodwardia virginica.*

***Galium tinctorium* L.** grows in fresh or brackish, tidal or nontidal bayous, bogs, carrs, depressions, ditches, dunes, floodplains, marshes, meadows, pools, prairies, sinkholes, sloughs, swales, swamps, and along the margins of lakes, ponds, rivers, and streams. Its habitats range from freshwater to nearly oligohaline (e.g., 0.18–0.44 ppt salinity) and in some instances can approach fairly saline conditions. Sites tend to be open areas that receive full sunlight, but the plants can tolerate partial shade. They usually occur in neutral or alkaline sites but are broadly adapted to a wide range of acidity (pH: 5.1–8.6). The substrates include clay, gravel, marl, muck, organic peat (quaking mats), and sand. Some plants have been found growing even on submerged logs. The pollination biology has not been studied. The plants form a seed bank and the seeds are known to germinate within 30–40 days (in light at 10°C) following a period (at least 2 weeks) of cold stratification. Although some treatments report this species to be an annual, it is a perennial from prostrate or subterranean rhizomes. The plants are known to recolonize burned areas fairly readily. **Reported associates:** *Abies balsamea, Acer negundo, A. rubrum, A. saccharinum, A. spicatum, Achillea millefolium, Acorus calamus, Agalinis paupercula, Agrimonia gryposepala, Agrostis alba, A. gigantea, A. scabra, Aletris, Alisma subcordatum, Alnus rugosa, A. serrulata, Alopecurus aequalis, Alternanthera philoxeroides, Amaranthus tuberculatus, Amelanchier, Amorpha fruticosa, Antennaria parlinii, Apios americana, Apocynum androsaemifolium, Aralia nudicaulis, Arctostaphylos uva-ursi, Arnoglossum plantagineum, Asclepias incarnata, Athyrium filix-femina, Azolla mexicana, Bacopa monnieri, Bartonia virginica, Betula lutea, B. papyrifera, Bidens cernuus, B. discoideus, B. frondosus, B. laevis, Boehmeria cylindrica, Botrychium matricariifolium, Bolboschoenus fluviatilis, Briza, Bromus, B. japonicus, Calamagrostis canadensis, Caltha palustris, Calystegia sepium, Campanula aparinoides, Cardamine, Carex aquatilis, C. atlantica, C. bebbii, C. bromoides, C. brunnescens, C. canescens, C. comosa, C. crawfordii, C. crinita, C. disperma, C. echinata, C. gigantea, C. granularis, C. grayi, C. hystricina, C. intumescens, C. lacustris, C. lasiocarpa, C. leptonervia, C. lupulina, C. oligosperma, C. pedunculata, C. pellita, C. projecta, C. retroflexa, C. retrorsa, C. rostrata, C. scoparia, C. stipata, C. stricta, C. trisperma, C. tuckermanii, C. vaginata, Carpinus caroliniana, Carya aquatica, Celtis laevigata, C. occidentalis, Cephalanthus occidentalis, Cerastium fontanum, Chamaedaphne calyculata, Chelone glabra, Cicuta bulbifera, C. maculata, Cinna latifolia, Cirsium nuttallii, Cladium mariscoides, Cladonia, Clintonia borealis, Comandra umbellata, Comarum palustre, Coptis trifolia, Cornus amomum, C. canadensis, C. foemina, Corylus cornuta, Crataegus, Cuscuta gronovii, Cyclospermum* *leptophyllum, Cyperus odoratus, C. rivularis, C. strigosus, Danthonia spicata, Decodon verticillatus, Decumaria barbara, Desmanthus, Digitaria, Distichlis spicata, Drosera intermedia, D. rotundifolia, Dryopteris carthusiana, D. cristata, Dulichium arundinaceum, Echinocystis lobata, Eleocharis acicularis, E. baldwinii, E. engelmannii, E. erythropoda, E. fallax, E. flavescens, E. intermedia, E. obtusa, E. olivacea, E. rostellata, E. smallii, Elymus canadensis, E. virginicus, Epigaea repens, Epilobium, E. coloratum, E. glandulosum, E. leptophyllum, Equisetum arvense, E. fluviatile, E. sylvaticum, Erechtites hieracifolia, Erigeron philadelphicus, Eriophorum virginicum, Eupatorium perfoliatum, Euphorbia corollata, Eurybia macrophylla, Eutrochium maculatum, Festuca, Fimbristylis autumnalis, Fragaria, Frangula alnus, Fraxinus nigra, F. pennsylvanica, Fuirena pumila, F. squarrosa, Galium asprellum, G. labradoricum, G. trifidum, G. triflorum, Gaultheria hispidula, Geranium, Geum rivale, Glyceria acutiflora, G. borealis, G. canadensis, G. grandis, G. melicaria, G. septentrionalis, Goodyera repens, Gymnocarpium disjunctum, Gynerium saccharoides, Hasteola suaveolens, Hedeoma hyssopifolia, Hibiscus moscheutos, Houstonia, Hydrocotyle umbellata, H. verticillata, Hypericum boreale, H. canadense, H. mutilum, Ilex decidua, I. mucronata, I. verticillata, Impatiens capensis, Iris versicolor, Itea virginica, Juncus acuminatus, J. alpinus, J. brevicaudatus, J. canadensis, J. effusus, J. gerardii, J. greenei, J. nodosus, Kalmia angustifolia, K. polifolia, Koeleria macrantha, Krigia, Larix laricina, Lathyrus graminifolius, L. palustris, Leersia oryzoides, Leptochloa, Leucospora, Lilium michiganense, Lindera benzoin, Liparis loeselii, Lipocarpha micrantha, Liquidambar styraciflua, Listera cordata, Lobelia kalmii, L. spicata, Lonicera villosa, Ludwigia alternifolia, L. leptocarpa, L. peploides, L. polycarpa, Lycopodiella inundata, Lycopus americanus, L. asper, L. rubellus, L. uniflorus, Lysimachia ciliata, L. terrestris, L. thyrsiflora, Lythrum lineare, L. salicaria, Maianthemum trifolium, Medicago lupulina, Mentha arvensis, M. spicata, Micranthemum umbrosum, Mimulus ringens, Monarda fistulosa, Moneses uniflora, Morus rubra, Muhlenbergia longiligula, M. mexicana, M. uniflora, Myosotis laxa, Myosoton aquaticum, Myrica gale, Onoclea sensibilis, Oryzopsis asperifolia, Osmunda claytoniana, O. regalis, Osmundastrum cinnamomeum, Panicum hemitomon, Paspalum urvillei, Peltandra virginica, Penthorum sedoides, Persea palustris, Persicaria amphibia, P. arifolia, P. coccinea, P. lapathifolia, P. maculosa, P. pensylvanica, P. punctata, P. sagittata, Petasites frigidus, Phalaris arundinacea, Phegopteris connectilis, Phleum pratense, Phragmites australis, Phlox, Phyla nodiflora, Physostegia virginiana, Picea mariana, Pilea fontana, P. pumila, Pinus arizonica, P. ponderosa, P. strobus, Plantago, Platanthera obtusata, P. psycodes, Pleurozium schreberi, Poa fendleriana, P. palustris, Pogonia ophioglossoides, Polytrichum, Populus tremuloides, Potentilla norvegica, Proserpinaca palustris, Pseudotsuga menziesii, Pteridium aquilinum, Ptilimnium capillaceum, Pycnanthemum virginicum, Quercus lyrata, Q. macrocarpa, Q. rugosa, Ranunculus lapponicus, R. pensylvanicus, R. pusillus, R. septentrionalis,*

Rhexia virginica, Rhododendron groenlandicum, Rhus glabra, R. typhina, Rhynchospora alba, R. capitellata, R. glomerata, Ribes americanum R. hirtellum, Rorippa palustris, Rosa arkansana, Rubus flagellaris, R. idaeus, R. setosus, Rumex altissimus, R. crispus, R. maritimus, R. verticillatus, Sacciolepis striata, Sagittaria brevirostra, S. lancifolia, S. latifolia, S. rigida, Salix ×glatfelteri, S. interior, S. myricoides, S. pedicellaris, S. petiolaris, S. pyrifolia, S. rigida, Sambucus nigra, Samolus valerandi, Sarracenia purpurea, Sassafras albidum, Saururus cernuus, Scheuchzeria palustris, Schoenoplectus pungens, S. smithii, S. tabernaemontani, Scirpus atrovirens, S. cyperinus, S. pedicellatus, Scutellaria galericulata, S. lateriflora, Senecio glabellus, S. neomexicanus, Sisyrinchium, Sium suave, Smilax rotundifolia, S. walteri, Solanum americanum, S. dulcamara, Solidago canadensis, S. gigantea, S. graminifolia, S. macrophylla, Sorbus americana, Sparganium americanum, S. eurycarpum, Spartina patens, S. pectinata, Sphagnum, Spiraea alba, S. tomentosa, Stachys hispida, S. palustris, Stellaria longifolia, S. prostrata, Streptopus roseus, Symphyotrichum lanceolatum, S. lateriflorum, S. puniceum, Symphoricarpos albus, Taxodium distichum, Taxus canadensis, Teucrium canadense, Thalictrum dasycarpum, T. fendleri, T. pubescens, Thelypteris palustris, Thuja occidentalis, Tilia americana, Tillandsia usneoides, Toxicodendron radicans, T. vernix, Triadenum fraseri, T. virginicum, Trientalis borealis, Trifolium resupinatum, Typha angustifolia, T. ×glauca, T. latifolia, Ulmus americana, Utricularia cornuta, U. gibba, Vaccinium angustifolium, V. macrocarpon, V. myrtilloides, V. oxycoccos, Valerianella, Verbena hastata, Veronica anagallis-aquatica, V. scutellata, Viburnum dentatum, V. edule, Vicia americana, V. cracca, Viola affinis, V. blanda, V. lanceolata, V. macloskeyi, V. renifolia, Vitis riparia, Woodwardia areolata, W. virginica, Xyris torta, Zanthoxylum americanum, Zizania aquatica, Zizaniopsis miliacea.

***Galium trifidum* L.** inhabits fresh or brackish bogs, carrs, depressions, ditches, dunes, fens, floodplains, marshes, meadows, ox-bows, prairies, seeps, swales, swamps, vernal pools and the margins of lakes, rivers, and streams at elevations of up to 3350 m. The plants occur from boreal or subarctic to temperate latitudes and are broadly adapted to a wide range of acidity (pH: 5.3–7.8). Their substrates include gravel, loam, marl, muck, sand, and peat. They tolerate low levels of salinity (occasionally at the margins of salt marshes or in temporary saline tidal pools) and grow well in full sun to partial shade. There is little information on the pollination biology or reproductive ecology of this species. The mechanism of seed dispersal also has not been reported. The plants do form a seed bank where as many as 130 viable seeds/m^2 have been retrieved. Vegetative reproduction occurs by means of a slender, horizontal, or underground rhizome. **Reported associates:** *Abies lasiocarpa, Acer rubrum A. saccharinum, Achillea millefolium, A. sibirica, Aconitum columbianum, Acorus americanus, Adenocaulon bicolor, Agrimonia gryposepala, Agrostis gigantea, A. scabra, A. stolonifera, A. tenuis, Aira caryophyllea, Alisma, Alnus rubra, A. rugosa, A. serrulata, A. sinuata, Alopecurus aequalis, Amaranthus tuberculatus, Amblystegium serpens, Amelanchier*

*alnifolia, Aneura pinguis, Angelica arguta, A. atropurpurea,
A. lucida, Antennaria racemosa, Anticlea elegans, Apocynum
androsaemifolium, Aralia nudicaulis, Arctostaphylos uva-ursi,
Asclepias incarnata, A. syriaca, Athyrium filix-femina, Atriplex
patula, Aulacomnium palustre, Betula glandulosa, B. lutea, B.
pumila, Bidens cernuus, B. connatus, B. coronatus, B. frondosus, Boehmeria cylindrica, Bolboschoenus fluviatilis, Bromus
ciliatus, B. vulgaris, Bryum pseudotriquetrum, Calamagrostis
canadensis, C. deschampsioides, C. stricta, Calla palustris, Calliergon giganteum, Caltha leptosepala, C. palustris,
Camassia quamash, Campanula aparinoides, C. rotundifolia,
Campylium stellatum, Cardamine cordifolia, Carex aquatilis,
C. buxbaumii, C. canescens, C. chordorrhiza, C. comosa, C.
deweyana, C. diandra, C. glareosa, C. hystericina, C. interior, C. lacustris, C. lanuginosa, C. lasiocarpa, C. leptalea,
C. limosa, C. lupuliformis, C. lyngbyei, C. nebrascensis, C.
obnupta, C. oligocarpa, C. ovalis, C. paupercula, C. pluriflora,
C. prairea, C. projecta, C. pseudocyperus, C. ramenskii, C. rariflora, C. rostrata, C. sartwellii, C. scoparia, C. scopulorum, C.
stricta, C. unilateralis, C. utriculata, C. vesicaria, C. viridula,
Chamerion angustifolium, Chara, Chelone glabra, Chimaphila
umbellata, Cicuta bulbifera, C. douglasii, C. mackenzieana, C.
maculata, C. virosa, Cirsium arvense, C. vulgare, Clintonia uniflora, Comarum palustre, Corallorrhiza striata, Cornus sericea,
Corylus cornuta, Cotula coronopifolia, Cratoneuron filicinum,
Cuscuta salina, Cyperus diandrus, C. strigosus, Dasiphora
floribunda, Decodon verticillatus, Deschampsia beringensis,
D. cespitosa, Desmodium, Disporum trachycarpum, Distichlis
spicata, Doellingeria umbellata, Drepanocladus aduncus,
Dryopteris cristata, Eleocharis acicularis, E. erythropoda,
E. fallax, E. palustris, E. parvula, Elymus glaucus, Elytrigia
repens, Empetrum nigrum, Epilobium ciliatum, E. hornemannii,
E. leptophyllum, E. palustre, Equisetum arvense, E. fluviatile,
E. hyemale, E. pratense, Erechtites hieracifolia, Eriophorum
scheuchzeri, Eupatorium perfoliatum, Eutrochium maculatum,
Festuca rubra, Fragaria vesca, F. virginiana, Galium aparine,
G. asprellum, G. brevipes, G. labradoricum, G. obtusum, G.
tinctorium, Geum macrophyllum, G. palustre, Glyceria elata,
G. grandis, G. striata, Goodyera oblongifolia, Grindelia stricta,
Helenium bigelovii, Helianthus strumosus, Heliopsis helianthoides, Heracleum lanatum, Hierochloe odorata, Hippuris
vulgaris, Holcus lanatus, Hordeum brachyantherum, Impatiens
capensis, Iris versicolor, I. virginica, Jaumea carnosa, Juncus
balticus, J. dudleyi, J. effusus, J. filiformis, Juniperus communis, Kelloggia galioides, Lactuca serriola, Laportea canadensis, Larix laricina, L. occidentalis, Lathyrus palustris, Leersia
oryzoides, Lemna minor, Leucanthemum vulgare, Ligusticum
scoticum, Lindera benzoin, Linnaea borealis, Listera convallarioides, Lobelia kalmii, L. siphilitica, Lonicera caerulea, L.
involucrata, Lotus corniculatus, Ludwigia palustris, Lupinus
argenteus, L. nootkatensis, Lycopus americanus, L. uniflorus, L. virginicus, Lysichiton americanus, Lysimachia ciliata,
L. quadriflora, L. terrestris, L. thyrsiflora, Lythrum salicaria,
Mahonia repens, Maianthemum stellatum, Marchantia polymorpha, Mentha arvensis, Menziesia ferruginea, Mertensia ciliata, Mimulus glabratus, M. guttatus, M. ringens, Muhlenbergia
glomerata, Myrica gale, Nasturtium officinale, Nyssa sylvatica,
Oenanthe sarmentosa, Onoclea sensibilis, Orthilia secunda,
Oryzopsis asperifolia, Osmorhiza berteroi, Oxypolis occidentalis, O. rigidior, Palustriella falcata, Parentucellia viscosa,
Parnassia fimbriata, Parthenocissus quinquefolia, Paxistima
myrsinites, Peltandra virginica, Penthorum sedoides, Persicaria
amphibia, P. coccinia, P. hydropiper, P. lapathifolia, P. maculosa, P. pensylvanica, P. punctata, P. sagittata, Petasites sagittatus, Phalaris arundinacea, Philonotis fontana, Phleum
alpinum, P. pratense, Phragmites australis, Phyllodoce empetriformis, Physostegia virginiana, Picea engelmannii, Pilea fontana, P. pumila, Pinus albicaulis, P. contorta, Plagiobothrys
figuratus, Plagiomnium ellipticum, Platanthera hyperborea,
Poa eminens, P. palustris, P. pratensis, Polemonium occidentale, Populus balsamifera, P. tremuloides, Potentilla pacifica,
Prenanthes alata, Primula cusickii, Pseudotsuga menziesii,
Pycnanthemum virginianum, Pyrola asarifolia, Ranunculus
gmelinii, R. pensylvanicus, R. repens, R. uncinatus, Rhamnus
alnifolia, Rhinanthus minor, Rhododendron glandulosum,
R. groenlandicum, Ribes, Rorippa alpina, R. palustris, R.
teres, Rosa palustris, R. woodsii, Rubus arcticus, R. idaeus,
R. pubescens, R. setosus, R. spectabilis, R. ursinus, Rudbeckia
californica, Rumex aquaticus, R. maritimus, R. orbiculatus,
R. salicifolius, R. verticillatus, Sagittaria cuneata, S. latifolia,
Salix candida, S. commutata, S. discolor, S. drummondiana,
S. fuscescens, S. geyeriana, S. interior, S. monticola, S. pedi-
cellaris, S. petiolaris, S. planifolia, S. rigida, S. scouleriana,
Sanguisorba canadensis, Sarcocornia perennis, Sarracenia
purpurea, Schizachne purpurascens, Schoenoplectus acutus,
S. pungens, S. tabernaemontani, S. torreyi, Scirpus atrovirens,
S. cyperinus, Scolochloa festucacea, Scorpidium cossonii,
Scutellaria galericulata, S. lateriflora, S. leonardii, Senecio
pseudaureus, S. triangularis, Shepherdia canadensis, Sidalcea
reptans, Sium suave, Solanum dulcamara, Solidago canaden-
sis, S. gigantea, Sparganium angustifolium, S. eurycarpum,
S. minimum, Spartina pectinata, Spergularia macrotheca,
Sphagnum, Spiranthes diluvialis, Spiraea alba, S. betulifolia,
S. tomentosa, Stachys albens, S. palustris, Stellaria humifusa,
S. longifolia, Symphoricarpos albus, Symphyotrichum boreale,
S. foliaceum, S. lanceolatum, S. novae-angliae, S. puniceum,
Symplocarpus foetidus, Thalictrum dasycarpum, Thelypteris
palustris, Tiarella trifoliata, Tolmiea menziesii, Tomentypnum
nitens, Toxicodendron vernix, Triadenum fraseri, T. virgini-
cum, Trientalis europaea, Triglochin maritimum, T. palustre,
Trillium ovatum, Typha angustifolia, T. ×glauca, T. latifolia,
Ulmus americana, Utricularia intermedia, U. macrorhiza,
Vaccinium globulare, V. membranaceum, V. myrtilloides, V.
scoparium, V. vitis-idaea, Veratrum californicum, Verbena
hastata, V. officinalis, Veronica anagallis-aquatica, Viburnum
dentatum, V. edule, V. lentago, V. nudum, V. opulus, Viola epipsila, V. macloskeyi, V. nephrophylla, V. orbiculata, Xerophyllum
tenax, Zanthoxylum americanum, Zizania aquatica.*

Use by wildlife: The mericarps of *Galium* species are consumed occasionally by waterfowl or UPL birds (Aves) and the
stems are eaten by muskrats (Mammalia: Rodentia: Muridae:
Ondatra zibethicus). Although *G. tinctorium* plants contain
8% protein (by dry weight), they also contain an extract that
deters feeding by crayfish (Crustacea: Decapoda: Cambaridae:

Procambarus acutus). *Galium* flowers provide nectar for a number of small flying insects. *Galium asprellum* provides cover and habitat for hatchling wood turtles (Reptilia: Testudines: Emydidae: *Glyptemys insculpta*) and is associated with a plant bug (Insecta: Hemiptera: Miridae: *Polymerus unifasciatus*). Stems of *G. obtusum* are eaten by swamp rabbits (Mammalia: Leporidae: *Sylvilagus aquaticus*). *Galium trifidum* constitutes a component of breeding pool habitat for the Oregon spotted frog (Amphibia: Ranidae: *Rana pretiosa*), a state-endangered species. It also is a host to several fungi (Ascomycota: Dothideomycetidae: *Cercospora punctoidea*, *Septoria cruciatae*; Erysiphomycetidae: *Golovinomyces cichoracearum*; Basidiomycota: Pucciniaceae: *Puccinia punctata*; Oomycota: Peronosporaceae: *Peronospora calotheca*). *Galium tinctorium* is a host of a rust fungus (Basidiomycota: Uropyxidaceae: *Aecidium sparsum*); both *G. asprellum* and *G. tinctorium* are hosts to leaf spot (Ascomycota: Dothideomycetidae: *Cercospora gallii*). The flowers of *G. trifidum* are visited by nectar-collecting insects (Insecta), which include short-tongued bees (Hymenoptera: Colletidae: *Hylaeus mesillae*) and flies (Diptera: Calliphoridae: *Helicobia rapax*; Muscidae: *Bithoracochaeta leucoprocta*; Syrphidae: *Chrysotoxum pubescens*, *Syritta pipiens*, *Toxomerus marginatus*; Tachinidae: *Chetogena claripennis*, *Gymnoclytia occidua*). Muskrats (Mammalia: Rodentia: Muridae: *Ondatra zibethica*) use plants of *G. trifidum* when building their mound-like lodges.

Economic importance: food: The roasted fruits of *Galium aparine* are said to make an excellent substitute for coffee (which also belongs to this family); however, none of the OBL *Galium* species reportedly is edible and *G. asprelleum* has even been described as being poisonous. Several *Galium* species contain an enzyme that is used to coagulate milk in the making of cheese, curds, and whey; **medicinal:** The Choctaw people used *Galium asprellum* to promote sweating, as a diuretic, and to treat measles. It also has been used as a medicine for kidney problems. Crude extracts are effective at inhibiting bacteria (Proteobacteria: Enterobacteriaceae: *Escherichia coli*). *Galium palustre* contains several iridoids (geniposidic acid, 10-deacetylasperulosidic acid, scandoside, monotropein, asperulosidic acid, deacetylasperuloside, asperuloside, 6-O-acetylscandoside), which represent a physiologically active class of compounds having anti-inflammatory, antimicrobial, antiviral, and enzyme inhibitory properties. *Galium tinctorium* was administered medicinally by several Native American tribes. The Ojibwa made an infusion from the plants to treat eczema, respiratory problems, ringworm and scrofula. The Cherokee used the same treatment for gallstones. The Iroquois applied a poultice made from the whole plant to relieve backaches, ruptures, and swelling. The Klallam, Makah, and Quinault tribes used a similar poultice as a hair tonic. The Kwakiutl rubbed plants on their skin to relieve chest pains. The Menominee and Miwok treated kidney ailments by an infusion or tea made from the whole plant. Several tribes (Iroquois, Karok, Quileute) believed that the plants possessed aphrodesiac properties; **cultivation:** *Galium palustre* and *G. tinctorium* are cultivated; **misc. products:** The resilient stems of *Galium* plants were once used as a mattress stuffing, hence their common name of "bedstraw."

Native Americans used various *Galium* species as an abrasive to remove pitch from their hands and (when dried) for starting fires. Crushed plants of *G. tinctorium* were used as a perfume (Makah, Omaha, Ponca) and as a hair wash (Nitinaht). A red dye used to color porcupine quills was derived from the roots of *G. tinctorium* by the Micmac. The French settlers of Canada obtained a red anthraquinone dye from the plants, which they used to color cloth. *Galium asprellum* is a plant recommended for wetland restoration. It has been shown to concentrate heavy metals (Cd, Mn, Pb, Zn) and may be suitable for use in remediation of metal-contaminated sites. *Galium trifidum* has been planted in the restoration of fens that have been impacted by peat mining; **weeds:** *Galium asprellum* is regarded as a weed of cranberry marshes in Wisconsin; **nonindigenous species:** *Galium palustre* has been introduced to Argentina, Australia (Tasmania), and New Zealand.

Systematics: Several phylogenetic studies have corroborated the placement of *Galium* within the monophyletic tribe Rubieae of subfamily Rubioideae (Rubiaceae). However, this rather large genus, which has been divided into 13 sections, is polyphyletic as traditionally circumscribed. Notably, several surveyed members of *Galium* section *Aparinoides* (which contains all but one of the species treated here) resolve as being more closely allied to section *Glabella* of the genus *Asperula*, with members of section *Platygalium* resolving among the genera *Cruciata* and *Valantia*. Within section *Aparinoides*, five informal "species groups" have been recognized, with *G. brevipes*, *G. tinctorium*, and *G. trifidum* placed in the "*G. trifidum* group," *G. palustre* in the "*G. palustre* group," and *G. labradoricum* and *G. obtusum* assigned to the "*G. obtusum* group." *Galium asprellum* is assigned to section *Trachygalium*, which also contains the North American *G. concinnum* (FACU, UPL). No recent taxonomic treatment has been prepared for this section. Additional sampling of *Galium* species in phylogenetic analyses will be necessary before the circumscription of the genus and its sections can be refined effectively. *Galium obtusum* has been merged with *G. tinctorium* by some authors, but this disposition is unfounded and based on the early confusion of these species. Both *G. brevipes* and *G. tinctorium* have been included as varieties (or subspecies) of *G. trifidum* by some authors; however, the latter has been shown to be isolated genetically from both of the former species. *Galium labradoricum* was regarded as a variety of *G. tinctorium* in older treatments; however, it is now considered to be closely related to *G. obtusum*. *Galium tinctorium* also has been placed in the genus *Asperula*. *Galium elongatum* has been regarded as an infraspecific taxon of *G. palustre*; however, it remains morphologically and cytologically distinct from that species. *Galium trifidum* is variable morphologically and ecologically with four recognized subspecies. The base chromosome number for *Galium* section *Aparinoides* is $x = 12$. Diploids ($2n = 24$) includes *G. labradoricum*, *G. tinctorium*, and *G. trifidum*. *Galium obtusum* ($2n = 48$) is a tetraploid. *Galium palustre* ($2n = 24$, 48, 96) occurs as diploid, tetraploid, or octoploid cytotypes, which arguably are of autoploid origin; however, only diploids have been reported from North America.

Comments: *Galium trifidum* extends through Eurasia and occurs throughout North America except for the southeast.

The other species are somewhat more restricted and occur mainly in eastern North America (*G. asprellum*, *G. obtusum*, *G. tinctorium* [also in north and central Europe]), northeastern North America (*G. brevipes*, *G. palustre* [also in Europe]), or northern North America (*G. labradoricum*).

References: Adamus, 2005; Allen et al., 2002; Anderson, 1943; Andersson & Nilsson, 2002; Andersson & Rova, 1999; Andreas & Bryan, 1990; Beal, 1977; Bergeron et al., 1996; Cellot et al., 1998; Cobbaert et al., 2004; Cooper & Jones, 2004; Falińska, 1999; Gilbert & Cooke, 2001; Hancock, 1942; Handa et al., 2002; Hanlon et al., 1998; Hanson, 1951; Harney et al., 1998; Jean & Bouchard, 1993; Jervis et al., 1993; Kangas & Hannan, 1985; Keller, 2000; Manen et al., 1994; McDonald et al., 1996; McKenzie et al., 2003; Meisel et al., 2002; Natali et al., 1995; Nilsson & Nilsson, 1978; Often et al., 2006; Peinado et al., 1998; Pellerin et al., 2006; Prusak et al., 2005; Puff, 1976, 1977; Rabinowitz & Rapp, 1985; Rabinowitz et al., 1981; Rivas-Martínez et al., 1999; Segadas-Vianna, 1951; Smith & Kadlec, 1983; Stewart, Jr. & Nilsen, 1993; Teppner et al., 1976; Terrel, 1972; Tuttle & Carroll, 2005; Watson et al., 2000; Wiegand, 1897.

4. Oldenlandia

Bluet, mille graines

Etymology: After Henrik Bernhard Oldenland (1663–1699)

Synonyms: *Anistelma*; *Edrastima*; *Eionitis*; *Gerontogea*; *Gonotheca*; *Hedyotis* (in part); *Karamyschewia*; *Listeria*; *Mitratheca*; *Stelmanis*; *Thecagonum*; *Theyodis*

Distribution: global: Africa; Asia; Australia; New World; **North America:** southeastern

Diversity: global: 100 species; **North America:** 4 species

Indicators (USA): OBL; FACW: *Oldenlandia boscii*

Habitat: freshwater; palustrine; **pH:** unknown; **depth:** <1 m; **life-form(s):** emergent herb

Key morphology: stems (to 3 dm) prostrate or spreading, much branched; leaves (to 25 mm) opposite, linear, sessile; flowers small, solitary, or in axillary clusters of 2–3, subsessile; corolla rotate, 4-merous; corolla lobes (to 1 mm) white, sometimes pink margined; ovary inferior; capsule (to 2.5 mm) campanulate, smooth; seeds numerous, angular, purple

Life history: duration: perennial (rhizomes); **asexual reproduction:** rhizomes; **pollination:** unknown; **sexual condition:** hermaphroditic; **fruit:** capsules (common); **local dispersal:** rhizomes, seeds; **long-distance dispersal:** seeds

Imperilment: (1) *Oldenlandia boscii* [G5]; S1 (MO, NC, VA); S3 (GA, KY)

Ecology: general: Ecological information for North American species of *Oldenlandia* is scarce. Several species are designated as FACW, FAC, or FACU, but only one is categorized as OBL (and only in a portion of its range). The pollination biology and seed ecology remain unreported for much of the genus.

Oldenlandia boscii (DC.) Chapm. grows in borrow pits, deltas, depressions, ditches, flatwoods, floodplains, lake terraces, mudflats, roadsides, sandbars, savannas, swales, swamps, and along the margins of lakes, ponds, rivers, and ephemeral streams at elevations of up to 620 m, mainly along the southeastern coastal plain of the United States. The plants occur on peat,

gravel, loamy sand, sand, or sandy loam in open areas or clearings. Habitats typically experience seasonal water and include dredged sites or exposed shores, pond bottoms and mudflats where drawdowns characteristically occur. The plants remain in flower and set fruit continuously throughout the year. The flowers are homostylous but their breeding system has not been determined. The seeds can represent a dominant component of the seed bank, even at sites where the species is rare in the standing vegetation, especially in areas that have experienced fires. The seeds germinate well after stratification for 40+ days at 6°C. Their method of dispersal is not known. **Reported associates:** *Betula nigra*, *Botrychium lunarioides*, *Cephalanthus occidentalis*, *Cladium jamaicense*, *Cyperus echinatus*, *C. erythrorhizos*, *C. esculentus*, *C. flavicomus*, *C. polystachyos*, *Digitaria*, *Diodia teres*, *Echinochloa*, *Eleocharis acicularis*, *E. melanocarpa*, *E. microcarpa*, *Eragrostis hypnoides*, *Erigeron quercifolius*, *Eupatorium capillifolium*, *Fimbristylis autumnalis*, *F. vahlii*, *Fuirena*, *Habenaria repens*, *Hydrolea quadrivalvis*, *Hypericum denticulatum*, *H. mutilum*, *Ilex*, *Juncus repens*, *Leersia hexandra*, *Leptochloa*, *Lindernia dubia*, *Lipocarpha maculata*, *L. micrantha*, *Liquidambar styraciflua*, *Ludwigia palustris*, *Micranthemum umbrosum*, *Mitreola petiolata*, *Mollugo verticillata*, *Nyssa ogeche*, *N. sylvatica*, *Oldenlandia uniflora*, *Panicum dichotomiflorum*, *P. hemitomon*, *P. wrightianum*, *Paspalum dissectum*, *P. laeve*, *P. notatum*, *P. urvillei*, *P. verrucosum*, *Planera aquatica*, *Quercus nigra*, *Rhexia mariana*, *Rhynchospora caduca*, *R. perplexa*, *Rotala ramosior*, *Sabatia calycina*, *S. campanulata*, *Salix nigra*, *Taxodium*, *Triadenum*, *Viola lanceolata*, *Xanthium strumarium*, *Xyris*.

Use by wildlife: none reported

Economic importance: food: not reported to be edible; **medicinal:** no reported uses; **cultivation:** *Oldenlandia boscii* is not cultivated; **misc. products:** none; **weeds:** none; **nonindigenous species:** *Oldenlandia boscii* is native to North America

Systematics: Phylogenetic studies using DNA sequence data resolve *Oldenlandia* within tribe Hedyotidae of subfamily Rubioideae (Rubiaceae); however, not as a monophyletic group. Although only a few species of this large genus have been included in phylogenetic surveys, it is evident that *Oldenlandia* is polyphyletic, and that the North American species do not share a common ancestry with the African species, which comprise the majority of the genus. Moreover, even the New World species of *Oldenlandia* are polyphyletic (Figure 5.47), and a thorough taxonomic reevaluation of the genus is needed. In particular, the OBL North American *O. boscii* does not associate with any species of *Oldenlandia* in studies conducted thus far (Figure 5.47). *Oldenlandia boscii* has been placed formerly in *Hedyotis* (as *H. boscii*); however, the polyphyly of *Hedyotis* (Figure 5.47) makes it difficult to evaluate the taxonomic merit of such a disposition at this time. It is evident that future studies eventually will result in the transfer of *O. boscii* to a different genus. The base chromosome number of *Oldenlandia* (as currently circumscribed) is $x = 9$, in which case *O. boscii* ($2n = 36$) would be tetraploid.

Comments: *Oldenlandia boscii* occurs in the southeastern United States

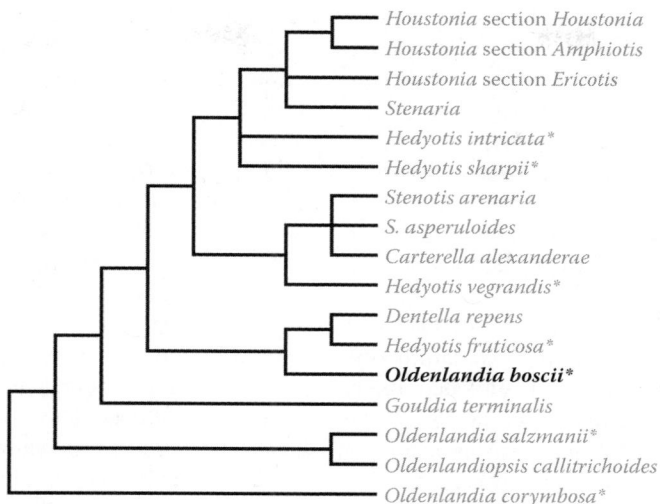

Houstonia section Houstonia
Houstonia section Amphiotis
Houstonia section Ericotis
Stenaria
Hedyotis intricata*
Hedyotis sharpii*
Stenotis arenaria
S. asperuloides
Carterella alexanderae
Hedyotis vegrandis*
Dentella repens
Hedyotis fruticosa*
Oldenlandia boscii*
Gouldia terminalis
Oldenlandia salzmanii*
Oldenlandiopsis callitrichoides
Oldenlandia corymbosa*

FIGURE 5.47 Relationships among representative taxa within tribe Hedyotidae of Rubiaceae (subfamily Rubioideae) as indicated by phylogenetic analysis of *trnL* sequence data. This analysis indicates the polyphyly of several genera (asterisked) including *Oldenlandia* and further studies that include additional taxa will be necessary to determine more realistic limits of these genera. The OBL North American *Oldenlandia boscii* (bold) does not group with other surveyed members of the genus, but resolves here with the Indian species *Dentella repens* and *Hedyotis fruticosa*. Analysis of nrITS sequence data (Church, 2003) differs by associating *O. boscii* with *Stenotis asperuloides* (however, based on a smaller assemblage of taxa) and also indicate the polyphyly of *Oldenlandia*. (Adapted from Church, S.A., *Mol. Phylogen. Evol.*, 27, 223–238, 2003.)

References: Andersson & Rova, 1999; Church, 2003; DeBerry & Perry, 2005; Kirkman & Sharitz, 1994; Lewis, 1962b; Terrell & Lewis, 1990.

5. *Pentodon*

Hale's pentodon

Etymology: from the Greek *penta odon* meaning "five toothed"

Synonyms: *Hedyotis* (in part); *Oldenlandia* (in part)

Distribution: global: Africa; New World; **North America:** southeastern

Diversity: global: 2 species; **North America:** 1 species

Indicators (USA): OBL: *Pentodon pentandrus*

Habitat: freshwater; palustrine; **pH:** alkaline; **depth:** <1 m; **life-form(s):** emergent herb

Key morphology: branches diffuse, prostrate, creeping or spreading, 4-angled; leaves (to 5 cm) opposite, lanceolate, tapering to short, decurrent petioles, stipules membranous; cymes short, axillary, or appearing terminal; pedicels (to 4 mm long) thick; flowers 5-merous; corolla campanulate, white, readily deciduous, the tube (to 3 mm) funnelform, the lobes (to 2.5 mm) triangular; ovary inferior, 2-locular, capsule (to 4 mm) 2-valved; seeds (20+) pitted, reddish brown

Life history: duration: annual (fruit/seeds); **asexual reproduction:** none; **pollination:** unknown; **sexual condition:** hermaphroditic; **fruit:** capsules (common); **local dispersal:** seeds; **long-distance dispersal:** seeds

Imperilment: (1) *Pentodon pentandrus* [G5]; S1 (GA)

Ecology: general: Little ecological information exists for this small genus.

Pentodon pentandrus **(Schumach. & Thonn.) Vatke** grows in ditches, flatwoods, floodplains, marshes, meadows, prairies, sloughs, swales, swamps, and along the margins of canals, lakes, ponds, rivers, and streams along the southeastern coastal plain. It occurs on substrates of limerock, marl, mud, silt, and sand. Although the plants sometimes are described as "light demanding" they also are reported in deep shade. They can occur in disturbed wetlands and in drying sites that experience seasonal inundation. Under suitable conditions, the plants can form fairly extensive sprawling mats. Nothing is known regarding the pollination biology or seed ecology of this species. **Reported associates:** *Acer rubrum, Amorpha fruticosa, Ampelopsis arborea, Andropogon glomeratus, A. virginicus, Asclepias perennis, Baccharis halimifolia, Berchemia scandens, Bidens alba, Blechnum serrulatum, Boehmeria cylindrica, Buchnera americana, Bumelia celastrina, Callicarpa americana, Chiococca parviflora, Cladium jamaicense, Conoclinium coelestinum, Crotalaria rotundifolia, Cynanchum scoparium, Desmodium paniculatum, Dichanthelium, Dichondra carolinensis, Echinochloa, Elephantopus alatus, Erechtites hieracifolia, Eryngium balduinii, Eupatorium capillifolium, E. mikanioides, Eustachys glauca, Galium hispidulum, Habenaria, Hypericum mutilum, Ilex glabra, Imperata cylindrica, Ipomoea sagittata, Iresine diffusa, Kosteletzkya pentacarpos, Lantana camara, Ludwigia repens, Lyonia fruticosa, Lythrum alatum, Malvastrum corchorifolium, Mitreola petiolata, M. sessilifolia, Muhlenbergia capillaris, Myrica cerifera, Nyssa, Oxalis, Parthenocissus quinquefolia, Paspalum ciliatifolium, P. monostachyum, Persicaria hydropiperoides, Phlebodium aureum, Physalis viscosa, Phytolacca americana, Piloblephis rigida, Pinus elliottii, Pityopsis graminifolia, Pluchea odorata,*

Polypremum procumbens, Pteridium aquilinum, Pterocaulon virgatum, Quercus laurifolia, Q. virginiana, Rhus copallinum, Rhynchospora pusilla, Rubus trivialis, Sabal palmetto, Salix caroliniana, Samolus parviflorus, Schinus terebinthifolius, Schizachyrium rhizomatum, Serenoa repens, Setaria magna, Sisyrinchium solstitiale, Smilax auriculata, Solidago, Sporobolus indicus, Stillingia sylvatica, Stipa, Taxodium, Tephrosia rugelii, Thelypteris kunthii, Toxicodendron radicans, Vaccinium myrsinites, Vitis munsoniana, Vittaria lineata.

Use by wildlife: none reported

Economic importance: food: *Pentodon pentandrus* is eaten as a vegetable in Ghana; **medicinal:** *Pentodon pentandrus* is used medicinally in Ghana; **cultivation:** not cultivated; **misc. products:** none; **weeds:** none; **nonindigenous species:** There is some debate whether *Pentodon pentandrus* is native to North America or if it was introduced from Africa.

Systematics: Preliminary phylogenetic investigations of Rubiaceae show *Pentodon* to be a relatively isolated genus within tribe Hedyotidae of subfamily Rubioideae (Figure 5.46). There are only two species in the genus, *Pentodon laurentioides* (from Somalia) and the wider ranging *P. pentandrus*; however, only the latter has been included in phylogenetic analyses. The base chromosome number of *Pentodon* is $x = 9$ with *P. pentandrus* ($2n = 18$) being diploid. Genetic studies to elucidate the geographical origin of New World populations of *P. pentandrus* might help to clarify its status as a native species in this region.

Comments: *Pentodon pentandrus* occurs along the coastal plain of the southeastern United States into Mexico, Central and South America as well as in Africa

References: Guarino, 1997; Lewis, 1965; Rogers, 2005.

6. *Pinckneya*

Bitterbark, fevertree

Etymology: after Charles Cotesworth Pinckney (1746–1825)
Synonyms: *Bartramia*; *Bignonia* (in part); *Eupinckneya*; *Mussaenda*; *Pinknea* (orth. var.)
Distribution: global: North America; **North America:** southeastern
Diversity: global: 1 species; **North America:** 1 species
Indicators (USA): OBL: *Pinckneya bracteata*
Habitat: freshwater; palustrine; **pH:** 3.9–5.0; **depth:** <1 m; **life-form(s):** emergent shrub, emergent tree
Key morphology: stems (to 8 m) woody, with white lenticels (to 3 mm) and conical buds; leaves (to 27 cm) opposite, ovate or elliptic, petioled (to 30 mm), clustered at branch tips, stipules (to 10 mm) narrowly triangular; panicles lax, with leaf-like bracts and opposite, decussate branchlets that terminate in cymes (to 15 cm) with up to 25 flowers; flowers 5-merous, pedicelled (to 5 mm); calyx with linear lobes (to 20 mm), lobes of some flowers expanded into showy, white, lavender, or lilac petal-like calycophylls (to 7.5 cm); corolla tubular, the lobes (to 5 cm) reflexed, fleshy, greenish, pinkish, white, yellow, or red/brown streaked; style (to 6.1 cm) exserted; capsules (to 2.5 cm) subglobose, bilobed, punctiform (with lenticels); seeds (to 3 mm) with a thin round wing (to 7 mm)

Life history: duration: perennial (buds); **asexual reproduction:** unknown; **pollination:** insect; **sexual condition:** hermaphroditic; **fruit:** capsules (common); **local dispersal:** seeds (wind); **long-distance dispersal:** seeds
Imperilment: (1) *Pinckneya bracteata* [G4]; S1 (SC); S3 (GA)
Ecology: general: *Pinckneya* is monotypic.

Pinckneya bracteata (Bartram) Raf. grows in bayheads, bays, bogs, ditches, flatwoods, floodplains, hammocks, sandridges, savannas, seeps, springheads, swamps, and along streams. Its substrates (muck, sand, sandy loam, sandy peat) typically are acidic (e.g., pH: 3.9–4.5); however, the plants also reportedly can survive on alkaline soils. The plants tolerate full sun to partial shade but cannot withstand deep shade or prolonged drought conditions. The plants can tolerate brief exposures to fairly cold temperatures (to −23°C). Flowering occurs in May–June and fruiting in July. An outcrossing breeding system is indicated by the protandrous flowers, which present pollen for 1–2 days before the anthers wither; the stigmas become receptive on the third day. The principal pollinators are cloudless sulfur butterflies (Lepidoptera: Pieridae: *Phoebis sennae*), which pollinate the flowers as they feed on the floral nectar. There is no information on seed ecology, but seeds available from commercial sources germinate in 2–8 weeks. The seeds are winged and presumably dispersed at least locally by the wind. The plants are propagated easily by softwood cuttings. Some occurrences have been described as clonal; however, there is no description of natural vegetative reproduction for this species. The plants occur commonly in bog ecotones. **Reported associates:** *Acer rubrum, Agalinis, Alnus serrulata, Andropogon virginicus, Aristida spiciformis, Athyrium filix-femina, Baccharis halimifolia, Balduina atropurpurea, Cacalia, Carex atlantica, Clethra alnifolia, Cliftonia monophylla, Coreopsis, Cyrilla racemiflora, Drosera intermedia, Erianthus gigantea, Eriocaulon decangulare, Eupatorium, Eurybia paludosa, Euthamia minor, Gordonia lasianthus, Helianthus, Hypericum, Ilex coriacea, I. glabra, I. opaca, Illicium floridanum, Itea virginica, Lachnanthes virginiana, Leucothoe axillaris, Liriodendron tulipifera, Lobelia glandulosa, Lyonia lucida, Magnolia virginiana, Myrica cerifera, M. heterophylla, Nyssa biflora, N. sylvatica, Osmunda regalis, Osmundastrum cinnamomeum, Pallavicinia lyellii, Panicum, Peltandra sagittifolia, P. virginica, Persea borbonia, P. palustris, Pinguicula pumila, Pinus elliottii, P. palustris, P. serotina, P. taeda, Polytrichum commune, Pyrrhobryum spiniforme, Quercus, Rhus copallinum, Rhynchospora, Rubus pensilvanicus, Sarracenia minor, S. purpurea, Schizachyrium, Sorbus, Sphagnum henryense, S. recurvum, Sporobolus, Taxodium ascendens, Toxicodendron vernix, Vaccinium arboreum, V. corymbosum, Viburnum nudum, Viola ×primulifolia, Woodwardia areolata, W. virginica, Xyris.*
Use by wildlife: *Pinckneya bracteata* is a host plant for several Lepidoptera including the silver-spotted skipper (Hesperiidae: *Epargyreus clarus*) and variable oakleaf caterpillar (Notodontidae: *Lochmaeus manteo*). The plants also are host to an Oomycota fungus (Pythiaceae: *Phytophthora nicotianae*) and thysanopteran (Insecta: Thysanoptera: Thripidae: *Thrips quinciensis*). The flowers provide nectar to butterflies such as the cloudless sulfur (see pollination above).

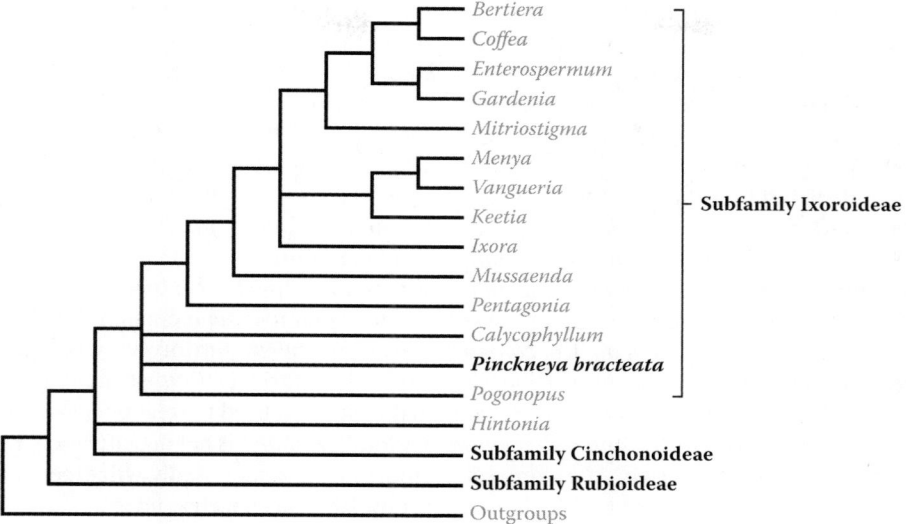

FIGURE 5.48 Phylogenetic position of *Pinckneya* (bold) as indicated by analysis of *rbcL* sequence data. These data indicate that the genus does not occur in Rubiaceae subfamily Cinchonoideae as once presumed; rather, it resolves near the base of subfamily Ixoroideae. (Adapted from Bremer, B. et al., *Ann. Missouri Bot. Gard.*, 82, 383–397, 1995.)

Economic importance: food: not edible; **medicinal:** The bark of *Pinckneya bracteata* was used as a substitute for quinine (*Cinchona*) to treat malaria and other fevers, especially during the American Civil War. The plants contain the physiologically active aribin (1-methyl-beta-carboline) and a glucoside called pinckneyin, but apparently are poor in alkaloids; however, their phytochemistry deserves a more detailed study; **cultivation:** *Pinckneya bracteata* is cultivated (sometimes as *P. pubens*) as an ornamental. Cultivars include 'Bostwick,' 'Carpenter Road Pink,' 'Larry's Party,' 'May's Pink,' 'Off-White,' 'Pink Fever,' 'Rose Pink,' and 'Savannah Pink'; **misc. products:** none; **weeds:** none; **nonindigenous species:** none

Systematics: *Pinckneya* is monotypic and therefore also monophyletic. Although morphological data indicate a fairly close relationship of *Pinckneya* to *Cinchona* (subfamily Cinchonoideae), *rbcL* sequence data (Figure 5.48) and chloroplast DNA [restriction fragment length polymorphism (RFLP)] data place the genus near *Calycophyllum* and *Pogonopus* within subfamily Ixoroideae. Combined morphological and cpDNA data resolve these three genera as a clade situated between the two subfamilies. The relatively recently discovered genus *Kerianthera* also is quite similar to *Pinckneya*, but the two genera have not yet been included in comprehensive phylogenetic analyses simultaneously to determine their degree of relatedness. *Pogonopus* has been hypothesized as the sister genus to *Pinckneya*, a result upheld by phylogenetic analysis of morphological data, but not cpDNA data, which resolve *Calycophyllum* and *Pogonopus* as a clade. However, combined cpDNA and morphological data do place *Pogonopus* and *Pinckneya* as sister genera. The chromosome number of *Pinckneya* is unknown. Population genetic studies should be undertaken to clarify the breeding system and to ascertain the extent of genetic variation residing in this rare and local species.

Comments: *Pinckneya bracteata* is narrowly restricted to the southeastern United States (Florida, Georgia, South Carolina).

References: Bremer, 1996; Bremer & Jansen, 1991; Bremer et al., 1995; Delprete, 1996; Elias, 1987; Godfrey & Wooten, 1981; Hasegawa & Hambrecht, 2003; Kirkbride, Jr., 1985; Lincicome, 1998; Naudain, 1885; Schornherst, 1943.

7. Strumpfia

Pride-of-Big-Pine
Etymology: after Christoph Karl Strumpff (1712–1754)
Synonyms: *Patsjotti*
Distribution: global: North America; **North America:** southeastern
Diversity: global: 1 species; **North America:** 1 species
Indicators (USA): OBL: *Strumpfia maritima*
Habitat: brackish/saline (coastal); palustrine; **pH:** alkaline; **depth:** <1 m; **life-form(s):** emergent shrub
Key morphology: stems (to 2 m) woody, ringed by stipule bases; leaves (to 3 cm) evergreen, linear, leathery, whorled, clustered at branch tips, petioled (to 1 mm), margins revolute, foliage pine-scented; racemes axillary, few-flowered; flowers 5-merous; corolla campanulate, tube (to 1 mm) short, lobes (to 6 mm across) spreading, white to pink; drupe (to 6 mm) fleshy, globose, white, with calyx persistent, 1- or 2-seeded
Life history: duration: perennial (buds); **asexual reproduction:** stem sprouts; **pollination:** insect; **sexual condition:** hermaphroditic; **fruit:** drupes (common); **local dispersal:** seeds; **long-distance dispersal:** seeds (birds)
Imperilment: (1) *Strumpfia maritima* [G4]; S1 (FL)
Ecology: general: *Strumpfia* is monotypic.

Strumpfia maritima **Jacq.** grows on coastal beaches, cliffs, flats, and swamps at low elevations (<10 m). Substrates can be saline and include pitted limestone, rocks, rubble, and sand. The plants are salt tolerant and can withstand heavy salt spray and seawater overwash. *Strumpfia maritima* blooms and fruits throughout the year. The flowers are proterogynous and the styles retract prior to the male phase. The poricidal anthers are fused together into a ring and facilitate "buzz pollination" of the flowers by insects. The seeds germinate (~6%) after 4–6 months when placed on moist peat. They are believed to be dispersed by birds. The stems will resprout from the stumps (coppice) when broken or cut. More detailed studies on the pollination biology, population genetics, and reproductive ecology would provide helpful information for managing this rare species. **Reported associates:** *Acacia farnesiana, Acanthocereus tetragonus, Andropogon ternarius, Aristida, Bourreria cassinifolia, Bursera simaruba, Byrsonima lucida, Caesalpinia pauciflora, Catesbaea parviflora, Chamaecrista lineata, Chamaesyce, Chrysobalanus icaco, Coccoloba uvifera, Coccothrinax argentata, Croton linearis, Dipholis salicifolia, Eugenia foetida, Evolvulus grisebachii, Ficus aurea, F. citrifolia, Forestiera segregata, Lantana involucrata, Linum arenicola, Metopium toxiferum, Myrica cerifera, Myrsine floridana, Opuntia stricta, Persea borbonia, Pinus elliottii, Piscidia piscipula, Psidium longipes, Randia aculeata, Rhachicallis americana, Rhus copallinum, Sabal palmetto, Savia bahamensis, Schizachyrium sanguineum, Serenoa repens, Sideroxylon salicifolium, Sophora tomentosa, Sorghastrum secundum, Spiranthes torta, Suriana maritima, Swietenia mahagoni, Thrinax morrisii, T. radiata.*

Use by wildlife: In the Caribbean islands, the flowers, fruits, leaves, and seeds of *Strumpfia maritima* are eaten by the endangered Bahaman Hutia (Mammalia: Rodentia: Capromyidae: *Geocapromys ingrahami*), Swan Island Hutia (Mammalia: Rodentia: *Geocapromys thoracatus*), Exuma Island iguana (Reptilia: Squamata: Iguanidae: *Cyclura cychlura figginsi*), the Cuban iguana (Reptilia: Squamata: Iguanidae: *Cyclura nubila nubila*), and undoubtedly by various other animals.

Economic importance: food: not edible; **medicinal:** Extracts of *Strumpfia maritima* contain the flavonol glycoside narcissin and exhibit antifertility activity in female rats (Mammalia: Rodentia: Muridae: *Rattus*). Leaf infusions are stimulating and have been used to treat poisonous bites and fevers; **cultivation:** *Strumpfia* is not in cultivation; **misc. products:** smoke from this plant supposedly will repel mosquitos; **weeds:** none; **nonindigenous species:** none

Systematics: Phylogenetic analyses of DNA sequence data resolve the monotypic *Strumpfia* as the sister group to a clade comprising tribes *Catesbaeeae* and *Chiococceae*, i.e., the "*Catesbaeeae-Chiococceae* complex" of Rubiaceae subfamily Cinchonoideae (Figure 5.49). Its consistent distinction from that complex by both molecular and morphological

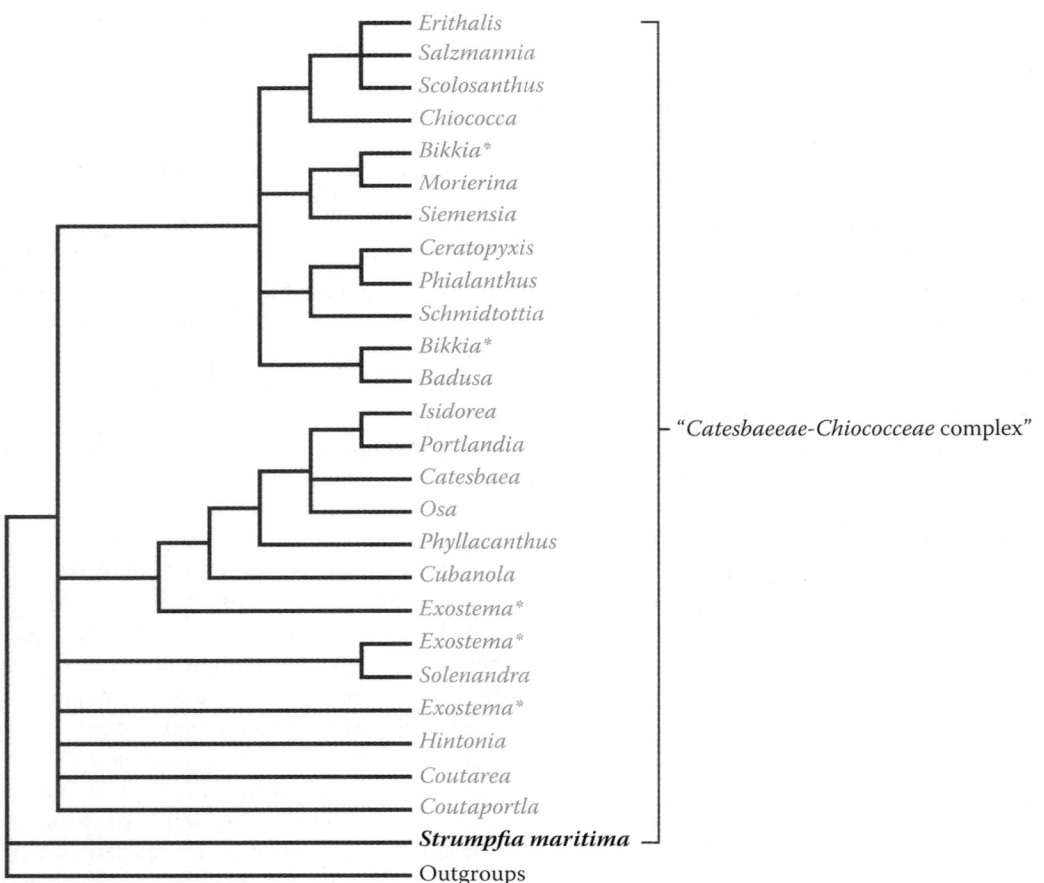

FIGURE 5.49 Relationships of *Strumpfia maritima* (bold) as indicated by phylogenetic analysis of combined *trnL-F* and nrITS sequence data. These results resolve *Strumpfia* as an isolated genus outside of the "CCC" (Catesbaeeae–Chiococceae complex) of Rubiaceae subfamily Cinchonoideae and also indicate the polyphyly of several larger genera (asterisked). (Adapted from Motley, T.J. et al., *Amer. J. Bot.*, 92, 316–329, 2005.)

data analysis has led to its current recognition as a monotypic tribe. The base chromosome number of *Strumpfia* is *x* = 11, with *S. maritima* (2*n* = 22) being a diploid.

Comments: *Strumpfia maritima* is restricted to Florida (where it is listed as endangered) and to several of the Caribbean islands

References: Hsu et al., 1981; Igersheim, 1993; Motley et al., 2005; Puff et al., 1995; Rova et al, 2002.

ORDER 4: LAMIALES [21]

The taxonomy of the order Lamiales, which contains roughly 18,000 species, has undergone extensive revision as a result of molecular phylogenetic research. Notably, this group now subsumes a number of former orders including Bignoniales, Callitrichales, Plantaginales, and Scrophulariales (Judd et al., 2002; Soltis et al., 2005). Many changes involve the family Scrophulariaceae, which is polyphyletic as formerly defined, and now has been subdivided to reflect more natural groups (Olmstead et al., 2001; Albach et al., 2005; Oxelman et al., 2005; Rahmanzadeh et al., 2005; Tank et al., 2006). A number of significant modifications (e.g., merger of Avicenniaceae and Acanthaceae; merger of Callitrichaceae and Hippuridaceae with Plantaginaceae; transfer of *Agalinis, Castilleja, Cordylanthus, Macranthera, Pedicularis* to Orobanchaceae; transfer of *Glossostigma, Mimetanthe, Mimulus*, to Phrymaceae; transfer of *Lindernia, Micranthemum* to Linderniaceae, transfer of many genera to Plantaginaceae, etc.) have been necessary to maintain taxa that are meaningful with respect to results from recent phylogenetic analyses. These dispositions have been followed here.

Molecular studies consistently resolve the Lamiales as monophyletic, and the clade possesses a number of defining characteristics including the replacement of starch by oligosaccharides for carbohydrate storage, the common occurrence of 6-oxygenated flavones, diacytic stomates, onagrad embryos, the extension of parenchyma into anther locules, and protein inclusions in the mesophyll nuclei (Judd et al., 2002). The flowers are mainly bilabiate (i.e., "two lipped") with two or four stamens, which are didynamous in the latter case.

Relationships among the families of Lamiales have been fairly well established as a result of combined analysis of multiple DNA sequence data sets (Olmstead et al., 2001; Oxelman et al., 2005). A current phylogenetic perspective of Lamiales indicates the repeated origin of OBL taxa across the entire order (Figure 5.50). Although it is quite likely that additional modification to the classification of Lamiales will be forthcoming as a consequence of continued phylogenetic investigations, these hypothesized relationships at least appear to reflect the major clades of diversification in the order with reasonable accuracy.

The Lamiales are important economically as the source of olives (*Olea*), ash lumber (*Fraxinus*), and medical digitalin (*Digitalis*). The group contains important ornamental plants such as African violets (*Saintpaulia*), shrimp plant (*Justicia*), and snapdragons (*Antirrhinum*). *Striga* (witchweed) is a major agricultural weed.

There are 10 families of Lamiales containing species that have been designated as obligate aquatics in North America:

1. **Acanthaceae** Juss.
2. **Lamiaceae** Martinov
3. **Linderniaceae** Borsch, K. Müll. & Eb. Fisch.
4. **Oleaceae** Hoffmanns. & Link
5. **Orobanchaceae** Vent.
6. **Phrymaceae** Schauer
7. **Plantaginaceae** Juss.
8. **Lentibulariaceae** Rich.

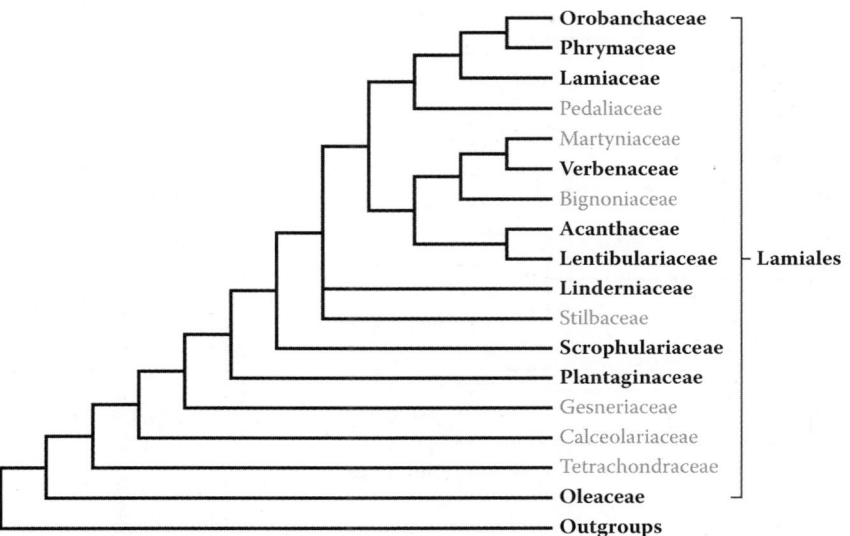

FIGURE 5.50 Hypothetical phylogenetic relationships in Lamiales as resolved by analysis of combined DNA sequence data. Families that contain obligate aquatics in North America are shown in bold. This cladogram would indicate that OBL taxa have evolved repeatedly within the order. (Adapted from Oxelman, B. et al., *Taxon*, 54, 411–425, 2005.)

9. **Scrophulariaceae** Juss.
10. **Verbenaceae** Adans.

Family 1: Acanthaceae [240]

The Acanthaceae are principally a herbaceous family (some shrubs and vines) consisting of nearly 3500 species that are distributed mainly in terrestrial habitats (rarely aquatic) ranging from tropical to warm temperate regions (McDade et al., 2000; Judd et al., 2002; Wasshausen, 2004). Phylogenetic analyses of various DNA sequence data (Figure 5.51) confirm that the family is monophyletic if merged with the former Avicenniaceae (Schwarzbach & McDade, 2002). By this broader circumscription, it is difficult to define the family morphologically, although the presence of articulated nodes, lack of endosperm, and flowers that are subtended by a bract and two bracteoles represent practical synapomorphies. The foliage often contains calcium carbonate inclusions known as cystoliths. It also is common for the ovules to be borne on modified stalks (retinacula), which forcefully discharge the seeds (up to 5 m) upon the explosive dehiscence of the capsular fruits.

The flowers are bilateral, 5-merous and bilabiate, nectariferous, usually protogynous and pollinated by insects or birds (Judd et al., 2002). Several popular ornamental genera (e.g., *Acanthus*, *Aphelandra*, *Barleria*, *Eranthemum*, *Fittonia*, *Hypoestes*, *Justicia*, *Pseuderanthemum*, *Ruellia*, *Strobilanthes*, and *Thunbergia*) occur in the family.

Acanthaceae predominantly are terrestrial; however, OBL North American species occur in four genera:

1. *Avicennia* L.
2. *Hygrophila* R. Br.
3. *Justicia* L.
4. *Stenandrium* Nees

Phylogenetic analyses (Figure 5.51) show all four genera to be relatively distantly related from one another, a result that indicates the OBL habit has arisen several independent times within the family, in both basal and relatively derived clades.

1. Avicennia

Black mangrove
Etymology: after Avicenna (Ibn Sina) (981–1037)
Synonyms: *Halodendrum*; *Hilairanthus*; *Horau*; *Sceura*; *Upata*
Distribution: **global:** pantropical; **North America:** southern
Diversity: global: 8 species; **North America:** 2 species
Indicators (USA): OBL: *Avicennia germinans*, *A. marina*
Habitat: marine; palustrine; **pH:** 5.8–8.5; **depth:** 0–2 m; **life-form(s):** emergent shrub, emergent tree
Key morphology: stems (to 25 m; usually <3 m) woody, producing dense, horizontally radiating, subterranean rows of emergent, woody, pencil-like roots (pneumatophores);

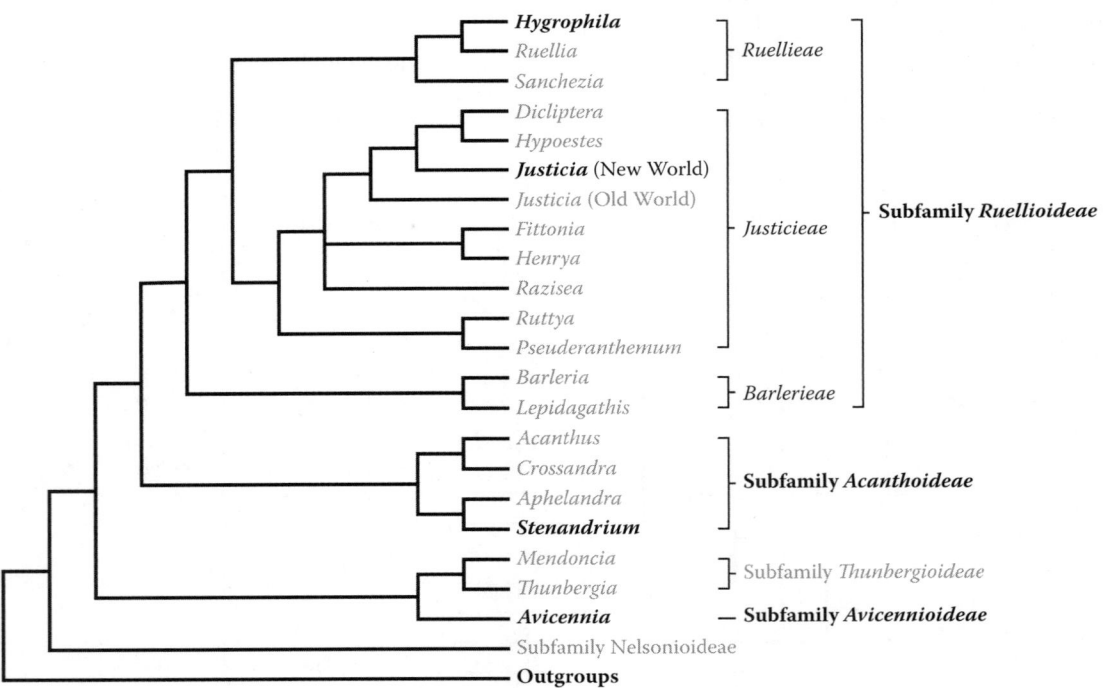

FIGURE 5.51 Phylogenetic relationships among selected genera of Acanthaceae as indicated by analysis of combined nuclear and cpDNA sequences. The distribution of taxa containing obligately aquatic species in North America (bold) suggests multiple, independent origins of the OBL habit in the family. The four subfamilies as well as tribes of subfamily Ruellioideae are shown for reference. (Adapted from Schwarzbach, A.E. & McDade, L.A., *Syst. Bot.*, 27, 84–98, 2002.)

leaves (to 15 cm) opposite, evergreen, leathery, short petioled (to 12 mm), pubescent beneath; cymes from upper axils, consisting of dense, spicate clusters (to 6.5 cm) of 2–7 sessile flower pairs; calyx 5-merous; corolla (to 2 cm) campanulate to rotate, divided approximately to middle, 4-merous, white; stamens 4; capsules (to 5 cm) flat, densely gray-hairy, dehiscence 2-valvate, 1-seeded; seeds large, lacking a seed coat

Life history: duration: perennial (buds); **asexual reproduction:** stem sprouts; **pollination:** insect; **sexual condition:** hermaphroditic; **fruit:** capsules (common); **local dispersal:** seeds (water); **long-distance dispersal:** seeds (water)

Imperilment: (1) *Avicennia germinans* [G5]; S3 (LA); (2) *A. marina* [GNR]

Ecology: general: Mangrove communities (comprising *Avicennia* and several other genera) dominate the coastal areas of warm, tropical regions and cover an estimated 175,000–220,000 hectares of coastal habitat in the southeastern United States. In North America, *Avicennia* species typically occupy more elevated (shoreward) sites in coastal salt marshes. All species are obligately aquatic halophiles; however, none can tolerate prolonged hypersalinity. Most species are fairly tolerant to a wide range of habitat conditions. Although they occur on sediments that tend to be circumneutral (e.g., pH: 7.0–7.2), acidity in some sites can drop markedly (e.g., pH: 2.5) if the soils dry out (due to oxygenation of sulfur). Seeds of *Avicennia* species are nondormant and cryptoviviparous, i.e., they germinate before dispersal; however, without emergence of the embryo. The seeds are also described as recalcitrant, i.e., they are shed with a high moisture content and lose their viability if the moisture content drops below 65%. The seeds of different species vary in their ability to float for prolonged periods, but all are water dispersed. The flowers are pollinated by insects, usually by bees (Hymenoptera). The leaves are adapted to saline conditions by their ability to extrude salt through specialized glands. The species all produce pneumatophores (aerially oriented roots), which provide aeration of the shoots through their lenticel system. Remarkably, although individual plants can produce prolific networks of radiating roots with pneumatophores, they do not reproduce vegetatively. However, cut or broken stems can resprout (coppice). These plants provide vital habitat for numerous marine organisms including amphibians, birds, crustaceans, fish, invertebrates, mammals, and reptiles that include endangered species such as the American crocodile (Reptilia: Crocodilia: Crocodylidae: *Crocodylus acutus*), Atlantic Ridley sea turtle (Reptilia: Testudines: Cheloniidae: *Lepidochelys kempii*), Florida manatee (Mammalia: Sirenia: Trichechidae: *Trichechus manatus latirostrus*), and hawksbill sea turtle (Reptilia: Testudines: Cheloniidae: *Eretmochelys imbricata*).

Avicennia germinans (**L.**) **L.** inhabits tidal or nontidal borrow pits, depressions, ditches, dunes, estuaries, flats, hammocks, salt marshes, shores, and swamps at or near sea level. The substrates are open and range from brackish to saline (0–90 ppt salinity) sediments (sand, marl, or shell middens) that tend to be circumneutral (pH: 5.8–7.9; optimum: 6.5–7.0). The sites are strongly reducing (at least −200 mV) and contain high levels of sulfides. Low levels of salinity have been observed to result in increased CO_2 assimilation by the plants. Although they are generally highly tolerant to inundation, they cannot withstand prolonged periods of pneumatophore submergence. The flowers are self-compatible but are protandrous and typically outcrossed by bees (Insecta: Hymenoptera). Following the dispersal of fruits in the water, the pericarp is shed from the buoyant embryo usually within 2 weeks. The two cotyledons expand into a wing-like structure, which quickly produces roots and facilitates the flotation and dispersal of the embryo. The embryos can remain afloat for more than 80 days and have a median viability of about 16 weeks. However, roots form within a week and the propagules must establish within 14 days. Once dispersed to a favorable site, seedlings will fully establish in less than 3 weeks. The plants often form extensive thickets. There is no true means of vegetative reproduction, but coppicing (resprouting) from broken or cut stems can occur. The seedlings can be seriously defoliated by larval mangrove buckeye moths (Insecta: Lepidoptera: Nymphalidae: *Junonia evarete*). **Reported associates:** *Acrostichum, Annona glabra, Atriplex pentandra, Baccharis halimifolia, Batis maritima, Blutaparon vermiculare, Bolboschoenus robustus, Borrichia arborescens, B. frutescens, Cladium jamaicense, Conocarpus erectus, Distichlis spicata, Encyclia tampensis, Ficus aurea, Heliotropium curassavicum, Ipomoea pes-caprae, I. violacea, Iva cheiranthifolia, I. frutescens, Jacquinia keyensis, Juncus roemerianus, Laguncularia racemosa, Lycium carolinianum, Metopium toxiferum, Monanthochloe littoralis, Persea palustris, Phlebodium aureum, Pinus elliottii, Pleopeltis polypodioides, Quercus virginiana, Rhabdadenia biflora, Rhizophora mangle, Sabal palmetto, Salicornia, Sambucus nigra, Sarcocornia perennis, Schinus terebinthifolius, Sesuvium maritimum, S. portulacastrum, Sideroxylon celastrinum, Spartina alterniflora, S. patens, S. spartinae, Sporobolus virginicus, Suaeda linearis, Taxodium ascendens, Tillandsia balbisiana, T. fasciculata, T. flexuosa, T. paucifolia, T. ×smalliana, T. utriculata, Vanilla barbellata.*

Avicennia marina (**Forssk.**) **Vierh.** was introduced to salt marshes and coastal shores in southern California. The substrates are saline mud or silt of various acidity (pH: 6.0–8.5). The plants are resistant to extremes of salinity (0–80 ppt), high sunlight, and hot and dry conditions and can also tolerate some shading. Sexual reproduction begins in plants that are 5–20 years old (depending on crowding conditions) and becomes more irregular in older plants. The flowers are self-compatible but are protandrous and pollinated by honeybees (Hymenoptera: Apidae: *Apis mellifera*). Fruits can be produced by self-pollination, but their progeny exhibit indications of inbreeding depression. Roughly 250 seeds/year are produced on average over the 100-year lifespan of a single plant. Unlike *A. germinans*, the seeds of *A. marina* sink once the pericarp has been shed, which occurs within a few (1–3) days. Flotation occurs longer in seawater than in 10% seawater (where pericarp loss occurs more quickly); however, propagules that sink initially in 10% seawater can regain their buoyancy and refloat after 3 days. Reflotation also has been observed in propagules that sink originally in seawater, but after a longer (1-week) duration. Despite their relatively poor flotation, propagules have been observed to become dispersed on shores more than 20 km from the nearest source. Yet, most propagules strand and establish close to the parental plants. The seedlings establish rapidly (often within 5 days) over a wide range of light, nutrients, and salinity. The seeds have been stored in dry air for up to 10 days without any loss of viability; however, they will die if they become excessively dehydrated (water content reduced by 35% or more). Sites with high salinity and limited phosphorous may induce dwarfed plants. *Avicennia marina* generally has a more temperate distribution than its congeners.

Reported associates (North America): *Arthrocnemum subterminale, Atriplex leucophylla, A. semibaccata, Batis maritima, Cressa truxillensis, Cuscuta salina, Distichlis spicata, Frankenia salina, Jaumea carnosa, Limonium californicum, Mesembryanthemum crystallinum, M. nodiflorum, Monanthochloe littoralis, Salicornia bigelovii, S. maritima, Salsola tragus, Sarcocornia perennis, Sesuvium verrucosum, Spartina foliosa, Spergularia macrotheca, Suaeda moquinii, Triglochin concinnum.*

Use by wildlife: *Avicennia* species are important ecologically by providing cover for waterfowl and other birds (Aves). *Avicennia germinans* is a host plant for crabs (Crustacea: Decapoda), snails (Mollusca: Gastropoda), and several Lepidoptera (Insecta: Nymphalidae: *Junonia evarete*; *J. genoveva*; Sphingidae: *Madoryx pseudothyreus*). The leaves contain iridoid glycosides, which are known to deter generalist herbivore feeding, but have been linked to host-specific feeding by some *Junonia* larvae. The flowers attract a number of insects (Insecta) including ants (Hymenoptera: Formicidae), honeybees (Hymenoptera: *Apis mellifera*), wasps (Hymenoptera: Vespidae), bugs (Hemiptera), flies (Diptera), bee flies (Diptera: Bombyliidae), soldier beetles (Coleoptera: Cantheridae), and moths (Lepidoptera). The plants provide important habitat for the Cuban yellow warbler (Aves: Parulidae: *Setophaga petechia gundlachi*), Florida prairie warbler (Aves: Parulidae: *Setophaga discolor paludicola*), little blue heron (Aves: Ardeidae: *Egretta caerulea*), reddish egret (Aves: Ardeidae: *Egretta rufescens*), white ibis (Aves: Threskiornithidae: *Eudocimus albus*) and sometimes furnish exclusive nesting sites for the American bald eagle (Aves: Accipitridae: *Haliaeetus leucocephalus*). They are critical habitat for a subspecies of the mangrove fox squirrel (Mammalia: Rodentia: Sciuridae: *Sciurus niger avicennia*). The herbage of *A. marina* has been used in other countries as fodder for camels (Mammalia: Artiodactyle: Camelidae: *Camelus*) and cattle (Mammalia: Bovidae: *Bos taurus*). In its introduced North American range, *A. marina* provides habitat for snails (Mollusca: Gastropoda), crabs (Crustacea: Decapoda: Brachyura), isopods (Crustacea: Isopoda), and the light-footed clapper rail (Aves: Rallidae: *Rallus longirostris levipes*). However, the nonindigenous California populations also provide roosting sites for various raptors, which prey upon young light-footed clapper rail chicks. Rich bacterial populations can occur on the leaves (Flavobacteriaceae: *Flavobacterium*) and roots (Proteobacteria: Vibrionaceae: *Vibrio*). In its native distributional range, the pneumatophores are known to support three distinct zones of epiphytic algae, i.e., an upper (Chlorophyta: Cladophoraceae: *Rhizoclonium*), middle (Rhodophyta: Rhodomelaceae: *Bostrychia*), and lower (Rhodophyta: Delesseriaceae: *Caloglossa*) zone. Many fungi (including wood-rotting species) also colonize plants in their native range.

Economic importance: food: The fruits of *Avicennia germinans* reportedly are edible once they are processed but should be avoided because they are toxic when raw. The tender leaves of *A. marina* have been eaten as a vegetable. The flowers of *A. germinans* are a major source of commercial nectar and important for the production of "mangrove honey." *Avicennia marina* is also an important honey plant in its native range; **medicinal:** *Avicennia germinans* has been used by various Caribbean people as an astringent and as a remedy for diarrhea, dysentery, edema, hemorrhoids, rheumatism, and sore throats. It is also used to control bleeding, treat wounds, and improve skin circulation. When gargled, bark decoctions are reportedly effective against tumors (especially of the larynx and throat ulcers). The leaves and twigs are known to possess a naphthoquinone (3-chlorodeoxylapachol), which exhibits antitumor activity in some human cancer lines. The plants have also been used in Africa to control lice, ringworm, and various skin parasites. The bark, roots, and wood of *A. marina* (and other species) contain tannins and lapachol, a compound that can cause contact dermatitis. The latter substance reportedly exhibits antitumor effects in rodents but apparently is ineffective on humans. In Africa and Asia, the acrid juice from *A. marina* has been used as an abortive, the root and bark as an aphrodisiac, the wood to treat snakebite, and an aqueous seed extract for sores. A poultice made from unripened fruits is used to treat wounds; a leaf poultice is applied to treat skin ailments. Extracts of *A. marina* are analgesic but to a lesser extent than morphine. The plant also has been taken to control ulcers; **cultivation:** *Avicennia marina* has been distributed in cultivation; **misc. products:** Wood from *Avicennia germinans* has been used for making crossties, piers, posts, utility poles, and for wharf construction. The ashes when added to water produce a soap substitute. The wood smoke supposedly repels mosquitos and is used for smoking fish. The bark has been used in leather tanning. The wood of *A. marina* has been used for firewood, as fuel for lime kilns, and to make poles and ribs for boats. The bark is the source of a brown dye; **weeds:** The nonindigenous *A. marina* is invasive in California saltmarshes (San Diego area) and has persisted there despite repeated attempts to eradicate the plants; **nonindigenous species:** *Avicennia marina* was planted intentionally (as research stock) in California in 1968 using plants originating from Auckland, New Zealand.

Systematics: Combined molecular data resolve *Avicennia* as a monophyletic genus within the basal elements of family Acanthaceae (Figures 5.51 and 5.52), where it is now recognized as a distinct subfamily (Avicennioideae). There has been debate whether intercontinental populations of *A. germinans* should be regarded as a single or multiple species (e.g., *A. africana*). Studies of cuticular hydrocarbons showed there to be substantial divergence between Old and New World populations; however, studies using amplified fragment length polymorphism (AFLP) markers indicate that African populations are more similar to western Atlantic populations (USA, Caribbean, South America) than any of these are to Pacific coast populations in Mexico and Central America. Furthermore, there is no detectable morphological divergence between even the Pacific coast and other populations. The AFLP markers also detected considerable variations in the pattern of genetic diversity among populations, with some populations showing much higher levels of inbreeding (possibly due to the entrapment of propagules by the rhizophores). East Atlantic populations retained the highest levels of genetic variation overall, with bottlenecks and founder effect implicated in the genetic erosion of several populations elsewhere in the range of the species. Microsatellite markers have revealed consistent patterns in preliminary population surveys. Similarly, microsatellite markers showed variable levels of genetic variation to exist in the widespread *A. marina*. Although most populations are predominantly outcrossed (e.g., average heterozygosity = 0.407), heightened inbreeding and reduced heterozygosity characterize those at the extremes of the species distributional range. Multiple cpDNA haplotypes have been reported for *A. marina* over a fairly small geographical range. *Avicennia germinans* is closely related to *A. bicolor* (Figure 5.52) and can occur sympatrically with that species; however, no hybrids have been reported between these species. Results of allozyme analysis confirm a close relationship between *A. marina* and *A. alba* and their distant relationship to *A. germinans*. The base chromosome number of *Avicennia* is *x* = 8 or 9. *Avicennia marina* (2*n* = 36,

64, 96) exists at several higher ploidy levels. No counts are reported for *A. germinans*.

Comments: *Avicennia germinans* occurs along the coastal southeastern United States. *Avicennia marina* is known only from coastal San Diego county, California.

References: Amusan & Adeniyi, 2005; Aronson, 1989; Aubé & Caron, 2001; Bandaranayake, 1998; Baskin & Baskin, 1998; Bowers, 1984; Cerón-Souza et al., 2006; Clarke, 1995; Clarke & Myerscough, 1991a,b; Dawson, 1989; Dodd et al., 2000, 2002; Duke, 1991; Duke & Wain, 1981; Duke et al., 1998; Feller & Sitnik, 1996; Garcia-Barriga, 1974–1975; Hartwell, 1967, 1971; Jones et al., 2005; Kado et al., 2004; Kathiresan & Bingham, 2001; Koprowski, 1994; Maguire et al., 2000; McMullen, 1987; Nettel et al., 2005; Rabinowitz, 1978; Schwarzbach & McDade, 2002; Schwarzbach & Ricklefs, 2001; Tomlinson, 1986.

2. *Hygrophila*

Hygro, Miramar weed, starhorn, water wisteria

Etymology: from the Greek *hygros philos* meaning "moisture loving"

Synonyms: *Asteracantha*; *Barleria* (in part); *Cardanthera*; *Heliadelphis*; *Hemidelphis*; *Justicia* (in part); *Kita*; *Nomaphila*; *Physichilus*; *Polyechma*; *Ruellia* (in part); *Synnema*

Distribution: global: pantropical; **North America:** southeastern

Diversity: global: 80 species; **North America:** 4 species

Indicators (USA): OBL: *Hygrophila corymbosa*, *H. difformis*, *H. lacustris*, *H. polysperma*

Habitat: freshwater; palustrine, riverine; **pH:** 5.0–9.0; **depth:** 0–4 m; **life-form(s):** emergent herb, submersed (vittate)

Key morphology: stems quadrangular, decumbent, rooting at nodes, submersed (to 50 cm), or emergent and creeping (to 10 cm) or with tips ascending (to 8 dm); leaves (to 15 cm) opposite, lanceolate, sessile, joined at their base by a flange; flowers sessile, in axillary (apical or verticillate) clusters; calyx (to 5 mm) 5-parted; corolla (to 8 mm) bilabiate, tubular (tube to 5 mm; limb to 2.5 mm), purple to blue in life, but drying to yellow; stamens 4 (didynamous) or 2 with 2 staminodes, filament bases joined by a membrane; capsules (to 12 mm) oblong, 2-celled, many seeded (to 30)

Life history: duration: perennial (rhizomes); **asexual reproduction:** rhizomes, leaf fragments, shoot fragments, shoot layering; **pollination:** insect or self; **sexual condition:** hermaphroditic; **fruit:** capsules (common); **local dispersal:** seeds (ballistic: water, wind), vegetative fragments (water); **long-distance dispersal:** seeds (animals), vegetative fragments (water)

Imperilment: (1) *Hygrophila corymbosa* [GNR]; (2) *H. difformis* [not ranked]; (3) *H. lacustris* [G5]; S1 (GA); (4) *H. polysperma* [GNR]

Ecology: general: *Hygrophila* is a large, predominantly Old World genus of herbaceous species that are found characteristically in wet habitats. Most species are emergent wetland plants, although some can also grow as fully submersed

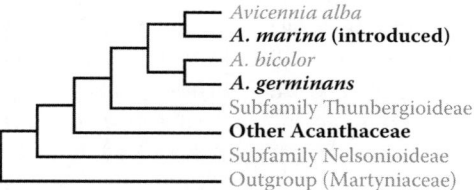

Avicennia alba
A. marina (introduced)
A. bicolor
A. germinans
Subfamily Thunbergioideae
Other Acanthaceae
Subfamily Nelsonioideae
Outgroup (Martyniaceae)

FIGURE 5.52 Phylogenetic relationships in *Avicennia* as indicated by analysis of combined molecular data. These results confirm the monophyly of *Avicennia* and also indicate that the two OBL North American species (in bold) are only distantly related. Because all *Avicennia* species are obligate aquatics, the OBL habit presumably arose only once in this clade. (Adapted from Schwarzbach, A.E. & McDade, L.A., *Syst. Bot.*, 27, 84–98, 2002.)

aquatics. The pollination biology and seed ecology are not well studied for most of the species. Some species require a period of warm seed stratification (e.g., 40°C for 90 days) for optimal germination. Several introductions of nonindigenous species (three of the four OBL North American species) have occurred as a consequence of escapes from cultivated, decorative aquarium plants.

Hygrophila corymbosa (**Blume**) **Lindau** occurs in springs, swamps, and other wet sites as an emergent or in shallow standing water. The substrates are described as muck. Currently, this nonindigenous species is known only from two North American localities (Florida); thus any ecological information from this region is minimal. The following information is derived mainly from cultivated material. The plants require fairly warm waters, with the minimum temperature for cultivation given as 22.2°C. This species thrives in high light but can tolerate low light conditions and is fast growing in culture. The plants are known to take up heavy metals from the water column and can tolerate up to 50 ppm cadmium without exhibiting adverse effects on their growth. They can easily be propagated vegetatively by "leaf disk" cultures. Details on the reproductive biology and seed ecology are scarce. The two voucher specimens for the nonindigenous North American populations document flowering plants in January–February, which suggests that sexual reproduction might occur year round. **Reported associates:** *Acer rubrum.*

Hygrophila difformis (**L.f.**) **Blume** occurs along pond margins on clay substrates. However, because this nonindigenous species is currently known from only one North American locality, there is little insight into its ecology in this region. The single voucher specimen collected in May possesses flowers. In aquariums, this species thrives in high light and nutrient conditions, but also tolerates low light levels. The plants can be propagated readily by vegetative cuttings. They are believed to release toxic phosphorus-containing allelochemicals that are highly inhibitory to cyanobacteria (Bacteria: Cyanophyceae: Chroococcaceae: *Microcystis aeruginosa*). **Reported associates:** none.

Hygrophila lacustris (**Schltdl. & Cham.**) **Nees** inhabits canals, depressions, ditches, drying ponds, flatwoods, floodplains, marshes, prairies, sandbars, sloughs, swamps, and the margins of lakes, ponds, and rivers. Its habitats generally are described as having acidic substrates (e.g., pH: 6.5–7.2) of sand, silt, loam, mud, or peat. The plants can tolerate partial shade but are not tolerant to salinity and decline at salinity levels greater than 3.5 ppt. Plants do not appear to reproduce vegetatively by fragmentation; however, vegetative reproduction by stem tip layering has been reported and extensive stands can occur. There is little information on the floral biology of this species. The seeds become adhesive when wet and likely adhere to animals, which potentially disperse them. Additional information on the floral and seed ecology of this species is needed. **Reported associates:** *Acer rubrum, Alternanthera philoxeroides, Ambrosia artemisiifolia, A. trifida, Ampelopsis arborea, Andropogon glomeratus, Arundinaria gigantea, Asclepias, Berchemia scandens, Bidens mitis, Boehmeria cylindrica, Boltonia, Brunnichia*

ovata, Campsis radicans, Carex comosa, C. crus-corvi, C. frankii, C. lupuliformis, C. lupulina, Carya aquatica, C. glabra, Cephalanthus occidentalis, Chasmanthium latifolium, C. laxum, C. sessiliflorum, Cocculus carolinus, Cornus foemina, Crataegus opaca, C. viridis, Cuphea carthagenensis, Cyperus virens, Diospyros virginiana, Echinodorus cordifolius, Eclipta prostrata, Eichhornia crassipes, Eleocharis, Eupatorium capillifolium, E. fistulosum, Forestiera acuminata, Fraxinus pennsylvanica, F. profunda, Gratiola virginiana, Helianthus simulans, Hydrocotyle verticillata, Hydrolea, Hypericum, Hyptis alata, Ilex decidua, Ilex opaca, Itea virginica, Juncus effusus, Justicia ovata, Leersia, Liquidambar styraciflua, Ludwigia decurrens, L. glandulosa, L. palustris, Lycopus rubellus, L. virginicus, Lysimachia radicans, Magnolia grandiflora, Mikania scandens, Mimulus, Myrica cerifera, Nyssa aquatica, N. biflora, N. sylvatica, Onoclea sensibilis, Osmunda regalis, Persicaria glabra, P. hydropiperoides, P. punctata, P. virginiana, Phanopyrum gymnocarpon, Pinus glabra, P. taeda, Pluchea camphorata, Pontederia cordata, Quercus laurifolia, Q. michauxii, Q. nigra, Q. pagoda, Q. phellos, Q. virginiana, Rhynchospora corniculata, Rubus argutus, Rumex verticillatus, Sabal minor, Sabatia calycina, Saccharum giganteum, Sagittaria lancifolia, S. latifolia, Salix nigra, Sambucus nigra, Saururus cernuus, Schoenoplectus, Scirpus cyperinus, Smilax glauca, S. rotundifolia, Solidago gigantea, Sparganium americanum, Sphenopholis pensylvanica, Spiranthes cernua, Styrax americanus, Symphyotrichum subulatum, Taxodium distichum, Thelypteris palustris, T. patens, Toxicodendron radicans, Tradescantia ohiensis, Triadenum walteri, Typha latifolia, Vernonia gigantea, Viburnum dentatum, Vitis rotundifolia, Woodwardia areolata, Xyris, Zizaniopsis miliacea.

Hygrophila polysperma (**Roxb.**) **T. Anderson** grows either as an emergent or as submersed (to 3+ m depths) in canals, floodplains, lakes, marshes, ponds, reservoirs, rivers, streams, swamps, and along river and streambanks at elevations of up to at least 6 m. This species is broadly tolerant to ecological conditions (e.g., pH: 5.0–9.0; optimally 5.0–7.0; water hardness: 30–140 ppm). Also, the extent of substrate nutrients does not appear to be critical because aquarium plants have been grown easily on pure sand. It also tolerates relatively low levels of light, with a photosynthetic compensation point that is deeper than that of many native submersed species. Although tropical in origin, *H. polysperma* can tolerate cold water temperatures (to 4°C) and will grow well in water from 10°C to 35°C (optimally at 22°C–28°C). These plants are highly plastic and can grow both as completely submersed or emergent lifeforms. Carbon dioxide uptake is facilitated through the aerial leaves of emergent plants. The plasticity in growth results in a higher allocation of structural shoot carbohydrates and lower leaf protein levels in plants growing as an emergent life-form. Sexual reproduction appears to be relatively uncommon in introduced North American populations and only has been observed on emergent plants. The flowers are adapted for insect-pollination; however, North American plants appear to be autogamous with high seed set observed even during cooler months when potential pollinators are not active. The seed ecology is poorly known with virtually no information for North American populations. These plants occur most commonly in canals or streams and thrive following disturbances caused by flooding events. The growth rate exhibited by the plants increases substantially (up to five times higher) in flowing water. The plants compete effectively (if not superiorly) with *Hydrilla* (another noxious nonindigenous aquatic weed), especially in waters of lower pH values. They also are superior competitors with the native *Ludwigia palustris*. During hot weather, the entire shoot can break off from the root crown to form large floating mats. Vegetative shoot fragmentation is the major mechanism of dispersal and fragment regeneration capacity is extremely high, exceeding that of *Hydrilla* and *Limnophila* in laboratory experiments. However, plants do not establish readily in deeper water. **Reported associates (North America):** *Acer rubrum, Alternanthera philoxeroides, Amaranthus australis, Ambrosia artemisiifolia, Bacopa, Bidens pilosa, Boehmeria cylindrica, Cabomba caroliniana, Caperonia palustris, Ceratophyllum demersum, Ceratopteris thalictroides, Colocasia esculenta, Commelina diffusa, Crinum americanum, Cryptocoryne beckettii, Cyperus iria, C. polystachyos, Dactyloctenium aegyptium, Desmodium tortuosum, Dichanthelium commutatum, Diodia virginiana, Echinochloa colona, Eclipta prostrata, Egeria densa, Eichhornia crassipes, Euphorbia heterophylla, Heteranthera dubia, Hydrilla verticillata, Hydrocotyle umbellata, Lemna, Limnophila sessiliflora, Ludwigia repens, Momordica charantia, Myriophyllum aquaticum, M. spicatum, Najas guadalupensis, Nuphar, Parietaria floridana, Persicaria punctatum, Pistia stratiotes, Pontederia cordata, Potamogeton illinoensis, Pouzolzia zeylanica, Sagittaria platyphylla, Salvinia minima, Samolus valerandi, Saururus cernuus, Sphagneticola trilobata, Vallisneria americana, Zizania texana.*

Use by wildlife: Dense stands of *Hygrophila lacustris* provide habitat for foraging gallinules (Aves: Rallidae: *Porphyrio*) and rails (Aves: Rallidae). The roots are used as microhabitat by various invertebrates. Rod-shaped virus-like particles occur in both variegated and nonvariegated forms of *H. difformis* and *H. polysperma*.

Economic importance: food: The leaves of *Hygrophila difformis* are eaten in India as a leafy vegetable; **medicinal:** Some *Hygrophila* species (e.g., *H. difformis*) have been found to contain effective antioxidants. *Hygrophila difformis* also contains iridoid glycosides (hygrophiloside), which have hepatoprotective activity. Ethanolic extracts have been shown to affect the sleep patterns of laboratory mice (Mammalia: Rodentia: Muridae: *Mus*). Benzene extracts contain cardiac glycosides, flavonoids, saponins, steroids, and tannins, and are highly lethal to helminths (Annelida: Oligochaeta: Megascolecidae: *Pheretima posthuma*), which are similar to human intestinal roundworm parasites. Whole-plant extracts effectively inhibit human pathogenic bacteria (Proteobacteria: Enterobacteriaceae: *Enterobacter faecalis, Escherichia coli, Klebsiella pneumoniae, Shigella dysenteriae*) and fungi (Ascomycota: Trichocomaceae: *Aspergillus niger*) in clinical tests; **cultivation:** Many of the commercial

aquarium plants sold as "*Hygrophila lacustris*" actually are *H. corymbosa*. The latter is also distributed under the cultivar names 'Greta,' 'Ruffle leaf,' 'Siamensis,' 'Siamensis Broadleaf,' and 'Stricta.' *Hygrophila difformis* also is cultivated as an ornamental aquarium plant. *Hygrophila polysperma* is another popular aquarium plant, which includes the narrow-leaved cultivar 'Ceylon' and the variegated cultivars 'Rosanervig' (also known as 'Marmor' and 'Sunset') and 'Tropic Sunset'; **misc. products:** Both *Hygrophila corymbosa* and *H. difformis* have been recommended for use in the removal of heavy metals (e.g., Cd, Pb) from water. *Hygrophila difformis* has also been considered for the removal of nutrients from wastewater due to its high aqueous uptake rate of nitrogen and phosphorous. Essential oils extracted from *H. difformis* are highly repellent to seed-eating beetles (Insecta: Coleoptera: Chrysomelidae: *Callosobruchus chinensis, C. maculatus*) and can be used to safeguard food products that often are damaged by them; **weeds:** *Hygrophila difformis* is a weed of rice (Poaceae: *Oryza sativa*) in Bangladesh and India. *Hygrophila polysperma* is listed as a Federal noxious weed in the United States. The plants can block culverts, foul water pumping stations, and interfere with navigation. Extensive stands of *H. polysperma* in the San Marcos River, Texas, threaten populations of the endangered Texas wild rice (Poaceae: *Zizania texana*). Studies have shown it to be quite resistant to herbicides, which favor its spread when applied to control co-occurring weeds such as *Hydrilla*. Other nonindigenous hygrophilas (*H. corymbosa, H. difformis*) exhibit similar weedy characteristics; **nonindigenous species:** Both *Hygrophila corymbosa* and *H. difformis* were introduced to Florida (from Asia) as escapes from cultivation. *Hygrophila corymbosa* was first reported from southern Florida in 1988 (the earliest voucher specimen from 1987), but it is believed to have been distributed to that state as early as 1917. It also has become naturalized in Taiwan, first appearing there in 2001. *Hygrophila difformis* was first reported from the United States (in Florida) in 2002. *Hygrophila lacustris* (as *H. costata*) has been introduced to Australia. *Hygrophila polysperma* (native to India and Malaysia) was introduced to North America also as an escape from cultivation. It was imported into Ohio and Florida as an aquarium plant prior to the 1950s, established in Florida by the late 1970s, and spread to 20 Florida counties by the 1990s. *Hygrophila polysperma* also was introduced to Hawaii and to the Erft River, Germany.

Systematics: Despite the small size of this genus, only one species (*Hygrophila corymbosa*) has been included routinely in any molecular phylogenetic survey. Consequently, the monophyly of *Hygrophila* has not been confirmed and its distinctness from some genera (e.g., *Nomaphila*) remains unsettled. With respect to the sole exemplar, several analyses of DNA sequence data have confirmed the placement of *Hygrophila* within tribe Ruellieae of Acanthaceae (Figure 5.53). In these studies, *Hygrophila* resolves close to *Hemigraphis, Strobilanthes*, and related genera (Figure 5.53). A comprehensive systematic study of *Hygrophila* is needed, as even estimates of the number of species in the genus vary considerably (25–100). Several authors recently have

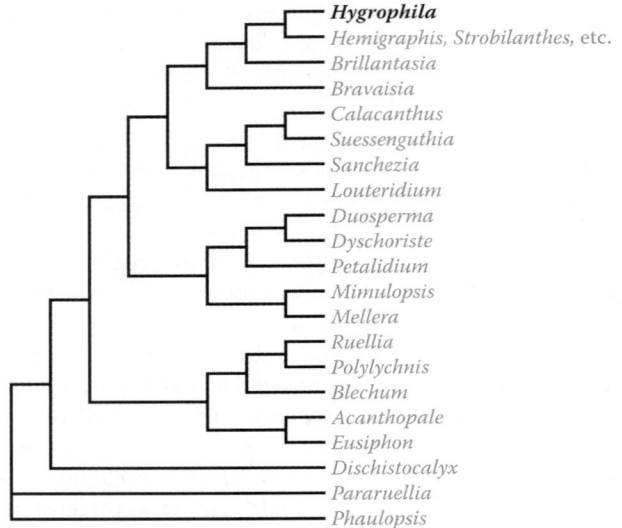

FIGURE 5.53 Phylogenetic relationships among genera of Acanthaceae tribe Ruellieae as indicated by analysis of nrITS sequence data. *Hygrophila* (bold), the only genus in the tribe with OBL North American species, is relatively derived in the group. The monophyly of large genus *Hygrophila* remains to be verified by the inclusion of additional species in similar analyses. (Adapted from Tripp, E.A., Ruellieae (Acanthaceae), Version 17 February 2007, *The Tree of Life Web Project*, Available online: http://tolweb.org/Ruellieae/, 2007.)

suggested that *Hygrophila lacustris* may be synonymous with *H. costata*, which extends into tropical America. However, such issues of synonymy should be reserved until a thorough study of the genus has been conducted. The basic chromosome number of *Hygrophila* is $x = 8$. Both *H. lacustris* and *H. polysperma* ($2n = 32$) are tetraploid; counts are unavailable for *H. corymbosa* or *H. difformis*. *Hygrophila corymbosa* contains several putative paralogs (*RCYC2A1/RCYC2A2*) of the *CYCLOIDEA2* (*CYC2*) gene.

Comments: *Hygrophila lacustris* occurs along the coastal plain of the southeastern United States; *H. polysperma* occurs in Florida, Texas, and Virginia

References: Angerstein & Lemke, 1994; Banerjee et al., 2012; Botts et al., 1990; Correll & Correll, 1975; De et al., 2014; Doyle et al., 2003; Godfrey & Wooten, 1981; Gordon & Gantz, 2011; Grant, 1955; Jensen & Nielsen, 1985; Kasselmann, 2003; Krishanu, 2012; La Starza et al., 2000; Les & Wunderlin, 1980; Lihua & Liu, 2009; Meyer & Lavergne, 2004; Moody, 1989; Moylan et al., 2004; Mühlberg, 1982; Owens et al., 2001; Oyedeji & Abowei, 2012; Pal & Samanta, 2011; Pal et al., 2010; Pandey et al., 2006; Penfound & Hathaway, 1938; Poole & Bowles, 1999; Proeseler et al., 1990; Reams, Jr., 1953; Samanta et al., 2012; Schmitz & Nall, 1984; Schwarzbach & McDade, 2002; Spencer & Bowes, 1985; Stutzenbaker, 1999; Sutton, 1995; Tripp, 2007; Tullock, 2006; Van Dijk et al., 1986; Vandiver, Jr., 1980; Vijayakumar et al., 2005; Wu et al., 2010; Wunderlin et al., 1988, 2002; Yaowakhan et al., 2005; Zhang et al., 2009; Zhong & Kellogg, 2014.

3. *Justicia*

Water-willow

Etymology: after James Justice (1698–1763)

Synonyms: *Acelica; Amphiscopia; Aulojusticia; Beloperone; Bentia; Calliaspidia; Calymmostachya; Centrilla; Chaetochlamys; Chaetothylax; Chaetothylopsis; Chiloglossa; Cyphisia; Cyrtanthera; Cyrtantherella; Dianthera; Dimanisa; Duvernoia; Dyspemptemorion; Emularia; Ethesia; Gendarussa; Glosarithys; Harnieria; Heinzelia; Hemichoriste; Heteraspidia; Ixtlania; Jacobinia; Kuestera; Libonia; Linocalyx; Lophothecium; Lustrinia; Mananthes; Nicoteba; Orthotactus; Petalanthera; Plagiacanthus; Plegmatolemma; Porphyrocoma; Psacadocalymma; Rhacodiscus; Rhaphidospora; Rhiphidosperma; Rhyticalymma; Rodatia; Rostellaria; Saglorithys; Salviacanthus; Sarojusticia; Sarotheca; Sericographis; Simonisia; Solenochasma; Stethoma; Tabascina; Thalestris; Thamnojusticia; Tyloglossa*

Distribution: global: cosmopolitan; **North America:** southern and eastern

Diversity: global: 400–600 species; **North America:** 17 species

Indicators (USA): OBL: *Justicia americana, J. crassifolia, J. lanceolata, J. ovata*

Habitat: freshwater; palustrine, riverine; **pH:** 5.1–7.8; **depth:** 0–2 m; **life-form(s):** emergent herb

Key morphology: stems (to 1 m) bluntly angular, ascending, rooting at lower nodes; leaves (to 20 cm) opposite, elliptic, lanceolate, linear or ovate, sessile; inflorescences terminal, axillary, long stalked (to 15 cm), dense or loosely flowered spikes; calyx (to 15 mm) 5-merous; corolla (to 3 cm) bilabiate, rose-lavender, purple, or white, the lower lipped 3-lobed, often with brown, purple, or white markings; stamens 2, weakly exserted from the corolla; capsule (to 2.3 cm) narrowed at base, 4-seeded; seeds (to 2 mm) discoid, the surface warty to papillose

Life history: duration: perennial (rhizomes); **asexual reproduction:** rhizomes, stolons, shoot fragments; **pollination:** insect; **sexual condition:** hermaphroditic; **fruit:** capsules (common); **local dispersal:** seeds (ballistic, water); rhizomes; **long-distance dispersal:** seeds (water); stem fragments (water)

Imperilment: (1) *Justicia americana* [G5]; SX (VT); S1 (IA, ON, QC); S2 (LA, MI); (2) *J. crassifolia* [G3]; S3 (FL); (3) *J. lanceolata* [G5]; S1 (OK); S2 (MO); S3 (KY); (4) *J. ovata* [G5]; S1 (IL, OK); S2 (MO); S3 (KY, NC, VA)

Ecology: general: Although most *Justicia* species are terrestrial plants, five of the North American taxa (29%) are ranked as wetland indicators. Four North American species (24%) are obligate aquatics (OBL), which grow as emergents on exposed, wet substrates, or in shallow water; however, none tolerates excessive inundation. All of the species have fairly showy, nectariferous flowers that attract large bees (Hymenoptera) and butterflies (Lepidoptera), which are the principal pollinators, but occasionally also are visited by hummingbirds (Aves: Trochilidae). Most of the species are genetically self-compatible but mainly are outcrossed due to protandry. The seeds lack the hygroscopic hairs found in other Acanthaceae and are dispersed locally by a ballistic mechanism.

***Justicia americana* (L.) Vahl** inhabits floodplains, seeps, swamps, and pond, reservoir, river, and stream margins at elevations of up to 300 m. The substrates are circumneutral (pH: 6.2–7.8) but tend to be alkaline (e.g., limestone), high in calcium, and coarse in texture. They include clay, cobble, gravel, muck, rock, sand, and silt. Plants become severely impacted at sites where acid mine discharge has occurred. The habitats are open (receiving full light) and the plants will be impacted negatively if they become shaded by the associated riparian vegetation. This is a river species and a good indicator of riverine littoral communities, but it does not occur in sites exposed to severe erosion. Although an obligate aquatic, this species can withstand up to 8 weeks of desiccation but declines rapidly if individuals are inundated for more than 4 weeks in duration. Optimum survival occurs at water depths of 15–50 cm, but the plants can tolerate depths to about 1 m once they have become established. Flowering can occur twice within a season. Although the breeding system needs to be evaluated (e.g., the degree, if any, of self-pollination has not been determined), the flowers are pollinated by insects including long-tongued bees (Megachilidae: *Hoplitis cylindricans, H. pilosifrons, Megachile addenda*) and short-tongued bees (Halictidae: *Augochlorella striata, Lasioglossum imitatus, L. pilosus, L. versatus*). The fruits ripen within 3 weeks and their seeds are ejected forcefully (about 1.2 m) upon capsule dehiscence. Although the mature fruits are capable of flotation (<7 days), the seeds will not germinate until they are liberated from the capsule. Most of the seeds will float for several hours (sinking completely within 2 days) and will germinate almost immediately whether they are stranded upon a shoreline or are submersed. In the latter case, most of the germinating seeds will regain buoyancy and float to the surface where they can be dispersed by water currents. Seedling densities can range from 100 to 2000+ per square meter on favorable, exposed sites. Because the seeds lack dormancy and germinate rapidly, overwintering in this species is entirely by means of its rhizomes/stolons (a distinction between these organs is difficult in this species). The plants are highly clonal and their survival is facilitated by their firm anchorage in the substrate and a rapid capacity for regrowth following flood events. As the plants are buried by silt and sediment, the number of shoots produced decreases substantially and concomitantly, a lower proportion is produced by the stems than by the rhizomes. Vegetative reproduction ceases in darkness. About 2.5 stolons are produced per plant on average, with each usually becoming about 1.3 m in length. Dense patches of plants can produce nearly 70 m of stolons per square meter. Vegetative reproduction also occurs by means of shoot fragments, which root at the nodes and can remain afloat for several days. The plants can be propagated artificially using apical stem cuttings. Leachates from the litter of *J. americana* are known to inhibit seed germination in cattail (*Typha latifolia*). **Reported associates:** *Acer negundo, A. rubrum, A. saccharinum, Alisma subcordatum, Alnus serrulata, Alternanthera philoxeroides, Amorpha fruticosa, Andropogon gerardii, Arisaema dracontium, Asclepias perennis, Athyrium, Bacopa monnieri, Baptisia australis, Betula nigra, Carex emoryi, C. gigantea,* *Celtis laevigata, Cephalanthus occidentalis, Colocasia esculenta, Commelina virginica, Cornus amomum, Cynodon dactylon, Cyperus, Decumaria barbara, Digitaria sanguinalis, Diodia teres, Eleocharis geniculata, E. montevidensis, E. obtusa, E. quadrangulata, Eragrostis hypnoides, Fallopia japonica, Fimbristylis autumnalis, F. miliacea, F. vahlii, Fraxinus caroliniana, Fuirena simplex, Glottidium vesicarium, Gratiola brevifolia, G. virginiana, Hydrocotyle ranunculoides, H. verticillata, Hymenocallis caroliniana, Ipomoea, Itea virginica, Jacquemontia tamnifolia, Juncus acuminatus, J. effusus, Lemna minor, Leersia lenticularis, L. oryzoides, Lindernia dubia, Lipocarpha micrantha, Liquidambar styraciflua, Lobelia cardinalis, Ludwigia octovalvis, L. palustris, L. peploides, Mikania scandens, Mollugo verticillata, Najas guadalupensis, Nyssa aquatica, N. biflora, Onoclea sensibilis, Orontium aquaticum, Osmunda regalis, Panicum agrostoides, Paspalum, Persicaria punctata, P. virginiana, Phanopyrum gymnocarpon, Phyla nodiflora, Pilea pumila, Platanus occidentalis, Podostemum ceratophyllum, Polypremum procumbens, Populus heterophylla, Portulaca oleracea, Potamogeton amplifolius, P. nodosus, Prunus, Quercus lyrata, Q. nigra, Rhexia, Rhynchospora colorata, R. nivea, Rotala ramosior, Sagittaria secundifolia, Salix caroliniana, S. exigua, S. interior, S. nigra, S. sericea, Saururus cernuus, Schoenoplectus pungens, S. tabernaemontani, Sisyrinchium angustifolium, Smilax, Solidago shortii, Sphagnum, Spiraea virginiana, Spiranthes cernua, Strophostyles helvula, Symphyotrichum lanceolatum, S. subulatum, Taxodium distichum, Triadenum, Ulmus americana, Xanthium strumarium, Xyris difformis, Zizaniopsis miliacea.*

***Justicia crassifolia* (Chapm.) Chapm. ex Small** grows either as an emergent or in shallow water in ditches, flatwoods, savannas, seeps, and swamps. Its substrates are described as acidic and include loamy sand and sand. The plants are regarded as indicators of seep communities and reportedly favor sites that are characterized by a mild soil disturbance. Virtually no information exists on site characteristics or the reproductive ecology of this rare species. The plants are known to form extensive clones in the water. Vegetative reproduction (and overwintering) occurs by means of rhizomes/stolons. **Reported associates:** *Agalinis, Aristida beyrichiana, Asclepias viridula, Balduina uniflora, Bartonia verna, Calopogon barbatus, Carphephorus pseudoliatris, Chamaecyparis thyoides, Chaptalia tomentosa, Ctenium aromaticum, Drosera capillaris, Eriocaulon compressum, Euphorbia inundata, Eurybia eryngiifolia, Lachnocaulon digynum, Lycopodium carolinianum, Parnassia caroliniana, Pinguicula ionantha, P. lutea, Pinus palustris, Platanthera integra, Pleea tenuifolia, Polygala lutea, Polygala polygama, Rhexia alifanus, Rhynchospora oligantha, Rhynchospora, Ruellia noctiflora, Sarracenia leucophylla, S. rubra, Serenoa repens, Syngonanthus flavidulus Taxodium, Utricularia subulata, Xyris ambigua, X. drummondii, X. scabrifolia.*

***Justicia lanceolata* (Chapm.) Small** inhabits depressions, ditches, flatwoods, floodplains, levees, marshes, ponds, river bottoms, sandbars, swamps, and the margins of lakes, rivers, and streams. The plants are emergents but can be found growing in standing water up to 0.5 m deep in open or shaded sites. They are well adapted to flooding and also are tolerant of saline storm surges, being able to survive for 6 months at salinity levels of 3 g/L; however, the plants will succumb after a 9-week exposure to salinities of 6 g/L. They are effective at trapping sediments, which include loam, loamy sand, muck, mucky sand, mud, and sandy loam. Few details have been provided on the reproductive biology of this species other than the plants being capable of producing thousands of seeds. **Reported associates:** *Acer rubrum, Aeschynomene indica, Alternanthera philoxeroides, Amaranthus cannabinus, Ammannia coccinea, Arundinaria gigantea, Asclepias perennis, Bacopa caroliniana, B. monnieri, Bidens, Callitriche, Carex louisianica, Carya aquatica, Cephalanthus occidentalis, Coleataenia longifolia, Colocasia esculenta, Cornus foemina, Crataegus viridis, Crinum americanum, Cyperus difformis, C. odoratus, Diospyros virginiana, Echinochloa walteri, Eclipta prostrata, Eichhornia crassipes, Eleocharis parvula, Fraxinus pensylvanica, F. profunda, Gleditsia aquatica, Gratiola neglecta, Heteranthera dubia, Hydrocotyle, Hymenocallis occidentalis, Iris fulva, I. virginica, Itea virginica, Juncus, Leersia oryzoides, Lemna valdiviana, Leptochloa, Lindera melissifolia, Lobelia cardinalis, Ludwigia leptocarpa, L. peploides, Micranthemum umbrosum, Mikania scandens, Myriophyllum, Najas guadalupensis, Nelumbo lutea, Nyssa aquatica, N. sylvatica, Panicum, Paspalum, Penthorum sedoides, Persicaria punctata, Phragmites australis, Planera aquatica, Pluchea, Populus heterophylla, Proserpinaca palustris, Quercus lyrata, Q. nigra, Q. texana, Rhynchospora corniculata, R. tracyi, Rosa palustris, Sabatia calycina, Sagittaria latifolia, S. platyphylla, Salix nigra, Schoenoplectus pungens, S. tabernaemontani, Sclerolepis uniflora, Smilax walteri, Sparganium, Sphenoclea zeylanica, Spirodela, Styrax americanus, Saururus cernuus, Symphyotrichum subulatum, S. tenuifolium, Taxodium distichum, Triadenum tubulosum, Typha domingensis, T. latifolia, Ulmus americana, Utricularia macrorhiza, Vallisneria americana, Vigna luteola, Zizaniopsis miliacea.*

Justicia ovata **(Walter) Lindau** occurs in bayous, deltas, depressions, ditches, floodplains, ponds, prairies, sloughs, swamps, and margins of ponds and rivers. The substrates tend to be acidic (pH: 5.1–7.4) and can consist of clay, loam, muck, mucky loam, sand, sandy loam, or silty clay. Some plants also have been reported growing on logs in rivers. The species is adapted to fairly deep shade. There are few accounts that provide much insight into the pollination biology (which occurs presumably by insects) or the seed ecology of this species. In new successional sites, this species tends to increase over time and to stabilize at an intermediate elevational zone (~35 cm). Once rooted, the plants form dense, circular, monospecific stands that effectively trap sediments. **Reported associates:** *Acer negundo, A. rubrum, Acmella oppositifolia, Allium vineale, Alnus serrulata, Alternanthera philoxeroides, Ammannia latifolia, Ampelopsis arborea, Amsonia rigida, Annona glabra, Asclepias perennis, Bacopa caroliniana, Berchemia scandens, Bidens laevis, Bignonia capreolata, Blechnum serrulatum, Boehmeria cylindrica,* *Brunnichia ovata, Callicarpa americana, Campsis radicans, Carex aquatilis, C. bromoides, C. crus-corvi, C. frankii, C. gigantea, C. glaucescens, C. intumescens, C. joorii, C. louisianica, C. lupulina, C. tribuloides, C. typhina, Carpinus caroliniana, Carya aquatica, Celtis laevigata, Cephalanthus occidentalis, Chasmanthium latifolium, C. laxum, C. sessiliflorum, Cladium jamaicense, Colocasia esculenta, Commelina diffusa, C. virginica, Cornus foemina, Crataegus opaca, Crinum americanum, Cynodon dactylon, Cyperus, Cyrilla racemiflora, Dichanthelium boscii, D. commutatum, Diodia virginiana, Diospyros virginiana, Dulichium arundinaceum, Echinochloa walteri, Eleocharis baldwinii, E. cellulosa, E. flavescens, E. microcarpa, Eragrostis hypnoides, Eupatorium leptophyllum, E. semiserratum, Euphorbia, Ficus aurea, Fimbristylis autumnalis, F. perpusilla, Fraxinus caroliniana, F. pennsylvanica, Fuirena squarrosa, Gratiola aurea, G. virginiana, Helenium flexuosum, Hydrocotyle umbellata, H. verticillata, Hydrolea ovata, Hygrophila lacustris, Hypericum mutilum, Hypoxis hirsuta, Hyptis alata, Ilex decidua, I. verticillata, I. vomitoria, Illicium floridanum, Itea virginica, Juncus coriaceus, J. polycephalus, J. repens, Juniperus virginiana, Laportea canadensis, Leersia lenticularis, L. oryzoides, Ligustrum sinense, Lindernia dubia, Lipocarpha micrantha, Liquidambar styraciflua, Lobelia cardinalis, Ludwigia alternifolia, L. leptocarpa, L. palustris, L. peploides, L. repens, Lycopus rubellus, L. virginicus, Lysimachia lanceolata, L. radicans, Magnolia virginiana, Micranthemum umbrosum, Mikania scandens, Myrica cerifera, Nyssa aquatica, N. biflora, N. ogeche, Onoclea sensibilis, Orontium aquaticum, Osmunda regalis, Panicum anceps, P. hemitomon, P. rigidulum, P. virgatum, Paspalum distichum, P. laeve, P. vaginatum, Parthenocissus quinquefolia, Peltandra virginica, Persicaria punctata, P. setaceum, Phanopyrum gymnocarpon, Phyla nodiflora, Physostegia, Pilea pumila, Pinus elliottii, P. taeda, Planera aquatica, Pleopeltis polypodioides, Pluchea camphorata, P. rosea, Pontederia cordata, Populus heterophylla, Proserpinaca palustris, P. pectinata, Rhynchospora colorata, R. corniculata, R. glomerata, R. inundata, R. microcarpa, R. tracyi, Rotala ramosior, Rudbeckia hirta, R. laciniata, Ruellia, Rumex, Quercus laurifolia, Q. lyrata, Q. phellos, Sabal minor, S. palmetto, Sabatia calycina, S. kennedyana, Saccharum baldwinii, S. giganteum, Sacciolepis striata, Sagittaria graminea, S. lancifolia, S. latifolia, S. platyphylla, Sambucus nigra, Saururus cernuus, Schoenoplectus pungens, S. tabernaemontani, Scirpus cyperinus, Scleria, Sebastiania fruticosa, Sesbania, Smilax tamnoides, S. walteri, Solanum carolinense, Solidago gigantea, S. leavenworthii, Spartina alterniflora, Spermacoce glabra, Sphagnum, Spiranthes cernua, Sporobolus poiretii, Styrax americanus, Symphyotrichum lateriflorum, Taxodium distichum, Thelypteris kunthii, T. palustris, Tillandsia usneoides, Toxicodendron radicans, T. pubescens, Tradescantia, Trepocarpus aethusae, Triadenum virginicum, T. walteri, Typha domingensis, Ulmus alata, U. americana, U. crassifolia, Utricularia foliosa, Vernonia gigantea, Viola affinis, Vitis rotundifolia, Wisteria frutescens, Woodwardia areolata, Xyris elliottii, Zizaniopsis miliacea.*

Use by wildlife: Seeds of *Justicia* species are eaten by pinnated grouse (Aves: Phasianidae: *Tympanuchus cupido*). The binding rhizomes of *Justicia americana* increase substrate stability, which is favorable to the establishment of sedentary mussels (e.g., Mollusca: Unionidae: *Villosa fabalis, V. lienosa*) and other invertebrates (e.g., Insecta: Plecoptera). The breakdown of *J. americana* foliage provides an important source of organic matter for river-dwelling invertebrates. The leaves are high in nitrogen (to 3.8%; >50% from protein) with nitrogen yields that can reach 590 kg/ha, potentially making them a relatively high quality animal feed. They have been used as a palatable feed for channel catfish (Vertebrata: Osteichthyes: Ictaluridae: *Ictalurus punctatus*). The plants provide habitat for crayfish (Crustacea: Cambaridae: *Orconectes luteus*), various fish (Vertebrata: Osteichthyes) such as the Cape Fear shiner (Cyprinidae: *Notropis mekistocholas*), Devils River minnow (Cyprinidae: *Dionda diaboli*), greater redhorse (Cyprinidae: *Moxostoma valenciennesi*), and Tippecanoe darter (Percidae: *Etheostoma tippecanoe*) and riverine map turtles (Reptilia: Emydidae: *Graptemys geographica*). A variety of insects (Insecta) forage for nectar on the flowers of *Justicia americana* including long-tongued bees (Hymenoptera: Anthophoridae: *Anthophora abrupta, Ceratina dupla, Epeolus bifasciatus, Florilegus condigna, Melissodes bimaculata, M. communis, Nomada articulata, Synhalonia dilecta, S. rosae, Triepeolus lunatus*; Apidae: *Apis mellifera, Bombus griseocallis, B. impatiens, B. pensylvanica*; Megachilidae: *Coelioxys octodentata, C. sayi, Hoplitis cylindricans, H. pilosifrons, Megachile brevis, M. montivaga, M. pugnatus, M. texana, Osmia distincta*), short-tongued bees (Hymenoptera: Andrenidae: *Calliopsis andreniformis*; Halictidae: *Agapostemon sericea, A. virescens, Augochlorella aurata, A. striata, Halictus confusus, H. ligatus, H. rubicunda, Lasioglossum imitatus, L. pilosus, L. versatus*), wasps (Scoliidae: *Campsomeris plumipes*; Sphecidae: *Ammophila nigricans, Bembix americana*), flies (Diptera: Bombyliidae: *Bombylius atriceps, B. helvus, Rhynchanthrax parvicornis*; Conopidae: *Physocephala tibialis, Zodion fulvifrons*; Syrphidae: *Allograpta obliqua, Eristalis tenax, Parhelophilus laetus, Sphaerophoria contiqua, Syritta pipiens, Tropidia quadrata*), and Lepidoptera (Hesperiidae: *Epargyreus clarus, Erynnis martialis, Euphyes vestris, Thorybes bathyllus*; Nymphalidae: *Chlosyne nycteis*; Lycaenidae: *Celastrina argiolus*; Pieridae: *Pieris rapae*). It also is a host plant for golden looper moth larvae (Lepidoptera: Noctuidae: *Argyrogramma verruca*). The roots are hosts to parasitic nematodes (Nematoda: Heteroderidae: *Meloidogyne*). *Justicia lanceolata* is resistant to herbivory by nutria (Mammalia: Rodentia: Myocastoridae: *Myocastor coypus*); whereas, *J. ovata* is grazed by nutria and also by swamp rabbits (Mammalia: Leporidae: *Sylvilagus aquaticus*). *Justicia ovata* is also a larval host plant of butterflies (Lepidoptera: Nymphalidae) including the Cuban crescent (*Anthanassa frisia*), phaon crescent (*Phyciodes phaon*), and Texan crescent (*Anthanassa texana*).

Economic importance: food: Although some *Justicia* species reportedly are edible, there are no accounts of any of the OBL species being used as food; **medicinal:** *Justicia americana* contains alkaloids, but these have not been characterized. The Seminoles used *Justicia crassifolia* as a treatment to restore virility; **cultivation:** Although there are numerous reports of "*Justicia ovata*" in cultivation, that name often has been misapplied to a different species (*Justicia candicans*) that is commonly cultivated as an ornamental; **misc. products:** *Justicia americana* has been planted to enhance fish habitat and to control erosion of streambanks; **weeds:** none; **nonindigenous species:** *Justicia americana* has been introduced to China, where it was planted intentionally.

Systematics: *Justicia* is the type genus of tribe Justicieae (Acanthaceae), which is demonstrably monophyletic by phylogenetic analysis of nucleotide sequence and indel data, and by their unusual tricolporate hexapseudocolpate pollen. Tribe Justicieae resolves as the sister clade to tribe Ruellieae (Figure 5.51). However, *Justicia* is itself problematical taxonomically. Although fewer than a dozen *Justicia* species have been included in phylogenetic analyses, the preliminary results clearly indicate that the genus is not monophyletic as currently circumscribed, but comprises distinct New World and Old World components (Figure 5.54). *Justicia lanceolata* (retained here as a species in compliance with the 2013 indicator list) is treated by many authors as a variety of *J. ovata*. Relationships among the obligate aquatic species have not been clarified. Of the four OBL North American species, only *J. americana* has been included in phylogenetic analyses; however, never with sufficient representation of taxa to determine even its proximity of relationship to New World vs. Old World species. Further systematic studies of this large and complex genus are needed. The basic chromosome number in *Justicia* is not readily interpretable, not only because the genus is polyphyletic, but also because of extensive variation in the haploid numbers reported ($n = 7, 9, 11–18, 22, 27, 28, 31$). The gametic number $n = 14$ occurs most commonly and $x = 7$ has been suggested as the base number for the genus. Chromosome numbers are not available for any the OBL North American species.

Comments: *Justicia americana* occurs in eastern North America, *J. lanceolata* and *J. ovata* are restricted to the southeastern United States, and *J. crassifolia* is endemic to Florida.

References: Baskin & Baskin, 1998; Beal, 1977; Daniel et al., 1984; Endress, 1994; Evers et al., 1998; Fassett, 1957; Fritz & Feminella, 2003; Fritz et al., 2004a,b; Garrett et al., 2004; Guo & Bradshaw, 1993; Hall & Penfound, 1939; Helm & Chabreck, 2006; Hill & Webster, 2004; Homoya & Abrell, 2005; Howard & Mendelssohn, 1999; Keiper et al., 1998; Koryak & Reilly, 1984; Lewis, 1980; Little, 1979; Llewellyn & Shaffer, 1993; Mathies et al., 1983; Penfound, 1940b; Piovano & Bernardello, 1991; Plunkett & Hall, 1995; Rapp et al., 2001; Scotland et al., 1995; Shaffer et al., 1992; Strakosh et al., 2005; Wahlberg, 2001; Wieland, 2000.

4. Stenandrium

Pineland pinklet, sweet shaggytuft

Etymology: from the Greek *stenos andros* meaning "narrow stamens"

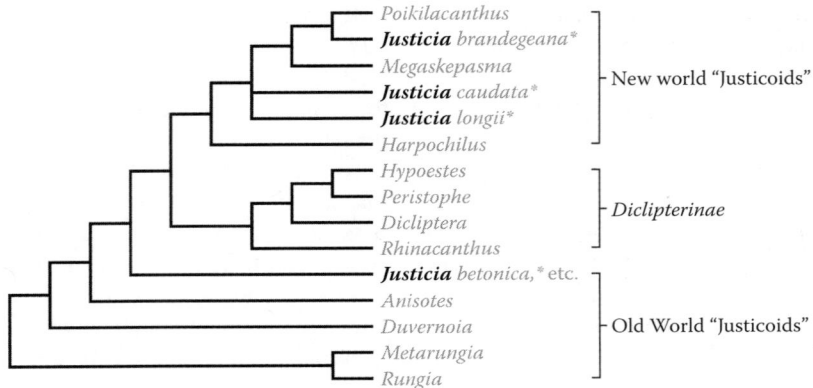

FIGURE 5.54 Intergeneric relationships in Acanthaceae tribe Justicieae as indicated by cladistic analysis of nrITS sequence data. Although only a few *Justicia* species were included in the analysis, it is evident that this genus (in bold) of roughly 400–600 species is polyphyletic (as indicated by asterisks), at least with respect to New World versus Old World components. None of the OBL North American species was included in the survey, and their possible relationship to other New World species requires verification. The New World clade is the sister group to subtribe Diclipterinae, which is monophyletic in this analysis. (Adapted from McDade, L.A. et al., *Syst. Bot.*, 25, 106–121, 2000.)

Synonyms: *Crossandra* (in part); *Gerardia*; *Stenandriopsis* (in part); *Synandra*

Distribution: global: New World; **North America:** southern

Diversity: global: 25 species; **North America:** 2 species

Indicators (USA): OBL; FACW: *Stenandrium dulce*

Habitat: freshwater; palustrine; **pH:** alkaline; **depth:** <1 m; **life-form(s):** emergent herb

Key morphology: acaulescent; leaves (to 10 cm) pubescent, petiolate (to 65 mm), ovate to elliptic, arranged in a rosette; inflorescence a head-like to elongate spike (to 85 mm), peduncles (to 20 cm) pubescent, the flowers sessile, 5-merous; calyx (to 11 mm) with lobes divided nearly to base; corolla (to 25 mm) tubular (to 16 mm), the limb (to 10 mm) spreading, bilabiate but appearing nearly radially symmetrical, pink to purple, whitish inside; androecium of 4 stamens and 1 staminode; style flared apically, funnel like; capsule (to 12 mm) ellipsoid, the surface barbed or bristly; seeds (to 4 mm) 4, laterally flattened

Life history: duration: perennial (rhizomes); **asexual reproduction:** rhizomes; **pollination:** insect; **sexual condition:** hermaphroditic; **fruit:** capsules (common); **local dispersal:** rhizomes; **long-distance dispersal:** seeds (animals)

Imperilment: (1) *Stenandrium dulce* [G3]; S1 (FL); S2 (TX)

Ecology: general: Most *Stenandrium* species are terrestrial and generally inhabit canyons, deserts, forests, grasslands, slopes, and dry arroyos, with a few growing along streams. Only one has wetland status in North America.

Stenandrium dulce **(Cav.) Nees** occurs in flatwoods, hammocks, prairies, and swamps of the southeastern coastal plain. The substrates typically lack loam, are low in nutrients, and have been characterized as alkaline sand, limestone, and Perrine marl. The plants occur in full sun to moderate shade. They are intolerant to salinity or desiccation. Although this species is categorized as an obligate aquatic, populations in Texas and Mexico reportedly have been observed to inhabit arid associations. There is scarce information on the reproductive biology of this species,

although it has been reported that the floral nectar occurs too deeply for bees (Hymenoptera) to obtain. No detailed study has been made on the pollination biology or seed ecology. The barbed/bristly capsule surface possibly facilitates adherence to animals for dispersal. The ecological information also is relatively minimal. Plants reportedly establish best in open sites and can spread like a groundcover when established. They are intolerant of competition with other plants and have been observed to grow rapidly and to flower in recently burned areas. The foliage contains 2-benzoxazolinone, which is phytotoxic to some weed seeds and 1,4-benzoxazin-3-ones, which are known to reduce survival of and deter feeding by aphids (Hemiptera: Aphididae: *Metopolophium dirhodum*). **Reported associates:** *Agalinis filifolia, Ambrosia artemisiifolia, Andropogon glomeratus, A. gyrans, A. virginicus, Aristida stricta, Astragalus cyaneus, Baccharis halimifolia, Bidens pilosus, Blechnum serrulatum, Callicarpa americana, Cassytha filiformis, Cephalanthus occidentalis, Chamaesyce porteriana, Cladium jamaicense, Conoclinium coelestinum, Conyza canadensis, Crinum americanum, Cynanchum scoparium, Cyperus haspan, Dichanthelium dichotomum, Dichondra carolinensis, Dyschoriste angusta, Elytraria caroliniensis, Erianthus giganteus, Eryngium balduinii, Eupatorium capillifolium, E. mikanioides, Euphorbia polyphylla, Eustachys glauca, Flaveria linearis, Fuirena scirpoidea, Galactia regularis, Gratiola hispida, Habenaria, Hedyotis nigricans, Helenium vernale, Heliotropium, Hymenocallis palmeri, Hypericum hypericoides, Ilex glabra, Ipomoea sagittata, Iresine diffusa, Justicia angusta, Liatris garberi, Linum carteri, Lippia nodiflora, Lobelia glandulosa, Lyonia fruticosa, Lythrum alatum, Melanthera angustifolia, Melochia corchorifolia, Mikania, Muhlenbergia capillaris, M. filipes, Myrica cerifera, Nephrolepis, Panicum hemitomon, P. tenerum, P. virgatum, Parthenocissus quinquefolia, Paspalum ciliatifolium, P. monostachyum, Passiflora suberosa, Penstemon fendleri, Phlebodium*

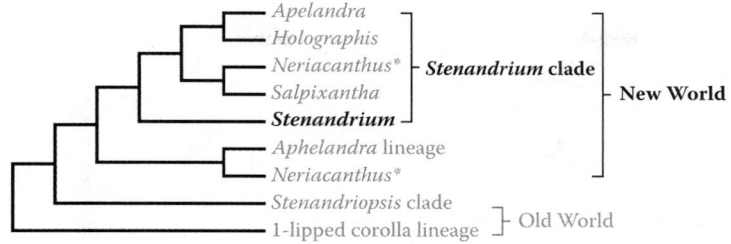

FIGURE 5.55 Phylogenetic relationships of the "2-lipped corolla lineage" (Acanthaceae tribe Acantheae) as resolved using combined nuclear and cpDNA sequence data. *Stenandrium* (bold) resolves as the basal genus in the monophyletic *Stenandrium* clade. This group, together with *Neriacanthus* and the *Aphelandra* lineage, are restricted to the New World; whereas, the *Stenandriopsis* clade and entire 1-lipped corolla lineage are Old World taxa. *Neriacanthus* (asterisked) is polyphyletic with respect to one species (*N. purdieanus*), which falls within the *Stenandrium* clade. Although *Stenandrium dulce* was not surveyed, it presumably falls within *Stenandrium* as indicated. (Adapted from McDade, L.A. et al., *Syst. Bot.*, 30, 834–862, 2005.)

aureum, Piloblephis rigida, Pinus elliottii, Piriqueta caroliniana, Pityopsis graminifolia, Pluchea odorata, P. rosea, Polygala grandiflora, Psilotum nudum, Pterocaulon virgatum, Quercus laurifolia, Rhus copallinum, Rhynchelytrum repens, Rhynchospora colorata, R. divergens, R. inundata, R. microcarpa, Rubus trivialis, Rudbeckia hirta, Ruellia caroliniensis, Sabal palmetto, Sacciolepis striata, Salix caroliniana, Schinus terebinthifolius, Schoenus nigricans, Schizachyrium rhizomatum, Scleria, Scoparia dulcis, Serenoa repens, Setaria geniculata, Silphium, Smilax auriculata, S. laurifolia, Solidago fistulosa, Sporobolus indicus, Symphyotrichum adnatum, Taxodium ascendens, Tephrosia rugelii, Teucrium canadense, Toxicodendron radicans, Vaccinium myrsinites, Vernonia blodgettii, Vitis munsoniana, Vittaria lineata, Woodwardia virginica.

Use by wildlife: *Stenandrium dulce* is a larval host plant of the definite patch butterfly (Insecta: Lepidoptera: Nymphalidae: *Chlosyne definita*).

Economic importance: food: not reported as edible; **medicinal:** The Florida Seminoles used *Stenandrium dulce* as both a sedative and a stimulant; **cultivation:** *Stenandrium dulce* is cultivated for natural wildflower gardens; **misc. products:** none; **weeds:** none; **nonindigenous species:** none

Systematics: Molecular data place *Stenandrium* within tribe Acantheae, subfamily *Acanthoidae* (Figure 5.51), which is the sister group to subfamily Ruellioideae of Acanthaceae. *Stenandrium* defines a well-supported clade of 2-lipped, New World genera (Figure 5.55); however, the analysis of additional taxa is necessary to evaluate the monophyly of the genus and to better elucidate interspecific relationships (*Stenandrium dulce* has not yet been included in any molecular phylogenetic survey). The lack of any other obligate aquatic taxa in *Stenandrium* indicates a single origin of the OBL habit in the genus. The base chromosome number of *Stenandrium* is $x = 13$. *Stenandrium dulce* ($2n = 52$) is a tetraploid.

Comments: *Stenandrium dulce* occurs in the southern United States (Florida, Georgia, Texas) but extends southward through Central America to Chile.

References: Bravo et al., 2004; Daniel, 1984; McDade et al., 2005; Piovano & Bernardello, 1991; SFWMD, 2003.

Family 2: Lamiaceae [258]

The "mint" family (Lamiaceae/Labiatae) is a large group consisting of approximately 7,000 species of herbs, shrubs, and trees, including many that are of considerable economic importance. Phylogenetic analyses of several cpDNA sequence data sets (Wagstaff & Olmstead, 1997; Wagstaff et al., 1998) have confirmed the monophyly of this fairly distinctive family, after incorporating some refinements such as the inclusion of many genera (e.g., *Callicarpa*) placed formerly in Verbenaceae. Morphologically, Lamiaceae are best defined by their ovules, which are attached laterally to false septa near the inrolled carpel margins (Judd et al., 2002). Many of the species have bilabiate corollas and gynobasic styles (which have evolved several times in the family). They frequently have square stems with opposite leaves (especially in the North American species) and contain various ethereal oils (e.g., terpenoids). The nectariferous flowers (often arranged in verticels) are pollinated by various insects and birds. Many of the species reportedly are gynodioecious. The fruits either are drupes (dispersed by birds and mammals) or nutlets (from a schizocarp), which are dispersed by gravity, wind, or water (Judd et al., 2002).

Because of their essential oils, Lamiaceae are the source of numerous culinary spices including basil (*Ocimum*), mint (*Mentha*), oregano (*Origanum*), rosemary (*Rosmarinus*), sage (*Salvia*), savory (*Satureja*), and thyme (*Thymus*). The family contains many ornamental genera such as *Agastache, Ajuga, Callicarpa, Lamium, Lavandula, Mentha, Monarda, Nepeta, Ocimum, Origanum, Phlomis, Plectranthus, Rosmarinus, Salvia, Scutellaria, Solenostemon, Stachys, Teucrium, Thymus*, and others. *Tectona* (formerly placed in Verbenaceae) is the source of the commercially valuable teak, which is used for lumber and as an ornamental wood.

Ecologically, mints usually occupy terrestrial habitats although at least a few representatives are found routinely in wetland floras. A fair number of species would be regarded as FACW inhabitants and 11 genera of Lamiaceae contain OBL species in North America:

1. *Callicarpa* L.
2. *Clinopodium* L.

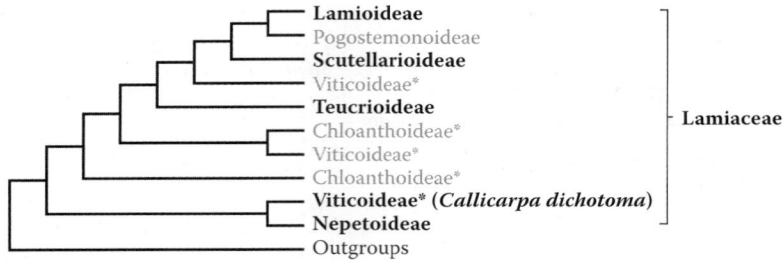

FIGURE 5.56 Subfamily relationships in Lamiaceae as indicated by phylogenetic analysis of combined *ndhF* and *rbcL* sequence data. Two subfamilies (asterisked) are polyphyletic in this analysis. The distribution of subfamilies containing OBL taxa (bold) indicates independent origins of the habit in the family. (Adapted from Wagstaff, S.J. et al., *Plant Syst. Evol.*, 209, 265–274, 1998.)

3. ***Hyptis*** Jacq.
4. ***Lycopus*** L.
5. ***Macbridea*** Elliott ex Nutt.
6. ***Mentha*** L.
7. ***Physostegia*** Benth.
8. ***Pogogyne*** Benth.
9. ***Scutellaria*** L.
10. ***Stachys*** L.
11. ***Trichostema*** L.

Currently, seven subfamilies are recognized in Lamiaceae (Figure 5.56), although two of them (Chloanthoideae, Viticoideae) are polyphyletic. Five subfamilies contain genera with OBL species: Lamioideae (*Macbridea*, *Physostegia*, *Stachys*); Nepetoideae (*Clinopodium*, *Hyptis*, *Lycopus*, *Mentha*, *Pogogyne*); Scutellarioideae (*Scutellaria*); Teucrioideae (*Trichostema*); Viticoideae (*Callicarpa*). Phylogenetic relationships among the subfamilies (Figure 5.56) indicate multiple origins of the OBL habit in the family.

1. *Callicarpa*

Beautyberry

Etymology: from the Greek *calo carpo* meaning "beautiful fruit"

Synonyms: *Amictonis*; *Burcardia*; *Geunsia*; *Illa*; *Johnsonia*; *Porphyra*; *Tomex*

Distribution: global: Africa; Asia; New World; **North America:** southeastern

Diversity: global: 140 species; **North America:** 3 species

Indicators (USA): OBL: *Callicarpa dichotoma*

Habitat: freshwater; palustrine; **pH:** 5.0–6.8; **depth:** <1 m; **life-form(s):** emergent shrub

Key morphology: stems (to 2 m) highly branched; leaves (to 10 cm) opposite, deciduous, subsessile, lanceolate to obovate, margins serrate; cymes (to 2.5 cm) axillary, paired, pedunculate (to 12 cm); flowers 4-merous, radially symmetrical; corolla (to 3 mm) tubular, purple; stamens exserted; fruit subglobose (to 2 mm), bright purple or white, in showy clusters

Life history: duration: perennial (buds); **asexual reproduction:** none; **pollination:** insect; **sexual condition:** hermaphroditic; **fruit:** drupes (common); **local dispersal:** drupes (gravity, water); **long-distance dispersal:** drupes (birds)

Imperilment: (1) *Callicarpa dichotoma* [GNR]

Ecology: general: *Callicarpa* mainly is a tropical and subtropical genus, which grows in mixed forests and on mountain slopes with only a few species occurring along rivers or streambanks. The native North American (and Puerto Rican) species are designated as FACW, FACU plants. Little information exists on the natural history of the genus. Most species appear to be entomophilous and are pollinated by various bees (Hymenoptera: Apidae: Halictidae). A few island species are functionally dioecious and produce nongerminable pollen in some of their flowers. The drupes are brightly colored and attract birds (Aves), which likely are their principal dispersal agents.

***Callicarpa dichotoma* (Lour.) K. Koch** grows in bogs, depressions and ditches at elevations of up to 600 m. Suitable substrates include loam or sandy loam with an acidic pH (optimal pH = 5.0–5.5; but up to pH = 6.8 is tolerated). The plants can survive on dry soil but thrive in wetter conditions and grow best in sites that receive full sunlight (4–5 h/day). They will grow in light shade and can withstand quite cold winters (to −29°C). The plants are self-compatible and some cultivars (e.g., "Issai") will self-pollinate and exhibit fairly high seed set. Reports of higher seed set in cross-pollinated plants are questionable, especially because many cultivars are derived clonally. The seeds require no stratification for germination, which can be enhanced using coir, i.e., coconut (Arecaceae: *Cocos nucifera*) fiber as a substrate. Germination also increases significantly after the fruit pulp has been removed. Germination of winter sown seeds occurs within 5 months (hardiness zone 6). The plants have low flammability and can live for up to 20 years. They can be propagated artificially by stem cuttings and by layering. Regeneration by micropropagation of axillary buds also has been successful. The OBL status of this species in North America is arguable, given that the plants inhabit mixed forests and mountain slopes in their native range. Further ecological studies of this species are necessary. **Reported associates:** none.

Use by wildlife: *Callicarpa dichotoma* is resistant to grazing by deer (Mammalia: Cervidae: *Odocoileus*) and to feeding by Japanese beetles (Insecta: Coleoptera: Scarabaeidae: *Popillia japonica*). The colorful drupes attract birds (Aves).

Economic importance: food: not edible; **medicinal:** Many *Callicarpa* species are described as having medicinal properties. *Callicarpa dichotoma* contains several phenylethanoid glycosides (which have been shown to be neuroprotective) and acteoside, which can alleviate some types of memory impairments. In Asia, the leaves (used as decoctions or powdered) are consumed to control internal bleeding and to heal wounds. Leaf powder is applied externally to control external bleeding. The leaves have been used to treat keratitis (inflammation of the cornea) and snakebites; **cultivation:** *Callicarpa dichotoma* is grown as an ornamental shrub and is regarded as the best and most attractive horticultural species in the genus. It received gold medals from the Pennsylvania Horticultural Society in 1989 and from the Georgia Green Industry in 2002. Cultivars include 'Albescens' (white drupes), 'Albifructus' (white drupes), 'Amethyst' (purple drupes), 'Duet' (variegated), 'Early Amethyst' (compact form), 'Early Profusion,' 'Issai' (compact form), 'Profussions' (purple drupes), 'Shirobana' (white drupes), 'Splashy' (variegated), 'Spring Gold' (golden young foliage), and 'Winterthur'; **misc. products:** none; **weeds:** *Callicarpa dichotoma* is regarded as invasive in North Carolina; **nonindigenous species:** *Callicarpa dichotoma* (Chinese beautyberry) was introduced from Asia as an ornamental garden plant and has become naturalized in several localities.

Systematics: Although *Callicarpa* was assigned formerly to Verbenaceae, molecular phylogenetic analyses (Figure 5.56) indicate that it should be transferred to Lamiaceae along with many other genera of subfamily Viticoideae, which is polyphyletic as formerly circumscribed. Specifically, the analysis of cpDNA sequence data resolve *Callicarpa dichotoma* as the sister group to subfamily Nepetoideae in a position that is distant from other Viticoideae. It will be necessary to include the other *Callicarpa* species in phylogenetic analyses before the monophyly of the genus can be verified. Also, the position of the genus is not firmly established with respect to other Lamiaceae and subsequent studies that include a broader sampling of taxa could result in different hypotheses of its relationship within the family; however, preliminary studies indicate a close alliance with Lamiaceae and a distant relationship to Verbenaceae. The base chromosome number of *Callicarpa* is $x = 9$. *Callicarpa dichotoma* ($2n = 36$) is a tetraploid.

Comments: *Callicarpa dichotoma* (native to China, Japan, Korea, and Vietnam) occurs in the eastern United States (North Carolina, South Carolina, Tennessee, Virginia)

References: AHPD, 1975; Baskin & Baskin, 1998; Held, 2004; Kärkönen et al., 1999; Kawakubo, 1990; Koo et al., 2005; Lee et al., 2006; Lewis, 1961; Shu, 1994; Sprinkles & Bachman, 1999; Tsukaya et al., 2003; Wagstaff & Olmstead, 1997; Wagstaff et al., 1998.

2. *Clinopodium*

Calamint, false pennyroyal, savory; calament, clinopode
Etymology: from the Greek *klineios podion* meaning "bed foot" in reference to the inflorescence shape of some species
Synonyms: *Calamintha*; *Micromeria* (in part); *Rizoa*; *Satureja* (in part); *Thymus* (in part)
Distribution: global: Africa; Asia; Europe; **North America:** widespread
Diversity: global: 25 species; **North America:** 13 species

Indicators (USA): OBL: *Clinopodium brownei*
Habitat: freshwater; palustrine; **pH:** unknown; **depth:** <1 m; **life-form(s):** emergent herb
Key morphology: stems (to 4 dm) weak, sprawling, square, herbage minty-aromatic; leaves (to 15 mm) opposite, suborbicular to ovate, petioled (to 5 mm); flowers bilabiate, solitary, axillary, pedicelled (to 15 mm); calyx (to 5 mm) turbinate; corolla (to 10 mm) lavender, pink, or whitish, the tube dilated above the calyx, the throat hairy, lower lip 3-lobed, upper lip hood-like, unlobed; stamens 4, didynamous, exserted; style gynobasic, ovary 4-lobed; nutlets minute, dark purple-red
Life history: duration: perennial (rhizomes); **asexual reproduction:** rhizomes, shoot fragments; **pollination:** insect; **sexual condition:** hermaphroditic; **fruit:** nutlets (common); **local dispersal:** rhizomes, stem fragments; **long-distance dispersal:** nutlets
Imperilment: (1) *Clinopodium brownei* [G5]; S1 (GA)
Ecology: general: *Clinopodium* predominantly is a terrestrial genus of fields, forests, grasslands, hillsides, and streamsides, with three North American species (23%) having some level of status as wetland indicators (OBL, FACW, FAC, FACU).

***Clinopodium brownei* (Sw.) Kuntze** grows in bayous, depressions, ditches, flatwoods, floodplains, marshes, meadows, seeps, springs, swamps, and along pond, river, and stream margins at elevations of up to 3300 m (outside of North America). Exposures are open to semishaded and the substrates include loamy sand, muck, mud, sand, sandy clay, and sandy loam. Apparently, the plants are well adapted to inundation as evidenced by their use in aquariums and in water gardens; however, there also are a number of reports of this species from drier waste areas. There are no detailed accounts on the pollination, floral biology, or seed ecology of this species. Vegetative reproduction occurs by fragmentation of the stems, which root at nodes and also by rhizomes, which result in a mat-forming habit. **Reported associates:** *Acer rubrum, Alternanthera philoxeroides, Baptisia alba, Berchemia scandens, Blechnum serrulatum, Briza, Bromus unioloides, Carex, Carpinus caroliniana, Centella erecta, Cephalanthus occidentalis, Cerastium, Cyperus retrorsus, C. surinamensis, C. virens, Epidendrum, Eryngium baldwinii, Forestiera acuminata, Hypoxis curtissii, Iris hexagona, Lindernia grandiflora, Liquidambar styraciflua, Lobelia cliffortiana, Magnolia, Medicago lupulina, Myrica inodora, Nuttallanthus canadensis, Persicaria hydropiperoides, Plantago virginica, Poa annua, Polypodium, Quercus lyrata, Rhynchospora colorata, Samolus valerandi, Spermolepis divaricata, Sphenopholis, Stellaria, Taxodium ascendens, Thelypteris patens, Tillandsia bartramii, T. fasciculata, T. utriculata, Toxicodendron radicans, Trifolium campestre, T. repens, Triodanis perfoliata, Ulmus americana, U. crassifolia, Valerianella radiata, Verbena, Vicia acutifolia, V. minutiflora, Vittaria.*
Use by wildlife: The nutlets of *Clinopodium brownei* are eaten by ground doves (Aves: Columbidae: *Columbina passerina*), where up to 22,000 seeds have been recovered from the crop of a single bird.

Economic importance: food: *Clinopodium brownei* is not reported to be edible; **medicinal:** Ethanolic extracts of *Clinopodium brownei* effectively inhibit several pathogenic bacteria (Bacteria: Staphylococcaceae: *Staphylococcus aureus*, *S. pyogenes*). The leaves have been used by the natives of Central America and Mexico to prepare a bath for treating headaches. South American natives have used a decoction made from the leaves as a cough medicine. In Jamaica, the leaves are brewed into a tea to treat stomach disorders and diarrhea; **cultivation:** *Clinopodium brownei* is sold as an oxygenating plant ("creeping charlie") for use in aquariums or ponds; **misc. products:** The essential oil of *Clinopodium brownei* primarily contains the mint-odored monoterpenes pulegone and menthone, which are used as flavorings and as fragrance agents for beverages, perfume, soap, and toiletry products. The plants also have been used as an insect repellant; **weeds:** none; **nonindigenous species:** none

Systematics: Over the years there has been much taxonomic confusion regarding the delimitation of *Calamintha*, *Clinopodium*, *Micromeria*, and *Satureja* within subfamily Nepetoideae, tribe Mentheae of Lamiaceae. Generally, the merger of *Calamintha* and *Clinopodium* has been accepted but contemporary authors have placed *Clinopodium brownei* in several other genera, notably *Micromeria* and *Satureja*. Analysis of cpDNA sequence data resolves *C. brownei* within a well-supported *Clinopodium* clade, which is distinct from (but closely related to) the clades that contain *Micromeria* and *Satureja* species (Figure 5.57). Various analyses cpDNA and nuclear DNA data resolve the genus as the sister group to

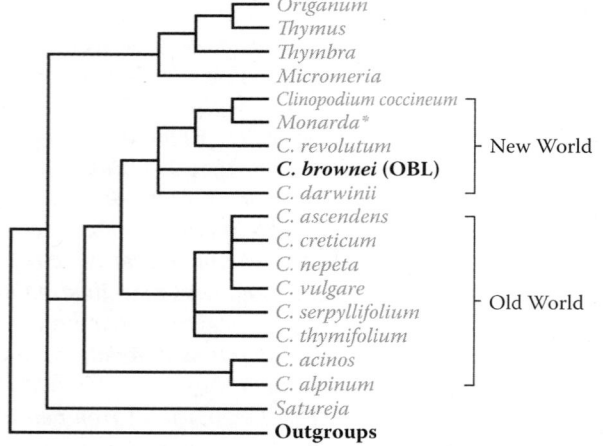

FIGURE 5.57 Phylogenetic relationships in Lamiaceae (subfamily Nepetoideae, tribe Mentheae) as indicated by analysis of cpDNA sequence data. These results resolve *Clinopodium* as a clade distinct from either *Micromeria* or *Satureja*, whose taxonomic limits have been confused in the past. Here the OBL North American *C. brownei* (in bold) nests deeply among the New World representatives of *Clinopodium*. Although the clade also includes *Monarda* (asterisked), other studies (e.g., Prather et al., 2002; Walker & Sytsma, 2007) show that genus to be quite distinct from *Clinopodium*; therefore, its placement here must be in error. (Adapted from Bräuchler, C. et al., *Taxon*, 54, 639–650, 2005; Bräuchler, C. et al., *Taxon*, 55, 977–981, 2006.)

Conradina (Figure 5.59), or at least closely associated with it. *Clinopodium brownei* occurs within a New World subclade along with one surveyed species of *Monarda*. However, the placement of *Monarda* in that analysis presumably is in error given that several other studies resolve *Clinopodium* and *Monarda* in distinct clades. Additional sampling of taxa and further analyses will be necessary to sort out the ultimate interrelationships of these genera. However, the inclusion of *C. brownei* in *Clinopodium* rather than either *Micromeria* or *Satureja* seems to be warranted based on the currently available evidence. The base chromosome number of *Clinopodium* probably is $x = 10$. Counts are not available for *C. brownei*.

Comments: *Clinopodium brownei* occurs along the Gulf coastal plain and extends into South America.

References: Ankli et al., 1999; Asprey & Thornton, 1953; Bermúdez & Velázquez, 2002; Bräuchler et al., 2005, 2006; Caceres et al., 1991; Cantino & Wagstaff, 1998; Edwards et al., 2006; Epling & Jativa, 1964, 1966; Garber, 1877; Harley & Paucar Granda, 2000; Mitchell & Ahmad, 2006; Morales Valverde, 1993; Passmore, 1981; Prather et al., 2002; Rojas & Usubillaga, 2000; Tucker et al., 1992.

3. *Hyptis*

Cluster bushmint, desert lavender, musky mint

Etymology: From the Greek *hyptios*, meaning "upturned," in reference to the lower lip of the flower

Synonyms: *Brotera*; *Broteroa*; *Condea*; *Gnoteris*; *Hypothronia*; *Mesosphaerum*; *Rhaphiodon*; *Schaueria*

Distribution: global: New World; **North America:** southern

Diversity: global: 400 species; **North America:** 6 species

Indicators (USA): OBL: *Hyptis alata*

Habitat: freshwater; palustrine; **pH:** broad range; **depth:** <1 m; **life-form(s):** emergent herb

Key morphology: stems (to 3 m) square, finely pubescent, the herbage strongly odorous; leaves (to 15 cm) opposite, lanceolate, tapering to base, margins coarsely serrate; flowers densely clustered in paired (to 12 pairs), axillary, long-peduncled (to 6 cm), globose heads (to 2.5 cm), each subtended by a leaflike involucre; flowers (to 12 mm) bilabiate, 5-lobed, white, purple-spotted, the lower lip with a saccate central lobe; stamens 4, didynamous; schizocarp dehiscing into 4, black, single-seeded nutlets (to 1.5 mm), which remain enclosed by the persistent calyx (to 8 mm)

Life history: duration: perennial (persistent rootstock); **asexual reproduction:** none; **pollination:** insect; **sexual condition:** hermaphroditic; **fruit:** nutlets (common); **local dispersal:** nutlets; **long-distance dispersal:** nutlets

Imperilment: (1) *Hyptis alata* [G5]; S3 (NC)

Ecology: general: *Hyptis* predominantly is a terrestrial South American genus, with only a few species found in wet sites. Only one North American species is categorized as an obligate aquatic. The reproductive biology has been studied in detail for only a few species. Most *Hyptis* appear to be self-compatible, but usually are outcrossed by insect or hummingbird (Aves: Trochilidae) pollinators. Bees or wasps (Hymenoptera) are the most common pollinators reported. In some species, the pollen is released explosively upon perturbation of the tensioned stamens by a pollinator (often by hummingbirds). Protandry appears to be common, with stigma receptivity occurring after the stamens have been tripped and most of the pollen has been dispersed. Little information on seed ecology exists for any of the species.

Hyptis alata (**Raf.**) **Shinners** inhabits bogs, borrow pits, depressions, ditches, flatwoods, marshes, meadows, prairies, roadsides, savannas, seasonal ponds, seeps, swales, swamps, and the margins of ponds and streams at elevations of up to 137 m. The substrates have been described both as acidic and as calcareous, and include clay, sand, and sandy loam. The plants occur where there is full sun to partial shade, and can tolerate some drying conditions. The pollination biology has not been described, but the flowers are visited by nectar seeking bees and wasps (Hymenoptera), which probably function as their pollinators. The mechanisms of seed dispersal and germination requirements also remain unknown. The seeds appear to represent a fairly minor component of wetland seed banks. A number of collections have been made from frequently burned sites, which indicates some degree of fire tolerance for this species. **Reported associates:** *Acmella oppositifolia, Agalinis obtusifolia, Agrostis hyemalis, Ambrosia artemisiifolia, Amphicarpum muhlenbergianum, Amsonia tabernaemontana, Andropogon brachystachyus, A. capillipes, A. glaucopsis, A. glomeratus, A. gyrans, n. mohrii, A. virginicus, Anemia wrightii, Aristida beyrichiana, A. palustris, A. purpurascens, A. spiciformis, Arnoglossum ovatum, Asclepias lanceolata, A. viridis, Asimina, Axonopus fissifolius, A. furcatus, Baccharis halimifolia, Bacopa caroliniana, B. rotundifolia, Bigelowia nudata, Boltonia asteroides, Buchnera longifolia, Carex glaucescens, Carphephorus, Centella erecta, Ceratiola ericoides, Chamaecrista deeringiana, C. nictitans, Chaptalia tomentosa, Chiococca parvifolia, Cirsium nuttallii, Cladium jamaicense, Mnesithea rugosa, Coreopsis linifolia, C. tinctoria, Ctenium aromaticum, Cyperus haspan, C. polystachyos, C. stenolepis, Dichanthelium aciculare, D. longiligulatum, D. scabriusculum, D. strigosum, Diodia virginiana, Drosera brevifolia, Eleocharis equisetoides, E. melanocarpa, E. microcarpa, E. quadrangulata, E. tuberculosa, Eragrostis elliottii, E. spectabilis, Erigeron strigosus, E. vernus, Eriocaulon compressum, E. decangulare, E. ravenelii, Eriochloa punctata, Eryngium aromaticum, E. yuccifolium, Eupatorium leptophyllum, E. leucolepis, E. mohrii, E. serotinum, Euthamia caroliniana, E. graminifolia, E. leptocephala, E. minor, Fimbristylis littoralis, Fuirena breviseta, F. bushii, Funastrum clausum, Gaura lindheimeri, Gratiola brevifolia, G. ramosa, G. virginiana, Helenium drummondii, H. pinnatifidum, Helianthus heterophyllus, H. radula, Hibiscus aculeatus, H. moscheutos, Hydrolea ovata, Hygrophila lacustris, Hymenocallis liriosme, Hypericum crux-andreae, H. galioides, H. hypericoides, H. nudiflorum, Ilex cassine, I. glabra, I. myrtifolia, Ipomoea sagittata, Iva angustifolia, Juncus effusus, J. marginatus, J. megacephalus, J. roemerianus, J. validus, Justicia ovata, Lachnocaulon anceps, L. beyrichianum, Leersia hexandra, Liatris acidota, Lilium catesbaei, Linum carteri, Lobelia flaccidifolia, Ludwigia linearis, L. microcarpa,* *L. pilosa, L. sphaerocarpa, L. suffruticosa, Lycopodiella alopecuroides, L. appressa, L. caroliniana, Lycopus rubellus, Lyonia ferruginea, L. fruticosa, L. lucida, Lythrum alatum, L. lineare, Marshallia tenuifolia, Mecardonia acuminata, Melaleuca quinquenervia, Mikania scandens, Mitreola sessilifolia, Muhlenbergia filipes, M. sericea, Myrica cerifera, Neptunia lutea, Odontosoria clavata, Oenothera drummondii, Oxypolis filiformis, Panicum boscianum, P. hemitomon, P. hians, P. rigidulum, P. scoparium, P. tenerum, P. virgatum, Paspalum floridanum, P. laeve, P. notatum, P. plicatulum, P. setaceum, P. urvillei, Persea palustris, Persicaria hydropiperoides, Phyla nodiflora, Piloblephis rigida, Pinus elliottii, P. taeda, Piriqueta caroliniana, Pityopsis graminifolia, Platanthera nivea, Pluchea rosea, Polygala grandiflora, P. leptocaulis, P. ramosa, P. rugelii, Pontederia cordata, Proserpinaca palustris, P. pectinata, Pterocaulon pycnostachyum, Ptilimnium capillaceum, Quercus minima, Q. pumila, Ratibida peduncularis, Rhexia lutea, R. mariana, Rhynchospora caduca, R. cephalantha, R. colorata, R. divergens, R. elliottii, R. fascicularis, R. filifolia, R. globularis, R. gracilenta, R. inexpansa, R. inundata, R. latifolia, R. macrostachya, R. microcarpa, R. nitens, R. perplexa, R. rariflora, Rubus trivialis, Rudbeckia texana, Sabal palmetto, Sabatia angularis, S. brevifolia, S. campanulata, Saccharum giganteum, Sarracenia, Saururus cernuus, Schinus terebinthifolius, Schizachyrium rhizomatum, S. scoparium, S. tenerum, Scirpus cyperinus, Scleria baldwinii, S. georgiana, S. muehlenbergii, S. reticularis, S. verticillata, Senna ligustrina, Serenoa repens, Setaria parviflora, Solidago sempervirens, S. stricta, Sorghastrum secundum, Spartina patens, Stillingia sylvatica, Stylisma aquatica, Symphyotrichum dumosum, Tradescantia hirsutiflora, Tridens strictus, Tripsacum dactyloides, Typha latifolia, Vaccinium myrsinites, Vernonia gigantea, Vitis rotundifolia, Xylorhiza tortifolia, Xyris fimbriata, X. jupicai, X. laxifolia, X. louisianica.*

Use by wildlife: The nectariferous flowers of *Hyptis alata* attract insects (Insecta) such as leaf-cutting bees (Hymenoptera: Megachilidae: *Megachile albitarsis*), roadside skippers (Lepidoptera: Hesperiidae: *Amblyscirtes aesculapius*), and thread-waisted wasps (Hymenoptera: Sphecidae).

Economic importance: food: not edible; **medicinal:** none reported; **cultivation:** *Hyptis alata* is distributed occasionally as an ornamental native wildflower; **misc. products:** none; **weeds:** none; **nonindigenous species:** none

Systematics: *Hyptis* is placed in tribe Ocimeae, subfamily Nepetoideae of Lamiaceae. Both the tribe and subfamily consistently resolve as monophyletic groups in phylogenetic analyses. The other four genera of Nepetoideae that contain OBL species (*Lycopus, Mentha, Micromeria, Pogogyne*) are placed within tribe Mentheae, which is also monophyletic and is distinct from *Ocimeae* (Figure 5.58). Consequently, the OBL habit appears to be derived independently in *H. alata*. Only a small representation of the rather large genus *Hyptis* has been studied phylogenetically, with the most comprehensive analysis containing only nine species (unfortunately not *H. alata*). Combined cpDNA sequence data resolve these *Hyptis* species as a well-supported clade, which is closely

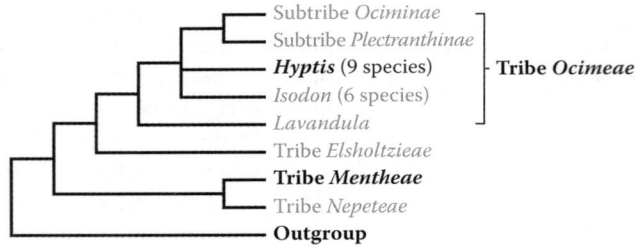

FIGURE 5.58 Phylogenetic analysis of Lamiaceae subfamily Nepetoideae, tribe Ocimeae showing the position of *Hyptis*. The cladogram, which is based on analysis of combined cpDNA sequence data, resolves *Hyptis* as a clade that is closely related to the genus *Isodon*. A similar result has been obtained by analysis of cpDNA restriction fragment data (Wagstaff et al., 1995), which resolved *H. alata* (not included in the above study) as the sister group to *Isodon*. Groups that contain OBL species in North America are indicated in bold. (Adapted from Paton, A.J. et al., *Mol. Phylogen. Evol.*, 31, 277–299, 2004.)

related to the principally Asian genus *Isodon* (Figure 5.58). A prior analysis (using cpDNA RFLP data) that included only *H. alata* resolved the species as the sister to *Isodon*. Therefore, it is reasonable to presume that *Hyptis* is monophyletic and is closely related to *Isodon* in tribe Ocimeae. However, more specific details on the relationship of *H. alata* will not be possible until additional *Hyptis* species are added to similar analyses. *Hyptis alata* sometimes has been subdivided into two subspecies, with subsp. *alata* in North America and subsp. *rugosula* in South America (the latter alternatively considered a distinct species). The base chromosome number of *Hyptis* is *x* = 7 or 8. The chromosome number of *H. alata* has not been reported.

Comments: *Hyptis alata* occurs along the southeastern coastal plain of the United States; however, the species is amphitropically disjunct in the New World, occurring also in South America.

References: Brantjes & De Vos, 1981; Correll & Correll, 1975; Godfrey & Wooten, 1981; Harley, 1983; Keller & Armbruster, 1989; Orzell & Bridges, 2006; Paton et al., 2004; Tobe et al., 1998; Wagstaff et al., 1995.

4. *Lycopus*

Bugleweed, water-horehound' chanvre d'eau, lycope
Etymology: from the Greek *lyco pous* meaning "wolf foot," with respect to the shape of the petal lobes
Synonyms: *Euhemus*; *Phytosalpinx*
Distribution: global: Asia; Australia; Europe; **North America:** widespread
Diversity: global: 15 species; **North America:** 8 species
Indicators (USA): OBL: *Lycopus americanus, L. amplectens, L. asper, L. cokeri, L. europaeus, L. rubellus, L. uniflorus, L. virginicus*
Habitat: brackish, freshwater; palustrine; **pH:** 3.5–7.8; **depth:** <1 m; **life-form(s):** emergent herb
Key morphology: stems (to 2 m) square, branched, or simple, herbage not fragrant; leaves (to 15 cm) opposite, linear to lanceolate, simple, or deeply pinnately lobed, sessile, or with

tapering petiolate bases; verticels (up to 30-flowered) dense, axillary; flowers 4- to 5-merous, radial to slightly bilateral; corolla (to 4.5 mm) tubular, funnelform to campanulate, white, or cream, sometimes purple spotted; stamens 4, didynamous, the posterior pair staminodial or absent, the anterior pair fertile, exserted; ovary deeply 4-lobed, the style gynobasic, ovary 4-lobed; schizocarp dehiscing into 4, black, single-seeded nutlets (to 2.2 mm), which remain enclosed by the persistent calyx (to 3.0 mm), which is subequal to or greater than the length of the nutlets

Life history: duration: perennial (stolons, tubers); **asexual reproduction:** stolons, tubers; **pollination:** insect; **sexual condition:** hermaphroditic; **fruit:** nutlets (common); **local dispersal:** nutlets (gravity), stolons, tubers; **long-distance dispersal:** nutlets (animals, water)

Imperilment: (1) *Lycopus americanus* [G5]; S1 (AK, GA); S2 (<u>AB</u>, NC, <u>NF</u>); S3 (WY); (2) *L. amplectens* [G5]; S1 (IN, MD, MS, NC); S2 (DE, NY); S3 (GA); (3) *L. asper* [G5]; SH (MO); S1 (KS); S2 (<u>ON</u>, <u>QC</u>); S3 (<u>AB</u>, <u>BC</u>, WY); (4) *L. cokeri* [G3]; S3 (NC); (5) *L. europaeus* [GNR] (6) *L. rubellus* [G5]; SH (WV); S1 (DC, KS, NY, PA, RI); S2 (MA, NJ, <u>ON</u>); (7) *L. uniflorus* [G5]; S1 (KS, WY); S2 (KY, MO); S3 (<u>AB</u>, AK, CA, NC); (8) *L. virginicus* [G5]; S2 (MI, <u>ON</u>, <u>QC</u>, VT)

Ecology: general: All *Lycopus* species grow in wet habitats and all of the North American species are categorized as OBL aquatics. This genus can be regarded as an excellent indicator of wetland habitat because it always occurs on hydric substrates although the plants seldom are found growing in more than a few cm of standing water. When growing in saturated substrates the plants avoid root anaerobiosis by translocating oxygen from their aerial shoot. Many of the species are thought to be gynodioecious, having both bisexual and functionally female (male sterile) flowers. However, only a few species have been investigated in enough detail to verify the presence of functionally male flowers. The pollen of all species is extremely similar and pollination of the relatively small flowers can be carried out by various insects (Insecta) including bees or wasps (Hymenoptera), beetles (Coleoptera), butterflies (Lepidoptera), and hoverflies (Diptera: Syrphidae). The nutlets are not mucilaginous but have either a nontuberculated or tuberculated corky crest that facilitates their flotation and water dispersal. One OBL taxon (*Lycopus* ×*sherardii* Steele) has been excluded from this treatment because of its uncertain taxonomic status (see *Systematics* below).

***Lycopus americanus* Muhl. ex W.P.C. Bartram** occurs in beaver ponds, bogs, borrow pits, brooks, canals, carrs, depressions, ditches, dunes, fens, floodplains, marshes, meadows, prairies, seeps, swales, swamps, vernal pools, and along margins of bayous, lakes, ponds, rivers, and streams at elevations of up to 1000 m. It occupies a wide range of habitats with basalt-, limestone- or sandstone-derived substrates (pH: 6.1–7.8) that include clay, clay loam, cobble, gravel, muck, mud, peat, rocks, sand, sandbars, sandy clay, sandy gravel, silt, and silty cobble. The plants occur in nontidal or tidal sites and can tolerate brackish conditions (0.5–10 ppt salinity) and partial shade. The flowers are insect pollinated and, like other *Lycopus* species, probably are protandrous and perhaps also gynodioecious (but this possibility remains to be verified). Several bees (Hymenoptera: Andrenidae: *Calliopsis andreniformis, Perdita octomaculata*; Apidae: *Ceratina dupla,* Halictidae: *Lasioglossum imitatum, L. versatum*) collect the pollen and probably serve as pollinators. Fruit production is high (some regard this species functionally as a FAC annual) with a single plant capable of producing up to 34,560 nutlets. The nutlets require cold stratification for 270 days to break their dormancy and will then germinate under a 30°C/20°C temperature regime. The plants are a significant element of the wetland seed bank, with high germination rates observed from core samples taken at 0–45 cm depths. Studies indicate frequent (up to 100%) colonization rates of the roots by arbuscular mycorrhizal fungi. The plants have a persistent stem base and freely produce stolons (but no tubers) as a means of vegetative reproduction. The stolons can spread laterally from the plants for a distance up to 2 dm. In cattail (Typhaceae: *Typha*) marshes, the plants occur more abundantly along paths made through the vegetation by mammals. **Reported associates:** *Acer rubrum, A. saccharinum, A. saccharum, Acorus calamus, Agalinis paupercula, A. purpurea, A. tenuifolia, Agastache, Agrimonia gryposepala, Agrostis alba, A. gigantea, A. scabra, Alisma subcordatum, A. triviale, Alnus rhombifolia, A. rugosa, Amaranthus tuberculatus, Ambrosia psilostachya, Amphicarpa bracteata, Anaphalis margaritacea, Andromeda glaucophylla, Andropogon gerardii, Anemone canadensis, Angelica atropurpurea, Apios americana, Apocynum cannabinum, Aralia nudicaulis, Aristida oligantha, Arnoglossum plantagineum, Artemisia ludoviciana, Asclepias incarnata, A. speciosa, A. syriaca, A. verticillata, Aulacomnium palustre, Baccharis halimifolia, Baptisia alba, Barbarea vulgaris, Berula erecta, Betula nigra, B. pumila, B. sandbergii, Bidens cernuus, B. connatus, B. coronatus, B. frondosus, Boehmeria cylindrica, Boisduvalia densiflora, Bolboschoenus fluviatilis, Boltonia decurrens, Brickellia californica, Bryum pseudotriquetrum, Calamagrostis canadensis, Calliergonella cuspidata, Caltha palustris, Calystegia sepium, Campanula aparinoides, Campylium stellatum, Cardamine bulbosa, Carex annectens, C. aquatilis, C. atherodes, C. bebbii, C. canescens, C. comosa, C. crawei, C. cristatella, C. eburnea, C. folliculata, C. flava, C. garberi, C. hystericina, C. interior, C. lacustris, C. lasiocarpa, C. longii, C. lurida, C. molesta, C. leptalea, C. microptera, C. nebrascensis, C. normalis, C. pallescens, C. pellita, C. praegracilis, C. prairea, C. retrorsa, C. rostrata, C. sartwellii, C. scoparia, C. sterilis, C. stipata, C. stricta, C. swanii, C. tribuloides, C. utriculata, a. vesicaria, C. viridula, C. vulpinoidea, Carya, Cassia fasciculata, Cephalanthus occidentalis, Chamaedaphne calyculata, Chasmanthium laxum, Chelone glabra, Cicuta bulbifera, C. maculata, Cirsium arvense, Comarum palustre, Coreopsis atkinsoniana, Cornus racemosa, C. sericea, Cuscuta pentagona, Cyperus bipartitus, C. diandrus, C. erythrorhizos, C. odoratus, C. rivularis, C. strigosus, Cypripedium reginae, Dasiphora floribunda, Datisca glomerata, Desmanthus illinoensis, Dichanthelium meridionale, Distichlis spicata, Doellingeria umbellata, Drepanocladus vernicosus, Drosera rotundifolia, Dulichium arundinaceum, Echinochloa crus-galli, Echinocystis lobata, Eleocharis acicularis, E. elliptica, E. erythropoda, E. obtusa, E. palustris, Elymus trachycaulus, E. villosus, E. virginicus, Epilobium ciliatum, E. coloratum, E. leptophyllum, Equisetum arvense, E. fluviatile, E. laevigatum, Erechtites hieraciifolia, Erigeron canadensis, E. philadelphicus, Eriophorum viridicarinatum, Eryngium yuccifolium, Eupatorium capillifolium, E. maculatum, E. perfoliatum, E. serotinum, Euthamia graminifolia, Fissidens adianthoides, Fragaria virginica, Fraxinus americana, F. nigra, Galium asprellum, G. brevipes, G. labradoricum, G. obtusum, G. tinctorium, G. trifidum, Gentiana andrewsii, G. linearis, G. rubricaulis, Gentianopsis virgata, Geum aleppicum, G. rivale, Glyceria canadensis, G. obtusa, G. striata, Helenium autumnale, Helianthus grosseserratus, Hibiscus moscheutos, Hordeum jubatum, Hydrocotyle umbellata, H. verticillata, Hypericum canadense, H. majus, H. perforatum, H. punctatum, H. sphaerocarpum, Impatiens capensis, Ipomoea lacunosa, Iris lacustris, I. versicolor, I. virginica, Iva annua, Juncus alpinus, J. balticus, J. brevicaudatus, J. canadensis, J. dudleyi, J. effusus, J. interior, J. longistylis, J. nodosus, J. torreyi, Kalmia angustifolia, K. polifolia, Lathyrus palustris, Leersia oryzoides, L. virginica, Liatris pycnostachya, L. spicata, Lilaeopsis chinensis, L. masonii, Lindernia dubia, Liparis loeselii, Lobelia cardinalis, L. inflata, L. kalmii, L. siphilitica, Ludwigia palustris, Lycopodiella inundata, Lycopus asper, L. uniflorus, L. virginicus, Lysimachia nummularia, L. quadriflora, L. terrestris, L. thyrsiflora, Lythrum alatum, L. salicaria, Melilotus alba, Mentha arvensis, Micranthes pensylvanica, Mikania scandens, Mimulus guttatus, M. ringens, Monarda fistulosa, Morus alba, Muhlenbergia asperifolia, M. glomerata, M. mexicana, Myrica cerifera, M. gale, Napaea dioica, Nemopanthus mucronata, Nyssa sylvatica, Onoclea sensibilis, Osmunda regalis, Osmundastrum cinnamomeum, Oxypolis rigidior, Oxytropis campestris, Packera aurea, Panicum capillare, P. occidentalis, P. virgatum, Parnassia glauca, Pedicularis lanceolata, Peltandra virginica, Penthorum sedoides, Persicaria amphibia, P. arifolia, P. coccinea, P. hydropiper, P. hydropiperoides, P. lapathifolia, P. pensylvanica, P. maculosa, P. sagittata, P. virginiana, Phalaris arundinacea, Phragmites australis, Phyla lanceolata, Physalis longifolia, Physostegia virginiana, Pilea fontana, Platanthera dilatata, P. huronensis, Pluchea odorata, Poa compressa, P. pratensis, Polemonium reptans, Polypogon monspeliensis, Pontederia cordata, Populus deltoides, P. fremontii, P. tremuloides, Potentilla anserina, P. arguta, P. simplex, Primula mistassinica, Prunella vulgaris, Prunus serotina, Ptilimnium capillaceum, Pycnanthemum virginianum, Quercus macrocarpa, Q. rubra, Ratibida pinnata, Rhamnus, Rhododendron groenlandicum, Rhynchospora capillacea, Ribes glandulosum, Rorippa columbiae, R. islandica, R. sylvestris, Rosa blanda, R. palustris, Rotala ramosior, Rubus idaeus, Rudbeckia laciniata, Rumex crispus, R. maritimus, u. orbiculatus, Sagittaria latifolia, Salix amygdaloides, S. candida, S. exigua, S. eriocephala, S. fragilis, S. interior, S. irrorata, S. myricoides, S. nigra, Sambucus nigra, Samolus valerandi, Saponaria* *officinalis, Sarracenia purpurea, Schizachyrium scoparium, Schoenoplectus americanus S. pungens, S. smithii, S. tabernaemontani, Scirpus atrocinctus, S. atrovirens, S. cyperinus, S. pendulus, Scleria verticillata, Scutellaria galericulata, Sium suave, Solanum dulcamara, Solidago altissima, S. canadensis, S. gigantea, S. graminifolia, Solidago occidentalis, S. ohioensis, S. uliginosa, S. patula, S. riddellii, Sorghastrum nutans, Sparganium eurycarpum, Spartina alterniflora, S. cynosuroides, S. patens, S. pectinata, Sphagnum warnstorfii, Sphenopholis intermedia, Spiraea alba, S. tomentosa, Spiranthes diluvialis, Sporobolus, Stachys palustris, Stellaria longifolia, Symphyotrichum boreale, S. ericoides, S. lanceolatum, S. novae-angliae, S. pilosum, S. puniceum, Symplocarpus foetidus, Teucrium canadense, Thalictrum dasycarpum, T. pubescens, Thelypteris palustris, Thuidium delicatulum, Thuja occidentalis, Tilia americana, Tomenthypnum nitens, Toxicodendron radicans, Tradescantia ohiensis, Triadenum fraseri, T. virginicum, Trollius laxus, Typha angustifolia, T. latifolia, Ulmus, Urtica dioica, Vaccinium macrocarpon, V. oxycoccos, Valeriana, Verbena hastata, V. urticifolia, Vernonia fasciculata, Veronica scutellata, Veronicastrum virginicum, Viola nephrophylla, Vitis riparia, Woodwardia areolata, Xanthium strumarium, Zigadenus.*

Lycopus amplectens Raf. grows in bays, bogs, depressions, ditches, floodplains, marshes, and savannas at elevations of up to 80 m. The substrates are clay, sand, or sandy peat. These plants often occur in naturally disturbed sites and common along the shorelines of newly created reservoirs. There is no information on the pollination or seed ecology of this relatively rare species. The plants reproduce vegetatively by means of long, branching runners, which bear fleshy, crescent-shaped, terminal tubers that can reach up to 5 cm in length. **Reported associates:** *Acer rubrum, A. saccharinum, Alternanthera philoxeroides, Asclepias incarnata, Bacopa caroliniana, Betula nigra, Bidens connatus, B. coronatus, B. frondosus, B. tripartita, Boehmeria cylindrica, Calamagrostis canadensis, Carex comosa, C. lupulina, Cephalanthus occidentalis, Cladium mariscoides, Clethra alnifolia, Conyza canadensis, Cornus obliqua, Cyperus, Dichanthelium scabriusculum, D. spretum, Drosera intermedia, Dulichium arundinaceum, Eleocharis acicularis, E. microcarpa, E. quadrangulata, E. tuberculosa, Erechtites hieracifolia, Eriocaulon aquaticum, E. decangulare, Eupatorium capillifolium, E. dubium, E. perfoliatum, Euthamia caroliniana, E. tenuifolia, Fraxinus pennsylvanica, Fuirena squarrosa, Galium tinctorium, Glyceria obtusa, Gratiola aurea, G. ramosa, Habenaria repens, Helenium autumnale, Hibiscus moscheutos, Hydrocotyle umbellata, H. verticillata, Hypericum canadense, H. mutilum, Juncus canadensis, J. militaris, J. polycephalus, Lactuca canadensis, Leersia oryzoides, Lindernia dubia, Liquidambar styraciflua, Lobelia canbyi, L. cardinalis, L. siphilitica, Lophiola aurea, Ludwigia alternifolia, L. decurrens, L. leptocarpa, L. palustris, Lycopus rubellus, L. virginicus, Lysimachia terrestris, Lythrum salicaria, Mikania scandens, Mimulus ringens, Monarda punctata, Muhlenbergia torreyana, Murdannia*

keisak, Nyssa biflora, N. sylvatica, Oclemena nemoralis, Panicum hemitomon, P. rigidulum, P. verrucosum, P. virgatum, Paspalum repens, Persicaria glabra, P. hydropiperoides, P. pensylvanica, P. sagittata, Pinus taeda, Plantago sparsiflora, Pluchea foetida, P. rosea, Polygala cymosa, Pontederia cordata, Populus deltoides, Proserpinaca palustris, Ptilimnium capillaceum, Quercus palustris, Rhexia virginica, Rhynchospora alba, R. careyana, Sabatia difformis, Sacciolepis striata, Sagittaria latifolia, Salix caroliniana, S. nigra, Schoenoplectus pungens, Scirpus cyperinus, Scutellaria lateriflora, Spiraea tomentosa, Taxodium ascendens, Triadenum virginicum, T. walteri, Typha latifolia, Vaccinium macrocarpon, Verbena hastata, Viola lanceolata, Woodwardia areolata, W. virginica, Xyris difformis.

Lycopus asper Greene grows in ditches, fens, marshes, meadows, sloughs, and along the margins of lakes, ponds, rivers, or streams at elevations of up to 1820 m. Its substrates are described as alkaline cobble, sand, silty clay or silty cobble with a fairly high organic matter content (20%). Some occurrences are described as being heavily shaded, but sites generally are open. The plants establish readily where wet, freshly deposited sand (e.g., along beaches) contacts areas of established sand. They often occur on waste ground and in polluted waters. Information on the reproductive ecology of this species is scarce. Vegetative reproduction is facilitated by short runners and stolons, which terminate in thickened tubers. The foliage supposedly has the odor of freshly sawn lumber. **Reported associates:** *Agalinis purpurea, Agrostis exarata, A. stolonifera, Allium canadense, Anemone canadensis, Anticlea elegans, Apocynum cannabinum, Artemisia tridentata, Atriplex, Berula erecta, Bidens cernuus, B. frondosus, Boehmeria cylindrica, Bromus tectorum, Calamagrostis stricta, Calystegia sepium, Carex aquatilis, C. bebbii, C. brunnescens, C. densa, C. diandra, C. interior, C. lacustris, C. lanuginosa, C. limosa, C. pellita, C. prairea, C. rosea, C. rostrata, C. sartwellii, C. tetanica, Catabrosa aquatica, Chamaedaphne calyculata, Chara, Chenopodium album, Chrysothamnus nauseosus, Cicuta douglasii, C. maculata, Cirsium altissimum, C. arvense, Cleomella, Comarum palustre, Coreopsis atkinsoniana, Cornus sericea, Cuscuta glomerata, Cypripedium candidum, Dipsacus fullonum, Distichlis spicata, Echinocystis lobata, Elaeagnus angustifolia, Eleocharis acicularis, E. compressa, E. palustris, Elymus canadensis, E. ×pseudorepens, E. repens, Epilobium ciliatum, E. leptophyllum, E. strictum, Equisetum arvense, E. hyemale, E. laevigatum, Eupatorium perfoliatum, Eutrochium maculatum, Fallopia convolvulus, Galium obtusum, G. trifidum, Gentianopsis virgata, Geum aleppicum, Glyceria grandis, G. striata, Glycyrrhiza lepidota, Helenium autumnale, Helianthus grosseserratus, Hierochloe odorata, Hordeum jubatum, Hypericum majus, Hypoxis hirsuta, Iris versicolor, Juncus balticus, J. brachycarpus, J. nodatus, J. nodosus, J. tenuis, u. torreyi, Lathyrus palustris, Leersia oryzoides, Lilium philadelphicum, Liparis loeselii, Lobelia kalmii, L. siphilitica, Lycopus americanus, L. uniflorus, Lysimachia quadriflora, L. thyrsiflora, Lythrum salicaria, Maianthemum stellatum, Mentha arvensis, Mimulus glabratus, M. ringens,*

Morus alba, Muhlenbergia racemosa, Nasturtium offici-nale, Oenothera elata, Packera aurea, Panicum capillare, P. occidentalis, P. virgatum, Parnassia glauca, Parthenocissus inserta, Pascopyrum smithii, Pedicularis lanceolata, Persicaria amphibia, P. hydropiperoides, P. lapathifolia, P. maculosa, Phalaris arundinacea, Phragmites australis, Platanthera dilatata, P. hyperborea, Polypogon monspelien-sis, Potentilla anserina, P. norvegica, Purshia tridentata, Pycnanthemum virginianum, Ranunculus circinatus, R. cym-balaria, R. sceleratus, Rhus trilobata, Rhynchospora capil-lacea, Ribes americanum, Ribes aureum, Rorippa columbiae, R. curvisiliqua, R. islandica, Rosa arkansana, R. woodsii, Rumex altissimus, R. crispus, R. maritimus, R. orbiculatus, R. salicifolius, R. verticillatus, Salix bebbiana, S. cordata, S. exigua, S. farriae, S. lucida, S. petiolaris, Sanicula canaden-sis, Sarcobatus, Scheuchzeria palustris, Schoenoplectus acutus, S. americanus, S. pungens, S. tabernaemontani, Scirpus cyperinus, S. pallidus, Scutellaria galericulata, S. lateriflora, Sium suave, Solanum dulcamara, Solidago gigan-tea, S. graminifolia, S. occidentalis, S. riddellii, Sonchus arvensis, Sparganium, Spartina pectinata, Sphenopholis obtusata, Spiraea alba, Spiranthes cernua, S. diluvialis, Stachys palustris, Symphyotrichum boreale, S. lanceola-tum, S. novae-angliae, S. puniceum, S. praealtum, Teucrium canadense, Thalictrum dasycarpum, Thaspium barbinode, Toxicodendron rydbergii, Triglochin maritimum, T. palustre, Typha angustifolia, T. latifolia, Urtica dioica, Utricularia minor, Veronica anagallis-aquatica, Viola nephrophylla, V. pratincola, Zizia aurea.

Lycopus cokeri H.E. Ahles ex Sorrie inhabits bays, savannas, seeps, swamps, and margins of ponds or lakes. The substrates are acidic sand. The habitats are quite specialized and include Carolina bays, sandhill seeps, and streamhead pocosins, which occur in areas of extensive sand deposits where the groundwater rises to the surface over impermeable layers of clay. In such areas, the herbaceous vegetation typi-cally is maintained by fire, which reduces the overstory cover in ecotones between the wet and UPL regions. Lake margins receiving groundwater seepage (including some artificial res-ervoirs) also provide similar, suitable habitat. No information exists on the reproductive ecology, seed ecology, or popula-tion genetics of this rare species, although such studies would provide valuable information for conservation management. Specific ecological information (e.g., substrate composition, pH, etc.) is quite limited and generally unavailable. **Reported associates:** *Acer rubrum, Allium, Andropogon virginicus, A. gyrans, Aristida palustris, Arundinaria gigantea, Asplenium platyneuron, Bacopa caroliniana, Bartonia capitata, Burmannia biflora, Calamovilfa brevipilis, Carex atlantica, C. glaucescens, C. leptalea, C. lonchocarpa, C. tenax, C. turgescens, Chamaecyparis thyoides, Dichanthelium sca-briusculum, Drosera capillaris, D. intermedia, D. rotundi-folia, Dulichium arundinaceum, Eleocharis quadrangulata, E. robbinsii, Erianthus giganteus, Eriocaulon compressum, Eryngium integrifolium, Eupatorium resinosum, Fuirena squarrosa, Glyceria obtusa, Goodyera pubescens, Huperzia selago, Hypericum canadensis, H. mutilum, H. perforatum,*

Ilex glabra, Iris virginica, Juncus biflorus, Lachnanthes car-oliana, Lachnocaulon anceps, Lilium pyrophilum, Lindera subcoriacea, Liquidambar styraciflua, Liriodendron tulip-ifera, Lobelia batsonii, L. pubera, Lycopodiella caroliniana, Lysimachia asperulifolia, Mayaca fluviatilis, Melanthium virginicum, Muhlenbergia capillaris, Myrica caroliniensis, M. cerifera, Nyssa biflora, Onoclea sensibilis, Osmunda regalis, Osmundastrum cinnamomeum, Oxypolis ternata, Panicum virgatum, Peltandra virginica, Pinus serotina, P. taeda, Platanthera clavellata, Pleea tenuifolia, Polygala lutea, Proserpinaca pectinata, Pteridium aquilinum, Rhexia, Rhynchospora latifolia, R. leptocarpa, R. scirpoi-des, Sagittaria macrocarpa, Sarracenia flava, S. purpurea, S. rubra, S. ×catesbyana, Schoenoplectus etuberculatus, S. subterminalis, Solidago patula, Sphagnum, Sporobolus pinetorum, Symphyotrichum novi-belgii, Taxodium ascen-dens, Tipularia discolor, Triadenum virginicum, Utricularia cornuta, U. juncea, Vaccinium crassifolium, Woodwardia areolata, Xyris.

Lycopus europaeus L. is reported from bogs, ditches, marshes, swales, swamps, and margins of lakes and rivers at elevations of up to 1000 m. This species occurs across a wide range of acidity (pH: 4.9–7.3), on substrates that include gravel, loam, peat, and sand. The sites typically are freshwater and nontidal; however, tidal sites with up to 2% salinity are toler-ated. The plants occupy the brackish supralittoral zone in the more saline sites. The species is gynodioecious with protan-drous hermaphroditic flowers. The female flowers tend to be larger than the hermaphrodites. The flowers are outcrossed by insects. Optimal seed germination requires light and alternat-ing temperatures. Germination is complex and occurs across a wide range of temperatures. Seeds stratified for 10 weeks at 5°C germinate (≤25%) from 12°C to 33°C. Germination occurs readily if there is a diurnal temperature fluctuation in excess of 7°C, and optimally (80%) with a 16°C fluctua-tion (i.e., 30°C/14°C). Germination at constant temperatures generally is much lower; however, high germination (90%) has been observed at 25°C and 15°C (in light) for seeds that have been stratified and buried. Primary dormancy is broken by exposure to low temperatures (≤12°C) and is induced at temperatures >14°C. High concentrations of gibberellic acid (1000 mg/L) or gibberellin A4 (10 mg/L) promote seed germi-nation at 25°C. The nutlets can remain viable after floating for up to 15 months. They are typically dispersed by water and, in riverine habitats, are transported over extended distances dur-ing flooding events. The plants yield a transient seed bank of 3–25 seeds/m². Many viable seeds have been recovered from the dung of cattle (Mammalia: Bovidae: *Bos taurus*) and horses (Mammalia: Equidae: *Equus ferus*), indicating that the plants also can be dispersed endozoically by such grazing animals. *Lycopus europaeus* is moderately mycotrophic and exhibits a correlated increase in biomass with root mycorrhiza density. Vegetative reproduction occurs by the production of stolonif-erous shoots, which terminate in very slender tubers. In one study, the plants exhibited an unusual correlation by grow-ing in the vicinity of asphalt-paved paths. There is extensive ecological information for this species in its native Old World range; however, specific data (e.g., associated species) for introduced North American populations are relatively scarce. **Reported associates (North America):** *Acer rubrum, A. saccharinum, Agalinis purpurea, Berchemia scandens, Bidens cernuus, B. comosus, Boehmeria cylindrica, Bolboschoenus fluviatilis, Carex bromoides, C. festucacea, C. haydenii, C. lacustris, C. laevivaginata, C. seorsa, Cicuta maculata, Commelina virginica, Cornus foemina, C. obliqua, Cyperus fuscus, C. odoratus, Decumaria barbara, Dichanthelium dichotomum, Echinochloa crus-galli, Fraxinus pennsylvanica, Glyceria striata, Helenium autumnale, Impatiens capensis, Iris pseudacorus, Justicia americana, Leersia oryzoides, Ligustrum sinense, Liquidambar styraciflua, Lonicera japonica, Lythrum salicaria, Microstegium vimineum, Mimulus alatus, M. ringens, Nyssa biflora, Osmunda regalis, Penthorum sedoides, Persicaria amphibia, Phalaris arundinacea, Ranunculus sceleratus, Rumex altissimus, Sagittaria latifolia, Salix interior, Saururus cernuus, Schoenoplectus tabernaemontani, Scutellaria lateriflora, Smilax, Solanum dulcamara, Taxodium distichum, Thelypteris palustris, Typha latifolia, T. ×glauca, Ulmus americana, Xanthium strumarium, Zephyranthes atamasca.*

Lycopus rubellus **Moench** inhabits beaver ponds, bogs, cliffs, depressions, ditches, flatwoods, floodplains, marshes, mudflats, prairies, seeps, sinkholes, sloughs, swamps, and margins of lakes, ponds, and streams. This is a broadly adapted species that grows in a variety of chert, limestone, and sandstone substrates (pH: 5.1–6.9) including clay, clay loam, muck, mucky peat, sand, sapric muck, and silty sand. The substrates often consist of a deep organic layer. The plants can occur in full sun to dense shade and have been observed growing on floating islands of vegetation and also on logs and stumps. There are no reports on the pollination biology or seed ecology, and it is not clear whether a seed bank develops for this species. Vegetative reproduction is accomplished by the production of branching, stoloniferous runners, which can extend up to 6 m in length, and terminate in elongate, cylindrical tubers that can be up to 10 cm. The plants typically grow clonally from the tubers, with reproduction from seed rarely observed at established sites. **Reported associates:** *Acalypha rhomboidea, Acer rubrum, Acmella oppositifolia, Agalinis tenuifolia, Agrostis hyemalis, A. stolonifera, Alnus glutinosa, A. serrulata, Alopecurus aequalis, Ambrosia, Amelanchier canadensis, Amorpha fruticosa, Ampelopsis arborea, Amphicarpum muehlenbergianum, Amsonia tabernaemontana, Andropogon capillipes, A. glomeratus, Apios americana, Aristida palustris, Arnoglossum ovatum, Asclepias lanceolata, A. perennis, A. viridis, Baccharis halimifolia, Bacopa caroliniana, Berchemia scandens, Bidens bipinnatus, B. comosus, B. connatus, B. discoideus, B. frondosus, B. vulgatus, Boehmeria cylindrica, Boltonia apalachicolensis, B. asteroides, Callicarpa americana, Campsis radicans, Cardamine pennsylvanica, Carex atlantica, C. comosa, C. crus-corvi, C. decomposita, C. glaucescens, C. lacustris, C. leptalea, C. lupulina, C. seorsa, C. stipata, C. tribuloides, C. tuckermanii, C. verrucosa, C. vulpinoidea, Carya, Celtis,*

Centella erecta, Cephalanthus occidentalis, Chasmanthium latifolium, C. laxum, C. sessiliflorum, Cicuta maculata, Cinna arundinacea, Circaea lutetiana, Cirsium arvense, Cladium jamaicense, Mnesithea rugosa, Commelina diffusa, C. virginica, Conoclinium coelestinum, Coreopsis tinctoria, Cornus amomum, C. florida, C. sericea, Cyperus erythrorhizos, C. ferruginescens, C. squarrosus, Decodon verticillatus, Decumaria barbara, Dichanthelium acuminatum, D. clandestinum, D. scabriusculum, Dulichium arundinaceum, Eleocharis acicularis, E. engelmannii, E. equisetoides, E. quadrangulata, E. tortilis, E. vivipara, Elephantopus, Epilobium coloratum, Eragrostis hypnoides, Erianthus giganteus, Erigeron annuus, Eriocaulon compressum, Eriochloa punctata, Eupatorium serotinum, Eryngium yuccifolium, Fagus grandifolia, Fimbristylis littoralis, Fraxinus caroliniana, F. pennsylvanica, Fuirena bushii, Galium obtusum, G. tinctorium, Gaura lindheimeri, Geum aleppicum, Glyceria acutiflora, G. septentrionalis, G. striata, Gratiola brevifolia, G. virginiana, Hackelia virginiana, Hedyotis, Heliopsis helianthoides, Heliotropium indicum, Hibiscus moscheutos, Hydrocotyle umbellata, H. verticillata, Hydrolea uniflora, H. ovata, Hygrophila lacustris, Hypericum fasciculatum, H. mutilum, H. perforatum, Hyptis alata, Hymenocallis liriosme, Ilex cassine, Illicium floridanum, Impatiens capensis, Ipomoea lacunosa, I. sagittata, Iris fulva, Itea virginica, Iva angustifolia, I. annua, Juglans, Juncus acuminatus, J. effusus, J. nodatus, u. polycephalus, J. roemerianus, Justicia ovata, Kyllinga brevifolia, Lachnanthes caroliana, Leersia hexandra, L. lenticularis, L. oryzoides, L. virginica, Leucothoe axillaris, L. racemosa, Lilium michiganense, Lindera benzoin, Lindernia dubia, Liquidambar styraciflua, Lobelia amoena, L. cardinalis, L. siphilitica, Lonicera tatarica, Ludwigia linearis, L. palustris, L. pilosa, L. polycarpa, L. sphaerocarpa, Lycopus americanus, L. amplectens, L. virginicus, Lyonia ligustrina, L. lucida, Lysimachia ciliata, L. radicans, Lythrum lineare, Magnolia virginiana, Mentha arvensis, M. spicata, Micranthemum umbrosum, Mikania scandens, Mimulus alatus, Muhlenbergia frondosa, M. schreberi, Murdannia keisak, Myosurus minimus, Myrica cerifera, Neptunia lutea, Nyssa aquatica, N. biflora, N. sylvatica, Oenothera drummondii, Onoclea sensibilis, Osmunda regalis, Osmundastrum cinnamomeum, Ostrya virginiana, Oxypolis filiformis, Panicum dichotomiflorum, P. gymnocarpon, P. hemitomon, P. rigidulum, P. virgatum, Parthenocissus quinquefolia, Paspalum floridanum, P. fluitans, P. plicatulum, P. pubiflorum, Peltandra virginica, Persea palustris, Persicaria hydropiper, P. hydropiperoides, P. lapathifolia, P. pensylvanica, P. sagittata, P. setacea, P. virginiana, Phanopyrum gymnocarpon, Phytolacca americana, Picea mariana, Pilea pumila, Pinus elliottii, P. glabra, P. palustris, Pluchea camphorata, Polygala leptocaulis, Pontederia cordata, Proserpinaca palustris, P. pectinata, Ptilimnium costatum, Quercus bicolor, Q. laurifolia, Q. nigra, Q. palustris, Q. phellos, Q. virginiana, Ranunculus flabellaris, R. sceleratus, Ratibida peduncularis, Rhapidophyllum hystrix, Rhexia mariana, R. virginica, Rhynchospora cephalantha, R. colorata, R. corniculata, R. elliottii, R. globularis, R. macrostachya, R.

microcarpa, R. miliacea, R. mixta, R. nitens, Rorippa palustris, Rubus trivialis, Rudbeckia texana, Rumex verticillatus, Sabal minor, Sabatia angularis, S. calycina, Saccharum giganteum, Sagittaria graminea, S. latifolia, S. rigida, Salix nigra, Sambucus nigra, Sarracenia, Saururus cernuus, Schoenoplectus tabernaemontani, Scirpus cyperinus, Scleria baldwinii, Scutellaria lateriflora, S. nervosa, Senecio glabellus, Sium suave, Smilax bona-nox, S. laurifolia, S. walteri, Solidago gigantea, S. sempervirens, Sparganium americanum, Spartina patens, Sphagnum, Sphenopholis pensylvanica, Spiranthes laciniata, Stachys tenuifolia, Symphyotrichum lanceolatum, S. lateriflorum, Taxodium ascendens, T. distichum, Teucrium canadense, Thelypteris palustris, Tilia americana, Tillandsia usneoides, Toxicodendron radicans, Tradescantia hirsutiflora, Triadenum walteri, Tridens strictus, Tripsacum dactyloides, Typha latifolia, Ulmus americana, Utricularia gibba, Vaccinium, Verbena hastata, Verbesina alternifolia, Vernonia gigantea, Viburnum nudum, Viola lanceolata, Vitis riparia, Wisteria frutescens, Woodwardia areolata, W. virginica, Xyris fimbriata, X. laxifolia.

***Lycopus uniflorus* Michx.** inhabits beaches, bogs, carrs, cliffs, depressions, ditches, dunes, fens, floodplains, marshes, meadows, pools, prairies, seeps, sloughs, swales, swamps, and the margins of brooks, lakes, ponds, rivers, and streams at elevations of up to 2000 m. This is an extremely broadly adapted species, which occurs on substrates derived from limestone, sandstone, or organic material (pH: 4.2–7.4) including clay, gravel, muck, mud, peat (including floating mats), sand, and sandy silt. The plants also have been observed growing on submerged logs and rotting stumps. Populations occur in both nontidal and tidal, freshwater sites. This species is regarded as a generalist with respect to light; however, some studies indicate that it favors medium to large gaps (60–190 m^2). The flowers presumably are insect pollinated, but few details are known regarding the floral biology or seed ecology. The seeds will germinate similarly when buried in any orientation and the plants do form a persistent seed bank. The roots are colonized (up to 73%) by arbuscular mycorrhizal Fungi. Vegetative reproduction occurs by means of runners or stolons (sometimes leafy), which produce conical or fusiform tubers. Clonal spread occurs by "guerrilla" type ramets. The plants are known to increase in cover significantly following burning. **Reported associates:** *Abies balsamea, Acer negundo, A. rubrum, A. saccharum, A. saccharinum, A. spicatum, Aconitum columbianum, Acorus calamus, Agalinis paupercula, A. purpurea, A. tenuifolia, Agrostis alba, A. scabra, A. stolonifera, Alisma subcordatum, A. triviale, Alnus rugosa, Alopecurus aequalis, Ambrosia artemisiifolia, Amelanchier, Amaranthus tuberculatus, Ammophila, Amphicarpa bracteata, Andromeda glaucophylla, A. polifolia, Andropogon gerardii, Anemone virginiana, Aneura pinguis, Angelica atropurpurea, Anticlea elegans, Apios americana, Apocynum cannabinum, Arethusa bulbosa, Asclepias incarnata, A. tuberosa, Asplenium trichomanes, Athyrium filix-femina, Aulacomnium palustre, Barbarea vulgaris, Bartonia virginica, Betula alleghaniensis, B. papyrifera, B. pumila, B. sandbergii, Bidens cernuus,*

B. connatus, B. coronatus, B. discoideus, B. frondosus, B. tripartita, Boehmeria cylindrica, Bolboschoenus fluviatilis, Botrychium multifidum, Bromus kalmii, Bryoxiphium norvegicum, Bryum pseudotriquetrum, Calamagrostis canadensis, C. pickeringii, C. stricta, Calamovilfa, Calla palustris, Calliergonella cuspidata, Calopogon tuberosus, Caltha palustris, Calystegia sepium, Campanula aparinoides, Campylium polygamum, C. stellatum, Cardamine pratensis, Carex aquatilis, C. atherodes, C. bebbii, C. buxbaumii, C. canescens, C. chordorrhiza, C. comosa, C. crawei, C. crinita, C. disperma, C. eburnea, C. echinata, C. exsiccata, C. flava, C. folliculata, C. haydenii, C. hystericina, C. interior, C. intumescens, C. lacustris, C. lasiocarpa, C. leptalea, C. limosa, C. livida, C. lupulina, C. magellanica, C. michauxiana, C. nebrascensis, C. oligosperma, C. pauciflora, C. pellita, C. prairea, C. pseudocyperus, C. retrorsa, C. rostrata, C. sartwellii, C. scoparia, C. sterilis, C. stipata, C. stricta, C. swanii, C. tenuiflora, C. tetanica, C. trisperma, C. vesicaria, C. viridula, C. wiegandii, Cephalanthus occidentalis, Chamaedaphne calyculata, Chamaesyce polygonifolia, Chamerion angustifolium, Chelone glabra, Cicuta bulbifera, C. douglasii, C. maculata, Cinna arundinacea, Circaea alpina, Cirsium arvense, C. muticum, C. vulgare, Cladium mariscoides, Clematis virginiana, Clethra alnifolia, Climacium dendroides, Clintonia borealis, Comarum palustre, Coptis trifolia, Coriospermum, Cornus canadensis, C. racemosa, C. rugosa, C. sericea, Cryptotaenia canadensis, Cuscuta gronovii, Cyperus bipartitus, C. erythrorhizos, C. odoratus, C. strigosus, Cypripedium acaule, C. reginae, Dasiphora floribunda, Decodon verticillatus, Dichanthelium acuminatum, Dicranella heteromalla, Doellingeria umbellata, Drepanocladus vernicosus, Drosera anglica, D. intermedia, D. rotundifolia, Dryopteris carthusiana, D. cristata, Dulichium arundinaceum, Echinochloa walteri, Eleocharis acicularis, E. atropurpurea, E. elliptica, E. engelmannii, E. intermedia, E. olivacea, E. palustris, E. rostellata, E. smallii, Elymus lanceolatus, E. riparius, E. virginicus, Epilobium ciliatum, E. coloratum, E. leptophyllum, Equisetum arvense, E. fluviatile, E. hyemale, E. sylvaticum, E. variegatum, Erigeron philadelphicus, Eriophorum angustifolium, E. gracile, E. tenellum, E. vaginatum, E. virginicum, E. viridicarinatum, Eupatorium perfoliatum, Euthamia graminifolia, E. tenuifolia, Eutrochium maculatum, Fallopia convolvulus, Fimbristylis autumnalis, Fissidens adianthoides, Fragaria virginiana, Frangula alnus, Fraxinus nigra, F. pennsylvanica, Galium brevipes, G. labradoricum, G. palustre, G. tinctorium, G. triflorum, Gaultheria hispidula, Gaylussacia baccata, G. frondosa, Gentiana linearis, Gentianopsis, Geum aleppicum, G. canadense, G. rivale, Glyceria canadensis, G. grandis, G. melicaria, G. nervata, G. occidentalis, G. striata, Glycyrrhiza lepidota, Gratiola aurea, Hamatocaulis vernicosus, Hasteola suaveolens, Helenium autumnale, Helianthemum canadense, Helianthus divaricatus, H. grosseserratus, Heracleum lanatum, Hordeum jubatum, Huperzia lucidula, Hydrocotyle americana, H. umbellata, Hypericum anagalloides, H. boreale, H. canadense, H. ellipticum, H. kalmianum, H. majus, H. mutilum, Hypoxis hirsuta, Ilex verticillata,

Impatiens capensis, I. pallida, Iris versicolor, I. virginica, Juncus articulatus, J. balticus, J. canadensis, J. coriaceus, J. effusus, J. nodosus, J. pelocarpus, J. torreyi, Juniperus horizontalis, J. virginiana, Kalmia angustifolia, K. polifolia, Lactuca canadensis, Laportea canadensis, Larix laricina, Lathyrus japonicus, L. latifolius, L. palustris, Leersia oryzoides, Leptodictyum riparium, Lespedeza capitata, Lindera benzoin, Lipocarpha micrantha, Liriodendron tulipifera, Lobelia cardinalis, L. inflata, L. kalmii, L. siphilitica, Lonicera oblongifolia, Ludwigia alternifolia, L. palustris, Lycopodiella inundata, Lycopodium annotinum, Lycopus americanus, L. asper, Lyonia ligustrina, Lysimachia ciliata, L. nummularia, L. quadriflora, L. terrestris, L. thyrsiflora, Lythrum salicaria, Maianthemum canadense, M. stellatum, M. trifolium, Meesia triquetra, Melampyrum lineare, Mentha arvensis, Mentha ×gracilis, Menyanthes trifoliata, Micranthes pensylvanica, Mimulus glabratus, M. ringens, Mitella diphylla, M. nuda, Mnium affine, Muhlenbergia glomerata, M. mexicana, M. uniflora, Myosotis scorpioides, Myrica gale, Nemopanthus mucronata, Nyssa sylvatica, Oclemena acuminata, Oenanthe sarmentosa, Onoclea sensibilis, Osmunda claytoniana, O. regalis, Osmundastrum cinnamomeum, Packera aurea, Palustriella falcata, Panicum flexile, P. virgatum, Parnassia glauca, Parthenocissus inserta, Pedicularis lanceolata, Peltandra virginica, Persicaria amphibia, P. hydropiper, P. hydropiperoides, P. lapathifolia, P. pensylvanica, P. punctata, P. sagittata, Petasites frigidus, Phalaris arundinacea, Philonotis fontana, Phleum pratense, Photinia floribunda, P. melanocarpa, Phragmites australis, Physostegia virginiana, Picea mariana, P. rubens, Pilea fontana, P. pumila, Pinus resinosa, P. strobus, Platanthera dilatata, Poa compressa, P. palustris, P. pratensis, Pogonia ophioglossoides, Polemonium occidentale, Polygala cruciata, Polymnia canadensis, Polytrichum commune, P. piliferum, Populus balsamifera, P. deltoides, P. tremuloides, Potentilla anserina, P. norvegica, P. simplex, Primula cusickii, Prunella vulgaris, Prunus pumila, Pycnanthemum virginianum, Quercus ellipsoidalis, Q. palustris, Q. rubra, Q. velutina, Ramalina intermedia, Ranunculus flabellaris, R. lapponicus, R. orthorhynchus, R. pensylvanicus, R. sceleratus, Rhamnus alnifolia, Rhexia virginica, Rhododendron groenlandicum, R. viscosum, Rhynchospora alba, R. capillacea, R. capitellata, R. fusca, Ribes americanum, R. hirtella, R. oxycanthoides, R. triste, Rorippa islandica, R. palustris, Rosa virginiana, R. woodsii, Rotala ramosior, Rubus canadensis, R. hispidus, R. idaeus, R. pubescens, Rudbeckia laciniata, Rumex aquaticus, R. orbiculatus, R. salicifolius, R. verticillatus, Sagittaria engelmanniana, S. latifolia, Salix alba, S. discolor, S. bebbiana, S. candida, S. eriocephala, S. pedicellaris, S. petiolaris, S. planifolia, S. pyrifolia, S. sericea, S. serissima, Sambucus nigra, Saponaria officinalis, Sarracenia purpurea, Scheuchzeria palustris, Schizachyrium scoparium, Schoenoplectus acutus, S. pungens, S. purshianus, S. subterminalis, S. tabernaemontani, S. torreyi, Scirpus atrovirens, S. cyperinus, S. microcarpus, S. pendulus, Scleria reticularis, S. triglomerata, S. verticillata, Scorpidium cossonii, S. scorpioides, Scutellaria galericulata, S. lateriflora,

Sium suave, Solanum dulcamara, S. ptychanthum, Solidago canadensis, S. gigantea, S. ohioensis, S. patula, S. riddellii, S. uliginosa, Sonchus arvensis, Sorghastrum nutans, Sparganium androcladum, S. eurycarpum, Spartina pectinata, Sphagnum angustifolium, S. fallax, S. fimbriatum, S. girgensohnii, S. magellanicum, S. rubellum, S. subsecundum, S. subtile, S. tenerum, S. viridum, S. warnstorfii, Spiraea alba, S. douglasii, S. tomentosa, Spiranthes romanzoffiana, Stachys hyssopifolia, S. palustris, Sullivantia sullivantii, Symphyotrichum boreale, S. cordifolium, S. lanceolatum, S. lateriflorum, S. novi-belgii, S. puniceum, Symphoricarpos occidentalis, Symplocarpus foetidus, Teucrium canadense, Thalictrum dasycarpum, T. pubescens, Thelypteris palustris, Thuidium delicatulum, Thuja occidentalis, Tomenthypnum nitens, Toxicodendron radicans, T. vernix, Triadenum fraseri, T. virginicum, Trichophorum alpinum, T. cespitosum, Triglochin palustre, Trientalis borealis, Trollius laxus, Tsuga canadensis, Typha angustifolia, T. latifolia, Ulmus americana, Urtica dioica, Utricularia cornuta, U. minor, Vaccinium angustifolium, V. corymbosum, V. macrocarpon, V. myrtilloides, V. oxycoccos, Verbena hastata, Veronica americana, V. anagallis-aquatica, Viburnum dentatum, V. nudum, Verbena hastata, Viola lanceolata, V. macloskeyi, V. nephrophylla, V. pallens, V. palustris, Vitis riparia, Woodwardia virginica, Xyris caroliniana, X. torta, Zanthoxylum americanum.

Lycopus virginicus **L.** occurs in beaver ponds, bogs, on cliffs, in deltas, depressions, ditches, floodplains, marshes, meadows, prairies, savannas, seeps, sloughs, swales, swamps, and along the margins of ponds, pools, rivers, and streams at elevations of up to 300 m. Like the previous species, this is a broadly adapted plant that inhabits substrates ranging from dolomite to sandstone (pH: 3.5–7.3) and including clay–loam, gravel, mud, organic debris, rock, sand, sand bars, and silty sand. Like several of its congeners, this species also occurs on rotting logs. The plants tolerate full sun to partial shade. Despite being a common species, there have been no detailed studies of its floral biology or seed ecology. Presumably it is insect pollinated. The plants are known to form seed banks in sites that are characterized by frequent to moderate inundation. The seeds also are found commonly as a component of river drift. Vegetative reproduction occurs by the development of slender stoloniferous runners, which only rarely form tubers. **Reported associates:** *Abies balsamea, Acer rubrum, A. negundo, A. saccharinum, A. saccharum, Aeschynomene, Agalinis maritima, Agrostis perennans, A. scabra, Alnus rugosa, A. serrulata, Alternanthera philoxeroides, Amelanchier, Amaranthus tuberculatus, Amphicarpaea bracteata, Amsonia tabernaemontana, Andropogon glomeratus, A. virginicus, Apios americana, Aralia nudicaulis, Arisaema dracontium, A. triphyllum, Arthraxon hispidus, Asclepias incarnata, Aulacomnium palustre, Baccharis halimifolia, Bartonia paniculata, B. virginica, Berchemia scandens, Betula alleghaniensis, B. nigra, Bidens aristosa, B. bipinnata, B. connata, B. coronata, B. discoidea, Boehmeria cylindrica, Boykinia aconitifolia, Brachyelytrum erectum, Calamagrostis canadensis, Calla palustris, Campsis radicans, Cardamine bulbosa, C. rotundifolia, Carex atlantica,*

C. buxbaumii, C. canescens, C. comosa, C. concatenata, C. crinita, C. debilis, C. frankii, C. glaucescens, C. gynandra, C. howei, C. intumescens, C. leptalea, C. lonchocarpa, C. lupulina, C. lurida, C. muskingumensis, C. pedunculata, C. scoparia, C. silicea, C. stipata, C. stricta, C. tenera, C. torta, C. trisperma, Carya, Cassia, Cephalanthus occidentalis, Cephalozia connivens, Chamaecyparis thyoides, Chamaedaphne calyculata, Chasmanthium latifolium, C. laxum, C. sessiliflorum, Chelone glabra, Cicuta bulbifera, Cinna arundinacea, Cladopodiella fluitans, Claytonia virginica, Coptis trifolia, Cornus canadensis, C. florida, C. racemosa, C. sericea, Crataegus opaca, Cryptotaenia canadensis, Cuscuta gronovii, Cyperus polystachyos, Decodon verticillatus, Dennstaedtia punctilobula, Dichanthelium clandestinum, D. sabulorum, D. scoparium, Diodia teres, Diospyros virginiana, Doellingeria umbellata, Drosera intermedia, D. rotundifolia, Dryopteris intermedia, Dulichium arundinaceum, Echinochloa, Eleocharis microcarpa, E. obtusa, E. olivacea, E. tortilis, Epilobium leptophyllum, Equisetum arvense, Eragrostis hypnoides, Eriophorum virginicum, Eupatorium fistulosum, E. semiserratum, E. serotinum, Euthamia tenuifolia, Fagus grandifolia, Frangula alnus, Fraxinus caroliniana, F. nigra, F. pensylvanica, Galium obtusum, G. palustre, G. tinctorium, G. triflorum, Gaultheria procumbens, Gaylussacia baccata, Geum virginianum, Gleditsia triacanthos, Glyceria melicaria, G. striata, Gratiola virginiana, Helenium autumnale, Hibiscus moscheutos, Holcus lanatus, Houstonia serpyllifolia, Hygrophila lacustris, Hypericum hypericoides, H. galioides, H. mutilum, Ilex mucronata, I. opaca, I. verticillata, Impatiens capensis, Ipomoea, Iris versicolor, Itea virginica, Juncus canadensis, J. coriaceus, J. effusus, J. greenei, J. gymnocarpus, J. subcaudatus, Justicia ovata, Kalmia latifolia, Laportea canadensis, Larix laricina, Leersia lenticularis, L. oryzoides, L. virginica, Leucobryum glaucum, Lindera benzoin, Liquidambar styraciflua, Liriodendron tulipifera, Lobelia cardinalis, Lonicera, Ludwigia alternifolia, L. decurrens, L. leptocarpa, L. palustris, Lycopus americanus, L. amplectens, L. rubellus, L. uniflorus, Lygodium japonicum, L. palmatum, Lyonia ligustrina, Lysimachia lanceolata, L. radicans, L. terrestris, L. thyrsiflora, Lythrum salicaria, Maianthemum canadense, Medeola virginiana, Mikania scandens, Mimulus alatus, M. ringens, Mitchella repens, Mitella nuda, Mitreola, Mylia anomala, Myrica caroliniensis, M. pensylvanica, Nyssa biflora, N. sylvatica, Onoclea sensibilis, Osmunda regalis, Osmundastrum cinnamomeum, Oxalis acetosella, Oxypolis rigidior, Packera aurea, Pallavicinia lyellii, Panicum capillare, P. rigidulum, P. virgatum, Parthenocissus quinquefolia, Paspalum urvillei, Peltandra virginica, Persicaria arifolia, P. hydropiperoides, P. sagittata, Phalaris arundinacea, Phanopyrum gymnocarpon, Photinia floribunda, Physalis angulata, Physostegia intermedia, Pilea pumila, Pinus glabra, Platanthera clavellata, P. flava, Platanus occidentalis, Pluchea camphorata, Poa compressa, Populus deltoides, Potentilla simplex, Proserpinaca palustris, Prunus serotina, P. virginiana, Quercus bicolor, Q. laurifolia, Q. lyrata, Q. nigra, Q. palustris, Q. phellos, Q. rubra, Q. stellata,
Ranunculus abortivus, Rhamnus alnifolia, Rhexia virginica, Rhynchospora alba, R. capitellata, R. corniculata, R. glomerata, R. macrostachya, R. miliacea, Rubus argutus, R. hispidus, R. idaeus, R. trivialis, Rudbeckia laciniata, Rumex altissimus, R. verticillatus, Sabal minor, Sabatia calycina, Saccharum giganteum, Sagittaria calycina, S. latifolia, Salix interior, S. nigra, S. sericea, Sanguisorba canadensis, Saponaria officinalis, Sarracenia purpurea, Saururus cernuus, Scirpus cyperinus, S. expansus, S. georgianus, Scrophularia marilandica, Scutellaria lateriflora, S. racemosa, Sida, Sium suave, Smilax rotundifolia, Solidago altissima, S. gigantea, S. patula, S. rugosa, Sparganium americanum, S. eurycarpum, Spartina patens, Sphagnum capillifolium, S. fimbriatum, S. magellanicum, S. palustre, S. papillosum, S. recurvum, Spiraea alba, Symphyotrichum dumosum, S. lateriflorum, Symplocarpus foetidus, Taxodium distichum, Tetraphis pellucida, Thalictrum pubescens, Thelypteris noveboracensis, T. palustris, Thuja occidentalis, Toxicodendron radicans, T. vernix, Trautvetteria caroliniensis, Triadenum virginicum, T. walteri, Triadica sebifera, Typha latifolia, Ulmus alata, U. americana, Urtica dioica, Vaccinium corymbosum, V. fuscatum, V. macrocarpon, V. oxycoccos, Verbena hastata, Vernonia fasciculata, V. gigantea, V. noveboracensis, Viburnum dentatum, V. nudum, V. prunifolium, V. recognitum, Viola blanda, V. cucullata, V. lanceolata, V. macloskeyi, V. ×primulifolia, Woodwardia areolata, W. virginica, Xanthium strumarium, Xanthorhiza simplicissima, Xyris difformis, X. laxifolia, Zanthoxylum americanum.

Use by wildlife: Tubers of *Lycopus asper* and *L. uniflorus* are an important food of muskrats (Mammalia: Muridae: *Ondatra zibethicus*) and the plants commonly occur on their mounds. *Lycopus europaeus* also is found on muskrat mounds and is readily eaten by cattle (Mammalia: Bovidae: *Bos taurus*) and occasionally by horses (Mammalia: Equidae: *Equus ferus*). Cattle also are known to graze on *L. americanus*. The nutlets of *L. uniflorus* occasionally are eaten by pintail ducks (Aves: Anatidae: *Anas acuta*) and those of *L. americanus, L. rubellus,* and *L. virginicus* also provide food for various waterfowl. The nutlets of *L. asper* are the largest of the North American species and contain 25% oil 17% protein by weight. Plants of *L. americanus* occur in the habitat associated with Le Conte's sparrow (Aves: Emberizidae: *Ammodramus caudacutus*). *Lycopus virginicus* contains no chemical defenses against herbivores, though the fresh foliage consists of 5.2% protein. *Lycopus* species are a host of larval hawkmoths (Insecta: Lepidoptera: Sphingidae: *Sphinx eremitus*). Flowers of *L. americanus* are visited by numerous insects (Insecta) including beetles (Coleoptera: Cantharidae: *Chauliognathus pennsylvanicus*), long-tongued bees (Hymenoptera: Anthophoridae (*Ceratina dupla, Melissodes comptoides, M. nivea, M. rustica, Triepeolus lunatus*; Apidae: *Apis mellifera, Bombus fervidus, B. griseocollis, B. impatiens, B. pensylvanica, B. rufocinctus, B. vagans, Psithyrus variabilis*; Megachilidae: *Coelioxys germana, Megachile brevis, M. mendica, M. petulans*), short-tongued bees (Andrenidae: *Calliopsis andreniformis, Perdita octomaculatus*; Colletidae: *Colletes americana, C. latitarsis, Hylaeus* *affinis, H. illinoisensis, H. mesillae*; Halictidae: *Agapostemon sericea, Augochlorella aurata, Augochloropsis metallica, Halictus confusus, H. ligatus, H. parallelus, H. rubicunda, Lasioglossum coriaceus, L. imitatus, L. versatus*), butterflies (Hesperiidae: *Ancyloxypha numitor, Pholisora catullus, Polites themistocles*; Lycaenidae: *Everes comyntas, Lycaeides melissa*; Nymphalidae: *Limenitus archippus, Phyciodes tharos*; Pieridae: *Colias philodice*), flies (Bombyliidae: *Exoprosopa fasciata, E. fascipennis*; Calliphoridae: *Cochliomyia macellaria, Helicobia rapax, Lucilia sericata*; Muscidae: *Limnophora narona, Neomyia cornicina*; Conopidae: *Physocephala texana, Physoconops brachyrrhinchus*; Empididae: *Empis clausa*; Sarcophagidae: *Sarcophaga sinuata*; Syrphidae: *Eristalis arbustorum, Orthonevra nitida, Paragus bicolor, P. tibialis, Sphaerophoria contiqua, Syritta pipiens, Toxomerus marginatus*; Tachinidae: *Acroglossa hesperidarum, Archytas analis, A. aterrima, Copecrypta ruficauda, Cylindromyia euchenor, C. fumipennis, Deopalpus hirsutus, Distichona auriceps, Gymnoclytia immaculata, Gymnosoma fuliginosum, Hystriciella pilosa, Linnaemya comta, Plagiomima spinosula, Scotiptera parvicornis, Senotainia rubriventris, Tachinomyia panaetius, Trichopoda pennipes*), moths (Ctenuchidae: *Cisseps fulvicollis*), and wasps (Chrysididae: *Hedychrum wiltii*; Pompilidae: *Anoplius lepidus, A. marginatus, Ceropales bipunctata, Entypus fulvicornis, Poecilopompilus interrupta*; Sapygidae: *Sapyga interrupta*; Scoliidae: *Scolia bicincta*; Sphecidae: *Ammophila nigricans, A. pictipennis, A. procera, Bicyrtes ventralis, Eremnophila aureonotata, Eucerceris zonata, Isodontia apicalis, Philanthus gibbosus, P. ventilabris, Prionyx atrata, P. thomae, Sceliphron caementaria, Sphex ichneumonea, S. pensylvanica, Stizoides renicinctus, Stizus brevipennis, Tachytes aurulenta, T. distinctus*; Tiphiidae: *Myzinum quinquecincta*; Vespidae: *Dolichovespula maculata, Eumenes fraterna, Euodynerus foraminatus, Parancistrocerus vagus, Polistes annularis, Polistes fuscata*). *Lycopus* species are highly susceptible to infection by rust fungi (Basidiomycota: Pucciniaceae: *Puccinia menthae*) and are hosts to leaf-spot fungi (Ascomycota: Mycosphaerellaceae: *Septoria lycopi*). A leaf blight fungus (Ascomycota: Pleosporomycetidae: *Ascochyta lophanthi*) occurs on *L. americanus* and *L. uniflorus*. *Lycopus rubellus* is a host of a leaf-spot fungus (Ascomycota: Dothideomycetidae: *Cercospora lycopi*). Parasitic chytrid fungi are hosted by *L. americanus* (Blastocladiomycota: Physodermataceae: *Physoderma lycopi*) and *L. uniflorus* (Chytridiomycota: Synchytriaceae: *Synchytrium cellulare*).

Economic importance: food: Although commonly called "water horehound" for its superficial resemblance to true horehound (*Marrubium vulgare*), *Lycopus* lacks the bitter properties of its namesake, which are used to flavor cough medicines and candy. However, the boiled, dried, or raw "roots" (stolons or tubers) of several species (*L. amplectens, L. asper, L. uniflorus*) are edible. Tubers of *L. amplectens* and *L. uniflorus* are prepared by boiling in salt water or by pickling in wine vinegar. Those of the latter species were eaten by the Okanagon and Thompson tribes. *Lycopus* tubers ("crow potatoes") also were eaten by the Ojibwa. Tubers of *L. uniflorus* can be peeled

and added to salads or made into relish; **medicinal:** *Lycopus* species contain various phenolic compounds, with some that inhibit gonadotropic hormones (e.g., luteinizing hormone) and others (e.g., lithospermic acid in *L. europaeus* and *L. virginicus*) that inhibit thyroid stimulating hormones (TSH). In particular, *L. virginicus* has been used as a medicinal plant since the early 19th century, to treat various disorders including cardiac ailments, indigestion, insomnia, pulmonary bleeding, and ulcers. Extracts of *L. americanus, L. europaeus* and *L. virginicus* have been used in traditional medicine to control minor hyperthyroid conditions. Ethanolic extracts of *L. europaeus* are known to be effective at controlling hyperthyroidism and to reduce symptoms of latent hyperthyreosis under controlled laboratory testing. *Lycopus americanus* has been used as an astringent, as a mild narcotic and sedative, and to control bleeding and coughs. It also can influence circulation, heart rate, and blood sugar levels. The Meskwaki people ate *L. americanus* to treat stomach cramps. The Iroquois administered decoctions of *L. asper* as a laxative. *Lycopus europaeus* extracts contain several isopimarane diterpenoids and demonstrably increase the efficacy of tetracycline and erythromycin against resistant strains of *Staphylococcus aureus* (Bacteria: Staphylococcaceae) bacteria. Methanolic extracts from leaves of *L. europaeus* are inhibitory to xanthine oxidase, which catalyzes reactions responsible for gout. Hydroalcoholic extracts from *L. europaeus* exhibit high antioxidative activity. The Cherokee applied chewed roots of *L. virginicus* to treat snakebite wounds. Tea made from *Lycopus* reportedly is strongly sedative and narcotic like; **cultivation:** Several species of *Lycopus* are cultivated including *L. americanus, L. europaeus, L. rubellus* and *L. virginicus*; **misc. products:** The juice from *L. americanus* can be used as a color-fast dye for linen, silk, and wool; **weeds:** *Lycopus europaeus* is invasive in eastern North America. *Lycopus americanus* and *L. uniflorus* are regarded as minor weeds of cranberry bogs in Wisconsin; **nonindigenous species:** *Lycopus asper*, a native of western North America, was introduced to eastern North America, likely as a seed contaminant of grain shipped by railway. *Lycopus europaeus* was introduced to eastern North America in the late 19th century, probably as a seed contaminant of shipping ballast. In North America it has been observed to spread at rates of up to 45 km/year and had spread to Illinois by 1978. Nutlets of *L. virginicus* have been found as a contaminant of forage and lawn mixtures.

Systematics: Molecular data (Figure 5.59) substantiate the placement of *Lycopus* within tribe Mentheae of Lamiaceae subfamily Nepetoideae. From the phylogenetic survey of about half of the genera of *Mentheae*, *Lycopus* resolves in a relatively isolated position that is distinct from other genera in the tribe that contain OBL North American species. The genus appears to be monophyletic by its highly uniform morphology; however, only a single species has been included in formal phylogenetic analyses and others should be added for confirmation. *Lycopus ×sherardi* is a putative hybrid between *L. uniflorus* and *L. virginicus*. Although it has been categorized as OBL, it has been excluded here until its taxonomic status can better be clarified. Some authors recognize *Lycopus*

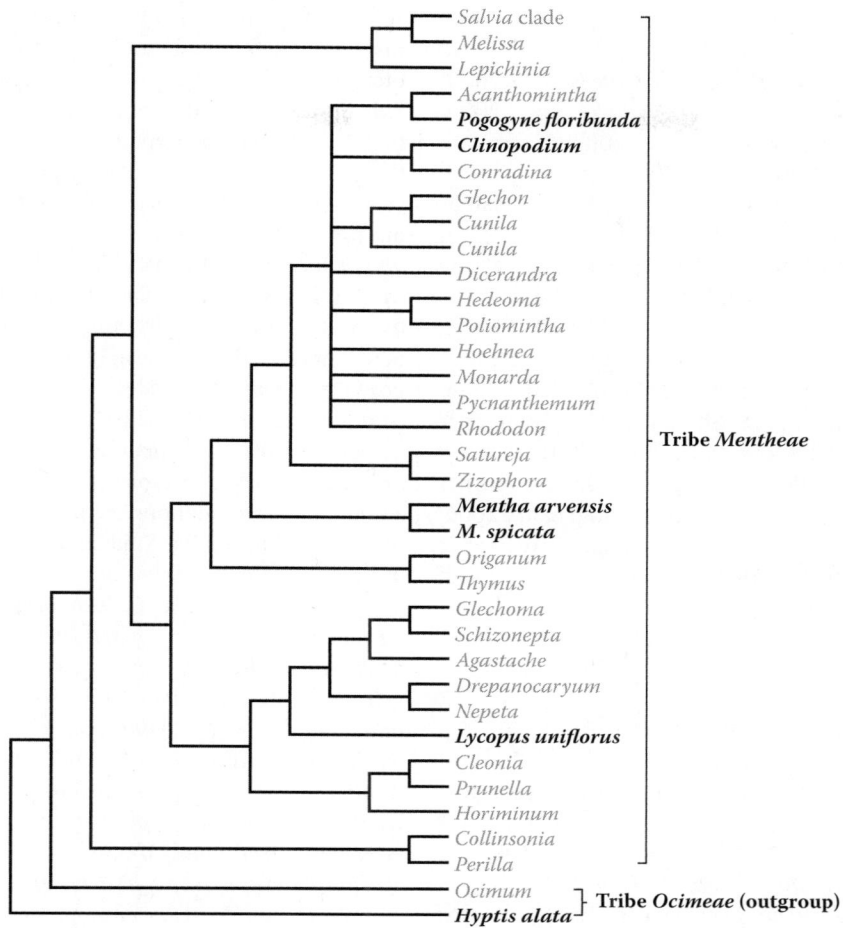

FIGURE 5.59 Phylogenetic relationships among genera of Lamiaceae tribe Mentheae (subfamily Nepetoideae) as indicated by analysis of combined cpDNA and nuclear DNA sequences. The resolution of taxa with OBL North American representatives (shown in bold) indicates multiple origins of the habit in the group. Roughly half of the genera of *Mentheae* were included in this analysis. (Adapted from Walker, J.B. & Sytsma, K.J., *Ann. Bot.* 100, 375–391, 2007.)

laurentianus as distinct from *L. americanus*, but both taxa are difficult to distinguish from one another. *Lycopus angustifolius* also has been recognized as a distinct species, but usually is regarded as being synonymous with *L. rubellus*, The name *L. lucidus* (an Old World species) has been misapplied to *L. asper* in older literature. *Lycopus cokeri* sometimes is merged with *L. virginicus* (as variety *pauciflorus*), but it is distinct morphologically and occupies a specialized habitat. The base chromosome number of *Lycopus* is $x = 11$. The genus is uniform cytologically, with *L. americanus*, *L. amplectens*, *L. asper*, *L. europaeus*, *L. rubellus*, *L. uniflorus*, and *L. virginicus* all being diploids ($2n = 22$). Fertile, introgressive hybrids between *L. americanus* and the introduced *L. europaeus* have been reported from disturbed shoreline localities. The putative hybrids have been well-documented morphologically, but should also be verified using genetic data. *Lycopus* would benefit from an extensive reexamination using genetic data to examine species boundaries, interrelationships, hybridization, etc. The relatively small size of the genus would facilitate a comprehensive study.

Comments: In North America, *Lycopus americanus* is widespread throughout, *L. asper* is widespread except for the southeastern United States and *L. uniflorus* is widespread except for the southern United States. The latter also extends into eastern Asia. *Lycopus amplectens* occurs in the eastern United States with *L. rubellus* and *L. virginicus* more broadly in eastern North America. *Lycopus cokeri* is endemic to North Carolina and South Carolina. The nonindigenous *L. europaeus* occurs throughout eastern North America and also in British Columbia.

References: Anderson, 1943, 1987; Anderson & Leopold, 2002; Andreas & Bryan, 1990; Auf'mkolk et al., 1984; Baskin & Baskin, 1998; Beal, 1977; Bonanno et al., 1998; Brändel, 2006; Braun, 1937; Brown, 1879; Cooper & Jones, 2004; Cosyns et al., 2005; El-Gazzar & Watson, 1968; Elias & Dykeman, 1990; Fassett, 1957; Felter & Lloyd, 1898; Fernald & Kinsey, 1943; Frolik, 1941; Gibbons et al., 2003; Girardin et al., 2001; Godefroid & Koedam, 2004; Hammer & Heseltine, 1988; Hanlon et al., 1998; Harder, 1985; Henderson, 1962; Hewitt & Miyanishi, 1997; Hong & Moon, 2003; Hooker &

Westbury, 1991; Hunt, 1943; Hussein et al., 1999; Johnson & Leopold, 1994; Kangas & Hannan, 1985; Karling, 1955a; Kelley et al., 1975; Kong et al., 2000; Kost & De Steven, 2000; Lachance & Lavoie, 2002; Lahring, 2003; Lake et al., 2000; Lamaison et al., 1991; Lamb & Mallik, 2003; Lieneman, 1929; Locky et al., 2005; Lovell, 1903; Lynn & Karlin, 1985; Mack & Boerner, 2004; Mackun et al., 1994; Mandossian & McIntosh, 1960; McAtee, 1925; Menges & Waller, 1983; Middleton, 2002b; Mohlenbrock, 1959a; Moon & Hong, 2003, 2006; Moorhead et al., 2000; Mukhin & Betekhtina, 2006; Nekola, 2004; Owens & Ubera-Jiménez, 1992; Perry & Uhler, 1981; Prusak et al., 2005; Redmond, 2007; Rodríguez-Riano & Dafni, 2007; Rubino et al., 2002; Schafale & Weakley, 1990; Schwintzer, 1978; Shi et al., 2002; Skinner & Sorrie, 2002; Sorrie, 1997; Sorrie et al., 2006; Sparrow, 1957; Stevens, 1932; Stuckey, 1969; Thompson, 1969a,b; Turner et al., 2000; Vartapetian & Andreeva, 1986; Vogt et al., 2006; Vonhoff & Winterhoff, 2006; Vonhoff et al., 2006; Walker & Sytsma, 2007; Webber & Ball, 1980; Weishampel & Bedford, 2006; Winterhoff et al., 1988.

5. Macbridea

Birds-in-a-nest, Carolina bogmint
Etymology: after Dr. James MacBride (1784–1817)
Synonyms: *Melittis* (in part); *Thymbra* (in part)
Distribution: global: North America; **North America:** southeastern
Diversity: global: 2 species; **North America:** 2 species
Indicators (USA): OBL: *Macbridea caroliniana*
Habitat: freshwater; palustrine, riverine; **pH:** 4.5–6.1; **depth:** <1 m; **life-form(s):** emergent herb
Key morphology: stems (to 9 dm) erect, square, gland dotted; leaves (to 12 cm) opposite, elliptic, the upper sessile or subsessile, the lower short petioled; flowers showy, bilabiate, in terminal, head-like clusters; corolla (to 3.5 cm) pink to lavender, purple, and white striped, cylindric at base, upper lip 2-lobed and hood-like, the lower 3-lobed and spreading, stamens didynamous, exserted; nutlets 4, the surface with irregular ribs
Life history: duration: perennial (rhizomes); **asexual reproduction:** rhizomes; **pollination:** insect; **sexual condition:** hermaphroditic; **fruit:** nutlets (common); **local dispersal:** nutlets (gravity), rhizomes; **long-distance dispersal:** nutlets (water)
Imperilment: (1) *Macbridea caroliniana* [G2]; S1 (GA); S2 (NC)
Ecology: general: Both species of *Macbridea* occur frequently in wet habitats and are designated either as OBL (*M. caroliniana*) or FACW (*M. alba*) wetland indicators. There have been fairly extensive studies on the pollination biology, population genetics, and seed ecology of the genus.

Macbridea caroliniana (Walter) S.F. Blake is a rare inhabitant of bogs, floodplains, forest seeps, marshes, and swamps along the southeastern coastal plain. Its substrates are acidic (e.g., pH: 4.5–6.1) and high in organic matter. This species most frequently occurs in blackwater stream swamps. Microsite studies have indicated that occurrences are associated with a rich herbaceous flora, high cover of mosses (18%), open canopies (89%), and a fairly high soil phosphorus content (6.5 ppm), and relatively low potassium levels. Flowering occurs from mid-July through early September. The flowers are self-compatible, but require insect pollination (Hymenoptera; Lepidoptera) to set seed because spontaneous selfing does not readily occur. The flowers likely are pollen limited. The seeds lack dormancy, germinate without pretreatment (often while still on the parent plant), and do not form a seed bank. The nutlets float and are dispersed by gravity and water. The genetic structure of populations is influenced by the particular river basin, with gap patches being more similar to one another genetically than individuals growing among sites under closed canopies. This species has been cultivated successfully using a planting mix consisting of perlite (50%), vermiculite (30%), and chopped or milled *Sphagnum* (20%). The plants reproduce vegetatively by rhizomes. Some populations are threatened by disturbance from feral pigs (Mammalia: Suidae: *Sus scrofa*). **Reported associates:** *Acer rubrum, Arundinaria gigantea, Boehmeria cylindrica, Carex atlantica, C. bromoides, C. leptalea, Carpinus caroliniana, Chamaecyparis thyoides, Clethra alnifolia, Decumaria barbara, Eupatorium resinosum, Fraxinus pennsylvanica, Helianthemum bicknelli, Hydrocotyle verticillata, Ilex coriacea, I. glabra, I. opaca, Isoetes riparia, Itea virginica, Lachnocaulon beyrichianum, Leucothoe axillaris, Liquidambar styraciflua, Liriodendron tulipifera, Lophiola americana, Ludwigia palustris, Lysimachia asperulifolia, Magnolia virginiana, Mitchella repens, Murdannia keisak, Myrica heterophylla, Nyssa biflora, N. sylvatica, Osmunda regalis, Osmundastrum cinnamomeum, Oxydendrum arboreum, Parnassia caroliniana, Persea palustris, Pinus palustris, P. taeda, Platanthera clavellata, Quercus laurifolia, Q. michauxii, Rhododendron flammeum, Rhynchospora decurrens, Sagittaria fasciculata, Saururus cernuus, Sphagnum, Toxicodendron radicans, T. vernix, Triadenum walteri, Ulmus rubra, Vitis, Woodwardia areolata, Xyris flabelliformis*.
Use by wildlife: The flowers of *Macbridea caroliniana* are visited by bumblebees (Insecta: Hymenoptera: Apidae: *Bombus impatiens*) and butterflies (Insecta: Lepidoptera: Hesperiidae: *Poanes zabulon*).
Economic importance: food: not edible; **medicinal:** none; **cultivation:** *Macbridea caroliniana* is cultivated occasionally as an ornamental for native bog gardens; **misc. products:** none; **weeds:** none; **nonindigenous species:** none
Systematics: *Macbridea*, with only two species, is assigned to tribe Lamieae subtribe Melittidinae (Lamiaceae subfamily Lamioideae). The genus should not be confused with *Macbridea* Raf. (=*Seutera* [Apocynaceae]), which does not have nomenclatural priority. Phylogenetic analysis of morphological data initially indicated *Macbridea* to be the sister group of a clade consisting of *Brazoria, Physostegia,* and *Warnockia*. That result was confirmed subsequently by phylogenetic analyses of the subtribe using cpDNA and nuclear sequence data. Those analyses demonstrated the monophyly of *Macbridea* and resolved the genus within a clade of North American endemics also comprising *Brazoria, Physostegia, Synandra, Warnockia* (Figure 5.60). These results are

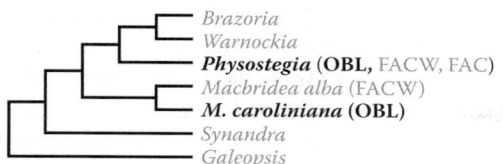

FIGURE 5.60 Relationships among several genera of Lamiaceae (Lamioideae; Lamiaeae; Melittidinae) as indicated by phylogenetic analyses of cpDNA and nuclear DNA sequences. These molecular data resolve the endemic North American genera (*Brazoria*, *Macbridea*, *Physostegia*, *Synandra*, and *Warnockia*) as a clade (Scheen & Albert, 2004). Phylogenetic analyses of morphological characters (Turner, 1996) and additional cpDNA sequence data (Chen et al., 2014) yield compatible topologies. Genera with OBL species (in bold) indicate independent origins of the habit in the subtribe. (Adapted from Scheen, A.-C. et al., *Cladistics*, 24, 299–314, 2008.)

consistent with those based on cytology, stomatal morphology and pollen sculpturing, which indicate that *Macbridea* forms a clade with *Brazoria* and *Physostegia* and is similar chromosomally to *Synandra*. The base chromosome number of *Macbridea* is *x* = 9 (apparently derived within subfamily Lamioideae). *Macbridea caroliniana* (2*n* = 18) is a diploid. Hybridization has not been reported in *Macbridea*.

Comments: *Macbridea caroliniana* is endemic to the southeastern United States.

References: Abu-Asab & Cantino, 1994; Adams et al., 2010; Beal, 1977; Cantino, 1985; Chafin, 2007; Chen et al., 2014; Fishbein & Stevens, 2005; Glitzenstein et al., 2001; Godfrey & Wooten, 1981; Pitts-Singer et al., 2002; Sanders & Cantino, 1984; Scheen & Albert, 2004; Scheen et al., 2008; Schulze et al., 2002; Turner, 1996; Weeks, 2009; Weeks & Walker, 2006.

6. *Mentha*

Mint, pennyroyal, peppermint, spearmint; menthe, menthe en épi, menthe poivrée, pouliot

Etymology: after Minthe, a mythical nymph who was changed into a mint plant by Persephone

Synonyms: *Audibertia*; *Menthella*; *Preslia*; *Pulegium*

Distribution: global: Africa; Asia; Australia; Europe; North America; **North America:** widespread

Diversity: global: 18 species; **North America:** 10 species

Indicators (USA): OBL: *Mentha aquatica*, *M.* ×*gracilis*, *M.* ×*piperita*, *M. pulegium*, *M. spicata*, *M.* ×*villosa*; **FACW:** *M. aquatica*, *M.* ×*gracilis*, *M.* ×*piperita*, *M. spicata*, *M.* ×*villosa*; **FAC:** *M.* ×*gracilis*, *M.* ×*villosa*

Habitat: brackish, freshwater; palustrine; **pH:** 4.5–8.5; **depth:** <1 m; **life-form(s):** emergent herb

Key morphology: stems (to 1 m) erect, square, the herbage glabrous or pubescent, sweet or spicy scented; leaves (to 8 cm) opposite, elliptic, lanceolate or ovate, sessile, or petiolate (to 15 mm), the margins serrate; flowers bilabiate, in dense, verticillate cymes that are axillary or condensed in terminal, spike-like, or globose clusters; corolla (to 8 mm) lavender, pink, violet, or white, tubular, with a 2-lobed upper and 3-lobed lower lip; stamens 4, equal, exserted (bisexual flowers) or included (pistillate flowers); style gynobasic, ovary 4-lobed; schizocarp dehiscing into four smooth nutlets (to 1 mm)

Life history: duration: perennial (rhizomes, stolons); **asexual reproduction:** rhizomes, stolons, vegetative fragments; **pollination:** insect; **sexual condition:** hermaphroditic; **fruit:** nutlets (common); **local dispersal:** rhizomes, stolons, seeds (gravity); **long-distance dispersal:** seeds (animals, water)

Imperilment: (1) *Mentha aquatica* [GNR]; (2) *M.* ×*gracilis* [GNA]; (3) *M.* ×*piperita* [GNA]; (4) *M. pulegium* [GNR]; (5) *M. spicata* [GNR]; (6) *M.* ×*villosa* [GNA];

Ecology: general: *Mentha* species (mints) occur quite characteristically in wetland habitats. Mints are similar ecologically and several species can be found coexisting in marshes. Nine species and hybrids (90%) are wetland indicators in North America; however, the only native species (*M. arvensis*) is designated as FACW. All of the OBL North American species are nonindigenous, and even these occur facultatively in wetlands and often can be found inhabiting fairly dry waste areas. Generally the species occupy open sites that are not permanently inundated, often where ice scour, wave action, or currents have created natural disturbances. Pollination typically is by insects (Insecta), mainly bumblebees (Hymenoptera: Apidae: *Bombus*) and honeybees (Hymenoptera: Apidae: *Apis mellifera*), but also by beetles (Coleoptera), butterflies (Lepidoptera) and hoverflies (Diptera: Syrphidae). The nutlets float and are dispersed by water; perhaps in some instances, they are dispersed by waterfowl. Vegetative reproduction occurs by the production of rhizomes or stolons.

Mentha aquatica **L.** grows emersed or in shallow water of fens, marshes, meadows, and along lake, pond, river, and stream margins at elevations of up to 975 m. The plants are adapted to a wide range of acidity (pH: 4.5–7.8) and tend to occur on sandy substrates (but also on muck). They are not salt-tolerant but can occur along brackish wetland margins or in tidal freshwater sites. They do withstand physical perturbation and tend to increase at sites where disturbance (e.g., trampling by livestock) occurs. This species is regarded as a wetland pioneer and occurs in sites receiving full sun to partial shade. The plants are gynomonoecious and the bisexual flowers are protandrous. Although self-compatible and capable of self-pollination, the flowers primarily are insect-pollinated. Individual plants produce about 200 nutlets in a season. The seeds require 30 days of cold stratification for optimal germination under a 20°C/15°C temperature regime but will germinate over a broad temperature range (9°C–36°C). High germination rates (>50%) require a temperature fluctuation of 8°C. Mature nutlets are dispersed by water and can represent a large proportion of viable propagules occurring in river drift. They form a transient to long-term persistent seed bank of up to 5360 seeds/m². The seed banks are best developed in mid- to late-succession stages of 9–80 years. Although total biomass can decrease by up to 57% when plants are inundated or flooded, the shoots elongate significantly as an adaptive response. Vegetative reproduction occurs mainly by the diffuse production of rhizomes. Plants also are dispersed by stem or rhizome fragments. The roots normally are colonized by arbuscular mycorrhizae. Grazing by large herbivores causes populations to decline. The plants are subject to higher levels of herbivory when growing out of the water, where herbivores (chrysomelid beetles) show no preference for the associated terrestrial plants. There is extensive ecological information regarding European populations of this species; however, few such studies exist with respect to those occurrences within its introduced North American range. Maximum yield of cultured plants is achieved using 12.5 × 16.5 cm pots containing Metro-Mix 500 or Pro-Mix BX growth media fertilized with 29.2 g Nutricote per pot. **Reported associates (North America):** *Agrostis, Alisma, Alnus rhombifolia, Bidens cernuus, B. frondosus, Callitriche, Caltha palustris, Carex bebbii, C. granularis, C. lacustris, C. lyngbyei, C. vulpinoidea, Deschampsia cespitosa, Eleocharis erythropoda, E. palustris, E. parvula, Epilobium ciliatum, Equisetum fluviatile, Eupatorium perfoliatum, Eutrochium maculatum, Festuca arundinacea, Galium trifidum, Glyceria elata, Gratiola neglecta, Hippuris vulgaris, Iris pseudacorus, Juncus articulatus, J. balticus, J. brachycephalus, J. effusus, J. oxymeris, Leersia oryzoides, Lilaea scilloides, Lilaeopsis occidentalis, Limosella aquatica, Lotus corniculatus, Lycopus americanus, L. europaeus, Lysichiton americanus, Lythrum salicaria, Mentha arvense, Menyanthes trifoliata, Mimulus guttatus, Myosotis scorpioides, Nasturtium officinale, Oenanthe sarmentosa, Persicaria amphibia, P. hydropiper, Phalaris arundinacea, Platanus racemosa, Platanthera dilatata, Populus fremontii, Potentilla anserina, Ribes hirtellum, Rubus bifrons, R. ursinus, Rumex occidentalis, Sagittaria latifolia, Salix lasiolepis, S. lucida, S. sitchensis, Scirpus atrovirens, S. cyperinus, S. lacustris, S. microcarpus, Sidalcea hendersonii, Sium suave, Solanum dulcamara, Sparganium eurycarpum, Symphyotrichum, Symplocarpus foetidus, Toxicodendron diversilobum, Typha latifolia, Veronica americana, Vitis girdiana.*

Mentha ×*gracilis* **Sole** inhabits cliffs, depressions, ditches, marshes, meadows, and the margins of lakes, ponds, rivers, and streams. North America sites typically have sandy substrates and range in exposure from full sun to shade. These hybrid plants generally are highly sterile; however, fertile backcrosses have been reported. Reproduction primarily is vegetative and occurs by means of rhizomes. There are few ecological accounts of this hybrid in its nonindigenous North American range. **Reported associates (North America):** *Acalypha, Acer rubrum, Agrostis gigantea, Alnus rugosa, Ambrosia trifida, Andropogon gerardii, Apocynum androsaemifolium, A. cannabinum, Artemisia serrata, Asclepias incarnata, Betula papyrifera, Bidens connatus, B. frondosus, Boehmeria cylindrica, Bolboschoenus fluviatilis, Calystegia sepium, Carex lacustris, C. torta, Chamaesyce maculata, Cirsium arvense, Coreopsis tinctoria, Cornus rugosa, Cyperus, Doellingeria umbellata, Echinochloa muricata, Eleocharis, Epilobium ciliatum, Equisetum arvense, Eupatorium perfoliatum, Euthamia graminifolia, Eutrochium maculatum, Galium, Helianthus grosseserratus, Heliopsis helianthoides, Hibiscus laevis, Hydrocotyle americana, Hypericum ellipticum, H. mutilum, Impatiens capensis, Juncus, Justicia americana, Leersia oryzoides,* *L. virginica, Lindernia dubia, Lobelia cardinalis, Lotus corniculatus, Ludwigia palustris, Lycopus uniflorus, Lysimachia ciliata, L. terrestris, L. vulgaris, Lythrum salicaria, Mentha arvensis, Mimulus glabratus, M. ringens, Mollugo verticillata, Napaea dioica, Nasturtium officinale, Oenothera biennis, Onoclea sensibilis, Packera aurea, Persicaria amphibia, P. hydropiper, P. maculosa, P. pensylvanica, P. posumbu, P. sagittata, Phalaris arundinacea, Phleum pratense, Pilea pumila, Pinus resinosa, Poa pratensis, Prunella vulgaris, Pycnanthemum virginianum, Rhexia virginica, Rorippa sylvestris, Sagittaria latifolia, Schoenoplectus pungens, Scirpus cyperinus, Setaria, Silphium perfoliatum, Solidago gigantea, S. rugosa, Sorghastrum nutans, Sullivantia, Symphyotrichum ciliolatum, S. lateriflorum, Tanacetum vulgare, Teucrium canadense, Triadenum virginicum, Typha latifolia, Urtica, Verbena hastata, V. urticifolia, Viola, Xanthium strumarium.*

Mentha ×*piperita* **L.** grows in bogs, floodplains, marshes, meadows, prairies, streams, swamps, and along the margins of brooks, lakes, ponds, rivers, and streams at elevations of up to 2500 m. The plants occur across a range of acidic substrates (pH: 5.5–6.5) including clay, gravel, sand, sandy loam, and silt. The sites often are open but the plants also can thrive in shade. This hybrid is highly pollen-sterile and its reproduction is almost entirely vegetative by means of stolons. Accordingly, allozyme analysis of natural populations indicates a lower level of genetic variation in this species than in its sexual congeners. The plants often are found growing in waste areas. Commercial propagation is by stem cuttings, stolon division, or the turion-like buds on the stolons. Plants also have been regenerated artificially by tissue culture. Long-day conditions enhance growth and will increase the monoterpene content. Commercial oil yields have been increased by applications of anhydrous ammonia at rates of 45–179 kg/ha. **Reported associates (North America):** *Acer saccharinum, Achillea millefolium, Acorus americanus, Agrostis, Alnus rhombifolia, Amaranthus tuberculatus, Ambrosia artemisiifolia, Artemisia biennis, Asclepias incarnata, Brassica nigra, Calamagrostis canadensis, Carex lurida, Carya ovata, Chelone glabra, Chenopodium album, Cirsium vulgare, Clinopodium arkansanum, Cornus sericea, Dipsacus sylvestris, Echinochloa walteri, Elymus virginicus, Epilobium, Eupatorium fistulosum, E. maculatum, E. perfoliatum, Euthamia graminifolia, Fraxinus pennsylvanica, Glechoma hederacea, Glyceria canadensis, Helenium autumnale, Impatiens capensis, Juglans nigra, Juniperus, Lactuca, Leonurus cardiaca, Lobelia siphilitica, Lolium, Lotus corniculatus, Lycopus americanus, L. virginicus, Lythrum salicaria, Melilotus alba, Mentha arvensis, Myosotis scorpioides, Oenothera biennis, Panicum virgatum, Pedicularis lanceolata, Peltandra virginica, Persicaria lapathifolia, P. pensylvanica, P. maculosa, P. punctata, Phleum pratense, Plantago lanceolata, Poa nemoralis, Prunus virginiana, Quercus agrifolia, Ranunculus pensylvanicus, R. repens, R. septentrionalis, Rhamnus, Rumex crispus, Salix, Sambucus nigra, Schoenoplectus pungens, Scutellaria lateriflora, Secale cereale, Solanum dulcamara, Solidago altissima, S. patula, Stachys palustris, S. tenuifolia, Symphyotrichum novae-angliae, S. puniceum, Symphytum* *officinale, Thalictrum, Trifolium hybridum, T. pratense, T. repens, Tripleurospermum maritima, Typha angustifolia, Ulmus americana, Verbena hastata, Vernonia noveboracensis, Vitis girdiana, V. riparia.*

***Mentha pulegium* L.** grows in depressions, ditches, floodplains, lagoons, marshes, meadows, prairies, sandbars, seeps, vernal pools, and along lake, pond, river, and stream margins at elevations of up to 1420 m. The plants can tolerate seasonal drought and brackish (alkaline) conditions and grow in full sun to partial shade. Substrates (pH: 5.0–8.5) are diverse and include clay, clay-loam, gravel, sand, serpentine, and silt. The flowers typically are hermaphroditic and protandrous, and the plants rarely are gynodioecious. Pollination is carried out by bees (Hymenoptera) and other insects. The rich genetic variation and high heterozygosity observed in natural populations (as determined by allozyme analysis) indicates a predominantly outcrossing breeding system. The nutlets (along with their enveloping calyx) are dispersed by the wind or by water and also by attaching to the fur of animals. Germination rates are high in natural populations. The seeds germinate throughout the year, but require light and fluctuating temperatures to break dormancy. They can germinate underwater and the seedlings are able to grow under conditions of shallow inundation. The largest seed banks occur in pastures (up to 176,000/m²) due to soil disturbance caused by trampling livestock. Vegetative reproduction occurs by means of rhizomes and stolons, which possess scale-like leaves. Vegetative reproduction also can occur by fragments of rhizomes or the aerial stems, which root at the nodes. The roots normally are associated with arbuscular mycorrhizae. In North America, plants have been found growing in both natural and artificially restored wetlands. Although this species currently is the least widely distributed of the OBL mints, it is becoming fairly prevalent in western North America where it is regarded as a weed. **Reported associates (North America):** *Achillea millefolium, Acmispon americanus, Agrostis stolonifera, Alnus rubra, Alopecurus, Amorpha fruticosa, Artemisia lindleyana, A. vulgaris, Beckmannia, Bidens frondosus, Callitriche, Carex aperta, C. feta, C. interrupta, C. obnupta, C. retrorsa, C. unilateralis, C. vulpinoidea, Centaurium erythraea, C. muehlenbergii, Chenopodium ambrosioides, Conyza canadensis, Coreopsis tinctoria, Cornus sericea, Cyperus acuminatus, C. erythrorhizos, C. esculentus, Daucus carota, Deschampsia cespitosa, Eleocharis palustris, Epilobium ciliatum, Equisetum arvense, Eragrostis, Eryngium petiolatum, Euthamia occidentalis, Fraxinus latifolia, Gaillardia aristata, Galium, Gnaphalium palustre, Helenium autumnale, Holcus, Impatiens capensis, Iris pseudacorus, Juncus effusus, Lathyrus sphaericus, Leersia oryzoides, Lilaeopsis occidentalis, Lindernia dubia, Ludwigia palustris, Lycopus americanus, Lysimachia nummularia, Lythrum salicaria, Medicago hispida, M. lupulina, Melilotus officinalis, Mentha arvensis, M. ×piperita, Mimulus guttatus, Mollugo verticillata, Myosotis laxa, Paspalum distichum, Persicaria hydropiperoides, Phalaris arundinacea, Plantago lanceolata, P. major, Poa pratensis, Polypodium glycyrrhiza, Portulaca oleracea, Prunella vulgaris, Psilocarphus elatior, Rorippa*

curvisiliqua, Rubus discolor, Rumex crispus, Salix commutata, S. fluviatilis, S. lucida, S. sitchensis, Sambucus racemosa, Schoenoplectus tabernaemontani, Scutellaria lateriflora, Sericocarpus rigidus, Solanum dulcamara, Sonchus asper, Spiraea douglasii, Trifolium arvense, T. repens, Typha latifolia, Urtica dioica, Veronica americana, Xanthium strumarium.

***Mentha spicata* L.** occurs in ditches, floodplains, marshes, meadows, prairies, ravines seeps, and along margins of ponds, rivers, and streams at elevations of up to 2320 m. The substrates (pH: 5.5–7.3) are derived from dolomite or granite and include adobe, clay, clay loam, humus, sand, sandy loam, sandy silt, and silt loam. The plants usually occur in open sites but can tolerate light shade. They are common in waste areas or drying sites. The flowers are self-compatible, but are protandrous and insect pollinated. Analyses of natural populations using allozyme data reveal a high level of genetic variation, which indicates a predominantly outcrossing breeding system. Vegetative reproduction occurs by the production of slender stolon-like rhizomes. A high yield of essential oils has been obtained by fertilizing commercial fields with 168 kg/ha of anhydrous NH_3. **Reported associates (North America):** *Acer negundo, Acorus calamus, Actaea alba, Adenostoma fasciculatum, Ageratina adenophora, Agropyron repens, Agrostis gigantea, A. stolonifera, Alnus rubra, Ambrosia artemisiifolia, Amphicarpaea bracteata, Aralia racemosa, Arctium minus, Armoracia rusticana, Artemisia douglasiana, Asclepias incarnata, A. perennis, Atriplex, Baccharis salicifolia, Bidens cernuus, Brassica nigra, Bromus inermis, Cardamine pennsylvanica, Carex trichocarpa, C. vulpinoidea, Caulophyllum thalictroides, Ceanothus crassifolius, Cenchrus longispinus, Centaurea, Chamaesyce maculata, Cichorium, Cirsium arvense, Commelina diffusa, Conium maculatum, Conoclinium coelestinum, Convolvulus arvensis, Cyperus eragrostis, C. erythrorhizos, C. ferruginescens, C. involucratus, C. squarrosus, Dasiphora floribunda, Digitaria sanguinalis, Dioscorea villosa, Dipsacus, Epilobium ciliatum, E. coloratum, Equisetum, Eragrostis hypnoides, Eupatorium altissimum, E. perfoliatum, Ficus carica, Foeniculum, Fragaria vesca, Fraxinus, Glechoma hederacea, Glyceria grandis, G. striata, Helianthus annuus, Heliotropium indicum, Hibiscus moscheutos, Hierochloe odorata, Hypericum perforatum, Impatiens capensis, Ipomoea lacunosa, Juncus balticus, Lactuca serriola, Laportea canadensis, Leersia lenticularis, L. oryzoides, Lepidium latifolium, Lindernia dubia, Lobelia cardinalis, Lotus corniculatus, Lycopus americanus, L. rubellus, Melica imperfecta, Menispermum canadense, Mentha arvensis, M. ×gracilis, Monarda fistulosa, Muhlenbergia frondosa, M. schreberi, Myosoton aquaticum, Myosurus minimus, Osmorhiza claytonii, Panicum rigidulum, Paspalum fluitans, Persicaria hydropiper, P. hydropiperoides, P. maculosa, P. punctata, P. sagittata, Phalaris arundinacea, Phleum pratense, Phryma leptostachya, Pilea pumila, Pinus ponderosa, Piptatherum miliaceum, Plantago lanceolata, Platanus racemosa, Pluchea camphorata, Poa palustris, P. pratensis, Populus, Ptilimnium costatum, Pycnanthemum virginianum, Quercus agrifolia, Ranunculus* *acris, R. pensylvanicus, Rosa, Rubus ursinus, Rumex aceto-sella, R. crispus, R. verticillatus, Salix exigua, S. lasiandra, S. lasiolepis, S. laevigata, S. monochroma, S. nigra, Salvia, Sanguinaria canadensis, Saururus cernuus, Schoenoplectus pungens, S. tabernaemontani, Scutellaria lateriflora, Setaria glauca, Sium suave, Solanum carolinense, S. dulcamara, Solidago canadensis, Spartina, Spiraea, Spiranthes diluvialis, Symphyotrichum lanceolatum, S. lateriflorum, S. pilosum, S. subspicatum, Toxicodendron diversilobum, Typha latifolia, Umbellularia californica, Urtica, Verbascum blattaria, V. thapsus, Verbena hastata, V. stricta, V. urticifolia, Vitis riparia, Xanthium strumarium.*

Mentha villosa Huds. grows in meadows, streams, and springs at elevations of up to 930 m. It occurs on substrates described as loam and sandy loam. The plants reproduce vegetatively by rhizomes. Other life-history information for this nonindigenous species is sparse. **Reported associates (North America):** *Anemopsis californica, Baccharis salicifolia, Carex, Juncus balticus, Poa annua, Salix exigua, S. laevigata, S. lasiolepis, Senecio vulgaris.*

Use by wildlife: The foliage or nutlets of *Mentha* species are consumed occasionally by green-winged teal ducks (Aves: Anatidae: *Anas crecca*), ruffed grouse (Aves: Phasianidae: *Bonasa umbellus*), and muskrats (Mammalia: Muridae: *Ondatra zibethicus*). *Mentha aquatica* is grazed by livestock. The flowers of *M. aquatica* and *M. pulegium* attract various butterflies (Insecta: Lepidoptera). In its North American range, *M. ×piperita* hosts a number of larval Lepidoptera (Noctuidae: *Amphipyra pyramidoides, Autographa californica, Euxoa ochrogaster, Peridroma saucia, Spodoptera eridania, Xylena nupera*; Nymphalidae: *Anartia jatrophae*; Pyralidae: *Pyrausta orphisalis, Synclita obliteralis*; Sphingidae: *Sphinx eremitus*). *Mentha pulegium* also is a larval host plant for the white peacock butterfly (Lepidoptera: Nymphalidae: *Anartia jatrophae*). In North America, *Mentha spicata* also hosts a variety of larval Lepidoptera (Arctiidae: *Haploa lecontei*; Geometridae: *Synchlora aerata, Xanthotype sospeta*; Lycaenidae: *Strymon melinus*; Noctuidae: *Autographa californica, Autoplusia egena*; Nymphalidae: *Vanessa cardui*; Pyralidae: *Synclita obliteralis*). *Mentha ×piperita* hosts various fungi including Ascomycota (Didymellaceae: *Boeremia strasseri*; Incertae sedis: *Trichothecium*; Nectriaceae: *Gibberella zeae*; Plectosphaerellaceae: *Verticillium tricorpus*; Pleosporaceae: *Alternaria*; Sclerotiniaceae: *Sclerotinia sclerotiorum*), Basidiomycota (Ceratobasidiaceae: *Thanatephorus cucumeris*), and Oomycota (Pythiaceae: *Pythium*). *Mentha ×piperita* and *M. spicata* are common hosts for several Ascomycota (Erysiphaceae: *Erysiphe cichoracearum*; Plectosphaerellaceae: *Verticillium albo-atrum*) and a rust (Basidiomycota: Pucciniaceae: *Puccinia menthae*); the latter also occurs on *M. ×gracilis. Mentha ×piperita* and *M. ×gracilis* are much more susceptibile to infection by wilt fungus (Plectosphaerellaceae: *Verticillium dahliae*) than *M. spicata.*

Economic importance: food: A nearly universal fondness for the taste of peppermint (*M. ×piperita*) and spearmint (*M. ×gracilis, M. spicata*) has led to their common use as flavorings for a variety of products including candy, chewing gum, jelly, medicine, tea, and toothpaste. Several popular alcoholic beverages (e.g., Crème de menthe, Mint Julep, Mojito) also contain peppermint or spearmint as an essential ingredient. Herbal teas made from peppermint leaves remain highly popular. The Cherokee and Kiowa tribes chewed spearmint leaves or added them to various foods. The Kawaiisu, Miwok, and Yuki people brewed the leaves as a tea. Leaves of *Mentha ×gracilis* can be eaten in salads or used to enhance the flavor of cooked foods. The leaves of water mint (*M. aquatica*) also are edible. *Mentha pulegium* is toxic and should never be eaten; **medicinal:** The mints have a long history of medicinal use that extends back to ancient Egypt (1000 BC). The Native Americans used peppermint and spearmint extensively in their medicines. The Cherokee relied on peppermint and spearmint to treat colic, cramps, fever, gas, headaches, indigestion, and nausea. They also used the plants as both a sedative and stimulant and applied peppermint or spearmint tinctures to relieve the discomfort of piles. The Delaware, Hoh, Oklahoma, and Quileute tribes made a general medicinal tonic from peppermint. The Iroquois administered an infusion of peppermint or spearmint for colds, fevers, and as a wash for general injuries. The Menominee used a peppermint poultice to relieve symptoms associated with pneumonia. Peppermint and spearmint were used by the Mohegans to rid children of intestinal parasites. The Iroquois ate spearmint to relieve stomach disorders. A leaf infusion of spearmint was taken by the Miwok people to ease diarrhea and relieve stomach disorders. *Mentha pulegium*, often distributed as a herbal remedy (pennyroyal), contains the monoterpene pulegone, a substance that is toxic to the liver and to other organs when metabolized into menthofuran. Products containing pennyroyal should be avoided because the oil is known to be fatal to humans, especially to children. The essential oils found in *M. aquatica* (water mint), *M. ×piperita* (peppermint), and *M. spicata* (spearmint) inhibit the growth of several pathogenic microbes in culture including *Escherichia coli* (Enterobacteriaceae), *Helicobacter pylori* (Helicobacteraceae), *Micrococcus flavus* (Micrococcaceae), *Salmonella enteritidis* (Enterobacteriaceae), multiresistant *Shigella sonnei* (Enterobacteriaceae), and methicillin sensitive and resistant strains of *Staphylococcus aureus* (Staphylococcaceae). Spearmint tea has been shown to inhibit the carcinogenic activation of heterocyclic amines. Spearmint extracts also are effective at suppressing benzoyl peroxide-induced tumor promotion in mice and have been used to treat gastric ulcers. Extracts of peppermint, spearmint, and water mint exhibit antioxidant properties, with peppermint shown to be effective at reducing induced lung tumors in laboratory mice. The active ingredient of peppermint oil is menthol, which is known to have appetite suppressant properties (thus its common use in after-dinner mints). Peppermint oil (standardized to 44% menthol) also is analgesic, antispasmodic (by blocking calcium channels), is somewhat antiviral, relieves colic and digestive gas, and stimulates bile production. It has been used clinically to treat arthritis, diverticulitis, gallstones, indigestion, irritable bowel syndrome, and to relieve postoperative nausea and menstrual cramps. Peppermint also contains perillyl alcohol, which causes regression of some animal tumors and may be therapeutic for colon cancer. In rare instances, menthol and other essential oils found in pennyroyal, peppermint, and spearmint can cause contact dermatitis or lip inflammation (when used as a toothpaste flavoring) in humans. Infusions of water mint are astringent and emetic and have been used to control diarrhea. *Mentha ×villosa* contains the monoterpene rotundifolone, which has known pain-relieving properties. Essential oils from the plant (notably piperitenone oxide) inhibit some bacteria (Eubacteria: Staphylococcaceae: *Staphylococcus aureus*) and have been shown to lower blood pressure in laboratory animals. Some fetal toxicity has been observed in the offspring of laboratory rats (Mammalia: Muridae: *Rattus*) that were given food containing the essential oils; **cultivation:** *Mentha aquatica* was cultivated widely as a condiment by the ancient Greeks and Romans and was used in Italy and Switzerland as far back as the Neolithic period. Today, *Mentha* species are grown commonly in herb gardens, as ornamentals, and for commercial production of mint oils. A few species occasionally are grown in aquariums. The midwestern United States produces 75% of the world's fresh peppermint supply. Mint cultivars include *M. aquatica*: 'Mandeliensis'; *M. ×gracilis*: 'Aurea,' 'Madeline Hill,' 'Neer-Kalka' [fertile], 'Variegata'; *M. ×piperita*: 'After Eight,' 'Basil,' 'Bergamot,' 'Black Mitcham,' 'Chocolate,' 'Citrata,' 'Eau de Cologne,' 'Extra Strong,' 'Frantsila,' 'Grapefruit,' 'Kukrail,' 'Lemon,' 'Lime,' 'Logee's,' 'Morocco,' 'Multimentha,' 'Murray Mitcham,' 'Native Wilmet,' 'Oma Streib,' 'Orange,' 'Reverchonii,' 'Reine Rouge,' 'Swiss,' 'Swiss Ricola'; *M. pulegium*: 'Upright'; *M. spicata*: 'Algerian Fruity,' 'Argentina,' 'Austrian,' 'Brundall,' 'Canaries,' 'Crispa,' 'Crispata,' 'Crispula,' 'Erospicata,' 'Guernsey,' 'Irish,' 'Kentucky Colonel,' 'Kerry,' 'Mexican,' 'Moroccan,' 'Nana,' 'Native Spearmint,' 'Newbourne,' 'Persian,' 'Pharaoh,' 'Rhodos,' 'Russian,' 'Small Dole,' 'Spanish Furry,' 'Spanish Pointed,' 'Tashkent,' 'Ukraine,' 'Variegata,' 'Verte Blanche,' 'Viridis,' and 'Westmeath,' and *M. ×villosa*: 'Cae Rhos Lligwy,' and 'Jack Green'; **misc. products:** The pleasant fragrance associated with mint oils has resulted in their widespread use in products ranging from deodorants to food additives. Six principal ketone compounds (controlled by the segregation of two genes) determine the specific fragrances of *Mentha* species and their hybrids, which range from sweet spearmint (carvone) to apple/pineapple (piperitenone oxide), and a musty odor (piperitone oxide). Interestingly, the chemical responsible for the pleasant aroma of spearmint (D-carvone) is simply the isomer of L-carvone, which lends the characteristic pungent odor to caraway (*Carum carvi*) and dill (*Anethum graveolens*), which both are members of the plant family Apiaceae. The leaves of water mint were smoked like tobacco in Europe during the Middle Ages. Oil derived from *M. pulegium* has been used as an insect repellant and peppermint oil is loathed by rats (Mammalia: Muridae: *Rattus*). Because of its readily discernable odor, peppermint oil has been used to detect leaks in pipes and by the United States Navy in shipboard gas-mask drills. Peppermint oil will inhibit mold when it is added to paste; **weeds:** *Mentha pulegium* is a weed of wet meadows in western North America. *Mentha ×piperita* also can become invasive; **nonindigenous species:** All of the OBL *Mentha* species and hybrids are nonindigenous to North America and were introduced (mainly from Europe) through cultivation.

Systematics: The systematic position of *Mentha* is well supported within several monophyletic groups: Lamiaceae subfamily Nepetoideae, tribe Mentheae (Figures 5.58 and 5.59). Phylogenetic analysis of combined cpDNA sequences for 15 *Mentha* species representing all currently recognized sections indicates that the genus also is monophyletic (Figure 5.61). The OBL North American species (and the hybrids) occur within two sections: *Mentha* (*M. aquatica*, *M. spicata* and

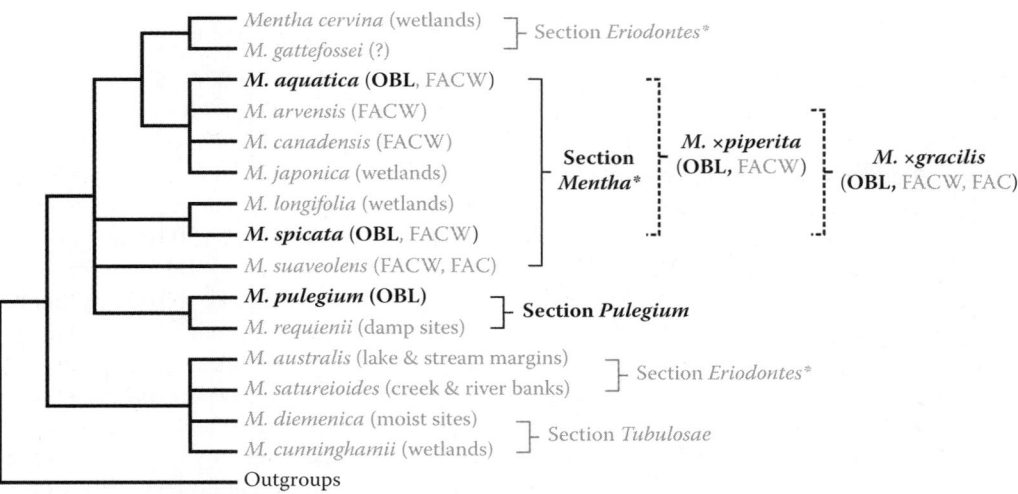

FIGURE 5.61 Analysis of combined cpDNA sequence data has substantiated the monophyly of *Mentha*, although two of the currently recognized sections (asterisked) may be polyphyletic. The distribution of the OBL taxa (bold) in two sections (each also containing facultative or nonwetland species) indicates that the habit likely has evolved independently in the genus. General habitat preferences are indicated for taxa not in North America. The hybrid parentages (dashed lines) have been confirmed by genetic analyses. (Adapted from Bunsawat, J. et al., *Syst. Bot.*, 29, 959–964, 2004.)

the hybrids) and *Pulegium* (*M. pulegium*), which differ by their base chromosome numbers (section *Pulegium*: $x = 10$; section *Mentha*: $x = 12$). North American populations of *M. pulegium* include diploids ($2n = 20$), triploids ($2n = 30$), and tetraploids ($2n = 40$); whereas, *M. spicata* mainly is tetraploid ($2n = 48$) and *M. aquatica* octaploid ($2n = 96$). If accurate, then the anomalous counts of $2n = 10$ reported for *M. pulegium* would indicate a lower base number ($x = 5$) for section *Pulegium*. A wide range of counts characterizes two of the hybrids, i.e., *M.* ×*gracilis* ($2n = 48, 54, 60, 72, 84, 96, 108, 120$) and *M.* ×*piperita* ($2n = 72, 96$). *Mentha* ×*villosa* reportedly is triploid ($2n = 36$). The hybrid nature of *M.* ×*gracilis* has been corroborated by artificial F_1 hybrids of *M. arvensis* ($2n = 96$) × *M. spicata* ($2n = 48$), which yielded a euploid series of $2n = 48, 60, 72, 84,$ and 96. Interspecific hybridization is a prominent feature of section *Mentha*, but is not known to occur elsewhere in the genus. Although hybrids routinely have been excluded from this book, three *Mentha* hybrids (*M.* ×*gracilis*, *M.* ×*piperita*, *M.* ×*villosa*) are included because of their economic importance. *Mentha spicata* itself presumably is also of hybrid origin (*M. longifolia* × *M. suaveolens*) and represents one parent of the natural hybrids *M.* ×*gracilis* (*M. arvensis* × *M. spicata*) and *M.* ×*piperita* (*M. aquatica* × *M. spicata*). Genetic analyses of AFLP and RAPD data are consistent with these presumed hybrid pedigrees and flavonoid data further corroborate the hybrid parentage of *M.* ×*piperita*. *Mentha* ×*villosa* is a putative hybrid between *M. spicata* × *M. suaveolens*. An allele inherited from *M. spicata* is known to cause spindle disruptions in its hybrids, which provides one explanation for their sterility. *Mentha canadensis* has been recognized as a distinct, endemic North American segregate of *M. arvensis* (FACW) and possibly is an amphidiploid hybrid involving *M. arvensis* and *M. longifolia*; however, additional genetic studies are necessary to confirm or refute this possibility.

Comments: *Mentha spicata* (Africa, Eurasia) is widespread in North America and *M.* ×*piperita* (Eurasia) is widespread except for central portions of North America; *M. aquatica* (Africa, Eurasia), *M.* ×*gracilis* (Europe), *M. pulegium* (Eurasia), and *M.* ×*villosa* (Africa, Europe) are disjunct in eastern and western North America.

References: Alkire & Simon, 1995; Anderson et al., 1996; Andrews, 1958; Bakerink et al., 1996; Baskin & Baskin, 1998; Bekker et al., 1999; Ben Fadhel & Boussaïd, 2004; Bonamonte et al., 2001; Brändel, 2006; Bunsawat et al., 2004; Burbott & Loomis, 1967; Cellot et al., 1998; Chambers & Hummer, 1994; Christy, 2004; Clayton & Orton, 2004; Darinot & Morand, 2001; da Silva Bezerra Guerra et al., 2012; De Carvalho & Da Fonseca, 2006; de Sousa et al., 2007; DiTomaso & Healy, 2003; Dorman et al., 2003; Douhan & Johnson, 2001; El-Zaher et al., 2005; Fassett, 1957; Földesi & Havas, 1979; Gobert et al., 2002; Grieve, 1980; Harley & Brighton, 1977; Hodkinson & Thompson, 1997; Hornok, 1992; Imai et al., 2001; Khanuja et al., 2000; Khayyal et al., 2001; Lahlou et al., 2002; Lawrence, 2006; Lenssen et al., 2000; Mimica-Dukić et al., 2003; Mohlenbrock, 1959a; Moore et al., 2006; Murray, 1960; Nedorostova et al., 2009;

Otto & Svensson, 1981; Panetta, 1985; Saleem et al., 2000; Samarth et al., 2006; Shasany et al., 2005a,b; Speichert & Speichert, 2004; Spirling & Daniels, 2001; Steyermark, 1931; Swanson & Nelson, 1942; Tucker & Fairbrothers, 1981, 1990; Voirin et al., 1999; Warren, 1993; Wilkinson & Beck, 1994; Yu et al., 2004.

7. Physostegia

False dragonhead, obedient plant; dracocéphale, physostégie

Etymology: from the Greek *physa stega* meaning "bladder chamber" in reference to the mature inflated calyx

Synonyms: *Dracocephalum* (in part)

Distribution: global: North America; **North America:** widespread

Diversity: global: 12 species; **North America:** 12 species

Indicators (USA): OBL: *Physostegia correllii*, *P. godfreyi*, *P. intermedia*, *P. ledinghamii*, *P. leptophylla*; **FACW:** *P. intermedia*, *P. parviflora*, *P. virginiana*

Habitat: brackish, freshwater; palustrine; **pH:** 5.0–7.7; **depth:** <1 m; **life-form(s):** emergent herb

Key morphology: stems (to 2.0 m) erect, unbranched, square; rosette leaves deciduous; cauline leaves (7.5–18.0 cm) opposite, sessile or the lower leaves petioled (3.0–6.5 cm); flowers bilabiate, subsessile or pedicelled (0.5–2.5 mm), in terminal racemes (1–20); corolla with flat upper lip and 3-lobed lower lip, white to deep reddish violet; stamens 4, didynamous, not exserted, the filaments coherent by interlocking hairs; style gynobasic, ovary 4-lobed; nutlets (to 4.2 mm) 4, triangular, smooth, or warty

Life history: duration: perennial (rhizomes, persistent rootstocks); **asexual reproduction:** rhizomes; **pollination:** insect; **sexual condition:** hermaphroditic; **fruit:** nutlets (common); **local dispersal:** nutlets (gravity), rhizomes; **long-distance dispersal:** nutlets (water), rhizome fragments (water)

Imperilment: (1) *Physostegia correllii* [G2]; S1 (LA); S2 (TX); (2) *P. godfreyi* [G3]; S3 (FL); (3) *P. intermedia* [G5]; SX (KY); S1 (MO); (4) *P. ledinghamii* [G4, G5]; S3 (AB); (5) *P. leptophylla* [G4]; S2 (GA, NC); S3 (VA); (6) *P. parviflora* [G4]; S1 (NE, UT); S2 (WY); S3 (BC); (7) *P. virginiana* [G5]; S1 (NH); S2 (NC, VT); S3 (QC)

Ecology: general: Most *Physostegia* species grow in wet sites (tolerating inundation up to 30 cm) and are at least capable of growing facultatively as wetland plants. They occur from sea level to elevations of 2300 m and grow in sites that span a wide range of acidity. The flowers are visited by more than 30 insect species as well as by hummingbirds (Aves: Trochilidae). Although they are self-compatible and capable of self-pollination, the species are protandrous and tend to be outcrossed by insect (Insecta) pollinators, mainly bumblebees (Hymenoptera: Apidae: *Bombus*). Yet, because of vegetative reproduction and extensive clonal growth, selfing due to pollination among different flowers of the same clone (geitonogamy) probably occurs frequently. Other bees (Hymenoptera: Apidae: *Apis*, *Anthophora*; Megachilidae: *Megachile*), soldier beetles (Coleoptera: Cantharidae: *Chauliognathus*), and wasps (Hymenoptera: Vespidae: *Polistes*, *Vespula*) function as pollinators when bumblebees are scarce. The seed ecology remains largely unstudied; however, low to good germination has been obtained for unstratified seeds in some species. The perennating buds are often not produced until after the flowering period has ceased and require a dormant period before bolting is induced. Two species (*Physostegia parviflora*, *P. virginiana*) were ranked as OBL in the 1996 list but have since been reclassified as FACW. They have been retained here because their treatments had been completed prior to the release of the 2013 list.

Physostegia correllii (**Lundell**) **Shinners** is a rare inhabitant of canals, creekbeds, ditches, marshes, prairies, and the margins of rivers and streams. Substrates are silty clay loams. The plants occur in disturbed, unstable sites, which include human-altered landscapes that contain large numbers of invasive species. The flowers are day-neutral and self-compatible but are pollinated primarily by bumblebees (Insecta: Hymenoptera: Apidae: *Bombus*). However, high seed set has been obtained when flowers are selfed artificially. Fairly good seed germination (30%) has been obtained after 6 months using air-dried seeds (dried at room temperature) that have been planted in potting soil. Enhanced germination can be achieved by soaking seeds for 24h in gibberellic acid (500ppm) and transferring them to paper that has been dampened with distilled water. The plants reproduce vegetatively by rhizomes (to 50cm), and populations can represent clones of one or few individuals. They are propagated easily from cuttings or by divisions of the rhizome. There are only 3–4 known extant sites for *P. correllii* in The United States. Additional ecological information including ecological associates, site characteristics, population genetic data and information on natural rates of sexual reproduction in this rare species would provide useful information for conservation programs and are encouraged. **Reported associates:** *Alternanthera philoxeroides, Arundo donax, Colocasia esculenta, Cyperus erythrorhizos, Diodia, Echinochloa colona, E. walteri, Eleocharis quadrangulata, Hydrocotyle, Leptochloa uninervia, Ludwigia, Mentha, Panicum, Phragmites australis, Persicaria, Pontederia cordata, Populus, Ricinus communis, Sagittaria, Scirpus, Sorghum, Tamarisk.*

Physostegia godfreyi **P.D. Cantino** grows in bogs, ditches, flatwoods, roadsides, savannas, seeps, and thickets, sometimes occurring in shallow standing water. The substrates are described as Plummer (grossarenic paleaqqults) or sand. Information regarding most aspects of the species biology and ecology of this plant is scarce. The nutlets differ from the smooth fruits of other *Physostegia* species by their warty surface. The plants reproduce vegetatively by the production of rhizomes (to 10cm). **Reported associates:** *Aletris, Aristida, Asclepias viridula, Balduina uniflora, Calopogon, Centella asiatica, Chaptalia tomentosa, Ctenium aromaticum, Diodia virginiana, Eriocaulon decangulare, Helianthus heterophyllus, Hypericum, Lachnanthes caroliniana, Lachnocaulon anceps, Linum westii, Ludwigia virgata, Muhlenbergia capillaris, Pinus palustris, Polygala chapmanii, P. cruciata, Rhexia alifanus, R. mariana, Rhynchospora colorata, R. compressa, R. filifolia, R. plumosa, Sabatia bartramii, S. stellaris, Sarracenia, Schoenolirion elliottii, Scleria hirtella,* *Scutellaria integrifolia, Serenoa repens, Stillingia aquatica, Triantha racemosa, Verbesina chapmanii, Xyris ambigua, X. baldwiniana, X. stricta.*

Physostegia intermedia (**Nutt.**) **Engelm. & A. Gray** occurs in depressions, ditches, intermittent ponds, marshes, meadows, oxbows, prairies, river, and stream beds, swamps, and along the margins of rivers and streams. The substrates are alkaline (pH: 7.3–7.7) and include clay, clay loam, loam, and sand. The plants tolerate full light to partial shade. Little information exists on the ecology of this species. Experimental pollinations indicate that the species is self-compatible and can produce high seed set when selfed. Details on natural pollination or other aspects of the floral biology are lacking. **Reported associates:** *Alternanthera philoxeroides, Amsonia tabernaemontana, Andropogon gerardii, Argythamnia humilis, Arnoglossum plantagineum, Asclepias perennis, Baptisia sphaerocarpa, Bidens aristosa, Bifora americana, Boltonia diffusa, Bouteloua curtipendula, Calamovilfa arcuata, Camassia scilloides, Cardamine bulbosa, Carex crinita, C. crus-corvi, C. lupulina, C. microdonta, Castilleja coccinea, Ceanothus herbaceus, Chasmanthium laxum, Cicuta mexicana, Colocasia esculenta, Cooperia drummondii, Coreopsis tripteris, Cyperus virens, Dichanthelium acuminatum, Diodia virginiana, Diospyros virginiana, Echinochloa colona, E. crus-galli, E. walteri, Echinodorus berteroi, Eleocharis palustris, E. quadrangulata, Elymus svensonii, Eryngium leavenworthii, Eupatorium coelestinum, E. serotinum, Euphorbia corollata, Fraxinus caroliniana, Gentiana alba, Hibiscus moscheutos, Houstonia purpurea, Hydrocotyle, Hymenocallis liriosme, H. occidentalis, Hypericum hypericoides, Iris virginica, Iva annua, Juncus coriaceus, Justicia americana, Liatris squarrosa, Liquidambar styraciflua, Ludwigia alternifolia, L. decurrens, L. peploides, Lycopus virginicus, Muhlenbergia cuspidata, Oenothera linifolia, Panicum dichotomiflorum, P. rigidulum, P. virgatum, Parthenium integrifolium, Paspalum floridanum, P. repens, Persicaria hydropiperoides, P. lapathifolia, Phragmites australis, Phyla lanceolata, Pluchea, Quercus lyrata, Q. nigra, Q. phellos, Q. stellata, Rhexia mariana, Rhynchospora glomerata, Rumex altissimus, Sagittaria graminea, S. lancifolia, Salix nigra, Schizachyrium scoparium, Schoenoplectus americanus, S. tabernaemontani, Sesbania drummondii, Silphium laciniatum, S. terebinthinaceum, Sorghastrum nutans, Spartina patens, Spiranthes magnicamporum, Sporobolus asper, S. clandestinus, Steinchisma hians, Stipa leucotricha, Symphyotrichum dumosum, S. subulatum, Teucrium canadense, Trachelospermum difforme, Tradescantia ohiensis, Tragia urticifolia, Trepocarpus aethusae, Trichostomum setaceum, Tridens flavus, Tripsacum dactyloides, Ulmus alata, U. americana, Valerianella radiata, Vernonia baldwinii.*

Physostegia ledinghamii (**Boivin**) **Cantino** inhabits ditches, flood plains, marshes, meadows, river flats, sloughs, thickets, woodlands, and the margins of lakes and streams. The plants typically occur within intermittently inundated sites that are situated just above the frequently flooded zone. Its substrates include mud and silt. The flowering period extends from late July through early August. The plants perennate and reproduce vegetatively by rhizomes, which can result in the development of clonal patches. **Reported associates:** *Anemone canadensis, Calamagrostis canadensis, Carex utriculata, Cicuta maculata, Cirsium, Eleocharis palustris, Geum aleppicum, Juncus filiformis, Mentha arvensis, Mimulus ringens, Persicaria amphibia, Poa palustris, Ranunculus aquatilis, R. macounii, Sagittaria cuneata, Veronica scutellata.*

Physostegia leptophylla **Small** occurs in ditches, floodplains, hammocks, marshes, river bottoms, seeps, sloughs, swamps, and along canal, river, or stream margins at elevations to at least 12 m. The plants can occur in full sun and also will tolerate shade. They often grow in shallow water and have a fairly high tolerance to inundation as indicated by their growing popularity as an aquarium plant (see below). Their habitats include fresh, non-tidal to brackish (tidal) sites with alluvial, muck, mud, peat, sandy clay, sandy loam, sandy peat, Surrency (arenic umbric paleaquults), or Winder (typic glossaqualfs) substrates. The reproductive biology of this species remains largely unstudied; however, artificial pollinations indicate that it is self-compatible. The plants are highly rhizomatous/stoloniferous, which can result in the development of large, clonal colonies. Populations of *P. leptophylla* are threatened by habitat drainage. More detailed ecological information is needed for this relatively rare species. **Reported associates:** *Acer rubrum, Asclepias lanceolata, Carex canescens, C. hyalinolepis, Centella asiatica, Cephalanthus occidentalis, Cladium mariscoides, Clethra alnifolia, Coreopsis falcata, Cyperus, Eleocharis fallax, E. rostellata, Eriocaulon decangulare, Eryngium aquaticum, Fraxinus, Itea virginica, Liquidambar styraciflua, Lyonia ligustrina, Myrica cerifera, Nyssa biflora, Orontium aquaticum, Osmunda regalis, Panicum virgatum, Peltandra virginica, Pogonia ophioglossoides, Pontederia cordata, Quercus laurifolia, Rhynchospora glomerata, Rosa palustris, Sabal palmetto, Salix caroliniana, Saururus cernuus, Schoenoplectus americanus, Taxodium distichum, Toxicodendron radicans, Triadenum walteri, Typha angustifolia, T. latifolia, Ulmus, Woodwardia areolata, Zizania aquatica.*

Physostegia parviflora **Nutt. ex A. Gray** is found in depressions, ditches, flats, marshes, meadows, mudflats, prairies, sloughs, swales, swamps, thickets, and along the shores of lakes, rivers, and streams, at elevations of up to 3121 m. Exposures include open to shaded sites. Reported substrates include gravel, gravelly clay, loam, muck, mud, sand, silt, and silt loam. The floral biology of this species has not been studied. The seeds require 1–2 months of cool (1°C–3°C), moist stratification to break their dormancy. The plants reproduce vegetatively by rhizomes (up to 12 cm). **Reported associates:** *Acer negundo, Acorus calamus, Allium schoenoprasum, Alnus rugosa, Apocynum cannabinum, Artemisia frigida, A. lindleyana, Calamagrostis stricta, Carex lanuginosa, C. rossii, C. stipata, C. vesicaria, Cicuta douglasii, Cirsium arvense, C. flodmanii, Coreopsis tinctoria, Cornus sericea, Deschampsia cespitosa, Dichanthelium acuminatum, Downingia elegans, Eleocharis acicularis, E. palustris,*

Epilobium watsonii, Equisetum arvense, Fraxinus pennsylvanica, Gaillardia aristata, Galium boreale, Helenium autumnale, Holcus lanatus, Juncus nevadensis, Ludwigia palustris, Lycopus asper, Lysimachia ciliata, L. hybrida, Medicago lupulina, Mentha arvensis, Menyanthes trifoliata, Mimulus guttatus, Persicaria amphibia, Petasites sagittatus, Phalaris arundinacea, Plantago major, Poa compressa, P. triflora, Populus tremuloides, P. trichocarpa, Prunella vulgaris, Rhamnus alnifolia, Rosa, Sagittaria cuneata, Salix bebbiana, S. exigua, S. luteosericea, S. maccalliana, S. prolixa, Schoenoplectus tabernaemontani, Senecio, Sium suave, Sparganium emersum, S. eurycarpum, Spiraea alba, Stachys palustris, Symphyotrichum chilense, Tanacetum vulgare, Toxicodendron radicans, Triglochin, Typha latifolia, Urtica dioica, Vicia cracca.

Physostegia virginiana **(L.) Benth.** grows in carrs, depressions, estuaries, fens, floodplains, glades, marshes, mudflats, prairies, savannas, seeps, swales, swamps, and along the margins of lakes, ponds, rivers, and streams, sometimes standing in shallow water. The plants are fairly broadly adapted to acidity (pH: 5.0–7.0), substrate composition (clay, gravel, limestone, rock, sand, sandy loam, and silt), and exposures (full sun to light shade). Typically they occur in freshwater and also found where conditions are somewhat brackish. Although categorized originally as an OBL species in some regions, *P. virginiana* has been reclassified as a FACW, FAC indicator, which more accurately reflects its ability to grow among species that are more characteristic of drier conditions. However, it thrives in wet habitats. This is a long-day plant, which requires a 16 h photoperiod to induce flowering. There is a high (but variable) incidence of anther sterility in populations, which indicates that the species is gynodioecious. The pollen-sterile plants produce significantly smaller flowers, exhibit more rapid stigma receptivity following anthesis, and produce fewer nutlets than do the hermaphrodites. The hermaphroditic flowers are self-compatible and exhibit high seed set when they are self-pollinated artificially. However, the flowers probably are outcrossed predominantly by bumblebees (Insecta: Hymenoptera: Apidae: *Bombus*), which are frequent insect visitors. A population genetic survey of this species would be informative and would be necessary to clarify the extent of outcrossing in natural populations. Seed production typically is fairly high, but there are no reports of seed bank formation. Six-month-old, unstratified seeds germinate variably (2%–60%) at 25°C. Good germination has been reported at 21°C–24°C for seeds that are damp stratified for 3 months at 4°C. The plants reproduce vegetatively by rhizomes, which can extend to 65 cm. The unusual common name of "obedient plant" refers to the tendency of the flowers to remain displaced when pushed to one side of the raceme, a phenomenon known as "catalepsy." Repeated pedicel movements are facilitated by a ring of specialized parenchyma cells at the pedicel base. The following list of associated species attempts to reflect natural occurrences rather than associations with planted cultivated material, which can include many terrestrial garden species. **Reported associates:** *Acalypha rhomboidea, Acer negundo, A. rubrum, A.*

saccharinum, *Achillea millefolium*, *Acorus calamus*, *Agastache nepetoides*, *Agrimonia parviflora*, *Agropyron repens*, *Agrostis*, *Alisma*, *Allium cernuum*, *Alnus rugosa*, *Ambrosia artemisiifolia*, *A. trifida*, *Amorpha canescens*, *Amphicarpa bracteata*, *Andropogon capillipes*, *A. gerardii*, *A. glaucopsis*, *A. virginicus*, *Antennaria neglecta*, *A. plantaginifolia*, *Apios americana*, *Apocynum sibiricum*, *Aquilegia canadensis*, *Arisaema dracontium*, *A. triphyllum*, *Arnoglossum ovatum*, *Artemisia serrata*, *Asclepias hirtella*, *A. incarnata*, *A. verticillata*, *A. viridiflora*, *Baptisia australis*, *B. leucantha*, *Betula nigra*, *Bidens cernuus*, *B. eatonii*, *B. hyperboreus*, *B. vulgatus*, *Bigelowia nuttallii*, *Boehmeria cylindrica*, *Bolboschoenus fluviatilis*, *Bouteloua curtipendula*, *Brickellia eupatorioides*, *Buchnera americana*, *Calamagrostis canadensis*, *Caltha palustris*, *Calystegia sepium*, *Campanula aparinoides*, *Carex crawei*, *C. eburnea*, *C. frankii*, *C. garberi*, *C. hyalinolepis*, *C. juniperorum*, *C. lacustris*, *C. leptalea*, *C. lurida*, *C. muskingumensis*, *C. projecta*, *C. stricta*, *C. torta*, *Carya ovata*, *Cassia fasciculata*, *Castilleja coccinea*, *Ceanothus americanus*, *Celtis occidentalis*, *Cephalanthus occidentalis*, *Chaetopappa asteroides*, *Chelone glabra*, *Cicuta maculata*, *Circaea lutetiana*, *Cirsium arvense*, *C. discolor*, *C. muticum*, *C. vulgare*, *Cinna arundinacea*, *Clematis ochroleuca*, *Clinopodium arkansanum*, *Comandra umbellata*, *Coreopsis major*, *C. tripteris*, *Cornus drummondii*, *C. obliqua*, *C. sericea*, *Coronilla varia*, *Crataegus*, *Croton monanthogynus*, *Cryptotaenia canadensis*, *Cuscuta gronovii*, *Cyperus odoratus*, *Cypripedium candidum*, *Dactylis glomerata*, *Dalea candida*, *Danthonia spicata*, *Desmodium glutinosum*, *D. rotundifolium*, *Dichanthelium depauperatum*, *D. oligosanthes*, *Echinacea pallida*, *E. sanguinea*, *E. simulata*, *Echinochloa*, *Eleocharis*, *Elymus canadensis*, *E. virginicus*, *Epilobium ciliatum*, *Equisetum arvense*, *Erigeron philadelphicus*, *Eriocaulon parkeri*, *Eryngium yuccifolium*, *Euonymus atropurpureus*, *Eupatorium altissimum*, *E. leucolepis*, *E. maculatum*, *E. perfoliatum*, *Euphorbia corollata*, *Eurybia hemispherica*, *Euthamia graminifolia*, *Evolvulus sericeus*, *Fallopia convolvulus*, *Fraxinus americana*, *F. nigra*, *F. pennsylvanica*, *Galactia regularis*, *Galium aparine*, *G. brevipes*, *G. circaezans*, *G. obtusum*, *G. trifidum*, *Gentiana andrewsii*, *Gentianella quinquefolia*, *Gentianopsis procera*, *Glyceria striata*, *Gratiola neglecta*, *Gaura biennis*, *G. filipes*, *Geum*, *Hasteola suaveolens*, *Hedyotis nigricans*, *Helenium autumnale*, *Helianthus divaricatus*, *H. grosseserratus*, *H. hirsutus*, *H. mollis*, *H. tuberosus*, *Helianthus* ×*verticillatus*, *Heliopsis helianthoides*, *Heliotropium tenellum*, *Hexalectris spicata*, *Hibiscus laevis*, *Hieracium*, *Houstonia canadensis*, *Hydrophyllum*, *Hymenocallis crassifolia*, *Hypericum dolabriforme*, *H. prolificum*, *H. sphaerocarpum*, *Ilex longipes*, *Impatiens capensis*, *Ionactis linariifolius*, *Iris versicolor*, *Isanthus brachiatus*, *Isoetes tuckermanii*, *Iva angustifolia*, *Juncus balticus*, *J. filipendulus*, *J. tenuis*, *Koeleria macrantha*, *Lactuca canadensis*, *L. floridana*, *Laportea canadensis*, *Lathyrus japonicus*, *L. latifolius*, *Leersia lenticularis*, *L. oryzoides*, *L. virginica*, *Lepidium virginicum*, *Lepraria*, *Lespedeza capitata*, *L. hirta*, *L. virginica*, *Liatris acidota*, *L.*

aspera, *L. cylindracea*, *L. punctata*, *L. pycnostachya*, *L. spicata*, *L. squarrosa*, *Linaria vulgaris*, *Liquidambar styraciflua*, *Lithospermum canescens*, *Lobelia cardinalis*, *L. siphilitica*, *L. spicata*, *Lolium perenne*, *Lonicera prolifera*, *Ludwigia microcarpa*, *Lychnis alba*, *Lycopus americanus*, *L. uniflorus*, *Lysimachia ciliata*, *L. nummularia*, *L. quadriflora*, *Lythrum salicaria*, *Maianthemum stellatum*, *Manfreda virginica*, *Marshallia caespitosa*, *M. mohrii*, *Matelea obliqua*, *Matteuccia struthiopteris*, *Mecardonia acuminata*, *Menispermum canadense*, *Mentha arvensis*, *Menyanthes trifoliata*, *Mitreola petiolata*, *Mnium punctatum*, *Muhlenbergia capillaris*, *M. frondosa*, *M. mexicana*, *Myosoton aquaticum*, *Napaea dioica*, *Neptunia lutea*, *Nyssa aquatica*, *N. sylvatica*, *Oligoneuron album*, *Onoclea sensibilis*, *Onosmodium virginianum*, *Ophioglossum engelmannii*, *Opuntia humifusa*, *Osmunda*, *Oxypolis rigidior*, *Packera aurea*, *Panicum virgatum*, *Parthenocissus*, *Paspalum floridanum*, *Pedicularis lanceolata*, *Persicaria amphibia*, *P. hydropiperoides*, *P. lapathifolia*, *P. maculosa*, *P. punctata*, *Phalaris arundinacea*, *Phlox divaricata*, *P. glaberrima*, *P. nivalis*, *P. pilosa*, *Phyla lanceolata*, *Physalis longifolia*, *Pilea pumila*, *Pinus banksiana*, *Plantago major*, *Poa compressa*, *P. palustris*, *P. pratensis*, *Pogonia ophioglossoides*, *Polygala senega*, *Populus deltoides*, *Porteranthus stipulatus*, *Potentilla anserina*, *P. simplex*, *Prenanthes alba*, *Prunus serotina*, *Ptilimnium costatum*, *Pycnanthemum virginianum*, *Quercus alba*, *Q. bicolor*, *Q. palustris*, *Q. rubra*, *Q. stellata*, *Ranunculus abortivus*, *R. fascicularis*, *R. septentrionalis*, *Ratibida pinnata*, *Rhexia lutea*, *R. virginica*, *Rhynchospora caduca*, *R. colorata*, *R. divergens*, *R. elliottii*, *R. glomerata*, *R. gracilenta*, *R. microcarpa*, *R. plumosa*, *R. thornei*, *Rorippa palustris*, *Rosa arkansana*, *R. carolina*, *R. setigera*, *Rubus flagellaris*, *Rudbeckia fulgida*, *R. hirta*, *R. laciniata*, *R. subtomentosa*, *R. texana*, *Ruellia humilis*, *Sabatia angularis*, *S. campanulata*, *Sagittaria latifolia*, *Salix fragilis*, *S. humilis*, *S. interior*, *S. nigra*, *Sambucus nigra*, *Samolus valerandi*, *Saururus cernuus*, *Schizachyrium scoparium*, *S. tenerum*, *Scirpus atrovirens*, *S. pendulus*, *Scrophularia marilandica*, *Scutellaria galericulata*, *S. lateriflora*, *S. parvula*, *Senna marilandica*, *Sicyos angulatus*, *Silphium integrifolium*, *S. perfoliatum*, *S. terebinthinaceum*, *S. trifoliatum*, *Sisymbrium altissimum*, *Sisyrinchium albidum*, *Sium suave*, *Smilax bonanox*, *Solanum dulcamara*, *Solidago canadensis*, *S. gigantea*, *S. nemoralis*, *S. ohioensis*, *S. ptarmicoides*, *S. riddellii*, *S. rigida*, *S. sempervirens*, *S. ulmifolia*, *Sorghastrum nutans*, *Sparganium*, *Spartina pectinata*, *S. spartinae*, *Spiraea alba*, *Spiranthes magnicamporum*, *Sporobolus asper*, *S. clandestinus*, *S. compositus*, *S. coromandelianus*, *S. heterolepis*, *S. neglectus*, *S. silveanus*, *S. vaginiflorus*, *Stachys tenuifolia*, *Symphoricarpos orbiculatus*, *Symphyotrichum concolor*, *S. cordifolium*, *S. ericoides*, *S. georgianum*, *S. laeve*, *S. lateriflorum*, *S. oblongifolium*, *S. ontarionis*, *S. patens*, *S. pratense*, *S. shortii*, *Taenidia integerrima*, *Talinum teretifolium*, *Taxodium distichum*, *Teucrium canadense*, *Toxicodendron radicans*, *Tradescantia ohiensis*, *Trantha glutinosa*, *Tripsacum dactyloides*, *Typha latifolia*, *Ulmus americana*, *U. rubra*, *Urtica dioica*, *Vaccinium arboreum*, *Verbena hastata*, *V. simplex*,

Verbesina helianthoides, V. virginica, Vernonia fasciculata, Veronicastrum virginicum, Viburnum rafinesqueanum, Viola sororia, Vitis aestivalis, V. riparia, Xanthium, Zizania aquatica, Zizia aptera.

Use by wildlife: *Physostegia* is a host genus for leaf beetles (Insecta: Coleoptera: Chrysomelidae: *Phyllobrotica physostegia*). Several species (e.g., *P. leptophylla, P. virginiana*) are planted to attract insects such as bumblebees (Hymenoptera: Apidae: *Bombus*) and butterflies (Lepidoptera). The nectar from several *Physostegia* species is "robbed" by halictid bees (Hymenoptera: Halictidae), which are too small to effect pollination, and by carpenter bees (Hymenoptera: Apidae: *Xylocopa virginica*), which reach the nectar by chewing into the corolla base. Such holes provide nectar access to small flies (Insecta: Diptera: Syrphidae), halictid bees, and honeybees (Hymenoptera: Apidae: *Apis mellifera*). *Physostegia virginiana* is a host plant of the stinkbug (Insecta: Hemiptera: Pentatomidae: *Cosmopepla bimaculata*), the black vine weevil (Insecta: Curculionidae: *Otiorhynchus sulcatus*), and larval dull-barred moths (Insecta: Tortricidae: *Endothenia hebesana*). The floral nectar is taken by butterflies (Insecta: Nymphalidae: *Danaus plexippus*; Pieridae: *Colias philodice*), long-tongued bees: (Apidae: *Bombus pensylvanica, B. vagans*; Anthophoridae: *Melissodes agilis, M. bimaculata, M. rustica*; Megachilidae: *Megachile brevis, M. latimanus*), and ruby-throated hummingbirds (Aves: Trochilidae: *Archilochus colubris*). *Physostegia parviflora* is a host of several phytopathogens including a powdery mildew fungus (Ascomycota: *Neoerysiphe galeopsidis*), alfalfa mosaic virus (Group IV ((+) ssRNA): Bromoviridae: *Alfamovirus*), and tobacco ringspot virus (Group IV ((+)ssRNA): Secoviridae: *Nepovirus*). Although some *Physostegia* species (e.g., *P. intermedia* and *P. virginiana*) are said to be resistant to grazing by deer (Mammalia: Cervidae: *Odocoileus*), there are reports that they are grazed. *Physostegia virginiana* (and probably other congeners) contains iridoid glycosides, which are substances known to deter feeding by generalist herbivores. The plants have low susceptibility to infection by the northern root-knot nematode (Nematoda: Heteroderidae: *Meloidogyne hapla*).

Economic importance: food: *Physostegia* is not edible; **medicinal:** The Meskwaki people used a leaf infusion of *Physostegia parviflora* as a cold remedy. *Phystostegia virginiana* was ineffective when tested for antibacterial activity; **cultivation:** *Physostegia* was cultivated in Europe as long ago as 1674. The rare *Physostegia correllii* is cultivated using material that has been propagated vegetatively from rhizome cuttings. *Physostegia intermedia* and *P. parviflora* are sold as native wildflowers. *Physostegia virginiana* is grown prolifically as an ornamental. In addition to native plants sold for natural wildflower gardens, there are numerous cultivars in circulation including 'Alba,' 'Bouquet Rose,' 'Crown of Snow,' 'Eyeful Tower,' 'Galadriel,' 'Grandiflora,' 'Ladhams' Variety,' 'Miss Manners,' 'Olympic Gold,' 'Pink Bouquet,' 'Red Beauty,' 'Rosea,' 'Rose Crown,' 'Rose Queen,' 'Snow Queen,' 'Summer Glow,' 'Summer Snow,' 'Summer Spire,' 'Variegata,' and 'Vivid.' Several of these cultivars have earned the Award of Garden Merit by the Royal Horticultural

Society. The vase life of cut *Physostegia* flowers can be extended by addition of a 2% sugar solution or by first dipping them for 1 h in a solution of sodium silver thiosulfate. *Physostegia virginiana* is recommended for use in gardens as a native alternative to *Lythrum salicaria* (Lythraceae), a noxious wetland weed. *Physostegia leptophylla* (distributed as "Florida Crypt") is used in aquariums and as an ornamental plant for shallow pond or water garden margins; **misc. products:** Seeds of *Physostegia virginiana* contain laballenic acid, which has a potential use as an industrial oil; **weeds:** Cultivars of *Physostegia virginiana* can become aggressive when planted in moist, open substrates; **nonindigenous species:** *Physostegia virginiana* is naturalized near Turin, Italy. In North America, naturalized populations are reported from Delaware, New Brunswick, and Nova Scotia.

Systematics: The common name "false dragonhead" alludes to the generic name *Dracocephalum* (literally "dragonhead"), which once was applied to this group, but has since been conserved nomenclaturally for an Old World genus of Lamiaceae. *Physostegia* is most similar morphologically to *Brazoria*, to which it is closely allied as the sister group to a *Brazoria* + *Warnockia* clade (Figure 5.60). Preliminary cladistic analyses of the genus using morphological data have yielded variable results, but have consistently indicated a close relationship between *P. godfreyi* and *P. purpurea*, and possibly, between *P. correllii* and *P. parviflora* as well. The application of DNA sequence data to phylogenetic studies of *Physostegia* might provide a better indication of interspecific relationships as well as a means of confirming the probable monophyly of the genus. The base chromosome number of *Physostegia* is $x = 19$, which indicates a possible amphidiploid origin from $x = 9$, 10 taxa. *Physostegia correllii, P. godfreyi, P. intermedia*, and *P. parviflora* are diploid ($2n = 38$); *P. ledinghamii* and *P. leptophylla* ($2n = 76$) are tetraploid; *P. virginiana* ($2n = 38, 76$) has both diploid and tetraploid counts reported. *Physostegia ledinghamii* (OBL) is believed to be a tetraploid hybrid derivative of *P. parviflora* (FACW) and *P. virginiana* (FACW), a possibility that should be investigated further using genetic data, especially given the wetland indictor status of the two putative parental species (although both had been classified as OBL in the 1996 list). Otherwise, interspecific hybridization in *Physostegia* is rare, and is believed to be prevented primarily by ecological and temporal isolating mechanisms. Two intergrading subspecies of *P. virginica* are recognized mainly on the basis of whether their perennating organs develop into primary and secondary, branching rhizomes (subspecies *virginica*), or into a nonrhizomatous, persistent rootstock (subspecies *praemorsa*). *Physostegia* species rarely coexist at a site. A few species are genetically crossincompatible. Additional molecular and genetic data should be collected to further investigate interspecific relationships and to better elucidate breeding systems within this genus.

Comments: *Physostegia ledinghamii* is the only OBL *Physostegia* species that is fairly widespread, and ranges from the northern central United States through central Canada (also widespread are the FACW *P. parviflora* in northwestern North America and *P. virginiana* in eastern North America).

The remaining species are more restricted, with *P. leptophylla* in the southeastern United States, *P. intermedia* narrowly distributed in the south central United States, *P. correllii* in Louisiana and Texas (but extending into Mexico), and *P. godfreyi*, which is endemic to Florida.

References: Beattie et al., 1989; Cantino, 1979, 1981, 1982; Chen et al., 2014; Coulter, 1882; Hagemann et al., 1967; Jones, 2005; Lockhart et al., 2002; Lowden, 1969; Mackey et al., 1987; Müller, 1933; Nass & Rimpler, 1996; Owens & Cole, 2003; Patchell, 2013; Phillippe et al., 2003; Sanders et al., 1945; Smeins & Diamond, 1983; Speichert & Speichert, 2004; Vujnovic et al., 2005.

8. *Pogogyne*

Beardstyle, mesa mint

Etymology: from the Greek *pogon gynê*, meaning "bearded style"

Synonyms: *Hedeomoides*

Distribution: global: Mexico, North America; **North America:** western

Diversity: global: 7 species; **North America:** 7 species

Indicators (USA): OBL: *Pogogyne abramsii, P. douglasii, P. floribunda, P. nudiuscula, P. ziziphoroides*

Habitat: freshwater; palustrine; **pH:** 5.6–7.8; **depth:** <1 m; **life-form(s):** emergent herb

Key morphology: stems (to 50 cm) square, erect or spreading, herbage aromatic; leaves (to 2 cm) opposite, obovate to spatulate; flowers bilabiate, in axial or terminal, head-like or spike-like verticels, or solitary, axillary, often enveloped by ciliate bracts; corolla (to 2 cm) tubular, lavender to purple, the upper lip entire, the lower lip 3-lobed, spreading, sometimes yellow-mottled; stamens 4, didynamous, all fertile or the upper (posterior) pair sterile; styles pubescent; nutlets (to 2.5 mm) 4, hairy at apex, persistent in calyx

Life history: duration: annual (fruit/seeds); **asexual reproduction:** none; **pollination:** insect; **sexual condition:** hermaphroditic; **fruit:** nutlets (common); **local dispersal:** nutlets (gravity, water, wind); **long-distance dispersal:** nutlets (birds, rabbits, water)

Imperilment: (1) *Pogogyne abramsii* [G2]; S2 (CA); (2) *P. douglasii* [G4]; (3) *P. floribunda* [G3]; S1 (ID, OR); S3 (CA); (4) *P. nudiuscula* [G1]; S1 (CA); (5) *P. ziziphoroides* [G5]

Ecology: general: *Pogogyne* is characteristic of vernal wetlands of the western coastal United States, with five species designated as OBL and two as FACW plants. No other Lamiaceae are found in vernal pool habitats. All of the species are annuals and, as in many other vernal pool species, they lack mechanisms for long-distance dispersal of seeds, a factor that has been attributed as an adaptation to the isolated, scattered distribution of suitable habitats. Other than their ability to tolerate inundation, the plants do not appear otherwise to be well-adapted to aquatic conditions. Their flowering and fruiting periods occur essentially under terrestrial conditions, only after the waters either have dried up or have receded from the habitat.

***Pogogyne abramsii* J.T. Howell** inhabits depressions, ditches, flats, streambeds, and vernal pools on coastal mesas at elevations of 90–200 m. The substrates usually are acidic (pH: 5.6–5.7) and are described as gravelly loam and sands of the Redding soil type, which are underlain by fine clay or by silica-cemented hardpan. The waters can be more alkaline (e.g., pH = 6.7–8.8; conductivity: 46–286 μmhos/cm). This species has poor drought tolerance, can withstand up to 4 months of inundation, but its vigor declines under conditions of prolonged inundation. Inundation results in reduced branching and elongation of the internodes. The plants thrive on moist soil, but benefit from inundation-induced mortality, which reduces both inter- and intraspecific plant densities. Growth is better in larger pools. Although this species is virtually confined to vernal pools, the plants occasionally have been observed growing in disturbed, UPL sites. Flowering occurs in the spring (April to May). The flowers are self-compatible but natural selfing has not been observed in the absence of pollinators. In experimental manipulations, seed set is much higher when the flowers are cross-pollinated. The principal pollinators are bees (Hymenoptera: Anthophoridae: *Exomalopsis nitens, E. torticornis*; Apidae: *Apis mellifera*). Pollinator visitation to plants increases proportionally with flower density; whereas, visits to individual flowers decrease. Plants in their natural habitats are larger, flower earlier, have higher flower densities, and yield higher levels of seed set than plants growing in artificially created vernal pools. However, plants in created sites experience relatively higher pollinator visits per flower and receive ample pollination for sufficient seed set. Fruiting occurs in the late spring (May to June) and the nutlets can be dispersed from July to September. A single plant produces 6–30 nutlets on average. *Pogogyne abramsii* is a winter annual, with seed germination initiating soon (usually within a week) after the first autumn rains commence (October to December). The inundated plants grow slowly until the spring when flowering begins (late April to early May). Seedling viability declines at sites that remain inundated for long periods of time. Greenhouse experiments using seed bank material indicate that a late-season germination pulse can occur in spring (April) due to seeds that float free of the soil. However, the seed bank often is likely to be depleted rapidly if sufficient moisture for germination is present. The seeds can retain up to 74% viability after 1 year of storage. About 80% of seeds germinate during the first moisture cycle; however, at least some seeds can remain viable and will germinate over four cycles. Laboratory germination of field-collected seeds has been successful under a 25°C/10°C temperature regime. Generally, germination rates are variable and are determined by seed size, with higher rates observed for larger seeds (>0.2 mg). Seed dormancy is physical and is determined by the seed coat, with higher dormancy observed in relatively more fecund plants. Exposure to continuous moisture for 48 h will break dormancy. There is no mechanism for long-distance seed dispersal. The nutlets are not shed from the calyx, even throughout the dry summer period, and often remain attached to the dried shoots until the stems eventually deteriorate and collapse. Local dispersal in water can occur by floating seeds, which may be transported among adjacent pools during periods of high water.

Occasionally, longer dispersal might occur by brush rabbits (Mammalia: Leporidae: *Sylvilagus bachmani*), which graze on the plants and can disperse the seeds in their fecal pellets; however, rabbits prefer green plants (before the seed is ripe) and only an extremely low frequency of eaten seeds (about 0.02%) is estimated to remain in a viable state. Birds (Aves), capable of transporting nutlets in mud on their feet, have been implicated as another possible biotic dispersal agent. **Reported associates:** *Acanthomintha ilicifolia, Alopecurus saccatus, Anagallis minima, Brodiaea insignis, B. jolonensis, B. orcuttii, Callitriche marginata, Centunculus minimus, Cotula coronopifolia, Crassula aquatica, C. drummondii, Deschampsia danthonioides, Downingia cuspidata, Elatine brachysperma, E. californica, californica, Eleocharis palustris, Eryngium aristulatum, Haplopappus venetus, Hemizonia fasciculata, Isoetes howellii, I. orcuttii, Juncus bufonius, J. dubius, J. triformis, Lasthenia californica, Lepidium nitidum, Lilaea scilloides, Lythrum hyssopifolia, Myosurus minimus, Navarretia fossalis, Orcuttia californica, Phalaris lemmonii, Pilularia americana, Plagiobothrys acanthocarpus, P. bracteatus, P. chorisianus, P. leptocladus, P. reticulatus, Plantago elongata, P. heterophylla, P. hookeriana, Psilocarphus brevissimus, P. tenellus, Veronica peregrina.*

Pogogyne douglasii **Benth.** grows in depressions, ditches, meadows, playas, prairies, sloughs, streambeds, swales, and vernal pools at elevations of up to 1040 m. Its substrates include adobe, black/heavy clay, muck, mud, sandy loam, and serpentine. The sites often are described as saline-alkaline, but specific values of acidity and salinity have not been reported. The floral biology has not been thoroughly studied. Flowering occurs late in summer to early fall. The seeds germinate in early summer once the waters have dried up. Additional research on various aspects of the population biology and ecology of this species is needed. **Reported associates:** *Acmispon wrangelianus, Alopecurus saccatus, Callitriche hermaphroditica, C. marginata, Castilleja campestris, Clarkia affinis, C. purpurea, Crassula aquatica, Cressa truxillensis, Deschampsia danthonioides, Dichelostemma capitatum, Downingia bella, D. bicornuta, D. cuspidata, D. insignis, D. ornatissima, D. pulchella, Eleocharis acicularis, E. macrostachya, Epilobium pygmaeum, E. torreyi, Eryngium castrense, E. alismifolium, E. aristulatum, Glyceria occidentalis, Gratiola ebracteata, Hemizonia congesta, Hesperevax acaulis, H. caulescens, Isoetes nuttallii, I. orcuttii, Juncus balticus, J. bufonius, J. capitatus, J. leiospermus, J. uncialis, Lasthenia ferrisiae, L. fremontii, L. glaberrima, L. glabrata, L. minor, Lilaea scilloides, Limnanthes alba, L. douglasii, Limosella aquatica, Lolium multiflorum, Lomatium californicum, Lythrum portula, Medicago polymorpha, Mimulus tricolor, Myosurus minimus, Navarretia leucocephala, Phalaris, Pilularia americana, Pinus sabiniana, Plagiobothrys greenei, P. leptocladus, P. stipitatus, P. undulatus, Pleuropogon californicus, Pogogyne ziziphoroides, Polygonum, Potamogeton diversifolius, Psilocarphus brevissimus, P. tenellus, Quercus douglasii, Q. wislizenii, Ranunculus aquatilis, R. bonariensis, R. lobbii, R. muricatus, Rumex, Sidalcea hirsuta, Spergularia*

salina, Toxicoscordion venenosum, Trifolium tridentatum, T. willdenovii, Veronica arvensis, V. peregrina.

Pogogyne floribunda **Jokerst** inhabits flats, intermittent lakes, playas, swales, and vernal pools at elevations from 1000 to 1750 m. Reported substrates include adobe, basalt cobble, clay, and peat. Because this species has been discovered only recently, there is a dearth of information regarding all aspects of its life history. **Reported associates:** *Agoseris heterophylla, Artemisia cana, A. ludoviciana, Bergia texana, Cryptantha intermedia, Deschampsia danthonioides, Downingia, Epilobium brachycarpum, E. torreyi, Eryngium mathiasiae, Iva axillaris, Juniperus, Lagophylla ramosissima, Leymus cinereus, Mimulus pygmaeus, Navarretia, Phacelia inundata, P. thermalis, Plagiobothrys hispidulus, P. stipitatus, Poa sandbergii, Polygonum polygaloides, Psilocarphus brevissimus, Rumex crispus, Taeniatherum caput-medusae, Verbena bracteata, Veronica peregrina.*

Pogogyne nudiuscula **A. Gray** is narrowly restricted to flats and vernal pools on coastal mesas at elevations from 100 to 250 m. The substrates are circumneutral (pH: 6.5–7.8) and consist of clay or gravelly clay loam of the Stockpen soil type. The plants can withstand some inundation, which results in reduced branching and elongated internodes. Flowering begins once the standing waters have dried up (May–July). The flowers are pollinated primarily by honey bees (Hymenoptera: Apidae: *Apis mellifera*). The seeds are not resistant to fire, which can damage them. Other life history information on this species is scarce but would be useful to assemble for assistance with conservation efforts. There are few remaining populations, and some literature accounts confuse this species with the closely related *P. abramsii*. **Reported associates:** *Acanthomintha ilicifolia, Adenostoma fasciculatum, Alopecurus saccatus, Anagallis minima, Atriplex semibaccata, Avena barbata, Brodiaea insignis, B. jolonensis, B. orcuttii, Bromus madritensis, Callitriche marginata, Centaurium davyi, Cotula coronopifolia, Crassula aquatica, C. drummondii, Deschampsia danthonioides, Downingia, Elatine brachysperma, E. californica, Eleocharis palustris, Eriogonum fasciculatum, Eryngium aristulatum, Haplopappus venetus, Hemizonia fasciculata, Hypochaeris glabra, Isoetes howellii, I. orcuttii, Juncus bufonius, J. dubius, J. triformis, Lasthenia californica, Lepidium nitidum, Lilaea scilloides, Lolium, Lythrum hyssopifolia, Myosurus minimus, Muilla clevelandii, Navarretia fossalis, N. intertexta, Ophioglossum californicum, Phalaris lemmonii, Pilularia americana, Plagiobothrys acanthocarpus, P. bracteatus, P. chorisianus, P. leptocladus, P. reticulatus, Plantago heterophylla, P. hookeriana, Polypogon monspeliensis, Psilocarphus brevissimus, P. tenellus, Rhus laurina, Stipa pulchra, Veronica peregrina.*

Pogogyne ziziphoroides **Benth.** grows in depressions, flats, meadows, swales, and vernal pools at elevations of up to 1075 m. The substrates are described as saline alkaline and include clay, clay loam, gravel, gravelly clay, gravelly loam, loam, sandy loam, serpentine, and stony hardpan. Other ecological information on this species is lacking. **Reported associates:** *Achyrachaena mollis, Allium amplectens, Alopecurus* *geniculatus, A. saccatus, Blennosperma nanum, Brodiaea californica, Brodiaea minor, Callitriche hermaphroditica, C. marginata, Castilleja campestris, Cerastium viscosum, Clarkia purpurea, Crassula aquatica, Deschampsia danthonioides, Downingia bicornuta, D. cuspidata, D. ornatissima, D. pulchella, D. yina, Elatine californica, Eleocharis acicularis, E. macrostachya, E. palustris, Epilobium torreyi, Erodium botrys, Eryngium alismifolium, E. armatum, E. castrense, E. petiolatum, E. vaseyi, Geranium dissectum, Glyceria occidentalis, Gratiola ebracteata, Hypochaeris glabra, Isoetes howellii, I. nuttallii, I. orcuttii, Juncus bufonius, J. capitatus, J. hemiendytus, J. leiospermus, J. uncialis, Lasthenia californica, L. chrysostoma, L. fremontii, L. glaberrima, L. platycarpha, Layia fremontii, Lepidium nitidum, Lilaea scilloides, Limnanthes alba, L. douglasii, L. floccosa, Limosella aquatica, Lolium multiflorum, Lomatium, Lupinus bicolor, Lythrum portula, Micropus californicus, Microseris acuminata, Mimulus angustatus, M. tricolor, Myosurus minimus, Navarretia filicaulis, N. heterandra, N. intertexta, N. leucocephala, N. myersii, N. pauciflora, Pilularia americana, Plagiobothrys greenei, P. leptocladus, P. scouleri, P. stipitatus, P. trachycarpus, P. undulatus, Pogogyne douglasii, Polygonum polygaloides, Potamogeton diversifolius, Psilocarphus brevissimus, P. tenellus, Ranunculus aquatilis, R. bonariensis, R. muricatus, Sedella pumila, Sidalcea hartwegii, S. hirsuta, Streptanthus diversifolius, Taeniatherum caput-medusae, Trichostema lanceolatum, Trifolium amplectens, T. tridentatum, T. variegatum, Triphysaria eriantha, Triteleia hyacinthina, Veronica arvensis, V. peregrina, Zigadenus fremontii.*

Use by wildlife: *Pogogyne abramsii* is eaten by brush rabbits (Mammalia: Leporidae: *Sylvilagus bachmani*) and pocket gophers (Mammalia: Geomyidae: *Thomomys bottae*). The pollen is gathered by bees (Insecta: Hymenoptera: Anthophoridae: *Exomalopsis nitens, E. torticornis*; Apidae: *Apis mellifera*) and flies (Insecta: Diptera: Bombyliidae: *Bombylius facialis*). The plants comprise an element of habitat for the endangered San Diego fairy shrimp (Crustacea: Branchinectidae: *Branchinecta sandiegonensis*).

Economic importance: food: The Concow tribe brewed a tea from the leaves of *Pogogyne douglasii*. The Numlaki and Yuki people seasoned wheat and barley pinole using the aromatic seeds of *P. douglasii*; **medicinal:** Leaves of *Pogogyne douglasii* were eaten by the Concow to relieve stomach and bowel pains; **cultivation:** *Pogogyne* is not cultivated, probably due to its demanding habitat requirements; **misc. products:** The Mendocino used plants of *P. douglasii* to repel fleas (Insecta: Siphonaptera); **weeds:** none; **nonindigenous species:** none

Systematics: Phylogenetic analysis of molecular data confirms the position of *Pogogyne* within tribe Mentheae of subfamily Nepetoideae of Lamiaceae, where it resolves as the sister genus to *Acanthomintha* (Figure 5.58). Although the monophyly of the morphologically and ecologically distinct *Pogogyne* has not been in question, it would be informative to include the remaining species of this small genus in phylogenetic analyses for confirmation and to provide much-needed

insights regarding interspecific relationships. The most recent systematic treatment of the genus recognized two subgenera primarily on the basis of whether the species possess either two (subg. *Pogogyne*) or four (subg. *Hedyomoides*) fertile stamens. The monophyly of these subgenera remains untested. *Pogogyne abramsii* and *P. nudiuscula* are believed to be closely related and once were treated as conspecific; however, morphometric analyses indicate that they are distinct. Preliminary cladistic analyses of morphological data indicate that *P. abramsii* and *P. nudiuscula* comprise a clade derived from *P. douglasii*. The Mexican populations of *P. nudiuscula* are distinct and may warrant recognition as a separate species. The base chromosome number of *Pogogyne* is $x = 19$. *Pogogyne douglasii* and *P. ziziphoroides* $(2n = 38)$ are diploid.

Comments: *Pogogyne abramsii* and *P. nudiuscula* are restricted to coastal mesas of San Diego County, California. Both are seriously imperiled (on U.S. Federal endangered species list), with the latter regarded as being on the verge of extinction. *Pogogyne douglasii* and *P. nudiuscula* are restricted to California, *P. ziziphoroides* occurs in California and southern Oregon, and *P. floribunda* ranges from northeastern California to Oregon and Idaho.

References: Baskin, 1994; Bauder, 1989, 2000; Bauder & McMillan, 1998; Bliss & Zedler, 1998; Borgias, 2004; Dole & Sun, 1992; Elam, 1998; Ertter, 2000; Howell, 1931; Hunt, 1992; Jokerst, 1992; McMillan, 1995; Purer, 1939; Ramaley, 1919; Schiller et al., 2000; Schleidlinger, 1981; USFWS, 1998c; Wacker & Kelly, 2004; Walker & Sytsma, 2007; Witham, 2006; Zammit & Zedler, 1990; Zedler & Black, 1992, 2004.

9. Scutellaria

Skullcap; casque, scutellaire

Etymology: from the Latin *scutella* meaning "flat dish," in reference to the ridged calyx

Synonyms: *Anaspis; Cassida; Cruzia; Harlanlewisia; Perilomia; Salazaria; Theresa*

Distribution: global: cosmopolitan; **North America:** widespread

Diversity: global: 300 species; **North America:** 45 species

Indicators (USA): OBL: *Scutellaria floridana, S. galericulata, S. lateriflora, S. racemosa*; **FACW:** *S. lateriflora, S. racemosa*

Habitat: freshwater; palustrine; **pH:** 3.4–7.5; **depth:** <1 m; **life-form(s):** emergent herb

Key morphology: stems (to 1 m) erect, square, herbage not aromatic; leaves (to 12 cm) opposite, sessile or petiolate (to 3 cm), the margins serrate or crenate; flowers axillary, bilabiate, stalked (to 4 mm), either solitary or in one-sided racemes (to 10 cm) bearing 2–44 flowers; calyx (to 6.5 mm) bilabiate, campanulate, splitting to base when mature, the upper segment with a protuberance or ridge-like projection; corolla (to 25 mm) tubular, curving, ascending, blue, pink, violet, or white, the lateral lobes connected to the entire, hood-like upper lip, lower lip spreading, convex, emarginate apically; stamens 4, didynamous, included under upper lip; nutlets (to 2 mm) 4, tuberculate

Life history: duration: perennial (rhizomes); **asexual reproduction:** rhizomes, stolons; **pollination:** insect, self; **sexual condition:** gynodioecious, hermaphroditic; **fruit:** nutlets (common); **local dispersal:** nutlets (water); **long-distance dispersal:** nutlets (water, animals)

Imperilment: (1) *Scutellaria floridana* [G2]; S2 (FL); (2) *S. galericulata* [G5]; SH (NC); S1 (DE, KS, MO, VA, WV); S2 (CA, MD, <u>YT</u>); S3 (WY); (3) *S. lateriflora* [G5]; S1 (CA); S2 (<u>NF</u>); S3 (<u>BC</u>, <u>SK</u>); (4) *S. racemosa* [GNR]

Ecology: general: Less than a third of the North American *Scutellaria* species have any type of wetland indicator status, and only four are designated as OBL. Although the genus occurs primarily in terrestrial habitats on a global basis, it is not uncommon to find at least one species growing in a North American wetland, because several of the more aquatic species are quite widespread. The flowers of most species are insect pollinated, but self-pollination and even cleistogamous flowers are known to occur in genus. Dispersal mechanisms for many species are not known. However, the nutlets of the aquatic species float and are dispersed by water. Seed dormancy in some species appears to involve an inhibitor, which can be inactivated by removing the pericarp or by soaking the nutlets in water for several days. Overall, seed longevity in the sediments is short, leading to the formation only of transient seed banks.

Scutellaria floridana **Chapm.** occurs in borrow pits, flatwoods, prairies, savannas, and seeps at low elevations (e.g., 4 m). The substrates are loamy/peaty sands or sand. These plants usually occur along the margins of wetlands where the water levels fluctuate. They depend on winter fires to maintain a suitable habitat by removal of competing vegetation. The plants have been observed to flower vigorously within 3 months after a site has burned. Flowering extends from April to October. The flowers are pollinated by bees (Hymenoptera: Halictidae; Megachilidae). Much of the basic life-history information (e.g., habitat information, reproductive biology, seed ecology, extent of vegetative reproduction, population genetics, etc.) remains unknown for this rare species, beyond a fairly basic level. **Reported associates:** *Aletris, Aristida stricta, Arnoglossum, Asclepias connivens, A. pedicellata, Clethra alnifolia, Cuphea aspera, Cyrilla, Drosera, Erechtites, Eriocaulon decangulare, Hypericum, Ilex, Justicia crassifolia, Lilium catesbaei, Ludwigia, Macbridea alba, Magnolia virginiana, Panicum, Pinguicula ionantha, Pleea tenuifolia, Polygala cruciata, P. lutea, Rhexia alifanus, Rhynchospora oligantha, Sarracenia flava, Scleria, Smilax laurifolia, Solidago, Sporobolus, Verbesina chapmanii.*

Scutellaria galericulata **L.** grows in bogs, canals, carrs, depressions, ditches, dunes, fens, floodplains, marshes, meadows, mudflats, prairies, sloughs, swales, swamps, and along the margins of lakes, ponds, rivers, and streams at elevations of up to 2100 m. It is one of the most widely distributed wetland plants in North America, and is adapted to a broad range of substrates (pH: 4.5–7.5) that include clay, gravel, loam, marly peat, muck, muddy cobble, peat, rock, sand, sandy loam, and silty sand. The plants can also be found growing on waterlogged logs and stumps. They tolerate full sun to considerable shade. Although the flowers are self-compatible, the plants normally are cross-pollinated by bees (Hymenoptera), flies (Diptera), and butterflies (Lepidoptera). Many plants are gynodioecious, having flowers that are male sterile. The seeds are dispersed by water and can remain afloat for up to a year. Although the seed surfaces are smooth, they adhere fairly well to various animal furs and it is likely that at least some degree of animal dispersal also occurs. The seeds apparently require a dormant period before germination. Ripe seeds that have been kept at ambient outdoor temperatures will germinate during the first spring. Wild seed placed at 10°C has been observed to germinate irregularly from 14 to 180 days. Seeds treated with a fairly high concentration of gibberellic acid (1000 mg/L) will germinate at 25°C. More precise germination requirements should be determined for this species. A transient seed bank of up to 73 seeds/m^2 has been observed. The seeds reach a higher frequency in the seed bank in later successional stages (10–20 years), which is proportional to the frequency of plants that are present in the standing vegetation. The plants usually are associated with arbuscular mycorrhizae. The plants reproduce vegetatively by slender rhizomes and somewhat thickened stolons. **Reported associates:** *Abies balsamea, Acer negundo, A. rubrum, A. saccharinum, Achillea millefolium, Acorus americanus, A. calamus, Agrostis scabra, A. stolonifera, Alisma subcordatum, Alnus rugosa, Amelanchier alnifolia, Ammophila breviligulata, Amorpha fruticosa, Andromeda glaucophylla, Andropogon gerardii, Angelica atropurpurea, Apocynum sibiricum, Arethusa bulbosa, Asclepias incarnata, A. speciosa, A. syriaca, Azolla mexicana, Betula lutea, B. nigra, B. papyrifera, B. ×sandbergii, Bidens cernuus, B. frondosus, Bistorta bistortoides, Boehmeria cylindrica, Calamagrostis canadensis, C. inexpansa, C. stricta, Calla palustris, Callitriche verna, Caltha palustris, Calystegia sepium, Campanula aparinoides, Carex alata, C. aquatilis, C. arcta, C. atherodes, C. athrostachya, C. aurea, C. brunnescens, C. buxbaumii, C. comosa, C. crawei, C. crinita, C. diandra, C. disperma, C. echinata, C. gynandra, C. hystericina, C. interior, C. lacustris, C. lasiocarpa, C. limosa, C. lurida, C. nebrascensis, C. pellita, C. praegracilis, C. prairea, C. retrorsa, C. rosea, C. rostrata, C. sartwellii, C. scoparia, C. stipata, C. tenera, C. tetanica, C. utriculata, C. vesicaria, C. vulpinoidea, Castilleja, Cephalanthus occidentalis, Chamaedaphne calyculata, Chelone glabra, Cicuta bulbifera, C. douglasii, C. maculata, Cirsium arvense, C. muticum, C. vulgare, Cladium mariscoides, Comarum palustre, Cornus canadensis, C. rugosa, C. sericea, C. unalaschkensis, Corylus, Dasiphora floribunda, Deschampsia cespitosa, Doellingeria umbellata, Drosera intermedia, D. rotundifolia, Dryopteris cristata, Dulichium arundinaceum, Echinocystis lobata, Eleocharis acicularis, E. atropurpurea, E. compressa, E. elliptica, E. obtusa, E. palustris, E. smalliana, Epilobium ciliatum, E. coloratum, E. glaberrimum, E. leptophyllum, Equisetum arvense, E. hyemale, E. laevigatum, Erechtites hieraciifolia, Eupatorium perfoliatum, Euthamia graminifolia, Eutrochium maculatum, Fraxinus nigra, Galium aparine, G. brevipes, G. labradoricum, G. obtusum, G. palustre, G. tinctorium, G. trifidum, G. triflorum,*

Gentianopsis, Geum macrophyllum, Glyceria canadensis, G. grandis, G. pallida, Gnaphalium obtusifolium, G. palustre, Gratiola aurea, Helianthus grosseserratus, H. occidentalis, Hesperostipa spartea, Hippuris vulgaris, Holodiscus discolor, Hordeum brachyantherum, Hypericum kalmianum, H. majus, H. scouleri, Hypoxis hirsuta, Impatiens capensis, Iris versicolor, I. virginica, Isoetes, Juncus balticus, J. brevicaudatus, J. canadensis, J. ensifolius, J. effusus, J. nevadensis, J. nodosus, J. orthophyllus, J. torreyi, Kalmia polifolia, Larix laricina, Lathyrus palustris, Lemna minor, Liatris aspera, L. pycnostachya, L. spicata, Lilium philadelphicum, Lobelia cardinalis, L. inflata, L. kalmii, Lonicera dioica, Ludwigia palustris, L. polycarpa, Lupinus, Lycopodium, Lycopus americanus, L. asper, L. uniflorus, Lysichiton americanus, Lysimachia ciliata, L. quadriflora, L. quadrifolia, L. terrestris, L. thyrsiflora, Lythrum alatum, L. salicaria, Mentha arvensis, Menyanthes trifoliata, Microseris laciniata, Mimulus glabratus, M. ringens, Mitchella repens, Mitella, Montia chamissoi, M. linearis, Muhlenbergia glomerata, M. richardsonis, Myosotis scorpioides, Myrica gale, Myriophyllum tenellum, Onoclea sensibilis, Osmorhiza berteroi, Osmunda regalis, Osmundastrum cinnamomeum, Panicum virgatum, Parnassia glauca, Parthenocissus inserta, Pedicularis lanceolata, Peltandra virginica, Penthorum sedoides, Persicaria amphibia, P. coccinea, P. hydropiper, P. hydropiperoides, P. punctata, P. sagittata, P. virginiana, Phragmites australis, Phyla lanceolata, Picea mariana, Pinus resinosa, P. strobus, Platanthera grandiflora, Poa palustris, P. pratensis, Phalaris arundinacea, Phleum pratense, Physalis virginiana, Pilea fontana, Platanthera dilatata, Polygonatum biflorum, Populus deltoides, P. tremuloides, Potentilla anserina, P. gracilis, Puccinellia pauciflora, Ranunculus alismifolius, R. flammula, R. pensylvanicus, Ratibida pinnata, Rhododendron, Rhynchospora alba, Ribes americanum, Rorippa palustris, Rosa gymnocarpa, R. nutkana, R. virginica, Rubus, Rumex aquaticus, R. orbiculatus, R. verticillatus, Sagittaria cuneata, S. latifolia, Salix bebbiana, S. candida, S. exigua, S. geyeriana, S. humilis, S. lutea, Sarracenia purpurea, Schizachyrium scoparium, Schoenoplectus acutus, S. pungens, S. tabernaemontani, Scirpus atrovirens, S. cyperinus, S. microcarpus, S. pedicellatus, Scutellaria lateriflora, Sium suave, Solidago gigantea, S. rugosa, Sparganium americanum, S. chlorocarpum, S. eurycarpum, Spartina pectinata, Sphagnum, Spiraea alba, Sporobolus, Stellaria longipes, Symphyotrichum boreale, S. puniceum, S. spathulatum, Symphoricarpos albus, Taraxacum officinale, Thalictrum pubescens, Thelypteris palustris, Thuja occidentalis, Toxicodendron radicans, T. vernix, Triadenum fraseri, T. virginicum, Trifolium longipes, T. pratense, Triteleia hyacinthina, Tsuga canadensis, Typha angustifolia, T. latifolia, T. ×glauca, Urtica dioica, Utricularia cornuta, Vaccinium macrocarpon, V. oxycoccos, Verbena hastata, Vernonia fasciculata, Veronica americana, V. scutellata, Vicia villosa, Viola biflora, Vitis riparia.

Scutellaria lateriflora L. inhabits bogs, canals, carrs, cliffs, depressions, ditches, dunes, fens, floodplains, marshes, meadows, oxbows, pools, prairies, ravines, sandbars, savannas, seeps, sloughs, swales, swamps, and the margins of lakes, ponds, rivers, and streams at elevations of up to 500 m. Like the preceding, this species also has very broad ecological tolerances and occurs on substrates derived from granite, limestone, or sandstone (pH: 3.4–7.3), which include gravel, loam, muck, mud, rock, sand, sandy loam, silt, and silt loam. *Scutellaria lateriflora* grows occasionally on waterlogged logs or on stumps and occurs in sites that vary in exposure from full sun to dense shade. Generally, this species is quite comparable ecologically to the preceding, except that it is not often reported from higher elevations. The two species can be found growing together at lower elevations. *Scutellaria lateriflora* also differs from *S. galericulata*, by flowers that usually are self-pollinating, although they are visited by bees (Hymenoptera) and flies (Diptera). The seeds require light for germination, which has been achieved at 22°C–30°C following 1–14 weeks of cold (4°C–10°C), moist stratification. Dried, freshly collected seeds kept at room temperature (20°C–25°C) germinate well (88%) within 11 days (after soaking in sterile water); however, their germination declines markedly (21% over 91 days) after 1 year of storage. A transient seed bank (up to 1475 seeds/m^2) is formed. The seeds float and are dispersed mainly by water. *Scutellaria lateriflora* is quite characteristic of wetland habitats; however, it often extends into drier sites as well and is more likely than the preceding species to be found in association with FACW vegetation, or with species that are more characteristic of UPL habitats. This species reproduces vegetatively by means of slender rhizomes or stolons. **Reported associates:** *Abies balsamea, Acalypha rhomboidea, Acer negundo, A. rubrum, A. saccharinum, A. saccharum, A. spicatum, Acorus calamus, Agrostis gigantea, A. stolonifera, Alisma subcordatum, Alnus rugosa, A. serrulata, Amaranthus tuberculatus, Ambrosia trifida, Amelanchier canadensis, Andromeda glaucophylla, Angelica atropurpurea, Apios americana, Apocynum cannabinum, A. sibiricum, Aquilegia canadensis, Arisaema, Asclepias incarnata, Atrichum oerstedianum, Berchemia scandens, Betula lutea, B. nigra, B. papyrifera, B. ×sandbergii, Bidens cernuus, B. coronatus, B. discoideus, B. frondosus, B. tripartita, B. vulgatus, Boehmeria cylindrica, Bolboschoenus fluviatilis, Calamagrostis canadensis, Calla palustris, Campanula aparinoides, Campsis radicans, Carex aquatilis, C. bebbii, C. granularis, C. grayi, C. gynandra, C. hystericina, C. intumescens, C. lacustris, C. lurida, C. muskingumensis, C. novae-angliae, C. pellita, C. rostrata, C. sartwellii, C. seorsa, C. stipata, C. stricta, C. torta, C. tuckermanii, C. utriculata, Cephalanthus occidentalis, Chamaedaphne calyculata, Chamaesyce polygonifolia, Chara, Chelone glabra, Cicuta bulbifera, C. maculata, Cinna arundinacea, Circaea alpina, Cirsium arvense, C. muticum, Clematis virginiana, Clintonia borealis, Coptis trifolia, Coriospermum, Cornus racemosa, C. sericea, Corylus, Cryptotaenia canadensis, Cuscuta gronovii, Cyperus bipartitus, C. odoratus, C. strigosus, Decodon verticillatus, Decumaria barbara, Dichanthelium acuminatum, D. clandestinum, Diervilla, Diospyros virginiana, Doellingeria umbellata, Drosera rotundifolia, Dryopteris carthusiana, Echinochloa crus-galli, E. walteri, Eleocharis acicularis, E. erythropoda, E. intermedia, E. palustris, Elymus* riparius, *E. virginicus, Epilobium coloratum, Equisetum arvense, E. hyemale, E. scirpoides, E. sylvaticum, Eragrostis hypnoides, Erigeron philadelphicus, Eryngium yuccifolium, Euonymus atropurpureus, Eupatorium fistulosum, E. maculatum, E. perfoliatum, E. serotinum, Euthamia graminifolia, Fallopia convolvulus, Fragaria virginiana, Fraxinus americana, F. caroliniana, F. pennsylvanica, Galium aparine, G. asprellum, G. brevipes, G. obtusum, G. trifidum, G. tinctorium, G. triflorum, Gentianopsis, Gleditsia triacanthos, Glyceria canadensis, G. septentrionalis, G. striata, Helenium autumnale, Hibiscus moscheutos, Huperzia lucidula, Hydrocotyle ranunculoides, H. verticillata, Hypericum mutilum, H. sphaerocarpum, Hypnum lindbergii, Ilex verticillata, Impatiens capensis, I. pallida, Iris pseudacorus, I. virginica, Itea virginica, Juncus alpinus, J. brevicaudatus, J. dudleyi, J. effusus, J. gymnocarpus, J. nodosus, J. torreyi, Justicia americana, Kalmia angustifolia, K. latifolia, K. polifolia, Lactuca canadensis, Laportea canadensis, Larix laricina, Leersia oryzoides, L. virginica, Leucothoe racemosa, Leymus arenarius, Lindera benzoin, Liquidambar styraciflua, Listera auriculata, Lobelia cardinalis, L. siphilitica, Ludwigia alternifolia, L. palustris, Lycopodium obscurum, Lycopus americanus, L. asper, L. rubellus, L. uniflorus, L. virginicus, Lyonia ligustrina, Lysimachia ciliata, L. nummularia, Lythrum alatum, L. salicaria, Maianthemum canadense, Menispermum canadense, Mentha arvensis, Mikania scandens, Mimulus ringens, Mitchella repens, Mitella diphylla, M. nuda, Monarda fistulosa, Muhlenbergia mexicana, Myosoton aquaticum, Myrica cerifera, M. gale, Nasturtium officinale, Nyssa aquatica, N. biflora, Onoclea sensibilis, Osmorhiza claytonii, Osmunda regalis, Osmundastrum cinnamomeum, Ostrya virginiana, Oxypolis rigidior, Panicum capillare, P. dichotomiflorum, Pedicularis lanceolata, Pellia epiphylla, Peltandra virginica, Penthorum sedoides, Persicaria amphibia, P. hydropiper, P. lapathifolia, P. posumbu, P. punctata, P. sagittata, P. setacea, P. virginiana, Phalaris arundinacea, Phragmites australis, Physalis longifolia, Physostegia virginiana, Picea mariana, Pilea pumila, Pinus banksiana, P. resinosa, Platanus occidentalis, Poa palustris, Polypodium appalachianum, Pontederia cordata, Populus deltoides, P. tremuloides, Potentilla anserina, P. arguta, P. norvegica, Prunella vulgaris, Pteridium aquilinum, Pycnanthemum virginianum, Pyrola minor, Quercus bicolor, Q. ellipsoidalis, Q. laurifolia, Q. velutina, Ranunculus acris, R. pensylvanicus, R. sceleratus, Ratibida pinnata, Rhexia virginica, Rubus occidentalis, R. pubescens, Rudbeckia laciniata, Rumex orbiculatus, R. salicifolius, R. verticillatus, Sagittaria latifolia, S. rigida, Salix amygdaloides, S. fragilis, S. interior, S. sericea, Sambucus nigra, Saponaria officinalis, Sarracenia purpurea, Sassafras albidum, Saururus cernuus, Schoenoplectus acutus, S. americanus, S. heterochaetus, S. pungens, S. purshianus, c. tabernaemontani, Scirpus atrovirens, S. cyperinus, S. expansus, S. georgianus, S. pendulus, Scutellaria galericulata, Sicyos angulatus, Silphium perfoliatum, Sium suave, Smilax lasioneura, S. tamnoides, S. walteri, Solanum dulcamara, S. nigrum, Solidago altissima, S. canadensis, S. gigantea, S. ohioensis, S. patula, S. riddellii,* S. rugosa, Sparganium eurycarpum, Spartina pectinata, Sphagnum, Sphenopholis pensylvanica, Spiraea alba, S. douglasii, S. tomentosa, Stachys tenuifolia, Streptopus amplexifolius, Symphoricarpos, Symphyotrichum boreale, S. cordifolium, S. dumosum, S. lanceolatum, S. lateriflorum, S. praealtum, S. puniceum, S. racemosum, Taxodium distichum, Taxus canadensis, Teucrium canadense, Thalictrum dasycarpum, T. pubescens, Thelypteris noveboracensis, T. palustris, Thuja occidentalis, Tilia americana, Toxicodendron radicans, Triadenum walteri, Trientalis borealis, Typha angustifolia, T. latifolia, T. ×glauca, Ulmus americana, Vaccinium angustifolium, V. corymbosum, V. fuscatum, V. macrocarpon, V. oxycoccos, Verbena hastata, Vernonia noveboracensis, Veronica americana, V. officinalis, Viola cucullata, Viola nephrophylla, Vitis riparia, Waldsteinia fragarioides, Xanthium strumarium, Zanthoxylum americanum.*

***Scutellaria racemosa* Pers.** inhabits bottoms, depressions, ditches, fields, floodplains, gardens, lawns, marshes, roadsides, slopes, swales, swamps, and the margins of lakes, ponds, and streams at low elevations. The plants sometimes occur in standing shallow water but often are found in quite dry sites. Collections consistently indicate full sunlight exposures. The substrates include muck, peaty muck, peaty sand, sand, sandy gravel, sandy loam, and sandy silt. Flowers and fruits are produced throughout the year. The plants perennate and reproduce vegetatively by rhizomes. The OBL status of this nonindigenous species is unusual, given its frequent association with ruderal species and numerous reports from such sites as agricultural fields, gardens, residential lawns, and commercial building landscapes. **Reported associates (North America):** *Acalypha gracilens, Alternanthera philoxeroides, A. sessilis, Amaranthus spinosus, Ambrosia trifida, Arenaria serpyllifolia, Baccharis halimifolia, Campsis radicans, Centella asiatica, Conyza bonariensis, Cynodon dactylon, Cyperus croceus, C. rotundus, Cyrilla racemiflora, Dichanthelium scabriusculum, Eleusine indica, Eryngium, Euphorbia heterophylla, Hypericum chapmanii, Hyptis mutabilis, Ilex cassine, Ipomoea hederifolia, I. cordatotriloba, I. quamoclit, Juncus, Mikania scandens, Nyssa biflora, Paspalum notatum, P. urvillei, Persicaria, Phalaris caroliniana, Phyllanthus urinaria, Pinus elliottii, Portulaca oleracea, Ranunculus muricatus, Rorippa, Sida rhombifolia, Sisyrinchium rosulatum, Smilax bona-nox, S. glauca, S. laurifolia, Sorghum halepense, Stachys floridana, Taxodium, Vaccinium, Veronica arvensis, Vitis rotundifolia.*

Use by wildlife: *Scutellaria galericulata* and *S. lateriflora* occasionally are grazed by elk (Mammalia: Cervidae: *Cervus canadensis*), but are not eaten by mule deer (Mammalia: Cervidae: *Odocoileus hemionus*). The seeds of *S. lateriflora* are eaten by pheasants (Aves: Phasianidae: *Phasianus colchicus*). The flowers of *S. floridana* are visited by crab spiders (Arthropoda: Arachnida: Thomisidae) and insects (Insecta) such as bumblebees (Hymenoptera: Apidae: *Bombus*), leafcutter bees (Hymenoptera: Megachilidae), sweat bees (Hymenoptera: Halictidae), butterflies (Lepidoptera), flies (Diptera), and katydids (Orthoptera: Tettigoniidae).

Those of *S. galericulata* are visited by bumblebees (Hymenoptera: Apidae: *Bombus*) and karner blue butterflies (Lepidoptera: Lycaenidae: *Lycaeides melissa*), which forage for the nectar. This species is the host of several larval Lepidoptera (Choreutidae: *Prochoreutis inflatella*; Gracillariidae: *Caloptilia scutellariella*; Pterophoridae: *Capperia evansi*). *Scutellatia lateriflora* hosts Lepidoptera larvae of *Capperia evansi* (Pterophoridae), *Prochoreutis inflatella* (Choreutidae), and *Scrobipalpa scutellariaeella* (Gelechiidae). Its flowers are visited by nectar-seeking anthophorid bees (Hymenoptera: Apidae: *Melissodes bimaculata*). Leaf beetles (Insecta: Chrysomelidae) are hosted by *S. galericulata* (*Phyllobrotica decorata*) and *S. lateriflora* (*Phyllobrotica limbata*). *Scutellaria galericulata* is a host of gray mold (Ascomycota: Sclerotiniaceae: *Botrytis cinerea*), which is a pest of various wine grapes (Vitiaceae: *Vitis*). *Scutellaria lateriflora* is the host of several fungi including leaf spot (Ascomycota: Mycosphaerellaceae: *Septoria scutellariae*) and white powdery mildew (Ascomycota: Erysiphaceae: *Erysiphe cruciferarum*).

Economic importance: food: *Scutellaria* is not edible; **medicinal:** *Scutellaria* species have a long history of use in both human and veterinary medicine. *Scutellaria galericulata* contains several phenylethanoid glycosides, which possess antioxidant properties that enable them to scavenge free radicals efficiently. Two of the glycosides isolated from *S. galericulata* (calceolarioside, osmanthuside) exhibit antimicrobial activity against several gram-positive bacteria, gram-negative bacteria, and yeast-like fungi. *Scutellaria lateriflora* has anticonvulsive and sedative properties and has been shown to reduce nervous tension. It has also been used to treat coughing, epilepsy, hiccoughs, and insomnia, to reduce symptoms of alcohol and tobacco withdrawal and as an anthelmintic. The activity has been attributed to several flavonoids (baicalin, baicalein), which bind to a receptor of gamma-aminobutyric acid (GABA), a main inhibitory neurotransmitter. The plants also contain flavonoids (scutellarin and ikonnikoside) that bind to the serotonin 5-hydroxytryptamine receptor 7 (5-HT7) in the brain, which has been linked to certain types of pain and inflammation. Extracts of *S. lateriflora* effectively suppress the onset of seizures in laboratory animals. Aqueous and alcoholic extracts of *S. lateriflora* exhibit anti-inflammatory activities. Dichloromethanolic and methanolic extracts from aerial tissues of *S. galericulata* and *S. lateriflora* exhibit antifungal (Ascomycota: Davidiellaceae: *Cladosporium cucumerinum*) and antimicrobial (Proteobacteria: Enterobacteriaceae: *Escherichia coli*) activity; those of *S. lateriflora* are active against some yeasts (Ascomycota: Saccharomycetidae: *Candida albicans*). *Scutellaria lateriflora* reportedly contains steroidal phytoestrogens. It allegedly contains hepatotoxic glycosides that can cause acute hepatitis and liver damage, but these effects also have been attributed to contamination of herbal preparations by germander (Lamiaceae: *Teucrium chamaedrys*). The common name of "mad dog skullcap" originated in the 18th century from its use as a treatment (albeit ineffective) for rabies. *Scutellaria lateriflora* has also been used to treat

anxiety, colic, and pain in racehorses (Mammalia: Equidae: *Equus ferus*). The Delaware and Oklahoma tribes used the tops of *S. galericulata* plants as a laxative. The Ojibwa made a heart medicine from the plants. The Cherokee included *S. lateriflora* (Gûnigwalĭ'skĭ) in a compound decoction used to control diarrhea or promote menstruation. They used the plants to calm the nerves. A decoction from the roots was administered as a kidney medicine and as an aid in childbirth. The Iroquois used the powdered roots of *S. lateriflora* as an antidiarrheal agent, a throat cleanser, and a smallpox preventative. *Scutellaria racemosa* contains baicalin (a prolyl endopeptidase inhibitor) and oroxylin A (a dopamine reuptake inhibitor). Hydroalcoholic extracts of *S. racemosa* effectively inhibit prolyl oligopeptidase, which has been implicated in several functions of the human central nervous system; **cultivation:** *Scutellaria galericulata* (e.g., 'Corinne Tremaine') and *S. lateriflora* are cultivated as water garden ornamentals. *Scutellaria lateriflora* is grown commercially in Australia and New Zealand as a source of medicinal herbs. Sometimes it is misrepresented in cultivation by *S. altissima*; **misc. products:** *Scutellaria galericulata* and *S. lateriflora* contain several neo-clerodane diterpenoids, which deter feeding by insects such as the Egyptian cotton leafworm (Insecta: Lepidoptera: Noctuidae: *Spodoptera littoralis*); **weeds:** *Scutellaria racemosa* is a weed of gardens, lawns, and disturbed sites; **nonindigenous species:** *Scutellaria racemosa* is native to South America and was introduced to the southeastern United States in the 1970s as a contaminant of nursery material or sod. Its spread has been facilitated by mowing equipment, which is used along highway right-of-ways.

Systematics: *Scutellaria* is the type genus of Lamiaceae subfamily Scutellarioideae, which resolves as the sister clade to subfamilies Lamioideae and Pogostemonoideae by analyses of cpDNA sequence data (Figure 5.56). Subfamily Scutellarioideae is distinguished morphologically by a bilabiate calyx having entire, rounded lips and by distinctively tuberculate fruits. The phylogenetic analysis of cpDNA sequence data, which included three of the four genera in the subfamily, placed *Tinnea* as most closely related to *Scutellaria*. Some authors have divided the genus into two subgenera that distinguish species having a one-sided inflorescence (subg. *Scutellaria*) from those with a four-sided inflorescence (subg. *Apeltanthus*). Phylogenetic studies of the large genus *Scutellaria* itself are highly incomplete due to the small number of species that have been analyzed simultaneously. Although *S. galericulata* (section *Galericularia*) and *S. lateriflora* (section *Lateriflorae*) are not regarded as being particularly closely related, cladistic analysis of morphological data resolves them as sister species. More comprehensive phylogenetic analyses of the genus, or at least of portions of it, will be necessary before any specific details of relationships, the naturalness of subgenera, sections, etc., can be evaluated with any certainty. Reliable species-specific markers have been obtained for *S. galericulata* and *S. lateriflora* using RAPD analysis and comparison of cpDNA (*rpl16*) sequences. These markers have proven to be useful for identifying the presence of the species in herbal drug preparations, which contain only fragmentary parts of the plants. The base chromosome number of *Scutellaria* is uncertain, with 2*n* counts ranging from 12 to 88. *Scutellaria galericulata* (2*n* = 32) and *S. lateriflora* (2*n* = 88) probably are polyploids. The sterile *S. churchilli* (2*n* = 60) is believed to represent an F$_1$ hybrid between *S. galericulata* and *S. lateriflora*; however, it cannot readily be synthesized by artificial crosses and probably arises rarely in nature, persisting vegetatively by rhizome fragmentation. *Scutellaria galericulata* was long known under the name of *S. epilobiifolia*, which occasionally appears in literature accounts. North American populations of *S. galericulata* reportedly differ from European populations by having nutlets with less-pronounced tubercles.

Comments: *Scutellaria galericulata* occurs nearly throughout all of North America (except the extreme Arctic) and extends into Eurasia; *S. lateriflora* is also widely distributed except for the central Rocky Mountains and the far north. *Scutellaria racemosa* occurs in the southern United States [also South America] and *S. floridana* is endemic to Florida.

References: Allen, 1983; Andersson et al., 2000; Awad et al., 2003; Bergeron et al., 1996; Bruno et al., 1998; Couvreur et al., 2004; Ersöz et al., 2002a,b; Falińska, 1999; Farrell & Mitter, 1990; Fernald, 1950; Gafner et al., 2003a,b; Gill & Morton, 1978; Hanlon et al., 1998; Harley et al., 2004; Heinrich, 1976; Hosokawa et al., 2000, 2005; Islam et al., 2011; Kral, 1981; Krings & Neal, 2001a,b; Lans et al., 2006; Leckie et al., 2000; Lynn & Karlin, 1985; Marques et al., 2010; Middleton, 2000; Miller, 1914; Mitchell, 1926; Mooney, 1891; Olmstead, 1989; Peredery & Persinger, 2004; Pitts-Singer et al., 2002; Rodríguez et al., 1993, 1996; Thompson, 1969b; USFWS, 1994; Van der Valk & Davis, 1978; Whiting et al., 2002; Wolfson & Hoffmann, 2003.

10. *Stachys*

Hedge-nettle, marsh betony, woundwort; Épiaire

Etymology: from the Greek *stachys* meaning "spike," in reference to the inflorescence

Synonyms: *Betonica*; *Phlomidoschema*; *Tetrahit*

Distribution: global: Africa; Asia; Europe; New World; **North America:** widespread

Diversity: global: 300 species; **North America:** 36 species

Indicators (USA): OBL: *Stachys ajugoides*, *S. albens*, *S. chamissonis*, *S. hyssopifolia*, *S. palustris*, *S. stricta*, *S. tenuifolia*; **FACW:** *S. chamissonis*, *S. hyssopifolia*, *S. palustris*, *S. tenuifolia*

Habitat: brackish (coastal), freshwater; palustrine, riverine; **pH:** 4.2–8.0; **depth:** <1 m; **life-form(s):** emergent herb

Key morphology: stems (to 2.5 m) erect, square, the herbage usually hairy (smooth in one species) and often malodorous; leaves (to 15 cm) opposite, simple, petioled (to 6 cm) to nearly sessile, the margins toothed; flowers bilabiate, in continuous or discontinuous, verticillate clusters arranged in 1-several terminal racemes or spikes (to 40 cm); corolla tubular (to 11–24 mm), magenta, pink, purple, red, or white, a ring of hairs inside above the base, the upper lip (to 5.5–10 mm) concave, entire, the lower lip (to 8.5–15 mm) 3-lobed; stamens 4, didynamous; style gynobasic; nutlets (to 2 mm) 4, oblong or ovoid

Life history: duration: perennial (rhizomes); **asexual reproduction:** rhizomes, tubers; **pollination:** insect; **sexual condition:** gynodioecious, hermaphroditic; **fruit:** nutlets (common); **local dispersal:** rhizomes, fragments; **long-distance dispersal:** nutlets (water)

Imperilment: (1) *Stachys ajugoides* [G4]; S1 (AK); S2 (BC); (2) *S. albens* [G4]; S1 (UT); (3) *S. chamissonis* [G4]; (4) *S. hyssopifolia* [G4]; S1 (DE, CT); S2 (NC, NJ, NY); S3 (VA); (5) *S. palustris* [G5]; S1 (KS, OK, VA); S2 (MA, PE); S3 (WY); (6) *S. stricta* [G4]; (7) *S. tenuifolia* [G5]; S1 (CT, DE, FL, NB, NC); S3 (NJ, WV)

Ecology: general: *Stachys* is fairly diverse ecologically, inhabiting grasslands, pastures, and woodlands as well as wetland habitats. About a third of the North American species have status as wetland indicators, but only seven of the species are regarded as obligate aquatics. The OBL *Stachys* species are insect-pollinated and share similar ecological characteristics. They occur frequently in ephemeral hydrological conditions at sites where the water has receded and the surface substrates are drying, such as along the margins of permanent wetlands.

Stachys ajugoides **Benth.** inhabits bluffs, canyons, chaparral, depressions, ditches, dunes, marshes, meadows, mudflats, savannas, seeps, sloughs, springs, swamps, vernal pools, washes, and the margins of bogs, lakes, ponds, rivers, and streams at elevations of up to 3600 m. The substrates span a broad range of acidity (pH: 4.6–8.0), are derived from basalt, granite, sandstone, serpentine, or schist, and include adobe, ash, clay, clay loam, cobble, gravel, loam, mud, rock, sand, sandy loam, silt, silty clay loam, and stony loam. Occurrences tend to be on nitrogen-rich soils with a few records reported from brackish soils and salt marshes. The plants will tolerate exposures ranging from full sun to deep shade. For an OBL species, *S. ajugoides* is unusual ecologically because it occurs infrequently in "typical" wetlands where permanent standing water supports associations of characteristic wetland plants. It is more common in wetland margins that are subject to periodic drying (especially riparian habitats) and in ephemeral or vernal sites where the water has receded. Consequently, many of the reported associates are plants of dry or arid conditions (e.g., Cactaceae), which occupy dry sites surrounding the temporary water, or that move into the sites once the water has receded. However, *S. ajugoides* does not persist in areas that permanently remain dry. There are also taxonomic problems (see below), which undoubtedly have influenced reports of associated species. *Stachys ajugoides* occurs often in disturbed sites, especially in areas that have burned recently. The flowers are pollinated by bumblebees (Hymenoptera: Apidae: *Bombus*) and honeybees (Hymenoptera: Apidae: *Ceratina pacifica*, *C. sequoiae*). The plants have a strong, musky odor. They are rhizomatous and can develop into fairly large clonal populations, which are able to persist among dense thickets of other vegetation. The roots are readily colonized by a vesicular–arbuscular mycorrhizal fungus (Glomeromycota: Glomeraceae: *Glomus macrocarpum*) in laboratory culture and presumably are mycorrhizal in nature. **Reported associates:** *Abies amabilis*,

A. concolor, A. grandis, Acer circinatum, A. macrophyllum, Achillea millefolium, Achlys triphylla, Acmispon glabrus, Aconogonon phytolaccifolium, Actaea rubra, Adenocaulon bicolor, Adenostoma fasciculatum, A. sparsifolium, Adiantum jordanii, A. pedatum, Agrostis exarata, A. gigantea, Aira caryophyllea, Allium validum, Alnus rhombifolia, A. rubra, A. rugosa, A. tenuifolia, A. viridis, Amaranthus californicus, Ambrosia chenopodiifolia, Amsinckia, Anaphalis margaritacea, Anemone deltoidea, Arbutus, Arctostaphylos glauca, A. pungens, Artemisia californica, A. douglasiana, A. palmeri, A. tridentata, Asarum caudatum, Asyneuma prenanthoides, Athyrium filix-femina, Avena, Baccharis salicifolia, B. sarothroides, Berberis nervosa, Betula papyrifera, Blechnum spicant, Brassica nigra, Brodiaea orcuttii, Bromus diandrus, B. hordeaceus, B. madritensis, B. vulgaris, Calocedrus decurrens, Calochortus albus, Calycanthus occidentalis, Calystegia macrostegia, Campanula scouleri, Cardamine occidentalis, Carduus pycnocephalus, Carex angustata, C. deweyana, C. hendersonii, C. jonesii, C. lanuginosa, C. obnupta, C. rossii, Castanopsis chrysophylla, Ceanothus cordulatus, C. leucodermis, Centaurea melitensis, Cephalanthus occidentalis, Chamerion angustifolium, Circaea alpina, Cistanthe umbellata, Claytonia perfoliata, C. sibirica, Cleome isomeris, Clintonia uniflora, Collinsia heterophylla, Conium maculatum, Cornus nuttallii, C. sericea, Corylus cornuta, Cryptantha simulans, Cupressus forbesii, Cylindropuntia californica, C. prolifera, Cynara cardunculus, Cynodon, Cyperus eragrostis, Cystopteris fragilis, Daucus carota, Dicentra formosa, Disporum hookeri, D. smithii, Dryopteris arguta, D. campyloptera, D. expansa, Eleocharis bella, E. macrostachya, Elymus glaucus, E. trachycaulus, Encelia californica, Epilobium canum, E. glaberrimum, E. minutum, Equisetum arvense, E. telmateia, Eriodictyon crassifolium, Eriogonum elongatum, E. fasciculatum, Erodium, Eucalyptus, Euphorbia misera, Ferocactus viridescens, Festuca californica, F. occidentalis, F. rubra, F. subuliflora, Frangula californica, Fraxinus latifolia, Galium andrewsii, G. aparine, G. nuttallii, G. triflorum, Gaultheria shallon, Glyceria striata, Gnaphalium palustre, Hemizonia fasciculata, Heracleum lanatum, Heteromeles arbutifolia, Hieracium albiflorum, Holcus lanatus, Holodiscus discolor, Hosackia crassifolia, Hydrophyllum tenuipes, Iris douglasiana, Isocoma menziesii, Juglans californica, Juncus ensifolius, J. nevadensis, J. phaeocephalus, J. xiphioides, Keckiella, Lactuca serriola, Lepechinia calycina, Lithophragma, Lolium multiflorum, Lupinus sellulus, Luzula campestris, L. parviflora, Lysichiton americanus, Maianthemum dilatatum, M. stellatum, Malacothamnus fasciculatus, Malosma laurina, Mammillaria dioica, Marah oreganus, Melica torreyana, Mentha arvensis, Menziesia ferruginea, Mimulus aurantiacus, M. cardinalis, M. dentatus, M. guttatus, M. torreyi, Monardella stoneana, M. villosa, Montia chamissoi, Nasturtium officinale, Nemophila parviflora, Oenanthe sarmentosa, Oenothera elata, Oplopanax horridum, Opuntia oricola, Osmorhiza chilensis, Oxalis oregana, Perideridia kelloggii, Persicaria lapathifolia, Phacelia mutabilis, Phleum pratense, Picea sitchensis, Pinus contorta, P. jeffreyi,
P. muricata, P. sabiniana, Plagiobothrys collinus, P. scouleri, Platanus racemosa, Plectritis macrocera, Poa palustris, P. pratensis, P. trivialis, Polypodium glycyrrhiza, P. hesperium, Polystichum munitum, Populus fremontii, Potentilla gracilis, Prunella vulgaris, Prunus ilicifolia, Pseudotsuga menziesii, Pseudotsuga menziesii, Pteridium aquilinum, Pycnanthemum californicum, Quercus agrifolia, Q. chrysolepis, Q. dumosa, Q. engelmannii, Q. kelloggii, Q. lobatum, Ranunculus californicus, R. repens, R. sceleratus, Rhamnus californica, R. crocea, R. ilicifolia, R. purshiana, Rhus integrifolia, R. trilobata, Ribes bracteosum, R. californicum, Rorippa, Rosa gymnocarpa, R. pisocarpa, Rubus discolor, R. laciniatus, R. parviflorus, R. spectabilis, R. ursinus, Rudbeckia occidentalis, Rumex crispus, Salix commutata, S. exigua, S. laevigata, S. lasiolepis, S. lucida, S. sitchensis, Salvia mellifera, Sambucus nigra, S. racemosa, Sanicula bipinnatifida, Satureja douglasii, Schoenoplectus, Scirpus microcarpus, Scrophularia floribunda, Selaginella cinerascens, Senecio triangularis, Sequoia sempervirens, Sidalcea oregana, Simmondsia chinensis, Sisyrinchium bellum, Solidago californica, Spergularia, Spiraea douglasii, S. splendens, Stellaria crispa, Stephanomeria, Streptopus amplexifolius, Symphoricarpos albus, S. mollis, Symphyotrichum foliaceum, Tamarix ramosissima, Taxus brevifolia, Thuja plicata, Tiarella trifoliata, Tolmiea menziesia, Toxicodendron diversilobum, Trientalis borealis, T. latifolia, Trifolium cyathiferum, Trillium ovatum, Trisetum, Tsuga heterophylla, Typha latifolia, Umbellularia californica, Urtica dioica, Vaccinium alaskaense, V. ovatum, V. parvifolium, Vancouveria hexandra, Veratrum californicum, Verbena lasiostachys, Veronica americana, Vicia, Viguiera laciniata, Viola glabella, V. macloskeyi, V. sempervirens, Vitis, Whipplea modesta, Yucca whipplei.

Stachys albens A. Gray grows in bogs, chaparral, ditches, marshes, meadows, seeps, sloughs, springs, swamps, washes, and along the margins of rivers and streams at elevations of up to 3020 m. The substrates are characterized as being somewhat alkaline (pH: data not available), and usually comprise granitic serpentine as cobble, gravel, loam, and sandy loam. The plants occur in full sun and also tolerate partial shade. Although the plants often are found in drier sites when mature, the localities indicate previously wet areas where water has receded. There are some stands reported from tidal areas and in recently burned sites. Neither the reproductive biology nor the seed ecology of this species has been studied formally. The flowers presumably are insect-pollinated. The herbage is densely wooly with a distinct fragrance that some have described as a "carpet factory smell." Vegetative reproduction occurs by the production of rhizomes, which can develop into fairly large, dense, clonal colonies. **Reported associates:** *Abies concolor, Acer macrophyllum, Achillea millefolium, Aconitum columbianum, Adenocaulon bicolor, Aesculus californica, Agastache urticifolia, Agrimonia gryposepala, Agropyron, Alnus rhombifolia, A. tenuifolia, Anemopsis californica, Angelica tomentosa, Aquilegia eximia, A. formosa, Arctostaphylos patula, Artemisia californica, A. douglasiana, A. dracunculus, A. ludoviciana, A. tridentata, Asclepias fascicularis, A. speciosa, Astragalus clevelandii, Baccharis* *glutinosa, B. salicifolia, Betula occidentalis, Brachypodium distachyon, Briza minor, Bromus diandrus, B. hordeaceus, B. madritensis, B. tectorum, Holcus lanatus, Calocedrus decurrens, Calycanthus occidentalis, Camassia quamash, Carduus pycnocephalus, Carex macloviana, C. nudata, C. praegracilis, C. serratodens, C. subfusca, Castilleja affinis, C. miniata, C. rubicundula, Ceanothus cuneatus, Chrysolepis sempervirens, Cirsium douglasii, Claytonia parviflora, Clematis ligusticifolia, Cornus sericea, Corylus cornuta, Cupressus sargentii, Cynosurus echinatus, Cyperus odoratus, Delphinium polycladon, D. uliginosum, Dicentra formosa, Distichlis spicata, Dryopteris arguta, Eleocharis macrostachya, E. parishii, Elymus glaucus, Epilobium glaberrimum, Epipactis gigantea, Equisetum arvense, Eriogonum fasciculatum, Euonymus occidentalis, Fragaria vesca, Galium aparine, G. trifidum, Geranium richardsonii, Glyceria striata, Helenium bolanderi, Helianthus exilis, Heracleum lanatum, H. maximum, Heteromeles arbutifolia, Hosackia oblongifolia, Hypericum anagalloides, H. formosum, Iris missouriensis, Isocoma menziesii, Juncus balticus, J. effusus, J. mexicanus, J. rugulosus, J. xiphioides, Keckiella cordifolia, Lilium kelleyanum, L. pardalinum, L. parvum, Lolium multiflorum, Lonicera involucrata, L. subspicata, Lupinus latifolius, Maianthemum stellatum, Malosma laurina, Medicago polymorpha, Mimetanthe pilosa, Mimulus cardinalis, M. guttatus, M. moschatus, M. nudatus, M. rubellus, Montia chamissoi, Nasturtium officinale, Oenanthe, Oenothera hookeri, Oxypolis occidentalis, Paspalum dilatatum, Phleum pratense, Phoradendron, Pinus coulteri, P. jeffreyi, P. lambertiana, P. ponderosa, Platanthera leucostachys, Platanus racemosa, Pluchea, Poa palustris, Polypogon monspeliensis, Populus fremontii, Potentilla glandulosa, Prosopis glandulosa, Prunus ilicifolia, Pseudotsuga macrocarpa, Pteridium aquilinum, Quercus agrifolia, Q. chrysolepis, Q. durata, Q. kelloggii, Q. wislizeni, Rhamnus californica, R. ilicifolia, Rhododendron occidentale, Ribes nevadense, R. roezlii, Rosa californica, Rubus parviflorus, R. ursinus, Rudbeckia californica, R. hirta, R. occidentalis, Rumex salicifolius, Salix breweri, S. caudata, S. gooddingii, S. laevigata, S. lasiolepis, S. lucida, S. melanopsis, Salvia apiana, Sambucus mexicana, Satureja mimuloides, Schoenoplectus acutus, S. pungens, Scirpus microcarpus, Scrophularia californica, Senecio clevelandii, S. triangularis, Sidalcea reptans, Sisymbrium, Solidago californica, S. canadensis, Sonchus oleraceus, Sphenosciadium capitellatum, Symphoricarpos, Taraxacum officinale, Toxicodendron diversilobum, Trichostema laxa, Typha angustifolia, T. latifolia, Umbellularia californica, Urtica dioica, U. holosericea, Veratrum californicum, Verbena litoralis, Vinca major, Viola adunca, Vitis californica.*

Stachys chamissonis Benth. occurs in coastal Pacific bogs, ditches, dunes, floodplains, gravel bars, gullies, marshes, seeps, swamps, and along lakeshores or stream margins at elevations of up to 1300 m. The substrates are circumneutral (pH: 6.0–8.0), siliceous or (mostly) organic, and consist of clay, humus, loam, muck, peat, or silt loam. The plants tolerate exposures ranging from full sun to dense shade. The flowers, with their distinctive rose-red to purple

corollas, are pollinated by black-chinned (Aves: Trochilidae: *Archilochus alexandri*) and other hummingbirds. Seeds collected in the spring and then cold-stratified (1°C) for 2–4 days have germinated at greenhouse temperatures (29°C/10°C). The seeds are not long lived and form a minor, transient seed bank. The plants reproduce vegetatively by rhizomes. They are tolerant of disturbance and are among the first plants to revegetate volcanic mudflows. Like several of the other species, *S. chamissonis* is characteristic of temporarily flooded habitats but is drought-tolerant and often forms dense thickets in sites that dry out as the waters recede. The foliage has a strong odor that is described as "disagreeable" or "musky." The foliage is highly susceptible to grazing by ungulate mammals. **Reported associates:** *Abies concolor, b. grandis, Acer circinatum, A. macrophyllum, Achlys triphylla, Aconitum columbianum, Actaea rubra, Adenocaulon bicolor, Adiantum pedatum, Alnus rubra, A. rugosa, Angelica arguta, Aralia californica, Artemisia tilesii, Aruncus sylvester, Asarum caudatum, Athyrium americanum, A. filix-femina, Blechnum spicant, Boykinia major, B. occidentalis, Bromus vulgaris, Calamagrostis canadensis, Caltha leptosepala, Canadanthus modestus, Cardamine angulata, Carex deweyana, C. hendersonii, C. laeviculmis, C. luzulina, C. mertensii, C. obnupta, C. stipata, Castilleja, Chamerion angustifolium, Cicuta douglasii, Cinna latifolia, Circaea alpina, Cirsium arvense, C. vulgare, Claytonia sibirica, Conocephalum conicum, Conyza canadensis, Cornus canadensis, C. sericea, C. unalaschkensis, Corydalis scouleri, Corylus cornuta, Crataegus monogyna, Delairea odorata, Dicentra formosa, Digitalis purpurea, Disporum smithii, Dryopteris expansa, Elymus glaucus, Epilobium anagallidifolium, E. ciliatum, E. glaberrimum, Equisetum arvense, Eurhynchium oreganum, Festuca subulata, Fraxinus latifolia, Frangula purshiana, Galium aparine, G. triflorum, Gaultheria shallon, Geum macrophyllum, Glyceria striata, Gnaphalium, Gymnocarpium dryopteris, Heracleum lanatum, Hieracium albiflorum, Holcus, Hydrophyllum fendleri, H. tenuipes, Hypericum anagalloides, Hypochaeris radicata, Ilex aquifolium, Juncus bufonius, J. ensifolius, Kindbergia praelonga, Lactuca muralis, Lamium maculatum, L. purpureum, Leucolepis acanthoneuron, Lithocarpus densiflorus, Lonicera involucrata, Lotus corniculatus, Lupinus latifolius, L. lepidus, Luzula melanocarpa, L. parviflora, Lysichiton americanus, Maianthemum dilatatum, M. stellatum, Malus fusca, Marah oreganus, Mentha arvensis, Menziesia ferruginea, Mertensia paniculata, Micranthes odontoloma, Mimulus guttatus, Mitella ovalis, M. pentandra, Montia parvifolia, Myosotis laxa, Myrica californica, Nemophila parviflora, Oemleria cerasiformis, Oenanthe sarmentosa, Oplopanax horridum, Osmorhiza berteroi, O. purpurea, Oxalis oregana, Penstemon, Petasites frigidus, Philadelphus lewisii, Physocarpus capitatus, Picea engelmannii, P. sitchensis, Plagiomnium insigne, Plagiothecium undulatum, Plantago lanceolata, Platanthera stricta, Pleuropogon refractus, Poa trivialis, Polystichum munitum, Populus balsamifera, Potentilla egedii, Prosartes hookeri, Prunella vulgaris, Pseudotsuga menziesii, Pteridium aquilinum, Pyrola asarifolia, Ranunculus repens, Rhamnus* *purshiana, Rhododendron macrophyllum, Ribes bracteosum, R. lacustre, Rorippa curvisiliqua, Rosa pisocarpa, R. woodsii, Rubus discolor, R. leucodermis, R. parviflorus, R. spectabilis, R. ursinus, Salix hookeriana, S. lucida, S. scouleriana, S. sitchensis, Sambucus racemosa, Scirpus microcarpus, Scrophularia lanceolata, Senecio pseudaureus, e. triangularis, Sequoia sempervirens, Spiraea douglasii, Streptopus amplexifolius, Symphoricarpos albus, Tanacetum vulgare, Taraxacum officinale, Tellima grandiflora, Thalictrum occidentale, Thuja plicata, Tiarella trifoliata, Tolmiea menziesii, Trautvetteria caroliniensis, Trientalis borealis, Trifolium, Trillium ovatum, Tsuga heterophylla, Typha latifolia, Urtica dioica, Vaccinium ovatum, V. ovalifolium, V. parvifolium, Valeriana sitchensis, Vancouveria hexandra, Veratrum viride, Veronica americana, V. scutellata, Viola glabella, Viola sempervirens.*

***Stachys hyssopifolia* Michx.** is found in bogs, depressions, ditches, dunes, marshes, meadows, prairies, savannas, swales, swamps, and along the margins of lakes and ponds. The substrates are acidic (pH: 4.2–7.0) and include gravel, sand, and sandy peat. The plants occur only in open sites, often in standing water up to 15 cm deep. Moreso than the other species, the habitats almost always are characterized by pronounced seasonal water level fluctuations (e.g., shallow pond shores) and mature plants often are observed on drying substrates after the standing waters have receded. This species is relatively rare throughout its range and there is very little known regarding most aspects of its life history. Presumably the flowers are insect pollinated. Many of the associated species that occur in these ephemeral water habitats are annuals; however, *S. hyssopifolia* is a perennial and no information on its seed ecology has been reported. Vegetative reproduction occurs by slender rhizomes that become thickened and tuberous. Despite the perennial habit, it can be difficult to relocate plants in consecutive years, perhaps due to a diminished production of the conspicuous flowering shoots.
Reported associates: *Acalypha gracilens, Achillea millefolium, Agalinis purpurea, Agrostis gigantea, A. hyemalis, Ambrosia artemisiifolia, Andropogon gerardii, A. virginicus, Aristida necopina, Asclepias rubra, Bulbostylis capillaris, Calamagrostis canadensis, Carex scoparia, Cladium mariscoides, Coreopsis rosea, Cuscuta corylii, Cyperus bipartitus, C. dentatus, C. erythrorhizos, Daucus carota, Dichanthelium spretum, Drosera filiformis, Dulichium arundinaceum, Echinodorus tenellus, Eleocharis acicularis, E. melanocarpa, E. microcarpa, E. minima, E. obtusa, E. olivacea, E. palustris, E. robbinsii, E. tricostata, Epilobium coloratum, Erechtites hieracifolia, Erigeron canadensis, Eriocaulon aquaticum, Eupatorium leucolepis, E. perfoliatum, E. resinosum, Euphorbia corollata, Fimbristylis autumnalis, Fuirena pumila, F. squarrosa, Gnaphalium obtusifolium, Gratiola aurea, Helianthus angustifolius, Hypericum boreale, H. canadense, H. majus, Juncus articulatus, J. biflorus, J. brachycarpus, J. debilis, J. effusus, J. pelocarpus, J. scirpoides, Krigia biflora, Lachnanthes caroliniana, Lactuca canadensis, Lechea mucronata, Lipocarpha micrantha, Lobelia canbyi, Ludwigia hirtella, L. sphaerocarpa, Lycopodiella*

appressa, L. inundata, Lycopus americanus, L. amplectens, L. uniflorus, Lysimachia hybrida, L. quadrifolia, L. terrestris, Maianthemum racemosum, Muhlenbergia torreyana, Nuttallanthus canadensis, Oenothera biennis, Oldenlandia uniflora, Panicum rigidulum, P. wrightianum, Paspalum dissectum, P. laeve, Persicaria hydropiper, P. hydropiperoides, P. maculosa, P. punctata, Poa compressa, P. pratensis, Polygala cruciata, Polytrichum, Proserpinaca palustris, Quercus coccinea, Q. velutina, Rhexia mariana, R. virginica, Rhus typhina, Rhynchospora capitellata, R. corniculata, R. glomerata, R. macrostachya, R. nitens, R. perplexa, R. scirpoides, Rotala ramosior, Rubus flagellaris, Sabatia kennedyana, Sagittaria graminea, S. teres, Salix discolor, S. nigra, Schizachyrium scoparium, Schoenoplectus hallii, S. smithii, Scirpus cyperinus, Scleria minor, S. reticularis, S. triglomerata, Setaria glauca, Solidago graminifolia, S. latissimifolia, S. remota, S. tenuifolia, Spartina pectinata, Spiraea alba, S. tomentosa, Spiranthes tuberosa, Strophostyles umbellata, Symphyotrichum dumosum, Taraxacum laevigatum, T. officinale, Tradescantia ohiensis, Triadenum virginicum, Typha angustifolia, Verbascum thapsus, Verbena hastata, Viola lanceolata, Xyris difformis, X. torta.

***Stachys palustris* L.** inhabits beaches, ditches, dunes, fens, flats, floodplains, marshes, meadows, prairies, sloughs, swales, and the margins of lakes, ponds, rivers, and streams at elevations of up to 2550 m. The plants occur on a wide range of substrates (pH: 5.3–7.5) that include clay, gravel, gravelly sand, mud, sand, sandy loam, silty clay, silty clay loam, and silt loam. Localities typically are exposed to full sun and represent sites with permanent water or that are at least inundated seasonally. The plants are gynodioecious with hermaphrodite flowers that are protandrous, self-compatible, and capable of spontaneous selfing. Ordinarily, the flowers are cross-pollinated (mean pollen:ovule ratio = 6182) by a variety of different bees (see *Use by wildlife*). A single plant will produce about 64 nutlets on average. The nutlets float quite well (for up to 229 days) and are dispersed by water; up to 20% of the seeds can germinate while they still are afloat. Dormancy of *S. palustris* seeds is induced at temperatures >8°C–10°C and is relieved at temperatures <5°C. Seeds stratified for 10 weeks at 5°C will germinate in the light at 12°C–36°C, but require a diurnal temperature fluctuation of at least 10°C to achieve high germination rates (>50%). Rates of 80% germination have been attained with a temperature fluctuation of 14°C (mean temperature = 22°C). Reduced germination (20%) occurs in darkness at 25°C. Extended stratification (>4 weeks) may increase seed mortality. The plants can form a transient or somewhat persistent seed bank of up to 312 nutlets/m². *Stachys palustris* is a strong competitor relative to other common wetland species but is sensitive to heavy metal pollution. It exhibits a slightly positive correlation with the burn frequency of habitats. The plants are associated with arbuscular mycorrhizae. They produce diffuse ramets that propagate vegetatively by means of subterranean stolons, which develop into fusiform tubers. **Reported associates:** *Acer saccharinum, A. saccharum, Agrostis gigantea, Agropyron repens, Alisma subcordatum, Alnus rugosa,*

A. tenuifolia, Ambrosia psilostachya, Ammophila breviligulata, Amorpha canescens, Andropogon gerardii, Anemone canadensis, Antennaria microphylla, A. neglecta, Anticlea elegans, Apocynum cannabinum, Arctostaphylos uva-ursi, Artemisia douglasiana, A. ludoviciana, A. serrata, Asclepias hirtella, A. incarnata, A. syriaca, Atriplex prostrata, Betula nigra, B. pumila, B. ×sandbergii, Bidens cernuus, B. vulgatus, Blephilia ciliata, Boehmeria cylindrica, Bolboschoenus maritimus, Boltonia asteroides, Brachyelytrum erectum, Calamagrostis canadensis, C. stricta, Calystegia sepium, Carex athrostachya, C. grayi, C. hyalinolepis, C. macloviana, C. pellita, C. praegracilis, C. sartwellii, C. stipata, C. stricta, C. tetanica, Carya ovata, Celtis occidentalis, Chenopodium album, C. glaucum, C. rubrum, Cicuta maculata, Cirsium arvense, C. flodmanii, Comandra umbellata, Comarum palustre, Conyza canadensis, Coreopsis palmata, C. tripteris, Cornus sericea, Corylus americana, Crataegus, Cuscuta pentagona, Cyperus bipartitus, C. schweinitzii, C. strigosus, Cypripedium candidum, Dalea purpurea, Deschampsia cespitosa, Desmodium adscendens, D. illinoense, Dichanthelium wilcoxianum, Eleocharis acicularis, E. compressa, E. palustris, Elymus alaskanus, E. canadensis, E. glaucus, E. trachycaulus, Epilobium ciliatum, Equisetum laevigatum, Eryngium yuccifolium, Euphorbia corollata, E. esula, Euthamia graminifolia, Eutrochium maculatum, Fimbristylis autumnalis, Fragaria virginiana, Fraxinus nigra, F. pennsylvanica, Fuirena squarrosa, Galium boreale, G. trifidum, Gentiana andrewsii, Gentianopsis, Geum aleppicum, Glyceria striata, Glycyrrhiza lepidota, Grindelia squarrosa, Helenium autumnale, Heracleum lanatum, Hibiscus moscheutos, Holcus lanatus, Hordeum jubatum, Hypericum perforatum, H. sphaerocarpum, Hypoxis hirsuta, Impatiens capensis, Iris virginica, Juncus alpinus, J. balticus, J. brevicaudatus, J. bufonius, J. canadensis, J. effusus, J. greenei, J. nodosus, Lactuca tatarica, Laportea canadensis, Larix laricina, Lathyrus venosus, Leersia oryzoides, Lespedeza capitata, Liatris pycnostachya, Lilium michiganense, Lipocarpha micrantha, Lobelia cardinalis, L. kalmii, L. spicata, Lycopus asper, L. uniflorus, Lysimachia ciliata, L. nummularia, Medicago sativa, Mentha arvensis, Mimulus ringens, Monarda fistulosa, Muhlenbergia asperifolia, Napaea dioica, Nasturtium officinale, Nepeta cataria, Oenothera pilosella, Onosmodium molle, Panicum virgatum, Parnassia glauca, Pedicularis canadensis, P. lanceolata, Pediomelum esculentum, Persicaria amphibia, P. coccinia, Phalaris arundinacea, Phlox pilosa, Phragmites australis, Physalis heterophylla, P. longifolia, Physostegia virginiana, Pinus banksiana, Plantago eriopoda, P. lanceolata, Platanthera leucophaea, P. praeclara, P. psycodes, Poa palustris, P. pratensis, Polygala verticillata, P. virginiana, Polygonum ramosissimum, Populus balsamifera, P. tremuloides, Potentilla anserina, P. arguta, Prenanthes racemosa, Primula meadia, Pseudotsuga menziesii, Pteridium aquilinum, Puccinellia nuttalliana, Pycnanthemum virginianum, Quercus ellipsoidalis, Ranunculus cymbalaria, Ratibida pinnata, Rhexia virginica, Rhynchospora capitellata, Rosa arkansana, Rubus, Rudbeckia hirta, R. laciniata, R. subtomentosa, Rumex crispus, Salicornia rubra, Salix lucida, S. scouleriana, Saponaria officinalis, Schizachyrium scoparium, Schoenoplectus pungens, S. smithii, S. tabernaemontani, Scirpus atrovirens, S. cyperinus, Scolochloa festucacea, Scutellaria lateriflora, Senecio congestus, S. hydrophilus, Setaria viridis, Silphium compositum, S. integrifolium, S. laciniatum, S. perfoliatum, Sium suave, Solanum dulcamara, Solidago canadensis, Sonchus arvensis, Sorghastrum nutans, Sparganium eurycarpum, Spartina pectinata, Spergularia salina, Sphagnum, Spiranthes diluvialis, Spiraea alba, Streptopus amplexifolius, Suaeda calceoliformis, Symphyotrichum eatonii, S. foliaceum, S. laeve, S. lanceolatum, S. novae-angliae, S. praealtum, S. puniceum, S. subspicatum, Taenidia integerrima, Taraxacum officinale, Teucrium canadense, Thalictrum dasycarpum, T. venulosum, Tilia americana, Toxicodendron radicans, Trichophorum cespitosum, Triglochin maritimum, Typha latifolia, Ulmus americana, U. rubra, Urtica dioica, Veronica scutellata, Veronicastrum virginicum, Viola lanceolata, Vitis riparia, Xyris torta, Zizia aptera.*

***Stachys stricta* Greene** grows in chapparal, marshes, meadows, mudflows, seeps, washes, and along the margins of rivers and streams at elevations of up to 650 m. The substrates (often serpentine) are described as clay, gravel, rock, sand, sandy clay loam, and sandy loam. The plants are found both in open or shady sites. Like its congeners, *S. stricta* has also an affinity for ephemeral water and occurs often in vernal seeps or on drying streambeds. Ecological life-history information for this species is extremely limited. There are no detailed studies of its reproductive biology or seed ecology. **Reported associates:** *Abies grandis, Achlys triphylla, Acmispon americanus, A. grandiflorus, Adenocaulon bicolor, Aesculus californica, Agrostis viridis, Ailanthus altissima, Aira caryophyllea, Alnus rubra, Asarum caudatum, Asclepias fascicularis, Atrichum selwynii, Avena barbata, Berberis nervosa, Blechnum spicant, Bloomeria, Brachypodium distachyon, Briza minor, Bromus diandrus, B. hordeaceus, B. madritensis, Calycanthus occidentalis, Calypogeia neesiana, Cardamine californica, Carduus pycnocephalus, Carex dudleyi, C. hendersonii, C. nudata, Castilleja, Centaurea melitensis, Centaurium, Cephalozia bicuspidata, Chiloscyphus pallescens, Cladonia, Claopodium whippleanum, Claytonia sibirica, Clintonia andrewsiana, Corallorhiza striata, Cynodon dactylon, Cynoglossum grande, Cynosurus echinatus, Cyperus eragrostis, Cystopteris fragilis, Daucus pusillus, Dendroalsia abietina, Dicranum fuscescens, D. tauricum, Disporum smithii, Eleocharis acicularis, E. macrostachya, Epilobium densiflorum, Eriodictyon californicum, Eurhynchium oreganum, E. praelongum, Ficus carica, Fissidens bryoides, Frullania tamarisci, Galium triflorum, Gastridium ventricosum, Gaultheria shallon, Geocalyx graveolens, Geranium bicknellii, Gnaphalium luteoalbum, Helenium puberulum, Helianthella californica, Hemizonia fitchii, Heteromeles arbutifolia, Hierochloe occidentalis, Hoita macrostachya, Hordeum brachyantherum, H. murinum, Hypericum anagalloides, Hypnum circinale, Hypogymnia enteromorpha, Iris macrosiphon, Isothecium myosuroides,*

Juncus bufonius, J. effusus, J. oxymeris, Lactuca serriola, Lemna, Lepidozia reptans, Lessingia nemaclada, Leucolepis acanthoneuron, Lithocarpus densiflorus, Lolium multiflorum, Lotus corniculatus, Lythrum californicum, L. hyssopifolia, Madia gracilis, Melilotus alba, Mimulus cardinalis, M. floribundus, M. guttatus, Mentha arvensis, Muhlenbergia, Myrica californica, Nassella pulchra, Neckera douglasii, Osmorhiza occidentalis, Oxalis oregana, Parmelia perlata, Paspalum dilatatum, P. distichum, Pellaea andromedaefolia, Pertusaria ambigens, Phacelia cicutaria, Pinus sabiniana, Plagiomnium insigne, Plagiothecium laetum, P. undulatum, Platismatia glauca, Poa, Polypodium scouleri, Polypogon monspeliensis, Polystichum munitum, Populus tremuloides, Porella navicularis, Porotrichum bigelovii, Pseudotsuga menziesii, Pteridium aquilinum, Pyrola picta, Quercus douglasii, Q. lobata, Q. wislizeni, Ranunculus muricatus, Rhamnus tomentella, Rhododendron macrophyllum, R. occidentale, Rhus diversiloba, Rorippa, Rosa, Rumex conglomeratus, R. crispus, R. pulcher, Salix laevigata, S. lasiolepis, S. lucida, Sambucus mexicana, Scapania bolanderi, Schoenoplectus, Scleropodium touretii, Sequoia sempervirens, Silybum marianum, Sisyrinchium bellum, Stephanomeria virgata, Symphoricarpos rivularis, Tetraphis pellucida, Tiarella unifoliata, Torilis arvensis, Torreya californica, Toxicodendron diversilobum, Trichostema rubisepalum, Trientalis latifolia, Trifolium willdenovii, Trillium ovatum, Triteleia hyacinthina, Tsuga heterophylla, Typha domingensis, T. latifolia, Umbellularia californica, Usnea, Vaccinium ovatum, V. parvifolium, Verbascum blattaria, Veronica anagallis-aquatica, Vicia villosa, Viola arvensis, V. sempervirens, Vulpia myuros, Xanthium strumarium.

Stachys tenuifolia Willd. grows in bayous, canals, carrs, depressions, ditches, fens, floodplains, gravel pits, marshes, meadows, oxbows, prairies, savannas, sloughs, swales, swamps, and along margins of brooks, lakes, ponds, rivers, and streams at elevations of up to 200 m. The substrates are derived mainly from sandstone (pH: 5.1–8.0) and include clay, gravel, loam, loamy muck, mud, peat, sand, sandy peat, silt, and silty clay loam. Exposures range from full sun to light shade. The plants can form thickets due to stems that root at the nodes and a creeping rhizomatous habit. The plants are found in relatively disturbed as well as undisturbed sites. This is another species where basic life-history information is sparse. **Reported associates:** *Acalypha rhomboidea, Acer negundo, A. rubrum, A. saccharinum, Agalinis tenuifolia, Alisma subcordatum, Allium canadense, Alternanthera philoxeroides, Amaranthus tuberculatus, Ambrosia artemisiifolia, Amphicarpa bracteata, Andropogon, Anemone canadensis, Angelica atropurpurea, Apios americana, Apocynum sibiricum, Artemisia serrata, Asclepias incarnata, A. perennis, A. sullivantii, Asparagus, Berteroa, Betula lutea, B. nigra, B. papyrifera, Bidens cernuus, B. comosus, B. connatus, B. discoideus, B. frondosus, B. vulgatus, Boehmeria cylindrica, Boltonia asteroides, Calamagrostis canadensis, Calystegia sepium, Campanula americana, Cardamine pensylvanica, Carex aquatilis, C. buxbaumii, C. granularis, C. grayi, C. hyalinolepis, C. haydenii, C. lacustris, C. lanuginosa,*

C. lupuliformis, C. lupulina, C. muskingumensis, C. normalis, C. squarrosa, C. stricta, C. tribuloides, C. trichocarpa, C. tuckermanii, C. typhina, C. vulpinoidea, Celtis occidentalis, Centaurea, Cephalanthus occidentalis, Cicuta bulbifera, C. maculata, Circaea lutetiana, Cirsium arvense, C. muticum, C. vulgare, Climacium americanum, Conocephalum conicum, Conyza, Cornus amomum, C. racemosa, C. sericea, Corylus, Crataegus, Cryptotaenia canadensis, Cynodon, Cyperus bipartitus, Dichanthelium acuminatum, D. clandestinum, Echinocystis lobata, Eleocharis elliptica, E. erythropoda, Elymus canadensis, E. repens, E. riparius, E. villosus, E. virginicus, Epilobium ciliatum, Equisetum arvense, E. hyemale, Eragrostis, Erechtites hieraciifolia, Erigeron canadensis, E. philadelphicus, Eryngium yuccifolium, Euonymus atropurpureus, Eupatorium fistulosum, E. maculatum, E. perfoliatum, E. rugosum, Eurybia macrophylla, Euthamia graminifolia, Fallopia scandens, Fraxinus pennsylvanica, Galium asprellum, G. boreale, G. brevipes, G. obtusum, G. trifidum, Geum canadense, Glechoma hederacea, Glyceria septentrionalis, G. striata, Gratiola neglecta, Hasteola suaveolens, Helenium autumnale, Helianthus grosseserratus, H. decapetalus, H. tuberosus, Heliopsis helianthoides, Hibiscus laevis, Hydrolea affinis, Hypericum perforatum, Hystrix patula, Impatiens capensis, I. pallida, Iodanthus pinnatifidus, Iris versicolor, I. virginicus, Juncus balticus, Laportea canadensis, Leersia lenticularis, L. oryzoides, L. virginica, Leonurus cardiaca, Liatris spicata, Lilium michiganense, Linaria, Lobelia cardinalis, L. siphilitica, Lychnis, Lycopus americanus, L. rubellus, Lysimachia ciliata, L. hybrida, L. nummularia, L. quadriflora, Lythrum alatum, Menispermum canadense, Mentha arvensis, M. spicata, Mikania scandens, Mimulus alatus, Mitchella repens, Monarda fistulosa, Muhlenbergia mexicana, Myosoton aquaticum, Napaea dioica, Nyssa biflora, Oenothera clelandii, Onoclea sensibilis, Oxypolis rigidior, Panicum repens, Parnassia glauca, Parthenocissus, Paspalum dilatatum, Pedicularis lanceolata, Penthorum sedoides, Persicaria coccinea, P. pensylvanica, P. posumbu, P. punctata, P. sagittata, P. virginiana, Phalaris arundinacea, Phlox stolonifera, Phragmites australis, Phyla lanceolata, Physalis longifolia, Physostegia virginiana, Pilea pumila, Platanthera leucophaea, Poa trivialis, Populus balsamifera, P. deltoides, P. tremuloides, Potentilla anserina, Prenanthes racemosa, Pteridium aquilinum, Ptilimnium costatum, Pycnanthemum flexuosum, P. virginianum, Quercus bicolor, Q. macrocarpa, Q. palustris, Q. rubra, Ranunculus abortivus, R. septentrionalis, Rhamnus cathartica, Rosa palustris, R. setigera, Rubus, Rudbeckia hirta, R. laciniata, Rumex orbiculatus, R. verticillatus, Sagittaria latifolia, Salix amygdaloides, S. nigra, Sambucus nigra, Samolus valerandi, Schoenoplectus tabernaemontani, Scirpus expansus, Scrophularia marilandica, Scutellaria lateriflora, S. nervosa, Packera aurea, P. glabella, Setaria, Sicyos angulatus, Silphium perfoliatum, Sium suave, Solanum dulcamara, S. nigrum, Solidago gigantea, S. rugosa, Sorghum halepense, Sparganium americanum, Spartina pectinata, Spilanthes, Spiraea alba, Sporobolus, Symphyotrichum cordifolium, S. dumosum, S. lanceolatum, S. lateriflorum, S. novae-angliae,

S. ontarionis, S. prenanthoides, S. puniceum, Taxodium distichum, Teucrium canadense, Thalictrum dasycarpum, Thelypteris palustris, Tilia americana, Toxicodendron radicans, Triadenum fraseri, T. walteri, Triadica sebifera, Trichophorum cespitosum, Trillium cernuum, Typha latifolia, Ulmus americana, Urtica dioica, Uvularia sessilifolia, Verbascum, Verbena hastata, V. urticifolia, Verbesina alternifolia, V. occidentalis, Vernonia missurica, Veronica scutellata, Viola affinis, V. cucullata, V. sororia, Vitis riparia, Zanthoxylum americanum, Zizania aquatica.

Use by wildlife: *Stachys ajugoides* is a host to several larval Lepidoptera (Insecta: Noctuidae: *Autographa pasiphaea, Megalographa biloba*; Tortricidae: *Aphelia alleniana*). The flowers are visited by nectar-seeking hummingbirds (Aves: Trochilidae), bees (Insecta: Hymenoptera: Megachilidae: *Osmia*), and butterflies (Lepidoptera). The nutlets are eaten by various wildlife. The plants are the hosts of several Fungi including powdery mildews (Ascomycota: Erysiphaceae: *Erysiphe galeopsidis, Podosphaera macularis*) and other Ascomycota (Mycosphaerellaceae: *Ramularia bullata, R. lamii*). The roots are eaten by deer (Mammalia: Cervidae: *Odocoileus*) and the foliage is eaten by grizzly bears (Mammalia: Ursidae: *Ursus arctos horribilis*). *Stachys albens* has been identified as a characteristic habitat component of the mountain beaver (Mammalia: Aplodontiidae: *Aplodontia rufa*). It is the host of a generalist predatory bug (Insecta: Heteroptera: Miridae: *Dicyphus hesperus*), which is being evaluated to control several agricultural and commercial pests. It is a host plant for several aphids (Insecta: Homoptera: Aphididae: *Amphorophora rubi, Aulacorthum solani*). The foliage of *S. chamissonis* is eaten by caterpillars (Lepidoptera: Noctuidae: *Euplexis benesimilis*), slugs (Mollusca: Arionidae: *Ariolimax, Arion*) and is grazed by deer (Mammalia: Cervidae: *Odocoileus*) and Roosevelt elk (Mammalia: Cervidae: *Cervus elaphus roosevelti*). It can comprise up to 1.6% of the diet of the latter. The juice is eaten by spittlebugs (Insecta: Homoptera: Cercopidae). The floral nectar is removed by butterflies (Lepidoptera) and hummingbirds (Aves: Trochilidae), which are the principal pollinators. Hoverflies (Insecta: Diptera: Syrphidae) feed on the pollen. The plants are a host of powdery mildew Fungi (Ascomycota: Erysiphaceae: *Erysiphe cichoracearum*). *Stachys palustris* and *S. tenuifolia* are other hosts of *Erysiphe galeopsidis* (Ascomycota: Erysiphaceae). The flowers of *S. palustris* are visited by many insects (Insecta), which remove the nectar. These include long-tongued bees (Hymenoptera: Anthophoridae: *Anthophora abrupta, A. terminalis, Ceratina dupla, Melissodes bimaculata, Nomada superba*; Apidae: *Bombus auricomus, B. bimaculatus, B. borealis, B. fervida, B. griseocallis, B. impatiens, B. pensylvanica, B. perplexus, B. vagans*; Megachilidae: *Hoplitis pilosifrons, Megachile brevis*); short-tongued bees (Andrenidae: *Calliopsis andreniformis*; Colletidae: *Hylaeus mesillae*; Halictidae: *Agapostemon virescens, Lasioglossum paradmirandus, L. versatus*), Lepidoptera (Hesperiidae: *Ancyloxypha numitor, Polites themistocles*; Noctuidae: *Anagrapha falcifera, Autographa precationis*; Papilionidae: *Papilio marcellus*), and flies

(Diptera: Bombyliidae: *Systoechus vulgaris*; Syrphidae: *Allograpta obliqua, Sphaerophoria, Syrphus ribesii, Toxomerus marginatus*). Only the bees (except *Lasioglossum versatus*) are functional pollinators. The plants host several larval Lepidoptera (Noctuidae: *Autographa pasiphaea*; Nymphalidae: *Euphydryas chalcedona*; Tortricidae: *Endothenia hebesana, E. quadrimaculana*). *Stachys palustris* can be grazed fairly heavily by white-tailed deer (Mammalia: Cervidae: *Odocoileus virginianus*), especially when the deer density is high.

Economic importance: food: The tuberous rhizomes of *Stachys hyssopifolia* are crisp and can be cooked or pickled, eaten raw like celery, or added to salads. Those of *S. palustris* reportedly are edible when boiled, dried, or made into bread. The Kawaiisu added *Stachy albens* to food as a spice. Haida children chewed the stems of *S. chamissonis* for the juice. The Makah and Quinault tribes used the plants to cover sprouts while steaming them. The Quinault would drink the sweet nectar from the flowers of *S. ajugoides*. Seeds of *S. palustris* were used as food by the Gosiute; **medicinal:** The Saanich tribe made a tonic from the rhizomes of *S. ajugoides*. People of the Green River Group and Puyallup tribes applied a poultice made from *S. ajugoides* plants to treat boils. The seeds of *S. ajugoides* contain beta-caryophyllene (14%), a substance having anesthetic activity like that of clove oil. *Stachys palustris* contains several iridoids (harpagide, acetylharpagide), which are substances attributed with analgesic and anti-inflammatory properties. The plants have been used as an antiseptic, to relieve gout, and to stop hemorrhages. An infusion of fresh or dried leaves of *S. palustris* was used by the Chippewa for colic. The Delaware and Oklahoma people added the root to a compound used for treating venereal disease; **cultivation:** *Stachys ajugoides, S. albens, S. hyssopifolia*, and *S. palustris* are cultivated infrequently. *Stachys chamissonis* is distributed more widely as an ornamental native wildflower for wet areas and water gardens. Cultivars include *S. chamissonis* var. *cooleyae* and 'Royal Cloak.' *Stachys stricta* includes the cultivar 'Dark Lilac'; **misc. products:** The Kawaiisu used the leaves of *S. albens* to cork their water bottles. The Haisla and Hanaksiala made fishing nets fashioned from fibers of *S. ajugoides*. *Stachys chamissonis* transplants well and has been considered as a useful plant for wetland restoration projects. Plants of *S. palustris* can be made into a yellow dye (when mixed with alum) or a black dye (when mixed with logwood). The coat of arms of Britain's Royal Society of Medicine features a hand clasping a bunch of *S. palustris* plants; **weeds:** The introduced hexaploid cytotype of *Stachys palustris* is regarded as a serious weed, especially in cultivated fields; **nonindigenous species:** Although *Stachys palustris* is native to the continent, hexaploid cytotypes have been introduced from Europe to some portions of northeastern North America. The plants probably originated as escapes from contaminated shipping ballast, which was disposed of during the early part of the 19th century.

Systematics: *Stachys* is placed in Lamiaceae subfamily Lamioideae, a position confirmed by the analysis of *rbcL* sequence data. However, the genus is not monophyletic in its current circumscription as indicated by species that associate with several other genera (e.g., *Phlomidoschema, Prasium, Sideritis*), and a clade of Hawaiian genera, which nests within *Stachys* (Figure 5.62). With only about 15% of the species surveyed phylogenetically at present, the systematics of this large genus undoubtedly will undergo substantial revision as further taxa are included in comparative phylogenetic analyses. The available evidence does indicate at least three independent origins of the OBL habit in North America and a relatively derived phylogenetic position of the aquatic taxa in the genus (Figure 5.62). The current state of taxonomy in *Stachys* is confusing and ambiguous species delimitations have made it extremely difficult to apply information in the literature to the correct taxon; undoubtedly some errors have been made. In the present treatment, *S. mexicana* and *S. rigida* are considered to be conspecific with *S. ajugoides* following recent taxonomic recommendations. However, whether one, two, or three taxa should be maintained in this instance deserves further study. *Stachys ajugoides* var. *rigida* sometimes is recognized as a distinct species (*S. rigida*), which is then designated as a FACW plant. Some authors have merged *S. mexicana* (FACW, UPL) with

S. rigida, while keeping the taxon distinct from *S. ajugoides*. *Stachys cooleyae* generally (but not universally) has been treated as a variety of *S. chamissonis* (var. *cooleyae*), which is the disposition followed here. Much confusion arises from the fact that the name "*Stachys ciliata*" has been applied as a synonym to two different species; i.e., to *S. chamissonis* (=*S. ciliata* Epling) and to *S. mexicana* [i.e., *S. ajugoides*] (=*S. ciliata* Dougl. ex Benth.), and that literature accounts seldom identify the author of the name used. There is no consensus whether *S. pilosa* should be merged with *S. palustris* (e.g., subsp./var. *pilosa*, subsp. *arenicola*) or maintained as distinct (as is done here). Although recent investigations have made some progress in elucidating the taxonomy of this large and difficult genus, additional systematic study is needed to further clarify species limits and relationships of the taxa treated herein. The chromosome numbers of *Stachys* are equally difficult to interpret. A wide range of variation ($2n = 10–102$) precludes an unambiguous assessment of base chromosome number for the genus as a whole. A base number of $x = 17$ has been suggested for the *S. palustris* "complex" and probably is applicable to all of the species treated here. Accordingly, *S. hyssopifolia* ($2n = 34$) would be diploid and *S. ajugoides*

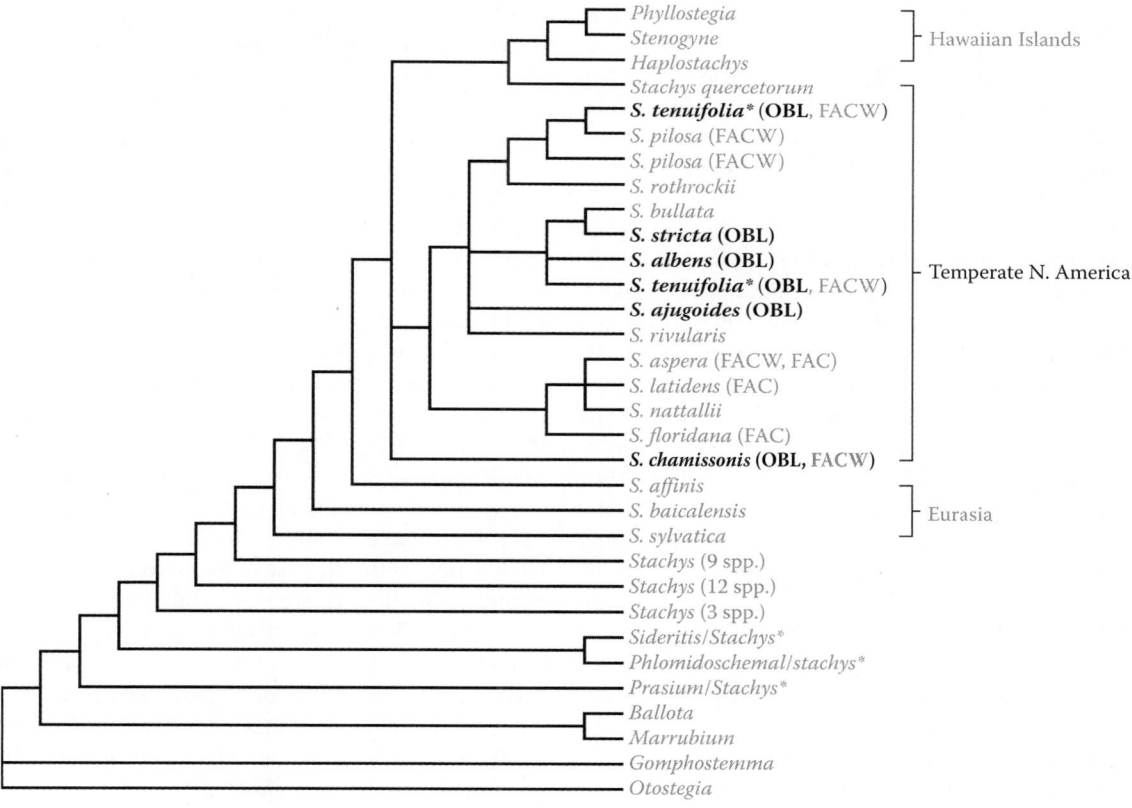

FIGURE 5.62 Phylogenetic relationships in *Stachys* based on analysis of 5S-NTS sequence data representing about 15% of the species. The result shown (one of 3783 equal minimum-length trees recovered using maximum parsimony) retains the major features of the strict consensus tree. This phylogenetic tree indicates at least three independent origins of OBL species (bold) from within a clade of temperate North American taxa. However, *Stachys* is not monophyletic (indicated by asterisks) as currently circumscribed unless several other groups (*Phlomidoschema, Prasium, Sideritis*, Hawaiian Islands genera) are included. The Hawaiian species are hypothesized to be of polyploid hybrid origin. The disparate placement of *S. tenuifolia* accessions (asterisked) might be due to a misidentification or perhaps to hybridization (e.g., involving *S. palustris*; not surveyed). Generally, the OBL species appear to be relatively derived within the genus. (Modified from Lindqvist, C. & Albert, V.A., *Amer. J. Bot.*, 89, 1709–1724, 2002.)

$(2n = 64, 66)$, *S. albens* $(2n = 66)$, *S. chamissonis* $(2n = 64)$ and *S. stricta* $(2n = 66)$ would be tetraploids. *Stachys tenuifolia* $(2n = 34, 68)$ consists of diploid and tetraploid cytotypes. *Stachys palustris* $(2n = 34, 68, 102)$ includes diploid, tetraploid, and hexaploid cytotypes. The chromosomes of some tetraploids are dimorphic with one or two pairs being about twice the length of the rest. The $2n = 66$ tetraploids have one longer chromosome pair and are believed to be derived from a $2n = 68$ ancestor by chromosomal fusion. The $2n = 64$ species have two pairs of longer chromosomes, which indicates a similar derivation from an ancestral $2n = 68$ tetraploid. Chromosomes of *S. tenuifolia* tetraploids $(2n = 68)$ are of nearly equal length. It is suspected that natural hybridization occurs between *S. palustris* and *S. tenuifolia*, but this possibility needs to be investigated experimentally.

Comments: *Stachys palustris* is widespread in North America except for the southeastern United States and also extends throughout Eurasia. More restricted ranges characterize *S. tenuifolia* (eastern North America), *S. hyssiopifolia* (eastern United States), *S. ajugoides* and *S. chamissonis* (western North America), *S. albens* (California, Nevada, Utah), and *S. stricta* (endemic to California).

References: Alexander, 2007; Allen-Diaz et al., 2004; Anderson et al., 2001; Bailey & Poulton, 1968; Biondi & Casavecchia, 2001; Bliss & Cox, 1964; Brändel, 2006; Cates & Orians, 1975; Christy, 2004; Cooke, 1997; Cruden, 1977; Edmonstone, 1841; Fernald & Kinsey, 1943; Fonda, 1974; Galatowitsch & van der Valk, 1996a; Gaudet & Keddy, 1995; Graham & Henry, 1933; Grant, 1949; Grant & Grant, 1968; Guard, 1995; Gunther, 1973; Halpern & Harmon, 1983; Hanley & Taber, 1980; Harmon & Franklin, 1995; Jenkins & Starkey, 1991; Klinka et al., 1995; Lawrence, 1993; Lindqvist & Albert, 2002; Löve & Löve, 1954; MacCracken, 2002; MacHutchon et al., 1993; Mac Nally, 1995; Michener, 1936a; Mohlenbrock, 1959a; Mulligan & Munro, 1983, 1989; Pabst & Spies, 1998; Price, 2001; Rivas-Martínez et al., 1999; Rustaiyan et al., 2006; Sheridan & Spies, 2005; Sluis & Tandarich, 2004; Smeins & Olsen, 1970; Speichert & Speichert, 2004; Stevens, 1932; Tester, 1996; Turner, 1995, 1998; USDA, 2005; van den Broek et al., 2005; Van der Valk & Davis, 1978; VanLaerhoven et al., 2006; Westman, 1975; Wilcock, 1974; Wolken et al., 2001.

11. *Trichostema*

Bluecurls

Etymology: from the Greek *trichos stemon* meaning "hair stamened" in reference to the capillary filaments

Synonyms: *Eplingia*; *Isanthus*

Distribution: global: North America; **North America:** widespread

Diversity: global: 17 species; **North America:** 17 species

Indicators (USA): OBL: *Trichostema austromontanum*

Habitat: freshwater; palustrine; **pH:** unknown; **depth:** <1 m; **life-form(s):** emergent herb

Key morphology: stems (to 5 dm) erect, hairy, the herbage with a strong odor; leaves (to 5 cm) opposite, hairy, elliptic, short-petioled (to 5 mm); cymes axillary, 1–3-flowered; flowers irregular; corolla tubular (to 3 mm), 5-lobed (to 3 mm), blue to purple, lower lobe white with purple blotches; stamens (to 5.5 mm) 4, purple, arching between upper two corolla lobes, much exserted; nutlets 4, joined at lower third, hairy, ridged

Life history: duration: annual (fruit/seeds); **asexual reproduction:** none; **pollination:** self; **sexual condition:** hermaphroditic; **fruit:** nutlets (common); **local dispersal:** nutlets (unknown); **long-distance dispersal:** none

Imperilment: (1) *Trichostema austromontanum* [G3]; S1 (CA)

Ecology: general: *Trichostema* species generally inhabit dry, open sites (e.g., chaparral, grassland, scrub), and also occur often along the dry margins of lakes and watercourses. Only four of the North American species (24%) are wetland indicators with one species designated as OBL. Most of the species are annuals with 4–5 being shrubby or subfruticose perennials. The species predominantly are cross-pollinated by bees (Insecta: Hymenoptera: Andrenidae, Anthophoridae, Apidae, Halictidae, Megachilidae), wasps (Insecta: Hymenoptera: Vespidae), or hummingbirds (Aves: Trochilidae); however, a few are autogamous. All of the species are self-compatible and set seed in the absence of pollinators. Some of the grassland species require prolonged burial and smoke exposure to trigger the germination of the dormant seeds. The seed ecology and dispersal mechanisms are unknown for most of the species. The narrow distributions of several species may be due to the adaptation of seeds to montane conditions or perhaps to reduced dispersability related to the basal fusion of the nutlets. Some species are known to be weakly allelopathic.

Trichostema austromontanum **F.H.** Lewis occurs in depressions, ditches, meadows, seeps, swales, and along the margins of lakes, pools, and streams at elevations from 500 to 2650 m. The substrates are granitic in origin and are described as clay loam, gravel, mud, sand, and sandy loam. Site exposures range from partial shade to full sun. Similar to the preceding genus, *Trichostema* usually occurs where the water is ephemeral or has receded, especially along the margins of lacustrine or riverine habitats. Depending on precipitation patterns, populations can vary by as much as from 11 to 10,000 individuals in any given year. Field observations have indicated that *T. austromontanum* has no functional pollinators. The flowers are self-compatible, produce no nectar, have low pollen:ovule ratios (mean = 233), and are autogamous. Similar levels of seed set occur whether flowers are visited by insects (90%) or are excluded from insects (87%). No information exists on the seed ecology. **Reported associates:** *Achillea millefolium, Acmispon americanus, Agrostis scabra, Ambrosia psilostachya, Artemisia dracunculus, A. tridentata, Atriplex coronata, Brodiaea terrestris, Bromus orcuttianus, B. tectorum, Carex douglasii, Castilleja lasiorhyncha, Collinsia childii, Cornus nuttallii, Cystopteris fragilis, Epilobium densiflorum, Epipactis gigantea, Erigeron philadelphicus, Eriodictyon californicum, Eriophyllum confertiflorum, Festuca, Heterocodon, Ipomopsis, Iris hartwegii, Juncus covillei, J. effusus, J. ensifolius, Lilium parryi, Lupinus, Mimulus brevipes, M. pilosus, Navarretia intertexta,*

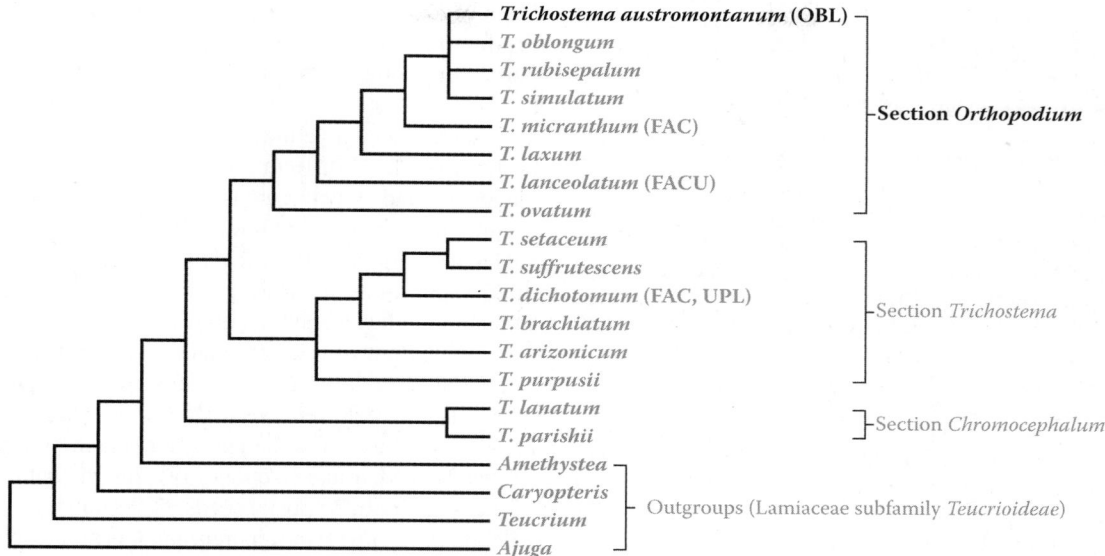

FIGURE 5.63 Phylogenetic relationships in *Trichostema* as indicated by analysis of combined nrITS and *ndhF* sequence data. *Trichostema austromontanum* (bold), the only OBL representative of the genus in North America, resolves within a relatively derived position close to *T. oblongum* and *T. micranthum*, the two putative progenitors of this alleged allotetraploid. Other analyses (e.g., Cantino et al., 1998; Wagstaff et al., 1998) indicate a sister group relationship of *Trichostema* and *Caryopteris*. (Adapted from Huang, M. et al., *Syst. Bot.*, 33, 437–446, 2008.)

Pinus coulteri, P. jeffreyi, P. ponderosa, Plectritis, Pteridium aquilinum, Quercus kelloggii, Q. wislizenii, Rosa woodsii, Salix exigua, Scutellaria bolanderi.

Use by wildlife: none reported

Economic importance: food: *Trichostema* is not edible; **medicinal:** none reported; **cultivation:** not cultivated; **misc. products:** none; **weeds:** none; **nonindigenous species:** none

Systematics: The placement of *Trichostema* in Lamiaceae subfamily Teucrioideae is supported by phylogenetic analysis of combined cpDNA sequences. Molecular data have also corroborated the monophyly of the genus, which is distinctive among mints in having nutlets that are fused for a third of their length. However, the relationships of the genus in the subfamily are not as straightforward. Analysis of nonmolecular data indicates an affinity to *Tetraclea, Oncinocalyx, Teucrium,* and *Teucridium*; however, several studies using DNA sequence data resolve *Caryopteris* as the nearest relative of *Trichostema*. Further complicating matters is the fact that nonmolecular and molecular data alike resolve *Caryopteris* as polyphyletic. The current evidence favors the retention of *Trichostema* with either *Amethystea* or *Caryopteris* sensu stricto (i.e., section *Caryopteris*) as its sister group. Earlier taxonomic treatments distinguished five sections within *Trichostema*; however analyses of combined nrITS and *ndhF* sequences resolve three subclades that correspond mainly to sections *Chromocephalum, Orthopodium,* and *Trichostema* (Figure 5.63). *Trichostema austromontanum* occurs in section *Orthopodium*, closely allied to *T. oblongum, T. rubisepalum,* and *T. simulatum*. The base chromosome number of *Trichostema* section *Orthopodium* is $x = 7$, which is believed to represent a derived base number reduction in the genus. All species in the section are diploid except for *T. austromontanum* ($2n = 28$), which is tetraploid. *Trichostema austromontana* is thought to be an allotetraploid derivative of *T. micranthum* and *T. oblongum*, a result that is supported by its combination of genetic (i.e., RAPD) and morphological markers that distinguish the putative ancestral species. Dwarfed, smaller-leaved plants are treated as *T. austromontanum* subsp. *compacta* and occur only within a single vernal pool in California.

Comments: *Trichostema austromontanum* is restricted to California and Nevada.

References: Armstrong & Crawford, 1996; Bauder & McMillan, 1998; Cantino et al., 1998; Heisey & Delwiche, 1985; Huang et al., 2000, 2008; Keeley & Fotheringham, 1998; Lewis, 1945, 1960; Spira, 1980; Wagstaff et al., 1998.

Family 3: Lentibulariaceae [3]

This cosmopolitan family of carnivorous plants contains roughly 320 aquatic, epiphytic, or terrestrial species. The family is distinctive by its combination of carnivorous habit, glandular hairs that secrete mucilage and digestive enzymes, absence of roots (except *Pinguicula*), bilabiate flowers with a lobed, saccate or spurred lower lip, and free-central placentation. The leaves are highly modified and include those with a flat, glandular flypaper-like surface and margins that inroll when in contact with prey (*Pinguicula*); forked, downward oriented, tubular, spirally twisted segments that lead to a passive bladder-like trap (*Genlisea*); or, highly dissected segments covered with valve-like bladders having a "trap-door" that closes actively as prey is ingested (*Utricularia*). Several species develop turion-like winter buds as hibernacula.

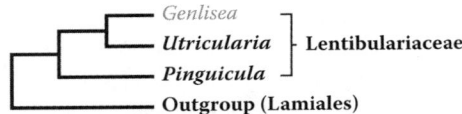

FIGURE 5.64 Phylogenetic relationships in Lentibulariaceae as indicated by analyses of cpDNA sequence data. These studies provide strong support for the monophyly of the family, but its nearest relative within Lamiales has not been ascertained with confidence. Taxa with OBL North American representatives are indicated in bold. Because *Genlisea* also is aquatic (but Old World), there appears to be a single origin of the OBL habit in the family. (Adapted from Jobson, R.W. et al., *Syst. Bot.*, 28, 157–171, 2003; Müller, K. et al., *Plant Biol.*, 6, 477–490, 2004; Cieslak, T. et al., *Amer. J. Bot.*, 92, 1723–1736, 2005.)

The flowers often are showy and typically are outcrossed by bees (Hymenoptera: Apidae) or wasps (Hymenoptera: Vespidae). However, they are self-compatible and autogamy can occur. In some *Utricularia* species, tactile sensitivity of the lower stigma lobe results in the nearly complete closure of the stigma when it has been pollinated (Newcombe, 1922). Some species produce cleistogamous flowers. The capsular fruits contain numerous seeds, which are dispersed by animals, water, and wind.

Ecologically, most Lentibulariaceae occur either in the water or on wet substrates. The habitats generally are characterized by nutrient-poor sediments (e.g., peat or sand), where low nitrogen levels are then offset by carnivory. The plants exhibit a diversity of habits including emergent, floating, submersed and suspended life-forms. Two genera occur in North America and both contain obligate aquatics:

1. *Pinguicula* L.
2. *Utricularia* L.

The monophyly of Lentibulariaceae has been confirmed consistently by several phylogenetic studies analyzing cpDNA sequence data (Chase et al., 1993; Jobson et al., 2003; Müller et al., 2004; Cieslak et al., 2005). Where all three genera have been included, *Genlisea* and *Utricularia* resolve as a well-supported clade, with *Pinguicula* as their sister group (Figure 5.64). However, there is no consensus on the sister group of Lentibulariaceae, with Bignoniaceae, and Lamiaceae (but not Byblidaceae or Martyniaceae as some had suspected) among the possible families suggested by recent analyses (e.g., Müller et al., 2004; Cieslak et al., 2005).

1. *Pinguicula*

Butterwort, pings; grassette
Etymology: from the Latin *pinguis*, meaning "fat" pertaining to the greasy texture of the leaves
Synonyms: *Isoloba*
Distribution: global: Asia; Europe; New World; **North America:** widespread
Diversity: global: 85 species; **North America:** 9 species
Indicators (USA): OBL: *Pinguicula caerulea, P. ionantha, P. lutea, P. macroceras, P. planifolia, P. primuliflora, P. pumila, P. villosa, P. vulgaris;* **FACW:** *P. lutea*

Habitat: freshwater; palustrine; **pH:** 3.1–8.2; **depth:** <1 m; **life-form(s):** emergent herb
Key morphology: stems compressed (<5 mm), with small (to 1 cm) or large (to 20 cm) rosettes of decumbent, sessile, succulent, greenish-yellow, or reddish leaves, the blades small (to 1.5 cm) or large (to 9 cm), elliptic to oblong, the upper surface glandular and mucilaginous, the margins inrolled; flowers showy, bilateral, nodding, solitary at apex of 1-numerous naked, glandular to villous scapes (to 30 cm); corolla (to 3.5 cm) tubular, pale to deep purple, pink, white, or yellow, sometimes streaked, the lower section contracted into a spur (to 9 mm), a pubescent white or yellow "palate" (to 6 mm) extending along the bottom of the throat, the limb spreading, with two upper lobes and three lower lobes, each lobe shallowly to deeply incised; stamens 2, the filaments thick, arching around ovary; stigma 2-lobed, a flap-like lobe covering the anthers; capsules (to 1 cm) globose, 2-valved; seeds (to 0.8 mm) numerous
Life history: duration: perennial (dormant apices, winter buds); **asexual reproduction:** adventitious plantlets, winter buds; **pollination:** insect; **sexual condition:** hermaphroditic; **fruit:** capsules (common); **local dispersal:** adventitious plantlets; gemmae (water); seed (water, wind); **long-distance dispersal:** seed (water, wind)
Imperilment: (1) *Pinguicula caerulea* [G4]; S3 (FL, NC); (2) *P. ionantha* [G2]; S2 (FL); (3) *P. lutea* [G4/G5]; S2 (LA, NC); S3 (FL); (4) *P. macroceras* [G5]; S3 (<u>BC</u>, CA, MT, OR); (5) *P. planifolia* [G3]; S1 (AL); S2 (MS); S3 (FL); (6) *P. primuliflora* [G3]; S1 (GA); S3 (AL, FL, MS); (7) *P. pumila* [G4]; S1 (AL, OK); S2 (NC); (8) *P. villosa* [G4]; S1 (<u>AB</u>); S2 (<u>BC</u>, <u>LB</u>, <u>ON</u>, <u>QC</u>, <u>SK</u>); S3 (<u>MB</u>); (9) *P. vulgaris* [G5]; S1 (ME, <u>NB</u>, NH, <u>NS</u>, VT, WI); S2 (NY, <u>SK</u>); S3 (<u>AB</u>, MI, MN)
Ecology: general: Nearly all of the North American *Pinguicula* species occur in wet habitats and are designated as OBL indicators. Only one (*P. lutea*) grows in somewhat drier habitats and is designated as OBL, FACW. The habitats often are acidic, but some species can occur at pH values that exceed 8.0. The self-compatible flowers predominantly are outcrossed by flying insects, typically small bees (Hymenoptera), although the progressive backward curvature of the stigma in unpollinated flowers eventually will result in self-pollination. Once open, the flowers remain receptive for 3–10 days. The flap-like stigma covers the anthers with the receptive surface exposed to the interior of the flower. Insects carrying pollen on their backs inevitably brush against the receptive stigma as they enter to collect nectar. They are forced against the top of the corolla as they crawl across a prominent, glandular structure known as the "palate." Insects exiting the flower force the stigma flap upward to expose the anthers, which dust the back of the insect with a fresh load of pollen. The fruits mature within 6–8 weeks when they then dehisce to release the numerous, dust-like seeds, which are wind dispersed. Generally, seeds of temperate *Pinguicula* species require cold stratification for germination and cold storage conditions to maintain their viability. Germination of surface-sown seed reportedly occurs within 1–4 months above 15°C. In more northerly distributed species, the root system dies back annually and the plants produce a

specialized turion-like winter bud, which is resistant to cold and desiccation and overwinters upon the decomposition of the normal foliage leaves. The buds often proliferate into clusters (the offshoots called "gemmae"), with each one capable of vegetative propagation. Details on the dispersal of the gemmae are lacking, but they reportedly can be dispersed in water following high periods of rain. *Pinguicula* species lack mycorrhizal associations and obtain their nutrients by carnivory. The plants emit a fungal odor, which helps to attract small insect prey such as aphids (Hemiptera: Aphididae), gnats and mosquitoes (Diptera), and springtails (Collembola) to their leaves, where they become entrapped upon the sticky, mucilaginous surface. Larger insects (>5 mm) tend to escape from the traps. Comprehensive lists of the insects trapped by these plants have been compiled for only a few of the species. The glandular leaf hairs secrete digestive enzymes such as acid phosphatase, amylase, esterase, and ribonuclease. Captured prey stimulates the leaf margins to roll inward, thereby increasing contact between the leaf surface and the prey. There is a surprising lack of detailed ecological and life-history information (e.g., habitat pH, pollinators, seed dormancy and germination requirements, prey species) for many of the North American species. Fresh seeds of some species reportedly germinate within days if they become submersed in water.

Pinguicula caerulea **Walter** inhabits bogs, depressions, ditches, flats, flatwoods, floodplains, prairies, savannas, seeps, and swamp margins at elevations of up to 46 m. The substrates include Albany (grossarenic paleudults), Basinger (spodic psammaquents), loamy sand, Meadowbrook (grossarenic ochraqualfs), muck, peat, Pomona (ultic Haplaquods), Pottsburg (grossarenic haplaquods), Ridgewood (aquic quartzipsamments), sand, sandy peat, and occasionally decaying logs. The plants tolerate full sun to light shade, but do not grow in sites where standing water persists. The corollas range from violet to white but usually have violet stripes, which serve as a "honey guide" to attract potential pollinators. The flowers are pollinated by long-tongued bees (Hymenoptera), which seek the nectar. Specific information on the reproductive biology and seed ecology of this species is otherwise quite incomplete and deserves detailed study. This species has been propagated successfully by tissue culture. **Reported associates:** *Aletris, Amianthium muscitoxicum, Amphicarpum purshii, Andropogon gerardii, A. glaucopsis, A. glomeratus, A. virginicus, Anthaenantia rufa, Aristida palustris, A. stricta, Arnoglossum ovatum, Asclepias rubra, Balduina atropurpurea, B. uniflora, Bigelowia nudata, Buchnera, Calopogon barbatus, Carex striata, Carphephorus odoratissimus, Centella erecta, Chaptalia tomentosa, Cirsium horridulum, Cleistes divaricata, Clethra alnifolia, Coreopsis falcata, C. linifolia, Ctenium aromaticum, Dichanthelium dichotomum, D. scabriusculum, Dionaea muscipula, Drosera brevifolia, D. capillaris, Erigeron vernus, Eriocaulon compressum, E. decangulare, Eupatorium leucolepis, Fimbristylis, Helenium pinnatifidum, H. vernale, Helianthus heterophyllus, Hypericum brachyphyllum, H. crux-andreae, Ilex glabra, I. myrtifolia, Iris tridentata,* *I. virginica, Isoetes, Lachnocaulon anceps, Lilium catesbaei, L. pyrophilum, Lobelia, Ludwigia pilosa, Lycopodium, Lysimachia loomisii, Marshallia graminifolia, Muhlenbergia expansa, Myrica cerifera, M. heterophylla, Nyssa, Oclemena reticulata, Osmundastrum cinnamomeum, Paspalum praecox, Persea palustris, Photinia pyrifolia, Pinguicula pumila, Pinus elliottii, P. palustris, Plantago sparsiflora, Platanthera ciliaris, P. integra, P. nivea, Pleea tenuifolia, Pluchea rosea, Pogonia ophioglossoides, Polygala brevifolia, P. hookeri, Pterocaulon pycnostachyum, Pyxidanthera barbulata, Quercus laevis, Q. laurifolia, Rhexia alifanus, R. lutea, R. mariana, Rhynchospora chapmanii, R. latifolia, Sabatia, Saccharum, Sarracenia flava, S. minor, S. purpurea, S. rubra, Scleria triglomerata, Scutellaria integrifolia, Serenoa repens, Sisyrinchium, Smilax laurifolia, Solidago verna, Sphagnum, Spiranthes, Sporobolus pinetorum, Styrax americanus, Taxodium ascendens, Triantha racemosa, Toxicodendron radicans, Triantha racemosa, Utricularia, Viburnum nudum, Viola lanceolata, Woodwardia virginica, Zephyranthes.*

Pinguicula ionantha **R.K. Godfrey** grows in bogs, canals, depressions, ditches, flatwoods, marshes, savannas, seeps, and along the margins of streams and swamps at elevations of at least 9 m. The substrates are described as acidic (pH: 6.1–6.5) and consist of clay, muck, muck over clay, peat, peaty sand Pelham (arenic paleaquults), Plummer (grossarenic paleaquults), sand, sandy peat, sandy peaty muck, and Scranton (humaqueptic psammaquents). This species is extremely tolerant of inundation and very wet conditions and will withstand submergence in shallow water for up to several days. It tolerates only partial shade, but commercial growers reportedly start the seedlings under full light to light shade (on a substrate of peat or 50%/50% peat/sand). Plants cultivated from seedlings (on peat or 1:1 peat/sand) produce their first flowers in about 3 years. The flowers have a pale violet or white corolla and emit a sweet, honey-like fragrance, which presumably helps to attract pollinating insects. However, details on natural pollinators and other life history information has not been published for this species. Low germination rates often are reported for commercially distributed seeds, but there is no way of knowing how seed was derived (e.g., selfed vs. outcrossed plants) in such instances. Successful germination (within 1 month) has been reported after thoroughly running water (until it is clear) through a prewashed 1:1 sand/peat mix (in a plastic pot), draining off the excess water, and sowing the seeds on the peat layer that forms on top; then sealing the entire pot in a plastic bag and placing it close to (about 15 cm) a light source. The plants have been successfully propagated artificially using tissue culture. The digestive pools that form on the leaves are known to absorb ultraviolet-light and are thought to provide a warning cue to alert potential insect visitors. The populations fluctuate in size from year to year and require prolonged, open conditions to persist. Natural disturbances (e.g., frequent fires) are needed to remove competing plants such as *Cliftonia monophylla, Cyrilla racemiflora,* and *Hypericum.* **Reported associates:** *Agalinis, Aletris lutea, Aristida beyrichiana, A.*

stricta, Calopogon barbatus, Chaptalia tomentosa, Cirsium lecontei, Cliftonia monophylla, Ctenium aromaticum, Cyrilla racemiflora, Drosera capillaris, D. tracyi, Eriocaulon compressum, E. decangulare, Eryngium aquaticum, Euphorbia inundata, Eurybia eryngiifolia, Gaylussacia dumosa, Helenium vernale, Helianthus heterophyllus, Huperzia selago, Hypericum chapmanii, H. microsepalum, Ilex myrtifolia, Lachnocaulon anceps, Lophiola americana, Lycopodiella alopecuroides, L. caroliniana, Macbridea alba, Magnolia virginiana, Marshallia tenuifolia, Myrica heterophylla, Nolina, Nyssa ursina, Parnassia caroliniana, Peltandra virginica, Pinguicula lutea, P. planifolia, P. primuliflora, P. pumila, Pinus elliottii, P. palustris, Pleea tenuifolia, Polygala cymosa, P. lutea, P. ramosa, Pteridium aquilinum, Rhexia alifanus, Rhynchospora chapmanii, Rudbeckia graminifolia, Sarracenia flava, S. psittacina, Sarracenia, Schoenus nigricans, Scleria baldwinii, Scutellaria floridana, Serenoa repens, Sporobolus floridanus, Stenanthium densum, Taxodium ascendens, TrianSAMEtha racemosa, Utricularia subulata, Woodwardia virginica, Xyris ambigua, Zigadenus glaberrimus.

***Pinguicula lutea* Walter** grows in barrens, bayheads, bogs, ditches, draws, flatwoods, marshes, meadows, poor fens, prairies, roadsides, savannas, scrub, seepage bogs, slopes, and along the margins of ponds and streams at elevations of up to 61 m. The plants occur in full sun to partial shade but do not tolerate prolonged periods of standing water. The substrates are described as acidic and are characterized as Albany (grossarenic paleudults), Eau Gallie (alfic haplaquods), Ellabelle (arenic umbric paleaquults), Escambia (plinthaquic paleudults), muck, Oldsmar (alfic arenich), Plummer (grossarenic paleaquults), Pomona (ultic haplaquods), Rains (typic paleaquults), sand, sandy loam (Ruston/Orangeburg), sandy peat, Stilson (arenic plinthic paludults), and Surrency (arenic umbric paleaquults). The flowers appear from late February to May and are pollinated by long-tongued bees (Hymenoptera), which forage for the nectar. The corollas normally are yellow but pure white-colored forms have been encountered. The seeds can be germinated as soon as they ripen, without any pretreatment. Seeds placed on moist *Sphagnum* in partial sunlight should germinate within 3 months. The plants require open sites with minimal competition to thrive and commonly occur in sites that are annually burned, grazed, or frequently mowed. A threshold of 5 mm has been determined experimentally as the size for prey that are able to escape the sticky leaf traps. **Reported associates:** *Acer rubrum, Agalinis, Aletris lutea, Andropogon arctatus, A. gyrans, A. liebmannii, A. virginicus, Aristida palustris, A. stricta, Aronia arbutifolia, Asclepias, Balduina uniflora, Bartonia verna, Bidens bipinnatus, Bigelowia nudata, Calopogon barbatus, Carphephorus odoratissimus, C. pseudoliatris, Chaptalia tomentosa, Cirsium, Clematis, Cliftonia monophylla, Coreopsis gladiata, Crotalaria purshii, Ctenium aromaticum, Cyrilla racemiflora, Dichanthelium aciculare, D. acuminatum, D. strigosum, Dionaea muscipula, Doellingeria sericocarpoides, Drosera brevifolia, D. capillaris, D. rotundifolia, D. tracyi, Elephantopus tomentosus, Erigeron vernus, Eriocaulon*

compressum, E. decangulare, E. texense, Eryngium aquaticum, Eupatorium, Euphorbia corollata, E. inundata, Eurybia eryngiifolia, Gamochaeta purpurea, Gaylussacia dumosa, Gelsemium sempervirens, Gymnadeniopsis integra, G. nivea, Hartwrightia floridana, Helenium vernale, Helianthus angustifolius, H. heterophyllus, H. radula, Heterotheca subaxillaris, Hypericum brachyphyllum, H. chapmanii, H. fasciculatum, H. microsepalum, H. suffruticosum, Hypoxis hirsuta, H. juncea Ilex coriacea,, I. glabra, I. myrtifolia, Iris verna, Justicia crassifolia, Lachnanthes caroliniana, Lachnocaulon anceps, L. digynum, Liatris spicata, Lilium catesbaei, Liquidambar styraciflua, Lobelia glandulosa, Lophiola aurea, Lycopodiella alopecuroides, L. appressa, Lycopodium caroliniana, Lyonia lucida, Macranthera flammea, Magnolia virginiana, Marshallia graminifolia, Muhlenbergia expansa, Myrica cerifera, M. heterophylla, Neottia bifolia, Nyssa biflora, Ophioglossum, Oxypolis filiformis, Parnassia caroliniana, Penstemon australis, Persea borbonia, Phlox pilosa, Pinguicula ionantha, P. planifolia, P. primuliflora, P. pumila, Pinus elliotii, P. palustris, P. pumila, P. serotina, Pityopsis graminifolia, Pleea tenuifolia, Pogonia ophioglossoides, Polygala cruciata, P. cymosa, P. lutea, P. nana, P. polygama, P. ramosa, Pteridium aquilinum, Quercus falcata, Q. geminata, Q. minima, Q. myrtifolia, Q. nigra, Q. pumila, Q. stellata, Q. virginiana, Rhexia alifanus, R. lutea, Rhus copallinum, Rhynchosia reniformis, Rhynchospora chapmanii, R. ciliaris, R. latifolia, R. oligantha, R. plumosa, Rubus flagellaris, Rudbeckia laciniata, Sabatia campanulata, S. decandra, S. difformis, S. macrophylla, Sagittaria, Sarracenia alata, S. flava, S. leucophylla, S. psittacina, S. purpurea, S. rubra, Schizachyrium sanguineum, S. scoparium, Scleria baldwinii, S. muehlenbergii, S. triglomerata, Serenoa repens, Sisyrinchium angustifolium, S. fuscatum, Smilax bona-nox, S. laurifolia, S. pumila, Solidago fistulosa, Sphagnum, Spiranthes brevilabris, Sporobolus floridanus, Stokesia laevis, Stylosanthes biflora, Symphyotrichum tradescantii, S. walteri, Symplocos tinctoria, Syngonanthus flavidulus, Taxodium ascendens, Tephrosia virginiana, Triantha racemosa, Utricularia subulata, Vaccinium corymbosum, V. darrowii, V. myrsinites, Valerianella radiata, Viburnum molle, V. nudum, Viola lanceolata, V. primulifolia, V. septemloba, Woodwardia virginica, Xyris ambigua, X. baldwiniana, X. brevifolia, X. caroliniana, Zigadenus glaberrimus.

***Pinguicula macroceras* Link** inhabits bogs, cliffs, flats, hummocks, marshes, meadows, muskeg, rock walls, seeps, slopes, springs, tidal flats, and lake, river, or stream margins at elevations of up to 3600 m. The reported substrates are acidic (pH: 5–6) or alkaline (no specific pH values given) and include gravel, limestone, rock, scree, serpentine, silt, and talus. The plants apparently favor northern exposures that experience partial to full shade and have also been found in southern exposures. They can withstand short cold periods down to −23°C, and can last up to a week at temperatures from 0°C to −7°C. Dormancy is induced when the temperature falls below 10°C, whereupon the leaves wither and the plants overwinter by means of specialized winter buds. The flowers have a purple corolla and are not fragrant. Specific natural pollinators (presumably bees) have not been identified. The seeds require an 8–12 week period of cold stratification (−10°C to 5°C) to germinate (placing seeds in a plastic bag with a few sprigs of damp peat is recommended while refrigerating them). Germination reportedly occurs at 12°C in the light after 30–120 days, but is erratic. Life-history information for the closely related *P. vulgaris* should be consulted as it is much better known and probably conspecific (see *Systematics* section below) or at least quite similar ecologically. **Reported associates:** *Abies amabilis, Achnatherum lemmonii, Agrostis hallii, Alnus viridis, Arnica spathulata, Blechnum spicant, Calamagrostis nutkaensis, Carex echinata, C. serratodens, Cassiope mertensiana, Castilleja miniata, Chamaecyparis lawsoniana, Cornus unalaschkensis, Danthonia californica, Darlingtonia californica, Delphinium nudicaule, Deschampsia cespitosa, Drosera rotundifolia, Elliottia pyroliflora, Empetrum nigrum, Epipactis gigantea, Erigeron foliosus, E. peregrinus, Eriophorum crinigerum, Galium ambiguum, Gentiana affinis, G. douglasiana, G. plurisetosa, Habenaria, Helenium bigelovii, Horkelia sericata, Huperzia selago, Iris innominata, Juncus orthophyllus, Larix laricina, Leucothoe davisiae, Lomatium californicum, L. howellii, Lonicera involucrata, Luzula, Myrica gale, Narthecium californicum, Parnassia palustris, Picea, Pinus contorta, P. monticola, Platanthera sparsiflora, Populus tremuloides, Potentilla anserina, Pyrrocoma racemosa, Rhododendron albiflorum, R. groenlandicum, R. occidentale, Rudbeckia californica, Salix bebbiana, Sanicula peckiana, Sanguisorba officinalis, Sphagnum, Symphyotrichum spathulatum, Trianta glutinosa, Trientalis borealis, T. europaea, Trifolium longipes, Tsuga heterophylla, T. mertensiana, Vaccinium oxycoccos, V. vitis-idaea, Xerophyllum tenax.*

Pinguicula planifolia **Chapm.** grows in bogs, depressions, ditches, flatwoods, savannas, seeps, swamps, and along pond or river margins. The substrates are acidic (pH: 4.7–5.9) and have been characterized as clay, Escambia (plinthaquic paleudults), Florala (plinthaquic paleudults), loamy sand, muck, organic sandy clay, peat, Pamlico-Dorovan (terric medisaprists), Pantego (umbric paleaquults), Pelham (arenic paleaquults), Plummer (grossarenic paleaquults), Rains (typic paleaquults), sand, sandy loam, sandy peat, and Scranton (humaqueptic psammaquents). The plants occur characteristically within a few cm of shallow water (where they can be difficult to detect unless they are flowering) or at least occupying open, extremely wet sites where they can form sizeable colonies. The flowers have pale violet corollas and are sweetly, honey-odored, which probably serves to attract pollinators (presumably bees, but not yet reported). The seeds require temperatures in excess of 15°C for adequate germination, which can be erratic. However, germination has been reported for seeds that are fairly green and not fully ripened. Variation in seed germination is probably related to their source, i.e., whether they originate from selfed flowers or from outcrossed plants. The plants can be cultured in a 1:1 mixture of peat/ sand but have a very slow growth rate and are not easy to grow. They will resprout following natural fires, which help to remove competing overstory vegetation. The plants have been propagated successfully by tissue culture. **Reported associates:** *Andropogon gyrans, A. mohrii, Aristida beyrichiana, A. palustris, Arnoglossum ovatum, Asclepias lanceolata, Balduina uniflora, Bartonia verna, Calopogon barbatus, Carphephorus pseudoliatris, Castanea pumila, Centella asiatica, Chaptalia tomentosa, Cirsium lecontei, Cliftonia monophylla, Coelorachis rugosa, Coreopsis nudata, Ctenium aromaticum, Cyrilla parvifolia, C. racemiflora, Dichanthelium erectifolium, D. scoparium, D. wrightianum, Doellingeria umbellata, Drosera capillaris, D. filiformis, D. tracyi, Eleocharis tuberculosa, Erigeron verna, Eriocaulon compressum, E. decangulare, E. lineare, E. nigrobracteatum, E. texense, Euphorbia inundata, Eurybia eryngiifolia, Helenium brevifolium, H. vernale, Helianthus heterophyllus, Hypericum brachyphyllum, H. chapmanii, H. fasciculatum, H. microsepalum, H. myrtifolium, Ilex decidua I. glabra, I. myrtifolia, I. vomitoria, Juncus polycephalus, Lachnocaulon digynum, Lachnanthes caroliana, Lilium catesbaei, Lobelia floridana, L. puberula, Lophiola americana, Ludwigia pilosa, Lycopodiella alopecuroides, L. caroliniana, Magnolia virginiana, Marshallia tenuifolia, Muhlenbergia expansa, Nyssa ursina, Oxypolis filiformis, Panicum nudicaule, Persea, Pinguicula ionantha, P. lutea, Pinus elliottii, P. palustris, Platanthera integra, Pleea tenuifolia, Pogonia ophioglossoides, Polygala cruciata, P. cymosa, P. lutea, P. nana, P. ramosa, Rhexia virginica, Rhynchospora careyana, R. chapmanii, R. corniculata, R. filifolia, R. harperi, R. macra, R. oligantha, R. stenophylla, Rudbeckia graminifolia, Sabatia bartramii, S. macrophylla, Sarracenia alata, S. flava, S. psittacina, S. rubra, Scleria baldwinii, S. muehlenbergii, Serenoa repens, Sporobolus floridanus, Stenanthium densum, Stillingia aquatica, Stokesia laevis, Syngonanthus flavidulus, Taxodium ascendens, Trianta racemosa, Utricularia subulata, Vaccinium darrowii, Woodwardia virginica, Xyris ambigua, X. baldwiniana, X. drummondii, X. platylepis, X. serotina, X. stricta.*

Pinguicula primuliflora **C.E. Wood & R.K. Godfrey** inhabits bogs, depressions, pond margins, savannas, seeps, springs, streams, and swamps. The substrates are described as Escambia (plinthaquic paleudults), loamy sand, muck, Pantego (umbric paleaquults), peat, Plummer (grossarenic paleaquults), Rutledge-Pamlico (terric medisaprists), and sand. This species occurs in shallow (sometimes flowing) water and even has been observed to form floating colonies on the water surface. The plants are fairly tolerant of shade. The flowers are unscented with pale purple to pink corollas, which resemble the blooms of a primrose (*Primula*). There have been no studies on the pollination biology or seed ecology of this species. These plants are distinct among the North American taxa in their ability to propagate vegetatively by the formation of vegetative plantlets at the tips of their leaves. The plantlets detach as the leaves decompose and are capable of independent establishment. These plants can be propagated by leaf cuttings or by tissue culture. Detailed studies have shown that larger insects (3.1–4.0 mm) are trapped near the edges of the leaves (where they are retained by the rolled margin); whereas, smaller insects (3.0–0.1 mm) are found progressively closer to the midrib. The plants occasionally are found growing in powerline easements, where the clearing of brush is thought to provide prolonged open conditions that mimic the effects of natural disturbances such as fire. **Reported associates:** *Aletris, Andropogon, Aristida beyrichiana, Arundinaria gigantea, Bigelowia nudata, Calamintha coccinea, Carex atlantica, C. collinsii, C. exilis, C. venusta, Chamaecyparis thyoides, Chaptalia tomentosa, Cirsium, Cliftonia monophylla, Cnidoscolus stimulosus, Conradina canescens, Crataegus lacrimata, Ctenium aromaticum, Cyrilla racemiflora, Drosera brevifolia, D. capillaris, D. rotundifolia, D. tracyi, Eriocaulon compressum, E. nigrobracteatum, Eriogonum tomentosum, Helenium brevifolium, Hypericum, Ilex glabra, Illicium floridanum, Lachnocaulon anceps, Liatris, Lupinus diffusus, Lycopodium, Lyonia lucida, Magnolia grandiflora, M. virginiana, Myrica cerifera, M. heterophylla, Orontium aquaticum, Pallavicinia lyellii, Panicum nudicaule, Persea palustris, Pieris phillyreifolia, Pinguicula ionantha, P. lutea, P. pumila, Pinus palustris, P. serotina, Pityopsis aspera, P. graminifolia, Pleea tenuifolia, Pogonia ophioglossoides, Pteridium aquilinum, Quercus laevis, Rhynchospora macra, R. stenophylla, Sarracenia alata, S. leucophylla, S. psittacina, S. rubra, Schoenoplectus etuberculatus, Serenoa repens, Smilax laurifolia, Sphagnum, Symphyotrichum walteri, Utricularia subulata, Vaccinium darrowii, Viola primulifolia, V. septemloba, Woodwardia areolata, Xyris caroliniana, X. chapmanii, X. drummondii.*

Pinguicula pumila Michx. is found in bogs, depressions, ditches, flatwoods, glades, hammocks, prairies, riverbottoms, roadsides, savannas, seeps, swales, swamps, and along the margins of streams at elevations of up to 70 m. The substrates have been described as ranging from strongly acidic to calcareous or marlaceous, but specific pH data for natural populations have not been compiled. The substrates have been characterized as clay, coral rock, Cudjoe complex, Cuthbert (typic hapludults), Leefield–Stilson complex, limerock, loamy sand, marl, Myakka (aeric haplaquods), Oldsmar (alfic arenic haplaquods), peaty sand, Pelham (arenic paleaquults), Pomona (ultic haplaquods), Pottsburg (grossarenic haplaquods), sand, sandy loam, and sandy peat. The flowers are variable with corollas that range from purple to yellow or white. Presumably, they are pollinated by bees (Hymenoptera); however, their pollination biology has not been described in any detail. The seeds (at least some of those produced) apparently are not dormant and do not require cold stratification. Variable germination rates have been obtained by floating seeds on distilled water in a petri dish or by sowing them in peat. Seeds collected from soil cores have germinated at 20°C under a 16-h photoperiod. Although reported to have short viability, fairly old seeds have germinated well after being retrieved from seed banks in both disturbed and non-disturbed sites where plants were absent from the standing vegetation. The specific germination requirements and conditions of dormancy for this species require further elucidation. Successful propagation has been achieved by tissue culture. These are minute plants, with rosettes that usually are less than 3.0 cm in diameter. Like the previous, this species has been observed to grow under

powerlines, which are habitats that are managed for overstory removal. **Reported associates:** *Acer rubrum, Agalinis, Aletris lutea, Alnus serrulata, Amphicarpum muhlenbergianum, Andropogon glomeratus, A. virginicus, Aristida beyrichiana, Athyrium filix-femina, Bacopa, Bartonia virginica, Blechnum serrulatum, Buchnera, Burmannia capitata, Calopogon tuberosus, Campanula, Carex amphibola, C. atlantica, C. glaucescens, C. tribuloides, Cassytha filiformis, Castanea pumila, Centella asiatica, Cephalanthus occidentalis, Chaptalia tomentosa, Chasmanthium laxum, Cladium jamaicense, Clethra alnifolia, Cliftonia monophylla, Coreopsis linifolia, C. tripteris, Ctenium aromaticum, Cyperus haspan, Cyrilla racemiflora, Desmodium lineatum, D. paniculatum, Dichanthelium dichotomum, D. ensifolium, D. scoparium, D. sphaerocarpon, Drosera brevifolia, D. capillaris, D. intermedia, D. rotundifolia, Dyschoriste, Eleocharis microcarpa, Elephantopus carolinianus, Elytraria caroliniensis, Erigeron strigosus, Eriocaulon decangulare, E. texense, Eryngium aquaticum, E. integrifolium, Eupatorium perfoliatum, E. rotundifolium, Euphorbia inundata, Eurybia eryngiifolia, Fagus, Fuirena breviseta, F. squarrosa, Gaylussacia dumosa, Gratiola pilosa, Hartwrightia floridana, Helenium pinnatifidium, Helianthus angustifolius, H. heterophylla, H. hirsutus, Heliotropium polyphyllum, Huperzia selago, Hypericum cistifolium, H. mutilum, H. brachyphyllum, H. tetrapetalum, Hypoxis, Ilex coriacea, I. glabra, Illicium floridanum, Iris hexagona, Itea virginica, Juncus coriaceus, J. scirpoides, Lachnanthes caroliniana, Lachnocaulon anceps, Liatris, Linaria canadensis, Lobelia puberula, Ludwigia alternifolia, Lycopodiella alopecuroides, L. caroliniana, Lycopodiella cernua, Lycopus virginicus, Lygodium japonicum, Lyonia lucida, Magnolia virginiana, Marshallia tenuifolia, Mitchella repens, Mitreola sessilifolia, Muhlenbergia capillaris, M. filipes, Myrica cerifera, M. heterophylla, Oclemena reticulata, Onoclea sensibilis, Osmunda regalis, Osmundastrum cinnamomeum, Panicum abscissum, P. tenerum, Parnassia caroliniana, Paspalum setaceum, P. urvillei, Peltandra sagittifolia, P. virginica, Persea palustris, Pinckneya bracteata, Pinguicula caerulea, P. ionantha, P. lutea, P. primuliflora, Pinus elliottii, P. palustris, P. taeda, Piriqueta caroliniana, Pityopsis graminifolia, P. oligantha, Pluchea camphorata, P. rosea, Pogonia ophioglossoides, Polygala grandiflora, P. lutea, P. mariana, P. rugelii, Pteridium aquilinum, Ptilimnium capillaceum, Rhexia alifanus, R. mariana, R. nuttallii, R. petiolata, R. virginica, Rhododendron canescens, Rhynchospora breviseta, R. cephalantha, R. ciliaris, R. divergens, R. fascicularis, R. oligantha, R. rariflora, Rudbeckia, Sabal palmetto, Sabatia difformis, S. grandiflora, Samolus ebracteatus, Sarracenia alata, S. flava, S. leucophylla, S. minor, S. purpurea, Schizachyrium rhizomatum, Scleria reticularis, Scutellaria integrifolia, Serenoa repens, Setaria geniculata, Sisyrinchium, Smilax auriculata, S. laurifolia, Solidago rugosa, Sphagnum, Spiranthes vernalis, Steinchisma hians, Symphyotrichum dumosum, S. lateriflorum, S. subulatum, Taxodium ascendens, Triantha racemosa, Toxicodendron radicans, Utricularia cornuta, U. juncea, U. subulata,*

Vaccinium elliottii, V. myrsinites, Viburnum nudum, Viola lanceolata, V. palmata, V. ×primulifolia, Vitis rotundifolia, Woodwardia areolata, W. virginica, Xyris ambigua, X. baldwiniana, X. difformis, X. elliottii, X. fimbriata, X. laxifolia, X. platylepis, X. torta, Zamia, Zigadenus glaberrimus.

***Pinguicula villosa* L.** occurs in subarctic bogs, depressions, fens, marshes, meadows, muskeg, slopes, strangmoor (flarks), tundra, and along the margins of creeks, subalpine lakes, and ponds at elevations of up to 170 m. The substrates reportedly are acidic or somewhat alkaline (e.g., pH: 7.4–7.5) and consist of shallow humus or peat underlain by gravel or silt. Studies have shown that *P. villosa* allocates about a third of its annual nitrogen to winter buds, another third to the seeds, and the remaining third to leaves and other structures. This species is small in stature (the rosettes range in size from 1.0 cm in full sun to 2.5 cm in deep shade) and expends many resources in producing sufficient vertical growth to extend above the surrounding *Sphagnum* canopy. Only 20%–40% of adult plants flower in a given year. The species has been described as having a mixed mating system. Insect pollinators have not been observed in wild populations, which are characterized by low flower production and seed set, and at least some self-pollination occurs. A correlation of flower production with average past summer temperature indicates that the floral primordia develop in the year prior to flowering. Plants that are large and vegetative have a greater probability of flowering in the subsequent year but flowering plants deplete about 20% more nitrogen than nonflowering plants in a season. The seeds require several months of cold stratification to break dormancy. Seedling emergence and other aspects of growth and reproduction are strongly influenced by microsite differences. Seed production significantly lowers survivorship and lowers reproduction in the subsequent year. The plants are not highly efficient at capturing prey but supplement about 50% of their nitrogen and 40% of their phosphorous requirements through carnivory (mostly Collembola). The amount of prey captured varies substantially from year to year, but does not appear to be correlated with site nutrient levels. This is a relatively late successional species in the arctic tundra where the growing season is less than 4 months in duration. Overwintering occurs by winter buds ("gemmae"), which can be separated to propagate the plants vegetatively. **Reported associates:** *Alnus viridis, Andromeda polifolia, Anemone multifida, A. narcissiflora, A. parviflora, Antennaria monocephala, Arnica frigida, Aulacomnium turgidum, Betula glandulosa, B. nana, Calamagrostis canadensis, Calliergon, Calypogeia sphagnicola, Cardamine pratensis, Carex aquatilis, C. capillaris, C. chordorrhiza, C. garberi, C. gynocrates, C. lenticularis, C. limosa, C. membranacea, C. microglochin, C. pauciflora, C. podocarpa, C. rostrata, C. rotundata, C. scirpoidea, C. trisperma, C. vaginata, Cassiope tetragona, Cetraria nivalis, Chamerion latifolium, Cladina mitis, Cladonia alpestris, C. rangiferina, Comarum palustre, Cypripedium, Dasiphora floribunda, Diapensia lapponica, Dicranum acutifolium, Drepanocladus, Drosera anglica, D. rotundifolia, Dryas integrifolia, Empetrum nigrum, Equisetum arvense, E. scirpoides, E. variegatum, Erigeron peregrinus, Eriophorum angustifolium, E. scheuchzeri,* *E. vaginatum, Festuca altaica, F. brachyphylla, Geocaulon lividum, Hierochloe odorata, Hypnum revolutum, Juncus, Kalmia angustifolia, Larix laricina, Leptarrhena pyrolifolia, Loiseleuria procumbens, Lycopodium complanatum, Maianthemum trifolium, Myrica gale, Nephroma, Pedicularis groenlandica, P. kanei, P. lapponica, P. sudetica, P. verticillata, Peltigera aphthosa, Picea mariana, Platanthera dilatata, Pleurozium schreberi, Pyrola asarifolia, Ranunculus gmelinii, Rhododendron groenlandicum, R. lapponicum, R. tomentosum, Rubus arcticus, R. chamaemorus, Salix arctophila, S. myrtillifolia, S. phlebophylla, S. reticulata, Sanguisorba canadensis, Sanionia uncinata, Saxifraga hirculus, Scheuchzeria palustris, Silene acaulis, Sphagnum fuscum, S. rubellum, S. warnstorfii, Spiranthes romanzoffiana, Splachnum luteum, Thalictrum alpinum, Tofieldia coccinea, T. pusilla, Tomenthypnum nitens, Trichophorum cespitosum, Triglochin palustre, Vaccinium oxycoccos, V. uliginosum, V. vitis-idaea, Valeriana capitata, Viola palustris.*

***Pinguicula vulgaris* L.** inhabits alpine, arctic, boreal or subarctic beaches, bogs, cliffs, depressions, dunes, meadows, muskeg, outcrops, perched meadows, scree, swales, talus, tundra, and the margins of lakes, ponds, rivers, and streams. Its habitats are extremely variable, with the only apparent common denominators being nutrient-poor conditions and a growing season of at least 60–90 days. The plants occur in open to partially shaded sites on substrates derived from limestone or sandstone and tolerate a broad acidity gradient (pH: 3.1–8.2); however, they tend to grow better on alkaline substrates. The reported substrates include clay, cobble, gravel, peat, rock, sand, schist, silt, and silty shale, but they typically are low in organic matter content. The floral primordia develop during the previous summer and only 20%–40% of adult plants typically will flower in a given year. The flowers are adichogamous, self-compatible, and can produce up to 140 seeds. Although they are adapted for outcrossing by bees (Hymenoptera), observations of potential pollinators are scarce. The species is regarded as being highly autogamous with some populations found to be largely inbreeding. Ovule production is directly proportional to the initial mass of the vegetative bud. Seed production does not decrease survivorship, but does result in reduced reproduction in the subsequent year. Seed viability generally is high. Imbibed seeds kept at 1°C–2°C, or seeds stored dry for 6 months, exhibited 100% germination under 18 h of light at temperatures above 10°C. Germination optimally requires white light and improves with (but does not require) cold stratification for 6 weeks at 3°C. High germination is reported for seeds that have been sterilized for 5 min in a 20% NaOCl solution, washed in deionized water, then treated with gibberellic acid (0.1% GA_3) and exposed to a constant temperature of 21°C. The minute seeds are wind dispersed (as the capsules are shaken) and have been found to develop into transient seed banks of up to 43 seeds m². The capsules do not open in wet weather. Due to their small size, the seeds contain few food reserves and probably do not survive often beyond a single year; however, delayed germination (2–3 years) has been observed in some sites. The root systems die back each year and the plants overwinter by a vegetative winter bud and its offshoots ("gemmae"), which typically are produced by more than 90% of plants. The winter buds (but not adult plants) will withstand freezing. Individual plants can live up to 10 years when grown under suitable conditions. Plants grown from the gemmae will begin to flower after 3 years. The gemmae can be dispersed by water following rains, snow melt or by other types of flooding. The plants allocate resources to growth in proportion to their size, with growth allocation decreasing as the plants become larger. Vegetative processes (growth and vegetative reproduction) are correlated positively. A threshold size must be attained to trigger sexual reproduction, but varies among sites. Larger plants have higher levels of both sexual and vegetative reproduction, but large, fruiting plants produce fewer winter buds ("gemmae"). Sexual reproduction depletes resources that can be restored to some degree by carnivory. The plants are highly efficient at capturing prey and can obtain 100% of their nitrogen and phosphorous required for flowering by carnivory; however, nitrogen and phosphorous uptake and content correlate with soil nitrogen levels rather than with prey capture. The captured prey includes mites (Acarina), small midges and gnats (Diptera: Nematocera), springtails (Collembola), and thrips (Thysanoptera). Within a site, the amount of prey captured remains fairly consistent from year to year, but can vary substantially among sites.

Reported associates: *Abies amabilis, Achillea borealis, Agrostis mertensii, Alectoria ochroleuca, Allium schoenoprasum, Alnus viridis, Andromeda polifolia, Androsace chamaejasme, Anemone drummondii, A. parviflora, Antennaria friesiana, A. pulcherrima, Anticlea elegans, Arctagrostis latifolia, Arctostaphylos alpina, A. canescens, A. columbiana, A. rubra, Arethusa bulbosa, Armeria maritima, Arnica frigida, Aruncus dioicus, Athyrium alpestre, Aulacomnium turgidum, Betula nana, Bistorta officinalis, B. vivipara, Bromus pumpellianus, Bryum, Calamagrostis hyperborea, C. purpurascens, Calliergon stramineum, C. trifarium, Calochortus tolmiei, Calopogon tuberosus, Caltha leptosepala, Calypogeia trichomanis, Calystegia polymorpha, Campanula uniflora, Cardamine purpurea, Carex aquatilis, C. atratiformis, C. atrofusca, C. aurea, C. bicolor, C. bigelowii, C. capillaris, C. capitata, C. chordorrhiza, C. concinna, C. dioica, C. flava, C. gynocrates, C. lachenalii, C. limosa, C. livida, C. lyngbyei, C. magellanica, C. meadii, C. microglochin, C. membranacea, C. misandra, C. pluriflora, C. podocarpa, C. rariflora, C. richardsonii, C. rotundata, C. rupestris, C. scirpoidea, C. vaginata, C. viridula, Cassiope tetragona, Castilleja cusickii, C. rhexiifolia, C. septentrionalis, Cetraria cucullata, C. islandica, C. juniperina, C. lacunosa, C. nivalis, Chamaecyparis, Chamerion angustifolium, C. latifolium, Cirsium pitcheri, Cladonia rangiferina, C. sylvatica, Conocephalum conicum, Cornicularia divergens, Cryptogramma acrostichoides, Cryptomnium hymenophylloides, Cypripedium passerinum, C. reginae, C. calceolus, Dactylina arctica, Danthonia intermedia, Darlingtonia californica, Dasiphora floribunda, Deschampsia, Diapensia lapponica, Dicranum scoparium, Doellingeria umbellata, Draba incana, Drosera anglica,*

D. linearis, D. rotundifolia, Dryas alaskensis, D. integrifolia, D. octopetala, Dupontia fisheri, Empetrum atropurpureum, E. nigrum, Encalypta ciliata, E. procera, Equisetum arvense, E. variegatum, Erigeron peregrinus, Eriodictyon californicum, Eriophorum angustifolium, E. chamissonis, E. russeolum, E. vaginatum, E. viridicarinatum, Eritrichium aretioides, Euphrasia, Eurybia sibirica, Eutrema edwardsii, Festuca brachyphylla, F. rubra, Fissidens osmundioides, Galearis rotundifolia, Garrya buxifolia, Gentiana douglasiana, G. prostrata, Gentianella propinqua, Geum calthifolium, G. peckii, Hedysarum alpinum, Hieracium robinsonii, Hierochloe alpina, Holodiscus discolor, Huperzia selago, Hylocomium splendens, Hypnum aduncum, Iris lacustris, Juncus arcticus, J. balticus, J. trifidus, J. triglumis, Juniperus communis, Kobresia myosuroides, K. simpliciuscula, Lagotis glauca, Larix laricina, Leptobryum pyriforme, Limprichtia revolvens, Liparis loeselii, Lobelia kalmii, Loiseleuria procumbens, Lonicera hispidula, L. involucrata, Lophozia rutheana, Luina hypoleuca, Luzula arcuata, L. confusa, L. multiflora, Madia minima, Micranthes nelsoniana, Mimulus aurantiacus, Minuartia arctica, M. rossii, Mnium thomsonii, Muhlenbergia racemosa, Myrica gale, Nephrophyllidium crista-galli, Ochrolechia frigida, Odontoschisma elongatum, Orobanche fasciculata, Oxytropis borealis, O. nigrescens, O. scammaniana, Papaver macounii, Parmelia omphalodes, Parnassia glauca, P. palustris, Parrya nudicaulis, Pedicularis capitata, P. groenlandica, P. kanei, P. langsdorffii, P. sudetica, Peltigera, Phlox sibirica, Picea mariana, Pinus contorta, P. monticola, Plantago macrocarpa, Platanthera aquilonis, P. hyperborea, P. dilatata, P. obtusata, P. psychodes, Pleurocladula albescens, Poa alpina, P. arctica, P. palustris, Pogonia ophioglossoides, Polytrichum, Populus balsamifera, Potentilla anserina, P. tridentata, P. vahliana, Preissia quadrata, Prenanthes racemosa, Primula egaliksensis, P. jeffreyi, P. mistassinica, Ptilidium, Pyrola, Rhododendron albiflorum, R. camtschaticum, R. groenlandicum, R. lapponicum, R. subarcticum, R. tomentosum, Rhytidium rugosum, Romanzoffia sitchensis, Rubus arcticus, R. chamaemorus, Sagina nodosa, Salix alaxensis, S. arbusculoides, S. arctophila, S. phlebophylla, S. pulchra, S. reticulata, S. richardsonii, S. rotundifolia, Sarracenia purpurea, Saussurea angustifolia, Saxifraga aizoides, S. flagellaris, S. oppositifolia, Schoenoplectus pungens, Sedum spathulifolium, Selaginella oregana, S. selaginoides, Shepherdia canadensis, Sibbaldiopsis tridentata, Silene acaulis, S. uralensis, Solidago decumbens, S. houghtonii, S. juncea, S. ptarmicoides, Sphagnum nitidum, S. papillosum, Sphaerophorus globosus, Spiranthes cernua, Swartzia montana, Tanacetum huronense, Thalictrum alpinum, T. occidentale, Thuja plicata, Tofieldia coccinea, T. pusilla, Tomentypnum nitens, Tortula ruralis, Trianthia glutinosa, Trichophorum alpinum, T. cespitosum, Triglochin maritimum, T. palustre, Trisetum spicatum, Tsuga mertensiana, Utricularia intermedia, Vaccinium oxycoccos, V. uliginosum, V. vitis-idaea, Viola glabella, Whipplea modesta, Woodsia glabella.

Use by wildlife: *Pinguicula pumila* is grazed by Key deer (Mammalia: Cervidae: *Odocoileus virginianus clavium*) and

P. vulgaris also is grazed by larger vertebrates. Not surprisingly, there are few insect (Insecta) predators reported for these carnivorous plants. *Pinguicula vulgaris* contains iridoid glucosides, which are bitter and are known to function generally as feeding deterrents.

Economic importance: food: *Pinguicula* leaves have been used in parts of Europe to coagulate milk into an allegedly palatable yogurt-like product that Linnaeus described as: "most delicious." In Norway it is available commercially as "Tjukkmjølk"; **medicinal:** The leaf secretion of *Pinguicula* species is said to be antiseptic and the leaves have often been applied to the skin to treat wounds and as a restorative poultice. A wash made from the leaves was used to remove hair lice from children and as a treatment for eczema. The leaves were used to treat ringworm by rubbing them on the afflicted part. European farmers used the plants to treat infected udders of dairy animals. The Seminoles administered a whole plant infusion of *P. pumila* to treat raw meat sickness and to alleviate severe abdominal pains. *Pinguicula vulgaris* has been used as a purgative, a cough remedy and treatment for throat

disorders. It was mixed with extracts of thyme (Lamiaceae: *Thymus*) to produce "diatussin," a concoction used to treat whooping-cough. Whole plants of *P. lutea* were used by the Seminoles in an analgesic infusion for treating abdominal pains; **cultivation:** *Pinguicula caerulea, P. ionantha, P. lutea, P. planifolia, P. primuliflora, P. villosa,* and *P. vulgaris* are commonly cultivated as novelty specimen plants. 'Rose' is a double-flowered cultivar of *P. primuliflora* and 'Pasco Giant' is a cultivar of *P. pumila*; **misc. products:** The dried roots of *P. vulgaris* were carried by the Oweekeno as a good luck charm. A floral essence of *P. villosa* is sold commercially. A dye has been obtained from the leaves of *P. vulgaris*; **weeds:** none; **nonindigenous species:** none

Systematics: Molecular systematic analyses clearly resolve *Pinguicula* as a monophyletic sister genus to a clade comprising *Genlisea* and *Utricularia* (Figure 5.64). However, the traditional concept of relationships within the genus is not well supported. Studies that include nearly half of the recognized species (Figure 5.65) indicate that all of the subgenera (*Isoloba, Pinguicula, Temnoceras*), several of the sections

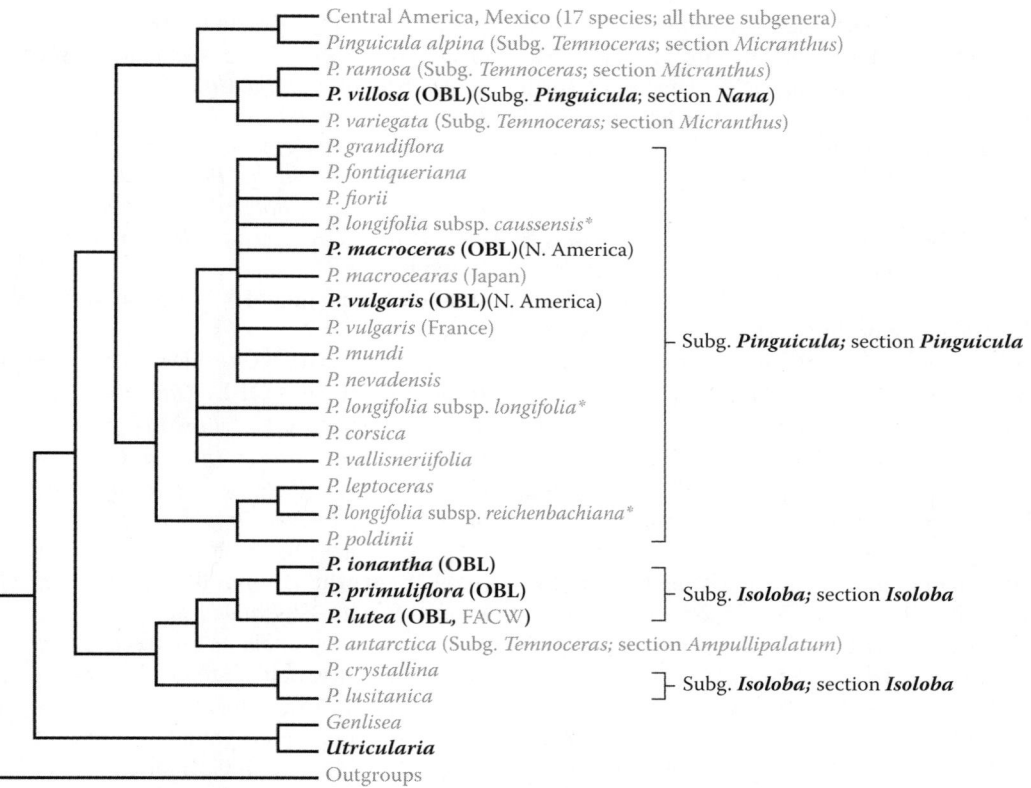

FIGURE 5.65 Interspecific relationships in *Pinguicula* based on phylogenetic analysis of cpDNA sequences for roughly half of the recognized species. The lack of monophyly indicated for any of the traditional subgenera (*Isoloba, Pinguicula,* and *Temnoceras*) warrants a thorough reevaluation of the existing classification. Several sections (e.g., *Isoloba*), as well as at least one species (asterisked), also are not monophyletic. The OBL North American taxa (in bold) do not occur within a unique clade, but are fairly dispersed throughout the genus. However, most *Pinguicula* species are aquatic and it is reasonable to assume a single origin of the habit for the entire genus. The DNA sequence data do not distinguish *P. macroceras* and *P. vulgaris* accessions phylogenetically, even when surveyed from different continents. The recognition of these morphologically similar taxa as distinct species should be reconsidered. The results are similar to analyses of nrITS data for a slightly different subset of taxa (Degtjareva et al., 2006; Shimai et al., 2007), where *P. planifolia* resolved as the sister group to a clade comprising *P. primuliflora, P. caerulea,* and *P. lutea,* with the latter two taxa associated as sister species. (Adapted from Cieslak, T. et al., *Amer. J. Bot.,* 92, 1723–1736, 2005.)

(e.g., *Isoloba*), and at least one species (*P. longifolia*), are not monophyletic. However, the species studied to date do exhibit some geographical integrity as present analyses generally have distinguished tropical, temperate Eurasian, and temperate New World clades (albeit, without high internal support in some instances). Six of the OBL North American species have been included in one phylogenetic analysis (Figure 5.65). Those results indicate a close relationship between *P. macroceras* and *P. vulgaris* (both Old and New World accessions of these species are unresolved as a polytomy) and between *P. ionantha* and *P. primuliflora* (which resolve as a clade with *P. lutea* as its sister species). The diminutive *P. villosa* resolves as the sister species to *P. ramosa*, a morphologically similar Japanese endemic. Other studies, which have analyzed nrITS sequence data for a slightly different subset of taxa, included several additional North American species and arrived at similar results; i.e., *P. planifolia* resolves as the sister species to a clade comprising (in respective order) *P. primuliflora*, *P. caerulea* and *P. lutea*. Together, these analyses indicate a distinct clade for the OBL southeastern North American species (*P. caerulea*, *P. ionantha*, *P. lutea*, *P. planifolia*, *P. primuliflora*; *P. pumila* has not yet been surveyed). Before a complete revision of the classification can be made for *Pinguicula*, it will be necessary to conduct further studies that include additional species and that combine multiple DNA data sets. North American *Pinguicula* species occur in two chromosomal series having a base number of either $x = 8$ or $x = 11$. The former series is more variable with diploids ($2n = 16$: *P. villosa*), tetraploids ($2n = 32$: *P. caerulea*, *P. lutea*, *P. planifolia*, *P. primuliflora*), and hexaploids ($2n = 64$: *P. macroceras*, *P. vulgaris*). The latter chromosomal series includes only diploids ($2n = 22$: *P. ionantha*, *P. pumila*). The endemic North American species do not hybridize, either naturally or experimentally. Their isolation is enforced by a combination of ecological, geographic, and chromosomal barriers as well as the tendency for some species to inbreed. However, hybrids have been reported for the more widespread *P. vulgaris*. *Pinguicula macroceras* is very similar to *P. vulgaris* and many authors have treated these taxa as conspecific (e.g., *P. vulgaris* subsp. *macroceras*). Such a disposition is reasonable given that the corolla and spur features that allegedly distinguish the two taxa overlap, they share the same chromosome number, and they are not divergent at the DNA level, even between geographically disparate (i.e., Old vs. New World) populations.

Comments: Distributions of *Pinguicula* species vary from widespread (circumboreal), e.g., *P. villosa* and *P. vulgaris* (northern North America; Eurasia), *P. macroceras* (northwestern North America; Japan; Russia) to increasingly restricted ranges as illustrated by *P. caerulea*, *P. lutea*, and *P. pumila* (southeastern United States), *P. planifolia* and *P. primuliflora* (Alabama, Florida, Mississippi), and *P. ionantha* (endemic to Florida). The latter is known only from 20 or so localities and is listed as a U.S. Federally threatened species. Nearly all *Pinguicula* species are threatened to some degree by overcollecting and by habitat destruction.

References: Alm, 2005; Avis et al., 1997; Baskin & Baskin, 1998; Blanchard, 1993; Calkins, 1879; Carroll, 1982; Casper, 1962; Cheers, 1992; Cieslak et al., 2005; Cohen et al., 2004; Cooper, 1913; Degtjareva1 et al., 2006; Drew & Shanks, 1965; Eckstein & Karlsson, 2001; Fernald & Kinsey, 1943; Folkerts & Freeman, 1989; Gano & McNeill, 1917; Garber, 1877; Gibson, 1991a; Gluch, 2005; Godfrey & Stripling, 1961; Hanson, 1951; Harper et al., 1998; Heslop-Harrison, 2004; Heslop-Harrison & Knox, 1971; Jobson et al., 2003; Joel et al., 1985; Karlsson, 1988; Karlsson & Carlsson, 1984; Karlsson et al., 1987, 1990, 1991, 1994; Kondo, 1969; Lamb, 1991; Larsen, 1965; Legendre, 2000; Lewis et al., 1928; Lloyd, 1976; Maas, 1989; Marco, 1985; Méndez & Karlsson, 2004, 2005; Mitchell, 2011; Molau, 1993; Müller et al., 2004; Neyland et al., 2004; Patzelt et al., 2001; Persson, 1952; Pessin, 1938; Pietropaolo & Pietropaolo, 1993; Raju, 1969; Remington et al., 1918; Ritchie, 1977, 1982; Sargent & Otto, 2004; Schlosser, 2009; Schnell, 2002; Shaver et al., 1996; Sheridan et al., 1997; Slack, 1981; Svensson et al., 1993; USFWS, 1994; Wells, 1996; Williams, 1990b; Worley & Harder, 1996.

2. *Utricularia*

Bladderwort; utriculaire

Etymology: from the Latin *utriculus* (diminutive of *uterus*), meaning "small belly," in reference to the foliar bladders

Synonyms: *Akentra*; *Aranella*; *Askofake*; *Avesicaria*; *Biovularia*; *Bucranion*; *Calpidisca*; *Cosmiza*; *Diurospermum*; *Enetophyton*; *Enskide*; *Hamulia*; *Lecticula*; *Lentibularia*; *Lepiactis*; *Megozipa*; *Meionula*; *Meloneura*; *Nelipus*; *Orchyllium*; *Pelidnia*; *Personula*; *Plectoma*; *Pleiochasia*; *Plesisa*; *Polypompholyx*; *Saccolaria*; *Sacculina*; *Setiscapella*; *Stomoisia*; *Tetralobus*; *Trilobulina*; *Trixapias*; *Vesiculina*; *Xananthes*

Distribution: global: cosmopolitan; **North America:** widespread

Diversity: global: 220 species; **North America:** 19 species

Indicators (USA): OBL: *Utricularia amethystina*, *U. cornuta*, *U. floridana*, *U. foliosa*, *U. geminiscapa*, *U. gibba*, *U. inflata*, *U. intermedia*, *U. juncea*, *U. macrorhiza*, *U. minor*, *U. ochroleuca*, *U. olivacea*, *U. purpurea*, *U. radiata*, *U. resupinata*, *U. simulans*, *U. striata*, *U. subulata*

Habitat: brackish, freshwater; lacustrine, palustrine; **pH:** 3.5–8.9; **depth:** 0–4 m; **life-form(s):** emergent herb, submersed (vittate), suspended

Key morphology: plants rootless, highly modified vegetatively, the distinction of stems and leaves difficult, the shoots (often referred to as stolons) variable, including highly dissected, foliose, photosynthetic organs (submersed or suspended), or inflated, radially-oriented, spongy-tissued floats, or subterranean, achlorotic, rhizoidal branches that anchor plants to the substrate, with some species dimorphic for different variants of these types; however, all species distinguished by the presence of pouch-like bladders (to 0.3–5.0 mm) that capture various invertebrates from the surrounding water or saturated substrates; tubers or turions (2.0–30+ mm) present or absent; inflorescence a one to many-flowered raceme (0.01–1.0 m), pedicels (to 3.0 cm) straight to recurved; calyx (0.6–5.0 mm) 2-lobed; corolla [chasmogamous flowers] (1.0 mm to 2.0 cm), bilabiate, with a conical spur (to 2.0 cm), white, pink, purple, or yellow, cleistogamous (when present) smaller and bud like; stamens 2; capsules (1.0–6.0 mm) one to many seeded

Life history: duration: annual (fruit/seeds); perennial (winter buds, turions); **asexual reproduction:** shoot fragments, stolons, tubers, turions (stem), winter buds; **pollination:** insect, cleistogamous, or self; **sexual condition:** hermaphroditic; **fruit:** capsules (common); **local dispersal:** fragments, turions, winter buds (water), seeds (water, wind); **long-distance dispersal:** turions (birds), fruits (water), seeds (birds, water, wind)

Imperilment: (1) *Utricularia amethystina* [G5]; (2) *U. cornuta* [G5]; SH (DE, MD); S1 (<u>AB</u>, AR, IL, NC, OH, <u>PE</u>, TN); S2 (IN, <u>LB</u>, PA, RI, <u>SK</u>); S3 (<u>MB</u>); (3) *U. floridana* [G3]; SH (NC); S1 (AL, SC); S2 (GA); (4) *U. foliosa* [G5]; S3 (GA); (5) *U. geminiscapa* [G4]; S1 (IN, OH, NC, RI, <u>PE</u>, WV); S2 (<u>NB</u>, <u>QC</u>); S3 (DE, <u>NF</u>, NY, <u>ON</u>, VA, VT, WI); (6) *U. gibba* [G5]; SH (IA); S1 (DC, <u>NB</u>, OR, RI); S2 (KS, <u>NS</u>, <u>QC</u>, WV); S3 (<u>BC</u>, DE, NJ, VT); (7) *U. inflata* [G5]; S1 (AL, DE, MD, MI, NJ, OK, PA, TN); S2 (NY); S3 (NC, VA); (8) *U. intermedia* [G5]; SH (IA); S1 (IL, MT, RI, UT, WY); S2 (CA, ND, OH, PA, WA); S3 (ID, IN, NJ); (9) *U. juncea* [G5]; S1 (OK); S2 (DE, NY, VA); S3 (LA, MD, NC); (10) *U. macrorhiza* [G5]; SH (DC); S1 (KY, NC, WV); S2 (KS, WY); S3 (VA); (11) *U. minor* [G5]; SX (DE); SH (NC, RI); S1 (IA, IL, IN, NE, NJ, <u>PE</u>, UT); S2 (CO, <u>NB</u>, ND, OR, <u>SK</u>, WA, WY, <u>YT</u>); S3 (CA, <u>MB</u>, NY, OH); (12) *U. ochroleuca* [G4]; S1 (AK, <u>BC</u>, CA, CO, NY, <u>NS</u>, <u>ON</u>, OR); S3 (<u>QC</u>); (13) *U. olivacea* [G4]; S1 (AL, GA, MS, NJ, SC, VA); S2 (NC); (14) *U. purpurea* [G5]; S1 (DE, MD, TX); S2 (IN, MS, <u>NB</u>, <u>NF</u>, VA); S3 (AL, LA, MN, NC, NJ, <u>QC</u>, RI, VT, WI); (15) *U. radiata* [G4]; SX (PA); SH (VT); S1 (IN, <u>NB</u>); S2 (NY, RI); S3 (DE, NC, NJ, <u>NS</u>, VA); (16) *U. resupinata* [G4]; SX (IN, <u>NB</u>, PA); S1 (AL, CT, DE, GA, MD, NC, NJ, <u>NS</u>, RI, TN, VT); S2 (MA, ME, NH, <u>QC</u>); S3 (MN, WI); (17) *U. simulans* [G4]; (18) *U. striata* [G4]; SH (CT); S1 (MD, VA); S2 (NY); S3 (NC); (19) *U. subulata* [G5]; SX (PA); S1 (MI, MO, OK, RI, TN); S2 (IN); S3 (DE, MA, MD, <u>NS</u>, NY, VA)

Ecology: general: All bladderwort species are associated with aquatic or wet conditions and inhabit diverse sites that include lakes, wet shores, and even moss-covered tree trunks as FAC epiphytes (the latter habit is absent in North American representatives). The North American species represent three distinct growth forms including: (1) the so-called "terrestrial" types, which actually occur on wet substrates or in very shallow water, but are anchored to the substrate and are not often inundated; (2) suspended or submersed types, which occur routinely below the surface of standing waters (either free or attached to the bottom); and (3) floating types, where the inflorescence is supported by radial vegetative floats that contain extensive air-space tissue (the submersed, supporting portion of the shoot can be anchored or free). All of the species lack true roots and when they become attached to the bottom, they are held in place by stolons or rhizoidal branches. The vegetative portions of the plants are so extremely modified, that distinctions between leaf and stem tissue are difficult to apply. All bladderworts produce small bladder-like structures that

trap and digest minute invertebrate prey for nutrient supplementation (although some develop mutualistic associations with various aquatic invertebrates). Enzyme-labeled fluorescence assays have documented the production of phosphatase by the glandular hairs within the traps. In any case, the presence of zooplankton correlates positively with bladderwort productivity. Animals are trapped as they touch any of the trigger hairs at the base of the bladder door, which causes it to spring open and to release the internal vacuum, resulting in the inflow of water along with any appropriately sized prey in the vicinity. The active ion transport process responsible for resetting the traps is demanding energetically and respiratory rates of the bladders far exceed those of the other vegetative structures. The higher respiratory rates have been attributed, at least in part, to a structural modification (contiguous cysteine residues in helix 3 of subunit I) of cytochrome c oxidase. The plants nonrandomly trap epiphytic organisms preferentially over planktonic ones, arguably because the epiphytes are more likely to use the trigger hairs as roosting platforms, or happen to swim closer to the traps. There is one report of diminutive aquatic angiosperm plants (Lemnaceae: *Wolffia*) being captured by a bladderwort (*U. macrorhiza*); however, it appears that they were captured incidentally as a consequence of the suction-like action of the traps and were not digested. The flowers of even the most aquatic bladderwort species do not function when they are wet and nearly all are held above the water surface by elongate scapes, which are supported by shoot systems that occasionally become modified as floats. Although the showy, bilabiate flowers with nectariferous spurs appear to be well adapted for insect pollination, no definitive pollinators have been identified. Most of the bladderwort species whose reproductive biology has been investigated are self-compatible and frequently (if not primarily) self-pollinating; however, those in the relatively primitive Australian section *Polypompholyx* are self-incompatible and are obligately outcrossed wetland annuals. A shift to inbreeding is understandable, especially in many of the derived aquatic species, given that the pollination of plants growing in deepwater habitats inevitably becomes impacted by a diminished pollinator pool over open water. Several self-compatible North American species (*U. geminiscapa, U. gibba, U. juncea, U. subulata*) produce bud-like cleistogamous flowers in addition to the more characteristic, showy chasmogamous type. There have been few detailed studies of seed germination in this genus; however, the seeds of the more northern distributed species generally require a period of cold storage and cold stratification to break dormancy. Most of the North American species are perennials although a few are annuals. Some of the perennials overwinter by producing winter buds or turions (categorized according to their degree of specialization). In some species (e.g., *U. macrorhiza*), the stems completely die back in the fall leaving behind only the compact turions; thus, the plants behave in essence like "vegetative annuals." It is not unusual to find several bladderwort species co-occurring in suitable habitats. Although *U. stygia* G. Thor was given OBL status in the 2013 wetland indicator list, it is not recognized here as being distinct from

U. ochroleuca (see *Systematics* below) and no individual treatment is provided.

Utricularia amethystina Salzm. ex A. St.-Hil. & Girard is a scarce annual or perennial, which grows in depressions, flatwoods, savannas, and swamps at elevations of up to 2800 m. This is a "terrestrial" bladderwort that is attached to the substrate (peat, rock, or sand) by means of rhizoidal branchlets. It is extremely shade tolerant and the plants generally do not persist beyond the flowering stage. There is virtually nothing else known of the life history of this species in North America. Plants have not been observed recently in Florida despite repeated attempts to relocate them. **Reported associates:** none reported in the Florida portion of the range.

Utricularia cornuta Michx. is a perennial inhabitant of bogs, borrow pits, depressions, ditches, dunes, fens, lakeshores, marshes, mudflats, pannes, pools, seeps, sloughs, swales, swamps, and pond margins. It grows either as a "terrestrial" or as a subaquatic form in shallow water (e.g., 0.5–100 cm). The plants are adapted to a wide range of water chemistry (pH: 4.1–8.1; alkalinity <100 mg/L $CaCO_3$; conductivity <210 µmhos/cm) and occur in full sunlight on substrates of clay, marl, mud, peat, sand, sandy muck, or silt. The plants thrive in open conditions and grow much taller in exposed sites than in sheltered areas. The flowers are entirely chasmogamous and self-compatible. Comparable, high seed set has been obtained from both selfed and outcrossed plants. Although the flowers are adapted for insect pollination, they primarily are self-pollinating. The seeds float and are dispersed either by the water or by wind. Seed germination is high (87%) when cultured in Moore's nutrient solution (pH: 4.5–5.0), but is substantially reduced (<11%) when the seeds are sown on filter paper moistened with water of the same pH (with or without gibberellic acid added). No winter buds are formed, but vegetative reproduction occurs by leaf fragments, which are dispersed by the water. The traps have been observed to contain soil nematodes (Nematoda). **Reported associates:** *Agalinis paupercula, A. purpurea, Agrostis perennans, Amerorchis rotundifolia, Andromeda glaucophylla, A. polifolia, Andropogon glomeratus, Arethusa bulbosa, Asclepias rubra, Symphyotrichum novi-belgii, Bartonia paniculata, Betula nana, B. pumila, Brasenia schreberi, Burmannia capitata, Calamagrostis canadensis, C. pickeringii, Calamovilfa longifolia, Calopogon tuberosus, Carex aquatilis, C. buxbaumii, C. canescens, C. chordorrhiza, C. crawei, C. eburnea, C. exilis, C. lasiocarpa, C. lenticularis, C. limosa, C. livida, C. lurida, C. oligosperma, C. pauciflora, C. stricta, C. trisperma, a. viridula, Chamaecyparis thyoides, Chamaedaphne calyculata, Chara, Cladium mariscoides, Cladopodiella fluitans, Cornus sericea, Cypripedium reginae, Danthonia epilis, Dasiphora floribunda, Dichanthelium acuminatum, D. dichotomum, Doellingeria umbellata, Drosera capillaris, D. filiformis, D. intermedia, D. rotundifolia, Dulichium arundinaceum, Elatine minima, Eleocharis acicularis, E. compressa, E. elliptica, E. melanocarpa, E. olivacea, E. quinqueflora, E. robbinsii, E. rostellata, E. tuberculosa, Empetrum nigrum, Eriocaulon aquaticum, E. compressum, E. decangulare, E. tuberculosa, Eriophorum vaginatum,* *E. virginicum, Eupatorium perfoliatum, Fimbristylis autumnalis, Fuirena scirpoidea, Gaylussacia baccata, G. dumosa, Gratiola aurea, Helianthus angustifolius, Hypericum kalmianum, H. lissophloeus, H. mutilum, Ilex glabra, Ionactis linariifolia, Iris versicolor, I. virginica, Juncus abortivus, J. balticus, J. canadensis, J. dichotomus, J. diffusissimus, J. effusus, J. militaris, J. nodatus, J. nodosus, J. pelocarpus, Kalmia angustifolia, Lachnanthes caroliana, Larix laricina, Linum medium, Lobelia canbyi, L. dortmanna, L. kalmii, L. nuttallii, Lophiola aurea, Ludwigia alternifolia, Lycopodiella alopecuroides, L. caroliniana, L. inundata, Mayaca fluviatilis, Menyanthes trifoliata, Muhlenbergia uniflora, Myrica gale, M. pensylvanica, Myriophyllum tenellum, Narthecium americanum, Nymphaea odorata, Nuphar variegata, Oclemena nemoralis, Orontium aquaticum, Panicum tenerum, Parnassia glauca, Photinia floribunda, P. melanocarpa, Picea mariana, Pinguicula, Pinus rigida, Platanthera blephariglottis, P. clavellata, P. dilatata, Pogonia ophioglossoides, Polygala cruciata, Rhexia virginica, Rhododendron groenlandicum, Rhynchospora alba, R. capillacea, R. cephalantha, R. chalarocephala, R. fusca, R. gracilenta, R. inundata, R. oligantha, Rubus chamaemorus, Sabatia difformis, Sagittaria engelmanniana, S. isoetiformis, Sarracenia alata, S. psittacina, S. purpurea, Scheuchzeria palustris, Schizachyrium scoparium, Schizaea pusilla, Schoenoplectus acutus, S. pungens, Scirpus cyperinus, Scleria reticularis, S. verticillata, Sclerolepis uniflora, Solidago ptarmicoides, S. uliginosa, Sphagnum bartlettianum, S. cuspidatum, S. fallax, S. magellanicum, S. majus, S. papillosum, S. portoricense, S. pulchrum, S. pylaesii, S. rubellum, Spiranthes cernua, Symphyotrichum dumosum, S. novi-belgii, Taxodium ascendens, Thelypteris palustris, Trianthu racemosa, Triadenum virginicum, Trichophorum alpinum, Triglochin maritimum, T. palustre, Utricularia gibba, U. intermedia, U. juncea, U. minor, U. purpurea, U. resupinata, U. striata, U. subulata, Vaccinium macrocarpon, V. oxycoccos, Woodwardia virginica, Xyris montana, X. smalliana.*

Utricularia floridana Nash is a perennial, which grows as an aquatic in bogs, lakes, and ponds at depths from 0.1 to 8.0 m. The waters are clear (secchi depth: 0.8–4.5 m), acidic (pH: 4.5–7.2; mean: 5.8), soft (alkalinity <25 mg/L $CaCO_3$; conductivity <151 µmhos/cm), and low in nutrients (total phosphorous <30 mg/L; total N: 600 mg/L (mean). Although dislodged plants can occur as suspended forms when dispersed over deeper water, the shoots usually are attached firmly to the bottom in water less than 1.5 m deep. Substrates include Chipley (aquic quartzipsamments), muck, mud, Pamlico-Dorovan (terric medisaprists), or sand. Shallower sites enable the plants to send up extremely long flowering scapes (up to 1 m) to raise the flowers (10 dm or more) above the water surface. There are no specific accounts of pollination, seed dispersal, or seed germination for this species. The plants produce two distinct shoot types: leafless shoots (attached to the substrate), which bear numerous bladders; and leafy shoots (extending into the water), which produce few or no bladders. The foliage secretes various polysaccharides, which serve to attract feeding bacteria and algae to the traps for capture. Nutrient supplementation by carnivory was indicated in one study where the plants contained 3.4% nitrogen, a level that was approximately twice the amount found in other co-occurring submersed species. Vegetative reproduction occurs by shoot fragmentation. This species often is found growing along with few others. **Reported associates:** Brasenia schreberi, Cabomba caroliniana, Eleocharis acicularis, E. baldwinii, E. robbinsii, Eriocaulon, Fuirena scirpoidea, Mayaca fluviatilis, Myriophyllum heterophyllum, M. laxum, M. pinnatum, Nelumbo lutea, Nuphar advena, Nymphaea odorata, Nymphoides aquatica, N. cordata, Potamogeton diversifolius, Rhynchospora tracyi, Sagittaria isoetiformis, S. subulata, Utricularia foliosa, U. gibba, U. inflata, U. purpurea, U. resupinata, Websteria confervoides.

Utricularia foliosa **L.** is a perennial, which occurs in bayous, canals, creeks, ditches, lakes, marshes, oxbows, ponds, pools, rivers, sloughs, and streams in waters spanning a fairly broad range of conditions (pH: 5.5–8.5, mean: 6.9; alkalinity <75 mg/L $CaCO_3$; conductivity <561 µmhos/cm; secchi depth = 0.5–3.7 m). The plants are somewhat salt tolerant. The shoots float or are suspended in water from 0.3 to 2.0 m in depth. The main vegetative axes are distinctively flattened. Some branches lack bladders and most bladders occur on the bushiest shoots. The vegetative organs generally are coated with mucilage. The showy flowers are self-compatible and attract few insect pollinators; however, genetic analyses using microsatellite markers indicate that populations are not entirely clonal but show some evidence of outcrossing. Experimentally selfed flowers have on average 68% fruit set, five seeds/fruit, and 43% seed set. The selfed seeds are viable and germinate well. These observations indicate a mixed mating system in this species. The berry-like capsules float and can be dispersed by the water; however, the seeds sink rapidly when they are shed. Vegetative reproduction occurs by shoot fragmentation. There is a positive correlation of prey capture with the content of carbohydrates, which may function as an attractant; however, the carbohydrate content actually is thought to be derived primarily from the associated periphyton communities. The periphyton facilitate phosphorous uptake to the shoots from the water. The plants optimize their resource allocation to carnivory by increasing the size and number of bladders when zooplankton abundance is low, and by decreasing them when water nutrient levels are high. Documented prey capture includes bugs (Hemiptera: Corixidae), flies (Diptera: Chironomidae; Ceratopogonidae), mayflies (Ephemeroptera: Caenidae), rotifers (Rotifera), and seed shrimp (Crustacea: Ostracoda). Trapping efficiency varies considerably with the bigger bladders capturing more, larger, and a greater taxonomic richness of prey. The optimal size of bladders has been determined as 1650 µm. The precise contribution of carnivory to the plants is difficult to elucidate. In some cases, the plants invest from 19% to 52% of their biomass in bladder production, but obtain most of their nitrogen and phosphorus through foliar uptake. The predominant uptake of nitrogen directly from the water as ammonium has been confirmed by studies using ^{15}N isotopic tracers. **Reported associates:** Alternanthera philoxeroides, Annona

glabra, Baccharis, Bacopa caroliniana, Blechnum serrulatum, Boehmeria cylindrica, Brasenia schreberi, Cabomba caroliniana, Cephalanthus occidentalis, Ceratophyllum demersum, Cladium jamaicense, Colocasia esculenta, Cyperus alternifolius, C. elegans, C. haspan, C. lecontei, Egeria densa, Eichhornia crassipes, Eleocharis baldwinii, E. cellulosa, E. elongata, Eriocaulon, Eupatorium leptophyllum, Ficus aurea, Fuirena scirpoidea, F. squarrosa, Heteranthera dubia, Hydrocotyle umbellata, Hypericum, Ipomoea aquatica, Juncus dichotomus, J. polycephalus, Justicia ovata, Lachnanthes caroliniana, Leersia hexandra, Lemna, Limnobium spongia, Ludwigia octovalvis, L. repens, Mayaca fluviatilis, Mikania scandens, Myrica cerifera, Myriophyllum heterophyllum, M. spicatum, Najas guadalupensis, Nitella, Nuphar, Nymphaea odorata, Nymphoides aquatica, Panicum hemitomon, P. repens, P. rigidulum, P. virgatum, Paspalidium geminatum, Paspalum distichum, Persea, Persicaria hydropiperoides, P. punctata, Pluchea rosea, Pontederia cordata, Potamogeton diversifolius, P. nodosus, Proserpinaca palustris, Rhizophora mangle, Rhynchospora inundata, R. microcarpa, R. tracyi, Ruppia maritima, Saccharum giganteum, Sagittaria graminea, S. lancifolia, S. subulata, Salix, Salvinia minima, Sambucus nigra, Taxodium distichum, Thelypteris kunthii, Typha domingensis, Utricularia floridana, U. gibba, U. inflata, U. purpurea, U. resupinata, Vallisneria americana, V. neotropicalis, Wolffia, Xyris elliottii.

Utricularia geminiscapa **Benj.** is a perennial found in bogs, creeks, depressions, ditches, lakes, meadows, ponds, pools, and swales in waters up to 2.3 m deep. The plants tolerate a fairly broad range of water conditions (pH: 3.5–8.6, mean: 6.7; alkalinity <230 mg/L $CaCO_3$; conductivity <430 µmhos/cm) and occur on muddy or sandy substrates. Most commonly this species grows as a suspended form in fairly shallow waters (mean depth: 1.3 m), but it has been found at a record depth of 18 m in one lake with extraordinarily clear water (secchi depth >18 m). Although the occurrences span a wide pH gradient, the species is most characteristic of fairly acidic sites (pH <5.5) and has disappeared quickly (within 5 years) from lakes that have been treated with calcite (stable $CaCO_3$). The plants are self-compatible and produce short-stalked, bud-like, self-pollinating, underwater cleistogamous flowers in addition to their normal, showy, aerial, chasmogamous flowers. Natural pollinators of the chasmogamous flowers have not been identified. Seeds stored in water at 1°C–3°C retained their viability (19%–38%) for up to 3 years. Vegetative reproduction occurs by shoot fragmentation and by production of loose, unspecialized winter buds. **Reported associates:** Andromeda glaucophylla, Andropogon glaucopsis, A. perangustatus, Brasenia schreberi, Calla palustris, Carex canescens, C. lacustris, C. lasiocarpa, C. oligosperma, C. pseudocyperus, C. rostrata, C. striata, C. stricta, Centella erecta, Ceratophyllum demersum, Chamaecyparis thyoides, Chamaedaphne calyculata, Cyperus bipartitus, Decodon verticillatus, Dichanthelium wrightianum, Drosera rotundifolia, Elatine minima, Eleocharis melanocarpa, E. palustris, E. robbinsii, Elodea canadensis, E. nuttallii, Eriocaulon aquaticum, E. compressum, Eriophorum vaginatum, Eupatorium leucolepis,

E. mohrii, E. resinosum, Galium triflorum, Heteranthera dubia, Hypericum denticulatum, H. fasciculatum, Ilex, Juncus brevicaudatus, Kalmia polifolia, Lachnanthes caroliniana, Leersia hexandra, Lemna minor, L. trisulca, Lobelia boykinii, L. dortmanna, Lysimachia loomisii, Megalodonta beckii, Myriophyllum farwellii, M. heterophyllum, M. laxum, M. sibiricum, M. spicatum, M. tenellum, M. verticillatum, Najas flexilis, Nitella, Nuphar advena, N. variegata, Nymphaea odorata, N. tuberosa, Panicum hemitomon, P. tenerum, P. verrucosum, P. virgatum, Persicaria amphibia, Photinia floribunda, Phragmites australis, Polygala cymosa, Potamogeton confervoides, P. foliosus, P. gramineus, P. natans, P. praelongus, P. pusillus, P. richardsonii, P. robbinsii, P. zosteriformis, Rhynchospora perplexa, R. pusilla, R. scirpoides, Sagittaria isoetiformis, S. latifolia, Sarracenia purpurea, Scheuchzeria palustris, Schoenoplectus acutus, S. subterminalis, S. tabernaemontani, S. torreyi, Scirpus cyperinus, Scleria reticularis, Sclerolepis uniflora, Sparganium angustifolium, Sphagnum, Spirodela polyrhiza, Stuckenia pectinata, Typha angustifolia, Utricularia gibba, U. intermedia, U. macrorhiza, U. minor, U. olivacea, U. purpurea, U. resupinata, U. striata, Vaccinium oxycoccos, Vallisneria americana, Viola blanda, V. lanceolata, Woodwardia, Zizania aquatica.

***Utricularia gibba* L.** is an annual or perennial, which inhabits bogs, canals, channels, depressions, ditches, dunes, fens, flatwoods, lakes, marshes, meadows, mudflats, ponds, pools, river backwaters, seeps, sinkholes, sloughs, streams, and swamp margins at elevations of up to 2500 m. The plants occur either rooted on moist shorelines, tangled in mats of floating vegetation, free-floating on the surface, or suspended in waters up to 3.8 m deep. They grow predominantly in open sites but otherwise are quite broadly tolerant ecologically (pH: 4.6–8.6; alkalinity <100 mg/L $CaCO_3$; conductivity <385 μmhos/cm) and will withstand some turbidity (secchi depth: 0.4–3.4 m) and brackish water. Reported substrates include loamy sand, muck, mud, sand, peat, peaty muck, and even floating or moss-covered logs (extending up to 30 cm above the water surface). Flowering occurs when plants are floating, entangled in floating mats or vegetation, or when attached to exposed substrates in water typically no more than 2–3 cm deep but almost never occurs in plants that are suspended in deeper water. The shoots can produce either erect inflorescences normally bearing 2–6 chasmogamous flowers, or reduced inflorescences with submersed cleistogamous flowers. Seed set occurs only on stranded plants or those in shallow water. The winged seeds are dispersed by the wind or by water. Germination requirements are not known with certainty; however, seed collected in field cores germinated without any pretreatment when placed under conditions ranging from drawdown to flooded states. Generally, germination has been observed to increase with the extent and duration of flooding. Vegetative reproduction (often resulting in dense floating mats) occurs by fragmentation of the stolons, which can reach 20 cm in length. Some authors describe this species as producing winter buds; however, such reports are likely due to misinterpretations of the cleistogamous flowers. Nitrogen supplementation results in the elongation of the stolons

because lateral branches do not develop in this species. An increase in internode growth and bladder number resulted when plants that were cultured under suboptimal nutrient conditions were fed paramecia (Chromalveolata: Ciliophora: Parameciidae). However, studies using isotopic tracers have indicated that *U. gibba* preferentially takes up nitrogen in the form of dissolved inorganic nitrogen rather than obtaining it from prey. This is a relatively easy plant to maintain in culture. **Reported associates:** *Acer rubrum, Agalinis purpurea, Alisma subcordatum, Alnus serrulata, Alternanthera philoxeroides, Ammannia coccinea, Andropogon virginicus, Arethusa bulbosa, Azolla caroliniana, Bacopa caroliniana, B. monnieri, Bidens cernuus, Brasenia schreberi, Cabomba caroliniana, Calopogon tuberosus, Carex bullata, C. comosa, C. crinita, C. exilis, C. glaucescens, C. lasiocarpa, C. lurida, C. rostrata, C. striata, C. tenuiflora, Cephalanthus occidentalis, Ceratophyllum demersum, C. echinatum, C. muricatum, Chamaecyparis thyoides, Chara, Cicuta bulbifera, Cladium jamaicense, C. mariscoides, Crinum americanum, Cyperus bipartitus, C. ferruginescens, C. strigosus, Decodon verticillatus, Drosera intermedia, D. rotundifolia, Dulichium arundinaceum, Eichhornia crassipes, Egeria densa, Elatine minima, Eleocharis acicularis, E. baldwinii, E. cellulosa, E. elongata, E. erythropoda, E. intermedia, E. olivacea, E. robbinsii, E. smallii, E. tuberculosa, Elodea nuttallii, Eriocaulon aquaticum, Eriophorum virginicum, Fimbristylis autumnalis, Fuirena squarrosa, Glossostigma cleistanthum, Glyceria canadensis, Gratiola aurea, Heteranthera dubia, Hymenocallis latifolia, Hypericum majus, Itea virginica, Isoetes echinospora, Juncus canadensis, J. effusus, J. coriaceus, J. pelocarpus, Landoltia punctata, Leersia oryzoides, Lemna minor, Lemna minuta, Leptochloa fusca, Limnobium spongia, Liparis loeselii, Lipocarpha micrantha, Liquidambar styraciflua, Lobelia dortmanna, L. kalmii, Lycopus, Ludwigia alternifolia, L. leptocarpa, L. palustris, L. repens, Luziola fluitans, Mayaca fluviatilis, Menyanthes trifoliata, Myriophyllum aquaticum, M. farwellii, M. heterophyllum, M. humile, M. sibiricum, M. tenellum, M. verticillatum, Najas flexilis, N. gracillima, N. guadalupensis, Nelumbo lutea, Nitella, Nuphar advena, N. variegata, Nymphaea odorata, Nyssa biflora, Orontium aquaticum, Panicum hemitomon, P. rigidulum, Paspalidium geminatum, Persicaria coccinea, Phyla lanceolata, Pinus taeda, Pistia stratiotes, Pogonia ophioglossoides, Pontederia cordata, Potamogeton amplifolius, P. bicupulatus, P. crispus, P. diversifolius, P. epihydrus, P. foliosus, P. friesii, P. gramineus, P. natans, P. oakesianus, P. pusillus, P. robbinsii, P. strictifolius, P. vaseyi, Proserpinaca palustris, Ranunculus flammula, Rhexia mariana, R. virginica, Rhynchospora alba, R. capillacea, R. capitellata, R. chalarocephala, R. fusca, R. inundata, R. scirpoides, R. tracyi, Rhus aromatica, Riccia fluitans, Sagittaria cuneata, S. graminea, S. lancifolia, S. latifolia, S. rigida, Salvinia, Sarracenia purpurea, Schoenoplectus acutus, S. pungens, S. smithii, S. subterminalis, Scirpus cyperinus, S. longii, Scleria, Sparganium americanum, S. androcladum, Sphagnum bartlettianum, S. fallax, S. flavicomans, S. magellanicum, S. recurvum, S. papillosum, S. torreyanum,*

S. henryense, Spirodela polyrhiza, Stuckenia pectinata, Taxodium, Thalia geniculata, Triadenum fraseri, T. virginicum, Trichophorum cespitosum, Triglochin palustre, Typha angustifolia, T. latifolia, Utricularia cornuta, U. floridana, U. foliosa, U. geminiscapa, U. intermedia, U. macrorhiza, U. minor, U. purpurea, U. radiata, U. resupinata, U. striata, U. subulata, Vallisneria americana, Xyris smalliana, X. torta.

***Utricularia inflata* Walter** is a perennial inhabitant of canals, cypress domes, depressions, ditches, dunes, flatwoods, floodplains, lagoons, lakes, marshes, ponds, pools, prairies, savannas, streams, and swamps. The plants occur in open sites where waters are fresh (<0.5 ppt salinity; cannot tolerate salinity >1 ppt), fairly clear (secchi depth: 0.5–3.6 m), circumneutral (pH: 5.5–8.6), relatively soft (alkalinity <100 mg/L CaCO$_3$), and of moderate conductivity (<615 µmhos/cm). They are almost always found floating or suspended in still or slow-moving waters less than 1.5 m deep. Neither the floral nor seed ecology of this species has been studied in detail. The plants do not always flower every year and some flowers have been observed to open while they are submersed under as much as a meter of clear water. Fruit set is indicated by the reflexing of the peduncles. A complete lack of fruit set has been observed in some populations that occur at the northern extent of the distributional range. Vegetative reproduction occurs by shoot fragmentation and by tubers, which are produced at ends of stolons, especially when in contact with the substrate (muck), or on plants that have become stranded on the shore when waters have receded. Tuber germination has not been observed in the field, but occurs readily when they are placed in water in the laboratory. The tubers facilitate survival over dry conditions. The plants can become excessively dense in still, protected waters and contribute importantly to the early development of floating marshland systems in the southern United States. Drawdowns result in reduced numbers of plants occurring in deeper waters but in increased numbers in shallow sites as competing species are eradicated. A general decline of rosulate species has been observed at sites where *U. inflata* proliferates, which in turn alters the sediment chemistry leading to increased pore water carbon dioxide and ammonium nitrogen levels. Studies using radioactive tracers confirm that a significant amount of phosphorous is taken up by the plants through carnivory. Foliar uptake of stable cesium ([133]Cs) from the water column has been documented. **Reported associates:** *Alternanthera philoxeroides, Azolla caroliniana, Brasenia schreberi, Cabomba caroliniana, Callitriche heterophylla, Carex decomposita, C. glaucescens, C. intumescens, C. longii, C. striata, Ceratophyllum demersum, Chara, Clethra alnifolia, Cyperus virens, Egeria densa, Eichhornia crassipes, Eleocharis obtusa, E. robbinsii, Elodea canadensis, Eriocaulon aquaticum, E. compressum, Hydrilla verticillata, Hydrocotyle umbellata, Hydrolea quadrivalvis, Lemna minor, L. valdiviana, Limnobium spongia, Lobelia dortmanna, Ludwigia peruviana, Luziola fluitans, Micranthemum umbrosum, Myriophyllum heterophyllum, M. hippuroides, M. spicatum, M. tenellum, Najas guadalupensis, Nelumbo lutea, Nitella, Nuphar advena, N. polysepala, Nymphaea odorata, Nymphoides aquatica, N. cordata,*

Nyssa, Orontium aquaticum, Panicum hemitomon, P. repens, Persicaria hydropiperoides Pieris phillyreifolia, Pinus elliottii, Potamogeton amplifolius, P. diversifolius, P. gramineus, P. illinoensis, P. praelongus, P. pusillus, P. richardsonii, P. robbinsii, P. zosteriformis, Proserpinaca palustris, Ranunculus longirostris, Riccia fluitans, Sagittaria graminea, S. subulata, Salvinia minima, Sarracenia flava, Sparganium angustifolium, Taxodium ascendens, Thelypteris, Tillandsia bartramii, Torreyochloa pallida, Utricularia foliosa, U. purpurea, U. radiata, Vallisneria americana, Woodwardia virginica.

***Utricularia intermedia* Hayne** is a perennial, which inhabits beaches, bogs, depressions, ditches, dunes, fens, flats, lakes, marshes, meadows, muskeg, ponds, pools, sloughs, swamps, and margins of rivers or streams at elevations of up to 2700 m. Generally, this species is a poor competitor and is intolerant of eutrophication. The plants occur in open (occasionally thickly vegetated) sites as a suspended life-form at depths up to 3.0 m, but thrive in shallower water (<1.0 m), where the stolons can extend (up to 15 cm) into the substrate. The foliage is dimorphic with photosynthetic leaves that lack bladders and achlorotic, leafless, bladder-bearing shoots, which typically are embedded in the substrate. A very high proportion of the total plant biomass (about 51%) is invested in the production of carnivorous shoots and bladders, with the latter representing roughly 55% of the carnivorous shoot biomass. The bladders are of medium size relative to other species and can capture prey from 1 to 4 mm in size. Although often regarded as an indicator of alkaline, minerotrophic conditions, this species occurs across a broad range of acidity (pH: 3.6–8.6), and actually is somewhat more common in acidic water (mean pH: 6.7) of moderate alkalinity (<230 mg/L CaCO$_3$) and conductivity (<400 µmhos/cm). Reported substrates include muck, mud, peat, peaty mud and sand. Flowering occurs infrequently and only when the plants reside in shallow water (<0.5 m). Although the plants are fertile, fruit production is variable and mature fruits often are not observed. One study in Japan reported fruit set of only 29.7% and an extremely low number of seeds per capsule (mean = 0.45). Field-collected seeds have germinated successfully in tap water. Reproduction primarily is vegetative, and occurs by means of the compact turions, which are produced in mid to late summer or earlier if in desiccating habitats. The turions are laden with starch and sink to the bottom as the shoots decompose. Their germination is coincident with the increase of temperature in early spring, although they will germinate earlier if they are brought into the laboratory and kept at room temperature. The turions contain polyamines (spermidine, homospermidine), which are believed to function as growth factors. Upon their germination, internal oxygen production provides buoyancy to the turions, which are then dispersed by water. **Reported associates:** *Abies balsamea, Acer rubrum, Acorus calamus, Agalinis purpurea, Alisma gramineum, Allionia gausapoides, Alnus rugosa, Alopecurus aequalis, Amerorchis rotundifolia, Andromeda glaucophylla, A. polifolia, Aneura pinguis, Arethusa bulbosa, Asclepias incarnata, Aulacomnium palustre, Betula papyrifera, B. pumila, Bidens cernuus, Botrychium simplex, Brachythecium, Brasenia schreberi,*

Bromus ciliatus, Bryum pseudotriquetrum, Cabomba caroliniana, Calamagrostis canadensis, Calliergon richardsonii, Calliergonella cuspidata, Callitriche hermaphroditica, Calopogon tuberosus, Campanula aparinoides, Campylium stellatum, Carex aquatilis, C. atherodes, C. brunnescens, C. buxbaumii, C. canescens, C. chordorrhiza, C. comosa, C. crawei, C. diandra, C. disperma, C. eburnea, C. exilis, C. flava, C. hystericina, C. interior, C. lacustris, C. lasiocarpa, C. leptalea, C. limosa, C. livida, C. magellanica, C. microglochin, C. oligosperma, C. ormostachya, C. pauciflora, C. pellita, C. prairea, C. rostrata, C. scirpoidea, C. sterilis, C. stricta, C. tenuiflora, C. trisperma, C. utriculata, C. vesicaria, C. viridula, Ceratophyllum demersum, Chamaedaphne calyculata, Chara vulgaris, Chelone glabra, Cicuta bulbifera, Cladium mariscoides, Clintonia borealis, Comarum palustre, Coptis trifolia, Cornus sericea, Cyperus bipartitus, Cypripedium reginae, Dasiphora floribunda, Decodon verticillatus, Doellingeria umbellata, Drepanocladus aduncus, D. crassicostatus, Drosera anglica, D. intermedia, D. linearis, D. rotundifolia, Dulichium arundinaceum, Eleocharis acicularis, E. compressa, E. elliptica, E. erythropoda, E. olivacea, E. palustris, E. quinqueflora, E. robbinsii, E. rostellata, E. smallii, Elodea canadensis, E. nuttallii, Epilobium leptophyllum, Equisetum arvense, E. fluviatile, E. hyemale, Eriocaulon aquaticum, Eriophorum angustifolium, E. crinigerum, E. gracile, E. spissum, E. virginicum, E. viridicarinatum, Eupatorium perfoliatum, Eutrochium maculatum, Fissidens adianthoides, Frangula alnus, Galium boreale, G. labradoricum, Gaultheria hispidula, Gentianopsis procera, Glyceria canadensis, G. striata, Hippuris vulgaris, Hypericum, Iris versicolor, Isoetes echinospora, I. muricata, Juncus albescens, J. alpinus, J. canadensis, J. stygius, Kalmia polifolia, Kobresia simpliciuscula, Larix laricina, Leersia oryzoides, Lemna minor, L. trisulca, Limprichtia revolvens, Linnaea borealis, Lobelia kalmii, L. siphilitica, Lycopodium annotinum, Lycopus americanus, L. uniflorus, Lysimachia quadriflora, L. terrestris, L. thyrsiflora, Megalodonta beckii, Mentha arvensis, Menyanthes trifoliata, Muhlenbergia glomerata, Myrica gale, M. pensylvanica, Myriophyllum alterniflorum, M. farwellii, M. sibiricum, M. spicatum, M. verticillatum, Najas flexilis, N. guadalupensis, N. minor, Nymphaea odorata, N. tetragona, N. tuberosa, Nuphar polysepala, N. ×rubrodisca, N. variegata, Osmunda regalis, Packera aurea, Panicum flexile, Parnassia glauca, Pedicularis lanceolata, Peltandra virginica, Persicaria amphibia, P. hydropiper, Phragmites australis, Picea mariana, Platanthera clavellata, P. dilatata, Pogonia ophioglossoides, Pohlia wahlenbergii, Pontederia cordata, Potamogeton alpinus, P. amplifolius, P. epihydrus, P. friesii, P. foliosus, P. gramineus, P. natans, P. obtusifolius, P. praelongus, P. pulcher, P. pusillus, P. richardsonii, P. robbinsii, P. spirillus, P. subsibiricus, P. zosteriformis, Potentilla gracilis, Pycnanthemum virginianum, Ranunculus gmelinii, R. longirostris, R. occidentalis, Rhamnus alnifolia, Rhynchospora alba, R. capillacea, R. scirpoides, Rubus pubescens, Rumex, Sagittaria cuneata, Sa. latifolia, Salix candida, S. discolor, S. pedicellaris, S. pyrifolia, S. serissima, Sarracenia purpurea, Scheuchzeria palustris, Schoenoplectus acutus, S. smithii, S. tabernaemontani, Scirpus cyperinus, S. longii, Scleria verticillata, Scorpidium scorpioides, Scutellaria galericulata, Sium suave, Solidago ohioensis, S. patula, S. uliginosa, Sorbus decora, Sparganium angustifolium, S. chlorocarpum, S. eurycarpum, S. fluctuans, S. natans, Sphagnum centrale, S. cuspidatum, S. fallax, S. fimbriatum, S. lescurii, S. papillosum, S. teres, S. warnstorfii, Spiraea, Spirodela polyrhiza, Stuckenia filiformis, S. pectinata, S. vaginata, Symphyotrichum lanceolatum, Thelypteris palustris, Thuja occidentalis, Tofieldia pusilla, Tolypella intricata, Tomentypnum nitens, Toxicodendron vernix, Triadenum fraseri, T. virginicum, Triantha glutinosa, Trichophorum alpinum, T. cespitosum, Trientalis borealis, Triglochin maritimum, T. palustre, Typha angustifolia, T. latifolia, Utricularia cornuta, U. geminiscapa, U. gibba, U. macrorhiza, U. minor, U. purpurea, U. radiata, U. resupinata, Vaccinium macrocarpon, V. oxycoccos, V. uliginosum, Vallisneria americana, Viola, Wolffia columbiana, Xyris.

***Utricularia juncea* Vahl** is an annual species, which occurs in very shallow waters or on the drying shores of bogs, depressions, ditches, dunes, flatwoods, floodplains, marshes, pools, savannas, seeps, sloughs, swales, swamps, or along the margins of pools and streams at elevations of up to 1300 m. The sites are open with acidic (specific pH values not reported), granitic and sandstone substrates that include loamy sand, muck, mud, peat and sand (sometimes covered by a thin organic layer). The plants are self-compatible and produce both chasmogamous and cleistogamous (65%–80%) flowers. Although the flowers are adapted for insect pollination, the plants primarily are self-pollinating and good seed set has been obtained whether the flowers are selfed or outcrossed experimentally. Germination of unstratified seed is high (98%) when cultured in Moore's nutrient solution (pH: 4.5–5.0), but is reduced (<13%) when the seeds are sown on filter paper moistened with water having the same pH (with or without gibberellic acid added). The seeds float and are dispersed by the water or by wind. A persistent seed bank (but of uncertain duration) is produced. Some vegetative reproduction occurs by means of leaf fragments, which are dispersed by the water. *Utricularia juncea* has been described as a fugitive species whose establishment depends on disturbance, or on sites that are free of litter (e.g., crayfish mounds). Fairly dense stands of plants develop occasionally but this species is highly intolerant of competition. Presumably (but not yet demonstrated conclusively) an annual, a large proportion of its biomass (>90%) is allocated to reproductive structures. There is little photosynthetic tissue, and the plants probably obtain significant quantities of nutrients via carnivory.

Reported associates: *Acer rubrum, Acrostichum, Agalinis purpurea, Agrostis perennans, Aletris lutea, Amphicarpum muhlenbergianum, Andropogon glomeratus, A. gyrans, A. virginicus, Annona glabra, Aristida beyrichiana, A. palustris, A. patula, Asimina reticulata, Bartonia paniculata, Bigelowia nudata, Blechnum serrulatum, Brasenia schreberi, Bryum pseudotriquetrum, Calamagrostis pickeringii, Calopogon tuberosus, Carex exilis, C. livida, Carphephorus carnosus, Centella asiatica, Chamaecyparis thyoides, Chloris,*

Cladium jamaicensis, C. mariscoides, Climacium americanum, Ctenium aromaticum, Cyperus haspan, C. odoratus, Danthonia epilis, Dichanthelium dichotomum, D. ensifolium, Drosera capillaris, D. filiformis, D. intermedia, Eleocharis obtusa, E. minima, E. olivacea, E. robbinsii, Erigeron vernus, Eriocaulon aquaticum, E. decangulare, Eriophorum virginicum, Eupatorium recurvans, Fimbristylis castanea, Fuirena pumila, F. scirpoidea, Gaylussacia dumosa, Gratiola aurea, Hartwrightia floridana, Hypericum cistifolium, Hypericum fasciculatum, H. myrtifolium, Ilex cassine, I. coriacea, I. glabra, Juncus acuminatus, J. biflorus, J. canadensis, J. dichotomus, J. megacephalus, J. militaris, J. pelocarpus, J. scirpoides, J. trigonocarpus, Kalmia angustifolia, Lachnanthes caroliana, Lachnocaulon anceps, Leersia hexandra, Liatris spicata, Linum medium, Lobelia brevifolia, L. canbyi, L. nuttallii, Lophiola aurea, Ludwigia, Lycopodiella alopecuroides, L. appressa, L. caroliniana, Lyonia lucida, Magnolia virginiana, Marshallia tenuifolia, Muhlenbergia uniflora, Myrica cerifera, M. pensylvanica, Narthecium americanum, Nymphaea odorata, Nymphoides cordata, Oclemena nemoralis, O. reticulata, Osmundastrum cinnamomeum, Oxypolis filiformis, Panicum abscissum, P. hemitomon, Persea borbonia, Philonotis fontana, Pinguicula, Pinus elliottii, P. palustris, P. rigida, Platanthera blephariglottis, Pluchea, Pogonia ophioglossoides, Polygala brevifolia, P. cruciata, P. ramosa, P. rugelii, P. setacea, Pontederia cordata, Proserpinaca pectinata, Pterocaulon pycnostachyum, Rhexia cubensis, R. nuttallii, R. mariana, R. virginica, Rhynchospora alba, R. breviseta, R. cephalantha, R. chalarocephala, R. ciliaris, R. colorata, R. fascicularis, R. filifolia, R. glomerata, R. gracilenta, R. inundata, R. macrostachya, R. oligantha, R. scirpoides, Sabal palmetto, Sabatia difformis, S. grandiflora, Sagittaria teres, Salix caroliniana, Sarracenia minor, S. purpurea, Saururus cernuus, Schizachyrium scoparium, S. stoloniferum, Schizaea pusilla, Scleria reticularis, S. verticillata, Serenoa repens, Smilax, Solidago sempervirens, Spartina bakeri, Sphagnum bartlettianum, S. magellanicum, S. portoricense, S. pulchrum, Stillingia aquatica, Symphyotrichum novi-belgii, Syngonanthus flavidulus, Taxodium distichum, Tillandsia utriculata, Triantha racemosa, Triadenum virginicum, Utricularia cornuta, U. fibrosa, U. purpurea, U. striata, U. subulata, Vaccinium macrocarpon., Woodwardia virginica, Xyris ambigua, X. caroliniana, X. elliottii, X. fimbriata, X. jupicai, X. smalliana, X. platylepis, X. torta.

***Utricularia macrorhiza* Leconte** is a perennial inhabitant of fresh to slightly brackish bayous, bogs, canals, channels, depressions, ditches, dunes, fens, flats, floating mats, floodplains, flowages, lakes, marshes, oxbows, ponds, pools, potholes, prairies, reservoirs, rivers, sloughs, streams and swales at elevations of up to 2700 m. This species is extremely versatile ecologically and is found in still to slow-moving hard and soft waters (alkalinity <265 mg/L CaCO$_3$) spanning a wide range of acidity (pH: 4.8–8.9) and nutrient levels (conductivity <400 µmhos/cm). The suspended plants grow mainly in shallow waters and also can occur at depths up to 2.5 m. Reported substrates typically are organic and include clay, gravel, loam, muck, peaty muck, sand, sandy muck, and silt.

The plants are tolerant of turbidity and eutrophic conditions but also occur in clear, oligotrophic waters. The reproductive biology of this species is enigmatic. Although the large, showy flowers are adapted for insect pollination, few potential pollinators have been reported. The flowers open only for a brief duration, usually 1 day. Presumably, they are self-compatible and capable of autogamy; however, these assumptions require confirmation. After pollination, the pedicels recurve and the peduncles reorient, pulling the flowers beneath the water where the capsules develop; however, seed maturation has been observed infrequently in many populations. This species rarely is represented in seed bank studies and the propagules probably do not persist for any significant duration. In one study, germination of *U. macrorhiza* seeds recovered from seed bank samples was only 3.3%. In another study, *U. macrorhiza* occurred commonly in both natural and restored wetlands, but did not germinate from seed bank samples. Abnormalities in ovule production and embryo sac development have been documented in some plants. Together, these observations indicate a low level of successful sexual reproduction in this species, although more critical studies of its reproductive biology are needed. A predominant mode of vegetative reproduction has been indicated by genetic studies in Japan using AFLP markers, which found most populations to consist of a single genotype. During their growing season, the plants can reproduce vegetatively by fragmentation. However, the shoots die back annually and overwintering is achieved by means of round or elongate, and often highly mucilaginous turions. In the fall, the turions detach from the decaying shoots and sink. Turion dormancy is maintained by inhibitory phytohormones (innate dormancy) and by low temperatures (imposed dormancy). As water temperatures increase in the spring, the turions become buoyant from photosynthetic oxygen that is generated during their germination. The turions are dispersed by water; however, due to their fairly large size, it is unlikely that they often are transported between bodies of water. Mechanisms of long-distance dispersal are difficult to elucidate in this species. Broader dispersal must be facilitated in large part by waterfowl vectors, which transport seeds (e.g., in mud on their feet) or vegetative fragments over greater distances. The number of bladders on plants varies among sites, but is consistent within a site, and correlates positively with the specific conductivity of the water. Bladder number is not related to prey availability. The plants are not effective bicarbonate users and their growth can be limited by low levels of dissolved inorganic carbon in the water column. Marl encrustation of the foliage can occur in highly alkaline lakes. The long-standing misapplication of the name "*U. vulgaris*" to North American plants of *U. macrorhiza* makes it extremely difficult to distinguish literature accounts, which specifically refer to the latter species. **Reported associates:** *Acorus calamus, Agalinis paupercula, Agrostis scabra, Alisma subcordatum, Arethusa bulbosa, Asclepias incarnata, Azolla caroliniana, Bacopa rotundifolia, Bartonia virginica, Bidens cernuus, Bolboschoenus fluviatilis, Brasenia schreberi, Butomus umbellatus, Cabomba, Calamagrostis canadensis, C. stricta, Calla palustris, Callitriche verna,*

Calopogon tuberosus, Carex aquatilis, C. atherodes, C. canescens, C. chordorrhiza, C. comosa, C. exilis, C. exsiccata, C. lasiocarpa, C. limosa, C. lupulina, C. rostrata, C. viridula, C. utriculata, Cephalanthus occidentalis, Ceratophyllum demersum, C. echinatum, Chamaedaphne calyculata, Chara deliculata, Cicuta virosa, Cladium jamaicense, Comarum palustre, Crassula aquatica, Decodon verticillatus, Drosera anglica, D. intermedia, D. rotundifolia, Dulichium arundinaceum, Eleocharis acicularis, E. erythropoda, E. olivacea, E. palustris, E. robbinsii, Elodea canadensis, E. nuttallii, Epilobium palustre, Equisetum fluviatile, Eriocaulon aquaticum, Eriophorum virginicum, Glyceria borealis, G. grandis, G. striata, Heteranthera dubia, H. limosa, Hippuris vulgaris, Hydrilla verticillata, Iris pseudacorus, I. versicolor, I. virginica, Isoetes tenella, Juncus alpinus, J. arcticus, J. balticus, J. effusus. J. marginatus, J. nodosus, J. pelocarpus, J. tenuis, Leersia oryzoides, Lemna minor, L. trisulca, Lobelia dortmanna, L. siphilitica, Ludwigia palustris, Lycopodiella inundata, Lycopus uniflorus, Lysichiton americanus, Megalodonta beckii, Mentha arvensis, Menyanthes trifoliata, Muhlenbergia uniflora, Myrica gale, Myriophyllum alterniflorum, M. farwellii, M. sibiricum, M. spicatum, M. verticillatum, Najas flexilis, N. guadalupensis, N. marina, Nelumbo lutea, Nitella, Nuphar advena, N. polysepala, N. ×rubrodisca, N. variegata, Nymphaea odorata, N. tetragona, Oenanthe sarmentosa, Panicum hemitomon, Peltandra virginica, Persicaria amphibia, P. coccinea, Phalaris arundinacea, Phragmites australis, Pogonia ophioglossoides, Polygala sanguinea, Pontederia cordata, Potamogeton alpinus, P. amplifolius, P. confervoides, P. crispus, P. epihydrus, P. foliosus, P. friesii, P. gramineus, P. natans, P. obtusifolius, P. praelongus, P. pulcher, P. pusillus, P. richardsonii, P. robbinsii, P. spirillus, P. zosteriformis, Proserpinaca palustris, Ranunculus aquatilis, R. longirostris, R. subrigidus, R. trichophyllus, Rhynchospora alba, R. capitellata, R. macrostachya, R. scirpoides, Riccia fluitans, Ricciocarpus natans, Rorippa, Rumex verticillatus, Ruppia cirrhosa, R. maritima, Sagittaria cuneata, S. falcata, S. graminea, S. latifolia, S. rigida, Sanionia uncinata, Schoenoplectus acutus, S. americanus, S. pungens, S. smithii, S. subterminalis, S. tabernaemontani, Sium suave, Sparganium angustifolium, S. chlorocarpum, S. eurycarpum, S. fluctuans, S. natans, Sphagnum magellanicum, S. pulchrum, S. recurvum, Spiraea douglasii, Spirodela polyrhiza, Stuckenia filiformis, S. pectinata, Triglochin maritimum, Typha angustifolia, T. latifolia, Utricularia geminiscapa, U. gibba, U. intermedia, U. minor, U. ochroleuca, U. purpurea, U. radiata, U. resupinata, Vaccinium, Vallisneria americana, Veronica catenata, Viola lanceolata, Warnstorfia fluitans, W. trichophylla, W. tundrae, Wolffia, Wolffiella gladiata, Xyris montana, X. torta, Zannichellia palustris, Zizania aquatica, Z. palustris.

Utricularia minor L. is a perennial, which grows in bogs, canals, depressions, ditches, dunes, fens, floating mats, gravel pits, inlets, lakes, marshes, meadows, muskeg, ponds, pools, prairies, seeps, sloughs, streams, swales, and swamps at elevations of up to 3350 m. It grows in open sites and tolerates a wide range of acidity (pH: 4.1–8.7) as well as a fairly broad range of alkalinity (<210 mg/L CaCO$_3$) and trophic conditions (conductivity <390 μmhos/cm); however, it is found primarily in nutrient-poor sites with low alkalinity and a basic pH. Occasionally it is reported from hot springs. The plants are known to decline when water quality deteriorates, possibly as a consequence of increased turbidity. They occur typically in shallow (<0.5 m) water where they are attached to the substrate by buried portions of the shoot, which have higher numbers of bladders than the free segments that extend into the water column. Less commonly, the plants can occur (also in fairly shallow water) as freely suspended or floating life forms. Substrates are of calcareous or volcanic derivation and include organic muck, mud, peat, sand, and sandy loam. Often a thin organic layer is present. The reproductive biology of this species is poorly known. Flowering occurs only in attached plants growing in shallow water. The flowers are chasmogamous and are adapted for insect pollinators; however, they produce very small quantities of pollen and pollinators have not been observed. It is likely that most seed production is due to autogamy, but this possibility needs to be confirmed experimentally. The seeds are essentially unwinged and presumably dispersed by waterfowl. Although a transient seed bank may develop, one study failed to detect germination of the species from seedbank samples taken where the plants were represented fairly commonly among the standing vegetation. Reproduction primarily is vegetative, either by fragmentation (during the growing season) or by means of the perennating turions. The turions are buoyant but remain attached to the plants by a basket-like enclosure of foliage. The turion/leaf units detach as the plants decompose in the autumn, but will float beneath the ice. As the surrounding foliar enclosure decomposes, the turions eventually are released to float freely at the water surface. The minute turions (<3 mm) are dispersed by the water and probably also on occasion by waterfowl if they become entangled in their plumage. Turion germination is promoted by warming water temperatures. In northern Europe, this species has been found to occur most often in small isolated bodies of water with low interconnectivity. The plants produce relatively small traps and capture prey items that are mainly under 1 mm in size. *Utricularia minor* often grows intermixed with the similar *U. intermedia*. **Reported associates:** *Acer rubrum, Agalinis paupercula, Alnus rugosa, Andromeda, Arethusa bulbosa, Aulacomnium palustre, Bidens coronatus, B. frondosus, Botrychium simplex, Brasenia schreberi, Calamagrostis canadensis, Calla palustris, Campylium stellatum, Carex aquatilis, C. buxbaumii, C. canescens, C. crawei, C. eburnea, C. exilis, C. flava, C. granularis, C. howei, C. interior, C. lacustris, C. lasiocarpa, C. limosa, C. livida, C. lyngbyei, C. rostrata, C. sartwellii, C. saxatilis, C. sterilis, C. tenuiflora, C. tetanica, C. trisperma, C. utriculata, C. vesicaria, C. viridula, Chamaedaphne calyculata, Cephalozia connivens, Ceratophyllum demersum, Chara vulgaris, Cicuta bulbifera, Cladium mariscoides, Cladopodiella fluitans, Comarum palustre, Cornus sericea, Cyperus engelmannii, Cypripedium reginae, Dasiphora floribunda, Decodon verticillatus, Deschampsia cespitosa, Drosera intermedia, D. rotundifolia,*

Dulichium arundinaceum, Eleocharis acicularis, E. elliptica, E. pauciflora, E. quinqueflora, E. robbinsii, E. rostellata, Elodea canadensis, Epilobium watsonii, Equisetum fluviatile, E. variegatum, Eriophorum angustifolium, E. gracile, E. virginicum, Gaylussacia baccata, Gentianopsis crinita, G. procera, Glyceria canadensis, Heteranthera dubia, Iris pseudacorus, I. versicolor, Isoetes, Juncus alpinus, J. articulatus, J. balticus, J. effusus, J. stygius, Kalmia, Kobresia myosuroides, K. simpliciuscula, Larix laricina, Lemna minor, L. trisulca, L. turionifera, Leucobryum glaucum, Liparis loeselii, Listera cordata, Lobelia kalmii, Lycopodiella inundata, Lycopus uniflorus, L. virginicus, Lysimachia quadriflora, L. thyrsiflora, Megalodonta beckii, Menyanthes trifoliata, Muhlenbergia glomerata, Mylia anomala, Myrica gale, Myriophyllum sibiricum, M. spicatum, Najas flexilis, Nuphar polysepala, N. variegata, Nymphaea odorata, N. tuberosa, Osmunda regalis, Panicum flexile, P. spretum, Parnassia glauca, Pedicularis groenlandica, Persicaria amphibia, P. sagittata, Phalaris arundinacea, Picea mariana, Platanthera dilatata, P. hyperborea, Polytrichum strictum, Pontederia cordata, Potamogeton amplifolius, P. crispus, P. epihydrus, P. friesii, P. gramineus, P. illinoensis, P. natans, P. praelongus, P. pulcher, P. pusillus, P. robbinsii, P. spirillus, P. zosteriformis, Ranunculus aquatilis, Rhodiola integrifolia, Rhododendron, Rhynchospora alba, R. capillacea, R. capitellata, R. fusca, Riccia fluitans, Sagittaria latifolia, Salix bebbiana, S. candida, S. gracilis, S. drummondii, S. pyrifolia, Sarracenia purpurea, Saxifraga, Schoenoplectus acutus, S. tabernaemontani, Scirpus cyperinus, S. microcarpus, Scleria verticillata, Sium suave, Solidago ohioensis, S. riddellii, Sparganium glomeratum, Sphagnum capillifolium, S. centrale, S. contortum, S. fimbriatum, S. flexuosum, S. magellanicum, S. papillosum, S. pulchrum, S. recurvum, S. squarrosum, Spiraea tomentosa, Spiranthes romanzoffiana, Spirodela, Stuckenia pectinata, Swertia perennis, Symphyotrichum boreale, S. lanceolatum, Tetraphis pellucida, Thalictrum alpinum, Thelypteris palustris, Triadenum fraseri, T. virginicum, Trichophorum alpinum, T. pumilum, Triglochin maritimum, T. palustre, Typha angustifolia, T. latifolia, Utricularia cornuta, U. geminiscapa, U. gibba, U. intermedia, U. macrorhiza, U. purpurea, U. resupinata, Vaccinium corymbosum, V. macrocarpon, Veronica, Warnstorfia exannulata, Wolffia columbiana, Woodwardia virginica, Xyris difformis, X. montana, X. torta, Viola lanceolata.

***Utricularia ochroleuca* R.W. Hartm.** inhabits bogs, depressions, fens, lakes, marshes, and streams at elevations of up to 2800 m. Depending on the region, the sites are quite variable in acidity (pH: 6.2–8.3) but characteristically are nutrient poor. The foliage is dimorphic with bladderless, photosynthetic shoots extending into the water column, and achlorotic subterranean shoots that bear the bladders. Typically, the plants occur attached to the bottom in shallow water. They will not flower if in deeper water. Although showy flowers are produced, they produce deformed pollen grains and never yield seed. Reproduction is exclusively vegetative by means of fragmentation, or by the production of small turions (to 7 mm), which are the only means of overwintering.

The sterility of this taxon indicates that it is of hybrid origin (see *Systematics* section below). One study estimated its total biomass investment in carnivorous shoot production to be 40%–59%, with 18%–29% allocated to the production of traps. **Reported associates:** *Anticlea elegans, Carex livida, C. microglochin, C. scirpoidea, C. viridula, Chara, Dasiphora floribunda, Eriophorum gracile, Isoetes lacustris, Juncus balticus, Kobresia myosuroides, K. simpliciuscula, Littorella, Lobelia dortmanna, Myriophyllum alterniflorum, Nitella, Packera pauciflora, Pedicularis groenlandica, Picea mariana, Potamogeton alpinus, P. gramineus, P. natans, P. pusillus, Primula egaliksensis, Ptilagrostis mongholica, Salix brachycarpa, S. candida, S. myrtillifolia, S. planifolia, S. serissima, Scorpidium scorpioides, Sisyrinchium pallidum, Sparganium angustifolium, Stuckenia pectinata, Thalictrum alpinum, Trichophorum pumilum, Triglochin maritimum, T. palustre, Utricularia macrorhiza, Valeriana edulis.*

***Utricularia olivacea* C. Wright** is a diminutive annual, which occurs in borrow pits, depressions, ditches, lakes, ponds, streams, and swamps. Some sites are acidic (e.g., pH: 4.9) but the full range of pH has not been surveyed adequately for this species. The plants grow either suspended above the bottom in water (from a few cm to 6 m deep) or in dense floating mats. The reported substrates include fibrous mat bottoms and sand. The species can tolerate eutrophication when growing as a floating life-form where it can develop into large, pure floating mats. Its reproductive biology has not been studied. The flowers are extremely reduced (<3.5 mm) and produce only a single, minute (0.5 mm) seed. Even though they are visited occasionally by midges and gnats (Diptera), the flowers must be principally self-pollinating because they are in close proximity to the water surface and do not open if inundated. A comprehensive life-history study is needed for this species. Ecological information also is scarce because the minute plants often are overlooked in field surveys. **Reported associates:** *Acer rubrum, Andropogon brachystachyus, A. glomeratus, A. virginicus, Carex canescens, Centella asiatica, Chamaecyparis thyoides, Cyperus erythrorhizos, C. haspan, Dichanthelium erectifolium, D. portoricense, Dionaea muscipula, Eleocharis baldwinii, E. minima, E. nana, E. robbinsii, Eriocaulon compressum, Eupatorium resinosum, Euthamia caroliniana, Fuirena scirpoidea, Hydrocotyle umbellata, Hypericum edisonianum, H. fasciculatum, Juncus megacephalus, Lachnanthes caroliniana, Lachnocaulon engleri, Leersia oryzoides, Lemna minor, Lilium catesbaei, Lindernia grandiflora, Ludwigia alternifolia, Mayaca aubletii, Myriophyllum laxum, Nymphaea odorata, Nyssa biflora, Orontium aquaticum, Panicum abscissum, P. hemitomon, P. verrucosum, Persicaria hirsuta, Polygala cymosa, Polygonella gracilis, Potamogeton confervoides, P. natans, Rhexia mariana, Rhynchospora cephalantha, R. fascicularis, R. filifolia, R. inundata, R. microcarpa, R. scirpoides, R. wrightiana, Sagittaria subulata, Schoenoplectus subterminalis, Scleria reticularis, S. triglomerata, Spartina bakeri, Sphagnum macrophyllum, Stenanthium densum, Thelypteris kunthii, Utricularia geminiscapa, U. purpurea, U. striata, Vaccinium macrocarpon, Xyris brevifolia, X. chapmanii, X. elliottii, X. fimbriata, X. flabelliformis.*

Utricularia purpurea **Walter** is a perennial, which inhabits depressions, ditches, lakes, marshes, ponds, pools, prairies, sinks, sloughs, streams, and swamps in still to slow-moving waters from a few centimeters up to 6.5 m deep (but ordinarily 2.0 m). The sites range from open to shaded exposures. The plants occur in softwaters that are acidic (pH: 4.5–7.5; mean: 6.0), very low in alkalinity (<45 mg/L $CaCO_3$), low in conductivity (<185 µmhos/cm), and nutrient-poor (total phosphorous <33 mg/L; mean total nitrogen: 680 mg/L). This is a freshwater species, but it has been reported from a few oligohaline sites. The substrates include marl, muck, mud, sand, and silt. The ecology of this species is somewhat anomalous. It cannot assimilate bicarbonate and has been observed to decline substantially in several studies where acidic lakes were treated with calcite (crystalline calcium carbonate). However, it also is a major substrate (as a floating mat) for calcareous periphyton, particularly in the Florida Everglades. Populations vary from being completely sterile to profusely flowering, but there have been few studies of the reproductive biology. It is likely that much of the seed set results from self-pollination, despite the showy flowers. Seeds that have been stored in water at 1°C–3°C retain their viability (42%–53%) for up to 3 years and might persist longer in seed banks. This species does not form turions but can reproduce vegetatively by shoot fragmentation. Despite the inability of its shoots to produce prolific branches, the plants can develop into fairly dense stands where conditions are optimal. The growing season is temperature and depth dependent and can extend to more than twice as long for plants growing at 2 m than at 6 m. The carnivorous habit of this species is enigmatic. The plants invest roughly 26% of their biomass in bladders, but trap very few invertebrates and obtain less than 1% of their total nitrogen from the captured prey. Studies using ^{15}N tracers indicate that the plants actually obtain most of their nitrogen by direct uptake of ammonium from the water. However, the bladders support living communities of microorganisms, which produce nutritive respiratory by-products (i.e., CO_2) and detritus, a further indication that mutualism rather than carnivory may be the primary function of bladders in this species. The captured periphyton facilitate phosphorous uptake by the plants. This species is regarded as an indicator of waters that are not phosphorous enriched. The older shoots typically become colonized by dense populations of epiphytic algae including Cyanobacteria (e.g., Oscillatoriaceae: *Schizothrix hofmanni*; Scytonemataceae: *Scytonema calcicola*), desmids (Desmidiales) and diatoms (Bacillariophyceae). **Reported associates:** *Alternanthera philoxeroides, Asclepias incarnata, Bacopa caroliniana, Bidens beckii, Brasenia schreberi, Cabomba caroliniana, C. pulcherrima, Callitriche, Carex rostrata, C. stricta, Cephalanthus occidentalis, Ceratophyllum demersum, Chara, Cladium jamaicense, C. mariscoides, Crinum americanum, Cyperus haspan, Cyrilla racemiflora, Drepanocladus, Drosera intermedia, Dulichium arundinaceum, Eichhornia crassipes, Eleocharis acicularis, E. baldwinii, E. cellulosa, E. elongata, E. equisetoides, E. obtusa, E. olivacea, E. palustris, E. robbinsii, Elodea canadensis, E. nuttallii, Equisetum fluviatile, Eriocaulon aquaticum, Fuirena scirpoidea, Glyceria canadensis, Gratiola aurea, Hydrilla verticillata, Hydrochloa caroliniensis, Hydrocotyle umbellata, Hymenocallis latifolia, H. palmeri, Hypericum, Ilex, Ipomoea aquatica, Isoetes braunii, I. prototypus, Juncus dichotomus, J. effusus, J. megacephalus, J. militaris, J. pelocarpus, J. subcaudatus, Justicia ovata, Lachnanthes caroliniana, Leersia hexandra, Lemna minor, Littorella americana, Lobelia dortmanna, Ludwigia octovalvis, L. repens, Lycopodiella alopecuroides, Mayaca fluviatilis, Myrica cerifera, Myriophyllum farwellii, M. heterophyllum, M. humile, M. sibiricum, M. spicatum, M. tenellum, M. verticillatum, Najas flexilis, N. gracillima, N. guadalupensis, Nelumbo lutea, Nitella prolonga, Nuphar variegata, Nymphaea leibergii, N. odorata, Nymphoides aquaticum, N. cordata, Nyssa, Panicum hemitomon, P. repens, Paspalidium geminatum, Paspalum distichum, Peltandra virginica, Persicaria hydropiperoides Pinus, Pistia stratiotes, Pontederia cordata, Potamogeton amplifolius, P. bicupulatus, P. confervoides, P. crispus, P. epihydrus, P. foliosus, P. gramineus, P. illinoensis, P. natans, P. oakesianus, P. obtusifolius, P. perfoliatus, P. praelongus, P. pulcher, P. richardsonii, P. robbinsii, P. zosteriformis, Proserpinaca pectinata, Ranunculus flammula, Rhizophora mangle, Rhynchospora inundata, R. macrostachya, R. scirpoides, R. tracyi, Ruppia maritima, Sagittaria graminea, S. lancifolia, S. subulata, S. teres, Salvinia rotundifolia, Sarracenia psittacina, Schoenoplectus pungens, S. subterminalis, S. torreyi, Solidago, Sparganium americanum, S. androcladum, S. angustifolium, S. fluctuans, Sphagnum macrophyllum, Spirodela polyrhiza, Stuckenia pectinata, Taxodium distichum, Typha latifolia, Utricularia cornuta, U. fibrosa, U. floridana, U. foliosa, U. geminiscapa, U. gibba, U. inflata, U. intermedia, U. juncea, U. macrorhiza, U. minor, U. olivacea, U. radiata, U. resupinata, U. striata, Vallisneria americana, Websteria confervoides, Wolffia columbiana, Wolffiella gladiata, Xyris fimbriata, X. smalliana, X. torta, Zizania aquatica.*

Utricularia radiata **Small** is an annual species (occasionally a perennial in warmer areas), which grows in backwaters, canals, depressions, ditches, dunes, flatwoods, lakes, marshes, ponds, pools, prairies, sinkholes, streams, and swamps. The plants typically occur in shallow (<1 m), still or slow-moving fresh waters, where salinity is <0.5 ppt (they cannot tolerate salinity >1.0 ppt). The waters are acidic (pH: 5.6–7.3; mean: 6.6) with low alkalinity (<33 mg/L $CaCO_3$), and the substrates often are organic muck. Like *U. inflata*, the plants are heterophyllous, with dissected, bladder-bearing submersed leaves and a radiating whorl of inflated, spongy leaves that floats on the water surface and supports the floral stalk. Plants growing in deeper water usually lack the floating leaves and are sterile. Some populations can produce dense stands of flowering plants. However, there is little known about the reproductive biology of this species. The peduncles ascend (or at least spread) when the fruits are set. The seeds have been germinated successfully when kept in water at room temperature (25°C) in a petri dishes. Vegetative reproduction during the growing season can be achieved by stem fragmentation. However, unlike *U. inflata*, the plants do not produce tubers if they become stranded on drying shores, and their persistence at a site is due primarily to seed. The seed dispersal mechanisms have not been elucidated. **Reported associates:** *Brasenia schreberi, Cabomba caroliniana, Calamagrostis canadensis, Callitriche, Chara vulgaris, Cladium jamaicense, Cyperus haspan, Decodon verticillatus, Drosera intermedia, Elatine minima, Eleocharis acicularis, E. olivacea, E. robbinsii, E. tricostata, Elodea canadensis, E. nuttallii, Eriocaulon aquaticum, Gratiola aurea, Hydrilla verticillata, Hypericum adpressum, Isoetes engelmannii, Juncus canadensis, Lemna minor, Lysimachia terrestris, Myriophyllum alterniflorum, M. heterophyllum, M. humile, M. spicatum, Najas flexilis, Nitella, Nuphar variegata, Nymphaea odorata, Nymphoides cordata, Nyssa biflora, Panicum hemitomon, P. repens, P. spretum, P. verrucosum, Pontederia cordata, Potamogeton amplifolius, P. bicupulatus, P. confervoides, P. diversifolius, P. epihydrus, P. foliosus, P. gramineus, P. natans, P. pusillus, P. robbinsii, Proserpinaca palustris, Sagittaria graminea, S. latifolia, Solidago tenuifolia, Sparganium, Sphagnum, Taxodium ascendens, T. distichum, Utricularia gibba, U. inflata, U. intermedia, U. macrorhiza, U. purpurea, Vallisneria americana, Viola primulifolia, V. sagittata, Xyris montana.*

Utricularia resupinata **B.D. Greene ex Bigelow** is a perennial species, which occurs in borrow pits, canals, depressions, ditches, dunes, flatwoods, lagoons, lakes, lakeshores, marshes, pools, savannas, and sloughs. The waters span a broad range of acidity (pH: 4.6–8.8; mean 6.1) and typically are quite clear (secchi depth: 1.3–5.7 m). The sites are characterized by low alkalinity (<60 mg/L CaCO$_3$), low specific conductivity (<230 μmhos/cm), and relatively nutrient-poor conditions (total phosphorous <18 mg/L; mean total N: 450 mg/L). The plants are attached to the substrate, which can be muck, mud, peat, sand, or sandy peat. Typically they grow in shallow waters less than 15 cm deep; however, under optimal clarity they can be found at depths up to 3 m. In Lake George, New York, much greater plant densities have been documented at depths of 1–2 m (500–2500 plants/m^2) than at 3 m (<400 plants/m^2). Despite the species name, the flowers are not resupinate. Flowering can be prolific, but occurs sporadically, especially in the northern portion of its range. This species is observed most frequently after the waters have receded and when temperatures are higher than average. Little is known about its pollination system or other aspects of its reproductive biology and seed ecology. It has not been observed to form a seed bank. The plants survive high water levels vegetatively by forming large mats, which can dislodge and float on the surface. There are no winter buds produced, but the plants can reproduce vegetatively by fragmentation. **Reported associates:** *Agalinis harperi, Baccharis, Bacopa caroliniana, Bidens beckii, B. cernuus, Brasenia schreberi, Carex lasiocarpa, C. lenticularis, C. rostrata, C. stricta, C. viridula, Cephalanthus occidentalis, Chara, Cladium jamaicense, C. mariscoides, Colocasia esculenta, Cyperus haspan, C. lecontei, Drosera intermedia, D. rotundifolia, Dulichium arundinaceum, Elatine minima, Eleocharis acicularis, E. baldwinii, E. cellulosa, E. elongata, E. robbinsii, E. smallii, Elodea canadensis, E. nuttallii, Equisetum variegatum, Eriocaulon aquaticum, E. lineare, Fontinalis, Fuirena scirpoidea, Gratiola aurea, Heteranthera*

dubia, Hydrocotyle umbellata, Hypericum, Isoetes braunii, Juncus balticus, J. dichotomus, J. militaris, J. pelocarpus, Justicia angusta, Lachnanthes caroliniana, Leersia hexandra, Littorella americana, Lobelia dortmanna, Ludwigia octovalvis, L. repens, Mayaca fluviatilis, Myrica cerifera, Myriophyllum alterniflorum, M. heterophyllum, M. tenellum, Najas flexilis, N. guadalupensis, Nitella, Nymphaea leibergii, N. odorata, Nuphar variegata, Nymphaea odorata, Nymphoides aquatica, N. cordata, Oxypolis filiformis, Panicum hemitomon, P. repens, P. tenerum, P. virgatum, Paspalum distichum, Parnassia palustris, Paspalidium geminatum, Persicaria hydropiperoides, Pinus, Pluchea rosea, Pontederia cordata, Potamogeton amplifolius, P. confervoides, P. diversifolius, P. gramineus, P. perfoliatus, P. praelongus, P. pusillus, P. robbinsii, P. spirillus, P. vaseyi, P. zosteriformis, Ranunculus longirostris, Rhynchospora fusca, R. tracyi, Sagittaria cristata, S. graminea, S. lancifolia, S. subulata, Sambucus nigra, Schoenoplectus subterminalis, Sparganium americanum, S. androcladum, S. angustifolium, S. chlorocarpum, Spiranthes cernua, Stuckenia pectinata, Subularia aquatica, Taxodium ascendens, T. distichum, Typha, Utricularia cornuta, U. floridana, U. foliosa, U. geminiscapa, U. gibba, U. intermedia, U. macrorhiza, U. minor, U. purpurea, U. subulata, Vallisneria americana, Viola lanceolata, Xyris.

Utricularia simulans **Pilg.** is a perennial found in canals, ditches, flatwoods, prairies, savannas, seeps, and sloughs. There is only minimal, nonspecific ecological information available for the habitats, which are described as acidic or calcareous substrates of peaty sand or sand that are low in nutrients. Life history information on this species is virtually nonexistent and a thorough study of the plant is needed. **Reported associates:** *Drosera brevifolia, Pinguicula pumila, Pinus elliottii, Serenoa repens, Taxodium ascendens, Utricularia juncea, U. subulata.*

Utricularia striata **Leconte ex Torr.** is a perennial species of bogs, ditches, flatwoods, floodplains, marshes, pools, ponds, shores, seeps, streams, and swamps. Typically, the habitat consists of exposed substrates that are overlain only by very shallow water (<10 cm); however, the plants also can occur in floating mats over deeper areas. The sites are acidic (pH: 4.9–6.6) with substrates of loamy sand, muck, peat, sand, or silt. The plants grow frequently in loose organic material (e.g., leaves) and sometimes occur in deeply shaded areas. In some places the plants appear only during low water periods (e.g., droughts), which expose ample quantities of suitable shoreline habitat. They will develop into dense mats under optimal conditions. The foliage is dimorphic. The traps occur mainly on modified stolons that penetrate the substrate, but are infrequent along the foliose shoots that remain above the substrate in the water. Like so many bladderworts, the reproductive biology of this species has not been studied formally and the method of pollination, seed dispersal, etc. remains to be elucidated. **Reported associates:** *Arethusa bulbosa, Cladium mariscoides, Coreopsis rosea, Drosera capillaris, D. intermedia, Eleocharis obtusa, E. olivacea, E. robbinsii, E. vivipara, Eriocaulon aquaticum, Gaylussacia dumosa, Gratiola aurea, Iris prismatica, Juncus militaris, Lobelia nuttallii, Ludwigia sphaerocarpa, Mayaca fluviatilis, Nyssa, Pinguicula caerulea, Pinus, Potamogeton diversifolius, Proserpinaca pectinata, Rhynchospora inundata,*

R. macrostachya, R. nitens, R. scirpoides, Sagittaria teres, Sarracenia minor, Sphagnum, Utricularia cornuta, U. fibrosa, U. gibba, U. juncea, U. olivacea, U. purpurea, U. subulata, Xyris smalliana, X. torta.

Utricularia stygia G. Thor. Excluded (see *Systematics* below).

Utricularia subulata L. is an annual plant of bogs, depressions, ditches, dunes, fens, flatwoods, pocosins, lake and pond margins, savannas, seeps, springs, and swales at elevations of up to 2600 m. The sites generally tend to be fairly acidic (pH: 3.7–5.6) but these figures are based only on a few published values. However, the ecological amplitude of these plants must be fairly broad given that this is the most widely distributed species in the genus. The plants are also reported from fens and reportedly can form calcareous mats that support filamentous cyanobacteria (e.g., Oscillatoriaceae: *Schizothrix calcicola*; Scytonemataceae: *Scytonema hofmanni*), which precipitate marl (calcium carbonate). The shoots are dimorphic with flat, unbranched photosynthetic leaf-like structures that develop above the sediments, and whitish, branching, stolon-like projections that penetrate the substrate for anchorage; both produce bladders. The substrates are loamy sand, peat, sand, and sandy peat and are more or less continually saturated and sometimes shallowly inundated. The plants also can grow in an unusual habitat along the edge of *Sphagnum* mats by establishing on bare granitic rock. Both chasmogamous and cleistogamous (sometimes entirely) flowers are produced and generate a large quantity of seed. Thus, the plants are at least highly (if not-predominantly) self-pollinating, which is advantageous for their annual habit. The seeds are easy to germinate and a successful procedure is to sow them in flats of sterilized soil, which are kept moist by a fine mist applied for two 20-min periods daily. Even without special care, the seeds are notorious for germinating as a contaminant in pots of other carnivorous plants that are grown in the greenhouse. Seed germination usually occurs within 4–6 weeks. Propagules remain dormant for a prolonged (but undetermined) time in the seed bank. In one study, the plants did not occur among the standing vegetation at a number of drying sites that had been burned from 1 to 30+ years prior, yet numerous seeds germinated from the seed bank samples. In another case, 1140 germinating seeds/m^2 were observed in samples taken from wet swale sediments. Experiments have indicated that plant establishment is greatly facilitated by mowing to remove the overstory vegetation, a task that normally would be accomplished by natural fires. The small stems often produce small dew-like droplets in the angles of the inflorescence, which may function to trap harmful insects.

Reported associates: *Acer rubrum, Agalinis fasciculata, A. neoscotica, A. purpurea, Agrostis perennans, Aletris farinosa, Amphicarpum muehlenbergianum, Andromeda glaucophylla, Andropogon glomeratus, A. virginicus, Aristida beyrichiana, A. palustris, Bacopa monnieri, Balduina uniflora, Bartonia paniculata, B. verna, B. virginica, Bidens frondosus, Buchnera americana, Calamagrostis pickeringii, Calamovilfa brevipilis, C. pickeringii, Calopogon barbatus, C. tuberosus, Carex canescens, C. exilis, C. limosa, C. livida, C. stricta, C. striata, C. trisperma, Carphephorus pseudoliatris, Castanea pumila, Centella asiatica, Cephalanthus occidentalis, Chamaecyparis thyoides, Chamaedaphne calyculata, Chaptalia tomentosa, Cirsium lecontei, Cladopodiella fluitans, Cladium mariscoides, Cleistes bifaria, Cliftonia monophylla, Conium maculatum, Coreopsis gladiata, C. rosea, Ctenium aromaticum, Cyperus bipartitus, C. odoratus, C. oxylepis, C. polystachyos, Cyrilla racemiflora, Danthonia epilis, Decodon verticillatus, Dichanthelium dichotomum, Dionaea muscipula, Drosera brevifolia, D. capillaris, D. filiformis, D. intermedia, D. rotundifolia, D. tracyi, Eleocharis acicularis, E. baldwinii, E. elliptica, E. flavescens, E. geniculata, E. melanocarpa, E. microcarpa, E. montevidensis, E. olivacea, E. parvula, E. pauciflora, E. robbinsii, E. tortilis, E. tuberculosa, Eriocaulon aquaticum, E. compressum, E. decangulare, E. nigrobracteatum, Eriophorum virginicum, Eupatorium compositifolium, Euphorbia inundata, Eurybia eryngiifolia, Fimbristylis autumnalis, F. castanea, Fuirena longa, F. pumila, F. scirpoidea, Gaylussacia baccata, G. dumosa, G. frondosa, G. mosieri, Helenium vernale, Helianthus heterophyllus, Hypericum adpressum, H. brachyphyllum, H. canadense, H. kalmianum, Ilex coriacea, I. glabra, I. myrtifolia, I. vomitoria, Impatiens capensis, Juncus acuminatus, J. balticus, J. biflorus, J. canadensis, J. dichotomus, J. effusus, J. greenei, J. megacephalus, J. pelocarpus, J. scirpoides, Justicia crassifolia, Kalmia angustifolia, K. polifolia, Lachnanthes caroliana, Lachnocaulon anceps, L. digynum, Larix laricina, Linum medium, L. striatum, Lobelia canbyi, L. kalmii, L. nuttallii, Lophiola aurea, Ludwigia decurrens, L. leptocarpa, L. sphaerocarpa, Lycopodiella alopecuroides, L. appressa, L. caroliniana, Lyonia lucida, Lysimachia asperulifolia, Magnolia virginiana, Muhlenbergia uniflora, Myrica gale, M. heterophylla, M. inodora, M. pensylvanica, Narthecium americanum, Nymphaea odorata, Nyssa biflora, Oclemena nemoralis, Oldenlandia uniflora, Oxalis stricta, Oxypolis filiformis, Panicum implicatum, P. nudicaule, P. verrucosum, P. virgatum, P. wrightianum, Parnassia caroliniarum, Peltandra sagittifolia, P. virginica, Persea, Persicaria hirsuta, P. puritanorum, Picea mariana, Pinguicula ionantha, P. lutea, P. planifolia, Pinus clausa, P. elliottii, P. palustris, P. rigida, Pleea tenuifolia, Pogonia ophioglossoides, Polygala brevifolia, P. cruciata, P. cymosa, P. lutea, P. nana, P. polygama, P. ramosa, Potentilla anserina, Rhexia alifanus, R. virginica, Rhododendron viscosum, Rhynchospora alba, R. capillacea, R. capitellata, R. cephalantha, R. chalarocephala, R. chapmanii, R. colorata, R. divergens, R. gracilenta, R. macra, R. microcarpa, R. nitens, R. oligantha, R. scirpoides, R. stenophylla, Rudbeckia graminifolia, Sabatia difformis, S. macrophylla, Sagittaria latifolia, Salix nigra, Sarracenia alata, S. flava, S. purpurea, S. psittacina, S. rubra, Scapania nemorosa, Schizachyrium scoparium, Schizaea pusilla, Schoenoplectus pungens, Scleria baldwinii, S. reticularis, S. verticillata, Sclerolepis uniflora, Sematophyllum demissum, Serenoa repens, Smilax, Solidago ohioensis, S. sempervirens, Sphagnum bartlettianum, S. compactum, S. magellanicum, S. portoricense, S. pulchrum, S. rubellum, S. subsecundum, Spiranthes vernalis, Stenanthium densum, Symphyotrichum novi-belgii, Syngonanthus flavidulus, Taxodium, Thalictrum revolutum, Thuidium delicatulum, Triantha racemosa, Triadenum virginicum, T. walteri, Utricularia cornuta, U. gibba, U. juncea, U.*

resupinata, *U. striata, Vaccinium corymbosum, V. darrowii, V. elliottii, V. macrocarpon, V. oxycoccos, Viola primulifolia, Vitis rotundifolia, Woodwardia virginica, Xyris ambigua, X. baldwiniana, X. caroliniana, X. difformis, X. drummondii, X. jupicai, X. montana, X. platylepis, X. scabrifolia, X. serotina, X. torta, Zenobia pulverulenta.*

Use by wildlife: The tangled mats of *Utricularia gibba* and underwater foliage of *U. floridana, U. foliosa, U. inflata, U. macrorhiza, U. purpurea,* and *U. resupinata* provide cover and shelter for small fish (Osteichthyes) and aquatic invertebrates. Recent studies have shown that in some bladderwort species, a fairly diverse living invertebrate community (e.g., Rotifera) is supported within the bladders, which provide the organisms with protection from fish and other predators. *Utricularia foliosa* is associated with a number of invertebrates including various flies (Insecta: Diptera: Chironomidae; Ceratopogonidae; Culicidae), mayflies (Insecta: Ephemeroptera: Caenidae; Baetidae), beetles (Insecta: Coleoptera: Ditiscidae), bugs (Insecta: Hemiptera: Corixidae; Pleidae), odonates (Insecta: Odonata), mites (Acari), seed shrimp (Crustacea: Ostracoda), horsehair worms (Nematomorpha), and gastropods (Gastropoda). Capsules of *U. gibba* are eaten by ducks (Aves: Anatidae) and its foliage is a minor but frequent diet constituent of the Texas river cooter turtle (Reptilia: Emydidae: *Pseudemys texana*). *Utricularia gibba* is a preferred food of the grass carp (Osteichthyes: Cyprinidae: *Ctenopharyngodon idella*), a fish often used for the biological control of aquatic plants. *Utricularia inflata* is an important cover plant for larval golden shiners (Osteichthyes: Cyprinidae: *Notemigonus crysoleucas*). The foliage of *U. macrorhiza* is a minor food of moose (Mammalia: Cervidae: *Alces alces*), muskrats (Mammalia: Muridae: *Ondatra zibethicus*) and some waterfowl (Aves: Anatidae). The plants are used as a nesting site by eared grebes (Aves: Podicipedidae: *Podiceps nigricollis*). *Utricularia minor* occurs in floating mats of grasses that are used for nesting by black terns (Aves: Laridae: *Chlidonias niger*). It is eaten as an autumn food by caribou (Mammalia: Cervidae: *Rangifer tarandus*). The trap doors of *U. macrorhiza* bladders are a preferred substrate for attachment by some sessile rotifers (Rotifera: Flosculariidae: *Ptygura beauchampi*), which are attracted to the structures by a chemical stimulus. Although not usually considered as important to wildlife, some *U. macrorhiza* plants have been found to contain as many as 37 invertebrate taxa (e.g., Diptera: Chironomidae; Gastropoda: *Gyraulus, Helisoma, Physa*; Hemiptera: *Neoplea striola*; Trichoptera: *Triaenodes abus*) averaging up to 80 animals/100 g, which surely provide an indirect source of food for waterfowl. The bladders of this species are known to contain some organisms, e.g., *Euglena* (Protozoa: Euglenaceae), *Colpidium colpoda* (Ciliophora: Turaniellidae), *Heteronema acus* (Protozoa: Euglenaceae), Nematoda, *Phacus longicauda* (Protozoa: Euglenaceae), and Rotifera, which not only survive but can reproduce within them; however, several captured protozoans including *Centropyxis aculeata* (Amoebozoa; Centropyxidae), *Paramecium* (Chromalveolata: Parameciidae), *Stentor polymorphus* (Chromalveolata: Stentoridae), and *Stylonychia*

pustulatus (Chromalveolata: Oxytrichidae) are readily killed and are digested in the bladders. *Utricularia purpurea* is used as habitat by the red-spotted newt (Amphibia: Salamandridae: *Notophthalmus viridescens*) and the banded killifish (Osteichthyes: Fundulidae: *Fundulus diaphanus*). Seeds of *U. subulata* are eaten by the Florida bobwhite (Aves: Odontophoridae: *Colinus virginianus floridanus*) and can make up from 14% to 23% of their spring diets.

Economic importance: food: not edible; **medicinal:** no uses reported; **cultivation:** Many *Utricularia* species are grown as specimens by carnivorous plant enthusiasts. *Utricularia cornuta, U. floridana, U. gibba, U. inflata, U. intermedia, U. juncea, U. macrorhiza, U. minor, U. ochroleuca, U. purpurea, U. radiata,* and *U. subulata* all reportedly are cultivated as ornamental water garden plants. *Utricularia gibba* is grown in aquariums; **misc. products:** none; **weeds:** Several of the North American *Utricularia* species can grow in profusion; however, they are native and seldom regarded as weedy. *Utricularia subulata* can become a greenhouse weed establishing in pots of other carnivorous plants. *Utricularia purpurea* occasionally is reported as a weed due to its prolific growth, sometimes even in regions where it is listed as imperiled. The densely spreading *U. inflata* is regarded as weedy in Massachusetts, New York and Washington states; **nonindigenous species:** All North American *Utricularia* species are regarded as native to the region. Although native to the southeastern United States, *U. inflata* recently has spread to Adirondack lakes of New York State (where it has been observed to suppress the growth rate of submersed rosulate species), as well as to Massachusetts, Pennsylvania and Washington state (the latter as an escape from cultivation). It is naturalized in Japan. *Utricularia subulata* is native to eastern North America and has recently naturalized in California, presumably as a contaminant of horticultural carnivorous plant shipments. *Utricularia geminiscapa* was introduced to New Zealand sometime before 1975, possibly through seed-contaminated machinery imported from the United States to drain wetlands. *Utricularia gibba* has been introduced to Europe and New Zealand. It is often inconspicuous occurrence in mats of vegetation facilitates its introduction along with other aquatic plants.

Systematics: Several studies (e.g., Figures 5.64 and 5.66) have confirmed that *Genlisea* (also aquatic, carnivorous) is the sister group of *Utricularia*. Although numerous synonyms exist for *Utricularia*, phylogenetic analyses (e.g., Figure 5.66) do not support the generic subdivision of the group, which is monophyletic as currently defined. Although most authors have distinguished *Polypompholyx* from *Utricularia* either at the generic or subgeneric level, phylogenetic analyses resolve the former taxon within *Utricularia* as the sister group to section *Pleiochasia* (e.g., Figure 5.66). Although only about a quarter of the extant *Utricularia* species have been included simultaneously in phylogenetic analyses (e.g., Figure 5.66), most of the currently accepted sections appear to natural, monophyletic units with a few exceptions. *Utricularia olivacea* (section *Utricularia*) is distinct from other members of that section and resolves with members of section *Vesiculina* (Figure 5.66). Also, *U. amethystina* (section *Foliosa*) resolves as closely related to both *U. huntii* (section *Psyllosperma*) and *U. tricolor* (section

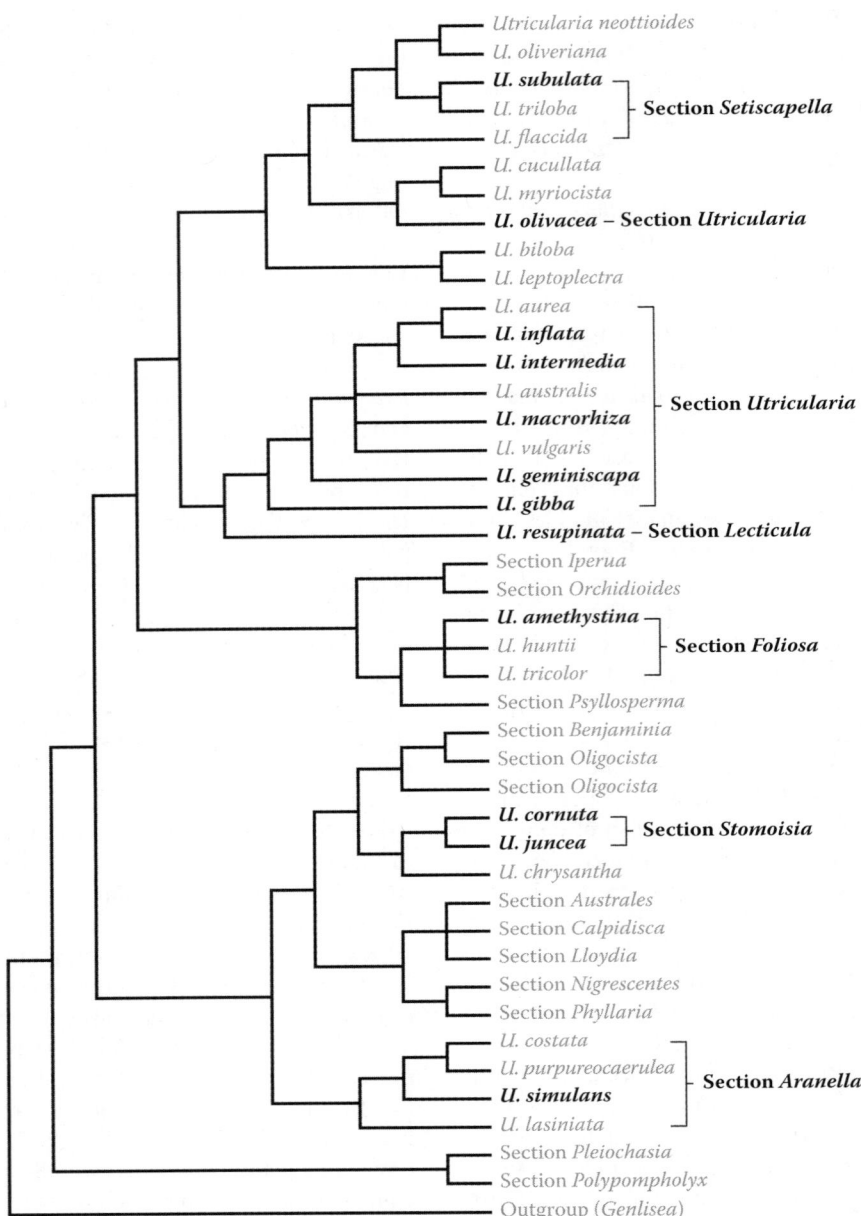

FIGURE 5.66 Phylogenetic relationships in *Utricularia* as indicated by analysis of combined cpDNA sequences for 58 species. All *Utricularia* species should be regarded as obligate (i.e., OBL) aquatics, which indicates a single origin of the habit in the genus. The North American representatives (all OBL) are indicated in bold. These results (based on approximately 25% of the total number of species) agree fairly well with the currently accepted sectional classification although a few problems (e.g., misplacement of *U. olivacea*; paraphyly of section *Setiscapella*) are evident. (Adapted from Jobson, R.W. et al., *Syst. Bot.* 28, 157–171, 2003.)

Foliosa). Further sampling of taxa in these groups may help to clarify the species relationships. One analysis (Figure 5.66), where twelve of the 19 North American species were included simultaneously, shows North America to be a phylogenetic mosaic of species from at least six separate radiations, with most of them originating from the diverse section *Utricularia*. Of the nine species not included, eight are assigned to section *Utricularia* and one (*U. purpurea*) to section *Vesiculina*. *Utricularia purpurea* was included in an analysis of *trnK* intron sequence data (with fewer taxa), where it resolved as the sister group to section *Setiscapella*. *Utricularia* is difficult

taxonomically due to the large number of variable species, many with widespread (intercontinental) distributions. In North America, a great deal of nomenclatural confusion has existed with respect to several taxa. The name "*Utricularia fibrosa*" has been particularly problematic because of nomenclatural homonyms. The authorship of the name is critical for the correct taxonomic assignment. *Utricularia fibrosa* Walter is now regarded as a synonym of *U. gibba*; whereas, *U. fibrosa* Britt. is a synonym of *U. striata* Le Conte ex Torrey. *Utricularia macrorhiza* long been known in North America as *U. vulgaris* (Old World), although these species are distinct. A close relationship of these taxa to one another, as well as to *U. australis* (Old World) is evident from analyses of DNA sequence data (Figure 5.66). Although all three species are quite similar morphologically, they have been distinguished genetically by *trnK* intron sequences, which resolve *U. australis* and *U. vulgaris* as sister species and *U. macrorhiza* as their closest relative. *Utricularia cornuta* and *U. juncea* (section *Stomoisia*) are also very similar morphologically and at times have been recognized as conspecific. Indeed, DNA sequence data indicate that they are sister taxa (Figure 5.66); however, they are seasonally isolated and are not interfertile when crossed experimentally. The distinctness of *U. intermedia* and *U. ochroleuca* remains in question. The latter is sterile and widely believed to represent a hybrid between *U. minor* × *U. intermedia*, which frequently grow together. Otherwise, there are no suspected instances of hybridization among North American *Utricularia* species. Certainly, the putative hybrid status of *U. ochroleuca* should be confirmed or refuted using genetic markers. Also, the Old World *U. stygia* (another sterile taxon) has been split from *U. ochroleuca* by some authors. Recognition of these sterile taxa at the rank of species does not appear to be prudent, and if they are proven to be of hybrid origin, they should be reclassified accordingly. Several European taxa, first thought to be hybrids, are now believed to represent dysploid vegetative apomicts. This possibility also should be considered with respect to the presumed North American hybrids. *Utricularia radiata* was once considered to be a variety of *U. inflata* (*U. inflata* var. *minor*); however, these taxa differ chromosomally and remain distinct morphologically, even in rare instances where they are found growing together. The original chromosome number of North American *Utricularia* cannot readily be elucidated because of the considerable variation in reported base numbers ($x = 7$–$11, 15$). *Utricularia cornuta*, *U. inflata* and *U. juncea* (all $2n = 18$) are $x = 9$ diploids [$2n = 36$ tetraploids also are known for *U. inflata*]; *U. subulata* ($2n = 30$) is an $x = 15$ diploid; *U. gibba* and *U. radiata* ($2n = 28$) are $x = 7$ tetraploids; *U. intermedia*, *U. macrorhiza*, *U. minor*, *U. ochroleuca*, *U. resupinata* (all $2n = 44$) are $x = 11$ tetraploids; *U. foliosa* ($2n = 42$) is an $x = 7$ hexaploid. The breeding systems of *Utricularia* species require intensive study. Despite having flowers that appear to be well-adapted for insect pollination, potential pollinators are seldom observed to visit any of the bladderwort species, which is remarkable given the considerable attention paid to these plants. Furthermore, genetic studies using AFLP markers have found very low levels of genetic variability within populations of *U. macrorhiza* (studied in Japan), indicating that they reproduce extensively by vegetative propagation. Similar studies of other species in the genus are strongly encouraged. The genome size of *U. gibba* (88 Mbp) is among the smallest known for any angiosperm.

Comments: In North America the species vary from being widespread throughout (*Utricularia macrorhiza* [also Central America and northern Asia]) to having progressively narrower distributions in northern North America (*U. intermedia*, *U. minor* and *U. ochroleuca* [circumboreal]), northeastern North America (*U. geminiscapa*), southern and eastern North America (*U. radiata, U. subulata* [worldwide], *U. striata*); eastern North America (*U. cornuta* [Central America], *U. purpurea* [Central America], *U. resupinata* [Central America]), eastern and western (disjunct) North America (*U. gibba* [worldwide]); northeastern and northwestern North America (*U. ochroleuca* [Eurasia]), southern and eastern United States (*U. inflata*; disjunct in Washington State), southeastern United States (*U. floriidana, U. foliosa* [Central & South America; Africa], *U. juncea* [Central & South America, Africa], *U. olivacea* [Central & South America]) and Florida (*U. amethystina* [Central & South America], *U. simulans* [Central & South America; Africa]). *Utricularia amethystina* is known in Florida only historically (two populations) with no recent observations.

References: Adamec, 1997, 1999, 2007; Adamec & Lev, 2002; Andreas & Bryan, 1990; Araki, 2002; Baskin & Baskin, 1998; Beal, 1977; Bergerud, 1972; Bern, 1997; Boe, 1994; Boylen et al., 1999; Brewer, 1999a; Brock, 1970; Brown, 1987; Bukaveckas, 1988; Burton, 1977; Cheers, 1992; Cherry & Gough, 2006; Christy, 2004; Cooper, 1996; Crow & Hellquist, 1985; Dahlgren & Ehrlén, 2005; Davis et al., 2005a; Dítě et al., 2006; Dolan & Sharitz, 1984; Dörrstock et al., 1996; Farooq & Siddiqui, 1965; Fields et al., 2003; Foster & Glaser, 1986; Fraser et al., 1980; Galatowitsch & van der Valk, 1996a,b; Gates, 1929; Glaser et al., 1981; Goldsby & Sanders, Sr., 1977; Greene, 1998; Greilhuber et al., 2006; Guisande et al., 2000; Haber, 1979; Harms, 1999; Heenan et al., 2004; Hegner, 1926; Hiebert et al., 1986; Hill et al., 1998; Hoyer et al., 1996; Hutchinson, 1975; Jobson & Albert, 2002; Jobson et al., 2003; Johnson & Steingraeber, 2003; Kameyama & Ohara, 2006; Kasselmann, 2003; Keddy, 1983; Keddy & Reznicek, 1982; Kitamura, 1991; Knight & Frost, 1991; Kondo, 1971, 1972; Kondo et al., 1978a; Kosiba, 1992; Krull, 1970; Laakkonen et al., 2006; Lahring, 2003; Liston & Trexler, 2005; Locky et al., 2005; Lollar et al., 1971; Looney & Gibson, 1995; Luken, 2005; Lynn & Karlin, 1985; Maliakal et al., 2000; McVaugh, 1943a; Müller & Borsch, 2005; Meyers & Strickler, 1979; Mitchell, 1994; Mitchell et al., 1994; Moeller, 1978a, 1980; Muenscher, 1944; Neid, 2006; Nekola, 2004; Nichols, 1931, 1999a; Noe et al., 2003; Oosting & Anderson, 1937; Orzell & Bridges, 2006; Pagano & Titus, 2004; Philbrick & Les, 1996; Pietropaolo & Pietropaolo, 1993; Pinder et al., 2006; Płachno et al., 2006; Poiani & Johnson, 1988; Preston & Croft, 1997; Reinert & Godfrey, 1962; Richards, 2001; Richards & Fry, 1996; Richards & Kuhn, 1996; Richardson, 1967; Roberts, 1972; Roberts et al., 1985; Ross et al., 2000; Russell, 1942; Rutishauser & Isler, 2001; Sanabria-Aranda et al., 2006; Schnell, 1980, 2002; Shannon, 1953; Sheldon & Boylen, 1977; Singer et al., 1983; Sorenson & Jackson, 1968; Sorrie et al., 2006; Squires & Lesack, 2003; Taylor, 1989; Thor, 1988; Ulanowicz, 1997; Urban et al., 2006; Vaithiyanathan & Richardson, 1999; Van der Valk & Davis, 1978; Villanueva et al., 1985; Wallace, 1978; Weiher et al., 1994; Whicker et al., 1990; Williams, 1990a; Wilson et al., 1993; Winston & Gorham, 1979a,b; York, 1905.

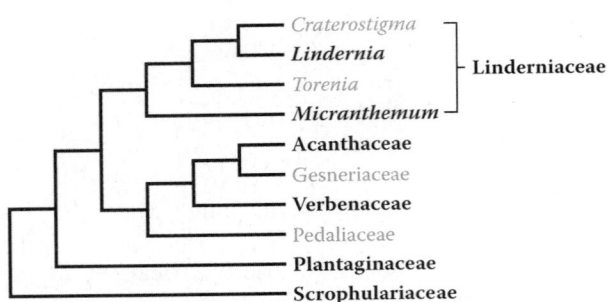

FIGURE 5.67 Relationships of Linderniaceae as indicated by analysis of combined DNA sequence data. Taxa with OBL North American representatives are indicated in bold. The single species of *Lindernia* surveyed in this analysis does not adequately represent the entire genus, which other studies demonstrate to be polyphyletic (see Figure 5.68). (Redrawn from Albach, D.C. et al., *Amer. J. Bot.*, 92, 297–315, 2005.)

Family 4: Linderniaceae [11]

The Linderniaceae were initially included as a part of the former Scrophulariaceae. Subsequently, they were transferred to a redefined Plantaginaceae and eventually became recognized as a distinct segregate family as a result of continuing phylogenetic studies (Figure 5.67). In its current circumscription, the family comprises roughly 200 species (Oxelman et al., 2005; Rahmanzadeh et al., 2005). Vegetatively, many species resemble Lamiaceae by their square stems and opposite leaves. However, the flowers, although tubular and often bilabiate, are otherwise dissimilar to Lamiaceae in not being arranged in verticels, having terminal (not gynobasic) styles, and fruits that are capsules rather than nutlets. Morphologically, the family is distinguished primarily by its geniculate anterior (abaxial) filaments with a basal swelling

(Oxelman et al., 2005; Rahmanzadeh et al., 2005). Usually, there are four stamens (or two reduced to staminodes); however, there are only two stamens in the minute flowers of *Micranthemum*. The filaments are often ornamented with colorful hairs and protuberances that mimic pollen-laden anthers. Pollination occurs mainly by insects, commonly by bees (Hymenoptera); however, cleistogamous, self-compatible flowers occur in *Lindernia Micranthemum* and *Vandellia* (East, 1940; Fernald, 1950; Endress, 1992; Fischer et al., 2013).

There have been several independent phylogenetic studies of Linderniaceae. An analysis of combined chloroplast DNA sequence data (Albach et al., 2005) originally indicated that *Lindernia* was distantly related to *Micranthemum* based on the sampling of only one species each from each of those genera (Figure 5.67). Although initial studies with a few representative taxa resolved *Torenia* and *Micranthemum* as being closely related (Albach et al., 2005; Oxelman et al., 2005), a more comprehensive analysis of taxa based on *matK* sequences (Rahmanzadeh et al., 2005) showed *Lindernia* to be polyphyletic, with *Torenia* nested among one subset of the species. In the most comprehensive molecular phylogenetic survey of the family (based on *trnK/matK* sequence data), Fischer et al. (2013) confirmed that *Lindernia* was highly polyphyletic and that the genus required substantial taxonomic refinement (Figure 5.68). In their study, the sensu stricto clade of *Lindernia* (which contained the type species) also included species of *Bryodes* and *Psammetes* (which they subsumed within the former). The remaining surveyed *Lindernia* species associated with several other clades/genera and were transferred accordingly to *Bonnaya*, *Chamaegigas*, *Craterostigma*, *Linderniella*, and *Torenia*. *Stemodiopsis* resolved as sister to the remainder of the family. Furthermore, that study resolved

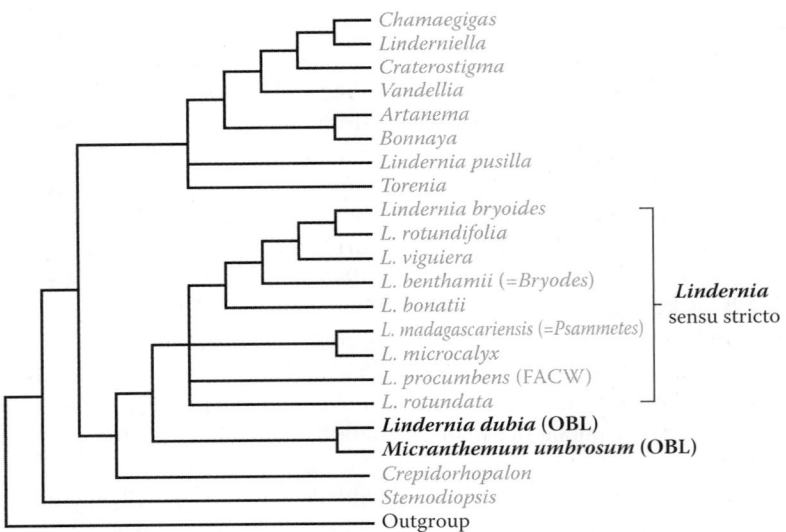

FIGURE 5.68 Phylogenetic relationships in Linderniaceae based on the analysis of *matK/trnK* intron sequences indicated the polyphyly of *Lindernia* and necessitated the transfer of many species to different genera. Taxa with OBL North American representatives (shown in bold) resolve as a sister clade to *Lindernia* sensu lato. Additional sampling of *Lindernia* and *Micranthemum* will be necessary to determine the appropriate taxonomic disposition of these two species, as well as the proper generic assignment of *L. pusilla*, whose relationship remains unsettled. (Redrawn from Fischer, E. et al., *Willdenowia*, 43, 209–238, 2013.)

Lindernia dubia and *Micranthemum umbrosum* as a sister clade to the *Lindernia* sensu stricto group, which would indicate the need for additional nomenclatural changes to either one or both taxa (Figure 5.68). However, the generic distinction between *Lindernia* and *Micranthemum* has been retained here pending the outcome of further taxonomic revision of this family as well as the retention of these names in the latest wetland indicator list.

Linderniaceae primarily are tropical and the species occur mostly in the Old World (Africa, Asia, Australia) except for the largest genus *Lindernia*, which nearly is cosmopolitan. Most of the species are terrestrial. *Craterostigma*, *Lindernia*, and *Torenia* are found in cultivation. *Torenia* (FACW) has been introduced to the southeastern United States. Two North American genera contain obligate wetland (OBL) aquatics:

1. ***Lindernia*** All.
2. ***Micranthemum*** Michx.

1. *Lindernia*

False pimpernel; fausse gratiole, lindernie
Etymology: after Franz Balthasar von Lindern (1682–1755)
Synonyms: *Anagalloides*; *Bazina*; *Bryodes*; *Ilysanthes*; *Psammetes*; *Pyxidaria*; *Tittmannia*
Distribution: global: Africa; Asia; Europe; New World; **North America:** E/W disjunct
Diversity: global: 100 species; **North America:** 8 species
Indicators (USA): OBL: *Lindernia dubia, L. grandiflora, L. monticola*
Habitat: freshwater; palustrine; **pH:** 5.1–9.1; **depth:** <1 m; **life-form(s):** emergent herb
Key morphology: stems (to 4 dm) square, rooting at nodes, creeping to weakly ascending; basal leaf rosette present or absent, cauline leaves (to 3 cm) opposite, sessile or subsessile; flowers bilabiate, solitary in leaf axils, stalked (to 4 cm); corolla (to 10 mm) tubular, blue, purple, violet, or white, the throat/tube purple streaked, upper lip shallowly 2-lobed, lower lip spreading and 3-lobed; fertile stamens 2, staminodia 2; capsules (to 7 mm) with numerous, minute seeds; seeds (to 0.5 mm) ribbed or winged, gold or reddish
Life history: duration: annual (fruit/seeds), perennial (persistent rootstock); **asexual reproduction:** none; **pollination:** insect or cleistogamous; **sexual condition:** hermaphroditic; **fruit:** capsules (common); **local dispersal:** seeds; **long-distance dispersal:** seeds
Imperilment: (1) *Lindernia dubia* [G5]; SH (DE, ME); S1 (AZ, NH, ON); S2 (BC, MI, NS, WV); S3 (MA, NC, QC, WA); (2) *L. grandiflora* [G5]; (3) *L. monticola* [G4]; S2 (FL, NC); S3 (AL)
Ecology: general: *Lindernia* is a frequent wetland inhabitant, with more than half of the temperate species worldwide found in wet sites. Six of the eight North American *Lindernia* species (75%) occur at least occasionally in wetlands with three designated as OBL aquatics. Often, the aquatic species occur on wetland margins, which dry out as the waters recede seasonally. The flowers are adapted for insect pollination by bees

(Hymenoptera) and perhaps flies (Diptera); however, some are cleistogamous and self-pollinating. Most species are annual, although a few can persist as short-lived perennials. Dispersal of the minute seeds has not been well-studied; however, they appear to be disseminated mainly by the wind and by water.

***Lindernia dubia* (L.) Pennell** is an annual, which inhabits bayous, canals, deltas, depressions, ditches, floodplains, gravel pits, marshes, meadows, mudflats, pools, prairies, sandbars, sloughs, swales, swamps, and margins of lakes, ponds, reservoirs, rivers, streams, and vernal pools at elevations of up to 1670 m. It is broadly adapted to granitic or limestone substrates (pH: 5.1–9.1), which include clay, cobble, gravel, muck, mud, muddy gravel, rock, sand, sandy clay, sandy peat, and stony gravel. This species has been described as thermophilous and "tolerant to all aquatic conditions." It is most characteristic of exposed drying substrates where the standing waters have receded. The plants can tolerate some shade but are susceptible to shading by other vegetation. The flowers are self-compatible but are cross-pollinated by short-tongued bees (Halictidae: *Lasioglossum imitatus*, *L. versatus*). Sometimes a large proportion of the flowers that develop early or late in the season is cleistogamous and self-pollinating. The plants produce numerous "dust-like" (0.4 mm) seeds that are dispersed by the wind or as drift in the water. A small number has also been recovered from the dung of horses (Mammalia: Equidae: *Equus ferus*). The seeds represent an important (often dominant) component of seed banks, reaching densities upward of 60,000 seeds/m^2. They persist in the seed bank, but their maximum longevity has not been determined. Specific germination requirements (e.g., dormancy) have also not been reported, but the seeds germinate readily either when exposed on the surface or while inundated. An average of 54 seedlings/m^2 can be produced under suitable conditions. Increased plant densities are known to occur in response to atmospheric CO_2 enrichment. The plants are intolerant to extreme pollution such as oil spills. **Reported associates:** *Acer rubrum, A. saccharinum, Alisma subcordatum, Alnus rhombifolia, Alternanthera philoxeroides, Amaranthus tuberculatus, Ambrosia artemisiifolia, Ammannia coccinea, A. robusta, Anagallis minima, Apocynum sibiricum, Asclepias incarnata, Bacopa rotundifolia, Betula nigra, Betula ×sandbergii, Bidens cernuus, B. connatus, B. discoideus, B. eatonii, B. frondosus, Bolboschoenus fluviatilis, Boltonia asteroides, B. decurrens, Bulbostylis capillaris, Butomus umbellatus, Calamagrostis canadensis, Cardamine longii, C. pensylvanica, Carex atherodes, C. crinita, C. echinata, C. lacustris, C. lupulina, C. stricta, C. pellita, C. scoparia, C. vesicaria, Chamaecrista fasciculata, C. nictitans, Chenopodium glaucum, Cicuta bulbifera, C. maculata, Clethra alnifolia, Coreopsis tinctoria, Cornus, Crassula aquatica, Cuscuta compacta, Cynodon dactylon, Cyperus acuminatus, C. bipartitus, C. difformis, C. erythrorhizos, C. ferruginescens, C. odoratus, C. polystachyos, C. squarrosus, C. strigosus, C. surinamensis, C. virens, Digitaria sanguinalis, Drosera intermedia, Dulichium arundinaceum, Echinochloa crus-galli, E. muricata, Eclipta prostrata, Elatine rubella, Eleocharis acicularis, E. aestuum,*

E. calva, E. engelmannii, E. intermedia, E. obtusa, E. palustris, E. quadrangulata, E. tuberculosa, Epilobium coloratum, Equisetum hyemale, Eragrostis hypnoides, E. pectinacea, Eriocaulon parkeri, Eupatorium dubium, E. perfoliatum, Euthamia tenuifolia, Fimbristylis autumnalis, F. miliacea, F. perpusilla, F. vahlii, Fraxinus profunda, Fuirena pumila, Galium tinctorium, Gentianopsis, Geocarpon minimum, Glottidium vesicarium, Glyceria, Gratiola aurea, G. neglecta, G. virginiana, Helenium autumnale, H. flexuosum, Heliotropium, Heteranthera dubia, H. limosa, Hibiscus moscheutos, Hordeum jubatum, Hypericum canadense, H. majus, H. mutilum, Ilex glabra, Impatiens capensis, Ipomoea, Iris versicolor, Isoetes butleri, I. melanopoda, I. riparia, Isolepis carinata, Iva annua, Jacquemontia tamnifolia, Juncus effusus, J. nodosus, J. pelocarpus, J. repens, J. torreyi, Justicia americana, Lactuca canadensis, Laportea canadensis, Lathyrus palustris, Leersia oryzoides, L. virginica, Leptochloa fusca, Leucospora multifida, Leucothoe racemosa, Limosella acaulis, L. aquatica, Lindera subcoriacea, Lindernia crustacea, Lipocarpha aristulata, L. micrantha, Lobelia cardinalis, L. siphilitica, Ludwigia alternifolia, L. glandulosa, L. palustris, L. peploides, L. polycarpa, L. sphaerocarpa, Lycopus americanus, L. amplectens, L. uniflorus, Lysimachia hybrida, L. nummularia, L. terrestris, Lythrum salicaria, Magnolia virginiana, Mentha arvensis, Micranthemum umbrosum, Mikania scandens, Mimulus alatus, M. cardinalis, M. ringens, Mitreola, Mollugo verticillata, Muhlenbergia mexicana, Nyssa biflora, Oldenlandia uniflora, Onoclea sensibilis, Oryza, Osmundastrum cinnamomeum, Panicum capillare, P. clandestinum, P. dichotomiflorum, P. gymnocarpon, P. hemitomon, P. verrucosum, Paspalum fluitans, Penthorum sedoides, Persicaria amphibia, P. hydropiperoides, P. lapathifolia, P. pensylvanica, P. punctata, Phalaris arundinacea, Phyla lanceolata, P. nodiflora, Pilularia americana, Pinus serotina, Plantago cordata, Pluchea camphorata, Polygonum aviculare, Polypremum procumbens, Pontederia cordata, Populus fremontii, Portulaca oleracea, Proserpinaca palustris, Puccinellia, Pycnanthemum virginianum, Quercus bicolor, Ranunculus abortivus, R. flabellaris, R. sceleratus, Rhexia virginica, Rhynchospora macrostachya, Rorippa palustris, R. sylvestris, Rotala ramosior, Rumex altissimus, R. maritimus, R. stenophyllus, R. verticillatus, Sabatia kennedyana, Sagittaria calycina, S. latifolia, Salix interior, S. nigra, Samolus valerandi, Saururus cernuus, Schoenoplectus acutus, S. hallii, S. pungens, S. smithii, S. tabernaemontani, Scirpus atrovirens, S. cyperinus, Scleria reticularis, Sclerolepis uniflora, Scutellaria lateriflora, Selaginella rupestris, Senna obtusifolia, Sium suave, Smilax laurifolia, Sparganium americanum, Sporobolus cryptandrus, Stachys pilosa, S. pycnantha, Strophostyles helvula, Symphyotrichum ontarionis, Taxodium distichum, Thelypteris palustris, Toxicodendron radicans, Tradescantia ohiensis, Triadenum virginicum, T. walteri, Typha latifolia, Verbena hastata, Viola, Woodwardia virginica, Xanthium strumarium, Xyris difformis.

***Lindernia grandiflora* Nutt.** occurs in canals, depressions, ditches, drains, flatwoods, floodplains, gravel pits, marshes, meadows, prairies, roadsides, savannas, scrub, seeps, sloughs, swamps, and along the margins of lakes, ponds, rivers, and streams. The substrates are characterized as muck, peat, peaty sand, and sand. The plants can tolerate full sun to partial shade. The life history of this species is largely unstudied and it has been described both as an annual and a perennial. Because seed sets have not been observed in greenhouse plants, it has been suggested that this species may be obligately outcrossed. The seeds are unique in the genus by having wing-like ribs, which might facilitate their dispersal by wind or water. The stems root at their nodes and the plants can form dense mats when growing in shallow water. **Reported associates:** *Acer rubrum, Alternanthera philoxeroides, Andropogon brachystachyus, A. glomeratus, A. virginicus, Aristida spiciformis, Bacopa caroliniana, B. monnieri, Carex vexans, Carya, Centella asiatica, Cephalanthus occidentalis, Commelina, Cyperus odoratus, Dichanthelium portoricense, Diodia virginiana, Eleocharis baldwinii, E. interstincta, Erigeron quercifolius, Eriocaulon compressum, Eupatorium leptophyllum, Euthamia caroliniana, Fuirena scirpoidea, Hibiscus grandiflorus, Hydrocotyle bonariensis, H. umbellata, Hypericum, Ilex cassine, Indigofera hirsuta, Juglans, Juncus marginatus, J. megacephalus, Lachnanthes caroliniana, Lachnocaulon, Leersia hexandra, Lilium catesbaei, Lobelia feayana, Ludwigia hirtella, L. peruviana, L. repens, Luziola fluitans, Magnolia virginiana, Mikania scandens, Myrica cerifera, Nyssa sylvatica, Oxypolis filiformis, Panicum hemitomon, P. repens, Persea borbonia, Persicaria, Phragmites australis, Phyla nodiflora, Pinus, Pluchea, Polygala nana, Pontederia cordata, Ptilimnium capillaceum, Quercus, Rhynchospora cephalantha, R. pusilla, Sabatia, Salix caroliniana, Samolus valerandi, Sapium sebiferum, Schoenoplectus californicus, Serenoa, Setaria, Smilax, Spartina bakeri, Sphagnum, Taxodium ascendens, T. distichum, Thelypteris kunthii, Typha, Urochloa mutica, Woodwardia virginica, Xyris jupicai, Zeuxine strateumatica.*

***Lindernia monticola* Nutt.** is restricted to bogs, depressions, meadows, pools, or seeps on granitic outcrops at elevations of up to 365 m. The soils typically comprise a shallow veneer (e.g., mucky sand) that overlays the granitic rock. The habitats typically are characterized by intermittent water. This species is described as an annual, biennial, or perennial. Presumably, it is an annual where the shallow habitats dry up completely during the hot summer season. A basal rosette of leaves occurs often in sites where water levels fluctuate, but is absent on plants that grow in more hydrologically stable sites. Elongation of the rosette internodes has been observed in plants that have been cultured on continuously moist soil under greenhouse conditions. The ribbed seeds probably are dispersed by the wind or by water. They occur within seed banks (up to 86 seeds m²) of uncertain duration. Seed germination has been achieved under a 25°C/15°C temperature regime following a 3-month period of cold stratification (5°C). Otherwise, there is very little life-history information available for this highly specialized species. Additional research on the floral and seed biology as well as the factors that influence perennation is needed. The plants sometimes occur in shallow, mud-bottomed pools that are devoid of other vegetation, or in association with a number of species that are restricted similarly to these specialized outcrop habitats. **Reported associates:** *Agrostis elliottiana, Allium cuthbertii, Amphianthus pusillus, Andropogon virginicus, Baptisia tinctoria, Botrychium lunarioides, Bulbostylis capillaris, Callitriche heterophylla, Cheilanthes lanosa, C. tomentosa, Chionanthus virginicus, Coreopsis major, Corydalis sempervirens, Croton willldenowii, Cyperus aristatus, Danthonia sericea, Diamorpha smallii, Diodia teres, Gnaphalium, Gratiola neglecta, Helianthus porteri, Houstonia caerulea, H. pusilla, Hypericum gentianoides, Isoetes melanopoda, I. melanospora, I. piedmontana, Juncus georgianus, Krigia virginica, Lepuropetalon spathulatum, Micranthes petiolaris, M. virginiensis, Minuartia glabra, M. uniflora, Nothoscordum bivalve, Nuttallanthus canadensis, Oenothera fruticosa, Ophioglossum crotalophoroides, Opuntia humifusa, Packera tomentosa, Pellaea wrightiana, Phacelia dubia, Phlox nivalis, Plantago virginica, Polygala curtissii, Portulaca smallii, Potentilla, Riccia dictyospora, Schizachyrium scoparium, Schoenolirion croceum, Scleria triglomerata, Sedum pusillum, Selaginella rupestris, Senecio smallii, S. tomentosus, Talinum teretifolium, Trifolium, Utricularia juncea, Woodsia obtusa.*

Use by wildlife: The flowers of *Lindernia dubia* are a source of nectar for short-tongued bees (Insecta: Hymenoptera: Andrenidae: *Calliopsis andreniformis*; Halictidae: *Lasioglossum imitatus, L. obscurus, L. pilosus, L. versatus*) and butterflies (Insecta: Lepidoptera: Hesperiidae: *Ancyloxypha numitor, Staphylus hayhurstii*; Lycaenidae: *Everes comyntas*; Pieridae: *Colias philodice, Eurema lisa*). It is a host plant for larvae of the white peacock butterfly (Lepidoptera: Nymphalidae: *Anartia jatrophae*).

Economic importance: food: none reported; **medicinal:** None of the OBL species of *Lindernia* has been used medicinally; **cultivation:** *Lindernia dubia* is cultivated occasionally. *Lindernia grandiflora* has become fairly prevalent commercially (both in North America and Europe) as a ground cover for planting along water garden margins, for use in hanging baskets, and sometimes for decorative planting in aquariums. It is commonly sold under the name of 'Angel's Tears'; **misc. products:** none; **weeds:** *Lindernia dubia* is a dominant rice paddy weed in Korea and Japan, where strains resistant to acetolactate synthase (ALS) inhibitor herbicides (bensulfuron-methyl, pyrazosulfuron-ethyl) have evolved. It is reported to be invasive in Belgium and a weed of ricefields in France and Spain. *Lindernia grandiflora* is regarded as a greenhouse weed in Finland; **nonindigenous species:** *Lindernia dubia* was introduced to Europe around 1851, reaching the Czech Republic by 1989 and Belgium by 1993. It was introduced to Japan around 1933 and also has been reported from Guangdong, China, and Taiwan. The minute seeds are frequent contaminants of seed mixes and soil. In one case, seedlings were observed to emerge from the soil of potted *Thuja occidentalis* specimens that had been shipped from Georgia to California. *Lindernia grandiflora* has appeared in Finnish gardens as an escape from greenhouse stock.

Systematics: Several phylogenetic analyses (based on cpDNA sequences) have demonstrated that *Lindernia* is not monophyletic and, in particular, separates those species having nonalveolate seeds (which include *L. procumbens*, the nomenclatural type of the genus) from those with alveolate seeds (e.g., most North American species; Figure 5.68). Consequently, a number of *Lindernia* species have been transferred to different genera (e.g., *Bonnaya, Ceratostigma, Linderniella, Torenia, Vandellia*). Although *Lindernia* has long been allied with *Torenia*, its polyphyletic nature now makes such associations uninterpretable. In its current circumscription, *Lindernia* (sensu stricto) is the sister group of the African/Malagasy genus *Crepidorhopalon* (Figure 5.68). The sensu stricto *Lindernia* clade also contains *Micranthemum* (Figure 5.68); however, additional taxon sampling will be necessary to determine the most appropriate taxonomic disposition for the latter group. *Lindernia dubia* is the only OBL species included in molecular phylogenetic analyses to date (Figure 5.68). *Lindernia grandiflora* has been segregated as a distinct section (section *Bazina*) because of its winged seeds. However, a more recent taxonomic treatment assigned all three of the OBL species (*L. dubia, L. grandiflora, L. monticola*) to section *Brachycarpae*. Relationships among the North American *Lindernia* taxa remain poorly understood. Morphometric analyses of *L. dubia* have failed to support the recognition of infraspecific taxa; however, AFLP markers reportedly can readily distinguish material identified as *L. dubia* subsp. *major* from typical *L. dubia*. Some authors continue to recognize *L. dubia* var. *anagallidea* as a distinct species (*L. anagallidea*). A thorough systematic survey of this complex is needed. The base chromosome number of *Lindernia* is difficult to determine because there is a wide range in reported $2n$ numbers ($2n = 14, 16, 18, 24, 28, 42$) and because the genus is polyphyletic as currently circumscribed. *Lindernia dubia* ($2n = 18, 32$) has both diploid and tetraploid cytotypes. Chromosome numbers are unreported for *L. grandiflora* and *L. monticola*.

Comments: *Lindernia dubia* is widespread (except for the north and Rocky Mountain region) and extends into South America. More restricted distributions characterize *L. grandiflora* (Florida, Georgia) and *L. monticola* (southeastern United States).

References: Abernethy & Willby, 1999; Baldwin & Derico, 1999; Beal, 1977; Berger & Elisens, 2005; Burk, 1977; Campbell & Gibson, 2001; Casper & Krausch, 1981; Chaw & Kao, 1989; D'Arcy, 1979; Fischer, 2004; Fischer et al., 2013; Houle & Phillips, 1988; Ikeda & Miura, 1994; Koizumi et al., 2004; Kuitunen & Lahtonen, 1994; Kurka, 1990; Leck, 2003; Lewis, 2000; Matthews & Murdy, 1969; McFarland & Rogers, 1998; Mohlenbrock et al., 1961; Neff & Baldwin, 2005; Pennell, 1935; Pyšek et al., 2002; Rahmanzadeh et al., 2005; Small, 1896; Stucky & Coxe, 1999; Tooker et al., 2002; Wyatt & Fowler, 1977; Yamaguchi et al., 2005; Yoshino et al., 2006.

2. *Micranthemum*

Baby's-tears, mudflower

Etymology: from the Greek *mikros anthemon* meaning "small flowered"

Synonyms: *Amphiolanthus*; *Globifera*; *Hemianthus*; *Hemisiphonia*; *Pinarda*

Distribution: global: New World; **North America:** eastern and southern

Diversity: global: 20 species; **North America:** 3 species

Indicators (USA): OBL: *Micranthemum glomeratum*, *M. micranthemoides*, *M. umbrosum*

Habitat: freshwater, saline (tidal); lacustrine, palustrine, riverine; **pH:** 5.3–8.5; **depth:** <1 m; **life-form(s):** emergent herb, submersed (vittate)

Key morphology: stems (to 30 cm) creeping or floating, glabrous, succulent; leaves (to 15 mm) opposite, entire, round to oblanceolate, essentially sessile; flowers (to 2 mm) minute, solitary, axillary, 4-merous, subsessile pedicels to 0.5 mm); corolla (to 2 mm) campanulate, the lobes imbricate; stamens 2, filaments geniculate; capsules (to 1 mm) subglobose, septicidal, irregularly dehiscent; seeds numerous, minute (to 0.3 mm), reticulate, ribbed

Life history: duration: annual (fruit/seeds); **asexual reproduction:** shoot fragments; **pollination:** self; **sexual condition:** hermaphroditic; **fruit:** capsules (common); **local dispersal:** fragments (water); **long-distance dispersal:** seeds

Imperilment: (1) *Micranthemum glomeratum* [G3]; (2) *M. micranthemoides* [GH]; SX (DE, NY, PA); SH (DC, MD, NJ, VA); (3) *M. umbrosum* [G5]; S1 (VA); S3 (NC)

Ecology: general: *Micranthemum* is a genus of amphibious annuals that is more common in warmer parts of the New World. The species grow either submersed in shallow water or as emergents along shorelines and on moist or drying substrates. The self-compatible flowers appear to be largely autogamous, but there have been no adequate investigations of the reproductive biology for any of the species. Cross-pollination by flies (Insecta: Diptera) probably occurs in some species. The minute seeds are thought to be dispersed in mud that becomes attached to waterbirds (Aves).

Micranthemum glomeratum **(Chapm.) Shinners** inhabits ditches, drains, swamps, and the shores of lakes, ponds, and rivers, often growing submersed in shallow water (to 1 m). Although found mainly in freshwater, the plants are euryhaline with a high salinity tolerance (up to 117 ppm sodium; 208 ppm chloride). Most occurrences are found in hard waters (mean $CaCO_3$ alkalinity: 30 ppm [2.8–55.2]; mean pH: 7.4 [6.3–8.5]). The substrates include gravel, mud, and sand. Optimal growth in water occurs at temperatures of 24°C–26°C. The reproductive biology of the minute (<2 mm) flowers has not been described, but they are either self-pollinated or pollinated by flies (Insecta: Diptera). The seeds probably are dispersed by water, but little information on seed ecology (e.g., germination requirements) exists for this species. The stems, which produce roots (up to 5 cm) along their nodes, can become dislodged during floods. The plants are annual, but can reproduce vegetatively by layering or by fragmentation of the stems during the growing season. **Reported associates:** *Acrostichum danaeifolium, Alternanthera philoxeroides, Amaranthus australis, Apios americana, Bacopa monnieri, Blechnum serrulatum, Boehmeria cylindrica, Canna flaccida, Centella asiatica, Cephalanthus occidentalis, Chamaecrista fasciculata, Chara,*

Colocasia esculenta, Commelina diffusa, Crinum americanum, Cyperus odoratus, Dichanthelium, Diodia virginiana, Dryopteris ludoviciana, Eclipta prostrata, Eleocharis baldwinii, Erechtites hieraciifolia, Eupatorium capillifolium, Fuirena, Galactia, Hydrocotyle umbellata, Hypericum tetrapetalum, Hyptis alata, Ipomoea indica, Juncus, Ludwigia peruviana, L. repens, Lygodium microphyllum, Mikania scandens, Mimosa quadrivalvis, Nuphar advena, Nymphaea odorata, Osmunda regalis, Panicum hemitomon, P. repens, P. rigidulum, Paspalum distichum, Pennisetum purpureum, Persicaria, Phyla nodiflora, Pluchea, Pontederia cordata, Psychotria nervosa, P. sulzneri, Ptilimnium capillaceum, Rapanea punctata, Rhynchospora inundata, R. rariflora, Rubus trivialis, Sabal palmetto, Sagittaria lancifolia, S. latifolia, Salix, Sambucus nigra, Sarcostemma clausum, Saururus cernuus, Sesbania punicea, Smilax bona-nox, Sphagneticola trilobata, Thelypteris interrupta, T. palustris, T. serrata, Toxicodendron radicans, Tripsacum dactyloides, Typha, Urena lobata, Vitis rotundifolia.

Micranthemum micranthemoides **(Nutt.)** Wettst. occurred historically along the banks of freshwater tidal rivers or streams (with 0.6–1.0 m of tidal influence) and on substrates of gravel, mud, or sand. The flowers are cleistogamous and self-pollinating, shedding their corollas while still unopened. No living populations of this species are known to survive (see cultivation and comments sections later). The following list of species represents historical or potential associates. **Reported associates:** *Crassula aquatica, Cyperus bipartitus, Elatine americana, E. minima, E. triandra, Eleocharis obtusa, E. parvula, Eriocaulon parkeri, Gratiola virginiana, Heteranthera reniformis, Isoetes mattaponica, I. riparia, I. saccharata, Lilaeopsis chinensis, Limosella aquatica, Orontium aquaticum, Peltandra virginica, Pontederia cordata, Sagittaria calycina, S. graminea, S. montevidensis, S. subulata, Schoenoplectus smithii, Zizania aquatica.*

Micranthemum umbrosum **(J.F. Gmel.) S.F. Blake** is an amphibious species that occurs in canals, depressions, ditches, marshes, pools, sandbars, streams, swamps, and along the margins of lakes, ponds, and rivers. The plants can grow entirely submersed in shallow water (typically to 1 m) or they can survive when exposed along drying margins of waterbodies. This species thrives at 20–26°C across a broad range of water conditions (pH: 5.3–8.4; mean 7.0) in sun or under partial shade. It occurs on clay loam, muck, sand, or silt, and also grows on logs or cypress "knees" (pneumatophores). Dislodged, floating plants are often sterile. There is no detailed information on the pollination biology, but the flowers are either autogamous or are pollinated by flies (Insecta: Diptera). Seed banks of up to 76 seeds/m² have been reported, but details on seed germination or longevity are lacking. This species will often colonize "snags" of vegetation that float in channels, or that become trapped within the debris that accumulates along the margins of streams or other watercourses. In open water it can form dense, circular mats that can extend for more than 1 m in diameter. The stem fragments can propagate vegetatively. **Reported associates:** *Acer rubrum, Alternanthera philoxeroides, Bacopa monnieri, Bidens cernuus, B. discoidea,*

B. frondosus, Boehmeria cylindrica, Callitriche heterophylla, Cardamine pensylvanica, Centella asiatica, Cephalanthus occidentalis, Chamaesyce, Cicuta mexicana, Cirsium, Colocasia esculenta, Commelina diffusa, Cyperus erythrorhizos, C. haspan, C. polystachyos, C. strigosus, Dichanthelium dichotomum, Digitaria sanguinalis, Diodia virginiana, Echinochloa muricata, Echinodorus cordifolius, Eclipta prostrata, Egeria densa, Eichhornia crassipes, Eleocharis baldwinii, E. obtusa, E. olivacea, E. radicans, E. vivipara, Eragrostis hypnoides, Erechtites hieraciifolia, Eryngium baldwinii, Fimbristylis autumnalis, F. perpusilla, Fuirena pumila, F. scirpoidea, Galium, Gratiola aurea, Habenaria repens, Helenium flexuosum, Heliotropium indicum, Hibiscus moscheutos, Hydrochloa caroliniensis, Hydrocotyle umbellata, H. verticillata, Hymenocallis caroliniana, Hypericum mutilum, Hypoxis curtissii, Ilex cassine, Isoetes louisianensis, Juncus coriaceus, J. effusus, J. filipendulus, J. marginatus, J. pelocarpus, J. repens, Justicia ovata, Lemna minor, Limnobium spongia, Lindernia dubia, Lipocarpha micrantha, Ludwigia decurrens, L. octovalvis, L. peploides, L. palustris, L. repens, L. sphaerocarpa, L. uruguayensis, Lycopus rubellus, L. virginicus, Mitreola petiolata, Mnium affine, Myrica cerifera, Myriophyllum heterophyllum, Nuphar, Nymphoides aquatica, Nyssa, Oldenlandia boscii, O. uniflora, Pallavicinia lyellii, Panicum dichotomiflorum, P. hemitomon, P. repens, P. rigidulum, P. verrucosum, Paspalum fluitans, P. notatum, Penthorum sedoides, Persicaria amphibia, P. hydropiperoides, P. pensylvanica, P. punctata, Phyla lanceolata, P. nodiflora, Phyllanthus urinaria, Pinus elliottii, Pluchea camphorata, Pontederia cordata, Proserpinaca palustris, Ranunculus laxicaulis, Rhynchospora corniculata, Rosa palustris, Rotala ramosior, Rubus, Sabatia calycina, S. kennedyana, Sagittaria lancifolia, S. latifolia, S. subulata, Salvinia rotundifolia, Sambucus nigra, Samolus valerandi, Saururus cernuus, Schoenoplectus californicus, S. pungens, Scirpus cyperinus, S. divaricatus, Taxodium distichum, Triadenum walteri, Typha, Utricularia gibba, U. inflata, U. macrorhiza, Viola primulifolia, Woodwardia areolata, Youngia japonica.

Use by wildlife: *Micranthemum umbrosum* provides cover for aquatic invertebrates and fish and is a host for several larval moths (Insecta: Lepidoptera: Pyralidae: *Parapoynx obscuralis, Langessa nomophilalis*). However, the plants contain several feeding deterrent chemicals (3,4,5-trimethoxyallylbenzene and various lignoids), which make them unpalatable to grass carp (Chordata: Cyprinidae: *Ctenopharyngodon idella*) and crayfish (Crustacea: Cambaridae: *Procambarus spiculifer, P. acutus*).

Economic importance: food: *Micranthemum umbrosum* is highly acrid and unpalatable; **medicinal:** none reported; **cultivation:** Aquarium plants distributed as "*Micranthemum micranthemoides*" or "*Hemianthus micranthemoides*" (a synonym) obviously are misidentified, given that living plants of this species have not been observed for nearly 70 years (see comments later). Most of the commercial material sold under these names probably is *M. glomeratum*, but its identity should be confirmed using genetic markers. *Micranthemum umbrosum* is also cultivated as an aquarium plant; however, it

is fairly difficult to culture unless strong light and supplemental CO_2 are provided; **misc. products:** none; **weeds:** none; **nonindigenous species:** none

Systematics: Molecular phylogenetic analyses (which include only *M. umbrosum*) resolve *Micranthemum* within Linderniaceae close to *Torenia* and *Lindernia* (Figure 5.67). A comprehensive molecular systematic study of *Micranthemum* is needed to test the monophyly and to elucidate relationships of the genus. Such an effort could be problematic given that the type of *Hemianthus* (*H. micranthemoides*) presumably is extinct. Moreover, there is a dearth of even basic systematic information (e.g., chromosome number reports) for *Micranthemum* and no comprehensive taxonomic revisions or monographs exist. Authors also disagree whether to merge or segregate *Hemianthus* (divided style, fused calyx, and long corolla tube) from *Micranthemum* (short style, free calyx lobes, short corolla tube), which results in confusing estimates of diversity (4–20 species) for the latter genus. The genera are treated here as synonymous.

Comments: *Micranthemum glomeratum* is endemic to Florida and occurs mainly in the central portion of that state. Once narrowly endemic to the eastern United States, living plants of *M. micranthemoides* have not been observed since 1941 despite more than 20 years of intensive searches. The species was declared to be extinct by the Center for Biological Diversity in 2001. *Micranthemum umbrosum* occurs throughout the southeastern United States and extends into Central and South America.

References: Aulbach-Smith & de Kozlowski, 1990; Cook, 1996a; D'Arcy, 1979; Fassett, 1928; Fernald, 1950; Fischer, 2004; Hoyer et al., 1996; Lane & Kubanek, 2006; Milius, 2002; Musselman et al., 2001; Nolfo-Clements, 2006; Parker et al., 2006; Pennell, 1935; Schmidt, 2005; Shull, 1903; Stoops et al., 1998; Tarver et al., 1978; Williams, 1972; Woodson et al., 1979.

Family 5: Oleaceae [25]

Although a relatively small family of only 600 species, Oleaceae are widely distributed and are highly valued as the source of ash lumber (*Fraxinus*), olives (*Olea*), and the sweetly fragrant jasmine (*Jasminum*) and lilac (*Syringa*). The family consists entirely of woody species (lianas, trees, shrubs), which are distinguished by their opposite leaves and two or more superposed buds. Phytochemically, the family lacks hydrolyzed or condensed tannins, but contains other phenylpropanoids (Green, 2004). Most species have showy, often fragrant flowers with only two stamens and are pollinated by nectar-seeking insects; however, both *Forestiera* and *Fraxinus* have reduced, mainly wind-pollinated flowers (Judd et al., 2002). A few genera are heterostylous and most species predominantly are outcrossed (Green, 2004). The fruits include berries, capsules, drupes, and samaras and are dispersed by birds or by the wind.

Some authors (e.g., Takhtajan, 1997) place Oleaceae within a distinct order (Oleales); however, most authors now recognize the family within the Lamiales, where molecular data resolve it as the sister group to the remainder of that order (Figure 5.50; Wagstaff & Olmstead, 1997; Wallander &

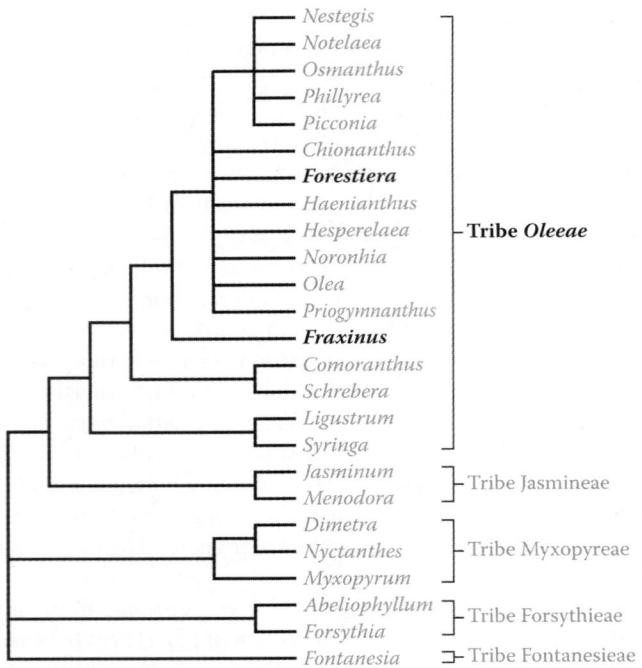

FIGURE 5.69 Phylogenetic relationships in Oleaceae as indicated by analysis of cpDNA sequence data. These results indicate that the OBL habit (taxa in bold) has evolved independently in the family. All 25 currently recognized genera were surveyed in this analysis. (Adapted from Wallander, E. & Albert, V.A., *Amer. J. Bot.* 87, 1827–1841, 2000.)

Albert, 2000). A similar iridoid and phenolic glycoside chemistry also indicates a relationship to Lamiales (Jensen et al., 2002). A recent phylogenetic survey, which included all 25 genera of Oleaceae (Figure 5.69), has led to the adoption of an infrafamilial classification that abandons the traditional subfamilies Oleoideae and Jasminoideae because the latter is paraphyletic (Wallander & Albert, 2000). Instead, five tribes are recognized (*Myxopyreae, Fontanesieae, Forsythieae, Jasmineae, Oleeae*), with the latter divided into four subtribes (*Ligustrinae, Schreberinae, Fraxininae, Oleinae*).

Oleaceae are primarily a terrestrial family, with OBL aquatics found in only two of the North American genera:

1. *Forestiera* Poir.
2. *Fraxinus* L.

Phylogenetic analysis of molecular (cpDNA sequence) data indicates that the OBL habit was acquired independently in these mainly terrestrial genera (Figure 5.69). However, both genera are fairly closely related and assigned to the same tribe (Oleeae). They differ essentially by their fruit types, a drupe in *Forestiera* and samara in *Fraxinus*. The basic haploid number of tribe Oleeae ($n = 23$) is believed to have been derived by allopolyploidy involving $x = 11$ and 12 species.

1. *Forestiera*
Swamp-privet
Etymology: after Charles Le Forestier (?–1820)

Synonyms: *Adelia; Bigelovia; Borya; Carpoxis; Geisarina; Nudilus; Piptolepis*
Distribution: global: New World; **North America:** southern
Diversity: global: 19 species; **North America:** 8 species
Indicators (USA): OBL: *Forestiera acuminata*
Habitat: freshwater; palustrine, riverine; **pH:** broad range; **depth:** <1 m; **life-form(s):** emergent shrub, emergent tree
Key morphology: stems (to 10 m) with spreading branches; leaves (to 12 cm) opposite, simple, deciduous, sparsely toothed above middle, petioles (to 2 cm) slender; flowers appearing before the leaves, bisexual or unisexual; the ♂ flowers numerous, in dense cymose clusters along branchlets; stamens 4; pistillode present or absent; the ♀ flowers several in a cluster; corolla absent; staminodes 4; drupes (to 3 cm) oblong, slightly compressed, 1-seeded, dark purple, wrinkled
Life history: duration: perennial (buds); **asexual reproduction:** stem sprouts; **pollination:** wind; **sexual condition:** polygamodioecious; **fruit:** drupes (common); **local dispersal:** drupes (fish, water); **long-distance dispersal:** drupes (birds)
Imperilment: (1) *Forestiera acuminata* [G5]; S1 (KS); S2 (OK); S3 (GA)
Ecology: general: Only one North American species of *Forestiera* is an obligate aquatic but a few others occur facultatively (FAC, FACU) in wetlands. The flowers are wind pollinated and are either unisexual and dioecious or polygamodioecious with mixtures of hermaphroditic and unisexual flowers. The hermaphroditic flowers are self-compatible. The fruits are eaten and dispersed by birds, fish, or small mammals.

Forestiera acuminata **(Michx.) Poir.** inhabits bayous, depressions, floodplains, sloughs, swales, swamps, and margins of lakes, rivers, and streams at elevations of up to 200 m. This species often occurs on calcareous sediments, but it is described as broadly tolerant to soil pH (specific pH values have not been reported). The substrates include clay, clay loam, loam, mud, sand, and silt loam. The plants are highly flood tolerant and can withstand periods of inundation lasting from 1 to 3 months. Where shoreline water levels vary, *F. acuminata* tends to occupy the zone between the average high and low water marks. This species occurs in full sun or in partial shade and is among the most common species to colonize light gaps (both interior and edge) in floodplain forests. Adult plants are fairly resistant to damage by hurricanes. *Forestiera acuminata* is polygamodioecious, that is, individuals generally are dioecious with some trees also producing mixtures of unisexual and hermaphroditic flowers, the latter apparently being self-compatible. The flowers are precocious (appearing before the leaves), lack a corolla, and are pollinated by the wind. The drupes have been described as miniature, wrinkled sausages. They have a thin endocarp. Aside from physical dormancy conferred by the enveloping stone (endocarp), the seeds germinate well when extracted from the fruit if held first in water for 1 week before planting. The seeds remain viable and exhibit germination rates that are nearly 40% higher after the fruits have been eaten by channel catfish (Chordata: Siluriformes: Ictaluridae: *Ictalurus punctatus*), which represent an unusual dispersal agent. New stems can develop from the base of senescing individuals in response to inundation; however, cuttings do not root well and cannot easily establish when planted. **Reported associates:** *Acalypha virginica, Acer negundo, A. rubrum, A. saccharinum, A. saccharum, Adiantum pedatum, Alnus serrulata, Alopecurus carolinianus, Alternanthera philoxeroides, Amelanchier arborea, Amorpha fruticosa, Ampelopsis arborea, A. cordata, Asimina triloba, Asplenium platyneuron, A. rhizophyllum, Azolla caroliniana, Berchemia scandens, Betula nigra, Bidens discoidea, Boehmeria cylindrica, Botrychium virginianum, Brunnichia ovata, Callitriche heterophylla, Campsis radicans, Carpinus caroliniana, Carex cherokeensis, C. crus-corvi, C. grayi, C. intumescens, C. joorii, C. lupuliformis, C. lupulina, C. typhina, Carya aquatica, C. illinoinensis, Celtis laevigata, Cephalanthus occidentalis, Chasmanthium sessiliflorum, Cicuta maculata, Cinna arundinacea, Cocculus carolinus, Commelina virginica, Conoclinium coelestinum, Cornus drummondii, C. florida, C. foemina, o. obliqua, Crataegus opaca, C. viridis, Cyperus pseudovegetus, Cystopteris fragilis, Dichanthelium dichotomum, Dioclea multiflora, Diodia virginiana, Diospyros virginiana, Echinochloa crus-galli, Echinodorus cordifolius, Eleocharis palustris, Fraxinus americana, F. caroliniana, F. pennsylvanica, F. profunda, Gleditsia aquatica, G. triacanthos, Glyceria striata, Gratiola neglecta, G. virginiana, Hibiscus moscheutos, Hygrophila lacustris, Hypericum hypericoides, Ilex decidua, I. opaca, I. vomitoria, Impatiens capensis, Itea virginica, Iva annua, Juglans nigra, Juncus effusus, Justicia ovata, Laportea canadensis, Leersia lenticularis, L. virginica, Lemna minor, Leucothoe axillaris, Lindera benzoin, Liquidambar styraciflua, Liriodendron tulipifera, Lobelia cardinalis, Ludwigia grandiflora, L. palustris, L. peploides, Lyonia lucida, Magnolia virginiana, Malvaviscus arboreus, Matteuccia struthiopteris, Mikania scandens, Mimulus alatus, Morus rubra, Myrica cerifera, Nelumbo lutea, Nyssa aquatica, N. biflora, N. ogeche, N. sylvatica, Onoclea sensibilis, Packera glabella, Penthorum sedoides, Persea borbonia, P. palustris, Persicaria amphibia, P. hydropiperoides, P. lapathifolia, P. virginiana, Phanopyrum gymnocarpon, Phegopteris hexagonoptera, Physocarpus opulifolius, Pilea pumila, Pinus serotina, Planera aquatica, Platanus occidentalis, Pleopeltis polypodioides, Polystichum acrostichoides, Populus deltoides, P. heterophylla, Proserpinaca palustris, Quercus alba, Q. bicolor, Q. lyrata, Q. nigra, Q. nuttallii, Q. palustris, Q. phellos, Q. texana, Q. virginiana, Rhododendron canescens, Rhynchospora corniculata, Rubus, Sabal minor, S. palmetto, Saccharum baldwinii, Sagittaria latifolia, Salix caroliniana, S. nigra, Sambucus nigra, Smilax tamnoides, Styrax americanus, Saururus cernuus, Scirpus atrovirens, Sisyrinchium angustifolium, Styrax americanum, Symphyotrichum lateriflorum, Taxodium ascendens, T. distichum, Toxicodendron radicans, Trachelospermum difforme, Triadenum walteri, Triadica sebifera, Typha latifolia, Ulmus americana, U. crassifolia, U. rubra, Urtica dioica, Viburnum rufidulum, Vitis aestivalis, V. cinerea, V. palmata, Wisteria frutescens, Woodsia obtusa, Xanthium strumarium.*

Use by wildlife: The twigs of *Forestiera acuminata* are browsed by white-tail deer (Mammalia: Cervidae: *Odocoileus virginianus*) and swamp rabbits (Mammalia: Leporidae: *Sylvilagus aquaticus*). However, dense thickets are less valuable to wildlife. The drupes are eaten by channel catfish (Chordata: Siluriformes: Ictaluridae: *Ictalurus punctatus*), mallards (Aves: Anatidae: *Anas platyrhynchos*), quail (Aves: Odontophoridae: *Colinus virginianus*), wild turkeys (Aves: Phasianidae: *Meleagris gallopavo*), and wood ducks (Aves: Anatidae: *Aix sponsa*). The plants are not susceptible to damage by gypsy moths (Insecta: Lepidoptera: *Lymantria dispar*) but they are eaten to some degree by the nonindigenous emerald ash borer (Insecta: Coleoptera: *Agrilus planipennis*). The seeds are eaten by weevil larva and pupae (Insecta: Coleoptera: Curculionidae: *Neotylopterus pallidus*). **Economic importance: food:** The drupes of *F. acuminata* reportedly are edible; **medicinal:** The Houma people prepared a decoction from the roots and bark of *F. acuminata* to use as a medicinal tea; **cultivation:** *Forestiera acuminata* is cultivated as a native shrub for wet sites and as an alternative planting in lieu of invasive privets (*Ligustrum*); **misc. products:** *Forestiera acuminata* has been considered for use in efforts to control shoreline erosion, but it cannot be established from cuttings; **weeds:** none; **nonindigenous species:** none

Systematics: Analyses of DNA sequence data place *Forestiera* species within a clade corresponding to Oleaceae tribe Oleeae, subtribe Oleinae; however, many of the generic and species relationships in that group are not resolved well and will require further study to elucidate. Of the four species included in those analyses, *F. acuminata* was most closely related to *F. neo-mexicana*, which resolved as its sister species; however, the placement of two other species (*F. eggersiana, F. segregata*) was ambiguous (Figure 5.70). The close relationship that has been suggested for *Forestiera* and *Chionanthus* is supported in a general sense by a molecular phylogenetic analysis (Figures 5.69 and 5.70) and also more specifically by a phenetic analysis of seed protein serological data. All genera assigned to subtribe Oleinae have a derived chromosomal base number of $x = 23$, which is believed to have arisen by allopolyploidy involving $x = 11$ and $x = 12$ taxa (presumably extinct). *Forestiera acuminata* ($2n = 46$) is regarded as a diploid.

Comments: *Forestiera acuminata* occurs in the southeastern United States, especially in the Mississippi embayment region.

References: Brown, 1943; Busbee et al., 2003; Chick et al., 2003; Clark, 1978; Correll & Correll, 1975; Dale, 1984b; East, 1940; Elias, 1987; Environmental Laboratory, 1987; Fischer et al., 1999; Godfrey & Wooten, 1981; Haack et al., 2002; Helm & Chabreck, 2006; Hoagland & Johnson, 2005; King & Antrobus, 2005; Leininger et al., 1997; McGregor, 1960; Piechura & Fairbrothers, 1983; Robertson, 2006; Speck, 1941; Taylor, 1945; Wallander & Albert, 2000.

2. Fraxinus

Pop ash, pumpkin ash, water ash; frêne

Etymology: the ancient Latin name used by Ovid, Virgil, and others

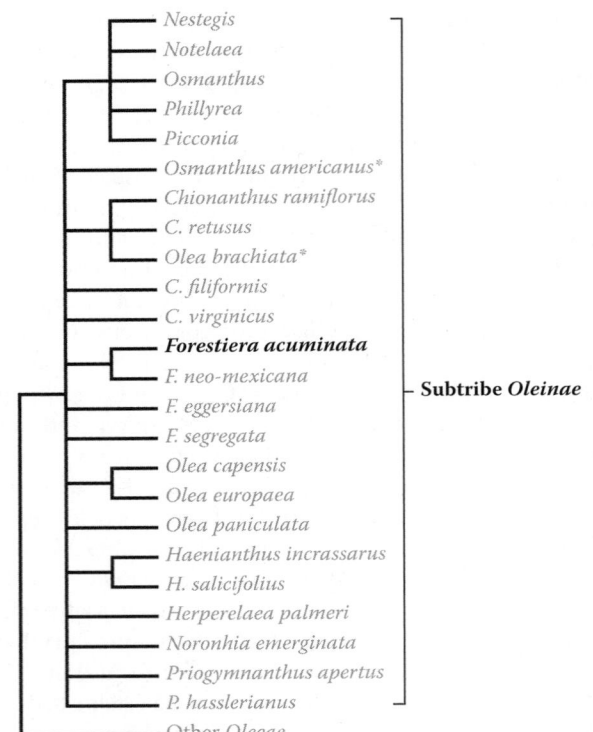

Nestegis
Notelaea
Osmanthus
Phillyrea
Picconia
*Osmanthus americanus**
Chionanthus ramiflorus
C. retusus
*Olea brachiata**
C. filiformis
C. virginicus
Forestiera acuminata
F. neo-mexicana
F. eggersiana
F. segregata
Olea capensis
Olea europaea
Olea paniculata
Haenianthus incrassarus
H. salicifolius
Herperelaea palmeri
Noronhia emerginata
Priogymnanthus apertus
P. hasslerianus
Other *Oleeae*

— Subtribe *Oleinae*

FIGURE 5.70 Phylogenetic relationships of four *Forestiera* species within Oleaceae tribe Oleeae, subtribe Oleinae as indicated by the analysis of cpDNA sequence data. Although *Forestiera* does not resolve as a clade, and several species of *Olea* and *Osmanthus* are misplaced (asterisks), this particular data set provided only weak support for the groups indicated and the relationships depicted should be regarded only as tentative. Nevertheless, the monophyly of *Forestiera* and its close relationship to *Chionanthus* remain plausible and are not contradicted by the results. (Adapted from Wallander, E. & Albert, V.A., *Amer. J. Bot.* 87, 1827–1841, 2000.)

Synonyms: *Apilia*; *Aplilia*; *Calycomelia*; *Fraxinoides*; *Leptalix*; *Mannaphorus*; *Meliopsis*; *Ornanthes*; *Ornus*; *Petlomelia*; *Samarpsea*; *Samarpses*
Distribution: global: Northern Hemisphere (temperate, subtropical); **North America:** widespread
Diversity: global: 50 species; **North America:** 18 species
Indicators (USA): OBL: *Fraxinus caroliniana*, *F. profunda*
Habitat: brackish (coastal), freshwater; palustrine, riverine; **pH:** 3.8–6.8; **depth:** <1 m; **life-form(s):** emergent shrub, emergent tree
Key morphology: trunk single (to 30 m) or multiple (to 12 m) with furrowed or scaly bark and a swollen or dilated base; leaves (to 45 cm) opposite, odd pinnate (5–9 leaflets), leaflets (to 25 cm) entire or toothed, stalked (to 2 cm); flowers unisexual (rarely bisexual), numerous, in dense, elongate, cymose clusters; corolla absent; the ♂ flowers with a minute calyx and 2 or 4 stamens, pistil absent; the ♀ flowers with a cup-like calyx and elongate pistil, stamens absent; samaras (to 8 cm), 1-seeded, with a dry, flattened wing (to 2 cm), occasionally 3-winged
Life history: duration: perennial (buds); **asexual reproduction:** none; **pollination:** wind; **sexual condition:** dioecious,

polygamous; **fruit:** samaras (common); **local dispersal:** samaras (gravity); **long-distance dispersal:** samaras (water)
Imperilment: (1) *Fraxinus caroliniana* [G4/G5]; (2) *F. profunda* [G4]; S1 (DC, NJ, PA); S2 (MD, MI, <u>ON</u>); S3 (MS)
Ecology: general: Ash trees generally are associated with UPL habitats, but nearly half of the North American species occur at least occasionally in wetlands. The flowers are mostly unisexual and dioecious; however, bisexual flowers and polygamous individuals occur occasionally. Species are both insect and wind pollinated and the flowers range from having the perianth completely present to entirely absent. The flowers of OBL North American species uniformly have a reduced calyx, no corolla, and are wind pollinated. The elongate fruits possess a flat, marginal wing, which facilitates their dispersal by wind in terrestrial habitats, and by water in aquatic habitats. The seeds of most ashes can be dried and stored at low temperatures. The life history of many *Fraxinus* species (especially European) has been studied extensively. However, the OBL North American species have been studied mainly in a broad ecological context.

Fraxinus caroliniana **Mill.** is a small tree that grows in bayous, canals, depressions, ditches, floodplains, hammocks, lagoons, marshes, savannas, sloughs, swamps, and along the margins of rivers and streams at elevations of up to at least 49 m. Although some substrates reportedly are derived from limestone, the sites usually are acidic (pH: 3.8–6.8) and include loamy sand, muck, mucky peat, and sandy loam, often with fairly high (e.g., 40%) organic matter content. The sites mostly are freshwater and nontidal but plants also have been reported from tidal localities in water that is of considerable salinity (e.g., 3.5% seawater). The species is described as shade intolerant but specimens have been collected at sites receiving full sun to dense shade. Experimentally, the seedlings grow optimally at 53% of full sunlight. The plants are extremely tolerant of flooding and often produce multiple trunks (sometimes having dilated bases) for support. The flowers are wind pollinated and occur in a dioecious arrangement. The seeds germinate at 30°C after 50–90 days of cold stratification (5°C). They tend to be poorly represented in floodplain seed banks, even where the adult plants occur in the standing vegetation. The seeds are dispersed mainly by the transport of the broadly winged samaras (occasionally 3-winged) by water. The seedlings require exposed sites for their initial establishment, but once established, they can tolerate up to 11 months of continuous, shallow inundation, or moist soil conditions, regardless of the soil type. **Reported associates:** *Acer rubrum, Acrostichum danaeifolium, Alnus serrulata, Amelanchier canadensis, Amorpha fruticosa, Ampelopsis arborea, Amsonia tabernaemontana, Annona glabra, Arundinaria gigantea, Asclepias perennis, Baccharis halimifolia, Bacopa caroliniana, Berchemia scandens, Betula nigra, Bidens aristosus, B. discoideus, Bignonia capreolata, Blechnum serrulatum, Boehmeria cylindrica, Brunnichia ovata, Callicarpa americana, Campsis radicans, Cardamine bulbosa, Carex alata, C. bromoides, C. cruscorvi, C. gigantea, C. glaucescens, C. hyalinolepis, C. intumescens, C. joorii, C. lonchocarpa, C. louisianica, C. lupulina, C. seorsa, C. stipata, C. tribuloides, C. venusta, Carpinus* caroliniana, Carya aquatica, Celtis laevigata, Cephalanthus occidentalis, Chasmanthium laxum, C. ornithorhynchum, C. sessiliflorum, Chionanthus virginicus, Chrysobalanus icaco, Cicuta maculata, Cinna arundinacea, Clematis virginiana, Clethra alnifolia, Cliftonia monophylla, Commelina virginica, Conocarpus erectus, Cornus foemina, Crataegus marshallii, C. opaca, Crinum americanum, Cyrilla racemiflora, Decodon verticillatus, Dichanthelium dichotomum, Diospyros virginiana, Dulichium arundinaceum, Eupatorium coelestinum, Ficus aurea, Forestiera acuminata, Fraxinus pennsylvanica, Frullania cobrensis, Galium obtusum, Gelsemium rankinii, Gleditsia aquatica, Glyceria septentrionalis, Hibiscus moscheutos, Hordeum pusillum, Hydrocotyle verticillata, Hypericum galioides, H. hypericoides, H. prolificum, Hypoxis curtissii, Ilex cassine, I. coriacea, I. decidua, I. opaca, I. verticillata, I. vomitoria, Ipomoea sagittata, Itea virginica, Juncus coriaceus, Justicia ovata, Kosteletzkya pentacarpos, Leersia lenticularis, Leucothoe axillaris, L. racemosa, Liquidambar styraciflua, Lobelia cardinalis, L. inflata, L. puberula, Ludwigia repens, Lycopus rubellus, L. virginicus, Lyonia ligustrina, L. lucida, Lysimachia lanceolata, Magnolia virginiana, Metopium toxiferum, Microstegium vimineum, Mikania scandens, Mitreola petiolata, Myrica cerifera, Myrsine floridana, Nephrolepis, Nyssa aquatica, N. biflora, N. ogeche, N. sylvatica, Onoclea sensibilis, Orontium aquaticum, Osmunda regalis, Osmundastrum cinnamomeum, Ostrya virginiana, Panicum rigidulum, Parthenocissus quinquefolia, Peltandra virginica, Persea borbonia, P. palustris, Persicaria hydropiperoides, P. setacea, Phanopyrum gymnocarpon, Physostegia intermedia, Pilea pumila, Pinus elliottii, P. palustris, P. taeda, Pistia stratiotes, Planera aquatica, Platanus occidentalis, Platanthera flava, Pleopeltis polypodioides, Pluchea camphorata, P. odorata, Polypogon, Pontederia cordata, Populus heterophylla, Proserpinaca pectinata, Psychotria nervosa, Quercus falcata, Q. laurifolia, Q. lyrata, Q. michauxii, Q. nigra, Q. pagoda, Q. phellos, Q. laevis, Q. shumardii, Q. stellata, Q. texana, Q. virginiana, Ranunculus sceleratus, Rhizophora mangle, Rhododendron viscosum, Rhynchospora cephalantha, R. colorata, R. corniculata, R. glomerata, R. inundata, R. miliacea, Roystonea elata, Rudbeckia laciniata, Sabal minor, S. palmetto, Saccharum baldwinii, Sagittaria lancifolia, S. latifolia, Salix caroliniana, S. nigra, Sambucus nigra, S. nigra, Saururus cernuus, Scirpus, Scutellaria lateriflora, Sebastiania fruticosa, Serenoa repens, Sideroxylon salicifolium, Smilax laurifolia, S. tamnoides, S. walteri, Spartina bakeri, Sphagnum macrophyllum, Sphenopholis pensylvanica, Spiranthes cernua, S. odorata, Styrax, Symplocos tinctoria, Taxodium ascendens, T. distichum, Thelypteris kunthii, Tillandsia usneoides, Toxicodendron radicans, Trachelospermum difforme, Triadenum walteri, Triadica sebifera, Ulmus alata, U. americana, Uniola latifolia, Vaccinium arboreum, V. elliottii, Viburnum nudum, V. obovatum, Vicia floridana, Viola, Vitis rotundifolia, Wisteria frutescens, Woodwardia areolata, W. virginica, Xyris laxifolia.

Fraxinus profunda (**Bush**) **Bush** is a large tree found in canals, depressions, floodplains, marshes, sloughs, swales, swamps, and along margins of ponds. The substrates are primarily of mineral composition (but with an organic content up to 16% or more), acidic (pH: 4.5–6.8), and include clay, clay loam, loam, muck, sand, silt, silty clay, and silt loam. The plants develop into a canopy or subcanopy species in nontidal, subtidal, or tidal sites and can tolerate cold temperatures down to −25°C. They typically grow where salinity levels are below 1 ppt, but apparently can withstand short exposures of up to 5 ppt during storm surges. The flowers are wind pollinated, occur in a dioecious arrangement, and begin to bear seed when individuals reach 10 years of age. The seeds are not produced prolifically and are wind or water dispersed, often being deposited in fairly low densities (<1/m²/year). The seeds reportedly require a period of cold stratification, and will germinate more readily under flooded rather than nonflooded conditions. Seedling establishment is most successful in bare, moist openings; however, the seedlings are somewhat shade tolerant. Adult plants have intermediate shade tolerance. This species is highly tolerant to shallow water inundation (average of 0.5 m depth), and occupies sites that are characterized by median spring/annual inundation periods of 35 and 103 days, respectively. However, faster growth occurs under more well-drained conditions. The swollen, "pumpkin-like" trunk serves as a buttress for added support in saturated soils. The plants are extremely susceptible to fire and also will die back during drought. **Reported associates:** *Acer negundo, A. rubrum, A. saccharinum, Alnus maritima, A. serrulata, Alternanthera philoxeroides, Ampelopsis arborea, A. cordata, Apios americana, Arisaema triphyllum, Arundinaria gigantea, Asclepias perennis, Azolla caroliniana, Berchemia scandens, Bidens discoidea, Bignonia capreolata, Boehmeria cylindrica, Brunnichia ovata, Campsis radicans, Carex atlantica, C. bromoides, C. crinita, C. crus-corvi, C. folliculata, C. glaucescens, C. hyalinolepis, C. intumescens, C. joorii, C. lonchocarpa, C. louisianica, C. lupulina, C. seorsa, C. stipata, C. stricta, Carpinus caroliniana, Carya aquatica, C. glabra, Celtis laevigata, Cephalanthus occidentalis, Chamaecyparis thyoides, Cicuta maculata, Cinna arundinacea, Cirsium muticum, Clematis virginiana, Clethra alnifolia, Commelina virginica, Cornus amomum, C. foemina, Crataegus viridis, Crinum americanum, Cyrilla racemiflora, Decumaria barbara, Diodia virginiana, Dioscorea villosa, Diospyros virginiana, Dulichium arundinaceum, Eupatorium coelestinum, Euphorbia purpurea, Forestiera acuminata, Fraxinus caroliniana, F. pennsylvanica, Gelsemium rankinii, Gleditsia aquatica, Gordonia lasianthus, Gratiola virginiana, Hydrocotyle, Ilex cassine, I. decidua, I. laevigata, I. opaca, I. verticillata, I. vomitoria, Impatiens capensis, Itea virginica, Iris fulva, Justicia ovata, Laportea canadensis, Leersia lenticularis, L. oryzoides, L. virginica, Lemna minor, Leucothoe racemosa, Lindera benzoin, Liquidambar styraciflua, Liriodendron tulipifera, Lobelia cardinalis, Ludwigia glandulosa, L. grandiflora, L. palustris, Lyonia lucida, Lysimachia radicans, Magnolia virginiana, Mikania scandens, Mitreola petiolata, Murdannia keisak, Myrica cerifera, Nyssa aquatica, N. biflora, N. sylvatica, Onoclea sensibilis, Orontium aquaticum, Osmunda regalis, Osmundastrum cinnamomeum, Parthenocissus quinquefolia, Peltandra virginica, Persea* palustris, Persicaria arifolia, P. hydropiperoides, P. sagittata, P. virginiana, Pilea pumila, Pinus serotina, P. taeda, Planera aquatica, Platanthera flava, Pluchea camphorata, Populus heterophylla, Peltandra virginica, Phanopyrum gymnocarpon, Proserpinaca palustris, P. pectinata, Quercus laurifolia, Q. lyrata, Q. michauxii, Q. nigra, Q. nuttallii, Q. palustris, Q. phellos, Q. texana, Rhododendron viscosum, Rhynchospora inundata, R. miliacea, R. mixta, Rosa palustris, Rudbeckia laciniata, Sabal minor, S. palmetto, Sagittaria latifolia, Salix nigra, Sassafras albidum, Saururus cernuus, Sideroxylon reclinatum, Sium suave, Smilax laurifolia, S. rotundifolia, S. walteri, Sphagnum, Spiranthes odorata, Symphyotrichum lateriflorum Taxodium ascendens, T. distichum, Thalictrum pubescens, Thelypteris palustris, Tillandsia usneoides, Toxicodendron radicans, Triadenum tubulosum, T. walteri, Ulmus americana, Vaccinium corymbosum, V. fuscatum, Viburnum dentatum, V. nudum, Vitis, Woodwardia areolata, W. virginica, Zizania aquatica.

Use by wildlife: Fraxinus caroliniana is used as nesting habitat for wood storks (Aves: Ciconiidae: Mycteria

americana). The seeds of F. caroliniana are eaten by birds (Aves) and mammals (Mammalia). The plants are host to fungi (Basidiomycota: Schizoporaceae: Schizopora paradoxa), larval Lepidoptera (Papilionidae: Papilio glaucus; Sesiidae: Podosesia syringae), several nematode root parasites (Nematoda: Heteroderidae: Meloidogyne incognita; Tylenchulidae: Tylenchulus palustris, T. semipenetrans), the introduced South African pit scale (Insecta: Hemiptera: Planchonia stentae), and a whitefly (Hemiptera: Aleyrodidae: Massilieurodes americanus). Seeds of F. profunda are eaten by wood ducks (Aves: Anatidae: Aix sponsa) and other birds. The foliage and twigs are browsed (but not preferably) by white-tailed deer (Mammalia: Cervidae: Odocoileus virginianus) and occasionally by swamp rabbits (Mammalia: Leporidae: Sylvilagus aquaticus). The plants are host to a spider mite (Arachnida: Acari: Tenuipalpidae: Brevipalpus fraxini). Older trees are susceptible to heartrot fungi (Basidiomycota: Polyporaceae: Lentinus tigrinus). All ash species are susceptible to infection by larvae of the nonindigenous emerald ash borer beetle (Insecta: Coleoptera: Buprestidae: Agrilus

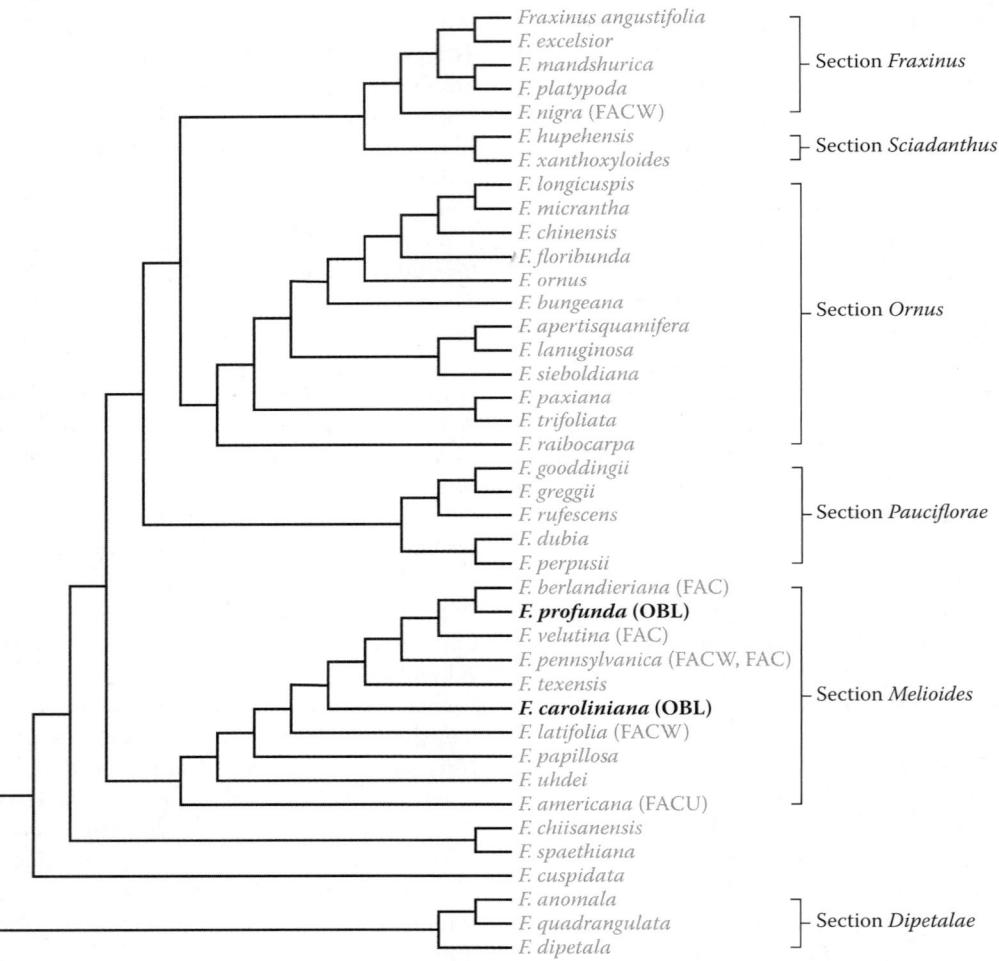

FIGURE 5.71 Phylogenetic relationships among sections of Fraxinus based on analysis of nrDNA sequences. Although the two OBL North American species (in bold) are not shown as sister species, a lack of resolution within sections Melioides and Pauciflorae resulted in some ambiguous relationships to be resolved arbitrarily. Consequently, additional data are needed to clarify relationships within sections Melioides and Pauciflorae more confidently. The wetland indicator status is provided (in parentheses) for those applicable North American taxa. (Adapted from Wallander, E., Plant Syst. Evol., 273, 25–49, 2008.)

planipennis), which feed on the phloem and disrupt nutrient transport.

Economic importance: food: not edible; **medicinal:** none; **cultivation:** *Fraxinus caroliniana* and *F. profunda* both are cultivated, with the latter more widely available; **misc. products:** The wood of *F. caroliniana* is soft and unsuitable as timber. The wood of *F. profunda* is of high quality and is used commercially for boxes, fuel, lumber, paper pulp, and tool handles; **weeds:** none; **nonindigenous species:** none

Systematics: Some confusion in evaluating literature reports is attributable to the fact that various authors have adopted the synonym *F. tomentosa* in place of *F. profunda*, a taxonomic issue that has not yet been resolved satisfactorily. The latter name has been retained here. Phylogenetic analyses of DNA sequence data (Figure 5.69) resolve *Fraxinus* within Oleaceae (tribe Oleeae) as the sister group to subtribe Oleinae. A molecular phylogenetic analysis of 27 *Fraxinus* taxa initially demonstrated the monophyly of the genus as well as that of the North American section *Melioides*, to which *F. caroliniana* and *F. profunda* are assigned (however, neither species was included in that analysis). A subsequent study of 40 *Fraxinus* species, which included both OBL taxa, predictably confirmed their placement in the clade representing section *Melioides*. However, because resolution was poor within section *Melioides* (as well as in section *Pauciflorae*), any ambiguous relationships within those sections were resolved arbitrarily (Figure 5.71). As a consequence, it cannot be estimated confidently whether the OBL habit was derived once or twice within the genus. A number of other species in the entirely North American section *Melioides* are FACW indicators (Figure 5.71). Because of polyploidy and hybridization, phylogenetic relationships in *Fraxinus* should be corroborated by analysis of cpDNA data. In particular, some authors have hypothesized a hybrid origin for *F. profunda*. A recent phylogenetic study of *Fraxinus* has analyzed combined nuclear and cpDNA data, but included only one of the OBL taxa (*F. caroliniana*). That study also resolved section *Melioides* as monophyletic, but depicted relationships within the section that were quite different from those deduced from nuclear ribosomal DNA (nrDNA) sequences (e.g., Figure 5.71). Furthermore, it is suspected that at least some of the material from section *Melioides* that was sampled for molecular analyses originated from arboretum specimens that were misidentified. Hopefully, further refinements to the systematic study of this group will be forthcoming. The base chromosome number of *Fraxinus* is $x = 23$ (see discussion under "*Systematics*" for *Forestiera*). *Fraxinus profunda* ($2n = 138$) is a hexaploid, which is suggested to have arisen either by hybridization involving tetraploid *F. americana* ($4x = 92$) and diploid *F. pennsylvanica* ($2n = 46$), or as an autopolyploid derivative of *F. pennsylvanica*. The latter hypothesis is supported by patterns of cuticle morphology and circumstantially by failed attempts to hybridize *F. americana* and *F. pennsylvanica*. However, it would be more insightful to investigate the origin of this species using appropriate genetic markers.

Comments: *Fraxinus caroliniana* occurs in the southeastern United States; whereas, *F. profunda* is distributed discontinuously throughout eastern North America.

References: Baskin & Baskin, 1998; Carroll, 2003; De Leon, 1961; Dow et al., 1990; DuBarry, Jr., 1963; Elias, 1987; Grand & Vernia, 2004; Griffin, III & Breil, 1982; Hardin & Beckmann, 1982; Harms, 1990; Hinsinger et al., 2013; Inserra et al., 1987; Jeandroz et al., 1997; Jones & McLeod, 1990; Light et al., 2002; Middleton, 2000; Monk, 1966; Nesom, 2014; Schneider & Sharitz, 1986, 1988; Stumpf & Lambdin, 2000; Titus, 1990, 1991; Townsend, 2001; Waldron et al., 1996; Wallace et al., 1996; Wallander, 2008; White, 1983.

Family 6: Orobanchaceae [61]

Some authors still do not recognize Orobanchaceae as a distinct family, but include the taxa within a broadly defined Scrophulariaceae (e.g., Fischer, 2004). However, molecular data have helped considerably to clarify the circumscription of Orobanchaceae by resolving the family as a clade that is distinct phylogenetically from Scrophulariaceae (Olmstead et al., 2001; Oxelman et al., 2005; Bennett & Mathews, 2006). This group consists of hemiparasitic or holoparasitic herbs, with the latter habit having been derived independently within the family on several occasions (Young et al., 1999). The few morphologically defining features of the group include specialized hairs that lack vertical septae, and a determinate, racemose inflorescence (Judd et al., 2002). Pollination in Orobanchaceae is biotic and is facilitated by insects and birds. Seed dispersal occurs by the wind. Overall, the family is not important economically, but contains the notorious witchweed (*Striga*), which is responsible for extensive crop damage.

Orobanchaceae consist of approximately 1700 species and have a nearly cosmopolitan distribution (Judd et al., 2002). The majority of species are terrestrial, with OBL aquatics occurring in only five North American genera:

1. ***Agalinis*** Raf.
2. ***Castilleja*** Mutis ex L. f.
3. ***Chloropyron*** Behr
4. ***Macranthera*** Nutt. ex Benth.
5. ***Pedicularis*** L.

The most detailed phylogenetic study of Orobanchaceae available was carried out using DNA sequences from the nuclear-encoded photoreceptor phytochrome A (*PHYA*) gene (Bennett & Matthews, 2006). This study of 98 species in 43 genera resolved six-well-supported clades (Figure 5.72). The four OBL North American genera included in that study (*Agalinis*, *Castilleja*, *Chloropyron*, *Pedicularis*) resolved in three different subclades, indicating that the aquatic habit has originated independently at least several times within the family (Figure 5.72). Phylogenetic studies using cpDNA sequences (Young et al., 1999) are not as well-resolved, but indicate that *Macranthera* also is closely related to *Agalinis* and *Castilleja*.

Systematic studies of tribe Pedicularideae (subtribe Castillejinae) by Tank & Olmstead (2008) and Tank et al.

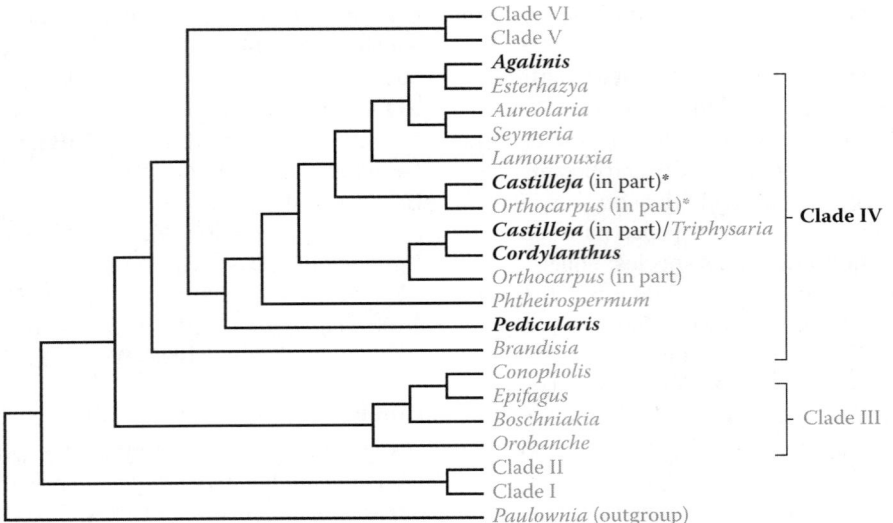

FIGURE 5.72 Phylogenetic relationships in Orobanchaceae as indicated by *PHYA* sequence data. All of the OBL North American taxa (bold) occur within one clade (clade IV) as resolved by this analysis; however, they are dispersed among three distinct subclades. A second clade containing *Castilleja* and *Orthocarpus* (asterisked) is likely due to paralogous sequences via a gene duplication. All members of clade IV are hemiparasites. (Adapted from Bennett, J.R. & Matthews, S., *Amer. J. Bot.*, 93, 1039–1051, 2006.)

(2009) have resulted in several realignments of genera. These modifications included the inclusion of *Clevelandia* and *Ophiocephalus* within *Castilleja*, and the elevation of two *Cordylanthus* subgenera to generic level as *Dicranostegia* and *Chloropyron*. The latter includes all of the OBL species assigned formerly to *Cordylanthus*. These proposed changes have been followed here.

1. *Agalinis*

False foxglove, gerardia; gérardie

Etymology: from the Greek *aga linos* meaning "quite linen" for the resemblance of some species to flax (*Linum*)

Synonyms: *Anisantherina*; *Dasystoma*; *Gerardia*; *Tomanthera*

Distribution: global: North America; South America; **North America:** eastern

Diversity: global: 40 species; **North America:** 33 species

Indicators (USA): OBL: *Agalinis calycina*, *A. linifolia*, *A. maritima*, *A. paupercula*, *A. tenella*; **FACW:** *A. linifolia*, *A. maritima*, *A. paupercula*, *A. tenella*; **FAC:** *A. paupercula*

Habitat: freshwater, saline; palustrine; **pH:** 5.8–7.5; **depth:** <1 m; **life-form(s):** emergent herb

Key morphology: stems (to 50 cm) erect, weakly to strongly four angled, simple or with opposite branches (or fascicles) apically; leaves (to 5 cm) opposite, linear (to 3 mm wide) or linear spatulate, succulent or dry, sessile; inflorescence a raceme (up to 20 flowers) or the opposite flowers appearing axillary; flowers 5-merous, pedicelled (to 15 mm), somewhat irregularly two-lipped (the two upper lobes reduced); corolla (to 2.5 cm) tubular, lavender pink, pink, purple, rose-purple, or violet, the throat swollen above with the lobes spreading; stamens (to 9 mm) 4, didynamous, glabrous, or pubescent; capsules (to 8 mm) globose with numerous seeds (to 1.5 mm)

Life history: duration: annual (fruit/seeds); **asexual reproduction:** none; **pollination:** insect or self; **sexual condition:** hermaphroditic; **fruit:** capsules (common); **local dispersal:** seeds (water, wind); **long-distance dispersal:** seeds (water, wind)

Imperilment: (1) *Agalinis calycina* [G1]; S1 (NM, TX); (2) *A. linifolia* [G4]; S2 (AL, LA); S3 (GA, NC); (3) *A. maritima* [G5]; SX (<u>NB</u>); S1 (<u>NS</u>); S2 (DE, NH, RI); S3 (NC, NY, ME, MS); (4) *A. paupercula* [G5]; SH (<u>MB</u>); S1 (<u>NB</u>, PA, VA); S3 (IA, NJ, NY, VT); (5) *A. tenella* [G4/G5]; SH (DE, MD); S2 (NC); S3 (LA)

Ecology: general: *Agalinis* species are broad ecologically and occur over an impressive range of habitats including dry sand barrens, grasslands, woodlands, and various fresh to saline wetlands. About half of the North American species have some type of wetland indicator status, but only five (15%) are designated as obligate aquatics. The genus contains annuals and perennials with both durations occurring among the OBL North American species. All of the species appear to be hemiparasites. Many *Agalinis* species are known to be self-compatible and are capable of self-pollination in the absence of pollinators. However, most also are herkogamous, with the spatial separation between the stigma and anthers greatly reducing the probability of selfing under usual conditions. Thus, seed set tends to be higher when pollinators are present. Pollinators include bees (Hymenoptera: Apidae: *Apis*, *Bombus*, *Melissodes*) and other insects. The OBL North American species are FAC hemiparasites and develop haustoria, which are induced by chemical exudates, for penetrating the root systems of their host plants. Although plants can grow to maturity in the absence of a host, they fare much better when a host is present. The seeds do not require a stimulus from a host plant to initiate their germination. The seeds of some species have physiological dormancy, but the seed ecology of the OBL species has not been investigated in detail. Some *Agalinis* species germinate at 20°C–24°C following 3–6 weeks of cold stratification (at 3°C–4°C). The dispersal mechanism is not known for the OBL species, but the small seeds probably are transported by the wind. Seeds of some species have been observed to float, which indicates potential for water dispersal. The plants often turn black when they are dried.

***Agalinis calycina* Pennell** is a rare annual, which grows in cienega, inland salt marshes, and seeps at elevations of up to 1555 m. The soils are described as alkaline, saline, and calcareous silty clays and loams that are derived from gypsum and limestone. The plants are in flower from August to October. The flowers are ephemeral and remain open for only 1 day from approximately 9:00 am until 5:00 pm. They then wither and fall from the plant. Pollinators have not been observed but are suspected to include bees (Hymenoptera) and butterflies (Lepidoptera). Seed set in these rare plants is high, which could indicate an autogamous breeding system as well. The capsules mature and dehisce in October, releasing large quantities of tiny seeds. The few remaining populations of this species are threatened by hydrological changes to the cienega environment and by potential exploitation of the habits for resource extraction. **Reported associates:** *Bolboschoenus maritimus, Cirsium wrightii, Distichlis spicata, Eleocharis palustris, E. rostellata, Fimbristylis puberula, Flaveria chlorifolia, Helianthus paradoxus, Heliotropium curassavicum, Juncus balticus, Limonium limbatum, Muhlenbergia asperifolia, Nesaea longipes, Phragmites australis, Pluchea odorata, Samolus ebracteatus, Schoenoplectus americanus, Sesuvium verrucosum, Spartina pectinata, Sporobolus airoides, S. texanus, Suaeda calceoliformis, Symphyotrichum ericoides, S. subulatum, Tamarix chinensis, Triglochin maritimum, Typha domingensis,*

***Agalinis linifolia* (Nutt.) Britt.** is a perennial inhabitant of barrens, bogs, canals, depressions, ditches, flatwoods, glades, marshes, ponds, prairies, savannas, seasonal wetlands, seepage slopes, shores, swamps, and the margins of ponds, sinks, and streams along the southeastern coastal plain at low elevations of up to 46 m. The plants often occur in standing, shallow water (to 20 cm) in sites that receive full sun or are partially shaded. Although primarily found in freshwater, some sites are characterized by mildly brackish conditions. The substrates are described as acidic (no specific pH values reported) and have been characterized as Basinger, clay, Immokalee, loamy sand, Placid (typic humaquepts), Pompano, Pottsburg (grossarenic haplaquods), sand, and sandy peat. The flowers are produced from September to October and attract bees (see *Use by wildlife*), which presumably pollinate them. The plants have been collected in sites that are burned annually, which indicates that they are fire resistant. This is the only perennial OBL *Agalinis* species. The plants reproduce vegetatively by rhizomes and can develop into thick clumps. Although hemiparasitic, the plants produce relatively few haustorial hairs, with no haustorial attachments observed in one study. **Reported associates:** *Amphicarpum muhlenbergianum, Andropogon virginicus, Annona glabra, Aristida purpurascens, A. stricta,*

Bacopa monnieri, Bigelowia nudata, Cassytha filiformis, Centella asiatica, Cirsium horridulum, Cladium jamaicense, Cliftonia monophylla, Coleataenia tenera, Conocarpus erectus, Coreopsis gladiata, Crinum americanum, Cyrilla racemiflora, Eleocharis cellulosa, E. interstincta, Eriocaulon compressum, E. decangulare, Eupatorium leptophyllum, E. mohrii, E. paludicola, Hymenocallis palmeri, Hypericum brachyphyllum, H. fasciculatum, Ilex glabra, I. myrtifolia, Leersia hexandra, Liatris spicata, Lobelia floridana, Ludwigia linearis, L. pilosa, L. simpsonii, Lycopus angustifolius, Muhlenbergia expansa, Nyssa biflora, Oxypolis filiformis, Panicum hemitomon, P. virgatum, Paspalum monostachyum, P. praecox, Persea, Persicaria hydropiperoides, Pinus elliottii, P. palustris, Pluchea baccharis, Polygala cymosa, Proserpinaca palustris, Rhexia aristosa, R. lutea, Rhizophora mangle, Rhynchospora careyana, R. colorata, R. divergens, R. latifolia, R. microcarpa, R. pleiantha, R. tracyi, Sabatia decandra, S. difformis, Sagittaria lancifolia, Sarracenia psittacina, Schizachyrium rhizomatum, S. scoparium, Schoenus nigricans, Serenoa repens, Solidago stricta, Sorbus, Symphyotrichum chapmanii, S. dumosum, S. tenuifolium, Taxodium ascendens, T. distichum, Tillandsia paucifolia, Utricularia cornuta, U. foliosa, U. purpurea, Xyris difformis.

***Agalinis maritima* (Raf.) Raf.** is an annual halophyte that inhabits beaches, dunes, flatwoods, saltmarshes, strands, and swales. The full range of acidity has not been characterized for this species; however, it is reported from acidic (e.g., pH: 5.8) to alkaline substrates (often sand) in brackish to saline sites and tolerates a range of salinity from 0.5 to 80 ppt (but generally under 18 ppt). The plants grow typically in the upper marsh, often in slight depressions where other vegetation is sparse. Prime habitats are inundated regularly by tidal waters and have substrates that are covered by an organic layer. This species tends to occur fairly sporadically throughout its range but it can become locally abundant, reaching densities of several hundred plants over a few m² of marsh. The overall size of the plants generally increases southward over its distributional range. The plants flower only on sunny days and the flowers remain open for only a short period of time (mid-morning until noon) and drop their corollas within a few hours afterwards. As the flowers mature, the anthers rotate from 45° to 90° relative to the filaments. Although the plants are self-compatible, pollination is facilitated by small insects (Hymenoptera) that forcibly enter and exit the flowers. Seed set is generally high. The plants are hemiparasitic on the rhizomes of *Borrichia frutescens, Distichlis spicata, Spartina alterniflora*, and *S. patens*. **Reported associates:** *Agropyron pungens, Agrostis scabra, Amaranthus australis, Andropogon virginicus, Aphanostephus skirrhobasis, Aristida, Artemisia campestris, Ascophyllum nodosum, Atriplex arenaria A. patula, A. prostrata, Avicennia germinans, Baccharis halimifolia, Bassia hirsuta, Batis maritima, Blutaparon vermiculare, Bolboschoenus robustus, Borrichia frutescens, Cakile, Carex silicea, Cenchrus, Centrosema virginianum, Chloris petraea, Conocarpus erectus, Conoclinium betonicifolium, Croton punctatus, Cynanchum angustifolium, Cynodon* *dactylon, Cyperus polystachyos, Dichanthelium sabulorum, Digitaria filiformis, Distichlis spicata, Eleocharis parvula, E. rostellata, Eragrostis oxylepis, Fimbristylis castanea, F. ferruginea, Heliotropium curassavicum, Heterotheca subaxillaris, Hibiscus moscheutos, Hydrocotyle bonariensis, Ipomoea pes-caprae, I. sagittata, I. stolonifera, Iva angustifolia, I. frutescens, I. imbricata, Jacquinia keyensis, Juncus canadensis, J. gerardii, J. greenei, J. roemerianus, Kosteletzkya pentacarpos, Liatris scariosa, Limonium carolinianum, L. nashii, Lycium carolinianum, Lysimachia maritima, Lythrum lineare, Monanthochloe littoralis, Myrica cerifera, M. pensylvanica, Oenothera parviflora, Panicum amarum, P. capillarioides, P. virgatum, Paspalum setaceum, P. vaginatum, Phragmites australis, Phyla nodiflora, Physalis cinerascens, Pinus taeda, Plantago maritima, Pluchea camphorata, P. odorata, P. purpurascens, Polygala verticillata, Portulaca rubricaulis, Potentilla anserina, Prunus maritima, Puccinellia maritima, Quercus, Rayjacksonia phyllocephala, Rhachicallis americana, Rhizophora mangle, Rhynchospora colorata, Sabatia stellaris, Salicornia bigelovii, Sarcocornia perennis, Schizachyrium maritimum, Schoenoplectus americanus, S. pungens, Serenoa repens, Sesuvium portulacastrum, Setaria magna, Solidago sempervirens, Spartina alterniflora, S. cynosuroides, S. patens, Spergularia marina, Sporobolus virginicus, Strophostyles helvola, Suaeda linearis, S. maritima, Symphyotrichum subulatum, S. tenuifolium, Teucrium canadense, Toxicodendron radicans, Triglochin maritimum, Triplasis purpurea, Typha angustifolia, Uniola paniculata, Vaccinium corymbosum, Viola lanceolata.*

Agalinis paupercula (A. Gray) Britton is an annual, which grows in barrens, beaches, bogs, borrow pits, deltas, depressions, ditches, dunes, fens, marshes, meadows, prairies, seeps, sloughs, swales, and along the margins of lakes, ponds, rivers, and streams. The plants can occur on exposed substrates or in water that is up to 0.5 m deep. Although sometimes regarded as an indicator of alkaline fens, this species also grows in more acidic sites (pH: 6.1–7.5). It is found not only on peat substrates, but also on those derived from dolomite or sandstone and described variously as clay, gravel, loam, marl, muck, mud, rock, sand, sandy loam, and silt. Exposures can range from full sun to shade. The plants are self-compatible and the floral morphology and development indicates a mixed mating system. As the flowers mature, the anthers rotate 45° to 90° relative to the filaments. The pollen is released within the closed flowers either 1 day prior to anthesis or as the flowers open. The pollen begins to germinate within 2 h after anthesis. Other details on the reproductive ecology of this species remain uncertain. **Reported associates:** *Agalinis purpurea, Agrostis gigantea, A. hyemalis, Anemone virginiana, Astragalus canadensis, Bartonia virginica, Betula glandulifera, Bidens cernuus, B. connatus, B. discoidea, B. frondosus, Calamagrostis canadensis, Calopogon tuberosus, Campanula aparinoides, Carex bebbii, C. crawfordii, C. lasiocarpa, C. prairea, C. scoparia, C. sterilis, Chamaedaphne calyculata, Cicuta bulbifera, Cirsium muticum, Cladium mariscoides, Cornus sericea, Cyperus bipartitus, C. engelmannii, C. strigosus, Decodon*

verticillatus, Desmodium canadense, Dichanthelium acuminatum, D. meridionale, Drosera intermedia, D. rotundifolia, Dulichium arundinaceum, Echinochloa muricata, E. walteri, Eleocharis robbinsii, Equisetum, Erigeron strigosus, Eriophorum angustifolium, E. virginicum, Eupatorium leucolepis, E. maculatum, E. perfoliatum, Fimbristylis autumnalis, Fragaria virginiana, Gentianopsis virgata, Glyceria canadensis, Gnaphalium obtusifolium, Helianthemum bicknellii, Hypericum majus, Impatiens capensis, Juncus balticus, J. brevicaudatus, J. canadensis, J. effusus, J. greenei, J. pelocarpus, Kalmia, Larix laricina, Liatris spicata, Liparis loeselii, Lipocarpha micrantha, Lobelia kalmii, L. spicata, Lycopodiella inundata, Lycopodium hickeyi, Lycopus americanus, L. uniflorus, Lysimachia hybrida, L. terrestris, Menyanthes trifoliata, Muhlenbergia glomerata, M. uniflora, M. mexicana, Parnassia glauca, P. palustris, Peltandra virginica, Persicaria lapathifolia, P. maculosa, Phragmites australis, Poa palustris, Polygala cruciata, Populus balsamifera, P. tremuloides, Potentilla anserina, Primula, Pycnanthemum verticillatum, Rhexia virginiana, Rhynchospora alba, R. capillacea, R. capitellata, R. glomerata, Rorippa palustris, Salix candida, S. interior, Sarracenia purpurea, Schoenoplectus pungens, S. smithii, Scirpus atrocinctus, S. cyperinus, Scleria verticillata, Scutellaria galericulata, Selaginella eclipes, Solidago gigantea, S. graminifolia, S. patula, S. riddellii, S. uliginosa, Sorghastrum nutans, Sphagnum, Spirea tomentosa, Spiranthes cernua, Stachys palustris, Symphyotrichum oolentangiense, Thuja occidentalis, Thelypteris palustris, Toxicodendron vernix, Triantha glutinosa, Trichophorum cespitosum, Utricularia cornuta, U. gibba, U. intermedia, U. purpurea, U. resupinata, Vaccinium macrocarpon, Verbena hastata, Viola lanceolata, Xyris caroliniana, X. torta.

Agalinis tenella Pennell is an annual species that occurs in barrens, bogs, flats, flatwoods, glades, hammocks, marshes, meadows, pocosins, prairies, roadsides, savannas, seepage bogs, swales, and along the margins of ponds and streams on the southeastern coastal plain at low elevations of up to 91 m. Occurrences are generally exposed to full sun. The substrates have been described as clay, Dothan (plinthic paleudults), gravel, oolite limestone, sand, sandy clay, sandy loam, and sandy peat. Populations have been reported both from recently burned as well as long-standing unburned sites. Flowering occurs from September to October. Although the pollination biology of this species has not been studied in any detail, the plants appear to be highly outcrossed with genetic analyses of microsatellite loci documenting large numbers of alleles, and high levels of heterozygosity in populations. **Reported associates:** *Agalinis aphylla, A. divaricata, A. filicaulis, Amphicarpum muhlenbergianum, Aristida palustris, A. spiciformis, A. stricta, Bigelowia nudata, Carphephorus odoratissimus, C. pseudoliatris, Cirsium virginianum, Ctenium aromaticum, Dichanthelium, Eragrostis elliottii, Eupatorium rotundifolium, Helianthus heterophyllus, Hypericum, Ilex glabra, Liatris, Muhlenbergia expansa, Pinus elliottii, P. palustris, Rhynchospora divergens, R. plumosa, Schizachyrium tenerum, Schoenus nigricans, Scleria muehlenbergii, Serenoa repens, Seymeria cassioides, Sorghastrum.*

Use by wildlife: *Agalinis maritima* is a host for larval insecta (Insecta: Lepidoptera: Nymphalidae) including the common buckeye (*Junonia coenia*) and the West Indian buckeye (*J. evarete*). *Agalinis calycina* also is a larval host of the common buckeye (*Junonia coenia*). *Agalinis linifolia* is eaten by deer (Mammalia: Cervidae: *Odocoileus*). Its flowers attract various bees (Insecta: Hymenoptera: Colletidae: *Hylaeus confluens*; Halictidae: *Lasioglossum creberrimum*; Megachilidae: *Megachile brevis*, *M. mendica*), which presumably function as pollinators.

Economic importance: food: not edible; **medicinal:** none; **cultivation:** A few *Agalinis* species are grown horticulturally, but none of the OBL aquatics; **misc. products:** none; **weeds:** none; **nonindigenous species:** none

Systematics: Phylogenetic analysis of combined cpDNA data indicate that *Agalinis* is monophyletic and is the sister group to a clade comprising *Aureolaria*, *Brachystigma*, *Dasistoma*, and *Seymeria*, which all are genera to which it has been allied in past taxonomic treatments. *Agalinis* species were placed formerly in the genus *Gerardia*; however, that name was found to have nomenclatural priority in the family Acanthaceae. Much taxonomic uncertainty remains in the genus. *Agalinis tenella* has been treated by many authors as a synonym of *A. obtusifolia*; however, phylogenetic analyses (Figure 5.74) and microsatellite data clearly distinguish the two taxa as distinct species. Unfortunately, the improper association of *A. tenella* with *A. obtusifolia* has made it extremely difficult to extract pertinent information for either taxon from the literature. Consequently, care is advised when evaluating the above information provided for *A. tenella*, which likely contains related inaccuracies. *Agalinis neoscotica* has been treated as a variety of either *A. paupercula* or *A. purpurea* and *A. paupercula* has been treated as a variety of *A. purpurea*. The distinctness of these taxa warrants a more critical evaluation. The OBL species have been assigned to three different sections within the genus: section *Erectae* (*A. tenella*), section *Heterophyllae* (*A. calycina*), and section *Purpureae* (*A. maritima*, *A. paupercula*). Both *Agalinis maritima* and *A. paupercula* often are assumed to be derived from (or at least closely related to) *A. purpurea*, with which they have been classified in section *Purpurea*, subsection *Purpureae*. However, several phylogenetic studies (e.g., Figures 5.73 and 5.74) have indicated that neither section *Purpurea* nor subsection *Purpureae* is monophyletic, and that *A. maritima* is not closely related either to *A. paupercula* or to *A. purpurea* (Figure 5.74). A fairly comprehensive assessment of interspecific relationships in *Agalinis* has been provided by a phylogenetic analysis of combined nuclear and cpDNA sequences for 29 of the 40 species, which included all four OBL taxa (Figure 5.74). That analysis indicated that the OBL habit has arisen repeatedly in the genus, in which numerous wetland indicators are scattered throughout. The molecular studies also call for a need to reevaluate the sectional classification of *Agalinis*, with a number of the traditional sections and subsections not resolved as clades as currently circumscribed. Two base chromosome number series occur in *Agalinis* with $x = 13$ conceivably being derived by a single evolutionary reduction from the $x = 14$ series (Figure 5.73). *Agalinis linifolia*, *A. maritima*, *A. paupercula* (all $2n = 28$), and *A. tenella* ($2n = 26$) are diploids. Counts are unavailable for *A. calycina*. Hybrids involving the OBL species have not been reported.

Comments: *Agalinis maritima* occurs along the coastal regions of eastern North America; *A. paupercula* is distributed in interior northeastern North America; *A. linifolia* and *A. tenella* occur primarily along the coastal plain of the southeastern United States. *Agalinis calycina* is restricted to New Mexico and Texas, but extends into Mexico.

References: Aronson, 1989; Baird & Riopel, 1985; Bartholomew et al., 2006; Baskin & Baskin, 1998; Canne, 1984; McLaughlin, 1932; Musselman & Mann, Jr., 1977; Musselman et al., 1978; Neel, 2002; Neel & Cummings, 2004; Nelson et al., 2000; Pennell, 1929, 1935; Pettengill & Neel, 2008, 2011; Poole et al., 2007; Ross et al., 2000; Sivinski, 2011; Stewart & Canne-Hilliker, 1998; Stuckey & Gould, 2000; Theodose & Roths, 1999; Van Auken et al., 2007; Vickery & Vickery, 1983; Visser et al., 1998.

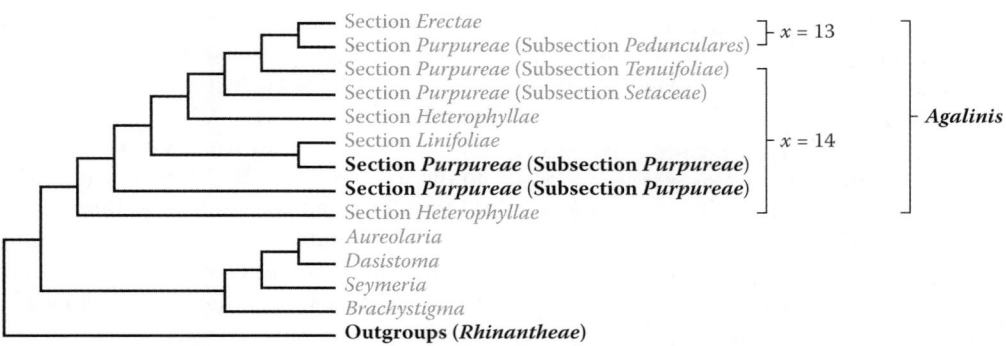

FIGURE 5.73 Relationships in *Agalinis* as indicated by phylogenetic analysis of combined cpDNA. The molecular data resolve *Agalinis* as monophyletic, but do not support the monophyly of sections *Heterophyllae* or *Purpureae*, or subsection *Purpureae*. These results indicate that the $x = 13$ chromosomal base number has been derived only once in the genus. Neither OBL North American species was included in this analysis; however, both are classified in section *Purpureae* and subsection *Purpureae* (bold), which is paraphyletic as sampled. (Adapted from Neel, M.C. & Cummings, M.P., *BMC Evol. Biol.*, 4, 15, 2004.)

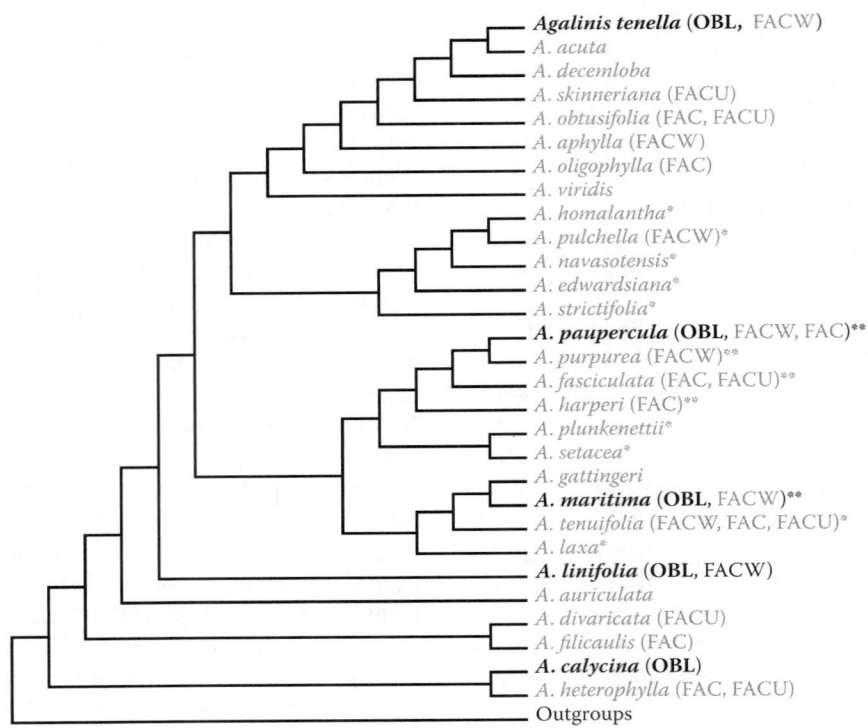

FIGURE 5.74 Interspecific relationships in *Agalinis* as reconstructed using combined nuclear and cpDNA sequence data. The distribution of the North American OBL species (bold) indicates five independent origins of the habit in the genus (other wetland indicator codes are provided in parentheses). Asterisked taxa represent those assigned to section *Purpureae*, with double asterisks denoting members of subsection *Purpureae*; neither group is monophyletic in this analysis. (Adapted from Pettengill, J.B. & Neel, M.C., *BMC Evol. Biol.*, 8(1), 264, 2008.)

2. *Castilleja*

Paintbrush, owl's-clover; castilléjie

Etymology: after Domingo Castillejo (1744–1793)

Synonyms: *Clevelandia; Euchroma; Gentrya; Ophiocephalus; Orthocarpus* (in part); *Triphysaria*

Distribution: global: Asia; Europe; New World; **North America:** widespread

Diversity: global: 180 species; **North America:** 105 species

Indicators (USA): OBL: *Castilleja minor;* **FACW:** *C. ambigua, C. campestris, C. miniata;* **FACU:** *C. miniata;* **unranked:** *C. hyetophila*

Habitat: brackish (coastal), freshwater; palustrine; **pH:** 4.3–8.0; **depth:** <1 m; **life-form(s):** emergent herb

Key morphology: stems (to 30–150 cm) erect; leaves (to 40–100), alternate, lanceolate, linear lanceolate or ovate, 0–5–lobed; inflorescence (to 12–40 cm) spike-like, the bracts (to 25–50 mm) green or colorful with tips ranging from white to yellow, red or rose purple; calyx 4-lobed, colored like the bracts; corolla (to 25–40 mm) tubular, orange, yellow, yellow green, red or rose purple, 2-lipped, the upper lip (to 5–20 mm) beak-like, the lower lip (to 5 mm) reduced to a small tridentate structure or pouch; stamens 4; styles exserted; capsules (to 15 mm) ovoid, asymmetric, dehiscence loculicidal; seeds (to 2 mm) brown, reticulate

Life history: duration: annual (fruit/seeds); perennial (rhizomes, tap roots); **asexual reproduction:** rhizomes; **pollination:** bird or insect; **sexual condition:** hermaphroditic; **fruit:** capsules (common); **local dispersal:** seeds (wind); **long-distance dispersal:** seeds (wind)

Imperilment: (1) *Castilleja ambigua* [G4]; S2 (<u>BC</u>, CA); (2) *C. campestris* [G4]; S2 (CA); (3) *C. hyetophila* [G4]; S3 (AK); (4) *C. miniata* [G5]; S1 (<u>YT</u>); S2 (CA); S3 (AK, OR); (5) *C. minor* [G5]; S1 (<u>BC</u>); S2 (AZ, MT, WY)

Ecology: general: *Castilleja* species occupy a variety of habitats ranging from dry chaparral to freshwater and saline wetlands. About 20% of the North American species are designated as wetland indicators, but only one as OBL. The 1996 wetland indicator list ranked five species as OBL indicators in at least a portion of their range. However, four of those species have since been removed from the 2013 list (*C. hyetophila*) or have been reclassified only as facultative (FACW, FAC) indicators (*C. ambigua, C. campestris, C. miniata*). Because the treatments for all five species had been completed prior to the release of the 2013 list, they have been retained here. Most *Castilleja* species are pollinated by hummingbirds (Aves: Trochilidae) with the remainder pollinated principally by short-tongued bees (Hymenoptera: Andrenidae). All species in the genus are hemiparasites, although most are not host specific but parasitize a variety of grasses (Poaceae), legumes (Fabaceae), sagebrush (*Artemisia*), and other herbaceous flowering plants. The full range of host species has not been elucidated for any of the species treated here.

***Castilleja ambigua* Hook. & Arn.** is a coastal annual, which inhabits beaches, bluffs, meadows, mudflats, prairies, saltmarshes, sloughs, strands, swales, swamps, terraces, vernal pools, and margins of ponds at elevations of up to 150 m. This species is typical of the high marsh and grows best between the mean higher tide water and high tide levels. The sites are open and sunny and categorized as alkaline, brackish or saline with substrates described as alluvium, clay, or sand. The flowers are pollinated by bees (Insecta: Hymenoptera). The maximum seed germination has been reported following 2 weeks of moist stratification at 5°C. High germination (61%) also has been achieved for seeds soaked in freshwater for 24 h followed by a 24-day stratification period. Plants will establish without a host if they are provided with supplemental water. Documented host plants include *Plantago maritima*, *Salicornia*, and *Spergularia macrotheca*. **Reported associates:** *Acmispon americanus*, *Arctostaphylos*, *Armeria maritima*, *Atriplex patula*, *Baccharis*, *Briza minor*, *Bolboschoenus maritimus*, *Carex lyngbyei*, *Claytonia perfoliata*, *Convolvulus*, *Cordylanthus maritimus*, *C. mollis*, *Corethrogyne californica*, *Cuscuta salina*, *Cynosurus echinatus*, *Dasiphora floribunda*, *Distichlis spicata*, *Eleocharis*, *Epilobium brachycarpum*, *Eriogonum latifolium*, *Euthamia occidentalis*, *Festuca rubra*, *Fragaria chiloensis*, *Galium trifidum*, *Geranium molle*, *Grindelia integrifolia*, *Helenium bolanderi*, *Holcus lanatus*, *Hosackia gracilis*, *Hypochaeris radicata*, *Iris douglasii*, *Isoetes nuttallii*, *Isolepis cernua*, *Jaumea carnosa*, *Juncus balticus*, *J. lesueurii*, *J. mexicana*, *Lepidium latifolium*, *Limnanthes macounii*, *Limonium californicum*, *Lolium perenne*, *Lycopus asper*, *Lysimachia maritima*, *Lythrum californicum*, *Mentha arvensis*, *Mimulus guttatus*, *Orthocarpus bracteosus*, *Parapholis incurva*, *Persicaria punctata*, *Plantago elongata*, *P. lanceolata*, *P. maritima*, *Pluchea odorata*, *Poa annua*, *Polygonum marinense*, *Potentilla anserina*, *Primula pauciflora*, *Prunella vulgaris*, *Psilocarphus elatior*, *Puccinellia pumila*, *Rosa nutkana*, *Rumex*, *Salicornia depressa*, *Salix sitchensis*, *Sarcocornia pacifica*, *Schoenoplectus acutus*, *S. pungens*, *Sedum lanceolatum*, *Sisyrinchium bellum*, *Sium*, *Spartina densiflora*, *S. foliosa*, *Spergularia canadensis*, *S. macrotheca*, *S. rubra*, *Symphyotrichum lentum*, *Trifolium*, *Triglochin maritimum*, *Triphysaria versicolor*, *Typha angustifolia*, *T. latifolia*, *Vulpia bromoides*.

Castilleja campestris (Benth.) **T.I. Chuang & Heckard** is an annual, which grows in depressions, ditches, meadows, vernal pools, swales, and along lake margins at elevations of up to 2300 m. This species usually is dominant in large (0.01–0.05 hectares) but shallow vernal pools where the plants can occur on exposed substrates or in shallow standing water (to 25 cm). The substrates tend to be acidic (pH: 4.3–6.2) and are described as adobe, clay, gravelly loam, gravelly sandy loam, mud, sandy loam, or vertisols. The reproductive biology of this species needs to be clarified; however, circumstantial evidence (e.g., diffuse spacing of individuals in populations) has led some to conclude that the species might largely be self-pollinating. FAC outcrossing by bees (Insecta: Hymenoptera) is also believed to occur but pollinators rarely have been observed at field sites. Conditions for seed germination have not been elucidated. Because of its vernal pool habitat, there

has been some suggestion that this species is not parasitic; however, a detailed investigation of its potential host plants remains to be conducted. **Reported associates:** *Alopecurus howellii*, *Cuscuta howelliana*, *Deschampsia danthonioides*, *Downingia bella*, *D. bicolor*, *D. bicornuta*, *D. cuspidata*, *D. ornatissima*, *D. pulchella*, *Elatine californica*, *E. chilensis*, *Eleocharis macrostachya*, *Epilobium torreyi*, *Eryngium aristulatum*, *E. spinosepalum*, *E. vaseyi*, *Gnaphalium palustre*, *Gratiola heterosepala*, *Juncus bufonius*, *J. hemiendytus*, *J. leiospermus*, *J. uncialis*, *Lasthenia fremontii*, *L. glaberrima*, *Legenere limosa*, *Leontodon taraxacoides*, *Limnanthes alba*, *Limosella aquatica*, *Linanthus*, *Mimulus guttatus*, *M. tricolor*, *Navarretia intertexta*, *N. leucocephala*, *N. myersii*, *Neostapfia colusana*, *Orcuttia inaequalis*, *O. pilosa*, *Pilularia americana*, *Plagiobothrys greenei*, *P. leptocladus*, *P. scouleri*, *P. stipitatus*, *P. undulatus*, *Pleuropogon californicus*, *Pogogyne ziziphoroides*, *Psilocarphus brevissimus*, *P. oregonus*, *Sidalcea*, *Tillaea aquatica*, *Veronica arvensis*, *V. peregrina*.

Castilleja hyetophila **Pennell** is a perennial found in tidal coastal meadows and saltmarshes. Virtually no ecological information exists for this species; however, it is quite similar to, and doubtfully distinct from, *C. miniata* (see *Systematics* later). **Reported associates:** none.

Castilleja miniata **Douglas ex Hook.** is a perennial, which occurs on beaches, in bogs, depressions, ditches, dunes, floodplains, glades, marshes, montane meadows, seeps, swales, swamps, and along the margins of lakes, rivers, and streams at elevations of up to 3500 m. It inhabits diverse sites ranging from alpine tundra to estuaries. The substrates are circumneutral (pH: 6.1–7.8) and include basaltic alluvium, clay loam, cobble, decomposed granite, gravel, humus loam, lava outcrops, limestone, loam, peat, rock, rocky sand, sand, sandy loam, serpentine, silt, and silty clay–loam. The plants occur in exposures ranging from shade to full sun. The flowers are almost entirely self-incompatible. Each plant produces an average of 4–5 inflorescences, with 13–14 fruits per inflorescence and 65 seeds per fruit. The flowers are pollinated by several species of hummingbirds (see *Use by wildlife*, later). Although the plants primarily are outcrossed, outbreeding depression (approximately 70%) has been detected in seed set for crosses made over 30 m distances relative to two meter distances. The seeds become dormant almost immediately when they mature and require no chilling to induce dormancy. The seeds often require between 14 and 16 weeks (sometimes up to 6 months) of chilling for germination. However, seed germination has been achieved following 3 months of moist stratification at 2°C–5°C, and good germination (14.7%) has been obtained after 4 weeks at 2°C followed by incubation for 4 weeks under a 10°C/20°C temperature regime. *Castilleja miniata* is known to be parasitic on the roots of *Alnus rubra*, *Artemisia californica* and *Lupinus argenteus*. Herbivory of *C. miniata* is reduced when growing in the presence of the latter species as a result of several quinolizidine alkaloids, which are transferred from the host plant to the foliage (but not to the nectar). Although *C. miniata* can establish without a host plant, it grows much better with a host. Experiments have shown that leguminous hosts (*Medicago sativa*) are more beneficial to the plants than are grass hosts (*Lolium perenne*). The plants are on average 47% larger and produce more than twice as many seeds when parasitizing *Lupinus* species (legumes) than other hosts. *Castilleja miniata* is regarded as fire resistant due to its long tap root, which can split and grow into cracks within the substrate. This species is quite widely distributed and grows in association with an impressive diversity of species. **Reported associates:** *Abies amabilis, A. concolor, A. grandis, A. lasiocarpa, A. magnifica, A. procera, Abronia latifolia, Acer glabrum, Achillea lanulosa, A. millefolium, Achnatherum lettermanii, Aconitum columbianum, Aconogonon davisiae, Agastache urticifolia, Agoseris aurantiaca, A. glauca, Agropyron dasystachyum, A. spicatum, A. trachycaulum, Agrostis diegoensis, A. exarata, A. scabra, A. stolonifera, A. variabilis, Allium cernuum, A. schoenoprasum, A. validum, Alnus rhombifolia, A. rugosa, A. sinuata, A. tenuifolia, A. viridus, Alopecurus, Amelanchier alnifolia, Anaphalis margaritacea, Anemone multifida, A. occidentalis, Angelica breweri, A. grayi, A. arguta, Antennaria alpina, A. lanata, Anticlea elegans, Apocynum androsaemifolium, A. cannabinum, Aquilegia caerulea, A. flavescens, A. formosa, Arabis glabra, A. drummondii, A. furcata, A. hirsuta, A. holboellii, A. sparsiflora, Arctostaphylos nevadensis, A. patula, A. uva-ursi, Arenaria capillaris, A. rubella, Arnica ×diversifolia, A. latifolia, A. longifolia, A. mollis, Artemisia californica, A. dracunculus, A. ludoviciana, A. rothrockii, A. tridentata, Aruncus sylvester, Asarum wagneri, Astragalus tenellus, A. vexilliflexus, Athyrium filix-femina, Atriplex confertifolia, Baccharis salicifolia, Balsamorhiza sagittata, Betula occidentalis, Bistorta bistortoides, Brachythecium, Bromopsis suksdorfii, Bromus carinatus, B. polyanthus, Calamagrostis breweri, C. canadensis, C. rubescens, Calocedrus decurrens, Calochortus invenustus, C. subalpinus, Caltha leptosepala, Camassia quamash, Campanula rotundifolia, Cardamine breweri, Carex angustata, C. aquatilis, C. buxbaumii, C. fracta, C. geyeri, C. hassei, C. heteroneura, C. hoodii, C. illota, C. jonesii, C. lanuginosa, C. lasiocarpa, C. lenticularis, C. limnophila, C. limosa, C. macloviana, C. mertensiana, C. microptera, C. nigricans, C. pellita, C. podocarpa, C. rostrata, C. scirpoidea, C. scopulorum, C. senta, C. spectabilis, C. straminiformis, C. subfusca, C. tenuiflora, C. utriculata, Castilleja fraterna, C. sulphurea, Ceanothus cordulatus, C. cuneatus, C. integerrimus, C. leucodermis, C. sanguineus, C. velutinus, Ceratoides lanata, Cercocarpus ledifolius, Chaenactis douglasii, Chimaphila umbellata, Chrysothamnus viscidiflorus, Cicuta douglasii, Cirsium drummondii, C. hookerianum, C. undulatum, Clematis occidentalis, Clintonia uniflora, Collinsia tenella, Collomia grandiflora, C. linearis, Comarum palustre, Conioselinum, Conium maculatum, Cornus, Crataegus rivularis, Crepis modocensis, Cupressus sargentii, Dactylis glomerata, Darlingtonia californica, Dasiphora floribunda, Delphinium barbeyi, D. glaucum, D. nuttallianum, Deschampsia atropurpurea, D. cespitosa, Dicranoweisia crispula, Dicranum, Distichlis, Draba aurea, Drosera rotundifolia, Eleocharis, Elymus elymoides, E. glaucus, E. lanceolatus, E. trachycaulus, Ephedra, Epilobium alpinum, E.* *anagallidifolium, E. angustifolium, E. glaberrimum, Epipactis gigantea, Equisetum arvense, E. fluviatile, E. laevigatum, E. variegatum, Erigeron aureus, E. coulteri, E. peregrinus, E. speciosus, Eriogonum flavum, E. wrightii, Eriogonum microthecum, E. nudum, E. pyrolifolium, E. strictum, E. umbellatum, Eriophorum crinigerum, E. polystachion, Erysimum capitatum, Erythronium grandiflorum, Eucephalus elegans, E. ledophyllus, Eurybia conspicua, E. integrifolia, Festuca idahoensis, F. rubra, F. thurberi, F. viridula, Fragaria chiloensis, F. virginiana, Frasera speciosa, Galium boreale, G. oreganum, G. triflorum, Gentiana calycosa, G. newberryi, Gentianella amarella, Gentianopsis simplex, Geranium richardsonii, G. viscosissimum, Geum triflorum, Gilia aggregata, Glyceria elata, Glycosma occidentalis, Gnaphalium chilense, Grayia, Grindelia, Hackelia californica, H. floribunda, Hedysarum sulphurescens, Helenium bigelovii, H. decurrens, Helianthella quinquenervis, Helianthus uniflorus, Heracleum lanatum, H. maximum, Heuchera cylindrica, Hieracium albiflorum, H. gracile, Holodiscus discolor, Hordeum brachyantherum, Horkelia rydbergii, Hosackia oblongifolia, Hydrophyllum capitatum, H. fendleri, Hypericum anagalloides, Hypochaeris radicata, Iris missouriensis, Juncus balticus, J. drummondii, J. effusus, J. mertensianus, J. parryi, J. xiphioides, Juniperus communis, J. occidentalis, Kalmia microphylla, K. polifolia, Koeleria macrantha, Lathyrus japonicus, L. nevadensis, Leymus cinereus, L. mollis, Ligusticum grayi, Lilium columbianum, Linanthus nuttallii, Linum lewisii, L. perenne, Listera cauriana, Lomatium dissectum, L. martindalei, L. triternatum, Lonicera involucrata, Luetkea pectinata, Luina stricta, Lupinus andersonii, L. arcticus, L. argenteus, L. latifolius, L. lepidus, L. polyphyllus, L. sericeus, Luzula comosa, L. hitchcockii, L. parviflora, Lycium, Machaeranthera canescens, Madia elegans, M. glomerata, Mahonia repens, Maianthemum racemosum, M. stellatum, Marchantia, Medicago lupulina, Melica bulbosa, M. spectabilis, Melilotus officinalis, Menyanthes trifoliata, Mertensia ciliata, Micranthes aprica, M. ferruginea, M. lyallii, M. nelsoniana, M. nidifica, Microseris alpestris, Mimulus cardinalis, M. guttatus, M. lewisii, M. primuloides, Mitella breweri, M. pentandra, Monarda fistulosa, Monardella odoratissima, Muhlenbergia filiformis, M. richardsonis, Myosotis scorpioides, Oenothera, Oreostemma alpigenum, Orobanche uniflora, Oryza, Osmorhiza occidentalis, Parnassia fimbriata, Paxistima myrsinites, Pedicularis bracteosa, P. contorta, P. groenlandica, P. ornithorhyncha, Penstemon albidus, P. attenuatus, P. cardwellii, P. confertus, P. davidsonii, P. fruticosus, P. humilis, P. leiophyllus, P. lyallii, P. procerus, P. rydbergii, Penstemon rupicola, P. shastensis, P. whippleanus, Perideridia gairdneri, P. parishii, Phacelia hastata, P. sericea, Phleum alpinum, P. pratense, Phlox diffusa, P. longifolia, Phyllodoce breweri, P. empetriformis, Picea engelmannii, Picrothamnus desertorum, Pinus albicaulis, P. balfouriana, P. contorta, P. flexilis, P. jeffreyi, P. monophylla, P. monticola, Plantago tweedyi, Platanthera dilatata, P. hyperborea, P. sparsiflora, P. stricta, Platanus racemosa, Poa alpina, P. compressa, P. cusickii, P. fendleriana, P. palustris, P. pratensis, P. secunda, P. wheeleri, Polemonium californicum,*

P. pulcherrimum, Polygonum douglasii, Polystichum lonchitis, Populus angustifolia, P. balsamifera, P. fremontii, P. tremuloides, P. trichocarpa, Potentilla diversifolia, P. drummondii, P. egedii, P. flabellifolia, P. glandulosa, P. gracilis, Primula jeffreyi, P. tetrandra, Prunella vulgaris, Prunus emarginata, P. fasciculata, P. virginiana, Pseudocymopterus montanus, Pseudoroegneria spicata, Pseudotsuga menziesii, Pteridium aquilinum, Purshia tridentata, Pycnanthemum californicum, Pyrola asarifolia, Pyrrocoma lanceolata, Ranunculus eschscholtzii, Rhododendron albiflorum, R. neoglandulosum, R. occidentale, Ribes bracteosum, R. cereum, R. howellii, R. lacustre, R. nevadense, Rosa woodsii, Rubus lasiococcus, R. parviflorus, R. pedatus, R. spectabilis, Salix arctica, S. barclayi, S. brachystachys, S. commutata, S. drummondiana, S. eastwoodiae, S. exigua, S. farriae, S. glauca, S. lasiolepis, S. lemmonii, S. ligulifolia, S. lucida, S. lutea, S. nivalis, Saussurea americana, Saxifraga bronchialis, S. oppositifolia, Scirpus congdonii, Scutellaria, Sedum divergens, S. lanceolatum, Selaginella densa, S. scopulorum, Senecio canus, S. cymbalarioides, S. crassulus, S. integerrimus, S. megacephalus, S. pseudaureus, S. serra, S. triangularis, Shepherdia canadensis, Sibbaldia procumbens, Sidalcea oregana, Silene parryi, Sisyrinchium bellum, S. idahoense, Sitanion hystrix, Solidago canadensis, S. multiradiata, Sorbus scopulina, S. sitchensis, Sphaeralcea, Sphenosciadium capitellatum, Spiraea betulifolia, S. densiflora, S. douglasii, Spiranthes porrifolia, Spiranthes romanzoffiana, Spraguea umbellata, Stachys albens, Stellaria crispa, S. jamesiana, S. longipes, Stenanthium occidentale, Stipa, Symphoricarpos albus, S. oreophilus, S. rotundifolius, Symphyotrichum foliaceum, S. laeve, S. spathulatum, Synthyris pinnatifida, Taraxacum officinale, Tetradymia axillaris, Thalictrum fendleri, T. occidentale, Thermopsis montana, Tragopogon porrifolius, Triantha glutinosa, Trientalis arctica, Trifolium latifolium, T. longipes, Trisetum spicatum, Triteleia hyacinthina, Tsuga mertensiana, Urtica dioica, Vaccinium deliciosum, V. globulare, V. membranaceum, V. nivictum, V. scoparium, Valeriana dioica, V. occidentalis, V. sitchensis, Veratrum californicum, V. viride, Veronica cusickii, V. serpyllifolia, V. wormskjoldii, Vicia americana, Viola macloskeyi, V. nuttallii, V. pallens, Wyethia amplexicaulis, W. ×magna, Xerophyllum tenax, Yucca brevifolia.

***Castilleja minor* (A. Gray) A. Gray** is an annual found in ditches, flats, floodplains, marshes, meadows, pools, seeps, swales, swamps, thickets, and washes or along the margins of canals, lakes, ponds, rivers, springs, and streams at elevations of up to 2560 m. The plants ordinarily occur on alkaline (pH: 6.9–8.0) or saline sites that often have ephemeral water or are in the process of drying out. The substrates are derived from limestone (occasionally from granite, schist, or serpentine) and are described as boulders, clay loam, gravel, limestone, loam, muck, rock, sand, sandy clay, sandy gravel, sandy loam, silt, and travertine. The plants tolerate exposures from shade to full sun. The flowers are pollinated by hummingbirds (Aves: Trochilidae). The seeds require a 30-day period of cold stratification (3°C) for germination, which occurs well while under a 21°C–25°C/10°C–16°C temperature

regime. Seeds collected from cooler sites germinate better at relatively cooler temperatures. The plants grow rapidly and will begin to produce flowers in about 15 weeks after germination. This species is known to be parasitic on *Packera clevelandii* but also has been established successfully using *Achnatherum hymenoides, Festuca idahoensis,* and *Penstemon pinifolius* as hosts. **Reported associates:** *Abies concolor, Acmispon heermannii, Adenostoma fasciculatum, A. sparsifolium, Agrostis exarata, A. gigantea, A. stolonifera, Allophyllum divaricatum, Amorpha californica, Antennaria microphylla, Aquilegia, Arctostaphylos glandulosa, A. glauca, A. pungens, Artemisia douglasiana, A. ludoviciana, A. tridentata, Asclepias speciosa, Atriplex patula, A. truncata, Baccharis glutinosa, B. salicifolia, B. viminea, Berula erecta, Bolboschoenus robustus, Bromus japonicus, B. tectorum, Bolboschoenus, Calamagrostis stricta, Calocedrus decurrens, Carex alma, C. hassei, C. hystericina, C. nebrascensis, C. pellita, C. scirpoidea, C. simulata, Ceanothus leucodermis, Centaurium exaltatum, C. venustum, Cercocarpus betuloides, Cicuta douglasii, Cirsium arvense, Cupressus, Dasiphora floribunda, Deschampsia cespitosa, D. elongata, Dipsacus fullonum, Distichlis spicata, Dryopteris arguta, Elaeagnus angustifolia, Eleocharis palustris, E. parishii, E. rostellata, Elymus caninus, E. trachycaulus, Epilobium adenocaulon, Epipactis gigantea, Erigeron philadelphicus, Eriodictyon trichocalyx, Eriogonum fasciculatum, Euthamia occidentalis, Fraxinus velutina, Glycyrrhiza lepidota, Helianthus nuttallii, Hordeum jubatum, Hosackia oblongifolia, Hypericum formosum, Juncus balticus, J. rugulosus, J. torreyi, J. xiphioides, Kochia scoparia, Lemna minor, Lepidospartum squamatum, Leymus cinereus, Lupinus, Lysimachia maritima, Machaeranthera canescens, Melilotus albus, Mentha spicata, Mimulus cardinalis, M. guttatus, M. pilosus, Muhlenbergia asperifolia, Nasturtium officinale, Oenothera hookeri, Packera clevelandii, Panicum pacificum, Penstemon centranthifolius, P. clevelandii, P. spectabilis, Persicaria lapathifolia, Phoradendron, Pinus jeffreyi, Platanus racemosa, Polypogon monspeliensis, Populus fremontii, Potentilla anserina, P. pensylvanica, Pseudotsuga macrocarpa, Pteridium aquilinum, Puccinellia nuttalliana, Pycnanthemum californicum, Pyrrocoma uniflora, Ranunculus cymbalaria, Rhamnus tomentellus, Quercus agrifolia, Q. kelloggii, Ribes, Rosa woodsii, Rubus ursinus, Salicornia rubra, Salix exigua, S. laevigata, S. lasiandra, S. lasiolepis, Sarcobatus vermiculatus, Schoenoplectus acutus, S. americanus, S. pungens, Scirpus saximontanus, Solidago, Sonchus asper, Spartina gracilis, Spiranthes diluvialis, Stipa, Symphyotrichum ascendens, Thelypodium integrifolium, Toxicodendron diversilobum, Trichostema austromontanum, Triglochin maritimum, Typha angustifolia, T. latifolia, Urtica, Verbena, Vitis girdiana, Zeltnera exaltata.*

Use by wildlife: *Castilleja miniata* is grazed by cows (Mammalia: Bovidae: *Bos taurus*), horses (Mammalia: Equidae: *Equus ferus*), and sheep (Mammalia: Bovidae: *Ovis aries*), and the fruits are eaten by marmots (Mammalia: Sciuridae: *Marmota vancouverensis*). The flowers and foliage are eaten by elk (Mammalia: Cervidae:

Cervus canadensis) and mule deer (Mammalia: Cervidae: *Odocoileus hemionus*). It is a larval host plant for several checkerspot butterflies (Insecta: Lepidoptera: Nymphalidae: *Euphydryas anicia*; *E. editha augusta*; *E. gillettii*). The seeds are eaten by flies (Insecta: Diptera: Agromyzidae: *Phytomyza subtenella*) and plume moths (Insecta: Lepidoptera: Pterophoridae: *Amblyptilia pica*). *Castilleja miniata* is associated with a variety of fungi (Ascomycota: Diademaceae: *Clathrospora permunda*; Erysiphaceae: *Sphaerotheca fuliginea*; Mycosphaerellaceae: *Mycosphaerella*, *Ramularia castilleyae*; Phaeosphaeriaceae: *Leptosphaeria concinna*; Pleosporaceae: *Lewia scrophulariae*, *Pleospora comata*, *P. njegusensis*; Sclerotiniaceae: *Botrytis cinerea*; family uncertain: *Didymella castillejae*; Basidiomycota: Cronartiaceae:

Cronartium coleosporioides; Entylomataceae: *Entyloma*). It is an alternate host of the white pine blister rust (Basidiomycota: Cronartiaceae: *Cronartium ribicola*). *Castilleja miniata* is a major food plant (nectar source) for several hummingbirds (Aves: Trochilidae) including broadtailed hummingbirds (*Selasphorus platycercus*), calliope hummingbirds (*Stellula* calliope), and rufous hummingbirds (*Selasphorus rufus*). The flowers contain a large amount of nectar (about 1.9 μL) with a sugar content of roughly 36%. These birds are known to feed nearly exclusively on the nectar of *C. miniata*, even when other suitable species are present. The different species avoid competition by feeding on flowers at different levels. *Castilleja minor* is grazed by deer (Mammalia: Cervidae: *Odocoileus*).

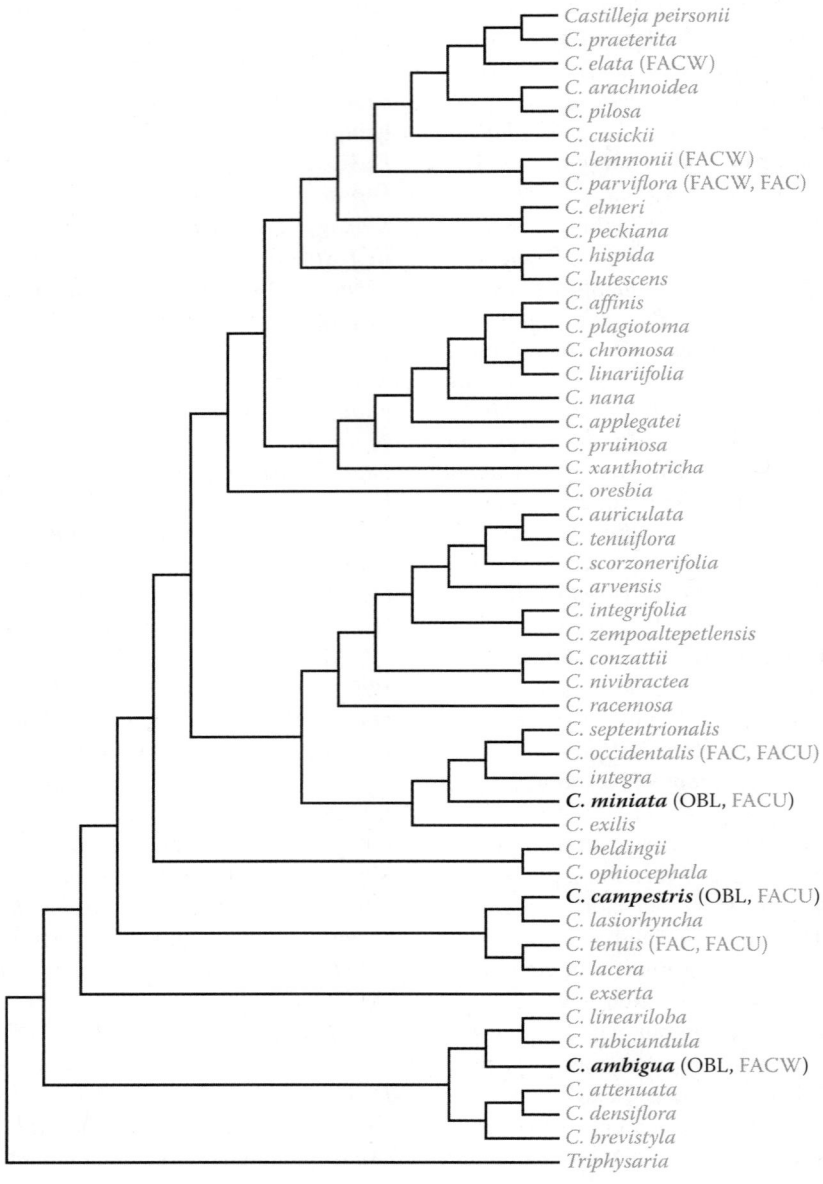

FIGURE 5.75 Phylogenetic relationships in *Castilleja* as indicated by analysis of combined DNA sequence data. Several independent origins of the OBL taxa (bold) are indicated by these phylogenetic relationships. (Adapted from Tank, D.C. & Olmstead, R.G., *Amer. J. Bot.*, 95, 608–625, 2008; nomenclature updated in accordance with Tank, D.C. et al., *Syst. Bot.*, 34, 182–197, 2009.)

Economic importance: food: Children of the Kwakiutl, Nitinaht, and Southern tribes sipped nectar from the flowers of *Castilleja miniata* as a candy-like treat; **medicinal:** The Gitksan tribe used a whole-plant decoction of *Castilleja miniata* as a diuretic, a purgative, to stop bleeding and to treat sore backs, eyes, and lungs. They also used a seed decoction as a cough suppressant; **cultivation:** *Castilleja miniata* is cultivated infrequently as an ornamental due to difficulties of establishing it with an appropriate host; **misc. products:** The Apache and White Mountain tribes made a dye from the root bark of *Castilleja miniata* and *C. minor* to color various animal skins. Bella Coola girls used the flowers of *C. miniata* in games. The Thompson tribe used *C. miniata* as a household decoration and to feed hummingbirds. The Nitinaht covered its flowers with snail slime as a means of trapping hummingbirds. Women of the Keres and Western tribes would hold the plants during ceremonial dances. It also was used by the Navajo and Ramah people as a medicine to protect hunters from malevolent spirits; **weeds:** none; **nonindigenous species:** none

Systematics: Some phylogenetic analyses of molecular data show *Castilleja* to be related most closely to *Cordylanthus* and *Orthocarpus* (Figure 5.72), which both are comprised entirely of annual species. However, studies that evaluate combined molecular data resolve *Castilleja* as paraphyletic unless circumscribed to include *Clevelandia* and *Ophiocephalus*, with *Triphysaria* placed as the immediate sister group (Figure 5.75). Molecular analyses also confirm that the monotypic *Gentrya* (once regarded as a distinct genus) is nested within *Castilleja*. Three subgenera have been recognized within *Castilleja*: *Castilleja*, *Colacus*, and *Gentrya*. Subgenera *Colacus* and *Gentrya* contain bee-pollinated species; whereas, subg. *Castilleja* is predominantly bird-pollinated. Analysis of molecular data demonstrate that subgenus *Colacus*, together with representatives of *Clevelandia* and *Ophiocephalus*, forms a basal grade of annual species. Subgenus *Colacus* is divided into three sections: *Oncorhynchus* (which contains *C. ambigua* and *C. campestris*), *Pilosae*, and *Pallescentes*. Subgenus *Castilleja* contains only a few annuals (e.g., *C. minor*) and groups the perennial taxa (e.g., *C. hyetophila*, *C. miniata*) within a single clade. Thus, the perennial habit appears to be derived in the genus. *Castilleja miniata* is believed to be closely related to *C. kaibabensis*, an open meadow species. Some authors have recognized *C. uliginosa* as distinct; however, it generally is regarded as a synonym of *C. miniata* subsp. *miniata*. In any event, this taxon was confined to two localities and the last known remaining individual reportedly died in 1987 despite efforts to conserve the plants. *Castilleja hyetophila* has been distinguished from *C. miniata* based on relatively minor characteristics (e.g., the shape of the bract apex), which are inconsistent. Some authors have merged this species with *C. miniata* and its taxonomic distinctness warrants further investigation (it has not been included in the 2013 wetland indicator list). Initially, *C. minor* was part of a complex containing *C. exilis*, *C. spiralis* and *C. stenantha*; however, recent treatments regard *C. exilis* as a subspecies or variety of *C. minor*, and *C.*

stenantha as a synonym for *C. minor* subsp. *spiralis* (=*C. spiralis*). The taxonomic difficulties in *Castilleja* are exacerbated by confusing patterns of morphological variation attributable to natural hybridization and polyploidy (especially in subg. *Castilleja*). Roughly half of the surveyed species of *Castilleja* (with a chromosomal base number of $x = 12$) are polyploids. *Castilleja ambigua*, *C. campestris*, and *C. minor* are diploid ($2n = 24$). *Castilleja miniata* contains a breadth of euploid cytotypes ($2n = 24, 48, 72, 96, 120, 144$) as well as some anomalous counts ($2n = 18$). Hybridization in *Castilleja* may be facilitated by the fact that reduced pollinator specificity has been observed at sites where different species coexist with morphologically intermediate plants. Although many taxa in *Castilleja* are thought to be allopolyploid derivatives, chromosomal, genetic, and morphological evidence suggests that homoploid reticulation also can explain existing patterns of interspecific variation in the genus. *Castilleja miniata* reportedly hybridizes with *C. crista-galli*, *C. hispida*, *C. peckiana*, *C. purpurascens*, *C. rhexijolia*, and *C. sulphurea*. *Castilleja peckiana* has been regarded as a morphologically intermediate hybrid involving *C. miniata* and *C. hispida*. It also has been hypothesized that *C. hispida* and *C. peckiana* are alloploids containing the genome of *C. miniata*. *Castilleja crista-galli* was thought to be an allopolyploid hybrid derivative of *C. linariifolia* and *C. miniata*; however, morphological and cpDNA RFLP data analysis does not support that hypothesis. The derivation of *C. crista-galli* as a variant of *C. miniata* resulting from introgression with *C. chromosa* is supported phenetically. It is apparent that additional studies of hybridization, particularly those incorporating genetic markers, would be quite helpful in interpreting the taxonomy and evolution of the genus.

Comments: Three species (*C. ambigua*, *C. miniata*, *C. minor*) occur generally throughout western North America. *Castilleja hyetophila* occurs only in Alaska and British Columbia and *C. campestris* is restricted to California and Oregon.

References: Adler, 2002; Adler & Wink, 2001; Bolen, 1964; Borland, 1994; Campbell et al., 1997; Chuang & Heckard, 1991, 1992a; Goman, 2001; Grant & Grant, 1968; Heckard, 1968; Heckard & Chuang, 1977; Heimbinder, 2001; Hersch & Roy, 2007; Hiratsuka & Maruyama, 1976; Holmgren, 1973; Lawrence, 2005; Lawrence & Kaye, 2005; Lertzman, 1981; Luna, 2005; Manson, 2005; Martin, 1988; Mathews & Lavin, 1998; Matthies, 1997; McCoy & Stermitz, 1983; McDonald, 2006; Meyer & Carlson, 2004; Nagorsen, 1987; Platenkamp, 1998; Skinner, 1928; Smith, 1929; St. John & Courtney, 1924; Tank, 2002, 2006; Welsh, 1974; Wilkins, 1957; Witham, 2006; Zambino et al., 2005; Zika, 1996.

3. *Chloropyron*

Bird's-beak

Etymology: from the Greek *chloro pyro* meaning "green wheat" in reference to the appearance of the plants

Synonyms: *Adenostegia*; *Cordylanthis* (in part); *Dicranostegia*

Distribution: global: North America; **North America:** western

Diversity: global: 18 species; **North America:** 18 species

Indicators (USA): OBL: *Chloropyron maritimum*, *C. molle*, *C. palmatum*

Habitat: brackish, saline; palustrine; **pH:** 7.0–9.8; **depth:** <1 m; **life-form(s):** emergent herb

Key morphology: Stems (to 60 cm) erect, herbage grayish-green to glaucous, marked with red or purple; leaves (to 2.5 cm) entire, oblong to lanceolate; spikes (to 15 cm) subtended by a tongue-like bract, the floral bracts (to 3 cm) boat-shaped, lanceolate, 3–7-lobed apically; calyx (to 2.5 cm) spathe-like; corolla (to 2.5 cm) tubular, club-shaped, the throat inflated (to 5 mm), creamy or white, 2-lipped, corolla lobes (to 7 mm), greenish, yellow, or reddish purple; stamens 2 or 4; capsules (to 10 mm) oblong, many-seeded (to 40); seeds (to 3 mm) reticulate

Life history: duration: annual (fruit/seeds); **asexual reproduction:** none; **pollination:** insect; **sexual condition:** hermaphroditic; **fruit:** capsules (common); **local dispersal:** seeds (water, wind); **long-distance dispersal:** seeds (water, wind)

Imperilment: (1) *Chloropyron maritimum* [G4]; S2 (CA, OR, UT); (2) *C. molle* [G2]; S1 (CA); S2 (CA); (3) *C. palmatum* [G1]; S1 (CA)

Ecology: general: *Chloropyron* is diverse ecologically, with species inhabiting chaparral, desert scrub, woodlands, and serpentine outcrops. Many of the species are well adapted to arid conditions. Four species are designated as wetland indicators with three ranked as OBL aquatics. All of the OBL taxa occur in the halophytic subgenus *Hemistegia. Chloropyron* is hemiparasitic with fairly nonspecific hosts. All three OBL species have been reared successfully using the common sunflower (Asteraceae: *Helianthus annuus*) as a host. The plants are provided with salt secreting glands, and exuded salt crystals often can be seen on their leaves and floral bracts.

Chloropyron maritimum **Nutt. ex Benth.** inhabits depressions, ditches, dunes, flats, hot springs, meadows, salt marshes, seeps, sloughs, swales, and the margins of saline lakes and streams at elevations of up to 1676 m. The habitats typically are brackish or saline (5–132 ppt NaCl) and are quite alkaline (pH: 7.0–9.8) with substrates principally consisting of sandy loam (63%–94% sand). This species usually occurs in sites with periodic tidal inundation, where it occupies the zone between the mean and extreme high water levels. It is intolerant to prolonged inundation and also occurs in nontidal sites. Despite being fairly salinity tolerant, the plants tend to occur at lower salinity sites in the field due to interactions with soil moisture. They often grow at low elevations (e.g., 20 cm) in the salt marsh, where episodes of tidal inundation are frequent; however, they also can be prevalent on well-drained soils (5%–33% moisture content). The flowers are unusual for the genus in having four (rather than two) functional stamens. Contrary to some reports, the flowers are self-compatible and can achieve high seed set (>70%) when they are hand pollinated. Yet, the flowers are only partially autogamous with a spontaneous seed set of only 3%–4%. A small proportion of the flowers can mature their anthers and stigmas simultaneously while both organs are enclosed within the corolla, arguably as a means of facilitating self-pollination when pollinators are scarce. High sediment nitrogen levels can result in large increases in plant biomass and flower production.

The principal pollinators are large-bodied bees (Insecta: Hymenoptera) that are capable of making physical contact with the ventrally protruding stigmas as they enter the flower to forage. These include honeybees (Apidae: *Bombus californicus, B. crotchii, B. sonorus; Melissodes tepida*) and leafcutter bees (Megachilidae: *Anthidium edwardsii, A. palliventre*). Less effective (smaller) pollinators include plasterer bees (Colletidae: *Colletes*) and sweat bees (Halictidae: *Dialictus*) as well as flies (Diptera: Bombyliidae) and moths (Lepidoptera: Pyralidae: *Lipographis fenestrella*). Populations can become pollinator limited and the seed set in some populations increased by 52%–89% when the flowers were hand pollinated. Fruit set also can be limited by low sediment nitrogen levels. Capsules each mature an average of about seven seeds and single plants typically produce from 15 to 30 seeds. The annual seed production can vary substantially (e.g., 75,000 vs. 750,000 seeds) in some populations. Natural seed germination occurs from February to March. Fresh seeds can germinate immediately if they are well-illuminated. Their germination is inhibited at temperatures above 27°C. Seeds stored dry (at 16°C–37°C) will remain viable for at least 14 months and will germinate well after being exposed to a brief cold period (15 days at 5°C). Seed germination is highly influenced by soil moisture and the germination speed by salinity levels. Under laboratory conditions, germination is high (70%–80%) at salinities from 2 to 23 ppt when soil moisture is high, but can drop to under 10% when salinity is high (23 ppt) and soil moisture is low (37%). The speed of germination also slows at higher salinity, especially when soil moisture is low. Seeds soaked in saltwater (44 ppt NaCl) and cold stratified (4°C) for 14 days showed fairly high germination rates (59%). Germination in fresh water is also good (40% after 2 weeks), which indicates that rainfall could substantially influence germination levels in the field. The longevity of seed bank propagules has not been determined. The seeds are dispersed by water. Drier soils will impede seedling establishment. The seedlings can experience a mortality rate of more than 50% by the end of the summer. An isozyme survey of 21 loci for four coastal saltmarsh populations revealed low levels of genetic variation (mean number of alleles per locus: 1.00–1.14; polymorphic loci: 0%–14.3%; observed heterozygosity: 0%–2%), which was expressed primarily as rare alleles at a few loci. It would be interesting to obtain comparable data from the more inland sites where the plants are more abundant. The plants are associated with sparse to moderate cover and respond positively to disturbances that reduce vegetative cover. This hemiparasite has been grown successfully using *Distichlis spicata* (its principal natural host), *Frankenia grandifolia, Monanthochloe littoralis, Salicornia depressa*, and *Schoenoplectus subterminalis* as hosts. The hosts are believed to be especially important by providing water during dry periods. Flower production was higher (13 vs. 1/plant) when plants were grown in association with a native grass (e.g., *Distichlis spicata*) than with a nonindigenous grass (*Parapholis incurva*) as a host. The presence of *C. maritimus* and other parasitic plants has been found to improve sediment salinity, redox potential, and habitat heterogeneity, resulting in higher overall plant species richness. **Reported associates:** *Acacia cyclops, Agropyron trachycaulum, Agrostis stolonifera, Alisma, Allenrolfea occidentalis, Amblyopappus pusillus, Anemopsis californica, Apium graveolens, Armeria maritima, Asclepias speciosa, Atriplex canescens, A. lentiformis, A. patula, A. phyllostegia, A. semibaccata, A. watsonii, Baccharis sarothroides, Batis maritima, Bolboschoenus maritimus, Bromus tectorum, Cakile edentula, Carex lanuginosa, C. microptera, C. nebrascensis, C. praegracilis, C. simulata, Castilleja ambigua, Centaurium exaltatum, Chrysothamnus albidus, Cicuta douglasii, Cleomella, Comandra umbellata, Cressa truxillensis, Cuscuta salina, Distichlis spicata, D. stricta, Eleocharis palustris, E. rostellata, Elymus triticoides, Epilobium ciliatum, Frankenia grandifolia, F. palmeri, F. salina, Glycyrrhiza lepidota, Haplopappus paniculatus, H. racemosus, Hordeum glaucum, H. jubatum, H. marinum, H. murinum, Hutchinsia procumbens, Iris missouriensis, Jaumea carnosa, Juncus acutus, J. arcticus, J. balticus, J. bufonius, J. ensifolius, J. torreyi, Kochia scoparia, Lactuca serriola, Lasthenia glabrata, Limonium californicum, Lycopus asper, Mentha arvensis, Mesembryanthemum nodiflorum, Mimulus guttatus, Monanthochloe littoralis, Nasturtium officinale, Nitrophila occidentalis, Parapholis incurva, Phragmites australis, Poa palustris, P. pratensis, Polygonum paronychia, Polypogon interruptus, P. monspeliensis, Potamogeton, Potentilla anserina, Puccinellia nuttalliana, Ranunculus cymbalaria, R. sceleratus, Rosa woodsii, Rumex crispus, Sagittaria cuneata, Salicornia bigelovii, S. depressa, Salsola, Sarcobatus vermiculatus, Schoenoplectus acutus, S. americanus, Senecio hydrophilus, Sium suave, Solidago missouriensis, Spartina densiflora, S. foliosa, S. gracilis, Spergularia marina, Sphenopholis obtusata, Sporobolus airoides, Suaeda depressa, S. esteroa, S. taxifolia, Symphyotrichum chilense, Tamarix, Trifolium campestre, T. repens, Triglochin concinnum, T. maritimum, T. palustre, Typha, Veronica americana, Xanthium strumarium.*

Chloropyron molle **(A. Gray) A. Heller** occurs in flats, floodplains, meadows, playas, salt marshes, sinks, sloughs, and along the margins of ephemeral lakes, pools, and streams at elevations of up to 155 m. The sites are described as brackish, saline or alkaline but the specific salinity and pH values have not been reported; however, 91% of populations are said to occur where substrates are slightly to moderately saline within the first meter of depth. When growing in a salt marsh, the plants inhabit the upper region at the limit of tidal influence. Characteristic substrates are described as adobe or heavy clay. The flowers have two functional stamens and are produced in proportion to the size and vigor of the plant. Peak flowering occurs during midsummer and declines in August. Although the flowers are self-compatible and are capable of self-fertilization, they normally are outcrossed by insects. Bumblebees (Hymenoptera: Apidae: *Bombus californicus*) are the most frequent pollinators with other potential pollinators including honeybees (Hymenoptera: Apidae: *Melissodes*), leafcutter bees (Hymenoptera: Megachilidae: *Anthidium*), and sweat bees (Hymenoptera: Halictidae: *Halictus, Lasioglossum*). Seed production varies considerably

but single fruits will average about 24 seeds and individual plants are capable of producing up to 32,000 seeds. The seeds remain afloat in fresh or salt water for extended periods of time and can be dispersed by water. However, most capsules remain closed on the parental plants and release their seeds directly beneath as the plants eventually collapse and decay at the end of the growing season. The plants regenerate from a seed bank that persists for at least several years. Maximum seed germination occurs on exposed substrates following winter rains (which reduce soil salinity) or when tidal inundation is frequent. Seedlings grow rapidly when tidal inundations reach their annual low (usually March to April). During periods of drought, seedlings can emerge as early as December. Seedling mortality is high when the seeds germinate near unsuitable hosts or in sites where there are few suitable hosts. Known angiosperm hosts of this hemiparasite include *Distichlis spicata* (Poaceae), *Salicornia depressa* (Amaranthaceae) and *Jaumea carnosa* (Asteraceae). Winter annuals such as *Hainardia cylindrica* (Poaceae), *Juncus bufonius* (Juncaceae), or *Polypogon monspeliensis* (Poaceae) are not good hosts because they die before the plants can reproduce. The occurrence of parasitic dodder (Convolvulaceae: *Cuscuta salina*) on associated species has been shown to positively influence *C. molle*. There are few remaining sites for *C. molle* but some of them contain between 1000–6000 individuals. **Reported associates:** *Achillea millefolium, Allenrolfea occidentalis, Apium graveolens, Atriplex cordulata, A. polycarpa, A. prostrata, Bolboschoenus maritimus, B. robustus, Cirsium hydrophilum, Chloropyron palmatum, Cotula coronopifolia, Cressa truxillensis, Cuscuta salina, Distichlis spicata, Frankenia salina, Grindelia stricta, Hainardia cylindrica, Isolepis cernua, Jaumea carnosa, Juncus balticus, J. gerardii, Lathyrus jepsonii, Lepidium latifolium, Lilaeopsis masonii, Limonium californicum, Lysimachia maritima, Nitrophila occidentalis, Phragmites australis, Plantago maritima, P. subnuda, Polygonum aviculare, Polypogon monspeliensis, Rumex crispus, Salicornia depressa, Schoenoplectus americanus, Sonchus oleraceus, Spartina patens, Sporobolus airoides, Suaeda suffrutescens, Symphyotrichum lentum, S. subulatum, Triglochin maritimum.*

Chloropyron palmatum **(Ferris) Tank & J.M. Egger** grows in depressions, marshes, meadows, sinks, sloughs, and along the margins of seasonal ponds at elevations of up to 85 m. This species occupies seasonally flooded alkaline (pH: 7.2–9.8 [optimum: 7.2–8.6]) sites that typically are saline. The flowers, which are produced from June to October, have two functional stamens. The plants are self-compatible and capable of self-pollination; however, outcrossing by bumblebees (Hymenoptera: Apidae: *Bombus californicus, B. occidentalis, B. vosnesenskii*) is common. The fruits mature in late summer and early fall with seed germination occurring from March to April. Typically, five seeds per fruit and 85 seeds per plant are produced on average; however, an individual plant can produce upward of 1000 seeds under optimal conditions. The seeds float and are dispersed by water and by wind. Seasonal flooding promotes seed germination (by reducing salinity) and seed dispersal. Dispersal by ants (Insecta: Hymenoptera: Formicidae) also has been suggested. Seeds that are cold stratified for 4–5 months at 4°C–5°C germinate well (94%) at 10°C but poorly (12%) at 27°C. Germination is higher at low salinity than at high salinity conditions. The seeds remain viable in the seed bank for at least 3 years, but annual fluctuations in the size of standing vegetation populations can be considerable. The survival rate of seedlings observed in some populations is fairly low (18%). An allozyme survey of 14 loci detected higher levels of genetic variation (mean number of alleles/locus: 1.1–1.6; polymorphic loci: 7.1%–35.7%; observed heterozygosity: 1.4%–2.8%) than in *Cordylanthus maritimum* (see earlier). Overall, the extent of genetic variation observed was low in some populations but in others was comparable to many annual, insect-pollinated species. Allozyme data indicated that larger populations did not harbor a greater level of genetic variation than smaller populations and that among-population variation represented only 2% of the total genetic variation observed. Genetic studies using ISSR markers indicated considerable genetic structure, with individual vernal pools containing genetically distinct subpopulations. Plants of *C. palmatum* survive poorly when grown without a host or with *Phleum*, *Pinus*, or *Quercus* as hosts, but they can be reared successfully using *Helianthus*. *Distichlis spicata*, *Kochia californica*, and *Suaeda moquinii* are believed to be the primary natural hosts. **Reported associates:** *Allenrolfea occidentalis*, *Arthrocnemum subterminale*, *Astragalus tener*, *Atriplex coronata*, *A. depressa*, *A. joaquiniana*, *Castilleja attenuata*, *Chloropyron molle*, *Cressa truxillensis*, *Deinandra bacigalupii*, *Distichlis spicata*, *Downingia pulchella*, *Eryngium aristulatum*, *Frankenia salina*, *Hemizonia parryi*, *H. pungens*, *Holocarpha obconica*, *Kochia californica*, *Lasthenia*, *Lepidium lasiocarpum*, *L. latipes*, *Lolium multiflorum*, *Salicornia depressa*, *Spergularia macrotheca*, *Suaeda moquinii*.

Use by wildlife: The capsules and ovules of *Chloropyron maritimum* are eaten by larvae of the salt marsh snout moth (Insecta: Lepidoptera: Pyralidae: *Lipographis fenestrella*). Developing fruits and seeds of *C. molle* also are consumed by several Lepidoptera (Pyralidae: *Lipographis fenestrella*; Tortricidae: *Saphenista*) and the mature seeds are fed on by the savannah sparrow (Aves: Emberizidae: *Passerculus sandwichensis*), western meadowlark (Aves: Icteridae: *Sturnella neglecta*) and the salt marsh harvest mouse (Mammalia: Muridae: *Reithrodontomys raviventris*). The foliage of *C. molle* is grazed by cattle (Mammalia: Bovidae: *Bos*) and feral hogs (Mammalia: Suidae: *Sus scrofa*).

Economic importance: food: not edible; **medicinal:** A few *Cordylanthus* species (but none of the OBL *Chloropyron* species) have been used medicinally by native North American tribes; **cultivation:** not cultivated; **misc. products:** none; **weeds:** none; **nonindigenous species:** none

Systematics: Generally, *Cordylanthus* is related most closely to *Castilleja* (Figure 5.72). However, molecular phylogenetic analyses demonstrate that *Cordylanthus* is at least biphyletic with respect to subgenera *Cordylanthus* and *Hemistegia*, which do not resolve as a clade. Yet, both

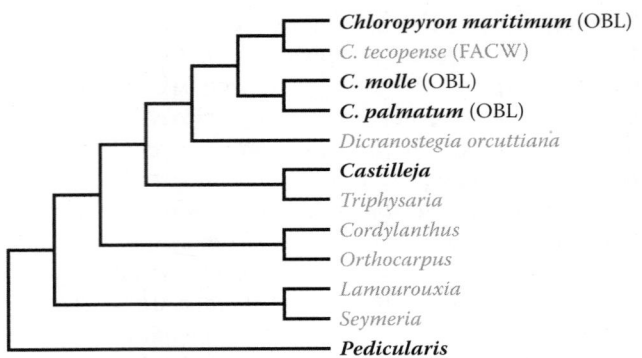

FIGURE 5.76 Phylogenetic relationships in Orobanchaceae subtribe Castillejinae based on DNA sequence data derived from *rps16*, *trnL/F*, nrETS, and nrITS regions. Taxa with OBL North American representatives are shown in bold. (Adapted from Tank, D.C. & Olmstead, R.G., *Amer. J. Bot.*, 95, 608–625, 2008; with nomenclature adapted from Tank, D.C. et al., *Syst. Bot.*, 34, 182–197, 2009.)

subgenera *Cordylanthus* and *Hemistegia* are each monophyletic, with the latter associating with the monotypic subgenus *Dicranostegia*. To resolve these discrepancies, *Cordylanthus* has been subdivided into two additional genera: *Chloropyron* (=subg. *Hemistegia*) and *Dicranostegia* (=subg. *Dicranostegia*). All three OBL *Chloropyron* species (*C. maritimum*, *C. molle*, and *C. palmatum*) occur within a clade of halophytes (Figure 5.76) that also contains *C. tecopense* (FACW). The three subspecies of *C. maritimum* include subsp. *canescens*, which occurs in inland localities and subsp. *maritimum* and subsp. *palustre*, which grow mainly in coastal sites. Two subspecies have been recognized for *C. molle*, with *C. molle* subsp. *molle* in coastal salt marshes and subsp. *hispidum* occurring in inland saline flats. The base chromosome number of *Chloropyron* is $x = 7$, but the diploid ancestors of the genus are believed to be extinct. Accordingly, *C. maritimum* ($2n = 30$) and *C. molle* ($2n = 28$) would be tetraploids and *C. palmatum* ($2n = 42$) a hexaploid. Morphology and cytology have indicated that *C. maritimum* and *C. molle* are closely related and that *C. palmatum* may be an allopolyploid involving *C. molle* as one parent. This interpretation generally seems compatible with molecular phylogenetic relationships, which resolve *C. maritimum* and *C. tecopense* in one clade and *C. molle* and *C. palmatum* as their sister clade (Figure 5.76).

Comments: *Chloropyron maritimum* occurs in the western United States; *C. molle* and *C. palmatum* are California endemics. The latter two species are restricted to few populations, with only four natural occurrences extant for *C. palmatum*. All three species are listed as federally endangered in the United States.

References: Ayres et al., 2007; Bauer, 1930; Chuang & Heckard, 1971, 1973, 1986; Coats et al., 1993; Fellows & Zedler, 2005; Fischer, 2004; Fleishman et al., 2001; Grewell, 2004, 2005; Grewell et al., 2003; Heimbinder, 2001; Helenurm & Parsons, 1997; Noe & Zedler, 2000, 2001; Parsons & Zedler, 1997; Ruygt, 1994; Tank & Olmstead, 2008; Tank et al., 2009; Vanderwier & Newman, 1984.

4. Macranthera

Flameflower, hummingbird flower, orange blackherb

Etymology: from the Greek *makros anthos* meaning "large flowered"

Synonyms: *Conradia*; *Dasystoma*; *Gerardia* (in part); *Russelia*; *Tomilix*; *Toxopus*

Distribution: global: North America; **North America:** southeastern

Diversity: global: 1 species; **North America:** 1 species

Indicators (USA): OBL: *Macranthera flammea*

Habitat: freshwater; palustrine; **pH:** 4.5–6.0; **depth:** <1 m; **life-form(s):** emergent herb

Key morphology: stems (to 30 dm) erect, obscurely 4-angled, usually unbranched, the herbage turning black when dry; leaves (to 33 cm), up to 46 in rosettes, or cauline (up to 16 cm), opposite, deeply pinnately lobed (grading to entire apically), sessile or with short, winged petioles; inflorescence a panicle of racemes with numerous flowers, pedicels (to 2 cm) recurved in fruit; calyx (to 1.5 cm) tubular, 5-parted, green; corolla (to 2.5 cm) tubular, 2-lipped (the lobes short), fleshy, orange (turning brown to black with age); stamens 4, exserted from corolla by long (to 46 mm), orange, filaments; capsule (to 1.8 cm) with short hairs; seeds (to 3 mm) irregular, with 2–3 membranous wings

Life history: duration: annual, biennial (fruit/seeds); **asexual reproduction:** none; **pollination:** hummingbird; **sexual condition:** hermaphroditic; **fruit:** capsules (common); **local dispersal:** seeds (water, wind); **long-distance dispersal:** seeds (water, wind)

Imperilment: (1) *Macranthera flammea* [G3]; S1 (GA); S2 (AL, LA, FL); S3 (MS)

Ecology: general: *Macranthera* is monotypic.

Macranthera flammea (W. Bartram) Pennell occurs on wet ground or in shallow waters of bays, bogs, depressions, flatwoods, floodplains, marshes, ponds, savannas, seeps, stream margins, and swamps. The substrates are acidic (pH: 4.5–6.0) and sandy. The plants are intolerant of standing water and occur mainly in sites with adequate water flow. The showy flowers are pollinated exclusively by hummingbirds (Aves: Trochilidae: *Archilochus colubris*). The flowers reportedly are protandrous, with the stamens withering before the style elongates and the stigma becomes receptive. The orange corolla is believed to assist the pollinators with finding the flowers in the dark swamps, which typically are inhabited by the plants. This species usually grows as a biennial, which produces a rosette of leaves the first year and a flowering shoot the second. Seeds stratified at 3°C for 150 days will germinate well at 25°C–29°C. The seeds have 2–3 membranous wings, which probably facilitates their flotation and dispersal by wind. Surprisingly, there has been no detailed study on the pollination biology of this conspicuously flowered species and few details also exist on other aspects of its life history. Plants of *M. flammea* provided with a host and fertilizer produced more haustoria and grew taller than those without a host or those deprived of fertilizer. In greenhouse culture the plants are able to reach maturity without a host; however, substantial numbers of haustoria are produced when they are grown on various angiosperm timber hosts including *Carya illinoensis*, *Celtis laevigata*, *Fraxinus pensylvanica*, *Liquidambar styraciflua*, *Nyssa sylvatica*, *Pinus elliotii*, *P. strobus*, *Platanus occidentalis*, *Quercus alba* and *Q. shumardii*; thus it does not appear to be very host specific. *Nyssa biflora* is a natural host. The blackening of tissues upon drying is due to the presence of orobanchin (a phenolic glycoside derivative of caffeic acid) and iridoids, which occur in significant amounts. The habitat of this species is threatened by damage from feral swine (Mammalia: Suidae: *Sus scrofa*). **Reported associates:** *Acer rubrum*, *Aletris lutea*, *Andropogon gyrans*, *A. mohrii*, *A. virginicus*, *Aristida beyrichiana*, *A. palustris*, *A. stricta*, *Arnoglossum sulcatum*, *Aronia arbutifolia*, *Arundinaria gigantea*, *Burmannia capitata*, *Calopogon*, *Carphephorus pseudoliatris*, *Clethra alnifolia*, *Cliftonia monophylla*, *Coreopsis gladiata*, *C. linifolia*, *Ctenium aromaticum*, *Cyrilla racemiflora*, *Dichanthelium*, *Doellingeria umbellata*, *Drosera capillaris*, *D. filiformis*, *D. intermedia*, *Elephantopus nudatus*, *Erianthus giganteus*, *Eriocaulon decangulare*, *E. texense*, *Eryngium integrifolium*, *Eupatorium*, *Eurybia eryngiifolia*, *Fimbristylis*, *Fraxinus caroliniana*, *Fuirena*, *Helianthus angustifolius*, *H. heterophyllus*, *Hypericum cistifolium*, *H. galioides*, *H. tetrapetalum*, *Hyptis alata*, *Ilex coriacea*, *I. glabra*, *Isoetes louisianensis*, *Itea virginica*, *Juncus trigonocarpus*, *J. validus*, *Lachnocaulon digynum*, *Liatris*, *Lilium catesbaei*, *Liriodendron tulipifera*, *Lophiola aurea*, *Ludwigia hirtella*, *Lycopodiella alopecuroides*, *Lyonia lucida*, *Magnolia virginiana*, *Melanthium virginicum*, *Muhlenbergia capillaris*, *M. expansa*, *Myrica cerifera*, *M. heterophylla*, *Nyssa biflora*, *Osmundastrum cinnamomeum*, *Panicum scoparium*, *Peltandra sagittifolia*, *Persea palustris*, *Pinguicula lutea*, *P. planifolia*, *P. primuliflora*, *Pinus elliottii*, *P. palustris*, *P. serotina*, *Platanthera integra*, *Polygala*, *Pteridium aquilinum*, *Rhexia*, *Rhododendron viscosum*, *Rhynchospora chapmanii*, *R. latifolia*, *R. oligantha*, *R. stenophylla*, *Sabatia macrophylla*, *Sarracenia alata*, *S. leucophylla*, *S. psittacina*, *S. rubra*, *Scleria muehlenbergii*, *Serenoa repens*, *Sphagnum*, *Stokesia laevis*, *Trantha racemosa*, *Viburnum nudum*, *Xyris*.

Use by wildlife: *Macranthera flammea* is a nectar source for ruby-throated hummingbirds (Aves: Trochilidae: *Archilochus colubris*).

Economic importance: food: not edible; **medicinal:** none; **cultivation:** not cultivated; **misc. products:** none; **weeds:** none; **nonindigenous species:** none

Systematics: Phylogenetic analyses of *rps*2 sequence data (Figure 5.77) place the monotypic *Macranthera* within a clade of New World taxa containing *Agalinis* and *Castilleja*. Within this clade is *Lamourouxia*, which resembles *Macranthera* by its pinnately incised leaves and similar floral morphology. However, the group is not resolved well enough to indicate any further details of relationships among these genera (Figure 5.77) and requires further study. The base chromosome number of *Macranthera* probably is $x = 7$. The chromosome number of *M. flammea* ($2n = 26$) is believed to be an aneuploid derivative of $2n = 28$ (tetraploid). The chromosomes of *M. flammea* are very large and the karyotype is asymmetric.

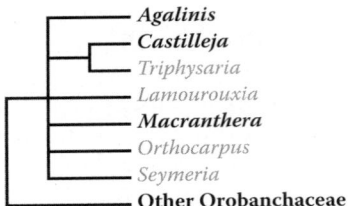

FIGURE 5.77 Relationships of *Macranthera* to other genera of Orobanchaceae as indicated by phylogenetic analysis of *rps*2 sequence data. Resolution is poor within this clade, in which all taxa have biogeographical affinities with the Americas. Taxa that include OBL indicators are indicated in bold. (Adapted from Young, N.D. et al., *Ann. Missouri Bot. Gard.*, 86, 876–893, 1999.)

Comments: *Macranthera flammea* is restricted to the coastal plain of the southeastern United States

References: Alford & Anderson, 2002; Baskin & Baskin, 1998; Determann et al., 1997; Engeman et al., 2007; Fischer, 2004; Godfrey & Wooten, 1981; Kondo et al., 1978b; Mann, Jr. & Musselman, 1981; Musselman, 1972; Musselman & Mann, Jr., 1977; Pennell, 1935; Pickens, 1927, 1930; Young et al., 1999.

5. *Pedicularis*

Fen/swamp betony, lousewort, elephant's head; pédiculaire

Etymology: from the Latin *pediculus* meaning "louse" owing to the belief that animals grazing near the plants would contract lice

Synonyms: *Elephantella*; *Pediculariopsis*

Distribution: global: Asia; Europe; New World; **North America:** widespread

Diversity: global: 600 species; **North America:** 36 species

Indicators (USA): OBL: *Pedicularis attollens, P. chamissonis, P. groenlandica, P. lanceolata*; **FACW:** *P. crenulata, P. groenlandica, P. lanceolata*; **unranked:** *P. palustris* [regarded here as OBL]

Habitat: freshwater; palustrine; **pH:** 3.8–8.5; **depth:** <1 m; **life-form(s):** emergent herb

Key morphology: stems (to 80 cm) erect, tomentose above; leaves arranged in a basal rosette (to 25 cm) or cauline (to 11 cm) and alternate or opposite, the blades pinnatifid with crenate or toothed lobes, or divided in up to 50 linear to oblong segments; racemes (to 12–30 cm) spike-like, pedicels (to 6 mm) short; calyx (to 12 mm) 2–5-lobed; corolla (to 7–26 mm) purple, red, white or yellow, 2-lipped, the upper lip or galea (to 15 mm) beak-, hood- or trunk-like, the lower lip (to 12 mm) 3-lobed and narrow to spreading; stamens 4, didynamous, anthers included; capsules (to 20 mm) lanceolate to ovoid; seeds (to 4 mm), smooth or reticulate

Life history: duration: biennial, perennial, summer annual (persistent rootstocks); **asexual reproduction:** crown/root division; **pollination:** insect; **sexual condition:** hermaphroditic; **fruit:** capsules (common); **local dispersal:** seeds (water, wind); **long-distance dispersal:** seeds (water, wind)

Imperilment: (1) *Pedicularis attollens* [G4]; (2) *P. chamissonis* [G4]; (3) *P. crenulata* [G4]; S1 (CA, NE, MT); S2 (WY);

(4) *P. groenlandica* [G4]; S1 (AK, <u>SK</u>); S2 (<u>YT</u>); S3 (<u>QC</u>); (5) *P. lanceolata* [G5]; SH (DE, KY); S1 (AR, GA, MA, MD, PA, NC, TN); S2 (CT, NY, WV); S3 (NE, NJ, VA); (6) *P. palustris* [G4]; S1 (<u>NS</u>); S2 (<u>NF</u>, <u>QC</u>)

Ecology: general: *Pedicularis* is a large and diverse genus of annual, biennial, and perennial hemiparasites. By and large, most of the species occupy UPL sites; however, over 60% of the North American species are wetland indicators, with four designated as OBL in at least a portion of their range. *Pedicularis crenulata* was classified as OBL, FACW in the 1996 indicator list, but has since been reclassified as FACW in the 2013 list. Because its treatment was prepared prior to the release of the 2013 revision, it has been retained here. *Pedicularis palustris*, which in North America occurs only in Canada (thus is unranked in the indicator list), should also be treated as an OBL species as has been done here. Many *Pedicularis* species are self-compatible, but require insects (Insecta) such as bumblebees (Hymenoptera: Apidae: *Bombus*) to achieve successful pollination. A few arctic species can self-pollinate and a few nectar producing species are pollinated by hummingbirds (Aves: Trochilidae). Generally, the nectar-producing species lack a beaked corolla, flower earlier in the season, and are pollinated by long-tongued queen bumble bees (Hymenoptera: Apidae: Bombus), which enter the flower in an upright (nototribic) position. The nectarless species have a beaked corolla, flower later in the season, and are pollinated by worker bumblebees (Hymenoptera: Apidae: Bombus), which enter the flower in an inverted (sternotribic) position and remove the pollen using wing vibrations. The seeds typically are characterized by physiological dormancy. Vegetative reproduction can occur by root or crown branching, but has been inadequately described for most species. In North America, the genus perhaps is best known for the Furbish lousewort (*P. furbishiae*), a protected species whose rediscovery caused the demise of a major hydroelectric construction project.

***Pedicularis attollens* A. Gray** is a perennial, which occurs in bogs, depressions, marshes, meadows, seeps, swales, and along the margins of lakes and streams at elevations from 335 to 3962 m. The plants grow mainly in alpine and subalpine sites on substrates that are described as granitic sand, gravel, humus, loam, peat, pumice, pumice sand, and silty pumice. The specific pH range has not been reported. The flowers are nectarless, have a pink to purple corolla, and are pollinated principally by worker bumblebees (Hymenoptera: Apidae: *Bombus bifarius, B. flavifrons, B. rufocinctus, B. vosnesenskii*), which remove the pollen entirely using wing vibrations ("buzz" pollination). Details on seed germination, dispersal, and other life history information are lacking for this species. **Reported associates:** *Abies concolor, A. lasiocarpa, Achillea lanulosa, Acmispon americanus, Aconitum columbianum, Agrostis oregonensis, Allium validum, Alnus tenuifolia, Antennaria media, A. rosea, Aquilegia truncata, Aulacomnium palustre, Bistorta bistortoides, Botrychium simplex, Cirsium nidulum, Calamagrostis breweri, Caltha leptosepala, Carex abrupta, C. aquatilis, C. echinata, Ca. exserta, C. heteroneura, C. hoodii, C. incurviformis, C. luzulifolia, C. nebrascensis, C.*

nigricans, *C. scopulorum, C. senta, C. simulata, C. sub-nigricans, C. utriculata, Cassiope mertensiana, Castilleja lemmonii, Corallorhiza maculata, Danthonia intermedia, D. unispicata, Darlingtonia californica, Dasiphora floribunda, Deschampsia cespitosa, Drepanocladus, Drosera rotundifolia, Eleocharis pauciflora, E. quinqueflora, Elymus glaucus, Epilobium ciliatum, E. minutum, Eriophorum gracile, Galium bifolium, Gentianopsis holopetala, G. simplex, Glyceria elata, Gaultheria humifusa, Heuchera rubescens, Holodiscus, Hosackia pinnata, Hypericum anagalloides, Ivesia lycopodioides, Juncus balticus, J. mertensianus, J. nevadensis, J. orthophyllus, Kalmia microphylla, K. polifolia, Ligusticum grayi, Lomatium grayi, Lonicera conjugialis, Luzula congesta, L. orestera, L. subcongesta, Madia bolanderi, Meesia triquetra, Menyanthes trifoliata, Micranthes californica, Mimulus guttatus, M. jepsonii, M. primuloides, Monardella odoratissima, Montia chamissoi, Muhlenbergia filiformis, Oreostemma alpigenum, Packera cymbalarioides, Pedicularis groenlandica, Penstemon confertus, P. rydbergii, Perideridia parishii, Phleum alpinum, Phyllodoce breweri, P. empetriformis, Pinguicula vulgaris, Pinus albicaulis, P. contorta, P. monticola, Platanthera leucostachys, Poa pratensis, Polemonium caeruleum, Potentilla drummondii, P. flabelliformis, P. gracilis, P. grayi, P. pennsylvanica, Primula fragrans, P. suffrutescens, P. tetrandra, Ptilagrostis kingii, Pyrola picta, Quercus vacciniifolia, Ranunculus occidentalis, Rhododendron, Rubus lasiococcus, Salix arctica, S. commutata, S. orestera, S. planifolia, Senecio triangularis, Solidago multiradiata, Sphagnum, Spiraea densiflora, S. douglasii, Spiranthes porrifolia, Stellaria longipes, Symphyotrichum spathulatum, Triantha glutinosa, Trifolium longipes, T. monanthum, Trisetum spicatum, T. wolfii, Tsuga mertensiana, Vaccinium cespitosum, V. uliginosum, Veronica americana, V. scutellata, V. serpyllifolia, Viola adunca, Wyethia mollis.*

***Pedicularis chamissonis* Steven** is a perennial, which grows in ditches, marshes, meadows, seeps, tundra, and along the margins of lakes, ponds, and streams at elevations of up to 2450 m. The substrates are acidic (pH: 5.0–6.0) and include deep loam, gravel, and volcanic soil. The flowers are nectarless, have a pink to reddish corolla, and typically are outcrossed by bumblebees (Hymenoptera: Apidae: *Bombus*), which enter the flowers upside down (sternotribic) and remove the pollen by vibrating their wings or by scraping the anthers. The seeds are dispersed by their attachment to animals, but their germination requirements have not been reported. High elevation plants studied in Japan required an average of 35 days before growth was initiated. **Reported associates:** *Aconitum maximum, Agrostis exarata, Anaphalis, Anemone narcissiflora, Angelica lucida, Arnica unalaschcensis, Athyrium filix-femina, Barbarea orthoceras, Bistorta vivipara, Cacalia auriculata, Calamagrostis langsdorffii, Carex macrochaeta, Castilleja unalaschcensis, Chamerion angustifolium, Cirsium kamtschaticum, Comarum palustre, Conioselinum pacificum, Cornus suecica, Dactylorhiza aristata, Deschampsia beringensis, Dryopteris, Empetrum, Epilobium hornemannii, Erigeron peregrinus, Festuca rubra, Fritillaria camschatcensis, Geranium erianthum, Geum calthifolium, Heracleum*

lanatum, Hierochloe odorata, Lathyrus palustris, Linnaea borealis, Lupinus nootkatensis, Lycopodium, Maianthemum dilatatum, Poa arctica, P. macrocalyx, Ranunculus occidentalis, Salix arctica, Trientalis borealis, T. europaea, Trisetum spicatum.

***Pedicularis crenulata* Benth.** is a perennial, which inhabits fens, meadows, and the margins of streams at elevations from 1525 to 2300 m. The substrates are circumneutral to alkaline (pH: 6.6–8.5) and commonly comprise loam or peat that is high in organic matter (e.g., 62%). The flowers have a magenta or white corolla, open in spring, and are in full bloom by early July. They are unable to set seed in the absence of pollinators. Pollination is achieved almost exclusively by bumblebee queens (Hymenoptera: Apidae: *Bombus appositus, B. californicus, B. flavifrons, B. frigidus, B. rufocinctus*), which enter the flowers in an upright (nototribic) position to gather the copius nectar. The pollen is transferred in their head-thorax crevice. Details on seed germination or dispersal mechanisms are unknown. These hemiparasitic plants are known to take up alkaloids (anagyrine) from their host (Fabaceae: *Thermopsis montana*). **Reported associates:** *Abies lasiocarpa, Achillea millefolium, Agrostis gigantea, A. scabra, Alopecurus borealis, Antennaria microphylla, Aquilegia caerulea, Arabis drummondii, Arctostaphylos uva-ursi, Arnica mollis, Artemisia frigida, Astragalus leptaleus, Bistorta bistortoides, B. vivipara, Bromus ciliatus, Calamagrostis canadensis, C. stricta, Caltha leptosepala, Cardamine cordifolia, Carex aquatilis, C. capillaris, C. illota, C. microptera, C. nebrascensis, C. parryana, C. pellita, C. praegracilis, C. scirpoidea, C. scopulorum, C. simulata, C. utriculata, Castilleja rhexiifolia, Chamerion angustifolium, Cirsium arvense, C. coloradense, Clementsia rhodantha, Conioselinum scopulorum, Crepis runcinata, Danthonia, Dasiphora floribunda, Delphinium barbeyi, Deschampsia cespitosa, Draba nivalis, Eleocharis quinqueflora, Elymus trachycaulus, Epilobium ciliatum, Erigeron speciosus, Festuca brachyphylla, Gentiana, Hierochloë hirta, Hippochaete variegata, Hordeum jubatum, Hymenoxys hoopesii, Hypericum scouleri, Iris missouriensis, Juncus arcticus, J. balticus, J. castaneus, J. longistylis, J. mertensianus, Kobresia myosuroides, K. simpliciuscula, Luzula parviflora, Maianthemum stellatum, Mentha arvensis, Mertensia ciliata, Micranthes odontoloma, Mimulus glabratus, Moehringia lateriflora, Muhlenbergia richardsonis, Oreochrysum parryi, Oxypolis fendleri, Oxytropis sericea, Packera pauciflora, Parnassia parviflora, Pedicularis groenlandica, Phleum pratense, Picea engelmannii, Plantago eriopoda, Poa compressa, P. leptocoma, P. pratensis, P. secunda, Polemonium foliosissimum, Potentilla anserina, P. gracilis, P. hippiana, P. pulcherrima, P. subjuga, Primula egaliksensis, P. incana, P. parryi, P. pauciflora, Ptilagrostis porteri, Ranunculus alismifolius, R. cardiophyllus, Rhodiola integrifolia, R. rhodantha, Rumex crispus, Salix brachycarpa, S. myrtillifolia, S. planifolia, Senecio taraxacoides, S. triangularis, Silene drummondii, Sisyrinchium pallidum, Spiranthes diluvialis, Stellaria longipes, Symphyotrichum spathulatum, Taraxacum officinale, Thalictrum alpinum, Thermopsis montana, Thlaspi montanum, Trichophorum pumilum, Trifolium* *longipes, Triglochin maritimum, Trisetum wolfii, Urtica dioica, Zizia aptera.*

Pedicularis groenlandica Retz. is a perennial found in bogs, marshes, meadows, muskeg, seeps, and along the margins of lakes, pools, and streams at elevations from 750 to 3600 m. The plants grow in sunny or shaded sites on a wide range of organic and mineral substrates (pH: 5.8–7.8) that include granite, humus, loam, muck, peat, rich muck, rock, sand, and talus. The plants can withstand cold temperatures down to −46°C. The flowers are nectarless, have a pink to reddish-purple corolla, and are pollinated by worker bumblebees (Hymenoptera: Apidae: *Bombus appositus, B. bifarius, B. centralis, B. flavifrons, B. lapponicus, B. melanopygus, B. occidentalis, B. rufocinctus*), which enter the flowers in an upright (nototribic) position, but remove the pollen entirely using wing vibration ("buzz" pollination). Extremely low seed set has (0.5%) been observed when pollinators are excluded, indicating that the flowers are at least partly self-compatible. The corollas normally are purple, but white-flowered forms also have been observed. Reduced seed set can occur when the plants grow along with *Lupinus latifolius* (Fabaceae) or *Bistorta bistortoides* (Polygonaceae) because of pollinator competition. The seeds have been described as being nondormant, with germination occurring at 22°C. However, they also are described as being physiologically dormant, with germination occurring over a 3-week period under greenhouse conditions following their exposure to 120 days of cold, moist stratification, but not during the first year when they have been stratified outdoors. Seeds placed in the dark at 18°C for 28 days exhibited very low (0.5%) germination. Germination reportedly can be enhanced by soaking the seeds for 2 weeks in cold water (4°C), or by treating them with gibberellic acid. Seedlings develop their first leaves after 4 weeks, when it then becomes necessary to provide them with a host plant. The plants are known to take up the pyrrolizidine alkaloid senecionine from their host (Asteraceae: *Senecio triangularis*).

Reported associates: *Abies lasiocarpa, Achillea millefolium, Aconitum columbianum, Agoseris, Agrostis gigantea, A. humilis, A. rossiae, A. scabra, A. variabilis, Allium schoenoprasum, A. validum, Alnus tenuifolia, A. viridis, Alopecurus pratensis, Andromeda polifolia, Anemone narcissiflora, Antennaria alpina, A. corymbosa, A. microphylla, Arctostaphylos alpina, Arnica chamissonis, A. cordifolia, A. longifolia, Artemisia cana, A. scopulorum, Astragalus alpinus, A. americanus, Betula nana, Bistorta bistortoides, Botrychium lanceolatum, Bromus vulgaris, Calamagrostis breweri, C. canadensis, Caltha leptosepala, Camassia quamash, Cardamine cordifolia, Carex aquatilis, C. arcta, C. bigelowii, C. buxbaumii, C. canescens, C. capillaris, C. capitata, C. illota, C. lenticularis, C. limosa, C. luzulina, C. muricata, C. nardina, C. nebrascensis, C. nigricans, C. ovalis, C. rostrata, C. rupestris, C. saxatilis, C. scopulorum, C. simulata, C. spectabilis, C. subnigricans, C. vesicaria, Cassiope mertensiana, Castilleja lemmonii, C. miniata, C. occidentalis, C. parviflora, Chamerion angustifolium, Cirsium neomexicanum, Cladonia pyxidata, Claytonia nevadensis, Clementsia rhodantha, Climacium dendroides, Comarum palustre,* *Corydalis caseana, Dasiphora floribunda, Deschampsia cespitosa, Drosera anglica, Dryas octopetala, Eleocharis palustris, E. pauciflora, E. quinqueflora, Empetrum nigrum, Epilobium alpinum, E. anagallidifolium, E. hornemannii, E. latifolium, Equisetum arvense, Erigeron melanocephalus, E. peregrinus, Eriophorum altaicum, E. angustifolium, E. crinigerum, E. gracile, E. vaginatum, Erythronium grandiflorum, Festuca brachyphylla, F. ovina, Fragaria virginiana, Galium boreale, G. trifidum, Gaultheria humifusa, Gentiana algida, Gentianopsis detonsa, G. simplex, Gentianella amarella, Geranium richardsonii, Geum macrophyllum, G. rossii, Hedysarum sulphurescens, Helenium bigelovii, Hierochloë hirta, Hypericum anagalloides, H. scouleri, Hypnum lindbergii, H. revolutum, Ivesia lycopodioides, Juncus albescens, J. alpinoarticulatus, J. arcticus, J. drummondii, J. longistylis, J. mertensianus, J. orthophyllus, J. parryi, Juniperus communis, Kalmia microphylla, Kobresia myosuroides, K. simpliciuscula, Larix lyallii, Lewisia sierrae, Ligusticum canbyi, L. grayi, L. tenuifolium, Listera cordata, Lloydia serotina, Loiseleuria procumbens, Lonicera caerulea, Luetkea pectinata, Lupinus argenteus, L. latifolius, Luzula campestris, L. comosa, L. piperi, L. subcapitata, Maianthemum amplexicaule, M. racemosum, M. stellatum, Menyanthes trifoliata, Menziesia ferruginea, Mertensia ciliata, Micranthes ferruginea, M. nelsoniana, M. odontoloma, M. oregana, M. subapetala, M. tolmiei, Mimulus guttatus, M. lewisii, M. primuloides, M. tilingii, Minuartia rossii, M. rubella, Mitella breweri, M. pentandra, Mnium, Montia, Muhlenbergia filiformis, Nuphar, Oreostemma alpigenum, Orthocarpus lacerus, Oryzopsis asperifolia, Osmorhiza depauperata, Oxypolis fendleri, O. occidentalis, Oxytropis sericea, Parnassia parviflora, Pedicularis attollens, P. crenulata, P. oederi, P. racemosa, Penstemon heterodoxus, Phalaris arundinacea, Phleum alpinum, P. pratense, Phyllodoce empetriformis, P. glanduliflora, Picea engelmannii, P. glauca, Pinus albicaulis, P. contorta, P. murrayana, Platanthera dilatata, P. hyperborea, P. obtusata, P. stricta, Poa abbreviata, P. alpina, P. compressa, P. pratensis, P. reflexa, Polemonium foliosissimum, Populus tremuloides, Potamogeton gramineus, Potentilla breweri, P. drummondii, P. diversifolia, P, gracilis, P. grayi, P. grandis, P. subjuga, Primula conjugens, P. egaliksensis, P. incana, P. jeffreyi, P. parryi, P. pauciflora, P. tetrandra, Ptilagrostis kingii, P. porteri, Pyrola minor, P. secunda, Ranunculus alismifolius, R. cymbalaria, R. eschscholtzii, Rhodiola integrifolia, R. rhodantha, Rhododendron albiflorum, R. neoglandulosum, R. tomentosum, Ribes montigenum, Rorippa, Rosa sayi, R. woodsii, Rubus acaulis, R. idaeus, Rudbeckia occidentalis, Rumex, Salix arctica, S. barclayi, S. bebbiana, S. boothii, S. brachycarpa, S. candida, S. commutata, S. drummondiana, S. eastwoodiae, S. geyeriana, S. monticola, S. myrtillifolia, S. orestera, S. petrophila, S. planifolia, S. phylicifolia, S. scouleriana, Saxifraga reticulata, Scutellaria, Selaginella selaginoides, Senecio cymbalarioides, S. triangularis, Sibbaldia procumbens, Silene acaulis, Sisyrinchium pallidum, Solidago multiradiata, Sorbus, Sphagnum, Spiraea douglasii, S. splendens, Spiranthes romanzoffiana, Stellaria longipes, Stereocaulon glareosum, S. nivale, Stipa, Streptopus* *amplexifolius, Stuckenia pectinata, Swertia perennis, S. radiata, S. perennis, Symphyotrichum foliaceum, S. spathulatum, Taraxacum officinale, Thalictrum alpinum, T. fendleri, T. sparsiflorum, T. venulosum, Tortula norvegica, Trichophorum pumilum, Trifolium hybridum, T. repens, Triglochin maritimum, T. palustre, Trisetum spicatum, Trollius laxus, Tsuga mertensiana, Utricularia ochroleuca, Vaccinium cespitosum, V. deliciosum, V. membranaceum, V. oxycoccos, V. scoparium, V. uliginosum, Valeriana edulis, V. sitchensis, Veratrum californicum, V. viride, Veronica alpina, V. cusickii, V. wormskjoldii, Viola adunca, V. macloskeyi, V. nephrophylla, V. palustris, Xerophyllum tenax, Zigadenus elegans.*

Pedicularis lanceolata Michx. is a short-lived perennial, which occurs in bogs, carrs, ditches, dunes, fens, floodplains, marshes, meadows, prairies, savannas, seeps, swales, swamps, and along the margins of lakes, ponds, pools, rivers, and streams. The plants thrive in open or shaded sites on alkaline (pH >7.0) limestone or peat substrates, but also can occur under acidic conditions. Their substrates can include clay, muck, sand, silt loam, and tufa. The plants normally grow for 2 years before reaching the flowering stage. Their flowers have a yellow corolla and are nectarless, but their structure is similar otherwise to that of the nectar-producing species. They have an intermediate pollination system achieved by nonvibrating, inverted (sternotribic), pollen-foraging, worker bumblebees (Hymenoptera: Apidae: *Bombus affinis, B. impatiens, B. vagans*). The seeds are winged and probably are wind dispersed. They germinate after being frozen for 5 weeks at 0°C, or when they have been exposed to alternating warm (20°C) and cold (4°C) periods followed by a treatment with gibberellic acid (2000 ppm). Initially, the plants will grow well without a host, but require a host in order to survive beyond 2 months. Growth in the presence of a host also will result in larger plants. The plants can be reared successfully using either clover (Fabaceae: *Trifolium incarnatum*) or wheat (Poaceae: *Triticum aestivum*) as an artificial host. Documented natural hosts are asterisked in the subsequent list of reported associates. There also is evidence of conspecific parasitism. Although the adults typically inhabit alkaline sites, the seedlings grow best at a low pH (6.2) where nutrients are more readily available. **Reported associates:** *Abies balsamea, Acer rubrum, A. saccharinum, Achillea millefolium, Acorus, Agalinis paupercula, A. tenuifolia, Ageratina altissima, Agrimonia parviflora, Agrostis alba, A. capillaris, A. gigantea, A. stolonifera, Alisma subcordatum, Alliaria petiolata, Alnus rugosa, A. serrulata, Alopecurus carolinianus, Ambrosia artemisiifolia, A. trifida, Amphicarpa bracteata, Andromeda polifolia, Andropogon gerardii, A. virginicus, Anemone canadensis, Angelica atropurpurea, Antennaria neglecta, Anticlea elegans, Apios americana, Apocynum cannabinum, A. sibiricum, Arnoglossum plantagineum, Asclepias incarnata, A. syriaca, A. tuberosa, Baptisia tinctoria, Berberis thunbergii, Betula glandulifera, B. nigra, B. papyrifera, B. pumila, B. ×sandbergii, B. cernuus, B. connatus, B. frondosus, B. laevis, Boehmeria cylindrica, Bromus ciliatus, B. inermis, Calamagrostis canadensis, C. stricta, Caltha palustris, Calystegia sepium, Campanula aparinoides,*

Campsis radicans, Cardamine bulbosa, C. pratensis, Carduus arvensis, Carex aquatilis, C. atlantica, C. bicknellii, C. blanda, C. bromoides, C. buxbaumii, C. communis, C. comosa, C. crawei, C. crinita, C. cryptolepis, C. flava, C. granularis, C. hystericina, C. interior, C. lacustris, C. lasiocarpa, C. leptalea, C. limosa, C. livida, C. lurida, C. pellita, C. prairea, C. retrorsa, C. sartwellii, C. schweinitzii, C. sterilis, C. stipata, C. stricta, C. suberecta, C. tetanica, C. tribuloides, C. trichocarpa, C. viridula, C. vulpinoidea, Carpinus caroliniana, Celastrus scandens, Cephalanthus occidentalis, Cerastium fontanum, Chara, Chelone glabra, C. lyonii, Cicuta bulbifera, C. maculata, Cinna arundinacea, Circaea lutetiana, Cirsium altissimum, C. discolor, C. lanceolatum, C. muticum, Cladium mariscoides, Clematis virginiana, Climacium americanum, Clinopodium vulgare, Comarum palustre, Cornus amomum, C. foemina, C. racemosa, C. rugosa, C. sericea, Corylus americana, Cuscuta, Cynanchum louiseae, Cyperus, Cypripedium kentuckiense, C. reginae, Dalea purpurea, Danthonia spicata, Dasiphora floribunda, Daucus carota, Decodon verticillatus, Deschampsia cespitosa, Desmodium cuspidatum, Dichanthelium acuminatum, D. clandestinum, Diodella teres, Diospyros virginiana, Doellingeria umbellata, Drosera anglica, D. rotundifolia, Dulichium arundinaceum, Echinochloa muricata, Elaeagnus commutata, Elaeodendron xylocarpum, Eleocharis acicularis, E. compressa, E. elliptica, E. erythropoda, E. palustris, E. pauciflora, E. rostellata, Elymus lanceolatus, E. trachycaulus, Epilobium coloratum, E. leptophyllum, E. strictum, Equisetum arvense, E. fluviatile, E. hyemale, E. laevigatum, E. palustre, E. variegatum, Erigeron pulchellus, Eriophorum angustifolium, E. gracile, E. perfoliatum, Euthamia graminifolia, E. tenuifolia, Eutrochium fistulosum, E. maculatum, Filipendula rubra, Fragaria vesca, F. virginiana, Frangula alnus, Fraxinus nigra, F. pennsylvanica, Galium aparine, G. boreale, G. brevipes, G. labradoricum, G. palustre, G. tinctorium, G. trifidum, Gentiana andrewsii, G. clausa, G. puberulenta, G. rubricaulis, Gentianopsis crinita, G. virgata, Geum canadense, G. rivale, Glechoma hederacea, Glyceria canadensis, G. grandis, G. septentrionalis, G. striata, Glycyrrhiza lepidota, Gomphrena globosa, Gratiola aurea, Helenium autumnale, Helianthus decapetalus, H. giganteus, H. grosseserratus, H. maximiliani, Hibiscus moscheutos, Hieracium caespitosum, H. odorata, Hierochloe odorata, Hydrocotyle americana, Hydrophyllum appendiculatum, Hypericum ascyron, H. kalmianum, H. mutilum, H. perforatum, Hypoxis hirsuta, Impatiens capensis, Iris pseudacorus, I. versicolor, I. virginica, Juncus alpinus, J. balticus, J. biflorus, J. brachycephalus, J. brevicaudatus, J. canadensis, J. dudleyi, J. effusus, J. nodosus, J. tenuis, J. torreyi, Juniperus virginiana, Lactuca, Larix laricina, Lathyrus palustris, Leersia oryzoides, Lespedeza procumbens, Leucanthemum vulgare, Liatris ligulistylis, L. pycnostachya, L. scariosa, Ligustrum vulgare, Lilium philadelphicum, Lindera benzoin, Linum sulcatum, Liparis loeselii, Liquidambar styraciflua, Liriodendron tulipifera, Lobelia cardinalis, L. kalmii, L. puberula, L. siphilitica, Lonicera japonica, L. morrowii, L. tatarica, Lotus corniculatus, Ludwigia alternifolia, Lycopus americanus, L. uniflorus, Lysimachia ciliata, L. quadriflora, L. quadrifolia, L. terrestris, L. thyrsiflora, Lythrum alatum, L. salicaria, Machaeranthera parviflora, Maianthemum canadense, M. racemosum, Malaxis spicata, Medicago lupulina, Melilotus alba, M. officinalis, Mentha aquatica, M. arvensis, Menyanthes trifoliata, Micranthes pensylvanica, Microstegium vimineum, Mimulus glabratus, M. ringens, Monarda media, Muhlenbergia asperifolia, M. glomerata, M. mexicanum, M. racemosa, Myosotis scorpioides, Myrica cerifera, M. gale, Nasturtium officinale, Oenothera parviflora, Onoclea sensibilis, Osmunda regalis, Oxalis corniculata, O. stricta, Oxypolis rigidior, Packera aurea, P. schweinitziana, Panicum flexile, P. virgatum, Parnassia caroliniana, P. glauca, P. grandifolia, P. palustris, Parthenocissus quinquefolia, Paspalum dilatatum, Peltandra virginica, Persicaria amphibia, P. coccinea, P. posumbu, P. sagittata, P. virginiana, Petalostemum purpureum, Petasites, Phalaris arundinacea, Phleum pratense, Phlox glaberrima, P. pilosa, Photinia floribunda, P. pyrifolia, Physocarpus opulifolius, Physostegia, Pilea fontana, P. pumila, Pinus strobus, Plantago lanceolata, P. major, P. rugelii, Platanthera clavellata, P. praeclara, Poa palustris, P. pratensis, Polygala cruciata, Ponthieva racemosa, Populus balsamifera, P. deltoides, P. grandidentata, P. tremuloides, Potentilla anserina, P. recta, P. simplex, Prunella vulgaris, Prunus serotina, Pteridium aquilinum, Pycnanthemum tenuifolium, P. virginianum, Quercus macrocarpa, Q. rubra, Ranunculus acris, R. hispidus, R. septentrionalis, Rhamnus alnifolia, R. frangula, Rhynchospora capillacea, R. capitellata, Ribes hirtellum, Rosa acicularis, R. carolina, R. multiflora, R. palustris, R. virginiana, Rubus allegheniensis, R. flagellaris, R. hispidus, R. idaeus, R. pubescens, Rudbeckia fulgida, R. hirta, R. laciniata, Rumex crispus, Sagittaria latifolia, Salix bebbiana, S. bicolor, S. candida, S. discolor, S. exigua, S. glaucophylloides, S. pedicellaris, S. petiolaris, S. sericea, Sanicula marilandica, S. odorata, Saururus cernuus, Schizachyrium scoparium, Schoenoplectus acutus, S. subterminalis, S. tabernaemontani, Scirpus atrovirens, S. cyperinus, S. hattorianus, S. lineatus, S. polyphyllus, Scleria verticillata, Scutellaria galericulata, Selaginella eclipes, Silphium integrifolium, S. terebinthinaceum, Sium suave, Smilax herbacea, Solanum carolinense, S. dulcamara, Solidago canadensis, S. gigantea, S. graminifolia, S. nemoralis, S. ohioensis, S. patula, S. ptarmicoides, S. riddellii, S. rigida, S. rugosa, S. uliginosa, Sonchus arvensis, Sorghastrum nutans, Sparganium androcladum, Spartina pectinata, S. gracilis, Sphagnum, Sphenopholis obtusata, Spiraea alba, S. tomentosa, Spiranthes cernua, S. lucida, Stellaria longifolia, Sullivantia sullivantii, Symphyotrichum boreale, S. ericoides, S. firmum, S. laeve, S. lanceolatum, S. lateriflorus, S. novae-angliae, S. pilosum, S. praealtum, S. prenanthoides, S. puniceum, Symplocarpus foetidus, Taraxacum officinale, Taxodium distichum, Thalictrum dasycarpum, T. dioicum, T. pubescens, Thelypteris palustris, Thuidium delicatulum, Thuja occidentalis, Toxicodendron radicans, T. vernix, Triadenum fraseri, Trantha glutinosa, Trichophorum cespitosum, Trifolium campestre, T. dubium, T. incarnatum, T. pratense, T. repens, *Triglochin maritimum, T. palustre, Triticum aestivum, Tsuga canadensis, Typha angustifolia, T. ×glauca, T. latifolia, Ulmus rubra, Valeriana edulis, Veratrum viride, Verbena hastata, V. urticifolia, Vernonia fasciculata, V. gigantea, V. noveboracensis, Viburnum acerifolium, V. dentatum, V. lentago, V. nudum, V. opulus, V. recognitum, Viola cucullata, V. lanceolata, V. nephrophylla, Vitis labrusca, V. riparia, Zizia aptera, Z. aurea.*

Pedicularis palustris L. is a biennial or summer annual, which inhabits coastal bogs, ditches, fens, marshes, and meadows at elevations of up to 400 m. It tolerates a broad range of acidity (pH: 3.8–8.3) but tends to occur in more acidic sites (pH: 5.1–6.9). The plants occur on organic to sandy substrates and are highly tolerant to sediment iron levels. This species is monocarpic (a biennial in North America). The flowers have a pink to purple corolla, are self-compatible, but produce nectar (average sugar content of 13.9%) and normally are pollinated nototribically by bumblebee (Hymenoptera: Apidae: *Bombus*) queens foraging for nectar, and sternotribically by bumblebee worker species that scrape pollen from the concealed anthers. A lack of pollinators can limit seed production. In some cases, self-pollination can lead to notable seed set; however, the seed number per capsule is reduced significantly in such cases. There usually are 20 seeds/flower and more than 1000 seeds/plant. The seeds are buoyant (remaining afloat for up to 557 days) and are dispersed by water. They have conditional primary dormancy, which is overcome when imbibed seeds are buried and then exposed to cold temperatures (4°C). As a consequence, maximum germination occurs in the spring and is reduced substantially in late summer and autumn. Fresh, untreated seeds exhibit low germination (1%–14%) and dry storage (28 days) induces secondary dormancy. Germination of dry-stored seeds is not strongly light dependent and occurs well under a 15°C/25°C temperature regime in the light. A long-term persistent seed bank of 60–140 seeds per m² is formed with a seed mortality of 75%, and germination rate greater than 58%, observed after 5 years. Genetic studies using AFLP markers have detected relatively high levels of genetic variation in both large and small populations. A significantly higher number of capsules and seeds per plant occurs in more genetically variable populations. Most of the ecological information on this species is extracted from European accounts given that few studies of North American populations are available. The following list of associates is compiled from European accounts that include taxa also found in the North American range of the species. **Reported associates:** *Agrostis canina, A. gigantea, A. stolonifera, Aneura pinguis, Angelica sylvestris, Anthoxanthum odoratum, Aulacomnium palustre, Betula nana, Bryum pseudotriquetrum, Calamagrostis stricta, Caltha palustris, Campylium stellatum, Carex canescens, C. diandra, C. echinata, C. gynocrates, C. lasiocarpa, C. limosa, C. mackenziei, C. magellanica, C. nigra, C. panicea, C. rostrata, C. vaginata, C. viridula, Cirsium palustre, Cladium mariscoides, Comarum palustre, Corallorhiza trifida, Drosera anglica, D. rotundifolia, Eleocharis quinqueflora, Empetrum nigrum, Epilobium palustre, Eriophorum angustifolium, E. vaginatum, Equisetum fluviatile, Festuca* *rubra, Filipendula ulmaria, Fissidens adianthoidesm, Galium palustre, Geranium sylvaticum, Holcus lanatus, Hylocomium splendens, Iris pseudacorus, Juncus articulatus, J. effusus, J. gerardii, Juniperus communis, Lathyrus palustris, Limprichtia revolvens, Lotus corniculatus, Lysimachia thyrsiflora, Mentha aquatica, Menyanthes trifoliata, Palustriella commutata, Parnassia palustris, Phragmites australis, Pinguicula vulgaris, Plantago maritima, Pleurozium schreberi, Potentilla anserina, P. erecta, Prunella vulgaris, Ranunculus flammula, R. repens, Rhizomnium pseudopunctatum, Rubus chamaemorus, Salix, Samolus valerandi, Scorpidium scorpioides, Sphagnum fuscum, Stellaria palustris, Thalictrum, Thelypteris palustris, Tomentypnum nitens, Triglochin maritimum, T. palustre, Vaccinium uliginosum, Valeriana dioica, Vicia cracca.*

Use by wildlife: *Pedicularis attollens* provides important breeding and postbreeding habitat for the white-tailed ptarmigan (Aves: Phasianidae: *Lagopus leucura*). *Pedicularis chamissonis* is an important element of nesting habitat for the Lapland bunting (Aves: Emberizidae: *Calcarius lapponicus*). It is found as cover on dune and tundra dens of the arctic fox (Carnivora: Canidae: *Alopex lagopus beringensis*). *Pedicularis chamissonis* is a host of the arctic blister rust (Basidiomycota: Cronartiaceae: *Cronartium kamtschaticum*) and powdery mildew (Ascomycota: Erysiphaceae: *Podosphaera macularis*). Nectar from *P. crenulata* is taken by bumblebees (Insecta: Hymenoptera: Apidae: *Bombus*) and leafcutting bees (Hymenoptera: Megachilidae). The plants are host to a sac fungus (Ascomycota: Bionectriaceae: *Nectriella pedicularis*) and a rust fungus (Basidiomycota: Cronartiaceae: *Cronartium coleosporioides*). *Pedicularis groenlandica* is highly digestible and contains from 8% to 20% crude protein (with highest levels occurring in the preflowering stage). The plants are eaten by the Rocky Mountain elk (Mammalia: Cervidae: *Cervus canadensis nelsoni*), Rocky Mountain goat (Mammalia: Bovidae: *Oreamnos americanus*), and sheep (Mammalia: Bovidae: *Ovis aries*). They also comprise nesting habitat of the yellow rail (Aves: Rallidae: *Coturnicops noveboracensis*). *Pedicularis groenlandica* is a host of the Gillett's checkerspot butterfly (Insecta: Lepidoptera: Nymphalidae: *Euphydryas gillettii*) as well as several fungi (Ascomycota: Sclerotiniaceae: *Sclerotinia coloradensis*; family uncertain: *Phoma pedicularis*; Basidiomycota: Cronartiaceae: *Cronartium coleosporioides*; Pucciniaceae: *Puccinia clintonii*). *Pedicularis lanceolata* is grazed by white-tailed deer (Mammalia: Cervidae: *Odocoileus virginianus*). It also is a host of the Baltimore checkerspot butterfly (Lepidoptera: Nymphalidae: *Euphydryas phaeton*). *Pedicularis attollens* and *P. groenlandica* usually are not mycorrhizal; however, some plants of *P. groenlandica* are associated with arbuscular mycorrhizae and dark septate endophytes. Short-tongued worker bumblebees (Hymenoptera: Apidae: *Bombus terricola*) rob the nectar from the flowers of *P. palustris* by perforating the base of the corolla, sometimes effecting pollination. The foliage of *P. palustris* is eaten by reindeer (Mammalia: Cervidae: *Rangifer tarandus*) and birds (Aves).

Economic importance: food: The Iroquois cooked the leaves of *Pedicularis lanceolata* as a vegetable; however, the plants also are considered to be poisonous. Because *P. groenlandica* can accumulate the toxic alkaloid senecionine when using *Senecio triangularis* as a host, it is best to avoid ingesting this plant; **medicinal:** *Pedicularis* extracts are sold as herbal remedies purported to have muscle relaxant and sedative properties. Native Americans reportedly smoked the flowers of some species for their soothing effect. The Washo people used a decoction of leaves from *Pedicularis attollens* as a tonic and applied a poultice made from the plant to treat wounds and swellings. The Cheyenne made a cough medicine from an infusion of the powdered leaves and stems from *P. groenlandica*; **cultivation:** Seeds of *Pedicularis chamissonis, P. crenulata, P. groenlandica* and *P. lanceolata* are distributed occasionally to plant as garden ornamentals; **misc. products:** *Pedicularis groenlandica* seeds are included as an ingredient of the Yellowstone National Park (YNP) wetlands seed mix. Seedlings or plugs of *P. lanceolata* also are planted for wetland restoration; **weeds:** none; **nonindigenous species:** none

Systematics: Phylogenetic studies of Orobanchaceae using *PHYA* sequence data (Figure 5.72) resolve *Pedicularis* at the base of a clade ("clade IV"), which contains several other genera with OBL North American taxa. The monophyly of *Pedicularis* has been confirmed by a phylogenetic analysis of combined nrITS and cpDNA (*matK*) sequence data for 71 species representing 20 sections and 43 series. However, only one of the OBL North American taxa (*P. chamissonis*; subgenus *Cyclophyllum*) was included in that analysis. A more recent study that evaluated the same two loci for more than 80 *Pedicularis* taxa (26 from North America) included all of the OBL species except *P. palustris*. That study corroborated the monophyly of *Pedicularis* and also indicated that in North America, the OBL habit has arisen independently at least three times (Figure 5.78). *Pedicularis lanceolata* is believed to be related to the Asian *P. resupinata* (subgenus *Allophyllum*). Unfortunately, both species have not yet been included simultaneously in any of the molecular phylogenetic analyses as a means of testing that hypothesis. *Pedicularis palustris* comprises two subspecies: *P. pedicularis* subsp. *palustris* (biennial) and *P. pedicularis* subsp. *karoi* (annual); the latter occurs only in Asia. Molecular data indicate the nearest relatives of *P. chamissonis* to be *P. kansuensis* and *P. nodosa* (subgenus *Sigmantha*), which resolve in a clade that is quite distant from other North American species. Studies of *Pedicularis chamissonis* in Japan have disclosed two taxa that differ by their cpDNA haplotypes, ISSR markers, and nrITS sequences. No evidence of gene flow between the taxa is evident and they also exhibit several morphological differences. Preliminary DNA evidence indicates that *P. chamissonis* should be subdivided into two similar but genetically distinct species. The remaining OBL species occur within a clade that is predominantly North American, but also includes several species with Asian and European distributions (Figure 5.78). The base chromosome number of *Pedicularis* is $x = 8$, with reports for *P. attollens, P. groenlandica, P. lanceolata,* and *P. palustris* indicating that they all are diploid ($2n = 16$).

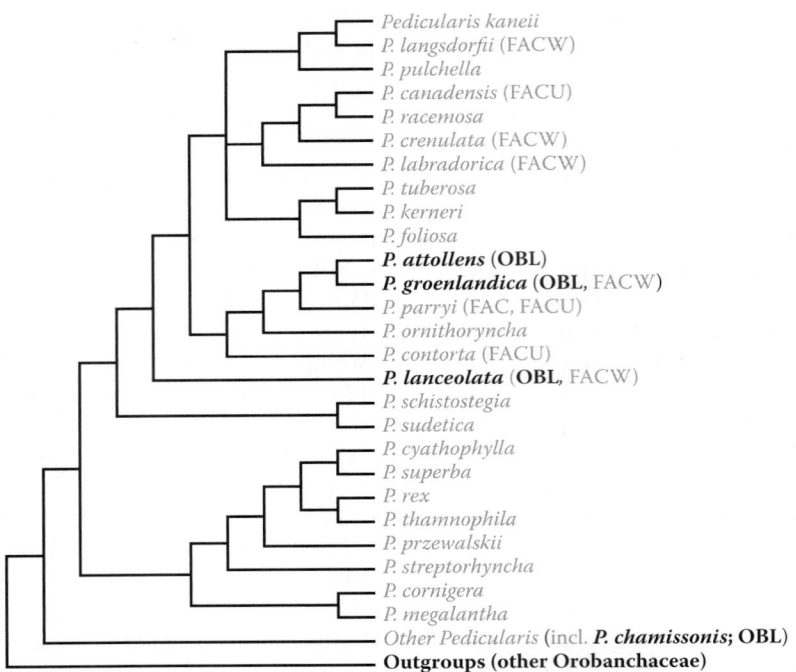

FIGURE 5.78 Phylogenetic relationships in *Pedicularis* as indicated by analyses of combined *matK* and nrITS sequence data. At least three independent origins of the OBL habit (bold) are indicated for the genus. Wetland indicator status is also provided, where applicable, for other North American species. (Adapted from strict consensus tree of Robart, B.W., *Syst. Bot.*, 30, 644–656, 2005.)

Hybridization between sympatric *Pedicularis* species is rare and often is limited by several types of isolating mechanisms. Although *P. groenlandica* and *P. attollens* are sister species, and are pollinated by the same bumblebee (Hymenoptera: Apidae: *Bombus*) species, their pollen is distributed dorsally by the former and on the forehead by the latter (resulting in so-called "*Pedicularis* type" of mechanical isolation). The pollinators also readily recognize the different species and appear to be highly constant in their floral visitation patterns (ethological isolation).

Comments: *Pedicularis* species with fairly broad distributions include *P. groenlandica*, which occurs in northern and western North America, and *P. lanceolata*, which occurs throughout eastern North America. More limited distributions characterize *P. crenulata* (western United States), *P. attollens* (California, Nevada, Oregon), *P. palustris* (eastern Canada, but extending into northern Europe), and *P. chamissonis* (Alaska, but extending into Eurasia).

References: Allard, 2001; Anderson et al., 1999; Andersson et al., 2000; Bamberg & Major, 1968; Bank II, 1953; Baskin & Baskin, 2002a; Benedict, 1983; Bonde, 1965; Borg et al., 1980; Cázares et al., 2005; Cholewa & Henderson, 1984b; Christy, 2004; Cooper & Sanderson, 1997; Costelloe, 1988; Doolittle et al., 1990; Doran, 1943; Duminil et al., 2007; Evans et al., 2001b; Fischer, 2004; Frederick & Gutiérrez, 1992; Frolik, 1941; Fryday & Glew, 2003; Fujii et al., 1997, 2001; Grant, 1994; Helm, 1982; Heusser, 1978; Jensen, 2004; Johnson & Steingraeber, 2003; Johnston et al., 1968; Jumpponen et al., 1998; Koeman-Kwak, 1973; Kufeld, 1973; Lackney, 1981; Leppik, 1967; Macior, 1968, 1969, 1970, 1973, 1977, 1983,

1993; Mehringer, Jr. et al., 1977; Peinado et al., 2005a; Piehl, 1965; Record, 2011; Ree, 2005; Robart, 2010; Robart et al., 2015; Saunders, Jr., 1955; Schaffner, 1904; Schmidt & Jensen, 2000; Schneider & Stermitz, 1990; Seaver, 1909; Snowden & Wheeler, 1993; Stern et al., 1993; Wang & Li, 2005; Wilson, 1948; Yoshie, 2008; Zagrebel'nyi, 2003.

Family 7: Phrymaceae [9–15]

Several molecular phylogenetic investigations (Beardsley & Olmstead, 2002; Beardsley et al., 2004; Beardsley & Barker, 2005) have warranted the transfer of several former genera of Scrophulariaceae tribe Mimuleae (e.g., *Glossostigma*, *Mimulus*) to Phyrmaceae, whose circumscription remains in a state of flux. Some analyses (Beardsley & Olmstead, 2002; Beardsley & Barker, 2005) suggest the inclusion of eleven genera (*Berendtiella*, *Elacholoma*, *Glossostigma*, *Hemichaena*, *Lancea*, *Leucocarpus*, *Mazus*, *Microcarpaea*, *Mimulus*, *Peplidium*, *Phryma*) within Phrymaceae; whereas, others (e.g., Oxelman et al., 2005) do not necessarily support the inclusion of *Mazus* and *Lancea* in the family. Barker et al. (2012) provided a prospectus, which circumscribed the family as having 13 genera comprising 188 species; however, that proposal recognized several segregate genera of *Mimulus*, which are not widely accepted. As a group, Phrymaceae possess a tubular, toothed calyx, loculicidal capsules, and tactiley sensitive, bilamellate stigmas. The family is relatively small in any case, with most estimates in the range of 175–230 species. Yet, there is substantial diversity in the group, including annual and perennial life histories and chasmogamous or cleistogamous flowers that are self-, insect-, or hummingbird

(Aves: Trochilidae) pollinated. Seed dispersal can occur by rain (splash-cup method), wind, adherence to mud on the feet of waterfowl (Aves: Anatidae), or by the dispersal of dislodged, capsule-bearing plants by water.

The nearest relatives of Phrymaceae remain uncertain because of poor resolution in phylogenetic analyses. There is weak support for a sister group relationship with *Paulownia*; however, it also is possible that *Paulownia* could be included within Phrymaceae (Beardsley & Olmstead, 2002). More recent analyses indicate a close (but unresolved) relationship among Phrymaceae, *Paulownia*, *Rehmannia*, *Mazus*, *Lancea*, and Orobanchaceae (Figure 5.50; but see details in Oxelman et al., 2005). Regardless, it does appear that the closest relatives of Phyrmaceae would be found somewhere within this group of genera.

Taxonomic problems also abound with respect to the circumscription of genera within family. Notably, *Mimulus* is not monophyletic as once circumscribed and has required major taxonomic realignments of species in order to reflect more natural genera that are consistent with results obtained from phylogenetic analyses. A number of these taxonomic reassignments remain unsettled and are not widely accepted. The present treatment follows conservative recommendations (Beardsley et al., 2004) to include the monotypic *Mimetanthe* (categorized as an OBL aquatic in the 1996 list, but as FACW in the 2013 revised list) within *Mimulus* and not to split the latter into several weakly defined segregate genera as Barker et al. (2012) suggested. *Glossostigma* is a relative newcomer to North America, having been first introduced to the continent in the late 20th century (Les et al., 2006). It was not assigned status as a wetland indicator in the 1996 list, but has been included among the obligate aquatics in the 2013 list. Consequently, there currently are two genera of Phyrmaceae that contain OBL species in North America:

1. *Glossostigma* Wight. & Arn.
2. *Mimulus* L.

1. *Glossostigma*

Mud mat

Etymology: from the Greek *glôssa stigmo*, meaning "tongue stigma" to describe the ligulate stigma
Synonyms: *Peltimela*
Distribution: global: Africa; Australia; Asia; North America; **North America:** eastern
Diversity: global: 7 species; **North America:** 1 species
Indicators (USA): OBL: *Glossostigma cleistanthum*
Habitat: freshwater; lacustrine, riverine; **pH:** 5.5–6.3; **depth:** 0–4 m; **life-form(s):** emergent herb, submersed (rosulate)
Key morphology: stems (to 20 cm) creeping, rooting at the nodes, the internodes variable (1–18 mm); leaves emersed (to 11 mm) or submersed (to 57 mm), opposite, spatulate, with two parallel, longitudinal lacunae basally; flowers solitary, cleistogamous, and short pedicelled (to 2 mm) when submersed, or chasmogamous and long pedicelled (>4 mm) when emersed; calyx (to 3 mm) 3-lobed, urceolate; corolla rudimentary in cleistogamous flowers, 5-lobed, bilabiate

and whitish in chasmogamous flowers; stamens 2; capsules (to 1.8 mm) many-seeded (to 73); seeds (to 0.5 mm) reticulate, flattened
Life history: duration: annual (fruit/seeds); perennial (rhizomes); **asexual reproduction:** rhizomes, shoot fragments; **pollination:** cleistogamous (self), insect; **sexual condition:** hermaphroditic; **fruit:** capsules (common); **local dispersal:** rhizomes, seeds (water, wind); **long-distance dispersal:** shoot fragments, seeds (waterfowl)
Imperilment: (1) *Glossostigma cleistanthum* [GNR]
Ecology: general: All *Glossostigma* species are obligate aquatics, which grow in shallow water or along the margins of lakes, ponds and watercourses. They are common at sites where water levels fluctuate, and where shorelines are characterized by a gentle slope. The genus contains annuals and perennials, which reproduce sexually by autogamy or are pollinated by insects. Local seed dispersal occurs by wind or a "splash cup" method, whereby the seeds are dislodged from the fruits by raindrops.

Glossostigma cleistanthum **W. R. Barker** occurs in shallow to deep waters of lakes, ponds and watercourses or their margins. The plants thrive in acidic (pH: 5.5–6.3), oligotrophic (10–34 ppb phosphorous) waters that are low in conductivity (29–97 μS/m^2) and alkalinity (4–17 mg/L). The substrates primarily are sand, sometimes mixed with silt and gravel. Unlike many wetland plants, this species grows well under conditions of permanent inundation. The plants behave as annuals when they are emersed (e.g., along shorelines), but perennate when they are submersed and will remain green throughout much of the winter when they occur at deeper sites. Emergent plants produce erect, chasmogamous flowers (mainly autogamous); whereas submersed plants produce prodigious quantities (up to 72% of all nodes) of spherical, cleistogamous flowers that self-pollinate. Roughly 2.5% of the flowers eventually produce mature fruits. The plants grow in thick patches that can exceed 25,000 plants m^{-2}. Thick beds of plants can yield upward of 23,000 seeds m^{-2}. Low germination (1.4%) has been observed for unstratified seeds (30 days at 15°C; 16 h light) placed on moist filter paper; however, much higher levels occur (>50%) after 60 days if brought to 24°C under greenhouse conditions. Seeds stratified at 4°C germinated somewhat better. However, seed germination apparently requires no pretreatment but simply a sufficient amount of time, perhaps to leach away inhibitory substances. Nearly 100% germination has been obtained for seeds after 6–8 months of storage in water at room temperature. The longevity of seeds in the seed bank is not known with certainty. Seeds can be dispersed locally from chasmogamous flowers by the wind or by rain (splash cup). Fruits derived from cleistogamous flowers sink immediately if they become dislodged; however, if they remain attached to a portion of shoot that contains the lacunate leaves, they will float with the plant fragment and ultimately will be deposited along the shoreline where they eventually germinate on the wet substrate. Long-distance dispersal occurs when the seeds are transported in mud that adheres to the feet of waterfowl (Aves: Anatidae). **Reported associates (North America):** *Brasenia*

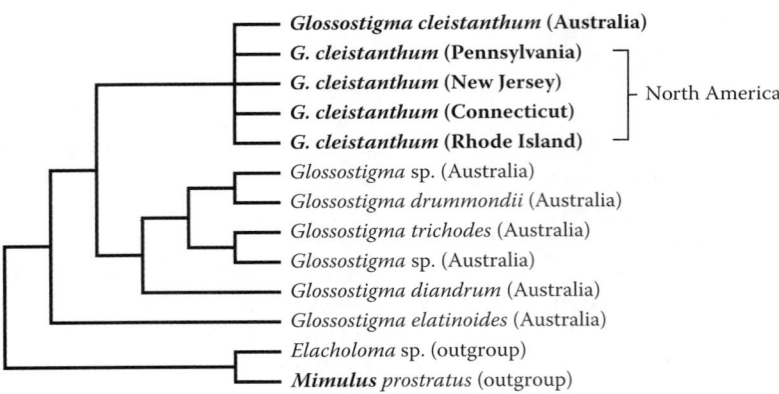

FIGURE 5.79 Phylogenetic relationships in *Glossostigma* as indicated by analysis of combined nrITS/ETS sequence data. All North American populations surveyed to date are identical to Australian material of *G. cleistanthum* for these loci but differ substantially from *G. diandrum*, the name originally applied to North American plants. North American taxa with OBL representatives are indicated in bold; however, the entire genus *Glossostigma* consists of aquatic species. (Adapted from Les, D.H. et al., *Amer. J. Bot.*, 93, 927–939, 2006.)

schreberi, Cabomba caroliniana, Callitriche heterophylla, Ceratophyllum demersum, C. echinatum, Crassula aquatica, Eleocharis acicularis, Elatine minima, Eriocaulon aquaticum, E. parkeri, Gratiola aurea, Heteranthera multiflora, Isoetes, Juncus pelocarpus, Lilaeopsis chinensis, Limosella subulata, Ludwigia palustris, Myriophyllum heterophyllum, M. heterophyllum × M. laxum, M. humile, M. tenellum, Najas flexilis, N. guadalupensis, Nymphoides cordata, Potamogeton bicupulatus, P. epihydrus, P. gramineus, Riccia fluitans, Sagittaria graminea, Sparganium, Utricularia gibba, U. striata, U. macrorhiza, Vallisneria americana.

Use by wildlife: The plants are eaten readily by ducks (Aves: Anatidae: *Anas*) and by Canada geese (Aves: Anatidae: *Branta canadensis*).

Economic importance: food: not edible; **medicinal:** none; **cultivation:** *Glossostigma* species are cultivated widely as freshwater aquarium plants. Being difficult taxonomically, it is not easy to identify the particular species that are being distributed by aquarium retailers; **misc. products:** none; **weeds:** *Glossostigma cleistanthum* is a fast-spreading weed that inhabits fairly pristine sites where it can exclude rare native species; **nonindigenous species:** *Glossostigma cleistanthum* was introduced to North America before 1992 most likely as a result of the careless disposal of aquarium specimens imported from Australia or New Zealand.

Systematics: An analysis of combined cpDNA (*trn*L/F spacer) and nrITS/external transcribed spacer (ETS) spacer sequence data for all currently known taxa in the genus resolved *Glossostigma* as monophyletic. However, the clade is nested along with several other genera (*Elacholoma, Microcarpaea, Peplidium*), within a larger clade comprising various species of *Mimulus*. From the taxa surveyed thus far, *Elacholoma* and *Mimulus prostratus* comprise the sister clade to *Glossostigma* (Figure 5.79). Molecular data place all North American *Glossostigma* populations surveyed to date with *G. cleistanthum* (Figure 5.79). Earlier reports of *G. diandrum* in North America were erroneous and resulted from taxonomic misidentification. The base chromosome number of *Glossostigma* is *x* = 5, with reported counts ranging from

diploid (2*n* = 10) to 12-ploid (2*n* = 60). *Glossostigma cleistanthum* (2*n* = 50) is a decaploid.

Comments: *Glossostigma cleistanthum* occurs in Connecticut, Delaware, Maryland, New Jersey, Pennsylvania, and Rhode Island.

References: Barker, 1982, 1992; Beardsley & Barker, 2005; Beardsley & Olmstead, 2002; Beuzenberg & Hair, 1983; Carter, 1973; Knapp et al., 2011; Lamont & Fitzgerald, 2001; Les et al., 2006; USDA, 1998; Willis, 1970.

2. *Mimulus*

Monkeyflower; mimule

Etymology: from the Greek *mimos*, meaning "actor," in reference to the mask-like corolla

Synonyms: *Cynorrhynchium*; *Diplacus*; *Erythranthe*; *Eunanus*; *Hemichaena*: *Mimetanthe*; *Monavia*

Distribution: global: Africa; Asia; Australia; North America; **North America:** widespread

Diversity: global: 100 species; **North America:** 95 species

Indicators (USA): OBL: *Mimulus alatus, M. alsinoides, M. angustatus, M. dentatus, M. dentilobus, M. eastwoodiae, M. floribundus, M. glabratus, M. glaucescens, M. guttatus, M. latidens, M. moschatus, M. pilosus, M. nudatus, M. primuloides, M. pygmaeus, M. ringens, M. tilingii, M. tricolor, M. washingtonensis*; **FACW:** *M. cardinalis, M. lewisii, M. moschatus, M. primuloides*

Habitat: freshwater; palustrine; **pH:** 4.9–8.5; **depth:** <1 m; **life-form(s):** emergent herb

Key morphology: stems (to 1–150 cm) erect or tufted; leaves (to 18–125 mm) opposite, sessile or petioled (to 12–95 mm), the blades lanceolate, oblong, ovate or round, the margins entire or toothed; flowers pedicelled (to 1–120 mm), bilabiate, axillary or in terminal racemes; corolla (to 11–60 mm) gold, maroon, purple, red, white or yellow, the throat with paired longitudinal folds; capsules (to 4–18 mm) many seeded

Life history: duration: annual (fruit/seeds) or perennial (bulblets, rhizomes, stolons); **asexual reproduction:** rhizomes; **pollination:** insect; **sexual condition:** hermaphroditic; **fruit:** capsules (common); **local dispersal:** bulblets, rhizomes, stolons; seeds (water, wind); **long-distance dispersal:** rhizome/stem fragments (water); seeds (birds, mammals) **Imperilment:** (1) *Mimulus alatus* [G5]; SX (MI); S1 (MA, NE); S2 (<u>ON</u>); S3 (CT, IA, NJ, NY); (2) *M. alsinoides* [G5]; S1 (ID); S3 (<u>BC</u>); (3) *M. angustatus* [G3]; (4) *M. cardinalis* [G5]; S1 (UT); S3 (AZ); (5) *M. dentatus* [G5]; S2 (<u>BC</u>); (6) *M. dentilobus* [G2]; S1 (TX); (7) *M. eastwoodiae* [G3]; S1 (AZ, CO); S3 (UT); (8) *M. floribundus* [G5]; S1 (<u>AB</u>); S2 (AR, WY); S3 (<u>BC</u>, CO); (9) *M. glabratus* [G5]; SH (MO); S1 (<u>AB</u>, CA, IL, <u>MB</u>, MI, <u>ON</u>, <u>QC</u>, <u>SK</u>, WY); S2 (IA); S3 (UT); (10) *M. glaucescens* [G3]; S3 (CA); (11) *M. guttatus* [G5]; S1 (MI, NE, ND); S2 (<u>SK</u>); S3 (NM); (12) *M. latidens* [G4]; S1 (OR); (13) *M. lewisii* [G5]; S2 (AK); S3 (<u>AB</u>, WY); (14) *M. moschatus* [G5]; S1 (MA, NH, VA); S2 (NJ, <u>ON</u>, VT, WY); (15) *M. pilosus* [G5]; S3 (UT); (16) *M. nudatus* [G3]; S3 (CA); (17) *M. primuloides* [G4]; S1 (NM, UT); S2 (MT); (18) *M. pygmaeus* [G4]; S2 (CA); (19) *M. ringens* [G5]; SH (CO); S1 (<u>AB</u>, AR, ID, MS, MT); S2 (LA); S3 (<u>PE</u>); (20) *M. tilingii* [G5]; S1 (<u>AB</u>, WY); (21) *M. tricolor* [G4]; S2 (OR); (22) *M. washingtonensis* [G4]; SX (WA)

Ecology: general: *Mimulus* is most diverse in western North America, where the species occur in a wide range of habitats including deserts, forests, and roadsides. Even in drier areas, they typically grow where soil conditions remain moist, often on sloping terrain. Generally, the species are quite common in disturbed areas and frequently are found on rocky, gravelly, or sandy substrates. Many of the species appear quickly after an area has been burned. More than a third of the North American *Mimulus* species are designated as wetland indicators, with just under a quarter of them categorized as OBL in at least some portion of their range. The flowers are showy and typically are pollinated by insects. Most of the species are self-compatible and are capable of self-pollination due to the simultaneous receptivity of the pollen and stigma; however, outcrossing is facilitated by the position of the stigma, which typically is elevated positionally above the stamens. The stigma lobes respond tactilely and reduce inadvertent self-pollinations by closing quickly once they have been contacted by a pollinator. Generally, the species that produce the most nectar also have the highest pollinator visitation rates, fruit set, and seed production levels. Microsatellite analyses have shown that both selfing and outcrossing taxa retain considerable genetic variation, but that the relative level in some selfing species has been reduced by historical population bottlenecks. Although both selfing and outcrossing species occur, the pollen to ovule ratios are relatively low for all *Mimulus* species. Their small seeds are dispersed by the wind, by water, in mud that adheres to the feet of waterfowl (Aves: Anatidae), and by mammals (Mammalia), which ingest them and transport them endozoically. *Mimulus cardinalis*, *M. lewisii*, and *M. pilosus* were ranked as OBL, FACW indicators in the 1996 list, but all three were reclassified as FACW in the 2013 list. The latter species is treated as *Mimetanthe pilosa* in the 2013 indicator list but has been retained here within *Mimulus* following taxonomic preferences (see *Systematics* later). Because their treatments had already been completed prior to the release of the 2013 revision, all three of these species have been retained here despite their revised indicator status. *Mimulus michiganensis* (OBL) is maintained as a distinct species in the 2013 indicator list, but is retained here as a variety of *M. glabra*; a separate treatment has not been provided (see *Systematics* later).

Mimulus alatus **Ait.** is a perennial, which occurs in bayous, deltas, ditches, floodplains, marshes, meadows, mudflats, prairies, sandbars, swales, swamps, and along the margins of ponds, rivers and streams. The sites range from open to shady and consist of circumneutral substrates (pH: 6.1–7.9) composed mainly of mud or sand. The plants can withstand temperatures as low as −31°C. The flowers are pollinated primarily by bumblebees (Apidae: *Bombus*). The seeds do not require stratification and will germinate at 21°C within 2 weeks when fresh. The seeds do not persist long in a seed bank and their viability has been observed to decrease (from 80% to 20%) after 2 years in storage. Vegetative reproduction occurs by the proliferation of stolons. **Reported associates:** *Acer negundo, A. nigrum, A. rubrum, A. saccharum, Acorus calamus, Agalinis purpurea, A. tenuifolia, Alisma subcordatum, Alnus rugosa, A. serrulata, Alternanthera philoxeroides, Amaranthus cruentus, A. spinosus, Ambrosia trifida, Ammannia coccinea, A. latifolia, Amorpha fruticosa, Amphicarpaea bracteata, Apios americana, Arundinaria gigantea, Asclepias incarnata, Bacopa, Berchemia scandens, Betula nigra, Bidens cernuus, B. comosus, B. connatus, B. discoidea, B. frondosus, B. laevis, B. polylepis, B. tripartitus, B. vulgatus, Boehmeria cylindrica, Bolboschoenus fluviatilis, Calamagrostis canadensis, Campanula americana, Campsis radicans, Cardamine pennsylvanica, Carex atlantica, C. baileyi, C. bromoides, C. crinita, C. debilis, C. festucacea, C. gracillima, C. grayi, C. gynandra, C. interior, C. intumescens, C. lacustris, C. laevivaginata, C. leptalea, C. lupulina, C. lurida, C. prasina, C. scoparia, C. seorsa, C. squarrosa, C. stipata, C. stricta, C. swanii, C. torta, C. tribuloides, C. typhina, C. vulpinoidea, Carpinus caroliniana, Carya ovata, Cephalanthus occidentalis, Chelone glabra, Cicuta maculata, Cinna arundinacea, Circaea lutetiana, Collinsonia canadensis, Commelina virginica, Conoclinium coelestinum, Cornus alternifolia, C. amomum, C. foemina, C. obliqua, C. racemosa, Crassula aquatica, Croton capitatus, Cyperus aristatus, C. flavescens, C. strigosus, Decumaria barbara, Desmodium cuspidatum, Dichanthelium clandestinum, D. dichotomum, Diodia teres, D. virginiana, Echinochloa crusgalli, E. walteri, Elatine americana, Eleocharis obtusa, E. ovata, E. palustris, Elymus virginicus, Epilobium coloratum, Equisetum fluviatile, E. hyemale, Eragrostis hypnoides, Eupatorium coelestinum, E. fistulosum, E. perfoliatum, E. rugosum, Fraxinus americana, F. nigra, F. pennsylvanica, Galium obtusum, Geum virginianum, Glyceria septentrionalis, G. striata, Habenaria, Helenium autumnale, Helianthus decapetalus, Heteranthera multiflora, Houstonia caerulea, Hydrangea arborescens, Hydrolea uniflora, Hypericum densiflorum, H. mutilum, H. perforatum, Ilex decidua, I. verticillata, Impatiens capensis, I. pallida, Iris pseudacorus, Itea virginica, Juglans nigra, Juncus coriaceus, J. effusus,*

J. tenuis, Laportea canadensis, Leersia oryzoides, L. virginica, Leptochloa, Leucospora multifida, Limosella subulata, Liatris spicata, Lindera benzoin, Lindernia dubia, Liquidambar styraciflua, Liriodendron tulipifera, Lobelia cardinalis, L. siphilitica, Lonicera morrowii, Lotus corniculatus, Ludwigia alternifolia, L. palustris, Lycopus americanus, L. rubellus, L. virginicus, Lysimachia ciliata, L. quadrifolia, Lythrum salicaria, Mecardonia, Mentha arvensis, Mikania scandens, Mimulus moschatus, M. ringens, Mollugo verticillata, Nemopanthus mucronatus, Nyssa biflora, N. sylvatica, Oldenlandia uniflora, Onoclea sensibilis, Osmunda regalis, Osmundastrum cinnamomeum, Ostrya virginiana, Oxypolis rigidior, Packera aurea, P. glabella, Panicum dichotomiflorum, P. rigidulum, Parthenocissus quinquefolia, Paspalum repens, Passiflora incarnata, Peltandra virginica, Penthorum sedoides, Persicaria arifolia, P. densiflora, P. hydropiper, P. hydropiperoides, P. lapathifolia, P. maculosa, P. pensylvanica, P. punctata, P. sagittata, Phlox maculata, Phragmites australis, Phryma leptostachya, Pilea pumila, Plantago major, Platanus occidentalis, Pluchea, Poa compressa, Polymnia canadensis, Pontederia cordata, Populus deltoides, P. heterophylla, Prunella vulgaris, Ptilimnium costatum, Quercus alba, Q. nigra, Q. palustris, Q. prinus, Q. rubra, Rhexia, Rhynchospora corniculata, R. glomerata, Rorippa islandica, Rosa multiflora, R. palustris, Rotala ramosior, Rubus argutus, R. hispidus, Rudbeckia hirta, R. laciniata, Ruellia caroliniensis, Rumex conglomeratus, R. verticillatus, Saururus cernuus, Sabal minor, Sabatia angularis, Sagittaria australis, S. latifolia, S. montevidensis, S. subulata, Salix alba, S. exigua, S. interior, S. nigra, S. sericea, Sambucus nigra, Saururus cernuus, Schoenoplectus pungens, S. purshianus, S. tabernaemontani, Scirpus atrovirens, S. cyperinus, Scutellaria lateriflora, S. nervosa, Setaria viridis, Sicyos angulatus, Sium suave, Smilax, Solidago caesia, S. rugosa, Sparganium eurycarpum, Spartina pectinata, Sphagnum, Stachys tenuifolia, Symphyotrichum lanceolatum, S. lateriflorum, S. prenanthoides, Symplocarpus foetidus, Taxodium distichum, Thelypteris palustris, Tilia americana, Toxicodendron radicans, T. vernix, Triadenum virginicum, T. walteri, Trifolium hybridum, T. stoloniferum, Typha angustifolia, T. latifolia, Ulmus americana, U. rubra, U. thomasii, Urtica dioica, Vaccinium corymbosum, V. fuscatum, Verbena hastata, Verbesina alternifolia, Vernonia noveboracensis, Viburnum dentatum, V. lentago, V. nudum, V. recognitum, Viola cucullata, Vitis, Zephyranthes atamasca, Zizania aquatica.

Mimulus alsinoides Dougl. ex Benth. is an annual, which grows in ditches, on cliffs or slopes, and along the banks of rivers and streams at elevations of up to 1375 m. The plants often occur in shady sites on cliffs, outcrops, in rocky crevices, or under ledges on substrates that include gravel, loam, shaley clay, or on thin soil over basaltic rock or sandstone. The plants are described as inhabiting acidic sites, although their specific pH tolerance has not been determined. Flowering occurs in spring (May to June). There is virtually no information on the reproductive biology of this species. The capsules produce numerous seeds, which reportedly will germinate in the light at 18°C–21°C after 1–3 months of cold stratification.

Reported associates: *Abies grandis, Acmispon parviflorus, Adiantum pedatum, Agrostis humilis, Aira caryophyllea, A. praecox, Allium cernuum, Amelanchier alnifolia, Anemone deltoidea, Arabis hirsuta, Arbutus menziesii, Arctostaphylos columbiana, A. nevadensis, A. uva-ursi, Arnica amplexicaulis, Aspidotis densa, Athyrium filix-femina, Barbarea vulgaris, Bolandra oregana, Calamagrostis canadensis, Caltha leptosepala, Camassia quamash, Cardamine oligosperma, Carex aquatilis, C. vesicaria, Ceanothus velutinus, Cheilanthes gracillima, C. siliquosa, Cicuta douglasii, Claytonia parviflora, Collinsia parviflora, Corylus cornuta, Dryopteris campyloptera, Elymus glaucus, Erysimum asperum, Festuca occidentalis, Fragaria chiloensis, Gaultheria shallon, Glyceria grandis, Heuchera micrantha, Holodiscus discolor, Hypericum anagalloides, H. scouleri, Juncus, Juniperus scopulorum, Luzula campestris, Mahonia aquifolium, Maianthemum dilatatum, Micranthes rufidula, Mimulus guttatus, Montia parvifolia, Nemophila parviflora, Oxalis oregana, Packera bolanderi, Paxistima myrsinites, Pedicularis groenlandica, Penstemon fruticosus, Pentagramma triangularis, Phacelia verna, Primula latiloba, Prunus emarginata, Pseudotsuga menziesii, Quercus, Rubus leucodermis, Rubus spectabilis, Salix lucida, Scirpus microcarpus, Selaginella douglasii, Spiraea douglasii, Stenanthium occidentale, Sullivantia oregana, Synthyris reniformis, Tellima grandiflora, Thuja plicata, Tsuga heterophylla, Vaccinium uliginosum, Viola glabella, Vulpia myuros.*

***Mimulus angustatus* (A. Gray) A. Gray** is an annual, which occurs in chaparral, depressions, flats, meadows, seeps, swales, vernal pools, and along the dried bottoms or margins of streams at elevations of up to 1525 m. The plants grow in sunny sites on substrates that include gravel, gravelly sand, obsidian rubble, red clay, rock, volcanic soil, and serpentine. The range of pH tolerance has not been reported. This is regarded as a fugitive species where optimal growth and reproduction occur in disturbed sites such as mounds that are created by pocket gophers (Rodentia: Geomyidae). Such disturbances limit competition with annual grasses and other herbs, resulting in higher germination and growth rates. Plant cover is highest in recently disturbed sites but declines rapidly, with a 50% reduction in cover observed within 2 years following a disturbance. Larger flowers also are produced on plants that grow in disturbed sites. Seed production is pollen limited and the amount of pollen deposited on the stigma increases with flower size. The plants are self-compatible, but higher seed set occurs when they are outcrossed by insects. Specific pollinators have not yet been identified. The seeds remain viable in the seed bank for at least several years, but little information on their germination requirements, etc. is available. **Reported associates:** *Adenostoma fasciculatum, Arctostaphylos canescens, Ceanothus cuneatus, Cupressus sargentii, Heteromeles arbutifolia, Isoetes orcuttii, Lasthenia californica, Lomatium, Mimulus guttatus, M. montioides, M. tricolor, Pinus sabiniana, Plagiobothrys glyptocarpus, Psilocarphus brevissimus, P. oregonus, P. tenellus, Quercus douglasii, Q. dumosa, Q. garryana, Ranunculus muricatus, Thysanocarpus radians, Toxicodendron diversilobum, Umbellularia californica.*

Mimulus cardinalis **Dougl. ex Benth.** is a perennial, which occurs in chaparral, meadows, oxbows, on sandbars, in seeps, sloughs, washes, and along the margins of lakes, ponds, rivers, and streams at elevations of up to 2625 m. The plants tolerate full sun to shade and occur on a fairly broad range of substrates (pH: 5.5–7.2) that include ash, calcareous alluvium, cobble, clay loam, gabbro/clay, granite, granitic loam, granitic sand, gravel, mud, sand, sandstone, sandy loam, schist, serpentine, and silty clay. They can withstand temperatures down to −33°C. The red-petalled flowers are self-compatible but normally do not set seed unless they are pollinated by hummingbirds (Aves: Trochilidae: *Calypte anna*). A single plant can produce several hundred fruits, with each containing up to 2500 seeds. Germination at 25°C is reported for seeds that have been stratified at 5°C–18°C for 2 weeks, then at 4°C for 6 weeks. Seeds not germinating within 10 weeks may require an additional 6 weeks of stratification at 4°C. The seeds are known to persist in the seed bank for at least 1 year. Although often categorized as an annual, the plants are perennial and reproduce vegetatively by means of a rhizome. Demographic studies have shown that marginal populations exhibit higher growth rates and survival than central populations. Annual survival is much higher for reproductive plants (72%–91%) than for vegetative plants (11%–22%). **Reported associates:** *Abies magnifica, Acer circinatum, A. macrophyllum, Acmispon glabrus, Adenocaulon bicolor, Adenostoma fasciculatum, Adiantum capillus-veneris, A. pedatum, Aesculus californica, Ageratina adenophora, Allionia gausapoides, Allium validum, Alnus rhombifolia, Aquilegia flavescens, A. formosa, Aralia californica, Arbutus menziesii, Artemisia californica, A. douglasiana, A. tridentata, Arundo donax, Asarum caudatum, Asclepias eriocarpa, Athyrium filix-femina, Atriplex canescens, Baccharis salicifolia, Bidens laevis, Boykinia elata, Brachythecium frigidum, Brickellia californica, Calycanthus occidentalis, Campanula prenanthoides, Carex hendersonii, C. lanuginosa, C. nudata, C. senta, Castilleja linariifolia, Ceanothus leucodermis, Centaurium venustum, Cercis canadensis, Chamaecyparis, Clematis lasiantha, C. ligusticifolia, Conocephalum conicum, Cornus glabrata, C. nuttallii, C. sericea, Corylus cornuta, Cyperus eragrostis, C. involucratus, Datisca glomerata, Dicentra formosa, Disporum hookeri, Distichlis spicata, Eleocharis parishii, Epipactis gigantea, Eriastrum densifolium, Erigeron philadelphicus, Eriodictyon crassifolium, E. trichocalyx, Eriogonum elongatum, E. fasciculatum, Frangula californica, Fraxinus latifolia, Galium triflorum, Glyceria elata, Helenium bigelovii, H. puberulum, Heteromeles arbutifolia, Juncus covillei, J. effusus, J. orthophyllus, J. rugulosus, Keckiella breviflora, Larrea tridentata, Ligusticum apiifolium, Lilium humboldtii, Lobelia cardinalis, Lysichiton americanus, Maianthemum stellatum, Malacothamnus, Melilotus alba, Mentha spicata, Mimulus floribundus, M. glaucescens, M. guttatus, M. lewisii, M. moschatus, M. pilosus, M. tilingii, Mitella ovalis, Montia perfoliata, Muhlenbergia rigens, Nasturtium officinale, Oenothera elata, Panicum pacificum, Parthenocissus vitacea, Peltiphyllum peltatum, Persicaria hydropiperoides,* *P. lapathifolia, Petasites palmatus, Phacelia, Philadelphus lewisii, Physocarpus capitatus, Pinus jeffreyi, Piptatherum miliaceum, Platanus racemosa, Polypogon monspeliensis, P. viridis, Populus fremontii, P. tremuloides, P. trichocarpa, Primula pauciflora, Prunella vulgaris, Prunus ilicifolia, Pseudotsuga menziesii, Quercus agrifolia, Q. engelmannii, Q. kelloggii, Q. wislizeni, Rhamnus ilicifolia, Rhus integrifolia, Ribes, Rorippa, Rosa gymnocarpa, Rubus leucodermis, R. ursinus, Rumex, Salix exigua, S. gooddingii, S. laevigata, S. lasiolepis, S. lucida, S. lutea, Salvia apiana, Sambucus nigra, Schoenoplectus acutus, Scoliopus hallii, Senecio triangularis, Solidago confinis, Stachys stricta, Stenanthium occidentale, Symphoricarpos rivularis, Symphyotrichum subulatum, Tamarix, Taxus brevifolia, Tellima grandiflora, Tiarella trifoliata, Toxicodendron diversilobum, T. rydbergii, Trifolium longipes, Typha domingensis, T. latifolia, Umbellularia californica, Urtica holosericea, Verbena lasiostachys, Veronica anagallis-aquatica, Viola glabella, Vitis californica, V. girdiana, Vulpia microstachys, Woodwardia fimbriata, Xanthium strumarium, Xylococcus bicolor, Yucca whipplei.*

Mimulus dentatus **Nutt. ex Benth.** is a perennial, which grows in coastal ditches, seeps, springs, swamps, and along the margins of rivers and streams at elevations of up to 400 m. The substrates are circumneutral (pH: 6.0–8.0) and include gravel, gravelly loam, sand, sandstone, and sandy loam. The plants can survive in full sun but occur mainly in partial shade and will tolerate cold temperatures down to −16°C. Little information exists on the pollination biology or seed ecology of this species. Vegetative reproduction occurs by extension of rhizomes. **Reported associates:** *Abies grandis, Acer circinatum, A. macrophyllum, Adenocaulon bicolor, Adiantum pedatum, Agrostis alba, Alnus oregana, A. rubra, Angelica arguta, Athyrium filix-femina, Bistorta bistortoides, Blechnum spicant, Boykinia elata, Bromus sitchensis, B. vulgaris, Cardamine angulata, C. occidentalis, C. oligosperma, Carex deweyana, Chamaecyparis lawsoniana, Chamerion angustifolium, Chrysosplenium glechomifolium, Circaea alpina, Cirsium arvense, Claytonia sibirica, Corydalis scouleri, Dicentra formosa, Digitalis purpurea, Dryopteris arguta, Elymus glaucus, Epilobium ciliatum, Equisetum laevigatum, Festuca, Filipendula occidentalis, Galium aparine, G. oreganum, G. triflorum, Glyceria striata, Holcus lanatus, Hydrophyllum, Juncus effusus, Lactuca serriola, Luzula parviflora, Lysichiton americanus, Maianthemum dilatatum, Marah oreganus, Melica subulata, Menziesia ferruginea, Micranthes occidentalis, Mimulus guttatus, M. moschatus, Mitella caulescens, M. ovalis, Montia parvifolia, Myrica californica, Oenanthe sarmentosa, Oplopanax horridum, Osmorhiza chilensis, Oxalis oregana, O. trilliifolia, Petasites frigidus, Picea sitchensis, Platanthera stricta, Pleuropogon refractus, Poa trivialis, Polypodium glycyrrhiza, P. scouleri, Polystichum kruckebergii, P. munitum, Prunella vulgaris, Ranunculus repens, R. uncinatus, Ribes bracteosum, Rosa nutkana, Rubus leucodermis, R. parviflorus, R. spectabilis, R. ursinus, Rumex crispus, R. obtusifolius, Salix lasiolepis, Sambucus callicarpa, S. racemosa, Saxifraga mertensiana, Scirpus microcarpus, Senecio jacobaea, S. triangularis,*

Sequoia sempervirens, Stachys mexicana, Stellaria calycantha, S. media, S. crispa, Streptopus amplexifolius, Thalictrum occidentale, Thuja plicata, Tiarella trifoliata, T. unifoliata, Tolmiea menziesii, Trautvetteria caroliniensis, Trifolium repens, Trisetum cernuum, Tsuga heterophylla, Urtica dioica, Vaccinium ovalifolium, V. parvifolium, Veratrum californicum, Veronica americana, Viola glabella.

Mimulus dentilobus B.L. Robins. & Fern. is a perennial found on ledges by waterfalls, on mud flats, seepage cliffs, springs, and along streambeds at elevations from 488 to 3108 m. The substrates have been described as gravel, silty loam, and travertine. The creeping stems root at their nodes and can form dense mats. Ecological information is extremely scarce for this species. **Reported associates:** *Acacia, Bothriochloa barbinodis, Carex, Castilleja linariifolia, Celtis, Fraxinus velutina, Juncus, Micranthes odontoloma, Mimulus guttatus, Parkinsonia, Polypogon monspeliensis, Prosopis, Veronica.*

Mimulus eastwoodiae Rydb. is a perennial, which inhabits shallow caves, cliffs, seeps, springs, and stream bottoms at elevations from 940 to 2180 m. The plants often occur in shade. The substrates typically are made up of igneous gravel, sand (Wingate-Keyanta) or sandstone (Cedar Mesa, Navajo) that is infused with calcareous to saline water. The specific range of pH tolerance has not been determined. The flowers have bright red corollas and are pollinated by hummingbirds (Aves: Trochilidae). There is little additional information available on the life history of this species. The plants are specialists in "hanging garden" habitats, which develop on wet cliff walls or alcoves, and support a distinct assemblage of suitably adapted species. The plants spread vegetatively by their stolons, which can root at the nodes. **Reported associates:** *Adiantum capillus-veneris, A. pedatum, Agrostis gigantea, A. semiverticillata, Alnus rhombifolia, Amelanchier utahensis, Andropogon hallii, Anticlea vaginata, Aquilegia micrantha, Arctostaphylos patula, Artemisia tridentata, Asclepias asperula, A. rusbyi, Athyrium filix-femina, Baccharis salicina, Berberis fremontii, Betula occidentalis, Brickellia longifolia, Calamagrostis scopulorum, Carex aurea, C. bolanderi, C. curatorum, C. specuicola, Castilleja, Celtis laevigata, Cercis occidentalis, Cercocarpus montanus, Circaea alpina, Cirsium arizonicum, C. rydbergii, Claytonia perfoliata, Coleogyne ramosissima, Collinsia childii, Corylus cornuta, Dalea candida, Dichanthelium acuminatum, Draperia systyla, Elaeagnus angustifolia, Ephedra, Epipactis gigantea, Equisetum hyemale, E. laevigatum, Ericameria nauseosa Erigeron kachinensis, E. sionis, E. zothecinus, Euphorbia brachycera, Fallugia paradoxa, Fendlera rupicola, Forestiera pubescens, Fraxinus anomala, Galium aparine, Glyceria elata, Gutierrezia sarothrae, Herrickia glauca, H. wasatchensis, Heterotheca villosa, Hordeum jubatum, Ipomopsis aggregata, Juncus effusus, J. ensifolius, Juniperus, Lobelia cardinalis, Maianthemum stellatum, Mentha arvensis, Mimulus cardinalis, M. guttatus, Nolina microcarpa, Opuntia engelmannii, O. phaeacantha, O. polyacantha, Osmorhiza berteroi, Packera multilobata, Panicum, Pellaea, Penstemon, Perityle specuicola, Petrophyton caespitosum, Phragmites australis, Pinus edulis,* *Platanthera zothecina, Populus fremontii, Primula pauciflora, P. specuicola, Prosopis velutina, Prunus virginiana, Pseudognaphalium, Pseudotsuga menziesii, Psorothamnus fremontii, Purshia mexicana, Quercus gambelii, Q. turbinella, Rhamnus betulaefolia, Rhizocarpon superficiale, Rhus trilobata, Rubus neomexicana, R. parviflorus, Salix exigua, Schizachyrium scoparium, Senna bicapsularis, Shepherdia, Solidago velutina, Stephanomeria, Sullivantia hapemanii, Viola glabella, Woodsia oregana, Yucca angustissima.*

Mimulus floribundus Lindl. is an annual plant reported from chaparral, crevices, depressions, ditches, hot springs, ledges, meadows, pools, sandbars, seeps, sloughs, springs, streams, swales, washes, and the margins of lakes, rivers, and streams at elevations of up to 2970 m. The substrates range from acidic to calcareous (the specific pH range has not been determined) and include basalt, clay, granite, gravel, limestone, loam, obsidian, rock, sand, schist, serpentine, shale, silt, and silt loam. The plants occur in sun or shade, often in flowing water, and often in areas that recently have been burned. They also are common in drying stream and riverbeds. The yellow-petalled flowers are self-compatible and highly self-pollinating, but there are few details on other aspects of the life history. The plants are glandular and coated with a mucilaginous slime of uncertain function. **Reported associates:** *Achillea millefolium, Acmispon americanus, A. glabrus, A. heermannii, Aconitum, Adenostoma fasciculatum, A. sparsifolium, Ageratina adenophora, Agropyron, Agrostis idahoensis, A. scabra, Allium acuminatum, Amorpha, Antennaria umbrinella, Aquilegia formosa, Arabidopsis thaliana, Arctostaphylos uva-ursi, Aristida, Artemisia californica, A. douglasiana, A. dracunculus, A. tridentata, Aspidotis californica, A. densa, Astragalus douglasii, A. lentiginosus, Avena barbata, Baccharis salicifolia, Bergerocactus emoryi, Boykinia rotundifolia, Brickellia californica, Bromus laevipes, B. madritensis, B. sterilis, Calamagrostis breweri, Calocedrus decurrens, Calochortus plummerae, C. venustus, Calystegia macrostegia, Camassia cusickii, Camissonia californica, C. hirtella, Campanula rotundifolia, Carex alma, C. aquatilis, C. douglasii, C. fracta, C. luzulaefolia, C. nebrascensis, C. rossii, C. scopulorum, C. spectabilis, Ceanothus crassifolius, C. leucodermis, C. palmeri, Centaurea melitensis, Ceanothus cuneatus, Cercocarpus betuloides, C. minutiflorus, Chaenactis artemisiifolia, Chenopodium botrys, Chorizanthe fimbriata, Cirsium occidentale, Clarkia purpurea, Claytonia lanceolata, Collinsia parviflora, Conium maculatum, Cryptantha micromeres, C. muricata, Cystopteris fragilis, Danthonia spicata, Datisca glomerata, Deinandra fasciculata, Delphinium cardinale, Deschampsia elongata, Dicentra chrysantha, Diplacus aurantiacus, Draba stenoloba, Dryopteris arguta, Dudleya lanceolata, Eleocharis, Elymus canadensis, E. glaucus, Emmenanthe penduliflora, Encelia farinosa, Ephedra, Epilobium ciliatum, E. oregonense, Equisetum laevigatum, Ericameria nauseosa, Erigeron coulteri, E. formosissimus, E. peregrinus, Eriogonum elongatum, E. fasciculatum, E. wrightii, Eriophyllum confertiflorum, Erodium botrys, E. cicutarium, Eucrypta chrysanthemifolia, Forestiera, Galium parishii, Garrya,*

Gilia leptalea, Gnaphalium californicum, G. purpureum, Gutierrezia sarothrae, Hemizonia, Heteromeles arbutifolia, Heterotheca grandiflora, Heuchera richardsonii, Hieracium albiflorum, Hirschfeldia incana, Hosackia crassifolia, H. oblongifolia, Jamesia americana, Juncus chlorocephalus, J. macrophyllus, J. mertensianus, Juniperus communis, J. occidentalis, Keckiella cordifolia, Lamarckia aurea, Lasthenia californica, Lathyrus vestitus, Lepidospartum squamatum, Lessingia filaginifolia, Lilium humboldtii, Linanthus ciliatus, Lonicera subspicata, Lupinus bicolor, L. hirsutissimus, L. latifolius, L. polyphyllus, Luzula multiflora, Lyonothamnus floribundus, Lythrum, Madia elegans, Maianthemum stellatum, Malacothamnus fasciculatus, Malacothrix saxatilis, Malosma laurina, Marah macrocarpus, Meconella denticulata, Melica imperfecta, Micranthes aprica, M. bryophora, M. nelsoniana, Mimulus breweri, M. cardinalis, M. guttatus, M. moschatus, M. parishii, M. pilosus, M. primuloides, Monarda fistulosa, Muhlenbergia racemosa, M. rigens, Nicotiana glauca, Opuntia phaeacantha, Orthocarpus hispidus, Oryzopsis asperifolia, Pediomelum argophyllum, Pellaea andromedifolia, P. mucronata, Penstemon gracilis, P. grinnellii, P. labrosus, Perideridia parishii, Phacelia brachyloba, P. cicutaria, P. minor, P. grandiflora, P. parryi, Phleum alpinum, Pinus contorta, P. jeffreyi, P. ponderosa, Piptatherum micranthum, Platanthera sparsiflora, Platanus racemosa, Poa epilis, P. nevadensis, P. rupicola, P. secunda, Polygonum douglasii, Polypodium californicum, Polypogon, Populus fremontii, Potentilla glandulosa, Primula tetrandra, Prunus ilicifolia, Pseudognaphalium viscosum, Pseudotsuga macrocarpa, Psoralea orbicularis, Pterostegia, Quercus agrifolia, Q. berberidifolia, Q. chrysolepis, Q. engelmannii, Q. kelloggii, Rafinesquia californica, Ranunculus alismifolius, Rhamnus pilosa, Rhododendron occidentale, Rhus integrifolia, R. ovata, R. trilobata, Ribes nevadense, Ricinus communis, Rosa californica, R. woodsii, Rubus ursinus, Rumex angiocarpus, Sagina, Salix exigua, S. laevigata, S. lasiolepis, S. scouleriana, Salvia apiana, S. columbariae, S. mellifera, Schizachyrium scoparium, Scirpus, Selaginella hansenii, Senecio flaccidus, Sequoia sempervirens, Sibbaldia procumbens, Silene laciniata, Sisymbrium orientale, Sisyrinchium, Solanum nigrum, Solidago missouriensis, Sonchus oleraceus, Stachys albens, Stellaria media, S. nitens, Stipa californica, Suksdorfia violacea, Tamarix, Taraxacum officinale, Toxicodendron diversilobum, Toxicoscordion fremontii, Trifolium longipes, T. obtusiflorum, T. repens, Trisetum spicatum, Typha, Urtica dioica, Vaccinium cespitosum, Venegasia carpesioides, Verbascum thapsus, Veronica wormskjoldii, Viola adunca, Vitis girdiana, Yucca whipplei.

Mimulus glabratus Kunth is a perennial inhabitant of brooks, ditches, fens, meadows, ponds, seeps, springs, streams, and the margins of lakes, ponds, rivers, and streams. It occurs mainly in sites that are calcareous (pH: 6.0–8.0), often in cold, flowing water, and in sunny to shaded exposures. The plants will grow with their stems floating in shallow water or creeping on exposed substrates that include gravel, limestone, muck, mud, rock, rotting logs, sand, and sandy humus. The flowers are self-compatible and readily self-pollinate. The

average pollen viability is 75% and seed set generally is high. Vegetative reproduction can occur by means of stoloniferous shoots. **Reported associates:** *Agrostis exarata, A. stolonifera, Ammannia coccinea, Amorpha fruticosa, Angelica atropurpurea, Batrachospermum, Berula erecta, Betula pumila, Bidens, Calamagrostis canadensis, C. stricta, Calliergon, Callitriche verna, Caltha palustris, Cardamine bulbosa, Carex aurea, C. bebbii, C. hystericina, C. nebrascensis, C. pellita, C. praegracilis, C. utriculata, C. vulpinoidea, Catabrosa aquatica, Chara, Cicuta douglasii, Cirsium muticum, C. vinaceum, Cornus sericea, Deschampsia cespitosa, Dryopteris cristata, Elaeagnus angustifolia, Eleocharis acicularis, E. palustris, Elodea canadensis, Epilobium ciliatum, Equisetum fluviatile, E. hyemale, Eriophorum angustifolium, Eupatorium perfoliatum, Eurybia integrifolia, Eutrochium maculatum, Galium labradoricum, Gentianopsis crinita, Geum aleppicum, G. macrophyllum, Glyceria grandis, G. striata, Hordeum jubatum, Houstonia humifusa, Impatiens capensis, Juncus balticus, J. nodatus, J. nodosus, J. tenuis, J. torreyi, Juniperus deppeana, Justicia americana, Leersia oryzoides, Lemna minor, Linaria canadensis, Liparis loeselii, Lobelia kalmii, Ludwigia palustris, Lycopus asper, Marchantia polymorpha, Melilotus officinalis, Mentha arvensis, Menyanthes trifoliata, Micranthes pensylvanica, Mimosa aculeaticarpa, Muhlenbergia emersleyi, Nasturtium officinale, Nolina microcarpa, Packera aurea, Panicum virgatum, Parnassia glauca, Pascopyrum smithii, Pedicularis lanceolata, Persicaria punctata, Pilea fontana, Platanthera hyperborea, Poa paludigena, P. palustris, P. pratensis, Polypogon monspeliensis, Potamogeton foliosus, Quercus emoryi, Ranunculus circinatus, R. cymbalaria, R. longirostris, Rosa arkansana, Rudbeckia hirta, Rumex orbiculatus, R. salicifolius, Salix discolor, S. exigua, S. interior, S. pedicellaris, Schoenoplectus acutus, S. pungens, S. tabernaemontani, Scirpus pallidus, Scleria verticillata, Scutellaria galericulata, Sparganium eurycarpum, Spartina pectinata, Spiranthes diluvialis, S. romanzoffiana, S. lucida, Sullivantia hapemanii, S. sullivantii, Symphyotrichum puniceum, Symplocarpus foetidus, Tamarix chinensis, Triglochin maritimum, Typha domingensis, T. latifolia, Vernonia sericea, Veronica americana, V. anagallis-aquatica, V. peregrina, Zannichellia palustris.*

Mimulus glaucescens Greene is an annual, which grows in chaparral, ditches, meadows, sandbars, seeps, washes, and along the margins of streams at elevations of up to 1525 m. It occurs in open, sunny sites on various substrates including basalt, clay, gravel, loam, mud, rock, sand, sandy loam, scree, serpentine, and talus. The flowers are self-compatible but are strongly herkogamous (with widely separated stigmas and anthers) and are outcrossed by insects. The plants exhibit high levels of inbreeding depression when they are self-pollinated. The seeds germinate under 16 h of light at 18°C and 8 h of darkness at 10°C. **Reported associates:** *Abies concolor, Adiantum aleuticum, Aesculus californica, Allium amplectens, A. campanulatum, Andropogon glomeratus, Avena, Briza minor, Bromus, Calocedrus decurrens, Ceanothus cuneatus, Cercis canadensis, Chenopodium botrys, Claytonia parviflora, Collinsia sparsiflora, Cyperus* *eragrostis, Darmera peltata, Dichelostemma capitatum, Epilobium brachycarpum, E. torreyi, Epipactis gigantea, Eriodictyon californicum, Erodium botrys, Erythronium multiscapoideum, Heteromeles arbutifolia, Heterotheca oregona, Holcus lanatus, Hypericum anagalloides, Juncus bufonius, Lepechinia calycina, Limnanthes alba, Linanthus ciliatus, Lithocarpus densiflorus, Lithophragma affinis, Lolium multiflorum, Lomatium, Lotus corniculatus, Lupinus bicolor, Malva nicaeensis, Marah fabaceus, Mimulus bicolor, M. cardinalis, M. guttatus, M. layneae, M. moschatus, M. torreyi, Muhlenbergia rigens, Onychium densum, Petrorhagia dubia, Pinus ponderosa, P. sabiniana, Platanthera dilatata, Polypogon monspeliensis, Potentilla glandulosa, Quercus douglasii, Q. kelloggii, Q. wislizenii, Rumex crispus, Scleranthus annuus, Scutellaria siphocamyploides, Senecio eurycephalus, Silene gallica, Sisyrinchium bellum, Thysanocarpus curvipes, Toxicodendron diversilobum, Toxicoscordion venenosum, Trifolium cyathiferum, T. tridentatum, T. variegatum, T. wormskioldii, Triteleia hyacinthina, Umbellularia californica, Vulpia.*

Mimulus guttatus DC. is an annual or perennial, which inhabits beaches, bogs, brooks, chaparral, cliffs, depressions, ditches, dunes, fens, floodplains, gravel bars, marshes, meadows, pools, rivulets, sandbars, seeps, sloughs, streams, springs, swales, vernal pools, washes, and the beds or margins of lakes, rivers, and streams at elevations of up to 4100 m. The plants often occupy cold, flowing water and occur on a wide range of substrates (pH: 6.0–8.0) that include basalt, clay, clay loam, cobble, granite, gravel, gravelly sandy loam, humus, limestone, loam, loamy clay, mudstone, peat, rock, rocky clay, sand, sandstone, sandy loam, schist, scree, serpentine, shale, silt, silt loam, and silty clay loam. These broadly adapted plants can tolerate exposures ranging from full sun to deep shade. The flowers occur in pairs at the nodes, usually with only one pair open at any given time. The number of flowers can vary considerably, and averages from 3 to 50 per plant. Flower production is resource dependent. Smaller flowers with less pollen occur under drier conditions than when water is readily available. Larger later-blooming flowers develop on plants that have not set seed than on those that already are producing seeds. The flowers are self-compatible and often are self-pollinating (20%–60%), but usually are outcrossed (40%–80%) and are pollinated by insects (Insecta) including bumblebees and small solitary bees (Hymenoptera: Apidae: *Bombus californicus*; *Apis mellifera*), butterflies (Lepidoptera), and flies (Diptera). The bees can distinguish and forage preferentially on plants with larger quantities of viable pollen. The stigmas are tactilely sensitive, with the lobes closing together within 6 s after any contact is made. Studies have found that most selfing (50%) occurs within single flowers, with the remainder attributable to pollination among flowers of a single individual (geitonogamy) or as a result of biparental inbreeding among close relatives. In unpollinated plants, selfing can be promoted upon contact of anthers and stigmas resulting from the curvature of stigma lobes (to contact anthers), or as the corollas are shed, which occurs within 9 days of flowering. The corolla captures and retains pollen

that is shed from the anthers as well as from visiting pollinators, which can facilitate self-pollination or even delay cross-pollination. However, most seed set usually occurs within the first day of flowering. Self-pollination in normally outcrossing populations results in reduced fitness (inbreeding depression) for many traits, including higher susceptibility to feeding by spittlebugs (Insecta: Homoptera: Cercopidae: *Philaenus spumarius*), but apparently does not alter extent of successful seed germination. The small (<0.02 mg) seeds can germinate in water or on sand and exhibit 33% germination within 1 day. Successful germination has been reported under a 12/12-h, 17°C–18°C/12°C–13°C temperature regime. Although germination has no special requirements, it is dependent to some degree on ploidy, as it occurs across a wider temperature range in tetraploids and aneuploids than in diploids. Most of the seeds remain buoyant for less than 1 h. The seeds disperse in water following a downstream direction (averaging a distance of about 275 m), by wind (averaging less than 0.5 meters), or in scat dispersed by grazing animals (up to 1 km). The plants often behave as annuals but can persist (as short-lived perennials) by the production of creeping rhizomes or stolons. Vegetative reproduction occurs through the dispersal of stem and rhizome fragments by the water. The plants form fragments readily, especially during conditions of high water flow, and the fragments retain high survival and colonization capacity. Some populations grow in substrates that are highly contaminated with copper, which they can tolerate (maintain good growth and seed germination) at sites where the total soil copper levels normally would be phytotoxic (up to 9500 ppm). Their copper tolerance is due at least in part to the presence of metal-binding metallothionein peptides. The plants growing in association with sedges (*Carex*) have higher survivorship during the winter and experience as much as a 75% reduction in herbivory. A 25-year observation of populations showed that they are highly dynamic, often dying out and reestablishing, with some exhibiting more than a 100-fold difference in size over a period of several years. Local extinctions often arise due to seasonal desiccation, which has been exacerbated by global climate change. **Reported associates:** *Abies amabilis, A. magnifica, Acer glabrum, A. macrophyllum, Achillea millefolium, Acmispon glabrus, Aconitum columbianum, Actaea rubra, Adenostoma fasciculatum, Adiantum aleuticum, A. capillus-veneris, Aesculus californica, Agastache urticifolia, Ageratina adenophora, Agropyron albicans, Agrostis densiflora, A. exarata, A. stolonifera, A. thurberiana, A. variabilis, A. viridis, Allium validum, Alnus oblongifolia, A. rhombifolia, A. sinuata, A. tenuifolia, Alopecurus aequalis, A. pratensis, Ambrosia psilostachya, Andropogon glomeratus, Androsace filiformis, Anemone piperi, Anemopsis californica, Angelica kingii, Antennaria parvifolia, A. rosea, Anticlea elegans, Aquilegia chrysantha, Arabis drummondii, Arbutus menziesii, Arctostaphylos patula, A. pungens, A. viscida, Aristida fendleriana, Arnica ×diversifolia, A. latifolia, A. mollis, Artemisia arbuscula, A. californica, A. douglasiana, A. ludoviciana, A. nova, r. rothrockii, A. tridentata, Arundo donax, Astragalus didymocarpus, Athyrium filix-femina, Atriplex canescens,*

Baccharis glutinosa, B. salicifolia, B. sarothroides, Berula erecta, Bidens aureus, B. ferulifolius, B. laevis, Bistorta bistortoides, B. vivipara, Boykinia major, Brassica campestris, B. geniculata, B. tournefortii, Brodiaea hyacinthina, B. orcuttii, B. terrestris, Bromus carinatus, B. laevipes, B. tectorum, Calamagrostis canadensis, C. rubescens, Calocedrus decurrens, Calochortus splendens, Camassia leichtlinii, C. quamash, Capsella bursa-pastoris, Cardamine cordifolia, Carduus nutans, Carex alma, C. amplifolia, C. aquatilis, C. athrostachya, C. aurea, C. canescens, C. cusickii, C. deweyana, C. exsiccata, C. gigas, C. haydeniana, C. hoodii, C. illota, C. lenticularis, C. leptopoda, Carex luzulina, C. microptera, C. nebrascensis, C. nervina, C. nigricans, C. norvegica, C. obnupta, C. pachystachya, C. pellita, C. praegracilis, C. rostrata, C. scirpoidea, C. simulata, C. specifica, C. utriculata, C. vesicaria, Castilleja attenuata, C. lineariiloba, C. miniata, C. minor, Catabrosa aquatica, Catalpa, Ceanothus crassifolius, C. cuneatus, C. greggii, C. leucodermis, C. palmeri, C. prostratus, Celtis, Centaurium exaltatum, Cerastium arvense, C. viscosum, Cercocarpus betuloides, Cheilanthes covillei, Chenopodium murale, Cicuta douglasii, Cinna latifolia, Circaea alpina, Cirsium arvense, C. foliosum, C. fontinale, C. scariosum, Claytonia cordifolia, Clematis ligusticifolia, Collinsia parviflora, C. sparsiflora, Conioselinum scopulorum, Conyza canadensis, Cotula coronopifolia, Cornus sericea, Crassula aquatica, Crataegus, Croton setigerus, Cryptantha barbigera, C. intermedia, C. micrantha, C. muricata, Cupressus forbesii, Cyperus involucratus, Cystopteris fragilis, Danthonia californica, a. intermedia, Darlingtonia californica, Dasiphora floribunda, Datisca glomerata, Daucus pusillus, Deschampsia cespitosa, D. elongata, Descurainia californica, D. incisa, D. pinnata, D. sophia, Dicentra formosa, Diplacus aurantiacus, Downingia bella, D. bicornuta, D. insignis, D. ornatissima, Drosera rotundifolia, Dudleya variegata, Eleocharis macrostachya, E. montevidensis, E. palustris, E. parishii, E. quinqueflora, E. rostellata, Elymus glaucus, E. trachycaulus, Encelia farinosa, Epilobium alpinum, E. angustifolium, E. ciliatum, E. glaberrimum, E. hornemannii, E. palustre, E. watsonii, Equisetum arvense, E. laevigatum, E. pratense, Erigeron algidus, E. coulteri, Eriodictyon crassifolium, Erigeron peregrinus, Eriogonum fasciculatum, E. nudum, E. thurberi, E. wrightii, Eriophorum, Eryngium castrense, Eschscholzia minutiflora, Festuca idahoensis, F. subulata, Fragaria vesca, F. virginiana, Frasera speciosa, Galium asperulum, G. bifolium, G. boreale, G. trifidum, G. triflorum, Garrya flavescens, Gayophytum diffusum, Gentiana calycosa, Gentianella amarella, Geranium richardsonii, Geum macrophyllum, Gilia capitata, Glyceria elata, G. grandis, G. striata, Hackelia floribunda, Haplopappus cuneatus, Helianthus nuttallii, Heliotropium, Heteromeles arbutifolia, Heracleum lanatum, H. maximum, Holcus lanatus, Hordeum brachyantherum, H. jubatum, H. murinum, Hosackia oblongifolia, Hydrocotyle, Hydrophyllum fendleri, Hypochaeris glabra, Iris missouriensis, Isoetes howellii, Iva axillaris, Juncus balticus, J. bufonius, J. conglomeratus, J. ensifolius, J. falcatus, J. longistylis, J. macrophyllus, J. mertensianus, J. nevadensis, J. phaeocephalus, J. rugulosus, J. tenuis, J. torreyi, J. xiphioides, Juniperus osteosperma, Kalmia microphylla, Keckiella antirrhinoides, Lactuca serriola, Lamium amplexicaule, Lasthenia californica, L. fremontii, Lemna minor, Lepechinia calycina, Lepidium perfoliatum, Lepidospartum squamatum, Ligusticum canbyi, Lilaea scilloides, Lilium kelleyanum, Limnanthes gracilis, Lolium, Lonicera interrupta, L. involucrata, Lupinus bicolor, L. burkei, L. concinnus, L. latifolius, Luzula campestris, L. parviflora, Lycopodium sitchense, Lycopus americanus, Lythrum californicum, L. hyssopifolia, Madia glomerata, Maianthemum stellatum, Malacothamnus fasciculatus, Malosma laurina, Marah, Marrubium vulgare, Marsilea mucronata, Mentha arvensis, Mertensia ciliata, M. franciscana, M. paniculata, M. perplexa, Micranthes odontoloma, Mimulus alsinoides, M. angustatus, M. cardinalis, M. dentatus, M. dentilobus, M. eastwoodiae, M. floribundus, M. glaucescens, M. lewisii, M. moschatus, M. nudatus, M. pilosus, M. primuloides, M. tilingii, Mitella pentandra, Monardella odoratissima, M. stoneana, Montia chamissoi, M. fontana, Muhlenbergia asperifolia, M. filiformis, Myosotis discolor, Nasturtium officinale, Navarretia leucocephala, Nemophila, Nuphar polysepala, Odontostomum hartwegii, Oenothera elata, O. speciosa, Orobanche uniflora, Oryzopsis hymenoides, Osmorhiza chilensis, O. occidentalis, Oxypolis fendleri, Panicum bulbosum, Parietaria, Parnassia fimbriata, P. palustris, Pedicularis groenlandica, Pellaea mucronata, Penstemon procerus, P. rydbergii, Perideridia lemmonii, Pectocarya linearis, Persicaria hydropiperoides, P. lapathifolia, Phacelia distans, P. hastata, Phalaris arundinacea, Philadelphus lewisii, Phleum alpinum, P. pratense, Picea engelmannii, Pinus contorta, P. edulis, P. jeffreyi, P. monophylla, P. ponderosa, P. sabiniana, Piptatherum miliaceum, Plagiobothrys collinus, P. leptocladus, P. parishii, P. stipitatus, Plantago lanceolata, P. major, Platanthera dilatata, P. sparsiflora, Platanus racemosa, P. wrightii, Plectritis congesta, Poa annua, P. pratensis, P. secunda, P. triflora, Polemonium occidentale, P. pulcherrimum, Polypogon monspeliensis, Populus fremontii, P. tremuloides, Potamogeton gramineus, Potentilla glandulosa, P. gracilis, P. pacifica, Primula tetrandra, Prunella vulgaris, Prunus armeniaca, P. emarginata, P. ilicifolia, Pseudotsuga menziesii, Psilocarphus brevissimus, Psorothamnus emoryi, Pteridium aquilinum, Puccinellia pauciflora, Purshia, Quercus agrifolia, Q. berberidifolia, Q. chrysolepis, Q. douglasii, Q. engelmannii, Q. kelloggii, Q. tomentella, Q. turbinella, Q. wislizenii, Ranunculus acriformis, R. aquatilis, R. bonariensis, R. circinatus, R. cymbalaria, R. macranthus, R. muricatus, R. occidentalis, R. pygmaeus, R. uncinatus, Rhamnus purshiana, Rhododendron occidentale, Rhus ovata, R. trilobata, Ribes aureum, R. hudsonianum, R. inerme, R. lacustre, R. nevadense, R. roezlii, Rorippa curvisiliqua, Rosa californica, R. woodsii, Rubus parviflorus, Rudbeckia laciniata, R. occidentalis, Rumex crispus, R. salicifolius, Sagittaria, Salix drummondiana, S. exigua, S. geyeriana, S. gooddingii, S. irrorata, S. lasiolepis, S. lemmonii, S. lucida, S. pseudomyrsinites, Salvia apiana, S. mellifera, Sambucus mexicana,

S. racemosa, Sanguisorba sitchensis, Sarcostemma cynan-choides, Schismus, Schoenoplectus acutus, S. pungens, Scirpus congdonii, S. microcarpus, Selaginella bigelovii, Senecio cymbalarioides, S. hydrophilus, S. triangularis, S. serra, Sidalcea malviflora, S. neomexicana, Silene gallica, Sisymbrium altissimum, Sisyrinchium bellum, S. californi-cum, S. elmeri, Sonchus oleraceus, Spergularia rubra, Sphenopholis obtusata, Sphenosciadium capitellatum, Sporobolus airoides, Stachys stricta, Stellaria longifolia, S. longipes, S. media, Stenanthium occidentale, Stipa coronata, S. occidentalis, Streptopus amplexifolius, Swertia perennis, Symphyotrichum ascendens, S. eatonii, S. spathulatum, S. subspicatum, Tamarix ramosissima, Taraxacum officinale, Thalictrum occidentale, Thlaspi arvense, Thysanocarpus curvipes, Tiarella trifoliata, Tofieldia, Toxicodendron diver-silobum, Toxicoscordion venenosum, Trichophorum, Trifolium longipes, T. repens, T. variegatum, T. wormskioldii, Triglochin maritimum, Triphysaria eriantha, Trisetum spica-tum, Tsuga mertensiana, Typha latifolia, Umbellularia cali-fornica, Uropappus lindleyi, Urtica dioica, Vaccinium myrtillus, Veratrum californicum, V. viride, Verbena verna, Veronica americana, Veronica anagallis-aquatica, Viola canadensis, V. glabella, V. macloskeyi, V. orbiculata, V. palustris, Vulpia, Woodwardia fimbriata, Xanthium strumar-ium, Xylococcus bicolor, Yucca whipplei.

Mimulus latidens (A. Gray) Greene is an annual, which grows in chaparral, depressions, meadows, mudflats, playas, vernal pools, and along the margins of lakes, ponds, and streams at elevations of up to 1711 m. The sites reportedly are alkaline (specific pH range not determined) with sub-strates consisting of adobe, clay, cobbley clay, mud, muddy clay, serpentine, and silty clay. There is very little life-history information available for this species. **Reported associates:** *Arabis, Artemisia cana, Bergia texana, Brodiaea jolonensis, Bromus hordeaceus, Centunculus minimus, Crassula aquat-ica, Deinandra fasciculata, Deschampsia danthonioides, Downingia cuspidata, Elatine brachysperma, Eleocharis macrostachya, E. palustris, Epilobium pygmaeum, Ericameria nauseosa, Eryngium aristulatum, Isoetes howel-lii, I. orcuttii, Juncus bufonius, Lepidium latipes, L. nitidum, Lilaea scilloides, Lythrum hyssopifolia, Malvella leprosa, Marsilea vestita, Muhlenbergia richardsonis, Myosurus min-imus, Nama stenocarpum, Navarretia fossalis, N. hamata, Ophioglossum californicum, Orcuttia californica, Phalaris lemmonii, Pilularia americana, Plagiobothrys acanthocar-pus, Plantago elongata, Plantago erecta, Pogogyne nudi-uscula, Psilocarphus brevissimus, Psilocarphus tenellus, Rotala ramosior, Sida Taraxia tanacetifolia, Verbena brac-teata, Veronica.*

Mimulus lewisii Pursh is a perennial, which is found grow-ing on cliffs, in ditches, flats, gravel bars, meadows, ravines, sandbars, seeps, springs, talus slopes, and along the margins of lakes and streams at elevations from 800 to 3100 m. The substrates are acidic (pH: 5.8–6.6) and can consist of basalt, clay, granite, gravel, humic loam, loam, pumice, rock, sand, or sandy gravel. The plants will withstand partial to fairly dense shade and can survive cold temperatures of less than −20°C.

Although self-compatible, the plants require cross pollination to achieve optimal seed set. The flowers have magenta-red corollas and are pollinated by insects (Insecta), primarily bees (Hymenoptera: Apidae: *Apis mellifera*; *Bombus vosnesenskii*; Megachilidae: *Osmia*); however, they are occasionally visited by hummingbirds (Aves: Trochilidae). A single plant can pro-duce several hundred fruits, with each one containing up to 2500 seeds. The seeds persist in the seed bank for at least 1 year with about 3% of the seeds remaining dormant during each season. Seed germination occurs readily at greenhouse temperatures. The seedlings typically establish in primary or early successional sites such as along the exposed grav-elly margins of lakes and streams. Seedling establishment is reduced at sites with high moss cover. Demographic studies have shown that central populations exhibit higher growth rates and survival than marginal populations. Annual survival is substantially higher for reproductive plants (81%–97%) than it is for vegetative plants (7%–26%). Vegetative reproduction occurs by means of rhizome production. The plants are a frequent component of the alpine "snow patch" flora. Field-surveyed plants have exhibited only a low incidence of arbus-cular mycorrhizae. **Reported associates:** *Abies lasiocarpa, A. magnifica, Acer glabrum, Achillea, Aconitum columbia-num, Aconogonon davisiae, Actaea rubra, Agrostis exarata, A. thurberiana, Alnus rhombifolia, A. tenuifolia, A. viridis, Alopecurus aequalis, Anaphalis margaritacea, Anemone parviflora, Angelica arguta, Antennaria media, Aquilegia caerulea, Arnica chamissonis, A. cordifolia, A. latifolia, A. longifolia, A. mollis, Athyrium distentifolium, A. filix-femina, Betula occidentalis, Bistorta bistortoides, Bolandra oregana, Bromus vulgaris, Boykinia major, Calamagrostis canadensis, Caltha leptosepala, Carex aquatilis, C. disperma, C. merten-sii, C. microptera, C. nigricans, C. paysonii, C. praegraci-lis, C. scopulorum, C. spectabilis, C. utriculata, Cassiope stelleriana, Castilleja elmeri, C. oreopola, C. parviflora, Catabrosa aquatica, Cinna latifolia, Claytonia lanceolata, Eleocharis quinqueflora, Elymus glaucus, Epilobium alpi-num, E. angustifolium, E. ciliatum, E. latifolium, Equisetum arvense, Erigeron peregrinus, Eriophorum angustifolium, Erythronium grandiflorum, Eucephalus engelmannii, Galium triflorum, Gentiana calycosa, Geum macrophyllum, Glyceria grandis, G. striata, Gaultheria humifusa, Hackelia floribunda, Heracleum lanatum, H. maximum, Hieracium gracile, Hosackia oblongifolia, Hypericum anagalloides, H. formosum, Hypnum, Ipomopsis aggregata, Juncus drum-mondii, J. ensifolius, J. longistylus, J. mertensianus, J. par-ryi, Kalmia microphylla, Larix lyallii, Leucothoe davisiae, Ligusticum grayi, Lomatium martindalei, Lupinus latifolius, L. polyphyllus, Luetkea pectinata, Luzula campestris, L. par-viflora, Marchantia, Menziesia ferruginea, Mertensia cili-ata, M. paniculata, Micranthes ferruginea, M. odontoloma, M. tolmiei, Mimulus cardinalis, M. guttatus, M. moschatus, M. primuloides, M. tilingii, Mitella nuda, M. pentandra, Oreostemma alpigenum, Osmorhiza occidentalis, Parnassia fimbriata, Pedicularis attollens, P. groenlandica, Penstemon procerus, Petasites, Philadelphus lewisii, Phleum alpi-num, Phyllodoce empetriformis, Picea engelmannii, Pinus* *albicaulis, P. contorta, Platanthera dilatata, P. stricta, Poa alpina, Polystichum lonchitis, Populus trichocarpa, Potentilla flabellifolia, Primula jeffreyi, Prunus virginiana, Pseudotsuga menziesii, Pulsatilla occidentalis, Ranunculus eschscholtzii, Ribes lacustre, Rubus parviflorus, Rudbeckia occidentalis, Salix commutata, S. drummondiana, S. planifolia, Sambucus racemosa, Scirpus microcarpus, Senecio cymbalarioides, S. triangularis, Sorbus scopulina, Spiraea betulifolia, S. densiflora, S. splendens, Stachys ajugoides, Stellaria crispa, S. obtusa, S. umbellata, Symphyotrichum foliaceum, S. subspicatum, Thalictrum occidentale, Tiarella unifoliata, Triantha glutinosa, T. occidentalis, Trisetum spicatum, Trollius laxus, Tsuga mertensiana, Urtica dioica, Vaccinium membranaceum, V. uliginosum, Vahlodea atropurpurea, Valeriana sitchensis, Veratrum viride, Veronica wormskjoldii, Viola glabella.*

Mimulus moschatus Dougl. ex Lindl. is a perennial, which occurs on cliffs and in bogs, depressions, ditches, gullies, marshes, meadows, pools, prairies, seeps, springs, streams, swales, swamps, washes, and along the margins of ponds, rivers, and streams at elevations of up to 3150 m. The plants are quite versatile ecologically and can tolerate water depths up to 1 m, full sun to deep shade, and cold temperatures to at least −15°C. They occur on a wide range of substrates (pH: 5.0–7.0) that include basalt, clay, granite, gravel, humus, limestone, loamy sand, muck, mud, mudstone, pumice, rock, rocky gravel, rocky loam, sand, sandy loam, serpentine, and shale. The flowers are pollinated by bees (Hymenoptera: Apidae) and bee flies (Diptera: Bombyliidae) and produce numerous seeds. Most of the seeds lack dormancy and are capable of germination immediately after ripening. Only about 3% of the seeds will persist to the following season (especially those produced by late-flowering plants), so that a long-term seed bank does not develop. The seeds are dispersed by the wind and water. Drought-stressed plants suffer high mortality and do not produce fruit. Populations can vary from a few to several thousand individuals. The plants are short-lived perennials that propagate by the development of very fine rhizomes. Generally, they are poor competitors and do well in disturbed areas. They often grow on sedge tussocks (e.g., *Carex nudata*), which provide a protective, stable substrate. Their glandular foliage produces a musk-like odor, which can be lost after periods of prolonged cultivation. **Reported associates:** *Abies concolor, A. lasiocarpa, Acer circinatum, A. macrophyllum, A. rubrum, Achlys triphylla, Aesculus californica, Agrostis stolonifera, Allium campanulatum, A. validum, Allotropa virgata, Alnus rhombifolia, A. rubra, Amelanchier alnifolia, Anaphalis margaritacea, Aquilegia formosa, Arbutus menziesii, Arctostaphylos patula, Aristolochia californica, Arnica mollis, Artemisia douglasiana, Asarum caudatum, Astragalus robbinsii, Athyrium filixfemina, Baccharis douglasii, B. pilularis, Berberis aquifolium, B. nervosa, B. thunbergii, Botrychium multifidum, Boykinia elata, B. occidentalis, Brachythecium frigidum, Calocedrus decurrens, Calochortus nudus, Calypogeia muelleriana, Campanula aparinoides, Carex crawfordii, C. garberi, C. gynandra, C. gynodynama, C. hendersonii, C. hysterica, C. lanuginosa, C. lenticularis, C. lurida, C. nudata, C. obnupta,* *C. scabrata, C. senta, Castilleja, Ceanothus foliosus, C. velutinus, Chamaecyparis lawsoniana, Chamaesyce serpyllifolia, Chrysolepis chrysophylla, Circaea alpina, C. pacifica, Cirsium hydrophilum, Claytonia parviflora, C. sibirica, Collinsia parviflora, Conocephalum conicum, Cornus canadensis, C. nuttallii, C. sericea, Corylus cornuta, Crataegus, Danthonia californica, Darlingtonia californica, Delphinium glaucum, Dennstaedtia punctilobula, Dicentra formosa, Eleocharis compressa, E. rostellata, Elymus glaucus, Epilobium canum, E. torreyi, Epipactis gigantea, Equisetum arvense, E. fluviatile, E. variegatum, Erechtites minima, Erythronium, Eutrochium maculatum, Fallopia japonica, Fraxinus dipetala, Galium tinctorium, G. trifidum, G. triflorum, Gaultheria ovatifolium, G. shallon, Gentianella quinquefolia, Geum macrophyllum, Glyceria elata, Gnaphalium palustre, Goodyera oblongifolia, Gratiola ebracteata, Habenaria, Helenium, Heracleum lanatum, H. maximum, Hoita, Holcus lanatus, Hosackia gracilis, Hypericum anagalloides, H. formosum, H. perforatum, Impatiens capensis, Juncus articulatus, J. bufonius, J. covillei, J. effusus, J. mexicanus, Leucothoe davisiae, Lilium humboldtii, L. pardalinum, L. washingtonianum, Linnaea borealis, Lupinus latifolius, Lychnis coronaria, Lyonia ligustrina, Lysichiton americanus, Lythrum salicaria, Marchantia polymorpha, Mentha arvensis, M. pulegium, Mertensia, Mimulus alatus, M. cardinalis, M. dentatus, M. floribundus, M. glaucescens, M. guttatus, M. lewisii, M. pilosus, M. primuloides, M. tilingii, Mitella breweri, M. ovalis, Monarda didyma, Myosotis laxa, M. scorpioides, Myrica californica, M. gale, Navarretia, Onoclea sensibilis, Osmundastrum cinnamomeum, Perideridia parishii, Picea engelmannii, Pinguicula vulgaris, Pinus contorta, P. jeffreyi, P. lambertiana, P. ponderosa, P. strobus, Plantago major, P. subnuda, Poa pratensis, Pohlia wahlenbergii, Persicaria punctata, Potentilla anserina, P. drummondii, P. glandulosa, P. gracilis, Primula jeffreyi, Prunella vulgaris, Pseudotsuga macrocarpa, P. menziesii, Pteridium aquilinum, Pycnanthemum californicum, Quercus chrysolepis, Q. kelloggii, Ranunculus alismifolius, R. trichophyllus, Rhamnus californica, Rhododendron macrophyllum, R. occidentale, Ribes bracteosum, R. cereum, R. nevadense, R. sanguineum, Rosa gymnocarpa, Rubus leucodermis, R. parviflorus, R. ursinus, Rumex crispus, R. obtusifolius, Salix cordata, S. lasiolepis, Sanicula tuberosa, Scirpus cyperinus, S. microcarpus, Scoliopus hallii, Scutellaria galericulata, Senecio triangularis, Sequoiadendron giganteum, Sidalcea oregana, Silene nivea, Sisyrinchium bellum, S. elmeri, Solidago spectabilis, Sphagnum, Spiraea alba, S. densiflora, Spiranthes lucida, S. romanzoffiana, Stachys ciliata, S. palustris, S. pycnantha, Stellaria longipes, S. media, Symphoricarpos albus, Symphyotrichum foliaceum, S. prenanthoides, S. spathulatum, S. tradescantii, Taxus brevifolia, Tellima grandiflora, Thalictrum, Thelypteris palustris, Tiarella trifoliata, T. unifoliata, Toxicodendron diversilobum, Triantha glutinosa, Trientalis latifolia, Trillium ovatum, T. rivale, Tsuga canadensis, T. heterophylla, Tussilago farfara, Vaccinium parviflorum, Veratrum californicum, V. viride, Veronica americana, Vicia americana, Viola glabella, Woodwardia fimbriata, Xanthium strumarium.*

Mimulus nudatus **Curran ex Greene** is an annual, which inhabits depressions, ditches, draws, glades, seeps, streambeds, and the margins of streams at elevations of up to 700 m. The plants occur in open sites and are restricted to serpentine substrates that include clay, gravel, and rocky clay. They also grow in the crevices of wet serpentine outcrops. The flowers primarily are outcrossed and pollinated by small sweat bees (Hymenoptera: Halictidae: *Dialictus*) and to a lesser extent by honey bees (Hymenoptera: Apidae: *Apis mellifera*). An outcrossing breeding system is indicated by relatively high levels of heterozygosity, which have been documented genetically using microsatellite markers. Viable seed production is reduced when plants co-occur along with the closely related *M. guttatus* as a consequence of asymmetric deposition of foreign pollen on the stigmas. The seeds are dispersed by water and apparently form a seed bank. The populations often establish in disturbed sites and exhibit large annual fluctuations with respect to precipitation patterns. The fluctuations occur synchronously among populations that are separated by as much as 300–400 m. Analysis of genetic segregation in hybrids between *M. nudatus* and the nonserpentine tolerant *M. marmoratus* have found that serpentine tolerance in the former is correlated positively with drought tolerance but not with Ni (nickel) tolerance. **Reported associates:** *Adenostoma fasciculatum, Allium falcifolium, A. fimbriatum, Anagallis arvensis, Ancistrocarphus filagineus, Aspidotis densa, Astragalus clevelandii, Bromus hordeaceus, B. madritensis, Calycadenia pauciflora, Carex serratodens, Centaurea melitensis, Chaenactis glabriuscula, Cirsium fontinale, C. hydrophilum, Cryptantha hispidula, Cupressus sargentii, Cypripedium californicum, Delphinium uliginosum, Epilobium minutum, Eriodictyon californicum, Helianthus exilis, Hesperolinon disjunctum, H. serpentinum, Lagophylla minor, Lessingia glandulifera, Linanthus latisectus, Lolium multiflorum, Lomatium macrocarpum, L. marginatum, Malacothrix clevelandii, M. floccifera, Mimulus guttatus, M. tricolor, Pinus sabiniana, Quercus berberidifolia, Q. durata, Senecio clevelandii, Streptanthus barbiger, S. breweri.*

Mimulus pilosus **(Benth.) S. Watson** is an annual, which is found on sand or gravel bars, in ditches, drying pools, floodplains, meadows, mudflats, pond or stream beds, seeps, sloughs, swales, washes, and along the margins of lakes, rivers, and streams at elevations of up to 2950 m. It is reported from various substrates including ash, clay, cobble, gabbro, granite, gravel, obsidian, rock, sand, sandstone, sandy loam, schist, serpentine, shale, and silt. The plants occur in exposures ranging from full sun to partial shade. This species produces a large seed bank in young shrub stands of chaparral, where it is associated with canopy gaps. It frequently recolonizes burned areas by its persistent seeds. Few other details have been reported concerning the pollination biology, seed ecology, or other life history aspects of these plants. **Reported associates:** *Abies concolor, A. magnifica, Acmispon americanus, A. prostratus, A. strigosus, Adenostoma fasciculatum, A. sparsifolium, Ageratina adenophora, Agrostis idahoensis, Allium amplectens, A. lacunosum, Allophyllum*

divaricatum, Amaranthus californicus, Ambrosia psilostachya, Antirrhinum cornutum, A. nuttallianum, Aquilegia formosa, Arctostaphylos patula, Aristida fendleriana, Artemisia californica, A. douglasiana, A. ludoviciana, A. tridentata, Baccharis salicifolia, Bromus madritensis, Cakile maritima, Calandrinia breweri, Calochortus albus, Camissonia californica, C. cheiranthifolia, C. hirtella, Carex douglasii, C. fracta, C. subbracteata, Ceanothus greggii, C. leucodermis, Centaurea melitensis, Centaurium venustum, Ceanothus cuneatus, C. incanus, C. papillosus, Cercocarpus betuloides, Chamaesyce glyptosperma, Cheilanthes covillei, Chorizanthe polygonoides, Chrysothamnus nauseosus, Cirsium vulgare, Clarkia purpurea, Cynodon dactylon, Cyperus eragrostis, Cystopteris fragilis, Danthonia californica, Deinandra fasciculata, D. floribunda, Deschampsia elongata, Dicentra chrysantha, Eleocharis bella, Elymus glaucus, Epilobium ciliatum, Eremocarpus setigerus, Ericameria cuneata, E. linearifolia, Eriogonum fasciculatum, Erodium cicutarium, Euphorbia ocellata, Forestiera neomexicana, Fremontodendron californicum, Gayophytum humile, Geranium richardsonii, Gnaphalium canescens, G. chilense, G. palustre, Haplopappus cuneatus, Helianthus gracilentus, Heliotropium curassavicum, H. europaeum, Heterocodon rariflorum, Heteromeles arbutifolia, Holocarpha, Hordeum hystrix, Horkelia elata, Hosackia stipularis, Juncus bryoides, J. bufonius, J. macrophyllus, J. mexicanus, Juniperus osteosperma, Keckiella antirrhinoides, Lamarckia aurea, Lepidospartum squamatum, Lessingia nemaclada, Leymus condensatus, Lupinus albifrons, L. latifolius, L. luteolus, L. polyphyllus, Maianthemum stellatum, Malosma laurina, Marsilea vestita, Medicago polymorpha, Melilotus indica, Mentzelia laevicaulis, Mimulus bolanderi, M. cardinalis, M. floribundus, M. guttatus, M. moschatus, M. palmeri, M. parishii, M. rattanii, M. rubellus, Mollugo verticillata, Monardella lanceolata, Muhlenbergia rigens, Nasturtium officinale, Navarretia hamata, N. intertexta, Panicum acuminatum, Penstemon grinnellii, Persicaria lapathifolia, Phacelia brachyloba, P. cicutaria, P. grandiflora, P. parryi, Pinus monophylla, P. sabiniana, Platanus racemosa, Poa pratensis, Polypogon monspeliensis, Populus fremontii, Potentilla glandulosa, Quercus agrifolia, Q. berberidifolia, Q. douglasii, Q. engelmannii, Q. lobata, Q. turbinella, Ranunculus sceleratus, Raphanus sativus, Rhamnus ilicifolia, R. pilosa, Rhododendron occidentale, Ribes nevadense, Robinia neomexicana, Romneya trichocalyx, Rhus ovata, R. trilobata, Rosa woodsii, Salix exigua, S. lasiolepis, S. scouleriana, Salvia apiana, Sanicula crassicaulis, Scrophularia californica, Scutellaria tuberosa, Senecio vulgaris, Solidago californica, Sporobolus airoides, Stachys ajugoides, S. albens, Stylocline gnaphalioides, Tamarix, Taraxacum officinale, Thalictrum fendleri, Toxicodendron diversilobum, Toxicoscordion fremontii, Trichostema lanceolatum, T. micranthum, Trifolium cyathiferum, T. microcephalum, Typha latifolia, Veronica peregrina, Vulpia, Xanthium strumarium, Xylococcus bicolor, Yucca whipplei.

Mimulus primuloides **Benth.** is a perennial, which occurs in bogs, depressions, ditches, flats, floodplains, marshes, meadows, springs, and along the margins of lakes, rivers, and streams at elevations of up to 3810 m. This species exists on a wide range of sites from acidic bogs to alkaline meadows (e.g., pH: 4.9–8.5). It is reported from substrates that include clay loam, granite, gravel, humus, loam, mud, organic loam, peat, pumice, rock, sand, and serpentine. The plants can be found on exposed substrates or in shallow water, and will tolerate full sun to partial shade, and cold temperatures down to −36°C. The flowers are pollinated by bees (probably *Bombus*), but their pollination biology has not been studied thoroughly. Field studies have shown that populations produce more than twice the density of plants at lower elevations (e.g., 1400 m) than at higher elevations (e.g., 3400 m). Populations growing at the highest elevations (e.g., 3400 m) allocate their highest relative proportion of resources to sexual reproduction but produce few seeds (3 seeds on average). Those growing at mid-elevations (e.g., 2400 m) have the highest relative resource allocation to vegetative structures and also produce the highest mean number of seeds per capsule (average of 69 seeds/capsule). The seeds require light but not cold stratification for germination, which will occur within 1–3 weeks when incubated at 22°C. Seed germination also is reportedly enhanced when the seeds are exposed to smoke that contains butenolides. Larger plants generally allocate a larger proportion of their biomass to vegetative reproduction, which occurs by the formation of vegetative bulblets from the rhizome and stolons. The seeds occur at low frequency and densities in closed forest seed banks. The plants rapidly colonize disturbed areas and can form dense patches within a year after an area has been burned, drained, or overgrazed. **Reported associates:** *Abies amabilis, A. concolor, A. grandis, A. lasiocarpa, A. magnifica, Achillea millefolium, Achlys triphylla, Achnatherum lettermanii, A. nelsonii, A. occidentalis, Aconitum columbianum, Agrostis aurantiaca, A. capillaris, A. exarata, A. humilis, A. idahoensis, A. lepida, A. pallens, A. scabra, A. stolonifera, A. thurberiana, A. variabilis, Allium validum, Alnus rugosa, A. sinuata, Alopecurus alpinus, Anacolia, Anemone deltoidea, A. oregana, Antennaria luzuloides, A. microphylla, A. rosea, Aquilegia formosa, Arctostaphylos patula, Arnica chamissonis, A. fulgens, A. mollis, Artemisia arbuscula, A. cana, A. ludoviciana, Aulacomnium palustre, Betula, Bistorta bistortoides, Botrychium simplex, Boykinia occidentalis, Bromus sitchensis, Bryum, Calamagrostis breweri, C. canadensis, Calocedrus decurrens, Calochortus nudus, Callitriche hermaphroditica, C. verna, Caltha leptosepala, Calystegia malacophylla, Camassia leichtlinii, C. quamash, Cardamine breweri, C. californica, Carex abrupta, C. angustata, C. angustior, C. aquatilis, C. athrostachya, C. aurea, C. brunnescens, C. buxbaumii, C. canescens, C. echinata, C. fissuricola, C. gymnoclada, C. heteroneura, C. hoodii, C. illota, C. incurviformis, C. integra, C. interrupta, C. jonesii, C. laeviculmis, C. lenticularis, C. leporinella, C. limosa, C. luzulina, C. mariposana, C. microptera, C. muricata, C. nebrascensis, C. nervina, C. nigricans, C. obnupta, C. pachystachya, C. phaeocephala, C. praeceptorium, C. praegracilis, C. raynoldsii, C. rostrata, C. scopulorum, C. simulata, C. spectabilis, C. subfusca, C. subnigricans, C. utriculata, C. vesicaria, Castilleja lemmonii, C. miniata, C. nana, C. senta, C. suksdorfii, C. tenuis, C. tumulicola, Ceanothus cordulatus, C. velutinus, Ceratodon, Chamerion angustifolium, Cirsium callilepis, Claytonia exigua, Clintonia uniflora, Collinsia parviflora, Comarum palustre, Corydalis caseana, Crassula aquatica, Cratoneuron filicinum, Danthonia californica, D. intermedia, D. unispicata, Darlingtonia californica, Delphinium glaucum, D. menziesii, Deschampsia cespitosa, D. elongata, Drepanocladus aduncus, Drosera anglica, D. rotundifolia, Eleocharis acicularis, E. bella, E. macrostachya, E. montevidensis, E. palustris, E. quinqueflora, E. rostellata, E. torticulmis, Elymus glaucus, E. trachycaulus, Epilobium alpinum, E. brevistylum, E. ciliatum, E. glaberrimum, E. halleanum, E. hornemannii, E. lactiflorum, E. oregonense, Equisetum arvense, E. telmateia, Eragrostis hypnoides, Erigeron foliosus, E. peregrinus, Eriophorum crinigerum, E. gracile, Erythronium grandiflorum, Euthamia occidentalis, Festuca idahoensis, F. rubra, Fragaria virginiana, Galium bifolium, G. trifidum, Gaultheria humifusa, Gentiana calycosa, G. newberryi, G. prostrata, G. sceptrum, Gentianella amarella, Gentianopsis holopetala, G. simplex, Glyceria borealis, G. striata, Gnaphalium palustre, Hastingsia alba, Helenium bigelovii, Hordeum brachyantherum, Hosackia pinnata, Hymenoxys hoopesii, Hypericum anagalloides, H. formosum, H. scouleri, Iris missouriensis, Juncus acuminatus, J. balticus, J. covillei, J. drummondii, J. effusus, J. howellii, J. macrandrus, J. mertensianus, J. mexicanus, J. nevadensis, J. orthophyllus, J. oxymeris, J. tenuis, J. xiphoides, Juniperus communis, Kalmia microphylla, K. polifolia, Koeleria macrantha, Larix occidentalis, Leersia oryzoides, Leptarrhena pyrolifolia, Leptobryum pyriforme, Ligusticum canbyi, L. grayi, Lilaeopsis occidentalis, Lilium pardalinum, Lindernia dubia, Linnaea borealis, Lomatium grayi, Lonicera caerulea, L. conjugialis, L. involucrata, Lophozia, Lupinus argenteus, L. burkei, L. confertus, L. latifolius, L. polyphyllus, Luzula campestris, L. comosa, L. multiflora, Lysichiton americanus, Madia bolanderi, M. glomerata, Marchantia, Meesia triquetra, Menyanthes trifoliata, Menziesia ferruginea, Micranthes ferruginea, M. nelsoniana, M. odontoloma, M. oregana, Microseris borealis, M. laciniata, M. nutans, Mimulus floribundus, M. guttatus, M. lewisii, M. moschatus, M. tilingii, Mitella pentandra, Montia chamissoi, M. linearis, Muhlenbergia asperifolia, M. filiformis, M. richardsonis, Myosotis laxus, Myriophyllum aquaticum, Navarretia, Nothocalais alpestris, Oreostemma alpigenum, Orobanche californica, Orthilia secunda, Oxypolis occidentalis, Packera cymbalarioides, Parnassia californica, P. fimbriata, P. palustris, Paspalum distichum, Pedicularis attollens, P. groenlandica, Pellaea breweri, Penstemon oreocharis, P. procerus, P. rydbergii, Perideridia bolanderi, P. oregana, P. parishii, Persicaria amphibia, P. hydropiperoides, Phalacroseris bolanderi, Phalaris arundinacea, Philonotis fontana, Phleum alpinum, Phlox gracilis, Phyllodoce breweri, P. empetriformis, Physocarpus capitatus, Picea engelmannii, Pinus albicaulis, P. contorta, P. monticola, Plagiobothrys figuratus, Platanthera dilatata,*

P. hyperborea, P. leucostachys, P. sparsiflora, P. stricta, Poa cusickii, P. epilis, P. palustris, P. pratensis, P. secunda, Polygonum cascadense, Polytrichum juniperinum, Populus tremuloides, Potamogeton gramineus, Potentilla brevifolia, P. drummondii, P. egedii, P. flabellifolia, P. glandulosa, P. gracilis, Primula conjugens, P. jeffreyi, P. tetrandra, Prunella vulgaris, Ptilagrostis kingii, Puccinellia simplex, Pyrrocoma apargioides, Quercus garryana, Raillardella pringlei, Ranunculus alismifolius, R. flammula, R. gormanii, R. occidentalis, R. populago, R. uncinatus, Rhododendron neoglandulosum, R. occidentale, Ribes lacustre, R. sanguineum, Rosa gymnocarpa, Rubus lasiococcus, R. ursinus, Rumex paucifolius, R. salicifolius, Sagina saginoides, Sagittaria latifolia, Salix commutata, S. drummondiana, S. eastwoodiae, S. exigua, S. geyeriana, S. jepsonii, S. lemmonii, S. myrtillifolia, S. orestera, S. planifolia, S. scouleriana, S. sitchensis, Sanguisorba occidentalis, Scirpus congdonii, S. criniger, S. pungens, Senecio clarkianus, S. cymbalarioides, S. integerrimus, S. scorzonella, S. triangularis, Sibbaldia procumbens, Sidalcea oregana, S. pedata, Sisyrinchium bellum, S. californicum, S. elmeri, S. idahoense, Solidago canadensis, S. multiradiata, Sorbus, Sphagnum squarrosum, Sphenosciadium capitellatum, Spiraea densiflora, S. douglasii, Spiranthes porrifolia, S. romanzoffiana, Stachys ajugoides, Stellaria calycantha, S. crispa, S. longipes, Stenanthium occidentale, Streptopus roseus, Symphoricarpos albus, Symphyotrichum foliaceum, S. hendersonii, S. spathulatum, Taraxacum officinale, Thalictrum fendleri, T. occidentale, Tofieldia glutinosa, Trautvetteria caroliniensis, Triantha occidentalis, Trifolium cyathiferum, T. longipes, T. monanthum, T. wormskjoldii, Triglochin palustre, Trisetum spicatum, Triteleia hyacinthina, Tsuga heterophylla, T. mertensiana, Utricularia intermedia, U. macrorhiza, Vaccinium cespitosum, V. globulare, V. membranaceum, V. nivictum, V. occidentale, V. scoparium, Vaccinium oxycoccos, V. uliginosum, Valeriana sitchensis, Veratrum californicum, Veronica americana, V. scutellata, V. serpyllifolia, V. wormskjoldii, Viola adunca, V. macloskeyi, V. orbiculata, V. palustris, Xerophyllum tenax.

Mimulus pygmaeus A.L. Grant is an annual, which grows in vernally wet depressions, flats, meadows, playas, swales, and along the margins of ephemeral pools and streams at elevations from 1100 to 1800 m. The substrates include clay, gravel, gravelly loam, mud, rock, and silt. Life history information is scarce for this rare species. **Reported associates:** *Acmispon americanus, Artemisia, Evax, Juncus, Linanthus liniflorus, Perideridia erythrorhiza, P. gairdneri, P. howellii.*

Mimulus ringens L. is a perennial, which grows on dikes, in bogs, carrs, depressions, ditches, dunes, floodplains, flowages, marshes, meadows, mudflats, oxbows, prairies, sandbars, seeps, sloughs, swales, swamps, and along the margins of lakes, ponds, rivers, and streams at elevations of up to 200 m. A wide range of acidity (pH: 5.3–8.3) is tolerated, but occurrences are more common on alkaline sites (mean pH: 7.1). The substrates consist of organic or inorganic clay, gravel, loam, muck, mucky peat, mud, sand, and sandy loam. The plants also have been found growing on stumps. They occur in open or shaded sites (more typically open) in water depths from −6.0 to +19.5 cm (relative to the substrate; mean: 11.5 cm), and will flower across conditions ranging from −6 to +6 cm of inundation; however, biomass generally declines with increasing water depths. The flowers are self-compatible and pollinated primarily by worker bumblebees (Hymenoptera: Apidae: *Bombus*; in order of visitation: *B. fervidus, B. impatiens, B. griseocollis, B. nevadensis, B. bimaculatus*). The visitation frequency of the bees is proportional to the size of the floral display with more flowers being probed in succession in the larger displays. A mixed mating system occurs, where the level of outcrossing correlates with the extent of anther/stigma separation (herkogamy), which varies widely in populations. The probability of geitonogamy is greater as more flowers are probed within an inflorescence, until the stigmas eventually close (15–90 min after pollination). Analysis of experimental populations using genetic markers indicated that higher plant densities resulted in a greater proportion of pollinator flights between plants and a higher frequency of outcrossing. Pollen carryover can extend gene flow by nearly 50% over the average distances traversed by pollinators. Pollinator competition with *Lobelia siphilitica* (Campanulaceae) can reduce seed set (by 37%) and outcrossing rates (by 20%) when the species co-occur. Multiple paternity of fruits is common, with one study finding that over 95% of the fruits had at least two pollen donors. When stored dry at 10°C–15°C, the seeds will germinate without stratification; however, light is essential and higher germination rates are attained under brighter conditions. The rates are high in water and seeds that have been stratified for 4–5 months at 4°C also germinate well, but not in the dark. Seed germination typically occurs within 4 weeks at 21°C. Higher ambient temperatures (32°C) can promote germination under diffuse light conditions. The small seeds persist for an undetermined time in the seed bank. They are dispersed by the wind or by water. The plants often occur in disturbed sites. They can spread vegetatively by stolon-like rhizomes but are not aggressive colonizers. This species is highly recruitment limited and may not be capable of establishment even when seeded intentionally into a site. The plants show a high incidence (80%–100%) of arbuscular mycorrhizae. The growth of the plants has been observed to decline when placed in competition with *Cyperus bipartitus, Juncus bufonius* or *Lythrum salicaria*. **Reported associates:** *Acalypha rhomboidea, Acer negundo, A. rubrum, A. saccharinum, Achillea millefolium, Acorus calamus, Agropyron repens, Agrostis alba, A. scabra, A. stolonifera, A. tenuis, Alisma subcordatum, Allium vineale, Alnus rugosa, A. serrulata, Ambrosia artemisiifolia, Amphicarpaea bracteata, Apios americana, Apocynum cannabinum, Arisaema triphyllum, A. dracontium, Asclepias incarnata, A. purpurascens, Symphyotrichum puniceum, Betula alleghaniensis, B. lenta, B. nigra, B. papyrifera, B. ×sandbergii, Bidens cernuus, B. connatus, B. frondosus, B. laevis, Boehmeria cylindrica, Brachythecium rivulare, Brassica, Bryhnia novae-angliae, Bolboschoenus fluviatilis, Calamagrostis canadensis, C. inexpansa, Caltha palustris, Calystegia sepium, Campanula aparinoides, Cardamine bulbosa, C. pensylvanica, Carex aquatilis, C. aurea, C. bebbii,*

C. canescens, C. grayi, C. gynandra, C. hystericina, C. lacustris, C. lanuginosa, C. lasiocarpa, C. pellita, C. projecta, C. scabrata, C. squarrosa, C. tenera, C. vaginata, C. vesicaria, C. viridula, C. vulpinoidea, Cephalanthus occidentalis, Chamaedaphne calyculata, Chamerion angustifolium, Chelone glabra, Chrysanthemum leucanthemum, Chrysosplenium americanum, Cicuta bulbifera, C. maculata, Cinna arundinacea, Circaea alpina, Cirsium arvense, Comarum palustre, Cornus amomum, C. racemosa, C. sericea, Corylus, Crataegus, Cuscuta arvensis, Cyperus aristatus, Cyperus bipartitus, C. diandrus, C. esculentus, C. ferruginescens, C. strigosus, Daucus carota, Desmodium glutinosum, Dioscorea villosa, Dryopteris cristata, D. spinulosa, Echinochloa crus-galli, E. walteri, Eleocharis acicularis, E. erythropoda, E. obtusa, E. smallii, Epilobium coloratum, Equisetum arvense, Eragrostis cilianensis, E. hypnoides, E. pectinacea, Erigeron annuus, E. philadelphicus, Eupatorium perfoliatum, E. serotinum, Eutrochium maculatum, Fallopia convolvulus, F. scandens, Fimbristylis autumnalis, Fraxinus americana, F. nigra, Galium aparine, G. obtusum, G. tinctorium, G. trifidum, G. triflorum, Gentianopsis virgata, Geum canadense, G. rivale, Glyceria borealis, G. melicaria, G. striata, Gratiola, Helenium autumnale, H. flexuosum, Hibiscus moscheutos, Hippuris vulgaris, Hydrocotyle americana, Hypericum canadensis, H. majus, H. mutilum, H. virginicum, Impatiens capensis, Iris versicolor, Juncus balticus, J. brevicaudatus, J. canadensis, J. dudleyi, J. effusus, J. gymnocarpus, J. nodosus, J. tenuis, J. torreyi, Justicia americana, Kalmia latifolia, Laportea canadensis, Lappula squarrosa, Leersia oryzoides, L. virginica, Lemna minor, Leymus arenarius, Lindera benzoin, Lipocarpha micrantha, Lobelia cardinalis, L. siphilitica, Ludwigia alternifolia, L. palustris, Lycopus americanus, L. asper, L. uniflorus, L. virginicus, Lyonia ligustrina, Lysimachia nummularia, L. terrestris, L. thyrsiflora, Lythrum alatum, L. salicaria, Magnolia tripetala, Melilotus alba, Mentha arvensis, Micranthes pensylvanica, Mimulus alatus, Muhlenbergia frondosa, Myosotis scorpioides, Myosoton aquaticum, Myrica caroliniensis, Nuphar variegatum, Nyssa sylvatica, Oenothera biennis, O. fruticosa, Onoclea sensibilis, Osmunda regalis, Osmundastrum cinnamomeum, Oxypolis rigidior, Panicum capillare, P. dichotomiflorum, Parnassia glauca, Pedicularis lanceolata, Penthorum sedoides, Persicaria amphibia, P. hydropiper, P. hydropiperoides, P. lapathifolia, P. pensylvanica, P. sagittata, Phalaris arundinacea, Phragmites australis, Phyla lanceolata, Physostegia virginiana, Picea, Pilea pumila, Pinus banksiana, Platanthera psycodes, Poa compressa, P. paludigena, Populus balsamifera, P. deltoides, Potentilla anserina, P. norvegica, Prunus serotina, Pycnanthemum muticum, P. virginianum, Quercus, Ranunculus pensylvanicus, R. septentrionalis, R. trichophyllus, Rhamnus, Rhexia virginica, Rhizomnium appalachianum, R. punctatum, Rhynchostegium serrulatum, Rosa multiflora, R. palustris, Rubus allegheniensis, Rudbeckia hirta, R. laciniata, Rumex crispus, R. orbiculatus, R. verticillatus, Sagittaria latifolia, Salix alba, S. babylonica, S. candida, S. eriocephala, S. exigua, S. glaucophylloides, S. gracilis, S. interior, S. lucida, S. nigra, S. petiolaris, S. planifolia, S. serissima, Sambucus nigra, Samolus parviflorus, Saponaria officinalis, Saururus cernuus, Schoenoplectus pungens, S. tabernaemontani, Scirpus atrovirens, S. georgianus, S. cyperinus, Scutellaria galericulata, S. lateriflora, Setaria viridis, Sium suave, Solanum carolinense, S. dulcamara, S. rostratum, Solidago altissima, S. canadensis, S. graminifolia, S. juncea, S. ohioensis, S. patula, S. riddellii, Sonchus, Sparganium americanum, S. eurycarpum, Spartina pectinata, Spiraea alba, S. tomentosa, Stachys palustris, Symphyotrichum lanceolatum, S. lateriflorum, S. puniceum, Symplocarpus foetidus, Thelypteris noveboracensis, T. palustris, Thuidium delicatulum, Tiarella cordifolia, Toxicodendron radicans, Trifolium pratense, Typha angustifolia, T. latifolia, Ulmus, Urtica dioica, U. procera, Utricularia macrorhiza, Vaccinium fuscatum, Vallisneria americana, Vernonia noveboracensis, Veratrum viride, Verbena hastata, V. urticifolia, Viola cucullata, V. macloskeyi, Xanthium strumarium.

***Mimulus tilingii* Regel** is a perennial inhabitant of cold alpine or subalpine brooks, ditches, flats, marshes, meadows, snowfields, springs, streams, or the margins of bogs, lakes, ponds, and smaller streams at elevations from 1400 to 3780 m. The substrates range from calcareous to volcanic and include clay, granite, granite loam, gravel, humus, limestone, loam, pumice, rock, sand, scree, and talus. Exposures from full sun to partial shade are tolerated. The plants principally are outcrossed by large bees (Insecta: Hymenoptera: Apidae: *Bombus*), hummingbirds (Aves: Trochilidae) or rarely by flies (Insecta: Diptera). The flowers, with <50% of the floral biomass allocated to corollas and stamens, have intermediate male function compared to congeners. Self-pollination is reduced by the movement of the stigma lobes, which close slowly (in about 12 seconds on average) after contact is made with a pollinator. A mixed mating system is indicated by intermediate pollen:ovule ratios and inbreeding coefficients (determined from allozyme data) relative to other congeners. The seeds germinate readily in 1–3 weeks at 21°C–22°C when sown on the surface in the light. They are dispersed by flowing water. The plants are short-lived perennials with a creeping habit and can form large mats by the proliferation of rhizomes or stolons. Vegetative reproduction occurs by the fragmentation of the rhizomes and their subsequent dispersal in the water. The plants are both nonmycorrhizal or associated with arbuscular mycorrhizae or dark, septate endophytes. The herbage is coated by a polysaccharide slime of uncertain function, which is secreted by glandular epidermal hairs. **Reported associates:** *Abies grandis, A. lasiocarpa, Agrostis diegoensis, A. idahoensis, Alnus, Antennaria media, Athyrium distentifolium, Barbarea, Bistorta bistortoides, Calamagrostis breweri, Caltha howellii, Carex heteroneura, C. nebrascensis, C. nigricans, C. spectabilis, C. vernacula, Cassiope, Castilleja lemmonii, C. miniata, C. parviflora, C. peirsonii, Cerastium fontanum, Cirsium vulgare, Claytonia megarhiza, Cystopteris fragilis, Dasiphora floribunda, Deschampsia atropurpurea, D. cespitosa, D. elongata, Elmera racemosa, Epilobium alpinum, E. ciliatum, E. glaberrimum, E. latifolium, Erigeron peregrinus, Eriogonum* *ovalifolium, Gaultheria shallon, Geranium californicum, G. richardsonii, Glyceria elata, Hypericum formosum, Juncus drummondii, J. mertensianus, J. parryi, Kalmia microphylla, Lewisia pygmaea, L. sierrae, Ligusticum grayi, Luetkea pectinata, Lupinus latifolius, Luzula hitchcockii, Micranthes ferruginea, M. nelsoniana, M. odontoloma, M. tolmiei, Mimulus cardinalis, M. guttatus, M. lewisii, M. moschatus, M. primuloides, Minuartia rubella, Monardella, Montia chamissoi, Muhlenbergia filiformis, Oreostemma alpigenum, Pedicularis attollens, P. groenlandica, Penstemon davidsonii, P. heterodoxus, Phleum alpinum, Phyllodoce breweri, P. empetriformis, Picea, Pinus albicaulis, P. contorta, Plantago lanceolata, Poa fendleriana, P. gracillima, P. secunda, P. wheeleri, Polygonum, Populus, Potentilla breweri, P. drummondii, P. flabellifolia, P. flabelliformis, Primula tetrandra, Pteridium aquilinum, Ranunculus populago, Rhododendron, Salix arctica, S. commutata, S. eastwoodiae, S. orestera, S. sitchensis, Senecio triangularis, Sibbaldia procumbens, Stellaria calycantha, Trifolium monanthum, Trisetum spicatum, Tsuga mertensiana, Vaccinium cespitosum, Vahlodea atropurpurea, Valeriana sitchensis, Veratrum californicum, V. viride, Veronica americana, Veronica wormskjoldii, Viola.*

Mimulus tricolor Hartw. ex Lindl. is an annual plant of depressions, ditches, gravel bars, meadows, mudflats, swales, vernal pools, and drying lake/pond margins or streambeds at elevations of up to 3140 m. The sites are characterized by full sunlight and occur where water recedes to expose the drying substrate. The substrates include adobe, clay, lava, mud, obsidian gravel, gravelly clay, rock, rocky clay, sand, serpentine, and stony loam. The flowers are pollinated and outcrossed by insects, which include bumblebees (Hymenoptera: Apidae: *Bombus*), solitary bees (Hymenoptera: Megachilidae), and occasional beeflies (Diptera: Bombyliidae). The fruits are hardened capsules, which remain indehiscent at maturity. They open only when exposed to water for at least 24 h; thus, a period of pooled water is essential for capsule dehiscence to commence. The seeds usually remain within the opened capsules, which are dispersed locally by water, or over greater distances if the sites are flooded extensively. Seed germination occurs after the spring waters recede and is relatively simultaneous, typically resulting in a clumped distribution of seedlings. An extensive seed bank is maintained, with some seeds known to persist for at least 8 years. Although populations may reappear consistently from year to year at a site, they can exhibit extensive fluctuations in size and area. The plants thrive in lightly disturbed sites and occur frequently on abandoned gopher (Mammalia: Rodentia: Geomyidae) mounds. **Reported associates:** *Abies, Achyrachaena mollis, Agrostis avenacea, Alopecurus californicus, A. geniculatus, A. saccatus, Amsinckia, Anagallis minima, Anthemis cotula, Arenaria californica, Blennosperma nanum, Brodiaea, Bromus hordeaceus, Callitriche heterophylla, C. marginata, Cardamine oligosperma, Castilleja attenuata, C. exserta, Cicendia quadrangularis, Clarkia purpurea, Claytonia dichotoma, Crassula aquatica, Croton setigerus, Crypsis vaginiflora, Cryptantha, Cuscuta howelliana, Damasonium californicum, Deschampsia danthonioides,*

Downingia bacigalupii, D. bella, D. bicornuta, D. cuspidata, D. elegans, D. insignis, D. ornatissima, Elatine ambigua, E. brachysperma, E. californica, Eleocharis acicularis, E. palustris, Epilobium cleistogamum, E. pallidum, E. pygmaeum, E. torreyi, Eremocarpus setigerus, Eryngium alismifolium, E. castrense, E. mathiasiae, E. vaseyi, Eschscholzia, Floerkea douglasii, Gilia intertexta, Gilia leucocephala, Glyceria, Gratiola ebracteata, Grindelia camporum, Hemizonia parryi, H. pungens, Hordeum marinum, Hypochaeris glabra, Isoetes orcuttii, Juncus bufonius, J. leiospermus, J. tenuis, J. uncialis, J. xiphioides, Lactuca serriola, Lasthenia fremontii, L. glabrata, Legenere limosa, Leontodon taraxacoides, Lepidium latifolium, L. latipes, Libocedrus, Limnanthes douglasii, Linanthus bicolor, Lolium perenne, L. multiflorum, L. temulentum, Lupinus bicolor, Lythrum californicum, L. hyssopifolia, Machaerocarpus californicus, Malvella leprosa, Marsilea vestita, Matricaria discoidea, Mimulus angustatus, M. bicolor, M. nudatus, Montia fontana, Myosurus minimus, Navarretia leucocephala, N. minima, N. myersii, N. prostrata, Orcuttia californica, O. inaequalis, O. tenuis, O. viscida, Persicaria amphibia, Phalaris lemmonii, Phyla nodiflora, Pilularia americana, Pinus ponderosa, Plagiobothrys bracteatus, P. leptocladus, P. scouleri, P. stipitatus, P. trachycarpus, Plantago elongata, Poa annua, Pogogyne ziziphoroides, Polygonum aviculare, P. douglasii, Polypogon maritimus, P. monspeliensis, Psilocarphus brevissimus, P. tenellus, Ranunculus aquatilis, Rorippa curvisiliqua, Rumex crispus, R. pulcher, Trifolium hirtum, Triteleia laxa, Tuctoria greenei, Veronica peregrina, Xanthium strumarium.

Mimulus washingtonensis Gandog. is an annual, which grows in draws, meadows, sandbars, seeps, on slopes, in wagon ruts, and along the margins of ponds, rivers, and springs at elevations of up to 1219 m. The plants occupy sites with vernal water and occur on substrates that are thin veneers of basaltic crust, gravel or sand, which overlay the granitic bedrock. Exposures can include sun or shade. These habitat conditions develop only in areas of low disturbance. The flowers have very low levels of self-pollination and principally are outcrossed by ground-nesting, solitary bees (Insecta: Hymenoptera: Megachilidae). Other insect pollinators can include bumblebees (Hymenoptera: Apidae), flies (Diptera: Syrphidae), and moths (Lepidoptera: Sphingidae). Competition for pollinators can occur with *M. guttatus* when the two species grow in close proximity. The seeds are dispersed by wind or by water during periods of spring runoff. An extensive seed bank develops in most cases. Generally, the sites are extremely nutrient poor and the plants grow in association with a nitrogen fixing Cyanobacteria (Nostocales: Nostocaceae: *Nostoc commune*). Few associated species reportedly co-occur with this species on the highly sterile sites. **Reported associates:** *Allium brandegeei, Alnus, Arenaria serpyllifolia, Asclepias, Astragalus diaphanus, Bromus tectorum, Carex, Cercocarpus ledifolius, Danthonia intermedia, Delphinium, Epilobium minutum, Heterocodon rariflorus, Juncus, Juniperus occidentalis, Lewisia pygmaea, L. triphylla, Mimulus breweri, M. guttatus, Montia linearis, Navarretia, Nostoc commune, Picea engelmannii, Pinus* *contorta, P. ponderosa, Plagiobothrys, Primula conjugens, Pseudoroegneria spicata, Purshia tridentata, Ribes cereum, Rudbeckia occidentalis, Senecio triangularis, Streptanthus, Thelypodium eucosmum, Triteleia hyacinthina,*

Use by wildlife: *Mimulus alatus* is a larval food plant of the buckeye butterfly (Insecta: Lepidoptera: Nymphalidae: *Junonia coenia*). Its nectar is taken by bumblebees (Insecta: Hymenoptera: Apidae: *Bombus pensylvanicus*) and its foliage is resistant to browsing by deer (Mammalia: Cervidae: *Odocoileus*). Other hosts for larval Lepidoptera include *M. cardinalis* (Noctuidae: *Abagrotis variata, Megalographa biloba*), *M. guttatus* (Noctuidae: *Amphipyra tragopoginis, Annaphila lithosina*; Nymphalidae: *Euphydryas chalcedona, Junonia coenia, Phyciodes mylitta*), *M. moschatus* (Noctuidae: *Annaphila casta, A. miona*; Nymphalidae: *Junonia coenia*), and *M. ringens* (Nymphalidae: *Euphydryas phaeton*). Fungal hosts include *M. alatus* (Ascomycota: Mycosphaerellaceae: *Cercospora mimuli*), *M. guttatus* (Ascomycota: Mycosphaerellaceae: *Ramularia mimuli*; Basidiomycota: Entylomataceae: *Entyloma clintonianum*; Oomycota: Peronosporaceae: *Peronospora jacksonii*), *M. lewisii* (Ascomycota: Didymosphaeriaceae: *Apiosporella mimuli*; Mycosphaerellaceae: *Ramularia lewisii, R. mimuli*), *M. moschatus* (Ascomycota: Mycosphaerellaceae: *Septoria mimuli*), *M. ringens* (Ascomycota: Mycosphaerellaceae: *Ramularia mimuli*; Blastocladiomycota: Physodermataceae: *Physoderma*), and *M. tilingii* (Ascomycota: Mycosphaerellaceae: *Ramularia mimuli, Septoria mimuli*; Basidiomycota: Entylomataceae: *Entyloma clintonianum*). *Mimulus alsinoides* occurs as a habitat component of two salamanders (Chordata: Amphibia: Plethodontidae: *Plethodon dunni, P. vehiculum*). The plants are not grazed by mule deer (Mammalia: Cervidae: *Odocoileus hemionus*). *Mimulus cardinalis* is a nectar source (average sugar content of 17%) for hummingbirds (Aves: Trochilidae). It is associated with the habitat of the panamint alligator lizard (Reptilia: Anguidae: *Elgaria panamintina*) and the Kanab ambersnail (Mollusca: Succineidae: *Oxyloma haydeni kanabensis*). *Mimulus guttatus* is eaten occasionally by deer (Mammalia: Cervidae: *Odocoileus*), which act as seed dispersal agents. *Mimulus lewisii* grows within the habitat of the lemming mouse (Mammalia: Rodentia: Muridae: *Synaptomys borealis*). It is categorized as highly tolerant to grazing by deer (Mammalia: Cervidae: *Odocoileus*). The roots and stems of *M. moschatus* are eaten by the Columbian black-tailed deer (Mammalia: Cervidae: *Odocoileus hemionus columbianus*), especially in early spring. The plants are susceptible to a number of pathogens including tobacco necrosis, tobacco and tomato ringspot, and alfalfa and cucumber mosaic viruses. *Mimulus primuloides* constitutes a portion of the habitat of the mountain pocket gopher (Mammalia: Rodentia: Geomyidae: *Thomomys monticola*). It is eaten by the devastating grasshopper (Insecta: Orthoptera: Acrididae: *Melanoplus devastator*). The foliage of *M. ringens* is fed on by larvae of the chalcedony midget moth (Insecta: Lepidoptera: Noctuidae: *Elaphria chalcedona*). The nectar is a source of food for the pipevine and spicebush swallowtail butterflies (Lepidoptera: Papilionidae: *Battus philenor, Papilio troilus*)

and skipper butterflies (Lepidoptera: Hesperiidae: *Epargyreus tityrus*). Other insect floral visitors of *M. ringens* include flies (Diptera: Syrphidae: *Allograpta obliqua, Sphaerophoria cylindrica, Syritta pipiens, Syrphus americana*), bees (Hymenoptera: Apidae: *Bombus americanorum, B. bimaculatus, B. fervidus, B. griseocollis, B. impatiens, B. nevadensis, B. virginicus, Ceratina dupla*; Halictidae: *Agapostemon radiatus, A. splendens*; Megachilidae: *Megachile latimanus*), and wasps (Hymenoptera: Sphecidae: *Microbembex monodonta*). In addition to its flowers, which provide food for pollinating insects, the seeds, and foliage of *M. washingtonensis* are eaten by various phytophagous insects.

Economic importance: food: Native North American tribes used several *Mimulus* species as foods. These included the raw stalks of *M. cardinalis* (Kawaiisu), fruits of *M. eastwoodiae* (Navajo, Kayenta), leaves, and shoots of *M. glabratus* for salad (Isleta), leaves of *M. guttatus* for salad (Mendocino), or boiled as a vegetable (Miwok), boiled shoots of *M. moschatus* as a vegetable (Miwok), and leaves of *M. tilingii* as greens (Neeshenam). The leaves of *M. ringens* also were eaten as salad greens by early North American settlers and by native tribes; **medicinal:** Various species of *Mimulus* were used by the Miwok to control diarrhea. The Karok washed their newborn babies with an infusion derived from *M. cardinalis*. The Navajo and Kayenta ingested *M. eastwoodiae* for hiccoughs. The Potawatomi prepared a general medicinal from *M. glabratus*. A decoction of stems and leaves from *M. guttatus* was used by the Kawaiisu as a steambath to treat back and chest pains. The Shoshoni applied a poultice of the crushed leaves to treat wounds and rope burns. The Yavapai made a tea from the plants for treating stomachaches. A root decoction of *M. ringens* was used as an anticonvulsive to treat epilepsy in Iroquois women and in a wash administered as a poison antidote; **cultivation:** *Mimulus* became a horticultural sensation in Europe during the mid-19th century due to the ease of hybridizing several species such as *M. cardinalis* and *M. lewisii*. Cultivated *Mimulus* species include *M. alatus, M. alsinoides, M. cardinalis* (including cultivars 'Dark Throat,' 'Red Dragon' and 'Santa Cruz Island Gold'), *M. eastwoodiae, M. guttatus* (including cultivar 'Richard Bish'), *M. lewisii* (including cultivar 'Sunset'), *M. moschatus* (including cultivar 'Variegatus'), *M. primuloides, M. ringens* and *M. tilingii*, which are grown mainly as water garden ornamentals. *Mimulus cardinalis* and *M. lewisii* have received the Award of Garden Merit by the Royal Horticultural Society. *Mimulus moschatus* was introduced to England in 1826, where the plants completely lost their musky fragrance within 80 years of cultivation; **misc. products:** *Mimulus* has become a model system for ecological genetics research. *Mimulus alatus* often is planted in wetland restoration projects. Essence of *M. lewisii* is sold for "aroma therapy"; **weeds:** *Mimulus guttatus* is invasive in the United Kingdom as is *M. moschatus* in Australia; **nonindigenous species:** *Mimulus guttatus* has been introduced to eastern North America, Europe (early 19th century), Japan and New Zealand. *Mimulus moschatus* was introduced to eastern North America, Australia, Europe (in 1826), New Zealand and South America (Chile). More research is necessary to determine whether all populations in eastern North America are escapes from cultivation, or if some are naturally disjunct.

Systematics: Phylogenetic evidence from DNA sequence data shows that *Mimulus* is not monophyletic as traditionally circumscribed. Notably, combined molecular data resolve a clade that includes *Berendtiella, Hemichaena, Leucocarpus*, and members of *Mimulus* subgenus *Schizoplacus*. The genus *Leucocarpus* associates with *Mimulus* sections *Erythranthe, Simiolus*, and the non-Australian members of section *Paradanthus* (Figure 5.80). Many of the sections (e.g., *Eunanus, Oenoe, Paradanthus*) also are polyphyletic as currently circumscribed (Figure 5.80). Although once assigned to the monotypic genus *Mimetanthe, Mimulus pilosus* resolves as the sister to a clade of species comprising *Mimulus* sections *Diplacus, Eunanus*, and *Oenoe* (Figure 5.80). However, depending on how *Mimulus* eventually becomes circumscribed, it is conceivable that *Mimetanthe* (i.e., section *Mimuloides*) could be reinstated as a distinct genus. It has been retained here within *Mimulus*. A cpDNA/nrDNA phylogeny (Figure 5.80) of the western North American species depicts relationships of all the OBL North American species except for *M. alatus* (eastern) and *M. dentilobus* (southwestern). *Mimulus alatus* is very similar to, and arguably is closely related to *M. ringens* (both section *Mimulus*); *M. dentilobus* is placed in section *Simiolus*. The phylogeny of *Mimulus* indicates repeated derivations of the OBL habit in the genus. AFLP and DNA sequence data support the monophyly of *Mimulus* section *Eurythanthe* but differ in their resolution of taxa. The relationships in section *Eurythanthe* remain somewhat ambiguous because different results are obtained using different genetic markers. *Mimulus cardinalis* resolves variously as the sister to *M. verbenaceus* (DNA sequence data; Figure 5.80) or *M. lewisii* (AFLP & allozyme data; Figure 5.81); *M. eastwoodiae* resolves either as closest to *M. rupestris* (allozymes), *M. nelsonii* (DNA sequence data; Figure 5.80), or *M. verbenaceus* (AFLP data; Figure 5.81). However, those phylogenetic analyses consistently indicate that both the OBL habit and hummingbird (Aves: Trochilidae) pollination are derived independently in *M. cardinalis* and *M. eastwoodiae*, which never resolve as sister taxa. The putative sister species *M. lewisii* (bee pollinated; higher elevations) and *M. cardinalis* (hummingbird pollinated; lower elevations) can be hybridized, but they are isolated in nature by having different elevational range centers as well as their discrete pollinator preferences. Flowers higher in anthocyanin and carotenoid content dissuade bee visitors and those high in nectar obtain greater numbers of hummingbird visits. Further isolation results from the lower vigor of their F_1 hybrids, which are characterized by reduced germination rates and low pollen and ovule fertility. *Mimulus lewisii* comprises two races that are distinct morphologically, genetically and geographically (northern Rocky Mountains vs. California Sierra Nevada). Allozyme data indicate low levels of intrapopulational genetic variation in section *Erythranthe*, a result attributed to rapid fixation of alleles in the typically small, isolated populations. The presumed close relationship of *M. alatus* and

M. ringens is evidenced by their ability to hybridize when sympatric. The hybrids are intermediate morphologically, have reduced pollen fertility (~37%), and exhibit biochemical (flavonoid) profiles that are additive relative to the parental species. Although both species have a fairly broad range of habitat tolerances, they tend to partition their occurrences ecologically between open marshland (*M. ringens*) and shady treamsides (*M. alatus*). An estuarine ecotype of *M. ringens* (recognized as var. *colpophilus*) remains morphologically distinct from var. *ringens* when the taxa are grown under common garden conditions. *Mimulus alsinoides* was once believed to be most closely related to *M. pulsiferae* (UPL); however, it is apparent that these two species are quite distantly related (Figure 5.80). The close relationship assumed between *M. dentatus* and *M. sessilifolius* is supported by molecular data (Figure 5.80). *Mimulus glabratus* was regarded as a parent of *M. michiganensis*, which had long been recognized as a variety of the former species; however, genetic data from RAPD markers, do not substantiate the hybrid origin of the latter, despite its low pollen fertility (<3%). Instead, *M. michiganensis* is suspected to be an aneuploid or otherwise genomically rearranged derivative of *M. glabratus*. Attempts at experimental hybridization between these taxa have failed, indicating that the taxa are well-isolated genetically. *Mimulus glabratus* previously was believed to be closely related to *M. guttatus* and *M. pilosiusculus*; however, DNA data indicate that it is rather distant from at least the former (Figure 5.80). The *Mimulus guttatus* complex consists of several closely related taxa (*M. cupriphilus, M. glaucescens, M. guttatus, M. laciniatus, M. marmoratus, M. nasutus, M. nudatus, M. pardalis, M. tilingii*), which have been difficult to evaluate taxonomically. *Mimulus glabratus* itself embodies a mixoploid polyploid series as indicated by allozyme data, which correlates well with morphology and ploidy levels. *Mimulus glaucescens, M. nudatus* and *M. tilingii* have been treated as varieties of *M. guttatus* and DNA data confirm that the taxa are at least closely related. *Mimulus glaucescens* is only weakly interfertile with *M. guttatus*, and few natural hybrids have been reported; furthermore, experimental hybridizations have shown that their F_2 progeny are highly sterile and it seems reasonable to maintain these as distinct species. Many authors do not accept *M. nasutus* as being distinct taxonomically from *M. guttatus*. However, genetic mapping of F_2 arrays indicates that their genomes are extensively divergent, resulting in unequal genetic transmission characterized by an excess of *M. guttatus* alleles in the hybrids. Nuclear DNA polymorphism analysis indicates that natural hybridization of these species can be asymmetrically introgressive from *M. nasutus* to *M. guttatus*. Analysis of nuclear DNA sequences indicates that there have been at least several origins of allotetraploid hybrids involving these species, which are highly selfing and are isolated reproductively from both parents. The presumed derivation of the serpentine endemic *M. nudatus* from *M. guttatus* is supported by their close relationship as indicated by analysis of DNA sequence data (Figure 5.80). The species are isolated by postzygotic incompatibility (low seed viability of 1.7%–12.0%) and by the greater drought and

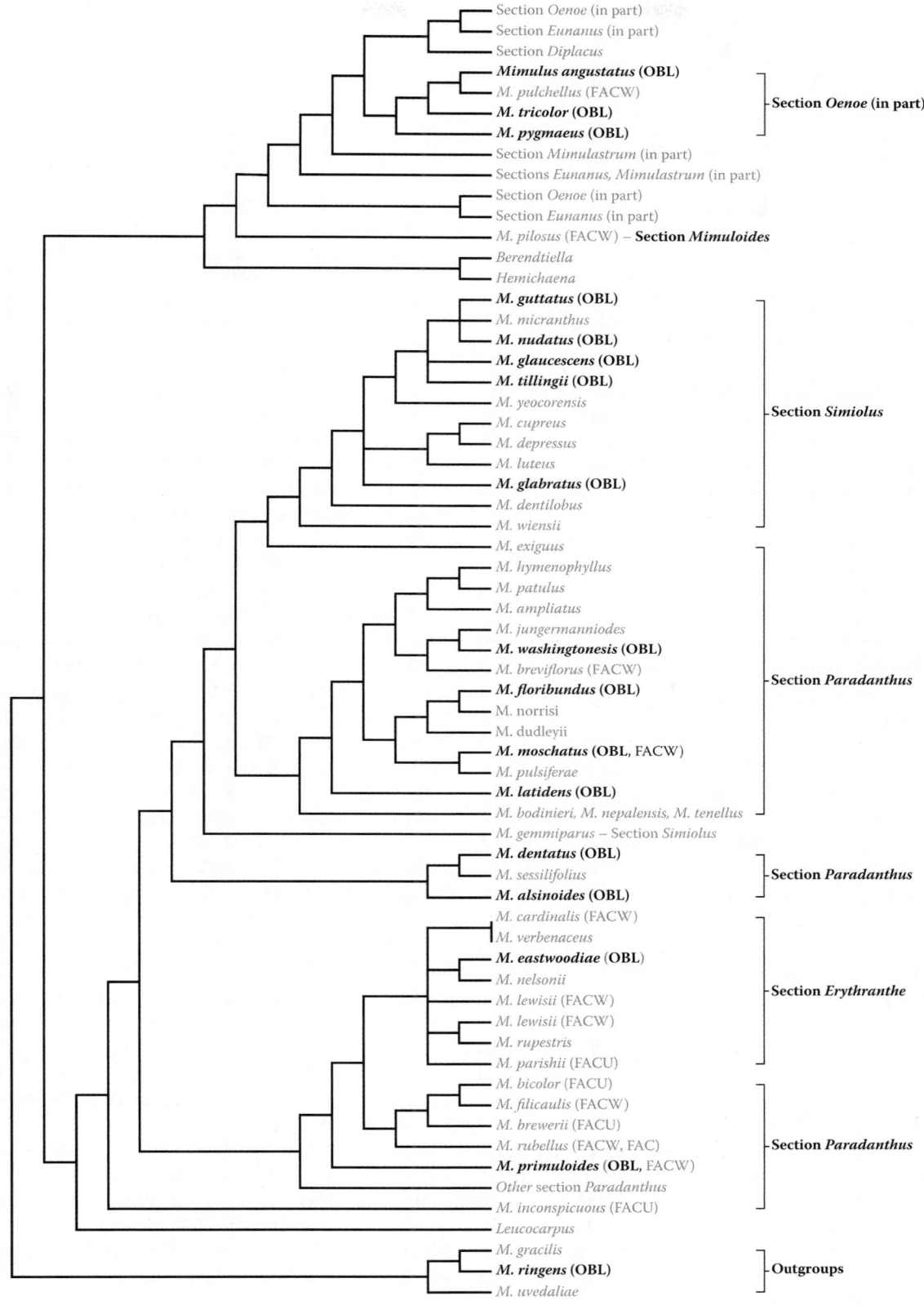

FIGURE 5.80 This phylogeny of western *Mimulus* species is based on combined cpDNA and nrDNA sequence data and includes all of the OBL North American taxa (indicated in bold) except *M. alatus* and *M. dentilobus*. The relationships rendered here indicate repeated origins of the OBL habit in the genus. These results show that not only are many sections (e.g., *Eunanthus, Oenoe, Paradanthus*) not monophyletic, but the genus itself is not monophyletic as currently circumscribed (e.g., the inclusion here of *Berendtiella, Hemichaena* and *Leucocarpus*). Section *Erythranthe* is one section that resolves as a clade (see also Figure 5.81). (Adapted from Beardsley, P.M. et al., *Amer. J. Bot.*, 91, 474–489, 2004.)

FIGURE 5.81 Phylogenetic relationships of *Mimulus* section *Erythranthe* as indicated by an analysis of AFLP data. These results, which are somewhat inconsistent with the analysis of DNA sequence data, are better resolved and also support the monophyly of *Erythranthe*. Allozyme data place *M. nelsonii* with *M. verbenaceus* and *M. eastwoodiae* with *M. rupestris* (Vickery, Jr. et al., 1989). (Adapted from Beardsley, P.M. et al., *Evolution*, 57, 1397–1410, 2003.)

calcium deficiency tolerance of the former. *Mimulus guttatus* and *M. glabratus* are hypothesized as the parents of *M. gemmiparus*, but further study (incorporating genetic evidence) is needed to substantiate this possibility. Phylogenetic analysis (Figure 5.80) has failed to support a close relationship between *M. latidens* and *M. pulsiferae*, which had been suggested formerly. *Mimulus evanescens* is a morphologically intermediate hybrid derivative of *M. latidens* and *M. breviflorus*. *Mimulus moschatus* occurs within a monophyletic "alliance" of 13 morphologically similar but genetically distinct species, many of which are quite rare. *Mimulus pygmaeus* was once placed within a monotypic section (*Microphyton*) due to its unusual fruit symmetry and anther development. However, analyses of DNA sequence data indicate that it is closely related to *M. tricolor* and other species in section *Oenoe* (Figure 5.80). The base chromosome number of *Mimulus* is difficult to ascertain because of extensive variability as well as the fact that the genus does not appear to be monophyletic as circumscribed. Chromosome numbers available for OBL species include: *M. cardinalis*, *M. eastwoodiae*, *M. lewisii* ($2n = 16$); *M. angustatus*, *M. tricolor* ($2n = 18$), *M. primuloides* ($2n = 18, 36$), *M. pygmaeus* ($2n = 20$), *M. glaucescens* ($2n = 28$), *M. glabratus* ($2n = 30, 60, 62$); *M. dentilobus*, *M. floribundus*, *M. latidens*, *M. moschatus* ($2n = 32$); *M. guttatus* ($2n = 28, 30, 32, 48, 56$); *M. ringens* ($2n = 16, 22, 24$); and *M. tilingii* ($2n = 28, 30, 48, 50, 56$).

Comments: *Mimulus* species vary in their distributions from being widespread (*M. ringens*) or occurring principally in eastern and western North America (*M. moschatus*), western North America (*M. cardinalis*, *M. dentatus*, *M. floribundus*, *M. guttatus*, *M. lewisii*, *M. pilosus*, *M. tilingii*), central North America (*M. glabratus*; extending to South America), eastern North America (*M. alatus*), northwestern United States (*M. washingtonensis*), western United States (*M. alsinoides*, *M. primuloides*), southwestern United States (*M. eastwoodiae*), or having more restricted distributions such as *M. dentilobus* (Arizona, New Mexico, Texas), *M. latidens*, *M. pygmaeus* and *M. tricolor* (California, Oregon), and *M. angustatus*, *M. glaucescens*, and *M. nudatus* (California).

References: Allen & Sheppard, 1971; Angert, 2005, 2006; Arathi & Kelly, 2004; Awadalla & Ritland, 1997; Barbour et al., 2007; Baskin & Baskin, 1998; Bauder & McMillan,

1998; Beardsley et al., 2003, 2004; Beaven & Oosting, 1939; Bell et al., 2005; Benedict, 1983; Bliss, 1986; Carr & Eubanks, 2002; Cázares et al., 2005; Charlesworth, 1992; Christy, 2004; Chuang & Heckard, 1992b; Conger, 1912; Correll & Correll, 1975; Cowan, 1945; Dole, 1990, 1992; Douglas, 1981; Dudash & Ritland, 1991; Dumas, 1956; Ewing, 2001; Fenster & Carr, 1997; Fishman et al., 2001; Flanagan et al., 1997; Fonda, 1974; Fraser & Karnezis, 2005; Galloway, 1995; Gardner & Macnair, 2000; Grant, 1924; Grant & Grant, 1968; Halpern et al., 1999; Harrison, 1999; Harrison et al., 2000; Harshberger, 1929; Holden, 1999; Howard & Berlocher, 1998; Hughes et al., 2001; Hutchings, 1932; Ingles, 1952; Inouye & Underwood, 2004; Jackson & Bliss, 1982; Johansson & Keddy, 1991; Karron et al., 1995a,b, 1997, 2004; Kelly & Willis, 1998; Kiang, 1971; Kirkpatrick & Barnes, 2000; Klyver, 1931; Latta & Ritland, 1994; Lawrence, 1939; Leck, 2003; Leclerc-Potvin & Ritland, 1994; Levine, 2000; McNair, 1992; Mehringer, Jr. et al., 1977; Meinke, 1995a,b, 2001; Mitchell, 1926; Mitchell et al., 2004, 2005; Mohlenbrock, 1959a; Neiland, 1958; Pabst & Spies, 1998; Peinado et al., 2005a; Pojar, 1975; Posto & Prather, 2003; Price, 1940; Ramaley, 1919; Ramsey et al., 2003; Richter & Stromberg, 2005; Ritland, 1989; Ritland & Leblanc, 2004; Ritland & Ritland, 1989; Robertson et al., 1999; Runkel & Roosa, 1999; Schemske & Bradshaw Jr., 1999; Schnepf & Busch, 1976; Seabloom et al., 1998; Shaw, 1930; Sheridan & Spies, 2005; Sparrow, 1975; Speichert & Speichert, 2004; Sperry, 1993; Stone & Drummond, 2006; Sutherland & Vickery Jr., 1988; Sweigart & Willis, 2003; Sweigart et al., 2008; Tilman, 1997; Truscott et al., 2006; Turner et al., 2000; USDA, 2007; Vickery, Jr., 1964, 1990a,b, 1991, 1999; Vickery, Jr. et al., 1986, 1989; Waser et al., 1982; Wells, 2006; Welsh, 1989; Welsh & Toft, 1981; Wheeler, 1942; Wheeler et al., 1983; Whittall et al., 2006; Willis, 1993; Windler, 1965; Wu et al., 2008; Zammit & Zedler, 1994.

Family 8: Plantaginaceae [110]

One of the more substantial taxonomic upheavals to result from ongoing phylogenetic research has been the drastic recircumscription of the former Scrophulariaceae, a family that long had been regarded as natural (Olmstead et al., 2001;

Albach et al., 2005; Oxelman et al., 2005; Rahmanzadeh et al., 2005; Tank et al., 2006). Several existing families (e.g., Orobanchaceae, Phrymaceae, Plantaginaceae) have required considerable conceptual modifications; whereas, others either have been abandoned (Callitrichaceae, Hippuridaceae) or newly established (Linderniaceae). Furthermore, compliance with the rules of botanical nomenclature has necessitated the transfer of most familiar "scroph" genera (except for the type genus *Scrophularia*) to the present family Plantaginaceae, which formerly contained only three genera (*Bougueria*, *Littorella*, and *Plantago*). However, the phylogenetic evidence in support of these taxonomic realignments (e.g., Figure 5.50) is compelling, leaving little choice other than to adopt a revised classification scheme to accommodate the clades indicated (see also the discussion under Lamiales).

As currently recognized, most Plantaginaceae are autotrophic herbaceous plants with simple leaves and bisexual, bilabiate flowers having sagittate anthers, a superior ovary, and loculicidal capsules with winged or angular seeds. Pollination is mainly biotic (birds, insects), but a few species are pollinated abiotically (water, wind). The family is a source of many popular ornamental plants including angelonia (*Angelonia*), beardtongue (*Penstemon*), foxglove (*Digitalis*), hebe (*Hebe*), snapdragon (*Antirrhinum*), speedwell (*Veronica*), and toad-flax (*Linaria*). Some members (e.g., *Digitalis*) contain cardiac glycosides and are important medicinally.

Although primarily terrestrial, Plantaginaceae are fairly rich in aquatic and wetland plants. The North American flora includes 14 different genera that contain species designated as OBL in some portion of their range:

1. ***Bacopa*** Aubl.
2. ***Callitriche*** L.
3. ***Chelone*** L.
4. ***Dopatrium*** Buch.-Ham. ex Benth.
5. ***Gratiola*** L.
6. ***Hippuris*** L.
7. ***Leucospora*** Nutt.
8. ***Limnophila*** R. Br.
9. ***Littorella*** P. J. Bergius
10. ***Mecardonia*** Ruiz & Pav.
11. ***Plantago*** L.
12. ***Stemodia*** L.
13. ***Synthyris*** Benth.
14. ***Veronica*** L.

The phylogenetic relationships depicted in Figure 5.82 include nine of the 14 genera that contain species categorized as OBL in North America. The figure also indicates a concentration of aquatics within tribe Gratioleae (see also Figure 5.86) and also that the OBL taxa have evolved throughout the family, and that they have arisen multiple times within it. Regarding the five OBL genera excluded from that analysis, *Dopatrium* and *Limnophila* are assigned to tribe Gratioleae; whereas, *Leucospora* is placed within tribe Stemodieae. Relationships also can be estimated reasonably for *Littorella*, which some authors merge with *Plantago*, and for *Synthyris*, which some

authors merge with *Veronica* (tribe Veroniceae). Although *Callitriche* and *Hippuris* have been segregated as distinct families (Callitrichaceae, Hippuridaceae) in the past, they clearly resolve as a clade embedded within Plantaginaceae. The monotypic genus *Amphianthus* (OBL) also has been problematic. Olmstead et al. (2001) verified the placement of *Amphianthus* in tribe Gratioleae close to *Gratiola*; however, a more recent study that included additional taxa indicated that *Amphianthus* is nested within *Gratiola* and should be merged within it (Estes & Small, 2008). That recommendation has been followed here. The segregation of tribe Gratioleae as a distinct family also has been proposed (Rahmanzadeh et al., 2005), but such a disposition is unnecessary from a phylogenetic standpoint and would simply add to unnecessary nomenclatural complexity.

1. *Bacopa*

Water-hyssop; hyssope d'eau

Etymology: derived from an aboriginal name used in French Guiana

Synonyms: *Allocalyx*; *Ancistrostylis*; *Anisocalyx*; *Brami*; *Bramia*; *Caconapea*; *Calytriplex*; *Cardiolophus*; *Geochorda*; *Habershamia*; *Heinzelmannia*; *Heptas*; *Herpestes*; *Herpestis*; *Hydranthelium*; *Hydrotrida*; *Ildefonsia*; *Macuillamia*; *Mecardonia* (in part); *Mella*; *Moniera*; *Monocardia*; *Obolaria* (in part); *Pagesia*; *Ranaria*; *Ranapalus*; *Septas*; *Septilia*; *Sinobacopa*

Distribution: global: Africa; Asia; Australia; New World; **North America:** widespread (except northeast and northwest)

Diversity: global: 60 species; **North America:** 7 species

Indicators (USA): OBL: *Bacopa caroliniana*, *B. egensis*, *B. eisenii*, *B. innominata*, *B. monnieri*, *B. repens*, *B. rotundifolia*

Habitat: brackish, freshwater; palustrine; **pH:** 4.9–8.7; **depth:** <1 m; **life-form(s):** emergent herb, submersed (vittate)

Key morphology: stems (from 5 to 60 cm) succulent, ascending in shallow water, decumbent or prostrate, glabrous or pubescent; leaves (to 10–40 mm) opposite, sessile, clustered apically in deeper water, glandular punctate, obovate to round, margins entire to crenate, bases tapering or clasping; flowers solitary or paired in upper axils, sessile or stalked (to 51 mm), paired bractlets present or absent; calyx (to 7 mm) 4–5-parted, lobed to base, 3 outer lobes larger and enclosing inner 2; corolla (to 14 mm) bilabiate, tubular campanulate, 3–5-lobed, blue, lavender, pink, violet or white; stamens 2, or 4 and didynamous; capsules (to 5 mm) with >30 seeds

Life history: duration: annual (fruit/seeds); perennial (rhizomes); **asexual reproduction:** rhizomes, shoot fragments, stolons; **pollination:** cleistogamous, insect, self; **sexual condition:** hermaphroditic; **fruit:** capsules (common); **local dispersal:** seeds, fragments (water); **long-distance dispersal:** seeds (waterfowl)

Imperilment: (1) *Bacopa caroliniana* [G4]; SH (VA); S1 (NC, TN); (2) *B. egensis* [GNR]; (3) *B. eisenii* [G3]; (4) *B. innominata* [G3]; SH (MD, NC); S1 (SC); S2 (VA); (5) *B. monnieri* [G5]; S1 (OK); S2 (NC, VA); (6) *B. repens* [GNR]; (7) *B. rotundifolia* [G5]; SH (NC); S1 (AB, AZ, ID, IN, MT, WY, VA); S3 (IA, KY, MN)

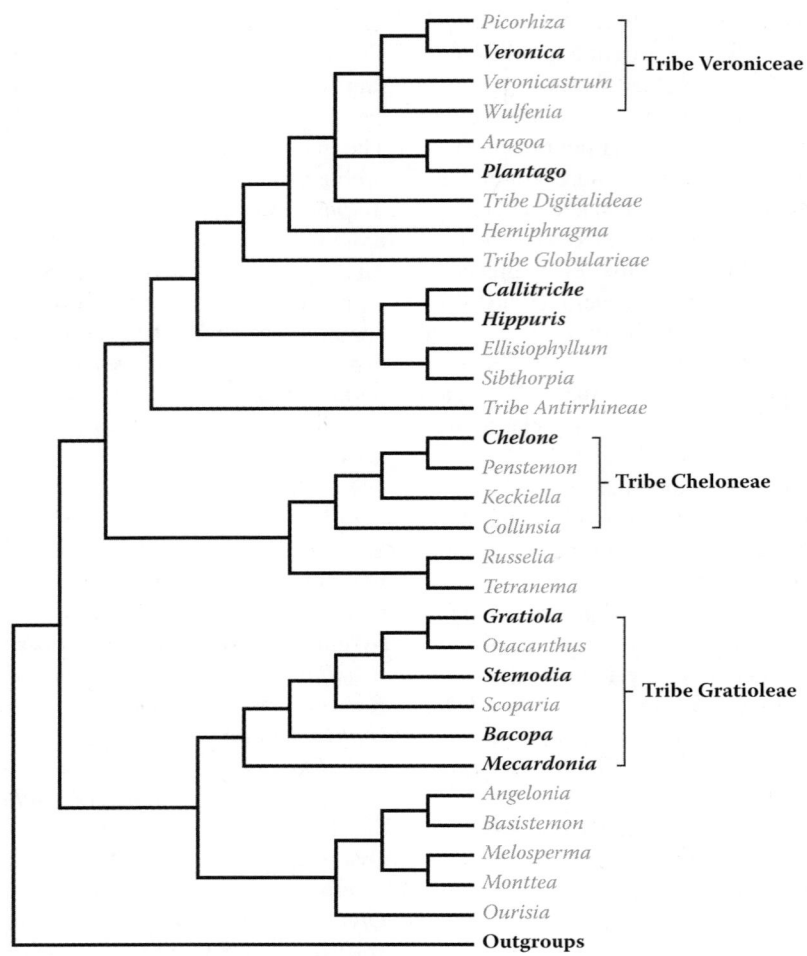

FIGURE 5.82 Intergeneric relationships in Plantaginaceae as indicated by analysis of combined DNA sequence data. Surveyed taxa include nine of the 15 genera containing OBL species in North America (in bold). These phylogenetic relationships illustrate that the OBL habit has arisen several times within the family and also indicate a concentration of aquatic genera occurring within tribe Gratioleae. (Adapted from Albach, D.C. et al., *Amer. J. Bot.*, 92, 297–315, 2005.)

Ecology: general: All North American *Bacopa* species are OBL aquatics, which share similar life-histories although they include both annuals and perennials. The species typically have broad ecological tolerances and inhabit shallow waters, shores, or sites that become exposed as the waters dry and recede. The plants essentially are emergents, but will grow in shallow water as vittate, submersed life-forms, which maintain their apices above the water by their buoyant terminal foliage. Specialized floating leaves do not develop per se. The flowers are self-incompatible, but breeding systems vary and include outcrossers, inbreeders, and species having mixed mating systems. Cleistogamous flowers are produced occasionally by some species; the chasmogamous flowers are pollinated mainly by a variety of bees (Insecta: Hymenoptera). Seed dispersal has been poorly documented, but probably occurs by water or by carriage on the feet or plumage of waterfowl. Seed banks often develop. The plants commonly form dense mats and can grow rather aggressively under suitable conditions.

Bacopa caroliniana (Walter) **B.L. Robins.** is a perennial, which inhabits canals, depressions, ditches, flatwoods, floodplains, hammocks, marshes, meadows, pools, ponds, prairies, sloughs, swamps, and the margins of ponds at elevations to at least 19 m. It can grow as a creeping form or as a submersed form with floating leaves in shallow, standing waters that are less than 1 m in depth. The plants occur predominantly in fresh water but also tolerate mildly brackish conditions (salinities to 3.0 ppt). They grow more commonly in acidic waters (mean pH: 6.6) but are found over a fairly wide range of habitats (pH: 4.9–8.4) with fairly average water chemistry (e.g., 23 μg/L mean total phosphorous; 740 μg/L mean total nitrogen; 18.8 mg/L mean $CaCO_3$ carbonate). In culture, the plants will grow well at water temperatures up to 25°C. In nature, they occur in full sun on substrates characterized as Duckston (typic psammaquents), karst, marl, muck, peaty muck, Placid (typic humaquepts), sandy loam, or silty clay loam. The reproductive biology has not been studied in detail; however, the flowers presumably are insect pollinated and reportedly do not set seed in the absence of pollinators. In any case, seed set often is poor. Specific conditions relating to seed dormancy have not been reported but germination has been noted to occur better under flooded than simply moist conditions. A persistent seed bank is formed and seed germination is observed commonly in seed bank studies. Vegetative reproduction occurs by spreading rhizomes. The plants are propagated easily by stem cuttings. This species often grows in floating mats in association with algae (Cyanobacteria and green algae), *Ludwigia*, and *Utricularia* species. The foliage is fragrant with a scent of lemon, which might act as a repellent to herbivores. **Reported associates:** *Acmella oppositifolia, Aeschynomene indica, Agrostis hyemalis, Alternanthera philoxeroides, Ampelopsis arborea, Amsonia tabernaemontana, Andropogon capillipes, A. glaucopsis, A. glomeratus, A. virginicus, Annona glabra, Aristida palustris, Arnoglossum ovatum, Aronia arbutifolia, Asclepias lanceolata, A. viridis, Axonopus furcatus, Azolla caroliniana, Baccharis halimifolia, Bacopa monnieri, Bidens laevis, B. mitis, Blechnum serrulatum, Boehmeria cylindrica, Boltonia asteroides, B. diffusa, Brasenia schreberi, Callicarpa americana, Carex decomposita, C. glaucescens, C. longii, C. verrucosa, Cassytha filiformis, Centella asiatica, C. erecta, Cephalanthus occidentalis, Chara, Cladium jamaicense, Coelorachis rugosa, Colocasia esculenta, Commelina diffusa, Coreopsis tinctoria, Crinum americanum, Cyperus articulatus, C. compressus, C. haspan, C. odoratus, C. polystachyos, C. surinamensis, Cyrilla racemiflora, Decodon verticillatus, Dichanthelium scabriusculum, Diodia virginiana, Distichlis spicata, Echinochloa walteri, Eichhornia crassipes, Eleocharis cellulosa, E. elongata, E. equisetoides, E. interstincta, E. quadrangulata, E. vivipara, Eragrostis elliottii, Erianthus giganteus, Eriocaulon compressum, Eriochloa punctata, Eryngium yuccifolium, Eupatorium leptophyllum, E. serotinum, Euthamia leptocephala, Ficus aurea, Fimbristylis autumnalis, F. caroliniana, F. dichotoma, F. littoralis, Fraxinus caroliniana, Fuirena bushii, F. pumila, F. squarrosa, Funastrum clausum, Galium tinctorium, Gaura lindheimeri, Gordonia lasianthus, Gratiola brevifolia, Gyrotheca tinctoria, Helianthus angustifolius, Hibiscus grandiflorus, H. moscheutos, Hydrochloa carolinensis, Hydrocotyle umbellata, Hydrolea ovata, Hymenocallis latifolia, H. liriosme, H. palmeri, Hypericum mutilum, H. virginicum, Hyptis alata, Ilex glabra, I. vomitoria, Ipomoea alba, I. sagittata, Iris, Iva angustifolia, Juncus effusus, J. marginatus, J. polycephalus, J. roemerianus, Justicia angusta, J. ovata, Kosteletzkya pentacarpos, Kyllinga brevifolia, Lachnanthes caroliana, Leersia hexandra, Lemna minor, Leptochloa fusca, Limnobium spongia, Ludwigia alternifolia, L. linearis, L. microcarpa, L. peruviana, L. pilosa, L. repens, L. sphaerocarpa, Luziola fluitans, Lycopus rubellus, Lyonia, Lythrum lineare, Magnolia virginiana, Mayaca fluviatilis, Mikania cordifolia, M. scandens, Muhlenbergia filipes, Myrica cerifera, Myriophyllum aquaticum, M. pinnatum, Neptunia lutea, Nuphar advena, Nymphaea odorata, Nymphoides aquatica, Nyssa biflora, Oenothera drummondii, Osmunda regalis, Oxypolis filiformis, Panicum hemitomon, P. repens, P. rigidulum, P. tenerum, P. virgatum, Paspalidium geminatum, Paspalum dissectum, P. distichum, P. floridanum, P. laeve, P. monostachyum, P. plicatulum, P. vaginatum, Peltandra virginica, Persea palustris, Persicaria hirsuta, P. hydropiperoides, P. punctata, Phragmites australis, Pinus palustris, Pluchea rosea, Polygala leptocaulis, Pontederia cordata, Proserpinaca palustris, P. pectinata, Ptilimnium capillaceum, Ratibida peduncularis, Rhexia mariana, R. virginica, Rhynchospora capitellata, R. cephalantha, R. colorata, R. corniculata, R. crinipes, R. divergens, R. elliottii, R. fascicularis, R. globularis, R. inundata, R. macrostachya, R. microcarpa, R. microcephala, R. mixta, R. nitens, R. tracyi, Rubus trivialis, Rudbeckia texana, Sabatia angularis, Saccharum giganteum, Sacciolepis striata, Sagittaria graminea, S. lancifolia, S. latifolia, Salix caroliniana, Salvinia minima, Sarracenia alata, Saururus cernuus, Schizachyrium rhizomatum, Schoenoplectus pungens, S. californicus, Schoenus nigricans, Scirpus cyperinus, Scleria baldwinii, S. reticularis, Serenoa repens, Sesbania drummondii, Setaria magna, S. parviflora, Solidago sempervirens, Spartina bakeri, S. cynosuroides, S. patens, S. spartinae, Sphagnum cuspidatum, Spiranthes laciniata, Stachys, Symphyotrichum elliotii, Taxodium ascendens, T. distichum, Teucrium canadense, Thalia geniculata, Thelypteris kunthii, T. thelypterioides, Tillandsia fasciculata, T. utriculata, T. variabilis, Tradescantia hirsutiflora, Triadenum walteri, Tridens strictus, Tripsacum dactyloides, Typha domingensis, T. latifolia, Utricularia gibba, U. foliosa, U. inflata, U. purpurea, Vallisneria americana, Vernonia gigantea, Vigna luteola, Woodwardia areolata, W. virginica, Xyris elliottii, X. fimbriata, X. laxifolia, X. smalliana, Zizaniopsis miliacea.*

***Bacopa egensis* (Poepp.) Pennell** is a perennial, which occurs in quiet, shallow waters of bayous and lakes, or along exposed lakebeds and the margins of pools. In standing waters the stems become spongy in texture and the leaves develop as a floating rosette. Although it is regarded as nonindigenous, it is believed to have been introduced naturally to North America from seeds transported by waterfowl; thus, its status as a native species should be reconsidered. Other life-history information is scarce for this species and only a few associates have been compiled. **Reported associates:** *Bacopa rotundifolia, Cyperus, Sagittaria latifolia.*

***Bacopa eisenii* (Kellogg) Pennell** is an annual (potentially perennial), which inhabits ditches, marshes, ponds, rice (*Oryza*) fields, vernal pools, and desiccating pond margins at elevations of up to 1372 m. The plants occur in full sun, in shallow (<30 cm) fresh or brackish water, on alkaline clay or muddy substrates. Although self-compatible, the large and showy flowers are primarily outcrossed. The level of spontaneous autogamy is quite low because the stamens do not recurve against the stigma when the flowers senesce as is the case with the selfing species. Honeybees (Hymenoptera: Apidae: *Apis mellifera*) are frequently observed pollinators. This species mainly is an annual, but it can perennate if suitable conditions exist; the plants have survived for more than a year under greenhouse culture. The seeds have a bladder-like coat and are dispersed by water. They also are dispersed by adherence to agricultural equipment. Vegetative reproduction can occur by shoot fragmentation, and can result in the formation of large, clonal mats. The plants are relatively resistant to 2-methyl-4-chlorophenoxyacetic acid (MCPA) herbicides, which are used to control rice field weeds, and they are spreading in rice fields as a consequence. Life-history information is scarce for this relatively restricted species. **Reported associates:** *Bacopa rotundifolia, Frankenia, Ludwigia, Marsilea, Salicornia.*

***Bacopa innominata* (G. Maza) Alain** is a perennial, which grows in tidal or nontidal freshwater depressions, ditches, flatwoods, floodplains, hammocks, marshes, mud-flats, prairies, roadsides, savannas, and along the margins of canals, ponds, rivers, and streams at elevations to at least 18 m. The plants are shade sensitive and can occur in wet woodlands, but only in light gaps. The reported substrates are calcareous and include Felda (arenic ochraqualfs), mud, and sand. Despite the rarity of this species, there is little pertinent ecological information available in the literature. The plants are known to develop into rather thick mats by the production of fleshy, horizontal stems (rhizomes). **Reported associates:** *Acer, Carpinus, Commelina diffusa, Cyperus croceus, Drymaria cordata, Elatine americana, Eriocaulon parkeri, Hydrocotyle verticillata, Hypericum mutilum, Justicia americana, Lilaeopsis chinensis, Liquidambar styraciflua, Magnolia virginiana, Nyssa, Osmundastrum cinnamomeum, Panicum repens, Paspalum notatum, Pinus elliottii, Quercus laurifolia, Sagittaria subulata, Ulmus.*

***Bacopa monnieri* (L.) Pennell** is a perennial, which occurs in bayous, beaches, bottoms, canals, depressions, ditches, dunes, flats, floodplains, marshes, meadows, ponds, pools, prairies, salt marshes, shores, swales, swamps, and along the margins of lakes, rivers, and streams at elevations of up to 1320 m (achieving this altitude in Asia). The plants also occur on floating "flotant" marshes. They grow across a wide range of tidal or nontidal, fresh to brackish (salinity to 10 ppt) habitats (e.g., pH: 5.4–8.7; water temperatures of 15°C–30°C), but occur mostly in hard water, alkaline, high-nutrient lakes (mean alkalinity: 38 mg/L $CaCO_3$; mean conductivity: 233 μS/cm; mean calcium: 33 mg/L; mean chloride: 41 mg/L; mean pH: 7.4). The shoots grow either as a floating form in shallow water or are prostrate and creep along exposed substrates, which include clay, limestone, muck, mud, sand, and silty clay. The flowers are self-compatible but typically are pollinated by insects such as honeybees (Hymenoptera: Apidae: *Apis*). Seed production is high with an average of 62 seeds produced per capsule. The seeds are consumed by ducks (Aves: Anatidae: *Anas*); however, their subsequent viability has not been verified and the role of waterfowl as potential dispersal agents for this species requires further study. The seeds initially are dormant for 2–3 weeks and require light for germination. Excellent germination (within 16 days) has been obtained for seeds that have been stored for 2 days at 4°C and then placed under a 14/10 h light/dark and 30°C/25°C day/night temperature regime. Field-collected seeds stored for 4 weeks at 21.0°C–35.7°C also have exhibited high germination. The seeds will germinate in salinities up to 8 ppt. Germination can be enhanced by treating the seeds with giberellic acid. This species forms persistent seed banks at densities that can surpass 2000 seeds/m². The plants grow into dense mats and reproduce vegetatively by fragmentation. Populations are highly resistant to damage by hurricanes, but plant biomass is reduced substantially by

fires. The plants are poor competitors and occur mainly in open sites that are not grazed. A genetic analysis of natural populations in India using RAPDs markers indicated fairly low levels of genetic variability, which reflects a combination of sexual and asexual reproduction. **Reported associates:** *Acer rubrum, Acrostichum danaeifolium, Adiantum capillus-veneris, Agalinis fasciculata, Aletris, Alternanthera philoxeroides, Ammannia latifolia, Amphicarpum, Andropogon, Annona glabra, Aristida, Avicennia germinans, Axonopus furcatus, Baccharis halimifolia, Bacopa caroliniana, Batis maritima, Bidens mitis, Bolboschoenus robustus, Boltonia diffusa, Borrichia frutescens, Buchnera americana, Carex longii, Carphephorus odoratissimus, Centella asiatica, Chamaesyce maculata, Chasmanthium, Cladium jamaicense, Commelina diffusa, Conocarpus erectus, Cynanchum angustifolium, Cyperus articulatus, C. compressus, C. haspan, C. odoratus, C. oxylepis, C. polystachyos, C. retrorsus, C. surinamensis, Diodia virginiana, Distichlis spicata, Drosera brevifolia, Echinochloa walteri, Eichhornia crassipes, Eleocharis albida, E. cellulosa, E. flavescens, E. geniculata, E. interstincta, E. montevidensis, E. parvula, E. vivipara, Eupatorium serotinum, Euthamia graminifolia, Fimbristylis autumnalis, F. caroliniana, F. castanea, F. dichotoma, Fuirena longa, F. pumila, F. scirpoidea, F. simplex, Funastrum clausum, Galium tinctorium, Helianthus angustifolius, Hibiscus moscheutos, Hydrocotyle umbellata, H. verticillata, Hypericum mutilum, Ipomoea sagittata, Iva angustifolia, Iva frutescens, Juncus effusus, J. marginatus, J. megacephalus, J. roemerianus, Justicia americana, Kosteletzkya pentacarpos, Kyllinga brevifolia, Laguncularia racemosa, Leersia hexandra, Lemna, Leptochloa fascicularis, L. fusca, Lobelia cardinalis, L. feayi, Ludwigia leptocarpa, L. octovalvis, L. palustris, L. peploides, L. repens, Luziola fluitans, Lythrum lineare, Marsilea vestita, Mayaca fluviatilis, Mikania cordifolia, M. scandens, Mitreola, Myrica cerifera, Najas guadalupensis, Oldenlandia uniflora, Onoclea sensibilis, Oxypolis, Panicum dichotomiflorum, P. hemitomon, P. repens, P. rigidulum, P. virgatum, Paspalidium geminatum, Paspalum dissectum, P. distichum, P. laeve, P. vaginatum, Phragmites australis, Phyla lanceolata, P. nodiflora, Pilea microphylla, Pluchea camphorata, P. odorata, P. rosea, Persicaria hirsuta, P. punctata, Pinus palustris, Pontederia cordata, Proserpinaca palustris, Ptilimnium capillaceum, Rhexia mariana, Rhynchospora colorata, R. divergens, R. fascicularis, R. globularis, R. inundata, R. microcarpa, R. microcephala, R. nitens, R. nivea, Rubus argutus, Sabal minor, S. palmetto, Sabatia, Sacciolepis striata, Sagittaria lancifolia, Salicornia, Salix caroliniana, S. nigra, Salvinia minima, Samolus ebracteatus, S. valerandi, Saururus cernuus, Schoenoplectus americanus, S. californicus, S. pungens, Scleria reticularis, Serenoa repens, Sesbania drummondii, Sesuvium maritimum, S. portulacastrum, Setaria geniculata, Solidago fistulosa. sempervirens, S. stricta, Spartina bakeri, S. cynosuroides, S. patens, S. spartinae, Spiranthes vernalis, Spirodela, Stuckenia pectinata, Symphyotrichum elliotii, S. subulatum, Teucrium canadense, Thelypteris palustris, Toxicodendron radicans, Triglochin striata, Typha* *angustifolia, T. domingensis, T. latifolia, Utricularia gibba, U. subulata, Vigna luteola, Xyris jupicai.*

Bacopa repens (Sw.) Wettst. is an introduced annual found in canals, ricefields, and pools at elevations of up to 100 m. It grows in clear or muddy shallow water (to 0.5 m deep) and usually roots in clay substrates. The flowers are self-compatible and are either chasmogamous or cleistogamous. The species most likely is self-pollinating because natural pollinators have not been observed (at least in North America), and the fruits often develop on underwater portions of the plants that occur as much as 20 cm below the water surface. Apparently the capsules do not dehisce while they remain on the plants, but are dispersed in their entirety as they detach from the plants. This is a mat-forming species and densities up to 50 plants per m² have been recorded in some rice field ponds. **Reported associates (North America):** *Ammannia auriculata, Bacopa rotundifolia, Dopatrium junceum, Eriocaulon cinereum, Heteranthera limosa, Rotala ramosior, Sagittaria guayanensis, Sphenoclea zeylanica.*

Bacopa rotundifolia (Michx.) Wettst. is an annual or perennial plant of buffalo wallows, depressions, ditches, flats, lakebeds, marshes, mudflats, playas, ponds, pools, ricefields (*Oryza*), sloughs, streams, swamps, and vernal pools at elevations of up to 1600 m. It can tolerate a fairly broad range of acidity (pH: 5.4–7.8) but tends to occur in localities with somewhat acidic conditions (e.g., pH: 6.6). The sites range from permanent (up to 40 cm deep) fresh to slightly brackish waters (salinity <3.0 ppt), to vernal or drying habitats. Reported substrates include alluvium, clay, granite, and mud. The plants are intolerant to shade but can withstand fire and cold temperatures to −36°C. They are self-compatible and the reduced flowers principally are autogamous as the anthers recurve to eventually contact the stigmas. The seeds germinate while they are submersed. Populations exhibit considerable annual fluctuations and develop a persistent seed bank. This species contributes substantially to the formation of detritus, which provides food for many grazing organisms. The net primary production of these plants has been estimated at 204 kcal/m²/year. The life-history of this species warrants further investigation. The plants apparently are annuals when grown under greenhouse conditions and live for only a few months after germination; however, they reportedly overwinter as rosettes in the field when they are growing in standing water. Vegetative reproduction occurs by stolons. **Reported associates:** *Alisma subcordatum, A. triviale, Alopecurus carolinianus, Ammannia coccinea, Azolla caroliniana, Bacopa repens, Bidens, Bolboschoenus fluviatilis, Brachiaria platyphylla, Carex conjuncta, C. vesicaria, Ceratophyllum demersum, Chara, Cicuta maculata, Coreopsis tinctoria, Cyperus acuminatus, C. esculentus, C. squarrosus, Diodia virginiana, Distichlis spicata, Echinochloa colona, E. crus-galli, E. muricata, Echinodorus cordifolius, E. rostratus, Eclipta prostrata, Elatine rubella, Eleocharis acicularis, E. engelmannii, E. obtusa, E. ovata, E. palustris, Equisetum hyemale, Eragrostis hypnoides, Eurystemon mexicanum, Fimbristylis autumnalis, Heteranthera limosa, Impatiens capensis, Isoetes melanopoda, Juncus interior, Leersia oryzoides, Lemna minor,* *L. perpusilla, Lepidium densiflorum, Leptochloa fascicularis, L. fusca, L. panicoides, Leucospora multifida, Limosella acaulis, L. aquatica, Lindernia anagallidea, L. dubia, Lipocarpha drummondii, Lobelia cardinalis, L. siphilitica, Ludwigia palustris, L. peploides, Malvella leprosa, Marsilea mucronata, M. vestita, Mimulus ringens, Mollugo verticillata, Myosurus minimus, Oenothera canescens, Oryza sativa, Packera aurea, Pascopyrum smithii, Persicaria amphibia, P. coccinea, P. lapathifolia, P. pensylvanica, Phyla, Pilularia americana, Plantago elongata, Polygonum ramosissimum, Pontederia cordata, Potamogeton diversifolius, P. gramineus, P. nodosus, Ranunculus abortivus, Rorippa palustris, R. sessiliflora, R. sinuata, R. teres, Rotala ramosior, Rumex crispus, R. stenophyllus, R. verticillatus, Sagittaria brevirostra, S. calycina, S. latifolia, S. longiloba, Salix gooddingii, Schoenoplectus acutus, S. tabernaemontani, Setaria glauca, Sium suave, Sparganium eurycarpum, Spermacoce glabra, Sphenoclea, Spirodela polyrrhiza, Stachys palustris, Tridens albescens, Typha angustifolia, T. latifolia, Utricularia macrorhiza, Veronica peregrina.*

Use by wildlife: *Bacopa caroliniana* provides habitat for the green-winged teal (Aves: Anatidae: *Anas carolinensis*) and also the alligator snapping turtle (Reptilia: Chelydridae: *Macrochelys temminckii*), especially when associated in floating mat communities. It is an important food of the pond-slider turtle (Reptilia: Emydidae: *Trachemys scripta*) and a natural host plant for several larval moths (Insecta: Lepidoptera: Pyralidae: *Parapoynx allionialis, P. maculalis, P. obscuralis*). *Bacopa caroliniana* is known to support epiphytic Cyanobacteria (Stignemotales) that cause avian vacuolar myelinopathy (AVM), a serious disease of herbivorous waterfowl and their predators such as the bald eagle (Aves: Accipitridae: *Haliaeetus leucocephalus*) and American coot (Aves: Rallidae: *Fulica americana*). *Bacopa monnieri* is a larval host plant of the white peacock butterfly (Insecta: Lepidoptera: Nymphalidae: *Anartia jatrophae*), a species that also has been reared successfully on plants of *B. caroliniana*. *Bacopa monnieri* occasionally is eaten by waterfowl but the seeds can be consumed in quantity, e.g., 1015 were recovered from the digestive tract of a single green-winged teal (Aves: Anatidae: *Anas carolinensis*). *Bacopa monnieri* often is a dominant vegetational component of nesting sites for various birds including black-crowned night herons (Aves: Ardeidae: *Nycticorax nycticorax*), cattle egrets (Aves: Ardeidae: *Bubulcus ibis*), great blue herons (Aves: Ardeidae: *Ardea herodias*), great egrets (Aves: Ardeidae: *Ardea alba*), snowy egrets (Aves: Ardeidae: *Egretta thula*), laughing gulls (Aves: Laridae: *Leucophaeus atricilla*), Louisiana herons (Aves: Ardeidae: *Egretta tricolor*), and roseate spoonbills (Aves: Threskiornithidae: *Platalea ajaja*). It also provides habitat for the dusky seaside sparrow (Aves: Emberizidae: *Ammodramus maritimus nigrescens*). *Bacopa rotundifolia* is eaten by white-tailed deer (Mammalia: Cervidae: *Odocoileus virginianus*). The seeds and shoots (with a caloric content of 4.3 kcal/g) also provide food for various waterfowl including green-winged teal (Aves: Anatidae: *Anas carolinensis*), pintail (Aves: Anatidae: *Anas acuta*) and ring-necked ducks (Aves: Anatidae: *Aythya collaris*). The dying foliage is eaten by snails (Gastropoda: Lymnaeidae: *Lymnaea palustris*; Physidae: *Physa integra*) and fly larvae (Insecta: Diptera: Chironomidae: *Cardiocladius*). The plants occasionally are used as oviposition sites by familiar blue damselflies (Insecta: Odonata: Coenagrionidae: *Enallagma civile*).

Economic importance: food: The lemon-scented leaves of *Bacopa caroliniana* have been used as a seasoning; **medicinal:** The Seminoles used leaf infusions containing *Bacopa caroliniana* as a sedative to treat coughs, trembles and shortness of breath. *Bacopa monnieri* (in Asia known as Brahmi) is a highly regarded medicinal plant and extracts have been shown to improve memory, reduce stress, and stimulate thyroid activity in laboratory studies. It is used in traditional Indian (Ayurvedic) medicine to treat asthma, bronchitis, diarrhea, epilepsy, insomnia, mental retardation, rheumatism, and as a diuretic. Alcoholic extracts exhibit antitumor activity against cancerous sarcoma cells. The extracts also are antibacterial, have antioxidant properties, can chelate metals in the blood, and reduce free radical production. Notably, the plants contain bacosides A and B, which are triterpene saponins with analgesic properties. These bacosides reportedly can repair damaged neurons by stimulating the production of proteins involved in neural-cell synapse regeneration. The shoots contain betulinic acid, oroxindin, and wogonin, which effectively inhibit the growth of some fungi (Ascomycota: Clavicipitaceae: *Claviceps fusiformis*; Pleosporaceae: *Alternaria alternata*); **cultivation:** *Bacopa caroliniana* is grown as an ornamental plant in aquariums and water gardens. *Bacopa monnieri* is widely grown as an aquarium plant; **misc. products:** Commercially distributed, contaminated rooted plant plugs of *Bacopa caroliniana* are a potential source of Collembola, fungus gnats (Diptera: Sciaridae: *Bradysia*), flower thrips (Thysanoptera: Thripidae: *Frankliniella occidentalis*), and other damaging arthropods in greenhouses. *Bacopa monnieri* can accumulate substantial quantities of heavy metals (Cd, Cr, Cu, Mn, Pb) and has been considered for use in metal removal from polluted water; **weeds:** *Bacopa eisenii*, *B. monnieri* and *B. rotundifolia* occasionally are reported as rice field weeds. Some plants of *B. rotundifolia* that were introduced to Malaysia have evolved resistance to ALS (acetolactate synthase) inhibitors including bensulfuron-methyl, metsulfuron-methyl, and pyrazosulfuron-ethyl herbicides; **nonindigenous species:** *Bacopa egensis*, a native of Africa and South America, was introduced to the southern United States in the early 19th century, but probably as a natural consequence of seed transported by waterfowl. Other introductions include *B. monnieri* (to Spain and Portugal), and *B. repens* (to the southern United States from South America). Although native to North America, *B. rotundifolia* probably was introduced to California via rice culture sometime before 1923. It also has been introduced to rice fields in Japan and Malaysia and was first reported from Europe in 1999.

Systematics: Although some authors have advocated combining *Mecardonia* with *Bacopa*, the genera appear to be closely related but distinct within tribe Gratioleae (Figures 5.82 through 5.86). Other authors have subdivided *Bacopa* into segregate genera including *Bramia* (*Bacopa monnieri*), *Hydrotrida* (*B. caroliniana*), and *Herpestris* (*B. rotundifolia*); however, a thorough systematic study of *Bacopa* (using morphological and molecular data) will be necessary before the presumed monophyly of the genus can be evaluated. Interspecific phylogenetic relationships remain poorly known in *Bacopa* and the genus would benefit from additional systematic study. The Mexican/South American *Bacopa repens* closely resembles *B. caroliniana* and might be closely related. *Bacopa eisenii* and *B. rotundifolia* also appear to be closely related and DNA sequence data (Figure 5.85) indicate a close relationship between *B. eisenii* and *B. repens*. Artificial crosses between the latter pair have been successful; however, their synthetic F_1 hybrids are sterile. *Bacopa stragula* is believed to represent simply an intertidal form of *B. innominata*; however, this assumption should be verified using genetic analyses. The basic chromosome number of *Bacopa* is uncertain. The genus contains many polyploids, with counts up to $2n = 80$ reported. *Bacopa repens* ($2n = 28$) presumably is a diploid; whereas, *B. eisenii* and *B. rotundifolia* (both $2n = 56$) as well as *Bacopa monnieri* ($2n = 64, 68$) appear to be tetraploid. An artificial higher autoploid of *B. monnieri* ($2n = 128$) exhibits reduced fertility.

Comments: *Bacopa rotundifolia* is the most widespread species in our range and occurs throughout much of North America, especially the central region. More limited distributions characterize *B. monnieri* (southern United States), *B. caroliniana* and *B. innominata* (southeastern United States), and *B. eisenii* (California, Nevada). Two species are nonindigenous: *Bacopa egensis* (currently known only from Arkansas and Louisiana) and *B. repens* (reported from California, South Carolina, Texas, and Virginia).

References: Austin, 2004b; Baldwin et al., 1996; Barrett & Strother, 1978; Bick & Bick, 1963; Busch et al., 1998; Chabreck & Palmisano, 1973; Chaudhuri et al., 2002; Chiscano, 1999; Chowdhuri et al., 2002; Cloyd & Zaborski, 2004; Darokar et al., 2001; DePoe, 1969; DiTomaso & Healy, 2003; Dolan, 2004; Elangovan et al., 1995; Fischer, 2004; Fritsch et al., 2007; Ghosh et al., 2007; Haukos & Smith, 1997; Herlong, 1979; Hill & Harris, 1943; Holder et al., 1980; Hoyer et al., 1996; Kadono, 2004; Kar et al., 2002; Kasselmann, 2003; Keeley & Zedler, 1998; Lane & Mitchell, 1997; Lederhouse et al., 1992; Lodha & Bagga, 2000; Martin & Uhler, 1951; Mason, 1957; Mathur & Kumar, 2001; McAtee, 1939; Olofsdotter et al., 2000; Parmenter, 1980; Penfound, 1953; Pennell, 1935; Poiani & Dixon, 1995; Polley & Wallace, 1986; Powers et al., 1978; Rave & Baldassarre, 1989; Roodenrys et al., 2002; Sasser et al., 1996; Schubauer & Parmenter, 1981; Schuyler, 1989; Sharma, 2005; Singh et al., 2005; Sinha, 1999; Sinha & Chandra, 1990; Speichert & Speichert, 2004; Srivastava et al., 2002; Stutzenbaker, 1999; Taylor et al., 1994; Thieret, 1970; Tilly, 1968; Visser et al., 2000; Vohora et al., 1997; Wetzel et al., 2001; White et al., 1982; Wilde et al., 2005; Woodson, Jr. et al., 1979; Zomlefer & Giannasi, 2005.

2. *Callitriche*

Water starwort; Callitriche

Etymology: from the Greek *kalli thrix* meaning "beautiful hair" in reference to the graceful stems

Synonyms: none

Distribution: global: cosmopolitan; **North America:** widespread

Diversity: global: 50 species; **North America:** 9 species

Indicators (USA): OBL: *Callitriche hermaphroditica, C. heterophylla, C. marginata, C. palustris, C. peploides, C. stagnalis, C. trochlearis*

Habitat: freshwater, brackish; lacustrine, palustrine, riverine; **pH:** 5.2–8.5; **depth:** 0–4 m; **life-form(s):** emergent herb, floating leaved, submersed (vittate)

Key morphology: stems (to 5 dm) slender and branching, submersed, floating or prostrate; leaves (to 5 cm) simple, opposite, linear to spatulate, decussate (cauline) or in terminal, floating rosettes, often notched apically; plants monoecious; flowers minute, 1–3 in axils, usually subtended by inflated bracts; perianth absent; stamen 1; schizocarps (to 1.6 mm) striate, pedicelled (to 25 mm), winged or wingless, splitting into 4 nutlet-like, 1-seeded segments (mericarps)

Life history: duration: annual (fruit/seeds); perennial (whole plants); **asexual reproduction:** shoot fragments; **pollination:** self (geitonogamous); water; **sexual condition:** monoecious; **fruit:** mericarps (common); **local dispersal:** seeds (animals, water), seedlings (water); **long-distance dispersal:** seeds (animals)

Imperilment: (1) *Callitriche hermaphroditica* [G5]; SH (VT); S1 (NF, NS, NY, PE); S2 (NB, NE, MI, WI, WY); S3 (QC); (2) *C. heterophylla* [G5]; S1 (MA, CO, MI, NE, UT, WI, YT); S2 (BC, IA, MB, ME, ON, WY); S3 (MN); (3) *C. marginata* [G4]; S1 (BC); S2 (OR); (4) *C. palustris* [G5]; S2 (NJ); S3 (DE, VA); (5) *C. peploides* [G4]; (6) *C. trochlearis* [G3]

Ecology: general: *Callitriche* principally is an aquatic genus, which contains both annual and perennial growth forms. All species are annuals when growing in desiccating habitats, but some plants can persist yearlong if water remains and conditions permit. All but two North American species (*C. terrestris*, FACW; *C. pedunculosa*, FACW) are designated as OBL indicators. Although the two FAC species inhabit moist sites, they also occur frequently in drier localities such as muddy agricultural fields. Three growth forms occur in the genus, ranging from submersed (inhabiting standing water) to amphibious (inhabiting moist soil or standing water) or "terrestrial" (inhabiting seasonally moist areas that dry out). When growing in standing waters, the stems either are completely submersed or produce a rosette of floating leaves. All species are monoecious and produce protogynous flowers that are devoid of a perianth. In most taxa, the reproductive structures extend above the water surface where pollen is transferred between the staminate and pistillate flowers of the same individual (aerial geitonogamy). Here the styles either come directly in contact with the dehiscing anthers, or the pollen grains drop from the anthers onto a subtending stigma. Some species with aerial geitonogamy also possess an unusual reproductive system where the pollen tubes germinate within the anthers of the staminate flowers,

grow through the vegetative tissues, and ultimately penetrate a pistillate flower at the same or adjacent node (internal geitonogamy). In this case, pollination is achieved without the pollen or stigma ever coming in contact with the water. One species (*C. hermaphroditica*) is water pollinated (hydrophilous). The seeds have variable dormancy requirements for germination. The light weight and flattened shape of the fruits facilitates their dispersal on water surfaces and some can remain afloat for several weeks. They also are transported by waterfowl as they adhere to mud that is attached to their feet. This treatment excludes *C. fassettii* and *C. longipedunculata* (both ranked as OBL in the 2013 indicator list), which are not recognized as distinct taxonomically (see *Systematics* later).

Callitriche hermaphroditica **L.** is an annual or perennial inhabitant of bogs, ditches, lakes, ponds, pools, rivers, sloughs, streams, and vernal pools at elevations of up to 3180 m. The plants occur in fresh to brackish waters (salinity up to 5.5 ppt) that often are less than 0.5 m in depth; however, they also have been reported from waters up to 4 m deep. The habitats are mesotrophic to eutrophic and principally softwater (pH: 5.6–7.9; total alkalinity <40 ppm; sulfates <5 ppm). The substrates include gravelly loam, mud, sand, sandy gravel, and silt. This species tolerates fairly cold temperatures and occurs northward to at least 64°30′N latitude. Temperatures can range from 10°C at germination to 22°C at the time of maximum seed development. Typically the plants are found in high disturbance but low stress sites. Aerial flowers are not produced and the fruits develop only from submersed flowers. During anthesis the styles extend to the subtending node in close proximity of the anthers. The pollen germinates precociously within the anthers, whereupon the pollen tubes grow to and penetrate the stigmatic tissue. Although this pollination system as described is not truly hypohydrophilous (the pollen itself is not transported within the water), there is some genetic evidence from RAPD markers that xenogamous pollen transfer also might occur in this species. The pollen is inaperturate and the exine is rudimentary, as is generally true for other hydrophilous plants. Light is required for seed germination. The seeds (sterilized in 0.5%–1.0% sodium hypochlorite) germinate within 3 weeks in distilled water kept at 10°C (to 92%) or 18°C (64%), but not at 24°C. They will withstand freezing (−15°C) for at least 3 weeks. The seeds are dispersed as they float (for up to several weeks) on the water surface. The plants persist as short-lived perennials in deeper water (surviving even beneath ice cover), but behave as annuals in sites where the waters have receded. Unlike most of its congeners, *C. hermaphroditica* can use bicarbonate ions in addition to CO_2 as a photosynthetic carbon source. **Reported associates:** *Allionia gausapoides, Alopecurus howellii, Bidens beckii, Brasenia schreberi, Calla palustris, Callitriche heterophylla, C. marginata, C. palustris, C. stagnalis, C. trochlearis, Cephalanthus occidentalis, Ceratophyllum demersum, Chara braunii, C. globularis, C. vulgaris, Cotula coronopifolia, Distichlis spicata, Downingia bacigalupii, Eleocharis acicularis, E. palustris, Elodea canadensis, Equisetum fluviatile, Heteranthera dubia, Hippuris vulgaris, Hydrocharis morsus-ranae, Isoetes echinospora, Lemanea, Lemna minor, L. trisulca, Lepidium oxycarpum, Lythrum hyssopifolia, Marsilea,*

Myosurus minimus, Myriophyllum sibiricum, M. spicatum, M. ussuriense, M. verticillatum, Najas flexilis, N. guadalupensis, Nasturtium officinale, Nitella flexilis, N. tenuissima, Nuphar variegatum, Nymphaea odorata, Persicaria, Pilularia americana, Plagiobothrys stipitatus, Plantago heterophylla, Potamogeton amplifolius, P. berchtoldii, P. crispus, P. diversifolius, P. epihydrus, P. foliosus, P. friesii, P. gramineus, P. illinoensis, P. natans, P. nodosus, P. perfoliatus, P. praelongus, P. pusillus, P. richardsonii, P. robbinsii, P. spirillus, P. zosteriformis, Porterella carnosula, Ranunculus aquatilis, R. longirostris, Ruppia maritima, Sagittaria cuneata, S. latifolia, Sarcocornia pacifica, Schoenoplectus acutus, Sparganium emersum, S. eurycarpum, Spirodela, Stuckenia filiformis, S. pectinata, S. vaginata, Tolypella intricata, Typha latifolia, Utricularia intermedia, U. macrorhiza, U. minor, Vallisneria americana, Wolffia borealis, W. columbiana, Zannichellia palustris, Zostera marina.

***Callitriche heterophylla* Pursh** is an annual or perennial reported from bogs, brooks, depressions, ditches, floodplains, gravel pits, marshes, meadows, mudflats, oxbows, pools, prairies, puddles, rivers, sandbars, seeps, sinkholes, streams, swales, swamps, vernal pools, washes, and the margins of pools, rivers, and streams at elevations of up to 3350 m. The habitats are circumneutral (pH: 5.2–7.2) with substrates that include clay, clay loam, granite, humus, mud, organic loam, peat, sand, sandy muck, and silt loam. The plant are tolerant of cold waters and shade. Generally, they grow as emergents or in shallow waters (to 1.2 m), but have been reported at depths up to 4.6 m. As the species name implies, the plants are heterophyllous, with ovate "terrestrial" leaves or linear "aquatic" leaves produced depending on the relative position of the shoot tip to the water surface. Sexual reproduction occurs by aerial or internal geitonogamy, and the fruits are produced by both the aerial and submersed flowers. This species can represent more than 28% of the propagules present in freshwater tidal marsh seed banks, even when the plants are not dominant in the standing vegetation. Over 250 seedlings/m² have been recorded in some field plots. Good germination has been reported within 6 weeks for seeds receiving 50% of full sunlight under standard greenhouse conditions (maintaining a minimum temperature of 18°C). One study reported significantly higher natural seed germination at sites that had been farmed previously than seeds recovered from unfarmed sites. The seeds remain viable when eaten by some animals, and likely are dispersed by endozoic transport. The plants can develop into fairly large mats but do not occur in dense stands of other species (e.g., *Typha*).
Reported associates: *Agrostis alba, Alisma subcordatum, Allium haematochiton, Alnus rubra, Alopecurus saccatus, Amaranthus cannabinus, Ammannia coccinea, Amphianthus pusillus, Arnoglossum plantagineum, Athyrium filix-femina, Atriplex patula, Azolla caroliniana, Bacopa caroliniana, Beckmannia syzigachne, Bidens cernuus, B. connatus, B. frondosus, B. laevis, Boehmeria cylindrica, Boldia erythrosiphon, Boykinia major, Brasenia schreberi, Callitriche hermaphroditica, C. palustris, C. terrestris, Cardamine longii, Carex annectens, C. atlantica, C. bromoides, C. deweyana,*

C. feta, C. intumescens, C. leptalea, C. lyngbyei, C. obnupta, C. oxylepis, C. pellita, C. radiata, C. rostrata, C. unilateralis, Castilleja attenuata, C. campestris, Cephalanthus occidentalis, Ceratophyllum demersum, Chara, Chasmanthium laxum, C. sessiliflorum, Cicuta douglasii, Cinna arundinacea, Cotula coronopifolia, Cyperus eragrostis, Cypripedium kentuckiense, Decodon verticillatus, Deschampsia cespitosa, Diamorpha smallii, Downingia bicornuta, Dulichium arundinaceum, Egeria densa, Eichhornia crassipes, Elatine brachysperma, Eleocharis acicularis, E. baldwinii, E. interstincta, E. palustris, E. parvula, Elodea canadensis, E. nuttallii, Epilobium ciliatum, Equisetum, Erechtites hieracifolia, Eriocaulon aquaticum, Fontinalis, Fraxinus latifolia, Galium parisiense, G. tinctorium, G. trifolium, Gilia, Glyceria acutiflora, G. canadensis, G. elata, G. striata, Gratiola aurea, G. virginiana, Heteranthera dubia, Howellia aquatilis, Hydrilla verticillata, Hydrochloa caroliniensis, Hydrocotyle umbellata, Hypericum mutilum, Impatiens capensis, Isoetes bolanderi, I. engelmannii, I. howellii, I. lacustris, I. melanopoda, I. melanospora, I. nuttallii, I. occidentalis, I. riparia, Juncus balticus, J. bufonius, J. coriaceus, J. effusus, J. ensifolius, J. tenuis, Justicia americana, Lasthenia fremontii, Leersia virginica, Lemna minor, Lilaea scilloides, Limnanthes, Limnobium spongia, Lindernia dubia, Lobelia cardinalis, Lomatium utriculatum, Ludwigia alternifolia, L. palustris, L. peploides, Lycopus virginicus, Lysichiton americanus, Lythrum portula, L. salicaria, Marsilea vestita, Mayaca fluviatilis, Micranthemum umbrosum, Mimulus glabratus, M. guttatus, Montia fontana, Murdannia keisak, Myosotis laxa, Myrica cerifera, Myriophyllum heterophyllum, Nasturtium officinale, Navarretia myersii, Nitella, Nuphar advena, N. macrophyllum, N. polysepala, N. variegata, Nymphaea odorata, Nymphoides aquatica, Nyssa, Oenanthe sarmentosa, Osmunda regalis, Osmundastrum cinnamomeum, Packera aurea, Panicum dichotomiflorum, P. hemitomon, P. tuckermanii, Peltandra virginica, Persicaria arifolia, P. pensylvanica, P. punctata, Phalaris arundinacea, Phegopteris hexagonoptera, Phyllanthus caroliniensis, Pilularia americana, Plagiobothrys fulvus, P. stipitatus, Platanthera ciliaris, P. clavellata, Poa trivialis, Podostemum ceratophyllum, Polypogon monspeliensis, Pontederia cordata, Potamogeton bicupulatus, P. crispus, P. diversifolius, P. epihydrus, P. foliosus, P. natans, P. nodosus, P. oakesianus, P. robbinsii, P. tennesseensis, Proserpinaca palustris, Quercus agrifolia, Q. engelmannii, Ranunculus aquatilis, R. muricatus, R. uncinatus, Rhamnus ilicifolia, Rhus trilobata, Rhynchospora capitellata, R. corniculata, Rudbeckia fulgida, Rumex crispus, Sagina decumbens, Sagittaria latifolia, S. sanfordii, Salix hookeriana, S. lucida, S. sitchensis, Schoenoplectus californicus, S. pungens, S. smithii, S. tabernaemontani, Scirpus atrovirens, S. cyperinus, S. microcarpus, Sium suave, Solidago rugosa, Sparganium americanum, S. angustifolium, S. emersum, Sphagnum cuspidatum, Spiraea douglasii, Spirodela polyrhiza, Stachys ajugoides, Stebbinsoseris heterocarpa, Taxodium distichum, Thelypteris noveboracensis, Torreyochloa erecta, T. pallida, Toxicodendron diversilobum, Trifolium willdenovii, Triglochin maritimum, Typha latifolia,

Utricularia gibba, U. minor, U. resupinata, Vallisneria americana, Veronica americana, V. peregrina, V. scutellata, Woodwardia areolata, W. virginica, Zannichellia palustris, Zizania aquatica.

Callitriche marginata Torr. is an annual, which grows in depressions, ditches, flats, marshes, mudflats, ponds, pools, streams, swales, vernal pools/ponds, and along their margins at elevations of up to 1890 m. Typical sites form as winter rains produce temporary water bodies, which dry up by early summer. The plants produce floating leaves in shallow water, which can range from clear to turbid conditions. They can tolerate partial to dense shade and can withstand from 25 to 45 days of inundation at depths up to 6 dm. The pools are circumneutral (pH: 5.6–7.2) with substrates that include basalt, clay, claypan, cobbley clay, granite, gravelly loam, mud, rock, and stony loam. This is a spring annual, which typically disappears by late summer. Site temperatures can range from 10°C at the time of seed germination (February and March) to 20°C when seed set occurs. Flowering occurs only under a 16-h photoperiod. Reproduction occurs by aerial geitonogamy with fruits produced only by aerial flowers. The seeds (sterilized in 0.5%–1.0% sodium hypochlorite) will germinate rapidly (typically within 1 week) after being placed in distilled water at 10°C (100%) or 18°C (100%), but not at 24°C. The seeds also have been observed to germinate within 18 days after being exposed to a period of warm stratification (dry storage at room temperature). Light is required for germination, but 25% germination will occur in seeds that have been kept in the dark. The seeds can withstand freezing (−15°C) for at least 3 weeks, and dried seeds germinate well after 2 weeks of exposure to a range of cold temperatures (−17°C to 1°C). The fruits can remain afloat on the water surface for up to several weeks. The seedlings also float and perhaps are dispersed similarly. The plants exhibit high carbonic anhydrase activity, which enables them to concentrate CO_2 following active inorganic carbon uptake. **Reported associates:** *Agrostis avenacea, Alopecurus saccatus, Anagallis arvensis, Artemisia cana, Astragalus pauperculus, Atriplex, Baccharis douglasii, Blennosperma nanum, Callitriche hermaphroditica, Carex tumulicola, Cicendia quadrangularis, Cotula coronopifolia, Crassula aquatica, C. connata, C. tillaea, Cressa truxillensis, Deschampsia gracilis, Dichelostemma capitatum, D. multiflorum, Deschampsia danthonioides, D. nubigena, Downingia bella, D. bicornuta, D. cuspidata, Elatine brachysperma, E. californica, E. chilensis, Eleocharis acicularis, E. palustris, Epilobium densiflorum, E. pygmaeum, Eryngium aristulatum, E. vaseyi, Gilia tricolor, Gnaphalium palustre, Gratiola heterosepala, Hordeum marinum, Isoetes howellii, I. nuttallii, I. orcuttii, Juncus bufonius, J. patens, Lasthenia fremontii, Lepidium, Lilaea scilloides, Limnanthes douglasii, Linanthus ciliatus, Lolium, Lotus corniculatus, Lupinus bicolor, Lythrum hyssopifolia, Malvella leprosa, Marsilea mucronata, M. vestita, Micropus californicus, Mimulus guttatus, Montia fontana, Muilla maritima, Myosurus minimus, Navarretia fossalis, N. leucocephala, N. prostrata, Orcuttia californica, Orthocarpus, Phalaris lemmonii, Pilularia americana, Plagiobothrys acanthocarpus, P. leptocladus, P. stipitatus,* *Plantago bigelovii, P. elongata, Pogogyne abramsii, P. nudiuscula, P. ziziphoroides, Polygonum aviculare, Polypogon monspeliensis, Psilocarphus brevissimus, P. oregonus, P. tenellus, Ranunculus aquatilis, R. bonariensis, Rumex crispus, Sedella pumila, Sisyrinchium californicum, Trifolium subterraneum, Triteleia hyacinthina, Veronica peregrina.*

Callitriche palustris L. is annual or perennial of bogs, brooks, canals, ditches, fens, gravel/sand bars, lakes, marshes, meadows, mudflats, oxbows, ponds, pools, potholes, rivers, sloughs, streams, swamps, and the margins of lakes, rivers, and streams at elevations of up to 3550 m. The habitats are open, fresh, or brackish (salinity up to 3 ppt.), circumneutral (pH: 5.2–7.4), still or flowing waters, with substrates of clay, gravel, gravel sand, sandy mud, muck, mud, sand, silt, or silty muck. The plants will tolerate very cold temperatures (to −36°C) and extend northward to at least 68° N latitude. The species is described as amphibious, with the ability to grow both as a submersed life form (at depths up to 1 m), or as an emergent on exposed, wet substrates. It is heterophyllous, producing a rosette of floating leaves when submersed. Mistakenly reported by some accounts as being apomictic (agamospermous), sexual reproduction occurs either by aerial geitonogamy or by internal geitonogamy and the plants will produce fruits in both the aerial and submersed flowers. Seed germination occurs typically at about 10°C with peak flowering and fruiting reached at 27°C. Light is necessary for seed germination. Slightly higher germination rates have been reported for seeds kept under flooded conditions; however, other studies have indicated that germination occurs only on noninundated substrates. Seeds (sterilized in 0.5%–1.0% sodium hypochlorite) germinate within 3 weeks of being placed in distilled water at 10°C (70%), 18°C (95%), or 24°C (42%). Faster germination has been observed for seeds pretreated for 11 days in 0.2 M NaCl. The seeds will withstand freezing (−15°C) for at least 3 weeks. The fruits can remain afloat for up to several weeks and are dispersed on the water surface. The seedlings also float and probably are dispersed in a similar fashion. Dispersal of fruits clinging to the fur of vertebrates also has been documented in some European populations. The plants are early colonizers (within 10 years) of newly formed aquatic habitats (e.g., beaver ponds, oxbows). They exhibit high peroxidase activity, which confers tolerance to pulp and paper mill effluents. **Reported associates:** *Agrostis variabilis, Alisma subcordatum, A. triviale, Alnus, Alopecurus aequalis, Azolla, Bidens beckii, B. cernuus Bistorta bistortoides, Calla palustris, Callitriche hermaphroditica, C. heterophylla, Caltha natans, Camassia, Cardamine parviflora, C. pensylvanica, Carex aquatilis, C. lenticularis, C. pellita, C. rostrata, C. utriculata, Celtis occidentalis, Cephalanthus occidentalis, Ceratophyllum demersum, Chrysosplenium americanum, Cicuta virosa, Comarum palustre, Deschampsia cespitosa, Drepanocladus aduncus, Echinochloa crus-galli, Egeria densa, Eleocharis acicularis, E. bella, E. compressa, E. palustris, Elodea canadensis, E. nuttallii, Equisetum arvense, E. fluviatile, Eragrostis hypnoides, Fraxinus nigra, Glyceria borealis, G. striata, Gnaphalium palustre, G. uliginosum, Gratiola neglecta,*

Heteranthera dubia, Hippuris vulgaris, Isoetes bolanderi, I. occidentalis, Juncus alpinus, J. arcticus, Leersia oryzoides, Lemna minor, L. trisulca, Limosella acaulis, Lindernia dubia, Lobelia cardinalis, Marsilea mucronata, M. vestita, Menispermum canadense, Menyanthes trifoliata, Mimulus glabratus, M. primuloides, Mougeotia, Myriophyllum heterophyllum, M. sibiricum, Najas flexilis, Nitella, Nuphar polysepala, N. variegata, N. ×rubrodisca, Phalaris arundinacea, Persicaria amphibia, P. hydropiper, P. sagittata, P. virginiana, Porterella carnosula, Potamogeton alpinus, P. diversifolius, P. epihydrus, P. foliosus, P. friesii, P. gramineus, P. illinoensis, P. natans, P. nodosus, P. obtusifolius, P. pusillus, P. richardsonii, P. zosteriformis, Ranunculus aquatilis, R. circinatus, R. gmelinii, R. longirostris, R. sceleratus, R. trichophyllus, Rhododendron, Riccia fluitans, Ricciocarpus natans, Sagittaria graminea, S. latifolia, Sanionia uncinata, Schoenoplectus pungens, S. tabernaemontani, Sparganium angustifolium, S. chlorocarpum, S. emersum, S. eurycarpum, S. minimum, S. natans, Spirodela polyrhiza, Stuckenia filiformis, S. pectinata, Typha latifolia, Utricularia macrorhiza, Vallisneria americana, Veronica americana, V. peregrina, Warnstorfia trichophylla, W. tundrae, Zannichellia palustris.

***Callitriche peploides** Nutt.* grows in depressions, ditches, floodplains, marshes, meadows, mudflats, pools, swamps, and along the margins of rivers and streams at elevations of up to 137 m. Most exposures are described as full sun. The substrates consist of granite, loam, muck, mud, sand, sandy clay, or silt. Sexual reproduction occurs by aerial geitonogamy, with fruits produced only by aerial flowers. The plants can develop into mats and often represent an early successional stage, especially in pools as sediment depth increases and water depth declines. A number of reports are from abandoned and current rice fields and from agricultural sites that are heavily disturbed by livestock. There is relatively little ecological information available for this species. **Reported associates:** *Anthoceros, Crassula aquatica, C. drummondii, Elatine brachysperma, Eleocharis obtusa, Isoetes lithophila, I. melanopoda, Juncus diffusissimus, Leptochloa fusca, Lobelia feayana, Parietaria, Riccia nigrella, Rorippa teres, Sedum nuttallii, Spartina, Sphaerocarpos, Taxodium, Veronica peregrina.*

***Callitriche stagnalis** Scop.* is an annual or perennial, which inhabits tidal or nontidal beaches, channels, depressions, ditches, dunes, marshes, mudflats, oxbows, ponds, puddles, sloughs, streams, swamps, and the margins of lakes and watercourses at elevations of up to 1760 m. This species is heterophyllous and produces a rosette of floating leaves (often in dense mats) when growing submersed in water (up to 9 dm depths). It also forms dense mats when growing on exposed substrates. The plants occur in circumneutral waters (pH: 6.4–8.5), but tend to inhabit more alkaline sites and do not grow in highly acidic sites (e.g., pH: 4.5). Reported substrates include gravel, mud, sand, and silt. In North America, the flowering period is prolonged from April to November. Sexual reproduction occurs by aerial geitonogamy, which produces fruits in aerial flowers only. Although the plants will grow luxuriantly in flowing water, they will not set seed

unless they are situated in calm waters or on exposed mudflats, where 100% seed set can be attained. More seeds have been observed in disturbed sites, where the seed banks can attain a propagule density of more than 700 fruits/m^2. Germination rates are high on exposed substrates. The mechanism of natural seed dispersal is uncertain, but recreational vehicles are believed to transport seeds in the mud that adheres to them. The plants cannot use bicarbonate for photosynthesis, but use free CO_2 efficiently, with a compensation point 2.5 μM (relative to the equilibrium concentration of 15 μM) and a low saturation point (250–500 μM CO_2). Their photosynthetic rate increases with water current velocity, reaching a maximum at 8–12 mm/s, but declining markedly above 20 mm/s. An annual life form dominates in colder climates; otherwise the plants more commonly are perennial. Vegetative reproduction can occur by stem fragments, which root at their nodes. This species frequently is an early colonizer of exposed mudflats. It often is found in disturbed sites and in North America is associated with a number of other nonindigenous invasive species. **Reported associates:** *Alnus rubra, Bidens, Brasenia schreberi, Butomus umbellatus, Callitriche hermaphroditica, Carex obnupta, C. utriculata, Ceratophyllum demersum, Chara, Crassula aquatica, Cyperus bipartitus, Egeria densa, Elatine americana, Eleocharis, Elodea canadensis, E. nuttallii, Epilobium ciliatum, Equisetum, Eriocaulon parkeri, Fontinalis antipyretica, Fraxinus latifolia, Galium cymosum, Gentianella victorinii, Glyceria, Howellia aquatilis, Hydrodictyon reticulatum, Iris pseudacorus, Isoetes, Leersia oryzoides, Lemna minor, L. trisulca, Lilaea scilloides, Lotus, Ludwigia palustris, Lythrum salicaria, Myriophyllum aquaticum, M. hippuroides, M. spicatum, Najas flexilis, N. guadalupensis, Nasturtium officinale, Nitella, Nuphar polysepala, Nymphaea odorata, Persicaria amphibia, P. hydropiperoides, Phalaris arundinacea, Picea sitchensis, Potamogeton amplifolius, P. crispus, P. epihydrus, P. foliosus, P. obtusifolius, P. pusillus, P. zosteriformis, Ranunculus aquatilis, R. repens, Salix, Schoenoplectus smithii, Sparganium angustifolium, Spiraea douglasii, Stuckenia filiformis, S. pectinata, Thuja plicata, Tsuga heterophylla, Typha latifolia, Utricularia macrorhiza, Vallisneria americana, Wolffia.*

***Callitriche trochlearis** Fassett* grows in ditches, marshes, meadows, ponds, potholes, rivers, streams, swales, vernal pools, washes, and along the margins of ponds, rivers, and streams at elevations of up to 3350 m. It is amphibious and heterophyllous, producing rosettes of floating leaves when growing in shallow water (up to 6 dm) but also occurring on exposed substrates of clay, mud, or sand. The plants can occur in watercourses where there are moderate currents. Sexual reproduction occurs by aerial and internal geitonogamy, whereby fruits can be found on both aerial and submersed flowers. Otherwise, there is sparse ecological information for this species and further study is needed. **Reported associates:** *Callitriche hermaphroditica, C. stagnalis, Eleocharis acicularis, Persicaria lapathifolia, Populus trichocarpa, Rorippa sinuata, Rumex salicifolius.*

Use by wildlife: The shoots and fruits of *Callitriche* species are eaten by various waterfowl, including black ducks

(Aves: Anatidae: *Anas rubripes*), buffleheads (Aves: Anatidae: *Bucephala albeola*), canvasbacks (Aves: Anatidae: *Aythya valisineria*), gadwells (Aves: Anatidae: *Anas strepera*), mallards (Aves: Anatidae: *Anas platyrhynchos*), redheads (Aves: Anatidae: *Aythya americana*), and wood ducks (Aves: Anatidae: *Aix sponsa*). Germinable seeds of *C. heterophylla* have been recovered from horse (Mammalia: Equidae: *Equus*) dung samples, indicating that the plants are grazed at least inadvertently by equines. *Callitriche marginata* is the host of the fungus *Heterodoassansia callitrichis* (Fungi: Basidiomycota: Doassansiaceae). Plants of *C. peploides* are eaten by the Florida marsh rabbit (Mammalia: Leporidae: *Sylvilagus palustris paludicola*). *Callitriche stagnalis* is abundant in the postbreeding habitat of the Oregon spotted frog (Amphibia: Ranidae: *Rana pretiosa*).

Economic importance: food: The non-North American *Callitriche antarctica* is eaten raw as a vegetable in a fashion similar to watercress. The plants are highly hydrated with only 7%–9% of their fresh weight as dry matter; **medicinal:** *Callitriche stagnalis* contains verbascocide, an anticarcinogenic antioxidant that can repair DNA that becomes damaged by free oxygen radicals. The plants also contain several iridoid glycosides including aucubin, which can be toxic due to its ability to bind proteins, and catalpol, which allegedly has neuroprotective properties. Also, this is the only species in the genus known to produce a positive test for alkaloids. In Europe, the plants were added as an ingredient of plasters, which were used to promote the discharge of fluid from infections; **cultivation:** *Callitriche hermaphroditica* is used as an oxygenating plant for aquariums and water gardens.

Callitriche palustris has been promoted as an aquarium plant since the mid-19th century. *Callitriche stagnalis* is grown occasionally in aquariums; **misc. products:** Some *Callitriche* species are used in Europe as indicators of environmental pollution; **weeds:** *Callitriche palustris* has been reported as a weed in Finnish agricultural fields and in Japanese rice paddies; **nonindigenous species:** *Callitriche stagnalis* was introduced to North America during the mid-19th century, presumably by ballast disposal at shipping ports, or as an escaped aquarium plant. *Callitriche heterophylla* and *C. stagnalis* were introduced to New Zealand and *C. peploides* was introduced to France and Taiwan sometime before 1997.

Systematics: Although *Callitriche* has often been segregated within a distinct family (Callitrichaceae), phylogenetic analyses clearly resolve the genus as the sister to *Hippuris* in a clade nested well within the limits of the redefined Plantaginaceae (Figure 5.82). The genus itself traditionally has been subdivided into three sections (*Callitriche*, *Microcallitriche*, *Pseudocallitriche*) of which only the latter (with *C. hermaphroditica*, *C. lusitanica*, *C. truncata*) is supported phylogenetically (Figure 5.83); whereas species segregated as section *Microcallitriche* (e.g., *C. deflexa*, *C. marginata*, *C. nuttallii*, *C. peploides*, *C. terrestris*) are interspersed among members assigned to section *Callitriche*. A more reasonable subdivision would be to segregate the Old and New World representatives of species outside of the *Pseudocallitriche* group (Figure 5.83), but this taxonomic modification has not yet been formally proposed. *Callitriche verna* is a synonym of *C. palustris* and *C. longipedunculata* is a synonym of *C. marginata*; the data reported for these taxa have been merged

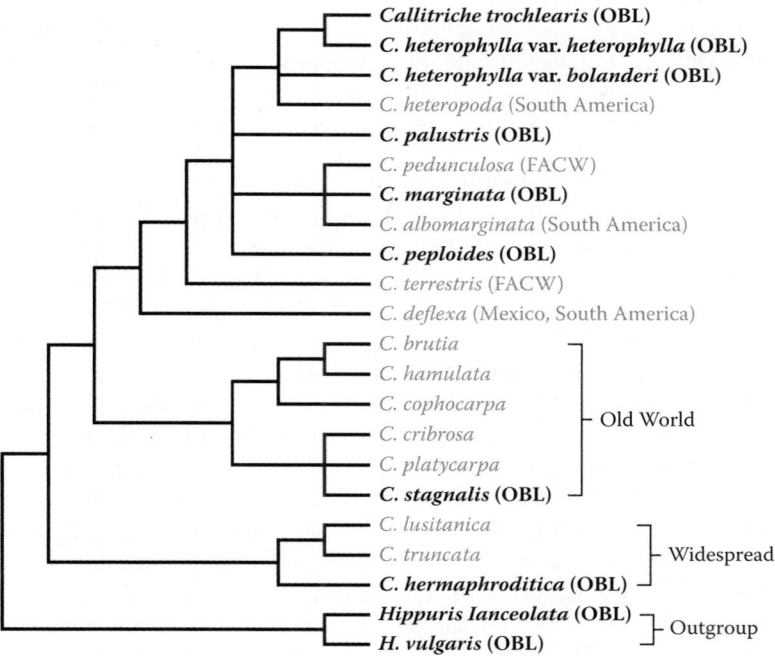

FIGURE 5.83 Phylogenetic relationships in *Callitriche* as indicated by analysis of *rbcL* sequence data. The relationships resolved by this study do not support the sectional subdivisions based on morphology, but suggest that the genus has diverged as more geographically cohesive groups (*C. stagnalis* is nonindigenous to the New World). The obligate (OBL) North American taxa are indicated in bold. The wetland indicator status is provided otherwise for North American species. (Adapted from Philbrick, C.T. & Les, D.H., *Aquat. Bot.*, 68, 123–141, 2000.)

under the latter names. Plants recognized as *C. intermedia* by some authors apparently are no more than higher elevation forms of *C. palustris*. Minor variants of *C. hermaphroditica* have been called *C. fassettii* (ranked as OBL in the 2013 indicator list) but these do not appear to warrant recognition as a distinct species and no separate treatment has been provided here. Two varieties of *C. heterophylla* are regularly distinguished (var. *heterophylla*; var. *bolanderi*), and *rbcL* sequence data indicate that these taxa might represent distinct species (Figure 5.83); however, additional morphological and molecular data are needed to address this question more critically. The base chromosome number of *Callitriche* is *x* = 5, with aneuploid (*x* = 4, 3) and polyploid series evolving on several occasions. Diploids (2*n* = 10) include *C. peploides* and *C. stagnalis* and tetraploids (2*n* = 20) include *C. heterophylla* (both varieties), *C. marginata*, and *C. palustris*. *Callitriche trochlearis* (2*n* = 40) is an octoploid and *C. hermaphroditica* (2*n* = 6) is a diploid derived from an *x* = 3 aneuploid series. In *C. hermaphroditica*, chromosome chains occur with the homologous chromosomes adjacent to one another, which facilitates synapsis. *Callitriche* can hybridize but hybrids have not been reported in North America.

Comments: *Callitriche* is distributed broadly with some species widespread essentially throughout North America (*C. heterophylla*; extending into the Caribbean and South America) and others widespread except for the southern United States (*C. palustris*; extending into Australia and Eurasia), or throughout northern and western North America (*C. hermaphroditica*; extending into Eurasia). More restricted distributions characterize *C. peploides* (southeastern United States; extending into Mexico and Costa Rica), *C. marginata* (British Columbia, California, Idaho, Oregon), and *C. trochlearis* (California, Idaho, Oregon). The nonindigenous *C. stagnalis* (native to Africa and Eurasia) is disjunct in eastern and western North America and also occurs in South America.

References: Baskin & Baskin, 1998; Blair, 1936; Bliss & Zedler, 1998; Bouchard et al., 1987; Campbell & Gibson, 2001; Christy, 2004; Cooper et al., 2000; Crowder et al., 1977; Damtoft et al., 1994; Dennis et al., 1979; Deschamp & Cooke, 1985; Erbar & Leins, 2004; Fassett, 1951; Humphreys, 1857; Kautsky, 1988; Keeley, 1999; Lahring, 2003; Lansdown, 2009; Leck & Graveline, 1979; Les & Mehrhoff, 1999; Linder, 1938; Liu et al., 2005; Maberly & Madsen, 2002; Madsen & Soendergaard, 1983; McCain & Christy, 2005; McLaughlin, 1974; Middleton, 2003; Mitchell, 1974; Moyle, 1945; Nishihiro et al., 2004; Palmer et al., 1992; Parker & Leck, 1985; Persson & Shacklette, 1959; Philbrick, 1984b, 1993; Philbrick & Anderson, 1992; Philbrick & Bernardello, 1992; Philbrick & Les, 1996, 2000; Philbrick et al., 1998; Purer, 1939; Raffauf, 1996; Ray et al., 2001; Robbins, 1918; Roy et al., 1992; Salonen et al., 2001; Sand-Jensen, 1983; Schmidt et al., 2004; Schotsman, 1954, 1967; Spencer & Ksander, 1998; Sperduto & Nichols, 2004; Vickery, 1995; Wacker & Kelly, 2004; Wagenaar, 1969; Walters & Wyatt, 1982; Whitehouse, 1933; Williams, 2006; Yamada et al., 2007; Zika, 1996.

3. *Chelone*

Balmony, turtlehead; tête de tortue

Etymology: from the Greek *chelone*, meaning "tortoise," with respect to the turtle-like shape of the corolla

Synonyms: none

Distribution: global: North America; **North America:** eastern

Diversity: global: 4 species; **North America:** 4 species

Indicators (USA): OBL: *Chelone cuthbertii*, *C. glabra*, *C. obliqua*

Habitat: freshwater; palustrine; **pH:** 5.6–8.4; **depth:** <1 m; **life-form(s):** emergent herb

Key morphology: stems (to 2 m) erect, bluntly 4-angled; leaves (to 18 cm) opposite, lanceolate, sessile or petioled (to 4 cm), the margins dentate or serrate; flowers subsessile, in condensed, terminal, spike-like racemes; calyx (to 1 cm) 5-lobed; corollas (to 3.7 cm) testudiform (turtle-like), white or greenish-yellow to purple or purple rose, the lower lip projecting as a palate that closes the corolla throat, the lower throat bearded internally with wooly white to yellow hairs; stamens 4, didynamous, their anthers confluent dorsally; staminode 1, shorter than the stamens; syles exserted at maturity; capsules (to 1.5 cm) ovoid; seeds numerous, flattened, winged

Life history: duration: perennial (rhizomes); **asexual reproduction:** rhizomes; **pollination:** insect; **sexual condition:** hermaphroditic; **fruit:** capsules (common); **local dispersal:** rhizomes; **long-distance dispersal:** seeds (water, wind)

Imperilment: (1) *Chelone cuthbertii* [G3]; S1 (GA); S2 (VA); S3 (NC); (2) *C. glabra* [G5]; S1 (AR); S2 (MB); S3 (IA, MS, NC, RI); (3) *C. obliqua* [G4]; SH (AR, GA, MS); S1 (IA, KY, MD, MI, TN, VA); S2 (KY, MO, NC); S3 (IL, IN, KY)

Ecology: general: All *Chelone* species are designated as OBL aquatics except for *C. lyonii* Pursh (FACW), which nevertheless can occur in their association. The flowers of *Chelone* are pollinated by large bumblebees (Hymenoptera: Apidae: *Bombus*), which clasp the hairs on the palate of the lower corolla lip while their weight forces the corolla open far enough for them to gain entry. As the bee enters the flower, it displaces the stamen filaments, causing the grouped anthers to release their pollen upon the insect's thorax. The seeds require cold stratification for optimal germination and are winged to facilitate water and wind dispersal. Allozyme data indicate that *Chelone* species possess a level of genetic variation comparable to other perennials with selfing to mixed breeding systems. The species reproduce vegetatively by means of short rhizomes but are not highly clonal.

***Chelone cuthbertii* Small** inhabits bogs, depressions, meadows, and seeps at elevations of up to 2011 m. The plants are shade tolerant; otherwise, there is very little ecological information available for this rare plant. The flowers are produced from August to September and have purplish corollas. The reproductive ecology of this species has not been investigated but probably is similar to that of the other species. **Reported associates:** *Aconitum reclinatum, Alnus serrulata, Arethusa bulbosa, Aulacomnium palustre, Bazzania trilobata, Boykinia aconitifolia, Calamagrostis coarctata, Calopogon tuberosus,*

Cardamine clematitis, Carex atlantica, C. echinata, C. folliculata, C. gynandra, C. intumescens, C. joorii, C. leptalea, C. trisperma, C. ruthii, Chasmanthium laxum, Chelone lyonii, Chrysosplenium americanum, Cicuta maculata, Diphylleia cymosa, Doellingeria umbellata, Drosera rotundifolia, Epilobium leptophyllum, Eriophorum virginicum, Fuirena breviseta, Helenium autumnale, Helonias bullata, Hottonia inflata, Houstonia serpyllifolia, Hydrocotyle americana, Hydrophyllum canadense, Hypericum buckleyi, H. densiflorum, H. graveolens, H. mitchellianum, H. prolificum, Ilex verticillata, Impatiens capensis, Juncus brevicaudatus, J. caesariensis, J. effusus, J. gymnocarpus, J. subcaudatus, Kalmia angustifolia, K. carolina, K. latifolia, Lilium grayi, Lycopus uniflorus, Lyonia ligustrina, Lysimachia terrestris, Melanthium virginicum, Menziesia pilosa, Micranthes careyana, M. micranthidifolia, Monarda didyma, Orontium aquaticum, Osmunda regalis, Osmundastrum cinnamomeum, Oxypolis rigidior, Packera aurea, Parnassia asarifolia, Persicaria sagittata, Photinia melanocarpa, P. pyrifolia, Platanthera grandiflora, Poa paludigena, Polytrichum commune, Rhizomnium appalachianum, Rhododendron catawbiense, R. maximum, R. viscosum, Rhynchospora capitellata, Rosa palustris, Rudbeckia laciniata, Sabatia campanulata, Sagittaria latifolia, Salix sericea, Sanguisorba canadensis, Sarracenia flava, S. purpurea, Schoenoplectus purshianus, Scirpus atrovirens, S. cyperinus, S. expansus, S. polyphyllus, Solidago elliottii, S. uliginosa, S. patula, Sorbus americana, Sparganium chlorocarpum, Sphagnum affine, S. bartlettianum, S. fallax, S. palustre, S. recurvum, Spiraea alba, S. tomentosa, Stenanthium gramineum, Symphyotrichum puniceum, Thalictrum clavatum, T. dioicum, Thelypteris palustris, T. simulata, Trautvetteria caroliniensis, Triantha glutinosa, Vaccinium macrocarpon, Veratrum viride, Veronica americana, Viburnum cassinoides, Viola cucullata, V. macloskeyi, V. ×primulifolia, Woodwardia areolata.

***Chelone glabra** L.* grows in bogs, depressions, ditches, fens, floodplains, marshes, meadows, prairies, river bottoms, seeps, sloughs, swales, swamps, and along the margins of lakes, ponds, rivers, and streams at elevations of up to 1150 m. It can occur in open sites or in shade, across a fairly wide range of substrates (pH: 6.0–8.4) that include clay, granite, loam, loamy muck, muck, peat, rocky loam, sand, sandstone, and sandy loam. The plants usually are found in no more than a few cm of standing water. Their showy white flowers, which provide good contrast for pollinators when growing in shady habitats, occur from July to October. The flowers produce copious amounts of nectar (averaging 3.3 mg/floret) but are pollinated only by a few select species of bumblebees (Hymenoptera: Apidae: *Bombus fervidus, B. terricola, B. vagans*), which primarily visit them during the morning (7:00–9:00 am). Floral visitation rates are low because it can take up to 30 seconds for an insect to enter a single flower successfully. The mouth of the corolla is "closed" and excludes other insects, which cannot learn how to enter the flower; however, corollas that are damaged by herbivores occasionally facilitate nectar theft. The four fertile anthers connect to form a single, dorsal pollen presentation unit. As

the flower is forced open, its filaments spread, allowing the pollen to be deposited on the thorax of the entering insect. The style elongates above the anthers, eventually reaching the corolla mouth. Consequently, self-pollination would be highly improbable in the absence of insects. Hummingbirds (Aves: Trochilidae: *Archilochus colubris*) also visit the flowers for nectar, but doubtfully function as pollinators. The seeds require a period of cold, moist stratification (2–4 months at 3°C–5°C) and high light levels (surface sowing) for effective germination, which occurs within 2–6 weeks when incubated subsequently at 13°C–18°C. The seeds are broadly winged and are dispersed both by water and the wind. *Chelone glabra* has reappeared dramatically in areas after they have been burned, sometimes following more than a 20-year absence from the site. Excessive insect herbivory by larval butterflies (Lepidoptera: Nymphalidae: *Euphydryas phaeton*) and sawflies (Hymenoptera: Tenthredinidae: *Macrophya nigra, Tenthredo grandis*) can severely reduce the growth of adult plants along with their flower and seed production. The roots are colonized by arbuscular mycorrhizal and dark septate endophytic fungi. **Reported associates:** *Abies balsamea, Acer rubrum, A. pensylvanicum, A. saccharinum, Acorus americanus, Adiantum capillus-veneris, Agalinis purpurea, Agastache nepetoides, Agrostis perennans, Aletris farinosa, Alisma triviale, Alnus rugosa, A. serrulata, Ambrosia artemisiifolia, Amphicarpaea bracteata, Angelica atropurpurea, A. triquinata, Apios americana, Apocynum cannabinum, Aralia nudicaulis, Arctium lappa, Arisaema triphyllum, Artemisia serrata, Aruncus dioicus, Arundinaria gigantea, Asclepias incarnata, A. hirtella, Atrichum oerstedianum, Bartonia paniculata, Betula alleghaniensis, B. nigra, B. papyrifera, B. ×sandbergii, Bidens cernuus, B. connatus, B. frondosus, B. vulgatus, Bignonia capreolata, Boehmeria cylindrica, Boykinia aconitifolia, Brachyelytrum erectum, Brachythecium rivulare, Bryhnia novae-angliae, Calamagrostis canadensis, C. stricta, Callitriche, Calopogon tuberosus, Caltha palustris, Calystegia sepium, Campanula aparinoides, C. divaricata, Campylium chrysophyllum, Cardamine bulbosa, C. clematitis, Carex aestivalis, C. alata, C. aquatilis, C. atherodes, C. atlantica, C. bromoides, C. collinsii, C. comosa, C. crinita, C. flava, C. folliculata, C. frankii, C. grayi, C. gynandra, C. haydenii, C. hystericina, C. interior, C. intumescens, C. lacustris, C. leptalea, C. lonchocarpa, C. lupulina, C. lurida, C. pedunculata, C. pellita, C. prairea, C. prasina, C. rostrata, C. sartwellii, C. scabrata, C. scoparia, C. sterilis, C. stipata, C. stricta, C. trichocarpa, C. vulpinoidea, Castilleja coccinea, Cephalanthus occidentalis, Cercis canadensis, Chamaedaphne calyculata, Chelone lyonii, Chrysosplenium americanum, Cicuta bulbifera, C. maculata, Cinna arundinacea, Circaea alpina, Cirsium arvense, Clematis virginiana, Clethra alnifolia, Comarum palustre, Conocephalum conicum, Coptis trifolia, Cornus amomum, C. canadensis, C. foemina, C. racemosa, C. sericea, Corylus, Crataegus, Cuscuta gronovii, C. rostrata, Cypripedium reginae, Dasiphora floribunda, Decumaria barbara, Deschampsia flexuosa, Dichanthelium clandestinum, D. dichotomum, D. sphaerocarpon, D. spretum, Diphylleia cymosa, Doellingeria* umbellata, Drosera rotundifolia, Dryopteris carthusiana, D. goldiana, Dulichium arundinaceum, Dumortiera hirsuta, Echinocystis lobata, Eleocharis obtusa, E. smalliana, Elymus virginicus, Epilobium coloratum, E. leptophyllum, Equisetum arvense, Eriophorum angustifolium, E. virginicum, Eupatorium fistulosum, E. maculatum, E. perfoliatum, E. purpureum, Eurybia divaricata, E. schreberi, Euthamia graminifolia, Eriophorum viridicarinatum, Fallopia scandens, Fragaria virginiana, Fraxinus americana, F. nigra, F. pennsylvanica, Fuirena squarrosa, Galium asprellum, G. palustre, G. tinctorium, G. triflorum, Gaultheria, Gentiana andrewsii, Gentianopsis virgata, Geum aleppicum, G. canadense, G. geniculatum, G. laciniatum, G. rivale, Glyceria canadensis, G. maxima, G. melicaria, G. pallida, G. striata, Hamamelis virginiana, Hasteola suaveolens, Helenium autumnale, Helianthus angustifolius, H. giganteus, H. grosseserratus, Helonias bullata, Heuchera villosa, Hibiscus moscheutos, Hookeria acutifolia, Houstonia serpyllifolia, Hydrangea arborescens, H. quercifolia, Hydrocotyle americana, Hylotelephium telephioides, Hypericum ascyron, H. densiflorum, H. frondosum, H. mutilum, H. virginicum, Ilex decidua, I. opaca, I. verticillata, Impatiens capensis, Iris versicolor, I. virginica, Itea virginica, Jamesianthus alabamensis, Juncus acuminatus, J. canadensis, J. coriaceus, J. effusus, J. gymnocarpus, Juniperus virginiana, Lactuca canadensis, Laportea canadensis, Lathyrus palustris, Larix laricina, Leersia oryzoides, Liatris microcephala, Lilium grayi, Lindera benzoin, Linum striatum, Liparis loeselii, Liquidambar styraciflua, Liriodendron tulipifera, Lobelia amoena, L. cardinalis, L. kalmii, L. siphilitica, Lonicera, Ludwigia palustris, Lycopus americanus, L. uniflorus, L. virginicus, Lygodium palmatum, Lysimachia ciliata, L. terrestris, L. thyrsiflora, Lythrum alatum, L. lineare, L. salicaria, Magnolia acuminata, M. virginiana, Maianthemum canadense, Marshallia trinervia, Menispermum canadense, Mentha arvensis, Micranthes micranthidifolia, M. pensylvanica, M. petiolaris, Microstegium vimineum, Mikania scandens, Mimulus ringens, Mitchella repens, Mitella nuda, Mitreola petiolata, Mnium hornum, Muhlenbergia glomerata, M. mexicana, Myrica cerifera, Napaea dioica, Nasturtium officinale, Nemopanthus mucronatus, Nyssa sylvatica, Oclemena acuminata, Oenothera fruticosa, Onoclea sensibilis, Osmunda regalis, Osmundastrum cinnamomeum, Oxalis acetosella, O. stricta, Oxypolis rigidior, Packera aurea, Panicum rigidulum, Parnassia asarifolia, P. glauca, P. grandifolia, P. palustris, Pedicularis lanceolata, Pellia appalachiana, Peltandra virginica, Penthorum sedoides, Persea palustris, Persicaria amphibia, P. arifolia, P. coccinea, P. hydropiperoides, P. pensylvanica, P. punctata, P. sagittata, Phalaris arundinacea, Philonotis fontana, Phlox glaberrima, P. maculata, Physocarpus opulifolius, Phytolacca americana, Picea mariana, Pilea fontana, P. pumila, Pinus strobus, Pityopsis ruthii, Plagiomnium ciliare, Platanthera clavellata, P. dilatata, P. flava, P. peramoena, P. psycodes, Poa paludigena, P. palustris, Proserpinaca palustris, Prunus serotina, Pteridium aquilinum, Pycnanthemum virginianum, Quercus bicolor, Q. palustris, Radula sullivantii, Ranunculus recurvatus, R. septentrionalis, Rhamnus alnifolia,

Rhizomnium appalachianum, R. punctatum, Rhododendron groenlandicum, R. viscosum, Rhus copallina, R. glabra, Rhynchospora alba, R. capillacea, R. capitellata, R. globularis, Rhynchostegium serrulatum, Rubus alleghieniensis, R. hispidus, Rudbeckia fulgida, R. laciniata, Rumex crispus, R. obtusifolius, R. orbiculatus, R. verticillatus, Sagittaria latifolia, Salix bebbiana, S. candida, S. discolor, S. nigra, S. petiolaris, Scapania nemorosa, Schoenoplectus acutus, S. tabernaemontani, Scirpus atrovirens, S. cyperinus, Scleria verticillata, Scutellaria galericulata, S. integrifolia, S. lateriflora, Selaginella apoda, Silphium perfoliatum, Sium suave, Solanum dulcamara, Solidago flexicaulis, S. gigantea, S. graminifolia, S. ohioensis, S. patula, S. riddellii, S. rugosa, S. uliginosa, Sorghastrum nutans, Sparganium americanum, S. eurycarpum, S. glomeratum, Spartina pectinata, Sphagnum lescurii, S. palustre, S. recurvum, Sphenopholis pensylvanica, Spiraea alba, S. tomentosa, Spiranthes cernua, Splachnum ampullaceum, Stellaria borealis, S. corei, Symphyotrichum boreale, S. cordifolium, S. lanceolatum, S. lateriflorum, S. novae-angliae, S. pilosum, S. puniceum, Symplocarpus foetidus, Thalictrum clavatum, T. dasycarpum, T. pubescens, Thelypteris noveboracensis, T. palustris, Thuidium delicatulum, Thuja occidentalis, Tiarella cordifolia, Toxicodendron vernix, Trautvetteria caroliniensis, Triadenum virginicum, Trientalis, Triglochin palustre, Trollius laxus, Tsuga canadensis, Typha angustifolia, T. latifolia, T. ×glauca, Ulmus americana, U. rubra, Utricularia cornuta, Vaccinium corymbosum, Valeriana ciliata, V. edulis, Vallisneria americana, Veratrum viride, Verbena hastata, Verbesina alternifolia, Vernonia gigantea, Viburnum dentatum, V. nudum, V. opulus, V. recognitum, Vicia cracca, Viola blanda, V. cucullata, V. lanceolata, V. macloskeyi, V. nephrophylla, V. renifolia, Vitis aestivalis, Woodwardia areolata, Xyris tennesseensis, X. torta, Zizia aurea.

***Chelone obliqua* L.** occurs in bogs, cliffs, fens, floodplains, gravel bars, marshes, meadows, outcrops, sloughs, swamps, and along the margins of lakes and streams at elevations of up to 1725 m. The plants tolerate full sun to partial shade and occur on gneiss and other acidic substrates (pH: 5.6–6.8). The flowers have purple-pink corollas and are produced from August to September. They reportedly are pollinated by bumblebees (Hymenoptera: Apidae: *Bombus*). The seeds must be cold (4°C) and moist stratified for optimal germination, which occurs within 2–4 weeks when incubated subsequently at 18°C–20°C. **Reported associates:** *Acalypha rhomboidea, Acer negundo, A. nigrum, A. rubrum, A. saccharinum, Aconitum uncinatum, Aesculus glabra, Ageratina altissima, Agrostis perennans, Alliaria petiolata, Ambrosia trifida, Angelica triquinata, Asarum canadense, Asclepias incarnata, Asimina triloba, Bidens frondosus, Boehmeria cylindrica, Boltonia asteroides, Calystegia sepium, Camassia scilloides, Campsis radicans, Carex folliculata, C. gynandra, C. ruthii, Carya illinoensis, Celtis occidentalis, Cercis canadensis, Chasmanthium latifolium, Cinna arundinacea, C. latifolia, Circaea lutetiana, Cryptotaenia, Dennstaedtia punctilobula, Diervilla sessilifolia, Elymus virginicus, Erythronium albidum, Eupatorium* *coelestinum, E. maculatum, E. rugosum, Eurybia divaricata, Filipendula rubra, Floerkea proserpinacoides, Forestiera acuminata, Frangula alnus, Fraxinus nigra, F. pennsylvanica, Galium asprellum, Gentiana clausa, Glyceria nubigena, Gymnocladus dioica, Helenium autumnale, Houstonia serpyllifolia, Huperzia porophila, Impatiens capensis, Ipomoea lacunosa, Iris virginica, Juglans nigra, Laportea canadensis, Leersia virginica, Lindera benzoin, Lobelia cardinalis, Lycopus americanus, Lysimachia nummularia, Menispermum canadense, Mertensia virginica, Micranthes petiolaris, Morus rubra, Muhlenbergia frondosa, M. schreberi, Myosotis scorpioides, Oclemena acuminata, Osmunda regalis, Osmundastrum cinnamomeum, Oxalis montana, Oxypolis rigidior, Persicaria pensylvanica, Phalaris arundinacea, Pilea pumila, Platanus occidentalis, Populus deltoides, Quercus bicolor, Ranunculus abortivus, Rhynchospora, Rubus occidentalis, Salix nigra, Sanicula odorata, Saururus cernuus, Scirpus cyperinus, Scutellaria lateriflora, Sicyos angulatus, Smilax hispida, Solidago canadensis, S. glomerata, Spermacoce glabra, Sphagnum, Symphyotrichum cordifolium, S. lanceolatum, Symplocarpus foetidus, Toxicodendron radicans, Trillium erectum, T. grandiflorum, Ulmus americana, Urtica dioica, Verbena urticifolia, Verbesina alternifolia, Viburnum dentatum, Viola striata, Vitis riparia.*

Use by wildlife: The foliage of *Chelone glabra* commonly is eaten by moose (Mammalia: Cervidae: *Alces alces*), the Ouachita map turtle (Reptilia: Emydidae: *Graptemys ouachitensis*), and white-tailed deer (Mammalia: Cervidae: *Odocoileus virginianus*). It is a larval host plant for several insects (Insecta: Lepidoptera Geometridae: *Eupithecia satyrata*; Noctuidae: *Papaipema nepheleptena*; Nymphalidae: *Euphydryas phaeton*; Torticidae: *Endothenia hebesana*). Larvae of *E. phaeton* often occur on the plants in nests made from drawn together leaves, comprising broods that can exceed forty individuals. The flowers are eaten by larvae of the satyr pug moth (Insecta: Lepidoptera: Geometridae: *Eupithecia satyrata fumata*). As the larvae and prepupae of sawflies (Insecta: Hymenoptera: Tenthredinidae: *Tenthredo grandis*) feed on the foliage of *C. glabra*, they ingest several iridoid glycosides: catalpol (a chemical deterrent to predators) is sequestered in the body of the insect with an aucubin eliminated in the frass. Another larval sawfly (Hymenoptera: Tenthredinidae: *Macrophya nigra*) also feeds on *C. glabra*. Larvae of the checkerspot butterfly *Euphydryas phaeton* (Lepidoptera: Nymphalidae) similarly acquire unpalatability from feeding on the foliage of this plant. The flowers provide nectar for several bumblebee (Insecta: Hymenoptera: Apidae: *Bombus*) pollinators, and occasionally to ruby-throated hummingbirds (Aves: Trochilidae: *Archilochus colubris*). Smaller insects like bee wolfs (Insecta: Hymenoptera: Sphecidae: *Philanthus solivagus*) similarly seek nectar from the corolla lips, but are unable to enter the flowers. *Chelone glabra* is host to leaf-spot Fungi (Ascomycota: Mycosphaerellaceae: *Septoria wilsonii*), powdery mildews (Ascomycota: Erysiphaceae: *Erysiphe cichoracearum, Neoerysiphe galeopsidis*), and rust Fungi (Basidiomycota: Pucciniaceae: *Puccinia andropogonis*). *Chelone obliqua* can be an early spring food of beavers

(Mammalia: Rodentia: Castoridae: *Castor canadensis*). The seeds of *C. obliqua* are eaten by larval bell moths (Insecta: Lepidoptera: Tortricidae: *Endothenia hebesana*) and leafminer flies (Insecta: Diptera: Agromyzidae: *Phytomyza cheloniae*), which can reduce the seed production of adult plants by more than 20%. The plants also are known to contain catalpol and reportedly are resistant to the northern root-knot nematode (Nematoda: Heteroderidae: *Meloidogyne hapla*).

Economic importance: food: The Cherokee boiled or fried the young shoots of *Chelone glabra* to use as a food; **medicinal:** The Cherokee used *Chelone glabra* to reduce fever and as a dietary aid, laxative, treatment for skin disorders, and vermifuge. The Iroquois employed the plants as a liver medicine. The Malecite and Micmac tribes used the plants as a contraceptive. This species also has been used as a treatment for herpes, indigestion, and jaundice; **cultivation:** All *Chelone* species are cultivated as ornamentals for wet areas. Cultivars of *C. glabra* include 'Black Ace,' 'Fall White,' and 'Montana'; those of *C. obliqua* include 'Forncett Foremost,' 'Forncett Poppet,' 'Ieniemienie,' 'Pink Sensation,' 'Pink Temptation,' and 'Praecox Nana.' Plants distributed as *C. obliqua* 'Alba' actually are *C. glabra*; **misc. products:** *Chelone glabra* is recommended as a plant for use in streamside restoration; **weeds:** none; **nonindigenous species:** *Chelone obliqua* occasionally escapes from cultivation from beyond its native range.

Systematics: Authors have disagreed on the circumscription of tribe Cheloneae with respect to the placement of *Collinsia* and *Tonella* (alternatively placed in tribe Collinseae) as well as *Tetranaema*, which traditionally has been included in the tribe. However, combined DNA sequence data support the monophyly of tribe Cheloneae only if *Collinsia* and *Tonella* are included and *Tetranema* is excluded (Figure 5.84). *Chelone* resolves consistently as the sister genus to *Nothochelone* and *Chionophila*, respectively; this clade is the sister group to *Penstemon* (Figure 5.84). The phylogenetic relationships among the North American *Cheloneae* are comparable with those estimated by RFLP analysis of PCR-amplified regions of cpDNA and other analyses of DNA sequence data (e.g., Figure 5.82). *Chelone* itself clearly is monophyletic as indicated by analyses of DNA and allozyme data. Phylogenetic analyses indicate that the genus (and OBL habit) is relatively derived in the tribe. The leaves of *Chelone glabra* tend to become narrower in the northwestern portion of its range and up to seven varieties have been named; however, none is supported by analyses of morphological or allozyme data. The base chromosome number of *Chelone* is $x = 14$. *Chelone cuthbertii* and *C. glabra* are diploids ($2n = 28$), with a tetraploid ($2n = 56$) race also known for the latter. The diploids are distinct morphologically. *Chelone obliqua* is a polyploid with tetraploid ($2n = 56$) and hexaploid ($2n = 84$) cytotypes. Currently, it is believed that hybrids and allopolyploids have arisen independently on several occasions in the genus. Some hexaploids and the tetraploids resemble *C. glabra* morphologically; whereas, other hexaploids resemble *C. lyonii*. Because of the recent divergence of taxa, allozymes exhibit insufficient variability to reconcile species relationships with certainty or

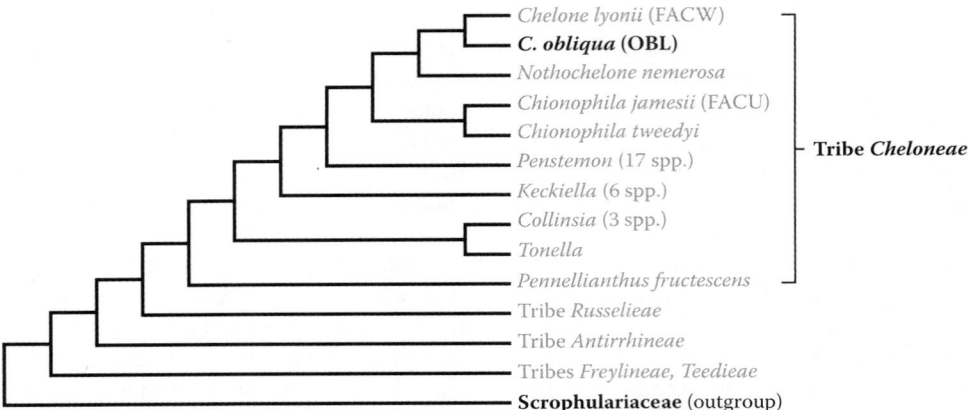

FIGURE 5.84   Relationships among genera of Plantaginaceae tribe Cheloneae as indicated by analysis of combined ITS and *matK* sequence data. The position of *Tonella* (not surveyed in the study) was inferred by its placement in a similar analysis of cpDNA RFLP data (Wolfe et al., 1997). Together, these results support the transfer of *Collinsia* (and *Tonella*) to tribe Cheloneae (Wolfe et al., 2002). Phylogenetic analyses consistently indicate a derived position of *Chelone* in *Cheloneae*, and an independent origin of the OBL habit in the genus (OBL North American taxa are indicated in bold). (Adapted from Wolfe, A.D. et al., *Syst. Bot.*, 27, 138–148, 2002.)

to test effectively for hybridization, which likely has occurred between *C. glabra* and *C. lyonii*, and perhaps, among other species. Allozyme analysis also indicates that gene flow occurs at relatively low levels in the genus.

Comments: *Chelone glabra* is distributed throughout eastern North America, *C. obliqua* occurs in the central and southeastern United States, and *C. cuthbertii* is restricted to isolated occurrences across the southeastern United States.

References: Bergeron et al., 1996; Bowers, 1980; Bowers et al., 1993; Cooperrider & McCready, 1970; Crosswhite, 1965; Crum et al., 1972; Emerton, 1888; Fischer, 2004; Franzyk et al., 2004; Heinrich, 1975a,b, 1976, 1979; Hunt, 1927; James, 1948; Keller, 2000; Lamondia, 1995; Lovell, 1898, 1899; Martin, 1887; Middleton, 2002a; Moll, 1976; Murie, 1934; Nelson et al., 1998; Nelson & Elisens, 1999; Pennell, 1935; Reed, 1913; Reese & Lubinski, 1983; Rindge, 1952; Roberts & Arner, 1984; Schmidt, Jr., 1957; Stamp, 1984, 1987; Weakley & Schafale, 1994; Weishampel & Bedford, 2006; Williams et al., 1999a, 2000; Wiser et al., 1996; Wolfe et al., 1997, 2002, 2006a; Wood et al., 1940; Zartman & Pittillo, 1998.

4. *Dopatrium*

Abunome, horsefly's eye

Etymology: derived possibly from *dopatta*, the Hindi name for a silk or muslin shawl

Synonyms: *Gratiola* (in part); *Ilygethos*; *Kyrtandra* (in part); *Lindernia* (in part); *Vandellia*

Distribution: global: Africa; Asia; Australia; **North America:** southern

Diversity: global: 12 species; **North America:** 1 species

Indicators (USA): OBL: *Dopatrium junceum*

Habitat: freshwater; palustrine; **pH:** alkaline (pH: 7.2–8.4); **depth:** <1 m; **life-form(s):** emergent herb, submersed (vittate)

Key morphology: stems (to 50 cm) erect, succulent, striate, profusely branched from base; leaves opposite, sessile, scale-like, or (to 2 cm) subspatulate to lanceolate, decreasing

in size apically, base clasping, margin entire; flowers pedicelled (to 1 cm), solitary, axillary; calyx (to 2 mm) campanulate, 5-lobed; corolla (to 7 mm) bilabiate, white, pale purple or rose, upper lip 2-lobed, lower lip 3-lobed and longer, with a white blotch at the base of the middle lobe; stamens (fertile) 2, included; staminodes 2, clavate, yellow; capsule (to 3.5 mm) ellipsoid-globose, loculicidal; seeds (to 0.5 mm) numerous, reticulately ribbed

Life history: duration: annual (fruit/seeds); **asexual reproduction:** none; **pollination:** chasmogamous (self), cleistogamous (self); **sexual condition:** hermaphroditic; **fruit:** capsules (common); **local dispersal:** seeds (water, wind); **long-distance dispersal:** seeds (mud adhering to waterfowl)

Imperilment: (1) *Dopatrium junceum* [GNR]

Ecology: general: *Dopatrium* species are annuals, which are adapted to grow in shallow ephemeral marshes or in seasonal rock (granite) or hardpan pools that fill during the rainy period and subsequently dry out. The seeds remain dormant in the bottom mud during the dry season and germinate as the rains return. The small seeds are dispersed passively by the wind or water. In the native range of the genus, it is not unusual for associates to be absent, but can include genera such as *Isoetes*, *Marsilea* and *Rotala*.

Dopatrium junceum (Roxb.) **Buch.-Ham. ex Benth.** grows in ditches, marshes, and rice fields at elevations of up to 100 m (but up to 2300 m in some portions of its native range). Although the seedlings are submersed, the adult plants can grow either submersed in shallow (to 30 cm) standing or slow-moving water, or emergent on exposed, desiccating substrates. The substrates characteristically are alkaline (pH: 7.2–8.4) muds. The plants probably are well-adapted to warm waters because some congeners reportedly can tolerate diurnal temperatures that fluctuate from 21°C to 46°C. The normally chasmogamous are flowers are induced as the waters recede; however, the lower flowers are cleistogamous when submersed. Pollination of the chasmogamous flowers is likely to occur by insects in the native range; however, introduced populations probably are primarily self-pollinating. Some fruits can develop while underwater. The seeds remain dormant unless they become inundated, and their germination occurs under water. Seedling emergence is strongly reduced by shading. The plants are not mycorrhizal. When growing in rice culture, *Dopatrium* can become highly infected by *Ralstonia solanacearum* (Bacteria: Burkholderiaceae), a gram negative pseudomonad bacterium that is the cause of southern wilt disease. The growth of *Dopatrium* plants in ricefields can be controlled effectively by the application of buckwheat (*Fagopyrum*) pellets at the rate of two tons per hectare.

Reported associates (North America): *Ammannia auriculata, Bacopa repens, Chara, Crassula aquatica, Eriocaulon cinereum, Heteranthera limosa, Lythrum hyssopifolia, Rotala ramosior, Sagittaria guayanensis, Sphenoclea zeylanica.*

Use by wildlife: The plants are host to the brown ring patch fungus (Basidiomycota: Ceratobasidiaceae: *Waitea circinata*), which causes sheath spot disease of rice (*Oryza*).

Economic importance: food: not eaten; **medicinal:** *Dopatrium junceum* is used in traditional folk medicine in India; **cultivation:** not cultivated; **misc. products:** none; **weeds:** *Dopatrium junceum* is a ricefield weed in California, Louisiana, Argentina, and throughout many parts of Asia; **nonindigenous species:** *Dopatrium junceum* was introduced to North America (in Louisiana by 1969) and to the Hawaiian Islands in association with the development of rice culture.

Systematics: Traditionally, *Dopatrium* has been placed within tribe Gratioleae of Plantaginaceae together with several other predominantly aquatic genera (*Deinostema, Hydrotriche, Limnophila*) in subtribe Dopatriinae. Phylogenetic analysis of combined DNA sequence data substantiates this disposition and resolves this group as a clade, with the exception of *Philcoxia*, which is not aquatic and is more distantly related (Figure 5.85). Furthermore, the tribes Gratioleae and Stemodieae are polyphyletic as currently delimited and their circumscriptions require a reevaluation. These analyses identify the sister group of *Dopatrium* as *Hydrotriche*, an aquatic Malagasy genus (not in North America), which also frequently occurs in rice fields. Both genera are related to *Limnophila*, which also is nonindigenous to North America. A phylogenetic survey of *Dopatrium* species has not yet been undertaken. The base chromosome number of *Dopatrium* is $x = 7$. *Dopatrium junceum* ($2n = 14$) is diploid.

Comments: At this time *Dopatrium junceum* is known in North America only from ricefields in California, Hawaii and Louisiana.

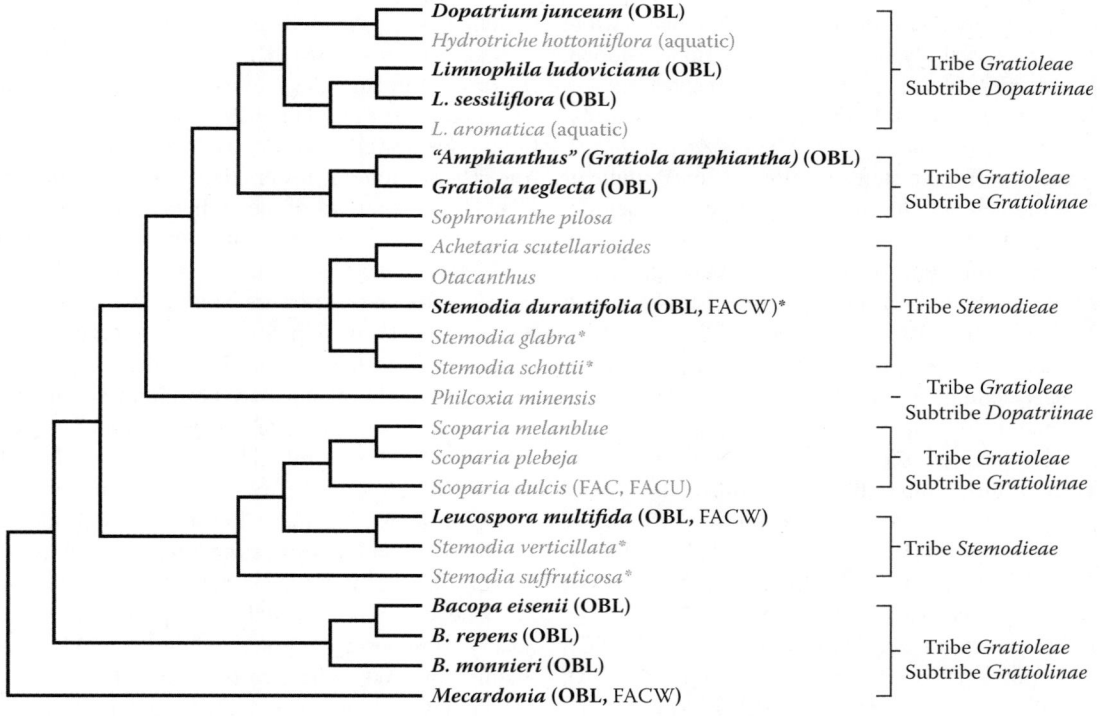

FIGURE 5.85 Phylogenetic relationships of Plantaginaceae tribes Gratioleae and Stemodieae (based on analyses of combined DNA sequence data) indicate that these tribes are polyphyletic as currently circumscribed. The aquatic Gratioleae subtribe Dopatriinae resolves as a clade in exclusion of the anomalous genus *Philcoxia*. The genus *Stemodia* (asterisked) is clearly polyphyletic and requires further study. Tribe Gratioleae contains the largest concentration of aquatic genera in Plantaginaceae, with multiple origins of the OBL habit (shown in bold) evident. Other analyses (Figure 5.86) confirm that *Amphianthus* is nested within *Gratiola* and should be transferred to that genus (see Estes & Small, 2008). (Adapted from Fischer, E., *The Families and Genera of Vascular Plants, Vol. VII. Flowering Plants, Dicotyledons: Lamiales (except Acanthaceae including Avicenniaceae)*. Springer-Verlag, Berlin, Germany, 2004; Fritsch, P.W. et al., *Proc. Calif. Acad. Sci.*, 58, 447–467, 2007.)

References: Chandran & Bhavanandan, 1981; Cook, 1996a,b; Cook et al., 1974; Fischer, 1997, 2004; Fritsch et al., 2007; Hiroomi, 2001; Khanh et al., 2005; Pradhanang & Momol, 2001; Ragupathy & Mahadevan, 1993; Razi, 1950; Soerjani et al., 1987; Thieret, 1970.

5. *Gratiola*

Golden-pert, hedgehyssop; gratiole

Etymology: from the Latin *gratia* meaning "grace" with reference to its medicinal utility

Synonyms: *Amphianthus*; *Fonkia*; *Nibora*; *Sophronanthe*; *Tragiola*

Distribution: global: Asia; Europe; North America; **North America:** widespread

Diversity: global: 20 species; **North America:** 12 species

Indicators (USA): OBL: *Gratiola amphiantha*, *G. aurea*, *G. ebracteata*, *G. floridana*, *G. heterosepala*, *G. neglecta*, *G. virginiana*, *G. viscidula*

Habitat: brackish, freshwater; lacustrine, palustrine, riverine; **pH:** 4.2–8.9; **depth:** 0–4 m; **life-form(s):** emergent herb, floating leaved, submersed (vittate)

Key morphology: stems (to 6.5 dm) erect to decumbent (dwarfed when submersed), obscurely 4-angled, often fleshy or succulent; leaves (to 7.0 cm) opposite, sessile, oblong to lanceolate, sometimes clasping at base, dentate, or glandular; flowers axillary, solitary, pedicelled (to 2.5 cm), often with paired bractlets subtending calyx; corolla (to 12 mm) bilabiate, white or yellow, the tube obscurely 4-angled, upper lip 1–2-lobed, lower lip 3-lobed; stamens 2; capsules (to 5 mm) spherical

The distinctive *Gratiola amphiantha*, which has been transferred only recently from the segregate genus *Amphianthus*, merits a separate description: stems small (to 15 cm), single or branched from the caudex, arising from a rosette of sessile, lanceolate leaves (to 8 mm) and terminating in a pair of ovate, floating leaves (to 8 mm), their petioles thin; foliage glandular punctate; flowers stalked (to 2 mm), either cleistogamous in submersed leaf axils or chasmogamous and solitary in floating leaf axils; corolla (to 4 mm) white to lightly purplish, slightly irregularly campanulate, the short lobes spreading; stamens 2, included; capsules (to 3 mm) lobed apically and basally, somewhat flattened; seeds (to 1 mm) brown, reticulate

Life history: duration: annual, biennial (fruit/seeds); perennial (rhizomes, stolons); **asexual reproduction:** rhizomes; **pollination:** insect, self or cleistogamous; **sexual condition:** hermaphroditic; **fruit:** capsules (common); **local dispersal:** seeds; **long-distance dispersal:** seeds

Imperilment: (1) *Gratiola amphiantha* [G2]; S1 (AL, SC); S2 (GA); (2) *G. aurea* [G5]; SH (NF); S1 (MI, NB, NC, PA); S2 (QC); S3 (DE); (3) *G. ebracteata* [G4]; S1 (MT); (4) *G. floridana* [G4]; S1 (TN); S3 (GA); (5) *G. heterosepala* [G3]; S1 (OR); S3 (CA); (6) *G. neglecta* [G5]; S1 (NE, NS, UT); S2 (AB, KS, WY); S3 (AZ, BC, NB, NC); (7) *G. virginiana* [G5]; S1 (KS, MI, RI); S2 (NJ, OH); (8) *G. viscidula* [G4]; SH (DC, DE); S1 (MD, MO); S2 (NC); S3 (KY, OH)

Ecology: general: All of the North American *Gratiola* species have some designation as wetland indicators, with just over half being categorized as OBL aquatics. The species

vary in their life histories (annual, biennial, or perennial) but they tend to occupy similar habitats, especially sites with wet, exposed substrates. The species are self-compatible and either self-pollinating or are pollinated by bees (Insecta: Hymenoptera). Self-pollination seems to be the prevalent mode of reproduction in the genus, with a specific pollinator [short-tongued bees (Insecta: Hymenoptera)] recorded only for *Gratiola virginiana*. The seeds are small (<0.8 mm) and probably are dispersed by water, wind, or by transport in mud attached to the feet of waterfowl (Aves: Anatidae).

Gratiola amphiantha **D. Estes & R. L. Small** is a winter annual restricted to shallow (0.3 m) depressions on granitic Piedmont outcrops, where rainwater collects. The sites are acidic. Typical susbstrates develop within the depressions as a thin (0.4–5 cm) layer of sandy-silty soil, which typically is low in organic matter (e.g., 4%–8%), nitrogen (0.6%), phosphorous (35 ppm), potassium (38 ppm), and other nutrients (150 ppm). The plants can grow where the water depth is between 0.1–10 cm, but occur most often at depths less than 3 cm. The entire life cycle can complete in only three to 4 weeks, which is an adaptation to the rapid desiccation that characterizes the shallow depressions. However, the plants also can live for several months but eventually die as a consequence of the spring droughts that occur from March to May. The seeds remain dormant throughout the warm summer months and germinate in late fall or early winter (December to January). A submerged rosette develops initially and eventually produces one or more stems, each with two floating leaves that support a flower. Flowering occurs from mid-February to April and fruiting from April to May. The plants also produce submerged cleistogamous flowers (axillary to the rosette leaves) in addition to the chasmogamous flowers that arise in the floating leaf axils. The close proximity of the anthers to the stigma in all flowers promotes self-pollination, and allozyme surveys have detected little genetic variation either within or between populations. The cleistogamous flowers can become chasmogamous if the waters recede enough for them to emerge, but no pollinating organisms have been observed. There are unconfirmed reports that the plants may be apomictic (agamospermous), but these have not been substantiated. The plants average just over one (1.1) fruit and 22 seeds per individual, although up to 100 seeds per plant have been observed. Intraspecific competition is high and the average fruit production per plant declines as plant densities increase. The seeds are unspecialized for dispersal, which occurs from May to June. Dispersal is believed to be facilitated by outwash during periods of heavy rains, or by animals, when the seeds or seed-bearing plant fragments adhere to their feet. The seeds are dormant when they first are shed and they remain dormant throughout the hot summer months. There is a seed bank formed with some seeds remaining dormant for at least several years. Light and standing water are the only known requirements for germination, which can occur within 3–4 days when the seeds become exposed to cooler weather and heavy rains. The plants are among the first colonizers of open outcrop depressions. They are adapted to the open outcrop sites by having a high light compensation point and a photosynthetic apparatus that saturates at very high light intensities. The plants are poor competitors and are acutely susceptible to drought. **Reported associates:** *Agrostis, Andropogon virginicus, Bulbostylis capillaris, Callitriche heterophylla, Cladonia, Croton willdenowii, Cyperus granitophilus, Diamorpha smallii, Dichanthelium, Erythronium americanum, Eupatorium capillifolium, Hypericum gentianoides, H. lloydii, Isoetes melanospora, I. piedmontana, I. tegetiformans, Juncus georgianus, Krigia virginica, Lindernia monticola, Minuartia glabra, M. godfreyi, M. groenlandica, Nothoscordum bivalve, Phacelia maculata, Polytrichum commune, P. juniperinum, Rumex acetosella, R. hastatulus, Schizachyrium scoparium, Sedum pusillum, Selaginella tortipila, Senecio tomentosus, Viola pedata, Yucca filamentosa.*

Gratiola aurea **Pursh** is a perennial, which inhabits beaches, bogs, depressions, ditches, lakes, ponds, shores, and the margins of rivers and streams. The waters span a broad range of acidity (pH: 5.5–8.9) but typically are oligotrophic with low conductivity (<110 μmhos/cm), low alkalinity (<50 mg/L CaCO$_3$) and low Mg (<10 mg/L) levels. The substrates include gravel, muck, mucky sand, mud, peat, rock, sand, and sandy loam. Although this species has been reported from brackish coastal pond habitats, its tolerance to salinity ordinarily appears to be quite low. The plants exist as two relatively discrete life-forms, growing either as an emergent on wet substrates (these plants are fertile with yellow-petalled flowers and broader leaves) or completely submersed (these plants are dwarfed, sterile, and possess narrow, awl-like leaves). Although sometimes described as a rosulate plant in the literature, the life-form is distinctly vittate, even when growing under submersed conditions. The submersed form is produced when the water levels fall temporarily to expose substrates that are suitable for seed germination. As the waters rise again, the developing seedlings transform into the dwarfed form and remain in a vegetative state. Although other *Gratiola* species typically are found only in fairly shallow waters (<0.5 m), the submersed form of this species has been reported at depths of up to 4 m. The reproductive biology has not been studied, but the flowers (of emergent plants) probably are either self-pollinated or are pollinated by small bees (Hymenoptera). The flowers produce many seeds, which can reach densities of more than 60 seeds/m^2 in the seed bank. Good seed germination (68%) has been reported for seeds that are placed under a 20°C/30°C temperature regime following 9 months of cold (4°C) stratification. *Gratiola aurea* is similar physiologically to the rosulate flora of oligotrophic lakes in acquiring from 60% to 100% of its total CO$_2$ through root uptake. The plants are categorized as stress tolerant but are poor competitors with other wetland vegetation. They often grow on ground where trampling has occurred and are frequent elements of early beach succession and lakeshore succession on leached sand and silt substrates. Vegetative reproduction occurs by means of a fleshy rhizome and stolons. **Reported associates:** *Agalinis purpurea, Agrostis scabra, Ambrosia artemisiifolia, Asclepias incarnata, Bartonia paniculata, B. virginica, Bidens connatus, B. discoidea, B. frondosus, Brasenia schreberi, Calamagrostis*

canadensis, Calluna vulgaris, Carex rostrata, C. scoparia, C. walteriana, Cephalanthus occidentalis, Ceratophyllum demersum, Chamaedaphne calyculata, Cladium mariscoides, Clethra alnifolia, Conyza canadensis, Coreopsis rosea, Crassula aquatica, Cuscuta coryli, Cyperus bipartitus, C. dentatus, C. diandrus, C. erythrorhizos, C. polystachyos, C. strigosus, Drosera filiformis, D. intermedia, D. rotundifolia, Dulichium arundinaceum, Echinodorus tenellus, Eclipta prostrata, Elatine americana, E. minima, Eleocharis acicularis, E. intermedia, E. melanocarpa, E. microcarpa, E. obtusa, E. olivacea, E. ovata, E. palustris, E. quadrangulata, E. robbinsii, E. tenuis, E. tricostata, E. tuberculosa, Elodea canadensis, E. nuttallii, Epilobium coloratum, Eragrostis hypnoides, Erechtites hieracifolia, Eriocaulon aquaticum, Eupatorium dubium, E. hyssopifolium, E. leucolepis, E. perfoliatum, Fimbristylis autumnalis, F. perpusilla, Fuirena pumila, F. squarrosa, Galium palustre, G. tinctorium, Glossostigma cleistanthum, Glyceria acutiflora, G. borealis, G. canadensis, Gnaphalium uliginosum, Gratiola virginiana, Helenium autumnale, H. flexuosum, Helianthus angustifolius, Heteranthera dubia, Hibiscus moscheutos, Hydrocotyle umbellata, Hypericum adpressum, H. boreale, H. canadense, H. majus, H. mutilum, H. perforatum, Iris virginica, Isoetes lacustris, I. tenella, I. tuckermanii, Juncus articulatus, J. brachycephalus, J. canadensis, J. debilis, J. militaris, J. pelocarpus, J. repens, Kalmia angustifolia, Lachnanthes caroliana, Lactuca canadensis, Lindernia dubia, Lipocarpha micrantha, Littorella americana, Lobelia dortmanna, Ludwigia alternifolia, L. palustris, L. sphaerocarpa, Lycopus amplectens, L. uniflorus, L. virginicus, Lysimachia terrestris, Lythrum salicaria, Melampyrum lineare, Micranthemum umbrosum, Mikania scandens, Mimulus ringens, Muhlenbergia uniflora, Myriophyllum humile, M. spicatum, M. tenellum, Najas flexilis, N. gracillima, Nitella, Nymphaea odorata, Nymphoides cordata, Oldenlandia uniflora, Panicum dichotomiflorum, P. longifolium, P. philadelphicum, P. rigidulum, P. spretum, P. verrucosum, P. virgatum, P. wrightianum, Paspalum laeve, Penthorum sedoides, Persicaria amphibia, P. hydropiper, P. hydropiperoides, P. maculosa, P. pensylvanica, P. puritanorum, Polygala cruciata, Pontederia cordata, Potamogeton bicupulatus, P. epihydrus, P. pusillus, Proserpinaca palustris, P. pectinata, Ranunculus flammula, R. reptans, Rhexia virginica, Rhododendron viscosum, Rhynchospora capillacea, R. capitellata, R. fusca, R. inundata, R. macrostachya, R. nitens, R. scirpoides, Rotala ramosior, Rubus hispidus, Sabatia kennedyana, Sacciolepis striata, Sagittaria graminea, S. latifolia, S. teres, Schoenoplectus pungens, S. purshianus, S. smithii, Scirpus cyperinus, Scleria reticularis, Sclerolepis uniflora, Scutellaria galericulata, S. lateriflora, Sium suave, Solidago caroliniana, S. graminifolia, S. tenuifolia, Sparganium americanum, S. angustifolium, S. chlorocarpum, Spartina pectinata, Sphagnum cuspidatum, Spiraea tomentosa, Stachys hyssopifolia, Subularia aquatica, Symphyotrichum racemosum, Triadenum virginicum, Utricularia cornuta, U. fibrosa, U. gibba, U. juncea, U. macrorhiza, U. purpurea, U. resupinata, U. striata, Vaccinium macrocarpon, Vallisneria americana, Verbena hastata, Viola lanceolata, Xyris difformis, X. smalliana, X. torta.

***Gratiola ebracteata* Benth. ex A. DC.** is an annual species, which is reported from beaches, bogs, channels, depressions, ditches, drying ponds, flats, floodplains, gravel pits, meadows, mudflats, roadsides, scour pools, seeps, sloughs, streams, swales, vernal pools, washes, and the shorelines of lakes, ponds, and rivers at elevations of up to 2080 m. It is found in tidal or non-tidal sites on substrates that include clay, granite, gravelly loam, hardpan, mud, sandstone, sandy silt, and volcanic rock. The plants most often grow in small but moderately deep pools (to about 30 cm depth); however, they are not found in very deep water. Other life-history information is lacking for this species. **Reported associates:** *Achlys triphylla, Aira caryophyllea, Alisma, Allium amplectens, Alopecurus aequalis, A. saccatus, Anagallis minima, Beckmannia syzigachne, Berberis nervosa, Blennosperma nanum, Briza minor, Bromus mollis, Callitriche marginata, Carex athrostachya, C. lenticularis, C. microptera, C. vesicaria, Castilleja campestris, Ceanothus cuneatus, Chara, Cicendia quadrangularis, Crassula aquatica, Damasonium californicum, Deschampsia cespitosa, D. danthonioides, Downingia bella, D. bicornuta, D. concolor, D. cuspidata, D. elegans, D. ornatissima, D. pusilla, Eleocharis acicularis, E. bella, E. macrostachya, E. obtusa, Elodea canadensis, Epilobium pygmaeum, Equisetum fluviatile, E. variegatum, Erodium botrys, Eryngium alismifolium, E. aristulatum, E. castrense, E. vaseyi, Fontinalis, Galium trifidum, Glyceria declinata, Gnaphalium, Gratiola heterosepala, G. neglecta, Hordeum depressum, Hosackia pinnata, Hypochaeris glabra, Isoetes howellii, I. nuttallii, I. orcuttii, Juncus acuminatus, J. bufonius, J. canadensis, J. leiospermus, J. nevadensis, J. supiniformis, J. tenuis, J. uncialis, J. xiphioides, Lasthenia burkei, L. fremontii, L. glaberrima, Legenere limosa, Lilaea scilloides, Limnanthes douglasii, Limosella aquatica, Lolium multiflorum, Lythrum hyssopifolia, L. portula, Mentha pulegium, Mimulus guttatus, M. tricolor, Montia dichotoma, M. fontana, Myosurus minimus, Myriophyllum hippuroides, M. ussuriense, Navarretia intertexta, N. leucocephala, N. minima, N. myersii, N. prostrata, Orcuttia pilosa, O. tenuis, O. viscida, Persicaria lapathifolia, Pilularia americana, Plagiobothrys austiniae, P. bracteatus, P. chorisianus, P. leptocladus, P. scouleri, P. stipitatus, Pleuropogon californicus, Poa annua, P. palustris, Pogogyne douglasii, P. ziziphoroides, Polygonum bidwelliae, P. douglasii, P. polygaloides, Polypogon monspeliensis, Populus, Porterella carnosula, Potentilla flabellifolia, P. glandulosa, Psilocarphus brevissimus, Pseudotsuga menziesii, Ranunculus aquatilis, R. bonariensis, R. flammula, R. lobbii, R. reptans, Ribes viscosissimum, Rorippa islandica, R. sinuata, Rumex salicifolius, Salix scouleriana, Scirpus microcarpus, Sericocarpus rigidus, Sidalcea hartwegii, S. hirsuta, Sparganium natans, Subularia aquatica, Trifolium depauperatum, T. hirtum, Triphysaria eriantha, Velezia rigida, Veronica peregrina.*

***Gratiola floridana* Nutt.** is an annual, which grows in ditches, floodplains, marshes, river bottoms, seeps, spring runs, swamps, and along the margin of brooks, ponds, and streams. The plants occur often in shallow waters in shady coastal plain habitats. The substrates have been characterized as circumneutral loam, mud, or Rutlege (typic humaquepts). The ecology and life history of this relatively rare plant deserves detailed study. **Reported associates:** *Clematis socialis, Phlox glaberrima, Ranunculus.*

***Gratiola heterosepala* Mason & Bacig.** is an annual, which occurs in borrow pits, marshes, swales, vernal pools, and along the margins of lakes, pools, and reservoirs at elevations of up to 1650 m. The sites tend to be acidic (e.g., pH: 5.0) but the complete range of acidity has not been determined. The principal substrates are clay (adobe), but also can include basalt, gravel, gravelly clay, gravelly loam, loam, loamy sand, mud, rocky gravel, and volcanic mud. The plants occur more commonly in pools that are relatively large ($3000–283,000 \text{ m}^2$) and deep (depth of 5–70 cm; average 24 cm). They produce 1–2 flowers when the vernal pools are inundated with <5 cm of water (mainly April to June). The flowers are self-compatible and the species predominantly is self-pollinating. Natural pollinators have not been observed in field surveys. The plants complete a rapid life cycle during the period when the vernal pools have begun to dry. Each fruit produces an average of 150 small seeds, which mature within 1–2 weeks of anthesis. The plants die soon afterwards. A large seed bank develops in which the seeds can remain dormant for at least 3 years. Seed germination occurs when the pools have become inundated following the onset of fall or winter rains. Seedling growth begins underwater. Dispersal is believed to be mainly by animals (livestock and waterfowl), probably by seeds that adhere to them in mud. Populations can exhibit considerable annual fluctuations in size depending on the amount of rainfall, but can reach upward of 1 million individuals under optimal conditions. In some localities, the plants seem to favor compacted substrates such as cattle hoofprints. Typically, site cover (herbaceous) varies from 5% to 85%, with an average cover of 57%. **Reported associates:** *Astragalus tener, Castilleja campestris, Chamaesyce hooveri, Distichlis spicata, Downingia bicornuta, D. cuspidata, D. pusilla, Eleocharis macrostachya, E. palustris, Epilobium pygmaeum, E. torreyi, Eryngium castrense, E. mathiasiae, E. vaseyi, Gratiola ebracteata, Isoetes howellii, I. nuttallii, I. orcuttii, Lasthenia ferrisiae, L. fremontii, L. glaberrima, Legenere limosa, Lepidium latipes, Limnanthes douglasii, Marsilea vestita, Montia fontana, Myosurus minimus, Navarretia leucocephala, N. myersii, N. prostrata, Neostapfia colusana, Orcuttia inaequalis, O. pilosa, O. tenuis, O. viscida, Plagiobothrys chorisianus, P. stipitatus, Pleuropogon californicus, Psilocarphus brevissimus, Sidalcea calycosa, Taeniatherum caput-medusae, Tuctoria greenei, T. mucronata.*

***Gratiola neglecta* Torr.** is an annual, which inhabits bogs, canals, depressions, ditches, floodplains, gravel pits, marshes, meadows, mudflats, pools, prairies, sand bars, savannas, seeps, sloughs, springs, swales, swamps, and the margins of lakes, ponds, rivers, and streams at elevations of up to 1950 m. The plants occur under tidal or nontidal conditions, but grow only as emergents or in shallow water. When the plants grow in tidal conditions they are less pubescent and more highly branched. The site exposures range from open to partially shaded conditions. The substrates are acidic to slightly alkaline (pH: 4.9–7.6) and can include clay, clay loam, gravel, gravelly mud, muck, mud, sand, silt, and silt loam. The tolerance to salinity has not been determined but the plants have been reported from salt marshes. Flowering occurs early in the season and does not persist long once seed has been set. The reproductive biology of this species has not been studied in detail; however, it is likely that the plants are self-pollinating. The seeds are dormant and can develop into large seed banks yielding as many as 175 seeds/m². The seed banks are known to persist even after the land has been farmed. The species can be well-represented in the seed bank even when it has been absent from the standing vegetation for intervals in excess of 6 years. Seed dormancy can be broken using a 30°C/15°C temperature germination regime, following a 3–4 month period of warm stratification (dry storage). The seeds also germinate when cold-stratified at 5°C for 30 days and then incubated at temperatures of 18°C–35°C. Seed germination occurs under either flooded or non-flooded conditions. This species can become weedy in agricultural fields and often is prominent in cropland after standing waters have been drawn down. **Reported associates:** *Abutilon theophrasti, Acalypha rhomboidea, Acer negundo, A. rubrum, A. saccharinum, Agrostis gigantea, Alisma subcordatum, A. triviale, Alnus, Alopecurus aequalis, A. carolinianus, Amaranthus cannabinus, A. hybridus, A. retroflexus, Ambrosia artemisiifolia, A. trifida, Amorpha fruticosa, Amphianthus pusillus, Amsonia tabernaemontana, Aristida purpurea, Arundinaria tecta, Atriplex, Beckmannia syzigachne, Bidens cernuus, B. laevis, B. tripartitus, Boehmeria cylindrica, Bolboschoenus maritimus, B. robustus, Boltonia decurrens, Bothriochloa laguroides, Bouteloua gracilis, Bromus hordeaceus, Bryum cyclophyllum, Calamagrostis canadensis, Callitriche heterophylla, C. palustris, C. peploides, C. terrestris, Caltha natans, Cardamine pensylvanica, Carex crebriflora, C. crus-corvi, C. frankii, C. gravida, C. lanuginosa, C. lurida, C. scoparia, C. stricta, C. sychnocephala, C. vulpinoidea, Centunculus minimus, Cephalanthus occidentalis, Cerastium glomeratum, Ceratophyllum demersum, Chamerion angustifolium, Commelina virginica, Conyza canadensis, Cornus, Crassula aquatica, Cuscuta, Cynodon dactylon, Cyperus bipartitus, C. diandrus, C. erythrorhizos, C. iria, C. odoratus, C. pseudovegetus, Diamorpha smallii, Dichanthelium acuminatum, D. polyanthes, D. yadkinense, Distichlis spicata, Echinochloa crus-galli, E. muricata, E. walteri, Elephantopus carolinianus, Elatine rubella, E. triandra, Eleocharis acicularis, E. engelmannii, E. ovata, E. palustris, E. parvula, Elodea canadensis, Elymus canadensis, Epilobium ciliatum, E. coloratum, Equisetum arvense, E. fluviatile, Festuca rubra, Fimbristylis autumnalis, F. miliacea, F. vahlii, Fraxinus caroliniana, F. nigra, F. pennsylvanica, Geranium carolinianum, Glyceria declinata, G. striata, Gnaphalium palustre,* *G. purpureum, G. uliginosum, Gratiola ebracteata, G. virginiana, Helianthus annuus, Houstonia purpurea, Hypericum boreale, H. tubulosum, Hypochaeris radicata, Hypoxis curtissii, Impatiens capensis, Isoetes melanopoda, I. melanospora, I. piedmontana, Isolepis carinata, Itea virginica, Juncus acuminatus, J. biflorus, J. bufonius, J. diffusissimus, J. dudleyi, J. effusus, J. gerardii, J. nevadensis, J. oxymeris, J. scirpoides, J. validus, Juniperus monosperma, Justicia ovata, Leersia oryzoides, Lemna minor, Lilaeopsis occidentalis, Limonium carolinianum, Limosella acaulis, L. aquatica, Lindernia dubia, L. monticola, Liquidambar styraciflua, Lobelia cardinalis, L. elongata, Lolium perenne, Ludwigia palustris, Lycopus, Lysimachia maritima, Lythrum tribracteatum, Magnolia virginiana, Marsilea vestita, Mazus pumilus, Mecardonia acuminata, Mentha arvensis, M. piperita, M. spicata, Mimulus ringens, Murdannia keisak, Myrica gale, M. pensylvanica, Myriophyllum spicatum, M. ussuriense, Nuttallanthus canadensis, Nyssa aquatica, N. biflora, N. sylvatica, Onoclea sensibilis, Oryza, Oxalis europaea, Panicum obtusum, P. rigidulum, P. virgatum, Pascopyrum smithii, Pedicularis lanceolata, Peltandra virginica, Persea palustris, Persicaria amphibia, P. arifolia, P. coccinea, P. hydropiper, P. hydropiperoides, P. maculosa, P. pensylvanica, P. punctata, P. sagittata, Phalaris angusta, P. arundinacea, Phyla lanceolata, Physalis pubescens, Pilea pumila, Pinus taeda, Plagiobothrys scouleri, Plantago maritima, P. pusilla, P. rugelii, P. virginica, Pluchea camphorata, Poa palustris, Polygala sanguinea, Polygonum ramosissimum, Populus deltoides, Quercus bicolor, Q. macrocarpa, Q. rubra, Ranunculus aquatilis, R. pusillus, R. reptans, R. sceleratus, Rhynchospora macrostachya, Rorippa islandica, R. palustris, Rumex crispus, Saccharum baldwinii, Sagittaria calycina, S. cuneata, S. latifolia, Salicornia europaea, Salix nigra, Saururus cernuus, Schoenoplectus pungens, S. tabernaemontani, Sisyrinchium rosulatum, Sium suave, Smilax laurifolia, S. walteri, Solanum dulcamara, Solidago canadensis, S. sempervirens, Soliva sessilis, Spartina alterniflora, S. patens, S. pectinata, Spiraea, Sporobolus airoides, Sporobolus cryptandrus, Stachys tenuifolia, Subularia aquatica, Symphyotrichum tenuifolium, Taraxacum officinale, Taxodium distichum, Thalictrum pubescens, Tilia americana, Toxicodendron radicans, Tradescantia, Triadenum tubulosum, Trifolium stoloniferum, Triglochin maritimum, Typha angustifolia, T. ×glauca, T. latifolia, Ulmus americana, Verbena hastata, Veronica americana, V. peregrina, Xanthium spinosum, Zizania aquatica.*

***Gratiola virginiana* L.** is an annual or biennial inhabitant of bogs, depressions, ditches, dunes, floodplains, marshes, meadows, mudflats, ponds, pools, ravines, ricefields, sloughs, springs, streams, swales, swamps, and the margins of lakes, ponds, rivers, and streams. The sites tend to be acidic (pH: 4.2–6.6) but occurrences have also been reported from calcareous localities. Reported substrates include clay, gravel, loam, muck, mud, muddy clay, muddy loam, peat, peaty muck, rock, sand, and sandy loam. The water conditions are tidal or non-tidal and range from fresh to brackish. When submersed in shallow (<30 cm) water, the plants often grow with only their apical foliage emergent; otherwise they are entirely emergent in sites with exposures ranging from full sun to deep shade. Occurrences are reported more frequently from shaded localities. The flowers are pollinated by small-tongued bees (Insecta: Hymenoptera), but are self-compatible and also capable of self-pollination, especially in areas subject to frequent inundation. Those flowers produced late in the season are cleistogamous and self-pollinating. Biennial plants flower early in the season (May); whereas, the small summer seedlings flower in late summer and early autumn. A single plant can produce upward of 46,300 seeds. This species is a common component of the seed bank, especially in wooded (swamp) habitats. The seed bank is known to persist at sites where the species is absent from the standing vegetation, and a considerable seed bank can remain even after a site has been farmed (although seed densities are reduced in such cases). When growing prostrate on soft tidal mud, the plants can become buried along with their mature fruits. As their capsules decay, the seeds are situated for optimum germination and development. The seeds germinate under a 30°C/15°C temperature regime following 2 months of warm stratification (dry storage). As in *G. neglecta*, the morphology of the plants differs when growing in tidal sites. In that case they have broader, shorter and thicker leaves with less-toothed margins, and greater numbers of closed flowers with shorter pedicels. Further study is needed to determine whether this observed variational pattern is genetically based and possibly worthy of taxonomic distinction. **Reported associates:** *Acer rubrum, Alnus serrulata, Amelanchier canadensis, Amaranthus cannabinus, Ampelopsis arborea, Aralia spinosa, Arisaema triphyllum, Aronia arbutifolia, Arundinaria gigantea, Asclepias perennis, Athyrium asplenioides, Berchemia scandens, Betula nigra, Bidens comosus, Boehmeria cylindrica, Bolboschoenus fluviatilis, Brunnichia ovata, Calamagrostis canadensis, Callitriche heterophylla, Campsis radicans, Carex albolutescens, C. atlantica, C. crinita, C. debilis, C. digitalis, C. folliculata, C. gracillima, C. incomperta, C. intumescens, C. joorii, C. lacustris, C. leptalea, C. lupulina, C. lurida, C. scoparia, C. seorsa, C. stipata, C. straminea, C. stricta, C. styloflexa, Carpinus caroliniana, Carya aquatica, Celtis laevigata, Cephalanthus occidentalis, Chasmanthium latifolium, C. laxum, C. sessiliflorum, Chelone glabra, Chionanthus virginicus, Cinna arundinacea, Commelina virginica, Crassula aquatica, Crataegus opaca, Cuscuta gronovii, Cyperus bipartitus, C. strigosus, Dichanthelium dichotomum, D. laxiflorum, Diodia virginiana, Dioscorea villosa, Diospyros virginiana, Echinochloa, Elatine americana, Eleocharis obtusa, E. palustris, E. tortilis, Eriocaulon parkeri, Fagus grandifolia, Fraxinus caroliniana, F. pennsylvanica, F. profunda, Galium tinctorium, Gaylussacia frondosa, Glyceria obtusa, G. striata, Gratiola neglecta, Hamamelis virginiana, Heteranthera reniformis, Hexastylis arifolia, Hygrophila lacustris, Hypericum hypericoides, Ilex laevigata, I. opaca, I. verticillata, I. vomitoria, Impatiens capensis, Isoetes melanopoda, I. riparia, Itea virginica, Juncus canadensis, J. coriaceus, J. effusus, J. repens, J. subcaudatus, Justicia ovata, Kyllinga gracillima, Leersia lenticularis, L. oryzoides, L. virginica, Lemna minor, Leucothoe racemosa, Lilium superbum, Limnobium spongia, Lindernia dubia, Liquidambar styraciflua, Liriodendron tulipifera, Lobelia cardinalis, Ludwigia grandiflora, L. palustris, L. peploides, Lycopus rubellus, L. virginicus, Lyonia ligustrina, Lysimachia radicans, Magnolia grandiflora, M. virginiana, Maianthemum canadense, Micranthemum micranthemoides, Mikania scandens, Mimulus alatus, M. ringens, Mitchella repens, Morus rubra, Myrica cerifera, Nuphar advena, Nyssa aquatica, N. biflora, N. sylvatica, Onoclea sensibilis, Orontium aquaticum, Osmunda regalis, Osmundastrum cinnamomeum, Oxydendrum arboreum, Oxypolis rigidior, Panicum spretum, Peltandra virginica, Penthorum sedoides, Persea palustris, Phanopyrum gymnocarpon, Pilea pumila, Pinus elliotii, P. glabra, P. taeda, Planera aquatica, Platanthera clavellata, Pluchea camphorata, Persicaria coccinea, P. hydropiperoides, P. punctata, P. sagittata, P. setacea, Pontederia cordata, Populus heterophylla, Potamogeton diversifolius, Proserpinaca palustris, Pteridium aquilinum, Quercus laurifolia, Q. lyrata, Q. michauxii, Q. nigra, Q. phellos, Q. virginiana, Rhexia virginica, Rhododendron viscosum, Rhynchospora capitellata, R. gracilenta, R. miliacea, Rosa palustris, Rotala ramosior, Rubus cuneifolius, Sabal minor, S. palmetto, Sabatia calycina, Saccharum giganteum, Sagittaria brevirostra, S. calycina, S. fasciculata, S. graminea, S. latifolia, S. montevidensis, S. subulata, Salix nigra, Sarracenia, Sassafras albidum, Saururus cernuus, Schoenoplectus pungens, S. purshianus, S. smithii, Scirpus cyperinus, S. polyphyllus, Scleria triglomerata, Selaginella, Senecio glabellus, Serenoa repens, Smilax glauca, S. pseudochina, S. rotundifolia, S. tamnoides, S. walteri, Solidago gigantea, Sparganium eurycarpum, Spartina alterniflora, S. cynosuroides, Sphagnum, Symphyotrichum lateriflorum, Symplocarpus foetidus, Symplocos tinctoria, Taxodium distichum, Thelypteris palustris, Toxicodendron radicans, T. vernix, Triadica sebifera, Triadenum fraseri, T. walteri, Ulmus americana, Vaccinium arboreum, V. atrococcum, V. corymbosum, Vernonia gigantea, Viburnum nudum, Viola ×primulifolia, Vitis cinerea, V. rotundifolia, Wisteria frutescens, Woodwardia areolata, W. virginica, Zizania aquatica.*

***Gratiola viscidula* Pennell** is a perennial, which is found in ditches, floodplains, marshes, swales, swamps, and along the margins of lakes, ponds, and streams. The sites are circumneutral (pH: 6.0–7.4) with exposures ranging from full sun to partial shade. There have been no studies of the reproductive biology of this species, but it is likely that the plants are self-compatible and are capable of self-pollination. Seed germination is light dependent (which prevents germination in buried seeds) and occurs maximally under a 30°C/15°C temperature regime following 3 months of cold (4°C) stratification. In the field, the seeds undergo several annual conditional dormancy/nondormancy cycles, with their maximum germination occurring in the spring (March to April) but declining by summer (June). The seeds survive for at least 30 months and can form a seed bank. Vegetative reproduction occurs by the production of creeping, weakly branching rhizomes. **Reported associates:** *Bartonia virginica, Calopogon tuberosus, Calycanthus floridus, Carex joorii, Cephalanthus* *occidentalis, Chrysogonum virginianum, Cleistes bifaria, Cypripedium kentuckiense, Eurybia saxicastellii, Gratiola pilosa, Hydrocotyle americana, Hypericum crux-andreae, Lathyrus palustris, Lobelia nuttallii, Lycopodiella appressa, Parnassia asarifolia, Platanthera ciliaris, P. clavellata, P. cristata, P. integrilabia, Polygala cruciata, P. nuttallii, Rhynchospora globularis, Sambucus racemosa, Scleria ciliata, Sphenopholis pensylvanica, Veratrum parviflorum, Vernonia noveboracensis, Viburnum lentago, Xyris caroliniana, X. torta.*

Use by wildlife: The flowers of *Gratiola* species are eaten by cucumber beetles (Insecta: Coleoptera: Chrysomelidae: *Acalymma, Diabrotica*). *Gratiola aurea* is eaten by larvae of the European corn borer (Insecta: Lepidoptera: Pyralidae: *Ostrinia nubilalis*).

Economic importance: food: Although there are no specific reports regarding the edibility of any of the OBL *Gratiola* species, none should be eaten because other members of the genus (*G. officinalis, G. peruviana*) can be quite poisonous; **medicinal:** Of the OBL North American species, only *Gratiola virginiana* (entire plants) reportedly has been used medicinally; however, the European *G. officinalis* (known as *Gratia Dei* for its medicinal properties) contains the active glucosides gratiolin and gratiosolin, and has long been employed as a cathartic, diuretic, emetic and vermifuge; **cultivation:** *Gratiola aurea* and *G. virginiana* are cultivated as ornamental water garden plants; **misc. products:** none;

weeds: *Gratiola neglecta* has been reported as a weed in agricultural fields; **nonindigenous species:** none
Systematics: The genus *Amphianthus* once was considered to be distinct from *Gratiola* and of uncertain affinity to other genera of Plantaginaceae. Although the fruit resembles *Veronica* (tribe Veroniceae), the corolla morphology and punctate foliage strongly indicate an alliance to tribe Gratioleae where it traditionally has been classified. Phylogenetic analyses of combined DNA sequence data not only confirm the placement of *Amphianthus* within tribe Gratioleae of Plantaginaceae, but also resolve the taxon as being deeply nested within the genus *Gratiola* as a relatively derived species (Figure 5.86). Thus, despite years of traditional taxonomic segregation, it becomes necessary to merge *Amphianthus* with *Gratiola*, where the taxon is recognized under the name *G. amphiantha*. The transfer of *Amphianthus* to *Gratiola* also is supported chromosomally (see later). Phylogenetic analyses of seven *Gratiola* species (including *G. amphiantha*) indicate that the genus is monophyletic as redefined, although inclusion of additional species is needed for further confirmation. Section *Nibora* includes the annual species *G. ebracteata, G. flava, G. floridana, G. heterosepala, G. neglecta,* and *G. virginiana*. Because *G. amphiantha* (also annual) is closely related to *G. neglecta* and *G. ebracteata* (Figure 5.86) it is appropriate to include it within section *Nibora. Gratiola aurea* forma *pusilla* is the name applied to the dwarfed, submerged forms of that species, which are phenotypically plastic rather than genetically

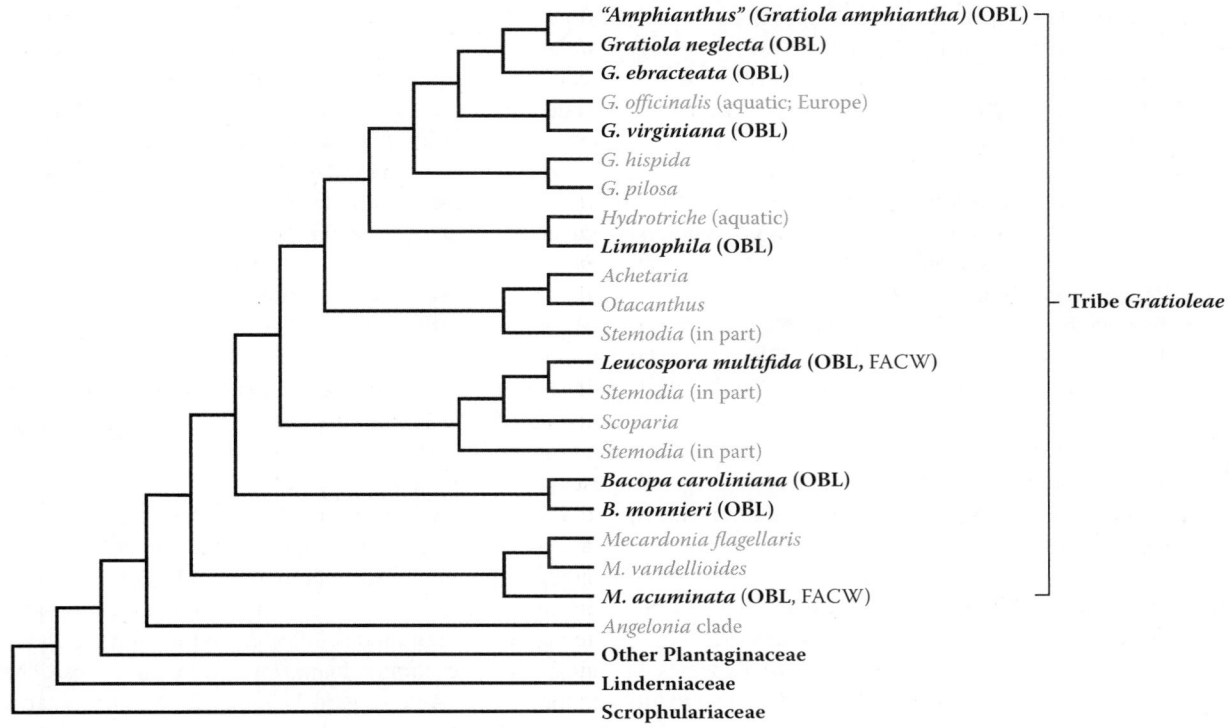

FIGURE 5.86 Phylogeny of Plantaginaceae tribe Gratioleae as indicated by analysis of cpDNA sequence data. From the species included in this analysis, *Gratiola* is monophyletic but only if the segregate genus *Amphianthus* (which nests within) is included. The aquatic habit probably arose once in *Gratiola* as indicated by the OBL North American taxa (bold), which resolve as a clade with the European *G. officinalis*, another aquatic. These results further indicate that *Stemodia* is not monophyletic as currently circumscribed. (Adapted from Estes, D. & Small, R.L., *Syst. Bot.*, 33, 176–182, 2008.)

distinct variants. *Gratiola virginiana* var. *aestuariorum* is the name given to estuarine forms of that species, but further study is needed to determine whether this taxon is distinct genetically from the typical variety. Some authors recognize *G. viscidula* var./subsp. *shortii* as an infraspecific taxon; however, its distinctness was not supported by a numerical analysis of key morphological characters. The base chromosome number of *Gratiola* is $x = 8$ ($x = 7$, 9 as an aneuploid series). The closely related *Gratiola amphiantha* and *G. neglecta* ($2n = 18$) are $x = 9$ diploids; *G. virginiana* ($2n = 16$) is an $x = 8$ diploid, and *G. aurea* ($2n = 28$) is an $x = 7$ tetraploid. Natural hybrids have not been reported for any of the North American species.

Comments: *Gratiola neglecta* is widespread in North America, with *G. aurea* in eastern North America, and *G. ebracteata* in western North America; *G. floridana*, *G. virginiana*, and *G. viscidula* occur in the southeastern United States; *G. heterosepala* is quite rare, with 82 extant sites in California and one site in Oregon. *Gratiola amphiantha* is an even rarer species that is endemic to Alabama, Georgia and South Carolina, where only 57 extant populations were located during a 1994 survey for the plants.

References: Adams et al., 2007; Barbour et al., 2007; Baskin & Baskin, 1988, 1998; Baskin et al., 1989; Ceska et al., 1986; Christy, 2004; Clarkson, 1958; Conard, 1935; Detmers, 1910a,b; Enser, 2001; Estes & Small, 2008; Ferren Jr. & Schuyler, 1980; Fischer, 2004; Fritsch et al., 2007; Funk & Fuller, 1978; Gibson et al., 2005; Graham & Henry, 1933; Hanlon et al., 1998; Hawkins & Richards, 1995; Hilton, 1993; Hilton & Boyd, 1996; Hopkins, 1969; Kaye et al., 1990; Keddy et al., 1994, 1998; Keeley & Zedler, 1998; Koning, 2005; Lam, 1998; Lammers, 1958; Leck & Graveline, 1979; Lahring, 2003; Les et al., 2006; Lewis et al., 1962; Lunsford, 1938; Matthews & Murdy, 1969; Mathies et al., 1983; McCanny et al., 1990; Middleton, 2003; Nielsen et al., 1991; Olmstead et al., 2001; Ornduff, 1969c; Parker & Leck, 1985; Pennell, 1919, 1935; Platenkamp, 1998; Schlising & Sanders, 1982; Schneider & Sharitz, 1986; Seabloom et al., 1998; Shipley & Parent, 1991; Smith et al., 1998; Sorrie et al., 2006; Speichert & Speichert, 2004; Sperduto & Nichols, 2004; Spooner, 1984; Stevens, 1932; Thomas, 1985; USFWS, 1993a, 2004; Wall et al., 2003; Wilson, 1935; Wisheu & Keddy, 1991.

6. *Hippuris*

Bottle-brush, mare's tail; hippuride

Etymology: from the Greek *hippos oura*, meaning "horse tail," in reference to the tail-like stem

Synonyms: none

Distribution: global: Asia; Australia; Europe; New World; **North America:** widespread, except for the southeast

Diversity: global: 3–4 species; **North America:** 3 species

Indicators (USA): OBL: *Hippuris montana*, *H. tetraphylla*, *H. vulgaris*; **FACW:** *H. montana*

Habitat: brackish (coastal) or freshwater; lacustrine, palustrine, riverine; **pH:** 3.0–7.5; **depth:** 0–3 m; **life-form(s):** emergent herb, submersed (vittate)

Key morphology: stems (to 0.1–1.5 m), submersed or emersed, erect, lacunae anastamosing; leaves (to 1–8 cm), entire, linear,

acute or obtuse apically, in whorls of 2–12, flaccid (submersed) or stiff (emergent); emergent herbage with peltate hairs; flowers solitary, axillary (upper), uni- or bisexual, sessile (upper) or short-stalked (lower), epigynous; calyx rim-like, adherent to ovary; corolla absent; stamen solitary, sessile or stalked (to 1 mm), arising from apex of calyx, anther large, often reddish; style single, lying in the groove between the anther lobes, stigmatic throughout; nutlet (to 3 mm) indehiscent, single-seeded

Life history: duration: perennial (rhizomes); **asexual reproduction:** rhizomes; **pollination:** wind; **sexual condition:** hermaphroditic; monoecious; **fruit:** nutlets (common); **local dispersal:** fruits/seeds (ice, water, wind); rhizome fragments (water); **long-distance dispersal:** seeds (waterfowl)

Imperilment: (1) *Hippuris montana* [G4]; S1 (AB, YT); S3 (BC); (2) *H. tetraphylla* [G5]; S1 (YT); S2 (BC); S3 (MB, QC); (3) *H. vulgaris* [G5]; SX (IN); SH (IA); S1 (NE, NH, NY, VT); S3 (AZ, ME, WY)

Ecology: general: All *Hippuris* species in North America are OBL aquatics that grow in standing water or on wet, exposed sediments. The genus is known for its heterophylly, which is regulated by abscisic acid. When growing in standing water, the emergent shoots produce foliage that is stiff and waxy and contains stomata; whereas, the underwater foliage is limp, membranaceous, and devoid of stomata. The flowers are either bisexual or unisexual. Occasionally the plants are polygamous, possessing both bisexual and unisexual flowers. The bisexual flowers often are described as wind-pollinated and protogynous, with the filaments elongating only after the stigmas shrivel; however, they also are described with the style commonly situated in between the anther sacs, which would facilitate selfing. In plants with unisexual flowers the males are situated beneath the females, which would indicate an adaptation for wind-mediated outcrossing. Certainly, the reproductive ecology of the genus requires further study. Dispersal of the buoyant fruits occurs by water, in mud adhering to the feet of water birds (Aves: Anatidae), or by endozoic transport within the gut of feeding birds (Aves). There also are reports that the fruits can become frozen in ice and subsequently are dispersed upon its melting and breakup. Vegetative reproduction in all species occurs by means of rhizomes. Care must be taken when evaluating literature accounts because there are numerous instances where aquatic members of the genus *Equisetum* (horsetails) have been confused with *Hippuris*.

***Hippuris montana* Ledeb.** grows in freshwater brooks, depressions, meadows, mires, muskeg, pools, runnels, seeps, slopes, streambeds, and the margins of lakes and streams at elevations of up to 1829 m. The plants are shade-intolerant and occur on saturated acidic soils of medium nitrogen levels, which include humus, loam, mud, peat, and rock. The populations extend northward beyond 62° N latitude and often develop in alpine areas of snowmelt. The flowers are polygamomonoecious (mainly unisexual and monoecious, but also with some hermaphrodites), with the staminate flowers occurring below the pistillate flowers. These are diminutive plants (stems <10 cm) with a moss-like habit. They can become well camouflaged among the various mosses with which they commonly are associated; thus, becoming less conspicuous to herbivores. Unlike the following species, these plants typically do not inhabit inundated sites, but occur instead on wet, mossy substrates. The ecological associates all are emergents. **Reported associates:** *Abies lasiocarpa, Adiantum aleuticum, Agrostis scabra, Artemisia norvegica, Asterella lindenbergiana, Athyrium filix-femina, Blechnum spicant, Blepharostoma trichophyllum, Calamagrostis canadensis, C. stricta, Callitropsis nootkatensis, Caltha biflora, C. leptosepala, Calypogeia trichomanis, Carex macrochaeta, C. nigricans, C. saxatilis, C. spectabilis, C. utriculata, Cassiope mertensiana, Chamaecyparis nootkatensis, Chamerion angustifolium, Diplophyllum albicans, Dryopteris dilatata, Eleocharis palustris, Empetrum nigrum, Enemion savilei, Epilobium ciliatum, E. palustre, Erigeron, Eriophorum angustifolium, E. scheuchzeri, Festuca, Geranium, Glyceria grandis, Gymnomitrion, Haplomitrium hookeri, Harrimanella stelleriana, Heracleum lanatum, Hieracium triste, Huperzia selago, Juncus drummondii, Kiaeria starkei, Larix lyallii, Leptarrhena pyrifolia, Ligusticum calderi, Lophozia ventricosa, Luetkea pectinata, Luzula parviflora, Lycopodium sitchense, Mimulus, Mitella, Moerckia blyttii, Nardia geoscyphus, Nephrophyllidium crista-galli, Oligotrichum parallelum, Pedicularis, Petasites frigidus, Phyllodoce aleutica, P. empetriformis, Pohlia wahlenbergii, Polystichum andersonii, P. lemmonii, Polytrichastrum alpinum, Potentilla, Primula cuneifolia, Pseudoleskea baileyi, Ranunculus eschscholtzii, R. gmelinii, Racomitrium canescens, Rhodiola integrifolia, Rhytidiadelphus squarrosus, Rhytidiopsis robusta, Rorippa palustris, Rubus arcticus, Salix interior, S. niphoclada, Sanguisorba, Saxifraga rivularis, Scapania paludosa, Senecio triangularis, Sinosenecio newcombei, Sorbus sitchensis, Sphagnum, Trichophorum pumilum, Tsuga mertensiana, Vaccinium, Vahlodea atropurpurea, Valeriana sitchensis, Veratrum viride, Veronica.*

Hippuris tetraphylla **L.f.** inhabits brackish bays, beaches, deltas, depressions, dunes, flats, floodplains, lagoons, marshes, meadows, mudflats, ponds, pools, sloughs, swales, and the margins of rivers and streams at low, coastal elevations. The plants occur in shallow conditions (depths: 15–40 cm) on circumneutral substrates (e.g., pH: 6.6–7.0) of fine sand, humus, silt, muck, or mud. This species is an oligohaline indicator (salinity: 0.5–5.0 ppt), which tolerates inundation above the tidal limit. It is cold tolerant with occurrences that extend northward to at least 67° N latitude. The flowers are perfect and either self- or wind pollinated. The seeds persist and occur commonly in the seed bank with densities of up to 1160 seeds/m² reported. These plants often are the initial colonizers of bare or soft, unconsolidated sediments of estuarine marshes and river deltas. They show higher abundance in sites that have been grazed by snow geese (Aves: Anatidae: *Chen caerulescens*). Local dispersal can occur vegetatively by the redistribution of floating rhizome fragments. **Reported associates:** *Arctophila fulva, Angelica lucida, Atriplex alaskensis, Bolboschoenus maritimus, Calamagrostis stricta, Caltha natans, Carex aquatilis, C. glareosa, C. limosa, C. lyngbyei, C. mackenziei, C. norvegica, C. paleacea, C. pluriflora, C.*

ramenskii, C. rostrata, C. saxatilis, C. subspathacea, C. ursina, Comarum palustre, Dupontia fisheri, Eleocharis acicularis, E. halophila, E. parvula, Equisetum, Eriophorum angustifolium, E. russeolum, Festuca rubra, Galium trifidum, Hierochloe odorata, Isoetes tenella, Leymus arenarius, Menyanthes trifoliata, Myriophyllum sibiricum, M. spicatum, Plantago maritima, Potentilla anserina, P. egedii, Puccinellia andersonii, P. grandis, P. lucida, P. phryganodes, Ranunculus cymbalaria, R. gmelinii, Salicornia europaea, Salix brachycarpa, S. ovalifolia, Senecio congestus, Sparganium hyperboreum, Spergularia canadensis, Stellaria humifusa, Stuckenia filiformis, S. pectinata, S. vaginata, Subularia aquatica, Triglochin maritimum, T. palustre, Utricularia minor, Zannichellia palustris.

Hippuris vulgaris **L.** is found in freshwater bogs, brooks, ditches, dunes, fens, flats, floodplains, lakes, marshes, meadows, mudflats, ponds, pools, shores, sloughs, streams, swamps, and along the margins of lakes, ponds, rivers, and streams at elevations of up to 2700 m. This species often occurs in cold-water habitats and extends to at least 69° N latitude where it is a prominent element of tundra pools. It usually grows in shallow waters, but has been reported to occur at depths of 2–3 m. Plants of a dwarfed stature can develop on drying shorelines, but they do not withstand prolonged desiccation. The plants tolerate only partial shade and typically occur where the sulfate ion (SO_4^{2-}) concentration exceeds 50 ppm. Although fairly acid conditions are tolerated (pH: 3.0–7.5), most occurrences tend to be in more alkaline habitats (mean pH: 7.0). Alkalinity normally ranges from 12 to 49.5 mg/L but the plants usually are absent from waters with a total alkalinity below 30 ppm. This species is known to survive fairly drastic reductions in site pH (e.g., from 7.0 to 3.0). The substrates can include gravel, humus, marl, marly muck, muck, mud, pebbles, sand, and silt. Flowering commences in late summer. The flowers reportedly are protogynous and wind pollinated, but further studies on the breeding system and reproductive ecology would be beneficial. Some accounts indicate that fruit production is rare; whereas, others indicate it to be fairly prolific. The nutlets are dispersed by water and also endozoically by waterfowl (Aves: Anatidae), which can consume sizeable quantities of the seed. Fresh seed is dormant (throughout at least the winter) and buoyant, remaining afloat for at least a short time (but reportedly up to a year). The seeds have been stored for several years in distilled water without germination, which typically must be induced by the abrasion or decay of the surrounding pericarp. Thoroughly dried fruits that are devoid of their exocarps (which are abraded in the gizzard of waterfowl) are viable and retain the ability to float for up to a week before sinking. Such fruits germinate within 1–2 weeks when allowed to settle on a mud layer in a container of distilled water. Germinating embryos develop a flange-like collar of hairs above the elongating root, which presumably anchors the seedlings and holds them above the surface of the loose mud as they extend. Although the seeds contain a relatively low percentage of ortho-dihydroxyphenols (which are thought to increase seed bank persistence), they are known to persist (at densities up to 13 seeds/m²) at sites where the species is absent from the standing vegetation. The seed bank is described as transient. The plants recolonize areas rapidly following flood disturbances and also have been observed to naturally colonize lands that have been reclaimed from coal-mining activity. They have been categorized ecologically as ruderals, which establish quickly following a disturbance, and as r-strategists, which are characterized by rapid, prolific propagation. Although the plants are capable of vegetative reproduction by fragmentation of the stout rhizomes, the floating fragments do not establish readily unless they are deposited on disturbed, exposed substrates. Rooted stem fragments must become anchored in the substrate and eventually will die if they remain afloat for extended periods. Detached buds also float, root quickly (but only in the autumn), and are dispersed readily on the water surface by currents, with an average survival rate of roughly 30%. The survival rates of rhizomes (about 45%) and stem fragments (about 25%) are highest during the spring. In the field, both vegetative regeneration and the likelihood of establishment are relatively higher in spring. Submersed shoots can remain green throughout the winter, which provides them with a competitive advantage in the following season. The aerial leaf morphology is induced by high light levels, low red to far-red light ratios, and temperatures above 10°C, and can occur on plants growing in as much as 1.5 m of water when these conditions are present. The leaves not only are distinct morphologically but also physiologically; e.g., the activity of ribulose bisphosphate carboxylase oxygenase is greater in the emergent leaves than in the submersed foliage of these C-3 plants.

Reported associates: *Acorus americanus, Alnus rugosa, Alopecurus aequalis, Apocynum cannabinum, Arctagrostis latifolia, Arctophila fulva, Artemisia biennis, Betula, Bidens cernuus, Bistorta vivipara, Bolboschoenus robustus, Brachythecium frigidum, Brasenia schreberi, Calamagrostis canadensis, C. stricta, Calla palustris, Calliergon giganteum, Callitriche hermaphroditica, C. palustris, Caltha palustris, Cardamine pratensis, Carex aquatilis, C. aurea, C. bebbii, C. canescens, C. cusickii, C. diandra, C. flava, C. lacustris, C. lanuginosa, C. limosa, C. microptera, C. nebrascensis, C. pellita, C. prairea, C. projecta, C. pseudocyperus, C. rossii, C. rostrata, C. spectabilis, C. stipata, C. tenera, C. utriculata, C. vaginata, C. vesicaria, C. viridula, Catabrosa aquatica, Ceratophyllum demersum, Chamerion angustifolium, Chara, Cicuta douglasii, C. virosa, Comarum palustre, Deschampsia cespitosa, Drepanocladus aduncus, Dupontia fisheri, Egeria densa, Eleocharis acicularis, E. palustris, Elodea canadensis, Epilobium ciliatum, E. palustre, Equisetum fluviatile, E. variegatum, Eriocaulon aquaticum, Eriophorum angustifolium, E. chamissonis, E. scheuchzeri, Fontinalis neomexicana, Galium trifidum, G. triflorum, Gentiana rubricaulis, Gentianella amarella, Glyceria borealis, G. grandis, G. striata, Heteranthera dubia, Hordeum jubatum, Juncus acuminatus, J. balticus, J. brevicaudatus, J. dudleyi, J. filiformis, J. nodosus, J. supiniformis, Lemna minor, L. trisulca, Lilaeopsis occidentalis, Lobelia dortmanna, Lyngbya, Lysichiton americanus, Lysimachia thyrsiflora, Megalodonta beckii, Mentha arvensis, Menyanthes trifoliata, Mimulus ringens, Myriophyllum alterniflorum, M. farwellii, M. heterophyllum, M. sibiricum, M. spicatum, M. verticillatum, Najas flexilis, Nephrophyllidium crista-galli, Nuphar advena, N. polysepala, N. ×rubrodisca, N. variegata, Nymphaea odorata, Paludella squarrosa, Parnassia palustris, Persicaria amphibia, Phalaris arundinacea, Phragmites australis, Plagiomnium insigne, Platanthera dilatata, P. hyperborea, Poa pratensis, Potamogeton alpinus, P. amplifolius, P. crispus, P. epihydrus, P. friesii, P. gramineus, P. illinoensis, P. natans, P. nodosus, P. obtusifolius, P. praelongus, P. richardsonii, P. robbinsii, P. zosteriformis, Ranunculus aquatilis, R. cymbalaria, R. gmelinii, R. longirostris, R. pallasii, R. sceleratus, R. trichophyllus, Riccia fluitans, Rorippa aquatica, R. palustris, R. teres, Rubus arcticus, Rumex aquaticus, R. maritimus, Ruppia maritima, Sagittaria cuneata, S. graminea, S. latifolia, Salix candida, Schoenoplectus acutus, S. tabernaemontani, Scorpidium scorpioides, Scutellaria galericulata, Sium suave, Sparganium angustifolium, S. chlorocarpum, S. eurycarpum, S. hyperboreum, S. minimum, Sphagnum lindbergii, S. squarrosum, Spiraea alba, Spiranthes romanzoffiana, Spirodela polyrhiza, Stachys palustris, Stellaria longifolia, Stuckenia pectinata, S. filiformis, Triantha glutinosa, Triglochin maritimum, Typha latifolia, Urtica dioica, Utricularia cornuta, U. intermedia, U. macrorhiza, Vallisneria americana, Veronica americana, Wolffia arrhiza, Zannichellia palustris, Zizania palustris.*

Use by wildlife: *Hippuris* plants provide important habitat structure for aquatic invertebrates. The seeds and leaves of *Hippuris tetraphylla* are eaten by the cackling Canada goose (Aves: Anatidae: *Branta canadensis minima*) and the spectacled eider (Aves: Anatidae: *Somateria fischeri*). The seeds and foliage of *H. vulgaris* are eaten by 25 species of waterfowl (Aves: Anatidae) with as many as 1090 seeds having been recovered from the stomach of a single bird. Notably, the plants are eaten by various waterfowl (Aves: Anatidae) including the American wigeon (*Anas americana*), black duck (*Anas rubripes*), blue-winged teal (*Anas discors*), cinnamon teal (*Anas cyanoptera*), emperor goose (*Chen canagica*), gadwell (*Anas strepera*), greater scaup (*Aythya marila*), green-winged teal (*Anas carolinensis*), lesser scaup (*Aythya affinis*), mallard (*Anas platyrhynchos*), pintail (*Anas acuta*), and wood duck (*Aix sponsa*). It is especially important as a food plant in western North America. The seeds also are eaten by more terrestrial avifauna such as the Wilson snipe (Aves: Scolopacidae: *Gallinago gallinago*), with up to 50 seeds recovered from a single bird. The winter-green foliage is grazed on by caribou (Mammalia: Cervidae: *Rangifer tarandus*) and probably to some degree by moose (Mammalia: Cervidae: *Alces alces*). The foliage also is reportedly eaten by goats (Mammalia: Bovidae: *Capra aegagrus*). The plants are a component of nesting habitat for the trumpeter swan (Aves: Anatidae: *Cygnus buccinator*). They are colonized by "Ingoldian" Fungi, i.e., those species normally associated with the decay of leaves that fall into streams.

Economic importance: food: The Alaskan Eskimo eat the young leaves of *Hippuris tetraphylla* as greens. The Inupiats and Yupiks prepare a soup by adding *Hippuris vulgaris* to water, cod liver oil, seal oil, and seal blood. The Inuktitut make a type of ice cream using the plants. They also use the plants as a soup condiment or eat them raw or mixed with salmon eggs and seal oil; **medicinal:** *Hippuris vulgaris* contains verbascoside (an antioxidant), and the iridoids aucubin (a hepatoprotectant) and catalpol, which reportedly has neuroprotective properties. In one study, its antioxidant activity ranked in the top 20% of plants tested, which included 22 species that have been used in traditional medicine. The juice of the plant (ingested or applied externally) has been used to reduce bleeding, to alleviate skin inflammation, and as a treatment for ulcers; **cultivation:** *Hippuris vulgaris* is grown as an ornamental water garden plant; **misc. products:** From this genus originated the term "*Hippuris* syndrome," which is used to describe plants with elongate stems having whorls of simple leaves; **weeds:** *Hippuris vulgaris* has been reported as a weed of no-till meadows in Alaska and Wyoming; however, many weed reports have confused this genus with horsetail (*Equisetum*). It also grows profusely in some English rivers; **nonindigenous species:** none

Systematics: Historically, *Hippuris* most often was placed either in Haloragaceae, or was treated as a distinct family (Hippuridaceae). However, recent phylogenetic studies resolve the genus firmly within Plantaginaceae as the sister group of *Callitriche* (Figure 5.82). Authors have disagreed on the number of species, with some recognizing a single cosmopolitan species (*Hippuris vulgaris*) comprising "ecological races" in the Arctic and Baltic regions. However, a more reasonable assessment is to recognize three species in North America alone, given that they are quite distinctive both ecologically and morphologically. *Hippuris montana* is a diminutive diploid (see later) of subalpine habitats, where it grows in dense patches among mosses on wet substrates but normally not under inundated conditions. It principally is monoecious; whereas, the other species mainly are hermaphroditic (although *H. tetraphylla* also has been reported as unisexual). *Hippuris vulgaris* occurs most commonly in standing water. It has bisexual flowers and lanceolate leaves in whorls of 8–12. Unlike the preceding taxa, *H. tetraphylla* inhabits brackish or saline sites and possesses distinctive blunt, ovate to oblong leaves in whorls of 4–6. The latter two species are tetraploid. Some authors regard the various differences as plastic environmental responses, but the distinctive features are maintained across their range. An Eurasian taxon (*H. lanceolata*; $2n = 32$) is regarded variously as conspecific with *H. vulgaris*, as a distinct species, or as a hybrid (then as *H. ×lanceolata*), involving *H. tetraphylla* and *H. vulgaris*. This taxon is distinct genetically from *H. vulgaris* (see Figure 5.83) and requires further study to clarify its taxonomic status. Certainly the taxonomy of *Hippuris* deserves reconsideration on a global basis. The base chromosome number of *Hippuris* is $x = 8$. *Hippuris montana* is diploid ($2n = 16$); whereas, *H. tetraphylla* and *H. vulgaris* are tetraploid ($2n = 32$). Hybrids have not been reported in North America.

Comments: *Hippuris vulgaris* is the most widespread taxon in the genus, occurring in north and western North America.

It extends into Eurasia, is disjunct in Pategonia, and fossils indicate that it occurred formerly in Antarctica. *Hippuris montana* is restricted to northwestern North America and *H. tetraphylla* occurs in northern North America (extending into Eurasia).

References: Aiken et al., 2007; Allen et al., 2002; Anderson, 1939; Andersson et al., 2000; Arber, 1920; Ashworth et al., 2004; Bärlocher, 1982; Barrat-Segretain & Bornette, 2000; Barrat-Segretain et al., 1998; Bodkin et al., 1980; Bouchard et al., 1987; Chandler, 1937; Chang et al., 2001; Christy, 2004; Clausen et al., 2006; Cody, 2000; Cook, 1978, 1988; Cooper, 1939b; Cottam & Knappen, 1939; Cowardin et al., 1979; Cronquist, 1981; Crow & Hellquist, 1983; Damtoft et al., 1994; Dirschl, 1969; Farmer et al., 1986; Fraser et al., 1980; Fraser et al., 2007; Goliber & Feldman, 1990; Good, 1924; Gosling & Baker, 1980; Greulich & Bornette, 2003; Grieve, 1980; Handa et al., 2002; Hazlett, 1989; Hazlett, 1998; Hendry et al., 1994; Jefferies, 1977; Jefferies et al., 1979; Jorgenson, 2000; Jutila b. Erkkilä, 1998; Kerbes et al., 1990; Klein, 1982; Klinka et al., 1995; Lacoul & Freedman, 2006a; Leins & Erbar, 2004; Mabbott, 1920; McAtee, 1918, 1939; McAvoy, 1931; McKelvey et al., 1983; Moore, 1973; Moyle, 1945; Pollock et al., 1998; Preston & Croft, 1997; Racine & Walters, 1994; Robert et al., 2004; Safford, 1888; Sedinger, 1986; Sedinger & Raveling, 1984; Sharratt et al., 2006; Speichert & Speichert, 2004; Stehn et al., 1993; Stoudt, 1944; Strong, 2000; Taylor, 1981; Tiner, 1999; Viereck et al., 1992; Vincent, 1958; Walker et al., 1994; Wheeler et al., 1983; White & Harris, 1966; Wilson, 1978; Worley, 1969.

7. Leucospora

Paleseed

Etymology: from the Greek *leukos spora*, meaning "white seed"

Synonyms: *Capraria* (in part); *Conobea* (in part)

Distribution: global: North America; **North America:** eastern

Diversity: global: 1–4 species; **North America:** 1 species

Indicators (USA): OBL; FACW: *Leucospora multifida*

Habitat: freshwater; palustrine; **pH:** circumneutral; **depth:** <1 m; **life-form(s):** emergent herb

Key morphology: stems (to 20 cm) quadrangular, much branched, prostrate to ascending; leaves (to 3 cm) opposite, petioled, pinnate (3–7 divisions), the segments pinnatifid; herbage glandular pubescent; flowers solitary, axillary, stalked (to 10 mm); calyx (to 5 mm) mainly free, linear, spreading; corolla (to 6 mm) bilabiate, the lobes much shorter than the tube, white to greenish white, often tinged with pink to lavender; stamens 4, inserted, didynamous; capsule ovoid; seeds minute, numerous, ridged

Life history: duration: annual (fruit/seeds); **asexual reproduction:** none; **pollination:** unknown; **sexual condition:** hermaphroditic; **fruit:** capsules (common); **local dispersal:** seeds (water, wind); **long-distance dispersal:** seeds (birds)

Imperilment: (1) *Leucospora multifida* [G5]; S1 (IA, NE); (ON); S3 (GA)

Ecology: general: *Leucospora* is monotypic.

Leucospora multifida (**Michx.**) **Nutt.** grows in depressions, ditches, floodplains, glades, levees, marshes, meadows, mudflats, prairies, sandbars, seeps, shores, sinkholes, springs, swales, swamps, and along the margins of ponds, pools, reservoirs, rivers, and streams. The plants tolerate only temporary periods of inundation and occur on open to lightly shaded sites on circumneutral to alkaline substrates (the specific pH range has not been reported), which include clay, gravel, limestone, loam, marl, mud, rock, sand, and silt. Although the pollination biology has not been studied, the flower structure indicates that they either are self-pollinating or are pollinated by bees (Hymenoptera). Numerous seeds are produced and are present in the seed bank of early forest successional stages (<10 years) and are known to persist at sites where the plants are absent from the standing vegetation. They also occur (at densities averaging 34 seeds/m²) in farmed swampland sites. The minute seeds float and are dispersed by the water or by wind; viable seeds also are known to be transported in dung from horses (Mammalia: Equidae: *Equus*). The seeds probably adhere also to mud on the feet of waterfowl (Aves: Anatidae). Several early occurrence records are from ballast disposal sites, which indicates that the species might have been introduced unintentionally as a consequence of shipping activity. The dormancy requirements of the seeds are complex. Dormancy is broken when the seeds are buried at 5°C or 15°C/6°C for 12 weeks; however, if they are kept under 20°C/10°C, 25°C/15°C, or 30°C/15°C temperature regimes, they will germinate only under 30°C/15°C or 35°C/20°C day/night temperature cycles. If they are not flooded, naturally buried seeds exhibit an annual conditional dormancy (summer–fall)/nondormancy (spring-early summer) cycle. Light is required for the germination of nondormant seeds; however, the seeds will not germinate at 15°C/6°C under any treatment. The breaking of winter dormancy and the onset of summer conditional dormancy are prevented by flooding. High temperatures induce conditional dormancy and enable those seeds that retain dormancy (because of winter flooding) to germinate when they experience warm summer temperatures (e.g., on exposed mudflats or shores). This species often grows in disturbed sites, but also occurs in natural, "high-quality" communities. It is a frequent pioneer species of desiccating sites such as drying lakeshores, pond bottoms, mudflats and streambeds. The plants compete poorly with larger vegetation. **Reported associates:** *Acalypha virginica, Acer negundo, A. saccharinum, Agalinis tenuifolia, Agrostis gigantea, Alisma subcordatum, Alopecurus carolinianus, Amaranthus tuberculatus, A. rudis, Ambrosia psilostachya, A. trifida, Ammannia coccinea, A. robusta, Apocynum cannabinum, Arenaria patula, Arthraxon hispidus, Atriplex patula, Bidens bipinnatus, B. cernuus, B. comosus, B. frondosus, B. tripartitus, Boehmeria cylindrica, Boltonia asteroides, Brachyelytrum erectum, Calystegia sepium, Carex frankii, C. lupulina, C. lurida, C. molesta, C. squarrosa, C. tribuloides, C. viridula, C. vulpinoidea, Chaiturus marrubiastrum, Chamaesyce maculata, C. nutans, C. prostrata, Chasmanthium, Chenopodium berlandieri, Cicuta maculata, Cinna arundinacea, Cirsium vulgare, Clinopodium*

arkansanum, Commelina communis, Conoclinium coelestinum, Croton capitatus, C. glandulosus, C. monanthogynus, Cynodon, Cyperus acuminatus, C. aristatus, C. bipartitus, C. diandrus, C. erythrorhizos, C. ferruginescens, C. flavescens, C. odoratus, C. squarrosus, C. strigosus, Dichanthelium clandestinum, Digitaria sanguinalis, Diospyros virginiana, Echinochloa crus-galli, Eclipta prostrata, Elaeagnus umbellata, Eleocharis erythropoda, E. geniculata, E. montevidensis, E. obtusa, E. ovata, Epilobium coloratum, Equisetum arvense, E. hyemale, Eragrostis frankii, E. hypnoides, E. pectinacea, E. reptans, Erigeron strigosus, Eupatorium perfoliatum, E. serotinum, E. urticifolium, Euphorbia dentata, Fallopia scandens, Fimbristylis autumnalis, F. vahlii, Galium parisiense, Glyceria striata, Gratiola neglecta, Helenium autumnale, Helianthus tuberosus, Heliotropium indicum, Hypericum canadense, H. mutilum, Impatiens capensis, Juncus acuminatus, J. coriaceus, J. diffusissimus, J. effusus, J. filipendulus, J. marginatus, J. tenuis, Juniperus virginiana, Justicia americana, Kyllinga pumila, Leersia oryzoides, L. virginica, Leptochloa fusca, Ligustrum sinense, Lindernia dubia, Liparis loeselii, Lipocarpha drummondii, L. micrantha, Lobelia siphilitica, Ludwigia glandulosa, L. palustris, Lycopus americanus, Mazus japonicus, Mecardonia acuminata, M. procumbens, Melochia pyramidata, Microstegium vimineum, Mimulus alatus, Mollugo verticillata, Muhlenbergia frondosa, M. sobolifera, Myosotis laxa, Myosurus minimus, Nasturtium officinale, Oxalis corniculata, Panicum capillare, P. dichotomiflorum, P. repens, P. scoparium, Paspalum dilatatum, Pastinaca sativa, Penthorum sedoides, Perilla frutescens, Persicaria lapathifolia, P. maculosa, P. pensylvanica, P. posumbu, P. punctata, Phyla lanceolata, P. nodiflora, Pilea pumila, Plantago virginica, Pluchea camphorata, Polypremum procumbens, Populus deltoides, Proserpinaca palustris, Prunella vulgaris, Quercus lyrata, Ranunculus abortivus, R. micranthus, R. parviflorus, R. pusillus, R. sardous, Rhynchospora capitellata, Rorippa sessiliflora, Rotala ramosior, Rumex conglomeratus, R. crispus, R. obtusifolius, Sabal palmetto, Sagittaria graminea, Salix caroliniana, S. interior, S. nigra, Samolus parviflorus, Satureja glabella, Schoenoplectus smithii, S. torreyi, Scirpus pendulus, Scutellaria lateriflora, S. parvula, Sedum pulchellum, Senecio glabellus, Sesbania, Setaria faberi, S. leucopila, S. parviflora, S. pumila, Sida, Solidago gigantea, Sonchus arvensis, Sorghum halepense, Sporobolus cryptandrus, S. vaginiflorus, Strophostyles, Symphyotrichum ontarionis, S. praealtum, Trifolium repens, Teucrium canadense, Triglochin palustre, Typha latifolia, Verbena simplex, V. urticifolia, Verbesina alternifolia, Veronica arvensis, Vitis aestivalis, V. cinerea, Xanthium strumarium.

Use by wildlife: none reported

Economic importance: food: none reported; **medicinal:** *Leucospora multifida* was used by the indigenous people of eastern Kansas as a restorative tonic; **cultivation:** not cultivated; **misc. products:** none; **weeds:** *Leucospora multifida* is considered to be a weed in Kansas; **nonindigenous species:** *Leucospora multifida* probably was introduced to some localities (e.g., Alabama, New Jersey, Pennsylvania) during the 19th century when it was found to be growing on mounds of discarded shipping ballast.

Systematics: *Leucospora* usually is regarded as monotypic, but a second species (*L. coahuilensis*) has been described from Mexico. In older treatments, *L. multifida* was placed in the genus *Conobea* (7 species). More recently, several authors have advocated the merger of *Leucospora* with the morphologically similar *Schistophragma* (two species) or *Stemodia* (56 species). Molecular data (Figures 5.85 and 5.86) resolve *L. multifida* within Plantaginaceae as the sister to *Stemodia verticillata*. However, both species occur within a heterogeneous clade along with some species of *Stemodia* (tribe Stemodieae), which is polyphyletic, and *Scoparia* (tribe Gratioleae). Neither *Conobea* nor *Schistophragma* were included in those analyses. Certainly, a more intensive phylogenetic study of *Leucospora* and related genera is needed to ascertain more appropriate taxonomic delimitations. *Leucospora* has not been investigated cytologically.

Comments: *Leucospora multifida* occurs in eastern North America.

References: Barringer & Burger, 2000; Baskin & Baskin, 1977, 1994, 1996; Campbell & Gibson, 2001; Gates, 1940; Harper, 1910; Henrickson, 1989; Johnson & Anderson, 1986; Joyner & Chester, 1994; Lindsey et al., 1961; Middleton, 2003; Mohlenbrock et al., 1961; Moore, 1976; Pennell, 1935; Roberts & Vankat, 1991; Smith, 1867; Smyth, 1903; Steinauer & Rolfkmeier, 2003; Thompson & McKinney, 2006; Turner, 1936; Turner & Cowan, 1993.

8. *Limnophila*

Ambulia, marshweed, rice–paddy herb

Etymology: from the Greek *limnê philos* meaning "pond friend"

Synonyms: *Ambulia*; *Ambuli*; *Bonnayodes*; *Cybbanthera*; *Diceros*; *Gratiola* (in part); *Hottonia* (in part); *Hydropityon*; *Stemodia* (in part); *Tala*; *Terebinthina*; *Stemodiacra*

Distribution: global: Africa; Asia; Australia; Europe; North America*; **North America:** southeastern

Diversity: global: 36 species; **North America:** 3 species

Indicators (USA): OBL: *Limnophila aromatica*, *L. indica*, *L. ×ludoviciana*, *L. sessiliflora*

Habitat: freshwater; lacustrine, palustrine, riverine; **pH:** 5.0–8.3; **depth:** 0–4 m; **life-form(s):** emergent herb, submersed (vittate)

Key morphology: stems round to quadrangular, erect or creeping in the air (to 1 m) or flexuous in the water (to 3.7 m); leaves opposite or whorled, either homophyllous, lanceolate/ovate (to 9 cm) and serrate, or heterophyllous with the submersed leaves (to 6 cm) laciniate to compound pinnate and grading transitionally upward to fleshy, entire to deeply pinnatifid emersed leaves (to 6.5 cm); flowers solitary or racemose, axillary, sessile or pedicelled (to 3 cm), often paired at nodes, sometimes cleistogamous and submersed; corollas (to 18 mm) tubular, funnelform to weakly bilabiate, varying (even within a species) in color from blue, purple, or violet to cream, mauve, pale yellow, pink, or white; stamens 4, didynamous, included; capsules (to 6 mm) broadly elliptical; seeds (to 0.7 mm) numerous

Life history: duration: annual (fruit/seeds); perennial (rhizomes); **asexual reproduction:** rhizomes, shoot fragments; **pollination:** insect, self or cleistogamous; **sexual condition:** hermaphroditic; **fruit:** capsules (common); **local dispersal:** fragments, seeds (water); **long-distance dispersal:** seeds (waterbirds)

Imperilment: (1) *Limnophila aromatica* [unranked]; (2) *L. indica* [GNR]; (3) *L. ×ludoviciana* [GNR]; (4) *L. sessiliflora* [GNR]

Ecology: general: All *Limnophila* species are aquatic or wetland plants that occur mainly in tropical and subtropical regions of the Old World where the genus is native. The plants typically occupy low elevation riverine marshes and floodplains but also are widespread in ricefields. Both annual and perennial habits are reported, which may reflect the local growing conditions. The reproductive biology of this genus is poorly known. Like other members of the family, the flowers appear to be adapted for pollination by insects (Insecta); however, they are self-compatible and might commonly self-pollinate. Several species (all within section *Limnophila*) produce submersed, cleistogamous flowers in addition to the typically emergent, chasmogamous ones. The species that occur in deeper waters exhibit heterophylly with respect to their submersed and emergent foliage. Homophyllous emergent species also occur. Many species are amphibious and grow equally well when they are submersed in up to several meters of water, or when they grow as emersed either in shallow water or on wet, exposed substrates.

Limnophila aromatica **(Lam.) Merr.** is an annual or perennial, which grows in ditches, marshes, ponds, pools, prairies, ricefields, and along the margins of rivers. Optimum growth occurs where light levels are high and the water is shallow (<1 m), acidic (pH: 5.0–7.0), of soft or medium hardness, and with temperatures that range from 22°C to 28°C. There are conflicting literature accounts indicating that the flowers are rarely produced in the wild but are prolifically produced in cultivation. Reliable information on the reproductive biology and ecology of this species is scarce. The seeds but can be a minor component of the seed bank in native habitats. This species is homophyllous and does not grow in deep water. The plants can be propagated vegetatively by stem cuttings, which root in water within 2 weeks. The foliage has a pungent, turpentine-like odor. **Reported associates (North America):** none reported.

Limnophila indica **(L.) Druce** is an amphibious perennial of ricefields, riverbanks, and swamps. It grows in still or flowing standing waters where the light levels are high. The water conditions are circumneutral (pH: 6.5–8.3) and range in temperature from 20°C to 25°C (optimum at 22°C–23°C). The substrates commonly are mud. The species is heterophyllous, having finely dissected submersed foliage and fleshy, undivided emergent foliage. The aerial foliage (and precocious flowering) can be induced by the application of abscisic acid. The seeds reportedly have a low germination rate. The plants are propagated readily by shoot cuttings and can be tissue cultured using 1 cm root tip sections that are floated on a 2% sucrose medium for 4 weeks. **Reported associates (North**

America): *Hydrocotyle umbellata, Juncus, Ludwigia repens, Proserpinaca pectinata, Salix, Taxodium.*

Limnophila ×ludoviciana **Thieret** inhabits ditches, lakeshores, and rice fields. It is a hybrid between the previous and the next species. **Reported associates (North America):** none.

Limnophila sessiliflora (**Vahl**) **Blume** is an amphibious perennial, which occurs in canals, channels, depressions, ditches, floodplains, lakes, marshes, ponds, sandbars, sloughs, streams, swamps, and along the margins of rivers and streams. It can inhabit still or flowing waters (to 4 m depth) with a pH ranging from 5.0 to 8.0, total phosphorous to 227 µg/L, and temperatures from 15°C to 28°C; optimal growth occurs at 20°C–26°C. The substrates include Felda (arenic ochraqualfs), sand, and sandy loam. Aerial chasmogamous flowers and submersed cleistogamous flowers are produced, but nothing is known of their pollination biology. Seed production is very high, typically yielding 200–300 seeds/capsule. The seeds require light for germination, which occurs optimally (to 96%) within 5–12 days when kept under submersed conditions at 30°C. Good germination also has been reported for dried seeds that are stored at 15°C–18°C and then incubated at 25°C. The seeds are dispersed locally when large masses of plants break free from the substrate and drift (along with fruits) across the water surface. Studies of propagule banks indicate that the seeds are produced in quantity (averaging more than 400/m^2) and can remain viable in the sediments for more than 30 years. This species is shade adapted and has a low CO_2 compensation point when submersed, which enables it to grow at relatively low light levels (water transparency ranging from 0.9 to 2.2 m). Biotypes that are resistant to ALS inhibiting sulfonylurea herbicides have been reported from Japan. The herbicide resistance has been attributed to an amino acid substitution (Gln for Pro), which induces a conformational change to the secondary structure of the acetolactate synthase protein. The plants often form dense mats by producing a root system that can become up to 1 m long. Vegetative reproduction occurs by leaf fragments and by shoot fragments, which will root readily if sufficient stem material (e.g., 6+ nodes) remains. The vegetative fragments cannot establish in deep water. **Reported associates (North America):** *Hydrilla verticillata, Nuphar, Nyssa sylvatica, Taxodium ascendens.*

Use by wildlife: Few wildlife uses are reported. *Limnophila sessiliflora* is the host plant for several insects (Insecta) including the plume moth (Lepidoptera: Pterophoridae: *Stenoptilodes taprobanes*) in the Old World and a microlepidopteran moth (Lepidoptera: Pyralidae: *Parapoynx*) in North America. A nematode (Nematoda: Aphelenchoididae: *Aphelenchoides fragariae*) can severely damage the foliage of *Limnophila* species. The sweetpotato whitefly (Insecta: Hemiptera: Aleyrodidae: *Bemisia tabaci*) also is a pest of the genus. *Limnophila indica* reportedly produces a fish toxin that is released when the foliage is damaged.

Economic importance: food: The raw or steamed young shoots of *Limnophila aromatica* are edible, have a sour and slightly bitter flavor, and have become a popular vegetable (similar to spinach) in Vietnamese cuisine. The plants have a lemony citrus fragrance and are used as an ingredient in *canh chua*, a sweet and sour Vietnamese soup. They frequently are prepared in a curry sauce. *Limnophila indica* also is eaten as a vegetable; **medicinal:** Leaf extracts of *Limnophila aromatica* are thought to reduce the risk of cardiovascular disease by suppressing the generation of superoxide and nitric oxide free radicals. Essential oils from the plants contain limonene and perillaldehyde, which have antibacterial and antioxidant properties. In traditional Asian medicine the plant has been administered to reduce stomach disorders, stimulate lactation, suppress fever, and as a poultice for leg sores. *Limnophila indica* also is regarded as having antiseptic properties and is mixed with coconut oil as a liniment to treat elephantiasis. It is used in India and the Philippines as a remedy for dysentery and indigestion; **cultivation:** *Limnophila aromatica* is grown as an aquarium or water garden plant and also in "homegardens" for use in local food preparation; *L. indica* and *L. sessiliflora* also are cultivated widely as aquarium and water garden specimens; **misc. products:** *Limnophila* has been considered for use as a water purifying species, to incorporate in the treatment of industrial wastewater effluents; **weeds:** *Limnophila aromatica, L. indica,* and *L. sessiliflora* all are common ricefield weeds. *Limnophila sessiliflora* is listed as a federal noxious weed by the U. S. Department of Agriculture (list of February 1, 2012); **nonindigenous species:** All *Limnophila* species are nonindigenous to the New World and were introduced to North America sometime before 1971 as escapes from cultivation. The first North American record of *L. aromatica* was in 1994 as a local escape from a nursery in Hillsborough County, Florida. The plants are believed to have been introduced to North America initially after the end of the Vietnam War in the 1970s, by Vietnamese immigrants who use it as a key ingredient of traditional Cambodian soup dishes.

Systematics: The monophyly of *Limnophila* is supported by phylogenetic analysis (Figure 5.87) and by distinctive morphological features that include a valvate capsule with a winged, central placental column. At one time the genus was believed to be related to the morphologically similar genera *Stemodia* and *Morgania*; however, it is related most closely to *Dopatrium* and *Hydrotriche*, which resolve as a sister clade (Figures 5.85 through 5.87). Currently, the genus is divided into four sections. Section *Limnophila* contains all species with finely dissected underwater foliage, including *L. indica* and *L. sessiliflora*. *Limnophila aromatica* is placed in section *Striatae*, which includes species with striate calyces. Some accounts (e.g., the 2013 wetland indicator list) treat *L. aromatica* as a subspecies of *L. chinensis* (i.e., *L. chinensis* subsp. *aromatica*), but most authors consider these taxa to represent distinct species, which is the practice followed here. A thorough systematic study of *Limnophila* is needed, as many of the taxonomic distinctions are not compelling and only a few taxa have been investigated phylogenetically. The base chromosome number of *Limnophila* is $x = 17$. *Limnophila indica* and *L. aromatica* consist of morphologically similar diploids and (presumably autoploid) tetraploids ($2n = 34, 68$); whereas only tetraploids have been reported for *L. sessiliflora* ($2n = 68$). Counts of $2n = 51$ also are reported for latter but indicate triploid hybrids, which are not necessarily interspecific. The triploids are sterile and propagate vegetatively. *Limnophila ×ludoviciana* Thieret is regarded as a fertile hybrid

FIGURE 5.87 Phylogenetic relationships of *Limnophila* as inferred by analysis of combined DNA sequence data. The upper clade (subtribe Dopatriinae) resolves with *Dopatrium* and *Hydrotriche* as the sister group to the genus. *Limnophila* ×*ludoviciana* is a putative hybrid between *L. sessiliflora* and *L. indica* and shares the cpDNA of the former. The distribution of OBL North American taxa (in bold) illustrates that this subtribe probably represents a common origin of the habit (*Hydrotriche* is an Old World aquatic). (Adapted from Fritsch, P.W. et al., *Proc. Calif. Acad. Sci.*, 58, 447–467, 2007.)

between *L. indica* and *L. sessiliflora*, which arose when the two species were grown together in cultivation. One surveyed population was tetraploid ($2n = 68$). Some care must be exercised when searching the literature because the name *Limnophila* also applies to a genus of crane flies (Insecta: Diptera: Tipulidae).

Comments: *Limnophila aromatica* (native to southeast Asia and Australia) occurs in Florida; *L. indica* (native to tropical Africa, Asia and Australia) occurs in Florida and Louisiana; *L. sessiliflora* (native to eastern tropical Asia) is reported from California, Florida, Georgia, and Texas. The hybrid *L.* ×*ludoviciana* is found in California and Louisiana.

References: Cook, 1996a,b; Fischer, 2004; Gielis, 2003; Godfrey & Wooten, 1981; Gorai et al., 2014; Imbruce, 2007; Jiwajinda et al., 2002; Johnson, 1999; Kasselmann, 2003; Kukongviriyapan et al., 2007; Les, 2002; Lin et al., 2004; Liu et al., 2005, 2006; Mahler, 1980; Mohan Ram & Rao, 1982; Nguyen, 2006; Nishihiro et al., 2006; Philcox, 1970; Pieroni & Price, 2006; Rao & Mohan Ram, 1981; Sculthorpe, 1967; Speichert & Speichert, 2004; Spencer & Bowes, 1985; Staples & Kristiansen, 1999; Suksri et al., 2005; Thabrew, 1981; Venkata Rao et al., 1989; Venkateswarlu, 1984; Wang et al., 2000; Windeløv, 1998; Wit, 1964; Yang & Yen, 1997; Zettler & Freeman, 1972.

9. *Littorella*

Shore-grass, shore-plantain, shore-weed; littorelle

Etymology: diminutive of the Latin *lītus* meaning "lakeshore" for its habitat

Synonyms: *Plantago* (in part)

Distribution: global: Europe; North America; South America; **North America:** northeastern

Diversity: global: 3 species; **North America:** 1 species

Indicators (USA): OBL: *Littorella americana*

Habitat: freshwater; lacustrine; **pH:** 5.8–8.8; **depth:** 0–4 m; **life-form(s):** emergent herb, submersed (rosulate)

Key morphology: stem stoloniferous; leaves (to 6 cm) rosulate, curved or straight, linear-subulate, flat; flowers unisexual (monoecious), 3–4-merous; ♂ flowers single, peduncled (to 4 cm), the stamen filaments (to 2 cm) strongly exserted, ♀ flowers 2–4, sessile; calyx (to 4 mm) unequal; corolla urceolate; achenes (to 2 mm), slender, oblong

Life history: duration: perennial (stolons); **asexual reproduction:** stolons; **pollination:** wind; **sexual condition:** monoecious; **fruit:** capsules (infrequent); **local dispersal:** seeds; stolons; **long-distance dispersal:** seeds (birds)

Imperilment: (1) *Littorella americana* [G5]; S1 (NY); S2 (MI, NB, VT, WI); S3 (ME, MN, NF, QC)

Ecology: general: All *Littorella* species are obligate aquatics, growing either as completely submersed life forms or as emergents when they become stranded on shorelines. The habitats usually are acidic, oligotrophic, softwater lakes in cold, temperate climates. The plants flower only when they are emersed from the water, as the flowers have no adaptations for functioning when wet. It is difficult to compile accurate life-history information on the North American species (*L. americana*) because of taxonomic confusion with its European congener (*L. uniflora*); however, the biology and ecology of both species appears to be quite similar. Some of the ecological information reported here is based on studies of *L. uniflora* and further research may be necessary to confirm the validity of information as it applies to the North American plants.

***Littorella americana* Fern.** inhabits beaches, shores, and the margins of lakes and ponds. These are plants of oligotrophic softwater habitats (e.g., conductivity <110 µmhos/cm; total alkalinity <50 mg/L) that typically range in acidity from pH: 5.8–8.8. In rare instances, the plants have been observed to occur in highly acidic conditions (pH: 3.9). The plants can survive at depths up to 4 m, but generally occur in much shallower water (e.g., 1.5 m). The substrates consist of clay, gravel, muck, mud, peat (rarely), sand, and sandy muck. Flowering reportedly is quite rare in this species, although sexual reproduction occurs commonly in the European congener. Flowering is initiated in summer (July) and occurs only if plants have become stranded on wet shores; whereas, submersed plants will remain vegetative. The terrestrial plants rapidly develop leaves that have reduced air–space tissues and higher numbers of stomata. Leaf growth also is more rapid under terrestrial conditions and flowering commences within 3–4 weeks. The plants are monoecious and are wind pollinated. When fertile, a male flower (with anthers extended on very long filaments) occurs singly at the tip of the floral scape, with 1–4 sessile female flowers produced at the base. Each fertile plant produces about 8.5 seeds on average. Seed germination is low (averaging <14%) and occurs optimally at 20°C in light, on a moist, bicarbonate-free substrate. However, a much higher germination rate (76%) can be achieved if the seeds initially are dried for 2–4 weeks, which appears to result in a loss of dormancy. The seed bank is long persistent, with the seeds being able to remain dormant for several decades. Specific field studies have estimated their longevity at 9–80 years, and probably more than 40 years. Long-distance dispersal is believed to occur as a consequence of endo- or exozoic seed transport by birds (Aves). The plants are known to be uprooted by ducks (Aves: Anatidae), which arguably are foraging for the edible fruits. *Littorella* is categorized ecologically as an "isoetid" aquatic, i.e., a rosulate life form that is characteristic of oligotrophic habitats. Like others in this ecological group, the plants are adapted to take up much of their CO_2 from the sediments, where the concentrations are substantially higher than in the surrounding waters. The land forms also rely heavily on sediment uptake of CO_2. Leaf uptake also occurs in either case. The plants take up carbon in a complex fashion. At night, they induce a C_4 CAM photosynthetic pathway (resulting in malate fixation); whereas, during the day, they shift to a C_3 mechanism, which decarboxylates the malate and also refixes the CO_2 that is produced as a consequence of their daytime respiration. Generally, higher levels of ambient CO_2 result in a proportionally higher C_3 activity. Photosynthesis is limited by ambient CO_2 concentrations, with relatively higher levels of O_2 being released to the sediments as CO_2 concentrations are increased. Higher concentrations of CO_2 in the water also result in higher water uptake by the plants. Generally, higher CO_2 concentrations yield larger plants with longer and more numerous leaves. Although typically regarded as a species of acidic water, it is interesting to note that the plants not only can tolerate a fairly alkaline pH, but also that comparative studies in North American lakes indicate that individual plant biomass is higher (>50%) in alkaline waters (pH: 7.8) than in acidic waters (pH: 6.0). However, the annual productivity rate per plant is somewhat higher and the plants grow much more densely under acidic, nutrient-poor conditions. The fewer but larger plants characteristic of the more eutrophic sites would possess an advantage in their sexual reproductive capacity because preliminary data indicate that the larger plants are more likely to flower than the smaller ones. Higher productivity has been observed for plants colonized by arbuscular and vesicular mycorrhizae, which thrive in the oxygen released by the roots of the plants. Vegetative reproduction is prominent and occurs by the proliferation of stolons. The vegetative growth of plants is important in stabilizing sand and gravel substrates for colonization by other species. Eutrophication has been identified as a principal factor accounting for the disappearance of *Littorella* from historical localities. **Reported associates:** *Asclepias incarnata, Brasenia schreberi, Carex lacustris, C. limosa, Ceratophyllum, Chara, Dulichium arundinaceum, Elatine minima, Eleocharis acicularis, E. ovata, E. palustris, Elodea canadensis, E. nuttallii, Eriocaulon aquaticum, Gratiola aurea, Heteranthera dubia, Hypericum boreale, Isoetes macrospora, I. tenella, Juncus militaris, J. pelocarpus, Lemna trisulca, Limosella aquatica, Lobelia dortmanna, Lycopus uniflorus, Lysimachia terrestris, Myriophyllum alterniflorum, M. farwellii, M. humile, M. sibiricum, M. spicatum, M. tenellum, Najas flexilis, Nitella, Nuphar advena, N. variegata, Nymphaea odorata, Persicaria amphibia, Potamogeton amplifolius, P. epihydrus, P. gramineus, P. natans, P. perfoliatus, P. praelongus, P. pusillus, P. richardsonii, P. robbinsii, P. spirillus, P. zosteriformis, Ranunculus reptans, Sagittaria graminea, Schizaea pusilla, Schoenoplectus acutus, Sparganium angustifolium, Spiraea alba, Stuckenia pectinata, Subularia aquatica, Typha latifolia, Utricularia cornuta, U. intermedia, U. purpurea, U. resupinata, Vallisneria americana.*

Use by wildlife: *Littorella americana* provides habitat for Oligochaete worms (Annelida: Oligochaeta: Tubificidae: *Ilyodrilus templetoni*) and it is an indicator of fine silts, which harbor burrowing species. Substantial losses of plants have been observed due to the nesting activity of fishes (Osteichthyes).

Economic importance: food: none; **medicinal:** *Littorella* (*L. uniflora*) contains iridoid glycosides, which are attributed with various analgesic, anti-inflammatory, antimicrobial, antitumor, purgative, sedative and vasoconstrictive properties; it also contains phenylethanoid glycosides, which can have antioxidant and antitumor properties; **cultivation:** *Littorella americana* is cultivated occasionally as an aquarium plant; **misc. products:** none; **weeds:** none; **nonindigenous species:** none

Systematics: Perhaps because it is such a small plant, *Littorella* has been picked on extensively by taxonomists! Over the years, *L. americana* has been demoted to a variety (=*L. uniflora* var. *americana*), transferred to a different species (=*L. uniflora*), and even moved to a different genus (=*Plantago uniflora*). However, recent molecular phylogenetic studies indicate that the genus remains distinct from *Plantago*, and resolves as its sister group (Figure 5.88). Although in such cases it would be compatible phylogenetically either to merge the genera or to retain their distinction, the latter is preferable given their morphological and ecological distinctness. Furthermore, retaining the generic names, which have been used in the literature for years, also facilitates retrieval of information for each genus. Because all *Littorella* species are aquatic, there appears to be a single origin of the OBL habit in the genus. Phylogenetic analysis also provides evidence for the distinctness of *L. americana* from *L. uniflora*; they are not sister taxa, despite the widespread practice of treating the former as a synonym or variety of the latter (Figure 5.88). The chromosome number of *Littorella* is $2n = 24$; however, counts are available only for the European *L. uniflora*. Additional cytological work on the genus might be informative.

Comments: *Littorella americana* occurs sporadically throughout northeastern North America

References: Andersen et al., 2006; Arts & van der Heijden, 1990; Aulio, 1986; Bagger & Madsen, 2004; Bekker et al., 1999; Bellemakers et al., 1996; Boston & Adams, 1986, 1987; Farmer & Spence, 1987; Fernald, 1918; Hoggard et al., 2003;

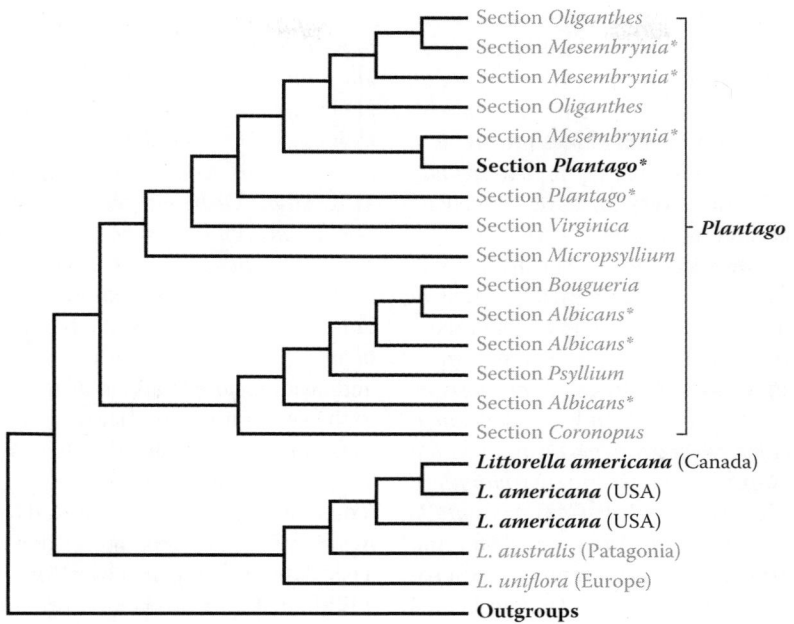

FIGURE 5.88 Relationships of *Plantago* and *Littorella* as indicated by analysis of ITS sequence data. Although some recent authors have advocated the merger of these two genera, the phylogenetic analyses maintain them as distinct sister clades and it would seem more appropriate to preserve nomenclature that has been in common use for centuries. These results also are consistent with those based on analysis of combined ITS and cpDNA data (e.g., Rønsted et al., 2002). Despite the common practice of merging *L. americana* with *L. uniflora*, these results clearly indicate the distinctness of those species. Asterisks indicate sections of *Plantago* that are not monophyletic in the analysis shown. Taxa with OBL North American representatives are indicated in bold. All three *Littorella* species would be categorized as OBL aquatics; however, only *L. americana* is North American. (Adapted from Hoggard, R.K. et al., *Amer. J. Bot.*, 90, 429–435, 2003.)

Ismailoglu et al., 2002; Keeley, 1998; Newmaster et al., 1997; Nichols, 1999a; Nielsen et al., 1991; Nielsen et al., 2004; Pearsall, 1917; Pedersen et al., 2006; Rahn, 1996; Robe & Griffiths, 1998; Rønsted et al., 2000, 2002; Salisbury, 1970; Schoof-van Pelt, 1973; Seddon, 1972; Zika, 1988.

10. *Mecardonia*

Axilflower, baby jump-up

Etymology: after Antonia de Meca y Cardona (18th century)

Synonyms: *Bacopa* (in part); *Erinus* (in part); *Herpestris*; *Lindernia* (in part); *Pagesia*

Distribution: global: New World; **North America:** southern

Diversity: global: 15 species; **North America:** 2 species

Indicators (USA): OBL: *Mecardonia acuminata, M. procumbens*; **FACW:** *M. acuminata, M. procumbens*

Habitat: brackish; freshwater; palustrine, riverine; **pH:** 5.8–7.6; **depth:** <1 m; **life-form(s):** emergent herb

Key morphology: stems square, much branched and prostrate (to 15 cm) or erect (to 6 dm); leaves (to 5 cm) opposite, indistinctly petioled, narrowed basally, serrate apically, glandular punctate; inflorescence solitary, axillary, the pedicels (to 30 mm) subtended by a pair of leaf-like bracts (to 4 mm); calyx (to 9 mm) divided nearly to base; corolla (to 12 mm) bilabiate, tubular, white (purple veined) or lemon yellow; stamens 4, included; capsules (to 6 mm) elliptical; seeds (to 0.3 mm) numerous, reticulate

Life history: duration: perennial (persistent stem bases); **asexual reproduction:** none; **pollination:** unknown; **sexual condition:** hermaphroditic; **fruit:** capsules (common); **local dispersal:** seeds; **long-distance dispersal:** seeds

Imperilment: (1) *Mecardonia acuminata* [G5]; SX (DC, IL); SH (KS); S1 (DE, GA, IN, MD, MO); S3 (KY); (2) *M. procumbens* [G5]

Ecology: general: Very little is known on the life history of *Mecardonia* species. Both of the North American species normally occupy wet (but not inundated) sites, but also can occur in drier waste areas. The flowers of some species are visited by bees (e.g., Hymenoptera: Apidae: *Euglossa*), but specific information on the reproductive ecology of the genus is scarce. Because highly successful seed production occurs in nonindigenous populations of some species, the plants possibly are self-compatible and self-pollinating despite their showy flowers, which appear ideally adapted for insect pollination. Preliminary genetic analyses are consistent with this interpretation. The foliage turns black as it dies back in the fall, causing some to think that the plants are diseased. Although perennial, some of the plants that are sold in the trade are categorized as annuals. Propagation by cuttings has been successful for some of the species.

Mecardonia acuminata **(Walter) Small** grows in freshwater bogs, borrow pits, canals, depressions, ditches, flatwoods, floodplains, glades, gravel bars, marshes, prairies, sand bars, savannas, swamps, and along the margins of ponds and streams at elevations of at least 9 m. It occurs in full sun to partial shade on wet ground or in very shallow standing water. The sites often are only seasonally wet (the moisture primarily from rainwater) and dry out during the summer. The substrates often are thin soil layers that develop over clay loam, gravel, limestone, loam, mud, rock, sand, sandy loam, silty loam, or silty sand. This species has been associated with acidic sites; however, the few reported pH values are alkaline (pH: 7.0–7.6). The plants are self-compatible and frequently are self-pollinating. The detection of very low levels of genetic variation within populations and most genetic variation residing among populations (ISSR markers), is consistent with inbreeding being the primary breeding system. The seeds of *M. acuminata* are a known component of some commercially distributed peat, and probably have been dispersed unwittingly as a soil contaminant. **Reported associates:** *Acer rubrum, Acmella oppositifolia, Allium cernuum, Ambrosia artemisiifolia, Ampelopsis arborea, Andropogon gerardii, Aristida palustris, A. virgata, Axonopus furcatus, Bacopa monnieri, Callicarpa americana, Carex glaucescens, Cassia, Centella erecta, Cirsium muticum, Cladium jamaicense, Colocasia esculenta, Commelina, Cuphea carthagenensis, Cyperus pseudovegetus, C. strigosus, Dalea foliosa, Dichanthelium aciculare, Diodia virginiana, Diospyros virginiana, Drosera intermedia, Eclipta prostrata, Eleocharis microcarpa, E. obtusa, Eragrostis elliottii, Eriocaulon ravenelii, Eupatorium capillifolium, E. coelestinum, E. compositifolium, Euthamia tenuifolia, Fagus grandifolia, Fimbristylis, Flaveria linearis, Fuirena pumila, F. squarrosa, Galactia volubilis, Gelsemium sempervirens, Helenium autumnale, H. pinnatifidum, Helianthus angustifolius, H. ×verticillatus, Heliotropium indicum, Heterotheca subaxillaris, Hymenocallis palmeri, Hypericum denticulatum, H. gymnanthum, H. mutilum, Hyptis alata, Ilex glabra, I. opaca, Iva annua, Juncus filipendulus, J. pelocarpus, J. repens, Juniperus, Justicia americana, Kalmia hirsuta, Lachnanthes caroliana, Leersia, Leucothoe racemosa, Liatris garberi, L. spicata, Ligustrum sinense, Lindera melissifolia, Linum intercursum, Liquidambar styraciflua, Lobelia puberula, Ludwigia linearis, L. linifolia, L. microcarpa, Lycopodiella alopecuroides, Lycopus, Lysimachia quadriflora, Marshallia mohrii, Melanthera angustifolia, Mitreola petiolata, Muhlenbergia capillaris, M. sericea, Myrica, Nyssa biflora, Onoclea sensibilis, Osmanthus americanus, Osmunda regalis, Ostrya virginiana, Oxypolis filiformis, O. rigidior, Panicum aciculare, P. dichotomiflorum, P. tenerum, P. verrucosum, P. virgatum, Parnassia grandifolia, Paspalum dilatatum, P. monostachyum, P. urvillei, Persea palustris, Persicaria hydropiperoides Phlox glaberrima, Photinia pyrifolia, Phyla nodiflora, Physostegia virginiana, Pinus echinata, P. elliottii, P. palustris, P. taeda, Pluchea camphorata, P. foetida, Polygala nana, Polypremum procumbens, Proserpinaca pectinata, P. ×intermedia, Ptilimnium costatum, Pycnanthemum virginianum, Quercus laurifolia, Q. virginiana, Rhexia mariana, R. virginica, Rhus copallina, Rhynchospora caduca, R. corniculata, R. divergens, R. floridensis, R. globularis, R. mixta, R. thornei, R. torreyana, R. tracyi, Rubus, Rudbeckia grandiflora, R. laciniata, R. triloba, Sabal minor, S. palmetto, Sabatia angularis, S. brevifolia, Sarracenia, Schizachyrium rhizomatum, S. scoparium, Scirpus cyperinus, Scleria reticularis, S. triglomerata,* *S. verticillata, Scoparia dulcis, Scutellaria integrifolia, Serenoa repens, Setaria parviflora, Silphium gracile, S. terebinthinaceum, Smilax, Solidago riddellii, S. rigida, S. sempervirens, S. tortifolia, Sorghastrum nutans, Spermacoce terminalis, Stylosanthes biflora, Symphyotrichum dumosum, S. laeve, Ulmus americana, Utricularia biflora, U. purpurea, Vaccinium, Verbena, Vernonia texana, Veronica, Viburnum obovatum, Viola ×primulifolia, Vitis, Xylorhiza tortifolia, Xyris laxifolia, X. tennesseensis.*

***Mecardonia procumbens* (P. Mill.) Small** inhabits fresh or brackish depressions, ditches, flats, floodplains, hammocks, lagoons, lawns, marshes, meadows, ponds, prairies, streams, and swamps at elevations of up to 1200 m. The plants occur in the sun or under partial shade on substrates of gravel, marl, sand, and sandy loam that tend to be acidic (pH: 5.8–6.5). The optimal temperature range is from 21°C to 24°C, with temperatures above 16°C needed to induce prolifically flowering plants. Cold temperatures down to −4°C can be tolerated. The plants remain in flower for an extended period of time and their flower production often is prolific. The pollination biology is unknown; however, large numbers of seeds are produced, even where plants are nonindigenous, which indicates either self-pollination or pollination by highly unspecialized pollinators. Germination of excavated seeds can occur within 1.5 months after they have been soaked in water for 24 h, dried, and provided with a 12 h photoperiod while incubating under a 35°C/25°C temperature regime. A seed bank is formed, with densities of 400 seeds/m² reported. The plants are quite durable and can grow into circular mats reaching up to 2 dm in diameter. **Reported associates:** *Alnus serrulata, Baccharis halimifolia, Byrsonima, Croton, Cynodon dactylon, Cyperus pseudovegetus, Dichanthelium dichotomum, D. sphaerocarpon, Elephantopus carolinianus, Eragrostis hypnoides, Euphorbia, Fimbristylis autumnalis, Fuirena simplex, Hypericum brachyphyllum, H. mutilum, Juncus nodatus, J. tenuis, Leucospora multifida, Ludwigia alternifolia, Lycopus virginicus, Osmunda regalis, Passiflora, Pectis, Pinus taeda, Pluchea camphorata, Rhynchospora glomerata, R. inexpansa, Salix nigra, Solidago rugosa, Sorghum halepense, Taxodium, Verbena, Viola ×primulifolia.*

Use by wildlife: *Mecardonia acuminata* is eaten by cattle (Mammalia: Bovidae: *Bos taurus*) and white-tailed deer (Mammalia: Cervidae: *Odocoileus virginianus*), especially in late fall and winter. It is a host plant for larval plume moths (Insecta: Lepidoptera: Pterophoridae: *Stenoptilodes brevipennis, S. taprobanes*) and for whiteflies (Insecta: Homoptera: Aleyrodidae: *Bemisia tabaci*).

Economic importance: food: no uses reported; **medicinal:** *Mecardonia procumbens* was used by the Nahua people of Central America and Mexico for dermatological problems. It also has been used to treat abscesses; **cultivation:** *Mecardonia procumbens* is a cultivated ornamental distributed mainly as the popular cultivar 'Gold Flake' [or 'USMECA67'], which is a hybrid between *M. caespitosa* (female) and *M. procumbens* (male). 'Sunmecareki' (also known as 'Prima Lemon Yellow') is another hybrid cultivar that is attributed to *M. procumbens*; **misc. products:** The 'Gold Flake' cultivar of *M. procumbens* is widely planted as a groundcover; **weeds:** *Mecardonia procumbens* has become a serious nuisance in parts of India. It also is reported as invasive in many other sites where it has been introduced; **nonindigenous species:** *Mecardonia procumbens* is naturalized in Africa, Australia (Queensland), India, Japan (Bonin Islands), Java, Nepal, Papua New Guinea, Taiwan (by 1999) and on the Pacific Islands of Babeldaob, Guam, Namoluk, Nauru, Peleliu, Pohnpei, Puluwat, Rota, Saipan, Tinian, Weno and Yap, where it often occurs in disturbed or cultivated sites.

Systematics: Although *Mecardonia* and *Bacopa* have been merged by some authors, analyses of various molecular datasets indicate that the genera are closely related but distinct phylogenetically (Figures 5.82, 5.85, and 5.86). *Mecardonia* is coherent morphologically and appears to be monophyletic as currently circumscribed (e.g., Figure 5.86); however, a comprehensive phylogenetic survey of the genus has not been conducted and is needed to test this assumption. The base chromosome number of *Mecardonia* is x = 11. *Mecardonia procumbens* is diploid (2n = 22); whereas, *M. acuminata* is tetraploid (2n = 44). Artificially induced tetraploids of *M. procumbens* have been synthesized and these grow slower, delay flowering, but flower longer than the diploids. A supernumerary ("B") chromosome has been found in *M. procumbens*. Several infraspecific taxa have been described for *Mecardonia acuminata*. The integrity of these taxa has been evaluated by morphological and genetic (ISSR markers) analyses, which fail to uphold the distinctness of *M. acuminata* var. *microphylla*, but do indicate the integrity of *M. acuminata* var. *peninsularis*. Natural hybrids are not reported in North America, but several hybrid interspecific cultivars have been synthesized, and these appear to retain at least some fertility.

Comments: *Mecardonia acuminata* occurs in the southeastern United States with *M. procumbens* ranging through the southern United States and extending into South America.

References: Ahedor & Elisens, 2006; Beal, 1977; Bostick, 1977; Chen & Wu, 2001; Clewell, 1985; D'Arcy, 1979; Fischer, 2004; Fosberg et al., 1979; Gielis, 2003; Godfrey & Wooten, 1981; Heinrich, 2000; Kaul, 1975; Kellman, 1978; Lewis et al., 1962; Long & Lakela, 1971; Neyland et al., 2004; Orzell & Bridges, 2006; Pennell, 1935; Porter, Jr., 1967; Quattrocchi, 2000; Rossow, 1987; Sinha, 1987; Takashi, 1999; Thill, 1984.

11. *Plantago*

Plantain; queue de rat

Etymology: from the Latin *planta*, meaning "sole of the foot" and referring to the appearance of the leaves when flattened on the ground

Synonyms: *Arnoglossum*; *Asterogeum*; *Bougueria*; *Coronopus*; *Lagopus*; *Psyllium*

Distribution: global: cosmopolitan; **North America:** widespread

Diversity: global: 270 species; **North America:** 30 species

Indicators (USA): OBL: *Plantago bigelovii, P. cordata, P. macrocarpa*; **FACW:** *P. australis, P. macrocarpa, P. sparsiflora*

Habitat: freshwater; riverine; **pH:** 4.5–7.9; **depth:** <1 m; **life-form(s):** emergent herb

Key morphology: stem a short caudex; leaves (to 50 cm) in a basal rosette, attenuate to cordate at base, sessile or staked by a pseudopetiole (to 15 cm), venation arcuate and palmate or parallel; inflorescence of 1-many elongate, axillary spikes (to 60 cm); flowers 4-merous; sepals (to 3.5 mm) dry, imbricate, persistent, the margins membraneous; petals (to 3.5 mm) whitish, connate basally, withering in fruit; stamens (to 2.3 cm) 2–4, strongly exserted; capsules (to 11 mm) circumscissile; seeds (to 3.8 mm) 1–9 per capsule

Life history: duration: annual (fruit/seeds); perennial (persistent caudex); **asexual reproduction:** rhizomes, rosette division; **pollination:** self, wind (chasmogamous); self (cleistogamous); **sexual condition:** hermaphroditic; **fruit:** capsules (common); **local dispersal:** seeds (gravity, water); **long-distance dispersal:** seeds (animals)

Imperilment: (1) *Plantago australis* [GNR]; (2) *P. bigelovii* [G4]; (3) *P. cordata* [G4]; SH (DC, FL, IA, KY, MD, VA); S1 (AL, IL, IN, MI, MS, NC, OH, <u>ON</u>, TN, WI); S2 (AR); S3 (GA, MO, NY); (4) *P. macrocarpa* [G4]; S1 (OR); S2 (WA); S3 (<u>BC</u>); (5) *P. sparsiflora* [G3]; S1 (NC); S2 (GA); S3 (FL)

Ecology: general: *Plantago* is diverse ecologically and can be found in habitats ranging from bogs to deserts, mountains, savannas, sea cliffs, and woodlands. Many wet sites typically have at least one *Plantago* species present. Of the 30 North American species, 22 are designated as wetland indicators, three categorized as OBL in some portion of their range. *Plantago australis* and *P. sparsiflora* (FACW) formerly were included among the OBL taxa in the 1996 indicator list. (Because their treatments were completed prior to the release of the 2013 list, they have been retained here.) Members of the genus also occur frequently in disturbed areas and often grow in naturally disturbed sites. The species include both annual and perennial life histories. All species are self- or wind-pollinated, long-day flowering plants with hermaphroditic flowers. Female plants have been reported occasionally in one species. Although most species are self-incompatible, several annuals and some of the perennials have derived self-compatibility. Some of the annual, self-compatible species also bear cleistogamous flowers, which produce relatively low quantities of pollen that is shed directly on the anthers before the flowers open. More pollen is produced in those species that possess chasmogamous flowers. The annuals have a higher reproductive output, which averages five times more fruits and seeds, and over four times the weight of seeds for a given leaf area, than the perennials. Most *Plantago* species produce a circumscissile capsule with seeds that become mucilaginous (and adhesive) when they are wet. Dispersal of seeds by adhesion is common in the genus.

Plantago australis **Lam.** is an annual, which inhabits bottoms, cienega, depressions, marshes, meadows, springs, and river margins at elevations of up to 2347 m. The substrates include sandy loam. The plants are self-compatible and produce cleistogamous, self-pollinating flowers. Seed production is high, with plants averaging 166 capsules per inflorescence. The seeds generally are categorized as gravity dispersed;

however, they also are known to be eaten and dispersed by mammals. Little additional ecological information exists for the nonindigenous North American portion of its range. **Reported associates (North America):** *Alnus tenuifolia, Ambrosia monogyra, Baccharis salicifolia, Carex, Celtis reticulata, Clematis drummondii, Condalia, Funastrum cynanchoides, Juncus bufonius, Lycium, Mimulus guttatus, Nasturtium officinale, Pinus ponderosa, Polypogon monspeliensis, Populus fremontii, Prosopis juliflora, P. velutina, Salix gooddingii, Salsola, Senegalia greggii, Tamarix, Yucca elata.*

Plantago bigelovii **A. Gray** is an annual plant of coastal bluffs, depressions, dunes, flats, ledges, marshes, meadows, mudflats, salt marshes, seeps, slopes, swales, vernal pools/lakeshores, and wallows at elevations of up to 700 m. The substrates are alkaline or saline and include clay, gravel, loam, mud, rock, sand, sandstone, and silty clay. Cleistogamous, self-pollinating flowers are produced in the early spring and have a high seed output, which averages 10.2 capsules per inflorescence and yields on average 2.26 capsules with 2.92 seeds (2.31 mg) per 10 cm^2 of leaf tissue. The seeds are dispersed by animals (endozoically) and are known to remain viable in the pellets of rabbits (Mammalia: Leporidae). **Reported associates:** *Amblyopappus pusillus, Anagallis minima, Arthrocnemum subterminale, Atriplex coronata, A. coulteri, A. parishii, A. semibaccata, Blennosperma nanum, Bromus hordeaceus, Centaurium muehlenbergii, Centromadia parryi, Cicendia quadrangularis, Cordylanthus palmatus, Crassula aquatica, C. connata, Cressa truxillensis, Deschampsia danthonioides, Dudleya greenei, Eleocharis macrostachya, Eryngium vaseyi, Entosthodon fascicularis, Fragaria, Frankenia salina, Hemizonia kelloggii, Hordeum depressum, Juncus bufonius, J. uncialis, Lasthenia californica, L. fremontii, Lepidium latipes, Lupinus, Lythrum hyssopifolia, Marsilea, Mesembryanthemum nodiflorum, Micranthes occidentalis, Microseris acuminata, Myosurus minimus, Parapholis incurva, Plagiobothrys austiniae, P. glyptocarpus, P. humistratus, P. leptocladus, P. scouleri, P. scriptus, P. stipitatus, Plantago elongata, P. erecta, Platanthera elegans, Pogogyne abramsii, Psilocarphus brevissimus, P. elatior, Senecio vulgaris, Spergularia macrotheca, S. platensis, S. salina, Trifolium depauperatum.*

Plantago cordata **Lam.** is a perennial inhabitant of floodplains and shallow (e.g., 10 cm) woodland streams and rivers. The substrates are circumneutral and include cobble, gravel, limestone, rock, sand, and slate. The plants can tolerate complete inundation for up to 4 months and sometimes occur under tidal, slightly brackish (salinity to 5 ppm) conditions. Occurrences typically are exposed to full sunlight early in the season before leaves are produced by the overstory vegetation, but the sites eventually become shady when the canopy closes. The young inflorescences are produced in the fall and overwinter in the protection of the petiolar leaf bases. Flowering begins in March to April, before canopy closure; however a second flowering event can occur occasionally in September to October. Anywhere from 10% to 74% of plants will in flower in a given year. An average of six spikes is produced

per plant. The flowers are chasmogamous and protogynous but are self-compatible, which facilitates geitonogamy. Allozyme data indicate a mixed breeding system. Although up to 130 flowers can be produced on a single spike, the species has a low reproductive output, producing on average 0.47 capsules and 0.76 seeds (0.98 mg) per 10 cm^2 of leaf tissue. Each capsule normally yields 2–3 seeds, which typically are short lived and survive only for a month when stored at 4°C. Seeds that are dried to 4%–5% moisture or that are exposed to liquid nitrogen remain viable and germinate best under an alternating temperature/light regime of 30°C (16 h light) and 15°C (8 h darkness). Flowering occurs 1 year after the plants first establish. This species has a specialized water dispersal mechanism. The seeds are attached to a spongy, buoyant placenta when shed, and are dispersed on the water surface. Upon contact with the water, the seed coat becomes mucilaginous and swells, which liberates the seeds from the placenta. The freed seeds float and will attach readily to most surfaces with which they come into contact due to the adhesiveness of the mucilage. Seed germination normally commences within 6–14 days, even if the seeds remain afloat. The seeds will not germinate under water. The seedlings remain afloat but must be deposited on a favorable substrate within a week in order to establish or they will sink and die. The plants are adapted to the stable stream environments that exist within climax hardwood forests. They cannot tolerate highly unstable, man-made environments and are quite susceptible to hydrological alterations, erosion, and excessive deposition of gravel bars where their reproductive threshold cannot be maintained. The corky caudex can become displaced horizontally as a rhizome-like rootstock. Studies of morphological characters have indicated that populations retain substantial genetic variability; however, electrophoretic data indicate that little genetic variation exists within populations despite the observation of extensive genetic differentiation among populations (e.g., $G_{ST} = 0.865$). The pattern of genetic variation is highly structured geographically, with populations north of the glacial boundary being indistinguishable by allozyme analysis. The characteristic broad-bladed foliage does not develop until summer, once flowering has concluded and the canopy has closed. **Reported associates:** *Acer negundo, A. rubrum, A. saccharum, Actaea, Alisma, Allium canadense, Ambrosia trifida, Amelanchier, Anemone quinquefolia, Arisaema triphyllum, Asarum canadense, Asclepias incarnata, Bidens, Boehmeria cylindrica, Calamagrostis, Campanula americana, Cardamine bulbosa, Carex albursina, C. laxiflora, C. retroflexa, Carpinus caroliniana, Carya glabra, C. ovata, Cephalanthus occidentalis, Chasmanthium latifolium, Cicuta, Circaea canadensis, Claytonia virginica, Cornus alternifolia, C. florida, C. racemosa, Corylus americana, Crataegus, Cuscuta, Cyperus, Decodon verticillatus, Desmodium illinoense, Dichanthelium boscii, Dryopteris marginalis, Equisetum arvense, E. hyemale, Erythronium americanum, Euonymus atropurpureus, E. obovatus, Eupatorium rugosum, E. serotinum, Fagus grandifolia, Floerkea proserpinacoides, Fragaria virginiana, Fraxinus americana, F. nigra, Galium aparine, G. circaezans, Geranium maculatum, Gleditsia triacanthos, Glyceria* *striata, Geum rivale, Hamamelis virginiana, Helianthus, Hepatica nobilis, Hydrophyllum virginianum, Hystrix patula, Impatiens, Iris, Lindera benzoin, Liriodendron tulipifera, Lobelia siphilitica, Lonicera dioica, Luzula campestris, Lycopus americanus, Lysimachia ciliata, L. nummularia, L. quadrifolia, Mertensia virginica, Mimulus alatus, Mitella diphylla, Muhlenbergia sobolifera, Nasturtium officinale, Onoclea sensibilis, Osmorhiza longistylis, Ostrya virginiana, Panax trifolius, Parthenocissus quinquefolia, Pellaea atropurpurea, Peltandra virginica, Phaseolus polystachios, Phlox divaricata, Phryma leptostachya, Physocarpus opulifolius, Pilea pumila, Plantago lanceolata, P. rugelii, Platanus occidentalis, Podophyllum peltatum, Polemonium reptans, Polygonatum pubescens, Persicaria virginiana, Polystichum acrostichoides, Pontederia cordata, Populus deltoides, P. tremuloides, Prenanthes alba, Prunella vulgaris, Prunus virginiana, Quercus alba, Q. muhlenbergii, Q. rubra, Ranunculus abortivus, R. recurvatus, R. septentrionalis, Rhamnus cathartica, Ribes cynosbati, Rumex obtusifolius, R. verticillatus, Sagittaria, Salix interior, S. rigida, Sambucus nigra, Sanguinaria canadensis, Sanicula marilandica, Scirpus atrovirens, Smallanthus uvedalius, Smilax bonanox, Solanum dulcamara, Solidago flexicaulis, Sorghastrum nutans, Staphylea trifolia, Symphyotrichum cordifolium, S. lateriflorum, S. novae-angliae, Thalictrum dioicum, T. revolutum, Thaspium barbinode, Thuja occidentalis, Tiarella cordifolia, Tilia americana, Toxicodendron radicans, Trillium grandiflorum, T. recurvatum, Typha latifolia, Ulmus americana, U. rubra, Viburnum lentago, Viola pubescens, V. renifolia, V. rostrata, V. sororia, Verbena hastata, Vitis riparia, Zanthoxylum americanum, Zizia aurea.*

***Plantago macrocarpa* Cham. & Schlecht.** is a perennial, which inhabits beaches, bogs, depressions, dunes, estuaries, fens, marshes, meadows, mires, mudflats, ponds, shores, slopes, sloughs, tidal flats, tundra, and the margins of lakes, rivers, and streams at elevations of up to 120 m. The plants can tolerate shallow standing water and saline sites, just above high tide limit. The habitats are open. Substrates are categorized as fibrisols, gleysols, humic gleysols, or ultramafic, and include mud, organic peat (pH: 4.5), organic silt, rock, sand, sandy muck, serpentine, silt, and stones. This species grows in cold regions to 60.7° N latitude and frequently occurs along snowbed margins. On average the plants produce extremely few seeds (1.1/capsule) but also extremely large seeds (avg. 2.9 mg). Unlike other *Plantago* species, the two-seeded fruits are indehiscent. The pollen:ovule ratio is extremely high (about 1200 pollen grains per seed set), which is indicative of an outcrossing breeding system. **Reported associates:** *Achillea millefolium, Agrostis aequivalvis, A. alba, A. exarata, Alnus, Anemone narcissiflora, A. oregana, Angelica lucida, Athyrium filix-femina, Bistorta vivipara, Blechnum spicant, Boykinia elata, B. intermedia, B. major, Brachythecium, Calamagrostis canadensis, Camassia quamash, Carex anthoxanthea, C. aquatilis, C. echinata, C. interior, C. lenticularis, C. livida, C. lyngbyei, C. macrochaeta, C. muricata, C. obnupta, C. phyllomanica, C. pluriflora, C. rostrata, C. utriculata, C. vesicaria, Castilleja miniata,*

Cetraria islandica, Cladina mitis, C. stellaris, C. rangiferina, C. maxima, Comarum palustre, Conioselinum chinense, C. gmelinii, C. pacificum, Coptis trifoliata, Cornus canadensis, C. suecica, Daucus carota, Deschampsia cespitosa, D. elongata, Dicranum scoparium, Drosera rotundifolia, Eleocharis palustris, Empetrum nigrum, Equisetum arvense, Erigeron peregrinus, Eriophorum chamissonis, Festuca rubra, Fritillaria camschatcensis, Galium aparine, G. trifidum, Gaultheria shallon, Gentiana sceptrum, Geranium erianthum, Geum calthifolium, G. macrophyllum, Heracleum lanatum, Hierochloe odorata, Hippuris vulgaris, Hordeum brachyantherum, Huperzia selago, Hylocomium splendens, Hypericum anagalloides, H. formosum, Iris setosa, Juncus arcticus, J. balticus, J. effusus, J. nevadensis, J. supiniformis, Kalmia microphylla, K. occidentalis, Lathyrus palustris, Leymus arenarius, L. mollis, Lilaeopsis occidentalis, Linnaea borealis, Loiseleuria procumbens, Lophocolea cuspidata, Luzula campestris, L. parviflora, Lycopus uniflorus, Lysichiton americanus, Lysimachia maritima, Maianthemum dilatatum, Malus fusca, Mentha arvensis, Menziesia ferruginea, Myrica gale, Nephroma arcticum, Nephrophyllidium crista-galli, Peltigera canina, Phalaris arundinacea, Picea sitchensis, Pinguicula vulgaris, Platanthera dilatata, P. greenei, P. stricta, Pleurozium schreberi, Poa komarowii, P. pratense, Polygonum aviculare, Potentilla anserina, P. egedii, P. pacifica, Primula jeffreyi, P. pauciflora, Prunella vulgaris, Ptilidium ciliare, Puccinellia, Racomitrium lanuginosum, Ranunculus cymbalaria, R. flammula, R. occidentalis, R. orthorhynchus, Rhamnus purshiana, Rhododendron groenlandicum, Rhytidiadelphus squarrosus, Rubus pedatus, R. spectabilis, Rumex acetosella, Salix, Schoenoplectus pungens, Senecio triangularis, Sidalcea hendersonii, Sisyrinchium angustifolium, Sium suave, Sphagnum, Spiraea douglasii, Stellaria humifusa, Streptopus amplexifolius, Symphyotrichum foliaceum, S. subspicatum, Thuja plicata, Torreyochloa pauciflora, Trianthella glutinosa, Trichophorum cespitosum, Trientalis europaea, Trifolium wormskjoldii, Triglochin maritimum, Trisetum canescens, Tsuga heterophylla, Vaccinium alaskense, V. ovalifolium, V. oxycoccos, V. parvifolium, V. uliginosum, Veratrum californicum, Vicia americana, Viola palustris, Xerophyllum tenax.

***Plantago sparsiflora* Michx.** is a perennial of coastal plain ditches, flatwoods, roadsides, savannas, and seeps at elevations to at least 6 m. The typical habitats are seasonally wet, open sites including light gaps. The substrates are circumneutral and include loam, marl, muck, mud, and sand. Typically the surface soils are acidic but are underlain by limestone. Only about 8% of the plants in a population will be in flower simultaneously and the reproductive output of the species is low, averaging just 1.9 inflorescences per plant and producing 0.80 capsules with 1.02 seeds (0.96 mg) per 10 cm² of leaf tissue. Generally there are only two seeds produced per capsule. A survival rate of up to 83% has been reported for populations that have been transplanted to favorable microsites. This species occurs often on mowed, wet roadsides and sometimes on lawns. It is regarded as an indicator species for fire-maintained flatwoods-savanna communities and requires frequent fires (1–10 year return intervals) to maintain the open character of the savannas necessary for its continued survival. **Reported associates:** *Acer, Agalinis purpurea, Allium, Andropogon capillipes, A. gerardii, A. glomeratus, A. mohrii, Apocynum cannabinum, Arnoglossum ovatum, Bigelowia nudata, Carex lutea, Centella erecta, Chaptalia tomentosa, Coreopsis linifolia, Ctenium aromaticum, Dichanthelium dichotomum, D. strigosum, Erigeron vernus, Eryngium yuccifolium, Eupatorium leucolepis, E. mohrii, Fimbristylis puberula, Gordonia, Helenium vernale, Helianthus angustifolius, Hyptis alata, Ilex glabra, Liatris spicata, Lyonia, Marshallia graminifolia, Muhlenbergia expansa, Myrica cerifera, Nyssa biflora, Panicum anceps, P. verrucosum, P. virgatum, Parnassia caroliniana, Paspalum praecox, P. urvillei, Pinus elliottii, P. serotina, P. palustris, P. taeda, Pityopsis graminifolia, Pluchea rosea, Polygala hookeri, Quercus pumila, Rhexia alifanus, Rhynchospora colorata, R. debilis, R. elliottii, R. oligantha, R. rariflora, R. thornei, Schizachyrium scoparium, Scleria muehlenbergii, Serenoa repens, Setaria parviflora, Solidago stricta, Sporobolus teretifolius, Symphyotrichum dumosum, S. walteri, Thalictrum cooleyi, Tridens ambiguus, Vaccinium.*

Use by wildlife: The fruits and seeds of *Plantago australis* are eaten by wild mountain tapirs (Mammalia: Tapiridae: *Tapirus pinchaque*) in South America. The seeds of *P. bigelovii* are eaten by brush rabbits (Mammalia: Leporidae: *Sylvilagus bachmani*), which act as their dispersal agents. *Plantago cordata* is grazed by white-tailed deer (Mammalia: Cervidae: *Odocoileus virginianus*), which can negatively impact populations considerably. *Plantago macrocarpa* is grazed havily by bear (Mammalia: Ursidae: *Ursus*) and also by deer (Mammalia: Cervidae: *Odocoileus*).

Economic importance: food: Despite their mutual common names, *Plantago* should not be confused with the banana-like plantain (Musaceae: *Musa*), which is a staple food in many tropical areas. Several species of *Plantago* also are edible. The leaves of *P. australis* are eaten either when raw or cooked. Native Alaskans eat the young, raw leaves of *P. macrocarpa* in salads or cook them as a spinach-like vegetable. The foliage of *P. cordata* is edible and is said to be quite tender when it has been cooked; **medicinal:** In tropical America, *Plantago australis* is used as an anti-inflammatory, antiseptic, and for treating amebic dysentery, general infections, malaria, ulcers, and urinary tract infections. The Tolewa and Yurok of North America have made a poultice from the steamed leaves of the plant to treat boils and cuts. Assays verify that the plant extracts exhibit high antiviral activity, but not against human immunodeficiency virus (HIV). However, chronic oral ingestion of the crude extracts can have mildly toxic effects on liver function. Infusions of the foliage reportedly are hypoglycaemic and are used in Mexico to treat diabetes. The Houma mixed grease or oil with the raw leaves of *P. cordata* to make a curative poultice for application to boils, burns, cuts, and sores. The Potawatomi used the dried or fresh roots of *P. cordata* to treat burns or as an ingredient in tea to remedy sore feet or upset stomachs. *Plantago cordata* and other species reportedly are anticarcinogenic. Testing of several species has demonstrated cytotoxic activity, arguably due to the occurrence of the flavonoid luteolin-7-O-β-glucoside, which is known to inhibit the proliferation of human cancer cells. The Aleuts prepared a boiled root decoction from *P. macrocarpa* for use as a curative tonic; **cultivation:** *Plantago cordata* has been cultivated as an ornamental water garden plant; **misc. products:** *Plantago macrocarpa* has been used in Alaska to dye yarn; **weeds:** *Plantago australis* is reported as a weed in Australia, Hawaii, and New Zealand as well as in many parts of its native range; **nonindigenous species:** *Plantago australis* is naturalized in New Zealand and Australia (arriving in Tasmania sometime after 1970).

Systematics: Several molecular systematic studies verify that *Plantago* is monophyletic, but only if the monotypic segregate genus *Bougueria* is included. It is the sister genus to *Littorella* (Figure 5.88), which some authors prefer to include within *Plantago* as a section. Few OBL *Plantago* species have been included in phylogenetic analyses, making it difficult to evaluate the distribution of the habit in the genus. The condition would seem to have evolved multiple times, given that the OBL species have been classified in different sections of subgenus *Plantago*, e.g., section *Micropsyllium* (*P. bigelovii*), section *Palaeopsyllium* (*P. cordata*), section *Plantago* (*P. sparsiflora*), and section *Virginica* (*P. australis*); however, phylogenetic analyses (e.g., Figure 5.88) also indicate that several sections (notably section. *Plantago*) are not monophyletic and further clarification of sectional limits is necessary. Combined DNA sequence data weakly support the association of *P. australis* and *P. myosuros*. Both are reasonably supported within a clade that represents section *Virginica* (subgenus *Plantago*). Analysis of nrITS data resolve *P. sparsiflora* (section *Plantago*) and *P. camtschatica* (section *Mesembrynia*) as sister species, but again only with weak support. The remaining OBL species have not yet been included in any comprehensive phylogenetic survey, although *P. bigelovii* sometimes is treated as a synonym of *P. elongata* (FACW; also in section *Micropsyllium*), which resolves as the sister to *P. heterophylla* (FAC, FACW). However, *P. bigelovii* ($x = 5$) and *P. elongata* ($x = 6$) represent different chromosomal series and the question of their distinctness will require further study for clarification. Additional phylogenetic insights will be possible only when a much denser survey of taxa within this large genus has been carried out. The basic chromosome numbers of *Plantago* include $x = 4$, 5, or 6. *Plantago bigelovii* ($2n = 20$) and *P. cordata* ($2n = 24$) are tetraploids. Natural hybrids have not been reported for any of the OBL species. Experimental crosses with *P. cordata* have yielded few, mainly inviable seeds (e.g., with *P. sparsiflora* as the maternal parent), and a single, highly sterile F_1 hybrid, with *P. eriopoda* (also $2n = 24$) as the paternal parent.

Comments: The OBL species are distributed rather distinctly, with *Plantago macrocarpa* in northwestern North America, *P. bigelovii* in western North America, *P. cordata* in eastern North America, and *P. sparsiflora* in the southeastern United States. *Plantago australis* is a native of Mexico and South America, which has been introduced to Arizona, Montana, and New Mexico.

References: Abad et al., 1999; Andrade-Cetto & Heinrich, 2005; Bank II, 1953; Bowles & Apfelbaum, 1989; Caraballo et al., 2004; Chambers, 2001; Coelho de Souza et al., 2004; D'Arcy, 1971; Dawe et al., 2000; Downer, 2001; Downing, 1996; Gálvez et al., 2003; Glitzenstein et al., 2001; Godfrey, 1961; Godfrey & Wooten, 1981; Harper, 1903, 1944, 1945; Harvey, 2000; Hoggard et al., 2003; Johnson, 1982; MacKenzie et al., 2000; Meagher et al., 1978; Miller et al., 1992; Mohlenbrock,1959b; Mymudes, 1991; Mymudes & Les, 1993; Palmeiro et al., 2002, 2003; Palomino et al., 2002; Pence & Clark, 2005; Persson, 1952; Primack, 1978a,b, 1979; Rahn, 1996; Rønsted et al., 2000, 2002; Rozefelds et al., 1999; Smith, 1973; Speck, 1941; Talbot & Talbot, 1994; Tessene, 1969; Zedler & Black, 1992.

12. *Stemodia*

Goat-weed, twintip

Etymology: from the Latin *stemo diacra* meaning "two-peaked stamen," in reference to the shape of some anthers

Synonyms: *Anamaria*; *Angervillea*; *Choadophyton*; *Dickia*; *Lendneria*; *Matourea*; *Morgania*; *Phaelypea*; *Poarium*; *Stemodiacra*; *Unanuea*; *Unannea*; *Valeria*; *Verena*

Distribution: global: Africa; Asia; Australia; neotropics; **North America:** southwestern

Diversity: global: 56 species; **North America:** 3 species

Indicators (USA): OBL; FACW: *Stemodia durantifolia*

Habitat: freshwater; riverine; **pH:** unknown; **depth:** <1 m; **life-form(s):** emergent herb

Key morphology: stems (to 1 m) erect, glandular hairy, somewhat woody at base; leaves (to 6 cm) opposite or whorled (in 3s), lanceolate, becoming smaller upward, bases auriculate, clasping, the margins serrate; inflorescence solitary, axillary, becoming racemose apically; flowers short pedicelled (to 0.5 mm); calyx (to 2 mm) 5-parted, nearly free to base; corolla (to 6 mm) blue to purple, bilabiate, tubular, the throat hairy; stamens 4, didynamous, anther thecae separated by a globose connective; capsule (to 3 mm) conical or ovoid, loculicidal and septicidal; seeds (to 0.3 mm) numerous, blackish brown, tuberculate

Life history: duration: annual (fruit/seeds); perennial (persistent stem bases); **asexual reproduction:** none; **pollination:** insect; **sexual condition:** hermaphroditic; **fruit:** capsules (common); **local dispersal:** seeds; **long-distance dispersal:** seeds

Imperilment: (1) *Stemodia durantifolia* [G5]; S2 (CA)

Ecology: general: Many *Stemodia* species grow in temporarily inundated habitats such as damp depressions, riverbanks, and the margins of mangrove swamps. Among the few North American species, only *S. durantifolia* occurs regularly in association with wet conditions. Although it is designated as an OBL wetland indicator in parts of North America, it also is ranked as FACW and can occur in association with Saguaro cacti and other desert species (see later). The floral morphology of *Stemodia* species is adapted for insect pollination and field observations in Australia indicate that the flowers are visited by a number of bees (Insecta: Hymenoptera: Apidae: *Apis*; Halictidae: *Lasioglossum*; Megachilidae). However, the ecology of most *Stemodia* species has not been studied thoroughly and few other aspects of their life history are known in any detail.

Stemodia durantifolia **(L.) Sw.** inhabits brooks, channels, crevices, drying riverbeds, floodplains, meadows, outflows, pools, seeps, springs, streams, washes, and the margins of pools and streams at elevations of up to 1463 m. The plants occur often in shallow water and grow on substrates that are described as granite, gravel, mud, rock, sand, sandy alluvium, silty sand, and stones. The life history of this species has not been investigated, even though it is quite common in the neotropics. Flowering occurs throughout the year, but other aspects of the reproductive ecology have not been studied. Different accounts describe the plants both as annuals and perennials, but without specific documentation. Most reports indicate a perennial habit and it is conceivable that the duration might vary depending on specific habitat conditions. The stems can become partly woody at the base, which probably allows the plants to persist. **Reported associates:** *Abutilon*, *Acacia angustissima*, *Adiantum*, *Agave chrysantha*, *A. schottii*, *Alhagi maurorum*, *Ambrosia ambrosioides*, *A. cordifolia*, *A. deltoidea*, *A. monogyra*, *Andropogon glomeratus*, *Aristida purpurea*, *Atriplex lentiformis*, *Avena fatua*, *Baccharis salicifolia*, *B. sarothroides*, *Bebbia juncea*, *Bidens*, *Boerhavia spicata*, *B. wrightii*, *Bothriochloa barbinodis*, *B. ischaemum*, *B. laguroides*, *Brickellia coulteri*, *Brodiaea orcuttii*, *Calibrachoa parviflora*, *Calliandra*, *Camissonia*, *Capsella bursa-pastoris*, *Carnegiea gigantea*, *Celtis reticulata*, *Cephalanthus occidentalis*, *Cheilanthes*, *Chilopsis linearis*, *Cirsium*, *Cupressus forbesii*, *Cylindropuntia acanthocarpa*, *C. bigelovii*, *C. fulgida* *Cynodon dactylon*, *Cyperus difformis*, *C. odoratus*, *Datura wrightii*, *Dodonaea viscosa*, *Echinocereus triglochidiatus*, *Encelia farinosa*, *Ephedra aspera*, *Equisetum*, *Eragrostis cilianensis*, *E. lehmanniana*, *Ericameria laricifolia*, *Eriogonum polycladon*, *Erodium cicutarium*, *Eucalyptus microtheca*, *Eysenhardtia orthocarpa*, *Ferocactus cylindraceus*, *Fouquieria splendens*, *Fraxinus velutina*, *Funastrum*, *Gossypium thurberi*, *Gutierrezia microcephala*, *Haplopappus*, *Helianthus annuus*, *Hordeum murinum*, *Hymenothrix wrightii*, *Hyptis emoryi*, *Iva hayesiana*, *Juncus acutus*, *Krameria*, *Larrea tridentata*, *Ludwigia peploides*, *Lycium parishii*, *Lythrum californicum*, *Machaeranthera*, *Mimosa aculeaticarpa*, *Mimulus guttatus*, *Mecardonia procumbens*, *Melilotus albus*, *Melinis repens*, *Mentzelia multiflora*, *Mimulus cardinalis*, *Monardella linoides*, *M. stoneana*, *Muhlenbergia rigens*, *Nicotiana obtusifolia*, *Olneya tesota*, *Opuntia engelmannii*, *Panicum capillare*, *Parkinsonia aculeata*, *P. microphylla*, *Parietaria hespera*, *Penstemon pseudospectabilis*, *Perityle emoryi*, *Persicaria lapathifolia*, *Peucephyllum schottii*, *Phragmites australis*, *Physalis angulata*, *P. longifolia*, *Plantago rhodosperma*, *Platanus racemosa*, *P. wrightii*, *Pleurocoronis pluriseta*, *Pluchea odorata*, *P. sericea*, *Polanisia dodecandra*, *Polypogon monspeliensis*, *Populus fremontii*, *Prosopis velutina*, *Punica granatum*, *Quercus agrifolia*, *Q. arizonica*, *Q. turbinella*, *Rhus ovata*, *Rumex*, *Salix exigua*, *S. gooddingii*, *S. laevigata*, *Schoenoplectus*, *Senegalia greggii*, *Senna*,

Simmondsia chinensis, Solanum americanum, Sorghum halepense, Sphaeralcea laxa, Tamarix ramosissima, Toxicodendron, Tribulus terrestris, Trixis californica, Typha domingensis, Vauquelinia californica, Veronica anagallis-aquatica, V. peregrina, Vitex agnus-castus, Vulpia myuros, Xanthium strumarium, Yucca baccata.

Use by wildlife: *Stemodia durantifolia* is a larval host plant for the tropical buckeye butterfly (Insecta: Lepidoptera: Nymphalidae: *Junonia genoveva*).

Economic importance: food: The boiled roots of some *Stemodia* species are eaten by the Australian Aborigines; however, none of the North American species is reported to be edible; **medicinal:** Several *Stemodia* species are used as medicinals, a factor that has been attributed to the occurrence of certain diterpenes (e.g., stemarin, stemodin, stemodinone), which can exhibit antiviral and anticancer activity. *Stemodia durantifolia* is regarded as an antifertility plant in Haiti; however, laboratory tests (on rats) have failed to indicate any fetal toxic activity. A decoction of the plant also is used in the Carribean region as a general tonic or as a menstrual stimulant; **cultivation:** *Stemodia durantifolia* is not cultivated; **misc. products:** none; **weeds:** *Stemodia durantifolia* is regarded as a widespread weed in lowland Panama and elsewhere in the neotropics; **nonindigenous species:** *Stemodia durantifolia* has been introduced to the Miami, Florida area.

Systematics: Several phylogenetic studies (e.g., Figure 5.86) indicate that *Stemodia* is polyphyletic as currently circumscribed. *Stemodia durantifolia* associates in a clade with other New World taxa (*Acharetia, Otacanthus, Stemodia glabra, S. schottii*); however, other New World *Stemodia* resolve within a distant clade containing *Leucospora* and *Scoparia* species. The systematics and circumscription of *Stemodia* require further study and clarification. The base chromosome number of *Stemodia* also is uncertain. Counts for South American species indicate $x = 11$; whereas, a base number of $x = 7$ is indicated for at least one Old World (Australian) species.

Comments: *Stemodia durantifolia* occurs in Arizona, California and Florida, and extends southward through South America.

References: Correll & Correll, 1975; D'Arcy, 1979; Elvin & Sanders, 2003; Fischer, 2004; Halberstein, 2005; Sosa & Seijo, 2002; Turner & Cowan, 1993; Weniger et al., 1982.

13. *Synthyris*

Kittentails

Etymology: from the Greek *syn thyra* meaning "united door" in reference to the adherent fruit valves

Synonyms: *Besseya*; *Veronica* (in part); *Wulfenia* (in part)

Distribution: global: North America; **North America:** western

Diversity: global: 19 species; **North America:** 19 species

Indicators (USA): OBL: *Synthyris ranunculina*

Habitat: freshwater; palustrine; **pH:** alkaline; **depth:** <1 m; **life-form(s):** emergent herb

Key morphology: stems reduced; leaves (to 2.0 cm) in basal rosettes, petioled (to 4.0 cm), reniform, shallowly palmately 5–7–lobed, the margins deeply crenate; racemes (to 8.0 cm) one or two, curved, with 1–3 alternate leaf-like bracts; flowers pedicelled (to 3.0 mm), congested apically; sepals 2, fused basally; corolla (to 4 mm) 3–4–parted, linear, nearly regular, essentially free, blue or violet; stamens 2, adnate to corolla; capsule ovate, compressed, septicidal; seeds few, flattened

Life history: duration: perennial (persistent rosette); **asexual reproduction:** rhizomes; **pollination:** insect; **sexual condition:** hermaphroditic; **fruit:** capsules (common); **local dispersal:** seeds; **long-distance dispersal:** seeds

Imperilment: (1) *Synthyris ranunculina* [G2]; S2 (NV)

Ecology: general: Most *Synthyris* species occupy cold sites such as alpine or subalpine tundra or tundra ecotones, montane grasslands, and conifer savannas. They also occur in lower elevation forests. Some species resist wind desiccation by their relatively low leaf water potential, higher leaf temperatures, and low transpirational rates. Only one species is regarded as an OBL indicator, but several grow near snow-melt areas where the ground is seasonally wet. The flowers are self-compatible but also are strongly protogynous and are pollinated primarily by insects (Hymenoptera). Pollinators in section *Besseya* include short-tongued bees (Insecta: Hymenoptera: Halictidae: *Dialictus anomalus*) and wasps (Insecta: Hymenoptera: Vespidae: *Pseudomasaris zonalis*). Bees also have been observed as floral visitors of species in subgenus *Missurica*. Flowering often occurs during the early spring when water is most plentiful. At least several species are cross-compatible, enabling hybridization to occur among species from different sections. Seed germination requirements vary, with some requiring 2–3 months of cold stratification and others germinating after storage at room temperature. Reported germination temperatures range from 16°C to 22°C.

Synthyris ranunculina **Pennell** grows in meadows, seeps, slopes, springs, along stream margins, and near snow banks at subalpine to alpine elevations from 2620 to 3590 m. The substrates have been described as silty clay loam or organic and/or gravelly soils that develop on carbonate (limestone) outcrops. The plants also occur on moss-covered rocks and in cliff crevices where sufficient moisture exists. They tend to occupy sites with a sheltered (e.g., northerly) exposure. Flowering occurs from late June to August. The species probably is pollinated by bees (Hymenoptera) but its reproductive ecology is poorly documented. Seeds that are planted in the spring reportedly will germinate after 1 month when incubated at 16°C–21°C. Vegetative reproduction occurs by means of a slender rhizome. Little additional life-history information is available and a thorough study of this rare species should be undertaken. The plants are subjected to minor threats such as trampling by hikers and horses (Mammalia: Equidae: *Equus*). **Reported associates:** *Antennaria soliceps, Aquilegia formosa, Botrychium crenulatum, Carex, Castilleja martinii, Chamerion angustifolium, Cirsium clokeyi, Cystopteris fragilis, Draba jaegeri, D. paucifructa, Equisetum arvense, Heuchera rubescens, Ivesia cryptocaulis, Pinus aristata, P. flexilis, P. longaeva, Platanthera sparsiflora, Populus tremuloides, Primula fragrans, Ribes montigenum, Silene clokeyi, Sphaeromeria compacta, Telesonix jamesii, Trisetum spicatum, Valeriana acutiloba.*

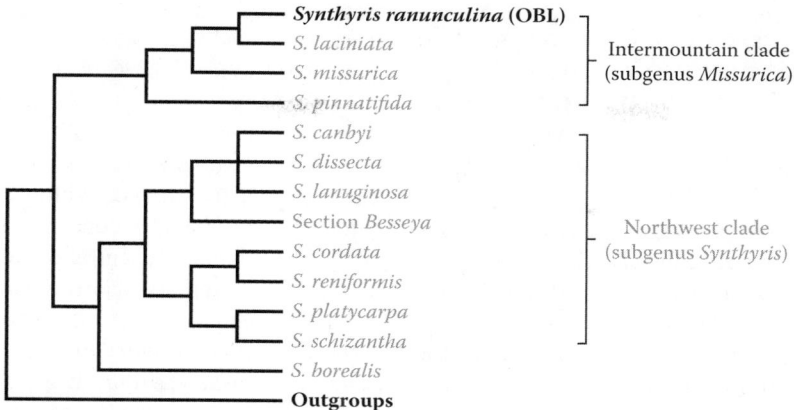

FIGURE 5.89 Phylogenetic relationships in *Synthyris* as reconstructed by analysis of combined morphological and DNA sequence data for all recognized species. Section *Besseya* includes seven species assigned formerly to the genus *Besseya*, which resolves as a clade imbedded within *Synthyris*. *Synthyris ranunculina* appears in a relatively derived position in the genus (OBL North American taxa are indicated in bold). (Adapted from Hufford, L. & McMahon, M., *Syst. Bot.*, 29, 716–736, 2004.)

Use by wildlife: Some *Synthyris* species contain toxic substances and are avoided by livestock.

Economic importance: food: not edible; **medicinal:** Shoot extracts from at least one species (*S. rubra*) are toxic to small vertebrates, but *S. ranunculina* has not been similarly evaluated; **cultivation:** Nearly two-thirds of *Synthyris* species are cultivated mainly as alpine or rock-garden plants; however, *S. ranunculina* is not commonly cultivated, presumably due to its rarity, or perhaps, because of its recalcitrance to flowering in cultivation (see *Systematics* later); **misc. products:** none; **weeds:** none; **nonindigenous species:** none

Systematics: Although some authors have argued for the inclusion of *Synthyris* within *Veronica*, the molecular phylogenetic evidence in support of this disposition is not persuasive. Because the analyses lack resolution at critical nodes of the phylogeny they do not preclude the possibility that the genera are distinct, as other (also molecular) analyses have indicated. Consequently, it is preferable to maintain the genera as distinct unless more compelling evidence is forthcoming. On the other hand, there is strong evidence that *Besseya* is derived from within *Synthyris* (Figure 5.89) and these genera should be merged. The former genus *Besseya* comprises a clade that is maintained as a separate section. *Synthyris ranunculina* occupies a relatively derived position in the genus (Figure 5.89), resolving as the sister species to *S. laciniata* within the subgenus *Missurica* (also known as the "intermountain clade"). The base chromosome number of *Synthyris* is *x* = 12. Eight species with reported counts are diploid (2*n* = 24) and one is tetraploid (2*n* = 48). *Synthyris ranunculina* has not been investigated cytologically. Hybridization has been reported in the genus and there are natural hybrids that involve members of different sections (i.e., *Besseya* and *Missurica*). Several synthetic hybrids also have been produced, but none involving *S. ranunculina*, which produces flowers with difficulty in cultivation. The number of floral parts (petals, sepals, stamens) in *S. ranunculina* has been found to vary when the plants are grown under different environmental conditions.

Comments: *Synthyris ranunculina* is a rare endemic restricted to roughly 36 populations along the eastern flank of the Spring Mountains, Nevada.

References: Albach & Chase, 2001; Albach et al., 2004a,b; Bliss, 1971; Chesnut, & Wilcox, 1901; Clokey, 1951; Cooper, 1952; Cooper & Bequaert, 1950; Hufford, 1992; Hufford & McMahon, 2004; Kruckeberg & Hedglin, 1963; McKone et al., 1995; Pennell, 1933; Rogers, 1996; Went, 1958.

14. *Veronica*

Brooklime, water speedwell; mouron d'eau, véronique

Etymology: after Saint Veronica, who wiped the face of Jesus with her veil

Synonyms: *Beccabunga*; *Diplophyllum*; *Cochlidospermum*; *Macrostemon*; *Odicardis*; *Oligospermum*

Distribution: global: Africa; Asia; Australia; Europe; New World; **North America:** widespread

Diversity: global: 250 species; **North America:** 34 species

Indicators (USA): OBL: *Veronica americana*, *V. anagallis-aquatica*, *V. beccabunga*, *V. peregrina*, *V. scutellata*, *V. serpyllifolia*; **FACW:** *V. peregrina*, *V. serpyllifolia*; **FAC:** *V. peregrina*, *V. serpyllifolia*

Habitat: freshwater; palustrine, riverine; **pH:** 4.4–10.0; **depth:** <1 m; **life-form(s):** emergent herb, submersed (vittate)

Key morphology: stems erect (to 1 m) or prostrate (to 55 cm), the foliage glabrous or hairy; leaves opposite, orbicular (to 5 cm) to narrowly lanceolate (to 9 cm), sessile or petioled, the margins smooth to serrate; racemes many flowered (to >30), terminal or axillary, the flowers pedicelled (to 16 mm); corolla (to 10 mm) rotate, 4-lobed, blue, violet or white; stamens 2, exserted, widely divergent; capsule (to 4 mm) flattened, obcordate; seeds (to 0.5–1.5 mm) flat

Life history: duration: annual (fruit/seeds) or perennial (rhizomes); **asexual reproduction:** shoot fragments, rhizomes; **pollination:** insect or self; **sexual condition:** hermaphroditic; **fruit:** capsules (common); **local dispersal:** seeds (rain), vegetative fragments (water); **long-distance dispersal:** seeds (water, animals)

Imperilment: (1) *Veronica americana* [G5]; SX (IN); SH (KY, MO); S1 (DE, IA, IL, TN, TX); S2 (KS, NC); S3 (YT); (2) *V. anagallis-aquatica* [G5]; S1 (NC, QC, VT); (3) *V. beccabunga* [GNR]; (4) *V. peregrina* [G5]; S2 (NB); S3 (NJ, NY, QC); (5) *V. scutellata* [G5]; SX (DC); S1 (IL, MD, TN, VA); S2 (LB, WV, WY); S3 (AB, BC, NF, YT); (6) *V. serpyllifolia* [G5]; S1 (SK); S2 (NB); S3 (AB, NF)

Ecology: general: Most *Veronica* species are more characteristic of UPL communities, but there also is a small concentration of aquatic and wetland species in the genus. The species occur throughout the Northern Hemisphere and at montane elevations in the tropics. Many are common roadside weeds. Six species are categorized as OBL in some portion of their North American range; however, *V. peregrina* and *V. serpyllifolia* are the least "aquatic" given that both often occupy UPL sites. In general, the OBL species tend to be broadly distributed geographically, extremely versatile ecologically, and occur across a wide variety of wetland habitats, which is evident by their extensive and diverse rosters of associated species (see later). *Veronica* species have been described variously as protogynous, weakly protogynous, and even as protandrous. Of those species that have been studied in detail, most are self-compatible, but normally are outcrossed due to herkogamy (widely separated stigma and anthers). Plants growing at higher elevations often are pollinator limited, and attain higher seed set when the flowers are pollinated manually. Reported insect (Insecta) pollinators include flies (Diptera: Muscoidea; Syrphidae) and bees (Hymenoptera: Apidae: *Bombus*; Halictidae). The seeds are dispersed over short distances by the rain (by means of a splash-cup mechanism) and over longer distances by water, wind, or through endozoic animal transport (facilitated especially by grazing agricultural animals).

Veronica americana **Schwein. ex Benth.** is a perennial, which inhabits beaches, bogs, brooks, depressions, ditches, fens, flats, gravel bars, marshes, meadows, sandbars, seeps, springs, streams, swales, swamps, and the margins of lakes, ponds, pools, rivers, and streams at elevations of up to 3200 m. The plants occur up to 64° N latitude and can tolerate cold temperatures down to −39°C. They occur in sites where the exposures range from full sun to fairly deep shade; however, they are much more common where light gaps occur and generally are regarded as being shade intolerant. Although extremely adaptable to diverse wetland conditions, this species is associated most commonly with cool, flowing, shallow (to 20 cm) watercourses. It has a broad pH tolerance (pH: 5.7–7.8), but is more common in hard waters (alkalinity >100 ppm) where the sulfate ion concentration is low (<50 ppm) and nitrogen levels are high. Specific conductance varies from 20 to 340 µS/cm. Site temperatures are relatively cool during the growing season (9°C–18°C). The substrates are of alkaline, granitic, or volcanic origin and include calcareous marl, clay, gravel, gravel loam, humus, loam, muck, mud, peat, pumice, rock, sand, sandy loam, serpentine, silt, silty loam, and silty muck. The plants also have been observed growing on floating logs. The flowers reportedly are insect pollinated (Diptera, Hymenoptera); however, detailed pollination studies have not

been made. Seeds that have been stratified in moist sand for 6 weeks at 3°C have exhibited high germination rates (86%) under a 29°C/10°C–20°C (8/16 h) temperature regime. A viable seed bank (from 3 to 20 seeds/m^2) has been documented. The plants reproduce vegetatively by rhizomes. If their leaf or stem tissues become wounded, the plants will generate periderm to seal the wounds within 2–3 weeks. The plants are known to take up and to concentrate manganese (Mn) at ratios of 7330:1 (tissue concentration relative to the surrounding water). **Reported associates:** *Abies lasiocarpa, Acer glabrum, A. macrophyllum, A. negundo, Achillea millefolium, Aconitum columbianum, Actaea rubra, Adiantum capillus-veneris, Agrostis exarata, A. gigantea, A. scabra, A. stolonifera, Alisma, Allium, Alnus oblongifolia, A. rugosa, A. tenuifolia, Alopecurus aequalis, A. pratensis, Amelanchier, Anagallis, Anaphalis margaritacea, Anemopsis californica, Angelica ampla, A. arguta, A. lucida, Aquilegia chrysantha, A. formosa, Arabis hirsuta, Arbutus, Arctophila fulva, Arnica cordifolia, Artemisia cana, A. douglasiana, Athyrium filix-femina, Atriplex, Baccharis salicifolia, Beckmannia syzigachne, Berula erecta, Betula occidentalis, Bidens cernuus, Bouteloua curtipendula, B. gracilis, Boykinia major, Brickellia californica, Bromopsis ciliata, Bromus carinatus, B. inermis, Calamagrostis canadensis, C. rubescens, Callitriche heterophylla, C. palustris, Caltha leptosepala, C. palustris, Capsella bursa-pastoris, Cardamine cordifolia, C. pensylvanica, C. umbellata, Carduus nutans, Carex angustata, C. anthoxanthea, C. aquatilis, C. athrostachya, C. aurea, C. barbarae, C. canescens, C. crawfordii, C. densa, C. ebenea, C. egglestonii, C. eurycarpa, C. flava, C. geyeri, C. hystericina, C. interior, C. jonesii, C. lanuginosa, C. leptalea, C. lyngbyei, C. microptera, C. nebrascensis, C. obnupta, C. pachystachya, C. pellita, C. phaeocephala, C. praegracilis, C. rostrata, C. scabrata, C. stipata, C. utriculata, C. vesicaria, Catabrosa aquatica, Cerastium arvense, C. fontanum, C. nutans, Chamaecyparis, Chamerion angustifolium, Chelone glabra, Chenopodium, Chrysosplenium americanum, Chrysothamnus viscidiflorus, Cicuta douglasii, C. virosa, Cinna latifolia, Cirsium arvense, C. muticum, C. vulgare, Claytonia cordifolia, Clintonia uniflora, Conioselinum chinense, Conyza canadensis, Coptis trifolia, Cornus sericea, Corydalis aquae-gelidae, Crupina vulgaris, Cryptantha simulans, Cyperus eragrostis, C. esculentus, Dasiphora floribunda, Deschampsia beringensis, D. cespitosa, D. elongata, Descurainia californica, Drosera rotundifolia, Echinochloa crus-galli, Egeria densa, Eleocharis bella, E. palustris, E. quinqueflora, Elymus canadensis, E. glaucus, E. trachycaulus, Epilobium ciliatum, E. hornemannii, E. minutum, E. oregonense, E. watsonii, Equisetum arvense, E. hyemale, E. laevigatum, E. pratense, E. scirpoides, E. sylvaticum, Ericameria nauseosa, Erigeron speciosus, Eriogonum umbellatum, Eupatorium perfoliatum, Eurybia conspicua, Festuca ovina, F. rubra, Filago arvensis, Fragaria virginiana, Frasera speciosa, Fraxinus latifolia, Galium aparine, G. boreale, G. trifidum, G. triflorum, Gaultheria shallon, Geranium bicknellii, G. erianthum, G. maculatum, G. richardsonii, Geum macrophyllum, Glyceria elata, G. grandis, G. striata,*

Gnaphalium palustre, Heracleum lanatum, H. maximum, H. sphondylium, Hippuris vulgaris, Holodiscus microphyllus, Hordeum brachyantherum, H. jubatum, Hydrophyllum fendleri, Hypericum anagalloides, Impatiens capensis, Iris missouriensis, Juncus balticus, J. bufonius, J. castaneus, J. effusus, J. ensifolius, J. nevadensis, J. phaeocephalus, J. saximontanus, J. supinus, J. tenuis, J. tracyi, J. xiphioides, Kochia scoparia, Lactuca serriola, Lemna minor, Leersia oryzoides, Leymus cinereus, Lilium pardalinum, Lomatium cous, Lonicera involucrata, Lotus corniculatus, Lupinus arbustus, L. latifolius, L. polyphyllus, Luzula parviflora, Lysichiton americanus, Lysimachia nummularia, Maianthemum racemosum, M. stellatum, M. trifolium, Marah fabaceus, Marchantia polymorpha, Melandrium album, Melilotus officinalis, Mentha arvensis, M. canadensis, M. ×piperita, Mertensia ciliata, M. franciscana, M. paniculata, M. perplexa, Micranthes odontoloma, Mimulus breviflorus, M. filicaulis, M. glabratus, M. guttatus, M. lewisii, M. moschatus, M. primuloides, Mitella pentandra, Monardella, Montia chamissoi, Muhlenbergia filiformis, Myosotis laxa, Nasturtium officinale, Navarretia intertexta, Oenanthe sarmentosa, Osmorhiza berteroi, O. depauperata, Oxypolis fendleri, Pascopyrum smithii, Pedicularis groenlandica, Persicaria hydropiperoides, P. lapathifolia, P. punctata, Petasites sagittatus, Phalaris arundinacea, Phegopteris connectilis, Philadelphus lewisii, Phleum pratense, Phyllodoce breweri, Picea engelmannii, Pilea fontana, Pinus contorta, Plagiobothrys cusickii, P. scouleri, Plantago major, Platanthera dilatata, Platanus wrightii, Poa arctica, P. palustris, P. pratensis, P. triflora, Podistera eastwoodiae, Pohlia wahlenbergii, Polemonium occidentale, Polypogon monspeliensis, P. viridis, Populus angustifolia, P. balsamifera, P. fremontii, P. tremuloides, Potentilla anserina, P. gracilis, Primula jeffreyi, Prunus virginiana, Pseudotsuga menziesii, Purshia, Pycnanthemum californicum, Pyrola secunda, Ranunculus aquatilis, R. californicus, R. cymbalaria, R. hyperboreus, R. longirostris, R. occidentalis, R. repens, R. reptans, R. trichophyllus, R. uncinatus, Rhamnus alnifolia, Rhododendron, Ribes aureum, R. bracteosum, R. cereum, R. hudsonianum, R. inerme, R. lacustre, R. oxyacanthoides, Rorippa islandica, R. sphaerocarpa, R. teres, Rosa woodsii, Rubus idaeus, R. pedatus, Rudbeckia laciniata, Rumex aquaticus, R. crispus, R. paucifolius, Sagina saginoides, Sagittaria latifolia, Salix bebbiana, S. boothii, S. candida, S. drummondiana, S. eastwoodiae, S. exigua, S. geyeriana, S. gooddingii, S. irrorata, S. lasiolepis, S. lemmonii, S. lutea, S. monticola, S. pseudomonticola, S. pseudomyrsinites, Sambucus racemosa, Sanicula marilandica, Schoenoplectus acutus, S. pungens, S. tabernaemontani, Scirpus microcarpus, Scutellaria galericulata, S. siphocamyploides, Senecio eremophilus, S. hydrophilus, S. triangularis, Sidalcea neomexicana, S. oregana, Silybum marianum, Solanum dulcamara, Sparganium angustifolium, S. eurycarpum, Sphagnum russowii, Sphenosciadium capitellatum, Spiraea betulifolia, S. douglasii, Stachys ajugoides, S. albens, S. bullata, S. pycnantha, Stellaria calycantha, S. longifolia, S. media, Streptopus amplexifolius, Swertia perennis,
Symphoricarpos rotundifolius, Symphyotrichum eatonii, S. foliaceum, S. spathulatum, Taraxacum officinale, Thalictrum occidentale, Thermopsis divaricarpa, Thlaspi arvense, Thuja occidentalis, Tragopogon dubius, Trifolium cyathiferum, T. longipes, T. repens, T. variegatum, Triticum cereale, Tsuga heterophylla, Typha latifolia, Urtica dioica, Vaccinium scoparium, Veratrum californicum, V. viride, Veronica anagallis-aquatica, V. officinalis, V. serpyllifolia, Vicia americana, Viola adunca, V. glabella, V. macloskeyi, V. pallens.

***Veronica anagallis-aquatica* L.** is an perennial plant of beaches, brooks, channels, depressions, ditches, drying riverbeds, dunes, flats, floodplains, gravel bars, marl lakes, marshes, meadows, mudflats, quarries, sand bars, seeps, sloughs, springs, streams, swales, washes, and the margins of lakes, ponds, pools, and rivers at elevations of up to 3050 m. The plants occur in full sun to partial shade, typically in shallow (e.g., 10–60 cm), alkaline (pH: 6.2–10.0), oligotrophic–mesotrophic to eutrophic waters of considerable hardness. Occasionally they have been reported from brackish sites. Their abundance at sites declines proportionally as ammonium concentrations increase above 0.5 mg/L. The substrates tend to be fairly fertile and are derived from dolomite, gneiss, granite, or sandstone parent materials, and include clay, clay loam, gravel, marl, muck, mud, rock, sand, sandy loam, silt, and silty loam. The flowers produced by emergent plants are self-compatible but are pollinated by flies (Insecta: Diptera). However, when submersed, the plants often produce cleistogamous flowers, which yield seed through self-pollination. The reproductive ecology of this species in North America is poorly documented and a detailed study would be useful. In Australia (where nonindigenous), substantial numbers of female "blue ants" (Hymenoptera: Tiphiidae: *Diamma bicolor*), which are actually wasps, have been observed to forage on the flowers. Thus, pollination does not appear to require specialist insect vectors. The plants are prolific seed producers. The seeds float and are water dispersed in drift, becoming deposited along the banks of riverine systems. The seedlings establish within a 2 m elevational zone from the water. A persistent seed bank is formed, which has been attributed in part to a relatively high concentration of *ortho*-dihydroxyphenols that occur in the seeds. Seed densities are highest within the first 5 cm of sediment and can surpass 600/m^2 (averaging 162/m^2 in one study). The seeds often do not occur where there is standing water, but they can be extremely abundant in river bed sediments. A considerable number of propagules can remain in the seed bank even after a flood event has occurred. Seed germination has been reported to occur at 15°C–19°C after 5 months of cold stratification. Seeds that have been recovered from animal (pony) dung remained viable and germinated after 4 weeks of cold stratification (5°C). Alternating temperatures are necessary to break dormancy; whereas, a combination of light and alternating temperatures (28°C/13°C) is necessary to induce optimal germination. These requirements prevent germination from occurring when the seeds are buried too deeply, or when a canopy of competitive overstory vegetation is present. The plants are known to become much more prevalent following disturbances that are created by flooding and also by livestock grazing. The seedlings quickly produce lateral branches, which root rapidly as a means of facilitating plant establishment in shifting substrates. They quickly will colonize newly exposed habitats along river channels where flooding has occurred. The roots usually are associated with arbuscular mycorrhizae. The foliage can concentrate manganese (Mn) from the surrounding waters at a ratio of 75:1. Emergent leaves are produced when the plants occur in low velocity currents; whereas, submersed leaves develop at high velocity currents. The emergent foliage resemble typical "sun leaves" in being larger, thicker denser, and heavier than the submersed foliage, which more closely resembles "shade leaves." The emergent leaves also have higher photosynthetic rates; however, the submersed leaf morphology provides the plants with a necessary flexibility for withstanding the forces associated with faster water currents. Although this species frequently has been described as an annual, it actually is a rhizomatous perennial. The indigenous/nonindigenous status of this species remains under some contention. **Reported associates:** *Acer saccharinum, Acmispon americanus, Adenostoma, Agrostis gigantea, Alisma triviale, Alnus rhombifolia, Alopecurus carolinianus, Amaranthus tuberculatus, Ammannia coccinea, Amorpha fruticosa, Anemopsis californica, Anthemis, Artemisia douglasiana, A. tridentata, Atriplex, Azolla, Baccharis glutinosa, B. salicifolia, Berula erecta, Betula nigra, Bidens cernuus, B. frondosus, B. laevis, Boltonia asteroides, Calamagrostis canadensis, Callitriche heterophylla, Carex crawei, C. garberi, C. lacustris, C. nebrascensis, C. praegracilis, C. stricta, C. subfusca, C. viridula, Ceanothus, Cephalanthus occidentalis, Ceratophyllum demersum, Chara, Chenopodium ambrosioides, Chrysothamnus, Cicuta bulbifera, C. maculata, Clematis ligusticifolia, Clinopodium glabellum, Conium maculatum, Cornus sericea, Cotula coronopifolia, Crypsis, Cyperus aristatus, C. eragrostis, Cyperus odoratus, Dasiphora floribunda, Deschampsia cespitosa, Descurainia incana, Eleocharis elliptica, E. erythropoda, E. montevidensis, E. palustris, E. parishii, Elodea bifoliata, E. nuttallii, Elymus trachycaulus, Epilobium ciliatum, E. glaberrimum, Epipactis gigantea, Eragrostis hypnoides, E. pectinacea, Erigeron divergens, Eriodictyon, Erodium, Eupatorium perfoliatum, Fraxinus latifolia, F. velutina, Gentianopsis virgata, Geranium carolinianum, Gnaphalium palustre, Hosackia oblongifolia, Howellia aquatilis, Hydrocotyle verticillata, Hymenoclea monogyra, Impatiens capensis, Juncus bufonius, J. effusus, J. rugulosus, Leersia oryzoides, Lemna minor, L. minuta, Leptochloa fusca, Lindernia dubia, Lobelia cardinalis, L. kalmii, Lolium multiflorum, Ludwigia palustris, L. polycarpa, Lupinus confertus, Lycopus americanus, L. asper, L. uniflorus, Lysimachia nummularia, L. quadriflora, Lythrum hyssopifolia, Melilotus indicus, Mentha arvensis, M. pulegium, Mimulus cardinalis, M. floribundus, M. glabratus, M. guttatus, M. pilosus, M. ringens, Myosotis discolor, Myriophyllum sibiricum, Najas guadalupensis, N. marina, Nasturtium officinale, Nicotiana glauca, Nitrophila occidentalis, Oscillatoria rubescens, Penthorum sedoides, Phalaris arundinacea, Phragmites australis, Pilea fontana, Platanus racemosa, Pluchea, Poa annua, Persicaria amphibia, P. lapathifolia, P. maculosa, P. punctata, Polypogon interruptus, P. monspeliensis, P. viridis, Populus fremontii, Potamogeton foliosus, P. gramineus, P. pusillus, Potentilla anserina, Primula mistassinica, Pseudognaphalium luteoalbum, Quercus agrifolia, Q. bicolor, Ranunculus cymbalaria, R. longirostris, R. pensylvanicus, R. sceleratus, R. trichophyllus, Raphanus sativus, Rhamnus ilicifolia, Rhizoclonium hieroglyphicum, Rorippa palustris, Rosa woodsii, Rubus discolor, R. ursinus, Rumex conglomeratus, R. crispus, R. salicifolius, R. verticillatus, Sagittaria latifolia, Salix candida, S. eriocephala, S. exigua, S. gooddingii, S. laevigata, S. lasiolepis, Salvia, Sambucus mexicana, Schoenoplectus acutus, S. tabernaemontani, Scirpus atrovirens, Scleria verticillata, Scutellaria galericulata, S. lateriflora, Selaginella eclipes, Sidalcea oregana, Sium suave, Solidago occidentalis, S. ohioensis, S. riddellii, Sonchus asper, Sparganium eurycarpum, Spirogyra, Sporobolus, Stuckenia pectinata, Symphyotrichum boreale, Thinopyrum ponticum, Toxicodendron diversilobum, Triadenum fraseri, Triantha glutinosa, Trifolium pratense, Typha latifolia, Urtica dioica, U. urens, Utricularia macrorhiza, Verbena hastata, V. robusta, Veronica americana, V. scutellata, Vitis californica, V. girdiana, Xanthium strumarium, Zigadenus.*

Veronica beccabunga L. is a perennial, which grows on beaches and in brooks, depressions, ditches, fens, lakeshores, meadows, mudflats, streams, swamps, and along the margins of watercourses at elevations of at least 65 m. It occurs in circumneutral sites (pH: 5.5–7.4) on substrates of gravel, rock, and sand. The flowers are self-compatible but are protogynous and insect pollinated, which results in a mixed (selfing/outcrossing) breeding system. The principal pollinators appear to be hover flies (Insecta: Diptera: Syrphidae: *Syritta pipiens*). Self-pollination occurs during unfavorable weather conditions, which prevents the flowers from fully opening. The capsules do not dehisce until they have become thoroughly wetted by rain, a trait which is believed to maximize the probability that standing water will be present at the appropriate time for seed dispersal. Numerous seeds are produced. The seeds are not buoyant but become mucilaginous when wet and are believed to be dispersed mainly by adhesion. They do appear in samples collected from water drift, perhaps by adhering to other floating material. In any case, the observation that they are deposited regularly within flood-deposited sediments indicates that some form of water dispersal (either directly or by adhesion to other floating materials) must be an effective means of dissemination. The seeds occur more frequently in surface sediments. They germinate well (100%) when stored dry for 3–12 months at 5°C, but nearly as well (98%) when stored at 20°C for more than a month. Seed germination (94%–100%) occurs in light, shade, or in darkness. Despite this apparent ease of germination, some seed dormancy must exist, given that a long-term, persistent seed bank (achieving densities of up to 2067 seeds/m²) also can develop. Entire plants or their fragments are buoyant and can become dislodged during floods. The plant fragments root readily and promote the vegetative reproduction of plants along banks, which can result in rapid colonization. This nonindigenous species has not spread rapidly in North America, perhaps due to niche preemption by the closely related *V. americana*. The shoots produce much shorter internodes when growing in cold water (e.g., 4°C), and the submersed leaves develop much thinner blades than those of the emergent foliage. The plants are not mycorrhizal. The North American associates are poorly known. **Reported associates (North America):** *Juncus effusus, Lysimachia ciliata, Lythrum salicaria, Myosotis scorpioides, Nasturtium officinale, Potamogeton crispus, Salix nigra.*

Veronica peregrina **L.** is a winter annual, which grows in bayous, canals, channels, deltas, depressions, ditches, dunes, flats, floodplains, gravel bars, gravel pits, marshes, meadows, mudflats, playas, sloughs, vernal pools, washes, and along the margins of lakes, ponds, prairies, reservoirs, rivers, seeps, streams, and swales at elevations of up to 3110 m. The plants tolerate sun to light shade and occur on alkaline or granitic substrates, often underlain by hardpan. These include adobe, clay, clay loam, cobbley clay, gravel, lava, loam, mud, rock, sand, sandy loam, sandy mud, schist, shells, silt, and silty clay. The plants are cold hardy and occur northward in localities up to 64° N latitude. The flowers remain open for less than 1.5 days, but flowering occurs earlier and for a prolonged time if the plants are grown at higher temperatures (22°C–32°C). The plants are self-compatible, usually self-pollinating (estimated 95% selfing rate), and occasionally cleistogamous. A highly autogamous breeding system also is indicated by an extremely low pollen to ovule ratio (average: 6.4:1). However, the flowers are protogynous and attract some bees (Hymenoptera: Halictidae), flies (Diptera: Syrphidae), and spiders (Arachnida), which may result in occasional episodes of outcrossing. Allozyme data have revealed that despite its breeding system, a high level of genetic variability and heterozygosity are maintained in the species because of hexasomic inheritance that is attributable to hexaploidy. An individual plant can produce 20 or more capsules (averaging 70 seeds each) and upward of 62,000 seeds, which are shed from the mature capsules easily, particularly during periods of rain. The seeds are adapted for splash-cup dispersal by raindrops, which distribute them at short distances of up to 85 cm. The seeds undergo a conditional/nondormancy cycle, where fresh seeds initially are dormant, but eventually lose their dormancy within 2 months. Subsequent germination rates up to 99% can be obtained under 20°C–30°C/10°C–15°C temperature regimes. Reduced germination occurs when seeds are incubated under a 15°C/6°C temperature cycle. Poor germination occurs in the dark (32% maximum), or when sufficient soil moisture is lacking. The seeds germinate under both flooded and nonflooded conditions, with somewhat higher rates observed for lesser-inundated sites. In some cases, cold stratification (3 weeks at 5°C) can result in higher germination rates (76% at 20°C–30°C) than those observed for untreated seeds (40%). Large seed banks (e.g., >4500/m²) can develop, with the highest densities occurring from 3 to 9 cm below the substrate surface. Higher seed densities have been recorded from disturbed sites than in comparable but undisturbed habitats. The plants are amphibious and exhibit extensive phenotypic plasticity, even within microhabitats. Those individuals growing at the periphery of pools are taller and bear more flowers and produce lighter seeds than those growing near the pool center. Allozyme data have indicated that plants growing at the periphery of pools are more variable genetically as well as phenotypically. The roots usually are associated with arbuscular mycorrhizae. This is an early successional species that colonizes pools as soon as a mud substrate begins to develop. The plants also appear quickly along the exposed mud of receding lakeshores. This species thrives on bare, wet soil and in disturbances. It often appears in agricultural fields, particularly if they are well irrigated. Occurrences also are reported from gardens, lawns, orchards, and even cracks in concrete driveways and sidewalks. Ruderal occurrences (cultivated land, gardens, fields, roadsides, railroad rights-of-way, waste areas) dominate the eastern localities; whereas, natural communities (e.g., vernal pools) are prevalent in western North America. It is likely that cultural disturbance of habitats has facilitated the eastward spread of the species. Large seed banks have been recorded in areas that have been planted with various forage seed mixtures, indicating that the species can occur as a seed contaminant. Its spread in Scandinavia has been attributed to seed contamination of garden stock in plant nurseries. **Reported associates:** *Achillea millefolium, Adiantum pedatum, Agropyron repens, Agrostis alba, A. hyemalis, A. idahoensis, A. scabra, Aira caryophyllea, Alisma triviale, Alnus, Alopecurus aequalis, A. carolinianus, A. geniculatus, A. pratensis, Amaranthus, Ammannia robusta, Amsinckia, Anagallis arvensis, Androsace, Andropogon gerardii, Antennaria neglecta, A. parvifolia, Arabidopsis thaliana, Arabis, Arisaema dracontium, Artemisia arbuscula, Atriplex polycarpa, Avena barbata, A. fatua, Baccharis salicifolia, Bassia hyssopifolia, Bidens, Boehmeria cylindrica, Brassica geniculata, Bromus hordeaceus, B. rubens, Buchloe dactyloides, Calamagrostis canadensis, Calandrinia ciliata, Callitriche marginata, C. palustris, C. peploides, Capsella bursa-pastoris, Cardamine oligosperma, C. parviflora, C. pensylvanica, Carex blanda, C. bromoides, C. caroliniana, C. cherokeensis, C. granularis, C. grisea, C. lupulina, C. nebrascensis, C. oxylepis, C. stipata, Castilleja attenuata, C. campestris, C. densiflora, Cerastium glomeratum, Chenopodium album, Chorizanthe polygonoides, Chrysanthemum leucanthemum, Cicuta maculata, Cirsium muticum, Clematis virginiana, Conyza canadensis, Coronopus didymus, Crassula aquatica, C. connata, C. drummondii, C. solierii, Cressa truxillensis, Crypsis schoenoides, Cryptotaenia canadensis, Cynara cardunculus, Cynodon, Cynosciadium digitatum, Cyperus strigosus, Danthonia spicata, Deschampsia danthonioides, Descurainia incana, Dichondra micrantha, Digitaria, Distichlis spicata, Downingia bella, D. cuspidata, D. insignis, D. ornatissima, D. pulchella, Dracocephalum parviflorum, Echinocystis lobata, Elatine brachysperma, E. californica, E. chilensis, E. rubella, Eleocharis acicularis, E. bella, E. erythropoda, E. macrostachya, E. obtusa, E. palustris, Ellisia nyctelea, Elymus virginicus, Epilobium leptophyllum, E. pygmaeum,*

Equisetum arvense, Erechtites hieraciifolia, Erigeron flagellaris, Eriogonum wrightii, Erodium botrys, E. brachycarpum, Eryngium aristulatum, E. vaseyi, Eupatorium, Euphorbia corollata, Frankenia salina, Galium aparine, G. tinctorium, Gamochaeta purpurea, Geranium carolinianum, Geum canadense, Glechoma hederacea, Glyceria striata, Gnaphalium palustre, Gratiola neglecta, Grayia spinosa, Hedeoma hispida, Hemizonia fasciculata, Hordeum jubatum, Horkelia, Hosackia oblongifolia, Houstonia caerulea, Hydrocotyle verticillata, Hymenoxys hoopesii, Hypericum majus, Isoetes howellii, I. orcuttii, Iva frutescens, Ivesia, Juncus balticus, J. brevicaudatus, J. bufonius, J. hemiendytus, J. leiospermus, Lamium, Lasthenia californica, L. ferrisiae, L. fremontii, L. glaberrima, L. glabrata, L. minor, Lepidium dictyotum, Leptochloa fusca, Limosella acaulis, L. aquatica, Lindera benzoin, Lipocarpha occidentalis, Lolium temulentum, Lupinus bicolor, Lycopus americanus, Lysimachia nummularia, Lythrum hyssopifolium, L. tribracteatum, Malvella leprosa, Marsilea vestita, Medicago polymorpha, Mentha arvensis, Mimulus alatus, M. guttatus, M. primuloides, Monolepis nuttalliana, Montia, Myosotis discolor, M. macrosperma, M. verna, Myosurus minimus, Nama hispidum, Nasturtium officinale, Navarretia fossalis, N. intertexta, N. leucocephala, Oenothera biennis, O. laciniata, Oplismenus hirtellus, Oxytropis deflexa, Panicum capillare, Pascopyrum smithii, Paspalum, Penstemon gracilis, Persicaria lapathifolia, P. maculosa, P. virginiana, Phalaris arundinacea, Phragmites australis, Phyla lanceolata, P. nodiflora, Phytolacca americana, Pilularia americana, Plagiobothrys acanthocarpus, P. bracteatus, P. chorisianus, P. leptocladus, P. scouleri, P. stipitatus, Plantago elongata, P. major, Poa annua, P. secunda, P. sylvestris, Pogogyne douglasii, Polygonum aviculare, Polypogon maritimus, Potentilla anserina, P. norvegica, P. rivalis, Prosopis glandulosa, Psilocarphus brevissimus, P. oregonus, Ranunculus abortivus, R. aquatilis, R. recurvatus, R. sceleratus, Rhynchospora, Riccia nigrella, Rorippa curvipes, R. teres, Rubus alleghheniensis, Rumex acetosella, R. crispus, R. maritimus, Sagina decumbens, Salix rigida, Samolus parviflorus, Saponaria officinalis, Schoenoplectus pungens, Senecio glabellus, S. pauperculus, S. vulgaris, Sibara virginica, Sida, Sidalcea oregana, Silene antirrhina, Sisymbrium irio, Sisyrinchium bellum, Sonchus asper, Spergularia marina, Stachys hispida, Stellaria longifolia, S. media, S. prostrata, Stipa pulchra, Suaeda moquinii, Symphyotrichum eatonii, Taraxacum officinale, Thelypteris palustris, Thlaspi arvense, Tradescantia ohiensis, Trichostema lanceolatum, Trifolium dubium, T. obtusiflorum, T. repens, Triodanis perfoliata, Trisetum pensylvanicum, Typha latifolia, Urtica chamaedryoides, Vaccinium, Veratrum, Veronica arvensis, V. serpyllifolia, Vicia ludoviciana, Viola sororia, Vitis riparia, Waldsteinia fragarioides, Xanthium strumarium.

Veronica scutellata L. is a perennial species of bogs, depressions, ditches, dunes, fens, marshes, meadows, pools, prairies, puddles, sloughs, swales, swamps, vernal pools, and margins of lakes, ponds, rivers, and streams at elevations of up to 2260 m. The habitats are characterized by waters that are acidic (pH: 4.4–6.3) and shallow (15–30 cm), and include sites with exposures ranging from full sun to dense shade. The plants commonly occur where the standing waters have receded (e.g., reservoir margins, dry creek beds). Reported substrates include clay, clay loam, clay mud, gravel, loam, muck, mud, peat, peaty muck, sand, and silty loam. This species is cold tolerant and inhabits sites up to 65° N latitude. The flowers are self-compatible, but are pollinated by flies (Insecta: Diptera). The capsules do not dehisce until they have become thoroughly wetted by rain. The seeds float, but only for a short period of time (<30 min). A long-term, persistent seed bank (up to 4640 seeds/m^2) develops and reaches its highest density at intermediate (2–6 cm) soil depths. The seeds require light for germination. Good germination (40%) occurs within 6 weeks if the seed is first stratified (3°C) in moist sand for 6 months and then exposed to a 28°C–29°C/10°C–20°C temperature regime (8/16 h). Viable seeds have been recovered from cattle (Mammalia: Bovidae: *Bos taurus*) dung, which indicates the potential for endozoic dispersal by vertebrates. Natural burning or other removal of woody overstory vegetation significantly increases habitat for the plants and facilitates their spread; however, the species itself is a poor survivor of fires. The plants also have been documented as occurring in the later successional stages following fire, appearing only after an adequate layer of humus and litter has built up. This species also has been categorized as a component of the reed (*Typha*) swamp successional stage, where it can develop into a dense undergrowth following the desiccation of standing winter waters. In one study, the plants were not adversely affected at a site that had been impacted by an oil spill. Vegetative reproduction occurs by means of creeping, stolon-like rhizomes. Clonally developed mats of the plants can persist within fairly dense stands of other rhizomatous species (e.g., *Phalaris*). This species often is found at sites that are rich in sedges (*Carex*), which the relatively frail plants sometimes will grow against for support. **Reported associates:** *Acer saccharinum, Acorus calamus, Agrostis alba, A. exarata, A. gigantea, Aira caryophyllea, Alisma, Alnus rugosa, A. tenuifolia, Alopecurus aequalis, A. geniculatus, Amphicarpaea bracteata, Artemisia campestris, Asclepias incarnata, A. tuberosa, Athyrium filix-femina, Beckmannia syzigachne, Betula nigra, Bidens cernuus, Bistorta bistortoides, Boehmeria cylindrica, Boisduvalia, Brodiaea coronaria, Calamagrostis canadensis, C. stricta, Calliergon giganteum, Callitriche heterophylla, C. palustris, Calopogon tuberosus, Camassia leichtlinii, C. quamash, Campanula aparinoides, Cardamine penduliflora, C. pensylvanica, Carex angustata, C. aquatilis, C. arcta, C. athrostachya, C. atherodes, C. bebbii, C. densa, C. deweyana, C. exsiccata, C. feta, C. laeviculmis, C. lacustris, C. lanuginosa, C. lemmonii, C. lenticularis, C. nebrascensis, C. obnupta, C. pansa, C. pellita, C. retrorsa, C. rostrata, C. saxatilis, C. scoparia, C. simulata, C. stipata, C. stricta, C. subfusca, C. unilateralis, C. utriculata, C. vulpinoidea, Cephalanthus occidentalis, Cicuta douglasii, C. maculata, Cirsium arvense, Cladium mariscoides, Comarum palustre, Convolvulus arvensis, Cornus amomum, C. obliqua, C. sericea, Crepis tectorum, Danthonia californica,*

Deschampsia cespitosa, D. danthonioides, Downingia elegans, D. yina, Dryopteris carthusiana, Eleocharis acicularis, E. bolanderi, E. palustris, E. quinqueflora, Elymus glaucus, Epilobium ciliatum, E. palustris, Equisetum arvense, E. fluviatile, E. sylvaticum, Eryngium alismifolium, E. petiolatum, Eupatorium perfoliatum, Eutrochium maculatum, E. purpureum, Fallopia convolvulus, Frangula alnus, F. purshiana, Fraxinus latifolia, F. nigra, F. oregana, F. pennsylvanica, Galeopsis tetrahit, Galium brevipes, G. obtusum, G. parisiense, G. tinctorium, G. trifidum, G. triflorum, Gaultheria shallon, Gentianopsis crinita, Geum aleppicum, G. macrophyllum, Glyceria grandis, G. occidentalis, G. septentrionalis, G. striata, Gratiola, Grindelia nana, Helenium bigelovii, Holcus lanatus, Hordeum brachyantherum, Hosackia pinnata, Howellia aquatilis, Hypericum anagalloides, Hypochaeris radicata, Impatiens capensis, Iris pseudacorus, I. versicolor, I. virginica, Isoetes nuttallii, Ivesia aperta, I. sericoleuca, Juncus acuminatus, J. balticus, J. effusus, J. ensifolius, J. lesueurii, J. macrandrus, J. nevadensis, J. orthophyllus, J. oxymeris, J. patens, J. tenuis, Lemna minor, Lilium, Limnanthes douglasii, Ludwigia palustris, L. polycarpa, Lupinus latifolius, Lycopus americanus, L. uniflorus, Lysichiton americanus, Lysimachia nummularia, L. quadrifolia, Lythrum alatum, Malus fusca, Melica, Mentha arvensis, M. pulegium, Menyanthes trifoliata, Microsteris gracilis, Mimulus clivicola, M. guttatus, M. primuloides, M. ringens, Montia linearis, Myosotis laxa, Myosurus minimus, Myrica californica, Navarretia leucocephala, Nuphar polysepala, Oenanthe sarmentosa, Oenothera perennis, Onoclea sensibilis, Oxypolis rigidior, Parnassia, Paspalum distichum, Penthorum sedoides, Perideridia bolanderi, P. gairdneri, P. parishii, Petasites frigidus, Phalaris arundinacea, Phragmites australis, Phyla lanceolata, Physostegia ledinghamii, Picea mariana, P. sitchensis, Pilea, Pinus contorta, Plagiobothrys hirtus, P. hispidulus, Platanthera blephariglottis, P. leucostachys, P. psycodes, Poa palustris, P. pratensis, Polygala polygama, Persicaria amphibia, P. hydropiper, P. hydropiperoides, P. maculosa, P. punctata, Potamogeton gramineus, Potentilla anserina, P. canadensis, P. egedii, P. norvegica, P. pacifica, P. polygaloides, Primula tetrandra, Proserpinaca palustris, Pseudotsuga menziesii, Quercus alba, Q. bicolor, Q. rubra, Ranunculus alismifolius, R. aquatilis, R. flabellaris, R. flammula, R. gmelinii, R. macounii, R. repens, Ribes americanum, Rorippa palustris, Rosa carolina, R. nutkana, Rubus ursinus, Rumex crispus, R. salicifolius, Sagittaria latifolia, Salix geyeriana, S. hookeriana, S. interior, Sanguisorba menziesii, Schoenoplectus pungens, S. tabernaemontani, Scirpus atrovirens, S. microcarpus, Scutellaria galericulata, S. lateriflora, Sisyrinchium sarmentosum, Sium suave, Solanum dulcamara, Solidago gigantea, Sonchus arvensis, Sparganium angustifolium, S. chlorocarpum, S. eurycarpum, Sphagnum, Spiraea alba, S. douglasii, Spirodela polyrrhiza, Stachys ciliata, Stellaria calycantha, S. crispa, S. longifolia, S. longipes, S. media, Symphyotrichum chilense, S. puniceum, S. spathulatum, Thalictrum fendleri, Thelypteris palustris, Thuja plicata, Torreyochloa pallida, Triadenum fraseri, T. virginicum, Trifolium longipes, Triteleia hyacinthina, Typha
latifolia, Urtica dioica, Utricularia cornuta, U. intermedia, Vaccinium ovatum, V. uliginosum, Veratrum californicum, Verbena hastata, Veronica americana, V. anagallis-aquatica, Viola palustris.

***Veronica serpyllifolia* L.** is a perennial, which occurs on beaches, in bogs, brooks, depressions, ditches, floodplains, gravel bars, marshes, meadows, sand bars, seeps, springs, swales, swamps, tundra, and along the margins of lakes, ponds, and streams at elevations of up to 3140 m. The sites are circumneutral (pH: 6.0–7.5) with substrates often derived from granitic parental material. They are described as clay, clay humus, clay loam, gravel, gravel loam, humus, mud, rock, sand, sandy clay, shale, or talus. The plants can withstand cold conditions and occur to 64° N latitude. Exposures range from full sun to deep shade. When growing in open habitats, a plant can flower during its first year of growth, but often will remain vegetative until its second year after establishment. Although the flowers are self-compatible, the incurved anthers limit the extent of autogamy; however, both self- and cross-pollinations are facilitated by insects. The relative maturation of sexes is variable and the flowers have been described as protandrous, protogynous, and homogamous. If insufficient light is available, the flowers do not open and instead will self-pollinate in a cleistogamous fashion. The anthers are extremely small relative to other congeners and produce much less pollen, which translates into a relatively low pollen:ovule ratio (32:1) that is indicative of a facultatively selfing breeding system. The fruits average 58 very small seeds (0.05 mg) per capsule, with typical fecundity being about 1000 seeds/plant. Annual seed mortality is extremely high (up to 86%). Other than light, which is essential, there are no special requirements for seed germination. The small seeds are dispersed easily in the air and also float for a long period of time (up to 15 days). They are dispersed by raindrops over short distances (<1 m). Many animal dispersal vectors also exist. In Europe, the seeds (not tested for viability) have been recovered from bird feces (e.g., Aves: Fringillidae: *Acanthis cannabina*), and viable seeds have been removed from the dung of cattle (Mammalia: Bovidae: *Bos*), deer (Mammalia: Cervidae: *Capreolus*; *Cervus*), rabbits (Mammalia: Leporidae) and sheep (Mammalia: Bovidae: *Ovis aries*). Birds also are believed to disperse the small seeds that become lodged in adhering bits of mud. The estimated longevity of seed bank propagules is 1–5 years; however, viable seeds have been recovered from moderate seed bank depths (15 cm) at sites that had been farmed (with moderate tillage) continuously for a quarter of a century. Seed densities in excess of 2670 seeds m² have been reported. The highest densities have been observed at 4–8 cm soil depths, where flooding was of greatest duration (presumably leading to higher rates of deposition). The seeds are extremely vagile and almost ubiquitous, being reported from the seed banks of a broad assortment of habitats including many temperate forests. Seed germination frequently is observed at sites where the plants are absent from the standing vegetation. The seedlings can emerge nearly year round with biannual peaks typically occurring in the spring (March) and fall (September). This is an early successional species that is highly tolerant to heavily compacted soils. It commonly is found growing in gardens and lawns. The plants even have been found in sites that were bulldozed following a forest fire. Although this species can behave like an annual, it is regarded as a FAC perennial. It possesses a rhizomatous habit that produces clones with densely packed shoots. The leaves are known to remain green throughout the winter in some localities; however, frost and herbivory have been identified as the major sources of plant mortality. The roots usually are associated with arbuscular mycorrhizae. Accounts regarding the nativity of this species can be confusing because North America comprises both a native and an introduced subspecies. **Reported associates:** *Abies balsamea, Acer macrophyllum, A. negundo, A. saccharinum, Achillea millefolium, Agrostis hyemalis, A. variabilis, Alnus, Alopecurus aequalis, Amelanchier alnifolia, Androsace septentrionalis, Antennaria parvifolia, Anthoxanthum, Arabis holboellii, Arctagrostis, Arnica longifolia, Artemisia ludoviciana, Asclepias syriaca, A. incarnata, Athyrium filix-femina, Betula papyrifera, Blepharoneuron tricholepis, Calamagrostis canadensis, Caltha leptosepala, C. palustris, Calystegia sepium, Camassia leichtlinii, Cardamine breweri, C. aquatilis, C. engelmannii, C. fracta, C. lanuginosa, C. lemmonii, Carex macloviana, C. nebrascensis, C. pensylvanica, C. stricta, Castilleja applegatei, C. miniata, Cerastium beeringianum, C. fontanum, Chamerion angustifolium, Chelone glabra, Cicuta, Circaea alpina, Cirsium muticum, Claytonia cordifolia, Collinsia parviflora, Cornus canadensis, C. racemosa, C. sericea, Cotula coronopifolia, Crataegus monogyna, Cruciata pedemontana, Cuscuta gronovii, Cytisus scoparius, Dactylis glomerata, Darlingtonia californica, Deschampsia cespitosa, Drosera rotundifolia, Echinocystis lobata, Eleocharis acicularis, Elymus glaucus, Epilobium hornemannii, Equisetum, Erigeron annuus, E. flagellaris, Eriophorum, Eupatorium perfoliatum, Eutrochium maculatum, Festuca brachyphylla, Fragaria platypetala, F. virginiana, Fraxinus, Galium aparine, G. trifidum, Gaultheria, Geranium carolinianum, Geum canadense, G. macrophyllum, Glechoma hederacea, Heracleum lanatum, Heuchera, Hieracium aurantiacum, Hypericum majus, H. perforatum, Hypoxis hirsuta, Iris missouriensis, Juncus effusus, J. mexicanus, J. orthophyllus, J. tenuis, Kelloggia galioides, Leptarrhena pyrolifolia, Leucanthemum vulgare, Lilium kelleyanum, Lobelia spicata, Lolium, Lonicera caerulea, L. conjugialis, L. morrowii, L. tatarica, Luetkea pectinata, Lupinus latifolius, L. lepidus, L. nootkatensis, L. polyphyllus, Luzula comosa, L. spicata, Lycopus uniflorus, Lysimachia nummularia, L. quadriflora, Marchantia, Menziesia ferruginea, Mertensia ciliata, Micranthes nidifica, M. odontoloma, Mimulus guttatus, M. lewisii, M. moschatus, Monarda fistulosa, Monardella villosa, Montia chamissoi, M. linearis, Nemophila spatulata, Orobanche uniflora, Ostrya virginiana, Oxalis oregana, Oxypolis occidentalis, Penstemon procerus, Petasites frigidus, Phleum alpinum, Picea sitchensis, Pinus contorta, Plagiobothrys scouleri, Plantago lanceolata, P. major, P. rugelii, Platanthera leucostachys, P. sparsiflora, Poa annua, P. compressa, P. leptocoma, P. pratensis, P. secunda, Polemonium elegans, Polygonum polygaloides, Polypodium, Polystichum, Populus deltoides, Potentilla simplex, Prunella vulgaris, Prunus serotina, Pseudotsuga menziesii, Pteridium aquilinum, Pterospora andromedea, Quercus borealis, Q. rubra, Ranunculus acris, R. orthorhynchus, R. repens, R. uncinatus, Rhododendron neoglandulosum, Ribes cereum, Rorippa palustris, Rosa, Rubus idaeus, R. parviflorus, R. procera, R. spectabilis, R. ursinus, Rumex acetosella, Salix discolor, S. interior, S. jepsonii, S. lasiolepis, S. nigra, S. scouleriana, S. sitchensis, Schoenoplectus, Senecio triangularis, Sisymbrium officinale, Sisyrinchium, Solidago ulmifolia, Sparganium angustifolium, Spergularia rubra, Sphagnum, Spiraea douglasii, S. splendens, Sporobolus, Stachys, Stellaria calycantha, S. longifolia, S. media, Symphyotrichum cordifolium, Taraxacum officinale, Tiarella trifoliata, Trifolium longipes, T. repens, Triosteum perfoliatum, Triphysaria pusilla, Typha latifolia, Vaccinium globulare, Veratrum californicum, Veronica americana, V. arvensis, V. peregrina, V. wormskjoldii, Viola macloskeyi, V. sororia, Woodsia scopulina, Zizia aurea.*

Use by wildlife: *Veronica* species generally provide habitat and cover for amphipods (Crustacea: *Gammarus*) and other freshwater invertebrates. *Veronica americana* and *V. anagallis-aquatica* are host plants of the common buckeye butterfly (Insecta: Lepidoptera: Nymphalidae: *Junonia coenia*). *Veronica americana* is fed on by larval caddisflies (Insecta: Trichoptera: Limnephilidae: *Cryptochia, Desmona bethula*). The foliage of *V. americana* is highly palatable to slugs (Gastropoda: Arionidae: *Ariolimax columbianus, Arion ater*). It also is a highly nutritious forage plant used by elk (Mammalia: Cervidae: *Cervus elaphus*) during the winter and can make up to 1% of the diet of mule deer (Mammalia: Cervidae: *Odocoileus hemionus hemionus*) in spring. However, neither *V. americana* nor *V. scutellata* are eaten by the Columbian black-tailed deer (Mammalia: Cervidae: *Odocoileus hemionus columbianus*). *Veronica americana* is the preferred habitat for a freshwater shrimp (Crustacea: Decapoda: Atyidae: *Paratya curvirostris*) in New Zealand. *Veronica anagallis-aquatica* is a host plant for the variable checkerspot butterfly (Insecta: Lepidoptera: Nymphalidae: *Euphydryas chalcedona*). The developing fruits are eaten by moth larvae (Insecta: Lepidoptera: Pterophoridae: *Mariana taprobanes*). The leaves have been considered as a source of livestock feed due to their high palatability and 47%–89% leaf protein content. In Europe, *V. beccabunga* is a secondary host plant to aphids (Insecta: Hemiptera: Aphididae: *Aphis triglochinis*). It is used by a wide variety of aquatic invertebrates and vertebrates in its native distributional range; however, a comparable ecological assessment has not been made for North American plants. Several *Veronica* species are hosts to Fungi including *V. americana* (Basidiomycota: Entylomataceae: *Entyloma linariae*), *V. anagallis-aquatica* (Oomycota: Peronosporaceae: *Peronospora grisea*), *V. peregrina* (Basidiomycota: Entylomataceae: *Entyloma linariae*; Oomycota: Peronosporaceae: *Peronospora grisea*), *V. scutellata* (Ascomycota: Mycosphaerellaceae: *Cercospora tortipes*; Basidiomycota: Pucciniaceae: *Puccinia albulensis*), and *V. serpyllifolia* (Basidiomycota: Entylomataceae: *Entyloma linariae*; Chytridiomycota: Synchytriaceae: *Synchytrium* *globosum*; Oomycota: Peronosporaceae: *Peronospora grisea*). In Europe, the roots of *V. beccabunga* are parasitized by a chytrid fungus (Chytridiomycota: Hyphochytriaceae: *Cystochytrium radicale*). *Veronica peregrina* is the host of sac Fungi (Ascomycota: *Thallospora aspera*), smut fungi (Basidiomycota: Entylomataceae: *Entyloma linariae*, *E. veronicae*), and a water mold (Oomycota: Peronosporaceae: *Peronospora grisea*). *Veronica peregrina* is a larval host plant of the tiger moth (Insecta: Lepidoptera: Arctiidae: *Grammia geneura*). Nectar-foraging insects (Insecta) on *V. peregrina* include bees (Hymenoptera: Andrenidae: *Andrena cressonii*; Halictidae: *Augochlorella aurata*, *Halictus confusus*, *Lasioglossum imitatus*, *L. pilosus*, *L. pruinosus*, *L. versatus*, *L. zephyrus*), flies (Diptera: Anthomyiidae: *Anthomyia acra*; Chloropidae: *Rhopalopterum soror*; Syrphidae: *Sphaerophoria contiqua*, *Toxomerus marginatus*), and bugs (Hemiptera: Thyreocoridae: *Corimelaena pulicarius*). The seeds of *V. peregrina* are eaten by the greater prairie chicken (Aves: Phasianidae: *Tympanuchus cupido*), and can be an important component of their diet, especially during spring (May to June). *Veronica scutellata* is susceptible to infection by a barley deformation phytoplasma (Firmicutes: Acholeplasmatales: Acholeplasmataceae: "*Candidatus* Phytoplasma asteris"). The foliage of *V. scutellata* is eaten by the larvae of watercress leaf beetles (Insecta: Coleoptera: Chrysomelidae: *Phaedon viridis*). Viable seeds recovered from cattle (Mammalia: Bovidae: *Bos*) dung indicate that the plants also are grazed to some extent by livestock. *Veronica serpyllifolia* is grazed by cattle (Mammalia: Bovidae: *Bos*) rabbits (Mammalia: Leporidae) and by sheep (Mammalia: Bovidae: *Ovis aries*); however, it is not eaten by meadow voles (Mammalia: Rodentia: Muridae: *Microtus pennsylvanicus*).

Economic importance: food: Several *Veronica* species (*V. americana*, *V. anagallis-aquatica*, *V. beccabunga*) have been eaten as a preventative for scurvy. When thoroughly washed, *V. americana* can be eaten in salads or used as a potherb, but only if gathered from unpolluted sites (the plants are known to concentrate heavy metals). The plants were gathered and eaten as a spring vegetable by the Nez Perce tribe. In Italy, *V. anagallis-aquatica* and *V. beccabunga* are eaten raw in salads. Raw or cooked leaves of *V. scutellata* also are eaten; **medicinal:** *Veronica americana* was used by the Navajo and Ramah as an emetic and as a gastrointestinal remedy. *Veronica anagallis-aquatica* contains iridoid glucosides and sesquiterpenes. In Turkey, the plants are boiled in milk to make a poultice for treating abdominal pain and aqueous extracts are used in a soothing bath for alleviating pain associated with rheumatism. These medicinal applications have been corroborated by the documented occurrence of catapol derived iridoid glucosides (catalposide, verproside), which are highly active as pain relief and anti-inflammatory agents. *Veronica beccabunga* commonly was administered as an early medicinal plant (known as *Herba Beccabungae*) and is still sold as a homeopathic remedy. In Ireland, it was boiled and sweetened with candy, and used as an expectorant. It is used as a diuretic in southern Spain. The Cherokee used *V. serpyllifolia* to treat coughs and fevers. They also applied a poultice of the plant to boils and used the warm juice for treating earaches. In Spain, the flowering shoots are used to provide an ocular antiseptic; **cultivation:** All of the OBL North American *Veronica* species are cultivated as water garden ornamentals. *Veronica beccabunga* includes the cultivar 'Don's Dyke.' *Veronica scutellata* is known as ornamental Hebe; **misc. products:** The presence of well-established mats of *Veronica americana* has been used as an indicator of stable habitat conditions for aquatic invertebrates. Dense patches also serve as a soil binder along muddy water margins; **weeds:** *Veronica anagallis-aquatica* is reported as a rice field weed in Egypt and other parts of the world. *Veronica peregrina* can become weedy along damp roadsides or on cultivated land. It also is a weed of plant nurseries, which can serve as an avenue for widespread introductions. It is reported as a weed in China, Europe and Japan. *Veronica serpyllifolia* is a weed of gardens, lawns, roadsides and waste areas; **nonindigenous species:** *Veronica americana* was introduced sometime before 1940 to New Zealand where it is thoroughly naturalized. The widespread *V. anagallis-aquatica* has been introduced to Australia, Japan, and to New Zealand; however, the precise native range of this widespread species has been difficult to determine. It is regarded both as a native and as a nonindigenous species in North America, where specimen records date back to the 19th century in California (1881) and Wisconsin (1882). Some reported occurrences are from ballast disposal sites, and it is possible that at least some populations of the species were introduced during the early days of commercial shipping to the New World. The native status of this species deserves further investigation, perhaps ideally using genetic markers. *Veronica beccabunga* (Eurasian) was introduced to North America before 1876, most likely as a contaminant of discharged shipping ballast, but perhaps also along with imported shipments of sport fish. *Veronica peregrina* was introduced to Europe in the 17th century, reaching Portugal before 1877, Scandinavia before 1971, Slovenia in 1992 and Croatia in 2003. The species also has been introduced to Australia, China, Japan, Korea, Mongolia, and Russia. It is naturalized in Hawaii where it was introduced before 1984. It is apparent that the species often is spread as a seed contaminant in forage mixtures and in commercially distributed garden plant stocks. Some western populations are believed to have been introduced from eastern North America. *Veronica scutellata* was introduced to New Zealand sometime before 1924 and to Australia (Tasmania) before 1974. The plants are spread, at least in part, by the transport of seed-contaminated commercial pulpwood. *Veronica serpyllifolia* comprises a nonindigenous North American taxon (subsp. *serpyllifolia*) as well as an indigenous one (subsp. *humifusa*). The former is believed to have been introduced from Europe as a seed contaminant. *Veronica serpyllifolia* is extremely vagile and has even been introduced to remote islands including Hawaii and sub-Antarctic isles.

Systematics: Phylogenetic studies incorporating various molecular data indicate that *Veronica* is not monophyletic as traditionally circumscribed and should include several formerly segregated genera (e.g., *Hebe*, *Parahebe*). The status of *Synthyris* is more controversial. Although combined DNA sequence data show it to nest within *Veronica* (Figure 5.90), the branch support provided by those data is weak with respect to critical nodes. A number of anomalies exist. Nuclear DNA data resolve *Paederota* within *Veronica* and subgenus *Beccabunga* as the first branching subclade; whereas, cpDNA (and the combined data) resolve *Paederota* outside of *Veronica* and subgenus *Veronica* as the first branching subclade. Because *Synthyris* resolves in the vicinity of these groups, which are flanked by weak branch support with respect to the various data sets, it seems premature to merge the genus with *Veronica* until more conclusive evidence is forthcoming. Sectional subdivisions have been used widely in many past classifications of *Veronica*, but the recent trend has been to delimit various clades as subgenera (e.g., Figure 5.90). In this respect (see Figure 5.90), all of the OBL North American taxa would occur within two subgenera: subg. *Beccabunga* (*Veronica americana*, *V. anagallis-aquatica*, *V. beccabunga*, *V. peregrina*, *V. serpyllifolia*) and subg. *Veronica*: (*V. scutellata*). Although excluded from the analysis depicted in Figure 5.90, *V. anagallis-aquatica* resolves as the sister species to *V. beccabunga* from analysis of nrITS sequences; however, *V. americana* (also regarded as closely related to the latter) has not yet been included in phylogenetic analyses. A number of infraspecific taxa are delimited in *Veronica*. Of the three recognized subspecies of *V. beccabunga* (subsp. *abscondita*, subsp. *beccabunga*, subsp. *muscosa*), the North American introductions appear to represent subsp. *beccabunga*. *Veronica peregrina* has two subspecies that are distinguished by their glabrous or pubescent foliage. Subspecies *peregrina* occurs mainly in eastern and southeastern United States, with subsp. *xalapensis* being more western and northern in its distribution. *Veronica serpyllifolia* also has two subspecies, one being native to North America (subsp. *humifusa*) and the other nonindigenous (subsp. *serpyllifolia*). The former has been treated by some authors as a distinct species. The basic chromosome number of the genus probably is $x = 9$; however, several modifications (e.g., $x = 7$, $x = 9$) occur, sometimes varying in counts reported for a single species. *Veronica serpyllifolia* ($2n = 14$; both subspecies) and *Veronica scutellata* ($2n = 18$) are diploid; *V. americana* ($2n = 36$) is tetraploid; *V. anagallis-aquatica* ($2n = 18, 36$) and *V. beccabunga* ($2n = 18, 36$) possess both diploid and tetraploid cytotypes; *V. peregrina* ($2n = 54$) is hexaploid. Hybridization reportedly is common among members of subgenus *Beccabunga*; however, there have been no reports of hybrids in North America.

Comments: *Veronica americana*, *V. anagallis-aquatica* (cosmopolitan), *V. peregrina* (New World) and *V. serpyllifolia* (also in Eurasia) are widespread throughout North America;

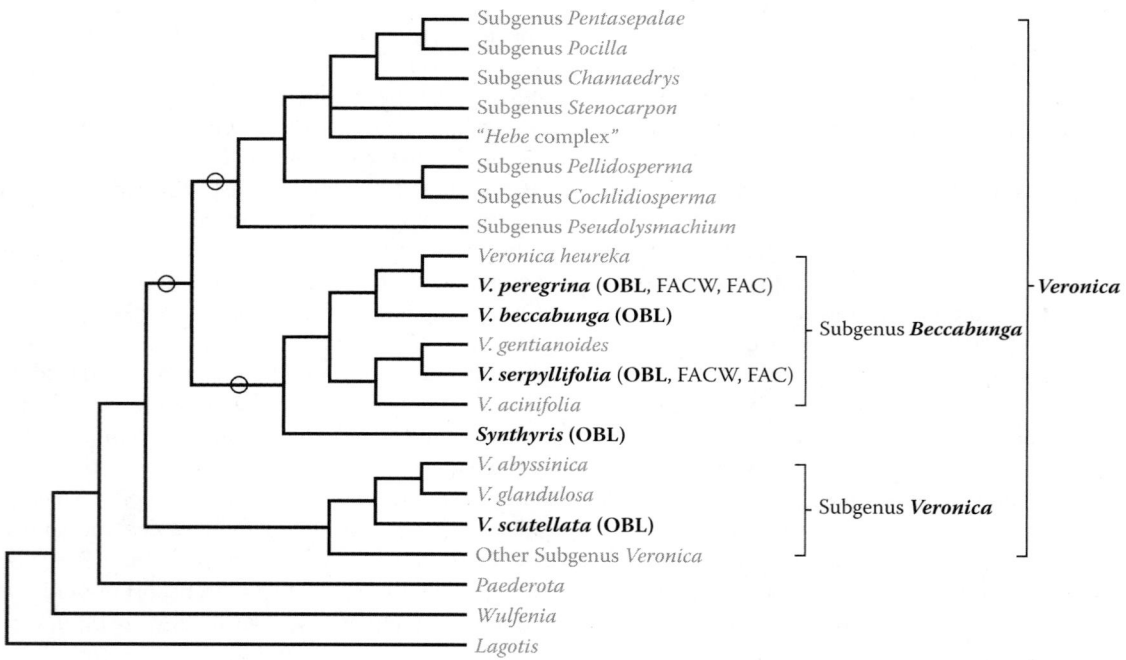

FIGURE 5.90 Phylogenetic relationships in *Veronica* as indicated by analysis of combined DNA sequence data. The aquatic North American species (OBL taxa indicated in bold) are restricted to two subgenera (*Beccabunga*, *Veronica*). The non-North American species of subg. *Beccabunga* shown (*V. acinifolia*, *V. gentianoides*, *V. heureka*) are also strongly associated with wet habitats and probably should be regarded as obligate aquatics; the two species related to *V. scutellata* in subg. *Veronica* (*V. abyssinica*, *V. glandulosa*) also associate with wetlands, but are more facultative indicators. In any case, the OBL habit appears to have arisen at least twice in the genus. Although these results show *Synthyris* to be nested within *Veronica*, the support for several critical branches (indicated by open circles) provided by the available DNA sequence data is low. Consequently, *Synthyris* has been maintained as distinct in this treatment pending the presentation of more definitive phylogenetic evidence. A separate analysis of nrITS data (Albach & Chase, 2001) included *V. anagallis-aquatica*, which resolved as the sister species to *V. beccabunga*. (Adapted from Albach, D.C. et al., *Ann. Missouri Bot. Gard.*, *Gard.*, 91, 275–320, 2004b.)

whereas, *V. scutellata* (also in Eurasia) is widespread except for the southern United States. *Veronica beccabunga* (Eurasia) has been introduced to northeastern and southwestern North America.

References: Albach & Chase, 2001; Albach et al., 2004a,b; Anderson & Leopold, 2002; Andersson et al., 2000; Ausden et al., 2005; Barbour, 1897; Bartgis, 1983; Baskin & Baskin, 1974, 1983, 1998; Beatley, 1956; Benson & Kurth, 1996; Blanchan, 1901; Boedeltje et al., 2003b; Boeger & Poulson, 2003; Boutin & Harper, 1991; Burk, 1977; Burrill, 1888; Byrd, 1984; Campbell, 1987; Cardina et al., 1991; Carpenter, 1978; Casper & Krausch, 1981; Cates & Orians, 1975; Cellot et al., 1998; Chippindale & Milton, 1934; Clark & Wilson, 1998, 2001; Cody, 2000; Combes, 1965; Cowan, 1945; Cruden, 1977; Dean et al., 1994; DiTomaso & Healy, 2003; Doyle, 2003; Drabble & Drabble, 1927; Drezner et al., 2001; Elias & Dykeman, 1990; Erman, 1984, 2002; Eybert & Constant, 1998; Fernald, 1950; Fernald & Kinsey, 1943; Frenot et al., 2005; Fricke & Steubing, 1984; Friedman et al., 1996; Godefroid & Koedam, 2004; Goodson et al., 2003; Grime et al., 1981; Gross, 1990; Grøstad et al., 1999; Guard, 1995; Guarrera, 2003; Gurnell et al., 2007a; Halsted, 1890; Harrod, 1964; Healy, 1944; Hendry et al., 1994; Hölzel & Otte, 2001; Jenkins & Starkey, 1993; Johnson & Altman, 1999; Kadono, 2004; Kampny & Dengler, 1997; Karling, 1943; Keeler, 1978; Kellman, 1974; Kerner von Marilaun, 1895; Klinka et al., 1995; Korschgen, 1962; Küpeli et al., 2005; Lang, 2006; Larsen, 1929; Leck & Graveline, 1979; Leck & Leck, 1998; Lentini & Venza, 2007; Les & Mehrhoff, 1999; Les & Stuckey, 1985; Linhart, 1974, 1976; Lippert & Jameson, 1964; Liu et al., 2005; Long et al., 2003; Marble et al., 1999; Matthews et al., 1990; McCain & Christy, 2005; McGregor, 1948; Mouissie, 2004; Moyle, 1945; Nakanishi, 2002; Neiland, 1958; Newman, 1948; Often et al., 2003, 2006; Olive, 1948; Öztürk & Fischer, 1982; Pakeman et al., 2002; Pandey & Srivastava, 1989; Parrish & Bazzaz, 1979; Pennell, 1935; Pieroni et al., 2005; Preston & Croft, 1997; Rabinowitz, 1978; Renne & Tracy, 2007; Rigat et al., 2007; Rivera et al., 2005; Robbins, 1918; Roberts, 1986; Robertson & James, 2007; Robinson, 2005; Rodger, 1933; Ruthven, 1911; Saeidi Mehrvarz & Kharabian, 2005; Saidak & Nelson, 1962; Sainty & Jacobs, 1981; Sanderson et al., 2007; Savile, 1956; Schneider & Melzer, 2003; Singer & Stireman III, 2001; Swingle, 1889; Sykes, 1981; Teketay, 1998; Ter Heerdt et al., 1996; Thieret, 1971; Topić & Ilijanić, 2003; Tracy & Sanderson, 2000; Turesson, 1916; Turki & Sheded, 2002; Vittoz & Engler, 2007; Von Oheimb et al., 2005; Wagner et al., 2003; Whitehouse, 1933; Wilkins, 1957; Wilson, 1908; Wolcott, 1937; Wolff et al., 2005; Wu et al., 2004; Zika, 1996.

Family 9: Scrophulariaceae [53]

As already noted in the introductions to the preceding several families, the circumscription of Scrophulariaceae has been altered extensively as a result of molecular phylogenetic studies (Olmstead et al., 2001; Albach et al., 2005; Oxelman et al., 2005; Rahmanzadeh et al., 2005; Tank et al., 2006). By the current circumscription, the constituents of Scrophulariaceae essentially represent the type genus *Scrophularia* along with most members of the former tribes *Aptosimeae, Hemimerideae, Leucophylleae, Limoselleae, Manuleae, Selagineae, Scrophularieae,* and *Teedieae,* along with the former families Buddlejaceae and Myoporaceae (Olmstead et al., 2001; Kornhall & Bremer, 2004; Oxelman et al., 2005). As a consequence, the taxonomic scope of this once very large and diverse family has diminished from more than 300 genera with 5850 species to a group consisting now of approximately 53 genera and roughly 1500 species. The number of tribes (redefined to represent major clades) has been reduced to eight (Figure 5.91).

Given their past taxonomic history and high degree of morphological similarity, it is not surprising that a distinction between the newly redefined Scrophulariaceae and Plantaginaceae can be made only with great difficulty. Judd et al. (2002) provide a few characters that potentially distinguish Scrophulariaceae with some efficacy including: nonsagittate anthers (sagittate in Plantaginaceae), anthers dehiscing by a single, distal, transverse slit (dehiscing by two longitudinal slits or by a single, inverted U- or V-shaped slit in Plantaginaceae), and a thick-walled megasporangium (thin walled in Plantaginaceae).

Pollination in Scrophulariaceae is facilitated primarily by insects, which frequent the flowers for their nectar; however, cleistogamy and chasmogamous autogamy also occur in the family. The seeds are dispersed mainly by the wind. *Buddleja, Diascia, Nemesia, Scrophularia, Sutera,* and *Verbascum* contain a large number of ornamental cultivars, and many of the other genera (e.g., *Alonsoa, Eremophila, Hebenstretia, Jamesbrittenia, Leucophyllum, Manulea, Myoporum, Selago*) contain at least a few cultivated representatives. The family is of minor economic importance otherwise.

Scrophulariaceae are almost entirely terrestrial. There is only a single North American genus with an obligate wetland (OBL) designation:

1. ***Limosella*** L.

Scrophulariaceae occupy a relatively central position within the Lamiales clade (Figure 5.50). A phylogenetic survey of the family is fairly complete, with most of the assigned genera having now been sampled in systematic studies (e.g., Figure 5.91). The dense degree of taxon sampling in the family has facilitated the delimitation of tribal limits that are meaningful phylogenetically. Because aquatic species often represent phylogenetically derived taxa, it is not surprising that the tribe containing *Limosella* (Limoselleae) resolves in a relatively derived position in the family (Figure 5.91). However, *Limosella* itself occurs in a relatively basal position within that tribe (Figure 5.91). All *Limosella* species should be regarded as aquatics and they grow only rarely under terrestrial conditions. This genus appears to represent the only aquatic radiation in Scrophulariaceae, with the remainder of the family populated primarily by terrestrial taxa (in both the New and Old World).

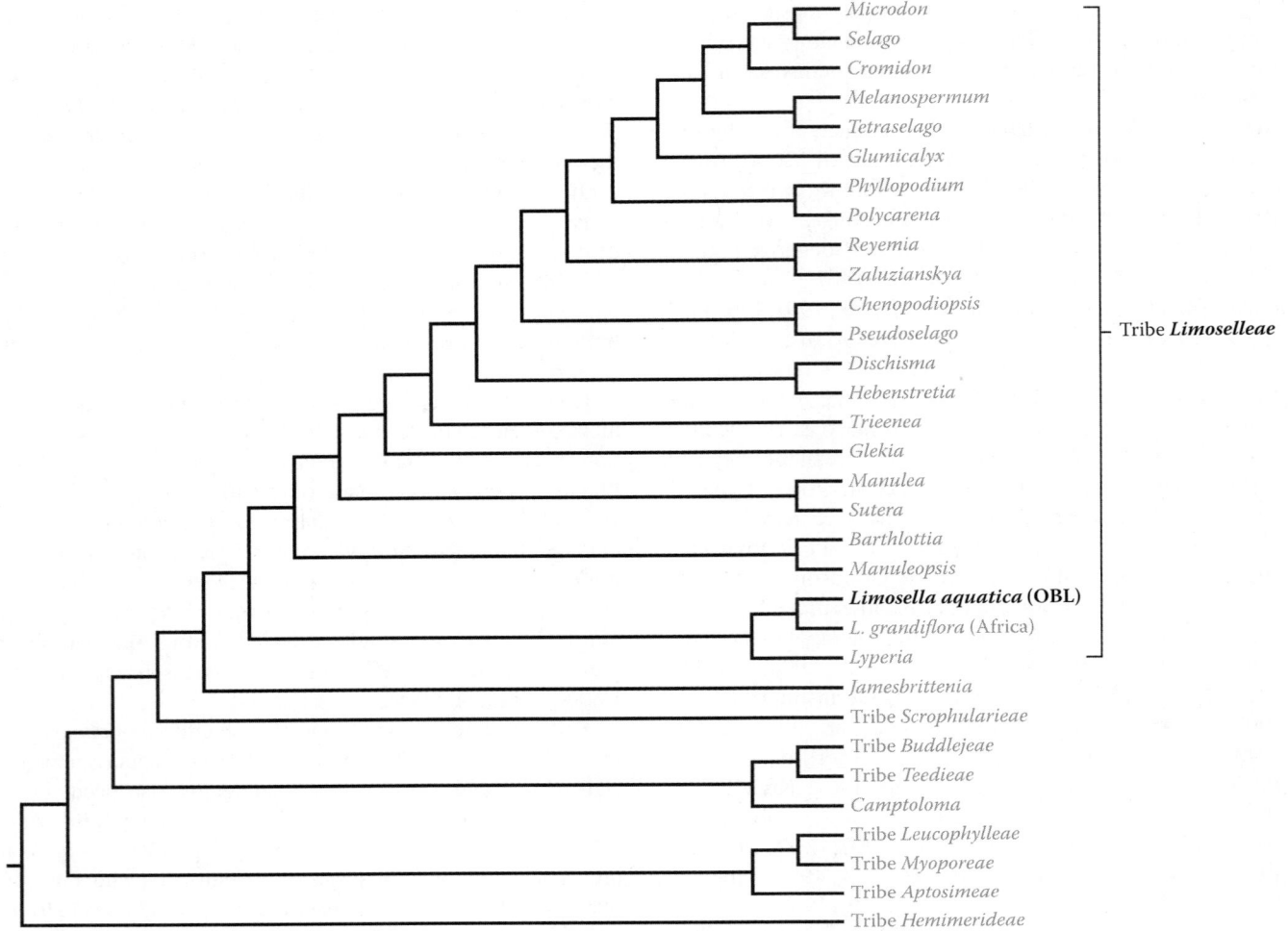

FIGURE 5.91 Phylogenetic relationships in Scrophulariaceae as indicated by analysis of combined DNA sequence data. *Limosella grandiflora* is African, and therefore not designated as an OBL North American taxon (indicated in bold); however, all *Limosella* species are aquatic, which indicates a single derivation of the OBL habit in this family. The single aquatic genus (*Limosella*) is an early diverging group in the relatively derived tribe Limoselleae. (Modified from Oxelman, B. et al., *Taxon*, 54, 411–425, 2005.)

1. *Limosella*

Mudwort; limoselle

Etymology: from the Latin *limosus* ("muddy") in reference to its habitat

Synonyms: *Ygramela*

Distribution: global: Africa; Asia; Europe; New World; **North America:** widespread (except central)

Diversity: global: 15 species; **North America:** 4 species

Indicators (USA): OBL: *Limosella acaulis, L. aquatica, L. australis, L. pubiflora*

Habitat: brackish (coastal) or freshwater; lacustrine, palustrine; **pH:** 6.0–8.2; **depth:** <1 m; **life-form(s):** emergent herb, floating leaved, submersed (rosulate)

Key morphology: stems (to 3 cm) horizontal, stoloniferous; leaves (to 12 cm) tufted in rosettes (to 20-leaved), awl-like or long-petioled (to 10 cm) and terminating in a spatulate blade (to 3 cm), stipules hyaline, ear-like, fused to lower leaf base; flowers 5-merous, solitary on naked peduncles (to 4 cm); corolla (to 4 mm) regular, campanulate, blue, pink or white;

stamens 4; capsules (to 5 mm) elliptical to globose; seeds numerous, the surface reticulate, stellate in cross-section

Life history: duration: annual (fruit/seeds); perennial (stolons); **asexual reproduction:** rhizomes, stolons; **pollination:** self; **sexual condition:** hermaphroditic; **fruit:** capsules (common); **local dispersal:** stolons, seeds (water); **long-distance dispersal:** seeds (birds)

Imperilment: (1) *Limosella acaulis* [G5]; S1 (<u>BC</u>); S2 (ID); S3 (WA); (2) *L. aquatica* [G5]; SH (MO, <u>NF</u>); S1 (<u>NB</u>); S2 (<u>ON</u>, <u>YT</u>); S3 (<u>AB</u>, AK, <u>BC</u>, MN, <u>QC</u>, WY); (3) *L. australis* [G4]; SX (DE, PA); SH (VA); S1 (<u>LB</u>, NC, NH, <u>NF</u>, NJ); S2 (MD, <u>NB</u>, <u>NS</u>); S3 (ME, NY, <u>PE</u>, <u>QC</u>); (4) *L. pubiflora* [G1Q]; SX (AZ); S1 (NM)

Ecology: general: The genus *Limosella* is entirely aquatic and all of the North American species are designated as OBL throughout their ranges. The genus occurs in temperate areas or in montane tropical habitats. The habit of the plants is very much reminiscent of *Glossostigma*, with which they easily can be confused. They mainly occupy muddy banks and shores or very shallow water where the water levels fluctuate. Pollination has been studied only in a few species, but the flowers generally appear to be self-compatible and self-pollinating. Like *Glossostigma*, the species technically are stolon-producing perennials, but often behave as annuals due to the difficulty of surviving the winter conditions (e.g., ice scour) associated with their exposed shoreline habitats. Perennation is most likely to occur where at least pockets of shallow water persist. Most species are well-represented in the seed bank. The reticulate seed coats trap air bubbles, which facilitate their flotation for water dispersal, and also increases their adherence to bits of mud that attach to the feet of migrating birds.

Limosella acaulis Sessé & Moc. grows as an annual or perennial in depressions, ditches, drying vernal pools, flats, marshes, mudflats, pools, seeps, swales, and along the margins of lakes, ponds, pools, reservoirs, and streams at elevations of up to 3200 m. The plants occur mainly on exposed substrates in full sunlight but sometimes also grow in very shallow water (<30 cm). The substrates often are described as alkaline but include adobe, clay, gravel, mud, peat, sand, sandy loam, sandy mud, shale, and volcanic soil. The plants can become mat forming by proliferation of the stolons. There is little additional life history information available for this species. **Reported associates:** *Agrostis idahoensis, Alisma, Alopecurus aequalis, Amaranthus californicus, Ammannia coccinea, A. robusta, Anagallis minima, Asclepias fascicularis, Bergia texana, Bidens, Callitriche palustris, Carex alma, C. densa, C. nebrascensis, Castilleja campestris, Chara, Chenopodium chenopodioides, Conyza canadensis, Crassula aquatica, Crypsis, Cyperus acuminatus, C. aristatus, C. bipartitus, Downingia bella, Elatine brachysperma, E. chilensis, Eleocharis acicularis, E. bella, E. ovata, E. palustris, E. parishii, E. parvula, Geum triflorum, Gnaphalium chilense, G. palustre, Gratiola neglecta, Hordeum brachyantherum, Hydrocotyle ranunculoides, Iris missouriensis, Isolepis cernua, Juncus balticus, J. bufonius, J. xiphioides, Limosella aquatica, Lindernia dubia, Lipocarpha aristulata, Marsilea vestita, Mimulus floribundus, M. guttatus, M. primuloides, Mollugo verticillata, Myosurus apetalus, Persicaria lapathifolia, Pilularia americana, Pinus ponderosa, Plagiobothrys hispidulus, P. stipitatus, Poa pratensis, Porterella carnosula, Potamogeton crispus, P. illinoensis, Ranunculus cardiophyllus, R. cymbalaria, Rorippa curvipes, Rotala ramosior, Rumex maritimus, R. salicifolius, Sagittaria sanfordii, Sparganium angustifolium, Stuckenia pectinata, Symphyotrichum frondosum, Typha, Veronica peregrina, Xanthium strumarium.*

Limosella aquatica L. is an annual or perennial resident of tidal or nontidally influenced beaches, borrow pits, depressions, ditches, flats, floodplains, gravel bars, lakes, marshes, meadows, ponds, pools, salt marshes, sand bars, seeps, sloughs, streams, swales, and the fluctuating margins of lakes, ponds, rivers, sinkholes, and streams at elevations of up to 3200 m. The plants can grow at water depths up to 30 cm and produce floating leaves when they are inundated. They can tolerate a small amount of salinity but occur most often in clear, oligotrophic lakes. Although they generally are associated with acidic conditions (pH: 6.0–7.0) the plants also have been observed to grow on sites (e.g., sandbars) that are alkaline (e.g., pH: 8.2). The substrates typically are infertile and include adobe, clay, gravel, mud, peat, rock, sand, silt, silty clay, silt loam, and volcanic material. Fairly cold temperatures are tolerated and the plants are found as far north as 64° N latitude. The plants are self-compatible and mainly produce cleistogamous flowers, especially when growing under submersed conditions. Charles Darwin reported that flowering has even been observed to occur on plants that were growing under the ice. Self-pollination of cleistogamous flowers has been attributed to a system of bubble pollination; however, the precise mechanism needs to be verified. The plants are well adapted to autogamy and selfed offspring are vigorous. Up to 200 seeds are produced per flower, with annual fecundity averaging 3600 seeds/plant, but reaching up to 12,950 seeds/plant for particularly robust individuals. Fruiting plants can occur at densities of up to 5/cm². The capsules can dehisce above or beneath the surface of the water. Minor dispersal can occur by water (however, seeds reportedly sink within a minute) and by wind. Longer dispersal is achieved by birds, which transport the seeds in mud that adheres to their feet (this method has been documented). Site occurrences are highly correlated with the extent of river connectivity, which emphasizes the efficacy of water dispersal. The propagules develop into persistent, long-term seed banks. Seed germination has been verified to occur after 27 months of burial, but seed viability certainly is retained for far longer periods. Rich seed banks can exist even where the plants are not represented in the standing vegetation. Seed banks often average only 4–7 seeds/m² but up to an estimated 5400/m² have been recovered within a 6 cm depth of sediment in some instances. Higher seed densities occur at sites that are ungrazed. The seeds have no special after-ripening requirements and will germinate within 3 days at room temperature; however, their germination is highly light dependent and can be reduced from 96% (in full light) to 2% (in dim light). Even in exposed field sites, seed germination can decline to 6% on cloudy days. The seeds will germinate equally well under inundated or noninundated conditions and treatments with brackish seawater have been found to enhance the germination of cold-treated seeds. The seedlings produce "prop hairs" at the base of the radicle, which help to anchor them in the mud. The plants typically occur in patchy sites and survive annually by their high seed production and recruitment from the seed banks. They commonly inhabit temporally disturbed shorelines. In Europe, they also have been observed to persist in artificially disturbed sites such as pig pastures, agricultural fields, and in wagon ruts. The plants most often are described as annuals, which probably applies in most cases; however, the plants are stoloniferous and have been reported to grow under ice cover (see earlier), which would indicate their ability to perennate in shallow water. **Reported associates:** *Alisma triviale, Alopecurus carolinianus, Amaranthus albus, Ambrosia, Ammannia robusta, Amorpha, Artemisia tilesii, Bacopa rotundifolia, Bidens cernuus, B. frondosus,*

Buchloe dactyloides, Callitriche heterophylla, Carex athrostachya, C. scoparia, C. scopulorum, Castilleja campestris, Ceratophyllum demersum, Chamaesyce glyptosperma, Chenopodium, Cirsium arvense, Coreopsis tinctoria, Crassula aquatica, Crypsis schoenoides, Cyperus acuminatus, C. aristatus, Deschampsia beringensis, D. cespitosa, D. danthonioides, Downingia bicornuta, D. cuspidata, D. pulchella, Echinochloa crus-galli, E. muricata, Elatine californica, E. chilensis, E. engelmannii, E. rubella, E. triandra, Eleocharis acicularis, E. ovata, E. palustris, Elodea canadensis, Epilobium ciliatum, E. densiflorum, E. torreyi, Equisetum arvense, E. fluviatile, E. variegatum, Eragrostis hypnoides, Festuca richardsonii, Gnaphalium palustre, G. uliginosum, Gratiola ebracteata, G. neglecta, Hackelia floribunda, Heteranthera dubia, H. limosa, Hordeum brachyantherum, H. jubatum, Isoetes melanopoda, I. muricata, Juncus balticus, J. bufonius, J. castaneus, J. hemiendytus, J. leiospermus, J. nevadensis, J. oxymeris, J. uncialis, Koenigia islandica, Lemna minor, Leptochloa fascicularis, Lilaea scilloides, Lilaeopsis chinensis, L. occidentalis, Limnanthes alba, Limosella acaulis, Lindernia dubia, Ludwigia palustris, Lythrum hyssopifolium, Marsilea vestita, Mentha arvensis, Mimulus guttatus, Mollugo verticillata, Montia fontana, Myosurus minimus, Myriophyllum spicatum, M. ussuriense, Navarretia intertexta, N. leucocephala, Oenothera canescens, Panicum capillare, Pascopyrum smithii, Persicaria amphibia, P. hydropiperoides, P. lapathifolia, P. pensylvanica, Phalaris arundinacea, Phyla cuneifolia, Pilularia americana, Plagiobothrys leptocladus, P. orientalis, P. scouleri, P. stipitatus, P. undulatus, Pogogyne ziziphoroides, Polemonium acutiflorum, Polygonum aviculare, Polypogon monspeliensis, Populus, Potamogeton richardsonii, Portulaca oleracea, Psilocarphus brevissimus, P. oregonus, Ranunculus cymbalaria, R. hyperboreus, R. reptans, Ribes inerme, Rorippa curvisiliqua, R. sinuata, R. teres, Rosa woodsii, Rumex crispus, R. maritimus, R. stenophyllus, Sagittaria calycina, S. latifolia, Salix exigua, S. geyeriana, Schoenoplectus americanus, S. triqueter, Scutellaria lateriflora, Senecio vulgaris, Sium suave, Solanum dulcamara, Spiranthes diluvialis, Stellaria fontinalis, Subularia aquatica, Symphyotrichum, Talinum parviflorum, Tamarix, Trichocoronis wrightii, Verbena bracteata, Vernonia fasciculata, Veronica arvensis, V. peregrina, V. persica, Xanthium strumarium.

Limosella australis R. Br. is a perennial of tidal habitats including ditches, estuaries, marshes, mudflats, pools, and the shores of lakes, ponds, and rivers. It can tolerate fresh to brackish conditions (0.5–18.0 ppt salinity) but usually is found in the mid to lower intertidal zone where the plants are exposed at low tide and submersed at high tide. The sediments are infertile and include gravel, mud, sand, and sandy gravel. The plants are self-pollinating, but flower rarely while they are inundated. Up to 100 seeds can be produced per flower. After seed set, the pedicels eventually recurve to facilitate the deposition of seeds in the substrate near the parental plant. Some dispersal by water probably also occurs, but longer dispersal distances would be most likely be achieved by seeds that become embedded in mud that adheres to waterfowl.

Seed germination and subsequent establishment are higher when the seeds are inundated by 5 cm of water than by 15 cm of water, with no establishment occurring at depths of 30 cm. Establishment is poor at noninundated sites. Thus, plants become more common, and their biomass increases as water depth declines, unless dry conditions are encountered. Seed germination and seedling establishment are reduced at drawdown rates of 5 cm/day compared to rates of only 1 cm/day. Individual plants live for an estimated 4–20 years. The plants are fairly tolerant to disturbances because of their stoloniferous habit. They exhibit less than 30% mortality when the leaves are 100% scorched by fire. Also, clipped plants (which simulates grazing) produce higher above-ground biomass. The vegetative spread by the stolons can yield dense tufts or highly scattered plants. This species is a pioneer of exposed sites and is not a very good competitor. The plants performed poorly in competition experiments with *Lilaeopsis masonii*, with their biomass decreasing in fresh water and declining substantially at a salinity of 7 ppt. **Reported associates:** *Bidens bidentoides, B. cernuus, B. laevis, Cirsium hydrophilum, Cordylanthus mollis, Crassula aquatica, Cyperus bipartita, Elatine minima, Eleocharis parvula, Equisetum fluviatile, Eriocaulon parkeri, Festuca filiformis, Glossostigma cleistanthum, Helenium bigelovii, Heteranthera reniformis, Hydrocotyle verticillata, Isoetes riparia, Isolepis cernua, Juncus acuminatus, Lathyrus jepsonii, Lilaeopsis carolinensis, L. chinensis, L. masonii, Lindernia dubia, Micranthemum micranthemoides, Mimulus guttatus, Myriophyllum tenellum, Najas guadalupensis, Nuphar advena, Persicaria arifolia, P. hydropiperoides, Pluchea carolinensis, Potamogeton crispus, Ptilimnium capillaceum, Sagittaria calycina, S. graminea, S. rigida, S. subulata, Samolus valerandi, Schoenoplectus acutus, S. pungens, S. tabernaemontani, Sium suave, Sparganium, Symphyotrichum lentum Triglochin striata, Vallisneria americana, Zannichellia palustris, Zizania aquatica.*

Limosella pubiflora Pennell is a stoloniferous perennial, which forms mats around ponds and cattle tanks. No adequate life history treatment can be provided because of the dearth of information available for this rare taxon. Furthermore, this species likely represents only a minor variant of *L. aquatica* (see *Systematics* later).

Reported associates: None.

Use by wildlife: *Limosella acaulis* is grazed by cattle (Mammalia: Bovidae: *Bos taurus*). *Limosella aquatica* is a host of the fungus *Doassansia limosellae* (Fungi: Basidiomycota: Doassansiaceae). *Limosella australis* is grazed by greater snow geese (Aves: Anatidae: *Chen caerulescens atlantica*).

Economic importance: food: not edible; **medicinal:** The Navajo and Ramah tribes used the rolled, washed leaves of *Limosella aquatica* to plug arrow and bullet wounds; **cultivation:** *Limosella aquatica* and *L. australis* are cultivated infrequently as aquarium plants; **misc. products:** The Ramah rubbed the leaves of *Limosella aquatica* on their bodies as a ceremonial medicine; **weeds:** *Limosella aquatica* is a minor weed of maize (Poaceae: *Zea mays*) fields in Mexico; **nonindigenous species:** *Limosella aquatica* is regarded as a cryptogenic, nonindigenous species in New Mexico. *Limosella* *australis* possibly was introduced to California, where it was first recorded in 1957. The native and nonindigenous ranges of most *Limosella* species require a thorough reevaluation, which will not be possible until the taxonomy of the genus has been critically revised.

Systematics: Phylogenetic analyses of DNA sequence data (Figures 5.91 and 5.92) confirm that *Limosella* is monophyletic. The *Limosella* clade nests among a number of genera assigned previously to Scrophulariaceae tribe Manuleeae; however, the tribal name Limoselleae has nomenclatural priority. The sister genus to *Limosella* remains somewhat uncertain, although some analyses support *Lyperia* in that position (Figure 5.91). A revision of *Limosella* has not been made since the early 20th century and is much needed in order to clarify species limits as well as their relationships. Only four taxa (about a fourth of the species) have been included in contemporary phylogenetic analyses. The African *L. macrantha* possibly is synonymous with the wider ranging *L. australis*. In North America, *L. pubiflora* has been recognized as a segregate of *L. aquatica*; however, it is not clearly distinct and its taxonomic status requires reevaluation. (such clarification might be difficult given that the taxon presumably is nearly extirpated in North America). A North American taxon known for a long time as *L. subulata* is now regarded as being synonymous with the widely distributed *L. australis*. Because several species apparently have much broader geographical distributions than thought previously, a thorough reexamination of the group is needed on a worldwide basis. The basic chromosome number for *Limosella* is $x = 10$. *Limosella australis* is diploid ($2n = 20$) and *L. aquatica* is tetraploid ($2n = 40$). Hybrids involving *Limosella aquatica × L. australis* have been reported; however, these are not well documented and much more intensive study in this area is needed.

Comments: *Limosella acaulis* occurs in the western United States and also in Mexico; *L. aquatica* is distributed in

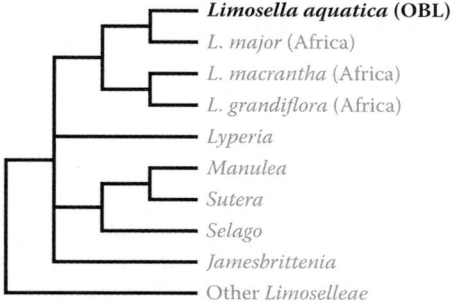

FIGURE 5.92 Phylogenetic relationships in *Limosella* as indicated by analysis of nrITS sequence data. Although only a quarter of the currently recognized species of *Limosella* are represented, it is reasonable to conclude that the genus is monophyletic. Because the entire genus is aquatic, a single origin of the aquatic habit is indicated for Scrophulariaceae (see also Figure 5.91). *Limosella aquatica* is the only species shown that is designated as OBL in North America (bold). However, some regard *L. macrantha* as synonymous with *L. australis*, which also is recognized as an OBL North American species. (Adapted from Kornhall, P. & Bremer, B., *Bot. J. Linn. Soc.*, 146, 453–467, 2004.)

northern and western North America with a wider range extending into Africa, Eurasia, and South America. *Limosella australis* occurs in eastern North America, but is disjunct in British Columbia and California (perhaps due to introductions). The species also occurs in Africa, Australia, South America, and Wales. *Limosella pubiflora* is restricted to Arizona and New Mexico.

References: Aronson, 1989; Baskin & Baskin, 1998; Ceska et al., 1986; Christy, 2004; Cox, 2001; Crosslé & Brock, 2002; Fassett, 1928; Ferren, Jr. & Schuyler, 1980; Giroux & Bedard, 1987; Glück, 1934; Jutila b. Erkkilä, 1998; Kerner von Marilaun, 1895; Melcher et al., 2000; Merritt & Cooper, 2000; Milberg & Stridh, 1994; Moore & Tryon, Jr., 1946; Mühlberg, 1982; Murphy, 2002; Nichols, 1920; Nicol et al., 2007; Philcox, 1990; Robertson & James, 2007; Salisbury, 1967, 1970; Stauffer, 1987; Vibrans, 1998; Wacker & Kelly, 2004; Wright, Jr. & Bent, 1968; Zebell & Fiedler, 1996; Zika, 1996.

Family 10: Verbenaceae [36]

The current circumscription of Verbenaceae that has emerged from a synthesis of various phylogenetic studies (e.g., Olmstead et al., 1993; Wagstaff & Olmstead, 1997; Schwarzbach & McDade, 2002; Oxelman et al., 2005; Marx et al., 2010) includes just over 1000 species. In order to achieve a monophyletic family, it has been necessary to reassign to other families, roughly two-thirds of the genera assigned formerly to Verbenaceae (e.g., *Avicennia* to Acanthaceae and *Callicarpa*, *Clerodendum*, *Tectona*, and *Vitex* to Lamiaceae). In the current circumscription, Verbenaceae are defined by their entirely indeterminate inflorescences, ovules that are attached marginally to false septa, thickened pollen apertures, a bilobed stigma, and a simple style that always is terminal (never gynobasic) in position; the flowers are showy but are only weakly bilabiate in symmetry (Judd et al., 2002). The family is widespread throughout temperate and tropical regions.

Verbenaceae are pollinated by bees and wasps (Insecta: Hymenoptera) or flies (Insecta: Diptera), which seek the floral nectar. The fleshy drupes are dispersed through ingestion by birds (Aves) and the nutlets (dispersed with the calyx attached) by a combination of abiotic (e.g., rain, wind) or biotic (birds) agents (Atkins, 2004). Popular ornamental plants include various species of *Lantana* and *Verbena* (the latter with several hundred cultivars). *Aloysia triphylla* ("lemon verbena") is used as a tasty lemony condiment for foods or in teas.

This family principally is terrestrial and includes only two genera that contain OBL North American species:

1. ***Phyla*** Lour.
2. ***Verbena*** L.

As currently indicated, Verbenaceae are a clade that is deeply nested within Lamiales in a position closely related to Martyniaceae and Bignoniaceae (Figure 5.50). Although past analyses included few representative genera of Verbenaceae as it is currently defined (e.g., Olmstead et al., 2007), more

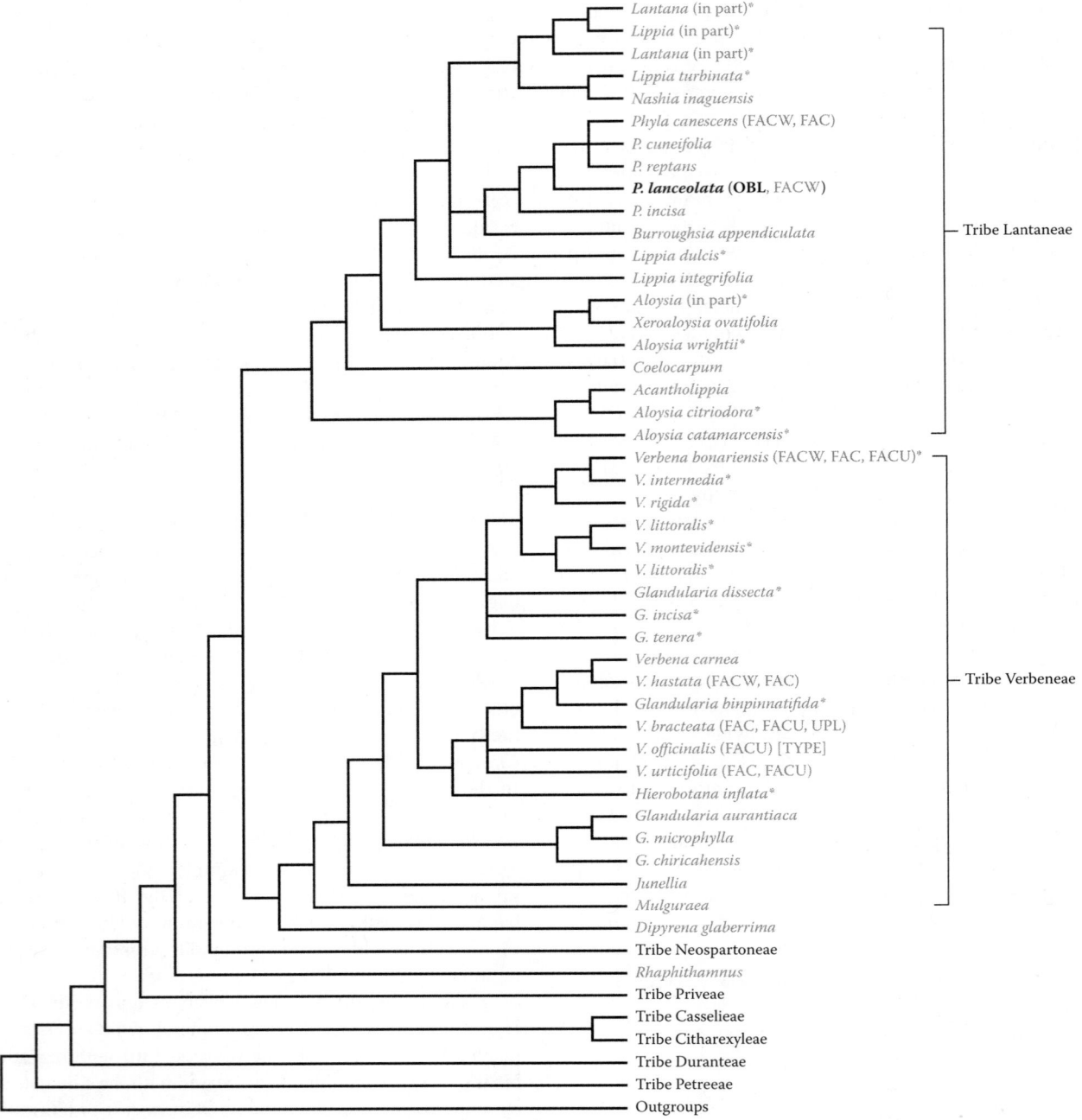

FIGURE 5.93 Relationships in Verbenaceae as indicated by results of combined DNA sequence data. Following the most recent circumscription of Verbenaceae, the included genera now resolve as a clade. However, the phylogenetic relationships also reveal that a number of genera are not monophyletic (asterisks indicate misplaced taxa). Taxonomic revisions of *Aloysia*, *Glandularia*, *Lantana*, *Lippia*, and *Verbena* are necessary to accommodate the relationships indicated. It is evident that the OBL habit (bold) likely has arisen twice in the family; even though *V. scabra* (OBL) has not been included among those *Verbena* species surveyed, it and *Phyla lanceolata* are placed in different tribes. (Adapted from Marx, H.E. et al., *Amer. J. Bot.* 97, 1647–1663, 2010.)

recent studies have provided a fairly comprehensive molecular systematic survey of the family as well as an improved tribal classification (Marx et al., 2010). However, these studies have also shown that a number of currently recognized genera, including *Verbena* itself, are not monophyletic as currently defined (Figure 5.93).

1. *Phyla*

Fog-fruit, frog-fruit; fraise de mer

Etymology: from the Greek *phulê* ("tribe"), in reference to the closely clustered flowers

Synonyms: *Cryptocalyx*; *Diototheca*; *Lippia* (in part); *Piarimula*; *Pilopus*; *Zapania* (in part)

Distribution: global: New World; paleotropics; **North America:** southern

Diversity: global: 15 species; **North America:** 7 species

Indicators (USA): OBL; **FACW:** *Phyla lanceolata*

Habitat: brackish, freshwater; palustrine; **pH:** 4.6–7.8; **depth:** <1 m; **life-form(s):** emergent herb

Key morphology: stems creeping, rooting at nodes, fertile branches ascending (to 6 dm), pubescence fugacious; leaves (to 7.5 cm) opposite, simple, elliptic, oblong, ovate or lanceolate, the margins serrate beyond the middle, the petioles (to 10 mm) decurrent; inflorescences of dense, many flowered, long-stalked (to 10 cm) spikes, which arise from terminal leaf axils, and elongate cylindrically (to 3.5 cm) at maturity; calyx 2-lobed, triangular, arching over fruit; corolla (to 2.5 mm) bilabiate, the petals white when open, lavender to purplish when immature (central apex of spike); stamens 4; schizocarp (to 1.5 mm) orbicular, splitting into 2 nutlets

Life history: duration: perennial (rhizomes); **asexual reproduction:** rhizomes, shoot fragments; **pollination:** insect; **sexual condition:** hermaphroditic; **fruit:** schizocarps (common); **local dispersal:** fragments (water), seeds; **long-distance dispersal:** seeds

Imperilment: (1) *Phyla lanceolata* [G5]; SH (NJ); S1 (DE, GA, UT); S2 (NC, ON)

Ecology: general: *Phyla* species often occur in disturbed habitats that include floodplains or comparable sites of ephemeral water, and are also common on cultivated land. Several of the species tolerate wet conditions but are not dependent on them for their prolonged survival. Five of the North American species are FACW indicators and even the one obligate species is categorized as FAC in a portion of its range. In those species studied, the flowers are self-compatible, but normally do not set seed unless they are pollinated by insects (Diptera, Hymenoptera). Pollen:ovule ratios indicate a facultatively autogamous breeding system. When viewed from above, the top of the spikes presents a "bulls-eye" pattern to pollinators by contrast of the dark, unopened central flowers (which provide a landing platform) and the lighter opened flowers that surround the central portion; the inflorescence appears much like a small capitulum. Dispersal can occur as vegetative fragments are displaced by flowing water. Endozoic transport of seeds by birds (Aves) and livestock (Mammalia: Bovidae) is suspected; however, mechanisms of seed dispersal (as well as seed germination requirements) are not well studied in the genus.

Phyla lanceolata **(Michx.) Greene** inhabits beaches, canals, depressions, desiccated lake bottoms and pools, ditches, dunes, flats, floodplains, lakeshores, levees, marshes, meadows, mudflats, potholes, prairies, ricefields, riverbanks, sandbars, sloughs, swales, and swamps at elevations of up to 1524 m. The plants can tolerate shade but are found most often growing in open sites where the canopy cover is sparse. The sites range from freshwater to oligohaline conditions and span a fairly broad range of acidity (pH: 4.6–7.8). The substrates include gravel, mud, muddy peat, rock, sand, sandy clay, sandy loam, and silty sandy loam. The flowers are pollinated by short-tongued bees (Insecta:

Hymenoptera: Andrenidae; Halictidae) and are described as being facultatively outcrossed. Persistence at sites depends strongly on recruitment from the seed bank, especially where disturbance is high. Requirements for seed germination have not been reported, but do not appear to be very complex. There is little information regarding seed dispersal. Vegetative fragments have been found to occur as a minor component of river drift. The ability of fragments to root readily at the nodes facilitates their establishment. This species is common on disturbed ground including recently abandoned agricultural fields and surface coal mines. The plants will tolerate burning, grazing, mowing, and trampling except at excessive levels. Vegetative propagation can be achieved effectively by stem cuttings, which root readily at their nodes. **Reported associates:** *Acalypha rhomboidea, Acer saccharinum, Acmella oppositifolia, Agropyron repens, Agrostis gigantea, A. stolonifera, Alisma subcordatum, A. triviale, Alternanthera philoxeroides, Amaranthus australis, A. cannabinus, A. tuberculatus, Ambrosia artemisiifolia, Ammannia coccinia, Amorpha fruticosa, Andropogon glomeratus, A. virginicus, Anthemis cotula, Apocynum cannabinum, Aristida oligantha, Asclepias incarnata, Axonopus fissifolius, Baccharis emoryi, B. halimifolia, B. salicifolia, B. salicina, Bacopa monnieri, Betula nigra, Bidens aristosus, B. cernuus, B. frondosus, B. laevis, Boehmeria cylindrica, Bolboschoenus maritimus, B. robustus, Boltonia decurrens, Bothriochloa longipaniculata, Bromus, Callitriche heterophylla, Caperonia palustris, Capsella bursa-pastoris, Carex annectens, C. brevior, C. buxbaumii, C. emoryi, C. frankii, C. grayi, C. lanuginosa, C. meadii, C. pellita, C. praegracilis, C. shortiana, C. stipata, Cassia fasciculata, Centella asiatica, Centrosema virginianum, Cephalanthus occidentalis, Chamaecrista fasciculata, C. nictitans, Cicuta maculata, Cinna arundinacea, Cirsium undulatum, Cladium jamaicense, Cornus drummondii, C. racemosa, C. sericea, Cuscuta glomerata, Cynodon dactylon, Cyperus aristatus, C. diandrus, C. erythrorhizos, C. esculentus, C. filicinus, C. haspan, C. iria, C. odoratus, C. strigosus, C. virens, Decodon verticillatus, Desmanthus illinoensis, Dichanthelium acuminatum, D. oligosanthes, D. sphaerocarpon, Digitaria sanguinalis, Diodia virginiana, Diospyros virginiana, Distichlis spicata, Echinochloa crus-galli, E. walteri, Echinocystis lobata, Eclipta prostrata, Eichhornia crassipes, Eleocharis acicularis, E. albida, E. baldwinii, E. calva, E. cellulosa, E. erythropoda, E. fallax, E. montana, E. obtusa, E. palustris, E. parvula, E. rostellata, E. tenuis, Epilobium coloratum, Equisetum hyemale, E. laevigatum, Eragrostis hypnoides, E. pectinacea, Ericameria nauseosa, Erigeron strigosus, Erechtites hieraciifolia, Erianthus giganteus, Eryngium yuccifolium, Eupatorium coelestinum, E. perfoliatum, E. serotinum, Euphorbia marginata, Eurybia divaricata, Eustoma exaltatum, Euthamia leptocephala, Fallopia convolvulus, Fimbristylis autumnalis, F. castanea, F. vahlii, Flaveria campestris, Fuirena pumila, Galactia volubilis, Galium aparine, Glechoma hederacea, Gleditsia triacanthos, Glottidium vesicarium, Glycyrrhiza lepidota, Habenaria repens, Helenium autumnale, Helianthus angustifolius,*

H. maximiliani, Hibiscus moscheutos, Hordeum jubatum, Hydrocotyle umbellata, H. verticillata, Hypericum galioides, H. mutilum, Impatiens capensis, Ipomoea lacunosa, I. sagittata, Iris versicolor, I. virginica, Iva annua, I. frutescens, Jacquemontia tamnifolia, Juncus balticus, J. filipendulus, J. interior, J. marginatus, J. roemerianus, J. saximontanus, J. tenuis, J. torreyi, Justicia ovata, Kosteletzkya pentacarpos, Lamium, Lappula squarrosa, Lathyrus palustris, Leersia hexandra, L. oryzoides, L. virginica, Leptochloa fascicularis, L. uninervia, Leucospora, Liatris lancifolia, Limnobium spongia, Lindernia dubia, Linum medium, Lipocarpha micrantha, Lobelia cardinalis, L. siphilitica, Ludwigia glandulosa, L. leptocarpa, L. linearis, L. repens, L. uruguayensis, Lycopus americanus, Lysimachia ciliata, L. nummularia, L. terrestris, Lythrum alatum, L. lineare, Medicago lupulina, Melilotus, Mentha arvensis, Micranthemum umbrosum, Mikania scandens, Mimosa, Mimulus glabratus, M. ringens, Mollugo verticillata, Muhlenbergia asperifolia, Myosoton aquaticum, Myosurus minimus, Myrica cerifera, Nama, Neptunia pubescens, Oenothera rhombipetala, Panicum capillare, P. dichotomiflorum, P. hemitomon, P. repens, P. rigidulum, P. vaginatum, P. virgatum, Parthenocissus quinquefolia, Pascopyrum smithii, Paspalum dilatatum, P. distichum, P. notatum, P. plicatulum, P. urvillei, Pennisetum glaucum, Penthorum sedoides, Persicaria amphibia, P. densiflora, P. hydropiper, P. hydropiperoides, P. lapathifolia, P. pensylvanica, P. punctata, P. setacea, P. virginiana, Phalaris arundinacea, Phleum pratense, Physostegia virginiana, Phyla nodiflora, Pilea pumila, Pinus elliottii, Plantago hirtella, P. major, Pluchea camphorata, P. odorata, Poa arida, P. palustris, P. pratensis, Polypogon monspeliensis, Polypremum procumbens, Pontederia cordata, Populus deltoides, P. fremontii, Portulaca oleracea, Psilocarya nitens, Ptilimnium, Quercus laurifolia, Q. virginiana, Ranunculus sceleratus, Rhexia mariana, Rhus trilobata, Rhynchospora caduca, R. corniculata, R. globularis, R. macrostachya, Rorippa islandica, R. palustris, R. sinuata, R. sylvestris, Rosa arkansana, R. palustris, Rotala ramosior, Rubus trivialis, Rumex crispus, R. maritimus, R. obtusifolius, R. stenophyllus, Sacciolepis striata, Sagittaria cuneata, S. falcata, S. lancifolia, S. latifolia, Salix amygdaloides, S. eriocephala, S. exigua, S. gooddingii, S. interior, S. nigra, Salvinia molesta, Sambucus nigra, Samolus valerandi, Schizachyrium scoparium, Schoenoplectus americanus, S. californicus, S. pungens, S. tabernaemontani, Scirpus atrovirens, S. pendulus, Scutellaria, Senna obtusifolia, Sesbania drummondii, S. herbacea, S. macrocarpa, Setaria faberi, S. geniculata, S. glauca, S. magna, Sisyrinchium angustifolium, Sium suave, Smilax, Solanum carolinense, Solidago, Sparganium, Spartina alterniflora, S. cynosuroides, S. patens, S. pectinata, Sphenopholis obtusata, Spiraea alba, Spiranthes cernua, Spirodela polyrhiza, Sporobolus asper, S. indicus, Stachys, Strophostyles helvula, Stuckenia pectinata, Suaeda calceoliformis, Symphyotrichum ericoides, S. lanceolatum, S. subulatum, Taraxacum officinale, Teucrium canadense, Thelypteris palustris, Toxicodendron radicans, Triadica sebifera, Trifolium repens, Typha angustifolia, T. domingensis, T. latifolia, Urtica dioica, Vaccinium, Verbena brasiliensis, V. hastata, V. scabra, Vernonia fasciculata, Veronica peregrina, Vigna luteola, Viola sororia, Vitis riparia, Xanthium strumarium.

Use by wildlife: The seeds of *Phyla* species are eaten by various waterfowl (Aves: Anatidae) including baldpate (*Anas americana*), blue-winged teal (*Anas discors*), green-winged teal (*Anas carolinensis*), mallard (*Anas platyrhynchos*), and wood ducks (*Aix sponsa*). They are a minor food of migrating soras (Aves: Rallidae: *Porzana carolina*). The foliage is a dietary component of stream-dwelling muskrats (Mammalia: Muridae: *Ondatra zibethicus*). *Phyla lanceolata* is a host plant for larval Lepidoptera (Insecta: Nymphalidae: *Charidryas gorgone, Junonia coenia, Phyciodes phaon*; Pterophoridae: *Lantanophaga pusilladactyla; Stenoptilodes taprobanes*). Normal growth of *Junonia coenia* larvae depends on their intake of iridoid glycosides, which occur in the foliage. The plants are host to several fungi (Ascomycota: Mycosphaerellaceae: *Cercospora lippiae*; Basidiomycota: Botryobasidiaceae: *Allescheriella crocea*) and a spider mite (Arthropoda: Acari: Tetranychidae: *Tetranychina lippiae*). More than 45 different insects (Insecta; mainly bees and flies) have been collected from the plants. The flowers are visited by short-tongued bees (Hymenoptera: Andrenidae: *Calliopsis andreniformis*; Halictidae: *Augochlorella aurata, Lasioglossum pectoralis, L. tegularis*), which collect the pollen and nectar. Other nectar-seeking insects include long-tongued bees (Hymenoptera: Apidae: Bombus *pensylvanica*; Anthophoridae: *Triepeolus lunatus, Florilegus condigna, Melissodes dentiventris*; Megachilidae: *Coelioxys octodentata; Megachile brevis*), short-tongued bees (Hymenoptera: Halictidae: *Agapostemon virescens, Augochlora purus, Augochlorella striata, Augochloropsis metallica, Halictus confusus, H. ligatus, Lasioglossum nymphaearum, L. pilosus*), wasps (Hymenoptera: Sphecidae: *Ammophila kennedyi, A. nigricans, A. pictipennis, Bicyrtes ventralis, Prionyx atrata*; Vespidae: *Stenodynerus ammonia, S. anormis*), flies (Diptera: Bombyliidae: *Exoprosopa fasciata, E. fascipennis, Systoechus vulgaris*; Conopidae: *Physoconops brachyrrhinchus*; Sarcophagidae: *Amobia floridensis*; Syrphidae: *Eristalis stipator, Helophilus fasciatus, H. latifrons, Orthonevra nitida, Paragus bicolor, Sphaerophoria contiqua, Syritta pipiens, Toxomerus marginatus, Tropidia quadrata*; Tachinidae: *Acroglossa hesperidarum, Archytas analis, Copecrypta ruficauda, Cylindromyia dosiades, C. euchenor, Periscepsia laevigata, Plagiomima spinosula*), and butterflies (Lepidoptera: Hesperiidae: *Ancyloxypha numitor, Hylephila phyleus, Polites peckius, P. themistocles*; Lycaenidae: *Everes comyntas, Lycaena hyllus*; Nymphalidae: *Phyciodes tharos*; Pieridae: *Colias philodice, Eurema lisa*).

Economic importance: food: not reported as edible; **medicinal:** The Mahuna used *Phyla lanceolata* as a treatment for rheumatism; **cultivation:** *Phyla lanceolata* is cultivated as an ornamental ground cover for wet areas; **misc. products:** *Phyla lanceolata* is recommended as a planting to stabilize shorelines and levees; **weeds:** *Phyla lanceolata* is not reported as weedy; **nonindigenous species:** Only *Phyla fruticosa* (FAC) is nonindigenous

Systematics: Although some previous analyses of DNA sequence data indicated that *Phyla* might not be entirely monophyletic, more comprehensive analyses (Figure 5.93) do resolve the group as a clade. The precise relationship to *Lippia*, which once subsumed the genus, is not clear; however both genera certainly are closely related. Further taxon sampling is necessary to assist with the delimitation of *Aloysia*, *Lantana*, *Lippia*, and *Phyla* and to provide for a better understanding of interspecific relationships within the latter genus. Currently, *P. lanceolata* resolves in a central position among the few congeners surveyed (Figure 5.93). There is weak support for the placement of *Burroughsia* as the sister genus to *Phyla* (Figure 5.93). The base chromosome number of *Phyla* is x = 9. *Phyla lanceolata* (2n = 36) is a tetraploid. Hybridization has been suspected in *Phyla* because of the similar floral morphology among species and complex patterns of morphological variation in certain taxa; however, adequate documentation has not been forthcoming.

Comments: *Phyla lanceolata* occurs throughout the southern half of North America and extends into Mexico.

References: Atkins, 2004; Battaglia et al., 2007; Beal, 1977; Bowers, 1984; Cole, 1992; Cooke, 1969; Cruden, 1977; Cruden & Lyon, 1985; Cruzan et al., 1988; Errington, 1941; Evans, 1979; Grace et al., 2000; Hoagland & Buthod, 2005; Hopkins, 1969; Lewis, 1961; Logan, 2003; Mabbott, 1920; Matthews et al., 1990; McAtee, 1918, 1925, 1939; Moldenke, 1958; Nolfo-Clements, 2006; Olmstead et al., 2007; Penfound, 1953; Quarterman, 1957; Robertson, 1924a; Rundle & Sayre, 1983; Sanders, 1987; Smith et al., 1998; Speichert & Speichert, 2004; Stutzenbaker, 1999; Ungar, 1964; Vincent et al., 2003; Visser et al., 1998, 2002.

2. *Verbena*

Vervain; verveine

Etymology: derived from the Greek *hiera botanê* meaning "sacred herb"

Synonyms: *Billardiera; Glandularia; Obletia; Shuttleworthia; Uwarowia*

Distribution: global: Africa; Asia; Europe; New World; **North America:** widespread

Diversity: global: 250 species; **North America:** 31 species

Indicators (USA): OBL; FACW: *Verbena scabra*

Habitat: brackish or freshwater; palustrine; **pH:** alkaline; **depth:** <1 m; **life-form(s):** emergent herb

Key morphology: stems (to 1.5 m) erect, quadrangular, herbage scabrous to strigose; leaves (to 13 cm) opposite, lanceolate, ovate or elongate ovate, petiolate (to 2.5 cm), the margins serrate dentate; spikes (to 20 cm) 1–5, secondarily arranged in diffuse axillary panicles, the flowers ascending or spreading, not overlapping; calyx (to 3 mm) 5-lobed, hispid, the acute apices connivent; corolla (to 5 mm) salverform, nearly regular, blue, lavender or pink; stamens 4, didynamous, included; style terminal, schizocarp (to 1.5 mm) enclosed by calyx, splitting into 4 trigonal nutlets

Life history: duration: biennial (tap root); perennial (tap root); **asexual reproduction:** stolons; **pollination:** insect; **sexual condition:** hermaphroditic; **fruit:** schizocarps (common); **local dispersal:** seeds; **long-distance dispersal:** seeds

Imperilment: (1) *Verbena scabra* [G5]; S1 (WV); S2 (AR, NC, VA)

Ecology: general: *Verbena* primarily is a terrestrial genus. There are 12 North American taxa with wetland indicator status, but only one species (*V. scabra*) is ranked as OBL in some parts of its range. *Verbena hastata* is encountered commonly in wetlands, but it is ranked as facultative (FAC to FACW) throughout its range. *Verbena* flowers are visited by nectar and pollen-collecting insects. In some species seed germination can be irregular and a period of cold stratification often is required to break dormancy. Seed dispersal has not been studied well enough in this genus to develop any broad generalizations.

Verbena scabra **Vahl** grows in bayheads, depressions, ditches, flats, floodplains, lakebottoms, marshes, meadows, seeps, shell mounds, swales, swamps, and along the margins of borrow pits, canals, ponds, rivers, and streams at elevations of up to 1158 m. It is found in brackish to fresh water on exposed substrates or in standing water up to 3 dm deep. Exposures ranging from shade to full sun are tolerated. The substrates are alkaline (e.g., pH: 8.6) and include clay, clay loam, limestone, muck, sand, sandy loam, shell mounds, and silt. The life history of this species, especially its reproductive ecology, is poorly known and deserves further study. The nutlets are coated with waxy granules, which may facilitate flotation. Occurrences of this species are concentrated along the coastal plain and the plants can be found on the unusual tree islands of the Everglades. They typically inhabit disturbed areas such as path margins, roadsides, and cleared or mowed areas. Although the plants have been described as being stoloniferous, they actually produce very short shoot clusters that arise from the base of their main crown. The roots are colonized by arbuscular mycorrhizae. **Reported associates:** *Acer rubrum, Agalinis linifolia, Andropogon glomeratus, Anemopsis californica, Baccharis emoryi, B. salicifolia, Callicarpa americana, Cephalanthus occidentalis, Chamaecrista, Chasmanthium nitidum, Cirsium, Cladium jamaicense, Commelina diffusa, Cynodon dactylon, Cyperus ligularis, C. surinamensis, Cyrilla racemiflora, Dichanthelium acuminatum, Dichanthium annulatum, Digitaria ciliaris, Diospyros virginiana, Eclipta prostrata, Eleocharis atropurpurea, Equisetum laevigatum, Erigeron quercifolius, Eryngium, Eupatorium, Fimbristylis autumnalis, Flaveria linearis, Hydrocotyle ranunculoides, Ilex glabra, Ipomoea, Itea virginica, Juncus balticus, J. effusus, J. saximontanus, Lemna, Leptochloa panicea, Liquidambar styraciflua, Ludwigia, Lythrum alatum, Medicago lupulina, Melothria pendula, Mikania scandens, Mitracarpus hirtus, Muhlenbergia asperifolia, Murdannia nudiflora, Myrica cerifera, Oxalis, Panicum capillare, Paspalum urvillei, Phyla lanceolata, P. nodiflora, Phyllanthus urinaria, Pinus taeda, Plantago hirtella, Platanus occidentalis, Pluchea odorata, Polygala grandiflora, Polypogon monspeliensis, Populus fremontii, Prosopis velutina, Quercus, Rhynchospora colorata, R. microcarpa, Rubus argutus, Sabal palmetto, Sabatia grandiflora, Saccharum giganteum, Salix caroliniana, S. exigua, S. gooddingii, Salvinia rotundifolia, Schoenoplectus* *americanus, S. pungens, Senna obtusifolia, Sesbania herbacea, Solidago canadensis, Sorghum halepense, Toxicodendron radicans, Typha latifolia, Vaccinium, Vigna luteola, Wolffiella gladiata, Zizaniopsis miliacea.*

Use by wildlife: *Verbena scabra* is a host plant of whiteflies (Insecta: Homoptera: Aleyrodidae: *Aleurotrachelus*; *Trialeurodes abutiloneus*). It is an occasional host of the nonindigenous light brown apple moth (Insecta: Lepidoptera: Tortricidae: *Epiphyas postvittana*) in California.

Economic importance: food: not edible; **medicinal:** none; **cultivation:** *Verbena scabra* is cultivated occasionally; **misc. products:** none; **weeds:** *Verbena scabra* has not been reported as weedy; **nonindigenous species:** *Verbena scabra* is indigenous.

Systematics: DNA sequence data analysis indicates that *Verbena* is not monophyletic, with numerous misplaced taxa. Because current analyses resolve a clade of species (including *V. officinalis*, the type) along with species of *Hierobotana* and *Glandularia* (Figure 5.93), a number of nomenclatural modifications will be necessary before the phylogenetic results can be rationalized taxonomically. More comprehensive phylogenetic surveys of this large genus will be necessary to better determine the relationship of *V. scabra*, which has not yet been included in such analyses. *Verbena scabra* is placed in section *Verbenaca* where it presumably is closely related to the morphologically similar *V. urticifolia* and somewhat more distantly to *V. carolina* (all three species are assigned to series *Leptostachyae*). Hybridization is relatively common in the genus and hybrids involving *V. scabra* and *V. urticifolia* are indicated by a number of specimens that combine the traits of both taxa. Further documentation of such hybrids would help to establish the presumed close relationship between these two species. A narrow-leaved variant of *V. scabra* has been described as forma *angustifolia*. The base chromosome number of *Verbena* is $x = 7$; *V. scabra* has not been counted chromosomally.

Comments: *Verbena scabra* occurs across the southern United States and extends into South America and the West Indies

References: Atkins, 2004; Aziz et al., 1995; Beal, 1977; Correll & Correll, 1975; Godfrey & Wooten, 1981; Lewis & Oliver, 1961; Mason, 1957; Perry, 1933; Tan & Judd, 1995.

Order 5: Solanales [6]

Even the acquisition of a considerable amount of DNA sequence data has not been able to fully solidify the phylogenetic circumscription of the order Solanales. Although molecular data consistently have grouped Solanales in a clade together with Lamiales and Gentianales, the placement of Boraginaceae and Lennoaceae among these groups has been controversial. Using data coded from the secondary structure of nrITS-1 sequences, Gottschling et al. (2001) resolved Lennoaceae (parasitic plants) within a "Boraginales" clade, which included Boraginaceae, Hydrophyllaceae, and several other smaller segregate families; whereas, other authors (e.g., Långström & Chase, 2002) have preferred to combine these

taxa into single, broadly defined Boraginaceae. Regardless of how this Boraginales/Boraginaceae clade is defined, it is becoming clear that the group may not belong to Solanales as had been postulated on the basis of morphology and some molecular data (see Soltis et al., 2005). Some analyses have placed it as the sister group to Lamiales (Albach et al., 2001); however, without topological support. Chloroplast *atpB* sequence data (Långström & Chase, 2002) resolve Boraginaceae (in the broad sense) as the sister group to a clade comprising Gentianales, Lamiales and Solanales (Figure 5.94) with improved (but still low) support. An analysis of six molecular data sets (Figure 5.95) provides yet another weakly supported result placing Boraginaceae as the sister group to a clade comprising Solanales and Lamiales. Yet, combined cpDNA and morphological data resolve Boraginaceae as the sister group to Solanaceae (which would permit its inclusion in the order), but still without support (Bremer et al., 2001).

Because the specific phylogenetic position of Boraginaceae has not been settled convincingly, it has been retained in Solanales in this treatment mostly as a matter of convenience; however, some recent studies (Weigend et al., 2014) indicate that it might be better to treat the group as a distinct order (Boraginales). Multiple sequence analyses (Figure 5.95) indicate a core Solanales as comprising five families (Convolvulaceae, Hydroleaceae, Montiniaceae,

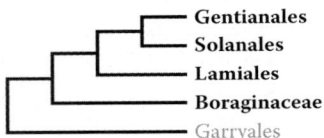

FIGURE 5.94 Phylogenetic position of Boraginaceae as indicated by analysis of *atpB* sequence data. The topology does not agree with results from multisequence analysis (Figure 5.95). Taxa containing OBL North American species are indicated in bold. (Adapted from Långström, E. & Chase, M.W., *Plant Syst. Evol.*, 234, 137–153, 2002.)

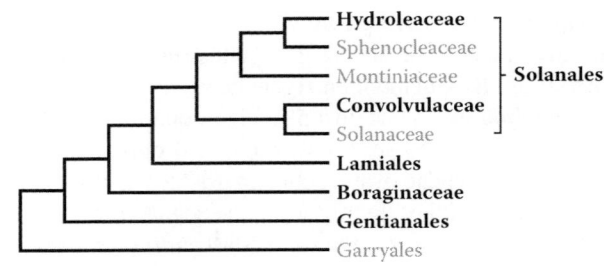

FIGURE 5.95 Component families of core Solanales as indicated by phylogenetic analysis of six DNA sequence datasets. This topology indicates a different placement of Boraginaceae (see Figure 5.94). The distribution of taxa containing OBL North American species (in bold) indicate at least two derivations of the aquatic habit in the order. Other recent studies (Weigend et al., 2014) have resolved a "Boraginales" clade, which is sister to Solanales in some analyses. (Adapted from Bremer, B. et al., *Mol. Phylogen. Evol.*, 24, 274–301, 2002.)

Solanaceae [including Duckeodendraceae, Nolanaceae], and Sphenocleaceae), which represent 140 genera and roughly 7000 species (Bremer et al., 2002; Soltis et al., 2005; Olmstead & Bohs, 2007).

In the circumscription followed here, Solanales include plants with plicate, radially symmetric, sympetalous corollas, an equal number of stamens and petals, and a lack of iridoid glycosides (Soltis et al., 2005). Most of the species are alternately leaved.

Four families of Solanales contain species designated as OBL indicators in North America:

1. **Boraginaceae** Juss.
2. **Convolvulaceae** Juss.
3. **Hydroleaceae** R. Br. ex Edwards
4. **Sphenocleaceae** T.Baskerv.

There have been several origins of the OBL habit in Solanales, at least with respect to these North American representatives (Figure 5.95). The genus *Solanum* includes several species that often grow in wetlands (e.g., *S. dulcamara*, *S. tampicense*), but none has been categorized as an OBL indicator. Two species (*S. campechiense*, *S. nudum*) are designated as OBL in the Caribbean region. Although the former occurs in North America (Texas), it has been excluded from this treatment because of its FACU ranking in that region.

Solanales are the source of many economic products and include important edible plants such as *Capsicum* (bell and cayenne peppers), *Ipomoea* (sweet potato), and *Solanum* (eggplant, potatoes, tomatoes). The group includes various ornamentals (*Cynoglossum, Daturea, Dichondra, Heliotropium, Ipomoea, Mertensia, Myosotis, Petunia, Physalis, Solanum, Symphytum*), tobacco (*Nicotiana*), and potent narcotics (*Atropa*—belladonna; *Datura*—jimsonweed). Many of the species are poisonous due to the presence of complex alkaloids. A few have been used as medicinals (*Borago, Lithospermum, Symphytum*).

Family 1: Boraginaceae [134]

The disputed phylogenetic placement of Boraginaceae already has been discussed earlier, although it is at least evident that their affinities lie somewhere in the proximity of Gentianales, Lamiales, and Solanales. The precise circumscription of Boraginaceae is subject to varying opinions, which reflect the different outcomes obtained from different analyses of DNA data. Analyses using a single cpDNA gene sequence (*atpB*) for a large sample of genera have argued for a narrower concept of the family, which includes several smaller segregates such as Cordiaceae, Ehretiaceae, Heliotropiaceae, and Hydrophyllaceae (Långström & Chase, 2002). However, many of the critical nodes, which provide essential structure to the topology regarding the distinctness of these putative families, are inadequately supported. A study that analyzed secondary structure data from the nrITS1 region for a smaller sample of genera (Gottschling et al., 2001) was able to resolve six fairly well-supported clades, which again, conceivably could be recognized as separate families. An analysis incorporating data from four cpDNA loci (Weigend et al., 2014) recommended the recognition of eight distinct clades within the group, with each clade corresponding approximately to a distinct family taxonomically (Figure 5.96). A comprehensive analysis that surveys over 70% of the genera using combined data from nuclear and plastid loci recovered five major clades that correspond well to subfamilies, and several subclades that correspond to tribes (Nazaire & Hufford, 2012).

It is not clear why some authors prefer to subdivide Boraginaceae into smaller taxonomic units. If a narrow concept is adopted, then at least *Heliotropium* and *Phacelia* would require transfers to different families (Heliotropiaceae, Hydrophyllaceae, respectively). Moreover, it seems impractical to accept a narrow family concept for several reasons. Either concept can be accommodated phylogenetically by recognizing each of the 6–8 clades as distinct families, or by recognizing them as subdivisions (e.g., subfamilies or tribes) within one family (Figure 5.96). Because much of the literature has followed a broad concept (e.g., Nazaire & Hufford, 2012), less confusion would result in retaining that perspective. Also, it seems of questionable merit to subdivide this group into an assemblage of families where several would contain only a handful of genera. For these reasons, a broad familial concept of Boraginaceae has been retained in this treatment.

The circumscription of Boraginaceae followed here includes roughly 2650 species of herbaceous and woody plants (Judd et al., 2002), which share several defining characters. These features include a rough vestiture of hairs with a basal cystolith, a cymose (usually scorpioid) inflorescence, and a plicate corolla. The style is either terminal or gynobasic in orientation and the fruits range from drupes to capsules or schizocarps (Judd et al., 2002).

The flowers usually are pollinated biotically by insects that include bees and wasps (Hymenoptera), butterflies (Lepidoptera), and flies (Diptera). Occasional pollinators include birds (Aves), bats (Mammalia: Chiroptera), and beetles (Insecta: Coleoptera). Breeding systems include self-pollinating species and a few distylous taxa (Judd et al., 2002). The propagules are dispersed biotically (drupes: birds, mammals; nutlets: mammals) and abiotically (nutlets: water, wind; capsules [seeds]: water, wind). Some fruits have structures that facilitate their attachment to fur or to clothing; a few possess arils and are ant dispersed (Judd et al., 2002).

Economically important members include several ornamentals (e.g., *Mertensia, Myosotis, Phacelia, Pulmonaria*) and a few medicinal plants (*Borago, Lithospermum, Symphytum*). Many species contain alkaloids and potentially are poisonous.

The majority of Boraginaceae are terrestrial. North America has five genera with species that are designated as OBL:

1. *Heliotropium* L.
2. *Mertensia* Roth
3. *Myosotis* L.
4. *Phacelia* Juss.
5. *Plagiobothrys* Fisch. & C. A. Mey.

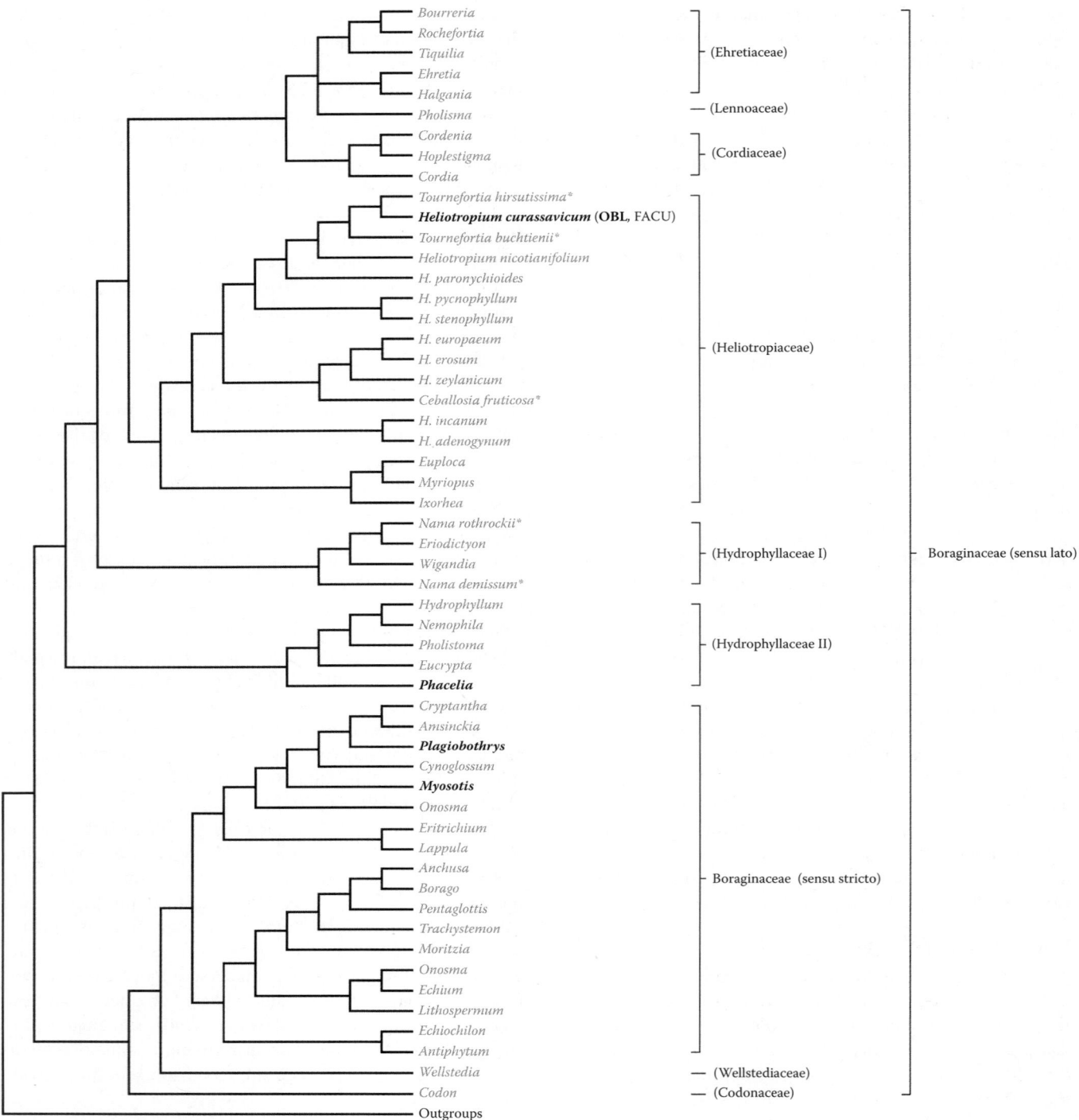

FIGURE 5.96 Phylogenetic relationships of Boraginaceae as reconstructed by combined cpDNA sequence data. Although the present treatment follows a broad taxonomic concept of the family, other authors have recommended its subdivision into additional families (indicated in parentheses) that correspond with the major clades recovered. Results from analysis of nrITS secondary structural data (Gottschling et al., 2001) differ by resolving six distinct clades, and also recommend a much narrower circumscription of Boraginaceae along with recognition of several additional families (e.g., Hydrophyllaceae). Although *Myosotis*, *Phacelia*, and *Plagiobothrys* all contain OBL species as indicated, none of them specifically was sampled in this analysis. However, if these genera are monophyletic, then the OBL habit (in bold) would have originated at least several times in the group. *Mertensia* recently has been included in a detailed phylogenetic analysis of the family, which places it near *Eritrichium* and *Lappula* (Nazaire & Hufford, 2012). A number of misplaced taxa (asterisked) indicate the need for further taxonomic refinements of those genera. (Adapted from Weigend, M. et al., *Cladistics*, 30, 508–518, 2014.)

1. *Heliotropium*

Heliotrope, turnsole; héliotrope

Etymology: from the Greek *hêlios tropeô* meaning "sun turning," with respect to the sunward movement of the leaves and flowers. The name commemorates the mythical water nymph Clytie, who in her unrequited love of the Greek sun god Apollo, became rooted to the earth and forever gazed after him as he moved across the sky

Synonyms: *Beruniella; Bourjotia; Bucanion; Ceballosia; Cochranea; Euploca; Lithococca; Meladendron; Parabouchetia; Schleidenia; Tournefortia; Valentina*

Distribution: global: pantropical; warm temperate; **North America:** southern and north central

Diversity: global: 250 species; **North America:** 21 species

Indicators (USA): OBL; FACU: *Heliotropium curassavicum*

Habitat: brackish, saline, freshwater; palustrine; **pH:** 6.6–10.3; **depth:** <1 m; **life-form(s):** emergent herb

Key morphology: stems (to 6 dm) decumbent, fleshy, glabrous, the foliage glaucous but turning black when dry; leaves (to 6 cm) alternate, linear to oblanceolate, succulent, sessile, or subpetiolate; cymes (to 10 cm) axillary, dividing into paired scorpoid spikes; flowers sessile; corolla (to 5 mm) tubular, funnelform, white with a yellow or violet "eyespot," the lobes plicate; schizocarp (to 3 mm), splitting into 4 nutlets, the calyx persistent

Life history: duration: perennial (rhizomes); **asexual reproduction:** rhizomes; **pollination:** insect; **sexual condition:** hermaphroditic; **fruit:** nutlets (common); **local dispersal:** seeds (water); **long-distance dispersal:** seeds (birds)

Imperilment: (1) *Heliotropium curassavicum* [G5]; SH (MB, NC); S1 (AB, NE, MO, TX, VA); S2 (OR, WY); S3 (UT)

Ecology: general: Most members of *Heliotropium* occur in UPL habitats. In North America, there are nine wetland indicator species, but only one (*H. curassavicum*) is categorized as OBL. There is some debate regarding the OBL designation of this species due to its occurrence in arid areas and in drier sites; however, this heterogeneity is reflected by its varied status (OBL, FACU) throughout different portions of its range. Members of this genus generally are regarded as being xenogamous (i.e., essentially outcrossed); however, the pollen/ovule ratios of some species indicate a facultatively autogamous breeding system. The pollen is binucleate, which has been associated with gametophytic self-incompatibility, but most species have not been surveyed for compatibility mechanisms. In any case, insect pollination appears to be the prevalent means of reproduction. In many *Heliotropium* species the pubescent foliage fluoresces, providing sharp contrast to the superimposed, nonfluorescent corollas, which facilitates pollinator attraction. However, the glaucous foliage of the OBL *H. curassavicum* does not fluoresce in this way. Some species have physiologically dormant seeds, which require warm stratification for germination. Others come out of dormancy when their seeds have been flooded.

Heliotropium curassavicum **L.** inhabits beaches, deltas, depressions, ditches, dunes, flats, floodplains, gravel bars, lake bottoms, marshes, meadows, ravines, sandbars, seeps, sloughs, swales, vernal pools, washes, and the exposed margins of alluvial fans, canals, lagoons, lakes, levees, playas, ponds, rivers, and streams at elevations of up to 2250 m. This species does not occur in standing water, but occupies sites where standing waters have receded. The habitats often are saline and tidal but freshwater inland sites are common. The plants occur on alkaline (pH: 6.6–10.3), brackish or saline (to 16 dS/m) mineral soils, which include adobe, clay, granite, gravel, gypsum, loam, mud, rock, sand, sandstone, sandy loam, schist, shells, silt, silty clay, and stony clay. Exposures range from full sun to partial shade. Cold temperatures to −33°C reportedly can be tolerated. The plants are regarded as short lived but are known to survive for at least 4 years. Sexual reproduction and seedling establishment predominate in open sites (~10% cover); whereas, vegetative reproduction is dominant in closed sites (90%–100% cover). The flowers are pollinated by various insects including Diptera (Stratiomyidae), Hymenoptera (Apidae; Megachilidae; Sphecidae; Vespidae), and Lepidoptera (Hesperiidae; Nymphalidae), which are known to visit the plants primarily to collect nectar (see *Use by wildlife* later). The white corollas have a yellow "eyespot" in the center, which changes to purple after a flower has been pollinated and serves as a cue to alert potential pollinators. A single plant can produce more than 141,000 seeds and 9800 vegetative root buds by its fourth year of age. Seedling emergence in either open or closed sites is highest for seeds buried at depths of 0–4 cm, with an emergence rate of 330 seedlings/m^2 observed from 2 to 4 cm soil depths. More than 62 million seeds can be produced within a 100 m^2 area. A seed bank is formed with densities of up to 440 seeds/m^2 observed. The seeds reportedly do not require cold stratification but do need high light levels for maximum germination. The germination rate declines as salinity levels increase. The thick fruit exocarp functions as a float to facilitate local dispersal by water. Seed dispersal is believed to occur by their adherence to the feet of various seabirds, shorebirds, and water birds. Survivorship of plants follows a "Deevey type III" curve, which represents high mortality in the early phases of growth and reproduction. Although this species sometimes is reported as an annual it is actually a perennial. Like annuals, the aerial stems die back after flowering and fruiting the first year, but the underground parts persist vegetatively by means of rhizomes (about 60% of the total biomass is allocated to underground tissues). The emergence of new aerial shoots is highest from soil depths of 6–10 cm, but continues even at depths up to 50 cm. Shoots can form sprawling clumps that emerge from a perennial crown or taproot and spread from lateral roots. The species is common in grazed and otherwise disturbed areas. Removal of competing vegetation by fire can result in an increased spread of the plants. The plants can survive in habitats varying widely in annual precipitation owing in part to their stomatal plasticity, which enables them to achieve higher photosynthetic rates at negative water potentials. Plants originating from cooler climates exhibit greater temperature acclimation ability. The roots are colonized by arbuscular–vesicular mycorrhizae.

Reported associates: *Abronia villosa, Acacia cyclops, Achillea millefolium, Achnatherum speciosum, Acmispon americanus, A. glabrus, A. heermannii, Adenostoma fasciculata, Agalinis calycina, Alternanthera, Amaranthus albus, Ambrosia acanthicarpa, A. dumosa, A. psilostachya, Andropogon glomeratus, Anemopsis californica, Apium graveolens, Artemisia californica, A. douglasiana, A. tridentata, Arthrocnemum subterminale, Asclepias fascicularis, Atriplex californica, A. canescens, A. cristata, A. lentiformis, A. minuscula, A. patula, A. phyllostegia, A. polycarpa, A. prostrata, A. rosea, Avena fatua, Avicennia germinans, Baccharis emoryi, B. halimifolia, B. pilularis, B. salicifolia, B. sarothroides, Bacopa monnieri, Bassia hyssopifolia, Batis maritima, Berula erecta, Bolboschoenus robustus, Borrichia frutescens, Brassica geniculata, B. nigra, Brickellia californica, Bromus diandrus, B. hordeaceus, B. tectorum, Cakile geniculata, C. maritima, Calochortus striatus, Camissonia cheiranthifolia, Carduus pycnocephalus, Carex praegracilis, Carpobrotus, Ceanothus tomentosus, Centaurea solstitialis, Centaurium exaltatum, Cercidium, Chamaesyce, Chenopodium rubrum, Chrysothamnus albidus, Cichorium intybus, Chrysothamnus nauseosus, Cirsium vulgare, Cleomella obtusifolia, C. parviflora, Conium maculatum, Convolvulus sepium, Conyza canadensis, Cordylanthus canescens, Coreopsis, Cotula coronopifolia, Crataegus berberifolia, Cressa truxillensis, Croton punctatus, Crypsis schoenoides, C. vaginiflora, Cucurbita foetidissima, Cynodon dactylon, Cyperus filicinus, C. odoratus, Datura wrightii, Deinandra floribunda, D. kelloggii, Digitaria, Distichlis spicata, Eclipta prostrata, Elaeagnus angustifolia, Eleocharis albida, E. montevidensis, E. palustris, Elymus elymoides, E. triticoides, Encelia californica, E. frutescens, Epilobium canum, E. ciliatum, Epipactis gigantea, Ericameria pinifolia, Erigeron procumbens, Eriogonum fasciculatum, E. parvifolium, Erodium cicutarium, Eucalyptus camaldulensis, Eupatorium serotinum, Eustachys glauca, E. petraea, Eustoma exaltatum, Euthamia leptocephala, E. occidentalis, Fimbristylis castanea, Frankenia salina, F. grandifolia, Gaura brachycarpa, Geraea canescens, Gilia, Glinus lotoides, G. radiatus, Gnaphalium luteoalbum, G. palustre, G. stramineum, Gutierrezia microcephala, Helianthus annuus, H. paradoxus, Heterotheca grandiflora, Hibiscus moscheutos, Hirschfeldia incana, Hydrocotyle umbellata, Ipomoea imperati, I. pes-caprae, Isocoma acradenia, I. menziesii, Iva angustifolia, I. cheiranthifolia, Jaumea carnosa, Juncus bufonius, J. mexicanus, J. patens, J. roemerianus, J. rugulosus, Kochia, Lactuca sativa, Lepidium latifolium, L. perfoliatum, L. virginicum, Lepidospartum squamatum, Leptochloa, Limonium californicum, Lolium multiflorum, Lupinus succulentus, Lycium carolinianum, L. cooperi, Lythrum californicum, L. hyssopifolium, Machaeranthera canescens, Malosma laurina, Malva parviflora, Malvella leprosa, Melilotus indica, Mesembryanthemum, Mimosa, Mimulus cardinalis, Monolepis nuttalliana, Muhlenbergia asperifolia, Nasturtium officinale, Nicotiana attenuata, N. glauca, Nitrophila occidentalis, Nothoscordum bivalve, Oenothera, Opuntia, Orcuttia, Oxystylis lutea, Palafoxia, Panicum amarum, Paspalum distichum, Pennisetum setaceum, Penstemon cordifolius, Petunia parviflora, Phacelia lutea, Phragmites australis, Phyla, Physalis viscosa, Plagiobothrys parishii, Plantago coronopus, Platanus racemosa, Pluchea purpurascens, Poa secunda, Polygonum aviculare, Polypogon monspeliensis, Populus balsamifera, P. fremontii, Potentilla egedii, Prosopis glandulosa, P. pubescens, Prunus armeniaca, P. fremontii, Ptilimnium, Purshia glandulosa, Quercus agrifolia, Ranunculus cymbalaria, Raphanus sativa, Rayjacksonia phyllocephala, Rhamnus ilicifolia, Rhus integrifolia, R. ovata, Rorippa sinuata, Rosa, Rumex crispus, Sabatia stellaris, Salix exigua, S. gooddingii, S. laevigata, S. lasiolepis, Salicornia bigelovii, Salsola, Salvia mellifera, Sambucus mexicana, Samolus cuneatus, Sarcobatus vermiculatus, Sarcocornia perennis, Schoenoplectus americanus, S. californicus, S. tabernaemontani, Senna obtusifolia, Sesuvium portulacastrum, S. verrucosum, Sicyos, Sideroxylon lanuginosum, Solanum carolinense, Solidago sempervirens, Sonchus asper, S. oleraceus, Sorghum halepense, Spartina alterniflora, S. foliosa, S. patens, S. pectinata, S. spartinae, Spergularia bocconii, S. salina, Sporobolus airoides, S. texanus, S. virginicus, Stanleya pinnata, Stillingia, Suaeda calceoliformis, S. conferta, S. linearis, S. moquinii, S. suffrutescens, Symphyotrichum subulatum, S. tenuifolium, Tamarix parviflora, T. ramosissima, Tidestromia lanuginosa, Toxicodendron diversilobum, Tradescantia occidentalis, Tridens strictus, Typha angustifolia, T. domingensis, T. latifolia, Uniola paniculata, Urtica, Veronica anagallis-aquatica, Vicia sativa, Xanthium strumarium, Yucca schidigera.*

Use by wildlife: *Heliotropium curassavicum* occurs in the habitat of the Gambel's quail (Aves: Odontophoridae: *Callipepla gambelii*), laughing gull (Aves: Laridae: *Leucophaeus atricilla*), least tern (Aves: Laridae: *Sternula antillarum*), and Texas nighthawk (Aves: Caprimulgidae: *Chordeiles acutipennis texensis*). In Hawaii, the plants are used as nesting sites by the Laysan teal (Aves: Anatidae: *Anas laysanensis*). The seeds can represent a substantial dietary fraction of coastal populations of the Florida duck (Aves: Anatidae: *Anas fulvigula fulvigula*) and are eaten by quail (Aves: Odontophoridae: *Colinus virginianus*). The floral nectar is taken by a number of insects (Insecta) including digger wasps (Hymenoptera: Sphecidae: *Steniolia duplicata*), mason wasps (Hymenoptera: Vespidae: *Euodynerus*), the MacNeill's desert sootywing (Lepidoptera: Hesperiidae: *Hesperopsis gracielae*), the painted crescent butterfly (Lepidoptera: Nymphalidae: *Phyciodes picta*), and the Carson wandering skipper butterfly (Lepidoptera: Hesperiidae: *Pseudocopaeodes eunus obscurus*); the latter is an endangered species in the United States. The flowers are visited (probably for nectar) by ground nesting or small carpenter bees (Hymenoptera: Apidae: *Anthophora abroniae*; *Ceratina arizonensis*), leafcutting bees (Hymenoptera: Megachilidae: *Ashmeadiella bigeloviae*; *A. cactorum*) and soldier flies (Diptera: Stratiomyidae: *Odontomyia inaequalis*). The shoots and inflorescences are used for roosting by the rubyspot damselfly (Odonata: Calopterygidae: *Hetaerina americana*). The leaves are fed on by larval leafminer flies (Diptera: Agromyzidae: *Liriomyza sorosis*), which themselves can harbor parasitic wasps (Hymenoptera: Pteromalidae: *Heteroschema*). The plants are a host to clover mites (Acari: Tetranychidae: *Bryobia praetiosa*), stinkbugs (Hemiptera: Pentatomidae), and a sac fungus (Ascomycota: Mycosphaerellaceae: *Cercospora heliotropii*). The leaves have a high nitrogen content (2.2% dry weight but also a high sodium content (3.1%). The proportion of free amino acids can increase by as much as 35% when growing in saline conditions. The foliar pyrrolizidine alkaloids are ingested by feeding male monarch butterflies (Lepidoptera: Nymphalidae: *Danaus plexippus*), and are passed to the females during mating, and ultimately to the eggs, where they function as an antipredatory chemical defense.

Economic importance: food: The seeds of *Heliotropium curassavicum* were used by the Tubatulabal people as a food source. The plants are eaten in Belize. However, due to the presence of toxic alkaloids (see later), this plant should be regarded as unfit for consumption and incidences of human poisoning have been reported from ingestion of herbal teas that contain it; **medicinal:** The foliage of *Heliotropium curassavicum* contains a number of pyrrolizidine alkaloids (average concentration of 2.5%) including coromandaline, curassavine, heliovicine, and a trachelanthamidine ester. The highest alkaloid concentrations occur at the onset of flowering and decline in older plants. Pyrrolizidine alkaloids are known to cause liver toxicity in humans, and extracts of this species can be genotoxic, resulting in abnormal metaphase during cell division. Consequently, all parts of the plant should be regarded as poisonous. The Paiute and Tubatulabal prepared medicinal decoctions from the leaves of *H. curassavicum* to treat diarrhea. The Paiute and Shoshoni administered a gargle made from the roots to remedy sore throats. The plants were used as a diuretic or emetic by the Paiute and Shoshoni. The Pima treated sores and wounds using a poultice made from the dried root and brewed the roots as a tea used for soothing soreness in the eyes. The Shoshoni treated measles and venereal diseases with a foliar decoction made from the plants. The Tubatulabal ingested a decoction of the foliage to relieve dysentery. This species still is used occasionally to treat a number of illnesses including arteriosclerosis, gout, neuralgia, phlebitis, and rheumatism. The extracts exhibit high antifungal activity against several Ascomycota (Candidaceae: *Candida albicans*; Trichocomaceae: *Penicillium citrinum*). The ethanolic extracts have been used to treat cancerous ulcers and reportedly exhibit antitumor activity against type P3888 leukemia. In Amazonia, a leaf decoction is applied to wounds as an antiseptic and is taken internally as a treatment for rheumatism. A powder made from the branches is used as a remedy for hemorrhoids. Other uses in that region include treatments for anemia, asthma, coughing, eczema, gonorrhea, and kidney stones. Some traditional uses included treatment of scorpion stings, but such applications can be attributed to the resemblance of the inflorescence to a scorpion tail rather than as any indication of potential efficacy in that regard; **cultivation:** *Heliotropium curassavicum* is not cultivated, perhaps because of its affinity for more saline soils; **misc. products:** *Heliotropium curassavicum* has been considered as a planting for the restoration of native grasslands. Native North

Americans made a purple dye from the plants. In Costa Rica, the ashes are used as salt. In Madagascar, the plants are burned to provide ash for agricultural fields; **weeds:** *Heliotropium curassavicum* is a common weed of disturbed areas in North America and other parts of the world; **nonindigenous species:** *Heliotropium curassavicum* probably is native only to the New World. It was introduced to and is regarded as an aggressive invasive species in the western Mediterranean region. It has been introduced to Africa (e.g., Canary Islands), Asia (e.g., Japan), and to Australia. It probably spread to some degree as a seed contaminant of shipping ballast, which is evidenced by its common occurrence in early ballast disposal sites.

Systematics: The closest relative of *Heliotropium* cannot be ascertained with certainty because a comprehensive phylogenetic analysis of all genera in Boraginaceae has not yet been performed; however, potential sister genera include *Euploca*, *Myriopus*, and *Ixorhea* (Figures 5.96 and 5.97). *Heliotropium* itself appears to be monophyletic, but only if members of *Ceballosia*, *Schleidenia*, and *Tournefortia* (except section *Cyphocyema*) are included (Figure 5.97). Some authors previously had placed several *Tournefortia* species in synonymy with *Heliotropium*. Because the type species of *Tournefortia* (*T. hirsutissima*) nests within *Heliotropium*, a different generic name (e.g., *Myriopus*) should be applied to the clade of "*Tournefortia*" species (i.e., section *Cyphocyema*) that falls outside the genus (Figure 5.97). Several clades exist within *Heliotropium* that might be suitable for recognition at the sectional level. There is an Old World clade that contains the type of the genus, two clades corresponding to the previously recognized sections *Heliothamnus* and *Orthostachys* (+ *Schleidenia*), and a weakly supported New World clade containing *Ceballosia* and many species assigned formerly to *Tournefortia*; the latter includes the OBL *H. curassavicum* (Figure 5.97). Although some authors favor the recognition of the "*Schleidenia*" clade as a distinct genus (i.e., *Schleidenia*), such a disposition seems unreasonable given that the retention of the clade in *Heliotropium* is compatible with the phylogenetic results and because most of the species in the clade already are commonly recognized under the generic name *Heliotropium*. By analysis of nrITS sequence data (Figure 5.97), *Heliotropium curassavicum* (two subspecies) resolves as a clade that is sister to *Tournefortia argentea*, a Hawaiian tree; however, the association is not well supported and additional species assigned to both genera need to be included in subsequent analyses for further clarification of interspecific relationships. Incorporation of additional molecular data could potentially provide better resolution of relationships within the genus. The basic chromosome number hypothesized for Boraginaceae is $x = 4$ and subfamily Heliotropoideae contains indicated base numbers of $x = 9$, 10, 11, and 13, which presumably are derived. In this interpretation, *Heliotropium curassavicum* ($2n = 26$) is a diploid with respect to a derived base number. In some introduced portions of its range, the plants exhibit various meiotic irregularities, have low pollen fertility, and reproduce primarily by vegetative means. The meiotic aberrations occur are higher frequencies in plants found growing in polluted sites.

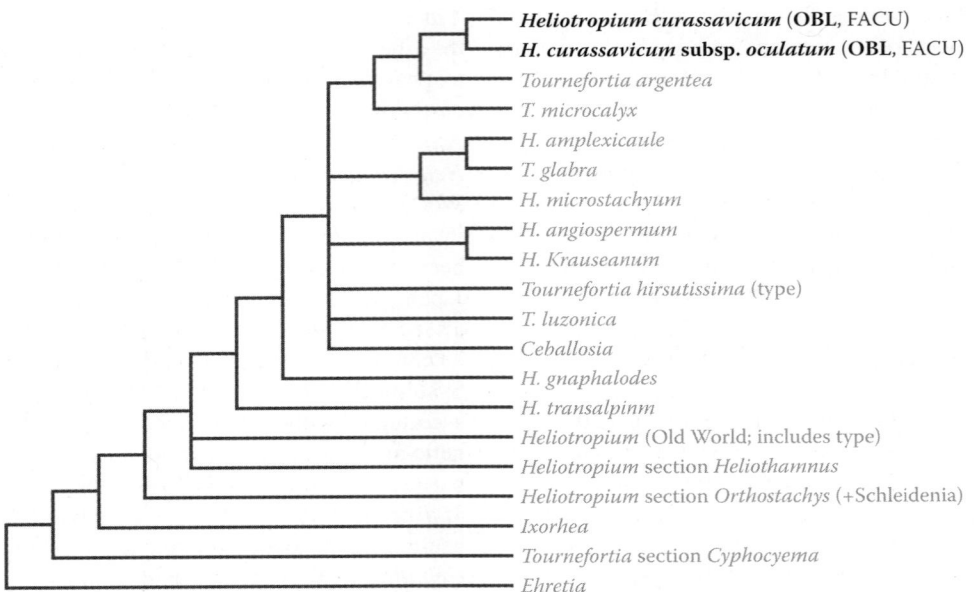

FIGURE 5.97 Phylogenetic relationships in *Heliotropium* as indicated by analysis of nrITS sequence data. These results indicate that *Tournefortia* is polyphyletic, with many species (including the type) nested within *Heliotropium*. The genus *Ceballosia* also nests within *Heliotropium*. Some authors favor the segregation of *Heliotropium* section *Orthostachys* (including *Schleidenia*) as a distinct genus; however, a broader concept of *Heliotropium* that includes this group is suggested here. The sole OBL North American representative (in bold) indicates an isolated origin of the aquatic habit in the genus and occurs in a relatively derived position in the topology. (Adapted from Nadja, D. et al., *Amer. J. Bot.*, 89, 287–295, 2002.)

Comments: *Heliotropium curassavicum* occurs throughout the United States (except the east central region) and extends into central Canada, and South America.

References: Aronson, 1989; Austin, 2004b; Aziz & Khan, 1996; Baskin & Baskin, 1998; Bolland et al., 1998; Brewbaker, 1967; Britton, 1951; Brown, 1879; Brown & Bugg, 2001; Carballo et al., 1992; Catalfamo et al., 1982; Cruden, 1977; Dhankar et al., 1998; Di Castri et al., 1990; Duke & Vásquez, 1994; Evans & Gillaspy, 1964; Faruqi & Holt, 1961; Feldman & Lewis, 2005; Frohlich, 1976; Gullion, 1962; Hegazy, 1994; Hegazy et al., 1994; Heiser, Jr. & Whitaker, 1948; Huxtable, 1980, 1989; James, 1960; Jurado et al., 2001; Kendall, 1964; Khan & Weber, 2006; Kloot, 1984; Mandeel & Taha, 2005; Michener, 1936a,b, 1939; Miller, 1988; Mooney, 1980; Mulligan, 2008; Nadja et al., 2002; Pickwell & Smith, 1938; Rahimi & Khatoon, 2004; Roy & Mooney, 1982; Sánchez, 1999; Schmidt & Scow, 1986; Stegmaier, Jr., 1972; Stieglitz, 1972; Stutzenbaker, 1999; Subramanian et al., 1980; Switzer & Grether, 2000; Thebeau & Chapman, 1984; Thompson & Slack, 1982; Timberlake, 1937; Tiner, 1993; USFWS, 2006a; Van Auken et al., 2007; Weaver, Jr. & Anderson, 2007; Weiss, 1995; Woodson, Jr. et al., 1969.

2. *Mertensia*

Bluebells; mertensia

Etymology: after Franz Karl Mertens (1764–1831)

Synonyms: *Casselia*; *Mertensianthe*; *Pneumaria*; *Pulmonaria* (in part); *Steenhammera*

Distribution: global: Asia; Europe; North America; **North America:** widespread

Diversity: global: 50 species; **North America:** 18 species

Indicators (USA): OBL: *Mertensia franciscana*; **FACW:** *M. ciliata*, *M. franciscana*

Habitat: freshwater; palustrine; **pH:** 5.4–6.8; **depth:** <1 m; **life-form(s):** emergent herb

Key morphology: stems (to 12 dm) erect, 1-many from the rootstock, herbage glabrous or pubescent; leaves (to 15 cm) alternate, elliptic, lanceolate or oblong, lower leaves petioled (to 15 cm), becoming sessile toward apex; cymes scorpoid and paniculate or umbellate from an axillary peduncle; flowers pedicelled (to 2 cm), regular, 5-merous; calyx (to 5 mm) divided nearly to base; corolla tubular (to 9 mm) plicate, dehiscent, blue (sometimes pink or white), the throat expanded distally into a 5-lobed limb (to 10 mm), 5 small appendages (fornices) present internally where tube and limb meet; style (to 20 mm) gynobasic, usually exserted; schizocarp dehiscing into 4, single-seeded, rugose nutlets (mericarps)

Life history: duration: perennial (caudex, rhizomes); **asexual reproduction:** rhizomes; **pollination:** insect, self; **sexual condition:** hermaphroditic; **fruit:** nutlets (infrequent); **local dispersal:** seeds (gravity, water), rhizome fragments; **long-distance dispersal:** seeds (birds, mammals)

Imperilment: (1) *Mertensia ciliata* [G5]; S1 (SD); S2 (NM); (2) *M. franciscana* [G3]

Ecology: general: *Mertensia* species grow under fairly diverse ecological conditions ranging from high montane elevations to seashores and woodlands. Most species are found in dry habitats. Although nine North American species (50%) are wetland indicators, they are found more commonly at sites where water is readily available, but not under conditions of permanent inundation. Even the OBL species (*M. franciscana*) is a FACW indicator in parts of its range. *Mertensia ciliata* was designated as OBL in the 1996 indicator list but has been reclassified as FACW in the 2013 list. Its treatment is included here because it was completed prior to the release of the updated list. The species are pollinated mainly by long-tongued bumblebees (Insecta: Hymenoptera: Apidae: *Bombus*) or by solitary bees (Insecta: Hymenoptera: Andrenidae: *Andrena*; Colletidae: *Hylaeus*; Halictidae: *Halictus*; Megachilidae: *Osmia*), with a few pollinated by hummingbirds (Aves: Trochilidae), and some that are self-pollinating. The fruits have no specialized dispersal mechanisms, but can float to be transported by fresh water, or in some cases, by salt water.

***Mertensia ciliata* (James ex Torr.) G. Don** grows in bogs, meadows, ravines, seeps, and along the margins of lakes and streams at higher elevations from 1300 to 3960 m. The plants require a constant moisture supply to maintain moist to wet soils, but do not occur in extremely saturated sites. The sites are relatively cool with a mean maximum temperature of 17°C–24°C and a mean minimum temperature of 3°C–7°C. The substrates tend to be acidic (pH: 5.4–6.8) and include clay loam, glacial till, granite, igneous talus, loam, shale alluvium, silt loam, volcanic rock, and decaying logs. Some substrates are fairly organic (6%–57% organic matter). The plants thrive and reproduce in open sites or in shaded sites receiving 10%–20% of full sunlight. They are known to occur as a late successional stage on fallen logs in mesic woodland sites. The overall size of plants decreases at higher elevations. Flowering commences after the snow has receded; there is a strong linear relationship between the first day of flowering and the first day of bare ground. The proportion of open flowers/time is positively skewed. The flowers are nectariferous and are mildly fragrant. Nectar production is highest in the morning and can reach levels as high as 17,200 µL/ha. The pollen is shed in about 10 days from the initial development of the floral bud (which is pink colored). The flowers open in the morning and the pollen is gone by the afternoon unless the weather is inclement (signaled by high humidity), which can delay pollen release by another 1–2 days. The mature corollas (which are blue in contrast to the buds) often dehisce in about 2 weeks. Clones can remain in flower for 3–5 weeks. The flowers are self-compatible and produce similar mean levels of seed set (16%–19%) whether they are self- or cross-pollinated. They are not apomictic. Levels of spontaneous self-pollination are extremely low and the plants will exhibit inbreeding depression if they are not outcrossed. Although pollinators often are scarce, there is significant pollen carryover (to the 11th or 12th successive flower), resulting in potentially high levels of outcrossing. Plant size is not related to inbreeding in this species. Pollination occurs mainly on calm, sunny days and is achieved by nectar-seeking insects that are large enough to force aside the anthers, which can occlude the floral tube. The major pollinators are bees (Hymenoptera: Apidae; Colletidae; Megachilidae). The most common pollinators are bumblebees (e.g., Hymenoptera: Apidae: *Bombus balteatus*, *B. bifarius*, *B. flavifrons*, *B. frigidus*, *B. insularis*, *B. lapponicus*,

B. melanopygus, *B. mixtus*, *B. occidentalis*). The corolla tube has narrow and wide morphs, which are used by different bumblebee castes where workers and queens use the wide morphs and workers solely visit the narrow morphs. The short-tongued bee species release pollen by vibrating the anthers. Occasional pollinators may include bee flies (Diptera: Bombyliidae: *Systoechus oreas*) and self-pollination may be facilitated by beetles (Insecta: Coleoptera: Staphylinidae). The flowers can produce both normal and abnormal anthers, the latter containing up to 85% defective pollen. Fruiting is rare and populations often are barren despite large productions of flowers. Seed set tends to be somewhat higher at higher elevations. When fruiting, the plants average about 24 nutlets/stem but only about 72% are viable. The nutlets are dispersed locally by gravity and by floodwaters. They probably are dispersed longer distances by birds such as pine siskins (Aves: Fringillidae: *Carduelis pinus*) or by adhering to the fur of mammals. The seeds are dormant when shed. In one study, storage at room temperature for several months yielded 2%–34% germination at 18°C; however, cold stratification at 5°C for 5–11 months resulted in low (0%–6%) germination. Higher germination (77%) was achieved within 12 days after first soaking seeds in concentrated sulfuric acid. Enhanced germination was achieved by nicking the pericarp corner with a razor blade, or by removing both the outer and inner seed coats (52% and 90% germination, respectively). Seeds do not germinate after a year of burial under natural conditions. Several years of burial (for decomposition of the pericarp) or some type of mechanical abrasion seems to be necessary for germination to occur in natural populations. Seed germination rarely is observed in the field. In the laboratory, it occurs better at 18°C than at 5°C. Some type of natural disturbance (e.g., sediment exposure from flooding or rodent activity) appears to be necessary for successful seedling establishment, and the species is characterized as being tolerant of intermediate disturbance levels. Local spread occurs mainly by the proliferation of rhizomes (which live for at least 9 years) and by subsequent ramet formation. Some vegetative reproduction may occur by the dispersal of rhizome fragments during floods, or by the eventual detachment of rhizomes from the parental caudex. Overall productivity (biomass) has been estimated to reach 4 g/m²/year. **Reported associates:** *Abies concolor, A. lasiocarpa, Acer glabrum, A. negundo, Achillea millefolium, Achnatherum nelsonii, Aconitum columbianum, Actaea rubra, Agastache urticifolia, Agoseris aurantiaca, A. glauca, Agrostis exarata, A. scabra, A. stolonifera, Allium cernuum, Alnus tenuifolia, Amelanchier alnifolia, A. utahensis, Androsace filiformis, Anemone narcissiflora, Angelica ampla, A. breweri, A. grayi, A. pinnata, Aquilegia caerulea, A. elegantula, A. flavescens, A. formosa, A. saximontana, Arnica cordifolia, A. longifolia, A. mollis, A. rydbergii, A. sororia, Artemisia tridentata, Athyrium americanum, A. filix-femina, Besseya wyomingensis, Bistorta bistortoides, B. vivipara, Bromopsis inermis, Bromus carinatus, B. ciliatus, B. hordeaceus, B. inermis, B. marginatus, B. porteri, Calamagrostis canadensis, Calochortus invenustus, Collomia linearis, Caltha leptosepala, Cardamine breweri,*

C. cordifolia, Carex aquatilis, C. athrostachya, C. canescens, C. disperma, C. ebenea, C. fracta, C. geyeri, C. heteroneura, C. illota, C. microptera, C. nigricans, C. norvegica, C. nova, C. pachystachya, C. pellita, C. rostrata, C. scopulorum, C. utriculata, Castilleja linariifolia, C. miniata, C. rhexiifolia, C. sulphurea, Cerastium arvense, Chaenactis douglasii, Chamerion angustifolium, Cirsium andersonii, C. arvense, C. drummondii, C. eatonii, C. foliosum, C. scopulorum, C. tioganum, C. vulgare, Claytonia lanceolata, Clematis hirsutissima, Comarum palustre, Conioselinum scopulorum, Conium maculatum, Cornus sericea, Corydalis aurea, C. caseana, Cryptogramma acrostichoides, Cynoglossum officinale, Cystopteris fragilis, C. reevesiana, Dasiphora floribunda, Delphinium barbeyi, D. glaucescens, D. nuttallianum, Deschampsia cespitosa, Disporum trachycarpum, Draba spectabilis, Eleocharis palustris, Elymus canadensis, E. glaucus, E. repens, E. trachycaulus, Epilobium alpinum, E. anagallidifolium, E. ciliatum, E. glaberrimum, E. hornemannii, E. saximontanum, Equisetum arvense, E. pratense, Ericameria nauseosa, Erigeron coulteri, E. elatior, E. peregrinus, E. speciosus, Erythronium grandiflorum, Eurybia conspicua, E. integrifolia, E. sibirica, Festuca brachyphylla, Fragaria vesca, F. virginiana, Galeopsis bifida, Galium aparine, G. asperulum, G. boreale, G. spurium, G. trifidum, G. triflorum, Gayophytum ramosissimum, Gentianella amarella, Gentianopsis thermalis, Geranium richardsonii, Geranium viscosissimum, Geum macrophyllum, G. rossii, Glyceria elata, G. striata, Glycosma occidentalis, Hackelia floribunda, H. micrantha, Hedysarum sulphurescens, Heracleum maximum, Hordeum brachyantherum, Hydrophyllum fendleri, Hymenoxys hoopesii, Juncus balticus, J. compressus, J. drummondii, J. mertensianus, J. saximontanus, J. tracyi, Juniperus communis, Kalmia microphylla, Ligusticum filicinum, L. porteri, L. tenuifolium, Lilium parvum, Lloydia serotina, Lonicera involucrata, Lupinus argenteus, L. polyphyllus, Luzula parviflora, Maianthemum racemosum, M. stellatum, Medicago polymorpha, Melica spectabilis, Mentha arvensis, Mertensia franciscana, M. lanceolata, Micranthes nidifica, M. odontoloma, Mimulus guttatus, Mitella pentandra, Moehringia lateriflora, Moneses uniflora, Muhlenbergia filiformis, Oreochrysum parryi, Oreoxis alpina, Orobanche uniflora, Orthilia secunda, Osmorhiza chilensis, O. depauperata, O. occidentalis, Oxypolis fendleri, Oxytropis riparia, Oxyria digyna, Parnassia fimbriata, P. parviflora, Paxistima myrsinites, Pedicularis groenlandica, Penstemon rydbergii, Phacelia heterophylla, Phalaris arundinacea, Phleum alpinum, P. pratense, Phyllodoce glanduliflora, Picea engelmannii, P. pungens, Pinus albicaulis, P. aristata, P. contorta, P. flexilis, Platanthera dilatata, P. sparsiflora, P. stricta, Poa alpina, P. arctica, P. compressa, P. leptocoma, P. nervosa, P. palustris, P. pratensis, P. reflexa, Polemonium foliosissimum, P. pulcherrimum, P. viscosum, Polygonum douglasii, Populus angustifolia, P. tremuloides, Potamogeton gramineus, Potentilla glandulosa, P. gracilis, P. pulcherrima, Primula parryi, P. pauciflora, Prunella vulgaris, Pseudocymopterus montanus, Pseudotsuga menziesii, Pyrola asarifolia, P. minor, Ranunculus alismifolius, R. pygmaeus,
R. uncinatus, Rhodiola integrifolia, R. rhodantha, Rhododendron neoglandulosum, Ribes cereum, R. hudsonianum, R. inerme, R. irriguum, R. lacustre, R. laxiflorum, R. montigenum, R. wolfii, Rorippa teres, Rosa acicularis, R. woodsii, Rubus idaeus, Rudbeckia californica, R. laciniata, R. occidentalis, Rumex aquaticus, R. crispus, Salix barclayi, S. bebbiana, S. boothii, S. brachycarpa, S. drummondiana, S. eastwoodiae, S. geyeriana, S. lasiolepis, S. lemmonii, S. lutea, S. monticola, S. planifolia, S. wolfii, Sambucus racemosa, Sarcodes sanguinea, Senecio amplectens, S. atratus, S. bigelovii, S. crassulus, S. eremophilus, S. fremontii, S. serra, S. sphaerocephalus, S. triangularis, Shepherdia canadensis, Sibbaldia procumbens, Sidalcea multifida, S. oregana, Silene drummondii, Sium suave, Solidago canadensis, S. multiradiata, Sorbus scopulina, Sphenosciadium capitellatum, Spiraea betulifolia, Stachys albens, Stellaria calycantha, S. longifolia, S. umbellata, Stipa occidentalis, Streptopus amplexifolius, Swertia perennis, Symphoricarpos albus, S. oreophilus, Symphyotrichum ascendens, S. ciliolatum, S. foliaceum, S. lanceolatum, S. spathulatum, Taraxacum officinale, Thalictrum alpinum, T. fendleri, T. occidentale, T. sparsiflorum, Thermopsis montana, T. rhombifolia, Thinopyrum intermedium, Thlaspi montanum, Torreyochloa pallida, Trifolium longipes, T. pratense, T. repens, Triglochin maritimum, Trisetum spicatum, T. wolfii, Trollius laxus, Urtica dioica, Vaccinium myrtillus, V. scoparium, Valeriana acutiloba, V. edulis, Verbascum thapsus, Veratrum californicum, V. tenuipetalum, Veronica americana, V. serpyllifolia, V. wormskjoldii, Vicia americana, Viola adunca, V. biflora, V. canadensis, V. lobata, V. renifolia.

***Mertensia franciscana* A. Heller** inhabits montane and subalpine cliffs, meadows, seeps, and the margins of lakes, ponds, rivers, and streams at elevations from 1700 to 3500 m. In sharp contrast to the preceding species, published information on the life history of this plant is scarce. However, its characteristics probably are quite similar given that the two species often are observed growing together in the same habitat and have similar associates. The plants tolerate sun or partial shade and are regarded as indicators of the bristlecone pine forest type. **Reported associates:** *Abies lasiocarpa, Acer glabrum, Achillea millefolium, Aconitum columbianum, Actaea rubra, Agastache pallidiflora, Agrostis gigantea, A. scabra, Allium geyeri, Alnus tenuifolia, Amelanchier utahensis, Anaphalis margaritacea, Androsace septentrionalis, Angelica pinnata, Aquilegia caerulea, Arenaria rubella, Arnica cordifolia, Bistorta vivipara, Botrychium lunaria, Bromus ciliatus, B. inermis, B. lanatipes, Calamagrostis canadensis, Campanula rotundifolia, Cardamine cordifolia, Carex aurea, C. deweyana, C. geyeri, C. lenticularis, C. microptera, C. norvegica, C. siccata, C. stipata, C. utriculata, Chamerion angustifolium, C. latifolium, Cicuta douglasii, Cinna latifolia, Cirsium arvense, C. parryi, C. vinaceum, Cornus sericea, Corydalis caseana, Dasiphora floribunda, Delphinium barbeyi, Deschampsia cespitosa, Draba spectabilis, Elymus canadensis, E. glaucus, Epilobium ciliatum, E. hornemannii, E. saximontanum, Equisetum arvense, E. laevigatum, E. pratense, Erigeron coulteri, E. elatior, E.*

speciosus, Euphorbia brachycera, Festuca ovina, F. thurberi, Fragaria vesca, F. virginiana, Galium boreale, G. trifidum, G. triflorum, Geranium richardsonii, Geum macrophyllum, G. turbinatum, Glyceria striata, Heracleum maximum, Hydrophyllum capitatum, H. fendleri, Hymenoxys hoopesii, Juncus balticus, J. saximontanus, Lathyrus lanszwertii, Ligusticum porteri, Lolium pratense, Lonicera involucrata, Luzula parviflora, Maianthemum stellatum, Mentha arvensis, Mertensia ciliata, Micranthes odontoloma, M. rhomboidea, Mimulus glabratus, M. guttatus, Mitella pentandra, Monardella glauca, Oreochrysum parryi, Orthilia secunda, Osmorhiza depauperata, O. occidentalis, Oxypolis fendleri, Pascopyrum smithii, Paxistima myrsinites, Penstemon whippleanus, Picea engelmannii, P. pungens, Pinus ponderosa, Poa fendleriana, P. leptocoma, P. pratensis, Polemonium viscosum, Populus tremuloides, Potentilla pensylvanica, Primula pauciflora, Prunella vulgaris, Pseudocymopterus montanus, Pseudostellaria jamesiana, Pseudotsuga menziesii, Pteridium aquilinum, Ranunculus inamoenus, R. uncinatus, Rhodiola rhodantha, Ribes inerme, R. montigenum, R. wolfii, Rosa woodsii, Rubus deliciosus, R. idaeus, R. parviflorus, Rudbeckia laciniata, Salix bebbiana, S. boothii, S. brachycarpa, S. drummondiana, S. ligulifolia, S. lutea, S. monticola, Sambucus racemosa, Saxifraga flagellaris, Senecio bigelovii, S. triangularis, Sibbaldia procumbens, Silene acaulis, Solidago multiradiata, Sorbus scopulina, Stellaria umbellata, Streptopus amplexifolius, Symphoricarpos albus, S. oreophilus, Symphyotrichum eatonii, S. foliaceum, Taraxacum officinale, Thalictrum fendleri, Thlaspi montanum, Trifolium longipes, Trisetum wolfii, Urtica dioica, Vaccinium scoparium, Valeriana occidentalis, Veratrum tenuipetalum, Veronica americana, Vicia americana, Viola canadensis.

Use by wildlife: The flowers of *Mertensia ciliata* are visited by various insects (Insecta) including nectar-gathering bees (Hymenoptera: Apidae: *Apis mellifera, Bombus balteatus, B. bifarius, B. flavifrons, B. frigidus, B. insularis, B. lapponicus, B. melanopygus, B. mixtus, B. occidentalis*; Colletidae: *Colletes paniscus*; Megachilidae: *Osmia*) and flies (Diptera: Bombyliidae: *Systoechus oreas*; Muscidae: *Hylemya, Paregle*). The smaller bees and flies often are nectar thieves and do not effect pollination. The floral ovaries are eaten by mites (Arachnida: Acari: Eriophyidae: *Aceria*). Fly (Insecta: Diptera) larvae have been observed feeding on various floral parts. The nutlets are eaten by pine siskins (Aves: Fringillidae: *Carduelis pinus*). The plants are palatable, nutritious (averaging 5022 calories/g leaf tissue), and are eaten by domestic cattle (Mammalia: Bovidae: *Bos taurus*) and sheep (Mammalia: Bovidae: *Ovis aries*), golden-mantled ground squirrels (Mammalia: Sciuridae: *Spermophilus lateralis*), grizzly bears (Mammalia: Ursidae: *Ursus arctos horribilis*), mule deer (Mammalia: Cervidae: *Odocoileus hemionus*), Rocky Mountain elk (Mammalia: Cervidae: *Cervus canadensis nelsoni*), Rocky Mountain goats (Mammalia: Bovidae: *Oreamnos americanus missoulae*), and yellow-bellied marmots (Mammalia: Sciuridae: *Marmota flaviventris*). The plants are hosts to butterfly larvae (Insecta: Lepidoptera:

Nymphalidae: *Euphydryas anicia, E. chalcedona*) and tiger moth larvae (Lepidoptera: Arctiidae: *Gnophaela latipennis*). They host several fungi (Ascomycota: Erysiphaceae: *Erysiphe cichoracearum*; Basidiomycota: Entylomataceae: *Entyloma serotinum*; Pucciniaceae: *Puccinia hydrophylli, P. mertensiae*). *Mertenisa ciliata* is used by the American pika (Mammalia: Ochotonidae: *Ochotona princeps*) for "haystacks," which refers to their cached piles of stored food. Elk (Mammalia: Cervidae: *Cervus*) use dense stands of the plants for bedding and for nursery sites.

Economic importance: food: The flowers of *Mertensia ciliata* reportedly are edible raw or as added to salads. The foliage should be avoided because of the presence of potentially toxic alkaloids; **medicinal:** *Mertensia ciliata* contains the pyrrolizidine alkaloids intermedine and lycopsamine, which can be hepatotoxic. The Cheyenne used an infusion of *M. ciliata* to increase lactation and administered infusions made from the leaves and roots to relieve itching associated with measles and smallpox; **cultivation:** *Mertensia ciliata* and *M. franciscana* are cultivated as a rock garden ornamentals; **misc. products:** A brown/yellow dye is obtained from the stem of *Mertensia ciliata*; **weeds:** none; **nonindigenous species:** none

Systematics: Care should be taken not to confuse this genus (*Mertensia* Roth) with the name *Mertensia* Willd., which is an illegitimate nomenclatural synonym of the fern genus *Dicranopteris*. The systematic position of *Mertensia* in Boraginaceae was uncertain for a long time. It has been placed previously in tribe Lithospermeae or Trigonotidae, with the latter designation favored by some contemporary authors. Although the genus resembles the Old World *Pulmonaria* superficially, the latter usually is placed in tribe Boragineae. *Mertensia* has been allied with *Anoplocaryum*, which is assigned to yet a different tribe: Eritrichieae. However, a recent and comprehensive analysis of combined DNA sequence data for Boraginaceae recovered *Mertensia* as a well-supported clade (after segregation of the genus *Pseudomertensia*), which resolved with strong internal support as the sister group of the genus *Asperugo*. Consequently, the proper placement of both genera appears to be within tribe Cynoglosseae of subfamily Boraginoideae. The systematic relationships within the genus itself were not understood well until recently. A recent phylogenetic analysis of 11 cpDNA regions for 55 *Mertensia* species has greatly helped to clarify the taxonomy of the group and recommended the recognition of two sections: section *Stenhammaria* and section *Mertensia*. The latter includes both of the OBL species. Several varieties of *M. ciliata* have been recognized, but these are based on overlapping, variable characters and are in need of further evaluation. *Mertensia ciliata* and *M. franciscana* are very similar morphologically and ecologically but are not closely related. *Mertensia ciliata* associates with accessions of *M. alpina, M. cana*, and *M. perplexa*; whereas, *M. franciscana* resolves in a distant clade as the sister species of *M. bakeri*. The base chromosome number of *Mertensia* is $x = 6$. *Mertensia ciliata* ($2n = 24, 48$) has both tetraploid and octoploid cytotypes; *M. franciscana* ($2n = 24$) is tetraploid. Natural hybrids have been reported between *M. ciliata* and *M. alpina* but are poorly documented. However, various accessions of these species that have been surveyed in phylogenetic analyses associate in a pattern that could indicate hybridization. Other hybrids include *M. ciliata* × *M. bakeri*, and *M. ciliata* × *M. perplexa*.

Comments: *Mertensia ciliata* occurs throughout the western United States with *M. franciscana* restricted to the southwestern United States.

References: Andersen & Armitage, 1976; Andersen et al., 1979; Armitage, 1979; Bauer, 1983; Bennett, 2003; Bliss, 1976; Britton, 1951; Carleton, 1966; Cronquist et al., 1984; Fisher & Fulé, 2004; Gerber, 1985; Inouye et al., 2000; Kazuo, 1994; Kufeld, 1973; Långström & Chase, 2002; Li & Stermitz, 1988; Macior, 1983; McCullough, 1948; Mealey, 1980; Moore, 1930; Nazaire & Hufford, 2012, 2014; Pelton, 1961; Pyke, 1982; Reeves, 1977; Rigg, 1942; Saunders, Jr., 1955; Skinner, 1928; Thomson, 1980; Vizgirdas & Rey-Vizgirdas, 2005; Wallmo et al., 1972; Williams, 1937; Wright, 1988.

3. *Myosotis*

Forget-me-not, scorpion-grass; ne m'oubliez pas

Etymology: from the Greek *muos ôta* meaning "mouse ear" in reference to the leaf shape

Synonyms: *Brunnera* (in part); *Eritrichium* (in part); *Myosotidium* (in part)

Distribution: global: Africa; Asia; Australia; Europe; New World; New Zealand; **North America:** widespread

Diversity: global: 100 species; **North America:** 11 species

Indicators (USA): OBL: *Myosotis laxa*, *M. scorpioides*; **FACW:** *M. scorpioides*; **FAC:** *M. scorpioides*

Habitat: freshwater; palustrine, riverine; **pH:** 4.5–7.4; **depth:** <1 m; **life-form(s):** emergent herb

Key morphology: shoots (to 6 dm) ascending or decumbent, rooting at lower nodes, herbage strigose; leaves (to 8 cm) sessile (upper) or short petioled (lower), alternate, oblanceolate, oblong, or spatulate; cymes (to 2 dm) one sided, scorpoid, racemose, loosely flowered, flowers regular, 5-merous; calyx (to 5 mm) hairy; corolla (to 10 mm) short salverform, the lobes rounded, abruptly spreading, blue (rarely white; turning pink with age), the center crested with a yellow "eye"; ovary deeply 4-parted, the style gynobasic, included; schizocarp dehiscing into 4 single-seeded, smooth nutlets (to 2 mm)

Life history: duration: perennial (dormant apices, stolons); **asexual reproduction:** shoot fragments, stolons; **pollination:** insect or self; **sexual condition:** hermaphroditic; **fruit:** nutlets (common); **local dispersal:** fragments, nutlets (water); **long-distance dispersal:** nutlets (animals)

Imperilment: (1) *Myosotis laxa* [G5]; S1 (IN, KY, NE); S2 (NC, RI, VT, WI); S3 (DE); (2) *M. scorpioides* [G5]

Ecology: general: *Myosotis* includes annuals, biennials, and perennials of temperate or higher elevational tropical regions. The species grow mainly in full sunlight and are found in seashore to montane communities including forests, wetlands, and cultivated sites. Although eight of the eleven North American species are designated as wetland indicators, most occur in UPL sites or only facultatively in wetlands. The flowers usually are self-compatible but are pollinated by assorted insects (Insecta) including honeybees (Hymenoptera: Apidae: *Apis*), solitary bees (Hymenoptera: Andrenidae: *Andrena*,), leaf-cutter bees (Hymenoptera: Megachilidae: *Megachile*, *Osmia*), beetles (Coleoptera), butterflies (Lepidoptera: Lycaenidae: *Lycaena*, *Polyommatus*; Pieridae: *Pieris*), and flies (Diptera: Muscidae; Syrphidae). The fruits generally are unspecialized for dispersal, but a few are dispersed by ants (Insecta: Hymenoptera: Formicidae) or are water dispersed. Most seeds germinate fairly readily (within 4 weeks) at 20°C. Some require a period of warm stratification.

Myosotis laxa **Lehm.** is a perennial, which grows in bogs, channels, depressions, ditches, dunes, fens, floodplains, marshes, meadows, mudflats, prairies, sandbars, seeps, sloughs, swales, swamps, and in shallow water (e.g., 0.5 m), or on the exposed/drying margins of lakes, ponds, rivers, and streams at elevations of up to 2050 m. It occupies relatively infertile sites on acidic to alkaline substrates (pH: 4.5–7.4) consisting of gravel, loam, muck, mud, sand, silt, or silt loam. The plants occur occasionally on bog mats or on decomposing wood. The stems lack aerenchyma tissue and are larger when growing in aerated substrates (e.g., near flowing water) or along with substrate aerating species (e.g., *Typha*); however, competition with the latter can occur unless soil temperatures remain relatively cool (11°C–12°C). The plants have been reported in exposures ranging from open to partially shaded sites and are known to establish readily in light gaps. The flowers are self-compatible and are either self- or insect pollinated. The single-seeded nutlets are buoyant, will remain afloat for up to 3.5 days, and are dispersed by the water. Seed germination reportedly will occur in the dark or in light under drained conditions but not when the seeds are waterlogged or submerged. High seed germination (95%) has been obtained within 2 weeks for untreated seeds placed under a 21°C/10°C temperature regime, when planted in a soil-less peat medium supplemented with micronutrients and a slow release fertilizer. Seed germination has been reported under a 25°C/15°C temperature regime. A fairly large seed bank can be formed with up to 260 seeds/m². In one study, 235 seedlings emerged from a 10 cm core taken from a site where the plants were absent from the standing vegetation. The life history of this species requires further clarification. The North American subspecies (subsp. *laxa*) appears to be a perennial (at least biennial); however, the Eurasian subspecies (subsp. *caespitosa*) is mainly an annual. In any case, the North American plants are weak perennials with a fibrous-rooted stem and they exhibit little or no vegetative reproduction. The roots are associated with arbuscular mycorrhizae. The plants have been observed to bioaccumulate arsenic at a level that is 4758 times higher than the levels in the ambient environment. **Reported associates:** *Acer circinatum*, *A. rubrum*, *A. saccharinum*, *Agrostis alba*, *A. depressa*, *A. stolonifera*, *Alnus rubra*, *A. rugosa*, *Amorpha fruticosa*, *Angelica genuflexa*, *Anthoxanthum odoratum*, *Apios americana*, *Asclepias incarnata*, *Beckmannia syzigachne*, *Betula nigra*, *Bidens cernuus*, *Boehmeria cylindrica*, *Calamagrostis canadensis*, *Callitriche heterophylla*, *Campanula americana*, *Cardamine bulbosa*, *Carex aquatilis*, *C. aurea*, *C. canescens*, *C. densa*, *C. deweyana*,

C. feta, C. gynandra, C. lacustris, C. lupulina, C. lyngbyei, C. obnupta, C. pellita, C. stipata, C. stricta, C. suberecta, C. unilateralis, C. vesicaria, Cephalanthus occidentalis, Cicuta bulbifera, C. douglasii, C. maculata, Cirsium arvense, C. muticum, Comarum palustre, Cornus obliqua, Corylus cornuta, Crataegus douglasii, Cyperus, Danthonia californica, Deschampsia cespitosa, Distichlis spicata, Drosera rotundifolia, Dulichium arundinaceum, Echinochloa crusgalli, Eleocharis acicularis, E. erythropoda, E. palustris, Epilobium ciliatum, E. densiflorum, E. glandulosum, E. leptophyllum, Equisetum arvense, E. fluviatile, Erigeron annuus, Eupatorium perfoliatum, E. purpureum, Frangula purshiana, Fraxinus pennsylvanica, Elymus glaucus, Galium obtusum, G. parisiense, G. tinctorium, G. trifidum, Geum canadense, Glyceria striata, Hippuris vulgaris, Holcus lanatus, Howellia aquatilis, Hypericum mutilum, Impatiens capensis, Iris pseudacorus, I. virginica, Juncus acuminatus, J. articulatus, J. effusus, J. gerardii, J. nevadensis, J. oxymeris, J. salicifolius, J. supiniformis, J. tenuis, Laportea canadensis, Leersia oryzoides, Lemna minor, Lobelia cardinalis, Lotus corniculatus, Ludwigia palustris, Lycopus americanus, L. uniflorus, Lysichiton americanus, Lysimachia ciliata, L. nummularia, L. terrestris, Lythrum salicaria, Mentha arvensis, Mimulus guttatus, Muhlenbergia asperifolia, Myosotis scorpioides, Nasturtium officinale, Oenanthe sarmentosa, Oenothera strigosa, Onoclea sensibilis, Osmunda regalis, Peltandra virginica, Persicaria hydropiper, P. hydropiperoides, P. pensylvanica, P. punctata, P. sagittata, Phalaris arundinacea, Phragmites australis, Phyla lanceolata, Pilea fontana, Plagiobothrys figuratus, Plantago lanceolata, P. maritima, Platanthera flava, Poa trivialis, Polypogon monspeliensis, Pontederia cordata, Potamogeton natans, Potentilla egedii, Pycnanthemum virginianum, Quercus palustris, Ranunculus alismifolius, R. flammula, Rhamnus alnifolia, Rorippa islandica, Rosa arkansana, R. woodsii, Rubus procerus, R. pubescens, R. ursinus, Rumex crispus, Sagittaria latifolia, Salix alba, S. hookeriana, S. interior, S. petiolaris, S. rigida, S. scouleriana, S. sitchensis, Schoenoplectus acutus, S. pungens, Scirpus cyperinus, S. microcarpus, S. pedicellatus, Scutellaria galericulata, Setaria, Sium suave, Solanum dulcamara, Solidago graminifolia, Sparganium angustifolium, S. eurycarpum, Spartina pectinata, Sphagnum, Spiraea alba, S. douglasii, Spirodela polyrrhiza, Stachys ciliata, Stellaria calycantha, S. media, Symphoricarpos albus, Thelypteris palustris, Torreyochloa pallida, Toxicodendron vernix, Trifolium repens, Typha angustifolia, T. latifolia, Ulmus americana, Urtica dioica, Utricularia, Verbascum thapsus, Verbena hastata, Veronica americana, V. anagallis-aquatica, V. scutellata, Viola palustris.

***Myosotis scorpioides* L.** is a perennial, which inhabits beaches, brooks, ditches, fens, flats, floodplains, marshes, meadows, mudflats, seeps, sloughs, springs, swamps, and the margins of lakes, rivers, and streams at elevations of up to 2900 m. The plants usually occur in cool, shallow (e.g., 15–30 cm) water with slow to moderate flow. They tolerate cold climates and extend onto tundra habitats up to 64.9° N latitude. Exposures ranging from full sun to deep shade are tolerated. The sites tend to be more mesotrophic or eutrophic than those of the previous species and often are described as alkaline; however, the plants occur over a broad range of acidity (pH: 4.7–7.1). The substrates include clay, gravel, loam, muck, mud, rock, sand, sandy loam, and silt. Newly emerging seedlings reach the flowering stage in about 95 days, but can be induced to flower in as little as 43 days if they are exposed to longer periods of light (15–24 h). The plants have a fairly long flowering period (i.e., May to September) in North America. The corollas are blue but turn to pink as they age, perhaps as a cue to pollinators. Unlike its congeners, this species possesses a multilocus self-incompatibility mechanism; individuals can be entirely self-incompatible but become increasingly self-compatible as heterozygosity increases. Consequently, the flowers either self-pollinate or are cross-pollinated by insects such as bees (Hymenoptera), flies (Diptera) and to a lesser extent, butterflies (Lepidoptera). The nutlets are water dispersed, mainly during late summer to fall. The small fruits float only until they become submerged, where after they sink; however, viable fruits can represent as much as 9% of the propagules that are recovered from river drift samples. They have been found in relatively high abundance in samples taken from irrigation water. Despite their lack of appendages, the nutlets have been found to adhere quite well to the fur of various vertebrates, and so they likely are dispersed to some degree by mammals; in Europe they are known to be dispersed by some ant (Hymenoptera: Formicidae) species. Viable seeds have been recovered from the dung of rabbits (Mammalia: Lagomorpha: Leporidae), which potentially transport the propagules endozoically. Seed germination has been reported after receiving no special treatment or after being exposed to a 4 week period of cold stratification at 4°C; a 25°C/15°C temperature regime commonly is used. A germination rate of 73% has been reported for seeds that have been dried at 32°C–34°C, refrigerated at 0°C–2°C, and then placed under a 30°C/20°C temperature regime (with 12 h light). The germination rate is reduced if seeds occur under waterlogged or submerged conditions. A seed bank is produced and the seeds are recovered commonly in seed bank samples (e.g., 7%–24% in some surveys). The density often is quite low, but can reach up to 1104 seeds/m². Shed seeds occur primarily in the surface sediments (<5 cm). They are long lived as evidenced by viable seeds retrieved from samples after 15–20 years despite a nearly 10-year absence from the standing vegetation. This is a relatively short-lived perennial (reproducing vegetatively by stolons), which typically lives for 2–3 years. Vegetative reproduction occurs by the dispersal of stolon or shoot fragments (mainly in the fall to winter period) and usually is the more prominent dispersal mechanism. This amphibious species exhibits few morphological or physiological differences between its submersed and emergent foliage, although the emergent leaves tend to have higher numbers of stomata. The plants are less stressed physiologically when growing under noninundated conditions and a 60% reduction in growth has been observed for plants growing in saturated sediments. The roots are colonized by arbuscular and dark septate endophyte mycorrhizae. The plants are common in disturbed sites,

especially where the upper sediment surface has been disturbed or removed. **Reported associates:** *Abies balsamea, A. lasiocarpa, Acalypha rhomboidea, Acer glabrum, A. rubrum, A. saccharinum, A. saccharum, Achillea millefolium, Actaea rubra, Agrostis stolonifera, Alisma, Alnus rubra, A. rugosa, Ambrosia trifida, Amelanchier alnifolia, Anemone canadensis, Apocynum cannabinum, Arabis drummondii, Asclepias incarnata, Barbarea vulgaris, Betula alleghaniensis, B. nigra, B. papyrifera, B. ×sandbergii, Bidens cernuus, Boehmeria cylindrica, Botrychium virginianum, Bromus carinatus, Calamagrostis canadensis, Calla palustris, Callitriche palustris, Caltha palustris, Campanula aparinoides, Cardamine pensylvanica, Carex aurea, C. bromoides, C. gynandra, C. muskingumensis, C. obnupta, C. stipata, Calystegia sepium, Celtis occidentalis, Cephalanthus occidentalis, Chamerion angustifolium, Chara, Cicuta douglasii, Clematis columbiana, Clintonia uniflora, Cornus sericea, Cryptotaenia canadensis, Cypripedium parviflorum, Dactylis glomerata, Eleocharis acicularis, Elymus glaucus, E. virginicus, Epilobium ciliatum, Equisetum fluviatile, E. hyemale, E. scirpoides, Eragrostis hypnoides, Eriogonum spergulinum, Eucephalus engelmannii, Eurybia macrophylla, Euthamia, Eutrochium maculatum, Fissidens fontanus, Fragaria virginiana, Fraxinus nigra, F. pennsylvanica, Galium triflorum, Gayophytum, Gentianopsis, Geum canadensis, G. macrophyllum, Glechoma hederacea, Glyceria, Heracleum maximum, Hieracium, Howellia aquatilis, Impatiens capensis, Iris versicolor, I. virginica, Juncus, Larix laricina, Lathyrus palustris, Leersia oryzoides, L. virginica, Lemna minor, Lobelia cardinalis, L. kalmii, Lonicera morrowii, Lycopus uniflorus, Lysimachia ciliata, L. nummularia, Maianthemum racemosum, Melica smithii, Menispermum, Mentha arvensis, Mertensia, Mimulus glabratus, M. ringens, Mitella breweri, Myosotis laxa, Myrica gale, Nasturtium officinale, Onoclea sensibilis, Osmunda regalis, Parnassia glauca, Persicaria maculosa, P. sagittata, P. virginiana, Phalaris arundinacea, Phyla lanceolata, Picea glauca, Plantago major, P. rugelii, Poa trivialis, Polemonium occidentalis, Populus angustifolia, P. balsamifera, Quercus macrocarpa, Q. rubra, Ranunculus alismifolius, R. flammula, R. hispidus, R. orthorhynchus, Ribes lacustre, Rosa palustris, Rubus parviflorus, Rumex occidentalis, R. salicifolius, Sagittaria latifolia, Salix drummondiana, Sambucus nigra, S. racemosa, Schoenoplectus tabernaemontani, Scirpus atrovirens, Scutellaria galericulata, S. lateriflora, Shepherdia canadensis, Smilax tamnoides, Solanum dulcamara, Solidago uliginosus, Sparganium, Sphagnum, Spiraea betulifolia, S. douglasii, Spiranthes diluvialis, Symphyotrichum ontarionis, Tanacetum vulgare, Taraxacum officinale, Thalictrum occidentale, Thelypteris palustris, Thuja occidentalis, T. plicata, Tiarella trifoliata, Tilia americana, Toxicodendron radicans, Tsuga canadensis, Typha ×glauca, Ulmus americana, Urtica dioica, Veratrum viride, Verbena hastata, Veronica americana, V. scutellata, Viola glabella, Vitis riparia, Wyethia, Zannichellia palustris.*
Use by wildlife: The seeds of *Myosotis* species are eaten by various birds (Aves). *Myosotis laxa* is a host to Fungi (Oomycota: Peronosporaceae: *Peronospora myosotidis*).

Myosotis scorpioides is a source of nectar for silver-spotted skippers (Insecta: Lepidoptera: Hesperiidae: *Epargyreus clarus*). It is eaten by rabbits (Mammalia: Lagomorpha: Leporidae) and crayfish (Crustacea: Decapoda: Cambaridae: *Orconectes rusticus*), but is resistant to grazing by deer (Mammalia: Cervidae: *Odocoileus virginianus*). It is a host to snails (Gastropoda: Succineidae). In the United Kingdom, it is the sole food of an uncommon lace bug (Insecta: Heteroptera: Tingidae: *Dictyla convergens*), which is not found in North America.

Economic importance: food: The plants should not be eaten due to the presence of potentially toxic alkaloids. *Myosotis scorpioides* reportedly can cause liver cancer if it is eaten in quantity; **medicinal:** *Myosotis scorpioides* contains several pyrrolizidine alkaloids (7-acetylscorpioidine, scorpioidine, symphytine, myoscorpine), which are lethal to mice (Mammalia: Rodentia: Muridae: *Mus*) in laboratory studies. Europeans once used the plant as a sedative and tonic, or applied extracts as a soothing eye bath. It has been used as a treatment for bronchitis and whooping cough; **cultivation:** *Myosotis laxa* and *M. scorpioides* are cultivated along with their interspecific hybrid *Myosotis ×suzae*. *Myosotis scorpioides* has a number of cultivars including 'Alba' (white flowered), 'Blaqua,' 'Ice Pearl,' 'John Beaty,' 'Mayfair,' 'Mermaid,' 'Oglethorpe,' 'Pinkie,' 'Semperflorens,' 'Snowflakes,' 'Unforgettable,' and 'Wisconsin'; **misc. products:** The Makah people rubbed plants of *Myosotis laxa* on their hair as a form of hairspray. *Vase with Myosotis and Peonies* is a notable painting by Vincent Van Gogh; **weeds:** *Myosotis laxa* and *M. scorpioides* are listed as weeds in Tasmania (Australia) and *M. scorpioides* is regarded as being invasive in many parts of North America; **nonindigenous species:** *Myosotis laxa* became naturalized in New Zealand around 1977. *Myosotis scorpioides* was introduced to North America, Mexico, and South America from Eurasia; it is nonindigenous to Japan. Introductions occur mainly as escapes from cultivation.

Systematics: Phylogenetic analyses place *Myosotis* within Boraginaceae tribe Cynoglosseae in the vicinity of *Cynoglossum* and *Plagiobothrys* (Figure 5.96). More comprehensive analyses have resolved *Myosotis* as the sister group of *Pseudomertensia*, a genus once recognized as a section within *Mertensia*. Analyses using combined DNA sequence data indicate that *Myosotis* is monophyletic, at least with respect to the taxa surveyed (roughly 1/3 of the total species), which represent both previously recognized sections and all relevant geographical regions. Those results indicate that the two sections (*Exarrhena*, *Myosotis*) do not correspond to the major clades recovered, and that a revised sectional classification is necessary. Of the two OBL species, only *M. laxa* (European material) was included in those phylogenetic analyses, where it resolves in a clade with the European *M. rehsteineri* and *M. debilis* (Figure 5.98). Consequently, the North American subspecies of *M. laxa* probably diverged from a European ancestor. Additional systematic research is needed to better elucidate patterns of variation associated with both of the OBL species. There are several subspecies

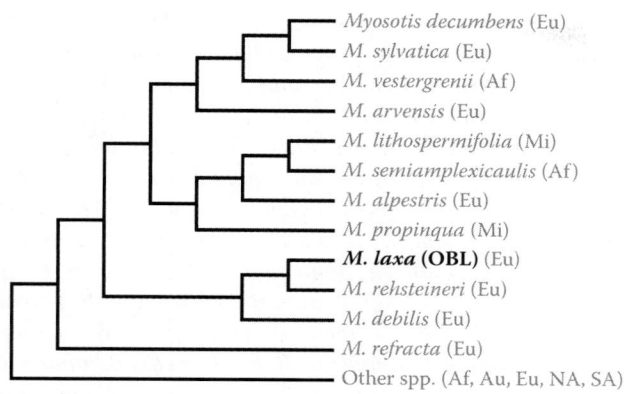

Myosotis decumbens (Eu)
M. sylvatica (Eu)
M. vestergrenii (Af)
M. arvensis (Eu)
M. lithospermifolia (Mi)
M. semiamplexicaulis (Af)
M. alpestris (Eu)
M. propinqua (Mi)
M. laxa (OBL) (Eu)
M. rehsteineri (Eu)
M. debilis (Eu)
M. refracta (Eu)
Other spp. (Af, Au, Eu, NA, SA)

FIGURE 5.98 Phylogenetic relationships in *Myosotis* as indicated by analysis of nrITS sequence data. Geographical affinities are indicated by Af (Africa), Au (Australasia), Eu (Europe), Mi (Middle East), NA (North America), and SA (South America). European material of the OBL *M. laxa* (in bold) resolves in a clade with other European species, which also is supported by cpDNA sequence data (not shown); consequently, the North American subspecies of *M. laxa* probably originated from Europe. The European *M. scorpioides* (not included earlier) presumably is closely related to *M. laxa* because of its ability to hybridize with that species. (Adapted from Winkworth, R.C. et al., *Mol. Phylogen. Evol.*, 24, 180–193, 2002.)

of *M. laxa* with subsp. *laxa* representing the North American plants and subspecies *baltica* and *cespitosa* assigned to Eurasian plants. Several varieties have been named for *M. scorpioides*. The base chromosome number for the genus presumably is *x* = 8. Both *Myosotis laxa* (2*n* = 80) and *M. scorpoides* (2*n* = 64) are high polyploids. These two species are able to hybridize and probably are fairly closely related. Their partially fertile, interspecific hybrid (involving the Old World subspecies of *M. laxa*) is known as *Myosotis ×suzae*. It has not been determined whether natural hybridization has yet occurred in North America between the native *M. laxa* and the introduced *M. scorpioides*. They appear to have similar ecological tolerances and associates; however, despite their nearly completely overlapping North American distributional ranges, it is surprising that they seldom have been reported as growing together. Controlled hybridization experiments between these taxa could provide additional insights.

Comments: *Myosotis laxa* and *M. scorpioides* both occur mainly in eastern and western (but not commonly in central) North America. *Myosotis laxa* extends into Eurasia and northern Africa; *M. scorpioides* is native to Eurasia.

References: Andersson et al., 2000; Bartow, 2003; Bennett, 2003; Boedeltje et al., 2003b, 2004; Britton, 1951; Callaway & King, 1996; Cellot et al., 1998; Cooke, 1997; Cosyns, 2004; Couvreur et al., 2004; Delatte & Chabrerie, 2008; Esler, 1987; Falińska, 1999; Fargione et al., 1991; Fassett, 1957; Germ & Gaberščik, 2003; Gurnell et al., 2007b; Hanlon et al., 1998; Hazlett, 1989; Hitchmough & Fieldhouse, 2004; Hoagland & Davis, 1987; Kelley & Bruns, 1975; Lenssen et al., 1998, 1999; Mason, 1957; Matus et al., 2003; McCain & Christy, 2005; Ogren, 2003; Preston & Croft, 1997; Prins, 1968; Ramaley,

1934; Resch et al., 1982; Robinson et al., 2006; Rozefelds et al., 1999; Rubino et al., 2002; Shaw, 1951; Speichert & Speichert, 2004; Šraj-Kržič et al., 2006; Swink, 1952; Varopoulos, 1979; Venables & Barrows, 1985; Vogt et al., 2007; Winkworth et al., 2002.

4. *Phacelia*

Scorpionweed

Etymology: from the Greek *phakelos* meaning "bundle" in reference to the dense inflorescence

Synonyms: *Eutoca*; *Miltitzia*; *Whitlavia*

Distribution: global: New World; **North America:** widespread

Diversity: global: 207 species; **North America:** 176 species

Indicators (USA): OBL; UPL: *Phacelia distans*

Habitat: freshwater; palustrine; **pH:** 7.4–8.4; **depth:** <1 m; **life-form(s):** emergent herb

Key morphology: stems (to 80 cm) decumbent or erect, pubescent; leaves (to 15 cm) alternate, 1–2 times pinnately compound, the segments rounded or toothed; cymes coiled, one sided, densely flowered, the flowers subsessile, 5-merous; calyx (to 5 mm) 5-lobed, densely hairy; corolla (to 9 mm) regular, 5-lobed, tubular, bell shaped to funnelform, blue or whitish, deciduous, with ovate scales at base of tube; capsules (to 3 mm) spherical, 2–4 seeded, puberulent; seeds (to 3 mm) brown, pitted

Life history: duration: annual; (fruits/seeds); **asexual reproduction:** none; **pollination:** insect; **sexual condition:** hermaphroditic; **fruit:** capsules (common); **local dispersal:** seeds (gravity); **long-distance dispersal:** seeds (animals)

Imperilment: (1) *Phacelia distans* [G5]

Ecology: general: *Phacelia* includes a diverse assemblage of annuals (both summer and winter) and perennial species that can occupy a variety of habitats including bluffs, chaparral, meadows, sand dunes, scrub, washes, and woodlands. Fourteen of the North American species (9%) are wetland indicators, but only one species is ranked as OBL. The flowers are self-compatible and are capable of autogamy, but usually are protandrous and are outcrossed by insects. In some species, the styles are deflected at an angle nearly perpendicular to the stamens, which mechanically reduces the probability of self-pollination. Although most species are hermaphroditic, at least some of them are gynodioecious. Seed dispersal mechanisms have been reported for few *Phacelia* species; however, their seeds are known to adhere to the fur of larger mammalian herbivores, which disperse them in that fashion. In other cases, they simply are dispersed by gravity and fall very close (<0.5 m) to the parental plants. Seed germination requirements vary. Species growing under milder conditions can lack dormancy altogether; whereas, those inhabiting various altitudinal zones can require proportionally increasing periods of cold stratification and higher thermoperiods as the elevation increases. Yet in other cases, nondormant seeds that fail to germinate in the autumn can acquire secondary dormancy and must undergo an after-ripening period during the following summer. Large, persistent seed banks have been reported for at least some species.

Phacelia distans **Benth.** is an annual, which inhabits alluvial fans, bluffs, canyons, chaparral, drainage channels, desert pavement, ditches, dunes, flats, floodplains, puddles, ravines, riverbanks, roadsides, sand bars, scalds, seeps, slopes, stream banks, stream beds, washes, and woodlands at elevations of up to 2682 m. The plants occur in exposures that range from full sun to light shade but produce higher specific leaf weights when growing under sunny conditions. The substrates have been described as alkaline (pH: 7.4–8.4) or saline and include alluvium, basalt cobble, boulders, clay, clay loam, gneiss, granite, granite loam, gravel, gravelly loam, gravelly sand, gravelly serpentine, limestone, loamy sand, mudstone, rock, rocky sand, sand, sandy gravel, sandy loam, selenite, shale, and talus. From the list of xerophytic associated species, it is difficult to reconcile this species as an OBL indicator. However, this species grows primarily in previously flooded sites that have since dried out. These weak-stemmed plants often are found growing beneath shrubs, which provide them with protection, shade, and support. Occurrences reportedly favor locations where there is a low diversity of other native or nonindigenous species; however, from the extensive list of reported associated species, this generalization is difficult to apply in a broad sense. The flowers appear from February to May, presumably are self-compatible, and are pollinated primarily by bees (Hymenoptera: Megachilidae). The seeds are dispersed by gravity or by animals. The plants produce a large seed bank (e.g., averaging 486 seeds/m^2); however, the deposition of seeds on the soil surface can reach substantially higher densities (e.g., 11,600 seeds/m^2). The seeds germinate in February. Field-collected seeds do not appear to be dormant and have germinated under a 10°C/6°C day/night temperature regime in flats that are watered daily. Although the plants often are reported in sites that have experienced recent fire, they are more prevalent at unburned sites. One study failed to recover seed germinants from soil samples that had been taken from burned areas. **Reported associates:** *Abronia maritima, A. umbellata, A. villosa, Abutilon incanum, Acalypha californica, Acamptopappus shockleyi, A. sphaerocephalus, Acer negundo, Achnatherum hymenoides, A. parishii, Achyronychia cooperi, Acmispon glabrus, A. humistratus, A. maritimus, A. rigidus, A. strigosus, Acourtia nana, A. wrightii, Adenophyllum cooperi, A. porophylloides, Adenostoma fasciculatum, A. sparsifolium, Agave chrysantha, A. deserti, A. mckelveyana, A. palmeri, A. toumeyana, Agoseris retrorsa, Allium burlewii, Allophyllum gilioides, Alnus oblongifolia, Aloysia wrightii, Ambrosia acanthicarpa, A. ambrosioides, A. chamissonis, A. chenopodiifolia, A. confertiflora, A. deltoidea, A. dumosa, A. eriocentra, A. monogyra, A. psilostachya, A. salsola, Amorpha californica, Amsinckia intermedia, A. menziesii, A. spectabilis, A. tessellata, Amsonia, Anemone tuberosa, Anisacanthus thurberi, Anisocoma acaulis, Anthriscus caucalis, A. luciana, Arctostaphylos glandulosa, A. glauca, A. manzanita, A. pungens, Argemone corymbosa, Argythamnia neomexicana, Aristida adscensionis, A. purpurea, A. ternipes, Artemisia californica, A. dracunculus, A. ludoviciana, A. nova, A. tridentata, Asclepias subulata, Astragalus lentiginosus,*

A. nothoxys, A. nuttallianus, Astrolepis cochisensis, A. sinuata, Atriplex canescens, A. cordulata, A. confertifolia, A. coronata, A. hymenelytra, A. lentiformis, A. polycarpa, A. serena, Avena barbata, A. fatua, Baccharis brachyphylla, B. salicifolia, B. sarothroides, B. sergiloides, Bahiopsis laciniata, B. parishii, Baileya multiradiata, B. pleniradiata, Bebbia juncea, Berberis haematocarpa, B. trifoliata, Bergerocactus emoryi, Bernardia myricifolia, Boechera perennans, B. sparsiflora, Bouteloua curtipendula, B. eriopoda, B. gracilis, Bowlesia incana, Brassica nigra, B. tournefortii, Brickellia baccharidea, B. californica, B. coulteri, Bromus carinatus, B. diandrus, B. hordeaceus, B. japonicus, B. rubens, B. tectorum, Bursera microphylla, Calandrinia ciliata, Calliandra eriophylla, Calocedrus decurrens, Camissoniopsis bistorta, C. cheiranthifolia, C. hirtella, Canotia holacantha, Capsella bursa-pastoris, Carex pansa, Carlowrightia arizonica, Carnegiea gigantea, Cassia, Castilleja angustifolia, C. applegatei, C. exserta, C. lanata, Caulanthus coulteri, C. lasiophyllus, Ceanothus crassifolius, C. cuneatus, C. greggii, C. leucodermis, C. oliganthus, C. palmeri, Celtis pallida, C. reticulata, Cenchrus ciliaris, C. setaceus, Centromadia pungens, Cercocarpus betuloides, C. ledifolius, C. montanus, Chaenactis carphoclinia, C. fremontii, C. stevioides, Chaetopappa ericoides, Cheilanthes covillei, C. newberryi, Chenopodium watsonii, Chilopsis linearis, Chorispora tenella, Chorizanthe brevicornu, Chrysothamnus viscidiflorus, Chylismia claviformis, Cirsium neomexicanum, Clarkia purpurea, C. unguiculata, Claytonia perfoliata, Clematis pauciflora, Coleogyne ramosissima, Coreopsis bigelovii, Corethrogyne filaginifolia, Corydalis, Cottea pappophoroides, Cottsia gracilis, Coursetia glandulosa, Crassula connata, C. tillaea, Crossosoma bigelovii, Croton californicus, C. sonorae, Cryptantha angustifolia, C. barbigera, C. clevelandii, C. gracilis, C. intermedia, C. maritima, C. muricata, C. nevadensis, C. pterocarya, C. utahensis, Cucurbita foetidissima, Cuscuta californica, Cylindropuntia acanthocarpa, C. bigelovii, C. californica, C. echinocarpa, C. fulgida, C. ganderi, C. leptocaulis, C. munzii, C. spinosior, C. versicolor, C. wolfii, Dalea searlsiae, Dasylirion wheeleri, Datura, Daucus pusillus, Delphinium parishii, D. scaposum, D. umbraculorum, Descurainia brevisiliqua, D. pinnata, Dichelostemma capitatum, Dieteria canescens, Digitaria californica, Diplacus aurantiacus, Distichlis spicata, Ditaxis lanceolata, Dodonaea viscosa, Draba cuneifolia, Dudleya blochmaniae, D. saxosa, Echinocereus engelmannii, E. fasciculatus, E. nicholii, Elymus elymoides, E. multisetus, Emmenanthe penduliflora, Encelia actonii, E. californica, E. farinosa, E. frutescens, E. virginensis, Ephedra aspera, E. californica, E. nevadensis, E. trifurca, E. viridis, Epilobium canum, Eragrostis curvula, E. intermedia, E. lehmanniana, Eremalche exilis, E. parryi, Eremothera boothii, E. chamaenerioides, Eriastrum densifolium, E. diffusum, E. eremicum, Ericameria brachylepis, E. cooperi, E. cuneata, E. laricifolia, E. linearifolia, E. nauseosa, Erigeron divergens, E. foliosus, E. lobatus, E. parishii, Eriodictyon crassifolium, E. trichocalyx, Eriogonum abertianum, E. deflexum, E. elongatum, E. fasciculatum, E. inflatum, E. nudum, E. wrightii,

Eriophyllum ambiguum, E. confertiflorum, E. lanosum, Erodium cicutarium, E. texanum, Eschscholzia californica, E. minutiflora, Escobaria alversonii, E. vivipara, Eucrypta chrysanthemifolia, Eulobus californicus, Euphorbia albomarginata, E. eriantha, E. pediculifera, Eurotia, Fagonia laevis, Ferocactus cylindraceus, F. viridescens, F. wislizeni, Forestiera, Fouquieria splendens, Frasera albomarginata, Fraxinus velutina, Fremontodendron californicum, Funastrum cynanchoides, Galium angustifolium, G. aparine, G. hilendiae, G. stellatum, Garrya flavescens, G. veatchii, Geraea canescens, Gilia angelensis, G. brecciarum, G. cana, G. capitata, G. sinuata, G. stellata, G. flavocincta, G. transmontana, Glandularia gooddingii, Gnaphalium, Grayia spinosa, Gutierrezia californica, G. sarothrae, Haplopappus linearifolius, Hazardia squarrosa, Helianthus gracilentus, Herissantia crispa, Herniaria hirsuta, Hesperocnide tenella, Heteromeles arbutifolia, Heterotheca villosa, Hibiscus denudatus, Hilaria rigida, Hoffmannseggia microphylla, Hordeum murinum, Horkelia rydbergii, Hyptis emoryi, Isocoma acradenia, I. menziesii, I. tenuisecta, Isomeris, Jatropha cardiophylla, Juglans major, Juncus bufonius, Juniperus californica, J. coahuilensis, J. deppeana, J. monosperma, J. osteosperma, Justicia californica, Keckiella antirrhinoides, K. ternata, Krameria bicolor, K. erecta, Krascheninnikovia lanata, Lappula occidentalis, Larrea tridentata, Lastarriaea coriacea, Lasthenia californica, L. chrysostoma, Layia glandulosa, L. pentachaeta, L. septentrionalis, Lepidium fremontii, L. lasiocarpum, L. virginicum, Lepidospartum squamatum, Leptosiphon aureus, Lessingia filaginifolia, Leymus condensatus, Linanthus bigelovii, Logfia filaginoides, L. gallica, Lomatium mohavense, L. nevadense, Lonicera subspicata, Acmispon glabrus, Lupinus arizonicus, L. benthamii, L. bicolor, L. concinnus, L. excubitus, L. hirsutissimus, L. shockleyi, L. sparsiflorus, L. subvexus, L. succulentus, Lycium andersonii, L. berlandieri, L. cooperi, L. exsertum, L. pallidum, Malacothamnus fasciculatus, Malacothrix californica, M. coulteri, M. glabrata, M. saxatilis, Malosma laurina, Malva parviflora, Mammillaria dioica, M. grahamii, Marah gilensis, M. macrocarpus, Marina parryi, Marrubium vulgare, Matelea producta, Medicago, Melampodium leucanthum, Melica californica, M. imperfecta, Melilotus indicus, Menodora scabra, Mentzelia involucrata, M. veatchiana, Mimosa aculeaticarpa, Mimulus aurantiacus, M. bigelovii, M. guttatus, M. kelloggii, M. pilosus, M. rubellus, Mirabilis laevis, M. multiflora, Monardella, Monolepis nuttalliana, Monoptilon bellioides, Mortonia utahensis, Morus microphylla, Muhlenbergia appressa, M. microsperma, M. porteri, M. rigens, Myosurus cupulatus, Nama hispida, N. pusilla, Nassella cernua, Nemacladus sigmoideus, Nemophila heterophylla, Nicotiana glauca, N. obtusifolia, Nolina microcarpa, Notholaena standleyi, Oenothera primiveris, Olneya tesota, Opuntia basilaris, O. bigelovii, O. chlorotica, O. engelmannii, O. littoralis, O. phaeacantha, O. ramosissima, Orthocarpus, Palafoxia linearis, Pappophorum vaginatum, Pappostipa speciosa, Parietaria floridana, P. pensylvanica, Parkinsonia florida, P. microphylla, P. praecox, Parthenium incanum, Pectocarya linearis, P. platycarpa, P. recurvata,
P. setosa, Pellaea truncata, Penstemon centranthifolius, P. eatonii, P. palmeri, P. parryi, P. spectabilis, P. subulatus, Peritoma arborea, Perityle emoryi, P. gilensis, Peucephyllum schottii, Phacelia campanularia, P. cicutaria, P. crenulata, P. fremontii, Phalaris minor, Phlox stansburyi, P. tenuifolia, Pholistoma auritum, Phoradendron californicum, Physaria gordonii, P. purpurea, P. tenella, Picrothamnus desertorum, Pinus edulis, P. jeffreyi, P. monophylla, P. ponderosa, P. sabiniana, Plagiobothrys arizonicus, P. jonesii, Plantago ovata, P. patagonica, Platanus racemosa, P. wrightii, Pleurocoronis laphamioides, P. pluriseta, Poa bigelovii, P. secunda, Populus fremontii, P. tremuloides, Porophyllum gracile, Proboscidea althaeifolia, Prosopis juliflora, P. velutina, Prunus andersonii, P. fasciculata, P. fremontii, P. ilicifolia, P. tenuifolia, P. virginiana, Pseudognaphalium biolettii, P. californicum, P. stramineum, Pseudotsuga macrocarpa, Psilostrophe cooperi, Psorothamnus arborescens, P. schottii, P. spinosus, Pteridium aquilinum, Pterostegia drymarioides, Purshia tridentata, Quercus agrifolia, Q. ajoensis, Q. berberidifolia, Q. chrysolepis, Q. cornelius-mulleri, Q. douglasii, Q. john-tuckeri, Q. kelloggii, Q. lobata, Q. turbinella, Q. wislizeni, Rafinesquia neomexicana, Ranunculus testiculatus, Rhamnus californica, R. crocea, R. ilicifolia, Rhus aromatica, R. integrifolia, R. ovata, Rhynchosia physocalyx, Ribes aureum, R. indecorum, R. quercetorum, Rosa californica, R. woodsii, Rubus ulmifolius, Rumex hymenosepalus, Salazaria mexicana, Salix exigua, S. gooddingii, S. laevigata, S. lasiolepis, Salvia apiana, S. columbariae, S. dorrii, S. mellifera, S. mohavensis, S. munzii, S. pachyphylla, Sambucus nigra, Schedonorus arundinaceus, Schismus arabicus, S. barbatus, Scrophularia californica, Sebastiania bilocularis, Selaginella arizonica, Senecio flaccidus, S. scorzonella, Senegalia greggii, Senna armata, Silene antirrhina, S. laciniata, Simmondsia chinensis, Sisymbrium altissimum, S. irio, Solanum parishii, Sonchus oleraceus, Sorghum halepense, Sphaeralcea ambigua, S. coulteri, S. grossulariifolia, S. laxa, Sporobolus contractus, Stenocereus thurberi, Stephanomeria exigua, S. pauciflora, Stillingia linearifolia, Stipa speciosa, Stutzia covillei, Suaeda nigra, Stylocline gnaphalioides, Tamarix aphylla, T. ramosissima, Tetracoccus fasciculatus, T. hallii, Tetradymia comosa, T. stenolepis, Thamnosma montana, Thysanocarpus curvipes, Tiquilia canescens, Toxicodendron diversilobum, Tradescantia occidentalis, Tridens muticus, Trifolium mucronatum, Trixis californica, Turritis glabra, Typha, Umbellularia californica, Uropappus lindleyi, Vachellia constricta, Vauquelinia californica, Verbascum thapsus, Verbena neomexicana, Veronica anagallis-aquatica, V. peregrina, Vulpia microstachys, V. myuros, V. octoflora, Xanthisma spinulosum, Xanthium strumarium, Xylococcus bicolor, Xylorhiza tortifolia, Yucca baccata, Y. brevifolia, Y. schidigera, Y. whipplei, Zinnia acerosa, Ziziphus obtusifolia, Z. parryi.

Use by wildlife: *Phacelia distans* is a highly preferred food of grazing sheep (Mammalia: Bovidae: *Ovis aries*) and land snails (Mollusca: Gastropoda: Helminthoglyptidae: *Helminthoglypta arrosa*). It is found occasionally in the diets of pronghorn antelope (Mammalia: Antilocapridae:

Antilocapra americana). The herbage is used as nesting material by the cactus wren (Aves: Troglodytidae: *Campylorhynchus brunneicapillus couesi*), which is the state bird of Arizona. The flowers attract a variety of foraging bees (insecta: Hymenoptera: Andrenidae: *Andrena macrocepha*, *A. sagittagalea*; Apidae: *Anthophora vannigera*, *Habropoda pallida*, *Xeromelecta californica*; Halictidae: *Dufourea tularensis*, *D. vernalis*, *Protodufourea eickworti*, *P. parca*, *P. zavortinki*; Megachilidae: *Anthidium angelarum*, *A. collectum*, *A. palmarum*, *Ashmeadiella gillettei*), with at least *Anthidium palmarum* serving as a primary pollinator. The plants host the larvae of ethmiid moths (Insecta: Lepidoptera: Ethmiidae: *Ethmia brevistriga*), jumping sand dune moths (Insecta: Lepidoptera: Scythrididae: *Areniscythris brachypteris*), and tiger moths (Insecta: Lepidoptera: Arctiidae: *Grammia geneura*). They attract adult leafhoppers (Insecta: Hemiptera: Cicadellidae: *Gloridonus atridorsum defectus*) and plant bugs (Insecta: Hemiptera: Miridae: *Dicyphus hesperus*, *Irbisia californica*, *I. silvosa*, *Melanotrichus vestitus*). The foliage is host to powdery mildew Fungi (Ascomycota: Erysiphaceae: *Alphitomorpha fuliginea*).

Economic importance: food: The Kawaiisu people steam cooked the leaves of *Phacelia distans* to eat as greens; **medicinal:** no medicinal uses have been reported; **cultivation:** *Phacelia distans* is not cultivated; **misc. products:** The seeds of *Phacelia distans* have been included in native plant mixes used for restoration projects of abandoned agricultural lands; **weeds:** *Phacelia distans* is not weedy; **nonindigenous species:** *Phacelia distans* was collected once in Wisconsin (in 1951) where it was regarded as an accidental introduction. It has appeared in Massachusetts on several occasions as an introduction. The sources of these introductions have not been identified.

Systematics: Phylogenetic analyses of DNA sequence data have confirmed that *Phacelia* is monophyletic and occurs within subfamily Hydrophylloideae of Boraginaceae (Figure 5.96) as the sister group of the genus *Romanzoffia* (Figure 5.99). *Phacelia distans*, the only OBL member of the genus, occupies a relatively derived position (Figure 5.99) within section *Ramosissimae* of subgenus *Phacelia*. *Phacelia ramosissima* (FACU) resolves as its most closely related species. Molecular analyses indicate that *P. distans* might comprise more than one species and requires additional taxonomic inquiry. The recognition of different taxa under that name could explain why such contrary wetland indicator categories (OBL, UPL) have been assigned to the taxon in different areas, but it is more likely that the UPL ranking from the northeast is a consequence of its introduction to that area in cultivated, terrestrial habitats. The base chromosome number of *Phacelia* is $x = 9$, 11, or 12. *Phacelia distans* ($2n = 22$) is diploid. No hybrids involving *P. distans* have been reported.

Comments: *Phacelia distans* is native to the southwestern United States but has been collected occasionally in Massachusetts and Wisconsin as the result of introductions, which do not seem to have persisted.

References: Bailey, 1922; Baskin & Baskin, 1973; Baskin et al., 1993; Bohart & Griswold, 1996; Cassis, 1984; Cavieres & Arroyo, 2000; Couvreur et al., 2004; Cruden, 1972; Drezner et al., 2001; Gillett, 1961; Hamilton, 2014; Huntzinger et al., 2011; Levy, 1988; Levy & Neal, 1999; Marushia & Allen, 2011; Michener, 1939; Muller, 1953; Murphy & Allen, 2009; Phillips et al., 1996; Ribble, 1968; Rubino, 2002; Schwartz, 1984; Schwartz & Scudder, 2000; Schwarz, 1927; Shields, 1966; Singer & Stireman III, 2001; Steers & Allen, 2012; Timberlake, 1937, 1939, 1941, 1951; Tluczek, 2012; Wainwright, 1978; Walden et al., 2014; Wallace & Szarek, 1981; Wei et al., 2007; Went et al., 1952.

5. *Plagiobothrys*

Popcornflower

Etymology: from the Greek *plagios bothros* ("oblique scar") referring to the scar marking the attachment point of the nutlets

Synonyms: *Allocarya*; *Allocaryastrum*; *Echinoglochin*; *Glyptocaryopsis*

Distribution: global: New World; **North America:** widespread except southeastern

Diversity: global: 65 species; **North America:** 42 species

Indicators (USA): OBL: *Plagiobothrys acanthocarpus*, *P. austiniae*, *P. chorisianus*, *P. distantiflorus*, *P. figuratus*, *P. glaber*, *P. hirtus*, *P. humistratus*, *P. leptocladus*, *P. parishii*, *P. strictus*, *P. tener*, *P. undulatus*; **FACW:** *P. bracteatus*, *P. figuratus*, *P. leptocladus*, *P. scouleri*, *P. stipitatus*

Habitat: freshwater to saline; palustrine; **pH:** 6.1–7.9; **depth:** <1 m; **life-form(s):** submersed (rosulate) herbs; emergent herbs

Key morphology: shoots (to 5 dm) erect to prostrate, the herbage usually strigose; leaves (to 10 cm) cauline (becoming smaller apically) or in basal rosettes, the lower opposite, the upper alternate, linear, entire; flowers regular, in racemes or spikes (coiled in bud), pedicels short (to 1 mm); corolla (to 12 mm) salverform with a short tube, the lobes spreading, white, inside of tube sometimes yellow, throat terminating in small crests; ovary deeply 4-lobed, style gynobasic; schizocarps dehiscing into four 1-seeded nutlets (to 2.5 mm), the nutlet surfaces wrinkly, bristly, or prickly, the attachment scar prominent

Life history: duration: annual (fruit/seeds); perennial (persistent crowns); **asexual reproduction:** shoot tips (layering); **pollination:** insect; **sexual condition:** hermaphroditic; **fruit:** nutlets (common); **local dispersal:** nutlets (gravity, water); **long-distance dispersal:** nutlets (animals)

Imperilment: (1) *Plagiobothrys acanthocarpus* [G4]; (2) *P. austiniae* [G4]; S2 (OR); (3) *P. bracteatus* [G4]; (4) *P. chorisianus* [G3]; S2 (CA); S3 (CA); (5) *P. distantiflorus* [G3]; (6) *P. figuratus* [G4]; S1 (BC, OR); (7) *P. glaber* [GH]; SH (CA); (8) *P. hirtus* [G1]; S1 (OR); (9) *P. humistratus* [G2]; (10) *P. leptocladus* [G4]; S1 (MT, WY); (11) *P. parishii* [G1]; S1 (CA); (12) *P. scouleri* [G5]; S1 (MB); S3 (AB); (13) *P. stipitatus* [G4]; (14) *P. strictus* [G1]; S1 (CA); (15) *P. tener* [G4]; (16) *P. undulatus* [G3]

Ecology: general: Over half of the North American *Plagiobothrys* species (67%) are at least FACW indicators, which grow commonly in moist grasslands or at

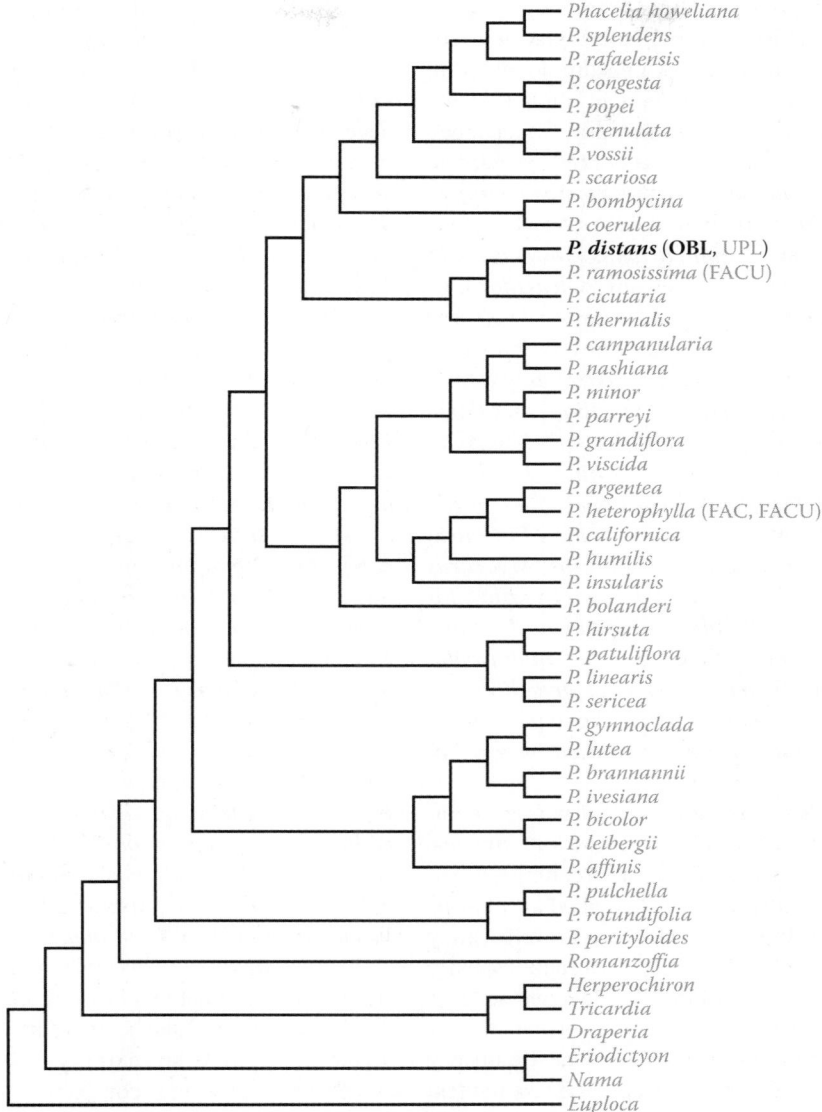

FIGURE 5.99 Phylogenetic relationships in *Phacelia* as derived from analysis of combined nrITS and *ndhF* sequence data. These reults place the OBL *P. distans* (bold) in a relatively derived position within the genus. The wetland indicator status is shown in parentheses where relevant. Most *Phacelia* species inhabit terrestrial sites. (Adapted from Walden, G.K. et al., *Madroño*, 61, 16–47 2014.)

sites characterized by seasonal moisture. The OBL species are almost entirely associated with vernal pools or similarly ephemeral or desiccating sites. Other species can inhabit various dry habitats including chaparral, scrub, and woodland communities that range from coastal to montane elevations. Treatments for *P. bracteatus*, *P. scouleri*, and *P. stipitatus* (all with OBL rankings in the 1996 list) have been retained here, as they were completed prior to their reclassification as FACW in the 2013 indicator list. *Plagiobothrys undulatus* was excluded from the 1996 list, when it was considered to be synonymous with *P. chorisianus*. However, it was ranked as OBL in the 2013 indicator list and has been included here. The life histories of most *Plagiobothrys* species have been poorly studied. The flowers appear to be self-compatible but are mostly outcrossed by insects. The extent of self-pollination

is unknown for most species. The seed ecology has been examined for only a few of the species and precise germination requirements, seed longevity data, etc., are not available for most of them. Mechanisms of seed dispersal are poorly known. A few species have barbed or hooked appendages on the nutlets, which probably facilitates their adhesion to fur and transport by mammals. Basic ecological information (e.g., soil pH, salinity, etc.) is wanting for many of the species. Habitats often are described as alkaline or saline, but with no corresponding values provided. Nearly all of the species are annuals, which enables the plants to survive in sites that dry out completely during the summer months.

***Plagiobothrys acanthocarpus* (Piper) I.M. Johnst.** is an annual found in depressions, drainage channels, flats, lake margins, meadows, savanna, seeps, sinks, stream sides, swales, and drying/vernal pools at elevations of up to 1436 m. The sediments are alkaline (specific pH values not reported) and include adobe, clay, granite, gravel, gypsum, loam, mud, and serpentine. There have been no detailed studies of the life history of this species. The nutlets are spiny appendaged and probably are dispersed by animals. **Reported associates:** *Allenrolfea occidentalis, Allium munzii, Artemisia californica, Asclepias fascicularis, Atriplex spinifera, Avena barbata, A. fatua, Blennosperma nanum, Bromus hordeaceus, B. madritensis, Callitriche marginata, Castilleja densiflora, C. exserta, Crassula aquatica, C. connata, Cyperus, Distichlis spicata, Elatine californica, Eleocharis palustris, Ephedra, Eriogonum fasciculatum, Erodium brachycarpum, E. moschatum, Eryngium aristulatum, Frankenia salina, Fritillaria biflora, Haplopappus, Harpagonella palmeri, Hemizonia fasciculata, Isoetes orcuttii, Juncus bufonius, Lasthenia californica, Lepidium, Lilaea scilloides, Limnanthes gracilis, Lythrum hyssopifolium, L. tribracteatum, Matricaria discoidea, Melica imperfecta, Mimulus guttatus, Myosurus minimus, Nassella pulchra, Navarretia fossalis, Orcuttia californica, Pilularia americana, Plagiobothrys fulvus, Plantago bigelovii, P. elongata, Pogogyne abramsii, P. nudiuscula, Polypogon monspeliensis, Primula, Psilocarphus brevissimus, Quercus engelmannii, Rumex crispus, Sambucus mexicana, Scirpus, Sisyrinchium bellum, Veronica peregrina, Vulpia myuros.*

***Plagiobothrys austiniae* (Greene) I.M. Johnst.** is a summer annual, which grows in alluvial fans, depressions, ditches, flats, swales, vernal pools, and margins of streams at elevations of up to 990 m. The substrates are described as alkaline and consist of clay, cobble, gravel, gravelly clay, gravelly sandy loam, loam, rock, and sand. The seeds germinate in early February, the plants bloom from March to May, and senesce by July. There are small prickles on the ribs of the nutlets, which probably facilitate animal dispersal. Other life history information on this species is scarce. **Reported associates:** *Achyrachaena mollis, Alopecurus saccatus, Amsinckia intermedia, Atriplex fruticulosa, Blennosperma nanum, Bromus hordeaceus, B. mollis, Callitriche, Camissonia graciliflora, Crassula aquatica, C. connata, C. tillaea, Deschampsia danthonioides, Distichlis spicata, Eleocharis palustris, Erodium botrys, Eschscholzia lobbii, Festuca megalura, Gratiola ebracteata, Juncus bufonius, Lasthenia californica, L. fremontii, Layia fremontii, Lepidium dictyotum, L. nitidum, Lolium multiflorum, Lupinus bicolor, Micranthes californica, Navarretia leucocephala, Pilularia americana, Plagiobothrys fulvus, P. greenei, P. nothofulvus, P. shastensis, Plantago bigelovii, P. coronopus, P. erecta, Pogogyne ziziphoroides, Quercus douglasii, Trifolium depauperatum, T. variegatum, Triphysaria eriantha.*

***Plagiobothrys bracteatus* (T.J. Howell) I.M. Johnst.** is an annual, which grows in channels, depressions, ditches, dunes, flats, floodplains, marshes, meadows, ruts, seeps, sloughs, swales, vernal pools, and along the drying margins of lakes, ponds, and streams at elevations of up to 2000 m. Although this species typically inhabits sites characterized by vernal water, the plants can withstand periods of inundation

lasting for up to 3–4 months. The substrates are described as alkaline or saline and range from adobe, basalt rubble, clay, gravel, gravelly clay, loamy clay, mud, mudstone, sand, sandy loam, sandy silt, schist, to volcanic ash. The seedlings develop into small rosettes that eventually produce branching stems, which can grow through the foliage of neighboring vegetation. There is no specific information available on the reproductive ecology of this species. **Reported associates:** *Alopecurus saccatus, Blennosperma bakeri, Bromus mollis, Callitriche, Calochortus uniflorus, Carex, Ceanothus cuneatus, Cerastium glomeratum, Crassula aquatica, Cuscuta howelliana, Danthonia californica, Deschampsia danthonioides, Downingia bacigalupii, D. bella, D. bicornuta, D. concolor, D. cuspidata, D. pulchella, D. yina, Draba verna, Eleocharis palustris, Epilobium ciliatum, Eremocarpus setigerus, Eriogonum wrightii, Eryngium aristulatum, E. armatum, E. mathiasiae, Erythronium howellii, Geranium dissectum, Glinus lotoides, Gratiola ebracteata, Grindelia camporum, Horkelia congesta, Isoetes nuttallii, Juncus bufonius, J. effusus, J. phaeocephalus, J. uncialis, J. xiphioides, Lasthenia fremontii, L. glaberrima, Limnanthes douglasii, L. floccosa, Lomatium cookii, Lotus, Lythrum hyssopifolium, Malvella leprosa, Mentha pulegium, Mimulus guttatus, Navarretia fossalis, N. leucocephala, N. minima, Pinus jeffreyi, P. ponderosa, Plagiobothrys glyptocarpus, P. greenei, P. stipitatus, Plantago coronopus, Pleuropogon californicus, Polypogon maritimus, Psilocarphus brevissimus, Quercus kelloggii, Ranunculus lobbii, Rhamnus tomentella, Rorippa curvisiliqua, Rumex crispus, Sambucus mexicana, Sidalcea calycosa, Trifolium obtusiflorum, Triteleia hyacinthina, Veronica peregrina, Xanthium strumarium.*

***Plagiobothrys chorisianus* (Cham.) I.M. Johnst.** is an annual, which inhabits bluffs, chaparral, depressions, dunes, flats, marshes, meadows, mudflats, prairies, seeps, swales, vernal pools, and the margins of ponds at elevations of up to 1350 m. Substrates consist of clay, gravel, sand, sandy peat, or mudstone. The plants flower from February to May and are known to persist in the seed bank for a fairly long (but undetermined) period of time. Other life history information is lacking for this species. **Reported associates:** *Acaena pinnatifida, Achillea millefolium, Alopecurus saccatus, Aphanes arvensis, Artemisia californica, Baccharis pilularis, Bromus hordeaceus, Callitriche marginata, Carex barbarae, C. obnupta, Castilleja campestris, C. latifolia, C. subinclusa, Ceanothus thyrsiflorus, Cerastium glomeratum, Cotula coronopifolia, Crassula aquatica, Cuscuta howelliana, Deschampsia danthonioides, Downingia bicornuta, Elatine californica, Eleocharis palustris, Epilobium cleistogamum, Erechtites glomerata, Erigeron glaucus, Eriogonum, Eriophyllum stoechadifolium, Erodium botrys, E. cicutarium, Eryngium armatum, Frangula californica, Gratiola ebracteata, Hemizonia fitchii, Heracleum maximum, Holocarpha macradenia, H. virgata, Hordeum marinum, Iris douglasiana, Isoetes howellii, I. orcuttii, Juncus bufonius, J. capitatus, J. uncialis, Lasthenia californica, L. fremontii, L. glaberrima, Layia chrysanthemoides, Lepidium latipes, L. nitidum, Limosella, Lolium perenne, Lythrum hyssopifolia,*

Oxalis pilosa, Pilularia americana, Plagiobothrys greenei, P. humistratus, P. leptocladus, P. stipitatus, Plantago elongata, P. erecta, Poa annua, Pogogyne ziziphoroides, Polypogon monspeliensis, Potentilla glandulosa, Psilocarphus brevissimus, P. tenellus, Pteridium aquilinum, Ranunculus bonariensis, R. lobbii, Rumex acetosella, Sagina procumbens, Salix lasiolepis, Salvia spathacea, Satureja douglasii, Scrophularia californica, Sedum spathulifolium, Senecio aronicoides, S. sylvaticus, Sidalcea malviflora, Spergularia rubra, Toxicodendron diversilobum, Trifolium barbigerum, Veronica peregrina.

***Plagiobothrys distantiflorus* (Piper) I.M. Johnst. ex M.E. Peck** is an annual, which occurs in depressions, seeps, swales, and vernal pools at elevations of up to 1040 m. It is found on substrates of Auburn series, clay loam, granite loam, granitic sand adobe, loam, and sandy loam. There is virtually nothing known regarding the biology or ecology of this species. **Reported associates:** *Alopecurus saccatus, Avena barbata, Bromus hordeaceus, Calandrinia ciliata, Castilleja campestris, Cerastium glomeratum, Crassula connata, Epilobium cleistogamum, Erodium botrys, Eryngium castrense, E. spinosepalum, Juncus bufonius, Lasthenia fremontii, Lolium multiflorum, Marsilea vestita, Plagiobothrys acanthocarpus, P. nothofulvus, P. stipitatus, Poa annua, Psilocarphus tenellus, Quercus douglasii, Ranunculus muricatus, Taraxacum officinale, Trifolium variegatum, Triteleia hyacinthina, Urtica urens, Vulpia bromoides.*

***Plagiobothrys figuratus* (Piper) I.M. Johnst. ex M.E. Peck** is an annual, which grows in depressions, ditches, meadows, prairies, and vernal pools at elevations of up to 1890 m. The plants sometimes occur in shallow standing water and tolerate full sun to partial shade. The substrates include clay, loam, sand, or silty clay loam. The plants flower in May to June but little information exists on their pollination biology. The flowers are visited by honeybees (Insecta: Hymenoptera: *Apis*), which might function as pollinators. The seeds germinate well (96%) without stratification or other treatment. Their method of dispersal is unknown. Sowing experiments show the seeds to be highly viable (87%). The seedlings emerge quickly and are capable of rapidly covering open sites. However, the species does persist well beyond the first year of establishment, probably due to competition. The roots are colonized by vesicular–arbuscular mycorrhizal fungi. **Reported associates:** *Acmispon americanus, Agrostis capillaris, A. exarata, A. stolonifera, Aira caryophyllea, Alopecurus, Anthoxanthum odoratum, Balsamorhiza deltoidea, Beckmannia syzigachne, Bistorta bistortoides, Briza minor, Brodiaea coronaria, Bromus mollis, Camassia leichtlinii, C. quamash, Cardamine penduliflora, Carex arcta, C. athrostachya, C. aurea, C. densa, C. feta, C. ovalis, C. tumulicola, C. unilateralis, Centaurium erythraea, C. muhlenbergii, C. umbellatum, Centaurium umbellatum, Centunculus minimus, Cerastium viscosum, Cirsium vulgare, Crataegus douglasii, Cynosurus cristatus, Danthonia californica, Daucus carota, Deschampsia cespitosa, Dipsacus fullonum, Downingia elegans, D. yina, Eleocharis acicularis, E. obtusa, E. palustris, Epilobium paniculatum, Equisetum hyemale,*

Erigeron decumbens, Eriophyllum lanatum, Eryngium petiolatum, Festuca arundinacea, F. rubra, Fraxinus latifolia, Galium aparine, G. cymosum, G. parisiense, Gaultheria shallon, Gentiana sceptrum, Glyceria occidentalis, Gnaphalium palustre, G. uliginosum, Grindelia integrifolia, Heterocodon rariflorum, Holcus lanatus, Hordeum brachyantherum, Hosackia gracilis, Hypericum perforatum, Hypochaeris radicata, Juncus acuminatus, J. articulatus, J. bufonius, J. effusus, J. ensifolius, J. nevadensis, J. oxymeris, J. patens, J. tenuis, Leontodon nudicaulis, Leucanthemum vulgare, Lithophragma parviflorum, Lolium perenne, Lomatium bradshawii, Ludwigia palustris, Lysimachia nummularia, Madia glomerata, Matricaria, Mentha arvensis, M. pulegium, Micranthes oregana, Microseris laciniata, Microsteris gracilis, Mimulus guttatus, Myosotis discolor, M. laxa, Orthocarpus bracteosus, Panicum capillare, P. occidentale, Parentucellia viscosa, Phalaris arundinacea, Pinus ponderosa, Plagiobothrys scouleri, Plantago lanceolata, Plectritis, Polygonum douglasii, Populus tremuloides, Potentilla gracilis, Prunella vulgaris, Pseudotsuga menziesii, Psilocarphus elatior, Quercus garryana, Ranunculus orthorhynchus, R. repens, Rorippa islandica, Rosa eglanteria, R. nutkana, R. pisocarpa, Rubus discolor, R. laciniatus, R. ursinus, Rumex acetosella, Salix geyeriana, S. piperi, S. sitchensis, S. lasiandra, Senecio jacobaea, Sericocarpus rigidus, Sidalcea virgata, Sisyrinchium angustifolium, Spiraea douglasii, Stellaria borealis, Symphyotrichum hallii, Toxicoscordion venenosum, Trifolium dubium, T. repens, Typha latifolia, Urtica dioica, Vaccinium myrtillus, Veronica scutellata, V. serpyllifolia, Vicia sativa, V. tetrasperma, V. villosa, Viola adunca.

***Plagiobothrys glaber* (A. Gray) I.M. Johnst.** is an annual of coastal, alkaline, or subsaline tidal marshes at elevations of up to 100 m. The plant is believed to have been extinct for many years (last collected in 1938) and nothing is known of its life history. Some have suggested that invasion of the nonindigenous *Cotula coronopifolia* might have facilitated its demise. **Reported associates:** none.

***Plagiobothrys hirtus* (Greene) I.M. Johnst.** is a winter annual or perennial of prairies and seasonal wet meadows at elevations of up to 270 m. The sediments are clay or silty clay loam (Conser or Ruch–Medford–Takilma series) that attain a depth of 160+ cm. The plants are shade intolerant and exhibit reductions in vigor, reproduction, and seedling establishment when growing under shaded conditions. They often form monospecific stands when growing in deeper pools. Natural disturbances (seasonal flooding and/or fire) are necessary to retain open habitat and to limit competition by species such as *Centaurea, Dipsacus fullonum, Fraxinus latifolia, Mentha pulegium,* and *Rubus discolor.* Flowering occurs from June to July. Larger plants can produce over 50 stems, each bearing up to 100 flowers. The flowers are self-compatible, but mainly are insect pollinated by bees (Hymenoptera: Apidae), butterflies (Lepidoptera), hover flies (Diptera: Syrphidae), and moths (Lepidoptera: Ctenuchidae: *Ctenucha*). The plants typically are highly fertile, with 73%–85% of the ovules developing into nutlets. However, seed set can be reduced indirectly in areas where pesticides (which eliminate susceptible pollinators) are used. Some degree of self-pollination reportedly occurs, but inbreeding depression has been indicated for seed viability, with 67% germination (at 21°C) observed for open-pollinated seeds vs. only 17% for selfed seeds. Generally, higher resources will yield larger seeds having higher germination rates, and the rates for selfed seeds will increase (up to 60%) if the plants are fertilized and watered. The seeds mature in summer–fall and germinate when the fall rains commence. Up to 95% of the seeds will germinate within 5 days if they are kept moist. Cold-stratified seeds (8 weeks at 4°C) germinate better (67%) under constant temperature (20°C) than under an alternating temperature regime (27% at 10°C/20°C). Unstratified seeds do not germinate under constant temperatures (20°C), but germinate well (67%) under an alternating temperature regime (10°C/20°C). The seedlings exhibit a mean survivorship of 66% after 1 month, and densities up to 78 seedlings/10 cm² have been observed in the field. New seedlings overwinter as submerged rosettes. The elongate stems can root at the nodes and reproduce new rosettes vegetatively by layering. In sites that dry out completely during the summer, the species behaves as a winter annual. Where sufficient moisture persists, the plants will perennate as rosettes for at least another winter, with connections among the ramets severing during that time. Common greenhouse experiments indicate that substantial genetic variation for various life-history traits exists among the populations. *Plagiobothrys hirtus* is extremely rare, with only an estimated 7000 plants remaining. **Reported associates:** *Agrostis alba, Alopecurus geniculatus, Beckmannia syzigachne, Camassia leichtlinii, Carex arcta, C. feta, C. unilateralis, Deschampsia cespitosa, Dipsacus sylvestris, Downingia yina, Glyceria occidentalis, Hordeum brachyantherum, Juncus acuminatus, J. balticus, J. effusus, J. oxymeris, J. patens, Limnanthes douglasii, Mentha arvensis, M. piperita, M. pulegium, Veronica scutellata.*

Plagiobothrys humistratus (Greene) I.M. Johnst. is an annual species of depressions, ditches, meadows, prairies, roadsides, seeps, sinks, swales, and vernal pools at elevations of up to 415 m. The substrates are described as alkaline or saline and consist of basalt, clay, cobbles, mud, sand, serpentine, or silty loam. Particulars on the life history of this species are lacking. **Reported associates:** *Allenrolfea occidentalis, Alopecurus saccatus, Amsinckia intermedia, Atriplex persistens, A. spinescens, Bromus hordeaceus, B. mollis, Callitriche marginata, Claytonia parviflora, Crassula aquatica, Cressa truxillensis, Distichlis spicata, Downingia pulchella, Eryngium castrense, Extriplex joaquiniana, Frankenia, Hemizonia congesta, Hordeum depressum, H. marinum, Kochia californica, Lasthenia californica, L. conjugens, L. fremontii, Layia fremontii, Lepidium dictyotum, Limnanthes douglasii, L. floccosa, Lolium multiflorum, L. perenne, Lythrum hyssopifolia, Myosurus, Nassella pulchra, Navarretia prostrata, Pilularia americana, Plagiobothrys acanthocarpus, P. chorisianus, P. stipitatus, P. trachycarpus, P. undulatus, Plectritis ciliosa, Psilocarphus brevissimus, Puccinellia simplex, Uropappus lindleyi, Vulpia bromoides, V. microstachys.*

Plagiobothrys leptocladus (Greene) I.M. Johnst. is a winter annual, which occurs in canals, depressions, ditches,

flats, marshes, meadows, mudflats, prairies, riverbeds, sandbars, sinks, swales, vernal pools, and along the desiccating margins of lakes, pools, reservoirs, and streams at elevations of up to 2500 m. The sites are open with circumneutral to alkaline (pH: 6.1–7.9) substrates described as adobe, clay, clay loam, loam, mud, sandy clay, silt, silty clay, silty loam, and stony clay. The seeds germinate in winter as ponding begins (December) and the seedlings develop in water even under ice cover (February to March). Ponding ceases in April. The plants flower in March–May and die shortly afterwards. Flowering as late as July to August has been reported for some localities, perhaps where water persists for a longer period. There is no information available on the pollination biology or seed ecology of this species. **Reported associates:** *Achnatherum hymenoides, Allium geyeri, A. nevii, Alopecurus carolinianus, A. saccatus, Alyssum, Artemisia cana, Atriplex argentea, A. coronata, A. parishii, A. polycarpa, A. serenana, Bassia hyssopifolia, Beckmannia syzigachne, Brassica geniculata, Brodiaea filifolia, B. orcuttii, B. terrestris, Bromus tectorum, Callitriche, Camassia quamash, Castilleja campestris, C. lineariiloba, Chrysothamnus viscidiflorus, Cirsium vulgare, Crassula aquatica, C. solierii, Cressa truxillensis, Crypsis schoenoides, Cryptantha torreyana, Damasonium californicum, Deschampsia danthonioides, D. gracilis, Distichlis spicata, Downingia bella, D. elegans, D. insignis, D. ornatissima, Eleocharis acicularis, E. palustris, Elymus elymoides, E. triticoides, Epilobium cleistogamum, E. pygmaeum, Eryngium aristulatum, E. castrense, E. vaseyi, Erysimum repandum, Frankenia salina, F. grandifolia, Grayia spinosa, Grindelia, Hemizonia pungens, Hesperostipa comata, Heterocodon rariflorus, Hordeum depressum, H. intercedens, H. marinum, Ipomopsis, Isoetes howellii, Iva axillaris, Juncus balticus, J. bufonius, Lasthenia californica, L. ferrisiae, L. fremontii, L. glaberrima, L. glabrata, Lepidium dictyotum, L. latifolium, L. latipes, Leymus triticoides, Limnanthes douglasii, Limosella aquatica, Lolium, Lomatium, Lupinus pusillus, Madia, Marsilea vestita, Mentzelia dispersa, Microseris platycarpha, Mimulus guttatus, Muhlenbergia richardsonis, Myosurus apetalus, M. minimus, Navarretia fossalis, N. intertexta, N. leucocephala, Neoholmgrenia andina, Orcuttia californica, Phacelia inundata, Plagiobothrys chorisianus, Plantago elongata, Pleuropogon californicus, Pogogyne douglasii, Polyctenium fremontii, Polygonum aviculare, P. polygaloides, Polypogon monspeliensis, Potentilla argentea, Psilocarphus brevissimus, P. globiferus, P. oregonus, Rhynchospora, Rorippa curvisiliqua, Rumex venosus, Sarcobatus vermiculatus, Spergularia bocconii, S. salina, Suaeda moquinii, Taraxia tanacetifolia, Trichostema lanceolatum, Veronica peregrina.*

Plagiobothrys parishii I.M. Johnst. is an annual, which grows in meadows, playas, seeps, springs, and along margins of vernal pools at elevations from 875 to 1400 m. It occupies substrates that consist of alkaline/saline clay, mud, or sandy silt. The plants flower from April to June. Other life history information is lacking. The species is quite rare and threatened by local removal of groundwater. **Reported associates:** *Anemopsis californica, Atriplex polycarpa,*

Bolboschoenus robustus, Calochortus striatus, Carex prae-gracilis, Centaurium exaltatum, Chrysothamnus albidus, Cressa truxillensis, Distichlis spicata, Eleocharis parishii, Eriastrum eremicum, Grindelia fraxinopratensis, Helianthus annuus, Heliotropium curassavicum, Hesperochiron pumilus, Hordeum jubatum, Iva axillaris, Juncus balticus, J. bufonius, J. mexicana, Lactuca sativa, Lepidium appelianum, L. flavum, Lysimachia maritima, Lythrum californica, Mimulus guttatus, Polypogon monspeliensis, Potentilla gracilis, Puccinellia parishii, Sarcobatus vermiculatus, Schoenoplectus pungens, Sidalcea neomexicana, Sporobolus airoides, Suaeda nigra, Tamarix ramosissima, Thelypodium crispum, Typha domingensis.

***Plagiobothrys scouleri* (Hook. & Arn.) I.M. Johnst.** is an annual, which inhabits beaches, bogs, borrow pits, ditches, flats, floodplains, gullies, meadows, playas, seeps, sloughs, springs, swales, vernal pools, and the margins of rivers and streams at elevations of up to 3140 m. The plants grow in open sites to partial shade and occur on alkaline or saline substrates of mineral or organic origin, which include clay, clay loam, granite, gravel, loam, mud, peat, sand, sandstone, sandy gravel, and silt clay. Details on the life history of this species have not been reported and the ecology is poorly understood. The plants often are found in disturbed sites and have become invasive in some parts of the world. This species has wider environmental tolerances relative to its OBL congeners. It is more FAC to wetland habitats, is not closely restricted to vernal pool habitats, and is the only one of its congeners that is reported to occur on organic substrates. Additional research on the biology and ecology of this species is needed. **Reported associates:** *Abies grandis, Achillea millefolium, Acmispon americanus, Agropyron, Agrostis hyemalis, A. microphylla, A. scabra, Aira praecox, Allium amplectens, A. douglasii, A. geyeri, Alnus rhombifolia, A. rubra, Alopecurus carolinianus, A. geniculatus, A. pratensis, Amelanchier, Amsinckia menziesii, Anagallis minima, Anthemis, Anthoxanthum odoratum, Armeria maritima, Artemisia arbuscula, A. cana, A. ludoviciana, A. tridentata, Asperugo procumbens, Balsamorhiza, Beckmannia syzigachne, Bidens cernuus, B. frondosus, Bistorta bistortoides, Bromus tectorum, Bryum miniatum, Buchloe dactyloides, Calamagrostis rubescens, Camassia quamash, Carex athrostachya, C. microptera, C. utriculata, Castilleja campestris, C. tenuis, C. victoriae, Cerastium glomeratum, C. viscosum, Cercocarpus, Claytonia cordifolia, Collinsia, Cornus sericea, Corrigiola litoralis, Crassula aquatica, C. connata, Cynosurus echinatus, Dactylis glomerata, Danthonia unispicata, Delphinium depauperatum, D. nuttallianum, Deschampsia cespitosa, D. danthonioides, D. elongata, Descurainia incana, Dianthus armeria, Downingia bacigalupii, D. elegans, D. laeta, Elatine californica, E. chilensis, E. rubella, Eleocharis acicularis, E. bella, E. palustris, Elymus virginicus, Epilobium ciliatum, Equisetum arvense, Erigeron, Eriogonum douglasii, E. heracleoides, Erodium cicutarium, Eryngium, Euphorbia serpyllifolia, Euphrasia nemorosa, Festuca idahoensis, F. rubra, Geranium, Glyceria grandis, Gnaphalium palustre, Gratiola ebracteata, G. heterosepala, G. neglecta, Grindelia, Holodiscus discolor,* *Hordeum brachyantherum, H. jubatum, Hosackia pinnata, Hypochaeris radicata, Ipomopsis, Isoetes howellii, Iva axillaris, Juncus bufonius, J. hemiendytus, J. tenuis, J. torreyi, J. uncialis, Lathyrus japonicus, Limosella aquatica, Lindernia dubia, Lolium, Lotus corniculatus, Lupinus sericeus, Marsilea vestita, Matricaria discoidea, Mentha pulegium, Mertensia, Microsteris gracilis, Mimulus guttatus, M. moschatus, Minuartia pusilla, Montia chamissoi, M. fontana, M. howellii, M. linearis, M. parvifolia, Myosurus minimus, Navarretia intertexta, N. squarrosa, Oenothera canescens, Pascopyrum smithii, Persicaria lapathifolia, Phalaris arundinacea, Phleum pratense, Phyla cuneifolia, Pinus contorta, P. ponderosa, Plagiobothrys figuratus, P. hispidulus, Plantago bigelovii, P. elongata, P. lanceolata, Plectritis congestus, Poa annua, P. bulbosa, P. confinis, P. cusickii, P. secunda, Polygonum aviculare, P. polygaloides, Populus tremuloides, Porterella carnosula, Potamogeton, Potentilla pacifica, Primula pauciflora, Prunella vulgaris, Prunus emarginata, P. serotina, Pseudoroegneria spicata, Pseudotsuga menziesii, Psilocarphus brevissimus, P. oregonus, Quercus garryana, Ranunculus flammula, R. orthorhynchus, R. repens, Rorippa, Rosa nutkana, Rumex obtusifolius, Sagina, Salix exigua, S. scouleriana, Salvia, Sarcobatus vermiculatus. Senecio, Spergularia rubra, Stellaria nitens, Suckleya suckleyana, Taraxacum officinale, Thlaspi, Trifolium dubium, T. repens, T. striatum, Triphysaria pusilla, T. versicolor, Triteleia hyacinthina, Vaccinium, Veratrum viride, Veronica americana, V. arvensis, V. biloba, V. peregrina, Viola praemorsa, Vulpia bromoides, V. myuros.*

***Plagiobothrys stipitatus* (Greene) I.M. Johnst.** is an annual species of depressions, ditches, flats, floodplains, lake bottoms, marshes, meadows, mudflats, oxbows, seeps, sloughs, swales, vernal pools, washes, and the margins of lakes, ponds, and streams at elevations of up to 3080 m. The plants occur in temporary standing water or in open, desiccating sites. The substrates are described as alkaline and include adobe, clay, clay loam, granite, gravelly clay, gravelly loam, humus, loam, loamy clay, muck, rock, sand, sandy loam, serpentine, and silty clay. There have been no life history studies of this species, which can become dominant in vernal pools. Little else has been reported on its ecology other than its associates. **Reported associates:** *Achyrachaena mollis, Acmispon wrangelianus, Aira caryophyllea, Allium amplectens, Alopecurus aequalis, A. geniculatus, A. saccatus, Amsinckia menziesii, Arabis, Aristida oliganthum, Atriplex joaquiniana, Avena barbata, Beckmannia syzigachne, Blennosperma nanum, Brassica geniculata, Briza minor, Brodiaea minor, Bromus diandrus, B. hordeaceus, B. mollis, B. rubens, B. tectorum, Callitriche marginata, Camassia quamash, Castilleja attenuata, C. campestris, Cerastium glomeratum, Cicendia quadrangularis, Clarkia purpurea, Crassula aquatica, C. connata, C. tillaea, Cressa truxillensis, Croton setigerus, Crypsis schoenoides, Cryptantha, Cyperus squarrosus, Deschampsia danthonioides, Dichelostemma multiflorum, Distichlis spicata, Downingia bacigalupii, D. bella, D. bicornuta, D. cuspidata, D. elegans, D. insignis, D. ornatissima, D. pulchella, Elatine californica, Eleocharis* acicularis, E. palustris, Epilobium pygmaeum, E. torreyi, Erodium botrys, Eryngium aristulatum, E. castrense, E. vaseyi, Eschscholzia, Festuca, Filago californica, Frankenia salina, Gastridium phleoides, Glyceria declinata, Gratiola ebracteata, Hemizonia congesta, Hesperevax caulescens, Hordeum depressum, H. glaucum, H. hystrix, H. leporinum, H. marinum, Hypochoeris glabra, Isoetes howellii, I. orcuttii, Juncus bufonius, J. capitatus, J. nevadensis, J. uncialis, Lasthenia californica, L. conjugens, L. ferrisiae, L. fremontii, L. glaberrima, L. glabrata, L. minor, Layia fremontii, Lepidium nitidum, Leymus cinereus, Lilaea scilloides, Limnanthes alba, Limosella aquatica, Lindernia dubia, Lolium multiflorum, Lythrum hyssopifolia, Malvella leprosa, Marsilea vestita, Medicago polymorpha, Micropus californicus, Microseris douglasii, Mimulus guttatus, M. nudatus, M. tricolor, Myosurus minimus, Navarretia leucocephala, N. myersii, N. pubescens, N. tagetina, Neostapfia colusana, Oenothera, Orcuttia inaequalis, O. tenuis, Phacelia, Pilularia americana, Plagiobothrys austiniae, P. bracteatus, P. chorisianus, P. fulvus, P. glyptocarpus, P. greenei, P. humistratus, P. undulatus, Pleuropogon californicus, Poa secunda, Pogogyne douglasii, P. ziziphoroides, Polypogon maritimus, P. monspeliensis, Primula clevelandii, Psilocarphus brevissimus, P. tenellus, Quercus lobata, Ranunculus bonariensis, Rotala ramosior, Sanicula bipinnatifida, Sida, Spergularia marina, Taeniatherum caput-medusae, Toxicodendron diversilobum, Trichostema lanceolatum, Trifolium depauperatum, T. longipes, T. microcephalum, T. willdenovii, Triphysaria eriantha, Triteleia laxa, Veronica arvensis, V. peregrina, Vulpia.

***Plagiobothrys strictus* (Greene) I.M. Johnst.** is an annual known from depressions, flats, meadows, marshes, meadows, playas, vernal pools, swales and the margins of streams at elevations of up to 182 m. The plants occur almost exclusively in alkaline/saline sites that are fed by sulfur hot springs. The substrates are a gravelly loam and clay admixture, which contains high concentrations of arsenic, boron, and sulfates. The plants flower in March–April. Other life history information is lacking. This biology and ecology of this severely imperiled species (U. S. Federally endangered) should be studied in detail as it currently is on the verge of extinction. An estimated 70% of its original populations have been extirpated and only two populations survive with less than 6000 plants remaining within a total combined area of 80 m². **Reported associates:** *Agrostis microphylla, Poa napensis, Puccinellia simplex.*

***Plagiobothrys tener* (Greene) I.M. Johnst.** is an annual occurring in depressions, ditches, meadows, seeps, vernal pools, and along the margins of streams at elevations of up to 1340 m. The reported substrates include alluvium, sand, and serpentine. This is yet another species in the genus for which ecological life history information is nearly nonexistent. **Reported associates:** *Juncus.*

***Plagiobothrys undulatus* (Piper) I.M. Johnst.** is an annual, which inhabits depressions, flats, gullies, meadows, mudflats, ponds, slopes, swales, vernal pools, and the receding margins of lakes and reservoirs at elevations of up to 400 m. These plants are C₃, nonbicarbonate users that grow as submersed rosulates when the water levels are high and

as emergents when the standing waters recede. Carbon fixation rates are much higher when the plants are emergent. The aqueous pH levels can range from 6.8 to 10.2 during the course of a day. The substrates are characterized as adobe, Bonsall sandy loam, clay, mud, sand, sandy peat, and silt. They include Laguna, Modesto, Riverbank, Turlock Lake, and Tuscan geomorphic surfaces and Alamo, Anita, Clear Lake, Cometa, Corning, Fiddyment, Redding, San Joaquin, San Ysidro, Toomes, and Tuscan soil series. Flowering occurs from March to June, with reproductive levels increasing as the water levels drop. The seeds germinate while under water and develop into submersed rosettes. In addition to natural occurrences, the plants appear commonly in restored and newly created pools, regardless of whether they have been seeded there or not. **Reported associates:** *Aira caryophyllea, Alisma, Alopecurus saccatus, Ambrosia psilostachya, Amsinckia menziesii, Anagallis minima, Artemisia tridentata, Avena barbata, A. fatua, Baccharis pilularis, Bromus mollis, Callitriche marginata, Capsella bursa-pastoris, Cotula coronopifolia, Crassula aquatica, Cressa truxillensis, Cryptantha nevadensis, Distichlis spicata, Downingia bella, D. bicornuta, D. cuspidata, D. pulchella, Echinodorus berteroi, Elatine brachysperma, E. californica, Eleocharis acicularis, E. macrostachya, E. palustris, Epilobium cleistogamum, Erodium botrys, Eryngium aristulatum, E. castrense, E. spinosepalum, E. vaseyi, Gnaphalium palustre, Gratiola ebracteata, Grindelia hirsutula, Heliotropium curassavicum, Hordeum brachyantherum, H. marinum, Horkelia, Isoetes howellii, I. orcuttii, Juncus bufonius, J. occidentalis, J. phaeocephalus, Lasthenia californica, L. glaberrima, Legenere valdiviana, Lepidium nitidum, Leptosiphon bicolor, Lilaea scilloides, Limosella acaulis, Lolium multiflorum, Lomatium utriculatum, Lythrum hyssopifolia, Marsilea vestita, Medicago polymorpha, Mimulus guttatus, Montia fontana, Myosurus, Navarretia leucocephala, N. prostrata, Orcuttia inaequalis, Phalaris lemmonii, Pilularia americana, Plagiobothrys acanthocarpus, P. humistratus, P. leptocladus, P. stipitatus, P. trachycarpus, Plantago lanceolata, Polypogon monspeliensis, Psilocarphus brevissimus, Ranunculus aquatilis, R. bonariensis, Rumex acetosella, R. crispus, Spergula arvensis, Trifolium monanthum, Veronica peregrina, V. serpyllifolia, Vicia benghalensis, V. sativa.*

Use by wildlife: *Plagiobothrys figuratus* is a minor source of nectar for honeybees (Hymenoptera: Apidae: *Apis*) during the mid to late summer. The foliage of *P. hirtus* is eaten by aphids (insecta: Hemiptera: Aphididae), caterpillars (Insecta: Lepidoptera), deer (Cervidae: Odocoileinae), and small rodents (Rodentia). The plants are grazed by domestic sheep (Bovidae: *Ovis aries*) and by cattle (Bovidae: *Bos taurus*). The floral nectar is taken by moths (Insecta: Lepidoptera: Ctenuchidae: *Ctenucha*). A weevil (Insecta: Coleoptera: Curculionidae: *Mogulones cruciger*), under consideration for biological control of houndstongue (*Cynoglossum officinale*), did not attack *P. hirtus* plants in field tests and exhibited low attack rates (and poor survival) in "no choice" experiments; it did not attack plants of *P. stipitatus* in any of the experiments. *Plagiobothrys stipitatus* is a host for nematodes

(Nemata: Pratylenchidae: *Pratylenchus morettoi*). Several fungi are hosted by *Plagiobothrys hirtus* (Chytridiomycota: Synchytriaceae: *Synchytrium myosotidis*) and *P. scouleri* (Ascomycota: Erysiphaceae: *Erysiphe cichoracearum*; Oomycota: Peronosporaceae: *Peronospora myosotidis*). *Plagiobothrys species* have been used as food for captive redfoot tortoises (Reptilia: Testudinidae: *Geochelone carbonaria*) and Russian tortoises (Reptilia: Testudinidae: *Agrionemys horsfieldii*).

Economic importance: food: *Plagiobothrys* species generally are regarded as inedible, but the flowers and shoots of some species have been used for food; **medicinal:** Several *Plagiobothrys* species contain various amounts of the naphthoquinone alkannin, which has astringent properties. The plants reportedly can contain toxic nitrate–nitrite levels; **cultivation:** *Plagiobothrys figuratus*, *P. hirtus* and *P. scouleri* are cultivated as ornamentals. The seed of *P. figuratus* is produced commercially and sold in bulk quantities; **misc. products:** *Plagiobothrys figuratus* often is seeded in wetland restoration projects. At least some species contain alkannin, a pH indicating naphthoquinone, from which an intense red dye can be made; **weeds:** *Plagiobothrys scouleri* is invasive in Japan and northeast Scotland and occasionally is reported as weedy in the western United States. It forms large stands on disturbed ground in New Brunswick; **nonindigenous species:** *Plagiobothrys figuratus* was introduced to Alaska before 1941 but is now believed to have been extirpated there. It has been introduced to Arkansas, Illinois, Michigan, and North Carolina. Commercial seed production of this species may account for these introductions because propagules of the species are known to contaminate commercial seed mixes. *Plagiobothrys scouleri* was introduced to Massachusetts before 1901 as a contaminant of wool waste that was used for fertilizer, but it has not persisted there. It is nonindigenous in Alaska, Maine, New Brunswick, Ontario, Pennsylvania, and Wisconsin. *Plagiobothrys scouleri* was introduced to Ireland, Scotland, and to other parts of Britain before 1974, probably as a contaminant of grass seed that was used in habitat restoration projects. It was introduced to Sweden (before 1994) as well as to Norway. Both *P. scouleri* and *P. stipitatus* have been introduced to Japan.

Systematics: Systematic relationships of *Plagiobothrys* remain poorly understood and the genus has been included in few molecular phylogenetic surveys at the time of this writing. A study incorporating two species (*P. arizonicus*, *P. humilis*) resolved *Amsinckia*, *Cryptantha* and *Plagiobothrys* as a well supported but polytomous clade. An analysis of five species (none OBL) resolved the genus as monophyletic and corroborated that close relationship among *Amsinckia*, *Cryptantha* and *Plagiobothrys*, but without more concise resolution (Figure 5.100). However, a preliminary phylogenetic analysis of *Plagiobothrys* based on morphological characters indicated that the genus might not be monophyletic as currently circumscribed. The same analyses indicated that the *Allocarya* group (once segregated as a distinct genus) includes the *Plagiobothrys* species with the latter name having nomenclatural priority. Most contemporary authors have merged the

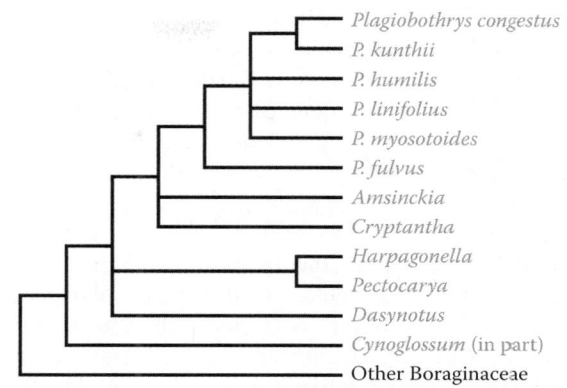

FIGURE 5.100 Phylogenetic relationships of *Plagiobothrys* as reconstructed by analysis of combined *rps16* and *trnL–trnF* sequence data. This analysis indicates either *Amsinckia* or *Cryptantha* as the sister group of *Plagiobothrys*, which resolves as a clade; however, no OBL species were surveyed in that study. (Adapted from Weigend, M. et al., *Mol. Phylogen. Evol.*, 68, 604–618, 2013.)

two genera. A more comprehensive molecular systematic study of this genus is needed to better determine the generic limits and its precise relationship to other genera of Boraginaceae. The base chromosome number of *Plagiobothrys* probably is $x = 6$, but counts have been reported for only a few of the species. *Plagiobothrys scouleri* ($2n = 54$) is a higher polyploid. *Plagiobothrys glaber* has been considered to be a variant of *P. stipitatus*; however, its likely extinction now precludes any genetic analyses that could test this possibility. *Plagiobothrys figuratus* has been recognized as a variety of *P. hirtus* and there seems to be considerable confusion in the literature regarding the distinction of these taxa. Several taxa in the *P. scouleri* "complex" recently have been reevaluated. There is still some question whether to retain *P. bracteatus* as a distinct species or to treat it as a variety of *P. scouleri*. *Plagiobothrys undulatus* is regarded by some authors as indistinct from *P. chorisianus*, under which it has been treated as a variety (*P. chorisianus* var. *undulatus*). A number of species are suspected of hybridization (e.g., *P. acanthocarpus* and *P. greenei*; *P. chorisianus* and *P. undulatus*; *P. distantiflorus* and *P. glyptocarpus*; *P. humistratus* and *P. scriptus*); however, these reports mostly are anecdotal and each case needs to be confirmed by careful genetic analyses.

Comments: *Plagiobothrys scouleri* is widespread except in the southeastern United States and extreme northern North America; *P. figuratus* occurs in northwestern North America and is sporadically introduced in the eastern United States; the remaining species occur in the western United States (*P. acanthocarpus*, *P. austiniae*, *P. bracteatus*, *P. chorisianus*, *P. distantiflorus*, *P. glaber*, *P. hirtus*, *P. humistratus*, *P. leptocladus* [also Saskatchewan], *P. parishii*, *P. stipitatus*, *P. strictus*, *P. tener*, *P. undulatus*). *Plagiobothrys glaber* is thought to be extinct.

References: Amsberry, 2001; Barbour et al., 2007; Carlson & Shephard, 2007; Chambers, 2007; Clark et al., 1997a, 2000; Clausnitzer & Huddleston, 2002; Dellavalle-Sanvictores, 2000; Ferren Jr. et al., 1998; Gimingham et al., 2002; Goltz,

2001; Ingham & Wilson, 1999; Ito & Esugi, 2004; Keeley, 1999; Keeley & Sandquist, 1991; Luc et al., 1986; MacKay, 2003; Olofson, 2000; Schlising & Sanders, 1982; Sorrie, 2005; Stenberg, 1995; Taylor, 2004; USFWS, 2002, 2003, 2006b; Weigend et al., 2010, 2013; Wilson et al., 2004; Winkworth et al., 2002; Zika, 1996.

Family 2: Convolvulaceae [60]

Phylogenetic analyses incorporating multiple cpDNA sequence data sets have determined that Convolvulaceae are monophyletic if the family is circumscribed to include several genera (*Cuscuta*, *Dichondra*) that have been segregated formerly as separate families (Stefanović et al., 2002). The monotypic genus *Humbertia* (*H. madagascariensis*) is the sister to the remainder of the family, a phylogenetic position that is equally consistent with treatments that recognize it as a distinct family (Humbertiaceae) or as a subfamily (Humbertioideae) of Convolvulaceae (Figure 5.101). The family Solanaceae represents the sister group of Convolvulaceae if *Humbertia* is recognized as a subfamily. Analysis of DNA sequence data has indicated that half of the traditionally recognized tribes (*Convolvuleae*, *Cresseae*, *Erycibeae*, *Merremieae*, *Poraneae*) are polyphyletic (Figure 5.101) and need to be redefined (Stefanović et al., 2002). Stefanović et al. (2003) have proposed a revised classification, which includes the merger of tribe Argyreieae with Ipomoeeae, and tribe Hildebrandtieae with Cressaeae, establishes several new tribes (Aniseieae, Cardiochlamyeae, Jacquemontieae) and resurrects a former tribe (Maripeae), to accommodate the phylogenetic results.

In the resulting circumscription, Convolvulaceae include close to 1700 species that consist mainly of twining herbs with laticifers and milky sap. Although the family is widespread, most species are tropical or subtropical. Some species are achlorotic parasites. The flowers are showy, radially symmetrical, with plicate, funnelform corollas, which often are convolute or twisting clockwise in bud (Judd et al., 2002). The ovary is superior, 2- or 4-lobed and consists of two carpels with axile placentation. The style is terminal or gynobasic and there are usually 2 ovules in each locule (Judd et al., 2002). The fruit develops into a papery capsule.

The flowers usually open only for a few hours and are pollinated by insects or hummingbirds (Aves: Trochilidae). The large seeds normally are dispersed by the wind, either individually or within the bladder-like capsule (Judd et al., 2002). The family includes the edible sweet potato (*Ipomoea batatas*) and several ornamental vines such as morning glory (*Ipomoea*), Christmas vine (*Porana*), and ponyfoot (*Dichondra*). Some species (e.g., *Convolvulus arvensis*) are serious agricultural weeds. Some *Ipomoea* species contain ergot alkaloids (e.g., ergine, ergonovine), which are hallucinogenic when they have been ingested.

Convolvulaceae predominantly are terrestrial and only two North American genera contain OBL species within the family (A third genus, *Convolvulus*, was included formerly among the OBL taxa in the 1996 indicator list):

1. ***Convolvulus*** L. (formerly OBL)
2. ***Cressa*** L.
3. ***Ipomoea*** L.

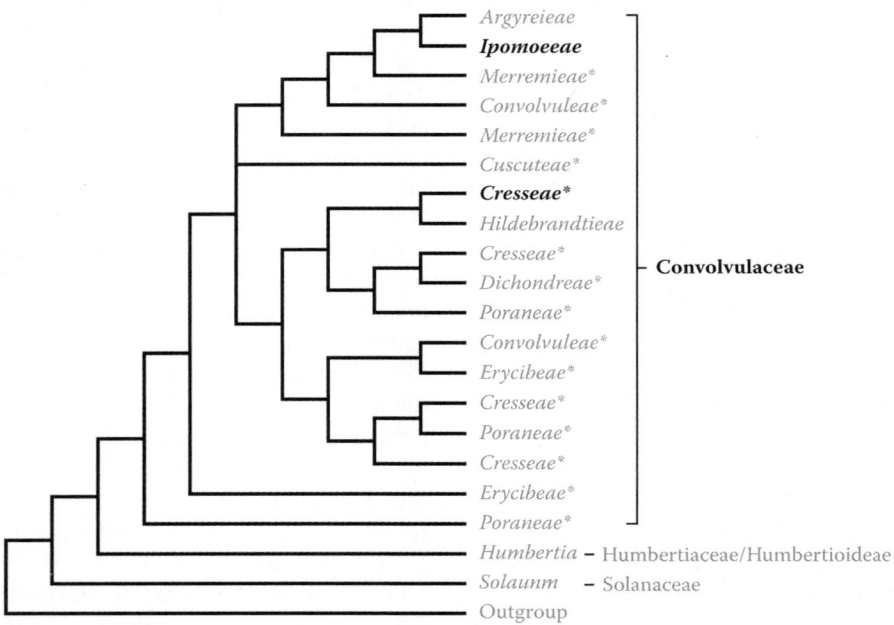

FIGURE 5.101 Tribal relationships in Convolvulaceae as indicated by analyses of combined cpDNA sequence data. The two genera with OBL North American species (*Cressa*, *Ipomoea*) occur in the respective tribes or portions of polyphyletic tribes (asterisked) that are indicated by bold type. The phylogenetic distribution of these taxa indicates that the OBL habit evolved independently within the family. *Humbertia* is recognized either as a distinct family or as a subfamily within Convolvulaceae. (Adapted from Stefanović, S. et al., *Amer. J. Bot.*, 89, 1510–1522, 2002.)

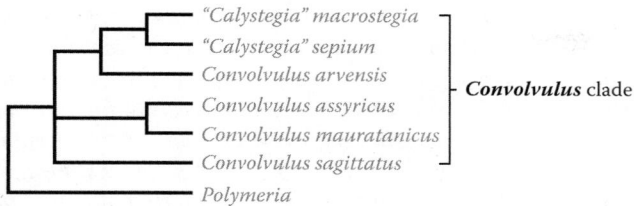

FIGURE 5.102 Phylogenetic relationships of *Convolvulus* as depicted from analysis of combined cpDNA sequence data. Although many authors continue to distinguish *Calystegia* as a distinct genus, doing so renders *Convolvulus* as paraphyletic, a result also found in a more comprehensive, recent study of the group (Williams et al., 2014). *Calystegia sepium* was classified as an OBL species in the 1996 indicator list, but was reclassified as FAC in the 2013 list. (Adapted from Stefanović, S. et al., *Amer. J. Bot.*, 89, 1510–1522, 2002.)

The phylogenetic distribution of these genera (Figure 5.101) indicates that the OBL habit originated independently in each. Although many authors continue to distinguish *Calystegia* from *Convolvulus*, these genera are regarded here as synonymous given that phylogenetic analyses (e.g., Figure 5.102) fail to distinguish them as separate clades (Stefanović et al., 2002, 2003).

1. *Convolvulus*

Hedge-bindweed; liseron des haies
Etymology: from the Latin *convolvo* meaning "to entwine"
Synonyms: *Bucharea*; *Calystegia*; *Idalia*; *Nemostima*; *Pantocsekia*; *Periphas*; *Podaletra*; *Rhodorhiza*; *Rhodoxylon*; *Strophocaulos*; *Symethus*; *Volvulus*
Distribution: global: cosmopolitan; **North America:** widespread
Diversity: global: 230 species; **North America:** 22 species
Indicators (USA): FAC: *Convolvulus sepium*
Habitat: brackish (coastal), freshwater; palustrine; **pH:** 5.3–7.8; **depth:** <1 m; **life-form(s):** emergent herb (vine)
Key morphology: stems (to 3 m) twining clockwise; leaves (to 10 cm) alternate, deltoid to lanceolate, the bases deeply cordate to sagittate, long petioled (to 8 cm); flowers solitary, axillary, closely subtended by a pair of sepaloid bracts (to 3.5 cm) that enclose the calyx, peduncles (to 15 cm) 4-angled; corolla showy (to 8 cm), regular, funnelform, white, sometimes with alternating pink or rose sectors; ovary 1-locular, stigma lobes flat; capsules (to 1 cm) spherical, with 2–4 large (to 5 mm), angular seeds
Life history: duration: perennial (rhizomes); **asexual reproduction:** rhizomes; **pollination:** insect; **sexual condition:** hermaphroditic; **fruit:** capsules (common); **local dispersal:** seeds (gravity, water); **long-distance dispersal:** seeds (water)
Imperilment: (1) *Convolvulus sepium* [G5]; SH (CA, NJ); S2 (WY); S3 (AB, NC, NJ, NY, WV)
Ecology: general: *Convolvulus* (including *Calystegia*) mainly is terrestrial with only two North American species designated as wetland indicators (FAC, FACU). In the 1996 indicator list, one species (*C. sepium*) was ranked as OBL in a

portion of its range. The revised (2013) list now categorizes *C. sepium* as FAC throughout its range, leaving no OBL species in the genus. However, because a rather extensive treatment for *C. sepium* had been completed prior to the release of the 2013 list, the account for the genus and this species has been retained here. Most of the *Convolvulus* species examined are self-incompatible (by a sporophytic system) and are outcrossed by insects (Insecta). Pollen removal by insect visitors often is extremely high but stigma deposition can be very low (e.g., 1%), which can result in low seed set. Because most species are perennials, clonal growth can limit the genetic variability of pollen, which is necessary for efficient reproduction in the self-incompatible species. All members of the genus are twining, herbaceous vines that grow upon other vegetation for support. The vines coil only in a clockwise (right-handed) direction. They are common species of open, disturbed sites.

Convolvulus sepium **L.** inhabits bayous, beaches, brooks, carrs, depressions, deltas, dikes, ditches, dunes, fens, flats, floodplains, lagoons, levees, marshes, meadows, prairies, river bottoms, sand bars, sloughs, springs, swales, swamps, and the margins of lakes, rivers, and streams at elevations of up to 500 m. This is a widely adapted species that grows in tidal (intertidal zone) or nontidal conditions, either on wet substrates or occasionally in shallow, standing water, which can range from fresh to oligohaline or mesohaline (to 3–10 ppt salinity). The plants can tolerate exposures ranging from full sun to shade. They grow on diverse substrates (pH: 5.3–7.8), which include clay, clay loam, gravel, limestone, loam, mud, peat, peaty sand, sand, sandy clayey loam, sandy gravel, sandy loam, sandy muck, and shells. The large flowers open as early as 3–4 A.M. to offer pollen and nectar to potential insect pollinators, which mainly include bees (e.g., Hymenoptera: Apidae: *Apis mellifera*; Bombus) and flies (e.g., Diptera: Syrphidae). They will remain closed under adverse weather conditions. The UV-absorbing properties of the white corolla provide a much sharper contrast to the green foliage than a UV-reflecting flower, which helps to attract the insect pollinators. Unlike its mainly self-incompatible congeners, this species has populations that are self-incompatible (sporophytic) or self-compatible. Because anthers and stigmas are separated spatially (herkogamy), spontaneous self-pollination (autogamy) does not occur readily (<5% seed set); however, pollinator-mediated selfing (including geitonogamy) is common in self-compatible populations. In a study of self-compatible plants, a mixed mating system was indicated by a somewhat lower pollen:ovule ratio (2264) than its outcrossing congeners (p:o ratio: 2704–3759). The ability of this species to self-pollinate results in higher median levels of fruit and seed set (67%, 46.3%, respectively) than its self-incompatible congeners, as the plants often are pollinator limited. The self-compatible plants are adapted for inbreeding and do not exhibit detectable differences (i.e., inbreeding depression) between self-pollinated and outcrossed flowers in terms of their fruit or seed production, or seed germination rates. Clonal growth in self-incompatible populations results in a high degree of sterility due to the lack of genetically distinct pollen that is necessary to achieve successful reproduction. The seeds are buoyant and will remain afloat for as long as 4 months; they conceivably are dispersed readily by water. The seeds reportedly require no pretreatment and will germinate (up to 59%) within 2 weeks of their planting when incubated at 15°C. Seed germination occurs in drained or saturated substrates, but not under inundated conditions. Some accounts indicate that light is required for germination. This factor could explain why the seeds sometimes are described as fragile or short-lived, but also reportedly develop into a persistent, long term seed bank where they can achieve densities as high as 280/m^2 and remain viable for up to 40 years. The seed bank formation is inconsistent as a number of seed bank studies have found this species to be relatively uncommon. The roots are fairly shallow but extensive, reaching up to 3 m in length. Tropane alkaloid derivatives released by the roots selectively promote the growth of certain soil bacteria. Studies in Europe have found the highest biomass to be produced under shoot competition, which occurs when several plants twine upon a single host stem. Low nutrient conditions induce the development of runners, which might facilitate the detection of better resource sites. Although this species can appear frequently in terrestrial habitats (e.g., cultivated land or disturbed sites), most collection records throughout its range are from wetlands. **Reported associates:** *Acalypha gracilens, A. rhomboidea, Acer rubrum, A. saccharinum, Acorus calamus, Aeschynomene virginica, Agalinis purpurea, A. tenuifolia, Agropyron repens, Agrostis gigantea, A. perennans, A. scabra, A. stolonifera, Alnus rubra, A. tenuifolia, Amaranthus cannabinus, A. tuberculatus, Ambrosia artemisiifolia, Amorpha fruticosa, Ampelamus albidus, Andropogon gerardii, A. virginicus, Anemonella thalictroides, Apios americana, Apocynum cannabinum, Arctium minus, Artemisia serrata, Arundo donax, Asclepias incarnata, Atriplex prostrata, Baccharis glutinosa, B. halimifolia, Barbarea vulgaris, Bidens cernuus, B. comosus, B. discoideus, B. frondosus, B. polylepis, Bignonia capreolata, Boehmeria cylindrica, Bolboschoenus novae-angliae, Bromus, Calamagrostis canadensis, C. stricta, Campsis radicans, Carex abscondita, C. albolutescens, C. bebbii, C. caroliniana, C. complanata, C. hormathodes, C. hyalinolepis, C. longii, C. lupulina, C. nebrascensis, C. obnupta, C. paleacea, C. pellita, C. scoparia, C. squarrosa, C. tribuloides, C. typhina, C. uberior, Castilleja coccinea, Cephalanthus occidentalis, Cerastium fontanum, Chamaecrista fasciculata, C. nictitans, Chasmanthium latifolium, C. laxum, Chenopodium album, C. rubrum, C. salinum, C. simplex, Cichorium intybus, Cicuta maculata, Cirsium arvense, Clematis ligusticifolia, C. terniflora, Commelina virginica, Conyza canadensis, Cornus sericea, Coronilla varia, Corylus cornuta, Cunila origanoides, Cuphea viscosissima, Cuscuta gronovii, Cyperus erythrorhizos, C. esculentus, C. odoratus, C. strigosus, Dactylis glomerata, Danthonia spicata, Datura stramonium, Decodon verticillatus, Descurainia sophia, Desmodium ciliare, D. paniculatum, Dichanthelium dichotomum, Digitaria ischaemum, Diodia teres, D. virginiana, Dioscorea quaternata, D. villosa, Distichlis spicata, Echinochloa crusgalli, E. walteri, Echinocystis lobata, Eclipta prostrata, Eleocharis acicularis, E. caribaea, E. erythropoda, E. ovata,* *Elymus canadensis, E. virginicus, Epilobium, Equisetum arvense, E. fluviatile, E. hyemale, Eragrostis, Erechtites hieracifolia, Erigeron strigosus, Eryngium prostratum, Erysimum cheiranthoides, Eucalyptus camaldulensis, Eupatorium coelestinum, E. perfoliatum, E. rugosum, E. serotinum, Euphorbia maculata, E. nutans, Fallopia scandens, Festuca elatior, F. rubra, Fimbristylis, Fraxinus nigra, Galium obtusum, G. tinctorium, G. triflorum, Geranium bicknellii, Geum canadense, Gleditsia triacanthos, Glyceria striata, Glycyrrhiza lepidota, Gnaphalium obtusifolium, Gypsophila paniculata, Hedeoma pulegioides, Hedyotis caerulea, H. purpurea, Helenium autumnale, H. flexuosum, Helianthus annuus, H. microcephalus, H. occidentalis, H. tuberosus, Heliotropium curassavicum, Hesperis matronalis, Hibiscus moscheutos, Hierochloe odorata, Hordeum jubatum, Humulus lupulus, Hydrocotyle umbellata, H. verticillata, Hypericum hypericoides, H. mutilum, H. perforatum, H. punctatum, H. stans, Ilex opaca, Impatiens capensis, I. ecalcarata, Ipomoea coccinea, I. hederacea, I. lacunosa, I. pandurata, Iris pseudacorus, I. versicolor, Iva frutescens, Juncus alpinus, J. balticus, J. canadensis, J. dichotomus, J. gerardii, J. nodosus, J. scirpoides, J. tenuis, Juniperus virginiana, Kochia scoparia, Kosteletzkya pentacarpos, Krigia biflora, Lathyrus palustris, Leersia oryzoides, L. virginica, Lespedeza cuneata, L. procumbens, L. striata, Liatris aspera, L. pycnostachya, Lindernia dubia, Linum striatum, Lobelia cardinalis, L. inflata, L. puberula, Lonicera japonica, Ludwigia alternifolia, L. decurrens, L. glandulosa, L. leptocarpa, L. palustris, Lycopus americanus, L. asper, L. rubellus, L. virginicus, Lysimachia lanceolata, L. nummularia, L. quadrifolia, L. terrestris, Lythrum salicaria, Maianthemum stellatum, Menispermum canadense, Mentha arvensis, M. ×gracilis, Microstegium vimineum, Mikania scandens, Mimulus alatus, Mitchella repens, Mollugo verticillata, Monarda fistulosa, Myosotis scorpioides, Myrica cerifera, M. pensylvanica, Napaea dioica, Oenothera biennis, Onoclea sensibilis, Osmunda claytoniana, O. regalis, Oxalis stricta, Panicum anceps, P. clandestinum, P. latifolium, P. polyanthes, P. rigidulum, P. sphaerocarpon, P. virgatum, Paronychia fastigiata, Parthenocissus quinquefolia, Paspalum laeve, Passiflora lutea, Pastinaca sativa, Penthorum sedoides, Persicaria arifolia, P. hydropiperoides, P. lapathifolia, P. pensylvanica, P. posumbu, P. punctata, P. sagittata, Phalaris arundinacea, Phragmites australis, Phyla lanceolata, Phyllanthus caroliniensis, Physalis angustifolia, P. pubescens, Physostegia virginiana, Pilea pumila, Planera aquatica, Plantago rugelii, Platanthera peramoena, P. psycodes, Pluchea odorata, Poa compressa, P. pratensis, Populus angustifolia, P. deltoides, P. fremontii, Potentilla anserina, P. norvegica, P. simplex, Polypremum procumbens, Prenanthes racemosa, Prunus americana, P. virginiana, Ptilimnium capillaceum, Ranunculus sceleratus, Ratibida pinnata, Rhexia mariana, R. virginica, Rhus glabra, R. typhina, Rhynchospora macrostachya, Rorippa islandica, R. palustris, Rosa arkansana, R. palustris, R. woodsii, Rotala ramosior, Rubus armeniacus, Rudbeckia hirta, R. laciniata, Rumex verticillatus, Sabatia angularis, Sacciolepis striata, Sagittaria australis, Salix* *amygdaloides, S. exigua, S. goodingii, S. interior, S. laevigata, S. ligulifolia, S. lucida, S. nigra, S. rigida, S. serissima, Sambucus nigra, Samolus floribundus, S. valerandi, Sanguisorba canadensis, Saururus cernuus, Schizachyrium scoparium, Schoenoplectus americanus, S. californicus, S. pungens, S. robustus, S. tabernaemontani, Scirpus cyperinus, Scutellaria galericulata, S. lateriflora, S. nervosa, Setaria faberi, S. glauca, S. viridis, Sida spinosa, Silene vulgaris, Silphium perfoliatum, Sinapis arvensis, Sisymbrium altissimum, Sisyrinchium angustifolium, Smilax glauca, S. hispida, S. rotundifolia, Solanum carolinense, S. dulcamara, S. nigrum, Solidago caesia, S. canadensis, S. gigantea, S. graminifolia, S. sempervirens, Sorghastrum, Sorghum halepense, Sparganium eurycarpum, Spartina cynosuroides, S. patens, S. pectinata, Spergularia salina, Spiraea alba, Spiranthes vernalis, Sporobolus, Stachys palustris, S. tenuifolia, Stellaria longifolia, Stipa spartea, Streptopus amplexifolius, Strophostyles helvola, Symphoricarpos occidentalis, Symphyotrichum cordifolium, S. dumosum, S. ericoides, S. lanceolatum, S. novae-angliae, S. novi-belgii, S. subspicatum, Taraxacum officinale, Teucrium canadense, Thalictrum dasycarpum, Thelypteris noveboracensis, T. palustris, Thinopyrum pycnanthum, Thlaspi arvense, Toxicodendron radicans, Triadenum tubulosum, T. virginicum, T. walteri, Trifolium hybridum, Triglochin maritimum, Triodanis perfoliata, Tripsacum dactyloides, Typha angustifolia, T. latifolia, Urtica dioica, Vaccinium formosum, Verbascum thapsus, Verbena hastata, Vernonia gigantea, V. noveboracensis, Veronica peregrina, Viburnum rafinesqueanum, Viola lanceolata, V. sagittata, Vitis girdiana, V. riparia, Woodwardia areolata, Xanthium strumarium, Zanthoxylum americanum.*

Use by wildlife: The foliage of *Convolvulus sepium* is fed upon by a variety of insect (Insecta) herbivores and is host to a number of moths (Lepidoptera: Bedelliidae: *Bedellia somnulentella*; Crambidae: *Hahncappsia marculenta, Herpetogramma pertextalis*; Pterophoridae: *Emmelina monodactyla, Oidaematophorus monodactylus*; Pyralidae: *Eurrhypara hortulata*), leaf beetles (Coleoptera: Chrysomelidae: *Chelymorpha cassidea, Deloyala guttata, Epithrix nitens, Megacerus discoidus, Metriona bicolor, Plagiometriona clavata*), sap beetles (Coleoptera: Nitidulidae: *Conotelus obscurus*), and a lace bug (Heteroptera: Tingidae: *Corythucha marmorata*). It is host to a few Fungi (Ascomycota: Mycosphaerellaceae: *Septoria convolvuli*; Basidiomycota: Pucciniaceae: *Puccinia convolvuli*; Oomycota: Albuginaceae; *Albugo ipomoeae-panduratae*) including several strains of yeast (Ascomycota: Metschnikowiaceae: *Metschnikowia continentalis*), which have been isolated from the flowers. The plants are browsed occasionally by white-tailed deer (Mammalia: Cervidae: *Odocoileus virginianus*). The flowers provide nectar to hummingbirds (Aves: Trochilidae). In Italy the plants are often used as fodder for rabbits (Mammalia: Leporidae). Tropane alkaloids released by the roots provide carbon and nitrogen to specific soil microbes (e.g., Bacteria: Rhizobiaceae: *Sinorhizobium meliloti*) that are capable of processing the compounds. The rhizomes store large amounts

of a jacalin-related mannose-specific lectin, which may function as an antiherbivory compound.

Economic importance: food: The stems and roots of *Convolvulus sepium* are eaten in China, India, New Zealand and elsewhere; however, they should be avoided due to the presence of potentially toxic alkaloids (see next) and their strong purgative properties. Chronic ingestion reportedly leads to stomach distress; **medicinal:** Extracts of *Convolvulus sepium* (sometimes misapplied medicinally as a form of "scammony") have been used as a laxative, to treat fevers, and to remedy liver dysfunctions. The roots contain calystegines, which are tropane alkaloid derivatives. The B2 calystegines can be toxic due to their glycosidase inhibitory properties. The plants contain lectins, which induce cell division (mitogenic) of immunologically functional T-cells; **cultivation:** *Convolvulus sepium* (especially subsp. *spectabilis*) is grown as an ornamental vine; **misc. products:** The flexible stems of *Convolvulus sepium* can be used as makeshift cordage; **weeds:** *Convolvulus sepium* often is reported as an agricultural or garden weed and has been increasingly problematic since the 1970s. A moth (Lepidoptera: Noctuidae: *Tyta luctuosa*) has been released as a biocontrol agent, but it has not been particularly effective. Escaped ornamental plants have become noxious weeds in some parts of the Middle East; **nonindigenous species:** *Convolvulus sepium* is indigenous to North America with exception of subsp. *sepium*, which is regarded as nonindigenous and invasive.

Systematics: Phylogenetic analysis of combined cpDNA sequence data resolves *Convolvulus* as a sister clade to *Polymeria*, which together comprise tribe Convolvuleae (Figures 5.101 and 5.102). Imbedded within *Convolvulus* is the morphologically distinctive section *Calystegia*, which is distinguished readily by a suite of consistent ovary, pollen, and stigma characters; however, its recognition as a separate genus (as has been commonplace for more than a decade) would render *Convolvus* as paraphyletic (Figure 5.102). A more recent study, which surveyed 62% of the accepted *Convolvulus* species and 81% of *Calystegia* species corroborated this finding. Consequently, *Calystegia* is not distinguished from *Convolvulus* in this treatment, although it probably should be retained as some lower-level taxon (e.g., section). *Convolvulus sepium* superficially resembles *C. arvensis* in its habit and habitat, and the plants should be examined carefully to avoid misidentifications. The cosmopolitan *C. sepium* is complex morphologically and should be reevaluated taxonomically on a worldwide basis. Currently, there are seven infraspecific taxa (subspecies) recognized in North America alone, but these deserve more critical study to assess their taxonomic distinctness, geographical distributions, indigenous status, etc. Hybridization among species or subspecies, including indigenous and nonindigenous taxa, is suspected to occur fairly often, but this possibility has not been investigated adequately in North America and warrants a thorough investigation. The basic chromosome number of *Convolvulus* is difficult to determine as a consequence of widespread aneuploidy, but is probably $x = 10$ or 11. *Convolvulus sepium* ($2n = 22$) is diploid with aneuploid counts ($2n = 20, 24$) reported (possibly as a consequence of hybridization).

Comments: *Convolvulus sepium* is nearly cosmopolitan. It is widespread throughout North America except in the far north and extends into Africa, Australia, Eurasia, New Zealand and South America.

References: Baker, 1961; Baskin & Baskin, 2006; Boedeltje et al., 2003a,b; Boldt et al., 1998; Boulger, 1875; Chittka, 1999; Duke & Bogenschutz-Godwin, 2002; Goldmann et al., 1990; Guntli et al., 1999; Gurnell et al., 2007b; Hashimoto, 2002; Johnson, 1999; Klimeš & Klimešová, 1994; Lachance et al., 1998; Leck & Leck, 1998; Lewis & Oliver, 1965; Luken & Thieret, 2001; Majka & Cline, 2006; Ogden, 1978; Stefanović et al., 2002, 2003; Strong & Kelloff, 1994; Stutzenbaker, 1999; Tepfer et al., 1988; Tipping & Campobasso, 1997; Ushimaru & Kikuzawa, 1999; Van Damme et al., 2004; Wheeler, Jr., 1987; Williams et al., 2014; Wilson, 1908.

2. *Cressa*

Alkali weed

Etymology: from the Greek *Krêssa* meaning "a woman of Crete"

Synonyms: *Carpentia*

Distribution: global: Africa; Asia; Australia; Europe; North America; **North America:** southwestern

Diversity: global: 4 species; **North America:** 2 species

Indicators (USA): OBL: *Cressa nudicaulis*

Habitat: saline (coastal); palustrine; **pH:** alkaline; **depth:** <1 m; **life-form(s):** emergent herb

Key morphology: stems (to 20 cm); leaves reduced, scale-like (to 4 mm); flowers solitary axillary; corolla (to 1 cm) white, funnelform; stamens 5, exserted; capsules 1–4–seeded

Life history: duration: perennial (rhizomes); **asexual reproduction:** rhizomes; **pollination:** unknown; **sexual condition:** hermaphroditic; **fruit:** capsules unknown; **local dispersal:** unknown; **long-distance dispersal:** unknown

Imperilment: (1) *Cressa nudicaulis* [G3]; S2 (TX)

Ecology: general: All *Cressa* species are halophytes and occupy similar saline habitats such as coastal lagoons, salt-marshes or alkaline flats. The other North American species (*C. truxillensis* [=*C. depressa*]) is a FACW indicator. The life history and ecology of the North American species are poorly understood.

Cressa nudicaulis Griseb. grows along saline coastal beaches, and in saline flats, salt marshes, and salt pans. The substrates are described as brackish, sandy clay. This species needs intensive study with even basic ecological or life history information sorely lacking. **Reported associates:** *Batis maritima, Borrichia frutescens, Cyperus articulatus, Distichlis, Panicum dichotomiflorum, P. obtusum, Salicornia, Spartina spartinae, Suaeda conferta, Xanthium strumarium.*

Use by wildlife: *Cressa nudicaulis* is eaten by endemic viscacha rats (Rodentia: Octodontidae: *Octomys mimax*) in western Argentina.

Economic importance: food: unknown; **medicinal:** unknown; **cultivation:** *Cressa nudicaulis* is not cultivated, probably because of its saline requirements; **misc. products:**

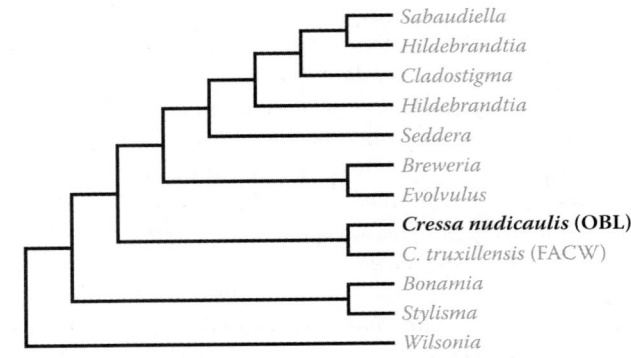

FIGURE 5.103 Phylogenetic relationships in Convolvulaceae tribe Cresseae (sensu lato) as indicated by analysis of combined cpDNA sequence data. The OBL North American species (in bold) forms a clade with the other (FACW) North American species; the two Old World *Cressa* species (both wetland taxa) were not included in the analysis. (Adapted from Stefanović, S. et al., *Amer. J. Bot.*, 89, 1510–1522, 2002; Stefanović, S. et al., *Syst. Bot.*, 28, 791–806, 2003.)

none; **weeds:** none; **nonindigenous species:** *Cressa nudicaulis* is indigenous.

Systematics: DNA sequence analysis indicates that *Cressa* is monophyletic (Figure 5.103); however, the two Old World species remain to be surveyed phylogenetically. The genus occupies a central position within tribe Cresseae, and all four species have an affinity for wetland habitats; thus, the OBL habit is considered as having a single derivation. The base chromosome number of the genus is $x = 7$; however, *Cressa nudicaulis* ($2n = 28$) is regarded as a diploid.

Comments: *Cressa nudicaulis* occurs in coastal, southern Texas, and extends into adjacent Mexico and into South America.

References: Austin, 1973, 1992, 2000; Sobrero et al., 2010.

3. *Ipomoea*

Water spinach

Etymology: from the Greek *ipos homoios* ("worm-like"), with respect to the twining habit

Synonyms: *Acmostemon; Batatas; Bonanox; Calonyction; Calycantherum; Diatremis; Dimerodisus; Exogonium; Mina; Parasitipomoea; Pharbitis; Quamoclit*

Distribution: global: cosmopolitan; **North America:** widespread except in north

Diversity: global: 600 species; **North America:** 49 species

Indicators (USA): OBL: *Ipomoea aquatica; I. asarifolia*

Habitat: freshwater; lacustrine, palustrine, riverine; **pH:** 6.8–8.2; **depth:** 0–2 m; **life-form(s):** emergent herb (vine), floating-leaved herb (vine)

Key morphology: stems (to 21 m) green or red, hollow or spongy, usually floating, with a milky sap, rooting at the nodes; leaves (to 12 cm) alternate, lanceolate to hastate or sagittate cordate, petioled (to 15 cm), margins entire; flowers pedicelled (to 7 cm), in long-stalked (to 9 cm) axillary clusters (1–5); corolla (to 6 cm) tubular, funnelform, lilac, pink, purple, or white, the throat often purple; stamens (to 1 cm) five, filaments hairy at base, attached to floral tube; ovary superior, 2-celled; capsule (to 16 mm) globose, 1–4-seeded; seeds (to 6 mm) brown, globular, velvety or smooth

Life history: duration: annual (fruit/seeds); perennial (rhizomes); **asexual reproduction:** rhizomes, shoot fragments; **pollination:** insect or self; **sexual condition:** hermaphroditic; **fruit:** capsules (infrequent); **local dispersal:** capsules, fragments (water); **long-distance dispersal:** seeds (water)

Imperilment: (1) *Ipomoea aquatica* [G5]; *I. asarifolia* (2) [unranked]

Ecology: general: *Ipomoea* species occur predominantly in terrestrial habitats. There are 22 North American species listed as wetland indicators, but most are ranked as FAC–UPL species. In the 1996 wetland indicator list, *Ipomoea aquatica* was designated as OBL only in the Caribbean region and Hawaii; however, that designation did not take into account its naturalization in continental North America, where it exclusively occupies wetland habitats. It is now regarded as an OBL North American species along with *I. asarifolia* (also introduced), which was excluded from the 1996 indicator list. Most of the species are herbaceous vines with large, showy flowers; a few species are shrubs or trees (to 15 m tall). Many of the species have sporophytic self-incompatibility and are outcrossed mainly by bees (Insecta: Hymenoptera). A few species are pollinated by bats (Mammalia: Chiroptera) or moths (Insecta: Lepidoptera). Other species are self-compatible and exhibit mixed mating systems. Only 1–4 seeds are produced by most species and are dispersed by animals or by the wind. Some species produce "labyrinth" seeds (i.e., containing air cavities), which are adapted for water dispersal. Some seeds contain hallucinogenic ergoline alkaloids. The genus is mainly tropical, extending into temperate areas except for colder climates.

Ipomoea aquatica **Forsk.** grows in canals, ditches, marshes, along the margins of rivers and streams, or on the surface of borrow pits, lakes, ponds, and streams at elevations of up to 500 m. The plants occur in sites with still or flowing water where they develop creeping or floating stems. They occupy a broad circumneutral range of substrates (pH: 6.8–8.2), which include clay loam, loam, and mud. This species has an affinity for tropical conditions and thrives where growing season temperatures are above 25°C. It is a short-day plant, with flowering induced at day lengths less than 12 h. Flowering is stimulated by drought conditions. The white corollas open in the morning (9 A.M.) and help dissipate heat to cool the temperature of the gynoecium when exposed to intense sunlight. The flowers mainly are self-pollinated (60%–65%) but also are outcrossed by insects (Insecta) including bees (Hymenoptera) and butterflies (Lepidoptera). Seed set is comparable between selfed plants (60%–84%) and open-pollinated plants (80%–88%). The seed set, seed weight, and seed number do not differ significantly between selfed and outcrossed flowers, an indication that the plants are adapted for inbreeding. Seed set in North American populations generally appears to be low. A single plant can produce up to 245 fruits, which normally are 1-seeded. The capsules are indehiscent and retain air pockets, which confer long-term

floatation. They are water-dispersed. The seeds mature in 33 (34% germination) to 61 days (75% germination) after anthesis. They are not dormant, but require mechanical scarification of the seed coat for germination. Once scarified, the seeds germinate within 24 h. The seeds germinate poorly under water or at temperatures below 25°C. They can be stored in sealed bags for up to 18 months at 10°C–20°C. Vegetative reproduction by fragmentation is the most common means of reproduction. The shoot fragments root readily at the nodes and develop quickly into new plants. The shoots can grow at rates of up to 10 cm/day and will reach lengths of more than 21 m. The plants do not survive when clipped underwater, which indicates that they are susceptible to grazing activity. Under ideal conditions, the plants can achieve a biomass (fresh weight) of 190,000 kg/ha within 9 months. Both UPL and aquatic cultivars are known. In China the roots are colonized by arbuscular mycorrhizae. Although the plants usually are perennial, an aneuploid strain is known that behaves like an annual. These plants easily are transformed genetically, facilitating the development of various genetically modified (GM) strains. **Reported associates (North America):** *Alternanthera philoxeroides, Azolla, Blechnum serrulatum, Boehmeria cylindrica, Cephalanthus occidentalis, Ceratopteris thalictroides, Eichhornia crassipes, Eleocharis baldwinii, Melaleuca quinquenervia, Myrica cerifera, Oxycaryum cubense, Panicum hemitomon, Pistia, Quercus laurifolia, Salvinia, Taxodium ascendens, Utricularia.*

Ipomoea asarifolia **(Desr.) Roemer & J.A. Schultes** inhabits beaches, canals, ditches, marshes, pools, ricefields, roadsides, swamps, and the margins of ponds at low elevations. The shoots sprawl along the shores of wet habitats, or they can float and extend along the surface of the water. The substrates have been described as muck or sand over limestone. Limited habitat data are available that are specific to North America due to the small number of known populations. The flowers are self-incompatible (through a late-acting system) and produce viable seeds only after they have been cross-pollinated. The flowers occur from January to July and remain open for about 6 h. They are pollinated primarily by bees (Insecta: Hymenoptera: Apidae; Megachilidae), which are guided to the flowers by magenta-colored "mesopetals" that function as honey guides. Nectar secretion is less than 1 μL. The seeds are dormant but will germinate well if they first are subjected to a 20-min period of scarification in concentrated sulfuric acid followed by their incubation under a 33°C/24°C temperature regime. Seed germination is insensitive to light. The highest seed germination rate occurs in distilled water and decreases with increasing osmotic stress. Seed germination declines with increasing nitrate levels. Comparable seedling emergence occurs at all planting depths from 1 to 10 cm, but is slightly reduced for surface-sown seeds. However, seedlings germinating at greater depths exhibit reduced biomass allocation to their roots. No persistent seed bank is formed, with 80%–90% mortality occurring within 18 months for seeds planted from the surface to 10 cm depths. Accordingly, it may be possible to eradicate this species by preventing the production of new flowers (e.g., by repeated mowing). The plants are sensitive to frost but are highly tolerant to artificial and natural disturbance. They are cultivated easily by shoot cuttings or by seed. **Reported associates (North America):** *Eupatorium capillifolium*, *Myrica cerifera*, *Setaria parviflora*.

Use by wildlife: A number of species feed on *Ipomoea aquatica* in its native range, but there is little information on its use by native wildlife in its nonindigenous North American range. Several families of insects (Insecta: Lepidoptera: Arctiidae; Bucculatricidae; Gelechiidae; Lyonetiidae; Noctuidae; Nymphalidae; Pyralidae; Sphingidae) use the plants as a larval food source in Asia. *Ipomoea aquatica* can provide a complete high-protein food for domestic rabbits (Mammalia: Leporidae). The plants are fed to cattle (Mammalia: Bovidae: *Bos taurus*), chickens (Aves: Phasianidae: *Gallus gallus*), ducks (Aves: Anatidae), fish (Chordata: Teleostomi), and pigs (Mammalia: Suidae: *Sus*). Plants that have been grown in North American greenhouses have suffered minor to moderate damage from green peach aphids (Insecta: Hemiptera: *Myzus persicae*) and Western flower thrips (Insecta: Thysanoptera: Thripidae: *Frankliniella occidentalis*) and are attacked more seriously by several Fungi (Ascomycota: Botryosphaeriaceae: *Phyllosticta ipomoeae*; Mycosphaerellaceae: *Phaeoisariopsis bati- cola*), and a bacterium (Proteobacteria: Pseudomonadaceae: *Pseudomonas syringae*). *Ipomoea asarifolia* is toxic to cattle (Mammalia: Bovidae: *Bos taurus*), goats (Mammalia: Bovidae: *Capra aegagrus*), and sheep (Mammalia: Bovidae: *Ovis aries*). The plants host symbiotic Fungi (Ascomycota: Clavicipitaceae: *Periglandula ipomoeae*).

Economic importance: food: *Ipomoea aquatica* has been cultivated for food in Asia at least since the 3rd century AD. The foliage contains 2% oxalate, up to 6.3% protein, 11% lipids, 18% crude fiber and an average carbohydrate content of roughly 4.3%. The seeds have a low oil content (8.3% total lipids). The raw or cooked leaves usually are eaten as a vegetable similar to spinach, or the shoot tips are pickled or added to soups or mixed with a variety of other vegetables. The plants are a major ingredient of Cantonese *furu*, Philippine *adobong kangkong* and *sinigang*, Thai *pak bung*, and Vietnamese *giàu muông*. The leaves contain vitamins A-C, β-carotene, and nearly three times the amount of vitamin E (as α-tocopherol) than spinach (*Spinacia oleracea*). They are extremely high in potassium and iron. Caution should be taken when eating plants of unknown provenance because they are known to concentrate heavy metals (e.g., chromium and lead) at high levels and also can contain dangerous levels of methyl mercury. Plants grown in wastewater have been observed to harbor thermotolerant coliform bacteria and parasites such as *Cryptosporidium*, *Cyclospora*, *Giardia*, and helminths; **medicinal:** *Ipomoea aquatica* contains insulin-like compounds. It has been used to treat diabetes in several countries and diets of the plants have been shown to induce hypoglycemic activity in laboratory animals. Aqueous stem extracts have antioxidant properties and exhibit antiproliferative activity on human lymphoma NB4 cells. Aqueous and ethanolic extracts exhibit moderate inhibitory activity of HIV-1 reverse transcriptase. Several pyrrolidine amides have been isolated from the foliage. The plants contain *N*-feruloyltyramines

(which are prostaglandin inhibitors) and bisphenol A, which mimics estrogen and can function as an endocrine disruptor. Because of its high iron content the foliage has been used to treat anemia. The seeds reportedly contain a strong pesticide that is capable of killing earthworms as well as leeches, tapeworms, and other intestinal parasites. A mouthwash made from the plants is effective against *Escherichia coli* and *Staphylococcus aureus*. The juice has been used as an emetic and to treat gastric or intestinal disorders. It is used as a diuretic and to remedy biliousness, hypertension, jaundice and nosebleeds. A poultice made from the foliage has been used to reduce fever, to relieve hemorrhoids and to treat boils or ringworm lesions. Despite its alleged medicinal virtues, excess ingestion of this species has been linked to leucorrhoea and other disorders. *Ipomoea asarifolia* is quite toxic and contains acylated anthocyanins and the ergoline alkaloids chanoclavine I, ergine, ergobalansinine, and lysergic acid α-hydroxyethylamide (LSH). Chanoclacine I and ergine induce hallucinogenic and psychotomimetic effects. LSH is similar chemically to the hallucinogenic drug LSD, and although it remains clinically untested, it is suspected to have similar effects on humans. However, the plants have strong antioxidant properties and the aqueous extracts exhibit significant analgesic and anti-inflammatory properties. The ergoline alkaloids are believed to be derived from symbiotic clavicipitaceous fungi, which colonize the plants; **cultivation:** *Ipomoea aquatica* is cultivated widely as a vegetable, especially in Southeast Asia. The cultivars (e.g., 'Ching Quat,' 'Pak Quat') include red-stemmed and green-stemmed forms. There is an UPL cultivar, which has been known as *Ipomoea reptans*. *Ipomoea asarifolia* is eaten as a vegetable, particularly in the New World tropics; **misc. products:** *Ipomoea aquatica* sometimes is grown to provide cover for fish production ponds. *Ipomoea asarifolia* is used as a hallucinogen in southern Mexico in magic and religious rituals; **weeds:** *Ipomoea aquatica* is listed as a Federal noxious weed in the United States where it obstructs water flow in canals, displaces native vegetation, and provide suitable breeding habitat for mosquitoes. *Ipomoea asarifolia* is reported as a weed in South America; **nonindigenous species:** *Ipomoea aquatica* is native to Asia and was introduced to North America (California, Florida) prior to 1950 as an escape from cultivation. It was introduced to Hawaii before 1871. Elsewhere it has been introduced to Africa, Australia, Central and South America, France, and to many islands including Cuba, Hispaniola, Honduras, Jamaica, Puerto Rico and others. *Ipomoea asarifolia* was introduced to Florida sometime before 1994, when the first population was discovered in Broward County. Believed to be originally native to the Old World, it has since spread throughout tropical America as well as Africa, India, and Southeast Asia.

Systematics: A large amount of DNA sequence data (including nuclear loci and complete cpDNA genomes) has been analyzed phylogenetically for Convolvulaceae, but has resulted in different interpretations regarding relationships in and of the large genus *Ipomoea*. These studies have indicated the monophyly of tribe Ipomoeeae, which includes *Argyreia*, *Astripomoea*, *Blinkworthia*, *Ipomoea*, *Lepistemon*,

Lepistemonopsis, *Paralepistemon*, *Rivea*, *Stictocardia*, and *Turbiina*. However, in all cases, molecular data have indicated that *Ipomoea* is not entirely monophyletic as currently circumscribed, with at least some species consistently falling outside of a core clade of species and associating with various other genera including *Argyreia*, *Lepistemon*, *Stictocardia*, and *Turbina*. An essential question is whether to merge several smaller genera with *Ipomoea*, or to remove the anomalous species and transfer them to other taxa. Further confounding the matter is that one of the anomalous species (*I. pes-tigridis*) is the type of the genus, to which the name is attached; if the latter approach is followed, then most current *Ipomoea* species would have to be renamed and transferred to a different genus. One problem with these analyses (particularly with the more variable nuclear DNA data) is that fairly distant taxa (as indicated by analyses of the more conserved cpDNA data) appear to have been selected as out-groups, a factor that certainly could influence the topology of the resulting cladograms and should be addressed in subsequent studies. Because the phylogeny of *Ipomoea* and allied genera cannot be evaluated with certainty at this time, the genus name is retained here in the broad sense at least until further clarification of generic boundaries can be achieved. Analyses of molecular and morphological data have consistently resolved *I. asarifolia* and *I. pes-caprae* as sister species. However, there have been contrasting indications of relationship for *I. aquatica* among the various data sets analyzed. Analyses of two nuclear loci weakly support the alliance of *I. aquatica* in a clade with *I. cairica* and *I. polymorpha* (using maximum parsimony) or *I. cairica* and *I. sepiaria* (Bayesian analysis); however, other analyses of these same loci [internal transcribed spacer (ITS), *waxy*] resolve *I. aquatica* as the sister species to *I. diamantinensis* (Figure 5.104). Unfortunately, there has not been an analysis of these same loci that includes *I. cairica*, *I. diamantinensis*, *I. polymorpha* and *I. sepiaria* simultaneously along with *I. aquatica*; thus it is impossible to conclude which of those species might be most closely allied with the latter. Three genetically distinct races of *I. aquatica* have been identified in Florida using RAPD markers and indicate multiple introductions of plants in that area. The basic chromosome number of *Ipomoea* is $x = 15$. *Ipomoea asarifolia* ($2n = 30$) is diploid. Although *I. aquatica* normally is diploid ($2n = 30$), a tetraploid race has been reported. The tetraploids have thicker and wider leaves with longer petioles, wider (but shorter) stems, longer and wider flowers, and larger seeds than diploids. Pollen is larger in the tetraploids, but viability is 100% in both races. Seed set is somewhat higher in the diploids. There is an aneuploid cytotype ($2n = 28$), which reportedly behaves as an annual. Hybrids involving *I. aquatica* or *I. asarifolia* have not been reported.

Comments: *Ipomoea aquatica* occurs in California and Florida and throughout the world in warm tropical or semi-tropical environments; *I. asarifolia* occurs in Florida and is common throughout the Old and New World tropics.

References: Anh et al., 2007; Austin, 2004a, b, 2007; Burks & Austin, 1999; Cerkauskas et al., 2006; Ching & Mohamed, 2001; Dias Filho, 1996, 1999; Ene-OjoAtawodi & Onaolapo, 2010; Erickson & Puttock, 2006; Eserman et al., 2014; Gai et al., 2006; Göthberg & Greger, 2006; Göthberg et al., 2004; Grubben, 2004; Harwood & Sytsma, 2003; Huang et al., 2005; Imbs & Pham, 1995; Jegede et al., 2009; Khamwan et al., 2003; Kiill & Ranga, 2003; King & Bamford, 1937; Lako et al., 2007; Lawal et al., 2010; Libert & Franceschi, 1987; Malaipan, 1983; Malalavidhane et al., 2000; Manos et al., 2001a; Markert et al., 2008; Meira et al., 2012; Middleton, 1990; Miller et al., 1999, 2002; National Academy of Sciences, 1976; Noureddin et al., 2004; Ogunwenmo, 2006; Panchaphong, 1987; Patiño & Grace, 2002; Patnaik, 1976; Phimmasan et al., 2004; Rai & Sinha, 2001; Stefanović et al., 2002; Steiner et al., 2011; Tofern et al., 1999; Tseng et al., 1992; Umar et al., 2007; Vaidhayakarn, 1987; Van & Madeira, 1998; Woodson, Jr. et al., 1975; Woradulayapinij et al., 2005.

Family 3: Hydroleaceae [1]

Once regarded as a segregate of Hydrophyllaceae (which itself has been merged with Boraginaceae), the relationships of Hydroleaceae have been clarified as a result of molecular phylogenetic analyses (e.g., Fay et al., 1998; Savolainen et al., 2000; Bremer et al., 2002). As a result of these and similar studies, the family now is recognized as a clade that is related only distantly to Boraginaceae/Hydrophyllaceae, but is allied as the sister group of Sphenocleaceae (Figure 5.105), a small family (1 genus; 2 species) that is characteristic of flooded habitats. Hydroleaceae are a small family, consisting solely of the genus *Hydrolea* with 11 species worldwide. The family is distinguished by having two styles, two oblique carpels, and glandular stamen filaments that are swollen to a deltoid base; like other members of Solanales, the corollas are plicate (Judd et al., 2002). Some species are thorny. All members are herbs or subshrubs that grow in, or at least very near to wet habitats. Most species inhabit tropical regions.

The broad filament bases restrict access to the nectar and probably are adaptations related to insect pollination (Erbar et al., 2005). However, even though the flowers are fairly showy and produce nectar, several species have low pollen:ovule ratios, anthers and stigmas that virtually touch, and are at least facultatively autogamous. The species have only minor economic importance as ornamentals and folk medicines.

There is one genus with OBL species in North America:

1. *Hydrolea* L.

1. *Hydrolea*

False fiddle leaf, sky-flower, water-olive, water-pod

Etymology: from the Greek *hudro elaia* ("water olive"), with respect to its habitat and appearance

Synonyms: *Ascleia*; *Beloanthera*; *Hydrolaea*; *Hydrolia*; *Lycium* (in part); *Nama*; *Reichelia*; *Sagonea*; *Steris*

Distribution: global: Africa; Asia; Australia; New World; **North America:** southeastern

Diversity: global: 11 species; **North America:** 5–6 species

Indicators (USA): OBL: *Hydrolea capsularis*, *H. corymbosa*, *H. ovata*, *H. quadrivalvis*, *H. spinosa*, *H. uniflora*

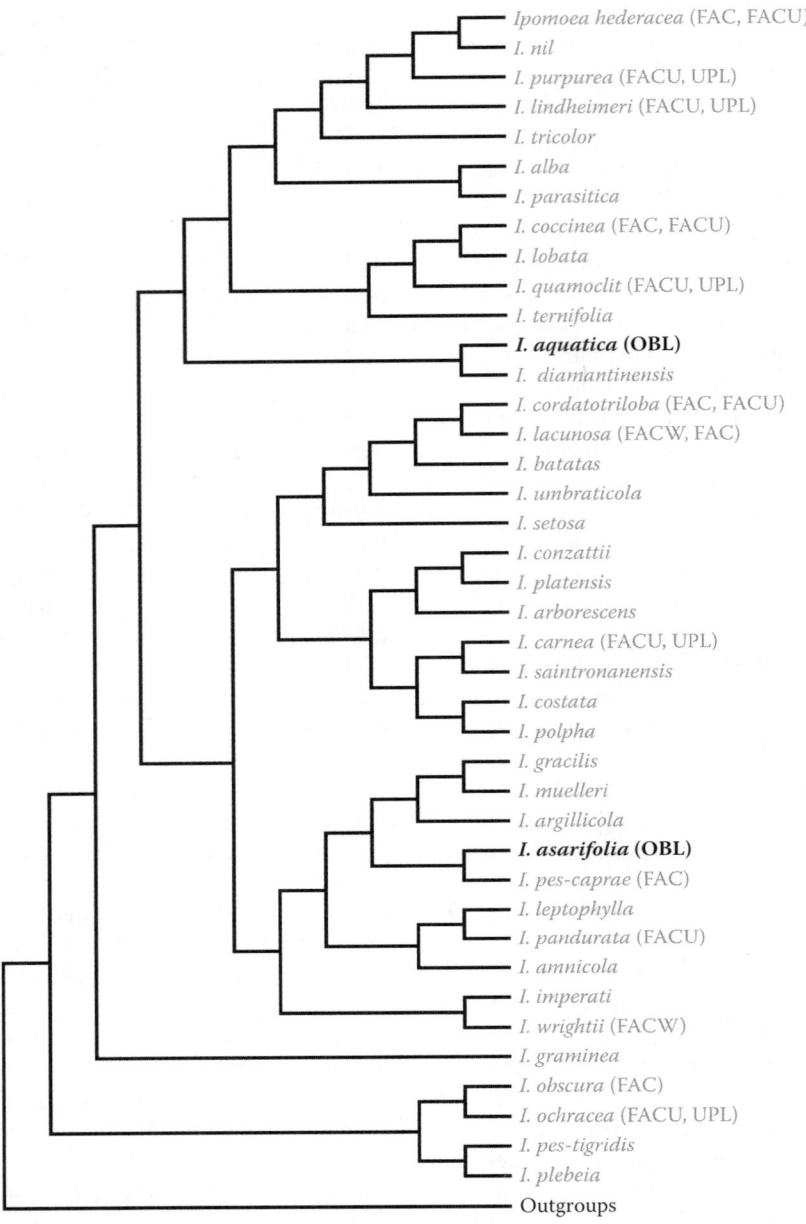

FIGURE 5.104 The relationships of OBL North American *Ipomoea* species (in bold) as estimated by phylogenetic analysis of combined nuclear DNA sequence data indicate their independent origins in the genus (wetland indicator designations are provided in parentheses). Other studies Manos et al., 2001a; Miller et al., 2002) have indicated that *Ipomoea* is not monophyletic as formerly circumscribed, but includes species from several other genera. (Adapted from Miller, R.E. et al., *Syst. Bot.*, 24, 209–227, 1999.)

Habitat: freshwater; palustrine; **pH:** 4.5–8.4; **depth:** <1 m; **life-form(s):** emergent herb, emergent shrub

Key morphology: stems (to 2 m) glabrous or hairy, succulent to woody, the bases often swollen with spongy aerenchyma tissue, 1–2 nodal thorns (to 30 mm) present or absent; leaves (to 12 cm) alternate, glabrous or hairy, ovate to lanceolate, the margins entire, undulate, or serrulate, sessile or petioled (to 1 cm); flowers 5-merous, in terminal clusters or corymbs, in leafy panicles, or in peduncled (to 4 cm) axillary clusters; corolla campanulate, the petals (to 12 mm) deep blue or white, connate at base; stamens free, inserted at base of the short corolla tube, the filaments abruptly dilated at base; ovary superior, styles (to 15 mm) 2–4; capsule (to 8 mm) globose to cylindrical, many seeded; seeds (to 0.7 mm) longitudinally ribbed, the surfaces reticulate

Life history: duration: perennial (persistent stem bases); **asexual reproduction:** rhizomes, stolons; **pollination:** autogamous or insect; **sexual condition:** hermaphroditic; **fruit:** capsules (common); **local dispersal:** seeds (water); **long-distance dispersal:** seeds (waterfowl)

Imperilment: (1) *Hydrolea capsularis* [unranked]; (2) H. *corymbosa* [G5]; S1 (SC); S3 (GA); (3) *H. ovata* [G5]; S1 (KY,

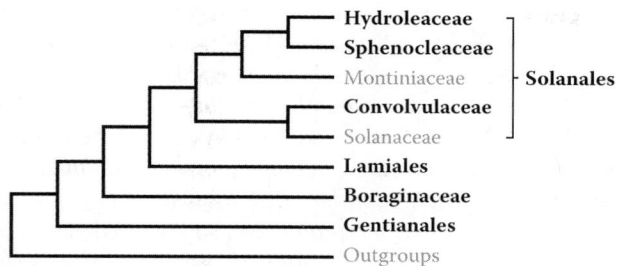

FIGURE 5.105 Arrangement of families in the Solanales, as indicated by phylogenetic analysis of combined DNA sequence data from six loci. Families containing OBL North American representatives are shown in bold. (Adapted from Bremer, B. et al., *Mol. Phylogen. Evol.*, 24, 274–301, 2002.)

TN); S2 (MO); (4) *H. quadrivalvis* [G5]; S3 (NC, VA); (5) *H. spinosa* [G5]; S1 (TX); (6) *H. uniflora* [G5]; S1 (IL, KY)

Ecology: general: All *Hydrolea* species are aquatic or wetland plants of tropical, subtropical or warm temperate regions. All of the North American species are designated uniformly as OBL. They have showy flowers that are adapted for insect pollination (e.g., producing nectar), but the few species studied for their reproductive ecology are self-compatible, have low pollen to ovule ratios, and primarily are autogamous. The life histories of most of the species remain poorly studied and there are few details published on their seed ecology, germination requirements, etc. Some seeds are known to float and are dispersed by water and waterfowl (Aves: Anatidae). The species are perennial and overwinter by their persistent stem bases. They usually are herbs but some can become small shrubs. In some cases, the plants will die back completely and have been described as annuals. The stems can develop into rhizomatous or stoloniferous structures that spread laterally and root at the nodes. The stem bases frequently exhibit a proliferation of air–space tissue. Many of the New World species possess nodal thorns, which probably deter herbivores.

Hydrolea capsularis **(L.) Druce** generally has been placed in synonymy with *H. spinosa* (see *Systematics* later). However, insufficient information is available at this time to determine its appropriate taxonomic status, and there is no effective means of compiling pertinent life history information for this taxon from the literature. Unavoidably, it has been excluded from further discussion.

Hydrolea corymbosa **J.F. Macbr. ex Ell.** occurs in depressions, ditches, flatwoods, marshes, meadows, prairies, savannas, swamps, and along the margins of borrow pits. The plants usually grow in shallow, standing water in open sites or in partial shade. The substrates are described as calcareous. The plants flower from July to September; otherwise, there is virtually nothing known regarding their life history. This species has been grown successfully using either Metro Mix 500 or Pro-Mix BX substrates to which 29.2 g of Nutricote fertilizer were added per pot, and irrigated thrice daily for 15 min intervals. Thorns are produced rarely in this species. **Reported associates:** *Agalinis linifolia, Asclepias perennis, Bacopa caroliniana, Carex glaucescens, Cephalanthus occidentalis,*

Coreopsis falcata, Crinum americanum, Cyperus haspan, Dichanthelium strigosum, Diodia virginiana, Eleocharis, Eragrostis elliottii, Eriocaulon decangulare, E. ravenelii, Eriochloa michauxii, Eryngium yuccifolium, Eupatorium, Eustachys glauca, Euthamia tenuifolia, Fimbristylis breviseta, Galactia elliottii, Gratiola ramosa, Helenium pinnatifidum, Hymenocallis, Hypericum fasciculatum, H. limosum, H. cistifolium, H. hypericoides, Hyptis alata, Ilex myrtifolia, Ipomoea sagittata, Iris tridentata, Iva microcephala, Juncus megacephalus, Lachnanthes caroliana, Lachnocaulon anceps, Leersia hexandra, Lobelia boykinii, L. cardinalis, L. glandulosa, Mecardonia acuminata, Mikania scandens, Mitreola petiolata, Muhlenbergia sericea, Myrica cerifera, Nemastylis floridana, Oxypolis filiformis, Panicum rigidulum, P. tenerum, P. verrucosum, P. virgatum, Parnassia caroliniana, Paspalum floridanum, P. praecox, Persicaria hydropiperoides Physostegia purpurea, Pinus elliottii, Piriqueta caroliniana, Pluchea rosea, Polygala cymosa, P. grandiflora, Pontederia, Proserpinaca pectinata, Quercus minima, Rhexia aristosa, mariana, Rhynchospora colorata, R. fascicularis, R. inundata, R. latifolia, R. odorata, Sabatia stellaris, Sagittaria lancifolia, Spiranthes laciniata, Steinchisma hians, Taxodium ascendens, Woodwardia virginica, Xyris caroliniana, X. fimbriata.

Hydrolea ovata **Nutt. ex Choisy** grows in bogs, depressions, ditches, flatwoods, meadows, pools, ponds, prairies, ricefields, savannas, swamps, and along the margins of gravel pits and streams. It is found in sunny to semishaded sites either in shallow water or where ponding has occurred for some duration. The substrates can consist of clay, sand, sandy loam, silty clay loam, or silty loam. The flowers are visited by various bees (Insecta: Apidae: Apis *mellifera, Bombus impatiens*; Halictidae: *Augochlora pura, Halictus ligatus, Lasioglossum imitatum*; Megachilidae: *Megachile mendica*) and presumably are pollinated and outcrossed by them. The extent of self-pollination is not known. The plants are thorny. Vegetative reproduction occurs by rhizomatous shoots. Other life history information is unavailable. **Reported associates:** *Acmella oppositifolia, Agrostis hyemalis, Albizia julibrissin, Alternanthera philoxeroides, Amsonia tabernaemontana, Andropogon capillipes, A. glomeratus, Aristida palustris, Arnoglossum ovatum, Asclepias lanceolata, A. viridis, Bacopa caroliniana, B. monnieri, Bidens laevis, Boltonia asteroides, B. diffusa, Brunnichia ovata, Carex glaucescens, Ca. joorii, C. verrucosa, Carya, Centella erecta, Cephalanthus occidentalis, Chara, Coelorachis rugosa, Coreopsis tinctoria, Cornus florida, Croton, Cynodon dactylon, Dichanthelium scabriusculum, D. spretum, Diospyros virginiana, Eleocharis baldwinii, E. equisetoides, E. flavescens, E. quadrangulata, E. quinqueflora, Eriocaulon compressum, Eriochloa punctata, Eryngium yuccifolium, Eupatorium serotinum, Euthamia leptocephala, Fagus grandifolia, Fimbristylis littoralis, Fuirena bushii, Gaura lindheimeri, Gratiola brevifolia, Hedyotis, Helenium, Helianthus angustifolius, Hibiscus moscheutos, Hydrocotyle umbellata, Hyptis alata, Hymenocallis liriosme, Ipomoea sagittata, Iva angustifolia, Juncus effusus, J. polycephalus, J. roemerianus, Justicia ovata, Leersia* *hexandra, L. oryzoides, Liquidambar styraciflua, Ludwigia leptocarpa, L. linearis, L. octovalvis, L. peploides, L. pilosa, L. sphaerocarpa, Lycopus rubellus, Lythrum lineare, Neptunia lutea, Oenothera drummondii, Oxypolis filiformis, Panicum hemitomon, P. virgatum, Paspalum floridanum, P. plicatulum, P. urvillei, P. vaginatum, Phyla nodiflora, Pinus elliottii, Pluchea rosea, Polygala leptocaulis, Persicaria, Proserpinaca palustris, P. pectinata, Ratibida peduncularis, Rhexia mariana, R. virginica, Rhynchospora capitellata, R. cephalantha, R. colorata, R. corniculata, R. elliottii, R. globularis, R. macrostachya, R. microcarpa, R. mixta, R. nitens, Rubus trivialis, Rudbeckia hirta, R. texana, Sabatia angularis, Saccharum, Sacciolepis striata, Sagittaria graminea, S. lancifolia, Sarracenia, Scleria baldwinii, Sesbania exaltata, Solidago sempervirens, Spartina patens, Spiranthes laciniata, Toxicodendron, Tradescantia hirsutiflora, Tridens strictus, Tripsacum dactyloides, Typha latifolia, Utricularia, Vaccinium, Vernonia gigantea, Xyris fimbriata, X. laxifolia.*

Hydrolea quadrivalvis Walter. inhabits bogs, ditches, flatwoods, marshes, sloughs, swamps, and the margins of borrow pits, ponds, and rivers. The sites usually are open and range from wet ground to shallow water (0.3–0.5 m deep). The plants tolerate a broad range of acidity (pH: 5.7–8.4) and usually occur on muddy substrates that are fairly high in organic matter. Flowering and fruiting begin in June, but the pollination biology has not been studied further in this species. The stems are thorny. The plants can form thick stands of intertangled, rhizomatous stems, which function in vegetative reproduction. **Reported associates:** *Ammannia, Callitriche, Cephalanthus occidentalis, Hibiscus laevis, Ludwigia leptocarpa, L. peploides, Lycopus, Nelumbo lutea, Persicaria hydropiperoides, Proserpinaca, Sagittaria latifolia, Veronica.*

Hydrolea spinosa L. is a herb or small shrub found in ditches, pools, and along the margins of rivers and canals, in full sun or light shade on marl or acidic (pH: 4.5–5.0) clay and silt substrates. The site conditions range from having permanent to seasonal water. The plants are adapted to self-fertilization because the stamens are in contact with the stigmas as the flowers open. The flowers are autogamous and readily set seed in the absence of pollinators. Some outcrossing probably occurs whenever appropriate insect (Insecta) agents are present. The seeds can float for several days and are dispersed by water and waterfowl (Aves: Anatidae). Seed germination requires light and can occur when the seeds are floating or when they are covered by the substrate. The shoots bear nodal thorns. Because of its extremely limited distribution in North America, there is little ecological information available on this species for the region. **Reported associates:** *Ludwigia peruviana.*

Hydrolea uniflora Raf. inhabits bogs, depressions, ditches, marshes, meadows, swamps, and the margins of borrow pits, lakes, ponds, rivers, sloughs, and streams. It grows in full sun in shallow water (to 0.6 m) or on wet substrates (e.g., pH: 6.3), which can consist of mud, sand, silty clay loam, silt, silty loam, or silty sand. The plants are thorny. There have been no detailed life history studies of this species. **Reported associates:** *Acer rubrum, Aeschynomene, Alternanthera philoxeroides, Andropogon virginicus, Asclepias, Carex*

crus-corvi, Carya, Celtis laevigata, Cephalanthus occidentalis, Chasmanthium latifolium, Crataegus, Cuscuta indecora, Cyperus pseudovegetus, Dichanthelium, Diodia virginiana, Diospyros virginiana, Echinochloa, Echinodorus cordifolius, Eleocharis obtusa, Hypericum, Iris fulva, Juncus elliottii, J. nodatus, Leersia oryzoides, Lemna valdiviana, Liquidambar styraciflua, Ludwigia glandulosa, L. palustris, L. peploides, Lygodium japonicum, Magnolia, Nyssa sylvatica, Panicum rigidulum, Paspalum dilatatum, P. notatum, Penthorum sedoides, Persicaria hydropiper, P. hydropiperoides, P. lapathifolia, Pinus taeda, Quercus lyrata, Q. nigra, Q. phellos, Q. shumardii, Rhynchospora corniculata, Sabal minor, Saccharum giganteum, Sagittaria, Salix nigra, Saururus cernuus, Smilax, Sphagnum, Steinchisma hians, Taxodium distichum, Toxicodendron radicans, Triadica sebifera, Typha latifolia, Ulmus alata, U. rubra, Vaccinium, Vitis.

Use by wildlife: Dense stands of *Hydrolea ovata* provide habitat for planktonic Crustacea (Cladocera: Chydoridae: *Alona affinis, A. setulosa, Chydorus brevilabris, Disparalona hamata*; Cyclopoida: Cyclopidae; Ostracoda), fly larvae (Insecta: Diptera: Chironomidae; Heleidae), mayfly nymphs (Insecta: Ephemeroptera), worms (Annelida: Oligochaeta; Nematoda), and protozoa colonies. Some of these organisms are eaten from the plants by top-feeding fish (Chordata: *Gambusia*). The flowers of *H. ovata* are visited by bumblebees and honeybees (Insecta: Hymenoptera: Apidae), leafcutting bees (Hymenoptera: Megachilidae), and sweat bees (Hymenoptera: Halictidae). They reportedly are attractive to butterflies (Insecta: Lepidoptera). *Hydrolea ovata* is a host to sac fungi (Ascomycota: Mycosphaerellaceae: *Cercospora namae*). Extracts of *H. quadrivalvis* have been shown to significantly deter feeding by crayfish (Crustacea: *Procambarus acutus*). *Hydrolea uniflora* is used for perching by the rainpool spreadwing damselfly (Insecta: Odonata: Lestidae: *Lestes forficula*).

Economic importance: food: *Hydrolea* species occasionally are eaten as a potherb. Analysis of *H. quadrivalvis* indicates that it is relatively low in protein (4.8%–11.1%) and contains 3.9% crude fat, 22.8% cellulose, 2.9% tannins and a caloric content of 4.00 Kcal/g. The foliage of most species is acrid and *H. spinosa* has been described as "intensively bitter" in taste. Leaves of the Old World *H. zeylanica* are eaten with rice in Thailand and Indonesia; **medicinal:** The foliage of native *Hydrolea* species has been pounded into pulp and used as a poultice in the southern United States. *Hydrolea spinosa* has been used in Venezuela as a treatment for tumors. Leaves of the Old World *H. zeylanica* reportedly are antiseptic; **cultivation:** *Hydrolea corymbosa* has appeared in cultivation relatively recently as a water garden marginal; **misc. products:** *Hydrolea ovata* has been planted in wetland restoration projects; **weeds:** *Hydrolea spinosa* is a rice field weed in Indonesia; **nonindigenous species:** *Hydrolea spinosa* is naturalized in Java and Sri Lanka.

Systematics: Although *Hydrolea* was once placed in the family Hydrophyllaceae, analysis of combined DNA sequence data indicates that the genus should be recognized as a distinct family that is the sister group to Sphenocleaceae

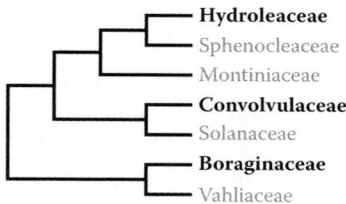

FIGURE 5.106 Sphenocleaceae and Hydroleaceae resolve strongly as a clade of Solanales by analysis of combined cpDNA and mitochondrial DNA (mtDNA) sequences. The placement of *Sphenoclea* (represented here by *S. zeylanica*) in Campanulaceae, as some authors maintain, clearly is unfounded as evidenced by these results. Taxa containing OBL North American species are highlighted in bold. (Adapted from Refulio-Rodriguez, N.F. & Olmstead, R.G. *Amer. J. Bot.*, 101, 287–299, 2014.)

(Figures 5.105 and 5.106). Two sections have been recognized in this small family: section *Attaleria* (Eastern Hemisphere) and section *Hydrolea* (Western Hemisphere). The sections differ slightly by their placental morphology (unwinged in section *Attaleria*; 2-winged in section *Hydrolea*) and chromosome numbers. The basic chromosome number for section *Hydrolea* is $x = 10$. *Hydrolea ovata*, *H. quadrivalvis*, and *H. uniflora* are diploid ($2n = 20$); whereas, *H. spinosa* is tetraploid ($2n = 40$); a diploid, thornless race of *H. spinosa* ($2n = 20$) has been reported. The basic chromosome number of section *Attaleria* is $x = 9$ or 12. *Hydrolea spinosa* is the most widespread New World species and the most variable morphologically. Some treatments (including the 2013 wetland indicator list) have recognized *H. capsularis* as being distinct taxonomically from *H. spinosa*, but there is insufficient evidence at this time to determine an appropriate disposition of this taxon. Three varieties are recognized for *H. spinosa*; whereas, no infraspecific taxa are recognized for any of the other species. The varieties overlap morphologically and a genetic analysis would be helpful to evaluate their distinctness. Despite its small size, a comprehensive phylogenetic analysis has not been conducted for *Hydrolea* and is needed to evaluate the monophyly of the genus and its sections. Interspecific hybrids are reported for *H. ovata* × *H. uniflora*.

Comments: *Hydrolea uniflora* and *H. ovata* occur in the central southern United States, with the latter extending into Mexico. *Hydrolea corymbosa* and *H. quadrivalvis* are found in the southeastern United States. *Hydrolea spinosa* (cited also as *H. capsularis*) occurs only in Texas, but extends into Mexico and South America. It was collected once from the Florida Everglades in 1996.

References: Bartholomew et al., 2006; Beal, 1977; Bried & Krotzer, 2005; Burrell, 1997; Campbell et al., 1982; Davenport, 1988; Erbar et al., 2005; Fernald & Kinsey, 1943; Hammer, 2002; Hardy & Raymond, 1991; Johnson, 1999; Kainradl, 1927; Lieneman, 1929; Little, 1979; Moody, 1989; Moore et al., 2006; Orzell & Bridges, 2006; Penfound et al., 1945; Porcher & Rayner, 2001; Prusak et al., 2005; Soerjani et al., 1987.

Family 4: Sphenocleaceae [1]

For years this family was allied incorrectly with Asterales or Campanulales (e.g., Cronquist, 1981). However, its correct phylogenetic placement within Solanales has been confirmed conclusively by phylogenetic analysis of DNA sequence data from coding and noncoding mitochondrial and chloroplast loci (Ferguson, 1998; Bremer et al., 2002; Refulio-Rodriguez & Olmstead, 2014). Those analyses consistently support Sphenocleaceae and Hydroleaceae as sister groups, which together with Montiniaceae, form a sister clade to Convolvulaceae and Solanaceae (e.g., Figure 5.106). This is a monogeneric family, comprising only two species originally of African affinity but now introduced broadly across the globe (Cook et al., 1974; Cook, 1996b).

Both species are annuals, with simple, spirally arranged leaves and inflorescences of numerous dense, terminal spikes. The flowers are bisexual and epigynous with a radially symmetrical, campanulate corolla; the bilocular ovary is partially inferior (Cronquist, 1981). The fruits are circumscissile capsules. The basic chromosome number of the family is $x = 12$.

These are herbs of wet, tropical environments. Although some insect (Insecta) pollination may occur, the plants are believed to be primarily autogamous (Cook, 1996b). The family is of little economic importance, but the species can become serious ricefield pests.

There is one genus with OBL species in North America:

1. ***Sphenoclea*** Gaertn.

1. *Sphenoclea*

Chickenspike, goosegrass, gooseweed, piefruit
Etymology: from the Greek *sphênos kleiô* ("enclosed wedge"), probably in reference to the cuneate fruit
Synonyms: *Gaertnera* (in part); *Pongati*; *Pongatium* (in part); *Rapinia* (in part); *Reichelia* (in part)
Distribution: global: Africa; Asia*; Australia*, New World*; **North America:** southern United States
Diversity: global: 2 species; **North America:** 1 species
Indicators (USA): OBL; FACW: *Sphenoclea zeylanica*
Habitat: brackish, freshwater; palustrine; **pH:** 6.9–8.0; **depth:** <1 m; **life-form(s):** emergent herb
Key morphology: Stems (to 18.6 dm) erect, hollow, proximally spongy and rooting at lower nodes; leaves (to 13.8 cm) alternate, elliptic to oblanceolate, petiolate (to 2.7 cm), margins entire, stipules lacking; inflorescences dense terminal or lateral spikes (to 10.5 cm), peduncled (to 10.3 cm); flowers sessile, bisexual, epigynous; corollas radial, 5-lobed, petals white, short tubular (to 2.3 mm), connate proximally; stamens 5, adnate to corolla tube and alternating with lobes; gynoecium syncarpous (2 carpels), inferior; fruit a sessile, circumscissile, wedge-shaped capsule (to 4 mm), seeds (to 0.5 mm) numerous.
Life history: duration: annual (seed); **asexual reproduction:** none; **pollination:** self; insect?; **sexual condition:** hermaphroditic; **fruit:** capsules (prolific); **local dispersal:** seeds (water); **long-distance dispersal:** seeds (birds and other animals, water)

Imperilment: (1) *Sphenoclea zeylanica* [G4/G5]; exotic

Ecology: general: *Sphenoclea* species can occur in standing water or as emergents on nearly any type of exposed, wet substrate. The plants tolerate extreme water level fluctuations by becoming submersed seasonally and also will thrive under drying, more terrestrial conditions. They must be emergent during their flowering period, which can persist throughout the year in some tropical areas. The plants are characterized as being principally self-pollinating, but might also be insect pollinated. Both species are annuals, which produce numerous seeds that are hydrochorous (dispersed by water) or disseminated in mud, by adhering perhaps to birds and other animals. The plants occur commonly along watercourses or with other annuals in the drawdown zone of lakes during low water level conditions.

Sphenoclea zeylanica **Gaertn.** inhabits freshwater to brackish (salinity ~18 ppt) bayous, borrow pits, canals, deltas, ditches, dredge spoil, floating mats, floodplains, floodways, gravel pits, lake beds, levees, marshes, rice fields, roadsides, sandbars, sandpits, sloughs, swamps, and the margins of lakes, ponds, reservoirs, rivers, streams and woodlands at elevations of up to 300 m [to 1500 m in native range]. Although the plants can tolerate a broad range of moisture conditions, flooding is essential and its delay or absence results in plants having shorter root and shoot lengths. The plants persist well where a level of standing water (up to 30–50 cm) is maintained throughout the year. Plant numbers increase proportionally with water depths from 0 to 8 cm and also when the duration of flooding is extended. Their biomass is highest at 2 cm depths when the water level is maintained constantly. Substrates can include alluvium, muck, mud, and sand. The full extent of acidity tolerance is unknown; however pH values of 6.2–8.0 have been reported from ricefields and from other sites. The plants thrive in full sun and are highly heat tolerant, and even are known to occur in thermal effluents. Flowering extends from late May to December. The plants produce prodigious quantities of small (~10 μg) seeds. Vast numbers (173,576/m² at 15 cm depth) have been recovered from Malaysian rice fields, where they can represent from 18% to 24% of the total seed pool. Local seed dispersal occurs by water; however, epizoic transport by waterfowl and other birds is implicated in their long distance dispersal. Seeds that have been sun dried for 3 days and then stored at 45°C for 4 months, germinate well at 25°C. Successful germination (in 6–8 weeks) also has been achieved following 3 months of cold stratification (7°C), when the seeds are sown in flats covered by a constant water depth of 2 cm and temperatures are maintained below 35°C. Although total immersion can reduce seed germination, it still can exceed 25% at depths of 20 cm. The plants can occur in brackish sites. However, their seed germination is inversely proportional to salinity, with reduced rates of 76% at a salinity of 4 dS/m, 65% at 8 dS/m, <17% at 16 dS m⁻¹ and complete inhibition at 24 dS m⁻¹. Shoot growth, root growth, and seedling vigor decline as salinity increases. The time required for germination increases with salinity levels up to 16 dS m⁻¹, above which germination is inhibited. This is a colonizing species. Studies conducted after a major

hurricane found that the plants significantly colonized a disturbed marsh within 1 year after the storm. In the United States, the species has been observed to reach its maximum cover in September. The roots are not known to be colonized by arbuscular mycorrhizal Fungi. **Reported associates (North America):** *Acalypha rhomboidea, Acer saccharinum, Aeschynomene indica, Alternanthera philoxeroides, Ammannia auriculata, A. coccinea, Ampelopsis arborea, Baccharis halimifolia, Bacopa egensis, B. monnieri, B. repens, B. rotundifolia, Bidens aristosus, Boehmeria cylindrica, Boltonia diffusa, Brunnichia ovata, Cardamine pensylvanica, Carex, Carya aquatica, Celtis laevigata, Cephalanthus occidentalis, Chara, Chasmanthium latifolium, Coleataenia longifolia, Colocasia esculenta, Commelina diffusa, C. virginica, Conoclinium coelestinum, Crinum americanum, Cynodon dactylon, Cyperus difformis, C. erythrorhizos, C. iria, C. odoratus, Dicliptera brachiata, Diodella teres, Diospyros virginiana, Distichlis spicata, Dopatrium junceum, Echinochloa crus-galli, E. walteri, Echinodorus cordifolius, Eclipta prostrata, Eichhornia crassipes, Eleocharis atropurpurea, E. microcarpa, E. obtusa, E. parvula, Elymus virginicus, Eriocaulon cinereum, Eupatorium serotinum, Euphorbia humistrata, Fimbristylis littoralis, Fraxinus pennsylvanica, Gleditsia aquatica, G. triacanthos, Heliotropium curassavicum, H. indicum, Heteranthera limosa, H. reniformis, Hibiscus laevis, H. moscheutos, Hydrocotyle umbellata, Hydrolea quadrivalvis, H. uniflora, Ilex decidua, Ipomoea lacunosa, Iva annua, Juncus, Justicia americana, Leersia oryzoides, Leptochloa fusca, Lindernia dubia, Lobelia cardinalis, Lonicera japonica, Ludwigia alternifolia, L. decurrens, L. grandiflora, L. leptocarpa, L. palustris, L. peploides, L. sphaerocarpa, Lycopus, Mikania scandens, Mimosa strigillosa, Mimulus alatus, Nyssa aquatica, Onoclea sensibilis, Oryza sativa, Packera glabella, Panicum repens, Paspalum repens, P. vaginatum, Pennisetum glaucum, Persicaria punctata, Phanopyrum gymnocarpon, Phragmites australis, Phyla lanceolata, Physalis angulata, Pilea pumila, Planera aquatica, Pluchea camphorata, P. foetida, P. odorata, Poa annua, P. autumnalis, Ranunculus hispidus, Rhynchospora macrostachya, Rotala indica, R. ramosior, Rubus trivialis, Sacciolepis striata, Sagittaria graminea, S. guayanensis, S. lancifolia, S. latifolia, S. platyphylla, Salix nigra, Samolus valerandi, Saururus cernuus, Schoenoplectus americanus, S. tabernaemontani, Sesbania drummondii, S. herbacea, Smilax bona-nox, Spartina alterniflora, S. patens, Sphenopholis obtusata, Symphyotrichum lateriflorum, S. subulatum, Taxodium distichum, Triadenum walteri, Typha angustifolia, T. latifolia, Vigna luteola, Viola sororia, Vitis rotundifolia, Xanthium strumarium, Zizaniopsis miliacea.*

Use by wildlife: *Sphenoclea zeylanica* reportedly is eaten by various mammalian herbivores such as cattle (Mammalia: Bovidae: *Bos*) despite several (but unconfirmed) reports of their toxicity to livestock.

Economic importance: food: Steamed young plants of *Sphenoclea zeylanica* are eaten with rice as a vegetable in Southeast Asia. The plants are fairly rich in boron (57 mg/100 g dry matter). However, their use as a food is cautioned given that fresh, uncooked plants reportedly are toxic to cattle (Mammalia: Bovidae: *Bos*) and to other mammals [see next]; **medicinal:** *Sphenoclea zeylanica*, which is categorized as a general medicinal plant in Thailand, contains small amounts of alkaloids, triterpenes and triterpene saponins. The plant extracts have extremely high superoxide scavenging activity. Interperitoneal injections of alcoholic whole-plant extracts are known to cause paralysis and other serious neurological disorders when administered to mice (Mammalia: Rodentia: *Mus*) and are lethal to them at concentrations exceeding 250 mg/kg; **cultivation:** *Sphenoclea zeylanica* is not cultivated intentionally; **misc. products:** Nine allelopathic growth inhibitors have been extracted from *Sphenoclea zeylanica*. Freshly harvested plants of *S. zeylanica* have been applied effectively as a nitrogen source (administered at 27 kg N/ha) in Asian rice paddies; **weeds:** Introduced through contaminated seed stock sometime prior to 1850, *Sphenoclea zeylanica* became well established in North American rice fields by 1903 and has persisted in those habitats ever since. Strains of *S. zeylanica* resistant to 2,4-D herbicides have evolved in several southeast Asian ricefields. The plants reportedly tolerate the herbicides molinate and propanil. However, they are susceptible to conidial suspensions (3.5×10^5 conidia/mL) of a leaf blight strain (Fungi: Ascomycota: Pleosporaceae: *Alternaria alternata* f.sp. *sphenocleae*), which has been considered for use as a biological control method for the species; **nonindigenous species:** *Sphenoclea zeylanica* is native to Africa but has been introduced extensively throughout southeast Asia and to Australia, Mexico, North America, and South America mainly as a consequence of contaminated rice seed stocks distributed for agricultural production.

Systematics: The phylogenetic position of *Sphenoclea* has proved to be difficult to determine and has resulted in recommendations for quite disparate taxonomic affiliations of the genus including Campanulaceae, Lythraceae, Phytolaccaceae, Portulacaceae, and Primulaceae. Although many contemporary treatments assign the genus to Campanulaceae, a number of molecular phylogenetic studies have clarified the placement of the genus (as a distinct family) within the order Solanales, in a position sister to Hydroleaceae (e.g., Figure 5.106). Furthermore, both Campanulaceae and Lobeliaceae exhibit an expansion of the inverted repeat into the small single-copy region of the chloroplast genome, which provides compelling evidence that *Sphenoclea* (which lacks the expansion), is not closely related to either family. *Sphenoclea* contains the storage carbohydrate inulin, a feature that had suggested a link to the inulin-containing families of Asterales and Campanulales; however, given the firm phylogenetic evidence to the contrary (Figure 5.106), the distribution of this substance (which also occurs in Boraginaceae and various monocotyledons) is not critically diagnostic in a taxonomic sense. Anatomical studies have indicated numerous differences between the floral vasculature of *Sphenoclea* and either Campanulaceae or Lobeliaceae, which further exemplifies a distant relationship to either family. On the other hand, the floral development of *Sphenoclea* differs in several respects to that of Hydroleaceae, which molecular data strongly support as its sister family. *Sphenoclea* arguably is monophyletic, although

only one of the species (*S. zeylanica*) has been included in formal phylogenetic analysis (Figure 5.106). The basic chromosome number of *Sphenoclea* is $x = 12$; *S. zeylanica* ($2n = 24$) is a diploid. No hybrids involving this genus have been reported.

Comments: *Sphenoclea zeylanica* has been introduced to (and persists in) Alabama, Arkansas, Florida, Georgia, Louisiana, Mississippi, Missouri, North Carolina, Oklahoma, South Carolina, and Texas.

References: Abulude et al., 2010; Boro & Sarma, 2013; Bremer et al., 2002; Carter et al., 2014; Chabreck & Palmisano, 1973; Civico & Moody, 1979; Cook, 2004; Cronquist, 1981; Dubey, 1986; Erbar et al., 2005; Ferguson, 1998; Fuller et al., 1985; Gupta, 1959; Hakim et al., 2011; Hall et al., 1971; Hirai et al., 2000; Howard & Wells, 2009; Itoh, 1994; Juraimi et al., 2012; Kent & Johnson, 2001; Knox & Palmer, 1999; Mabbayad & Watson, 1995; Masangkay et al., 1999; Prayoonrat, 2005; Raffauf & Higurashi, 1988; Rodrigues & Naik, 2009; Sahid et al., 1995; Smith, Jr., 1988; Soerjani et al., 1987; Start & Handasyde, 2002; Thieret, 1970, 1972; Tidestrom, 1913; Toriyama et al., 2005; Tumlison & Serviss, 2007; Vijaya, 2010; Yang et al., 2006.

DICOTYLEDONS II ("ASTERID" TRICOLPATES)—"CORE" ASTERIDS: EUASTERIDS II ("CAMPANULIDS")

The informal clade that has become known as "euasterids II" or "campanulids" (Figure 5.107) is the sister group of the euasterid I clade (Figure 5.1) and includes 42 families distributed mostly among four subclades that represent the major orders: Apiales, Aquifoliales, Asterales, and Dipsacales. Additional families (Bruniaceae, Columelliaceae, Eremosynaceae, Escalloniaceae, Paracryphiaceae, Polyosmaceae, Sphenostemonacae, and Tribelaceae) occur among a number of more poorly supported clades and have not been assigned conclusively to specific orders (Figure 5.107). Although Aquifoliales are well supported as the sister clade to the remainder of the campanulids, the relationships among the other orders and families have remained weakly resolved despite the analysis of a substantial amount of DNA sequence data (Figure 5.107). Aquatic plants occur only within the four major orders, where the OBL habit has arisen independently (Figure 5.107).

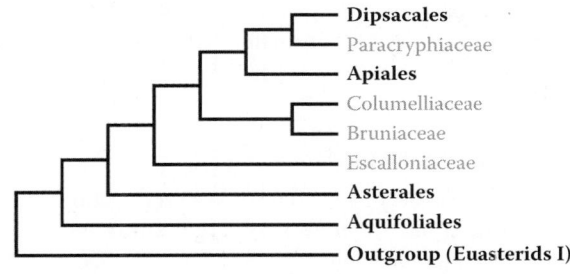

FIGURE 5.107 Relationships of major orders and families in the Euasterid II (campanulid) clade based on analysis of 10 chloroplast gene sequences. Groups containing OBL taxa in North America are indicated in bold. The OBL habit appears to have evolved several times within the campanulids (see also Soltis et al., 2005). (Adapted from Tank, D.C. & Donoghue, M.J., *Syst. Bot.*, 35, 425–441, 2010.)

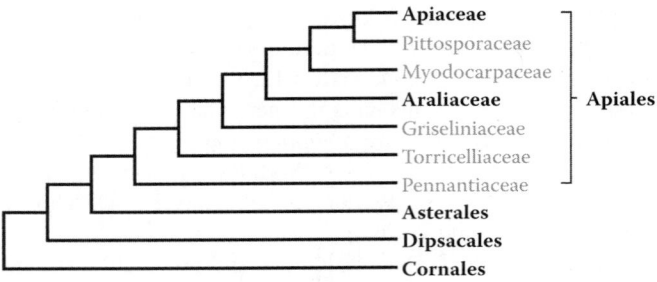

FIGURE 5.108 Phylogeny of Apiales as indicated by analysis of combined DNA sequence data. The seven families indicated comprise the order as circumscribed in the most recent classification (Plunkett et al., 2004b). These results provide compelling evidence to retain Apiaceae and Araliaceae as distinct families. The distribution of taxa with OBL species (bold) indicates two separate origins of the habit in the order. (Adapted from Chandler, G. & Plunkett, G.M., *Bot. J. Linn. Soc.*, 144, 123–147, 2004.)

ORDER 6: APIALES [7]

Apiales are a large order including more than 500 genera and over 5000 species (Plunkett et al., 2004b). Once thought to be related to rosids, the association of the order with the asterids is indicated convincingly by phylogenetic analyses of molecular data (e.g., Figures 5.1 and 5.107). Morphological similarities between the asterids and Apiales include a single integument, tenuinucellate ovules, S-type sieve-tube plastids, similar triterpenic sapogenins, polyacetylenes, alkaloids, flavonols, acetate-derived arthroquinones, and isopentenyl-substituted coumarins and absence of iridoids and tannins (Plunkett, 2001). The basic chromosome number of the order is $x = 6$ or 12 (Plunkett et al., 2004b).

Analyses of combined DNA sequence data have clarified phylogenetic relationships within the order substantially, with the recent classification of the group now recognizing seven distinct families (Figure 5.108). Among the most significant resolutions has been a refined circumscription of Apiaceae and Araliaceae, which have been treated inconsistently as distinct families or as subfamilies of Apiaceae. Recent studies resolve the taxa as distinct clades, separated by two subclades, which represent the families Myodocarpaceae and Pittosporaceae; the former family Mackinlayaceae has been transferred to Apiaceae as subfamily Mackinlayoideae (Figure 5.108; Plunkett et al., 2004b).

Family 1: Apiaceae [455]

Alternatively known by the conserved family name Umbelliferae, Apiaceae are a very large family with an estimated 455 genera and as many as 3750 species (Downie et al., 2000, 2001; Constance & Affolter, 2004). The family has been the subject of intensive phylogenetic study, which has incorporated a large amount of DNA sequence data (e.g., Katz-Downie et al., 1999; Downie et al., 2000; Chandler & Plunkett, 2004; Hardway et al., 2004; Plunkett et al., 2004b). These studies have revised the circumscription of the family to represent

a soundly monophyletic group, which can be subdivided into four clades that correspond to subfamilies: Apioideae, Azorelloideae, Mackinlayoideae, and Saniculoideae (Figure 5.109; Plunkett et al., 2004b). Most species occur within subfamily Apioideae, which contains roughly 400 of the genera and more than 3000 species (Plunkett et al., 2004b). In North America, the genera with OBL species occur within two subfamilies: *Eryngium* in subfamily Saniculoideae and the remainder in subfamily Apioideae. The OBL habit has evolved independently in each of the subfamilies. Within the latter subfamily, most of the aquatic genera are found in tribe Oenantheae, and it is fortunate that this group has now been studied fairly thoroughly (Hardway et al., 2004; Spalik & Downie, 2006; Feist & Downie, 2008; Spalik et al., 2009). These studies have helped to clarify the limits of several genera as well as their interrelationships. In particular, there is compelling evidence to remove the unusual "rachis-leaved" species from the genera *Ptilimnium* and *Oxypolis* and transfer them to the genera *Harperella* and *Tiedemannia*, respectively (Feist & Downie, 2008). That recommendation has been followed here.

Another important achievement has been the improved phylogenetic placement of taxa comprising the former subfamily Hydrocotyloideae (also recognized by some as a distinct family), which was found to be polyphyletic as previously circumscribed. Most of the species are resolved within the two clades that now are recognized as subfamilies Azorelloideae and Mackinlayoideae (Plunkett et al., 2004b). Molecular phylogenetic studies consistently have indicated that *Hydrocotyle* itself (the former type genus of the subfamily) should be included in Araliaceae (Chandler & Plunkett, 2004; Plunkett et al., 2004b).

Apiaceae have a cosmopolitan distribution and the species, though most of them terrestrial, occupy extremely diverse habitats ranging from arctic to tropical climates. Apiaceae usually are recognized easily by their suite of distinctive features. Although their habit varies to include trees, shrubs, vines, and herbs, the plants typically are aromatic (volatile terpenoids), with hollow stems, and often possess compound leaves having a characteristic sheathing petiolar base (a prominent feature of the vegetable celery). Some species have tubular, hollow, septate "rachis-leaves," where the

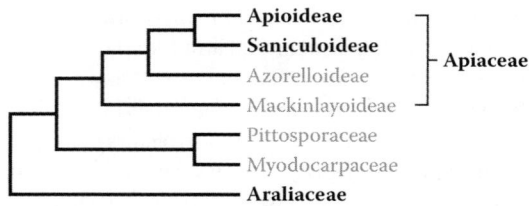

FIGURE 5.109 Relationships of subfamilies in Apiaceae as indicated by phylogenetic analysis of combined DNA sequence data. Taxa containing OBL species are indicated in bold. (Adapted from Downie, S.R. et al., *Amer. J. Bot.*, 87, 273–292, 2000; Chandler, G. & Plunkett, G.M. *Bot. J. Linn. Soc.*, 144, 123–147, 2004; Plunkett, G.M. et al., *S. Afr. J. Bot.*, 70, 371–381, 2004b.)

pinnae are reduced to hydathodes (Feist & Downie, 2008). A conspicuous trait of Apiaceae is their determinate umbellate inflorescence, which can be arranged secondarily into panicles, racemes, spikes or compound umbels, and usually is subtended by an involucre. The flowers are bisexual (except in rare cases) and reduced in size, usually with minute sepals and five small, monotonously white or yellow petals. The ovary is inferior and possesses a style that is swollen at its base as a nectariferous stylopodium. The fruit is a schizocarp that dehisces as two mericarps, which often remain attached by a fine filament (the carpophore). The carbohydrate umbelliferose is sequestered as a storage product and the endosperm contains petroselenic acid (Judd et al., 2002; Constance & Affolter, 2004).

For a family of this magnitude, there is surprisingly little information available on breeding systems or seed dispersal mechanisms (Constance & Affolter, 2004). The species frequently exhibit at least partial self-compatibility and presumably are either self-pollinating, or are pollinated by generalist insects. Interestingly, the type genus of the family (*Apium*) is derived from the Latin *apis*, meaning "bee" because the flowers attract these insects; some more restrictive pollinator systems also occur (Constance & Affolter, 2004). However, a number of studies (e.g., Gaudeul & Till-Bottraud, 2003) indicate that protandry, coupled with alternating sex expression within individual plants, can promote high levels of outcrossing. Species with winged or spiny mericarps generally are adapted for dispersal by wind or animals, respectively; however, overall structural and ecological factors ultimately regulate propagule dispersal in nature (Putz & Schmidt, 1999).

Many economically important plants occur here including popular culinary spices such as anise (*Pimpinella*), caraway (*Carum*), coriander (*Coriandrum*), cumin (*Cuminum*), dill (*Anethum*), fennel (*Foeniculum*), parsley (*Petroselinum*), and vegetables: carrot (*Daucus*), celery (*Apium*), and parsnip (*Pastinaca*, *Peucedanum*). Several of these same plants have been used as medicinal herbs and as ornamental plants. More than 160 genera currently are in cultivation, with the largest numbers of cultivars occurring in *Astrantia*, *Daucus* and *Eryngium*. A few taxa are notoriously toxic; e.g., *Cicuta* (regarded as the most toxic plants in North America) and *Conium* (poison hemlock), which led to Socrates' demise (Sullivan, 2001). Giant hogweed (*Heracleum mantegazzianum*) is a nonindigenous species with phototoxic sap, which causes severe dermatitis when coming into contact with skin that has been exposed to the sun.

There are 15 North American genera that contain OBL species within the family. *Conium maculatum* originally was regarded as OBL in the 1996 indicator list and is retained here, but has since been reclassified as a FACW, FAC species:

1. *Angelica* L.
2. *Berula* W. D. J. Koch
3. *Cicuta* L.
4. *Conium* L. [formerly OBL]
5. *Eryngium* L.
6. *Harperella* Rose
7. *Helosciadium* W. D. J. Koch
8. *Heracleum* L.
9. *Lilaeopsis* Greene
10. *Limnosciadium* Mathias & Constance
11. *Oenanthe* L.
12. *Oxypolis* Raf.
13. *Ptilimnium* Raf.
14. *Sium* L.
15. *Tiedmannia* Torr. & A. Gray
16. *Trepocarpus* Nutt. ex DC

1. *Angelica*

Alexanders, swamp whiteheads, woollyhead parsnip; angélique

Etymology: from the Latin *angĕlĭcus* ("angel-like"), with respect to its healing properties

Synonyms: *Angelocarpa*; *Archangelica*; *Callisace*; *Coelopleurum*; *Czernaevia*; *Epikeros*; *Gomphopetalum*; *Ostericum*; *Physolophium*; *Porphyroscias*; *Selinum* (in part); *Sphenosciadium*

Distribution: global: Asia; Europe; North America; **North America:** widespread except central

Diversity: global: 110 species; **North America:** 23 species

Indicators (USA): OBL: *Angelica atropurpurea*; **FACW:** *A. capitellata*

Habitat: freshwater; palustrine, riverine; **pH:** alkaline; **depth:** <1 m; **life-form(s):** emergent herb

Key morphology: stems (to 3 m) hollow, purple, glabrous, the herbage aromatic; leaves (to 4 dm) once, bi- or tri-ternately compound, the leaflets (to 1.5 dm) serrate to lobed, the petioles (to 4 dm) with inflated sheaths (to 6.5 cm); umbels (to 2 dm) compound, smooth to tomentose, peduncled (to 40 cm), with up to 45 smooth or tomentose rays (to 10 cm), umbellets spherical; flowers minute, 5-merous, pedicels reduced to a disk; petals greenish, purplish, or white; ovary inferior; mericarps (to 8 mm) compressed, smooth or tomentose, with a thin, marginal wing and unequally winged ribs

Life history: duration: perennial (persistent rootstock); **asexual reproduction:** none; **pollination:** insect; **sexual condition:** hermaphroditic; **fruit:** schizocarps (common); **local dispersal:** mericarps (water, wind); **long-distance dispersal:** mericarps (water, wind)

Imperilment: (1) *Angelica atropurpurea* [G5]; SH (MD); S1 (DE, KY, RI, TN); S2 (NC, PE); S3 (IA, QC); (2) *A. capitellata* [G5]

Ecology: general: Although *Angelica* species are associated most often with UPL meadows or woodlands, nine of the North American species occur at least occasionally in wetlands. *Angelica capitellata* (as *Sphenosciadium capitellatum* A. Gray) originally was categorized as OBL, FACW in the 1996 indicator list; however, it has been reclassified as FACW in the 2013 revision. Having been completed prior to the release of the revised list, its treatment has been retained here. *Angelica* plants are pollinated by generalist insects (Insecta), but some species may attract specific pollinators by unique combinations of their volatile floral scent components (various benzenoids and monoterpenes). The nectar of some species appears to have an intoxicating effect on some scavenging insects. The flowers are self-compatible and perfect but entire plants exhibit "temporal dioecism," where selfing is reduced by synchronized sex expression. All flowers on a plant first enter a protandrous phase for several days, and the stigmas do not become receptive until all of the stamens have withered. The fruits are flat, often winged, and usually buoyant. They are dispersed by water or wind.

Angelica atropurpurea **L.** inhabits bogs, depressions, ditches, fens, flats, floodplains, marshes, meadows, prairies, sandbars, seeps, sloughs, springs, swales, swamps, and the margins of lakes, rivers and streams. The plants grow on wet ground to shallow (<8 cm), often cold water, in exposures that vary from full sun to partial or deep shade. They usually are reported from alkaline sites (e.g., pH: 6.9–8.0) in calcareous substrates described as gravel, muck, peat, sand, or silty muck. This is a short lived, monocarpic perennial, which overwinters by means of a large (to 15 cm), branched, woody rootstock. It takes 2–4 years before reaching flowering stage, but the plants flower only once and then die. The flowers are self-compatible but temporally dioecious and outcrossed by generalist bees and wasps (Hymenoptera), beetles (Coleoptera) and flies (Diptera). The mericarps are slightly winged and appear to be dispersed mainly by wind and water. The seeds reportedly have double dormancy and may require 2 years for natural germination to occur. Light is required. Although germination is difficult, it can be enhanced by soaking seeds for 5 min in 7.6 M potassium hydroxide. The seeds normally are short lived (<1 year), but can remain viable for several years if stored frozen. Further studies on the pollination biology and seed ecology of this species are needed. **Reported associates:** *Acer rubrum, Achillea millefolium, Alnus rugosa, Amphicarpa bracteata, Andropogon gerardii, Apios americana, Apocynum cannabinum, Artemisia serrata, Asclepias incarnata, A. syriaca, Barbarea vulgaris, Boehmeria cylindrica, Bromus ciliatus, Calamagrostis canadensis, Caltha palustris, Campanula aparinoides, Carex aquatilis, C. bebbii, C. bromoides, C. canescens, C. comosa, C. hystericina, C. lacustris, C. scoparia, C. stipata, C. stricta, C. trichocarpa, C. utriculata, C. vesicaria, C. vulpinoidea, Cephalanthus occidentalis, Chelone glabra, Cichorium intybus, Cicuta maculata, Cirsium vulgare, Clematis virginiana, Comarum palustre, Cornus amomum, C. racemosa, C. sericea, Cryptotaenia canadensis, Cypripedium parviflorum, Dactylis glomerata, Doellingeria umbellata, Dulichium arundinaceum, Echinocystis lobata, Elymus canadensis, Epilobium coloratum, Equisetum arvense, Euonymus atropurpureus, Eupatorium perfoliatum, Eutrochium maculatum, E. purpureum, Fraxinus nigra, F. pennsylvanica, Galium obtusum, Gentiana andrewsii, Glyceria canadensis, G. striata, Hamamelis, Hasteola suaveolens, Helianthus giganteus, Heracleum maximum, Heuchera richardsonii, Hydrocotyle americana, Hypoxis hirsuta, Impatiens capensis, Iris versicolor, Larix laricina, Leersia oryzoides, Leonurus cardiaca, Ligusticum scoticum, Lilium michiganense, Lobelia siphilitica, Lycopus americanus, Lysimachia ciliata, L. terrestris, L. thyrsiflora, Lythrum salicaria, Maianthemum stellatum,*

Micranthes pensylvanica, Monarda fistulosa, Muhlenbergia mexicana, Myosotis scorpioides, Napaea dioica, Nasturtium officinale, Nepeta cataria, Onoclea sensibilis, Osmunda claytoniana, O. regalis, Oxypolis rigidior, Pedicularis lanceolata, Peltandra virginica, Phalaris arundinacea, Phleum, Pilea pumila, Plantago lanceolata, Poa paludigena, Pontederia cordata, Populus, Quercus macrocarpa, Ranunculus septentrionalis, Rosa palustris, Rudbeckia hirta, R. laciniata, Salix interior, Sambucus nigra, Schoenoplectus tabernaemontani, Scirpus atrovirens, S. cyperinus, Silphium perfoliatum, Sium suave, Solidago gigantea, S. ohioensis, S. patula, S. riddellii, S. sempervirens, Spartina pectinata, Spiraea alba, Symphyotrichum lanceolatum, S. puniceum, Symplocarpus foetidus, Thalictrum dasycarpum, T. thalictroides, Thelypteris palustris, Thuja occidentalis, Tilia americana, Toxicodendron vernix, Tradescantia ohiensis, Trollius laxus, Typha latifolia, Ulmus americana, U. rubra, Urtica dioica, Verbena hastata, Viola, Zizia aurea

Angelica capitellata **(A. Gray) Spalik, Reduron & S. R. Downie** grows in bogs, ditches, marshes, meadows, seeps, springs, swales, thickets, and along the margins of brooks, lakes and streams at elevations of up to 3415 m. It often is found in forest openings that receive full sun to shade. The substrates include granitic rock, granitic sand, loam, loam–shale, organic humus, pumice, silt, and silty loam. The flowers are pollinated by a host of insects (Insecta) including Coleoptera, Diptera, Hymenoptera and Lepidoptera (see *Use by wildlife*). Otherwise, little information on the life history of this species exists. Presumably, its reproductive and seed ecology are similar to the previous species. **Reported associates:** *Abies concolor, A. magnifica, Achillea millefolium, Aconitum columbianum, Agrostis rossiae, Allium validum, Alnus tenuifolia, Angelica arguta, Aquilegia formosa, Arctostaphylos, Arenaria kingii, Arnica fulgens, A. longifolia, A. mollis, Artemisia ludoviciana, A. tridentata, Aulacomnium palustre, Betula occidentalis, Bistorta bistortoides, Bryum, Calamagrostis canadensis, Calocedrus decurrens, Caltha leptosepala, Cardamine breweri, Carex abrupta, C. alma, C. aquatilis, C. echinata, C. illota, C. jonesii, C. nebrascensis, C. nervina, C. nigricans, C. rostrata, C. scopulorum, Castilleja miniata, Ceanothus leucodermis, Cercocarpus, Chamerion angustifolium, Cornus, Delphinium glaucum, Deschampsia cespitosa, Drepanocladus, Drosera rotundifolia, Eleocharis quinqueflora, Equisetum, Epilobium hornemannii, Ericameria nauseosa, Erigeron coulteri, Eriogonum umbellatum, Frasera speciosa, Gentiana calycosa, Geum macrophyllum, Glyceria elata, Hastingsia alba, Helenium bigelovii, Heracleum maximum, H. sphondylium, Hordeum brachyantherum, Hymenoxys hoopesii, Hypericum anagalloides, H. formosum, Ipomopsis aggregata, Iris missouriensis, Juncus macrandrus, J. nevadensis, Lilium kelleyanum, Linanthus nuttallii, Lonicera conjugialis, L. involucrata, Lupinus arbustus, L. confertus, L. latifolius, L. leucophyllus, Luzula subcongesta, Maianthemum stellatum, Meesia triquetra, Mertensia ciliata, Micranthes oregana, Mimulus guttatus, M. primuloides, Monardella odoratissima, Muhlenbergia filiformis, Oreostemma alpigenum, Orobanche californica,*

Osmorhiza occidentalis, Oxypolis occidentalis, Parnassia, Pedicularis groenlandica, Perideridia parishii, Philonotis, Phleum alpinum, Pinus contorta, P. jeffreyi, P. lambertiana, P. monticola, P. ponderosa, Platanthera dilatata, P. leucostachys, P. sparsiflora, Poa pratensis, P. secunda, Polemonium, Polygonum douglasii, Populus tremuloides, Potentilla drummondii, Primula tetrandra, Purshia tridentata, Ribes cereum, Rosa woodsii, Rudbeckia, Rumex, Salix eastwoodiae, S. lasiolepis, S. lutea, S. scouleriana, Scirpus congdonii, S. diffusus, Senecio triangularis, Sidalcea, Sisyrinchium bellum, Solidago, Spiraea splendens, Stellaria longipes, Swertia perennis, Symphyotrichum foliaceum, S. spathulatum, Taraxacum officinale, Thalictrum fendleri, Triantha glutinosa, Trifolium longipes, T. monanthum, Vaccinium uliginosum, Veratrum californicum, Veronica americana, Vicia, Viola adunca, V. macloskeyi.

Use by wildlife: The seeds of *Angelica atropurpurea* contain 24% protein, 19% fiber, 13% fat, and are eaten by a variety of wildlife. The foliage is grazed by white-tailed deer (Mammalia: Cervidae: *Odocoileus virginianus*). The herbage contains angular and linear furanocoumarins, which are toxic to many insects (Insecta) and result in an assemblage of specialized herbivores resistant to the compounds. These include larval Lepidoptera (Noctuidae: *Idia americalis, Papaipema harrisii*; Oecophoridae: *Agonopterix clemensella*; Papilionidae: *Papilio brevicauda, P. polyxenes*) and Diptera (Tephritidae: *Euleia fratria*). The plants are hosts of several sac fungi (Ascomycota: Dermateaceae: *Mollisia angelicae*; Mycosphaerellaceae: *Mycosphaerella angelicae, Passalora angelicae*; Valsaceae: *Diaporthopsis angelicae*; Venturiaceae: *Piggotia depressa*). Large numbers of soldier beetles (Insecta: Coleoptera: Cantharidae: *Chauliognathus marginatus*) have been reported from the inflorescences. The nectariferous flowers attract short-tongued bees (Insecta: Andrenidae: *Andrena hippotes, A. imitatrix, A. miranda, A. persimulata, A. spiraeana, A. thaspii*; Colletidae: *Hylaeus modestus*; Halictidae: *Sphecodes illinoensis*) and flies (Insecta: Diptera: Bombyliidae: *Villa alternata*), which probably serve as pollinators. This species is useful for conservation habitat management programs because it attracts many natural arthropod (Insecta) enemies including balloon flies (Diptera: Empididae), beetles (Coleoptera: Cantharidae; Coccinellidae), flower bugs (Heteroptera: *Orius insidiosus*), and wasps (Hymenoptera: Braconidae; Chalcidoidea; Cynipoidea) but relatively few herbivores such as aphids and leafhoppers (Hemiptera: Aphidoidea; Cicadellidae), lygus bugs (Hemiptera: Miridae: *Lygus*), root-maggot flies (Diptera: Anthomyiidae), thrips (Thysanoptera), and weevils (Coleoptera: Curculionidae). *Angelica capitellata* is poisonous to (but is eaten rarely by) livestock. Cattle calves (Mammalia: Bovidae: *Bos*) that are fed the green herbage (at 10–16 g/kg body weight) can exhibit respiratory distress within 4 h and die within 60 h. Consumption of the plants by horses (Mammalia: Equidae: *Equus*) results in photosensitization, which can lead to nasal and skin lesions as well as temporary (but long-lasting) blindness. The plants contain several coumarins (isoimperatorin, phellopterin,

imperatorin, oxypeucedanin, isopimpinellin) as well as a linear furanocoumarin. The flowers are visited by numerous generalist pollinators (Insecta) including many Diptera (Anthomyiidae; Bombyliidae: *Bombylius, Hemipenthes, Systoechus*; Conopidae; Milichiidae; Syrphidae: *Cheilosia. Eristalis*; Tachinidae: *Tachina*), Coleoptera (Buprestidae: *Anthaxia*; Cerambycidae: *Leptura propinqua, Pachyta armata*; Melyridae; Mordellidae; Scraptiidae), Hymenoptera (Apidae: *Apis mellifera, Bombus bifarius, B. fernaldae, B. insularis, B. silvicola, B. vosnesenskii*; Colletidae: *Hylaeus*; Halictidae: *Lasioglossum*; Megachilidae: *Chelostoma, Heriades, Megachile, Osmia*; Nyssonidae; Pompilidae: *Aporus luxus*; Pteromalidae: *Thinodytes petiolatus*; Sphecidae: *Ammophila*), and some Lepidoptera (Lycaenidae: *Plebejus saepiolus*). The flowers can shelter large numbers of predatory wasps (Insecta: Hymenoptera: Eumenidae: *Symmorphus*). The plants are used by leaf-mining flies (Insecta: Diptera: Agromyzidae: *Phytomyza*) and are larval hosts to butterflies (Insecta: Lepidoptera: Papilionidae *Papilio indra, P. zelicaon*). They host various fungi (Ascomycota: Diademaceae: *Comoclathris magna*; Pleosporaceae: *Clathrospora diplospora*; Venturiaceae: *Fusicladium peucedani, Pollaccia peucedani*).

Economic importance: food: The roots and seeds of *Angelica* species are used to flavor several alcoholic beverages including benedictine, chartreuse, gin, and vermouth. The leaf stalks and shoots of *Angelica atropurpurea* are edible and are eaten raw in salad or boiled as a stewed vegetable. The dried roots and young shoots have been candied and made into a confection since early colonial days (the fresh roots are poisonous); **medicinal:** *Angelica purpurea* is believed to be only mildly therapeutic. The dried roots (known pharmaceutically as *Angelica Radix*) contain angelic acid, coumarins, phellandrene, sesquiterpenes, and valeric acid. They were used by the Cherokee, Delaware, Iroquois, and Oklahoma tribes as an emetic, to induce perspiration, and to treat bronchitis, colds, colic, fever, flatulence, flu, gout, headaches, intestinal cramps, menstrual problems, pneumonia, rheumatism, sore throats, and urinary disorders. The Iroquois applied a root poultice to promote healing of broken bones and used a root decoction for treating frostbite. The Menominee applied a poultice of cooked, mashed roots as a general analgesic and to relieve swelling. The Meskwaki boiled the foliage to make a tea, which they used as a remedy for hayfever. Herbal teas containing this plant are available commercially and are said to relieve upset stomachs. The Paiute used root preparations of *A. capitellata* to repel head lice and to treat venereal sores; **cultivation:** *Angelica atropurpurea* is distributed as potted plants or as seed for growing as a native ornamental species. The seeds are contained in some commercial wildflower mixes; **misc. products:** The Delaware and Oklahoma tribes mixed the seeds of *Angelica atropurpurea* with tobacco for smoking. The Iroquois made a poison from the fresh roots. Essential oils from various *Angelica* species have been used as aromatic ingredients in perfume, soap, and shampoo; **weeds:** *Angelica atropurpurea* occasionally is reported as a weed, but this designation seems unwarranted and arises, perhaps, because of misidentifications with other species; **nonindigenous species:** *Angelica atropurpurea* is native to North America

Systematics: Due to the enormity of Apiaceae, it is difficult to evaluate the phylogenetic placement of *Angelica* with precision, given that many genera have not yet been surveyed. There have been several phylogenetic studies of the family that have included *Angelica* with the most inclusive containing 44 species, or roughly 40% of the total estimated number in the genus. In that analysis, *Peucedanum* resolved as the sister genus to *Angelica*. However, some analyses have indicated an endemic North American clade comprising numerous, poorly defined genera (e.g., *Aletes, Lomatium, Musineon, Oreoxis, Pseudocymopterus, Pteryxia, Tauschia*) as the sister clade to *Angelica*, and it is evident that further sampling and analyses will be necessary before a clearer picture of generic circumscriptions and relationships will emerge. The gist of most existing molecular studies is that *Angelica* consistently resolves among a group of similar genera that has been referred to informally as the "*Angelica* clade," which occupies a relatively derived position within subfamily Apioideae (Figure 5.110). Although phylogenetic resolution among many genera in the Apioideae is poor, fairly comprehensive phylogenetic surveys now justify the recognition of a monophyletic tribe Selineae, which is defined to include those genera formerly assigned to the "*Angelica*" and "*Arracacia*" clades. However, further complicating matters are the results of several analyses indicating that *Angelica* was not monophyletic as previously circumscribed. The heterogeneity of *Angelica* has been indicated by results of comparative serological analysis of seed proteins. One analysis of ITS data resolved a large clade of *Angelica* species, which included *Sphenosciadium capitellatum* and *Selinum pyrenaeum* as well but placed several *Angelica* species with other genera. Further corroborative analyses have resulted in the recommended transfer of both species to *Angelica*, which is the disposition accepted here. Several analyses have shown *Angelica capitellata* (formerly *Sphenosciadium*) to resolve with the North American species *A. arguta* and *A. breweri*, which comprised a clade distinct from Old World species. A more inclusive survey of ITS data for 44 Angelica species further refined the circumscription of *Angelica* as a monophyletic genus by incorporating several genera (*Coelopleurum, Czernaevia*), and one species of *Ostericum* (*O. koreanum*), which nested within a clade containing the type species *A. sylvestris*, along with the exclusion of several *Angelica* species, which associated with different genera. Unfortunately, neither *Angelica atropurpurea* nor *A. capitellata* was included in the broadest survey of the genus; thus, the relationships of the OBL North American species remain uncertain. Further clarification of generic limits for *Angelica* will require more extensive surveys of taxa in the Apioideae as well as additional data. No DNA sequence has yet been analyzed phylogenetically for *A. atropurpurea*, so it can only be presumed to fall within the clade of North American *Angelica* species along with *A. capitellata*. In any case, *Angelica* (the only genus of tribe Selineae to contain OBL species) represents at least one independent origin of the habit in subfamily Apioideae (Figure 5.110). The basic chromosome number of *Angelica* is $x = 11$. Based on other reported counts, the genus (as well as both *A. atropurpurea* and *A. capitellata*) is nearly uniformly diploid ($2n = 22$). Hybrids involving *A. atropurpurea* or *A. capitellata* have not been reported.

Comments: *Angelica atropurpurea* occurs in northeastern North America and *A. capitellata* is restricted to the western United States (CA, ID, NV, OR).

References: Alarcón et al., 2008; Bell & Constance, 1957; Berenbaum, 1981, 1983; Castlebury et al., 2003; Crête et al., 2001; Cruden & Hermann-Parker, 1977; Dearness, 1924; Downie et al., 2000, 2001, 2002; East, 1940; Ellis & Everhart, 1890; Feng et al., 2009; Fernald & Kinsey, 1943; Fiedler & Landis, 2007; Fiedler et al., 2007a; Fuller & McClintock, 1986; Gao et al., 1998; Katz-Downie et al., 1999; Lee & Soine, 2006; Runkel & Roosa, 1999; Sarker & Nahar, 2004; Shneyer et al., 2003; Smiley & Rank, 1986; Spalik et al., 2004; Spencer, 1981; Spinner & Bishop, 1950; Sun & Downie, 2004; Sun et al., 2004; Swift, 2008; Thorp et al., 1983; Tollsten et al., 1994; Valiejo-Roma et al., 2002; Wheeler, Jr., 1988; Wood et al., 1940.

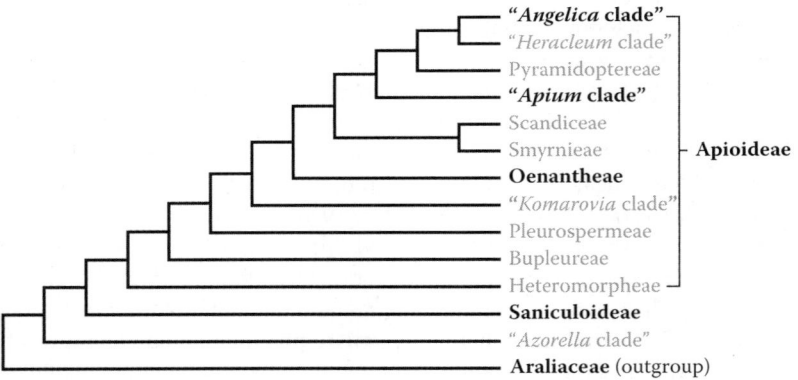

FIGURE 5.110 Major lineages in Apiaceae subfamily Apioideae as indicated by parsimony analysis of five combined molecular data sets. The scattered distribution of groups containing OBL North American species indicates at least three independent origins of the habit in the subfamily, with a fourth in subfamily Saniculoideae. *Angelica* is the only genus in the "*Angelica* clade" that contains an OBL species. (Adapted from Downie, S.R. et al., *Edinb. J. Bot.*, 58, 301–330, 2001.)

2. Berula

Cutleaf waterparsnip

Etymology: from the Greek *bêrullos*, a precious sea-green gem

Synonyms: *Afrocarum*; *Siella*; *Sium* (in part)

Distribution: global: Africa; Asia; Europe; North America; **North America:** central, western

Diversity: global: 7 species; **North America:** 1 species

Indicators (USA): OBL: *Berula incisa*

Habitat: freshwater, saline lacustrine, palustrine, riverine; **pH:** 5.3–8.7; **depth:** 0–2 m; **life-form(s):** emergent herb, submersed (rosulate)

Key morphology: roots fibrous, in fascicles; stems (to 8 dm) stoloniferous; leaves (to 3 dm), pinnately compound (decompound when submersed), the leaflets (to 4 cm) entire, lobed or serrate, in 5–12 pairs, petioles (to 12 cm) narrowly sheathing; umbels (to 2 dm) compound, with up to 15 rays (to 3 cm), pedicels to 5 mm; flowers small, white; stylopodium conical; mericarps (to 2 mm) orbicular, flattened laterally, ribs inconspicuous, pericarp thick, corky

Life history: duration: perennial (rhizomes, stolons); **asexual reproduction:** rhizomes, shoot fragments, stolons; **pollination:** insect; **sexual condition:** hermaphroditic; **fruit:** schizocarps (common); **local dispersal:** water (dislodged plantlets, fragments, fruits); **long-distance dispersal:** fruits (birds)

Imperilment: (1) *Berula incisa* [G4]; SX (NY); S1 (BC); S2 (IA, MI, WY)

Ecology: general: *Berula* species occupy aquatic or wetland habitats, often characterized by flowing water or in riverine wetlands, where dispersal by fragmentation is most effective. The habitats typically are alkaline (even somewhat saline) with fairly high conductivity. The flowers reportedly are insect pollinated (entomophilous), but the reproductive biology of this genus is poorly known. The corky fruits are buoyant, and are dispersed locally by water.

***Berula incisa* (Torr.) G. N. Jones** is a submersed or emergent aquatic, which grows in still to rapidly flowing waters of creek/river beds, ditches, dunes, fens, flats, floodplains, gullies, hot springs, marshes, meadows, oxbows, ponds, pools, prairies, rivers, saltmarshes, seeps, sloughs, springs, streams, swamps, washes, and along the margins of canals, ditches, lagoons, lakes, ponds, reservoirs, rivers and streams at elevations of up to 2347 m. It occurs in clear, shallow, mesotrophic or nutrient poor, hard waters (e.g., total alkalinity: 233 mg/L as $CaCO_3$; 575–800 μmhos/cm^{-1}; pH = 5.3–8.7; 7.5 = photosynthetic optimum) having mean N-NH_4 of 3.5–11.0 μg/L and mean P-PO_4 levels of 7.0–12.5 μg/L. The plants occupy open, sunny sites or partial shade on substrates of basalt, calcareous gravel, muck, mud, obsidian, peat, granitic sand, sandy clay, silt, or silty loam. Sometimes they occur in saline or subsaline substrates. The flowers are insect-pollinated, but submersed plants typically remain sterile. In the Old World, substantial pollen loads have been observed on some beetles (Insecta: Coleoptera: Cerambycidae: *Acmaeops collaris*). Otherwise, little information on pollination biology exists for this species. Observed seed densities range from 26 to 2260 seeds/m^2 (mean = 373 seeds/m^2). The seeds reportedly have no after-ripening requirement for germination and can develop a transient or short-term persistent seed bank. Plants often are abundant even at sites where no propagule bank is maintained. The flattened, corky mericarps are dispersed by water. They are believed to adhere to birds, which would provide wider dispersal. The plants are evergreen and persist throughout the winter. They are clonal and fragment irregularly, and the fragments can be well represented in river drift, especially after floods. Establishment of small, detached plants can occur during autumn. Exposure to stress induces higher biomass allocation to shoots, which can result in higher survivorship during subsequent stressful periods. This species can grow into dense masses with highly intertangled roots. The rhizomes and stolons are short lived. The roots are infected by vesicular–arbuscular mycorrhizae, but only when the plants grow under emergent conditions. The plants are tolerant to disturbance and effectively colonize sites undergoing restoration. They are vegetatively plastic, with size-dependent morphological traits and patch variability decreasing as nutrients decrease and flow velocity increases. The chlorophyll content of emersed and submersed leaves is similar; however, the latter are characterized by a slightly higher photosynthetic rate. Carbon assimilation from the sediments increases as water flow velocity decreases. Biomass production is estimated as 0.5 kg/m^2, yielding 8.5 tons/ha/year dry weight and 6.3 tons/ha/year organic matter. Field studies have shown that nutrient retention of a stream bottom covered by *Berula* plants, is about three times higher than that of barren sand. **Reported associates:** *Acer negundo, Agalinis purpurea, Ageratina adenophora, Agoseris aurantiaca, Agrostis exarata, A. gigantea, A. scabra, A. stolonifera, Alisma gramineum, Allium canadense, Alnus oblongifolia, A. rhombifolia, Ambrosia acanthicarpa, A. confertiflora, A. psilostachya, Ammannia coccinea, Anemopsis californica, Angelica atropurpurea, Anticlea elegans, Apium graveolens, Arundo donax, Asclepias fascicularis, A. speciosa, Atriplex phyllostegia, A. prostrata, A. rosea, A. semibaccata, Baccharis salicifolia, B. sarothroides, B. sergiloides, Bassia hyssopifolia, Betula pumila, Bidens aureus, B. cernuus, B. connatus, B. ferulifolius, B. frondosus, B. laevis, B. tripartitus, Boehmeria cylindrica, Bromus rubens, Calamagrostis stricta, Caltha leptosepala, C. palustris, Cardamine bulbosa, Carex aquatilis, C. hystericina, C. lurida, C. nebrascensis, C. pellita, C. praegracilis, C. rosea, C. rostrata, C. sartwellii, C. scirpoidea, C. simulata, C. stricta, Castilleja angustifolia, C. linariifolia, C. minor, Centaurium exaltatum, Chara vulgaris, Chenopodium rubrum, Cicuta bulbifera, C. maculata, Cirsium altissimum, C. arvense, C. muticum, C. wrightii, Cladium mariscoides, Clematis ligusticifolia, Cleome serrulata, Cleomella obtusifolia, Conyza canadensis, Cordylanthus maritimus, Cornus sericea, Cotula coronopifolia, Cressa erecta, C. runcinata, Cuscuta glomerata, Cypripedium candidum, Dasiphora floribunda, Deschampsia cespitosa, Descurainia pinnata, Dichanthelium acuminatum, Distichlis spicata, Echinochloa* crus-galli, Eleocharis acicularis, E. palustris, E. parishii, E. rostellata, Elodea canadensis, E. nuttallii, Elymus canadensis, E. trachycaulus, Epilobium ciliatum, E. strictum, Epipactis gigantea, Equisetum fluviatile, E. hyemale, Eriogonum fasciculatum, Eriophorum angustifolium, Eupatorium perfoliatum, Euthamia graminifolia, E. occidentalis, Eutrochium maculatum, Fissidens grandifrons, Forestiera pubescens, Fraxinus velutina, Galium labradoricum, G. obtusum, G. trifidum, Gentianopsis crinita, G. virgata, Geum aleppicum, Glaux maritima, Glyceria borealis, G. grandis, G. striata, Glycyrrhiza lepidota, Grindelia columbiana, G. squarrosa, Gutierrezia microcephala, Helenium autumnale, H. puberulum, Helianthus annuus, H. nuttallii, Heterotheca grandiflora, Hippuris vulgaris, Hordeum brachyantherum, H. jubatum, Hierochloe odorata, Hosackia oblongifolia, Hydrocotyle verticillata, Hygroamblystegium fluviatile, Hypoxis hirsuta, Impatiens capensis, Ipomoea, Iris missouriensis, Iva axillaris, Juglans major, Juncus acuminatus, J. balticus, J. brachycarpus, J. dudleyi, J. mexicanus, J. nodosus, J. rugulosus, J. torreyi, J. xiphioides, Lactuca serriola, Leersia oryzoides, Lemna gibba, L. minor, L. minuta, Lepidium latifolium, Liparis loeselii, Lobelia cardinalis, L. kalmii, L. siphilitica, Lobularia maritima, Ludwigia palustris, Lycopus americanus, L. asper, L. rubellus, Lysimachia quadriflora, L. thyrsiflora, Lythrum californicum, Maianthemum stellatum, Malosma laurina, Marchantia polymorpha, Marrubium vulgare, Melilotus alba, M. officinalis, Mentha arvensis, Menyanthes trifoliata, Mimulus cardinalis, M. glabratus, M. guttatus, M. ringens, Muhlenbergia asperifolia, M. rigens, Myosotis scorpioides, Myriophyllum sibiricum, Nasturtium officinale, Nitrophila occidentalis, Oenothera elata, O. speciosa, Oligoneuron riddellii, Oscillatoria rubescens, Packera aurea, Panicum bulbosum, P. capillare, P. virgatum, Parnassia glauca, Parthenocissus, Pedicularis groenlandica, P. lanceolata, Persicaria hydropiper, P. hydropiperoides, Petasites sagittatus, Phalaris arundinacea, Phragmites australis, Phyla lanceolata, Physalis longifolia, Pilea fontana, Pinus edulis, P. ponderosa, Piptatherum miliaceum, Plantago australis, P. subnuda, Platanthera hyperborea, Platanus, Pluchea carolinensis, P. sericea, Polypogon interruptus, P. monspeliensis, P. viridis, Populus angustifolia, P. deltoides, P. fremontii, Potamogeton alpinus, P. foliosus, P. gramineus, P. natans, P. strictifolius, Potentilla anserina, Prunella vulgaris, Pseudognaphalium luteoalbum, Puccinellia nuttalliana, Pycnanthemum virginianum, Quercus agrifolia, Q. chrysolepis, Q. dunnii, Q. wislizeni, Ranunculus cymbalaria, R. longirostris, R. macranthus, R. sceleratus, Rhizoclonium hieroglyphicum, Rhodiola integrifolia, Rhus trilobata, Rhynchospora capillacea, Ribes americanum, Ricinus communis, Rosa arkansana, R. californica, R. woodsii, Rubia tinctoria, Rumex crispus, R. maritimus, R. orbiculatus, R. verticillatus, Sagittaria cuneata, Salazaria mexicana, Salix eriocephala, S. exigua, S. geyeriana, S. gooddingii, S. laevigata, S. lasiolepis, S. pedicellaris, Sambucus nigra, Samolus valerandi, Sanicula canadensis, Saururus cernuus, Schoenoplectus acutus, S. americanus, S. pungens, S. tabernaemontani, Scirpus microcarpus, Scleria verticillata, Scutellaria galericulata, Shepherdia, Sisymbrium altissimum, Sium suave, Solidago canadensis, S. spectabilis, S. uliginosa, Sonchus asper, Sorghastrum nutans, Sparganium angustifolium, S. eurycarpum, Spartina gracilis, S. pectinata, Sphagnum, Spiranthes cernua, S. lucida, S. romanzoffiana, Stuckenia filiformis, Symphyotrichum ascendens, S. boreale, S. divaricatum, S. eatonii, S. laeve, S. novae-angliae, S. puniceum, Symplocarpus foetidus, Tamarix ramosissima, Thalictrum dasycarpum, Thaspium barbinode, Torreyochloa pallida, Toxicodendron diversilobum, Trifolium fragiferum, T. repens, Triglochin maritimum, Typha angustifolia, T. domingensis, T. latifolia, Urtica dioica, Utricularia macrorhiza, Verbena bracteata, Veronica americana, V. anagallis-aquatica, Viola nephrophylla, V. sororia, Vitis girdiana, Wolffia borealis, W. columbiana, Xanthium strumarium, Zannichellia palustris.

Use by wildlife: *Berula incisa* produces coumarins, which are chemical deterrents to herbivory. One amphipod species (Crustacea: Amphipoda: Dogielinotidae: *Hyalella azteca*) consumes the roots of the plants, gaining protection from predation by ingestion of their defensive chemicals. *Berula incisa* is a host plant for insect larvae (Insecta: Lepidoptera: Papilionidae: *Papilio polyxenes*) and a mite (Acari: Tenuipalpidae: *Brevipalpus obovatus*). The latter is Asian, but related species are known to occur in North America. The plants are a host to fungi (Ascomycota: Mycosphaerellaceae: *Septoria sii*). *Berula incisa* is an element of habitat for the Page springsnail (Gastropoda: Hydrobiidae: *Pyrgulopsis morrisoni*) in Arizona, for the endangered least chub (Osteichthyes: Cyprinidae: *Iotichthys phlegethontis*) in Utah, and for the San Cristóbal pupfish (Osteichthyes: Profundulidae: *Profundulus hildebrandi*), which is endemic to Chiapas, Mexico. The plants are highly palatable to (and are readily consumed by) pond snails (Gastropoda: Lymnaeidae: *Lymnaea stagnalis*). The foliage contains falcarindiol and falcarindol, which are algicidal. These plants reportedly are eaten readily when fed to captive leopard tortoises (Reptilia: Testudinidae: *Stigmochelys pardalis*). Various inconsistencies exist in the literature. In Africa, plants of *B. incisa* reportedly have killed cattle within an hour after consumption; however in Italy, they are used as a food supplement for cattle.

Economic importance: food: The leaves and flowers of *Berula incisa* were eaten by the Apache and White Mountain tribes. The flowers are used as food in Asia by the people of central Turkey. However, due to the presence of coumarins, the plants (at least their roots) should be regarded as toxic; **medicinal:** The foliage and blooms of *B. incisa* were used as a general medicinal by the Apache and White Mountain people. The Zuni treated athlete's foot and rashes using an infusion made from the whole plant; it was an ingredient in "schumaakwe cakes," which were applied externally to treat rheumatism and swelling. This species is used medicinally in KwaZulu-Natal for treating wounds. Cold dichloromethane/methanol extracts of the foliage exhibit antiplasmodial activity against strain D10 of *Plasmodium falciparum*, one of the protozoan parasites responsible for human malaria; **cultivation:** *Berula incisa* occasionally is sold as an ornamental water garden plant; **misc. products:** none; **weeds:** *Berula incisa* is reported as a weed in Australia, Mexico and South Africa; **nonindigenous species:** *Berula incisa* has been introduced to Australia where it is adventive in New South Wales, South Australia, Victoria, and Western Australia. It was introduced unintentionally to the Kashmir region of Asia, where it has become naturalized.

Systematics: The systematics of *Berula* have been in a state of flux. Once recognized as containing only two species (*B. erecta, B. thunbergii*), *Berula* has been shown by recent molecular phylogenetic investigations to include the monotypic *Afrocarum* (*A. imbricatum*) and several species of *Sium* (*S. bracteatum, S. burchellii, S. repandum*), which were misplaced from the core members of that latter genus (Figure 5.112). Consequently, molecular studies support a current circumscription of the genus as containing eight species (Figure 5.113). *Berula erecta*, which comprises several distinct clades as indicated by major DNA sequence divergence (Figure 5.113), has been split into three distinct species (*B. erecta, B. incisa, B. thunbergii*). In particular, the North American plants are distinct from Old World material and their restoration to species status as *B. incisa* is appropriate. *Berula* sensu lato clearly is a member of the tribe/clade *Oenantheae*, where it resolves phylogenetically as a group

that is closely related to *Sium* and *Helosciadium* (Figures 5.111 through 5.113). The base number of *Berula* probably is $x = 6$. The diploid chromosome number of *B. incisa* is variable and includes counts of $2n = 12, 18, 20$.

Comments: *Berula incisa* occurs in central and western North America. Some care should be taken when checking literature accounts, because the genus has been applied mistakenly to the similarly spelled genus for birch (*Betula*).

References: Anderson, 1943; Beck-Nielsen & Madsen, 2001; Bell & Constance, 1957; Berenbaum, 2009; Bolen, 1964; Bornette et al., 1994; Cellot et al., 1998; Childers et al., 2003b; Clarkson et al., 2004; Combroux et al., 2002; Crawford & Hartman, 1972; Elger et al., 2004; Ertuğ, 2000; Flowers, 1934; Frampton, 1995; Gross, 1999; Hardway et al., 2004; Haslam, 1978; Khuroo et al., 2007; Kirsten & Blinn, 2003; Lewis, & Elvin-Lewis, 2003; Martinez, 2006; Mitchell, 1974; Moran et al., 2002; Moseley, 1995; Nekola, 2004; Osmond et al., 1981; Pearson & Leoschke, 1992; Preston & Croft, 1997; Puijalon & Bornette, 2004; Puijalon et al., 2008; Spalik & Downie, 2006; Spalik et al., 2009; Thompson & Grime, 1979; Tremolieres et al., 1994; Velázquez-Velázquez & Schmitter-Soto, 2004; Viegi et al., 2003; Vogt et al., 2004; Willemstein, 1987; Wright & Mills, 1967; Zalewski et al., 1998.

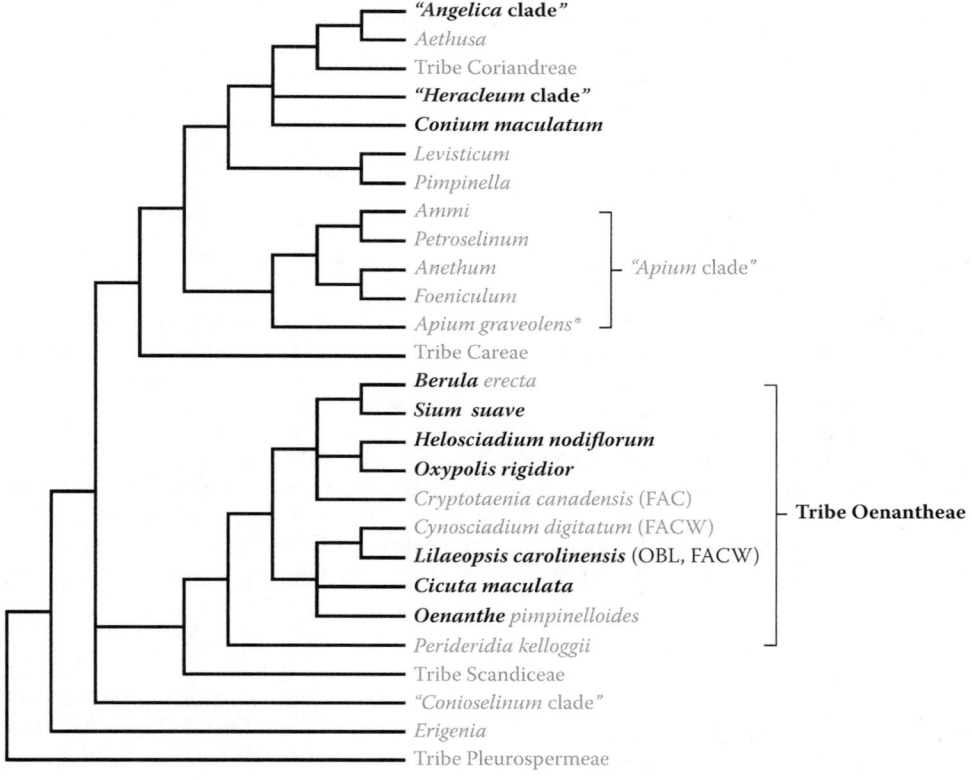

FIGURE 5.111 Broad relationships in Apiaceae subfamily Apioideae as indicated by analysis of nrITS DNA sequence data. Taxa with OBL North American representatives are indicated in bold. Although *Helosciadium* had been treated formerly as a section of *Apium*, these results clearly indicate its distinctness from the latter (asterisked), which occurs in the distantly related "*Apium* clade." Most of the aquatic Apiaceae occur in tribe Oenantheae, where the habit could have had a single origin; 2–3 other derivations of OBL taxa are indicated. (Adapted from Downie, S.R. et al., *Can. J. Bot.*, 80, 1295–1324, 2002; Valiejo-Roman, C.M. et al., *Taxon*, 51, 685–701, 2002.)

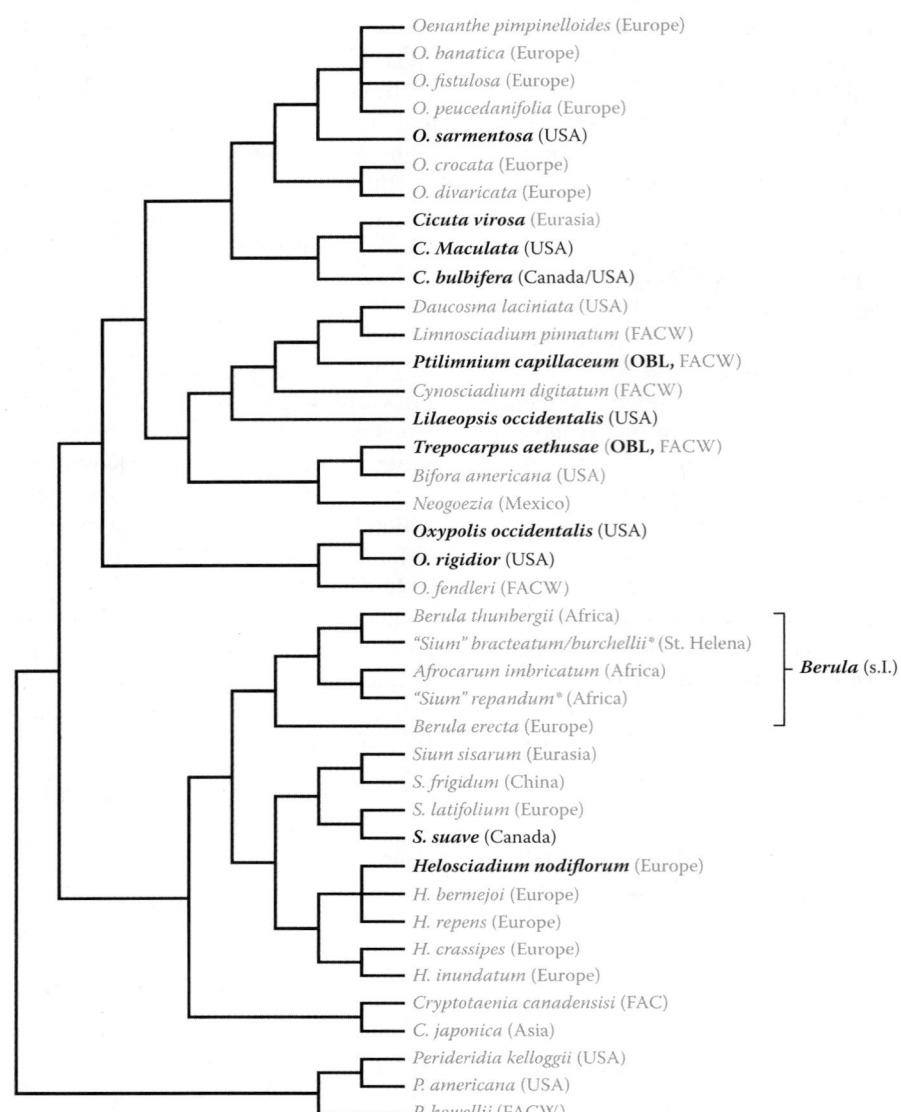

FIGURE 5.112 Phylogenetic relationships in Apiaceae tribe Oenantheae as resolved by analysis of nrITS sequence data. OBL North American taxa are indicated in bold. Most of the genera resolve as monophyletic except *Berula*, which includes a few wayward species assigned previously to *Sium* (asterisked) and the monotypic African genus *Afrocarum*; the remainder of *Sium* resolves as a clade sister to *Helosciadium*. The origin of material is indicated except where wetland indicator designations are provided (these are all from North America). The number of origins of the aquatic habit is difficult to ascertain in this tribe. The entire *Berula/Helosciadium/Sium* clade (where nearly all species are aquatic or wetland plants), might represent only one origin; separate origins are indicated for *Cicuta* and *Oxypolis*, with another origin in *Oenanthe* (where only about six of the 40 species are aquatic); as many as three separate origins could have occurred in the *Limnosciadium* group. Conversely, the entire tribe could be regarded as having an aquatic origin, with several reversals back to the terrestrial habit. Some intergeneric relationships differ slightly from those depicted in other analyses with more DNA sequence data but fewer taxa (see Figure 5.111). (Adapted from Hardway, T.M. et al., *S. Afr. J. Bot.*, 70, 393–406, 2004.)

3. Cicuta

Cowbane, water hemlock; carotte à Moreau, cicutaire

Etymology: from the Latin *cicuta*, meaning "pipe" in reference to the hollow internodes, which were used by shepherds to make flutes

Synonyms: none

Distribution: global: Asia; Europe; North America; **North America:** widespread

Diversity: global: 4 species; **North America:** 4 species

Indicators (USA): OBL: *Cicuta bulbifera, C. douglasii, C. maculata, C. virosa*

Habitat: freshwater; palustrine; **pH:** 3.7–8.5; **depth:** <1 m; **life-form(s):** emergent herb

Key morphology: rootstock chambered, sap yellow to reddish brown, roots fibrous or tuberous; shoots (to 25 dm) erect, hollow; leaves (to 40 cm) alternate, 1–3 pinnate, the divisions

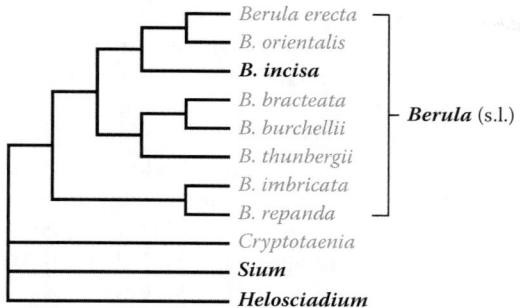

Berula (s.l.)

— Berula erecta
— B. orientalis
— **B. incisa**
— B. bracteata
— B. burchellii
— B. thunbergii
— B. imbricata
— B. repanda
— Cryptotaenia
— **Sium**
— **Helosciadium**

FIGURE 5.113 Interspecific relationships in the redefined *Berula* as indicated by phylogenetic analyses. In this circumscription, the genus includes eight distinct species, all essentially aquatic. OBL North American taxa are indicated in bold. (Adapted from Spalik, K. et al., *Taxon*, 58, 735–748, 2009.)

linear or broad (to 4 cm), margins dentate or serrate, petioles (to 30 cm) sheathed at base, axillary bulbils sometimes present; umbels (to 18 cm) axillary and terminal, twice compound, with up to 30 spreading rays (to 6 cm), involucre absent or deciduous; flowers (to 3 mm) epigynous, 5-merous, pedicelled (to 10 mm); calyx of minute (to 0.5 mm) deltoid lobes; petals (to 1.3 mm) white, short clawed at base; mericarps (to 4 mm) ovoid to round, slightly compressed, corky-ribbed; seeds (to 3.3 mm) flat or concave

Life history: duration: perennial (rhizomes; tubers); **asexual reproduction:** bulbils, rhizomes, tubers; **pollination:** insect; **sexual condition:** hermaphroditic; **fruit:** schizocarps (common); **local dispersal:** bulbils, fruits, tuberous rootstocks (water); **long-distance dispersal:** fruits (birds, mammals)

Imperilment: (1) *Cicuta bulbifera* [G5]; SH (DC, OR, VA); S1 (DE, MD, NC, WV, WY); S2 (AK, ID, WA, <u>YT</u>); S3 (<u>BC</u>, MT, NE); (2) *C. douglasii* [G5]; (3) *C. maculata* [G5]; S1 (<u>BC</u>, FL, KS, NC); S2 (<u>QC</u>, <u>YT</u>); S3 (WY); (4) *Cicuta virosa* [G4]; S2 (<u>BC</u>); S3 (<u>AB</u>, <u>QC</u>, <u>YT</u>)

Ecology: general: All *Cicuta* species occur in wetlands and are regarded as OBL aquatics. The plants commonly contain insecticidal furanocoumarins, which are associated ecologically with species of wetlands and other open habitats. *Cicuta* is self-compatible but protandrous and "temporally dioecious," where there is little overlap between the male and female flowering phases within individual umbels. Consequently, the breeding system is presumed to be predominantly outcrossing. The fruits are surrounded by a corky coat, which enables them to float and they are dispersed by water. The seeds must be in water to germinate, which in all species is promoted by a 12/12 h alternating temperature regime of 15°C/21°C for 2 weeks, followed by a constant temperature of 21°C. Germination markedly decreases in dried seeds. The plants (except *C. bulbifera*) develop rosettes during their first year from seed and require a vernalization period in order to flower. All species overwinter by their persistent rootstocks. The repeated replacement of rootstocks each fall results in clonal growth that can persist for decades. The primary root is short lived in *Cicuta*, and is replaced by a secondary root system during the seedling stage.

Cicuta bulbifera **L.** grows on beaches, in bogs, carrs, ditches, fens, flats, floodplains, marshes, meadows, mudflats, oxbows, prairies, sandbars, seeps, sloughs, springs, streams, swales, swamps, and along the margins of flowages, lakes, ponds, and rivers. This species extends northward to 65.5° N latitude. The plants occur in shallow (to 30 cm; average: 13–19 cm) soft water or hard water in open to lightly shaded sites. The substrates can be chemically diverse (pH: 3.7–8.5) but typically are acidic (usually pH <7.0) and include clay, cobble, gravel, humus, marl, muck, mucky peat, mud, peat, sand, and silt. Plants grow on floating organic mats and even can occur on mossy logs. The foliage contains several furanocoumarins (which can be insecticidal) and the plants are associated with a more specialized insect fauna; those Apiaceae lacking the compounds have more "generalist" insect associates. The minute flowers are visited by small insects (Insecta) including bees (Hymenoptera), beetles (Coleoptera), flies (Diptera) and wasps (Hymenoptera); however, the plants frequently are found without mature fruits. As in other congeners, the flowers presumably are protandrous and the umbels temporally dioecious, but these assumptions require verification for this species. The seeds are reported to be poisonous and are avoided by birds; however, viable seeds have been recovered from the scat of the common raccoon (Mammalia: Procyonidae: *Procyon lotor*). Thus, at least some dispersal must occur by endozoic transport (although there is no indication of how long the raccoons remain viable afterwards). Despite the ability to carry out normal sexual reproduction, the flowers often abort and reproduction predominantly is vegetative, occurring by axillary bulbils. The bulbils are produced on the stems but eventually detach and float on the water surface where they are dispersed widely. Plants show a high ecological association with habitats that have been invaded by purple loosestrife (*Lythrum salicaria*). However, unlike that species, they are more likely to occur where low fire frequency has resulted in a fairly high accumulation of litter. Plants have been found to occur much more densely on overwintering mounds of the muskrat (Mammalia: Muridae: *Ondatra zibethicus*). **Reported associates:** *Abies balsamea, Acer rubrum, A. saccharinum, A. spicatum, Acorus calamus, Agalinis paupercula, Agrimonia, Alisma subcordatum, A. triviale, Alnus rugosa, A. serrulata, Amaranthus rudis, A. tuberculatus, Amerorchis rotundifolia, Andromeda polifolia, Andropogon gerardii, Asclepias incarnata, Betula lutea, B. papyrifera, B. pumila, Bidens cernuus, B. coronatus, B. frondosus, B. tripartitus, Boehmeria cylindrica, Bolboschoenus fluviatilis, Brachythecium rivulare, Calamagrostis canadensis, C. stricta, Calla palustris, Caltha palustris, Campanula aparinoides, Cardamine pensylvanica, Carex aquatilis, C. buxbaumii, C. comosa, C. crawei, C. diandra, C. eburnea, C. hystericina, C. lacustris, C. lanuginosa, C. lasiocarpa, C. limosa, C. pellita, C. prairea, C. retrorsa, C. rossii, C. rostrata, C. sartwellii, C. stipata, C. stricta, C. utriculata, Centaurea, Cephalanthus occidentalis, Chamaedaphne calyculata, Chara, Chelone glabra, Cicuta douglasii, C. virosa, Cirsium muticum, Cladium mariscoides, Clintonia borealis, Comarum palustre, Coptis trifolia,*

Cornus amomum, C. obliqua, C. sericea, Cyperus ferruginescens, C. strigosus, Cypripedium reginae, Dasiphora floribunda, Decodon verticillatus, Doellingeria umbellata, Drepanocladus, Drosera anglica, D. rotundifolia, Dulichium arundinaceum, Echinochloa walteri, Eleocharis acicularis, E. erythropoda, E. palustris, E. tenuis, Epilobium palustre, E. strictum, Equisetum arvense, E. fluviatile, E. variegatum, Eragrostis hypnoides, Erechtites hieracifolia, Eriophorum angustifolium, E. gracile, Eupatorium perfoliatum, Eutrochium maculatum, Fraxinus nigra, F. pennsylvanica, Galium asprellum, G. circaezans, G. labradoricum, G. tinctorium, G. trifidum, Geum canadense, Glyceria canadensis, G. striata, Helenium autumnale, Helianthus giganteus, Hibiscus moscheutos, Hippuris montana, Huperzia selago, Hypericum boreale, H. canadense, H. majus, Ilex verticillata, Impatiens capensis, Iris versicolor, I. virginica, Juncus arcticus, J. brevicaudatus, J. canadensis, J. stygius, Larix laricina, Lathyrus palustris, Leersia oryzoides, Lemna minor, L. trisulca, Limnobium spongia, Lobelia kalmii, L. siphilitica, L. spicata, Ludwigia palustris, Lycopodium clavatum, L. complanatum, Lycopus americanus, L. uniflorus, Lysimachia quadriflora, L. terrestris, L. thyrsiflora, Lythrum salicaria, Mentha arvensis, Menyanthes trifoliata, Muhlenbergia glomerata, M. mexicana, Myriophyllum sibiricum, Nasturtium officinale, Nymphaea, Oligoneuron ohioense, Onoclea sensibilis, Osmunda regalis, Parnassia palustris, Pedicularis macrodonta, P. parviflora, Penthorum sedoides, Persicaria amphibia, P. coccinea, P. hydropiper, P. hydropiperoides, P. lapathifolia, P. pensylvanica, P. punctata, Phalaris arundinacea, Phragmites australis, Phryma leptostachya, Picea mariana, Pilea fontana, P. pumila, Pinus strobus, Platanthera clavellata, P. dilatata, P. psychodes, Potamogeton natans, P. richardsonii, P. zosteriformis, Potentilla norvegica, Pycnanthemum virginianum, Ranunculus abortivus, R. macounii, Rhynchospora alba, R. capillacea, Riccia fluitans, Rorippa palustris, Rumex orbiculatus, Rumex verticillatus, Sagittaria latifolia, Salix alaxensis, S. bebbiana, S. candida, S. nigra, S. niphoclada, S. pedicellaris, S. petiolaris, S. pyrifolia, Sarracenia purpurea, Scheuchzeria palustris, Schoenoplectus heterochaetus, S. tabernaemontani, Scirpus cyperinus, Scutellaria galericulata, S. incana, S. lateriflora, Sium suave, Solanum dulcamara, Solidago gigantea, Sparganium erectum, S. eurycarpum, Spartina pectinata, Sphagnum teres, Spiraea alba, S. douglasii, S. tomentosa, Spirodela polyrhiza, Symphyotrichum boreale, S. lanceolatum, S. lateriflorum, S. novae-angliae, S. puniceum, Thuja occidentalis, Thelypteris palustris, Toxicodendron vernix, Triadenum fraseri, T. virginicum, Trichophorum alpinum, T. pumilum, Triglochin maritimum, T. palustre, Typha angustifolia, T. latifolia, Ulmus americana, Urtica dioica, Utricularia cornuta, U. intermedia, U. macrorhiza, U. minor, Vaccinium corymbosum, V. oxycoccos, Vallisneria americana, Verbena hastata, V. urticifolia, Vernonia fasciculata, Veronica anagallis-aquatica, Viburnum, Viola cucullata, Wolffia, Xanthium strumarium, Zanthoxylum americanum.

Cicuta douglasii (DC.) **Coult. & Rose** occurs in still or flowing fresh to saline waters of bogs, ditches, depressions,

fens, flats, marshes, meadows, mudflats, ponds, salt marshes, seeps, sloughs, swamps, and along the margins of rivers and streams at elevations of up to 2740 m. Waters tend to be shallow but can reach depths of up to 65 cm. The plants can tolerate full sun to deep shade. Reported substrates include clay, clay loam, granite, gravelly loam, muddy clay, rocks, and sandy loam. Some plants have been observed growing on the top of floating logs. The seeds do not retain viability when they have been dried but germinate well after 12–24 months of storage in water (0.5°C–25°C). Successful germination has been achieved by placing the seeds under an alternating 12/12 h, 30°C/20°C light and temperature regime. The seeds are not long lived and do not germinate after 2–3 years of storage. The seeds float and are dispersed by water. They have been recovered from traps placed in irrigation ditches in quantities reaching 0.82 seeds per 254 kL of water. As the fall season approaches, roundish rootstocks form at the plant base but above the ground level. These structures can detach during periods of flooding and serve as propagules. Data reported for this species should be interpreted cautiously, because this species apparently is an allotetraploid, and is extremely difficult to distinguish morphologically from *C. maculata*. **Reported associates:** *Abies lasiocarpa, Achillea millefolium, Acorus calamus, Actaea rubra, Agastache urticifolia, Agoseris glauca, Agropyron spicatum, Agrostis humilis, A. stolonifera, Alisma triviale, Allium brevistylum, Alnus rhombifolia, A. rugosa, Alopecurus aequalis, Alyssum desertorum, Amelanchier alnifolia, Antennaria parvifolia, Arenaria congesta, Athyrium filix-femina, Baccharis salicifolia, Balsamorhiza sagittata, Beckmannia syzigachne, Betula occidentalis, Bidens cernuus, Bistorta bistortoides, Bouteloua curtipendula, B. gracilis, Bromus carinatus, Calamagrostis canadensis, C. rubescens, C. stricta, Caltha leptosepala, Carex aquatilis, C. bebbii, C. geyeri, C. illota, C. lenticularis, C. microptera, C. nebrascensis, C. praegracilis, C. retrorsa, C. rossii, C. simulata, C. stipata, C. utriculata, C. vesicaria, Castilleja miniata, Ceanothus velutinus, Cerastium arvense, Chamerion angustifolium, Cicuta bulbifera, Cirsium arvense, Collinsia parviflora, Comarum palustre, Corchorus aestuans, Cornus sericea, Crataegus douglasii, Cyperus esculentus, Dactylis glomerata, Dasiphora floribunda, Deschampsia cespitosa, Drosera rotundifolia, Dryopteris campyloptera, Eleocharis acicularis, E. palustris, E. parishii, Elymus albicans, E. glaucus, E. virginicus, Epilobium ciliatum, E. watsonii, Equisetum arvense, E. hyemale, E. laevigatum, E. sylvaticum, Erigeron subtrinervis, Eriophorum gracile, Eucephalus engelmannii, Euthamia occidentalis, Eutrochium maculatum, Festuca idahoensis, Fragaria vesca, Galium boreale, G. trifidum, G. triflorum, Geranium richardsonii, Geum macrophyllum, Glyceria elata, G. grandis, G. occidentalis, G. striata, Helianthus nuttallii, Heracleum lanatum, H. maximum, Heterotheca villosa, Hippuris vulgaris, Howellia aquatilis, Hypericum anagalloides, H. scouleri, Juncus balticus, J. bufonius, J. nodosus, J. saximontanus, J. torreyi, Juniperus communis, Lactuca, Lathyrus ochroleucus, Lemna minor, Lepidium campestre, Lupinus latifolius, Luzula parviflora, Lycopodium annotinum, Lycopus asper,*

Lysimachia ciliata, Mahonia nervosa, Maianthemum stellatum, Medicago lupulina, Melica bulbosa, Melilotus officinalis, Mentha arvensis, Mertensia franciscana, M. lanceolata, Micranthes oregana, Mimulus alsinoides, M. floribundus, M. guttatus, M. primuloides, Monarda fistulosa, Muhlenbergia filiformis, Myosotis scorpioides, Osmorhiza chilensis, O. occidentalis, Oxalis, Oxypolis fendleri, Oxytropis riparia, O. splendens, Paxistima myrsinites, Pedicularis groenlandica, P. parryi, Persicaria amphibia, P. lapathifolia, Petasites sagittatus, Phalaris arundinacea, Phleum alpinum, P. pratense, Physocarpus capitatus, Picea engelmannii, Pinus contorta, Plantago major, Platanthera dilatata, P. sparsiflora, Poa cusickii, P. palustris, P. pratensis, Polemonium occidentalis, Polygonum polygaloides, Populus angustifolia, P. balsamifera, P. tremuloides, Potamogeton gramineus, Potentilla anserina, Primula jeffreyi, P. tetrandra, Pseudotsuga menziesii, Pyrola asarifolia, P. chlorantha, Ranunculus flabellaris, R. gmelinii, R. longirostris, R. macounii, R. orthorhynchus, Ribes inerme, Ricciocarpus natans, Rorippa palustris, R. sphaerocarpa, Rosa acicularis, Rubus idaeus, Rudbeckia laciniata, Rumex aquaticus, R. crispus, R. salicifolius, Salix barclayi, S. bebbiana, S. drummondiana, S. eastwoodiae, S. exigua, S. geyeriana, S. lutea, S. monica, S. monochroma, Schoenoplectus tabernaemontani, Scirpus microcarpus, S. pallidus, Senecio hydrophilus, S. serra, S. triangularis, Sium suave, Solanum dulcamara, Solidago canadensis, S. gigantea, Sparganium angustifolium, Sphagnum, Symphoricarpos albus, Symphyotrichum foliaceum, S. laeve, S. lanceolatum, S. spathulatum, Taraxacum officinale, Thalictrum occidentale, Thelypodiopsis sagittata, Toxicodendron rydbergii, Trifolium longipes, Tripleurospermum maritima, Trisetum spicatum, Typha latifolia, Urtica dioica, Verbena bracteata, V. stricta, Veronica americana, V. anagallis-aquatica, Viola canadensis, V. nuttallii.

***Cicuta maculata* L.** inhabits tidal or nontidal, freshwater or brackish (0.5%–1.5% salinity), shallow (to 25 cm) waters of brooks, canals, channels, depressions, ditches, fens, flats, floodplains, marshes, meadows, prairies, river bottoms, sandbars, seeps, sloughs, springs, swales, swamps, thickets, vernal pools, and the margins of lakes, ponds, rivers, springs, and streams at elevations of up to 2590 m. Site exposures vary from open to shaded. The substrates are alkaline (pH: 7.2–8.5) and can consist of clay, loam, muck, mucky peat, peat, sand, sandy clay silt, sandy gravel, sandy loam, or silty loam. The plants have been observed growing on stumps, on floating logs, in mats of floating vegetation, or on hummocks formed by *Decodon* (Lythraceae) plants. Flowering will not occur unless the tuberous rootstocks have developed adequate food reserves. The plants produce five inflorescences on average. The flowers (produced May to August) are visited by various insects (Insecta) including bees (Hymenoptera), butterflies (Lepidoptera), and flies (Diptera), which serve as pollinators. There is evidence that the plants might be apomictic, at least on occasion. A single plant can produce upwards of 5500 seeds (each averaging 2.4 mg) in a season. The seeds have morphological dormancy and are highly viable (100%) with 37% germination (in 1–9 days at 21°C) reported after receiving 3 months of moist-cold treatment; i.e., stored on moist filter paper at 3.3°C. Good germination has been observed for stratified seeds (5°C) brought to a 25°C/15°C temperature regime. The seeds form a minor seed bank and their germination has been observed (albeit at a small level) even from forested soils. Typically, these plants are short lived, monocarpic perennials, where the stem and old rootstocks die back in fall, leaving the new tuberous rootstocks to overwinter. The rootstocks develop underground and provide no effective means of vegetative reproduction. However, some plants have been observed to produce slender, shallow rhizomes, which can facilitate asexual reproduction. As in other members of the genus, the mericarps float and are dispersed mainly by water. **Reported associates:** *Acer negundo, A. rubrum, A. saccharinum, Aconitum reclinatum, Acorus calamus, Actaea rubra, Agoseris glauca, Agrostis gigantea, A. hyemalis, A. stolonifera, Alisma subcordatum, Allium cernuum, Alnus oblongifolia, A. rugosa, A. serrulata, Alternanthera philoxeroides, Amelanchier canadensis, Amaranthus cannabinus, Ambrosia psilostachya, A. trifida, Amphicarpaea bracteata, Andropogon gerardii, Anemone canadensis, Anticlea elegans, Apios americana, Apocynum cannabinum, Arisaema triphyllum, Arnica fulgens, Arnoglossum plantagineum, Artemisia serrata, A. tridentata, Asclepias incarnata, A. speciosa, A. sullivantii, Athyrium filix-femina, Baccharis halimifolia, Berberis haematocarpa, B. repens, Berchemia scandens, Betula lutea, Bidens bipinnatus, B. cernuus, B. discoideus, B. frondosus, B. laevis, B. tripartitus, Boehmeria cylindrica, Boltonia asteroides, Bouteloua curtipendula, Boykinia aconitifolia, Brachythecium rivulare, Bromus inermis, Calamagrostis canadensis, C. inexpansa, C. stricta, Callicarpa americana, Calliergon, Caltha palustris, Calystegia sepium, Camassia scilloides, Campanula rotundifolia, Campanulastrum americanum, Cardamine clematitis, Carex alata, C. atherodes, C. atlantica, C. aurea, C. bromoides, C. chordorrhiza, C. communis, C. comosa, C. crawei, C. debilis, C. diandra, C. hyalinolepis, C. hystericina, C. intumescens, C. lacustris, C. leptonervia, C. lupuliformis, C. lupulina, C. lurida, C. magellanica, C. muskingumensis, C. nebrascensis, C. occidentalis, C. pellita, C. praegracilis, C. ruthii, C. sartwellii, C. scoparia, C. seorsa, C. stipata, C. stricta, C. tetanica, C. utriculata, C. vesicaria, C. vulpinoidea, Carpinus caroliniana, Ceanothus greggii, Celtis laevigata, Cephalanthus occidentalis, Ceratophyllum demersum, Chamerion angustifolium, Chara, Chelone glabra, C. lyonii, Chenopodium rubrum, Chrysosplenium americanum, Cinna arundinacea, Cirsium arvense, C. undulatum, Cladium jamaicense, Claytonia caroliniana, Clematis crispa, Clethra alnifolia, Comandra umbellata, Comarum palustre, Commelina diffusa, C. virginica, Conoclinium coelestinum, Coreopsis tripteris, Cornus amomum, C. foemina, C. sericea, Cryptotaenia canadensis, Cuscuta gronovii, C. pentagona, Cyperus aristatus, C. bipartitus, C. flavescens, C. strigosus, Cypripedium candidum, Dactylis glomerata, Dalea candida, D. purpurea, Dasiphora floribunda, Decodon verticillatus, Decumaria barbara, Deschampsia cespitosa, Desmanthus illinoensis, Desmodium* *canadense, Dichanthelium dichotomum, Diphylleia cymosa, Doellingeria sericocarpoides, Dracocephalum parviflorum, Dryopteris spinulosa, Dulichium arundinaceum, Echinochloa, Eclipta prostrata, Eichhornia crassipes, Eleocharis acicularis, E. atropurpurea, E. elliptica, E. ovata, E. palustris, Elymus caninus, E. glaucus, E. trachycaulus, E. villosus, Ephedra, Epilobium californicum, E. coloratum, E. leptophyllum, Equisetum arvense, E. hyemale, E. laevigatum, Eriophorum chamissonis, E. viridicarinatum, Eryngium yuccifolium, Euonymus obovata, Eupatorium perfoliatum, Euthamia graminifolium, Eutrochium fistulosum, E. maculatum, Fallopia convolvulus, Forestiera, Frangula californica, Fraxinus caroliniana, F. nigra, F. pennsylvanica, F. profunda, F. velutina, Galium aparine, G. asprellum, G. boreale, G. obtusum, G. tinctorium, Gentiana andrewsii, Geum aleppicum, G. canadense, G. geniculatum, Glottidium vesicarium, Glyceria canadensis, G. grandis, G. septentrionalis, G. striata, Glycyrrhiza lepidota, Hamamelis virginiana, Hasteola suaveolens, Helenium autumnale, Helianthus annuus, H. giganteus, Heliopsis helianthoides, Hesperostipa spartea, Heracleum lanatum, Heuchera rubescens, Hibiscus laevis, H. moscheutos, Hierochloe odorata, Hordeum jubatum, Houstonia serpyllifolia, Hydrocotyle ranunculoides, H. umbellata, H. verticillata, Hymenocallis floridana, Hypericum formosum, H. majus, H. mutilum, H. sphaerocarpum, Hypnum, Hypoxis hirsuta, Ilex opaca, I. verticillata, Impatiens capensis, Iris pseudacorus, I. versicolor, I. virginica, Itea virginica, Iva frutescens, Juglans major, Juncus balticus, J. brevicaudatus, J. dudleyi, J. effusus, J. longistylus, J. nodosus, J. roemerianus, J. tenuis, J. torreyi, Koeleria macrantha, Kosteletzkya pentacarpos, Laportea canadensis, Larix laricina, Lathyrus palustris, L. venosus, Leersia oryzoides, L. virginica, Lemna minor, Leucothoe racemosa, Liatris pycnostachya, L. spicata, Lilaeopsis chinensis, Lilium grayi, L. michiganense, L. philadelphicum, L. superbum, Lindera benzoin, Linum rigidum, Liquidambar styraciflua, Lobelia cardinalis, L. siphilitica, L. spicata, Lolium arundinaceum, L. pratense, Lonicera japonica, Ludwigia palustris, Lupinus, Lycopus americanus, L. asper, L. rubellus, L. uniflorus, Lyonia ligustrina, Lysimachia nummularia, L. quadriflora, L. quadrifolia, L. terrestris, L. thyrsiflora, Lythrum lineare, Magnolia virginiana, Mahonia haematocarpa, Melilotus alba, M. officinalis, Mentha arvensis, Menyanthes trifoliata, Micranthes micranthidifolia, Mimosa aculeaticarpa, Mimulus alatus, M. ringens, Monarda didyma, M. fistulosa, Muhlenbergia racemosa, Murdannia keisak, Myrica caroliniensis, M. cerifera, Napaea dioica, Nasturtium officinale, Nuphar advena, Nyssa biflora, N. sylvatica, Oenothera pallida, Oligoneuron ohioense, O. riddellii, Onoclea sensibilis, Osmunda regalis, Osmundastrum cinnamomeum, Oxydendrum arboreum, Oxypolis rigidior, Pachystima myrsinites, Packera aurea, P. glabella, P. paupercula, Panicum capillare, P. virgatum, Parnassia asarifolia, P. glauca, Parthenium integrifolium, Parthenocissus quinquefolia, Paspalum vaginatum, Pedicularis lanceolata, Peltandra virginica, Penstemon eatonii, Persea palustris, Persicaria arifolia, P. hydropiper, P. hydropiperoides, P. punctata, P.*

sagittata, P. setacea, P. virginiana, Phalaris arundinacea, Phanopyrum gymnocarpon, Phleum pratense, Phlox pilosa, Phragmites australis, Physostegia virginiana, Pilea pumila, Pinus taeda, Pluchea, Poa palustris, P. pratensis, Polygala verticillata, Polygonum achoreum, Pontederia cordata, Potamogeton illinoensis, Potentilla arguta, P. norvegica, Prenanthes racemosa, Primula meadia, Prunella vulgaris, Ptilimnium capillaceum, Pycnanthemum virginianum, Quercus bicolor, Q. gambelii, Q. michauxii, Q. turbinella, Ranunculus acris, R. macounii, R. sceleratus, Ratibida pinnata, Rhynchospora corniculata, Ribes americanum, Rosa blanda, R. palustris, R. woodsii, Rudbeckia hirta, R. laciniata, Rumex crispus, R. conglomeratus, R. orbiculatus, R. verticillatus, Saccharum giganteum, Sagittaria australis, S. lancifolia, S. latifolia, Salix bebbiana, S. boothii, S. exigua, S. interior, S. lutea, S. nigra, S. petiolaris, Sambucus nigra, Saururus cernuus, Schizachyrium scoparium, Schoenocrambe linearifolia, Schoenoplectus pungens, S. tabernaemontani, Scirpus atrovirens, S. cyperinus, S. pendulus, Scolochloa festucacea, Scrophularia lanceolata, Scutellaria galericulata, S. lateriflora, Sidalcea, Silphium compositum, S. laciniatum, S. perfoliatum, Sisymbrium altissimum, Sisyrinchium angustifolium, S. demissum, Sium suave, Smilax bona-nox, Solidago canadensis, S. patula, S. sempervirens, Sonchus arvensis, Sorghastrum nutans, Sparganium americanum, S. natans, Spartina cynosuroides, S. pectinata, Sphagnum, Sphenopholis pensylvanica, Spiraea alba, Stachys palustris, S. tenuifolia, Stellaria corei, S. longifolia, Symphoricarpos occidentalis, Symphyotrichum ericoides, S. lanceolatum, S. lateriflorum, S. novae-angliae, S. puniceum, Tanacetum vulgare, Taxodium distichum, Teucrium canadense, Thalictrum alpinum, T. clavatum, T. dasycarpum, T. pubescens, Thaspium trifoliatum, Thelesperma megapotamicum, Thelypteris noveboracensis, T. palustris, Thuja occidentalis, Tiarella cordifolia, Toxicodendron radicans, Tradescantia occidentalis, Trautvetteria caroliniensis, Triadenum fraseri, T. walteri, Triadica sebifera, Trifolium, Triglochin maritimum, T. palustre, Typha angustifolia, T. latifolia, Ulmus americana, Urtica dioica, Vaccinium corymbosum, Valeriana ciliata, V. edulis, Vallisneria americana, Veratrum viride, Verbascum thapsus, Verbena hastata, V. urticifolia, Vernonia fasciculata, Veronica anagallis-aquatica, Veronicastrum virginicum, Viburnum recognitum, Viola cucullata, V. macloskeyi, Vitis arizonica, Woodwardia areolata, Zanthoxylum americanum, Zizania aquatica, Zizaniopsis miliacea, Zizia aptera.

Cicuta virosa L. grows in still to slightly flowing fresh to brackish waters of bogs, fens, marshes, meadows and along the margins of bays, lakes, ponds, and streams in boreal portions of North America (to 67° N latitude) at elevations of up to 800 m. The plants grow in open to lightly shaded sites and often occur on floating mats. The substrates are somewhat acidic (pH: 5.1–6.6; extremes = 4.2–7.4) and include mineral soils and peat. The flowers (typically opening from July to August) are pollinated by unspecialized insects including bumblebees (Hymenoptera: *Bombus*), flies (Diptera: Syrphidae) and wasps (Hymenoptera: Vespidae). The bright coloration of the stylopodium may function in attracting insects (Insecta). This species forms a transient seed bank with densities of up to 52 seeds/m² reported. The seeds have innate dormancy, which is readily broken a by a moist-chilling treatment and a fluctuating (12/12 h; 15°C/26°C) temperature regime. The floating fruits are dispersed on the water surface. In the fall season, new above-ground rootstocks form at the base of the plants; these can detach and serve as propagules, especially during periods of high water. **Reported associates:** *Acorus americanus, Alnus tenuifolia, Arctophila fulva, Bidens cernuus, Bistorta vivipara, Calamagrostis canadensis, Calla palustris, Calliergon, Caltha natans, C. palustris, Cardamine pratensis, Carex aquatilis, C. canescens, C. diandra, C. lasiocarpa, C. lyngbyei, C. pseudocyperus, C. rostrata, C. utriculata, Cicuta bulbifera, Cinclidium, Comarum palustre, Conioselinum gmelinii, Drepanocladus aduncus, Equisetum fluviatile, E. palustris, Festuca rubra, Galium trifidum, Hippuris vulgaris, Hordeum brachyantherum, Hydrocotyle vulgaris, Juncus bufonius, Lemna minor, Luzula multiflora, Lysimachia thyrsiflora, Menyanthes trifoliata, Myrica gale, Myriophyllum, Parnassia palustris, Pedicularis macrodonta, Petasites sagittatus, Phragmites australis, Poa palustris, Ranunculus cymbalaria, R. gmelini, R. thyrsiflora, Rhinanthus minor, Rumex arctica, Salix candida, S. planifolia, Schoenoplectus tabernaemontani, Scutellaria galericulata, Sium suave, Sparganium angustifolium, S. natans, Sphagnum, Stellaria longifolia, Typha angustifolia, T. latifolia.*

Use by wildlife: In general, *Cicuta* species are highly poisonous to livestock and other animals and rarely are eaten. As little as 1.2–2.7 g of fresh *C. douglasii* tubers per kg body weight can be lethal to domestic sheep (Mammalia: Bovidae: *Ovis aries*). However, some organisms are adapted for feeding on these plants. *Cicuta bulbifera* is a larval food plant of the eastern black swallowtail butterfly (Insecta: Lepidoptera: Papilionidae: *Papilio polyxenes*). The plants contribute to habitat of the jumping mouse (Mammalia: Dipodidae: *Zapus hudsonius*). The flowers, roots and stems of *Cicuta douglasii* are known to be foraged by grizzly bears (Mammalia: Ursidae: *Ursus arctos horribilis*) and the seeds are eaten harmlessly by quail (Aves: Phasianidae: *Coturnix*). It is a larval host of several insects (Insecta: Lepidoptera: Oecophoridae: *Depressaria angustati, D. daucella*; Papilionidae: *Papilio polyxenes*). It is a host of rust Fungi (Basidiomycota: Pucciniaceae: *Uromyces americanus, U. lineolatus*) and a parasitic leaf-spot fungus (Ascomycota: Mycosphaerellaceae: *Septoria sii*) as well as other Fungi including Ascomycota (Mycosphaerellaceae: *Mycosphaerella sagedioides*; Venturiaceae: *Fusicladium peucedani*), Basidiomycota (Pucciniaceae: *Puccinia cicutae*), and Blastocladiomycota (Physodermataceae: *Physoderma*). Among its congeners *C. maculata* hosts the most diverse larval lepidopteran fauna (Insecta: Lepidoptera), which includes species of Epermeniidae (*Epermenia cicutaella*), Noctuidae (*Papaipema birdi*), Oecophoridae (*Agonopterix clemensella, Depressaria cinereocostella, D. juliella*), and Papilionidae (*Papilio polyxenes, P. zelicaon*). This species is a known host of at least 47 species of flies (Insecta: Diptera: Syrphidae; Tachinidae) and 37 species of parasitic wasps in six families (Insecta: Hymenoptera: Braconidae; Chrysididae; Eucoilidae;

Ichneumonidae; Pteromalidae; Tiphiidae). Tumbling flower beetles (Insecta: Coleoptera: Mordellidae: *Mordella quadripunctata*) and short-winged flower beetles (Kateretidae: *Heterhelus abdominalis*) have been collected from the flowers, which also are visited by pollinating bees (e.g., Hymenoptera: Apoidea: *Hylaeus illinoisensis*), snout butterflies (e.g., Lepidoptera: Nymphalidae: *Libytheana carinenta*), and flies (e.g., Diptera: Tabanidae: *Tabanus calens*). Weevil larvae (Insecta: Coleoptera: Curculionidae: *Apion pennsylvanicum*), undergo their development within the shoots. *Cicuta maculata* also hosts numerous fungi including Ascomycota (Incertae sedis: *Ascochyta thaspii, Neottiospora caricum, Nigredo scirpi*; Botryosphaeriaceae: *Phyllosticta abietis*; Leptosphaeriaceae: *Coniothyrium*; Mycosphaerellaceae: *Ramularia cicutae, Septoria sii*; Pleosporaceae: *Alternaria, Dendryphiella vinosa, Pleospora herbarum*; Protomycetaceae: *Protomyces macrosporus*; Rhizinaceae: *Phymatotrichopsis omnivora*; Sclerotiniaceae: *Botrytis cinerea*), Basidiomycota (Pucciniaceae: *Puccinia cicutae, Uromyces lineolatus*), and Oomycota (Peronosporaceae: *Plasmopara nivea*). Despite the toxicity of *C. maculata*, the stems (but not the leaves) are eaten by muskrats (Mammalia: Muridae: *Ondatra zibethicus*) as a summer food. Blue winged warblers (Aves: Parulidae: *Setophaga pinus*) have been observed to feed on various insects taken from the umbels. *Cicuta virosa* is a larval host of some Lepidoptera (Oecophoridae: *Depressaria daucella*) and several aphids (Insecta: Hemiptera: Aphididae: *Aphis*; *Rhopalosiphum*; *Siphocoryne*). The plants are distasteful to and are avoided by the phytophagous grass carp (Osteichthyes: Cyprinidae: *Ctenopharyngodon idella*).

Economic importance: food: *Cicuta* plants contain highly poisonous oenanthotoxin and cicutoxin, which are unsaturated polyacetylene alcohols that act as potent central nervous system receptor blockers. As many as 11 different polyacetylenes have been reported from the aerial and underground tissues of *C. virosa*. In this species, the physical form of cicutoxin (i.e., a rigid specific-length hydrocarbon chain separating two hydroxy groups) correlates with prolonged neuronal action potential, which has been postulated as the basis of its toxicity to vertebrates. Species in this genus are regarded as the most poisonous plants in all of North America and their ingestion induces nausea and abdominal cramps within minutes, eventually leading to seizures and death. Mortality rates approaching 30% have been reported. *Cicuta maculata* and *C. douglasii* are the most highly (i.e., deadly) poisonous (0.75–1.01 mg/g cicutoxins); whereas, *C. bulbifera* and *C. virosa* contain much smaller amounts (0.01–0.07 mg/g) of cicutoxins. The toxins are concentrated most highly in the root/rhizome of the plant. No part of any of these plants ever should be eaten and even a single bite of the more toxic species could be fatal. The consumption of wildfowl, which have fed on *C. virosa* plants, also can result in neurotoxic effects and renal failure. Similarly, honey derived from the pollen of *C. virosa* can convey its toxic properties to those who ingest it. Although *C. virosa* was well known as poisonous by various native tribes (e.g., Haisla, Hanaksiala; Inupiat, Kuskokwagmiut, and Western Eskimos), the

Inuktitut Eskimos would wrap fresh fish in the leaves of *C. virosa* before boiling. There is no effective antidote to cicutoxin poisoning; however, an unsubstantiated report suggests that a tea made from *Eupatorium perfoliatum* might alleviate symptoms; **medicinal:** The potent properties of these plants led to their widespread use in traditional medicine. The Carrier people of British Columbia applied the heated roots of *Cicuta douglasii* as a "plaster" to treat arthritis and rheumatism. The dried roots of *C. maculata* were used in a similar fashion by the Cree, Iroquois, and Woodlands tribes. The Idaho Kootenai treated sores by applying a paste from the pounded roots of *C. douglasii*. The Iroquois made a decoction of *C. maculata* to treat bruises, sprained joints, and broken bones. The Paiute and Shoshoni used a similar preparation to alleviate muscle pains and rheumatism. They used the root pulp to reduce swelling from snake bites. The Shoshoni made an eye wash from the roots. The Seminoles formulated a decoction from the roots stems and leaves of *C. maculata* to treat high fevers. The roots were ingested as a means of suicide by the Iroquois and Montana tribes, given as a poison by the Onondaga, and used as a contraceptive by the Cherokee. Cicutoxin isolated from *Cicuta maculata* exhibits antileukemial activity. Seizures induced by ingestion of cicutoxin reportedly can be alleviated substantially by intravenous administration of the benzodiazepine derivative diazepam ("valium"). Aqueous and ethanolic extracts of *C. maculata* exhibit slight antimicrobial activity to *Staphylococcus aureus*; **cultivation:** *Cicuta* rarely is available commercially and all species are discouraged from plantings because of their extreme toxicity; **misc. products:** The powdered root of *Cicuta douglasii* was used as an arrow poison by some North American tribes. The Klamath mixed the mashed roots of *C. maculata* with rattlesnake venom or decomposed deer liver to make arrow poisons. Ingestion of *C. maculata* was used ceremonially by the Cherokee. The Cherokee treated corn seeds with a root infusion of *C. maculata* for use as an insecticide during planting. The Iroquois applied a decoction of this plant as a disinfectant floor wash. The Ojibwa used the roots in preparing a hunting medicine. The Chippewa mixed seeds of *C. maculata* with tobacco before smoking it; **weeds:** The highly toxic *Cicuta douglasii* and *C. maculata* often are regarded as serious weeds especially in areas where cattle are grazed; **nonindigenous species:** none

Systematics: *Cicuta* is a sister clade to *Oenanthe* in tribe Oenantheae of Apiaceae (Figures 5.112 and 5.114). Other than by molecular data, the monophyly of the genus is indicated by several shared characters including a thickened stem base with transverse partitions, primary lateral leaflet veins directed toward the margin sinuses, and presence of the poisonous substances cicutoxin and cicutol. Molecular data and cytological analyses have indicated that *C. douglasii* may be a polyphyletic allotetraploid derived maternally from *C. maculata* and also that *C. bulbifera* may be of hybrid origin (Figure 5.114). The same analyses indicate a possible fifth species in the genus, although this question currently remains

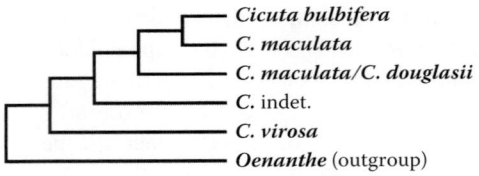

FIGURE 5.114 Phylogenetic relationships in the genus *Cicuta*, as indicated by phylogenetic analysis of maternal (cpDNA) markers. The nuclear DNA markers (not shown) yield a discordant phylogeny, with the position of *C. bulbifera* altered to comprise a sister clade to a mixed *C. maculata/C. douglasii* clade (possibly indicating the hybrid origin of *C. bulbifera*). *Cicuta douglasii* is believed to be an allotetraploid derived maternally from *C. maculata*. The unidentified taxon (indet.) was identified previously as *C. douglasii*, but may represent a novel species. All taxa shown (bold) are categorized as OBL in North America. (Adapted from Lee, C.-S. & Downie, S.R., *Can. J. Bot.*, 84, 453–468, 2006.)

under study. The base chromosome number of *Cicuta* is $x = 11$; *Cicuta bulbifera*, *C. maculata* and *C. virosa* are diploids ($2n = 22$); whereas, *C. douglasii* ($2n = 44$) apparently is an allotetraploid having a complement of 22 smaller and 22 larger chromosomes.

Comments: *Cicuta maculata* is widespread in North America. *Cicuta bulbosa* occurs in the northern half of North America, *C. virosa* only in the far north (Alaska and Canada; also in Eurasia), and *C. douglasii* only in western North America.

References: Aarvik et al., 1994; Ajima et al., 1999; Allen et al., 2002; Auclair et al., 1976; Baten, 1936; Beal, 1898; Beauchamp, 1900; Bell & Kane, 1981; Berenbaum, 1981; Borchardt et al., 2008; Buell & Buell, 1975; Christensen & Brandt, 2006; Comes et al., 1978; COSEWIC, 2004b; Coulter & Rose, 1889; Coville, 1897; Crawford & Hartman, 1972; Crevecœur, 1905; Cronin et al., 1978; Cruden & Hermann-Parker, 1977; Detmers, 1910b; Eleuterius & Caldwell, 1984; Enders, 1932; Fenton, 1986; Finke & Scriber, 1988; Fleckenstein, 2007; Graenicher, 1911; Hager & Vinebrooke, 2004; Hamilton & Bunnell, 1987; Hapner & Reinartz, 2005; Heidel, 1983; Hendrix & Sun, 1989; Hestand et al., 1973; Hewitt & Miyanishi, 1997; Hines & Ruiz, 2000; Holm, 1925; Johnston et al., 2007; Jones, 2007; Jutila, 2002; Kangas & Hannan, 1985; Karlin & Lynn, 1988; Karling, 1956; Kawahara, 2006; Kellerman & Carleton, 1885; Kelley & Bruns, 1975; Kercher et al., 2007; Konoshima & Lee, 1986; Landenberger & McGraw, 2004; Leck, 1996; Leck & Simpson, 1987; Lee & Downie, 2006; Lisberg & Young, 2003; Majka et al., 2008; Maurya et al., 2004; McAtee, 1920; McDonald, 1976; McGill, 1973; Mulé, 1983; Mulligan & Munro, 1981; North & Nelson, 1985; Panter et al., 1996; Plitmann, 2002; Quimby, 1951; Racine et al., 1998a, 1998b; Reed, 1936; Ritch-Krc et al., 1996; Robertson, 1924b; Rüdiger, 1974; Salsbury, 1984; Scatizzi et al., 1993; Shay, 1986; Sorensen & Holden, 1974; Stalter, 1973; Stevens, 1932; Talbot et al., 1988; Thompson & McKinney, 2006; Tooker & Hanks, 2000; Tooker et al., 2006; Vestal, 1914; Whitehouse & Bayley, 2005; Whiteside, J. E. 1887; Wittstock et al., 1995, 1997.

4. Conium

Poison hemlock; ciguë tachée, ciguë maculée

Etymology: from the Greek *konê* ("murder"), alluding to a nefarious use of this highly poisonous plant

Synonyms: none

Distribution: global: Africa; Asia; Australia*; Europe; North America*; South America*; **North America:** widespread

Diversity: global: 2–5 species; **North America:** 1 species

Indicators (USA): FACW; FAC: *Conium maculatum*

Habitat: freshwater; lacustrine, palustrine, riverine; **pH:** 5.6–8.7; **depth:** <1 m; **life-form(s):** emergent herb

Key morphology: Stems (to 3 m) erect, hollow, with purple spots or streaks, herbage odorous; leaves (to 4 dm) 3–4 times pinnately compound, blades broadly triangular in outline, the leaflets pinnately incised, petioles (to 25 cm) dilated, sheathing; umbels compound, lateral and terminal, arranged in compound cymes, peduncled (to 8 cm), with 4–6 scarious bracts (to 5 mm), 5–6 bractlets (to 3 mm), 10–20 rays (to 5 cm), ascending, spreading; flowers (to 3 mm) pedicelled (to 6 mm), 5-merous; calyx absent; petals (to 1.5 mm) white to yellowish, broadly obovate, narrowed at base; fruit (to 3 mm) ovate, slightly compressed, prominently waxy ribbed; seeds grooved

Life history: duration: annual, biennial, winter annual (taproot); **asexual reproduction:** none; **pollination:** insect; **sexual condition:** hermaphroditic; **fruit:** schizocarps (prolific); **local dispersal:** seeds (animals, water, wind); **long-distance dispersal:** seeds (animals, water)

Imperilment: (1) *Conium maculatum* [G5]

Ecology: general: *Conium* is a common plant of North American wetlands and also can occupy a broad range of dry habitats. Although the 1996 wetland indicator list included *Conium maculatum* among the OBL taxa (as OBL, FACW, FAC), the revised lists now categorize it only as FACW, FAC. It has been retained here because the treatment was completed prior to release of the revised indicator lists.

Conium maculatum **L.** is a biennial, which colonizes canals, canyons, channels, creekbeds, depressions, ditches, flats, floodplains, gullies, levees, marshes, meadows, pastures, prairies, ravines, salt marshes, sand bars, seeps, springs, stream bottoms, swales, thickets, washes, and the margins of lakes, rivers, and streams at elevations of up to 3354 m. The plants occur in open areas or more commonly in shaded sites on a diversity of substrates described as clay, clay loam, cobbly loam, gravel, gravel loam, limestone, loam, pea gravel, rock, salt crust, sand, sandy loam, silt, and silt loam. As demonstrated by its wide range of wetland indicator categories, described habitats, and substrate types, this is primarily a FACW species, which often occurs in more UPL habitats. Although these plants can behave as annuals, they typically grow as biennials, especially when situated in moist sites. Once flowering has occurred, the plants die by summer's end. One plant can produce upwards of 39,000 seeds, with the highest numbers produced by the tallest plants. Potential seed densities can reach 540,000/m². The seeds are dispersed in water, by adhering to farm equipment or clothing, or by birds and rodents. The seeds are morphologically

dormant, which requires growth and differentiation of the embryo to stimulate germination. Mature seeds will germinate within 13–18 days if dispersed during summer or early autumn; however, later dispersal normally delays germination until the following spring. Germination rates typically are high (83%–87%) but vary with respect to seed maturity. Seeds that are retained on plants during late fall and early winter will acquire physiological dormancy, which is induced by cold stratification, and results in a more complex morphophysiological dormancy. This condition is broken by summer conditions, which convert the seeds back to simple morphological dormancy. Optimal germination occurs under a 30°C/15°C temperature regime and typically is higher on soil substrates than on sand. A transient seed bank can develop with seeds surviving for up to 5 years. Seedlings appear mainly in late winter and early spring (January to April), but appreciable emergence also occurs from late summer through autumn (July to September). Seedling recruitment is highly dependent on the extent of cover, with maximum emergence occurring at sites having the lowest extent of vegetational cover and litter. Seedling survival is not clearly associated with particular cover values, although better survival has been reported for shaded sites. Generally, the plants favor disturbed sites for establishment, where many seedlings occur initially; however, density-dependent mortality eventually reduces considerably the number of surviving plants. Nitrogen additions have been shown to favor the growth of *Conium* over several co-occurring species. **Reported associates:** *Abies concolor, Acer negundo, Achillea millefolium, Acmispon maritimus, Aconitum columbianum, Adenostoma fasciculatum, Aesculus californica, Ageratina adenophora, Agropyron repens, Agrostis gigantea, A. viridis, Allium canadense, Alnus rhombifolia, Ambrosia artemisiifolia, Anemopsis californica, Anthemis, Apium graveolens, Apocynum cannabinum, Arabis glabra, Arctium minus, Artemisia californica, A. douglasiana, A. dracunculus, A. tridentata, Arundo donax, Asclepias fascicularis, A. syriaca, A. tuberosa, Atriplex, Avena fatua, Baccharis douglasii, B. emoryi, B. pilularis, B. salicifolia, B. sarothroides, Bassia, Brassica geniculata, B. nigra, Bromopsis inermis, Bromus catharticus, B. carinatus, B. diandrus, B. inermis, B. rubens, B. tectorum, Calocedrus decurrens, Calochortus splendens, Capsella bursa-pastoris, Carduus nutans, C. pycnocephalus, Carex athrostachya, C. barbarae, C. hystricina, C. microptera, C. praegracilis, C. scopulorum, C. spissa, Castilleja miniata, Ceanothus crassifolius, Centaurea melitensis, C. solstitialis, Cephalanthus occidentalis, Cerastium glomeratum, Cercocarpus, Chamerion angustifolium, Chenopodium californicum, Chrysanthemum coronarium, Cirsium arvense, C. vinaceum, Conioselinum scopulorum, Convolvulus arvensis, Conyza canadensis, Cornus sericea, Cynodon dactylon, Cryptantha barbigera, Cucurbita foetidissima, Cyperus eragrostis, C. involucratus, Dasiphora floribunda, Daucus carota, Desmodium, Dipsacus fullonum, Dryopteris arguta, Dudleya lanceolata, D. variegata, Echinochloa crus-galli, Eleocharis, Ellisia nyctelea, Elymus glaucus, Epilobium* ciliatum, Equisetum telmateia, Ericameria pinifolia, Erigeron foliosus, E. rybius, Eriodictyon crassifolium, Eriogonum fasciculatum, Eryngium aquaticum, Eucalyptus camaldulensis, Euphorbia peplus, Eurybia integrifolia, Ferocactus viridescens, Festuca elatior, F. idahoensis, Ficus carica, Foeniculum vulgare, Fragaria americana, Fraxinus pennsylvanica, Galium angustifolium, G. aparine, G. boreale, Gaura biennis, Geranium carolinianum, G. caespitosum, G. richardsonii, G. viscosissimum, Geum aleppicum, Gnaphalium californicum, Hackelia micrantha, Hedeoma pulcherrima, Hedypnois cretica, Helianthus grosseserratus, Heliopsis helianthoides, Heterotheca, Hirschfeldia incana, Holodiscus dumosus, Hordeum brachyantherum, H. murinum, H. pusillum, Hymenoclea monogyra, Hymenoxys hoopesii, Impatiens, Iris missouriensis, Jamesia americana, Juncus acutus, J. balticus, J. longistylis, J. xiphioides, Juniperus monosperma, J. deppeana, Keckiella cordifolia, Lactuca serriola, Lamium purpureum, Leonurus cardiaca, Leymus cinereus, Lilium humboldtii, Lolium multiflorum, Lomatium triternatum, Lonicera subspicata, Ludwigia hexapetala, Lupinus arboreus, L. argenteus, L. polyphyllus, Lycium californicum, Maianthemum racemosum, M. stellatum, Malosma laurina, Malva neglecta, M. parviflora, M. sylvestris, Marrubium vulgare, Matricaria matricarioides, Melica bulbosa, M. imperfecta, Melilotus alba, M. officinalis, Mentha pulegium, Mertensia ciliata, Mimulus aurantiacus, M. cardinalis, M. glabratus, M. guttatus, Mirabilis nyctaginea, M. oxybaphoides, Muhlenbergia rigens, Myosotis, Nama hispidum, Nassella, Nasturtium officinale, Nepeta cataria, Nicotiana glauca, N. trigonophylla, Oenothera biennis, Osmorhiza obtusa, O. occidentalis, Oxalis corniculata, Parkinsonia aculeata, Pascopyrum smithii, Pastinaca sativa, Persicaria coccinea, P. maculosa, Phacelia cicutaria, Phalaris arundinacea, Phleum alpinum, P. pratense, Phoenix canariensis, Pholistoma auritum, Picea pungens, Pinus ponderosa, P. strobiformis, Piptatherum miliaceum, Plantago coronopus, P. lanceolata, P. rugelii, Platanus acerifolia, P. racemosa, Pluchea odorata, Poa compressa, P. pratensis, P. secunda, Polanisia dodecandra, Polypogon, Populus angustifolia, P. ×canescens, P. fremontii, P. tremuloides, Potamogeton gramineus, Potentilla pulcherrima, P. recta, Prosopis, Prunella vulgaris, Prunus ilicifolia, Pseudotsuga menziesii, Quercus gambelii, Ranunculus abortivus, Rhamnus ilicifolia, Rhus integrifolia, R. laurina, R. trilobata, Ribes speciosum, Ricinus communis, Robinia neomexicana, Rosa californica, R. woodsii, Rubus discolor, R. ursinus, Rudbeckia laciniata, R. occidentalis, Rumex crispus, Salix exigua, S. gooddingii, S. irrorata, S. laevigata, S. lasiandra, S. lasiolepis, S. lutea, S. luteosericea, Salsola, Salvia apiana, S. mellifera, Sambucus californica, S. nigra, Saponaria officinalis, Sarcocornia perennis, Schinus, Schismus, Schoenoplectus americanus, Scirpus, Scrophularia californica, S. lanceolata, Senecio triangularis, S. wootonii, Silene laciniata, Silybum marianum, Sisymbrium irio, S. officinale, Solanum douglasii, Solidago californica, S. canadensis, S. occidentalis, Sonchus, Sorghum halepense, Stachys albens, Stipa robusta, Symphyotrichum ascendens, S. pilosum, Tamarix, Tanacetum parthenium, Taraxacum officinale, Thalictrum fendleri, T. occidentale, Thinopyrum ponticum, Toxicodendron diversilobum, T. radicans, Tradescantia ohiensis, Tragopogon dubius, Trifolium fendleri, T. longipes, Typha, Ulmus, Umbellularia californica, Urtica dioica, Veratrum californicum, Verbascum thapsus, Verbena macdougalii, Verbesina alternifolia, Veronica anagallis-aquatica, V. peregrina, Vicia sativa, Vinca major, Vitis girdiana, V. riparia, Wyethia amplexicaulis, Yucca whipplei, Xanthium strumarium.

Use by wildlife: Conium maculatum is acutely and chronically toxic to livestock, which may feed intentionally or inadvertently on the plants. The coniine and γ-coniceine alkaloids are teratogenic, causing a form of "crooked calf disease" in cattle (Mammalia: Bovidae: Bos taurus) as well as similar deformities in swine (Mammalia: Suidae: Sus scrofa) and sheep (Mammalia: Bovidae: Ovis aries). The lethal doses for these animals are 5.3, 8.0, and 10.0 g/kg live weight respectively. Toxicity of coniine is roughly eightfold that of γ-coniceine. Nevertheless, the foliage or seeds often are eaten readily by these livestock as well as by goats (Mammalia: Bovidae: Capra aegagrus), chickens (Aves: Phasianidae: Gallus gallus), elk (Mammalia: Cervidae: Cervus canadensis), and wild turkeys (Aves: Phasianidae: Meleagris gallopavo). Symptoms of severe poisoning can result from acute doses ranging from as low as 3.3 mg/kg (cattle) to 100 mg/kg (turkey chicks; Meleagris). Conium also contains linear furanocoumarins, which typically are toxic or repellant to generalist (Insecta: Lepidoptera) larvae, and in combination with the piperidine alkaloids (found in few other Apiaceae), often results in a highly specialized faunal association. The poison hemlock moth (Lepidoptera: Oecophoridae: Agonopterix alstroemeriana) feeds exclusively on C. maculatum. However, in some nonindigenous areas (e.g., California, Illinois), the faunal association mainly comprises generalist insect (Insecta) herbivores, including other introduced species such as cabbage loopers (Lepidoptera: Noctuidae: Trichoplusia ni), parsnip aphids (Homoptera: Aphididae: Hyadaphis foeniculi), and leafhoppers (Homoptera: Cicadellidae: Euscelidius variegatus), as well as native Lepidoptera like the anise swallowtail butterfly (Papilionidae: Papilio zelicaon), loop moths (Geometridae: Eupithecia miserulata), Ranchman's tiger moth (Arctiidae: Platyprepia virginalis), and Virginian tiger moth (Arctiidae: Spilosoma virginica), which apparently have adapted to feeding on the introduced plants. Though a generalist, the cabbage looper suffers no ill effects from feeding other than experiencing a somewhat longer development time. Likewise, generalist corn earworm moths (Lepidoptera: Noctuidae: Helicoverpa zea) exhibit no harmful symptoms after feeding on the plants. However, Conium is a nonpreferred host to at least the anise swallowtail butterfly, which exhibits relatively higher larval and pupal mortality and increased diapause frequency in comparison to the other host plants. These plants are a host of stink bugs (Insecta: Hemiptera: Pentatomidae: Euschistus conspersus) and are eaten by several Orthoptera (Insecta) including the Lakin, two-striped, and spur-throated grasshoppers (Acrididae: Melanoplus lakinus, M. bivittatus, M. differentialis). The honeysuckle aphid

(Insecta: Hemiptera: Aphididae: *Rhopalosiphum conii*) is a vector for the transmission of the poison hemlock ringspot virus. Other viral pathogens infecting *Conium* include alfalfa mosaic virus (Bromoviridae: *Alfamovirus*), carrot thin leaf virus (Potyviridae: *Potyvirus*), and celery mosaic virus (Potyviridae: *Potyvirus*). The plants are readily infected by *Xylella fastidiosa* (Proteobacteria: Xanthomonadaceae), the causative bacterial agent of Pierce's disease in grapes. *Conium maculatum* is strongly associated with the habitat of the white-footed mouse (Mammalia: Muridae: *Peromyscus leucopus*), but it is uncertain whether it is eaten by the rodent. The seeds are eaten by bobwhite quail (Aves: Odontophoridae: *Colinus virginianus*).

Economic importance: food: All parts of *Conium maculatum* (especially the seeds) are highly poisonous and never should be eaten under any circumstances. It is the infamous hemlock that not only led to Socrates demise but also is responsible occasionally for the accidental poisoning of other humans. The toxicity of *Conium* is less than that of *Cicuta*, but still it potentially is deadly. A 100 mg dose of coniine generally is regarded as being lethal to humans. Some children have survived ingestion of the plants when only small amounts were consumed and where medical treatment was administered rapidly; only a single child's death was attributed to hemlock poisoning during one 8-year study period. Although still debated, it is plausible that animals, which have eaten seeds of this plant, can pass along the toxic properties to humans when used as a source of meat or milk. Such an effect has been demonstrated in other vertebrates, where domestic dogs (Mammalia: Canidae: *Canis*) were fatally poisoned after consuming the meat of quail (Aves: Phasianidae: *Coturnix coturnix*), which had been fed a diet of hemlock seeds; **medicinal:** The toxic properties of *Conium* were recognized by various North American tribes including the Klallam, Lakota, and Snohomish people. The herbage contains a number of piperidine alkaloids and their derivatives (described as having a "mousy" smell), of which coniine and γ-coniceine exhibit the most toxic properties. These compounds act as nondepolarizing blockers of the neuromuscular junction, which can lead to death by respiratory failure. There is no antidote to the toxins and prompt medical treatment should be sought if ingestion of this plant is suspected. Despite its potency, the plant has been used in various poultices or ointments to treat inflammation, ulcers, and joint pain. Tinctures and extracts of the extracted juice have been administered pharmaceutically for use as sedatives and antispasmodics, or to treat angina, asthma, chorea, epilepsy, whooping cough, and stomach pains. The furanocoumarins of *C. maculatum* have antifungal properties; **cultivation:** Poisonous plants such as *Conium* seldom are cultivated, but *C. maculatum* contains at least one cultivar, which is traded as 'Golden Nemesis'; **misc. products:** Plant extracts of *C. maculatum* were used by the ancient Greeks for executions. The hollow stems have been used as pea shooters and whistles; however, both uses have resulted in human poisoning. Klallam women rubbed the roots on their body as a form of love potion. Several North American tribes used the plants as a source of arrow poison; **weeds:** Because of its toxicity to humans and livestock, *Conium maculatum* generally is regarded as a noxious weed throughout North America. The nonindigenous poison hemlock moth (Lepidoptera: Oecophoridae: *Agonopterix alstroemeriana*) defoliates the plants and has been used in a number of eradication programs, especially in western regions; **nonindigenous species:** Although many erroneously regard it as indigenous, *Conium maculatum* was introduced to North America as a garden plant. It also has been introduced to Australia (known there as "carrot fern"), The British Isles, New Zealand, and South America.

Systematics: *Conium* is a problematic genus in a systematic context. It contains at least two species (including *C. divaricatum*, a Greek endemic); however, an additional three taxa from southern Africa have been treated variously either as distinct species (i.e., *C. chaerophylloides*, *C. fontanum*, *C. sphaerocarpum*), or as varieties of *C. maculatum*. Also unresolved is the phylogenetic placement of *Conium* itself. Although the genus clearly falls within subfamily Apioideae (as indicated by analysis of multiple DNA sequences), its tribal placement has not yet been established confidently. DNA sequences from the nrITS region weakly support an association with *Heracleum* (Figure 5.111), which has been assigned to tribe Tordylieae; however, a definitive tribal placement of *Conium* has not been made at this time. A study of flavonoid dioxygenases has found that the enzyme flavone synthase I is unique to Apiaceae and resulted from a duplication of the flavanone 3β-hydroxylase gene within the family. Because *Conium* resolves phylogenetically within the subgroup of genera that possesses the enzyme, additional work in this area could help to further elucidate the relationships of the genus. The chromosome number of $2n = 22$ for both *C. maculatum* and *C. divaricatum* indicates that they both are diploids with a base number of $x = 11$.

Comments: *Conium maculatum* is widespread throughout North America, except in the far north.

References: Al-Barwani & Eltayeb, 2004; Baskin & Baskin, 1990, 1998; Berenbaum, 1981; Berenbaum & Harrison, 1994; Caplan, 1966; Castells & Berenbaum, 2008; Constantinidis et al., 1997; Cordes & Downie, 2007, 2008; Downie et al., 2001; Frank & Reed, 1987; Frank et al., 1995; Freitag & Severin, 1945; Gebhardt et al., 2005; Goeden & Ricker, 1982; Jennings, 1941; Keeler & Balls, 1978; López et al., 1999; Mamolos & Veresoglou, 2000; McKenna et al., 2001; Mitich, 1998; Nitao, 1987; Parsons, 1973; Reynolds, 2005; Roberts, 1979; Silvertown & Tremlett, 1989; Sims, 1980; Tremlett et al., 1984; Van Deusen & Kaufman, 1977; Wistrom & Purcell, 2005; Woodard, 2008; Zhou et al., 2008.

5. *Eryngium*

Eryngo, rattlesnake master; panicaut

Etymology: from the Greek *erunganô* meaning "to belch," a response supposedly elicited by ingestion of the plant

Synonyms: none

Distribution: global: cosmopolitan; **North America:** widespread

Diversity: global: 230 species; **North America:** 35 species

Indicators (USA): OBL: *Eryngium aquaticum*, *E. aristulatum*, *E. articulatum*, *E. castrense*, *E. constancei*, *E. jepsonii*,

E. mathiasiae, E. petiolatum, E. phyteumae, E. pinnatisectum, E. prostratum, E. racemosum, E. sparganophyllum; **FACW:** *E. prostratum*
Habitat: brackish (coastal); freshwater; lacustrine, palustrine, riverine; **pH:** 5.0–9.0; **depth:** <1 m; **life-form(s):** emergent herb
Key morphology: Stems prostrate (0.6–0.8 m) or erect (0.4–1.8 m), branched or unbranched, sometimes rooting at nodes; leaves basal and cauline, petioles (4–10 cm) septate or not, bladeless or with membranous or coriaceous, linear, ovate or round, entire to pinnately or palmately lobed or dissected blades (0.7–4 dm); flowers solitary and arising from nodes, or in cymes, umbels or racemes consisting of several to numerous heads (4–25 mm), with 4–17 spiny bracts (0.7–2.7 cm); flowers sessile, each subtended by one bracteole (to 15 mm), sepals (0.8–4.0 mm) persistent on fruit; petals (to 1.5 mm) blue, purple or white; fruit (2.0–4.0 mm) obovate or round, not ridged, the surface covered by tubercles or scales (to 1 mm).
Life history: duration: annual; biennial (taproot); perennial (rhizomes); **asexual reproduction:** rhizomes; **pollination:** insect; **sexual condition:** hermaphroditic; **fruit:** schizocarps (common); **local dispersal:** water; **long-distance dispersal:** animals; water
Imperilment: (1) *Eryngium aquaticum* [G4]; SX (NY, PA); S1 (MS, NC, SC); S2 (DE); S3 (NC, NJ); (2) *E. aristulatum* [G5]; S2 (CA); (3) *E. articulatum* [G5]; SH (WA); (4) *E. castrense* [G4]; (5) *E. constancei* [G1]; S1 (CA); (6) *E. jepsonii* [unranked]; (7) *E. mathiasiae* [G3]; (8) *E. petiolatum* [G4]; S1 (WA); (9) *E. phyteumae* [unranked]; (10) *E. pinnatisectum* [G3]; S3 (CA); (11) *E. prostratum* [G5]; S1 (IL, KS); S3 (NC); (12) *E. racemosum* [G2]; S2 (CA); (13) *E. sparganophyllum* [G2]; SH (NM); S1 (AZ)
Ecology: general: *Eryngium* species are broadly adapted ecologically. Although many species occur in UPLs about a third of the North American species are obligately aquatic, with two-thirds of the species occurring at least facultatively in wetlands. Most of the wetland species occur in vernal habitats. The concentration of aquatic species is partitioned phylogenetically, at least in a general sense. A predominantly New World clade comprises species that are most inclined toward aquatic or semiaquatic habitats, and most of the Old World species represent a clade that is associated with arid or semiarid habitats. The nectar-producing flowers and showy involucres of many species appear to be adapted for insect pollination; however, details on the reproductive biology for most members of this large genus are incomplete. Generally, the flowers appear to be self-compatible but protandrous. Pollination typically involves insects (Insecta), which visit the plants for nectar, because reported rates of autogamy are quite low. Pollinators mainly are generalist butterflies (Lepidoptera), bees (Hymenoptera), and flies (Diptera), with some species reportedly visited by more than 100 different insects. Exceptions surely occur. At least one isolated island species is protogynous and pollinated by hummingbirds (Aves: Trochilidae). Although geitonogamous pollen transfers occur frequently, outcrossing also appears to occur regularly, at least in the few species that have been studied in detail. It likely is

facilitated by the temporal partitioning of male and female phases ("temporal dioecism") within the inflorescence. The seed ecology of *Eryngium* species also has been poorly studied; however, three nonwetland species that have been studied all uniformly exhibited morphophysiological dormancy. Most species possess fruits that are somewhat scaly, which enables them to cling to animal fur for potential dispersal. Otherwise, observed local seed dispersal is quite restricted, occurring on the order of a few hundred centimeters.

***Eryngium aquaticum* L.** inhabits tidal and nontidal channels, depressions, ditches, flatwoods, marshes, meadows, prairies, savannas, seeps, swales, swamps, and the margins of ponds and rivers. This species appears to have fairly broad ecological tolerances. Habitats are freshwater (0–0.2 ppt salinity) or sometimes brackish (0.2–0.5 ppt salinity), and include sediments that are acidic (pH: 5.0–6.0) with 24%–30% organic matter and high amounts of silt (52%–58%; <25% clay or sand). Other reported substrates are sandy loam over sandy clay loam. This species also grows on calcareous floodplains where the calcium content is high, but the pH is variable from low to high values. The plants flower from July to August. The northward limit of this species is determined in part by its seed ecology. Seed germination is markedly reduced (5%) when they are refrigerated for 71 days compared to those that are not (38%). Germination time also was observed to be long in refrigerated seeds (42–49 days) compared to unrefrigerated seeds (27–90 days). The seeds are extremely scaly and are likely dispersed by adhering to animal fur. The plants are perennial and produce short basal stolons or "offshoots." Some of the early literature can be misleading because several authors regarded *E. aquaticum* as a synonym of *E. yuccifolium*. **Reported associates:** *Acer, Aeschynomene virginica, Agalinis purpurea, Aletris lutea, Amaranthus cannabinus, Andropogon, Apios americana, Arundo donax, Baccharis glomeruliflora, Bacopa monnieri, Bidens laevis, Boehmeria cylindrica, Bolboschoenus robustus, Boltonia asteroides, Carex canescens, C. comosa, C. emoryi, C. lacustris, Cassia fasciculata, Castilleja coccinea, Centella erecta, Chasmanthium latifolium, C. laxum, Chenopodium album, Cinna arundinacea, Cladium jamaicense, C. mariscoides, Coelorachis rugosa, Collinsonia serotina, Conium maculatum, Crinum americanum, Ctenium aromaticum, Cuscuta gronovii, Cyperus haspan, C. pseudovegetus, Cypripedium reginae, Desmodium, Dichanthelium hirstii, D. ovale, D. spretum, Echinochloa crus-galli, Eleocharis engelmannii, E. fallax, E. quadrangulata, E. rostellata, E. tuberculosa, Erechtites hieracifolia, Eriocaulon decangulare, Erythronium albidum, Euphorbia inundata, Eurybia eryngiifolia, Fimbristylis autumnalis, F. thermalis, Fuirena, Gaylussacia dumosa, Gratiola, Helenium autumnale, Helianthus angustifolius, Hibiscus moscheutos, Hydrocotyle, Hymenocallis floridana, Hyptis alata, Impatiens capensis, Iris virginica, Juncus acuminatus, J. biflorus, J. coriaceus, J. effusus, J. megacephalus, J. roemerianus, Kosteletzkya pentacarpos, Lilium canadense, L. michauxii, L. philadelphicum, Limosella australis, Lindernia monticola, Lipocarpha maculata, Lobelia cardinalis, L. glandulosa, Ludwigia alata,*

Lycopodium appressum, L. carolinianum, Lysimachia lanceolata, Lythrum lineare, Micranthemum micranthemoides, Mikania scandens, Myrica heterophylla, Oligoneuron rigidum, Orontium aquaticum, Osmunda regalis, Panicum hemitomon, P. rigidulum, P. virgatum, Parnassia caroliniana, Paspalum distichum, P. laeve, Peltandra virginica, Persicaria glabra, P. hydropiperoides, P. sagittata, Pinguicula ionantha, P. lutea, P. pumila, Pinus palustris, Polygala brevifolia, P. lutea, Pontederia cordata, Proserpinaca palustris, P. pectinata, Pteridium aquilinum, Ptilimnium capillaceum, Rhexia alifanus, R. aristosa, R. virginica, Rhynchospora chalarocephala, R. colorata, Rudbeckia fulgida, Rumex verticillatus, Sabatia angularis, S. difformis, S. dodecandra, Saccharum giganteum, Sagittaria falcata, S. graminea, S. lancifolia, Salix caroliniana, Saururus cernuus, Schoenoplectus pungens, S. tabernaemontani, Scirpus cyperinus, Scleria reticularis, Sclerolepis uniflora, Serenoa repens, Setaria magna, Silphium asteriscus, Solidago sempervirens, Spartina cynosuroides, Sporobolus teretifolius, Symphyotrichum novi-belgii, Taxodium distichum, Tradescantia virginiana, Triantha racemosa, Trillium recurvatum, T. viride, Tripsacum dactyloides, Typha angustifolia, Typha domingensis, T. latifolia, Verbesina occidentalis, Woodwardia virginica, Xyris caroliniana, X. smalliana, Zigadenus glaberrimus, Zizania aquatica, Zizaniopsis miliacea.

Eryngium aristulatum Jeps. is a biennial, which occurs on saturated ground or in standing waters from 10 to 45 cm deep (average depth: 17 cm). Its habitats include channels, deltas, depressions, ditches, dry lakebeds, flats, floodplains, marshes, meadows, playas, salt marshes, seeps, sloughs, swales, vernal pools, and the margins of creeks, lakes, ponds, and reservoirs at elevations of up to 1130 m. Although this species spans a fairly wide range of wetland habitats, it is found primarily in vernal pools or in other seasonally desiccated sites, and often is seen only after the water has receded. It also has been observed to grow in some artificially drained areas. It is known to occur in alkaline (pH 7.9–9.0), fresh to saline sites (in full sun to partial shade) on adobe, basalt, clay (Pescadero), mud, peaty soil, sand, sandy loam (Solano), serpentine, Stockpen soils, and vertisol substrates. The plants occur mainly in freshwater sites but are categorized as halophytes because of their salt tolerance. These are C_3 plants, but exhibit higher ribulose-1,5-bisphosphate carboxylase/oxygenase (RUBISCO) activity when growing under terrestrial conditions and greater phosphoenolpyruvate (PEP) carboxylase activity when growing as aquatics. They do not use bicarbonate as a carbon source. Populations retain a low but fairly stable frequency of individuals, despite the marked variation in annual precipitation. The sharply edged fruits remain attached to fur even after significant agitation, thereby indicating their potential for long-distance dispersal (epizoic). **Reported associates:** *Acmispon glabrus, A. wrangelianus, Adenostoma fasciculatum, Alopecurus, Amaranthus, Anagallis arvensis, Asclepias fascicularis, Astragalus didymocarpus, Atriplex, Avena barbata, Brodiaea filifolia, Bromus hordeaceus, Calochortus, Centromadia fitchii, C. parryi, Cirsium fontinale, Clematis ligusticifolia, Convolvulus arvensis, Cressa truxillensis, Croton*

setiger, Crypsis schoenoides, Cuscuta howellii (parasitic on), *Cyperus squarrosus, Deinandra fasciculata, Deschampsia danthonioides, Distichlis spicata, Downingia concolor, D. cuspidata, Elatine heterandra, Eleocharis macrostachya, Epilobium densiflorum, E. pygmaeum, Erodium botrys, E. cicutarium, Eryngium constancei, Festuca, Frankenia salina, Heliotropium, Hemizonia congesta, H. fitchii, Hesperevax caulescens, Heterocodon rariflorum, Hordeum murinum, Juncus xiphioides, Lamarckia aurea, Lasthenia burkei, L. californica, L. glaberrima, L. glabrata, L. macrantha, Lilaea scilloides, Limnanthes vinculans, Lindernia dubia, Lolium multiflorum, Lupinus bicolor, Lythrum hyssopifolia, Malvella leprosa, Nassella pulchra, Navarretia fossalis, N. plieantha, N. prostrata, Neostapfia colusana, Orcuttia californica, O. tenuis, Persicaria coccinea, Phyla nodiflora, Plagiobothrys chorisianus, P. leptocladus, Plantago lanceolata, Pogogyne abramsii, P. nudiuscula, Polygonum aviculare, Polypogon monspeliensis, Psilocarphus brevissimus, Ranunculus californicus, Rorippa curvisiliqua, Rotala ramosior, Rumex crispus, Schoenoplectus acutus, Sida, Sisyrinchium bellum, Spergula arvensis, Stachys ajugoides, Trifolium variegatum, T. willdenovii, Tuctoria mucronata, Veronica peregrina, Xanthium spinosum, X. strumarium.*

Eryngium articulatum Hook. is reported from ditches, flats, floodplains, gullies, marshes, meadows, outwash plains, salt marshes, sloughs, streambeds, swales, vernal pools, washes, and the margins of lakes, rivers, and streams at elevations of up to 1830 m. It inhabits fresh to brackish tidal or nontidal waters up to 1 m deep and grows in full sun on substrates reported as adobe, basalt, clay, gravel, muck, or mud. These salinity-tolerant plants are categorized as halophytes and have been observed growing in the high brackish marsh zone of wetlands in the San Francisco Bay area. **Reported associates:** *Agrostis exarata, Alisma triviale, Asclepias fascicularis, Beckmannia syzigachne, Briza maxima, Brodiaea californica, B. coronaria, Callitriche verna, Calystegia sepium, Carex comosa, Cephalanthus occidentalis, Clarkia, Deschampsia danthonioides, Eleocharis macrostachya, Epilobium, Erigeron, Eryngium mathiasiae, Euthamia occidentalis, Geranium, Helenium bigelovii, Hydrocotyle verticillata, Isolepis cernua, Juncus balticus, u. dubius, J. oxymeris, Lilaeopsis masonii, Mimulus guttatus, Navarretia leucocephala, Perideridia erythrorhiza, Phacelia, Phragmites australis, Plantago, Polygonum aviculare, P. polygaloides, Potentilla anserina, Rosa californica, Rubus, Rumex crispus, Sagittaria cuneata, Schoenoplectus acutus, S. americanus, S. californicus, Symphyotrichum lentum, Taeniatherum caput-medusae, Triglochin striatum, Typha latifolia.*

Eryngium castrense Jeps. inhabits creekbeds, depressions, ditches, flats, gullies, ravines, riverbeds, seeps, sloughs, swales, vernal pools, wallows, washes, and the margins of lakes, ponds, reservoirs, and streams at elevations of up to 900 m. Habitats typically are open and sunny with substrates that include adobe, alluvium, basalt, clay (Peters), clay–gravel, clay–loam, gravelly loam, muck, mud, Redding soil, rock, sand, and Tuscan mudflow. The hollow stems maintain aeration of the roots during winter months. Flowering occurs into early summer, after other vernal pool species have died back. Additional life history information on this species is lacking. **Reported associates:** *Achyrachaena mollis, Acmispon americanus, Aira caryophyllea, Allium amplectens, Alopecurus saccatus, Amsinckia menziesii, Anagallis arvensis, Asclepias fascicularis, Avena barbata, Blennosperma nanum, Briza minor, Brodiaea californica, B. elegans, B. terrestris, Bromus hordeaceus, Callitriche heterophylla, C. marginata, Calycadenia multiglandulosa, Castilleja attenuata, C. campestris, C. lineariiloba, Centaurium muehlenbergii, C. venustum, Centromadia fitchii, Centunculus minimus, Cerastium glomeratum, Chamaesyce ocellata, Chlorogalum angustifolium, Cicendia quadrangularis, Clarkia purpurea, Crassula aquatica, Croton setigerus, Crypsis schoenoides, C. vaginiflora, Cuscuta howelliana, Cynodon dactylon, Deschampsia danthonioides, Downingia bicornuta, D. cuspidata, D. insignis, D. ornatissima, D. pusilla, Elatine californica, Eleocharis acicularis, E. macrostachya, E. palustris, Elymus, Epilobium cleistogamum, E. densiflorum, E. pallidum, E. pygmaeum, E. torreyi, Erodium botrys, Eryngium pinnatisectum, E. vaseyi, Galium parisiense, Gastridium ventricosum, Geranium dissectum, Glyceria occidentalis, Gratiola ebracteata, Hemizonia fitchii, Heterocodon rariflorum, Holocarpha virgata, Hordeum marinum, Hosackia oblongifolia, Hypochaeris glabra, Isoetes howellii, I. nuttallii, I. orcuttii, Juncus bufonius, J. capitatus, J. leiospermus, J. tenuis, J. uncialis, J. xiphioides, Lasthenia californica, L. fremontii, L. glaberrima, Layia fremontii, Leontodon taraxacoides, Lepidium nitidum, Lilaea scilloides, Limnanthes douglasii, L. floccosa, Linanthus bicolor, Lolium multiflorum, Lomatium caruifolium, Lupinus bicolor, Lythrum hyssopifolia, L. tribracteatum, Marsilea vestita, Medicago polymorpha, Micropus californicus, Microseris acuminata, Mimulus guttatus, M. tricolor, Minuartia californica, Montia fontana, Myosurus minimus, Nasturtium officinale, Navarretia filicaulis, N. intertexta, N. leucocephala, N. tagetina, Neostapfia colusana, Odontostomum hartwegii, Orcuttia pilosa, O. tenuis, Parentucellia viscosa, Parvisedum pumilum, Paspalum dilatatum, Phyla nodiflora, Pilularia americana, Plagiobothrys austiniae, P. bracteatus, P. glyptocarpus, P. leptocladus, P. stipitatus, Plantago erecta, Poa annua, Pogogyne ziziphoroides, Polypogon monspeliensis, Populus, Potamogeton diversifolius, Proboscidea louisianica, Psilocarphus brevissimus, Quercus douglasii, Ranunculus aquatilis, R. bonariensis, R. muricatus, Raphanus raphanistrum, Rumex acetosella, R. crispus, R. pulcher, Salix, Scribneria bolanderi, Silene gallica, Soliva sessilis, Taeniatherum caput-medusae, Trichostema lanceolatum, Trifolium depauperatum, T. dubium, r. fucatum, T. hirtum, T. variegatum, Triteleia hyacinthina, Triphysaria eriantha, Typha, Veronica anagallis-aquatica, V. peregrina, Vicia villosa, Vulpia bromoides, V. microstachys, V. myuros.*

***Eryngium constancei* M.Y. Sheikh** grows typically in more than 10 cm of water in meadows and vernal pools at elevations of up to 853 m. The substrates are described as fine, powdery, volcanic silty clay. Flowering occurs after the waters evaporate from the pools (mainly in June to August).

The fruits possess tooth-like scales, which probably promote their dispersal on animal fur. Other information on the reproductive biology or life history of this species is scarce. The plants have been described as annuals, biennials, and perennials. This rare species is restricted to only a few sites in Lake and Sonoma counties, California. **Reported associates:** *Centaurium davyi, Clarkia purpurea, Cornus nuttallii, Cuscuta howelliana, Downingia concolor, Eleocharis, Eryngium aristulatum, Gratiola ebracteata, Juncus, Lilaea scilloides, Madia elegans, Mentha pulegium, Mimulus tricolor, Navarretia leucocephala, N. pauciflora, N. plieantha, Perideridia gairdneri, P. howellii, P. kelloggii, Plagiobothrys stipitatus, P. tener, Pogogyne douglasii, Quercus kelloggii.*

***Eryngium jepsonii* J.M. Coult. & Rose** inhabits ditches, flats, and river channels in areas near the margins of chaparral at elevations of up to 161 m. Habitats are open sites, often along north-facing slopes. The substrates are described as adobe, clay (dense), and vertisols. The plants flower in April and set fruits from August to September. Additional life-history information is scarce. **Reported associates:** *Acmispon wrangelianus, Brassica nigra, Bromus madritensis, Calochortus argillosus, Hemizonia congesta, Hesperevax caulescens, Lasthenia glabrata, Lolium multiflorum, Phalaris brachystachys, Trifolium willdenovii.*

***Eryngium mathiasiae* M.Y. Sheikh** grows in standing (to 40 cm) or receding waters of depressions, ditches, flats, floodplains, lakebeds, lakes, meadows, ponds, swales, vernal pools, and along the margins of streams and swamps at elevations from 975 to 1580 m. Habitats are open and receive full sun. They are often completely dry by early July. Reported substrates include basalt, clay, clay loam, mud, rock, and rocky clay. The plants are categorized as perennials and persist by means of a basal rosette of leaves and taproot. Otherwise, little life history information exists for this species. **Reported associates:** *Acmispon denticulatus, Agoseris heterophylla, Alopecurus, Artemisia cana, Beckmannia syzigachne, Brodiaea coronaria, Bromus japonicus, B. tectorum, Calycadenia fremontii, Camissonia tanacetifolia, Claytonia dichotoma, Croton setigerus, Cuscuta howelliana, Deschampsia danthonioides, Downingia bacigalupii, D. bicornuta, D. cuspidata, Eleocharis palustris, Epilobium brachycarpum, E. cleistogamum, E. pallidum, E. pygmaeum, E. torreyi, Eremocarpus setigerus, Eryngium alismifolium, E. articulatum, Gnaphalium palustre, Gratiola ebracteata, G. heterosepala, Grindelia camporum, Helianthus annuus, Iva axillaris, Juncus nevadensis, Lactuca serriola, Lupinus bicolor, Lythrum hyssopifolia, Machaerocarpus californicus, Mimulus tricolor, Montia linearis, Myosurus minimus, Navarretia intertexta, N. leucocephala, N. minima, Osmorhiza occidentalis, Perideridia, Persicaria amphibia, Phlox gracilis, Plagiobothrys bracteatus, P. scouleri, Poa sandbergii, P. secunda, Pogogyne floribunda, P. ziziphoroides, Polygonum douglasii, P. polygaloides, Psilocarphus brevissimus, Rumex crispus, Tragopogon dubius, Triteleia hyacinthina, Vulpia microstachys.*

***Eryngium petiolatum* Hook.** is found in ditches, meadows, prairies, swales, and in vernal pools or streams at elevations of up to 556 m. This species tends to occupy sites where standing water persists for a relatively long time period. The substrates have been described as adobe, clay and hardpan. The plants appear in June, flower in August and set fruit in September. The flowers attract bees (Insecta: Hymenoptera), which probably function as their pollinators. The plants perennate from a taproot. The scaly fruits probably are dispersed by animals. **Reported associates:** *Achillea millefolium, Agrostis alba, A. stolonifera, Aira caryophyllea, Alisma, Alopecurus geniculatus, A. saccatus, Apocynum cannabinum, Beckmannia syzigachne, Bidens cernuus, Callitriche, Calochortus longebarbatus, Camassia quamash, Carex unilateralis, Castilleja tenuis, Cyperus erythrorhizos, Danthonia californica, Daucus carota, Deschampsia cespitosa, D. danthonioides, Downingia elegans, D. yina, Echinochloa crus-galli, Eleocharis acicularis, E. obtusa, E. palustris, Epilobium densiflorum, E. minutum, Eriophyllum lanatum, Gastridium phleoides, Gnaphalium, Gratiola ebracteata, Grindelia integrifolia, G. nana, Holcus lanatus, Hypericum perforatum, Hypochaeris radicata, Isoetes nuttallii, Juncus acuminatus, J. bufonius, J. effusus, J. hemiendytus, Lasthenia californica, L. glaberrima, Limnanthes floccosa, Lolium perenne, Lomatium cookii, Ludwigia palustris, Lysimachia nummularia, Lythrum salicaria, Melica, Mentha pulegium, Microseris laciniata, Myosurus minimus, Navarretia leucocephala, Panicum capillare, Paspalum distichum, Perideridia, Persicaria maculosa, Phalaris arundinacea, Phlox gracilis, Plagiobothrys bracteatus, P. nothofulvus, P. stipitatus, Plantago lanceolata, Pogogyne ziziphoroides, Prunella vulgaris, Psilocarphus brevissimus, Perideridia gairdneri, Rosa eglanteria, R. nutkana, Rubus, Rumex acetosella, R. crispus, Schoenoplectus tabernaemontani, Scirpus microcarpus, Trichostema lanceolatum, Trifolium depauperatum, Triteleia hyacinthina, Typha latifolia, Veronica peregrina, V. scutellata, Vicia villosa.*

Eryngium phyteumae **Delar.** inhabits ciénagas (small marshes) and meadows at elevations of up to 2150 m. Virtually nothing is known about the life history of this species. **Reported associates:** none.

Eryngium pinnatisectum **Jeps.** grows in depressions, ditches, flats, flatwoods, limestone barrens, swales, vernal pools, and on the ephemeral margins of streams and lakes at elevations of up to 915 m. The plants occur on clay or granitic clay substrates where they flower from June to August. The fruits are densely scaled and probably are dispersed by adhering to animal fur. The life history of this species is poorly known otherwise. **Reported associates:** *Avena, Brodiaea, Daucus, Eremocarpus, Eryngium castrense, Holocarpha, Lolium multiflorum, Navarretia, Polypogon monspeliensis.*

Eryngium prostratum **Nutt. ex DC.** lives in depressions, ditches, floodplains, meadows, mudflats, seeps, swamps, thickets, and along the margins of lakes, ponds, and rivers at elevations below 100 m. The habitats typically are open (to partially shaded) and are characterized by temporary waters associated with seasonal flooding. Substrates include bare soil, clay, clay loam, muck, mud, rock, sand, sandy clay, sandy clay loam, silt, and silty clay loam. The plants are low-growing (prostrate) perennials (also described as

annuals), which often occur in tangled mats (resembling vines) by rooting at the nodes. This species is associated with disturbance and it commonly invades bottomland forest patch openings where suitable habitat exists. It is a component of the "*Ludwigia–Rhexia*" ecological group. Flowering occurs from May to October. The fruits are papillose or tuberculate (but not scaly) and their means of dispersal is unknown; however, they are likely to be dispersed by water. **Reported associates:** *Acer rubrum, Ageratina altissima, Agrostis hyemalis, Allium, Ammannia coccinea, Anthemis cotula, Axonopus furcatus, Bidens aristosus, B. frondosus, Boehmeria cylindrica, Boltonia diffusa, Botrychium biternatum, B. dissectum, Briza minor, Callicarpa americana, Carex intumescens, C. lupulina, C. tribuloides, C. typhina, Cephalanthus occidentalis, Chamaecrista nictitans, Coelorachis rugosa, Commelina diffusa, C. virginica, Conoclinium coelestinum, Cornus florida, Crotalaria sagittalis, Cynodon dactylon, Cyperus pseudovegetus, C. squarrosus, Diodia teres, Echinochloa muricata, Eclipta prostrata, Eleocharis acicularis, E. obtusa, E. palustris, Elephantopus carolinianus, Elymus virginicus, Eragrostis hypnoides, Erigeron, Eriocaulon decangulare, Eupatorium perfoliatum, E. serotinum, Fimbristylis autumnalis, F. vahlii, Forestiera acuminata, Fraxinus pennsylvanica, Gamochaeta purpurea, Helenium flexuosum, H. pinnatifidum, Hydrocotyle umbellata, Hypericum brachyphyllum, H. drummondii, H. gentianoides, H. mutilum, Ipomoea lacunosa, Iva annua, Juncus acuminatus, J. tenuis, Justicia americana, Kummerowia striata, Laportea canadensis, Leersia oryzoides, L. virginica, Lespedeza cuneata, Lindernia dubia, Linum striatum, Lipocarpha micrantha, Liquidambar styraciflua, Lobelia cardinalis, Ludwigia alternifolia, L. decurrens, L. glandulosa, L. linearis, L. linifolia, L. palustris, L. peploides, L. polycarpa, Lysimachia ciliata, Mimulus alatus, Mitreola sessilifolia, Oldenlandia boscii, Panicum dichotomiflorum, P. rigidulum, P. tenerum, P. virgatum, Paspalum laeve, P. notatum, P. urvillei, Penthorum sedoides, Persicaria hydropiper, P. lapathifolia, P. maculosa, P. pensylvanica, P. punctata, Phyla lanceolata, Pilea pumila, Pinus echinata, P. elliottii, P. taeda, Platanus occidentalis, Pluchea camphorata, Polygala cymosa, Polypremum procumbens, Populus deltoides, Quercus palustris, Ranunculus laxicaulis, R. sardous, Rhexia aristosa, R. mariana, R. virginica, Rhus copallina, Rhynchospora perplexa, R. pleiantha, R. tracyi, Rotala ramosior, Rubus argutus, Rudbeckia laciniata, R. mohrii, Rumex crispus, Saccharum giganteum, Sacciolepis striata, Salix nigra, Sassafras albidum, Schoenoplectus pungens, Scirpus cyperinus, Scleria georgiana, Sisyrinchium angustifolium, Smilax, Solanum carolinense, Sonchus, Spigelia marilandica, Stachys tenuifolia, Stenotaphrum secundatum, Trachelospermum difforme, Triadenum tubulosum, T. walteri, Vitis rotundifolia.*

Eryngium racemosum **Jeps.** inhabits ephemeral depressions, flats, pools, river bottoms, and swales mainly along floodplains, or pond and reservoir margins at elevations less than 25 m. Substrates are described as alkaline and include

Marcuse clay, sandy gravel, and silty clay. Flowering occurs from June to August. The plants are weak perennials, producing prostrate stems that root at the nodes. However, they also grow as annuals during exceptionally dry years. The numbers of populations and individuals have been observed to vary considerably, depending on the extent and pattern of annual precipitation. The plants require a seasonal flooding regime to maintain wetland conditions and associated scouring to reduce competition. An estimated quarter of all historical populations is believed to have been extirpated due to various flood control projects. **Reported associates:** *Astragalus tener, Atriplex cordulata, A. depressa, A. joaquiniana, A. minuscula, Cirsium crassicaule, Gratiola heterosepala, Hibiscus californicus, Isocoma arguta, Lathyrus jepsonii, Legenere limosa, Lepidium latipes, Lilaeopsis masonii, Populus fremontii, Rosa californica, Rubus vitifolius, Salix, Symphyotrichum lentum.*

Eryngium sparganophyllum **Hemsl.** occurs in ciénaga (small marshes), meadows, and springs at elevations of up to 1525 m. It grows on substrates of organic muck or silty clay loam and flowers from March to June. The fruits are scaly and probably are dispersed by adhering to animal fur. The plants are described as perennials. **Reported associates:** *Anemopsis californica, Arbutus, Asclepias subverticillata, Carex praegracilis, Eleocharis, Helianthus annuus, Juncus balticus, Lobelia cardinalis, Lythrum californicum, Schoenoplectus pungens, Sisyrinchium.*

Use by wildlife: *Eryngium aquaticum* is a host for larval moths (Insecta: Lepidoptera: Gelechiidae: *Coleotechnites eryngiella*; Noctuidae: *Papaipema eryngii*). It also is eaten by several cray fish (Decapoda: Cambaridae: *Procambarus spiculifer*; *P. acutus*) and by grass carp (Cypriniformes: Cyprinidae: *Ctenopharyngodon idella*). *Eryngium aristulatum* is eaten by the western pocket gopher (Mammalia: Geomyidae: *Thomomys bottae*) especially during the early spring season. Its flowers are used as a nectar source by the Acmon Blue butterfly (Insecta: Lepidoptera: Lycaenidae: *Plebejus acmon*). Several Fungi are reported from *E. aquaticum* including *Cylindrosporium eryngii* (Ascomycota: Leotiomycetidae), *Entyloma eryngii* (Basidiomycota: Entylomataceae), and *Septoria eryngicola* (Ascomycota: Mycosphaerellaceae).

Economic importance: food: *Eryngium articulatum* is a honey plant and provides a source of good-tasting dark honey; **medicinal:** Extracts from the flowers and leaves of *Eryngium prostratum* can partially inhibit the growth of *Staphylococcus aureus* (Eubacteria: Staphylococcaceae). Although their efficacy has not been evaluated clinically, tinctures or infusions of dried *E. aquaticum* roots have been used widely in traditional medicine and remain in use to the present day as a treatment for urinary infections and muscle spasms. Similar preparations were used by various native American tribes as an emetic (Koasati), as a diuretic, expectorant, or stimulant (Choctaw), to relieve nausea (Alabama, Cherokee), to treat intestinal worms (Delaware), and as a treatment for gonorrhea (Choctaw, Delaware, Oklahoma). The Choctaw also used the root of *E. aquaticum* as an antitoxin to treat snakebite. The putative antivenom properties of this plant (which are

unsubstantiated) persist in many folklore accounts and account for the common name of "rattlesnake master"; **cultivation:** *Eryngium aquaticum* and *E. petiolatum* are distributed commercially as ornamental garden specimens; **misc. products:** Passage of the type strain of tobacco mosaic virus (TMV) through *Eryngium aquaticum* is used to prepare the U2 strain of TMV, which is modified as a result; **weeds:** *Eryngium prostratum* can be troublesome as a garden weed; **nonindigenous species:** *Eryngium aquaticum* is regarded as introduced to the Canadian provinces of British Columbia and Ontario.

Systematics: The systematics of *Eryngium*, the largest genus in Apiacae, have been well studied. The genus occurs within subfamily Saniculoideae (Figure 5.109) and phylogenetic analyses of nearly half the known species, using several molecular datasets, confirm that *Eryngium* is monophyletic, with *Sanicula* as its sister group. One analysis using only nrITS data weakly supports the Old and New World clades as including *Sanicula*; however, that result probably is an artifact of the particular outgroup used. The genus is characterized by several morphological synapomorphies including nonpalmate leaves, capitate inflorescence, showy involucre, and hermaphroditic flowers that are subtended by a bract. All of the New World species resolve within a clade that corresponds with subgenus Monocotyloidea. Unfortunately, as comprehensive as these phylogenetic studies have been, only two OBL species (*E. articulatum, E. prostatum*) have been surveyed simultaneously (a different study included only *E. petiolatum*), leaving the relationships of most of the aquatic species unresolved. Yet, even this small sample indicates the independent derivation of the OBL habit in these two species (Figure 5.115). It also is interesting that the vernal pool specialist *E. articulatum* is closely related to (and apparently derived from) several Australian species (*E. ovinum, E. rostratum, E. vesiculosum*), which also share affinities for vernal habitats (Figure 5.115). Inclusion of other North American vernal pool species of *Eryngium* would be interesting to determine whether they have originated from a single or multiple radiations into this specialist habitat. A number of hybrids have been reported in the genus and may account for some of the variability associated with several of the OBL taxa. The base chromosome number of *Eryngium* ranges from $x = 5–8$; the latter is found most commonly and is characteristic of all OBL species surveyed: *Eryngium aquaticum* ($2n = 16, 96$) has diploid and unusual 12-ploid cytotypes; *E. articulatum, E. constancei, E. pinnatisectum*, and *E. racemosum* ($2n = 32$) are tetraploid; *E. aristulatum* ($2n = 32, 64$) contains tetraploid and octoploid cytotypes; *E. castrense* and *E. mathiasiae* ($2n = 64$) are octoploids.

Comments: The North American distribution of *Eryngium* includes Eastern coastal North America (*E. aquaticum*), the southern United States (*E. prostratum*), western United States (*E. articulatum*), and the northwestern United States (*E. petiolatum*). Species with limited distributions include *E. sparganophyllum* (Arizona and New Mexico; also extending into Mexico); *E. phyteumae* (Arizona; extending into Mexico), and a large number of species restricted to California (*E. aristulatum, E. castrense, E. constancei, E. jepsonii, E. mathiasiae,*

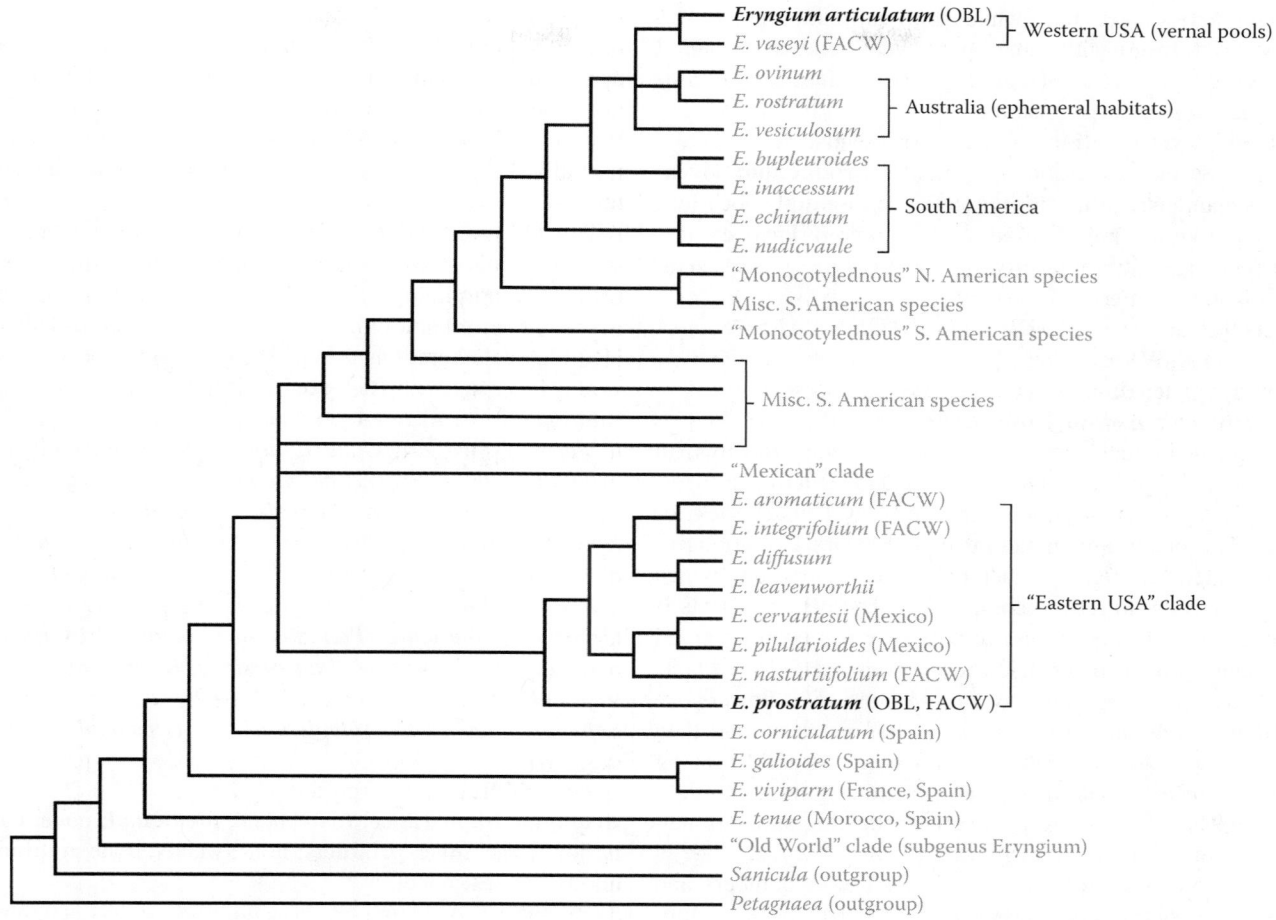

FIGURE 5.115 Phylogenetic relationships in *Eryngium* based on analysis of combined *trnQ–trnK* and ITS sequence data. OBL North American taxa are shown in bold (other wetland categories also are indicated for North American species). Although only two OBL taxa were included in this analysis of 118 species, it is evident that the OBL habit has arisen independently at least twice in the genus. These results indicate that the vernal pool species in the western United States may be derived from plants of similar habitat conditions in Australia. (Adapted from Calviño, C.I. et al., *Mol. Phylogen. Evol.*, 46, 1129–1150, 2008.)

E. pinnatisectum, E. racemosum). Eryngium aristulatum is of particular conservation concern, with one 1999 survey able to locate only approximately 1000 surviving individuals among four extant populations.

References: Aronson, 1989; Baden et al., 1975; Barbour et al., 2007; Baskin & Baskin, 1998; Bauder, 2000; Baye et al., 2000; Bell & Constance, 1960; Bernardello et al., 2001; Borgias, 2004; Bowden, 1945; Bray, 1957; Breden et al., 2001; Calviño et al., 2008; Carlquist & Pauly, 1985; Crouch & Golden, 1997; Evans et al., 2003; Gaudeul & Till-Bottraud, 2003, 2004; Harper, 1900; Herbold & Moyle, 1989; Herndon, 1968; Hill, 1992; Hunt, 1992; Kadereit et al., 2008; Keeley, 1999; Kelch & Murdock, 2012; Kirkman et al., 2000; Klein et al., 2007; Knuth, 1906, 1908; Krochmal, 1968; Lippert & Jameson, 1964; Lyder, 2009; Mathias, 1994; Mathias & Constance, 1941a; Molano-Flores, 2001; Moore et al., 2001; Mundry & Priess, 1971; Nichols, 1934; Parker & Hay, 2005; Pates & Madsen, 1955; Pellett, 1920; Schuyler et al., 1993; SFWO, 2009; Uphof, 1922; Witham, 2006; Witsell & Baker, 2006.

6. *Harperella*

Bishop's weed

Etymology: after Roland McMillan Harper (1878–1966)

Synonyms: *Ptilimnium* (in part)

Distribution: global: North America; **North America:** southeastern

Diversity: global: 1[–3] species; **North America:** 1[–3] species

Indicators (USA): OBL: *Harperella nodosa*

Habitat: freshwater; palustrine, riverine; **pH:** 7.0; **depth:** <1 m; **life-form(s):** emergent herb

Key morphology: Stems (to 1.2 m) erect, purplish, ribbed, the lower nodes often rooting and occasionally producing rosettes; leaves (to 4 dm) hollow, quill-like, regularly septate, clasping at base; herbage weakly dill scented; umbels compound, peduncled (to 1 dm), with 5–15 primary rays (to 2.0 cm) and lanceolate bracts (to 5 mm); 5–15 pedicels (to 5 mm) arise from a small involucre (<0.5 mm); flowers regular, bisexual; perianth campanulate, 5-merous; sepals (to 2 mm) persistent in fruit; petals (to 0.5 mm) white, short clawed, roundish; stamens (to 1.0 mm) with tapering filaments and rose-colored anthers; ovary inferior; schizocarps (to 2 mm) broadly elliptical, back of mericarps 3-ribbed

Life history: duration: annual (fruit/seeds) or perennial (vegetative rosette offshoots); **asexual reproduction:** adventitious plantlets; **pollination:** insect or self; **sexual condition:** hermaphroditic; **fruit:** schizocarps (common); **local dispersal:** seeds (gravity); **long-distance dispersal:** seeds, vegetative plantlets (water)

Imperilment: (1) *Harperella nodosa* [G2]; S1 (AL, GA, MD, NC, SC, VA, WV); S2 (AR)

Ecology: general: monotypic (see *Systematics* section).

Harperella nodosa Rose occurs on sand/gravel bars or in shallow, protected areas of swiftly flowing streams, in ephemerally flooded meadows and pond margins, or in seeps. Substrates consist of cobble, granite, gravel, mud, rock, or sand. The plants are intolerant of either dry conditions or deep water but readily tolerate periodic flooding, which seems essential to reduce competition. The pH of habitats is neutral (pH: 7.0) and excessive mortality has been observed for plants grown under acidic conditions (pH <5.0). Plants reach the flowering stage by July–August. The flowers are self-compatible, but protandrous; selfing results in less than 1% seed set. Overall, seed set is low but comparable among open-pollinated, outcrossed, and geitonogamous pollinations (25%, 26%, 21%, respectively), indicating a mixed mating system that is insect mediated. Seed set decreases late in the flowering season because of lower flower numbers and fewer pollinators. The life cycle is complete by late summer or early fall, when the fruiting stems topple and drop their seeds into the soil at the site of contact with the inflorescence. As a consequence, most seeds germinate close to the parental plant. Longer dispersal distances are facilitated by water due to the ability of the seeds to float. The seeds germinate (sometimes while on the inflorescence) by the end of September and the plants will overwinter in shallow water as rosettes. Germination rates are high. Estimates of population size vary, with some having fewer than 100 and others with more than a million individuals. Some plants are at least partially frost tolerant and are capable of surviving intermittent cold periods as low as −10°C. Adult plants sometimes will produce vegetative rosettes as basal offshoots. These can detach and become dispersed by water in which they survive well. Two morphologically and genetically distinct ecotypes occur, which inhabit either stream or pond margin habitats. Hydrological alterations present the greatest threat to the survival of this species. **Reported associates:** *Alnus serrulata, Andropogon gerardii, Arthraxon hispidus, Carex lupulina, C. torta, C. walteri, Conoclinium coelestinum, Dulichium arundinaceum, Eleocharis melanocarpa, E. tricostata, Eutrochium fistulosum, Fimbristylis, Gratiola brevifolia, Hydrolea ovata, Hypericum denticulatum, H. fasciculatum, H. myrtifolium, Isoetes riparia, Juncus repens, Justicia americana, Lobelia cardinalis, Ludwigia, Lysimachia terrestris, Orontium aquaticum, Panicum hemitomon, Proserpinaca pectinata, Psilocarya, Rhexia aristosa, R. mariana, R. virginica, Rhynchospora colorata, R. perplexa, R. microcarpa, Scirpus expansus, Sclerolepis uniflora, Tripsacum dactyloides, Xyris.*

Use by wildlife: none reported.

Economic importance: food: none; **medicinal:** none; **cultivation:** none; **misc. products:** none; **weeds:** none; **nonindigenous species:** none

Systematics: Although *Harperella nodosa* has long been placed in *Ptilimnium*, its phylogenetic distinctness necessitates that it now be regarded as a distinct genus (Figure 5.116). Molecular data resolve *Harperella* in a position as the sister group to a *Limnosciadium/Daucosma* clade, which in turn is sister to a *Tiedemannia/Ptilimnium* clade (Figure 5.116).

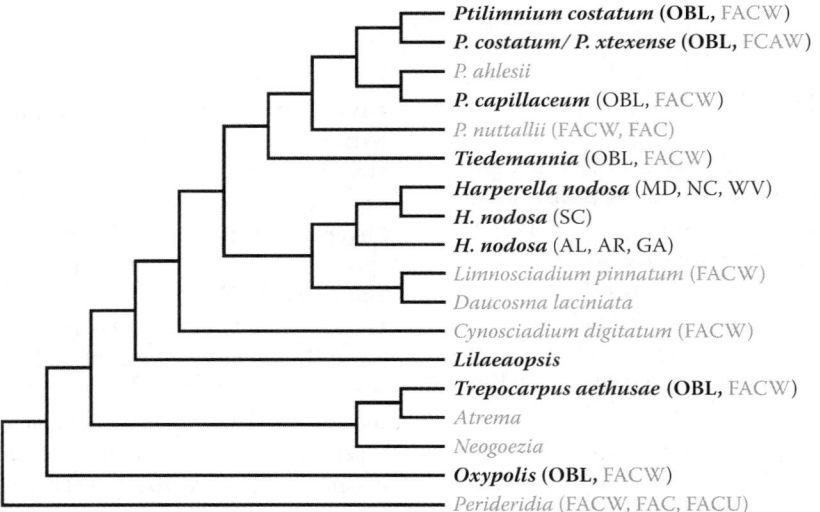

FIGURE 5.116 Relationships of *Harperella* as indicated by likelihood analysis of nrITS data. OBL North American taxa (or genera containing OBL species) are shown in bold with other wetland categories provided for North American species. Geographical segregates (by U.S. state) also are indicated for *Harperella*. (Adapted from Feist, M.A.E. & Downie, S.R., *Syst. Bot.*, 33, 447–458, 2008.)

Although regarded here as monotypic, genetic analysis of most known extant populations of *Harperella* confirm that the genus is monophyletic. However, *Harperella nodosa* itself has been subdivided into three taxa (sometimes distinguishing *H. fluviatilis* Rose and *H. vivipara* Rose as separate species) that are distinct genetically and ecologically. Populations corresponding to at least two of these taxa have been indicated consistently by DNA sequence analysis (Figure 5.116) and by allozyme data. Yet, even though *H. fluviatilis* and *H. vivipara* have been categorized as discrete OBL species, most authors do not recognize them as distinct from *H. nodosa*, although there may be merit to adopting some level of intraspecific taxonomic recognition. They have not been distinguished in the present treatment pending further study, and because of the impracticality of attempting to correctly parse the vague existing literature accounts to the appropriate taxon. Allozyme analysis indicated further that populations are characterized by low levels of genetic variation (H_T <0.09), which is partitioned mainly among populations (G_{ST} = 0.68). The base chromosome number of *Harperella* is $x = 6$; *H. nodosa* ($2n = 12$) is a diploid.

Comments: *Herperella nodosa* occurs sporadically in the southeastern United States (AR, AL, GA, MD, NC, SC, WV). It is listed by the U.S. Federal Government as an endangered species and currently survives in fewer than 15 extant populations.

References: Bartgis, 1997; Easterly, 1957; Feist & Downie, 2008; Hardcastle & Williams, 2001; Kress et al., 1994; Maddox & Bartgis, 1990; Marcinko & Randall, 2008; Rose, 1911; USFWS 1988.

7. Helosciadium

Fool's watercress, marsh umbrella, water celery

Etymology: the Greek *helos skiadeion* ("marsh umbrella") referring to the habitat and inflorescence shape

Synonyms: *Apium* (in part); *Ciclospermum* (in part); *Heliosciadium*; *Sium* (in part)

Distribution: global: Europe; North America* **North America:** southern

Diversity: global: 5 species; **North America:** 1 species

Indicators (USA): OBL: *Helosciadium nodiflorum*

Habitat: freshwater; palustrine, riverine; **pH:** alkaline; **depth:** <1 m; **life-form(s):** emergent herb, submersed (vittate)

Key morphology: shoots (to 11 dm) rooting at lower nodes; the leaves (to 20 cm) pinnately compound (odd) with up to 4 pairs of leaflets (to 6 cm), the lower crenate, 3-lobed apically, the upper serrate, the petioles (to 3 dm) conspicuously sheathed (to 10 cm); umbels compound, with up to 20 unequal rays (to 2 cm); flowers small, the corolla (to 0.5 mm) white or greenish white; schizocarps with 5 slightly raised ridges, the mericarps (to 2 mm) ovate oblong, brownish

Life history: duration: perennial (persistent rootstocks); **asexual reproduction:** shoot fragments; **pollination:** insect; **sexual condition:** hermaphroditic; **fruit:** schizocarps (common); **local dispersal:** fragments, mericarps (water); **long-distance dispersal:** mericarps (water)

Imperilment: (1) *Helosciadium nodiflorum* [GU]

Ecology: general: *Helosciadium* should be regarded as an aquatic genus, with all of the species growing directly in water or at least consistently being associated with aquatic habitats. The species occur most frequently in watercourses, but also grow in marshes and other wetlands. One species (*H. bermejoi*) inhabits drying streambeds. The long association of this genus with *Apium* makes it difficult to elucidate distinctive ecological characteristics. Generally, the flowers are probably self-compatible but protandrous and insect pollinated. The few available data indicate that the seeds have no after-ripening requirement for germination, and will germinate either when submersed or emersed. Presumably, all of the seeds are water dispersed; however, they supposedly do not float well (at least when fresh) and are characteristically absent from river drift samples.

***Helosciadium nodiflorum* W.D.J. Koch** grows in canals, ditches, marshes, springs and along the margins of lakes, ponds, rivers and streams at elevations of up to 350 m. It is particularly common in clear, shallow water (up to 1.2 m) along the margins of high order streams, which typically are from 2.5 to 8 m in width. These are moderately shade-tolerant plants that are highly tolerant of turbulence (withstanding up to 6 h of intense agitation), but occur most often in sites with moderate flow rates. This species grows mainly in alkaline waters (pH: 7.7–8.0; alkalinity 170–250 ppm), which are low in nutrients (3–6 ppm nitrate; <0.3 ppm ammonia nitrogen; <0.3 ppm phosphate phosphorous); however, the substrates (most often sand or gravel) are mesotrophic and contain moderate levels of calcium (100–150 ppm) and high amounts of nutrients (e.g., >10 ppm ammonia nitrogen; 93 ppm phosphate). The plants are quite tolerant of pollution and often flourish in eutrophic waters. Flowers and fruits are produced readily. The flowers are protandrous and are pollinated by insects (Insecta). The fruits lack a dormancy requirement and are dispersed by the water. They remain afloat for less than 2 days when ripe, but for more than 90 days when dry (and still retain their germinability). The seeds germinate on wet substrates or in shallow water. Seedling establishment requires open conditions and they can appear quickly in sites that have been dredged. However, seedling survivorship usually is low and continual disturbance is necessary to alleviate competition with other species. A seed bank of up to 50 seeds per m² can develop, usually in the mid-to-upper reaches of the floodplain; however, seed deposition often is undetected on substrates where the plants occur in the standing vegetation. The populations persist principally by means of overwintering shoots rather than as seedlings. The plants are perennial from a persistent root crown. Vegetative propagation occurs by shoot fragments, which develop new roots within a few days. The stems can become fairly persistent when rooting firmly into gravel substrates. The foliage has high resistance to flow and the plants dislodge with relatively little force and often are uprooted. The plants tend to be shallowly rooted (22–29 cm) but are fairly tolerant to substrate desiccation if not for prolonged duration. Somewhat longer roots are produced in drained sites. In watercourses, the plants grow in clumps where the silt accumulates. The stature of the plants becomes shorter if they are submersed. The plants can oxygenate the substrate when growing under hypoxic conditions. The rooting of shoots can be enhanced by treatment with β-cyclodextrin/naphthyloxyacetic acid. **Reported associates (North America):** none reported.

Use by wildlife: *Helosciadium nodiflorum* is highly palatable to grazing livestock. It is a host to a number of invertebrates in its native range; however, its wildlife value in North America is unknown. The plants have shown resistance to leaf mining flies (Insecta: Diptera: *Liriomyza trifolii*) in feeding tests.

Economic importance: food: In Crete, Cyprus, Italy, Spain, and Tunisia the leaves of *Helosciadium nodiflorum* are eaten as a boiled vegetable or raw in salads. If eaten, the plants must be collected from unpolluted localities because they are resistant to heavy metals and are able to uptake significant quantities of As, Cd, Cr, Cu, Hg, Mn, Ni, Pb, Sn, U, and Zn; **medicinal:** *Helosciadium nodiflorum* has been used in Europe as a digestive remedy, diuretic, stimulant, to reduce inflammation, and as a treatment for lymphatic tumors. The plants contain phototoxic linear furanocoumarins, but at a relatively low foliar concentration (11.8 μg/g fresh weight), which is below the threshold for inducing contact dermatitis (18 μg/g). The aerial portions exhibit a high (30%) free radical scavenging activity. The plants also have been used to treat a lack of appetite and inflammation of mucous membranes in domestic animals; **cultivation:** *Helosciadium nodiflorum* is grown infrequently as a water garden ornamental; **misc. products:** none; **weeds:** *Helosciadium nodiflorum* is regarded as a weed in Portugal and Spain, where it is native. It has not yet been reported as invasive anywhere in North America; **nonindigenous species:** *Helosciadium nodiflorum* was introduced to North America (reportedly before 1788), and also to Chile (before 1878), Mexico and New Zealand (before 1947). Several older North American records are from shipping ballast disposal sites, which provides one possible means of introduction. It also has been sold in markets mistakenly as watercress (*Nasturtium*).

Systematics: *Helosciadium* had been regarded as a section of *Apium* for many years until molecular data demonstrated conclusively that these genera are distinct and are not closely related (Figure 5.111). Unfortunately, the name *Apium nodiflorum* appears for *H. nodiflorum* in most literature accounts. The distant relationship of these genera also is evidenced by their extensive genetic distance (AFLP markers) and explains why attempts to hybridize *Helosciadium* species with cultivated celery (*Apium graveolens*) have failed in crop improvement studies. Molecular data analyses also have confirmed that *Helosciadium* is monophyletic and contains five species (Figures 5.112 and 5.117). However, there is still uncertainty regarding the specific sister group of *Helosciadium*. Analysis of combined DNA sequence data indicates *Oxypolis* as the sister genus (Figure 5.111), but analysis of nrITS data (for a larger number of species and genera) resolves *Helosciadium* as the sister genus to *Sium* (Figure 5.112). *Helosciadium nodiflora* is related closely to *H. bermejoi* (a rare endemic of Minorca) and to *H. repens* (Figure 5.112). The basic chromosome number of *Helosciadium* is *x* = 11. *Helosciadium nodiflora* (2*n* = 22) is a

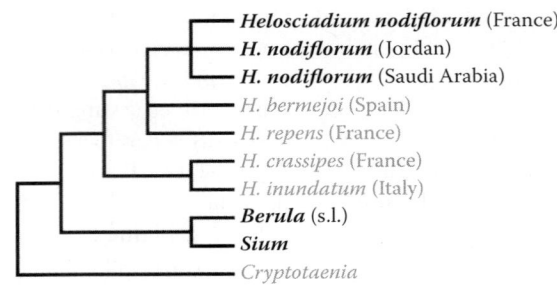

FIGURE 5.117 Phylogenetic relationships in *Helosciadium* as resolved by analysis of nrITS sequence data. OBL North American taxa are shown in bold. Accessions of *H. nodiflorum* are similar genetically despite their geographical origin (indicated in parentheses); it has not been determined where the North American plants originated. All species of *Helosciadium*, *Berula* and *Sium* are aquatic and wetland plants, which indicates a single origin of the aquatic habit in this clade of genera. (Adapted from Spalik, K. & Downie, S.R., *Amer. J. Bot.*, 93, 747–761, 2006.)

diploid; however, an aneuploid count (2*n* = 20) also has been reported for this species. Natural hybrids have been reported between *H. inundatum* × *H. nodiflorum* (known as *Apium ×moorei*) and between *H. nodiflorum* × *H. repens*; neither has been documented genetically.

Comments: *Helosciadium nodiflorum* has been reported sporadically from California, New Jersey, Pennsylvania, and South Carolina, but doubtfully has persisted in some of these localities

References: Bullitta et al., 2007; Champion & Tanner, 2000; Chandra, & Kulshreshtha, 2004; Chorianopoulou et al., 2001; Cook, 1996a; Coulter & Rose, 1887; Darwin, 1859; Della et al., 2006; Esler, 1987; Goodson et al., 2002, 2003; Guppy, 1906; Gurnell et al., 2007a; Hardway et al., 2004; Haslam, 1978; Haslam & Wolseley, 1981; Haslam et al., 1975; Johnson, 1999; Johnson & Brooke, 1998; Justin & Armstrong, 1987; Kokkinou et al., 2001; Lentini & Venza, 2007; Muminović et al., 2004; Onaindia et al., 1996; Pieroni et al., 2002; Preston & Croft, 1997; Rivera et al., 2005; Thommen & Westlake, 1981; Trumble et al., 1990; Valiejo-Roman et al., 2002; Villaseñor & Espinosa-Garcia, 2004; Vlyssides et al., 2005.

8. *Heracleum*

Cow parsnip; Grande Berce

Etymology: after Hercules (mythical) for the large size attained by some species

Synonyms: *Sphondylium*

Distribution: global: Africa; Asia; Europe; North America; **North America:** widespread

Diversity: global: 90–120 species; **North America:** 3 species

Indicators (USA): FACW; FAC; FACU: *Heracleum maximum*

Habitat: freshwater; palustrine, riverine; **pH:** 5.4–7.4; **depth:** <1 m; **life-form(s):** emergent herb

Key morphology: Stems (to 2.8 m) hollow, coarse, ribbed, pubescent; leaves (to 0.6 m) ternately compound to lobed, irregularly toothed, petioles (to 0.5 m) stout with decurrent sheaths (to 5 cm); herbage pungent; umbels (to 20 cm) compound (1–4 secondary umbels), peduncles (to 20 cm) pubescent, bracts (1–6) dilated basally; umbellets (15–40) on unequal stalks (to 10 cm), pedicels (to 30 mm) 10–50; flowers bisexual or staminate; sepals absent, petals white or purplish, occasionally cleft apically; schizocarps (to 12 mm) obovate, strongly flattened, inconspicuously ribbed

Life history: duration: perennial (persistent roots); **asexual reproduction:** none; **pollination:** insect; **sexual condition:** andromonoecious; **fruit:** schizocarps (common); **local dispersal:** seeds (wind); **long-distance dispersal:** seeds (water?)

Imperilment: (1) *Heracleum maximum* [G5]; SH (KY); S1 (GA, KS); S2 (TN); S3 (DE, MD, NC)

Ecology: general: *Heracleum* species are diverse and occur mainly throughout the temperate parts of the Northern Hemisphere, but extend into higher elevations at tropical latitudes. They occupy many types of habitat including montane forest, grassland and scrub. Most of the species are terrestrial, and even *Heracleum maximum* occurs across an extremely wide range of soil moisture gradients as indicated by its variable wetland indicator status and the eclectic assemblage of associated species across its range. It was regarded formerly as an OBL indicator (1996 list) only in the southwestern United States, but generally tends to grow in denser patches when occurring in wet, shaded sites. Its treatment (completed prior to the release of the revised indicator list) has been retained here. Most *Heracleum* species are perennial (occasionally biennial), and persist from the stem bases, which produce clusters of thickened cylindrical or fusiform roots. The plants are andromonoecious with self-compatible, protandrous flowers that are pollinated by unspecialized insects (Insecta: Coleoptera; Diptera; Hemiptera; Hymenoptera), at least in those species whose reproductive biology has been studied in any detail. The fruits are broadly winged, which allows for some degree of wind dispersal. Fruits of some species also are known to float for up to 3 days, which facilitates water dispersal.

Heracleum maximum W. Bartram inhabits beaches, bogs, bottomland, ditches, estuaries, floodplains, gravel bars, gullies, marshes, meadows, prairies, ravines, riverbeds, seeps, shores, sloughs, springs, swamps, thickets, woodlands, and the margins of lakes, rivers, and streams at elevations of up to 3750 m. The plants extend northward to at least to 65.9° N latitude. This species can grow in full sun to deep shade and also tolerates some salinity. It occurs on substrates described as black clay, black loam, clay loam, gravel, humus, loam, muck, organic, rock, sand, sandy humus, sandy loam and silt loam. The andromonoecious umbels contain a mix of hermaphroditic and staminate flowers, which are insect pollinated (Insecta: Coleoptera: Cleridae: *Trichodes nutalli*; Diptera [primarily], Hymenoptera). The plants are polycarpic perennials with a flowering peak that usually occurs from mid-to-late June (however, flowering ranges from April to August, depending on the locality). The hermaphroditic flowers are protandrous and the plants generally are temporally dioecious. Geitonogamous pollinations are rare owing to the slight extent of overlap between the flowering phases. The primary umbels exhibit much higher seed set (>80%) than the secondary umbels (<20%), which function primarily as males. In one study, reproductive output was estimated at 590 flowers/umbel, 14.7% hermaphroditic flowers and 8 seeds/umbel. However, when the primary umbel was damaged by herbivory, the output increased to 661 flowers/umbel, 23.7% hermaphroditic flowers and 192 seeds/umbel. The altered sex expression is believed to be a response that helps to restore fecundity otherwise lost to floral herbivory. Single plants are known to produce as many as 2280 seeds. The seeds retain longevity for up to 3 years when stored at 1°C. They have weak morphophysiological dormancy, but require only cold stratification to break dormancy (112 days under a 22°C/17°C temperature regime). Like related species, the seeds probably float and are dispersed by water. The plants occupy both seral and climax communities in disturbed or undisturbed sites. The plants respond variably to canopy removal. They are most susceptible to fire damage during periods of drought.

Reported associates: *Abies amabilis, A. balsamea, A. concolor, A. lasiocarpa, Acer glabrum, A. negundo, A. rubrum, A. saccharinum, A. saccharum, Achillea millefolium, Achnatherum nelsonii, Aconitum columbianum, Actaea rubra, Aesculus glabra, Agastache pallidiflora, A. urticifolia, Agoseris aurantiaca, Agrimonia gryposepala, Agrostis exarata, A. gigantea, A. scabra, A. stolonifera, Allium dichlamydeum, A. geyeri, A. textile, Alnus rugosa, A. tenuifolia, A. viridis, Ambrosia artemisiifolia, Amelanchier alnifolia, A. pallida, A. pumila, A. utahensis, Amsinckia lycopsoides, Anaphalis margaritacea, Androsace filiformis, A. septentrionalis, Angelica ampla, A. arguta, A. dawsonii, A. genuflexa, A. grayi, A. lucida, A. pinnata, Apios americana, Aquilegia caerulea, A. flavescens, A. formosa, Arabis drummondii, A. glabra, Aralia nudicaulis, Arctagrostis latifolia, Arctium lappa, A. minus, Arctostaphylos uva-ursi, Arenaria capillaris, Arnica cordifolia, A. mollis, Artemisia ludoviciana, A. serrata, A. tilesii, Asclepias syriaca, Athyrium filix-femina, Barbarea vulgaris, Betula alleghaniensis, B. nana, B. occidentalis, B. papyrifera, Bistorta bistortoides, Brachythecium starkei, Bromus anomalus, B. carinatus, B. ciliatus, B. inermis, B. japonicus, B. lanatipes, B. tectorum, B. vulgaris, Calamagrostis canadensis, C. nutkaensis, C. rubescens, C. stricta, Calochortus apiculatus, C. invenustus, Caltha leptosepala, C. palustris, Campanula rotundifolia, Campanulastrum americanum, Campylium stellatum, Cardamine bulbosa, C. cordifolia, C. douglassii, C. oligosperma, Carex aquatilis, C. aurea, C. athrostachya, C. bebbii, C. deweyana, C. ebenea, C. geyeri, C. heteroneura, C. hoodii, C. intumescens, C. jonesii, C. lenticularis, C. limosa, C. lyngbyei, C. macloviana, C. microptera, C. norvegica, C. nova, C. obnupta, C. occidentalis, C. pellita, C. preslii, C. straminiformis, C. stricta, C. utriculata, C. vesicaria, Carya ovata, Castilleja miniata, C. sulphurea, Cephalanthus occidentalis, Cercocarpus montanus, Chaenactis douglasii, Chaiturus marrubiastrum, Chamerion angustifolium, C. latifolium, Cichorium intybus, Cicuta douglasii, C. maculata, Cinna latifolia, Circaea alpina, Cirsium arvense, C. eatonii, C. hookerianum, C. parryi, C. vulgare, Claytonia lanceolata, C. perfoliata,*

C. sibirica, Clematis ligusticifolia, C. occidentalis, C. borealis, Clintonia uniflora, Collomia linearis, Conioselinum scopulorum, Cornus obliqua, C. sericea, Coronilla varia, Corydalis aurea, C. caseana, Corylus cornuta, Crataegus douglasii, Cryptotaenia canadensis, Cynoglossum officinale, Cystopteris fragilis, Dactylis glomerata, Dasiphora floribunda, Delphinium barbeyi, D. glaucum, D. nuttallianum, D. ×occidentale, Deschampsia beringensis, D. cespitosa, Descurainia incana, Dipsacus fullonum, D. laciniatus, Disporum hookeri, D. trachycarpum, Draba albertina, D. spectabilis, Dryopteris campyloptera, Dudleya farinosa, Eleocharis, Elymus glaucus, E. lanceolatus, E. repens, E. trachycaulus, Elytrigia intermedia, Epilobium ciliatum, E. hornemannii, E. saximontanum, Equisetum arvense, E. hyemale, E. laevigatum, E. pratense, E. telmateia, Erigeron coulteri, E. elatior, E. eximius, E. flagellaris, E. glaucus, E. peregrinus, E. speciosus, E. subtrinervis, Eriogonum latifolium, Eriophyllum confertiflorum, Erythronium americanum, E. grandiflorum, Eucephalus engelmannii, E. ledophyllus, Eurhynchium pulchellum, Eurybia conspicua, E. glauca, E. macrophylla, Festuca rubra, F. subulata, F. thurberi, Fragaria vesca, F. virginiana, Fraxinus nigra, F. pennsylvanica, Fritillaria camschatcensis, Galeopsis bifida, Galium aparine, G. boreale, G. trifidum, G. triflorum, Gayophytum ramosissimum, Gentianella amarella, Gentianopsis thermalis, Geranium caespitosum, G. maculatum, G. richardsonii, G. viscosissimum, Geum canadense, G. macrophyllum, Glechoma hederacea, Glyceria elata, G. grandis, G. striata, Goodyera oblongifolia, Gymnocarpium dryopteris, Gymnocladus dioicus, Hackelia floribunda, H. micrantha, Heliomeris multiflora, Hepatica nobilis, Heuchera cylindrica, Hieracium aurantiacum, Holcus lanatus, Horkelia, Humulus lupulus, Hydrophyllum capitatum, H. fendleri, H. virginianum, Hymenoxys hoopesii, Hypericum anagalloides, Impatiens, Ipomopsis aggregata, Iris douglasiana, I. missouriensis, Jamesia americana, Juncus balticus, J. compressus, J. drummondii, J. dudleyi, J. mertensianus, J. orthophyllus, J. saximontanus, J. tracyi, Juniperus scopulorum, Lappula occidentalis, Lathyrus ochroleucus, Leonurus cardiaca, Lepidium virginicum, Leymus arenarius, L. cinereus, L. mollis, Ligusticum porteri, L. tenuifolium, Linaria vulgaris, Linnaea borealis, Listera cordata, Lomatium dissectum, Lonicera canadensis, L. involucrata, L. utahensis, Lupinus argenteus, L. latifolius, Luzula comosa, L. parviflora, Mahonia repens, Maianthemum dilatatum, M. racemosum, M. stellatum, Marchantia, Matteuccia struthiopteris, Medicago lupulina, Melica spectabilis, M. subulata, Menispermum canadense, Mentha arvensis, Mertensia ciliata, M. franciscana, M. paniculata, Micranthes odontoloma, Mimulus guttatus, Mitella pentandra, Mnium spinulosum, Moehringia lateriflora, M. macrophylla, Moneses uniflora, Montia chamissoi, Myosotis scorpioides, Napaea dioica, Nassella viridula, Oenanthe sarmentosa, Oenothera elata, Orthilia secunda, Osmorhiza berteroi, O. depauperata, O. longistylis, O. occidentalis, Osmunda claytoniana, Ostrya virginiana, Oxalis violacea, Oxypolis fendleri, O. occidentalis, Panax trifolius, Parthenocissus quinquefolia, Pastinaca sativa, Paxistima myrsinites, Pedicularis bracteosa, P. groenlandica, P. procera, Penstemon barbatus, P. fruticosus, P. virgatus, Persicaria punctata, Petasites sagittatus, Phacelia heterophylla, Phalaris arundinacea, Philadelphus lewisii, Phleum alpinum, P. pratense, Phragmites australis, Physocarpus capitatus, Picea engelmannii, P. engelmannii, P. glauca, P. pungens, Pinus contorta, P. ponderosa, P. resinosa, P. strobiformis, Plantago major, Platanthera dilatata, P. stricta, Poa compressa, P. leptocoma, P. palustris, P. pratensis, P. reflexa, Polemonium foliosissimum, Polygonatum biflorum, Polygonum douglasii, Populus angustifolia, P. balsamifera, P. tremuloides, P. pulcherrima, Populus ×canescens, Potentilla anserina, P. gracilis, P. pacifica, P. plattensis, P. pulcherrima, Primula pauciflora, Prunella vulgaris, Prunus serotina, P. virginiana, Pseudocymopterus montanus, Pseudoroegneria spicata, Pseudostellaria jamesiana, Pseudotsuga menziesii, Pteridium aquilinum, Pyrola asarifolia, P. minor, Quercus alba, Q. macrocarpa, Ranunculus abortivus, R. inamoenus, R. occidentalis, R. uncinatus, Rhamnus alnifolia, Rhodiola integrifolia, R. rhodantha, Ribes americanum, R. cynosbati, R. inerme, R. lacustre, R. laxiflorum, R. missouriense, R. montigenum, R. triste, R. wolfii, Rosa acicularis, R. carolina, R. eglanteria, R. multiflora, R. nutkana, R. woodsii, Rubus deliciosus, R. idaeus, R. parviflorus, R. spectabilis, Rudbeckia californica, R. laciniata, Rumex acetosella, R. crispus, R. salicifolius, Salix barclayi, S. bebbiana, S. boothii, S. brachycarpa, S. drummondiana, S. exigua, S. geyeriana, S. glauca, S. lasiolepis, S. ligulifolia, S. lucida, S. lutea, S. monticola, S. planifolia, S. scouleriana, Sambucus nigra, S. racemosa, Sanguinaria canadensis, Sanicula marilandica, Schoenoplectus acutus, Scirpus microcarpus, Senecio bigelovii, S. hydrophiloides, S. serra, S. triangularis, Setaria, Shepherdia canadensis, Sibbaldia procumbens, Sidalcea candida, S. multifida, S. reptans, Silene parryi, Silphium perfoliatum, Solanum dulcamara, Solidago canadensis, S. simplex, Sonchus, Sorbus decora, S. scopulina, Sphenosciadium capitellatum, Spiraea betulifolia, S. douglasii, Stachys albens, Stellaria crassifolia, S. longifolia, S. obtusa, Streptopus amplexifolius, S. lanceolatus, Symphoricarpos albus, S. occidentalis, S. oreophilus, Symphyotrichum ciliolatum, S. eatonii, S. foliaceum, S. laeve, S. lanceolatum, Symplocarpus foetidus, Tanacetum vulgare, Taraxacum officinale, Teucrium canadense, Thalictrum dasycarpum, T. fendleri, T. occidentale, T. sparsiflorum, Thermopsis montana, T. rhombifolia, Thlaspi montanum, Thuja occidentalis, T. plicata, Tilia americana, Tragopogon pratensis, Trautvetteria caroliniensis, Trifolium hybridum, T. pratense, T. repens, Trillium cernuum, T. grandiflorum, T. ovatum, Triteleia ixioides, Trollius laxus, Tsuga canadensis, Tsuga heterophylla, Typha angustifolia, T. latifolia, Ulmus americana, Urtica dioica, Vaccinium myrtillus, Valeriana occidentalis, V. sitchensis, Veratrum californicum, V. tenuipetalum, V. viride, Veronica americana, V. serpyllifolia, V. wormskjoldii, Vicia americana, Viola adunca, V. canadensis, V. glabella, V. macloskeyi, V. renifolia, V. sororia, Vitis riparia, Xerophyllum tenax, Zanthoxylum americana, Zigadenus.

Use by wildlife: The Karok Indians regarded the roots of *Heracleum maximum* as poisonous to cattle but the Anticosti, Hesquiat, and Thompson tribes used the plants as forage for their cows (Mammalia: Bovidae: *Bos primigenius*). The roots are fed on by grizzly bears (Mammalia: Ursidae: *Ursus arctos*). The herbage is eaten by black bears (Mammalia: Ursidae: *Ursus americanus*), beavers (Mammalia: Castoridae: *Castor canadensis*) and to a minor extent by the California quail (Aves: Odontophoridae: *Callipepla californica*). It also is eaten by deer, elk, moose (Mammalia: Cervidae), and various small mammals. The plants are known to provide cover for several birds and successful nesting sites for the American eider (Aves: Anatidae: *Somateria mollissima dresseri*). The flowers attract a huge arthropod (Insecta) fauna of more than 274 species including seven species of long-tongued bees (Hymenoptera: Apidae, Megachilidae), 38 species of short-tongued bees (Hymenoptera: Colletidae, Halictidae), and a large diversity of Diptera (137 species) including various Agromyzidae, Anthomyiidae, Bibionidae, Bombyliidae, Chloropidae, Conopidae, Empididae, Muscidae, Sarcophagidae, Sciaridae, Sepsidae, Stratiomyidae, Syrphidae, Tabanidae, and Tachinidae. However, only three bee species have been observed to regularly visit the blooms. Other incidental floral visitors include beetles (Coleoptera) in the families Anthicidae (*Corphyra labiata*), Cerambycidae (*Euderces*), Cetoniidae (*Euphoria sepulcralis, Trichiotinus affinis, T. piger*), Chrysomelidae (*Diabrotica 12-punctata, D. vittata*), Cleridae (*Trichodes nutalli*), Curculionidae (*Centrinites*), Dermestidae (*Anthrenus castaneae, Cryptorhopalum haemorrhoidale, Orphilus*), Lampyridae (*Telephorus flavipes*), Meloidae (*Epicauta*), Melyridae (*Anthocomus*), Mordellidae (*Mordella marginata, Mordellistena biplagiata, Mordellochroa scapularis*), and a few bugs (Hemiptera): Thyreocoridae (*Corimelaena pulicaria*); Miridae (*Phytocoris*). The plants are also host to numerous species of parasitic wasps (Hymenoptera) in at least 14 families (Braconidae, Chalcididae, Chrysididae, Crabronidae, Eucoilidae, Eumenidae, Gasteruptiidae, Ichneumonidae, Pompilidae, Pteromalidae, Sphecidae, Tenthredinidae, Tiphiidae, Vcspidae). They also host various Lepidoptera larvae including species of Geometridae (*Eupithecia tripunctaria*), Hesperiidae (*Polites*), Noctuidae (*Papaipema harrisii*), Nymphalidae (*Basilarchia archippus, Cercyonis, Charidryas, Phyciodes*), Oecophoridae (*Agonopterix clemensella, A. flavicomella, Depressaria pastinacella*), and Papilionidae (*Iphiclides, Papilio brevicauda, P. kahli, P. machaon, P. polyxenes, P. zelicaon*). In addition to insects, this species also is a host to a great diversity of fungi including Ascomycota (*Anaphysmene heraclei, Colletotrichum dematium, Crocicreas cyathoideum, C. nigrofuscum, Cylindrosporium umbelliferarum, Davidiella macrocarpa, Diaporthe arctii, Didymella exigua, Epicoccum nigrum, Heteropatella umbilicata, Hyponectria sceptri, Leptosphaeria asparagina, L. doliolum, L. maculans, L. modesta, L. simmonsii, Leptospora rubella, Naevia stenospora, Nectriella pedicularis, Ophiobolus acuminatus, Ostracoderma, Passalora angelicae, Phoma asteriscus, P. complanata, P. heraclei,*

P. minuta, P. pedicularis, Phyllachora heraclei, Phyllosticta heraclei, Phymatotrichopsis omnivora, Plagiosphaera umbelliferarum, Pleospora herbarum, Ramularia heraclei, Rhabdospora aristata, R. heraclei, R. pastinacina, R. pleosporoides, Sirococcus americanus, Stellothyriella graminis, Stemphylium botryosum, Trichopezizella relicina) and Basiodiomycota (*Puccinia bakeriana*).

Economic importance: food: The young petioles and stems of *Heracleum maximum* have been eaten like celery once their hairy surface has been removed. When boiled twice, the first water is discarded to remove any bitter components. However, young, developing leaves of *H. maximum* probably should not be eaten because they are strongly cyanogenic, a property that gradually weakens as the foliage matures. This plant was eaten by many native North Americans. Various flower, stem and leaf preparations (e.g., mixed with blood, grease, honey, seal oil or sugar; boiled; raw; roasted; added to soups) provided food for the Alaska Native, Bella Coola, Blackfoot, California Indian, Carrier, Coeur d'Alene, Costanoan, Cree (Woodlands), Gitksan, Haisla, Hanaksiala, Hesquiat, Hoh, Karok, Kitasoo, Klamath, Kwakiutl (Southern), Makah, Mendocino, Meskwaki, Mewuk, Montana, Nitinaht, Ojibwa, Okanagan, Okanagan-Colville, Oweekeno, Pomo (Kashaya), Quileute, Quinault, Salish (Coast), Shuswap, Spokan, Thompson, Tolowa, Wet'suwet'en, Yuki, and Yurok tribes; **medicinal:** The roots of *Heracleum maximum* contain antibiotic and immunostimulant properties. Twenty mg of dried plant material can inhibit the growth of *Mycobacterium tuberculosis* and *M. avium* (Actinobacteria: Mycobacteriaceae). Plant extracts also exhibit antifungal activity and completely inhibit the growth of several Ascomycota: *Trichophyton tonsurans, T. rubrum*, and *Microsporum canis*. Aqueous root extracts are immunostimulative, indicating that they might also be antiviral. The seeds are mildly antioxidant. Contact with the foliage is known to cause phytophotodermatitis due to the presence of furanocoumarins. *Heracleum maximum* was utilized routinely as a food and medicinal by more than 20 native North American tribes. The leaves, roots or flowers were used similarly by many tribes (Aleut, Bella Coola, Blackfoot, California Indian, Northern Carrier, Chippewa, Cree, Gitksan, Haisla, Iroquois, Kashaya, Makah, Malecite, Meskwaki, Mewuk, Micmac, Ojibwa, Okanagan-Colville, Omaha, Paiute, Pawnee, Pomo, Quinault, Sanpoil, Shoshoni, Shuswap, Sikani, Tanaina, Washo, Woodlands) to prepare an analgesic poultice or decoction for treating assorted ailments such as back pains, boils, bruises, cholera, colds, cuts, erysipelas, eye sores, flesh worms, headaches, influenza, intestinal pains, mumps, muscle aches, neuralgia, rheumatism, smallpox, sore throats, stomach cramps, swellings, tuberculosis, urinary disorders, and wounds as well as lung, hip or limb pains. The Blackfoot used a stem infusion to remedy diarrhea and to aid in the removal of warts. The Cree, Shoshoni, and Washo tribes treated toothaches by placing pieces of the root in cavities. Powdered roots were used by the Cree and Thompson Indians to treat venereal diseases. The Meskwaki relieved severe headaches by ingesting the seeds of *H. maximum*. The Okanagon and Thompson tribes used the roots in a cathartic decoction. The Winnebago inhaled smoke from burned plant tops as a remedy for convulsions and fainting; **cultivation:** *Heracleum maximum* generally is not cultivated; **misc. products:** The Southern Carrier tribe of British Columbia used an infusion of *Heracleum maximum* flowers as an insect repellant. The Cree tipped their arrows with a poison prepared from root extracts. The Iroquois washed their rifles using a root decoction as a hunting medicine. The Kwakiutl and Coast Salish tribes made a hair ointment by mixing the dried roots with oil. A scalp cleanser made from a decoction of roots or stems was used by the Okanagan-Colville and Sanpoil tribes. The hollow stems were fashioned into flutes or used as makeshift straws by the Blackfoot and Cheyenne. The Haisla and Hanaksiala crafted stems into whistles. The Karok obtained a yellow dye from the roots. Flower stalks were fashioned into baskets by the Makah and Quileute tribes. The hollow stems were used as water containers by the Pomo (Kashaya). The Blackfoot and Pomo (Kashaya) children fashioned the stems into toy blowguns; **weeds:** *Heracleum maximum* occasionally is regarded as a field weed, especially in western North America; **nonindigenous species:** *Heracleum mantegazzianum* and *Heracleum sphondylium* are introduced to North America, but neither is a wetland plant.

Systematics: There is some discrepancy whether *Heracleum* is monophyletic as currently circumscribed. An analysis of nrITS sequences for relatively few taxa initially indicated that *Heracleum* might not be monophyletic but possibly includes *Pastinaca* (with which it often is confused). However, an expanded analysis of nrITS data for a fairly large number of taxa (Figure 5.118) generally resolved *Heracleum* as a distinct sister clade of *Pastinaca* (etc.), with the exception of a few misplaced taxa (e.g., *Heracleum marashicum*, *Mandenovia*, *Symphyoloma*). However, section *Heracleum*, which contains *H. maximum*, does not appear to be monophyletic in its present circumscription (Figure 5.118). There also is confusion regarding which name should correctly be applied to the native North American species. Although the name *H. lanatum* (1803) has been retained in much of the literature, the name *H. maximum* (1791) clearly has nomenclatural priority, and has been adopted here. Occasional tolerance to hydric conditions by *H. maximum* appears to be an unusual feature of the genus given that none of its closely related species is indicative of wetland habitats. *Heracleum sphondylium* is known to hybridize with *H. mantegazzianum* and both species are introduced to North America. These species are related to each other approximately as closely as they are to *H. maximum* (Figure 5.118). Consequently, hybridization with *H. maximum* could be a concern even though no instances have been reported. The basic chromosome number of *Heracleum* is $x = 11$. *Heracleum maximum* ($2n = 22$) is a diploid.

Comments: *Heracleum maximum* occurs throughout most of North America except for the southernmost United States.

References: Aikman et al., 1996; Baldwin & Bender, 2009; Baskin & Baskin, 1998; Bell & Constance, 1957; Borchardt et al., 2008a,b; Choate, 1967; Crispens, Jr. et al., 1960; Downie & Katz-Downie, 1996; Esser, 1995b; Fernald & Kinsey, 1943; Hendrix, 1984; Konuma & Yahara, 1997;

Linke et al., 2005; Logacheva et al., 2008; Mawdsley, 2002; McClellan et al., 2003; McCutcheon et al., 1997; Milligan, 2008; Moerman, 1996; Robertson, 1928; Stevens, 1957; Stoner & Rasmussen, 1983; Tooker & Hanks, 2000; Tooker et al., 2006; Tuell et al., 2008; Viereck et al., 1992; Webster et al., 2006, 2008.

9. *Lilaeopsis*

Grasswort; Liléopsis

Etymology: resembling *Lilaea*, a genus named for Alire Raffeneau-Delile (1778–1850)

Synonyms: *Crantzia*; *Crantziola*; *Hallomuellera*; *Hydrocotyle* (in part)

Distribution: global: Africa; Australia; Europe*; New Zealand; New World; **North America:** coastal (Atlantic, Gulf, Pacific)

Diversity: global: 13 species; **North America:** 5 species

Indicators (USA): OBL: *Lilaeopsis carolinensis*, *L. chinensis*, *L. masonii*, *L. occidentalis*, *L. schaffneriana*

Habitat: brackish (coastal), freshwater; lacustrine, palustrine; **pH:** 6.1–8.7; **depth:** 0–3 m; **life-form(s):** emergent herb, submersed (rosulate)

Key morphology: stems (to 30+ cm) horizontal (occasionally short ascending), rhizomatous, rooting at nodes; leaves (to 5–52 cm; longer in submersed plants) simple, entire, phylloidal (spatulate to quill-like), hollow, terete to strongly flattened, transversely septate (to 20 septae), arising from rhizome, sheathing (to 2.7 cm); herbage glabrous; umbels simple, axillary, peduncled (to 90 mm) or rarely sessile, 2–15-flowered; involucre of 2–5 inconspicuous bracts (to 3.5 mm); flowers regular, bisexual, 5-merous, pedicellate (to 30 mm); calyx minute, tooth-like; petals (to 1.8 mm) ovate, reflexed to spreading, maroon, pale green, pink, white or yellowish white; stamens alternate; ovary bicarpellate, inferior; 2-seeded, stylopodium flattened to conical; schizocarps (to 2.8 mm) 5-ribbed, some or all of the ribs with spongy tissue, carpophore absent; seeds circular to hemispherical

Life history: duration: perennial (rhizomes); **asexual reproduction:** rhizomes; **pollination:** self; **sexual condition:** hermaphroditic; **fruit:** schizocarps (common); **local dispersal:** rhizomes, seeds (water); **long-distance dispersal:** rhizomes, seeds (birds)

Imperilment: (1) *Lilaeopsis carolinensis* [G3]; SH (AR, GA); S1 (AL, VA); S2 (MS, NC, SC); S3 (FL); (2) *L. chinensis* [G5]; S1 (NS, RI); S2 (MA, ME, NH, NY); S3 (CT, NC); (3) *L. masonii* [G3]; S3 (CA); (4) *L. occidentalis* [G4]; S3 (BC); (5) *L. schaffneriana* [G4]; S2 (AZ)

Ecology: general: All *Lilaeopsis* species occur in aquatic or wetland habitats, with all of the North American taxa designated as OBL throughout their ranges. Even though some flowers can produce nectar and brightly colored perianths, they are inconspicuous, self-compatible, and most of the species mainly are self-pollinating (or possibly apomictic). Potentially pollinating insects are observed rarely on flowers in the field. Submersed plants tend to be sterile, but fruit production has been observed rarely in some submersed umbels. The spongy mericarps are buoyant and float on the water surface when

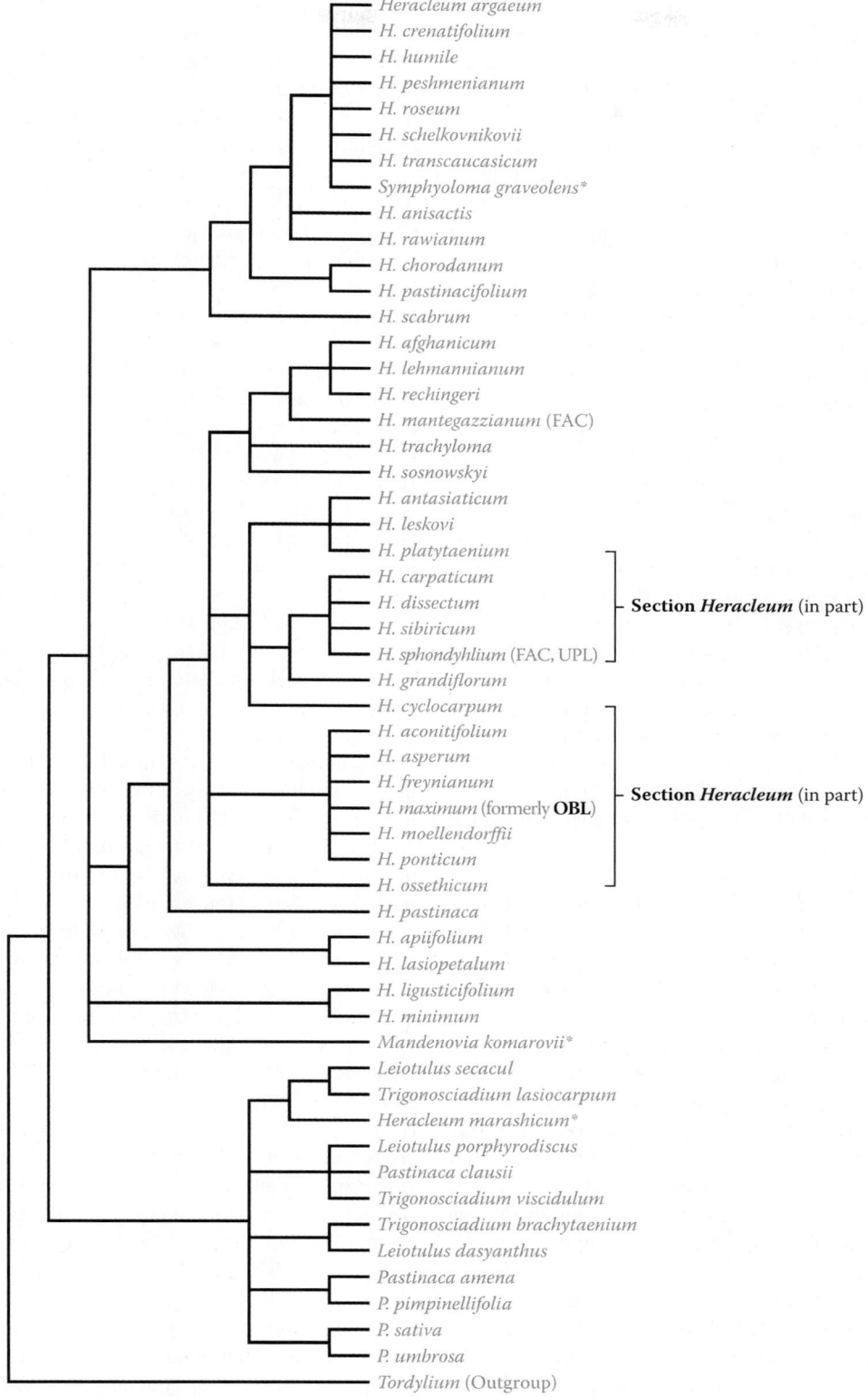

FIGURE 5.118 Phylogenetic relationships in *Heracleum* as depicted by nrITS sequence data analysis. The formerly OBL North American taxon *H. maximum* (now FACW, FAC, FACU) has no closely related wetland indicators. This analysis indicates that *Heracleum* is essentially monophyletic (except asterisked taxa), and resolves as the sister clade to one containing *Pastinaca*, which seems highly problematic. *Heracleum maximum* is placed in section *Heracleum*, which does not resolve as a clade in its current circumscription. (Adapted from Logacheva, M.D. et al., *Plant Syst. Evol.*, 270, 139–157, 2008.)

shed for dispersal. The seeds (and perhaps rhizome fragments) are transported further distances by birds. However, the pedicels of most species recurve in fruit, which tends to push many of the mericarps into the sand or mud, in this way ensuring a local seed bank. The leaves of most species exhibit considerable morphological plasticity and grow much shorter on emergent plants than when submersed. Although often found growing in dense patches, *Lilaeopsis* species are poor competitors and rely on site disturbances (e.g., fluctuating water levels) to reduce local interspecific competition.

Lilaeopsis carolinensis **J.M. Coult. & Rose** grows in brackish or fresh waters of canals, ditches, pools, streams, margins of lakes and ponds, or in floating mats of vegetation at elevations of up to 600 m. The plants are found rooted to the substrate (e.g., clay, mud, stumps) or tangled within floating mats of vegetation (e.g., *Hydrilla*). The plants can grow as emergents or as totally submersed life-forms, thriving when in 40 cm of warm (22°C–26°C), circumneutral (pH: 6.2–7.8) waters of relatively low calcium carbonate hardness (35–250 ppm $CaCO_3$). When grown under full sunlight, the leaves on emergent plants are shorter and narrower. Flowering and fruit production is high and occurs from March to August. The flowers are self-compatible and are self-pollinating. Although rare, a small amount of fruit production has been observed on umbels that develop completely underwater. It is not known whether such observations indicate cleistogamy or apomixis. The mericarps can remain afloat for more than 8 months after they are shed, due to the presence of spongy tissue in all of the ribs. Vegetative reproduction is prolific, with rhizome extension occurring more rapidly than in the other congeners. **Reported associates:** *Alternanthera philoxeroides, Cardamine, Cyperus, Eleocharis, Elymus virginicus, Fimbristylis castanea, Hydrilla verticillata, Hydrocotyle, Iva imbricata, Ludwigia, Marsilea, Myrica cerifera, Osmunda regalis, Packera glabella, Panicum amarum, P. repens, Peltandra virginica, Rosa palustris, Sagittaria platyphylla, Salvinia minima, Sesuvium portulacastrum, Thelypteris palustris.*

Lilaeopsis chinensis **(L.) Kuntze** occurs in the brackish tidal waters of bayous, canals, deltas, depressions, ditches, marshes, mudflats, river margins, saltmarshes, and shores near sea level. Plants typically occupy the intertidal zone (surviving periodic inundations of up to 3 m) where salinity remains fairly low (0.8‰–14.7‰), and usually grow at sites where at least some freshwater intrusion occurs. The species is highly tolerant of natural disturbance. The plants are found growing frequently in turf mats comprising *Spartina alterniflora* or *Schoenoplectus americanus*. Emergent plants grow in full sun to deep shade on substrates described as gravel, loamy mud, mud, peaty muck/mud, sand, sandy peat, and silt. They thrive in sunny and better drained sites, which tend to produce higher numbers of flowers and leaves. Densities of 20–81 leaves/100 cm² have been observed with some populations estimated to produce up to 2.3 million leaves. Flowering is prolific and occurs from March to June in the south of the range and from August to September in the north. Some populations produce upwards of 110,000 flowers at densities averaging 14 flowers/100 cm². Despite the high level of flowering, fruit production observed in cultivated plants

is low, perhaps indicating the absence of suitable pollinators or low levels of self-pollination. Spongy tissue is confined to the lateral ribs of the mericarps. Vegetative reproduction occurs by rhizomes, which expand several decimeters each season and result in densities as high as 1250 plants/25 cm². **Reported associates:** *Alternanthera philoxeroides, Amaranthus cannabinus, Atriplex prostrata, Bidens, Bolboschoenus maritimus, B. robustus, Boltonia asteroides, Carex, Cicuta maculata, Cladium jamaicense, Crassula aquatica, Cyperus tetragonus, Distichlis spicata, Eleocharis halophila, E. palustris, E. parvula, Eriocaulon parkeri, Eupatorium capillifolium, Festuca filiformis, Fimbristylis castanea, F. spadicea, Glaux maritima, Glottidium vesicarium, Hibiscus moscheutos, Hydrocotyle umbellata, Iva frutescens, Juncus acuminatus, J. roemerianus, Limonium carolinianum, Limosella australis, Lycopus americanus, Lythrum lineare, Panicum virgatum, Peltandra virginica, Phragmites australis, Plantago maritima, Pluchea odorata, Persicaria arifolia, P. hydropiperoides, P. punctata, Polypogon monspeliensis, Pontederia cordata, Ptilimnium capillaceum, Sagittaria calycina, S. falcata, S. graminea, S. lancifolia, S. subulata, Samolus valerandi, Schoenoplectus americanus, S. pungens, S. tabernaemontani, Sesbania punicea, Sesuvium maritimum, Solidago sempervirens, Spartina alterniflora, S. patens, S. cynosuroides, Spergularia salina, Sporobolus virginicus, Symphyotrichum tenuifolium, Triglochin striatum, Typha angustifolia, Zannichellia palustris, Zizania aquatica, Zizaniopsis miliacea.*

Lilaeopsis masonii **Mathias & Constance** is found in the brackish, tidally influenced waters of coves, deltas, flats, marshes, sloughs and river margins at elevations of up to 36 m. The plants are most common in the intertidal zone of brackish marshes where they grow on substrates ranging from clay, clay loam, loam, loamy sand, mud, sand, sandy clay loam, silt, and silty clay loam, to logs and stumps. Measurements of salinity in the habitats range from 0.25‰ to 8.5‰. Greenhouse studies indicate that growth decreases at salinities of 3, 6, 9–12, and 24‰ relative to fresh water; however, the extent of decrease is much less than that exhibited by some potentially competing species. The substrates are circumneutral (pH: 6.1–7.2) and typically contain moderate to high nutrient levels (22–88 ppm ammonia nitrogen; 5–45 ppm phosphorous, 30–789 ppm potassium). Flowering occurs from April to November, but flower and fruit production are low compared to congeners. The seeds will germinate in salinities up to 12‰, but germination steadily decreases as the salinity increases. The plants form extensive clonal colonies and persist primarily by vegetative reproduction, which occurs by the extension of creeping rhizomes. Competition experiments indicate that interactions with *Isolepis cernua* and *Hydrocotyle verticillata* (but not *Limosella subulata*) result in significantly reduced growth. Populations also are threatened by the growth of the invasive water hyacinth (*Eichhornia crassipes*). Biomass is suppressed when plants are exposed to as little as 0.2 L/m² of crude oil, especially at sites with higher salinities. **Reported associates:** *Cyperus eragrostis, Distichlis spicata, Eichhornia crassipes, Foeniculum vulgare, Hydrocotyle umbellata, H. verticillata, Isolepis cernua,*

Juncus, Lathyrus jepsonii, Limosella subulata, Polygonum aviculare, Potentilla anserina, Schoenoplectus acutus, S. americanus, S. californicus, Symphyotrichum chilense, S. lentum, Triglochin maritimum, T. striatum.

***Lilaeopsis occidentalis* J.M. Coult. & Rose** inhabits brackish or saltwater habitats such as beaches, canals, depressions, dunes, flats, lagoons, marshes, meadows, mudflats, ponds, prairies, salt marshes, sloughs, and the margins of lakes and rivers at elevations of up to 765 m. This species is designated as a halophyte but it might be FAC in this respect. There are a few, rare inland sites, and the species has been cultivated successfully in fresh water (300 ppm, total dissolved solids). The usual habitat is the intertidal zone, most commonly in the region occurring below the level of high tides. Substrates include cobble mud, mud, muddy ooze, rocks, sand, and sandy gravel. The plants survive in full sun to partial shade, but produce smaller leaves under the former relative to the latter conditions; the longest leaves occur on fully submersed plants (e.g., in 20–25 cm water). Flowering and fruiting occur from June to October. The mericarps possess spongy tissue but only on their lateral ribs. Vegetative reproduction occurs by rhizomes. **Reported associates:** *Alnus, Amorpha, Brasenia schreberi, Callitriche, Carex aperta, C. lyngbyei, Coreopsis atkinsoniana, Cotula coronopifolia, Crassula aquatica, Cyperus bipartitus, C. strigosus, Echinochloa, Eleocharis palustris, E. parvula, Elodea, Equisetum, Eragrostis hypnoides, Euthamia, Gratiola, Helenium, Isoetes, Isolepis cernua, Juncus balticus, Leersia, Lindernia, Limosella, Lindernia dubia, Lythrum salicaria, Mimulus, Najas, Oenanthe sarmentosa, Phalaris arundinacea, Philonotis fontana, Polygonum, Polypogon, Potentilla pacifica, Salix, Sarcocornia perennis, Schoenoplectus triqueter, Spartina foliosa, Symphyotrichum, Triglochin maritimum, T. concinnum, T. striatum, Veronica.*

***Lilaeopsis schaffneriana* (Schlecht.) J.M. Coult. & Rose** occupies perennial, slow-flowing freshwater habitats that include ditches, lake margins, marshes, pools, springs, and the floodplains of rivers and streams at elevations of up to 2135 m. The plants grow submersed in shallow water (5–25 cm deep) or as emergents, usually in open sunlight, but sometimes in shaded woodlands. The substrates consist of clay or silt, contain some organic matter, and tend to be alkaline (pH: 7.9–8.7). This is one of only four species in the genus to occupy exclusively freshwater, noncoastal habitats. The plants are in flower from March to October, producing 3–10 flowers/umbel. Flowering and fruiting is less prolific (often completely absent) where competing vegetation occurs, but several hundred flowers/m² can be produced in open sites. The water-dispersed mericarps are very buoyant due to the presence of spongy tissue, which is produced on all of the ribs. A seed bank is indicated by the recovery of viable seeds in the soil after 2 years following drought conditions. Once mature, newly produced seeds will germinate well (~90%) without stratification (or other treatments) in as little as 1–2 weeks. Seedling survivorship is high and growth can proceed throughout the year in the absence of frost. This species is a poor competitor, and requires low-level disturbances (e.g., drying/flooding cycles, scouring, sedimentation, trampling)

to remove leaf litter and to reduce competition with other vegetation. However, the plants are sensitive to excessive flood scouring, livestock trampling, dredging, and other hydrological alterations. Field studies have shown that plants grow much better (developing flowers and longer leaves) in plots where potentially competing vegetation has been clipped back. The plants are highly clonal, persisting, and spreading vegetatively by slender rhizomes. Populations of this rare species (known from only 16 extant sites) have been transplanted successfully. North American plants are recognized taxonomically as var. *recurva*, which differs from the typical variety by having fruits that are longer than broad rather than broader than long. **Reported associates:** *Anemopsis californica, Baccharis emoryi, B. salicifolia, Berula erecta, Carex lanuginosa, Chara, Cirsium neomexicanum, Cyperus odoratus, Eleocharis macrostachya, Equisetum laevigatum, Juncus bufonius, J. interior, J. mexicanus, J. saximontanus, J. torreyi, Lemna minor, Lobelia cardinalis, Medicago sativa, Muhlenbergia, Nasturtium officinale, Persicaria punctata, Polypogon monspeliensis, Populus fremontii, Rumex crispus, R. salicifolius, Salix gooddingii, Samolus vagans, Schoenoplectus pungens, Sisymbrium irio, Sonchus asper, Spiranthes delitescens, Symphyotrichum subulatum, Typha domingensis, Xanthium strumarium, Zannichellia palustris.*

Use by wildlife: *Lilaeopsis carolinensis* is used as shelter by small invertebrates. *Lilaeopsis chinensis* is associated to a minor degree as habitat for the salt marsh snail (Mollusca: Gastropoda: Ellobiidae: *Melampus bidentatus*).

Economic importance: food: none; **medicinal:** none; **cultivation:** *Lilaeopsis carolinensis* (as the oxymoronic "giant micro sword") is sold as an aquarium plant or as a "marginal" water garden ornamental; **misc. products:** none; **weeds:** none; **nonindigenous species:** *Lilaeopsis carolinensis* was introduced to Portugal and Spain sometime before 1956. It also has been suggested (but not verified) that North American plants might have been introduced from South America early in the 19th century.

Systematcs: The phylogenetic affinities of *Lilaeopsis* have been obscured by its simplified morphology. However, DNA sequence data analyses (for 13/15 species) consistently confirm that the genus is monophyletic and place it within tribe Oenantheae (subfamily Apiodieae), where it resolves most closely as the sister genus to *Cynosciadium* (Figures 5.111, 5.112, 5.116, and 5.119). All of the species in the genus are aquatic, so that the clade represents a single origin of the habit. The North American taxa do not comprise a single clade (Figure 5.119) due to the disparate position of *L. occidentalis* and *L. masonii*. Furthermore, phylogenetic analyses also indicate that *L. occidentalis*, *L. macloviana*, and *L. masonii* should best be recognized as a single, amphitropic species. *Lilaeopsis* has a base chromosome number of $x = 11$. *Lilaeopsis chinensis* and *L. carolinensis* are diploid ($2n = 22$); whereas, *L. masonii* and *L. occidentalis* are tetraploid ($2n = 44$).

Comments: *Lilaeopsis carolinensis* occurs in the southeast United States (also South America and Europe [introduced]); *L. chinensis* occurs in eastern North America

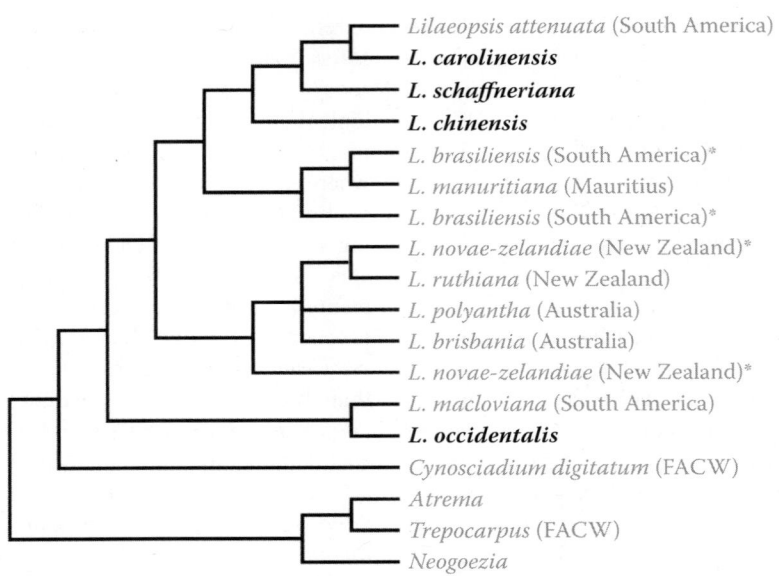

FIGURE 5.119 Phylogenetic relationships of *Lilaeopsis* as reconstructed from combined DNA sequence data. The OBL North American species of *Lilaeopsis* (bold) are not sister taxa, which indicates a complex biogeographical history for the group. *Lilaeopsis masonii* (excluded from the earlier analysis) resolves with *L. occidentalis* and *L. macloviana* by nrITS data (Bone et al., 2011), an indication that all three taxa might represent a single amphitropic species. Narrower concepts of species also have been suggested for the clades containing discordant taxa (asterisked). (Adapted from Bone, T.A. phylogenetic and biogeographical study of the umbellifer genus *Lilaeopsis*. MS thesis. Department of Plant Biology, University of Illinois at Urbana-Champaign, Urbana, IL, 342 pp, 2007; Bone, T.S. et al., *Syst. Bot.*, 36, 789–805, 2011.)

and *L. occidentalis* in western North America; endemic taxa include *L. masonii* (California) and *L. schaffneriana* (Arizona).

References: Affolter, 1985; Aronson, 1989; Bone, 2007; Bone et al., 2011; COSEWIC, 2004c; de Almeida & Freitas, 2006; Downie et al., 2002, 2008; Feist & Downie, 2008; Ferren Jr. & Schuyler, 1980; Gerst, 1996; Hardway et al., 2004; Hatch & Slack, 2008; Kasselmann, 2003; Kerwin, 1972; Malcom & Radke, 2008; Nelson, 1986; Petersen et al., 2002; Schuyler et al., 1993; Titus & Titus, 2008a,b,c; Valiejo-Roman et al., 2002; Wells, 1928; Zebell & Fiedler, 1996.

10. *Limnosciadium*

Dogshade

Etymology: from the Greek *limnê skiadeion* meaning "swamp umbrella," after its umbellate inflorescence

Synonyms: *Aethusa* (in part); *Cynosciadium* (in part); *Oenanthe* (in part)

Distribution: global: North America; **North America:** south-central

Diversity: global: 2 species; **North America:** 2 species

Indicators (USA): OBL: *Limnosciadium pumilum*; **FACW:** *L. pinnatum, L. pumilum*

Habitat: freshwater; lacustrine, palustrine; **pH:** unknown; **depth:** <1 m; **life-form(s):** emergent herb

Key morphology: stems erect (to 8 dm) or diffuse (to 4 dm), glabrous; roots fascicled; leaves (to 8–20 cm), simple and septate or pinnate (up to 9 filiform to lanceolate leaflets), petioles sheathing; umbels compound (to 20–flowered), terminal, axillary, up to 12 rays (to 5 cm), pedicels (to 8 mm) with

bractlets (to 5 mm); calyx (to 0.5–1.5 mm) tooth-like; petals (to 1.5 mm) ovate/obovate, white; mericarps (to 4 mm) oblong to orbicular, corky winged, stylopodium conical, carpophore bifid apically; seeds flattened

Life history: duration: annual (fruit/seeds); **asexual reproduction:** none; **pollination:** unknown; **sexual condition:** hermaphroditic; **fruit:** schizocarps (common); **local dispersal:** mericarps; **long-distance dispersal:** mericarps

Imperilment: (1) *Limnosciadium pinnatum* [G5]; S1 (KS, MO); (2) *L. pumilum* [G5]

Ecology: general: In the 1996 list, both *Limnosciadium* species were categorized as OBL as well as FACW indicators in various portions of their range, an implication of primarily FACW habitation. In the 2013 list, only *L. pumilum* retained partial OBL status, with *L. pinnatum* being ranked consistently as FACW. Because of their prior completion, the treatments for both species have been retained here. Regardless of their wetland indicator status, both *Limnosciadium* species can appear occasionally in atypical sites such as lawns or in UPL fields. Presumably the flowers are insect pollinated and the corky fruits are dispersed by water. However, nearly nothing has been reported on the life history of either species, and the information summarized later is sparse and admittedly incomplete.

***Limnosciadium pinnatum* (DC.) Mathias & Constance** grows on beaches, in depressions, ditches, lawns, marshes, meadows, prairies, and along the margins of lakes, ponds, rivers, and streams. The plants have been reported from polluted areas and also from saline sites. They grow in areas that receive full sunlight and occur on substrates that include clay loam, mud, and rock. Flowering has been observed from May to June. **Reported associates:** *Acer rubrum, Bromus japonicus, Chloracantha spinosa, Nyssa sylvatica, Paspalum plicatulum, P. texanum, Quercus, Schoenoplectus hallii, Stenotaphrum secundatum.*

Limnosciadium pumilum (Engelm. & A. Gray) Mathias & Constance occurs in bogs, depressions, ditches, marshes, meadows, prairies, ricefields, and swales. Some occurrences are reported from deeply shaded sites. Substrates include clay sand, sand, and sandy loam. Flowering has been observed from March to June. **Reported associates:** *Apium graveolens, Astranthium integrifolium, Hydrocotyle verticillata, Prosopis, Ptilimnium, Quercus.*

Use by wildlife: *Limnosciadium pinnatum* provides habitat for birds (Aves) in wet, coastal prairies. This species also is a host plant for the cotton fleahopper (Insecta: Hemiptera: Miridae). *Limnosciadium pumilum* has been identified as a minor dietary component of cattle (Mammalia: Bovidae: *Bos taurus*) during the spring season in some areas.

Economic importance: food: none; **medicinal:** none; **cultivation:** none; **misc. products:** none; **weeds:** none; **nonindigenous species:** *Limnosciadium pinnatum* was first reported from Illinois between 1956 and 1978 and is regarded as a nonnative species in that state

Systematics: *Limnosciadium* was distinguished from *Cynosciadium* (now monotypic) on the basis of several morphological incongruencies. Phylogenetic analyses of various DNA sequence data (Figures 5.112, 5.116 and 5.120) confirm that the two species of *Limnosciadium* are monophyletic, that they are distinct from *Cynosciadium*, and that they resolve as the sister clade to the monotypic *Daucosma laciniatum*, an endemic, UPL species. The basic chromosome number of *Limnosciadium* is $x = 6$. *Limnosciadium pumilum* ($2n = 12$) is a diploid.

Comments: Both *Limnosciadium pinnatum* and *L. pumilum* occur in the south-central United States.

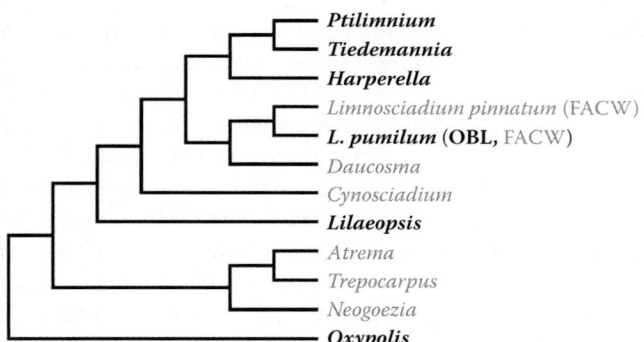

FIGURE 5.120 Phylogenetic relationships of *Limnosciadium* and other members of the "North American endemics" clade of Apiaceae tribe Oenantheae as indicated by analysis of combined DNA sequence data. Taxa containing OBL wetland indicators are shown in bold. The OBL habit apparently has arisen several times within this clade. (Adapted from Downie, S.R. et al., *Can. J. Bot.*, 86, 1039–1064, 2008.)

References: Bell & Constance, 1960; Cook, 1996a; Downie et al., 2008; Frasure, 1979; Henry & Scott, 1981; Mathias & Constance, 1941b; Oberholser, 1925.

11. Oenanthe

Water dropwort, water parsley; Oenanthe

Etymology: from the Greek *oinos antheô* meaning "wine flower," in reference to its fragrance

Synonyms: *Actinanthus; Oenosciadium; Phellandrium; Stephanorossia; Volkensiella*

Distribution: global: Africa; Asia; Europe; North America; **North America:** western

Diversity: global: 25–30 species; **North America:** 5 species

Indicators (USA): OBL: *Oenanthe aquatica, O. javanica, O. sarmentosa*

Habitat: brackish (coastal), freshwater; lacustrine, palustrine; **pH:** 4.0–8.3; **depth:** <1 m; **life-form(s):** emergent herb submersed (vittate)

Key morphology: stems (to 1.5 m) erect or decumbent, rooting at nodes, rhizomes/roots tuberous or lacking; leaves (to 3–6 dm) to 2–5 times pinnately compound, leaflets (to 6 cm) serrate or pinnately lobed, petioles (to 3.5 dm) sheathing; umbels compound, terminal, peduncled (to 13 cm), rays (to 3 cm) up to 20, bracts inconspicuous; flowers (to 3.8 mm) pedicelled (to 6 mm), radial, the outermost occasionally bilateral and/or staminate; calyx lobes (to 1 mm) lanceolate; petals (to 2 mm) broad, white to reddish; schizocarps (to 4.0 mm) cylindrical, the ribs broad, corky

Life history: duration: annual, biennial (fruit/seeds; rosettes), perennial (rhizomes, tuberous roots); **asexual reproduction:** none; **pollination:** insect; **sexual condition:** hermaphroditic; **fruit:** schizocarps (common); **local dispersal:** mericarps (water); **long-distance dispersal:** mericarps (birds)

Imperilment: (1) *Oenanthe aquatica* [GNR]; (2) *O. javanica* [GNR]; (3) *O. sarmentosa* [G5]

Ecology: general: *Oenanthe* is ecologically versatile, with many species growing in UPLs but also having an affinity for wetter sites. About six of the 30 or so species regularly occur in aquatic habitats. The plants can grow either as annuals, biennials, or perennials. The flowers are insect pollinated and protandrous. The corky-ridged fruits of the OBL species are dispersed locally by the water and over greater distances by adhering to the feet or plumage of birds (Aves).

Oenanthe aquatica (L.) Poir. occurs in the still or slow-moving waters of canals, depressions, ditches, marshes or along the margins of lakes, ponds, rivers, and streams. It grows in shallow wetlands where a dense vegetational cover is lacking, and also in shallow littoral, acidic waters (pH: 4.0–8.3; mean: 5.6). The plants thrive on wet, exposed sediments of gravel, mud, sand, or silt, which are high in carbon and nitrogen. When submersed, the plants rarely exceed a maximum water depth of 0.5–1.0 m, but occasionally have been found at depths of up to 2.5 m. Flowering does not occur in submersed plants, and for emergents is higher in biennials than in annuals; flowering also is reduced when the plants are shaded. The flowers are insect pollinated, protandrous, and the plants functionally are andromonoecious. Specific pollinators have not been reported. Seed production generally ranges from 500 to 17,000 seeds/individual; however, extremely robust plants are capable of producing upwards of 40,000 seeds. The fruits are buoyant and remain afloat from several hours until about 2 days (some as long as 5 days). Most freshly ripened fruits remain dormant, but up to 80% germination within 30 days has been reported for some seeds exposed to the air. The germination rate is lower in water. Salinities of 3.3‰ are tolerated fairly well, but seed germination is greatly reduced at 10‰. Optimal germination for dormant seeds occurs after they have been stratified in fresh water, then incubated in the air at 20°C under an 8/16h light/dark regime. Overall germination can vary from 40% to 90% for seeds originating from different localities, and always is higher under aerobic conditions. A small percentage (~1%) of sunken seeds germinates quickly, which produces buoyant seedlings that rise to the water surface where they are dispersed. Fluctuating water levels are important for the persistence of this species and masses of seedlings often are observed emerging from the bottom sediments that become exposed as a result of receding water levels. The foliage of submersed plants produces much more delicately divided leaves, with thinner segments. Stands can achieve densities of 500 plants/m² and can produce up to 1369 g/m² of biomass annually. Overall, the annual forms are rather poor competitors and the highest biomass (and flower production) is achieved by biennial plants growing in the littoral zone. **Reported associates (North America): none.**

Oenanthe javanica **(Blume) DC.** inhabits shallow water or saturated substrates of ditches, fens, marshes, oxbows, swamps, and the margins of lakes and streams at elevations of up to 250m. The reported substrates include mud, sand, and silt bars. Flowering occurs from July to August in North America. This is a stoloniferous perennial, which can form tubers and often grows in large, clonal clumps. However, allozyme data indicate that sexual reproduction occurs frequently, and leads to outcrossing (*tm* = 0.358) and high levels of genetic variability (e.g., 55% polymorphic loci; 2.83 alleles per polymorphic locus; 0.236 expected heterozygosity). Lower levels of genetic variation and reduced outcrossing rates are characteristic of cultivated populations. Although the seeds of *O. javanica* germinate with difficulty, their dormancy can be broken effectively by soaking them in a solution of 0.1%–3% K_2NO_3, 49% H_2SO_4, and 200 mg/L gibberellin. A daily period of illumination for at least 6h is required for optimal germination. Seeds will germinate from 10°C to 30°C, with an optimum temperature of 20°C. Good germination also has been achieved following a change in temperature from 10°C to 25°C. The seeds will not germinate in waters over 60°C. Their storage in sand can be optimized by treating them first in a solution of 0.2% K_2NO_3, 0.1% H_2O_2, and 10% PEG. The roots are fairly short (<15 cm), can live for 24 days, and attain a biomass of 1.2 g plant⁻¹. They are colonized by vesicular arbuscular mycorrhizae. **Reported associates (North America):** *Acer negundo, A. saccharinum, Apios americana, Bidens cernuus, Boehmeria cylindrica, Calystegia sepium, Carya cordiformis, Celastrus orbiculatus, Celtis occidentalis, Cirsium arvense, C. vulgare, Corylus*

americana, Cryptotaenia canadensis, Elymus virginicus, Eutrochium maculatum, Impatiens capensis, Lycopus virginicus, Persicaria hydropiperoides, P. sagittata, Phalaris arundinacea, Phytolacca americana, Quercus bicolor, Rhamnus cathartica, Rudbeckia laciniata, Scirpus atrovirens, Symphyotrichum lateriflorum, Typha, Viola sororia.

Oenanthe sarmentosa **C. Presl ex DC.** grows in brackish and fresh waters of beaches, bogs, brooks, cienega, ditches, dunes, estuaries, flats, floodplains, lagoons, marshes, meadows, muskeg, ponds, river bottoms, salt marshes, sloughs, springs, swales, swamps, and along the margins of lakes, reservoirs, rivers and streams at elevations of up to 1745 m. This broadly adapted species occurs in purely freshwater sites to tidally influenced estuarine habitats or salt marsh, extending northward to at least 60° N latitude. It tolerates full sun but often grows in shaded areas or openings of swamp forests. It normally is found in damp sites or in standing (or slowly flowing) water less than 15 cm in depth, but also appears frequently in drying sites. The usual substrates are acidic (pH: 4.5–6.0) and nitrogen rich. They include basalt, clay loam, granitic sand, gravel, loam, loamy sand, muck, mud, organic loam, peat, sand, sandy ash, sandy clay loam, sandy loam, sandstone, serpentine, and silt loam. Flowering occurs from June to August. Pollination is achieved by bumblebees (Hymenoptera: Apidae: *Bombus occidentalis*) and other insects such as flies (Diptera) and wasps (Hymenoptera). Plants flower commonly and produce large numbers of fruits. The mericarps float (for at least 30 days) and are dispersed on the water surface to suitable sites along the shore, or are transported longer distances by adhering to mud on the feet of waterfowl. Floating seeds eventually sink and lie dormant on the bottom until the sediments are exposed, which initiates germination. A 50% germination rate has been reported by first soaking seeds for 24h and then providing 21–30 days of cold stratification. Germination should occur within 30 days. This species is a significant competitor and the plants can achieve high densities and biomass. Vegetative reproduction occurs by means of fleshy, tuberous roots and rhizomes. **Reported associates:** *Abies concolor, Acer circinatum, A. macrophyllum, Achillea millefolium, Adenostoma fasciculatum, Adiantum pedatum, Agrostis alba, A. exarata, A. hyemalis, A. stolonifera, Alisma, Alnus rhombifolia, A. rubra, A. rugosa, A. viridis, Amelanchier alnifolia, Anemopsis californica, Angelica genuflexa, A. lucida, Anthoxanthum odoratum, Aquilegia formosa, Artemisia californica, A. douglasiana, A. dracunculus, Athyrium filix-femina, Baccharis, Beckmannia syzigachne, Bidens cernuus, Blechnum spicant, Bromus, Calamagrostis canadensis, Callitriche heterophylla, Calamagrostis canadensis, Calocedrus decurrens, Camassia quamash, Carex aquatilis, C. bolanderi, C. brunnescens, C. cusickii, C. deweyana, C. echinata, C. exsiccata, C. feta, C. interior, C. leptalea, C. lyngbyei, C. obnupta, C. pellita, C. stipata, C. unilateralis, C. utriculata, Castilleja unalaschcensis, Ceanothus leucodermis, C. palmeri, Cercocarpus montanus, Chrysosplenium glechomifolium, Cicuta douglasii, Cirsium hydrophilum, C. vulgare, Claytonia sibirica, Comarum palustre, Conium maculatum, Cornus sericea,*

Crataegus douglasii, C. monogyna, Danthonia californica, Deschampsia cespitosa, Digitalis purpurea, Distichlis spicata, Drepanocladus, Drosera rotundifolia, Dulichium arundinaceum, Eleocharis obtusa, E. palustris, Epilobium brachycarpum, E. ciliatum, Equisetum arvense, E. fluviatile, E. hyemale, Eriogonum fasciculatum, Eurhynchium praelongum, Festuca subulata, Frangula californica, F. purshiana, Fraxinus latifolia, Fritillaria biflora, Galium aparine, G. parisiense, G. trifidum, G. triflorum, Gaultheria shallon, Gentiana sceptrum, Geranium erianthum, Geum macrophyllum, Glaux maritima, Glyceria grandis, G. striata, Grindelia, Heliotropium, Heracleum lanatum, Heteromeles arbutifolia, Hippuris vulgaris, Holcus lanatus, Hosackia oblongifolia, Howellia aquatilis, Hydrocotyle verticillata, Hypericum anagalloides, H. bryophytum, Impatiens capensis, Iris pseudacorus, Jaumea carnosa, Juncus balticus, J. bufonius, J. effusus, J. lesueurii, J. tenuis, J. xiphoides, Keckiella antirrhinoides, Lathyrus palustris, Lilaeopsis occidentalis, Lolium arundinaceum, Lonicera involucrata, Lotus corniculatus, Ludwigia peploides, Lupinus latifolius, L. nootkatensis, Lycopus uniflorus, Lysichiton americanus, Lysimachia terrestris, Lythrum salicaria, Mahonia aquifolium, Maianthemum dilatatum, Malosma laurina, Malus fusca, Marrubium vulgare, Melilotus alba, Mentha arvensis, Menyanthes trifoliata, Mimulus cardinalis, M. guttatus, M. moschatus, Mitella ovalis, Myosotis laxa, Myrica californica, Myriophyllum aquaticum, Nasturtium officinale, Nuphar polysepala, Oemleria cerasiformis, Oxalis trilliifolia, Persicaria amphibia, P. hydropiperoides, Phacelia imbricata, Phalaris arundinacea, Phragmites australis, Physocarpus capitatus, Picea engelmannii, P. sitchensis, Pinus contorta, P. coulteri, P. jeffreyi, P. monticola, P. ponderosa, Platanthera dilatata, P. stricta, Platanus racemosa, Poa palustris, P. trivialis, Polypodium glycyrrhiza, Polypogon viridis, Polystichum munitum, Populus balsamifera, P. fremontii, Potentilla egedii, P. pacifica, Prunella vulgaris, Prunus, Pteridium aquilinum, Quercus agrifolia, Q. chrysolepis, Q. kelloggii, Ranunculus alismifolius, R. lobbii, R. repens, R. uncinatus, Rhinanthus minor, Rhododendron neoglandulosum, Rhus ovata, Ribes bracteosum, R. nevadense, R. spectabilis, Rosa californica, R. nutkana, R. pisocarpa, Rubus discolor, R. laciniatus, R. parviflorus, R. spectabilis, R. ursinus, Rumex crispus, R. obtusifolius, Sagittaria latifolia, Salix commutata, S. exigua, S. hookeriana, S. laevigata, S. lasiandra, S. lucida, S. nigra, S. sitchensis, Sambucus racemosa, Sanguisorba officinalis, Sarcocornia perennis, Schoenoplectus acutus, S. pungens, Scirpus microcarpus, Scrophularia californica, Scutellaria galericulata, Sium suave, Solidago spectabilis, Sparganium androcladum, S. angustifolium, Spartina foliosa, Sphagnum henryense, S. palustre, Spiraea douglasii, Stachys ajugoides, S. albens, S. ciliata, S. mexicana, Stellaria calycantha, S. crispa, Symphoricarpos albus, Symphyotrichum subspicatum, Tellima grandiflora, Thuja plicata, Tolmiea menziesii, Torreyochloa pallida, Toxicodendron diversilobum, Trientalis europaea, Trifolium wormskjoldii, Triglochin concinnum, T. maritimum, Tsuga heterophylla, Typha latifolia, Urtica dioica, Vaccinium ovatum, V. parvifolium, V. *uliginosum, Veratrum californicum, Veronica americana, V. scutellata, Vicia gigantea, Viola palustris.*

Use by wildlife: Ingestion of *Oenanthe aquatica* reportedly causes poisoning in young cattle (Mammalia: Bovidae: *Bos*). *Oenanthe sarmentosa* contains from 16% to 24% crude protein and is eaten by black bears (Mammalia: Ursidae: *Ursus americanus*), black-tailed deer (Mammalia: Cervidae: *Odocoileus hemionus*), elk (Mammalia: Cervidae: *Cervus elaphus*), and grizzly bears (Mammalia: Ursidae: *Ursus arctos*). The plants also are eaten by banana slugs (Mollusca: Gastropoda: Arionidae: *Arion, Ariolimax*). This is a host plant for several larval insects (Insecta: Lepidoptera: Oecophoridae: *Agonopterix oregonensis, A. rosaciliella, Depressaria daucella*; Papilionidae: *Papilio zelicaon*). The flowers attract several adult butterflies (Lepidoptera) such as the great copper (Lycaenidae: *Lycaena xanthoides*) and Lorquin's admiral (Nymphalidae: *Limenitis lorquini*), which visit them for nectar. This species also hosts various Fungi including Ascomycota (Mycosphaerellaceae: *Septoria oenanthis*) and Basidiomycota (Pucciniaceae: *Uromyces americanus, U. lineolatus, U. scirpi*; Raveneliaceae: *Nyssopsora echinata*). A nonindigenous moth (Insecta: Lepidoptera: Noctuidae: *Chrysodeixis chalcites*) was discovered by U.S. Customs officials on *O. javanici* plants being imported from Hawaii.

Economic importance: food: The raw or cooked stems of *Oenanthe sarmentosa* were used as food by the Costanoan, Cowlitz, Hesquiat, Miwok, Skokomish, and Snuqualmie tribes. *Oenanthe javanica* often is eaten as a vegetable in India, Vietnam, and other parts of Asia. The petioles of wild plants contain 29.70 mg/g soluble sugar and 0.77 mg/g phosphorus and the leaves contain up to 0.19 mg/g vitamin C and 0.35 mg/g iron. However, some *Oenanthe* species (especially the stems and roots) contain oenanthotoxin, a highly poisonous polyacetylene, which affects the central nervous system as a gamma-aminobutyric acid antagonist and can induce seizures when ingested even in small amounts. Consequently, it would be prudent to avoid eating any of the species; **medicinal:** The fruits of *Oenanthe aquatica* contain β-phellandrene, a monoterpene that induces expectorant activity. Although the Haisla and Hanaksiala regarded *O. sarmentosa* to be highly toxic, it was used medicinally by the Kitasoo, Kwakiutl, Kwakwaka'wakw, Nuxalkmc and Tsimshian tribes as an emetic or purgative, by the Makah as a laxative, and by the Nitinaht to speed up childbirth. *Oenanthe javanica* has long been used as a medicinal plant in Asia. The foliage contains flavones and other phenolics, which have been shown to inhibit replication of the hepatitis B virus and to exhibit antidiabetic activity in laboratory animals. The plant extracts contain persicarin, which is believed responsible for their ability to detoxify bromobenzene-induced hepatic lipid peroxidation and for their protective properties against glutamate-induced neurotoxicity. However, the extracts also are known to cause DNA damage in human lymphocytes. Dried root and shoot powder is inhibitory (13%–16%) to the mycelial growth of some fungi (Ascomycota: Nectriaceae: *Fusarium oxysporum*), which can be pathogenic to humans. The plant extracts are known to accelerate ethanol metabolism in laboratory animals. There has been at least one documented case of human contact dermatitis caused by exposure to *O. javanica*; **cultivation:** *Oenanthe aquatica* is sold as an ornamental water garden plant, with at least one cultivar distributed as 'Variegata'; *O. sarmentosa* also is planted as a wetland ornamental, and has a dwarf cultivar known as 'Flamingo'; *Oenanthe javanici* has been considered for use as a plant in Everglades farmlands production; **misc. products:** *Oenanthe sarmentosa* readily survives transplantation and is used in wetland restoration projects. It also is planted to reduce water flow and to facilitate sediment deposition. Makah and Quileute children fashioned toy whistles out of the cut stems. The seeds of *Oenanthe aquatica* yield β-phellandrene, a fragrant terpenoid that imparts a spicy, citrus aroma. *Oenanthe javanica* is effective at removing nitrogen and phosphorous from eutrophic and wastewater systems. Transgenic kanamycin and hygromycin resistant plants have been developed successfully. *Oenanthe javanica* also has been considered as a biological agent to facilitate removal of various metals from water. Plants grown for 5 days in wastewater containing 2 ppm gold can accumulate up to 12.2 kg of the metal per ton of root ash; **weeds:** *Oenanthe aquatica* and *O. javanica* should be regarded as at least potentially weedy due to their high potential reproductive capacity; **nonindigenous species:** The Eurasian *Oenanthe aquatica* has been introduced to Maryland, Ohio, and the District of Columbia; *O. javanica* has been introduced from Asia to Illinois, Missouri, Virginia, and Wisconsin. The popularity of both species as vegetables in Asia has contributed to their introduction to North America. Asian immigrants have been observed planting the latter along streams in the Washington, DC area as a food source.

Systematics: Some care should be taken when consulting the taxonomic literature, because *Oenanthe* also is a genus of passerine birds (Aves). Phylogenetic studies including 5–6 *Oenanthe* species indicate that the genus is monophyletic and comprises the sister clade to *Cicuta* (Figures 5.111, 5.112, 5.114 and 5.121). These studies also confirm the placement of the genus within the clade representing tribe Oenantheae. Fortunately, the limited taxon sampling has included two of the OBL North American species, which appear to be very closely related (Figure 5.121). However, the precise relationship of these species to one another and to the remainder of the genus will not be possible until more of the ~30 species have been surveyed phylogenetically. The basic chromosome number of *Oenanthe* is $x = 11$. *Oenanthe aquatica* ($2n = 22$) is a diploid and *O. sarmentosa* ($2n = 44$) a tetraploid. *Oenanthe javanica* ($2n = 20, 42, 63$) also is very similar to *O. sarmentosa*, and some authors have suggested that it be recognized as a subspecies of the latter. However, the different cytological data raise additional questions regarding the nature of *O. javanica* and its distinctness from *O. sarmentosa*.

Comments: *Oenanthe aquatica* currently is known only in Maryland, Ohio, and the District of Columbia where it has been introduced; *O. javanica* is a relatively recent introduction to Illinois, Missouri, Virginia, and Wisconsin. *Oenanthe sarmentosa* occurs throughout coastal western North America.

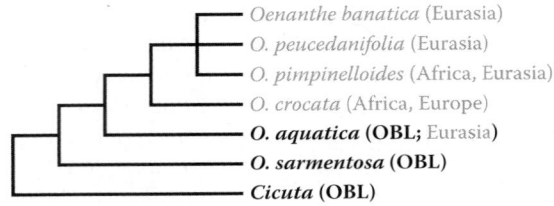

FIGURE 5.121 Phylogenetic relationships in *Oenanthe* as indicated by a likelihood analysis of combined cpDNA + nrITS sequence data. The OBL North American species are indicated in bold (*O. aquatica* is nonindigenous to North America). Parsimony analysis of the same data resolved *O. aquatica* and *O. sarmentosa* as sister species. (Adapted from Downie, S.R. et al., *Can. J. Bot.*, 86, 1039–1064, 2008.)

References: Bajwa et al., 2001; Bell & Constance, 1957; Casper & Krausch, 1981; Cates & Orians, 1975; Choi & Huh, 2007; Christy, 1993, 2004; Cooke, 1997; Cowan, 1945; Denoth & Myers, 2007; Downie et al., 2008; Endo et al., 2002; Frenkel et al., 1981; Fry, 1920; Guard, 1995; Han et al., 2008; Hroudová et al., 1992; Huh et al., 2002; Jain et al., 2011; Jenkins & Starkey, 1993; Jensch & Poschlod, 2008; Kim et al., 2009; Kuebel & Tucker, 1988; Lai et al., 2011; Ma et al., 2010; MacHutchon et al., 1993; McBain & Trush, Inc., 2004; McCain & Christy, 2005; McMinn, 1951; Miedzobrodski, 1960; Morton & Snyder, 1978; Murillo et al., 2013; Naruhashi & Iwatsubo,1998; Park et al., 1996; Pellmyr, 1987; Pojar, 1973; Preston & Croft, 1997; Qiu & Li, 2000; Qiyan et al., 2006; Quanyu et al., 1998; Rejmánková, 1992; Rigg, 1922; Schep et al., 2009; Sekine et al., 2007; Simmons et al., 2008; Speichert & Speichert, 2004; Stephens & Dowling, 2001; Vincieri et al., 1985; Wan-Ibrahima et al., 2010; Yang, 2009; Yang et al., 2000; Yatskievych & Raveill, 2001; Young, 2001; Zhou et al., 2005.

12. *Oxypolis*

Cowbane, hog-fennel; oxypolis

Etymology: from the Greek *oxys polos* meaning "sharp axis" to describe the lance-like leaves of some species [which, ironically, have now been transferred to *Tiedemannia*]

Synonyms: *Archemora*; *Neurophyllum*; *Peucedanum* (in part); *Sataria*; *Sium* (in part); *Tiedemannia* (in part)

Distribution: global: North America; **North America:** widespread (except central, northern)

Diversity: global: 4 species; **North America:** 4 species

Indicators (USA): OBL: *Oxypolis occidentalis*, *O. rigidior*, *O. ternata*; **FACW:** *O. fendleri*, *O. ternata*

Habitat: freshwater; lacustrine, riverine; **pH:** 4.5–7.0; **depth:** 0–2 m; **life-form(s):** emergent herb

Key morphology: roots tuberous, fascicled; stems (to 1.5 m) erect, glabrous; leaves (to 30 cm) mainly basal, pinnate (upper leaves sometimes undivided), with 5–13 broad to linearly lanceolate, entire to serrate leaflets (to 9.5 cm), petioles (to 5 dm) dilated, sheathing; umbels compound, flat-topped, peduncled (to 35 cm), rays 12–45 (to 12 cm); flowers (to 3 mm) pedicelled (to 15 mm); calyx conspicuously toothed; petals white or purple; schizocarps (to 7 mm) oblong to ovate, flattened, prominently winged, corky ribbed; stylopodium conical

Life history: duration: perennial (tubers); **asexual reproduction:** tubers; **pollination:** insect; **sexual condition:** hermaphroditic; **fruit:** schizocarps (common); **local dispersal:** mericarps (water, wind); **long-distance dispersal:** mericarps (animals?, water, wind)

Imperilment: (1) *Oxypolis fendleri* [G4]; S1 (UT); S2 (WY); (2) *O. occidentalis* [G4]; S2 (BC, OR); (3) *O. rigidior* [G5]; SH (NY); S2 (ON, PA); S3 (IA); (4) *O. ternata* [G3]; SH (VA); S1 (SC); S2 (FL, GA); S3 (NC)

Ecology: general: In the 1996 indicator list, all North American *Oxypolis* species were designated as OBL plants at least in some portion of their North American range, where they occur most often in fens or wet meadows. The treatment for *O. fendleri* (now categorized entirely as FACW) is included here because it had been completed prior to the release of the revised list. Although often viewed as an indicative genus of alkaline sites, most of the pH values reported for the habitats are circumneutral or acidic, including many that are associated with fens (which have been termed "acidic fens" by some). Occasionally, these plants can be found in waters up to 2 m in depth, but mostly they occur on moist substrates or in shallow water. The flowers are insect pollinated (entomophilous) and are known to be visited by various bees (Hymenoptera: Colletidae; Halictidae), beetles (Coleoptera: Cerambycidae), flies (Diptera), and other insects. Short-tongued bees are most often implicated as their principal pollinators. The seed ecology is poorly known, and details regarding germination and dispersal are scarce. The flattened, winged fruits probably float and are likely to be dispersed by both wind and water. All species are tuberous perennials.

Oxypolis fendleri **(A. Gray) A. Heller** grows in ditches, fens, marshes, meadows, seeps, streams, and along the margins of lakes and streams at higher elevations from 2100 to 3960 m. It occurs on moist ground or in shallow water in partially to deeply shaded sites. The substrates are circumneutral (e.g., pH: 6.5–7.0) and are described as clay, gravel, gravel loam, humic silt, igneous alluvium, fallen logs, loam, sand, sandy clay loam, and silty sand. Flowering occurs from June to August. Vegetative reproduction occurs by means of fascicled tubers. Other details on the life history of this species are lacking. **Reported associates:** *Abies concolor, A. lasiocarpa, Acer glabrum, Achillea millefolium, Aconitum columbianum, Actaea rubra, Agoseris aurantiaca, Agrostis exarata, A. gigantea, A. scabra, A. stolonifera, Allium geyeri, Alnus rugosa, A. tenuifolia, Amelanchier pumila, Anaphalis margaritacea, Anemone narcissiflora, Angelica grayi, A. ampla, A. pinnata, Anticlea elegans, Aquilegia caerulea, Androsace septentrionalis, Arnica cordifolia, A. mollis, Artemisia scopulorum, Betula, Bistorta bistortoides, B. viviparum, Bouteloua curtipendula, B. gracilis, Brickellia californica, Bromus ciliatus, B. porteri, Calamagrostis canadensis, Caltha leptosepala, Campanula rotundifolia, Cardamine cordifolia, Carex aquatilis, C. aurea, C. capillaris, C. disperma, C. illota, C. interior, C. jonesii, C. microglochin, C. microptera, C. norvegica, C. nova, C. pellita, C. pelocarpa, C. saxatilis, C. scopulorum,* *C. stipata, C. utriculata, Castilleja miniata, C. rhexiifolia, C. sulphurea, Cerastium brachypodum, Chaenactis douglasii, Chamerion angustifolium, Cicuta douglasii, Cinna latifolia, Cirsium perplexans, Coeloglossum viride, Comarum palustre, Conioselinum scopulorum, Corallorhiza maculata, Cornus sericea, Corydalis caseana, Cystopteris fragilis, Danthonia intermedia, Dasiphora floribunda, Delphinium barbeyi, D. occidentale, Deschampsia cespitosa, Descurainia incana, Draba aurea, D. spectabilis, Eleocharis quinqueflora, Elymus canadensis, E. trachycaulus, Epilobium ciliatum, E. hornemannii, E. palustre, E. saximontanum, Equisetum arvense, E. laevigatum, E. pratense, E. variegatum, Erigeron coulteri, E. elatior, E. flagellaris, E. glabellus, E. peregrinus, E. speciosus, Eriophorum altaicum, E. angustifolium, Festuca brachyphylla, F. pratensis, Fragaria vesca, F. virginiana, Galeopsis bifida, Galium boreale, Gaultheria humifusa, Gentianopsis thermalis, Geranium caespitosum, G. richardsonii, Geum aleppicum, G. macrophyllum, Glyceria striata, Helianthella quinquenervis, Heracleum maximum, H. sphondylium, Heuchera parvifolia, Hieracium fendleri, Hydrophyllum fendleri, Hymenoxys helenioides, H. hoopesii, Juncus balticus, J. castaneus, J. compressus, J. drummondii, J. mertensianus, J. saximontanus, J. traceyi, Juniperus scopulorum, Lepidium latifolium, Lewisia pygmaea, Ligusticum porteri, Linnaea borealis, Listera cordata, Lomatogonium rotatum, Lonicera involucrata, Lupinus argenteus, Luzula parviflora, Maianthemum stellatum, Mentha arvensis, Mertensia ciliata, M. franciscana, Micranthes odontoloma, Mimulus guttatus, M. primuloides, Mitella pentandra, M. stauropetala, Moehringia lateriflora, M. macrophylla, Moneses uniflora, Montia chamissoi, Oreochrysum parryi, Orthilia secunda, Osmorhiza berteroi, O. depauperata, Oxalis violacea, Oxytropis deflexa, Packera neomexicana, Parnassia fimbriata, Pedicularis crenulata, P. groenlandica, Penstemon strictus, P. virgatus, P. whippleanus, Petasites sagittatus, Phacelia heterophylla, Phleum alpinum, P. pratense, Picea engelmannii, P. pungens, Pinus contorta, P. ponderosa, Plantago tweedyi, Platanthera dilatata, P. hyperborea, P. obtusata, P. sparsiflora, P. stricta, Poa alpina, P. annua, P. arctica, P. compressa, P. leptocoma, P. palustris, P. pratensis, P. reflexa, Podistera eastwoodiae, Polemonium pulcherrimum, Populus angustifolia, P. tremuloides, Potentilla anserina, P. pensylvanica, P. pulcherrima, Primula parryi, P. pauciflora, Prunella vulgaris, Pseudocymopterus montanus, Pseudotsuga menziesii, Pyrola asarifolia, P. chlorantha, P. minor, P. picta, Ranunculus alismifolius, R. inamoenus, R. uncinatus, Rhodiola integrifolia, R. rhodantha, Ribes inerme, R. lacustre, R. montigenum, Robinia hispida, Rosa woodsii, Rubus deliciosus, R. parviflorus, R. strigosus, Rudbeckia laciniata, Rumex crispus, Salix arizonica, S. bebbiana, S. brachycarpa, S. drummondiana, S. lucida, S. monticola, S. planifolia, S. wolfii, Sambucus racemosa, Scirpus microcarpus, Senecio atratus, S. bigelovii, S. taraxacoides, S. triangularis, Sibbaldia procumbens, Silene drummondii, Sorbus scopulina, Stellaria calycantha, S. crassifolia, S. longifolia, S. umbellata, Streptopus amplexifolius, Swertia perennis, Symphoricarpos albus, Symphyotrichum boreale, S. foliaceum,*

S. lanceolatum, Taraxacum officinale, Thalictrum alpinum, T. fendleri, Thermopsis montana, T. rhombifolia, Thlaspi montanum, Torreyochloa pallida, Trautvetteria caroliniensis, Trifolium brandegeei, T. kingii, T. longipes, T. parryi, T. pratense, T. repens, Trisetum wolfii, Trollius laxus, Ulmus, Urtica dioica, Vaccinium myrtillus, V. scoparium, Veratrum californicum, V. tenuipetalum, Veronica americana, V. anagallis-aquatica, V. wormskjoldii, Vicia americana, Viola adunca, V. canadensis, Viola macloskeyi.

Oxypolis occidentalis J.M. Coult. & Rose occurs in bogs, cienaga, creekbeds, fens, marshes, meadows, seeps, springs, and along the margins of ponds and streams at elevations of up to 2700 m. The plants grow in wet substrates or in shallow, still or flowing waters, occasionally at depths to 2 m. They occupy sites ranging from open sun to dense shade and occur on substrates that can be acidic (e.g., pH: 5.2) and contain relatively high amounts of organic matter (e.g., 29%). More alkaline sites are implied by many collection localities, but supporting ecological data are lacking. Substrate types include dark humus, granite sand, muck, peat, quartz gravel, and sandy loam. Flowering occurs from July to August. The seeds reportedly require a period of cold stratification for germination, but specific requirements have not been detailed. In areas where the plants are grazed by deer (Mammalia: Cervidae: *Odocoileus*), they do not flower. Individuals reproduce vegetatively by means of their fascicled tubers. **Reported associates:** *Abies concolor, A. magnifica, Achillea millefolium, Aconitum columbianum, Agrostis exarata, A. idahoensis, A. oregonensis, A. stolonifera, Allium validum, Angelica capitellata, A. genuflexa, Aquilegia formosa, Aulacomnium palustre, Bistorta bistortoides, Bryum, Calamagrostis canadensis, Callitriche hermaphroditica, Calocedrus decurrens, Caltha leptosepala, Camassia quamash, Canadanthus modestus, Carex aquatilis, C. echinata, C. illota, C. lemmonii, C. limosa, C. luzulina, C. nebrascensis, C. nervina, C. nudata, C. rossii, C. scopulorum, C. utriculata, C. vesicaria, Castilleja minor, Circaea alpina, Cirsium douglasii, Claytonia saxosa, Corydalis caseana, Crassula aquatica, Deschampsia cespitosa, D. elongata, Drepanocladus aduncus, Drosera rotundifolia, Eleocharis decumbens, E. montevidensis, E. palustris, E. pauciflora, E. quinqueflora, E. rostellata, Epilobium ciliatum, Equisetum arvense, Eriophorum, Galium trifidum, G. triflorum, Gentiana newberryi, Gentianopsis simplex, Glyceria striata, Hastingsia alba, Helenium bigelovii, Heracleum maximum, Hypericum anagalloides, Isoetes bolanderi, Kelloggia galioides, Juncus chlorocephalus, J. drummondii, J. macrandrus, J. mexicanus, J. nevadensis, J. rugulosus, J. xiphioides, Ligusticum grayi, Lilium parvum, Listera, Lonicera conjugialis, Lophozia, Hosackia oblongifolia, Lupinus burkei, L. latifolius, Madia bolanderi, Marchantia polymorpha, Mentha arvensis, Micranthes oregana, Mimulus guttatus, M. moschatus, M. primuloides, Mitella breweri, Muhlenbergia asperifolia, M. filiformis, M. richardsonis, Oreostemma alpigenum, O. elatum, Orobanche californica, Orthilia secunda, Parnassia californica, Pedicularis attollens, P. groenlandica, Perideridia bacigalupii, P. bolanderi, P. gairdneri, P. parishii, Phalacroseris bolanderi, Philonotis*

fontana, Phleum alpinum, P. pratense, Pinus contorta, P. jeffreyi, Platanthera dilatata, P. leucostachys, P. sparsiflora, Potentilla drummondii, P. gracilis, Primula fragrans, P. jeffreyi, Prunella vulgaris, Pseudotsuga menziesii, Pteridium aquilinum, Pycnanthemum californicum, Ranunculus alismifolius, Rhododendron neoglandulosum, Ribes cereum, Rudbeckia californica, R. hirta, Salix drummondiana, S. lasiolepis, S. lemmonii, S. planifolia, Schoenoplectus pungens, Scirpus congdonii, S. microcarpus, Senecio clarkianus, S. scorzonella, S. triangularis, Sidalcea oregana, S. reptans, Sisyrinchium elmeri, Solidago californica, S. canadensis, Sphagnum subsecundum, Spiraea ×hitchcockii, S. douglasii, Spiranthes romanzoffiana, S. stellata, Stachys ajugoides, S. albens, S. chamissonis, Stellaria, Symphyotrichum spathulatum, Triantha glutinosa, Trifolium longipes, T. monanthum, T. wormskjoldii, Triglochin palustre, Vaccinium uliginosum, Veratrum californicum, Viola adunca, V. macloskeyi.

Oxypolis rigidior (L.) Raf. grows in bogs, creek bottoms, ditches, fens, floodplains, marshes, meadows, prairies, seeps, spray cliffs, swales, swamps, and along the margins of lakes, rivers, and streams at elevations of up to 1725 m. Exposures can vary from full sun to substantial shade. The substrates often are described as alkaline (e.g., alkalinity 190 mg/L; total hardness 212 mg/L) or as mafic, and span a broad range of acidity (pH: 4.2–8.2), with many reports indicating circumneutral values. They include clay, gravel, loam, marl, muck, mud, peat, peaty loam, sand, sandy loam, and silt. The plants flower from late July to September. They are pollinated predominantly by short-tongued bees (e.g., Hymenoptera: Colletidae, Halictidae), but possibly also by other insects (Insecta) such as flies (Diptera) and wasps (Hymenoptera), which mainly seek nectar or are predatory. Plant size is correlated positively with total seed production and with average seed mass, which is relatively low (mean: 1.8 mg). The seeds may require special conditions for germination, because some attempts using typical stratified and unstratified conditions were unsuccessful. However, other reports indicate that seeds sown at 20°C will germinate in less than 2 weeks. Germination reportedly is enhanced by smoke, and the plants are characterized as being fire tolerant. Consequently, it is likely that natural fire cycles are important in the life history of this species. The plants persist principally by means of vegetative reproduction, which relies on thickened tubers that are found at substrate depths from 3 to 12 cm. The primary root thickens by the end of the first year of growth and 5–7 secondary, fusiform roots typically are produced during 4–5 subsequent growing years and thicken similarly as the plants mature. Once the plants fully mature, the shoots die-back to the clustered roots each fall, and then regenerate from a renewed apical bud. **Reported associates:** *Acer rubrum, Adiantum capillus-veneris, Agalinis purpurea, Agrostis gigantea, A. perennans, Aletris farinosa, Allium cernuum, Alnus serrulata, Amorpha canescens, Amphicarpaea bracteata, Angelica atropurpurea, Andropogon gerardii, A. glomeratus, A. mohrii, A. virginicus, Anemone virginiana, Antennaria plantaginifolia, Apios americana, Apocynum cannabinum, Arethusa bulbosa, Arisaema triphyllum, Arnoglossum plantagineum,*

Aruncus dioicus, Asclepias hirtella, A. incarnata, A. sullivantii, A. tuberosa, Athyrium filix-femina, Atrichum oerstedianum, Aulacomnium palustre, Bazzania trilobata, Betula nigra, B. ×sandbergii, Bignonia capreolata, Bidens connatus, Boehmeria cylindrica, Boykinia aconitifolia, Bromus ciliatus, B. kalmii, Calamagrostis canadensis, C. stricta, Calopogon tuberosus, Caltha palustris, Calystegia sepium, Campylium chrysophyllum, Cardamine bulbosa, C. clematitis, Carex atherodes, C. atlantica, C. baileyi, C. bebbii, C. bromoides, C. buxbaumii, C. crawei, C. crinita, C. debilis, C. echinata, C. exilis, C. folliculata, C. gracilescens, C. gracillima, C. granularis, C. gynandra, C. haydenii, C. hystericina, C. interior, C. intumescens, C. lacustris, C. leptalea, C. lurida, C. pellita, C. sartwellii, C. sterilis, C. stricta, C. tetanica, C. trichocarpa, C. trisperma, C. viridula, Carpinus caroliniana, Castilleja coccinea, Catalpa bignonioides, Cephalanthus occidentalis, Cercis canadensis, Chamaecyparis thyoides, Chasmanthium laxum, Chelone cuthbertii, C. glabra, C. lyonii, C. obliqua, Chrysosplenium americanum, Cicuta maculata, Cinna arundinacea, Cirsium arvense, C. hillii, C. muticum, Cladium mariscoides, Clematis virginiana, Clethra acuminata, Clinopodium arkansanum, C. glabellum, Commelina virginica, Conoclinium coelestinum, Coreopsis gladiata, C. tripteris, Cornus amomum, C. foemina, C. obliqua, C. racemosa, C. rugosa, C. sericea, Corylus americana, Crataegus, Ctenium aromaticum, Cuscuta compacta, Cyperus strigosus, Cypripedium candidum, C. reginae, Dasiphora floribunda, Decumaria barbara, Dichanthelium dichotomum, D. oligosanthes, Diospyros virginiana, Diphylleia cymosa, Doellingeria umbellata, Drosera rotundifolia, Dryopteris cristata, Dulichium arundinaceum, Dumortiera hirsuta, Eleocharis obtusa, E. tenuis, E. tortilis, Elephantopus carolinianus, Elymus canadensis, E. virginicus, Epilobium coloratum, E. leptophyllum, Equisetum arvense, Erechtites hieracifolia, Erigeron strigosus, Eriocaulon decangulare, Eriophorum virginicum, Eryngium integrifolium, E. yuccifolium, Eupatorium perfoliatum, Euphorbia purpurea, Eurybia divaricata, Euthamia tenuifolia, Eutrochium fistulosum, E. maculatum, Fallopia scandens, Filipendula rubra, Fontinalis sphagnifolia, Fragaria virginiana, Fraxinus americana, F. nigra, Fuirena simplex, F. squarrosa, Galium asprellum, G. boreale, Gentiana andrewsii, Gentianopsis crinita, G. virgata, Geum geniculatum, G. obtusum, G. rivale, Glyceria melicaria, G. striata, G. obtusa, Hasteola suaveolens, Helenium autumnale, H. brevifolium, Helianthus angustifolius, H. grosseserratus, H. mollis, Hesperostipa spartea, Heuchera americana, H. villosa, Hookeria acutifolia, Houstonia caerulea, H. serpyllifolia, Huperzia porophila, Hydrangea quercifolia, Hydrocotyle, Hydrophyllum canadense, Hypericum densiflorum H. kalmianum, H. mutilum, H. prolificum, Hypnum imponens, Hypoxis hirsuta, Impatiens capensis, Ilex collina, I. opaca, I. verticillata, Iris versicolor, I. virginica, Isoetes engelmannii, I. valida, Itea virginica, Jamesianthus alabamensis, Juncus acuminatus, J. brachycephalus, J. coriaceus, J. dudleyi, J. effusus, J. gymnocarpus, J. militaris, J. nodosus, J. subcaudatus, Juniperus virginiana, Justicia americana, Kalmia carolina, K. latifolia, Koeleria macrantha, Krigia montana, Larix laricina, Laportea canadensis, Lathyrus palustris, Leersia virginica, Liatris pycnostachya, L. spicata, Lilium canadense, L. grayi, L. michiganense, L. superbum, Linum striatum, Liparis loeselii, Liquidambar styraciflua, Liriodendron tulipifera, Lobelia amoena, L. cardinalis, L. kalmii, L. nuttallii, L. puberula, L. siphilitica, L. spicata, Ludwigia alternifolia, L. leptocarpa, L. palustris, Lycopus americanus, L. virginicus, Lygodium palmatum, Lysimachia terrestris, Lyonia ligustrina, Lysimachia ciliata, L. quadriflora, L. quadrifolia, Lythrum alatum, Maianthemum canadense, M. stellatum, Marshallia grandiflora, M. trinervia, Medeola virginiana, Melanthium virginicum, Menyanthes trifoliata, Menziesia pilosa, Micranthes micranthidifolia, M. pensylvanica, M. petiolaris, Microstegium vimineum, Mikania scandens, Milium effusum, Mimulus ringens, Mitreola petiolata, Mnium hornum, Monarda fistulosa, Muhlenbergia glomerata, M. mexicana, M. racemosa, M. richardsonis, Myrica cerifera, Nasturtium officinale, Nyssa sylvatica, Oenothera fruticosa, Oligoneuron album, O. ohioense, O. riddellii, O. rigidum, Onoclea sensibilis, Orontium aquaticum, Osmunda regalis, Osmundastrum cinnamomeum, Packera aurea, P. paupercula, Panicum virgatum, Parnassia asarifolia, P. glauca, P. grandifolia, Parthenium integrifolium, Pedicularis canadensis, P. lanceolata, Peltandra virginica, Persicaria sagittata, Phalaris arundinacea, Philonotis fontana, Phlox carolina, P. glaberrima, P. maculata, P. pilosa, Photinia melanocarpa, P. pyrifolia, Physocarpus opulifolius, Physostegia virginiana, Picea rubens, Pinus taeda, Plagiomnium ciliare, Platanthera clavellata, P. flava, P. grandiflora, P. integrilabia, P. peramoena, Platanus occidentalis, Poa alsodes, P. compressa, P. paludigena, Pogonia ophioglossoides, Polemonium vanbruntiae, Polytrichum commune, Potentilla anserina, Prenanthes racemosa, Primula meadia, Ptilimnium costatum, Pycnanthemum tenuifolium, P. virginianum, Quercus macrocarpa, Rhamnus alnifolia, Rhexia virginica, Rhizomnium appalachianum, Rhododendron catawbiense, R. maximum, R. minus, R. viscosum, Rhus glabra, Rhynchospora alba, R. capillacea, R. capitellata, R. colorata, R. globularis, R. glomerata, R. gracilenta, Rosa blanda, R. palustris, Rubus allegheniensis, R. hispidus, Rudbeckia fulgida, R. hirta, R. laciniata, R. subtomentosa, Rumex orbiculatus, Sagittaria latifolia, Salix bebbiana, S. candida, S. caroliniana, S. discolor, S. humilis, S. sericea, S. serissima, Sanguisorba canadensis, Sarracenia flava, Scapina nemorosa, Schizachyrium scoparium, Schoenoplectus pungens, S. purshianus, Scirpus atrovirens, S. expansus, S. cyperinus, S. georgianus, S. pendulus, S. polyphyllus, Scleria muehlenbergii, S. verticillata, Scutellaria galericulata, S. lateriflora, Selaginella apoda, Silphium compositum, S. integrifolium, S. trifoliatum, Smilax tamnoides, Solidago canadensis, S. gigantea, S. patula, S. rugosa, S. uliginosa, Sorghastrum nutans, Spartina pectinata, Sphagnum affine, S. bartlettianum, S. fallax, S. fuscum, S. lescurii, S. palustre, S. recurvum, Spiraea alba, S. tomentosa, Spiranthes cernua, Sporobolus heterolepis, Stellaria corei, Stenanthium gramineum, Symphyotrichum boreale, S. divaricatum, S. ericoides,

S. laeve, S. lanceolatum, S. novae-angliae, S. novi-belgii, S. pilosum, S. puniceum, Thalictrum clavatum, T. dasycarpum, T. pubescens, Thaspium trifoliatum, Thelypteris noveboracensis, T. palustris, Thuidium delicatulum, Thuja occidentalis, Tiarella cordifolia, Tiedemannia teretifolia, Tilia americana, Toxicodendron radicans, Tradescantia ohiensis, Trautvetteria caroliniensis, Triadenum virginicum, Trichophorum cespitosum, Triglochin maritimum, Tripsacum dactyloides, Tsuga canadensis, Typha latifolia, Ulmus rubra, Vaccinium corymbosum, V. macrocarpon, V. simulatum, Valeriana edulis, Vernonia fasciculata, V. gigantea, V. noveboracensis, Veronicastrum virginicum, Viburnum nudum, Viola cucullata, V. macloskeyi, V. nephrophylla, V. pedatifida, V. ×primulifolia, V. walteri, Woodwardia areolata, Xanthorhiza simplicissima, Xyris tennesseensis, X. torta, Zizia aurea.

Oxypolis ternata (Nutt.) Heller inhabits bogs, depressions, ditches, flatwoods (and their ecotones), pocosins, savannas, seeps and swamps. Plants occur commonly in naturally disturbed areas. They are found on substrates (Bladen, Plummer types) that are acidic (e.g., pH: 4.5) and low in calcium (125 ppm) and organic matter (3.5%); they usually are described as loamy sand, or sand (84%), with low amounts of silt (13%) and clay (3%). The flowers are produced from September to October and are pollinated by a variety of insects (Insecta) including flies (Diptera), bees (Hymenoptera), beetles (Coleoptera), and butterflies (Lepidoptera). Fruiting extends from October to November. The plants can reproduce vegetatively by sprouting from their clustered tubers. The mixed OBL/FACW ranking of this species reflects that it is more common in mesic than in wet sites, but it never occurs in dry sites. **Reported associates:** *Aletris aurea, Andropogon arctatus, A. gyrans, A. liebmannii, A. virginicus, Aristida beyrichiana, A. palustris, A. stricta, Arnoglossum ovatum, Bigelowia nudata, Calamovilfa brevipilis, Carphephorus paniculatus, C. pseudoliatris, C. tomentosus, Chaptalia tomentosa, Cleistes divaricata, C. bifaria, Coreopsis floridana, C. linifolia, Ctenium aromaticum, Erigeron vernus, Eriocaulon decangulare, Eryngium integrifolium, Eurybia chapmanii, Gaylussacia frondosa, Helianthus heterophyllus, Ilex glabra, Juncus trigonocarpus, Lachnanthes caroliana, Liquidambar styraciflua, Marshallia graminifolia, Muhlenbergia capillaris, M. expansa, Myrica caroliniensis, M. cerifera, Panicum virgatum, Pinguicula caerulea, Pinus palustris, P. serotina, Pleea tenuifolia, Rhexia alifanus, Rhynchospora baldwinii, R. chapmanii, R. latifolia, R. plumosa, Sarracenia flava, Schwalbea americana, Smilax laurifolia, Sphagnum, Sporobolus pinetorum, Taxodium ascendens, Xyris ambigua, Zigadenus glaberrimus.*

Use by wildlife: *Oxypolis rigidior* supposedly is poisonous to domestic cattle (Mammalia: Bovidae: *Bos taurus*), hence the common name of "cowbane." However, *O. fendleri* and *O. occidentalis* are eaten quite regularly by mule deer (Mammalia: Cervidae: *Odocoileus hemionus*) and *Oxypolis occidentalis* also is grazed by Sitka black-tailed deer (*O. hemionus sitkensis*). *Oxypolis rigidior* is a host plant of several Lepidoptera (Insecta) including tawny-edged skippers

(Hesperiidae: *Polites themistocles*), white m-hairstreaks (Lycaenidae: *Parrhasius m-album*), viceroys (Nymphalidae: *Limenitis archippus*), larval concealer moths (Oecophoridae: *Depressaria cinereocostella*), and black swallowtail butterflies (Papilionidae: *Papilio polyxenes*); *O. ternata* also is thought to host black swallowtail larvae. *Oxypolis rigidior* is eaten readily by swarming grasshoppers (Insecta: Orthoptera: Acrididae: *Schistocerca emarginata, S. obscura*). Several Fungi are hosted by *Oxypolis fendleri* and *O. occidentalis* (Ascomycota: Mycosphaerellaceae: *Ramularia heraclei*; Basidiomycota: Pucciniaceae: *Puccinia ligustici*), and also by *O. rigidior* (Ascomycota: Incertae sedis: *Ascochyta thaspii, Dilophospora geranii, Neottiospora umbelliferarum*; Leptosphaeriaceae: *Ophiobolus nigroclypeatus*; Mycosphaerellaceae: *Cercospora, Passalora depressa, P. platyspora, Septoria umbelliferarum*). The flowers of *O. occidentalis* attract various bees (Hymenoptera: Colletidae: *Hylaeus basalis, H. coloradensis, H. episcopalis*), beetles (Coleoptera: Cerambycidae: *Cosmosalia chrysocoma, Stenostrophia tribalteata*), flies (Diptera) and even stoneflies (Plecoptera). The flowers of *O. rigidior* have an immense associated fauna and attract various bees (Hymenoptera: Apidae: *Apis mellifera*; Colletidae: *Hylaeus affinis, H. annulatus, H. mesillae*; Halictidae: *Halictus rubicundus, Lasioglossum admirandum, L. imitatum, L. versatum, Sphecodes dichrous*) and flies (Diptera: Anthomyiidae: *Calythea pratincola, Delia platura*; Bombyliidae: *Anthrax alternata*; Calliphoridae: *Lucilia sericata*; Chamaemyiidae: *Leucopis minor*; Chloropidae: *Apallates coxendix, Hippelates plebejus, Liohippelates flavipes, L. pusio, Olcella cinerea, O. trigramma*; Conopidae: *Physoconops brachyrhynchus, Zodion americanum*; Empidae: *Empis clausa*; Ephydridae: *Ochthera lauta*; Milichiidae: *Leptometopa latipes*; Muscidae: *Limnophora narona, Lispe consanguinea, Morellia micans, Musca domestica, Neomyia cornicina*; Pipunculidae: *Chalarus spurius*; Sarcophagidae: *Amobia aurifrons, Helicobia rapax, Ravinia anxia, R. derelicta, R. stimulans, Senotainia rubriventris, Sphixapata trilineata*; Sciaridae: *Sciara atrata*; Syrphidae: *Allograpta obliqua, Didea fuscipes, Eristalis arbustorum, Melangyna umbellatarum, Paragus bicolor, Platycheirus hyperboreus, Spilomyia longicornis, Syritta pipiens, Syrphus ribesii, Toxomerus geminatus, T. marginatus, T. politus*; Tachinidae: *Belvosia unifasciata, Chetogena claripennis, Clausicella geniculata, Cylindromyia dosiades, Gymnoclytia occidua, Gymnosoma fuliginosum, Euclytia flava, Myiopharus ancilla, Nemorilla pyste, Phasia aeneoventris, Phasia purpurascens, Spallanzania hesperidarum, Trichopoda pennipes*; Tephritidae: *Euaresta bella*). The flowers also are visited by numerous wasps (Hymenoptera: Braconidae: *Agathis simillimus, Apanteles crassicornis, Bracon apicatus, Cardiochiles abdominale, Chelonus sericeus, Meteorus areolatus, Microplites croceipes, M. gortynae, Monogonogastra fuscipennis, Opius gahani, Urosigalphus femoratus, Vipio rugator*; Chalcididae: *Bruchophagus gibba*; Chrysididae: *Chrysis venusta, Hedychrum violaceum, Holopyga ventrale*; Eucoilidae: *Pseudeucoila stigmata*; Figitidae: *Figites impatiens, Neralsia armatus, Prosaspicera albihirta*; Ichneumonidae: *Ceratogastra ornata*, *Ichneumon ambulatorius*; Leucospididae: *Leucospis affinis*; Mutillidae: *Dasymutilla macra*, *Myrmosa unicolor*, *Timulla vagans*; Perilampidae: *Perilampus fulvicornis*, *Perilampus hyalinus*; Pompilidae: *Ageniella fulgifrons*, *Anoplius lepidus*, *A. nigritus*, *Entypus fulvicornis*, *Evagetes ingenuus*, *Tachypompilus ferruginea*; Pteromalidae: *Eutrichosoma mirabile*; Scoliidae: *Scolia bicincta*; Sphecidae: *Ancistromma distincta*, *Astata unicolor*, *Cerceris clypeata*, *C. compacta*; *Chalybion californicus*, *Chlorion aerarius*, *Ectemnius rufipes*, *E. trifasciatus*, *Gorytes simillimus*, *Isodontia apicalis*, *Lestica confluentus*, *Lindenius columbianus*, *Lyroda subita*, *Oxybelus emarginatus*, *O. mexicanus*, *O. packardii*, *O. uniglumis*, *Prionyx atrata*, *Pseudoplisus phaleratus*, *Sceliphron caementaria*, *Stizoides renicinctus*, *Tachysphex belfragei*, *T. distinctus*; Tiphiidae: *Myzinum quinquecincta*, *Tiphia intermedia*, *T. letalis*, *T. transversa*; Vespidae: *Euodynerus foraminatus*, *Leionotus ziziae*, *Polistes carolina*, *P. fuscata*; *Stenodynerus fundatiformis*, *S. histrionalis*). Other visitors to the flowers of *O. rigidior* include beetles (Coleoptera: Cantharidae: *Chauliognathus pennsylvanicus*; Chrysomelidae: *Diabrotica cristata*, *D. undecimpunctata*, *Luperaltica nigripalpis*, *Sennius cruentatus*; Coccinellidae: *Cycloneda sanguinea*, *Hippodamia convergens*; Curculionidae: *Centrinaspis picumna*; Latridiidae: *Melanophthalma distinguenda*; Meloidae: *Epicauta cinereus*, *E. pensylvanica*, *Pyrota germari*; Mordellidae: *Mordella melaena*, *Mordellistena limbalis*; Rhipiphoridae: *Macrosiagon limbata*; Scarabaeidae: *Euphoria sepulcralis*), plant bugs (Hemiptera: Anthocoridae: *Orius insidiosus*; Lygaeidae: *Oncopeltus fasciatus*; Miridae: *Lygus lineolaris*, *Plagiognathus obscurus*), and lacewings (Neuroptera: Chrysopidae: *Chrysoperla plorabunda*).

Economic importance: food: The roots of *Oxypolis rigidior* were baked and eaten by the Cherokee. However, several sources characterize *O. fendleri* and *O. rigidior* as poisonous (especially the roots), although no cases of human fatalities apparently have been documented; **medicinal:** none; **cultivation:** *Oxypolis rigidior* is in cultivation as a water garden ornamental; **misc. products:** none; **weeds:** none; **nonindigenous species:** none

Systematics: Recent phylogenetic studies incorporating DNA sequence data have demonstrated the traditional generic concept of *Oxypolis* to be polyphyletic, necessitating the transfer of all the "rachis-leaved" species (for which the genus originally was named) to a different genus (*Tiedemannia*). These two genera also differ by their base chromosome number. As redefined, *Oxypolis* resolves within tribe Oenantheae as the sister genus to a clade containing several groups of "rachis-leaved" species (*Harperella*, *Tiedemannia*), the similar *Lilaeopsis*, and several broad-leaved genera (Figures 5.112 and 5.116). Three of the four species in the newly-defined genus are ranked as OBL, which indicates a single origin of the habit. Relationships within *Oxypolis* (Figure 5.122) indicate an initial radiation of species from the western United States (*O. fendleri*) both westward (*O. occidentalis*) and eastward (*O. rigidior*, *O. ternata*). Three subclades occur within *O. occidentalis*, which correspond roughly to Oregon/British Columbian populations, northern Cascade California populations, and other California localities (Figure 5.122). The base chromosome number of *Oxypolis* is $x = 8$ or 9. All counted species are apparently tetraploids based either on $2n = 32$ (*O. rigidior*) or $2n = 36$ (*Oxypolis fendleri*, *O. occidentalis*).

Comments: The species nearly are allopatric (*Oxypolis occidentalis* in far western North America, *O. fendleri* in the western United States, and *O. rigidior* in eastern North America), with the exception of *O. ternata* (southeastern United States), which overlaps with the preceding; *O. ternata* has been attributed erroneously to Texas.

References: Austin et al., 2008; Bell & Constance, 1960; Bowles et al., 2005; Carr et al., 2010; Cheney & Marr, 2007;

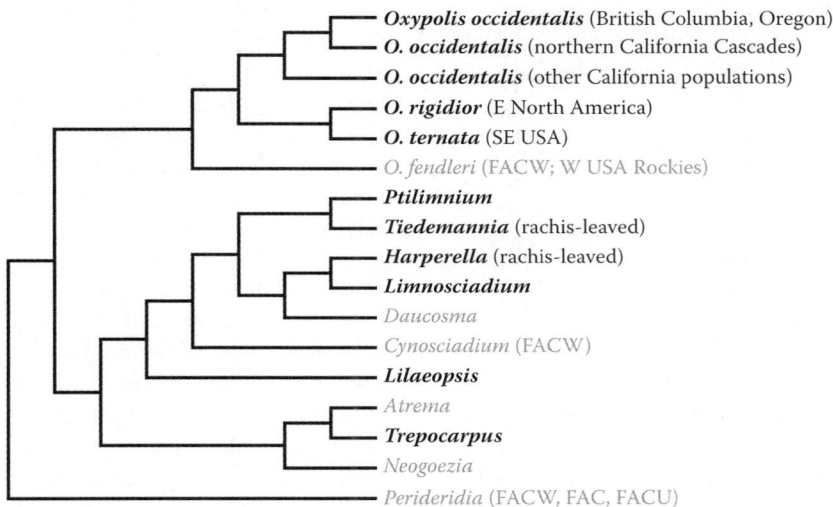

FIGURE 5.122 Phylogenetic relationships of *Oxypolis* as indicated by analysis of nrITS sequence data. Taxa with OBL North American representatives are highlighted in bold; other wetland designations are as indicated. *Oxypolis* is distant phylogenetically from *Tiedemannia*, which contains the "rachis-leaved" species formerly assigned to the genus. Three subclades occur in *O. occidentalis*, with their approximate geographical distinctions indicated. (Adapted from Feist, M.A.E. & Downie, S.R., *Syst. Bot.*, 33, 447–458, 2008.)

Cook, 1996; Cooper & Wolf, 2006; Crawford & Hartman, 1972; Davis, 1914; Davis et al., 2008; Dillingham, 2005; Feist, 2009; Feist & Downie, 2008; Fernald & Kinsey, 1943; Greene & Curtis, 1950; Hendrix & Sun, 1989; Holm, 1925; Knight & Walker, 2001; Lawrence & Biswell, 1972; MacRae & Rice, 2007; McCullough, 1948; Muldavin et al., 2000; Otte, 1975; Peterson, 1977; Ratliff, 1985; Robertson, 1924a,b, 1928; Rodger, 1998; Sherff, 1912; Singhurst & Bridges, 2007; Sorrie et al., 2003, 2006; Stubbs, 1992; Tooker & Hanks, 2000; Tooker et al., 2006; Vestal, 1914; Walker & Peet, 1984; Wallmo et al., 1972; Weems, Jr., 1953; Williams, 2003; Wiser et al., 1996; Zartman & Pittillo, 1998.

13. *Ptilimnium*

Herbwilliam, mock bishopweed

Etymology: from the Greek *ptilon mnion* meaning "feather moss" in reference to the finely dissected leaves

Synonyms: *Aethusa* (in part); *Ammi*; *Discopleura*; *Harperella* (in part); *Harperia*; *Sison*

Distribution: global: North America; **North America:** eastern

Diversity: global: 5 species; **North America:** 5 species

Indicators (USA): OBL: *Ptilimnium ahlesii*, *P. capillaceum*, *P. costatum*, *P. ×texense*; **FACW:** *P. capillaceum*, *P. costatum*

Habitat: brackish (coastal), freshwater; palustrine; **pH:** 4.2–8.0; **depth:** <1 m; **life-form(s):** emergent herb

Key morphology: Stems (to 15 dm) erect, branching; leaves (to 4 cm) pinnately decompound, with filiform, decompound divisions, petioles (to 3 cm) sheathing; umbels compound, axillary or terminal, peduncled (to 9.5 cm), 5–24-rayed (to 4.5 cm); flowers (5–20/umbel) pedicelled (to 6 mm); calyx of small teeth; petals white or pink; anthers purple to rose colored; styles (to 0.5–3 mm) spreading, styolopodium conical; schizocarp (to 4 mm) orbicular or nearly so, with conspicuous dorsal ribs and a longitudinal band of corky tissue

Life history: duration: annual (fruit/seeds); **asexual reproduction:** none; **pollination:** insect or self; **sexual condition:** hermaphroditic; andromonoecious; **fruit:** schizocarps (infrequent); **local dispersal:** seeds (water, wind); **long-distance dispersal:** seeds (water, wind)

Imperilment: (1) *Ptilimnium ahlesii* [G1]; SH (GA); S1 (NC); (2) *P. capillaceum* [G5]; SX (PA); SH (MO); S1 (KY); S2 (RI); S3 (NY); (3) *P. costatum* [G4]; SH (KY); S1 (AL, AR, NC); S2 (GA); S3 (IL, TX); (4) *P. ×texense* [GNA]

Ecology: general: *Ptilimnium* principally is a wetland genus, with three of its five species ranked as OBL and one as FAC, FACW. The remaining species (*P. ahlesii*) was recently described and has not yet formally been assigned a wetland indicator status. It is included here as OBL due to the consistent wetland conditions reported for all of its known occurrences. The flowers of all *Ptilimnium* species are either autogamous or insect pollinated (Insecta), then presumably by small flies (Diptera) and Hymenoptera (e.g., Megachilidae: *Anthidiellum perplexum*). All species of *Ptilimnium* are annuals, but their seed ecology has been described in detail only for *P. nuttallii*, which unfortunately, is the single member of the genus that is not ranked as OBL. In that species, the seeds

have morphological dormancy (up to 29% germinating in 2 weeks at 15°C/6°C, 20°C/10°C, 25°C/15°C) or a more complex morphophysiological dormancy, which requires higher temperatures (25°C/15°C, 30°C/15°C) to initiate germination. The seed dispersal mechanism has not been described for this genus, but the corky mericarps presumably float to some extent and are capable of water dispersal.

***Ptilimnium ahlesii* Weakley & G.L. Neasom** is a rare plant of tidal fresh or slightly brackish waters in marshes, sloughs, and swamps. Virtually nothing is known of the life history of this species other than its flowering time (May–early June) and fruiting period (late May–July). **Reported associates:** *Nyssa, Taxodium, Typha.*

***Ptilimnium capillaceum* (Michx.) Raf.** grows in brackish (0.5–18 ppt salinity) or fresh waters of bogs, creekbeds, cypress domes, deltas, depressions, ditches, flatwoods, hammocks, interdunal swales, marshes, meadows, prairies, salt marshes, seeps, springs, swales, swamps, thickets, and along the margins of canals, lakes, ponds, rivers, and streams at elevations of up to 40 m. The sites range in exposure from open sunlight to light shade and often are characterized by some degree of disturbance. Substrates span a wide range of acidity (pH: 4.8–8.0) and comprise clay, clay loam, loamy sand, muck, mud, peat, sand, sandstone, sandy clay, sandy clay loam, silty loam, or silty clay loam, which can contain as much as 55% organic matter. Occasionally the plants can be found growing on logs, stumps, or in mats of vegetation. Tolerance to severe flooding seems to be minimal, although many occurrences of this species are reported from sites that are described as disturbed. A seed bank is formed and the seeds have germinated within 6–8 weeks (at greenhouse temperatures held below 35°C) from soil cores that had been stored for 3 months at 7°C. Experimental manipulations show that this species responds variously to simulated saltwater flooding disturbances but does well in moderately saline sites after they are drained. In one study, the plants and their seed bank were restricted to the area extending from the waterline to 1 m above the waterline. Seedlings are not produced densely (4–8/m²) and the seeds do not germinate well under flooded conditions. Other studies also indicate quite low germination from seed bank samples, perhaps due to complex dormancy requirements. The plants are winter annuals in the southern portion of their range (germinating in autumn or early winter; dying back by late spring), but grow as spring annuals in the north (germinating in early spring and persisting through summer). In the south they occur most frequently in the spring. Plants flower from June to October. The floral or seed biology has not been described in detail. However, the species is believed to be primarily autogamous as evidenced in part by the low proportions of staminate flowers in the umbels. The plants have successfully established from floating coastal wrack, due to the presence of viable propagules (presumably seeds). Some studies report up to 30% colonization of the roots by arbuscular mycorrhizae. **Reported associates:** *Acer negundo, A. rubrum, Acmella oppositifolia, Aeschynomene, Agalinis maritima, A. purpurea, Agrostis stolonifera, Alternanthera philoxeroides, Amaranthus cannabinus, Ambrosia trifida,*

Amorpha, Andropogon capillipes, A. glaucopsis, A. glomeratus, A. virginicus, Aralia spinosa, Arundinaria, Asclepias incarnata, A. lanceolata, Baccharis halimifolia, Bacopa monnieri, Bidens connatus, B. coronatus, Briza minor, Boehmeria cylindrica, Bolboschoenus maritimus, B. robustus, Callicarpa americana, Calystegia sepium, Carex atlantica, C. hormathodes, C. hyalinolepis, C. intumescens, C. longii, C. lupulina, C. seorsa, C. straminea, C. stricta, Carpinus caroliniana, Carya, Chasmanthium laxum, C. sessiliflorum, Chenopodium rubrum, Cicuta maculata, Cinna arundinacea, Cladium jamaicense, C. mariscoides, Clematis crispa, Commelina diffusa, C. virginica, Crinum, Cuscuta gronovii, Cynodon dactylon, Cyperus erythrorhizos, C. filicinus, C. polystachyos, C. virens, Diodia virginiana, Diospyros virginiana, Distichlis spicata, Drosera brevifolia, Echinochloa crus-galli, E. walteri, Eclipta prostrata, Elatine americana, Eleocharis acicularis, E. albida, E. fallax, E. halophila, E. palustris, E. parvula, E. rostellata, Eriocaulon decangulare, Eryngium aquaticum, Eupatorium capillifolium, E. compositifolium, E. mohrii, E. rotundifolium, Euthamia caroliniana, Fagus, Festuca filiformis, Fimbristylis autumnalis, F. castanea, Fuirena pumila, Galium obtusum, G. palustre, Glottidium vesicarium, Glyceria septentrionalis, Gratiola aurea, Hibiscus grandiflorus, H. moscheutos, Hierochloe odorata, Honckenya peploides, Hydrocotyle ranunculoides, H. umbellata, H. verticillata, Hypericum canadense, H. hypericoides, Ilex cassine, I. vomitoria, Impatiens capensis, Ipomoea sagittata, Iris giganticaerulea, I. prismatica, I. versicolor, Iva frutescens, Juncus ambiguus, J. articulatus, J. canadensis, J. coriaceus, J. effusus, J. gerardi, J. marginatus, J. scirpoides, J. validus, Kosteletzkya pentacarpos, Leersia oryzoides, L. virginica, Leptochloa fusca, Limosella australis, Liquidambar styraciflua, Lilaeopsis chinensis, Lilium superbum, Limonium carolinianum, Liriodendron tulipifera, Lobelia cardinalis, L. puberula, Lonicera, Ludwigia alternifolia, L. glandulosa, L. microcarpa, L. palustris, L. suffruticosa, Lycopus americanus, L. rubellus, Lyonia ligustrina, Lythrum lineare, Macrothelypteris torresiana, Magnolia virginiana, Melothria pendula, Mikania scandens, Myrica cerifera, Nyssa sylvatica, Oenothera fruticosa, Onoclea sensibilis, Osmunda regalis, Osmundastrum cinnamomeum, Panicum amarum, P. dichotomiflorum, P. hemitomon, P. repens, P. virgatum, Paspalum dilatatum, P. notatum, P. repens, P. urvillei, Peltandra virginica, Persea palustris, Persicaria arifolia, P. hydropiperoides, P. lapathifolia, P. pensylvanica, P. punctata, P. virginiana, Petiveria alliacea, Phragmites australis, Phyla nodiflora, Physostegia, Pinus elliottii, P. taeda, Pluchea carolinensis, P. foetida, Polygala grandiflora, Polygonum ramosissimum, Pontederia cordata, Potentilla anserina, Pteridium aqulinum, Ptilimnium costatum, Pyrrhopappus, Quercus nigra, u. virginiana, Rhexia mariana, R. petiolata, Rhododendron viscosum, Rhus copallinum, Rhynchospora caduca, R. fascicularis, R. glomerata, R. microcarpa, Rosa palustris, Rubus argutus, Rudbeckia, Rumex maritimus, R. verticillatus, Sabal minor, Sabatia stellaris, Sacciolepis striata, Sagittaria calycina, S. lancifolia, S. latifolia, S. subulata, Salix, Salvinia minima, Samolus *valerandi, Sanguisorba canadensis, Sanicula canadensis, Sarracenia flava, Sassafras albidum, Saururus cernuus, Schinus terebinthifolius, Schoenoplectus americanus, S. heterochaetus, S. pungens, S. tabernaemontani, Scirpus cyperinus, Sesbania, emerus, Sesuvium portulacastrum, Setaria, Sisyrinchium, Sium suave, Smilax bona-nox, S. laurifolia, Solidago sempervirens, Spartina alterniflora, S. cynosuroides, S. patens, S. pectinata, Spergularia salina, Sphagnum, Spiranthes cernua, Suaeda maritima, Symphyotrichum divaricatum, S. novi-belgii, S. subulatum, Taxodium ascendens, T. distichum, Teucrium canadense, Thalictrum pubescens, Thelypteris hispidula, T. interrupta, T. kunthii, Tillandsia bartramii, Toxicodendron radicans, T. vernix, Triadica sebifera, Typha angustifolia, T. latifolia, Ulmus alata, Urochloa reptans, Vaccinium corymbosum, V. elliottii, V. stamineum, V. virgatum, Verbena, Verbesina virginica, Vicia ludoviciana, Vigna luteola, Viola, Vitis, Woodwardia areolata, Xanthium strumarium, Xyris difformis, Zannichellia palustris.*

***Ptilimnium costatum* (Ell.) Raf.** inhabits bogs, bottomlands, flatwoods, meadows, prairies, savannas, seeps, swamps, and the margins of ponds and streams. Occurrences usually are reported as shaded sites that are on acidic (e.g., pH: 4.2–4.8; 1.9%–2.4% organic matter) substrates (mud, sand), but calcareous conditions also have been attributed to some localities. Flowering occurs from June to October. Although other life history information is minimal for this species, a compilation of various reports provides a fairly substantial list of associated species. **Reported associates:** *Acer negundo, A. rubrum, Agalinis fasciculata, A. purpurea, A. tenuifolia, Aletris aurea, Alnus serrulata, Alternanthera philoxeroides, Aristida purpurascens, Asclepias perennis, A. rubra, Bidens laevis, B. mitis, Boltonia diffusa, Callicarpa americana, Calopogon tuberosus, Cardamine pensylvanica, Carex comosa, Chamaecrista fasciculata, Cicuta maculata, Circaea lutetiana, Clematis socialis, Coreopsis linifolia, C. tripteris, Cornus obliqua, Cyperus haspan, Dichanthelium acuminatum, D. dichotomum, D. scoparium, Diodia virginiana, Doellingeria umbellata, Drosera brevifolia, D. capillaris, Dulichium arundinaceum, Eleocharis montevidensis, E. quadrangulata, E. tuberculosa, Eragrostis spectabilis, Eriocaulon decangulare, Eryngium integrifolium, Eupatorium leucolepis, E. perfoliatum, E. rotundifolium, Eutrochium fistulosum, Fimbristylis castanea, Fuirena breviseta, F. bushii, Galium obtusum, G. tinctorium, Gelsemium sempervirens, Helianthus angustifolius, Hydrocotyle umbellata, Hydrolea uniflora, Hymenocallis caroliniana, Hypericum crux-andreae, Hypoxis hirsuta, Ilex opaca, I. vomitoria, Ipomoea cordatotriloba, Iris hexagona, I. virginica, Itea virginica, Juncus marginatus, J. scirpoides, Leersia, Liatris pycnostachya, Lilaeopsis chinensis, Linum medium, Liquidambar styraciflua, Lobelia puberula, Ludwigia alata, L. alternifolia, L. hirtella, L. palustris, Lycopus rubellus, Lyonia ligustrina, Magnolia virginiana, Marshallia graminifolia, Melanthium virginicum, Mikania scandens, Mimulus alatus, Mitreola sessilifolia, Murdannia keisak, Myrica caroliniensis, M. cerifera, Nyssa biflora, Onoclea sensibilis, Osmunda regalis, Osmundastrum cinnamomeum, Oxypolis rigidior, Packera glabella, Panicum*

anceps, *P. verrucosum, P. virgatum, Parnassia grandifolia, Paspalum floridanum, P. laeve, P. praecox, Peltandra
virginica, Persea palustris, Persicaria hydropiperoides, P.
punctata, P. sagittata, Phanopyrum gymnocarpon, Phlox
carolina, Photinia pyrifolia, Phyla lanceolata, P. nodiflora,
Physostegia, Pinus palustris, P. taeda, Pityopsis graminifolia, Platanthera ciliaris, P. cristata, Pluchea, Pogonia ophioglossoides, Polygala cruciata, Pontederia cordata, Pteridium
aquilinum, Pterocaulon virgatum, Ptilimnium capillaceum,
Quercus alba, Q. michauxii, Q. palustris, Rhexia mariana,
R. petiolata, Rhododendron oblongifolium, Rhynchospora
chalarocephala, R. globularis R. glomerata, R. gracilenta,
R. inexpansa, R. microcarpa, R. mixta, R. rariflora, Rubus,
Sabatia gentianoides, Saccharum giganteum, Sagittaria
lancifolia, S. latifolia, Sarracenia alata, Saururus cernuus,
Schizachyrium scoparium, Schoenoplectus tabernaemon-
tani, Scirpus, Scleria ciliata, S. reticularis, S. triglomerata,
Scutellaria integrifolia, S. nervosa, Sium suave, Smilax laurifolia, Solidago patula, S. rugosa, Spartina patens, Sphagnum,
Spiranthes cernua, Stenanthium densum, Symphyotrichum
elliottii, S. lateriflorum, S. tenuifolium, Tephrosia onobrychoides, Toxicodendron vernix, Triadenum virginicum, T. walteri,
Typha angustifolia, T. latifolia, Ulmus rubra, Vaccinium fuscatum, Viburnum nudum, Vigna luteola, Viola primulifolia,
Woodwardia areolata, W. virginica, Xyris ambigua, X. baldwiniana, X. difformis, X. scabrifolia, X. torta, Zizania aquatica, Zizaniopsis milacea.

Ptilimnium ×texense J.M. **Coult. & Rose** occurs in bogs,
marshes, and swamps. The habitats are described as acidic.
Flowering occurs from July to August. Detailed life history
information is lacking for this taxon. **Reported associates:**
none reported, but presumably similar to those reported for
Ptilimnium costatum.

Use by wildlife: *Ptelimnium capillaceum* is a common food
of the Florida duck (Aves: Anatidae: *Anas fulvigula fulvigula*). It is a host plant of larval insects (Insecta) including
the black swallowtail butterfly (Lepidoptera: Papilionidae:
Papilio polyxenes), the celery leaftier moth (Lepidoptera:
Crambidae: *Udea rubigalis*), and water weevils (Coleoptera:
Curculionidae: *Lixellus filiformis, Pnigodes buchanani*).
It also is host to Basidiomycota fungi (Ceratobasidiaceae:
Thanatephorus cucumeris). The plants are known to harbor
the celery mosaic potyvirus (Potyviridae: *Potyvirus*). The
nectariferous flowers attract small flies (Insecta: Diptera;

e.g., Conopidae: *Physoconops weemsi*) and wasps (Insecta:
Hymenoptera).

Economic importance: food: none; **medicinal:** The leaves
of *P. capillaceum* are used in Santería medicine as a tea for
treating upset stomachs or gas pains; **cultivation:** Seeds of
Ptilimnium capillaceum and *P. costatum* occasionally are sold
to be grown as native garden ornamentals; **misc. products:**
none; **weeds:** *Ptilimnium capillaceum* has been reported as a
weed of celery seedbeds, moist cropland and wet waste areas;
nonindigenous species: none

Systematics: Transfer of the former *Ptilimnium nodosum* to
the genus *Harperella* (see earlier; Figures 5.116 and 5.122)
now renders *Ptilimnium* as monophyletic. DNA sequence
data analyses clearly place *Ptilimnium* in tribe Oenantheae
but only weakly support *Tiedemannia* as its sister group
(Figure 5.123). Other analyses indicate the sister group as
a clade comprising the two monotypic genera *Daucosma*
and *Limnosciadium*. All analyses indicate that the OBL
habit arose only once within *Ptilimnium* (Figure 5.123). The
recently described *P. ahlesii* is indistinguishable from *P. capillaceum* by a comparison of nrITS sequences; however, it is
maintained as a recently diverged species because of its consistently larger fruits and nonoverlapping reproductive phenology with the latter. Although *P. ×texense* was presumed to
have arisen as a hybrid between *P. capillaceum* and *P. nuttallii*, evidence from DNA analysis (Figure 5.123) resolves the
taxon with neither species, but within *P. costatum* as some
authors have suggested. There is some degree of genetic subdivision within *P. costatum* and the species (as well as the
status of *P. ×texense*) requires further study. The base chromosome number of *Ptilimnium* is $x = 7$ or 8. *Ptilimnium ahlesii* ($2n = 14$) is diploid; whereas, *P. capillaceum* ($2n = 14, 16,
28$) and *P. costatum* ($2n = 22, 32$) have multiple diploid and
tetraploid cytotypes.

Comments: *Ptilimnium capillaceum* and *P. costatum* occur
broadly throughout southeastern North America; whereas,
more restricted distributions characterize *P. ahlesii* (GA, NC,
SC) and *P. ×texense* (LA, TX).

References: Aziz & Sylvia, 1995; Baldwin & Mendelssohn,
1998a; Baskin et al., 1999b; Beckwith & Hosford, 1957; Bell,
1971; Burke, 1963; Camras, 2007; Collins & Wien, 1995; Cook,
1996a; Easley & Judd, 1990; Easterly, 1957; Feist & Downie,
2008; Flynn et al., 1995; Fox, 1914; Harshberger, 1909;
Howard & Wells, 2009; Landman et al., 2007; MacRoberts

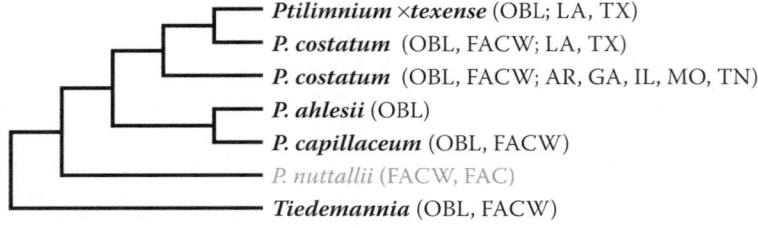

FIGURE 5.123 Phylogenetic relationships of *Ptilimnium* as indicated by nrITS sequence analysis. A single origin of the OBL habit (bold)
is indicated in *Ptilimnium*. These results refute the proposed origin of *P. ×texense* as a hybrid of *P. capillaceum* and *P. nuttallii*, but rather
support it as an element of *P. costatum* as suggested by Mathias & Constance (1961). (Adapted from Feist, M.A.E. & Downie, S.R., *Syst.
Bot.*, 33, 447–458, 2008.)

et al., 2001; Miller, 1931; Minchinton, 2006; Mohlenbrock, 1959a; Nichols, 1920; Orsenigo & Zitter, 1971; Osorio, 2009; Penfound & Hathaway, 1938; Porter-Utley, 1997; Toth, 2005; Usinger, 1956; Warner, 1926; Weakley, 2005; Weakley & Nesom, 2004; Wetzel et al., 2001; White & Simmons, 1988.

14. *Sium*

Waterparsnip; berle douce

Etymology: from *sion*, the Greek name for water parsnip

Synonyms: *Apium* (in part); *Cicuta* (in part); *Drepanophyllum* (in part)

Distribution: global: Asia; Europe; North America; **North America:** widespread

Diversity: global: 8 species; **North America:** 1 species

Indicators (USA): OBL: *Sium suave*

Habitat: brackish (coastal); freshwater; lacustrine, palustrine; **pH:** 4.6–8.0; **depth:** <1 m; **life-form(s):** submersed (rosulate), emergent herb

Key morphology: roots fusiform, clustered; stems (to 12 dm) erect, branched, corrugated, hollow; leaves (to 25 cm) 2–3-pinnate with narrow lobes or divisions, or once odd pinnate with 3–9 pairs of broad, serrate leaflets (to 4 cm), petioles (to 8 dm), segmented, sheathing; umbels (to 8 cm) compound, terminal, peduncled (to 10 cm), usually opposite a leaf, with up to 20 unequal rays (to 3 cm), the bracts (to 15 mm) and bractlets (to 3 mm) reflexed, leaf-like; flowers (10–20 per umbellet) pedicelled (to 5 mm), occasionally bilateral; calyx (to 2 mm) tooth-like, petals white, broad, narrowed apically; schizocarps (to 3 mm) ovoid, somewhat compressed, ribbed, corky, narrowly winged

Life history: duration: perennial (corms, rootstocks); **asexual reproduction:** roots; **pollination:** insect or geitonogamous; **sexual condition:** andromonoecious; **fruit:** schizocarps (common); **local dispersal:** water (mericarps); **long-distance dispersal:** water, animals (mericarps)

Imperilment: (1) *Sium suave* [G5]; S1 (AZ, KS, LA, TX); S2 (AZ, WY, <u>YT</u>); S3 (GA, KY, NC, WV)

Ecology: general: *Sium* is a characteristic genus of moist habitats worldwide. *Sium suave* occurs throughout North American wetlands where it displays an extremely broad range of ecological tolerances. The floral biology of *Sium* has been studied in detail for only a few species, but most appear to be temporally dioecious. The plants are andromonoecious, with the first umbels to open producing protandrous hermaphroditic flowers and the later umbels being mostly or entirely staminate. As the stigmas of the hermaphroditic flowers become receptive, they can be pollinated geitonogamously by pollen from the staminate flowers if any overlap occurs between sexual phases among umbels; otherwise the flowers are entomophilous. The corky mericarps are dispersed by water and also probably by their attachment to animal vectors.

Sium suave **Walter** inhabits bogs, borrow pits, brooks, channels, depressions, ditches, fens, flats, floodplains, gravel pits, marshes, meadows, mudflats, ponds, prairies, puddles, river bottoms, rivers, sandbars, savannas, sloughs, swales, swamps, vernal pools, and the margins of lakes, ponds, rivers, and streams at elevations of up to 2621 m. It occurs well into colder climates extending northward to at least latitude 66.7° N. The plants grow on saturated soils or in shallow (0.1–0.5 m deep) standing waters; they tolerate regular or permanent inundation at depths up to 0.2 m. Plants often are found in drying sites or where the waters are receding. Although occasionally reported from brackish sites (e.g., 1–9 ppt salinity), this species is found more often in freshwater habitats with low conductivity (92–1863 μmhos/ cm; mean: 672 μmhos/cm). Although plants can grow under brackish conditions, they exhibit reduced biomass when propagated under elevated sediment salt (NaCl) concentrations (0.25–1 ppt). Exposures range from full sun to shade. Substrates span a wide range of acidity (pH: 4.6–8.0) and include clay, clay loam, dolomite, humus, humus silt, loam, muck, mucky loamy sand, organic muck, peat, sand, sandy loam, silt, and silty mud of moderate nutrient content (e.g., 10.7 ppm NH_4 nitrogen). The growth of plants increases continually with the addition of NO_3–N, which also makes them more susceptible to displacement by competitors such as *Phalaris arundinacea* (Poaceae). The plants also occur in bog mats or in mats of floating vegetation. Sexual reproduction occurs primarily by entomophily. The umbels are protandrous, but the staminate flowers produce nectar several hours before the hermaphroditic flowers. The overlap between staminate and pistillate phases varies among umbels of different order, with the highest degree of overlap occurring among the secondary umbels. However, such overlap normally is quite low and the level geitonogamy is very limited. A fairly extensive seed bank with up to 150 seeds/m^2 can develop in natural wetlands; however, the plants do not appear to develop a seed bank in restored wetlands and occur much more frequently in natural wetlands. Although seed germination is not particularly high, the species reportedly is propagated readily by seed. In one study, seeds stratified at 4°C for 9 months and placed under a 20°C/30°C temperature regime, began to germinate after 6 days, achieving 18% total germination with a maximum rate of 0.15/day. The embryos contain high concentrations of furanocoumarins, which are known to have dormancy, germination deterrent, and antimicrobial properties. Higher seed germination occurs during drawdown conditions than when under submersed conditions, and populations ordinarily increase following natural drawdowns. Occurrences are highly associated with disturbance. One 4-year analysis indicated that a wetland oil spill had little effect on the plants. The foliage of this species can vary considerably, with early spring plants (which often occur in standing water) first producing highly dissected (bi-tripinnate) leaves, which then grade successively to pinnate leaves with fairly broad leaflets. Natural vegetative reproduction has not been described, but probably occurs at least occasionally by the fragmentation or dislodgement and dispersal of the clustered roots, which can be divided successfully to propagate cultivated plants. This species is potentially suitable for propagation by tissue culture. **Reported associates:** *Abies concolor, A. lasiocarpa, Acalypha rhomboidea, Acer negundo, A. rubrum, A. saccharinum, Achillea millefolium, Aconitum columbianum, Acorus calamus, Agrostis alba, A. oregonensis, A. scabra,*

A. stolonifera, Alisma subcordatum, A. triviale, Alnus rugosa, A. tenuifolia, Alopecurus aequalis, Amaranthus cannabinus, A. tuberculatus, Amblystegium serpens, Ambrosia artemisiifolia, A. trifida, Amorpha fruticosa, Anemone canadensis, Angelica atropurpurea, Apios americana, Arisaema dracontium, Arnoglossum plantagineum, Asclepias incarnata, Atriplex patula, Azolla mexicana, Beckmannia syzigachne, Betula lutea, B. nigra, Bidens cernuus, B. coronatus, B. frondosus, B. laevis, B. tripartitus, Bistorta bistortoides, Boehmeria cylindrica, Bolboschoenus fluviatilis, B. maritimus, B. robustus, Boltonia asteroides, Brachythecium mildeanum, Brasenia schreberi, Bryum, Calamagrostis canadensis, C. stricta, Calla palustris, Callitriche verna, Caltha palustris, Campanula aparinoides, Campylium stellatum, Cardamine bulbosa, C. pennsylvanica, Carex alata, C. aquatilis, C. atherodes, C. athrostachya, C. canescens, C. comosa, C. crinita, C. cryptolepis, C. diandra, C. grayi, C. gynandra, C. hormathodes, C. hyalinolepis, C. hystricina, C. intumescens, C. jonesii, C. lacustris, C. laeviconica, C. lasiocarpa, C. lenticularis, C. leptalea, C. lupuliformis, C. lupulina, C. lurida, C. lyngbyei, C. microptera, C. muskingumensis, C. nebrascensis, C. obnupta, C. oligosperma, C. paleacea, C. projecta, C. pseudocyperus, C. rostrata, C. salina, C. scoparia, C. stipata, C. stricta, C. tribuloides, C. tuckermanii, C. typhina, C. utriculata, C. viridula, C. vulpinoidea, Celtis occidentalis, Cephalanthus occidentalis, Ceratophyllum, Chamerion angustifolium, Chara, Chelone glabra, Cicuta bulbifera, C. maculata, Cinna arundinacea, Circaea lutetiana, Cirsium arvense, C. muticum, Cladium jamaicense, Clethra alnifolia, Climacium dendroides, Comarum palustre, Coptis trifolia, Cornus amomum, C. foemina, C. sericea, Corydalis caseana, Crataegus, Cryptotaenia canadensis, Cuphea viscosissima, Cuscuta compacta, C. cuspidata, Cyperus bipartitus, C. ferruginescens, C. filicinus, C. haspan, C. strigosus, Dactylis glomerata, Daucus carota, Decodon verticillatus, Deschampsia cespitosa, Dichanthelium acuminatum, Dichelyma uncinatum, Doellingeria umbellata, Drepanocladus aduncus, Dryopteris intermedia, Dulichium arundinaceum, Echinochloa walteri, Eleocharis acicularis, E. erythropoda, E. fallax, E. obtusa, E. olivacea, E. palustris, E. parishii, E. parvula, E. robbinsii, E. rostellata, E. uniglumis, Elymus trachycaulus, E. virginicus, Epilobium ciliatum, E. coloratum, E. leptophyllum, E. palustre, Equisetum arvense, E. fluviatale, Eragrostis frankii, E. hypnoides, Erechtites hieracifolius, Erigeron philadelphicus, Eriocaulon aquaticum, Eupatorium perfoliatum, E. serotinum, Euthamia graminifolia, Eutrochium maculatum, E. purpureum, Fagus grandifolia, Festuca rubra, Fragaria virginiana, Frangula alnus, Fraxinus nigra, F. pennsylvanica, F. profunda, Fritillaria camschatcensis, Galium aparine, G. boreale, G. brevipes, G. labradoricum, G. palustre, G. tinctorium, G. trifidum, G. triflorum, Geranium richardsonii, Geum macrophyllum, G. rivale, Glyceria borealis, G. grandis, G. elata, G. septentrionalis, G. striata, Gratiola aurea, Gymnocarpium dryopteris, Hamamelis, Helenium autumnale, H. flexuosum, Helianthus nuttallii, H. tuberosus, Heracleum sphondylium, Hibiscus moscheutos, Hippuris vulgaris, Hordeum jubatum, Hosackia oblongifolia, Hottonia inflata, Howellia aquatilis, Hydrocotyle umbellata, Hymenoxys hoopesii, Hypericum anagalloides, H. formosum, Hypnum lindbergii, Ilex verticillata, Impatiens capensis, Iris versicolor, I. virginica, Juncus balticus, J. brevicaudatus, J. canadensis, J. effusus, J. militaris, J. nevadensis, J. nodosus, J. pelocarpus, J. roemerianus, Kosteletzkya pentacarpos, Lactuca floridana, Laportea canadensis, Larix, Leersia lenticularis, L. oryzoides, Lemna minor, Leptobryum pyriforme, Leymus arenarius, Lobelia cardinalis, L. dortmanna, Lonicera involucrata, Ludwigia palustris, Lycopus americanus, L. asper, L. uniflorus, Lysimachia ciliata, L. lanceolata, L. nummularia, L. terrestris, L. thyrsiflora, Lythrum alatum, L. salicaria, Maianthemum canadense, M. stellatum, M. trifolium, Matteuccia struthiopteris, Mentha arvensis, M. piperita, Menyanthes trifoliata, Mertensia ciliata, Mikania scandens, Mimulus guttatus, M. ringens, Mitchella repens, Mitella nuda, Muhlenbergia asperifolia, M. sylvatica, Myosotis, Myrica cerifera, M. gale, Najas flexilis, Nitella flexilis, Nuphar polysepalum, N. variegata, Nymphaea odorata, Nymphoides cordata, Nyssa aquatica, Oenanthe sarmentosa, Onoclea sensibilis, Osmunda regalis, Osmundastrum cinnamomeum, Panicum capillare, P. dichotomiflorum, P. rigidulum, P. virgatum, Parthenocissus quinquefolia, Peltandra virginica, Penthorum sedoides, Persicaria amphibia, P. arifolia, P. coccinea, P. hydropiper, P. pensylvanica, P. maculosa, P. punctata, P. sagittata, Phalaris arundinacea, Phlox divaricatus, Phragmites australis, Phyla lanceolata, Physostegia virginiana, Picea engelmannii, P. pungens, Pilea fontana, P. pumila, Pinus strobus, Piperia elegans, Platanthera hyperborea, P. psycodes, P. sparsiflora, Pluchea carolinensis, Poa pratensis, Polygala polygama, Polypogon monspeliensis, Polystichum munitum, Pontederia cordata, Populus deltoides, P. fremontii, P. tremuloides, Portulaca oleracea, Potamogeton gramineus, P. natans, Potentilla anserina, P. egedii, P. norvegica, Primula pauciflora, Prunella vulgaris, Pseudotsuga, Pteridium aquilinum, Ptilimnium capillaceum, Pycnanthemum virginianum, Pyrola elliptica, Quercus bicolor, Ranunculus abortivus, R. flabellaris, R. gmelinii, R. lobbi, R. occidentalis, R. pensylvanicus, R. sceleratus, R. septentrionalis, Rhynchospora alba, R. macrostachya, Ribes lacustre, Ricciocarpus natans, Rorippa islandica, Rosa palustris, Rubus idaeus, R. parviflorus, R. pubescens, R. spectabilis, Rudbeckia laciniata, Rumex aquaticus, R. orbiculatus, R. salicifolius, R. verticillatus, Sagittaria lancifolia, S. latifolia, S. teres, Salix boothii, S. discolor, S. drummondiana, S. interior, S. lucida, S. nigra, S. petiolaris, S. planifolia, Sambucus nigra, S. racemosa, Samolus valerandi, Saururus cernuus, Schoenoplectus acutus, S. heterochaetus, S. pungens, S. subterminalis, S. tabernaemontani, S. torreyi, Scolochloa festucacea, Scirpus cyperinus, S. pedicellatus, S. pendulus, Scutellaria galericulata, S. lateriflora, Senecio congestus, S. triangularis, Sidalcea neomexicana, Silene nivea, Sisyrinchium angustifolium, Solanum dulcamara, Sparganium americanum, S. angustifolium, S. erectum, S. eurycarpum, S. glomeratum, S. natans, Spartina alterniflora, S. pectinata, Sphaerophysa salsula, Sphagnum, Spiraea alba, S. douglasii, Spiranthes diluvialis, Stachys albens, S.

palustris, S. tenuifolia, Stellaria longifolia, Symphyotrichum boreale, S. foliaceum, S. lanceolatum, S. lateriflorum, S. ontarione, S. puniceum, S. racemosum, S. subulatum, Symplocarpus foetidus, Taraxacum officinale, Thelypteris palustris, Thermopsis montana, Thuja occidentalis, Torreyochloa pallida, Toxicodendron radicans, Triadenum virginicum, Triantha glutinosa, Triglochin maritimum, T. palustre, Tsuga canadensis, Typha angustifolia, T. latifolia, Ulmus americana, U. rubra, Urtica dioica, Utricularia macrorhiza, U. minor, Veratrum californicum, V. tenuipetalum, Veronica americana, V. scutellata, Viburnum dentatum, Vicia americana, Viola cucullata, V. macloskeyi, Vitis riparia, Woodwardia virginica, Xanthium strumarium, Zanthoxylum americanum, Zizania aquatica.

Use by wildlife: *Sium suave* is eaten by muskrats (Mammalia: Cricetidae: *Ondatra zibethicus*), particularly during summer, and can be a preferred food item of white-tailed deer (Mammalia: Cervidae: *Odocoileus virginianus*). It is found in trace amounts in the diets of various shoal-water ducks. When growing in floating mats (e.g., with *Menyanthes*), the plants can provide nesting habitat for waterfowl such as the lesser scaup (Aves: Anatidae: *Aythya affinis*) and greater scaup (*A. marila*). The foliage contains linear furanocoumarins, which can be phototoxic to nonspecialized insect (Insecta) herbivores. However, most insects associated with the plants appear to be generalists. The flowers are visited by a variety of flies (Diptera: Tabanidae: *Stonemyia rasa, S. tranquilla, Tabanus novaescotiae, T. quinquevittatus, T. trispilus, T. typhus*) and Lepidoptera (e.g., Nymphalidae: *Libythea carinenta*), which forage for nectar. At least 38 species of wasps (Hymenoptera: Braconidae; Chrysididae; Eucoilidae; Ichneumonidae; Tiphiidae; Pteromalidae) also have been collected from the plants. *Sium suave* is a larval host plant for Lepidoptera in three families including Noctuidae (*Papaipema birdi*), Oecophoridae (*Agonopterix clemensella, A. nervosa, Depressaria cinereocostella*) and Papilionidae (*Papilio polyxenes*). The latter is known to be unaffected by the presence of the linear furanocoumarins in the foliage. *Sium suave* is a host of the northern root-knot nematode (Nemata: Heteroderidae: *Meloidogyne hapla*) and also for the carrot weevil (Insecta: Coleoptera: Curculionidae: *Listronotus oregonensis*), which is a major pest of carrots in northeastern North America. The plants also host various leaf and stem spot fungi including Ascomycota (Botryosphaeriaceae: *Phyllosticta*; Mycosphaerellaceae: *Cercospora sii, Septoria sii*; Phaeosphaeriaceae: *Stagonospora*), Basidiomycota (Pucciniaceae: *Uromyces americanus, U. lineolatus*), and Blastocladiomycota (Physodermataceae: *Physoderma palustris, P. vagans*). The Thompson people fed the rootstocks of *Sium suave* to their cattle as fodder. The plants contain 7.5%–10.7% crude protein and 49.3%–50.9% digestible dry matter.

Economic importance: food: The Shuswap considered the flowers of *Sium suave* to be poisonous, but the roots and tubers were eaten by the Algonquin, Bella Coola, Carrier, Cree, Interior Salish, Okanagan-Colville, Quebec, Shuswap, Thompson, Wet'suwet'en, and Woodlands communities. The Klamath and Montana Indian tribes made a relish from

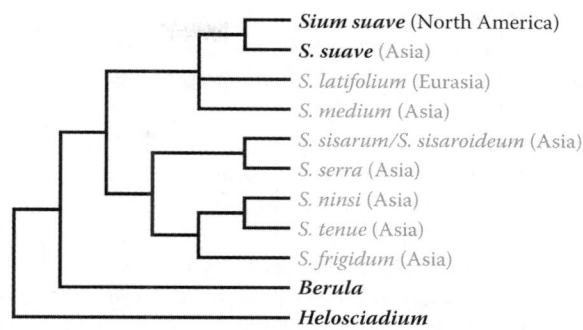

FIGURE 5.124 Phylogenetic relationships of *Sium* based on analyses of nrITS sequences. North American taxa with OBL representatives are indicated in bold; however, all *Sium* species inhabit wet sites and probably represent a single origin of the habit for the genus. (Adapted from Spalik, K. & Downie, S.R., *Amer. J. Bot.*, 93, 747–761, 2006.)

the foliage. The plants are eaten in parts of Mexico, where they are known as *Berro palmita*. The related, Old World skirret (*Sium sisarum*) is cultivated widely as a root crop. Although *Sium suave* generally is regarded as nonpoisonous, the roots do contain small quantities of toxic polyacetylenes, and it might be prudent to avoid them as a foodstuff, especially given that it can be confused with the highly toxic *Cicuta*; **medicinal:** The roots of *Sium suave* were used by the Iroquois to prepare a poultice for treating pain and in a compound decoction used to alleviate epilepsy. The Lakota administered the roots for stomach ailments; **cultivation:** *Sium suave* is planted occasionally as a water garden ornamental; **misc. products:** Toy whistles were made from the stems of *Sium suave* by Lakota children; **weeds:** none; **nonindigenous species:** none

Systematics: Molecular phylogenetic analyses have compelled the transfer of several former *Sium* species (*S. bracteatum, S. burchellii, S. repandum*) to *Berula* (Figures 5.112, 5.113), in order to circumscribe both *Berula* and *Sium* as monophyletic sister genera. As a result, *Sium* currently contains 8–9 species, which occur throughout the Nearctic and Palearctic regions (Figure 5.124). Because all *Sium* species are associated with wetlands, a single origin of the OBL habit is assumed for the group. *Sium suave* is the only New World native but it is related most closely to the central Asian *S. medium* and Eurasian *S. latifolium* (Figure 5.124). Evidently, the North American populations of *S. suave* are derived from Asian sources and a genetic survey of this widespread species might provide an interesting perspective of its biogeographical history. *Sium carsonii* (also ranked as OBL) has been distinguished from *S. suave* on the basis of its overall weaker habit, but it is doubtfully more than just an aquatic growth form of the latter and is not treated here as a distinct species. The base chromosome number of *Sium* is $x = 5$–6. *Sium suave* ($2n = 12$) is a diploid with corroborating counts obtained across North America. An anomalous report of $2n = 22$ is due to the misidentification of the source material, which actually was *Cicuta bulbifera*.

Comments: *Sium suave* occurs throughout North America and extends throughout northern portions of Asia.

References: Bélair & Benoit, 1996; Bell, 2004b; Berenbaum, 1981, 2009; Blickle, 1955; Boivin, 1999; Brockett, 1970; Burk, 1977; Crawford & Hartman, 1972; Crowden et al., 1969; Cruden & Hermann-Parker, 1977; Dawe et al., 2000; Dionne, 1976; Fassett, 1921; Field & Philipp, 2000; Galatowitsch & van der Valk, 1996a,b; Girardin et al., 2001; Gottesfeld, 1994; Green & Galatowitsch, 2002; Hubbard & Boe, 1988; Kawahara, 2006; Knuth, 1906, 1908, 1909; Leck & Simpson, 1987; Locky et al., 2005; Majak et al., 2008; Miklovic & Galatowitsch, 2005; Mullin et al., 1997; Murashige, 1974; Nichols, 1975; Roman et al., 2001; Shipley & Parent, 1991; Shull, 1905; Skinner & Telfer, 1974; Sletten & Larson, 1984; Spalik & Downie, 2006; Speichert & Speichert, 2004; Stollberg, 1950; Takos, 1947; Tooker & Hanks, 2000; Van der Valk & Davis, 1978; Walker & Wehrhahn, 1971; Walker et al., 2005; Walton & Lakela, 1995; Wilken, 1970; Zobel & March, 1993.

15. *Tiedemannia*

Water dropwort

Etymology: after Friedrich Tiedemann (1781–1861)

Synonyms: *Oenanthe* (in part); *Oxypolis* (in part); *Peucedanum* (in part)

Distribution: global: North America; **North America:** southeastern

Diversity: global: 2 species; **North America:** 2 species

Indicators (USA): OBL: *Tiedemannia canbyi*, *T. filiformis*; **FACW:** *T. filiformis*

Habitat: freshwater; lacustrine, palustrine; **pH:** acidic; **depth:** <1 m; **life-form(s):** emergent herb

Key morphology: stems (to 2 m) erect, fluted or striate, purplish to green; leaves (to 12.9 cm) linear subulate, terete, septate, petiolar base (to 6.1 cm) clasping; umbels (to 15 cm) compound, peduncled (to 10 cm), 7–12 rays (to 3 cm), bracts (to 2.5 cm) linear setaceous or lance attenuate; flowers (to 1.5 mm) 25-numerous, bisexual or unisexual, pedicelled (to 1 cm), bractlets (to 10 mm) linear attenuate/subulate; sepals (to 0.5 mm) triangular, green to reddish; petals (to 1.3 mm) white or garnet maroon, short clawed, incurved, concave; hypanthium (to 0.6 mm) greenish pink; filaments (to 1.5 mm) white or garnet maroon; schizocarp (to 9 mm) ovate or elliptic, compressed, margins corky winged

Life history: duration: perennial (rhizomes); **asexual reproduction:** rhizomes; **pollination:** insect or geitonogamous; **sexual condition:** andromonoecious; **fruit:** schizocarps (common); **local dispersal:** rhizomes; fruits (water); **long-distance dispersal:** fruits (water)

Imperilment: (1) *Tiedemannia canbyi* [G2]; SX (DE); S1 (MD, NC); S2 (GA, SC); (2) *T. filiformis* [G5]; S3 (NC)

Ecology: general: Both members of this small genus inhabit wetlands, with one species (*T. filiformis*) more sporadically so (OBL, FACW). The life history of these plants is poorly known and much work is needed to elucidate accurately their reproductive and seed ecology. The inflorescences are trimonoecious (polygamous), with individual flowers being either bisexual, pistillate or staminate. Their pollination system

(presumably insect) has not been studied in any detail. The flattened fruits have a corky wing and presumably float, which would allow for at least some degree of water dispersal.

***Tiedemannia canbyi* (J.M. Coult. & Rose) Feist & S.R. Downie** grows in Carolina bays, ponds, savannahs, sloughs, and swamps. Ironically, the type locality was a moist "UPL" meadow, a habitat not associated with extant occurrences. The plants thrive under conditions of infrequent and shallow (5–30 cm) inundation, and in sites with sparse canopy cover. Regular fire frequency appears to be a critical element of the habitat with individuals in some populations exhibiting a marked increase following wildfires. The substrates are either loam (e.g., Grady, McColl, Rembert, Portsmouth) or sandy loam (e.g., Coxville, Rains) underlain by clay. They are described as acidic and contain medium to high amounts of organic matter. The bisexual flowers (appearing from late June to October) are protandrous and in some umbels may produce inner florets that are strictly male or outer florets that are strictly female. Although a mixed mating system is indicated by this unusual sexual condition, the extent of outcrossing has not been determined empirically. Sexual reproduction can be greatly diminished (by up to 17%) as a result of herbivory from larval insects (Insecta: Lepidoptera: Papilionidae: *Papilio polyxenes*), which regularly sever the stems directly below the umbels. This is a highly clonal species and most reproduction occurs vegetatively, either by extension of the stolon-like rhizomes (which can attain 1 m in length), or by more local rooting of decumbent stems. Other aspects of the reproductive and seed ecology of this species require further study. **Reported associates:** *Agalinis linifolia, Andropogon, Aristida palustris, Carex striata, Centella asiatica, Cephalanthus occidentalis, Cladium mariscoides, Clethra alnifolia, Diospyros virginiana, Eleocharis melanocarpa, E. microcarpa, Harperella nodosa, Hypericum adpressum, H. denticulataum, Ilex amelanchier, I. myrtifolia, Iris, Lachnanthes caroliana, Liquidambar styraciflua, Lobelia boykinii, Ludwigia sphaerocarpa, Manisuris rugosa, Myrica cerifera, Nymphaea odorata, Nyssa biflora, Panicum hemitomon, P. rigidulum, P. verrucosum, Pinus serotina, Pluchea rosea, Polygala cymosa, Pontederia cordata, Proserpinaca pectinata, Rhexia aristosa, Rhynchospora inundata, R. tracyi, Saccharum baldwinii, S. giganteum, Sagittaria engelmanniana, Sarracenia flava, Stillingia aquatica, Taxodium ascendens, Triadenum virginicum, Utricularia geminiscapa, Woodwardia virginica.*

***Tiedemannia filiformis* (Walter) Feist & S.R. Downie** inhabits shallow bogs, borrow pits, canals, depressions, ditches, flats, flatwoods, glades, marshes, meadows, ponds, pools, prairies, roadsides, savannahs, seeps, sinkholes, swamps, and the margins of lakes, ponds, rivers, and streams at low elevations (e.g., 5–17 m). Occurrences include freshwater to brackish sites with exposures ranging from full sublight to moderate shade. The substrates usually are described as acidic (pH: 5.4–5.6) and include clay, clay loam, muck, mud, peat, sand, sandy loam, or silt loam (Atmore series); however, alkaline marl substrates also have been reported. Flowering extends from late May to September. The seeds are dispersed by water, in which they reach their maximum entrainment in June and September. Vegetative reproduction occurs by means of caudex offsets. **Reported associates:** *Acer rubrum, Agalinis linifolia, A. purpurea, Aletris aurea, Amphicarpum muhlenbergianum, Amsonia tabernaemontana, Andropogon arctatus, A. brachystachyus, A. glomeratus, A. gyrans, A. virginicus, Anthaenantia rufa, Aristida palustris, A. purpurascens, A. stricta, Arnoglossum ovatum, Arundinaria tecta, Asclepias lanceolata, Axonopus furcatus, Bacopa caroliniana, Balduina uniflora, Bidens mitis, Bigelowia nudata, Calamovilfa curtissii, Carex exilis, C. glaucescens, C. verrucosa, Carphephorus carnosus, C. pseudoliatris, Cassytha filiformis, Centella asiatica, Cephalanthus occidentalis, Chaptalia tomentosa, Cirsium horridulum, Cladium jamaicense, C. mariscoides, Cleistesiopsis divaricata, Cliftonia monophylla, Coleataenia longifolia, C. tenera, Coreopsis floridana, C. gladiata, C. nudata, Crataegus opaca, Crinum americanum, Ctenium aromaticum, Cyperus haspan, C. sanguinolentus, Cyrilla racemiflora, Dichanthelium acuminatum, D. nudicaule, D. scabriusculum, D. scoparium, D. sphaerocarpon, Diodella teres, Diodia virginiana, Diospyros virginiana, Drosera brevifolia, D. intermedia, Eleocharis confervoides, E. elongata, E. equisetoides, E. quadrangulata, E. tuberculosa, Eragrostis elliottii, Erigeron vernus, Eriocaulon compressum, E. decangulare, E. nigrobracteatum, Eupatorium mohrii, Eurybia eryngiifolia, Euthamia graminifolia, Fuirena bushii, F. scirpoidea, F. squarrosa, Gratiola brevifolia, G. pilosa, Gymnadeniopsis integra, Helianthus angustifolius, H. heterophyllus, Hibiscus moscheutos, Hydrolea ovata, Hymenocallis palmeri, H. tridentata, Hypericum brachyphyllum, H. fasciculatum, H. myrtifolium, Hypoxis juncea, Ilex glabra, I. myrtifolia, Iris tridentata, Juncus effusus, J. polycephalus, Lachnanthes caroliniana, Lachnocaulon digynum, Leersia hexandra, Liatris spicata, Lilium catesbaei, Linum medium, Lobelia floridana, L. glandulosa, Lophiola aurea, Ludwigia linearis, L. pilosa, L. simpsonii, L. sphaerocarpa, L. suffruticosa, Lycopodiella appressa, Lycopus rubellus, Magnolia virginiana, Marshallia graminifolia, Mikania batatifolia, Mnesithea rugosa, Muhlenbergia capillaris, M. expansa, Myrica heterophylla, Nyssa biflora, Panicum hemitomon, P. virgatum, Paspalum monostachyum, P. praecox, P. urvillei, Persea palustris, Pinguicula planifolia, Pinus elliottii, P. palustris, P. taeda, Piriqueta cistoides, Pleea tenuifolia, Pluchea baccharis, Pogonia ophioglossoides, Polygala cymosa, P. lutea, P. ramosa, P. rugelii, Proserpinaca palustris, P. pectinata, Quercus, Rhexia alifanus, R. lutea, R. mariana, R. nashii, R. nuttallii, R. petiolata, R. virginica, Rhynchospora careyana, R. cephalantha, R. chalarocephala, R. chapmanii, R. ciliaris, R. colorata, R. corniculata, R. divergens, R. elliottii, R. fascicularis, R. filifolia, R. galeana, R. globularis, R. harperi, R. inundata, R. latifolia, R. macra, R. microcarpa, R. mixta, R. nitens, R. oligantha, R. perplexa, R. pineticola, R. plumosa, R. pusilla, R. tracyi, Rubus, Rudbeckia texana, Sabatia decandra, S. grandiflora, Sacciolepis indica, S. striata, Sagittaria graminea, Sarracenia alata, S. flava, S. leucophylla, S. psittacina, Schizachyrium scoparium, Schoenus nigricans, Scleria pauciflora, S. verticillata, Sisyrinchium angustifolium, S. rhizomatum, S. scoparium, Scleria baldwinii, S. muehlenbergii, Setaria parviflora, Smilax laurifolia, Solidago fistulosa, S. stricta, Spermacoce prostrata, Sphagneticola trilobata, Sphagnum, Spiranthes laciniata, S. longilabris, Sporobolus teretifolius, Stillingia, Styrax americanus, Symphyotrichum dumosum, S. subulatum, S. tenuifolium, Taxodium ascendens, Toxicodendron, Trianthia racemosa, Tridens strictus, Tripsacum dactyloides, Utricularia purpurea, U. subulata, Vaccinium corymbosum, Viola lanceolata, Woodwardia virginica, Xyris ambigua, X. baldwiniana, X. difformis, X. elliottii, X. fimbriata, X. flabelliformis, X. laxifolia, X. platylepis, X. serotina, X. stricta, Zigadenus glaberrimus.*

Use by wildlife: *Tiedemannia canbyi* is a host plant for insect larvae (Insecta: Lepidoptera: Papillionidae: *Papilio polyxenes*) and is eaten by scale insects (Hemiptera: Coccoidea) and grasshoppers (Arthropoda: Orthoptera). The foliage is browsed on by deer (Mammalia: Cervidae: *Odocoileus*). *Tiedemannia filiformis* is a preferred larval host plant of the black swallowtail butterfly (Insecta: Lepidoptera: Papillionidae: *Papilio polyxenes*). Its flowers are visited by plasterer bees (Insecta: Hymenoptera: Colletidae: *Hylaeus confluens*).

Economic importance: food: not edible; **medicinal:** *Tiedemannia filiformis* was used as a medicinal plant by the native people of Florida; **cultivation:** none; **misc. products:** none; **weeds:** none; **nonindigenous species:** none

Systematics: Refined phylogenetic insights resulting from analyses of DNA sequence data have warranted the split of *Tiedemannia* from *Oxypolis*, where both species of the former genus had been assigned previously. The current generic circumscriptions now separate those species with compound leaves (*Oxypolis*) from those that are strictly "rachis-leaved" (*Tiedemannia*), where the foliage develops in a fashion similar to that of *Lilaeopsis*. The two resulting groups (*Oxypolis, Tiedemannia*) are surprisingly distant phylogenetically (Figure 5.125). The sister genus of *Tiedemannia* is *Ptilimnium*, which consists entirely of OBL species (see earlier); thus, the habit probably has arisen only once in these clades. *Tiedemannia greenmanii* previously was treated as a distinct species, but molecular and other data (Figure 5.125) now indicate that it should be recognized as a subspecies of *T. filiformis*, which is the convention followed here. The basic chromosome number of *Tiedemannia* is $x = 7$. Both *T. canbyi* and *T. filiformis* are tetraploids ($2n = 28$), with *T. filiformis* subsp. *greenmanii* (Mathias & Constance) Feist & S.R. Downie having larger chromosomes than *T. filiformis* subsp. *filiformis*.

Comments: Both *Tiedemannia canbyi* and *T. filiformis* are restricted to the southeastern United States.

References: Bartholomew et al., 2006; Belden et al., 1988; Bell & Constance, 1960; Boone et al., 1984; Chafin, 2007; Downie et al., 2008; Eleuterius & Caldwell, 1984; Feist & Downie, 2008; Feist et al., 2012; Godfrey & Wooten, 1981; Hessl & Spakman, 1995; Hinman & Brewer, 2007; Hutton, 2010; Judd, 1982; Kaplan, 1970; Loveless, 1959; Monette &

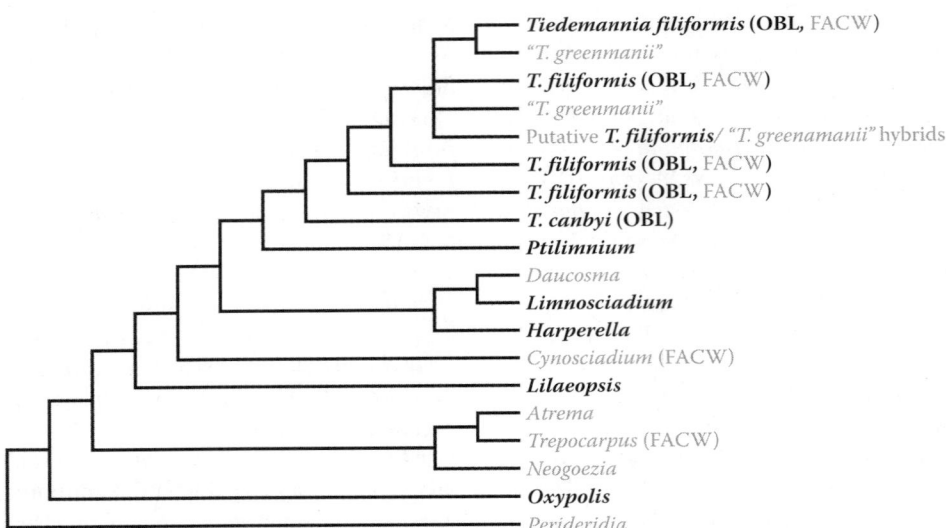

FIGURE 5.125 Relationships of *Tiedemannia* based on analysis of nrITS sequence data. Taxa containing OBL North American species are indicated in bold. (Adapted from Feist, M.A.E. & Downie, S.R., *Syst. Bot.*, 33, 447–458, 2008.)

Markwith, 2012; Murdock & Rayner, 1990; Orzell & Bridges, 1993; Palmer, 1919; Sims, 2007; Tucker et al., 1983; Walker & Peet, 1984.

16. *Trepocarpus*

Whitenymph

Etymology: from the Greek *trepo karpus* meaning "turning fruit" (of uncertain reference)

Synonyms: none

Distribution: global: North America; **North America:** south central United States

Diversity: global: 1 species; **North America:** 1 species

Indicators (USA): OBL; FACW: *Trepocarpus aethusae*

Habitat: freshwater; palustrine; **pH:** alkaline; **depth:** <1 m; **life-form(s):** emergent herb

Key morphology: stems (to 7 dm) erect, with few ascending branches, taproot slender; leaves (to 10 cm) short petioled, pinnately decompound, the divisions (<1 mm wide) flat, linear; umbels compound, peduncled (to 6 cm), 2–4 rays (to 1.5 cm), bracts (to 10 mm) linear subulate; flowers (to 2 mm) 2–8 per umbelette, bisexual; sepals (to 2 mm) subulate, persistent; petals white, broadly obovate; anthers yellow; styles very short, the stylopodium short-conic; schizocarp (to 10 mm) linear oblong, glabrous, flattened laterally, strong ribbed.

Life history: duration: annual (seeds); **asexual reproduction:** none; **pollination:** insect; **sexual condition:** hermaphroditic; **fruit:** schizocarps (common); **local dispersal:** fruits (water); **long-distance dispersal:** fruits (water)

Imperilment: (1) *Trepocarpus aethusae* [G4/G5]; S1 (MO, SC); S2 (GA); S3 (KY)

Ecology: general: *Trepocarpus* is monotypic.

Trepocarpus aethusae **Nutt. ex DC** is a winter annual, which grows in borrow pits, bottomlands, clearings, deltas, depressions, ditches, floodplains, hammocks, levees, meadows, prairies, river bottoms, roadsides, and along the margins of lakes, rivers, streams, swales, swamps, and thickets at low elevations. Most occurrences are found in partial shade (just above the zone of frequent flooding), but the plants also can tolerate full sunlight. The substrates typically are calcareous and include clay, loamy sand, and sand. The plants flower under both short and long days, but flower more rapidly under the latter conditions. Flowering occurs from April to August, with the fruits produced mainly in August. The flowers are protandrous and likely are pollinated by insects (Insecta), although no specific pollinators have been identified. As a winter annual, the seeds germinate in autumn but the plants do not bolt, flower, or set seed until the following growing season. The seeds are morphophysiologically dormant. Their physiological dormancy can be broken by warm stratification, which is followed by embryo growth. Subsequent germination will occur in light (68%–73%) or in darkness (52%–65%) under 15°C/6°C, 20°C/10°C, or 25°C/15°C temperature regimes. Fewer than 12% of the seeds normally will reenter secondary (physiological) dormancy. However, *T. aethusae* forms a long-lived persistent seed bank in which the seeds can live for at least 8 years. Delayed germination occurs due to delays in the loss of physiological dormancy. The seeds mature in mid to late August and early September. The mericarps float in the water and remain buoyant for at least 1 week. They are relatively heavy (~13.7 mg) and principally dispersed by water. This species has become rare in the flora and is adapted to periodic disturbances, such as the conditions that are characteristic of floodplains. **Reported associates:** *Acer rubrum, A. saccharinum, Allium vineale, Amorpha fruticosa, Andropogon virginicus, Aristolochia tomentosa, Asclepias perennis, Boehmeria cylindrica, Campsis radicans, Carex caroliniana, C. muskingumensis, C. typhina, Carya aquatica, Celtis laevigata, Chaerophyllum procumbens, Chasmanthium, Cicuta maculata, Clematis crispa, Crataegus mollis, Crinum americanum, Danthonia spicata, Daucus carota, Dioclea multiflora, Diospyros virginiana, Elymus virginicus, Fraxinus pennsylvanica, Geranium* *carolinianum, Helianthus tuberosus, Hydrocotyle, Ilex vomitoria, Iris hexagona, Juniperus virginiana, Kummerowia striata, Leitneria floridana, Lespedeza stuevei, Liquidambar styraciflua, Lonicera japonica, Panicum, Parthenocissus quinquefolia, Platanus occidentalis, Populus, Prunus serotina, Quercus lyrata, Q. nigra, Q. pagoda, Q. palustris, Q. phellos, Q. shumardii, Q. texana, Rhus copallina, Rhynchospora colorata, Robinia pseudoacacia, Rosa multiflora, Rubus allegheniensis, Sabal palmetto, Salix interior, Samolus, Schedonorus pratensis, Solidago canadensis, S. gigantea, S. nemoralis, Symphoricarpos orbiculatus, Taxodium distichum, Toxicodendron radicans, Tridens flavus, Ulmus alata, U. americana, Valeriana radiata, Vitis.*

Use by wildlife: none reported.

Economic importance: food: none reported; **medicinal:** none reported; **cultivation:** none; **misc. products:** *Trepocarpus* plants emit an odor that has been described as turpentine-like; **weeds:** *Trepocarpus aethusae* has been described as weedy in disturbed sites in prairies; **nonindigenous species:** none

Systematics: The monotypic *Trepocarpus* is assigned to tribe Oenantheae, where it resolves phylogenetically as the sister group to *Bifora* (Figure 5.112). There are no known reports of chromosome counts or hybrids for the genus.

Comments: *Trepocarpus aethusae* is endemic to the south-central United States.

References: Baskin et al., 2003; Harper, 1932; Schlessman & Barrie, 2004; Warner, 1926; Wilm & Taft, 1998.

Family 2: Araliaceae [43]

The circumscription of Araliaceae (1450 species) has been amended substantially as ongoing phylogenetic investigations continue to propose refinements to the classification of the family (Judd et al., 2008). The alteration most relevant to this treatment is the inclusion of *Hydrocotyle* (and *Trachymene*), which formerly had been treated either as members of Apiaceae subfamily Hydrocotyloideae or were placed within a distinct family (Hydrocotylaceae Bercht. & J. Presl). The transfer of *Hydrocotyle* and *Trachymene* is warranted by repeated studies (Plunkett et al., 1996a,b, 1997; Downie et al., 1998, 2000; Plunkett & Lowry, 2001; Chandler & Plunkett, 2004; Plunkett et al., 2004a,b; Nicolas & Plunkett, 2009), which place these genera within the Araliaceae clade in contrast to other genera of the former subfamily, which resolve primarily within the subfamilies Azorelloideae or Mackinlayoideae of Apiaceae (Plunkett et al., 2004b). Some authors now recognize subfamily Hydrocotyloideae (with *Hydrocotyle* and *Trachymene*) as a subdivision of Araliaceae, with the remainder of species assigned to subfamily Aralioideae (Judd et al., 2008). The family itself remains closely related to Apiaceae, Myodocarpaceae, and Pittosporaceae (Figures 5.108 and 5.126).

Araliaceae are distributed mainly throughout the humid tropics, but also extend into the temperate regions as well as into seasonally dry areas (Frodin, 2004). Despite its considerable size and diversity (trees, shrubs, and herbs), there are no aquatic plants in the family other than in *Hydrocotyle*.

Features of "core" Araliaceae include stipules, bisexual and radial epigynous flowers with broadly inserted petals (without

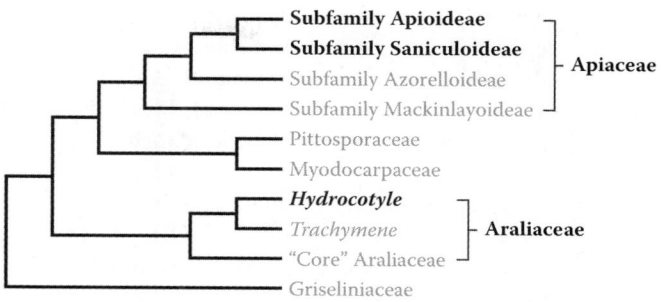

FIGURE 5.126 Relationships of Araliaceae and Apiaceae based on combined 26S and cpDNA data. North American taxa containing OBL species are indicated in bold. (Adapted from Chandler, G. & Plunkett, G.M., *Bot. J. Linn. Soc.*, 144, 123–147, 2004.)

inflexed apices) and globose drupes or berries consisting of 2–5 carpels; *Hydrocotyle* also lacks features commonly associated with Apiaceae such as a carpophore, compound umbels, and vittae and possesses valvate (rather than imbricate) aestivation (Chandler & Plunkett, 2004). The flowers of Araliaceae are small, usually aggregated, and are pollinated by various insects (Insecta: e.g., Coleoptera, Diptera, Hymenoptera, and Lepidoptera). The fleshy fruits most often are dispersed by birds (Aves) (Judd et al., 2008).

The family includes several medicinally noteworthy plants such as ginseng (*Panax ginseng*) and sarsaparilla (*Aralia*) as well as the ornamental English ivy (*Hedera*) and umbrella plant (*Schefflera*).

There is only one North American genus within the family in which OBL species occur:

1. *Hydrocotyle* L.

1. Hydrocotyle

Marsh pennywort; Écuelle d'eau

Etymology: from the Greek *hudro kotulê* ("water cup"), in reference to the leaf morphology

Synonyms: *Centella* (in part)

Distribution: global: cosmopolitan; **North America:** widespread (except far north and mountains)

Diversity: global: 130 species; **North America:** 11 species

Indicators (USA): OBL: *Hydrocotyle americana, H. ranunculoides, H. umbellata, H. verticillata*

Habitat: freshwater; lacustrine, palustrine, riverine; **pH:** 4.1–9.4; **depth:** <1 m; **life-form(s):** emergent herb, floating leaved

Key morphology: stems (<1 to several meters) delicate or coarse, creeping or floating, stoloniferous, rooting at nodes; leaves (to 8 cm) reniform or suborbicular, shallowly to deeply 3–18-cleft, the margins crenate or crenulate, petioles (1–40 cm) attached at leaf base (in sinus) or centrally peltate, stipules scale-like; umbels (to 3 cm) axillary, essentially sessile or peduncled (to 35 cm), spherical or hemispherical, simple, or in interrupted spikes (to 17 cm), or in 2–6 verticillate clusters; flowers (few–100) small, 5-merous, sessile or pedicelled (to 25 mm); calyx absent; petals (to 1.8 mm) ovate, whitish; schizocarps (to 3 mm) flattened laterally, broader than long, ribbed, or smooth

Life history: duration: perennial (stolons, rhizomes, tubers); **asexual reproduction:** stolons, rhizomes, shoot fragments, tubers; **pollination:** insect or self; **sexual condition:** hermaphroditic; **fruit:** schizocarps (common); **local dispersal:** rhizomes/stolons; seeds (water); **long-distance dispersal:** seeds (animals, water)

Imperilment: (1) *Hydrocotyle americana* [G5]; SH (AR, NF); S1 (IN, KY, SC, TN); S3 (MN, NC, OH, PE, QC); (2) *H. ranunculoides* [G5]; SH (BC); S1 (IL, KY, MO, NJ, NY); S2 (KS, WA, WV); S3 (GA, NC); (3) *H. umbellata* [G5]; SH (PA); S1 (CT, NS, OH); S3 (NY); (4) *H. verticillata* [G5]; SH (RI); S1 (CT, MO, NJ, NY, OR, UT); S2 (MA)

Ecology: general: All *Hydrocotyle* species inhabit wet or at least very moist sites (*H. bonariensis* Comm. ex Lam. and *H. pusilla* A. Rich. are categorized as OBL but only in the Carribean region; consequently they have been excluded from this treatment). The floral biology of this genus has been elucidated from a relatively limited number of observations. The flowers are minute and are very weakly protandrous, which results predominantly in autogamous self-pollination as the stigmas become receptive by the time the last-maturing anthers discharge their pollen. Otherwise, pollination occurs by "promiscuous" unspecialized insects such as flies (Insecta: Diptera: Muscidae). Occasionally, the flowers can become unisexual (female) by degradation of the anthers. The seeds are dispersed by water (some remaining buoyant for weeks), or are carried along with floating, fruit-laden plants that are transported by water currents. The seeds cannot withstand prolonged immersion in saltwater. Most of the species reproduce vegetatively by stoloniferous stems (sometimes described as rhizomes), which root along the nodes. The degree of stolon elongation is directly proportional to the fertility of the substrate. The plants also disperse by floating vegetative fragments, which are transported along with river drift. A few species are monocarpic ephemerals that persist from a seed bank. Some seeds are known to remain viable for at least 12 years. The seeds often constitute a large proportion of agricultural soil seed banks. They are eaten by livestock and can be dispersed along with their dung.

Hydrocotyle americana L. occurs in bogs, brooks, carrs, depressions, ditches, fens, floodplains, marshes, meadows, seeps, spray cliffs, springs, swamps, and along the margins of lakes, ponds, rivers, streams, and waterfalls at elevations of up to 610 m. The plants grow in deep shade to full sun but are associated mostly with damp, shady localities and often with disturbance. Substrates include muck, rock, sand, and sandstone. They are relatively acidic (pH: 5.7–7.2; 112–218 ppm alkalinity), fairly infertile (0.05–0.13 ppm phosphorous) and contain varied amounts of organic matter (19%–85%). The pollination biology has not been investigated. Seeds kept at 4°C for 14 weeks will germinate at greenhouse temperatures of 22°C–30°C. Germination also has been successful using seeds stratified for 4–8 weeks and then sown (moist) at 21°C. The seeds occur in some seed banks at fairly high frequency (e.g., 12% of plots in one study). Vegetative reproduction occurs by means of elongate stolons, which produce terminal tubers. The roots are colonized by arbuscular mycorrhizal and dark septate endophytic fungi. The plants are known to occur in sites contaminated by heavy metals (Cu, Pb, Zn). Some occurrences (notably seepages) can be particularly rich in bryophytes and ferns. **Reported associates:** *Abies balsamea, Acer rubrum, A. saccharinum, A. saccharum, Adiantum pedatum, Alnus rugosa, A. serrulata, Angelica atropurpurea, Apocynum androsaemifolium, Aralia nudicaulis, A. racemosa, Arisaema triphyllum, Asarum canadense, Asclepias incarnata, A. syriaca, Asplenium monanthes, A. montanum, A. trichomanes, Athyrium filix-femina, Atrichum oerstedianum, A. undulatum, Bazzania denudata, Berberis thunbergii, Betula alleghaniensis, B. papyrifera, B. pumila, Bidens connatus, B. pilosus, Boykinia aconitifolia, Brachythecium rivulare, Bryhnia novae-angliae, Bryocrumia vivicolor, Bryum, Calamagrostis canadensis, Calla palustris, Calycanthus floridus, Caltha palustris, Campanula aparinoides, Cardamine bulbosa, C. pensylvanica, Carex biltmoreana, C. bromoides, C. crinita, C. folliculata, C. gynandra, C. leptalea, C. prasina, C. retrorsa, C. scabrata, C. stricta, C. torta, Chamaedaphne calyculata, Chelone glabra, Chrysogonum virginianum, Chrysosplenium americanum, Cicuta bulbifera, Cinna arundinacea, C. latifolia, Circaea alpina, C. lutetiana, Clematis virginiana, Climacium americanum, Clintonia borealis, Conocephalum conicum, Coptis trifolia, Cornus alternifolia, C. canadensis, Cynodon dactylon, Cyperus esculentus, Cypripedium acaule, C. kentuckiense, Cystopteris protrusa, Dalibarda repens, Dasiphora floribunda, Dennstaedtia punctilobula, Deparia acrostichoides, Desmodium paniculatum, Dichanthelium ensifolium, Dichodontium pellucidum, Dicranum fulvum, Diphyscium cumberlandianum, Drosera rotundifolia, Dryopteris carthusiana, D. intermedia, Dumortiera hirsuta, Epilobium ciliatum, Equisetum arvense, E. fluviatile, Eupatorium perfoliatum, Eurybia saxicastellii, Eutrochium fistulosum, E. purpureum, Fagus grandifolia, Fissidens osmundioides, Floerkea proserpinacoides, Fontinalis sphagnifolia, Fragaria virginiana, Fraxinus nigra, Galax urceolata, Galium asprellum, G. kamtschaticum, G. palustre, G. triflorum, Gaultheria procumbens, Gentiana andrewsii, G. pennelliana, Geum aleppicum, G. canadense, G. rivale, Glyceria canadensis, G. melicaria, G. striata, Grammitis nimbata, Gratiola viscidula, Gymnocarpium disjunctum, Helenium autumnale, Heuchera parviflora, H. villosa, Houstonia serpyllifolia, Huperzia lucidula, H. porophila, Hymenophyllum tayloriae, Hyophila involuta, Hypericum mutilum, H. perforatum, Hypnum, Ilex verticillata, Impatiens capensis, I. pallida, Iris versicolor, Isoetes engelmannii, Juglans cinerea, Laportea canadensis, Larix laricina, Lathyrus palustris, Leersia oryzoides, Lobelia amoena, Lonicera canadensis, Ludwigia palustris, Lycopodium annotinum, Lycopus uniflorus, L. virginicus, Maianthemum canadense, Mentha arvensis, Menyanthes trifoliata, Micranthes careyana, M. caroliniana, M. pensylvanica, Microstegium vimineum, Mimulus ringens, Mitchella repens, Mitella nuda, Mnium hornum, M. marginatum, Myosotis scorpioides, Myrica gale, Oncophorus raui, Onoclea sensibilis, Osmorhiza claytonii, Osmunda claytoniana, Osmundastrum cinnamomeum, Oxalis montana, O.*

stricta, Packera aurea, P. schweinitziana, Palamocladium leskeoides, Paspalum notatum, Pedicularis lanceolata, Pellia appalachiana, P. epiphylla, P. neesiana, Persicaria sagittata, Phalaris arundinacea, Phegopteris connectilis, Philonotis fontana, Phyla nodiflora, Physocarpus opulifolius, Picea rubens, Pilea pumila, Pinus strobus, Plagiochila austini, P. caduciloba, P. sharpii, P. sullivantii, Plagiomnium affine, P. carolinianum, P. ciliare, Plantago major, Platanthera psycodes, Platylomella lescurii, Poa paludigena, Polemonium vanbruntiae, Polypodium virginianum, Polystichum acrostichoides, Populus tremuloides, Prunella vulgaris, Pseudotaxiphyllum distichaceum, Quercus rubra, Ranunculus abortivus, R. hispidus, Rhexia virginica, Rhizomnium appalachianum, R. punctatum, Ribes americanum, Riccardia multifida, Rosa carolina, Rubus fruticosus, R. hispidus, R. pubescens, Rudbeckia laciniata, Rumex verticillatus, Salix alba, S. lucida, Sambucus racemosa, Scirpus atrovirens, Scutellaria lateriflora, Selaginella apoda, Sesbania herbacea, Solanum dulcamara, Solidago canadensis, S. patula, Sonchus asper, Sphagnum angustifolium, S. girgensohnii, S. quinquefarium, Sphenopholis pensylvanica, Spiraea alba, Steerecleus serrulatus, Stenotaphrum secundatum, Streptopus lanceolatus, Sullivantia sullivantii, Symphyotrichum prenanthoides, Symplocarpus foetidus, Thalictrum clavatum, T. pubescens, Thamnobryum alleghaniense, Thelypteris noveboracensis, T. palustris, Thuidium delicatulum, Thuja occidenatlis, Tiarella cordifolia, Tilia americana, Toxicodendron vernix, Tradescantia ohiensis, Trautvetteria carolinensis, Triadenum virginicum, Trientalis borealis, Trichomanes boschianum, T. intricatum, T. petersii, Tsuga canadensis, Ulmus americana, Veratrum viride, Verbena rigida, Veronica americana, Viburnum lantanoides, V. lentago, Viburnum nudum, Viola blanda, V. cucullata, V. lanceolata, V. rotundifolia, V. sororia, Vittaria appalachiana, Zanthoxylum americanum.

Hydrocotyle ranunculoides L. f. grows in freshwater arroyos, brooks, canals, ciénagas, ditches, dunes, fens, flats, floodplains, marshes, meadows, ponds, pools, river bottoms, sinks, sloughs, streams, swales, swamps, and along the margins of lakes, ponds, reservoirs, and rivers at elevations of up to 1440 m. The plants do not tolerate salinity and decline rapidly at levels of 3.5‰ or higher. They often root along shorelines and then extend their floating stems across shallow water areas (up to 1.6 m but usually <50 cm deep), often forming dense mats by proliferation of the stolons; they dominate in the "pioneer zone" of the mat community. The plants can maintain their growth under drained conditions, but grow optimally in waterlogged sites; they readily recover from complete submergence. The substrates are somewhat alkaline (pH: 6.2–8.6; e.g., 280 ppm total alkalinity; 18 ppm $CaCO_3$) and can consist of clay, gravel, mud, peaty mud, or sand. The plants can grow nearly year round (tolerating temperatures from 0°C to 30°C) and produce flowers from March to November. Their pollination biology has not been studied. The peduncles recurve when fruits are mature, directing the seeds toward the substrate. Higher seed production occurs when plants are rooted on shore, but the species is not well represented in the seed bank. Seed germination occurs

in June; the specific requirements for germination have not been determined. Floating plants frequently reproduce vegetatively. Floating seeds and plant fragments are dispersed by water, animals, or by movement of substrates containing such propagules. Floating, fruit-bearing plant fragments are known to disperse substantial quantities of seeds. The plants grow quickly under optimal conditions and can achieve a biomass as high as 532 g of dry weight per m^2 at growth rates of up to 0.132 g g^{-1} dry wt d^{-1}. **Reported associates:** *Acer saccharinum, Ageratina jucunda, Alisma subcordatum, A. triviale, Alnus rubra, Alternanthera philoxeroides, Amaranthus australis, Andropogon glomeratus, Angelica genuflexa, Astragalus pomonensis, Athyrium filix-femina, Azolla caroliniana, Baccharis salicifolia, Bidens cernuus, B. connatus, B. discoideus, B. laevis, Boehmeria cylindrica, Cabomba caroliniana, Calamagrostis canadensis, Callitriche stagnalis, Carex alata, C. aquatilis, C. atlantica, C. buxbaumii, C. comosa, C. conoidea, C. hyalinolepis, C. interrupta, C. lacustris, C. lanuginosa, C. lasiocarpa, C. lyngbyei, C. obnupta, C. prairea, C. seorsa, C. suberecta, C. tetanica, Ceratophyllum demersum, Cicuta maculata, Cirsium arvense, C. tioganum, Comarum palustre, Commelina diffusa, Cotula coronopifolia, Cyperus esculentus, C. haspan, C. iria, C. niger, C. odoratus, C. strigosus, Darlingtonia californica, Deschampsia cespitosa, Digitaria bicornis, D. ciliaris, Dulichium arundinaceum, Echinochloa crus-galli, Echinodorus berteroi, E. cordifolius, Eclipta prostrata, Egeria densa, Eichhornia crassipes, Eleocharis acicularis, E. baldwinii, E. erythropoda, E. intermedia, E. obtusa, Eupatorium capillifolium, E. compositifolium, Fimbristylis autumnalis, Frangula purshiana, Fraxinus velutina, Fuirena pumila, Galium aparine, G. trifidum, Gamochaeta falcata, Glyceria septentrionalis, Gnaphalium palustre, Habenaria repens, Hibiscus moscheutos, Hierochloe odorata, Hydrilla verticillata, Hydrocotyle umbellata, Hypericum anagalloides, Ipomoea trifida, Iris pseudacorus, Juncus balticus, J. effusus, J. falcatus, J. marginatus, J. megacephalus, J. scirpoides, Landoltia punctata, Leersia hexandra, L. oryzoides, Lemna minor, L. perpusilla, L. valdiviana, Lilaeopsis, Limnobium spongia, Liquidambar styraciflua, Lobelia kalmii, Lonicera involucrata, L. japonica, Lotus corniculatus, Ludwigia glandulosa, L. leptocarpa, L. microcarpa, L. palustris, L. peploides, L. peruviana, L. repens, L. uruguayensis, Luziola fluitans, Lycopus americanus, L. uniflorus, L. virginicus, Lysichiton americanus, Lysimachia quadriflora, L. terrestris, L. thyrsiflora, Lythrum alatum, L. portula, L. salicaria, Malus fusca, Melothria pendula, Mentha arvensis, Mikania scandens, Mimulus guttatus, Muhlenbergia rigens, Murdannia keisak, M. nudiflora, Myriophyllum aquaticum, M. pinnatum, M. spicatum, Najas guadalupensis, Nasturtium officinale, Nelumbo lutea, Nicotiana glauca, Nuphar polysepala, Nuttallanthus canadensis, Nymphaea odorata, Nyssa biflora, Oenanthe sarmentosa, Osmunda regalis, Oxalis, Oxycaryum cubense, Packera glabella, Panicum anceps, P. capillare, P. hemitomon, P. repens, Parnassia grandifolia, Paspalum distichum, P. fluitans, P. repens, Pedicularis lanceolata, Pentodon pentandrus, Persicaria amphibia,*

P. arifolia, P. glabra, P. hydropiperoides, P. lapathifolia, P. maculosa, P. pennsylvanica, P. punctata, Phalaris arundinacea, Phyllanthus urinaria, Picea sitchensis, Pinus contorta, Polystichum munitum, Pontederia cordata, Populus deltoides, P. fremontii, Potamogeton natans, P. nodosus, Potentilla pacifica, Proserpinaca palustris, Ptilimnium capillaceum, Rhynchospora, Ricciocarpus natans, Rubus argutus, R. discolor, Rumex pulcher, Saccharum giganteum, Sacciolepis striata, Sagittaria graminea, S. lancifolia, S. latifolia, S. subulata, Salix gooddingii, S. lasiolepis, Salvinia minima, S. molesta, S. rotundifolia, Sambucus nigra, Saururus cernuus, Schoenoplectus acutus, S. americanus, S. tabernaemontani, Scirpus cyperinus, Scutellaria galericulata, S. lateriflora, Senna obtusifolia, Sesbania herbacea, Sorghum halepense, Sparganium americanum, S. angustifolium, Spiraea douglasii, Spirodela polyrhiza, Stachys palustris, S. tenuifolia, Symphyotrichum subulatum, Taxodium distichum, Torreyochloa pallida, Triadenum walteri, Typha domingensis, T. latifolia, Ulmus rubra, U. serotina, Urtica dioica, Utricularia gibba, Verbena hastata, V. scabra, Veronica americana, V. anagallis-aquatica, Vitis, Wolffia columbiana, Wolffiella gladiata, Xyris, Zizaniopsis miliacea.

***Hydrocotyle umbellata* L.** inhabits beaches, bogs, canals, deltas, depressions, ditches, dunes, flatwoods, levees, marshes, meadows, oxbows, savannahs, seeps, sinkholes, sloughs, swales, swamps, and the margins of lakes, ponds, rivers, and streams at elevations of up to 1371 m. The plants originate on shorelines where they creep along the ground and often dominate on newly exposed substrates. Eventually they can form mats by extending their stoloniferous stems across the surface of adjacent waters, by as much as 15 m during a single season. The plants are adapted to a broad range of freshwater habitats and also to brackish sites (to 18 ppt salinity) that are under some tidal influence. They can survive in full to partial sunlight and can colonize waters at depths to 1 m. A broad ecological tolerance is indicated by the wide reported range of acidity (pH: 4.1–9.4; reported means: 6.4–6.9). Mean values for other variables such as alkalinity (27 ppm $CaCO_3$), conductance (174 µS/cm), total phosphorous (45 µg/L) and total nitrogen (880 µg/L) indicate a preference for low alkalinity and nutrient levels. The range of substrate types is diverse and includes gravel, muck, mud, sand, sandy loam, sandy peat, and peat (which can contain up to 95% organic matter). The flowers are protandrous and are produced from April to November. There is virtually nothing known regarding the pollination biology; however, the flowers probably are highly autogamous. Viable seeds are not produced in the north of the range (e.g., Nova Scotia). Otherwise, the seeds germinate in June. The plants do persist in the seed bank (except in northern areas) and their germination success is higher for seeds under flooded conditions than simply moist conditions. Despite a very high rate of oxygen transport (3.49 g O_2 kg^{-1} dry root mass per hour), the plants do not tolerate submergence well and occur preferentially where water temperatures are not excessively warm (<42°C). Under optimal conditions, the plants can attain a biomass of 41 tons (dry wt)/ha year (at densities of 250–650 g dry wt/m^2) and growth rates of up to 18.3 g/dry wt/m^2. Increased

aqueous nitrogen concentrations (up to 20 ppm) will result in increased biomass. Growth rates are nearly twice as high on dry ground than under flooded conditions. Flooded plants allocate extra resources to petiole growth, which enables them to maintain their leaf blades on the water surface. Flooding (15–30 cm depth) also results in reduced biomass, leaf number, and reproductive resource allocation as well as increased adaxial stomate density. Unseasonable flooding in natural populations and late establishment can result in reduced biomass, and the plants thrive where the shorelines are exposed from 49 to 71 days or more. The plants also achieve a higher biomass (but lower root-to-shoot ratio) when grown experimentally with water hyacinth (*Eichhornia crassipes*) compared to a monoculture. However, under these conditions their shorter roots were less efficient as storage organs for nitrogen and phosphorus. This species mainly reproduces vegetatively by its rhizomatous stolons. When the plants are grown on dry soils they can produce underground tubers. The foliage contains triterpenoid glycosides, which are thought to be allelopathic to some species. **Reported associates:** *Acer rubrum, Acorus calamus, Agalinis purpurea, Agrostis perennans, Alternanthera philoxeroides, Ammania, Ampelopsis arborea, Andropogon virginicus, Annona glabra, Baccharis halimifolia, B. salicifolia, Bacopa caroliniana, B. monnieri, Bidens coronatus, B. laevis, Boehmeria cylindrica, Brasenia schreberi, Briza minor, Callitriche heterophylla, Calystegia sepium, Carex cryptolepis, C. glaucescens, C. longii, C. lupulina, C. pumila, C. stricta, C. verrucosa, Centella asiatica, Cephalanthus occidentalis, Chara, Chasmanthium laxum, Chrysopsis godfreyi, Cicuta maculata, Cirsium arvense, Cladium jamaicense, C. mariscoides, Coreopsis rosea, Cryptocoryne, Cyperus bipartitus, C. esculentus, C. odoratus, C. pseudovegetus, C. retrorsus, C. strigosus, C. virens, Decodon verticillatus, Diodia virginiana, Distichlis spicata, Dulichium arundinaceum, Eclipta prostrata, Eichhornia crassipes, Eleocharis acicularis, E. albida, E. baldwinii, E. cellulosa, E. elongata, E. equisetoides, E. melanocarpa, E. obtusa, E. olivacea, E. palustris, E. parvula, E. quadrangulata, E. robbinsii, E. rostellata, Erianthus compactus, Eriocaulon aquaticum, E. compressum, Eryngium prostratum, Eupatorium capillifolium, E. perfoliatum, Euthamia tenuifolia, Fimbristylis autumnalis, F. castanea, Fraxinus velutina, Fuirena pumila, F. squarrosa, Galactia microphylla, Galium obtusum, Gratiola aurea, Habenaria repens, Heterotheca subaxillaris, Hibiscus moscheutos, Hydrocotyle ranunculoides, H. verticillata, Hydrolea quadrivalvis, Hypericum adpressum, Ilex, Impatiens pallida, Ipomoea imperati, Iris prismatica, Isoetes tuckermanii, Itea virginica, Iva imbricata, Juncus canadensis, J. coriaceus, J. effusus, J. elliottii, J. filiformis, J. marginatus, J. militaris, J. pelocarpus, J. repens, J. scirpoides, J. tenuis, Juniperus virginiana, Leersia hexandra, L. oryzoides, L. virginica, Lemna minor, L. valdiviana, Leptochloa, Lilaeopsis chinensis, Lindernia dubia, Lipocarpha micrantha, Lobelia dortmanna, Ludwigia arcuata, L. decurrens, L. palustris, L. peploides, L. peruviana, L. repens, Luziola fluitans, Lycopus americanus, L. amplectens, L. rubellus, L. uniflorus, Lysimachia terrestris,*

Mikania scandens, Monarda punctata, Myrica cerifera, Myriophyllum aquaticum, M. heterophyllum, M. pinnatum, Nuphar advena, N. orbiculata, Nymphaea odorata, Nymphoides aquatica, N. cordata, Nyssa aquatica, N. sylvatica, Osmunda regalis, Panicum amarum, P. hemitomon, P. repens, P. verrucosum, Paspalidium geminatum, Paspalum, Peltandra virginica, Persea borbonia, Persicaria coccinea, P. glabra, P. hydropiperoides, P. pensylvanica, P. punctata, Phragmites australis, Phyla lanceolata, P. nodiflora, Pinus echinata, P. elliottii, P. palustris, Plantago, Platanus racemosa, Pluchea carolinensis, P. foetida, Poa annua, Pontederia cordata, Populus fremontii, Potamogeton diversifolius, P. gramineus, P. illinoensis, P. natans, P. nodosus, P. robbinsii, Prunus serotina, Pteridium aquilinum, Ptilimnium capillaceum, Quercus nigra, Q. virginiana, Ranunculus reptans, Rhynchospora corniculata, R. glomerata, R. inexpansa, R. scirpoides, Riccia fluitans, Ricciocarpus natans, Rorippa palustris, Sabal palmetto, Sabatia kennedyana, S. stellaris, Saccharum giganteum, Sacciolepis striata, Sagittaria lancifolia, S. latifolia, S. subulata, S. teres, Salix caroliniana, S. laevigata, S. lasiolepis, S. nigra, Samolus valerandi, Sarracenia, Saururus cernuus, Schizachyrium maritimum, Schoenoplectus americanus, S. heterochaetus, S. pungens, S. smithii, Scirpus cyperinus, Sesuvium portulacastrum, Sium suave, Smilax walteri, Sonchus oleraceus, Sparganium americanum, Spartina alterniflora, S. cynosuroides, S. patens, Spiraea alba, Sporobolus virginicus, Stachys hyssopifolia, Stuckenia pectinata, Symphyotrichum, Taraxacum officinale, Taxodium ascendens, T. distichum, Thelypteris palustris, Toxicodendron radicans, Triadenum virginicum, T. walteri, Trifolium dubium, Typha angustifolia, Uniola paniculata, Urena lobata, Utricularia foliosa, U. inflata, U. purpurea, Vallisneria americana, Verbena, Wisteria frutescens, Woodwardia areolata, Xyris smalliana, Zizania aquatica, Zizaniopsis miliacea.

***Hydrocotyle verticillata* Thunb.** is found in backwaters, bayheads, beaches, bogs, canals, cliffs, depressions, ditches, dunes, flats, flatwoods, floodplains, levees, marshes, pools, prairies, seeps, sloughs, springs, streambeds, swales, swamps, and along the margins of lakes, ponds, rivers, sinkholes, and streams at elevations of up to 3205 m. Colonized sites span the spectrum from full sun to shade and range from fresh to brackish (to 12 ppt salinity) waters of fairly broad acidity (pH: 5.0–8.0) at temperatures below 25°C; lower temperatures (10°C–15°C) are tolerated for short times. Typically, the plants grow on exposed, wet substrates but spread readily across open water surfaces by the formation of mats, which can reach 60 cm in thickness, and easily support the weight of an adult human. The shoots do not tolerate prolonged submergence. Reported substrates include clay, clay loam, gravel, loam, muck, mucky peat, mud, peat, rock, sand (granitic), sandy clay, sandy loam, sandy peat, silt, silty loam, and silty sand. The reproductive biology has not been studied in any detail and is not well characterized. Cultivated material is difficult to propagate sexually. These plants are dispersed by their buoyant seeds, which can remain afloat for several weeks. Small seed banks (1 seed/m²) have been reported,

with germination rates reportedly higher when located in drained sites than under flooded conditions. Reproduction is predominantly vegetative and occurs by means of the stoloniferous stems, which can be propagated using relatively small fragments. This species has a poor tolerance for "overwash" disturbance, i.e., the deposition of sand due to storm activity. **Reported associates:** *Acacia greggii, Acalypha rhomboidea, Acer rubrum, Agrostis exarata, A. gigantea, Alnus serrulata, Alternanthera philoxeroides, Amelanchier canadensis, Amaranthus cannabinus, Ambrosia artemisiifolia, A. psilostachya, Amorpha fruticosa, Ampelopsis arborea, Andropogon gerardii, A. glomeratus, Anemopsis californica, Apios americana, Apium graveolens, Arisaema dracontium, Artemisia douglasiana, Asclepias incarnata, A. perennis, Asplenium platyneuron, Atriplex, Baccharis halimifolia, B. salicifolia, Bacopa monnieri, Berchemia scandens, Berula erecta, Bidens discoideus, Boehmeria cylindrica, Brunnichia ovata, Callitriche heterophylla, Campsis radicans, Cardamine pensylvanica, Cardiospermum halicacabum, Carex atlantica, C. bromoides, C. caroliniana, C. cherokeensis, C. comosa, C. grayi, C. hormathodes, C. intumescens, C. joorii, C. leptalea, C. lupulina, C. oxylepis, C. retroflexa, C. seorsa, C. stipata, C. stricta, Carya, Celtis laevigata, Cenchrus tribuloides, Centella asiatica, C. erecta, Cephalanthus occidentalis, Chasmanthium laxum, Chilopsis linearis, Chloracantha spinosa, Chrysothamnus, Cicuta maculata, Cinna arundinacea, Cirsium muticum, C. vulgare, Cladium jamaicense, Clematis crispa, Commelina communis, C. diffusa, C. virginica, Conium maculatum, Cornus foemina, Crataegus viridis, Crinum americanum, Cryptotaenia canadensis, Cynodon dactylon, Cynosciadium digitatum, Cyperus erythrorhizos, C. filicinus, C. haspan, C. odoratus, Decodon verticillatus, Dichanthelium acuminatum, D. dichotomum, Dichondra micrantha, Diodia virginiana, Distichlis spicata, Dryopteris celsa, Dulichium arundinaceum, Echinochloa walteri, Eichhornia crassipes, Eleocharis acicularis, E. compressa, E. fallax, E. geniculata, E. montevidensis, E. obtusa, E. rostellata, E. tortilis, Elymus virginicus, Eragrostis hypnoides, Eriocaulon koernickianum, Erodium cicutarium, Eupatorium capillifolium, E. serotinum, Fimbristylis autumnalis, F. perpusilla, F. puberula, Fraxinus caroliniana, F. pennsylvanica, F. velutina, Fuirena simplex, Galium aparine, G. obtusum, G. tinctorium, G. trifidum, Geranium carolinianum, Geum canadense, Glyceria septentrionalis, Gratiola aurea, G. virginiana, Helenium flexuosum, Helianthus annuus, Hibiscus moscheutos, Hydrilla verticillata, Hydrocotyle umbellata, Hymenoclea monogyra, Hymenocallis rotata, Hypericum mutilum, Ilex decidua, Illicium, Impatiens capensis, Iris hexagona, Isoetes engelmannii, Isolepis cernua, Itea virginica, Iva frutescens, Juncus balticus, J. biflorus, J. bufonius, J. coriaceus, J. diffusissimus, J. effusus, J. polycephalus, J. repens, J. roemerianus, J. scirpoides, J. texanus, J. torreyi, Justicia ovata, Kosteletzkya pentacarpos, Leersia lenticularis, L. oryzoides, Leitneria floridana, Lemna minor, Leucothoe racemosa, Lilaeopsis masonii, L. occidentalis, Limnobium spongia, Limosella subulata, Lindera benzoin, Lindernia dubia, Lipocarpha* *micrantha, Liquidambar styraciflua, Lobelia cardinalis, Lonicera japonica, Ludwigia leptocarpa, L. palustris, L. peploides, Luziola fluitans, Lycopus rubellus, L. virginicus, Lygodium japonicum, Lyonia lucida, L. ligustrina, Lythrum lineare, Macbridea caroliniana, Malvaviscus penduliflorus, Marsilea mutica, M. vestita, Melia azedarach, Melothria pendula, Mentha spicata, Mentzelia, Micranthemum umbrosum, Mikania scandens, Mimulus glabratus, M. guttatus, Mitchella repens, Mitreola petiolata, Morus alba, M. microphylla, Muhlenbergia asperifolia, M. lindheimeri, Murdannia keisak, Myosotis macrosperma, Myrica cerifera, Myriophyllum aquaticum, M. pinnatum, Najas marina, Nasturtium officinale, Nicotiana glauca, Nuphar, Nymphaea odorata, Nyssa aquatica, N. biflora, N. sylvatica, Oenanthe sarmentosa, Onoclea sensibilis, Oplismenus hirtellus, Osmunda regalis, Osmundastrum cinnamomeum, Packera glabella, Palafoxia arida, Panicum hemitomon, P. repens, P. virgatum, Parthenocissus quinquefolia, Paspalum urvillei, Peltandra virginica, Persicaria hydropiperoides, P. pensylvanica, P. punctata, P. sagittata, P. setacea, P. virginiana, Phanopyrum gymnocarpon, Phragmites australis, Phyla lanceolata, Physostegia leptophylla, Pilea fontana, P. pumila, Pinus elliottii, Planera aquatica, Platanthera clavellata, Platanus racemosa, Pluchea camphorata, P. foetida, P. odorata, Poa annua, Polypogon monspeliensis, Polypremum procumbens, Pontederia cordata, Populus fremontii, Proserpinaca palustris, Prosopis velutina, Prunella vulgaris, Ptilimnium capillaceum, Quercus agrifolia, Q. lyrata, Q. texana, Ranunculus recurvatus, R. sceleratus, Rhynchospora caduca, R. colorata, R. corniculata, R. miliacea, R. nivea, Rubus argutus, R. trivialis, Rumex crispus, R. verticillatus, Sabal minor, S. palmetto, Sabatia kennedyana, S. stellaris, Saccharum giganteum, Sagittaria fasciculata, S. lancifolia, Salix caroliniana, S. gooddingii, S. nigra, Salsola, Salvia lyrata, Salvinia molesta, Sambucus nigra, Samolus valerandi, Sapindus saponaria, Saururus cernuus, Schoenoplectus acutus, S. americanus, S. californicus, S. pungens, Scutellaria lateriflora, Serenoa repens, Silybum marianum, Smilax bona-nox, S. laurifolia, S. tamnoides, Solidago patula, Sparganium eurycarpum, Spartina alterniflora, S. cynosuroides, S. patens, Sphagnum, Sphenopholis pensylvanica, Spiranthes cernua, Spirodela polyrhiza, Steinchisma hians, Stellaria prostrata, Styrax americanus, Symphyotrichum lentum, Tamarix chinensis, Taxodium distichum, Tillandsia usneoides, Toxicodendron radicans, Triadenum walteri, Triadica sebifera, Tripsacum dactyloides, Typha angustifolia, T. domingensis, T. latifolia, Ulmus americana, U. crassifolia, Urtica chamaedryoides, Utricularia, Vaccinium, Veronica anagallis-aquatica, V. peregrina, Viola affinis, Vitis arizonica, V. cinerea, V. mustangensis, Woodsia obtusa, Woodwardia areolata, W. fimbriata, W. virginica, Xanthium strumarium, Zannichellia palustris, Zizaniopsis miliacea.*

Use by wildlife: Experiments have shown that larvae of the black swallow butterfly (Insecta: Lepidoptera: Papilionidae: *Papilio polyxenes*) will starve when provided only with *Hydrocotyle americana* as a food source. *Hydrocotyle ranunculoides* contains oleanane triterpane and ursane glycosides,

which are known to deter feeding by some insects (Insecta). However, it is fed on readily by a host-specific weevil (Coleoptera: Curculionidae: *Lixellus elongatus*). The plants also are hosts of bacterial wilt (Proteobacteria: Ralstoniaceae: *Ralstonia solanacearum*), which is a serious disease of tomato (*Solanum lycopersicum*). *Hydrocotyle ranunculoides* is an associate of planktonic Crustacea (Isopoda: Sphaeromatidae: *Gnorimosphaeroma oregonensis*; Amphipoda: Corophiidae: *Corophium spinicorne*). The seeds and foliage of *Hydrocotyle umbellata* are eaten by ducks (Aves: Anatidae) and the foliage by Florida marsh rabbits (Mammalia: Leporidae: *Sylvilagus palustris paludicola*). The tubers contain 8.6% crude protein, 0.8% fat, and 88% carbohydrates. The herbage of the plants contains 9.4%–9.8% crude protein. *Hydrocotyle umbellata* is a host of larval insects (Insecta: Lepidoptera: Noctuidae: *Enigmogramma basigera*; Papilionidae: *Papilio polyxenes*), serpentine leaf miner flies (Insecta: Diptera: Agromyzidae: *Liriomyza sativae*), spider mites (Acarina: Tetranychidae: *Tetranychus tumidus*), and wasps (Insecta: Hymenoptera: Sphecidae: *Oxybelus fulvipes*). The foliage is fed upon somewhat by the glassy winged locust (Insecta: Orthoptera: Acrididae: *Stenacris vitreipennis*). The roots provide habitat for amphipods (Crustacea: Amphipoda: *Crangonyx floridanus*) and isopods (Crustacea: Isopoda: *Caecidotea racovitzai*). This species also is a host of the parasitic angiosperm *Orobanche ramosa* (Orobanchaceae). The plants are used as nesting materials by a South American catfish (Osteichthyes: Callichthyidae: *Hoplosternum littorale*), which has been introduced recently to Florida. *Hydrocotyle verticillata* is the host of a leaf miner fly (Insecta: Diptera: Agromyzidae: *Liriomyza minutiseta*). *Hydrocotyle* species are hosts to a fair number of leaf spot and gall Fungi. All four of the OBL species are fungal hosts for Ascomycota (Mycosphaerellaceae: *Cercospora hydrocotyles*) and Basidiomycota (Pucciniaceae: *Puccinia hydrocotyles*). *Hydrocotyle ranunculoides* and *H. umbellata* also host a member of Blastocladiomycota (Physodermataceae: *Physoderma hydrocotylidis*). In addition, *H. americana* hosts several other Ascomycota (Erysiphaceae: *Erysiphe heraclei, E. pisi*; *Septoria pallidula, S. spegazzinii*; Sclerotiniaceae: *Sclerotinia minor*), and Chytridiomycota (Synchytriaceae: *Synchytrium aureum*); *H. ranunculoides* other Basidiomycota (Entylomataceae: *Entyloma fimbriatum, E. hydrocotyles*); and *H. umbellata* other Ascomycota (Incertae sedis: *Ascochyta*), Basidiomycota (*Uromyces americanus, U. lineolatus*), and Chytridiomycota (Synchytriaceae: *Synchytrium aureum, S. bonaerense*).

Economic importance: food: Plants of *Hydrocotyle americana* were eaten as greens by the indigenous Canadian people. *Hydrocotyle verticillata* also is regarded as edible and comprises 11.8% dry matter by weight. *Hydrocotyle ranunculoides* has a high moisture content (95%) but contains 24% crude protein, 2% fat, 12% fiber, 29% low acid detergent fiber, and 45% total carbohydrate. It was eaten as a minor vegetable in Mesoamerica and has a caloric value (cal/g dry wt) of 3300 and ash-free caloric value of 4059; **medicinal:** The Seminole used various parts of *Hydrocotyle umbellata* to treat asthma, pneumonia, and "turtle sickness" (cough, shortness of breath, trembles). Outside of North America, the root and juice is used to treat leprosy (Venezuela) and as a diuretic or treatment for rheumatism and intestinal disorders (Cuba). *Hydrocotyle verticillata* has been used by the Hawaiians to remedy respiratory disorders. Various *Hydrocotyle* species have been used to treat renal disorders in different parts of the world, but such applications probably are due to the resemblance of the leaves to kidneys; **cultivation:** *Hydrocotyle americana* is planted around the margins of ornamental ponds and in waterfalls; *H. umbellata* and *H. ranunculoides* also are planted as aquarium plants or water garden ornamentals, the latter sometimes distributed as the cultivar 'Ruffles.' *Hydrocotyle verticillata* is another aquarium and water garden ornamental, sometimes distributed under the cultivar name 'Little Umbrellas'; **misc. products:** *Hydrocotyle ranunculoides* has been used to generate methane through fermentation and to treat wastewater. It also is planted in holding ponds to facilitate denitrification and pesticide degradation. *Hydrocotyle umbellata* also has been investigated for sewage treatment and effectively reduces the biological oxygen demand of wastewater; **weeds:** *Hydrocotyle americana* is cited as a weed in several parts of the world, but the taxonomic integrity of many reports is questionable. The numerous reports from New Zealand actually represent *H. heteromeria* and reports from Europe are based on the incorrect synonymy of *H. americana* with the clearly distinct *H. ranunculoides*. *Hydrocotyle ranunculoides* is invasive in Africa, Asia, Australia, and Europe, where it often is in cultivation. *Hydrocotyle umbellata* is regarded as a serious weed of St. Augustine turf grass (Poaceae: *Stenotaphrum secundatum*); *H. verticillata* also grows in well-watered lawns; **nonindigenous species:** *Hydrocotyle ranunculoides* has been introduced to Africa, Australia, and Europe. The introductions of *H. ranunculoides* to Australia and Europe have been confirmed by DNA "barcoding" markers. *Hydrocotyle umbellata* was introduced to Thailand as an escape from cultivation and also is naturalized in New Zealand. *Hydrocotyle verticillata* was introduced to Hawaii in the early 19th century. The reports of *H. verticillata* in Japan represent occurrences of *H. verticillata* var. *triradiata* (A. Rich.) Fernald, which some authors recognize as a distinct species (=*H. prolifera* Kellogg).

Systematics: Although an extensive phylogenetic survey of the ~130 *Hydrocotyle* species has not yet been conducted, those investigations that have sampled smaller numbers of species indicate that the genus is monophyletic (e.g., Figure 5.127). In the surveys conducted thus far, the closest sister genus to *Hydrocotyle* is the Australian *Neosciadium*; however, this result might be an artifact of incomplete taxon sampling. Both genera resolve with *Harmsiopanax*, *Trachymene* and *Uldinia* as the sister clade to the remaining Araliaceae (Figures 5.126 and 5.127). The OBL North American species have not been included in any comprehensive phylogenetic analysis and further clarification of their relationships in *Hydrocotyle* awaits more thorough taxonomic sampling of this group. However, one analysis that includes two OBL taxa (Figure 5.128) indicates the independent origin of their aquatic habit. The current wetland indicator list recognizes *H. prolifera* as a distinct

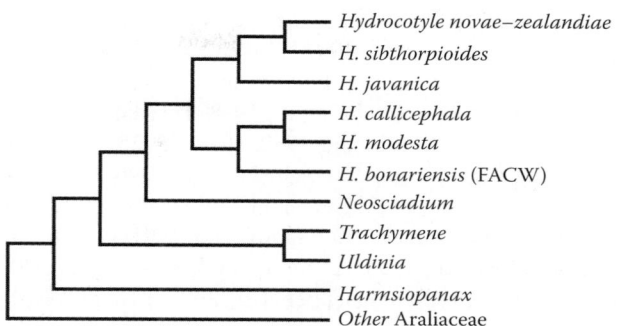

FIGURE 5.127 Relationships of *Hydrocotyle* in Araliaceae as indicated by analysis of combined cpDNA sequence data. None of the OBL North American species was included in the analysis. (Adapted from Nicolas, A.N. & Plunkett, G.M., *Mol. Phylogen. Evol.*, 53, 134–151, 2009.)

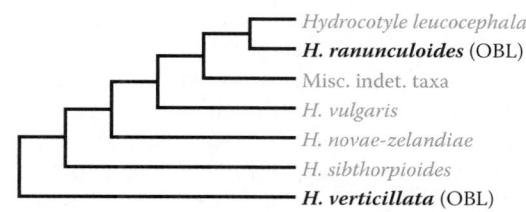

FIGURE 5.128 Interspecific relationships in *Hydrocotyle* as indicated by analysis of *trnH–psbA* sequence data. Although representing only a small number of *Hydrocotyle* species, this result indicates the independent derivation of the OBL habit (bold) in the two North American species surveyed. (Adapted from Van De Wiel, C.C.M. et al., *Mol. Ecol. Resour.*, 9, 1086–1091, 2009.)

OBL species; however, it is included here as a variety of *H. verticillata*; e.g., var. *triradiata* (A.Rich.) Fern., which is in accord with most contemporary treatments. The basic chromosome number of *Hydrocotyle* is $x = 9$ or 12. *Hydrocotyle ranunculoides* ($2n = 24, 48$) contains diploid and tetraploid cytotypes, *H. umbellata* ($2n = 48$) is tetraploid, and *H. verticillata* ($2n = 176 \pm 8$) is a high-order polyploid, which is among the highest chromosome numbers known in the genus. Despite these variable and high ploidy series, few hybrids have been suspected in the genus and none has yet been adequately documented.

Comments: *Hydrocotyle americana* occurs in northeastern North America; *H. ranunculoides*, *H. umbellata* and *H. verticillata* occur throughout the United States except for the central portions and mountainous regions. The ranges of *H. ranunculoides*, *H. umbellata* and *H. verticillata* extend into Central and South America.

References: Agami & Reddy, 1991; Austin, 2004b; Bartgis & Lang, 1984; Bazzaz, 1991; Beal, 1977; Bell et al., 1993; Bermingham, 2010; Blair, 1936; Boyd & McGinty, 1981; Brown & Tighe, 1992; Bryson et al., 2008; Busey & Johnson, 2006; Castellanos & Rijo, 1982; Chavasiri et al., 2005; Christy, 2004; Cordo et al., 1982; Cosyns et al., 2005; Coulter, 1904; Dawe & Reekie, 2007; De Lange et al., 2005;

DiTomaso & Healy, 2003; Eriksen, 1968; Everitt et al., 2008; Eyles, 1941; Fahrig et al., 1993; Finke & Scriber, 1988; Flinn et al., 2008; Glenn et al., 1995; Grant, 1949; Greca et al., 1994; Guppy, 1906; Hanlon et al., 1998; Heaven et al., 2003; Heyligers, 2007; Hilliard, Jr., 1982; Holm, 1925; Hong et al., 2008; Howard & Chege, 2007; Hoyer et al., 1996; Huffman & Lonard, 1983; Hunt, 1943; Hunt et al., 2008; Hussner & Lösch, 2007; Hussner & Meyer, 2009; Jantrarotai, 1991; Janzen, 1984; Johnson & Brooke, 1998; Kadono, 2004; Kasselmann, 2003; Keddy & Wisheu, 1989; Knepper et al., 2002; Knight & Miller, 2004; Knuth, 1906; Kuhnlein & Turner, 1991; Lallana, 1990; Lee & Berenbaum, 1992; Lusk & Reekie, 2007; Mandossian & McIntosh, 1960; Melcher et al., 2000; Middleton, 2009; Moorhead & Reddy, 1990; Morrison & Smith, 2007; Mulhouse et al., 2005; Murray-Gulde et al., 2005; Nagata, 1971, 1985; Nico & Muench, 2004; Nicolas & Plunkett, 2009; Owens et al., 2001; Parker et al., 1973; Patton & Judd, 1988; Penfound & Hathaway, 1938; Picking & Veneman, 2004; Picó & Nuez, 2000; Plunkett et al., 2004a; Porcher, 1981; Reddy & DeBusk, 1984; Reddy & Tucker, 1985; Robertson, 1888, 1889a; Ruiz Avila & Klemm, 1996; Russell, 1942; Saba, 1974; Schumacher, 1956; Singhurst et al., 2007; Speichert & Speichert, 2004; Stegmaier, Jr., 1966; Steubing et al., 1980; Stutzenbaker, 1999; Taylor et al., 1995; Thieret, 1971; Thomas et al., 1973; Toft et al., 1999; Van De Wiel et al., 2009; Walley, 2007; Watson et al., 1994b; Weishampel & Bedford, 2006; Webb & Beuzenberg, 1987; Wells, 1928; Wetzel et al., 2001; Wisheu & Keddy, 1989, 1991; Wolverton & McDonald, 1981; Yang et al., 2008; Yi et al., 2004b; Yoon et al., 2006; Zartman & Pittillo, 1998; Zungsontiporn, 2002.

Order 7: Aquifoliales [5]

Aqufoliales have been redefined based on the results obtained from asterid families, which consistently resolve the order as the sister clade to the remainder of the euasterid II group (campanulids), i.e., Apiales+Asterales+Dipsacales (Bremer et al., 2002). Most recently, the order has been defined to include five families: Aquifoliaceae Bercht. & J. Presl, Cardiopteridaceae Blume, Helwingiaceae Decne., Phyllonomaceae Small, and Stemonuraceae Kårehed (Bremer et al., 2002; APG, 2003). Within Aquifoliales, Aquifoliaceae are relatively derived and include the only OBL species within the order (Figure 5.129).

Family 1: Aquifoliaceae [1]

Although *Aquifolium* Mill. is a later synonym of *Ilex* L., the family name derived from the former genus has been conserved nomenclaturally. Several molecular phylogenetic studies have demonstrated that the genus known previously as *Nemopanthus* Raf. nests within *Ilex* (Cuénoud et al., 2000; Powell et al., 2000) and most recent taxonomic treatments appropriately have transferred the former genus to the latter (Manen et al., 2002; Gottlieb et al., 2005). As a result, Aquifoliaceae now comprise a single genus (*Ilex*), to which the common name of "holly" typically is applied. The family is most closely related to the much smaller Asian Helwingiaceae (<5 species) and Central American Phyllonomaceae (4 species) (Figure 5.129).

Hollies are woody plants (trees and shrubs) with simple, alternate leaves and minute stipules (Judd et al., 2008). The leaf margins range from smooth to deeply toothed and spinose. Some species are evergreen. The flowers principally are 4–6-merous and unisexual (the plants dioecious), with a superior ovary and short (or no) style (Elias, 1987; Judd et al., 2008). The fruits are red, orange, purple, or black drupes, which contain 4–6 pits (Elias, 1987; Judd et al., 2008). The flowers are insect pollinated and are outcrossed (mainly by bees: Hymenoptera); the fruits are bird dispersed (Judd et al., 2008).

The attractive foliage and bright berries of hollies have inspired their widespread use in ornamental wreaths and for other holiday decorations. They are also cultivated widely as decorative trees shrubs. The leaves (and twigs) of one species (*Ilex paraguariensis*) are high in caffeine and theobromine; in South America they are used to prepare a popular, stimulating beverage known as *yerba maté* (Reginatto et al., 1999). The berries of many holly species contain the bitter purine alkaloid ilicin and are highly toxic to humans, yet they are eaten extensively (and safely) by birds. The fine grain of holly wood makes it particularly suitable for use in fine decorative wood inlays. It also is a fairly heavy wood and has been used to manufacture various products ranging from chess pieces to piano keys. The American holly (*Ilex opaca*) is the state tree of Delaware.

Aquifoliaceae are distributed worldwide and contain more than 400 species. The species mostly are terrestrial and typically inhabit acidic sites (Judd et al., 2008). There is one North American genus with OBL species:

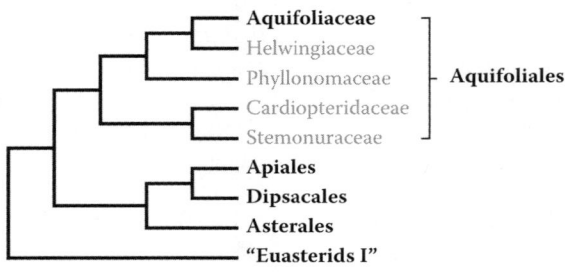

FIGURE 5.129 Phylogenetic relationships of Aquifoliales as indicated by combined morphological and DNA sequence data analysis. Taxa containing OBL North American species are highlighted in bold. (Adapted from Kårehed, J., *Amer. J. Bot.*, 88, 2259–2274, 2001.)

1. *Ilex* L.

1. *Ilex*

Holly; houx

Etymology: from the Latin *ilex* meaning "oak," in reference to the similarly incised leaves of some species

Synonyms: *Nemopanthus* Raf.; *Prinos* G. Don

Distribution: global: Africa; Australia; Eurasia; New World; **North America:** E/W disjunct

Diversity: global: 400 species; **North America:** 22 species

Indicators (USA): OBL: *Ilex amelanchier, I. laevigata, I. mucronata;* **FACW:** *I. laevigata, I. verticillata*

Habitat: freshwater; palustrine; **pH:** 3.5–6.5; **depth:** <1 m; **life-form(s):** emergent shrub, emergent tree

Key morphology: Shrubs or small trees (to 8 m); leaves (to 10 cm) deciduous, alternate, simple, lanceolate to obovate, tapering to base, lower surface glabrous or pubescent, the margins smooth to sharply serrate, petioles (to 20 mm) slender; plants dioecious; flowers 4-merous, the ♂ flowers stalked (to 16 mm), clustered (2–10) in axils, peduncled (to 2 cm), the ♀ flowers solitary, paired or in small clusters, sessile or stalked (to 5 mm); petals greenish, white or yellow; drupes (to 10 mm) berry-like, orange, red or scarlet (rarely yellow); seeds (4–10; to 4 mm) smooth or deeply furrowed.

Life history: duration: perennial (buds); **asexual reproduction:** stem clusters, layering, root suckers; **pollination:** insect; **sexual condition:** dioecious; **fruit:** drupes (common); **local dispersal:** drupes (water); **long-distance dispersal:** drupes (birds, small mammals)

Imperilment: (1) *Ilex amelanchier* [G4]; S2 (AL, FL, GA, LA); S3 (MS, NC, SC); (2) *I. laevigata* [G5]; S3 (DE, ME); (3) *I. mucronata* [G5]; S2 (RI); S3 (IN, MD, OH, WV); (4) *I. verticillata* [G5]; S1 (IA, TX); S2 (AR)

Ecology: general: Although only three North American species are classified as OBL, 16 of the 22 species (73%) are categorized as wetland indicators at some level. *Ilex verticillata* (FACW) formerly was categorized as OBL in a portion of its range. Its treatment has been retained here, having been completed prior to the release of the revised indicator list. Ecologically, *Ilex* is a highly versatile genus that is distributed nearly worldwide from 63°N to 35°S latitudes, and extends from sea level to elevations of 4000 m. Even though most *Ilex* species grow in terrestrial sites such as mountainous slopes, most of the species occur in humid habitats and many of the North American species are found frequently in fairly wet places. Most *Ilex* habitats are acidic. About half of the North American species can attain tree status; the remainder are shrubs. All species are dioecious and are pollinated by insects (ants, bees, flies, wasps), which forage for nectar secreted by glands at the petal bases. Bees (Insecta: Hymenoptera: Apidae: *Apis, Bombus*) are common pollinators. The sex ratios for progeny of open-pollinated plants are nearly equal but can vary widely in natural populations, tending to be male-biased in most of those studied. The orange or red drupes of the OBL species are dispersed mainly by birds (trasported endozooically), which use them as an important winter and spring food source (they become more palatable after overwintering).

Dispersal (endozooic transport) also occurs by black bears (Mammalia: Ursidae: *Ursus americanus*), coyotes (Mammalia: Canidae: *Canis latrans*), foxes (Mammalia: Canidae: *Vulpes vulpes*), fishers (Mammalia: Mustelidae: *Martes pennanti*), and raccoons (Mammalia: Procyonidae: *Procyon lotor*). The fruits of at least some species also float, so some dispersal by water is possible. The seeds are difficult to germinate and some have complex morphophysiological dormancy, resulting in delayed germination from 1 to 3 years, and a rate as minute as one in ten million. Some species reproduce vegetatively by root suckers or by the layering of branch tips that come in contact with the ground (e.g., when ladened with snow). Many hollies can be propagated vegetatively by shoot cuttings, which root most effectively at a low pH (~5.0).

***Ilex amelanchier* M.A. Curtis ex Chapman** grows in bayheads, depressions, flatwoods, floodplains, ponds, sandhills, savannahs, sloughs, swamps, thickets, and along the margins of canals, rivers, and streams at elevations of up to 350 m. The substrates are acidic (pH: 3.5–6.5) sands or silt (e.g., Guyton silt loam). The plants grow in full to partial sunlight. They are hardy from zones 6–10a and typically require a minimum of 200 frost-free days. Flowering occurs in April–May and the fruits mature from October to November. Single female plants grown in cultivation are known to produce low seed set, which is attributed to hybridization with other cultivated specimens (e.g., *I. opaca, I. verticillata*). Seed germination is highest when they are sowed immediately. The seeds can take years to germinate, but enhanced germination reportedly has been achieved by double stratification of 30–60 days at 20°C–30°C followed by 60–90 days at 5°C. **Reported associates:** *Acer rubrum, Alnus serrulata, Amsonia tabernaemontana, Aristida palustris, Berchemia scandens, Bignonia capreolata, Carex glaucescens, C. verrucosa, Cephalanthus occidentalis, Chamaecyparis thyoides, Clethra alnifolia, Coelorachis rugosa, Cornus foemina, Crataegus opaca, Cyrilla racemiflora, Dichanthelium, Diospyros virginiana, Eleocharis, Eriocaulon compressum, Fraxinus caroliniana, Gratiola ramosa, Ilex cassine, I. coriacea, I. decidua, I. myrtifolia, I. opaca, I. verticillata, Itea virginica, Lachnanthes caroliana, Leucothoe racemosa, L. recurva, Liquidambar styraciflua, Lobelia boykinii, Ludwigia pilosa, Lyonia lucida, Magnolia virginiana, Mikania scandens, Myrica cerifera, Nymphaea, Nyssa aquatica, N. biflora, N. ogeche, N. sylvatica, Onoclea sensibilis, Panicum hemitomon, P. rigidulum, P. tenerum, Pinus palustris, P. taeda, Polygala cymosa, Populus heterophylla, Proserpinaca pectinata, Quercus laurifolia, Rhexia virginica, Rhynchospora corniculata, R. filifolia, R. microcarpa, R. mixta, R. perplexa, R. tracyi, Sabal minor, Sagittaria, Salix nigra, Smilax laurifolia, S. walteri, Sphagnum, Stylisma aquatica, Taxodium ascendens, T. distichum, Toxicodendron radicans, Vaccinium formosum, V. fuscatum, Wisteria frutescens, Woodwardia areolata, W. virginica, Xyris laxifolia, Zenobia pulverulenta.*

***Ilex laevigata* (Pursh) A. Gray** occurs in bogs, Carolina bays, depressions, fens, floodplains, pocosins, swamps, thickets, and along the margins of ponds and streams at low elevations. It is hardy from zones 4b to 7b and occurs on acidic (pH: 4.4–6.5) substrates of clay, loam, muck (Dorovan), peat, and sand. The flowers are produced from April to June with fruits present from September to January. The seeds are notoriously difficult to germinate and no effective method has been reported. **Reported associates:** *Abies balsamea, Acer rubrum, Alnus maritima, A. rugosa, A. serrulata, Amelanchier canadensis, Apios americana, Aralia nudicaulis, Arisaema triphyllum, Arundinaria gigantea, Bazzania trilobata, Berberis thunbergii, Betula alleghaniensis, B. populifolia, Bignonia capreolata, Boehmeria cylindrica, Calamagrostis canadensis, C. coarctata, Calla palustris, Carex bromoides, C. crinita, C. folliculata, C. intumescens, C. lacustris, C. seorsa, C. stricta, C. trisperma, Cephalanthus occidentalis, Chamaecyparis thyoides, Chamaedaphne calyculata, Cicuta maculata, Cinna arundinacea, Clethra alnifolia, Commelina virginica, Coptis trifolia, Cornus amomum, C. foemina, Cyrilla racemiflora, Decodon verticillatus, Dioscorea villosa, Drosera rotundifolia, Dulichium arundinaceum, Eriophorum tenellum, E. vaginatum, E. virginicum, Fraxinus nigra, F. pennsylvanica, F. profunda, Gaultheria hispidula, G. procumbens, Gaylussacia baccata, G. dumosa, G. frondosa, Glyceria canadensis, Goodyera pubescens, Gordonia lasianthus, Helonias bullata, Hypnum imponens, Ilex coriacea, I. glabra, I. mucronata, I. opaca, I. verticillata, Iris versicolor, Itea virginica, Juncus brachycephalus, J. canadensis, Kalmia angustifolia, K. carolina, K. cuneata, K. latifolia, Larix laricina, Leersia oryzoides, Leucothoe racemosa, Lindera benzoin, Liquidambar styraciflua, Lonicera japonica, Lycopodiella appressa, Lyonia ligustrina, L. lucida, Lysimachia terrestris, Magnolia virginiana, Maianthemum canadense, M. trifolium, Myrica cerifera, M. heterophylla, Nyssa biflora, N. sylvatica, Oclemena acuminata, Onoclea sensibilis, Osmunda claytoniana, O. regalis, Osmundastrum cinnamomeum, Pallavicinia lyellii, Parthenocissus quinquefolia, Peltandra virginica, Persea palustris, Persicaria arifolia, P. sagittata, Photinia melanocarpa, P. pyrifolia, Picea mariana, P. rubens, Pinus rigida, P. serotina, P. strobus, P. taeda, Polytrichum, Pteridium aquilinum, Quercus bicolor, Q. palustris, Rhododendron canadense, R. canescens, R. groenlandicum, R. maximum, R. viscosum, Rhynchospora alba, Rosa palustris, Rubus hispidus, Sarracenia flava, S. purpurea, Saururus cernuus, Scirpus atrocinctus, Smilax glauca, S. laurifolia, S. pseudochina, S. rotundifolia, Sphagnum fallax, S. flavicomans, S. girgensohnii, S. magellanicum, S. pulchrum, S. palustre, Spiraea alba, S. latifolia, S. tomentosa, Symplocarpus foetidus, Symplocos tinctoria, Thalictrum pubescens, Thelypteris palustris, Tillandsia usneoides, Toxicodendron radicans, T. vernix, Trientalis borealis, Tsuga canadensis, Ulmus americana, Vaccinium corymbosum, V. crassifolium, V. formosum, V. fuscatum, V. fuscatum×V. corymbosum, V. macrocarpon, V. oxycoccos, Veratrum viride, Viburnum dentatum, V. nudum, Vitis rotundifolia, Woodwardia areolata, W. virginica, Zenobia pulverulenta, Zizania aquatica.*

Ilex mucronata (L.) **M. Powell, Savol. & S. Andrews** inhabits bogs, depressions, ditches, fens, marshes, meadows, river bottoms, sloughs, swamps, thickets, and the margins of lakes, ponds, rivers, and streams at elevations of up to 1430 m.

Exposures range from full sun to partial shade. The plants are hardy in zones 5a–8b. Their substrates are acidic (pH: 3.5–5.4), nutrient poor (e.g., 1.7–2.0 ppm Ca; 470–730 µg/L orthophosphate; 700–2000 µg/L NH_3-N), and include organic muck, peat, sand, and shallow peat over clay loam. This species is intolerant of fairly low sodium (112 mg/L) and chloride (168 mg/L) levels. Flowering occurs from May to June. The drupes ripen in July and contain an average of 2.6 filled seeds; they are dispersed in August but can persist on plants throughout much of the winter. Like most holly seeds, their germination is slow and difficult. In one study where seeds were cleaned and sowed outdoors, their germination rates varied from 16% to 66% over periods ranging from 146 to 626 days. In another study, refrigerated seeds exhibited somewhat lower (14%) and longer (338–565 days) germination rates than unrefrigerated seeds (24%; 252–511 days). In bogs, the plants are more common and denser on grounded *Sphagnum* mats than when growing on floating mats. Populations are known to decrease by as much as 17% in areas that become flooded due to the activity of beavers (Mammalia: Castoridae: *Castor canadensis*). The plants reproduce vegetatively by producing small clonal clusters of 10+ connected stems ("clonal fragments"). These stems are more numerous and more closely spaced in sunny localities, and are fewer but more widely spaced in shaded sites. The plants can live for up to 48 years. They are categorized as having intermediate shade tolerance and tolerance to moderate-severity disturbances. They are fire tolerant (even catastrophic fires) and new shoots (5 cm) have been observed to sprout from the roots within 3 weeks following a fire. They also appear in sites that have been commercially logged. The plants often become dominant (especially after disturbances such as fires) and effectively can exclude herbaceous and other woody species once they become established. This species accumulates zinc at levels (300–700 ppm) that are far higher than its associates. **Reported associates:** *Abies balsamea, Acer pensylvanicum, A. rubrum, Alnus rugosa, A. serrulata, A. viridis, Amelanchier bartramiana, A. canadensis, Andromeda polifolia, Aralia nudicaulis, Arctostaphylos uva-ursi, Arisaema triphyllum, Athyrium filix-femina, Aulacomnium palustre, Bartonia paniculata, Bazzania trilobata, Betula alleghaniensis, B. lenta, B. nana, B. papyrifera, B. populifolia, B. pumila, Botrychium simplex, Calamagrostis canadensis, Calla palustris, Calopogon tuberosus, Caltha palustris, Carex canescens, C. chordorrhiza, C. comosa, C. crinita, C. disperma, C. exilis, C. folliculata, C. intumescens, C. leptalea, C. lucorum, C. magellanica, C. michauxiana, C. oligosperma, C. pedunculata, C. pensylvanica, C. polymorpha, C. stricta, C. trisperma, C. utriculata, C. wiegandii, Carpinus caroliniana, Cavernularia hultenii, Cephalanthus occidentalis, Chamaedaphne calyculata, Chelone glabra, Chrysosplenium americanum, Cladina arbuscula, C. stellaris, Cladium mariscoides, Cladonia, Clethra alnifolia, Clintonia borealis, Comarum palustre, Comptonia peregrina, Coptis trifolia, Cornus canadensis, C. sericea, Corylus americana, C. cornuta, Cypripedium acaule, Dalibarda repens, Danthonia spicata, Dasiphora floribunda, Deschampsia flexuosa, Dicranum polysetum, D. scoparium, D. undulatum,*

Diervilla lonicera, Doellingeria umbellata, Drosera intermedia, D. rotundifolia, Dryopteris campyloptera, D. carthusiana, D. cristata, D. intermedia, Empetrum eamesii, E. nigrum, Epigaea repens, Eriocaulon aquaticum, Eriophorum angustifolium, E. virginicum, Eurybia macrophylla, Fallopia cilinodis, Festuca ovina, Fragaria virginiana, Fraxinus nigra, Galium triflorum, Gaultheria hispidula, G. procumbens, Gaylussacia baccata, G. dumosa, Gentiana linearis, Glyceria canadensis, Grimmia, Gymnocarpium dryopteris, Huperzia lucidula, Hylocomium splendens, Hypericum gentianoides, Hypnum curvifolium, H. imponens, Ilex laevigata, I. montana, I. verticillata, Impatiens capensis, Iris versicolor, Juncus, Juniperus communis, Kalmia angustifolia, K. polifolia, Larix laricina, Leucobryum glaucum, Lindera benzoin, Linnaea borealis, Listera auriculata, Lonicera canadensis, L. oblongifolia, L. villosa, Lycopodium annotinum, L. dendroideum, Lycopus uniflorus, L. virginicus, Lyonia ligustrina, Lysimachia terrestris, L. thyrsiflora, Maianthemum canadense, M. trifolium, Melampyrum lineare, Menyanthes trifoliata, Menziesia pilosa, Minuartia glabra, Mitchella repens, Mitella diphylla, M. nuda, Mnium, Moneses uniflora, Muhlenbergia uniflora, Myrica gale, M. pensylvanica, Myriophyllum tenellum, Nyssa sylvatica, Oclemena acuminata, O. nemoralis, Odontoschisma sphagni, Onoclea sensibilis, Orthilia secunda, Oryzopsis asperifolia, Osmunda claytoniana, O. regalis, Osmundastrum cinnamomeum, Oxalis montana, Pallavicinia lyellii, Peltandra virginica, Phegopteris connectilis, Photinia melanocarpa, P. pyrifolia, Picea glauca, P. mariana, P. rubens, Pinus banksiana, P. resinosa, P. rigida, P. strobus, Piptatherum pungens, Platanthera blephariglottis, Pleurozium schreberi, Pogonia ophioglossoides, Polypodium appalachianum, P. virginianum, Polytrichum commune, P. juniperinum, P. piliferum, Populus tremuloides, Prunus serotina, Pteridium aquilinum, Ptilidium ciliare, Ptilium crista-castrensis, Quercus ilicifolia, Q. prinoides, Q. rubra, Rhamnus alnifolia, Rhododendron canadense, R. groenlandicum, R. maximum, R. prinophyllum, R. viscosum, Rhynchospora alba, Rhytidiadelphus triquetrus, Rosa acicularis, R. carolina, Rubus chamaemorus, R. hispidus, R. idaeus, R. pubescens, Rumex acetosella, Salix bebbiana, S. discolor, Sarracenia purpurea, Scheuchzeria palustris, Schizachyrium scoparium, Scutellaria galericulata, Sibbaldiopsis tridentata, Solidago nemoralis, S. simplex, S. uliginosa, Sorbus americana, S. decora, Sphagnum angustifolium, S. capillifolium, S. centrale, S. fallax, S. fimbriatum, S. flavicomans, S. fuscum, S. girgensohnii, S. magellanicum, S. palustre, S. papillosum, S. pulchrum, S. pylaesii, S. recurvum, S. rubellum, S. russowii, S. tenellum, Spiraea alba, S. tomentosa, Symplocarpus foetidus, Thalictrum pubescens, Thuja occidentalis, Thelypteris palustris, T. simulata, Tiarella cordifolia, Toxicodendron vernix, Triadenum virginicum, Trichophorum cespitosum, Trientalis borealis, Trillium undulatum, Tsuga canadensis, Typha, Umbilicaria, Vaccinium angustifolium, V. corymbosum, V. macrocarpon, V. myrtilloides, V. oxycoccos, V. pallidum, V. vitis-idaea, Viburnum lantanoides, V. nudum, V. recognitum, Viola renifolia, Woodwardia virginica.

Ilex verticillata **(L.) A. Gray** is found on bluffs, dunes, in bayous, bogs, depressions, ditches, flats, flatwoods, floodplains, flowages, lagoons, marshes, meadows, ravines, river bottoms, sandbars, sandplains, seeps, sinkhole depressions, sloughs, string bogs, swales, swamps, thickets, and along the margins of lakes, ponds, pools, rivers, and streams at elevations of at least 1024 m. The plants occur in both tidal and nontidal sites. They are slightly salt tolerant (<5 ppt) and can endure temperatures as cold as −35°C (they are hardy in zones 3a–9a). Site exposures range from open to shaded conditions. Substrates are acidic (pH: 3.5–5.5) and include clay, clay loam, gravel–sand–peaty muck, humus clay, loam, muck, mucky loam, peat, sand, sandstone, sandy loam, and silt, typically with a high organic content. Flowering occurs from April to July. The flowers are pollinated by generalist insects (Insecta) such as beetles (Coleoptera: Elateridae; Mordellidae), bees (Hymenoptera: Apidae: *Bombus*), and flies (Diptera: Muscidae). Individual plants produce from 10 to 1000 drupes with seed set being significantly higher for plants when growing in connected sites relative to those in fragmented habitats. The fruits mature in about 88 days and average 4.1 viable seeds/fruit. Fruits ripen in late September–October but become wrinkled and brown by late January or early February. It is not unusual for 65% of the drupes to be removed from the plants by seed predators within 14–16 weeks. Seed germination is very low, reaching only about 2.3% after 1.8 years; however, seedlings often can occur in fair numbers (e.g., 0.4/m²). Untreated seeds that are sown outdoors typically will take 2–3 years to germinate. Good germination (52%–73%) has been achieved after subjecting seeds to warm stratification (30°C/20°C day/night) in moist sand for 2 months followed by cold stratification (5°C) for 60 days. In general, the plants are slow-growing and tend to be much more common in the standing vegetation than as represented by the seed bank. They often occur in single tree gaps. Densities as high as 1000 stems/hectare have been reported and the net aboveground primary production has been measured at 73 g/m²/year. The root system is quite superficial and is effective in surface soil competition. The plants produce suckers and can undergo vegetative reproduction by the layering of their shoot tips, which is facilitated by unusually heavy snowfalls. In some areas, plants appear to have been displaced by invading glossy buckthorn (*Rhamnus frangula*). **Reported associates:** *Abies balsamea, Acer negundo, A. rubrum, A. saccharum, A. saccharinum, Acorus calamus, Agrimonia gryposepala, A. parviflora, Agrostis gigantea, Alnus rugosa, A. serrulata, A. viridis, Amelanchier arborea, Amorpha fruticosa, Andropogon glomeratus, Aralia nudicaulis, Arisaema dracontium, A. triphyllum, Asclepias incarnata, A. perennis, Asimina triloba, Athyrium filix-femina, Aulacomnium palustre, Bartonia paniculata, B. virginica, Bazzania trilobata, Berchemia scandens, Betula alleghaniensis, B. nigra, B. papyrifera, B. populifolia, B. pumila, B. ×sandbergii, Bidens frondosus, Bignonia capreolata, Boehmeria cylindrica, Bryoxiphium norvegicum, Calamagrostis canadensis, Calla palustris, Calliergon cordifolium, C. giganteum, Caltha palustris, Campylium stellatum, Cardamine bulbosa, C.*

douglassii, Carex abscondita, C. atlantica, C. bromoides, C. brunnescens, C. buxbaumii, a. canescens, C. chordorrhiza, C. collinsii, C. crinita, C. crus-corvi, C. debilis, C. disperma, C. echinata, C. flava, C. folliculata, C. gigantea, C. glaucescens, C. granularis, C. grayi, C. gynocrates, C. interior, C. intumescens, C. joorii, C. lacustris, C. leptalea, C. lonchocarpa, C. lupuliformis, C. lupulina, C. lurida, C. oligosperma, C. pedunculata, C. retrorsa, C. scoparia, C. stipata, C. striata, C. stricta, C. trisperma, C. turgescens, C. typhina, C. utriculata, C. verrucosa, C. wiegandii, Carpinus caroliniana, Carya cordiformis, C. ovata, Cayaponia quinqueloba, Celtis laevigata, Cephalanthus occidentalis, Chamaedaphne calyculata, Chasmanthium laxum, Chelone glabra, Chionanthus virginicus, Chrysosplenium americanum, Cinna arundinacea, Cladina arbuscula, Cladium mariscoides, Clematis virginiana, Clethra alnifolia, Comarum palustre, Commelina virginica, Coptis trifolia, Cornus amomum, C. canadensis, C. foemina, C. obliqua, C. racemosa, C. sericea, Corylus americana, Crataegus spathulata, C. viridis, Crinum americanum, Cuscuta, Cypripedium parviflorum, C. reginae, Cyrilla racemiflora, Danthonia spicata, Dasiphora floribunda, Decodon verticillatus, Decumaria barbara, Deparia acrostichoides, Deschampsia flexuosa, Desmodium illinoense, Dichanthelium boscii, D. dichotomum, D. oligosanthes, Dicranum undulatum, Diervilla lonicera, Diospyros virginiana, Doellingeria umbellata, Drosera rotundifolia, Dryopteris carthusiana, D. cristata, Dulichium arundinaceum, Eleocharis tuberculosa, Epilobium leptophyllum, Equisetum fluviatile, Erechtites hieracifolia, Eriophorum virginicum, Euonymus americana, E. atropurpureus, Eupatorium fistulosum, E. perfoliatum, E. pilosum, E. rotundifolium, Fagus grandifolia, Festuca ovina, Frangula alnus, Fraxinus caroliniana, F. nigra, F. pennsylvanica, F. profunda, Galium obtusum, G. triflorum, Gaultheria hispidula, G. procumbens, Gaylussacia baccata, Gelsemium rankinii, Gentiana saponaria, Geum canadense, G. rivale, Glyceria canadensis, G. grandis, G. striata, Gordonia lasianthus, Gymnocarpium disjunctum, G. dryopteris, Habenaria repens, Hamamelis virginiana, Heliopsis helianthoides, Helonias bullata, Hibiscus moscheutos, Hieracium, Houstonia serpyllifolia, Hydrocotyle, Hylocomium splendens, Hypericum densiflorum, H. ellipticum, H. gentianoides, H. kalmianum, Hypnum imponens, Ilex collina, I. decidua, I. laevigata, I. montana, I. mucronata, I. opaca, Illicium floridanum, Impatiens capensis, Iris versicolor, Itea virginica, Juncus brevicaudatus, J. canadensis, J. effusus, J. filiformis, J. repens, J. subcaudatus, Justicia ovata, Kalmia angustifolia, K. carolina, K. latifolia, K. polifolia, Larix laricina, Leersia lenticularis, L. oryzoides, L. virginica, Leptodictyum riparium, Leucobryum glaucum, Leucothoe racemosa, Ligustrum sinense, Lilium michiganense, Lindera benzoin, Linnaea borealis, Liquidambar styraciflua, Liriodendron tulipifera, Lobelia amoena, L. cardinalis, Lonicera canadensis, L. dioica, L. japonica, L. oblongifolia, Lycopodiella appressa, Lycopodium obscurum, Lycopus virginicus, Lygodium palmatum, Lyonia ligustrina, L. lucida, Lysimachia ciliata, L. quadrifolia, L. terrestris, Lythrum lineare, Magnolia virginiana, Maianthemum canadense, M. stellatum, M. trifolium, Malaxis unifolia, Medeola virginiana, Menyanthes trifoliata, Menziesia pilosa, Micranthes pensylvanica, Microstegium vimineum, Mikania scandens, Mimulus, Minuartia glabra, Mitchella repens, Mitella diphylla, M. nuda, Monarda fistulosa, Moneses uniflora, Myrica gale, Nyssa aquatica, N. biflora, N. sylvatica, Oclemena acuminata, Onoclea sensibilis, Orontium aquaticum, Orthilia secunda, Oryzopsis asperifolia, Osmunda claytoniana, O. regalis, Osmundastrum cinnamomeum, Oxalis dillenii, Oxydendrum arboreum, Oxypolis rigidior, Packera aurea, Pallavicinia lyellii, Panicum virgatum, Parnassia asarifolia, P. glauca, Parthenocissus quinquefolia, Peltandra virginica, Persea borbonia, P. palustris, Persicaria arifolia, P. hydropiper, P. hydropiperoides, P. punctata, P. sagittata, P. virginiana, Phalaris arundinacea, Phanopyrum gymnocarpon, Phegopteris connectilis, Phlox glaberrima, Photinia melanocarpa, P. pyrifolia, Physocarpus opulifolius, Phytolacca americana, Picea glauca, P. mariana, P. rubens, Pilea pumila, Pinus banksiana, P. serotina, P. strobus, P. taeda, Platanthera clavellata, P. dilatata, P. integrilabia, Platanus occidentalis, Pleurozium schreberi, Pluchea camphorata, Poa compressa, P. palustris, P. trivialis, Polygonatum pubescens, Polymnia canadensis, Polypodium virginianum, Polytrichum commune, P. juniperinum, P. piliferum, Pontederia cordata, Populus balsamifera, P. grandidentata, P. heterophylla, P. tremuloides, Prunus pensylvanica, P. serotina, P. virginiana, Pteridium aquilinum, Quercus alba, Q. bicolor, Q. ellipsoidalis, Q. lyrata, Q. macrocarpa, Q. pagoda, Q. palustris, Q. phellos, Q. rubra, Q. velutina, Ranunculus hispidus, Rhamnus alnifolia, R. cathartica, Rhexia, Rhizomnium appalachianum, R. punctatum, Rhododendron arborescens, R. canadense, R. catawbiense, R. groenlandicum, R. maximum, R. periclymenoides, R. viscosum, Rhynchospora alba, R. capitellata, R. gracilenta, Rhytidiadelphus squarrosus, R. triquetrus, Ribes, Robinia pseudoacacia, Rosa blanda, R. palustris, R. virginiana, Rubus argutus, R. hispidus, R. idaeus, R. pubescens, Rumex acetosella, R. verticillatus, Sabal minor, Saccharum baldwinii, Sagittaria latifolia, Salix fragilis, S. humilis, S. nigra, S. pedicellaris, S. sericea, Sambucus nigra, Sanguinaria canadensis, Sanguisorba canadensis, Sarracenia oreophila, S. purpurea, Sassafras albidum, Saururus cernuus, Schoenoplectus tabernaemontani, Scirpus atrovirens, S. cyperinus, S. expansus, S. polyphyllus, Scleria triglomerata, Scutellaria integrifolia, S. lateriflora, Senecio glabellus, Sibbaldiopsis tridentata, Smilax glauca, S. laurifolia, S. rotundifolia, S. tamnoides, S. walteri, Solanum dulcamara, Solidago patula, S. rugosa, S. simplex, S. uliginosa, Sorbus americana, S. decora, Sparganium eurycarpum, Sphagnum affine, S. angustifolium, S. bartlettianum, S. capillifolium, S. centrale, S. compactum, S. fallax, S. fimbriatum, S. flavicomans, S. girgensohnii, S. magellanicum, S. palustre, S. papillosum, S. recurvum, S. subtile, S. wulfianum, Spiraea alba, S. tomentosa, Styrax americanus, Symphyotrichum lateriflorum, S. puniceum, S. racemosum, Symplocarpus foetidus, Taxodium ascendens, T. distichum, Taxus canadensis, Thalictrum dasycarpum, Thalictrum pubescens, Thelypteris noveboracensis, T. palustris, T. simulata, Thuidium delicatulum, Thuja occidentalis, Tiarella cordifolia, Tilia americana, Toxicodendron radicans, T. vernix, Triadenum fraseri, T. virginicum, T. walteri, Trientalis borealis, Tsuga canadensis, Typha angustifolia, T. latifolia, T. ×glauca, Ulmus alata, U. americana, Umbilicaria, Uvularia sessilifolia, Vaccinium angustifolium, V. corymbosum, V. fuscatum, V. macrocarpon, V. myrtilloides, V. oxycoccos, V. simulatum, V. vitis-idaea, Veratrum viride, Viburnum dentatum, V. nudum, V. prunifolium, Viola cucullata, V. macloskeyi, V. ×primulifolia, Vitis riparia, V. rotundifolia, Wisteria frutescens, Woodwardia areolata, W. virginica, Xanthorhiza simplicissima, Zanthoxylum americanum, Zizania aquatica.

Use by wildlife: The fruits of all the OBL *Ilex* species are relished by various songbirds (Aves), particularly in the spring when they are most palatable (e.g., in *I. verticillata*, phenolics and saponins decrease markedly after 5 months of ripening). It is believed that such compounds protect the developing fruits from insect and fungal predation. The fruits of *I. laevigata* are eaten by various birds (Aves) including the American robin (Turdidae: *Turdus migratorius*), brown thrasher (Mimidae: *Toxostoma rufum*), cedar waxwing (Bombycillidae: *Bombycilla cedrorum*), gray catbird (Mimidae: *Dumetella carolinensis*), hermit thrush (Turdidae: *Catharus guttatus*), northern bobwhite (Odontophoridae: *Colinus virginianus*), and northern flicker (Picidae: *Colaptes auratus*). *Ilex laevigata* is used to some extent by male white-tailed deer (Mammalia: Cervidae: *Odocoileus virginianus*) as a tree for scent "rubbing." It is a host to fungi (Ascomycota: Dermateaceae: *Cenangella ravenelii*) and also to several insects (Insecta: Lepidoptera: Lycaenidae: *Celastrina argiolus*; Lymantriidae: *Lymantria dispar*; Sphingidae: *Sphinx luscitiosa*). The plants also harbor Putnam's scale insects (Insecta: Hemiptera: Diaspididae: *Aspidiotus ancylus*). The yellow-bellied sapsucker (Aves: Picidae: *Sphyrapicus varius*) obtains sap from *I. mucronata*, and the fruits of the latter (which contain 7.7% crude fat) are eaten by many birds (Aves) including the American robin (Turdidae: *Turdus migratorius*), eastern bluebird (Turdidae: *Sialia sialis*), eastern phoebe (Tyrannidae: *Sayornis phoebe*), eastern towhee (Emberizidae: *Pipilo erythrophthalmus*), white-eyed vireo (Vireonidae: *Vireo griseus*), white-throated sparrow (Emberizidae: *Zonotrichia albicollis*), and wood thrush (Turdidae: *Hylocichla mustelina*). Foliage of *I. mucronata* is browsed by caribou (Mammalia: Cervidae: *Rangifer tarandus*), moose (Mammalia: Cervidae: *Alces alces*), snowshoe hare (Mammalia: Leporidae: *Lepus americanus*), and (less often in winter) white-tailed deer (Mammalia: Cervidae: *Odocoileus virginianus*). The plants are excavated by leaf mining flies (Insecta: Diptera: Agromyzidae: *Phytomyza nemopanthi*). They also are a minor forage item of beavers (Mammalia: Castoridae: *Castor canadensis*). The plants host various insects (Insecta: Lepidoptera) including Geometridae (*Sabulodes caberata*), Saturniidae (*Hyalophora columbia*) Sphingidae (*Sphinx kalmiae*), and Tortricidae (*Larisa subsolana, Rhopobota finitimana, R. naevana, R. unipunctana*). They also host numerous fungi including Ascomycota (Cucurbitariaceae: *Curreya peckiana*; Dermateaceae:

Dermea peckiana, Durandiella nemopanthis, Godroniopsis nemopanthi; Erysiphaceae: *Microsphaera nemopanthis, M. penicillata*; Rhytismataceae: *Rhytisma ilicis-canadensis*; Sarcosomataceae: *Conoplea sphaerica*; Venturiaceae: *Venturia curviseta*), and Basidiomycota (Hymenochaetaceae: *Phellinus inermis*; Meruliaceae: *Irpex lacteus*). This species is a component of nesting habitat for Bicknell's thrush (Aves: Turdidae: *Catharus bicknelli*) and yellow-bellied flycatcher (Aves: Tyrannidae: *Empidonax flaviventris*). Fruits of *I. verticillata* are highly preferred by various seed predators (birds, invertebrates, mammals). They contain phenolics and saponins in addition to soluble carbohydrates (88%) and are categorized as a low-quality bird food. Containing approximately 43% hexose, 3.7%–4% fat, 2.9%–6.1% protein, and 15.6% fiber, the fruits are eaten by more than a dozen bird (Aves) species including the American robin (Turdidae: *Turdus migratorius*), American woodcock (Scolopacidae: *Scolopax minor*), black-capped chickadee (Paridae: *Poecile atricapillus*), black duck (Anatidae: *Anas rubripes*), brown thrasher (Mimidae: *Toxostoma rufum*), cedar waxwing (Bombycillidae: *Bombycilla cedrorum*), eastern bluebird (Turdidae: *Sialia sialis*), evening grosbeak (Fringillidae: *Coccothraustes vespertinus*), gray catbird (Mimidae: *Dumetella carolinensis*), hermit thrush (Turdidae: *Catharus guttatus*), house finch (Fringillidae: *Carpodacus mexicanus*), mallard duck (Anatidae: *Anas platyrhynchos*), northern mockingbird (Mimidae: *Mimus polyglottos*), pileated woodpecker (Picidae: *Dryocopus pileatus*), ruffed grouse (Phasianidae: *Bonasa umbellus*), song sparrow (Emberizidae: *Melospiza melodia*), white-throated sparrow (Emberizidae: *Zonotrichia albicollis*), wild turkey (Meleagridinae: *Meleagris gallopavo*), and yellow-bellied sapsucker (Picidae: *Sphyrapicus varius*). They are an important food of black bears (Mammalia: Ursidae: *Ursus americanus*), particularly during winter (December), and also are eaten by deer mice (Mammalia: Muridae: *Peromyscus leucopus, P. maniculatus*) and bugs (Insecta: Hemiptera: Pentatomidae: *Banasa calva*). The plants are a minor food item for white-tailed deer (Mammalia: Cervidae: *Odocoileus virginianus*) and raccoons (Mammalia: Procyonidae: *Procyon lotor*). The shoots can be an important winter browse for the eastern cottontail rabbit (Mammalia: Leporidae: *Sylvilagus floridanus*). *Ilex verticillata* is a host for numerous fungi (Ascomycota: Botryosphaeriaceae: *Guignardia philoprina, Phyllosticta haynaldi*; Dermateaceae: *Cenangella ravenelii, Dermea peckiana, Gloeosporium niveum*; Diaporthaceae: *Diaporthe epimicta, D. ilicis, D. oxyspora*; Erysiphaceae: *Microsphaera penicillata*; Helotiaceae: *Godroniopsis nemopanthi*; Incertae sedis: *Dendrophoma nigrescens, Trichothecium roseum*; Myriangiaceae: *Myriangium asterinosporum*; Rhytismataceae: *Rhytisma concavum, R. ilicis-canadensis, R. prini*; Tubeufiaceae: *Rebentischia massalongii*). It is only mildly susceptible to inoculations of *Phytophthora ramorum* (Oomycota: Pythiaceae), which is a serious pathogen of oaks (*Quercus*) in western North America and also is fairly resistant to Florida wax scale (Insecta: Hemiptera: Coccidae: *Ceroplastes floridensis*). The plants also host many insects (Insecta) including Lepidoptera (Geometridae: *Cepphis armataria*; Limacodidae: *Prolimacodes badia*; Lycaenidae: *Callophrys henrici*; Mimallonidae: *Cicinnus melsheimeri*; Noctuidae: *Harrisimemna trisignata*; Notodontidae: *Schizura unicornis*; Saturniidae: *Automeris io, Hemileuca nevadensis*; Sphingidae: *Dolba hyloeus*; Tortricidae: *Archips fuscocupreanus, Rhopobota dietziana, R. finitimana, R. naevana, R. unipunctana, Sparganothis daphnana, S. reticulatana*) and plant bugs (Heteroptera: Miridae: *Cariniocoris ilicis*). The foliage is damaged by leaf-mining flies (Insecta: Diptera: Agromyzidae: *Phytomyza verticillatae*) but is fed on only occasionally by Japanese beetles (Insecta: Coleoptera: Scarabaeidae: *Popillia japonica*). The plants are hosts to rose aphids (Insecta: Hemiptera: Aphididae: *Macrosiphum rosae*), woolly alder aphids (Hemiptera: Aphididae: *Paraprociphilus tessellatus*), and occasionally are found with woolly elm aphids (Hemiptera: Aphididae: *Eriosoma americanum*). The presence of *I. verticillata* is correlated positively with the occurrence of deer ticks (Acari: Ixodidae: *Ixodes scapularis*), the principal transmitting agents of Lyme disease. The plants also are correlated positively with nesting habitat of the four-toed salamander (Amphibia: Plethodontidae: *Hemidactylium scutatum*). They are used as nesting material by mallard ducks (Aves: Anatidae: *Anas platyrhynchos*).

Economic importance: food: The leaves of 60 *Ilex* species (but none OBL) are used in beverages, with several of them having high caffeine and alkaloid contents. The Potawatomi kept the berries of *Ilex mucronata* as a sour, bitter food. However, *I. mucronata* and *I. verticillata* should be avoided as a food because at least the bark is known to be cyanogenic; **medicinal:** Several *Ilex* species are used medicinally in various parts of the world. In North America preparations of *Ilex mucronata* were taken by the Malecite, Ojibwa, and Potawatomi tribes to treat coughs, fevers, kidney ailments, tuberculosis, and as a general medicinal tonic. The Hocąk (Winnebago) brewed a tea from the bark to induce vomiting. The Iroquois used a bark decoction of *I. verticillata* as an emetic, cathartic, and treatment for biliousness. A root decoction was made to alleviate hay fever. The Ojibwa used the bark to prepare a remedy for diarrhea. The fruits of *I. mucronata* contain a lectin that causes agglutination of porcine (but not human) blood cells. The berries of *I. verticillata* contain cytotoxic ursolic acid, which is known to inhibit some cancer cells by interrupting the signal transducer and activator of transcription 3 (STAT3) activation pathway; the substance also has anti-inflammatory and antimicrobial properties; **cultivation:** *Ilex amelanchier* and *I. laevigata* occasionally are found in cultivation; however *I. verticillata* is grown extensively as an ornamental and is distributed as numerous cultivars such as: 'Afterglow', 'Aurantiaca', 'Berry Nice', 'Berry Heavy', 'Bright Horizon', 'Carolina Cardinal', 'Christmas Cheer', 'Chrysocarpa', 'Citronella', 'Compacta', 'Cresgold', 'Earlibright', 'Fructu Albo', 'Golden Male' (♂), 'Golden Rain', 'Golden Verboom' (♂), 'Jim Dandy' (♂), 'Jolly Red', 'Kolmber', 'Magical Berry', 'Magical Times', 'Maryland Beauty', 'Nana', 'Oosterwijk', 'Quansoo', 'Quitsa', 'Red Sprite', 'Scarlet O'Hara', 'Shaver', 'Southern Gentleman' (♂), 'Spravy', 'Spriber', 'Stop Light', 'Sunset', 'Sunsplash',

'Tiasquam', 'Vermeulen', 'Winter Gold', and 'Winter Red', and also as several *I. serrata*×*I. verticillata* hybrid cultivars: 'Apollo' (♂), 'Autumn Glow', 'Bonfire', 'Harvest Red', 'Raritan Chief' (♂), and 'Sparkleberry'; **misc. products:** The Menominee made an arrow poison from the berries of *Ilex mucronata*. The bark of *I. verticillata* can be used to produce a green dye. Refined extracts of *I. verticillata* have been recommended as a source of biodegradable surfactant because of their high saponin content (18.9% crude saponins by dry weight); **weeds:** none; **nonindigenous species:** none (OBL)

Systematics: *Ilex* is the sole genus of Aquifoliaceae and is situated phylogenetically in the order Aquifoliales as the sister group to the small Asian family Helwingiaceae (Figure 5.129). Molecular data indicate that "*Nemopanthus*" is nested within *Ilex* as the sister species of *I. amelanchier* (by both nuclear and cpDNA data), thus forcefully arguing for its inclusion within the latter genus as *I. mucronata*. Both species occur within one clade of Asian/North American species (Figure 5.130); whereas, *I. verticillata* resolves in a second, distinct clade of Asian/North American species (Figure 5.130). Although *I. laevigata* has not been included in phylogenetic analyses, its high degree of similarity to *I. verticillata* makes it likely that these species are quite closely related. If that is the case, then *I. laevigata* would represent a different origin of the OBL habit than indicated by the two OBL species sampled. The infrageneric classification of *Ilex* requires extensive revision as the subgeneric classification of the genus is inconsistent with the results of molecular data analyses. Although *I. verticillata* and *I. amelanchier* were both assigned to subgenus *Prinus*, their phylogenetic relationships place them in different clades where they each associate with species representing an amalgam of different subgenera (Figure 5.130). A proclivity for hybridization in *Ilex* has been indicated by the discordance of nuclear and chloroplast gene phylogenies in the genus. A number of interspecific hybrid cultivars have been developed between *Ilex serrulata* and *I. verticillata*, which molecular data show to be very closely related (Figure 5.130). There also is molecular evidence of intersectional hybridization in the genus. The basic chromosome number of *Ilex* is $x = 17$–20 and many species are high polyploids. *Ilex mucronata* ($2n = 40$) is a diploid and *I. verticillata* ($2n = 72$) is a tetraploid.

Comments: *Ilex amelanchier* occurs in the southeastern United States, *I. laevigata* and *I. verticillata* range through eastern North America; *I. mucronata* is distributed throughout northeastern North America.

References: Abbott et al., 1990; Adams, 1927; Ahti & Henssen, 1965; Ashe, 1897; Baker et al., 1945; Baldwin, 2007; Baskin & Baskin, 1998; Benner & Bowyer, 1988; Bergerud & Nolan, 1970; Blood et al., 2010; Boyle et al., 2007; Braun, 1936; Breden et al., 2001; Brown, 1946; Buell, 1946; Buell & Wistendahl, 1955; Byers et al., 2007; Cain & Penfound, 1938; Chalmers & Loftin, 2006; Chandler & Hooper, 1982; Coulter, 1953; Cousens et al., 1988; Cuénoud et al., 2000; Cullina, 2002; DeCoursey, 1963; DeGraaf, 2002; Dirr & Heuser, Jr., 2006; Dodds, 1960; Donahue, 1954; Elam, 2007; Elias, 1987; Fang & McLaughlin, 1990; Felt, 1901; Foley, 1974; Frappier et al., 2003; Gallant et al., 2004;

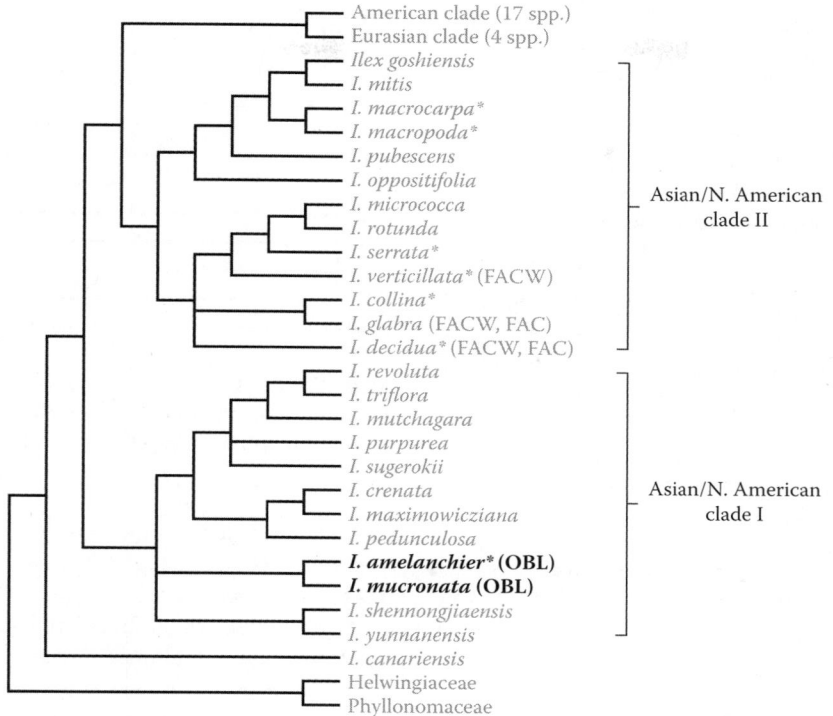

FIGURE 5.130 Phylogenetic relationships in *Ilex* as indicated by combined cpDNA data; the same relative relationships among the three included OBL species (shown in bold) also are indicated by nuclear data. Asterisked species are those assigned formerly to subgenus *Prinus* (the remainder represents various other subgenera). (Adapted from Manen, J.-F. et al., *Plant Syst. Evol.*, 235, 79–98, 2002.)

Gargiullo & Stiles, 1991, 1993; Geisler, 1926; Gerloff et al., 1966; Gervais & Wheelwright, 1994; Getz, 1961; Goldblatt, 1976a, 1979; Gorchov, 1990; Gottlieb et al., 2005; Griffiths & Piercey-Normore, 1995; Groves, 1937; Haegele, 2007; Hamerstrom & Blake, 1939b; Hamilton, Jr., 1941; Harper, 1935; Held, 2004; Henry, 1989; Hill, 1987; Huenneke, 1983; Ives, 1923; Jones, 2007; Kindscher & Hurlburt, 1998; Landers et al., 1979; LeBlanc & Leopold, 1992; Leonard, 1964; Little, Jr., 1944; Loizeau & Spichiger, 2004; Loizeau et al., in press; Lubelczyk et al., 2004; Lynn, 1984; Maier, 2003; Manen et al., 2002; Meiners & Stiles, 1997; Mendall, 1949; Mills et al., 2009; Mitchell & Niering, 1993; Nichols, 1934; Odum, 1942; Paratley & T. J. Fahey, 1986; Pettingill, Jr., 1939; Pringle, 1977; Quinby, 2000; Rathcke, 1988; Reyes et al., 2010; Richburg et al., 2001; Schwintzer, 1978; Seigler, 1976; Setoguchi & Watanabe, 2000; Sharitz, 2003; Sipple & Klockner, 1980; Skutch, 1929; Smith et al., 2007; Stiles, 1980; Sundue, 2006; Sweetman, 1944; Telfer, 1974; Tewksbury et al., 2002; Tooley & Browning, 2009; Torimaru & Tomaru, 2005; Tsang & Corlett, 2005; USDA, 1948; Wheelwright, 1986; Whitney & Moeller, 1982; Wieder et al., 1989; Willson, 1993; Witmer, 1996; Zimmerman & Hitchcock, 1929.

ORDER 8: ASTERALES [12]

The order Asterales (roughly 25,000 species) is dominated by the large family Asteraceae Martinov, which alone

accounts for about 23,000 species (Judd et al., 2008). Most of the remaining species occur within Campanulaceae Juss. The monophyly of Asterales has been demonstrated by many molecular phylogenetic studies employing cpDNA RFLP data and DNA sequence data from *atpB*, *matK*, *ndhF*, *rbcL*, 18S nrDNA, and other loci; it also is marked by several nonmolecular synapomorphies including valvate petals, lack of apotracheal parenchyma, presence of ellagic acid, and carbohydrates stored as inulin (Barkley et al., 2006; Judd et al., 2008).

The precise relationship of the order to other campanulids has remained unresolved (Figure 5.107); however, Asterales clearly associate with Dipsacales and Apiales along with a number of rather small and as of yet ordinally unplaced families (Figures 5.107, 5.129). A recent appraisal incorporating 17 kb of DNA sequence data for 122 taxa (Tank & Donoghue, 2010) has provided higher resolution of this group and more convincing internal support, although some relationships still remain uncertain. In this interpretation (Figure 5.131), Asterales resolve as the sister group of the clade containing Apiales, Dipsacales, and four unassigned families (Bruniaceae, Columelliaceae, Escalloniaceae, Paracryphiaceae). To date, this arrangement represents the best available hypothesis regarding the higher level relationships of Asterales. Three families with OBL North American taxa are found within the order, namely Asteraceae, Campanulaceae, and Menyanthaceae Dumort (Figure 5.132).

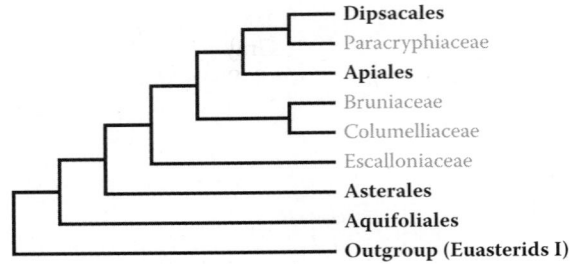

FIGURE 5.131 Relationships among campanulid taxa as indicated by phylogenetic analysis of ~17kb of DNA sequence data. Groups containing OBL taxa are indicated in bold. (Adapted from Tank, D.C. & Donoghue, M.J., *Syst. Bot.*, 35, 425–441, 2010.)

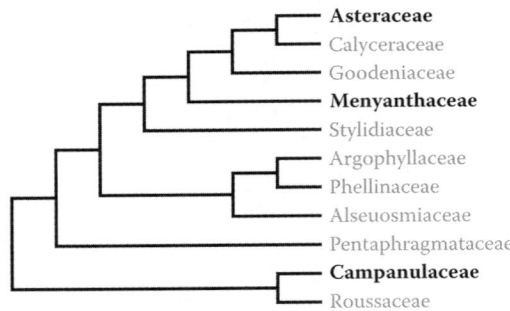

FIGURE 5.132 Relationships among the families of Asterales as indicated by phylogenetic analysis of ~17kb of DNA sequence data. Families containing OBL taxa (bold) indicate multiple origins of the aquatic habit within this order. (Adapted from Tank, D.C. & Donoghue, M.J., *Syst. Bot.*, 35, 425–441, 2010.)

Family 1: Asteraceae [1,535]

Also referenced widely by the conserved name Compositae, Asteraceae are generally regarded as the largest angiosperm family having an estimated 23,600+ species (Stevens, 2001; Judd et al., 2008; Panero & Funk, 2008). Arguably, this large and cosmopolitan group is the most intensively studied phylogenetically of any angiosperm family and its relationships have been the subject of rigorous inquiry for decades. Numerous studies have confirmed the monophyly of Asteraceae repeatedly, and there is little doubt regarding the phylogenetic integrity of the group as a clade. The most persuasive phylogenetic analyses resolve Asteraceae and Calyceraceae as sister groups (Figure 5.132). Both families contain tracheary elements with simple pits (Gustafsson & Bremer, 1995), which are likely to be synapomorphic.

The infrafamilial classification of Asteraceae has undergone countless revisions, which seemingly continue incessantly as the constant flow of information from phylogenetic analyses suggests additional refinements. The system used here follows Panero & Funk (2008), who recommended the recognition of 12 subfamilies to delimit the major clades of Asteraceae (Figure 5.133). Tribal designations continue to be evaluated with no strong consensus of accepted circumscriptions emerging at this time.

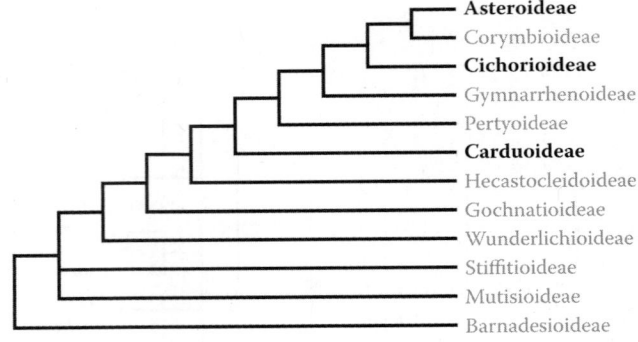

FIGURE 5.133 Major phylogenetic clades (subfamilies) of Asteraceae as indicated by combined analysis of sequence data from 10 cpDNA loci (Adapted from Panero, J.L. & Funk, V.A., *Mol. Phylogen. Evol.*, 47, 757–782, 2008).

Asteraceae are well-characterized morphologically and chemically. Broadly shared (synapomorphic) features include: epigynous flowers (reduced as florets) that are aggregated into heads (capitula), which are surrounded by specialized bracts called phyllaries; sepals modified as various structures (typically related to dispersal) known collectively as the pappus, which ultimately persists on the fruits (cypselae, but often referred to as achenes); anthers fused together, forming a cylindrical passage through which the style elongates and eventually protrudes (acting as a pollen presentation mechanism). The plants lack iridoids but typically contain sesquiterpene lactones (Judd et al., 2008), which can be allergenic and a cause of contact dermatitis.

Representatives of Asteraceae can be found in nearly every vegetated place on earth and also occur in most types of habitats. Although the species primarily inhabit open or dry sites, a few are adapted to survive in wetlands and some even under truly aquatic conditions. However, OBL Asteraceae are relatively uncommon and represent only 112/2413 (4.6%) of the North American species.

Asteraceae include herbaceous plants, shrubs, trees, and vines. Floral morphology is particularly diverse in Asteraceae and comprises numerous variants including asymmetric, strap-like bisexual ("ligulate") or pistillate/neuter ("ray") florets, and perfect, radially symmetrical "disc" florets. The florets can occur homogeneously in a head (e.g., ligulate or discoid), or in various combinations such as radiate heads, which are characterized by central disc florets and marginal ray florets. Many other floret variants and combinations also occur. As in Apiaceae, the heads commonly are arranged secondarily into even more complex inflorescences.

Most Asteraceae are genetically self-incompatible by means of a sporophytic ([sporophytic self-incompatibility (SSI)]) system (Hiscock & McInnis, 2003) and are outcrossed by biotic pollinators, which are attracted readily to the compound floral displays (Judd et al., 2008). Pollination occurs mainly by insects (Insecta), but a number of species are wind pollinated. Some species produce seeds apomictically. In most cases, the fruits are dispersed by the wind, which interacts with their capillary pappus bristles. In some species, the pappus is rigid and barbed, which facilitates epizooic dispersal by their ready attachment to the fur of animals. Fruits of the genus *Bidens* (so named for its toothed pappus) often are found clinging to the clothing (and to the fur of accompanying pets) of those who frequent wetland habitats.

Despite its enormity of species, Asteraceae provide surprisingly few economically important plants, but several of these are quite significant. Foodstuffs include endive (*Cichorium*), globe artichoke (*Cynara scolymus*), lettuce (*Lactuca sativa*), and Jerusalem artichoke (*Helianthus tuberosus*). The North American sunflower (*Helianthus annuus*) is grown as a source of seed oil and as an ornamental. Flavoring and cooking oils are also obtained from safflower (*Carthamus tinctorius*). Guayule (*Parthenium argentatum*) is grown commercially as an alternative source of rubber, which is used in the manufacture of hypoallergenic latex. Coneflower (*Echinacea*) is a popular medicinal plant and also is grown as an ornamental. The mysterious absinthe spirits owe their distinctive flavor to the flowers and foliage of grand wormwood (*Artemisia absinthium*). Other horticulturally important genera include asters (*Symphyotrichum*), blazingstar (*Liatris*), black-eyed Susan (*Rudbeckia*), dahlias (*Dahlia*), daisy (*Chrysanthemum*), goldenrod (*Solidago*), marigolds (*Calendula*), tickseed (*Coreopsis*), and zinnias (*Zinnia*). The common dandelion (*Taraxacum officinale*) is a ubiquitous lawn weed, and ragweeds (*Ambrosia* spp.) are legendary for their allergenic pollen, which is a common cause of "hayfever."

There are 45 North American genera that contain OBL species within the family. These occur mainly in subfamily Asteroideae (41 OBL genera) with the exception of *Cirsium* (Carduoideae) and *Microseris*, *Phalacroseris*, and *Vernonia* (Cichorioideae) (Figure 5.133).

1. *Ampelaster* G. L. Nesom
2. *Arctanthemum* (Tzvelev) Tzvelev
3. *Arida* (R. L. Hartman) D. R. Morgan & R. L. Hartman
4. *Arnoglossum* Raf.
5. *Artemisia* L.
6. *Baccharis* L.
7. *Bidens* L.
8. *Blennosperma* Less.
9. *Boltonia* L'Hér.
10. *Borrichia* Adans.
11. *Carphephorus* Cass.
12. *Cirsium* Mill.
13. *Coreopsis* L.
14. *Cotula* L.
15. *Doellingeria* (Mill.) Nees
16. *Eclipta* L.
17. *Erigeron* L.
18. *Eupatorium* L.
19. *Eurybia* (Cassini) Cassini
20. *Euthamia* (Nutt.) Cass.
21. *Eutrochium* Rafinesque
22. *Gnaphalium* L.
23. *Hartwrightia* A. Gray ex S. Watson
24. *Helenium* L.
25. *Helianthus* L.
26. *Hesperevax* (A. Gray) A. Gray
27. *Jamesianthus* S. F. Blake & Sherff
28. *Jaumea* Pers.
29. *Lasthenia* Cass.
30. *Marshallia* Schreb.
31. *Microseris* D. Don
32. *Mikania* Willd.
33. *Oclemena* Greene
34. *Packera* Á. Löve & D. Löve
35. *Phalacroseris* A. Gray
36. *Pluchea* Cass.
37. *Psilocarphus* Nutt.
38. *Raillardella* (A. Gray) Benth.
39. *Rudbeckia* L.
40. *Sclerolepis* Cass.
41. *Senecio* L.
42. *Shinnersia* R. M. King & H. Robinson
43. *Solidago* L.
44. *Symphyotrichum* Nees.
45. *Vernonia* Schreb.

1. *Ampelaster*

Climbing aster

Etymology: a combination of the Greek *ampelos* (vine) and genus *Aster*

Synonyms: *Aster* (in part); *Lasallea*; *Sitilias*; *Symphyotrichum* (in part); *Virgulus* (in part)

Distribution: global: North America; **North America:** southeastern United States

Diversity: global: 1 species; **North America:** 1 species

Indicators (USA): OBL: *Ampelaster carolinianus*

Habitat: freshwater; palustrine; **pH:** 5.1–6.5; **depth:** <1 m; **life form(s):** emergent shrub or vine

Key morphology: Stems (to 4 m) woody, shrubby, producing weak, vine-like branches that sprawl or climb over other plants; leaves deciduous or evergreen, alternate, sessile, the blades (to 70 mm) 1-nerved, elliptic to ovate or lanceolate, reduced distally, margins entire, bases clasping; heads (to 15/branch) radiate, peduncled (to 4 cm), hairy, arranged secondarily in panicles, the receptacles lacking pales; ray florets (to 70) pistillate, the rays (to 20 mm) pale rose to pink; disc florets (to 50), bisexual, the corollas (to 8 mm) yellow (turning to rose purple); cypselae (to 4.3 mm) brownish, 9–12-ribbed, pappus persistent, of orangish, barbed bristles (30–45)

Life history: duration: perennial (buds); **asexual reproduction:** shoot fragments; **pollination:** insect; **sexual condition:** gynomonoecious; **fruit:** cypsela common; **local dispersal:** stem fragments, cypselae (wind); **long-distance dispersal:** cypselae (wind)

Imperilment: (1) *Ampelaster carolinianus* [G5]

Ecology: general: *Ampelaster* is monotypic.

Ampelaster carolinianus (Walter) G. L. Nesom inhabits tidal or nontidal depressions, ditches, floodplains, roadsides, shores, swamps, thickets, woodlands, and the margins of rivers and streams at elevations of up to 30 m. Occurrences range from open sunny sites to shaded understory on substrates of loamy sand or mud, which reportedly are acidic (pH: 5.1–6.5). Flowering occurs late in the season, commencing in mid-October. Pollination is by bees (Insecta: Hymenoptera) and butterflies (Insecta: Lepidoptera). Clones can form by the production of adventitious roots on branches, which subsequently separate from the plant as ramets. Additional life-history information is unavailable. **Reported associates:** *Acer rubrum, Aeschynomene americana, Ambrosia artemisiifolia, Ampelopsis arborea, Andropogon virginicus, Annona glabra, Aristida purpurascens, Baccharis halimifolia, Bidens pilosus, Boehmeria cylindrica, Callicarpa americana, Cardamine pensylvanica, Carex, Centella asiatica, Cirsium horridulum, Clematis virginiana, Commelina diffusa, Conyza canadensis, Crinum americanum, Cynodon dactylon, Desmodium triflorum, Dichondra carolinensis, Digitaria ciliaris, Diospyros virginiana, Eclipta prostrata, Eleocharis, Eryngium baldwinii, Eupatorium capillifolium, Euphorbia hypericifolia, Galium tinctorium, Hypericum mutilum, Imperata cylindrica, Iris hexagona, Lantana camara, Leitneria floridana, Lepidium virginicum, Liquidambar styraciflua, Ludwigia peruviana, Lygodium japonicum, Lythrum alatum, Macroptilium lathyroides, Magnolia virginiana, Melothria pendula, Micranthemum umbrosum, Mikania scandens, Morrenia odorata, Myrica cerifera, Nyssa, Oxalis corniculata, Oxycaryum cubense, Panicum repens, Parthenocissus quinquefolia, Passiflora incarnata, Persea palustris, Persicaria hydropiperoides, Phlebodium aureum, Phyla nodiflora, Phytolacca americana, Pinus, Quercus laurifolia, Q. virginiana, Rhus copallinum, Rhynchosia cinerea, Rosa palustris, Rubus argutus, Sabal palmetto, Salix caroliniana, Salvia misella, Sambucus nigra, Samolus valerandi, S. ebracteatus, Schinus terebinthifolius, Setaria parviflora, Sida rhombifolia, Solanum diphyllum, Stylisma patens, Taxodium distichum, Thelypteris kunthii, Toxicodendron radicans, Typha latifolia, Ulmus americana, Urena lobata, Viola sororia, Vitis.*

Use by wildlife: The flowers of *Ampelaster carolinianus* are eaten by wild turkeys (Aves: Phasianidae: *Meleagris gallopavo*) and are known to attract bees (Insecta: Hymenoptera: Apidae: *Bombus*) and numerous butterflies (Insecta: Lepidoptera) such as the cloudless sulfur (Pieridae: *Phoebis sennae*), dainty sulfur (Pieridae: *Nathalis iole*), eastern tiger swallowtail (Papilionidae: *Papilio glaucus*), gulf fritillary (Nymphalidae: *Agraulis vanillae*), monarch (Nymphalidae: *Danaus plexippus*), and pearl crescent (Nymphalidae: *Phyciodes tharos*). The plants are also a host to painted lady caterpillars (Lepidoptera: Nymphalidae: *Vanessa virginiensis*), false spider mites (Acari: Tenuipalpidae: *Brevipalpus phoenicis*), fruit flies (Insecta: Diptera: Tephritidae: *Neaspilota achilleae, Trupanea actinobola*), and leaf-miner flies (Insecta: Diptera: Agromyzidae). The hosted leaf-miner flies are parasitized by wasps (Insecta: Hymenoptera: Eulophidae).

Economic importance: food: not reported to be edible; **medicinal:** The Seminole prepared a decoction from *Ampelaster carolinianus* leaves, which was taken to remedy itchy skin; **cultivation:** *Ampelaster carolinianus* is cultivated as an ornamental and is effectively grown in a 4:5:1 (per

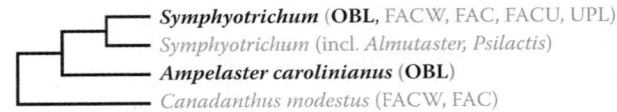

FIGURE 5.134 Phylogenetic position of *Ampelaster* in Asteraceae tribe Astereae, subtribe Symphyotrichinae, which summarizes analyses of four nuclear DNA regions. Although molecular data have not adequately resolved relationships within *Symphyotrichum*, it is evident that the OBL habit has arisen independently in each genus. (Adapted from Vaezi, J. & Brouillet, L., *Mol. Phylogen. Evol.*, 51, 540–553, 2009; Morgan, D.R. & Holland, B., *Syst. Bot.*, 37, 818–832, 2012.)

volume) mixture of compost:pine bark:sand; **misc. products:** none; **weeds:** none; **nonindigenous species:** none

Systematics: *Ampelaster* is placed within subtribe Symphyotrichinae (Asteraceae: tribe Astereae), where it resolves as the sister genus to *Symphyotrichum* (Figure 5.134). Some regard *Ampelaster* as a primitive, woody member of *Symphyotrichum* rather than as a monotypic entity, in which case it is recognized as *S. carolinianum* (Walter) Wunderlin & B. F. Hansen. The merger of these genera has merit, because DNA sequence data indicate that at least one *Symphyotrichum* species (*S. tenuifolium*) has introgressed with *Ampelaster* historically, an indication that the genera are ineffectively isolated. Furthermore, *S. gypsophilum* also associates with *Ampelaster* in molecular-based trees, which indicates that the two taxa either are relatively closely related or also have experienced historical hybridization. The genera are kept separate here pending the outcome of more definitive investigations. Although phylogenetic relationships in *Symphyotrichum* have been difficult to resolve (see Figure 5.171), it is evident that the OBL habit in *Ampelaster* has been derived independently (Figure 5.134). The base chromosome number of *Ampelaster* is *x* = 9, with *A. carolinianus* (2*n* = 18) representing a diploid.

Comments: *Ampelaster carolinianus* occurs along the coastal plain of the southeastern United States.

References: Childers et al., 2003b; Hammer, 2002; Morgan & Holland, 2012; Nesom, 1994, 2000; Schmidt, 2005; Semple, 2006; Stegmaier, Jr., 1968b; Tamang, 2005; Vaezi & Brouillet, 2009; Wilson & Stoffella, 2006; Wunderlin & Hensen, 2001.

2. Arctanthemum

Arctic daisy; chrysantheme arctique

Etymology: from the Greek *arktos anthemon*, meaning northern flower

Synonyms: *Chrysanthemum* (in part); *Dendranthema* (in part); *Leucanthemum* (in part)

Distribution: global: Asia; Eurasia; North America; **North America:** northern

Diversity: global: 1 species; **North America:** 1 species

Indicators (USA): OBL; FACW: *Arctanthemum arcticum*

Habitat: brackish, freshwater; palustrine; **pH:** unknown; **depth:** <1 m; **life form(s):** emergent herb

Key morphology: Shoots (to 40 cm) ascending, solitary from rosette, simple or branching distally; leaves alternate, fleshy, margins entire to dentate; the basal leaves (to 5 cm), wedge shaped to spatulate, 3–7-lobed, petioled (to 9.5 cm); the cauline leaves (to 3.5 cm), alternate, narrowly lanceolate, petioled (to 5.5 cm) to sessile; heads single or in groups of 2–3, radiate, long peduncled; involucre (to 2.9 cm) hemispherical, phyllaries (to 44) in 3–4 series, receptacle dome-like, pales lacking; ray florets (9–30) pistillate, the rays (to 31 mm) white; disc florets to 360+) bisexual, corollas campanulate, 5-lobed, yellow; cypselae (to 2.5 mm) cylindrical to obconical, 5–10-ribbed, pappus lacking.

Life history: duration: perennial (taproot, rhizomes); **asexual reproduction:** rhizomes; **pollination:** insect; **sexual condition:** gynomonoecious; **fruit:** cypsela (common); **local dispersal:** rhizomes, cypselae (gravity); **long-distance dispersal:** cypselae (animals?)

Imperilment: (1) *Arctanthemum arcticum* [G5]; SH (BC); S2 (YT); S3 (MB, ON, QC)

Ecology: general: *Arctanthemum* is monotypic.

 Arctanthemum arcticum **(L.) Tzvelev** inhabits fresh to brackish (salinity = 11.1–14.4 ppt) beaches, dunes, estuaries, flood plains, heath tundras, meadows, rocky coasts, salt marshes, sloughs, swards, tidal flats and marshes, and mouths of streams at elevations of up to 10 m. Plants occur in full sunlight at sites averaging 0.9 m from the mean high water level when growing in coastal environments. The substrates include clay, gravel, rock, sand, and silty loam. Flowering occurs from July to September. There have been no detailed studies of the reproductive biology or seed ecology of this species. The gynomonoecious heads contain flowers that are either pistillate (rays) or bisexual (disc), and likely self-incompatible. They are pollinated and presumably outcrossed by insects (Insecta) such as bees (Hymenoptera) and butterflies (Lepidoptera), which forage for pollen and nectar. The plants occur typically in rather harsh sites, where annual temperatures can range only from 0°C to 9°C, winds can reach up to 150 km/h, average relative humidity can exceed 90% and partial overcast can occur on more than 90% of the days throughout the year. However, disturbances from herbivory are less tolerated, with significantly higher abundances recorded in ungrazed sites. Where they do occur, the plants often represent a significant portion of the total site biomass. In one study the C:N:P ratios for aboveground biomass were determined as C:N = 17.46; C:P = 17.46; N:P = 12.64. The plants are rhizomatous and reproduce vegetatively by that means. **Reported associates:** *Achillea millefolium, Androsace chamaejasme, Angelica lucida, Arctagrostis latifolia, Arctophila fulva, Arctopoa eminens, Arnica unalaschcensis, Artemisia norvegica, Athyrium filix-femina, Atriplex gmelinii, A. patula, Aulacomnium palustre, Bistorta vivipara, Bolboschoenus maritimus, Calamagrostis canadensis, C. deschampsioides, C. lapponica, Carex circinata, C. glareosa, C. lyngbyei, C. pluriflora, C. ramenskii, C. rariflora, C. subspathacea, Cochlearia officinalis, Comarum palustre, Cornus suecica, Deschampsia cespitosa, Dupontia fisheri, Empetrum nigrum, Epilobium, Eriophorum angustifolium, E. russeolum, Festuca rubra, Fritillaria camschatcensis, Galium trifidum, Geum, Hordeum brachyantherum, Iris setosa, Juncus arcticus, J. alpinus, Lathyrus japonicus, Leymus arenarius, L. mollis, Ligusticum scoticum, Lupinus arcticus, Luzula parviflora, Lysimachia maritima, Matricaria discoidea, Montia fontana, Parnassia palustris, Petasites frigidus, Plantago major, P. maritima, Platanthera, Poa arctica, P. glauca, P. pratensis, Polemonium caeruleum, Potentilla anserina, Primula borealis, P. cuneifolia, P. pauciflora, Puccinellia borealis, P. nutkaensis, P. nutalliana, P. phryganodes, P. tenella, Ranunculus cymbalaria, R. flammula, R. hyperboreus, Rhinanthus minor, Rhodiola integrifolia, Rumex arcticus, Salicornia maritima, S. rubra, Salix brachycarpa, S. fuscescens, S. myrtillifolia, S. ovalifolia, Sanionia uncinata, Saussurea nuda, Sibbaldia procumbens, Silene involucrata, Spergularia canadensis, Stellaria humifusa, S. longipes, Suaeda calceoliformis, S. maritima, Taraxacum officinale, Tephroseris palustris, Trientalis europaea, Trifolium repens, Triglochin maritimum, T. palustre, Tripleurospermum maritima, Veronica, Viola langsdorffii, Wilhelmsia physodes.*

Use by wildlife: *Arctanthemum arcticum* is grazed by lesser snow geese (Aves: Anatidae: *Chen caerulescens*) and is strongly associated with their nesting sites. The flowers are visited by various insects (Insecta) including bumblebees (Hymenoptera: Apidae: *Bombus lucorum*) and are a nectar source for several butterflies (Lepidoptera) including the bog fritillary (Nymphalidae: *Boloria eunomia*), giant sulfur (Pieridae: *Colias gigantea*), and Palaeno sulfur (Pieridae: *Colias palaeno*), which likely function as pollinators. The plants also host a sac fungus (fungi: Ascomycota: Pleosporaceae: *Pleospora cerastii*).

Economic importance: food: not reported as edible; **medicinal:** no uses reported; **cultivation:** *Arctanthemum arcticum* is distributed under the cultivar names 'Polarstern', 'Roseum', and 'Schwefelglanz'; **misc. products:** none; **weeds:** none; **nonindigenous species:** none

Systematics: *Arctanthemum* is regarded as a monotypic genus within the "*Chrysanthemum* group" of Asteraceae tribe Anthemideae subtribe Artemisiinae. Phylogenetic analyses of internal transcribed spacer (ITS) and combined ITS/ETS or ITS/*trn*L-F intergenic spacer (IGS) sequence data consistently place *Arctanthemum* with *Ajania pacifica* in a clade, which is nested among numerous *Chrysanthemum* and *Ajania* species (e.g., Figure 5.135). These results clearly indicate that the circumscription of these genera should be reconsidered. *Arctanthemum* was segregated from *Chrsyanthemum* primarily on the basis of its short stature and rosulate (vs. leafy stemmed) habit, which represents a rather weak distinction for generic subdivision. In any case, there seems to be no compelling evidence to support the recognition of *Arctanthemum* as a distinct genus and it seems better treated as comprising a section of *Chrysanthemum*, as it formerly had been recognized; as such it is called *Chrysanthemum arcticum* L. Two subspecies have been distinguished (subsp. *arcticum*, subsp. *polare*), again primarily on the basis of plant stature. The basic chromosome number of *Arctanthemum* is $x = 9$, with *A. arcticum* ($2n = 18$) being diploid. Several anomalous counts ($2n = \sim64, 72, 90$) also have been reported and a comprehensive cytological survey is needed.

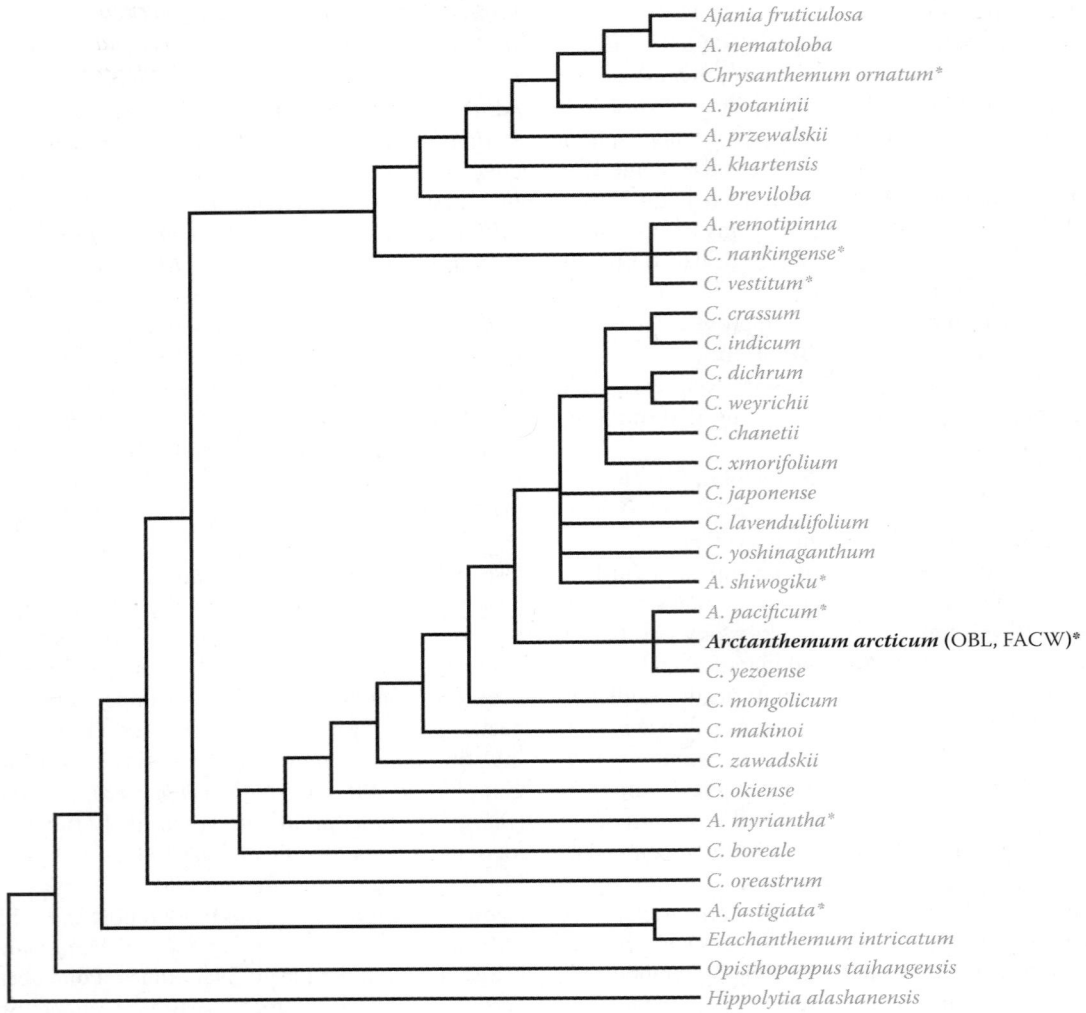

FIGURE 5.135 Phylogenetic position of *Arctanthemum* within Asteraceae tribe Anthemideae (subtribe Artemisiinae) based on analysis of combined ITS/*trn*L-F IGS sequence data. These results indicate that the current generic circumscriptions of *Ajania* and *Chrysanthemum* (together with *Arctanthemum*) represent highly polyphyletic groups (asterisks denote conspicuously misplaced taxa). It seems more reasonable to return to the classification of *A. arcticum* as a species of *Chrysanthemum*, given that it is embedded phylogenetically among most of the other species. Nevertheless, it is evident that the OBL habit of *Arctanthemum arcticum* (bold) is derived uniquely within this group. (Redrawn from Zhao, H.-B. et al., *Plant Syst. Evol.*, 284, 153–169, 2010.)

Comments: *Arctanthemum arcticum* occurs across the colder boreal regions of North America and extends into northern Eurasia.

References: Bennett et al., 2010; Blondeau, 1989; Brouillet, 2006a; Byrd, 1984; Ganter et al., 1996; Handa et al., 2002; Hanson, 1951; Hik et al., 1992; Hines & Ruiz, 2000; Jorgenson, 2000; Lelej et al., 2012; Macdonald & Barbour, 1974; Macoun, 1899; Masuda et al., 2009; Mulder et al., 1996; Ngai & Jefferies, 2004; Oosting & Parshall, 1978; Sellers, 1979; Troy & Wimber, 1968; Viereck et al., 1992; Watson et al., 2002; Zhao et al., 2010.

3. *Arida*

Desert tansy-aster

Etymology: from the Latin *aridus* (dry), in reference to the xeric habitats of many species

Synonyms: *Aster* (in part); *Bigelovia*; *Hazardia*; *Leucosyris*; *Linosyris*; *Machaeranthera* (in part)

Distribution: global: Mexico; North America; **North America:** southwestern United States

Diversity: global: 9 species; **North America:** 6 species

Indicators (USA): OBL: *Arida carnosa*

Habitat: saline; palustrine; **pH:** 8.2–10.1; **depth:** <1 m; **life form(s):** emergent subshrubs

Key morphology: Stems (to 90 cm) 1–10, much branched, partially upright to sprawling, weakly woody at base, green and often fleshy distally; leaves sessile, glabrous, scale-like (to 4 mm) or linear (to 2 cm) and fleshy, the margins entire; flower heads on broadly spreading branchlets (diffusely cymose), the phyllaries (to 7 mm) in 4–5 series; heads discoid, 5–40–flowered, pales lacking; corolla (to 7 mm) radial, yellow (sometimes reddish), 5-lobed; cypselae (to 4 mm) cylindric, ribbed, pappus of whitish, unequal bristles (to 6 mm)

Life history: duration: perennial (rhizomes); **asexual reproduction:** rhizomes; **pollination:** insect; **sexual condition:** hermaphroditic; **fruit:** cypselae (common); **local dispersal:** gravity/wind (cypselae); **long-distance dispersal:** wind (cypselae)

Imperilment: (1) *Arida carnosa* [G3–G4]; S2 (AZ)

Ecology: general: Although the name of this genus ironically reflects an association with dry, arid habitats, the species often occur in mudflats, swales, washes, and along the margins of playas and water sources in otherwise very dry desert regions. However, only two species are ranked as wetland indicators with just one designated as OBL. The flower heads of most of the species are radiate; however, the OBL *Arida carnosa* is discoid. Pollination for most species appears to occur by insects. The seeds possess a long-bristled pappus and are dispersed by wind. Details on germination requirements are not currently available.

Arida carnosa (A. Gray) **D.R. Morgan & R.L. Hartm.** grows in alluvial fans, depressions, ditches, dunes, flats (alkaline, ash, mud), meadows, oases, playas, seeps, sloughs, springs, scrub, washes, and along margins of canals and channels at elevations from −9 to 1554 m. The substrates are saline and alkaline (pH: 8.2–10.1) and include granite sand, sandy clay loam, and silt. The plants flower from August to September. The flowers are visited by bees (Insecta: Hymenoptera), which are probable pollinators. Whether the rayless condition of the heads correlates with any degree of self-compatibility and/or self-pollination has not been determined and more detailed information on the floral biology and seed ecology of this species is needed. **Reported associates:** *Allenrolfea occidentalis, Ambrosia salsola, Anemopsis californica, Apocynum cannabinum, Artemisia tridentata, Atriplex argentea, A. barclayana, A. canescens, A. confertifolia, A. elegans, A. hymenelytra, A. lentiformis, A. parryi, A. phyllostegia, A. polycarpa, A. pusilla, A. torreyi, Baccharis sarothroides, B. sergiloides, Bassia hyssopifolia, Bolboschoenus maritimus, B. robustus, Calochortus striatus, Carex praegracilis, Carsonia sparsifolia, Centromadia pungens, Chloropyron maritimum, Cleomella obtusifolia, C. parviflora, Cressa truxillensis, Crypsis schoenoides, Cyperus laevigatus, Distichlis spicata, Eleocharis rostellata, Ericameria albida, E. nauseosa, Euthamia occidentalis, Frankenia salina, Grayia spinosa, Grindelia fraxinipratensis, Gutierrezia microcephala, Helianthus annuus, H. nuttalli, Heliotropium curassavicum, Hordeum jubatum, Isocoma acradenius, Juncus acutus, J. balticus, J. bufonius, Kochia californica, Lactuca sativa, L. serriola, Larrea tridentata, Lepidium nitidum, Leucosyris arizonica, Lycium andersonii, L. cooperi, L. fremontii, Lythrum californicum, Nitrophila occidentalis, Oxytenia acerosa, Peritoma lutea, Phragmites australis, Plagiobothrys parishii, Pluchea odorata, P. sericea, Polypogon monspeliensis, Populus fremontii, Prosopis glandulosa, P. pubescens, P. velutina, Puccinellia rupestris, Rosa woodsii, Rumex crispus, Ruppia maritima, Sarcobatus vermiculatus, Salix exigua, Schoenoplectus americanus, Senegalia greggii, Sesuvium verrucosum, Solidago confinis, Spartina gracilis, Spergularia macrotheca, Sporobolus airoides, Suaeda nigra, S. occidentalis, Symphyotrichum ascendens, Tamarix aphylla, T. ramosissima, Tetradymia, Triglochin maritimum, Typha domingensis, Zeltnera exaltata.*

Use by wildlife: The flowers of *Arida carnosa* are visited by sand bees (Insecta: Hymenoptera: Andrenidae: *Perdita heterothecae*), which exhibit a similar geographical distribution and probably effect pollination in many cases. The plants are a host of the introduced Australian light brown apple moth (Insecta: Lepidoptera: Tortricidae: *Epiphyas postvittana*).

Economic importance: food: not reported to be edible; **medicinal:** none reported; **cultivation:** *Arida carnosa* is not in cultivation, perhaps due to its rather extreme habitat requirements; **misc. products:** none; **weeds:** none; **nonindigenous species:** none

Systematics: *Arida* is placed in tribe Astereae (subtribe Machaerantherinae) of Asteraceae. *Arida* was initially included within *Aster, Machaeranthera* or as a section of *Arida* or *Machaeranthera*. Phylogenetic studies of this group are difficult to interpret because of conflicting results and equally difficult to discuss based on the past taxonomical wanderings of this genus. For those species surveyed (*A. riparia, A. parviflora, A. turneri*), restriction site data from cpDNA strongly support a sister group relationship of *Pyrrocorna* (as *Machaeranthera* section *Arida*). However, nrITS data resolve *Arida* and *Pyrrocorna* as distant. Such inconsistent relationships between maternally inherited cpDNA and biparental nuclear DNA are difficult to explain without assuming that reticulate evolution has been involved in the origin of several taxa. Further systematic studies need to be conducted with *Arida carnosa*, which phylogenetic analyses place near *Machaeranthera* and *Triniteurybia* (Figure 5.136); however, a more inclusive analysis including other taxa has not been conducted, leaving its precise relationships unresolved. The base chromosome number of *Arida* is $x=5$. *Arida carnosa* ($2n=10$) is a diploid, as are the other species assigned to this genus.

Comments: *Arida carnosa* occurs in Arizona, California, and Nevada and southward into Mexico.

References: El-Lissy, 2007; Ezcurra et al., 1988; Hartman & Bogler, 2006; Holland, 1986; Morefield, 1988; Morgan, 2003; Morgan & Hartman, 2003; Morgan & Simpson, 1992; Morhardt & Morhardt, 2004; Pruski & Hartman, 2012; Selliah & Brouillet, 2008.

4. *Arnoglossum*

Indian-plantain, lamb's tongue

Etymology: from the Greek *arnos glôssa* ("lamb's tongue") alluding to the shape of the leaves

Synonyms: *Cacalia; Mesadenia*

Distribution: global: North America; **North America:** eastern

Diversity: global: 8 species; **North America:** 8 species

Indicators (USA): OBL: *Arnoglossum sulcatum*

Habitat: freshwater; palustrine, riverine; **pH:** unknown; **depth:** <1 m; **life form(s):** emergent herb

Key morphology: Caudex short; stems (to 15 dm) purplish, strongly grooved or ribbed; basal leaves (to 25 cm) lanceolate to ovate, margins crenate, sinuate or entire, petioles (to 45 cm) diminishing in length upward (uppermost leaves sessile); corymbs mostly diffuse; involucre (to 10 mm) cylindric, the phyllaries linear-oblong with strongly winged midribs and scarious margins; heads discoid, 5-flowered; corollas (to 9.5 mm) radial, creamy yellow, greenish or white, 5-lobed, lobes recurved; cypselae (to 4 mm) brown, fusiform, 6–8-ribbed, crowned by numerous (to 120+), capillary pappus bristles (to 7 mm).

Life history: duration: perennial (caudex); **asexual reproduction:** rhizomes; **pollination:** insect; **sexual condition:** hermaphroditic; **fruit:** cypselae (common); **local dispersal:** fruits (wind); **long-distance dispersal:** fruits (wind)

Imperilment: (1) *Arnoglossum sulcatum* [G3]; S1 (GA); S2 (AL); S3 (FL)

Ecology: general: *Arnoglossum* species occur in a range of habitats from dry sand ridges to wet, boggy lowlands. Several species have wetland indicator status (FACU–FACW); however, only *A. sulcatum* is designated as OBL. The presumably self-incompatible flowers are cross-pollinated by various Diptera (flies) Hymenoptera (bees, wasps), and other insects (Insecta). The seeds are dispersed by wind.

Arnoglossum sulcatum (Fernald) H. Robinson inhabits bayheads, bogs, ditches, flatwoods, floodplains, seeps, swales, swamps, thickets and margins of lakes, ponds, and streams at elevations from 1 to 100 m. The plants occur in exposures of full sun to shade or partial shade and on substrates that include loam, loamy sand, muck, peat, and sandy peat. The plants perennate by means of a short, thick caudex, from which small rosettes are produced by basal offshoots. These rosettes become detached as their connecting stems deteriorate and may take several years before flowering shoots are produced. Dispersal occurs primarily by the seeds, which are provided with a pappus. This species occurs more often where the duration of flooding is not excessive, and where occasional fires help to reduce competing vegetation. Nothing has been reported on the pollination biology or seed ecology of this species. **Reported associates:** *Andropogon glomeratus, Arundinaria gigantea, Ctenium aromaticum, Cyrilla racemiflora, Juncus trigonocarpus, Liriodendron tulipifera, Magnolia virginiana, Nyssa, Pinus elliottii, P. palustris, Sarracenia leucophylla, Sphagnum, Toxicodendron vernix.*

Use by wildlife: none reported

Economic importance: food: The powdered leaves of some *Arnoglossum* species have been used as a spice; **medicinal:** Antitumor activity is associated with highly oxygenated oplopanes derived from *Arnoglossum atriplicifolium*; however, *A. sulcatum* has not been screened for similar substances; **cultivation:** not in cultivation; **misc. products:** none; **weeds:** none; **nonindigenous species:** none

Systematics: *Arnoglossum* (subfamily Asteroideae) is placed within tribe Senecioneae and represents one group of "cacalioid" species that were associated formerly with the genus *Senecio*. Like other cacalioids, *Arnoglossum* is characterized by having an entire (vs. divided or cleft) stigma morphology.

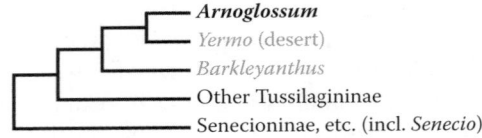

FIGURE 5.136 Phylogenetic relationships of *Eurybia* and *Arida*, two Astereae with OBL North American indicators (bold) based on combined nrETS/ITS sequence data. The OBL taxa are not particularly closely related, indicating independent origins of the habit in this group. (Adapted from Selliah, S. & Brouillet, L., *Botany*, 86, 901–915, 2008.)

Cacalia itself largely has been dismantled taxonomically into *Arnoglossum*, *Parasenecio*, and other segregate genera. The systematic position of *Arnoglossum* has been clarified much by phylogenetic analyses of DNA sequence data, which place the genus within the subtribe Tussilagininae, in a position not with traditional *Senecio* species, but instead in a distant clade as the sister group of the genus *Yermo*, a desert inhabitant (Figure 5.137). The OBL habit clearly is derived independently in *Arnoglossum* (Figure 5.137). The base chromosome number for *Arnoglossum* is $x=28$. The genus is highly polyploid with counts reported for all species representing a descending aneuploid series of $2n=56, 54, 52,$ or 50; *A. sulcatum* is $2n=50$.

Comments: *Arnoglossum sulcatum* occurs only in Alabama, Florida, and Georgia.

References: Anderson, 2006; Barkley, 1985; Flood, 2010; Godfrey & Wooten, 1981; Harper, 1903; McDearman, 1984; Pelser et al., 2007, 2010; Shaw, 2008; Wetter, 1983.

5. *Artemisia*

Wormwood; armoise

Etymology: after Artemis, the mythical Greek goddess

Synonyms: *Absinthium; Chamartemisia; Crossostephium; Dracunculus; Elachanthemum; Filifolium; Mausolea; Neopallasia; Oligosporus; Picrothamnus; Seriphidium; Sphaeromeria*

Distribution: global: Africa; Asia; Europe; New World; **North America:** widespread

Diversity: global: 350–500 species; **North America:** 50 species

Indicators (USA): OBL: *Artemisia lindleyana*

Habitat: freshwater; palustrine, riverine; **pH:** unknown; **depth:** <1 m; **life form(s):** emergent herb

Key morphology: Shoots (to 70 cm) erect, clustered, woody at base; leaves (to 5 cm) simple, linear-oblanceolate, entire or tipped by a few teeth or narrow divisions, green, and glabrous above, matted with dense white hairs below; heads with 5–9 ray and 10–30 disk florets, corollas yellow or somewhat reddish; involucre campanulate, of linear or lanceolate, white hairy phyllaries (to 3 mm); cypselae (to 0.5 mm) elliptical, glabrous.

Life history: duration: perennial (taproot); **asexual reproduction:** none; **pollination:** wind; **sexual condition:** hermaphroditic; **fruit:** cypselae (common); **local dispersal:**

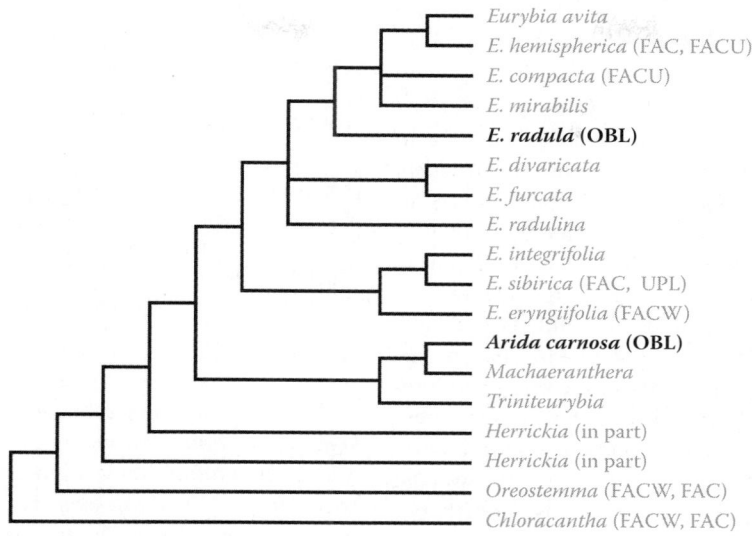

Eurybia avita
E. hemispherica (FAC, FACU)
E. compacta (FACU)
E. mirabilis
E. radula (OBL)
E. divaricata
E. furcata
E. radulina
E. integrifolia
E. sibirica (FAC, UPL)
E. eryngiifolia (FACW)
Arida carnosa (OBL)
Machaeranthera
Triniteurybia
Herrickia (in part)
Herrickia (in part)
Oreostemma (FACW, FAC)
Chloracantha (FACW, FAC)

FIGURE 5.137 Simplified summary showing proximate relationships of *Arnoglossum* as indicated by analyses of nrDNA and cpDNA sequences. *Arnoglossum* (bold) contains only one OBL species and its sister genus *Yermo* is a desert inhabitant. Consequently, it is evident that the OBL habit has been derived independently in the genus. These studies also illustrate the large phylogenetic separation of *Arnoglossum* and *Senecio*, with which it was once presumed to be closely related. (Adapted from Pelser, P.B. et al., *Amer. J. Bot.*, 97, 856–873, 2010.)

fruits (water, wind); **long-distance dispersal:** fruits (water, wind)

Imperilment: (1) *Artemisia lindleyana* [G5]; S3 (<u>BC</u>)

Ecology: general: *Artemisia* includes annual, biennial, and perennial herbs and shrubs with characteristically aromatic foliage. Most species typically inhabit dry, open areas and are not found in wetlands. The majority of species are wind pollinated. The plants contain various sesquiterpene lactones and terpenoids, which are believed to deter herbivores. The seeds of many of the temperate perennial species have physiological dormancy. Although the pappus is deciduous and lacking, dispersal of the very small and extremely light fruits can occur by water, wind, through shifting seed-bearing substrates or by animals.

Artemisia lindleyana **Besser.** occurs on floodplains, gravel bars, and along the margins of lakes, rivers, and streams at elevations from 275 to 4115 m. Plants are found in exposures ranging from open sun to light shade and on substrates of cobble, gravel, rock, sand, silt and talus. Flowering occurs from mid-summer to mid-fall and populations exhibit high pollen fertility (83%–100%). The small fruits probably are dispersed by both wind and water although there are no detailed studies of the reproductive biology or seed ecology of this species. Conditions for seed germination are unknown. Rhizomes apparently are not produced but the taproot can divide; however, vegetative reproduction has not been documented in this species. The uncertain distinction of this species from *A. ludoviciana* makes it difficult to compile accurate data; however, the list of associated species has been limited to those occurrences either specifically referring to *A. lindleyana* or at least to wetland sites. **Reported associates:** *Achillea millefolium, Agrostis, Allium brevistylum, A. cernuum, Amorpha fruticosa, Apocynum cannabinum, Aristida purpurea,* *Arnica, Artemisia campestris, A. dracunculus, A. tridentata, Bromus tectorum, Carex, Chrysothamnus, Coreopsis tinctoria, Cyperus esculentus, C. squarrosus, Deschampsia cespitosa, Dermatocarpon fluviatile, Dieteria canescens, Elaeagnus angustifolia, Eleocharis palustris, Elymus lanceolatus, Epilobium, Equisetum hyemale, Eragrostis hypnoides, E. pectinacea, Ericameria nauseosa, Eriogonum compositum, E. umbellatum, Euthamia occidentalis, Fraxinus latifolia, Gaillardia aristata, Helenium autumnale, Holodiscus discolor, Juncus, Leymus cinereus, L. racemosus, Lindernia dubia, Medicago lupulina, Melilotus alba, Mentha arvensis, M. pulegium, Morus alba, Pascopyrum smithii, Persicaria maculata, Phalaris arundinacea, Pinus contorta, Plantago lanceolata, P. major, Poa bulbosa, P. compressa, Populus balsamifera, P. tremuloides, Ribes, Rorippa columbiae, R. curvisiliqua, Rosa woodsii, Salix exigua, S. fluviatilis, S. lucida, Schoenoplectus acutus, Senecio triangularis, Sphenosciadium capitellatum, Sporobolus compositus, S. cryptandrus, Stellaria, Symphyotrichum subspicatum, Thalictrum sparsiflorum, Trifolium arvense, Ulmus pumila, Verbena bracteata, Veronica anagallis-aquatica, Xanthium strumarium.*

Use by wildlife: Most *Artemisia* species contain sesquiterpene lactones, which can be toxic if plants are grazed heavily by livestock. Lepidoptera (Insecta) larvae associated with *Artemisia ludoviciana* (but not necessarily *A. lindleyana*) include several Nymphalidae (*Vanessa cardui, V. virginiensis*) and Tortricidae (*Phaneta artemisiana, P. infimbriana*). Likewise, fungi hosted by *A. ludoviciana* (but not necessarily *A. lindleyana*) include Ascomycota (Botryosphaeriaceae: *Phyllosticta raui*; Erysiphaceae: *Golovinomyces orontii*; Mycosphaerellaceae: *Ramularia artemisiae*), Basidiomycota (Pucciniaceae: *Puccinia atrofusca, P. chrysanthemi, P.*

cnici-oleracei, *P. conferta*, *P. dracunculina*, *P. ludovicianae*), and Oomycota (Peronosporaceae: *Paraperonospora leptosperma*).

Economic importance: food: The Gosiute people once ate the seeds of *A. ludoviciana* subsp. *incompta* as a source of food; **medicinal:** The sesquiterpene lactones of *Artemisia* species are allergenic and exposure to *A. lindleyana* has induced dermatitis in some people. The pollen reportedly also is highly allergenic. Several ethnobotanical uses are reported for *A. ludoviciana* subsp. *incompta*. These include various tribal uses as general medicinals (Bella Coola), headache remedy (Carrier, Northern), and as a poultice to treat sprains and swelling (Carrier, Southern); **cultivation:** *Artemisia lindleyana* is grown occasionally as a native garden specimen; **misc. products:** *Artemisia lindleyana* contains a large number of methylated flavonoids of uncertain function; **weeds:** *Artemisia lindleyana* is not reported as weedy; **nonindigenous species:** none

Systematics: *Artemisia* is complex taxonomically. The genus typifies subtribe Artemisiinae of Anthemideae (subfamily Asteroideae), which is monophyletic when *Hippolytia* and *Nipponanthemum* are included. Although *Artemisia* has been subdivided variously into a dozen segregate genera, a broader generic concept is most consistent with clades resolved by molecular phylogenetic analyses. So interpreted, the genus would include species uniformly possessing the "*Artemisia*-type" pollen, and which resolve as the sister clade to *Kaschgaria* (Figure 5.138). Several of the segregate genera (*Absinthium*, *Dracunculus*, *Seriphidium*) occur as discrete clades within the genus and typically are recognized as subgenera. *Artemisia lindleyana* has been merged (with subsp. *incompta*) under *A. ludoviciana* (FACU–, FACU, UPL), a montane species that occurs rarely in wet areas and is found mainly at higher elevations (>1900 m). However, this subspecies is said to vary, with the component referred to as *A. lindleyana* possibly warranting discrete recognition. Consequently, its species status is retained here in order to include it among the OBL taxa. Because *Artemisia ludoviciana* is most closely related to nonwetland North American species (Figure 5.139), it is evident that the OBL habit was derived uniquely in *A. lindleyana*. The basic chromosome number of *Artemisia* is $x=9$. *Artemisia lindleyana* has diploid races ($2n=18$), but predominantly is tetraploid

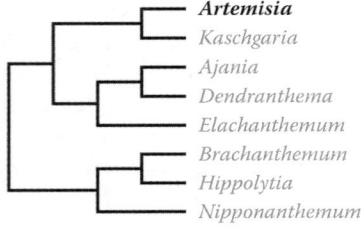

FIGURE 5.138 Phylogenetic placement of *Artemisia* as indicated by analysis of nuclear ETS and ITS sequence data. The OBL habit (bold) is uniquely derived in *Artemisia*. (Adapted from Sanz, M. et al., *Taxon*, 57, 66–78, 2008.)

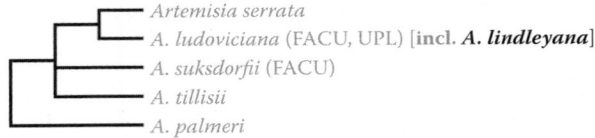

FIGURE 5.139 Phylogenetic relationships among several North American species of *Artemisia* as resolved by analysis of nrITS sequence data. The OBL habit (bold) is restricted to *A. lindleyana*, which has been merged with *A. ludoviciana* (as indicated here), but also is regarded as a potentially distinct segregate. (Adapted from Watson, L.E. et al., *BMC Evol. Biol.*, 2, 17, 2002.)

($2n=36$). Crosses made between tetraploid races of *A. lindleyana* (×*A. douglasiana*; ×*A. ludoviciana* subsp. *incompta*; ×*A. michauxiana*) show high pollen fertility (94%–100%); however, only *A. lindleyana* is not maintained as a distinct species among these. The taxonomic status of *A. lindleyana* and its degree of relationship to *A. ludoviciana* deserve further study.

Comments: *Artemisia lindleyana* is limited to western North America (British Columbia, Arizona, California, Idaho, Montana, New Mexico, Oregon, Washington)

References: Baskin & Baskin, 1998; Christy, 2004; Estes, 1969; McCain & Christy, 2005; McCormick & Bohm, 1986; Sanz et al., 2008; Shultz, 2006; Vallès & McArthur, 2001; Watson et al., 2002.

6. *Baccharis*

Groundsel-tree; baccaride

Etymology: from *baccaris*, the Latin name of a plant with fragrant roots

Synonyms: *Achyrobaccharis*; *Baccharidastrum*; *Molina*; *Neomolina*; *Pingraea*; *Polypappus*; *Pseudobaccharis*; *Psila*

Distribution: global: New World; **North America:** southern

Diversity: global: 320 species; **North America:** 21 species

Indicators (USA): OBL: *Baccharis douglasii*

Habitat: brackish (coastal); freshwater; palustrine; **pH:** 7.5; **depth:** <1 m; **life form(s):** emergent subshrub

Key morphology: Stems (to 210 cm) simple, erect, striate, glandular, resinous, woody at base; leaves (to 130 mm) lanceolate, entire or finely dentate, black gland dotted, 1–3-nerved, short petiolate; heads unisexual with either staminate (to 40) or pistillate (to 150) florets; involucre hemispherical with narrowly lanceolate, gland dotted, resinous phyllaries (to 4 mm); cypselae (to 1.5 mm), glandular, with a capillary, scabrous pappus (to 7 mm)

Life history: duration: perennial (rhizomes); **asexual reproduction:** rhizomes; **pollination:** insect; **sexual condition:** dioecious; **fruit:** cypselae (common); **local dispersal:** rhizomes, fruits (wind); **long-distance dispersal:** fruits (wind)

Imperilment: (1) *Baccharis douglasii* [G5]; S1 (OR)

Ecology: general: *Baccharis* is most diverse in the warmer portions of the New World, exhibiting the highest diversity in tropical and subtropical montane areas. Most species are woody and include shrubs to small trees. The flowers are pollinated by insects (Insecta) including small bees (Hymenoptera), flies

(Diptera), and beetles (Coleoptera). The species predominantly are dioecious and are necessarily outcrossed. Abiotic (wind) dispersal of the fruits is facilitated by the persistent pappus. Of several species studied, the fruits germinated at 20°C/30°C without requiring pretreatment. Germination often occurs in disturbed mineral soils. Twelve *Baccharis* species have some status as wetland indicators (FACW, FAC, FACU, UPL) but only one species is categorized as OBL.

Baccharis douglasii **DC.** inhabits fresh or brackish beaches, ditches, dry lake beds, dunes, flats, floodplains, marshes, meadows, mudflats, pools, river bottoms, salt marshes, seeps, sloughs, springs, streams, swales, swamps, thickets, washes, and margins of lakes, ponds, and streams at elevations of up to 1200 m. Most occurrences are reported in sites with shallow (<15 cm) water. The plants tolerate full sunlight to light shade and occur on various substrates described as diatomaceous earth, gravel, loam, sand, sandy gravel, sandy loam, serpentine, silt, silty clay loam, and silty loam. Neither the floral biology nor seed ecology of this species has been studied in detail. The plants flower from July to September and are pollinated primarily by insects (Insecta) including bees (Hymenoptera) and flies (Diptera). Vegetative reproduction occurs by means of rhizomes and can result in populations with extensive clonal structure. Although designated as OBL, one study found that plants commonly occurred in sites that did not experience periodic anaerobic conditions. This species also has been observed on cutover former stands of redwood (*Sequoia sempervirens*). **Reported associates:** *Acer macrophyllum, Aesculus californica, Alnus rhombifolia, Ambrosia chamissonis, Anemopsis californica, Anthemis cotula, Apium graveolens, Artemisia californica, A. douglasiana, A. dracunculus, Arundo donax, Atriplex leucophylla, A. patula, A. prostrata, A. semibaccata, Avena barbata, Baccharis pilularis, B. salicifolia, B. salicina, Bassia, Bolboschoenus maritimus, B. robustus, Brassica, Bromus diandrus, B. hordeaceus, B. madritensis, Cakile maritima, Carex barbarae, C. praegracilis, Carpobrotus edulis, Centaurea melitensis, C. solstitialis, Cirsium, Clematis ligusticifolia, Conium maculatum, Conyza canadensis, Coreopsis gigantea, Cortaderia, Cynodon dactylon, Cyperus eragrostis, Distichlis spicata, Eleocharis palustris, Epilobium canum, E. ciliatum, Equisetum hyemale, E, telmateia, Eucalyptus camaldulensis, Ficus carica, Foeniculum vulgare, Frangula californica, Frankenia salina, Fraxinus latifolia, Grindelia hirsutula, Helminthotheca echioides, Heteromeles arbutifolia, Heterotheca grandiflora, Hordeum murinum, Isocoma menziesii, I. veneta, Jaumea carnosa, Juglans californica, Juncus acutus, J. balticus, J. effusus, J. mexicanus, J. phaeocephalus, J. xiphioides, Limonium californicum, Lolium multiflorum, Lupinus albifrons, Lythrum, Marrubium vulgare, Melilotus officinalis, Mentha arvensis, M. pulegium, Mimulus cardinalis, M. guttatus, Myoporum laetum, Nasturtium officinale, Oxalis pes-caprae, Paspalum dilatatum, Persicaria maculata, P. punctata, Piptatherum miliaceum, Plantago major, Platanus racemosa, Pluchea carolinensis, P. sericea, Populus fremontii, Polypogon monspeliensis, Pseudognaphalium luteoalbum, Quercus agrifolia, Q. lobata, Rosa californica, Rubus armeniacus, R. discolor, R. ursinus, R. vitifolius, Rumex crispus,* *R. salicifolius, Salicornia, Salix babylonica, S. gooddingii, S. laevigata, S. lasiolepis, Sambucus, Schoenoplectus californicus, Silybum marianum, Solanum douglasii, Solidago, Stephanomeria, Symphyotrichum lanceolatum, S. subulatum, Toxicodendron diversilobum, Typha angustifolia, T. domingensis, T. latifolia, Umbellularia californica, Urtica dioica, Verbena lasiostachys, Vitis californica, V. girdiana, Vulpia bromoides.*

Use by wildlife: *Baccharis douglasii* provides habitat for brush mice (Mammalia: Cricetidae: *Peromyscus boylii*). It is eaten by desert cottontails (Mammalia: Leporidae: *Sylvilagus audubonii*) and riparian brush rabbits (Mammalia: Leporidae: *Sylvilagus bachmani riparius*). The flowers are foraged for their nectar by a number of bees (Insecta: Hymenoptera: Colletidae: *Hylaeus episcopalis, H. mesillae, H. modestus, H. nevadensis, H. polifolii, H. rudbeckiae*), and are also visited by soldier flies (Insecta: Diptera: Stratiomyidae: *Odontomyia hirtocculata*). The foliage is fed on by a leaf beetle (Insecta: Coleoptera: Chrysomelidae: *Trirhabda flavolimbata*) and by several leaf mining flies (Insecta: Diptera: Agromyzidae: *Liriomyza douglasii, L. togata*). This species is regarded as a good nectar source for various butterflies and moths (insecta: Lepidoptera) including hairstreaks (Lycaenidae) such as the tailed copper (*Lycaena arota*), and acmon blue (*Plebejus acmon*), and Nymphalidae such as the American painted lady (*Vanessa virginiensis*) and common buckeye (*Junonia coenia*). The plants also host fungi (Ascomycota: Mycosphaerellaceae: *Cercospora baccharidis*).

Economic importance: food: not edible; **medicinal:** The Costanoan Indians used *Baccharis douglasii* for cleansing wounds (decoction), as a disinfectant (powdered stems), to treat boils (leaf poultice), and for kidney problems (infusion). The Luiseño boiled whole plants to make an infusion for treating sores; **cultivation:** *Baccharis douglasii* is distributed by some native plant nurseries; **misc. products:** The Luiseño used the stems of *Baccharis douglasii* as the shaft for fire-starting tools. Some species (including the potentially conspecific *B. glutinosa*) have been planted for erosion control; **weeds:** none; **nonindigenous species:** none

Systematics: *Baccharis* (subfamily Asteroideae) has been placed in tribe Astereae subtribe Baccharidinae, where it resolves close to *Heterothalamus* (Figure 5.140), with which it has been merged by some authors. *Baccharis douglasii*

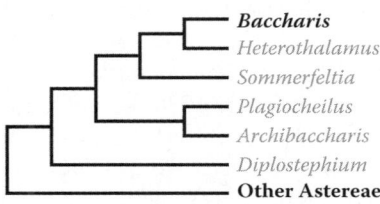

FIGURE 5.140 Phylogenetic placement of *Baccharis* as indicated by analysis of combined ETS and ITS sequence data. The absence of OBL taxa (indicated in bold) in the related genera *Heterothalamus* and *Sommerfeltia* indicates an independent origin of the habit in *Baccharis*. (Adapted from Karaman-Castro, V. & Urbatsch, L.E., *Syst. Bot.*, 34, 805–817, 2009.)

apparently has not yet been included in phylogenetic studies of this large genus, so its interspecific relationships remain uncertain. In a recent taxonomic (but not phylogenetic) overview of the genus, *B. douglasii* was placed in synonymy with the principally South American *B. glutinosa* (subgenus *Molina*) despite having a longer corolla (exceeding half the style length), than typical representatives of the species, which have corollas rarely reaching half the style length. Although it seems probable that *B. douglasii* and *B. glutinosa* are at least fairly closely related, additional study (preferably incorporating genetic data) should be undertaken to better evaluate the taxonomic status of these species. The base chromosome number of *Baccharis* is $x=9$; *Baccharis douglasii* ($2n=18$) is diploid.

Comments: *Baccharis douglasii* occurs in California and Oregon but extends southward into Mexico.

References: Baskin & Baskin, 1998; Byrd & Kelly, 2006; Chapman & Willner, 1978; Evens & San, 2004; Huffman, 2004; James, 1960; Karaman-Castro & Urbatsch, 2009; Lieneman, 1929; Müller, 2006; Spencer, 1981; USDA, 1948; Williams, 1988.

7. Bidens

Beggar-ticks, water-marigold; bident, fourchette

Etymology: from the Latin *bi dens* ("two teeth") for the paired pappus bristles of some cypselae

Synonyms: *Campylotheca; Coreopsis* (in part); *Diodonla; Megalodonta; Microlecane*

Distribution: global: Africa; North America; South America; **North America:** widespread

Diversity: global: 150–250 species; **North America:** 25 species

Indicators (USA): OBL: *Bidens aureus, B. beckii, B. bigelovii, B. cernuus, B. connatus, B. hyperboreus, B. laevis, B. mitis, B. trichospermus, B. tripartitus, B. vulgatus*; **FACW:** *B. bigelovii, B. tripartitus*

Habitat: brackish (coastal); freshwater (tidal/nontidal); lacustrine, palustrine, riverine; **pH:** 3.7–9.8; **depth:** 0–7 m; **life form(s):** emergent herb, submersed (vittate)

Key morphology: Stems (to 400 cm) usually single, round or 4-angled, commonly striate or grooved; leaves (to 4.5–22 cm) cauline, opposite, simple or pinnately/laciniately compound (submersed foliage highly dissected into capillary segments in *B. beckii*), blades/pinnae deltoid, lanceolate, linear, or ovate, margins entire to serrate, petioled (to 20–50 mm) or sessile; heads discoid (rayless) or radiate with to 6–13 ray florets (orange, white, or yellow) and to 25–150 disc florets (orange to yellow); involucre of 3–21 leafy bracts/bractlets (to 5–75 mm); phyllaries (4–30) biseriate (to 6–12 mm), their margins usually hyaline; cypselae (to 5–15 mm), compressed or flat, 3–4-angled, the margins sometimes corky winged, awnless or with a persistent pappus of 1–4 antrorsely or retrorsely barbed awns (to 1–7 mm).

Life history: duration: annual (fruit/seeds); perennial (rhizomes, whole plants); **asexual reproduction:** leaf (cotyledon) fragments, rhizomes, turions (stem); **pollination:** insect; **sexual condition:** hermaphroditic; **fruit:** cypselae

(common); **local dispersal:** fruits (water, wind); **long-distance dispersal:** fruits (animals, water, wind)

Imperilment: (1) *Bidens aureus* [G5]; (2) *B. beckii* [G4]; SH (OH); S1 (CT, IA, IL, IN, NJ, NH, PA, SK); S2 (MA, MT); S3 (BC, MB, NB, NS, NY, QC); (3) *B. bigelovii* [G5]; S1 (OK); (4) *B. cernuus* [G5]; S1 (AL); S2 (YT); S3 (GA, NC, WY); (5) *B. connatus* [G5]; SH (WV); S1 (RI); S2 (PE); S3 (NS); (6) *B. hyperboreus* [G4]; SH (NH); S1 (MA, NS, NY, ON); S3 (ME, NB); (7) *B. laevis* [G5]; S1 (MO, NH); S2 (NY); S3 (NC, PA); (8) *B. mitis* [G4]; S1 (MD, NJ); S2 (DE, TX); S3 (NC); (9) *B. trichospermus* [G5]; SH (NC, WV); S 1 (RI); S2 (MD, ON); S3 (DE); (10) *B. tripartitus* [G5]; S2 (WY); (11) *B. vulgatus* [G5]; S1 (BC, DE); S2 (WY); S3 (NC)

Ecology: general: *Bidens* includes herbaceous annuals and perennials. It is a common resident of wetlands and 75% of the North American taxa have some level of status as wetland indicators (at least as FACW); 11 of these are OBL, and one species (*B. beckii*) is a submersed aquatic. Many of the species are quite vigorous and can reach large numbers of individuals (especially the annuals) when growing in suitable sites. The genus includes both self-compatible and self-incompatible species; however, the plants are pollinated primarily by insects (Insecta) such as bees (Hymenoptera), butterflies (Lepidoptera), and flies (Diptera). Some species have mixed mating systems that include both selfing and outcrossing modes; some are principally autogamous. The capitula can be rayed or rayless. Interestingly, at least one rayed species (*B. cernuus*) appears to be normally sexual; whereas, several rayless taxa (e.g., *B. connatus, B. frondosus, B. vulgatus*) appear to be at least partially apomictic. Conceivably, apomixis would ensure seed set in cases where pollinator visitation was severely reduced because of the rayless floret condition. Dispersal of the fruits occurs mainly by their attachment to the fur or feathers of animals by their barbed pappus bristles; however many of the fruits also float and are dispersed on the surface of the water. Where the pappus is absent or reduced, the fruits are dispersed by wind or water. Species of colder temperate regions typically have physiologically dormant seeds, which must be cold-stratified to induce germination.

***Bidens aureus* (Aiton) Sherff** is an annual or perennial observed on floodplains, and in cienega, flats, marshes, meadows, pools, springs, stream bottoms, washes, and along the margins of channels, rivers, and streams at elevations from 900 to 2835 m. The plants occupy shaded to open sites on substrates described as clay, sand, fine silt, gneiss, granite, gravel, loam, muck, peat, and rock outcrops. They grow in warmer regions, tolerating low temperatures to −12°C. Flowering occurs from April to November. Higher biomass and greater flower production has been achieved when plants are grown under high nutrient conditions. Application of daminozide ($4.25\,g/dm^3$) will slow growth but it increases the production of flower buds. In most specimens the pappus is absent but in some it develops as several barbed awns. Reportedly the plants propagate readily from seeds (germinating in about 20 days), if collected in late autumn and stored at room temperature until spring (April). The perennial form overwinters by rhizomes. This species has been observed to be stable in burned sites. **Reported associates:** *Acacia angustissima, A. greggii, Acalypha neomexicana, Agave schottii, Ageratina herbacea, Amauriopsis dissecta, Ambrosia ambrosioides, A. psilostachya, Anisacanthus, Antennaria, Arbutus arizonica, Arundo donax, Astrolepis sinuata, Baccharis neglecta, B. salicifolia, Berberis wilcoxii, Berula erecta, Bidens aureus, B. bigelovii, B. laevis, Boerhavia scandens, Bothriochloa barbinodis, B. laguroides, Bouteloua aristidoides, B. chondrosioides, B. curtipendula, B. gracilis, Brachystigma wrightii, Brickellia amplexicaulis, B. coulteri, B. eupatorioides, B. floribunda, B. grandiflora, Carex pellita, C. praegracilis, Carminatia tenuiflora, Castilleja patriotica, Celtis laevigata, C. pallida, Cenchrus setaceus, Cercocarpus montanus, Cheilanthes bonariensis, C. lindheimeri, Chenopodium ambrosioides, Comandra umbellata, Coreocarpus arizonicus, Cottea pappophoroides, Cynodon dactylon, Cyperus, Dasylirion wheeleri, Digitaria sanguinalis, Dodonaea viscosa, Echinocereus coccineus, Echinochloa colona, Eleocharis palustris, Elymus canadensis, E. trachycaulus, Epilobium canum, Eragrostis intermedia, Ericameria nauseosa, Eriochloa acuminata, Eryngium lemmonii, Fouquieria splendens, Fraxinus velutina, Hilaria belangeri, Houstonia, Hydrocotyle, Hymenoxys hoopesii, Ipomoea costellata, I. cristulata, I. hederacea, Ipomopsis thurberi, Juglans major, Juncus balticus, Juniperus coahuilensis, J. deppeana, Lasianthaea podocephala, Leptochloa panicea, Linum, Lobelia cardinalis, Macroptilium, Melinis repens, Mentzelia isolata, Mimosa aculeaticarpa, Mimulus guttatus, Morus microphylla, Muhlenbergia asperifolia, M. emersleyi, M. microsperma, M. rigens, Oenothera speciosa, Opuntia engelmannii, Oxalis alpina, Panicum bulbosum, P. hirticaule, Pappophorum vaginatum, Parkinsonia aculeata, P. florida, Paspalum dilatatum, Pinus cembroides, P. discolor, P. engelmannii, P. leiophylla, P. ponderosa, P. strobiformis, Platanus wrightii, Polypogon monspeliensis, Populus fremontii, P. tremuloides, Porophyllum ruderale, Prosopis, Pseudognaphalium canescens, Quercus arizonica, Q. emoryi, Q. hypoleucoides, Q. oblongifolia, Q. rugosa, Q. turbinella, Ranunculus macranthus, Rhus trilobata, Rumex, Salix exigua, S. gooddingii, S. nigra, Schoenoplectus americanus, Setaria leucopila, Solidago velutina, Sorghum halepense, Spiranthes delitescens, Sporobolus wrightii, Tagetes lemmonii, Toxicodendron radicans, Typha, Viguiera cordifolia, Vilis arizonica, Yucca madrensis.*

***Bidens beckii* Torr. ex Spreng.** is a submersed perennial, which grows at depths to 7.0 m in still or slow-moving waters of bogs, lagoons, lakes, ponds, rivers, sloughs, and streams at elevations of up to 300 m. This heterophyllous species is emergent in shallow water but grows entirely submersed (with consistently dissected foliage) at greater depths. Waters mainly are alkaline (pH: 5.6–9.8; means=7.6; 8.0) with low conductivity (25–225 μmhos/cm) and alkalinity (<190 mg/l CaCO$_3$; means=48.8; 77; 90.6 mg/l). The plants usually occur on soft substrates but also are reported variously on marl, muck, mud, rock, sand, sandy mud or silt bottoms. They are known to overwinter at the bottom of ice-covered lakes in a physiologically active state. Categorized as a deep water species, the shoots can survive for 24 days under hydrostatic pressures comparable to those at a 21 m depth. However, plants flower and set seed (July–September) mostly when in quiet, shallow water. In one region (British Columbia) fewer than 7% of populations were observed to flower regularly. Pollination occurs by insects (Insecta), which include Coleoptera, Diptera, Hymenoptera, Lepidoptera, and Odonata. The lack of seed set observed on isolated heads indicates that plants are genetically self-incompatible. Seed set in natural populations typically is quite low, with only a few fruits maturing on each head. The seeds are denser than water and sink when shed from the plants; thus, dispersal must occur mainly by water currents. The seeds will germinate when soaked on filter paper after removal of the pericarp and seed coat. Vegetative reproduction occurs by rhizomes, shoot fragmentation, and submersed shoot turions. The rhizomes are produced extensively in shallow water, but become less abundant (and often deteriorate) in deeper water. The turions remain viable after 108 days of storage at 4°C, but are killed when frozen (e.g., −6°C) for prolonged periods. Viable turions quickly expand at 21°C when exposed to 16 h of light. Normally the species occurs at relatively low cover (1.5%–2.5%), but local densities can approach 400 plants/m^2. Increased abundance has been observed following successive periods of winter lake drawdown as the plants spread into shallower waters. *Bidens beckii* is categorized as intolerant of human disturbance and is known to have disappeared from several lakes as they became increasingly eutrophic. It is intolerant of turbidity. Introduction of Eurasian water milfoil (*Myriophyllum spicatum*) also has been correlated with severely reduced plant density. The plants reportedly withstand herbicide treatments of up to 5 g/l fluridone. **Reported associates:** *Brasenia schreberi, Butomus umbellatus, Calla palustris, Callitriche hermaphroditica, C. palustris, Ceratophyllum demersum, Chara aspera, C. contraria, C. globularis, C. haitensis, C. vulgaris, C. zeylanica, Comarum palustre, Elatine minima, Eleocharis acicularis, E. palustris, Elodea canadensis, Equisetum fluviatile, Eriocaulon aquaticum, Gratiola aurea, Heteranthera dubia, Hippuris vulgaris, Isoetes echinospora, I. lacustris, I. macrospora, Juncus pelocarpus, Lemna minor, L. trisulca, Littorella uniflora, Lobelia dortmanna, Myriophyllum alterniflorum, M. heterophyllum, M. sibiricum, M. spicatum, M. tenellum, M. verticillatum, Najas flexilis, N. guadalupensis, N. marina, Nelumbo lutea, Nitella flexilis, N. tenuissima, Nuphar advena, N. rubrodisca, N. variegata, Nymphaea odorata, N. tuberosa, Persicaria amphibia, Pontederia cordata, Potamogeton alpinus, P. amplifolius, P. crispus, P. epihydrus, P. friesii, P. foliosus, P. gramineus, P. illinoensis, P. natans, P. perfoliatus, P. praelongus, P. pusillus, P. richardsonii, P. robbinsii, P. spirillus, P. strictifolius, P. vaseyi, P. zosteriformis, Ranunculus flammula, R. longirostris, R. trichophyllus, Rorippa aquatica, Sagittaria cristata, S. graminea, S. latifolia, S. rigida, Schoenoplectus acutus, S. pungens, S. subterminalis, S. tabernaemontani, Sparganium angustifolium, S. emersum, S. eurycarpum, S. fluctuans, Spirodela polyrhiza, Stuckenia filiformis, S. pectinata, Subularia aquatica, Typha*

×glauca, *T. latifolia, Utricularia geminiscapa, U. gibba, U. intermedia, U. macrorhiza, U. minor, U. purpurea, Vallisneria americana, Wolffia columbiana, Zizania palustris.*

Bidens bigelovii A. Gray is an annual species of bottoms, draws, flats, floodplains, gulches, meadows, and margins of intermittent streams, pools, rivers, springs, streams, and washes at elevations from 900 to 2500 m. The plants typically grow in wet, shaded sites within the forested understory, but also have been reported from more open localities. The substrates are described as alluvial loam, basalt, clay, clay loam, granite, gravel, gravel loam, limestone cobble, limestone sand, loam, rhyolite, sand, and sandy loam. The heads occur in both rayed and rayless forms. Flowering occurs from June to November and good seed set (e.g., 20 fruits/head) has been observed in some populations (August–October). The cypselae possess barbed awns, which facilitate their attachment to animals for dispersal. Other information on the life history of this species (at least in North America) is scarce. **Reported associates:** *Acer negundo, Agave palmeri, Alnus oblongifolia, Amaranthus palmeri, Ambrosia psilostachya, Amorpha fruticosa, Anisacanthus thurberi, Anoda cristata, Arbutus arizonica, A. texana, Arctostaphylos pungens, Astragalus, Baccharis glutinosa, Bahia dissecta, Bidens aureus, Bouteloua curtipendula, B. gracilis, Bouvardia ternifolia, Brickellia floribunda, Bromus, Cercocarpus montanus, Chilopsis linearis, Chrysothamnus nauseosus, Crotalaria pumila, Cupressus, Cyperus, Dasylirion, Echinochloa crusgalli, Equisetum, Eragrostis intermedia, Erigeron canadensis, Eriochloa acuminata, Eryngium lemmonii, Erythrina flabelliformis, Fraxinus, Garrya wrightii, Geranium, Gnaphalium chilense, Gomphrena sonorae, Guardiola platyphylla, Gutierrezia sarothrae, Heliomeris longifolia, Heterotheca subaxillaris, Hilaria, Hymenothrix wrightii, Ipomoea hederifolia, Ipomopsis thurberi, Juglans major, J. microcarpa, Juniperus coahuilensis, J. deppeana, J. monosperma, J. scopulorum, Koeleria, Lasianthaea podocephala, Lobelia cardinalis, Lycurus, Pinus cembroides, P. discolor, P. edulis, P. leiophylla, P. ponderosa, Platanus wrightii, Populus fremontii, Prosopis, Prunus, Ptelea, Quercus arizonica, Q. emoryi, Q. grisea, Q. hypoleucoides, Q. oblongifolia, Q. rugosa, Q. toumeyi, Q. undulata, Rhamnus crocea, Rhus glabra, R. trilobata, R. virens, Roldana hartwegii, Salix gooddingii, S. lasiolepis, S. taxifolia, Senecio, Solanum douglasii, Tetramerium hispidum, Toxicodendron radicans, Viguiera cordifolia, Vitis arizonica.*

Bidens cernuus L. is a summer annual, which inhabits beaches, bogs, brooks, canals, creek beds, depressions, dikes, ditches, dunes, ephemeral ponds, fens, flats, flowages, interdunal swales, lake beds, marshes, meadows, mudflats, muskrat mounds, prairies, river beds, sand bars, sand plains, seeps, sloughs, springs, strands, swales, swamps, thickets, and margins of lakes, ponds, reservoirs, rivers, and streams at elevations of up to 3200 m. This is one of the region's most widely distributed wetland species, with localities extending as far north as 66.86° N latitude. Plants normally grow on exposed wet sediment but also tolerate shallow (i.e., <50 cm) standing fresh to slightly brackish water. They occur most often in full sun but also partial shade across a broad range of habitat conditions (e.g., pH = 5.1–8.8; alkalinity ($CaCO_3$) = 200–320 mg/l; hardness ($CaCO_3$) = 275–850 mg/l; conductivity = 212 μS cm^{-1}). Associated substrates also are diverse and include alluvium, bog mats, clay, clay loam, granitic sand, gravel, gravelly muck, gravelly rock, humus, loam, loamy sand, logs, marl, muck, mucky peat, mud, muddy cobble, muddy gravel, peat, rock, sand, sandstone, sandy clay, sandy loam, silt, silty clay and volcanic bedrock. These are long day plants with rayed, protandrous flowers pollinated primarily by insects (Insecta) such as bees (Hymenoptera) and flies (Diptera). Genetic studies indicate a relatively high level of variation, which is associated with outcrossing. Individual flowers can produce up to 190 seeds (averaging 104.7 mg), yielding more than 1,000 seeds/plant. Although seeds display up to 95% viability, they are short lived and survive for less than 20 months in persistent but short-term seed banks, which can attain densities of up to 10,648 seeds/m^2. The seed banks are important sources of propagules for recruitment during periods of drawdown and are effective at establishing plants in restored wetlands. Seedling recruitment is highest in open sites where water is 5 cm below the sediment surface and decreases steadily when inundated to 10 cm depth. Germination of inundated seeds also is lower in coarser sediments. Seeds are physiologically dormant and require 6–12 weeks of cold (2°C–8°C) stratification for germination in light under a fluctuating 30°C–35°C/20°C temperature regime. Secondary dormancy is induced in seeds kept at 12°C for 20 weeks. Scarification of the seed coat greatly enhances germination rates (in the absence of after ripening), especially for seeds treated with gibberellic acid or *N*-6-benzyladenine. In the field, seeds break dormancy in the spring when temperatures are <15°C; dormancy is induced in summer when temperatures elevate above this level. Seeds become completely dormant by late summer until decreasing temperatures (<7°C) relieve secondary dormancy. The barbed fruits are dispersed by clinging to birds and also by salamanders (Amphibia: *Ambystoma*), which passively transport them during their fall migrations. The fruits also have a buoyant coat, which facilitates local dispersal by water. Seedlings emerge at a mean air temperature of 17°C. Burial decreases seedling emergence, which is arrested when seeds are covered by 0.25–2 cm of sediment. There is no mechanism of vegetative reproduction. This species has a low relative growth rate compared to other wetland species. The highest net photosynthetic rate occurs during vegetative growth (May–August), and decreases by 50% during flowering (August–September) and fruiting (September–November). Plants survive for 202 days on average after achieving an average total biomass of 1.58 g. *Bidens cernuus* can become fairly aggressive and has shown to compete effectively against purple loosestrife (*Lythrum salicaria*) and other emergent wetland species. Plants can be controlled quite readily using the herbicide metsulfuron methyl. **Reported associates:** *Abies concolor, A. lasiocarpa, Acacia, Acalypha rhomboidea, Acer negundo, A. rubrum, A. saccharinum, Achillea millefolium, Acorus calamus, Achnatherum lemmonii, A. nelsonii, Agalinis purpurea,*

Agrostis gigantea, A. scabra, A. stolonifera, Alisma subcordatum, A. triviale, Alnus rugosa, Alopecurus aequalis, Amaranthus tuberculatus, Ambrosia artemisiifolia, A. trifida, Andromeda glaucophylla, Anticlea elegans, Arctium minus, Aristolochia serpentaria, Arnoglossum plantagineum, Artemisia biennis, A. ludoviciana, A. serrata, A. tridentata, Asclepias incarnata, A. syriaca, Balsamorhiza sagittata, Beckmannia syzigachne, Berula erecta, Betula alleghaniensis, B. nigra, B. occidentalis, B. papyrifera, B. pumila, Betula ×sandbergii, Bidens connatus, B. frondosus, B. polylepis, B. trichospermus, B. tripartitus, B. vulgatus, Bistorta bistortoides, Boehmeria cylindrica, Bolboschoenus fluviatilis, Brasenia schreberi, Bromus japonicus, B. latiglumis, B. marginatus, Calamagrostis canadensis, Calla palustris, Carduus acanthoides, Carex aperta, C. aquatilis, C. bebbii, C. canescens, C. chordorrhiza, C. crawei, C. crinita, C. diandra, C. garberi, C. grayi, C. hyalinolepis, C. lacustris, C. muskingumensis, C. nebrascensis, C. oligosperma, C. rostrata, C. scoparia, C. stricta, C. tribuloides, C. utriculata, C. viridula, C. vulpinoidea, Carya ovata, Cercocarpus montanus, Chamaedaphne calyculata, Chamaesyce polygonifolia, Chelone glabra, Chenopodium, Cicuta bulbifera, C. douglasii, C. maculata, C. virosa, Cirsium arvense, C. discolor, C. vulgare, Clematis virginiana, Comarum palustre, Conyza canadensis, Corispermum, Cornus racemosa, C. sericea, Crataegus, Cryptotaenia canadensis, Cyperus bipartitus, C. diandrus, C. erythrorhizos, C. squarrosus, C. strigosus, Dasiphora floribunda, Datura, Dipsacus fullonum, Doellingeria umbellata, Drosera rotundifolia, Dulichium arundinaceum, Echinochloa muricata, E. pungens, Elaeagnus angustifolia, Eleocharis acicularis, E. calva, E. engelmannii, E. erythropoda, E. intermedia, E. obtusa, E. olivacea, E. palustris, E. robbinsii, E. rostellata, E. smallii, Elodea canadensis, Elymus glaucus, E. repens, E. trachycaulus, E. virginicus, Epilobium campestre, E. ciliatum, E. leptophyllum, E. palustre, Equisetum arvense, E. fluviatile, Eragrostis hypnoides, Ericameria viscidiflora, Erigeron lonchophyllus, Erechtites hieraciifolius, Erysimum cheiranthoides, Euonymus atropurpureus, Eupatorium perfoliatum, Euthamia graminifolia, Eutrochium maculatum, Fallopia scandens, Fimbristylis autumnalis, Fraxinus nigra, Galium brevipes, G. tinctorium, G. trifidum, Gentiana andrewsii, G. rubricaulis, Gentianopsis virgata, Geranium, Glyceria canadensis, G. grandis, G. maxima, G. striata, Gnaphalium, Helenium autumnale, Helianthus giganteus, Hippuris vulgaris, Hordeum jubatum, Hydrophyllum, Hypericum canadensis, H. ellipticum, H. majus, Impatiens capensis, Iris missouriensis, I. virginica, Juncus alpinus, J. balticus, J. bufonius, J. effusus, J. ensifolius, J. filiformis, J. nodosus, J. pelocarpus, J. tenuis, J. torreyi, Juniperus, Kalmia polifolia, Lactuca, Laportea canadensis, Larix laricina, Leersia oryzoides, L. virginica, Lemna minor, L. minuscula, L. valdiviana, Leonurus cardiaca, Lilaeopsis, Lilium michiganense, Limosella acaulis, Lindernia dubia, Lipocarpha micrantha, Lobelia cardinalis, L. dortmanna, L. inflata, L. kalmii, L. spicata, L. siphilitica, Ludwigia palustris, Lupinus, Lychnis alba, Lycopodiella inundata, Lycopus americanus, L. asper,

L. uniflorus, Lysimachia ciliata, L. hybrida, L. nummularia; L. terrestris, L. thyrsiflora, Lythrum salicaria, Malva neglecta, Medicago lupulina, M. sativa, Melilotus alba, M. officinalis, Menispermum canadense, Mentha arvensis, Mertensia ciliata, Mimulus glabratus, M. guttatus, M. ringens, M. torreyi, Monarda fistulosa, Muhlenbergia asperifolia, Myosoton aquaticum, Myrica gale, Myriophyllum aquaticum, Napaea dioica, Nasturtium officinale, Nuphar polysepala, Nymphaea odorata, Onoclea sensibilis, Oxalis, Oxypolis rigidior, Panicum capillare, Parnassia, Pedicularis lanceolata, Penthorum sedoides, Persicaria amphibia, P. coccinea, P. lapathifolia, P. maculosa, P. pensylvanica, P. punctata, P. sagittata, P. virginiana, Phalaris arundinacea, Phleum pratense, Phragmites australis, Phyla lanceolata, Physalis longifolia, Physostegia virginiana, Picea mariana, Pilea fontana, P. pumila, Pinus banksiana, P. ponderosa, P. resinosa, P. strobus, Poa palustris, P. pratensis, Polygonum aviculare, Polypogon monspeliensis, Pontederia cordata, Populus angustifolia, P. deltoides, P. fremontii, P. tremuloides, Potentilla anserina, P. simplex, Primula mistassinica, Prosopis, Prunella vulgaris, Pseudognaphalium stramineum, Pycnanthemum, Quercus ellipsoidalis, Ranunculus aquatilis, R. circinatus, R. gmelini, R. macounii, R. pensylvanicus, R. reptans, R. sceleratus, R. trichophyllus, Rhamnus, Rhododendron groenlandicum, Rhynchospora scirpoides, Ribes cereum, R. montigenum, Rorippa palustris, Rosa woodsii, Rubus, Rudbeckia hirta, R. laciniata, R. occidentalis, Rumex altissimus, R. arctica, R. fueginus, R. maritimus, R. salicifolius, R. verticillatus, Sagittaria latifolia, Salix bebbiana, S. interior, Sambucus nigra, S. racemosa, Saponaria officinalis, Sarracenia purpurea, Schedonorus arundinaceus, Schoenoplectus acutus, S. pungens, S. purshianus, S. smithii, S. tabernaemontani, S. torreyi, S. triqueter, Scirpus atrovirens, S. cyperinus, S. pallidus, Scleria verticillata, Scrophularia marilandica, Scutellaria galericulata, S. lateriflora, Sericocarpus oregonensis, Sicyos angulatus, Silphium perfoliatum, Sium suave, Smilax lasioneura, Solanum dulcamara, Solidago canadensis, S. gigantea, S. riddellii, Sorghum halepense, Sparganium chlorocarpum, S. eurycarpum, Spartina pectinata, Sphagnum squarrosum, Sphenopholis intermedia, Spiraea alba, S. tomentosa, Spiranthes cernua, Spirodela polyrhiza, Stachys tenuifolia, Symphoricarpos oreophilus, Symphyotrichum boreale, S. firmum, S. lanceolatum, S. ontarionis, S. puniceum, Taraxacum officinale, Teucrium canadense, Thalictrum dasycarpum, T. fendleri, Thelypteris nevadensis, T. palustris, Thuja occidentalis, Tilia americana, Toxicodendron rydbergii, T. vernix, Triadenum fraseri, T. virginicum, Trifolium, Typha angustifolia, T. ×glauca, T. latifolia, Ulmus americana, Urtica dioica, Utricularia gibba, U. intermedia, U. macrorhiza, Vaccinium macrocarpon, V. oxycoccos, Verbascum thapsus, Verbena bracteata, V. hastata, V. stricta, Vernonia fasciculata, Veronica americana, V. anagallis-aquatica, V. scutellata, Veronicastrum virginicum, Vicia, Viola lanceolata, Xanthium strumarium, Zannichellia palustris, Zizania aquatica.

***Bidens connatus* Muhl. ex Willd.** is a summer annual, which occupies intertidal, tidal, or nontidal sites including bogs, channels, depressions, ditches, dunes, exposed bottoms, fens, floodplains, flowages, gravel bars, gullies, kettle holes, marshes, meadows, mudflats, oxbows, pools, prairies, sandbars, seeps, sloughs, swales, swamps, and margins of lakes, ponds, rivers streams, and waterfalls at elevations of up to 1700 m. The plants occur on moist soil or in standing water up to 15 cm in depth. Exposures vary from full sun to shade and substrates include gravel, gravelly clay, moss tufa, muck, mucky peat, muddy peat, peat, peaty muck, rock, sand, sandy muck, sandy peat, silt, and stones. Occurrences span a broad range of pH (4.8–7.6). The flowers are pollinated by bumblebees (Insecta: Hymenoptera: Apidae: *Bombus impatiens, B. pensylvanicus)*, but at least some populations are suspected to produce seeds apomictically. The plants are known to produce as many as 1803 fruits per shoot resulting in densities of 8169 cypselae/m². Lower densities (<150/m² have been reported in some seed bank studies. The seeds are physiologically dormant and require cold (4°C) stratification for germination in light under a fluctuating 35°C/20°C temperature regime. They will germinate when exposed to 16 h of light at 20°C after clipping the tops of cypselae. Germination rates are high (>98%) and seedling establishment averages 42%. The seedlings attain relative growth rates averaging 0.2370 g·g⁻¹·day⁻¹. The plants regularly colonize sedge (*Carex*) tussocks and populations are known to recover well after removal of weedy stands of reed (*Phragmites*). **Reported associates:** *Abutilon theophrasti, Acalypha gracilens, A. ostryifolia, A. rhomboidea, Acer rubrum, A. saccharinum, Acorus calamus, Agalinis purpurea, A. tenuifolia, Agrimonia parviflora, Agrostis alba, A. gigantea, A. scabra, A. stolonifera, Alisma subcordatum, A. triviale, Alnus rugosa, Amaranthus arenicola, A. graecizans, A. tuberculatus, Ambrosia artemisiifolia, Ammannia coccinea, Amorpha fruticosa, Ampelopsis aconitifolia, A. arborea, Andropogon virginicus, Apios americana, Apocynum androsaemifolium, A. cannabinum, Aralia nudicaulis, A. racemosa, Arisaema triphyllum, Artemisia biennis, Asclepias incarnata, Asparagus officinalis, Betula nigra, B. papyrifera, B. pumila, Bidens aristosus, B. bipinnatus, B. cernuus, B. discoideus, B. frondosus, B. laevis, B. trichospermus, B. tripartitus, B. vulgatus, Boehmeria cylindrica, Bolboschoenus fluviatilis, Brassica juncea, Bromus inermis, Calamagrostis canadensis, C. stricta, Calla palustris, Caltha palustris, Campanula aparinoides, Campsis radicans, Carex aquatilis, C. aurea, C. crinita, C. hyalinolepis, C. hystricina, C. interior, C. intumescens, C. lacustris, C. lupulina, C. lyngbyei, C. muskingumensis, C. obnupta, C. projecta, C. pseudocyperus, C. stipata, C. stricta, C. tribuloides, C. viridula, Cenchrus longispinus, Cephalanthus occidentalis, Chaerophyllum procumbens, Chamaecrista fasciculata, Chamaedaphne calyculata, Chamaesyce humistrata, C. nutans, Chenopodium album, Cicuta bulbifera, C. maculata, Cirsium, Cladium mariscoides, Coleataenia longifolia, Comarum palustre, Conocephalum conicum, Conoclinium coelestinum, Conyza canadensis, Cornus drummondii, C. obliqua, Crassula aquatica, Cuscuta gronovii, Cycloloma atriplicifolium, Cynanchum laeve, Cyperus dentatus, C. erythrorhizos, C.*

esculentus, C. ferruginescens, C. flavescens, C. odoratus, C. squarrosus, C. strigosus, Dalibarda repens, Datura stramonium, Decodon verticillatus, Desmanthus illinoensis, Desmodium cuspidatum, Dichanthelium clandestinum, D. dichotomum, Digitaria ischaemum, Diodia teres, D. virginiana, Drosera filiformis, D. intermedia, Dryopteris cristata, D. spinulosa, Dulichium arundinaceum, Dysphania ambrosioides, Echinochloa crus-galli, E. muricata, E. walteri, Echinodorus tenellus, Eclipta prostrata, Elatine americana, Eleocharis acicularis, E. ambigens, E. engelmannii, E. erythropoda, E. ovata, E. palustris, E. smallii, Eleusine indica, Epilobium ciliatum, E. coloratum, E. glandulosum, E. leptophyllum, Equisetum arvense, E. fluviatile, E. variegatum, Eragrostis cilianensis, E. hypnoides, E. pectinacea, Erechtites hieraciifolius, Erigeron annuus, E. philadelphicus, E. strigosus, Eriocaulon aquaticum, Eupatorium perfoliatum, E. pilosum, E. rugosum, E. serotinum, Euphorbia maculata, Euphrasia stricta, Euthamia graminifolia, E. tenuifolia, Eutrochium maculatum, Fallopia scandens, Fimbristylis autumnalis, Forestiera acuminata, Fragaria virginiana, Fraxinus nigra, Galium asprellum, G. palustre, G. tinctorium, G. trifidum, G. triflorum, Gamochaeta purpurea, Gaultheria procumbens, Gentiana andrewsii, G. rubricaulis, Glechoma hederacea, Glyceria canadensis, G. striata, Gratiola aurea, Hedeoma hispida, Helenium flexuosum, Helianthus tuberosus, Heliotropium indicum, Hemerocallis fulva, Hibiscus laevis, H. trionum, Hydrocotyle americana, Hypericum canadense, H. majus, H. mutilum, H. punctatum, Impatiens capensis, I. ecalcarata, Ipomoea lacunosa, Iris pseudacorus, I. versicolor, I. virginica, Isoetes tuckermanii, Juncus acuminatus, J. alpinus, J. arcticus, J. articulatus, J. brevicaudatus, J. bufonius, J. canadensis, J. diffusissimus, J. dudleyi, J. effusus, J. militaris, J. pelocarpus, J. tenuis, J. torreyi, Justicia americana, Kummerowia stipulacea, K. striata, Kyllinga pumila, Larix laricina, Lathyrus palustris, Leersia lenticularis, L. oryzoides, L. virginica, Lemna minor, L. trisulca, Leptochloa panicea, L. fusca, Lespedeza procumbens, L. thunbergii, Leucanthemum vulgare, Leucospora multifida, Limosella subulata, Lindera benzoin, Lindernia dubia, Lipocarpha micrantha, Lobelia cardinalis, L. dortmanna, L. inflata, L. kalmii, L. siphilitica, Ludwigia leptocarpa, L. palustris, L. peploides, Lycopus americanus, L. rubellus, L. uniflorus, Lysimachia terrestris, L. thyrsiflora, Lythrum salicaria, Matteuccia struthiopteris, Mentha arvensis, M. piperita, M. pulegium, Menyanthes trifoliata, Microstegium vimineum, Mikania scandens, Mimulus alatus, M. ringens, Mollugo verticillata, Morus rubra, Muhlenbergia schreberi, Myosotis scorpioides, Myrica gale, Nasturtium officinale, Nymphoides cordata, Oenothera laciniata, Onoclea sensibilis, Osmunda regalis, Osmundastrum cinnamomeum, Oxalis stricta, Oxypolis rigidior, Panicum capillare, P. dichotomiflorum, P. philadelphicum, Paronychia fastigiata, Paspalum pubiflorum, P. repens, Parthenocissus quinquefolia, Peltandra virginica, Penthorum sedoides, Persicaria amphibia, P. arifolia, P. coccinea, P. hydropiper, P. hydropiperoides, P. lapathifolia, P. longiseta, P. maculosa, P. pensylvanica, P. punctata, P. sagittata, P. virginiana,

Phalaris arundinacea, Phragmites australis, Phyla lanceolata, Physalis pubescens, Phytolacca americana, Picea mariana, Pilea pumila, Plantago cordata, P. lanceolata, P. rugelii, Platanus occidentalis, Pluchea camphorata, Poa annua, Pontederia cordata, Populus balsamifera, P. deltoides, P. tremuloides, Portulaca oleracea, Potamogeton, Potentilla anserina, P. norvegica, P. pacifica, P. simplex, P. supina, Prunella vulgaris, Pycnanthemum virginianum, Quercus bicolor, Q. palustris, Q. rubra, Ranunculus abortivus, R. repens, R. sceleratus, Rhexia mariana, Rhynchospora corniculata, Rorippa islandica, R. palustris, R. sessiliflora, Rosa carolina, R. palustris, Rotala ramosior, Rubus hispidus, R. idaeus, R. trivialis, Rumex altissimus, R. britannica, R. crispus, R. obtusifolius, R. verticillatus, Sagittaria brevirostra, S. latifolia, S. montevidensis, Salix amygdaloides, S. interior, S. lucida, S. nigra, S. pedicellaris, S. petiolaris, Sambucus nigra, Samolus valerandi, Schedonorus pratensis, Schoenoplectus acutus, S. pungens, S. tabernaemontani, Scirpus cyperinus, S. hattorianus, S. polyphyllus, Scrophularia marilandica, Scutellaria galericulata, S. lateriflora, Senna marilandica, Setaria faberi, S. pumila, S. viridis, Sida spinosa, Sisyrinchium californicum, Sium suave, Solanum americanum, S. carolinense, S. dulcamara, Solidago canadensis, S. ohioensis, S. patula, S. rugosa, Sonchus asper, Sorghum halepense, Sparganium americanum, S. eurycarpum, Spermacoce glabra, Sphagnum, Spiranthes cernua, Spiraea alba, S. tomentosa, Sporobolus cryptandrus, S. vaginiflorus, Stellaria media, Symphyotrichum lanceolatum, S. lateriflorum, S. ontarionis, S. racemosum, S. pilosum, S. praealtum, S. puniceum, S. subspicatum, Taraxacum erythrospermum, T. officinale, Thalictrum, Thelypteris palustris, Thuja occidentalis, Toxicodendron radicans, T. vernix, Triadenum fraseri, T. virginicum, Tridens flavus, Trientalis borealis, Trifolium repens, Triglochin concinnum, Tussilago farfara, Typha angustifolia, T. latifolia, Ulmus americana, U. rubra, Urochloa platyphylla, Urtica dioica, Vaccinium macrocarpon, Verbascum thapsus, Verbena hastata, V. urticifolia, Verbesina alternifolia, Vernonia gigantea, Viola blanda, V. lanceolata, Vitis riparia, Wolffia, Woodwardia virginica, Xanthium strumarium, Xyris difformis, Zizania aquatica.

***Bidens hyperboreus* Greene** is an annual, which occurs in tidal, fresh to brackish estuaries, marshes, mudflats, and margins of rivers at elevations of up to 10 m. The plants grow on gravel or sandy substrates. Flowering occurs from August to October, with fruiting in September and October. Other pertinent ecological information is scarce. Much of the habitat loss for this species is attributed to the damming of rivers. **Reported associates:** *Acorus calamus, Agrimonia rostellata, Alisma, Amaranthus tuberculatus, Arabis drummondii, Bidens bidentoides, B. eatonii, B. frondosus, B. laevis, Bolboschoenus fluviatilis, Calamagrostis canadensis, Campanula rotundifolia, Cardamine longii, Carex crawfordii, C. davisii, C. scoparia, C. stricta, Crassula aquatica, Cyperus lupulinus, C. strigosus, Elatine americana, Eleocharis diandra, E. obtusa, Equisetum palustre, Eriocaulon parkeri, Impatiens capensis, Isoetes riparia, Juncus brachycephalus, J. dudleyi, Limosella australis, Lindernia dubia, Lobelia cardinalis, Ludwigia palustris, Lythrum salicaria, Mentha arvensis, Micranthemum micranthemoides, Mikania scandens, Najas muenscheri, Nuphar advena, Penthorum sedoides, Persicaria arifolia, P. punctata, P. sagittata, Plantago cordata, Pontederia cordata, Sagittaria montevidensis, S. rigida, S. subulata, Samolus valerandi, Schoenoplectus pungens, S. tabernaemontani, Sium suave, Solidago simplex, Symphyotrichum lateriflorum, S. pilosum, S. puniceum, Typha angustifolia, T. latifolia, Zizania aquatica.*

***Bidens laevis* (L.) Britton, Sterns & Poggenb.** is a characteristic annual (occasionally perennial) of still to slowly moving, tidal, brackish (e.g., chlorinity=0.9%–2.8%) to fresh waters of bayous, bottoms, cienega, ditches, dunes, estuaries, floodplains, inlets, levees, marshes, mudflats, pools, sinkholes, sloughs, springs, swales, swamps, and margins of canals, lakes, ponds, rivers, and streams at elevations of up to 2800 m. The species also occurs frequently on "flotants," i.e., floating mats comprising water hyacinth (*Eichhornia crassipes*) and other vegetation. Exposures typically are open sunny sites but extend to shaded understory. Plants grow on wet substrates or in shallow water (<25 cm), typically at sites where average tidal flooding is 30–50 cm but <4 h duration. Reported substrates are acidic (pH: 5.0–6.4) and include alluvium, clay, gravel, loamy clay, muck, mud, peaty sand, sand, sandy gravel, serpentine, silt, and silty loam. The silt content typically is high (52%–60%) and sand content low (1%–25%), but plants also occur on sandy substrates, especially in western localities. Percent organic matter typically ranges from 24%–30% but can reach 78% in floating mats. Pollination is facilitated by insects (Insecta), especially honeybees (Hymenoptera: Apidae: *Apis mellifera*); *B. laevis* pollen has been found in more than 90% of honey samples evaluated in some areas. The flowers possess ultraviolet-absorbing floral anthochlor pigments and have a higher sucrose ratio (mean fructose:glucose:sucrose ratio of 33:31:36) than most Asteraceae. The seeds develop into type III persistent seed banks that retain some viability for at least 19 months. They can occur in up to 94% of seed bank samples taken at some localities and yield as many as 4600 seedlings/m^2. Although a correspondence between seed bank densities and seedling densities was observed over a 10-year seed bank study, neither factor correlated strongly with the percent cover achieved by plants in a given year. Successful germination also did not correspond closely to establishment success. Germination requires at least 8 weeks of cold stratification (5°C). A rate of 50% germination can be achieved under 35°C/20°C for seeds initially breaking dormancy (8–12 week stratification), or at 15°C/6°C for those having longer stratification (16 weeks). Seeds will not germinate at 5°C. Germination is inhibited in darkness and under hypoxic conditions; however, >80% germination can be achieved for inundated seeds not under hypoxia. Germination declines with depth of seed burial (to 5 cm) and can be as much as twice as high under nonflooded relative to flooded conditions in some cases. However, seeds that are not buried are susceptible to much higher rates of herbivory and more often are eaten. Neither relative humidity nor duration of storage substantially affects germination rates.

Dried seeds retain 92% of their water content after 1 month of storage and seeds stored at 5°C retained viability after 4 years of storage. The barbed, awned seeds are dispersed by attachment to animal fur or feathers and exhibit strong adherence to various fibers, especially those strongly curled. They also are dispersed effectively by water, with nearly 40% of the seed produced in one study remaining afloat for more than 30 days. Masses of floating seeds in tidal marshes often are dominated by those of this species. The number of seedlings peaks in May, gradually tapering off through October. Adult plants can withstand periodic flooding up to 100 cm, but eventually are restricted to high marsh areas because the seedlings are intolerant of prolonged inundation (<5 h). Plants growing along stream banks are relatively smaller and produce fewer flowers than those in the high marsh. This species is among the most persistent within its community. The roots are associated with arbuscular mycorrhizal fungi. Although annual, vegetative reproduction during the growing season has been observed through rooting of detached cotyledons. The insecticide endosulfan disrupts chromosomal spindle function at concentrations of 5 µg/l. Stem and leaf extracts of *B. laevis* exhibit some allelopathic activity. Simulated oil spill conditions resulted in a significant decline in seedling number and survival. Exclosure experiments indicate that herbivory significantly reduces the dominance of *B. laevis* in low marsh sites.

Reported associates: *Acer rubrum, Acmella oppositifolia, Acorus calamus, Aeschynomene indica, A. virginica, Agrostis viridis, Alnus serrulata, Alternanthera philoxeroides, Amaranthus cannabinus, Ambrosia acanthicarpa, A. psilostachya, A. trifida, Amorpha, Anemopsis californica, Apios americana, Arundo donax, Asclepias incarnata, Azolla caroliniana, Baccharis salicifolia, Berula erecta, Bidens aureus, B. connatus, B. frondosus, B. hyperboreus, B. mitis, B. trichospermus, Boehmeria cylindrica, Bolboschoenus fluviatilis, B. robustus, Brasenia schreberi, Brickellia floribunda, Calystegia sepium, Carex hyalinolepis, C. hystricina, C. lupulina, C. lurida, C. pellita, C. praegracilis, C. stipata, C. stricta, Celtis reticulata, Cephalanthus occidentalis, Chamaecrista fasciculata, Chasmanthium latifolium, C. laxum, Chenopodium album, Chilopsis linearis, Cicuta maculata, Cinna arundinacea, Cirsium, Cladium jamaicense, Conium maculatum, Conoclinium coelestinum, Conyza canadensis, Cornus amomum, C. foemina, Crinum americanum, Cuscuta gronovii, C. pentagona, Cynodon dactylon, Cyperus diandrus, C. erythrorhizos, C. esculentus, C. iria, C. odoratus, C. polystachyos, C. strigosus, C. refractus, Decodon verticillatus, Decumaria barbara, Dichanthelium scoparium, Dieteria canescens, Dulichium arundinaceum, Echinochloa crus-galli, E. walteri, Eclipta prostrata, Eichhornia crassipes, Eleocharis baldwinii, E. engelmannii, E. fallax, E. obtusa, E. palustris, E. parvula, E. quadrangulata, Elymus trachycaulus, Epilobium ciliatum, Equisetum laevigatum, Eriochloa villosa, Eryngium aquaticum, Eupatorium perfoliatum, Festuca pratensis, Fimbristylis thermalis, Fraxinus pennsylvanica, F. velutina, Galium obtusum, G. tinctorium, Glyceria striata, Gratiola virginica, Habenaria repens, Heteranthera multiflora, Heterotheca grandiflora, Hibiscus moscheutos,* *Hydrilla verticillata, Hydrocotyle ranunculoides, H. umbellata, Hymenocallis crassifolia, Hypericum ×dissimulatum, H. mutilum, Impatiens capensis, Iris versicolor, I. virginica, Iva frutescens, Juglans major, Juncus acuminatus, J. balticus, J. biflorus, J. coriaceus, J. effusus, J. megacephalus, J. repens, J. roemerianus, J. torreyi, Leersia oryzoides, Lemna gibba, L. minor, Lilaeopsis chinensis, Limnobium spongia, Lindera benzoin, Liquidambar styraciflua, Lobelia cardinalis, Lonicera japonica, Ludwigia alternifolia, L. decurrens, L. grandiflora, L. leptocarpa, L. palustris, L. peploides, L. repens, Luziola fluitans, Lycopus americanus, L. virginicus, Lythrum lineare, Marrubium vulgare, Melilotus officinalis, Mentha arvensis, Mentha spicata, Micranthemum umbrosum, Mikania scandens, Mimulus guttatus, Muhlenbergia asperifolia, Murdannia keisak, Myrica cerifera, Myriophyllum, Nasturtium officinale, Nelumbo lutea, Nuphar advena, Nymphaea odorata, Nyssa aquatica, Oenothera speciosa, Orontium aquaticum, Osmorhiza longistylis, Osmunda regalis, Panicum anceps, P. hemitomon, P. virgatum, Paspalum distichum, P. vaginatum, Peltandra virginica, Persicaria arifolia, P. glabra, P. hydropiperoides, P. meisneriana, P. punctata, P. sagittata, P. setacea, Phalaris arundinacea, Phragmites australis, Phyla nodiflora, Physocarpus opulifolius, Pilea fontana, P. pumila, Pluchea carolinensis, Polypogon monspeliensis, Pontederia cordata, Populus fremontii, P. ×hinckleyana, Potamogeton, Prosopis velutina, Ptilimnium capillaceum, Quercus lobata, Ranunculus macranthus, R. sceleratus, Rhynchospora colorata, R. corniculata, Ribes aureum, Ricciocarpus natans, Rosa palustris, Rumex crispus, R. verticillatus, Saccharum giganteum, Sacciolepis striata, Sagittaria lancifolia, S. latifolia, S. subulata, Salix exigua, S. gooddingii, S. laevigata, S. lasiolepis, S. nigra, Salvinia minima, Saururus cernuus, Schoenoplectus acutus, S. americanus, S. californicus, S. pungens, S. tabernaemontani, Scirpus cyperinus, Scrophularia marilandica, Senegalia greggii, Sesbania drummondii, S. herbacea, Setaria magna, Sidalcea neomexicana, Sium suave, Sparganium americanum, S. eurycarpum, Spartina alterniflora, S. cynosuroides, S. patens, Symphyotrichum frondosum, S. puniceum, S. racemosum, S. subulatum, S. tenuifolium, Taraxacum officinale, Taxodium distichum, Teucrium canadense, Thalictrum pubescens, Thelypteris palustris, Thinopyrum elongatum, Triadenum virginicum, T. walteri, Triodia flava, Tripsacum dactyloides, Typha angustifolia, T. domingensis, T. latifolia, Ulmus serotina, Urtica dioica, Vachellia constricta, Verbesina occidentalis, Vernonia noveboracensis, Viburnum dentatum, Veronica anagallisaquatica, Vitis, Woodwardia areolata, Xanthium, Zizania aquatica, Zizaniopsis miliacea, Zuloagaea bulbosa.*

***Bidens mitis* (Michx.) Sherff** is a summer annual, which resides in fresh to saline, often tidal sites including baygalls, bayous, beaches, bogs, cypress domes, depressions, ditches, estuaries, flatwoods, floodplains, hammocks, levees, marshes, meadows, mudflats, ponds, pools, prairies, ravines, salt marsh, savanna, seeps, sloughs, strands, swales, swamps, and margins of canals, lakes, and streams at elevations of up to 300 m. The plants occur in open, sunny to partially shaded sites on exposed, wet substrates or in up to 30–60 cm standing water. Reported substrates are diverse and include acidic clay (Yorktown series: typic fluvaquent), siliceous types (Basinger: spodic psammaquents; Meadowbrook: grossarenic ochraqualfs; Plummer: grossarenic paleaquults; Pompano: typic psammaquents; Waveland: sand, sandy loam (Dothan: plinthic paleudults), muck (Terra Ceia: typic medisaprists), limestone, peat, sandy muck, sandy peat, and even mossy logs. The substrates can be quite acidic (pH: 4.0–5.0) but sometimes are described as calcareous. The plants also occur frequently on bulging or floating peat beds called "batteries." There is little specific information available on the reproductive ecology of this species. The prominent yellow ray florets produce ultraviolet-absorbing anthochlor pigments, which provide "honey guides" to attract insect pollinators such as bees (Insecta: Hymenoptera). The plants come into flower relatively late in the season (September–November). Details regarding seed germination have not been reported. When growing in deeper water, plants can become spindly and obtain their support from other aquatic vegetation such as pickerel weed (*Pontederia*). **Reported associates:** *Acer rubrum, Agalinis purpurea, Aletris aurea, Alternanthera philoxeroides, Amaranthus cannabinus, Amelanchier, Amphicarpum muhlenbergianum, Andropogon arctatus, A. glomeratus, A. virginicus, Aristida palustris, A. purpurascens, A. stricta, Aronia arbutifolia, Arundinaria gigantea, Asclepias incarnata, Baccharis glomeruliflora, Bacopa caroliniana, Balduina uniflora, Bejaria racemosa, Bidens laevis, B. trichospermus, Bigelowia nudata, Blechnum serrulatum, Boehmeria cylindrica, Brasenia schreberi, Calamagrostis canadensis, Callicarpa americana, Campsis radicans, Carex albolutescens, C. atlantica, C. comosa, C. joorii, C. intumescens, C. tribuloides, C. flaccosperma, C. hyalinolepis, C. louisianica, C. lupulina, Carphephorus odoratissimus, C. pseudoliatris, Centella asiatica, Cephalanthus occidentalis, Ceratophyllum muricatum, Chamaecrista fasciculata, Chamaecyparis thyoides, Cicuta maculata, Cirsium nuttallii, Cladium jamaicense, C. mariscoides, Clethra alnifolia, Cliftonia monophylla, Coleataenia longifolia, C. tenera, Commelina virginica, Conyza canadensis, Coreopsis gladiata, Ctenium aromaticum, Cuscuta, Cyperus. flavescens, C. haspan, C. lecontei, C. odoratus, C. polystachyos, C. virens, Cyrilla racemiflora, Decodon verticillatus, Desmodium, Dichanthelium ensifolium, D. nudicaule, Drosera intermedia, Dulichium arundinaceum, Dysphania ambrosioides, Eleocharis baldwinii, E. elongata, E. montevidensis, E. olivacea, E. quadrangulata, Enterolobium cyclocarpum, Erechtites hieraciifolius, Eriocaulon compressum, E. decangulare, Eryngium aquaticum, Eupatorium capillifolium, E. leptophyllum, Eurybia eryngiifolia, Euthamia graminifolia, Fimbristylis autumnalis, F. castanea, Fuirena breviseta, F. squarrosa, Galium tinctorium, Gelsemium sempervirens, Gordonia lasianthus, Gratiola virginiana, Habenaria repens, Helianthus angustifolius, H. heterophyllus, Heterotheca subaxillaris, Hibiscus moscheutos, Hydrocotyle umbellata, H. verticillata, Hypericum brachyphyllum, H. cistifolium, H. fasciculatum, H. gentianoides,* *H. hypericoides, H. lissophloeus, H. mutilum, H. tetrapetalum, Hyptis alata, Ilex cassine, I. coriacea, I. glabra, I. myrtifolia, I. vomitoria, Illicium parviflorum, Iris hexagona, I. versicolor, Isopterygium tenerum, Itea virginica, Iva microcephala, Juncus canadensis, J. effusus, J. marginatus, J. megacephalus, J. pelocarpus, J. roemerianus, Justicia ovata, Lachnanthes caroliana, Lachnocaulon anceps, L. digynum, Lantana camara, Leersia hexandra, L. lenticularis, L. oryzoides, L. virginica, Leitneria floridana, Leucobryum albidum, Leucothoe racemosa, Liatris spicata, Lilaeopsis chinensis, Lindernia grandiflora, Linum medium, Liquidambar styraciflua, Lobelia glandulosa, Lophiola aurea, Ludwigia alata, L. alternifolia, L. curtissii, L. glandulosa, L. lanceolata, L. maritima, L. palustris, L. peruviana, L. suffruticosa, Lycopodiella, Lycopodium alopecuroides, Lycopus rubellus, Lyonia ligustrina, L. lucida, L. mariana, Lysimachia radicans, Magnolia virginiana, Mayaca fluviatilis, Mecardonia acuminata, Megathyrsus maximus, Melinis repens, Mikania scandens, Mitreola sessilifolia, Mnesithea tuberculosa, Muhlenbergia capillaris, Murdannia keisak, Myrica cerifera, M. heterophylla, M. pensylvanica, Myriophyllum, Nuphar advena, Nymphaea odorata, Nymphoides aquatica, Nyssa biflora, N. sylvatica, Oenothera fruticosa, Onoclea sensibilis, Orontium aquaticum, Osmunda regalis, Osmundastrum cinnamomeum, Oxycaryum cubense, Oxypolis filiformis, O. rigidior, Panicum dichotomiflorum, P. hemitomon, P. repens, P. verrucosum, P. virgatum, Parthenocissus quinquefolia, Paspalum notatum, Passiflora incarnata, Peltandra sagittifolia, P. virginica, Penthorum sedoides, Persea palustris, Persicaria hirsuta, P. hydropiperoides, P. punctata, P. sagittata, Phragmites australis, Phyla lanceolata, P. nodiflora, Physostegia, Pinus elliottii, P. palustris, P. rigida, P. serotina, P. taeda, Platanthera integra, Pleea tenuifolia, Pluchea camphorata, P. foetida, P. rosea, Pogonia ophioglossoides, Polygala ramosa, Polypremum procumbens, Pontederia cordata, Proserpinaca palustris, P. pectinata, Prunus serotina, Ptilimnium costatum, Pycnanthemum, Quercus elliottii, Q. hemisphaerica, Q. laurifolia, Q. minima, Q. myrtifolia, Q. virginiana, Rhexia alifanus, R. lutea, R. mariana, R. petiolata, R. virginica, Rhus copallinum, Rhynchospora alba, R. chalarocephala, R. corniculata, R. divergens, R. fascicularis, R. filifolia, R. globularis, R. inundata, R. macra, R. microcarpa, R. microcephala, R. oligantha, R. perplexa, R. plumosa, Riccia fluitans, Richardia scabra, Rosa palustris, Rudbeckia fulgida, Rubus, Sabal palmetto, Sabatia brevifolia, Saccharum giganteum, Sacciolepis indica, S. striata, Sagittaria australis, S. graminea, S. lancifolia, S. latifolia, Salvinia rotundifolia, Sambucus nigra, Sarracenia flava, S. minor, S. psittacina, Saururus cernuus, Schizachyrium scoparium, Schoenoplectus tabernaemontani, Scleria ciliata, S. muehlenbergii, S. triglomerata, Senna obtusifolia, Serenoa repens, Setaria parviflora, Sida acuta, Silphium, Smilax laurifolia, Solidago fistulosa, Sorghastrum secundum, Sphagnum, Spiranthes cernua, Stachys, Stenotaphrum secundatum, Symphyotrichum dumosum, S. elliotii, S. lateriflorum, S. tenuifolium, Taxodium ascendens, T. distichum, Thelypteris, Toxicodendron radicans, Triadenum tubulosum,*

T. virginicum, Triantha racemosa, Typha angustifolia, T. domingensis, T. latifolia, Utricularia floridana, U. gibba, U. subulata, Vaccinium corymbosum, V. macrocarpon, Vigna luteola, Vitis rotundifolia, Woodwardia virginica, Xyris ambigua, X. baldwiniana, X. fimbriata, X. jupicai, X. smalliana, X. stricta, Zigadenus glaberrimus, Zizania aquatica, Zizaniopsis miliacea.

Bidens trichospermus (Michx.) Britton (=***B. coronatus***) is a summer annual, which inhabits fresh to brackish sites (salinity <0.8 ppt) in bogs, carrs, depressions, dikes, ditches, estuaries, fens, floodplains, flowages, kettle holes, marshes, meadows, mudflats, ponds, prairies, river bottoms, sandbars, sand plains, sloughs, springs, string bogs, swales, swamps, and margins of lagoons, lakes rivers, and streams at elevations of up to 300 m. Plants occur in open sites and tolerate a relatively narrow (intermediate) soil moisture gradient ranging from wet substrates to shallow (<12 cm) standing waters. Reported substrates include alkaline histosols, gravel, humus, loamy muck, loamy sand, marl, muck, mud, peat, peaty muck, sand, sandy gravel, sandy muck, silt, silty loam, *Sphagnum* mats and sedge tussocks. Despite a number of accounts implying alkaline conditions, directly measured occurrences indicate more acidic reactions (pH 3.7–6.9). The yellow-rayed flowers open from July to October and attract insect pollinators (Insecta); however, prolific seed production in garden plantings indicates some degree of self-compatibility. Seed germination is inhibited by sedimentation of both fine and coarse materials at higher loadings. The seeds normally require stratification for germination, with rates of 68% obtained at 21°C/32°C when stored for 6 months under cool (4°C), moist conditions. Stratification in moist soil produces better germination than does moist sand. Lower germination (10%–20%) has been observed within 31–68 days with mature, freshly collected or dried seeds. Plants can tolerate moderate disturbance and establish quickly in recently burned or otherwise disturbed localities. **Reported associates:** *Acer rubrum, A. saccharinum, Achillea millefolium, Acorus calamus, Agalinis purpurea, Agrostis hyemalis, Alisma, Alnus rugosa, A. serrulata, Ambrosia bidentata, Amelanchier canadensis, Amaranthus cannabinus, A. tuberculatus, Ammannia latifolia, Amphicarpa bracteata, Andropogon, Anemone cylindrica, Apios americana, Apocynum androsaemifolium, A. cannabinum, Aronia ×prunifolia, Asclepias incarnata, Aspidium angustifolium, Aulacomnium palustre, Baccharis halimifolia, Bartonia virginica, Betula alleghaniensis, B. papyrifera, B. pumila, B. ×sandbergii, Bidens cernuus, B. comosus, B. connatus, B. frondosus, B. laevis, B. mitis, Boehmeria cylindrica, Bolboschoenus fluviatilis, B. robustus, Calamagrostis canadensis, C. inexpansa, Calla palustris, Calopogon tuberosus, Caltha palustris, Calystegia sepium, Campanula americana, C. aparinoides, Carex aquatilis, C. atlantica, C. crinita, C. canescens, C. chordorrhiza, C. comosa, C. disperma, C. howei, C. hyalinolepis, C. hystericina, C. lacustris, C. lanuginosa, C. lasiocarpa, C. lupulina, C. oligosperma, C. prairea, C. rostrata, C. sterilis, C. stipata, C. stricta, C. tenera, C. trisperma, C. viridula, Carpinus caroliniana, Chamaecrista fasciculata, Celastrus scandens,* *Cephalanthus occidentalis, Cephalozia connivens, Chamaedaphne calyculata, Chara, Chelone glabra, Cicuta bulbifera, C. maculata, Cinna arundinacea, Circaea alpina, Cirsium muticum, Cladium mariscoides, Cladopodiella fluitans, Clethra alnifolia, Comandra umbellata, Comarum palustre, Commelina virginica, Comptonia peregrina, Conyza, Coptis trifolia, Coreopsis tripteris, Cornus amomum, C. obliqua, C. sericea, Cuscuta gronovii, Cyperus bipartitus, C. erythrorhizos, C. esculentus, C. filicinus, C. strigosus, Dasiphora floribunda, Daucus carota, Decodon verticillatus, Dichanthelium acuminatum, D. scoparium, Diodia virginiana, Dirca palustris, Distichlis spicata, Doellingeria umbellata, Drosera intermedia, D. rotundifolia, Dryopteris cristata, Dulichium arundinaceum, Echinochloa crus-galli, E. walteri, Echinocystis lobata, Eleocharis fallax, E. obtusa, E. olivacea, E. palustris, E. parvula, Elymus virginicus, Epilobium coloratum, E. leptophyllum, Equisetum arvense, E. fluviatile, Eragrostis ciliaris, Erechtites hieraciifolius, Erigeron, Eriophorum angustifolium, E. virginicum, Eryngium yuccifolium, Eupatorium capillifolium, E. maculatum, E. perfoliatum, E. serotinum, Euthamia graminifolia, Fragaria virginiana, Frangula alnus, Fraxinus pennsylvanica, Galium labradoricum, G. obtusum, G. tinctorium, G. palustre, G. trifidum, Gaultheria procumbens, Gaylussacia baccata, Gentiana andrewsii, Gentianopsis crinita, G. virgata, Glyceria canadensis, G. striata, Gordonia lasianthus, Habenaria repens, Helenium autumnale, Helianthus giganteus, H. hirsutus, Hibiscus moscheutos, Hieracium aurantiacum, Hypericum gentianoides, H. majus, H. mutilum, H. punctatum, Ilex coriacea, I. glabra, I. verticillata, Impatiens capensis, Ipomoea purpurea, Iris versicolor, I. virginica, Itea virginica, Iva annua, I. frutescens, Juncus brachycephalus, J. canadensis, J. dudleyi, J. effusus, J. tenuis, J. torreyi, Kosteletzkya pentacarpos, Lachnanthes caroliana, Lactuca canadensis, Larix laricina, Leersia oryzoides, Lemna minor, Leucobryum glaucum, Leucothoe racemosa, Liatris spicata, Lindera benzoin, Liparis loeselii, Liriodendron tulipifera, Lobelia cardinalis, L. kalmii, L. siphilitica, Lonicera canadensis, L. japonica, Ludwigia alternifolia, L. decurrens, L. lanceolata, L. leptocarpa, L. linearis, L. palustris, L. pilosa, L. sphaerocarpa, Lycopodiella inundata, Lycopus americanus, L. uniflorus, L. virginicus, Lyonia ligustrina, L. lucida, Lysimachia nummularia, L. terrestris, L. thyrsiflora, Lythrum alatum, L. lineare, L. salicaria, Magnolia virginiana, Maianthemum canadense, Medeola virginiana, Menispermum canadense, Mentha arvensis, M. spicata, Menyanthes trifoliata, Mikania scandens, Mimulus ringens, Mitchella repens, Monarda fistulosa, Muhlenbergia glomerata, M. mexicana, M. racemosa, M. uniflora, Murdannia keisak, Mylia anomala, Myrica cerifera, Nasturtium officinale, Nemopanthus mucronatus, Nuphar, Nyssa sylvatica, Onoclea sensibilis, Orontium aquaticum, Osmunda claytoniana, O. regalis, Osmundastrum cinnamomeum, Oxypolis rigidior, Packera aurea, Pallavicinia lyellii, Panicum flexile, P. virgatum, Parnassia glauca, Parthenocissus quinquefolia, Pedicularis lanceolata, Peltandra virginica, Penthorum sedoides, Persea borbonia, Persicaria amphibia, P. arifolia,*

P. hydropiper, P. hydropiperoides, P. punctata, P. sagittata, Phalaris arundinacea, Phragmites australis, Physocarpus opulifolius, Picea mariana, Pilea fontana, P. pumila, Pinus elliottii, Plantago, Pluchea carolinensis, P. foetida, Poa nemoralis, P. palustris, P. pratensis, Polygonum ramosissimum, Pontederia cordata, Populus tremuloides, Potentilla norvegica, P. simplex, Proserpinaca palustris, Prunella vulgaris, Pteridium aquilinum, Ptilimnium capillaceum, Pycnanthemum virginianum, Quercus, Rhamnus alnifolia, Rhexia virginica, Rhododendron groenlandicum, Rhynchospora alba, R. fascicularis, R. glomerata, R. macrostachya, Ribes americanum, R. cynosbati, Rorippa islandica, Rosa carolina, R. palustris, Rubus hispidus, R. idaeus, R. pubescens, Rudbeckia fulgida, R. hirta, Rumex britannica, R. crispus, R. mexicanus, R. verticillatus, Sabatia stellaris, Saccharum giganteum, Sagittaria graminea, S. latifolia, Salix amygdaloides, S. bebbiana, S. candida, S. discolor, S. interior, S. nigra, S. pedicellaris, S. petiolaris, Sambucus nigra, Samolus valerandi, Sanicula marilandica, Sarracenia purpurea, Saururus cernuus, Schizachyrium scoparium, Schoenoplectus acutus, S. acutus×S. tabernaemontani, S. americanus, S. pungens, S. tabernaemontani, Scirpus atrovirens, S. cyperinus, Scleria verticillata, Scutellaria galericulata, S. lateriflora, Sium suave, Smilax laurifolia, S. walteri, Solanum dulcamara, Solidago canadensis, S. fistulosa, S. gigantea, S. juncea, S. patula, S. riddellii, S. uliginosa, Sonchus, Sorbus americana, Sparganium americanum, S. eurycarpum, Spartina alterniflora, S. cynosuroides, S. patens, S. pectinata, Sphagnum capillifolium, S. fimbriatum, S. magellanicum, S. palustre, S. papillosum, S. recurvum, Sphenopholis obtusata, S. pensylvanica, Spiraea alba, S. salicifolia, S. tomentosa, Spiranthes cernua, S. praecox, Stachys palustris, Stellaria longifolia, Symphyotrichum boreale, S. firmum, S. lanceolatum, S. laterifiorum, S. novae-angliae, S. puniceum, S. racemosum, S. subulatum, Symplocarpus foetidus, Taraxacum officinale, Taxodium distichum, Tetraphis pellucida, Teucrium canadense, Thalictrum pubescens, Thelypteris palustris, Thuja occidentalis, Tilia, Toxicodendron radicans, T. vernix, Triadenum fraseri, T. virginicum, Tridens flavus, Trifolium pratense, Typha angustifolia, T. latifolia, Ulmus americana, Utricularia macrorhiza, U. minor, Uvularia perfoliata, Vaccinium corymbosum, V. fuscatum, V. macrocarpon, V. oxycoccos, Verbena hastata, V. urticifolia, Vernonia gigantea, V. noveboracensis, Veronica scutellata, Viburnum dentatum, V. nudum, V. recognitum, Viola blanda, V. cucullata, V. lanceolata, V. macloskeyi, Vitis vulpina, Woodwardia areolata, W. virginica, Xyris difformis, X torta, Zanthoxylum americanum, Zizania aquatica.

Bidens tripartitus L. inhabits bogs, bottoms, ditches, flats, floodplains, marshes, meadows, mudflats, sand bars, sloughs, swamps, and margins of lakes, ponds, pools, reservoirs, rivers, and streams at elevations of up to 1804 m. Plants occur in open exposures but tolerate light shade. They grow on wet substrates or in shallowly inundated (e.g., 3 cm) sites. The reported substrates tend to be acidic (pH 5.9–7.1) and include basalt, gravel, mud, muddy sand, organic muck, sand, sandy clay, and silt. This is a long-day plant with protandrous flowers

that either self-pollinate or are insect pollinated by various bees (Insecta: Hymenoptera: *Andrena; Bombus*) or flies (Insecta: Diptera). This rayless species is believed primarily to be self-pollinating and genetic studies have documented relatively low levels of genetic variation in populations, especially compared to *B. cernuus,* a rayed, outcrossed congener. Aposporous apomixis has been reported for *B. tripatitus,* but the evidence is not compelling. The plants produce on average about 83 fruits/capitulum. The seeds are adapted to germinate upon processing cues associated with shallow, flooded conditions; i.e., fluctuating temperatures and sufficient light. They are capable of germinating (>40%) throughout the year under high (>21°C) or fluctuating (22°C±4°C) temperatures. When stratified across a wide temperature range (3°C–18°C), the seeds exhibit more than 60% germination, but stratification at 4°C for 8–12 weeks effectively breaks dormancy. Dormancy is induced in spring and summer as temperatures rise above 7°C, and is broken when autumn temperatures drop below 12°C. Seeds from a single plant can give rise to different ecomorphs, which emerge in the subsequent spring and autumn periods. Primary dormancy is broken after 1 year of soil burial, with germination occurring within 4–5 days. Secondary dormancy induction increases relative to the length of complete seed submergence (100% after 5 days); however, secondary dormancy is not induced in floating seeds. The seeds become dormant regardless of oxygen levels and whether fresh or dry and require light (red most effectively) for germination. Highest germination is achieved at shallow burial depths (0.5–1.0 cm) and under low oxygen levels (5%–10%). A short-term persistent seed bank can develop, with observed densities ranging from 3 to 9096 seeds/m^2. The use of 1–100 mM solutions of thiourea promotes germination (25°C). Treatment with potassium cyanide (KCN) effectively breaks dormancy. The barbed cypselae are dispersed easily on animal fur (e.g., cattle, donkeys, sheep, and wild boars). Lower dispersal potential exists for some plants, which possess a genetic polymorphism that produces awnless cypselae. The cypselae also remain buoyant even when soaked, which facilitates their water dispersal. Water dispersal is typical along rivers where it has been well documented. The seeds contain a relatively high amount of minerals (high ash content) with respect to their mass (2.6–2.8 mg). They also contain fairly high levels (~16%) of unsaturated fatty acids (oils). The plants are characterized by a high relative growth rate and can become aggressive. Their roots occasionally are associated with vesicular–arbuscular mycorrhizae. **Reported associates:** *Abutilon theophrasti, Acer negundo, A. rubrum, Agalinis purpurea, Agrostis gigantea, A. scabra, Alisma subcordatum, Alnus serrulata, Amaranthus cannabinus, Ambrosia artemisiifolia, Ammannia coccinea, Apios americana, Arctophila fulva, Asclepias incarnata, Bidens cernuus, B. frondosus, B. vulgatus, Boehmeria cylindrica, Bolboschoenus robustus, Calamagrostis canadensis, Carex athrostachya, C. conjuncta, C. frankii, C. granularis, C. lasiocarpa, C. lurida, C. rostrata, C. stricta, C. vulpinoidea, Cephalanthus occidentalis, Chenopodium glaucum, Cicuta maculata, Conoclinium coelestinum, Cuscuta gronovii, Cynodon dactylon, Cyperus aristatus, C. bipartitus,*

C. esculentus, C. flavescens, C. strigosus, Cuscuta cephalanthi, Dichanthelium acuminatum, Drosera intermedia, Dulichium arundinaceum, Echinochloa muricata, E. walteri, Eclipta prostrata, Elaeagnus angustifolia, Eleocharis acicularis, E. calva, E. erythropoda, E. obtusa, E. ovata, E. parvula, Elodea nuttallii, Epilobium coloratum, Ericameria nauseosa, Eriocaulon aquaticum, Eupatorium serotinum, Euthamia graminifolia, Eutrochium fistulosum, Fallopia scandens, Fallugia paradoxa, Galium asprellum, G. trifidum, Geum rivale, Glyceria striata, Gratiola neglecta, Hibiscus moscheutos, H. trionum, Hypericum boreale, H. canadense, H. majus, H. mutilum, Impatiens capensis, Juncus acuminatus, J. canadensis, J. debilis, J. diffusissimus, J. effusus, J. pelocarpus, J. subcaudatus, J. tenuis, J. torreyi, Kalmia polifolia, Kummerowia striata, Leersia oryzoides, Lindernia dubia, Linum striatum, Lobelia siphilitica, Ludwigia alternifolia, L. palustris, Lycopus uniflorus, Lysimachia terrestris, Lythrum salicaria, Mimulus alatus, M. ringens, Morus alba, Muhlenbergia asperifolia, M. uniflora, Myriophyllum tenellum, Nasturtium officinale, Nuphar polysepala, Oxalis stricta, Peltandra virginica, Penthorum sedoides, Persicaria careyi, P. lapathifolia, P. longiseta, P. maculosa, P. pensylvanica, P. punctata, P. sagittata, Plantago rugelii, Platanus occidentalis, Pontederia cordata, Populus deltoides, Porterella carnosula, Potamogeton crispus, Ranunculus cymbalaria, R. sardous, Rhexia mariana, R. virginica, Rhynchospora capitellata, R. fusca, Rotala ramosior, Rubus argutus, Rumex conglomeratus, Sagittaria australis, S. cuneata, S. graminea, S. latifolia, S. montevidensis, S. subulata, Salix exigua, S. geyeriana, S. gooddingii, S. nigra, Sambucus nigra, Samolus valerandi, Sarracenia purpurea, Scheuchzeria palustris, Schoenoplectus pungens, S. olneyi, S. purshianus, S. tabernaemontani, Scirpus atrovirens, S. cyperinus, S. polyphyllus, Scutellaria lateriflora, Setaria faberi, Solidago rugosa, Sparganium americanum, Spiraea alba, S. tomentosa, Sium suave, Symphyotrichum ciliatum, S. pilosum, S. puniceum, Tamarix, Triadenum fraseri, Typha angustifolia, T. latifolia, Utricularia cornuta, Veronica anagallis-aquatica, Viola lanceolata, Xanthium, Xyris difformis, X. jupicai.

Bidens vulgatus Greene is an annual species found on fresh to slightly brackish substrates (conductivity: 40–2000 μS cm^{-1}) of bogs, ditches, draws, fens, floodplains, marshes, meadows, outwashes, prairies, river bottoms, sand bars, swamps, washes, and the margins of brooks, canals, lakes, ponds, reservoirs, rivers, streams, and thickets at elevations of up to 1614 m. Exposures include open and shaded sites characterized by intermediate moisture levels. The substrates are circumneutral (e.g., pH 7.0) and include clay, clay loam, gravel, loam, marl, muck, mud, sand, sandy clay, sandy loam, and silt. Belknap, Karnak, Tradewater Karnak, and Tradewater Stendal soil types have been reported. The primary means of pollination has not been established, although the rayless capitula might indicate some degree of self-pollination. A single plant can produce up to 1940 relatively large (up to 14 mm) cypselae. Their barbed awns adhere to the fur of various mammals (Mammalia) including bears (Ursidae: *Ursus americanus*), bison (Bovidae: *Bison bison*),

deer (Cervidae: *Odocoileus virginianus*), mice (Cricetidae: *Peromyscus maniculatus*), and raccoons (Procyonidae: *Procyon lotor*), which disperse them. The fruits often are flattened and probably also water dispersed to some extent. Reasonable germination has been obtained for seeds that have been dried at 22°C for 5 days, stored at 5°C in the dark for 2 months, and then stored at 2°C for 5 months under dark, moist conditions. The percent germination decreases relative to substrate particle size. The plants often occur in disturbed sites but are highly sensitive to atmospheric sulfur dioxide. Their roots are associated with endomycorrhizae. **Reported associates:** *Acalypha rhomboidea, Acer negundo, A. rubrum, A. saccharinum, Achillea millefolium, Acmispon americanus, Agalinis purpurea, A. tenuifolia, Ageratina altissima, Agrostis alba, A. scabra, A. stolonifera, Alisma subcordatum, A. triviale, Amaranthus tuberculatus, Ambrosia artemisiifolia, A. trifida, Amorpha fruticosa, Amphicarpaea bracteata, Andropogon gerardii, A. virginicus, Anemopsis californica, Antennaria microphylla, Anthemis cotula, Apios americana, Apocynum cannabinum, Arida parviflora, Arnica cordifolia, Artemisia absinthium, A. biennis, A. vulgaris, Asclepias incarnata, A. ovalifolia, A. speciosa, Atriplex argentea, Beckmannia syzigachne, Berula erecta, Betula nigra, Bidens cernuus, B. connatus, B. frondosus, B. tripartitus, Boehmeria cylindrica, Bolboschoenus fluviatilis, B. maritimus, Boltonia asteroides, Brickellia eupatorioides, Bromus ciliatus, B. inermis, Calamagrostis canadensis, C. stricta, Calopogon tuberosus, Caltha palustris, Campanula americana, C. aparinoides, Campsis radicans, Carex aquatilis, C. cristatella, C. crus-corvi, C. emoryi, C. grayi, C. hystericina, C. interior, C. lacustris, C. lasiocarpa, C. limosa, C. louisianica, C. lupuliformis, C. lupulina, C. lurida, C. muskingumensis, C. nebrascensis, C. pellita, C. prairea, C. sartwellii, C. scoparia, C. stipata, C. stricta, C. tetanica, C. typhina, C. vulpinoidea, Carya ovata, Celtis occidentalis, Cephalanthus occidentalis, Cerastium fontanum, Chelone glabra, Cicuta bulbifera, C. maculata, Cinna arundinacea, Cirsium discolor, Conyza canadensis, Cornus amomum, C. racemosa, C. sericea, Crataegus, Cryptotaenia canadensis, Cuscuta glomerata, C. gronovii, C. indecora, C. pentagona, Cyperus erythrorhizos, C. esculentus, C. ferruginescens, C. odoratus, C. strigosus, Cypripedium reginae, Dichanthelium dichotomum, Doellingeria umbellata, Dulichium arundinaceum, Dyssodia papposa, Echinochloa crus-galli, E. muricata, Echinocystis lobata, Eleocharis acicularis, E. elliptica, E. erythropoda, E. obtusa, E. palustris, Elymus trachycaulus, E. virginicus, Epilobium leptophyllum, Equisetum arvense, E. hyemale, Eragrostis cilianensis, Eriophorum angustifolium, Euonymus atropurpureus, Eupatorium perfoliatum, Euphorbia corollata, Euthamia graminifolia, Eutrochium maculatum, E. purpureum, Fallopia dumetorum, Forestiera acuminata, Fraxinus nigra, F. pennsylvanica, Galium trifidum, Gaura podocarpa, Gentiana andrewsii, Gentianopsis crinita, Geum aleppicum, G. laciniatum, Glyceria grandis, G. striata, Glycyrrhiza lepidota, Grindelia squarrosa, Helenium autumnale, H. microcephalum, Helianthus hirsutus, H. nuttallii, H. strumosus, Heliopsis helianthoides, Hibiscus moscheutos,*

Hieracium scabrum, Hordeum jubatum, Hypericum gentianoides, Hypoxis hirsuta, Ilex verticillatus, Impatiens capensis, Iris versicolor, Isocoma pluriflora, Juncus balticus, J. canadensis, J. dudleyi, J. effusus, J. nodosus, Laportea canadensis, Leersia oryzoides, L. virginica, Lemna minor, Lilium philadelphicum, Linum sulcatum, Liquidambar styraciflua, Lobelia siphilitica, Ludwigia palustris, Lycopodium tristachyum, L. habereri, Lycopus americanus, L. asper, L. uniflorus, L. virginicus, Lysimachia ciliata, L. nummularia, L. thyrsiflora, Lythrum salicaria, Machaeranthera tanacetifolia, Maianthemum stellatum, Medicago lupulina, M officinalis, Menispermum canadense, Mentha arvensis, Menyanthes trifoliata, Mimulus glabratus, M. ringens, Monarda fistulosa, Muhlenbergia glomerata, M. mexicana, Myosoton aquaticum, Nepeta cataria, Oenothera biennis, O. villosa, Onoclea sensibilis, Osmunda regalis, Oxypolis rigidior, Panicum capillare, Parthenocissus vitacea, Paspalum distichum, Pedicularis lanceolata, Penthorum sedoides, Persicaria amphibia, P. coccinea, P. hydropiperoides, P. lapathifolia, P. maculosa, P. pensylvanica, P. punctata, P. sagittata, Phalaris arundinacea, Phragmites australis, Phleum pratense, Physalis longifolia, Physostegia virginiana, Peltandra virginica, Pilea fontana, P. pumila, Plantago major, Poa palustris, P. pratensis, Populus deltoides, P. tremuloides, Potentilla anserina, P. norvegica, Psilactis asteroides, Pseudognaphalium stramineum, Quercus alba, Q. phellos, Ranunculus cymbalaria, R. pensylvanicus, R. sceleratus, Rhynchospora capitellata, Ribes americanum, Rorippa amphibia, R. aquatica, Rubus idaeus, Rudbeckia hirta, Rumex britannica, R. crispus, R. mexicanus, R. salicifolius, R. verticillatus, Sagittaria brevirostra, S. latifolia, Salix alba, S. amygdaloides, S. exigua, S. interior, S. myricoides, S. nigra, S. petiolaris, Sambucus nigra, Saururus cernuus, Schoenoplectus acutus, S. pungens, S. tabernaemontani, Scirpus cyperinus, S. pallidus, S. sylvaticus, Scrophularia marilandica, Scutellaria galericulata, S. lateriflora, Senecio viscosus, Setaria pumila, Sicyos angulatus, Sium suave, Solidago canadensis, S. gigantea, S. nemoralis, S. rigida, Sorghastrum nutans, Sparganium eurycarpum, Spartina pectinata, Spiranthes cernua, Sporobolus contractus, S. flexuosus, Stachys palustris, S. tenuifolia, Stellaria longifolia, Strophostyles helvola, Symphyotrichum boreale, S. ciliatum, S. ericoides, S. frondosum, S. laeve, S. lanceolatum, S. novae-angliae, S. praealtum, S. racemosum, S. sericeum, S. subulatum, Taraxacum officinale, Thalictrum dasycarpum, Thaspium barbinode, Thelypteris palustris, Toxicodendron radicans, Triadenum fraseri, Trifolium hybridum, T. pratense, Typha angustifolia, T. latifolia, Ulmus americana, Urtica dioica, Vaccinium corymbosum, Verbena hastata, V. urticifolia, Vernonia gigantea, Veronica scutellata, Vicia, Viola blanda, V. nephrophylla, V. sagittata, Xanthium strumarium, Zeltnera calycosa, Zizia aurea.

Use by wildlife: *Bidens beckii* has been observed to support as many as 28 macroinvertebrates per mm² of tissue. These include caddis flies (Insecta: Trichoptera: *Hydroptila, Leptocerus americanus, Oecetis, Orthotrichia, Oxyethira, Polycentropus, Setodes, Triaenodes*), crustaceans (Amphipoda: *Hyalella*; Cladocera: *Daphnia*; Poecilostomatoida: *Ergasilus*),

damselflies and dragonflies (Insecta: Odonata: *Argia, Chromagrion, Erythemis*), isopods (Isopoda: *Caecidotea*), mayflies (Insecta: Ephemeroptera: *Attenella, Caenis, Callibaetis*), midges (Insecta: Diptera: Ceratopogonidae: *Bezzia, Dasyhelea, Probezzia*), molluscs (Gastropoda: *Aplexa elongata, Ferrissia, Gyraulus parvus, Laevapex, Menetus dilatatus, Pisidium, Planorbella trivolvis, Valvata sincera*), weevils (Insecta: Coleoptera: Curculionidae: *Euhrychiopsis lecontei*), and worms (Oligochaeta: *Stylaria*). The weevil *Euhrychiopsis lecontei* is known to oviposit on *B. beckii*. The seeds are eaten by various waterfowl (Aves: Anatidae). Although its foliage contains highly toxic phenolic compounds, *B. cernuus* is a host plant of Nearctic moths (Insecta: Lepidoptera: Tortricidae: *Epiblema otiosana*) and weevils (Insecta: Curculionidae: Coleoptera: *Apion*). Other associated invertebrates include aphids (Insecta: Hemiptera: Aphididae: *Rhopalosiphum nymphaeae*) and several flies (Insecta: Diptera: Tephritidae: *Campiglossa absinthii, Paroxyna absinthii*). Predatory stinkbugs (Insecta: Hemiptera: Pentatomidae: *Perillus circumcinctus*) reproduce in stands of *B. cernuus*. These plants also host a variety of fungi including Ascomycota (Erysiphaceae: *Podosphaera fuliginea, P. fusca, P. macularis, P. xanthii*; Mycosphaerellaceae: *Cercospora bidentis, C. umbrata, Pseudocercospora megalopotamica, Ramularia concomitans*; Protomycetaceae: *Protomyces gravidus*), and Basidiomycota (Entolomataceae: *Entyloma compositarum*; Pucciniaceae: *Puccinia obtecta*). Flowers of *B. cernuus* are visited by numerous insects (Insecta) including long-tongued bees (Hymenoptera: Anthophoridae: *Ceratina dupla, Epeolus autumnalis, Melissodes agilis, M. boltoniae, M. druriella*; Apidae: *Apis mellifera*; *Bombus fraternus, B. griseocollis, B. impatiens, B. pensylvanicus*; Megachilidae: *Megachile brevis, M. latimanus, M. mendica, M. petulans*), short-tongued bees (Hymenoptera: Andrenidae: *Andrena aliciae*; Colletidae: *Colletes compactus*; Halictidae: *Halictus ligatus, Lasioglossum versatus*), beetles (Coleoptera: Cantharidae: *Chauliognathus pennsylvanicus*; Chrysomelidae: *Diabrotica undecimpunctata*), butterflies (Lepidoptera: Hesperiidae: *Atalopedes campestris*; Lycaenidae: *Lycaena hyllus*; Nymphalidae: *Danaus plexippus, Limenitis archippus, Phyciodes tharos, Vanessa atalanta, V. virginiensis*; Pieridae: *Colias philodice*), flies (Diptera: Anthomyiidae: *Zaphne ambigua*; Bombyliidae: *Poecilanthrax alcyon, Sparnopolius confusus, Systoechus vulgaris*; Calliphoridae: *Cochliomyia macellaria*; Muscidae: *Neomyia cornicina*; Syrphidae: *Eristalis dimidiatus, E. transversus, Helophilus fasciatus, Toxomerus marginatus*; Tachinidae: *Archytas aterrima, Plagiomima spinosula*; Tephritidae: *Icterica circinata*), moths (Lepidoptera: Arctiidae: *Cisseps fulvicollis*; Noctuidae: *Anagrapha falcifera, Feltia jaculifera, Helicoverpa zea, Heliothis*), and wasps (Hymenoptera: Ichneumonidae: *Exetastes suaveolens*; Scoliidae: *Scolia bicincta*; Vespidae: *Polistes annularis, P. fuscata, Eumenesfraterna*). The floral nectar is consumed by various bees and beetles and the pollen collected and/or eaten by both as well. *Bidens connatus* is a host to several fungi (Ascomycota: *Cercospora nubecula, C. umbrata, Podosphaera aphanis, P. fuliginea, P. fusca, Protomyces andinus, P.*

gravidus, Pseudocercospora megalopotamica). The plants also are hosts to aphids (Insecta: Hemiptera: *Aphis coreopsidis*) and the European corn borer (Insecta: Lepidoptera: Crambidae: *Ostrinia nubilalis*). Pollen of *B. connatus* is collected by bumblebees (Insecta: Hymenoptera: *Bombus impatiens, B. pensylvanicus*). *Bidens laevis* is eaten and damaged by nutria (Mammalia: Rodentia: Myocastoridae: *Myocastor coypus*). However, when the barbed achenes of the plants penetrate the skin of nutria, secondary bacterial and fungal infections can result in dermatitis and skin lesions, which devalue the pelts for use as fur. *Bidens laevis* provides roosting habitat for the red-winged blackbird (Aves: Icteridae: *Agelaius phoeniceus*). Its seeds are eaten by American wigeon (Aves: Anatidae: *Anas americana*) and redhead ducks (Aves: Anatidae: *Aythya americana*). The leaves are eaten by beetles (Insecta: Coleoptera: Chrysomelidae *Coreopsomela elegans*) and are fed upon by aphids (Insecta: Hemiptera: *Macrosiphum rudbeckia*). They also are parasitized by Ascomycota fungi (Erysiphaceae: *Podosphaera macularis*; Mycosphaerellaceae: *Cercospora umbrata*). *Bidens laevis* is associated with *Iris* whitefly (Insecta: Heteroptera: Aleurodidae: *Aleyrodes spiraeoides*). The flowers are visited by various butterflies (Insecta: Lepidoptera) including blues and hairstreaks (Lycaenidae), dusky wing (Hesperiidae: *Erynnis*), monarch and painted lady (Nymphalidae: *Danaus plexippus; Vanessa*), Mormon metalmark (Riodinidae: *Apodemia mormo*), orange sulphurs, and California dog face (Pieridae: *Colias eurytheme; Zerene eurydice*). *Bidens mitis* is a host plant for bacterial wilt (Bacteria: Burkholderiaceae: *Ralstonia solanacearum*). The flowers of *B. mitis* are visited by bees (Insecta: Hymenoptera: Apidae: *Ceratina dupla*) and Monarch butterflies (Insecta: Lepidoptera: Nymphalidae: *Danaus plexippus*). *Bidens trichospermus* is a host of aphids (Insecta: Hemiptera: Aphididae: *Uroleucon chrysanthemi*) and rust fungi (Basidiomycota: Pucciniaceae: *Puccinia obtecta*). The plants also host larvae of the tickseed moth (Insecta: Lepidoptera: Noctuidae: *Cirrhophanus triangulifer*). Several fungi (Ascomycota: *Cercospora bidentis, C. umbrata; Podosphaera fuliginea*) parasitize the plants. *Bidens tripartitus* is associated with several flies (Insecta: Diptera: Agromyzidae: *Napomyza lateralis, Phytomyza horticola*; Tephritidae: *Campiglossa absinthii, Dioxyna bidentis, Icterica seriata, Paroxyna absinthii*). Bumblebees (Insecta: Hymenoptera: *Bombus*) are known to collect pollen and nectar from the flowers. The roots are readily infected by nematodes (Nematoda: Longidoridae: *Longidorus africanus*), which cause galls characterized by elevated cell-wall protein and free amino acid levels. The species is a known host of the cucumber mosaic virus (Bromoviridae). *Bidens vulgatus* provides a component of nesting habitat for the sedge wren (Aves: Troglodytidae: *Cistothorus platensis*). Occurrence of this species is correlated strongly with the presence of northern short-tailed shrews (Mammalia: Soricidae: *Blarina brevicauda*) and cottontail rabbits (Mammalia: Leporidae: *Sylvilagus floridanus*). It is a host to TSWV, the tomato spotted wilt virus (Bunyaviridae: *Tospovirus*), which is transmitted by thrips insects (Insecta: Thysanoptera: Thripidae: *Frankliniella occidentalis*). An aphid (Insecta: Hemiptera: Aphididae: *Aphis coreopsidis*) is known to infest the foliage and inflorescences. The plants also are associated with a number of fungi including Ascomycota (Erysiphaceae: *Sphaerotheca castagnei, S. fuliginea, S. fusea, S. humuli, S. macularis*; Mycosphaerellaceae: *Septocylindrium concomitans*), Basidiomycota (Entylomataceae: *Entyloma compositarum*), and Oomycota (Peronosporaceae: *Plasmopara halstedii*). *Bidens cernuus, B. frondosus,* and *B. vulgatus* are hosts of the northern root-knot nematode (Nematoda: Heteroderidae: *Meloidogyne hapla*). *Bidens cernuus, B. connatus* and *B. laevis* can become seriously infested with an Oomycota fungus (fungi: Peronosporales: *Plasmopara halstedii*). Larvae of the bidens borer (Insecta: Lepidoptera: Tortricidae: *Epiblema otiosana*) excavate the stems of *B. cernuus, B. connatus* and *B. vulgatus*. Ingestion of the barbed achenes of various *Bidens* species can cause damage to the gill arches of young salmon (Vertebrata: Osteichthyes: Salmonidae: *Oncorhynchus kisutch*).

Economic importance: food: A tea has been made from the leaves of *Bidens aureus*. Native American tribes in the Texas area used the heads of *B. bigelovii* to prepare a beverage. The young leaves of *B. connatus* allegedly are edible when cooked. Leaves of *B. laevis* and *B. tripartitus* also are reported to be edible. Raw or boiled leaves of *B. tripartitus* are eaten by the Daffla people of northern India. *Bidens laevis* was used as food by the Paiute; **medicinal:** Flavonoid extracts of *Bidens aureus* reportedly are antiulcerogenic. A decoction made from the leaves was taken to relieve diarrhea and as a digestive aid and sedative. *Bidens cernuus* has been used in folk medicine as an expectorant, emenogogue, and remedy for croup. It also has been used as a treatment for urinary tract infections. The plants contain phenylheptatriyne (1-phenylhepta-1,3,5-triyne), which is an effective antibiotic against Gram-positive bacteria, fungi and yeasts. Plant extracts are a source of cerbiden, which shows high antifungal activity against various strains of yeast (Ascomycota: Candidaceae: *Candida*). Essential oils extracted from *B. cernuus* also include (E,E)-1,3,11-tridecatriene-5,7,9-triyne, and sesquiterpenes (caryophyllene oxide; humulene-1,2-epoxide; a-selinene), with the latter group also exhibiting antimicrobial properties. Acetone, alcohol and ether extracts of *B. cernuus* inhibit gram+ bacteria. Shoot isolates of *B. cernuus* exhibit scavenging activity for DPPH (1,1-diphenyl-2-picrylhydrazyl) radicals. The root and seeds of *B. connatus* were used as an expectorant and to treat various menstrual problems. An infusion made from the root was prepared for alleviating severe coughs, heart palpitations, and croup. A poultice made from heated leaves was applied to the throat and chest to treat bronchial and throat ailments. The juice of *B. laevis* inhibits growth of Mycobacteria. The Florida Seminoles used *B. mitis* to treat diarrhea, eye problems, fevers and headaches. The Mikasuki and Seminole Seminoles treated fire, mist, and sun sicknesses using root or whole plant infusions of *B. trichospermus* ("hasabi"). This species also was used by the Seminoles as an eye medicine, to reduce fever, relieve pain, and to treat diarrhea and rheumatism (the latter by external application). *Bidens tripartitus* plants (especially the flowers) are a source of natural antioxidants, which exhibit free radical scavenging activity. Extracts from this species exhibit both antifungal and antibacterial properties and have been used to treat various ailments in folk medicine. Ethanolic extracts (50 μg/ml) contain 1-phenyl-3,5,7-heptatriyn, which inhibits (>80%), a protozoan parasite (Chromalveolata: Plasmodiidae: *Plasmodium falciparum*) that causes human malaria. Analgesic and antipyretic properties of *B. tripartitus* infusions make them potentially useful in treatment of inflammations. These properties are due at least in part to the presence of okanins, which inhibit production of nitric oxide, a causative agent in many inflammatory diseases. In Russia, the aerial portions of the plant are used to treat dermatitis and allergic reactions. Methylene chloride extracts of *B. tripartitus* exhibit high antithrombin activity. An essential oil derived from the roots of *B. tripartitus* possesses strong antifungal properties; **cultivation:** *Bidens aureus* is widely grown as an ornamental and contains a number of cultivars including: 'All Gold', 'Blacksmith's Flame', 'Cream Streaked Yellow', 'Golden Drop', 'Hannay's Lemon Drop', 'Julia's Gold', 'Lemon Queen', 'Rising Sun', 'Soft Yellow', and 'Super Nova'. Seeds of *Bidens cernuus*, *B. connatus*, *B. laevis*, *B. mitis* and *B. trichospermus* are marketed for native wildflower gardens; **misc. products:** *Bidens cernuus* and *B. trichospermus* are plants recommended for attracting butterflies. *Bidens laevis* has been considered as an organism for aquatic toxicity bioassays because of its high sensitivity to mutagenic compounds; **weeds:** *Bidens bigelovii* is regarded as a weed of milpa agricultural systems in Mexico. *Bidens cernuus* is a weed of carrot farms in Quebec. *Bidens connatus* and *B. trichospermus* are regarded as weeds of commercial North American cranberry bogs. *Bidens aureus*, *B. connatus* and *B. vulgatus* all are regarded as weeds in Europe. *Bidens mitis* plants produce germination stimulants for witchweed (*Striga lutea*). *Bidens trichospermus* is a common weed in the Eastern Hemisphere. *Bidens tripartitus* is a weed of maize, soybeans and sugar beet crops in Europe and rice fields in Asia. Triazine-resistant biotypes have been reported from Austria. *Bidens vulgatus* is a common weed in North American agricultural fields, gardens, waysides, yards, and other UPL sites; **nonindigenous species:** *Bidens aureus* is adventive in Argentina and naturalized in Asia (Japan) and Europe; *B. connatus* and *B. vulgatus* are naturalized in Asia and Europe. *Bidens connatus* was introduced in the western United States as a seed contaminant of cranberry stocks. *Bidens tripartitus* is regarded as nonindigenous to the Hudson River Basin of New York. *Bidens vulgatus* has been introduced to Belgium, Finland, France, Great Britain, New Zealand, Romania, and Yugoslavia. It also is believed to have been introduced to the western United States.

Systematics: *Bidens* presents many taxonomic problems. Until the advent of molecular data, the genus widely was regarded as monophyletic, although there was some debate whether to include the submersed aquatic species *Megalodonta beckii*, which possessed an anomalous chromosome number. Morphologically, this species is more similar to *Cosmos* than to *Bidens* or *Coreopsis*; however, various molecular phylogenetic analyses incorporating chloroplast and nuclear DNA sequences indicate that *Bidens* is not monophyletic, with many of the species (including *B. beckii*, *B. cernuus*, and *B. coronatus*) resolving within a clade containing both *Coreopsis* and *Cosmos* (Figure 5.141). The preliminary phylogenetic studies of *Bidens* indicate that the OBL habit has evolved at least twice in the genus, with a major radiation of aquatic and wetland species centered on members of section *Bidens*. There also are problems with the taxonomic delimitation of some species. Morphological and cytological data indicate that at least some plants referred to as *B. connatus* likely represent hybrids involving *B. cernuus* and *B. frondosus*. The putative hybrids possess a combination of morphological characteristics both intermediate and similar to the presumed parents. Some authors place *B. comosus* and *B. connatus* in synonymy with *B. tripartitus*, making it difficult to extract information accurately from the literature; however, these taxa are quite dissimilar morphologically. The distinctness of *B. cernuus* and *B. laevis* is debateable and deserves further study. The base chromosome number of *Bidens* is $x=13$. Presumed diploid counts occur in *Bidens beckii* ($2n=26$) and in *B. aureus*, *B. connatus*, *B. hyperboreus*, *B. laevis*, *B. mitis*, *B. trichospermus*, and *B. vulgatus* (all $2n=24$). Aneuploid and/or polyploid variation has been reported for *B. bigelovii* ($2n=48$), *B. cernuus* ($2n=24, 48$), *B. connatus* ($2n=24, 48, 60, 72$), *B. hyperboreus* ($2n=24, 48$), *B. laevis* ($2n=22, 24$), *B. tripartitus* ($2n=48$), and *B. vulgatus* ($2n=12, 24, 48$). Counts for *B. aureus* from Mexico exhibit extreme variation ($n=17–18, 23, ~35$). Failure to observe meiotic divisions in pollen mother cells suggests that *B. connatus* and *B. vulgatus* may be apomictic, at least in some populations.

Comments: The distributions of the OBL *Bidens* species are quite varied. *Bidens cernuus* and *B. tripartitus* are widespread in North America (and also in Eurasia); whereas, *B. aureus* is restricted to Arizona (but extends into Mexico and Central America). The other OBL species occur in northern North America (*B. beckii*), northeastern North America (*B. hyperboreus*), central North America (*B. vulgatus*), eastern North America (*B. connatus*, *B. trichospermus*), southern North America (*B. laevis*; extending into South America), the southeastern United States (*B. mitis*), and the southwestern United States (*B. bigelovii*).

References: Aldrich, 1943; Allison, 1967; Almeida-Cortez et al., 1999; Anderson et al., 1968; Andreas & Bryan, 1990; Austin, 2004b; Bačić, 1983; Baden III et al., 1975; Baldwin & Derico, 1999; Baldwin & Pendleton, 2003; Baldwin & Sharpe, 2009; Baldwin et al., 2001; Ballard, 1988; Baskin & Baskin, 1998; Beard, 1973; Beaven & Oosting, 1939; Beck et al., 2010; Bélair & Benoit, 1996; Benvenuti, 2007; Benvenuti & Macchia, 1997; Blanckaert et al., 2007; Bonasera et al., 1979; Bondarenko et al., 1968; Boot, 1914; Boutin et al., 2000; Boylen et al., 1999; Brandão et al., 1997; Brändel, 2004; Brenner et al., 1982; Brown, 1878; Caccavari & Fagúndez, 2010; Cain, 1928; Calero et al., 1996; Carlquist & Pauly, 1985; Ceska & Ceska, 1986; Chabreck et al., 1977; Chalchat et al., 2009; Coulter, 1904; Couvreur et al., 2005; Crawford & Mort, 2005; Crowe & Parker, 1981; Cypert, 1961, 1972; Dale, 1984b; Dale & Egley, 1971; Dana et al., 1965; D'Angelo et al., 2005; Davis, 1903,

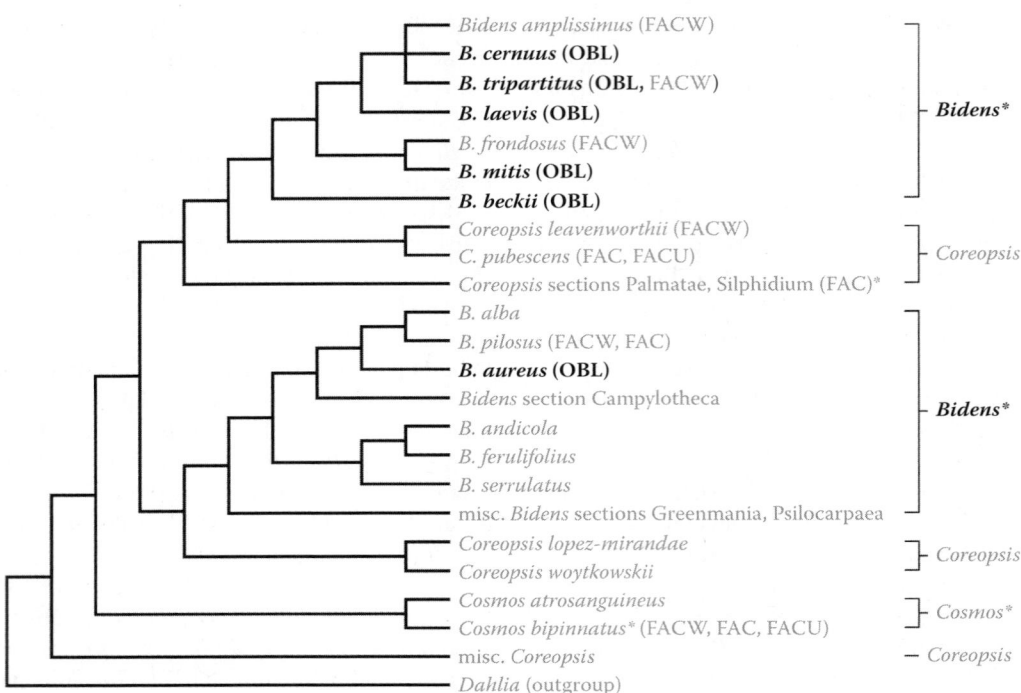

FIGURE 5.141 Phylogenetic relationships of *Bidens* as estimated from ITS sequence data. The relationships depicted indicate that the OBL taxa (indicated in bold) have evolved at least twice in *Bidens*, which with *Cosmos*, is nested polyphyletically within *Coreopsis* (shown by asterisks). These results are concordant with those obtained from a smaller subset of taxa (Ganders et al., 2000). *Bidens amplissimus* has not been assigned a wetlands designation, but also is characteristic of wetland habitats (Strother & Weedon, 2006). (Adapted from Sayre, C., A new species of *Bidens* (Asteraceae: Heliantheae) from Starbuck Island provides evidence for a second colonization of Pacific Islands by the genus, MSc Thesis, The University of British Columbia, Vancouver, BC, 61 pp, 2001.)

1910; Dearness, 1929; Decker, 1932; Dikova, 1989; Dittmar & Neely, 1999; Doumlele et al., 1985; Dovrat et al., 2012; Eddy, 1925; Ehrenfeld, 2005; Eichler & Boylen, 2009; Epstein & Cohn, 1971; Evans, 1979, 1982; Farnsworth & Meyerson, 1999; Fenner, 1983; Fishbein et al., 1997; Fisher et al., 1996; Foote, 1967; Friauf, 1953; Frolik, 1941; Galatowitsch & van der Valk, 1996a; Ganders et al., 2000; Gargiullo, 2010; Gates, 1911; Gaudet & Keddy, 1995; Gleason et al., 2009; Goun et al., 2002; Gratani et al., 2008; Grime & Hunt, 1975; Grombone-Guaratini et al., 2004; Gruberová et al., 2001; Guppy, 1906; Haines, 2003; Hall, 2010; Halsted, 1888; Harley & Harley, 1987; Harrison, Jr. & Chabreck, 1988; Haworth & McPherson, 1994; Hazlett, 1989; Healy, 1946; Hladun et al., 2002; Hogue, 1976; Hong et al., 2008; Hopfensperger & Baldwin, 2009; Howard, 1915; Howell, 1937; Hunt, 1943; Hutton, 2010; Isaak et al., 1959; Jackson, 1921; Jog et al., 2005; Jones, 1941, 2005; Jurik et al., 1994; Jutila, 2002; Kangas & Hannan, 1985; Keddy & Constabel, 1986; Keddy & Ellis, 1985; Keddy & Reznicek, 1982; Keller, 2000; Kellerman, 1908; Kellogg & Bridgham, 2002; Khan & Brush, 1994; Khan et al., 2002; Kil et al., 2012; Kim et al., 1999; Kimball & Crawford, 2004; Knuth, 1908; Ku et al., 1993; Kulbaba, 2004; Lacoursière et al., 1976; Larson, 1992; Leck, 1996; Leck & Brock, 2000; Leck & Graveline, 1979; Leck & Simpson, 1987, 1992, 1995; Leck et al., 1994; Leonard, 1959, 1967; Leroux et al., 1996; Leyer, 2006; Lieneman, 1929; Lindsey et al., 1961; Linzon et al., 1973; Lowcock & Murphy, 1990; Luken & Thieret, 2001; Madsen,

2006; Mamedov et al., 2005; McAtee, 1925; McCormac, 1993; McCoy & Hales, 1974; McDougall & Glasgow, 1929; Meanley, 1965; Melchert, 2010; Mills et al., 1996; Mitich, 1994; Miura et al., 2007; Montes et al., 1993; Moravcová et al., 2010; Nekola, 2004; Nekola & Lammers, 1989; Nelson & Anderson, 1983; Nelson et al., 2002; Nickell, 1959; Nowak, 2003; Noyes, 2007; Ortega et al., 2000; Padgett et al., 2004; Papp, 1959; Parker & Leck, 1985; Parrella et al., 2003; Peach & Zedler, 2006; Pérez et al., 2008, 2011; Perry & Atkinson, 1997; Perry & Hershner, 1999; Perry & Uhler, 1981; Pieroni & Price, 2006; Plante, 2000; Pozharitskaya et al., 2010; Racine et al., 1998b; Reed, 1913; Rho & Lee, 2004; Ristich et al., 1976; Robertson, 1928; Roman et al., 2001; Römermann et al., 2005; Rossell et al., 2008; Runkel & Roosa, 1999; Sanger & Gorham, 1973; Sasser et al., 1995, 1996; Sayre, 2001; Scogin & Zakar, 1976; Segadas-Vianna, 1951; Seybold et al., 2002; Sharpe & Baldwin, 2009; Sheldon & Boylen, 1977; Shemluck, 1982; Silvani et al., 2008; Simpson et al., 1983; Sipple & Klockner, 1980; Skroch & Dana, 1965; Slater, 2008; Smirnov et al., 1995; Solarz & Newman, 1996; Solbrig et al., 1972; Srivastava et al., 2010; Stalter & Baden, 1994; Steinauer et al., 1996; Stevens, 1932; Stotz, 2000; Strimaitis & Sheldon, 2011; Strother & Weedon, 2006; Stuckey, 1971; Sun & Ganders, 1988; Sundue, 2005, 2006; Svenson, 1935; Tadesse et al., 2001; Tanahara & Maki, 2010; Tatic & Zukowski, 1973; Thompson & Green, 2010; Thompson & Grime, 1983; Thompson & McKinney, 2006; Tomczykowa et al., 2008, 2011; Torres &

Galetto, 2002; Transeau, 1905; Tryon & Easterly, 1975; Turner, 1934; Tuttle, 1954; Van der Valk & Davis, 1978; Verloove, 2003; Visser et al., 1999; Voss, 1996; Walkinshaw, 1935; Wallentinus, 2002; Ware, 1915; Warrington, 1986; Warwick, 1991; Weiher & Keddy, 1995; Whitson, 1905; Wilson, 1908, 1935; Wolniak et al., 2007; Wright & Wright, 1932; Zhu et al., 2009; Zika, 2003; Zomlefer et al., 2007.

8. *Blennosperma*
stickyseed

Etymology: from the Greek *blennos sperma* meaning "slimy seed"

Synonyms: none

Distribution: global: North America; South America; **North America:** western

Diversity: global: 3 species; **North America:** 2 species

Indicators (USA): OBL: *Blennosperma bakeri*

Habitat: freshwater; palustrine; **pH:** unknown; **depth:** <1 m; **life·form(s):** emergent herb

Key morphology: Stems (to 30 cm) glabrous, branched upward from the base; leaves (to 15 cm) linear or 1–3-lobed (lobes to 3 cm), becoming gradually reduced and 0–5-lobed upward, margins entire; involucre of 6–8 ovate phyllaries (to 8 mm), which curve over the mature cypselae; 6–13 ray florets (whitish to yellow) with red stigmas, 30–70 disk florets (yellow) with white stigmas; cypselae (to 4 mm) papillate, strongly 4–6 angled; pappus lacking

Life history: duration: annual (fruit/seeds); **asexual reproduction:** none; **pollination:** insect; self; **sexual condition:** monoecious; **fruit:** cypselae (common); **local dispersal:** water; **long-distance dispersal:** animals; water

Imperilment: (1) *Blennosperma bakeri* [G1]; S1 (CA)

Ecology: general: This small genus of just three species occupies a wide range of habitats from dry grasslands to wetlands, but only one species is OBL (the other North American species is ranked FACW). The flowers are cross-pollinated by various insects (Insecta) but include both self-incompatible and self-compatible reproductive modes. The cypselae become mucilaginous when wet, hence the common name. This adhesive property facilitates their dispersal via attachment to various animals.

Blennosperma bakeri **Heiser** grows in shallow (<1 m) water of flats, meadows, swales, vernal pools, wallows, and along the margins of intermittent streams at elevations of up to 40 m. The substrates are characterized as loam. The self-incompatible flowers produced in early spring (March) are outcrossed and pollinated by a variety of insects (see *Use by wildlife* later). Cypselae develop only in the ray florets. When wet, the tuberculate cypselae secrete adhesive mucilage. This substance adheres strongly to most materials presumably including the plumage of waterfowl, which potentially could achieve long-distance dispersal. The tuberculate pericarps also provide buoyancy, which facilitates local water dispersal. Historical data indicate that populations have experienced substantial size fluctuations. Populations are particularly susceptible to grazing by livestock, but appear to have persisted in some areas that had been tilled for agricultural

applications. Several sites have been extirpated due to housing construction, viticulture, and damage from off-road vehicles. **Reported associates:** *Achyrachaena mollis, Anagallis arvensis, Avena barbata, Briza minor, Bromus hordeaceus, Carex, Castilleja densiflora, Convolvulus, Crassula connata, Deschampsia danthonioides, Downingia concolor, Erodium brachycarpum, Eryngium aristulatum, Hordeum pusillum, Hypochaeris glabra, Juncus bufonius, J. phaeocephalus, Limnanthes douglasii, Lasthenia glaberrima, Layia chrysanthemoides, Lupinus nanus, Lilaea scilloides, Madia sativa, Cicendia quadrangularis, Oenothera, Phalaris lemmonii, Plagiobothrys greenei, P. undulatus, Plagiobothrys bracteatus, Pleuropogon californicus, Poa annua, Ranunculus californicus, R. muricatus, Rumex acetosella, R. crispus, Spergula arvensis, Trifolium depauperatum, T. willdenovii, Triphysaria versicolor, Triteleia hyacinthina, Vulpia bromoides, V. octoflora, Veronica peregrina.*

Use by wildlife: The flowers of *Blennosperma bakeri* are visited by a variety of insects (Insecta) including bees (Hymenoptera: Andrenidae: *Andrena*; Apidae: *Nomada*), bee flies (Diptera: Bombyliidae), dance flies (Diptera: Empididae), drone flies (Diptera: Syrphidae), and beetles (Coleoptera: Chrysomelidae, Dermestidae, Melyridae). The plants also are associated with thrips insects (Thysanoptera: Thripidae: *Frankliniella minuta*).

Economic importance: food: The Neeshenam tribes ground the parched seeds of *Blennosperma nanum* into flour for bread; conceivably, the seeds of *B. bakeri* also could have been used in this way; **medicinal:** none; **cultivation:** none; **misc. products:** none; **weeds:** none; **nonindigenous species:** none

Systematics: Phylogenetic analyses resolve *Blennosperma* within Senecioneae subtribe Tussilagininae in a subclade with *Crocidium* (North America) and *Ischnea* (New Guinea). A close relationship between *Blennosperma* and *Crocidium* was first elucidated from their similar pollen cell wall ultrastructure and later by their comparable flavonoid profiles. However, cladistic analysis of morphological data placed *Blennosperma* closest to *Ischnea*, with *Crocidium* as the sister group to both genera. Studies using nrITS and combined nrITS/*trnK/matK* sequence data confirmed the monophyly of the subclade comprising these genera and also that *Blennosperma* is more closely related to *Ischnea* by resolving *Crocidium* as their sister genus. A subsequent analysis combining ETS, ITS, and cpDNA data provided additional confirmation for these relationships (Figure 5.142). Interspecific crosses made between *B. bakeri* and its two congeners yielded from 0–30% seed set; however, pollen fertility of the F$_1$'s was quite different, being 0% in *B. bakeri*×*B. chilense* crosses but 0%–67% in *B. bakeri*×*B. nanum* crosses. Because a large proportion of stained pollen in the latter instance was deformed, it is likely that these hybrids also were much less fertile. Cytological observations indicate that the South American *B. chilense* is an amphiploid whose genome was derived from the two North American species. This conclusion is supported by observed chromosome numbers of $n=9$ (*B. bakeri*), $n=7$ (*B. nanum*) and $n=16$ (*B. chilense*). Genetic surveys of *B. bakeri*

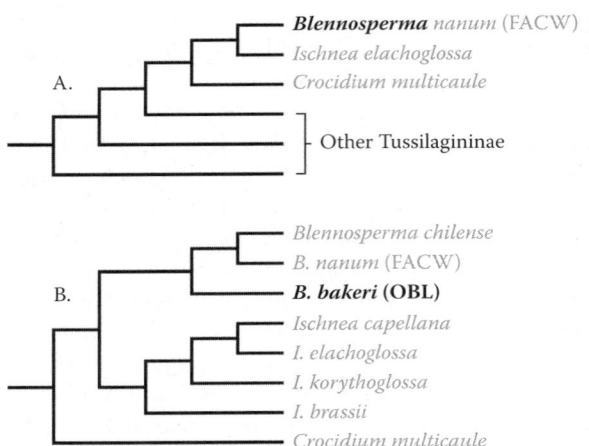

FIGURE 5.142 Phylogenetic relationships of *Blennosperma*. A. Intergeneric relationships indicated by analysis of combined DNA sequence data from eight nuclear and cpDNA regions. B. Interspecific relationships in *Blennosperma*, *Crocidium* and *Ischnea* derived from parsimony analysis of 62 morphological characters (adapted from Swenson & Bremer, 1997). Phylogenetic relationships among the three genera were congruent among all data sets analyzed. Taxa including OBL species are indicated in bold. (Adapted from Pelser, P.B. et al., *Amer. J. Bot.*, 97, 856–873, 2010.)

populations using allozymes, ISSR and RAPDS markers indicates that they are divergent and structured spatially, presumably as a consequence of ineffective gene flow.

Comments: *Blennosperma bakeri* is narrowly endemic and survives in an estimated 22 populations restricted to northern California.

References: Ayres & Sloop, 2008; Eakins, 1994; Elam, 1998; Leong & Bailey, 2000; Ornduff, 1963, 1964; Pelser et al., 2007, 2010; Skvarla & Turner, 1966; Smith, 1991; Strother, 2006a; Swenson & Bremer, 1997, 1999; Wagstaff et al., 2006.

9. *Boltonia*

Doll's-daisy, false chamomile

Etymology: after James Bolton (1758–1799)

Synonyms: *Matricaria* (in part)

Distribution: global: North America; **North America:** eastern

Diversity: global: 5 species; **North America:** 5 species

Indicators (USA): OBL; FACW: *Boltonia asteroides*

Habitat: brackish (coastal), freshwater (tidal/nontidal); palustrine, riverine; **pH:** 7.4–8.2; **depth:** <1 m; **life-form(s):** emergent herb

Key morphology: Stems (to 200 mm) erect, stoloniferous; leaves (to 22 cm) linear or lanceolate/oblanceolate; heads (10–100) radiate, peduncled (to 22 cm), arranged in ascending/spreading corymbs or panicles; bracts (to 12 cm) leaflike; phyllaries (to 4 mm) in 3–5 series; ray florets (20–60) with white or lilac petals (to 13 mm), disk florets (65–170) yellow; cypselae (to 2.0 mm) obovoid, winged (to 0.5 mm), 2–3-awned (to 2 mm).

Life history: duration: perennial (rhizomes, stolons); **asexual reproduction:** rhizomes, stolons; **pollination:** insect; **sexual condition:** gynomonoecious; **fruit:** cypselae (common); **local dispersal:** cypselae (water, wind); **long-distance dispersal:** cypselae (animals?, water, wind)

Imperilment: (1) *Boltonia asteroides* [G5]; S1 (MD, PA); S2 (DE, KS, <u>MB</u>, MI, NC, <u>SK</u>); S3 (GA, OH, VA)

Ecology: general: All *Boltonia* species have an affinity for wet habitats. Although not ranked OBL, two species are found occasionally in wetlands: *B. caroliniana* (FACW) and *B. diffusa* (FACW, FAC). *Boltonia apalachicolensis* and *B. decurrens* (unranked) occur mainly on floodplains and their wetland status should be reevaluated. The capitula contain pistillate ray florets and hermaphroditic disk florets, making the plants functionally gynomonoecious. Presumably, all of the species are predominantly outcrossed. The cypselae are flat and winged, which enhances flotation and facilitates water dispersal. The characteristic presence of several rigid awns may assist with animal dispersal. All species are perennial.

***Boltonia asteroides* (L.) L'Her.** occurs in tidal or nontidal, brackish, freshwater or saline sites (conductivity: 40–6800 µS cm^{-1}) including bogs, Carolina bays, channels, depressions, dikes, ditches, dunes, flood plains, marshes, meadows, mudflats, prairies, river bottoms, saltmarsh, sandbars, savannas, sloughs, streams, swales, swamps, thickets, and the margins of ponds, pools, rivers, sinkholes, and streams at elevations of up to 500 m. The plants usually are found in exposures of full sun but also tolerate partial shade. Although often reported to require acidic substrates (pH <6.8), sites more typically are circumneutral to somewhat alkaline (pH 7.4–8.2). The substrates include clay, clay loam, dolomite, loam, mud, sand, silty loam, silty clay, silty clay loam, and even waterlogged tree stumps. The flowers are arranged in showy heads and are pollinated primarily by bees and wasps (Insecta: Hymenoptera); siphonate flies (Insecta: Diptera) also visit the flowers but are less effective as pollinators. Single capitula can produce more than 200 florets and individual plants are known to yield upward of 10,000 seeds. The seeds require cold, moist stratification for germination (placement of seeds in moist sand kept at 1°C–4°C for 60–120 days). Dispersal occurs mainly through transport along the surface of water. Seeds will persist in seed banks, but for an undetermined duration. Despite its OBL wetland status, *B. asteroides* is fairly desiccation-resistant, with growth essentially unaffected by fall or spring drought periods. It also is quite frost resistant. The plants reproduce vegetatively by a radiating series of long, creeping rhizomes produced from the stem base. **Reported associates:** *Abutilon theophrasti, Acalypha rhomboidea, Acer negundo, A. rubrum, A. saccharinum, Acmella oppositifolia, Ageratina altissima, Agrostis hyemalis, A. perennans, Alisma subcordatum, Alternanthera philoxeroides, Amaranthus tuberculatus, Amblystegium serpens, Ambrosia artemisiifolia, A. trifida, Ammannia coccinea, Amsonia tabernaemontana, Andropogon gerardii, A. glomeratus, A. virginicus, Anemone canadensis, Apocynum cannabinum, Asclepias incarnata, A. lanceolata, A. syriaca, A. verticillata, A. viridis, Aulacomnium palustre, Axonopus fissifolius, Bacopa* caroliniana, *Betula nigra, Bidens aristosus, B. cernuus, B. frondosus, B. vulgatus, Boehmeria cylindrica, Bolboschoenus fluviatilis, Bothriochloa longipaniculata, Bouteloua curtipendula, Bromus inermis, Calamagrostis canadensis, C. stricta, Campsis radicans, Campylium stellatum, Cardiospermum halicacabum, Carex annectens, C. cristatella, C. crus-corvi, C. cryptolepis, C. frankii, C. grayi, C. hyalinolepis, C. lupuliformis, C. lupulina, C. lurida, C. muskingumensis, C. pellita, C. praegracilis, C. sartwellii, C. stricta, C. tribuloides, C. tuckermanii, C. typhina, C. viridula, Carya illinoinensis, Celtis occidentalis, Centella asiatica, Centrosema virginianum, Cephalanthus occidentalis, Chamaecrista fasciculata, Chara, Chasmanthium latifolium, Chelone obliqua, Cicuta maculata, Cinna arundinacea, Cirsium arvense, C. discolor, C. vulgare, Cladium jamaicense, C. mariscoides, Coleataenia longifolia, Conoclinium coelestinum, Conyza canadensis, Coreopsis tinctoria, Cornus drummondii, C. racemosa, Crinum americanum, Cuphea viscosissima, Cyperus aristatus, C. cephalanthus, C. erythrorhizos, C. haspan, C. odoratus, C. strigosus, Daucus carota, Desmanthus illinoensis, Dichanthelium acuminatum, D. hirstii, D. oligosanthes, D. sphaerocarpon, Diodia virginiana, Diospyros virginiana, Distichlis spicata, Echinochloa muricata, Echinocystis lobata, Eleocharis erythropoda, E. intermedia, E. obtusa, E. quadrangulata, E. olivacea, E. palustris, E. tuberculosa, Elymus virginicus, Equisetum arvense, Eragrostis frankii, E. pectinacea, Erechtites hieraciifolius, Eriocaulon compressum, Eriochloa punctata, Eryngium aquaticum, E. yuccifolium, Eupatorium altissimum, E. leucolepis, E. perfoliatum, E. serotinum, Euthamia graminifolia, E. leptocephala, Festuca, Fimbristylis autumnalis, F. benghalensis, F. thermalis, Forestiera acuminata, Fraxinus pennsylvanica, Fuirena pumila, Galactia volubilis, Gymnadeniopsis nivea, Gymnopogon ambiguus, Helenium autumnale, Helianthus angustifolius, H. grosseserratus, H. tuberosus, Hibiscus laevis, H. lasiocarpos, H. moscheutos, Hydrolea ovata, Hymenocallis liriosme, H. occidentalis, Hypericum galioides, Hypnum lindbergii, Hyptis alata, Ipomoea, I. lacunosa, I. nil, I. sagittata, Iris versicolor, I. virginica, Isolepis carinata, Iva angustifolia, I. annua, Juglans nigra, Juncus balticus, J. marginatus, J. megacephalus, J. repens, J. roemerianus, J. tenuis, J. torreyi, Juniperus virginiana, Lachnanthes caroliana, Lactuca canadensis, Laportea canadensis, Leersia hexandra, L. lenticularis, L. oryzoides, L. virginica, Lemna minor, Liatris pycnostachya, Lilaeopsis chinensis, Lindernia dubia, Linum medium, Lipocarpha micrantha, Liquidambar styraciflua, Lobelia cardinalis, L. puberula, L. siphilitica, Lotus corniculatus, Ludwigia alternifolia, L. grandiflora, L. palustris, L. peploides, L. polycarpa, L. sphaerocarpa, Lycopodiella appressa, L. prostrata, L. ×copelandii, Lycopus americanus, L. rubellus, L. virginicus, Lysimachia nummularia, Lythrum alatum, L. lineare, Melilotus alba, Mentha arvensis, Mimulus alatus, M. ringens, Mnesithea rugosa, Morus alba, Muhlenbergia frondosa, M. schreberi, Neptunia lutea, N. pubescens, Oenothera drummondii, O. lindheimeri, Osmunda regalis, Panicum capillare, P. dichotomiflorum, P. hemitomon, P. verrucosum, P. virgatum, Paspalum floridanum, P. plicatulum, Pastinaca sativa, Peltandra virginica, Penthorum sedoides, Persicaria amphibia, P. coccinea, P. hydropiperoides, P. lapathifolia, P. pennsylvanica, P. punctata, P. setacea, P. virginiana, Phalaris arundinacea, Phragmites australis, Phyla lanceolata, Physalis virginiana, Physostegia virginiana, Pilea pumila, P. lanceolata, Platanus occidentalis, Pluchea baccharis, P. foetida, P. odorata, Poa nemoralis, P. palustris, Polygala cymosa, P. leptocaulis, Pontederia cordata, Populus deltoides, Proserpinaca palustris, P. pectinata, Prunella vulgaris, Ptilimnium capillaceum, Quercus bicolor, Q. palustris, Ranunculus sceleratus, Ratibida peduncularis, Rhexia aristosa, R. mariana, R. virginica, Rhynchospora caduca, R. careyana, R. chalarocephala, R. colorata, R. corniculata, R. filifolia, R. globularis, R. macrostachya, R. microcarpa, R. nitens, R. perplexa, Rorippa aquatica, R. islandica, Rubus flagellaris, R. trivialis, Rumex altissimus, R. crispus, R. verticillatus, Sabatia angularis, S. difformis, Saccharum giganteum, S. villosum, Sagittaria lancifolia, S. latifolia, Salix amygdaloides, S. interior, S. nigra, Saururus cernuus, Schizachyrium scoparium, Schoenoplectus pungens, S. tabernaemontani, Scirpus atrovirens, Scleria georgiana, S. reticularis, Sclerolepis uniflora, Scutellaria galericulata, Sesbania punicea, S. vesicaria, Setaria faberi, S. glauca, S. parviflora, Sicyos angulatus, Sium suave, Smilax tamnoides, Solidago altissima, S. canadensis, S. gigantea, S. sempervirens, Sonchus arvensis, Sorghastrum nutans, Sparganium eurycarpum, Spartina cynosuroides, S. patens, S. pectinata, Spermacoce glabra, Spiranthes odorata, Sporobolus compositus, Stachys palustris, S. tenuifolia, Strophostyles umbellata, Symphyotrichum ericoides, S. ontarionis, S. lanceolatum, S. pilosum, S. racemosum, S. tradescantii, Taxodium ascendens, T. distichum, Teucrium canadense, Toxicodendron radicans, Tradescantia hirsutiflora, Tridens flavus, T. strictus, Typha angustifolia, T. latifolia, Ulmus americana, Urtica dioica, Vaccinium corymbosum, Verbena brasiliensis, Verbesina alternifolia, V. hastata, V. urticifolia, Vernonia fasciculata, V. gigantea, Vitis riparia, Woodwardia areolata, Xanthium strumarium, Xyris smalliana, X. torta, Zizania aquatica, Zizaniopsis miliacea.*

Use by wildlife: The flowers of *Boltonia asteroides* are visited by many nectar and/or pollen collecting insects (Insecta) including bees (Hymenoptera: Andrenidae: *Andrena simplex, Calliopsis coloradensis, Perdita boltoniae, P. octomaculata, Pseudopanurgus compositarum, P. solidaginis*; Apidae: *Bombus pensylvanicus, Melissodes confusa, Megachile brevis, M. mendica*; Colletidae: *Colletes americanus, Hylaeus affinis, H. mesillae;* Halictidae: *Agapostemon sericeus, Halictus confusus, H. ligatus, Sphecodes mandibularis*), beetles (Coleoptera: Cantharidae: *Chauliognathus pennsylvanicus*; Chrysomelidae: *Diabrotica longicornis, D. undecimpunctata; Microrhopala xerene*; Coccinellidae: *Hippodamia parenthesis*); bugs (Hemiptera: Miridae: *Lygus pratensis*); butterflies and moths (Lepidoptera: Erebidae: *Cisseps fulvicollis, Utetheisa ornatrix*; Hesperiidae: *Hylephila phyleus*; Noctuidae: *Helicoverpa armigera*; Nymphalidae: *Phyciodes tharos*; Pieridae: *Colias philodice*), flies (Diptera: Anthomyiidae: *Anthomyia*; Bombyliidae: *Sparnopolius*

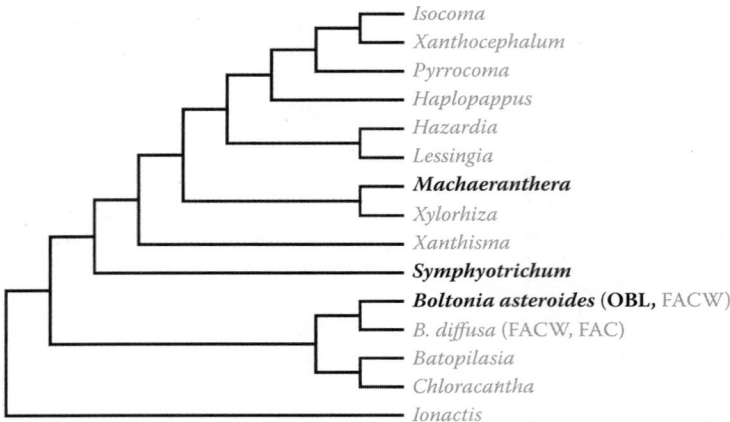

FIGURE 5.143 Phylogenetic relationships among some genera of Asteraceae tribe Astereae based on combined nrITS/ETS data. Taxa containing OBL species are indicated in bold. (Adapted from Urbatsch, L.E. et al., *Amer. J. Bot.*, 90, 634–649, 2003.)

fulvus; Calliphoridae: *Cochliomyia macellaria*; Conopidae: *Thecophora occidensis*; Sarcophagidae: *Amobia aurifrons*, *Sarcophaga*, *Senotainia flavicornis*, *Spallanzania hesperidarum*, *Sphixapata trilineata*; Sepsidae: *Sepsis*; Syrphidae: *Eristalis dimidiata*, *Eupeodes americanus*, *Mesograpta marginata*, *Paragus tibialis*, *Sphaerophoria cylindrica*, *Syritta pipiens*; Tachinidae: *Gymnoclytia immaculata*, *G. occidua*, *Gymnosoma fuliginosum*, *Leucostoma simplex*, *Phasia purpurascens*; Tephritidae: *Dioxyna picciola*), and wasps (Hymenoptera: Eumenidae: *Stenodynerus anormis*; Sphecidae: *Prionyx thomae*; Vespidae: *Polistesfuscata*). Fungi hosted by the plants include Ascomycota (Erysiphaceae: *Erysiphe cichoracearum*; Mycosphaerellaceae: *Didymaria*, *Septoria erigerontis*), and Basidiomycota (Entylomataceae: *Entyloma compositarum*; Pucciniaceae: *Aecidium boltoniae*, *Puccinia asterum*, *P. dioicae*, *P. extensicola*).

Economic importance: food: not edible; **medicinal:** Crude seed extracts of *Boltonia asteroides* exhibit fairly high antioxidant activity; **cultivation:** *Boltonia asteroides* performs well as a garden plant and cultivars include 'Diva', 'Jim Crockett', 'Masbolimket', 'Nana', 'Pink Beauty', and 'Snowbank'; **misc. products:** none; **weeds:** none; **nonindigenous species:** *Boltonia asteroides* has been introduced to France, Germany and Japan, as an escape from cultivation.

Systematics: Although phylogenetic analyses of *Boltonia* thus far have included only two of the five species, preliminary results indicate that the genus is monophyletic and closely related to *Batopilasia* and *Chloracantha* within tribe Astereae (Figure 5.143). Morphological and allozyme markers have indicated a low level of introgression in sympatric populations of *Boltonia asteroides* and *B. decurrens*, which were regarded formerly as conspecific varieties. The base chromosome number of the genus is $x=9$. Currently, three infraspecific taxa are recognized in *B. asteroides*: var. *latisquama* ($2n=18$) var. *asteroides* ($2n=36$), and var. *recognita* ($2n=36$). An anomalous count of $2n=27$ indicates the existence of triploids, perhaps in cultivated material resulting from intervarietal crosses involving var. *latisquama*.

Comments: *Boltonia asteroides* occurs in the eastern half of North America, with disjunct localities in Idaho and Oregon.
References: Borchardt et al., 2008a; Clark, 1983; Cranfill, 1981; DeWoody et al., 2011; Eleuterius, 1972; Eleuterius & McDaniel, 1978; Gleason et al., 2009; Grace et al., 2000; Karaman-Castro & Urbatsch, 2006; Kessler, Jr. et al., 2000; Lehmann & Jage, 2004; McCormac, 1993; Mito & Uesugi, 2004; Poiani & Dixon, 1995; Prevete et al., 2000; Reese & Lubinski, 1983; Robertson, 1891, 1894, 1922, 1924b, 1928; Sherff, 1912; Steffen, 1997; Stevens, 1924, 1932; Urbatsch et al., 2003; Yukio, 1965.

10. *Borrichia*

Sea oxeye daisy, seaside tansy
Etymology: after Ole Borrich (1626–1690)
Synonyms: *Buphthalmum*; *Diomedea*
Distribution: global: North America; Mexico; West Indies*;
North America: southeastern
Diversity: global: 3 species; **North America:** 3 species
Indicators (USA): OBL; FACW: *Borrichia frutescens*
Habitat: saline (coastal); palustrine; **pH:** 7.8–8.7; **depth:** <1 m; **life·form(s):** emergent shrub
Key morphology: Stems (to 1.5 m) woody, branched, erect or decumbent, with resinous sap, rhizomatous; leaves (to 10 cm) opposite, simple, coriaceous or succulent, linear to ovate, sessile or petioled, the margins entire or with spine-tipped teeth; inflorescence terminal with 1–10 cymose heads, heads radiate; involucre (to 18 mm) globose or broader; 10–45 phyllaries (to 4 mm) arranged in 2–3 series; receptacle with lanceolate to ovate pales (to 6 mm), partially enclosing the cypselae; ray florets (7–30), pistillate, corollas yellow; disc florets (20–75), bisexual, corollas yellow; cypselae (to 5 mm) gray or black, compressed or pyramidal, 3–4-angled, faces finely reticulate; pappus crown-like, 3–4-angled, persistent or late falling
Life history: duration: perennial (buds, rhizomes); **asexual reproduction:** rhizomes; **pollination:** insect; **sexual condition:** gynomonoecious; **fruit:** cypselae (common); **local dispersal:** water; **long-distance dispersal:** water

Imperilment: (1) *Borrichia frutescens* [G5]; SH (MD)

Ecology: general: *Borrichia* is a small genus of OBL or FACW shrubby or suffrutescent perennials, which inhabit warmer, coastal areas. Plants can develop into large clonal stands by rhizomatous growth. The flowers are insect pollinated (entomophilous) and gynomonoeciously arranged in each head with pistillate rays florets and bisexual disk florets. The smooth seeds float and primarily are water dispersed.

Borrichia frutescens (**L.) DC.** inhabits brackish to saline beaches, ditches, dunes, estuaries, flats, hammocks, mangrove swamps, marshes, meadows, mudflats, prairies, shell middens, tidal ponds, and margins of bayous, canals, channels, driftlines, lakes, tidal rivers, and streams at elevations from 2 to 10 m. This is a highly salt-tolerant shrub, which can withstand salinities from 3 to 100 ppt (conductivity: >56 dS/m) and repeated exposure to saltwater spray. Substrates are alkaline (pH 7.8–8.7) and include limestone, loam, loamy sand, marl, mud, rock, sand, sandy loam, sandy marl, and silty clay loam. Typical substrates consist of 0.6%–3.0% organic matter, 21–100 ppm phosphorous, and 2.1–6.7 ppm calcium. Exposures are in full sunlight. Salt tolerance is facilitated physiologically by an adapted malate dehydrogenase enzyme, which exhibits optimal activity in high ambient salt concentrations and by tolerance to high levels of salinity (0.7 M salt) without accumulation of proline, an osmoregulatory amino acid. Plants typically occupy an intermediate zone (>2 m in elevation; 5–20 m in extent) between the low marsh (*Spartina alterniflora*) and high marsh (*Iva frutescens*) vegetation and do not thrive in areas that are flooded regularly (water levels generally are <10 cm). They most often are described as extending to, but not beyond, the limit of normal tide influence. Movement of *Borrichia* into the low marsh also is inhibited by the higher levels (>0.25 mM) of sulfide, which significantly inhibits its productivity at concentrations of 0.5–1.0 mM. Competitive interactions prevent *Spartina* plants from encroaching into the *Borrichia* zone. Flowers are obligately outcrossed and pollinated by bees (Hymenoptera) and butterflies (Lepidoptera). The smooth seeds are dispersed by water. The duration of their flotation varies with respect to salinity, ranging from 18 days (freshwater) up to 238 days for either fresh seeds (15 ppt salinity) or those stratified at 6°C (36 ppt salinity). Viability is high (~97%). Seeds collected in August–September will germinate after stratification at 4°C for several months when placed in a 10 ppt salt solution. As salinity levels increase, plants transition from sexual to asexual reproduction. Spread of plants into hypersaline microsites is facilitated by clonally integrated ramets, which redistribute water. Although highly clonal, allozyme studies have indicated higher than expected levels of genetic variation among clones within populations, which emphasizes the importance of seed production for initial clonal establishment. Pollen-mediated gene flow persists across salinity gradients, but at a reduced (10%) level. Pollen is dispersed typically within a 0.26 m radius, with a potential paternity pool existing only for 1.91 m. Although the leaves are succulent, this is a C3 plant that acclimates well to temperature. Productivity (net daily photosynthesis) ranges from 0.052 to 0.079 g CO_2 g dry weight (DW)$^{-1}$ day^{-1} and is optimal at 25°C (summer; 20°C–25°C during winter). Populations are highly susceptible to disturbance by recreational vehicles. This is a secondary successional species, which is fairly slow to recolonize disturbed habitats. Some transplants reportedly fail in low marsh sites and persist poorly even in higher marsh habitat; however, densely planted material can spread quickly to exclude other species. Addition of treated wastewater can increase growth of plants by as much as 55%. The plants are known to become heavily parasitized by dodder (*Cuscuta*). **Reported associates:** *Acrostichum danaeifolium, Agalinis maritima, Allowissadula lozanii, Alternanthera philoxeroides, Amaranthus cannabinus, Ambrosia psilostachya, Andropogon glomeratus, Ardisia escallonoides, Atriplex cristata, A. matamorensis, A. patula, A. prostrata, Avicennia germinans, Baccharis halimifolia, Bacopa monnieri, Batis maritima, Billieturnera helleri, Blutaparon vermiculare, Bolboschoenus robustus, Borrichia arborescens, Bothriochloa laguroides, Bouteloua dactyloides, Byrsonima lucida, Casuarina, Celtis, Cenchrus spinifex, Chenopodium album, Chiococca alba, Chromolaena odorata, Cissus trifoliata, Citharexylum berlandieri, Conocarpus erectus, Cressa nudicaulis, Croton linearis, Cuscuta indecora, Cynodon dactylon, Cyperus articulatus, C. grayi, Dalea scandens, Desmanthus virgatus, Dichanthium aristatum, Digitaria ciliaris, Distichlis littoralis, D. spicata, Echinocactus texensis, Echinocereus, Eleocharis albida, E. obtusa, E. parvula, Eragrostis, Eustachys petraea, Eustoma exaltatum, Exothea paniculata, Festuca rubra, Fimbristylis caroliniana, F. castanea, F. puberula, Flaveria linearis, Forestiera angustifolia, F. segregata, Fuirena simplex, Gaillardia pulchella, Geranium carolinianum, Havardia pallens, Heliotropium angiospermum, H. curassavicum, Heterotheca subaxillaris, Hibiscus moscheutos, Hordeum pusillum, Hydrocotyle bonariensis, Ilex vomitoria, Ipomoea alba, I. sagittata, Iresine rhizomatosa, Isocoma drummondii, Iva annua, I. frutescens, I. imbricata, Juncus roemerianus, Juniperus virginiana, Karwinskia humboldtiana, Kosteletzkya pentacarpos, Laguncularia racemosa, Lepidium virginicum, Leucophyllum frutescens, Lilaeopsis chinensis, Limonium carolinianum, L. nashi, Lycium carolinianum, Lythrum alatum, L. lineare, Maytenus phyllanthoides, Melothria pendula, Metastelma barbigerum, Metopium toxiferum, Myrica cerifera, Myrsine floridana, Oenothera laciniata, O. patriciae, Opuntia engelmannii, O. humifusa, O. pusilla, Orthosia scoparia, Panicum virgatum, Pappophorum vaginatum, Parthenocissus quinquefolia, Paspalum distichum, P. monostachyum, Persea borbonia, P. palustris, Phragmites australis, Phyla nodiflora, Physalis walteri, Pinus taeda, Pluchea odorata, Plumbago scandens, Polypogon monspeliensis, Portulaca pilosa, Prosopis glandulosa, P. reptans, Psychotria nervosa, Quercus geminata, Rayjacksonia phyllocephala, Rhizophora mangle, Rhynchosia americana, R. senna, Ruellia nudiflora, Rumex verticillatus, Sabal palmetto, Sabatia dodecandra, S. stellaris, Sagittaria lancifolia, Salicornia bigelovii, S. depressa, S. maritima, Sarcocornia perennis, Schizachyrium scoparium, Schoenoplectus americanus, S. californicus, S. pungens,*

S. tabernaemontani, Sesuvium portulacastrum, Setaria leucopila, S. parviflora, Seutera angustifolia, Sideroxylon, Smilax auriculata, Solanum americanum, Solidago sempervirens, Spartina alterniflora, S. cynosuroides, S. patens, S. spartinae, Spergularia salina, Sporobolus coromandelianus, S. pyramidatus, S. virginicus, S. wrightii, Strophostyles helvola, Suaeda conferta, S. linearis, Symphyotrichum subulatum, S. tenuifolium, Taxodium ascendens, Teucrium canadense, Thespesia populnea, Thrinax morrisii, Tillandsia usneoides, Toxicodendron radicans, Typha angustifolia, Uniola paniculata, Vigna luteola, Yucca aloifolia, Y. treculeana, Zanthoxylum fagara.

Use by wildlife: Marsh rice rats (Mammalia: Cricetidae: *Oryzomys palustris*) feed on the flowers and leaves of *Borrichia frutescens* and it comprises an important habitat for cotton rats (Mammalia: Cricetidae: *Sigmodon hispidus exputus*). These plants are the preferred diet and source of cover for the endangered Lower Keys marsh rabbit (Mammalia: Leporidae: *Sylvilagus palustris hefneri*). *Borrichia frutescens* constitutes the dominant species in habitats supporting many wetland bird (Aves) colonies including great blue herons (Ardeidae: *Ardea herodias*), little blue herons (Ardeidae: *Egretta caerulea*), Louisiana herons (Ardeidae: *Egretta tricolor*) [which nest in *Borrichia*], reddish egrets (Ardeidae: *Egretta rufescens*), snowy egrets (Ardeidae: *Egretta thula*), and white-faced ibis (Threskiornithidae: *Plegadis chihi*). It also is an important element of habitat selected by the least tern (Aves: Laridae: *Sternula antillarum*) and is a preferred nesting habitat for immature brown pelicans (Aves: Pelecanidae: *Pelecanus occidentalis*) and laughing gulls (Aves: Laridae: *Leucophaeus atricilla*). This species comprises foraging and roosting habitat for the Aplomado falcon (Aves: Falconidae: *Falco femoralis*) and is fed upon by Canada geese (Aves: Anatidae: *Branta canadensis*). Pollen is collected from the flowers by bees (Insecta: Hymenoptera: Megachilidae: *Megachile brevis pseudobrevis*). It is the larval host plant for several moths (Insecta: Lepidoptera: Tortricidae: *Epiblema praesumptiosum, E. separationis*) and also supports adult beetles (Coleoptera: Meloidae: *Nemognatha punctulata*), which preferentially oviposit on the plants, as well as gall-making flies (Insecta: Diptera: Cecidomyiidae:

Asphondylia borrichiae), leaf-mining flies (Diptera: Agromyzidae: *Melanagromyza minimoides* [and a parasitic wasp: Insecta: Hymenoptera: Pteromalidae: *Heteroschema punctata*]), leafhoppers (Insecta: Homoptera: Cicadellidae: *Draeculacephala jloridana*), planthoppers (Homoptera: Delphacidae: *Pissonotus quadripustulatus*), scale insects (Homoptera: Diaspididae: *Aonidomytilus concolor*), and weevils (Insecta: Coleoptera: Curculionidae: *Listronotus borrichiae*). The inflorescences can become infested with larval and pupal fruit flies (Insecta: Diptera: Tephritidae: *Paracanthaforficula, Cecidocharella borrichia*). The foliage is fed upon by crabs (Crustacea: Ocypodidae: *Uca pugnax;* Sesarmidae: *Armases cinereum*) and marsh snails (Gastropoda: Ellobiidae: *Melampus bidentata*). The plants are hosts to a nematode (Nematoda: Tylenchulidae: *Tylenchulus palustris*). Hosted fungi include Ascomycota (*Cercosporidium*) and Basidiomycota (*Aecidium borrichiae, Puccinia mirifica, P. triannulata*).

Economic importance: food: not edible; **medicinal:** The leaves of *Borrichia frutescens* are brewed into a medicinal tea in the Bahamas. The plants contain cycloartanes, which inhibit the growth of *Mycobacterium tuberculosis*; **cultivation:** not cultivated; **misc. products:** none; **weeds:** none; **nonindigenous species:** *Borrichia frutescens* has been introduced to Bermuda, the West Indies, and to coastal areas of southeastern Spain.

Systematics: *Borrichia* resolves near the base of subtribe Engelmanniinae (tribe Heliantheae) close to *Vigethia*, the only other shrubby member of the otherwise herbaceous subtribe (Figure 5.144). The genus comprises two species (*B. arborescens, B. frutescens*) and their fertile interspecific F_1 hybrid (*B. ×cubansa*), which is intermediate morphologically to the parental species. The F_1 hybrids exhibit complete bivalent pairing, indicating a lack of chromosomal isolating barriers. Hybrid zones are maintained by positive assortative mating and clonal reproduction, and comprise mainly parental genotypes and later generation (F_2 or backcross) progeny. Backcrosses occur unidirectionally to the *B. frutescens* parent. The base chromosome number of *Borrichia* is $x=14$. All species are diploid ($2n=28$) but triploids ($2n=42$) occur occasionally in *B. frutescens*.

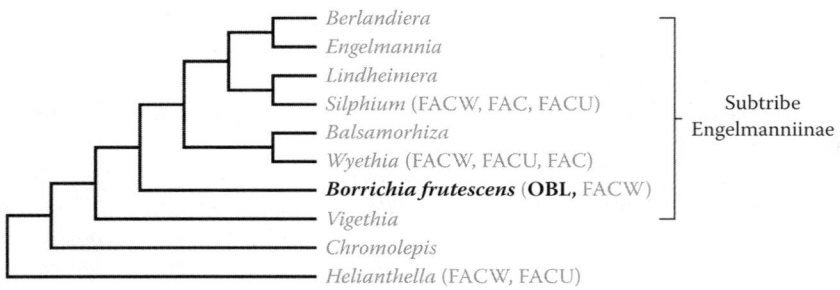

FIGURE 5.144 Intergeneric relationships within Asteraceae tribe Heliantheae subtribe Engelmanniinae as resolved from analysis of combined ETS, ITS, and 3'*trnK* sequence data. *Borrichia frutescens* is the only OBL indicator (shown in bold) within this subtribe, indicating an independent origin of the habit there. (Adapted from Moore, A.J. & Bohs, L., *Syst. Bot.*, 32, 682–691, 2007.)

Comments: *Borrichia frutescens* occurs in the southeastern United States.

References: Antlfinger, 1981, 1982; Antlfinger & Dunn, 1979; Aronson, 1989; Blus & Keahey, 1978; Bradley & Dunn, 1989; Burger, 1979; Bush & Huettel, 1969; Cantrell et al., 1996; Cavalieri & Huang, 1977, 1979; Cattell & Karl, 2004; Crespo et al., 2002; Dow et al., 1990; Elsey-Quirk et al., 2009; Forbes et al., 2008; Forys & Humphrey, 1999; Glazener, 1946; Haines & Montague, 1979; Hosier & Eaton, 1980; Jackson, 1922; Judd & Lonard, 2002; Judd et al., 1977; Kerwin & Pedigo, 1971; Kruchek, 2004; Levin, 1988; Marquardt & Pennings, 2010; McCaffrey & Dueser, 1990; McDaniel, 1971a; Moon et al., 2000; Moore & Bohs, 2007; O'Brien, 1981; Oosting, 1945; Packer, 1987; Pennings & Callaway, 2000; Pennings & Moore, 2001; Perez et al., 1996; Renda & Rogers, 1995; Richards et al., 2004; Rossi et al., 1996; Schwartz, 1954; Scifres et al., 1980; Semple, 1977, 1978; Semple & Semple, 1977; Stalter, 1973; Stalter & Batson, 1969, 1973; Stegmaier, Jr., 1967, 1973; Stiling et al., 1992; Thebeau & Chapman, 1984; Thompson & Slack, 1982; Tunnell et al., 1995; Zomlefer et al., 2007.

11. *Carphephorus*

Chaffhead

Etymology: from the Greek *karphos phora* meaning "bearing chaff"

Synonyms: *Litrisa; Trilisa*

Distribution: global: North America; **North America:** southeastern

Diversity: global: 7 species; **North America:** 7 species

Indicators (USA): OBL: *Carphephorus pseudoliatris*

Habitat: freshwater; palustrine; **pH:** 4.3–7.0; **depth:** <1 m; **life·form(s):** emergent herb

Key morphology: Stems (to 100 cm) erect, unbranched (scopiform), softly villous to hirsute; leaves (to 40 cm) alternate, mainly basal, narrowly linear, surfaces gland dotted, margins involute; heads discoid with 12–35 florets, in flat corymbs, peduncles villous; involucre (to 9 mm); phyllaries (15–40) triangular lanceolate, in several series; receptacle with pales; cypsela (to 3 mm) prismatic, 10-ribbed; pappus of 35–40 barbate, biseriate bristles (to 7 mm)

Life history: duration: perennial (caudex); **asexual reproduction:** basal stem offshoots; **pollination:** insect; **sexual condition:** hermaphroditic; **fruit:** cypselae (common); **local dispersal:** wind (cypselae); **long-distance dispersal:** wind (cypselae)

Imperilment: (1) *Carphephorus pseudoliatris* [G4]; SH (GA)

Ecology: general: *Carphephorus* species occupy habitats ranging from dry dunes and sand flats to low, wetland sites. Six of the seven species occur at least occasionally in wetlands (FACU, FACW) but only one is ranked as OBL. The flowers are pollinated by insects (Insecta); primarily bees (Hymenoptera) and butterflies (Lepidoptera). The bristle-plumed seeds are dispersed by the wind. Little is known regarding specific details of the floral biology or seed ecology of any of the species.

Carphephorus pseudoliatris Cassini grows in barrens, bogs, depressions, ditches, flatwoods, marshes, meadows, prairies, savannas, seeps, swales, and along the margins of swamps at elevations of up to 90 m. Typical sites are exposed to full sun. The substrates frequently are acidic (e.g., pH 4.3–4.4), but plants apparently also tolerate near neutral conditions. The soils are described as loamy sand, muck, peaty sand, sand, and sandy peat. They are classified technically as Compass (plinthic paleudults), Cuthbert (typic hapludults), Leon (aerie haplaquods), Plummer (grossarenic paleaquults), Rutledge (typic humaquepts), and Scranton (humaqueptic psammaquents). *Carphephorus pseudoliatris* is a long-lived perennial, which benefits from extended fire-free periods. Flowering occurs from August to October and fruiting from September to November. The flowering frequency initially increases significantly following a fire. Flowers reportedly are insect pollinated (Insecta), but details on the floral biology are lacking in the literature. The long-bristled cypselae are wind dispersed, but information regarding germination conditions is unavailable. Root growth extends to an average 27 cm depth, which probably facilitates survival through periodic fire episodes. **Reported associates:** *Agalinis aphylla, A. filicaulis, A. obtusifolia, Aletris aurea, A. lutea, Andropogon arctatus, A. glomeratus, A. gyrans, A. liebmannii, Anthaenantia rufa, Aristida palustris, A. purpurascens, A. simpliciflora, A. stricta, Arnoglossum ovatum, Arundinaria gigantea, Asclepias connivens, A. rubra, Balduina uniflora, Bidens mitis, Bigelowia nudata, Burmannia capitata, Calopogon pallidus, C. tuberosus, Carex turgescens, Carphephorus corymbosus, C. odoratissimus, C. paniculatus, Chaptalia tomentosa, Clethra alnifolia, Cliftonia monophylla, Coreopsis gladiata, Croton michauxii, Ctenium aromaticum, Dichanthelium acuminatum, D. dichotomum, D. ensifolium, D. nudicaule, Doellingeria sericocarpoides, Drosera capillaris, D. filiformis, D. tracyi, Erigeron vernus, Eriocaulon compressum, E. decangulare, E. texense, Eryngium integrifolium, Eupatorium leucolepis, E. rotundifolium, Eurybia eryngiifolia, Euthamia graminifolia, Fuirena squarrosa, Gaylussacia mosieri, Gymnadeniopsis integra, Helianthus heterophyllus, H. radula, Hypericum brachyphyllum, Hypoxis wrightii, Ilex glabra, I. myrtifolia, I. vomitoria, Iva microcephala, Juncus trigonocarpus, Kalmia hirsuta, Lachnanthes caroliana, Lachnocaulon anceps, L. digynum, Liatris gracilis, L. pilosa, L. spicata, Lilium catesbaei, Linum floridanum, Lobelia brevifolia, Lophiola aurea, Lycopodiella alopecuroides, L. caroliniana, Lycopodium, Magnolia virginiana, Muhlenbergia capillaris, M. expansa, Myrica heterophylla, Nolina atopocarpa, Nyssa biflora, N. ursina, Oxypolis denticulata, O. filiformis, Panicum virgatum, Paspalum praecox, Pinguicula lutea, P. planifolia, Pinus elliottii, P. palustris, Pityopsis graminifolia, P. oligantha, Pleea tenuifolia, Polygala chapmanii, P. cruciata, P. lutea, P. nana, P. ramosa, Quercus marilandica, Q. pumila, Rhexia alifanus, R. lutea, R. petiolata, Rhynchospora baldwinii, R. chalarocephala, R. chapmanii, R. latifolia, R. macra, R. oligantha, R. plumosa, Sabatia brevifolia, S. macrophylla, Sarracenia alata, S. flava, S. leucophylla, S. psittacina, S. purpurea, Schizachyrium scoparium, S. tenerum, Scleria baldwinii, S. georgiana, S. muehlenbergii, S. pauciflora, S. reticularis,*

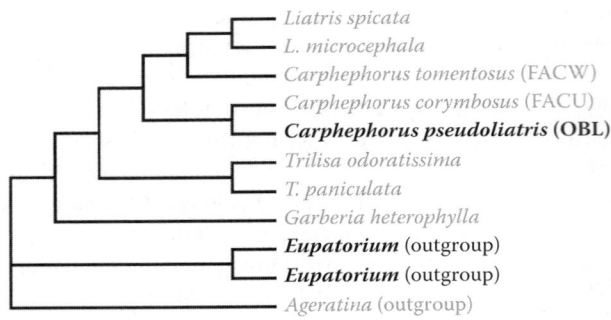

FIGURE 5.145 Cladogram of Eupatorieae subtribe Liatrinae generated from combined nrITS/ETS and cpDNA sequence data indicate that *Carphephorus* is not monophyletic, but instead represents a basal grade of *Liatris*. The OBL habit (taxa in bold) of *Carphephorus pseudoliatris* indicates a separate evolution of the habit. (Adapted from Schilling, E.E., *Mol. Phylogen. Evol.*, 59, 158–167, 2011a.)

Serenoa repens, Seymeria cassioides, Smilax laurifolia, Sorghastrum secundum, Stokesia laevis, Symphyotrichum adnatum, S. chapmanii, S. dumosum, Triantha racemosa, Utricularia juncea, U. subulata,, Verbesina chapmanii, Xyris ambigua, X. baldwiniana, X drummondii, X. scabrifolia, X. stricta, X. torta, Zigadenus glaberrimus.

Use by wildlife: none reported

Economic importance: food: reported; **medicinal:** not reported; **cultivation:** not cultivated; **misc. products:** none; **weeds:** none; **nonindigenous species:** none

Systematics: *Carphephorus* is a member of subtribe Liatrinae in tribe Eupatorieae. As its name implies, *C. pseudoliatris* was placed formerly in *Liatris* do to its similarity to members of that genus. A close relationship to *Liatris* is supported by analysis of nrITS sequence data, which in one analysis resolved *C. pseudoliatris* with *Liatris* species rather than a congener (*C. corymbosus*). Combined nrITS/ETS and cpDNA resolve *Carphephorus* as a basal grade to *Liatris* (Figure 5.145) and indicate that the genus is not monophyletic as circumscribed. The generic distinction of *Carphephorus* warrants reconsideration. Phylogenetic relationships might better be reflected by assigning all species to the genus *Liatris*. The base chromosome number of the genus is $x=10$; *Carphephorus pseudoliatris* and other species in the genus are uniformly diploid ($2n=20$).

Comments: *Carphephorus pseudoliatris* is a regional endemic of the southeastern United States.

References: Brewer et al., 2011; Carr, 2007; Hinman & Brewer, 2007; Nesom, 2006a; Schilling, 2011a; Schmidt & Schilling, 2000; Walker & Silletti, 2005.

12. *Cirsium*

Thistle; chardon

Etymology: from the Greek *kirsos* meaning "swollen vein," a malady reputedly remedied by this plant

Synonyms: *Breea*; *Cnicus* (in part)

Distribution: global: Africa; Asia; Europe; North America; **North America:** widespread

Diversity: global: 200 species; **North America:** 62 species

Indicators (USA): OBL: *Cirsium crassicaule, C. douglasii, C. fontinale, C. hydrophilum, C. muticum, C. vinaceum, C. wrightii;* **FACW:** *C. muticum*

Habitat: brackish; freshwater; palustrine; **pH:** 5.0–8.6; **depth:** <1 m; **life form(s):** emergent herb

Key morphology: Stems (to 3 m) single to several, much branched, glabrous to tomentose, spiny, sometimes hollow; leaves (to 7.0 dm) cauline and in basal rosettes, unlobed to deeply pinnatifid or pinnate, bristle-tipped, slender-spined (to 3.0 mm) to stoutly spined (to 3.0 cm), petioles slender, more or less winged or absent; heads single to numerous, discoid, corymbose or paniculate, peduncled (to 1.5 dm), with 25+ florets; involucre (to 5.0 cm) with imbricate (to 5–12 seriate) phyllaries (to 2.0 cm), minutely spinulose or slender spined (to 9.0 mm); receptacle without pales; corollas (to 3.2 cm) white to pinkish purple, lavender, or rose-purple; cypselae (to 6.0 mm) smooth, compressed, apically rimmed (to 0.3 mm), brown to black, with persistent or falling, flattened, plumose pappus bristles (to 2.0 cm)

Life history: duration: biennial (basal rosettes); perennial (basal rosettes); **asexual reproduction:** adventitious plantlets, rhizomatous lateral roots; **pollination:** insect; **sexual condition:** hermaphroditic; **fruit:** cypselae (common); **local dispersal:** seeds (wind); **long-distance dispersal:** seeds (animals, water, wind)

Imperilment: (1) *Cirsium crassicaule* [G2]; S2 (CA); (2) *C. douglasii* [G4]; (3) *C. fontinale* [G2]: S1-S2 (CA); (4) *C. hydrophilum* [G1]; S1 (CA); (5) *C. muticum* [G5]; SH (DE); S1 (AL, AR, FL,GA, OK); S2 (IA, KY, NC, ND, PE, SK); S3 (MD, VT); (6) *C. vinaceum* [G2]; S2 (NM); (7) *C. wrightii* [G2]: S1 (AZ, TX); S2 (NM).

Ecology: general: *Cirsium* is diverse ecologically, occupying sites from sea level to alpine elevations and from boreal to tropical climates. Habitats include deserts, dunes, forests, and meadows in addition to wetlands. Most North American species are characteristic of terrestrial communities but 26 species (42%) possess some status as wetland indicators. Only seven North American OBL species (11%) are OBL indicators. Although *C. loncholepis* and *C. scariosum* were assigned OBL indicator status in the 1996 list, most recent authors now treat *C. loncholepis* as a variety of *C. scariosum*, which itself has been downgraded to FACW status. *Cirsium virginianum* also has been downgraded from OBL, FACW status in the 1996 list to FACW in the 2013 list. Consequently, all three taxa have been excluded from the present treatment. *Cirsium wrightii* was added as an OBL indicator in the 2013 list and is included here. *Cirsium* contains perfect flowers (except *C. arvense*) that include self-compatible and self-incompatible genetic systems. They are pollinated primarily by insects (Insecta). Even though these plants are known for their "thistle down," which actually is the plumose pappus, the seeds of some species are heavy and are not dispersed far by the wind. Many of these seeds also float or can become entangled in animal fur to achieve greater dispersal distances. Germination generally requires some period of cold stratification to break dormancy. It is difficult to distinguish the duration of thistle plants as biennial, given that they often persist for more than 2 years as a nonflowering rosette. However, the plants are strictly monocarpous and will die subsequent to flowering. All species grow from a taproot, but some produce branching, woody rhizomatous root sprouts capable of generating adventitious rosettes from the maternal individuals. Consequently, one individual might actually flower in several subsequent seasons due to a delayed flowering response associated with these younger and smaller vegetatively generated plantlets.

Cirsium crassicaule (Greene) Jeps. is an annual or biennial resident of freshwater canals, channels, deltas, ditches, levees, marshes, sloughs, streams, washes, and riverbank scrub at elevations of up to 100 m. Substrates are described as sandy alkaline soils. The flowers are insect pollinated and produced from May to August. Local vegetative reproduction can occur by formation of basal rosette offshoots. Additional life history information is needed for this rare species. **Reported associates:** *Eryngium racemosum, Hibiscus californicus, Lathyrus jepsonii, Lilaeopsis masonii, Populus fremontii, Rosa californica, Rubus vitifolius, Salix.*

Cirsium douglasii DC. is a biennial or perennial inhabitant of bogs, draws, flats, marshes, meadows, seeps, swamps, and margins of ponds and streams at elevations of up to 2200 m. Sites often are vernally wet and sunny or in partial shade. The plants occur on serpentine substrates having textures of gravel, muck, sand, or silty clay loam. Flowering occurs from March to October. Good germination (>50%) has been obtained for seeds that are washed for 2 h in tap water, immersed in a 200 ppm solution of gibberellic acid for 24 h, surface sterilized for 20 min in 10% sodium hypochlorite (7% activated chlorine), and then given three consecutive 5 min washed in sterile distilled water. Germination (on nutrient agar) should occur within 8 weeks at 19°C under a 16 h photoperiod. **Reported associates:** *Abies concolor, Achillea millefolium, Acmispon americanus, Agrimonia gryposepala, Agrostis viridis, Alnus rhombifolia, Angelica tomentosa, Aquilegia eximia, Arctostaphylos, Artemisia douglasiana, Asclepias speciosa, Astragalus clevelandii, Calocedrus decurrens, Camassia quamash, Carex barbarae, C. ovalis, C. praegracilis, C. serratodens, Castilleja lacera, C. miniata, C. minor, Ceanothus, Chamerion angustifolium, Circaea alpina, Cordylanthus tenuis, Cupressus, Cypripedium, Darlingtonia californica, Delphinium uliginosum, Deschampsia cespitosa, Dieteria canescens, Eleocharis macrostachya, E. parishii, Epilobium ciliatum, Epipactis gigantea, Eriogonum nudum, E. umbellatum, Galium bifolium, Hastingsia alba, Helenium bigelovii, Helianthus exilis, Heterocodon rariflorus, Hoita macrostachya, Horkelia californica, Hypericum anagalloides, H. formosum, Juncus balticus, J. bufonius, J. effusus, J. hesperius, J. oxymeris, J. xiphioides, Lilium pardalinum, Madia elegans, Maianthemum stellatum, Melilotus alba, Mentha pulegium, Mimulus guttatus, M. layneae, M. nudatus, Nasturtium officinale, Neottia convallarioides, Packera clevelandii, P. streptanthifolia, Pinus jeffreyi, Plagiobothrys tener, Plantago subnuda, Polygonum polygaloides, Populus tremuloides, Potentilla gracilis, Prunella vulgaris, Quercus vacciniifolia, Rhamnus californica, Rudbeckia occidentalis,* *Sairocarpus cornutus, Salix breweri, S. lasiolepis, Scutellaria siphocampyloides, Sidalcea oregana, Sisyrinchium californicum, Solidago velutina, Spiraea douglasii, Stachys ajugoides, S. albens, Symphyotrichum spathulatum, Trichostema laxum, Trifolium fragiferum, T. glomeratum, T. obtusiflorum, T. subterraneum, T. variegatum, T. wormskioldii, Toxicoscordion venenosum, Triteleia peduncularis, Viola adunca, Zeltnera trichantha.*

Cirsium fontinale (Greene) Jeps. is a short lived (2–3 years), monocarpic perennial, which grows in bogs, canals, channels, creekbeds, gullies, meadows, seeps, springs, swales, and along the margins of streams at elevations of up to 860 m. Plants occur exclusively on perennially wet serpentine substrates, which have been described as adobe, clay, gravel, rock, sand, or sandy clay loam. Occurrences vary from boggy soils several decimeters thick (which produces the most robust plants), to rocky deposits interspersed with finer grained sediments. Substrates are alkaline (pH 7.6–7.8), low in organic matter (4%–14%), and contain high levels of Ca (505–810 mg/L) and Mg (2280–5540 mg/L). Phosphorous levels range from 14 to 23 mg/L. Concentration ranges of other constituents are reported as: Fe (36–131 mg/L), K (165–282 mg/L), Mn (9–31 mg/L), Na (52–182 mg/L), and Zn (0.8–2.4 mg/L). Plants can tolerate shade but occur more commonly in sunnier sites that seldom freeze during the winter or exceed 27°C during the summer. A large rosette is formed after the first year of growth and flowering normally occurs during the second year. Most plants die after flowering, but more vigorous specimens can survive a third year and flower again before they die. Flowering occurs from May to July, sometimes extending through October. The extent of flowering is proportional to the amount of water available, with stressed plants producing fewer flowers; however, flowers develop more rapidly under drought conditions. Each plant typically produces a single stalk bearing about 30–50 capitula, which individually contain 193 florets on average; however, seed set is much lower, averaging between 7–49 seeds (typically 15–25) per capitulum. The flowers remain open for 24 h and can be produced over a 3 month period. The degree of autogamous pollination is very low, and pollinators (Insecta) are required to achieve normal seed set. These include bees (Hymenoptera), flies (Diptera), moths (Lepidoptera: Sphingidae) and butterflies (Lepidoptera: Nymphalidae: *Danaus plexippus*; Pieridae: *Pieris rapae*). *Bombus vosnesenskii* (Hymenoptera: Apidae) is the principal pollinator. Each seed weighs approximately 4.5 mg. Germination of seeds often occurs within the leaf shadow of the maternal rosette, especially where dense competing vegetation is present. Germination rates of collected seeds are fairly high (56%–92%). Successful treatments placed cleaned seeds in dry storage (14% relative humidity) for a minimum of 3 weeks with subsequent storage at −18°C. High germination is achieved under an 11/13 h 20°C/12°C regime. The maximum longevity of buried seeds remains undetermined. Wind dispersal of the heavy seeds seldom exceeds several meters and greater dispersal distances are achieved when seeds are dispersed in water or become attached to animal fur. Vegetative reproduction occurs by the production of plantlets that develop from horizontally spreading rootstocks. Although restricted to nutrient-poor serpentine substrates, *C. fontinale* has been shown to compete well with congeneric species even under high nutrient conditions. **Reported associates:** *Achillea millefolium, Acmispon glabrus, A. humistratus, Aesculus californica, Agrostis exarata, A. viridis, Anagallis arvensis, Aquilegia eximia, Artemisia californica, Arctostaphylos glauca, Astragalus curtipes, Avena barbata, Baccharis pilularis, Bromus diandrus, B. hordeaceus, B. madritensis, Calochortus clavatus, Calystegia macrostegia, C. purpurata, Carex obispoensis, C. serratodens, Castilleja minor, Ceanothus cuneatus, C. leucodermis, Centaurea solstitialis, Chorizanthe palmeri, Cirsium vulgare, Clarkia purpurea, Claytonia exigua, Conium maculatum, Corethrogyne filaginifolia, Cortaderia jubata, Cryptantha flaccida, Cynodon dactylon, Deinandra fasciculata, Distichlis spicata, Dudleya lanceolata, Eleocharis, Elymus glaucus, Epilobium ciliatum, E. minutum, Eriogonum nudum, Eriophyllum confertiflorum, Eschscholzia californica, Euphorbia spathulata, Galium aparine, G. porrigens, Gilia achilleifolia, Gnaphalium, Hazardia squarrosa, Helenium bigelovii, Heliotropium curassavicum, Helminthotheca echioides, Hemizonia congesta, Hesperoyucca whipplei, Heteromeles arbutifolia, Hoita strobilina, Hordeum brachyantherum, H. marinum, H. murinum, Juncus balticus, J. phaeocephalus, J. xiphioides, Koeleria macrantha, Lactuca saligna, L. serriola, Lessingia micradenia, Leymus triticoides, Lolium multiflorum, L. perenne, Lomatium, Lotus corniculatus, Lupinus succulentus, Lythrum californicum, Medicago polymorpha, Melica torreyana, Melilotus indicus, Mimulus guttatus, Monolopia gracilens, Nassella pulchra, Persicaria lapathifolia, Phacelia, Phalaris, Pickeringia montana, Pinus sabiniana, Plantago, Platanus, Poa secunda, Polypogon monspeliensis, Potentilla anserina, Quercus agrifolia, Q. durata, Q. lobata, Ranunculus californicus, Rhamnus californica, Rubus ulmifolius, Rumex acetosella, R. crispus, Salix breweri, Sambucus nigra, Scirpus, Scrophularia californica, Silene laciniata, Sisyrinchium bellum, Solanum umbelliferum, Sonchus asper, S. oleraceus, Stachys ajugoides, S. pycnantha, Streptanthus glandulosus, Taeniatherum caput-medusae, Toxicodendron diversilobum, Triteleia peduncularis, Typha latifolia, Umbellularia californica, Verbena lasiostachys, Vulpia microstachys, Zeltnera davyi.*

***Cirsium hydrophilum* (Greene) Jeps.** is a biennial or monocarpic, short-lived perennial (2–3 years), which grows in brackish or fresh, tidal, or nontidal sites including ditches, flats, marshes, meadows, seeps, sloughs, springs, streams, and the margins of channels and streams at elevations of up to 457 m. Occurrences in tidal marshes occupy the upper intertidal marsh plain, usually along small stream banks that drain the surface peat layer. The substrates (peat or serpentine) are muds or silty clays that are high in organic matter in the upper horizon, with the mineral content increasing with depth (Joice soil series). The upper layers of these soils are strongly acidic (pH 5.0); whereas, the deeper layers (1.5–3.0 m) become progressively more alkaline (pH 8.2–8.5). The acidity increases when the soils are drained. There is limited information on

the reproductive ecology of this species. Flowers are produced successively throughout the summer (July–September). Larger, branched plants can produce several hundred capitula throughout the season. Pollination apparently occurs by insects (Insecta), primarily bumblebees (Hymenoptera: Apidae: *Bombus vosnesenskii*) and several other bee species. Reproductive output is low, with only 3–15 seeds produced per capitulum and many bearing no seeds at all. Despite the presence of a pappus (normally associated with wind dispersal), the fairly heavy seeds detach from the pappus before maturity and are dropped close to the maternal plants. Because they float and remain viable when soaked, the seeds conceivably could be dispersed more widely by water; however, long-distance dispersal has not been observed for this species. Most colonization occurs where bare soil or gaps have been exposed by disturbance. Precise requirements for germination have not been determined but optimal levels are believed to occur under lower salinity conditions. Seeds occur densely in the soil below the producing plants and are known to retain high viability for at least 5 years in artificial storage; thus, a persistent natural seed bank probably exists. Adult plants grow better during periods of lower water levels and higher salinity, presumably due to an associated decrease in competition. Introduced, nonindigenous pepperweed plants (*Lepidium latifolium*) occupy similar habitats and pose a particularly serious competitive threat. **Reported associates:** *Achillea millefolium, Apium graveolens, Arthrocnemum subterminale, Atriplex prostrata, Baccharis douglasii, Calystegia sepium, Cicuta maculata, Cirsium vulgare, Cordylanthus mollis, Cuscuta salina, C. subinclusa, Distichlis spicata, Eleocharis macrostachya, Epilobium ciliatum, Eurybia radulina, Euthamia occidentalis, Frankenia salina, Galium triflorum, Grindelia hirsutula, Helminthotheca echioides, Jaumea carnosa, Juncus balticus, Lepidium latifolium, Lotus corniculatus, Lysimachia maritima, Lythrum californicum, Oenanthe sarmentosa, Persicaria punctata, Phragmites australis, Pluchea odorata, Polypogon monspeliensis, Potentilla anserina, Rumex californicus, R. crispus, Salicornia depressa, Samolus valerandi, Sarcocornia pacifica, Schoenoplectus acutus, S. americanus, S. californicus, Sium suave, Solanum douglasii, Sonchus oleraceus, Symphyotrichum lentum, S. subulatum, Triglochin maritimum, Typha angustifolia, T. domingensis, T. latifolia.*

***Cirsium muticum* Michx.** is a biennial resident of bogs, carrs, depressions, dikes, ditches, fens, floodplains, low woodlands, marshes, meadows, prairies, seeps, springs, swamps, thickets, and margins of lakes, ponds, and streams at elevations of up to 1500 m. Individuals occur across a broad gradient of exposures (open to densely shaded sites) and have a high shade tolerance index (7.7/9). Substrates are calcareous (neutral to alkaline; pH: 7.1–8.6) and include clay, Genesee silt loam, gravel, humus, marl, marly silt, muck, mucky sand, peat, peaty loam, sand, sandy clay, and sandy loam. In patterned peatlands, the plants occupy the hummocks rather than the hollows. The flowers appear to be pollinated mainly by bumblebees (Insecta: Hymenoptera: Apidae: *Bombus fervidus, B. impatiens, B. pensylvanicus, B. vagans*); however, various butterflies (Insecta: Lepidoptera) may also contribute to pollination. Birds (Aves) also visit the flowers, but there is no evidence that they function as pollinators. The seeds are dispersed by the wind, facilitated by their long (12–20 mm) pappus bristles. Low seed germination (8%) has been achieved with or without a cold treatment (4°C for 83 days); however, germination occurs more rapidly in cold treated seeds (21–42 days) than in untreated seeds (29–299 days). A persistent seed bank has been reported for this species. Plants in the northern portion of their range are threatened by competition from the nonindigenous European marsh thistle (*Cirsium palustre*). **Reported associates:** *Abies balsamea, Acer rubrum, A. saccharum, A. spicatum, Achillea millefolium, Aconitum uncinatum, Acorus calamus, Actaea rubra, Agrostis scabra, Allium cernuum, Alnus rugosa, A. serrulata, Andropogon gerardii, A. glomeratus, A. virginicus, Angelica atropurpurea, Anthoxanthum odoratum, Anticlea elegans, Apios americana, Apocynum cannabinum, Aralia nudicaulis, Arethusa bulbosa, Arisaema dracontium, Arnoglossum ovatum, A. plantagineum, Asclepias incarnata, A. syriaca, Aureolaria patula, Barbarea vulgaris, Betula alleghaniensis, B. papyrifera, B. pumila, B. ×sandbergii, Bidens cernuus, B. discoideus, Boehmeria cylindrica, Bromus ciliatus, Calamagrostis canadensis, C. stricta, Calopogon tuberosus, Caltha palustris, Campanula aparinoides, Cardamine bulbosa, C. diphylla, Carex atlantica, C. bromoides, C. buxbaumii, C. canescens, C. caroliniana, C. cherokeensis, C. echinata, C. emoryi, C. exilis, C. haydenii, C. hystericina, C. lasiocarpa, C. leptalea, C. lupulina, C. lurida, C. oxylepis, C. pedunculata, C. pellita, C. prairea, C. sterilis, C. stipata, C. stricta, C. tenuiflora, C. tetanica, Carpinus caroliniana, Castilleja coccinea, Cerastium fontanum, Chamaedaphne calyculata, Chelone glabra, Cicuta maculata, Cinna arundinacea, Circaea alpina, Cirsium arvense, C. palustre, Clethra alnifolia, Clintonia borealis, Comarum palustre, Conium maculatum, Coreopsis tripteris, Cornus amomum, C. canadensis, C. rugosa, C. sericea, Cryptotaenia canadensis, Cuscuta gronovii, Cynosciadium digitatum, Cypripedium reginae, Cystopteris bulbifera, Dasiphora floribunda, Daucus carota, Deschampsia cespitosa, Desmodium canadense, Dichanthelium clandestinum, D. dichotomum, Dichondra repens, Dipsacus fullonum, Doellingeria umbellata, Dryopteris cristata, Eleocharis elliptica, E. tenuis, E. tortilis, Epilobium coloratum, E. leptophyllum, Equisetum arvense, E. fluviatile, Erigeron pulchellus, E. strigosus, Eriophorum chamissonis, Eryngium integrifolium, Eupatorium perfoliatum, E. rotundifolium, Euphorbia purpurea, Euthamia graminifolia, Eutrochium maculatum, E. purpureum, Filipendula rubra, Fraxinus nigra, F. profunda, Fuirena squarrosa, Galium aparine, G. asprellum, G. boreale, G. labradoricum, G. tinctorium, Gentiana andrewsii, G. saponaria, Gentianopsis crinita, G. virgata, Geranium carolinianum, Geum canadense, Glechoma hederacea, Glyceria striata, Hamamelis virginiana, Helenium autumnale, H. brevifolium, Helianthus giganteus, H. grosseserratus, H. ×verticillatus, Houstonia caerulea, Hydrocotyle verticillata, Hypericum ascyron, H. kalmianum, Hypoxis hirsuta, Ilex opaca,*

Impatiens capensis, Iris versicolor, I. virginica, Itea virginica, Juncus filiformis, J. filipendulus, J. nodosus, J. subcaudatus, Kalmia polifolia, Lactuca canadensis Larix laricina, Lathyrus palustris, Leersia oryzoides, Liatris ligulistylis, L. pycnostachya, L. spicata, Lilium michiganense, L. philadelphicum, Lindera benzoin, Liparis loeselii, Liquidambar styraciflua, Lobelia kalmii, L. siphilitica, Ludwigia microcarpa, Lycopus americanus, L. uniflorus, L. virginicus, Lyonia ligustrina, Lysimachia quadriflora, L. thyrsiflora, Lythrum alatum, Magnolia virginiana, Maianthemum stellatum, Malaxis unifolia, Marshallia mohrii, Matteuccia struthiopteris, Mecardonia acuminata, Mentha arvensis, Mertensia paniculata, Micranthes pensylvanica, Mikania scandens, Mimulus glabratus, Mitchella repens, Mitreola petiolata, Monarda, Muhlenbergia glomerata, M. racemosa, M. richardsonis, Myosotis macrosperma, Neottia cordata, Nyssa sylvatica, Oenothera biennis, O. perennis, Onoclea sensibilis, Oplismenus hirtellus, Osmunda regalis, Osmundastrum cinnamomeum, Oxypolis rigidior, Packera aurea, P. glabella, P. paupercula, Panicum virgatum, Parnassia caroliniana, P. glauca, P. grandifolia, P. palustris, Parthenocissus quinquefolia, Pedicularis lanceolata, Persicaria sagittata, P. virginiana, Phalaris arundinacea, Phleum pratense, Phlox pilosa, Physocarpus opulifolius, Physostegia virginiana, Picea glauca, Pilea fontana, P. pumila, Pinus rigida, P. strobus, Plantago cordata, Platanthera flava, P. lacera, P. psycodes, Poa annua, P. nemoralis, P. palustris, P. pratensis, Populus deltoids, P. heterophylla, P. tremuloides, Pteridium aquilinum, Ptilimnium costatum, Pycnanthemum tenuifolium, P. virginianum, Quercus michauxii, Q. nigra, Q. phellos, Q. rubra, Ranunculus recurvatus, Rhamnus alnifolia, Rhododendron groenlandicum, R. viscosum, Rhynchospora alba, R. caduca, R. capillacea, R. capitellata, R. glomerata, R. gracilenta, R. thornei, Rosa carolina, Rubus arcticus, R. hispidus, R. parviflorus, R. pubescens, Rudbeckia fulgida, R. hirta, R. subtomentosa, Rumex britannica, R. mexicanus, R. orbiculatus, Salix bebbiana, S. candida, S. discolor, S. pedicellaris, S. petiolaris, S. ×bebbii, Samolus valerandi, Sanguisorba canadensis, Schizachyrium scoparium, Schoenoplectus acutus, S. americanus, S. pungens, Scirpus atrovirens, S. expansus, Scleria muehlenbergii, S. verticillata, Scutellaria galericulata, S. lateriflora, Selaginella apoda, S. eclipes, Silene stellata, Silphium compositum, S. trifoliatum, Sium suave, Smilax rotundifolia, Solanum dulcamara, Solidago canadensis, S. gigantea, S. nemoralis, S. ohioensis, S. patula, S. ptarmicoides, S. riddellii, S. rigida, S. rugosa S. uliginosa, Sorghastrum nutans, Spartina pectinata, Sphagnum, Sphenopholis intermedia, S. obtusata, S. pensylvanica, Spiraea alba, S. tomentosa, Spiranthes cernua, Stellaria longifolia, S. prostrata, Symphyotrichum boreale, S. firmum, S. laeve, S. lanceolatum, S. lateriflorum, S. novae-angliae, S. novi-belgii, S. puniceum, Taraxacum officinale, Thalictrum dasycarpum, T. pubescens, Thelypteris palustris, Thuja occidentalis, Tilia americana, Triadenum fraseri, T. virginicum, T. walteri, Trichophorum cespitosum, Trientalis borealis, Triglochin maritimum, T. palustre, Trollius laxus, Tsuga canadensis, Typha angustifolia, T. latifolia, Ulmus americana, Urtica chamaedryoides,

U. dioica, Vaccinium oxycoccos, Valeriana ciliata, Veratrum virginicum, Verbena hastata, Vernonia fasciculata, V. noveboracensis, Veronica americana, V. peregrina, Veronicastrum virginicum, Viburnum lentago, V. trilobum, Viola cucullata, V. macloskeyi, V. nephrophylla, V. sororia, V. walteri, Xyris torta, Zizia aurea.

***Cirsium vinaceum* Wooton & Standl.** is a short-lived perennial, which inhabits forest edges, meadows, seeps, springs, and margins of streams at elevations from 2460 to 3020 m. Substrates are wet, calcareous (travertine) deposits with an average pH=8.1. They contain high levels of nitrogen and sulfate but low levels of phosphorous. Plants remain vegetative for 1-several years, during which time they can propagate vegetatively by producing adventitiously rooting rosette offshoots along short rhizomes. The plants die after flowering, which occurs from late June through August. Populations have exhibited a continued annual decline in the number of flowering stems since monitoring began in 1998. The flowers are partially self-compatible but primarily are outcrossed by hummingbirds (Aves: Trochilidae), or various insects (Insecta) such as bees (Hymenoptera), beetles (Coleoptera), flies (Diptera) and moths (Lepidoptera), which presumably act as pollen vectors. Native bee pollinators are less active in smaller patches of plants. Seed set can be limited by the presence of heterospecific pollen, which hinders reproduction and promotes interspecific hybridization. The seeds are large and heavy and possess a pappus for wind dispersal, but also are dispersed by water. Wind-mediated transport of seeds across the smooth surface of snowpack might facilitate their dispersal during the winter. Hundreds of viable seeds have been recovered from floating seed traps placed as far as 280 m from parental plants. In some instances entire capitula are dispersed by the water, which may result in the deposition of numerous seeds at a distant site, depending on how many are shed en route. Effective dispersal distances of up to 0.8 km have been postulated. Germination has been achieved after refrigerating seed (4°C) for 5 months, then treating with a 5% bleach solution. The highest germination (54%) was achieved under light (9/15 h) at room temperature. This species is threatened by invasion of nonindigenous teasel (*Dipsacus fullonum*) into habitats, which results in interference such as reduced rosette size, by grazing from domestic cattle, and destruction associated with human foot traffic. **Reported associates:** *Carduus nutans, Conium maculatum, Cirsium arvense, C. vulgare, Daucus carota, Dipsacus fullonum, Taraxacum officinale, Nasturtium officinale, Tragopogon pratensis, Verbascum thapsus.*

***Cirsium wrightii* A. Gray** is a biennial or monocarpic perennial, which occurs in cienegas, ditches, marshes, meadows, seeps, springs, and along the margins of ponds and streams at elevations from 1152 to 2393 m. Preferred substrates comprise perennially wet alkaline (travertine) gravel, rock, and sand. Virtually nothing is known about the reproductive biology or seed ecology of this species. **Reported associates:** *Baccharis salicifolia, B. salicina, Berula erecta, Distichlis spicata, Elaeagnus angustifolia, Eleocharis rostellata, Flaveria chlorifolia, Geranium, Helianthus paradoxus,*

Juncus balticus, Juniperus, Lobelia cardinalis, Lythrum californicum, Monotropa, Muhlenbergia asperifolia, Phragmites australis, Pinus monophylla, P. ponderosa, Populus fremontii, Quercus gambelii, Salix, Schoenoplectus acutus, S. americanus, S. pungens, Solidago, Symphyotrichum laeve, Tamarix ramosissima, Typha angustifolia, T. latifolia.

Use by wildlife: *Cirsium* (thistle) seeds are well known as a favored food of many birds, particularly goldfinches (Aves: Fringillidae: *Carduelis* spp.), which also use the pappus bristles in their nest construction. In laboratory trials, *Cirsium crassicaule* is a marginal host for a fly (Insecta: Diptera: Syrphidae: *Cheilosia corydon*) that has been used for biological control of several invasive thistles. A lengthy list of insect pollinators (Insecta) has been compiled for *C. crassicaule*, which includes beetles (Coleoptera: Cetoniidae: *Trichius piger*; Cerambycidae: *Typocerus sinuatus*; Chrysomelidae: *Diabrotica longicornis*), bugs (Hemiptera: Lygaeidae: *Lygaeus turcicus*), flies (Diptera: Bombyliidae: *Exoprosopa fasciata, Systoechus vulgaris*; Conopidae: *Physocephala tibialis*; Tabanidae: *Scepsis fulvicollis*), various Hymenoptera such as bumblebees and honey bees (Apidae: *Anthophora walshii, Apis mellifera, Bombus auricomus, B. fraternus, B. griseocollis, B. impatiens, B. pensylvanicus, B. vagans, B. variabilis, Ceratina dupla, Habropoda laboriosa, Melissodes agilis, M. bimaculata, M. coloradensis, M. communis, M. comptoides, M. desponsa, M. druriella, M. trinodis, Melitoma taurea, Ptilothrix bombiformis, Svastra obliqua, Triepeolus concavus, T. donatus, T. lunatus, T. nevadensis, T. remigatus*), ground-nesting bees (Colletidae: *Colletes eulophi*), solitary bees (Megachilidae: *Heriades leavitti, Megachile latimanus, M. montivaga, M. parallela, M. pugnata*), and sweat bees (Halictidae: *Agapostemon radiatus, A. texanus, A. viridulus, Halictus ligatus, Lasioglossum imitatum, L. pectorale, L. pilosum, L. pruinosum, L. versatum, Nomia nortoni*), and Lepidoptera including butterflies (Nymphalidae: *Anosia plexippus, Argynnis cybele, Limenitis archippus, Phyciodes tharos, Speyeria idalia, Vanessa atalanta, V. cardui*); Papilionidae: *Battus philenor, Papilio cresphontes, P. glaucus, P. polyxenes, P. troilus*; Pieridae: *Abaeis nicippe, Colias philodice, Phoebis eubule, Pontia protodice*), skipper moths (Hesperiidae: *Atrytone zabulon, Polites peckius, P. taumas, Thorybes bathyllus*), and sphinx moths (Sphingidae: *Hemaris thysbe, Hyles lineata*). It also is a host of larval artichoke plume moths (Insecta: Lepidoptera: Pterophoridae: *Platyptilia carduidactyla*). *Cirsium douglasii* also is a host plant for larval butterflies and moths (Insecta: Lepidoptera: Nymphalidae: *Phyciodes mylitta, Vanessa cardui*; Pterophoridae: *Platyptilia carduidactyla*). Rust fungi (Basidiomycota: Pucciniaceae: *Puccinia calcitrapae*) have been reported from the plants. *Cirsium fontinale* flowers attract insects (Insecta) such as bumblebees and honey bees (Hymenoptera: Apidae: *Apis mellifera, Bombus vosnesenskii*), sweat bees (Halictidae), and flies (Diptera), which serve as pollinators. They also are visited by several birds (Aves) including the American bushtit (Aegithalidae: *Psaltriparus minimus*), blackbirds (Icteridae), hummingbirds (Trochilidae), and sparrows (Passeridae). The inflorescences host ants (Insecta: Hymenoptera: Formicidae), aphids (Insecta: Hemiptera: Aphidoidea), and occasionally a small but usually insignificant number of seedhead weevils (Insecta: Coleoptera: Curculionidae: *Rhynocyllus conicus*). The plants also host larval artichoke plume moths (Insecta: Lepidopticta: Pterophoridae: *Platyptilia carduidactyla*). Fruit flies also occur on the plants (Insecta: Diptera: Tephritidae: *Paracantha gentilis*) and can cause minor damage to them. The young leaves of *C. fontinale* are fed on by domestic cattle (Bovidae: *Bos bos*). *Cirsium hydrophilum* hosts several larval Lepidoptera (Insecta: Nymphalidae: *Phyciodes mylitta*, *Vanessa cardui*; Pterophoridae: *Platyptilia carduidactyla*) and rust fungi (Basidiomycota: Pucciniaceae: *Puccinia laschii*). Introduced weevils (Insecta: Coleoptera: Curculionidae: *Rhinocyllus conicus*) have been reared from collected flower heads and pose a threat to the survival of this rare species. Flowers of this species are visited by bumblebees (Insecta: Hymenoptera: Apidae: *Bombus vosnesenskii*), which probably represent their principal pollinators. *Cirsium muticum* is used as forage by the eastern wild turkey (Aves: Phasianidae: *Meleagris gallopavo*). It is a host plant for several larval Lepidoptera (Insecta: Lycaenidae: *Calephelis muticum*; Nymphalidae: *Vanessa cardui*; Pterophoridae: *Platyptilia carduidactyla*) and aphids (Hemiptera: Aphididae: *Uroleucon pepperi*). It is also a host to a variety of fungi including Ascomycota (Botryosphaeriaceae: *Phyllosticta cirsii*; Erysiphaceae: *Erysiphe cichoracearum*; Leptosphaeriaceae: *Leptosphaeria mesoedema*; Pseudoperisporiaceae: *Chaetosticta perforata*), Basidiomycota (Pucciniaceae: *Puccinia calcitrapae*), and Oomycota (Albuginaceae: *Albugo tragopogonis*, *Pustula tragopogonis*). The flowers are visited by ruby-throated hummingbirds (Aves: Trochilidae: *Archilochus colubris*), which forage for small insects amongst them and spicebush swallowtail butterflies (Insecta: Lepidoptera: Papilionidae: *Papilio troilus*). Fruit flies (Insecta: Diptera: Tephritidae: *Terellia palposa*) also have been collected from the flowers. The seeds are an important source of food for the American goldfinch (Aves: Fringillidae: *Carduelis tristis*). The plants also provide habitat for the prairie deer mouse (Rodentia: Cricetidae: *Peromyscus maniculatus*) and reflexed Indiangrass leafhopper (Insecta: Homoptera: Cicadellidae: *Flexamia reflexa*). *Cirsium vinaceum* seeds are eaten by a number of invertebrate predators including artichoke plume moths (Insecta: Lepidoptera: Pterophoridae: *Platyptilia carduidactyla*), bumble flower beetles (Insecta: Coleoptera: Scarabaeidae: *Euphoria inda*), fruit flies (Insecta: Diptera: Tephritidae: *Paracantha gentilis*), seed-head weevils (Insecta: Coleoptera: Curculionidae: *Rhinocyllus conicus*), and stem borer weevils (Insecta: Coleoptera: Curculionidae: *Lixus pervestitus*). Together, these predators have been known to reduce seed production by as much as 98% in some populations. The flowers also are visited by hummingbirds (Aves: Trochilidae) and bees (Hymenoptera). *Cirsium wrightii* is a host of rust fungi (Basidiomycota: Pucciniaceae: *Puccinia californica*; *P. cirsii*). The flowers attract hummingbirds (Aves: Trochilidae).

Economic importance: food: The roots of many thistle species are eaten when raw, boiled or roasted. The immature inflorescences also are eaten raw or cooked like asparagus. Young leaves can be eaten raw (after spines are removed) or brewed into tea. The young stems can be peeled and cooked. Once boiled, thistle seeds can be eaten or ground into baking flour; **medicinal:** Thistle shoots have been eaten as a stomach remedy, or made into a salve for treating infections. A decoction from thistle roots has been used to treat asthma; **cultivation:** *Cirsium douglasii* has been tissue cultured successfully; **misc. products:** The thistle plant (*Cirsium*) is the national flower and symbol of Scotland; **weeds:** Many *Cirsium* species are regarded as noxious in North America because of their competitive vigor and spiny habit; **nonindigenous species:** All of the OBL *Cirsium* species are indigenous to North America.

Systematics: *Cirsium* is assigned to tribe Cardueae of subfamily Carduoideae (Figure 5.133), where it includes a major lineage of indigenous North American species. Molecular data resolve *Cirsium* as clade sister to the morphologically similar *Carduus*, which differs primarily by its nonplumose pappus. It is conceivable that further sampling within the latter may disclose inconsistencies in the classification of both genera. The North American *Cirsium* species resolve as a clade within the genus, although it is only weakly supported phylogenetically. Stronger support identifies a clade of species restricted to the California floristic province. The OBL North American *Cirsium* species are not monophyletic, with the habit originating at least twice: in a group of western North American species (*C. douglasii*, *C. fontinale*, *C. hydrophilum*), and in the Eastern North American *C. muticum* (Figure 5.146). These groups are fairly distantly related. Hybridization is common in *Cirsium* and is associated with wide chromosomal variation. *Cirsium* has a base chromosome number of $x = 17$. The native North American *Cirsium* species are functionally diploid, but represent an extensive reduction series ranging from $2n = 34$ to $2n = 18$. Intraspecific variation still requires further investigation in a number of species. Three varieties of *C. fontinale* are recognized: (var. *fontinale*, var. *campylon*, var. *obispoense*); however, the latter does not resolve phylogenetically close to type (see Figure 5.146). All three varieties are $2n = 34$ or $34 + 1B$ [accessory chromosome]. Two varieties have been recognized in *C. hydrophilum*, and these differ ecologically. Although both are wetland denizens, *C. hydrophilum* var. *hydrophilum* occurs in tidal marshes, whereas var. *vaseyi* is restricted to serpentine seeps. Both taxa are $2n = 32$. Hybridization between *C. hydrophilum* and *C. vulgare* has been postulated on the basis of some morphological anomalies, but has not been confirmed otherwise. *Cirsium crassicaule* is $2n = 32$, *C. douglasii* is $2n = 30$, 34, and *C. fontinale* is $2n = 34 + 1B$. The widespread *C. muticum* is more variable cytologically ($2n = 20$ [21, 22, 23, 30]) and hybridizes with *C. discolor* ($2n = 20$ [21, 22]; FACU, UPL), especially in areas where wet and dry habitats are juxtaposed. The parental species differ by 2–3 translocations resulting in extremely low (6%) pollen fertility in their F_1 hybrids. *Cirsium vinaceum* hybridizes with *C. parryi* and *C. wrightii*. The hybrids derived from *C. parryi* have been observed to be more common during wetter-than-average years, which

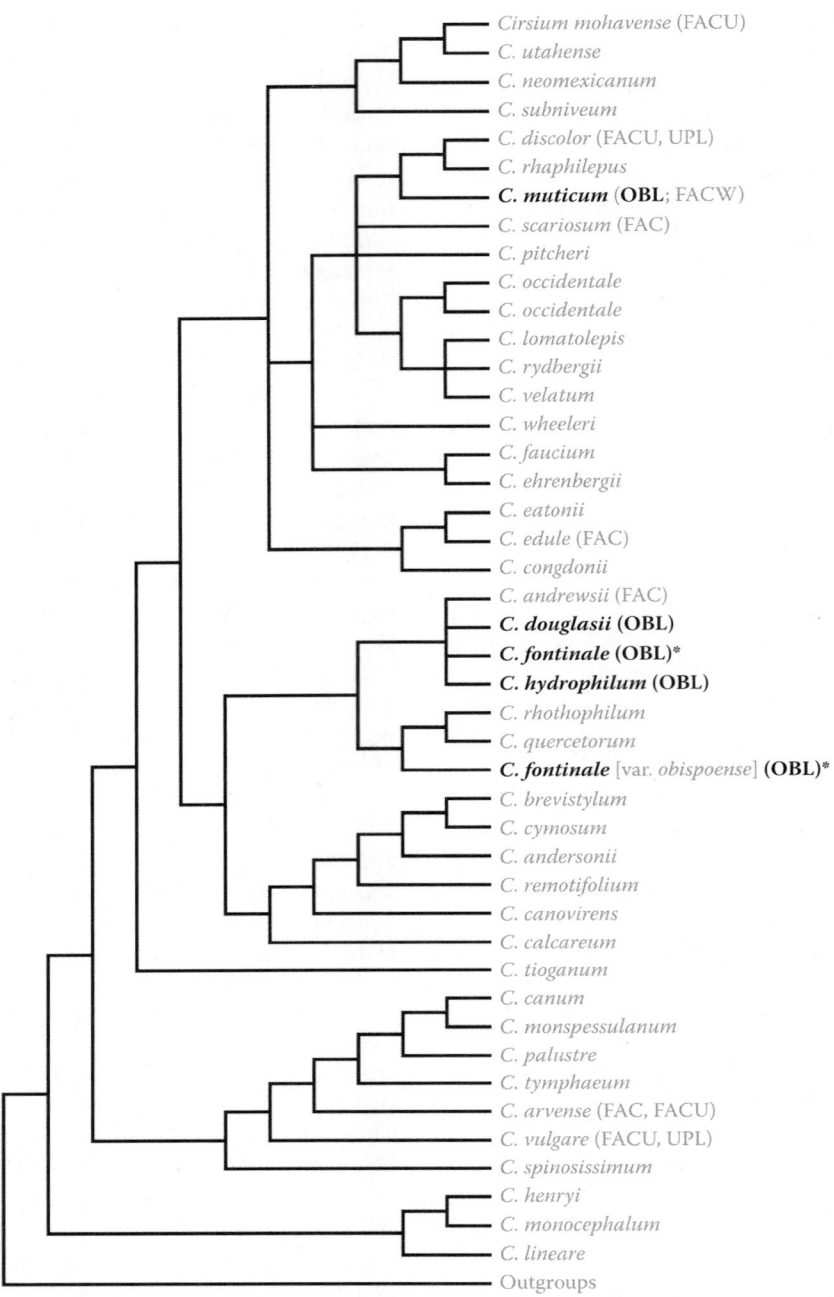

FIGURE 5.146 Phylogenetic relationships among *Cirsium* species reconstructed using nrITS and ETS data. Clearly there have been at least two independent origins of the OBL habit (taxa in bold). However, the different placement of *C. fontinale* accessions (asterisked) requires further evaluation. These results agree closely with those of a later analysis by Bodo Slotta et al. (2012). (Adapted from Kelch, D.G. & Baldwin, B.G., *Mol. Ecol.*, 12, 141–151, 2003.)

perhaps provide better germination conditions for the hybrid offspring. Hybridization between *C. wrightii* and C. *texanum* is suspected from observations of morphologically intermediate plants. Although detailed morphological analyses have provided some compelling evidence of hybridization, better documentation of hybrids, preferably incorporating genetic data, would be useful.

Comments: *Cirsium muticum* (throughout eastern North America) is the only widespread OBL thistle in North America. *Cirsium douglasii* has a more limited distribution in the western United States (California, Nevada, and Oregon). Three species (*C. crassicaule C. fontinale, C. hydrophilum*) occur only in California, one species (*C. wrightii*) occurs in Arizona and New Mexico (erroneously reported from Texas), and one species (*C. vinaceum*) is restricted to New Mexico.

References: Amon et al., 2002; Ashe, 2010; Beattie, 1995; Bess, 2005; Biswas & Mallik, 2010; Blair, 1940; Bloom, 1977; Buchmann et al., 2010; Chipping, 1994; Clark, 1997b;

Conners, 1967; Costelloe, 1988; Craddock & Huenneke, 1997; Ellett, 1970; Evens & San, 2004; Farr & Rossman, 2013a,b; Frankton & Moore, 1963; Freestone & Inouye, 2006; Frolik, 1941; Gordon, 1933; Han & McPheron, 1994; Huenneke & Thomson, 1995; Humbert et al., 2007; Kelch & Baldwin, 2003; Kuhn et al., 2011; Lucas, 1893; Moran, 1981, 1984; Nekola, 2004; Nguyen et al., 2004; Nichols, 1934; Nickell, 1951; Ownbey, 1951; Ownbey et al., 1975; Pasqualetto & Dunn, 1990; Powell & Knight, 2009; Powell et al., 2011; Reed, 1913; Rizza et al., 1988; Rusk, 1942; Shapiro, 1974b; Showers, 2010; Singhurst et al., 2007; Sivinski & Lightfoot, 1993; Standley, 1916; Thieret, 1971; Turner, 1928; Turner et al., 1987a,b; USFWS, 1993b, 2007, 2009, 2010; Verhey, 2007; Vizgirdas & Rey-Vizgirdas, 2005; Werner & Zedler, 2002; Wheeler et al., 1983.

13. *Coreopsis*

Tickseed; Coréope

Etymology: from the Greek *koris opsis* ("bug-like") for the tick-like achenes

Synonyms: *Agarista*; *Anacis*; *Calliopsis*; *Campylotheca*; *Chrysomelea*; *Chrysostemma*; *Coreopsoides*; *Diodonta*; *Diplosastera*; *Electra*; *Epilepis*; *Lechea*; *Leptosyne*; *Peramibus*; *Pugiopappus*; *Selleophytum*; *Tuckermannia*

Distribution: global: North America; Old World tropics; **North America:** widespread

Diversity: global: 35 species; **North America:** 34 species

Indicators (USA): OBL: *Coreopsis floridana, C. nudata*

Habitat: freshwater; lacustrine, palustrine; **pH:** unspecified; **depth:** <1 m; **life form(s):** emergent herb

Key morphology: Stems (to 12 dm) erect, glabrous, simple or moderately branched; upper cauline leaves alternate or opposite, reduced, linear, awliform or filiform; basal leaves awliform (to 40 cm) and attenuate or narrowly elliptic, lanceolate or oblong (to 15 cm) and petiolate (to 20 cm); capitula (1–10) radiate, involucres campanulate, outer phyllaries (to 6 mm) shorter than inner (to 15 mm); ray florets neuter, the rays (to 3.0 cm) lavender, pink or yellow, broadly 3-lobed apically; disc florets (to 4.8 mm) 40–120, the corollas 4-lobed, yellowish to purplish brown; stamens 4; receptacle chaffy, pales (to 10 mm) linear; cypselae (to 3–5 mm) oblong, purplish or grayish brown, the margin winged by irregular pectinate teeth or lobes (to 0.4 mm), the pappus of paired, membranous or anteriorly setose awns or scales (to 2.5 mm)

Life history: duration: short-lived perennial (rhizomes); **asexual reproduction:** rhizomes; **pollination:** insect; **sexual condition:** hermaphroditic; **fruit:** cypselae (common); **local dispersal:** seeds (gravity); **long-distance dispersal:** seeds (water, wind?)

Imperilment: (1) *Coreopsis floridana* [G3–G4]; (2) *C. nudata* [G3]; S1 (AL, MS); S2 (LA); S3 (GA)

Ecology: general: *Coreopsis* species primarily occur in temperate regions, but extend into the New and Old World tropics. They occupy a wide range of habitats from dry, rocky sites to low, wet areas. Roughly 35% of the species are designated at some level as wetland indicators; however, only two are categorized as OBL. Sporophytic self-incompatibility, which requires outcrossing to achieve seed set, is widespread in the genus (e.g., sections *Calliopsis*, *Coreopsis*; *Eublepharis*), and occurs in both OBL species. The flowers are insect pollinated (Insecta), typically by generalist bees (Hymenoptera: Halictidae) or bee flies (Diptera: Bombyliidae). Flowers of all North American species contain anthochlor pigments, which presumably function in the attraction of pollinators due to their ultraviolet spectral properties. Seeds of the few species investigated in detail exhibit either physiological dormancy or essentially are nondormant. Integrated phylogenetic/life history data indicate that the pattern of speciation in *Coreopsis* exhibits a general correlation between genetic distance and degree of reproductive isolation. However, shifts from a perennial to annual life history have greatly accelerated the acquisition of reproductive isolation in the genus.

***Coreopsis floridana* E. B. Smith** is a short-lived perennial inhabitant of bogs, depressions, ditches, flatwoods, glades, hammocks, marshes, prairies, savannas, seeps, sloughs, swales, and swamps at elevations of up to 50 m. The plants are hardy in zones 8–10. They are intolerant of salinity and can be harmed by salt spray. The plants will thrive in full sun but also occur in localities with partial shade. The substrates associated with specific occurrences are described as Basinger (spodic psammaquents), Bladen (typic albaquults), Holopaw (grossarenic ochraqualfs), Immokalee, loamy sand, peat, peaty sand, Plummer (grossarenic paleaquults), Pompano, sand, sandy peat, and Valkaria (spodic psammquents). Flowering occurs from October to November. Plants will produce 6–23 capitula per stem, and up to 15 flowering stems per plant. Herbivory can reduce head production by as much as 50% and plant height by 40%. The flowers are self-incompatible and are pollinated by insects. Specific pollinators have not been enumerated but likely are dominated by small bees (Insecta: Hymenoptera). The mechanism of seed dispersal has not been elucidated, but the marginally winged cypselae probably facilitate their dispersal by wind or water currents. Seeds germinate best when dried quickly and deteriorate if exposed to moisture. Fresh seed is highly viable (82%–91%) and no treatment is necessary for germination (96% when sown immediately). Dry storage for 2 weeks (15°C or 32°C; RH = 33%) may promote optimal after ripening and has yielded 100% germination. Generally the seeds have a fairly short storage life. A cool (20°C–25°C), dry (relative humidity: 24%–34%) place should be used if short-term storage is necessary; colder temperatures (15°C) optimize longer storage times. **Reported associates:** *Aletris lutea, Amphicarpum muhlenbergianum, Andropogon brachystachyus, A. gyrans, A. liebmannii, A. virginicus, Anthaenantia rufa, Aristida palustris, A. spiciformis, A. stricta, Arnoglossum ovatum, Arundinaria tecta, Balduina uniflora, Bigelowia nudata, Carphephorus pseudoliatris, Centella asiatica, Chaptalia tomentosa, Clethra alnifolia, Cliftonia monophylla, Coleataenia longifolia, Coleataenia tenera, Coreopsis gladiata, Ctenium aromaticum, Cyrilla racemiflora, Dichanthelium sphaerocarpon, D. strigosum, Eleusine indica, Eragrostis elliottii, Erigeron vernus, Eriocaulon compressum, E. decangulare, Eryngium yuccifolium, Eupatorium mohrii, Euthamia graminifolia,*

Fuirena scirpoidea, Gaylussacia mosieri, Gratiola ramosa, Helenium pinnatifidum, Helianthus angustifolius, Hypericum cistifolium, H. fasciculatum, H. gentianoides, H. myrtifolium, H. tetrapetalum, Hyptis alata, Ilex coriacea, I. glabra, Iva microcephala, Juncus trigonocarpus, Juniperus virginiana, Lachnocaulon anceps, Liatris spicata, Linum medium, Liquidambar styraciflua, Lobelia glandulosa, Ludwigia linearis, L. linifolia, L. maritima, Lycopodiella prostrata, Mnesithea rugosa, Myrica cerifera, Oldenlandia uniflora, Oxypolis denticulata, O. filiformis, Panicum verrucosum, Paspalum praecox, Persicaria setacea, Pinus elliottii, P. serotina, P. taeda, Pluchea baccharis, P. foetida, Polygala lutea, P. ramosa, Proserpinaca pectinata, Quercus minima, Rhexia mariana, R. nuttallii, Rhynchospora filifolia, R. inundata, R. latifolia, R. oligantha, R. tracyi, Richardia scabra, Sabal palmetto, Sabatia grandiflora, Sarracenia leucophylla, Schizachyrium rhizomatum, S. scoparium, Schoenus, Scleria baldwinii, S. georgiana, S. muehlenbergii, Serenoa repens, Seymeria cassioides, Smilax laurifolia, Symphyotrichum chapmanii, Taxodium ascendens, Tradescantia ohiensis, Utricularia subulata, Viola lanceolata, Vitis rotundifolia, Xyris ambigua, X. difformis, X. elliottii, X. flabelliformis.

***Coreopsis nudata* Nuttall** is a perennial, which occupies standing shallow water (<15 cm) or saturated sites in barrens, bogs, borrow pits, depressions, ditches, flatwoods, marshes, prairies, savannas, seepage bogs, swamps, and along pond, pool, or stream margins at elevations of up to 50 m. Plants tolerate exposures from full sun to partial shade. Reported substrates include loamy sand, muck, mucky clay, mud, peat, peaty muck, Pelham (arenic paleaquults), sand, sandy loam, and sandy peat. Flowering occurs from March to June. The plants are self-incompatible and are pollinated by insects. The reproductive biology is not well-documented, but the principal pollinators are likely to be bees (Insecta: Hymenoptera) as in most other *Coreopsis* species. Although the rays of the inflorescence are pink rather than yellow (most other congeners), no specialized pollinators have been reported. The mechanism of seed dispersal has not been determined, but probably occurs to some degree by wind and water, as facilitated by the irregularly winged margins of the cypselae. Seeds stored at 5°C for 0–8 months exhibited their highest germination (19%–50%) after 38 days under 16/8 h, 30°C/25°C or 10°C/5°C regimes. However, mature, untreated seed also reportedly germinate readily simply when sown in a pan of moist soil kept under warm conditions. Vegetative reproduction occurs by rhizomes, which arise from a fleshy root. The plants benefit from periodic fires, which remove competing woody vegetation. **Reported associates:** *Agalinis linfolia, Aletris aurea, Amphicarpum muhlenbergianum, Andropogon, Aristida purpurascens, A. stricta, Arnoglossum ovatum, Asclepias lanceolata, A. tuberosa, Bigelowia nudata, Burmannia capitata, Calamovilfa curtissii, Calopogon tuberosus, Carex glaucescens, Carphephorus odoratissimus, Centella asiatica, Cladium mariscoides, Clethra alnifolia, Ctenium aromaticum, Cyrilla racemiflora, Dichanthelium acuminatum, D. scoparium, D. sphaerocarpon, Eleocharis tuberculosa, Erigeron vernus, Eriocaulon compressum, E.*

decangulare, Fuirena scirpoidea, Gymnadeniopsis nivea. Helenium vernale, Helianthus radula, Hypericum brachyphyllum, H. chapmanii, H. fasciculatum, H. myrtifolium, Ilex myrtifolia, Juncus diffusissimus, J. polycephalus, J. repens, Lachnanthes caroliniana, Lobelia floridana, Lophiola aurea, Ludwigia pilosa, L. sphaerocarpa, Lycopodiella alopecuroides, Lyonia lucida, Magnolia virginiana, Mnesithea rugosa, Nyssa biflora, Oxypolis filiformis, Persea palustris, Pinguicula planifolia, Pinus elliottii, P. palustris, Pluchea foetida, Polygala cymosa, Pontederia cordata, Rhexia alifanus, R. virginica, Rhynchospora careyana, R. cephalantha, R. corniculata, R. filifolia, R. latifolia, Rudbeckia mohrii, Sabatia brevifolia, S. decandra, Sarracenia minor, Scleria baldwinii, Smilax laurifolia, S. walteri, Stylisma aquatica, Symphyotrichum adnatum, Taxodium ascendens, T. distichum, Tillandsia bartramii, Triantha racemosa, Utricularia purpurea, Woodwardia virginica, Xyris serotina.

Use by wildlife: *Coreopsis floridana* is browsed by white-tailed deer (Mammalia: Cervidae: *Odocoileus virginianus*), with up to 60% of plants showing associated damage. It is a host to plant pathogenic fungi (Ascomycota: Pleosporaceae: *Alternaria*). Butterflies (Insecta: Lepidoptera) forage for nectar from the flowers. The flowers of *Coreopsis nudata* are hosts to larval fruit flies (Insecta: Diptera: Tephritidae: *Dioxyna picciola*).

Economic importance: food: not edible; **medicinal:** no medicinal uses; **cultivation:** *Coreopsis floridana* (Florida Coreopsis) has been seed cultivated as a native wildflower garden plant since 2004 and is available sporadically from specialty nurseries; **misc. products:** *Coreopsis* is the state wildflower of Florida; **weeds:** no OBL weeds; **nonindigenous species:** none.

Systematics: Traditionally, *Coreopsis* has been placed in subfamily Asteroideae, tribe Coreopsideae of Asteraceae. However, molecular phylogenetic analyses have indicated repeatedly that *Coreopsis* is not monophyletic as currently circumscribed, but includes a number of other genera (*Bidens, Coreocarpus, Cosmos, Thelesperma*) that nest within a larger clade containing most of the species (Figure 5.147). It is apparent that extensive taxonomic realignments must eventually be made to reflect a circumscription of *Coreopsis* that is more consistent with phylogenetic relationships. Both *Coreopsis floridana* and *C. nudata* are assigned to section *Eublepharis*. *Coreopsis floridana* has been merged with *C. gladiata* by some authors. It is retained here as a distinct species because it is dissimilar genetically from the latter (nrITS sequences) and also differs chromosomally. The basic chromosome number of *Coreopsis* section *Eublepharis* is $x=13$. *Coreopsis gladiata* ($2n=26$) is a diploid; whereas, *C. floridana* ($2n=156$) is a dodecaploid. It is suspected to be a polyploid hybrid derivative of *C. gladiata* or *C. falcata* and *C. linifolia*; artificial hybridizations involving *C. floridana* and *C. gladiata* and *C. linifolia* have been unsuccessful. Generally, hybridization is much more common among the perennial *Coreopsis* species than annuals and rarely is successful when involving polyploids. *Coreopsis nudata* ($2n=26$) is diploid. *Coreopsis floridana* represents one of the most derived species in the

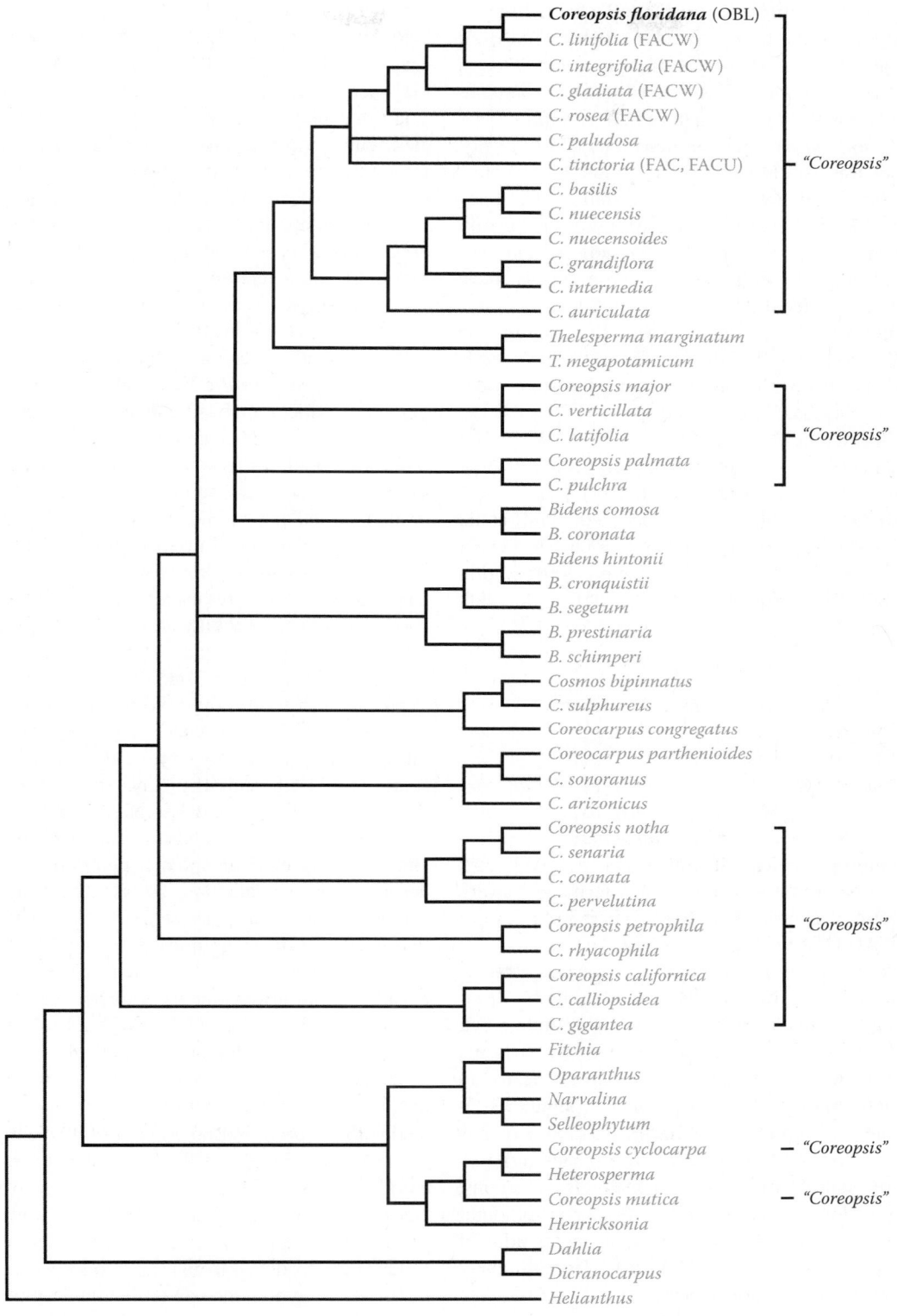

FIGURE 5.147 Maximum likelihood phylogeny of *Coreopsis* and associated genera based on combined nrITS and cpDNA sequences. It is apparent from this and other studies (e.g., Kim et al., 1999; Sayre, 2001; Crawford & Mort, 2005) that the current circumscription of *Coreopsis* is not phylogenetically robust and requires further refinement. This phylogenetic tree indicates that the OBL (bold) *C. floridana* originated from a highly derived clade dominated by FACW species. (Redrawn from Mort, M.E. et al., *Taxon*, 57, 109–120, 2008.)

genus (Figure 5.147), confirming that the OBL habit also is a highly derived trait in the group where it has emerged from a grade dominated by FACW species. *Coreopsis floridana* is closely related to *C. linifolia* (FACW), which some authors place in synonymy with *C. gladiata* (and *C. floridana*!). DNA sequences of *C. nudata* have not yet been included in phylogenetic analyses, but would represent desirable "new data." *Coreopsis nudata* probably is closely related to *C. rosea* (FACW), with which it shares anomalous (for the genus) pink-petaled ray florets; both are members of section *Eublepharis*. However, interspecific hybrids between these species have reduced (<10%) pollen fertility.

Comments: Both OBL *Coreopsis* species occur in the Southeastern United States. The distribution of *C. floridana* is more restricted (Florida) with *C. nudata* somewhat more widespread (Alabama, Florida, Georgia, Louisiana, Mississippi).

References: Aldrich et al., 2006, 2007; Anon, 1879; Archibald et al., 2005; Barkley et al., 2006; Baskin & Baskin, 1998; Benjamin, 1934; Crawford & Mort, 2005; Crawford & Smith, 1983; De Groote et al., 2011; Harper, 1900; Huff et al., n.d.; Kim et al., 1999; Loehrlein & Siqueira, 2005; Mort et al., 2008; Norcini & Aldrich, 2007, 2008; Sayre, 2001; Smith, 1975, 1976, 1982, 1983; Strother, 2006b; Tadesse et al., 1996, 2001; Tobe et al., 1998.

14. *Cotula*

Brassbuttons, waterbuttons; cotule pied-de-corbeau

Etymology: from the Greek *kotulê* ("small cup"), with respect to the sheathing bases of some leaves

Synonyms: *Brocchia*; *Ctenosperma*; *Gymnogyne*; *Lancisia*; *Leptinella*; *Leptogyne*; *Machlis*; *Otoglyphis*; *Pleiogyne*; *Sphaeroclinium*; *Strongylosperma*; *Symphyomera*

Distribution: global: Africa; Australia; Europe*; Mexico*; South America*; **North America*:** northeastern and western

Diversity: global: 55 species; **North America:** 3 species

Indicators (USA): OBL: *Cotula coronopifolia*

Habitat: brackish (coastal), freshwater; palustrine, riverine

pH: 5.2–7.7; **depth:** <1 m; **life form(s):** emergent herb

Key morphology: Stems (to 30 cm) fleshy, prostrate to erect, rooting at nodes; leaves (to 7 cm) alternate, sessile, sheating at base, blades gland dotted, entire to pinnatifid; heads (to 12.0 mm) eradiate, axillary, on naked, curving peduncles (to 10 cm); involucre with up to 30 phyllaries (to 5 mm) in 2–3 series; peripheral florets (to 40) rayless (corolla lacking), pistillate, distinctly stalked (to 1 mm); disk florets numerous, bisexual, with yellow, 4-lobed corollas (to 1.5 mm); outer cypselae (to 2.0 mm) corky winged along margin, the adaxial (inner) surface papillate; inner cypselae (to 1.5 mm), minutely stalked, essentially wingless; pappus absent.

Life history: duration: annual (seeds); perennial (persistent shoots); **asexual reproduction:** persistent shoots, shoot fragments; **pollination:** insect; **sexual condition:** gynomonoecious; **fruit:** cypselae (common); **local dispersal:** cypselae (water); **long-distance dispersal:** cypselae (birds)

Imperilment: (1) *Cotula coronopifolia* [GNR] – nonindigenous.

Ecology: general: Most *Cotula* species occur in nonwetland habitats with only *C. cornopifolia* regarded as OBL among the three species introduced to North America. The South African *C. myriophylloides* is the only other truly aquatic species in the genus. *Cotula* contains both annual and perennial species, which can vary within a species among regions. The floral biology is diverse with the capitula of some species being entirely hermaphroditic and others possessing unisexual florets in dioecious, monoecious, or gynomonoecious sexual conditions. Most species examined are self-compatible but self-incompatible species also are reported. The florets are either wind or insect pollinated (Insecta), with the former system more characteristic of the dioecious taxa. Because most species originate from mild climates, the seeds of many have no dormancy requirements. Several of the species are weedy and can effectively colonize disturbed sites.

Cotula coronopifolia **L.** is a summer annual or short-lived perennial, which inhabits fresh to saline backwaters, beaches, depressions, ditches, dunes, flats, floodplains, lagoons, marshes, meadows, mudflats, playas, pools, river bottoms, saltmarshes, sandbars, scrub, seeps, sloughs, springs, strands, swales, vernal pools, and the margins of lakes, ponds, and streams at elevations of up to 1200 m. Sites range from saturated ground to flowing water with exposures from full sun to partial shade. Substrates usually are described as alkaline (pH 7.4–7.7), but some brackish sites are acidic (pH 5.2). Substrates often are brackish (mean conductivity = 3.4–20.3 dS/m; salinity = 1.8–12.1 ppt; Na = 40.4 meq/l; Mg = 13.3 meq/l; B = 1 ppm). Types of substrates are diverse and include adobe, bare nutrient-rich mud, clay, clay loam, clay mud, cobble, gravel, muck, mucky clay, rocky clay, sand, sandy clay, sandy gravel, sandy loam, silt, and silty clay. Specific soil types include Reyes, Joice, Omni, and Pescadero series. The life history varies among regions with European plants reportedly annual and North American plants perennial. This distinction deserves further evaluation. The plants are capable of forming extensive mats, with the stems rooting at the nodes. They thrive under tidal conditions but generally occur above the high tide mark at an elevational range from 1.8 to 2.7 m MLLW (mean lower low water). They can tolerate pulses of up to 0.8% salinity for as long as 6 weeks. They are intolerant of frost but highly tolerant to periodic flooding; however, they do cannot withstand prolonged deep inundation. Submergence results in increased growth (root and shoot biomass) associated with the proliferation of photosynthetically active adventitious roots, increased Na+ content in roots, increased chlorophyll A and B content, and improved water use efficiency. Flooding also increases the amount of root aerenchyma tissue and the extent of stem xylem lignification. Flowers are produced from March to December. *Cotula coronopifolia* is gynomonoecious, having capitula with female outer florets (up to 9% of all florets) and hermaphrodite inner florets. Both the outer and inner florets are self-compatible and produce abundant seed when self-pollinated. Generally, the florets are adapted morphologically for self-pollination (e.g., low pollen production of 300 grains/floret); however, the outer (female) florets are receptive several days before the central hermaphroditic florets open, which would promote outcrossing when pollen vectors are present. The average seed set and germination rate are slightly higher for outcrossed seed. The flowers contain a nectary and are pollinated by unspecialized insects (Insecta), which can include stinkbugs (Hemiptera: Pentantomidae) and blowflies (Diptera: Calliphoridae). The outer cypselae possess corky wings and are dispersed by water; the inner cypselae are wingless and fall near the maternal plants. Fruit dispersal over greater distances occurs from transport in mud or by migratory birds. Numerous, germinable seeds have been recovered from the gizzards of various duck species (Aves: *Anas*), which readily consume them. Seeds are nondormant, and exhibit high germination rates (at 5, 30°C/15°C) even when freshly collected. Under saline conditions, salt-induced dormancy occurs as an osmotic effect, by slowing the rate of water uptake by the seed. Seeds retain viability with germination rates of up to 71% after storage in 2% NaCl for 43 days. Germination usually occurs in late fall or winter following the onset of rains. Seeds survive for up to 2 years in field conditions (short-term, persistent seed bank). They can reach high densities (up to 6600/m²) in saline marsh sediments. Because the plants often occupy highly disturbed habitats where native species do not thrive (e.g., dredge spoils), they pose a lesser threat of competition. The plants often colonize bare mud flats, but die back as other vegetation increases in density. **Reported associates (North America):** *Abies grandis, Acer glabrum, Achillea millefolium, Achlys triphylla, Achnatherum occidentale, Acmispon parviflorus, Adenostoma fasciculatum, Agrostis stolonifera, Alisma, Allium, Alnus rhombifolia, A. rugosa, Alopecurus aequalis, A. geniculatus, Ambrosia chamissonis, A. psilostachya, Amelanchier alnifolia, Ammophila arenaria, Anagallis arvensis, Anemopsis californica, Antennaria argentea, Apium graveolens, Apocynum androsaemifolium, Artemisia californica, A. douglasiana, A. pycnocephala, Arthrocnemum subterminale, Atriplex dioica, A. patula, A. prostrata, A. semibaccata, Avena barbata, A. fatua, Azolla, Baccharis pilularis, B. salicifolia, Barbarea orthoceras, Berberis aquifolium, Bidens cernuus, Bolboschoenus maritimus, Brassica nigra, Briza, Brodiaea terrestris, Bromus hordeaceus, B. rubens, B. tectorum, Cakile edentula, C. maritima, Calamagrostis rubescens, Callitriche hermaphroditica, C. heterophylla, C. marginata, Carex amplifolia, C. athrostachya, C. geyeri, C. lyngbyei, C. obnupta, C. praegracilis, C. unilateralis, Carpobrotus chilensis, Castilleja ambigua, C. lineariiloba, Camissoniopsis cheiranthifolia, Ceanothus, Centromadia pungens, Chenopodium foliosum, C. rubrum, Chrysanthemum, Cinna latifolia, Claytonia perfoliata, Clematis ligusticifolia, Collomia linearis, Comarum palustre, Conium maculatum, Conyza, Corallorhiza maculata, Cornus sericea, Cotula australis, Crassula aquatica, C. connata, Cressa truxillensis, Crypsis schoenoides, C. vaginiflora, Cryptantha, Cuscuta, Cyperus, Cypripedium montanum, Dactylis glomerata, Deschampsia cespitosa, D. elongata, Deinandra fasciculata, Dittrichia graveolens, Distichlis spicata, Downingia cuspidata, D. insignis, Drymocallis glandulosa, Eclipta prostrata, Eleocharis acicularis, E. macrostachya, E. montevidensis, E. palustris, E. parvula, Elymus glaucus, Epilobium ciliatum, Equisetum arvense, E. laevigatum, E.* telmateia, Eriogonum fasciculatum, Erodium botrys, Eryngium castrense, Eschscholzia californica, Foeniculum vulgare, Frankenia salina, Galium aparine, G. triflorum, Geum macrophyllum, Gnaphalium palustre, Goodyera oblongifolia, Grindelia hirsutula, Hainardia cylindrica, Heliotropium curassavicum, Helminthotheca echioides, Heracleum sphondylium, Heuchera cylindrica, Holcus lanatus, Holodiscus discolor, Hordeum brachyantherum, H. depressum, Hydrocotyle ranunculoides, Hydrophyllum capitatum, Hypochaeris, Isolepis cernua, Iva hayesiana, Jaumea carnosa, Juncus acutus, J. balticus, J. bufonius, J. effusus, J. lesueurii, J. phaeocephalus, J. xiphioides, Kickxia, Lasthenia californica, L. conjugens, L. glaberrima, Lemna minor, L. minuta, Lepidium latifolium, L. oxycarpum, Lilaea scilloides, Lilaeopsis occidentalis, Lilium columbianum, Limnanthes, Limonium, Lobelia cardinalis, Lolium multiflorum, L. perenne, Lonicera involucrata, Lotus corniculatus, L. uliginosus, Ludwigia peploides, Lupinus arboreus, Lysimachia maritima, Lythrum californicum, L. hyssopifolia, Maianthemum racemosum, M. stellatum, Malacothamnus, Malvella leprosa, Matricaria discoidea, Medicago polymorpha, Melica subulata, Melilotus albus, Mentha arvensis, Mesembryanthemum, Mimulus floribundus, M. guttatus, Muhlenbergia rigens, Myosurus minimus, Myrica californica, Najas marina, Nasturtium officinale, Navarretia leucocephala, N. squarrosa, Oenanthe sarmentosa, Orthocarpus, Oryza, Osmorhiza purpurea, Parapholis strigosa, Paxistima myrsinites, Persicaria, Phacelia, Phalaris arundinacea, Phragmites australis, Pinus contorta, P. ponderosa, Plagiobothrys bracteatus, P. leptocladus, P. stipitatus, P. undulatus, Plantago elongata, P. heterophylla, P. lanceolata, P. major, P. maritima, Platanthera dilatata, Platanus racemosa, Plecostachys serpyllifolia, Poa palustris, P. nemoralis, P. pratensis, Pogogyne abramsii, Polypogon monspeliensis, Populus balsamifera, P. fremontii, P. tremuloides, Potentilla anserina, Prosartes hookeri, Prunus emarginata, P. virginiana, Pseudognaphalium microcephalum, P. stramineum, Pseudotsuga menziesii, Psilocarphus, Pteridium aquilinum, Pseudostellaria jamesiana, Psilocarphus brevissimus, P. elatior, Pterospora andromedea, Quercus agrifolia, Q. douglasii, Q. garryana, Ranunculus cymbalaria, R. lobbii, R. sceleratus, Raphanus, Ribes lacustre, Rosa gymnocarpa, R. nutkana, R. woodsii, Rubus leucodermis, R. parviflorus, Rumex conglomeratus, R. crispus, Ruppia maritima, Salicornia depressa, S. rubra, Salix gooddingii, S. hookeriana, S. laevigata, S. lasiolepis, Sambucus nigra, Sarcocornia pacifica, Schedonorus arundinaceus, Schoenoplectus acutus, S. americanus, S. californicus, S. pungens, Scrophularia californica, Senecio vulgaris, Silybum marianum, Sisyrinchium bellum, Solidago canadensis, Sonchus asper, S. oleraceus, Spartina densiflora, Spergula arvensis, Spergularia bocconii, S. canadensis, S. macrotheca, S. salina, Stuckenia pectinata, Suaeda, Symphoricarpos albus, Tamarix, Taraxacum officinale, Toxicodendron diversilobum, Tragopogon dubius, Trifolium depauperatum, T. dubium, T. repens, T. wormskioldii, Triglochin maritimum, T. striata, Triphysaria versicolor, Typha angustifolia, T. domingensis, T. latifolia, Urtica dioica,*

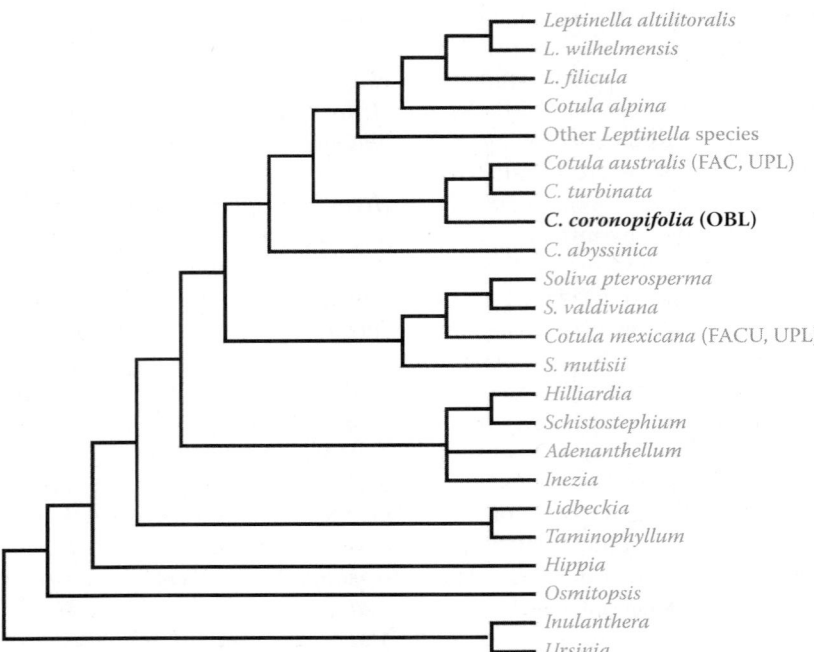

FIGURE 5.148 Phylogeny of Asteraceae tribe Anthemideae generated from combined nrITS and cpDNA sequence data shows a nested placement of *Leptinella* and *Soliva* species within *Cotula*. *Cotula coronopifolia* represents a single derivation of the OBL habit (bold) from the species sampled. (Redrawn from Himmelreich, S. et al., *Mol. Phylogenet. Evol.*, 65, 464–481, 2012.)

Veronica americana, *V. peregrina*, *Vicia*, *Woodsia oregana*, *Xanthium*, *Zannichellia palustris*, *Zostera japonica*.

Use by wildlife: Caterpillars (Insecta: Lepidoptera) have been reported to feed on the capitula of *Cotula coronopifolia*. The plants are eaten readily by domestic livestock, but have been suspected of causing poisoning in grazing animals. Various ducks (Aves: Anatidae: *Anas*) consume the seeds and pintail ducks (*Anas acuta*) have been observed to select habitats rich in *C. coronopifolia* in California. Pure stands of *C. coronopifolia* are associated negatively with occurrences of the rare salt marsh harvest mouse (Mammalia: Rodentia: Cricetidae: *Reithrodontomys raviventris*).

Economic importance: food: not edible; **medicinal:** Aerial portions of plants contain several alkaloids (cotuzine A and B). Extracts also contain 6-methoxy-1-benzofuran-4-ol, a substance with antibacterial and antifungal properties; **cultivation:** 'Cream Buttons' is one cultivar of *Cotula coronopifolia*; **misc. products:** *Cotula coronopifolia* has been considered for use in phosphorous removal from irrigation drainage water. A golden dye has been obtained from the plants; **weeds:** *Cotula coronopifolia* is regarded as a weed in natural wetland systems. Severe growths can cause blockages in canals; **nonindigenous species:** *Cotula coronopifolia* was introduced to North America from South Africa in the 19th century with records dating back to 1862 (West Coast) and 1879 (East Coast). By the early 20th century it had become recognized as a "ballast weed," which implicates accidental disposal of shipping ballast as the original source of propagules.

Systematics: *Cotula* is assigned to tribe Anthemideae of Asteraceae (subfamily Asteroideae), where it is placed within section *Cotula*. Opinions have varied whether to treat *Leptinella* as a section within *Cotula*, or as a distinct genus. However, recent phylogenetic analyses indicate that *Leptinella* species are nested within *Cotula*, along with *Soliva* (Figure 5.148); *Cotula* is not monophyletic unless these other genera are included. Most Anthemideae are $x=9$, with *Cotula* species having $x=8$, 9 or 10. The count of $2n=20$ for *Cotula coronopifolia* indicates that it is diploid.

Comments: *Cotula coronopifolia* occurs mainly along coastal regions throughout western and northeastern North America.

References: Baskin & Baskin, 1998; Casazza et al., 2012; Churchill, 1902; Cook, 1996a; DiTomaso & Healy, 2003; Douglas & Illingworth, 2004; Edgar, 1958; Eicher, 1988; Fritz et al., 2009; Gerhardt & Collinge, 2003; Goodman et al., 2010; Himmelreich et al., 2012; Houghton & Uhlig, 2003; Jefferson, 1974; Kether et al., 2012; Li & Recknagel, 2002; Liouane et al., 2012; Lloyd, 1972; Lloyd & Webb, 1987; McLaughlin, 1974; Partridge & Wilson, 1987; Pegtel, 1998; Pickart, 2006; Powell et al., 1974; Raulings et al., 2011; Rich et al., 2012; Smaoui et al., 2011; Ungar, 1987; Van der Toorn, 1980; Van der Toorn & ten Hove, 1982; Watson, 2006a; Webb, 1998; Widgren, 2003; Woo et al., 2008.

15. Doellingeria

Tall flat-topped aster, white-top

Etymology: after Ignatz Doellinger (1770–1841)

Synonyms: *Aster* (in part); *Diplopappus*; *Triplopappus*

Distribution: global: North America; **North America:** eastern

Diversity: global: 3 species; **North America:** 3 species

Indicators (USA): OBL; FACW: *Doellingeria umbellata*
Habitat: freshwater; palustrine; **pH:** 4.1–7.0; **depth:** <1 m;
life form(s): emergent herb
Key morphology: Stems (to 2 m) erect, single or up to 20;
leaves basal and cauline, alternate, sessile; basal leaves oblanceolate, withering; cauline leaves (to 15 cm) reduced distally,
elliptic, margins entire; heads (3–300+) radiate, peduncled
(to 10 mm), arranged secondarily in corymbs, involucre (to
5 mm) with phyllaries in 3–4 series; ray florets (2–16) pistillate, the rays (to 12.3 mm) white; disc florets (5–50) bisexual,
corollas (to 6 mm) 5-lobed, pale yellow; cypselae (to 3.2 mm)
4–6-ribbed, strigose, outer pappus of subulate scales (to
0.8 mm), inner pappus of 60–90 whitish, barbed bristles (to
6.2 mm).
Life history: duration: perennial (caudex, rhizomes); **asexual reproduction:** rhizomes; **pollination:** insect; **sexual
condition:** gynomonoecious; **fruit:** cypselae (common); **local
dispersal:** rhizomes, cypselae (gravity); **long-distance dispersal:** cypselae (wind)
Imperilment: (1) *Doellingeria umbellata* [G5]; S1 (MO, OK);
S2 (AB, NE, SD); S3 (DE, IA, NC, SK)
Ecology: general: *Doellingeria* species occur on dry or wet
sites ranging from bogs to woodlands and mountainous terrain. Only one species (*D. umbellata*) is a wetland indicator,
and even it is ranked more commonly as FACW than OBL. All
species presumably are self-incompatible and are outcrossed
by insects. Like many radiate composites, *Doellingeria* species are gynomonoecious with pistillate rays florets and
bisexual disc florets. In addition to promoting outcrossing,
this arrangement also has been shown to reduce herbivory
to the ray florets, which is significantly lower than the level
experienced by the disk florets. The dispersal of the cypselae
by wind is facilitated by their prominent pappus bristles. The
seeds are dormant are require a period of cold stratification to
break dormancy in most cases. The plants provide an important source of forage for grazing mammals.

Doellingeria umbellata **(Mill.) Nees** inhabits barrens,
beaver dams, bluffs, bogs, carrs, copses, dikes, ditches, fens,
floodplains, marshes, meadows, mudflats, prairie fens, prairies, roadsides, seeps, shores, sloughs, stream bottoms,
swales, swamps, thickets, and margins of baygalls, flowages,
gravel pits, lakes, rivers, streams, thickets, and woodlands at
elevations of up to 1800 m. The plants grow close to sources
of water, with one study finding them to occur no further than
18 m from watercourse margins. Exposures can vary from
sun to partial shade. The substrates are acidic to neutral
(pH 4.1–7.0) and are described as clay, clay loam, felspathic
granite, gravel, gravelly humus, humus, limestone, loam,
loamy sand, loamy till, mafic, Miami silt loam, muck, mud,
peat, quartzite talus, rock, sandy gravel, sandy leaf mold,
sandy loam, sandy muck, and sandy peat. The plants reportedly are self-incompatible, with less than 0.1% seed set resulting from self-pollinations. Flowering proceeds from late
summer into fall, with an average production of 34–38 flowers per head. Generally, larger numbers of flowers are produced by heads that develop earlier in the season. Plants
growing in darker sites produce significantly higher mean

numbers of flowers (32–36.5) than those in better-illuminated
sites (29.5). Disc florets are produced in a roughly 2.5–3.8:1
ratio to ray florets. The flowers are pollinated by various
insects (see *Use by wildlife*), which typically yields 11 (6 ray,
5 disc; rarely as many as 20) cypselae per head. The cypselae
are dispersed mainly by gravity or the wind. They remain
viable for more than 135 days, with natural germination
occurring in early spring. Cypselae have germinated successfully after 1 week of air drying followed by 20 weeks of dark,
cold (−4°C) stratification and subsequent exposure to 24°C
under a 12 h photoperiod. Observed germination rates are
relatively low (7%–23%) but are 2.5–3 times higher if the
pappus is removed from the fruit. Relative to other Asteraceae,
the cypselae are fairly heavy, weighing an average of 0.6–
0.8 mg and sometimes reaching 1.1 mg. The overall germination rate is higher for heavier cypselae, but weight differences
among germinable fruits are not correlated with germination
success. Plants can be common in the standing vegetation, yet
absent from the seed bank. They have been observed to attain
a mean above-ground biomass of 8.6 g/m^2. Although not
known to be particularly fire tolerant, the plants occur often
in areas of frequent fires or recent burns. The plants perennate by an enlarged caudex and can form moderate clones by
the formation of short to long rhizomes. Up to 58% of the
roots have been observed to be colonized by arbuscular
mycorrhizal and dark septate endophytic fungi. Foliar extracts
can be allelopathic, and are known to inhibit seed germination and seedling shoot/root growth of black cherries
(Rosaceae: *Prunus serotina*). **Reported associates:** *Abies
balsamea, Acer pensylvanicum, A. rubrum, A. saccharum,
Achillea millefolium, Agalinis purpurea, Agrimonia parviflora, Agrostis perennans, A. scabra, Alisma, Alnus rugosa,
A. viridis, Ambrosia trifida, Amianthium muscitoxicum,
Andromeda polifolia, Andropogon gerardii, Anticlea elegans, Apocynum cannabinum, Aralia racemosa, Arisaema
triphyllum, Arnoglossum plantagineum, Aronia arbutifolia,
A. melanocarpa, Asclepias incarnata, Athyrium filix-femina,
Avenella flexuosa, Betula alleghaniensis, B. papyrifera, B.
populifolia, B. pumila, B. ×sandbergii, Bidens cernuus, B.
trichospermus, Boehmeria cylindrica, Brachyelytrum erectum, Bryum pseudotriquetrum, Calamagrostis canadensis,
C. coarctata, C. stricta, Calla palustris, Caltha palustris,
Campanula aparinoides, Campylium stellatum, Cardamine
bulbosa, Carex aquatilis, C. atlantica, C. buxbaumii, C.
cryptolepis, C. debilis, C. digitalis, C. echinata, C. flava, C.
garberi, C. gynandra, C. haydenii, C. hystericina, C. interior,
C. intumescens, C. lacustris, C. lasiocarpa, C. leptalea, C.
livida, C. lucorum, C. lurida, C. magellanica, C. nebrascensis,
C. oligosperma, C. pellita, C. pensylvanica, C. polymorpha, C.
prairea, C. scoparia, C. sterilis, C. stricta, C. torta, C. trisperma, C. utriculata, C. viridula, C. vulpinoidea, Carpinus
caroliniana, Cephalanthus occidentalis, Chamaedaphne
calyculata, Chelone glabra, Cicuta bulbifera, C. maculata,
Cinna latifolia, Cirsium muticum, Cladium mariscoides,
Cladonia, Clematis virginiana, Clethra alnifolia, Climacium
dendroides, Comarum palustre, Comptonia peregrina,
Cornus amomum, C. canadensis, C. foemina, C. sericea,*

Corylus, Cuscuta compacta, C. pentagona, Cyperus strigosus, Danthonia compressa, D. spicata, Dasiphora floribunda, Dennstaedtia punctilobula, Dichanthelium acuminatum, D. clandestinum, Dicranum, Drepanocladus, Drosera rotundifolia, Dulichium arundinaceum, Eleocharis acicularis, E. elliptica, E. melanocarpa, E. tuberculosa, E. wolfii, Elymus lanceolatus, E. riparius, E. virginicus, Epilobium leptophyllum, E. strictum, Equisetum arvense, E. fluviatile, Eriophorum angustifolium, E. virginicum, Eupatorium perfoliatum, Eurybia macrophylla, Euthamia graminifolia, Eutrochium fistulosum, E. maculatum, Fagus grandifolia, Fallopia japonica, F. scandens, Festuca ovina, Filipendula rubra, Fragaria virginiana, Fraxinus americana, F. nigra, Galium asprellum, G. trifidum, Gaultheria hispidula, Gaylussacia baccata, G. frondosa, Gentiana andrewsii, Gentianopsis virgata, Geum aleppicum, Glyceria canadensis, G. grandis, G. melicaria, G. septentrionalis, Gymnadeniopsis clavellata, Hamamelis virginiana, Helenium autumnale, Helianthus angustifolius, H. giganteus, H. grosseserratus, Hieracium robinsonii, Holcus lanatus, Hydrocotyle americana, Hypericum ascyron, H. kalmianum, H. mutilum, H. nudiflorum, Ilex mucronata, I. opaca, I. verticillata, Impatiens capensis, Iris versicolor, I. virginica, Itea virginica, Juncus canadensis, J. dichotomus, J. diffusissimus, J. effusus, J. marginatus, J. nodatus, J. secundus, Kalmia angustifolia, K. latifolia, Larix laricina, Leersia oryzoides, Liatris aspera, L. ligulistylis, L. pycnostachya, L. spicata, Lilium philadelphicum, Limprichtia revolvens, Liparis loeselii, Liquidambar styraciflua, Lobelia cardinalis, L. kalmii, L. siphilitica, Ludwigia, Luzula campestris, Lycopodium obscurum, Lycopus americanus, L. uniflorus, L. virginicus, Lygodium palmatum, Lyonia ligustrina, L. mariana, Lysimachia asperulifolia, L. ciliata, L. quadriflora, L. terrestris, L. thyrsiflora, Lythrum alatum, L. salicaria, Magnolia virginiana, Maianthemum canadense, M. trifolium, Matteuccia struthiopteris, Medeola virginiana, Mentha arvensis, Menyanthes trifoliata, Micranthes pensylvanica, Mimulus ringens, Mitchella repens, Muhlenbergia glomerata, M. mexicana, M. richardsonis, Myrica gale, Oclemena acuminata, Onoclea sensibilis, Osmunda claytoniana, O. regalis, Osmundastrum cinnamomeum, Oxalis montana, Oxypolis rigidior, Packera paupercula, Panicum capillare, Parnassia glauca, P. palustris, Pedicularis lanceolata, Persicaria lapathifolia, P. maculosa, P. punctata, P. sagittata, Phalaris arundinacea, Philonotis fontana, Phlox maculata, Phragmites australis, Picea glauca, P. mariana, P. rubens, Pilea pumila, Pinus banksiana, P. resinosa, P. rigida, P. strobus, Piptatherum racemosum, Platanthera dilatata, P. flava, Platanus occidentalis, Pogonia ophioglossoides, Polytrichum commune, P. juniperinum, Populus balsamifera, P. deltoides, P. tremuloides, P. grandidentata, Potentilla simplex, Prunus pensylvanica, P. serotina, Pteridium aquilinum, Pycnanthemum virginianum, Quercus ilicifolia, Ranunculus hispidus, Rhamnus alnifolia, R. frangula, Rhododendron canadense, R. groenlandicum, R. maximum, R. viscosum, Rhynchospora alba, R. capillacea, R. capitellata, Rosa palustris, Rubus allegheniensis, R. hispidus, R. idaeus, R. pubescens, R. repens, Rudbeckia laciniata, Rumex acetosella, R. altissimus, Sagittaria latifolia, Salix ×bebbii, S. candida, S. cordata, S. discolor, S. eriocephala, S. petiolaris, S. sericea, S. serissima, Sanguisorba canadensis, Sarracenia purpurea, Schizachyrium scoparium, Schoenoplectus acutus, S. pungens, Scirpus atrovirens, S. cyperinus, S. expansus, Scleria verticillata, Scorpidium scorpioides, Scutellaria galericulata, S. lateriflora, Silphium terebinthinaceum, Smilax glauca, Solidago altissima, S. canadensis, S. gigantea, S. nemoralis, S. ohioensis, S. patula, S. riddellii, S. rugosa, S. uliginosa, Sorbus americana, Spartina pectinata, Sphagnum fallax, S. girgensohnii, S. magellanicum, S. palustre, S. recurvum, Spiraea alba, S. tomentosa, Spiranthes cernua, S. lucida, S. romanzoffiana, Stachys, Streptopus lanceolatus, Symphyotrichum boreale, S. cordifolium, S. dumosum, S. firmum, S. lanceolatum, S. novae-angliae, S. prenanthoides, S. puniceum, Symplocarpus foetidus, Thalictrum dasycarpum, T. pubescens, Thelypteris noveboracensis, T. palustris, Thuja occidentalis, Tomentypnum nitens, Toxicodendron radicans, T. vernix, Triadenum fraseri, T. virginicum, Triantha glutinosa, Trichophorum alpinum, Trientalis borealis, Triglochin maritimum, T. palustre, Trientalis borealis, Tsuga canadensis, Tussilago farfara, Typha angustifolia, T. ×glauca, T. latifolia, Ulmus americana, Urtica dioica, Uvularia grandiflora, Vaccinium angustifolium, V. corymbosum, V. oxycoccos, Valeriana edulis, Verbena hastata, Vernonia fasciculata, V. noveboracensis, Veronicastrum virginicum, Viburnum dentatum, V. nudum, Viola cucullata, V. sororia, Vitis labrusca, Woodwardia areolata, Zigadenus glaberrimus, Zizia aurea.

Use by wildlife: The cypselae of *Doellingeria umbellata* are eaten by moth larvae (Insecta: Lepidoptera: Coleophoridae: *Coleophora bidens*) and by various birds (Aves) including the eastern goldfinch (Fringillidae: *Carduelis tristis*), ruffed grouse (Phasianidae: *Bonasa umbellus*), swamp sparrow (Emberizidae: *Melospiza georgiana*), and wild turkey (Phasianidae: *Meleagris gallopavo*). The plants are eaten by beetles (Insecta: Coleoptera: Curculionidae: *Anthonomus lecontei*; Chrysomelidae: *Microrhopala excavata*) and are grazed by caribou (Mammalia: Cervidae: *Rangifer tarandus*) and Rocky Mountain elk (Mammalia: Cervidae: *Cervus elaphus nelsoni*). They are a minor food of meadow voles (Mammalia: Rodentia: Muridae: *Microtus pennsylvanicus*) and also provide food for larval cutworms (Insecta: Lepidoptera: Noctuidae: *Allagrapha aerea, Cucullia florea, C. postera, Papaipema impecuniosa*), Harris'/silvery checkerspot, and northern/pearl crescent butterflies (Insecta: Lepidoptera: Nymphalidae: *Chlosyne harrisii* [occurring only on *D. umbellata*], *C. nycteis*; *Phyciodes cocyta, P. tharos*). The roots are fed upon by larval clearwing moths (Insecta: Lepidoptera: Sesiidae: *Carmenta corni*). The plants also are a host of aphids (Insecta: Hemiptera: Aphididae: *Macrosiphum ambrosiae, M. rudbeckiae; Uroleucon olivei, U. paucosensoriatum*) and to several fungi including powdery mildew (Ascomycota: Erysiphaceae: *Golovinomyces asterum*) and other plant pathogens (Ascomycota: Botryosphaeriaceae: *Phyllosticta astericola*; Mycosphaerellaceae: *Cercosporella virgaureae, Pseudocercospora quarta*; Pleosporales:

Ascochyta compositarum). The flowers contain nectar with an average of 0.003 mg of sugar and are visited by various bees (Insecta: Hymenoptera: Apidae: *Apis mellifera*; *Bombus citrinus*; Colletidae: *Hylaeus mesillae*; Halictidae: *Augochlora aurata*, *A. pura*, *Lasioglossum cressonii*, *L. fuscipenne*; Megachilidae: *Megachile relativa*), beetles (Coleoptera: Nitidulidae: *Carpophilus brachypterus*), butterflies and moths (Insecta: Lepidoptera: Noctuidae: *Oncocnemis piffardi*; Nymphalidae: *Danaus plexippus*), and small flies (Insecta: Diptera), which all are likely pollinators as they forage for pollen and nectar.

Economic importance: food: *Doellingeria umbellata* is not reported to be edible; **medicinal:** The plants contain several sesquiterpene lactones (e.g., eudesmanolides and selinene and tremetone derivatives), which have been credited with a variety of anti-inflammatory, antirheumatic, expectorant, perspiration inducing and menstruation easing properties. The Mohegans used a leaf infusion of *Doellingeria umbellata* to treat stomach disorders; **cultivation:** *Doellingeria umbellata* is cultivated as a garden ornamental and sometimes is distributed under the cultivar name 'Weisser Schirm'; **misc. products:** The Potawatomi smudged flowers of *Doellingeria umbellata* to protect the sick from evil spirits. The seeds are included in some mixes used for revegetation of woodland margins and temporarily flooded roadside habitats. It is regarded as an important honey plant. The foliage contains large amounts of hydrocarbons (0.8%) such as polyisoprenes (rubber). This species is represented by one of the remarkable Blaschka glass floral models of the Ware collection displayed at Harvard University; **weeds:** none; **nonindigenous species:** none

Systematics: In Asteraceae, *Doellingeria* is placed within the large tribe Astereae, where it occupies a position between the Southern Hemisphere grade and the North American clade, being the sister genus to the latter. Formerly, distinctive leaf and inflorescence characters distinguished *Doellingeria* species as a section (section *Doellingeria*) of *Aster* in the broad, former circumscription of that genus. Molecular systematic studies eventually showed the group to occupy a discrete phylogenetic position at the base of the North American Astereae clade (Figure 5.149), and it is now widely recognized as a distinct genus. The OBL habit of *D. umbellata* represents an independent origin in the group (Figure 5.149). Two of the three *Doellingeria* species have been included in molecular phylogenetic analyses, which support the monophyly of the genus. Intergeneric hybrids between *Doellingeria umbellata* and *Oclemena nemoralis* have been reported, but are doubtful. The basic chromosome number of *Doellingeria* is $x=9$; *D. umbellata* ($2n=18$) is a diploid.

Comments: *Doellingeria umbellata* occurs across eastern North America.

References: Anderson, 1919; Andrus, 1988; Barney et al., 2007; Bartram, 1913; Bergerud, 1972; Bertin & Kerwin, 1998; Bertin et al., 2010; Bohlmann et al., 1980; Braun, 1940; Bucyanayandi & Bergeron, 1990; Carr et al., 1986; Chmielewski, 1999; Chmielewski & Ruit, 2002; Chmielewski & Strain, 2007; Feist et al., 2008; Ferguson, 1975; Greene,

1945, 1956; Greenfield & Karandinos, 1979; Grundel et al., 2011; Hagan et al., 2006; Heinrich, 1976; Horsley, 1977; Jones, 1978; Jost et al., 1999; Kent, 1908; Leonard, 1936; MacDonagh, 2010; Majka et al., 2007; Marr et al., 2009; McCabe, 1985; McKenzie et al., 2009; Moran, 1984; Morse, 2005; Neck, 1976; Nesom, 2001; Noyes & Rieseberg, 1999; Pellett, 1920; Pogue, 2005; Quattrocchi, 2012; Robinson et al., 2002; Selliah & Brouillet, 2008; Semple & Brouillet, 1980; Semple & Chmielewski, 2006; Strickler et al., 1996; Tolstead, 1942; Weishampel & Bedford, 2006; Williams et al., 1999a; Wiser, 1998; Zavitkovski, 1976.

16. *Eclipta*

False daisy, Yerba de Tajo; éclipte

Etymology: from the Greek *eklipês* ("omitted"), pertaining to the lack of a pappus

Synonyms: *Verbesina* (in part)

Distribution: global: Africa*; Australia; Eurasia*; New World; **North America:** eastern and southern

Diversity: global: 6 species; **North America:** 1 species

Indicators (USA): OBL; FACW; FAC: *Eclipta prostrata*

Habitat: freshwater or brackish (coastal); palustrine, riverine; **pH:** 4.2–6.8; **depth:** <1 m; **life form(s):** emergent herb

Key morphology: Stems (to 80 cm) weak, erect or decumbent, rooting adventitiously at the nodes; leaves (to 10 cm) opposite, lanceolate, sessile or short-petioled, margins entire to irregularly serrate, surfaces pilose (hairs to 0.6 mm), tapering at base; peduncles (to 7 cm) axillary or terminal, solitary or in groups (to 3); capitula radiate, involucre (to 5 mm) hemispherical, phyllaries (to 6 mm) 5–12 in two rows, spreading in fruit; ray florets (to 3 mm) numerous (to 70), in 2–3 series, pistillate or neuter, ligules white, 0–2-lobed; disc florets (to 2.0 mm) numerous (to 30+), bisexual, corolla white, 4–5-lobed, pales (to 2.6 mm) linear, pectinate; cypselae (to 2.3 mm) brown or black, elongate and wedge shaped, weakly 3–4-angled, flattened, margins thickened and cartilaginous, surfaces tuberculate or warty; pappus absent or cup-like, ciliate with 2–3 teeth.

Life history: duration: annual (fruits/seeds); perennial (persistent stems); **asexual reproduction:** shoot fragments; **pollination:** insect; **sexual condition:** gynomonoecious; **fruit:** cypselae (common); **local dispersal:** seeds (water, wind); **long-distance dispersal:** seeds (water, wind)

Imperilment: (1) *Eclipta prostrata* [G5]; SH (WI); S1 (MI, NY) S2 (ON)

Ecology: general: *Eclipta* is a small genus of annuals and weak perennials that grow in damp grasslands or in wet or intermittently flooded sites including the margins of lakes and watercourses. Localities include both freshwater inland and saline coastal sites. Substrates range from heavy clay soils to loam. Some species are better adapted to aquatic conditions where they are able to extend their shoots through the water column to produce emergent flowering branches. The stems root adventitiously at the nodes. The seeds are not dormant. Most species occur in Australia and South America. The globally distributed *E. prostrata* is the only weedy and the only OBL member of the genus.

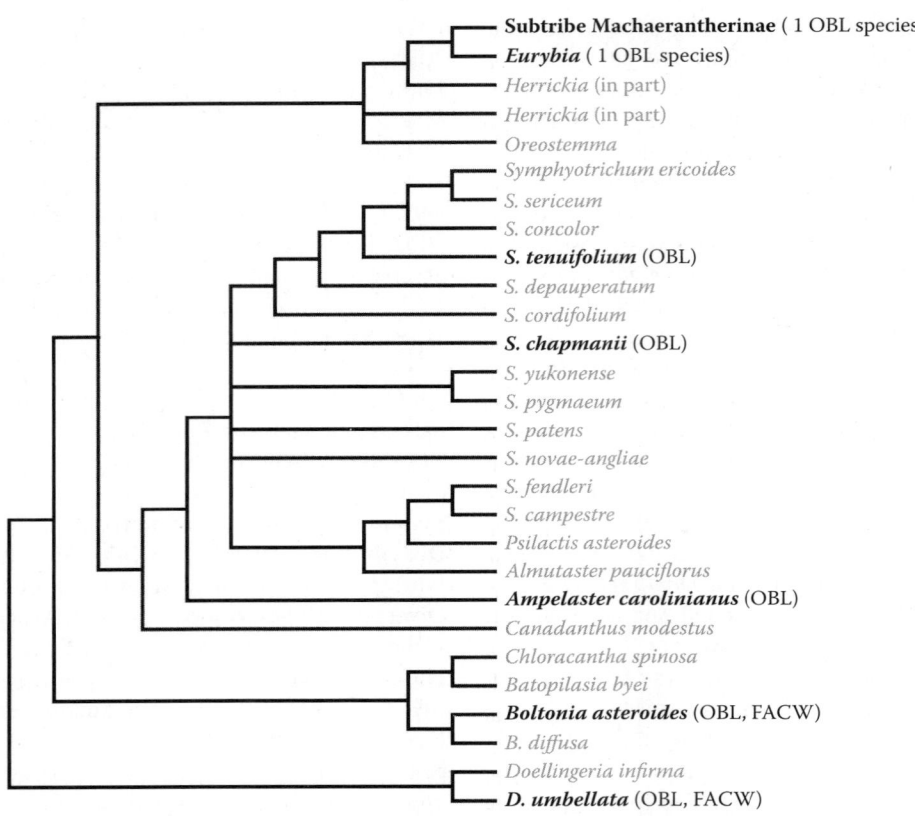

FIGURE 5.149 Phylogenetic tree constructed from ITS sequence data showing position of *Doellingeria* as the sister genus to the North American clade of Asteraceae tribe Astereae. OBL taxa (shown in bold) have evolved repeatedly within this clade of Astereae. (Adapted from Selliah, S. & Brouillet, L., *Botany*, 86, 901–915, 2008.)

Eclipta prostrata **(L.) L.** is an annual or short-lived perennial, which colonizes alluvial fans, basins, beaches, borrow pits, canals, channels, cypress domes, depressions, dikes, ditches, dunes, flats, flatwoods, floodplains, hammocks, levees, low woods, marshes, mudflats, pools, prairies, rice fields, river bottoms, sandbars, scrub, seeps, shoals, streams, swales, swamps, washes, and the margins of canals, lakes, pools, rivers, sloughs, streams, and tide pools at elevations of up to 2600 m. Plants can grow on wet sediments or in shallow (e.g., 15 cm) standing water. They also are common in disturbed sites where water has receded. Exposures range from full sun to partial shade. The substrates are described as adobe, adobe clay, calcareous talus, clay, clay loam, cobble, granite, gravel, loamy sand, muck, mud, rock, sand, sandy clay, sandy loam, sandy peat, silt, and silty clay. The plants are tolerant to salinity, with individuals originating from saline sites exhibiting higher tolerance to salinity-induced wilting than plants from nonsaline sites. They grow well across a wide range of acidity (pH 4.2–6.8), with higher pH conditions generally producing somewhat larger individuals. Photoperiodically, *E. prostrata* essentially is a qualitative short-day plant, with an absolute requirement of short days to induce flowering. However, some plants originating from different geographical regions exhibit quantitative short day or day-neutral flowering responses. Flowering initiates in the

5th week after seedlings emerge and fruits begin to set in the 6th week. Despite the voluminous literature on this species, the only potential pollinators reported have been small insects (Insecta) such as thrips (Thysanoptera) and ants (Hymenoptera: Formicidae). The "type II" pollen of *E. prostrata* has a spiny, echinate, outer surface (tectum), which presumably is adapted for insect pollination. *Eclipta prostrata* is both self- and cross-compatible, but higher seed production occurs when flowers are outcrossed by insects. Several adaptations promote outcrossing. The florets become receptive asynchronously with the outer (pistillate) ray florets opening before the protandrous hermaphroditic disc florets. Single capitula can produce 100–200 cypselae. Seed production varies among three main morphotypes, with erect plants producing the highest average number per capitulum (83±6.55), semierect plants having intermediate cypselae production (72±7.54), and prostrate plants producing the fewest cypselae (59±3.0). An individual plant can produce as many as 14,000 cypselae, with fruit production increasing from the 6th to 10th week after seedling emergence. Seed rains of 13–534/m² have been observed. Cypselae mature fully in 10–14 days. The seeds of *E. prostrata* are particularly adapted to germinate in warm, wet soils. They can germinate under salt stress (<3% seawater; 0–120 mM NaCl) but are highly sensitive to water stress (<−0.4 MPa). Rapidly dehydrated seeds retain high germination rates, but slow dehydration reduces their viability substantially. They have no dormancy or after ripening requirements but retain viability when stored at room temperature (for more than 5 months) and at 4.4°C (50% humidity). Seed germination is inhibited completely by the dark. Generally, germination occurs in the light at levels from 6% to 100% of full sunlight at temperatures from 10°C to 35°C. There is an inverse relationship between cypsela weight and time to germination. Optimal germination temperatures have been reported as 15°C and 25°C. It is highest (83%) at the upper extreme of the range and can reach 93% under alternating 30°C/20°C day/night temperatures. High germination rates (87%–93%) have been observed from pH 2.2–10, with an optimum at pH 5–7. Up to 89% germination occurs in distilled water (pH 5.5) or in solutions buffered from pH 6 to 7. Up to 80% germination can occur when seed is heated in an oven for 5 min at140°C then followed by a 14-day incubation at 30°C/20°C. Germination is highest at the soil surface and does not occur in seed buried at 2.5cm. In field conditions, seedling emergence also is highest (83%) for seeds on the surface and declines to zero at depths of only 0.5cm. Seed germination is unaffected following a 28 day exposure to the antimicrobial triclosan at 0.4–1000 ppb. Inoculation by arbuscular mycorrhizal fungi (which occurs maximum on drier sediments) results in higher shoot/root ratios; uninoculated plants form longer roots with greater surface area. The seeds often germinate in seedbank studies and densities up to 566/m² have been reported. Their maximum duration in the seed bank has not been determined, but presumably is short. Shallow flooding (10 cm water) reduces growth in young seedlings, but stimulates growth in older seedlings. The plants often form mats, which can reach 1m in extent. They have been micropropagated successfully by culturing cotyledonary nodes on Murashige and Skoog (MS) medium combined with 4.4 mM benzyl adenine, 4.6mM kinetin, 4.9mM 2-isopentenyladenine, 1.4mM gibberellic acid, 5% coconut water, and 3% sucrose. Optimal rooting (94.3%) occurred on full strength MS medium containing 9.8mM indole butyric acid. The tissues accumulate heavy metals (Fe, Mn, Zn), which should be taken into consideration when selecting collecting sites for medicinal uses. Kaur (2011) provided a comprehensive summary of life history information for this species. **Reported associates:** *Abutilon theophrasti, Acacia, Acer negundo, A. rubrum, A. saccharinum, Acmispon americanus, Ageratina adenophora, Ailanthus altissima, Alnus rhombifolia, Alternanthera philoxeroides, Amaranthus californicus, Ambrosia monogyra, Ammannia coccinea, Artemisia californica, Arundo donax, Atriplex argentea, A. canescens, A. hymenelytra, Azolla caroliniana, Baccharis salicifolia, B. sarothroides, Bebbia, Betula nigra, Bidens frondosus, B. laevis, Boehmeria cylindrica, Bolboschoenus maritimus, Caesalpinia pulcherrima, Calamagrostis canadensis, Calibrachoa parviflora, Calyptocarpus vialis, Cardamine pensylvanica, Carex longii, Carya illinoinensis, Celtis laevigata, C. reticulata, Cenchrus setaceus, Cephalanthus occidentalis, Chenopodium album, Cirsium vulgare, Clinopodium brownei, Coleataenia anceps, Colocasia esculenta, Conium maculatum, Conyza canadensis, Cressa truxillensis, Crypsis schoenoides, Cryptantha, Cynodon dactylon, Cyperus difformis, C. eragrostis, C. erythrorhizos, C. involucratus, C. odoratus, C. polystachyos, C. pygmaeus, C. surinamensis, C. virens, Daucus pusillus, Dicoria canescens, Digitaria sanguinalis, Diodia virginiana, Diospyros virginiana, Dysphania ambrosioides, Echinochloa colona, E. crus-galli, Eichhornia crassipes, Eleocharis geniculata, E. parishii, Encelia farinosa, Epilobium canum, E. ciliatum, Equisetum, Eragrostis cilianensis, Erechtites hieraciifolius, Eriochloa villosa, Erodium texanum, Eucalyptus, Eupatorium compositifolium, Euthamia occidentalis, Ficus carica, Foeniculum vulgare, Fraxinus pennsylvanica, Fraxinus velutina, Galium tinctorum, Glandularia pulchella, Glinus lotoides, Gnaphalium palustre, Hedyotis, Helianthus annuus, Heliotropium curassavicum, Heteranthera dubia, Heterotheca subaxillaris, Hirschfeldia incana, Hydrocotyle ranunculoides, H. umbellata, Hymenoclea sandersonii, Hypericum mutilum, Ilex glabra, Impatiens, Ipomoea coccinea, Juncus acutus, J. diffusissimus, J. dubius, J. effusus, J. tenuis, Justicia americana, Lactuca saligna, Larrea tridentata, Leersia oryzoides, Lemna minor, Lepidium, Lepidospartum squamatum, Leptochloa fusca, L. viscida, Leucophyllum frutescens, Leucospora multifida, Ligustrum sinense, Limnobium spongia, Lindernia dubia, Liquidambar styraciflua, Lobelia cardinalis, Ludwigia peploides, Lycium, Lysimachia nummularia, Lythrum californicum, Malosma laurina, Malva parviflora, Malvella leprosa, Melilotus alba, Mentha arvensis, Mentzelia laevicaulis, Microstegium vimineum, Mikania scandens, Mimulus alatus, M. cardinalis, Modiola caroliniana, Mollugo verticillata, Morus alba, Murdannia keisak, Myrica cerifera, Nerium oleander, Nicotiana glauca, N. obtusifolia, Nyssa, Onoclea sensibilis, Oxypolis filiformis, Panicum capillare, P. dichotomiflorum, Parkinsonia, Paspalum dilatatum, P. distichum, Penthorum sedoides, Perityle, Persicaria lapathifolia, P. punctata, P. virginiana, Peucephyllum schottii, Phyla lanceolata, P. nodiflora, Physalis angulata, Pilea pumila, Pinus edulis, Pistia stratiotes, Platanus occidentalis, P. wrightii, Pluchea odorata, P. sericea, Poa annua, Polanisia dodecandra, Polygonum aviculare, Polypogon monspeliensis, P. viridis, Populus fremontii, Potamogeton, Prosopis glandulosa, P. velutina, Pulicaria paludosa, Quercus agrifolia, Q. phellos, Ranunculus sceleratus, Rhus integrifolia, Rhynchospora, Ricinus communis, Rorippa, Rumex, Sabal palmetto, Sagittaria filiformis, S. latifolia, Salix caroliniana, S. exigua, S. gooddingii, S. lasiolepis, S. lucida, S. nigra, Salsola, Saururus cernuus, Schoenoplectus californicus, S. pungens, S. tabernaemontani, Scirpus cyperinus, Sesbania, Sesuvium verrucosum, Setaria magna, Sisymbrium irio, Solanum americanum, Solidago caesia, S. gigantea, Sorghum halepense, Spirodela polyrrhiza, Stemodia durantifolia, Stuckenia pectinata, Symphyotrichum subulatum, Tamarix aphylla, T. chinensis, T. ramosissima, Taxodium distichum, Toxicodendron radicans, Tribulus terrestris, Typha domingensis, T. latifolia, Ulmus parvifolia, U. rubra, U. serotina, Urtica dioica, Utricularia, Verbena* *brasiliensis, Veronica anagallis-aquatica, Viburnum prunifolium, Vitex agnus-castus, V. vulpina, Washingtonia filifera, Xanthium spinosum, X. strumarium, Zizaniopsis miliacea.*

Use by wildlife: Perhaps because of its potent chemical composition, reports of phytophagous invertebrates associated with *Eclipta prostrata* are scarce. However, despite having antifungal properties, it does host a number of pathogenic plant fungi including several Ascomycota (Botryosphaeriaceae: *Phyllosticta*; Diaporthaceae: *Phomopsis*; Glomerellaceae: *Colletotrichum*; Mycosphaerellaceae: *Cercospora*; Pleosporaceae: *Alternaria*; Sclerotiniaceae: *Sclerotinia minor*) and Basidiomycota (Cystofilobasidiaceae: *Itersonilia perplexans*). The plants also can become infected by the *Alternanthera* yellow vein virus (AlYVV). *Eclipta prostrata* has been used as a food additive for domestic livestock because it stimulates their growth.

Economic importance: food: The leaves of *Eclipta prostrata* have been cooked as a vegetable; however, ingestion of this plant is not recommended given its potent medicinal properties; **medicinal:** *Eclipta prostrata* has been used extensively as a source of medicinals, although much of the following information has been obtained from tests conducted on laboratory rodents and may not apply similarly to humans (except where noted). Alcoholic extracts exhibit lipid-lowering activity. Human patients taking 3 grams of powdered plant material daily for 60 days showed a significant decline in blood lipids and triglycerides and lower blood pressure. Aqueous and hydroalcoholic extracts have activity as stress-reducing agents. Alkaloids extracted from the plants are analgesic. Leaf suspensions have potent antihyperglycemic activity. Ethyl acetate extracts of *E. prostrata* contain the coumestans wedelolactone and demethyl-wedelolactone, which exhibit antihepatotoxic activity and stimulate regeneration of liver cells. Wedelolactone is a potent 5-lipoxygenase-inhibitor (which acts to reduce inflammation) and also inhibits HIV-1 integrase. Because the plants also contain 5-hydroxymethyl-(2,2′:5′,2″)-terthienyl tiglate, which inhibits HIV-1 protease, they show promise for the treatment of AIDS patients. Ethanolic extracts exhibit antidiabetic activity by inhibiting alpha-glucosidase and aldose reductase. Extracts even have shown to be effective at promoting hair growth. Fractionated ethanolic extracts contain effective antibacterial properties against *Salmonella typhi*. Plant extracts are strongly inhibitory to various yeast fungi (Ascomycota: Saccharomycetaceae: *Candida albicans, C. tropicalis*; Basidiomycota: *Rhodotorula glutinis*). Aqueous and ethanolic extracts inhibit growth of pathogenic spirochetes (Spirochaetes: Leptospiraceae: *Leptospira interrogans*). Ethanolic extracts of the aerial foliage contain sitosterol, stigmasterol and wedelolactone, which effectively neutralize the lethal properties of rattlesnake (Reptilia: Viperidae: *Crotalus*) venom and inhibit muscular release of creatine kinase; extracts also are effective at neutralizing venom toxicity from other crotalid snakes (Reptilia: Viperidae: *Bothrops jararaca, B. jararacussu; Lachesis muta*). Methanolic extracts generally have been shown to elevate immune responses and aqueous extracts also enhance immune responses and disease resistance in Tilapia

(Osteichthyes: Cichlidae: *Oreochromis mossambicus*) when administered as a food supplement. Avian coccidiosis, a parasitic disease of poultry can be prevented by administering a 120 ppm dose of *E. prostrata* coumestans. Dasyscyphin C extracted from the leaves effectively inhibits the proliferation of fish nodavirus (FNV) at 20 µg/mL. Shoot and root extracts are toxic to root-knot nematodes (Nematoda: Heteroderidae: *Meloidogyne graminicola, M. javanica*); **cultivation:** *Eclipta prostrata* is not in horticultural cultivation; **misc. products:** A variety of cotton dyes has been obtained from the leaves of *Eclipta prostrata. Eclipta* species also produce a permanent golden-yellow dye used to color hair. Larvae of mosquitoes (Diptera: Culicidae: *Culex quinquefasciatus*) exposed for 3 days to 1%–5% aqueous extracts of *E. prostrata* are completely killed, making the plants a potential source of natural insecticide. *Eclipta* leaves have been used in the biosynthesis of silver (Ag) nanoparticles. The nanoparticles are larvicidal with the potential to act as a natural control agent for several mosquitoes (Diptera: Culicidae: *Anopheles subpictus; Culex tritaeniorhynchus*). Extracts from the plants have been used as an additive to inhibit corrosion on steel exposed to hydrochloric acid; **weeds:** *Eclipta prostrata* is reported as a major weed of 17 other crops (e.g., cotton, peanuts, rice) in 35 countries, especially southeast Asia and other regions of high rainfall. It also interferes with the production of container-grown landscape plants; **nonindigenous species:** *Eclipta prostrata* was introduced to Japan around 1948. It was introduced to France as a rice field weed and was first discovered in Bulgaria in 2006. Introductions also have occurred in Italy, Portugal, and Spain and throughout Asia, often through rice cultivation. It has been naturalized in Hawaii since 1871.

Systematics: *Eclipta* is assigned to tribe Heliantheae subtribe Ecliptinae, which cpDNA and nrITS data demonstrate is polyphyletic. Specifically, *Eclipta* associates in a clade comprising most Mesoamerican lowland genera of the subtribe, where it resolves fairly closely to *Tilesia* (Figure 5.150). There it is basal to the "wedelioid group," a clade that contains genera having winged cypselae and a constricted, crownlike pappus from which bristles arise. In any case, *Eclipta* occupies a relatively isolated position in the subtribe. *Eclipta prostrata* formerly was known by the name *E. alba* (L.) Hassk., to which much of the literature still refers. Although most authors have merged these taxa, some authors retain the two species as distinct, and morphologically distinguish *E. prostrata* (multi-cellular hairs; obovate achenes) from *E. alba* (four-celled hairs; angular obovate achenes). There also have been three distinct phenotypes ("morphotypes") recognized for *E. alba* and RAPD markers have characterized some populations as substantially divergent genetically from others. A further taxonomic evaluation of this species (and additional genetic data) seems necessary to determine the correct number and circumscription of taxa assigned to this name. The chiasma frequency remains higher yearlong for the prostrate morphotype than for the erect or intermediate morphotypes. All morphotypes share the same chromosome number ($2n = 22$), which indicates that they are uniformly diploid. However, aneuploid derivatives ($2n = 18, 20, 24$) occasionally

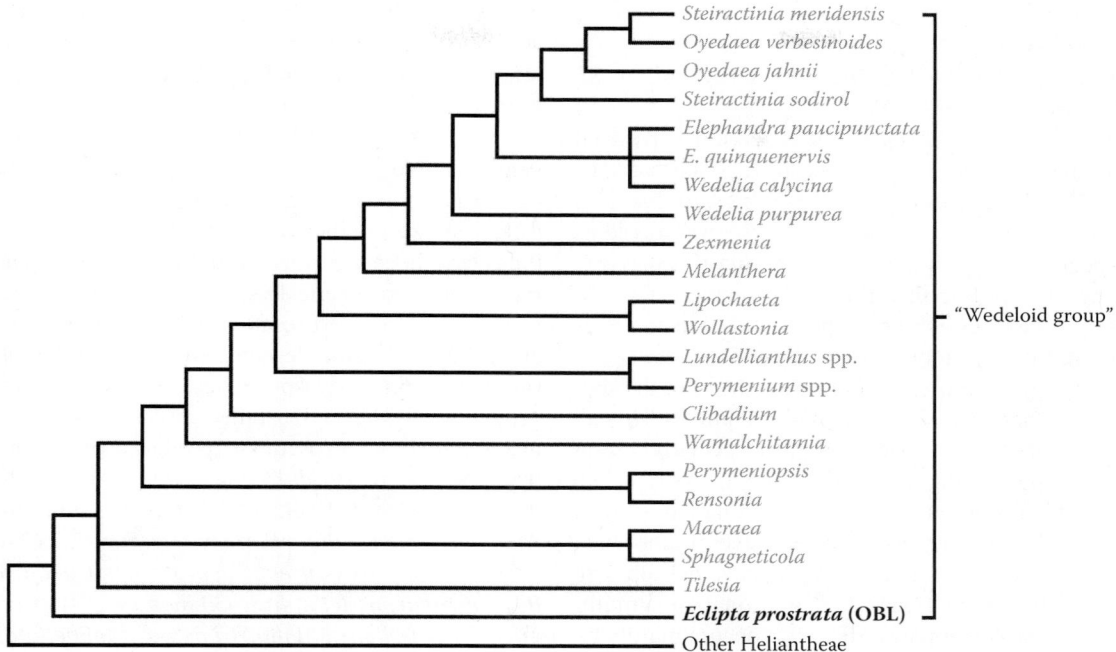

FIGURE 5.150 Phylogenetic position of *Eclipta prostrata* among the "wedeloid group" of subtribe Ecliptinae (Asteraceae: Heliantheae) as indicated by cpDNA RFLP data analysis. The OBL *Eclipta prostrata* (bold) occupies a rather isolated position at the base of this clade. (Redrawn from Panero, J.L. et al., *Amer. J. Bot.*, 86, 413–427, 1999.)

are reported. Interspecific hybrids have been obtained, at least artificially, for other *Eclipta* species; however, no incidences of hybridization have been reported that involve *E. prostrata*.

Comments: *Eclipta prostrata* is distributed throughout eastern and southern North America, being more common at lower latitudes.

References: Altom & Murray, 1996; Ananthi et al., 2003; Baskaran & Jayabalan, 2005; Begum et al., 2003; Chauhan & Johnson, 2008; Chmielewski, 2005; Christybapita et al., 2007; Dhawan, 2007; Ge & Wan, 1990; Gupta, 1992; He et al., 2008; Hussain & Kahn, 2010; Jaiswal et al., 2012; Jauzein, 1991; Jayathirthaa & Mishraa, 2004; Jha et al., 2009; Karthikumar et al., 2007; Kaur, 2011; Khanna & Kannabiran, 2007; Krishnan et al., 2010; Kumaria et al., 2006; Kunkel, 1984; Lee & Moody, 1988; Lu et al., 2010; Melo et al., 1994; Melouk et al., 1992; Michels et al., 2011; Moody, 1981; Mors et al., 1989; Nakatani & Kusanagi, 1991; Orchard & Cross, 2013; Panero et al., 1999; Pereira et al., 1998; Perveen, 1999; Prabhu et al., 2008; Prasad & Rao, 1979; Rajakumar & Rahuman, 2011; Rangineni et al., 2007; Roy et al., 2008; Saggoo et al., 2010; Sawant et al., 2004; Shao et al., 2011; Soerjani et al., 1987; Stevens et al., 2009; Strother, 2006c; Tewtrakul et al., 2007; Thakur & Mengi, 2005; Tzonev, 2007; Umemoto & Koyama, 2007; Umemoto & Yamaguchi, 1999; Umemoto et al., 1987; Urbatsch, 2000; Vankar et al., 2007; Varshney & Sharma, 1979; Venkatesan & Ravi, 2004; Wagner & Fessler, 1986; Wagner et al., 1986; Wasule, 2011; Wehtje et al., 2006.

17. *Erigeron*

Daisy fleabane, fleabane; vergerette

Etymology: from the Greek *eri gerôn* ("early old man"), referring to the graying flowers that appear in spring

Synonyms: *Achaetogeron*; *Aphanostephus*; *Apopyros*; *Aster* (in part); *Conyza*; *Darwiniothamnus*; *Hysterionica*; *Iotasperma*; *Leptostelma*; *Neja*; *Pappochroma*; *Stenactis*; *Trimorpha*

Distribution: global: cosmopolitan; **North America:** widespread

Diversity: global: 390 species; **North America:** 177 species

Indicators (USA): OBL: *Erigeron kachinensis, E. vernus*

Habitat: freshwater to slightly saline; palustrine; **pH:** 4.5–9.1; **depth:** <1 m; **life form(s):** emergent herb

Key morphology: Stems (to 6 dm) decumbent, prostrate or erect; leaves in basal rosettes (to 15 cm) or cauline (reduced or bract-like), blades oblanceolate to spatulate, sometimes fleshy or folded, the margins entire or denticulate; heads radiate, few (1–4/stem) or in loose corymbs (to 25/stem); involucre (to 11 mm) hemispherical, phyllaries (to 5 mm) in 2–4 series; pales absent; ray florets (to 15–40) pistillate, with white or pinkish corollas (to 8 mm); disc florets (to 3.8 mm) numerous, bisexual, corollas yellow; cypselae (to 2.0 mm), 2–4–nerved, sparsely strigose; outer pappus setiform (to 0.4 mm), the inner of 12–25 capillary bristles.

Life history: duration: biennial, perennial (caudex, taproot); **asexual reproduction:** rhizomes; **pollination:** insect; **sexual condition:** gynomonoecious; **fruit:** cypselae (common); **local dispersal:** seeds (gravity); **long-distance dispersal:** seeds (wind)

Imperilment: (1) *Erigeron kachinensis* [G2]; S1 (CO); S2 (UT); (2) *E. vernus* [G5]; S2 (VA)

Ecology: general: Most *Erigeron* species are terrestrial and occur in dry sites within montane or grassland habitats. Fewer than 16% of the North American species are listed as wetland indicators and only two species (~1%) are ranked as OBL. All species are gynomonoecious, having pistillate ray florets and hermaphroditic disc florets. Pollination is mediated by insects (Insecta). Observed pollinators include bees (Hymenoptera), flies (Diptera), and to a lesser degree, wasps (Hymenoptera) and butterflies (Lepidoptera). Most of the species are sexual, genetically self-incompatible, and outcrossed. However, some species are self-compatible (especially during later floral development stages), and some (mainly polyploids) are known to be diplosporous apomicts. Germination data are unavailable for seeds of the OBL species. Several other taxa have physiologically dormant seeds, a few have no apparent dormancy and others have seeds where germination is inhibited by filtered light. Presumably, dispersal occurs mainly by wind, facilitated by the pappus bristles. However the maximum, empirically derived dispersal distance for cypselae of one *Erigeron* species was only 1–4 m.

Erigeron kachinensis **S.L. Welsh & Glen Moore** occupies specialized habitats of alcoves, crevices, hanging gardens, seeps, springs, streambeds, and margins of washes at elevations from 1400 to 3597 m. The plants characteristically occur in shaded sites that maintain high levels of soil moisture (5.8%–25.6%). The sandstone-derived substrates (Cedar Mesa, Navajo) are sandy loams with 70.4%–89.6% sand content and low organic matter content (0.8%–2.2%). They are highly alkaline (pH 7.8–9.1) and fairly saline (conductivity=6.0–31.1 mmhos/cm) with high concentrations of calcium (6758–10884 ppm), phosphorous (7–11 ppm), potassium (80–229 ppm), and sodium (127–2535 ppm). June soil temperatures range from 13.2°C to 15.8°C. Flowering occurs from May to July. Plants sometimes flower twice a year (early spring and late summer). Individual plants produce up to 2.3 capitula on average with each producing an average of up to 29.8 florets. The number of heads/plant correlates proportionally with the extent of resources (e.g., water, phosphorous) available. The principal pollinators are generalist bee flies and syrphid flies (Insecta: Diptera: Bombyliidae; Syrphidae). Sawflies (Insecta: Hymenoptera: Tenthredinidae) and March flies (Insecta: Diptera: Bibionidae) visit the flowers less frequently, but may function as secondary pollinators. The plants are self-incompatible, obligate outcrossers. Coupled with the characteristically small population sizes, this breeding system leads to low fertilization rates (48%) and high losses (56%) of developing ovules due to inbreeding depression. Field studies indicate that only 42%–53% of flowers are pollinated, leading to the production of 19–25 mature seeds/capitulum. Generally, fecundity is higher for larger plants, but 80% of the annual seed production in a population can result from smaller plants due to their larger numbers. The seeds (cypselae) are dispersed primarily by gravity. Seeds appear to lack

dormancy and have been germinated on moist filter paper in petri dishes under greenhouse conditions. Seed-grown plants reach maturity in 6 months. Allozyme markers indicate that only 22.8% of the observed genetic variation is distributed among populations. Most populations are characterized by significant deviation of fixation indices and appear to be experiencing genetic drift. The mean observed heterozygosity (0.166) showed an increase with size and/or age. Three genetic races have been identified, with high elevation plants exhibiting the greatest genetic divergence from the others. The plants grow from taprooted rosettes that form loose mats or small clumps up to 7.3 cm; the stems are substoloniferous. Densities range from 6.1 to 27.4 plants/m². Smaller plants have slower growth rates, higher mortality and lower survival rates. An average of 1.5–2.5 rosettes is produced per plant. **Reported associates:** *Acer glabrum, Adiantum capillus-veneris, Aliciella subnuda, Amelanchier utahensis, Anticlea vaginata, Aquilegia micrantha, Berberis fremontii, Calamagrostis scopulorum, Carex aurea, C. curatorum, Castilleja linariifolia, Cercocarpus ledifolius, Cirsium arizonicum, C. rydbergii, Clematis ligusticifolia, Comandra umbellata, Epipactis gigantea, Erigeron sionis, E. concinnus, Fendlera rupicola, Frasera speciosa, Galium multiflorum, Hedeoma drummondii, Helianthella microcephala, Heterotheca villosa, Hymenopappus filifolius, Ipomopsis congesta, Juncus arcticus, Juniperus osteosperma, Linanthus pungens, Mimulus eastwoodiae, Oenothera longissima, Packera multilobata, Penstemon watsonii, Perityle specuicola, Philadelphus microphyllus, Pinus edulis, P. ponderosa, Platanthera sparsiflora, P. zothecina, Populus fremontii, Primula specuicola, Ranunculus cymbalaria, Rhamnus betulifolia, Salix exigua, Solidago velutina, Symphyotrichum ascendens, S. chilense.*

Erigeron vernus **(L.) Torr. & A. Gray** is a biennial or short-lived perennial, which inhabits barrens, bayous, bogs, borrow pits, ditches, flatwoods, floodplains, hammocks, meadows, prairies, savannas, sea-level fens, seeps, swales, swamps, and margins of lakes and ponds at elevations of up to 50 m. It grows in full sun to partially shaded sites that are acidic (pH 4.5–5.5). Substrates are described as loamy sand, peat, peaty sand, sand, sandy clay, sandy loam, and sandy peat or more specifically as Basinger (spodic sammaquents), Ona and Smyrna (typic and aeric haplaquods), Pottsburg (grossarenic haplaquods), Rutledge (typic humaquepts), and Vero (alfic haplaquods) soil types. Flowering occurs from March to October. Further details on the reproductive ecology of this species have not been determined. The plants persist from a thick caudex, from which elongate rhizomes can arise to produce additional rosettes. Nothing is known about the seed ecology, but the propagules must be able to withstand fires. The plants frequently are reported in disturbed sites such as those that have been recently burned or cut over. Additional life history information is needed for this species. **Reported associates:** *Acer rubrum, Aletris farinosa, Amphicarpum muhlenbergianum, Andropogon arctatus, A. glomeratus, A. gyrans, A. liebmannii, A. longiberbis, A. virginicus, Anthaenantia rufa, Aristida longespica, A. palustris, A. purpurascens, A. stricta, Arnoglossum ovatum, Aronia arbutifolia, Arundinaria* *gigantea, A. tecta, Asimina reticulata, Axonopus compressus, Balduina atropurpurea, B. uniflora, Bidens trichospermus, Bigelowia nudata, Buchnera floridana, Bulbostylis ciliatifolia, Carex glaucescens, C. lonchocarpa, C. striata, C. venusta, C. verrucosa, Carphephorus odoratissimus, C. paniculatus, C. pseudoliatris, Centella asiatica, Chamaecrista nictitans, Chaptalia tomentosa, Chasmanthium, Chloris, Chrysopsis gossypina, Cleistes, Clethra alnifolia, Coleataenia longifolia, Coreopsis gladiata, C. nudata, Crotalaria rotundifolia, Ctenium aromaticum, Cyrilla racemiflora, Dichanthelium aciculare, D. acuminatum, D. dichotomum, D. ensifolium, D. ovale, D. scabriusculum, D. sphaerocarpon, D. stigmosum, D. strigosum, Digitaria ischaemum, D. longiflora, Diodia virginiana, Drosera brevifolia, D. capillaris, D. intermedia, D. rotundifolia, Eleocharis baldwinii, E. melanocarpa, E. rostellata, E. tuberculosa, E. uniglumis, Elephantopus nudatus, Eragrostis refracta, Erigeron quercifolius, Eriocaulon compressum, E. decangulare, Eryngium integrifolium, E. yuccifolium, Eupatorium leucolepis, E. semiserratum, Fimbristylis castanea, F. thermalis, Fragaria virginiana, Fuirena breviseta, F. pumila, F. scirpoidea, Galactia elliottii, Galium obtusum, Gaylussacia dumosa, G. mosieri, Gelsemium rankinii, G. sempervirens, Geum canadense, Gordonia lasianthus, Gymnopogon brevifolius, Helenium amarum, Helianthus heterophyllus, Houstonia caerulea, Hypericum brachyphyllum, H. chapmanii, H. cistifolium, H. crux-andreae, H. fasciculatum, H. galioides, H. hypericoides, H. mutilum, H. myrtifolium, H. tenuifolium, Hyptis alata, Ilex coriacea, I. glabra, I. myrtifolia, I. opaca, Itea virginica, Juncus abortivus, J. acuminatus, J. debilis, J. dichotomus, J. megacephalus, J. pelocarpus, J. repens, J. trigonocarpus, Krigia dandelion, Kummerowia striata, Lachnanthes caroliniana, Lachnocaulon anceps, L. compressum, Liatris pilosa, L. spicata, Lilium catesbaei, Linum medium, Liquidambar styraciflua, Lobelia glandulosa, Lophiola aurea, Ludwigia linearis, L. pilosa, Lycopodiella alopecuroides, L. appressa, L. caroliniana, Lycopus rubellus, Lyonia fruticosa, L. lucida, L. mariana, Magnolia virginiana, Marshallia graminifolia, Mitchella repens, Mnesithea tessellata, Muhlenbergia capillaris, M. expansa, Myrica cariniensis, M. cerifera, M. heterophylla, Nuttallanthus canadensis, Nyssa biflora, Oclemena reticulata, Oldenlandia uniflora, Onoclea sensibilis, Osmunda regalis, Osmundastrum cinnamomeum, Oxypolis filiformis, Panicum verrucosum, P. virgatum, Parnassia caroliniana, Parthenocissus quinquefolia, Paspalum floridanum, P. laeve, P. praecox, Persea borbonia, P. palustris, Pinguicula caerulea, P. lutea, P. primuliflora, Pinus caribaea, P. elliottii, P. palustris, P. serotina, P. taeda, Pityopsis graminifolia, Plantago lanceolata, Pleea tenuifolia, Pluchea baccharis, P. foetida, P. odorata, Polygala brevifolia, P. cymosa, P. ramosa, Proserpinaca palustris, P. pectinata, Pteridium aquilinum, Pterocaulon pycnostachyum, Quercus laevis, Q. laurifolia, Q. minima, Q. nigra, Rhexia alifanus, R. lutea, R. nashii, R. nuttallii, R. petiolata, R. virginica, Rhus copallina, Rhynchospora alba, R. baldwinii, R. cephalantha, R. chapmanii, R. colorata, R. compressa, R. corniculata, R. decurrens, R. fascicularis, R. gracilenta, R. inexpansa, R. inundata, R. latifolia,* *R. megalocarpa, R. microcephala, R. oligantha, R. plumosa, Rubus trivialis, Sabal palmetto, Sabatia brevifolia, S. decandra, S. dodecandra, S. macrophylla, Saccharum giganteum, S. villosum, Sagittaria graminea, Sarracenia alata, S. flava, S. minor, Saururus cernuus, Schizachyrium condensatum, S. scoparium, Scleria ciliata, S. muehlenbergii, S. pauciflora, S. verticillata, Sclerolepis uniflora, Scutellaria integrifolia, Serenoa repens, Sisyrinchium mucronatum, Smilax auriculata, S. laurifolia, Smilax laurifolia, S. walteri, Solidago fistulosa, S. rugosa, Sorghastrum secundum, Spartina patens, Sphagnum, Spiranthes, Sporobolus pinetorum, Stenanthium densum, Stillingia, Symphyotrichum dumosum, S. patens, Syngonanthus flavidulus, Taxodium ascendens, Tillandsia bartramii, T. usneoides, Toxicodendron radicans, Triadenum virginicum, Triantha racemosa, Tridens flavus, Utricularia juncea, Vaccinium crassifolium, V. darrowii, V. myrsinites, V. stamineum, V. tenellum, Viburnum nudum, Viola lanceolata, V. primulifolia, Woodwardia areolata, W. virginica, Xyris ambigua, X. brevifolia, X. elliottii, X. fimbriata, X. torta.*

Use by wildlife: The leaves of *Erigeron kachinensis* can become infected by plant pathogenic fungi. The basal leaves host gall-forming Lepidoptera (Insecta). The flower heads are eaten by small rodents (Mammalia: Rodentia). *Erigeron vernus* is a host of seed-eating fruit flies (Insecta: Diptera: Tephritidae: *Neaspilota achilleae, Trupanea actinobola*), which can infest the capitula. It also hosts larvae of the celery leaftier moths (Insecta: Lepidoptera: Crambidae: *Udea rubigalis*), which are common on plants from February through June. The plants occur within the summer habitat of Henslow's sparrow (Aves: Emberizidae: *Ammodramus henslowii*).

Economic importance: food: not edible; **medicinal:** The common name "fleabane" supposedly originated from the perceived ability of these plants to repel fleas (Insecta: Siphonaptera); however, there seems to be no empirical evidence to support this claim (other sources attribute the name to the flea-like seeds). Aqueous leaf and stem extracts of *Erigeron vernus* do provide complete inhibition of *Pseudomonas aeruginosa*, a Gram-negative bacterium associated with wounds; **cultivation:** neither OBL species is in cultivation; **misc. products:** none; **weeds:** none; **nonindigenous species:** both OBL species are indigenous.

Systematics: Within Asteraceae *Erigeron* is classified in tribe Astereae subtribe Conyzinae (500 species), which also includes *Aphanostephus, Apopyros, Conyza Darwiniothamnus, Hysterionica, Leptostelma,* and *Neja.* It is the only one of these genera to have a distribution extending into the Old World. The taxonomy of *Erigeron* and these allied genera is extremely problematic. The incorporation of DNA sequence data shows *Erigeron* to be highly polyphyletic and has presented a hypothetical phylogeny (Figure 5.151) that is difficult to reconcile with any current taxonomic scheme. Far more study will be necessary before a reasonable solution will be reached as to how an acceptable taxonomic circumscription of genera in the Conyzinae might be devised. Based on current information (Figure 5.151) it seems difficult not to advocate the merger of several genera (*Aphanostephus, Apopyros, Conyza, Hysterionica, Neja*) with *Erigeron* in an

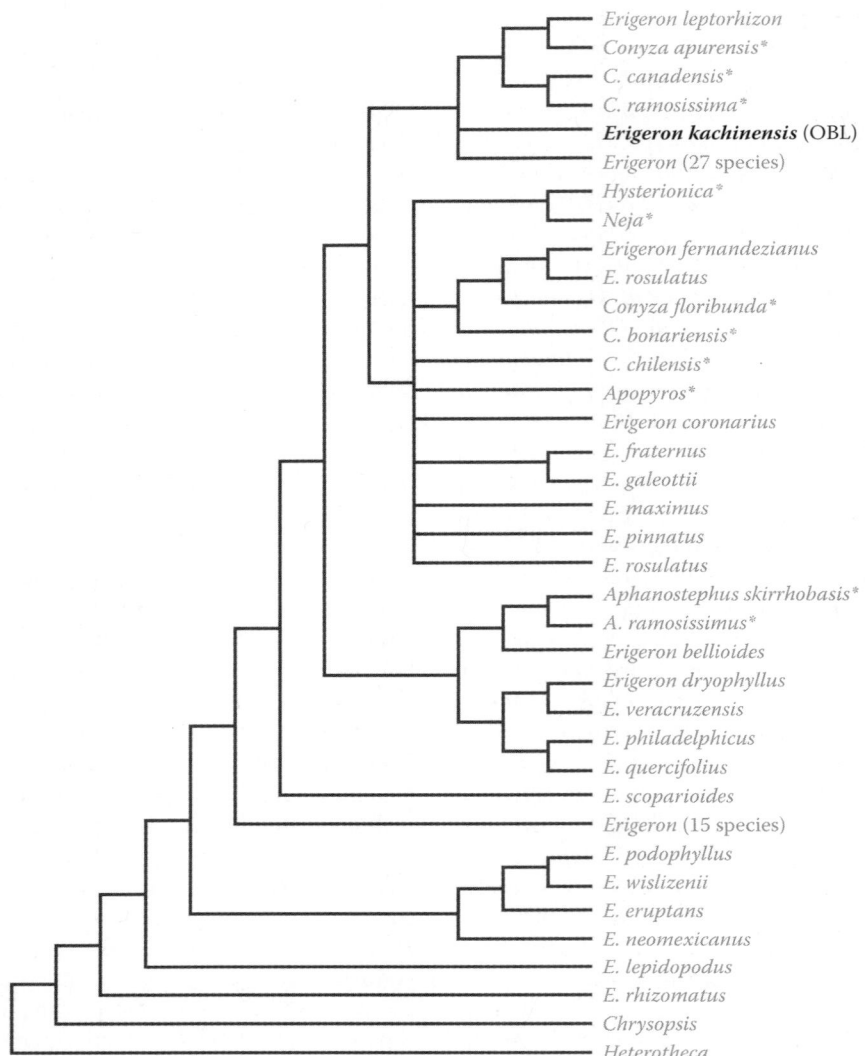

FIGURE 5.151 Phylogenetic relationships of Asteraceae tribe Astereae subtribe Conyzinae reconstructed from nrITS sequence data show *Erigeron* (*sensu lato*) to be highly polyphyletic if several other subordinate genera (asterisked) are to be maintained. The OBL species *E. kachinensis* is shown in bold. *Erigeron vernus* (OBL) was not sampled in this analysis. (Redrawn from Noyes, R.D., *Plant Syst. Evol.*, 220, 93–114,, 2000b.)

expanded concept of the genus. The relationship between the two OBL species does not appear to be close as they are assigned to different sections: *Scopulincola* (*E. katchinensis*) and *Erigeridium* (*E. vernus*). However, no members of the latter section have yet been included in phylogenetic analyses of the genus. The base chromosome number of *Erigeron* is $x=9$. *Erigeron vernus* ($2n=18$) is diploid; *E. kachinensis* has not been counted.

Comments: *Erigeron kachinensis* is endemic to Colorado and Utah. *Erigeron vernus* is endemic to the coastal plain of the southeastern United States and does not extend northward of the Salisbury Embayment in eastern Virginia.

References: Allphin & Harper, 1994, 1997; Allphin & Windham, 2002; Allphin et al., 1996, 2002; Armbruster & McGuire, 1991; Baskin & Baskin, 1998; Benjamin, 1934; Carr, 2007; Gaddy, 1982; Grieve, 1980; Huffman & Judd, 1998; Miller,

1931; Morris, 2007; Murrill, 1940a; Nesom, 2006b, 2008; Noyes, 2000a,b; Noyes & Allison, 2005; Noyes & Rieseberg, 1999; Noyes et al., 1995; Pates & Madsen, 1955; Pearson, 1954; Penfound & Watkins, 1937; Pranty & Scheuerell, 1997; Sheldon & Burrows, 1973; Smith, 1996; Sorrie & Weakley, 2001; Stegmaier, Jr., 1968; Uphof, 1922; Wells, 1928.

18. *Eupatorium*

Boneset, thoroughwort; Eupatoire

Etymology: after Mithridates VI Eupator (132–63 BC)

Synonyms: *Ageratina*; *Chone*; *Chromolaena*; *Critonia*; *Conoclinium*; *Cunigunda*; *Fleischmannia*; *Koanophyllon*; *Tamaulipa*; *Traganthes*; *Uncasia*

Distribution: global: Africa; Eastern Asia; Europe: North America; **North America:** eastern

Diversity: global: 41 species; **North America:** 27 species

Indicators (USA): OBL: *Eupatorium perfoliatum, E. resinosum*; **FACW:** *E. perfoliatum*

Habitat: freshwater; palustrine; **pH:** 4.4–8.0; **depth:** <1 m; **life form(s):** emergent herb

Key morphology: Stems (to 1.5 m) single, from caudices or rhizomes, branched distally, puberulent; leaves (to 15 cm) usually opposite (occasionally whorled, or the distal alternate), sessile, narrowly to broadly lanceolate, pilose or puberulent below, the margins serrate, the bases connate perfoliate or cuneate; heads discoid, arranged in corymbs, phyllaries (to 4.5 mm) 1–12 in 1–3 series, pales lacking; florets (7–14), 5-merous, perfect, corollas (to 3.5 mm), white to pinkish; cypselae (to 3.0 mm) 5-ribbed, pappus (to 3.5 mm) of 20–30 diminutively barbed bristles.

Life history: duration: perennial (overwintering stem buds); **asexual reproduction:** short rhizomes; **pollination:** insect; **sexual condition:** hermaphroditic; **fruit:** cypselae (common); **local dispersal:** seeds (gravity, wind); **long-distance dispersal:** seeds (wind)

Imperilment: (1) *Eupatorium perfoliatum* [G5]; S2 (PE); (2) *E. resinosum* [G3]; SX (DE, NY); S1 (SC); S2 (NJ); S3 (NC)

Ecology: general: *Euptorium* species occupy a fairly wide range of fresh to saline habitats including dunes, dry, open sites, grasslands, and low, wet areas. They occur commonly in wetlands (especially along pond margins) with about half of the North American species assigned some wetland indicator status. All *Eupatorium* species produce discoid heads, which contain bisexual florets. Most species are sexual but a few are agamospermous. Insect (Insecta) pollination characterizes most of the taxa including the OBL species; however, some dissected-leaved species ("dog fennels") possess features associated with wind-pollination such as pollen with reduced exine spines, elongate stigmas, reduced flower sizes, lax inflorescences, and self-compatibility. The insect-pollinated species have upright inflorescences with showy flowers, short stigmas, spiny pollen, and sporophytic self-incompatibility. Most species disperse their cypselae primarily by wind, facilitated by a bristled pappus. Seeds are either nondormant or possess physiological dormancy. Both OBL species perennate by the development of one or more overwintering buds at the base of each stem.

Eupatorium perfoliatum **L.** inhabits baygalls, beaches, bogs, carrs, crevices, depressions, ditches, dunes, fens, flats, flatwoods, floodplains, hammocks, lagoons, marshes, meadows, mudflats, oak openings, prairies, seeps, shores, sloughs, springs, swales, swamps, thickets, and the margins of canals, lakes, ponds, pools, rivers, and streams at elevations of up to 500 m. Exposures range widely from full sun to various degrees of shade. The plants occur in shallow water (<20 cm) or on exposed alkaline soils (e.g., pH: 6.5–8.0; alkalinity 112–425 mg $CaCO_3$/l;), which are described as clay, cobble, dolomite, gravel, loam, loamy muck, muck, mud, peat, peaty muck, rock, sand, sandy alluvium, sandy clay, sandy loam, silt, and silt loam. This species can tolerate a fair degree of substrate desiccation and frequently occurs in drying sites, where water has receded. Although perennial, some authors categorize this species as a "FAC annual," i.e., a perennial

capable of setting seed in the first year. Demographic studies observed 36%–75% of plants to be in flower across three populations studied for three consecutive years. The plants averaged 9–10 cypselae/head and 173–684 heads/plant across these populations. Asexual (ramet) reproduction ranged from 0.30 to 0.92 new ramets/year. From 60% to 100% of the new ramets will flower within their first year. Plants primarily are self-incompatible with populations averaging 7–10 seeds/head; however, some self-compatibility (yielding 1–4 seeds/head on average) also occurs. Outcrossing appears to dominate and crossing studies have documented hybrid vigor in interpopulationally outcrossed progeny. Populations are structured genetically as numerous subpopulations. The flowers are pollinated by various insects (Insecta) such as bees and wasps (Hymenoptera), flies (Diptera), and moths (Lepidoptera). In particular, honey bees (Hymenoptera: Apidae: *Apis mellifera*) are strongly attracted to the flowers, which also are visited by andrenid bees (Hymenoptera: Andrenidae: *Andrena flavipes*) and sweat bees (Hymenoptera: Halictidae: *Lasioglossum cinctipes, L. cressonii, L. zonulum*). One introduction of purple loosestrife (*Lythrum salicaria*) resulted in a 19% reduction in seed set due to increased foreign pollen loads. Seed mass averages 0.12–0.15 g and correlates positively with germination rate. The cypselae commonly are represented in seed banks; however, seed bank densities in draw down pools of the Mississippi River were low (3%) with an average of 2.2 propagules per m². Dormancy is broken after 112 days of dry stratification, with germination (21°C–32°C/4°C–21°C; 40%–70% relative humidity) initiating in 8 days and peaking at 14 days. Germination rates of 70% have been achieved after 270 days of cold (4°C) stratification followed by a 30°C/20°C temperature regime, with germination initiating in 4 days. Once stratified, seeds germinate well at 25°C (relative humidity = 70%) under both flooded and drawdown conditions. However, some banked seeds germinated only under nonflooded conditions. Laboratory germination rates of 80% have been reported. Higher germination rates (relative to hand-processed seed) have been obtained for seeds passed through a conditioning machine, which is described as a "brush-type thresher." Seedlings grow rapidly and can easily be transplanted to the field with high survivorship. Seedling density generally is lower in sediments of increasingly coarser texture. Seedling recruitment is somewhat higher where water is standing (1–15 cm depth), than in drawn down sediments. The plants contain an average of 1.0% nitrogen but only ~0.2% phosphorous by dry weight. The shoots become nearly horizontal at the substrate surface where they produce adventitious roots that penetrate 4–6 cm in peat, but reach 5–10 cm in mineral soils. Up to 59% of the root length can be colonized by arbuscular mycorrhizal fungi. Vegetative reproduction can occur by production of large whitish rhizomes. Adult plants exhibit symptoms (chlorotic spots, premature leaf drop) when exposed to 90 or 120 ppb ozone; however, seedlings are relatively resistant to similar ozone exposures. The plants occur disproportionately on grazed sites and often occur in areas of light disturbance. They exhibited a 50% decrease in biomass when grown in swards of other species compared to monospecific stands. The plants have been implicated as potential hosts for the hemiparasitic *Pedicularis lanceolata* (Orobanchaceae). **Reported associates:** *Acer rubrum, Achillea millefolium, Acorus calamus, Agalinis purpurea, Ageratina altissima, Agrostis hyemalis, A. scabra, A. stolonifera, Alisma, Allium canadense, A. cernuum, Alnus rugosa, A. serrulata, Ambrosia artemisiifolia, Amphicarpaea bracteata, Andropogon gerardii, A. virginicus, Anemone canadensis, Angelica atropurpurea, Anthoxanthum odoratum, Anticlea elegans, Apios americana, Apocynum cannabinum, Arctostaphylos uva-ursi, Arnoglossum plantagineum, Artemisia serrata, Asclepias incarnata, A. sullivantii, A. syriaca, A. verticillata, Aulacomnium palustre, Baptisia lactea, Betula alleghaniensis, B. populifolia, B. pumila, Bidens cernuus, B. connatus, B. frondosus, B. laevis, B. trichospermus, Bolboschoenus fluviatilis, Boltonia asteroides, Brachythecium rutabulum, Bromus, Bryum pseudotriquetrum, Calamagrostis canadensis, C. stricta, Calliergonella cuspidata, Callitriche heterophylla, Caltha palustris, Calystegia sepium, Campanula aparinoides, Campylium stellatum, Cardamine bulbosa, C. pratensis, Carex albolutescens, C. annectens, C. aurea, C. bromoides, C. buxbaumii, C. crawei, C. cristatella, C. crus-corvi, C. cryptolepis, C. disperma, C. eburnea, C. frankii, C. garberi, C. hyalinolepis, C. hystericina, C. interior, C. lacustris, C. leptalea, C. longii, C. lupulina, C. lurida, C. muricata, C. oligosperma, C. pellita, C. prairea, C. rosea, C. scoparia, C. stipata, C. stricta, C. suberecta, C. tomentosa, C. trichocarpa, C. viridula, C. vulpinoidea, Castilleja coccinea, Cephalanthus occidentalis, Ceratophyllum demersum, Chamaecrista, Chamerion angustifolium, Chelone glabra, Chenopodium album, Cicuta bulbifera, C. maculata, Circaea lutetiana, Cirsium arvense, C. muticum, Cladonia chlorophaea, C. cristatella, Clematis virginiana, Climacium americanum, Clinopodium acinos, C. glabellum, Coleataenia longifolia, Convolvulus arvensis, Cornus amomum, C. racemosa, C. sericea, Crataegus, Cuscuta gronovii, Cyperus bipartitus, C. erythrorhizos, C. esculentus, C. flavescens, C. odoratus, C. strigosus, Cypripedium reginae, Dasiphora floribunda, Decodon verticillatus, Desmodium glutinosum, Dichanthelium acuminatum, D. polyanthes, D. scoparium, Digitaria filiformis, Diodella teres, Dipsacus laciniatus, Distichlis spicata, Doellingeria umbellata, Drepanocladus aduncus, Drosera intermedia, D. rotundifolia, Dryopteris cristata, Dulichium arundinaceum, Echinochloa crus-galli, E. muricata, Eleocharis acicularis, E. elliptica, E. obtusa, E. olivacea, E. palustris, E. tenuis, Elephantopus, Elymus canadensis, Epilobium ciliatum, E. coloratum, E. leptophyllum, E. palustre, E. strictum, Equisetum arvense, E. fluviatile, E. pratense, E. sylvaticum, E. variegatum, Erechtites hieraciifolius, Erigeron annuus, Eriophorum virginicum, E. viridicarinatum, Eryngium yuccifolium, Eupatorium capillifolium, E. rotundifolium, E. semiserratum, E. serotinum, Eurhynchium hians, Euthamia graminifolia, Eutrochium dubium, E. maculatum, E. purpureum, Fagus, Fallopia convolvulus, Filipendula rubra, Fimbristylis autumnalis, Fissidens adianthoides, Fragaria virginiana, Fraxinus americana, F. nigra, Galium asprellum, G. labradoricum, G. palustre, G. tinctorium, Gentiana andrewsii, G. clausa, Gentianopsis crinita, G. virgata, Geum rivale, Glyceria canadensis, G. obtusa, G. striata, Gratiola neglecta, Habenaria repens, Helenium autumnale, Helianthus giganteus, H. grosseserratus, Hibiscus moscheutos, Hieracium, Hydrocotyle umbellata, Hypericum boreale, H. canadense, H. hypericoides, H. kalmianum, H. mutilum, Hypnum lindbergii, Hypoxis hirsuta, Impatiens capensis, Iris versicolor, I. virginica, Iva microcephala, Juncus acuminatus, J. arcticus, J. balticus, J. brachycephalus, J. bufonius, J. canadensis, J. coriaceus, J. effusus, J. marginatus, J. militaris, J. scirpoides, J. secundus, J. tenuis, J. torreyi, Lactuca canadensis, Leersia oryzoides, Lemna minor, Liatris pycnostachya, L. spicata, Lilium michiganense, Linum striatum, Lipocarpha micrantha, Lobelia kalmii, L. puberula, L. siphilitica, L. spicata, Lonicera japonica, L. oblongifolia, L. villosa, Ludwigia alternifolia, L. leptocarpa, L. palustris, L. peploides, Lycopus americanus, L. uniflorus, L. virginicus, Lysimachia ciliata, L. quadriflora, L. terrestris, L. thrysiflora, Lythrum alatum, L. salicaria, Magnolia virginiana, Maianthemum canadense, M. racemosum, M. trifolium, Medicago lupulina, Melilotus alba, Mentha arvensis, Micranthes pensylvanica, Mikania scandens, Mimulus glabratus, M. ringens, Monarda fistulosa, Muhlenbergia glomerata, M. mexicana, M. racemosa, M. richardsonis, M. sylvatica, M. uniflora, Myrica gale, Napaea dioica, Nuphar advena, Nyssa, Oclemena nemoralis, Oenothera biennis, O. fruticosa, Onoclea sensibilis, Osmunda regalis, Osmundastrum cinnamomeum, Oxalis montana, O. stricta, Oxypolis rigidior, Oxytropis campestris, Packera aurea, P. paupercula, Panicum capillare, P. dichotomiflorum, Parnassia glauca, Parthenium integrifolium, Parthenocissus vitacea, Paspalum, Pedicularis lanceolata, Peltandra virginica, Pennisetum glaucum, Penstemon digitalis, Penthorum sedoides, Persicaria amphibia, P. arifolia, P. coccinea, P. hydropiperoides, P. lapathifolia, P. maculata, P. pensylvanica, P. punctata, P. sagittata, Phalaris arundinacea, Phleum pratense, Phlox maculata, P. pilosa, Phragmites australis, Phyla lanceolata, P. nodiflora, Physcomitrium pyriforme, Physostegia virginiana, Picea mariana, Pilea pumila, Pinus strobus, P. taeda, Pityopsis, Plagiomnium ellipticum, Platanthera dilatata, Pluchea camphorata, Poa compressa, P. nemoralis, P. pratensis, Pogonia ophioglossoides, Polygala sanguinea, Polygonum ramosissimum, Pontederia cordata, Potamogeton epihydrus, P. natans, Potentilla anserina, P. norvegica, Prenanthes racemosa, Primula mistassinica, Proserpinaca palustris, Prunella vulgaris, Prunus serotina, Pteridium, Ptilimnium capillaceum, Pueraria montana, Pycnanthemum flexuosum, P. muticum, P. tenuifolium, P. virginianum, Quercus alba, Q. pagoda, Q. palustris, Q. velutina, Ranunculus sceleratus, R. septentrionalis, Rhexia mariana, R. virginica, Rhynchospora alba, R. capillacea, R. capitellata, R. glomerata, Ribes americanum, Rorippa palustris, Rosa multiflora, R. palustris, R. setigera, Rubus allegheniensis, R. sachalinensis, Rudbeckia fulgida, R. hirta, R. laciniata, Rumex orbiculatus, Sacciolepis striata, Sagittaria latifolia, Salix bebbiana, S. candida, S. discolor, S. exigua, S.*

interior, S. nigra, S. petiolaris, S. pyrifolia, S. serissima, Sambucus nigra, Sanguisorba canadensis, Schizachyrium scoparium, Schoenoplectus acutus, S. pungens, S. purshianus, S. tabernaemontani, Scirpus atrocinctus, S. atrovirens, S. cyperinus, S. polyphyllus, Scleria verticillata, Scutellaria galericulata, S. integrifolia, S. lateriflora, Selaginella apoda, Senecio suaveolens, Setaria faberi, S. parviflora, Silphium perfoliatum, S. terebinthinaceum, Sinapis arvensis, Sisyrinchium angustifolium, Sium suave, Smilax tamnoides, Solidago canadensis, S. gigantea, S. juncea, S. ohioensis, S. patula, S. puberula, S. riddellii, S. uliginosa, Solanum carolinense, S. dulcamara, Sonchus, Sorghastrum nutans, Sparganium eurycarpum, Spartina patens, S. pectinata, Sphagnum capillifolium, S. warnstorfii, Spiraea alba, S. tomentosa, Stachys palustris, Stuckenia filiformis, Sullivantia sullivantii, Symphyotrichum dumosum, S. ericoides, S. firmum, S. lanceolatum, S. novae-angliae, S. novi-belgii, S. pilosum, S. praealtum, S. puniceum, Symplocarpus foetidus, Taraxacum officinale, Thalictrum dasycarpum, T. pubescens, Thaspium trifoliatum, Thelypteris palustris, Thuidium delicatulum, Thuja occidentalis, Torreyochloa pallida, Toxicodendron radicans, T. vernix, Triadenum virginicum, Trichophorum alpinum, Trichostema, Trifolium hybridum, T. pratense, T. repens, Trollius laxus, Typha angustifolia, T. latifolia, Ulmus americana, U. rubra, Urtica, Vaccinium macrocarpon, Valeriana edulis, Verbascum blattaria, V. thapsus, Verbena hastata, V. stricta, V. urticifolia, Vernonia gigantea, V. noveboracensis, Veronicastrum virginicum, Viburnum dentatum, Viola cucullata, V. lanceolata, V. macloskeyi, V. primulifolia, V. sagittata, V. sororia, Vitis riparia, Wolffia columbiana, Xylorhiza tortifolia, Zizia aurea.

***Eupatorium resinosum* Torr. ex DC.** grows in bogs, borrow pits, marshes, pine barrens, pocosins, seeps, swamps, and along the margins of streams at elevations of up to 100 m. The plants are shade intolerant and occur on peat and other acidic substrates (pH = 4.4–7.0; specific conductance 45–77 μS cm^{-1}). Flowering occurs from August to September and fruits are set in October. Demographic studies observed just 2%–27% of plants to be in flower across three populations studied for three consecutive years. The plants studied averaged 3–9 cypselae/head and 30–91 heads/plant across populations. Asexual (ramet) reproduction ranged from 0.12–0.65 new ramets/year. Only 11%–31% of new ramets flower within the first year. The flowers are completely self-incompatible and produce on average three seeds/head in small populations to nine seeds/head in larger populations. Populations are structured as numerous subpopulations and crossing studies have documented inbreeding depression from spatially close crosses and hybrid vigor in interpopulationally outcrossed progeny. The flowers are pollinated by various insects (Insecta) including bees and wasps (Hymenoptera), flies (Diptera), and moths (Lepidoptera). The seeds will germinate (in a greenhouse mist room) after 6 months of dry/cold storage followed by a 30-day moist/cold treatment. This species has a lower reproductive output compared to the closely related *E. perfoliatum* (earlier), which is thought to be a consequence of resource limitations imposed by the nutrient-poor conditions of its habitat.

Its cypselae mass averages 0.22 g and does not correlate with germination rate. Vegetative reproduction occurs by means of short rhizomes.

Reported associates: *Acer rubrum, Agrostis perennans, Andropogon virginicus, Apios americana, Aronia arbutifolia, Asclepias incarnata, Bartonia paniculata, B. virginica, Bidens frondosus, B. trichospermus, Calamagrostis canadensis, Calamovilfa brevipilis, Carex albolutescens, C. atlantica, C. canescens, C. exilis, C. folliculata, C. lonchocarpa, C. striata, C. turgescens, Cephalanthus occidentalis, Chamaecyparis thyoides, Chamaedaphne calyculata, Cladium mariscoides, Clethra alnifolia, Coleataenia longifolia, Cuscuta, Cyperus dentatus, Danthonia sericea, Decodon verticillatus, Dichanthelium acuminatum, D. scabriusculum, Drosera intermedia, D. rotundifolia, Dulichium arundinaceum, Eleocharis acicularis, E. flavescens, E. robbinsii, E. tenuis, E. tuberculosa, Eriocaulon aquaticum, E. compressum, Eriophorum virginicum, Eubotrys racemosa, Eupatorium pilosum, Euthamia graminifolia, Fuirena squarrosa, Galium tinctorium, Gaylussacia baccata, G. dumosa, G. frondosa, Glyceria obtusa, Gymnadeniopsis clavellata, Hypericum canadense, H. densiflorum, H. denticulatum, H. fasciculatum, H. mutilum, Ilex glabra, I. verticillata, Iris versicolor, Juncus canadensis, J. effusus, J. militaris, Kalmia angustifolia, Lachnanthes caroliniana, Leersia oryzoides, Lindera subcoriacea, Liriodendron tulipifera, Lobelia nuttallii, Ludwigia alternifolia, L. palustris, Lycopodiella, Lycopus cokeri, L. uniflorus, Lyonia mariana, Lysimachia terrestris, Muhlenbergia uniflora, Nuphar variegata, Nymphaea odorata, Nyssa biflora, N. sylvatica, Oclemena nemoralis, Onoclea sensibilis, Orontium aquaticum, Osmunda regalis, Osmundastrum cinnamomeum, Oxypolis denticulata, O. rigidior, Panicum verrucosum, P. virgatum, Peltandra virginica, Persicaria hydropiperoides, P. sagittata, Pinus rigida, P. serotina, P. taeda, Pogonia ophioglossoides, Polygala cruciata, Pontederia cordata, Potamogeton confervoides, P. epihydrus, P. oakesianus, Rhexia virginica, Rhododendron viscosum, Rhynchospora alba, R. capitellata, R. chalarocephala, R. fusca, R. macra, R. scirpoides, R. stenophylla, Rosa palustris, Rubus hispidus, Sabatia difformis, Saccharum giganteum, Sagittaria engelmanniana, S. macrocarpa, Sarracenia flava, S. purpurea, Sassafras albidum, Schoenoplectus etuberculatus, S. subterminalis, Scirpus cyperinus, Scleria reticularis, Smilax glauca, S. pseudochina, S. rotundifolia, Solidago patula, Sparganium americanum, Sphagnum, Spiraea tomentosa, Symphyotrichum novi-belgii, Thelypteris palustris, Toxicodendron radicans, Triadenum virginicum, Utricularia geminiscapa, U. gibba, U. olivacea, U. purpurea, Vaccinium corymbosum, V. macrocarpon, Veratrum virginicum, Viola lanceolata, Woodwardia areolata, W. virginica, Xyris difformis, X. fimbriata, X. scabrifolia, Zizania aquatica.*

Use by wildlife: The stems of Eupatorium perfoliatum are a summer food of the muskrat (Rodentia: Cricetidae: *Ondatra zibethicus*). The plants provide habitat for crawfish frogs (Anura: Ranidae: *Lithobates areolatus*) and serve as nesting sites for the mourning warbler (Aves: *Geothlypis* *philadelphia*) and red-winged blackbird (*Agelaius phoeniceus*). They are hosts to numerous insects (Insecta) including aphids (Hemiptera: Aphididae: *Aphis spiraecola, Macrosiphum pallidum, Uroleucon ambrosiae, U. eupatorifoliae*), leaf beetles (Coleoptera: Chrysomelidae: *Ophraella notata*), long-horned beetles (Coleoptera: Cerambycidae: *Brachyleptura champlaini, Typocerus velutinus*), tumbling flower beetles (Coleoptera: Mordellidae: *Mordella albosuturalis, M. melaena; Mordellina pustulata; Mordellistena convicta*), gall gnats (Diptera: Cecidomyiidae: *Brachyneura eupatorii, Clinorhyncha eupatoriflorae, Contarinia perfoliata, Neolasioptera perfoliata*), several larval Lepidoptera (Bucculatricidae: *Bucculatrix eupatoriella*; Geometridae: *Chlorochlamys chloroleucaria, Semiothisa continuata*; Lycaenidae: *Celastrina argiolus*; Noctuidae: *Schinia trifascia*; Pterophoridae: *Hellinsia lacteodactylus*; Sesiidae: *Carmenta pyralidiformis*; Tortricidae: *Endopiza slingerlandana*), and adult mites (Acari: Phytoseiidae: *Proprioseiopsis unicus*). Fungi hosted by plants include pathogenic blights (Ascomycota: *Ascochyta compositarum*), leaf molds (Ascomycota: Mycosphaerellaceae: *Passalora perfoliati, Septoria eupatorii*), leaf spot (Ascomycota: Botryosphaeriaceae: *Phyllosticta decidua*), powdery mildew (Ascomycota: Leotiomycetidae: *Erysiphe cichoracearum*), and rusts (Basidiomycota: Pucciniaceae: *Puccinia eleocharidis*). The flowers are visited by an impressive list of insects (Insecta) including long-tongued bees (Hymenoptera: Apidae: *Apis mellifera* [honey bee]; *Bombus affinis, B. fraternus, B. griseocollis, B. impatiens, B. pensylvanicus; Melissodes druriella*; Megachilidae: *Megachile brevis, M. centuncularis, M. mendica*), short-tongued bees (Hymenoptera: Andrenidae: *Andrena simplex*; Colletidae: *Colletes eulophi, Hylaeus affinis, H. mesillae, H. modestus*; Halictidae: *Augochlorella aurata, Augochloropsis metallica, Halictus confusus, H. ligatus* [sweat bee], *H. rubicundus, Lasioglossum albipenne, L. connexum, L. ellisiae, L. pilosum, L. versatum, L. vierecki, L. zephyrum; Sphecodes clematidis, S. cressonii, S. pimpinellae*), bean leaf beetles (Coleoptera: Chrysomelidae: *Cerotoma trifurcata*), blister beetles (Coleoptera: Meloidae: *Epicauta pensylvanica*), checkered beetles (Coleoptera: Cleridae: *Trichodes apivorus*), flower beetles (Coleoptera: Carabidae: *Lebia ornata; Microlestes maurus*), fruit and flower chafers (Coleoptera: Scarabaeidae: *Euphoria sepulcralis*), locust borers (Coleoptera: Cerambycidae: *Megacyllene robiniae*), snout beetles (Coleoptera: Curculionidae: *Cylindridia prolixa, Odontocorynus umbellae*), soldier beetles (Coleoptera: Cantharidae: *Chauliognathus pennsylvanicus*), spotted cucumber beetles (Coleoptera: *Diabrotica undecimpunctata*), wedge-shaped beetles (Coleoptera: Rhipiphoridae: *Macrosiagon limbata*), bugs (Hemiptera: Lygaeidae: *Lygaeus kalmii*; Miridae: *Adelphocoris rapidus, Lygus lineolaris*; Phymatidae: *Phymata fasciata*; Thyreocoridae: *Corimelaena pulicaria*), butterflies (Lepidoptera: Hesperiidae [skippers]: *Ancyloxypha numitor, Euphyes dion, Poanes hobomok, Polites orgenes*; Lycaenidae: *Lycaena hyllus*; Nymphalidae: *Boloria bellona, Danaus plexippus, Euphydryas phaeton, Limenitis archippus, Phyciodes tharos*; Pieridae: *Colias philodice, Pieris rapae*), flies (Diptera: Anthomyiidae: *Delia platura*; Bibionidae: *Dilophus stigmaterus*; Bombyliidae: *Exoprosopa fasciata, E. fascipennis, Poecilanthrax halcyon, Poecilognathus punctipennis, Sparnopolius confusus*; Calliphoridae: *Cochliomyia macellaria, Lucilia illustris, L. sericata, Phormia regina, Pollenia rudis*; Conopidae: *Physoconops brachyrhynchus, Thecophora occidensis*; Empididae: *Empis clausa*; Muscidae: *Lispe albitarsis, Musca domestica, Neomyia cornicina, Stomoxys calcitrans*; Sarcophagidae: *Helicobia rapax, Ravinia stimulans, Sarcophaga sarracenioides*; Sepsidae: *Sepsis violacea*; Stratiomyidae: *Hedriodiscus binotata, Nemotelus glaber*; Syrphidae: *Cheilosia punctulata, Eristalinus aeneus, Eristalis arbustorum, E. dimidiatus, E. tenax, E. transversus, Paragus bicolor, Sphaerophoria contiqua, Spilomyia longicornis, S. sayi, Syritta pipiens, Syrphus ribesii, Toxomerus marginata, T. polita, Tropidia quadrata*; Tabanidae: *Hybomitra lasiophthalmus*; Tachinidae: *Archytas analis, A. aterrima, Euclytia flava, Gnadochaeta globosa, Gymnoclytia immaculata, G. occidua, Hystriciella pilosa, Leucostoma simplex, Lydina areos, Opsidia gonioides, Phasia aeneoventris, Plagiomima spinosula, Senotainia rubiventris, Spallanzania hesperidarum, Sphixapata trilineata, Tachinomyia panaetius, Xanthomelanodes arcuatus*), moths (Lepidoptera: Arctiidae: *Cisseps fulvicollis, Utetheisa bella*; Noctuidae: *Caenurgina erechtea, Feltia jaculifera*), thrips (Thysanoptera: Phlaeothripidae: *Haplothrips niger*; Thripidae: *Thrips tabaci*), and wasps (Hymenoptera: Braconidae: *Apanteles crassicornis, Chelonus sericeus, Vipio haematodes*; Chrysididae: *Hedychrum violaceum, H. wiltii*; Crabronidae: *Cerceris clypeata, C. nigrescens, C. prominens, Crabro tumidus, Ectemnius atriceps, E. decemmaculatus, E. continuus, E. dilectus, E. rufifemur, E. trifasciatus, Eucerceris zonata, Larra analis, Larropsis distincta; Oxybelus emarginatus, O. mexicanus, O. uniglumis, Philanthus bilunatus, P. gibbosus, P. politus, P. ventilabris, Stizus brevipennis, Tachysphex acutus, T. pompiliformis, Tachytes distinctus, T. pepticus*; Cynipidae: *Lestica confluentus*; Eumenidae: *Ancistrocerus adiabatus, A. antilope, A. campestris, Eumenes fraterna, Euodynerus foraminatus, Stenodynerus anormis*; Ichneumonidae: *Ceratogastra ornata, Cremastus cressoni*; Perilampidae: *Perilampus hyalinus*; Pompilidae: *Anoplius aethiops, A. lepidus, Ceropales maculata, Poecilopompilus interrupta*; Sapygidae: *Sapyga interrupta*; Scoliidae: *Scolia bicincta*; Sphecidae: *Ammophila kennedyi, A. nigricans, A. procera, Anacrabro ocellatus, Eremnophila aureonotata, Isodontia apicalis, I. philadelphica, Sceliphron caementaria, Sphex ichneumonea, S. nudus*; Tiphiidae: *Myzinum quinquecincta*; Vespidae: *Dolichovespula maculata, Polistes annularis, P. carolina, P. fuscata, Vespula germanica*). The foliage of *Eupatorium resinosum* is fed upon by white-tailed deer (Mammalia: Cervidae: *Odocoileus virginianus*).

Economic importance: food: Native Americans cooked and ate the leaves of *Eupatorium perfoliatum*. A bitter tea or tonic has been made from the dried leaves and flower tops; **medicinal:** *Eupatorium perfoliatum* has a long history of medicinal use and is becoming more widespread in homeopathic medicine. Although some sources reference the use of the plant for healing fractured bones (hence "boneset"), the common name actually derives from its use in treating influenza, which by its severity, was known colloquially as "break-bone fever" in parts of the United States. Despite any alleged use, caution should be taken when ingesting any medication containing this plant, especially during pregnancy due to potential abortifacient effects. Medicinal use of *E. perfoliatum* (various parts of the plant) was widespread among native North American tribes. It was used widely as a treatment for fever, colds and the flu (Cherokee, Delaware, Iroquois, Menominee, Mohegan, Nanticoke, Oklahoma, Seminole, Shinnecock), as a gastrointestinal aid (Cherokee, Delaware, Iroquois, Mohegan, Ontario), and to treat urinary or other renal disorders (Iroquios, Kosati, Micmac, Penobscot). The Cherokee also used the plants to induce sweating (also the Shinnecock) and as an antiseptic, diuretic, emetic (also the Koasati and Seminole), purgative, and stimulant. The Iroquois used herbal preparations derived from this plant to treat bone fractures, dermatological problems, hemorrhoids, pleurisy, pneumonia, and venereal disease (also the Micmac and Penobscot) and as an analgesic and laxative. Other uses included treatments for hemorrhage (Penobscot), rheumatism (Chippewa), snakebite (Chippewa, Meskwaki), and worms (Meskwaki). The Chippewa prepared an abortifacient from the roots and the Iroquis made a poison from the herbage. During the American Civil War, the Confederate army considered *E. perfoliatum* as a possible substitute for the expensive quinine used to treat malaria. When administered at a high dosage, the extracts reportedly treated mild cases of malaria effectively. Acidic extracts are ineffective at inhibiting malarial parasites (Plasmodiidae: *Plasmodium berghei*) in laboratory mice (Rodentia: *Mus*); however, homeopathic preparations (30 cH centesimal dilution) reportedly inhibit the multiplication of this parasite significantly. Sesquiterpene lactones (especially guaianolide) in extracts of the aerial foliage also are reported to exhibit antiprotozoal activity against malarial parasites (Plasmodiidae: *Plasmodium falciparum*). The plants contain protocatechuic acid, which is a major antioxidant. Various extracts have exhibited potent anti-inflammatory activity. Ethanolic leaf extracts reputedly show potent cytotoxic activity and weak antibacterial activity against Gram-positive bacteria (*Bacillus megaterium*; *Staphylococcus aureus*). Similarly, polysaccharides (4-O-methyl-glucurono-xylans) from alkaline aqueous extracts are reported to enhance in vitro phagocytosis of human granulocytes by 30%. Although the phagocytic activity of extracts has not been demonstrated conclusively, the anti-inflammatory properties have been well-documented and corroborate prior folk medicinal uses of this plant. A homeopathic 30 cH dose administered to Brazilian residents, who were highly infected by Dengue fever, resulted in a substantial decline in the incidence of the disease. *Eupatorium perfoliatum* also has been a suspected agent of hayfever. *Eupatorium resinosum* contains unspecified alkaloids, but its medicinal properties have not been evaluated; **cultivation:** Plants of *Eupatorium perfoliatum* have been cultivated as long ago as 1831, when they were advertised for sale in a catalog of

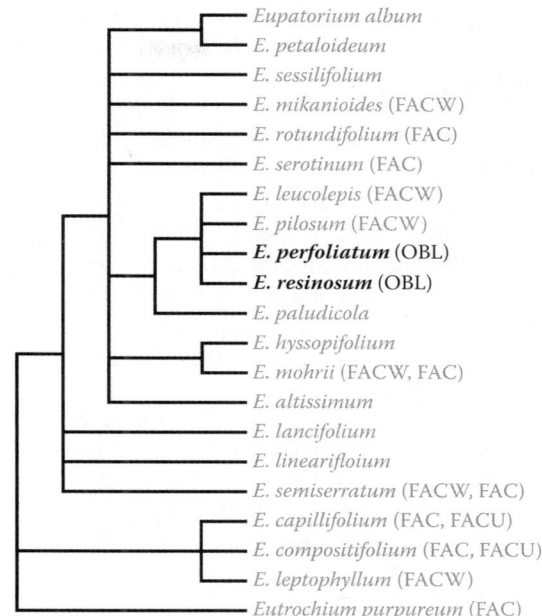

FIGURE 5.152 Phylogenetic relationships among members of the "*Eupatorium album* complex" as indicated by Bayesian analysis of nrITS sequence data. Although many interrelationships in this group are not resolved well, both OBL species (bold) are closely related and occur within a small clade of otherwise FACW species. (Adapted from Schilling, E. E., *Syst. Bot.*, 36, 1088–1100, 2011b. hybrid taxa excluded.)

the Bartram Botanic Garden in Philadelphia. Seeds of *E. perfoliatum* currently are distributed by a number of native wildflower nurseries; **misc. products:** *Eupatorium perfoliatum* has been considered as a plant for habitat management because it attracts high numbers of natural arthropod enemies and herbivores. *Eupatorium* species are planted to attract butterflies (Lepidoptera) to gardens; **weeds:** no (OBL species); **nonindigenous species:** Both OBL species are indigenous.

Systematics: *Eupatorium* is assigned to tribe Eupatorieae of Asteraceae. *Eupatorium perfoliatum* and *E. resinosum* both are $2n=20$, which is interpreted to be the diploid level in *Eupatorium* (basic number: $x=10$). However, extensive gene duplications evidenced by isozyme data indicate that $2n=10$ may have been the original diploid number of the genus. *Eupatorium perfoliatum* and *E. resinosum* are closely related species (Figure 5.152), and it is likely that the OBL habit evolved only once in their common ancestor. The two species hybridize where they co-occur but seem to be at least partially isolated by their different ecological affinities; *E. perfoliatum* is a broadly adapted species of alkaline habitats, and *E. resinosum* is a much more restricted specialist of open, acidic habitats. *Eupatorium perfoliatum* also is known to cross with *E. serotinum* to produce the hybrid *Eupatorium×truncatum*, to which some refer improperly as *E. resinosum* var. *kentuckiense* Fernald. *Eupatorium fernaldii* represents a more complex "trihybrid" derived from *E. petaloideum*, *E. sessilifolium* and *E. perfoliatum*. *Eupatorium resinosum* is believed to represent one parental lineage involved in the formation of the allopolyploid *Eupatorium leucolepis*.

Comments: *Eupatorium perfoliatum* is common throughout eastern North America. *Eupatorium resinosum* is rare and sporadic in the eastern United States (Delaware, New Jersey, New York, North Carolina, South Carolina).

References: Anderson, 1919; Archibald, 1957; Aviado & Reutter, 1969; Baldwin et al., 2001; Baskin & Baskin, 1998; Bjerknes et al., 2007; Brown & Goertz, 1978; Byers, 1995, 1998; Byers & Meagher, 1997; Campbell, 1933; Carr, 1831; Choesin & Boerner, 2000; Churchill et al., 1990; Conard, 1924; Den Breeÿen et al., 2006; Dittmar & Neely, 1999; Emerson, 1921; Ernst, 2002; Felt, 1908; Fiedler & Landis, 2007; Gosling, 1986; Grant, 1953; Greene, 1948; Grieve, 1980; Grundel et al., 2011; Habtemariam, 2008; Habtemariam & Macpherson, 2000; Hamerstrom, Jr. & Blake, 1939a; Hasegawa, 2007; Hensel et al., 2011; Horr & McGregor, 1948; Hotchkiss & Stewart, 1947; Isaacs et al., 2009; Johnson, 1974; Jones, 1924; Jones, Jr., 1968; Keddy & Ellis, 1985; Keddy et al., 1998; Kenow & Lyon, 2009; Kern, 1919; King & Robinson, 1970; Knisley & Denmark, 1978; LaDeau & Ellison, 1999; Laidig & Zampella, 1996; Leonard, 1936; LeSage, 1986; Lindig-Cisneros & Zedler, 2002; Lira-Salazar et al., 2006; Lisberg & Young, 2003; Lovell, 1908, 1915; Maas et al., 2011a,b; Mandossian & McIntosh, 1960; Marino, 2008; Matthews et al., 1990; McJannet et al., 1995; McKain et al., 2010; McVaugh, 1957; Michel & Henein, 2007; Nichols, 1915; Nielsen & Moyle, 1941; Nuzzo, 1978; Orendovici et al., 2003; Osvald, 1935; Paine & Ribic, 2002; Picking & Veneman, 2004; Price, 1960; Record, 2011; Robertson, 1929; Ruthven, 1911; Saunders, 2011; Schilling, 2011b; Schmidt & Schilling, 2000; Scholtens, 1991; Shipley & Parent, 1991; Shipley et al., 1991; Siripun & Schilling, 2006; Sorrie et al., 2006; Sullivan, 1975; Tucker & Dill, 1989; Tuell et al., 2008; Vestal, 1914; Vitt & Horton, 1990; Vollmar et al., 1986; Voss, 1950; Wagner et al., 1985; Weiher & Keddy, 1995; Weishampel & Bedford, 2006; Wheeler, 1983; White, 1965; Wiefenbach, 1993; Willaman & Schubert, 1961; Williams, 1891, 2000; Williams et al., 2012; Yahara et al., 1989; Zampella & Laidig, 1997; Zampella et al., 2001, 2006.

19. *Eurybia*
Wood-aster
Etymology: from the Greek *eurys baios* (wide, few), in reference to the rays
Synonyms: *Aster* (in part); *Biotia*; *Triniteurybia*
Distribution: global: Eurasia; North America; **North America:** widespread
Diversity: global: 23 species; **North America:** 23 species
Indicators (USA): OBL: *Eurybia radula*
Habitat: freshwater; palustrine; **pH:** 3.8–6.7; **depth:** <1 m; **life form(s):** emergent herb
Key morphology: Stems (to 1 m) erect, solitary, reddish; leaves (to 11.5 cm) alternate, lanceolate to elliptic, scabrous above (villous below), deeply serrate, sessile or winged petiolate (to 4.2 cm), bases tapering to clasping; heads (to 25) radiate, solitary or in corymbs, phyllaries (30–50) in 4–5 series, pales absent; ray florets 13–30 with pale blue or violet rays (to 1.5 cm), pistillate; disc florets 37–72, bisexual, radial,

corollas (to 6 mm) yellow; cypselae (to 3.5 mm) fusiform to cylindric, slightly compressed, 8–18-ribbed, light brown to straw-colored, pappus bristles (to 5.2 mm) barbed
Life history: duration: perennial (rhizomes); **asexual reproduction:** rhizomes; **pollination:** insect; **sexual condition:** gynomonoecious; **fruit:** cypselae (common); **local dispersal:** cypselae (gravity, wind); rhizomes; **long-distance dispersal:** cypselae (pappus)
Imperilment: (1) *Eurybia radula* [G5]; SH (DE, NY); S1 (CT, KY, MA, MD, NJ, ON, VA); S2 (PA); S3 (PE)
Ecology: general: *Eurybia* is diverse ecologically, with species ranging from low to high elevations and from dry, open habitats to shaded woodlands. Several species occur commonly in moist sites and nine (39%) are wetland indicators, but only one is recognized as OBL. The species are gynomonoecious, having pistillate ray florets and bisexual disk florets. At least several tested species are self-incompatible and are outcrossed by insects (Insecta) such as butterflies (Lepidoptera) and bees (Hymenoptera). Genetic bottlenecks can result in the loss of incompatibility alleles, causing reduced seed set in populations of some rarer species. The self-incompatibility system also is known to have broken down in some taxa, which are capable of inbreeding. The seeds can be nondormant; however, physiological dormancy occurs in at least some species. Accumulations of leaf litter can reduce germination rates.

Eurybia radula (Aiton) G.L. Nesom is a clonal perennial, which resides in or on bogs, cliffs, ditches, floodplains, heaths, meadows, patterned fens, swamps, and margins of lakes, streams, and waterfalls at elevations of up to 1385 m. In shorelines, plants have been observed to occupy the zone from 0.5 to 1.0 m above the water level elevation. The substrates are acidic (pH 3.8–6.7) and comprise peat, muck or podzolic soils. Flowering occurs from July to September. Pollination is carried out by insects (Insecta), which achieve higher rates of successful pollination in field sites than in forest populations. A higher rate of pollination also occurs in larger populations than in smaller ones. The seeds germinate best after cold stratification (4°C) on moist filter paper for at least 6 weeks. Seeds from more southern portions of the range (e.g., Pennsylvania) weigh more than those from the north (e.g., Maine). Their longevity in the seed bank has not been determined. The plants develop into lax clones, which are propagated vegetatively by rhizomes.
Reported associates: *Abies balsamea, Acer rubrum, Aconitum uncinatum, Agalinis purpurea, Ageratina altissima, Agrostis hyemalis, A. mertensii, Alnus rugosa, A. viridis, Amelanchier arborea, Anaphalis margaritacea, Andromeda polifolia, Arethusa bulbosa, Arnica chamissonis, Aronia melanocarpa, A. ×prunifolia, Athyrium filix-femina, Aulacomnium palustre, Avenella flexuosa, Bartonia paniculata, Betula michauxii, B. papyrifera, B. pumila, Bidens connatus, Brachyelytrum erectum, Bryum pseudotriquetrum, Calamagrostis canadensis, C. pickeringii, Calliergon stramineum, Campylium stellatum, Carex brunnescens, C. bullata, C. buxbaumii, C. deflexa, C. disperma,*

C. echinata, C. exilis, C. hormathodes, C. intumescens, C. lasiocarpa, C. leptalea, C. livida, C. michauxiana, C. muricata, C. oligosperma, C. pauciflora, C. rariflora, C. rostrata, C. scirpoidea, C. stylosa, C. tribuloides, C. trisperma, C. wiegandii, Cephalozia connivens, Chamaedaphne calyculata, Cladonia rangiferina, Cladopodiella fluitans, Coptis trifolia, Cornus canadensis, Crataegus macrosperma, Cuscuta rostrata, Dasiphora floribunda, Deschampsia cespitosa, Dichanthelium dichotomum, D. scoparium, Dicranum polysetum, Diervilla lonicera, Doellingeria umbellata, Drosera rotundifolia, Dryopteris campyloptera, D. intermedia, Empetrum nigrum, Epilobium palustre, E. strictum, Equisetum arvense, E. fluviatile, Eriophorum virginicum, E. viridicarinatum, Euthamia graminifolia, Galium kamtschaticum, G. tinctorium, Gaultheria hispidula, Gaylussacia dumosa, G. baccata, Geocaulon lividum, Glyceria canadensis, G. obtusa, G. striata, Gymnadeniopsis clavellata, Helonias bullata, Houstonia caerulea, Hydrangea arborescens, Hylocomium splendens, Hypericum densiflorum, Ilex glabra, I. mucronata, Juncus stygius, J. trifidus, Juniperus communis, Kalmia latifolia, K. polifolia, Larix laricina, Lilium superbum, Limprichtia revolvens, Linnaea borealis, Lobelia kalmii, L. nuttallii, Lonicera villosa, Lycopodiella alopecuroides, L. inundatum, Lycopodium alpinum, L. annotinum, L. sitchense, Lycopus uniflorus, Lysimachia terrestris, L. ×producta, Maianthemum canadense, M. trifolium, Marshallia grandiflora, Menyanthes trifoliata, Menziesia pilosa, Muhlenbergia glomerata, Mylia anomala, Myrica gale, M. pensylvanica, Oclemena acuminata, O. nemoralis, Onoclea sensibilis, Osmunda regalis, Osmundastrum cinnamomeum, Oxalis montana, Oxypolis rigidior, Packera indecora, Paronychia argyrocoma, Paspalum longipilum, P. setaceum, Pellia endiviifolia, Physocarpus opulifolius, Picea mariana, P. rubens, Pinus rigida, Platanthera blephariglottis, P. dilatata, P. peramoena, Pleurozium schreberi, Polygala nuttallii, P. sanguinea, Polytrichum strictum, Prenanthes trifoliolata, Primula mistassinica, Pteridium aquilinum, Rhamnus alnifolia, Rhododendron arborescens, R. canadense, R. groenlandicum, R. maximum, Rhus typhina, Rhynchospora alba, R. capitellata, R. fusca, Riccardia palmata, Rosa nitida, Rubus arcticus, R. chamaemorus, Salix pedicellaris, Sanguisorba canadensis, Sarracenia purpurea, Schizaea pusilla, Schoenoplectus torreyi, Selaginella selaginoides, Sibbaldiopsis tridentata, Smilax pseudochina, S. rotundifolia, Solidago bicolor, S. rugosa, S. macrophylla, S. uliginosa, Sphagnum angustifolium, S. capillifolium, S. fallax, S. fuscum, S. girgensohnii, S. lindbergii, S. magellanicum, S. palustre, S. papillosum, S. pulchrum, S. recurvum, S. rubellum, S. subnitens, S. subtile, S. teres, S. warnstorfii, Spiraea alba, Spiranthes cernua, Stenanthium leimanthoides, Thalictrum pubescens, Thelypteris noveboracensis, Thuja occidentalis, Tomenthypnum nitens, Trautvetteria caroliniensis, Trianthia glutinosa, Trichophorum alpinum, T. cespitosum, Triglochin maritimum, Vaccinium corymbosum, V. macrocarpon, V. myrtilloides, V. oxycoccos, V. uliginosum, Veratrum viride, Viburnum edule, V. nudum, Viola lanceolata, V. pedatifida, Woodsia ilvensis, Woodwardia virginica, Xyris montana, X. torta.

Use by wildlife: *Eurybia radula* is the host of a gall-forming midge (Insecta: Diptera: Cecidomyiidae: *Asteromyia laeviana*) and rust fungi (Basidiomycota: Pucciniaceae: *Puccinia asteris*). Case-building caterpillars (Insecta: Lepidoptera: Coleophoridae: *Coleophora*) attack the seeds and can excavate a high percentage of them. The plants also are susceptible to infection by TSWV, the tomato spotted wilt virus (Bunyaviridae: *Tospovirus*). Occurrences do not seem to be altered by the presence of browsing deer (Mammalia: Cevidae: *Odocoileus virginianus*), which are known to graze on other members of the genus.

Economic importance: food: not edible; **medicinal:** none reported; **cultivation:** There is a cultivar of *Eurybia radula* known as 'August Sky'; **misc. products:** none; **weeds:** none; **nonindigenous species:** none

Systematics: *Eurybia* represents one of the genera split from the North American *Aster* complex (where it was classified as section *Biotia*) as molecular phylogenetic studies clarified relationships within that group. The precise classification of *Eurybia* remains problematic. The genus had been included within tribe Astereae subtribe Solidagininae; however, initial phylogenetic analysis of nrITS data placed it instead with genera assigned to subtribe Machaerantherinae of Astereae. More thorough sampling indicated that *Eurybia* is sister to a clade that includes *Triniteurybia* and various genera of subtribe Machaerantherinae, but neither of the other "eurybioid" genera (i.e., *Herrickia, Oreostemma*) (Figure 5.136). The base chromosome number of *Eurybia* is $x=9$. The genus includes both diploids and polyploids, which seem to associate independently. *Eurybia radula* is a diploid ($2n=18$), and occupies a fairly solitary position amidst other diploid species (*E. avita, E. compacta, E. divaricata, E. hemispherica, E. mirabilis*). The chromosome morphology is believed to be a "derived" type, where a large satellite is attached to a short proximal portion of the short arm. There are no reports of hybrids involving *E. radula*.

Comments: *Eurybia radula* occurs in northeastern North America.

References: Bartram, 1913; Beck et al., 2004; Brouillet, 2006a; Brouillet et al., 1998; Fernald, 1901; Foster & King, 1984; Grondin & Ouzilleau, 1980; Harvey, 1903; Hutton, 1974; Keddy & Wisheu, 1989; Loeffler, 2013; Parrella et al., 2003; Pellerin et al., 2006; Reinartz & Les, 1994; Sabourin, 2006; Selliah, 2009; Selliah & Brouillet, 2008; Semple et al., 1983; Sperduto et al., 2000; Stireman III et al., 2010; Titus et al., 1995.

20. *Euthamia*

Goldentop; verge d'or

Etymology: from the Greek *eu thaminos* ("well-crowded") for the densely arranged flowers

Synonyms: *Solidago* (in part)

Distribution: global: Asia*; Europe*; Mexico; North America; **North America:** widespread

Diversity: global: 6 species; **North America:** 6 species
Indicators (USA): OBL; FACW: *Euthamia occidentalis*
Habitat: brackish; freshwater; palustrine, riverine; **pH:** 7.9–8.7; **depth:** <1 m; **life form(s):** emergent herb or subshrub
Key morphology: Stems (to 2 m) erect, herbaceous or somewhat woody; leaves (to 10 cm) linear, 3–5-nerved, margins entire but scabrous, apices sharply acute to acuminate, faces glandular dotted (to 56 dots/mm²), sessile; heads radiate, in elongate paniculate clusters, phyllaries (to 4.9 mm) in 3–5 series, pales absent; ray florets (to 28) pistillate, with yellow rays (to 2.5 mm); disc florets (to 18) bisexual, with yellow, 5-lobed, corollas (to 4.2 mm); cypselae (to 1 mm) fusiform, strigose, pappus of 25–45 barbellate bristles (to 1.5 mm)
Life history: duration: perennial (rhizomes); **asexual reproduction:** rhizomes; **pollination:** insect; **sexual condition:** gynomonoecious; **fruit:** cypselae (common); **local dispersal:** cypselae (water, wind); **long-distance dispersal:** cypselae (water, wind)
Imperilment: (1) *Euthamia occidentalis* [G5]; S2 (WY)
Ecology: general: All North American *Euthamia* are wetland indicators, which frequently occur on lake shores and in other moist, open areas. Only one species is regarded as OBL with the others more prevalent in mesic sites. All species are rhizomatous perennials that extend across a range of acidic to alkaline habitats, which often are characterized by sandy substrates. The plants are gynomonoecious, having pistillate ray florets and hermaphroditic disc florets. The self-incompatible flowers are pollinated by insects (Insecta). The cypselae have a bristly pappus and are dispersed by wind. The seeds germinate fairly readily across a wide range of conditions at temperatures above 18°C.

Euthamia occidentalis **Nutt.** grows in fresh to brackish/saline (0.14%–0.17% total salts), tidal or nontidal cienaga, depressions, ditches, flats, floodplains, gravelbars, marshes, meadows, river bottoms, sandbars, seeps, sinks, streambeds, swales, swamps, thickets, tinaja, washes, and along the margins of canals, lagoons, lakes, levees, ponds, rivers, and sloughs at elevations of up to 2591 m. Mainly riparian, the plants will tolerate shallow water and exposures ranging from fully open to shaded conditions. They also can tolerate average monthly temperatures ranging from −8°C to 33°C. Substrates are described as adobe, alluvium, clay, cobble, cobbly gravelly sand, gravel, loam, mud, rocks, ryolite, sand, sandy cobble, sandy clay, sandy gravel, sandy gravelly loam, sandy loam, serpentine, shale, silt, and silt loam. Specific soil types have been categorized as Dia, Dithold, Eastfork, haplaquents, Parran, typic psammaquents, and Umberland. They are alkaline (pH 7.9–8.7) with moderate levels of phosphorous (16–26 ppm), nitrate nitrogen (14–24 ppm) and specific conductivity (1.6–3.2 mmhos/cm). Substrate composition varies from high proportions of sand (43%–52%) and gravel (38%–66%) with small amounts of silt (2%–10%), to high proportions of silt (45%–50%) with lesser amounts of sand (35%–39%) and clay (15%–16%). The organic matter content typically is low (1.4%–2.3%). The flowers are produced from July to November and are pollinated by bees or other insects (see *Use by wildlife* later). The seeds are dispersed by wind,

or in water. Numerous viable seeds (34% germination rate under a 12/12, 30°C/20°C regime) have been retrieved from irrigation water, making this pathway an alternative dispersal route. Pretreatment of seeds is unnecessary and germination occurs at ambient greenhouse temperatures. Germination rates for 6–12 month old seeds are 37%–64% in shadehouse conditions. The plants can shed about 500 seeds/m² and seedbank samples have yielded 24–63 germinating seeds/962 cm³ of sediment sampled. The field longevity of seeds has not been fully determined, but seeds have germinated after 48 months of storage in water, or after 60 months of dry storage. The association of this plant with natural disturbance is due in part to the seedlings, which survive and persist better in newly opened habitats than in longer established open sites or in sites where a canopy has developed. Vegetative reproduction occurs by proliferation of rhizomes. **Reported associates:** *Acer negundo, Achillea millefolium, Achnatherum hymenoides, Adenostoma, Agrimonia gryposepala, Agrostis exarata, A. gigantea, A. stolonifera, Alhagi maurorum, Alisma triviale, Allium, Alnus rhombifolia, Alopecurus saccatus, Ambrosia psilostachya, A. trifida, Amelanchier alnifolia, Ammannia coccinea, Amorpha fruticosa, Andropogon glomeratus, Anemopsis californica, Apium graveolens, Apocynum cannabinum, Aristida purpurea, Artemisia californica, A. douglasiana, A. frigida, A. ludoviciana, A. tridentata, A. vulgaris, Arthrocnemum subterminale, Arundo donax, Asclepias speciosa, Astragalus pycnostachyus, Atriplex canescens, A. lentiformis, A. parryi, Baccharis douglasii, B. pilularis, B. salicifolia, B. salicina, B. sarothroides, Bacopa, Balsamorhiza sagittata, Bassia hyssopifolia, Berberis repens, Berula erecta, Betula papyrifera, Bidens frondosus, B. tripartitus, Boerhavia coulteri, Bolboschoenus maritimus, Botrychium multifidum, Bouteloua gracilis, Brickellia californica, B. longifolia, Bromus inermis, B. tectorum, Calamagrostis scopulorum, Callitriche heterophylla, Calystegia sepium, Campanula rotundifolia, Carex aperta, C. aquatilis, C. feta, C. interrupta, C. lenticularis, C. nebrascensis, C. obnupta, C. pellita, C. spissa, C. stipata, C. utriculata, C. vulpinoidea, Carpobrotus edulis, Castilleja angustifolia, C. linariifolia, C. minor, Celtis reticulata, Centaurea solstitialis, C. ×moncktonii, Cercis occidentalis, Cercocarpus montanus, Chenopodium macrospermum, Chloracantha spinosa, Chloropyron maritimum, Chrysothamnus greenei, C. viscidiflorus, Cicuta douglasii, C. maculata, Cirsium arvense, C. hydrophilum, C. vulgare, Clematis ligusticifolia, Conium maculatum, Conyza canadensis, Coreopsis tinctoria, Cornus sericea, Croton setiger, Crypsis schoenoides, Cuscuta salina, Cyperus acuminatus, C. eragrostis, C. erythrorhizos, C. esculentus, Cynodon dactylon, Cynosurus echinatus, Cytisus scoparius, Danthonia intermedia, Datisca glomerata, Daucus carota, Delairea odorata, Deschampsia cespitosa, Dichanthelium oligosanthes, Dieteria canescens, Dipsacus fullonum, Distichlis littoralis, D. spicata, Dysphania ambrosioides, Echinochloa crus-galli, Eclipta prostrata, Elaeagnus angustifolia, Eleocharis erythropoda, E. macrostachya, E. palustris, E. parishii, E. rostellata, Elymus glaucus, E. repens,*

Encelia resinifera, Epilobium brachycarpum, E. ciliatum, E. densiflorum, Epipactis, Equisetum arvense, E. hyemale, E. laevigatum, E. telmateia, Eragrostis hypnoides, Eremogone congesta, E. fendleri, Ericameria albida, E. nauseosa, E. parryi, E. pinifolia, Erigeron, Eriogonum fasciculatum, Eryngium heterophyllum, Eucalyptus camaldulensis, E. cladocalyx, Eustoma exaltatum, Festuca arizonica, F. idahoensis, Fimbristylis thermalis, Foeniculum vulgare, Frankenia salina, Fraxinus pennsylvanica, F. velutina, Galium trifidum, Gastridium ventricosum, Glyceria elata, G. striata, Glycyrrhiza lepidota, Gnaphalium palustre, Grindelia hirsutula, Helenium autumnale, H. puberulum, Helianthus californicus, H. nuttallii, H. paradoxus, Heliotropium curassavicum, H. europaeum, Helminthotheca echioides, Heterotheca oregona, H. villosa, Hibiscus moscheutos, Hirschfeldia incana, Hoita macrostachya, H. orbicularis, Holcus lanatus, Holodiscus discolor, Hordeum jubatum, H. murinum, Horkelia, Hypericum anagalloides, H. perforatum, Isocoma menziesii, Isolepis carinata, Jacobaea vulgaris, Jaumea carnosa, Juglans, Juncus acuminatus, J. acutus, J. arcticus, J. balticus, J. bufonius, J. dubius, J. effusus, J. falcatus, J. macrandrus, J. nevadensis, J. oxymeris, J. patens, J. phaeocephalus, J. tenuis, J. torreyi, J. xiphioides, Juniperus communis, J. occidentalis, J. scopulorum, Kickxia elatine, Kochia scoparia, Koeleria macrantha, Krascheninnikovia lanata, Lactuca serriola, Lasthenia glabrata, Lathyrus jepsonii, Leersia oryzoides, Lepidium draba, L. latifolium, Lepidospartum squamatum, Leptochloa fusca, Leymus cinereus, L. triticoides, Lilaeopsis masonii, L. occidentalis, Limonium californicum, Lindernia dubia, Lolium multiflorum, Lotus corniculatus, Ludwigia palustris, L. peploides, Lupinus, Luzula campestris, Lycopus americanus, L. asper, Lysimachia maritima, Lythrum californicum, L. salicaria, Malosma laurina, Medicago sativa, Melilotus albus, M. officinalis, Mentha arvensis, M. pulegium, Mentzelia, Mimulus cardinalis, M. guttatus, M. moschatus, M. primuloides, Mollugo verticillata, Morus alba, Muhlenbergia asperifolia, M. filiculmis, M. montana, M. racemosa, M. rigens, Myriophyllum aquaticum, Myosotis laxa, Nasturtium officinale, Oenanthe sarmentosa, Oenothera deltoides, O. elata, Opuntia polyacantha, Osmorhiza berteroi, Pascopyrum smithii, Paspalum distichum, Persicaria amphibia, P. coccinea, P. hydropiperoides, P. lapathifolia, P. maculosa, P. punctata, Phalaris arundinacea, Phlox hoodii, Phragmites australis, Phyla nodiflora, Picea engelmannii, P. glauca, Pinus edulis, P. flexilis, P. ponderosa, Plagiobothrys chorisianus, P. figuratus, P. mollis, Plantago coronopus, P. major, P. patagonica, Platanus wrightii, Pluchea carolinensis, P. odorata, P. sericea, Poa annua, P. bulbosa, P. compressa, P. pratensis, P. secunda, P. trivialis, Polypogon monspeliensis, Populus deltoides, P. fremontii, P. tremuloides, Portulaca oleracea, Potentilla anserina, P. gracilis, P. newberryi, Prosopis glandulosa, Prunus virginiana, Psathyrotes, Pseudotsuga menziesii, Pyrrocoma lanceolata, P. racemosa, Quercus agrifolia, Q. chrysolepis, Q. kelloggii, Q. macrocarpa, Q. wislizeni, Q. ×moreha, Ranunculus flammula, Raphanus sativus, Rhamnus californica, Rhus aromatica, R. integrifolia, Ribes aureum, Ricinus communis, Rorippa columbiae, R. curvisiliqua, Rosa californica, R. woodsii, Rubus ulmifolius, R. ursinus, Rumex conglomeratus, R. crispus, Sagittaria latifolia, Salicornia depressa, Salix amygdaloides, S. columbiana, S. exigua, S. gooddingii, S. hookeriana, S. laevigata, S. lasiolepis, S. lucida, S. lutea, S. melanopsis, S. nigra, S. sessilifolia, S. sitchensis, Salsola, Salvia apiana, Sarcobatus vermiculatus, Sarcocornia pacifica, Schedonorus arundinaceus, Schoenoplectus acutus, S. americanus, S. californicus, S. pungens, S. tabernaemontani, Senegalia greggii, Shepherdia canadensis, Sisymbrium altissimum, Sisyrinchium halophilum, Sium suave, Solanum dulcamara, Solidago canadensis, S. gigantea, S. velutina, Sonchus, Sorghum halepense, Spartina gracilis, Sphaeralcea, Spiraea betulifolia, S. douglasii, Spiranthes diluvialis, Sporobolus airoides, Stachys albens, Suaeda calceoliformis, S. californica, S. nigra, Symphoricarpos albus, Symphyotrichum chilense, S. lanceolatum, S. spathulatum, S. subulatum, Taeniatherum caput-medusae, Tamarix chinensis, T. ramosissima, Tetradymia canescens, Thinopyrum intermedium, Toxicodendron rydbergii, Trifolium repens, Triglochin maritimum, Triticum aestivum, Typha angustifolia, T. domingensis, T. latifolia, Ulmus pumila, Umbellularia californica, Urtica dioica, Vachellia constricta, Verbascum thapsus, Veronica americana, Vicia nigricans, Vinca major, Viola adunca, Vitis californica, V. girdiana, Xanthium strumarium, Zeltnera namophila.

Use by wildlife: *Euthamia occidentalis* is a host to several plant pathogenic fungi including powdery mildew (Ascomycota: Erysiphaceae: *Erysiphe cichoracearum*), leaf spot (Ascomycota: Mycosphaerellaceae: *Neoramularia spissa*) and rusts (Basidiomycota: Pucciniaceae: *Puccinia asterum, P. dioicae*). The flowers are visited by various bees (Insecta: Hymenoptera: Andrenidae: *Andrena aurihirta, A. citrinihirta, Perdita ciliata, P. interserta; Protoxaea gloriosa*; Apidae: *Melissodes paulula*; Colletidae: *Hylaeus leptocephala, H. mesillae, H. polifolii*; Halictidae: *Lasioglossum incompletum*; Megachilidae: *Megachile gentilis, M. perihirta*), which probably represent their major pollinators. Secondary pollinators include beetles (Insecta: Coleoptera: Melyridae: *Listrus senilis*), which eat the pollen and nectar. The flowers also provide nectar for a variety of butterflies and moths (Insecta: Lepidoptera: Hesperiidae; Nymphalidae: *Speyeria coronis*; Pieridae: *Colias eurytheme*) and are hosts to scorpion wasps (Insecta: Hymenoptera: Ichneumonidae: *Nothocremastus intermedius*) and soldier flies (Insecta: Stratiomyidae: *Hedriodiscus truquii, Odontomyia tumida*). A solitary wasp (Insecta: Hymenoptera: Crabronidae: *Rhopalum gracile*) makes its nests in the stems. The plants are dominant at nesting sites of the yellow-billed cuckoo (Aves: Cuculidae: *Coccyzus americanus*).

Economic importance: food: none reported; **medicinal:** none reported; **cultivation:** not in cultivation; **misc. products:** *Euthamia occidentalis* is commonly planted in marsh restoration projects. The plants are used as a source of dye. The Coloma people of California sometimes used *E. occidentalis* as a spindle plant in hand drills for fire starting, but

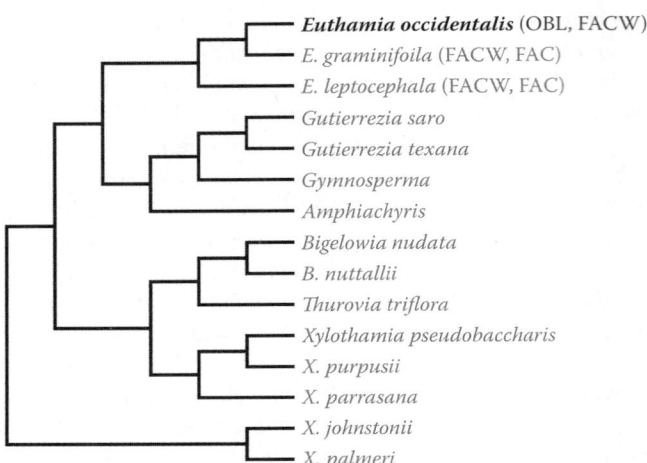

FIGURE 5.153 Relationships among selected genera of Asteraceae (Astereae) as resolved by phylogenetic analysis of nrITS (ETS, ITS, indel) data. The OBL *Euthamia occidentalis* (bold) occurs with other wetland indicators in a clade that is sister to *Amphiachyris*, *Gutierrezia*, and *Gymnosperma*. (Adapted from Urbatsch, L.E. et al., *Amer. J. Bot.*, 90, 634–649, 2003.)

it was not preferred for this purpose. Unlike some goldenrods (with which it was once allied), *E. occidentalis* does not contain rubber; **weeds:** none; **nonindigenous species:** The *Euthamia* species introduced to Eurasia is *E. graminifolia* not *E. occidentalis*.

Systematics: *Euthamia* is a member of tribe Astereae (subtribe Solidagininae) where it associates closely with *Amphiachyris*, *Gymnosperma* and *Gutierrezia* (Figure 5.153), but distantly from *Solidago*, where it had been assigned previously. Phylogenetic analysis including three of the species (Figure 5.153) indicates that the genus is monophyletic as currently circumscribed. The status of *E. galetorum* is unresolved as some regard it as a variety of *E. graminifolia*. The base chromosome number of *Euthamia* is $x = 9$. *Euthamia occidentalis* ($2n = 18$) is diploid. There are no reports of hybridization involving *E. occidentalis*.

Comments: *Euthamia occidentalis* occurs throughout western North America

References: Auble et al., 1994; Bollman et al., 2012; Christy, 2004; Comes et al., 1978; Dixon & Johnson, 1999; Haines, 2006; Hall & Long, 1921; Hurd, Jr. & Linsley, 1976; Hutchinson et al., 2007; Jackson, 1918; James, 1960; Jensen, 2007; Kelley & Bruns, 1975; Laymon, 1998; Lee & Scott, 2007; Macdonald & Barbour, 1974; Mawdsley, 1999, 2003; McCain & Christy, 2005; Moore et al., 2011; Morse, 2008; Nachlinger, 1988; Nesom, 2000; Odion et al., 1992; Pisula & Meiners, 2010; Saunders, 2011; Shapiro, 1974; Sieren, 1981; Spence & Henderson, 1993; Thompson, 2001; Trowbridge, 2007; Urbatsch et al., 2003; Vaghti & Keeler-Wolf, 2004; Whitcraft et al., 2011.

21. *Eutrochium*

Joe Pye weed; eupatoire

Etymology: from the Greek *eu trocho* (truly wheel-like), in reference to the leaf whorls

Synonyms: *Eupatoriadelphus*; *Eupatorium* (in part)

Distribution: global: North America; **North America:** widespread

Diversity: global: 5 species; **North America:** 5 species

Indicators (USA): OBL: *Eutrochium fistulosum*, *E. maculatum*; **FACW:** *E. fistulosum*, *E. maculatum*

Habitat: freshwater; palustrine; **pH:** 4.5–7.95; **depth:** <1 m; **life form(s):** emergent herbs

Key morphology: Stems (to 37 dm) solid or hollow, striate, purplish; leaves (to 28 cm) in whorls of 3–11, petioled (to 5 cm), elliptic to broadly/narrowly lanceolate or lance-ovate, pinnately veined, margins finely serrate or doubly serrate with blunt or sharp teeth, bases gradually or abruptly tapered; heads with 4–22 florets, arranged in dome-like or flat-topped compound corymbs; involucres (to 9 mm) purplish; florets (to 6 mm) with pink to purple corollas (to 7.5 mm); cypselae (to 5.5 mm) ribbed, yellow to black, with a rosy or purplish, barbellate pappus of up to 40 bristles.

Life history: duration: perennial (rhizomes); **asexual reproduction:** rhizomes; **pollination:** insect; **sexual condition:** hermaphroditic; **fruit:** cypselae (common); **local dispersal:** rhizomes, cypselae (wind); **long-distance dispersal:** cypselae (wind)

Imperilment: (1) *Eutrochium fistulosum* [G5]; S1 (MI, NH); S2 (ME); (2) *E. maculatum* [G5]; SH (KY); S1 (AB, ID, MT, WV); S2 (VA)

Ecology: general: *Eutrochium* occurs most commonly in meadow or woodland habitats that are characterized by either acidic or alkaline substrates. Four of the five species are wetland indicators. Both OBL species also are ranked as FACW in part of their range, indicating their broad range of habitat tolerance. All five species are perennial. *Eutrochium* reportedly includes both strongly self-incompatible as well as self-compatible species; some species are apomictic. The flowers typically are pollinated by various insects (Insecta). The cypselae are wind dispersed as facilitated by their bristly pappus. It is not known whether the fruits also are capable of water dispersal. The seeds are physiologically dormant and normally require a period of cold stratification to induce germination. They do not appear to persist for long duration in the seed bank.

***Eutrochium fistulosum* (Barratt) E.E. Lamont** is reported from bogs, deltas, depressions, ditches, flatwoods, floodplains, glades, meadows, pocosins, ravines, river bottoms, seeps, springs, swales, swamps, washes, and the margins of lakes, ponds, and streams at elevations of up to 1400 m. This species occupies sites where exposures vary from open (full sunlight) to various degrees of shade. The substrates are acidic to neutral (pH 4.5–7.1) and are described as gravel, loamy sand, muck, sand, sandstone and sandy clay and more specifically as Gilpin, Wellston and Muskingum soil series. Plants can grow quite tall, but their height ceases to increase once inflorescences are produced. The flowers are visited and pollinated by bumblebees and butterflies, which forage for pollen and nectar (see *Use by wildlife*). The seeds have physiological dormancy, which is broken by cold stratification at 5°C (up to 12 weeks). Optimal germination occurs under a 30°C/15°C temperature regime. The minimum germination temperature decreases as the after ripening of cypselae progresses. Low germination can be obtained using fresh seeds harvested in late August or early September. These should be planted on a soilless medium, lightly covered with medium, and watered regularly. Seed bank studies found higher numbers of propagules in farmed areas than intact communities. Exposure to silver nanoparticles enhances the germination rate of seeds, but also results in decreased root and shoot growth. Measured primary productivity is relatively low, yielding an annual above ground biomass of $44.4\,g\,m^{-2}$. The plants reproduce vegetatively by the production of rhizomes. **Reported associates:** *Acer negundo, A. pensylvanicum, A. rubrum, A. spicatum, Actaea podocarpa, Aesculus flava, Ageratina altissima, Agrostis perennans, Alliaria petiolata, Allium tricoccum, Alnus serrulata, Ambrosia, Amorpha fruticosa, A. ouachitensis, Amphicarpaea bracteata, Amsonia hubrichtii, A. illustris, Andropogon gerardii, A. glomeratus, Angelica triquinata, Apios americana, Apocynum cannabinum, Arisaema dracontium, A. triphyllum, Arnoglossum ovatum, Aronia arbutifolia, A. melanocarpa, Arundinaria gigantea, Asclepias verticillata, Athyrium filix-femina, Betula alleghaniensis, B. nigra, Bidens frondosus, Bignonia capreolata, Blephilia ciliata, Boehmeria cylindrica, Boltonia diffusa, Botrychium virginianum, Boykinia aconitifolia, Calamagrostis canadensis, Carex atlantica, C. bromoides, C. crinita, C. granularis, C. leptalea, C. lurida, C. misera, C. stricta, C. torta, Carpinus caroliniana, Carya glabra, C. ovata, Cercis canadensis, Chamaecrista fasciculata, Chasmanthium latifolium, C. laxum, Cinna arundinacea, C. latifolia, Clintonia borealis, Coleataenia anceps, Commelina erecta, Conoclinium coelestinum, Coreopsis tripteris, Cornus alternifolia, C. amomum, C. florida, Crataegus, Cuscuta rostrata, Cypripedium reginae, Deparia acrostichoides, Dichanthelium dichotomum, D. oligosanthes, D. polyanthes, Diervilla sessilifolia, Diodella teres, Diodia virginiana, Dioscorea villosa, Diospyros virginiana, Drosera capillaris, Dryopteris marginalis, Echinacea purpurea, Eleocharis tuberculosa, Elymus hystrix, Epilobium coloratum, E. hirsutum, Equisetum arvense, Eriocaulon decangulare, Eriophorum virginicum, Eryngium integrifolium, E. yuccifolium, Euonymus obovatus, Eupatorium capillifolium, E. perfoliatum, E. pilosum, E. rotundifolium, Euphorbia corollata, Eurybia chlorolepis, Fagus grandifolia, Frasera caroliniensis, Fuirena squarrosa, Gentianella quinquefolia, Geranium maculatum, Glyceria striata, Gratiola pilosa, Gymnadeniopsis clavellata, Halesia diptera, Hamamelis vernalis, Helianthus angustifolius, H. divaricatus, H. mollis, Holcus lanatus, Houstonia serpyllifolia, Huperzia lucidula, Hydrangea arborescens, Hygroamblystegium tenax, Hypericum mutilum, H. prolificum, Ilex decidua, Impatiens capensis, Isoetes riparia, Itea virginica, Juncus caesariensis, J. canadensis, J. coriaceus, J. effusus, Juniperus virginiana, Justicia americana, Lactuca canadensis, Laportea canadensis, Larix laricina, Leersia oryzoides, Lespedeza capitata, L. procumbens, Liatris aspera, L. squarrosa, Lilium superbum, Lindera benzoin, Linum medium, Liquidambar styraciflua, Lithospermum canescens, Lobelia cardinalis, L. floridana, L. siphilitica, L. spicata, Lonicera japonica, Ludwigia decurrens, Lycopodiella appressa, Lycopus uniflorus, L. virginicus, Lygodium palmatum, Lyonia ligustrina, Lysimachia fraseri, L. terrestris, Magnolia macrophylla, Maianthemum racemosum, Manfreda virginica, Mimulus, Monarda fistulosa, Oclemena acuminata, Onoclea sensibilis, Orontium aquaticum, Ostrya virginiana, Osmunda regalis, Osmundastrum cinnamomeum, Oxypolis rigidior, Packera aurea, Panicum virgatum, Parnassia grandifolia, Parthenocissus quinquefolia, Pedicularis lanceolata, Perilla frutescens, Persicaria arifolia, P. pensylvanica, P. sagittata, Phalaris arundinacea, Phlox pilosa, Physostegia virginiana, Picea rubens, Platanthera leucophaea, Platanus occidentalis, Pleopeltis polypodioides, Poa nemoralis, Polygala cruciata, Polygonatum pubescens, Polypodium appalachianum, Polystichum acrostichoides, Populus tremuloides, Potentilla simplex, Prenanthes altissima, Ptilimnium nodosum, Pycnanthemum tenuifolium, Quercus alba, Q. marilandica, Q. michauxii, Q. nigra, Q. shumardii, Ratibida pinnata, Rhamnus caroliniana, Rhexia mariana, R. virginica, Rhododendron arborescens, R. minus, Rhynchospora capitellata, R. gracilenta, R. rariflora, Rosa carolina, R. palustris, Rubus canadensis, Rudbeckia auriculata, R. hirta, R. laciniata, R. mohrii, Ruellia caroliniensis, Rugelia nudicaulis, Rumex crispus, Sabal minor, Sabatia angularis, Saccharum giganteum, Sagittaria latifolia, Salix nigra, Sambucus nigra, S. racemosa, Sanguisorba canadensis, Sarracenia oreophila, Sassafras albidum, Schizachyrium scoparium, Scirpus cyperinus, Scleria ciliata, S. muehlenbergii, Senecio vulgaris, Silphium compositum, S. integrifolium, S. trifoliatum, Solanum carolinense, Solidago caesia, S. glomerata, S. patula, S. rigida, S. rugosa, Sorghastrum nutans, Sorghum halepense, Sphagnum palustre, Stellaria corei, Stenanthium gramineum, Streptopus amplexifolius, Styrax grandifolius, Symphyotrichum dumosum, S. laeve, S. puniceum, S. undulatum, Symplocarpus foetidus, Thalictrum dioicum, T. mirabile, Thelypteris noveboracensis, T. palustris, Thyrsanthella difformis, Tilia americana, Toxicodendron radicans, Trautvetteria caroliniensis, Trillium decipiens, T. erectum, Typha angustifolia, Ulmus alata, U. americana, Vaccinium corymbosum, Vaucheria, Vernonia gigantea, V. lettermannii, V. noveboracensis, Viburnum dentatum, V. lantanoides, V. rufidulum, Viola blanda, V. cucullata, V. primulifolia, V. sororia, Woodwardia areolata, Xyris jupicai, Zizia aurea.*

***Eutrochium maculatum* L.** grows in tidal or nontidal sites within bogs, carrs, ditches, dune depressions, fens, floodplains, flowages, hedgerows, marshes, meadows, prairies, river bottoms, sand bars, seeps, sloughs, swales, swamps, thickets, and along the margins of brooks, canals, dikes, lakes, ponds, reservoirs, rivers, and streams at elevations of up to 2500 m. Although plants are reported to occur in sun or partial shade, they generally are highly shade intolerant (ranked as "9" in a 1–9 tolerance index). Sites usually occur on calcareous circumneutral to alkaline substrates (pH 6.8–7.95), which are described as clay, clay loam, dolomite, Dunkirk fine sandy loam, Dunkirk silt clay loam, Genesee silt loam, gravel, loam, loamy clay, loamy sand, marl, marly silt, muck, peat, peaty loam, rock, rocky clay, sand, sandy muck, sandy peat, Wabash silt–loam, and Withee–Freer–Marshfield soil association (the plants also have been observed on granitic rock and in acid, peaty soil). Although several reports indicate that the flowers are self-compatible, their high pollinator visitation rate surely promotes a fair level of outcrossing. The plants generally are among the most frequently visited by insects (Insecta) at a site of occurrence (32.9% of all insect visits among 10 species in one study). Bumblebees (Hymenoptera: Apidae: *Bombus*) characteristically pollinate the flowers, which often receive dominant visitation rates relative to other wetland species. Gene flow is strongly influenced by plant spacing, with pollinator foraging areas being inversely proportional to plant density. Individuals of the nonindigenous *Lythrum salicaria* (Lythraceae) that are introduced to populations can result in increased pollinator visitation rates to *E. maculatum*, but decreased seed set due to enhanced deposition of foreign pollen. The seeds weigh from 0.28–0.30 mg. They are physiologically dormant and require stratification at 0°C–5°C for 30–140 days to initiate germination. After stratification, germination typically occurs within 30 days (at 25°C) in shade or full sunlight (but better under low light). Optimal rates are achieved by keeping seeds at 25°C for 33 days, then at 15°C/25°C cycles for 14 days. Seeds stratified at 4°C for 9 months showed 40% germination when exposed to a 20°C/30°C daily temperature cycle. Seeds will remain viable for up to 3 years when stored under dry conditions (30% relative humidity) at 4°C. Although properly described as perennial, seed germination characteristics have led some to categorize this species as a "FAC annual" due to its ability to flower within the first growing season. Seedling biomass is suppressed by 61% under cold (16°C/6°C, 16/8 h) relative to warm (24°C/14°C, 16/8 h) conditions. Vegetative tissues contain intermediate levels of nutrients relative to other wetland species: 0.7%–1.3% nitrogen and 0.2%–0.3% phosphorous (as dry weight). Plants (and seedlings) are attributed with a relative growth rate of 0.24 g/g d^{-1}. The weight (g) of plants growing in monoculture swards of tall-growing species can be as much as 85% lower than those growing alone. An increase in abundance has been observed for plants in burned sites, perhaps due to the relaxation of competitive interactions. Root airspace tissue can increase by as much as 600% in flooded plants relative to drawdown conditions. Despite such adaptations, *E. maculatum* is highly sensitive to flooding, which results in considerably lower height and biomass compared to most other species studied. Consequently, although common in wetlands, this species is unlikely to persist in sites that become inundated or receive large volumes of runoff. This species is mycorrhizal, with up to 71% of root surfaces colonized by arbuscular mycorrhizal and dark septate endophytic fungi. The plants occur mostly in areas of very light disturbance; however, they are known to thrive on arsenic-rich mine tailings. Areas grazed by cattle are characterized by shorter plants relative to undisturbed sites. The plants produce a fibrous root system and propagate vegetatively by the production of rhizomes, which can result in the development of small, clonal colonies. **Reported associates:** *Abies balsamea, Acer negundo, A. rubrum, A. saccharinum, Achillea millefolium, Acorus calamus, Actaea rubra, Agalinis purpurea, Agastache scrophulariifolia, Ageratina altissima, Agrostis gigantea, A. scabra, A. stolonifera, Aletris farinosa, Alisma triviale, Alnus rugosa, Amaranthus albus, A. tuberculatus, Amelanchier canadensis, Ambrosia artemisiifolia, A. psilostachya, Amelanchier arborea, Andropogon gerardii, Anemone canadensis, A. cylindrica, Angelica genuflexa, Anthoxanthum odoratum, Apios americana, Apocynum androsaemifolium, A. cannabinum, Artemisia serrata, Asclepias incarnata, Athyrium filix-femina, Berula erecta, Betula alleghaniensis, B. occidentalis, B. papyrifera, B. pumila, B. ×sandbergii, Bidens cernuus, B. trichospermus, Blephilia hirsuta, Boehmeria cylindrica, Bolboschoenus fluviatilis, Bromus ciliatus, B. inermis, B. kalmii, B. latiglumis, Calamagrostis canadensis, C. stricta, Calla palustris, Caltha palustris, Campanula aparinoides, Campylium stellatum, Cardamine bulbosa, C. diphylla, C. pensylvanica, C. pratensis, Carex aquatilis, C. crinita, C. flava, C. folliculata, C. hystericina, C. lacustris, C. laevivaginata, C. lasiocarpa, C. laxiflora, C. leptalea, C. lupulina, C. lurida, C. pedunculata, C. prairea, C. rostrata, C. scoparia, C. shortiana, C. sterilis, C. stricta, C. tetanica, C. torta, C. trichocarpa, C. utriculata, C. vulpinoidea, Castilleja coccinea, Chamaedaphne calyculata, Chamerion angustifolium, Chelone glabra, Cicuta bulbifera, C. maculata, Cirsium altissimum, C. muticum, C. vulgare, Cladonia chlorophaea, C. cristatella, Cladium mariscoides, Clematis virginiana, Clethra alnifolia, Comandra umbellata, Conyza canadensis, Coreopsis tripteris, Cornus amomum, C. racemosa, C. sericea, Corylus cornuta, Crataegus douglasii, Crepis runcinata, Cryptotaenia canadensis, Cuscuta gronovii, Cyperus, Cypripedium parviflorum, Dasiphora floribunda, Daucus carota, Deschampsia cespitosa, Dichanthelium clandestinum, Doellingeria umbellata, Drosera intermedia, D. rotundifolia, Dryopteris intermedia, Echinocystis lobata, Elaeagnus umbellata, Eleocharis erythropoda, E. palustris, E. rostellata, Elymus canadensis, Epilobium ciliatum, E. coloratum, E. leptophyllum, Epipactis gigantea, Equisetum arvense, E. fluviatile, E. pratense, E. sylvaticum, Erechtites hieracifolia, Erigeron pulchellus, Eupatorium perfoliatum, Euthamia graminifolia, Eutrochium purpureum, Filipendula rubra, Fraxinus nigra, F. pennsylvanica, Funaria hygrometrica, Galium asprellum, G. boreale, G. labradoricum, G. palustre, G. tinctorium, G. trifidum, G. triflorum, Gentiana andrewsii, Gentianopsis virgata, Glechoma hederacea, Glyceria grandis, G. striata, Glycyrrhiza lepidota, Habenaria, Helenium autumnale, Helianthus grosseserratus, H. nuttallii, Heliopsis helianthoides, Heracleum sphondylium, Hieracium canadense, Hordeum jubatum, Hydrocotyle americana, Hypericum ascyron, H. majus, Hypoxis hirsuta, Ilex verticillata, Impatiens capensis, Iris pseudacorus, I. virginica, Juglans nigra, Juncus acuminatus, J. balticus, J. brachycephalus, J. dudleyi, J. effusus, J. interior, J. nodosus, J. tenuis, Juniperus virginiana, Lactuca biennis, Larix laricina, Lathyrus* *palustris, Leersia oryzoides, L. virginica, Leonurus cardiaca, Leucanthemum vulgare, Liatris pycnostachya, L. spicata, Lilium michiganense, Liparis loeselii, Lobelia cardinalis, L. kalmii, L. siphilitica, Lonicera hirsuta, L. involucrata, L. morrowii, Lotus uliginosus, Ludwigia palustris, Lycopus americanus, L. asper, L. europaeus, L. uniflorus, L. virginicus, Lysichiton americanus, Lysimachia ciliata, L. quadriflora, L. quadrifolia, L. thyrsiflora, Lythrum alatum, L. salicaria, L. virgatum, Maianthemum stellatum, Malva neglecta, Marchantia polymorpha, Matteuccia struthiopteris, Melilotus, Menispermum canadense, Mentha arvensis, Mertensia paniculata, Mimulus alatus, Monarda fistulosa, Muhlenbergia glomerata, M. mexicana, M. racemosa, Myrica gale, M. pensylvanica, Napaea dioica, Nasturtium officinale, Oclemena nemoralis, Onoclea sensibilis, Osmunda claytoniana, O. regalis, Osmundastrum cinnamomeum, Oxalis stricta, Oxypolis rigidior, Packera aurea, P. paupercula, P. pseudaurea, Panicum capillare, Parnassia caroliniana, P. glauca, P. palustris, Parthenocissus quinquefolia, Pedicularis canadensis, P. lanceolata, Persicaria amphibia, P. hydropiper, P. lapathifolia, P. maculosa, P. punctata, P. sagittata, P. virginiana, Phalaris arundinacea, Phleum pratense, Phlox maculata, P. pilosa, Phragmites australis, Physocarpus opulifolius, Physostegia virginiana, Picea glauca, P. mariana, Pilea fontana, P. pumila, Pinus ponderosa, P. sylvestris, Plantago lanceolata, P. major, Platanthera ciliaris, P. psycodes, Poa nemoralis, P. palustris, P. pratensis, P. saltuensis, Polemonium vanbruntiae, Polygala cruciata, Polygonatum biflorum, P. pubescens, Polygonum aviculare, Populus angustifolia, P. deltoides, P. tremuloides, Potentilla norvegica, Prenanthes altissima, Prunella vulgaris, Prunus americana, P. virginiana, Pteridium aquilinum, Pycnanthemum virginianum, Quercus bicolor, Rhamnus alnifolia, Rhododendron groenlandicum, Rhus aromatica, Rhynchospora capillacea, R. capitellata, R. fusca, Ribes americanum, R. hirtellum, Ribes americanum, Robinia pseudoacacia, Rorippa islandica, R. palustris, Rosa palustris, Rubus deliciosus, R. parviflorus, R. pubescens, R. sachalinensis, Rudbeckia fulgida, R. hirta, R. laciniata, Rumex britannica, R. crispus, R. obtusifolius, Sagittaria latifolia, Salix alba, S. bebbiana, S. candida, S. discolor, S. eriocephala, S. exigua, S. interior, S. lucida, S. petiolaris, S. pyrifolia, S. serissima, Sambucus nigra, Schedonorus arundinaceus, Schoenoplectus acutus, S. pungens, S. purshianus, S. tabernaemontani, Scirpus atrovirens, S. cyperinus, Scleria pauciflora, S. triglomerata, Scrophularia marilandica, Scutellaria galericulata, Selaginella apoda, Senecio suaveolens, Setaria glauca, Sidalcea hendersonii, Silphium perfoliatum, Sisymbrium altissimum, Sisyrinchium angustifolium, Sium suave, Solanum dulcamara, S. ptychanthum, Solidago altissima, S. canadensis, S. gigantea, S. juncea, S. missouriensis, S. ohioensis, S. patula, S. ptarmicoides, S. puberula, S. riddellii, S. rigida, S. rugosa, S. uliginosa, Sonchus arvensis, Sorghastrum nutans, Sparganium eurycarpum, Spartina pectinata, Sphagnum angustifolium, S. cuspidatum, S. teres, Sphenopholis obtusata, Spiraea alba, S. douglasii, S. tomentosa, Spiranthes cernua, S. lucida, S. vernalis, Stachys palustris, Stellaria longifolia, Symphyotrichum boreale, S. ericoides, S. firmum, S. lanceolatum, S. lateriflorum, S. novae-angliae, S. novi-belgii, S. prenanthoides, S. puniceum, S. racemosum, Symplocarpus foetidus, Taraxacum officinale, Teucrium canadense, Thalictrum dasycarpum, Thelypteris palustris, Thuidium delicatulum, Thuja occidentalis, Tilia americana, Toxicodendron radicans, T. vernix, Triadenum fraseri, Trifolium hybridum, T. pratense, T. repens, Triosteum perfoliatum, Trollius laxus, Tsuga canadensis, Typha angustifolia, T. latifolia, T. ×glauca, Ulmus americana, U. rubra, Urtica dioica, Vaccinium macrocarpon, V. oxycoccos, Veratrum viride, Verbena hastata, V. urticifolia, Verbesina alternifolia, Vernonia gigantea, Veronica americana, Veronicastrum virginicum, Viburnum dentatum, V. opulus, Viola cucullata, V. sororia, Vitis riparia, Xyris torta, Zanthoxylum americanum, Zizea aurea.*

Use by wildlife: *Eutrochium fistulosum* is a larval host of the three-lined flower moth (Insecta: Lepidoptera: Noctuidae: *Schinia trifascia*). It also is fed upon by adult leaf beetles (Insecta: Coleoptera: Chrysomelidae: *Sumitrosis inaequalis, Systena hudsonias*). The flowers provide pollen to bumblebees (Insecta: Hymenoptera: *Bombus impatiens*) and are a source of nectar for adult fritillary, skipper, snout, and swallowtail butterflies (Insecta: Lepidoptera: Hersperiidae: *Epargyreus clarus*; Papilionidae: *Battus philenor, Papilio glaucus, P. troilus*; Nymphalidae: *Libytheana fulvescens, Speyeria diana*). The plants host fungi that are pathogenic to insects (Zygomycota: Entomophthoraceae: *Eryniopsis lampyridarum*) as well as powdery mildews and other Ascomycota (Erysiphaceae: *Erysiphe cichoracearum*; Leptosphaeriaceae: *Leptosphaeria macrospora*). *Eutrochium maculatum* is eaten by various mammals (Mammalia) including domestic cattle (Bovidae: *Bos bos*), eastern cottontail rabbits (Leporidae: *Sylvilagus floridanus*), muskrats (Cricetidae: *Ondatra zibethicus obscurus*), white-tailed deer (Cervidae: *Odocoileus virginianus*), and wild elk (Cervidae: *Cervus elaphus nelsoni*). The latter browse the plants more frequently in grassland habitats than in mixed forest sites. The seeds are eaten by a number of birds (Aves) including mallard ducks (Anatidae: *Anas platyrhynchos*), swamp sparrows (Emberizidae: *Melospiza georgiana*), and wild turkeys (Phasianidae: *Meleagris gallopavo*) and also by white-footed mice (Mammalia: Rodentia: Cricetidae: *Peromyscus leucopus*). The plants are hosts to insects (Insecta) such as ambush bugs (Heteroptera: Phymatidae: *Phymata americana*), aphids (Hemiptera: Aphididae: *Macrosiphum ambrosiae, Uroleucon eupatoricolens*), the garden tiger moth (Lepidoptera: Arctiidae: *Arctia caja*), and also the northern root-knot nematode (Secernentea: Meloidogynidae: *Meloidogyne hapla*). They are used for oviposition by the endangered Mitchell's satyr butterfly (Lepidoptera: Nymphalidae: *Neonympha mitchellii mitchelli*). Slugs (Mollusca: Gastropoda) feed on the shoots, mainly as they emerge from the soil in the spring. The foliage is eaten by Japanese beetles (Insecta: Coleoptera: Scarabaeidae; *Popillia japonica*). The flowers are visited routinely by soldier beetles (Insecta: Coleoptera: Cantharidae), flower longhorn beetles (Insecta: Coleoptera: Cerambycidae:

Stictoleptura canadensis), and tumbling flower beetles (Insecta: Coleoptera: Mordellidae: *Mordella marginata*), various bees (Insecta: Hymenoptera) including bumble-bees, carpenter bees and honeybees (Apidae: *Apis mellifera, Bombus fervidus, Xylocopa*), and leaf-cutting bees (Insecta: Megachilidae), as well as silver-spotted skipper, monarch, checkerspot, great spangled fritillary and swallowtail butterflies (Insecta: Lepidoptera: Hesperiidae: *Epargyreus clarus*; Nymphalidae: *Danaus plexippus, Euphydryas phaeton, Speyeria cybele*; Papilionidae: *Papilio glaucus*), and hoverflies (Insecta: Diptera: Syrphidae: *Spilomyia fusca*). Beetle larvae (Insecta: Coleoptera: Mordellidae: *Mordellina pustulata*) have been reared from the stems. Damaged plants release volatile pyrrolizidine alkaloid derivatives. These compounds have been shown to attract several moths (Insecta: Lepidoptera: Arctiidae: *Cisseps fulvicollis, Ctenucha virginica, Halysidota tessellaris*), which then feed on the flowers, leaves, and roots. Other moths (Arctiidae: *Arctia caja, Haploa confusa*) also are hosted by the plants. A diversity of associated fungi occurs including Ascomycota (Diaporthaceae: *Diaporthe arctii*; Erysiphaceae: *Erysiphe cichoracearum*; Hyaloscyphaceae: *Solenopezia solenia, Trichopezizella relicina*; Leptosphaeriaceae: *Leptosphaeria jacksonii*; Mycosphaerellaceae: *Passalora perfoliati, Septoria eupatorii*) and Basidiomycota (Ceratobasidiaceae: *Ceratobasidium anceps*; Pucciniaceae: *Puccinia eleocharidis*). It also is a host plant of the parasitic angiosperms *Cuscuta gronovii* and *Pedicularis lanceolata* (Orobanchaceae).

Economic importance: food: *Eutrochium* species should not be eaten due to the presence of potent alkaloids; some species are suspected to cause livestock poisoning; **medicinal:** These plants are known to contain a number of alkaloids (echinatine, pyrrolizidine, trachelanthamidine), which can cause chronic liver poisoning and haemorrhage leading to cirrhosis. *Eutrochium fistulosum* was used in Florida to treat kidney disorders and as a tea to remove polyps. *Eutrochium maculatum* contains several pyrrolizidine alkaloids including echinatine, trachelanthamidine, and low (presumably subtoxic) levels (~0.06%) of lycopsamine; however, the plants also contain the guaianolide Cumambrin B, which can cause contact dermatitis. It was used extensively as a medicinal plant by Native American tribes. The Algonquin and Quebec used it to treat menstrual disorders and as a remedy for venereal disease (where white-corolla forms were administered to males and pink forms to females). The Cherokee administered various medicines as aerosols, using the hollow stems to aspirate them. They also used the root as a diuretic, for treating "dropsy," gout, rheumatism, to relieve nausea, and for various "female problems." An infusion was administered for kidney ailments. Root decoctions were prepared to alleviate painful urination. The Chippewa used root decoctions to prepare washes for treating joint inflammations and as a sedative to calm restless children. They also brewed a tea from the dried leaves and flower heads to induce perspiration. The Hocak burned the plants as a smudge to combat various illnesses. The Iroquois administered a compound decoction of crushed plants as a remedy for diarrhea, liver ailments, rheumatism, and stomach gas. A root infusion was taken to reduce chills and fever due to colds, for kidney problems, tuberculosis, and to relieve pain following childbirth. A root decoction was taken to allevaite symptoms of gonorrhea; **cultivation:** *Eutrochium fistulosum* and *E. maculatum* are cultivated as ornamentals in native wildflower gardens and to attract butterflies. Cultivars of *E. fistulosum* include: 'Bartered Bride', 'Berggarten,' 'Big Nate,' 'Carin,' 'Early Riser,' 'Gateway,' 'Ivory Towers,' 'Joe White,' and 'Massive White' (an Award of Garden Merit plant); cultivars of *E. maculatum* include: 'Atropurpureum', 'Augustrubin', and 'Gateway'; **misc. products:** The Cherokee used the stems of *E. maculatum* as a straw to extract water from low lying springs. *Eutrochium fistulosum* and *E. maculatum* are used in wetland restoration; seed of the latter is a component of "SubmerSeed™," a commercial aggregate used for wetland plant propagation. *Eutrochium fistulosum* and *E. maculatum* are an important source of honey; **weeds:** not weedy; **nonindigenous species:** none.

Systematics: Multiple DNA data sets confirm that *Eutrochium* is monophyletic and resolves within tribe Eupatorieae as a sister clade to other species of *Eupatorium*, within which it was formerly classified (Figure 5.154). Consequently, the decision to distinguish *Eutrochium* from *Eupatorium* taxonomically is more a matter of choice than necessity, as it would be phylogenetically compatible to retain the group as an infrageneric subdivision under a broader generic concept of *Eupatorium*. The latter option seems more reasonable, especially given the small size (five species) of the segregate genus. Admittedly, the extent of genetic divergence between *Eutrochium* and other *Eupatorium* species appears to be fairly high; however, the genetic distance separating *Eupatorium capillifolium* from *E. hyssopifolium* exceeds that level, yet they are retained within the same genus (Figure 5.154). The distinction of taxa primarily on the basis of molecular data is not an advisable practice. The genus is recognized here in the narrower sense only to conform with the usage followed by most recent floristic treatments. *Eutrochium fistulosum* and *E. purpureum* are not sister species, which indicates the independent derivation of the OBL habit in the genus (Figure 5.154). *Eutrochium fistulosum* and *E. purpureum* are scarcely distinguishable morphologically and genetically. Their taxonomic status as different species should be reconsidered, especially in light of known hybridization involving *E. purpureum* and all other species in the genus including *E. fistulosum*. All *Eutrochium* species (including *E. fistulosum* and *E. maculatum*) are $2n=20$ and are regarded as diploids; however, extensive gene duplications detected from allozyme analysis of *E. fistulosum* indicate that the original diploid level probably was $2n=10$ (i.e., $x=5$). The DNA content of *E. maculatum* has been calculated as 19.63 ± 2.66 feulgen absorption units (FAU). Populations of *E. maculatum* evaluated by AFLP analysis indicated lower than expected values of gene diversity ($H_T=0.21$; $H_S=0.17$) and moderate differentiation ($G_{ST}=0.18$), perhaps due to their small, fragmented populations or lower than expected outcrossing rates.

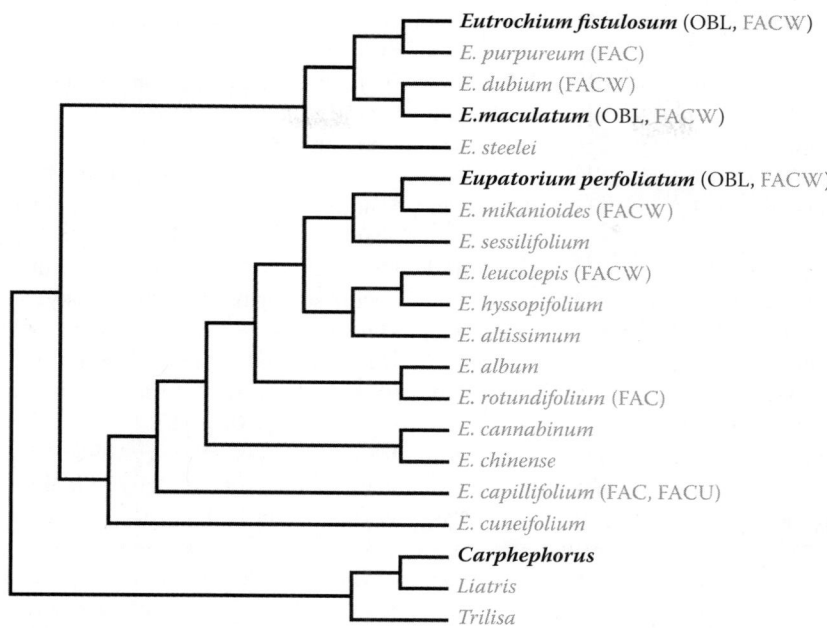

FIGURE 5.154 Phylogenetic relationships among selected representatives of Asteraceae tribe Eupatorieae showing placement of *Eutrochium* and related groups based on ITS sequence data. *Eutrochium* resolves as a sister clade to *Eupatorium*, a genus to which it was assigned formerly. Wetland indicator species are distributed across the tribe; however, the placement of OBL taxa (in bold) indicates several independent origins of the habit. (Modified from Schmidt, G.J. & Schilling, E.E. *Amer. J. Bot.*, 87, 716–726, 2000.)

Comments: *Eutrochium fistulosum* occurs in eastern North America; *E. maculatum* is widespread.

References: Ahmad & Hitchmough, 2007; Allbritton et al., 2002; Amon et al., 2002; Archibald, 1957; Atkinson et al., 2010; Austin, 2004b; Bacone et al., 1983; Barney et al., 2007; Bartgis, 1997; Bartgis & Lang, 1984; Basinger, 2003; Baskin et al., 1993; Bélair & Benoit, 1996; Belt & Kirk, 2009; Benzecry, 2012; Bermingham, 2010; Braun, 1935; Cain & Slater, 1948; Calhoun, 1985; Carleton, 1979; Darlow, 2000; Davidson et al., 1997; Densmore, 1974; Dziadyk & Clambey, 1980; Ebinger & Bacone, 1981; Edgin, 2004; El-Lakany & Dugle, 1972; Ellett, 1966; Frolik, 1941; Gargiullo, 2010; Gosling, 1986; Grabas & Laverty, 1999; Hitchmough & Wagner, 2011; Homoya et al., 1985; Hopkins, 1969; Humbert et al., 2007; Ito et al., 2000; Jennings, 1942; Jost et al., 1999; Kawahara, 2006; Keddy et al., 1998; Kercher & Zedler, 2004; Kindscher & Hurlburt, 1998; Krasnoff & Dussourd, 1989; Krischik & Zbinden, n.d.; Lamb et al., 2003; Levin & Kerster, 1969; Lisberg & Young, 2003; McJannet et al., 1995; Michel & Henein, 2007; Middleton, 2002b, 2003; Moncada et al., 2005; Moran, 1984; Norcini et al., 2012; Ortwine-Boes & Silbernagel, 2003; Punzalan et al., 2008; Raitviir et al., 1991; Record, 2011; Rudolph et al., 2006; Schilling et al., 1999; Schmidt & Schilling, 2000; Scholtens, 1991; Schoolmaster, Jr., 2005; Shipley, 1989; Shipley & Parent, 1991; Soper & Payne, 1997; Tryon & Easterly, 1975; Tsuda & Marion, 1963; Waldbauer & Ghent, 1984; Weishampel & Bedford, 2006; Wheeler, Jr., 1988; White, 1965; White et al., 2007; Whitehouse, 2007; Wiedenfeld et al., 2009; Williams, 1990a; Winsor, 1983; Yahara et al., 1989; Yin et al., 2012; Zettler et al., 1996.

22. *Gnaphalium*

Marsh cudweed; gnaphale des marais

Etymology: from the Greek *gnaphallon* ("lock of wool") in reference to the tomentose foliage

Synonyms: *Amphidoxa, Demidium, Filaginella, Homognaphalium*

Distribution: global: Africa, Asia, Australia, New World; **North America:** Widespread (except southeastern)

Diversity: global: 38 species; **North America:** 3 species

Indicators (USA): OBL; FACW: *Gnaphalium palustre*

Habitat: freshwater/brackish; palustrine; **pH:** alkaline; **depth:** <1 m; **life form(s):** emergent herb

Key morphology: Shoots (to 30 cm) clothed in woolly hairs, the branches decumbent, basal; leaves (to 3.5 cm) spoon-like or oblong, alternate, cauline, entire; capitula clustered in terminal heads, subtended by bracts (to 12 mm); capitula disc-like, phyllaries (to 4 mm) brown, their bases wooly, pales lacking; peripheral florets (to 80) pistillate, the inner florets (to 7) bisexual, corollas whitish; cypselae (to 0.55 mm) oblong, smooth or pipillate, pappus of 8–12 barbed bristles, readily dehiscent, falling singly.

Life history: duration: annual (fruit/seeds); **asexual reproduction:** none; **pollination:** insect; **sexual condition:** gyno-monoecious; **fruit:** cypselae (common); **local dispersal:** gravity, wind (cypselae); **long-distance dispersal:** animals [endozoic] (cypselae)

Imperilment: (1) *Gnaphalium palustre* [G5]; S2 (NE); S3 (AB, WY)

Ecology: general: *Gnaphalium* consists of annuals, biennials, or perennials, which occur from high to low elevations in forests, grasslands, and wetlands. Most of the species are associated with moist areas and many are regarded as weedy. All three of the North American species have some wetland status (FACW, FAC, UPL) and typically occur in wet, marginal habitats; however, only one is ranked as OBL. There is little information on the reproductive biology of the North American species. The inflorescences (capitula) of *Gnaphalium* are complex, consisting of a mixture of pistillate and bisexual florets. The plants presumably are self-incompatible and pollinated primarily by insects (Insecta); however, these assumptions are based mainly on circumstantial evidence and deserve further evaluation. Some *Gnaphalium* species are described as capable of selfing. The species also are generally described as wind dispersed. However, unlike many Asteraceae, the pappus is shed from the cypselae, which does not facilitate wind dispersal. The fruits survive ingestion by vertebrates, but the extent of their consumption by more vagile agents (e.g., waterfowl) has not been determined. In Europe, *Gnaphalium* plants are used as nesting material by magpies (Aves: Corvidae: *Pica pica*), which are believed to disperse the accompanying seeds secondarily.

Gnaphalium palustre **Nutt.** is a summer annual, which occurs in tidal or nontidal habitats including alluvial fans, arroyos, beaches, beaver ponds, chaparrel, depressions, ditches, drying brooks, dune pools, lake and pool bottoms, flats, floodplains, gravel bars, marshes, meadows, mudflats, pools, potholes, prairies, rain pools, rivulets, sand bars, seeps, slopes, springs, strands, streambeds, swales, vernal pools, washes, woodlands, and along the margins of canals, lakes, ponds, reservoirs, streams, and vernal pools at elevations of up to 3350 m. Occurrences range from fully open to shady sites. The substrates usually are described as alkaline (e.g., pH 7.8) but specific pH data are scarce. They are described as adobe, basalt, clay, clay loam, cobble, Dia, Dithod, granite, granitic loam, gravel, gravelly sand, humic sand, humus, Julian schist, loamy clay, loamy sand, muck, mud, Parmleed–Bidman association, pebble, quartzite, rocky basalt, rocky loam, Sagouspe, sand, sandy gravel, sandy loam, sandy rock, serpentine, Sespe formation, shale, silt, silty clay, silty clay loam, and silt loam. This species has remarkably broad ecological tolerances as evidenced by the extensive list of associated species provided. Although mainly observed at freshwater sites, there also have been several records from salt marshes or "slightly saline" river systems. The plants are most characteristic of sites where the water has receded, where they occur on drying mud. Ironically, although several studies have documented their sensitivity to fires and grazing, collection data commonly describe plants growing in disturbed sites including recently burned areas and heavily overgrazed locales. There is no definitive information on the reproductive biology of this species. The flowers are visited by butterflies (Insecta: Lepidoptera), but specific pollinators have not been identified. Seed production is common as is expected for an annual. The seeds have been germinated from seed banks sampled from several different wetland sites. In some cases, they were exposed to ambient local conditions for 1 year, then moved to a greenhouse where they were kept under continuously moist conditions for germination. Other seeds extracted from seed bank cores germinated after storage at 20°C for 5 months followed by a 10-day treatment at 5°C. These were lightly watered and kept under a 20°C/18°C day/night temperature regime. Under natural conditions, germination occurs somewhat later than many other species. It is uncertain how well the dehiscent pappus helps to disperse the small seeds by wind. The seeds do remain viable after ingestion by horses and passage through their digestive tract. Other potential means of dispersal have not been evaluated adequately.

Reported Associates: *Abies concolor, A. grandis, Acer negundo, Achillea millefolium, Acmispon americanus, A. glabrus, A. humistratus, Achnatherum lettermanii, A. parishii, Adenostoma fasciculatum, A. sparsifolium, Agoseris heterophylla, Agrostis capillaris, A. exarata, A. gigantea, A. idahoensis, A. scabra, A. stolonifera, Alisma triviale, Allophyllum integrifolium, Alopecurus saccatus, Alnus tenuifolia, Alopecurus aequalis, A. arundinaceus, A. carolinianus, A. pratensis, Amaranthus albus, A. californicus, A. powellii, A. watsonii, Ambrosia monogyra, Ammannia coccinea, Amorpha californica, A. fruticosa, Amsonia palmeri, Anagallis arvensis, A. minima, Anisacanthus thurberi, Antennaria dimorpha, A. racemosa, A. rosea, Anthemis cotula, Apocynum cannabinum, Arctostaphylos patula, A. pungens, A. viscida, Arnica chamissonis, Artemisia arbuscula, A. biennis, A. californica, A. douglasiana, A. dracunculus, A. frigida, A. ludoviciana, A. palmeri, A. tridentata, Arthrocnemum subterminale, Asclepias fascicularis, Astragalus miser, Avena barbata, A. fatua, Baccharis pilularis, B. salicifolia, B. sarothroides, B. sergiloides, Bahiopsis laciniata, Barbarea orthoceras, Bassia hyssopifolia, Beckmannia syzigachne, Bidens cernuus, B. frondosus, Bistorta bistortoides, Boechera parishii, Bolboschoenus fluviatilis, Brassica nigra, Briza minor, Brodiaea filifolia, B. terrestris, Bromus carinatus, B. diandrus, B. hordeaceus, B. japonicus, B. rubens, B. secalinus, B. sterilis, B. tectorum, Callitriche hermaphroditica, C. marginata, Callitropsis macnabiana, Calocedrus decurrens, Calochortus invenustus, C. palmeri, Calibrachoa parviflora, Camassia quamash, Camassia leichtlinii, Capsella bursa-pastoris, Carex abrupta, C. alma, C. athrostachya, C. douglasii, C. fracta, C. hassei, C. heteroneura, C. interrupta, C. jonesii, C. lenticularis, C. nebrascensis, C. praegracilis, C. retrorsa, C. rostrata, C. schottii, C. subfusca, C. unilateralis, Castilleja lasiorhyncha, C. miniata, C. tenuis, Ceanothus cordulatus, C. greggii, C. leucodermis, C. megacarpus, Celtis reticulata, Centaurea stoebe, Cerastium glomeratum, Cercocarpus betuloides, C. ledifolius, C. montanus, Chamaebatia foliolosa, Chenopodium album, C. berlandieri, C. fremontii, C. glaucum, C. leptophyllum, Cicuta, Cirsium arvense, C. scariosum, C. vulgare, Clematis ligusticifolia, Collinsia parviflora, C. torreyi, Collomia linearis, Coreopsis, Cornus sericea, Corydalis aurea, Conyza canadensis, Crassula aquatica, C. connata,*

Cressa truxillensis, Croton setiger, Crypsis alopecuroides, C. schoenoides, Cryptantha simulans, Cuscuta, Cylindropuntia, Cynodon dactylon, Cynosurus echinatus, Cyperus acuminatus, C. aristatus, C. eragrostis, C. esculentus, C. squarrosus, Cypselea humifusa, Dactylis glomerata, Damasonium californicum, Danthonia spicata, D. unispicata, Datisca glomerata, Deinandra fasciculata, D. kelloggii, Deschampsia cespitosa, D. danthonioides, D. elongata, Descurainia pinnata, D. sophia, Dicentra chrysantha, Dichanthelium acuminatum, Dichelostemma capitatum, Dieteria canescens, Digitaria sanguinalis, Downingia cuspidata, D. elegans, D. laeta, D. yina, Drymocallis glandulosa, D. lactea, Dryopteris arguta, Dudleya edulis, D. pulverulenta, Dysphania botrys, Echinocereus triglochidiatus, Echinochloa crus-galli, Eclipta prostrata, Elatine brachysperma, Eleocharis acicularis, E. bella, E. macrostachya, E. montevidensis, E. ovata, E. palustris, E. parvula, E. quinqueflora, Elodea canadensis, Elymus glaucus, E. trachycaulus, Encelia farinosa, Epilobium brachycarpum, E. campestre, E. ciliatum, E. densiflorum, E. oreganum, E. torreyi, Equisetum laevigatum, Eragrostis hypnoides, E. intermedia, Eremogone congesta, Ericameria nauseosa, Erigeron divergens, Eriochloa villosa, Eriodictyon trichocalyx, Eriogonum fasciculatum, E. parishii, E. wrightii, Eriophyllum confertiflorum, Erodium botrys, E. cicutarium, Eryngium aristulatum, E. castrense, Eucalyptus globulus, Euphorbia abramsiana, E. glyptosperma, E. heterophylla, E. serpyllifolia, Eurybia conspicua, E. integrifolia, Fallopia convolvulus, Festuca idahoensis, F. rubra, Forestiera pubescens, Frankenia salina, Fraxinus pennsylvanica, F. velutina, Fremontodendron, Galium, Garrya veatchii, Gayophytum decipiens, G. diffusum, Geranium carolinianum, Geum, Gilia modocensis, Glinus lotoides, Glyceria elata, G. leptostachya, Gnaphalium uliginosum, Gratiola, Grindelia, Gutierrezia sarothrae, Helianthus annuus, Heliotropium curassavicum, Hemizonella minima, Herniaria hirsuta, Hesperolinon micranthum, Hesperoyucca whipplei, Heterocodon rariflorus, Heteromeles arbutifolia, Hirschfeldia incana, Holcus lanatus, Hordeum brachyantherum, H. intercedens, H. jubatum, H. murinum, H. pusillum, Horkelia rydbergii, Hydrocotyle ranunculoides, Hypericum anagalloides, H. perforatum, H. scouleri, Hypochaeris glabra, Ipomoea cristulata, Iris hartwegii, Isoetes, Isolepis cernua, Ivesia unguiculata, Juglans major, Juncus arcticus, J. balticus, J. bryoides, J. bufonius, J. effusus, J. ensifolius, J. longistylis, J. macrandrus, J. macrophyllus, J. nevadensis, J. orthophyllus, J. phaeocephalus, J. saximontanus, J. tenuis, J. tiehmii, J. trilocularis, J. xiphioides, Juniperus californica, J. coahuilensis, J. occidentalis, J. osteosperma, Lactuca serriola, Lagophylla ramosissima, Lamarckia aurea, Larrea tridentata, Lasthenia californica, L. ferrisiae, Lathyrus japonicus, Leersia oryzoides, Lepidium thurberi, L. perfoliatum, L. virginicum, Lepidospartum squamatum, Leptochloa panicea, Lessingia nemaclada, Leucanthemum vulgare, Leymus cinereus, L. triticoides, Lilaea scilloides, Lilaeopsis, Lilium humboldtii, Limosella acaulis, L. aquatica, Lindernia dubia, Lithocarpus densiflorus, Lobularia maritima, Logfia filaginoides, Lolium multiflorum, L. perenne, L. temulentum,

Ludwigia palustris, Lupinus bicolor, L. breweri, L. lepidus, Lycium, Lythrum hyssopifolia, L. portula, Madia elegans, Malosma laurina, Malvella leprosa, Marah macrocarpa, Marchantia, Marsilea vestita, Matricaria discoidea, Melica, Melilotus albus, M. indicus, Mentha arvensis, Mentzelia veatchiana, Microsteris gracilis, Mimulus androsaceus, M. angustatus, M. bicolor, M. breweri, M. filicaulis, M. floribundus, M. guttatus, M. moschatus, M. parishii, M. pilosus, M. primuloides, M. suksdorfii, Montia chamissoi, Muhlenbergia filiformis, M. richardsonis, M. rigens, Myosurus minimus, Myriophyllum spicatum, Nama hispida, Nassella cernua, N. lepida, N. pulchra, N. viridula, Navarretia capillaris, N. divaricata, N. hamata, N. intertexta, N. leucocephala, N. prostrata, N. squarrosa, Nemophila, Nicotiana obtusifolia, Oenothera villosa, Opuntia basilaris, O. littoralis, O. ×vaseyi, Orcuttia californica, Panicum capillare, Parkinsonia microphylla, Penstemon rostriflorus, Perideridia gairdneri, P. parishii, Persicaria amphibia, P. hydropiperoides, P. lapathifolia, P. maculata, P. punctata, Phacelia davidsonii, P. hastata, Phalacroseris bolanderi, Phalaris arundinacea, P. lemmonii, P. paradoxa, Phaseolus angustissimus, Phleum pratense, Phyla nodiflora, Pilularia americana, Pinus contorta, P. coulteri, P. jeffreyi, P. lambertiana, P. monophylla, P. ponderosa, Plagiobothrys acanthocarpus, P. bracteatus, P. figuratus, P. hispidulus, P. leptocladus, P. trachycarpus, P. undulatus, Plantago lanceolata, P. major, P. ovata, Platanus racemosa, Plectritis congesta, Poa annua, P. atropurpurea, P. bulbosa, P. compressa, P. leibergii, P. nemoralis, P. pratensis, P. secunda, Polanisia dodecandra, Polemonium, Polygonum aviculare, P. douglasii, P. parryi, P. polygaloides, Polypogon monspeliensis, Populus angustifolia, P. fremontii, P. trichocarpa, Porterella carnosula, Potamogeton crispus, Potentilla anserina, P. biennis, P. gracilis, P. norvegica, P. rivalis, P. wheeleri, Primula hendersonii, Prosopis juliflora, P. velutina, Pseudognaphalium luteoalbum, P. stramineum, Pseudoroegneria spicata, Pseudotsuga menziesii, Psilocarphus brevissimus, P. oregonus, Purshia mexicana, P. tridentata, Quercus agrifolia, Q. berberidifolia, Q. corneliusmulleri, Q. dumosa, Q. emoryi, Q. grisea, Q. turbinella, Q. wislizeni, Rafinesquia californica, Ranunculus californicus, R. cymbalaria, R. flammula, R. orthorhynchus, R. sceleratus, Rhamnus californica, R. crocea, R. ilicifolia, Rhododendron occidentale, Rhus aromatica, R. integrifolia, R. ovata, Ribes cereum, R. nevadense, Rigiopappus leptocladus, Rorippa columbiae, R. curvipes, R. curvisiliqua, R. palustris, R. teres, Rosa californica, R. nutkana, R. woodsii, Rubus ulmifolius, R. ursinus, Rumex acetosella, R. crispus, R. fueginus, R. hymenosepalus, R. maritimus, R. salicifolius, Sagina saginoides, Sagittaria cuneata, S. latifolia, Sairocarpus cornutus, Salix boothii, S. exigua, S. gooddingii, S. laevigata, S. lasiolepis, S. lutea, Salsola tragus, Salvia apiana, S. leucophylla, S. mellifera, S. scouleriana, Sambucus nigra, Schedonorus arundinaceus, Schoenoplectus acutus, S. americanus, S. californicus, S. tabernaemontani, Schismus barbatus, Scirpus microcarpus, Sclerolinon digynum, Scribneria bolanderi, Scutellaria galericulata, S. lateriflora, Senecio aronicoides, S. crassulus, S. scorzonella, S. vulgaris, Sidalcea malviflora,

S. oregana, Sisymbrium altissimum, Sisyrinchium bellum, Solanum dulcamara, S. rostratum, Sonchus oleraceus, Sorghum halepense, Sparganium angustifolium, S. emersum, S. eurycarpum, Spartina pectinata, Spergularia bocconii, S. rubra, S. salina, Sphenosciadium capitellatum, Sporobolus airoides, S. wrightii, Stachys albens, S. rigida, Stellaria longipes, S. media, Stemodia durantifolia, Stephanomeria virgata, Stipa, Stylocline gnaphaloides, Suaeda nigra, Symphoricarpos albus, S. rotundifolius, Symphyotrichum foliaceum, S. frondosum, S. lanceolatum, S. spathulatum, Taeniatherum caput-medusae, Tamarix ramosissima, Taraxia tanacetifolia, Taraxacum californicum, T. officinale, Thalictrum fendleri, Thelypodium stenopetalum, Torreyochloa pallida, Toxicodendron diversilobum, Tragia, Tragopogon dubius, Trianta occidentalis, Trichostema lanceolatum, T. oblongum, Trifolium bolanderi, T. cyathiferum, T. longipes, T. monanthum, T. repens, T. variegatum, T. wormskioldii, Trisetum cernuum, Typha angustifolia, T. domingensis, Urtica dioica, Veratrum californicum, Verbascum thapsus, Verbena bracteata, Veronica americana, V. anagallis-aquatica, V. gracilis, V. peregrina, V. persica, Viola macloskeyi, Vulpia microstachys, V. myuros, V. octoflora, Wyethia angustifolia, Xanthium strumarium, Xylococcus bicolor, Yucca baccata, Zannichellia palustris, Zeltnera calycosa, Z. exaltata.

Use by wildlife: *Gnaphalium palustre* is a larval host of the American painted lady butterfly (Insecta: Lepidoptera: Nymphalidae: *Vanessa virginiensis*). The plants also are hosts to plant pathogenic fungi (Ascomycota: Pleosporaceae: *Alternaria*).

Economic importance: food: *Gnaphalium palustre* is not reported to be edible; **medicinal:** Decoctions and infusions prepared from *Gnaphalium palustre* plants exhibit very weak antioxidant activity; **cultivation:** *Gnaphalium palustre* is not currently in cultivation; **misc. products:** none; **weeds:** Because its seeds often occur in fallow fields, *G. palustre* sometimes is regarded as an agricultural weed; **nonindigenous species:** *Gnaphalium palustre* has been introduced to several sites in Alaska, where it is nonindigenous.

Systematics: *Gnaphalium* is the type genus of tribe Gnaphalieae (Asteraceae subfamily Asteroideae), which is most diverse in the Southern Hemisphere. Although relatively few *Gnaphalium* species have been surveyed in recent molecular systematic studies, there are indications that the genus is not monophyletic as currently circumscribed (Figure 5.155). Also, various data (cpDNA vs. nuclear DNA sequences) resolve different genera (e.g., *Syncarpha, Vellereophyton*) as the sister group to *Gnaphalium* "sensu stricto," i.e., the clade that contains the type species *G. uliginosum*. Additional study is needed to clarify relationships of and within this genus. However, it is evident that *Gnaphalium* is at least somewhat closely related to *Antennaria, Gamochaeta* and *Pseudognaphalium* (the latter to which several former species have been transferred). At this time *G. palustre* had not yet been sampled in any molecular phylogenetic analysis, leaving its precise relationships unresolved. *Gnaphalium* contains diploid and tetraploid taxa; *G. palustre* ($2n=14$) is diploid.

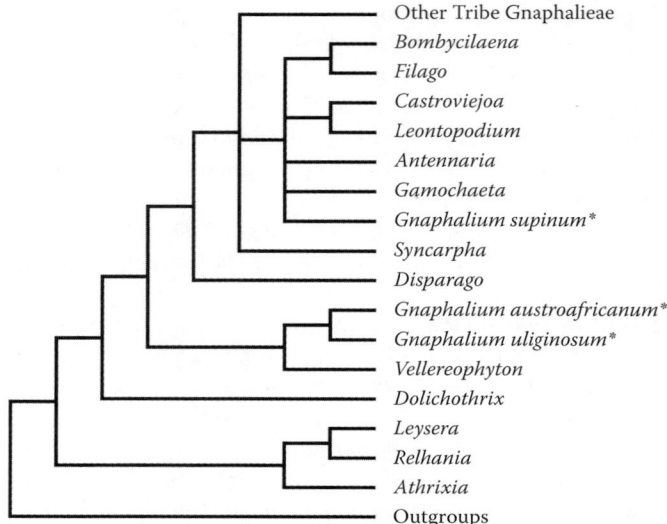

FIGURE 5.155 Phylogenetic relationships in Asteraceae tribe Gnaphalieae as indicated by phylogenetic analysis of cpDNA sequence data. These results indicate that *Gnaphalium* (asterisked) is not monophyletic as currently circumscribed. The OBL *Gnaphalium palustre* has not yet been included in phylogenetic analyses. (Adapted from Smissen, R.D. et al., *Taxon*, 60, 649–662, 2011.)

There have been no reports of hybridization involving *G. palustre*.

Comments: *Gnaphalium palustre* occurs in western North America.

References: Archibold & Hume, 1983; Arhangelsky, 2005; Christy, 2004; Cooke, 1997; Czarnecka et al., 2012; DeBenedetti & Parsons, 1984; Douglas et al., 2003; Guard, 1995; Hutchinson et al., 2007; Keil et al., 1988; Medeiros & Steiner, 2002; Nachlinger, 1988; Nesom, 2006c; Odion et al., 1988; Pospelova & Barnaulov, 2000; Quinn et al., 2008; Schilling & Floden, 2012; Shapiro, 1974b; Smissen et al., 2011.

23. *Hartwrightia*

Hartwrightia

Etymology: after Samuel Hart Wright (1825–1905)

Synonyms: none

Distribution: global: North America; **North America:** southeastern

Diversity: global: 1 species; **North America:** 1 species

Indicators (USA): OBL: *Hartwrightia floridana*

Habitat: freshwater; palustrine; **pH:** acidic; **depth:** <1 m; **life form(s):** emergent herb

Key morphology: Shoots (to 12 dm) erect, branched distally, with short, thick rhizomes, herbage glandular, viscid; leaves (to 25 cm) alternate, basal (elliptic; long petioled) and cauline (reduced, bract-like, nearly sessile), margins entire or wavy; heads discoid, arranged in corymbs, few-flowered (7–10), phyllaries (12–15) in 2–3 series, pales absent or peripheral; florets bisexual, corollas (to 3 mm) pinkish purple to white, radial; cypselae (to 3.5 mm) 5-angled or grooved, glandular dotted, pappus absent or of fragile, glandular hairs.

Life history: duration: perennial (rhizomes); **asexual reproduction:** rhizomes; **pollination:** insect; **sexual condition:** hermaphroditic; **fruit:** cypselae (common); **local dispersal:** animals [exozoic] (cypselae); **long-distance dispersal:** animals [exozoic] (cypselae)

Imperilment: (1) *Hartwrightia floridana* [G2]; S1 (GA); S2 (FL)

Ecology: general: *Hartwrightia* is monotypic.

Hartwrightia floridana **A. Gray ex S. Watson.** is perennial inhabitant of baygalls, bogs, depressions, flatwoods, marshes, meadows, prairies, seeps, strands, stream terraces, swales, and margins of depression marshes at elevations of up to 30 m. The plants occupy open sites on substrates that are described as loamy sand, peat, peat muck, and sandy peat. The soil pH associated with *Hartwrightia* occurrences has not been determined; however, the upper 40 cm of "Plantation" soil series associated with cutthroat grass (*Coleataenia longifolia*) communities (characteristic *Hartwrightia* habitat) typically is acidic (pH 5.3–5.8) and other occurrences are described as "acidic seeps." The reproductive biology of *Hartwrightia* remains unstudied. The flowers are believed to be predominantly outcrossed by insects (Insecta) based on field observations of flowers apparently being pollinated by medium to large-sized bees (Hymenoptera). The cypselae lack a pappus but are covered with sticky glands, which are thought to facilitate dispersal by attachment to insects (e.g., ants) or other animals. Nothing is known of the seed germination requirements or potential seed bank development. The plants are highly vulnerable to grazing by cattle but rely on periodic fire to reduce stands of competitive vegetation. A thorough study of the biology and ecology of this rare species is needed. **Reported associates:** *Acer rubrum, Amphicarpum muhlenbergianum, Andropogon glomeratus, A. gyrans, A. virginicus, Aristida purpurascens, A. stricta, Arundinaria, Balduina atropurpurea, Bartonia verna, Bidens, Bigelowia nudata, Burmannia biflora, Centella asiatica, Chaptalia tomentosa, Cliftonia monophylla, Coleataenia longifolia, Coreopsis floridana, Ctenium aromaticum, Cyrilla racemiflora, Dichanthelium ensifolium, Drosera capillaris, Eragrostis elliottii, Erigeron vernus, Eriocaulon compressum, E. decangulare, Eupatorium mohrii, Euthamia graminifolia, Fuirena scirpoidea, Gaylussacia dumosa, Gordonia lasianthus, Habenaria repens, Helianthus angustifolius, Hypericum cistifolium, H. fasciculatum, H. myrtifolium, Ilex coriacea, I. glabra, I. myrtifolia, Juncus, Lachnanthes caroliniana, Liatris spicata, Lobelia glandulosa, Lycopodiella alopecuroides, L. appressa, L. caroliniana, Lyonia fruticosa, L. lucida, Magnolia virginiana, Marshallia graminifolia, Myrica cerifera, M. heterophylla, Nyssa biflora, Oclemena reticulata, Osmundastrum cinnamomeum, Oxypolis filiformis, Panicum verrucosum, Paspalum praecox, Persea palustris, Pinus elliottii, P. palustris, Platanthera blephariglottis, Rhexia mariana, R. nuttallii, R. petiolata, Rhynchospora ciliaris, R. fascicularis, R. microcephala, R. oligantha, Sabatia difformis, S. grandiflora, S. macrophylla, Sarracenia minor, S. psittacina, Schizachyrium scoparium, Scleria muehlenbergii, S. reticularis, Serenoa repens, Smilax laurifolia, Solidago, Sorghastrum secundum, Sphagneticola trilobata, Sphagnum, Taxodium ascendens, Toxicodendron vernix, Utricularia subulata, Vaccinium myrsinites, Woodwardia areolata, Xyris ambigua, X. elliottii, X. platylepis.*

Use by wildlife: The flowers attract larger-sized bees (Insecta: Hymenoptera).

Economic importance: food: *Hartwrightia* is not reported to be edible; **medicinal:** none reported; **cultivation:** *Hartwrightia* is not currently in cultivation; **misc. products:** none; **weeds:** none; **nonindigenous species:** none

Systematics: *Hartwrightia* is assigned to the small subtribe Liatrinae of tribe Eupatorieae, which also contains *Carphephorus, Garberia, Liatris, Litrisa,* and *Trilisa*. Molecular phylogenetic studies of Liatrinae place *Hartwrightia* variously with *Trilisa* (nuclear DNA) or *Litrisa* (cpDNA), which indicates a possible hybrid origin of the genus in the subtribe (Figure 5.156). The secondary chemistry of *Hartwrightia* is similar to *Liatris*, which also occurs within the subtribe (Figure 5.156). *Hartwrightia floridana* (2n=20)

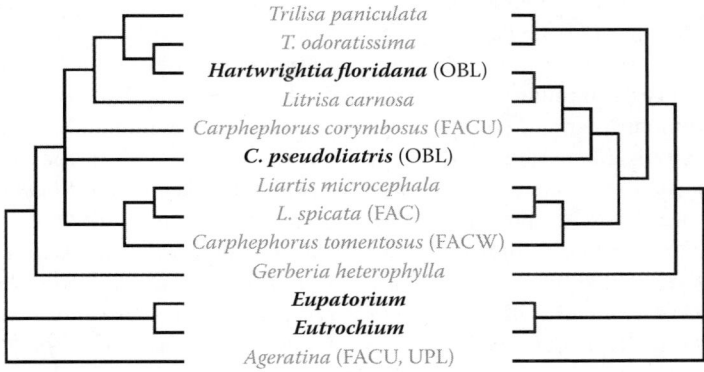

FIGURE 5.156 Discordant phylogenetic placement of *Hartwrightia floridana* as resolved by nuclear ETS/ITS sequence data (left) and cpDNA sequence (right). *Hartwrightia* is possibly of hybrid origin from ancestors similar to *Trilisa* and *Litrisa*. Neither data set places *Hartwrightia* with *Carphephorus pseudoliatris*, indicating their independent origin of the OBL habit (taxa with OBL species are indicated in bold). (Adapted from Schilling, E. E., *Mol. Phylogen. Evol.,* 59, 158–167, 2011a.)

has the same diploid chromosome number as other members of Liatrinae. There have been no contemporary reports of hybridization involving *Hartwrightia*.

Comments: *Hartwrightia floridana* is restricted to Georgia and Florida (United States).

References: Adams et al., 2010; Bohlmann et al., 1981; Carter et al., 2009; Chafin, 2008; Gage, 1985; Nesom, 2006d; Robinson, 1978; Robinson et al., 1989; Schilling, 2011a; Tobe et al., 1998; USFWS, 2013.

24. *Helenium*

Sneezeweed; Hélénie

Etymology: after Helen of Troy (mythical)

Synonyms: *Actinea*; *Actinella*; *Cephalophora*; *Heleniastrum*; *Leptopoda*; *Ptilepida*

Distribution: global: New World; **North America:** widespread

Diversity: global: 32 species; **North America:** 18 species

Indicators (USA): OBL: *Helenium brevifolium*, *H. drummondii*, *H. pinnatifidum*, *H. thurberi*, *H. virginicum*; **FACW:** *H. drummondii*

Habitat: freshwater; palustrine, riverine; **pH:** 4.5-alkaline; **depth:** <1 m; **life form(s):** emergent herb

Key morphology: Stems (to 130 cm) unbranched, slightly to strongly winged; leaves (to 20 cm) cauline, alternate, glabrous, lanceolate to lance-linear or obovate, entire, laciniate, lobed, pinnatifid, or undulate-serrate; heads single or up to 120+ per plant (in corymbs or panicles), radiate or discoid, pedunculed (to 30 cm); ray florets (0–34; to 25 mm) yellow (when present), neuter or pistillate; disc florets (to ~1000) bisexual, 4 or 5-lobed, with yellow, yellow-brown, yellow-green, purplish, or reddish-brown corollas (to 5.6 mm); cypselae (to 2.3 mm), with 5–12 entire or lacerate, awnless or aristate scales (to 3.7 mm).

Life history: duration: annual (seeds); perennial (taproot); **asexual reproduction:** root and stem sprouts; **pollination:** insect; **sexual condition:** hermaphroditic or gynomonoecious; **fruit:** cypselae (common); **local dispersal:** gravity (seeds); **long-distance dispersal:** water?, wind (seeds),

Imperilment: (1) *Helenium brevifolium* [G4]; S1 (AL, GA, LA, NC, SC, TN); S2 (VA); (2) *H. drummondii* [G4]; S3 (TX); (3) *H. pinnatifidum* [G4]; S2 (NC, SC); (4) *H. thurberi* [G4]; (5) *H. virginicum* [G3]; S2 (VA); S3 (MO)

Ecology: general: *Helenium* consists of annuals and perennials, with the latter more common. Nearly all of the North American species occur commonly along streams or near the edges of ponds and other standing water, with 15 species categorized as wetland indicators at some level (five as OBL). The heads are radiate in all of the OBL species except the annual *H. thurberi*, where they are discoid. Pollinators include various insects (Insecta) including bees and wasps (Hymenoptera), beetles (Coleoptera), butterflies (Lepidoptera), and flies (Diptera). The cypselae possess a pappus of scales (sometimes awned) and are dispersed by gravity, water or wind. Although the seeds of some species (e.g., *H. autumnale*) can float, floatation has not been documented for any of the OBL species. The seeds of different

species can lack dormancy completely or require cold stratification to break physiological dormancy.

***Helenium brevifolium* (Nutt.) Alph. Wood** is a perennial, which inhabits barrens, bayheads, bogs, bottoms, ditches, fens, flatwoods, glades, gravel bars, marshes, meadows, ravines, savannas, seeps, sloughs, swamps, and margins of rivers and streams at elevations of up to 1370 m. Exposures of full sun are tolerated. Reported substrates are acidic (e.g., pH 5.1–5.2) and include Escambia (plinthaquic paleudults), mafic, muck, Mulat (typic ochraquults), Porters, sandy clay, sandy loam and sandy peat. There is no reliable information on the reproductive biology or seed ecology of this species and additional life-history information is needed. **Reported associates:** *Acer rubrum*, *Aconitum uncinatum*, *Alnus serrulata*, *Amelanchier*, *Amianthium muscitoxicum*, *Andropogon*, *Anemone quinquefolia*, *Arisaema triphyllum*, *Aristida spiciformis*, *A. stricta*, *Aronia arbutifolia*, *Balduina uniflora*, *Betula*, *Calamagrostis canadensis*, *Calopogon tuberosus*, *Carex abscondita*, *C. atlantica*, *C. conoidea*, *C. debilis*, *C. leptalea*, *C. lurida*, *C. mitchelliana*, *C. schweinitzii*, *C. sterilis*, *C. torta*, *C. venusta*, *Castanea pumila*, *Castilleja coccinea*, *Centella asiatica*, *Chamaecyparis thyoides*, *Chaptalia tomentosa*, *Chelone glabra*, *C. obliqua*, *Cirsium muticum*, *Cladium mariscoides*, *Cleistesiopsis divaricata*, *Coreopsis gladiata*, *Crataegus flava*, *Ctenium aromaticum*, *Decumaria barbara*, *Dichanthelium clandestinum*, *D. dichotomum*, *Diospyros*, *Drosera filiformis*, *D. tracyi*, *Eleocharis tenuis*, *Eriocaulon compressum*, *E. lineare*, *Fimbristylis puberula*, *Fraxinus pennsylvanica*, *Glyceria striata*, *Gymnadeniopsis integra*, *Hexastylis*, *Houstonia caerulea*, *Hypericum brachyphyllum*, *Ilex glabra*, *I. opaca*, *I. verticillata*, *I. vomitoria*, *Itea virginica*, *Juncus gymnocarpus*, *J. subcaudatus*, *Juniperus virginiana*, *Kalmia angustifolia*, *K. latifolia*, *Leersia oryzoides*, *L. virginica*, *Liriodendron tulipifera*, *Lonicera japonica*, *Lycopus uniflorus*, *L. virginicus*, *Lyonia ligustrina*, *Magnolia tripetala*, *M. virginiana*, *Marshallia grandiflora*, *Medeola virginiana*, *Microstegium vimineum*, *Mitchella repens*, *Muhlenbergia expansa*, *M. glomerata*, *Myrica gale*, *Narthecium americanum*, *Nyssa biflora*, *N. sylvatica*, *Oenothera perennis*, *Osmunda regalis*, *Osmundastrum cinnamomeum*, *Oxydendrum arboreum*, *Oxypolis rigidior*, *Panicum virgatum*, *Parnassia grandifolia*, *Parthenocissus quinquefolia*, *Peltandra virginica*, *Persea*, *Physocarpus opulifolius*, *Pinguicula planifolia*, *Pinus palustris*, *P. rigida*, *P. strobus*, *P. virginiana*, *Platanthera integrilabia*, *Pleea tenuifolia*, *Pogonia ophioglossoides*, *Polygala lutea*, *P. nana*, *Prunus serotina*, *Pycnanthemum tenuifolium*, *Quercus rubra*, *Q. velutina*, *Rhamnus alnifolia*, *Rhododendron arborescens*, *R. viscosum*, *Rhynchospora alba*, *R. capitellata*, *R. oligantha*, *Rubus hispidus*, *Sagittaria fasciculata*, *Sanguinaria canadensis*, *Sanguisorba canadensis*, *Sarracenia flava*, *S. jonesii*, *S. psittacina*, *S. purpurea*, *Schizachyrium scoparium*, *Scirpus expansus*, *Scutellaria integrifolia*, *Selaginella apoda*, *Smilax laurifolia*, *S. rotundifolia*, *S. walteri*, *Solidago caesia*, *S. patula*, *S. rugosa*, *S. uliginosa*, *Sphagnum*, *Spiraea alba*, *S. tomentosa*, *Symphyotrichum novi-belgii*, *Thalictrum pubescens*, *Thuidium*, *Tofieldia glabra*, *Toxicodendron* *radicans, T. vernix, Triantha glutinosa, Utricularia subulata, Vaccinium corymbosum, V. darrowii, Viburnum nudum, Viola cucullata, V. primulifolia, V. walteri, Vitis rotundifolia, Woodwardia areolata, Xanthorhiza simplicissima, Xyris platylepis, X. torta.*

Helenium drummondii H. Rock is a perennial found in bogs, ditches, meadows, prairies, savannas, swamps, woodlands, and along the margins of ponds at elevations of up to 50 m. Substrates are acidic (e.g., pH 4.5) and include the following soil types: Caddo (typic glossaqualf), Guyton (typic glossaqualf), Kinder (typic glossaqualf), and Mabank/Wilson (oxyaquic vertic haplustalfs). Virtually no other life-history information exists for this species and additional study is needed. **Reported associates:** *Aletris aurea, Andropogon glomeratus, A. gyrans, A. liebmannii, A. virginicus, Anthaenantia rufa, Aristida palustris, A. purpurascens, Arnoglossum ovatum, Asclepias longifolia, Baptisia leucophaea, Carex glaucescens, Centella asiatica, Chaptalia tomentosa, Coleataenia longifolia, Coreopsis gladiata, Croton capitatus, Ctenium aromaticum, Dichanthelium acuminatum, D. dichotomum, D. scabriusculum, Echinacea purpurea, Eleocharis microcarpa, E. tortilis, E. tuberculosa, Eriocaulon decangulare, E. texense, Eryngium integrifolium, Eupatorium leucolepis, Euthamia leptocephala, Fuirena bushii, F. pumila, F. squarrosa, Gratiola brevifolia, Gymnadeniopsis nivea, Hypericum brachyphyllum, H. galioides, Hyptis alata, Juncus marginatus, J. scirpoides, J. trigonocarpus, J. validus, Lobelia flaccidifolia, Lycopodiella appressa, Marshallia caespitosa, M. graminifolia, Mimosa strigillosa, Muhlenbergia capillaris, M. expansa, Nyssa biflora, Panicum virgatum, Paspalum floridanum, P. notatum, Pluchea baccharis, Polygala ramosa, Proserpinaca pectinata, Prosopis glandulosa, Ratibida laciniata, Rhexia lutea, Rhynchospora berteroi, R. elliottii, R. filifolia, R. gracilenta, R. latifolia, R. macra, R. microcarpa, R. oligantha, R. perplexa, R. plumosa, R. rariflora, R. stenophylla, Rudbeckia hirta, Sabatia campanulata, Saccharum giganteum, Schoenolirion croceum, Scleria ciliata, S. georgiana, S. muehlenbergii, S. pauciflora, Silphium, Stylisma aquatica, Styrax americanus, Tridens ambiguus, Xyris stricta.*

Helenium pinnatifidum (Schwein. ex Nutt.) Rydb. occurs in canals, depressions, ditches, bogs, borrow pits, flatwoods, glades, hammocks, marshes, meadows, ponds, prairies, savannas, seepage slopes, scrub, sloughs, swales, swamps, woodlands, and along margins of cypress domes, lakes, pools, and swamps at elevations of up to 50 m. The plants tolerate full sun to partial shade. They occur exposed or in shallow water on alkaline substrates that include Basinger (Spodic Psammaquents), clay, EauGallie (Alfic Haplaquods), loam, loamy clay, loamy sand, marl, peat, Pomona (Ultic Haplaquods), Pottsburg (Grossarenic Haplaquods), Riviera (Arenic Glossaqualfs), sand, sandy loam, sandy peat, Smyrna and Myakka (Aeric Haplaquods), and Vero (Alfic Haplaquods). The plants grow only in sites with temporary water and are not found where permanent water conditions persist. Flowering occurs early in the year (January–April). The flowers are visited (and likely pollinated) by bees (Insecta:

Hymenoptera) and butterflies (Insecta: Lepidoptera). The seeds weigh about 0.37 mg and are dispersed by wind during the dry season (March–May). Seed germination occurs in wet depressions and in solution holes. Additional details on the reproductive biology and seed ecology are unavailable. **Reported associates:** *Acer rubrum, Agalinis aphylla, A. filicaulis, Aletris aurea, A. lutea, Amphicarpum amphicarpon, A. muhlenbergianum, Andropogon glomeratus, A. virginicus, Anthaenantia rufa, Aristida palustris, A. rhizomophora, A. spiciformis, A. stricta, Arnoglossum ovatum, Arundinaria tecta, Asclepias lanceolata, Asemeia grandiflora, Axonopus furcatus, Bacopa caroliniana, Balduina uniflora, Baptisia alba, Berchemia scandens, Bigelowia nudata, Boltonia asteroides, Buchnera floridana, Calamovilfa brevipilis, Carex gholsonii, C. granularis, Carphephorus odoratissimus, Centella asiatica, Chamaecrista nictitans, Chaptalia tomentosa, Cirsium horridulum, C. nuttallii, Cladium jamaicense, Coleataenia longifolia, C. tenera, Coreopsis gladiata, Ctenium aromaticum, Cyperus haspan, C. polystachyos, Dichanthelium aciculare, D. acuminatum, D. dichotomum, D. strigosum, Diodia virginiana, Dionaea muscipula, Diospyros virginiana, Drosera brevifolia, D. capillaris, Eleocharis flavescens, Eragrostis elliottii, Erigeron quercifolius, E. vernus, Eryngium integrifolium, E. yuccifolium, Eriocaulon compressum, E. decangulare, E. ravenelii, Eryngium prostratum, Eupatorium leptophyllum, E. leucolepis, E. rotundifolium, E. pilosum, Euphorbia inundata, Eustachys glauca, Euthamia graminifolia, Fothergilla gardenii, Fuirena breviseta, F. scirpoidea, F. squarrosa, Gaylussacia frondosa, Gentiana pennelliana, Gratiola ramosa, Gymnadeniopsis integra, G. nivea, Helenium vernale, Helianthus heterophyllus, Hypericum brachyphyllum, H. cistifolium, H. fasciculatum, H. hypericoides, H. mutilum, Hyptis alata, Ilex cassine, I. glabra, I. myrtifolia, Ipomoea sagittata, Juncus marginatus, J. megacephalus, J. polycephalus, J. repens, J. trigonocarpus, Lachnanthes caroliniana, Lachnocaulon anceps, Liatris garberi, Lilium catesbaei, Liquidambar styraciflua, Lobelia canbyi, L. glandulosa, Ludwigia linearis, L. linifolia, L. microcarpa, L. pilosa, L. virgata, Lycopus rubellus, Lyonia ligustrina, L. lucida, Lysimachia loomisii, Mecardonia acuminata, Mikania scandens, Mitreola petiolata, M. sessilifolia, Mnesithea rugosa, Muhlenbergia capillaris, M. expansa, M. sericea, Myrica cerifera, Nyssa biflora, Oxypolis canbyi, O. filiformis, Panicum hemitomon, P. virgatum, Parnassia caroliniana, Paspalum praecox, Persea palustris, Persicaria hydropiperoides, Phyla nodiflora, Pinguicula caerulea, P. pumila, Pinus elliottii, P. palustris, P. serotina, Piriqueta cistoides, Platanthera ciliaris, P. cristata, Pleea tenuifolia, Pluchea baccharis, Pogonia ophioglossoides, Polygala brevifolia, P. cruciata, P. cymosa, P. hookeri, P. lutea, P. nana, Pontederia cordata, Proserpinaca pectinata, Prunus caroliniana, Pteridium aquilinum, Pterocaulon pycnostachyum, Ptilimnium capillaceum, Pycnanthemum flexuosum, Pyxidanthera barbulata, Quercus minima, Q. phellos, Rhexia alifanus, R. aristosa, Rhynchospora cephalantha, R. chapmanii, R. chalarocephala, R. colorata, R. corniculata, R. decurrens, R. divergens, R. fascicularis, R.*

filifolia, R. galeana, R. globularis, R. gracilenta, R. inundata, R. latifolia, R. microcarpa, R. microcephala, R. perplexa, R. pleiantha, R. plumosa, R. tracyi, Rubus trivialis, Rudbeckia graminifolia, R. mohrii, Ruellia noctiflora, Sabal palmetto, Sabatia difformis, Saccharum giganteum, Sagittaria graminea, Sarracenia flava, S. minor, S. purpurea, Schizachyrium rhizomatum, Schoenolirion albiflorum, Scirpus pendulus, Scleria baldwinii, S. georgiana, S. muehlenbergii, S. pauciflora, S. reticularis, S. verticillata, Serenoa repens, Setaria parviflora, Seymeria cassioides, Smilax laurifolia, Solidago stricta, Sphagnum, Spiranthes laciniata, Steinchisma hians, Stillingia sylvatica, Symphyotrichum chapmanii, S. dumosum, Syngonanthus flavidulus, Taxodium ascendens, Thelypteris interrupta, Toxicodendron radicans, Utricularia inflata, U. purpurea, Vaccinium crassifolium, Verbesina heterophylla, Vitis rotundifolia, Woodwardia areolata, W. virginica, Xyris ambigua, X. caroliniana, X. elliottii, X. fimbriata, X. jupicai.

Helenium thurberi A. Gray is a riparian annual, which inhabits arroyos, bottoms, ditches, draws, floodplains, meadows, scrub, streambeds, washes, woodlands, and the margins of canals, ponds, rivers, and streams at elevations of up to 1600 m. The plants tolerate exposures of full sun to partial shade. They are found on substrates described as alluvium, limestone, mud, sand and silt. Life history information for this species is sparse. **Reported associates:** Acalypha neomexicana, Allophyllum gilioides, Amaranthus palmeri, Ambrosia acanthicarpa, A. confertiflora, A. monogyra, A. psilostachya, A. trifida, Argemone pleiacantha, Aristida adscensionis, A. ternipes, Artemisia ludoviciana, Astragalus nothoxys, Baccharis salicifolia, B. salicina, B. sarothroides, Bidens bigelovii, B. laevis, B. leptocephala, Boerhavia coccinea, Bouteloua aristidoides, B. barbata, B. curtipendula, B. gracilis, Bromus rubens, Calibrachoa parviflora, Celtis reticulata, Cenchrus incertus, Chamaecrista nictitans, Chilopsis linearis, Chloracantha spinosa, Chloris virgata, Comandra umbellata, Conyza canadensis, Corydalis aurea, Croton texensis, Cuphea wrightii, Cynanchum, Cynodon dactylon, Cyperus esculentus, C. niger, C. odoratus, Datura discolor, D. wrightii, Digitaria sanguinalis, Diodella teres, Echinochloa crus-galli, Eleocharis montevidensis, E. palustris, Elymus canadensis, Equisetum laevigatum, Eragrostis cilianensis, E. pectinacea, Erigeron divergens, Eriochloa acuminata, Eriogonum polycladon, Euphorbia hyssopifolia, E. indivisa, Fraxinus, Gomphrena sonorae, Helianthus annuus, H. petiolaris, Heterosperma pinnatum, Heterotheca subaxillaris, Heliomeris longifolia, Ipomoea barbatisepala, I. cristulata, I. hederacea, Juncus balticus, J. bufonius, J. torreyi, Laennecia coulter, Ludwigia palustris, Machaeranthera tagetina, M. tanacetifolia, Melampodium strigosum, Melia azedarach, Melilotus albus, M. indicus, Mentzelia pumila, Mimosa aculeaticarpa, Mimulus guttatus, Mirabilis longiflora, Mollugo verticillata, Muhlenbergia rigens, Nasturtium officinale, Nerium oleander, Nicotiana glauca, N. obtusifolia, Oenothera cespitosa, O. elata, Persicaria lapathifolia, P. pensylvanica, P. punctata, Phaseolus vulgaris, Physalis acutifolia, Piptatherum miliaceum, Plantago major, Polanisia dodecandra, Polypogon monspeliensis, P. viridis, Populus fremontii, Prosopis velutina, Pseudognaphalium canescens, P. stramineum, Quercus arizonica, Ranunculus sceleratus, Rumex crispus, Salix bonplandiana, S. exigua, S. gooddingii, S. laevigata, Sambucus nigra, Schoenoplectus acutus, S. americanus, Senecio flaccidus, Senegalia greggii, Senna hirsuta, Setaria leucopila, Sida abutifolia, Sisymbrium irio, Solanum americanum, S. nigrescens, S. rostratum, Sorghum bicolor, S. halepense, Sphaeralcea fendleri, Sporobolus contractus, S. cryptandrus, S. wrightii, Stephanomeria pauciflora, Symphyotrichum falcatum, S. subulatum, Tagetes lemmonii, Tamarix ramosissima, Tecoma stans, Typha domingensis, Ulmus parvifolia, Verbesina encelioides, Veronica anagallis-aquatica, Xanthisma gracile, X. spinulosum, Xanthium strumarium, Xanthocephalum gymnospermoides, Zinnia peruviana.

Helenium virginicum S.F. Blake is a perennial, which grows emersed or in shallow waters (to 50 cm) of bogs, ditches, meadows, prairies, swales, or borders of lakes and sinkhole ponds at elevations of up to 500 m. The substrates are acidic (pH 4.5–5.1) and comprise up to 2 m of gray clay mixed in a cobble, gravel, and sand matrix, which overlies dolomite or limestone. They can contain high concentrations of Al (1902 ppm) and As (22.3 ppm) but relatively low levels of B (0.017 ppm), Ca (364 ppm), K (100 ppm), Mg (80 ppm) and P (37 ppm). Some sites have more typical levels of Al, As, B, Ca, K, Mg and P, but very high levels of Cu. The habitats are characterized by variable durations of annual inundation; however, optimal conditions represent standing water in winter through mid-summer, with dry conditions for the remainder of the year. The plants are shade intolerant. They live for at least 4 years and are polycarpic. The stems bolt in mid-May with flowering commencing in early July and peaking in early August. The flowers are sporophytically self-incompatible with an observed seed set of 76.8% per outcrossed heads compared to 0.8% in selfed heads. The specific pollinators are unknown but are suspected to comprise various insects (Insecta) such as bees (Hymenoptera), butterflies (Lepidoptera), hoverflies (Diptera), and wasps (Hymenoptera). The germination rate of seeds is high (92%). Freshly dispersed seeds (which occurs in late fall) germinate poorly (<5%); however, once dormancy is broken (requiring 2 months of cold stratification), the seeds will germinate well at 19°C–32°C. No germination occurs in the dark or in standing water. The first year of growth develops into a rosette, with flowering shoots produced in subsequent years. A viable seed bank (80% germination) is maintained for at least 7 years. The persistent (type IV) seed bank is believed to be critical for survival of this species in face of extensive long-term environmental stochasticity. The seed bank enables plants to survive periods of excessive hydrological variability, which can eliminate potentially competitive vegetation. Vegetative reproduction occurs by the production of sprouts from fallen stems. Regeneration from root sprouts has been observed in greenhouse culture, but has not been documented in wild populations. Most individual plants in the field represent distinct genets. Seedling survival is poor. This species competes poorly with weedy plants but can persist in dryer sites once competitors have been removed. Transplanted individuals exhibit higher survival when placed near existing conspecific populations than in uncolonized sites. **Reported associates:** *Agrostis perennans, Bidens frondosus, Boltonia asteroides, Carex barrattii, C. lasiocarpa, Coleataenia longifolia, Cyperus dentatus, Dichanthelium acuminatum, Diodella teres, Echinodorus tenellus, Eleocharis acicularis, E. melanocarpa, Erechtites hieraciifolius, Fimbristylis autumnalis, Helenium flexuosum, Hypericum boreale, Isoetes virginica, Lysimachia hybrida, Panicum hemitomon, P. philadelphicum, P. verrucosum, Paspalum laeve, Quercus palustris, Rhexia mariana, R. virginica, Schoenoplectus torreyi, Sphagnum, Stachys hyssopifolia, Symphyotrichum dumosum, S. pilosum, Trichostema dichotomum, Viola lanceolata.*

Use by wildlife: *Helenium pinnatifidum* is a consistent feature of nesting habitat for the Cape Sable seaside sparrow (Aves: Emberizidae: *Ammodramus maritimus mirabilis*). It also provides a source of nectar for butterflies (Insecta: Lepidoptera). The flowers of *H. thurberi* are visited by an assortment of bees (Insecta: Hymenoptera) including andrenid bees (Andrenidae: *Calliopsis rozeni, Perdita callicerata*), honeybees (Apidae: *Anthophora maculifrons, Melissodes paroselae*), leafcutter bees (Megachilidae: *Ashmeadiella cactorum; Ashmeadiella occipitalis*), plasterer bees (Colletidae: *Colletes perileucus*), and sweat bees (Halictidae: *Halictus ligatus, Sphecodes mandibularis*).

Economic importance: food: not edible; **medicinal:** The lactones of several *Helenium* species are potentially allergenic and can cause contact sensitivity. *Helenium brevifolium* contains brevilins, which include sequiterpene lactones that have been shown to block signaling in cancer cells. The lactone pinnatifidin is derived from *H. pinnatifidum*; *H. thurberi* contains tenulin (I) and thurberilin and *H. virginicum* contains virginolide; **cultivation:** *Helenium brevifolium* is commercially available but uncommon in trade. It sometimes is distributed under the cultivar name 'Dunkel Pracht'. *Helenium pinnatifidum* also is reported to be in cultivation;

misc. products: none; **weeds:** despite the common name, none of the OBL species is considered to be a nuisance; **nonindigenous species:** all of the OBL species are indigenous.

Systematics: *Helenium* is the type genus of tribe Helenieae wherein it is closely related to *Gaillardia* and *Marshallia* (Figure 5.157). *Helenium brevifolium* is believed to be closely related to *H. campestre*. It also is suspected to be one hybrid parent, which gave rise to *H. flexuosum*. Although some have viewed *H. virginicum* as a segregate of *H. autumnale*, common garden experiments showed genetic differences in height, bolting date, blooming period, cauline leaf abundance/morphology, pappus length, and basal leaf length during flowering. Analyses of nuclear ITS sequences corroborate the distinction of these species and confirm that disjunct Missouri populations attributed to *H. virginicum* indeed are conspecific. DNA data also indicate *H. autumnale* as the sister species of *H. virginicum* (Figure 5.157). The base chromosome number of *Helenium* is given as $x = 17$. In this interpretation, the cytotypes of all OBL species would indicate derivations by aneuploid reduction: *H. brevifolium* ($2n = 26, 28$); *H. drummondii* ($2n = 32$); *H. pinnatifidum* ($2n = 32, 34$); *H. thurberi* ($2n = 26$); *H. virginicum* ($2n = 28$).

Comments: *Helenium brevifolium* and *H. pinnatifidum* occur in the southeastern United States. The other OBL species have more narrowly restricted ranges: *H. drummondii* (LA, TX), *H. virginicum* (MO, VA) and *H. thurberi* (AZ, but extending into Mexico).

References: Adams et al., 2005; Alexander, 1938; Avis et al., 1997; Baskin & Baskin, 1998; Bierner, 1974, 2006; Carr, 2007; Chamberlain et al., 2013; Chen et al., 2013; Clark, 1998; Ferguson & Wunderlin, 2006; Hallmark & Morgan, 2010; Herz & Lakshmikantham, 1965; Herz & Santhanam, 1967; Herz et al., 1962, 1968; Hilgard, 1873; Kirkman et al., 2000; Knox, 1987, 1997; Knox et al., 1995; Ley et al., n.d.; Lockwood et al., 1999; Marrs-Smith, 1983; Messmore & Knox, 1997; Mossman, 2009; Orzell & Bridges, 2006; Prenger, 2005; Rimer & Summers, 2006; Rock, 1957; Schilling & Floden,

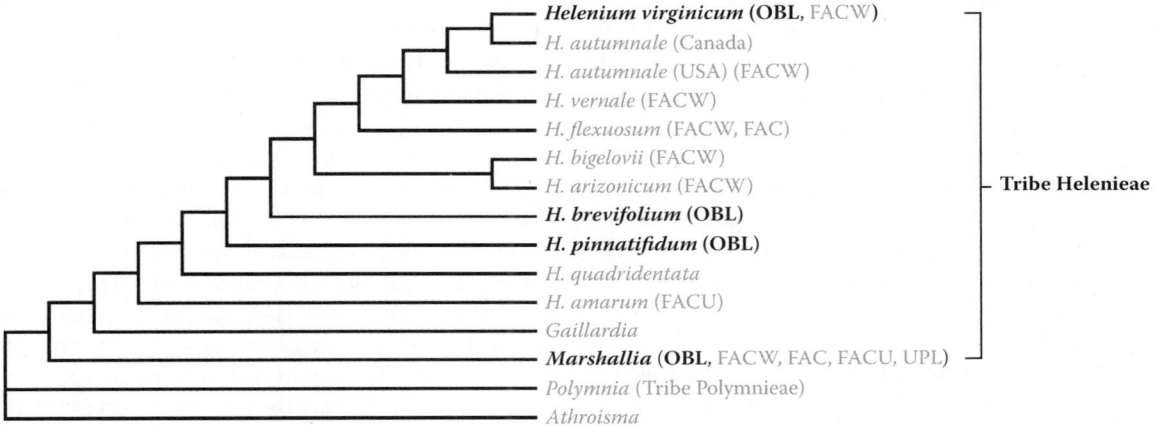

FIGURE 5.157 Interspecific relationships in Asteraceae tribe *Helenieae* as indicated by analysis of nrITS sequene data. The OBL habit (bold) has evolved several times within the genus *Helenium*. (Adapted from Schilling, E.E. & Floden, A., *Phytoneuron*, 81, 1–6, 2013.)

2013; Scott et al., 2001; Seymour, 2011; Simurda & Knox, 2000; Simurda et al., 2005; Solis-Garza & Jenkins, 1998; Stromberg et al., 2005; USDA, 1984; Van Alstine, 2000.

25. *Helianthus*

Sunflower; tournesol

Etymology: from the Greek *hêlios anthos* ("sun flower") for the sun-like inflorescence

Synonyms: *Harpalium*

Distribution: global: Mexico; North America; **North America:** widespread

Diversity: global: 52 species; **North America:** 52 species

Indicators (USA): OBL: *Helianthus californicus, H. heterophyllus, H. verticillatus*

Habitat: brackish, freshwater; palustrine, riverine; **pH:** acidic; **depth:** <1 mm; **life form(s):** emergent herb

Key morphology: Shoots (to 350 cm) erect, glabrous or hispid; leaves basal and opposite or cauline and alternate or whorled, sessile or with winged or wingless petioles (to 3 cm); blades (to 28 cm) ovate, linear, lanceolate, or spatulate, the surface with or without glandular dots, the margins entire, revolute or serrulate; heads (1–10) radiate, peduncled (to 15 cm), involucres (to 25 mm) campanulate to hemispheric, with 24–55 phyllaries (to 35 mm); receptacle with pales (to 11 mm); ray florets (10–21; to 26 mm), yellow, neuter; disk florets (to 150+) bisexual, corollas (to 8 mm) yellow or reddish lobed; cypselae (to 5 mm) glabrous, the pappus of 2 aristate scales (to 4 mm) with or without an additional 1–3 deltate scales (to 1.5 mm).

Life history: duration: perennial (crown buds, rhizomes); **asexual reproduction:** rhizomes; **pollination:** insect; **sexual condition:** hermaphroditic; **fruit:** cypselae (common); **local dispersal:** gravity, wind (seeds); **long-distance dispersal:** birds (seeds)

Imperilment: (1) *Helianthus californicus* [G4]; (2) *H. heterophyllus* [G4]; S1 (AL, GA); S3 (NC); (3) *H. verticillatus* [G1Q]; S1 (AL, GA, TN)

Ecology: general: *Helianthus* contains annuals and perennials, but all of the OBL species are perennial. Just under half of the species (23) have wetland indicator status including the three OBL taxa. Although some *Helianthus* cultivars are self-compatible, most wild species retain sporophytic self-incompatibility and predominantly are outcrossed by insects (Insecta). The principal pollinators are bees (Hymenoptera). Outcrossing is facilitated by protandry of the bisexual disc florets (the ray florets are neuter). Cold temperatures are thought to increase the potential for self-incompatibility. Seeds of most wild species investigated have physiological dormancy, which must be broken by a period of cold stratification. However, dormancy requirements also diminish in cultivated plants. Because the pappus (comprising readily deciduous scales) is not modified well for wind dispersal, seed dispersal is highly local unless ingestion by birds (or human intervention) occurs.

Helianthus californicus **DC.** is typically a riparian species that grows in freshwater or tidal brackish sites of canyons, channels, creekbeds, flats, floodplains, gulches, marshes, meadows, ravines, seeps, springs, streams, and along the margins of levees, sloughs, and perennial or

intermittent streams at elevations of up to 1828 m. Plants occur in open sun to partial shade but normally above the high water mark. Substrates have been described variously as alluvium, clay, decomposed granite, gneiss, gravel, muck, rock, sand, sandy loam, and serpentine. The flowering period is long and occurs from June to October. The flowers are pollinated by insects such as bees and butterflies (see *Use by wildlife*). The plants produce rhizomes and can be propagated vegetatively by the separation of dormant bare roots, which will establish well in streamside localities where individuals can persist from 8 to 10 years. There is no additional information on the reproductive biology or seed ecology of this species. **Reported associates:** *Adenostoma fasciculatum, A. sparsifolium, Aesculus californica, Ambrosia psilostachya, Apium graveolens, Arctostaphylos glauca, Artemisia douglasiana, A. ludoviciana, A. dracunculus, A. tridentata, A. vulgaris, Baccharis salicifolia, Calystegia sepium, Ceanothus greggii, Cercocarpus betuloides, Cheilanthes clevelandii, Clematis ligusticifolia, Datisca glomerata, Distichlis spicata, Epilobium ciliatum, Eriogonum apiculatum, E. fasciculatum, Euthamia occidentalis, Fraxinus velutina, Grindelia hirsutula, Hoita macrostachya, H. orbicularis, Juncus effusus, J. macrophyllus, Juniperus californica, Lathyrus jepsonii, Lemna, Leymus triticoides, Lilaeopsis masonii, Lonicera, Melilotus alba, Mimulus cardinalis, Muhlenbergia rigens, Pellaea mucronata, Persicaria, Pinus coulteri, P. jeffreyi, Platanus racemosa, Pluchea carolinensis, Quercus agrifolia, Q. berberidifolia, Q. dumosa, Ribes, Romneya coulteri, Rosa californica, Salix exigua, S. laevigata, S. lasiolepis, Schoenoplectus acutus, S. americanus, S. californicus, Solidago velutina, Symphoricarpos albus, Toxicodendron diversilobum, Typha domingensis, Urtica dioica.*

Helianthus heterophyllus **Nutt.** grows in bogs, depressions, ditches, flats, flatwoods, marshes, meadows, savanna, seepage bogs, and seepage slopes at elevations of up to 50 m. Exposures typically are in full sun. Substrates are acidic and described as aeric palequults, plummer (grossarenic palequults), sand, sandy clay, sandy peat, or fine loamy sand underlain by clay. Occurrences often represent sites that have been burned recently. The average maximum root depth (22 cm) exceeds that of most savanna species surveyed and likely affords protection during fire episodes. Life history information for this species is scanty. The flowers appear to be pollinated primarily by bees (see *Use by wildlife*). Seeds available from wildflower garden suppliers presumably have been stratified and reportedly germinate within a few weeks at temperatures above 27°C. Perennation occurs by means of crown buds and production of short rhizomes. The crown buds are not dormant and can produce daughter plants (occasionally flowers) during the same season. *Helianthus heterophyllus* has been described as one of the "more weedy" savanna species, which can persist and even thrive on soils that have been altered by conversion to pasture. **Reported associates:** *Acer rubrum, Agalinis aphylla, A. fasciculata, A. filicaulis, A. purpurea, Aletris aurea, A. lutea, A. obovata, Alternanthera sessilis, Amianthium muscitoxicum, Amphicarpum amphicarpon, Andropogon arctatus, A. glomeratus, A. gyrans, A.*

liebmannii, A. virginicus, Angelica dentata, Anthaenantia rufa, Aristida palustris, A. purpurascens, A. simpliciflora, A. stricta, Arnoglossum ovatum, Aronia arbutifolia, Arundinaria gigantea, Asclepias connivens, A. tuberosa, Balduina uniflora, Baptisia lanceolata, B. simplicifolia, Bartonia verna, Bidens bipinnatus, B. mitis, Bigelowia nudata, Calamovilfa brevipilis, Calopogon pallidus, C. tuberosus, Carex verrucosa, Carphephorus odoratissimus, C. paniculatus, C. pseudoliatris, C. tomentosus, Centella asiatica, Cephalanthus occidentalis, Chaptalia tomentosa, Cirsium horridulum, C. lecontei, Cleistesiopsis divaricata, Clethra alnifolia, Cliftonia monophylla, Cnidoscolus urens, Coleataenia longifolia, C. tenera, Coreopsis gladiata, Ctenium aromaticum, Cuphea carthagenensis, Cyrilla racemiflora, Dichanthelium acuminatum, D. ensifolium, D. nudicaule, D. scabriusculum, Diodia virginiana, Dionaea muscipula, Doellingeria sericocarpoides, Drosera brevifolia, D. capillaris, D. filiformis, D. intermedia, D. tracyi, Eclipta prostrata, Erigeron vernus, Eriocaulon compressum, E. decangulare, E. texense, Eryngium integrifolium, Eupatorium album, E. capillifolium, E. leucolepis, E. mohrii, E. rotundifolium, Euphorbia inundata, Eurybia eryngiifolia, Euthamia graminifolia, Fuirena breviseta, F. scirpoidea, F. squarrosa, Gaylussacia dumosa, G. mosieri, Gymnadeniopsis integra, G. nivea, Helenium pinnatifidum, H. vernale, Helianthus angustifolius, H. radula, Hieracium, Hydrocotyle umbellata, Hypericum brachyphyllum, H. crux-andreae, H. fasciculatum, H. galioides, H. nitidum, Hypoxis juncea, H. rigida, H. wrightii, Ilex coriacea, I. glabra, I. myrtifolia, I. vomitoria, Ionactis linariifolia, Juncus dichotomus, Lachnanthes caroliniana, Lachnocaulon anceps, L. digynum, Liatris pilosa, L. spicata, Lilium catesbaei, Linum floridanum, Liquidambar styraciflua, Lobelia brevifolia, L. floridana, L. glandulosa, L. nuttallii, Lophiola aurea, Ludwigia leptocarpa, L. linearis, L. pilosa, Lycopodiella alopecuroides, L. caroliniana, L. prostrata, Lyonia lucida, Lysimachia loomisii, Macranthera flammea, Magnolia virginiana, Marshallia graminifolia, Mitreola petiolata, Muhlenbergia expansa, Myrica heterophylla, Nolina atopocarpa, Nyssa biflora, Osmunda regalis, Osmundastrum cinnamomeum, Oxalis corniculata, Oxypolis denticulata, O. filiformis, Panicum verrucosum, Paspalum plicatulum, P. praecox, P. urvillei, Persea palustris, Persicaria hydropiperoides, Phoebanthus tenuifolius, Pinguicula caerulea, P. ionantha, P. lutea, P. planifolia, P. pumila, Pinus elliottii, P. palustris, P. serotina, Pityopsis graminifolia, P. oligantha, Pleea tenuifolia, Pogonia ophioglossoides, Polygala brevifolia, P. chapmanii, P. cruciata, P. cymosa, P. hookeri, P. lutea, P. ramosa, Proserpinaca pectinata, Pteridium aquilinum, Pterocaulon pycnostachyum, Pyxidanthera barbulata, Quercus geminata, Q. laurifolia, Q. margaretta, Q. marilandica, Q. minima, Q. pumila, Rhexia alifanus, R. lutea, R. mariana, R. nashii, R. nuttallii, R. petiolata, Rhynchosia, Rhynchospora baldwinii, R. cephalantha, R. chalarocephala, R. chapmanii, R. ciliaris, R. corniculata, R. latifolia, R. macra, R. oligantha, R. plumosa, R. rariflora, Rubus cuneifolius, R. trivialis, Rudbeckia graminifolia, Sabatia decandra, S. difformis, S. macrophylla, Sacciolepis

striata, Sarracenia alata, S. flava, S. leucophylla, S. minor, S. purpurea, S. psittacina, S. rubra, Schizachyrium scoparium, S. tenerum, Scleria baldwinii, S. georgiana, S. muehlenbergii, S. oligantha, S. pauciflora, S. reticularis, S. triglomerata, Serenoa repens, Seymeria cassioides, Sisyrinchium fuscatum, Smilax laurifolia, Solidago fistulosa, S. stricta, Sphagnum, Stenanthium densum, Stillingia aquatica, Stokesia laevis, Stylosanthes, Styrax americanus, Symphyotrichum adnatum, S. chapmanii, S. dumosum, Taxodium ascendens, T. distichum, Trianthema racemosa, Utricularia juncea, U. subulata, Vaccinium corymbosum, V. darrowii, V. elliottii, V. nudum, Verbesina chapmanii, V. heterophylla, Viola lanceolata, V. palmata, V. primulifolia, V. septemloba, Woodwardia virginica, Xyris ambigua, X. baldwiniana, X. caroliniana, X. difformis, X. laxifolia, X. scabrifolia, X. stricta, Zigadenus glaberrimus.

***Helianthus verticillatus* Small** occupies barrens, depressions, floodplains, prairies, terraces, and margins of streams at elevations of up to 300 m. It occurs on soils that are acidic, seasonally wet, high in organic matter and described as Ketona series (mixed, thermic, vertic ochraqualfs), sandy clay, sandy loam (Enville, Iuka), silt loam (Conasauga, Falaya, Lyerly, Townley, Wolftever), or silty clay loam (Dowellton, Gaylesville). Flowering occurs from late August to October. The flowers are self-incompatible and are believed to be pollinated principally by bees (Insecta: Hymenoptera) and butterflies (Insecta: Lepidoptera). The seeds are physiologically dormant. Good germination has been obtained after drying the seed for 1 week, placing it at 4°C for 1 month, then sowing it in a greenhouse under a 16/8 light regime at 25°C–30°C and 70% humidity. However, field germination rates appear to be relatively low. Seed of this species have been collected as a germplasm resource. The plants produce thickened, tuberous roots and rhizomes that can extend to 25 cm. Stems transplant easily and the rhizomes spread rapidly to form dense colonies. Population genetic studies have shown that clonal spread of plants is extremely localized. Management practices include prescribed fire cycles every 2–3 years to sustain the prairie-like conditions of its original habitat. **Reported associates:** *Agalinis purpurea, Andropogon gerardii, Bidens bipinnatus, Carex cherokeensis, Collinsonia canadensis, Helenium autumnale, Helianthus angustifolius, Hypericum sphaerocarpum, Jamesianthus alabamensis, Lobelia cardinalis, Marshallia mohrii, Oxypolis rigidior, Panicum virgatum, Physostegia virginiana, Prenanthes barbata, Pycnanthemum virginianum, Rudbeckia heliopsidis, Sabatia capitata, Saccharum giganteum, Schizachyrium scoparium, Silphium compositum, Solidago riddellii, Sorghastrum nutans, Sporobolus heterolepis, Symphyotrichum novae-angliae, Xyris tennesseensis.*

Use by wildlife: The flowers of *Helianthus californicus* attract bees (e.g., Insecta: Hymenoptera: Apidae: *Apis mellifera*) and butterflies (Insecta: Lepidoptera), which serve as pollinators. The plants also are a host of powdery mildew fungi (Ascomycota: Erysiphaceae: *Erysiphe cichoracearum, Golovinomyces ambrosiae*). Leaves of *H. heterophyllus* possess linear glandular trichomes, which sequester sequiterpene lactones for protection against herbivores and pathogens. The flowers of *H. heterophyllus* attract several pollinating bees (Insecta: Hymenoptera: Apidae: *Apis mellifera*, *Bombus griseocollis*, *Melissodes dentiventris*, *Xylocopa virginica*; Megachilidae: *Megachile xylocopoides*). The plants are hosts to aphids (Insecta: Hemiptera: Aphididae: *Uroleucon helianthicola*) and rust fungi (Basidiomycota: Pucciniaceae: *Puccinia helianthorum*). Roots of *H. heterophyllus* are fairly resistant to carrot beetles (Insecta: Coleoptera: Scarabaeidae: *Ligyrus gibbosus*), which are a serious pest of cultivated sunflowers. However, the plants are susceptible to *Alternaria helianthi* (fungi: Ascomycota: Pleosporaceae), a fungal disease of cultivated sunflowers. The seeds of *H. verticillatus* are eaten by insects and finches (Aves: Fringillidae).

Economic importance: food: *Helianthus* species include the edible Jerusalem artichoke (*H. tuberosus*) and commercial sunflower "seeds" (i.e., cypselae). None of the OBL species reportedly is edible; however, sunflower oil, which is derived from the cypselae, is edible; **medicinal:** The seeds of *Helianthus californicus*, *H. heterophyllus* and *H. verticillatus* contain phytosterols, which are known to lower serum cholesterol in humans by preventing intestinal cholesterol absorption. *Helianthus californicus* also contains sequiterpene lactones and diterpenes. *Helianthus heterophyllus* contains lactones as well as several trihydroxy C_{18} acids known to have antifungal properties; **cultivation:** *Helianthus californicus* was cultivated in European gardens by the mid-19th century, as evidenced by a plate of "*Helianthus californicus insignis*" (an unusual double-flowered form) drawn by Louis Van Houtte in 1845. It remains in bloom throughout the summer. *Helianthus heterophyllus* and *H. verticillatus* are cultivated for native wildflower gardens. Stocks of *H. verticillatus* have been preserved for breeding as a crop wild relative; **misc. products:** *Helianthus californicus* is highly resistant to stalk rot (fungi: Ascomycota: Sclerotiniaceae: *Sclerotinia sclerotiorum*) and has been used to hybridize resistant genes into cultivated sunflowers (*H. annuus*). Native Americans used the stalks of "jatbil" (*H. californicus*) to make arrow shafts. *Helianthus verticillatus* cypselae contain a reasonably high amount of oil (323.4 g/kg), which can be used in various food and industrial applications; **weeds:** Because of its ability to tolerate disturbance to savanna soils, *Helianthus heterophyllus* has been regarded as "weedy" in such areas; **nonindigenous species:** All of the OBL species are indigenous.

Systematics: *Helianthus* is the type genus of tribe Heliantheae in Asteraceae subfamily Asteroideae. The genus is fairly isolated in the tribe, but resolves as the sister genus to *Phoebanthus*, which some authors merge with *Helianthus* (Figure 5.158). All three OBL species have been classified within section *Divaricatus*; however, phylogenetic analyses of ETS data indicate that sectional divisions in the genus do not reflect natural groups (Figure 5.159). Despite incomplete resolution in molecular phylogenetic trees, the wide separation of the three OBL *Helianthus* species indicates that each has derived the habit independently (Figure 5.159). *Helianthus verticillatus* associates with several FACW species. The base chromosome number of *Helianthus* is $x = 17$. *Helianthus*

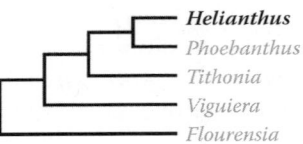

FIGURE 5.158 Relationships of *Helianthus* to other genera of subtribe Helianthinae based on analysis of nrITS sequence data. *Helianthus* is the only genus in this group to contain OBL species. (Adapted from Schilling, E.E. et al., *Syst. Bot.*, 23, 177–187, 1998.)

heterophyllus and *H. verticillatus* are diploids ($2n = 34$); whereas, *H. californicus* ($2n = 102$) is a hexaploid. Some populations of *H. heterophyllus* are regarded as triploids because of irregular meiosis and production of univalents and trivalents. The DNA content of *H. heterophyllus* (81.80 FAU) is intermediate among the $2n = 34$ species surveyed. Although *H. verticillatus* sometimes is designated as *H. ×verticillatus*, genetic data indicate that it did not arise as a hybrid involving *H. angustifolius* and *H. eggertii* or *H. grosseserratus* as has been hypothesized. The cpDNA is transmitted maternally in *Helianthus*; however hybrids involving *H. verticillatus* have shown a low level (1.86%) of paternal transmission and heteroplasmy in the offspring. Population genetic studies using simple sequence repeat (SSR) markers have disclosed high genetic and clonal diversity in *H. verticillatus*, despite the relatively few remaining individuals of this rare species. *Helianthus californicus* is very similar morphologically to *H. nuttallii*, which is believed to have contributed to its hexaploid genome. A recently described tetraploid (*H. inexpectatus*) is morphologically intermediate between *H. californicus* and *H. nuttallii*, and has been proposed as a possible connecting link between those hexaploid and diploid species. *Helianthus californicus* (hexaploid) × *H. annuus* (diploid) hybrids are partially fertile (37.8% pollen stainability) with 0.19% seed set in sib-pollinated offspring and 2.71% seed set from backcross progeny.

Comments: *Helianthus heterophyllus* and *H. verticillatus* occur in the southeastern United States; *H. californicus* occurs in California, but extends into Baja California, Mexico.

References: Aschenbrenner et al., 2013; Astiz et al., 2011; Avis et al., 1997; Bartholomew et al., 2006; Brewer et al., 2011; Campbell, 1999; Carr et al., 2010; Chafin, 2007; Elam & Cantrell, 2011; Ellis & McCauley, 2009; Ellis et al., 2006, 2008; Everett, 2012; Feng et al., 2006; Fernández-Cuesta et al., 2011; Gandhi et al., 2005; Gershenzon et al., 1984; Guertin, 2013; Gulya et al., 2007; Heiser & Smith, 1955; Herz & Bruno, 1986; Hurd, Jr. et al., 1980; Hurst, 2001; Jones, 1994; Kalk, 2011; Keil & Elvin, 2010; Kelsey, 1889; Khoury et al., 2013; Lobello et al., 2000; Long, 1966; Mandel, 2010; Matthews et al., 2002; Mitchell, 2011; Morris et al., 1983; Noss, 2013; Porter & Fraga, 2005; Rheinhardt & Brinson, 2002; Rogers et al., 1980; Schilling, 2001, 2006; Schilling et al., 1998; Seiler et al., 2010; Sims & Price, 1985; Stringer, 2013; Timme et al., 2007; USDA, 1984; Vaghti & Keeler-Wolf, 2004; Walker & Peet, 1984; Watson, 1928.

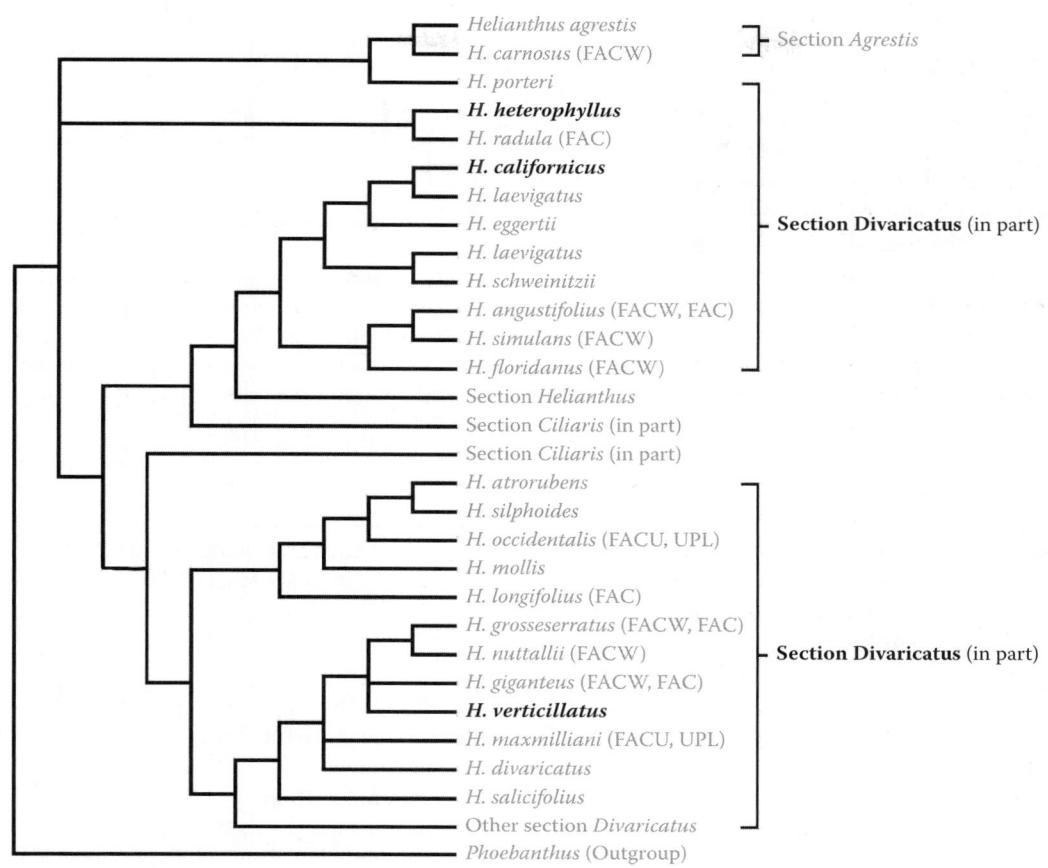

FIGURE 5.159 Interspecific relationships within *Helianthus* based on analysis of ETS sequence data. These relationships indicate three separate origins of OBL taxa (bold) in the genus. The sectional divisions currently in use do not reflect monophyletic groups. (Adapted from Timme, R.E. et al., *Amer. J. Bot.*, 94, 1837–1852, 2007.)

26. *Hesperevax*

Hogwallow starfish, involucrate evax

Etymology: a combination of the Greek *hesperos* (western) and the genus *Evax*

Synonyms: *Evax* (in part); *Psilocarphus* (in part)

Distribution: global: North America; **North America:** southwestern United States

Diversity: global: 3 species; **North America:** 3 species

Indicators (USA): OBL: *Hesperevax caulescens*

Habitat: freshwater; palustrine; **pH:** alkaline; **depth:** <1 m; **life form(s):** emergent herb

Key morphology: Stems typically absent or erect to decumbent (1–4; to 17 cm); leaves (to 90 mm) basal or cauline, congested distally, petioles thickened basally; inflorescence of terminal glomerules (to 25 mm) of 10–40 clustered, discoid heads, subtended by 12–20 whorled, spreading leaves (to 90 mm); adjacent receptacles connate proximally; pales (to 4.5 mm) vertically ranked; pistillate flowers 3–25, corollas obscure; staminate flowers 3–6, corollas to 1.6 mm. Cypselae (to 2 mm) smooth, glabrous, pappus lacking.

Life history: duration: annual (fruit/seeds); **asexual reproduction:** none; **pollination:** unknown; **sexual condition:** monoecious; **fruit:** cypselae (common); **local dispersal:** cypselae; **long-distance dispersal:** cypselae

Imperilment: (1) *Hesperevax caulescens* [G3]; S3 (CA)

Ecology: general: All three *Hesperevax* species are monoecious annuals. Only *H. caulescens* is an obligate aquatic, with *H. sparsiflora* (FACU) and *H. acaulis* (no indicator status) favoring dry, open habitats. There is virtually nothing reported on the reproductive or seed ecology of this genus, and only scarce information exists on other life history features.

Hesperevax caulescens (Benth.) A. Gray resides in depressions, flats, meadows, mudflats, roadsides, swales, and vernal pools at elevations of up to 502 m. Although sites typically possess water levels from 7 to 16 cm, the plants occur on the periphery of standing water where substrate desiccation is proceeding. The substrates are described as adobe, Anita clay, Bear Creek, crumbly clay shale, drying shrink-swell clay, gravel, Mehrten and Riverbank formations, muck, mud, serpentine, and Capay or Clear Lake vertisols. The plants flower and fruit from March to June. Occurrences tend to be more common on small (~1 m) clay mounds. Other life history information for this species is wanting. **Reported Associates:** *Achyrachaena mollis, Acmispon wrangelianus, Aira caryophyllea, Alopecurus saccatus, Amsinckia eastwoodiae, Anagallis minima, Ancistrocarphus filagineus, Androsace elongata, Anthemis cotula, Astragalus tener, Athysanus pusillus, Avena barbata, A. fatua, Blennosperma nanum, Briza minor,*

Bromus hordeaceus, B. madritensis, Calandrinia ciliata, Castilleja attenuata, Centromadia fitchii, Cerastium glomeratum, Chlorogalum angustifolium, Clarkia, Claytonia parviflora, Collinsia sparsiflora, Crassula aquatica, C. connata, C. tillaea, Croton setiger, Cuscuta howelliana, Deschampsia danthonioides, Downingia bicornuta, D. insignis, D. ornatissima, Eleocharis macrostachya, Epilobium cleistogamum, Erodium botrys, E. brachycarpum, E. cicutarium, E. moschatum, Eryngium aristulatum, E. castrense, E. vaseyi, Fritillaria agrestis, F. pluriflora, Gastridium phleoides, Geranium dissectum, Grindelia hirsutula, Hedypnois cretica, Hemizonia congesta, Hesperevax acaulis, Holocarpha virgata, Hordeum marinum, H. murinum, Hypochaeris glabra, Juncus bufonius, Lactuca serriola, Lasthenia californica, L. fremontii, L. glaberrima, L. glabrata, L. minor, Layia chrysanthemoides, L. fremontii, Lepidium latipes, L. nitidum, Leptosiphon bicolor, Lilaea scilloides, Limnanthes douglasii, Logfia gallica, Lolium multiflorum, Lotus, Lupinus bicolor, L. nanus, Lythrum hyssopifolia, Medicago polymorpha, Micropus californicus, Microseris acuminata, M. douglasii, M. elegans, Mimulus tricolor, Myosurus minimus, Navarretia heterandra, N. intertexta, N. leucocephala, Pilularia americana, Plagiobothrys acanthocarpus, P. fulvus, P. greenei, P. stipitatus, Plantago erecta, P. pusilla, Poa annua, P. bulbosa, P. secunda, Pogogyne douglasii, P. ziziphoroides, Psilocarphus brevissimus, P. oregonus, P. tenellus, Ranunculus hebecarpus, Rostraria cristata, Rumex crispus, Sagina decumbens, Senecio vulgaris, Soliva sessilis, Sonchus oleraceus, Taeniatherum caput-medusae, Taraxacum officinale, Tetrapteron graciliflorum, Toxicoscordion fremontii, Trifolium bifidum, T. dubium, T. gracilentum, T. hirtum, T. depauperatum, T. fucatum, T. jokerstii, T. microcephalum, T. variegatum, T. willdenovii, Triphysaria eriantha, Triteleia hyacinthina, T. laxa, Veronica peregrina, Vulpia bromoides, V. microstachys, V. myuros.

Use by wildlife: none reported

Economic importance: food: none reported; **medicinal:** unknown; **cultivation:** not currently in cultivation; **misc. products:** none; **weeds:** none; **nonindigenous species:** none

Systematics: *Hesperevax* was assigned formerly to tribe Inuleae, but currently is placed among the "*Filago*" group of tribe Gnaphalieae. Although the genus contains species once assigned to the Old World genus *Evax*, morphological cladograms indicated it to be distant from that genus and instead, likely allied to *Ancistrocarphus* as its sister group. The distant relationship of *Evax* and *Hesperevax* was confirmed by analysis of molecular data; however, that analysis included only one *Hesperevax* species (*H. sparsiflora*) and was unable to sample material of *Ancistrocarphus*. From the taxa surveyed, *Hesperevax* associated as the sister genus to a clade comprising *Psilocarphus* and one *Micropus* species (*M. californicus*) (Figure 5.160). The lack of monophyly indicated for several genera (*Evax, Filago, Gnaphalium, Micropus*) indicates the need for further taxonomic refinements within this group. Chromosome numbers have not been reported for *Hesperevax*.

Comments: *Hesperevax caulescens* is endemic to California, but possibly extends into Mexico.

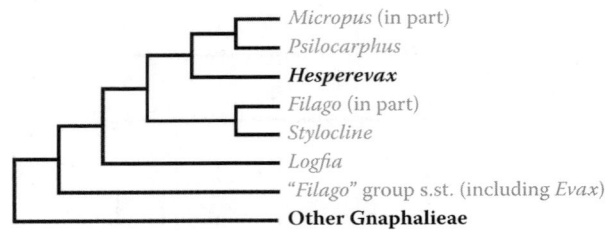

FIGURE 5.160 Phylogenetic relationships among genera of Asteraceae tribe Gnaphalieae as indicated by analysis of combined ITS/ETS sequence data. Taxa containing OBL species are indicated in bold. The lack of monophyly indicated for several genera indicates that the taxonomic subdivision of Gnaphalieae requires extensive revision. (Adapted from Galbany-Casals, M. et al., *Taxon*, 59, 1671–1689, 2010.)

References: Barbour et al., 2007; Buck, 2004; Buck-Diaz et al., 2012; Dittes and Guardino, 2002; Galbany-Casals et al., 2010; Lazar, 2006; Morefield, 1992a,b.

27. Jamesianthus

Alabama warbonnet

Etymology: after Robert Leslie James (1897–1977) combined with *anthos*, Greek for "flower"

Synonyms: none

Distribution: global: North America; **North America:** Alabama and Georgia (endemic)

Diversity: global: 1 species; **North America:** 1 species

Indicators (USA): OBL: *Jamesianthus alabamensis*

Habitat: freshwater; palustrine, riverine; **pH:** 7.3–8.2; **depth:** <1 m; **life form(s):** emergent herb

Key morphology: Stems (to 150+ cm) glandular above; leaves (to 90 mm) cauline, opposite, petiolate or sessile, lanceolate, margins entire or denticulate, bases auricled to produce squareish appearance; heads radiate, solitary or secondarily in corymbs, peduncled (to 8 cm); involucres (to 12+ mm) campanulate, phyllaries (14–18; to 4 mm) mostly in 3 series; ray florets (6–8), pistillate, fertile, laminae (to 15 mm) yellow; disc florets (20–30), bisexual or functionally staminate internally, corollas (to 6 mm) yellow, 5-lobed; cypselae (to 4 mm) obovoid (rays) or ellipsoid to clavate (disc), pappus lacking or of 6–8 barbellate bristles (to 5 mm), borne on crowns.

Life history: duration: perennial (fibrous rootstock buds); **asexual reproduction:** none; **pollination:** unknown (insect?); **sexual condition:** gynomonoecious; **fruit:** cypselae (common); **local dispersal:** cypselae (water); **long-distance dispersal:** cypselae (water)

Imperilment: (1) *Jamesianthus alabamensis* [G3]; S1 (GA); S3 (AL)

Ecology: general: *Jamesianthus* is monotypic.

***Jamesianthus alabamensis* S. F. Blake & Sherff** grows along the margins of first- and second-order streams at elevations of up to 200 m. The plants occur mainly in the shade. The substrates are described as Banger limestone outcrops sandwiched in Hartselle sandstone and covered by deposits of Tuscaloosa gravels and sand. The pH of the water is alkaline (7.3–8.2) with a specific conductivity of 225–270 μmhos. Flowering in this short-day plant initiates when day length drops below 13.5 h, which normally occurs from mid-August to October in its native range. The cypselae mature from late summer to fall. Plants that have been cultivated further north do not commence flowering until the end of October when frost destroys the flowers and kills the plants, which have not attained proper dormancy. No pollinators have been observed in the wild, but are suspected to be bees (Insecta: Hymenoptera). This species is believed to be an ecological endemic, closely adapted to the specific habitat it occupies and restricted there by limitated dispersal. Transplanted plants have been observed to grow well and reproduce in UPL habitats ranging from open pine forests to deciduous woods. However, dispersal of cypselae by wind or animals is not facilitated by the pappus, which is deciduous. The cypselae do float for a short time, which achieves their dispersal along the river banks. The cypselae reportedly are physiologically dormant and require cold stratification for germination. Optimal germination occurs in light under a 35°C/20°C temperature regime. Other reports indicate that cypselae germinate well (~30%) with or without stratification along a pH gradient from 4 to 9. **Reported associates:** *Acer negundo, A. rubrum, Adiantum capillus-veneris, Alnus serrulata, Aruncus dioicus, Bignonia capreolata, Carex torta, Cercis canadensis, Chelone glabra, Coreopsis tripteris, Cornus amomum, Elymus virginicus, Equisetum hyemale, Eutrochium fistulosum, Fagus grandifolia, Fraxinus pennsylvanica, Fuirena squarrosa, Hamamelis virginiana, Helenium autumnale, Helianthus angustifolius, H. microcephalus, Hydrangea quercifolia, Hypericum densiflorum, Iris verna, Itea virginica, Juncus coriaceus, Juniperus virginiana, Justicia*

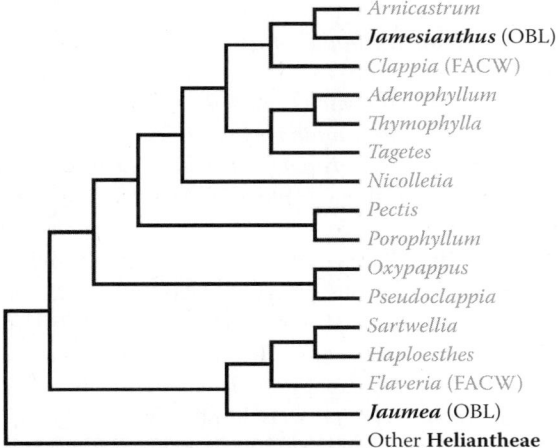

FIGURE 5.161 Phylogenetic placement of *Jamesianthus* among members of subtribe Pectidinae, tribe Tageteae of Asteraceae based on nrITS sequence data. The resolution of *Arnicastrum* as its sister genus confirms the relationship hypothesized by morphological characters. The OBL habit (bold taxa) is obviously derived independently in *Jamesianthus*. (Adapted from Baldwin, B.G. et al., *Syst. Bot.*, 27, 161–198, 2002.)

americana, Liriodendron tulipifera, Marshallia trinervia, Mitreola petiolata, Myrica cerifera, Nasturtium officinale, Oxypolis rigidior, Parnassia grandifolia, Phlox glaberrima, Plantago cordata, Platanus occidentalis, Rhus glabra, Rudbeckia fulgida, R. hirta, R. laciniata, R. triloba, Salix caroliniana, S. nigra, Symphyotrichum pilosum, Trautvetteria caroliniensis, Verbesina virginica, Xyris tennesseensis.

Use by wildlife: none reported
Economic importance: food: unknown; **medicinal:** unknown; **cultivation:** not in cultivation; **misc. products:** none; **weeds:** none; **nonindigenous species:** none
Systematics: *Jamesianthus* is placed within subtribe Pectidinae of tribe Tageteae, which resolves with the "Heleniod" members of the broadly circumscribed tribe Heliantheae. Its similar morphology (graduated multiseriate involucres of broad, rounded phyllaries; long pedunculate, large, solitary capitula; pappus of bristles) indicated a close relationship to *Arnicastrum* and *Clappia*, which eventually was confirmed by analysis of DNA sequence data (Figure 5.161). The chromosome number of *Jamesianthus* is 2n=32, a number shared with the closely related *Clappia*.
Comments: *Jamesianthus alabamensis* is restricted to Alabama and Georgia.
References: Baldwin et al., 2002; Baskin & Baskin, 1998, 2002b; Cullina, 2009; Dennis, 1982; Johnson, 1942; Sherff, 1940; Strother, 2006d.

28. Jaumea

Marsh jaumea, salty susan
Etymology: after J. H. Jaume St. Hilaire (1772–1845)
Synonyms: *Coinogyne*
Distribution: global: Central America; North America; Mexico; South America; **North America:** western
Diversity: global: 2 species; **North America:** 1 species
Indicators (USA): OBL: *Jaumea carnosa*
Habitat: saline; palustrine; **pH:** 6.4–7.4; **depth:** <1 m; **life form(s):** emergent herb
Key morphology: Stems (to 30 cm) ascending or sprawling, branched from base; leaves (to 35 mm) opposite, sessile, succulent, spatulate to linear, margins entire; heads radiate, solitary, involucre (to 8 mm) with 12–15 membranous or succulent purplish phyllaries in 3 or more series; receptacular pales absent; ray florets (3–10), pistillate, fertile; laminae (to 3 mm) yellow; disc florets (20–50), bisexual, fertile, corollas (to 7 mm) yellow, 5-lobed; cypselae (to 3 mm) obconic to columnar, 10-ribbed, glabrous, pappus lacking or of 1–5 subulate scales.
Life history: duration: perennial (rhizomes, stolons); **asexual reproduction:** plant fragments, rhizomes, stolons; **pollination:** insect; **sexual condition:** gynomonoecious; **fruit:** cypselae (common); **local dispersal:** cypselae (water); **long-distance dispersal:** cypselae (water)
Imperilment: (1) *Jaumea carnosa* [G4/G5]; S2 (BC)
Ecology: general: Both *Jaumea carnosa* (OBL) and the related *J. linearifolia* (Mexico southward) are rhizomatous perennial salt marsh plants with succulent leaves. The

photosynthetic pathway of *Jaumea* is unclear. *Jaumea carnosa* has been categorized as a C3 species; however, its congener *J. linearifolia* is known to possess Kranz anatomy, which is associated with C4 photosynthesis. *Jaumea* also is closely related to *Flaveria* (Figure 5.161), which is well-known for its inclusion of C4 species. Information on the reproductive biology and seed ecology of *Jaumea* has only been studied in for *J. carnosa* and is described later. Pollination occurs by insects (Insecta), but few specific pollinators have been described. The seeds are dispersed by water, either directly or indirectly in drifting wrack. No animal vectors have yet been identified, but this possibility requires further evaluation.

***Jaumea carnosa* (Less.) A. Gray** inhabits brackish to saline sites (salinity = 15–31 g/kg) including bases of sea cliffs, beaches, depressions, dunes, meadows, salt marshes, seeps, springs, swales, tidal mud flats, and the margins of lagoons, lakes, ponds, rivers, and sloughs at elevations of up to 71 m. Plants generally occur near or below (−0.05–0.42 m) the high tide limit or mean high water mark. They are tolerant to flooding and occur in areas ranging from 0 to 30% in flooding duration. Exposures are in full sun. The substrates include clay, gravel, rocky sand, sand, sandy loam, silt, and silty sand. Typical nutrient levels associated with stands are: Ca (3–11 meq/100 g), K (608–2301 ppm), Mg (7–27 meq/100 g), Na (51–254 meq/100 g), percent total N (0.16–0.82), NH_4 (1–28 ppm), and NO_3 (1–92 ppm), P (9–21 ppm). Production levels can reach 370 g/m^2 and can represent as much as 72% of the total biomass in some low mesohaline sites. Canopy cover can attain 34%. The plants regulate NaCl uptake when growing in saline habitats to maintain stable shoot salt concentrations. Leaf and stomate numbers decrease while leaf areas increase with respect to salt content. Increased nitrogen results in increased biomass but only at lower salinity levels. Growth can be depressed substantially by competition in low salinity sites. Removal of parasitic *Cuscuta* plants from habitats results in increased (4%) plant abundance. Overall, plant stature is least and leaves are thickest at 0.15 m above mean high water mark. The roots are colonized by arbuscular mycorrhizal (AM) fungi, with significantly lower levels of infection occurring in more UPL sites. Although sexual reproduction is less common than vegetative reproduction, high seed production can occur in conditions of prolonged flooding and lowered salinity. Flowering occurs from May to September, with seeds produced from June to September. The flowers are visited by various Hymenoptera (Insecta) including sand wasps (Crabronidae: *Bembix*) and sweat bees (Halictidae), which represent some of their pollinators. The highest reproductive output occurs at the lowest limit of the tidal range with 12.9% flowering in the lower zone and 0% in the upper marsh. The seeds, which average 0.09 mg, can achieve sediment densities averaging 14,164/m^2. They germinate readily without stratification, but can be stored at 5°C for up to a year. Germination has been achieved from seeds (best collected in August) stored at 5°C for 4–5 days, then placed under greenhouse conditions with daylight increasing from 9 to 13 hrs, and temperatures ranging from 13°C to 31°C. A low to moderate density seed bank is maintained

with up to 153 seedlings/m^2 germinating from sediment samples. Individuals in the standing crop typically are overrepresented with respect to the seed bank propagule pool. The seeds are buoyant and a high percent (98%) will remain afloat after 24 h, which facilitates local water dispersal. Seedling densities of 0.01/m^3 seawater have been recovered from tidal floodwater and 0.25/kg from wrack samples. Reproduction primarily is asexual (vegetative) and plants can form dense clonal patches or mats; they achieve their highest biomass when planted in monocultures. Plant fragments can root adventitiously and function as vegetative propagules. *Jaumea carnosa* is relatively sensitive to small anthropogenic disturbances, only able to recover from those occurring early in the growing season. **Reported associates:** *Abronia maritima, Acacia cyclops, Acmispon glabrus, Adenostoma fasciculatum, Artemisia californica, Arthrocnemum subterminale, Arundo donax, Astragalus pycnostachyus, Atriplex amnicola, A. lentiformis, A. patula, A. prostrata, A. semibaccata, A. watsonii, Baccharis douglasii, B. pilularis, B. salicifolia, B. sarothroides, Batis maritima, Bolboschoenus maritimus, Bromus madritensis, Cakile maritima, Camissoniopsis cheiranthifolia, Carex lyngbyei, Carpobrotus chilensis, C. edulis, Castilleja ambigua, Ceanothus, Chenopodium, Chloropyron maritimum, Conium maculatum, Cotula coronopifolia, Cuscuta pacifica, C. salina, Deschampsia cespitosa, Distichlis littoralis, D. spicata, Eschscholzia californica, Fragaria, Frankenia palmeri, F. salina, Fucus gardneri, Grindelia hirsutula, G. integrifolia, Heliotropium curassavicum, Heteromeles arbutifolia, Hordeum brachyantherum, Isocoma menziesii, I. veneta, Isolepis cernua, Juncus acutus, J. balticus, J. filiformis, J. lesueurii, Leymus innovatus, Limonium californicum, Lolium perenne, Lysimachia maritima, Mesembryanthemum crystallinum, M. nodiflorum, Myosurus minimus, Parapholis incurva, Plantago maritima, Platanus racemosa, Poa nemoralis, Populus fremontii, Potentilla anserina, Puccinellia nuttalliana, Ricinus communis, Rubus ursinus, Ruppia maritima, Salicornia depressa, Salix lasiolepis, Salsola soda, Sarcocornia perennis, Schoenoplectus californicus, S. olneyi, Solidago, Spartina foliosa, Spergularia canadensis, S. macrotheca, S. media, S. salina, Suaeda californica, S. taxifolia, Symphyotrichum chilense, S. subulatum, Tamarix, Triglochin maritimum, Typha domingensis, Vaucheria.*

Use by wildlife: Flowers of *Jaumea carnosa* provide nectar and pollen for various foraging insects (Insecta). *Jaumea carnosa* provides habitat for various birds (Aves) including California black rails (Rallidae: *Laterallus jamaicensis coturniculus*), light-footed clapper rails (Rallidae: *Rallus longirostris levipes*), Belding's Savannah sparrows (Emberizidae: *Passerculus sandwichensis beldingi*), and also for some mammals (Mammalia) including the salt marsh harvest mouse (Cricetidae: *Reithrodontomys raviventris*), and marsh-dwelling shrews (Mammalia: Soricidae: *Sorex ornatus californicus S. vagrans sonomae, S. trowbridgii*). The plants provide cover and act as tidal refugia during high tides for various marsh wildlife such as rodents (Mammalia: Rodentia). *Jaumea carnosa* is a natural host of the hemi-parasitic

Chloropyron maritimum (Orobanchaceae) and of the parasitic *Cuscuta pacifica* and *C. salina* (Convolvulaceae). The plants are hosts to club fungi (Basidiomycota: Pucciniaceae: *Uromyces junci*).

Economic importance: food: none reported; **medicinal:** none; **cultivation:** not cultivated; **misc. products:** *Jaumea carnosa* often is planted in salt marsh restoration projects; **weeds:** none; **nonindigenous species:** none

Systematics: *Jaumea* is placed within subtribe Jaumeinae of tribe Tageteae of Asteraceae. It resolves as the sister group to subtribe Flaveriinae, which includes *Flaveria*, *Hypoësthes* and *Sartwellia* (Figure 5.161). The subtribe once also included *Venegasia*, which has since been moved to tribe Madieae on the basis of molecular evidence. There is little other systematic information on *J. carnosa* or its relationship to *J. lineari-folia*. The base chromosome number of *Jaumea* is *x* = 19, with *J. carnosa* (2*n* = 38) representing a diploid. The OBL habit has arisen independently in the genus.

Comments: *Jaumea carnosa* occurs in coastal western North America.

References: Allison, 1995; Armitage et al., 2006; Baldwin et al., 2002; Barbour, 1978; Brown & Bledsoe, 1996; Burg et al., 1980; Callaway et al., 2003; Cordazzo et al., 2007; Costea et al., 2009; Diggory & Parker, 2011; Eilers, 1974; Evens et al., 1991; Flowers et al., 1986; Giles-Johnson et al., 2011; Grewell, 2008; Heatwole, 2004; Hopkins & Parker, 1984; Janousek & Folger, 2012; Jefferson, 1974; Keammerer & Hacker, 2013; Massey et al., 1984; Morzaria-Luna & Zedler, 2007; Nelson & Kashima, 1993; Page, 1997; Powell, 1993; Ryan & Boyer, 2012; Seliskar, 1985; Shellhammer, 2012; Shennan et al., 1998; Smith & Turner, 1975; St. Omer & Schlesinger, 1980; Strother, 2006e; Sullivan & Zedler, 1999; Sustaita et al., 2011; USFWS, 2013b; Zedler, 2000.

29. *Lasthenia*

Goldfields

Etymology: after the female Greek philosopher Lasthenia of Mantinea (ca. 300 BC)

Synonyms: *Baeria*; *Burrielia*

Distribution: global: Mexico; North America; South America; **North America:** western

Diversity: global: 18 species; **North America:** 17 species

Indicators (USA): OBL: *Lasthenia burkei, L. ferrisiae, L. fremontii, L. glaberrima*

Habitat: freshwater, saline; palustrine; **pH:** 5.5–9.0+; **depth:** <1 m; **life form(s):** emergent herb

Key morphology: Stems (to 40cm) erect or sprawling, branched proximally or distally, glabrous or hairy; leaves (to 50–100mm), cauline, opposite, petioled or sessile with bases nearly connate and sheathing, linear, simple or pinnately lobed, margins entire or toothed; heads radiate (occasionally disciform in *L. glaberrima*), solitary or arranged secondarily in corymbs; involucres (to 5–10mm) campanulate, hemispheric or obconic, the phyllaries (5–16) in 1–2 series; receptacles conic or dome-like, lacking pales; ray florets 6–13, pistillate, fertile, rays (to 2–10mm; rarely absent in *L. glaberrima*), oblong to ovate, gold, yellow or orangish; disc florets (5–100+), bisexual, corollas yellow to orangish, 4–5-lobed; cypselae (to 1.5–4.0mm) black to gray, flattened or clavate, pappus lacking or of 1–10 scales

Life history: duration: annual (fruit/seeds); **asexual reproduction:** none; **pollination:** insect; **sexual condition:** gynomonoecious; **fruit:** cypselae (common); **local dispersal:** cypselae (gravity); **long-distance dispersal:** cypselae (animals, water)

Imperilment: (1) *Lasthenia burkei* [G1]; S1 (CA); (2) *L. ferrisiae* [G3]; S3 (CA); (3) *L. fremontii* [G4]; (4) *L. glaberrima* [G5]; S1 (BC, WA)

Ecology: general: *Lasthenia* species occupy nearly every type of edaphic landscape including coastal bluffs, deserts, grasslands, guano deposits, saline/alkaline flats, serpentine, vernal pools and woodlands. Ten species (59%) are categorized as wetland indicators, but only four (24%) as OBL. All of the OBL *Lasthenia* species are annual but two non-OBL species are perennial. The OBL taxa inhabit both freshwater and saline habitats and typically are associated with vernal pools. Exposures are open and receive full sunlight. The flowers are frequented by various insects (Insecta) including bees (Hymenoptera), bee flies (Diptera: Bombyliidae), flower beetles (Coleoptera: Melyridae), moths (Lepidoptera), and fungus gnats (Diptera: Sciaridae), which forage for nectar and facilitate pollination. Most of the species are self-incompatible but a few (including *L. glaberrima*) are self-pollinating. *Lasthenia* species characteristically are gynomonoecious, having radiate heads that possess fertile pistillate ray florets and bisexual disc florets. Disc-like heads with reduced or absent rays can occur in *L. glaberrima*, which probably is a consequence of reduced pollinator reliance resulting from self-pollination. The degree of seed set is proportional to the density of flowers in populations, which is a function of pollinator attraction. The seeds germinate without pretreatment. The pappus is lacking or modified into scales. Some seeds become viscid when wet, which assists in their attachment to animal fur. Transport of seeds by waterfowl or water also has been postulated. However, seed dispersal of many taxa is quite limited and achieved mainly by gravity. Seeds of some species retain >50% germination after 50 years of storage.

Lasthenia burkei (Greene) Greene inhabits flats, marshes, meadows, swales, vernal pools, and margins of streams at elevations of up to 673 m. Plants occur mainly below the pool margins where flooded conditions persist from winter through spring. The substrates are described as adobe, gravel, and serpentine. This species occupies a zone characterized by a fairly narrow ecological range of temperatures (15°C–18°C) and annual rainfall (75–125 cm). Flowering occurs throughout June. The flowers are strongly self-incompatible and outcrossed by various insects (Insecta) including beetles (Coleoptera), flies (Diptera: Bombylidae; Syrphidae), moths (Lepidoptera) and bees (Hymenoptera: Andrenidae; Apidae). Pollinator visits are more prevalent in natural sites than in created habitats. Seed production can approach 96% in open-pollinated plants; however, enclosed plants set seed at lower rates (38%), which indicates a FAC ability to self-pollinate in at least some individuals. The average annual seed set ranges from 83 to 92 seeds/inflorescence. The maximum observed seed set is 281 seeds/inflorescence. Germination rates of 56%–64% have been observed for seeds kept for 4 weeks on sterile agar under greenhouse conditions. The bristly pappus is thought to facilitate attachment to feathers or fur. It is not known whether seeds are buoyant, but they are presumed to float for at least a short duration. **Reported associates:** *Downingia concolor, Plagiobothrys.*

Lasthenia ferrisiae **Ornduff** occupies alkaline or saline flats, meadows, rain pools, river bottoms, salt marsh, scalds, sinks, swales, vernal pools, and margins of ponds and pools at elevations of up to 940 m. These plants occur only in alkaline/saline sites with some described as hypersaline. Sediment descriptions include adobe, claypan, heavy clay, loam, Modesto geomorphic surface, sand, sandy clay, silt, and Solano soil series. Like the preceding, this species exists within a fairly narrow ecological range in temperature (15°C–18°C) and annual rainfall (20–60 cm). Several sources distinguish *L. ferrisiae* from other OBL species by its typical occurrence near (not below) the pond margins. However, many occurrences are associated with relatively large (avg. 66,800 m²) and deep pools (avg. 68 cm) with moderate cover (avg. 64%). Sites can be characterized by a cover of nonindigenous species exceeding 30%. Flowering occurs from February to May. The flowers are self-incompatible and predominantly outcrossed by insects (Insecta). Details on specific pollinators or seed germination and dispersal are wanting. **Reported associates:** *Achyrachaena mollis, Allenrolfea occidentalis, Arthrocnemum subterminale, Astragalus tener, Atriplex confertifolia, A. coronata, A. spinifera, Bromus hordeaceus, B. madritensis, B. rubens, Cordylanthus, Cotula coronopifolia, Cressa truxillensis, Crypsis schoenoides, Delphinium recurvatum, Deschampsia, Distichlis spicata, Downingia bella, D. insignis, D. pulchella, Epilobium campestre, Erodium cicutarium, Eryngium aristulatum, Frankenia salina, Hordeum depressum, H. marinum, Juncus bufonius, Lasthenia californica, L. conjugens, L. fremontii, L. glaberrima, L. glabrata, L. gracilis, L. minor, L. platycarpha, Layia platyglossa, Lepidium dictyotum, L. jaredii, L. nitidum, Lolium multiflorum, Lotus, Lythrum hyssopifolium, Malvella leprosa, Mentha pulegium, Microseris douglasii, Myosurus minimus, Navarretia, Parapholis incurva, Phyla nodiflora, Plagiobothrys humistratus, P. leptocladus, P. stipitatus, Plantago elongata, Pogogyne douglasii, Polypogon monspeliensis, Psilocarphus, Rumex, Salicornia, Spergularia platensis, S. salina, Suaeda, Trifolium depauperatum, T. variegatum, Veronica peregrina, Vulpia bromoides.*

Lasthenia fremontii **(Torr. ex A. Gray) Greene** is known from depressions, ditches, flats, low plains, meadows, prairie, rain pools, ricefields, scalds, swales, vernal pools, and margins of salt marsh at elevations of up to 735 m. Substrates are alkaline (pH>9.0) or slightly saline and are described as adobe, clay, claypan, lava beds, mud, Redding gravelly loam, rocky clay, sand, sandy loam, and silty clay. Most individuals occur below the pool margin with edge plants representing less than 20% of the total cover. Edge plants have lower seed production and their seed exhibits lower probability of emergence than plants oriented nearer to pool centers. Peak flowering

occurs in early April to early May, with deeper water plants incrementally flowering later than edge plants in proportion to the pollinator pool. The proportion of plants in flower also is higher earlier in the year for edge plants. The flowers are self-incompatible and are outcrossed by both generalist and oligolectic bees (Insecta: Hymenoptera: Andrenidae; Halictidae) and bee flies (Insecta: Diptera: Bombylidae). Pollinator limitation is higher in sites comprising many unrelated taxa than in those consisting primarily of more closely related species. Pollinators are attracted less to plants that flower earlier in the season than those flowering later. Occurrences of this species indicate a narrow temperature tolerance (14°C–16°C) but broader ecological amplitude in annual precipitation (20–140 cm). Because of competition with surrounding vegetation, populations occur mainly below the margins of vernal pools in areas of prolonged winter inundation. However, plants also grow well under noninundated greenhouse conditions, or at shallower sites when other vegetation has been removed. The plants contain flavonoid sulfates, which are believed to function by inactivating harmful plant waste products generated in aquatic habitats. Seeds have been germinated successfully (40%–53% germination) without pretreatment by placing them in the light on moist filter paper kept at 20°C and at 25%–40% relative humidity. In natural populations seeds germinate in late fall to early winter, following the onset of winter rains. Seeds are dispersed mainly locally by gravity. The plants are parasitized by dodder (*Cuscuta howelliana*). **Reported associates:** *Achyrachaena mollis, Acmispon wrangelianus, Allium amplectens, Alopecurus saccatus, Amsinckia, Anthemis cotula, Astragalus tener, Atriplex joaquiniana, Blennosperma nanum, Brodiaea minor, Bromus hordeaceus, Callitriche marginata, Castilleja attenuata, C. campestris, Centromadia fitchii, Cerastium glomeratum, Clarkia purpurea, Crassula aquatica, C. tillaea, Cressa truxillensis, Crypsis schoenoides, Cuscuta howelliana, Deschampsia danthonioides, Distichlis spicata, Downingia bacigalupii, D. bella, D. bicornuta, D. concolor, D. insignis, D. ornatissima, D. pusilla, Eleocharis acicularis, E. macrostachya, E. palustris, Epilobium cleistogamum, Erodium botrys, E. cicutarium, Eryngium aristulatum, E. castrense, E. vaseyi, Frankenia salina, Gratiola ebracteata, G. heterosepala, Hemizonia congesta, Hesperevax caulescens, Hordeum depressum, Hypochaeris glabra, Isoetes orcuttii, Jaumea carnosa, Juncus bufonius, J. leiospermus, Lactuca serriola, Lasthenia californica, L. glaberrima, L. glabrata, Layia chrysanthemoides, L. fremontii, Legenere valdiviana, Lepidium dictyotum, L. latipes, L. nitidum, L. oxycarpum, Limnanthes alba, L. douglasii, Lolium multiflorum, L. perenne, Lupinus bicolor, Lythrum hyssopifolium, Medicago polymorpha, Micropus californicus, Microseris, Mimulus tricolor, Mollugo verticillata, Montia fontana, Nassella pulchra, Navarretia leucocephala, N. tagetina, Orcuttia tenuis, Pilularia americana, Plagiobothrys leptocladus, P. stipitatus, Plantago bigelovii, P. coronopus, P. elongata, Poa annua, Pogogyne douglasii, P. ziziphoroides, Psilocarphus brevissimus, P. chilensis, P. oregonus, Ranunculus bonariensis, Rumex crispus, Saxifraga, Sedella pumila, Sidalcea calycosa, S. diploscypha, Spergularia rubra, S. salina, Taeniatherum caput-medusae, Taraxacum officinale, Trifolium bifidum, T.*

depauperatum, T. willdenovii, Triglochin, Triphysaria eriantha, Triteleia hyacinthina, Vulpia bromoides.

***Lasthenia glaberrima* DC.** grows in depressions, ditches, flats, lagoons, lake beds, meadows, prairies, sulfur flats, swales, vernal pools and along the margins of floodplains, ponds, pools, rivers, streams, and washes at elevations of up to 1402 m. Individuals dominate deeper pools and are distributed mainly below the pool margins over alkaline (pH>9.0), subalkaline, or saline substrates described as adobe, clay, granite, gravelly loam, muck, mucky clay, mud, silicate, and Tuscan mudflow. This self-compatible plant is the only OBL *Lasthenia* species that is strongly self-pollinating. Relative to the other OBL taxa, this species also is more tolerant ecologically and occurs within a relatively broad temperature (9°C–17°C) and precipitation (25–140 cm) range. Flowering occurs from March to July. The flowers have inconspicuous greenish corollas that are not attractive to insects. In the field, the ray florets set little seed, but are fully fertile when pollinated artificially. Seed dispersal initiates in mid-May. Dispersal occurs mainly by gravity, but some animal or bird dispersal has been suggested. Seed germination studies obtained the highest average germination rates (96%) by placing seed in a warm chamber for 6 weeks followed by transfer to colder conditions (5°C/15°C) for 6 weeks. Reduced rates (43%) were observed after cold stratification at 5°C for 6 weeks followed by 6 weeks under a 8/16 h, 10°C/20°C temperature regime. Only 4%–5% germination rates were obtained from fresh seeds or for seeds dried at 15% relative humidity then frozen. **Reported associates:** *Achyrachaena mollis, Alisma, Alopecurus geniculatus, A. saccatus, Anagallis arvensis, Anthemis cotula, Anthoxanthum odoratum, Blennosperma nanum, Bolboschoenus maritimus, Briza minor, Brodiaea, Bromus hordeaceus, Callitriche heterophylla, C. marginata, Capsella bursa-pastoris, Castilleja attenuata, C. campestris, C. densiflora, Centaurea calcitrapa, Centromadia fitchii, Cerastium glomeratum, Convolvulus arvensis, Cotula coronopifolia, Crassula aquatica, Cuscuta howelliana, Damasonium californicum, Deschampsia danthonioides, Distichlis spicata, Downingia bella, D. bicornuta, D. concolor, D. cuspidata, D. insignis, D. ornatissima, D. yina, Echinodorus berteroi, Eleocharis acicularis, E. macrostachya, E. palustris, Epilobium brachycarpum, E. campestre, E. cleistogamum, Erodium botrys, E. moschatum, Eryngium aristulatum, E. armatum, E. castrense, E. petiolatum, E. vaseyi, Frankenia salina, Geranium dissectum, Gratiola ebracteata, G. heterosepala, Grindelia, Hemizonia congesta, Hesperevax caulescens, Hordeum brachyantherum, H. depressum, H. marinum, Horkelia, Hypochaeris glabra, H. radicata, Isoetes howellii, I. nuttallii, Juncus bufonius, J. hemiendytus, J. phaeocephalus, J. tenuis, J. xiphioides, Lactuca serriola, Lasthenia californica, L. conjugens, L. fremontii, L. glabrata, Layia chrysanthemoides, Legenere valdiviana, Lepidium latipes, L. nitidum, L. oxycarpum, Lilaea scilloides, Limnanthes douglasii, Lotus corniculatus, Lolium perenne, Lupinus bicolor, Lythrum hyssopifolium, Malvella leprosa, Medicago polymorpha, Mentha pulegium, Micropus californicus, Mimulus tricolor, Muilla maritima, Myosurus minimus, Navarretia leucocephala, N. myersii, Orthocarpus, Phleum pratense, Pilularia americana, Plagiobothrys bracteatus, P. chorisianus, P. figuratus, P. stipitatus, P. undulatus, Plantago bigelovii, P. elongata, P. erecta, Pleuropogon californicus, Poa annua, P. bulbosa, Pogogyne douglasii, P. ziziphoroides, Polypogon maritimus, P. monspeliensis, Psilocarphus brevissimus, P. chilensis, P. oregonus, Ranunculus alismifolius, R. bonariensis, R. orthorhynchus, Rumex crispus, Salix exigua, Sonchus oleraceus, Spergularia rubra, Trichostema lanceolatum, Trifolium barbigerum, T. bifidum, T. depauperatum, T. dubium, T. willdenovii, Triphysaria eriantha, Triteleia hyacinthina, Veronica peregrina, Vulpia bromoides.*

Use by wildlife: *Lasthenia* flowers host a large diversity of bees (Insecta: Hymenoptera: Andrenidae: *Andrena; Calliopsis; Panurginus*; Apidae: *Apis, Bombus, Nomada*; Halictidae: *Halictus, Lasioglossum*; Megachilidae: *Osmia*), bugs (Insecta: Hemiptera: Miridae: *Coquillettia; Hoplomachidea*), flies (Insecta: Diptera), moths (Insecta: Lepidoptera: *Schinia*), and various other insects (e.g., Coleoptera, Orthoptera). The flowers of *Lasthenia burkei* are visited by various bees (Insecta: Hymenoptera: Andrenidae: *Andrena angustitarsata, A. pensilis, A. submoesta*; Apidae: *Apis mellifera* (occasional); Halictidae: *Halictus tripartitus; Lasioglossum titusi*) and bee flies (Diptera: Bombyliidae: *Conophorus cristatus*), which forage for nectar. Flowers of *L. fremontii* attract large numbers of bees (Insecta: Hymenoptera: Andrenidae: *Andrena submoesta; Calliopsis*; Halictidae: *Lasioglossum titusi*) and bee flies (Insecta: Diptera: Bombyliidae: *Bombylius*) as well as occasional muscoid flies (Insecta: Diptera: Anthomyiidae) and parasitoid wasps (Insecta: Hymenoptera: Braconidae).

Economic importance: food: Seeds of several *Lasthenia* species (but none of the OBL taxa) have been used as food (eaten dry or ground into flour); **medicinal:** none reported; **cultivation:** Some *Lasthenia* species are cultivated, but none of the OBL taxa; **misc. products:** *Lasthenia* is the name of the newsletter published by the Davis Herbaria Society; **weeds:** none; **nonindigenous species:** none

Systematics: Various molecular data sets support the monophyly of *Lasthenia*. The OBL *Lasthenia* species are assigned to three different sections, which also appear to be monophyletic: section *Hologymne* (*L. ferrisiae*), section *Lasthenia* (*L. glaberrima*), and section *Ornduffia* (*L. burkei, L. fremontii*). Results from phylogenetic analyses of combined ETS, ITS, and cpDNA data fully support these relationships (Figure 5.162). The base chromosome number of *Lasthenia* is $x = 8$, with several lineages derived via descending aneuploidy. *Lasthenia ferrisiae* is closely related to *L. chrysantha* (FAC) and *L. glabrata* (FACW) (all $2n = 14$) and is suspected as originating as a hybrid derived from those taxa. The ranges of all three species overlap. Crosses between *L. chrysantha* and *L. glabrata* are fertile and produce a cypsela morphology quite similar to that of *L. ferrisiae*. *Lasthenia burkei* and *L. fremontii* are closely related to *L. conjugens* (FACW) and all share a chromosome number of $2n = 12$ (Figure 5.162). The three species primarily are allopatric. *Lasthenia burkei* and *L. fremontii* are sister species (Figure 5.162); however, their

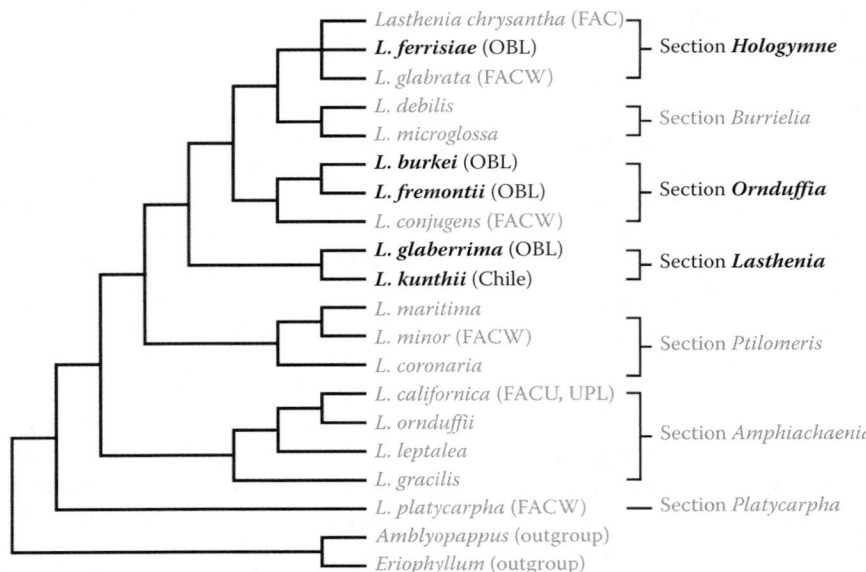

FIGURE 5.162 Phylogenetic relationships in *Lasthenia* as indicated by combined ETS, ITS, and cpDNA sequence data. The OBL habit (taxa in bold) has evolved several times within the genus but is concentrated among the more derived species. Because *L. kunthii* is not North American, it has not been evaluated as a wetland indicator; however, it shares virtually identical habitats with *L. glaberrima* and should be categorized as OBL. (Adapted from Emery, N.C. et al., *Research and Recovery in Vernal Pool Landscapes, Studies from the Herbarium*, Chico State University, Chico, CA, 2012a.)

artificial hybrids exhibit low fertility (20%). Hybrids between *L. conjugens* and *L. burkei* have higher fertility (75%) than hybrids involving either of those species and *L. fremontii* (20%), which differs chromosomally by a reciprocal translocation. Allozyme data have indicated higher genetic diversity in *L. burkei* and *L. conjugens* populations ($H_T=0.195$–0.209) than in *L. fremontii* ($H_T=0.168$), which also displays lower interpopulational differentiation. ISSR and RAPD markers indicate that 78% of genetic variation in *L. burkei* is partitioned within sites. Two distinct flavonoid races have been detected in *L. burkei*. *Lasthenia glaberrima* is related to the Chilean vernal pool species *L. kunthii* (Figure 5.162), which shares the same chromosome number ($2n=10$) and can be crossed successfully with the former to produce fully fertile hybrids.

Comments: *Lasthenia burkei*, *L. ferrisiae* and *L. fremontii* are restricted to California; *L. glaberrima* extends from California to British Columbia.

References: Barbour et al., 2007; Bohm & Banek, 1987; Borgias, 2004; Buck, 2004; Buck-Diaz & Evens, 2011; Buck-Diaz et al., 2012; Chan & Ornduff, 2006; Chan et al., 2001; COS et al., 2010; Crawford & Ornduff, 1989; Desrochers & Dodge, 2003; Emery, 2009; Emery et al., 2011, 2012a,b; Fairbarns, 2003; Graffis, 2013; Guerrant, Jr. & Raven, 1998; Harborne, 1975; Lazar, 2006; Linhart, 1976; Moore et al., 2003; Ornduff, 1963, 1969b; Platenkamp, 1998; Rajakaruna, 2003; Rajakaruna et al., 2003; Sargent et al., 2011; Sloop et al., 2012a; Thorp, 2014; Thorp & Leong, 1995.

30. *Marshallia*
Barbara's buttons

Etymology: after Moses Marshall (1758–1813)
Synonyms: *Athanasia*
Distribution: global: North America; **North America:** southeastern United States
Diversity: global: 7 species; **North America:** 7 species
Indicators (USA): OBL: *Marshallia graminifolia*
Habitat: freshwater; palustrine; **pH:** acidic; **depth:** <1 m; **life form(s):** emergent herb
Key morphology: Stems (to 120 cm) erect, simple or branched near or above middle; leaves (to 34 cm) alternate, margins entire, those basal petiolate and oblanceolate, 3-nerved, those cauline sessile, ascending and linear; heads (1–34; to 30 mm) discoid, solitary or arranged secondarily in corymbs, long-peduncled (to 50 cm); involucres (to 25 mm) with 12–24 persistent phyllaries (to 9.5 mm) in 2 series, receptacles with linear-attenuate pales; ray florets absent; disc florets (to 160) fertile, corollas 5-lobed (to 6 mm), pale lavender to purple; cypselae (to 4 mm) obconic, 5-angled, pappus of scales (to 2 mm), usually finely denticulate marginally.
Life history: duration: perennial (caudex; rhizomes); **asexual reproduction:** rhizomes; **pollination:** insect; **sexual condition:** hermaphroditic; **fruit:** cypselae (common; **local dispersal:** cypselae (gravity); **long-distance dispersal:** cypselae (animals)
Imperilment: (1) *Marshallia graminifolia* [G4]
Ecology: general: Six *Marshallia* species (86%) are wetland indicators, but only one is categorized as OBL. The species occur mainly in pine or hardwood forest understory or in open outcrops, but several occur frequently in wet areas near the margins of streams. The flowers are insect

pollinated (entomophilous) by bees (Hymenoptera), butterflies (Lepidoptera) and scarab beetles (Scarabaeidae: *Euphoria*). They are mainly outcrossed with low levels of autogamy (0%–2.7%) among the species. The seeds require no pretreatment for germination. The cypselae are believed to be dispersed by birds and other small animals, facilitated by their persistent, scale-like pappus.

Marshallia graminifolia (**Walter**) **Small** inhabits barrens, bogs, depressions, ditch banks, flatwoods, meadows, pocosins, prairies, savannah, seeps, swales, and margins of ponds and streams at elevations of up to 500 m. Plants often occur in shallow standing water (to 15 cm) on acidic substrate types described as Alapaha (arenic plinthic paleaquults), Allanton (grossarenic haplaquods), Basinger, Caddo (typic glossaqualf), Compass (plinthic paleudults), Cuthbert (typic hapludults), Floridana (arenic argiaquolls), gravel, Guyton (typic glossaqualf), Kinder (typic glossaqualf), Leon (aeric haplaquods), loam, loamy sand, muck, Mulat (typic ochraquults), Pelham (arenic paleaquults), Placid (spodic psammaquents-typic haplaquods-typic humaquepts), Pottsburg (grossarenic haplaquods), Rains, Rutlege (typic humaquepts), sand, sandy loam, sandy peat, St. Johns, Surrency (arenic umbric paleaquults), and Tocoi (ultic haplaquods). Flowering occurs from July to October. The flowers are protandrous, self-incompatible, outcrossed by insects (Insecta), and set no seed when artificially self-pollinated. The average seed size of the inner florets (3.61 mm) is somewhat larger than that of the outer florets (3.12 mm). The seeds require no pretreatment and germinate in about 10 days after being rinsed in 10% bleach and placed on water-moistened filter paper. **Reported associates:** *Acer rubrum, Agalinis filicaulis, Aletris aurea, A. lutea, Andropogon glomeratus, A. liebmannii, A. ternarius, A. virginicus, Anthaenantia rufa, Aristida palustris, A. purpurascens, A. spiciformis, A. stricta, Arnoglossum ovatum, Aronia arbutifolia, Asclepias longifolia, Axonopus fissifolius, Bacopa caroliniana, Balduina atropurpurea, B. uniflora, Bigelowia nudata, Calamovilfa brevipilis, Calydorea coelestina, Carex striata, Carphephorus paniculatus, C. pseudoliatris, C. tomentosus, Centella asiatica, Chaptalia tomentosa, Cleistesiopsis bifaria, C. divaricata, Clethra alnifolia, Cliftonia monophylla, Coleataenia longifolia, Coreopsis gladiata, Ctenium aromaticum, Cyrilla racemiflora, Dichanthelium acuminatum, D. ensifolium, D. scabriusculum, D. strigosum, Drosera capillaris, D. tracyi, Eleocharis tuberculosa, Eriocaulon compressum, E. decangulare, E. texense, Erigeron vernus, Eryngium aromaticum, E. integrifolium, Eupatorium leucolepis, Eurybia eryngiifolia, Fimbristylis, Fuirena breviseta, Gaylussacia frondosa, Gratiola hispida, Gymnadeniopsis nivea, Habenaria, Helenium drummondii, Helianthus angustifolius, H. floridanus, H. heterophyllus, Hypericum brachyphyllum, H. cistifolium, H. crux-andreae, Hyptis alata, Ilex coriacea, I. glabra, I. myrtifolia, Lachnocaulon anceps, Liatris spicata, Lilium, Linum medium, Liquidambar styraciflua, Lobelia brevifolia, L. flaccidifolia, Lophiola aurea, Ludwigia hirtella, L. pilosa, Lycopodiella alopecuroides, Magnolia virginiana, Marshallia caespitosa, Mnesithea tessellata, Muhlenbergia expansa, Myrica caroliniensis, M. cerifera, M. heterophylla, Nyssa biflora, Oclemena reticulata, Osmundastrum cinnamomeum, Oxypolis denticulata, O. filiformis, Panicum hemitomon, P. virgatum, Persea palustris, Pinckneya bracteata, Pinguicula caerulea, Pinus elliottii, P. palustris, P. serotina, Pityopsis oligantha, Pleea tenuifolia, Pluchea baccharis, Polygala cruciata, P. lutea, P. ramosa, Quercus, Rhexia alifanus, R. lutea, Rhynchospora cephalantha, R. elliottii, R. gracilenta, Rhynchospora inundata, R. latifolia, R. macra, R. microcarpa, R. oligantha, R. plumosa, R. pusilla, Rubus, Rudbeckia graminifolia, R. mohrii, Sabatia campanulata, S. macrophylla, Saccharum, Sarracenia flava, S. leucophylla, S. minor, S. psittacina, Schoenolirion croceum, Scleria georgiana, S. pauciflora, S. reticularis, Scutellaria integrifolia, Serenoa repens, Smilax laurifolia, Solidago stricta, Sphagnum, Sporobolus pinetorum, Styrax americanus, Symphyotrichum chapmanii, Taxodium ascendens, Toxicodendron radicans, Triantha racemosa, Vaccinium myrsinites, Verbesina chapmanii, Viburnum nudum, Vitis rotundifolia, Woodwardia virginica, Xyris caroliniana, Zigadenus.*

Use by wildlife: none reported.

Economic importance: food: not reported as edible; **medicinal:** none reported; **cultivation:** Several *Marshallia* species are cultivated as garden ornamentals; *M. graminifolia* occasionally is available from native garden plant suppliers; **misc. products:** none; **weeds:** none; **nonindigenous species:** none

Systematics: Within Asteraceae, *Marshallia* has been difficult to place phylogenetically, and has been classified variously in *Heliantheae* and *Eupatorieae*. The most comprehensive analysis is derived from nrITS data, which place *Marshallia* near the base of the tribe Heliantheae as the sister genus to *Pelucha* (Figure 5.163). Analysis of cpDNA data also places *Marshallia* within *Heliantheae*; however, with *Psilostrophe* as its sister genus (*Pelucha* was not sampled in that study). Thus it would appear that although the tribal placement of *Marshallia* has been resolved, further sampling of *Heliantheae* will be necessary to definitively settle the placement of *Marshallia* within the tribe. *Marshallia graminifolia* was once recognized as two species (*M. graminifolia* and *M. tenuifolia*) or subspecies until isozyme data showed populations of both taxa to comprise one, genetically uniform species. All populations similarly are relatively devoid of genetic variation with low heterozygosity, which suggests their derivation from a genetically depauperate ancestor arising from a bottleneck. The base chromosome number of *Marshallia* is $x=9$. *Marshallia graminifolia* ($2n=18$) is a diploid. It has not been reported to hybridize with other taxa. The OBL habit of *Marshallia graminifolia* is uniquely derived within the genus (Figure 5.163).

Comments: *Marshallia graminifolia* occurs throughout the southeastern United States.

References: Afonso, 2013; Cariaga et al., 2008; Coin, 2005; Goertzen et al., 2003; Sorrie et al., 2006; Watson, 2006b; Watson & Estes, 1990; Watson et al., 1994a.

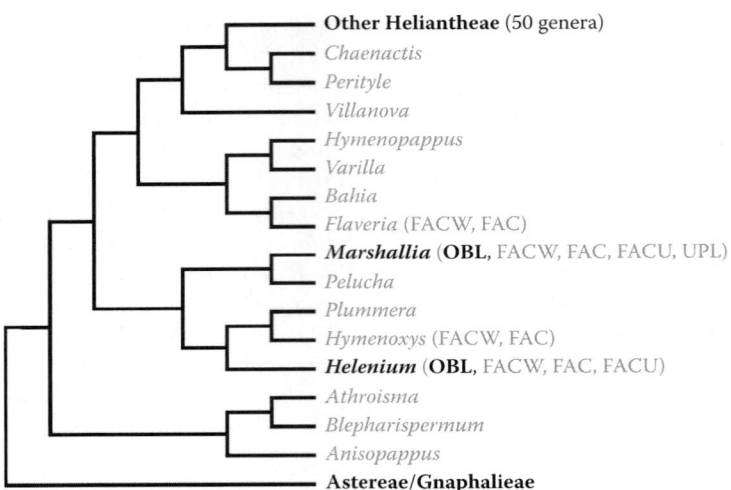

FIGURE 5.163 Phylogenetic relationships among genera of Heliantheae as indicated by analysis of nrITS sequence data. This tree indicates the independent origin of the OBL habit (bold) in *Marshallia*, which also contains species that are lesser wetland indicators. Other wetland indicator categories are provided where assigned. Sequence analysis of cpDNA resolves *Marshallia* as the sister genus to *Psilostrophe* (Cariaga et al., 2008), which was not included in the above study; however, that analysis excluded *Pelucha*, making a definitive appraisal of the closest relative to *Marshallia* problematic. (Adapted from Goertzen, L.R. et al., *Mol. Phylogen. Evol.*, 29, 216–234, 2003.)

31. *Microseris*

Silverpuffs

Etymology: from the Greek *mikros seris* ("small seris"), the latter being a plant similar to endive

Synonyms: *Apargidium*; *Calais*; *Ptilocalais*; *Scorzonella*

Distribution: global: Australia; New Zealand; North America; South America; **North America:** western

Diversity: global: 14 species; **North America:** 11 species

Indicators (USA): OBL: *Microseris borealis*

Habitat: freshwater; palustrine; **pH:** 4.0–7.0+; **depth:** <1 m; **life form(s):** emergent herb

Key morphology: Plants stemless, rhizomatous; leaves (to 30 cm) basal, oblanceolate, petioled, margins entire to sparsely denticulate; peduncle (to 70 cm) solitary, naked, erect; involucres (to 18 mm) ovoid, phyllaries (to 40) in 3–5 series; pales absent; florets (to 50) surpassing phyllaries (by 5+ mm), corollas yellow orange; cypselae (to 8 mm) columnar, pappus of up to 48 barbellate bristles (to 10 mm).

Life history: duration: perennial (rhizomes); **asexual reproduction:** rhizomes; **pollination:** insect; **sexual condition:** hermaphroditic; **fruit:** cypsela (common); **local dispersal:** rhizomes, cypselae (wind); **long-distance dispersal:** cypselae (wind)

Imperilment: (1) *Microseris borealis* [G5]; S1 (CA); S2 (WA)

Ecology: general: *Microseris* includes both annuals and perennials, which grow in forests, grasslands, hillsides, savannas, and woodlands as well as in wetlands. The various species occur over clay, rock, sand, serpentine and organic substrates. Three North American species (27%) are wetland indicators, but only one is categorized as OBL. The flowers are self-compatible (smaller-flowered annual species) or self-incompatible (larger-flowered perennial species); the anthers are either tetrasporangiate or bisporangiate. Although the self-compatible species frequently self-pollinate, details on the reproductive ecology of the self-incompatible species are scarce. Pollination has been reported by bees (Insecta: Hymenoptera) and hoverflies (Insecta: Diptera: Syrphidae). Little is known with respect to the seed ecology.

Microseris borealis (**Bong.**) **Schultz-Bip.** is a perennial, which inhabits bogs, fens, flats, forest clearings, marshes, meadows, muskeg, slopes, swamps, talus, and margins of streams at elevations of up to 1800 m. Sites vary from quite acidic pH (4.0–5.5) to alkaline (7.0+) and occur on granite outcrops, gravel, peat (predominantly), and rock substrates. Flowering occurs from June to September. The flowers are self-incompatible and outcrossed by insects (Insecta). Seed collections report 10% empty cypselae, which may be due to interspecific hybridization and subsequent embryo mortality. The seeds are dispersed by wind as facilitated by their pappus bristles. There is no information on germination requirements or seed bank potential. The plants spread vegetatively by their rhizomes, which produce fleshy adventitious roots. **Reported associates:** *Abies amabilis, A. concolor, A. lasiocarpa, A. procera, Aconitum, Aconogonon davisiae, Antennaria alpina, Callitropsis nootkatensis, Caltha leptosepala, Camassia, Carex aquatilis, C. nigricans, C. pauciflora, C. pluriflora, C. spectabilis, Cassiope mertensiana, Castilleja parviflora, Chamaecyparis, Chamerion angustifolium, Cladina, Cornus unalaschkensis, Deschampsia cespitosa, Drosera anglica, D. rotundifolia, Eleocharis palustris, Empetrum nigrum, Epilobium anagallidifolium, Eremogone capillaris, Eriogonum umbellatum, Eriophorum, Festuca, Gaultheria, Gentiana calycosa, G. douglasiana, Hieracium triste, Hypericum anagalloides, Juncus drummondii, J. mertensianus, Juniperus communis, Kalmia microphylla, Larix, Ligusticum calderi, L. grayi, Luetkea pectinata, Lupinus*

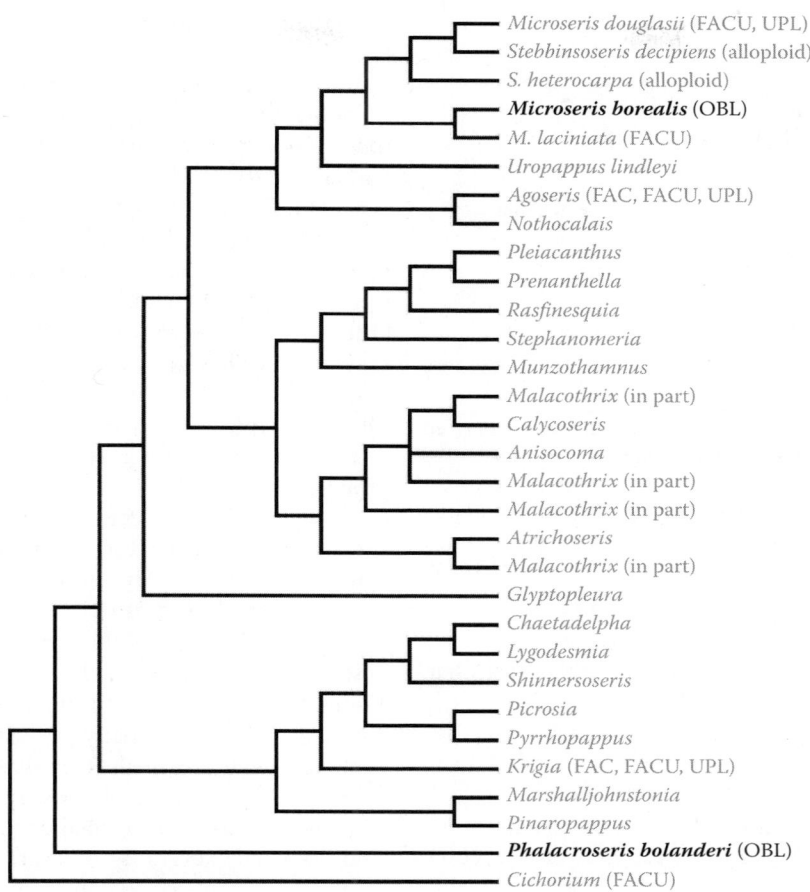

FIGURE 5.164 Phylogenetic relationships in Asteraceae tribe Cichorieae reconstructed from nuclear DNA sequence data. The tribe is relatively devoid of wetland indicators except for two OBL genera (in bold), which represent independent origins of the aquatic habit in the group. The nesting of *Stebbinsoseris* species within *Microseris* is a result of their allotetraploid derivation from *Microseris* and *Uropappus*. (Adapted from Lee, J. et al., *Syst. Bot.*, 28, 616–626, 2003.)

arcticus, Lycopodium clavatum, L. sitchense, Micranthes ferruginea, M. tolmiei, Myrica gale, Nephrophyllidium crista-galli, Nothocalais alpestris, Oreostemma alpigenum, Packera subnuda, Pedicularis bracteosa, Phyllodoce empetriformis, Picea sitchensis, Pinguicula vulgaris, Pinus albicaulis, P. contorta, Platanthera stricta, Poa, Potentilla flabellifolia, Primula jeffreyi, Pseudotsuga, Racomitrium lanuginosum, Rhododendron albiflorum, R. groenlandicum, Rubus chamaemorus, R. lasiococcus, Salix commutata, Sanguisorba officinalis, Senecio triangularis, Sorbus sitchensis, Sphagnum andersonianum, S. austinii, S. papillosum, S. rubellum, Symphyotrichum foliaceum, Thuja plicata, Triantha glutinosa, Trichophorum cespitosum, Trientalis europaea, Tsuga heterophylla, T. mertensiana, Vaccinium cespitosum, V. deliciosum, V. membranaceum, V. oxyccocos, V. uliginosum, Veratrum viride, Veronica wormskjoldii, Xerophyllum tenax.

Use by wildlife: Stems and leaves of *Microseris borealis* are eaten during summer and fall by black bears (Mammalia: Ursidae: *Ursus americanus*). Roughly 10% of cypselae reportedly are damaged by insects (Insecta) and fungi, but the specific taxa have not been identified.

Economic importance: food: not reported as edible; **medicinal:** none reported; **cultivation:** A few *Microseris* species (but not *M. borealis*) occasionally are in cultivation; **misc. products:** none; **weeds:** none; **nonindigenous species:** none

Systematics: *Microseris* is placed within tribe Cichorieae of Asteraceae subfamily Cichorioideae. Nuclear DNA sequence data indicate its relationships to be with *Stebbinsoseris* and *Uropappus* (Figure 5.164), the former being an allopolyploid derivative of *Microseris* and *Uropappus*. *Microseris* is monophyletic in exclusion of the allopolyploid taxa. The relationship of these three genera is supported by their shared cypselae ornamentation consisting of scurfy-scaled papillae or trichomes. Chloroplast RFLP data and nrITS sequence data (Figure 5.165) resolve *M. borealis* within a clade consisting entirely of perennial species. The chloroplast genome of *Microseris* contains from 2 to 4 tandemly repeated *trnF* genes: 2 or 4 in the perennial North American species (2 in *M. borealis*) and 2 to 3 in the annual or Old World species. The base chromosome number of *Microseris* is $x=9$. *Microseris borealis* ($2n=18$) is diploid, but has the highest DNA content among the eight members of the genus measured.

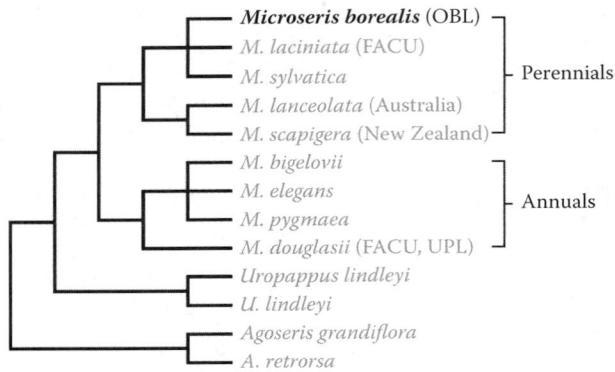

FIGURE 5.165 Phylogenetic relationships in *Microseris* derived from analysis of nrITS sequence data. *Microseris borealis*, the only OBL (bold) species in the genus, resolves within a clade of New and Old World perennial species. (Adapted from Vijverberg, C.A., Adaptive radiation of Australian and New Zealand *Microseris* (Asteraceae), PhD dissertation, University of Amsterdam, Amsterdam, Netherlands, 183, 2001.)

Experimental crosses between *M. borealis* and the Australian or New Zealand species (*M. lanceolata*, *M. scapigera*) have been unsuccessful.

Comments: *Microseris borealis* extends northward from California to Alaska in western North America.

References: Battjes et al., 1994; Chambers, 1963, 2006; Golumbia, 2001; Huntley, 2012; Lee et al., 2003; MacHutchon, 1999; Pak et al., 2001; Price & Bachmann, 1975; Price & Baranova, 1976; Sneddon, 1977; Vijverberg, 2001; Vijverberg & Bachmann, 1999; Vijverberg et al., 1999; Wallace & Jansen, 1990; Zald, 2010.

32. *Mikania*

Climbing hempweed

Etymology: after Joseph Gottfried Mikan (1743–1814)

Synonyms: *Cacalia* (in part); *Carelia*; *Eupatorium* (in part); *Kanimia*; *Willoughbya*

Distribution: global: North America; Paleotropics; South America; **North America:** eastern

Diversity: global: 450 species; **North America:** 3 species

Indicators (USA): OBL: *Mikania batatifolia*, *M. scandens*; **FACW:** *M. scandens*; **FAC:** *M. scandens*

Habitat: brackish, freshwater; palustrine; **pH:** 5.8–8.7; **depth:** <1 m; **life form(s):** emergent, herbaceous vines

Key morphology: Stems (to 300 cm+) twining, obscurely 6-angled, sometimes winged, internodes long (to 15 cm); leaves (to 15 cm) opposite, triangular to deltate-ovate, palmately veined, the bases cordate to hastate, apices acute to acuminate, petioled (to 50 mm); heads (to 7 mm) discoid, arranged secondarily in corymbs (to 12 cm), phyllaries (to 6 mm) 4, in 2 series, linear to narrowly ovate; receptacles flat, pales lacking; florets 4, corollas (to 5.4 mm) 5-lobed, white, pinkish, or purplish; cypselae (to 2.2 mm) brown or blackish, prismatically ribbed, densely gland dotted, pappus of 30–40 persistent, barbellate bristles (to 4.5 mm).

Life history: duration: perennial (dormant shoots); **asexual reproduction:** fragmentation (shoots); **pollination:** insect, self; **sexual condition:** hermaphroditic; **fruit:** cypsela (prolific); **local dispersal:** cypselae (water, wind); **long-distance dispersal:** cypselae (birds, wind)

Imperilment: (1) *Mikania batatifolia* [G3, G5]; (2) *M. scandens* [G3, G5]; SX (MI); SH (ME); S1 (IN); S2 (NH)

Ecology: general: Although *Mikania* includes an estimated 450 species, only three occur in North America. All three are wetland indicators, and all but *M. cordifolia* (FACW) have OBL status. *Mikania* species mainly are perennial, herbaceous vines. They grow in grasslands, salt marshes, swamps, woodlands, and in various wet, open areas. Their preferred substrates typically are calcareous, but can span a wide range of acidity. The flowers are protandrous but remain capable of autogamy in at least some species. Although the flowers are bisexual, some species are cryptically dioecious. The flowers are visited by various insects (Insecta: Coleoptera, Diptera, Hemiptera, Hymenoptera, Lepidoptera), but bees (Hymenoptera) and flies (Diptera) are the principal pollinators. Seed production is prolific and can average 94,500/m^2 in some species. Higher seed set (61%) is achieved under conditions of moderate rainfall, than under higher precipitation (51%). The cypselae initially possess a bristly pappus, which facilitates wind dispersal. Regardless of whether the pappus is present, the cypselae of some species can float in excess of 7 days, making water another important vector for local seed dispersal. Several species grow quite rapidly and easily can become weedy. Extensive growths of these vines can severely reduce the growth of shrubs and other vegetation as they twine over the foliage.

Mikania batatifolia **DC.** occurs in bayheads, flatwoods, prairies, roadsides, salt marshes, savannas, swamps, woodlands, and along the margins of borrow pits at elevations near sea level. It grows on alkaline substrates such as coral, oolite, sand, and sandy loam. Little information exists specifically for this species because much of the literature refers to the following species (*M. scandens*), which many consider to be synonymous taxonomically. **Reported associates:** *Annona glabra, Bacopa monnieri, Crinum americanum, Funastrum clausum, Muhlenbergia capillaris, Oxypolis filiformis, Panicum hemitomon, Paspalidium geminatum, Pontederia cordata, Proserpinaca palustris, Rhynchospora colorata, Sagittaria lancifolia, Salix caroliniana, Schoenus nigricans, Stillingia, Tillandsia fasciculata, T. utriculata, T. variabilis.*

Mikania scandens **(L.) Willdenow** inhabits bayous, bogs, canals, depressions, ditches, dunes, flats, flatwoods, floodplains, glades, hammocks, levees, marshes, meadows, potholes, prairies, ravines, roadsides, savannas, slopes, sloughs, streams, thickets, swales, swamps, woodlands, and the margins of lakes, ponds, rivers, and streams at elevations of up to 500 m. The plants tolerate shallow water and grow in full sun to dense shade over a wide diversity of sites from freshwater to brackish areas (0.5–18.0 ppt salinity). They often are represented in the transitional marsh region between the freshwater and saline extremes. Occurrences are characterized by a broad range of acidity (pH: 5.8–8.7; mean: 7.5), low mean

Secchi depth (1.5 m), and high mean values of alkalinity (mg/L as $CaCO_3$), conductance (204 μS/cm), calcium (30 mg/L), and magnesium (14 mg/L). The mean nutrient levels are relatively high (total N: 1,050 μg/L; total P: 55 μg/L; total K: 3.3 μg/L) with somewhat elevated levels of chloride (34 mg/L), sodium (19 mg/L) and sulfate (18 mg/L). Although this species can occur across a wide spectrum of environmental conditions, it is found most frequently in alkaline, hard water sites. The substrates have been described as Biscayne rock outcrop (fluvaquents), clay, compacted limerock, Copeland (typic argiaquolls), humus, limestone, loam, loamy sand, muck, oolite, Samsula–Myakka variant, peat, sand, and sandy loam. The plants also are found growing on floating mats of water hyacinth (Pontederiaceae: *Eichhornia crassipes*) and swamp loosestrife (Lythraceae: *Decodon verticillatus*). *Mikania scandens* is highly tolerant of flooding and achieves relative growths rates that are 50% higher when growing under flooded conditions relative to dry soil conditions. This tolerance is due in part to an extensive aerenchyma system and to heightened stem stomatal development that occurs under flooded conditions. Prolific adventitious roots also are produced, some of which extend to the surface and float as a means of enhancing internal aeration of tissues. Metabolically, the plants show no accumulation of ethanol under anaerobic conditions as well. These are vines that often climb upon other vegetation for support. The twining of the stems is believed to be influenced by geophysical factors. In one comparison most stems (93%) coiled clockwise in the Northern Hemisphere, while those of the Southern Hemisphere mainly coiled counterclockwise (73%); coiling at the equator was observed to be nearly equal (50:50) in directionality. This is an intermediate day plant requiring between 12.5 and 16 h of light daily; flowering will be inhibited under longer or shorter photoperiods. Nevertheless, the plants flower nearly continuously throughout the year in the southern portions of their range. The flowers are pollinated by a variety of insects (Insecta), which forage for nectar. Autogamous pollination also has been suggested by correlated patterns of morphological variation, but has not been confirmed in this species. The cypselae are dispersed by the wind (assisted by the bristly pappus), or by water (as evidenced by the recovery of viable fruits from drift line samples). Seed germination occurs after cold stratification followed by exposure to a 30°C/15°C temperature regime. Seven weeks of stratification at 5°C followed by 30°C temperatures under 12 h day lengths has induced reasonable germination. An after ripening period decreases the minimum germination temperature. An extensive seed bank is formed, which correlates with plant abundance. Both the standing vegetation and the seed bank are restricted to the region above the water line. The mean seed bank densities in a newly created wetland (initially lacking a seed bank) increased from 17/m² (after 4 years) to 13,533/m² (after 17 years) with percent cover increasing respectively from 9.4% to 41.1%. High densities of germinable seed (380–9,567/m²) are maintained early in the season (June), indicating at least some degree of persistence in the seed bank. A persistent seed bank also has been indicated in hardwood swamps, where germinable seeds have been recovered where the plants do not occur in the standing vegetation. Although many seeds are produced, they often have low viability (3%) with low germinability (6.1%), which results in low seedling production (6/m²). Seeds germinate well in saturated conditions (100%) but not in flooded conditions (0%). Total stem length also was observed to be greater under nonflooded (15.5 cm) than continuously flooded (2.4 cm) greenhouse conditions. Productivity is moderate to high, in one case reaching 1,123 g/m². Plant abundance varies across sites but responds positively following fire episodes. Studies of a Mississippi wetland found there to be a significant negative correlation in relative cover between *M. scandens* and *Leersia oryzoides* (Poaceae). Growth of *M. scandens* often can attain weedy proportions, especially in chronically flooded areas. However, this plant is regarded as an important "umbrella" species, whose protection conceivably results in the preservation of greater overall community biodiversity. Vegetative reproduction reportedly can occur by shoot fragmentation. **Reported associates:** *Abildgaardia ovata, Acalypha rhomboidea, Acer negundo, A. rubrum, A. saccharinum, Acorus calamus, Acrostichum danaeifolium, Aeschynomene virginica, Agalinis fasciculata, A. purpurea, Alnus rugosa, A. serrulata, Amaranthus cannabinus, Ambrosia artemisiifolia, Ammannia latifolia, Amorpha fruticosa, Ampelopsis arborea, Amphicarpaea bracteata, Andropogon arctatus, A. glomeratus, A. virginicus, Anemia adiantifolia, Annona glabra, Apios americana, Apocynum cannabinum, Arisaema triphyllum, Aristida stricta, Aristolochia serpentaria, Aronia arbutifolia, Arundinaria gigantea, Asclepias incarnata, A. perennis, Asimina triloba, Asplenium platyneuron, Athyrium filix-femina, Avicennia germinans, Azolla caroliniana, Baccharis angustifolia, B. halimifolia, Bacopa caroliniana, B. monnieri, Barbarea vulgaris, Batis maritima, Berchemia scandens, Betula nigra, Bidens frondosus, B. laevis, B. polylepis, Bigelowia nudata, Bignonia capreolata, Blechnum serrulatum, Blutaparon vermiculare, Boehmeria cylindrica, Bolboschoenus novae-angliae, B. robustus, Borrichia frutescens, Botrychium biternatum, Byrsonima lucida, Callicarpa americana, Calystegia sepium, Campsis radicans, Carphephorus odoratissimus, Carex abscondita, C. atlantica, C. bromoides, C. comosa, C. crinita, C. crus-corvi, C. debilis, C. emoryi, C. folliculata, C. frankii, C. gigantea, C. grayi, C. grisea, C. hyalinolepis, C. intumescens, C. joorii, C. leptalea, C. longii, C. lupuliformis, C. lupulina, C. lurida, C. retrorsa, C. scoparia, C. stipata, C. stricta, C. typhina, Carpinus caroliniana, Carya aquatica, C. cordiformis, C. tomentosa, C. illinoinensis, Celtis laevigata, Cenchrus tribuloides, Cephalanthus occidentalis, Chamaecrista fasciculata, C. nictitans, Chasmanthium latifolium, C. laxum, Chelone glabra, Chiococca alba, Cinna arundinacea, Cirsium, Cladium jamaicense, Clematis terniflora, Clethra alnifolia, Coccoloba, Coleataenia longifolia, Colocasia esculenta, Commelina virginica, Conocarpus erectus, Conoclinium coelestinum, Coreopsis leavenworthii, Cornus amomum, C. drummondii, C. foemina, Corydalis flavula, Crinum americanum, Ctenium aromaticum, Cuscuta gronovii, Cyperus erythrorhizos, C.*

esculentus, C. odoratus, C. polystachyos, C. pseudovegetus, C. strigosus, Cyrilla racemiflora, Decodon verticillatus, Dichanthelium aciculare, D. boscii, D. clandestinum, D. commutatum, D. scoparium, Digitaria ischaemum, Diodella teres, Diodia virginiana, Diospyros virginiana, Distichlis spicata, Drosera brevifolia, D. capillaris, Dysphania ambrosioides, Echinochloa muricata, E. walteri, Echinodorus cordifolius, Eclipta prostrata, Eichhornia crassipes, Eleocharis fallax, E. obtusa, Erechtites hieraciifolius, Erigeron quercifolius, E. vernus, Eriocaulon decangulare, Eriophorum virginicum, Eryngium aromaticum, E. baldwinii, Euonymus americanus, Eupatorium capillifolium, E. hyssopifolium, E. leptophyllum, E. perfoliatum, E. rotundifolium, E. serotinum, Euthamia graminifolia, Fagus grandifolia, Fallopia scandens, Ficus aurea, Fimbristylis autumnalis, F. caroliniana, F. spadicea, Forestiera acuminata, F. segregata, Fraxinus caroliniana, F. pennsylvanica, Fuirena breviseta, F. squarrosa, Funastrum clausum, Galium obtusum, G. tinctorium, Gaylussacia dumosa, G. mosieri, G. nana, Gratiola virginiana, Gymnadeniopsis clavellata, Habenaria repens, Hamamelis virginiana, Helenium autumnale, Helianthus angustifolius, H. radula, H. tuberosus, Hibiscus moscheutos, Hydrocotyle umbellata, H. verticillata, Hygrophila lacustris, Hypericum canadense, H. galioides, H. hypericoides, H. microsepalum, H. mutilum, Hyptis alata, Ilex cassine, I. decidua, I. glabra, I. myrtifolia, I. opaca, Impatiens capensis, Ipomoea alba, I. lacunosa, I. sagittata, Iris pseudacorus, Itea virginica, Iva frutescens, Juncus acuminatus, J. canadensis, J. coriaceus, J. dichotomus, J. effusus, J. megacephalus, J. polycephalus, J. roemerianus, J. scirpoides, J. tenuis, Juglans nigra, Juniperus, Justicia americana, J. ovata, Kosteletzkya pentacarpos, Kummerowia stipulacea, Lachnanthes caroliniana, Laguncularia racemosa, Leersia lenticularis, L. oryzoides, L. virginica, Lemna minor, Leptochloa panicea, Ligustrum sinense, Limonium carolinianum, Lindera benzoin, Lindernia dubia, Liquidambar styraciflua, Lobelia cardinalis, Lonicera japonica, Ludwigia alternifolia, L. decurrens, L. leptocarpa, L. linearis, L. palustris, L. peploides, L. repens, Lycium carolinianum, Lycopodiella alopecuroides, Lycopus americanus, L. rubellus, L. virginicus, Lyonia fruticosa, L. lucida, L. mariana, Lysimachia terrestris, Magnolia virginiana, Maytenus, Mentha cardiaca, Mimulus alatus, Mitchella repens, Mosiera longipes, Morus rubra, Muhlenbergia capillaris, Murdannia keisak, Myrica caroliniensis, M. cerifera, Nelumbo lutea, Nyssa aquatica, N. biflora, Oenothera biennis, Oldenlandia uniflora, Onoclea sensibilis, Opuntia stricta, Osmunda regalis, Osmundastrum cinnamomeum, Oxalis stricta, Oxypolis filiformis, Packera glabella, Panicum dichotomiflorum, P. hemitomon, P. verrucosum, P. virgatum, Parthenocissus quinquefolia, Paspalidium geminatum, Paspalum dilatatum, P. floridanum, P. laeve, P. notatum, P. urvillei, Passiflora lutea, Peltandra virginica, Penthorum sedoides, Persea palustris, Persicaria arifolia, P. glabra, P. hydropiperoides, P. lapathifolia, P. pensylvanica, P. punctata, P. sagittata, P. setacea, P. virginiana, Phanopyrum gymnocarpon, Phragmites australis, Phyla lanceolata, P. nodiflora, Physalis heterophylla, Phytolacca americana, Pilea fontana, P. pumila, Pinus elliott, P. palustris, P. serotina, P. taeda, Pisonia rotundata, Planera aquatica, Platanthera blephariglottis, Platanus occidentalis, Pleopeltis polypodioides, Pluchea baccharis, P. carolinensis, P. camphorata, Polygala brevifolia, P. lutea, P. nana, Polypodium virginianum, Polypremum procumbens, Pontederia cordata, Populus deltoides, Portulaca oleracea, Proserpinaca palustris, P. pectinata, Ptilimnium capillaceum, Quercus laurifolia, Q. lyrata, Q. michauxii, Q. nigra, Q. pagoda, Rhexia alifanus, R. mariana, R. petiolata, Rhizophora mangle, Rhynchospora capitellata, R. colorata, R. corniculata, R. globularis, R. inundata, R. macrostachya, R. microcarpa, R. tracyi, Rorippa palustris, Rosa palustris, Rotala ramosior, Rubus trivialis, Rudbeckia laciniata, Rumex verticillatus, Sabal minor, S. palmetto, Sabatia brevifolia, S. macrophylla, Saccharum giganteum, Sacciolepis striata, Sagittaria graminea, S. lancifolia, S. latifolia, Salicornia depressa, Salix caroliniana, S. nigra, Sambucus nigra, Samolus ebracteatus, S. valerandi, Sanicula canadensis, Sarracenia flava, Saururus cernuus, Schinus terebinthifolius, Schoenoplectus pungens, S. tabernaemontani, Scirpus atrovirens, S. cyperinus, Scleria triglomerata, Serenoa repens, Sesuvium portulacastrum, Setaria parviflora, Seutera angustifolia, Sida, Sium suave, Smilax bona-nox, S. laurifolia, S. rotundifolia, S. tamnoides, S. walteri, Solanum carolinense, Solidago leavenworthii, S. patula, S. sempervirens, Sophora, Spartina alterniflora, S. bakeri, S. cynosuroides, S. patens, Sphagnum, Spiranthes cernua, Sporobolus floridanus, S. virginicus, Strophostyles helvola, Stylosanthes hamata, Suriana maritima, Symphyotrichum lateriflorum, S. novi-belgii, S. racemosum, S. subulatum, S. tenuifolium, Symplocos tinctoria, Taxodium distichum, Teucrium canadense, Thalictrum pubescens, Thelypteris kunthii, T. palustris, Tillandsia fasciculata, T. usneoides, T. utriculata, T. variabilis, Toxicodendron radicans, Triadenum virginicum, T. walteri, Tripsacum dactyloides, Typha angustifolia, T. domingensis, T. latifolia, Ulmus alata, U. americana, Utricularia foliosa, Vaccinium corymbosum, Vallisneria americana, Verbena hastata, Vernonia noveboracensis, Veronica peregrina, Viburnum dentatum, V. nudum, Vicia floridana, Viola primulifolia, V. sororia, Vitis aestivalis, V. rotundifolia, Wisteria frutescens, Woodwardia areolata, Xanthium strumarium, Xyris ambigua, X. elliottii, X. platylepis, X. tennesseensis, X. torta, Zizania aquatica.

Use by wildlife: *Mikania batatifolia* is a floristic component of Florida Panther (Mammalia: Felidae: *Puma concolor couguar*) habitat. It is a host of mites (Arachnida: Acari: Phytoseiidae: *Proprioseius meridionalis*) and leaf-miner flies (Insecta: Diptera: Agromyzidae: *Phytobia*); the latter in turn are hosts to parasitic wasps (Insecta: Hymenoptera: Eulophidae: *Derostenus agromyzae*; *Tetrastichus*); Pteromalidae: *Heteroschema*). The plants themselves are parasitized by nematodes (Nematoda: Tylenchulidae: *Tylenchulus palustris, T. semipenetrans*). The flowers are visited by bees (Insecta: Hymenoptera: Halictidae: *Lasioglossum longifrons*). *Mikania* *scandens* provides habitat for the dusky seaside sparrow (Aves: Emberizidae: *Ammodramus maritimus nigrescens*) and nesting habitat for black rails (Aves: Rallidae: *Laterallus jamaicensis*) and other birds. Birds (Aves) also occasionally eat the seeds. The foliage is consumed by nutria (Mammalia: Rodentia: Myocastoridae: *Myocastor coypus*) and the stems and flowers by free-ranging ring-tailed lemurs (Mammalia: Lemuridae: *Lemur catta*), which were established for breeding on St. Catherines Island, Georgia in 1985. The plants are fed on by planthoppers (Insecta: Hemiptera: Flatidae: *Cyarda acutissima*) and soldier beetles (Insecta: Coleoptera: Cantharidae: *Chauliognathus pennsylvanicus*). They host melon thrips (Insecta: Thysanoptera: Thripidae: *Thrips palmi*), larval Lepidoptera (Insecta: Arctiidae: *Cosmosoma auge*, *C. myrodora*; Geometridae: *Semiothisa continuata*; Riodinidae: *Calephelis virginiensis*), and plant bugs (Insecta: Hemiptera: Miridae: *Taylorilygus pallidulus*). The plants are parasitized by nematodes (Nematoda: Tylenchulidae: *Tylenchulus palustris*). They also host several fungi including Ascomycota (Erysiphaceae: *Erysiphe cichoracearum*; Mycosphaerellaceae: *Cercospora mikaniicola*, *Passalora mikaniigena*, *Septoria mikaniae*) and Basidiomycota (Pucciniaceae: *Puccinia spegazzinii*). The flowers are visited by "lovebug" flies (Insecta: Diptera: Bibionidae: *Plecia nearctica*), carpenter-leafcutter bees (Insecta: Hymenoptera: Apidae: *Xylocopa virginica*; Megachilidae: *Megachile xylocopoides*), and various wasps (Insecta: Hymenoptera: Vespidae: *Polistes*). They provide a source of nectar for hairstreak (Insecta: Lepidoptera: Lycaenidae: *Eumaeus atala*), viceroy (Insecta: Nymphalidae: *Limenitis archippus*), and admiral (Insecta: Nymphalidae: *Limenitis arthemis*) butterflies.

Economic importance: food: Although not used as human food, the foliage of *Mikania* is a source of vitamins A and C and contains roughly 3.6% nitrogen, 50.9% crude protein, 607.3 μg/g β-carotene and has a relatively high in vitro digestibility (60.9%). Palatability can be reduced by alkaloids and phenolics, whose levels vary; **medicinal:** Various *Mikania* species are used worldwide as a source of natural therapeutic drugs. Both *M. batatifolia* and *M. scandens* contain mikanolide, which has antifungal (Ascomycota: Candidaceae: *Candida albicans*) properties. The Seminole used extracts of *M. scandens* as a treatment for dermatological disorders. The Mikasuki and Seminole derived a treatment for "snake sickness" from the plants. Leaf extracts of the plant are used in India to treat bruises, stomach ulcers, and wounds. Hydromethanolic foliar extracts are strongly antioxidant and exhibit analgesic properties shown to depress the central nervous system in mice (Rodentia: Muridae: *Mus*). The extracts also possess antioxidant properties, which are effective at reducing alcohol and isoniazid induced liver damage in laboratory animals; **cultivation:** *Mikania scandens* occasionally is cultivated; **misc. products:** Extracts from aerial portions of *M. scandens* exhibit strong allelopathic inhibitory effects on the germination of wheat (Poaceae: *Triticum aestivum*) and chickpea (Fabaceae: *Cicer arietinum*) seeds. *Mikania*

scandens contains mikanolide, which has been patented (US 6767561) as a DNA polymerase inhibitor; **weeds:** *Mikania scandens* is invasive in Asia. It is reported as a weed in the Malay Peninsula, where it can result in reduced fruit size of mangosteen crops. Plants were introduced as ornamentals to Guam, where they have become weeds. It also occasionally is viewed as a weed in the southern United States, especially in flooded communities; **nonindigenous species:** *Mikania scandens* has been introduced to southern Asia and several Pacific Islands. It is a prohibited species in South Africa.

Systematics: *Mikania* is classified as subtribe Mikaniinae within tribe Eupatorieae of Asteraceae. Few of the 400+ *Mikania* species have been sampled in phylogenetic studies with assorted molecular data currently available for only eight species. Analyses of these species indicate that the genus is monophyletic. It resolves in a relatively derived portion of the tribe in association with *Sclerolepis* and *Shinnersia*, which are two OBL genera (Figure 5.166). *Mikania batatifolia* and *M. scandens* are extremely similar, but arguably can be distinguished by various leaf and floral characters. *Mikania batatifolia* also uniquely possesses the flavone batatifolin; whereas, *M. scandens* produces the germacranolides desoxymikanolide, miscandenin and scandenolide, which have not been found in the former. Because many authors have merged the two species, it often

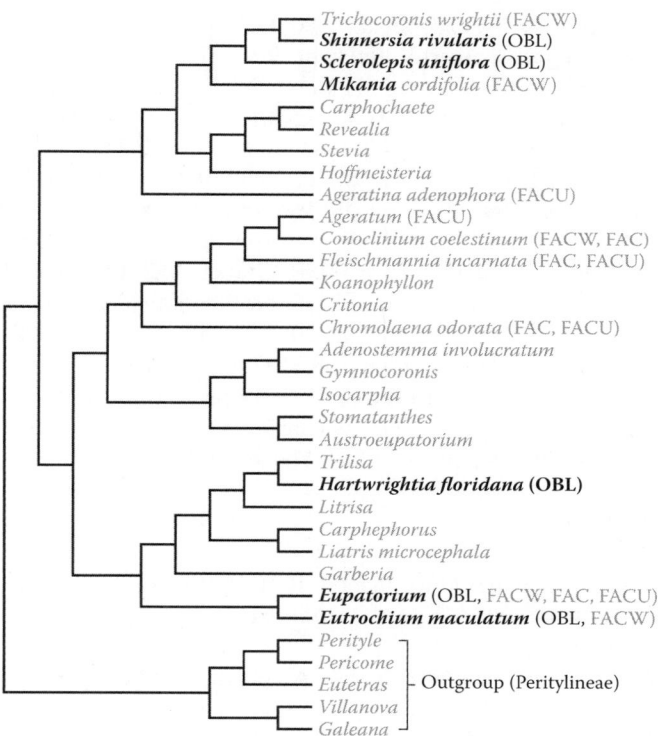

FIGURE 5.166 Intergeneric relationships in Asteraceae tribe Eupatorieae based on combined nrITS and *matK* sequence data showing relative positions of OBL taxa (in bold). *Mikania* (represented here by a non-OBL species) occupies a relatively derived phylogenetic placement, clustering with several OBL taxa. (Adapted from Tippery, N.P. et al., *Syst. Bot.*, 39, 1217–1225, 2014.)

is impossible to determine whether literature references have been correctly attributed. If these species truly are distinct, then it is evident that the names have been misapplied often in the literature. This problem should be kept in mind when interpreting the information provided in the preceding species accounts. Analysis of nrITS data for four species shows *M. scandens* to be much more closely related to the nonindigenous, invasive *M. cordata* and *M. micrantha* than to *M. cordifolia*. The base chromosome number of *Mikania* has been specified both as $x=17$ and $x=18$. The chromosome numbers of *M. batatifolia* and *M. scandens* (diploid) have been reported as $2n=34$, 36 and 38, but uncertain identification of material has made it difficult to determine whether the different counts reported associate with the different species, or simply represent intraspecific variation. Supranumerary (i.e., "B") chromosomes also have been observed in *M. scandens*.

Comments: *Mikania batatifolia* occurs from the West Indies into Florida; *M. scandens* extends northward from Florida to Maine (historically).

References: Altfeld, 2003; Atkinson et al., 1993; Baldwin et al., 2001; Baskin & Baskin, 1998; Baskin et al., 1993; Belden et al., 1988; Borthwick et al., 1950; Bried et al., 2007; Cerana, 2004; Coleman, 2007; Collins & Wien, 1995; Davis, 1974; Denmark & Muma, 1966; Dewanji et al., 1997; Dey et al., 2011, 2012; Dow et al., 1990; Ervin, 2007; Graenicher, 1927; Hall et al., 2007; Harborne & Baxter, 1993; Hasan et al., 2009; Hoffman & Dawes, 1997; Holder et al., 1980; Holmes, 1990, 1991, 1993, 2006; Hopfensperger & Baldwin, 2009; Hoyer et al., 1996; Huffman & Lonard, 1983; Hutton, 2010; Inserra et al., 1990; Keil et al., 1988; Koi, 2008; Leck, 2012; Legare & Eddleman, 2001; Macanawai et al., 2012; Maity & Ahmad, 2012; Maity et al., 2012; McConnell & Muniappan, 1991; McLain, 1986; Moon et al., 1993; Nauman, 1981; Neff et al., 2009; Nordman, 2004; Penfound & Earle, 1948; Ponzio et al., 2004; Rabb, 1960; Rowlee, 1894; Ruas et al., 2000; Rufatto et al., 2012; Schiefer, 1999; Schmitt, Jr., 1955; Schneider & Sharitz, 1986; Sharpe & Baldwin, 2009; Smith et al., 2012; Snodgrass et al., 1984b; Stegmaier, Jr., 1972; Strong & Kelloff, 1994; Taylor & Grace, 1995; Thomas & Vince-Prue, 1997; Tippery et al., 2014; Van der Valk & Rosburg, 1997; Watanabe et al., 1995; Yaacob & Tindall, 1995; Yang et al., 2005; Young, 1986.

33. *Oclemena*

Bog aster

Etymology: after Otto Clemen (1871–1946)

Synonyms: *Aster* (in part); *Calianthus*; *Eucephalus* (in part); *Galatella*

Distribution: global: North America; **North America:** eastern

Diversity: global: 4 species; **North America:** 4 species

Indicators (USA): OBL; FACW: *Oclemena nemoralis*

Habitat: freshwater; palustrine; **pH:** 4.0–6.7; **depth:** <1 m; **life-form(s):** emergent herb

Key morphology: Stems (to 90 cm) erect, single; leaves (to 100) alternate, cauline, crowded proximally, often withering at flowering, sessile or short petioled, the blades (to 6 cm) linear to linear lanceolate, margins recurved, minutely toothed;

heads (to 15) radiate, single or arranged secondarily in corymbs, slenderly peduncled (to 7 cm), involucre (to 7.5 mm) cylindic campanualte, the phyllaries (to 50) in 3–4 series, often pink margined, pales lacking; ray florets (to 25) pistillate, the rays (to 18 mm) pale to pink (rarely white); disc florets (to 35) bisexual, the corollas (to 7.5 mm) 5-lobed, pale to pinkish yellow; cypselae (to 3 mm) compressed, gland dotted, 5–8-ribbed, pappus of 40–55+ short (outer) or long (inner), bristles.

Life history: duration: perennial (rhizomes); **asexual reproduction:** rhizomes; **pollination:** insect; **sexual condition:** gynomonoecious; **fruit:** cypselae (common); **local dispersal:** rhizomes, cypselae (gravity, wind); **long-distance dispersal:** cypselae (wind)

Imperilment: (1) *Oclemena nemoralis* [G5]; SX (DE); S1 (CT, PA); S2 (PE, RI, VT); S3 (MI, NY)

Ecology: general: Only one *Oclemena* species is an OBL indicator, with the other three taxa designated as FACW, FAC, FACU, or UPL. The species usually are found on acidic substrates and occur in dry to wet areas at sites ranging from mesic woods to bogs or bog–forest ecotones. The radiate heads are gynomonoecious with pistillate ray florets and bisexual disc florets. As in many other radiate species, herbivory of the disc florets exceeds that of the rays. Little specific information exists on the reproductive biology or seed ecology of the species.

Oclemena nemoralis (Aiton) Greene occurs in barrens, bogs, heaths, patterned peatlands, poor fens, swamps, and along the margins of lakes, ponds, and streams at elevations of up to 900 m. Sites typically are acidic (pH 4.0–6.7) and low in calcium (<50 μeq/L Ca^{2+}) but can be weakly minerotrophic, especially in coastal areas influenced by salt spray (which is tolerated). In patterned landscapes, the plants occur evenly across the flarks and hummocks. Representative water chemistry features specific conductivity of 66–78 μs/cm; 342–410 μeq/L Cl–; 27–62 μeq/L SO_4^{2-}; no detectable NO_3^-; 1.53 μeq/L Ca^{2+}; and 1.2 μeq/L Mg^{2+}. The substrates are described as bog iron muck, peat (*Sphagnum*), or quartzite sand. Flowering occurs from the latter part of July through mid-September. The flowers presumably are self-incompatible and are outcrossed by unspecialized pollinators (Insecta: Coleoptera, Diptera, Homoptera, Hymenoptera, Lepidoptera). The bristly pappus facilitates dispersal of the cypselae by wind. The cypselae are fairly small (averaging about 0.4 mg), but are dispersed at a relatively low launch height (0.4 m), and achieve a terminal velocity of 0.42 m/s; consequently, they would settle to the ground in roughly 1 second under conditions of still air. Details of seed dormancy have not been reported, but a period of cold stratification is assumed. Wild gathered seeds germinated successfully (50%) under a 12-h photoperiod and 25°C/15°C temperature regime. Vegetative reproduction occurs by means of elongate rhizomes. **Reported associates:** *Acer rubrum, Agalinis purpurea, Agrostis perennans, A. stolonifera, Alnus rugosa, A. serrulata, Amelanchier canadensis, Andromeda polifolia, Andropogon glomeratus, A. virginicus, Anemone virginiana, Arethusa bulbosa, Aronia arbutifolia, A. melanocarpa, A. ×prunifolia, Aulacomnium*

palustre, Bartonia paniculata, B. virginica, Betula michauxii, B. populifolia, B. pumila, Calamagrostis canadensis,
C. pickeringii, C. stricta, Calopogon tuberosus, Campylium
stellatum, Carex albolutescens, C. atlantica, C. bullata, C.
canescens, C. chordorrhiza, C. collinsii, C. exilis, C. folliculata, C. lasiocarpa, C. limosa, C. livida, C. magellanica, C.
michauxiana, C. oligosperma, C. pensylvanica, C. rariflora,
C. rostrata, C. striata, C. stricta, C. trisperma, C. utriculata,
C. wiegandii, Cephalanthus occidentalis, Chamaecyparis
thyoides, Chamaedaphne calyculata, Cladium mariscoides,
Cladonia rangiferina, Clethra alnifolia, Coleataenia longifolia, Comarum palustre, Coptis trifolia, Cornus canadensis, Cuscuta, Cyperus dentatus, Danthonia sericea, Decodon
verticillatus, Dichanthelium acuminatum, D. ensifolium, D.
scabriusculum, Drosera filiformis, D. intermedia, D. rotundifolia, Dulichium arundinaceum, Eleocharis flavescens,
E. robbinsii, E. tenuis, E. tuberculosa, Empetrum nigrum,
Eriocaulon aquaticum, E. compressum, E. decangulare,
Eriophorum angustifolium, E. virginicum, Eubotrys racemosa, Eupatorium resinosum, E. pilosum, Eurybia radula,
Euthamia graminifolia, Gaylussacia baccata, G. dumosa, G.
frondosa, Geocaulon lividum, Glyceria obtusa, Gymnocolea
inflata, Hypericum canadense, H. densiflorum, H. denticu-
latum, H. mutilum, Ilex glabra, I. laevigata, I. mucronata,
I. verticillata, Iris setosa, I. versicolor, Juncus canadensis,
J. effusus, J. militaris, J. pelocarpus, Juniperus communis,
Kalmia angustifolia, K. buxifolia, K. latifolia, K. polifolia,
Lachnanthes caroliniana, Larix laricina, Leersia oryzoides,
Linnaea borealis, Lobelia canbyi, L. nuttallii, Lonicera vil-
losa, Lophiola aurea, Lycopodiella alopecuroides, L. caro-
liniana, Lycopus amplectens, L. virginicus, Lyonia ligustrina,
L. mariana, Lysimachia terrestris, Magnolia virginiana,
Maianthemum canadense, M. trifolium, Malaxis unifo-
lia, Melampyrum lineare, Menyanthes trifoliata, Mitchella
repens, Muhlenbergia torreyana, M. uniflora, Myrica gale,
M. pensylvanica, Narthecium americanum, Nyssa sylvatica,
Orontium aquaticum, Osmunda regalis, Osmundastrum cinnamomeum, Oxypolis rigidior, Panicum verrucosum, P. virgatum, Pellia, Peltandra virginica, Picea mariana, Pinus
rigida, P. strobus, Platanthera blephariglottis, P. dilatata,
Pleurozium schreberi, Pogonia ophioglossoides, Polygala
cruciata, Polytrichum juniperinum, Pontederia cordata,
Ptilidium ciliare, Quercus ilicifolia, Rhamnus alnifolia,
Rhexia virginica, Rhododendron canadense, R. groenlandicum, R. viscosum, Rhynchospora alba, R. capitellata, R.
cephalantha, R. chalarocephala, R. fusca, R. gracilenta, R.
oligantha, Rosa palustris, R. virginiana, Rubus chamaemorus,
R. hispidus, Sabatia difformis, Sagittaria engelmanniana,
Sanguisorba canadensis, Sarracenia purpurea, Sassafras
albidum, Scheuchzeria palustris, Schizachyrium scoparium, Schizaea pusilla, Schoenoplectus subterminalis,
Scirpus cyperinus, Scleria reticularis, Smilax glauca, S.
herbacea, S. laurifolia, S. rotundifolia, S. walteri, Solidago
erecta, S. nemoralis, S. puberula, S. uliginosa, Sparganium
americanum, Sphagnum angustifolium, S. bartlettianum,
S. capillifolium, S. cuspidatum, S. fallax, S. fimbriatum, S.
flavicomans, S. girgensohnii, S. magellanicum, S. majus,
S. palustre, S. papillosum, S. portoricense, S. pulchrum, S.
pylaesii, S. squarrosum, S. recurvum, S. rubellum, S. subsecundum, S. teres, S. warnstorfii, Spiraea alba, S. tomentosa,
Symphyotrichum boreale, S. novi-belgii, Symplocarpus foetidus, Thuja occidentalis, Toxicodendron radicans, Triadenum
fraseri, T. virginicum, Triantha racemosa, Trichophorum
alpinum, T. cespitosum, Trientalis borealis, Typha angustifolia, T. latifolia, Utricularia cornuta, U. juncea, U. subulata,
Vaccinium angustifolium, V. corymbosum, V. macrocarpon,
V. oxycoccos, V. uliginosum, Viburnum nudum, Viola lanceolata, Woodwardia areolata, W. virginica, Xyris difformis, X.
montana, Zizania aquatica.

Use by wildlife: *Oclemena nemoralis* is a host for larval case
moths (Insecta: Lepidoptera: Coleophoridae: *Coleophora
nemorella*), which feed on the seeds. The plants are also hosts
to rust fungi (Basidiomycota: Coleosporiaceae: *Coleosporium
solidaginis*; Pucciniaceae: *Puccinia asterum*).

Economic importance: food: There are no reliable reports
regarding the edibility of *Oclemena* species; **medicinal:**
The Chippewa administered a root decoction of *Oclemena
nemoralis* as drops to treat ear aches; **cultivation:** *Oclemena
nemoralis* is not cultivated commercially; **misc. products:**
none; **weeds:** none; **nonindigenous species:** none

Systematics: In their former circumscription, *Oclemena*
species were assigned to section *Acuminati* of *Aster* subgenus *Doellingeria*, the latter now elevated to generic rank. A
close relationship between *Oclemena* and *Doellingeria* has
been presumed on the basis of their similar phyllary morphology; however, at the time of this writing, no species of
Oclemena had yet been included in a molecular systematic
study. Consequently, it is not possible to determine whether
Oclemena and *Doellingeria* are closely or distantly related, or
even if they should be maintained as separate genera and further investigations concerning these questions are necessary.
The base chromosome number of *Oclemena* is $x=9$; *O.
nemoralis* ($2n=18$) is diploid. Morphological and cytological
data indicate that *Oclemena nemoralis* crosses with *O. acuminata* to yield an introgressive hybrid, which has been called
O. ×blakei. In contrast to the affinity of *O. nemoralis* for bogs,
O. acuminata inhabits forest understory. However, studies of
the malate dehydrogenase enzyme have found no evidence of
biochemical specialization in its thermal or kinetic properties, despite the contrasting habitats of the two species. An
intergeneric hybrid between *Doellingeria umbellata* and
Oclemena nemoralis has been reported, but never has been
verified genetically and is highly unlikely on the basis of morphological characteristics.

Comments: *Oclemena nemoralis* occurs mainly in glaciated
northeastern North America.

References: Almquist & Calhoun, 2003; Anderson & Davis,
1998; Bertin et al., 2010; Brouillet, 2006c; Brouillet & Simon,
1981, 1998; Brouillet et al., 1998; Chmielewski & Strain,
2007; Hill, 1976; Jones & Young, 1983; Knowlton, 1915;
Laidig & Zampella, 1996; Semple & Brouillet, 1980; Trudeau
et al., 2013; Weiher & Keddy, 1995; Wells, 1996.

34. *Packera*

Butterweed, groundsel, ragwort; Séneçon

Etymology: after John G. Packer (b. 1929)

Synonyms: *Senecio* (in part)

Distribution: global: Asia; North America; **North America:** widespread

Diversity: global: 64 species; **North America:** 54 species

Indicators (USA): OBL: *Packera glabella*, *P. subnuda*; **FACW:** *P. glabella*, *P. subnuda*

Habitat: freshwater; palustrine; **pH:** 4.9–8.4; **depth:** <1 m; **life-form(s):** emergent herb

Key morphology: stems (to 70+ cm) 1–3, erect, solid or hollow, striate or smooth; basal leaves distinctly or obscurely petioled, the blades (to 4–15+ cm) ovate, obovate, oblanceolate, elliptic, or lyrate, the margins crenate to undulate; cauline leaves sessile to weakly clasping, gradual to abruptly reduced upward; heads (1 or up to 30+) radiate, solitary or arranged secondarily in cymes or umbels, peduncled, phyllaries 13–21 (to 8 mm) in 1–2 series, receptacle flat, pales absent; ray florets (to 13) pistillate, the rays (to 12 mm) yellow; disc florets (to 55+) bisexual, the corollas (to 3.5 mm) 5-lobed, yellow; cypselae (to 2.5 mm) cylindric, 5–10-ribbed, a pappus of up to 60 bristles (to 6 mm) present.

Life history: duration: annual, biennial (seeds), perennial (caudex, rhizomes); **asexual reproduction:** rhizomes; **pollination:** insect; **sexual condition:** gynomonoecious; **fruit:** cypselae (common); **local dispersal:** cypselae (gravity, wind), rhizomes; **long-distance dispersal:** cypselae (animals)

Imperilment: (1) *Packera glabella* [G5]; S1 (NE); S2 (KS); S3 (NC); (2) *P. subnuda* [G5]; S2 (AB, AK); S3 (BC, WY)

Ecology: general: *Packera* is a fairly large and diverse genus, which can occur from sea-level to alpine elevations from subtropical to arctic regions. It includes annuals, biennials and perennials, and weedy plants of highly disturbed sites as well as rare, habitat specialists. Various species occur on dry rock or sand, serpentine soils, or organic peat across a spectrum of sites that includes grasslands, mountainous terrain, and woodlands. *Packera* generally is not regarded as being a characteristic wetland genus. Twenty-four North American species (44%) have some status as wetland indicators, but only two (3.7%) are categorized as OBL in at least a portion of their range; none is entirely OBL. The heads can be radiate or discoid with florets that are normally self-incompatible but occasionally self-compatible. The radiate heads are gynomonoecious with pistillate ray florets and bisexual disc florets. Both OBL species are radiate. Pollination generally occurs by means of bees (Hymenoptera) and other insects (Insecta). The cypselae possess a pappus of numerous bristles and are dispersed by animals, gravity, or wind.

Packera glabella **(Poir.) C. Jeffrey** is an annual or biennial found in nontidal or tidal reaches of depressions, ditches, dikes, flats, flatwoods, floodplains, glades, hammocks, marshes, meadows, pools, prairies, ravines, ricefields, roadsides, sandbars, seeps, sloughs, swamps, and along the margins of bayous, canals, lakes, ponds, rivers, streams, and woodlands at elevations of up to 600 m. Plants can tolerate shallow standing water (1–60 cm) and withstand a range of exposures from full sun to partial shade. The substrates are alkaline (pH 7.1–8.4) and are described as alluvium, calcareous, clay, limestone, loam, marl, muck, mud, *Rangia* shells (Mollusca: Bivalvia: Mactridae), sand, sandy clay, sandy loam, sandy silt, and silt. This is a later-emerging winter annual, which sometimes is described as a spring annual. Germination begins in the late fall with peak levels occurring in January and February. The adult plants reach their peak flowering from March to late May. The flowers are pollinated by various insects (Insecta), primarily bees (Hymenoptera). Specific germination requirements for the seeds have not been reported. The recovery of viable seeds from the dung of horses (Mammalia: Equidae: *Equus ferus*) indicates that endozoic mammalian dispersal occurs to some extent. This species often is described as weedy, especially in the early winter when it represents one of the few species in the vegetation and can attain significant densities. The plants respond favorably to vegetation thinning, in one case increasing seven-fold in frequency and eightfold in mean cover. **Reported associates:** *Acalypha rhomboidea, Acer rubrum, A. negundo, A. saccharinum, A. saccharum, Aesculus pavia, Agalinis tenuifolia, Alliaria petiolata, Alnus serrulata, Alternanthera philoxeroides, Amaranthus cannabinus, Ambrosia artemisiifolia, A. trifida, Amorpha fruticosa, Ampelopsis arborea, A. cordata, Apios americana, Apocynum cannabinum, Arisaema dracontium, Aristolochia serpentaria, Arnoglossum diversifolium, Arundinaria gigantea, Asclepias perennis, Asimina triloba, Asplenium platyneuron, Baccharis, Barbarea vulgaris, Berchemia scandens, Betula nigra, Bidens comosus, B. connatus, B. discoideus, B. frondosus, B. vulgatus, Bignonia capreolata, Boehmeria cylindrica, Brunnichia ovata, Callicarpa americana, Campanula americana, Campsis radicans, Canna flaccida, Capsicum annuum, Cardamine pensylvanica, Carex albida, C. amphibola, C. bromoides, C. caroliniana, C. cherokeensis, C. debilis, C. frankii, C. grayi, C. hyalinolepis, C. intumescens, C. lacustris, C. lupulina, C. lurida, C. oxylepis, C. retroflexa, C. rosea, C. socialis, C. tribuloides, C. typhina, C. vulpinoidea, Carpinus caroliniana, Carya aquatica, C. glabra, C. illinoinensis, Catalpa speciosa, Cayaponia quinqueloba, C. vulpinoidea, Celtis laevigata, C. occidentalis, Cephalanthus occidentalis, Cercis canadensis, Chaerophyllum procumbens, Chasmanthium laxum, Chionanthus virginicus, Cicuta maculata, Cinna arundinacea, Circaea canadensis, Cirsium arvense, C. muticum, C. vulgare, Cladium jamaicense, Clematis crispa, C. reticulata, Clethra alnifolia, Cocculus carolinus, Commelina communis, C. diffusa, C. virginica, Conoclinium coelestinum, Conyza canadensis, Cornus drummondii, C. foemina, Crataegus viridis, Crinum americanum, Cryptotaenia canadensis, Cuscuta campestris, Cynodon dactylon, Cynosciadium digitatum, Cyperus esculentus, C. haspan, C. odoratus, Cyrilla racemiflora, Dactylis glomerata, Daucus carota, Decodon verticillatus, Decumaria barbara, Dichanthelium commutatum, Dichondra repens, Digitaria bicornis, Diodia virginiana, Diospyros virginiana, Eclipta prostrata, Elymus virginicus, Elytraria, Equisetum arvense,*

E. hyemale, Erechtites hieraciifolius, Erigeron annuus, E. philadelphicus, E. quercifolius, Erythrina herbacea, Eubotrys racemosa, Euonymus americanus, Eupatorium capillifolium, E. perfoliatum, Euthamia graminifolia, Eutrochium purpureum, Forestiera, Fraxinus americana, F. caroliniana, F. pennsylvanica, F. profunda, Funastrum clausum, Galium aparine, G. bermudense, G. obtusum, G. tinctorium, G. triflorum, Gamochaeta pensylvanica, G. stagnalis, Gelsemium sempervirens, Geranium carolinianum, Geum canadense, G. vernum, Glechoma hederacea, Gleditsia aquatica, G. triacanthos, Glyceria septentrionalis, G. striata, Gonolobus suberosus, Habenaria repens, Hackelia virginiana, Halesia diptera, Hasteola robertiorum, Helenium flexuosum, Helianthus decapetalus, H. grosseserratus, Heliotropium indicum, Hibiscus, Hydrocotyle ranunculoides, H. verticillata, Hydrolea uniflora, Hypericum perforatum, Ilex decidua, I. opaca, I. vomitoria, Impatiens capensis, Iris hexagona, Itea virginica, Juglans nigra, Juncus effusus, J. marginatus, Juniperus virginiana, Kalmia latifolia, Lactuca serriola, Lamium purpureum, Laportea canadensis, Leersia lenticularis, L. oryzoides, Leitneria floridana, Ligustrum sinense, Lindera benzoin, Liquidambar styraciflua, Liriodendron tulipifera, Lobelia cardinalis, L. elongata, L. inflata, L. siphilitica, Lonicera japonica, Ludwigia bonariensis, L. palustris, Lycopus americanus, L. rubellus, Lyonia lucida, Lysimachia ciliata, Maclura pomifera, Magnolia grandiflora, M. virginiana, Melica mutica, Microstegium vimineum, Mikania cordifolia, M. scandens, Mimulus alatus, Morus rubra, Murdannia keisak, Myosurus minimus, Myosotis macrosperma, Myrica cerifera, Myrsine floridana, Nuttallanthus canadensis, Nyssa aquatica, N. biflora, N. ogeche, N. sylvatica, Onoclea sensibilis, Oplismenus hirtellus, Orontium aquaticum, Orthosia scoparia, Osmunda regalis, Osmundastrum cinnamomeum, Ostrya virginiana, Oxalis stricta, Oxydendrum arboreum, Panicum hemitomon, Parthenocissus quinquefolia, Passiflora lutea, P. suberosa, Paysonia lescurii, Peltandra virginica, Perilla frutescens, Persea borbonia, P. palustris, Persicaria arifolia, P. glabra, P. hydropiper, P. pensylvanica, P. virginiana, Phlebodium aureum, Phlox divaricata, Phoradendron serotinum, Physalis longifolia, Physostegia leptophylla, Phytolacca americana, Pilea pumila, Pinus elliottii, P. serotina, P. taeda, Planera aquatica, Platanus occidentalis, Poa annua, P. autumnalis, Polymnia canadensis, Polypodium vulgare, Pontederia cordata, Populus deltoides, P. heterophylla, Potentilla norvegica, Prunus angustifolia, P. caroliniana, P. serotina, Psychotria nervosa, Ptilimnium capillaceum, P. costatum, Pycreus flavidus, Quercus alba, Q. hemisphaerica, Q. laurifolia, Q. lyrata, Q. michauxii, Q. muehlenbergii, Q. nigra, Q. pagoda, Q. palustris, Q. phellos, Q. shumardii, Q. texana, Q. virginiana, Ranunculus abortivus, R. recurvatus, Rhododendron viscosum, Rhus copallinum, Rhynchospora, Rivina humilis, Rorippa palustris, Rosa palustris, Rubus, Ruellia caroliniensis, Sabal minor, S. palmetto, Sabatia calycina, Sagittaria lancifolia, Salix nigra, Sambucus nigra, Samolus valerandi, Sanicula canadensis, Saururus cernuus, Scleria pauciflora, Scutellaria lateriflora, S. nervosa, Senna obtusifolia, Serenoa repens, Sideroxylon lanuginosum, S. lycioides, S. reclinatum, S. tenax, Sisyrinchium angustifolium, Sium suave, Smilax auriculata, S. bona-nox, S. ecirrhata, S. laurifolia, S. pumila, S. tamnoides, Solanum nigrum, Solidago canadensis, S. flexicaulis, Sphenopholis nitida, Spiranthes cernua, S. vernalis, Stachys tenuifolia, Stellaria cuspidata, Symphyotrichum lanceolatum, S. lateriflorum, Taxodium ascendens, T. distichum, Teucrium canadense, Thelypteris kunthii, T. palustris, Tillandsia recurvata, T. usneoides, Toxicodendron radicans, Trepocarpus aethusae, Triadenum walteri, Trichostema dichotomum, Trillium, Triosteum aurantiacum, Typha latifolia, Ulmus alata, U. americana, U. crassifolia, U. rubra, Uniola latifolia, Urtica chamaedryoides, Verbascum blattaria, Verbena hastata, V. urticifolia, Vernonia gigantea, Veronica peregrina, Viburnum dentatum, V. obovatum, V. prunifolium, V. rufidulum, Vicia acutifolia, Viola lanceolata, V. pedata, V. sororia, Vitis aestivalis, V. cinerea, V. munsoniana, V. rotundifolia, Woodwardia areolata, Ximenia americana, Zephyranthes atamasco, Zizaniopsis miliacea.

Packera subnuda (**DC.**) **D.K. Trock & T.M. Barkl.** is a perennial species of beaches, bogs, depressions, fens, floodplains, gravel bars, hummocks, marshes, meadows, seeps, slopes, snowfields, swales, swamps, tundra, and the margins of brooks, lakes, pools, rivers and streams at elevations from 1036 to 3489 m. The plants typically occupy alpine or subalpine habitats with exposures varying from sun to shade. The substrates (pH 4.9–7.5) are characterized as calcareous, calc-hornfel, clay loam, granite, gravelly loam, hardpan, hornfel, humus, limestone, loam, muck, mud, organic loam, peat, rock, sand, sandy loam, scree, stony loam, and talus. Flowering proceeds from June to August. Additional details on the reproductive biology and seed ecology are unavailable. The roots are colonized by arbuscular mycorrhizal fungi and vegetative reproduction can occur by the production of rhizomes. When growing in patterned peatlands, the plants occur mainly in the troughs. **Reported associates:** *Abies concolor, A. grandis, A. lasiocarpa, Acer, Aconitum, Agastache, Allium validum, Alnus, Anemone, Antennaria corymbosa, Aquilegia formosa, A. jonesii, Arabis, Arnica, Artemisia cana, A. scopulorum, Athyrium filix-femina, Aulacomnium palustre, Betula glandulosa, B. occidentalis, Bistorta vivipara, Bryum pseudotriquetrum, Calamagrostis canadensis, Calliergon giganteum, C. stramineum, Caltha leptosepala, Campanula rotundifolia, Campylium stellatum, Cardamine, Carex aquatilis, C. buxbaumii, C. canescens, C. deweyana, C. heteroneura, C. illota, C. interior, C. lasiocarpa, C. limosa, C. livida, C. muricata, C. nigricans, C. rostrata, C. scopulorum, C. simulata, C. utriculata, C. viridula, Cercocarpus ledifolius, Chamerion latifolium, Comarum palustre, Danthonia intermedia, Dasiphora floribunda, Deschampsia cespitosa, Drepanocladus aduncus, Drosera anglica, Eleocharis quinqueflora, Epilobium ciliatum, E. palustre, Equisetum arvense, E. variegatum, Erigeron peregrinus, Eriophorum angustifolium, E. chamissonis, Fragaria virginiana, Galium trifidum, Gentiana calycosa, Geum rossii, Glyceria elata, G. striata, Helodium blandowii, Heracleum, Juncus drummondii, J.*

ensifolius, J. mertensianus, J. parryi, Juniperus occidentalis, Kalmia microphylla, K. polifolia, K. procumbens, Larix lyallii, Leptarrhena pyrolifolia, Lonicera caerulea, Luzula, Mertensia, Micranthes odontoloma, M. subapetala, Mimulus lewisii, Muhlenbergia filiformis, Packera contermina, Parnassia fimbriata, Pedicularis bracteosa, P. groenlandica, Petrophytum caespitosum, Philonotis fontana, Phleum alpinum, Picea engelmannii, Pinguicula vulgaris, Pinus albicaulis, P. contorta, P. flexilis, Plagiomnium cuspidatum, Poa leptocoma, P. palustris, Podagrostis thurberiana, Polemonium occidentale, Platanthera, Potentilla ×diversifolia, Primula, Pseudotsuga menziesii, Ranunculus, Rhodiola rhodantha, R. rosea, Rhododendron columbianum, Rubus arcticus, Rudbeckia, Salix arctica, S. glauca, S. orestera, S. planifolia, S. wolfii, Scirpus microcarpus, Scrophularia lanceolata, Senecio sphaerocephalus, S. triangularis, Sibbaldia procumbens, Smelowskia calycina, Solidago spathulata, Sphagnum riparium, S. squarrosum, S. teres, S. warnstorfii, Spiranthes romanzoffiana, Swertia perennis, Symphyotrichum foliaceum, Synthyris wyomingensis, Tomentypnum nitens, Trichophorum cespitosum, Trifolium parryi, Triglochin maritimum, Trollius albiflorus, T. laxus, Vaccinium scoparium, V. uliginosum, Valeriana, Viola macloskeyi, V. palustris.

Use by wildlife: *Packera glabella* provides food for swamp rabbits (Mammalia: Leporidae: *Sylvilagus aquaticus*) and habitat for wild turkeys (Aves: Phasianidae: *Meleagris gallopavo silvestris*). It is a breeding plant for seed bugs (Insecta: Hemiptera: Lygaeidae: *Neacoryphus bicrucis*). The decomposing plants are eaten by crayfish (Crustacea: Decapoda). The flowers are visited by bees (Insecta: Hymenoptera: Megachilidae: *Osmia georgica*), beetles (Insecta: Coleoptera: Coccinellidae: *Coccinella septempunctata*; *Harmonia axyridis*), and fruit flies (Insecta: Diptera: Tephritidae: *Trupanea actinobola*), which presumably act as pollinators. The plants carry BiMoV, the *Bidens* mottle virus (Potyviridae: *Potyvirus*) and are hosts to leafminer flies (Insecta: Diptera: Agromyzidae: *Agromyza maculosa*, *Liriomyza trifolii*) and plant bugs (Insecta: Hemiptera: Miridae: *Lygocorus semivittatus*, *Lygus lineolaris*, *Neurocolpus nubilus*, *Taylorilygus pallidulus*). The foliage contains toxic pyrrolizidine alkaloids, which have been linked to livestock poisoning. *Packera subnuda* is a larval food of the cinnabar moth (Insecta: Lepidoptera: Erebidae: *Tyria jacobaeae*). The plants also host several fungi including Basidiomycota (Entylomataceae: *Entyloma compositarum*; Pucciniaceae: *Puccinia subcircinata*) and Oomycota (Albuginaceae: *Albugo tragopogonis*; Synchytriaceae: *Synchytrium*).

Economic importance: food: *Packera* should never be eaten due to the presence of pyrrolizidine alkaloids, which are carcinogenic, hepatotoxic, and have been linked to cirrhosis; **medicinal:** *Packera glabella* contains the pyrrolizidine alkaloids florosenine (acetylotosenine), integerrimine, otosenine, senecionine, and senkirkine, which are associated with various disorders (see previous). Dried plant matter fed at 20% of initial body weight is lethal to rats (Mammalia: Rodentia: Muridae: *Rattus norvegicus*) due to the presence of senecionine

and other alkaloids; **cultivation:** *Packera glabella* is not cultivated; **misc. products:** none; **weeds:** *Packera glabella* often is described as weedy due to its abundance (especially in late winter and spring) and its potential for rapid growth. It is classified as a noxious weed in the state of Ohio, primarily with respect to agricultural activities; **nonindigenous species:** *Packera glabella* has been introduced to Iowa and Ontario.

Systematics: The former "Aureoid group" of *Senecio* essentially became the genus *Packera*, as molecular evidence increasingly indicated its distinctness from the former genus. Within Asteraceae, *Packera* currently is assigned to subtribe Senecioninae of tribe Senecioneae in subfamily Asteroideae. Molecular studies of *Packera* have indicated extensive historical hybridization in the genus, which complicates its phylogenetic evaluation. Nevertheless, the most recent phylogenetic appraisal indicates Old World members of *Pericallis* or *Senecio* as the closest relatives of *Packera*, which appears to be monophyletic as currently circumscribed (Figure 5.167). Within *Packera*, ITS sequence data place *P. glabella* within the basal section *Sanguisorboidei*, with *P. subnuda* resolving in a distantly related clade (Figure 5.167). This result indicates that the OBL habit has arisen independently at least twice in the genus. Nomenclatural problems abound in *Packera* and

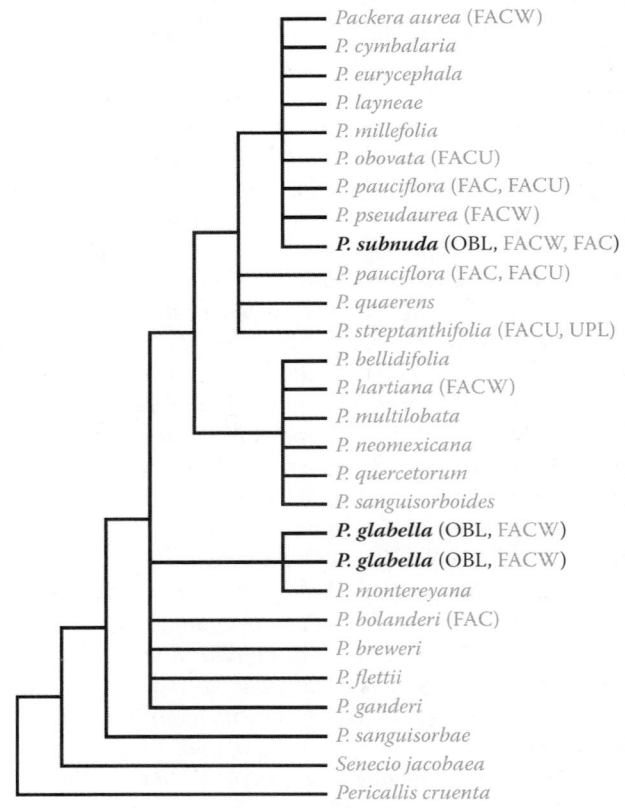

FIGURE 5.167 Phylogenetic relationships in *Packera* as resolved using ITS sequence data. The distant relationship of the two OBL species (bold) indicate the independent derivation of the habit in this genus. (Adapted from Bain, J.F. & Golden, J.L., *Mol. Phylogenet. Evol.*, 16, 331–338, 2000.)

can make it difficult to associate the correct taxon with literature accounts, which often list only the species name without the authority name. One example is *Senecio cymbalarioides* Buek, which is a synonym of *P. subnuda*; whereas, *Senecio cymbalarioides* Nuttall (=*Packera streptanthifolia*) is not. Because both species overlap geographically, those literature accounts listing only "*S. cymbalarioides*" cannot be assigned confidently to one or the other species. Although care has been taken to avoid such errors in assigning information properly, there is little that can be done to evaluate whether an original literature account had interpreted *P. subnuda* correctly in terms of its synonymy. The base number of *Packera* is $x=23$. *Packera glabella* ($2n=46$) is diploid. Two varieties of *P. subnuda* are recognized, which correlate cytologically; i.e., *P. subnuda* var. *subnuda* ($2n=46$) and *P. subnuda* var. *moresbiensis* ($2n=90$). The latter has been recognized at species rank as *Packera moresbiensis*. *Packera subnuda* populations possess diverse haplotypes, despite the lack of intrapopulational variation, with the greatest genetic subdivision occurring along either side of the continental divide. This pattern implicates historical hybridization as the initial source of haplotype diversity, which subsequently was reduced by drift. Hybridization occurs frequently among *Packera* species, but has not been reported (at least contemporaneously) for either of the OBL taxa.

Comments: *Packera glabella* occurs throughout the south-central United States; whereas, *P. subnuda* grows in northwestern North America.

References: Avis et al., 1997; Bain & Golden, 2000; Barkley, 1988; Brown & Tighe, 1992; Burke et al., 2012; Campbell & Gibson, 2001; Cázares et al., 2005; Cripps & Eddington, 2005; Daubenmire, 1990; Fleckenstein, 2007; Goeger et al., 1983; Golden & Bain, 2000; Han & McPheron, 1994; Helm & Chabreck, 2006; Homoya & Rayner, 1987; Hopkins, 1969; Horn & Somers, 1981; Johnson & Billings, 1962; Kapadia et al., 1990; Karacetin, 2007; Lemly, 2007; Mathies et al., 1983; Mohlenbrock, 1959a; Moseley & Pitner, 1996; Nelson, 1986; Orsenigo & Zitter, 1971; Purcifull & Zitter, 1971; Ray et al., 1987; Ruch et al., 1998; Smith et al., 2001; Snodgrass, 2003; Snodgrass et al., 1984b; Solbreck & Pehrson, 1979; Stegmaier, Jr., 1966, 1967b; Thieret, 1971; Trock, 2006; Zwank et al., 1988.

35. *Phalacroseris*

Mock dandelion

Etymology: from the Greek *phalakros seris* ("bald seris"), for its resemblance to endive (seris), but lacking a pappus

Synonyms: *Microseris* (in part)

Distribution: global: North America; **North America:** California

Diversity: global: 1 species; **North America:** 1 species

Indicators (USA): OBL: *Phalacroseris bolanderi*

Habitat: freshwater; palustrine; **pH:** alkaline; **depth:** <1 m; **life-form(s):** emergent herb

Key morphology: Stems (to 45 cm) erect, scapiform; leaves (to 20 cm) in basal rosettes, petioled, linear to oblanceolate, somewhat fleshy, the margins entire; heads solitary, involucre (to 13 mm) campanulate, phyllaries 8–25 in 2–4 series, connate at base; receptacles lacking pales; up to 35 florets (to 18 mm), corolla yellow; cypselae (to 4 mm) brown, obscurely 3-angled, lacking a pappus or with a short (<0.5 mm) crown.

Life history: duration: perennial (caudex; fleshy taproot); **asexual reproduction:** none; **pollination:** insect; **sexual condition:** hermaphroditic; **fruit:** cypsela (common); **local dispersal:** cypselae (gravity); **long-distance dispersal:** cypselae (gravity)

Imperilment: (1) *Phalacroseris bolanderi* [G3, G4]

Ecology: general: Monotypic (see later).

Phalacroseris bolanderi **A. Gray** inhabits bogs, cienega, fens, meadows, roadsides, seeps, slopes, swamps, and woodlands at elevations from 1600 to 2834 m. The plants occur in exposures ranging from full sun to mixed sun and shade. The substrates are described as Cryumbrepts, granite, muck, mud, peat, rocks, and ultramafic. The plants flower from June to August. Their principal pollinators are honeybees and leaf-cutting bees (Insecta: Hymenoptera: Apidae; Megachilidae). There is little additional information on the reproductive biology or seed ecology of this species. Because the cypselae lack a pappus, most dispersal presumably occurs by gravity. The stems perennate by a caudex. This species can replace sedges (Cyperaceae) in areas that are heavily trampled or grazed by livestock. **Reported associates:** *Abies concolor*, *Achnatherum occidentale*, *Agrostis idahoensis*, *A. scabra*, *Allium yosemitense*, *Alnus rugosa*, *Artemisia*, *Aulacomnium*, *Bistorta bistortoides*, *Caltha leptosepala*, *Camassia quamash*, *Carex echinata*, *C. feta*, *C. fissuricola*, *C. heteroneura*, *C. integra*, *C. jonesii*, *C. lemmonii*, *C. lenticularis*, *C. luzulina*, *C. scopulorum*, *C. utriculata*, *Castilleja miniata*, *Chaenactis douglasii*, *Cryptantha affinis*, *Danthonia californica*, *Darlingtonia californica*, *Deschampsia cespitosa*, *Drepanocladus*, *Drosera rotundifolia*, *Drymocallis glandulosa*, *Eleocharis decumbens*, *E. macrostachya*, *E. palustris*, *E. quinqueflora*, *Equisetum arvense*, *Eriophorum crinigerum*, *Eriophyllum nubigenum*, *Eucephalus breweri*, *Glyceria elata*, *Hastingsia alba*, *Helenium bigelovii*, *Heterocodon rariflorus*, *Hosackia oblongifolia*, *H. pinnata*, *Hulsea brevifolia*, *Hypericum anagalloides*, *Ivesia unguiculata*, *Juncus balticus*, *J. ensifolius*, *J. howellii*, *J. nevadensis*, *J. oxymeris*, *Kalmia microphylla*, *Lilium*, *Lonicera caerulea*, *Luzula subcongesta*, *Meesia triquetra*, *Micranthes oregana*, *Mimulus guttatus*, *M. moschatus*, *M. primuloides*, *Muhlenbergia andina*, *M. filiformis*, *Narthecium californicum*, *Oreostemma alpigenum*, *Oxypolis occidentalis*, *Parnassia palustris*, *Pedicularis attollens*, *Penstemon newberryi*, *Perideridia bolanderi*, *P. parishii*, *Philonotis fontana*, *Pinus contorta*, *P. jeffreyi*, *P. lambertiana*, *Platanthera dilatata*, *P. sparsiflora*, *P. yosemitensis*, *Poa pratensis*, *Populus*, *Potentilla gracilis*, *Primula jeffreyi*, *P. tetrandra*, *Ptychostomum pacificum*, *Rhododendron columbianum*, *Scirpus diffusus*, *S. microcarpus*, *Senecio clarkianus*, *S. scorzonella*, *S. triangularis*, *Sisyrinchium*, *Solidago elongata*, *Sorbus californica*, *Sphagnum subsecundum*, *S. teres*, *Spiraea douglasii*, *Spiranthes romanzoffiana*, *Stellaria longipes*, *Symphyotrichum spathulatum*, *Triantha occidentalis*, *Trifolium bolanderi*, *T. longipes*,

T. wormskioldii, Utricularia intermedia, Vaccinium uligino-sum, Viola macloskeyi, Zeltnera venusta.

Use by wildlife: *Phalacroseris bolanderi* provides habitat and cover for vagrant shrews (Mammalia: Soricidae: *Sorex vagrans*). It is eaten by grazing mammals, more prevalently in late summer (August) than earlier in the season. The flowers are visited by an assortment of bees (Insecta: Hymenoptera: Apidae: *Bombus melanopygus*; Megachilidae: *Osmia brevis, O. cobaltina, O. coloradensis, O. juxta, O. pusilla, O. texana*), which serve as their pollinators.

Economic importance: food: not reported as edible; **medicinal:** none reported; **cultivation:** not in cultivation; **misc. products:** none; **weeds:** none; **nonindigenous species:** none

Systematics: The monotypic *Phalacroseris* (tribe Cichorieae) is assigned to subtribe Microseridinae by some authors or to its own subtribe (Phalacroseridinae) by others. The latter disposition seems reasonable given that analysis of nuclear DNA sequence data resolves the genus in an isolated position near the base of tribe Cichorieae (Figure 5.164). The solitary position of the genus is indicated further by its distinctive pollen morphology. The chromosome number ($x=9$) is $2n=18$. Although the plants are diploid, they have a large genome (2C DNA content = 124.5).

Comments: *Phalacroseris bolanderi* is endemic to California.

References: Chambers, 2006a, 2012; Colwell et al., 2007; Feuer & Tomb, 1977; Ingles, 1961; Kie & Myler, 1987; Lee et al., 2003; Ratliff, 1985; Sikes et al., 2013; Spence & Shevock, 2012.

36. *Pluchea*

Camphorweed, marsh fleabane, sourbush

Etymology: after Abbé Noel-Antoine Pluche (1688–1761)

Synonyms: *Baccharis* (in part); *Berthelotia*; *Conyza* (in part); *Erigerodes*; *Erigeron* (in part); *Eremohylema*; *Eyrea*; *Gnaphalium* (in part); *Gymnostylis* (invalid); *Gynema*; *Pachythelia*; *Placus* (in part); *Polypappus*; *Spiropodium*; *Stylimnus*; *Tecmarsis*

Distribution: global: Africa, Asia, Australia, New World; **North America:** southern and eastern

Diversity: global: 40–60 species; **North America:** 9 species

Indicators (USA): OBL: *Pluchea foetida, P. longifolia, P. odorata, P. sagittalis*; **FACW:** *P. foetida, P. odorata*

Habitat: brackish, freshwater, or saline; palustrine; **pH:** 4.5–7.6; **depth:** <1 m; **life-form(s):** emergent herb

Key morphology: Stems (to 250 cm) erect, fibrous-rooted, winged or wingless, glandular, minutely to densely hairy, rhizomes present or absent; leaves cauline, sessile or petiolate, the blades (to 10–20 cm) lanceolate, lance-ovate, oblong, elliptic or ovate, bases clasping or not, margins denticulate to toothed; heads (to 12 mm) arranged secondarily in dense to loose corymbs, involucres (to 6–12 mm) campanulate or hemispheric, the phyllaries in 3–6+ series, the receptacles flat, lacking pales; outer florets pistillate, in 3–10+ series, inner florets (to 40+) functionally staminate, corollas 4–5-lobed, cream, pink, rose-purple, white, or yellow; cypselae oblong-cylindric, 4–8-ribbed, pappus a single series of persistent, free or basally connate bristles (to 4 mm).

Life history: duration: annual (seeds), perennial (persistent rootstocks); **asexual reproduction:** rhizomes, stolons; **pollination:** insect; **sexual condition:** hermaphroditic, gynomonoecious or cryptically monoecious; **fruit:** cypselae (common); **local dispersal:** cypselae (water, wind); rhizomes; **long-distance dispersal:** cypselae (wind)

Imperilment: (1) *Pluchea foetida* [G5]; SH (MO, NJ); S3 (DE); (2) *P. longifolia* [G3, G4]; S3 (FL); (3) *P. odorata* [G5]; S1 (PA); S2 (UT); S3 (KS, NC); (4) *P. sagittalis* [GNR]

Ecology: general: Because phylogenetic studies have not yet resolved whether *Pluchea* is monophyletic, any description of the genus should be considered as provisional with respect to summaries of specific features. As currently circumscribed, *Pluchea* is a morphologically diverse genus consisting of annual herbs or perennial herbs, woody shrubs and trees with aromatic foliage. Most of the species are tropical or warm temperate in distribution, typically growing in association with sites characterized by some degree of soil moisture. However, some inhabit dry cliffs, saline habitats or arid wasteland. The genus is found commonly along shores, drying lake or river beds, floodplains, and other sites of ephemeral water. All nine North American species are wetland indicators, with four of them (44%) categorized as OBL. The heads typically are heterogamous; i.e., they contain one to several rows of marginal female florets with various numbers (usually fewer) of central hermaphroditic disc florets, which often (~50% of the species) are functionally staminate. Thus the species tend to be cryptically monoecious although a few species produce entirely hermaphroditic florets. The heads are protogynous, with the outer female florets maturing before the staminate inner florets produce pollen. Consequently, the plants arguably are outcrossed, but it has not been determined whether any level of self-incompatibility (or selfing) also occurs. Pollination occurs by insects (Insecta), which forage for nectar and pollen. Although species often are present in seed banks, their seed ecology has not been studied adequately. Dispersal occurs mainly by the wind but likely to some extent by water also. Some species reproduce vegetatively by production of rhizomes or stolons.

Pluchea foetida (L.) DC. is an annual or perennial inhabitant of freshwater beaver ponds, bogs, borrow pits, cypress bays, depressions, ditches, dunes, flatwoods, floodplains, meadows, pine barrens, prairies, roadsides, savannah, swales, swamps, and margins of lakes, ponds, pools, and streams at elevations of up to 20 m. The plants have been observed to grow in as much as 30 cm of standing water. Exposures range from open sites to semi- or deep shade, but occurrences typically correlate with high light levels. The substrates are calcareous (pH 6.6–7.6) and have been described as Dorovan muck, loamy sand, muck, peat, placid, and myakka (typic humaquepts and aeric haplaquods), sand, sandy loam, and sandy peat. Flowering occurs from July to October, but nearly year-round in the southern United States. There have been no reports of self-compatibility in this species. The flowers produce copious nectar and are insect pollinated (entomophilous) primarily by bees (Hymenoptera) and butterflies (Lepidoptera). The cypselae possess a persistent pappus of

long bristles, which assists with their dissemination by wind. The barbellate pappus bristles may also facilitate some animal dispersal. It is unknown whether the cypselae are buoyant in water. A seed bank is formed, which persists for at least several years. It has been found at sites both with and without plants in the standing vegetation. Cores have produced an average germinable seed density of $27/m^2$. Untreated, field-collected seeds have germinated under greenhouse conditions with four watering cycles/day. The plants persist in annually burned pinelands. An average density of 1.1 germinable seeds/m^2 has been found at one site following a fire. The plants sometimes produce rhizomes, which provide a means of vegetative reproduction. **Reported associates:** *Acer rubrum, Ageratina altissima, Alternanthera philoxeroides, Amaranthus, Ambrosia artemisiifolia, Ampelopsis arborea, Andropogon arctatus, A. glomeratus, A. virginicus, Apios americana, Aristida palustris, A. stricta, Berchemia scandens, Boehmeria cylindrica, Bolboschoenus robustus, Boltonia asteroides, Borrichia frutescens, Buchnera, Carex alata, C. bullata, C. striata, Carpinus caroliniana, Carya aquatica, Centella asiatica, Cephalanthus occidentalis, Cicuta maculata, Cladium jamaicense, Clethra alnifolia, Cliftonia, Coleataenia anceps, C. tenera, Cuscuta, Cyperus, Cyrilla racemiflora, Decodon verticillatus, Dichanthelium aciculare, D. acuminatum, Diodia virginiana, Diospyros, Distichlis spicata, Drosera brevifolia, D. intermedia, Eleocharis flavescens, E. microcarpa, E. parvula, E. rostellata, Eragrostis elliottii, Erechtites hieraciifolius, Eriocaulon decangulare, Eubotrys racemosa, Eupatorium leptophyllum, E. perfoliatum, E. rotundifolium, Euthamia graminifolia, Eutrochium purpureum, Fimbristylis autumnalis, F. caroliniana, F. spadicea, Flaveria linearis, Galium bermudense, Gordonia lasianthus, Habenaria repens, Hibiscus dasycalyx, H. moscheutos, Hydrocotyle bonariensis, H. umbellata, Hymenocallis palmeri, Hypericum hypericoides, H. mutilum, Ilex coriacea, I. glabra, Ipomoea sagittata, Itea virginica, Iva frutescens, Juncus canadensis, J. dichotomus, J. repens, J. scirpoides, J. validus, Juniperus virginiana, Kosteletzkya pentacarpos, Lemna minor, Liquidambar styraciflua, Lobelia puberula, Ludwigia alata, L. alternifolia, L. glandulosa, L. palustris, Lycopodiella appressa, Lyonia ligustrina, Lythrum lineare, Magnolia virginiana, Mecardonia acuminata, Melanthera angustifolia, Mikania scandens, Mitreola, Muhlenbergia capillaris, Murdannia keisak, Myrica cerifera, M. pensylvanica, Nyssa sylvatica, Osmunda regalis, Osmundastrum cinnamomeum, Oxypolis, Panicum amarum, P. hemitomon, P. verrucosum, P. virgatum, Paspalum monostachyum, P. vaginatum, Persea borbonia, P. palustris, Persicaria hydropiperoides, P. lapathifolia, P. sagittata, Phragmites australis, Phyla lanceolata, P. nodiflora, Physostegia leptophylla, Pinus elliottii, P. taeda, Platanthera cristata, Pluchea camphorata, Polygala cruciata, P. lutea, Proserpinaca palustris, P. pectinata, P. ×intermedia, Ptilimnium capillaceum, Quercus alba, Q. michauxii, Q. pagoda, Q. palustris, Q. phellos, Rhexia mariana, R. virginica, Rhododendron viscosum, Rhynchospora colorata, R. corniculata, R. divergens, R. floridensis, R. globularis, R. macrostachya, R. microcephala, R. tracyi, Rosa palustris,*
Rubus hispidus, R. pensilvanicus, Rumex verticillatus, Sabal minor, S. palmetto, Sabatia decandra, S. stellaris, Saccharum giganteum, Sagittaria lancifolia, S. latifolia, Samolus ebracteatus, Sarracenia flava, Schizachyrium rhizomatum, S. scoparium, Schoenoplectus pungens, Scirpus cyperinus, Scleria, Sclerolepis uniflora, Serenoa repens, Setaria magna, S. parviflora, Smilax bona-nox, S. laurifolia, S. rotundifolia, Solidago fistulosa, S. sempervirens, Spartina alterniflora, S. patens, Spermacoce neoterminalis, Sphagnum, Spiranthes cernua, Spirodela polyrhiza, Sporobolus virginicus, Symphyotrichum subulatum, Taxodium, Thelypteris palustris, Toxicodendron radicans, T. vernix, Triadenum virginicum, T. walteri, Triadica sebifera, Typha angustifolia, T. latifolia, Utricularia gibba, Vaccinium corymbosum, V. stamineum, Vigna luteola, Vitis rotundifolia, Woodwardia areolata, W. virginica, Xyris difformis, X. jupicai, X. laxifolia, X. smalliana, X. torta.

***Pluchea longifolia* Nash** is a perennial, which occurs in brackish to tidal freshwater canals, depressions, ditches, flatwoods, floodplains, hammocks, marshes, prairies, seepage bogs, swamps, and along the margins of lakes at elevations of up to 17 m. The plants often are found growing in shallow water (e.g., 5 cm). The plants tolerate shade and occur on loamy sand, marl, peat, sandy loam, and sandy peat substrates. Flowering occurs from June to October. There is little other life-history information available for this species. **Reported associates:** *Acalypha gracilens, Acer, Acmella oppositifolia, Amorpha fruticosa, Asclepias perennis, Axonopus furcatus, Bacopa monnieri, Betula nigra, Boehmeria cylindrica, Carya aquatica, C. glabra, Celtis laevigata, Centella asiatica, Cephalanthus occidentalis, Cladium jamaicense, Clematis crispa, Coleataenia anceps, Conoclinium coelestinum, Cornus foemina, Crataegus aestivalis, C. marshallii, Cyperus croceus, C. polystachyos, Dichanthelium commutatum, Ditrysinia fruticosa, Dyschoriste humistrata, Elytraria caroliniensis, Eriochloa michauxii, Fraxinus, Hypericum galioides, Hypoxis curtissii, Hyptis alata, Ilex decidua, Justicia, Leitneria floridana, Liquidambar styraciflua, Mecardonia acuminata, Mitreola petiolata, Myrica cerifera, Nyssa, Oldenlandia uniflora, Panicum hemitomon, Penthorum sedoides, Persea palustris, Persicaria punctata, Pluchea camphorata, Polypremum procumbens, Quercus laurifolia, Q. nigra, Q. shumardii, Q. virginiana, Sabal palmetto, Saccharum baldwinii, Salix caroliniana, Samolus valerandi, Taxodium, Ulmus americana, Vernonia.*

***Pluchea odorata* (L.) Cass.** occupies fresh to brackish (salinity = 0.5–18 ppt) or saline sites including bayous, cienega, dikes, ditches, dunes, estuaries, flats, floodplains, hummocks, levees, marshes, meadows, ponds, prairies, ravines, riverbottoms, salt marshes, scrub, seeps, sinks, sloughs, springs, swamps, washes, and margins of canals, hot springs, lakes, rivers, and streams at elevations from −63 to 1430 m. The plants can occur on moist substrates or in standing water (to 30 cm) at sites ranging from open (optimal) to shaded exposures. The substrates are alkaline (pH 4.5–7.0) and have been described as alkali silt, clay, granite, gravel, limestone, loamy clay, muck, mud, muddy peat, peat, sand, sandy clay, silty clay, and silty loam. Plants thrive more on sediments of higher organic matter content. Flowering occurs from July to October, but nearly year-round in the southern United States. Although not reported, it is presumed that the flowers are self-incompatible. Regardless, they are likely to be principally outcrossed because of cryptic monoecy and protogynous floral maturation. Pollination occurs by nectar-foraging insects (Insecta) such as bees (Hymenoptera) and butterflies (Lepidoptera). The cypselae are equipped with a plumose pappus, which facilitates their dispersal by wind. Local dispersal likely occurs by means of floating cypselae as well. Efficient dispersal of *P. odorata* is evidenced by observations that the seeds can account for as much as 31%–42% of the total seed pool dispersed into a site. Seeds do persist to some degree in seed banks, but often at a fairly low frequency (<2%). Although stratification is not necessary in all cases, good germination has been reported for seeds stratified at 4°C for 6 weeks, then exposed to a 14/10h, 25°C/15°C light/temperature regime. Seedlings in the field germinate better on moist substrates (1–60 plants/m²) than on thoroughly saturated sites (0–0.4 plants/m²). When growing as perennials, the plants can reproduce vegetatively by rhizomes. The roots often are colonized to a large extent (e.g., 69.0%) by arbuscular mycorrhizae and dark septate endophyte fungi. **Reported associates:** *Abronia maritima, Acacia, Acer rubrum, Achillea millefolium, Acmispon americanus, Acrostichum danaeifolium, Adiantum, Aeschynomene virginica, Agalinis maritima, A. purpurea, Almutaster pauciflorus, Alopecurus saccatus, Alternanthera philoxeroides, Amaranthus albus, A. australis, A. cannabinus, A. palmeri, Amblyopappus pusillus, Ambrosia artemisiifolia, A. chamissonis, A. monogyra, A. psilostachya, A. trifida, Ammannia coccinea, A. latifolia, Andropogon virginicus, Anemopsis californica, Annona glabra, Apios americana, Apium graveolens, Apocynum cannabinum, Artemisia californica, Arthrocnemum subterminale, Arundo donax, Asclepias incarnata, Athyrium filix-femina, Atriplex lentiformis, A. patula, A. prostrata, Avicennia germinans, Azolla caroliniana, Baccharis douglasii, B. halimifolia, B. neglecta, B. pilularis, B. salicifolia, B. salicina, B. sarothroides, B. sergiloides, Bacopa monnieri, Bassia hyssopifolia, Batis maritima, Berula erecta, Bidens frondosus, B. laevis, B. trichospermus, Blechnum serrulatum, Blutaparon vermiculare, Boehmeria cylindrica, Boerhavia, Bolboschoenus maritimus, B. novae-angliae, B. robustus, Borrichia frutescens, Bothriochloa barbinodis, Brassica nigra, Brickellia californica, Bromus rubens, B. hordeaceus, Calamagrostis canadensis, Callicarpa americana, Calystegia sepium, Camissoniopsis cheiranthifolia, Carex annectens, C. blanda, C. longii, C. obnupta, C. pellita, C. praegracilis, Celtis laevigata, Cenchrus ciliaris, C. clandestinus, Cephalanthus occidentalis, Chasmanthium laxum, Chenopodium berlandieri, C. glaucum, C. pratericola, C. rubrum, Cicuta maculata, Cirsium horridulum, C. hydrophilum, C. vulgare, Cladium jamaicense, Clematis ligusticifolia, Comarum palustre, Conium maculatum, Conoclinium coelestinum, Conyza canadensis, C. floribunda, Cortaderia selloana, Cornus foemina, C. sericea, Cotula australis, Crypsis schoenoides, Cuscuta salina, C. subinclusa, Cycloloma atriplicifolium,* *Cynodon dactylon, Cyperus dentatus, C. difformis, C. erythrorhizos, C. filicinus, C. involucratus, C. laevigatus, C. odoratus, C. onerosus, C. polystachyos, C. pygmaeus, C. spectabilis, Datura wrightii, Decodon verticillatus, Deschampsia cespitosa, Dichanthelium acuminatum, Diodia virginiana, Diplacus aurantiacus, Distichlis spicata, Dudleya lanceolata, Echinochloa colona, E. crus-galli, E. walteri, Eclipta prostrata, Eichhornia crassipes, Eleocharis fallax, E. flavescens, E. geniculata, E. macrostachya, E. montevidensis, E. obtusa, E. palustris, E. parvula, E. rostellata, Elymus trachycaulus, Encelia farinosa, Epilobium ciliatum, Equisetum laevigatum, Erechtites hieraciifolius, Eriocaulon aquaticum, Eriochloa villosa, Eriogonum fasciculatum, Eucalyptus camaldulensis, Eupatorium capillifolium, E. perfoliatum, E. serotinum, Euphorbia prostrata, Eurybia radulina, Eustoma exaltatum, Euthamia graminifolia, E. occidentalis, Ficus aurea, Fimbristylis caroliniana, F. spadicea, Flaveria campestris, Forestiera pubescens, Fraxinus caroliniana, Fuirena simplex, Galium obtusum, G. triflorum, Glebionis coronaria, Gratiola aurea, Grindelia hirsutula, Helianthus annuus, Heliomeris multiflora, Heliotropium curassavicum, Helminthotheca echioides, Heteromeles arbutifolia, Hibiscus moscheutos, Hirschfeldia incana, Hydrocotyle ranunculoides, H. umbellata, H. verticillata, Hypericum mutilum, Ilex cassine, I. myrtifolia, Impatiens capensis, Imperata brevifolia, Ipomoea sagittata, Iris pseudacorus, Isocoma acradenia, I. menziesii, Iva annua, I. frutescens, Jaumea carnosa, Juncus acuminatus, J. acutus, J. articulatus, J. balticus, J. canadensis, J. cooperi, J. effusus, J. hybridus, J. interior, J. militaris, J. roemerianus, J. scirpoides, J. texanus, J. torreyi, J. xiphioides, Keckiella, Kochia scoparia, Kosteletzkya pentacarpos, Laguncularia racemosa, Larrea, Lathyrus jepsonii, Leersia hexandra, L. oryzoides, L. virginica, Lemna minor, Lepidium latifolium, Leptochloa fusca, L. viscida, Lilaeopsis chinensis, L. masonii, Limnobium spongia, Limonium californicum, L. carolinianum, Limosella australis, Lobelia cardinalis, Ludwigia alternifolia, L. grandiflora, L. hexapetala, L. octovalvis, L. palustris, L. peploides, L. peruviana, L. repens, Lycium californicum, L. carolinianum, Lycopus americanus, L. asper, Lyonia mariana, Lysimachia maritima, Lythrum alatum, L. californicum, L. lineare, Magnolia virginiana, Malosma laurina, Melilotus albus, Mentha arvensis, M. spicata, Mesembryanthemum nodiflorum, Mikania scandens, Mimulus cardinalis, M. guttatus, Morus alba, Muhlenbergia asperifolia, M. fragilis, Myoporum laetum, Myrica cerifera, M. pensylvanica, Myriophyllum spicatum, Myrsine floridana, Najas guadalupensis, Nelumbo lutea, Nephrolepis, Nicotiana glauca, Nitrophila occidentalis, Nymphaea odorata, Nymphoides cordata, Oenanthe sarmentosa, Oenothera jamesii, Osmunda regalis, Panicum capillare, P. hemitomon, P. virgatum, Parkinsonia florida, Parthenocissus quinquefolia, Paspalum laeve, P. minus, P. urvillei, P. vaginatum, Peltandra virginica, Peritoma arborea, Persea palustris, Persicaria amphibia, P. arifolia, P. hydropiperoides, P. maculosa, P. pensylvanica, P. punctata, P. sagittata, Phragmites australis, Phyla lanceolata, P. nodiflora, Pistia stratiotes, Plantago major, P. maritima, Platanus racemosa, Pleopeltis* *polypodioides, Pluchea sericea, Polygonum aviculare, P. ramosissimum, Polypogon interruptus, P. monspeliensis, P. viridis, Pontederia cordata, Populus fremontii, Portulaca oleracea, Potentilla anserina, Prosopis glandulosa, P. velutina, Pseudognaphalium luteoalbum, Psychotria nervosa, Ptilimnium capillaceum, Puccinellia fasciculata, Pyrrhopappus pauciflorus, Quercus agrifolia, Q. laurifolia, Ranunculus sceleratus, Raphanus, Rayjacksonia annua, Rhamnus crocea, Rhizophora mangle, Rhus integrifolia, Rhynchospora capitellata, R. colorata, R. corniculata, Ricinus communis, Rosa palustris, Roystonea regia, Rubus argutus, R. trivialis, R. ulmifolius, Rumex altissimus, R. californicus, R. crispus, R. dentatus, R. maritimus, Sabal palmetto, Sabatia grandiflora, S. stellaris, Sacciolepis striata, Sagittaria lancifolia, S. latifolia, S. montevidensis, Salicornia bigelovii, S. depressa, Salix caroliniana, S. exigua, S. gooddingii, S. laevigata, S. lasiolepis, Salsola collina, S. tragus, Salvia apiana, S. mellifera, Salvinia minima, S. molesta, Sambucus nigra, Samolus ebracteatus, S. valerandi, Sanguisorba minor, Sarcocornia pacifica, S. perennis, Saururus cernuus, Schoenoplectus acutus, S. americanus, S. californicus, S. pungens, S. tabernaemontani, Scirpus microcarpus, Scutellaria galericulata, Sesbania drummondii, S. herbacea, Sesuvium portulacastrum, S. verrucosum, S. maritimum, Setaria magna, Sisyrinchium demissum, Sium suave, Smilax bona-nox, Solanum douglasii, S. ptychanthum, Solidago gigantea, S. sempervirens, Sonchus asper, S. oleraceus, Sparganium eurycarpum, Spartina alterniflora, S. cynosuroides, S. patens, S. pectinata, S. spartinae, Spergularia salina, Sphagnum, Sphenopholis obtusata, Spirodela, Sporobolus airoides, S. texanus, S. wrightii, Stachys albens, Steinchisma hians, Strophostyles leiosperma, Stuckenia pectinata, Suaeda calceoliformis, S. linearis, Symphyotrichum lentum, S. subulatum, Tamarix aphylla, T. chinensis, T. ramosissima, Taxodium distichum, Thelypteris interrupta, T. palustris, Tillandsia, Toxicodendron radicans, Triadica sebifera, Triglochin maritimum, Typha angustifolia, T. domingensis, T. latifolia, Urtica dioica, Utricularia subulata, Vaccinium corymbosum, Verbena bonariensis, V. hastata, V. lasiostachys, Vernonia baldwinii, Veronica anagallis-aquatica, Vigna luteola, Vitex, Vitis girdiana, Washingtonia filifera, Woodwardia areolata, Xanthium strumarium, Zannichellia palustris, Zizania aquatica, Zizaniopsis miliacea.*

Pluchea sagittalis (Lam.) Cabrera has been reported occasionally from wet, open historical ballast disposal sites, riparian habitats and roadsides at elevations of up to 10 m. Soils typically are clay or sand. Flowering occurs from July to August. In its indigenous tropical range, the flowers are visited (and likely pollinated) by generalist bees and wasps (Insecta: Hymenoptera). The seeds require no stratification for germination. Studies of native sites in Argentina, Brazil, and Puerto Rico consistently have found only a minimal seed bank (0–9 seeds/m^2) to be present. The plants are highly lead tolerant with the roots able to accumulate up to 6730 µg Pb/g and the shoots up to 550 µg Pb/g dry weight. Studies in Argentina show that the plants can increase substantially following fires, which remove dense, competing grassland vegetation. This species is considered to be adventive in North America and should be monitored accordingly. **Reported associates (North America):** none

Use by wildlife: The inflorescences of *Pluchea foetida* and *P. odorata* are hosts to the sourbush seed fly (Insecta: Diptera: Tephritidae: *Acinia picturata*). *Pluchea foetida* is a host to assassin bugs (Insecta: Hemiptera: Reduviidae: *Apiomerus crassipes*), fruit flies (Insecta: Diptera: Tephritidae: *Dyseuaresta mexicana, Neaspilota punctistigma*), and predatory mites (Acari: Phytoseiidae: *Typhlodromalus peregrinus*). The flowers provide nectar and pollen for honeybees (Insecta: Hymenoptera: Apidae: *Apis mellifera*) as well as gray hairstreaks and tiger swallowtail butterflies (Insecta: Lepidoptera: Papilionidae: *Papilio glaucus*; Lycaenidae: *Strymon melinus*). *Pluchea longifolia* provides habitat for the cotton mouse (Mammalia: Cricetidae: *Peromyscus gossypinus*) and the plants are eaten by the eastern woodrat (Mammalia: Cricetidae: *Neotoma floridana*). *Pluchea odorata* provides habitat for seaside sparrows (Aves: Emberizidae: *Ammodramus maritimus*). The flowers are visited by small flies (Insecta: Diptera), katydids (Insecta: Orthoptera: Tettigoniidae), and butterflies (Insecta: Lepidoptera) such as blues (Lycaenidae), palmetto skippers (Hesperiidae: *Euphyes arpa*), and redbanded hairstreaks (Lycaenidae: *Calycopis cecrops*). They also attract a large assortment of bees (Insecta: Hymenoptera: Andrenidae: *Perdita octomaculata*; Apidae: *Bombus fervidus, B. griseocollis, B. impatiens, Ceratina calcarata, C. dupla, Epeolus lectoides, E. scutellaris, Melissodes bimaculata, Xylocopa virginica*; Colletidae: *Colletes americanus, C. speculiferus, Hylaeus affinis, H. mesillae, H. modestus, H. schwarzii*; Halictidae: *Agapostemon sericeus, A. splendens, A. texanus, A. virescens, Augochlora pura, Augochlorella aurata, Halictus ligatus, H. rubicundus, Lasioglossum oceanicum, L. nymphale, L. pilosum, L. zephyrum*; Megachilidae: *Anthidium maculifrons, Coelioxys germana, Megachile gemula, M. latimanus, M. mendica*). The plants also host various larval insects (Insecta: Lepidoptera: Arctiidae: *Tyria jacobaeae*; Bucculatricidae: *Bucculatrix plucheae*; Geometridae: *Synchlora frondaria*; Noctuidae: *Eublemma minima*; Pterophoridae: *Oidaematophorus lienigianus*; Pyralidae: *Unadilla floridensis*) and mites (Acari: Tenuipalpidae: *Brevipalpus obovatus*). The latter are capable of transmitting plant pathogenic viruses. The plants also harbor the phytopathogenic bacterium *Xylella fastidiosa* (Proteobacteria: Xanthomonadaceae), which is another vector of several plant diseases. *Pluchea sagittalis* contains eudesmanes, which have been shown to deter feeding by polyphagous insect larvae. In its native South American range, the flowers attract beetles (Insecta: Coleoptera: Cerambycidae: *Apagomerella versicolor*), predatory crab spiders (Arachnida: Araneae: Philodromidae; Thomisidae), stingless bees (Insecta: Hymenoptera: Apidae: Meliponini), and wasps (Insecta: Hymenoptera: Masaridae: *Trimeria rachiphorus*); the seeds are eaten by finches (Aves: Fringillidae: *Poospiza melanoleuca*). Plaster bees (Insecta: Hymenoptera: Colletidae: *Hylaeus polifolii*) reportedly have been collected from plants in California; however, this account probably is erroneous, given there are no other records of the plants in the state. **Economic importance: food:** *Pluchea* is not palatable for human consumption; **medicinal:** *Pluchea* species have been used widely in traditional medicine to treat numerous disorders. Several *Pluchea* species (*P. foetida*, *P. sagittalis*) contain eudesmane-type sesquiterpenoids, which exhibit antibacterial, antifungal, and antitumor properties. The Choctaw prepared a leaf decoction of *P. foetida* (*Hoshukkosona*) to alleviate fever. Dichloromethane extracts of *P. odorata* contain substances that exhibit anticancer activity and inhibit the inflammatory response. Leaf extracts of *P. odorata* exhibit strong antioxidant properties and are used in the Carribean, Central America and South America as a remedy for colds, digestive problems, migraines, and other ailments. Ethanolic and ethanolic/aqueous extracts of *P. odorata* inhibit growth of yeast (fungi: Ascomycota: Saccharomycetaceae: *Candida krusei*) at 200 µg/ml. *Pluchea sagittalis* is well known as a medicinal plant throughout tropical America. It has been used as an analgesic and to remedy digestive disorders and the shoot, leaf and flower extracts have been shown to alter the absorptive characteristics of the gastrointestinal mucosa. Dichloromethane and aqueous extracts of *P. sagittalis* have strong anti-inflammatory properties, which has been attributed to their potent antioxidant activity and ability to inhibit expression of hsp72 (a heat shock protein) in human neutrophil leukocytes. Leaf extracts also exhibit substantial antiproliferative activity, which has stimulated interest in their potential use for treatment of cancer cells; **cultivation:** *Pluchea odorata* ("sweet-scent") is cultivated as a garden plant in the United States and Cuba; **misc. products:** An aromatic and light amber honey is derived from *Pluchea foetida* and *P. odorata*. *Pluchea odorata* is known for the unusual frost ribbons ("crystallofolia"), which appear on the shoots during frost events. It is a source of cuauhtemone, which is a growth inhibitor of corn and bean seeds. Lead tolerance in *P. sagittalis* has made it a candidate for phytoremediation of sites contaminated by heavy metals. In Brazil, the plants are of religious significance to the Candomblé; **weeds:** *Pluchea odorata* is regarded as a weed in most of its nonindigenous range; **nonindigenous species:** The South American and West Indian *P. sagittalis* was introduced to Alabama and Florida in the 19th century as a result of disposed shipping ballast; however, a lack of recent records indicates that it has not spread in North America since its initial introduction. *Pluchea odorata* has become naturalized in Hawaii, where it is under biological control by a flower-head feeding fly (Diptera: Tephritidae: *Acinia picturata*). It also has become a naturalized weed in China and Taiwan. **Systematics:** *Pluchea* is assigned to tribe Inuleae, a placement consistent with phylogenetic analyses of chloroplast RFLP and *ndhF* sequence data. Morphological data (from 27 Old World and one New World species) indicate that *Pluchea* is monophyletic and represents the sister group to *Streptoglossa* and *Karelinia*. However, *ndhF* data resolve 14 sampled species within a clade interspersed among members of *Cylindrocline*, *Doellia*, *Epaltes*, *Karelinia*, *Laggera*, *Porphyrostemma*, *Pseudoconyza*, *Sphaeranthus*, *Tessaria*, *Streptogliossa*, and

Coleocoma (Figure 5.168). It is evident from the conflicting results between morphological and molecular data that the taxonomic circumscription of *Pluchea* deserves further evaluation. None of the OBL species has yet been included in a comprehensive molecular phylogenetic study, leaving their interrelationships unsettled. A comprehensive molecular phylogenetic analysis of *Pluchea* would help to evaluate the monophyly of the genus and its interspecific relationships. The base chromosome number for *Pluchea* is $x = 10$. *Pluchea odorata* and *P. sagittalis* ($2n = 20$) are diploids. No counts are reported for *P. foetida* or *P. longifolia*. Sterile, but vegetatively reproducing hybrids of *P. indica* × *P. odorata* occur spontaneously in Hawaii, where both species have been introduced. **Comments:** *Pluchea odorata* is distributed widely across the southern half of North America, extending through Mexico to

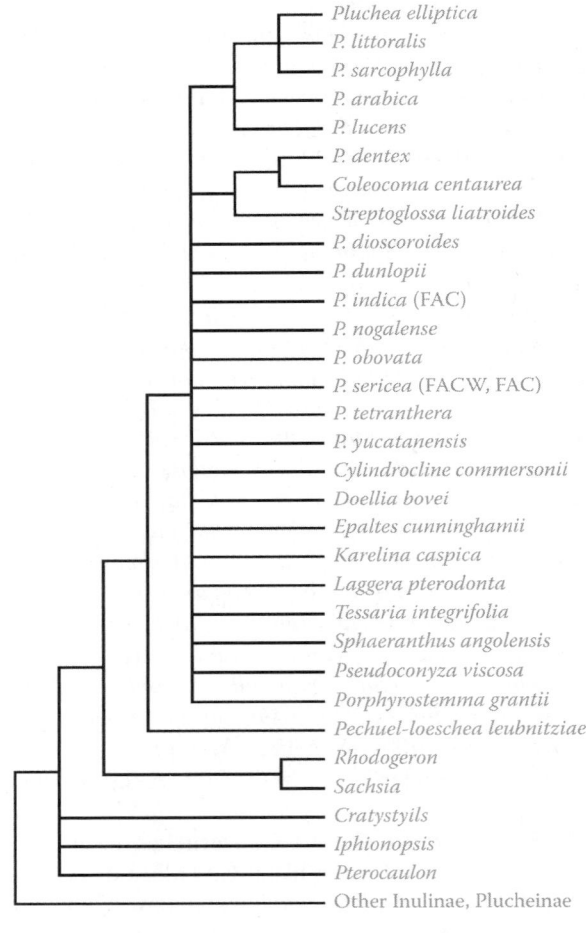

FIGURE 5.168 Phylogenetic relationships of *Pluchea* species and other members of Asteraceae tribe Inuleae based on analysis of *ndhF* sequence data. Although morphological data resolve *Pluchea* as monophyletic (King-Jones, 2001), the *ndhF* data do not. Rather, they reveal the extremely close relationship of *Pluchea* and a number of other genera in the tribe and the need to reevaluate the generic circumscriptions currently in use. None of the *Pluchea* taxa sampled in this analysis represented an OBL species, leaving their relationships unresolved. (Adapted from Anderberg, A.A. et al., *Org. Divers. Evol.*, 5, 135–146, 2005.)

South America; *P. foetida* is found throughout the southeastern portion of North America and extends to Mexico and the West Indies. *Pluchea longifolia*, is restricted to Florida. The non-indigenous *P. sagittalis* has been introduced to Alabama and Florida from South America or West Indies. There is an unconfirmed, indirect record from Riverside, California, as the host plant of a plaster bee specimen (see *Use by wildlife* earlier).

References: Ahemd & Kamel, 2013; Alyokhin et al., 2001; Anderberg, 1991; Anderberg et al., 2005; Baldwin et al., 1996, 2010; Battaglia et al., 2007; Benjamin, 1934; Biabiany et al., 2013; Bohlmann & Mahanta, 1978; Carpenter et al., 2006; Chapla & Campos, 2011; Childers & Denmark, 2011; Childers et al., 2003b; Cooperrider & Galang, 1965; Córdova et al., 2010; De la Peña, 2011; Deyrup et al., 2002; Elam et al., 2009; Eldridge, 1975; Evens & San, 2005; Feldman & Lewis, 2005; Feldman et al., 2007; Ferren, Jr. & Schuyler, 1980; Figueredo et al., 2011; Glenn et al., 2006; Godfrey & Wooten, 1981; Goyal & Aggarwal, 2013; Gridling et al., 2009; Hicks, 2003; Hopfensperger et al., 2009; Hunt, 1943; Ivie & Watson, 1974; Kalmbacher et al., 2005; Keeley & Jansen, 1991; Kern et al., 2012; King-Jones, 2001; Logan, 1973; Logarzo et al., 2010; Makings, 2013; Marrero et al., 2013; Monty & Emerson, 2003; Morton, 1964; Nesom, 2006e; Nesom & O'Kennon, 2008; Nichols et al., 2013; Nolfo-Clements, 2006; Nyman, 2011; Pearson, 1954; Peng et al., 1998; Pérez-García et al., 1996, 2001; Platt et al., 2006; Porter, Jr., 1967; Rasser et al., 2013; Rossato et al., 2010, 2012; Schmalzer, 1995; Singhurst & Holmes, 2012; Smith et al., 2002; Srivastava & Shanker, 2012; Stegmaier, Jr., 1967b; Steinberg, 1980; Stevens et al., 2010; Strong & Kelloff, 1994; Stubbs et al., 1992; Stuckey & Gould, 2000; Taylor, 2012; Vasey et al., 2012; Vera et al., 2008; Vincent et al., 2003; Vinha et al., 2011; Voeks, 1997; Vossler, 2012; Ward et al., 2013; Warner, 1926; Watanabe et al., 2007; Wilson, 2011; Yuan et al., 2010; Zhou et al., 2010.

37. *Psilocarphus*

Woolly marbles, woollyheads

Etymology: from the Greek *psilos karphos* ("smooth chaff"), with respect to the papery pales

Synonyms: none

Distribution: global: Mexico; North America; South America; **North America:** western

Diversity: global: 5 species; **North America:** 5 species

Indicators (USA): OBL: *Psilocarphus oregonus*, *P. tenellus*

Habitat: freshwater (or slightly brackish); palustrine; **pH:** 6.3–8.1; **depth:** <1 m; **life-form(s):** emergent herb

Key morphology: stems (to 20 cm) 2–10, ascending to prostrate, greenish to silvery, downy or silky; leaves (to 20 mm) cauline, opposite, linear to narrowly oblanceolate, obovate, or spatulate, the uppermost spreading around or appressed to heads; heads (to 6 mm) solitary, globose, involucre lacking; pistillate pales (to 2.7 mm) wooly, dehiscent, enclosing cypselae as a sac-like hood at maturity, staminate pales lacking; pistillate florets (8–100) with filiform corollas; staminate florets 2–10, their corollas (to 1.5 mm) 4- or 5-lobed; cypselae (to 1.2 mm) brown, cylindric or compressed, shiny, pappus absent.

Life history: duration: annual (fruit/seeds); **asexual reproduction:** none; **pollination:** self; **sexual condition:** gynomonoecious (cryptic); **fruit:** cypselae (common); **local dispersal:** cypselae (water); **long-distance dispersal:** cypselae (animals) **Imperilment:** (1) *Psilocarphus oregonus* [G4]; S3 (ID); (2) *P. tenellus* [G4]; S2 (ID); S3 (BC)

Ecology: general: All five *Psilocarphus* species are wetland indicators, with three rated as FACW and two as OBL. The breeding system of *Psilocarphus* deserves further study. The heads are cryptically gynomonoecious, with pistillate outer florets and a few functionally staminate inner florets. The heads are rayless and the corollas of the pistillate florets are filiform. The flowers presumably inbreed through self-pollination because they lack attractive floral displays and have a close stigma to pollen proximity. There also has been the suggestion that the plants might be apomictic. The abundance of plants has been correlated with annual precipitation patterns, but specific mechanistic details are few. The seeds are dispersed subsequent to the first fall rains. Germination initiates in spring, when evaporating waters begin to warm. The cypselae lack a pappus and likely are dispersed locally by water, assisted by the indumentum that clothes the pistillate pales, which eventually enclose the cypselae when mature. Long distance dispersal of cypselae by migrating shore birds has been hypothesized as an explanation for the amphitropical distribution of some species. Viable seeds also have been recovered from the droppings of rabbits (Mammalia: Leporidae), which also likely disperse the plants. These plants are among the most common species found in or near vernal pools, but they also occur in conspicuously dry sites where water has receded. Nonetheless, the species are highly tolerant of inundation, with concomitant stem elongation facilitating their adjustment to increasing water levels. The species often thrive on open, litter-free substrates that result from cattle grazing or other disturbances.

Psilocarphus oregonus **Nutt.** occurs in alluvial fans, depressions, ditches, drying pool beds, estuaries, flats, floodplains, hummocks, marshes, meadows, mudflats, playas, prairies, ruts, slopes, stream beds, swales, vernal pools, wallows, and along the margins of lakes, ponds, and reservoirs at elevations of up to 1700 m. They are most commonly associated with vernal pools, where the depths can range from 1 to 19 cm. Plants also occur commonly in ephemerally wet sites that are drying or have fully dried out. Exposures tend to receive full sun. The substrates are alkaline (often pH 7–8) with plants sometimes venturing into marginally saline sites. They include adobe, ashy silt, basalt, Capay silty clay, clay, clay loam, Cometa–Fiddyment series, Corning gravelly loam, gravel, lava, obsidian rubble, rock, silt, and vertisols. In one analysis, the plants represented 0.1%–8.4% of the total cover and occurred on substrates that were not saline (<2 dS/m), but were generally alkaline (pH 6.3–8.1), comprised 10%–15% clay, 60%–65% silt, and 25% sand, and contained 0.5%–0.8% organic carbon. In another case the percent cover in constructed vernal pools increased more than sevenfold over natural reference populations from which the seed was obtained. Flowering occurs from March to August.

All *Psilocarphus* species presumably self-pollinate; however, their breeding system has not been documented experimentally. The seed ecology also is poorly understood, with moisture being the only reported germination requirement. Seed dispersal also has not been investigated in any detail, but must be quite efficient. The cypselae are believed to be dispersed locally by water and over greater distances by birds (Aves) and possibly by small mammals. The plants also have been found within bales of hay, which suggests an inadvertent, alternative means of seed dispersal. **Reported associates:** *Achillea millefolium, Achyrachaena mollis, Acmispon wrangelianus, Agoseris, Allium geyeri, Alopecurus saccatus, Anthemis cotula, Arctostaphylos, Arnica, Artemisia cana, A. tridentata, A. tripartita, Astragalus tener, Avena fatua, Blennosperma nanum, Brodiaea terrestris, Bromus hordeaceus, Calandrinia ciliata, Callitriche marginata, Castilleja attenuata, C. campestris, C. exserta, Ceanothus cuneatus, Centromadia fitchii, Cerastium glomeratum, Chrysothamnus, Cicendia quadrangularis, Claytonia perfoliata, Cotula coronopifolia, Crassula aquatica, C. connata, Cressa truxillensis, Croton setiger, Cryptantha torreyana, Cuscuta howelliana, Deschampsia danthonioides, Distichlis spicata, Downingia bicornuta, D. concolor, D. cuspidata, D. insignis, D. ornatissima, D. pusilla, Elatine californica, E. chilensis, Eleocharis acicularis, E. bella, E. macrostachya, E. palustris, Epilobium brachycarpum, E. cleistogamum, E. minutum, Erodium botrys, E. cicutarium, Eryngium aristulatum, E. armatum, E. castrense, E. vaseyi, Eschscholzia californica, Euphorbia, Frankenia salina, Gratiola heterosepala, Grayia spinosa, Grindelia hirsutula, Hedypnois cretica, Hemizonia congesta, Hesperevax caulescens, Holocarpha virgata, Hordeum brachyantherum, H. depressum, H. marinum, H. murinum, Hypochaeris glabra, Ipomopsis, Isoetes howellii, I. orcuttii, Juncus balticus, J. bufonius, J. capitatus, J. uncialis, Lactuca serriola, Lasthenia californica, L. conjugens, L. glaberrima, L. glabrata, L. fremontii, Layia chrysanthemoides, Legenere valdiviana, Lepidium latipes, Leymus cinereus, Lilaea scilloides, Limnanthes douglasii, Lolium multiflorum, L. perenne, Lomatium farinosum, L. grayi, Lupinus bicolor, Lythrum hyssopifolia, Madia, Malvella leprosa, Marsilea vestita, Medicago polymorpha, Microseris douglasii, M. elegans, Mollugo verticillata, Montia fontana, Myosurus apetalus, M. minimus, Nassella pulchra, Navarretia intertexta, N. leucocephala, Neoholmgrenia andina, Phalaris lemmonii, Pilularia americana, Pinus jeffreyi, P. ponderosa, Plagiobothrys chorisianus, P. greenei, P. hispidulus, P. humistratus, P. leptocladus, P. stipitatus, Plantago aristata, P. bigelovii, P. coronopus, P. elongata, P. maritima, Pleuropogon californicus, Poa annua, P. secunda, Pogogyne douglasii, P. ziziphoroides, Polygonum polygaloides, Polypogon monspeliensis, Potamogeton, Psilocarphus brevissimus, P. tenellus, Quercus douglasii, Q. garryana, Q. wislizeni, Ranunculus muricatus, Rumex crispus, R. pulcher, Sagina apetala, Sarcobatus, Senecio vulgaris, Soliva sessilis, Sonchus oleraceus, Taeniatherum caput-medusae, Taraxacum officinale, Taraxia tanacetifolia, Tetradymia, Trichostema lanceolatum, Trifolium barbigerum, T. bifidum,* *T. depauperatum, T. fucatum, T. gracilentum, T. variegatum, T. willldenovii, T. wormskioldii, Triphysaria eriantha, Veronica peregrina, Vicia sativa, Vulpia bromoides.*

***Psilocarphus tenellus* Nutt.** inhabits barrens, chaparral, depressions, ditches, drying vernal pools, flats, gravel bars, meadows, prairies, puddles, roadbeds, roadsides, scrub, seeps, slopes, streambeds, swales, wallows, washes, woodlands, and margins of arroyos, dune slacks, pools, rivers, and streams at elevations of up to 2650 m. The plants have a relative low tolerance of inundation, being present mainly at sites that were inundated for less than 6–17 days in different studies. They occupy open to partially shaded exposures, typically occurring on bare, alkaline, often compacted substrates that include adobe, basalt, Blanca volcaniclastics, Capay silty clay, chert, clay, clay loam, Conejo volcanics, granite, gravel, gravelly clay, gravelly sand, gravelly sandy loam, Hanford coarse sandy loam, loamy clay, metavolcanic, mud, rock, sand, sandstone, sandy clay loam, sandy loam, serpentine, shale, shaly loam, silt, silty loam, stones, stony Tuscan loam, and Tuscan volcanics. It is not unusual to find plants in more UPL locations, especially in sites that are disturbed by fire or human influences (grazed areas, trails, vehicle ruts, etc.). Flowering occurs from March to August, with self-pollination likely predominating. Details on the reproductive and seed ecology are scarce but seed germination reportedly is suspended during dry years. Heat treated seeds (100°C for 7 min) exhibit no difference in germination compared to untreated seeds. **Reported associates:** *Achyrachaena mollis, Acmispon glabrus, A. junceus, A. strigosus, Adenostoma fasciculatum, Agrostis scabra, Aira caryophyllea, Alopecurus carolinianus, A. saccatus, Anagallis arvensis, A. minima, Aphanes occidentalis, Aralia californica, Artemisia californica, Astragalus gambelianus, Avena barbata, A. fatua, Baccharis pilularis, B. sarothroides, Blennosperma nanum, Bloomeria crocea, Brassica nigra, Bromus diandrus, B. hordeaceus, B. rubens, Callitriche marginata, Callitropsis forbesii, Camissonia campestris, Cardionema ramosissimum, Castilleja attenuata, C. densiflora, Ceanothus cuneatus, C. tomentosus, Centaurea melitensis, C. stoebe, Cerastium glomeratum, Chenopodium, Chlorogalum pomeridianum, Cicendia quadrangularis, Clarkia purpurea, Cneoridium dumosum, Collomia heterophylla, Cotula australis, C. coronopifolia, Crassula aquatica, C. connata, C. tillaea, Cryptantha microstachys, Cynosurus cristatus, C. echinatus, Dactylis glomerata, Deinandra fasciculata, Deschampsia danthonioides, D. elongata, Dicentra chrysantha, Dichelostemma capitatum, D. ida-maia, Distichlis spicata, Downingia bicornuta, D. cuspidata, D. ornatissima, Dryopteris arguta, Elatine brachysperma, Eleocharis acicularis, E. macrostachya, E. palustris, Eriogonum fasciculatum, Erodium botrys, E. cicutarium, E. moschatum, Eryngium armatum, E. castrense, E. spinosepalum, Festuca, Gastridium ventricosum, Gilia, Glinus, Gnaphalium palustre, Gratiola ebracteata, Helianthemum scoparium, Hemizonia congesta, Herniaria hirsuta, Heterocodon rariflorus, Heteromeles arbutifolia, Hirschfeldia incana, Holcus lanatus, Hordeum brachyantherum, H. marinum, H. murinum, Hypochaeris radicata,*

Isoetes orcuttii, Juncus bufonius, J. uncialis, Juniperus californica, Lamarckia, Lasthenia californica, L. fremontii, L. glaberrima, Layia fremontii, L. platyglossa, Lepidium nitidum, L. oblongum, L. virginicum, Leptosiphon bicolor, Lilaea scilloides, Limnanthes douglasii, Linanthus dichotomus, Logfia filaginoides, L. gallica, Lolium multiflorum, L. perenne, Lomatium caruifolium, L. lucidum, Lupinus bicolor, Lythrum hyssopifolia, Malacothamnus densiflorus, Malosma laurina, Marah macrocarpa, Matricaria discoidea, Melica imperfecta, Microseris acuminata, Mimulus guttatus, Mollugo verticillata, Montia fontana, Nassella pulchra, Navarretia intertexta, N. leucocephala, Nemophila maculata, Ophioglossum californicum, Opuntia prolifera, O. littoralis, Pedicularis, Pentachaeta lyonii, Phalaris lemmonii, Pilularia americana, Pinus jeffreyi, P. ponderosa, Plagiobothrys acanthocarpus, P. chorisianus, P. collinus, P. leptocladus, P. nothofulvus, P. scouleri, P. stipitatus, P. tenellus, P. trachycarpus, P. undulatus, Plantago coronopus, P. elongata, P. erecta, P. lanceolata, P. major, Platanus racemosa, Pleuropogon californicus, Poa annua, P. pratensis, Pogogyne douglasii, P. nudiuscula, P. ziziphoroides, Polygonum aviculare, Polypogon monspeliensis, Potentilla menziesii, P. recta, Pseudotsuga menziesii, Psilocarphus brevissimus, P. oregonus, Quercus agrifolia, Q. douglasii, Q. garryana, Q. engelmannii, Ranunculus bonariensis, R. californicus, Rhamnus californica, R. crocea, Rhus integrifolia, Rumex pulcher, Sagina apetala, S. decumbens, Salix, Salvia apiana, S. mellifera, Sanicula tuberosa, Selaginella cinerascens, Spergularia rubra, S. salina, S. villosa, Stylocline gnaphalioides, Taeniatherum caput-medusae, Taraxacum officinale, Thysanocarpus, Toxicodendron diversilobum, Trichostema lanceolatum, Trifolium depauperatum, T. dubium, T. gracilentum, T. hirtum, T. microcephalum, T. repens, T. subterraneum, T. variegatum, Triodanis perfoliata, Triphysaria eriantha, Uropappus lindleyi, Verbena lasiostachys, Veronica peregrina, V. verna, Viguiera laciniata, Vulpia bromoides, V. myuros, Wyethia angustifolia, Xylococcus bicolor.

Use by wildlife: *Psilocarphus tenellus* is a larval host plant of the painted lady butterfly (Insecta: Lepidoptera: Nymphalidae: *Vanessa cardu*).

Economic importance: food: not edible; **medicinal:** no uses reported; **cultivation:** not cultivated; **misc. products:** none; **weeds:** none; **nonindigenous species:** none

Systematics: Placed previously in tribe Inuleae, *Psilocarphus* currently is assigned to tribe Gnaphalieae where it resolves in a clade with *Micropus*; that clade in turn is sister to *Hesperevax* among those genera surveyed (Figure 5.160). Neither the monophyly of the genus nor interspecific relationships have yet been corroborated by DNA sequence data, which currently are available for only one species (*P. brevissimus*; FACW). A natural, intergeneric hybrid between *Hesperevax caulescens* and *Psilocarpha oregonus* has been reported from California. Although malformed and sterile, the hybrid would attest to the relatively close relationship of these genera that has been indicated by molecular phylogenetic analysis. However, this hybridization report bears further evaluation given that the presumed strongly self-pollinating reproductive biology of the

species would reduce the likelihood of such an event. The base chromosome number of *Psilocarphus* is $x=14$. Chromosome numbers have not been reported for *P. oregonus* or *P. tenellus*.

Comments: *Psilocarphus oregonus* and *P. tenellus* occur in western North America.

References: Barbour et al., 2007; Bauder, 2000; Buck, 2004; Collinge et al., 2013; Crowe et al., 1994; Douglas & Illingworth, 2004; Douglas et al., 2003; Keeley & Zedler, 1998; Lazar, 2006; Mason, 1957; Morefield, 1992b,c, 2006b; Odion, 2000; Powell et al., 1974; Pucci, 2007; Witham, 1992; Zedler, 1987.

38. *Raillardella*

Showy raillardella, silvermat

Etymology: a diminutive of the genus *Raillardia*, an orthographic variant named after Laurent Railliard (1792–1845)

Synonyms: *Railliardia*

Distribution: global: North America; **North America:** western United States

Diversity: global: 3 species; **North America:** 3 species

Indicators (USA): OBL: *Raillardella pringlei*

Habitat: freshwater; palustrine; **pH:** alkaline; **depth:** <1 m; **life-form(s):** emergent herb

Key morphology: Stems (to 50+ cm) erect, scapose; leaves in basal rosettes or becoming opposite, sessile, the blades (to 15 cm) linear to lanceolate or obovate, entire to minutely toothed; heads solitary or 2–3 in corymbs, peduncled (to 28 cm), phyllaries 6–13, receptacle with free or fused pales (to 14 mm); ray florets 6–13, pistillate, corolla rays (to 20 mm) orange to red-orange; disc florets 45–80, bisexual, corollas (to 11.5 mm) 5-lobed, orange to red-orange; cypselae (to 8 mm) club shaped, with a pappus of 8–17 bristle-like scales (to 11 mm).

Life history: duration: perennial (rhizomes); **asexual reproduction:** rhizomes; **pollination:** insect; **sexual condition:** gynomonoecious; **fruit:** cypselae (common); **local dispersal:** cypselae (wind); **long-distance dispersal:** cypselae (wind)

Imperilment: (1) *Raillardella pringlei* [G3]; S3 (CA)

Ecology: general: *Raillardella* is a small genus but it is found in a variety of habitats. One species is designated as OBL and another as FACU. A third species is not a wetland indicator but occurs in dry, gravelly sites. The species include both rayed and rayless flower heads, which would imply some differences in their pollination biology; however, information on the life-history of this genus is quite scarce. The ray florets (when present) are pistillate and fertile; the disc flowers are hermaphrodite and fertile. All of the species are self-incompatible. The cypselae are equipped with long pappus scales, which indicates that they are dispersed primarily by the wind. Conditions for seed germination have not been identified.

***Raillardella pringlei* Greene** grows in bogs, fens, meadows, seeps, and along the margins of streams at elevations from 1295 to 2285 m. The plants occur on serpentine-derived soils described as rock or sand. Flowering occurs from July to October. Pollinators have not been identified; however, the brightly colored rays would indicate insects (Insecta) such as bees (Hymenoptera) and butterflies (Lepidoptera) to be likely

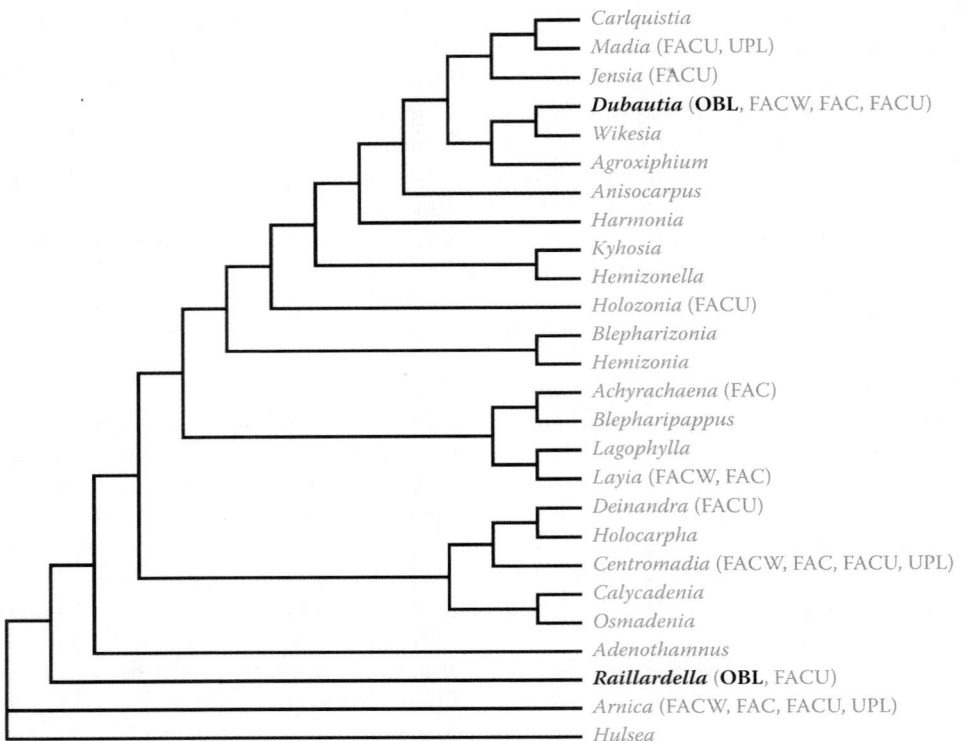

FIGURE 5.169 Phylogenetic relationships within Asteraceae tribe Madieae subtribe Madiinae reconstructed using combined cpDNA and nuclear DNA sequences. Wetland indicators (in parentheses) are rare in this group. OBL species occur only in two genera (bold): the relatively derived *Dubautia* (OBL only in Hawaii) and the more basal *Raillardella*, which indicates independent origins of the habit in the subtribe. (Redrawn from Baldwin, B.G., *Ann. Mo. Bot. Gard.*, 93, 64–93 2006.)

suspects. Vegetative reproduction can occur by the production of rhizomes. Additional life-history information is lacking. **Reported associates:** *Abies ×shastensis, Adiantum, Anemone drummondii, Caltha leptosepala, Carex echinata, C. scopulorum, Cypripedium californicum, Danthonia californica, Darlingtonia californica, Deschampsia cespitosa, Drosera rotundifolia, Eleocharis decumbens, Eriogonum nudum; Eucephalus ledophyllus, Gentianella amarella, Gentianopsis holopetala, Hastingsia alba, Helenium bigelovii, Juncus balticus, J. nevadensis, Narthecium californicum, Parnassia cirrata, P. palustris, Perideridia parishii, Pinus contorta, P. monticola, P. balfouriana, Polystichum lemmonii, Primula jeffreyi, Trianta occidentalis, Veratrum.*
Use by wildlife: none reported
Economic importance: food: not reported as edible; **medicinal:** none reported; **cultivation:** not cultivated; **misc. products:** none; **weeds:** none; **nonindigenous species:** none
Systematics: *Raillardella* belongs to the tribe Madieae, which alternatively has been recognized as a subtribe of tribe Heliantheae. Phylogenetically, it occupies a position near the base of the tribe, close to *Adenothamnus* (Figure 5.169) and other members of the group known as "tarweeds." The base chromosome number of the genus is $x = 17$ or 18; *Raillardella pringlei* ($2n = 34$) is a diploid. Artificial interspecific hybrids (*R. pringlei × R. argentea*; *R. pringlei × R. scaposa*) have been made, with the latter producing normal meiotic bivalents. One

intergeneric hybrid (*Raillardella pringlei × Raillardiopsis muirii* [$2n = 16$]) produced mainly univalents.
Comments: *Raillardella pringlei* is endemic to California.
References: Baldwin, 2006, 2012; Baldwin & Strother, 2006; Baldwin & Wessa, 2000; Ferlatte, 1978; Kyhos et al., 1990; Sikes et al., 2012; Stout, 2012.

39. Rudbeckia

Coneflower; Rudbeckie
Etymology: after Olaus Johannes Rudbeck (1630–1702) and Olaus Olai Rudbeck (1660–1740)
Synonyms: *Centrocarpha*; *Dracopis*
Distribution: global: Europe*; North America; **North America:** widespread
Diversity: global: 23 species; **North America:** 23 species
Indicators (USA): OBL: *Rudbeckia alpicola, R. fulgida, R. mohrii*; **FACW:** *R. alpicola*; **FAC:** *R. fulgida*
Habitat: freshwater; palustrine; **pH:** 5.6–7.9; **depth:** <1 m; **life-form(s):** emergent herb
Key morphology: Stems (to 150 cm) 1–15+, erect, branching distally; leaves alternate, basal and cauline, petioled or sessile, unlobed or pinnate to pinnatifid (3–9-lobed), the blades (to 25–70 mm) elliptic, linear, lanceolate, or ovate, the margins entire, dentate, or serrate; heads discoid or radiate, solitary or arranged secondarily (2–15) in corymbs, phyllaries (to 1–6 cm) persistent, in 1–3 series, receptacles (to 80 mm)

columnar, hemispherical, or ovoid, pales (to 7 mm) present; ray florets (7–15) neuter or absent, the rays (to 30 mm) elliptic to oblanceolate, yellow to yellow-orange; disc florets (100–500) bisexual, corollas (to 5.8 mm) 5-lobed, brownish purple or yellowish green to brownish purple or maroon; cypselae (to 5 mm) black, 4-angled, pappus crown-like or of scales (to 1.2 mm).

Life history: duration: perennial (rhizomes, stolons); **asexual reproduction:** rhizomes, stolons; **pollination:** insect, self; **sexual condition:** hermaphroditic; **fruit:** cypsela (common); **local dispersal:** cypselae (gravity, water), rhizomes, stolons; **long-distance dispersal:** cypselae (birds)

Imperilment: (1) *Rudbeckia alpicola* [G5T3T4]; (2) *R. fulgida* [G5]; S1 (DE, ON); S2 (DC); S3 (MD, PA, WV); (3) *R. mohrii* [G4]

Ecology: general: Most *Rudbeckia* species grow in meadows, prairies, or along streams, where moisture conditions can be fairly high. Seventeen of the species (74%) are wetland indicators with three (13%) ranked as OBL. The genus contains annuals, biennials and perennials, but the three OBL species are perennial. Both rayed and rayless species occur; however, the ray florets always are neuter. The flowers either can be self-compatible and self-pollinated, or self-incompatible and outcrossed. Cross-pollination occurs by insects (Insecta). *Rudbeckia* species also are known for agamospermous reproduction via meiotic diplospory. Their cypselae lack a pappus for wind dispersal and appear to be dispersed mainly by gravity. Transport of cypselae by water also is likely but has not yet been verified. The seeds are eaten readily by a variety of birds (Aves), which are believed to facilitate longer dispersal distances. The seeds also are a contaminant of grass and forage, which has resulted in their inadvertent dispersal through human agricultural activities.

Rudbeckia alpicola **Piper** grows in bogs, meadows, roadsides, seeps, thickets, and along the margins of streams at elevations of up to 1500 m. The inflorescences contain only disc florets, which are produced from mid-summer through fall. Thick rhizomes provide a means of vegetative reproduction. Despite its rarity, there is virtually no life-history information reported for this species. **Reported associates:** none.

Rudbeckia fulgida **Aiton** inhabits bogs, ditches, fens, flatwoods, glades, meadows, oak openings, prairies, prairie-fens, seeps, and woodlands at elevations of up to 700 m. This is a widespread species that occurs under a variety of conditions from dry to wet sites in full sun to shade. The substrates are characterized by a wide range of acidity (pH 5.6–7.9) but often are alkaline and high in calcium (e.g., 15,427 ppm). They are described as clay, clay loam, limestone, loamy sand, sand, sandy peat, Sumter, and Talbott. Much information is available on this species because of its many horticultural applications. The juvenile phase of growth ends at production of the 10th node. Flowering occurs from July to October. The plants become taller and produce more and larger flowers when grown at cooler temperatures (18°C–21°C), but take longer to flower than those grown at warmer temperatures (24°C–27°C). These are long-day plants with a critical daylength of 14 h. A

cold treatment is not necessary to promote flowering; however, pretreatment (15 weeks at 5°C) reduces the critical daylength to 13 h and can accelerate flowering by 25–30 days. Applications of 25–100 ppm of the plant growth regulator CYC (cyclanilide) are shown to increase shoot development, flowering, and plant width, but to delay flowering. Most varieties are self-incompatible, but some commercial varieties reportedly can be self-compatible or apomictic. Diplosporic agamospermy has been documented in the tetraploid *R. fulgida* var. *sullivantii*. Field studies have indicated plants to be mainly self-incompatible and outcrossed by insects (Insecta), which primarily are bees (Hymenoptera), butterflies (Lepidoptera), and flies (Diptera). The ray florets display a high degree of ultraviolet reflectance, which presumably functions as a guide for pollinators. There is a low level of self-compatibility (or apomixis) with 12.5%–14.3% seed set occurring in bagged heads or from geitonogamous pollinations respectively. The percent seed set in natural populations generally is high (26.1–69.5%) with most exceeding 45%. Presumably a seed bank is formed, but there are no specific details in the literature. Germination requirements can vary geographically. Seeds should be sown at 30°C, which is the optimal germination temperature. If no germination occurs within 4 weeks, they should be cold stratified (−4°C to +4°C) for 2–4 weeks. Germination of recalcitrant seeds can be improved significantly by "osmotic priming". Compared to unprimed seeds, the germination rate at 30°C was doubled and the time to germination halved for seeds primed in −1.3 MPa KNO$_3$. The plants propagate vegetatively by rhizomes or by stolons, which produce rosettes at their tips. Although categorized as OBL in some regions, this species is fairly drought resistant, displaying low leaf water potential at stomatal closure, low lethal water potential and a relatively high leaf osmotic adjustment during severe drought conditions. The plants also are salt-tolerant, surviving well for over 4 weeks at concentrations of 0.25 M NaCl. These physiological adaptations undoubtedly enable plants to colonize drier areas as well, accounting for its wide distribution and extensive list of plant associates. Mycorrhizae have been found on plants but only in scarce amounts. **Reported associates:** *Acer rubrum, Adiantum capillus-veneris, Aesculus pavia, Agalinis auriculata, A. tenuifolia, Allium cernuum, Alnus serrulata, Andropogon gerardii, A. glomeratus, A. virginicus, Anemone virginiana, Anemonella thalictroides, Angelica venenosa, Apios americana, Aquilegia canadensis, Aristida, Arnoglossum plantagineum, Aruncus dioicus, Arundinaria gigantea, Asclepias hirtella, A. incarnata, A. syriaca, A. variegata, A. verticillata, A. viridiflora, A. viridis, Asimina parviflora, Athyrium filix-femina, Baccharis glomeruliflora, Baptisia australis, Berchemia scandens, Betula pumila, Bidens, Bignonia capreolata, Blephilia ciliata, Boehmeria cylindrica, Bouteloua curtipendula, Brickellia eupatorioides, Brintonia discoidea, Bromus kalmii, Buchnera americana, Callirhoe papaver, Callitriche heterophylla, Campsis radicans, Cardamine bulbosa, Carex atlantica, C. blanda, C. bromoides, C. cherokeensis, C. crawei, C. eburnea, C. flava, C. hystericina, C. interior, C. intumescens,*

C. leptalea, C. lurida, C. oligocarpa, C. oxylepis, C. radiata, C. sterilis, C. stricta, C. swanii, C. tetanica, C. tribuloides, Carpinus caroliniana, Carya glabra, C. tomentosa, Castilleja coccinea, Cercis canadensis, Chamaecrista fasciculata, Chasmanthium laxum, Chelone glabra, Cinna arundinacea, Cirsium muticum, Cladium jamaicense, Climacium americanum, Clinopodium glabrum, Comandra umbellata, Conoclinium coelestinum, Coreopsis major, C. tripteris, Cornus amomum, C. florida, Cuscuta gronovii, Cyperus, Cypripedium kentuckiense, C. parviflorum, Danthonia spicata, Dasiphora floribunda, Daucus carota, Delphinium carolinianum, Desmodium canescens, D. paniculatum, Dichanthelium boscii, D. clandestinum, D. dichotomum, Diospyros virginiana, Echinacea purpurea, E. simulata, Eleocharis elliptica, E. flavescens, E. tortilis, Elephantopus tomentosus, Elymus virginicus, Equisetum arvense, E. fluviatile, E. hyemale, Eryngium aquaticum, E. integrifolium, E. yuccifolium, Euonymus americanus, Eupatoriadelphus maculatus, Eupatorium album, E. perfoliatum, E. serotinum, Euphorbia commutata, E. corollata, E. pubentissima, Eurybia hemispherica, Eutrochium fistulosum, E. maculatum, Fagus grandifolia, Filipendula rubra, Flaveria, Frasera caroliniensis, Fraxinus pennsylvanica, Fuirena simplex, F. squarrosa, Galactia volubilis, Galium triflorum, Gentiana andrewsii, Gentianella quinquefolia, Gillenia stipulata, Glyceria striata, Gymnadeniopsis clavellata, Hamamelis virginiana, Helenium autumnale, Helianthus angustifolius, H. atrorubens, H. divaricatus, H. grosseserratus, H. hirsutus, H. mollis, H. radula, Heliotropium tenellum, Hexalectris spicata, Hexastylis arifolia, Houstonia canadensis, H. purpurea, Hydrangea quercifolia, Hydrocotyle, Hypericum densiflorum, H. dolabriforme, H. hypericoides, H. sphaerocarpum, Hypoxis hirsuta, Ilex opaca, Impatiens capensis, Ionactis linariifolia, Iris verna, Itea virginica, Jamesianthus alabamensis, Juncus brachycephalus, J. coriaceus, J. dudleyi, J. effusus, J. nodosus, Juniperus virginiana, Leersia oryzoides, L. virginica, Lepuropetalon spathulatum, Lespedeza capitata, L. cuneata, L. hirta, L. violacea, Liatris aspera, L. hirsuta, L. pycnostachya, L. spicata, L. squarrosa, L. squarrulosa, Ligustrum, Lindera benzoin, Linum sulcatum, Liparis liliifolia, Liquidambar styraciflua, Liriodendron tulipifera, Lithospermum canescens, Lobelia cardinalis, L. glandulosa, L. kalmii, L. puberula, L. siphilitica, L. spicata, Ludwigia, Lycopus uniflorus, Lysimachia lanceolata, L. quadriflora, Lythrum alatum, Magnolia tripetala, Manfreda virginica, Marshallia trinervia, Matelea obliqua, Melica mutica, Menispermum canadense, Menyanthes trifoliata, Microstegium vimineum, Mikania scandens, Minuartia patula, Mitchella repens, Mitella diphylla, Mitreola petiolata, Mnesithea, Monarda fistulosa, Muhlenbergia capillaris, Myrica cerifera, Nasturtium officinale, Nyssa sylvatica, Oenothera filipes, Onoclea sensibilis, Onosmodium virginianum, Osmunda regalis, Osmundastrum cinnamomeum, Ostrya virginiana, Oxalis violacea, Oxydendrum arboreum, Oxypolis rigidior, Packera anonyma, P. aurea, Panicum flexile, P. virgatum, Parnassia grandifolia, Paspalum, Pedicularis canadensis, P. lanceolata, Pellaea atropurpurea, Phegopteris hexagonoptera,

Phlox glaberrima, P. pilosa, Physocarpus opulifolius, Physostegia virginiana, Pilea fontana, Pinus echinata, P. taeda, Piptochaetium avenaceum, Pityopsis graminifolia, Plantago rugelii, Platanthera ciliaris, Pogonia ophioglossoides, Polygala boykinii, Polygonatum biflorum, Ponthieva racemosa, Prenanthes serpentaria, Primula meadia, Pseudognaphalium obtusifolium, Pycnanthemum tenuifolium, P. virginianum, Pyracantha, Quercus alba, Q. falcata, Q. muehlenbergii, Q. stellata, Q. velutina, Ratibida pinnata, Rhamnus caroliniana, Rhododendron viscosum, Rhus aromatica, R. glabra, Rhynchospora capillacea, R. capitellata, R. gracilenta, R. glomerata, Rosa carolina, Rubus allegheniensis, Rudbeckia laciniata, R. subtomentosa, Ruellia caroliniensis, R. humilis, Sabatia angularis, Salix caroliniana, S. humilis, S. petiolaris, Salvia lyrata, Schizachyrium scoparium, Schoenoplectus americanus, Schoenus nigricans, Scirpus atrovirens, S. cyperinus, S. pendulus, S. polyphyllus, Scleria muehlenbergii, S. triglomerata, S. verticillata, Scutellaria parvula, Securigera varia, Selaginella apoda, S. ludoviciana, Setaria, Sideroxylon, Silphium asteriscus, S. compositum, S. integrifolium, S. trifoliatum, Sisyrinchium albidum, S. angustifolium, Smilax bona-nox, S. glauca, Solidago canadensis, S. gigantea, S. juncea, S. nemoralis, S. odora, S. patula, S. ptarmicoides, S. rigida, S. rugosa, S. sphacelata, S. uliginosa, Sorghastrum nutans, Spartina pectinata, Spiranthes lucida, S. magnicamporum, Sphagnum, Sporobolus clandestinus, S. compositus, S. heterolepis, S. junceus, S. vaginiflorus, Stachys crenata, Stenaria nigricans, Symphyotrichum dumosum, S. laeve, S. lateriflorum, S. novae-angliae, S. oblongifolium, S. patens, S. pilosum, S. pratense, S. prenanthoides, S. puniceum, Tephrosia virginiana, Thalictrum dasycarpum, T. revolutum, Thelypteris noveboracensis, T. palustris, Thuidium delicatulum, Toxicodendron radicans, Trautvetteria caroliniensis, Trichostema brachiatum, Tridens flavus, Trollius laxus, Ulmus rubra, Utricularia minor, Vaccinium corymbosum, V. elliottii, V. pallidum, Veratrum virginicum, Verbesina virginica, Vernonia missurica, Veronicastrum virginicum, Viburnum rufidulum, Viola cucullata, V. palmata, V. septemloba, Xyris tennesseensis, X. torta, Zizia aptera.

***Rudbeckia mohrii* A. Gray** occupies bayous, borrow pits, depressions, ditches, flatwoods, savannas, sloughs, swales, swamps, and margins of ponds at elevations of up to 50 m. The plants are reported most often growing in shallow, standing water and can occur in exposures ranging from full sun to shaded conditions. Substrates are described as loamy sandy loam, Lynchburg (aeric paleaquults), sand, and Surrency (umbric paleaquults). Flowering occurs from summer through fall with the heads producing both ray and disc florets. The plants are rhizomatous. Other life history information is lacking for this species. **Reported associates:** *Ammannia coccinea, Andropogon virginicus, Asclepias tuberosa, Bacopa, Coleataenia longifolia, C. tenera, Crataegus aestivalis, Cyperus iria, Digitaria ciliaris, Dichanthelium acuminatum, D. sphaerocarpon, Diodia teres, Diospyros virginiana, Drosera brevifolia, Echinochloa colona, Eclipta prostrata, Eriocaulon decangulare, Eryngium prostratum,*

Eutrochium fistulosum, Helenium pinnatifidum, Hydrolea ovata, Hypericum brachyphyllum, Juncus repens, Leersia hexandra, Lobelia floridana, Ludwigia linearis, L. linifolia, Mitreola sessilifolia, Mnesithea rugosa, Mollugo verticillata, Panicum hemitomon, Panicum verrucosum, Pluchea baccharis, Persicaria hydropiperoides, P. lapathifolia, Pityopsis graminifolia, Polygala cymosa, Polypremum procumbens, Quercus virginiana, Rhynchospora filifolia, R. perplexa, R. pleiantha, R. tracyi, Rhexia aristosa, R. mariana, Rumex crispus, Saccharum giganteum, Scleria georgiana, Sesbania herbacea, Sporobolus floridanus, Stylisma aquatica, Taxodium, Urochloa platyphylla, Viola lanceolata.

Use by wildlife: *Rudbecki fulgida* hosts several fungi including Ascomycota (Leotiomycetidae: *Golovinomyces ambrosiae*) and Oomycota (Peronosporaceae: *Basidiophora entospora, Plasmopara halstedii*). The latter can cause a severe leaf disease. The plants also host orb-web building spiders (Arachnida: Araneidae: *Argiope aurantia*) and nematodes (Nematoda: Aphelenchoididae: *Aphelenchoides fragariae, A. ritzemabosi*). The flowers are visited primarily (>80%) by small bees (Insecta: Hymenoptera: Apidae: *Ceratina*; Halictidae: *Lasioglossum*) and butterflies (Insecta: Lepidoptera: Hesperiidae: *Atalopedes campestris*; Nymphalidae; Pieridae), and to a lesser extent (<5%) by bumblebees (Insecta: Apidae) and flies (Insecta: Diptera: Syrphidae). Other reported floral visitors include various bees and wasps (Insecta: Hymenoptera: Andrenidae; Formicidae; Ichneumonidae; Megachilidae; Sphecidae; Vespidae), flies (Diptera: Bombyliidae; Calliphoridae; Lonchaeidae; Stratiomyidae; Tachinidae), beetles (Coleoptera: Buprestidae; Cantharidae; Cerambycidae; Chrysomelidae; Curculionidae; Meloidae; Mordelidae), bugs (Hemiptera: Cercopidae; Miridae; Nabidae; Phymatidae), and butterlies and moths (Insecta: Lepidoptera: Arctiidae; Lycaenidae; Pyralidae; Sesiidae). Flowerheads of the plants are eaten by several larval looper moths (Insecta: Lepidoptera: Geometridae: *Eupithecia miserulata, Chlorochlamys chloroleucaria, Synchlora aerata*) and sunflower moths (Insecta: Lepidoptera: Pyralidae: *Homoeosoma electellum*). The foliage is browsed by mammals (Mammalia) such as white-tailed deer (Cervidae: *Odocoileus virginianus*), rabbits (Lagomorpha: Leporidae), groundhogs (Rodentia: Sciuridae: *Marmota monax*), cattle (Bovidae: *Bos primigenius*), and other domestic farm animals. The plants serve as effective hosts for *Agalinis auriculata* (Plantaginaceae), a rare, parasitic angiosperm.

Economic importance: food: The flowers of *Rudbeckia fulgida* reportedly are edible; **medicinal:** The Cherokee used various preparations of *Rudbeckia fulgida* to treat dropsy, earaches, flux, snakebites, and worms. The plants contain a glucuronoxylan, which has a cough suppressant activity surpassing that of many common nonnarcotic medicines; **cultivation:** *Rudbeckia* species are grown commonly as garden ornamentals. Some nurseries are beginning to propagate *R. alpicola*. Cultivars of *R. fulgida* include 'Anthony Brooks', 'Blovi', 'City Garden', 'Early Bird Gold', 'Goldschirm', 'Goldsturm' [1999 perennial plant of the year award], 'Pot of Gold', 'Sun Baby', and 'Viette's Little Suzie'. *Rudbeckia*

mohrii seeds are available infrequently from wildflower nurseries; **misc. products:** *Rudbeckia fulgida* seeds are common in various "conservation" seed mixes sold for habitat restoration projects. This species also is being considered for use in nutrient removal from storm water and for beautification of urban demolition sites. The surface microstructure of its ray florets has been used to provide engineering insight into the development of artificial surfaces for water harvesting and passive surface transport; **weeds:** none; **nonindigenous species:** none

Systematics: *Rudbeckia* and other "coneflower" genera are assigned to subtribe Rudbeckiinae of tribe Heliantheae in the Asteraceae. *Rudbeckia* occupies a relatively derived position within the subtribe and is divided into three subgenera: *Rudbeckia, Macrocline,* and *Dracopis.* The latter is monotypic and is represented by the former genus *Dracopis.* Three independent origins of the aquatic habit are indicated by the phylogenetic distances separating all three OBL *Rudbeckia* species (Figure 5.170), which occur in two different subgenera. *Rudbeckia alpicola* and *R. mohrii* are placed within subgenus *Macrocline*; whereas, *R. fulgida* is assigned to subgenus *Rudbeckia.* The widespread occurrence of wetland indicators in *Rudbeckia* (especially FACW), would suggest that much of the genus has adapted well to aquatic habitat conditions (Figure 5.170). Taxonomists differ considerably in their interpretation of *R. fulgida.* Although seven varieties generally are recognized, some authors have retained most of those as distinct species (e.g., *R. deamii, R. palustris, R. speciosa, R. sullivantii, R. umbrosa*) or have subdivided some (e.g., *R. fulgida* var. *fulgida*) even further to recognize species such as *R. tenax.* In such interpretations, *R. fulgida* primarily occupies acidic habitats and the calcareous sites are associated with different species (e.g., *R. palustris, R. speciosa, R. sullivantii, R. tenax*). Without detailed genetic studies of infraspecific taxa within *R. fulgida,* it is difficult to address which system represents a better taxonomic scheme. Here, all mentioned varieties have been retained with all pertinent ecological information summarized under the single species *R. fulgida.* Obviously, these taxonomic issues must be considered when referring to the information reported earlier. If *R. fulgida* truly represents several distinct species, then the life history information and roster of associated species would not be reflected accurately. The level of genetic variation (% polymorphic loci assessed using RAPD markers) was found to be high in both natural (14%–69%) and cultivated (40%–63%) populations of *R. fulgida.* The basic chromosome number of *Rudbeckia* is $x = 16$, 18, or 19. *Rudbeckia fulgida* ($x = 19$) includes diploids ($2n = 38$) and tetraploids ($2n = 76$); *R. mohrii* ($x = 18$; $2n = 36$) is diploid. Tetraploid *R. fulgida* plants have a relative 1C DNA content of 8.5–8.9 pg. *Rudbeckia fulgida* can hybridize successfully with *R. hirta* and *R. missouriensis* although spontaneous natural hybrids have not been reported.

Comments: *Rudbeckia fulgida* occurs throughout eastern North America. *Rudbeckia alpicola* is endemic to Washington state and *R. mohrii* occurs only in Georgia and Florida.

References: Adrienne et al., 1985; Aldrich et al., 1981; Baskin et al., 1987; Campbell & Seymour, Jr., 2013; Chapman &

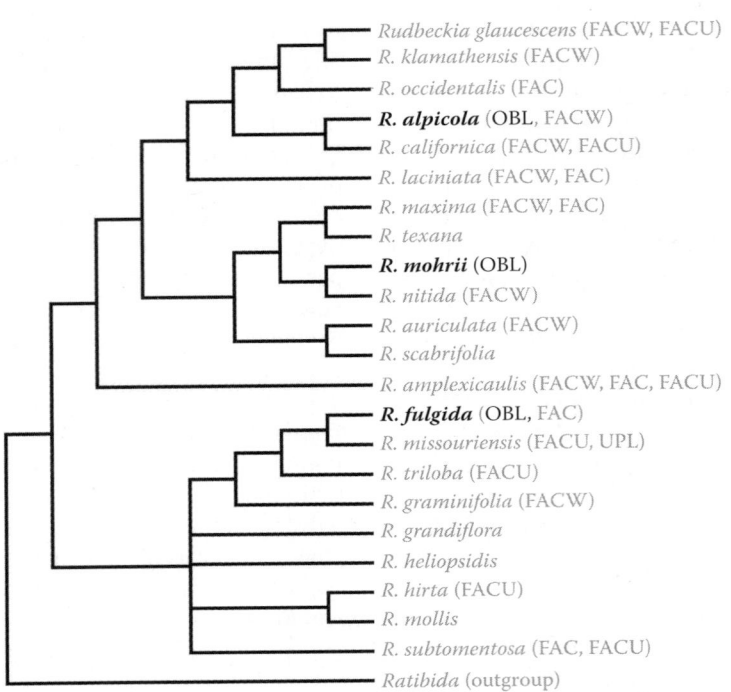

FIGURE 5.170 Interspecific relationships in *Rudbeckia* as reconstructed from combined nrITS sequences and cpDNA RFLP data. The three OBL species (bold) clearly have evolved independently in the genus, which contains numerous wetland indicators (codes in parentheses). These data also distinguish *R. alpicola* from *R. occidentalis*, within which it has been treated as a variety by some authors. The cp and nuclear topologies differ in the position of *R. nitida*, which may indicate ancestral hybridization; however, all three OBL species remain distinct in either tree. (Adapted from Urbatsch, L.E. & Jansen, R.K., *Syst. Bot.*, 20, 28–39, 1995; Urbatsch, L.E. et al., *Syst. Bot.*, 25, 539–565, 2000.)

Augé, 1994; Coletta, 2014; Cox & Urbatsch, 1994; De Groote et al., 2011; Diamond, Jr. et al., 2006; Echols & Zomlefer, 2010; Farris et al., 2011; Fay et al., 1993, 1994; Gerber et al., 2011; Harper, 1903; Hess, 2009; Hong, 2006; Kardošová et al., 1998; Kirkman et al., 2000; Koch et al., 2013; Kohl, 2008; Köppler et al., 2014; McDougall & Glasgow, 1929; Namestnik et al., 2012; Nosál'ová et al., 2000; Noyes, 2007; Palmer et al., 2009; Pill et al., 2000; Richards, 1986; Runkle et al., 1999; Scott & Molano-Flores, 2007; Scott et al., 2007; Singhurst et al., 2007; Stelzer et al., 2010; Stuber, 2013; Urbatsch & Cox, 2006; Urbatsch & Jansen, 1995; Urbatsch et al., 2000; Wu, 2012; Yuan et al., 1998a,b.

40. *Sclerolepis*

Bogbutton
Etymology: from the Greek *sklêros lepis* ("hard scale"), with respect to the pappus
Synonyms: *Ethulia*; *Sparganophorus* (in part)
Distribution: global: North America; **North America:** eastern United States
Diversity: global: 1 species; **North America:** 1 species
Indicators (USA): OBL: *Sclerolepis uniflora*
Habitat: freshwater; lacustrine, palustrine; **pH:** 5.6–6.7; **depth:** 0–3 m; **life-form(s):** emergent or submersed (vittate) herb
Key morphology: Stems (to 60 cm) creeping, erect, or floating; leaves cauline, whorled (3–6), sessile, the blades

(to 23 mm) linear, single nerved, the margins entire; heads discoid, solitary, peduncled (to 6 cm), phyllaries (to 4 mm) persistent, in 2 or more series, receptacles conical or hemisperical, pales lacking; florets (to ~50) bisexual, corollas (to 3 mm) 5-lobed, pink, lavender, or whitish; cypselae (to 3.5 mm) 5-ribbed, crowned by a persistent pappus of five broad scales.
Life history: duration: perennial (rhizomes); **asexual reproduction:** rhizomes, shoot fragments; **pollination:** unknown; **sexual condition:** hermaphroditic; **fruit:** cypselae (common to rare); **local dispersal:** cypselae (vector unknown), stem fragments; **long-distance dispersal:** unknown
Imperilment: (1) *Sclerolepis uniflora* [G4]; S1 (LA, MA, NH, RI, VA); S2 (DE, MD, NC, NJ)
Ecology: general: *Sclerolepis* is monotypic.

Sclerolepis uniflora **(Walter) Britton, Sterns & Poggenb.** is found in bogs, borrow pits, bottomlands, canals, depressions, ditches, flatwoods, levees, ponds, poor fens, savannas, seeps, swales, swamps, and along the margins of lakes, ponds, and rivers at elevations of up to 100+ m. The plants can occur as submersed aquatics or emergent helophytes. The plants occur mainly in the submersed form northward, but almost exclusively in the emergent form in the southern portion of their range. The submerged form is vegetative and has been reported from depths up to 3 m in oligotrophic or mesotrophic lakes. The emergent forms occur on clay, peat, or sand substrates. Flowering occurs from May to October,

and commences increasingly later northward in the range; the plants remain vegetative if they become inundated. There are no reports on pollinators, whether plants are self-incompatible, or other aspects of the reproductive biology. Seed set is abundant in the southern portion of the range, but is rare in the north. Germination requirements are unknown. The plants are rhizomatous and can develop into dense mats. Vegetative reproduction also occurs by plant fragments, which root after washing up on the shore. Plants rarely are able to overwinter in the terrestrial form. No additional life-history information is available. **Reported associates:** *Acer rubrum, Andropogon glomeratus, Aronia arbutifolia, Bacopa caroliniana, Boltonia asteroides, Carex lurida, Centella asiatica, Cephalanthus occidentalis, Cladium mariscoides, Clethra alnifolia, Coleataenia longifolia, Cyperus odoratus, C. polystachyos, Dichanthelium acuminatum, D. dichotomum, D. hirstii, D. sphaerocarpon, Diodia virginiana, Dulichium arundinaceum, Echinochloa muricata, Eleocharis microcarpa, E. olivacea, E. quadrangulata, E. tricostata, E. tuberculosa, Eriocaulon aquaticum, E. compressum, Eryngium aquaticum, Eubotrys racemosa, Eupatorium leptophyllum, E. leucolepis, E. pilosum, Euthamia graminifolia, Fimbristylis autumnalis, Fuirena pumila, Gratiola aurea, Hydrocotyle umbellata, Hypericum canadense, H. cistifolium, H. denticulatum, H. fasciculatum, Gratiola ramosa, Ilex glabra, Juncus canadensis, J. effusus, J. militaris, J. repens, Leersia hexandra, Lindernia dubia, Lobelia boykinii, L. canbyi, L. dortmanna, L. nuttallii, Ludwigia alternifolia, L. linearis, L. sphaerocarpa, L. suffruticosa, Luziola fluitans, Lycopus rubellus, Mnesithea rugosa, Myriophyllum heterophyllum, Nymphaea odorata, Nymphoides aquatica, Nyssa biflora, Oldenlandia uniflora, Panicum dichotomiflorum, P. hemitomon, P. verrucosum, Paspalum laeve, Persicaria hydropiperoides, Pinus taeda, Pluchea bacchais, P. foetida, Polygala cymosa, Pontederia cordata, Potamogeton epihydrus, P. natans, Proserpinaca intermedia, P. palustris, P. pectinata, Ptilimnium nodosum, Rhexia aristosa, R. mariana, R. virginica, Rhynchospora alba, R. filifolia, R. gracilenta, R. harperi, R. inundata, R. macrostachya, R. perplexa, Rubus hispidus, Sabatia difformis, Sagittaria engelmanniana, S. graminea, Sarracenia, Scirpus cyperinus, Scleria georgiana, S. reticularis, Smilax rotundifolia, Sparganium americanum, Sphagnum pylaesii, Symphyotrichum pilosum, Taxodium ascendens, T. distichum, Thelypteris palustris, Toxicodendron radicans, Triadenum virginicum, Utricularia cornuta, U. gibba, U. radiata, Vaccinium corymbosum, Viola lanceolata, Wolffia, Xyris ambigua, X. jupicai, X. smalliana.*
Use by wildlife: *Sclerolepis uniflora* occurs within the nesting habitat of the Florida sandhill crane (Aves: Gruidae: *Grus canadensis pratensis*).
Economic importance: food: unknown; **medicinal:** none reported; **cultivation:** not cultivated; **misc. products:** none; **weeds:** none; **nonindigenous species:** none.
Systematics: The monotypic *Sclerolepis* is assigned to subtribe Trichocoroninae (tribe Eupatorieae) of Asteraceae. Phylogenetic analyses of molecular data have confirmed the

monophyly of Trichocoroninae and its sister relationship to subtribe Mikaniinae. Within Trichocoroninae, *Sclerolepis* resolves consistently and strongly as the sister genus to a clade comprising *Shinnersia* (OBL) and *Trichocoronis* (FACW) (Figure 5.166). The base chromosome number of *Sclerolepis* is $x=15$. *Sclerolepis uniflora* ($2n=30$) is diploid.
Comments: *Sclerolepis uniflora* is endemic to the eastern coastal plain of the United States.
References: Dunlop, 2001; Fernald, 1912; Harrison & Knapp, 2010; King et al., 1976; Lamont, 2006; Ludwig et al., 1986; McAvoy & Wilson, 2014; Tippery et al., 2014; Watanabe et al., 1995.

41. *Senecio*
Groundsel, ragwort; séneçon
Etymology: from the Latin *sĕnex* (aged person), referring to the white pappus bristles
Synonyms: *Aetheolaena; Cadiscus; Culcitium; Dendrophorbium* (in part); *Hasteola* (in part); *Iocenes; Jacobaea* (in part); *Lasiocephalus; Packera; Pseudogynoxys* (in part); *Rhetinodendron; Robinsonia; Symphyochaeta; Synosma; Vendredia*
Distribution: global: cosmopolitan; **North America:** widespread
Diversity: global: 1000+ species; **North America:** 55 species
Indicators (USA): OBL: *Senecio crassulus, S. eremophilus, S. hydrophilus, S. triangularis;* **FACW:** *S. triangularis;* **FAC:** *S. eremophilus;* **FACU:** *S. crassulus, S. eremophilus*
Habitat: brackish, freshwater palustrine; **pH:** 4.3–7.6; **depth:** <1 m; **life-form(s):** emergent herb
Key morphology: Stems (to 70–200 cm) erect, single or loosely clustered (2–4); leaves alternate, basal or cauline and distributed evenly or progressively reduced distally, petioled or sessile, the blades (to 20 cm) elliptic, lanceolate, ovate or triangular, unlobed, pinnately lobed or lacerate, the margins dentate to entire; heads radiate or discoid, solitary or (to 12–80) arranged secondarily in corymbs or racemes, phyllaries (to 10 mm) with black or green tips, receptacles flat or convex, pales absent; ray florets absent or (5–13), pistillate, the rays (to 15 mm) yellow; disc florets (5–80+) bisexual, corollas 5-lobed, yellow; cypselae (to 3.5 mm) glabrous, 5-ribbed, pappus of 30–80 hair-like bristles.
Life history: duration: perennial (persistent caudex, rhizomes); **asexual reproduction:** rhizomes, stolons; **pollination:** insect, self; **sexual condition:** gynomonoecious, hermaphroditic; **fruit:** cypselae (common); **local dispersal:** cypselae (wind), rhizomes; **long-distance dispersal:** cypselae (animals)
Imperilment: (1) *Senecio crassulus* [G5]; S3 (WY); (2) *S. eremophilus* [G5]; S1 (MT); S2 (ND, ON); S3 (WY); (3) *S. hydrophilus* [G5]; SH (BC); S3 (MT, WY); (4) *S. triangularis* [G5]
Ecology: general: With more than 1000 species, *Senecio* ranks among the 10 largest flowering plant genera. This group is extraordinarily diverse and includes herbaceous annuals, biennials, and perennials, subshrubs, and shrubs. Understandably, the species occur across the face of the earth

and occupy a plethora of habitats representing a broad spectrum of arid to wet conditions. Yet, despite this great diversity, there are relatively few aquatics and the genus is not one typically associated with wetlands. Twenty-one of the 55 (38%) North American species are wetland indicators at some level, but only four (7%) are ranked as OBL, and none grows as a submersed life form. Two of the OBL species, *S. crassulus* (OBL, FACU) and *S. eremophilus* (OBL, FAC, FACU), can survive under quite contrasting conditions and actually are fairly common in drier sites as their different regional designations (and lists of associated species) indicate. Ironically, the latter species is commonly known as "desert ragwort". As its name implies, *S. hydrophilus* is the most highly adapted to wetland habitats in the genus. Both self-incompatible and self-compatible species occur, but the reproductive biology has been poorly studied for the OBL species. Their cross-pollination relies on insects (Insecta). Although apomixis has been suspected in a few cases, it has not yet been documented for any *Senecio* species. The heads typically are radiate, although sometimes (e.g., occasionally in *S. hydrophilus*) they lack ray florets. When present, the ray florets are pistillate and fertile while the disc florets are bisexual and fertile. The seeds are equipped with a pappus of numerous bristles, which facilitate wind dispersal over relatively short distances (mainly <5 m) depending primarily on their height of release. Greater dispersal distances can be achieved if the cypselae encounter turbulence induced convection, which raises them higher into the air. The cypselae also can attach to animal fur or to clothing, which potentially provides for greater dispersal distances. The seeds often require a cold stratification treatment to induce germination, except for some of the alpine taxa.

***Senecio crassulus* A. Gray** is a perennial, which colonizes alpine and subalpine depressions, forest openings, gravel bars, gullies, marshes, meadows, playas, seeps, slopes, swales, thickets, tundra, and margins of lakes, ponds, and streams at elevations from 2200 to 3962 m. The plants occur in full sun to partial shade, most often on southern exposures and rarely on western exposures. The substrates often are described as calcareous, but plants occur on substrates having pH values ranging from 4.8 to 7.6. These include Andesitic basalts, basalt, clay, clay loam, clay silt, clayey sand, dolomite, granite, gravel, gravelly rocky loam, limestone, loam, quartizite talus, rock, sandy clay, sandy–gravelly rocky loam, sandy loam, sandy rock, scree, silt, silt loam, stony loam, talus, tuff, and Wasatch limestone. Representative sites have an organic matter content of 14%–25%, mean total nitrogen of 0.31%–0.75%, and C:N ratio of 14:1–28:1. Copper concentrations exceeding 250 ppm are tolerated. Flowering commences within 28–52 days after snowmelt. The plants produce 1–6 heads per scape. The corolla tubes average 4.9–5.8 mm in length and the flowers are pollinated primarily by short- and medium-tongued bumblebees (Insecta: Hymenoptera), which have tongues of similar length (5.7–8.5 mm). Butterflies (Insecta: Lepidoptera) are secondary pollinators. The ratio of bee to butterfly visits varies among sites from about 3:1 to 25:1. Bees visit a greater mean number of plants during each foraging bout (23.75) than do butterflies (3.74). In one study, pollinator visitation correlated

with plant density and also with the occurrence of other bumblebee-pollinated species. However, another study found that pollinator visitation rates were not determined by plant density, but more likely reflected available nectar supplies. A small level of seed set observed in plants where inflorescences were bagged to exclude insects, indicates some degree of self-pollination (or perhaps apomixis). The plants have been categorized as self-compatible. Details on seed germination have not been reported and it is not known whether a seed bank persists. The plants reproduce vegetatively by lateral runners (branching rhizomes), which are 5–10 cm long and occur at 10–15 cm depth. This species is characteristic of naturally disturbed habitats (e.g., low burn severity) and occurrences are dissociated with areas having well-established stands of vegetation. The peak standing crop has been estimated at 2.5 g/m². **Reported associates:** *Abies lasiocarpa, Achillea millefolium, Aconogonon phytolaccifolium, Agoseris glauca, Agrostis variabilis, Alnus viridis, Amelanchier utahensis, Anaphalis margaritacea, Anchusa officinalis, Androsace filiformis, A. septentrionalis, Angelica roseana, Antennaria alpina, A. corymbosa, A. lanata, A. microphylla, A. racemosa, Anticlea elegans, Aquilegia flavescens, Arctostaphylos uva-ursi, Arnica cordifolia, A. latifolia, A. longifolia, A. parryi, A. rydbergii, Artemisia frigida, A. scopulorum, A. tridentata, Astragalus alpinus, Avenula hookeri, Balsamorhiza sagittata, Berberis repens, Bistorta bistortoides, B. vivipara, Boechera stricta, Bromus carinatus, Calamagrostis canadensis, C. rubescens, Caltha leptosepala, Calochortus eurycarpus, Camassia quamash, Carex aquatilis, C. athrostachya, C. douglasii, C. elynoides, C. geyeri, C. interior, C. luzulina, C. microptera, C. nebrascensis, C. nigricans, C. obtusata, C. paysonis, C. petasata, C. phaeocephala, C. raynoldsii, C. rossii, C. rupestris, C. scopulorum, Castilleja applegatei, C. cusickii, C. miniata, C. occidentalis, C. pulchella, C. rhexiifolia, C. sulphurea, Cerastium arvense, Chaenactis, Chamerion angustifolium, Cheilanthes gracillima, Chionophila tweedyi, Cirsium foliosum, Claytonia, Cryptogramma crispa, Cymopterus longipes, Cynoglossum, Danthonia intermedia, Dasiphora floribunda, Delphinium barbeyi, D. depauperatum, D. nuttallianum, D. ×occidentale, Deschampsia cespitosa, Draba crassa, D. spectabilis, Dryas hookeriana, D. octopetala, Eleocharis palustris, E. parishii, Elymus elymoides, E. glaucus, E. lanceolatus, E. ×pseudorepens, E. trachycaulus, Epilobium anagallidifolium, E. minutum, Equisetum arvense, E. laevigatum, Eremogone aculeata, E. congesta, Ericameria suffruticosa, Erigeron eatonii, E. elatior, E. formosissimus, E. grandiflorus, E. peregrinus, E. rydbergii, E. speciosus, Eriogonum flavum, E. pyrolifolium, E. umbellatum, Erythronium grandiflorum, Eucephalus engelmannii, Eurybia integrifolia, Festuca idahoensis, F. ovina, F. rubra, F. thurberi, Fragaria vesca, F. virginiana, Galium bifolium, G. boreale, Geranium richardsonii, G. viscosissimum, G. rossii, Geum rossii, G. triflorum, Glyceria grandis, Hackelia floribunda, Hedysarum sulphurescens, Helianthella quinquenervis, H. uniflora, Heliomeris multiflora, Hieracium scouleri, H. triste, Hordeum brachyantherum, Hulsea algida, Hymenoxys hoopesii, Hypericum formosum, Ipomopsis aggregata, Juncus drummondii, J.*

parryi, *Juniperus communis, Kalmia microphylla, Kobresia myosuroides, Larix, Lathyrus lanszwertii, Leptosiphon nuttallii, Lewisia triphylla, Ligusticum canbyi, L. filicinum, L. grayi, L. porteri, Linnaea borealis, Linum lewisii, Lloydia serotina, Lomatium, Lonicera utahensis, Lupinus argenteus, L. caudatus, L. parviflorus, L. wyethii, Luzula hitchcockii, L. spicata, Melica bulbosa, M. spectabilis, Mertensia alpina, M. ciliata, Micranthes bryophora, M. oregana, M. subapetala, Mimulus breweri, Minuartia obtusiloba, Noccaea fendleri, Oreostemma alpigenum, Orthilia secunda, Oryzopsis asperifolia, Osmorhiza berteroi, O. occidentalis, Oxytropis borealis, O. campestris, Paxistima myrsinites, Packera cana, P. streptanthifolia, Pedicularis bracteosa, P. contorta, P. cystopteridifolia, P. groenlandica, P. racemosa, Pellaea breweri, Penstemon globosus, P. whippleanus, Perideridia bolanderi, Phleum alpinum, Phlox longifolia, Phyllodoce empetriformis, Picea engelmannii, P. pungens, Pinus albicaulis, P. aristata, P. contorta, P. flexilis, Piptatherum exiguum, Plantago tweedyi, Poa alpina, P. bulbosa, P. pratensis, P. secunda, Polemonium pulcherrimum, P. viscosum, Polygonum douglasii, P. minimum, P. polygaloides, Populus tremuloides, Potentilla ×diversifolia, P. gracilis, Primula jeffreyi, P. pauciflora, Pseudotsuga menziesii, Quercus gambelii, Ranunculus alismifolius, R. eschscholtzii, Rhodiola rhodantha, Ribes lacustre, R. montigenum, Rosa acicularis, Rubus idaeus, R. sachalinensis, Rumex paucifolius, Salix arctica, S. exigua, S. geyeriana, S. glauca, S. planifolia, S. reticulata, Sambucus, Saxifraga adscendens, Sedum lanceolatum, Senecio amplectens, S. atratus, S. eremophilus, S. fremontii, S. hydrophilus, S. serra, S. triangularis, Sibbaldia procumbens, Sidalcea oregana, Smelowskia americana, S. calycina, Solidago multiradiata, Sorbus scopulina, Spiraea splendens, Stellaria calycantha, S. longifolia, S. longipes, S. umbellata, Stipa, Swertia perennis, Symphoricarpos, Symphyotrichum foliaceum, Synthyris pinnatifida, Taraxacum officinale, Thalictrum occidentale, Trifolium dasyphyllum, T. haydenii, T. longipes, T. parryi, Trisetum spicatum, Trollius, Tsuga mertensiana, Ulmus pumila, Urtica dioica, Vaccinium membranaceum, V. scoparium, Valeriana acutiloba, V. edulis, Vicia americana, Viola nuttallii, Wyethia amplexicaulis, Xerophyllum tenax.*

Senecio eremophilus Richardson is a perennial, which occurs in bogs, burns, depressions, ditches, drainages, draws, flats, forests, gullies, marshes, meadows, moors, plains, prairies, ravines, roadcuts, roadsides, seeps, slopes, stream bottoms, and the margins of lakes, ponds, rivers, and streams at elevations of up to 3700 m. Exposures range from fully open sites to moderate shade. The substrates reported include alluvium, aquie fluvent, basalt, basaltic loam, boulders, clay loam, decomposed shale, Dutton volcanics, granitic rock, gravel, gravelly loam, humus, igneous cobble, igneous gravel, limestone, loam, loamy volcanics, Mancos shale, Navajo blowsand, quartzite, rocky loam, rubble, sand, sandy clay, sandy clay loam, sandy loam, scree, silt, stony humus-loam, and talus. The plants occur commonly in eroded or disturbed areas such as burns, clearcuts, heavily trampled sites, and overgrazed areas, which they colonize readily. In meadows they tend to be more common in unburned sites. The caudex can branch and plants

often are observed growing in clumps. The reproductive and seed ecology has not been described. A seed bank of undetermined longevity can develop, yielding as many as 23 seedlings/m². **Reported associates:** *Abies concolor, A. lasiocarpa, Acer grandidentatum, Achillea millefolium, Achnatherum lemmonii, Acmispon wrightii, Agastache breviflora, A. pallidiflora, Agave palmeri, A. parryi, Ageratina rothrockii, Agrimonia striata, Agrostis gigantea, A. scabra, A. stolonifera, Allium cernuum, Alnus tenuifolia, Alopecurus aequalis, Amaranthus, Amauriopsis dissecta, Anaphalis, Antennaria parvifolia, Aquilegia chrysantha, A. elegantula, A. formosa, Arbutus arizonica, Arctostaphylos patula, A. pungens, Arenaria lanuginosa, Aristida purpurea, Arnica cordifolia, Artemisia arbuscula, A. cana, A. ludoviciana, A. michauxiana, A. tridentata, Astragalus rusbyi, Atriplex canescens, Balsamorhiza sagittata, Berberis repens, Betula, Bidens aureus, Boechera stricta, Bouteloua gracilis, Brickellia grandiflora, Brodoa oroarctica, Bromus carinatus, B. ciliatus, B. inermis, B. marginatus, B. sterilis, B. tectorum, Calamagrostis canadensis, Canotia holacantha, Carex aquatilis, C. diandra, C. inops, C. pellita, C. rostrata, C. siccata, C. wootonii, Ceanothus, Cerastium nutans, Cercocarpus ledifolius, Cercocarpus ×montanus, Chamerion angustifolium, Chimaphila umbellata, Chrysothamnus viscidiflorus, Cicuta bulbifera, C. maculata, Cirsium arvense, C. drummondii, C. eatonii, C. flodmanii, C. parryi, C. wheeleri, Comandra umbellata, Comarum palustre, Commelina dianthifolia, Conyza, Corallorhiza maculata, Cornus sericea, Cryptantha pterocarya, Cyperus fendlerianus, Dalea formosa, Danthonia parryi, Delphinium andesicola, D. scaposum, Desmodium, Dieteria canescens, Draba helleriana, Dracocephalum parviflorum, Echinocereus, Elymus arizonicus, E. elymoides, Epilobium ciliatum, E. strictum, Equisetum arvense, E. fluviatile, Eragrostis, Eremogone eastwoodiae, E. fendleri, Erigeron arizonicus, E. divergens, E. eximius, E. formosissimus, E. oreophilus, E. philadelphicus, E. rybius, E. speciosus, E. vreelandii, Eriogonum racemosum, Eriophorum chamissonis, Euphorbia spathulata, Fendlera rupicola, Festuca thurberi, Fragaria vesca, Frasera speciosa, Galinsoga parviflora, Galium proliferum, Gayophytum diffusum, Gentianella amarella, Geranium richardsonii, Geum macrophyllum, Glandularia bipinnatifida, Glossopetalon spinescens, Glyceria striata, Gnaphalium, Gutierrezia microcephala, G. sarothrae, Hackelia floribunda, Halenia rothrockii, Hedeoma hyssopifolia, Helianthella parryi, Heliomeris multiflora, Heterotheca fulcrata, Heuchera sanguinea, Hieracium abscissum, Holodiscus discolor, Hordeum brachyantherum, H. jubatum, Houstonia wrightii, Hymenoxys hoopesii, H. richardsonii, Hypnum, Iris missouriensis, Juncus balticus, J. longistylis, Juniperus communis, J. deppeana, J. osteosperma, Koeleria macrantha, Lappula occidentalis, Lathyrus lanszwertii, Leptosiphon aureus, Linanthus pungens, Linum puberulum, Lithospermum multiflorum, Lomatium foeniculaceum, Lonicera, Lupinus argenteus, L. aridus, L. palmeri, Madia glomerata, Melilotus alba, Menyanthes trifoliata, Mimulus guttatus, Mirabilis albida, M. longiflora, Monarda citriodora, Monardella, Muhlenbergia emersleyi, M. montana, Nemacladus glanduliferus, Nolina* *microcarpa, Oenothera elata, O. hartwegii, Opuntia phaeacantha, Oreochrysum parryi, Orthilia secunda, Osmorhiza depauperata, Oxalis, Oxytropis lambertii, Packera pauciflora, P. pseudaurea, Parthenium incanum, Paxistima myrsinites, Pedicularis grayi, Pennellia micrantha, Penstemon barbatus, P. virgatus, Pericome caudata, Perityle coronopifolia, Phacelia affinis, P. hastata, Phleum pratense, Physaria gordonii, Picea engelmannii, P. pungens, Pinus arizonica, P. cembroides, P. contorta, P. edulis, P. flexilis, P. leiophylla, P. monophylla, P. ponderosa, P. strobiformis, Piptochaetium pringlei, Plantago major, Pleiacanthus spinosus, Poa pratensis, P. secunda, Populus balsamifera, P. tremuloides, Potentilla anserina, Prunella vulgaris, Prunus, Psacalium decompositum, Pseudocymopterus montanus, Pseudognaphalium macounii, Pseudostellaria jamesiana, Pseudotsuga menziesii, Ptelea trifoliata, Pteridium aquilinum, Quercus gambelii, Q. grisea, Q. hypoleucoides, Q. oblongifolia, Q. rugosa, Q. turbinella, Ranunculus gmelinii, Rhamnus californica, R. crocea, Ribes cereum, Robinia neomexicana, Rosa woodsii, Rubus arcticus, R. leucodermis, R. neomexicanus, R. parviflorus, R. sachalinensis, Rudbeckia laciniata, Rumex obtusifolius, Salix bebbiana, S. geyeriana, S. lasiandra, Sambucus nigra, Sanguisorba minor, Scirpus microcarpus, Scrophularia lanceolata, Scutellaria galericulata, Senecio bigelovii, S. crassulus, S. flaccidus, S. sacramentanus, S. spartioides, Sidalcea, Silene laciniata, Solidago missouriensis, S. velutina, S. wrightii, Stachys, Stellaria longifolia, Stenotus, Stevia plummerae, Symphoricarpos oreophilus, Symphyotrichum ascendens, S. ericoides, S. puniceum, Thalictrum fendleri, Thelypodium wrightii, Thlaspi arvense, Tiquilia canescens, Tragia ramosa, Trautvetteria caroliniensis, Trifolium hybridum, T. pinetorum, T. repens, Triglochin maritimum, Typha latifolia, Urtica dioica, U. gracilenta, Vaccinium, Valeriana sorbifolia, Veratrum, Verbascum thapsus, Verbena, Veronica americana, V. anagallis-aquatica, Vicia americana, V. pulchella, Viola canadensis, Vitis arizonica, Yucca baccata, Y. schottii, Zuloagaea bulbosa.*

Senecio hydrophilus Nutt. is a biennial or short-lived perennial, which inhabits tidal or non-tidal beaches, bogs, diches, estuaries, fens, flats, floodplains, gravel bars, gulches, hummocks, marshes, meadows, mudflats, potholes, roadsides, seeps, slopes, sloughs, strands, stream bottoms, swales, swamps, thickets, vernal pools, and the margins of canals, cienega, lakes, ponds, reservoirs, rivers, and streams, at elevations of up to 3049 m. The plants often are reported in shallow standing water (to 10–20 cm deep), which can range from fresh to brackish or subsaline. Exposures usually receive full sunlight. The substrates tend to be alkaline and are described as alluvium, basalt, clay, gravel, Lickskillet silt loam, loam, mud, peat, rock, sandy gravel, sand, sandy loam, silt, and silt humus; some individuals also have been observed growing atop floating logs. Flowering occurs from May to June. Individual plants can vary from producing no ray florets to an average of about five per head. However, their level of self-incompatibility and other aspects of their breeding system and reproductive ecology have not been elucidated. The flowers presumably are pollinated by insects (Insecta). The seed ecology also has not been studied adequately. **Reported associates:** *Achillea millefolium, Agropyron, Agrostis stolonifera, Allium schoenoprasum, Alopecurus magellanicus, A. pratensis, Ambrosia artemisiifolia, Angelica ampla, A. kingii, Antennaria corymbosa, A. pulcherrima, Arnica chamissonis, A. mollis, Artemisia ludoviciana, Astragalus lemmonii, Atriplex calotheca, Barbarea orthoceras, Bassia hyssopifolia, Bistorta, Brassica nigra, Bromus inermis, Calamagrostis canadensis, C. stricta, Camassia, Carduus, Carex aquatilis, C. atherodes, C. athrostachya, C. buxbaumii, C. diandra, C. disperma, C. lanuginosa, C. lenticularis, C. microptera, C. nebrascensis, C. pellita, C. praegracilis, C. rossii, C. rostrata, C. simulata, C. utriculata, C. vesicaria, Castilleja miniata, Cicuta douglasii, C. maculata, Cirsium arvense, C. hydrophilum, C. scariosum, Comarum palustre, Crataegus, Dasiphora floribunda, Deschampsia cespitosa, Distichlis spicata, Downingia elegans, Eleocharis palustris, Epilobium ciliatum, Equisetum arvense, Erigeron lonchophyllus, E. peregrinus, Erysimum cheiranthoides, Gentianopsis detonsa, G. simplex, Geum aleppicum, G. macrophyllum, Glyceria, Grindelia squarrosa, Helianthus nuttallii, Heracleum sphondylium, Hordeum brachyantherum, H. jubatum, Impatiens capensis, Iris missouriensis, Iva axillaris, Juncus arcticus, J. balticus, J. ensifolius, J. longistylis, J. nevadensis, Lactuca serriola, Lemna minor, Lepidium latifolium, L. perfoliatum, Leymus triticoides, Linaria vulgaris, Lupinus lepidus, Lycopus asper, Lysimachia ciliata, L. maritima, Maianthemum stellatum, Medicago lupulina, Melilotus alba, Mentha arvensis, Menyanthes trifoliata, Mimulus guttatus, Muhlenbergia richardsonis, Myriophyllum sibiricum, Nasturtium officinale, Nepeta cataria, Oreostemma alpigenum, Packera debilis, Pascopyrum smithii, Pedicularis groenlandica, Penstemon pratensis, P. rydbergii, Perideridia parishii, Petasites frigidus, Phalaris arundinacea, Phleum pratense, Picea engelmannii, P. glauca, Pinus contorta, P. ponderosa, Platanthera obtusata, Pluchea odorata, Poa nemoralis, P. pratensis, P. secunda, Populus angustifolia, Porterella carnosula, Potamogeton gramineus, P. natans, Potentilla anserina, P. biennis, Pyrrocoma lanceolata, Ranunculus flammula, R. uncinatus, Ribes aureum, R. lacustre, Rosa nutkana, R. woodsii, Rumex occidentalis, Salix exigua, S. geyeriana, S. lutea, Sarcobatus, Sarcocornia pacifica, Schoenoplectus americanus, S. pungens, S. tabernaemontani, Scirpus microcarpus, Scutellaria galericulata, S. lateriflora, Senecio crassulus, S. sphaerocephalus, Sidalcea oregana, Sisyrinchium halophilum, Sium suave, Solanum dulcamara, Solidago canadensis, Spartina pectinata, Stachys palustris, Stellaria longifolia, S. longipes, Suaeda, Symphyotrichum eatonii, S. foliaceum, S. laeve, S. spathulatum, S. subspicatum, Taraxia breviflora, Thelypodium paniculatum, Thlaspi arvense, Toxicodendron radicans, Trifolium hybridum, T. longipes, Triglochin maritimum, Typha latifolia, Urtica dioica, Utricularia, Veronica americana, V. serpyllifolia, Wyethia.*

Senecio triangularis Hook. is a perennial, which inhabits bogs, carrs, cobble bars, depressions, draws, fens, floodplains, glades, gravel bars, marshes, meadows, ravines, roadsides, scrub, seeps, slopes, snowbanks, snow patches, swales, swamps, tundra, and the margins of lakes, ponds, and streams at elevations of up to 3658 m. Plants often occur in the shade of coniferous overstory but also extend into open, sunny areas, especially in riparian sites. The substrates generally are acidic (pH 4.3–6.7) but vary in composition to include alluvium, clay, clay loam, cobble, granite, gravel, humus, krummholz, loam, organic, rock, rocky sand, sand, sandstone, sandy loam, shale, and silt. The heads contain both ray and disk florets. The corolla tubes of the flowers are short, which best accommodates short-tongued bees (Insecta: Hymenoptera: Apidae) as their principal pollinators. The plants presumably are self-incompatible. The seeds are physiologically dormant and require 5 months of cold, moist stratification to induce their germination. Seeds collected in September have a 70%–90% germination rate and are estimated to survive for up to 5 years when stored at 3°C–5°C in sealed containers. Germination requires that seeds are sown on the surface, as buried seeds do not germinate. Natural germination occurs in the fluctuating temperatures of early spring. Poor germination rates have been reported for seeds recovered from seed banks, when cold stored at 1.1°C for several days, then placed under greenhouse conditions at day/night temperatures averaging 29.7°C/10.4°C. The shoots arise from a branched caudex, with some reports referring to rhizome production. The plants exhibit high resilience and tolerance to trampling. **Reported associates:** *Abies amabilis, A. concolor, A. grandis, A. lasiocarpa, A. magnifica, Abronia latifolia, Acer glabrum, Achillea millefolium, Achnatherum thurberianum, Achlys triphylla, Aconitum columbianum, A. delphiniifolium, Aconogonon davisiae, A. phytolaccifolium, Actaea rubra, Adenocaulon bicolor, Agastache urticifolia, Agoseris aurantiaca, A. glauca, Agrostis exarata, A. idahoensis, A. pallens, Allium brevistylum, A. geyeri, A. schoenoprasum, A. validum, Alnus rhombifolia, A. rubra, A. rugosa, A. viridis, Alopecurus magellanicus, Amelanchier alnifolia, Anaphalis margaritacea, Androsace septentrionalis, Anemone occidentalis, A. parviflora, A. piperi, Angelica ampla, A. arguta, A. dawsonii, A. grayi, A. lucida, A. pinnata, Antennaria corymbosa, A. lanata, A. media, A. neglecta, A. umbrinella, Anthoxanthum monticola, Anticlea elegans, A. occidentalis, Aquilegia coerulea, A. flavescens, A. formosa, Arabis nuttallii, Arctagrostis latifolia, Arnica cordifolia, A. latifolia, A. longifolia, A. mollis, A. ovata, Artemisia norvegica, A. tilesii, A. tridentata, Asarum caudatum, Asplenium trichomanes-ramosum, Astragalus miser, Athyrium filix-femina, Aulacomnium palustre, Barbilophozia lycopodioides, Betula occidentalis, Bistorta bistortoides, B. vivipara, Blechnum spicant, Boechera stricta, Boykinia major, Brachythecium albicans, B. groenlandicum, B. mildeanum, B. starkei, Bromus carinatus, B. ciliatus, B. inermis, B. lanatipes, B. porteri, B. vulgaris, Bryoria fuscescens, B. implexa, Bryum pseudotriquetrum, Calamagrostis canadensis, C. lapponica, C. purpurascens, C. rubescens, C. stricta, Calocedrus decurrens, Caltha leptosepala, Campanula rotundifolia, C. scabrella, Canadanthus modestus, Cardamine breweri, C. cordifolia, Carex aquatilis, C. athrostachya, C. aurea, C. bolanderi, C. brunnescens, C. canescens, C. concinnoides, C. disperma, C. echinata, C. exsiccata, C. fracta, C. geyeri, C. heteroneura, C. hoodii, C. illota, C. jonesii, C. laeviculmis, C. lemmonii, C. lenticularis, C. leptalea, C. leptopoda, C. limosa, C. livida, C. luzulina, C. macrochaeta, C. magellanica, C. mertensii, C. microchaeta, C. microptera, C. nebrascensis, C. nigricans, C. norvegica, C. nova, C. occidentalis, C. pachystachya, C. pellita, C. phaeocephala, C. podocarpa, C. pyrenaica, C. raynoldsii, C. rossii, C. rupestris, C. scirpoidea, C. scopulorum, C. spectabilis, C. subfusca, C. utriculata, C. vernacula, C. viridula, Cassiope mertensiana, C. tetragona, Castilleja lemmonii, C. miniata, C. parviflora, C. rhexiifolia, C. senta, C. suksdorfii, C. sulphurea, Catabrosa aquatica, Cerastium arvense, Ceratodon purpureus, Chaenactis douglasii, Chamaecyparis lawsoniana, Chamerion angustifolium, C. latifolium, Chimaphila umbellata, Cichorium intybus, Cicuta douglasii, Cinna latifolia, Circaea alpina, Cirsium scariosum, Cistanthe umbellata, Cladonia, Claytonia cordifolia, C. lanceolata, C. sibirica, Clintonia uniflora, Collomia debilis, Conioselinum scopulorum, Coptis occidentalis, Cornus canadensis, C. sericea, Corydalis caseana, Crataegus douglasii, Crepis runcinata, Cystopteris fragilis, Dactylis glomerata, Dactylorhiza viridis, Dasiphora floribunda, Delphinium barbeyi, D. glaucum, D. ×occidentale, Deschampsia cespitosa, D. elongata, Draba aurea, D. spectabilis, Drosera, Drymocallis glabrata, Dryopteris, Elymus elymoides, E. glaucus, E. lanceolatus, E. trachycaulus, Empetrum nigrum, Epilobium anagallidifolium, E. ciliatum, E. glaberrimum, E. halleanum, E. hornemannii, E. lactiflorum, E. minutum, Equisetum arvense, E. hyemale, E. sylvaticum, Erigeron coulteri, E. eatonii, E. elatior, E. formosissimus, E. peregrinus, E. speciosus, Erythronium grandiflorum, E. montanum, Eucephalus engelmannii, E. ledophyllus, Eurybia conspicua, E. sibirica, Festuca altaica, F. brachyphylla, F. rubra, F. thurberi, F. viridula, Fragaria vesca, F. virginiana, Frasera speciosa, Galeopsis bifida, Galium boreale, G. trifidum, G. triflorum, Gaultheria humifusa, Gentiana calycosa, G. parryi, Gentianella amarella, Gentianopsis detonsa, G. thermalis, Geranium erianthum, G. richardsonii, Geum macrophyllum, G. triflorum, Glandularia gooddingii, Glyceria borealis, G. elata, G. grandis, G. striata, Goodyera oblongifolia, Graphephorum wolfii, Gymnocarpium dryopteris, Hackelia diffusa, H. floribunda, H. micrantha, Helenium bigelovii, Heracleum lanatum, H. sphondylium, Hieracium albiflorum, H. scouleri, H. triste, Hordeum brachyantherum, Horkelia rydbergii, Hosackia oblongifolia, Huperzia selago, Hydrophyllum fendleri, Hymenoxys hoopesii, Hypericum anagalloides, Ipomopsis aggregata, Juncus articulatus, J. balticus, J. chlorocephalus, J. compressus, J. confusus, J. conglomeratus, J. drummondii, J. ensifolius, J. filiformis, J. mertensianus, J. parryi, J. regelii, J. saximontanus, Juniperus communis, Kalmia microphylla, Larix lyallii, Letharia vulpina, Leymus mollis, Ligusticum canbyi, L. grayi, L. porteri, L. tenuifolium, Lilium parryi, L. parvum, Linnaea borealis, Lloydia serotina, Lomatium martindalei, Lonicera involucrata, L. utahensis, Luetkea pectinata, Lupinus latifolius, L. nootkatensis, L. sericeus, Luzula campestris, L. hitchcockii,*

L. parviflora, L. subcongesta, Lysichiton americanus, Maianthemum racemosum, M. stellatum, Menyanthes trifoliata, Menziesia ferruginea, Mertensia ciliata, M. franciscana, M. paniculata, Micranthes ferruginea, M. lyallii, M. nelsoniana, M. odontoloma, M. oregana, M. razshivinii, Mimulus guttatus, M. lewisii, M. moschatus, M. primuloides, M. tilingii, Mitella breweri, M. pentandra, M. stauropetala, Monardella, Moneses uniflora, Myosotis arvensis, Nasturtium officinale, Neottia banksiana, N. cordata, Nothocalais alpestris, Oenanthe sarmentosa, Oplopanax horridus, Oreostemma alpigenum, Orthilia secunda, Osmorhiza berteroi, O. depauperata, O. occidentalis, O. purpurea, Oxypolis fendleri, O. occidentalis, Oxyria digyna, Packera cymbalaria, P. pseudaurea, P. subnuda, Parmeliopsis ambigua, Parnassia fimbriata, Pascopyrum smithii, Paxistima myrsinites, Pedicularis attollens, P. bracteosa, P. contorta, P. groenlandica, P. racemosa, Penstemon fruticosus, P. lyallii, P. rydbergii, P. whippleanus, Perideridia parishii, Petasites frigidus, Phacelia heterophylla, P. leptosepala, Philonotis fontana, Phleum alpinum, P. pratense, Phlox diffusa, Phyllodoce breweri, P. empetriformis, Picea engelmannii, P. glauca, Pinus albicaulis, P. contorta, P. jeffreyi, P. lambertiana, P. monticola, P. ponderosa, Platanthera dilatata, P. obtusata, P. sparsiflora, P. stricta, Poa alpina, P. fendleriana, P. leptocoma, P. palustris, P. pratensis, P. reflexa, P. stebbinsii, P. stenantha, P. wheeleri, Podagrostis aequivalvis, P. thurberiana, Pohlia nutans, Polemonium pulcherrimum, Polystichum lonchitis, P. munitum, Polytrichum commune, P. piliferum, Populus tremuloides, P. trichocarpa, Potamogeton gramineus, Potentilla flabellifolia, P. ×diversifolia, P. pulcherrima, Primula conjugens, P. frigida, P. jeffreyi, P. parryi, P. pauciflora, P. tetrandra, Prosartes hookeri, P. trachycarpa, Prunus emarginata, Pseudocymopterus montanus, Pseudoleskea radicosa, Pseudotsuga menziesii, Pteridium aquilinum, Ptilagrostis kingii, Pyrola asarifolia, P. minor, Quercus gambelii, Q. kelloggii, Ranunculus alismifolius, R. eschscholtzii, R. pygmaeus, R. uncinatus, Rhodiola integrifolia, R. rhodantha, R. rosea, Rhododendron albiflorum, R. columbianum, R. macrophyllum, Rhynchospora capitellata, Ribes acerifolium, R. bracteosum, R. erythrocarpum, R. hudsonianum, R. lacustre, R. laxiflorum, R. montigenum, R. nevadense, R. wolfii, Rosa acicularis, R. woodsii, Rubus idaeus, R. leucodermis, R. parviflorus, Rudbeckia californica, Rumex acetosella, R. fueginus, R. occidentalis, Salix arctica, S. barclayi, S. bebbiana, S. boothii, S. brachycarpa, S. commutata, S. drummondiana, S. eastwoodiae, S. farriae, S. jepsonii, S. lasiolepis, S. lutea, S. monticola, S. orestera, S. planifolia, S. pseudomonticola, S. reticulata, S. scouleriana, S. tweedyi, S. wolfii, Sambucus nigra, S. racemosa, Sanguisorba canadensis, S. officinalis, S. stipulata, Sanicula marilandica, Sasa, Saxifraga tricuspidata, Sedum lanceolatum, Senecio bigelovii, S. crassulus, S. hydrophiloides, S. integerrimus, S. lugens, S. scorzonella, S. serra, Sibbaldia procumbens, Sidalcea oregana, S. reptans, Silene parryi, Sisyrinchium californicum, Solidago confinis, S. multiradiata, Sorbus scopulina, S. sitchensis, Sphagnum henryense, Sphenosciadium capitellatum, Spiraea betulifolia, S. splendens, Stachys ajugoides, S. albens, S. chamissonis, *Stellaria calycantha, S. crassifolia, S. crispa, S. longifolia, S. obtusa, S. umbellata, Streptopus amplexifolius, Swertia perennis, Symphoricarpos albus, Symphyotrichum ciliolatum, S. eatonii, S. foliaceum, S. lanceolatum, Taraxacum ceratophorum, T. officinale, Taxus brevifolia, Thalictrum occidentale, Thlaspi montanum, Thuja plicata, Tiarella trifoliata, Tomentypnum nitens, Trautvetteria caroliniensis, Triantha glutinosa, Trichophorum, Trifolium haydenii, T. longipes, T. monanthum, T. repens, Trillium ovatum, Trisetum spicatum, Trollius albiflorus, T. laxus, Tsuga heterophylla, T. mertensiana, Urtica dioica, Vaccinium cespitosum, V. membranaceum, V. myrtillus, V. ovalifolium, V. scoparium, V. uliginosum, Vahlodea atropurpurea, Valeriana acutiloba, V. dioica, V. sitchensis, Verbascum thapsus, Veratrum californicum, V. viride, Veronica alpina, V. americana, V. cusickii, V. serpyllifolia, V. wormskjoldii, Vicia americana, Viola adunca, V. canadensis, V. glabella, V. macloskeyi, V. orbiculata, V. palustris, Xerophyllum tenax.*

Use by wildlife: Many *Senecio* species produce pyrrolizidine alkaloids, which generally are effective at reducing herbivory. The flowers of *Senecio crassulus* are visited and pollinated by bumblebees (Insecta: Hymenoptera: Apidae: *Bombus appositus, B. bifarius, B. flavifrons, B. frigidus, B. mixtus, B. rufocinctus, B. sitkensis, B. sylvicola, B. occidentalis*), leafcutting bees (Insecta: Megachilidae: *Osmia montana, O. subaustralis*), and butterflies (Insecta: Lepidoptera: Nymphalidae: *Aglais milberti, Polygonia gracilis*), which forage for nectar. *Bombus sylvicola* is the most frequent pollinator. Army cutworm moths (Insecta: Lepidoptera: Noctuidae: *Euxoa auxiliaris*) forage for the floral nectar nocturnally. The foliage is eaten by mule deer (Mammalia: Cervidae: *Odocoileus hemionus*), particularly along cuts and roadsides, by the Rocky Mountain goat (Mammalia: Bovidae: *Oreamnos americanus*), primarily in the fall, and by sheep (Mammalia: Bovidae: *Ovis aries*). The plants are hosts to several fungi (Ascomycota: Phaeosphaeriaceae: *Nodulosphaeria octoseptata, N. olivacea*; Pleosporaceae: *Pleospora richtophensis*; Basidiomycota: Pucciniaceae: *Puccinia senecionis, P. subcircinata*). They provide a habitat component of the yellow-bellied marmot (Mammalia: Rodentia: *Marmota flaviventris*). The flowers of *S. eremophilus* are visited by nectar-seeking dreamy duskywing butterflies (Insecta: Lepidoptera: Hesperiidae: *Erynnis icelus*) and green lattice moths (Insecta: Lepidoptera: Arctiidae: *Gnophaela vermiculata*) as well as plant bugs (Insecta: Hemiptera: Miridae: *Labopidea simplex*). *Senecio hydrophilus* is used as cover by nesting ducks (Aves: Anatidae) including gadwall (*Anas strepera*), cinnamon teal (*Anas cyanoptera*), shoveler (*Anas clypeata*), pintail (*Anas acuta*), and mallard (*Anas platyrhynchos*). It is a larval host of the ragwort stem borer moth (Insecta: Lepidoptera: Noctuidae: *Papaipema insulidens*). Such larvae provide food for nestling bobolinks (Aves: Icteridae: *Dolichonyx oryzivorus*). The plants also host several fungi (Ascomycota: Erysiphaceae: *Erysiphe cichoracearum*, Mycosphaerellaceae: *Ramularia senecionis*; Basidiomycota: Pucciniaceae: *Puccinia subcircinata*). The foliage is eaten by mule deer (Mammalia: Cervidae: *Odocoileus hemionus*).

Senecio triangularis provides habitat for the Canada lynx (Mammalia: Felidae: *Lynx canadensis*) and grizzly bears (Mammalia: Ursidae: *Ursus arctos*). The foliage is highly digestible (67.6%), contains 24.8% crude protein, and is eaten by mule deer (Mammalia: Cervidae: *Odocoileus hemionus*) and Rocky Mountain goats (Mammalia: Bovidae: *Oreamnos americanus*). It is an important food of the Rocky Mountain elk (Mammalia: Cervidae: *Cervus elaphus*) and comprises a small percentage of the diet for red-backed voles (Mammalia: Rodentia: Muridae: *Clethrionomys gapperi*) and Rocky Mountain moose (Mammalia: Cervidae: *Alces alces*). It is a host to aphids (Insecta: Hemiptera: Aphididae: *Aphis*), bees (Insecta: Hymenoptera: Apidae: *Bombus bifarius, B. fernaldae, B. flavifrons, B. frigidus, B. insularis, B. mixtus, B. occidentalis, B. sylvicola, Melissodes pallidisignata*; Halictidae: *Agapostemon femoratus, Halictus farinosus, H. rubicundus*; Megachilidae: *Hoplitis albifrons, Megachile pugnata, Osmia coloradensis, O. montana, O. pentstemonis, O. subaustralis, O. tristella*), larval Lepidoptera (Arctiidae: *Tyria jacobaeae*; Choreutidae: *Caloreas augustella*; Nymphalidae: *Charidryas palla*), and plant bugs (Insecta: Hemiptera: Miridae: *Labopidea nigrisetosa, Lygus aeratus*). Muscoid flies (Insecta: Diptera: Tephritidae: *Trypeta flaveola*) have been collected from plants and the adults reared from leaf mines of garden-grown plants. The flowers provide nectar to Gillett's checkerspot butterfly (Insecta: Lepidoptera: Nymphalidae: *Euphydryas gillettii*), but *Bombus frigidus* is the most important pollinator. Many fungi also are hosted by the plants including Ascomycota (Botryosphaeriaceae: *Phyllosticta garrettii*; Chaetothyriomycetidae: *Coniosporium microsporum*; Didymosphaeriaceae: *Apiosporella mimuli*; Dothideomycetidae: *Mycosphaerella tassiana, Ramularia filaris*; Erysiphaceae: *Alphitomorpha fuliginea, Podosphaera macularis, P. senecionis*; Leotiomycetidae: *Heteropatella umbilicata*; Pleosporomycetidae: *Leptosphaeria doliolum*), Basidiomycota (Coleosporiaceae: *Coleosporium occidentale*; Exobasidiomycetidae: *Entyloma calendulae*; Pucciniaceae: *Puccinia angustata, P. eriophori, P. senecionis, P. subcircinata*), and Chytridiomycota (Synchytriaceae: *Synchytrium*). The roots become heavily colonized by arbuscular mycorrhizal fungi.

Economic importance: food: *Senecio crassulus* should never be eaten because it contains toxic pyrrolizidine alkaloids (e.g., riddelline), which have anticholinergic and hepatotoxic properties and likely are carcinogenic. Similarly, *S. triangularis* contains numerous pyrrolizidine alkaloids including integerrimine, neotriangularine, retrorsine, rosmarinine, platy-phylline, senecionine, triangularine, 7-angelylretronecine, 7-senecioylretronecine, 7-angelyl-9-sarracinylretronecine, and 7-senecioyl-9-sarracinylretronecine; **medicinal:** *Senecio eremophilus* has been used as a herbal remedy in some parts of the world, but should not be taken as a medicinal due to its riddelline content (see previous). An infusion made from the pulverized roots and leaves of *S. triangularis* were used by the Cheyenne as a sedative and to relieve chest pain. The plants contain senecionine and senecionine N-oxide, which have tumor-suppressant properties. The alkaloid senecionine can be transferred from plants of *S. triangularis* to its root parasites *Pedicularis bracteosa* and *P. groenlandica* (Orobanchaceae); **cultivation:** Although numerous *Senecio* are popularly grown as ornamental garden plants, none of the OBL species has been cultivated; **misc. products:** Seeds of *Senecio triangularis* are used in some erosion control blankets; **weeds:** *Senecio eremophilus* has been categorized as weedy in its nonindigenous range; **nonindigenous species:** *Senecio eremophilus* was introduced to the Yukon Territory sometime before 1968, which is when it was first collected in that area. It is believed to have spread into the region through transportation corridors.

Systematics: *Senecio* is the type genus of Senecioneae, the largest tribe of Asteraceae with more than 150 genera and 3000 species. *Senecio* (subtribe Senecioninae) alone accounts for about 1000 of the species, which ranks it among the largest of angiosperm genera. The circumscription of this speciose genus has been changed repeatedly as phylogenetic analyses continually confirmed polyphyletic associations with other genera. A recent circumscription, which incorporated molecular data from 186 *Senecio* species, proposed delimiting the genus as one clade (*Senecio* sensu stricto), which is well supported by several molecular data sets. Included in that group are most species regarded previously as belonging to *Senecio*, as well as members of six other genera: *Aetheolaena, Culcitium, Hasteola, Iocenes, Lasiocephalus,* and *Robinsonia*. That disposition has been followed here. Unfortunately, *S. triangularis* was the only OBL species included in the molecular study, where it resolved among a cohesive group of North American *Senecio* species (and *Hasteola suaveolens*) within a New World clade (Figure 5.171). *Senecio eremophilus* has been included in a less comprehensive molecular phylogenetic study; however, the exclusion of other OBL species in both studies makes it impossible to evaluate the interrelationships of the aquatic-adapted *Senecio* as a whole. The base chromosome number of *Senecio* is $x=10$. *Senecio crassulus* and *S. hydrophilus* are tetraploid ($2n=40$); whereas, *S. eremophilus* (38, 40, 44) and *S. triangularis* ($2n=40$, 80) also contain aneuploids and polyploids as indicated. *Senecio eremophilus* reportedly hybridizes with *S. spartioides* ($2n=40$).

Comments: *Senecio crassulus* occurs in the western United States, *S. eremophilus* in northern and western North America, and *S. hydrophilus* and *S. triangularis* in western North America.

References: Allen et al., 1991; Austin et al., 2007; Barkley, 2006; Baskin & Baskin, 1998; Bennett & Mulder, 2009; Boggs et al., 2006; Bull et al., 1968; Christy, 2004, 2013; Cole, 1995; Cooper et al., 1999; Cripps & Eddington, 2005; Deschamp, 1977; Diehl & McEvoy, 1989; Doyle, 1972, 2003; Dungan & Wright, 2005; Ellison, 1949, 1954; Emmel et al., 1992; Fowler et al., 2012; French et al., 1994; Ghazoul, 2005; Hamer & Herrero, 1987; Han & Norrbom, 2005; Harmon & Franklin, 1995; Harshberger, 1929; Hooke, 2006; Jackson & Bliss, 1982; Johnston et al., 1968; Krueger, 1972; Kufeld, 1973; Kupchan & Suffness, 1967; Langel et al., 2011; Lewis et al., 1928; Mancuso, 2001; McCain & Christy, 2005; McEvoy & Cox, 1987; Merritt & Merritt, 1978; Monty et al., 2008;

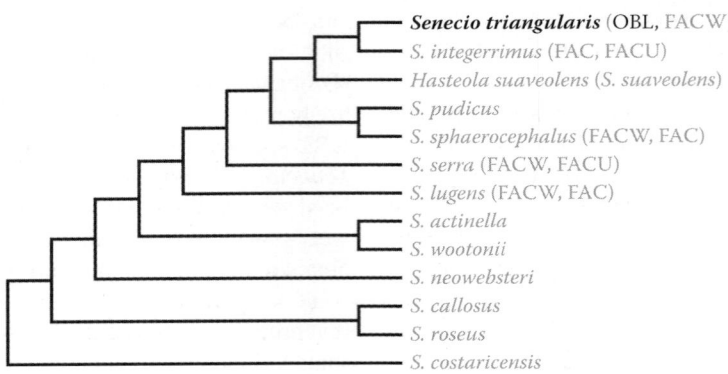

FIGURE 5.171 Phylogenetic position of *Senecio triangularis* within a clade of predominantly North American *Senecioninae*. Although other OBL species were excluded from this analysis, *S. triangularis* (and other wetland indicators) occupy a relatively derived position in the group, which also includes the former *Hasteola*. (Adapted from Pelser, P.B. et al., *Taxon*, 56, 1077–1104, 2007.)

Newman et al., 2014; Oosting & Billings, 1943; Pelser et al., 2002, 2007, 2010; Pickford & Reid, 1943; Pyke, 1982; Pyke et al., 2012; Quinlan et al., 2003; Roitman, 1983; Rüeger & Benn, 1983; Rust, 2002; Saunders, Jr., 1955; Schmitt, 1980, 1983; Schneider & Stermitz, 1990; Scianna & Lapp, 2005; Sheldon & Burrows, 1973; Smith, 1981a; Stoughton & Marcus, 2000; Svendsen, 1974; Thomson, 1980, 1981a, 1982; Turesson, 1914; USFWS, 2013b; Waddington, 1979; Wallmo et al., 1972; Wang & Kemball, 2005; Wells et al., 2012; West & Reese, 1991; Whitcraft et al., 2011; Williams & Marshall, 1938; Wilson et al., 2010; Wittenberger, 1978.

42. *Shinnersia*

Mexican oak leaf, Rio Grande bugheal
Etymology: after Lloyd Herbert Shinners (1918–1971)
Synonyms: *Trichocoronis* (in part)
Distribution: global: North America; **North America:** south central (Texas)
Diversity: global: 1 species; **North America:** 1 species
Indicators (USA): OBL: *Shinnersia rivularis*
Habitat: freshwater; palustrine, riverine; **pH:** 5.5–9.0; **depth:** <1 m; **life-form(s):** emergent herb; submersed (vittate)
Key morphology: Stems (to 1+ m) highly branched, creeping or submersed, rhizomatous, rooting at the nodes; leaves opposite, sessile, often flaring, the blades (to 6 cm) oblanceolate or 3–5-lobed; heads (to 9 mm) discoid, hemispherical, solitary (rarely paired), peduncled (to 8 cm), phyllaries (25–30; to 6 mm) in 2–3 series, pales lacking; disc florets (75–100) with white corollas (to 3 mm); cypselae (to 2.3 mm) prismatic (4–5-ribbed), brown, sparsely gland dotted and hairy, the pappus lacking.
Life history: duration: perennial (rhizomes); **asexual reproduction:** rhizomes, shoot fragments; **pollination:** unknown; **sexual condition:** hermaphroditic; **fruit:** cypselae (rare to common); **local dispersal:** cypselae (water); **long-distance dispersal:** cypselae (water)
Imperilment: (1) *Shinnersia rivularis* [G2/G3]; S1 (TX)
Ecology: general: *Shinnersia* is monotypic.

Shinnersia rivularis (A. Gray) King & H. Rob. grows in rivers, springs, streams and spring-fed swales at elevations of up to 3000 m. The plants typically are submersed (<10 cm aerial) and rooted in slow moving waters (from 0.3 to 1.0 m in depth), but also can survive as emergents when stranded on shorelines. They grow best under high light conditions and at water temperatures from 18°C to 30°C. Although its natural water sources are derived from calcareous outcrops, cultivated plants reportedly grow well across a broad range of acidity (pH = 5.5–9.0). The substrates are described as mucky gravel. Flowering occurs year-round but peaks in March–May. The failure of introduced European plants to set seed probably is due to self-incompatibility coupled with a lack of compatible genotypes. No information exists on the reproductive or seed ecology of this species. Vegetative reproduction occurs by means of long rhizomes and fragmentation of lower stem regions, which root at the nodes. **Reported associates:** *Eleocharis cellulosa*, *E. interstincta*, *Hydrocotyle umbellata*, *Justicia americana*, *Ludwigia*, *Myriophyllum*, *Nasturtium officinale*, *Nomaphila stricta*, *Nymphoides aquatica*, *Potamogeton illinoensis*.

Use by wildlife: *Shinnersia rivularis* can become infected with CMV, the cucumber mosaic virus (Bromoviridae: *Cucumovirus*), which causes vein clearing of the foliage (see *Cultivation* later).

Economic importance: food: unknown; **medicinal:** unknown; **cultivation:** *Shinnersia rivularis* is cultivated as a decorative aquarium plant and survives well under aquarium conditions. The cultivar 'White-Green' is a variegated leaf form caused by a viral infection; **misc. products:** none; **weeds:** *Shinnersia rivularis* is regarded as a weed in its nonindigenous range. The plants can be extremely fast-growing under suitable conditions; **nonindigenous species:** *Shinnersia rivularis* has been introduced to Austria, Germany, Hungary, and Slovakia, presumably as an escape from cultivation in aquaria.

Systematics: *Shinnersia* is placed within tribe Eupatorieae (subtribe Trichocoroninae) of Asteraceae. Analysis of DNA sequence data resolves *Shinnersia* as the sister genus to *Trichocoronis* (Figure 5.166), with which it has been merged by some authors. The two genera comprise a sister clade to *Sclerolepis*, which also is OBL. *Shinnersia* differs from *Trichocoronis* by its solitary axillary capitula, lack of a pappus, and several other features. The base chromosome number of *Shinnersia* is $x=15$. *Shinnersia rivularis* ($2n=60$) is tetraploid.

Comments: *Shinnersia rivularis* occurs in Texas and extends into Coahuila and Nuevo León, Mexico.

References: Cook, 1996; Correll & Correll, 1975; Kasselmann, 2003; Nesom, 2006f; Pavol et al., 2009; Poole et al., 2007; Proeseler et al., 1990; Tippery et al., 2014.

43. *Solidago*

Goldenrod; verge d'or

Etymology: from the Latin *solidus ago* ("become sound"), which is attributed to alleged healing properties

Synonyms: *Actipsis*; *Aster* (in part); *Brachychaeta*; *Brintonia*; *Leioligo*; *Oligoneuron*; *Oreochrysum*

Distribution: global: Eurasia; New World; **North America:** widespread

Diversity: global: 100 species; **North America:** 77 species

Indicators (USA): OBL: *Solidago confinis*, *S. houghtonii*, *S. latissimifolia*, *S. ohioensis*, *S. patula*, *S. riddellii*, *S. stricta*, *S. uliginosa*, *S. verna*; **FACW:** *S. confinis*, *S. stricta*

Habitat: brackish, freshwater; palustrine; **pH:** 3.7–8.3; **depth:** <1 m; **life-form(s):** emergent herb

Key morphology: Shoots (to 200 cm) 1–10, erect, rhizomatous or arising from a woody caudex; leaves basal and cauline, sessile or with tapering petioles (to 25 cm) that often are winged and clasping or auricled at their base, the blades (to 60 cm) elliptic, elliptic lanceolate, lanceolate, lanceolate elliptic, linear lanceolate/oblanceolate, ovate, or subcordate, the margins entire, serrate, or crenate serrate; heads (to 800) radiate, peduncled (to 10 mm), secund or not, arranged secondarily in corymbs, panicles, or thyrsiform panicles, involucres (to 9 mm) campanulate, phyllaries (to 18) in 3–5 series, receptacles lacking pales; ray florets (1–12) pistillate, rays (to 2.5–7.9 mm) yellow; disc florets (4–27) bisexual, 5-lobed, the corollas (to 5.2 mm) yellow; cypselae (to 2.5 mm) obconic, ribbed (8–10), glabrous to strigose, pappus persistent, of 2 series of 25–45 barbellate bristles (to 5.5 mm)

Life history: duration: perennial (woody caudex, rhizomes); **asexual reproduction:** branching caudex, rhizomes; **pollination:** insect; **sexual condition:** gynomonoecious; **fruit:** cypselae (common); **local dispersal:** cypselae (wind); **long-distance dispersal:** cypselae (wind)

Imperilment: (1) *Solidago confinis* [G4]; (2) *S. houghtonii* [G3]; S1 (NY); S2 (ON); S3 (MI); (3) *S. latissimifolia* [G5]; SX (DC); SH (WV); S1 (DE, NY, RI); S2 (NC, VA); S3 (MD, NJ, NS); (4) *S. ohioensis* [G4]; S2 (NY); S3 (OH); (5) *S. patula* [G5]; SH (NH); S1 (AR, IA, OK, VA, WV); S2 (NC); S3 (DE, KY, MD, NC, VA, VT); (6) *S. riddellii* [G5]; SH (ND); S1 (AR, GA, SD); S2 (KS, MB); S3 (IA, MO, ON); (7) *S. stricta* [G4/G5]; SX (DE); S1 (TN, WV); S2 (KY, VA); S3 (NC, NJ); (8) *S. uliginosa* [G4/G5]; SX (DC); SH (AL); S1 (DE, GA, IA, NC); S2 (PA, VA); S3 (MB, MD, NJ, NY, WV); (9) *S. verna* [G3]; S2 (SC); S3 (NC)

Ecology: general: *Solidago* is a relatively large genus of perennials, which occupies a wide diversity of habitats ranging from shaded woodlands to open rocky cliffs. It is a familiar genus of grassland habitats, and often occurs where high soil moisture is maintained. Twenty-nine of the species (38%) are ranked at some level as wetland indicators, with about 12% as OBL. From those species studied in detail, the flowers of *Solidago* are self-incompatible and rely on outcrossing by insects (Insecta) such as bees and wasps (Hymenoptera), flies (Diptera), or butterflies and moths (Lepidoptera). The heads are gynomonoecious with pistillate ray florets and bisexual disc florets. Because the ray florets lack pollen producing organs, they are damaged significantly less by herbivores. The seeds are equipped with a pappus of long bristles and are dispersed by the wind, but generally settle within 2.0 m of the maternal plant. Some species contribute extensively to the seed bank, with densities approaching 50,000 seeds/m². The duration of the seed bank varies among species from transient (1 year) to persistent (>4 years). In at least some cases, conditional dormancy prevents seeds from germinating in the first Autumn after dispersal (which prevents seedling death from freezing), then requiring a 16-month period for germination. Vegetative reproduction occurs by the formation of rhizomes, which can lead to extensive clonal development, or by the branching of the caudex, which results in clumps of stems.

***Solidago confinis* A. Gray** is a plant of chaparral, cienega, depressions, ditches, draws, dunes, flats, floodplains, hot springs, marshes, meadows, river beds, roadsides, salt marshes, seeps, slopes, springs, swamps, and the margins of lakes and streams at elevations of up to 2500 m. Exposures range from sun to shade. The substrates are alkaline and described as clay, granitic loam, limestone, loam, marl, mud, rock, sand, serpentine, shale, silt, and silty loam. Little life-history information is available for this species. 20°C is the temperature recommended for the germination of mature seeds. The plants perennate from a woody taproot, from which slender rhizome-like branches are produced. **Reported associates:** *Abies concolor, Achillea fremontii, A. millefolium, Achnatherum occidentale, Adenostoma fasciculatum, A. sparsifolium, Adiantum capillus-veneris, Ageratina adenophora, Agoseris heterophylla, Agropyron, Agrostis idahoensis, A. stolonifera, Alnus rhombifolia, Amorpha californica, Andropogon glomeratus, Anemopsis californica, Antennaria rosea, Aquilegia formosa, Arabidopsis thaliana, Arctostaphylos glauca, A. patula, Artemisia californica, A. douglasiana, A. dracunculus, A. ludoviciana, A. tridentata, Asclepias fascicularis, Aspidotis californica, Astragalus douglasii, A. lentiginosus, Atriplex canescens, A. polycarpa, Avena barbata, Baccharis salicifolia, B. salicina, B. sergiloides, Barbarea orthoceras, Berula erecta, Bouteloua gracilis, Brickellia californica, Bromus carinatus, B. ciliatus, B. diandrus, B. tectorum, Calocedrus decurrens, Calochortus plummerae, Calystegia macrostegia, C. occidentalis, Camissoniopsis hirtella, Carex alma, C. athrostachya, C. aurea, C. bolanderi, C. fracta, C. jonesii, C. occidentalis, C. pellita, C. richardsonii, C. schottii, C. senta, C. spissa, C. subfusca, Castilleja miniata, Ceanothus cordulatus, C. leucodermis, Centaurea melitensis,*

Ceanothus cordulatus, Cercocarpus betuloides, C. ledifolius, Chaenactis artemisiifolia, Cirsium occidentale, C. scariosum, Clematis ligusticifolia, Cornus sericea, Cotoneaster lacteus, Crotalus, Cryptantha micromeres, C. muricata, Datisca glomerata, Delphinium cardinale, Deschampsia elongata, Diplacus aurantiacus, Distichlis spicata, Drymocallis lactea, Dryopteris arguta, Dudleya lanceolata, Dysphania botrys, Eleocharis montevidensis, E. rostellata, Emmenanthe penduliflora, Epilobium canum, E. ciliatum, Epipactis gigantea, Equisetum hyemale, E. laevigatum, Ericameria nauseosa, E. pinifolia, Erigeron divergens, Eriogonum fasciculatum, E. wrightii, Eriophyllum confertiflorum, Erodium cicutarium, Erysimum capitatum, Eucrypta chrysanthemifolia, Eulobus californicus, Euthamia occidentalis, Festuca rubra, Fragaria, Fraxinus velutina, Fremontodendron californicum, Gamochaeta purpurea, Garrya flavescens, Gentiana fremontii, Gentianella amarella, Geranium richardsonii, Geum, Gilia capitata, Glyceria elata, Gutierrezia sarothrae, Hedeoma drummondii, Helenium bigelovii, H. puberulum, Helianthus nuttallii, Heracleum maximum, Hesperoyucca whipplei, Heteromeles arbutifolia, Hirschfeldia incana, Hoita orbicularis, Holcus lanatus, Horkelia rydbergii, Hosackia crassifolia, H. oblongifolia, Hypericum anagalloides, H. scouleri, Iris missouriensis, Jaumea carnosa, Juncus balticus, J. effusus, J. macrandrus, J. orthophyllus, J. xiphioides, Juniperus occidentalis, Keckiella cordifolia, K. ternata, Lathyrus vestitus, Lepidospartum squamatum, Leymus cinereus, L. triticoides, Lilium humboldtii, L. parryi, Lobelia cardinalis, Lonicera subspicata, Lupinus burkei, L. hirsutissimus, L. hyacinthinus, L. latifolius, Lythrum californicum, Madia elegans, Maianthemum stellatum, Malacothrix saxatilis, Malosma laurina, Marah macrocarpa, Meconella denticulata, Melica imperfecta, Melilotus albus, Mentha arvensis, Mentzelia laevicaulis, Mimulus cardinalis, M. floribundus, M. guttatus, M. tilingii, Muhlenbergia andina, M. asperifolia, M. richardsonis, M. rigens, Nasturtium officinale, Nicotiana glauca, Oenanthe sarmentosa, Oenothera elata, Oxypolis occidentalis, Pellaea mucronata, P. andromedifolia, Penstemon labrosus, Perideridia parishii, Phacelia cicutaria, P. grandiflora, P. minor, Phragmites australis, Pinus coulteri, P. edulis, P. jeffreyi, P. monophylla, P. ponderosa, P. sabiniana, Platanthera dilatata, Platanus racemosa, Pluchea sericea, Poa pratensis, Polypodium californicum, Populus balsamifera, P. fremontii, P. trichocarpa, Potentilla biennis, P. wheeleri, Prunus ilicifolia, Pseudognaphalium californicum, Pteridium aquilinum, Pterostegia drymarioides, Pycnanthemum californicum, Quercus agrifolia, Q. chrysolepis, Q. cornelius-mulleri, Q. durata, Q. turbinella, Rafinesquia californica, Rhamnus californica, Rhus integrifolia, R. ovata, Ribes cereum, R. nevadense, Ricinus communis, Rosa californica, R. woodsii, Rubus ursinus, Salix exigua, S. gooddingii, S. laevigata, S. lasiolepis, S. lutea, S. scouleriana, Salvia apiana, S. columbariae, S. pachyphylla, Schoenoplectus, Schoenoplectus pungens, Schoenus nigricans, Scirpus microcarpus, Senecio triangularis, Senegalia greggii, Setaria parviflora, Sidalcea malviflora, Silene laciniata, Sisymbrium orientale, Sisyrinchium bellum, Solidago *velutina, Sonchus oleraceus, Sphenosciadium capitellatum, Sporobolus airoides, Stachys albens, Stellaria longipes, S. media, S. nitens, Stephanomeria virgata, Symphoricarpos rotundifolius, Symphyotrichum spathulatum, Thalictrum fendleri, Toxicodendron diversilobum, Trifolium obtusiflorum, T. wormskioldii, Typha latifolia, Umbellularia californica, Urtica dioica, Venegasia carpesioides, Veratrum californicum, Veronica americana, V. anagallis-aquatica, Vitis, Yucca baccata.*

***Solidago houghtonii* Torr. & A. Gray ex A. Gray** grows on alvars, beaches, dunes, flats, shores, and along the margins of ponds at elevations of up to 400 m. Substrates consist of alkaline (pH 7.0–8.0) calcareous sands, cobble, marl, or rock. Flowering occurs from July to October with fruit formation from August to November. Rhizome size is the best predictor of flowering potential. The flowers are self-incompatible and rely on insects (Insecta) for cross-pollination. The seeds are physiologically dormant and require cold stratification (and light) for germination. Buried seeds do not germinate. Vegetative reproduction provides an important means of persistence and occurs by means of rhizome-like branches that extend from the caudex. **Reported associates:** *Agalinis purpurea, Ammophila breviligulata, Andropogon gerardii, Anticlea elegans, Artemisia campestris, Calamagrostis canadensis, Calamovilfa longifolia, Carex conoidea, C. crawei, C. garberi, C. viridula, Castilleja coccinea, Cirsium pitcheri, Cladium mariscoides, Clinopodium glabrum, Cypripedium calceolus, C. candidum, Dasiphora floribunda, Deschampsia cespitosa, Eleocharis quinqueflora, E. rostellata, Elymus canadensis, Equisetum laevigatum, Euthamia graminifolia, Gentianopsis virgata, Geocaulon lividum, Houstonia longifolia, Hypericum kalmianum, H. perforatum, Juncus balticus, J. canadensis, J. vaseyi, Juniperus horizontalis, Kalmia, Larix laricina, Lathyrus japonicus, Lobelia kalmii, L. spicata, Maianthemum stellatum, Muhlenbergia glomerata, Myrica gale, Packera paupercula, Parnassia glauca, Physocarpus opulifolius, Pinus banksiana, P. strobus, Potentilla anserina, Prunus pumila, Rhododendron groenlandicum, Salix cordata, S. exigua, S. myricoides, Schizachyrium scoparium, Scleria verticillata, S. uliginosa, Shepherdia canadensis, Solidago ohioensis, S. spathulata, Spiranthes casei, S. cernua, S. romanzoffiana, Sporobolus heterolepis, Tanacetum bipinnatum, Thuja occidentalis, Triantha glutinosa, Trichophorum cespitosum, T. clintonii, Triglochin maritimum, T. palustre, Valeriana sitchensis, Viola sororia.*

***Solidago latissimifolia* P. Mill.** occurs in brackish to freshwater coastal plain roadsides, swamps and thickets at elevations of up to 80 m. Flowering occurs from August to October, and essentially persists year-long in the southern portion of its range. Life history information for this species is scarce, with some reported under the synonymous name *S. elliottii*. Vegetative reproduction is facilitated by formation of elongate, creeping rhizomes. **Reported associates:** *Andromeda polifolia, Andropogon virginicus, Apios americana, Arethusa bulbosa, Asclepias rubra, Calopogon tuberosus, Carex venusta, Chamaedaphne calyculata, Cladonia, Eleocharis tortilis, Empetrum nigrum, Eriocaulon decangulare,*

Eriophorum virginicum, Eupatorium leptophyllum, Eurybia spectabilis, Fuirena squarrosa, Hypericum brachyphyllum, Ilex laevigata, Juncus diffusissimus, Lachnanthes caroliniana, Liatris scariosa, Lycopodiella appressa, Oclemena nemoralis, Panicum hemitomon, P. verrucosum, Pinus elliottii, Polygala lutea, Pontederia cordata, Pyxidanthera barbulata, Rhynchospora careyana, R. cephalantha, R. filifolia, R. gracilenta, R. inundata, Sarracenia purpurea, Scleria triglomerata, Sphagnum, Triadenum virginicum, Utricularia subulata, Vaccinium, Xyris difformis, X. fimbriata.

***Solidago ohioensis* Riddell** inhabits beaches, bluffs, bogs, depressions, dunes, fens, flats, marshes, meadows, pannes, prairies, ravines, ridges, roadsides, shores, springs, swales, swamps, and the margins of borrow pits, lakes, ponds, and rivers at elevations of up to 300 m. The habitats usually are alkaline and calcareous (e.g., pH: 6.0–7.8; 261–392 mg $CaCO_3$/L) with substrates consisting of clay, cobble, dolomite, gravel, loam, marl, sand, sandy loam, or silty clay. The seeds are physiologically dormant and normally require at least 30 days of cold, moist stratification to promote germination. Optimal germination occurs in the light (on the surface or in light cover), at 15°C–16°C. Seeds should be stored under cold and dry conditions. Seeds average about 0.28 mg in weight and number approximately 0.6–1.1 million/kg. An 8.3:1 ratio of seeds is needed to produce an equivalent number of seedlings in revegetated areas. Plants overwinter by their persistent caudex. Areas managed by prescribed burning and shrub removal showed a successive decrease in cover but not in the frequency of this species. **Reported associates:** *Acer rubrum, A. saccharum, Agalinis purpurea, A. skinneriana, Agrostis gigantea, Aletris farinosa, Alisma, Allium cernuum, Alnus rugosa, Amphicarpaea bracteata, Andropogon gerardii, Aneura pinguis, Apios americana, Apocynum cannabinum, Arnoglossum plantagineum, Asclepias incarnata, A. sullivantii, Aulacomnium palustre, Betula pumila, Bidens cernuus, B. trichospermus, Bromus ciliatus, Bryum pseudotriquetrum, Calamagrostis canadensis, Calopogon tuberosus, Caltha palustris, Campanula americana, C. rotundifolia, Campylium stellatum, Cardamine bulbosa, Carex aquatilis, C. bicknellii, C. buxbaumii, C. crawei, C. eburnea, C. flava, C. haydenii, C. hystericina, C. interior, C. lasiocarpa, C. leptalea, C. pellita, C. prairea, C. sartwellii, C. scoparia, C. sterilis, C. stricta, C. tenera, C. trichocarpa, C. viridula, Chara vulgaris, Chelone glabra, Cicuta maculata, Cirsium muticum, Cladium mariscoides, Clinopodium glabrum, Comandra umbellata, Comarum palustre, Coreopsis tripteris, Cornus amomum, C. foemina, C. sericea, Cuscuta, Cyperus, Cypripedium candidum, Dasiphora floribunda, Deschampsia cespitosa, Desmodium canadense, D. illinoense, D. paniculatum, Dichanthelium acuminatum, Doellingeria umbellata, Drosera rotundifolia, Eleocharis elliptica, E. palustris, E. quinqueflora, E. rostellata, Epipactis helleborine, Equisetum arvense, E. scirpoides, E. variegatum, Eriophorum angustifolium, E. viridicarinatum, Eryngium yuccifolium, Eupatorium altissimum, E. perfoliatum, Euthamia graminifolia, Eutrochium maculatum, Filipendula rubra, Fissidens adianthoides, Fragaria virginiana, Fraxinus pennsylvanica, Galium boreale, Gentiana andrewsii, Gentianopsis crinita, G. virgata, Geum canadense, G. rivale, Glyceria striata, Helianthus giganteus, H. grosseserratus, Hypericum kalmianum, H. prolificum, Impatiens capensis, I. pallida, Inula helenium, Iris versicolor, Juncus articulatus, J. balticus, J. brachycephalus, J. dudleyi, J. nodosus, Juniperus horizontalis, Larix laricina, Leersia oryzoides, Liatris pycnostachya, L. spicata, Lilium philadelphicum, Limprichtia revolvens, Liparis loeselii, Lobelia canbyi, L. kalmii, L. siphilitica, Lonicera oblongifolia, Lophiola aurea, Ludwigia palustris, Lycopus americanus, L. uniflorus, Lysimachia quadriflora, Lythrum alatum, Maianthemum stellatum, Mentha arvensis, Menyanthes trifoliata, Mimulus glabratus, Monarda fistulosa, Muhlenbergia glomerata, M. mexicana, M. richardsonis, M. sylvatica, Myrica gale, M. pensylvanica, Nasturtium officinale, Osmunda regalis, Oxypolis rigidior, Packera aurea, P. paupercula, Panicum flexile, P. virgatum, Parnassia glauca, Parthenium integrifolium, Pedicularis lanceolata, Phlox glaberrima, P. pilosa, Physocarpus opulifolius, Physostegia virginiana, Platanthera peramoena, Pogonia ophioglossoides, Populus balsamifera, P. deltoides, P. tremuloides, Potentilla anserina, Pycnanthemum virginianum, Rhamnus alnifolia, Rhynchospora alba, R. capillacea, Rosa palustris, R. setigera, Rubus pubescens, Rudbeckia hirta, Sagittaria latifolia, Salix candida, S. cordata, S. discolor, S. myricoides, S. nigra, S. pedicellaris, S. rostrata, S. serissima, Sanguisorba canadensis, Sarracenia purpurea, Schizachyrium scoparium, Schoenoplectus acutus, S. pungens, S. tabernaemontani, Scirpus atrovirens, Scleria verticillata, Scorpidium scorpioides, Selaginella apoda, Silphium integrifolium, S. terebinthinaceum, S. trifoliatum, Solidago altissima, S. canadensis, S. gigantea, S. patula, S. riddellii, S. rugosa, S. stricta, S. uliginosa, Sorghastrum nutans, Spartina pectinata, Sphagnum teres, S. warnstorfii, Sporobolus heterolepis, Stachys palustris, Symphyotrichum ericoides, S. firmum, S. lanceolatum, S. novae-angliae, S. oblongifolium, S. pilosum, S. puniceum, Symplocarpus foetidus, Thalictrum dasycarpum, Thaspium trifoliatum, Thelypteris palustris, Thuja occidentalis, Toxicodendron radicans, T. vernix, Triadenum virginicum, Triantha glutinosa, Trichophorum alpinum, T. cespitosum, Trifolium pratense, Triglochin maritimum, T. palustre, Tsuga canadensis, Typha angustifolia, T. latifolia, T. ×glauca, Utricularia cornuta, U. intermedia, U. minor, Valeriana edulis, Verbena hastata, Vernonia gigantea, Veronicastrum virginicum, Viburnum lentago, Zizia aurea.*

***Solidago patula* Muhl. ex Willd.** grows in baygalls, bogs, ditches, fens, floodplains, meadows, prairies, ravines, seeps, slopes, spray cliffs, swamps, thickets, woodlands, and along the margins of lakes rivers, and streams at elevations of up to 1800 m. Although exposures from open sites to partial and dense shade are tolerated, this is categorized as a gap species because the mean importance value is lowest (0.414) under closed canopies and increases substantially (1.03–1.26) where there are medium to large gaps. The substrates are acidic to alkaline (pH 5.8–8.0) and include gravel, Houghton muck, mud, peat, Plummer (grossarenic paleaquults), sand, sandy humus, and silt. The plants are in flower from August to

October. The flowers are pollinated by insects (Insecta), which seek their nectar and pollen. A seed bank of undetermined longevity is formed. Seeds taken from soil cores germinate in the light at 25°C. The roots (up to 84%) are colonized by arbuscular/vesicular mycorrhizal fungi and dark septate endophytes, which facilitate the uptake of phosphorous. Vegetative reproduction occurs by the formation of elongate, creeping rhizomes. **Reported associates:** *Acer rubrum, Achillea millefolium, Aconitum reclinatum, Agrostis perennans, Aletris farinosa, Alnus rugosa, A. serrulata, Amelanchier arborea, Amphicarpaea bracteata, Andropogon gerardii, Anemone virginiana, Aneura pinguis, Angelica atropurpurea, Apios americana, Arethusa bulbosa, Arisaema triphyllum, Aronia arbutifolia, A. melanocarpa, Asclepias incarnata, A. tuberosa, Athyrium filix-femina, Atrichum altecristatum, A. oerstedianum, Aulacomnium palustre, Bazzania trilobata, Betula alleghaniensis, B. lenta, B. pumila, B. ×sandbergii, Bignonia capreolata, Boehmeria cylindrica, Bromus ciliatus, B. kalmii, Bryhnia novae-angliae, Bryum pseudotriquetrum, Calamagrostis canadensis, Calliergonella cuspidata, Calopogon tuberosus, Caltha palustris, Campanula aparinoides, Campylium stellatum, Cardamine bulbosa, C. clematitis, Carex aquatilis, C. atlantica, C. baileyi, C. bromoides, C. buxbaumii, C. collinsii, C. comosa, C. crinita, C. debilis, C. echinata, C. flava, C. folliculata, C. gynandra, C. hystericina, C. interior, C. intumescens, C. lacustris, C. lasiocarpa, C. leptalea, C. leptonervia, C. lurida, C. pellita, C. prairea, C. ruthii, C. scabrata, C. sterilis, C. stricta, C. tenax, C. torta, C. trichocarpa, C. trisperma, Carpinus caroliniana, Castilleja coccinea, Cephalanthus occidentalis, Chamaedaphne calyculata, Chasmanthium laxum, Chelone cuthbertii, C. glabra, C. lyonii, Cicuta maculata, Cinna arundinacea, Cirsium muticum, Cladium mariscoides, Claytonia caroliniana, Clematis virginiana, Climacium, Collinsonia canadensis, Comarum palustre, Conioselinum chinense, Coreopsis tripteris, Cornus amomum, C. foemina, C. sericea, Cuscuta gronovii, Cyperus flavescens, Cypripedium reginae, Danthonia spicata, Dasiphora floribunda, Decodon verticillatus, Decumaria barbara, Deschampsia cespitosa, Desmodium canadense, Dichanthelium clandestinum, D. dichotomum, D. polyanthes, Dicranum fulvum, Doellingeria umbellata, Drosera rotundifolia, Dryopteris cristata, D. intermedia, Dulichium arundinaceum, Eleocharis elliptica, E. rostellata, Epilobium coloratum, Equisetum arvense, E. fluviatile, Eriophorum virginicum, E. viridicarinatum, Eubotrys racemosa, Euonymus americanus, E. obovatus, Eurybia divaricata, Euthamia graminifolia, Eutrochium fistulosum, E. maculatum, Eupatorium perfoliatum, Filipendula rubra, Fissidens adianthoides, Fontinalis sphagnifolia, Fragaria virginiana, Fraxinus americana, F. nigra, Galium asprellum, G. triflorum, G. obtusum, Gaylussacia bigeloviana, Gentiana andrewsii, Gentianopsis, Geum geniculatum, G. rivale, Gillenia trifoliata, Glyceria melicaria, G. striata, Gymnadeniopsis clavellata, Helenium autumnale, Helianthus giganteus, H. grosseserratus, H. schweinitzii, Heuchera villosa, Hookeria acutifolia, Houstonia serpyllifolia, Hypericum densiflorum, H. denticulatum, H. mutilum, Ilex collina, I. coriacea, I. montana, I. opaca, I. verticillata, Illicium floridanum, Impatiens capensis, I. pallida, Iris versicolor, Isoetes engelmannii, Itea virginica, Juncus acuminatus, J. brachycephalus, J. caesariensis, J. coriaceus, J. effusus, J. gymnocarpus, J. marginatus, J. subcaudatus, Juniperus virginiana, Kalmia angustifolia, K. latifolia, Krigia montana, Larix laricina, Leersia oryzoides, L. virginica, Leucothoe fontanesiana, Lilium grayi, L. superbum, Limprichtia revolvens, Lindera benzoin, Linum striatum, Liquidambar styraciflua, Liriodendron tulipifera, Lobelia amoena, L. cardinalis, L. kalmii, L. puberula, L. siphilitica, Lonicera oblongifolia, Ludwigia alternifolia, L. palustris, Luzula multiflora, Lycopus americanus, L. uniflorus, L. virginicus, Lyonia ligustrina, L. lucida, Lysimachia ciliata, L. fraseri, L. lanceolata, L. quadriflora, Magnolia grandiflora, M. tripetala, M. virginiana, Medeola virginiana, Mentha arvensis, Menyanthes trifoliata, Menziesia pilosa, Micranthes pensylvanica, M. petiolaris, Microstegium vimineum, Mimulus ringens, Mnium hornum, Monarda didyma, M. fistulosa, Muhlenbergia glomerata, Myrica cerifera, M. gale, M. heterophylla, M. pensylvanica, Nyssa aquatica, N. biflora, N. sylvatica, Oenothera fruticosa, O. gaura, Onoclea sensibilis, Onosmodium bejariense, Orontium aquaticum, Osmunda claytoniana, O. regalis, Osmundastrum cinnamomeum, Oxypolis rigidior, Packera aurea, Parnassia asarifolia, P. glauca, P. grandifolia, Pedicularis lanceolata, Peltandra sagittifolia, Persea palustris, Persicaria arifolia, P. sagittata, Phalaris arundinacea, Phanopyrum gymnocarpon, Philonotis fontana, Phlox glaberrima, Physocarpus opulifolius, Picea glauca, P. rubens, Pilea fontana, P. pumila, Pinus rigida, P. strobus, Plagiomnium ciliare, Platanthera dilatata, P. grandiflora, P. lacera, P. peramoena, Platanus occidentalis, Poa paludigena, P. palustris, P. pratensis, Pogonia ophioglossoides, Polypodium virginianum, Polystichum acrostichoides, Polytrichum commune, Prenanthes racemosa, Prunella vulgaris, Pycnanthemum virginianum, Ranunculus recurvatus, Rhamnus alnifolia, R. lanceolata, Rhexia virginica, Rhizomnium appalachianum, R. punctatum, Rhododendron arborescens, R. catawbiense, R. maximum, R. viscosum, Rhynchospora alba, R. capillacea, R. capitellata, R. globularis, Ribes hirtellum, Rosa palustris, Rubus hispidus, R. pubescens, R. repens, Rudbeckia fulgida, R. laciniata, Sagittaria latifolia, Salix bebbiana, S. candida, S. caroliniana, S. discolor, S. eriocephala, S. humilis, S. nigra, S. petiolaris, S. sericea, S. serissima, Sarracenia jonesii, S. oreophila, S. purpurea, Sanguisorba canadensis, Scapania nemorosa, Schizachyrium scoparium, Schoenoplectus acutus, S. tabernaemontani, Scirpus atrovirens, S. cyperinus, S. expansus, S. georgianus, S. polyphyllus, Scorpidium scorpioides, Scutellaria lateriflora, Scytonema, Selaginella apoda, S. arenicola, Sematophyllum marylandicum, Senecio suaveolens, Silphium asteriscus, S. trifoliatum, Smilax laurifolia, S. tamnoides, Solidago altissima, S. caesia, S. flexicaulis, S. gigantea, S. ohioensis, S. radula, S. rugosa, S. uliginosa, Sorghastrum nutans, Sphagnum affine, S. angustifolium, S. bartlettianum, S. fallax, S. flexuosum, S. fuscum, S. lescurii,*

S. palustre, S. recurvum, S. subsecundum, S. subtile, S. teres, S. warnstorfii, Sphenopholis pensylvanica, Spiraea alba, S. tomentosa, Spiranthes cernua, S. romanzoffiana, Stenanthium gramineum, Sullivantia sullivantii, Symphyotrichum boreale, S. dumosum, S. firmum, S. laeve, S. lateriflorum, S. novae-angliae, S. praealtum, S. prenanthoides, S. puniceum, Symplocarpus foetidus, Taxodium distichum, Tetragonotheca ludoviciana, Teucrium canadense, Thalictrum clavatum, T. dasycarpum, T. pubescens, Thelypteris noveboracensis, T. palustris, T. simulata, Thuidium delicatulum, Thuja occidentalis, Tomentypnum nitens, Toxicodendron radicans, T. vernix, Trautvetteria caroliniensis, Triadenum virginicum, Trichophorum alpinum, Triglochin maritimum, T. palustre, Trollius laxus, Tsuga canadensis, Typha angustifolia, T. ×glauca, T. latifolia, Utricularia intermedia, Vaccinium corymbosum, V. macrocarpon, Veratrum virginicum, V. viride, Verbena hastata, Vernonia noveboracensis, Viburnum dentatum, V. lentago, V. nudum, Viola cucullata, V. macloskeyi, V. primulifolia, Vitis aestivalis, Woodwardia areolata, W. virginica, Xanthorhiza simplicissima, Xyris tennesseensis, X. torta, Zizia aurea.

***Solidago riddellii* Frank ex Riddell** grows in bogs, depressions, ditches, fens, marshes, prairies, prairie-fens, roadsides, seeps, swales, and along the margins of lakes, rivers, and streams at elevations of up to 400 m. Despite their OBL status, plants occur most frequently in mesic to wet-mesic sites. They grow on acidic to alkaline (pH 5.6–8.3) substrates, which are described as dolomite, gravel, marl, muck, mucky sand, peat, sand, and sandy clay. Flowering occurs late in the season (August–October). The flowers are pollinated primarily by bees (see *Use by wildlife* later), which forage for nectar and pollen. The seeds are dispersed by the wind. The seeds are physiologically dormant and require a period of cold, moist stratification (at least 3 weeks at 4°C) for germination. Because light also is required, optimal germination is attained when seeds are surface sown at 20°C. In natural habitats, drawdown conditions are necessary for seed germination. Overwintering occurs by a persistent woody caudex, which can produce new shoots vegetatively from rhizomatous branches. **Reported associates:** *Acalypha rhomboidea, Achillea millefolium, Agalinis purpurea, A. tenuifolia, Agrimonia parviflora, Agrostis stolonifera, Allium cernuum, Andropogon gerardii, Anemone canadensis, Angelica atropurpurea, Anticlea elegans, Apocynum cannabinum, Asclepias incarnata, A. sullivantii, Betula pumila, B. ×sandbergii, Bidens, Bromus ciliatus, Calamagrostis canadensis, C. stricta, Calopogon tuberosus, Caltha palustris, Calystegia sepium, Campanula aparinoides, Cardamine bulbosa, Carex atherodes, C. buxbaumii, C. crawei, C. hystricina, C. interior, C. cryptolepis, C. lasiocarpa, C. pellita, C. prairea, C. sartwellii, C. sterilis, C. stricta, Chelone glabra, Cicuta maculata, Cirsium muticum, Cladium mariscoides, Comandra umbellata, Coreopsis tripteris, Cornus amomum, C. sericea, Cyperus bipartitus, C. strigosus, Cypripedium candidum, Dalea purpurea, Dasiphora floribunda, Dichanthelium acuminatum, D. leibergii, Doellingeria umbellata, Drosera rotundifolia, Echinacea pallida, Eleocharis elliptica, E.*

palustris, Epilobium coloratum, E. leptophyllum, E. strictum, Equisetum arvense, Eriophorum angustifolium, Eryngium yuccifolium, Eupatorium perfoliatum, Euphorbia corollata, Euthamia graminifolia, Eutrochium maculatum, Fragaria virginiana, Frangula alnus, Galium boreale, G. pilosum, Gentiana andrewsii, Gentianopsis crinita, G. virgata, Geum aleppicum, G. rivale, Glyceria striata, Helenium autumnale, Helianthemum bicknellii, Helianthus grosseserratus, Hemerocallis fulva, Heuchera richardsonii, Hypericum kalmianum, H. perforatum, Ilex verticillata, Juncus canadensis, J. marginatus, J. tenuis, J. torreyi, Larix laricina, Lathyrus palustris, Lespedeza capitata, Liatris aspera, L. ligulistylis, L. pycnostachya, L. spicata, Lilium philadelphicum, Liparis loeselii, Lithospermum canescens, Lobelia canbyi, L. kalmii, L. siphilitica, Lophiola aurea, Lycopus americanus, L. rubellus, Lysimachia quadriflora, Lythrum alatum, Maianthemum stellatum, Mentha arvensis, Micranthes pensylvanica, Muhlenbergia glomerata, M. racemosa, M. richardsonis, Onoclea sensibilis, Oxypolis rigidior, Panicum virgatum, Parnassia caroliniana, P. glauca, P. palustris, Pascopyrum smithii, Pedicularis canadensis, P. lanceolata, Phalaris arundinacea, Phlox glaberrima, P. maculata, Physocarpus opulifolius, Platanthera leucophaea, P. peramoena, Polygala senega, Potentilla norvegica, P. simplex, Prenanthes racemosa, Primula meadia, Prunella vulgaris, Pseudognaphalium obtusifolium, Pycnanthemum pilosum, P. virginianum, Ratibida pinnata, Rhynchospora alba, R. capillacea, Rorippa palustris, Rosa carolina, R. palustris, Rudbeckia fulgida, R. hirta, R. laciniata, R. triloba, Salix ×bebbii, S. candida, S. discolor, S. petiolaris, S. ×rubella, Schizachyrium scoparium, Schoenoplectus acutus, S. pungens, Scirpus atrovirens, Scleria verticillata, Scutellaria parvula, Silphium integrifolium, S. laciniatum, S. terebinthinaceum, Solidago canadensis, S. gigantea, S. nemoralis, S. ohioensis, S. rigida, S. uliginosa, Sorghastrum nutans, Spartina pectinata, Sphenopholis intermedia, Spiraea alba, Spiranthes cernua, Sporobolus compositus, S. heterolepis, Symphyotrichum boreale, S. ericoides, S. laeve, S. lateriflorum, S. novae-angliae, S. puniceum, Thalictrum dasycarpum, Thaspium trifoliatum, Thelypteris palustris, Tradescantia ohioensis, Triantha glutinosa, Triglochin maritimum, T. palustre, Typha angustifolia, T. ×glauca, T. latifolia, Utricularia minor, Valeriana edulis, Verbena hastata, Vernonia fasciculata, Veronicastrum virginicum, Viburnum lentago, Viola sagittata, V. sororia, Zanthoxylum americanum, Zizia aurea.

***Solidago stricta* Aiton** inhabits tidal or nontidal, freshwater to brackish sites including barrens, beaches, bogs, depressions, ditches, dunes, flats, flatwoods, floodplains, glades, lagoons, marshes, prairies, roadsides, salt flats, savannas, seeps, woodlands, and margins of canals, ephemeral ponds, hammocks, lakes, and salt marsh at elevations of up to 300 m. Exposures range from full sun to partial shade. The substrates are acidic to alkaline (pH 5.6–7.8) and include alluvial sand, Biscayne (typic fluvaquents), Dothan (plinthic paleudults), Hallandale (lithic psammaquents), Leefield (plinthaquic paleudults), limestone, loamy sand, marl, oolite, peaty sand,

Samsula (terric medisaprists), sand, and sandy peat. The flowers are produced from August to November and are pollinated by bees (Hymenoptera) and other insects (Insecta). Requirements for seed germination have not been determined. Vegetative reproduction occurs by means of slender, stolon-like rhizomes. The plants often appear in recently burned or otherwise disturbed sites and persist in areas characterized by light anthropogenic disturbance and in remediated sites. They exhibit significantly enhanced abundance in mowed sites. **Reported associates:** *Acer rubrum, Agalinis georgiana, A. linifolia, A. purpurea, Alnus serrulata, Ampelopsis arborea, Amphicarpum muhlenbergianum, Andropogon arctatus, A. glomeratus, A. gyrans, A. virginicus, Anemia wrightii, Aristida purpurascens, A. stricta, Aronia arbutifolia, Arundinaria tecta, Asemeia grandiflora, Baccharis glomeruliflora, B. halimifolia, Bidens, Bigelowia nudata, Buchnera longifolia, Calamovilfa brevipilis, Carex glaucescens, Carphephorus paniculatus, C. tomentosus, Cassytha filiformis, Centella asiatica, Chamaecrista deeringiana, C. nictitans, Chaptalia tomentosa, Chiococca alba, Cirsium horridulum, C. virginianum, Cladium jamaicense, Clinopodium dentatum, Coleataenia anceps, C. longifolia, C. tenera, Coreopsis gladiata, Crotalaria purshii, Ctenium aromaticum, Cyperus, Danthonia spicata, Desmodium lineatum, D. tenuifolium, Dichanthelium aciculare, D. acuminatum, D. dichotomum, D. longiligulatum, D. ovale, D. scabriusculum, D. sphaerocarpon, D. strigosum, Diodia virginiana, Diospyros virginiana, Dyschoriste oblongifolia, Elephantopus nudatus, Elionurus tripsacoides, Eragrostis elliottii, E. refracta, E. spectabilis, Erigeron vernus, Eriocaulon ravenelii, Eriochloa michauxii, Eryngium aquaticum, E. yuccifolium, Eupatorium leptophyllum, E. leucolepis, E. mohrii, E. rotundifolium, E. semiserratum, Euthamia graminifolia, Fuirena breviseta, Funastrum clausum, Gaylussacia frondosa, Gentiana autumnalis, Gratiola pilosa, Gymnopogon brevifolius, Helenium pinnatifidum, Helianthus angustifolius, Hibiscus aculeatus, H. coccineus, Hydrocotyle, Hymenocallis palmeri, Hypericum crux-andreae, H. microsepalum, H. setosum, Hyptis alata, Ilex coriacea, I. glabra, Juncus roemerianus, Lachnocaulon anceps, Lespedeza capitata, Liatris garberi, L. gracilis, L. spicata, Linum carteri, L. medium, Liquidambar styraciflua, Liriodendron tulipifera, Lobelia glandulosa, L. nuttallii, Ludwigia hirtella, L. microcarpa, L. simpsonii, Magnolia virginiana, Marshallia graminifolia, Mikania scandens, Mnesithea rugosa, Muhlenbergia sericea, Myrica cerifera, M. heterophylla, Muhlenbergia capillaris, M. expansa, Nyssa biflora, N. sylvatica, Osmundastrum cinnamomeum, Oxydendrum arboreum, Oxypolis filiformis, Panicum verrucosum, P. virgatum, Paspalum monostachyum, P. praecox, P. setaceum, P. urvillei, Pinus elliottii, P. palustris, P. serotina, P. taeda, Piriqueta cistoides, Pityopsis aspera, P. graminifolia, Pluchea baccharis, Pogonia ophioglossoides, Polygala lutea, P. nana, Pteridium aquilinum, Pycnanthemum flexuosum, Quercus laevis, Q. minima, Q. phellos, Q. pumila, Rhexia alifanus, R. mariana, R. nashii, Rhododendron canescens, Rhus copallinum, Rhynchospora chapmanii, R. colorata, R. debilis, R. divergens, R. elliottii, R. microcarpa, R. tracyi, R. stenophylla, R. torreyana, Rubus argutus, R. trivialis, Rudbeckia fulgida, Sabatia campanulata, S. decandra, Saccharum baldwinii, S. coarctatum, S. giganteum, Sarracenia flava, S. purpurea, S. ×catesbaei, Schizachyrium rhizomatum, S. scoparium, S. tenerum, Scleria muehlenbergii, S. reticularis, S. triglomerata, S. verticillata, Scutellaria integrifolia, Senecio, Senna ligustrina, Serenoa repens, Smilax glauca, S. laurifolia, S. walteri, Solidago arguta, S. ohioensis, S. odora, S. sempervirens, Sorghastrum nutans, Spartina alterniflora, Sphagnum, Sphenomeris clavata, Spiranthes cernua, Sporobolus curtissii, Symphyotrichum dumosum, S. lateriflorum, S. tenuifolium, S. walteri, Taxodium ascendens, Tephrosia spicata, Vaccinium corymbosum, Viburnum rufidulum, Vigna luteola, Viola lanceolata, Vitis rotundifolia, Xyris ambigua, X. caroliniana.*

***Solidago uliginosa* Nutt.** occurs in bogs, dikes, ditches, fens, flats, floodplains, grass balds, marshes, meadows, prairies, roadsides, ruts, savannas, seeps, shores, stream deltas, string bogs, swamps, thickets, and along the margins of lakes, rivers, and streams at elevations of up to 1500+ m. The plants are found in open sunny exposures to partial shade, on wet substrates or in shallow (e.g., 8 cm deep) standing water. The substrates represent a broad range of acidity (pH 3.7–8.2) and include clay, gravel, humus, muck, peat, peaty muck, sand, sandy peat, and silt. The flowers are insect-pollinated, principally by bees (Insecta: Hymenoptera). This is a short-day plant, with flowering commencing about July 28th and lasting for 43–46 days, with the peak occurring around 2 September. Elevated temperatures delay flowering significantly (by up to 6 days). The seeds require a period of cold stratification for germination. Leaf extracts of *S. uliginosa* have a strong allelopathic affect, which reduces germination of jack pine (Pinaceae: *Pinus banksiana*) seeds. In fen habitats, up to 73% of the roots are colonized by arbuscular mycorrhizal fungi and dark septate endophytes. Vegetative reproduction occurs by the development of elongate, branching rhizomes. Greater occurrences of this species are associated with sites where white-tailed deer (Mammalia: Cervidae: *Odocoileus virginianus*) are present. **Reported associates:** *Abies balsamea, Acer rubrum, A. saccharinum, A. saccharum, Aconitum uncinatum, Acorus calamus, Agrostis hyemalis, A. perennans, Alnus rugosa, A. serrulata, Amelanchier, Andromeda polifolia, Andropogon gerardii, Anemone canadensis, Aneura pinguis, Angelica atropurpurea, Anthoxanthum odoratum, Aralia racemosa, Arethusa bulbosa, Aronia arbutifolia, A. melanocarpa, A. ×prunifolia, Asclepias incarnata, Aulacomnium palustre, Avenella flexuosa, Bidens cernuus, B. trichospermus, Betula papyrifera, B. pumila, B. ×sandbergii, Boehmeria cylindrica, Bromus ciliatus, B. kalmii, Bryum pseudotriquetrum, Calamagrostis canadensis, C. stricta, Calliergon giganteum, C. trifarium, Calliergonella cuspidata, Calopogon tuberosus, Caltha palustris, Calystegia sepium, Campanula aparinoides, Campylium stellatum, Carex aquatilis, C. arcta, C. atlantica, C. buxbaumii, C. canescens, C. chordorrhiza, C. cumulata, C. debilis, C. echinata, C. exilis, C. flava, C. folliculata, C. haydenii, C. hystericina, C. interior, C. lacustris, C. lasiocarpa, C. leptalea, C.*

limosa, C. livida, C. oligosperma, C. pellita, C. prairea, C. rostrata, C. scoparia, C. sterilis, C. stricta, C. tenuiflora, C. tetanica, C. trisperma, C. wiegandii, Carpinus caroliniana, Castilleja coccinea, Cetraria islandica, Chamaedaphne calyculata, Chara vulgaris, Chelone glabra, Cicuta bulbifera, Cirsium muticum, C. palustre, Cladina arbuscula, C. mitis, C. rangiferina, C. terrae-novae, Cladium mariscoides, Cladonia alpestris, C. arbuscula, C. crispata, C. mitis, C. rangiferina, C. uncialis, Clematis virginiana, Comarum palustre, Comptonia peregrina, Coptis trifolia, Cornus canadensis, C. foemina, C. racemosa, C. sericea, Cypripedium acaule, C. parviflorum, C. reginae, Danthonia compressa, Dasiphora floribunda, Dennstaedtia punctilobula, Deschampsia cespitosa, Dichanthelium acuminatum, D. clandestinum, D. dichotomum, Dicranum polysetum, Doellingeria umbellata, Drosera intermedia, D. rotundifolia, Dryopteris carthusiana, D. cristata, Dulichium arundinaceum, Eleocharis elliptica, E. palustris, E. rostellata, E. tenuis, Elymus trachycaulus, Empetrum nigrum, Epilobium leptophyllum, Equisetum arvense, E. fluviatile, Erigeron philadelphicus, Eriophorum angustifolium, E. vaginatum, E. virginicum, E. viridicarinatum, Eupatorium perfoliatum, Eurybia radula, Euthamia graminifolia, Eutrochium maculatum, Fissidens adianthoides, Fragaria virginiana, Fraxinus nigra, Galium asprellum, G. boreale, G. labradoricum, G. tinctorium, G. triflorum, Gaultheria hispidula, G. procumbens, Gaylussacia baccata, G. dumosa, Gentiana linearis, Gentianopsis virgata, Geum aleppicum, G. rivale, Gillenia trifoliata, Glyceria canadensis, G. grandis, G. striata, Hamatocaulis vernicosus, Helenium autumnale, H. brevifolium, Helianthus giganteus, Hieracium aurantiacum, Houstonia caerulea, Hydrophyllum canadense, Hylocomium splendens, Hypericum boreale, H. densiflorum, H. ellipticum, H. kalmianum, Ilex verticillata, Impatiens capensis, Iris versicolor, Juncus brachycephalus, J. brevicaudatus, J. canadensis, J. dudleyi, J. effusus, J. filiformis, J. greenei, J. militaris, J. nodosus, J. stygius, J. subcaudatus, Juniperus communis, J. virginiana, Kalmia angustifolia, K. latifolia, K. polifolia, Kurzia setacea, Lactuca canadensis, Larix laricina, Leersia oryzoides, Liatris pycnostachya, L. spicata, Lilium michiganense, Limprichtia revolvens, Linum striatum, Liparis loeselii, Lobelia kalmii, Lonicera oblongifolia, L. villosa, Lycopodium clavatum, Lycopus americanus, L. uniflorus, Lyonia ligustrina, Lysimachia quadriflora, L. thyrsiflora, Lythrum alatum, Maianthemum trifolium, Meesia triquetra, Mentha arvensis, Menyanthes trifoliata, Muhlenbergia glomerata, M. mexicana, Mylia anomala, Myrica gale, M. pensylvanica, Nemopanthus mucronatus, Neottia bifolia, Oenothera perennis, Osmunda regalis, Osmundastrum cinnamomeum, Oxypolis rigidior, Packera aurea, P. paupercula, P. schweinitziana, Paludella squarrosa, Panicum virgatum, Parnassia glauca, P. grandifolia, Pedicularis canadensis, P. lanceolata, Peltandra virginica, Persicaria sagittata, Phalaris arundinacea, Philonotis fontana, Phragmites australis, Physocarpus opulifolius, Picea glauca, P. mariana, P. rubens, Pinus banksiana, P. rigida, P. strobus, Platanthera dilatata, P. hyperborea, P. psycodes, Pleurozium schreberi, Poa pratensis,

Pogonia ophioglossoides, Polygala sanguinea, Polytrichum strictum, Populus grandidentata, P. tremuloides, Potentilla canadensis, Pteridium aquilinum, Pycnanthemum tenuifolium, P. virginianum, Quercus rubra, Ranunculus gmelinii, Rhamnus alnifolia, R. frangula, Rhododendron canadense, R. groenlandicum, R. maximum, R. viscosum, Ribes hirtellum, Rhynchospora alba, R. capitellata, Rhytidium rugosum, Ribes hirtellum, Rosa palustris, Rubus arcticus, R. chamaemorus, R. hispidus, R. pubescens, Rudbeckia fulgida, R. hirta, Rumex acetosella, Sabatia campanulata, Salix bebbiana, S. candida, S. discolor, S. eriocephala, S. humilis, S. nigra, S. pedicellaris, S. petiolaris, S. sericea, S. serissima, Sanguinaria canadensis, Sanguisorba canadensis, Sanicula marilandica, Sarracenia purpurea, Scheuchzeria palustris, Schizachyrium scoparium, Schoenoplectus acutus, Scirpus atrovirens, S. cyperinus, S. expansus, Scleria verticillata, Scorpidium scorpioides, Scutellaria galericulata, Selaginella apoda, Silphium terebinthinaceum, Sisyrinchium angustifolium, Smilax tamnoides, Solidago canadensis, S. gigantea, S. juncea, S. ohioensis, S. patula, S. riddellii, S. rugosa, Sorghastrum nutans, Sparganium emersum, S. eurycarpum, Sphagnum affine, S. angustifolium, S. bartlettianum, S. capillifolium, S. contortum, S. flavicomans, S. fuscum, S. magellanicum, S. rubellum, S. subsecundum, S. teres, S. warnstorfii, Spiraea alba, S. tomentosa, Spiranthes lucida, S. romanzoffiana, Splachnum ampullaceum, Symphyotrichum boreale, S. firmum, S. laeve, S. lanceolatum, S. novae-angliae, S. novibelgii, S. puniceum, Symplocarpus foetidus, Thalictrum dasycarpum, T. pubescens, Thelypteris palustris, Thuidium delicatulum, Thuja occidentalis, Tomenthypnum nitens, Toxicodendron vernix, Triadenum virginicum, Triantha glutinosa, Trichophorum alpinum, T. cespitosum, Trientalis borealis, Triglochin maritimum, Trollius laxus, Tsuga canadensis, Typha angustifolia, T. ×glauca, T. latifolia, Utricularia cornuta, U. intermedia, U. minor, Uvularia grandiflora, Vaccinium angustifolium, V. corymbosum, V. macrocarpon, V. myrtilloides, V. oxycoccos, Valeriana ciliata, V. edulis, Viburnum dentatum, V. nudum, Viola cucullata, V. sororia, V. walteri, Xyris torta, Zizia aurea.

***Solidago verna** M.A. Curtis ex Torr. & A. Gray* is a rare plant of bogs, ditches, flatwoods, meadows, roadsides, sandhills, savannas, and seeps at elevations of up to 70 m. Although restricted to relatively poorly drained soils, the plants do not tolerate inundation or prolonged saturation of soils, which include Goldsboro and Lenoir (aeric paleaquult) series. Populations thrive under a fairly open canopy (usually *Pinus*) in areas where 2–3 year cycles of fire reduce shading by competing vegetation. Unlike most goldenrods, this species flowers in the spring (May–June). Little other information is available on the reproductive or seed ecology. Seedlings establish only in open sites cleared by fire or other disturbance. The plants perennate by a persistent underground rosette. **Reported associates:** *Agalinis aphylla, Aletris aurea, Aristida stricta, Bigelowia nudata, Calamovilfa brevipilis, Carphephorus paniculatus, C. tomentosus, Chaptalia tomentosa, Cleistesiopsis bifaria, C. divaricata, Ctenium aromaticum, Danthonia sericea, Dionaea muscipula, Erigeron* *vernus, Eriocaulon decangulare, Eryngium integrifolium, Gaylussacia frondosa, Gymnadeniopsis nivea, Ilex glabra, Lilium pyrophilum, Liquidambar styraciflua, Lysimachia asperulifolia, Marshallia graminifolia, Muhlenbergia expansa, Oxypolis denticulata, Parnassia caroliniana, Pinguicula caerulea, Pinus palustris, P. taeda, Pteridium aquilinum, Rhynchospora oligantha, Solidago pulchra, Xyris scabrifolia.*

Use by wildlife: Various insects (Insecta) are attracted to *Solidago* flowers to gather their nectar and pollen. The flowers of *S. confinis* are visited by numerous bees (Hymenoptera: Andrenidae: *Andrena aurihirta, Perdita colei, P. gutierreziae, P. media, P. zonalis*; Colletidae: *Hylaeus asininus, H. episcopalis, H. wootoni*; Megachilidae: *Anthidium placitum, A. tenuiflorae, Heriades cressoni, Megachile angelarum, M. gentilis, M. nevadensis, M. perihirta*; Miridae: *Lopidea taurina*), beetles (Coleoptera: Ripiphoridae: *Macrosiagon pavipenne*), and soldier flies (Diptera: Stratiomyidae: *Hedriodiscus truquii*). The flowers also are known to attract butterflies (Lepidoptera: Lycaenidae: *Plebejus acmon, Strymon melinus*). The plants are hosts to several fungi (Ascomycota: Botryosphaeriaceae: *Phyllosticta sphaeropsispora*; Basidiomycota: Agaricomycetidae: *Mycena clavata*). The flowers of *S. houghtoni* provide pollen and nectar to a variety of insects (Insecta), which serve as pollen vectors. These include bees (Hymenoptera: Halictidae), butterflies (Lepidoptera), flies (Diptera), moths (Lepidoptera), and wasps (Hymenoptera). Other insects associated with the plants include beetles (Coleoptera: Phalacridae), bugs (Hemiptera: Nabidae), butterflies (Lepidoptera: Coliophoridae), flies (Diptera: Bombyliidae; Syrphidae), and froghoppers (Orthoptera: Cercophidae). *Solidago ohioensis* is a host to aphids (Insecta: Hemiptera: Aphididae: *Uroleucon lanceolatum*), gall midges (Insecta: Diptera: Cecidomyiidae: *Asteromyia modesta*) and fungi (Basidiomycota: Pucciniaceae: *Puccinia dioicae*). The flowers attract buckeyes (Insecta: Lepidoptera: Nymphalidae: *Junonia coenia*) and other butterflies, particularly in late summer. The flowers of *S. patula* attract bees (Insecta: Hymenoptera), beetles (Insecta: Coleoptera), flies (Insecta: Diptera), and wasps (Insecta: Hymenoptera: Eulophidae). The foliage is a host to larval moths (Insecta: Lepidoptera: Gracillariidae: *Cremastobombycia solidaginis*) and a variety of fungi including Ascomycota (Botryosphaeriaceae: *Phyllosticta*; Erysiphaceae: *Golovinomyces asterum*; Mycosphaerellaceae: *Cercosporella virgaureae, Septoria solidaginicola*; Rhytismataceae: *Rhytisma solidaginis*) and Basidiomycota (Coleosporiaceae: *Coleosporium asterum*; Pucciniaceae: *Puccinia asterum, P. dioicae*). The plants commonly are parasitized by dodder (Convolvulaceae: *Cuscuta gronovii*). The herbage of *S. riddellii* contains 6.1% protein (about average for most plants) and 8.4% polyphenolics. The flowers are among those that are most highly attractive to bees (Insecta: Hymenoptera) in the standing vegetation and are visited, especially in late season, by various species including mining bees (Insecta: Hymenoptera: Andrenidae: *Andrena*), honey bees (Insecta: Hymenoptera: Apidae: *Bombus impatiens, Ceratina calcarata, Melissodes, Xylocopa*

virginica), plasterer bees (Insecta: Hymenoptera: Colletidae: *Hylaeus affinis, H. annulatus, H. modestus*), and sweat bees (Insecta: Hymenoptera: Halictidae: *Agapostemon virescens, Augochlorella aurata, Halictus confusus, H. ligatus, H. rubicundus, Lasioglossum admirandum, L. pilosum*). In addition they attract buckeye, pearl crescent, and orange sulfur butterflies (Insecta: Lepidoptera: Nymphalidae: *Junonia coenia, Phyciodes tharos*; Pieridae: *Colias eurytheme*), assorted flies (Insecta: Diptera), moths (Insecta: Lepidoptera) and wasps (Insecta: Hymenoptera). The inflorescences provide cover for predatory assassin bugs (Insecta: Hemiptera: Reduviidae: *Arilus cristatus, Phymata pennsylvanica*) and praying mantids (Mantodea: Mantidae). The flowers of *S. stricta* are visited by plasterer bees (Insecta: Hymenoptera: Colletidae: *Colletes simulans*) and are hosts to thick-headed flies (Insecta: Diptera: Conopidae: *Physoconops weemsi*). The plants are hosts to larval butterflies and moths (Insecta: Lepidoptera: Noctuidae: *Schinia nundina*) and red pine needle rust fungi (Basidiomycota: Coleosporiaceae: *Coleosporium asterum*). *Solidago uliginosa* is a host to aphids (Insecta: Hemiptera: Aphididae: *Uroleucon pieloui, U. pseudambrosiae*), leaf-folding caterpillars (Insecta: Lepidoptera: Gelechiidae: *Dichomeris leuconotella*; *Gnorimoschema gallaeasterella*), mold fungi (Ascomycota: Davidiellaceae: *Cladosporium astericola*), and powdery mildew fungi (species undetermined). The flowers are visited routinely by bumblebees and honeybees (Insecta: Hymenoptera: Apidae: *Apis, Bombus*). This species is regarded as an important honey plant. The foliage occasionally is clipped by muskrats (Mammalia: Cricetidae: *Ondatra zibethicus*). The plants are dominant at nesting sites of the bobolink (Aves: Icteridae: *Dolichonyx oryzivorus*) and occur to a lesser extent at nesting sites of the savannah sparrow (Aves: Emberizidae: *Passerculus sandwichensis*).

Economic importance: food: *Solidago* is not often used as food although the leaves of some species reportedly are edible; **medicinal:** Because goldenrod pollen is transported by insects, it is not a factor in human allergies as is often thought. This unfounded association occurred because goldenrods flower simultaneously with ragweed (Asteraceae: *Ambrosia*), whose wind-borne pollen is highly allergenic. The Potawatomi prepared a root poultice from *S. uliginosa* as a treatment for skin boils. *Soldago uliginosa* was used in sweat baths by the Hocąk of Wisconsin; **cultivation:** *Solidago confinis* is cultivated as a garden plant in California. *Solidago ohioensis* is grown as an ornamental species for native plant gardens. *Solidago riddellii* and *S. stricta* are readily available from seed catalogs; **misc. products:** *Solidago ohioensis* has been used in revegetation programs for urban parks in England. The leaves of *S. riddellii* are a source of natural rubber (cis-1,4-polyisoprene) with a 0.7% hydrocarbon content; **weeds:** Because of its ubiquity, *Solidago* is regarded inappropriately as a weedy genus by many. *Solidago latissimifolia* has been reported as a rare weed of cornfields and pastures in Wisconsin, which is a region far beyond its native range and warrants further documentation; **nonindigenous species:** *S. latissimifolia* (Wisconsin).

Systematics: *Solidago* is assigned to tribe Astereae where it is closely related to *Ericameria*, *Euthamia* and *Heterotheca* (Figure 5.172). *Solidago* presents numerous taxonomic problems and various interpretations have made it difficult to properly apply information gleaned from the literature to a particular species. Accordingly, this issue should be considered carefully when evaluating the information summarized in the species accounts. Naturally occurring hybrids between *S. rugosa* and *S. sempervirens* have been recognized at species rank as *S. ×asperula*, which has been categorized as OBL. However, because this taxon remains interfertile with the parental species, it is recognized here as an interspecific hybrid and has been excluded from the species accounts. The circumscription of *Solidago stricta* embodies much confusion. In one interpretation, *S. stricta* includes the taxon known formerly as *S. gracillima* as a subspecies, which included

material known as *S. perlonga*. More recently, arguments have been made not only to retain *S. gracillima* as a distinct species but also to subdivide *S. stricta* into two additional species, in which the name *S. stricta* is applied to the segregate known formerly as *S. perlonga* with the name *S. virgata* applied to the other segregate. In the present work, *S. stricta* is retained in the broader sense to include both *S. gracillima* and *S. perlonga*, primarily because it would be virtually impossible to parse the information in the literature so that accurate correspondence with each of the segregate taxa could be made. Moreover, a comprehensive systematic study of this group should be made to determine whether these taxa merit species status rather than infraspecific recognition under *S. stricta*. Consequently, all information pertaining to *S. gracillima* or *S. perlonga* has been summarized here under the name *S. stricta*. Based on a morphometric analysis, a subspecies of *S. patula* (*S. patula* subsp. *strictula*) recently has been elevated to species rank as *Solidago salicina*. However, because this question also has not been evaluated thoroughly, all information pertaining to both taxa has been combined here under *S. patula*. *Solidago houghtonii*, *S. ohioensis* and *S. riddellii* are placed by some authors in the genus *Oligoneuron*. It is difficult to resolve such taxonomic issues satisfactorily because no comprehensive molecular phylogenetic study of *Solidago* has been undertaken at this time. However, phylogenetic assessments of cpDNA and ETS/ITS sequence data for five *Oligoneuron* and 9–26 *Solidago* species resolve all species assigned to the former genus as nested among the latter (Figure 5.172); hence all have been retained here within *Solidago*. Taxonomically, the OBL species occur within two sections and various subsections of *Solidago*, which is consistent with their dispersed placements throughout the genus (Figure 5.172). Section *Solidago* contains subsection *Argutae* (*S. patula*, *S. verna*), subsection *Junceae* (*S. confinis*), subsection *Maritimae* (*S. stricta*, *S. uliginosa*), and subsection *Venosae* (*S. latissimifolia*); whereas, *S. houghtonii*, *S. ohioensis* and *S. riddellii* are assigned to section *Ptarmicoidei*. However, the monophyly of these infrageneric subdivisions is questionable, especially with respect to section *Ptarmicoidei* (Figure 5.172). Morphometric analyses place *S. confinis* with the western *S. guiradonis* and *S. spectabilis* in a group that is distinct from eastern North American and South American species. It is evident that much additional systematic research is needed in this genus. The base chromosome number of *Solidago* is $x = 9$. Most of the OBL species are diploids ($2n = 18$), which include *S. confinis*, *S. ohioensis*, *S. patula*, *S. riddellii*, and *S. verna*. Mixtures of diploid, tetraploid and hexaploid cytotypes occur in *S. latissimifolia* and *S. stricta* ($2n = 18, 36, 54$) and *S. uliginosa* ($2n = 18, 36$). The tetraploid cytotypes of *S. uliginosa* are uniquely restricted to limestone outcrops compared to the widespread (but not on limestone outcrops) diploids. From 1 to 2 supernumerary chromosomes have been observed in 3.4%–5.5% of root-tip preparations examined in *S. uliginosa*; however, they have not been found in the meiotic anther cells. *Solidago stricta* is believed to hybridize with *S. sempervirens* when in proximity of salt marshes. *Solidago houghtonii* ($2n = 54$) is an allohexaploid

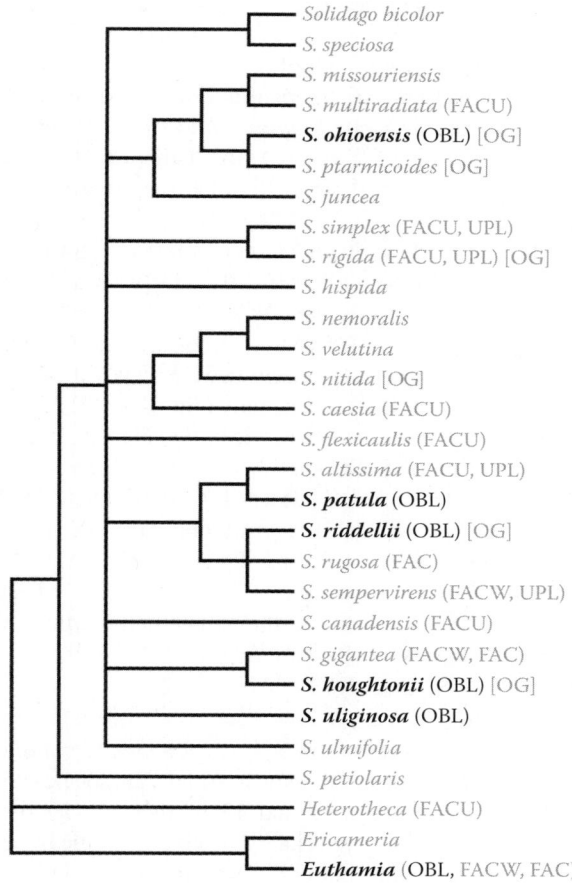

FIGURE 5.172 Phylogenetic relationships among *Solidago* species as indicated by analysis of combined cpDNA sequences. Species assigned formerly to the genus *Oligoneuron* [OG] do not resolve as a separate clade, but nest among *Soldago* species. Within *Solidago*, the [OG] species have been assigned to section *Ptarmicoidei*, which this analysis depicts as polyphyletic. These results also indicate that the OBL habit (taxa in bold) has evolved repeatedly in the genus. *Solidago houghtonii* is an allohexaploid, whose maternal parent was *S. gigantea* as indicated here. (Adapted from Laureto, P.J. & Barkman, T.J., *Syst. Bot.*, 36, 209–226, 2011.)

derived from complex introgressive hybridization involving *S. gigantea* as the maternal parent and *S. riddellii*, *S. ptarmicoides*, and *S. ohioensis* as paternal contributors.

Comments: *Solidago latissimifolia*, *S. patula* and *S. uliginosa* occur throughout eastern North America), *S. houghtonii*, *S. ohioensis* and *S. riddellii* in the midwestern region, and *S. stricta* in the southeastern United States. More restricted distributions occur in *S. verna* (North and South Carolina) and *S. confinis* (California and Nevada; extending into Baja California).

References: Amon et al., 2002; Anderson, 1919; Anderson & Leopold, 2002; Basinger, 2003; Bertin et al., 2010; Betz & Cole, 1969; Blasdale, 1919; Bonner et al., 2009; Bowles et al., 1996; Brotherson, 1983; Brown, 1967; Cain & Slater, 1948; Camras, 2007; Carr, 2007; Carr et al., 1986; Chmielewski et al., 1987; Choesin & Boerner, 2000; Cornwell et al., 2001; Dale et al., 2002; Davis, 1929; Dicks et al., 2010; Dölle & Schmidt, 2009; Faber-Langendoen & Maycock, 1994; Fell, 1957; Fleming et al., 2007; Genesis Nursery, 2013; Gilliam et al., 2006; Gleason, 1917; Goodwillie & Franch, 2006; Goodwin, 1937; Gordon, 1933; Greene, 1948; Gross & Werner, 1983; Heinrich, 1975a; Hitchmough et al., 2004; Homoya et al., 1985; James, 1960; Johnston & Reekie, 2008; Kapoor, 1978; Keil et al., 1988; Kindscher & Hurlburt, 1998; Klips, 2003; Lamont & Young, 2005; Laphitz et al., 2011; Laureto & Barkman, 2011; Loeffler, 1994; Markle, 1915; Middleton, 2002b; Moran, 1981; Morton, 1979; Nekola, 2004; Nelson, 2006; Nichols & Sperduto, 2012; Nolin & Runkle, 1985; Ogle, 1989; Orzell & Bridges, 2006; Pellerin et al., 2006; Penskar, 1997; Pringle, 1977, 1982; Rentch & Fortney, 1997; Richards, 1972, 1976; Sanderson et al., 2007; Schilling et al., 2008; Schoolmaster, Jr., 2005; Semple, 2012, 2013; Semple & Cook, 2006; Semple et al., 1984, 2012; Shamey, 2011; Simmons & Strong, 2002; Sladen, 1916; Soper & Payne, 1997; Sorrie et al., 2006; Stireman III et al., 2010; Stoynoff, 1993; Taylor & Raney, 2013; Tompkins et al., 2010; Tuell et al., 2008; Vestal, 1914; Walck et al., 1996, 1998; Warren, 2001; Weakley & Schafale, 1994; Weishampel & Bedford, 2006; Wells, 1996; Wieder et al., 1981, 1984; Zaremba, 2004; Zartman & Pittillo, 1998.

44. *Symphyotrichum*

Aster

Etymology: from the Greek *symphysis trichos* ("grown together hair"); of uncertain application

Synonyms: *Aster* (in part); *Eurybia* (in part); *Heleastrum*

Distribution: global: Asia, Europe*, New World; **North America:** widespread

Diversity: global: 90 species; **North America:** 77 species

Indicators (USA): OBL: *Symphyotrichum bahamense, S. boreale, S. chapmanii, S. defoliatum, S. divaricatum, S. elliottii, S. expansum, S. lanceolatum, S. lentum, S. novi-belgii, S. praealtum, S. puniceum, S. squamatum, S. subulatum, S. tenuifolium*; **FACW:** *S. lanceolatum, S. novi-belgii, S. praealtum, S. subulatum*

Habitat: brackish, freshwater, saline; palustrine; **pH:** 4.0–12.6; **depth:** <1 m; **life-form(s):** emergent herb

Key morphology: Shoots (to 80–300 cm) erect, annual or perennial and rhizomatous; leaves basal and cauline, the basal leaves petioled (to 15 cm), winged or wingless, the blades (to 3–25 cm) elliptic, lanceolate, linear, linear-lanceolate, linear-oblanceolate, oblanceolate, obovate, ovate, ovate-lanceolate, spatulate, or suborbiculate; cauline leaves short-petioled, sessile or clasping, the blades (to 9–22 cm) elliptic, lanceolate, lanceolate-ovate, linear, linear-lanceolate, oblanceolate, oblong, or subulate; heads (1 or to 50–150) radiate, peduncled (to 2–5 or to 20 cm), single or arranged secondarily in corymbose panicles or racemes, involucres (to 6.2–15 mm), phyllaries (to 62) in 4–6 series, receptacles lacking pales; ray florets (to 23–60) pistillate, rays (to 2–20 mm) blue, blue-lavender, blue-violet, lavender, light pink, pale purple, pale rose, pink, purple, rose-purple, violet, or white; disk florets (to 13–90) bisexual, 5-lobed, the corollas (to 4.7–7.5 mm) brown, brownish purple, cream, light yellow, pink, pinkish purple, purple, reddish-brown, or yellow; cypselae (to 2.0–4.5 mm) brown, gray, gray-brown, purple, or tan, pappus of persistent bristles (to 4.2–6.5 mm) in 1–3 series.

Life history: duration: annual (seeds); perennial (rhizomes; thickened caudex); **asexual reproduction:** rhizomes, stolons; **pollination:** insect, self; **sexual condition:** gynomonoecious; **fruit:** cypselae (common); **local dispersal:** cypselae (wind, water); rhizomes; **long-distance dispersal:** cypselae (animals, water)

Imperilment: (1) *Symphyotrichum bahamense* [not ranked]; (2) *S. boreale* [G5]; SH (VT); S1 (NF, NJ, PA, WA, WV); S2 (IA, ID, IN, NS, NY, PE, YT, WY); S3 (NB, OH, SD); (3) *S. chapmanii* [G2/G3]; SH (AL), S2 (FL); (4) *S. defoliatum* [G2]; S2 (CA); (5) *S. divaricatum* [G5]; S1 (WV); S2 (MO); (6) *S. elliottii* [G4]; S1 (VA); S2 (NC); S3 (SC); (7) *S. expansum* [GNR]; (8) *S. lanceolatum* [G5]; SH (NY, QC); S1 (NJ, TX); S2 (NC, NF); S3 (BC, GA, IA, WY); (9) *S. lentum* [G2]; S2 (CA); (10) *S. novi-belgii* [G5]; SH (GA); S2 (NB, PA, QC, WV); S3 (NJ, NY, QC); (11) *S. praealtum* [G5]; SX (DC); SH (MA); S1 (GA, KS, KY, MD, NJ, TN, VA); S2 (KY, ON); S3 (MI, PA, WV); (12) *S. puniceum* [G5]; S1 (MS); S2 (TX); S3 (BC, KY); (13) *S. squamatum* [G5]; (14) *S. subulatum* [G5]; S1 (ME, NV, PE); S2 (NB, NY); S3 (NC); (15) *S. tenuifolium* [G5]; S1 (NH); S3 (NC, NY)

Ecology: general: *Symphyotrichum* is a large and diverse genus of annuals and perennials. It is encountered commonly in wetlands, with 42 of the North American species (55%) listed as wetland indicators and 15 of them (19%) ranked as OBL. The genus is remarkably diverse ecologically as is well illustrated by the impressive spectrum of acidity (representing nearly 9 pH units) tolerated by the species. Most *Symphyotrichum* species are self-incompatible and require outcrossing by insects (Insecta), which predominantly are bees (Hymenoptera). A number of the annual species are self-compatible, mainly autogamous, and tend to have shorter ray florets. The radiate heads are gynomonoecious with outer pistillate ray florets and central bisexual disk florets. The bisexual florets are protandrous, maturing pollen prior to stigma receptivity. Pollen must be removed almost entirely before the stigmas are exposed. The cypselae mature in late summer to late fall and are dispersed locally by the wind, which interacts with the long pappus bristles, or by animals, which consume the cypselae and scatter viable propagules in their feces. The seeds of some species also are buoyant and can be dispersed along the surface of water. Cypselae also have been recovered from plumage, indicating the potential for their ectozoic long-distance dispersal by birds. Seeds of different species vary from being nondormant and capable of immediate germination to those having strong physiological dormancy. These are ubiquitous plants of open marsh and meadow, where many species achieve significant population sizes. Consequently, they often are characterized as "weedy". Their attractive floral displays and prolific seed output have resulted in numerous instances of nonidigenous introductions worldwide as a consequence of escapes from cultivation and as agricultural seed contaminants.

Symphyotrichum bahamense (Britton) G.L. Nesom is an annual, which resides in depressions, ditches, lawns, marshes, meadows, river channels, roadsides, salt marshes, seeps, stream beds, and margins of borrow pits, ponds, and woodlands at elevations of up to 1024 m (California). Reported substrates include marl, rock, and sand. The flowers are self-compatible. The plants characteristically occur in disturbed, ruderal areas. Otherwise, life history information for this species is scarce. **Reported associates:** *Aeschynomene indica, Baccharis salicifolia, Brickellia californica, Cynodon dactylon, Digitaria ciliaris, D. ischaemum, D. sanguinalis, Echinochloa crusgalli, Epilobium brachycarpum, Eriogonum fasciculatum, Juncus balticus, J. dubius, Lepidospartum squamatum, Nasturtium officinale, Rumex crispus, Schedonorus arundinaceus, Scirpus, Tamarix ramosissima, Typha, Xanthium strumarium.*

Symphyotrichum boreale (Torr. & A. Gray) A. Löve & D. Löve is a perennial, which inhabits bogs, carrs, dikes, ditches, dunes, fens, flats, marshes, meadows, patterned peatlands, prairies, roadsides, sandbars, seeps, sloughs, swales, swamps, woodlands, and the margins of lakes, ponds, reservoirs, rivers, and streams at elevations of up to 3240 m. Plants occur in full sun to shady exposures across a fairly broad range of acidity (pH 5.5–8.0), but generally on calcareous substrates (pH 7.0–8.0), which include gravel, histosols, loam, logs, marl, marly peat, muck, peat, rocky granite, sand, sandy gravel, sandy loam, sandy rock, and talus. The plants are in flower from August to October. Vegetative reproduction occurs by elongate rhizomes. A large proportion of roots (40%) is colonized by arbuscular mycorrhizal fungi and dark septate endophytes. Additional details on the reproductive biology and seed ecology are lacking. **Reported associates:** *Acer rubrum, Achillea millefolium, Agalinis purpurea, Agoseris glauca, Agrostis stolonifera, Alnus, Andromeda polifolia, Andropogon gerardii, Anemone canadensis, Antennaria parvifolia, Apocynum cannabinum, Arethusa bulbosa, Aronia melanocarpa, Asclepias incarnata, A. syriaca, A. tuberosa, Betula glandulosa, B. neoalaskana, B. occidentalis, B. pumila, B. ×sandbergii, Bidens trichospermus, Boehmeria cylindrica, Bromus ciliatus, Calamagrostis canadensis, C. stricta, Calamovilfa longifolia, Calopogon tuberosus, Campanula*

aparinoides, C. rotundifolia, Campylium stellatum, Carex aquatilis, C. aurea, C. buxbaumii, C. canescens, C. capillaris, C. chordorrhiza, C. crawei, C. cryptolepis, C. diandra, C. echinata, C. exilis, C. flava, C. gynocrates, C. interior, C. lacustris, C. lanuginosa, C. lasiocarpa, C. leptalea, C. limosa, C. livida, C. microglochin, C. muricata, C. nebrascensis, C. oligosperma, C. pellita, C. prairea, C. rostrata, C. sartwellii, C. scirpoidea, C. simulata, C. sterilis, C. tenuiflora, C. tetanica, C. trisperma, C. utriculata, C. viridula, Chamaedaphne calyculata, Chara, Chelone glabra, Cicuta bulbifera, C. maculata, Cirsium muticum, Cladium mariscoides, Comarum palustre, Comptonia peregrina, Cornus sericea, Crepis runcinata, Cypripedium parviflorum, Danthonia, Dasiphora floribunda, Decodon verticillatus, Deschampsia cespitosa, Doellingeria umbellata, Drepanocladus revolvens, Drosera linearis, D. rotundifolia, Eleocharis elliptica, E. palustris, E. quinqueflora, E. robbinsii, E. rostellata, Elymus trachycaulus, Epilobium leptophyllum, E. palustre, E. strictum, Equisetum arvense, E. fluviatile, Eriophorum angustifolium, E. gracile, E. viridicarinatum, Eupatorium perfoliatum, Euthamia graminifolia, Eutrochium maculatum, Festuca trachyphylla, Fragaria virginiana, Galium boreale, G. trifidum, G. labradoricum, Gentiana fremontii, Gentianella amarella, Gentianopsis crinita, G. virgata, Glyceria grandis, G. striata, Glycyrrhiza lepidota, Hippuris vulgaris, Hordeum jubatum, Hypericum kalmianum, Impatiens capensis, Iris versicolor, Juncus alpinoarticulatus, J. balticus, J. canadensis, J. greenei, J. nodosus, J. stygius, Kobresia simpliciuscula, Larix laricina, Liatris aspera, L. pycnostachya, Lobelia kalmii, Lonicera oblongifolia, L. villosa, Lycopus americanus, L. asper, L. uniflorus, Lysimachia maritima, L. quadriflora, L. thyrsiflora, Lythrum alatum, Mentha arvensis, Menyanthes trifoliata, Mimulus, Muhlenbergia glomerata, M. richardsonis, Myrica gale, Oclemena nemoralis, Onoclea sensibilis, Ophioglossum pusillum, Osmunda regalis, Oxypolis rigidior, Packera paupercula, Parnassia glauca, P. palustris, P. parviflora, Pedicularis groenlandica, P. lanceolata, Persicaria amphibia, P. lapathifolia, Petasites frigidus, Phleum pratense, Phragmites australis, Picea glauca, P. mariana, Physostegia, Pilea fontana, Pinus banksiana, P. contorta, Plantago major, Platanthera dilatata, P. hyperborea, P. obtusata, Pleurozium schreberi, Poa interior, P. pratensis, Pogonia ophioglossoides, Populus balsamifera, P. tremuloides, Potentilla anserina, Prenanthes racemosa, Primula egaliksensis, P. incana, P. pauciflora, Puccinellia nuttalliana, Pycnanthemum tenuifolium, P. virginianum, Pyrola asarifolia, Ranunculus gmelinii, Rhamnus alnifolia, Rhododendron groenlandicum, Rhynchospora alba, R. capillacea, R. glomerata, Ribes oxyacanthoides, Rosa acicularis, R. woodsii, Rubus pubescens, Salix arbusculoides, S. bebbiana, S. candida, S. discolor, S. glauca, S. maccalliana, S. monticola, S. pedicellaris, S. petiolaris, S. planifolia, S. pseudomonticola, S. sericea, S. serissima, Saponaria officinalis, Sarracenia purpurea, Scheuchzeria palustris, Schizachyrium scoparium, Schoenoplectus acutus, S. pungens, S. tabernaemontani, Scirpus microcarpus, Scleria verticillata, Scorpidium scorpioides, Scutellaria galericulata, S. lateriflora, Senecio lugens,

Sisyrinchium montanum, Solidago canadensis, S. gigantea, S. ohioensis, S. patula, S. riddellii, S. uliginosa, Sonchus arvensis, Sparganium eurycarpum, Sphagnum capillifolium, S. contortum, S. fuscum, S. magellanicum, S. teres, S. warnstorfii, Spiraea douglasii, S. tomentosa, Spiranthes cernua, Sporobolus heterolepis, Swertia perennis, Symphoricarpos occidentalis, Symphyotrichum ericoides, S. firmum, S. lanceolatum, S. novae-angliae, Taraxacum officinale, Thalictrum alpinum, T. venulosum, Thelypteris palustris, Thuja occidentalis, Tomentypnum nitens, Toxicodendron vernix, Triadenum fraseri, Triantha glutinosa, Trichophorum alpinum, T. cespitosum, T. pumilum, Trifolium pratense, Triglochin maritimum, T. palustre, Typha latifolia, Urtica dioica, Utricularia intermedia, Vaccinium macrocarpon, V. oxycoccos, V. uliginosum, Vicia americana, Viola lanceolata, V. sororia.

Symphyotrichum chapmanii (Torr. & A. Gray) Semple & Brouillet is a perennial, which grows in barrens, bogs, depressions, ditches, flatwoods, poor fens, prairies, savannas, seeps, and swamps at elevations of up to 30 m. Substrates are acidic (e.g., pH 5.4–5.6) and include loamy sand, Plummer (grossarenic paleaquults), sandy peat, and Pantego (umbric paleaquults). The plants are especially common in frequently burned sites. They are in flower from August to December and perennate from a cormose rootstock, from which stout rhizomes are produced. Other life-history information is wanting. **Reported associates:** *Agalinis aphylla, A. filicaulis, A. linifolia, A. purpurea, Aletris aurea, A. lutea, Amphicarpum muhlenbergianum, Andropogon arctatus, A. glomeratus, A. liebmannii, A. ternarius, Aristida palustris, A. purpurascens, A. stricta, Arnoglossum ovatum, Aronia arbutifolia, Balduina uniflora, Bartonia paniculata, Bigelowia nudata, Calopogon pallidus, Carex joorii, Carphephorus pseudoliatris, Chamaecyparis thyoides, Chaptalia tomentosa, Cladium mariscus, Cliftonia monophylla, Clethra alnifolia, Coleataenia longifolia, Coreopsis gladiata, C. leavenworthii, Ctenium aromaticum, Cuscuta compacta, Cyrilla racemiflora, Dichanthelium acuminatum, D. ensifolium, D. nudicaule, Drosera capillaris, D. tracyi, Eriocaulon compressum, E. decangulare, Eupatorium mohrii, E. semiserratum, Euphorbia inundata, Eurybia eryngiifolia, Euthamia graminifolia, Fuirena squarrosa, Gaylussacia mosieri, Gelsemium rankinii, Gymnadeniopsis nivea, Habenaria, Helenium pinnatifidum, Helianthus floridanus, H. heterophyllus, Hypericum brachyphyllum, H. fasciculatum, Hypoxis curtissii, Ilex coriacea, I. glabra, I. myrtifolia, Juncus marginatus, J. polycephalus, J. trigonocarpus, Justicia crassifolia, Lachnanthes caroliniana, Lachnocaulon anceps, Liatris spicata, Lilium catesbaei, Lobelia floridana, Lophiola aurea, Ludwigia linearis, L. pilosa, Lycopodiella alopecuroides, L. caroliniana, Lyonia fruticosa, L. lucida, Magnolia virginiana, Marshallia graminifolia, Mitreola sessilifolia, Muhlenbergia sericea, Myrica cerifera, M. inodora, Nyssa biflora, Osmanthus americanus, Oxypolis denticulata, O. filiformis, Panicum dichotomiflorum, Persea palustris, Pieris phillyreifolia, Pinguicula planifolia, Pinus elliottii, P. serotina, Pityopsis, Platanthera ciliaris, Pleea tenuifolia, Pluchea baccharis, P. camphorata, P. foetida, P. carolinensis, Pogonia ophioglossoides, Polygala cruciata, P. cymosa,*

P. lutea, Proserpinaca pectinata, Rhexia alifanus, R. virginica, Rhododendron viscosum, Rhynchospora baldwinii, R. cephalantha, R. chapmanii, R. corniculata, R. filifolia, R. gracilenta, R. latifolia, R. oligantha, R. plumosa, R. rariflora, Rudbeckia graminifolia, Sabatia macrophylla, S. quadrangula, Sagittaria graminea, Sarracenia flava, S. minor, S. psittacina, Scirpus cyperinus, Scleria baldwinii, S. muhlenbergii, S. reticularis, Seymeria cassioides, Smilax laurifolia, Sporobolus curtissii, Styrax americanus, Syngonanthus flavidulus, Taxodium ascendens, Utricularia cornuta, U. juncea, Vaccinium corymbosum, Verbesina heterophylla, Vitis, Woodwardia virginica, Xyris ambigua, X. baldwiniana, X. elliottii, X. stricta, Zigadenus glaberrimus.

Symphyotrichum defoliatum (Parish) G.L. Nesom is a cespitose perennial, which occurs in chaparral, cienaga, ditches, meadows, ravines, roadsides, slopes, springs, swales, and along the margins of streams at elevations of up to 2042 m. The plants usually occur in open sites but also tolerate partial shade. Reported substrates include adobe, clay loam, granite, schist, sand, and sandy loam. Flowering occurs from August to November. The plants can develop into small clumps by production of short rhizomes and occur commonly in disturbed areas. Additional life history information for this species is unavailable. However, given the list of reported associates and frequency of this species in dry grasslands, its OBL indicator designation seems unusual. **Reported associates:** *Achillea millifolium, Acmispon argophyllus, Adenostoma fasciculatum, A. sparsifolium, Agnorhiza ovata, Amaranthus, Amelanchier utahensis, Ambrosia psilostachya, Arctostaphylos glandulosa, Artemisia, Asclepias fascicularis, Avena barbata, Baccharis pilularis, Bromus carinatus, B. diandrus, B. hordeaceus, B. rubens, B. tectorum, Calochortus, Carex, Ceanothus greggii, C. spinosus, Cercocarpus betuloides, Cirsium vulgare, Conyza canadensis, Cordylanthus, Corethrogyne filaginifolia, Distichlis spicata, Drymocallis glandulosa, Epilobium brachycarpum, E. ciliatum, Eriogonum fasciculatum, E. wrightii, Euthamia occidentalis, Garrya, Hazardia, Gutierrezia californica, Heliotropium, Hirschfeldia incana, Hoita orbicularis, Juncus arcticus, J. balticus, J. dubius, Lagophylla ramosissima, Leptosiphon liniflorus, Leymus triticoides, Lolium, Madia elegans, Melilotus albus, Muhlenbergia rigens, Nassella cernua, Persicaria, Pinus jefferyi, Plantago major, Platanus racemosa, Poa pratensis, Prunus virginiana, Pseudognaphalium luteoalbum, Pycnanthemum californicum, Quercus agrifolia, Q. engelmannii, Q. kelloggii, Ranunculus californicus, Rhamnus, Rhus aromatica, Rosa californica, Rumex salicifolius, Salix lasiolepis, Salvia apiana, Sambucus nigra, Sysimbrium, Sisyrinchium bellum, Solidago velutina, Symphoricarpos albus, Thermopsis californica, Trichostema lanatum, Typha domingensis, T. latifolia.*

Symphyotrichum divaricatum (Nutt.) G.L. Nesom is an annual, which grows in fresh to brackish (0–10 ppt salinity) sites including ditches, dunes, flats, floodplains, levees, marshes, meadows, pools, prairies, ravines, roadsides, shores, stream bottoms, thickets, washes, and the margins of bayous, canals, lakes, playas, woodlands, rivers, salt marsh and streams at elevations of up to 1829 m. Sites usually are open with mineral or organic substrates described as clay, clay sand, cobble, loam, rock, sand, sandy alluvium, sandy clay, shells, silt, and Tapeats sandstone. The flowers are self-incompatible. The seed ecology have not been studied in any detail. This species commonly invades disturbed sites and tolerates drought, fire, flooding (periodic), grazing, and mowing. The plants occur frequently along railroad tracks, which probably represent effective dispersal corridors. They are affected negatively by pepperweed (Brassicaceae: *Lepidium latifolium*) and have been observed to increase in abundance at sites where glyphosate herbicide and hand-pulling treatments have been implemented to control such weedy species. A fair amount of life-history information in the literature refers not to this species but to *Aster divaricatus* L., which is synonymous with *Eurybia divaricata* (see *Systematics*) and care should be taken not to confuse these taxa. **Reported associates:** *Achillea millefolium, Acmispon americanus, Agalinis fasciculata, Allium canadense, A. drummondii, Amaranthus tuberculatus, Ambrosia confertiflora, A. psilostachya, A. trifida, Ampelopsis arborea, Amphiachyris dracunculoides, Andropogon glomeratus, Anemone berlandieri, Apium graveolens, Asclepias stenophylla, A. viridis, Atriplex canescens, A. cristata, A. dioica, A. prostrata, Baccharis salicifolia, B. salicina, Bacopa monnieri, Bolboschoenus maritimus, Borrichia frutescens, Bothriochloa ischaemum, B. laguroides, Bouteloua curtipendula, B. dactyloides, Bradburia pilosa, Brickellia eupatorioides, Briza minor, Bromus arvensis, Callirhoe involucrata, Cardamine hirsuta, Cardiospermum halicacabum, Carduus nutans, Carex arkansana, C. cherokeensis, C. crus-corvi, C. leavenworthii, Castilleja indivisa, Celtis laevigata, C. reticulata, Centaurium texense, Cerastium glomeratum, Chamaecrista fasciculata, Chenopodium berlandieri, C. glaucum, C. pratericola, Chloracantha spinosa, Chloris virgata, Cirsium texanum, Cissus trifoliata, Cocculus carolinus, Conoclinium betonicifolium, Conyza canadensis, Coreopsis tinctoria, Crataegus spathulata, Crepis pulchra, Cressa truxillensis, Croton monanthogynus, C. texensis, Cucumis melo, Cuscuta cuspidata, C. indecora, C. pentagona, C. salina, Cycloloma atriplicifolium, Cynodon dactylon, Cyperus articulatus, C. echinatus, C. eragrostis, C. erythrorhizos, C. ochraceus, C. odoratus, C. polystachyos, C. squarrosus, Daucus carota, Delphinium carolinianum, Desmanthus illinoensis, D. leptolobus, Dichanthelium oligosanthes, Dichanthium aristatum, Distichlis spicata, Echinochloa colona, E. crusgalli, Eclipta prostrata, Eleocharis interstincta, E. palustris, E. parvula, E. rostellata, Elymus canadensis, E. virginicus, Encelia farinosa, Erigeron philadelphicus, E. strigosus, Eriochloa punctata, Erodium cicutarium, Eryngium hookeri, Euphorbia bicolor, E. dentata, E. hexagona, E. serpens, Eustoma exaltatum, Forestiera acuminata, Fragaria vesca, Fraxinus pennsylvanica, F. velutina, Funastrum cynanchoides, Gaillardia pulchella, Galium aparine, Geranium carolinianum, Geum canadense, Glandularia bipinnatifida, Gleditsia triacanthos, Gonolobus suberosus, Grindelia ciliata, G. squarrosa, Helianthus annuus, H. maximiliani, H. petiolaris, Heliotropium curassavicum, Helminthotheca echioides, Hopia obtusa, Hordeum jubatum, H. pusillum, Hydrocotyle bonariensis, Hypericum anagalloides, Ipomoea amnicola, I. sagittata, Iva annua, Juglans major, Juncus dudleyi, J. kelloggii, J. torreyi, Juniperus virginiana, Justicia americana, Kochia scoparia, Koeleria macrantha, Lactuca serriola, Lamium amplexicaule, Lathyrus hirsutus, Lepidium latifolium, L. virginicum, Leptochloa fusca, L. panicea. Liatris punctata, Lindheimera texana, Lolium perenne, Ludwigia octovalvis, Lycium carolinianum, Lythrum alatum, Machaeranthera tanacetifolia, Maclura pomifera, Malachra capitata, Malvastrum coromandelianum, Marsilea vestita, Medicago orbicularis, Megathyrsus maximus, Melilotus albus, M. officinalis, Melothria pendula, Mikania scandens, Mimosa asperata, M. roemeriana, Mnesithea cylindrica, Monarda citriodora, Myosurus minimus, Myosotis macrosperma, Nassella leucotricha, Neptunia lutea, Nothoscordum bivalve, Oenothera curtiflora, O. glaucifolia, O. laciniata, O. speciosa, O. suffulta, Opuntia phaeacantha, Oxalis stricta, Packera tampicana, Panicum capillare, P. coloratum, P. millegrana, P. virgatum, Parkinsonia aculeata, Parthenocissus quinquefolia, Pascopyrum smithii, Paspalum denticulatum, P. dilatatum, Passiflora incarnata, P. floridanum, Persicaria amphibia, P. pensylvanica, Phalaris caroliniana, Phragmites australis, Phyla nodiflora, Physalis angulata, P. longifolia, P. mollis, Physaria gracilis, Plantago patagonica, P. rhodosperma, Platanus occidentalis, P. wrightii, Pluchea carolinensis, P. odorata, Poa compressa, Polygonum aviculare, P. ramosissimum, Polypogon interruptus, P. monspeliensis, Polytaenia nuttallii, Populus deltoides, P. fremontii, Prosopis glandulosa, P. velutina, Prunus angustifolia, Pyrrhopappus pauciflorus, Ptilimnium nuttallii, Quercus shumardii, Ratibida columnifera, Rubus flagellaris, Rudbeckia amplexicaulis, R. hirta, Rumex crispus, Sabatia campestris, Salix amygdaloides, S. exigua, S. gooddingii, S. interior, S. nigra, Salsola collina, S. tragus, Sapindus saponaria, Sarcocornia pacifica, Scandix pecten-veneris, Schizachyrium scoparium, Schoenoplectus americanus, S. californicus, S. pungens, S. tabernaemontani, Sesbania drummondii, S. herbacea, S. vesicaria, Sesuvium maritimum, S. portulacastrum, S. verrucosum, Setaria pumila, Sida, Sideroxylon lanuginosum, Sisyrinchium angustifolium, Smilax bonanox, Solanum campechiense, S. dimidiatum, Solidago canadensis, S. petiolaris, Sonchus asper, S. oleraceus, Sorghastrum nutans, Sorghum halepense, Sporobolus airoides, S. compositus, S. texanus, Stachys crenata, Stellaria media, Stenaria nigricans, Styphnolobium affine, Suaeda calceoliformis, S. linearis, S. nigra, Symphyotrichum ericoides, S. lanceolatum, Symphoricarpos orbiculatus, Tamarix ramosissima, Taraxacum officinale, Teucrium canadense, Thelesperma filifolium, Torilis arvensis, Toxicodendron radicans, Tragia betonicifolia, Tridens albescens, T. muticus, Trifolium campestre, Triodanis perfoliata, Tripsacum dactyloides, Typha domingensis, Ulmus americana, U. crassifolia, Urochloa fuscata, Urtica chamaedryoides, Valerianella radiata, Verbena halei, Veronica peregrina, Vicia minutiflora, V. sativa, Vigna luteola, Vitis mustangensis, Xanthium strumarium, Zanthoxylum clava-herculis.*

***Symphyotrichum elliotii* (Torr. & A. Gray) G.L. Nesom** is a perennial, which inhabits tidal or non-tidal freshwater to brackish (e.g., salinity = 2 ppt) baygalls, bays, bogs, deltas, depressions, ditches, flatwoods, floodplains, hammocks, marshes, prairies, savannas, swales, and margins of lakes, swamps and thickets at elevations of up to 50+ m. The plants occur where water levels usually are less than 5 cm, although depths can exceed 70 cm for short durations. Occurrences range from open sites to partial shade, but shade can result in more frail plants. Sites characteristically have nutrient-poor acidic substrates including clay, loamy sand, Samsula (terric medisaprists), and sand; however occurrences on marl also are reported. Flowering occurs from August to October. The flowers are pollinated by various insects (Insecta) including bees (Hymenoptera), butterflies (Lepidoptera) and flies (Diptera). The plants produce long rhizomes and can quickly fill vacant sites through prolific clonal reproduction. **Reported associates:** *Acer rubrum, Agalinis purpurea, Alnus serrulata, Ampelaster carolinianus, Amphicarpum muhlenbergianum, Andropogon virginicus, Aristida stricta, Baccharis halimifolia, Bejaria racemosa, Bidens laevis, B. mitis, Blechnum serrulatum, Carex, Cicuta maculata, Cliftonia monophylla, Coleataenia longifolia, Cyperus distinctus, Cyrilla racemiflora, Drosera brevifolia, Eleocharis montevidensis, E. quadrangulata, Erigeron quercifolius, Euthamia graminifolia, Galium tinctorium, Hydrocotyle umbellata, Hypericum brachyphyllum, Ilex coriacea, I. glabra, Iris hexagona, Itea virginica, Juncus marginatus, Kalmia hirsuta, Lilaeopsis chinensis, Liriodendron tulipifera, Lobelia elongata, Ludwigia maritima, L. repens, Lycopus rubellus, Lyonia ferruginea, L. fruiticosa, L. lucida, Magnolia virginiana, Mikania scandens, Mnesithea rugosa, Murdannia keisak, Myrica cerifera, Nyssa biflora, Oenothera biennis, Panicum hemitomon, Persea palustris, Persicaria sagittata, Phyla lanceolata, Pinus elliottii, P. palustris, P. taeda, Polygala lutea, Quercus, Rhexia petiolata, Rhus copallinum, Rhynchospora corniculata, R. microcarpa, R. miliacea, Rubus cuneifolius, Saccharum baldwinii, Sagittaria lancifolia, S. latifolia, Salix, Sambucus, Schoenoplectus tabernaemontani, Scirpus cyperinus, Serenoa repens, Solidago petiolaris, Syngonanthus flavidulus, Taxodium ascendens, Typha angustifolia, Ulmus, Vaccinium myrsinites, Viburnum nudum, Woodwardia virginica, Xyris brevifolia, X. jupicai, Zizaniopsis miliacea.*

***Symphyotrichum expansum* (Poepp. ex Spreng.) G.L. Nesom** is an annual plant of fresh to saline sites including arroyos, beaches, chapparal, creek and lake bottoms, ditches, dunes, flats, floodplains, gravel bars, levees, marshes, meadows, ricefields, riverbeds, saltmarsh, seeps, sinks, slopes, swales, swamps, vernal pools, washes, waste areas, and margins of canals, channels, lakes, ponds, reservoirs, rivers, sloughs, and streams at elevations of up to 2320 m. Exposures range from open sites to partial shade. The substrates usually are described as alkaline or slightly alkaline and include clay, granite, gravel, loam, Monterey shale, muck, mucky silt loam, mud, peaty silt, sand, sandstone, sandy loam, silt, silty clay, and silty loam. The flowers are self-compatible and the plants presumably autogamous; however, they are visited by bees (Insecta: Hymenoptera) and some outcrossing may occur. The small seeds weigh about 0.10 mg and occur mainly within the top 1–2 cm of substrate. Germination has been obtained under greenhouse conditions (with lighting extended to approximate local spring photoperiod) following 30+ days of cold stratification (4°C). This species often is characterized as weedy and became common in the California ricefields during the 1920s. **Reported associates:** *Abies concolor, Agropyron desertorum, Agrostis stolonifera, Alnus oblongifolia, Amauriopsis dissecta, Ambrosia acanthicarpa, A. psilostachya, Ammannia auriculata, A. coccinea, Anemopsis californica, Arbutus arizonica, Arctostaphylos, Arundo donax, Atriplex patula, A. rosea, Baccharis douglasii, B. neglecta, B. salicifolia, B. sarothroides, B. sergiloides, Bassia hyssopifolia, Bidens aureus, B. frondosus, B. leptocephalus, Bolboschoenus maritimus, Brassica, Brickelia atractyloides, B. eupatorioides, B. grandiflora, Bromus ciliatus, Calliandra eriophylla, Carex, Carya illinoinensis, Celtis reticulata, Cenchrus setaceus, Centromadia fitchii, Chilopsis linearis, Chloropyron molle, Cladium, Commelina dianthifolia, Conyza canadensis, Cortaderia, Croton texensis, Crypsis schoenoides, Cucurbita foetidissima, Cynodon dactylon, Cyperus eragrostis, C. odoratus, Distichlils spicata, Dysphania pumilio, Dyssodia papposa, Echinochloa crus-galli, Echinodorus berteroi, Eleocharis montevidensis, E. rostellata, Elymus elymoides, Encelia farinosa, Epilobium brachycarpum, E. ciliatum, Epipactis gigantea, Equisetum arvense, E. laevigatum, Erigeron neomexicanus, Eriogonum, Erodium cicutarium, Eryngium lemmoni, Eucalyptus camaldulensis, Euphorbia revoluta, Ferocactus, Ficus carica, Flaveria campestris, Foeniculum vulgare, Frankenia palmeri, Fraxinus velutina, Geranium caespitosum, Gossypium hisutum, Grindelia, Gutierrezia, Helenium microcephalum, Helianthus annuus, Heliotropium curassavicum, Helminthotheca echioides, Heterotheca grandiflora, Heuchera sanguinea, Hilaria, Holocarpha virgata, Hyptis emoryi, Ipomoea purpurea, Ipomopsis thurberi, Isocoma acradenia, I. menziesii, Juglans major, Juncus acutus, J. balticus, J. bufonius, J. torreyi, Juniperus coahuilensis, J. deppeana, Kallstroemia grandiflora, Kochia scoparia, Koeleria macrantha, Lactuca serriola, Laennecia coulteri, Larrea tridentata, Lasianthaea podocephala, Leersia oryzoides, Lepidium latifolium, Leptochloa, Lobelia cardinalis, Lythrum californicum, Malosma laurina, Malva parviflora, Melilotus albus, Mesembryanthemum crystallinum, M. nodiflorum, Mimulus guttatus, Mirabilis coccinea, Muhlenbergia rigens, Nolina microcarpa, Oenothera elata, Panicum capillare, P. hirticaule, P. virgatum, Parkinsonia microphylla, Paspalum distichum, Persicaria amphibia, Phoenix canariensis, Phragmites australis, Pinus cembroides, P. leiophylla, P. ponderosa, Platanus racemosa, P. wrightii, Plantago lanceolata, Pluchea odorata, P. sericea, Poa bigelovii, Polypogon interruptus, P. monspeliensis, P. viridis, Populus fremontii, Potamogeton, Prosopis juliflora, P. pubescens, P. velutina, Pseudotsuga menziesii, Puccinellia fasciculata, Quercus emoryi, Q. gambelii, Q. grisea, Q. hypoleucoides, Q. oblongifolia, Q. rugosa, Q. turbinella, Rayjacksonia annua, Rhus glabra,*

R. integrifolia, Ricinus communis, Robinia neomexicana, Rorippa, Rubus ulmifolius, Rumex crispus, Salix exigua, S. gooddingii, S. lasiolepis, Salsola tragus, Sambucus, Schinus molle, S. terebinthifolius, Schoenoplectus americanus, S. californicus, S. pungens, Scrophularia parviflora, Senegalia greggii, Sesbania, Sesuvium verrucosum, Sisymbrium irio, Solanum rostratum, Sorghum halepense, Sphaeralcea hastulata, Sphenopholis obtusa, Sporobolus contractus, S. texanus, S. wrightii, Suaeda calceoliformis, Tamarix chinensis, T. ramosissima, Thinopyrum intermedium, T. ponticum, Turricula parryi, Typha domingensis, Ulmus parvifolia, U. pumila, Verbesina longifolia, Veronica americana, Viguiera cordifolia, Vitis arizonica, Washingtonia filifera, W. robusta, Wislizenia refracta, Woodsia phillipsii, Xanthium strumarium, Zeltnera calycosa.

***Symphyotrichum lanceolatum* (Willd.) G.L. Nesom** is a perennial, which inhabits fresh or saline sites including backwashes, beaches, carrs, channels, cienega, copses, deltas, depressions, dikes, ditches, draws, fens, flats, floodplains, glades, hummocks, marshes, meadows, mudflats, prairies, ravines, river and stream bottoms, roadsides, salt marshes, sand bars, savanna, scrub, sloughs, swamps, thickets, and the margins of canals, gravel pits, lakes, ponds, reservoirs, rivers, springs, and streams at elevations of up to 2987 m. This species occupies sites spanning a very broad range of soil moisture as evidenced by its multiple wetland indicator designations and the eclectic assemblage of associated UPL and lowland species. Standing water levels can reach 34–42 cm. The plants occur in a wide range of exposures from full sun to shade. They normally are associated with alkaline (pH >7) substrates, but can occur along a pH gradient from 4.6 to 7.8. Substrates are described as basalt cobble, clay, clay loam, clay loam muck, Denver arapahoe, gravel, gravel loam, hermit shale, limestone, loam, loamy sand, marl, muck, mud, organic aquie fluvent, peat, rocks, sand, sandy clay, sandy loam, sandy peat, silt, silty loam, silty sand, and silty sandy loam. Many sites tend toward a high sand content (e.g., 54%). Flowering occurs from August to October. Flowering time can be influenced by the intensity of light once the required photoperiod has been achieved. Flower heads on the terminal branches open about 1 week before those on other parts of the plant. The florets mature centripetally, with the ray florets maturing 3–4 days prior to the disc florets. Approximately half of the ramets in a population will initiate flowering by September. Flowering is not significantly related to plant size; however, earlier flowering can occur in plants grown under short-day photoperiods. The plants are strongly self-incompatible (producing only few seeds when selfed) and are essentially obligately outcrossed. Individual flowers produce an average of 0.018 mg of sugar, which attracts numerous insect (Insecta) pollinators. In one study, the most effective pollinators were honeybees (Hymenoptera: Apidae: *Apis mellifera*), flies (Diptera: Syrphidae: *Eristalis*), and wasps (Hymenoptera: Vespidae: *Dolichovespula arenaria, Polistes fuscatus*). Pollinators visit the flowers throughout the day but are most prevalent at dawn and dusk. Observed pollen viability is high, ranging from 48% to 99% across the five principal cytotypes

examined. Cypselae mature in 3–4 weeks subsequent to pollination. Plants are capable of producing upward of 200,000 cypselae. However, the occurrence of mature seeds in a population can differ substantially (e.g., 20%–70%). Pollinator satiation within large clones has been observed to limit pollen transfer among clones (which is necessary in a self-incompatible species), and can reduce seed set to only 0.1%–2.8%. The cypselae are small, extremely light (0.127 mg), and predominantly are wind dispersed, with 21% of the mass allocated to the pappus (37 bristles of 4 mm length); occasionally the corollas remain attached to the disc florets when shed. Viable cypselae (3% incidence) have been recovered from the fecal pellets of white-tailed deer (Mammalia: Cervidae: *Odocoileus virginianus*), which demonstrates that endozoic animal dispersal also occurs. From 60% to 100% of cypselae can germinate without treatment. Germination rates (after 14 days at 14 h photoperiod; 25°C/15°C temperature regime) improve somewhat after 5 months of storage at room temperature (67%–98%) or after 5 months of cold (5°C), moist stratification (73%–100%). Good germination also is reported (for seed stratified at 4°C) using a 12 h photperiod under a 20°C/5°C temperature regime. Generally, fall germination is minimal with most seedlings emerging in the spring. Immediate germination can be induced by carefully rupturing both the pericarp and seed coat. Germination rates of 35%–65% are typical in natural populations. Germination of some cytotypes requires a critical weight to be achieved (0.075 mg for tetraploids; 0.100 mg for hexaploids). Time to germination increases proportionally with cytotype level, and is roughly 1.9 times longer for octoploids than tetraploids. Germinable seeds can achieve densities of 2.9/m² and represent 0.51% relative abundance in the seed bank. During the first year of growth, more resources are allocated to vegetative growth (e.g., rhizomes) rather than seed production. About 26% of the total dry weight biomass is allocated to rhizome production. Perennation occurs by dormant rhizome buds or rosettes of scale leaves. Most rhizomes occur with the top 0–5 cm of soil. Physiological integration among ramets is low. Although connections among rhizomes often sever from the main plant during the winter, many persist for several seasons, resulting in an intertwined network of rhizomes up to 8 mm thick and more than 4 m in length. New rhizome production (also rosettes and flowerheads) is substantially higher in weeded sites. Vescicular–arbuscular mycorrhizae occurred on 44%–47% of the roots in one population attributed to *S. lanceolatum*. This is an early successional species that often is characterized as weedy. The plants are found most frequently in 2–6 year old fields, but usually are absent in older stands of vegetation. The plants also occur most frequently at a distance of 1–2 m from trees but do not occur alongside them. *Symphyotrichum lanceolatum* has branching shoots with exponentially increasing leaf populations, which lead to high leaf turnover rates and apical leaf crowding. In one analysis the specific leaf area of the foliage was 205 cm²/g with a dry matter content of 0.28 g/g. Essential oils extracted from the flowers are alleopathic (exhibiting inhibition of germination and growth) against *Lactuca sativa* (Asteraceae) seeds.

Reported associates: *Abies concolor, Acer glabrum, A. negundo, A. rubrum, A. saccharinum, A. saccharum, Achillea millefolium, Acorus calamus, Adiantum capillus-veneris, Agalinis tenuifolia, Agave utahensis, Agrostis gigantea, A. stolonifera, Alisma triviale, Allium canadense, A. cernuum, A. schoenoprasum, Almutaster pauciflorus, Alnus oblongifolia, A. rugosa, A. tenuifolia, A. viridis, Alternanthera philoxeroides, Amelanchier utahensis, Amauriopsis dissecta, Ambrosia psilostachya, A. tomentosa, Amorpha fruticosa, Ampelopsis arborea, Amphicarpaea bracteata, Andropogon gerardii, Anemone canadensis, Anemopsis californica, Antennaria rosea, Apocynum cannabinum, Aquilegia, Aralia racemosa, A. spinosa, Arctium minus, Arctostaphylos uva-ursi, Arisaema dracontium, Aristida longespica, A. oligantha, Arnica chamissonis, Artemisia biennis, A. cana, A. ludoviciana, A. serrata, A. tridentata, Arundinaria gigantea, Asclepias incarnata, A. speciosa, Asimina triloba, Asparagus officinalis, Atriplex patula, Baccharis salicifolia, Baptisia lactea, Barbarea vulgaris, Berchemia scandens, Berula erecta, Betula alleghaniensis, B. nigra, B. occidentalis, B. papyrifera, B. pumila, Bidens cernuus, B. connatus, B. frondosus, Bignonia capreolata, Boehmeria cylindrica, Bolboschoenus fluviatilis, B. maritimus, Boltonia asteroides, Bothriochloa barbinodis, Bouteloua curtipendula, B. eriopoda, B. gracilis, B. hirsuta, B. simplex, Brickellia californica, B. floribunda, B. longifolia, Bromus inermis, B. japonicus, B. tectorum, Calamagrostis canadensis, C. rubescens, C. scopulorum, C. stricta, Caltha palustris, Campanula aparinoides, Campsis radicans, Carex aquatilis, C. atherodes, C. bebbii, C. buxbaumii, C. cherokeensis, C. comosa, C. complanata, C. conoidea, C. cristatella, C. curatorum, C. debilis, C. deweyana, C. disperma, C. geyeri, C. glaucodea, C. haydenii, C. hoodii, C. hystericina, C. lacustris, C. lasiocarpa, C. leptalea, C. lurida, C. molesta, C. nebrascensis, C. pellita, C. praegracilis, C. prairea, C. rossii, C. rostrata, C. sartwellii, C. scopulorum, C. secalina, C. siccata, C. simulata, C. stipata, C. stricta, C. tetanica, C. tribuloides, C. trisperma, C. utriculata, C. vesicaria, C. viridula, C. vulpinoidea, Carpinus caroliniana, Carya cordiformis, C. illinoinensis, C. ovata, Castilleja lineariiloba, C. miniata, Celtis laevigata, C. reticulata, Ceanothus fendleri, Cephalanthus occidentalis, Cercis canadensis, C. occidentalis, Cercocarpus ledifolius, Chamaedaphne calyculata, Chasmanthium latifolium, Chelone glabra, Chenopodium album, C. neomexicanum, Chicorum intybus, Chilopsis linearis, Chimaphila umbellata, Chylismia specicola, Cicuta douglasii, C. maculata, Cinna latifolia, Circaea canadensis, Cirsium arizonicum, C. arvense, C. rydbergii, C. wrightii, Clematis virginiana, Climacium dendroides, Clinopodium glabellum, Cocculus carolinus, Coleataenia longifolia, Comarum palustre, Conyza canadensis, Cordylanthus, Cornus amomum, C. florida, C. sericea, Corylus americana, Crataegus marshallii, C. pruinosa, C. viridis, Cryptotaenia canadensis, Cuscuta gronovii, Cynodon dactylon, Danthonia parryi, Dasiphora floribunda, Datura wrightii, Delphinium ×occidentale, Deschampsia cespitosa, Dicoria canescens, Dieteria canescens, Digitaria cognata, Dioscorea villosa, Diospyros virginiana, Distichlis, Ditrysinia fruticosa, Doellingeria umbellata, Drosera rotundifolia, Drymocallis arguta, Dysphania graveolens, Echinochloa crus-galli, Echinocystis lobata, Elaeagnus angustifolia, Eleocharis acicularis, E. compressa, E. erythropoda, E. macrostachya, E. obtusa, E. palustris, E. parvula, Elymus elymoides, E. glaucus, E. lanceolatus, E. repens, E. trachycaulus, E. virginicus, Encelia resinifera, Ephedra viridis, Epilobium ciliatum, E. coloratum, Epipactis gigantea, Equisetum arvense, E. fluviatile, E. hyemale, Eragrostis curvula, E. hypnoides, Ericameria nauseosa, Erechtites hieraciifolius, Erigeron flagellaris, E. formosissimus, E. lonchophyllus, Erodium cicutarium, Eryngium aquaticum, Eupatorium perfoliatum, Euthamia graminifolia, E. gymnospermoides, E. occidentalis, Eutrochium maculatum, E. purpureum, Fallopia scandens, Fallugia paradoxa, Fendlera rupicola, Festuca arizonica, F. thurberi, Fimbristylis autumnalis, F. littoralis, Forestiera pubescens, Fraxinus americana, F. anomala, F. nigra, F. pennsylvanica, F. velutina, Gaillardia aristata, Galinsoga parviflora, Galium obtusum, G. tinctorium, G. triflorum, Gaultheria, Gentiana, Gentianella amarella, Gentianopsis detonsa, Geranium viscosissimum, Geum canadense, G. macrophyllum, Glyceria canadensis, G. grandis, G. striata, Glycyrrhiza lepidota, Grindelia squarrosa, Gutierrezia sarothrae, Gymnadeniopsis clavellata, Hackelia virginiana, Helenium autumnale, Helianthus annuus, H. giganteus, H. grosseserratus, H. nuttallii, Heliomeris longifolia, Heracleum lanatum, Herrickia glauca, Hesperostipa spartea, Heterotheca fulcrata, H. villosa, Hibiscus moscheutos, Hieracium, Hordeum brachyantherum, H. jubatum, Hymenoxys hoopesii, Hypericum ascyron, Ilex decidua, I. opaca, Impatiens capensis, Ipomopsis aggregata, Iris versicolor, I. virginica, Iva annua, Juglans major, J. nigra, Juncus acuminatus, J. arcticus, J. balticus, J. dudleyi, J. ensifolius, J. interior, J. saximontanus, J. torreyi, Juniperus communis, J. deppeana, J. osteosperma, J. scopulorum, Justicia americana, Kalmia, Kochia scoparia, Lactuca serriola, Laportea canadensis, Leersia oryzoides, Leptochloa panicea, Leymus arenarius, L. cinereus, L. triticoides, Liatris pycnostachya, Lilium superbum, Lindera benzoin, Lindernia dubia, Linnaea borealis, Lipocarpha micrantha, Liquidambar styraciflua, Lobelia cardinalis, L. kalmii, L. siphilitica, Lupinus argenteus, Lycopodiella inundata, Lycopus americanus, L. asper, L. uniflorus, Lysimachia quadriflora, L. terrestris, Lythrum alatum, L. californicum, Maianthemum stellatum, Malva neglecta, Maurandella antirrhiniflora, Medicago lupulina, Melilotus albus, M. officinalis, Mentha arvensis, Mimulus cardinalis, M. guttatus, Mirabilis albida, Mitchella repens, Mitella pentandra, Monarda fistulosa, Moneses uniflora, Morus rubra, Muhlenbergia asperifolia, M. mexicana, M. richardsonis, M. rigens, M. tricholepis, M. wrightii, Myosotis scorpioides, Myosoton aquaticum, Napaea dioica, Nasturtium officinale, Nepeta, Nyssa aquatica, N. sylvatica, Oenothera biennis, O. elata, O. riparia, Onoclea sensibilis, Opuntia polyacantha, Orthilia secunda, Orthocarpus, Osmundastrum cinnamomeum, Ostrya knowltonii, Oxypolis fendleri, Panicum capillare, P. miliaceum, P. virgatum, Pappostipa speciosa, Parnassia, Parthenocissus* *quinquefolia, P. vitacea, Pascopyrum smithii, Paspalum distichum, Pastinaca sativa, Pedicularis groenlandica, P. lanceolata, Perideridia gairdneri, Persicaria amphibia, P. sagittata, P. virginiana, Petrophytum caespitosum, Phalaris arundinacea, Phaseolus angustissimus, Phleum pratense, Phragmites australis, Phyla lanceolata, Picea engelmannii, P. glauca, Pilea fontana, Pinus contorta, P. edulis, P. ponderosa, Plantago eriopoda, Platanthera dilatata, P. sparsiflora, Platanus occidentalis, P. wrightii, Poa nemoralis, P. palustris, P. pratensis, P. sylvestris, P. trivialis, Polanisia dodecandra, Polygonum aviculare, Polypogon interruptus, P. viridis, Populus angustifolia, P. balsamifera, P. deltoides, P. fremontii, P. tremuloides, P. trichocarpa, Primula specuicola, Proboscidea parviflora, Prunus serotina, P. virginiana, Pseudotsuga menziesii, Ptelea trifoliata, Pteridium aquilinum, Purshia mexicana, Pycnanthemum virginianum, Pyrola asarifolia, Quercus alba, Q. bicolor, Q. gambelii, Q. macrocarpa, Q. michauxii, Q. muehlenbergii, Q. nigra, Q. phellos, Q. shumardii, Ranunculus aquatilis, R. cymbalaria, Ratibida pinnata, Rhamnus betulifolia, R. californica, Rhododendron columbianum, Rhus aromatica, Ribes americanum, R. aureum, Rosa blanda, R. palustris, R. woodsii, Rotala ramosior, Rubus idaeus, R. pubescens, R. sachalinensis, R. ulmifolius, Rudbeckia laciniata, R. occidentalis, Ruellia strepens, Rumex densiflorus, Sabal minor, Sabatia calycina, Salix alaxensis, S. amygdaloides, S. bebbiana, S. discolor, S. drummondiana, S. eriocephala, S. exigua, S. fragilis, S. interior, S. fragilis, S. geyeriana, S. gooddingii, S. laevigata, S. lasiolepis, S. ligulifolia, S. lucida, S. lutea, S. maccalliana, S. petiolaris, S. wolfii, Sambucus nigra, Sanicula canadensis, Sanvitalia abertii, Saponaria officinalis, Schedonorus arundinaceus, Schoenoplectus acutus, S. americanus, S. pungens, S. tabernaemontani, Scirpus atrovirens, S. cyperinus, S. microcarpus, Scolochloa festucacea, Scrophularia marilandica, Senecio hydrophilus, S. suaveolens, S. triangularis, Setaria, Shepherdia rotundifolia, Silphium perfoliatum, Sium suave, Smilax bona-nox, S. rotundifolia, S. tamnoides, Solanum sarrachoides, Solidago auriculata, S. canadensis, S. confinis, S. gigantea, S. nana, S. riddellii, S. rigida, S. velutina, Sonchus arvensis, Sorghastrum nutans, Sorghum halepense, Sparganium eurycarpum, Spartina pectinata, Sphagnum, Sphenopholis intermedia, Spiraea alba, S. tomentosa, Spiranthes diluvialis, Sporobolus airoides, Stachys palustris, Streptopus amplexifolius, Symphoricarpos albus, S. oreophilus, Symphyotrichum cordifolium, S. drummondii, S. dumosum, S. ericoides, S. falcatum, S. foliaceum, S. laeve, S. lateriflorum, S. novae-angliae, S. oolentangiense, S. puniceum, S. spathulatum, Symplocarpus foetidus, Tamarix chinensis, Taxodium ascendens, T. distichum, Teucrium canadense, Thalictrum dasycarpum, T. fendleri, T. occidentale, Thelypodium integrifolium, Thelypteris palustris, Thuja occidentalis, Thyrsanthella difformis, Toxicodendron radicans, T. rydbergii, Tradescantia subaspera, Tragopogon, Trepocarpus aethusae, Tribulus terrestris, Tridens strictus, Trientalis borealis, Trifolium pratense, Triglochin maritimum, Trollius laxus, Typha angustifolia, T. domingensis, T. latifolia, Ulmus alata, U. americana, U. pumila, U. rubra,*

Urtica dioica, Vaccinium corymbosum, V. membranaceum, Verbascum thapsus, Verbena bracteata, V. hastata, Verbesina alternifolia, Vernonia gigantea, Veronica anagallis-aquatica, Viburnum lentago, Viola, Vitis arizonica, V. cinerea, V. riparia, V. rotundifolia, V. vulpina, Wisteria frutescens, Xanthisma spinulosum, Xanthium strumarium, Yucca baccata, Zanthoxylum americanum, Zeltnera calycosa, Z. namophila, Zizania aquatica.

***Symphyotrichum lentum* (Greene) G.L. Nesom** is a perennial, which inhabits fresh to brackish tidal or nontidal beaches, deltas, ditches, levees, marshes, meadows, salt marshes, sloughs, and margins of lakes, pools, rivers, and streams at elevations of up to 1402 m. Site exposures vary from full sun to partial shade. The substrates are described as "slightly alkaline" and include clay, mud, and sand. Flowers are produced from May to November. The plants propagate vegetatively by the production of long rhizomes. Additional life history information is scarce for this rare species. **Reported associates:** *Acmispon, Apium graveolens, Atriplex prostrata, Carpobrotus edulis, Distichlis spicata, Eleocharis macrostachya, Epilobium ciliatum, Eryngium articulatum, Euthamia occidentalis, Grindelia hirsutula, Hydrocotyle verticillata, Jaumea carnosa, Juncus balticus, Lepidum latifolium, Lilaeopsis masonii, Limosella australis, Lysimachia maritima, Lythrum californicum, Mentha, Persicaria amphibia, Schoenoplectus americanus, Triglochin maritimum, Typha, Verbena, Xanthium strumarium.*

***Symphyotrichum novi-belgii* (L.) G.L. Nesom** is a perennial, which inhabits fresh to brackish (salinity: 0–25 ppt), tidal or nontidal barrens, bogs, cliffs, depressions, dunes, fens, marshes, meadows, pools, roadsides, salt marshes, savannas, seeps, shores, swamps, thickets, and the margins of lakes, ponds, reservoirs, rivers, and streams at elevations of up to 800 m. The plants usually occupy open sites but also can tolerate partial shade. The substrates have been described as clay, gravel, loam, mafic, muck, peat, rock, sand, and stone. The plants also have been found growing on logs and stumps. Flowering occurs from August to October. The plants do not require vernalization for bolting and flowering, but bolting is promoted by long photoperiods in contrast to short days, which delay stem elongation. The plants are classified as long-day/short-day plants. The minimum time from sowing to anthesis is achieved by growing plants under long photoperiods until the main stem reaches 5 cm in length, followed by a subsequent shift to short photoperiods. The flowers are strongly self-incompatible and are outcrossed by various insects (Insecta) such as bees (Hymenoptera), beetles (Coleoptera), butterflies and moths (Lepidoptera), and flies (Diptera). Field observations have recorded from 2.75 to 14.0 bees comprising 6–12 species per plant as floral visitors. Late-blooming plants can host a disproportionate number of male bees (93%). The individual blossom area is large (averaging 6.02 cm²). However, despite having an extremely large total floral display area, one study showed that the plants were visited by a less than proportional number of bees relative to other plant species. The cypselae are dispersed by the wind, facilitated by long pappus bristles (5–6 mm). They can persist in the seed bank for at least a short period. Optimal seed germination requirements have not been determined; however, germination has been obtained after 2–6 weeks of cold stratification at 5°C followed by germination at 20°C under a 14-h photoperiod. Germination has been described as "slow." The plants can be highly clonal and reproduce vegetatively by the production of long rhizomes. Because of self-incompatibility, introduced plants that develop from single clones are essentially sterile. The clones are long lived, in one case still persisting after 150 years since its initial introduction to Great Britain. Although the plants grow well in oligohaline sites (<5 ppt NaCl), a substantial decline in cover (from 3.67% to 0.13%) was observed following a restoration project, which increased the salinity >25 ppt. Because of its ornamental desireability, effective procedures have been developed for adventitious shoot regeneration and transgenic transformation. **Reported associates:** *Acer rubrum, Achillea millefolium, Aconitum uncinatum, Agalinis purpurea, Ageratina altissima, Agrostis capillaris, A. hyemalis, A. perennans, A. scabra, A. stolonifera, Ailanthus altissima, Alnus serrulata, Alopecurus pratensis, Amelanchier canadensis, A. spicata, Amaranthus cannabinus, Ambrosia artemisiifolia, Amorpha fruticosa, Anagallis arvensis, Andropogon gerardii, A. glomeratus, A. virginicus, Anthoxanthum nitens, A. odoratum, Apios americana, Apocynum cannabinum, Aralia racemosa, Arethusa bulbosa, Aronia arbutifolia, A. melanocarpa, A. ×prunifolia, Asclepias incarnata, Astragalus alpinus, Athyrium filix-femina, Atriplex prostrata, Avenella flexuosa, Baccharis halimifolia, Baptisia tinctoria, Bartonia paniculata, Bidens connatus, B. frondosus, Boehmeria cylindrica, Bolboschoenus robustus, Borrichia frutescens, Bromus tectorum, Bryum, Cakile edentula, Calamagrostis pickeringii, Callitriche heterophylla, Calopogon tuberosus, Calystegia sepium, Campanula rotundifolia, Carex albolutescens, C. atlantica, C. bullata, C. canescens, C. collinsii, C. emoryi, C. exilis, C. folliculata, C. hormathodes, C. leptalea, C. livida, C. paleacea, C. pensylvanica, C. striata, C. stricta, C. torta, C. trisperma, Castilleja coccinea, Cephalanthus occidentalis, Chamaecyparis thyoides, Chamaedaphne calyculata, Chenopodium rubrum, Cirsium muticum, Cladium mariscoides, Clethra alnifolia, Coleataenia longifolia, Conium maculatum, Cornus amomum, C. sericea, Cuscuta gronovii, Cyperus dentatus, C. erythrorhizos, C. filicinus, Cyrilla, Danthonia sericea, D. spicata, Decodon verticillatus, Deschampsia cespitosa, Desmodium canadense, Dichanthelium acuminatum, D. clandestinum, D. dichotomum, D. ensifolium, D. scabriusculum, Drosera filiformis, D. intermedia, D. rotundifolia, Dulichium arundinaceum, Echinochloa walteri, Eleocharis acicularis, E. fallax, E. flavescens, E. parvula, E. robbinsii, E. rostellata, E. tenuis, E. tuberculosa, E. uniglumis, Elymus repens, E. virginicus, Epilobium coloratum, Erechtites hieracifolius, Erigeron annuus, Eriocaulon aquaticum, E. compressum, E. decangulare, Eriophorum virginicum, Eryngium aquaticum, Eubotrys racemosa, Eupatorium perfoliatum, E. resinosum, Euphorbia cyathophora, Euthamia graminifolia, Eutrochium dubium, E. purpureum, Fallopia scandens, Festuca rubra, Fragaria virginiana, Fraxinus pennsylvanica, Galium obtusum, G. tinctorium, Gaylussacia baccata, G. dumosa, G. frondosa, Gentianella amarella, Glyceria obtusa, G. striata, Gratiola, Gymnadeniopsis clavellata, Hedysarum alpinum, Helenium brevifolium, Heliopsis helianthoides, Hibiscus moscheutos, Hieracium, Hordeum jubatum, Houstonia caerulea, Hypericum boreale, H. canadense, H. densiflorum, H. denticulatum, H. mutilum, Ilex glabra, I. laevigata, I. verticillata, Impatiens capensis, Iris prismatica, I. versicolor, Iva frutescens, Juncus bufonius, J. canadensis, J. effusus, J. gerardii, J. militaris, J. pelocarpus, J. subcaudatus, J. tenuis, Juniperus virginiana, Kalmia angustifolia, Kosteletzkya pentacarpos, Lachnanthes caroliniana, Lathyrus japonicus, Leersia oryzoides, Lemna minor, Lepidium ruderale, L. virginicum, Leptodictyum riparium, Ligusticum scoticum, Lilaeopsis chinensis, Limonium carolinianum, Limosella australis, Lindera benzoin, Lindernia dubia, Lobelia canbyi, L. cardinalis, L. nuttallii, Lonicera morrowii, Lophiola aurea, Ludwigia palustris, Lycopodiella alopecuroides, L. caroliniana, Lycopus americanus, L. amplectens, L. uniflorus, L. virginicus, Lyonia ligustrina, Lysimachia terrestris, Lythrum hyssopifolia, L. salicaria, Magnolia virginiana, Maianthemum stellatum, Mikania scandens, Mitchella repens, Muhlenbergia glomerata, M. uniflora, Myrica caroliniensis, M. pensylvanica, Narthecium americanum, Nuphar variegata, Nymphaea odorata, Oclemena nemoralis, Oenothera biennis, O. perennis, Onoclea sensibilis, Orontium aquaticum, Osmunda regalis, Osmundastrum cinnamomeum, Oxypolis rigidior, Oxytropis campestris, Pallavicinia lyellii, Panicum verrucosum, P. virgatum, Parnassia grandifolia, Parthenocissus quinquefolia, Peltandra virginica, Persicaria arifolia, P. hydropiper, P. hydropiperoides, P. maculosa, P. pensylvanica, P. punctata, P. sagittata, Phalaris arundinacea, Phragmites australis, Physocarpus opulifolius, Pilea fontana, P. pumila, Pinus rigida, P. strobus, Plantago maritima, Platanus occidentalis, Pluchea camphorata, Poa compressa, P. pratensis, Pogonia ophioglossoides, Polygala cruciata, Polygonum aviculare, P. ramosissimum, Pontederia cordata, Portulaca oleracea, Potamogeton confervoides, Potentilla anserina, Prenanthes racemosa, Prunus maritima, P. pumila, P. serotina, Ptilimnium capillaceum, Pycnanthemum tenuifolium, Quercus ilicifolia, Q. palustris, Q. rubra, Ranunculus sceleratus, Rhamnus alnifolia, Rhexia virginica, Rhododendron viscosum, Rhus typhina, Rhynchospora alba, R. capitellata, R. cephalantha, R. chalarocephala, R. fusca, R. gracilenta, R. oligantha, Rosa blanda, R. palustris, R. rugosa, R. virginiana, Rubus allegheniensis, R. hispidus, R. idaeus, Rumex acetosella, R. crispus, R. verticillatus, Sabatia difformis, Sagittaria engelmanniana, S. graminea, S. latifolia, Salix nigra, Sambucus nigra, Samolus valerandi, Sanguisorba canadensis, Sarracenia purpurea, Schizachyrium scoparium, Schizaea pusilla, Schoenoplectus americanus, S. pungens, S. subterminalis, S. tabernaemontani, Scirpus cyperinus, S. expansus, Scleria reticularis, Scutellaria lateriflora, Sedum, Selaginella apoda, Sericocarpus asteroides, Sibbaldiopsis tridentata, Silene vulgaris, Sium suave, Smilax glauca, S. herbacea, S.*

rotundifolia, S. walteri, Solanum dulcamara, S. nigrum, Solidago juncea, S. rugosa, S. sempervirens, S. simplex, S. uliginosa, Sparganium americanum, Spartina alterniflora, S. cynosuroides, S. patens, S. pectinata, Spergularia rubra, S. salina, Sphagnum bartlettianum, S. magellanicum, S. portoricense, S. pulchrum, Spiraea alba, S. tomentosa, Symphyotrichum anticostense, S. subulatum, S. tenuifolium, Tanacetum bipinnatum, T. vulgare, Tetraphis pellucida, Teucrium canadense, Thalictrum pubescens, Thelypteris palustris, Thinopyrum pungens, Toxicodendron radicans, T. rydbergii, T. vernix, Triadenum virginicum, T. walteri, Triantha racemosa, Trientalis borealis, Triglochin maritimum, Trisetum spicatum, Typha angustifolia, T. domingensis, T. latifolia, Urtica dioica, Utricularia cornuta, U. gibba, U. juncea, U. striata, U. subulata, Vaccinium angustifolium, V. corymbosum, V. macrocarpon, V. myrtilloides, V. pallidum, Viburnum dentatum, V. nudum, V. recognitum, Viola cucullata, V. lanceolata, V. walteri, Vitis labrusca, Woodwardia areolata, W. virginica, Xyris difformis, X. torta, Zizania aquatica.

Symphyotrichum praealtum (Poir.) G.L. Nesom is a perennial, which occurs in bottoms, ditches, fens, marshes, meadows, prairies, railroad right-of-ways, ravines, roadsides, savannas, seeps, shores, swales, swamps, thickets, waste areas, woodlands, and along the margins of canals, lakes, ponds, reservoirs, rivers, springs, and streams at elevations of up to 2407 m. The full range of substrate acidity has not been determined but fairly low pH values (4.0–5.5) have been measured. Reported substrates include clay, clay loam, cobble, gravel, loam, and sand. The plants occur frequently along railroads, which probably provide an effective corridor for wind dispersal of the cypselae. Flowering occurs late in the season (August–November). The plants are strongly self-incompatible, with less than 0.1% seed set occurring from self-pollinations. The flowers are outcrossed by insects (Insecta), primarily butterflies and moths (Lepidoptera). The small seeds (0.22 mg) are dispersed primarily by the wind, facilitated by long (5.5 mm) pappus bristles. The plants can become highly clonal by the production of long stoloniferous rhizomes, which can spread the plant vegetatively through fragmentation. Open habitats created by natural or human disturbances (e.g., clearing, fire) are essential for the persistence of this species. Field surveys have found more than 67% of plants to be colonized by arbuscular mycorrhizal and dark septate endophytic fungi. **Reported associates:** *Acer rubrum, Aconitum uncinatum, Agalinis heterophylla, A. purpurea, A. tenuifolia, Agrostis gigantea, A. perennans, A. stolonifera, Alnus tenuifolia, Amaranthus tuberculatus, A. viridis, Ambrosia artemisiifolia, A. trifida, Amorpha canescens, Ampelopsis arborea, Andropogon gerardii, A. glomeratus, Anemone canadensis, A. virginiana, Apocynum cannabinum, Asclepias incarnata, A. sullivantii, A. syriaca, A. tuberosa, Baccharis halimifolia, B. salicifolia, Baptisia bracteata, Boehmeria cylindrica, Bolboschoenus fluviatilis, Boltonia asteroides, Bothriochloa saccharoides, Bouteloua curtipendula, B. eriopoda, B. gracilis, Brickellia eupatorioides, Bromus inermis, Caragana, Carex annectens, C. atlantica, C. granularis, C. gynandra, C. scoparia, C. squarrosa,*

C. viridula, Ceanothus herbaceus, Celtis laevigata, Cercis, Chloris canterae, Cicuta maculata, Cinna arundinacea, Cirsium arvense, C. muticum, C. wrightii, Clematis virginiana, Clinopodium glabrum, Conoclinium coelestinum, Conyza canadensis, Coreopsis tinctoria, C. tripteris, Cornus amomum, C. racemosa, Crataegus, Cucumis melo, Cynodon dactylon, Cyperus acuminatus, C. echinatus, C. erythrorhizos, C. odoratus, C. squarrosus, C. strigosus, Dactyloctenium aegyptium, Dasiphora floribunda, Dennstaedtia punctilobula, Deschampsia cespitosa, Dichanthelium clandestinum, D. dichotomum, D. oligosanthes, Digitaria ciliaris, Doellingeria umbellata, Dysphania ambrosioides, Echinochloa crus-galli, E. crus-pavonis, Eleocharis quinqueflora, E. tenuis, Elymus canadensis, E. repens, Equisetum arvense, Eragrostis hypnoides, Eryngium yuccifolium, Eupatorium altissimum, E. serotinum, Euphorbia corollata, Euthamia graminifolia, Eutrochium fistulosum, E. maculatum, Filipendula rubra, Fragaria virginiana, Galium aparine, G. tinctorium, Gamochaeta purpurea, Gentianella amarella, Gentianopsis virgata, Geranium texanum, Geum canadense, Glandularia bipinnatifida, Helenium amarum, H. autumnale, Helianthus annuus, H. giganteus, H. grosseserratus, H. mollis, H. pauciflorus, H. tuberosus, Heliopsis helianthoides, Heterotheca, Hymenoxys hoopesii, Hypericum densiflorum, H. mutilum, Ilex montana, Impatiens capensis, Juncus biflorus, J. effusus, J. greenei, J. torreyi, Juniperus scopulorum, Lactuca serriola, Lantana horrida, Lepidium virginicum, Leptochloa panicea, Lespedeza violacea, Leucospora multifida, Liatris spicata, Lobelia cardinalis, L. kalmii, Lycopus americanus, L. uniflorus, Lysimachia ciliata, Lythrum alatum, Medicago alba, Melilotus indicus, Mentha arvensis, Monarda fistulosa, Morus, Oenothera gaura, Onoclea sensibilis, Onosmodium bejariense, Osmundastrum cinnamomeum, Oxalis stricta, Panicum virgatum, Parnassia glauca, Paspalum urvillei, Pastinaca sativa, Pedicularis lanceolata, Persicaria amphibia, P. sagittata, Phalaris arundinacea, Physostegia virginiana, Phytolacca americana, Picea rubens, Pinus edulis, P. ponderosa, Pluchea odorata, P. sericea, Poa pratensis, Polypogon monspeliensis, Populus angustifolia, Prenanthes racemosa, Prunus serotina, Ptilimnium nuttallii, Pycnanthemum virginianum, Pyrrhopappus pauciflorus, Quercus gambelii, Q. macrocarpa, Q. palustris, Ranunculus sardous, Ratibida pinnata, Rhus glabra, Rhynchospora glomerata, Rorippa sessiliflora, Rubus hispidus, Rudbeckia hirta, Rumex crispus, Saccharum giganteum, Salix exigua, S. sericea, Sambucus nigra, Sanguisorba canadensis, Schedonorus pratensis, Schizachyrium scoparium, Scirpus cyperinus, S. pendulus, Securigera varia, Setaria parviflora, Silphium integrifolium, S. laciniatum, S. terebinthinaceum, S. trifoliatum, Solanum americanum, Solidago altissima, S. canadensis, S. gigantea, S. missouriensis, S. odora, S. patula, S. riddellii, S. rigida, S. rugosa, S. uliginosa, Sonchus asper, Sorghastrum nutans, Spartina pectinata, Sphagnum, Spiraea alba, Sporobolus compositus, S. heterolepis, Stellaria longifolia, Stenaria nigricans, Symphyotrichum ericoides, S. lanceolatum, S. novae-angliae, S. patens, S. pratense, S. subulatum, Teucrium canadense, Thelypteris noveboracensis,

Triadica sebifera, Triglochin maritimum, Triodanis perfoliata, Tripsacum dactyloides, Typha angustifolia, T. latifolia, Verbena hastata, V. litoralis, Verbesina alternifolia, Vernonia noveboracensis, Vicia sativa, Viola lanceolata, V. pedatifida, V. sororia.

Symphyotrichum puniceum (L.) **Á. Löve & D. Löve** is a perennial plant of beaches, bogs, carrs, depressions, ditches, fens, floodplains, levees, marshes, meadows, pinelands, pools, prairie fens, prairies, riparian snowbeds, savannas, seeps, sloughs, swales, swamps, thickets, tussocks (tops), and the margins of flowages, lakes, ponds, reservoirs, rivers, and streams at elevations of up to 2000 m. The plants are good indicators of tidal or nontidal freshwater marshes, which have an intermediate level of flooding. Tolerance to some salinity is (0–3 psu; practical salinity units) has resulted in their categorization as a fresh/oligohaline species. The species is broadly shade tolerant with exposures ranging from full sun to moderate shade. The plants occupy a wide range of acidic to calcareous habitats (pH 4.1–8.0), but occurrences often correlate with higher (more alkaline) pH. Although sometimes categorized as calciphiles, these plants actually are calcium generalists, which tolerate elevated calcium levels without influence on growth or accumulation rates. Tissue calcium levels range from 0.7% to 1.5% and correlate with levels present in the water but not in the sediment. Tissue magnesium levels range from 0.1% to 0.4%. The substrates include clay, clay loam, granite, gravel, histosols, loam, muck, peat, Potsdam sandstone, sand, sandy gravel, sandy loam, and silt. The plants can remain inconspicuous throughout most of the growing season, but attain dominance when flowering occurs late in the season. Flowering occurs from August to November. The flowers are pollinated by bees (Hymenoptera), butterflies (Lepidoptera) and other insects (Insecta). Observed seed set is relatively low (averaging 0.1–14.8 seeds set per inflorescence). The cypselae average 0.38 mg and are crowned by long pappus bristles (to 3.8 mm), which facilitate wind dispersal. The cypselae launch height is fairly high (~1.5 m), which results in greater dispersal distances when obstacles occur nearby. Seeds of the bisexual disk florets suffer a much higher rate of herbivory than the ray florets. Germinable seeds have been recovered from fecal pellets of white-tailed deer (Mammalia: Cervidae: *Odocoileus virginianus*), which indicates that some endozoic dispersal also occurs. The cypselae are found commonly in seed banks and can reach average denities of 10,300/m^2. Germination has been achieved following a period of winter stratification, but specific requirements have not been elucidated. They germinate only under nonflooded conditions. Shoots arise from a thick, branched caudex, from which short or long rhizomes are produced. The roots (6%–46%) are colonized by arbuscular mycorrhizal and dark septate endophytic fungi, which can result in reduced biomass under high water table conditions. **Reported associates:** *Abies balsamea, Acer pensylvanicum, A. rubrum, A. saccharum, Achillea millefolium, Acorus calamus, Actaea rubra, A. pachypoda, Agalinis purpurea, Ageratina altissima, Agrostis gigantea, Alnus rugosa, A. serrulata, A. viridis, Alisma, Ambrosia trifida, Amphicarpaea bracteata, Andropogon gerardii, A. glomeratus, Anemone canadensis, Anthoxanthum nitens, Apios americana, Apocynum androsaemifolium, A. cannabinum, Arctium minus, Arethusa bulbosa, Aronia melanocarpa, Artemisia biennis, A. serrata, Asclepias incarnata, A. rubra, A. syriaca, A. verticillata, Athyrium filix-femina, Avenella flexuosa, Barbarea vulgaris, Betula papyrifera, B. ×sandbergii, Bidens cernuus, B. frondosus, B. mitis, B. trichospermus, Bolboschoenus fluviatilis, B. robustus, Bromus ciliatus, B. latiglumis, Bryum, Burmannia capitata, Calamagrostis canadensis, C. stricta, Calla palustris, Callitriche verna, Calopogon tuberosus, Caltha palustris, Calystegia sepium, Campanula aparinoides, Cardamine pensylvanica, C. pratensis, Carex aquatilis, C. atherodes, C. atlantica, C. aurea, C. bebbii, C. buxbaumii, C. comosa, C. crinita, C. cristatella, C. cryptolepis, C. diandra, C. flava, C. folliculata, C. gynandra, C. haydenii, C. hystericina, C. interior, C. intumescens, C. lacustris, C. lasiocarpa, C. leptalea, C. limosa, C. lurida, C. oligosperma, C. pellita, C. prairea, C. rariflora, C. rostrata, C. sartwellii, C. scabrata, C. sterilis, C. stipata, C. stricta, C. torta, C. tribuloides, C. trichocarpa, C. trisperma, C. vaginata, C. vulpinoidea, Carya illinoinensis, Chamaedaphne calyculata, Chamerion angustifolium, Chelone glabra, Chrysosplenium americanum, Cicuta bulbifera, C. maculata, Cinna arundinacea, C. latifolia, Cirsium arvense, C. discolor, C. muticum, C. vulgare, Cladium mariscoides, Clematis virginiana, Climacium dendroides, Clintonia borealis, Coleataenia anceps, Comarum palustre, Cornus amomum, C. foemina, C. racemosa, C. sericea, Cryptotaenia canadensis, Cuscuta europaea, C. gronovii, Cyperus strigosus, Cypripedium reginae, Dasiphora floribunda, Desmodium, Dichanthelium acuminatum, D. clandestinum, Doellingeria umbellata, Drosera rotundifolia, Dryopteris campyloptera, D. carthusiana, D. cristata, D. intermedia, Dulichium arundinaceum, Eleocharis elliptica, E. erythropoda, E. obtusa, E. palustris, E. rostellata, Elymus lanceolatus, E. virginicus, Epilobium ciliatum, E. coloratum, E. leptophyllum, Equisetum arvense, E. fluviatile, E. scirpoides, E. sylvaticum, Eriocaulon decangulare, E. koernickianum, Eriophorum angustifolium, E. virginicum, E. viridicarinatum, Euonymus atropurpureus, Eupatorium perfoliatum, Eurybia macrophylla, Euthamia graminifolia, Eutrochium fistulosum, E. maculatum, E. purpureum, Fallopia japonica, F. scandens, Festuca rubra, Filipendula rubra, Fragaria virginiana, Fraxinus nigra, F. pennsylvanica, Galium aparine, G. asprellum, G. boreale, G. labradoricum, G. tinctorium, G. triflorum, Gaultheria hispidula, Gentiana andrewsii, Gentianopsis crinita, Geum aleppicum, G. canadense, G. palustre, G. peckii, G. rivale, Glyceria canadensis, G. grandis, G. melicaria, G. striata, Habenaria repens, Helenium autumnale, Helianthus giganteus, H. grosseserratus, Hibiscus moscheutos, Hieracium aurantiacum, Holcus lanatus, Houstonia caerulea, Hypericum mutilum, H. prolificum, Ilex montana, I. mucronata, I. verticillata, Impatiens capensis, Iris versicolor, I. virginica, Juncus arcticus, J. balticus, J. brachycephalus, J. brevicaudatus, J. canadensis, J. dudleyi, J. effusus, J. interior, J. nodosus, J. tenuis, Kalmia angustifolia, K. latifolia, Laportea canadensis, Lathyrus palustris, Larix laricina,*

Leersia oryzoides, L. virginica, Lemna minor, Leonurus cardiaca, Leucothoe fontanesiana, Liparis loeselii, Lobelia cardinalis, L. kalmii, L. siphilitica, Lomatogonium rotatum, Lonicera ×bella, Lonicera villosa, Ludwigia alternifolia, L. palustris, Lycopus americanus, L. uniflorus, Lysimachia ciliata, L. quadriflora, L. terrestris, L. thyrsiflora, L. vulgaris, Lythrum alatum, L. salicaria, Maianthemum trifolium, Mayaca fluviatilis, Melampyrum lineare, Mentha arvensis, Mentha piperita, M. spicata, Menyanthes trifoliata, Mertensia paniculata, Micranthes pensylvanica, Microstegium vimineum, Mimulus ringens, Mitchella repens, Mitella diphylla, Monarda fistulosa, Muhlenbergia glomerata, Murdannia keisak, Napaea dioica, Nasturtium officinale, Nuphar advena, Onoclea sensibilis, Osmundastrum cinnamomeum, Oxalis montana, Oxypolis rigidior, Packera aurea, P. paupercula, Panicum capillare, P. virgatum, Parnassia caroliniana, P. glauca, P. palustris, Parthenocissus, Pedicularis lanceolata, Peltandra virginica, Penthorum sedoides, Persicaria amphibia, P. arifolia, P. hydropiperoides, P. lapathifolia, P. maculosa, P. pensylvanica, P. punctata, P. sagittata, P. virginiana, Phalaris arundinacea, Phegopteris connectilis, Phlox maculata, Phragmites australis, Picea glauca, P. mariana, Pilea fontana, P. pumila, Pinus sylvestris, Plantago maritima, Platanthera dilatata, P. psycodes, Platanus occidentalis, Pogonia ophioglossoides, Polygala sanguinea, Polygonatum biflorum, Polystichum braunii, Populus deltoides, P. grandidentata, P. tremuloides, Potentilla anserina, Prenanthes alba, Pteridium aquilinum, Puccinellia pumila, Pycnanthemum virginianum, Quercus falcata, Ranunculus gmelinii, R. pensylvanicus, Rhamnus alnifolia, Rhododendron maximum, Rhynchospora capillacea, R. chalarocephala, R. glomerata, Ribes americanum, R. cynosbati, R. hirtellum, R. hudsonianum, R. lacustre, R. triste, Rorippa palustris, Rubus hispidus, R. idaeus, R. parviflorus, R. pubescens, R. repens, R. sachalinensis, Rudbeckia laciniata, Rumex britannica, Saccharum giganteum, Sagittaria latifolia, Salix bebbiana, S. ×bebbii, S. candida, S. discolor, S. eriocephala, S. lucida, S. nigra, S. petiolaris, S. sericea, Sambucus nigra, S. racemosa, Sanguisorba canadensis, Sanicula, Sarracenia alata, S. purpurea, Scheuchzeria palustris, Schizachyrium scoparium, Schoenoplectus acutus, S. tabernaemontani, Scirpus atrovirens, S. cyperinus, S. expansus, S. pendulus, Scrophularia marilandica, Scutellaria galericulata, S. lateriflora, Senecio suaveolens, Silphium perfoliatum, Sium suave, Smilax lasioneura, Solanum dulcamara, Solidago altissima, S. canadensis, S. flexicaulis, S. gigantea, S. macrophylla, S. patula, S. riddellii, S. rugosa, S. uliginosa, Sparganium americanum, S. chlorocarpum, S. eurycarpum, Spartina pectinata, Sphagnum magellanicum, S. recurvum, Sphenopholis intermedia, Spiraea alba, S. tomentosa, Spiranthes cernua, Stellaria borealis, Stenanthium densum, Symphyotrichum boreale, S. cordifolium, S. firmum, S. lanceolatum, S. lateriflorum, S. novae-angliae, S. prenanthoides, Symplocarpus foetidus, Teucrium canadense, Thalictrum dasycarpum, T. pubescens, Thelypteris palustris, Thuja occidentalis, Tiarella cordifolia, Tilia americana, Toxicodendron radicans, T. vernix, Tradescantia ohiensis,

Triadenum fraseri, T. virginicum, Trichophorum cespitosum, Trifolium, Triglochin maritimum, T. palustre, Trillium cernuum, Tripsacum dactyloides, Tsuga canadensis, Tussilago farfara, Typha angustifolia, T. latifolia, Urtica dioica, Utricularia cornuta, Uvularia sessilifolia, Vaccinium cespitosum, V. macrocarpon, Valeriana edulis, Veratrum viride, Verbena hastata, Vernonia fasciculata, V. noveboracensis, Veronicastrum virginicum, Viburnum lentago, V. nudum, Vinca, Viola cucullata, V. renifolia, V. sororia, Vitis riparia, Xyris torta, Zizania aquatica.

***Symphyotrichum squamatum* (Spreng.) G.L. Nesom** is a nonindigenous annual species of brackish and saline ballast disposal sites, beaches, ditches, floodplains, marshes, meadows, salt marshes, and waste areas at elevations of up to 36 m. There is little life-history information in the literature that pertains specifically to introduced North American populations of this species. However, some information can be gleaned from reports in other regions. North American occurrences usually occur on brackish, sandy substrates. Plants in Argentina occur on saline soils categorized as typic natraquolls, which also occur in the western United States. In Europe, the plants have been observed growing on highly alkaline sediments (pH: 11.8–12.6; electrical conductivity: 5980 µS/cm; total organic C: 4.12%; total N: 0.23%; total Ca: 1.27%; total Na: 664.2 mg/kg; P: 11 mg/kg). The plants are self-compatible and presumably autogamous. The cypselae possess a bristly pappus (to 5.3 mm), and certainly are wind-dispersed to some degree. In its native South American range, viable seed has been recovered from the fecal pellets of hares (Mammalia: Lagomorpha: Leporidae: *Lepus europaeus*), which also confirms the potential for animal dispersal of cypselae in this species. Arbuscular mycorrhizal fungi colonization (non-North American) has been observed in up to 41% of the root surfaces. The plant extracts are allelopathic, with 10% solutions completely (100%) inhibiting rootlet growth in wheat (Poaceae: *Triticum*). **Reported associates (North America):** *Asclepias fascicularis, Malvella leprosa.*

***Symphyotrichum subulatum* (Michx.) G.L. Nesom** is an annual inhabitant of tidal or non-tidal barrens, bayheads, beaches, bottomlands, chaparral, depressions, ditches, dunes, floodplains, lawns, marshes, meadows, roadsides (especially when salted), shores, sloughs, swamps, thickets, washes, and margins of canals, lakes, rivers, and salt marshes at elevations of up to 300 m. The plants tolerate a wide range of salinity including freshwater sites, but more often occur in oligohaline communities with brackish to saline conditions (0–10+ ppt salinity). They do not tolerate excessive flooding, which can reduce biomass significantly (e.g., from 0.37–0.44 g to 0–0.02 g), especially in disturbed sites (from 0.66–3.86 g to 0 g). Exposures are open with full sunlight. Substrates include clay, Copeland (typic argiaquolls), loam, marly sand, rock, sand, sandy clay, sandy loam, shell mounds, and spodic psammaquents. Flowering occurs from midsummer through early winter. The plants are self-compatible with the flowers being principally autogamous. About 40% seed set occurs in bagged (pollinator excluded) flowers compared to open-pollinated plants (60%) and no evidence of apomixis has been found.

Wind dispersal of seeds is facilitated by the bristly pappus (to 5.5 mm) and achieves the highest deposition rates in high elevation marshes. The fresh seeds also are buoyant and water dispersed. They remain afloat for an average of 38–62 days in salinities ranging from 0 to 36 ppt. Flotation decreases slightly with dry stratification, but some seeds can remain afloat for up to 140 days. Various studies have found mean densities of 170–2631 seeds/m² in seedbank samples. The greatest seed deposition occurs where mean water levels are less than 0.05 cm in depth. Germination has been achieved for dry-stored seeds following a period of cold stratification at 6°C for 109 days. The highest seed density (98%) occurs in dry sites and increases in plots that have been cleared of vegetation. The highest seedling emergence rates have been observed in low elevation, open canopy marshes. Arbuscular mycorrhizal and dark septate endophytic fungi are known to colonize up to 67% of the roots. The plants attain a typical mean biomass of 15.6 g dry wt. This is a "disturbance" or "fugitive annual" species, which often occurs in ruderal sites or areas of natural perturbations such as burned sites, where increased biomass has been observed relative to unburned sites. The plants are prolific colonizers, especially amidst *Spartina* (Poaceae) clones, but exhibit reduced growth where *Salicornia* (Amaranthaceae) dominates; however, the highest seed bank densities have been observed within the *Salicornia* zone. The plants show elevated "Kranz syndrome" $^{13}C/^{12}C$ isotope ratios, but have not been found to undergo C_4 photosynthesis.

Reported associates: *Achillea millefolium, Agrostis stolonifera, Alternanthera philoxeroides, Amaranthus australis, A. cannabinus, Ambrosia artemisiifolia, A. grayi, Ammannia coccinea, Ammophila breviligulata, Ampelopsis, Andropogon glomeratus, Arundo donax, Asclepias incarnata, A. perennis, Atriplex patula, A. prostrata, Avicennia germinans, Baccharis halimifolia, Bacopa monnieri, Batis maritima, Bidens laevis, Blutaparon vermiculare, Bolboschoenus maritimus, B. robustus, Borrichia frutescens, Carex hormathodes, C. silicea, Celtis, Centella asiatica, Chenopodium berlandieri, C. rubrum, Cirsium horridulum, Cladium jamaicense, C. mariscoides, Conoclinium coelestinum, Conyza canadensis, Coreopsis tinctoria, Crinum americanum, Cuscuta indecora, Cyperus grayi, C. haspan, C. odoratus, C. polystachyos, C. virens, Decodon verticillatus, Distichlis spicata, Echinochloa crus-galli, E. walteri, Eclipta prostrata, Eleocharis cellulosa, E. compressa, E. fallax, E. intermedia, E. macrostachya, E. parvula, E. quadrangulata, E. uniglumis, Erigeron quercifolius, Eupatorium capillifolium, Eustoma exaltatum, Euthamia, Festuca rubra, Fimbristylis caroliniana, F. spadicea, Fuirena squarrosa, Galium tinctorium, Helianthus annus, Heliotropium curassavicum, Hibiscus moscheutos, Hordeum jubatum, Hudsonia tomentosa, Hymenocallis crassifolia, Ipomoea sagittata, Iris versicolor, I. virginica, Iva annua, I. frutescens, Juglans microcarpa, Juncus arcticus, J. articulatus, J. bufonius, J. effusus, J. greenei, J. hybridus, J. gerardii, J. pelocarpus, J. roemerianus, J. scirpoides, Juniperus virginiana, Kosteletzkya pentacarpos, Lachnanthes caroliniana, Laguncularia racemosa, Lechea maritima, Leersia hexandra, Leitneria floridana, Lemna minor, Leptochloa fusca, Lilaeopsis chinensis, Limonium carolinianum, Lobelia elongata, Ludwigia glandulosa, L. linearis, Lycopus uniflorus, Lythrum hyssopifolia, L. lineare, Marsilea vestita, Mikania scandens, Myrica caroliniensis, M. pensylvanica, Oenothera parviflora, Orontium aquaticum, Osmunda regalis, Panicum amarum, P. dichotomiflorum, P. virgatum, Paspalum distichum, P. minus, P. urvillei, P. vaginatum, Peltandra virginica, Persicaria hydropiper, P. pensylvanica, P. punctata, Phragmites australis, Phyla lanceolata, P. nodiflora, Plantago maritima, Pluchea carolinensis, Polygala rugelii, Polygonella articulata, Polygonum ramosissimum, Polypremum procumbens, Pontederia cordata, Potentilla anserina, Prosopis, Prunus maritima, Ptilimnium capillaceum, Rhizophora mangle, Rhynchospora macrostachya, Rosa carolina, R. rugosa, Rumex maritimus, Sabatia, Sacciolepis striata, Sagittaria lancifolia, S. longiloba, S. papillosa, Salicornia bigelovii, S. rubra, Samolus valerandi, Sarcocornia perennis, Schoenoplectus americanus, S. californicus, S. pungens, S. tabernaemontani, Scutellaria galericulata, Sesbania herbacea, Sesuvium portulacastrum, Setaria, Sium suave, Smilax, Solidago rugosa, S. sempervirens, Spartina alterniflora, S. bakeri, S. cynosuroides, S. patens, Spergularia salina, Sporobolus texanus, S. virginicus, Stachys aspera, Steinchisma hians, Suaeda calceoliformis, S. linearis, Symphyotrichum novi-belgii, S. tenuifolium, Teucrium canadense, Toxicodendron radicans, Triglochin maritimum, Typha angustifolia, T. domingensis, T. latifolia, Vigna luteola, Xyris jupicai, Zizaniopsis miliacea, Ziziphus jujuba.*

***Symphyotrichum tenuifolium* (L.) G.L. Nesom** is a perennial inhabitant of tidal freshwater, brackish or saline bayous, coastal prairie, ditches, flats, levees, marshes, meadows, salt marshes, sandbars, sloughs, swales, woodlands, and the margins of tidal rivers and pools at elevations of up to 10 m. Although individuals can be found where salinity approaches 55 ppt, they occur mainly within the lower salinity (5–23 ppt) reaches of the high marsh zone. Plants originating from southern latitudes arguably are more salt-tolerant and attain higher shoot biomass and net photosynthetic rates than those from northern latitudes across a range of salinities from 0 to 60 ppt. Exposures typically are in full sun, or occasionally in partial shade. The substrates are circumneutral to alkaline (pH 6.7–7.6) and include clay, loamy sand, mud, and sand. Flowering occurs from July to February but some populations may flower both in the spring and in the fall. The reproductive ecology of this species has not been studied in detail, but the flowers presumably are pollinated by various insects (Insects), primarily bees (Hymenoptera). The rays are white and the disk yellow when the flowers fully open, with both turning pink to rosy pink or lavender as the flowers age. This color change probably cues pollinators in determining which florets remain unvisited. A long-bristled pappus (to 6.1 mm) facilitates short-distance dispersal of cypselae by wind currents. However, cypselae have been recovered from the plumage of Brant geese (Aves: Anatidae: *Branta bernicla*), which demonstrates the potential for their long-distance dispersal by birds. Specific germination requirements have not been reported. Observations indicate that roughly 25% of seeds emerge as seedlings, with 50%–90% of those surviving to the end of the first growing season (the higher numbers associated with sites cleared of competing vegetation). From 20% to 75% of the first year seedlings survive through the end of the second growing season, with higher survivorship generally occurring again in cleared sites. This is a "fugitive" species, where the seedlings occur more commonly in bare patches resulting from disturbance, than in the established vegetation. Increased biomass has been observed at sites following the thinning of dense stands of rush plants (Juncaceae: *Juncus*). Peak biomass measurements range from 1.9 to 27.8 g/m^2, with the lower values associated with freshwater sites relative to those of intermediate salinity. Plants do not persist in plots that have received extended nitrogen fertilization. The stems become flattened by the high tides of late summer and fall (August–September), with only shorter plants remaining erect. The plants can reproduce vegetatively and develop into cespitose clusters or larger colonies by the production of short (to 70 cm in northern populations) or long (to 1.2 m in southern populations) rhizomes or stolons. **Reported associates:** *Acrostichum danaeifolium, Agalinis linifolia, A. maritima, Alternanthera philoxeroides, Amaranthus australis, A. cannabinus, Andropogon, Aristida purpurascens, Atriplex calotheca, A. glabriuscula, A. patula, Avicennia germinans, Baccharis angustifolia, B. halimifolia, Bacopa monnieri, Batis maritima, Bidens, Blutaparon vermiculare, Bolboschoenus robustus, Borrichia frutescens, Cakile edentula, Calystegia sepium, Carex lupulina, Cassytha filiformis, Cenchrus tribuloides, Centella asiatica, Chenopodium berlandieri, Cicuta maculata, Cirsium horridulum, Cladium jamaicense, Coleataenia longifolia, C. tenera, Colocasia esculenta, Conocarpus erectus, Conyza canadensis, Cyperus erythrorhizos, C. filicinus, C. lupulinus, C. odoratus, C. strigosus, C. tetragonus, Diodia virginiana, Distichlis littoralis, D. spicata, Echinochloa walteri, Eclipta prostrata, Eleocharis cellulosa, E. fallax, E. parvula, E. quadrangulata, E. rostellata, Eriocaulon, Eupatorium capillifolium, Eustachys petraea, Fimbristylis autumnalis, F. caroliniana, F. spadicea, Forestiera segregata, Fuirena breviseta, Heterotheca subaxillaris, Hydrocotyle bonariensis, H. umbellata, H. verticillata, Hymenocallis palmeri, Ipomoea sagittata, Iva angustifolia, I. frutescens, I. imbricata, Juncus gerardii, J. roemerianus, Kosteletzkya pentacarpos, Laguncularia racemosa, Leersia hexandra, L. oryzoides, Leptochloa fusca, Limonium carolinianum, Ludwigia grandiflora, L. peruviana, L. repens, L. simpsonii, Lycium carolinianum, Lythrum lineare, Mikania scandens, Myrica cerifera, M. pensylvanica, Oenothera humifusa, Oldenlandia uniflora, Opuntia humifusa, Oxypolis filiformis, Packera glabella, Panicum amarum, P. dichotomiflorum, P. hemitomon, P. virgatum, Paspalum dissectum, P. monostachyum, Peltandra virginica, Persicaria hydropiperoides, P. punctata, Phragmites australis, Phyla lanceolata, Pinus elliottii, Plantago maritima, Pluchea baccharis, Polygonum ramosissimum, Polypogon monspeliensis, Pontederia cordata, Portulaca oleracea, Potentilla anserina, Pseudognaphalium stramineum, Quercus, Rhizophora mangle, Rhynchospora colorata, R. corniculata, R. divergens, R. microcarpa, R. tracyi, Rubus trivialis, Rumex verticillatus, Sabal palmetto, Sabatia decandra, S. stellaris, Saccharum giganteum, Sacciolepis striata, Sagittaria lancifolia, S. latifolia, S. subulata, Salicornia bigelovii, S. depressa, S. maritima, Salix caroliniana, Samolus valerandi, Sarcocornia perennis, Schinus terebinthifolius, Schizachyrium rhizomatum, Schoenoplectus americanus, S. californicus, S. pungens, Scleria reticularis, Serenoa repens, Sesuvium portulacastrum, Setaria magna, S. parviflora, Seutera angustifolia, Smilax auriculata, S. bona-nox, Solidago sempervirens, S. stricta, Spartina alterniflora, S. bakeri, S. cynosuroides, S. patens, S. spartinae, Spergularia canadensis, Sporobolus virginicus, Suaeda linearis, Symphyotrichum dumosum, S. novi-belgii, S. subulatum, Taxodium ascendens, Thelypteris palustris, Toxicodendron radicans, Triglochin maritimum, T. striata, Typha angustifolia, T. domingensis, T. latifolia, Uniola paniculata, Urena lobata, Utricularia, Vigna luteola, Zizania aquatica, Zizaniopsis miliacea.*

Use by wildlife: *Symphyotrichum boreale* is a host to a sac fungus (Fungi: Ascomycota: Leotiomycetidae: *Erysiphe cichoracearum*). *Symphyotrichum divaricatum* is a host of phytopathogenic bacteria (Proteobacteria: Xanthomonadaceae: *Xylella fastidiosa*), rusts (fungi: Basidiomycota: *Aecidium microsporum*; Coleosporiaceae: *Coleosporium asterum*; Pucciniaceae: *Puccinia asteris, P. cnici-oleracei*), smuts (fungi: Basidiomycota: Ustilaginaceae: *Ustilago dubiosa*), and other plant pathogenic fungi (Valsaceae: *Diaporthe linearis, Diaporthopsis*). The mature flowers are eaten by Canada geese (Aves: Anatidae: *Branta canadensis*). *Symphyotrichum elliottii* is a host of parasitic nematodes (Nematoda: Pratylenchidae: *Pratylenchus coffeae*; Tylenchulidae: *Tylenchulus palustris*). The flowers are visited in the fall by bees (Insecta: Hymenoptera: Apidae: *Bombus*), butterflies (Insecta: Lepidoptera), and fruit flies (Insecta: Diptera: Tephritidae: *Trupanea actinobola*). The flowers of *S. expansum* are visited by bees (Insecta: Hymenoptera). *Symphyotrichum lanceolatum* is fed upon by larval beetles (Insecta: Coleoptera: Chrysomelidae: *Microrhopala xerene, Sumitrosis inaequalis*), butterflies (Insecta: Lepidoptera: Nymphalidae: *Chlosyne gorgone, Phyciodes batesii, P. pratensis, P. pulchella, P. tharos*), flies (Insecta: Diptera: Agromyzidae: *Agromyza curvipalpis, A. reptans, Nemorimyza posticata, Phytomyza albiceps*), and white-tailed deer (Mammalia: Cervidae: *Odocoileus virginianus*). The phloem is fed upon by adult leafhoppers (Insecta: Hemiptera: Cicadellidae: *Macrosteles quadrilineatus*), plant bugs (Insecta: Hemiptera: Miridae: *Plagiognathus politis*), and treehoppers (Insecta: Hemiptera: Membracidae: *Publilia modesta*). The list of its documented floral visitors (all potential pollinators) is lengthy and includes long-tongued bees (Insecta: Hymenoptera: Apidae: *Apis mellifera, Bombus affinis, B. griseocallis, B. impatiens, B. pensylvanica, B. vagans, Triepeolus cressonii*; Anthophoridae: *Ceratina dupla, Melissodes bimaculata, M. boltoniae, M. rustica, M. subillata, Xylocopa virginica*; Megachilidae: *Coelioxys octodentata, Heriades leavitti, Megachile brevis,*

M. latimanus, M. mucida, M. relativa), short-tongued bees (Insecta: Hymenoptera: Andrenidae: *Andrena asteris, A. asteroides, A. hirticincta, A. nubecula, A. simplex, Heterosarus compositarum*; Colletidae: *Colletes americana, C. compactus, Hylaeus mesillae, H. modestus*; Halictidae: *Agapostemon sericea, A. splendens, A. texanus, A. virescens, Augochlora purus, Augochlorella aurata, A. striata, Augochloropsis metallica, Halictus confusus, H. ligatus, H. rubicunda, Lasioglossum albipennis, L. connexus, L. coriaceus, L. forbesii, L. fuscipennis, L. imitatus, L. paraforbesii, L. pectoralis, L. pilosus, L. tegularis, L. versatus, L. zephyrum, Paralictus platyparius, Sphecodes clematidis, S. cressonii, S. davisii, S. dichroa, S. ranunculi, S. stygius*), beetles (Coleoptera: Cantharidae: *Chauliognathus pennsylvanicus*; Cerambycidae: *Megacyllene robiniae*; Chrysomelidae: *Diabrotica undecimpunctata*; Meloidae: *Epicauta pensylvanica*; Scarabaeidae: *Euphoria sepulcralis*), bugs (Insecta: Hemiptera: Alydidae: *Alydus eurinus*; Miridae: *Lygus lineolaris*), butterflies, and moths (Insecta: Lepidoptera: Hesperiidae: *Atalopedes campestris, Erynnis martialis, Polites peckius*; Lycaenidae: *Celastrina argiolus, Lycaena phlaeas*; Noctuidae: *Anagrapha falcifera*; Nymphalidae: *Danaus plexippus, Limenitis arthemis, Nymphalis antiopa, Phyciodes tharos, Polygonia comma, Speyeria nokomis, Vanessa atalanta, V. cardui, V. virginiensis*; Pieridae: *Colias philodice, Pieris rapae, Pontia protodice*; Sesiidae: *Cisseps fulvicollis*), flies (Insecta: Diptera: Bombyliidae: *Sparnopolius confusus, Villa fulviana*; Calliphoridae: *Calliphora vicina, Cochliomyia macellaria, Lucilia illustris, L. sericata, Phormia regina, Pollenia rudis*; Conopidae: *Thecophora occidensis*; Muscidae: *Graphomya americana, G. maculata, Musca domestica, Neomyia cornicina, Stomoxys calcitrans*; Otitidae: *Chaetopsis aenea*; Sarcophagidae: *Helicobia rapax, Ravinia anxia, Sphixapata trilineata*; Syrphidae: *Allograpta obliqua, Eristalinus aeneus, E. anthophorina, E. arbustorum, E. dimidiatus, E. flavipes, E. stipator, E. tenax, E. transversus, Eristalis, Eupeodes americanus, Helophilus fasciatus, Paragus tibialis, Parhelophilus laetus, Sphaerophoria contiqua, Spilomyia longicornis, S. sayi, Syritta pipiens, Syrphus ribesii, Toxomerus geminatus, T. marginatus, T. politus, Tropidia quadrata*; Tachinidae: *Archytas analis, A. aterrima, Brachicoma sarcophagina, Copecrypta ruficauda, Estheria abdominalis, Gymnoclytia immaculata, Gymnosoma fuliginosum, Linnaemya comta, Leucostoma simplex, Phasia aeneoventris, Spallanzania hesperidarum, Tachinomyia panaetius, Trichopodes pennipes*), and wasps (Insecta: Hymenoptera: Braconidae: *Coeloides scolytivorus*; Ichneumonidae: *Ichneumon laetus, Lissonota clypeator*; Pompilidae: *Episyron biguttatus*; Sphecidae: *Ammophila kennedyi, Ammophila nigricans, A. procera, Bembix americana, Cerceris prominens, Ectemnius continuus, E. maculosus, E. trifasciatus, Eremnophila aureonotata, Eucerceris zonata, Isodontia apicalis, Liris argenta, Philanthus gibbosus, P. politus, P. ventilabris*; Tiphiidae: *Myzinum quinquecincta*; Vespidae: *Ancistrocerus adiabatus, A. antilope, A. catskill, Dolichovespula arenaria, D. maculata, Eumenes fraterna, Euodynerus annulatus, E. foraminatus, Leionotus scrophulariae, Polistes carolina, P. fuscata,*

Stenodynerus anormis, Vespula germanica). The plants host several fungi including powdery mildew (Ascomycota: Erysiphaceae: *Erysiphe cichoracearum*) and rusts (Basidiomycota: Coleosporiaceae: *Coleosporium asterum*; Pucciniaceae: *Erysiphe cichoracearum*). Also hosted are gall midges (Insecta: Diptera: Cecidomyiidae: *Rhopalomyia asteriflorae*), which cause flower and bud galls. Leaf blister is caused by another gall midge (Insecta: Diptera: Cecidomyiidae: *Asteromyia paniculata*), which establishes following a fungal infection (fungi: Ascomycota: Rhytismataceae: *Rhytisma asteris*). *Symphyotrichum novi-belgii* is a host of several fungi including leaf-spot (Ascomycota: Mycosphaerellaceae: *Septoria atropurpurea*), powdery mildew (Ascomycota: Erysiphaceae: *Erysiphe cichoracearum*), rusts (Basidiomycota: Coleosporiaceae: *Coleosporium asterum*; Pucciniaceae: *Puccinia asteris*), smuts (Basidiomycota: Entylomataceae: *Entyloma compositarum*), water molds (Oomycota: Peronosporaceae: *Basidiophora entospora*), wilt (Ascomycota: Chaetothyriomycetidae: *Phialophora asteris*), and other fungal plant pathenogens (Ascomycota: Corynesporascaceae: *Corynespora*). The plants also host leaf nematodes (Nematoda: Tylenchida: Aphelenchoididae: *Aphelenchoides ritzemabosi*), mites (Acari: Tenuipalpidae: *Brevipalpus obovatus*), and are susceptible to *Aster* chlorotic stunt virus (Betaflexiviridae: *Carlavirus*). The plants also have become infected experimentally with TSWV, the tomato spotted wilt virus (Bunyaviridae: *Tospovirus*). *Symphyotrichum praealtum* is a winter food source for swamp and cottontail rabbits (Mammalia: Leporidae: *Sylvilagus aquaticus, S. floridanus alacer*). It is a host of rust fungi (Basidiomycota: Coleosporiaceae: *Coleosporium asterum*) and is a larval food plant for pearl crescent butterflies (Insecta: Lepidoptera: Nymphalidae: *Phyciodes tharos*) and a larval host for fruit flies (Insecta: Diptera: Tephritidae: *Campiglossa albiceps*). The flowers are visited by bumblebees (Insecta: Hymenoptera: Apidae: *Bombus*), butterflies and moths (Insecta: Lepidoptera: Nymphalidae: *Phyciodes tharos*; Pieridae: *Colias eurytheme, C. philodice, Pyrisitia lisa, Pontia protodice*), and other insects. *Symphyotrichum puniceum* provides habitat for four-toed salamanders (Amphibia: Plethodontidae: *Hemidactylium scutatum*) and northern yellow throats (Aves: Parulidae: *Geothlypis trichas*). It is a larval food plant for the European corn borer (Insecta: Lepidoptera: Crambidae: *Ostrinia nubilalis*), pearl crescent butterfly (Insecta: Lepidoptera: Nymphalidae: *Phyciodes tharos*) and plume moth (Insecta: Lepidoptera: Pterophoridae: *Lioptilodes albistriolatus*), which feed on the foliage and flowers. It also provides a food source for the imperiled Chittenango ambersnail (Mollusca: Gastropoda: Succineidae: *Novisuccinea chittenangoensis*). The floral nectar and pollen attract various bees (Insecta: Hymenoptera: Andrenidae: *Perdita octomaculata*; Apidae: *Bombus perplexus, B. terricola, B. vagans*; Colletidae: *Colletes americanus, C. compactus*; Megachilidae: *Megachile relativa*) and butterflies (Insecta: Lepidoptera: Nymphalidae: *Boloria titania montinus, Neonympha mitchellii mitchellii*), which serve as pollinators. The plants host a number of fungi including powdery mildews (Ascomycota: Erysiphaceae: *Alphitomorpha fuliginea*, *Erysiphe cichoracearum*), rusts (Basidiomycota: Coleosporiaceae: *Coleosporium asterum*, *C. solidaginis*, *C. sonchi-arvensis*; Pucciniaceae: *Puccinia asteris*, *P. cnicioleracei*, *P. dioicae*), smuts (Basidiomycota: Entylomataceae: *Entyloma compositarum*), water molds (Oomycota: Peronosporaceae: *Basidiophora entospora*), and other plant pathogens (Ascomycota: Botryosphaeriaceae: *Phyllosticta*; Clavicipitaceae: *Guignardia haydenii*; Dermateaceae: *Mollisia cinerea*; Leptosphaeriaceae: *Ophiobolus crassus*; Mycosphaerellaceae: *Cercosporella virgaureae*, *Ramularia asteris*, *R. filaris*, *Septoria astericola*, *S. atropurpurea*; Valsaceae: *Diaporthe linearis*). Gall midges (Insecta: Diptera: Cecidomyiidae) produce galls on the inflorescences of the plants. *Symphyotrichum squamatum* reportedly is toxic to cattle (Mammalia: Bovidae: *Bos*). *Symphyotrichum subulatum* represents a major dietary component of some white-tailed deer (Mammalia: Cervidae: *Odocoileus virginianus*) populations. The plants host larval moths (Insecta: Lepidoptera: Pterophoridae: *Lioptilodes albistriolatus*). They also are eaten by nutria (Mammalia: Rodentia: *Myocastor coypus*), which can significantly reduce their biomass. The flowers are visited by various foraging bees (Insecta: Hymenoptera: Andrenidae: *Calliopsis coloradensis*; Colletidae: *Hylaeus affinis*), butterflies (Insecta: Lepidoptera: Nymphalidae: *Phyciodes tharos*), lace bugs (Insecta: Hemiptera: Tingidae: *Corythucha marmorata*), leafhoppers (Insecta: Hemiptera: Cicadellidae: *Empoasca bifurcata*, *E. delongi*, *Scaphoideus*), pirate bugs (Insecta: Hemiptera: Anthocoridae: *Cardiastethus assimilis*), plant bugs (Insecta: Hemiptera: Miridae: *Lygus lineolaris*), and planthoppers (Insecta: Hemiptera: Delphacidae: *Liburniella ornata*, *Sixeonotus insignis*, *Taylorilygus apicalis*). Foliage of *S. tenuifolium* is eaten by marsh crabs (Arthropoda: Crustacea: Sesarmidae: *Armases cinereum*), which prefer plants from northern over southern latitudes.

Economic importance: food: None of the OBL *Symphyotrichum* species reportedly is edible; **medicinal:** A root decoction of *Symphyotrichum divaricatum* has been used to alleviate headaches and a preparations of mashed roots are applied for toothaches. Ethanolic extracts of the flowers, stems and leaves of *S. lanceolatum* exhibit antibacterial activity against *Streptococcus pyogenes* (Eubacteria: Streptococcaceae) and *Salmonella enterica* (Eubacteria: Enterobacteriaceae) and fungicidal activity against *Cylindrocladium spathulatum* and *Fusarium oxysporum* (fungi: Ascomycota: Nectriaceae). Floral ethanolic extracts also effectively inhibit (by 91%) development of parasitic nematode larvae (Nematoda: Trichostrongylidae) in sheep (Mammalia: Bovidae: *Ovis aries*). The Zuni prepared a decoction from *S. lanceolatum* to treat arrow and bullet wounds and used the dried, pulverized plants for abrasions. They also inhaled smoke from the crushed flowers to stop nosebleeds. The Iroquois prepared an infusion of the plants for treating fevers. *Symphyotrichum novi-belgii* contains flavonoids and triterpenoid saponins. The extracts exhibit molluscicidal activity against snails (Mollusca: Gastropoda: Planorbidae: *Biomphalaria alexandrina*). Exposure to *S. novi-belgii* has been linked to contact dermititis in greenhouse workers. *Symphyotrichum praealtum* contains coumarins, which exhibit a number of medicinal properties. The Ramah Navajo tribe used a decoction of *S. praealtum* for snakebite and a cold infusion of plants as an eye medicine and to treat stomachaches or internal injuries. The Meskwaki prepared a stimulant drug from the plants for reviving unconscious persons. *Symphyotrichum puniceum* was listed as a pharmaceutical drug in 19th century references and was used as a medicinal plant by several North American tribes. The roots and rhizomes provide antispasmodic and nerve-calming remedies. The Cree and Woodlands people used a root decoction of *S. puniceum* to promote menstruation, to induce sweating for relief of fever, to promote recovery after childbirth, to reduce facial paralysis, and to treat teething pains in children; the roots were chewed to ease toothaches. The Iroquois administered a root infusion to treat colds, pneumonia, tuberculosis, and fevers such as typhoid. *Symphyotrichum squamatum* has been used medicinally as an antidiarrheal agent, tumor suppressant, and to expedite wound healing. Crude hydroalcoholic extracts reduce gastric acid secretion and have antiulcer properties. The roots of *S. subulatum* contain the diacetylene *(E)-4-(3-acetoxyprop-1-enyl)-2-methoxyphenyl (S)-2-methylbutanoate*, which has anti-inflammatory properties. The plants also contain a phloroglucinol glycoside (*1-[(butanoyl)phloroglucinyl]-β-D-gluco-pyranoside*), which exhibits moderate to strong free-radical scavenging activity; **cultivation:** Seeds of *Symphyotrichum elliottii* are available from some specialty nurseries; however, the plants can become overly aggressive in gardens. *Symphyotrichum lanceolatum* and *S. novi-belgii* are cultivated as ornamentals and for use in the cut-flower industry. Cultivars of the latter (which also is known as Michaelmas Daisy) are numerous (>240) and contain both dwarf and tall forms. They include: 'Ada Ballard', 'Alert', 'Alice Haslam', 'Alma Potschke', 'Andenken', 'Audrey', 'Barbados', 'Barr's Pink', 'Benary's Composition', 'Bkatrpu', 'Bkatrrv', 'Bkatrwt', 'Blue Eyes', 'Casablanca', 'Carnival', 'Celeste', 'Chelsea', 'Chequers', 'Crimson Viking', 'Dark Milka', 'Dasdebi', 'Dasdem', 'Dasfour', 'Dasgra', 'Dasing', 'Dasjes', 'Daskat', 'Dasmag', 'Dasone', 'Dukaster', 'Eventide', 'Fellowship', 'Freja Viking', 'Heinz Richard', 'Henry I Blue', 'Henry III', 'Herbstschnee', 'Island Bahamas', 'Jenny', 'Karmijn Milka', 'Kristina', 'Lady In Blue', 'Loke Viking', 'Magic Purple', 'Malba', 'Margrethe Viking', 'Marie Ballard', 'M.C. Snowy', 'Melody', 'Mittelmeer', 'Moercassino', 'Mt. Everest', 'Niobi', 'Odin Viking', 'Patricia Ballard', 'Patricia Viking', 'Peter Harrison', 'Peter III Blue', 'Professor Anton Kippenburg', 'Purple Dome', 'Purple Monarch', 'Purple Viking', 'Raspberry Swirl', 'Royal Ruby', 'Samoa', 'Sasha', 'Schneekissen', 'Schöne von Dietlikon', 'Snow Cushion', 'Thyra Viking', 'Tiny Tot', 'Twinkle', 'Victoria Celeste', 'Victoria Diana', 'Victoria Elisabeth', 'Victoria Fanny', 'Victoria Gaby', 'Victoria Ilona', 'Victoria Jane', 'Victoria Xenian', 'Violetta', 'Water Perry', 'White Ladies', 'White Swan', 'Winston Churchill', 'Woods Light Blue', 'Woods Pink', 'Woods Purple', and 'Wood's Series' among others. Many of the cultivars have U.S. patents pending. *Symphyotrichum puniceum* is sometimes distributed as the cultivar 'Firmus'; **misc. products:** *Symphyotrichum novi-belgii* has been highly recommended as a plant for use in "green roofs" in Nova Scotia. This species also is planted widely for prairie restoration. The Chippewa smoked root tendrils of *S. puniceum* with tobacco as a charm to lure game while hunting. The Ramah Navajo smoked dried leaves of *S. praealtum* to bring good luck in hunting. The name *Aster puniceus* (=*Symphyotrichum puniceum*) appears in the poem "Moose Hunting" by Carol Steinhagen; **weeds:** Although native to North America, *Symphyotrichum lanceolatum* is aggressive and can become particularly problematic in recently plowed, neglected, or poorly managed agricultural fields. Elsewhere, it is highly invasive in Hungary and Serbia, where it seriously threatens natural communities. *Symphyotrichum novi-belgii* is reported as an invasive in Austria, Czech Republic, and Romania. Resistance to imazapyr, an ALS-inhibiting herbicide, has occurred in nonindigenous plants of *S. squamatum* introduced to Spain. This species has become a nuisance agricultural weed throughout Europe; **nonindigenous species:** Several of the OBL taxa have been introduced to various regions including *Symphyotrichum expansum* (Hawaii, Japan), *S. squamatum* (from Latin America to the southeastern United States), and *S. subulatum* (Africa, Asia and salt mine areas of Michigan, Ohio, and Ontario). *Symphyotrichum lanceolatum* was brought to Europe for cultivation in the late 17th century, whereafter it escaped and became naturalized, especially along large, lowland rivers. It spread subsequently to Austria, Belgium, Czech Republic, France, Germany, Great Britain, Hungary, Italy, Poland, Romania, Serbia, Slovakia, Spain, and Ukraine. Similarly, *S. novi-belgii* was introduced to Europe as an ornamental, perhaps by the early 18th century, and also has become as widely naturalized there. It also is naturalized in Australia and South Korea. *Symphyotrichum squamatum* has been introduced to Africa (Egypt, Tunisia, Zimbabwe), Asia (Bahrain, Iraq, Kuwait, Japan, Qatar, Turkey), Australia, and Europe (Croatia, Cyprus, France, Greece, Italy, Malta, Portugal, Slovenia, Spain), possibly as a contaminant of rice stocks or other agricultural products. It has spread extensively throughout agricultural fields, particularly in the Mediterranean basin. **Systematics:** Major taxonomic realignments necessitated by evidence from molecular and morphological phylogenetic studies have resulted in the transfer of numerous, familiar North American *Aster* species to the genus *Symphyotrichum*. *Symphyotrichum* is the largest genus of subtribe Symphyotrichinae (Asteraceae: tribe Astereae), which currently, also includes *Almutaster*, *Ampelaster*, *Canadanthus*, and *Psilactis*. The genus *Aster* itself now refers almost entirely to Eurasian species. Molecular phylogenetic analyses (e.g., Figure 5.173) have cast considerable doubt on the distinctness of *Almutaster* and *Psilactis* from *Symphyotrichum*, and it seems advisable that those genera should be merged. Incomplete lineage sorting, hybridization and recent species radiations have made it difficult to resolve relationships within *Symphyotrichum* using nuclear DNA (*GAPDH*, ETS, ITS, 5S) sequence data, which exhibit numerous inconsistencies among their rendered phylogenies. In a relative sense, the nuclear ITS region has been identified as providing the

best estimate of phylogeny from among the nuclear regions evaluated (Figure 5.173). It is understandable that numerous taxonomic issues remain to be resolved within the genus. However, the molecular phylogenetic analyses consistently indicate multiple origins of the OBL habit in the genus. *Symphyotrichum firmum* (as *Aster firmus*) has been treated as a subspecies or variety of *S. puniceum*; however, the taxa are quite distinct and are retained here as separate species. Consequently, it is possible that some of the information compiled for *S. puniceum* (cited in the literature simply as *Aster puniceus*), may actually refer to *Symphyotrichum firmum*; care was taken to minimize such errors. *Symphyotrichum subulatum* is considerably problematic taxonomically, which makes it particularly difficult to associate literature information with the proper taxon. It has been recognized as a single species with five varieties or as five distinct species, all which have been ranked as OBL. Based on their generally allopatric ranges and high degree of reproductive isolation from one another, the five taxa have been retained here as distinct species: *S. bahamense* (=*S. subulatum* var. *elongatum*), *S. divaricatum* (=*S. subulatum* var. *ligulatum*), *S. expansum* (=*S. subulatum* var. *parviflorum*), *S. squamatum* (=*S. subulatum* var. *squamatum*) and *S. subulatum*. To further complicate matters, the name "*Aster divaricatus*" is a synonym for two different genera depending on the authorship: *Aster divaricatus* L. being the proper synonym for *Eurybia divaricata*; whereas, *Aster divaricatus* (Nutt.) Torr. & A. Gray non L. is the correct synonym applied to *Symphyotrichum divaricatum*. Consequently, literature accounts that simply specify *Aster divaricatus* without authorship cannot be assigned to the correct genus with certainty. Moreover, some accounts have misapplied the incorrect synonym, making it quite difficult to elucidate the taxon being referenced. Nevertheless, every effort has been made to attribute literature information properly to each species (or variety) as cited; however, much information simply is assigned to *S. subulatum* (without indication of variety), leaving little recourse but to compile it under that name. *Symphyotrichum lentum* is similar morphologically to *S. chilense*, and has been treated by some as a variety of that species. The basic chromosome numbers of *Symphyotrichum* are $x=5$, 7, 8, 13, 18, and 21. Based on this series, diploids would include *S. defoliatum* ($2n=36$), *S. divaricatum*, *S. expansum*, *S. subulatum*, and *S. tenuifolium* ($2n=10$), *S. chapmanii* ($2n=14$), and *S. elliottii* ($2n=16$); tetraploids would include *S. bahamense* and *S. squamatum* ($2n=20$); *S. novi-belgii* ($2n=48$) would be hexaploid. The karyotype of *S. chapmanii* indicates it to be an aneuploid derivative of an $x=9$ series. Multiple cytotypes are exhibited by *S. boreale* ($2n=16$, 32, 48, 64), *S. lanceolatum* ($2n=32$, 40, 48, 56, 64), *S. lentum* ($2n=16$, 64), *S. praealtum* ($2n=32$, 48, 64), and *S. puniceum* ($2n=16$, 32). The estimated genome size of *S. lanceolatum* (1Cx-value) is relatively small (0.69 pg). The tetraploid and octoploid cytotypes of *S. praealtum* correlate with the taxonomic varieties *S. praealtum* var. *praealtus* and *S. praealtum* var. *angustior* respectively. *Symphyotrichum divaricatum* has been implicated as one parental lineage involved in the hybrid origin of *S. tenuifolium*. *Symphyotrichum boreale* is

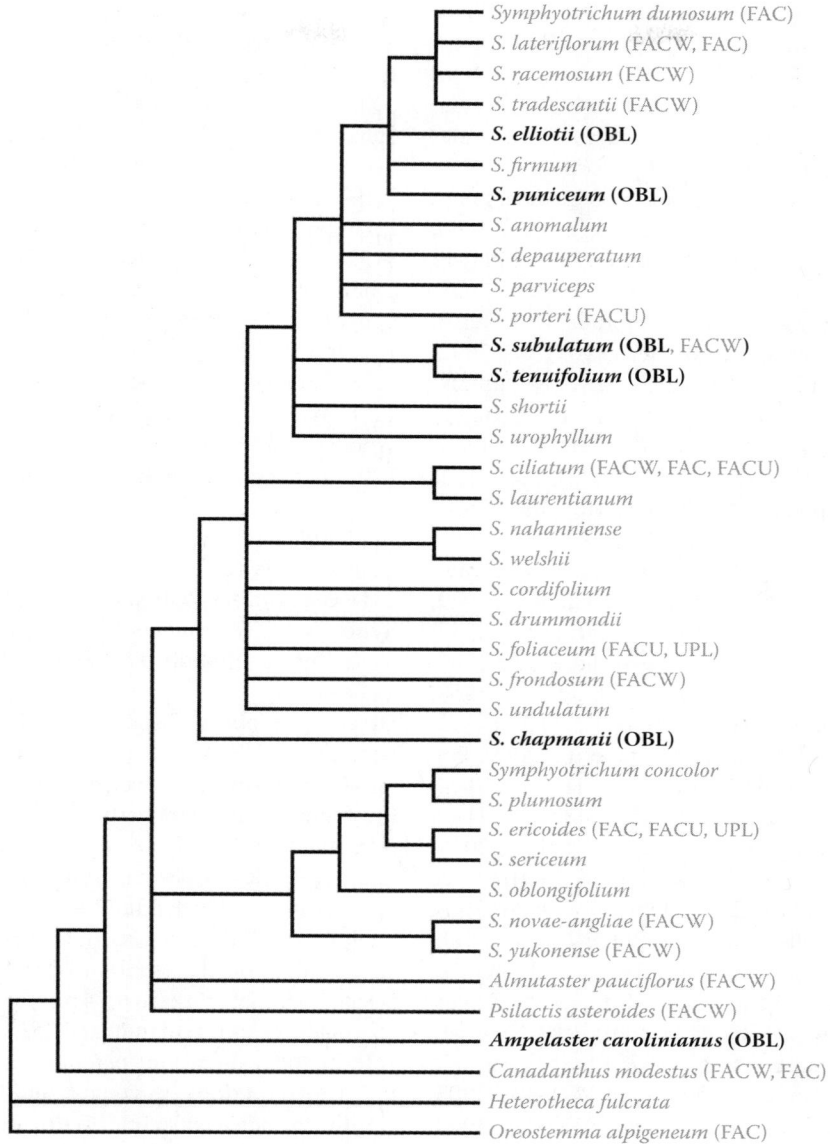

FIGURE 5.173 Phylogenetic relationships among selected *Symphyotrichum* species based on analysis of nrITS sequence data. Because of extensive hybridization, lineage sorting, etc., the phylogenetic structure of this genus has been resolved inconsistently using different molecular data sets; however, the results from the nrITS region are believed to provide a fairly accurate overview of interspecific relationships. It is evident at least that the OBL habit has arisen numerous times within *Symphyotrichum*, and also independently in the closely related genus *Ampelaster*. These results also argue for the inclusion of *Almutaster* and *Psilactis* within *Symphyotrichum*. (Adapted from Vaezi, J. & Brouillet, L., *Mol. Phylogenet. Evol.*, 51, 540–553, 2009.)

suspected to hybridize with *S. lanceolatum* and *S. puniceum*. *Symphyotrichum lanceolatum* hybridizes frequently with *S. laeve*, *S. lateriflorum*, and *S. puniceum*, but also occasionally with *S. boreale*, *S. ciliolatum*, *S. cordifolium*, *S. dumosum*, *S. ericoides*, *S. longifolium*, *S. novi-belgii*, *S. praealtum*, *S. sagittifolium*, and *S. undulatum*. Hybridization between *S. lanceolatum* and *S. lateriflorum* has been postulated in the origin of *S. ontarionis*. *Symphyotrichum ×subgeminatum* is a hybrid involving *S. novi-belgii* and *S. ciliolatum*. Hexaploid *S. novi-belgii* and tetraploid *S. boreale* have been implicated in the hybrid origin of the decaploid *S. anticostense*, a conclusion

that is supported only weakly by genetic data. Population genetic studies of *S. praealtum* have revealed high levels of clonal diversity and interpopulational subdivision.

Comments: *Symphyotrichum lanceolatum* is widespread in North America. More regional distributions characterize *S. boreale* (northern North America), *S. puniceum* (eastern and northern North America), *S. novi-belgii* (northeastern North America), *S. subulatum* (eastern North America), *S. praealtum*, *S. tenuifolium* (eastern United States), *S. elliottii* (southeastern United States), *S. squamatum* (introduced to southeastern United States, extending into South America),

S. expansum (southern United States, extending into South America), and *S. divaricatum* (south-central United States, extending into Mexico). Restricted distributions occur in *S. bahamense* (California [introduced], Florida, Georgia, extending into the West Indies), *S. chapmanii* (Florida, Alabama), and *S. defoliatum*, *S. lentum* (California).

References: Albert & Tepley, 2005; Allen, 1984; Anderson & Davis, 1998; Aronson & Galatowitsch, 2008; Arthur, 1898; Bainard et al., 2012; Baldwin & Mendelssohn, 1998b; Baldwin et al., 2001, 2010; Balogh, 2001; Baskin & Baskin, 1998; Beas et al., 2013; Beal, 1977; Bennett & Moran, 2013; Bertin & Kerwin, 1998; Bertin et al., 2010; Bertness et al., 1992; Boulos & Al-Hasan, 1986; Box et al., 1967; Boyer & Burdick, 2010; Brewer & Grace, 1990; Brewer et al., 1997; Briggs et al., 1989; Brodie, 1909; Brouillet et al., 2006; Buckallew, 2007; Burge & Isaac, 1977; Byers et al., 2007; Campbell & Seymour, Jr., 2011; Campbell et al., 2010; Carter et al., 2009; Casebere & Lodato, 2011; Chen et al., 2009; Childers et al., 2003; Chmielewski & Semple, 2001; Chmielewski & Strain, 2007; Clark, 2011; Cockel, 2010; Conard & Galligar, 1929; Crain, 2008; Cretini et al., 2012; Darlow, 2000; Davis, 1915; DeBarros, 2010; Decker et al., 2006; Desserud et al., 2006; Dias et al., 2006, 2009; Dow et al., 1990; Drawe, 1968; Dressler et al., 1987; Drohan et al., 2006; Egan & Ungar, 2000; Ellis & Everhart, 1885a; Elsey-Quirk et al., 2009; Fox et al., 2012; Freire et al., 2005; Gabriel & de la Cruz, 1974; Gladstar & Hirsch, 2000; Glattstein, 1991; Grant, 2013; Haag et al., 2005; Hao et al., 2011; Hejda et al., 2009; Hess, 2009; Hill et al., 2012; Hilty, 2014; Hodgson, 1928; Hoffman & Dawes, 1997; Horsburgh et al., 2011; Hubbard & Judd, 2013; Humbert et al., 2007; Hunt-Joshi et al., 2005; Inserra et al., 1990a; Jedlička & Prach, 2006; Jones, 1941, 2013; Judd & Lonard, 2004; Kandalepas et al., 2010; Kercher & Zedler, 2004; Kindscher, 1994; Kindscher & Tieszen, 1998; Ko et al., 2009; Kohl, 2011; Kristiansen et al., 1997; Krychak-Furtado et al., 2011; Laidig & Zampella, 1996; Leck & Leck, 2005; Leck et al., 2009; Lee et al., 2012; Le Page & Keddy, 1998; Lovell, 1907, 1913; Macdonald, 2003; Martin et al., 1982; Matthews et al., 1990; Meyer et al., 2008; Mohamed et al., 2011; Mohlenbrock et al., 1961; Mongelli et al., 1997; Moore et al., 2009; Morgan, 2008; Morgan & Holland, 2012; Mørk et al., 2012; Myers et al., 2004; Nelson, 1986; Nesom, 2005; Nesom & O'Kennon, 2008; Newmaster et al., 2012; Nichols & Nichols, 2008; Nichols et al., 2013; Nolfo-Clements, 2006; Nolin & Runkle, 1985; Novak & Foote, 1968; Oliveira et al., 2005; Olson & Fletcher, 2000; Orzell & Bridges, 1993, 2006; Osuna et al., 2003; Parrella et al., 2003; Paulsen et al., 1998; Payne, 2010; Peach & Zedler, 2006; Penfound & Hathaway, 1938; Pennings et al., 2001, 2003; Petrović et al., 2013; Proffitt et al., 2005; Quattrocchi, 2012; Rand, 2000; Reddoch & Reddoch, 2005; Renz et al., 2012; Rightmyer, 2008; Ritchie, 1957; Rosen et al., 2013; Rosenfeld, 2004; Ross et al., 2000; Russo et al., 2013; Schmalzer, 1995; Schmid & Bazzaz, 1994; Semple, 1982; Semple & Brammall, 1982; Semple et al., 1983; Sharpe & Baldwin, 2009, 2012; Shipley & Vu, 2002; S°rbu & Oprea, 2010; Sites & McPherson, 1981; Slocum & Mendelssohn, 2008; Smith, 1940; Smith & Turner, 1975; Smith et al., 2008; Snedden & Steyer, 2013; Spence, 2005; Sperduto & Nichols, 2012; Stalter, 2004; Stalter & Baden, 1994; Stevens et al., 2010; Stewart, 1953; St. Hilaire, 2003; Stromberg et al., 2009, 2011; Stubbs et al., 1992; Stuckey & Gould, 2000; Stutzenbaker, 1999; Sutton & Steck, 2005; Taboada et al., 1998; Taylor, 2012; Taylor & Grace, 1995; Tenorio & Drezner, 2006; Turki & Sheded, 2002; Urbatsch, 2013; USDA, 1984; Vaezi, 2008; Vasconcelos et al., 1999; Vasey et al., 2012; Vignolio & Fernández, 2006; Vincent et al., 2003; Vivian-Smith & Stiles, 1994; Walton, 1995; Ward & Clewell, 1989; Warners & Laughlin, 1999; Weaver, 1960; Weishampel & Bedford, 2006; Werier & Naczi, 2012; Wetzel et al., 2004; White & Simmons, 1988; Wigand et al., 2011; Williams, 1997; Williams et al., 2008; Willmer, 2011; Wilzer et al., 1989; Wolfe et al., 2006b; Woodcock & Pekkola, 2014; Woodcock et al., 2013; Yavorska, 2009; Zerbe et al., 2004; Zhen et al., 2009.

45. *Vernonia*

Ironweed; vernonie

Etymology: after William Vernon (1688–1711)

Synonyms: *Chrysocoma* (in part); *Serratula* (in part)

Distribution: global: New World; **North America:** central and eastern

Diversity: global: 20 species; **North America:** 17 species

Indicators (USA): OBL; FACW: *Vernonia lettermannii*

Habitat: freshwater; riverine; **pH:** unknown; **depth:** <1 m; **life-form(s):** emergent herb

Key morphology: Shoots (to 7 dm) erect; leaves mostly cauline, the blades (to 10 cm) filiform; heads (to 12 mm) numerous, discoid, peduncled (to 20+ mm), arranged secondarily in corymb-like inflorescences, involucre (to 10+ mm) narrowly campanulate, phyllaries (30–40+) in 5–6+ series; florets (to 14) 5-lobed, purple; cypselae (to 4 mm) with a persistent pappus of 25+ outer scales (to 0.5 mm) and 35+ inner bristles (to 7+ mm).

Life history: duration: perennial (caudex, rhizomes); **asexual reproduction:** rhizomes; **pollination:** insect; **sexual condition:** hermaphroditic; **fruit:** cypselae (common); **local dispersal:** cypselae (gravity, wind); **long-distance dispersal:** cypselae (animals?, wind)

Imperilment: (1) *Vernonia lettermannii* [G3]; S2 (OK); S3 (AR)

Ecology: general: *Vernonia* species often occur in dry meadows, grasslands and woodlands, but nearly every species can be found growing at least occasionally in wet sites such as bottomlands, floodplains, marshes or damp shores. Ten of the North American species (59%) are wetland indicators (FACW-UPL), including the single OBL species. Those species that have been investigated in detail are insect pollinated, and self-incompatible; however, they are highly cross-compatible, which often leads to the formation of numerous interspecific hybrids. Documented pollinators include various nectar and pollen-seeking insects (Insecta) including bee flies (Diptera: Bombyliidae), bees (Hymenoptera: Apidae), and butterflies (Lepidoptera: Hesperiidae; Nymphalidae; Papilionidae; Pieridae). Most of the seeds require cold stratification to induce germination; however, germination rates tend to be relatively low overall. The plants perennate from an enlarged caudex and reproduce vegetatively by the production of short rhizomes. The foliage of many species contains sequiterpene lactones, which are known to deter a variety of mammalian herbivores. It is a common site to see *Vernonia* plants standing as the only erect vegetation in meadows otherwise densely grazed by domestic cattle (Mammalia: Bovidae: *Bos*). Those *Vernonia* species lacking sequiterpene lactones are grazed heavily and have more restricted distributional ranges than their congeners.

***Vernonia lettermannii* Engelm. ex A. Gray** inhabits floodplains, terraces, washes, and the margins of lakes, rivers and streams at elevations of up to 200 m. The plants can tolerate full sun to partial shade. Interestingly, this plant also is regarded as drought tolerant. Occurrences are found on nutrient-poor substrates including cobble, gravel bars, rock crevices, and sandbars. The flowers open from July to September and are pollinated by insects (Insecta). The specific mechanism of seed dispersal has not been determined but likely involves a combination of wind (interacting with the long pappus bristles) and perhaps animals (by adhesion of cypselae to fur or feathers). The seeds will germinate at 21°C after receiving 90 days of moist, cold stratification at 4°C. The plants also can be propagated vegetatively by cuttings, which will develop numerous roots after 3 weeks under a misting water environment. **Reported associates:** *Acer rubrum, Alnus serrulata, Ambrosia, Amorpha fruticosa, A. ouachitensis, Amsonia hubrichtii, A. illustris, Andropogon gerardii, Apios americana, Betula nigra, Boltonia diffusa, Cephalanthus occidentalis, Chasmanthium latifolium, Coleataenia anceps, Commelina erecta, Cornus amomum, Crataegus, Diodia virginiana, Diospyros virginiana, Eutrochium fistulosum, Hamamelis vernalis, Hypericum prolificum, Ilex decidua, I. vomitoria, Juniperus virginiana, Liquidambar styraciflua, Ludwigia decurrens, Panicum virgatum, Perilla frutescens, Persicaria pensylvanica, Phyllanthopsis phyllanthoides, Platanus occidentalis, Salix caroliniana, Schizachyrium scoparium, Styrax grandifolius, Thyrsanthella difformis, Ulmus alata, Vaccinium corymbosum.*

Use by wildlife: The inflorescences of *Vernonia lettermannii* are highly attractive to butterflies (Insecta: Lepidoptera), and honeybees (Insecta: Hymenoptera), which seek the nectar and pollen and effect pollination. The foliage contains glaucolide-A, a sesquiterpene lactone that is known to deter feeding by several armyworms (Insecta: Lepidoptera: Noctuidae: *Spodoptera eridania, S. frugiperda, S. ornithogalli*) and saddleback caterpillars (Insecta: Lepidoptera: Limacodidae: *Acharia stimulea*). The plants also are relatively resistant to herbivory by deer (Mammalia: Cervidae) and rabbits (Mammalia: Leporidae).

Economic importance: food: not edible; **medicinal:** no medicinal uses known; **cultivation:** 'Iron Butterfly' is a very popular and widely distributed cultivar of *Vernonia lettermannii*, which was developed at the University of Georgia; **misc. products:** In 2007, the United States Postal Service issued a postage stamp featuring a *Vernonia* plant with a butterfly (Lepidoptera: Pieridae: *Zerene cesonia*) pollinator; **weeds:** Despite its ominous common name, *Vernonia lettermannii* is not regarded as weedy; **nonindigenous species:** none

Systematics: Phylogenetic analysis of combined DNA sequence data, which includes nine North American *Vernonia* species, indicates that the genus is monophyletic as currently recognized (Figure 5.174). This principally North American group (but extending into Mexico and the West Indies), is the

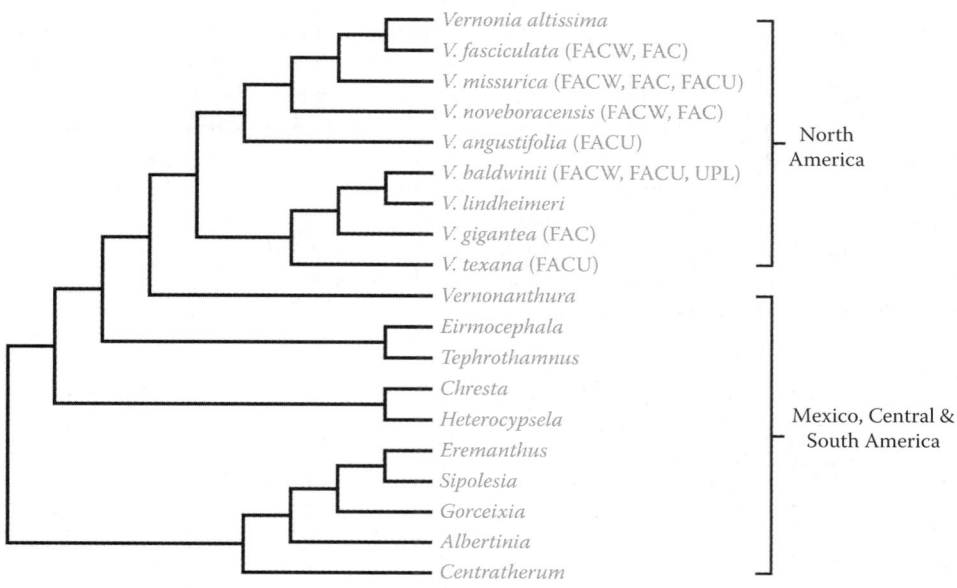

FIGURE 5.174 Phylogenetic relationships of *Vernonia* and other members of tribe Vernonieae (Asteraceae) as indicated by analysis of combined DNA sequence data. *Vernonia* is a principally North American genus, which contains many wetland indicators (the OBL species *V. lettermannii* was not sampled in this analysis). It resolves as a clade that descends from *Vernonanthura* and other Latin American members of the tribe. (Adapted from Keeley, S.C. et al., *Mol. Phylogenet. Evol.*, 44, 89–103, 2007.)

type genus of tribe Vernonieae where it resolves as the sister to *Vernonanthura* (Mexico to South America). *Vernonia* contains many wetland indicators. *Vernonia lettermannii* is classified within series *Verae* (subsection *Paniculatae*; section *Lepidoploa*) of the genus, where it is most similar morphologically to *V. fasciculata*. Although *V. lettermannii* has not yet been included in a molecular phylogenetic analysis, it presumably would occur in a position near *V. fasciculata*. The presence of awl-shaped hairs in pits of the upper leaf surface occur only among the North American species *V. fasciculata*, *V. lettermannii*, *V. marginata* and *V. texana* and has been interpreted as evidence of a close relationship among these species. However, phylogenetic relationships (Figure 5.174) resolve two of these species (*V. fasciculata*, *V. texana*) in different clades, which raises questions regarding the taxonomic utility of this feature. Consequently, the precise relationship of *V. lettermannii* remains undetermined. The basic chromosome number of *Vernonia* is *x*=17; *V. lettermannii* (2*n*=34) is a diploid. Isolating mechanisms among *Vernonia* species principally involve ecological and/or morphological divergence but only weak genetic barriers. Consequently, there are numerous possibilities for hybridization and nearly every species has been implicated in hybridization with another. *Vernonia lettermannii* forms natural hybrids with *V. baldwinii*. It also has been crossed successfully with *V. angustifolia*, *V. lindheimeri*, and *V. missurica*.

Comments: *Vernonia lettermannii* occurs only in Arkansas and Oklahoma.

References: Abdel-Baset et al., 1971; Burnett et al., 1977; Correll & Correll, 1975; Crandall & Tyrl, 2006; Cullina, 2009; Faust, 1972; Gleason, 1923; Jones, Jr., 1966, 1970, 1972, 1976; Keeley et al., 2007; Seigler, 1999; Strother, 2006f.

Family 2: Campanulaceae [90]

For many years Campanulaceae and Lobeliaceae were treated as related but distinct families, a taxonomic convention that slowly has given way to the merger of the families more recently. Whether or not these families should be combined appears to be a matter of preference more than necessity, given that the two groups resolve as sister clades (Figure 5.175) in phylogenetic analyses using DNA sequence data from chloroplast and nuclear loci (e.g., Eddie et al., 2003; Antonelli, 2008; Haberle et al., 2009). Consequently, whether they are recognized as two families or as a single family containing two subfamilies (Campanuloideae, Lobelioideae), either scheme is equally defensible by phylogenetic criteria. Because there seems to be no compelling argument for accepting one disposition over the other, the merger of the families has been followed here to reflect the taxonomic scheme used in recent compilations. When combined in this way, Campanulaceae represent roughly 2500 species with nearly 40% of those occurring within the genera *Campanula* and *Lobelia* (Judd et al., 2008; Morin, 2012a). Neither of these large genera is monophyletic as currently circumscribed.

Campanulaceae are united by a number of features including their generally herbaceous habit (with some secondarily derived, woody species), presence of lacticifers with milky sap, stamens attached to a nectary disk atop the inferior ovary, plunger pollination presentation mechanism, presence of polyacetylenes and absence of iridoids, and carbohydrates stored as inulin (Judd et al., 2008). Subfamily Campanuloideae is defined by its predominantly radial flowers, campanulate corollas, free anthers and filaments, and invaginating hairs on the upper style; Lobelioideae have bilateral flowers, connate filaments and anthers, and a longitudinal slit in the upper corolla lobe (Judd et al., 2008). The showy flowers are outcrossed by birds and insects and produce seeds, which are dispersed by wind (capsular fruits) or by birds (berries).

Representatives of Campanulaceae are distributed widely across temperate and subtropical, or montane tropical regions. The family is reasonably well-represented in wetlands, with OBL species known from 6 of the 14 genera (43%) found in North America. Four of the genera (*Downingia*, *Howellia*, *Legenere*, *Porterella*) are entirely aquatic.

Several genera (*Campanula*, *Codonopsis*, *Jasione*, *Lobelia*) are important economically as horticultural specimens. Medicinal drugs have been obtained from *Adenophora*, *Campanula*, *Codonopsis*, *Lobelia*, *Platycodon*, and a number of other genera. The roots of *Campanula rapunculus* are eaten as a vegetable in Europe.

Six North American genera contain OBL aquatics, with the majority (all but *Campanula*) occurring within subfamily Lobelioideae.

1. ***Campanula*** L.
2. ***Downingia*** Torr.
3. ***Howellia*** A. Gray

FIGURE 5.175 Phylogenetic relationships in Campanulaceae *sensu lato* (including Lobeliaceae) based on DNA sequence analysis of 680 accessions. In studies that include outgroups (e.g., Antonelli, 2008), subfamily Lobelioideae resolves as a sister clade to Campanuloideae, rather than a paraphyletic grade as depicted here. Groups containing OBL North American taxa (bold) indicate the independent derivation of the habit in the two subfamilies. (Adapted from Mansion, G. et al., *PLoS ONE*, 7(11), e50076, 2012.)

4. *Legenere* McVaugh
5. *Lobelia* L.
6. *Porterella* Torr.

1. *Campanula*

Bellflower, harebell; Campanule

Etymology: diminutive of the Latin *campana* ("bell"), which refers to the flower shape

Synonyms: *Annaea*; *Astrocodon*; *Brachycodon*; *Brachycodonia*; *Campanulastrum*; *Cenekia*; *Decaprisma*; *Depierrea*; *Diosphaera*; *Drymocodon*; *Echinocodonia*; *Erinia*; *Gadellia*; *Hemisphaera*; *Hyssaria*; *Lacara*; *Loreia*; *Marianthemum*; *Megalocalyx*; *Mzymtella*; *Nenningia*; *Neocodon*; *Pentropis*; *Petkovia*; *Popoviocodonia*; *Pseudocampanula*; *Quinquelocularia*; *Rapuntia*; *Rapuntium*; *Rotantha*; *Roucela*; *Sicyocodon*; *Symphiandra*; *Symphyandra*; *Talanelis*; *Tracheliopsis*; *Trachelioides*; *Wahlenbergia* (in part); *Weitenwebera*

Distribution: global: Africa; Asia; Europe; North America; **North America:** widespread (except south, central United States)

Diversity: global: 420 species; **North America:** 34 species

Indicators (USA): OBL: *Campanula aparinoides*, *C. californica*, *C. floridana*

Habitat: freshwater; palustrine; **pH:** 4.1–6.9; **depth:** <1 m; **life-form(s):** emergent herb

Key morphology: Stems (to 1 m) weak or reclining, angled or winged, the edges smooth or with stiff, recurved hairs; leaves sessile to short-petioled, the blades (to 6 cm) ovate, lanceolate, linear-elliptic or linear-lanceolate, the margins entire, crenate, or remotely or finely toothed, the surfaces smooth or with scabrid veins and edges; flowers (to 1.5 cm) solitary, pedicelled (to 20 mm), axillary or on on short, leafy terminal branches, corollas (to 6 mm) 5-merous, bluish, violet, or whitish, campanulate, funnelform or rotate; capsule (to 4 mm) hemispheric or obconic, weakly ribbed, dehiscent by basal or medial pores; seeds (to 0.5 mm) numerous.

Life history: duration: perennial (rhizomes); **asexual reproduction:** rhizomes; **pollination:** insect, self; **sexual condition:** hermaphroditic; **fruit:** capsule (common); **local dispersal:** water (seeds); **long-distance dispersal:** water (seeds)

Imperilment: (1) *Campanula aparinoides* [G5]; SH (CO); S1 (AB, GA, WY), S2 (DE, NC, ND, SK, TN); S3 (NS, WV); (2) *C. californica* [G3]; S3 (CA); (3) *C. floridana* [G3]

Ecology: general: This large and diverse genus of perennial (all OBL species) and annual herbs inhabits a diversity of primarily North Temperate sites ranging from deserts to wetlands and woodlands at sea level to alpine elevations. This is not a particularly characteristic group of wetland plants with less than a third of the species (29%) having any level of wetland indicator status and only three (9%) categorized as OBL. Outcrossing breeding systems facilitated by insect pollination predominate in the genus, which possesses an intricate "plunger" mechanism of pollen release (described under *C. aparinoides* later) to minimize self-pollination. Although most species that have been tested are self-incompatible, some are fully self-compatible such as *C. americana*, which is related to the OBL *C. floridana*. However, even in self-incompatible taxa, selfing and low levels of ensuing seed set have been observed to occur in those older flowers that remain unpollinated, thereby providing a mixed mating system. There remains a surprising lack of information on the specific reproductive biology of different species in this large and showy-flowered genus and additional studies are encouraged. All of the species produce multiseeded capsules. Seed germination typically is reported to occur at temperatures of 18°C–22°C, with stratification for 2–4 weeks at –4°C to 4°C recommended for recalcitrant germinants. The seeds of most *Campanula* species (including terrestrials) are adapted primarily for dispersal by water, or by gravity for those species inhabiting cliffs or ledges. Seed banks can be sparse or quite extensive (>1000/m²) depending on the species.

***Campanula aparinoides* Pursh** is a perennial, which inhabits bays, beaches, bogs, carrs, depressions, ditches, fens, floodplains, hummocks, lakeshores, marshes, meadows, prairie fens, prairies, ravines, roadsides, seeps, shores, sloughs, swales, swamps, thickets, and the margins of brooks, channels, lagoons, lakes, ponds, rivers, sloughs, and streams at elevations of up to 1676 m. It can occur in shaded or open sites. The plants tolerate acidic as well as alkaline, calcareous (261–392 mg $CaCO_3$/L) substrates (pH 4.1–6.9), which include clay, gravel, histosols, limestone cobble, marl, marly sand, muck, mucky peat, mud, peat, peat muck, rock, rotting stumps, sand, sandy gravelly peat, sandy loam, sandy silt, and silty clay. They have been observed on peat mats more than 45 cm in thickness. The weak stem of this species promotes a climbing habit, which relies on the presence of other vegetation for support. The plants are in flower from the end of June through mid-September, at least within the principal portion of their range. The flowers are outcrossed by insects, which are attracted to their nectar. Thrips (Insecta: Thysanoptera) are the most prevalent floral visitors with butterflies and moths (Insecta: Lepidoptera) and beeflies (Insecta: Diptera: Bombyliidae) observed in lesser numbers. Outcrossing of the protandrous flowers is facilitated by a "plunger" mechanism, whereby the anthers form an enveloping ring around the brush-like style, and dehisce inward. The elongating style sweeps the pollen grains out of the flower, while the receptive stigma surfaces remain folded together and inaccessible to the pollen. The receptive surfaces are exposed as the stigma lobes unfold and receive foreign pollen from visiting insects, which also pick up self-pollen (from the backs of the now expanded stigma lobes) to be transferred to other flowers. The prevalence and movement of the small thrips throughout the floral interior is believed to effect self-pollination in the event that the flowers are not visited by larger insects. However, no seed set has been observed in a population that was introduced to Europe, which indicates that the plants might represent a self-incompatible clone (or that appropriate pollinators are not present). Thus, the extent of self-compatibility in this species needs to be clarified. Seed set commences in late August. The seeds are dispersed primarily by water and possess epidermal cells with large lumens, which enhances their buoyancy. The seeds germinate optimally at 21°C, but other requirements have not been specified. A thick litter layer is believed to suppress seed germination in the wild. The plants occur commonly on tussocks, typically expanding their coverage from May to September. They are found both on hummocks and in hollows, with a somewhat higher frequency in the former. This species persists in sites characterized by light to heavy disturbance. It occurs more often in places where litter mass is relatively high and fire incidence relatively low; however, it is known to reappear after decades of absence in areas that have been burned. Vegetative reproduction occurs by rhizomes, which can displace those of *Galium* (Rubiaceae) species. Its spread by seed and vegetative growth is relatively slow. **Reported associates:** *Abies balsamea, Acer rubrum, A. saccharinum, A. saccharum, Acorus americanus, A. calamus, Agalinis purpurea, Agrimonia parviflora, Agrostis gigantea, A. hyemalis, A. perennans, Aletris farinosa, Alisma triviale, Alnus rugosa, A. serrulata, Alopecurus aequalis, A. carolinianus, Amaranthus tuberculatus, Ambrosia artemisiifolia, Ammannia auriculata, Amorpha fruticosa, Andropogon gerardii, Anemone canadensis, Angelica atropurpurea, Anthoxanthum nitens, Apios americana, Apocynum androsaemifolium, Arethusa bulbosa, Asclepias incarnata, A. purpurascens, A. rubra, Baptisia tinctoria, Barbarea vulgaris, Bartonia virginica, Beckmannia syzigachne, Betula alleghaniensis, B. papyrifera, B. pumila, Bidens cernuus, B. comosus, B. connatus, B. frondosus, B. trichospermus, Boehmeria cylindrica, Botrychium virginianum, Bromus ciliatus, Calamagrostis canadensis, Calla palustris, Callitriche heterophylla, Calopogon tuberosus, Caltha palustris, Calystegia sepium, Cardamine bulbosa, Carex alata, C. atlantica, C. aurea, C. aquatilis, C. bebbii, C. bicknellii, C. buxbaumii, C. canescens, C. chordorrhiza, C. comosa, C. crinita, C. diandra, C. garberi, C. gynandra, C. hystericina, C. interior, C. lacustris, C. laevivaginata, C. lasiocarpa, C. leptalea, C. limosa, C. livida, C. lurida, C. oligosperma, C. pauciflora, C. pellita, C. prairea, C. rostrata, C. sartwellii, C. scoparia, C. shortiana, C. sterilis, C. stipata, C. straminea, C. stricta, C. suberecta, C. tomentosa, C. torta, C. tribuloides, C. trichocarpa, C. utriculata, C. vesicaria, C. vulpinoidea, Carpinus caroliniana, Castilleja coccinea, Cephalanthus occidentalis, Chamaedaphne calyculata, Chamerion angustifolium, Chelone glabra, Cicuta bulbifera, C. maculata, Cirsium arvense, C. muticum, Cladina, Cladium mariscoides, Clinopodium glabrum, Clintonia borealis, Coleataenia longifolia, Comarum palustre, Cornus amomum, C. foemina, C. sericea, Corylus americana, Cuscuta gronovii, Cyperus strigosus, Cypripedium candidum, C. reginae, Dactylorhiza viridis, Dasiphora floribunda, Decodon verticillatus, Deschampsia cespitosa, Dichanthelium acuminatum, D. clandestinum, D. dichotomum, Doellingeria umbellata, Drepanocladus aduncus, Drosera rotundifolia, Dryopteris cristata, Dulichium arundinaceum, Elaeagnus umbellata, Eleocharis acicularis, E. elliptica, E. erythropoda, E. obtusa,* *E. palustris, E. quinqueflora, E. tenuis, E. tuberculosa, Epilobium ciliatum, E. coloratum, E. leptophyllum, E. strictum, Equisetum arvense, E. fluviatile, E. hyemale, E. laevigatum, E. variegatum, Erechtites hieraciifolius, Eriophorum gracile, E. vaginatum, E. virginicum, E. viridicarinatum, Eupatorium perfoliatum, E. rotundifolium, Euthamia graminifolia, Eutrochium dubium, E. maculatum, E. purpureum, Fallopia scandens, Filipendula rubra, Fragaria virginiana, Fraxinus nigra, F. pennsylvanica, Galium asprellum, G. boreale, G. labradoricum, G. obtusum, G. palustre, G. tinctorium, G. trifidum, G. triflorum, Gentiana andrewsii, G. saponaria, Gentianopsis crinita, G. virgata, Geocaulon lividum, Geum canadense, G. rivale, Glyceria borealis, G. canadensis, G. grandis, G. striata, Goodyera oblongifolia, Gymnadeniopsis clavellata, Helenium autumnale, Helianthus grosseserratus, Heracleum sphondylium, Heuchera richardsonii, Hibiscus moscheutos, Hieracium aurantiacum, H. piloselloides, Hydrocotyle americana, Hypericum kalmianum, H. perforatum, Ilex verticillata, Impatiens capensis, Iris versicolor, I. virginica, Juncus acuminatus, J. balticus, J. brachycephalus, J. brevicaudatus, J. canadensis, J. dichotomus, J. dudleyi, J. effusus, J. marginatus, J. scirpoides, J. tenuis, Juniperus communis, Lactuca biennis, L. canadensis, Larix laricina, Lathyrus palustris, Leersia oryzoides, Lemna minor, L. trisulca, Liatris pycnostachya, Lilium philadelphicum, L. superbum, Linnaea borealis, Lobelia cardinalis, L. kalmii, L. siphilitica, L. spicata, Lonicera oblongifolia, L. villosa, Ludwigia alternifolia, L. palustris, Lycopus americanus, L. uniflorus, Lysimachia ciliata, L. quadriflora, L. terrestris, L. thyrsiflora, Lythrum alatum, L. salicaria, Maianthemum canadense, M. stellatum, Melampyrum lineare, Mentha arvensis, M. ×piperita, Menyanthes trifoliata, Micranthes pensylvanica, Mimulus ringens, Mitella nuda, Muhlenbergia glomerata, M. mexicana, Myrica gale, Nuphar advena, Onoclea sensibilis, Orbexilum pedunculatum, Osmunda regalis, Osmundastrum cinnamomeum, Oxypolis rigidior, Packera aurea, P. paupercula, Panicum virgatum, Parnassia glauca, Parthenocissus quinquefolia, Paspalum laeve, Pedicularis lanceolata, Peltandra virginica, Penstemon digitalis, Penthorum sedoides, Persicaria amphibia, P. coccinea, P. hydropiper, P. hydropiperoides, P. lapathifolia, P. maculosa, P. pensylvanica, P. punctata, P. sagittata, Phalaris arundinacea, Phleum pratense, Phlox glaberrima, Picea glauca, P. mariana, Physostegia virginiana, Pilea fontana, P. pumila, Pinus banksiana, P. strobus, Platanthera hyperborea, P. leucophaea, P. obtusata, P. psycodes, Poa nemoralis, P. palustris, P. pratensis, Pogonia ophioglossoides, Polemonium reptans, Polygala cruciata, Pontederia cordata, Populus balsamifera, P. tremuloides, Potentilla anserina, Primula mistassinica, Proserpinaca palustris, Pycnanthemum tenuifolium, P. virginianum, Pyrola asarifolia, Ranunculus acris, R. pensylvanicus, R. septentrionalis, Rhamnus frangula, Rhododendron groenlandicum, Rhynchospora alba, R. glomerata, Ribes americanum, R. cynosbati, R. glandulosum, R. hirtellum, Rorippa palustris, Rosa palustris, Rubus hispidus, R. idaeus, R. pubescens, R. sachalinensis, Rudbeckia* *fulgida, R. hirta, R. subtomentosa, R. triloba, Rumex britannica, R. crispus, R. verticillatus, Sagittaria cuneata, S. latifolia, Salix bebbiana, S. caroliniana, S. candida, S. cordata, S. eriocephala, S. exigua, S. discolor, S. pedicellaris, S. petiolaris, S. sericea, Sambucus nigra, Sarracenia purpurea, Scheuchzeria palustris, Schizachyrium scoparium, Schoenoplectus acutus, S. heterochaetus, S. pungens, S. subterminalis, S. tabernaemontani, Scirpus atrovirens, S. cyperinus, S. pendulus, S. serissima, Scleria reticularis, Scutellaria galericulata, S. integrifolia, S. lateriflora, Selaginella eclipes, Senecio suaveolens, Silphium integrifolium, S. terebinthinaceum, Sium suave, Solanum dulcamara, Solidago altissima, S. gigantea, S. houghtonii, S. ohioensis, S. patula, S. ptarmicoides, S. riddellii, S. rigida, S. rugosa, S. uliginosa, Sorghastrum nutans, Sparganium americanum, S. androcladum, S. eurycarpum, Spartina pectinata, Sphagnum girgensohnii, S. magellanicum, S. palustre, Sphenopholis intermedia, Spiraea alba, Spiranthes cernua, Stachys appalachiana, S. eplingii, Stellaria alsine, Symphyotrichum boreale, S. ericoides, S. firmum, S. laeve, S. lanceolatum, S. lateriflorum, S. novae-angliae, S. novi-belgii, S. praealtum, S. puniceum, Symplocarpus foetidus, Thalictrum dasycarpum, T. pubescens, Thelypteris palustris, Thuja occidentalis, Torreyochloa pallida, Toxicodendron vernix, Tradescantia ohiensis, Triadenum fraseri, T. virginicum, Triantha glutinosa, Trichophorum alpinum, Triglochin maritimum, T. palustre, Typha angustifolia, T. latifolia, T. ×glauca, Ulmus americana, Urtica dioica, Usnea, Utricularia cornuta, U. intermedia, Vaccinium macrocarpon, V. oxycoccos, Valeriana edulis, Vernonia noveboracensis, Veratrum virginicum, Verbena hastata, Vernonia baldwinii, V. fasciculata, Veronica scutellata, Veronicastrum virginicum, Viburnum dentatum, V. lentago, V. nudum, V. opulus, Viola blanda, V. cucullata, V. primulifolia, V. sororia, Vitis riparia, Woodwardia virginica, Xyris torta, Zanthoxylum americanum.*

***Campanula californica* (Kellogg) A. Heller** is a weak perennial found in bogs, depressions, fens, lagoons, marshes, meadows, prairies, roadsides, seeps, slopes, swales, swamps, woodlands, and along the margins of watercourses at elevations of up to 1523 m. Exposures range from open sites to partial shade. Substrates include gravelly loam, humus, and sand. Flowering extends from June to October. Vegetative reproduction occurs by production of rhizomes. Because of their frail stems, the plants are particularly susceptible to anthropogenic disturbances associated with livestock grazing and logging operations. Other life history information for this species is scarce. **Reported associates:** *Alnus rubra, Anthoxanthum odoratum, Baccharis pilularis, Berberis, Bistorta bistortoides, Blechnum spicant, Calamagrostis nutkaensis, Camassia quamash, Carex obnupta, Castilleja chrymactis, Claytonia sibirica, Cynosurus echinatus, Deschampsia, Equisetum arvense, Gaultheria shallon, Helenium bolanderi, Helianthus, Holcus lanatus, Hosackia gracilis, Hypericum anagalloides, Hypochaeris radicata, Juncus effusus, Lupinus, Lysichiton americanus, Maianthemum dilatatum, Mimulus guttatus, M. moschatus, Perideridia gairdneri, Platanthera*

dilatata, Polystichum munitum, Potentilla anserina, Prunella vulgaris, Prunus, Pseudotsuga menziesii, Ranunculus orthorhynchus, Rhododendron columbianum, Rubus ursinus, Sidalcea calycosa, Sisyrinchium bellum, S. californicum, Sphagnum, Spiranthes romanzoffiana, Stachys ajugoides, S. chamissonis, Stellaria littoralis, Veronica scutellata, Viola sempervirens.

***Campanula floridana* S. Watson ex A. Gray** is a perennial inhabitant of depressions, ditches, flatwoods, floodplains, hammocks, lawns, marshes, meadows, roadsides, swamps and the margins of rivers and streams. The plants occur typically in sites receiving full sun exposure. The substrates are described as calcareous and include limerock, marl, and peat. The plants essentially remain in flower throughout the year. Like *C. aparinoides*, the seeds possess enlarged epidermal lumens, which enhances their buoyancy. Vegetative reproduction occurs by rhizomes (from which multiple stems extend) and can result in a mat-forming habit. **Reported associates:** *Acer negundo, Acmella oppositifolia, Agalinis purpurea, Aletris obovata, Allium canadense, Andropogon virginicus, Arnoglossum ovatum, Asclepias lanceolata, A. longifolia, A. michauxii, A. perennis, Asemeia grandiflora, Bidens mitis, B. pilosus, Boltonia diffusa, Buchnera americana, B. floridana, Calopogon tuberosus, Campsis radicans, Celtis laevigata, Cenchrus incertus, Centrosema virginianum, Cephalanthus occidentalis, Chamaecrista fasciculata, C. nictitans, Chaptalia tomentosa, Chasmanthium, Cirsium horridulum, C. nuttallii, Clematis crispa, Conoclinium coelestinum, Coreopsis basalis, C. gladiata, C. leavenworthii, Crotalaria rotundifolia, Dalea carnea, Dichanthelium acuminatum, Dichondra caroliniana, Diospyros virginiana, Drosera capillaris, Dyschoriste oblongifolia, Elephantopus elatus, Eremochloa ophiuroides, Erigeron quercifolius, E. strigosus, E. vernus, Eryngium aquaticum, E. baldwinii, Eupatorium mohrii, E. perfoliatum, Euphorbia cyathophora, Eustachys paspaloides, Evolvulus sericeus, Galium tinctorium, Glandularia pulchella, Helianthus angustifolius, H. radula, Helenium pinnatifidum, Heterotheca subaxillaris, Hibiscus aculeatus, Hydrocotyle umbellata, Hydrolea quadrivalvis, Hypericum mutilum, H. myrtifolium, Hyptis alata, H. mutabilis, Indigofera caroliniana, Iris hexagona, Juncus marginatus, Juniperus virginiana, Lachnanthes caroliniana, Lachnocaulon, Lepidium virginicum, Liatris gracilis, L. spicata, Linum medium, Lobelia glandulosa, Lonicera sempervirens, Ludwigia maritima, Lyonia lucida, Lythrum alatum, Melanthera nivea, Mikania scandens, Mimosa strigillosa, Mitreola petiolata, M. sessilifolia, Myrica cerifera, Nyssa sylvatica, Oenothera laciniata, O. simulans, Opuntia humifusa, Oxalis corniculata, Oxypolis filiformis, Panicum, Paspalum, Persicaria hydropiperoides, Phlox drummondii, Phyla nodiflora, Physalis walteri, Physostegia purpurea, Pinguicula pumila, Pinus elliottii, P. palustris, Pityopsis graminifolia, Pluchea baccharis, Polygala balduinii, P. boykinii, P. incarnata, P. lutea, P. nana, P. violacea, Polygonella gracilis, Pontederia cordata, Pterocaulon pycnostachyum, Ptilimnium capillaceum, Pyrrhopappus* carolinianus, *Quercus incana*, *Q. laevis*, *Rhexia mariana*, *R. petiolata*, *Rhynchospora colorata*, *Rosa palustris*, *Rubus cuneifolius*, *Rudbeckia fulgida*, *R. hirta*, *Ruellia caroliniensis*, *Rumex hastatulus*, *Sabal palmetto*, *Sabatia brevifolia*, *S. stellaris*, *Saccharum giganteum*, *Sageretia minutiflora*, *Sagittaria graminea*, *S. lancifolia*, *Salvia lyrata*, *Saururus cernuus*, *Scutellaria integrifolia*, *Serenoa repens*, *Silphium asteriscus*, *Sisyrinchium angustifolium*, *Smilax laurifolia*, *Solidago fistulosa*, *S. odora*, *Spermacoce verticillata*, *Spermolepis divaricata*, *Sphenopholis obtusata*,

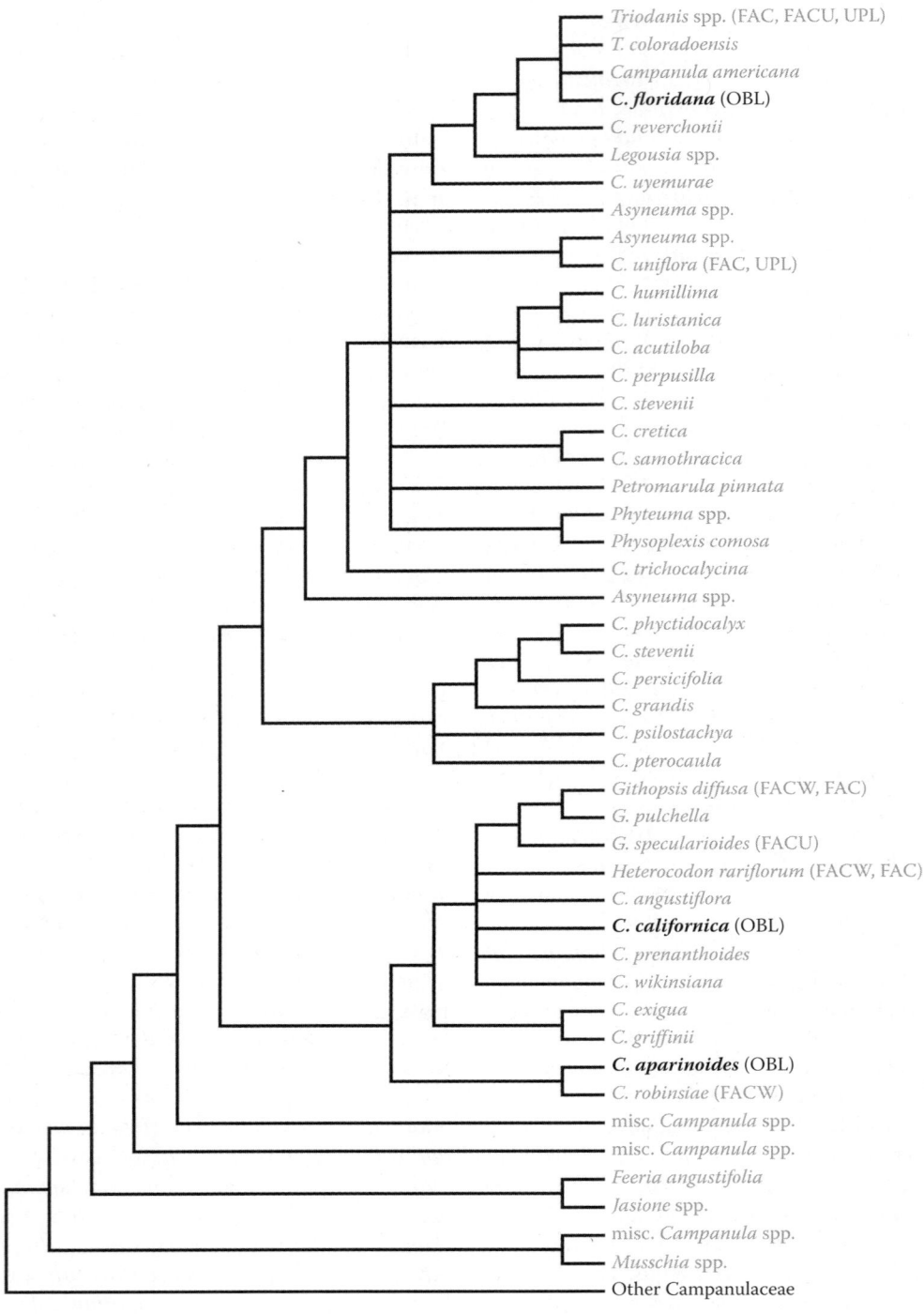

FIGURE 5.176 Phylogenetic relationships of *Campanula* based on analysis of *petD* group II intron sequences. These results indicate that the OBL habit has originated three separate times in the North American species (bold). Similar to results from analysis of nrITS sequence data (Roquet et al., 2008), this tree also demonstrates that the large genus *Campanula* is not monophyletic as currently circumscribed, but includes species from more than a dozen other genera. Much additional taxonomic refinement is necessary in this genus. (Adapted from Mansion, G. et al., *PLoS ONE*, 7(11), e50076, 2012.)

Spigelia loganioides, Spiranthes vernalis, Stachys floridana, Stenotaphrum secundatum, Strophostyles umbellata, Styrax americanus, Symphyotrichum dumosum, Toxicodendron radicans, Tradescantia ohiensis, Tridens flavus, Triodanis perfoliata, Utricularia foliosa, Verbena halei, Vicia acutifolia, Vigna luteola, Vitis shuttleworthii, Wahlenbergia, Xyris, Zephyranthes treatiae.

Use by wildlife: *Campanula aparinoides* commonly occurs within the brooding habitat of the ruffed grouse (Aves: Phasianidae: *Bonasa umbellus*). The flowers attract various insects (Insecta) to their nectar including pollinating butterflies and moths (Lepidoptera) and beeflies (Diptera: Bombyliidae) as well as flower thrips (Insecta: Thysanoptera: Thripidae: *Frankliniella tritici*), which are believed to facilitate secondary self-pollination. The plants are also hosts to several sac fungi (Ascomycota: Mycosphaerellaceae: *Pseudocercospora campanulae, Septoria campanulae*).

Economic importance: food: not reported as edible; **medicinal:** Young Iroquois women were given a decoction made from the stems of *Campanula aparinoides* to induce childbirth. The Hocąk made a medicinal steam from the plants, which was inhaled; **cultivation:** None of the OBL species is currently in cultivation; **misc. products:** none; **weeds:** Although a Florida endemic, *Campanula floridana* commonly occurs in ruderal habitats and often is regarded as a weed, especially of turfgrass; **nonindigenous species:** *Campanula aparinoides* was introduced to Pälkäne, Finland sometime before 1932, presumably as a contaminant of grass or clover seed.

Systematics: The taxonomy of *Campanula* is in a shambles. As currently circumscribed, the genus is extensively polyphyletic with the species interspersed among more than 15 other genera (Figure 5.176). Phylogenetic analyses have been difficult to conduct because of the sheer enormity of the genus; however, improvements in the classification continue to appear as a consequence of such work. A sister species relationship between *C. aparinoides* and *C. robinsiae* is evident not only by their placement in molecular analyses (e.g., Figure 5.176), but also by their distinctive seed morphology, which differs from other members of the genus. The taxa have also been considered as conspecific subspecies. *Campanula floridana* and *C. robinsiae* formerly were thought to be closely related and even to comprise the distinct genus *Rotantha*; however, it is now clear that they are unrelated, with *C. robinsiae* instead closely related to *C. aparinoides* (Figure 5.176). The seed morphology of *C. californica* and *C. prenanthoides* is distinctive and similar, and corresponds with their close phylogenetic placement (Figure 5.176). The small seeds of *C. floridana* (once thought to be closely related to *C. aparinoides*) also are distinctive; however, they indicate no obvious affinity with other *Campanula* species. *Campanula californica* possesses an unusual pantoporate pollen type, which is known in only four other species within the family; however, it does not appear to indicate any close relationship among those species. The base chromosome number of *Campanula* is regarded as *x* = 17, although this high number most likely is derived. *Campanula aparinoides* (2*n* = 34, 136, 170) contains diploid, octoploid

and decaploid cytotypes. The chromosome numbers of *C. californica* and *C. floridana* have not been determined. No hybrids involving any of the three OBL *Campanula* species have been reported.

Comments: *Campanula aparinoides* is widespread throughout eastern North America. Highly restricted ranges characterize *C. californica* (California) and *C. floridana* (Florida).

References: Albert & Tepley, 2005; Alsos et al., 2003; Auclair et al., 1976; Bowles, 1991; Burns, 1911; Choesin & Boerner, 2000; Eisto et al., 2000; Ferris, 1970; Galloway et al., 2003; Godfrey, 1975; Godfrey & Wooten, 1981; Harrington, 1986; Herring & Davis, 2004; Hill, 1891; Howell, 1970; Isaak et al., 1959; Kindscher & Hurlburt, 1998; Kost et al., 2007; Laanei et al., 1983; Mac Elwee, 1900; Mansion et al., 2012; Martin, 1981; McAtee, 1920; McCarty et al., 2001; Middleton, 2002a; Nekola, 2004; Nichols, 1915; Norcini, 2010; Nowicke et al., 1992; Orzell & Kurz, 1986; Pammel, 1908; Peach & Zedler, 2006; Poindexter & Nelson, 2011; Roquet et al., 2008; Rosatti, 1986; Ruthven, 1911; Sherff, 1912; Shetler & Morin, 1986; Stephenson et al., 2000; Swink, 1952; Tobe et al., 1998; Tolstead, 1942; Vaccaro et al., 2009; Vogler & Stephenson, 2001; Wherry, 1927; White, 1965.

2. *Downingia*

Calicoflower, skyblue

Etymology: after Andrew Jackson Downing (1815–1852)

Synonyms: *Bolelia; Clintonia; Gynampsis; Lobelia* (in part); *Wittea*

Distribution: global: Chile; North America; **North America:** western

Diversity: global: 14 species; **North America:** 14 species

Indicators (USA): OBL: *Downingia bacigalupii, D. bella, D. bicornuta, D. concolor, D. cuspidata, D. elegans, D. insignis, D. laeta, D. montana, D. ornatissima, D. pulchella, D. pusilla, D. willamettensis, D. yina*

Habitat: freshwater; palustrine; **pH:** 6.0–10.3; **depth:** <1 m; **life-form(s):** emergent or submersed (vittate) herb

Key morphology: Stems (to 40 cm) decumbent or erect; leaves (to 25 mm) cauline, sessile, awl-like to lanceolate, margins entire, typically deciduous before anthesis; flowers resupinate (usually) or nonresupinate, sessile (but appearing stalked due to their elongate ovary), arranged in spikes, bracts (to 28 mm) foliar; corollas (to 19 mm) blue (usually), pink or white, bilabiate, upper lip 2-lobed, lower lip 3-lobed with a medial white/yellow spot; filaments and anthers of stamens fused into a tube surrounding the style; ovary elongate (to 90 mm), inferior, twisted (in resupinate flowers), 1–2 loculed; capsules dehiscent along 3–5 lateral slits; seeds minute (<1 mm), numerous, fusiform, with longitudinal or spiral lines.

Life history: duration: annual (fruit/seeds); **asexual reproduction:** none; **pollination:** insect or self; **sexual condition:** hermaphroditic; **fruit:** capsules (common); **local dispersal:** seeds (water); **long-distance dispersal:** seeds (waterfowl)

Imperilment: (1) *Downingia bacigalupii* [G4]; S2 (ID); (2) *D. bella* [G2/3]; (3) *D. bicornuta* [G3/4]; (4) *D. concolor* [G4]; S1 (CA); (5) *D. cuspidata* [G3]; (6) *D. elegans* [G5];

SH (BC); (7) *D. insignis* [G4]; S1 (ID); (8) *D. laeta* [G5]; S1 (MT, SK, WY); S2 (AB, CA, UT); (9) *D. montana* [G3/4]; (10) *D. ornatissima* [G2/3]; (11) *D. pulchella* [G3]; (12) *D. pusilla* [G2]; S2 (CA); (13) *D. willamettensis* [undetermined]; (14) *D. yina* [G4]

Ecology: general: All *Downingia* species are annual, obligate aquatics of intermittent waters such as vernal pools. Flowers typically are produced within 45–56 days after germination. Most species have insect-pollinated (entomophilous) blue or purple, resupinate flowers (non-resupinate in *D. laeta*) and all are characterized by a staminal column comprising fused anthers and filaments. The species are genetically self-compatible, so all have at least some potential for self-pollination. However, self-pollination is greatly minimized by protandry. Within the genus occur both primarily outcrossing and self-pollinating species, which exhibit three basic floral morphs. "Short anther tube" flowers have a small distance (~1.0 mm) between the lower corolla lip and anthers, which is more suitable for pollen deposition on smaller insects. "Long anther tube" flowers exhibit a greater distance (~4.3 mm) between the lower corolla lip and anthers, which is more suited for pollen deposition on larger insects. Both types are associated with outcrossed flowers. "Reduced" flowers (associated with selfing species) are diminuitive and nonresupinate, with an included staminal column. In outcrossed flowers, the insects are attracted to the floral nectaries. Large epidermal hairs on the upper anther surface function as trigger hairs. When tripped, these result in the ejection of pollen from a flap and pollen deposition, usually nototribically (on the back of the insect). Pollen deposition on the stigma is facilitated by a brush-like structure of hairs beneath the stigma. In primarily self-pollinating species (*D. laeta, D. pusilla*), the flowers are reduced and their stigma expands while still within the anther column. Each capsule produces about 150 min (<1 mm) seeds, which can persist in large seed banks for several years while awaiting suitable germination conditions. Cool temperatures and soil saturation are regarded as essential germination requirements. Seed germination follows the winter rains and most will germinate while submersed. Successful germination has been reported for various species by sowing them in February in a standard sand–soil–compost mixture, which is firmly packed and then covered by a thin, fine soil layer. No stratification appears to be necessary. The seeds are then scattered on the soil, lightly dusted by sifted soil, and then then lightly tamped. The pots must be kept constantly wet and provided with approximately 16 h of illumination at 17°C–20°C for some species, but 7°C–10°C for others. Seed germination is inhibited at higher temperatures (25°C/15°C day/night regime). Germination usually occurs in 5–7 days. The seeds and seedlings of *Downingia* species float and they are dispersed locally by water. Long-distance dispersal is presumed to occur primarily through the dispersal of seeds by waterfowl (Aves: Anatidae). The transport of seed in dried mud that has been caked on to automobiles has been implicated in the dispersal of some species.

Downingia bacigalupii **Weiler** is reported from borrow pits, depressions, ditches, dried stream beds, flats, floodplains, lake and stream beds, marshes, meadows, mudflats, playas, ponds, pools, puddles, swales, vernal pools, wallows, washes, and the margins of lakes, ponds, reservoirs, and small streams at elevations of up to 2523 m. The plants can be found in standing water up to 10 cm deep, but occur often on what appears to be thoroughly dried ground, at sites where water was present formerly. Exposures typically occur in full sun. The substrates are described as adobe, basalt cobble, clay, clay loam, clay mud, gravel, loam, mud, peat, rock, sand, sandy loam, serpentine, silt, and volcanic. Flowering and seed set (April–August) occur after the waters recede and the substrates desiccate. The large flowers differ from most *Downingia* in having a unilocular (cf. bilocular) ovary. They are outcrossed by small bees (Insecta: Hymenoptera: Andrenidae: *Panurginus*), which, unlike most *Downingia*, collect the pollen sternotribically (on their undersides) while hanging upside down at the tip of the anther column. Seeds can germinate underwater or on moist soil from late spring to early summer. Those planted in February germinate in mid-February to early March with the seedlings exhibiting more extensive vegetative growth than plants developing from seeds sown later in the season. Depending on water levels, young plants can grow initially as submersed before eventually becoming emergent. Grazing and trampling by livestock can damage plants significantly. **Reported associates:** *Alisma, Allium, Alopecurus geniculatus, Ambrosia psilostachya, Arctostaphylos viscida, Artemisia cana, A. ludoviciana, A. tridentata, Beckmannia syzigachne, Camassia quamash, Carex ferruginea, Ceanothus cuneatus, Chrysothamnus viscidiflorus, Cyperus squarrosus, Damasonium californicum, Deschampsia danthonioides, D. elongata, Distichlis spicata, Downingia bicornuta, D. insignis, D. laeta, D. yina, Echinops sphaerocephalus, Eleocharis macrostachya, E. palustris, Epilobium campestre, Eryngium alismifolium, E. articulatum, Euphorbia serpyllifolia, Galium, Gnaphalium, Gratiola, Isoetes bolanderi, Iva axillaris, Juncus arcticus, J. balticus, J. bufonius, J. confusus, J. ensifolius, J. nevadensis, Limosella aquatica, Lolium, Marsilea vestita, Mimulus tricolor, Muhlenbergia richardsonis, Myosurus, Navarretia breweri, N. divaricata, N. leptalea, N. leucocephala, Oenothera flava, Orcuttia tenuis, Pascopyrum smithii, Perideridia, Phleum pratense, Pinus ponderosa, Plagiobothrys hispidulus, P. mollis, P. scouleri, Poa secunda, Polygonum polygaloides, Porterella carnosula, Psilocarphus brevissimus, P. oregonus, Quercus garryana, Rubus, Sagittaria, Schoenoplectus acutus, Taeniatherum caput-medusae, Trifolium, Triteleia, Vicia, Viola adunca.*

Downingia bella **Hoover** occupies depressions, lake beds, marshes, meadows, swales, vernal pools, wallows and the margins of lakes and ponds at elevations of up to 1615 m. Exposures occur in full sun. The plants normally grow in standing water (5–35 cm deep) or occasionally on drying substrates. Plants can cover the bottoms of deeper pools. The substrates include alkaline adobe, Capay clay, claypan, gravel, gravelly loam (San Joaquin series), mud, northern hardpan, sand, silty loam, and vina loam. Typically, the surrounding waters are poorly buffered (specific conductance = 329 S/cm) and can experience wide fluctuations in acidity (e.g., pH = 6.4–10.3) during a single diurnal period. Free CO_2 levels approximate $4.2\,mol/m^3$. Photosynthesis occurs primarily via a C_3 pathway but stomata are absent on the submersed foliage. Specific leaf mass averages $8.3\,g/m^2$ with 67% air space and total leaf chlorophyll content averages from 344 to $433\,\mu g/g$ of fresh tissue. Flowering occurs from 20 March to 15 May and correlates positively with water depth. The flowers are outcrossed primarily by bees (Insecta: Hymenoptera: Apidae: *Bombus californicus*, *B. sonorus*). Niche partitioning of pollinators and flowering time has been observed between *D. bella* and *D. cuspidata* when growing in sympatry. The seeds will germinate while under water. Disturbance by animals (hoofprints) can create microdepressions, which provide suitable habitat by creating deeper pockets in the substrate where the plants can flourish. **Reported associates:** *Alopecurus saccatus, Anagallis minima, Asclepias fascicularis, Astragalus tener, Avena barbata, Brodiaea filifolia, B. orcuttii, B. terrestris, Callitriche longipedunculata, C. marginata, Centromadia pungens, Crassula aquatica, C. connata, Cressa truxillensis, Deschampsia danthonioides, Downingia bicornuta, D. cuspidata, D. insignis, D. ornatissima, D. pulchella, D. yina, Elatine brachysperma, E. californica, Eleocharis acicularis, E. macrostachya, E. montevidensis, E. palustris, Epilobium brachycarpum, E. campestre, E. torreyi, Eryngium aristulatum, E. castrense, E. vaseyi, Gratiola ebracteata, Grindelia hirsutula, Isoetes howellii, I. orcuttii, Juncus bufonius, J. xiphioides, Lactuca serriola, Lasthenia californica, L. ferrisiae, L. fremontii, L. glaberrima, L. minor, Lepidium nitidum, Lilaea scilloides, Limnanthes alba, Lythrum hyssopifolia, Marsilea vestita, Medicago polymorpha, Mimulus guttatus, Minuartia californica, Montia fontana, Myosurus minimus, Navarretia hamata, N. intertexta, N. leucocephala, N. prostrata, Orcuttia californica, Phalaris lemmonii, Pilularia americana, Plagiobothrys humistratus, P. leptocladus, P. stipitatus, P. undulatus, Plantago bigelovii, P. elongata, Pogogyne douglasii, P. ziziphoroides, Psilocarphus brevissimus, P. tenellus, Ranunculus bonariensis, Senecio vulgaris, Sidalcea hartwegii, Spergularia salina, Trifolium depauperatum, T. variegatum, Veronica peregrina.*

***Downingia bicornuta* A. Gray** occurs in borrow pits, depressions, ditches, flats, marshes, meadows, mudflats, ponds, sloughs, streams, swales, vernal pools, wallows, and along the margins of lakes, ponds, reservoirs, rivers, and swamps at elevations of up to 1737 m. The plants are emergent or submersed in 5–61 cm of still or flowing water (average depth = 17 cm), but tend to dominate in the deeper pools where they are known to cover up to 35% of the bottom. Inhabited pools average $5000\,m^2$. The plants can withstand minimum temperatures of 8.3°C. Exposures occur in full sun. The substrates vary in acidity (pH 6.0–8.2), and include adobe, Alamo series, alluvium, Anita

series, basaltic clay, clay, Clear Lake series, Cometa series, Corning series, Fe and Si cemented hardpan, Fiddyment series, loam, Mehrten (Amador-Gillender), mud, Pentz series, Redding gravelly loam, sand, San Joaquin series, stony Tuscan loam, Toomes series, Tuscan series, and volcanic rock. The plants can take up to 70 days to produce flowers (mid-April through early August), which float on the surface if developing in deeper water. Pollinators include solitary bees (Insecta: Hymenoptera: Andrenidae: *Andrena, Panurginus atriceps*; Halictidae: *Lasioglossum*). The capsules dehisce almost immediately upon maturation. **Reported associates:** *Alisma triviale, Alopecurus pratensis, A. saccatus, Anagallis arvensis, Artemisia cana, A. tridentata, Beckmannia syzigachne, Callitriche heterophylla, C. marginata, C. verna, Castilleja campestris, C. tenuis, Convolvulus arvensis, Cotula, Crassula aquatica, Cuscuta howelliana, Cyperus, Deschampsia danthonioides, Downingia bacigalupii, D. bella, D. concolor, D. cuspidata, D. insignis, D. laeta, D. ornatissima, D. pulchella, D. pusilla, D. yina, Elatine californica, E. heterandra, Eleocharis macrostachya, E. palustris, E. quinqueflora, Epilobium torreyi, Eryngium castrense, E. vaseyi, Frankenia salina, Glyceria ×occidentalis, Gratiola ebracteata, G. heterosepala, Hordeum brachyantherum, H. jubatum, Hypochaeris glabra, Isoetes howellii, Juncus bufonius, J. leiospermus, J. uncialis, Juniperus occidentalis, Lasthenia fremontii, L. glaberrima, Legenere limosa, Leontodon saxatilis, Lepidium latifolium, Lilaea scilloides, Limnanthes alba, L. douglasii, Limosella aquatica, Lindernia dubia, Lythrum hyssopifolia, L. portula, Marsilea vestita, Navarretia leucocephala, N. tagetina, Orcuttia tenuis, Persicaria coccinea, Phalaris arundinacea, Plagiobothrys hispidulus, P. leptocladus, P. scouleri, P. stipitatus, P. undulatus, Pleuropogon californicus, Poa annua, Pogogyne douglasii, P. ziziphoroides, Porterella carnosula, Potamogeton foliosus, Potentilla gracilis, Psilocarphus brevissimus, Ranunculus aquatilis, R. bonariensis, Rotala ramosior, Sagittaria latifolia, Sidalcea diploscypha, Taraxacum officinale, Trifolium, Triphysaria, Triteleia hyacinthina, Utricularia macrorhiza, Veronica americana, V. anagallis-aquatica, V. arvensis, Viola.*

***Downingia concolor* Greene** inhabits depressions, ditches, dry lake beds, flats, marshes, meadows, mudflats, ponds, shores, vernal pools, wallows, and the margins of lakes and ponds at elevations of up to 1433 m. It occupies sites where flowing or standing waters reside only during winter and spring, but not in summer or fall. The plants most often are found as emergents on drying substrates that remain after temporary waters recede, but also can occur as submersed forms (to at least 25 cm) at sites where standing waters persist. They can form concentric rings around the periphery of ponds where they thrive mainly at shallow or intermediate depths. Flowering is noticeably reduced after only 2 weeks of seedling submergence. Adaptive elongation of plants occurs after 8 weeks of inundation; however, seedlings cannot survive prolonged inundation (4 months). The plants are most common early in the season (April–May), becoming less prevalent or even disappearing in later months (June–July). Exposures occur under full sun. The substrates are described as adobe, alkali, alluvium, clay, loamy clay, muck, mucky silty loam, mud, sand, serpentine, silt, and volcanic. Flowering and fruiting proceeds from 1 April to 15 July. The flowers are pollinated by bees (Insecta: Hymenoptera: Andrenidae: *Panurginus atriceps*; Halictidae: *Lasioglossum*), and the capsules do not dehisce until the seeds have fully matured. The seeds must be in an osmotic solution with positive water potential in order to germinate. Experimental germination rates are higher in the light than in the dark, and (in light) increase with seed density. Natural germination occurs at cool temperatures (7°C–10°C); experimentally, the highest rates of germination were attained at 10°C–15°C under 11 h of light. A large, persistent seed bank develops and seeds can remain viable for extended periods of time after floating on the water surface. Floating seedlings can remain dormant for at least a year, especially at lower temperatures (10°C). **Reported associates:** *Agrostis elliottiana, Aira, Alopecurus saccatus, Anagallis arvensis, Briza, Brodiaea jolonensis, B. orcuttii, Callitriche, Castilleja attenuata, Convolvulus arvensis, Cotula coronopifolia, Crassula aquatica, Deschampsia danthonioides, Downingia bicornuta, D. cuspidata, D. elegans, D. insignis, D. pulchella, D. pusilla, Elatine californica, Eleocharis macrostachya, Epilobium, Erodium botrys, Eryngium alismifolium, E. aristulatum, E. armatum, E. vaseyi, Grindelia hirsutula, Hordeum marinum, H. murinum, Isoetes howellii, Juncus bufonius, J. phaeocephalus, Lasthenia burkei, L. californica, L. conjugens, L. fremontii, L. glaberrima, L. glabrata, Layia chrysanthemoides, Lepidium nitidum, Lilaea scilloides, Limnanthes alba, L. douglasii, Lolium multiflorum, Lupinus bicolor, Lythrum hyssopifolia, Malvella leprosa, Mentha pulegium, Mimulus tricolor, Montia fontana, Muhlenbergia, Navarretia intertexta, Plagiobothrys stipitatus, P. undulatus, Pleuropogon californicus, Pogogyne douglasii, Polypogon monspeliensis, Psilocarphus brevissimus, Rumex crispus, Triphysaria eriantha.*

Downingia cuspidata **(Greene) Greene ex jeps.** grows in depressions, flats, marshes, meadows, mudflats, playas, puddles, roadsides, streambeds, swales, vernal pools, wallows, and along the margins of lakes, ponds, reservoirs, and streams at elevations of up to 1709 m. The plants can grow when completely submersed (3–46 cm depth) and will tolerate inundation for 40 days or more. However, they occur primarily on wet or drying substrates, which include acidic, alkaline or subalkaline adobe, alluvial clay, alluvium, Anita series, basalt, clay, clay loam, claypan, Fe and Si cemented hardpan, granitic sand, gravel, Hideaway series, Huerhuero loam, humus, muck, pebbly clay, Ramona clay loam, Red Buff loam, Redding series, San Joaquin series, serpentine, stony Tuscan loam, Toomes series, Tuscan (TuB), and vina loam. Flowering occurs from late March to mid-June and correlates negatively with water depth. The principal pollinators are bees (Insecta: Hymenoptera: Andrenidae; Halictidae— see *Use by wildlife*). Niche partitioning (different pollinators and flowering times) has been reported in sympatric

populations of *D. cuspidata* and *D. bella*. Seed germination is initiated at 17°C–20°C. A persistent seed bank presumably exists for this species, whose populations can fluctuate widely from year to year. **Reported associates:** *Alopecurus saccatus, Anagallis minima, Astragalus tener, Avena barbata, Blennosperma nanum, Brodiaea filifolia, B. jolonensis, B. orcuttii, Callitriche longipedunculata, C. marginata, Castilleja attenuata, C. campestris, C. tenuis, Centromadia fitchii, Cicendia quadrangularis, Crassula aquatica, Croton setiger, Crypsis schoenoides, C. vaginiflora, Cuscuta howelliana, Damasonium californicum, Deinandra fasciculata, Deschampsia danthonioides, Distichlis spicata, Downingia bella, D. bicornuta, D. concolor, D. ornatissima, D. pulchella, D. pusilla, Elatine brachysperma, E. californica, Eleocharis acicularis, E. macrostachya, E. palustris, Epilobium campestre, E. cleistogamum, E. densiflorum, E. torreyi, Erodium botrys, Eryngium aristulatum, E. castrense, E. spinosepalum, E. vaseyi, Frankenia salina, Gnaphalium palustre, Gratiola heterosepala, Hordeum marinum, Isoetes howellii, I. nuttallii, I. orcuttii, Juncus bufonius, J. effusus, J. leiospermus, J. sphaerocarpus, J. uncialis, Lasthenia fremontii, L. glaberrima, Layia, Legenere limosa, Lemna, Lepidium nitidum, Leptosiphon ciliatus, Lilaea scilloides, Limnanthes alba, L. douglasii, Limosella aquatica, Lomatium caruifolium, Lythrum hyssopifolia, Malvella leprosa, Marsilea vestita, Mimulus guttatus, M. tricolor, Minuartia californica, Montia fontana, M. linearis, Muilla maritima, Myosurus minimus, Navarretia fossalis, N. hamata, N. leucocephala, N. prostrata, Nemophila maculata, Ophioglossum californicum, Orcuttia californica, O. pilosa, O. tenuis, Phalaris lemmonii, Pilularia americana, Plagiobothrys acanthocarpus, P. greenei, P. humistratus, P. leptocladus, P. nothofulvus, P. stipitatus, P. undulatus, Plantago elongata, P. erecta, Pleuropogon californicus, Pogogyne abramsii, P. douglasii, P. ziziphoroides, Psilocarphus brevissimus, P. tenellus, Puccinellia simplex, Rumex, Sedella pumila, Sisyrinchium bellum, Trifolium depauperatum, T. variegatum, T. wormskioldii, Tryphysaria, Veronica arvensis, V. peregrina.*

Downingia elegans **(Douglas ex Lindl.) Torr.** inhabits depressions, ditches, flats, marshes, meadows, mudflats, prairies, riverbeds, roadsides, seeps, sloughs, swales, vernal pools, and the margins of lakes, ponds and streams at elevations of up to 1889 m. The plants sometimes are found growing partly immersed but usually occur in shallow beds. The substrates are circumneutral (pH: 6.1–7.8 depending on depth) and are described as adobe, basalt, gravel, loam, muck, mud, sand, and vertisols. Flowering continues from late May to early October. The flowers have a unilocular ovary and are pollinated by insects (Insecta) such as bees (Hymenoptera) or occasionally by butterflies (Lepidoptera). The seeds are produced prolifically and their biomass can account for up to 1.3% of the total seed mass in some seasonal wetlands. Germination (under 16 h light) is highest for fresh seeds (93%) or for warm/cold stratified seed (92%; seeds kept at 10°C/20°C for 6 weeks then at 5°C/15°C for 6 weeks). It is considerably reduced (74%) for cold-stratified seed. Germination also is high (90%) for dried, frozen seed. Fresh seed also can be sowed at 13°C–16°C. Successful germination has been reported at 10°C for seeds planted in a soil-less, peat-based medium, to which micronutrients and slow-release 14–14–14 fertilizer have been added. Plants in the standing crop can attain a mean percent cover of 0.25–0.42 and mean percent biomass of 0.7–1.9 in seasonal wetlands. Up to 96% of the root surfaces can be colonized by vesicular-arbuscular mycorrhizal fungi. The only known Canadian population of this species was eliminated following habitat alterations for waterfowl management. **Reported associates:** *Agrostis, Alopecurus geniculatus, A. saccatus, Artemisia cana, Beckmannia syzigachne, Brodiaea terrestris, Callitriche marginata, Camassia, Carex unilateralis, Castilleja campestris, C. lineariiloba, Ceanothus cuneatus, Centaurium, Coreopsis tinctoria, Damasonium californicum, Danthonia, Deschampsia cespitosa, D. danthonioides, Downingia concolor, D. insignis, Drymocallis arguta, Eleocharis acicularis, E. macrostachya, E. palustris, Epilobium densiflorum, Eryngium castrense, E. mathiasiae, E. petiolatum, Fritillaria affinis, Gratiola, Grindelia hirsutula, Helenium autumnale, Hordeum, Iva axillaris, Juncus acuminatus, J. effusus, Leymus triticoides, Lolium, Lysimachia ciliata, L. nummularia, Melica, Mentha arvensis, Mimulus guttatus, M. tricolor, Myosurus apetalus, Navarretia leucocephala, Olsynium douglasii, Pedicularis, Penstemon, Perideridia, Persicaria amphibia, Phalaris arundinacea, Physostegia parviflora, Pilularia americana, Plagiobothrys figuratus, P. hirtus, P. hispidulus, P. leptocladus, P. scouleri, Polyctenium fremontii, Polygonum polygaloides, Populus trichocarpa, Potentilla, Primula, Prunella vulgaris, Psilocarphus brevissimus, Ranunculus aquatilis, R. flammula, Sagittaria cuneata, Salix exigua, S. prolixa, Schoenoplectus tabernaemontani, Senecio, Siam suave, Taraxia tanacetifolia, Trifolium, Veratrum viride, Veronica scutellata.*

***Downingia insignis* Greene** is found in depressions, ditches, dried basins, flats, levees, marshes, meadows, mudflats, playas, roadsides, saline flats, sandbars, streams, swales, vernals pools, and along the margins of lakes, reservoirs, and streams at elevations of up to 1805 m. Typically, the plants grow at intermediate depths (12–15 cm) in the standing waters of temporary pools. The substrates are alkaline (pH >9.0) and are described as adobe, Capay silty clay, clay, claypan, clay silt, Clear Lake clay adobe, gravelly clay, Marine series, mud, Natrixeralfs great group, peat, peaty muck, Riz series, silt, volcanic sand, and Willows series. Flowering and fruiting occur from early March through mid-June (rarely to August). The flowers are of the long anther tube type. This conformation results in their pollen being deposited on the back surfaces (nototribically) of large (>8 mm) pollinating bees (Insecta: Hymenoptera), which hang upside down during the collection process. **Reported associates:** *Acmispon wrangelianus, Alopecurus saccatus, Arabis, Artemisia cana, A. tridentata, Astragalus tener, Callitriche marginata, Castilleja campestris, Cotula coronopifolia, Crassula aquatica, Cressa truxillensis, Crypsis schoenoides, Damasonium californicum,*

Deschampsia danthonioides, Distichlis spicata, Downingia bacigalupii, D. bella, D. bicornuta, D. concolor, D. elegans, D. ornatissima, D. pulchella, D. pusilla, Eleocharis acicularis, E. macrostachya, E. palustris, Epilobium campestre, E. densiflorum, E. torreyi, Ericameria nauseosa, Eryngium aristulatum, E. vaseyi, Frankenia salina, Grindelia hirsutula, Hemizonia congesta, Hesperevax caulescens, Hordeum marinum, Inula, Juncus bufonius, J. capitatus, Juniperus occidentalis, Lagophylla, Lasthenia californica, L. ferrisiae, L. fremontii, L. glaberrima, L. glabrata, Lepidium latipes, L. nitidum, Lilaea scilloides, Limnanthes douglasii, Limosella aquatica, Lolium multiflorum, L. perenne, Lythrum hyssopifolia, L. tribracteatum, Malvella leprosa, Mimulus tricolor, Myosurus minimus, M. sessilis, Navarretia leucocephala, Neostapfia colusana, Oenothera flava, Orcuttia pilosa, Phacelia thermalis, Plagiobothrys hispidulus, P. leptocladus, P. stipitatus, Plantago bigelovii, P. elongata, Pleuropogon californicus, Pogogyne douglasii, Polygonum aviculare, Polypogon maritimus, P. monspeliensis, Porterella carnosula, Psilocarphus brevissimus, Rumex, Sarcobatus vermiculatus, Schoenoplectus maritimus, Sida, Spergularia salina, Thinopyrum ponticum, Trifolium willdenovii, Veronica peregrina.

***Downingia laeta* (Greene) Greene** resides in borrow pits, bottoms, depressions, ditches, meadows, mudflats, playas, potholes, saline flats, seeps, shores, sloughs, vernal pools, and along the margins of lakes, ponds, reservoirs, and streams at elevations of up to 2207 m. The plants occur in shallow water (15–20 cm) or are exposed on drying substrates. Substrates include clay, clayey mud, mud, sand, and silt. Flowering occurs from May through July and fruiting from July to October. Unlike most *Downingia*, the flowers are not resupinate and primarily are self-pollinating. No pollinators have been reported and it is uncertain whether any degree of outcrossing occurs. This rare species occurs scattered throughout desert regions, which isolates it from most other *Downingia* species. Its primary means of dispersal is unknown. **Reported associates:** *Alisma gramineum, Beckmannia syzigachne, Bistorta bistortoides, Carex athrostachya, C. utriculata, Castilleja tenuis, Chrysothamnus, Distichlis spicata, Downingia bacigalupii, D. bicornuta, Eleocharis macrostachya, E. palustris, Epilobium campestre, Gnaphalium palustre, Gratiola heterosepala, Hordeum jubatum, Iris, Juncus bufonius, Lasthenia glaberrima, Limosella, Marsilea vestita, Plagiobothrys scouleri, Poa, Polypogon monspeliensis, Porterella carnosula, Potentilla anserina, Psilocarphus brevissimus, Rumex crispus, Sagittaria cuneata, Sarcobatus, Trifolium pratense, Veronica peregrina.*

***Downingia montana* Greene** occurs in ditches, meadows, vernal pools, washes, and along the margins of lakes and streams at elevations of up to 2255 m. Unlike other *Downingia* species, which commonly occur in vernal pools, this species is most common in open meadows within forests of *Abies, Calocedrus decurrens, Pinus ponderosa, Pseudotsuga,* and *Quercus kelloggii.* The substrates include clay, muck, mud, rock, sand, and serpentine. The plants are in flower and fruit from late May through early August. The flowers have a unilocular ovary. Other life history information for this species is scarce. **Reported associates:** *Carex, Castilleja lacera, Juncus bufonius, J. uncialis, Limnanthes, Mimulus biolettii, M. leptaleus, M. luteus, M. pulchellus, M. pulsiferae, Navarretia, Plagiobothrys hispidulus, Polygonum parryi, Triteleia hyacinthina.*

Downingia ornatissima **Greene** grows in depressions, ditches, draws, flats, floodplains, meadows, mudflats, roadsides, slopes, swales, streambeds, vernal pools, wallows, and along the margins of levees at relatively low elevations of up to 268 m. The plants grow as emergents or in shallow standing water (5–9 cm) of small pools (<1000 m^2). The substrates are characterized as adobe, alkali clay, Anita series, Arbuckle series, Capay clay, clay loam, Corning gravelly loam, hardpan, Mehrten formation (Amador-Gillender series), mud, Pentz series, Redding gravel, sand, San Joaquin sandy loam, Toomes series, Tuscan series, and vertisols. Flowering extends from about 10 April to 30 May. The flowers are of the short anther tube type and are pollinated by specialized bees (Insecta: Hymenoptera: Andrenidae). **Reported associates:** *Achyrachaena mollis, Acmispon wrangelianus, Alopecurus saccatus, Avena, Blennosperma nanum, Briza minor, Brodiaea, Bromus hordeaceus, Callitriche marginata, Castilleja campestris, C. exserta, Centromadia fitchii, Clarkia purpurea, Claytonia perfoliata, Crassula aquatica, Cuscuta howelliana* [parasitic], *Deschampsia danthonioides, Downingia bella, D. bicornuta, D. cuspidata, D. insignis, D. pulchella, D. pusilla, Elatine californica, Eleocharis palustris, Epilobium brachycarpum, E. cleistogamum, Eryngium castrense, E. vaseyi, Gratiola, Hemizonia congesta, Hesperevax caulescens, Hordeum marinum, Isoetes howellii, I. orcuttii, Juncus balticus, J. bufonius, Lasthenia californica, L. fremontii, L. glabrata, Layia chrysanthemoides, Myosurus minimus, Lepidium nitidum, Limnanthes douglasii, L. floccosa, Lolium multiflorum, Lupinus bicolor, Lythrum hyssopifolium, Medicago polymorpha, Mimulus tricolor, Minuartia douglasii, Navarretia intertexta, N. leucocephala, Orcuttia inaequalis, O. tenuis, Phalaris lemmonii, Pilularia americana, Plagiobothrys acanthocarpus, P. chorisianus, P. humistratus, P. leptocladus, P. stipitatus, Plantago elongata, Pogogyne douglasii, P. ziziphoroides, Populus, Potamogeton, Psilocarphus brevissimus, P. chilensis, P. oregonus, Rumex crispus, Sagina apetala, Taeniatherum caput-medusae, Taraxacum officinale, Trifolium depauperatum, T. hirtum, T. willdenovii, Triphysaria eriantha, Triteleia hyacinthina, Tuctoria greenei, Veronica peregrina, Vicia sativa.*

Downingia pulchella **(Lindl.) Torr.** occupies depressions, ditches meadows, plains, swales, vernals pools, wallows, and the margins of lagoons, lakes, and salt marshes at elevations of up to 1677 m. It occurs along shallow margins or in standing water (8–10 cm). Site exposures are open and substrates can be acidic (least typical), alkaline or saline. They are described as adobe, claypan, Edminster series, Fe–Si cemented hardpan, Hillmar series, Kesterson series, Modesto, mud, Redding series, sand, and San Joaquin series. The flowers have a short anther tube and are pollinated primarily by specialist bees

(Insecta: Hymenoptera: Andrenidae). The usual period of flowering and fruiting extends from 15 April to 7 June. **Reported associates:** *Achyrachaena mollis, Allenrolfea occidentalis, Alopecurus geniculatus, A. saccatus, Arthrocnemum subterminale, Astragalus tener, Atriplex parishii, Blennosperma nanum, Castilleja campestris, Centromadia fitchii, Cotula coronopifolia, Crassula aquatica, Cressa truxillensis, Crypsis schoenoides, Cuscuta howelliana, Damasonium californicum, Deschampsia danthonioides, Distichlis spicata, Downingia bella, D. bicornuta, D. concolor, D. cuspidata, D. insignis, D. ornatissima, Eleocharis acicularis, E. palustris, Epilobium torreyi, Eryngium aristulatum, E. vaseyi, Frankenia salina, Hordeum depressum, Hypochaeris glabra, Juncus leiospermus, J. tenuis, J. uncialis, Lasthenia californica, L. conjugens, L. ferrisiae, L. fremontii, L. glaberrima, L. glabrata, L. platycarpha, Layia gaillardioides, Legenere limosa, Lepidium acutidens, Lilaea scilloides, Limnanthes alba, L. douglasii, Limosella aquatica, Lythrum hyssopifolia, Malvella leprosa, Mimulus angustatus, M. tricolor, Myosurus minimus, M. sessilis, Navarretia intertexta, N. leucocephala, N. prostrata, Phalaris lemmonii, Plagiobothrys humistratus, P. leptocladus, P. scouleri, P. stipitatus, P. trachycarpus, P. undulatus, Plantago elongata, Pogogyne douglasii, P. nudiuscula, P. ziziphoroides, Psilocarphus brevissimus, Rumex, Salicornia, Sida, Toxicoscordion venenosum, Triphysaria, Veronica arvensis.*

Downingia pusilla **(G. Don) Torr.** inhabits alluvial terraces, ditches, flats, meadows, mud beds, stream beds, swales, vernal pools, wallows, and the margins of sloughs at relatively low elevations of up to 448 m. Characteristic habitats are small to large (220–56,500 m^2), shallow (average depth = 12 cm) pools with high plant cover (e.g., 66%–78%). Exposures range from full sunlight to partial shade. The substrates are characterized as alkaline or saline and include alluvium, Antioch, clay, Corning gravelly loam, gravel, Corning-Redding, gravelly loam, Hornitos, mud, Pescadero, Redding, San Ysidro, stony clay, and volcanic sediment. Typically, the plants are in flower and fruit from 27 March to 27 May. The flowers are of the reduced type and primarily are self-pollinating. The life cycle is rapid with flowering plants developing in as little as 15 days from germination. Populations exhibit large annual and year-to-year fluctuations, which are ameliorated by a large seed bank. Major anthropogenic threats to the plants include agricultural and urban development, forestry, grazing, and damage by off-road vehicles. The species also is known to have disappeared from sites due to competition by invasive grasses (Poaceae). **Reported associates:** *Achyrachaena mollis, Alopecurus saccatus, Astragalus tener, Avena fatua, Blennosperma nanum, Brodiaea, Bromus hordeaceus, Callitriche marginata, Castilleja campestris, Centromadia fitchii, Cerastium glomeratum, Cicendia quadrangularis, Cotula coronopifolia, Crassula aquatica, C. connata, Croton setiger, Deschampsia danthonioides, Distichlis spicata, Downingia bicornuta, D. concolor, D. cuspidata, D. insignis, D. ornatissima, Elatine californica, E. chilensis, Eleocharis acicularis, E. macrostachya, E. palustris, Epilobium brachycarpum, E. cleistogamum, Erodium botrys, Eryngium* aristulatum, E. castrense, Frankenia salina, Gratiola ebracteata, G. heterosepala, Hemizonia congesta, Holocarpha virgata, Hordeum depressum, H. marinum, Hypochaeris glabra, Isoetes howellii, I. orcuttii, Juncus bufonius, J. capitatus, J. uncialis, J. xiphioides, Lactuca serriola, Lasthenia conjugens, L. fremontii, L. glaberrima, L. glabrata, Layia chrysanthemoides, Legenere limosa, Lepidium latipes, Lilaea scilloides, Limnanthes douglasii, Lolium perenne, Lupinus bicolor, Lythrum hyssopifolia, Marsilea vestita, Medicago polymorpha, Microseris douglasii, Mimulus tricolor, Montia fontana, Myosurus minimus, Navarretia leucocephala, Phalaris lemmonii, Pilularia americana, Plagiobothrys greenei, P. humistratus, P. leptocladus, P. stipitatus, P. undulatus, Plantago elongata, Pleuropogon californicus, Pogogyne douglasii, P. ziziphoroides, Polypogon monspeliensis, Psilocarphus brevissimus, P. oregonus, P. tenellus, Ranunculus bonariensis, R. muricatus, Rumex crispus, Senecio vulgaris, Sidalcea diploscypha, Taeniatherum caput-medusae, Trifolium barbigerum, T. depauperatum, T. variegatum, Triphysaria eriantha, Veronica peregrina, Vulpia bromoides.

***Downingia willamettensis* M.E. Peck** occurs in ditches, meadows, ponds, pond beds, sloughs, streams, vernal pools, and along the margins of lakes and ponds at elevations of up to 1371 m. Substrates are reported as clay or silt. Flowering and fruiting occur from 1 June to 1 August. The scarce amount of literature information on this species likely is the result of incomplete parsing of information attributed to *Downingia yina* (see next) with which it has long been synonymized. Consequently, this account is highly preliminary and it will be necessary to reevaluate life history information for details that pertain specifically to *D. willamettensis*. **Reported associates:** *Amaranthus californicus, Eleocharis, Gnaphalium palustre, Gratiola ebracteata, Rorippa curvisiliqua.*

***Downingia yina* Applegate** is found in depressions, draws, flats, lake beds, meadows, mudflats, prairies, slopes, swales, vernal pools, and along the margins of lakes and ponds at elevations of up to 1982 m. The substrates are slightly acidic (pH 6.2–6.8) and include adobe, Agate-Winlo, basalt rubble, clay, loam, muddy clay, serpentine, and volcanic pumice and loam. Flowering and fruiting occur from 1 July to 15 August. The flowers are pollinated by bees (Insecta: Hymenoptera) and perhaps to some degree by butterflies (Insecta: Lepidoptera). Fresh seeds have germinated successfully following a 4°C–10°C/2°C–4°C day/night temperature regime when planted in a soil-less peat-based medium provided with micronutrients and a slow-release 14–14–14 fertilizer. Transplantable individuals develop in about 3 months. The plants reportedly persist more frequently in pools that are not grazed by cattle (Mammalia: Bovidae: *Bos*). **Reported associates:** *Abies concolor, Agoseris heterophylla, Alisma gramineum, Allium geyeri, Alopecurus saccatus, Apera interrupta, Artemisia cana, Beckmannia syzigachne, Brodiaea, Bromus japonicus, Callitriche longipedunculata, Calocedrus decurrens, Camassia leichtlini, C. quamash, Cardamine oligosperma, Carex douglasii, Castilleja tenuis, Centromadia fitchii, Cerastium glomeratum, Cladonia gracilis, Clarkia gracilis, Collomia linearis, Croton setiger, Danthonia californica,*

Dasiphora floribunda, Delphinium distichum, Deschampsia cespitosa, D. danthonioides, Downingia bacigalupii, D. bella, D. bicornuta, Echinops sphaerocephalus, Elatine heterandra, Eleocharis acicularis, E. palustris, Elymus elymoides, Epilobium brachycarpum, E. campestre, E. densiflorum, E. torreyi, Erigeron pumilus, Erodium cicutarium, Eryngium petiolatum, Geranium dissectum, Gnaphalium palustre, Gratiola heterosepala, Grindelia hirsutula, G. integrifolia, Hackelia bella, Helenium bigelovii, Hesperochiron pumilus, Hordeum marinum, Hydrophyllum capitatum, Hypochaeris radicata, Idahoa scapigera, Iris missouriensis, Isoetes howellii, I. nuttallii, Juncus acuminatus, J. balticus, J. bufonius, Lactuca serriola, Lagophylla ramosissima, Lasthenia californica, L. glaberrima, Leptosiphon bicolor, Lilaea scilloides, Limnanthes floccosa, Limosella aquatica, Lindernia dubia, Lolium perenne, Lomatium cookii, L. utriculatum, Ludwigia, Lupinus bicolor, Madia, Marsilea vestita, Mentha pulegium, Micropus californicus, Microsteris gracilis, Moenchia erecta, Montia dichotoma, M. linearis, Myosotis, Myosurus apetalus, M. minimus, Navarretia leucocephala, Olsynium douglasii, Orthocarpus bracteosus, Panicum, Phalaris, Pinus jeffreyi, P. ponderosa, Plagiobothrys austiniae, P. bracteatus, P. hirtus, P. leptocladus, P. undulatus, Poa bulbosa, P. nemoralis, Polygonum aviculare, P. polygaloides, Porterella carnosula, Primula jeffreyi, Psilocarphus brevissimus, P. leptocladus, Ranunculus aquatilis, Rorippa curvisiliqua, Rosa, Rotala ramosior, Rumex crispus, R. salicifolius, Sonchus asper, Taeniatherum caput-medusae, Trichostema lanceolatum, T. oblongum, Trifolium arvense, T. willdenovii, T. wormskioldii, Tripsacum lanceolatum, Triteleia hyacinthina, Valerianella locusta, Veratrum californicum, Veronica arvensis, V. peregrina, Vulpia myuros.

Use by wildlife: *Downingia* species are visited by a diverse insect (Insecta) fauna including bees (Hymenoptera: Andrenidae: *Panurginus*; Anthophoridae: *Anthophora urbana, Nomada*; Apidae: *Apis mellifera, Bombus huntii, B. vosnesenskii*; Colletidae: *Hylaeus*; Halictidae: *Halictus, Lasioglossum*; Megachilidae: *Osmia*), beetles (Coleoptera: Chrysomelidae; Coccinellidae; Dasytidae; Meloidae: *Meloe*), bugs (Hemiptera: Miridae), butterflies (Lepidoptera: Adelidae: *Adela*; Satyridae), flies (Diptera: Calliphoridae: *Phormia*; Scathophagidae: *Scatophaga*; Syrphidae), and wasps (Hymenoptera: Pompilidae). The flowers of several species (*D. bicornuta, D. concolor, D. cuspidata, D. ornatissima, D. pulchella*) are pollinated by female specialist bees (Insecta: Hymenoptera: Andrenidae: *Panurginus atriceps*); however, there is some debate whether the bee species visiting these *Downingia* species may represent an oligolege distinct from *P. atriceps*. Additional floral visitors of *D. concolor* include sweat bees (Insecta: Halictidae: *Lasioglossum*) and flies (Insecta: Diptera: Anthomyiidae; Calliphoridae; Syrphidae). Flowers of *D. bella* are visited by bees (Insecta: Hymenoptera: Anthophoridae; Apidae: *Bombus californicus, B. sonorus*), butterflies (Insecta: Lepidoptera: Hesperiidae: *Hesperia lindseyi*), and wasps (Insecta: Hymenoptera: Vespidae). The flowers of *D. cuspidata* attract wasps (Insecta: Hymenoptera: Vespidae) and those of *D. elegans* and *D. yina* are visited by nectaring great copper butterflies (Insecta: Lepidoptera: Lycaenidae: *Lycaena xanthoides*), and to a minor extent by honeybees (Insecta: Hymenoptera: Apidae: *Apis*). Livestock graze on *D. bacigalupii* and the seedlings of *D. bicornuta* are eaten by tadpole shrimp (Crustacea: Triopsidae: *Lepidurus packardi*). *Downingia elegans* is a host to a club fungus (Fungi: Basidiomycota: Doassansiaceae: *Doassansia downingiae*).

Economic importance: food: None of the *Downingia* species reportedly is edible; **medicinal:** none reported; **cultivation:** *Downingia concolor* and *D. cuspidata* have been recommended as potential border plants. In the later 19th-century, *D. elegans* was touted as a "half-hardy annual" summer bedding plant. However, despite their inherent beauty, *Downingia* species rarely are cultivated, mainly due to their fastidious habitat requirements; **misc. products:** The seed of *Downingia elegans* is sold for wetland restoration projects. *Downingia pulchella* is featured among the flowers in the astounding Ware collection of Blaschka glass flower models; **weeds:** none; **nonindigenous species:** *Downingia elegans* has been introduced to The United Kingdom and central Europe via contaminated grass seed originating from western North America. It also has been introduced to the Australian Alps.

Systematics: Phylogenetic analyses of DNA sequence data from several nuclear and plastid loci confirm the monophyly of *Downingia*. Depending on the taxa surveyed, its sister group resolves as the western North American genus *Porterella* (Figure 5.177), or as the Mediterranean genus *Solenopsis*. Although disparate biogeographically, *Solenopsis* contains annual species that inhabit vernal pools like *Downingia*. Traditionally, *Downingia* and *Howellia* have been assigned to subtribe Howelliinae of tribe Loebelieae, and a close relationship of these genera along with *Legenere* and *Porterella*

seems probable. Additional insight into intergeneric relationships should be gained once a more comprehensive analysis that includes all of these genera is conducted. *Downingia yina* recently has been subdivided into three morphologically cryptic species (*D. pulcherrima*, *D. willamettensis*, and *D. yina*). Consequently, it virtually has been impossible to determine whether some literature information attributed to *D. yina* should instead refer to *D. willamettensis* (*D. pulcherrima* has been excluded here pending more definitive investigations). Spiral (vs. longitudinally) lined seeds occur only in *D. cuspidata* and *D. pusilla* and represent a morphlogical synapomorphy for these species. Although the basic chromosome number of subfamily Lobelioideae is considered to be $x=7$, that of *Downingia* generally is interpreted as $x=11$. Subsequent aneuploid modifications have produced an array of diploid cytotypes, which conceivably enforce reproductive isolation among the species: *Downingia bacigalupii*, *D. ornatissima*, and *D. yina* ($2n=24$); *D. bella*, *D. bicornuta*, *D. cuspidata*, *D. insignis*, *D. laeta*, *D. montana*, *D. pulchella*, and *D. pusilla* ($2n=22$); *D. concolor* ($2n=16, 18$); *D. elegans* ($2n=20$); and *D. willamettensis* ($2n=12, 16, 20$). Few *Downingia* species can hybridize due to strong genetic isolating barriers. Hybrids have been reported between *D. bacigalupii* × *D. yina* (as *D. pulcherrima*) and *D. elegans* × *D. willamettensis*. Fertile synthetic hybrids have been derived from *D. insignis* × *D. pulchella*.

Comments: The broadest distributions occur in *Downingia laeta* (California to Alberta and Saskatchewan) and *D. elegans* (California to British Columbia). Narrower ranges characterize *D. bacigalupii*, *D. bicornuta*, and *D. insignis* (California, Idaho, Nevada, Oregon), *D. yina* and *D. willamettensis* (California, Oregon, Washington), and *D. montana* and *D. pulchella* (California, Oregon); four species are

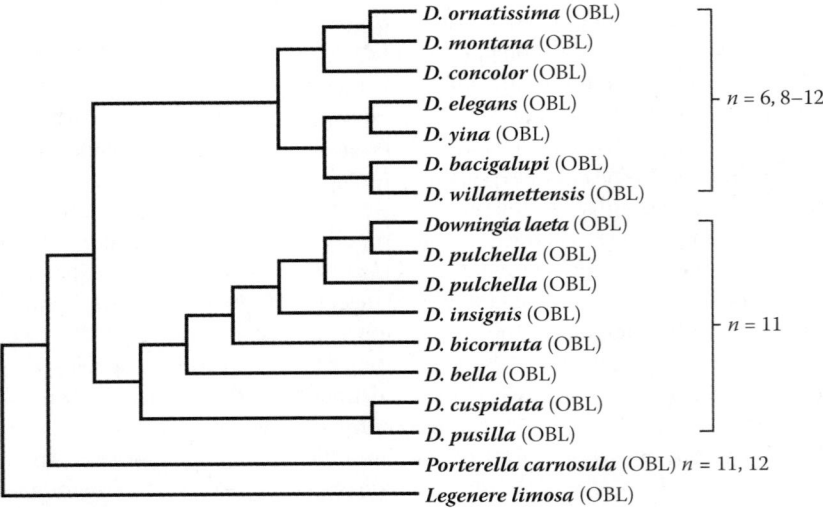

FIGURE 5.177 Phylogenetic relationships in *Downingia* as indicated by analysis of combined nrITS and cpDNA sequence data. The genus is characterized by two clades consisting of $n=11$ species (lower clade) and a mixture of derived chromosomal numbers (upper clade). All species (as well as their putative outgroups) are OBL taxa (in bold), which indicates a single origin of the habit in the genus and potentially its sister genera. (Adapted from Schultheis, L.M., *Syst. Bot.*, 26, 603–621, 2001.)

endemic to California: *D. bella*, *D. concolor*, *D. cuspidata*, and *D. ornatissima*. In North America, *D. pusilla* occurs only in California but also is known from Argentina and Chile.

References: Adams & Dawson, 2009; Antonelli, 2008; Barbour et al., 2007; Barry, 1998; Bartow, 2007a,b; Baskin, 1994; Bauder, 1992; Bauder & McMillan, 1998; Bauder et al., 1998; Briggs & Fitzgerald, 1986; Brown, 1999; Buck, 2004; Buck-Diaz et al., 2012; Cassell, 2010; Christy, 2013; Clausnitzer & Huddleston, 2002; Comer et al., 2005; Croel & Kneitel, 2011a; Egger et al., 2012; Elam, 1998; Evans-Peters, 2010; Evans-Peters et al., 2012; Foote, 1961; Fraser, 2000;Gimingham et al., 2002; Gleason, 2002; Gluesenkamp & Wirka, 2006; Graffis, 2013; Guerrant, Jr. & Raven, 1998; Holland, 1986; Ingham & Wilson, 1999; Kaplan, 1967; Kaye et al., 2011; Keeley & Sandquist, 1991; Keeley & Zedler, 1998; Kent, 1908; King, 1992; Lammers, 1993; Lazar, 2006; Lenski, 1986, 1987; Linhart, 1976; Lippert & Jameson, 1964; LSA, 2010; Martin & Lathrop, 1986; McDougall et al., 2005; Mc Vaugh, 1941; Moore et al., 2003; Orcutt, 1883; Platenkamp, 1998; Ramaley, 1919; Rymer, 2011; Schlising & Sanders, 1982; Schmidt, 1980, 1989; Schultheis, 2001, 2010, 2012; Severns et al., 2006; Silveira, 2000a,b; Thorp, 1979; Thorp & Leong, 1995, 1998; Wacker & Kelly, 2004; Wallis, 2001; Wildsmith, 1884; Wood, Jr., 1961.

3. *Howellia*

Water howellia

Etymology: after Thomas Howell (1842–1912) and Joseph Howell (1830–1912)

Synonyms: none

Distribution: global: North America; **North America:** Northwestern United States

Diversity: global: 1 species; **North America:** 1 species

Indicators (USA): OBL: *Howellia aquatilis*

Habitat: freshwater; lacustrine, palustrine; **pH:** 6.9–7.0; **depth:** to 2.5 m; **life-form(s):** submersed (vittate) herb

Key morphology: Shoots (to 60 cm) submersed or floating, much branched; leaves (to 5 cm) cauline, alternate, linear, sessile, the margins entire to minutely toothed; flowers (3–10) cleistogamous or chasmogamous, the latter resupinate, axillary, solitary, subsessile, or pedicellate (to 8 mm); corolla (to 3 mm; absent in cleistogamous flowers) white, bilabiate, upper lip (to 1.3 mm) 2-lobed and spreading, lower lip (to 1.3 mm) 3-lobed, and erect; stamens fused into a tube; ovary (to 13 mm) inferior; capsules (to 13 mm) 1-loculed, dehiscent irregularly by lateral slits or tears; seeds (to 4 mm) 1–5, cylindric, shiny brown.

Life history: duration: annual (fruit/seeds); **asexual reproduction:** none; **pollination:** cleistogamous/self; **sexual condition:** hermaphroditic; **fruit:** capsules (common); **local dispersal:** seeds (gravity); **long-distance dispersal:** seeds (waterfowl)

Imperilment: (1) *Howellia aquatilis* [G3]; S1 (ID, OR); S2 (CA, WA); S3 (MT)

Ecology: general: *Howellia* is monotypic.

Howellia aquatilis **A. Gray** inhabits shallow (20–250 cm) depressions, lakes, marshes, oxbows, ponds, pools, and potholes at elevations of up to 1377 m. The plants are restricted to ephemeral freshwater sites where the waters are fed in spring by snowmelt and then recede seasonally as evaporation and transpiration of standing vegetation exceed groundwater recharge rates. The habitats typically are surrounded by various trees and/or shrubs. The waters are neutral in acidity (pH 6.9–7.0) with low conductivity (100–120 μS/cm) and moderate dissolved oxygen (6.1–6.5 mg/L). Some diffuse shade by marginal woody plants is tolerated, but deeper shade by dense graminoid (Poaceae) populations can be highly detrimental. Although substrates have been described as clay, clay muck, muck, and mucky peat–clay, the plants grow best on a coarse-textured organic surface (peat) less than 25 cm in depth and exhibit markedly reduced growth on finer textured organic substrates or on mineral soils. Sampled substrates contained approximately 55% organic matter. The plants are obligate inbreeders. Flower production (May–August) initiates on the submersed stems, which produce self-pollinated, cleistogamous flowers, and proceeds to the emergent axes, which produce chasmogamous flowers that also self-pollinate. In the latter, the stigma pushes through the surrounding anther tube after the pollen has been discharged. Pollen viability (assessed via stainability) is high (93%). The capsules produce few (1–5) but large (2–4 mm) seeds, which are dispersed in mid-late summer and germinate in October. The seedlings persist throughout winter, with growth resuming in spring in up to 2 m of water. No after ripening period is necessary. Under experimental conditions, germination is highest (87%–99%) for fresh seeds kept aerobically for 50–100 days at 20°C in complete darkness or in equal (12/12 h) light and darkness, then brought under a 5°C/20°C (12/12 h) temperature regime. Reduced germination (53% after 60 days) was obtained for seeds kept in dry storage for 8 months. The seeds require a dry pond bed for germination; they do not germinate under water or under constant temperature (20°C). Fall germination ceases as waters recover the seed. Seedling growth is rapid and can exceed 1 cm in less than a week. Seed density in the seedbank is highest in late summer, immediately following site desiccation, and can be reduced by 82%–90% by the subsequent spring. Samples taken in May–June recovered seeds at densities of 673–1263/m²; densities of samples taken in September–October varied from 6861 to 2904/m², with the latter decrease attributed to germination. Deposition of thick litter layers by competing grasses (Poaceae) is believed to interfere with seedling growth. The large seeds are not buoyant but sink when released. Their long-distance dispersal is presumed to occur by means of migrating waterfowl (Aves: Anatidae). **Reported associates:** *Acorus calamus, Alisma triviale, Alnus rhombifolia, A. rugosa, Alopecurus aequalis, Amelanchier alnifolia, Betula papyrifera, Callitriche heterophylla, C. stagnalis, Carex atherodes, C. retrorsa, C. rostrata, C. stipata, C. vesicaria, Cicuta douglasii, Cornus sericea, Crataegus douglasii, Dactylis glomerata, Eleocharis palustris, Equisetum fluviatile, Fraxinus latifolia, Galium, Glyceria borealis, G. ×occidentalis, Juniperus communis, Lemna minor, L. trisulca, Ludwigia palustris, Myriophyllum spicatum, Myosotis scorpioides, Nuphar variegata,*

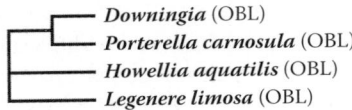

Downingia (OBL)
Porterella carnosula (OBL)
Howellia aquatilis (OBL)
Legenere limosa (OBL)

FIGURE 5.178 Phylogenetic tree constructed from nrITS sequence data showing close proximity of four entirely OBL (bold) genera of Campanulaceae. It is conceivable that this group represents a single aquatic clade; however, additional taxon sampling is necessary before this possibility can be evaluated adequately. (Adapted from Schultheis, L.M., *Syst. Bot.*, 26, 603–621, 2001.)

Persicaria coccinea, Phalaris arundinacea, Physocarpus capitatus, Pinus ponderosa, Populus tremuloides, P. trichocarpa, Potamogeton gramineus, P. natans, Pseudotsuga menziesii, Ranunculus aquatilis, R. flabellaris, R. flammula, R. gmelinii, Ricciocarpus natans, Rosa, Sagittaria cuneata, Salix bebbiana, S. drummondiana, Sium suave, Sparganium emersum, S. natans, Symphoricarpos albus, Typha latifolia, Utricularia macrorhiza, Veronica anagallis-aquatica, V. catenata.

Use by wildlife: The plants are grazed by dabbling ducks (Aves: Anatidae) and the flowers are visited infrequently by small flies (Insecta: Diptera). The underwater shoots support caddis fly (Insecta: Trichoptera) cases.

Economic importance: food: none reported; **medicinal:** no uses reported; **cultivation:** not cultivated; **misc. products:** none; **weeds:** none; **nonindigenous species:** none

Systematics: *Downingia* and *Howellia* have been assigned to subtribe Howelliinae of tribe Loebelieae and the most comprehensive phylogenetic studies to date place *Howellia* (monotypic) near the genera *Downingia* (14 species), *Legenere* (monotypic), and *Porterella* (monotypic), which all entirely comprise OBL species (Figure 5.178). Although a more definitive phylogenetic placement of *Howellia* will require additional sampling, it is likely to remain closely related to these genera. An allozyme survey of 18 genetic loci showed there to be no detectable variation within or among *H. aquatilis* populations, a result attributed to inbreeding. A subsequent study using higher resolution (AFLP) markers showed that 83.8% of the genetic variation detected in California individuals occurred within populations and 16.2% among populations. The pattern of genetic differentiation among distant sites strongly implicates migrating waterfowl as agents of long-distance dispersal for the species. The base chromosome number of *Howellia* and related genera is $x=11$; *H. aquatilis* ($2n=22$) is diploid.

Comments: *Howellia aquatilis* occurs only in California, Idaho, Montana, Oregon, and Washington.

References: Lesica, 1992, 1997; Lesica et al., 1988; Lichthardt & Gray, 2002; Lichthardt & Moseley, 2000; Morin, 2012b; Reeves & Woessner, 2004; Schierenbeck & Phipps, 2010; Shelly & Moseley, 1988.

4. *Legenere*
False Venus' looking glass
Etymology: an anagram of E. L. Greene (1843–1915)

Synonyms: *Howellia* (in part)
Distribution: global: North America; **North America:** California (endemic)
Diversity: global: 1 species; **North America:** 1 species
Indicators (USA): OBL: *Legenere limosa*
Habitat: freshwater; lacustrine, palustrine; **pH:** unknown; **depth:** <1 m; **life-form(s):** submersed (vittate) herb
Key morphology: Stems (to 30 cm) decumbent or erect, simple or branched; leaves (to 3 cm) cauline, sessile, linear (submersed) or oblong-lanceolate (emergent), readily deciduous, margins entire; flowers (to 12) pedicelled (to 31 mm), axillary or in terminal racemes (to 14 cm), cleistogamous or chasmogamous, not resupinate; calyx 4-merous (cleistogamous flowers) or 5-merous (chasmogamous flowers); corolla (absent in cleistogamous flowers) tubular (to 1.5 mm), white or yellowish, bilabiate, upper lip 3-lobed (to 2 mm), lower lip 2-lobed (to 2 mm); stamens fused into a tube; ovary (to 3.5 mm) inferior; capsule (to 10 mm) unilocular, dehiscent apically; seeds (to 20; to 1 mm), brown, shiny.

Life history: duration: annual (fruit/seeds); **asexual reproduction:** none; **pollination:** insect or self (cleistogamous); **sexual condition:** hermaphroditic; **fruit:** capsules (common); **local dispersal:** seed (gravity, water); **long-distance dispersal:** seed (waterfowl)
Imperilment: (1) *Legenere limosa* [G2]; S2 (CA)
Ecology: general: *Legenere* is monotypic.

Legenere limosa **(Greene) McVaugh** grows in ephemeral ditches, marshes, meadows, ponds, sloughs, and vernal pools at elevations of up to 883 m. The plants occur in relatively large (e.g., 27,600–141,400 m²) and deep (23–38 cm) pools with high herbaceous cover (60%–95%). They typically grow under submersed conditions (depths ~23 cm) until the water levels recede during the summer. Their substrates are described as Alamo clay, Cometa, gravel, gravelly clay, Hambright, mud, northern basalt flow, northern claypan, northern hardpan, northern volcanic ashflow, northern volcanic mudflow, Rocklin, San Joaquin, and Tuscan. Flowering takes place from April to June. Both apetalous cleistogamous and petaled chasmogamous flowers are produced. The former are self-pollinated and frequently are the most common type observed. The latter are visited by small bees (Insecta: Hymenoptera); however, they occur less frequently and their reproductive biology has not been elucidated. A persistent seedbank is assumed, inspired by observations that plants can suddenly reappear at a site after decades of absence. **Reported associates:** *Alopecurus saccatus, Astragalus, tener, Brodiaea terrestris, Callitriche marginata, Castilleja campestris, Convolvulus arvensis, Crassula aquatica, Deschampsia danthonioides, Distichlis spicata, Downingia bicornuta, D. cuspidata, D. pusilla, Eleocharis macrostachya, E. palustris, Epilobium cleistogamum, Eryngium castrense, Frankenia salina, Glyceria declinata, G. ×occidentalis, Gratiola ebracteata, G. heterosepala, Isoetes howellii, Juncus xiphioides, Lasthenia conjugens, L. fremontii, L. glaberrima, Lilaea scilloides, Lolium multiflorum, Lotus corniculatus, Marsilea vestita, Navarretia leucocephala, Neostapfia colusana, Orcuttia pilosa, O. viscida, O. tenuis, Phyla nodiflora, Ranunculus* *bonariensis, Rorippa curvisiliqua, Plagiobothrys chorisianus, P. stipitatus, P. undulatus, Pleuropogon californicus, Pogogyne serpylloides, Psilocarphus oregonus, P. tenellus, Schoenoplectus.*

Use by wildlife: none reported

Economic importance: food: not reported as edible; **medicinal:** no known uses; **cultivation:** *Legenere* is not cultivated; **misc. products:** none; **weeds:** not weedy; **nonindigenous species:** none

Systematics: *Legenere* (monotypic) is closely related to *Howellia* (Figure 5.178), the genus to which it originally was referred. Both genera are OBL annuals with an affinity for ephemeral waters, which are characteristics shared by yet another closely related genus *Porterella* (Figure 5.178). Additional taxon sampling is necessary to better indicate how these genera resolve phylogenetically among each other and with other Campanulaceae. The chromosome number of *Legenere* has not been determined.

Comments: *Legenere limosa* is endemic to California.

References: Abrams, 1960; Barbour et al., 2007; Buchmann et al., 2010; Dittes & Guardino, 2002; Lazar, 2006; Mason, 1957; McVaugh, 1943b; Morin, 2012c; Nakamura & Nelson, 2001; Platenkamp, 1998; Witham & Kareofelas, 1994.

5. *Lobelia*

Cardinal flower, lobelia; lobélie

Etymology: after Mathias de L'Obel (1538–1616)

Synonyms: *Dortmanna; Enchysia; Haynaldia; Isolobus; Laurentia; Mezleria; Neowimmeria; Parastranthus; Petromarula; Pratia; Rapuntium; Tupa*

Distribution: global: cosmopolitan; **North America:** widespread

Diversity: global: 415 species; **North America:** 29 species

Indicators (USA): OBL: *Lobelia amoena, L. boykinii, L. canbyi, L. cardinalis, L. dortmanna, L. elongata, L. flaccidifolia, L. floridana, L. glandulosa, L. kalmii, L. paludosa, L. siphilitica;* **FACW:** *L. cardinalis, L. flaccidifolia, L. siphilitica*

Habitat: brackish or freshwater; lacustrine, palustrine, riverine; **pH:** 4.1–9.0; **depth:** 0–3 m; **life-form(s):** emergent herb, submersed (rosulate) herb

Key morphology: Stems (to 6–18 dm) erect, simple or somewhat branched, sap milky; leaves few to 27, basal, the blades (to 3.5–40 cm), succulent or not, oblanceolate, lanceolate or linear; or leaves 15–50, cauline, alternate, the blades (to 2.5–18 cm), filiform, lanceolate, lance-ovate, linear, oblong, obovate, ovate, or subulate, the margins entire, crenate or serrate, sessile or petioled (to 3 cm); racemes (to 20–50 cm) terminal, to 20–100–flowered, secund or not; flowers (to 13–45 mm) perfect or pistillate, 5-merous, tubular, bilabiate, resupinate, lower lip (to 20 mm) 3-lobed, spreading, upper lip (to 20 mm) 2-lobed, erect or recurved, pedicels (to 5–17 mm) with 2 bracteoles (to 1 mm) near or at base; corolla bright, light or deep blue, lavender, violet or red, the tube fenestrate or not; filaments united into a tube (to 3.5–33 mm), the base split; ovary 2-celled, partially to wholly inferior; capsules (to 3–10 mm) globose to sub-globose, many-seeded, dehiscent

apically, calyx persistent; seeds (to 0.4–0.8 mm) elongate, amber brown, brown, or dark brown, pitted-reticulate.

Life history: duration: biennial, perennial (basal offshoots); **asexual reproduction:** rhizomes, rosettes; **pollination:** bird or insect; **sexual condition:** gynodioecious or hermaphroditic; **fruit:** capsules (common); **local dispersal:** seeds (gravity); **long-distance dispersal:** seeds (birds, mammals)

Imperilment: (1) *Lobelia amoena* [G4]; S1 (TN); S3 (NC); (2) *L. boykinii* [G2/G3]; SX (DE); S1 (AL, FL, MS, NJ); S2 (GA, NC); S3 (SC); (3) *L. canbyi* [G4]; S1 (MD); S2 (DE, TN); S3 (GA, NC, NJ); (4) *L. cardinalis* [G5]; SH (<u>MB</u>); S1 (NE, NV); S2 (CO); S3 (<u>NB</u>); (5). *L. dortmanna* [G4/G5]; SH (NJ); S1 (<u>AB</u>, AK, OR, <u>PE</u>); S2 (<u>MB</u>, PA, RI, <u>SK</u>, WA); (6) *L. elongata* [G4/G5]; S1 (DE, LA, VA); S3 (MD); (7) *L. flaccidifolia* [G5]; S3 (LA); (8) *L. floridana* [G5]; (9) *L. glandulosa* [G4/G5]; S3 (NC); (10) *L. kalmii* [G5]; S1 (PA, <u>NS</u>, SD, WA, WV); S2 (MA, MT, NH); S3 (<u>AB</u>, CT, IA, <u>NB</u>, <u>NF</u>, NJ); (11) *L. paludosa* [G4/G5]; (12) *L. siphilitica* [G5]; SX (ME); SH (<u>MB</u>); S1 (MA, NH, VT, WY); S3 (GA, LA)

Ecology: general: *Lobelia* is a large genus with an impressive diversity of species. Having nearly a worldwide distribution, these plants inhabit a spectrum of communities ranging from cold-temperate to tropical latitudes, sea-level to alpine elevations, and mesic to entirely aquatic habitats. However, all of the OBL North American species belong to the same section *Lobelia*, which is characteristic of wet habitats. Accordingly, this is a familiar genus in and around North American wetlands with 24 species (83%) having some wetland indicator designation and nearly half of those (41%) categorized as OBL in some portion of their range. Almost all species of section *Lobelia* are short-lived perennials (living no longer than 3–4 years), which produce flowering shoots that bolt from basal rosettes. One group consists of small-flowered (7–15 mm), diploid plants, to which *L. boykinii* and *L. canbyi* belong. Medium-sized flowers (11–33 mm) and narrow leaves characterize *L. flaccidifolia, L. floridana, L. glandulosa,* and *L. paludosa. Lobelia amoena, L. cardinalis, L. elongata,* and *L. siphilitica* are characterized by having larger flowers (15–50 mm). Many of the species (76%) in this section are self-fertile, but most are outcrossed by insects (Insecta), or in some cases, by nectar-foraging birds (Aves). As in other Campanulaceae, the flowers possess a tube of fused anthers surrounding a brush-like style, which pushes pollen out prior to the opening of the receptive stigmatic surfaces. Some species are gynodioecious. The tropical species typically have nondormant seeds; whereas, some of the temperate species (e.g., *L. kalmii*) have physiologically dormant seeds, which require cold or warm stratification in order to germinate. Seeds of most section *Lobelia* germinate when fresh (or up to 5 years of age) with no requirement for a cold treatment. Germination in some species is reliant on light (e.g., *L. cardinalis, L. siphilitica*) and the seeds of *L. dortmanna* germinate while under water. The seeds can be bird or wind dispersed, but often have no apparent specialized dispersal mechanisms and are passively dispersed by gravity or water, or are dispersed in the feces of mammals, which graze on the plants.

The plants yield a milky sap, which can contain toxic alkaloids and resins.

***Lobelia amoena* Michx.** is a biennial or perennial, which grows in backwaters, bottomlands, channels, depressions, ditches, flats, flatwoods, floodplains, hummocks, marshes, roadsides, seeps, slopes, spray cliffs, swamps, woodlands, and along the margins of ponds, rivers, and streams at elevations of up to 777 m. It is found typically in partial to deep shade and on calcareous (e.g., pH 6.8) substrates that include alluvium, clay, limestone, loam, loamy sand, muck, mucky loam, rock, sand, and sandy loam. Flowering occurs from July to October. The plants perennate by basal offshoots and often appear in recently burned areas. The seeds reportedly germinate when fresh (placed on top of moist soil) and remain viable for at least 5 years. Additional life history information on this species is wanting. **Reported associates:** *Acer rubrum, A. saccharum, Agalinis linifolia, Arisaema triphyllum, Aristida spiciformis, A. stricta, Athyrium filix-femina, Atrichum oerstedianum, Bidens mitis, Boltonia apalachicolensis, Bryhnia novae-angliae, Burmannia biflora, Camellia, Campsis, Campylium chrysophyllum, Carex debilis, Carya aquatica, C. glabra, Celtis laevigata, Chelone glabra, Coleataenia longifolia, Collinsonia canadensis, Commelina virginica, Conoclinium, Dichanthelium dichotomum, Diospyros, Diphyscium cumberlandianum, Dracaena, Dumortiera hirsuta, Eleocharis, Eriocaulon, Eupatorium serotinum, Eutrochium fistulosum, Eurybia divaricata, Fissidens dubius, Fuirena squarrosa, Heuchera villosa, Hookeria acutifolia, Houstonia serpyllifolia, Hydrangea quercifolia, H. radiata, Hypericum galioides, H. mutilum, Ilex, Juniperus, Leersia, Ligustrum lucidum, L. sinense, Liquidambar styraciflua, Lobelia cardinalis, L. glandulosa, Lonicera japonica, Ludwigia alternifolia, L. glandulosa, Luzula multiflora, Lycopus rubellus, L. virginicus, Lysimachia ciliata, Mayaca fluviatilis, Micranthes petiolaris, Microstegium vimineum, Mitchella repens, Mnium hornum, Nandina domestica, Onoclea sensibilis, Ostrya, Oxypolis rigidior, Oxystegus tenuirostris, Panicum hemitomon, Parnassia asarifolia, Penthorum sedoides, Phanopyrum gymnocarpon, Philonotis fontana, Pinus palustris, P. taeda, Plagiomnium ciliare, Planera aquatica, Pluchea, Polygala cymosa, P. ramosa, Polystichum acrostichoides, Quercus laurifolia, Q. lyrata, Q. virginiana, Rhamnus, Rhizomnium appalachianum, Rhododendron maximum, Rhynchospora, Rudbeckia laciniata, Sabal minor, Saccharum baldwinii, Sarracenia, Scapania nemorosa, Schizachyrium scoparium, S. tenerum, Schoenus, Selaginella apoda, Sematophyllum marylandicum, Sideroxylon reclinatum, Smilax bona-nox, S. rotundifolia, S. smallii, S. tamnoides, Solidago curtisii, S. fistulosa, S. patula, Sorghastrum, Sphagnum lescurii, Taxodium distichum, Thalictrum clavatum, Thuidium delicatulum, Thelypteris noveboracensis, Trautvetteria caroliniensis, Triadenum virginicum, T. walteri, Triadica sebifera, Tsuga canadensis, Ulmus americana, Vernonia noveboracensis, Viburnum obovatum, Viola cucullata, V. sagittata, Wisteria frutescens, Woodwardia areolata, Xanthorhiza simplicissima, Xyris tennesseensis.*

***Lobelia boykinii* Torr. & A. Gray ex A. DC.** is a biennial or perennial, which inhabits bays, bogs, borrow pits, depressions, ditches, flatwoods, meadows, ponds, pools, and savannas at elevations of up to 55 m. The plants occur under a semiopen broken canopy of tree cover in sites that are flooded seasonally. They often grow in shallow (to 43 cm) standing waters. The substrates are described as clay, loamy sand, muck, and peat. These plants are obligate outcrossers with pollen-limited sexual reproduction. All of the pollen is pushed out of the anther tube prior to stigma receptivity and the flowers are self-incompatible. The most prevalent pollinators are bees (see *Use by wildlife*). Flowering proceeds from May to July and the seeds are dispersed from July to September. Persistent standing water will delay flowering and fruiting by up to a month. The seeds fall beneath the maternal plants, but are buoyant and can be dispersed locally on the surface when standing water is present. Natural germination (from <2% to 8%) occurs during late summer and early fall if soils receive adequate rainfall to become saturated. The seeds require complete saturation for germination, but germinate well only when fresh and are thought to be short-lived (but also reportedly viable for as long as 5 years). It has been suggested that some seeds might persist longer in a seedbank, but this possibility has not been verified. Rosettes develop during the fall and the seedlings will continue to grow throughout winter, even when covered by water; seedling survival typically is around 60%–80%. The rosettes disappear by April, but are produced again in the fall by the rhizomes at a distance up to 5 cm from the original rosettes. After 19 months (April), the rosettes will either bolt or die back once more. Rhizomes that survive past the flowering stage will produce 1–2 new rosettes in the fall. Rosettes that detach also can be dispersed by water as ramets. Herbivory can destroy more than half (to 55%) of the flowering shoots in a population. The plants have been shoot-cultured and successfully cryopreserved using the encapsulation dehydration method. **Reported associates:** *Agalinis aphylla, A. linifolia, Andropogon glomeratus, A. gyrans, Aristida palustris, Boltonia asteroides, Carex striata, Centella asiatica, Coleataenia longifolia, C. tenera, Cyrilla racemiflora, Dichanthelium acuminatum, D. hirstii, D. sphaerocarpon, Eleocharis melanocarpa, E. robbinsii, Eriocaulon compressum, Eupatorium leucolepis, E. mohrii, E. paludicola, Euthamia graminifolia, Hypericum denticulatum, H. fasciculatum, H. harperi, H. myrtifolium, Ilex amelanchier, I. myrtifolia, Iris tridentata, Lachnanthes caroliniana, Leersia hexandra, Lilium catesbaei, Lobelia floridana, Ludwigia pilosa, L. suffruticosa, Lysimachia loomisii, Mnesithea rugosa, Muhlenbergia torreyana, Nyssa, Oxypolis canbyi, Panicum hemitomon, P. verrucosum, P. virgatum, Paspalum praecox, Pinguicula planifolia, Pinus taeda, Pluchea baccharis, P. foetida, Polygala cymosa, Rhexia aristosa, Rhynchospora brachychaeta, R. careyana, R. filifolia, R. galeana, R. harperi, R. inundata, R. microcarpa, R. perplexa, R. pusilla, R. tracyi, Sabatia difformis, Sagittaria isoetiformis, Sarracenia flava, Scleria baldwinii, S. reticularis, Sclerolepis uniflora, Spiranthes laciniata, Symphyotrichum chapmanii, Taxodium ascendens, Utricularia striata, Vaccinium* *corymbosum, Viola lanceolata, Woodwardia virginica, Xyris ambigua, X. smalliana, Zenobia pulverulenta.*

Lobelia canbyi A. Gray is an annual, which grows in barrens, bays, bogs, depression ponds, flatwoods, floodplains, marshes, pocosins, savannas, swales, vernal pools, and along the margins of lakes, ponds, rivers, and streams at elevations of up to 283 m. The substrates are acidic (pH 4.1–6.0) and include loamy sand, muck, Mullica sandy loam, and sand. Flowering commences in mid-July and extends through late September. The seeds reportedly germinate when fresh (placed on top of moist soil) and remain viable for 5 years or more. More detailed information on the reproductive and seed ecology of this species is unavailable. **Reported associates:** *Acer rubrum, Agalinis purpurea, Agrostis perennans, Aletris farinosa, Amphicarpum amphicarpon, Andropogon glomeratus, A. virginicus, Bartonia paniculata, Betula populifolia, Bulbostylis capillaris, Calamagrostis canadensis, C. coarctata, C. pickeringii, Calamovilfa brevipilis, Calopogon tuberosus, Cardamine bulbosa, Carex barrattii, C. bullata, C. exilis, C. livida, C. scoparia, Chamaecyparis thyoides, Chamaedaphne calyculata, Cladium mariscoides, Clethra alnifolia, Coleataenia longifolia, Corema conradii, Coreopsis rosea, Danthonia sericea, Dichanthelium acuminatum, D. dichotomum, D. ensifolium, D. scabriusculum, Drosera intermedia, D. rotundifolia, Dulichium arundinaceum, Eleocharis melanocarpa, E. microcarpa, E. obtusa, E. palustris, E. robbinsii, E. tricostata, E. tuberculosa, Equisetum arvense, Eriocaulon aquaticum, E. compressum, E. decangulare, Eriophorum virginicum, Eupatorium album, E. hyssopifolium, E. leucolepis, E. perfoliatum, E. rotundifolium, Eurybia compacta, Euthamia graminifolia, Fimbristylis autumnalis, Fuirena squarrosa, Geum rivale, Glyceria obtusa, Gratiola aurea, Helenium pinnatifidum, Hypericum boreale, H. canadense, H. crux-andreae, H. mutilum, Ilex glabra, Iris prismatica, Juncus acuminatus, J. canadensis, J. debilis, J. dichotomus, J. effusus, J. marginatus, J. militaris, J. pelocarpus, J. scirpoides, Kalmia angustifolia, Lachnanthes caroliniana, Liatris pilosa, Liparis loeselii, Lobelia nuttallii, Lophiola aurea, Ludwigia linearis, Lycopodiella alopecuroides, L. caroliniana, Lycopus amplectens, Lysimachia terrestris, Minuartia caroliniana, Muhlenbergia torreyana, M. uniflora, Nymphoides cordata, Oclemena nemoralis, Panicum verrucosum, P. virgatum, Parnassia caroliniana, Platanthera blephariglottis, P. peramoena, Pogonia ophioglossoides, Polygala cruciata, P. lutea, P. nuttallii, Proserpinaca pectinata, Rhexia mariana, R. virginica, Rhynchospora alba, R. capitellata, R. cephalantha, R. chalarocephala, R. corniculata, R. fusca, R. glomerata, R. gracilenta, R. knieskernii, R. macrostachya, R. oligantha, R. perplexa, R. scirpoides, R. torreyana, Rotala ramosior, Rubus hispidus, Sabatia difformis, Saccharum baldwinii, S. giganteum, Sarracenia purpurea, Schizachyrium scoparium, Schoenoplectus hallii, S. pungens, Scirpus cyperinus, Scleria reticularis, Sclerolepis uniflora, Solidago ohioensis, S. riddellii, Spiraea alba, S. tomentosa, Stachys hyssopifolia, Symphyotrichum dumosum, S. novi-belgii, Taxodium, Thelypteris palustris, Triadenum virginicum, Trianthema racemosa, Typha latifolia, Utricularia* *cornuta, U. juncea, U. minor, U. subulata, Vaccinium corymbosum, V. macrocarpon, Viola lanceolata, Woodwardia virginica, Xyris difformis, X. smalliana, X. torta.*

Lobelia cardinalis L. is a short-lived perennial, which inhabits backwaters, bottomlands, cienega, depressions, ditches, fens, floodplains, gravelbars, hammocks, hanging gardens, marshes, meadows, prairies, roadsides, rockbars, sandbars, seeps, slopes, sloughs, springs, swamps, washes, and the margins of borrow pits, ponds, rivers, and streams at elevations of up to 2469 m. This is a broadly adapted species, which is found in fresh to oligohaline or mesohaline sites. The plants occur most typically in riparian habitats or in places (e.g., floodplain meadows or oxbows) that were associated formerly with riverine conditions. They tolerate fair amounts of standing water (to 30 cm) but also grow well in mesic sites. In the southwest, they can be found in association even with xeric species in areas where the water is ephemeral or has receded. Exposures vary from full sun to dense shade. The substrates range from acidic to alkaline (pH 5.8–7.8) and include alluvial sand, Chestatee stony sandy loam, clay, clay humus, clay loam, cobble, gravel, gravelly loam, gravelly sand, histosols, humus, limestone, Muav limestone, loam, muck, mucky peat, mud, rock, sand, sandy loam, sandy peat, sandy silt, sandstone, siliceous rocks, silt, silty clay loam, silty mud, and Wayland silt loam. The plants also have been observed growing on stumps. Flowering occurs from July through mid-September with fruiting in August through October. The flowers are self-compatible (producing normal seed quantities when selfed), but are strongly protandrous and are almost entirely outcrossed in nature. High levels of inbreeding depression (54%–83% decrease in net fertility) also have been demonstrated. As in other Campanulaceae, the pollen is plunged through a fused stamen tube, which envelops the style and its unexposed stigmatic surfaces. The pollen is released for several days before the elongating style emerges from the tube and reflexes to expose the stigmatic surfaces, which are receptive for 2–3 days. The flowers are pollinated exclusively by hummingbirds (Aves: Trochilidae), which are attracted to the scarlet red blooms and their nectar. The nectar contains about 16% sugar, but some populations apparently do not produce any. This reproductive system has a high capacity for long-distance gene flow, and experiments have shown that the transfer of genes from horticulturally grown cultivars to wild populations is possible. Some (but not all) natural populations have been shown to be pollen limited, with significantly higher seed production observed in flowers provided with supplemental pollen. In natural populations, the duration of staminate and pistillate floral phases is regulated by pollinator activity. The length of the staminate phase is inversely proportional to the extent of pollen removal with the pistillate phase inversely proportional to the duration of the staminate phase. The extent of seed production is determined mainly by the genotype of the maternal plant. Stressful conditions (10% ambient light) result in lower seed numbers and nectar production, but do not affect the pollen number or seed weight. Floral nectar production varies with date, is greater in staminate phase flowers, and decreases as the flowers age. It also is higher in plants having larger numbers of flowers. The small seeds are believed to be dispersed by flotation, or by their adherence to mud on the feet (or lodged in the feathers) of migrating birds (Aves). In some studies, the seeds required 9–12 days of light (95% germination), but no cold treatment, in order to germinate. Others have categorized the seeds as physiologically dormant, requiring a period of cold stratification and a 30°C/15°C temperature regime to germinate. They will not germinate when kept in the dark. The seeds retain their viability for up to 25 years when stored in sealed glass vials, or for more than 12 years when stored at −5°C in open containers. However, they have not been recovered from seed-bank samples taken in floodplain swamps. Individual plants propagate vegetatively by means of short rhizomes, and their roots are colonized by vesicular-arbuscular mycorrhizal fungi. Plants treated with growth regulators (600 ppm benzyladenine) produce significantly more basal branches than control plants. Although this is a widespread and broadly adapted species, attempts to reintroduce populations in Massachusetts were unsuccessful. It is believed that short periods of grazing might help to reestablish plants in some areas. **Reported associates:** *Acacia angustissima, Acalypha virginica, Acer leucoderme, A. negundo, A. rubrum, A. saccharinum, A. saccharum, Achnatherum hymenoides, Acourtia wrightii, Acrostichum danaeifolium, Adiantum capillus-veneris, Agave palmeri, A. schottii, A. utahensis, Ageratina herbacea, Agrostis exarata, Alhagi maurorum, Alisma subcordatum, Alnus oblongifolia, A. serrulata, Amaranthus cannabinus, Amauriopsis dissecta, Ambrosia artemisiifolia, A. confertiflora, A. psilostachya, Amelanchier canadensis, A. utahensis, Amorpha fruticosa, Ampelopsis arborea, Amphicarpaea bracteata, Amsonia palmeri, Anaphalis margaritacea, Andropogon gerardii, A. glomeratus, A. virginicus, Anemopsis californica, Apocynum cannabinum, Aquilegia chrysantha, A. desertorum, A. formosa, A. micrantha, Arbutus arizonica, A. xalapensis, Arctostaphylos, Arisaema triphyllum, Arnoglossum plantagineum, Aronia melanocarpa, Artemisia bigelovii, A. ludoviciana, A. tridentata, Arundinaria gigantea, Asclepias incarnata, A. perennis, A. subverticillata, A. syriaca, Asimina triloba, Asplenium platyneuron, Astragalus preussii, Athyrium filix-femina, Atriplex canescens, Baccharis neglecta, B. salicifolia, B. salicina, B. sarothroides, B. sergiloides, Bacopa monnieri, Berberis fremontii, B. haematocarpa, Berchemia scandens, Berula erecta, Betula alleghaniensis, B. nigra, B. occidentalis, Bidens aureus, B. connatus, B. discoideus, B. trichospermus, B. vulgatus, Bignonia capreolata, Boehmeria cylindrica, Bolboschoenus robustus, Boltonia asteroides, Bothriochloa barbinodis, Botrychium dissectum, Bouteloua aristidoides, B. curtipendula, B. eriopoda, B. gracilis, Brickellia amplexicaulis, B. californica, B. eupatorioides, B. grandiflora, B. longifolia, Bromus, Brunnichia ovata, Bryum, Calamagrostis canadensis, C. scopulorum, Callitriche heterophylla, Calycanthus floridus, Campsis radicans, Canotia holacantha, Cardamine bulbosa, Carex abscondita, C. alata, C. alma, C. annectens, C. atlantica, C. bromoides, C. crinita, C. crus-corvi, C. curatorum, C. debilis, C. folliculata, C. frankii,*

C. hyalinolepis, C. hysterica, C. intumescens, C. joorii, C. leptalea, C. lonchocarpa, C. lupuliformis, C. lupulina, C. lurida, C. muskingumensis, C. oxylepis, C. praegracilis, C. radiata, C. retroflexa, C. senta, C. seorsa, C. stipata, C. torta, Carpinus caroliniana, Carya aquatica, C. cordiformis, C. glabra, C. ovata, C. tomentosa, Ceanothus, Celtis laevigata, C. occidentalis, C. pallida, C. reticulata, Cephalanthus occidentalis, Cercis occidentalis, Cercocarpus intricatus, C. montanus, Chasmanthium latifolium, C. laxum, Cheilanthes eatonii, C. fendleri, Choisya dumosa, Cicuta maculata, Cinna arundinacea, Cirsium arizonicum, C. rydbergii, C. wrightii, Cladium californicum, Clematis virginiana, Commelina virginica, Conocephalum conicum, Conyza canadensis, Cornus amomum, C. florida, C. foemina, Cosmos bipinnatus, Crataegus viridis, Crinum americanum, Crossosoma, Croton, Cryptantha confertiflora, Cuscuta indecora, Cynodon dactylon, Cyperus filicinus, C. strigosus, Cypripedium kentuckiense, Dasylirion wheeleri, Datura, Decodon verticillatus, Decumaria barbara, Desmodium batocaulon, Dichanthelium boscii, D. clandestinum, Diodia virginiana, Diospyros virginiana, Doellingeria umbellata, Dulichium arundinaceum, Dyschoriste schiedeana, Echinocereus coccineus, E. pectinatus, Echinochloa walteri, Eclipta prostrata, Eleocharis geniculata, E. macrostachya, E. montevidensis, E. palustris, E. parvula, E. rostellata, Elephantopus carolinianus, E. tomentosus, Elodea, Elymus canadensis, E. lanceolatus, E. virginicus, Ephedra, Epilobium canum, Epipactis gigantea, Equisetum arvense, E. hyemale, E. laevigatum, Eragrostis lehmanniana, Ericameria nauseosa, Eriochloa villosa, Eryngium lemmonii, Eubotrys racemosa, Euonymus americanus, Eupatorium perfoliatum, E. serotinum, Euphorbia arizonica, Euthamia occidentalis, Eutrochium fistulosum, E. maculatum, Evolvulus, Fagus grandifolia, Fallopia japonica, Fallugia paradoxa, Ferocactus wislizeni, Fimbristylis puberula, Fouquieria, Forestiera acuminata, F. pubescens, Fraxinus americana, F. anomala, F. caroliniana, F. nigra, F. pennsylvanica, F. profunda, F. velutina, Fuirena simplex, Galium obtusum, Garrya flavescens, Gelsemium rankinii, Gentiana andrewsii, Gentianella amarella, Gentianopsis detonsa, Geranium, Glandularia gooddingii, Gleditsia triacanthos, Glyceria canadensis, G. septentrionalis, G. striata, Gratiola virginiana, Gutierrezia sarothrae, Gymnadeniopsis clavellata, Halesia diptera, Hamamelis virginiana, Hedeoma, Helenium autumnale, Helianthus annuus, Hemerocallis fulva, Herrickia glauca, H. wasatchensis, Heterotheca subaxillaris, H. villosa, Hibiscus laevis, H. moscheutos, Hydrocotyle verticillata, Hymenoxys hoopesii, Hypericum mutilum, H. punctatum, Hyptis alata, Ilex cassine, I. decidua, I. opaca, I. verticillata, I. vomitoria, Impatiens capensis, Imperata brevifolia, Ipomoea, Ipomopsis thurberi, Iris versicolor, I. virginica, Iresine angustifolia, Isocoma acradenia, Itea virginica, Jatropha macrorhiza, Juglans cinerea, J. major, J. microcarpa, Juncus balticus, J. brachycephalus, J. coriaceus, J. dudleyi, J. ensifolius, J. effusus, J. pylaei, J. tenuis, J. torreyi, J. xiphioides, Juniperus coahuilensis, J. deppeana, J. monosperma, J. osteosperma, J. scopulorum, J. virginiana, Justicia americana, J. ovata, Kosteletzkya pentacarpos, Laportea, Lasianthaea podocephala, Leersia lenticularis, L. oryzoides, L. virginica, Liatris spicata, Ligusticum canadense, Limnobium spongia, Lindera benzoin, Lindernia dubia, Linum striatum, Liquidambar styraciflua, Liriodendron tulipifera, Lobelia amoena, L. anatina, L. elongata, L. inflata, L. puberula, L. siphilitica, Lonicera, Ludwigia alternifolia, L. grandiflora, L. octovalvis, L. palustris, L. repens, Lycopus americanus, L. rubellus, L. virginicus, Lyonia ligustrina, Lysimachia maritima, L. nummularia, L. quadriflora, Lythrum californicum, Macroptilium, Magnolia grandiflora, M. macrophylla, M. tripetala, Mammillaria heyderi, Mandevilla, Marrubium vulgare, Medeola virginiana, Menispermum canadense, Mentha arvensis, M. spicata, Mentzelia puberula, Microstegium vimineum, Mikania scandens, Mimosa dysocarpa, Mimulus alatus, M. cardinalis, M. ringens, Mitchella repens, Morus rubra, Muhlenbergia asperifolia, M. californica, M. emersleyi, M. rigens, M. schreberi, M. thurberi, Murdannia keisak, Myosoton aquaticum, Myrica cerifera, Najas, Nasturtium microphyllum, Nitrophila, Nolina microcarpa, Nuphar, Nyssa aquatica, N. biflora, N. sylvatica, Oenothera elata, O. longissima, Onoclea sensibilis, Opuntia basilaris, O. engelmannii, O. phaeacantha, O. polyacantha, Orontium aquaticum, Osmunda regalis, Osmundastrum cinnamomeum, Ostrya virginiana, Oxydendrum arboreum, Oxypolis rigidior, Oxytenia acerosa, Packera aurea, Panicum virgatum, Parnassia grandifolia, Parthenocissus quinquefolia, Paspalum distichum, Pellaea truncata, Peltandra virginica, Penstemon digitalis, Penthorum sedoides, Pericome caudata, Perityle ciliata, P. congesta, Persea palustris, Persicaria arifolia, P. hydropiper, P. hydropiperoides, P. lapathifolia, P. longiseta, P. maculosa, P. pensylvanica, P. punctata, P. sagittata, P. setacea, P. virginiana, Petrophytum caespitosum, Phalaris arundinacea, Phanopyrum gymnocarpon, Phegopteris hexagonoptera, Phlox glaberrima, Phragmites australis, Phyla lanceolata, Physalis hederifolia, Physostegia virginiana, Phytolacca americana, Pilea pumila, Pinus cembroides, P. discolor, P. echinata, P. edulis, P. leiophylla, P. ponderosa, P. taeda, Planera aquatica, Plantago, Platanthera ciliaris, P. flava, P. peramoena, Platanus occidentalis, P. wrightii, Pleopeltis polypodioides, Pluchea odorata, P. sericea, Poa nemoralis, Polystichum acrostichoides, Pontederia cordata, Populus balsamifera, P. deltoides, P. fremontii, P. heterophylla, Potamogeton nodosus, Prosopis velutina, Prunella vulgaris, Prunus serotina, Purshia mexicana, Pseudognaphalium canescens, Pteridium aquilinum, Ptilimnium capillaceum, Pycnanthemum virginianum, Quercus alba, Q. bicolor, Q. emoryi, Q. falcata, Q. gambelii, Q. grisea, Q. hypoleucoides, Q. lyrata, Q. laurifolia, Q. macrocarpa, Q. michauxii, Q. nigra, Q. oblongifolia, Q. pagoda, Q. palmeri, Q. palustris, Q. phellos, Q. rugosa, Q. shumardii, Q. turbinella, Ranunculus flabellaris, R. sceleratus, Rhamnus betulifolia, R. californica, R. cathartica, Rhododendron alabamense, R. canescens, R. maximum, R. viscosum, Rhus aromatica, R. microphylla, R. virens, Rhynchospora capitellata, R. colorata, R. corniculata, R. inundata, R. miliacea, R. nivea, Ribes americanum, Rubus

hispidus, Rudbeckia fulgida, Ruellia caroliniensis, Sabal minor, S. palmetto, Sabatia angularis, Sagittaria latifolia, Salix bonplandiana, S. eriocephala, S. exigua, S. gooddingii, S. humilis, S. laevigata, S. lasiolepis, S. nigra, S. sericea, S. taxifolia, Sambucus nigra, Samolus valerandi, Sanguisorba canadensis, Sanicula canadensis, Saponaria officinalis, Sassafras albidum, Saururus cernuus, Schizachyrium scoparium, Schoenoplectus americanus, S. tabernaemontani, Schoenus nigricans, Scirpus atrovirens, S. caroliniana, S. cyperinus, S. expansus, S. polyphyllus, Scutellaria galericulata, S. lateriflora, Senegalia greggii, Setaria parviflora, Shepherdia rotundifolia, Silene, Silphium asteriscus, Sium suave, Smilax bona-nox, S. walteri, Solanum americanum, S. douglasii, S. dulcamara, S. elaeagnifolium, Solidago altissima, S. canadensis, S. confinis, S. gigantea, S. nana, S. patula, S. rugosa, S. velutina, Sorghastrum nutans, Spartina alterniflora, S. pectinata, Sphaeralcea, Sphagnum, Sphenopholis pensylvanica, Spiranthes odorata, Stachys hispida, Staphylea trifolia, Stephanomeria, Symphyotrichum dumosum, S. ericoides, S. laeve, S. lanceolatum, S. lateriflorum, S. praealtum, S. puniceum, S. racemosum, Symplocos tinctoria, Tamarix chinensis, Taxodium distichum, Teucrium canadense, Thelypodium integrifolium, T. wrightii, Thelypteris palustris, T. noveboracensis, Torreyochloa pallida, Toxicodendron radicans, Triadenum walteri, Triglochin maritimum, Trixis californica, Tsuga canadensis, Tussilago farfara, Typha angustifolia, T. domingensis, T. latifolia, Ulmus alata, U. americana, U. rubra, Urtica dioica, Utricularia, Vaccinium corymbosum, V. elliottii, V. pallidum, Verbena hastata, Vernonia fasciculata, V. noveboracensis, Viburnum dentatum, V. lentago, V. nudum, Viguiera cordifolia, Viola primulifolia, Vitis arizonica, V. girdiana, V. riparia, V. rotundifolia, Wisteria frutescens, Woodwardia areolata, W. fimbriata, Xanthisma spinulosum, Xanthium, Xylorhiza tortifolia, Yucca baccata, Y. elata, Y. madrensis, Zanthoxylum americana, Zornia.

Lobelia dortmanna L. is a perennial, which grows submersed in fresh, soft water, oligotrophic lakes and ponds at depths from 0.1 to 3.0 m (average: 1.5 m) at elevations of up to 746 m. The waters lack perceptible flow, and have low alkalinity (<70 mg/L total alkalinity as $CaCO_3$), low conductivity (<130 μmhos/cm), and high light penetration (e.g., secchi disk = 6.5 m). Acidity can range from pH: 4.4–8.9. Nutrient ranges (all in μeq/L) recorded for several acidic sites (pH 4.4–6.0) in Nova Scotia include: Al (2–28); Ca (15–39); Cl (107–138); Fe (1–11); HCO_3 (0–12); K (0–8); Mg (25–39); Na (104–139); NH_4 (0.7–3.2); SO_4 (52–89). Increased alkalinity and SO_4^{2-} can affect populations negatively by increasing the extent of sediment mineralization, which causes anoxia. The plants are highly freeze-tolerant, but are intolerant to drought or eutrophic conditions. They occur normally along the shallow littoral zone of lakes, and grow poorly when exposed on shorelines. They colonize substrates of gravel, mucky gravel, mud, sand, sandy gravel, sandy peat, silt, or stones. The substrates sometimes are covered by a layer of organic detritus (to 2.5 cm). Most occurrences are on sand. The plants also have been seen growing on pieces of driftwood. Individuals persist as completely submersed, perennial, evergreen rosettes, and only their flower stalks (which can reach 2 m in length) become emergent during sexual reproduction. New plants take from 2 to 5 years to reach flowering, which occurs from May to October but peaks in July–August. Floral initiation correlates with a 19°C temperature. Both cleistogamous (submersed) and chasmogamous (emergent) flowers are produced. The latter are protandrous and possibly are insect-pollinated (entomophilous) on occasion, but both types predominately are self-pollinating. Unlike its congeners, the stigma remains enclosed within the anther tube. One flower typically produces 41–175 (max.=334; avg.=118) seeds, which are dispersed from mid-June through summer. The seeds float for only a short period and then sink rapidly, but they do not settle below 5 mm in silty sediments. Seed germination occurs from July through autumn. Some seeds require no pretreatment for germination, while others (even in the same population) require at least 1 month of cold stratification (at 1°C–3°C) in order to break dormancy. The seeds must receive at least minimal light for germination (e.g., 4 μmol photons/m²/sec), which will not occur in complete darkness. Germination rates are higher under flooded (28%) than nonflooded (10%) conditions; however, oxygen deficiency (caused by flooding) can induce secondary dormancy, requiring 30–180 days of cold stratification for germination at 18°C–25°C. One protocol reported 83% germination (21°C, 12 h light) for seeds pretreated in high humidity for 3 weeks at 2°C, then placed on 1% agar for 4 weeks at 6°C. Seeds can attain densities of 41/m² in seed banks, which have been described either as short-term or persisting for >20 years in duration. Dried and frozen seeds can be stored for at least several months. The plants undergo C_3 photosynthesis with maximum rates measured at 2.5 mg O_2/g fresh wt/hr, which is a very low level compared to other submersed aquatic plants. They do not undergo any crassulacean acid metabolism (CAM)-like metabolism. The plants grow slowly and have an extremely low biomass production ratio. They can use ammonium as a nitrogen source but have higher uptake rates for nitrate. Most of their CO_2 is taken up from the sediment by the roots, and is translocated through an extensive internal lacunal system to the leaf mesophyll. The plants cannot use bicarbonate as a carbon source. The mature leaves contain 20–30 mg/g total N and 1–2 mg/g total P. Phosphorous, and up to 83% of inorganic nitrogen, are taken up by the root system rather than from the water column. Nutrients are moved internally through the plant by acropetal water transport, which is driven by root pressure and guttation through specialized leaf hydathodes. In sediments high in reduced Fe and Mn and organic matter, Fe and Mn root plaques can form, which restrict P uptake and concomitantly reduce biomass production. The plants release virtually all of their photosynthetic O_2 to the sediments. This action promotes oxidation and may assist in nutrient uptake, but also presents the possibility of nighttime oxygen depletion. The condition is exacerbated in sediments having an increased organic matter content, which results in depleted sediment O_2 levels and NH_4^+, Fe^{2+}, and CO_2 accumulation. Consequently, levels of organic matter tend to be relatively low, with one study finding maximum

biomass accumulation across an organic matter gradient to occur at 4.23% organic matter content. About 60% of the maximum summer biomass overwinters, but little additional growth occurs from October to early May. Vegetative propagation occurs by axillary buds that develop at the base of old flowering shoots. In one study, 98% of the annual population increase resulted from vegetative reproduction. The plants also have been known to produce small vegetative plantlets instead of flowers. The roots can become colonized by arbuscular mycorrhizal fungi. **Reported associates:** *Brasenia schreberi, Callitriche heterophylla, Chara, Drosera intermedia, Elatine minima, Eleocharis acicularis, E. palustris, Elodea canadensis, E. nuttallii, Equisetum fluviatile, Eriocaulon aquaticum, Gratiola aurea, Heteranthera dubia, Isoetes ×eatoni, I. echinospora, I. lacustris, Juncus militaris, J. pelocarpus, J. supiniformis, Littorella uniflora, Ludwigia palustris, Myriophyllum elatinoides, M. tenellum, Najas flexilis, Nitella, Nuphar polysepala, Nymphaea odorata, Nymphoides cordata, Pontederia cordata, Potamogeton amplifolius, P. bicupulatus, P. epihydrus, P. gramineus, P. natans, P. robbinsii, P. spirillus, Ranunculus aquatilis, R. flammula, Rhynchospora fusca, Sagittaria graminea, Schoenoplectus pungens, S. subterminalis, S. torreyi, Sparganium americanum, S. angustifolium, S. emersum, S. fluctuans, Subularia aquatica, Utricularia cornuta, U. gibba, U. purpurea, U. radiata, U. resupinata.*

***Lobelia elongata* Small** is a perennial, which occurs primarily along the coastal plain in fresh to oligohaline, tidal or nontidal sites including bogs, bottomland, ditches, flatwoods, floodplains, marshes, roadsides, savannas, seeps, slopes, sloughs, swamps, and the margins of borrow pits, ponds, rivers, and streams at low elevations. Exposures can range from full sun to shade. The substrates (e.g., pH 6.2) include clay, marl, muck, mud, and sand. Flowering extends from July through October. Pollination has been attributed to butterflies (Insecta: Lepidoptera), but without adequate documentation. Additional life-history information should be obtained for this rare species, which often occurs at sites containing other rare plants. **Reported associates:** *Acer rubrum, Amaranthus cannabinus, Amsonia tabernaemontana, Arundinaria tecta, Asclepias incarnata, Bidens bipinnatus, Bignonia, Boehmeria cylindrica, Bolboschoenus robustus, Campsis, Carex crebriflora, C. decomposita, C. reniformis, Carpinus carolinianus, Chamaecyparis thyoides, Chrysopsis gossypina, Cladium jamaicense, C. mariscoides, Cleistesiopsis divaricata, Coleataenia longifolia, Cornus foemina, Cyperus haspan, Cyrilla racemiflora, Decodon verticillatus, Dichanthelium dichotomum, D. polyanthes, Eleocharis fallax, E. radicans, E. rostellata, E. uniglumis, Elephantopus carolinianus, Erigeron vernus, Eriocaulon decangulare, Eryngium aquaticum, Euphorbia bombensis, Fimbristylis caroliniana, Galium bermudense, Gratiola neglecta, Hibiscus moscheutos, Houstonia purpurea, Hypoxis curtissii, Ilex opaca, Iresine rhizomatosa, Itea virginica, Iva imbricata, Juncus elliottii, J. megacephalus, J. roemerianus, Kalmia angustifolia, Lilaeopsis carolinensis, Limosella australis, Lobelia cardinalis, Ludwigia alata, L. brevipes, Lythrum lineare, Mikania scandens,*

Myrica cerifera, Nyssa aquatica, N. sylvatica, Panicum virgatum, Paspalum distichum, Perilla frutescens, Persicaria, Phalaris caroliniana, Phyla lanceolata, Phragmites australis, Phyla nodiflora, Physalis viscosa, Physostegia leptophylla, Planera aquatica, Pontederia cordata, Ptilimnium capillaceum, Quercus hemisphaerica, Rhynchospora colorata, R. fascicularis, R. macrostachya, Rosa palustris, Saccharum baldwinii, Sagittaria graminea, S. lancifolia, Schoenoplectus americanus, S. pungens, Scleria ciliata, Sium suave, Smilax, Spartina alterniflora, S. cynosuroides, Spiranthes odorata, Stachys aspera, Stewartia malacodendron, Symphyotrichum elliotii, S. subulatum, Taxodium distichum, Teucrium canadense, Tillandsia usneoides, Trachelospermum jasminoides, Typha angustifolia, T. latifolia, Vaccinium elliottii, V. macrocarpon, Vitis.

***Lobelia flaccidifolia* Small** is an annual, which inhabits bottomlands, depressions, ditches, flatwoods, floodplains, prairies, roadsides, savannas, swales, swamps and the margins of ponds, rivers, and streams at low elevations along the Coastal Plain. The plants grow in full sun to shaded sites on substrates described as alluvium, Caddo (typic glossaqualf), Guyton (typic glossaqualf), Kinder (typic glossaqualf), loam, loamy sand, mud, sand, and sandy peat. Flowering occurs from June (earlier than most of its other congeners) through September. Little additional life-history information has been published on this species. **Reported associates:** *Aletris aurea, Andropogon glomeratus, A. liebmannii, Aristida palustris, Arnoglossum ovatum, Asclepias longifolia, Bidens mitis, Centella asiatica, Chaptalia tomentosa, Coleataenia longifolia, Coreopsis gladiata, Ctenium aromaticum, Dichanthelium acuminatum, D. scabriusculum, D. scoparium, Diospyros virginiana, Eleocharis microcarpa, E. tuberculosa, Eriocaulon decangulare, E. texense, Eryngium integrifolium, Eupatorium leucolepis, Euthamia leptocephala, Fuirena bushii, Gratiola brevifolia, G. pilosa, Gymnadeniopsis nivea, Helenium drummondii, Helianthus angustifolius, Hypericum brachyphyllum, H. galioides, Hyptis alata, Juncus validus, Lycopodiella appressa, Marshallia caespitosa, M. graminifolia, Mayaca fluviatilis, Muhlenbergia expansa, Panicum virgatum, Pinus palustris, Pluchea baccharis, Polygala lutea, P. ramosa, Polypremum procumbens, Proserpinaca pectinata, Rhexia lutea, Rhynchospora elliottii, R. fascicularis, R. filifolia, R. gracilenta, R. latifolia, R. microcarpa, R. perplexa, R. plumosa, R. pusilla, R. rariflora, Sabatia campanulata, Saccharum giganteum, Scleria georgiana, S. muehlenbergii, S. pauciflora, Schoenolirion croceum, Stylisma aquatica, Xyris stricta.*

***Lobelia floridana* Chapm.** is a perennial, which grows in barrens, bogs, borrow pits, depressions, ditches, flatwoods, meadows, mudflats, pocosins, roadsides, savannas, swamps, and along the margins of streams at low elevations along the coastal plain. The plants occur commonly in shallow standing water (to 15 cm deep) and grow on substrates described as loamy sand, muck, peaty sand, and sandy peat. The flowers appear from March through September and are pollinated by bees (see *Use by wildlife*). Perennation is by a short caudex. Other details on the life history of this species are

lacking. **Reported associates:** *Agalinis fasciculata, A. linifolia, Agrostis hyemalis, Aletris lutea, Amphicarpum muhlenbergianum, Andropogon gyrans, A. virginicus, Aristida palustris, A. purpurascens, A. stricta, Arnoglossum ovatum, Asclepias lanceolata, Balduina uniflora, Bidens mitis, Bigelowia nudata, Briza minor, Buchnera americana, Burmannia capitata, Carphephorus odoratissimus, Chaptalia tomentosa, Centella asiatica, Clethra alnifolia, Cliftonia monophylla, Coleataenia longifolia, C. tenera, Coreopsis gladiata, C. lanceolata, C. nudata, Ctenium aromaticum, Cyrilla racemiflora, Dichanthelium aciculare, D. acuminatum, D. dichotomum, D. scoparium, D. sphaerocarpon, Diodia virginiana, Drosera capillaris, Eleocharis tuberculosa, Elephantopus nudatus, Eragrostis elliottii, E. refracta, Erigeron vernus, Eriocaulon compressum, E. decangulare, Eryngium integrifolium, E. yuccifolium, Eupatorium album, E. mohrii, E. semiserratum, Euthamia graminifolia, Eutrochium fistulosum, Gymnopogon brevifolius, Helenium flexuosum, H. vernale, Helianthus angustifolius, H. heterophyllus, H. radula, Hydrocotyle umbellata, Hypericum brachyphyllum, H. chapmanii, H. cistifolium, H. fasciculatum, H. galioides, H. gymnanthum, H. microsepalum, H. myrtifolium, Hyptis alata, Ilex glabra, I. myrtifolia, Juncus biflorus, J. dichotomus, J. elliottii, J. polycephalus, Lachnanthes caroliniana, Lespedeza repens, Liatris chapmanii, L. spicata, Lobelia boykinii, L. glandulosa, Lophiola aurea, Ludwigia linearis, L. pilosa, L. virgata, Lyonia lucida, Macbridea, Mnesithea rugosa, Nyssa biflora, Oxypolis filiformis, O. greenmanii, Paspalum circulare, P. floridanum, P. setaceum, Pinguicula planifolia, Pinus elliottii, P. palustris, P. taeda, Pityopsis graminifolia, Pleea tenuifolia, Pluchea odorata, Polygala cruciata, P. cymosa, P. lutea, P. ramosa, Rhexia alifanus, R. lutea, R. mariana, R. nuttallii, R. virginica, Rhynchospora careyana, R. corniculata, R. filifolia, R. glomerata, R. gracilenta, R. harperi, R. inexpansa, R. plumosa, R. pusilla, R. wrightiana, Rudbeckia graminifolia, R. mohrii, Ruellia ciliosa, Sabatia brevifolia, S. decandra, S. stellaris, Sagittaria graminea, Salvia lyrata, Sarracenia flava, S. psittacina, S. rosea, Schizachyrium sanguineum, S. scoparium, Scleria baldwinii, S. ciliata, S. muehlenbergii, Sericocarpus tortifolius, Seymeria cassioides, Solidago fistulosa, S. odora, Sphagnum, Spiranthes longilabris, Sporobolus, Stenanthium densum, Stylosanthes biflora, Symphyotrichum chapmanii, Taxodium ascendens, Tradescantia ohiensis, Triantha racemosa, Trichostema dichotomum, Utricularia subulata, Vernonia gigantea, Viola lanceolata, Woodwardia virginica, Xyris ambigua, X. caroliniana, X. difformis, X. torta.*

***Lobelia glandulosa* Walter** is a perennial, which grows in freshwater tidal or nontidal sites including barrens, bogs, depressions, ditches, flatwoods, glades, hammocks, marshes, meadows, pocosins, prairies, roadsides, savannas, seeps, sloughs, swales, swamps, and the margins of bays, canals, ponds, and streams at low elevations (e.g., 29 m) along the coastal plain and outer piedmont. The plants sometimes are found growing with their bases in shallow standing water. Exposures vary from fully open to deeply shaded sites. The substrates generally are described as calcareous and include clay loam, loam, loamy sand, Okeelanta–Terra Ceia (medisaprists), Pelham (arenic paleaquults), peat, peaty sand, Perrine marl, sand, sandy alluvium, sandy loam, and sandy peat. This species flowers quite late in the season (September–November) but is reported to remain in bloom for most of the year. The flowers attract hummingbirds (see *Use by wildlife*); however, it is not certain if they function as pollinators. Other aspects of the reproductive biology and seed ecology of this species remain unknown. The plants perennate from a slender caudex. Their resistance to fire is uncertain. They have been collected in frequently burned areas but also have disappeared from sites in years subsequent to burns.

Reported associates: *Acer rubrum, Agalinis filifolia, A. linifolia, A. purpurea, Andropogon arctatus, A. gerardii, A. glomeratus, A. gyrans, A. longiberbis, A. virginicus, Aristida palustris, A. spiciformis, A. stricta, Arnoglossum ovatum, Asclepias lanceolata, Axonopus furcatus, Balduina uniflora, Bartonia, Berchemia scandens, Bidens aristosus, Bigelowia nudata, Burmannia biflora, Callitriche terrestris, Cassytha filiformis, Centella asiatica, Cirsium nuttallii, Cladium jamaicense, Coleataenia longifolia, C. tenera, Coreopsis gladiata, Crinum americanum, Ctenium aromaticum, Cyperus polystachyos, Cyrilla racemiflora, Dichanthelium acuminatum, D. dichotomum, D. strigosum, Diodia virginiana, Drosera brevifolia, D. capillaris, Dyschoriste angusta, Echinochloa muricata, Eleocharis flavescens, Elytraria caroliniensis, Eragrostis, Erigeron vernus, Eriocaulon decangulare, Eryngium integrifolium, Eupatorium mikanioides, Euphorbia polyphylla, E. porteriana, Eustachys glauca, Flaveria linearis, Fuirena breviseta, F. squarrosa, Helenium autumnale, H. pinnatifidum, H. vernale, Hydrocotyle umbellata, Hymenocallis palmeri, Hypericum cistifolium, H. fasciculatum, H. gentianoides, H. hypericoides, Hyptis alata, Ilex glabra, Juncus trigonocarpus, Justicia angusta, Lachnocaulon anceps, Leersia virginica, Liatris garberi, L. spicata, Linum carteri, Lipocarpha micrantha, Lobelia amoena, L. brevifolia, L. feayana, L. floridana, Lophiola aurea, Ludwigia alternifolia, L. glandulosa, L. microcarpa, Lycopodiella appressa, Lycopus rubellus, Lyonia, Magnolia virginiana, Mayaca fluviatilis, Melanthera angustifolia, Microstegium vimineum, Mikania scandens, Mitreola petiolata, M. sessilifolia, Mnesithea tessellata, Muhlenbergia expansa, M. sericea, Myrica cerifera, Nyssa biflora, N. sylvatica, Oxypolis filiformis, Panicum hemitomon, P. virgatum, Parietaria, Paspalum praecox, Phyla nodiflora, Pinguicula pumila, Pinus elliottii, P. palustris, Pleea tenuifolia, Pluchea baccharis, P. foetida, Polygala cymosa, P. lutea, P. ramosa, Polypremum procumbens, Proserpinaca pectinata, Pterocaulon pycnostachyum, Rhexia alifanus, R. lutea, R. petiolata, Rhus copallinum, Rhynchospora colorata, R. corniculata, R. divergens, R. globularis, R. inundata, R. latifolia, R. macra, R. microcarpa, R. oligantha, R. perplexa, Rubus trivialis, Ruellia caroliniensis, Sabal palmetto, Sabatia dodecandra, S. macrophylla, Saccharum giganteum, Sarracenia leucophylla, Schizachyrium rhizomatum, Schoenus nigricans, Scleria muehlenbergii, S. pauciflora, Serenoa repens, Seymeria cassioides, Smilax laurifolia, Spermacoce remota,* *Sphagnum, Stenandrium dulce, Stenaria nigricans, Stillingia, Symphyotrichum adnatum, S. chapmanii, S. dumosum, Syngonanthus flavidulus, Taxodium ascendens, T. distichum, Toxicodendron radicans, Triadenum virginicum, T. walteri, Utricularia subulata, Vernonia blodgettii, Vitis rotundifolia, Xyris caroliniana, X. elliottii, X. jupicai.*

Lobelia kalmii L. is a biennial, which inhabits alvar, beaches, bogs, borrow pits, carrs, cliffs, crevices, depressions, ditches, dunes, fens, flats, floating mats, floodplains, hot springs (to 32°C–36°C), ledges, marshes, meadows, mudflats, pannes, prairies, prairie fens, roadsides, sandbars, seeps, shores, sloughs, springs, streams, string bogs (in flarks), swales, swamps, thickets, and along the margins of lakes, ponds, rivers, and streams at elevations of up to 714 m. Although this calciphile is regarded widely as a fen indicator, it also occupies various habitats with mineral substrates, especially when calcareous. Most occurrences are in full sun but can extend to partial shade. Substrates typically are alkaline (pH: 6.0–9.0; ~300 mg/L $CaCO_3$) and include basalt, cobble, dolomite, gravel, humus, limestone, marl, marly clay, muck, mucky clay, mud, peat, rock, sand, silty gravel, and tufa. The plants are in flower from 31 July to 13 October. The flowers are protandrous and expel their pollen prior to the emergence and receptivity of the stigmas. They are pollinated primarily by bees (see *Use by wildlife*); however, high seed set, even where pollinators are scarce, have indicated that self-pollination may occur as well. The fruits contain several hundred seeds. Their dispersal mechanism is uncertain. The seeds reportedly will germinate in 15–20 days at 22°C after 3 months of cold stratification at 4°C–5°C. In one study, plants increased greatly over a 4-year period of sand accretion at the site; however, this response could have represented their initially low frequency at the study onset. A higher abundance of plants also has been observed at sites grazed by white-tailed deer (Mammalia: Cervidae: *Odocoileus virginianus*) relative to ungrazed sites. **Reported associates:** *Acer rubrum, Agalinis purpurea, A. tenuifolia, Agrostis capillaris, A. gigantea, Allium cernuum, A. schoenoprasum, Alnus rugosa, A. viridis, Amelanchier humilis, Andromeda polifolia, Andropogon gerardii, Anemone parviflora, Aneura pinguis, Antennaria, Anticlea elegans, Apocynum cannabinum, Aquilegia canadensis, Arnoglossum plantagineum, Aronia melanocarpa, Asclepias incarnata, Aulacomnium palustre, Berula erecta, Betula pumila, B. ×sandbergii, Bidens cernuus, B. trichospermus, Boehmeria cylindrica, Bromus ciliatus, B. kalmii, Bryum pseudotriquetrum, Calamagrostis canadensis, C. stricta, Calliergon giganteum, C. trifarium, Calliergonella cuspidata, Calopogon tuberosus, Calystegia sepium, Campanula aparinoides, C. rotundifolia, Campylium polygamum, C. stellatum, Cardamine pratensis, Carex aquatilis, C. atherodes, C. aurea, C. buxbaumii, C. castanea, C. conoidea, C. crawei, C. cryptolepis, C. diandra, C. eburnea, C. exilis, C. flava, C. garberi, C. granularis, C. haydenii, C. hystericina, C. interior, C. lacustris, C. lasiocarpa, C. leptalea, C. limosa, C. livida, C. pellita, C. prairea, C. rostrata, C. sartwellii, C. sterilis, C. stricta, C. tenuiflora, C. tetanica, C. trisperma, C. viridula, Castilleja coccinea, Chamaedaphne* *calyculata, Chara vulgaris, Chelone glabra, Cicuta bul-bifera, C. maculata, Cirsium muticum, Cladium maris-coides, Climacium, Clinopodium glabellum, C. glabrum, Collema fuscovirens, C. tenax, Comarum palustre, Coreopsis tripteris, Cornus amomum, C. foemina, C. sericea, Cyperus bipartitus, Cypripedium candidum, C. parviflorum, C. regi-nae, Cystopteris bulbifera, Dasiphora floribunda, Decodon verticillatus, Deschampsia cespitosa, Dichanthelium acu-minatum, Doellingeria umbellata, Drepanocladus, Drosera rotundifolia, Dryas integrifolia, Dulichium arundinaceum, Eleocharis elliptica, E. erythropoda, E. palustris, E. quin-queflora, E. rostellata, E. tenuis, Epilobium leptophyllum, E. strictum, Epipactis gigantea, Equisetum arvense, E. flu-viatile, E. variegatum, Erigeron hyssopifolius, Eriophorum angustifolium, E. virginicum, E. viridicarinatum, Eupatorium perfoliatum, Euthamia graminifolia, E. gymnospermoides, Eutrochium maculatum, Filipendula rubra, Fissidens adianthoides, Fragaria vesca, Galium asprellum, G. boreale, G. labradoricum, G. trifidum, Gentiana andrew-sii, Gentianopsis crinita, G. virgata, Geum rivale, Gillenia trifoliata, Glyceria grandis, G. septentrionalis, G. striata, Graphephorum melicoides, Grimmia, Helenium autumnale, Helianthus nuttallii, Houstonia longifolia, Hypericum kal-mianum, Ilex verticillata, Ionactis linariifolia, Iris versicolor, Juncus alpinus, J. articulatus, J. balticus, J. brachycepha-lus, J. dudleyi, J. nodosus, J. stygius, J. torreyi, J. triglumis, Juniperus horizontalis, J. virginiana, Kalmia polifolia, Larix laricina, Lathyrus palustris, Lechea, Leptodictyum riparium, Liatris ligulistylis, L. pycnostachya, L. spicata, Lilium phila-delphicum, Limprichtia revolvens, Linaria vulgaris, Linum virginianum, Liparis loeselii, Lobelia siphilitica, L. spicata, Lonicera oblongifolia, Lycopus americanus, L. uniflorus, Lysimachia quadriflora, L. thyrsiflora, Lythrum alatum, L. salicaria, Meesia triquetra, Melilotus albus, Mentha arven-sis, Menyanthes trifoliata, Mimulus glabratus, Monarda fistulosa, Muhlenbergia glomerata, M. racemosa, M. rich-ardsonis, Myrica gale, M. pensylvanica, Onoclea sensibilis, Osmunda regalis, Oxypolis rigidior, Oxytropis campestris, Packera aurea, P. indecora, P. paupercula, P. schweinitziana, Paludella squarrosa, Panicum flexile, Parnassia caroliniana, P. glauca, P. palustris, Pedicularis canadensis, P. lanceolata, Phalaris arundinacea, Philonotis fontana, Phlox glaber-rima, Phragmites australis, Physaria arctica, Physocarpus opulifolius, Physostegia virginiana, Picea mariana, Pilea fontana, P. pumila, Pinguicula vulgaris, Platanthera dila-tata, P. hyperborea, P. psycodes, Poa alpina, P. compressa, Pogonia ophioglossoides, Potentilla anserina, P. nivea, Prenanthes racemosa, Primula mistassinica, Proserpinaca palustris, Pycnanthemum virginianum, Ranunculus septen-trionalis, Rhamnus alnifolia, R. frangula, Rhynchospora alba, R. capillacea, R. capitellata, Rosa blanda, Rubus pubescens, Rudbeckia fulgida, R. hirta, Sabatia angula-ris, Sagittaria graminea, Salix ×bebbii, S. candida, S. dis-color, S. eriocephala, S. interior, S. lucida, S. pedicellaris, S. sericea, S. vestita, Sanguisorba canadensis, Sarracenia purpurea, Saxifraga oppositifolia, Schizachyrium scopar-ium, Schoenoplectus acutus, S. americanus, S. pungens,* *S. subterminalis, S. tabernaemontani, Scirpus atrovirens, S. cyperinus, S. lineatus, Scleria verticillata, Scorpidium scorpioides, Scutellaria galericulata, Selaginella apoda, S. selaginoides, Sibbaldiopsis tridentata, Silphium terebin-thinaceum, Sisyrinchium angustifolium, Smilax tamnoides, Solidago bicolor, S. canadensis, S. gigantea, S. nemoralis, S. ohioensis, S. patula, S. ptarmicoides, S. riddellii, S. uligi-nosa, Sorghastrum nutans, Spartina pectinata, Sphagnum contortum, S. teres, S. warnstorfii, Sphenopholis intermedia, S. obtusata, Spiraea alba, S. tomentosa, Spiranthes cernua, S. lucida, S. romanzoffiana, Sporobolus neglectus, Stachys palustris, Symphyotrichum boreale, S. dumosum, S. ericoi-des, S. firmum, S. laeve, S. lanceolatum, S. novae-angliae, S. pilosum, S. puniceum, S. spathulatum, Symplocarpus foetidus, Thalictrum dasycarpum, T. pubescens, Thelypteris palustris, Thuja occidentalis, Tomentypnum nitens, Tortella tortuosa, Tortula ruralis, Toxicodendron rydbergii, T. ver-nix, Triadenum fraseri, T. virginicum, Triantha glutinosa, Trichophorum alpinum, T. cespitosum, T. clintonii, Triglochin maritimum, T. palustre, Trisetum spicatum, Trollius laxus, Tussilago farfara, Typha latifolia, Utricularia cornuta, U. gibba, U. intermedia, U. minor, Vaccinium angustifolium, V. cespitosum, V. macrocarpon, V. oxycoccos, Valeriana edulis, Verbena hastata, Viburnum lentago, Viola cucullata, V. pal-mata, V. sororia, Vitis vulpina, Zizia aurea.*

Lobelia paludosa Nutt. occurs in barrens, depressions, ditches, dunes, flats, flatwoods, floodplains, marshes, prai-ries, roadsides, savannas, seeps, swales, swamps, and along the margins of woodlands and ponds at low elevations across the coastal plain. The plants often are found in shallow water and grow in open to shaded exposures. The substrates include loamy sand, peat, Pomona fine sand, sand, sandy peat, spodo-sol, and ultic haplaquods. The main flowering period occurs from February to May, but possibly can extend throughout the year. The plants are observed often in sites that have been burned recently (within 1–12 months), which indicates a requirement for reduced competition from other species in order to thrive. Additional details on the reproductive biol-ogy and seed ecology are unavailable. **Reported associates:** *Acrostichum danaeifolium, Agalinis filicaulis, A. obtusifo-lia, Agrostis hyemalis, Aletris lutea, A. obovata, A. ×totte-nii, Allium canadense, Ampelopsis arborea, Andropogon brachystachyus, Aristida palustris, A. stricta, Asclepias connivens, A. longifolia, A. michauxii, Balduina angusti-folia, Baptisia lanceolata, Berlandiera pumila, Bigelowia nudata, Boltonia diffusa, Buchnera americana, Calopogon multiflorus, C. pallidus, Carex glaucescens, C. lupuliformis, Carphephorus odoratissimus, C. pseudoliatris, Chaptalia tomentosa, Cirsium horridulum, Coreopsis gladiata, C. lan-ceolata, C. nudata, Croptilon divaricatum, Ctenium aromati-cum, Cyclosorus dentatus, Cyperus erythrorhizos, Cyrilla racemiflora, Dalea carnea, D. pinnata, Dichanthelium dichotomum, D. ensifolium, D. strigosum, Drosera brevifo-lia, D. tracyi, Eleocharis, Elephantopus elatus, Erigeron stri-gosus, E. vernus, Eriocaulon compressum, E. decangulare, Eryngium integrifolium, E. yuccifolium, Eupatorium album, E. leucolepis, E. mohrii, E. rotundifolium, E. semiserratum,*

Euphorbia inundata, Eurybia eryngiifolia, Gaylussacia dumosa, G. mosieri, Gelsemium sempervirens, Gratiola hispida, G. pilosa, Gymnadeniopsis nivea, Helenium vernale, Helianthus angustifolius, H. carnosus, H. heterophyllus, H. radula, Houstonia procumbens, Hydrocotyle ranunculoides, Hypericum brachyphyllum, H. chapmanii, H. fasciculatum, H. mutilum, H. suffruticosum, Hypoxis juncea, Ilex ambigua, I. glabra, Juncus marginatus, J. megacephalus, Lachnanthes caroliniana, Lachnocaulon anceps, Lechea minor, Lepidium virginicum, Liatris chapmanii, L. spicata, L. tenuifolia, Limnobium spongia, Lophiola aurea, Ludwigia maritima, L. virgata, Lyonia fruticosa, Macbridea alba, Mimosa quadrivalvis, Mitchella repens, Myrica, Nyssa, Oclemena reticulata, Oenothera laciniata, O. simulans, Oldenlandia uniflora, Oxycaryum cubense, Panicum virgatum, Paspalum setaceum, Phoebanthus tenuifolius, Physostegia godfreyi, Pinguicula pumila, Pinus elliottii, P. palustris, Piriqueta cistoides, Pityopsis graminifolia, P. oligantha, Polygala crenata, P. cruciata, P. lutea, P. nana, P. polygama, P. ramosa, Pteridium aquilinum, Pterocaulon pycnostachyum, Quercus minima, Q. pumila, Rhexia alifanus, R. mariana, Rhynchospora colorata, R. latifolia, R. macra, Rubus, Rudbeckia graminifolia, R. hirta, R. mohrii, Ruellia ciliosa, Sabal palmetto, Sabatia stellaris, Sagittaria graminea, Salvia lyrata, Sarracenia flava, S. minor, S. psittacina, Saururus cernuus, Schizachyrium tenerum, Scleria ciliata, Scutellaria integrifolia, Serenoa repens, Seymeria cassioides, Sisyrinchium angustifolium, Solidago odora, Stenanthium densum, Stillingia sylvatica, Symphyotrichum chapmanii, Taxodium ascendens, T. distichum, Tephrosia florida, Tradescantia hirsutiflora, T. ohiensis, Triantha racemosa, Urochloa mutica, Utricularia cornuta, Vaccinium myrsinites, Valerianella radiata, Verbesina chapmanii, Viburnum obovatum, Xyris ambigua, X. brevifolia, X. caroliniana.

***Lobelia siphilitica* L.** is common in tidal or nontidal sites including alluvial fans, beaches, bluffs, bogs, carrs, channels, deltas, depressions, ditches, draws, fens, flats, floodplains, gravel bars, marshes, meadows, oxbows, prairies, ravines, river bottoms, roadsides, sandbars, seeps, shores, slopes, sloughs, spray cliffs, springs, streambeds, streams, swales, swamps, thickets, and along the margins of lakes, ponds, rivers, and streams at elevations of up to 1676 m. The plants occur frequently (but not exclusively) in association with riverine habitats. They often appear in drying sites where the water levels have receded, but also can be found in shallow (e.g., 15 cm deep) standing water. Although this species can be found in a broad range of habitats, the plants commonly occur sporadically at a site and rarely in large numbers. Exposures range from full sun to shade. The substrates are described most often as calcareous and alkaline (pH >6.5; 300–400 mg/l $CaCO_3$) and include alluvium, clay, gravel, humus, limestone, loam, loamy clay, loamy sand, marl, mucky loam, mucky peat, mud, peaty muck, rock, rubble, sand, sandy clay, sandy loam, silt, and silty muck. The flowering period extends at least from 28 July through 15 October. Variation in bolting time and the extent of fruit production are under genetic, rather than plastic control. The corollas vary widely in color from light blue to purple, a polymorphism that does not appear to be maintained by pollinator or herbivore selection. The plants are self-compatible with normal seed quantities produced in artificially selfed flowers. However, they are almost entirely outcrossed because of strong protandry and high levels of inbreeding depression, which can lead to a 34%–71% decrease in net fertility. Crosses can vary with respect to whether a positive or negative effect occurs from inbreeding. High levels of outcrossing and heterozygosity have been verified in some populations using genetic (microsatellite) markers. The flowers are pollinated primarily by bumblebees (see *Use by wildlife*) and several natural populations have been shown to be strongly pollen-limited. Some (but not all) populations are gynodioecious. Gynodioecious populations can consist of from 3% to 93% female plants, but show only slightly elevated levels of inbreeding depression in the hermaphroditic compared to the female individuals. Hermaphroditic flowers generally are preferred by pollinators. Female plants have higher (14%–32%) photosynthetic rates, but produce smaller corollas (by 4%–8%) than the hermaphrodites. The highly variable frequency of females observed among populations probably occurs because different populations may lack some of the interacting mitochondrial cytoplasmic male sterility genes and nuclear restorer genes, which determine sex expression in this species. Females are more common where there is a higher annual temperature, lower soil moisture, and lower fruit predation, which may indicate sex-specific selection. However, females also are more prevalent in small populations, which are influenced by drift, inbreeding and founder effects. Smaller female display sizes have been observed in the absence of coflowering plant species, which compete for pollinators. Hermaphrodites that carry more restorer genes have lower pollen viability. Fruiting peaks from October to November, with individual plants capable of producing upward of 8100 seeds. Like the related *L. cardinalis*, the seeds are minute (~20 μg), and likely are dispersed locally by water or to greater distances by their adherence to various animals. Fresh seeds require light, but no cold treatment in order to germinate; however, good germination (86%) has been obtained within 2 weeks at ambient greenhouse temperatures, following an 8-week period of cold (5°C) stratification. Seedling survival is about 78% under greenhouse conditions. In another study, air-dried seeds that were stored at 4°C could be germinated by placing them between moist filter paper layers, cold stratifying them for 12 weeks at 4°C, and then incubating them at 24°C under a 12/12 h light regime. Seeds collected from smaller natural populations exhibited higher germination rates than those from larger populations, which may reflect the generally more southern localities associated with smaller populations, where relatively higher temperatures occur. Female flowers also produce larger seeds with higher germination rates. The roots (up to 74%) are colonized by arbuscular mycorrhizal fungi and dark septate endophytic fungi. The occurrence of plants at a site results in a consistent fungal rhizosphere community compared to adjacent habitat where the plants are absent. Individuals transplanted to a temperate

deciduous woodland understory site declined and then died after 2 years, presumably because of soil moisture levels that were inadequate to sustain the plants. **Reported associates:** *Acalypha, Acer negundo, A. nigrum, A. rubrum, A. saccharinum, A. saccharum, Adiantum capillus-veneris, Agalinis tenuifolia, Ageratina altissima, Agrimonia striata, Agrostis gigantea, A. stolonifera, Alisma subcordatum, Alliaria petiolata, Alnus rugosa, Amaranthus tuberculatus, Ambrosia, Amorpha fruticosa, Amphicarpaea bracteata, Andropogon gerardii, Angelica atropurpurea, Arnoglossum plantagineum, Artemisia, Asarum canadense, Asclepias incarnata, A. syriaca, Bacopa rotundifolia, Betula alleghaniensis, B. pumila, B. ×sandbergii, Bidens aristosus, B. cernuus, B. frondosus, B. laevis, B. trichospermus, B. vulgatus, Boehmeria cylindrica, Botrychium dissectum, Bouteloua curtipendula, Calamagrostis canadensis, Callitriche palustris, Caltha palustris, Campanula americana, Cardamine pensylvanica, Carex annectens, C. brevior, C. conjuncta, C. emoryi, C. frankii, C. grayi, C. hystericina, C. interior, C. leptalea, C. lurida, C. prairea, C. stricta, C. tribuloides, C. vesicaria, C. vulpinoidea, Carpinus, Carya cordiformis, C. glabra, C. ovata, Castilleja coccinea, Cephalanthus occidentalis, Ceratophyllum demersum, Cercis canadensis, Chara, Chelone glabra, Chenopodium, Cichorium intybus, Cicuta maculata, Cinna arundinacea, Cirsium arvense, C. muticum, Clematis virginiana, Clintonia borealis, Commelina communis, Conocephalum conicum, Conoclinium coelestinum, Coreopsis tripteris, Cornus alternifolia, C. amomum, C. rugosa, C. sericea, Crataegus, Cryptotaenia canadensis, Cyperus flavescens, C. strigosus, Cypripedium reginae, Dasiphora floribunda, Daucus carota, Decodon verticillatus, Diodia virginiana, Doellingeria umbellata, Drosera brevifolia, Echinochloa, Eleocharis acicularis, E. rostellata, Elymus virginicus, Epilobium coloratum, E. leptophyllum, Equisetum hyemale, Erigeron annuus, Eriophorum angustifolium, Eupatorium perfoliatum, E. serotinum, Eustoma exaltatum, Euthamia graminifolia, Eutrochium fistulosum, E. maculatum, E. purpureum, Fragaria virginiana, Fraxinus nigra, Fuirena simplex, Galium, Gentiana andrewsii, Gentianopsis virgata, Geum aleppicum, G. canadense, Gleditsia triacanthos, Glyceria, Glycyrrhiza lepidota, Gnaphalium, Helianthus grosseserratus, H. strumosus, Helenium autumnale, Hemerocallis fulva, Heuchera parviflora, Hibiscus moscheutos, Hydrangea arborescens, Hypericum ascyron, H. mutilum, H. perforatum, Impatiens capensis, I. pallida, Iris versicolor, Juglans cinerea, Juncus coriaceus, J. effusus, J. filipendulus, J. interior, J. nodosus, J. tenuis, J. torreyi, Lactuca canadensis, L. floridana, Larix laricina, Leersia oryzoides, L. virginica, Lemna turionifera, Liatris pycnostachya, Lilium michiganense, Lindera benzoin, Lindernia dubia, Liparis loeselii, Liquidambar styraciflua, Lobelia cardinalis, L. inflata, L. kalmii, L. spicata, Ludwigia alternifolia, L. palustris, L. peploides, Lycopus americanus, L. europaeus, L. rubellus, L. uniflorus, Lysimachia ciliata, L. lanceolata, Lythrum alatum, L. salicaria, Maianthemum stellatum, Mentha, Menyanthes trifoliata, Micranthes pensylvanica, Microstegium vimineum, Mimulus glabratus, M.*
ringens, Monarda fistulosa, Muhlenbergia glomerata, Myosotis scorpioides, Myrica gale, Nasturtium, Onoclea sensibilis, Osmunda regalis, Osmundastrum cinnamomeum, Ostrya virginiana, Oxalis stricta, Oxypolis rigidior, Packera aurea, Panicum virgatum, Parnassia glauca, P. grandifolia, P. palustris, Pascopyrum smithii, Pedicularis lanceolata, Penstemon digitalis, Persicaria hydropiperoides, P. lapathifolia, P. pensylvanica, P. punctata, P. sagittata, P. virginiana, Phalaris arundinacea, Phleum pratense, Phlox maculata, Physostegia virginiana, Picea, Pilea pumila, Pinus banksiana, Platanthera aquilonis, P. grandiflora, Pluchea camphorata, Poa pratensis, Pogonia ophioglossoides, Populus deltoides, P. tremuloides, Potentilla anserina, Prunella vulgaris, Prunus, Pteridium aquilinum, Pycnanthemum virginianum, Quercus alba, Q. bicolor, Q. ellipsoidalis, Q. rubra, Q. velutina, Ranunculus abortivus, Rhus typhina, Rhynchospora capillacea, R. capitellata, Ribes, Rosa blanda, R. multiflora, Rubus allegheniensis, Rudbeckia fulgida, R. laciniata, R. triloba, Ruellia strepens, Rumex crispus, R. verticillatus, Sagittaria latifolia, Salix amygdaloides, S. ×bebbii, S. candida, S. discolor, S. exigua, S. eriocephala, S. lutea, Sanicula, Schizachyrium scoparium, Scleria verticillata, Schoenoplectus acutus, S. americanus, S. pungens, Scirpus atrovirens, S. cyperinus, S. pendulus, Scytonema, Selaginella apoda, Silphium asteriscus, S. perfoliatum, S. terebinthinaceum, Sium suave, Smilax lasioneura, Solanum carolinense, S. dulcamara, Solidago altissima, S. gigantea, S. patula, S. riddellii, S. rigida, Sorghastrum nutans, Spartina pectinata, Sphagnum lescurii, Sphenopholis obtusata, Spiraea alba, Spiranthes cernua, S. diluvialis, Stachys palustris, S. tenuifolia, Staphylea trifolia, Stellaria fontinalis, Symphyotrichum boreale, S. firmum, S. lanceolatum, S. lateriflorum, S. novae-angliae, S. pilosum, S. puniceum, S. shortii, S. subspicatum, Symplocarpus foetidus, Teucrium, Thalictrum, Thelypteris palustris, Thuja occidentalis, Tilia americana, Toxicodendron radicans, Tragopogon, Trautvetteria caroliniensis, Triglochin palustre, Tripsacum dactyloides, Typha latifolia, Ulmus americana, Urtica, Vaccinium pallidum, Valeriana edulis, Verbena hastata, V. urticifolia, Verbesina alternifolia, Vernonia fasciculata, V. gigantea, V. missurica, Veronicastrum virginicum, Viburnum prunifolium, Viola cucullata, V. sororia, Vitis riparia, Xanthium strumarium, Xyris jupicai, Zannichellia palustris, Zanthoxylum americanum.

Use by wildlife: The alkaloids present in the latex of *Lobelia* species function to deter herbivores; reduced levels of herbivory (e.g., in late-flowering *L. siphilitica* individuals) can result in plants characterized by a lower latex content. *Lobelia amoena* is a host to several fungi (Ascomycota: Mycosphaerellaceae: *Cercospora lobeliae, Passalora effusa*) and its flowers provide nectar to southbound migrating monarch butterflies (Insecta: Lepidoptera: Nymphalidae: *Danaus plexippus*). The foliage of *L. boykinii* is grazed by deer (Mammalia: Cervidae: *Odocoileus virginianus*), insects (Insecta: *Orthoptera*), rabbits (Mammalia: Lagomorpha: Leporidae), and voles (Mammalia: Rodentia: Muridae: *Microtus*). The flowers are visited and pollinated primarily by bees (Insecta: Hymenoptera) including bumblebees (Apidae:

Bombus griseocollis, *B. impatiens*, *B. pensylvanicus*, *B. vagans*), carpenter bees (Apidae: *Ceratina dupla*), honey bees (Apidae: *Anthophora walshii*, *Apis mellifera*, *Eucera speciosa*, *Melissodes bimaculata*, *M. communis*, *Svastra obliqua*), leafcutter bees (Megachilidae: *Anthidium psoraleae*, *Coelioxys octodentata*, *C. rufitarsis*, *Hoplitis pilosifrons*, *Hoplitis spoliata*, *H. truncata*, *Megachile brevis*, *M. petulans*, *M. rugifrons*), plasterer bees (Colletidae: *Hylaeus confluens*, *H. modestus*), sweat bees (Halictidae: *Agapostemon viridulus*, *Augochlorella aurata*, *A. striata*, *Halictus confusus*, *Lasioglossum coriaceum*, *L. imitatum*, *L. pilosum*, *L. versatum*), and bee flies (Diptera: Bombyliidae: *Geron holosericeus*, *Systoechus vulgaris*), but occasionally by ruby-throated hummingbirds (Aves: Trochilidae: *Archilochus colubris*). They also are visited (but probably not often pollinated) by various butterflies (Insecta: Lepidoptera) including checkered whites (Pieridae: *Pontia protodice*), dusky wings (Hesperiidae: *Erynnis juvenalis*), eastern tailed-blues (Lycaenidae: *Cupido comyntas*), least skipperlings (Hesperiidae: *Ancyloxypha numitor*), monarchs (*Danaus plexippus*), cabbage butterflies (Pieridae: *Pieris rapae*), small coppers (Lycaenidae: *Lycaena phlaeas*), swallowtails (Papilionidae: *Battus philenor*, *Papilio troilus*), and skippers (Hesperiidae: *Pholisora catullus*, *Polites peckius*, *P. taumas*). The plants (especially flowers and fruits) are parasitized by several pathogenic fungi (Ascomycota: Pleosporaceae: *Alternaria*; Sclerotiniaceae: *Botrytis cinerea*). The anther tubes of *L. cardinalis* are eaten by striped garden caterpillars (Insecta: Lepidoptera: Noctuidae: *Trichordestra legitima*) and the ovules and seeds are eaten by larval weevils (Insecta: Coleoptera: Curculionidae: *Curculio hispidulus*). The inflorescences are consumed by slugs (Mollusca: Gastropoda: Arionidae: *Arion subfuscus*) and the leaves are trenched and grazed by pink-washed looper moth larvae (Insecta: Lepidoptera: Noctuidae: *Enigmogramma basigera*). The flowers attract moths and butterflies (e.g., Insecta: Lepidoptera: Noctuidae: *Autographa precationis*; Papilionidae: *Papilio troilus*) and ruby-throated hummingbirds (Aves: Trochilidae: *Archilochus colubris*), the latter being their sole pollinator. The plants are hosts to several fungi including chytrids (Chytridiomycota: Physodermataceae: *Physoderma*), leaf spot (Ascomycota: Mycosphaerellaceae: *Cercospora lobeliae*, *Passalora effusa*, *P. lobeliae-cardinalis*, *Septoria lobeliae*), and rust (Basidiomycota: Pucciniaceae: *Puccinia lobeliae*). They are also host plants of WTV, the wound tumor virus (Reoviridae: *Phytoreovirus*). The roots host several nematodes (Nematoda: Anatonchidae: *Miconchus exilis*; Cyatholaimidae: *Achromadora pseudomicoletzkyi*, *A. terricola*; Comesomatidae: *Vasostoma*; Diplogastridae: *Paroigolaimella coprophaga*; Diplopeltidae: *Cylindrolaimus obtusus*; Dorylaimidae: *Aporcelaimus*, *Discolaimium*, *Dorylaimellus parvulus*, *D. virginianus*, *Pungentus pungens*; Hoplolaimidae: *Helicotylenchus digonicus*; Monhysteridae: *Monhystera*; Mononchidae: *Mononchus tunbridgensis*; Plectidae: *Plectus armatus*; Tylenchidae: *Aglenchus thornei*, *Psilenchus gracilis*, *P. magnidens*). *Lobelia dortmanna* is grazed by caribou (Mammalia: Cervidae: *Rangifer tarandus*),

larval moths (Insecta: Lepidoptera: Crambidae: *Eoparargyractis plevie*), and caddis flies (Insecta: Trichoptera: Limnephilidae: *Limnephilus infernalis*). The flowers of *L. elongata* are visited by butterflies (e.g., Insecta: Lepidoptera: Hesperiidae: *Amblyscirtes aesculapius*), which have been suggested (but not demonstrated) to be their pollinators. The flowers of *L. floridana* are visited by several bees (Insecta: Hymenoptera) including bumblebees (Apidae: *Bombus griseocollis*, *B. impatiens*), leafcutting bees (Megachilidae: *Megachile georgica*), and sweat bees (Hymenoptera: Halictidae: *Lasioglossum coreopsis*), which all potentially function as pollinators. The flowers of *L. glandulosa* are visited by ruby-throated hummingbirds (Aves: Trochilidae: *Archilochus colubris*). The flowers of *L. kalmii* are pollinated by sweat bees (Insecta: Hymenoptera: Halictidae: *Lasioglossum*) and also are visited by cabbage butterflies (Insecta: Lepidoptera: Pieridae: *Pieris rapae*) and flies (Insecta: Diptera: Syrphidae). The flowering stalks are grazed by deer (Mammalia: Cervidae: *Odocoileus virginianus*). The plants are a habitat indicator of the dorcas copper butterfly (Insecta: Lepidoptera: Lycaenidae: *Lycaena dorcas*) in Nova Scotia, and of the bog conehead katydid (Insecta: Orthoptera: Tettigoniidae: *Neoconocephalus lyristes*) and dusky-faced meadow grasshopper (Insecta: Orthoptera: Tettigoniidae: *Orchelimum concinnum*) in Ohio. They also are an important habitat component for Hine's emerald dragonfly (Insecta: Odonata: Corduliidae: *Somatochlora hineana*) in Michigan and the Dakota skipper (Insecta: Lepidoptera: Hesperiidae: *Hesperia dacotae*) in Manitoba. The foliage hosts chytrids (fungi: Chytridiomycota: Physodermataceae: *Physoderma*) and leaf-spot fungi (Ascomycota: Mycosphaerellaceae: *Septoria lobeliae*). The flowers of *L. paludosa* provide nectar for various butterflies (Insecta: Lepidoptera). *Lobelia siphilitica* is eaten by white-tailed deer (Mammalia: Cervidae: *Odocoileous virginianus*) even though some have categorized it as "deer resistant". It is a larval host plant for dark-spotted palthis moths (Insecta: Lepidoptera: Noctuidae: *Palthis angulalis*) and the leaves are eaten by larval looper moths (Insecta: Lepidoptera: Noctuidae: *Enigmogramma basigera*). Numerous insects visit the flowers to gather nectar and pollen, the latter having a mean nutritional value of 5560 calories/gram. The flowers are pollinated primarily by bumblebees (Insecta: Hymenoptera: Apidae: *Anthophora terminalis*, *Bombus affinis*, *B. auricomis*, *B. fervidus*, *B. griseocollis*, *B. impatienis*, *B. pensylvianicus*, *B. vagans*, *Ceratina calcarata*). Other floral visitors include insects (Insecta) such as carpenter bees (Apidae: *Xylocopa*), honey bees (Apidae: *Apis mellifera*), plasterer bees (Colletidae: *Hylaeus affinis*), sweat bees (Hymnoptera: Halticidae: *Agapostemon virescens*, *Augochlorella aurata*, *Halictus rubicundus*, *Lasioglossum admirandum*, *L. coriaceum*), and syrphid flies (Diptera: Syrphidae), as well as hummingbirds (Aves: Trochilidae). The roots host leafminer flies (Insecta: Diptera: Agromyzidae: *Melanagromyza virens*), nematodes (Nematoda: Anguinidae: *Ditylenchus*; Dorylaimidae: *Dorylaimellus tenuidens*, *Eudorylaimus acuticauda*, *E. obtusicaudatus*, *Mesodorylaimus bastiani*; Mononchidae: *Mononchus papillatus*), and wasps

(Insecta: Hymenoptera: Braconidae: *Dacnusa*). The seeds are eaten by larval weevils (Insecta: Coleoptera: Curculionidae: *Miarus hispidulus*); the adult weevils feed on the pollen and nectar. The plants are also hosts to a variety of fungi including leaf spot (Ascomycota: Botryosphaeriaceae: *Phyllosticta bridgesii*; Mycosphaerellaceae: *Cercospora lobeliae, Passalora effusa, P. lobeliae-cardinalis, Septoria lobeliae, S. lobeliae-syphiliticae*; Sclerotiniaceae: *Cristulariella pyramidalis*), root rot (Ascomycota: Sclerotiniaceae: *Phymatotrichopsis omnivora*), and rust (Basidiomycota: Pucciniaceae: *Puccinia lobeliae*).

Economic importance: food: None of the *Lobelia* species has been used as a source of food and should ever be eaten due to the presence of pyridine alkaloids such as lobeline and lobinaline, which makes many of them (e.g., *L. cardinalis, L. floridana, L. kalmii, L. siphilitica*) toxic. *Lobelia siphilitica* has been implicated in some cases of livestock poisoning; **medicinal:** *Lobelia cardinalis* and *L. elongata* contain the alkaloid lobinaline, which is known to reduce blood pressure; however, home remedies made from these plants can be dangerously toxic if taken in moderate to high dosages. *Lobelia floridana* and *L. siphilitica* also can cause systemic poisoning in humans. Total alkaloid concentration in *L. cardinalis* leaves can reach 0.44% and their ingestion can induce convulsions, diarrhea, nausea, salivation, vomiting, and in some cases, coma. Nevertheless, this species had numerous medicinal uses by various Native American tribes. The Cherokee "snuffed" a cold infusion to stop nosebleeds, used a crushed leaf poultice to relieve headaches, a root infusion to treat worm infections, and internal infusions to remedy colds, fevers, rheumatism, stomach problems, and as a sore healing aid. They also administered preparations of the plants to treat croup and syphilis. The Delaware and Oklahoma used a root infusion for typhoid. The plant was regarded as a panacea by the Iroquois, who also believed that it strengthened the potency of all medicines. Specifically, they used a root infusion or poultice as an analgesic and a root decoction as a wash or poultice for chancres and fever sores. The plants also were used in medications for alleviating menstrual problems, for cramps and fevers, as a wash for injuries, in compounds used to treat consumption and epilepsy, and combined in medicines taken to ease grief. Plant decoctions or root infusions were taken by the Iroquois, Meskwaki, and Pawnee as a love potion. The Zuni included the plants as an ingredient in "schumaakwe cakes", which were used to control rheumatism and swelling. *Lobelia kalmii* was used as an emetic by the Cree, Hudson Bay and Iroquois tribes. The Iroquois also made an infusion from pulverized plants, which they used as drops for treating abscesses and earaches. *Lobelia siphilitica* contains several alkaloids (including lobeline) and the polyacetylene lobetyolin and appears in European medicinal herbals dating back to at least 1781. It was used extensively by the Cherokee as a treatment for croup, rheumatism, and syphilis. They applied a poultice of crushed leaves for headaches, used a leaf infusion for combatting colds and fevers, and administered a root infusion to treat slowly healing sores, stomach problems and worms. A root poultice was used for

controlling throat "risings" and a cold infusion "snuffed" to control nosebleeds. The Iroquois regarded the plants highly as a medicinal (second only to *Sassafras*) and used it for treating syphilis and as a purgative. They also gargled with the plant to suppress coughing; **cultivation:** Attempts to grow *Lobelia amoena* as a garden plant have been unsuccessful. Because of its brilliant red flowers and broad site adaptability, *L. cardinalis* is cultivated extensively and includes the cultivars: 'Bee's Flame,' 'Black Truffle,' 'Elmfeuer,' 'Eulalia Berridge,' 'Fan Scarlet,' 'Fried Green Tomatoes,' 'Frielings Ghost,' 'Gladys Lindley,' 'Golden Torch,' 'Huntsman,' 'Illumination,' 'Mrs Furnell,' 'Queen Victoria' (an Award of Garden Merit plant), 'Rose Beacon,' 'Russian Princess,' 'Shrimp Salad,' 'Small Form,' 'The Bishop,' and 'The Test'. The hybrid *L. ×speciosa* and its cultivars 'Fan Burgundy,' 'Hadspen Purple,' and 'Ruby Slippers' also are grown widely as ornamentals. Seeds of *L. kalmii* are available from some suppliers. Cultivars include 'Abby' and 'Blue Shadow'. *Lobelia siphilitica* is widely cultivated and its seed is readily available commercially. Cultivars include 'Alba,' 'Blue Cardinal,' 'Blue Selection,' 'Hillview Blue,' 'La Fresco,' 'Lilac Candles,' 'Nana,' 'Rosea,' and 'White Candles'; **misc. products:** The Iroquois used an infusion from the shoots and flowers of *Lobelia cardinalis* as a basket wash. They administered a root infusion or poultice to protect against problems caused by witchcraft. The plants were used ceremonially by the Meskwaki in funerals and to fend off storms. The Jemez included the flowers used in their ceremonial rain dance. Infusions from macerated plants of *L. siphilitica* were taken by the Iroquois to ward off witchcraft. The Meskwaki finely chopped and ingested its roots as a love potion. *Lobelia cardinalis* and *L. siphilitica* have been used as a model system for studying phenotypic plasticity in plants. Methanolic extracts of *L. siphilitica* have high larvicidal activity on mosquitoes (Diptera: Culicidae: *Culex quinquefasciatus*). *Lobelia siphilitica* is considered to be a suitable species for use in "green roof" construction, especially in shade applications. *Lobelia cardinalis* and *L. siphilitica* often are found in wetland seed mixes, which are used for habitat restoration; **weeds:** Several *Lobelia* species (e.g., *L. siphilitica*) occasionally have been described as "lawn weeds"; however, none is known to reach nuisance levels; **nonindigenous species:** none reported.

Systematics: *Lobelia* is placed within the subtribe Siphocampylinae of tribe Lobelieae in subfamily Lobelioideae of Campanulaceae. Preliminary phylogenetic analyses demonstrate that the genus is not monophyletic, with various species falling among many other genera including *Brighamia, Clermontia, Cyanea, Delissea, Diastatea, Downingia, Grammatotheca, Hippobroma, Solenopsis*, and *Trematolobelia* (Figure 5.179). All 12 OBL species have been placed consistently within one section of the genus, section *Lobelia*. However, in a phylogenetic analysis that included three OBL species (*L. cardinalis, L. dortmanna*, and *L. kalmii*), the result resolves a clade in which *L. rotundifolia* (assigned to section *Tylomium*) is nested among them (Figure 5.179). It is clear that much taxonomic refinement is needed for this large and complex group. A recent evaluation

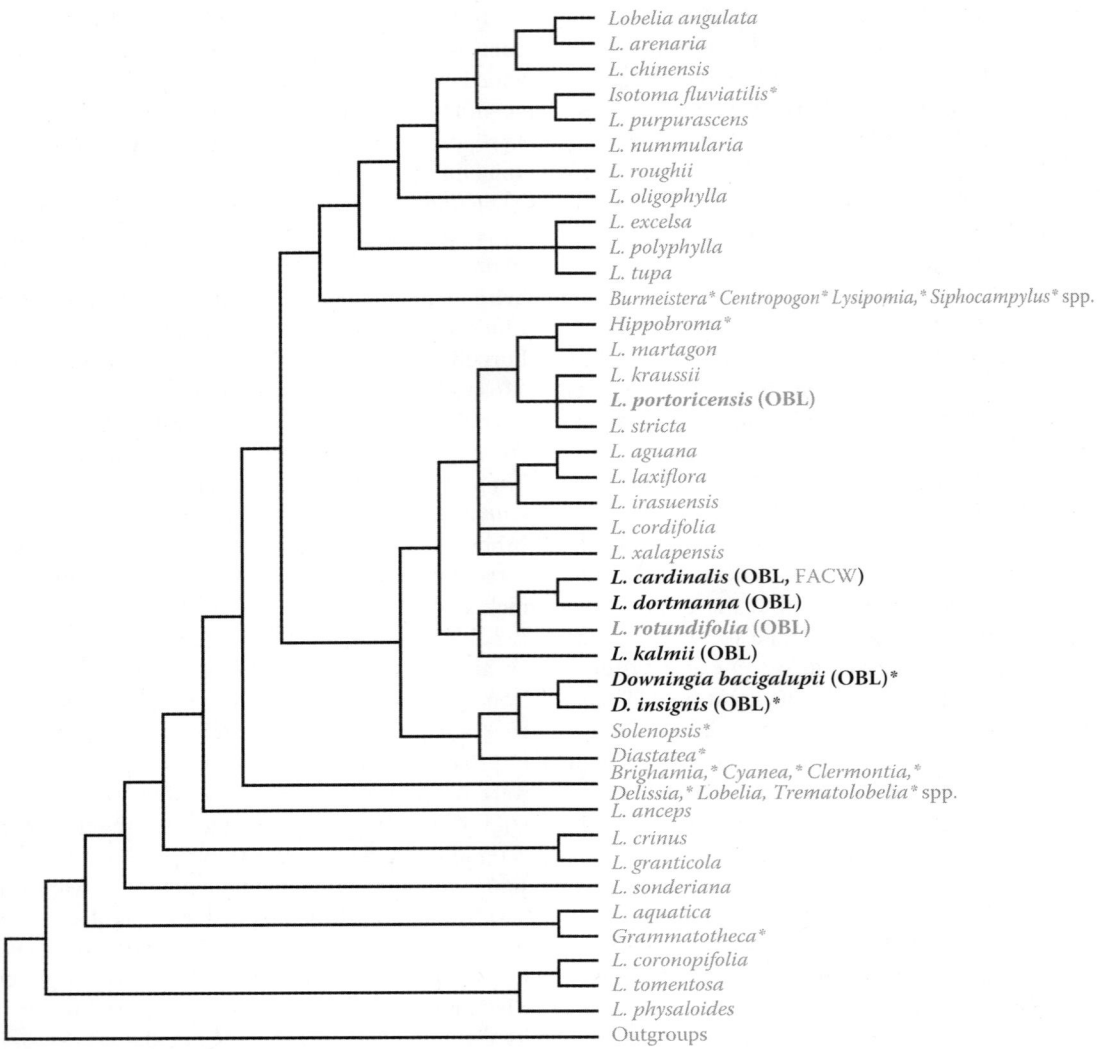

Lobelia angulata
L. arenaria
L. chinensis
*Isotoma fluviatilis**
L. purpurascens
L. nummularia
L. roughii
L. oligophylla
L. excelsa
L. polyphylla
L. tupa
Burmeistera Centropogon* Lysipomia,* Siphocampylus* spp.*
*Hippobroma**
L. martagon
L. kraussii
L. portoricensis (OBL)
L. stricta
L. aguana
L. laxiflora
L. irasuensis
L. cordifolia
L. xalapensis
L. cardinalis (OBL, FACW**)**
L. dortmanna (OBL)
L. rotundifolia (OBL)
L. kalmii (OBL)
Downingia bacigalupii* (OBL)
D. insignis* (OBL)
*Solenopsis**
*Diastatea**
Brighamia, Cyanea,* Clermontia,**
Delissia, Lobelia, Trematolobelia* spp.*
L. anceps
L. crinus
L. granticola
L. sonderiana
L. aquatica
*Grammatotheca**
L. coronopifolia
L. tomentosa
L. physaloides
Outgroups

FIGURE 5.179 Phylogenetic relationships of *Lobelia* species reconstructed by analysis of combined chloroplast DNA sequence data from the *rbcL, ndhF,* and *trnL–F* regions. This analysis clearly demonstrates that the current generic concept of *Lobelia* does not represent a monophyletic group (anomalous genera are indicated by asterisks). The OBL North American species included in this analysis (black bold) resolve in a clade with *L. rotundifolia,* which has been classified in a different section. The two OBL species in gray bold occur outside of the North American region considered in this treatment. (Adapted from Antonelli, A., *Mol. Phylogenet. Evol.,* 46, 1–18, 2008.)

has divided the genus into 18 sections, based primarily on comparative morphological characteristics. However, given the preliminary phylogenetic results, and the fact that such analyses still have not yet included many of the taxa, any classification of this complex genus should be regarded as tentative at best. The base chromosome number of *Lobelia* is $x=7$. Diploids ($2n=14$) include *L. boykinii, L. canbyi, L. cardinalis, L. dortmanna, L. flaccidifolia, L. kalmii* and *L. siphilitica; L. amoena, L. elongata,* and *L. glandulosa* ($2n=28$) are tetraploid; *L. floridana* and *L. paludosa* ($2n=42$) are hexaploid. Counts of $2n=24$, 28 and 32 have been reported for European populations of *L. dortmanna. Lobelia cardinalis, L. elongata,* and *L. glandulosa* occasionally exhibit supernumerary chromosomes. Too few species have been included in molecular phylogenetic investigations to provide definitive answers regarding species relationships; however,

several hypotheses have been proposed on the basis of comparative morphology and crossing studies. *Lobelia amoena* is believed to be of allotetraploid origin (with one parent similar to *L. puberula*) and *L. boykinii* and *L. canbyi* apparently are closely related. *Lobelia cardinalis* and *L. siphilitica* presumably are closely related because of their ability to hybridize; however, the occurrence of the alkaloid lobinaline (present in *L. cardinalis* and *L. elongata* but absent in *L. puberula* and *L. siphilitica*), indicates a closer relationship between *L. cardinalis* and *L. elongata.* The two hexaploids (*L. floridana* and *L. paludosa*) are believed to be sister taxa. Further insight into interspecific relationships in this genus will require much additional systematic research and increased taxon sampling in phylogenetic analyses. Interspecific hybrids occur rarely in natural communities, due to effective isolating mechanisms. However, synthetic interspecific hybrids are numerous and have been produced between the following species pairs: *L. amoena*×*L. elongata*, *L. amoena*×*L. glandulosa*, *L. amoena*×*L. speciosa*, *L. canbyi*×*L. gattingeri*, *L. canbyi*×*L. nuttallii*, *L. elongata*×*L. ×speciosa*, *L. georgiana*×*L. glandulosa*, *L. glandulosa*×*L. elongata*, *L. flaccidifolia*×*L. feayana*, *L.flaccidifolia*×*L. puberula*, *L. glandulosa*×*L. speciosa*, *L. paludosa*×*L. floridana*, *L. siphilitica*×*L. brevifolia*, *L. siphilitica*×*L. georgiana*, and *L. siphilitica*×*L. puberula*. *Lobelia ×speciosa* is the name given to hybrids between *L. siphilitica*×*L. cardinalis*. Chloroplast DNA primarily is inherited maternally in *L. siphilitica*; however, heteroplasmy and low levels of paternal inheritance have been observed in some populations.

Comments: *Lobelia cardinalis* is widespread in all but northwestern North America and extends into Mexico. Two species occur in northern North America (*L. dortmanna*, *L. kalmii*) and one in eastern North America (*L. siphilitica*). The range of *L. dortmanna* also extends into northwestern Europe. The remaining OBL species are distributed either in the southern United States (*L. flaccidifolia*, *L. floridana*) or southeastern United States (*Lobelia amoena*, *L. boykinii*, *L. canbyi*, *L. elongata*, *L. glandulosa*, *L. paludosa*).

References: Abel & Friesen, 1991; Adams et al., 2010; Alexander & Svenson, 1943; Anderson, 1987; Atwood, 1941; Avis et al., 1997; Ayers, 2012; Bach, 2001; Bailey, 2002; Baker, 1975; Bartholomew et al., 2006; Bartkowska & Johnston, 2012; Barton, 1960; Baskin & Baskin, 1998; Beal, 1977; Bergerud, 1972; Bertin, 1982; Bowden, 1959, 1960, 1961; Burrows & Tyrl, 2012; Campbell et al., 1989; Carlquist, 1981; Carr, 2007; Caruso, 2012; Caruso & Case, 2007; Caruso et al., 2003, 2005, 2006, 2010; Catling et al., 1986; Chaubal et al., 1962; Choesin & Boerner, 2000; Christensen & Sand-Jensen, 1998; Clark & Pence, 1999; Colin & Jones, 1980; Cooke & Lefor, 1998; Core, 1967; Costello, 1936; Cronin et al., 1998; Davis, 1993; Devlin, 1988; Devlin & Stephenson, 1984, 1985, 1987; Devlin et al., 1987; Drayton & Primack, 2012; Dudle et al., 2001; Durewicz, 2012; Dvorak & Volder, 2010; Farmer, 1989; Farmer & Spence, 1987; Fernald, 1899, 1916; Flint, 1882; Freeman et al., 1991; Gaddy, 1982; Gano, 1917; Geisler, 1926; Getter & Rowe, 2008; Gibbons et al., 1990; Godfrey & Wooten, 1981; Gordon, 1933, 1989, 2009, 2011; Grundel et al., 2011; Guttenberg, 1881; Hamm et al., 1986; Hammer, 2002; Harms & Grodowitz, 2009; Harper, 1904; Heus, 2003; Hirst, 1983; Home, 1781; Homoya et al., 1985; Hovatter et al., 2011; Johnson & Galloway, 2008; Johnston, 1991, 1992; Katsar et al., 2007; Keddy, 1981; Keddy et al., 2006; Kesting et al., 2009; Khare, 2007; Klymko et al., 2012; Lacey et al., 2001; Lammers, 2011; Latimer & Freeborn, 2009; LeBlond et al., 2007; Lee et al., 2006; Ley et al., n.d.; Lieneman, 1929; Lindsey et al., 1961; Little, Jr., 1939; Macior, 1967; Manske, 1938; Markle, 1915; McAvoy & Wilson, 2014; McMillan & Porcher, 2005; McVaugh, 1936; Miller & Stanton-Geddes, 2007; Minno & Slaughter, 2003; Mitchell, 1926; Moeller, 1978b; Mohlenbrock, 1959b; Mold, 2012; Møller & Sand-Jensen, 2011; Moran, 1981; Mottl et al., 2006; Moyle, 1945; Mutikainen & Delph, 1998; Myers et al., 2004; Nichols, 1999a; Nielsen & Sand-Jensen, 1989; Norcini

& Nelson, 2010; Oppel et al., 2009; Orr & Dickerson, 1966; Parachnowitsch et al., 2012; Pavela, 2009; Pearson, 1954; Pedersen, 1993; Pedersen & Sand-Jensen, 1992; Pellerin et al., 2006; Penfound & Watkins, 1937; Philbrick & Les, 1996; Pigliucci & Schlichting, 1995; Pigliucci et al., 1997; Proell, 2009; Pulido et al., 2012; Rawinski & Ludwig, 1992; Richardson et al., 1984; Riefner, Jr. & Hill, 1983; Rigney, 2013; Rosatti, 1986; Royo et al., 2008; Sah et al., 2007; Sand-Jensen et al., 1982; Schlichting & Devlin, 1992; Schmidt, 2005; Schneider & Sharitz, 1986; Schweitzer, 2007; Sinnott, 1912; Smith, 1996; Smolders et al., 1995; Sorrie et al., 2006; Sparrow, 1975; Stevens, 1957; Stewart & Kantrud, 1972; Sweetman, 1928; Swink, 1952; Tehon et al., 1946; Thomas, 1933; Thomas & Alexander, 1962; Thompson & McKinney, 2006; Tiner, 2009; Tuell et al., 2008; Uphof, 1922; Wassink & Caruso, 2013; Weber & Rooney, 1994; Weishampel & Bedford, 2006; Wherry, 1927; Williamson & Black, 1981; Wilson & Keddy, 1988; Wisheu & Keddy, 1991; Wium-Andersen, 1971; Wolfe et al., 2006; Zartman & Pittillo, 1998.

6. *Porterella*

Porterplant

Etymology: after Thomas C. Porter (1822–1901)

Synonyms: *Laurentia* (in part); *Lobelia* (in part)

Distribution: global: North America; **North America:** western United States

Diversity: global: 1 species; **North America:** 1 species

Indicators (USA): OBL: *Porterella carnosula*

Habitat: freshwater; palustrine; **pH:** 6.0; **depth:** <1 m; **life-form(s):** emergent or submersed (vittate) herb

Key morphology: stems (to 30 cm) erect, slightly succulent, the branches (if present) arising from base; leaves cauline, sessile, the blades (to 15 mm) narrowly ovate or triangular, margins entire or sparsely toothed; flowers bisexual, resupinate, arranged in racemes, pedicelled (to 3 cm); sepals (to 10 mm) spreading, linear lanceolate; corolla bilabiate, blue to white, hypanthium (to 6 mm) present, upper lip (to 2 mm), 2-lobed, erect, lower lip (to 8 mm) 3-lobed, spreading, with a yellow spot; stamens fused in a tube (to 10 mm); ovary inferior, 5-angled, 2-celled; capsules (to 14 mm) obconic to cylindrical, dehiscence valvate; seeds (to 1 mm) numerous, smooth, very finely striate.

Life history: duration: annual (fruit/seeds); **asexual reproduction:** none; **pollination:** unknown; **sexual condition:** hermaphroditic; **fruit:** capsules (common); **local dispersal:** seeds; **long-distance dispersal:** seeds

Imperilment: (1) *Porterella carnosula* [G4]; S1 (UT, WY); S2 (ID)

Ecology: general: *Porterella* is monotypic.

Porterella carnosula **(Hook. & Arn.) Torr.** is a montane annual, which inhabits brooks, channels, depressions, ditches, flats, floodplains, marshes, meadows, mudflats, playas, roadsides, ruts, seeps, sloughs, streambeds, swales, vernal pools, wallows, washes, and the margins of lakes, ponds, reservoirs, and streams at elevations from 1249 to 3241 m. Most occurrences are exposed to full sunlight. The plants typically grow as emergents along the drying margins of sites where temporary waters have receded; however, sterile plants can grow with their shoots submersed in shallow (e.g., 5 cm) water. The substrates are acidic (e.g., pH 6.0) and include adobe, basalt cobble, clay, clay loam, humus, loam, mud, rock, sand, sandy loam, sandy mud, and silty clay loam. The flowers are produced from May to August. Otherwise, there is no additional information available on the reproductive biology or seed ecology of this species. **Reported associates:** *Abies lasiocarpa, A. magnifica, Arnica longifolia, Artemisia tridentata, Bromus, Camassia quamash, Carex athrostachya, C. lenticularis, C. nebrascensis, C. utriculata, C. vesicaria, Castilleja pilosa, Chrysothamnus, Collinsia grandiflora, C. parviflora, Damasonium californicum, Deschampsia cespitosa, Downingia bacigalupii, D. bicornuta, D. insignis, D. laeta, D. yina, Drymocallis lactea, Eleocharis bella, E. macrostachya, E. ovata, E. palustris, Epilobium, Galium trifidum, Geum triflorum, Glyceria borealis, Gratiola ebracteata, G. neglecta, Heterocodon rariflorus, Hordeum brachyantherum, Isoetes bolanderi, Juncus balticus, J. bufonius, Juniperus occidentalis, Lasthenia glaberrima, Lilaea scilloides, Limosella acaulis, Lonicera, Marsilea vestita, Mentha arvensis, Mimulus evanescens, M. floribundus, M. leptaleus, M. pilosus, M. suksdorfii, Muhlenbergia, Myosurus apetalus, Nuphar polysepala, Orthocarpus, Pedicularis attollens, Perideridia, Persicaria amphibia, Picea engelmannii, Pilularia americana, Pinus contorta, P. ponderosa, Plagiobothrys hispidulus, P. scouleri, Poa bulbosa, P. pratensis, Polygonum polygaloides, Potamogeton, Potentilla flabellifolia, Ranunculus cardiophyllus, R. flammula, Rorippa curvisiliqua, Sagittaria cuneata, Trifolium, Vaccinium, Valeriana edulis, Veronica peregrina.*

Use by wildlife: none reported.

Economic importance: food: not reported as edible; **medicinal:** no uses reported; **cultivation:** not cultivated; **misc. products:** none; **weeds:** none; **nonindigenous species:** none

Systematics: *Porterella* is classified in subtribe Siphocampylinae of tribe Lobelieae in subfamily Lobelioideae of Campanulaceae. In some phylogenetic analyses (Figures 5.177 and 5.178), *Porterella* resolves near *Downingia, Howellia,* and *Legenere,* which share with it similar habitat and biogeographical characteristics. However, some authors have placed *Howellia* and *Porterella* in different subtribes, while others have assigned *Porterella* to a group (excluding *Downingia, Howellia,* and *Legenere*) that comprises the former segregates of *Laurentia* (*Diastatea, Enchysia, Hippobroma, Isotoma, Palmerella, Siphocampylus,* and *Solenopsis*), which possess an entire, rather than split, corolla tube. It is evident that a more comprehensive phylogenetic survey, which includes all of these genera, will be necessary in order to obtain a clearer understanding of their interrelationships. The base chromosome number of *Porterella* is reported as $x = 11$ or 12. Some uncertainty exists whether both numbers are reliable with several authors accepting $x = 11$ as the correct base number. In either case, *Porterella carnosula* would represent a diploid ($2n = 22$ or 24). The cpDNA genome of *Porterella* was found to contain an unusually large 696-bp insertion in the *ycf1* gene. No hybrids involving *Porterella* have been reported.

Comments: *Porterella carnosula* occurs in Arizona, California, Idaho, Nevada, Oregon, Utah, and Wyoming.

References: Blackwell, 2006; Correll & Correll, 1975; Crespo et al., 1998; Knox, 2014; Lammers, 1993; Meinke, 1995b; Morin, 2012d; Ratliff, 1985; Serra & Crespo, 1997; Stace & James, 1996; Wimmer, 1948; Zika, 1996.

Family 3: Menyanthaceae [6]

The Menyanthaceae (bogbeans) are a well-defined family consisting of about 80 species worldwide. All members are insect pollinated, aquatic, or semiaquatic herbs that inhabit freshwater lakes, ponds, and wetlands (Kadereit, 2007; Tippery et al., 2008). The three North American genera consist entirely of OBL species and the remaining species in the family also are helophytes or hydrophytes. Menyanthaceae once had been regarded as closely related to (or even part of) Gentianaceae (Lindsey, 1938); however, contemporary phylogenetic analyses consistently place the family firmly within the Asterales, in a position remote from Gentianaceae (Downie & Palmer, 1992; Judd & Olmstead, 2004; Judd et al., 2008; Tank & Donoghue, 2010). The chemistry of Menyanthaceae also differs from that of Gentianaceae, by its discordant flavonoid profile (Bohm et al., 1986) and lack of the cyclitol L-(+)-bornesitol, which is prevalent throughout the latter (Schilling, 1976). Menyanthaceae clearly are monophyletic and resolve as a clade when analyzed using *ndhF,* nrITS, *rbcL,* or *trnK/matK* sequence data in conjunction with numerous Asterales outgroup genera (Olmstead et al., 2000; Tippery et al., 2008). At a higher taxonomic level, the family consistently resolves within an "MCGA" clade (Figures 5.132 and 5.180) comprising Menyanthaceae, Goodeniaceae, Calyceraceae, and Asteraceae, based on analyses employing data from multiple nuclear and chloroplast loci as well as morphological characters (Olmstead et al., 2000; Albach et al., 2001; Lundberg & Bremer, 2003; Judd & Olmstead, 2004; Tank & Donoghue, 2010; Maia et al., 2014).

Species of Menyanthaceae are rhizomatous perennials, which characteristically possess sheathing petioles, a radially symmetrical, rotate or funnelform, sympetalous corolla that is imbricate or valvate in bud, five alternate, epipetalous stamens with sagittate anthers, and a superior to partially inferior ovary. A number of species possess fringed petals or interstaminal scales. The fruit is an irregularly dehiscent or valvate capsule. Different species can produce monomorphic hermaphroditic flowers, heterostylous flowers, or unisexual flowers in dioecious or gynodioecious arrangements (Ornduff, 1988; Tippery & Les, 2008). Although both self-compatible and self-incompatible species occur, Menyanthaceae lack the plunger type of pollen presentation mechanism found in many other Asterales. The flowers typically are showy and are pollinated by various insects. Vegetative habits include large to diminutive emergent wetland plants as well as strictly aquatic species with specialized floating leaves.

Menyanthes, Nephrophyllidium, Nymphoides, and *Villarsia* are cultivated as ornamentals for aquaria and water gardens and some taxa are used for food (*Menyanthes*) or

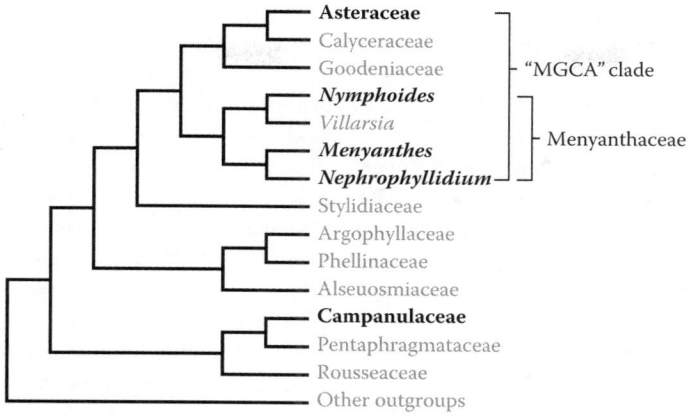

FIGURE 5.180 Phylogenetic placement of Menyanthaceae within the "MGCA" clade of Asterales based on a combined analysis of morphological and *atpB*, *ndhF*, and *rbcL* nucleotide sequence data. The Menyanthaceae are entirely aquatic with all three North American genera containing OBL taxa (shown in bold); *Villarsia* also is aquatic, but occurs outside of North America. (Adapted from Lundberg, J. & Bremer, K., *Int. J. Plant Sci.*, 164, 553–578, 2003.)

medicine (*Menyanthes*, *Nymphoides*) (Cook, 1996a; Tippery et al., 2008). Horticultural exchange has introduced a number of nonindigenous weeds to North America and elsewhere.

The largest genus is *Nymphoides*, with about 50 species distributed worldwide. Phylogenetic studies (Tippery & Les, 2009) have substantially modified the circumscriptions of *Liparophyllum* Hooker f. (7 spp.) and *Villarsia* Vetenat (3 spp.), along with the transfer of most species in the latter to *Ornduffia* Tippery & Les (7 spp.). *Liparophyllum*, *Ornduffia*, and *Villarsia* and are restricted to the Southern Hemisphere; *Menyanthes* and *Nephrophyllidium* (Northern Hemisphere) are monotypic.

Three North American genera contain OBL aquatics:

1. ***Menyanthes*** L.
2. ***Nephrophyllidium*** Gilg.
3. ***Nymphoides*** Ség.

1. *Menyanthes*

Bog-bean, buck-bean, marsh trefoil; trèfle d'eau
Etymology: from the Greek *mênuô anthos* ("disclosing flowers"), referring to the sequential, upward opening of flowers along the raceme
Synonyms: none
Distribution: global: Africa, Asia, Europe, North America;
North America: widespread (principally northern)
Diversity: global: 1 species; **North America:** 1 species
Indicators (USA): OBL: *Menyanthes trifoliata*
Habitat: freshwater; lacustrine, palustrine; **pH:** 2.2–8.0;
depth: 0–2.0 m; **life-form(s):** emergent/floating herb
Key morphology: Shoots (to 1.5 m) rhizomatous, natant, rooting adventitiously; leaves emersed, compound, trifoliate, petiolate (to 30 cm), bases sheathing, leaflets (to 12 cm) sessile, elliptical to ovate or obovate, margins entire; racemes (to 60 cm) 3–25-flowered, peduncled (to 30 cm); flowers pedicellate (to 20 mm), hermaphroditic, heteromorphic (distylous);

corolla (to 20 mm) 4–6-merous, the lobes spreading, inner surfaces white or tinged with purple, with dense, fringe-like hairs, outer surfaces pinkish, smooth; stamens dimorphic, $2\times$ longer in thrums than pins, anthers sagittate, reddish to purple; ovary partly inferior, ringed with glands, style long (to 8 mm; pins) or short (to 5.3 mm; thrums); capsules (to 12 mm) globose or ellipsoid, dehiscent by 2 valves, 10–30-seeded; seeds (2.5 mm), light brown, broadly ellipsoid, smooth, partially compressed
Life history: duration: perennial (rhizomes); **asexual reproduction:** rhizomes; **pollination:** insect; **sexual condition:** hermaphroditic (heterostylous); **fruit:** capsules (common); **local dispersal:** seeds and shoot fragments (water); **long-distance dispersal:** seeds (birds)
Imperilment: (1) *Menyanthes trifoliata* [G5]; SX (DE); S1 (AZ, MD, MO, NC, SD, VA, WV); S2 (IA, NE, NJ, OH, RI, UT); S3 (IN, ND)
Ecology: general: *Menyanthes* is monotypic.

Menyanthes trifoliata **L.** is a perennial, which occurs in tidal or nontidal sites including beaches, bogs, canals, carrs, depressions, ditches, dunes, fens, flats, floodplains, lakes, marshes, meadows, mires, mudflats, muskeg, oxbows, patterned peatlands, ponds, pools, potholes, seeps, shores, sloughs, streams (lentic), swamps, and along the margins of lakes, ponds, and rivers at elevations of up to 3557 m. The plants can produce emergent shoots on wet substrates, or extend floating shoots in water up to 2.0 m deep. Occurrences range from about 40°–70° N latitude and include sites where average air temperatures fall to −45°C in January or climb to 27°C in July and annual rainfall varies from <25 to 250 cm. More southerly localities tend to occur at higher elevations. The plants contain an extensive internal air-space system, which maintains a linear gradient of O_2 from the leaves to the rhizome and roots along an inverse CO_2 gradient. They are well known for their ability to translocate O_2 from the roots to the surrounding sediments, which provides a mechanism for sequestering phosphorus as well as for establishing favorable concentration gradients in the rhizosphere. Site exposures usually are open or receive only minimal to partial shade. A respectable range of habitats is tolerated with substrates spanning a spectrum of acidity from pH=2.2–8.0 and comprising substrates characterized as alluvium, clay, clay loam, granite, gravel, humus, hydric borofibrist, marl, muck, muck loam, mucky peat, mud, peat, sand, or typic cryofibrist. The content of organic matter in the substrate can range from 13 to 92%. The plants also grow occasionally on floating logs, or quite often in floating mats. This is a pioneer mat-forming species with its survival related to the water holding capacity of the mat. Flowering generally takes place from May to July, but has been observed to occur as late as October 23, especially following periods of drought. The mean density of flowering spikes has been estimated at 0.93/m². The plants are strongly self-incompatible and are obligately outcrossed ($r=1$). The flowers are heterostylous (distylous), weakly protogynous, and are pollinated by short-tongued insects (see *Use by wildlife*). Long and short style morphs typically occur in a 1:1 ratio in populations. The pollen of long style morphs is smaller (~50 µ) than that of short style morphs (~126 µ). Homostylous individuals occur rarely, but are characterized by considerably reduced seed set compared to dimorphic plants. Some populations exhibit substantial pollen inviability and high levels of illegitimate pollen on stigmas. Mean pollen dispersal has been estimated at 2.6 m and pollen limitation can be a major factor contributing to low seed set. The neighborhood area is small, with that of one population estimated at 10–15 m². Up to 25 capsules can develop on an inflorescence, each containing from 4 to 31 seeds, yielding an average reproductive output of 210 seeds/plant. Seed set typically exceeds 80% in natural populations; however seeds collected from an arctic portion of the range (69°22′N, 152°10′W; 107 m elevation) did not germinate in the light or in darkness at 22°C. The seeds (reaching up to 2.3 mm) will float and can remain viable for at least 15 months; however, studies in Europe have found large seed pools that consist entirely of inviable seeds. The average local seed dispersal distance can be as low as <0.2 m. They also are eaten (and presumably dispersed) by birds (Aves), fish (Osteichthyes), and reindeer (Mammalia: Cervidae: *Rangifer*). The seeds also are believed to be transported in mud that adheres to the feet of migrating waterfowl (Aves: Anatidae). The seeds are highly viable but will remain dormant for at least 6 months when in water. If the seed coat is scarified to allow water imbibition, germination normally will occur within 14 days. The seeds are categorized as being morphophysiologically dormant; however, the highest experimental germination (80%) has been obtained for scarified seed placed in the light but receiving no cold stratification. Ecologically, seedling recruitment appears to be of far less importance than clonal reproduction. Vegetative reproduction occurs by proliferation and branching of the rhizomes as well as by their fragmentation and dispersal. Sections of rhizome have remained afloat for 187 days (100%), with nearly 70% of them retaining their ability to resprout. The percentage of generative buds on rhizomes increases proportionally with rhizome size. Lateral buds form primarily on rhizomes 2–32 cm in length but rarely on those longer than 32 cm. Studies have indicated that the plants can alter their morphology via phenotypic plasticity in order to persist in populations invaded by competing vegetation. A reduced number of flower buds and lower seed production can occur when plants are shaded by competing vegetation; however, such inflorescences also are characterized by a higher survival rate than those produced in completely open sites. The plants decompose rapidly, losing about 80% of their biomass after 3 years. The roots reportedly lack mycorrhizae. **Reported associates:** *Abies concolor, Acer macrophyllum, A. rubrum, Agrostis scabra, Allium schoenoprasum, Alisma, Alnus rugosa, A. serrulata, A. viridis, Amelanchier utahensis, Andromeda polifolia, Andropogon gerardii, Anemone parviflora, Aneura pinguis, Arethusa bulbosa, Aronia melanocarpa, Athyrium filixfemina, Aulacomnium palustre, Betula cordifolia, B. glandulosa, B. lenta, B. michauxii, B. nana, B. occidentalis, B. papyrifera, B. populifolia, B. pumila, Bidens connatus, B. trichospermus, B. vulgatus, Bolboschoenus fluviatilis, Brasenia schreberi, Bromus ciliatus, Bryum pseudotriquetrum, Calamagrostis canadensis, C. stricta, Calla palustris, Calliergon giganteum, C. stramineum, Calliergonella cuspidata, Calopogon tuberosus, Calla palustris, Caltha palustris, Campanula aparinoides, Campylium polygamum, C. stellatum, Canadanthus modestus, Cardamine bulbosa, Cardamine pratensis, Carex aquatilis, C. aurea, C. brunnescens, C. buxbaumii, C. canescens, C. capillaris, C. chordorrhiza, C. conoidea, C. cryptolepis, C. cusickii, C. diandra, C. dioica, C. echinata, C. exilis, C. exsiccata, C. flava, C. granularis, C. gynandra, C. hystericina, C. illota, C. interior, C. lacustris, C. lasiocarpa, C. leptalea, C. limosa, C. livida, C. lurida, C. lyngbyei, C. magellanica, C. microptera, C. nebrascensis, C. oligosperma, C. pauciflora, C. pellita, C. pluriflora, C. prairea, C. pseudocyperus, C. rossii, C. rostrata, C. rotundata, C. sartwellii, C. scoparia, C. scopulorum, C. simulata, C. sterilis, C. stipata, C. stricta, C. suberecta, C. tenuiflora, C. trisperma, C. utriculata, C. vesicaria, C. viridula, Carpinus caroliniana, Castilleja coccinea, Cerastium, Chamaecyparis thyoides, Chamaedaphne calyculata, Cicuta bulbifera, C. douglasii, Cirsium vulgare, Cladium mariscoides, Cladopodiella fluitans, Comarum palustre, Coptis trifolia, Cornus amomum, C. racemosa, C. sericea, Cypripedium acaule, Darlingtonia californica, Dasiphora floribunda, Decodon verticillatus, Deschampsia cespitosa, Dicranum polysetum, Doellingeria umbellata, Drepanocladus sendtneri, Drosera anglica, D. intermedia, D. linearis, D. rotundifolia, Dryopteris carthusiana, D. cristata, Dulichium arundinaceum, Eleocharis acicularis, E. elliptica, E. palustris, E. quinqueflora, E. rostellata, E. tenuis, Epilobium anagallidifolium, E. leptophyllum, E. palustre, Equisetum arvense, E. fluviatile, E. hyemale, E. palustre, Eriophorum angustifolium, E. callitrix, E. chamissonis, E. gracile, E. scheuchzeri, E. tenellum, E. vaginatum, E. virginicum, E. viridicarinatum, Eupatorium perfoliatum, Eurybia radula, Euthamia graminifolia, Eutrochium maculatum, Fissidens adianthoides, Fragaria virginiana, Fraxinus nigra, Fuirena simplex, Galium labradoricum, G. palustre, G. trifidum, G.*

triflorum, Gaultheria hispidula, G. shallon, Gaylussacia baccata, Gentianopsis virgata, G. thermalis, Glyceria grandis, G. striata, Gymnadeniopsis clavellata, Gymnocolea inflata, Hamatocaulis lapponicus, H. vernicosus, Helenium autumnale, Hippuris vulgaris, Hudsonia, Hylocomium splendens, Ilex mucronata, I. verticillata, Impatiens capensis, Iris versicolor, Isoetes engelmannii, Juncus balticus, J. drummondii, J. effusus, J. ensifolius, J. stygius, J. subcaudatus, Kalmia microphylla, K. polifolia, Larix laricina, Lemna trisulca, L. valdiviana, Leptodictyum riparium, Limprichtia revolvens, Lobelia kalmii, L. siphilitica, Lonicera involucrata, L. oblongifolia, L. villosa, Lycopus asper, L. uniflorus, Lyonia ligustrina, Lysichiton americanus, Lysimachia quadriflora, L. terrestris, L. thyrsiflora, Maianthemum trifolium, Meesia triquetra, Mentha arvensis, Micranthes pensylvanica, Mimulus primuloides, Mitella nuda, Mnium, Muhlenbergia glomerata, Myrica gale, M. pensylvanica, Myriophyllum verticillatum, Nephrophyllidium crista-galli, Nitella, Nuphar advena, N. polysepala, N. variegata, Nymphaea tetragona, Oclemena nemoralis, Oreostemma alpigenum, Osmunda regalis, Osmundastrum cinnamomeum, Oxypolis rigidior, Packera aurea, P. streptanthifolia, Parnassia glauca, P. palustris, Pedicularis groenlandica, P. labradorica, P. lanceolata, Peltandra virginica, Persicaria amphibia, Petasites frigidus, Phalaris arundinacea, Phragmites australis, Physocarpus opulifolius, Picea engelmannii, P. glauca, P. mariana, Pilea pumila, Pinus contorta, P. ponderosa, P. resinosa, P. strobus, Plantago major, Platanthera blephariglottis, P. dilatata, P. hyperborea, Pleurozium schreberi, Poa pratensis, Pogonia ophioglossoides, Polytrichum strictum, Populus trichocarpa, Potamogeton alpinus, P. epihydrus, P. gramineus, P. natans, Potentilla breweri, Primula jeffreyi, P. tetrandra, Pyrola asarifolia, Quercus vacciniifolia, Ranunculus pallasii, Rhamnus alnifolia, R. frangula, R. purshiana, Rhododendron canadense, R. columbianum, R. groenlandicum, Rhynchospora alba, R. capillacea, Ribes, Rosa palustris, Rubus arcticus, R. chamaemorus, R. pubescens, Rudbeckia fulgida, Rumex occidentalis, Salix arbusculoides, S. bebbiana, S. candida, S. caroliniana, S. discolor, S. drummondiana, S. exigua, S. farriae, S. humilis, S. maccalliana, S. niphoclada, S. pedicellaris, S. planifolia, S. serissima, S. wolfii, Sanguisorba canadensis, Sarracenia purpurea, Scheuchzeria palustris, Schoenoplectus acutus, S. americanus, S. subterminalis, S. tabernaemontani, Scirpus atrovirens, S. cyperinus, S. microcarpus, S. pendulus, Scleria verticillata, Scorpidium scorpioides, Selaginella apoda, S. eclipes, Senecio eremophilus, S. hydrophilus, S. triangularis, Shepherdia canadensis, Silphium terebinthinaceum, Sisyrinchium, Solidago ohioensis, S. patula, S. rigida, S. uliginosa, Sparganium angustifolium, S. eurycarpum, S. fluctuans, Sphagnum angustifolium, S. capillifolium, S. contortum, S. cuspidatum, S. fallax, S. fuscum, S. magellanicum, S. majus, S. nitidum, S. papillosum, S. pulchrum, S. recurvum, S. rubellum, S. subsecundum, S. subtile, S. teres, S. warnstorfii, Spiraea alba, S. douglasii, S. tomentosa, Spiranthes romanzoffiana, Splachnum ampullaceum, Symphyotrichum boreale, S. foliaceum, S. puniceum,

Symplocarpus foetidus, Taraxacum officinale, Taxus canadensis, Thelypteris palustris, Thuja occidentalis, Tomenthypnum nitens, Toxicodendron vernix, Triadenum fraseri, T. virginicum, Triantha glutinosa, Trichophorum alpinum, T. cespitosum, Trientalis borealis, T. europaea, Trifolium repens, Triglochin maritimum, T. palustre, Trollius albiflorus, Typha angustifolia, T. latifolia, Utricularia cornuta, U. gibba, U. intermedia, U. macrorhiza, U. minor, Vaccinium corymbosum, V. macrocarpon, V. oxycoccos, V. uliginosum, Verbascum thapsus, Veronica americana, Viola cucullata, V. macloskeyi, Warnstorfia exannulata, Xyris montana.

Use by wildlife: *Menyanthes* plants and rootstocks are eaten by deer (Mammalia: Cervidae: *Odocoileus*) and moose (Mammalia: Cervidae: *Alces alces*). They also are grazed occasionally by cattle (Mammaila: Bovidae: *Bos*) and sheep (Mammalia: Bovidae: *Ovis*). The foliage is fed upon by several larval moths (Insecta: Lepidoptera: Pyralidae: *Munroessa icciusalis*; Saturniidae: *Hemileuca lucina, H. maia, H. nevadensis*; Tortricidae: *Aphelia alleniana*). The herbage also hosts a variety of fungi including chytrids (Blastocladiomycota: Physodermataceae: *Cladochytrium menyanthis*), gray mold (Ascomycota: Sclerotiniaceae: *Botrytis*), and leaf spot (Ascomycota: Botryosphaeriaceae: *Phyllosticta*; Pleosporomycetidae: *Septoria menyanthis*). Leaf extracts are strongly attractive to bean weevils (Insecta: Coleoptera: Chrysomelidae: *Acanthoscelides obtectus*). The roots host endophytic fungi (Ascomycota: Vibrisseaceae: *Phialocephala fortinii*). The flowers are visited by bees (Insecta: Hymenoptera: Apidae: *Apis mellifera, Bombus borealis, B. sandersoni, B. terricola*), beetles (Insecta: Coleoptera: Nitidulidae: *Meligethes viridescens*), butterflies (Insecta: Lepidoptera: Pieridae: *Pieris napi, P. rapae*), and syrphid flies (Insecta: Diptera: Syrphidae: *Eristalis anthophorina, Melanostoma mellinum, Neoascia globosa*). Frogs (Amphibia: Ranidae: *Lithobates sylvaticus*) use the plants as a substrate for depositing their egg masses. The plants provide habitat for damselflies (Insecta: Odonata: *Lestes disjunctus, L. forcipatus*) and are used at least by the latter species for oviposition. The seeds are eaten by ducks (Aves: Anatidae: *Anas*), red-necked phalaropes (Aves: Scolopacidae: *Phalaropus lobatus*), and other shorebirds.

Economic importance: food: Alaskan Natives used the dried, ground, and leached rootstocks of *Menyanthes* for food or ground them into flour to make bread. The rootstocks also were kept as an emergency food. The Hesquiat used the plants as forage for deer (Mammalia: Cervidae); **medicinal:** *Menyanthes trifoliata* contains several physiologically active glycosides including loganin, menyanthin, sweroside and the seco-cyclopentane glucosides dihydrofoliamenthin, foliamenthin, and menthiafolin. Loganin reportedly inhibits Cox-1 and suppresses a tumor necrosis factor (TNF-α), which is associated with autoimmune disorders. It also has demonstrated a protective effect against hepatic oxidative stress under type 2 diabetes. Sweroside has been shown to be a potential therapeutic agent for treating osteoporosis. The plants also contain several lupane triterpenes that significantly inhibit the

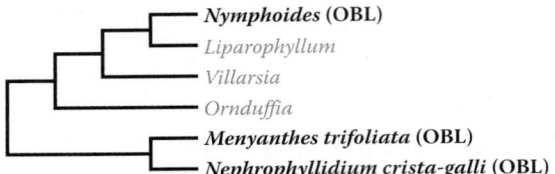

FIGURE 5.181 Phylogenetic relationships among the six genera of Menyanthaceae as indicated by analysis of combined nuclear and chloroplast DNA sequences and DNA secondary structure data. The OBL North American taxa are indicated in bold. Although wetland rankings have not been assigned to the Old World genera (*Liparophyllum*, *Ornduffia* and *Villarsia*), they are completely aquatic as well, which indicates a single origin of the aquatic habit in the family. (Adapted from Tippery, N.P. & Les, D.H., *Novon*, 19, 404–411, 2009.)

synthesis of prostaglandins, which regulate numerous physiological activities. Polysaccharides isolated from the plants stimulate immune cells and act as potent suppressive and anti-inflammatory agents. Decoctions of *M. trifoliata* have been shown to be anti-inflammatory and effective in treating acute renal failure in laboratory animals. Extracts of *M. trifoliata* are highly cytotoxic and exhibit high antitumor potential. Other secondary constituents include coumarins (scoparone, braylin, scopoletin) and the terpenoid lactone loliolide in aerial parts and the triterpenoid saponin menyanthoside in the rhizomes. *Menyanthes* was used as a general medicinal by the Menominee, Micmac, and Tlingit people. The Aleuts prepared an infusion of the roots as a tonic to alleviate constipation, gas pains, and rheumatism. The Kwakiutl used a root or leaf decoction for treating stomach sickness. They prepared a root and stem decoction to assuage blood spitting and administered a root or leaf decoction to promote weight gain when sick with the flu; **cultivation:** *Menyanthes trifoliata* is distributed widely as an ornamental water garden plant. Several successful procedures have been developed for regenerating the plants by tissue culture; **misc. products:** *Menyanthes* leaves have been used as a substitute for hops in beer brewing; **weeds:** *Menyanthes trifoliata* is regarded as a weed in New Zealand; **nonindigenous species:** *Menyanthes trifoliata* was introduced to (and allegedly has been eradicated from) New Zealand. It is introduced as an escape from cultivation in water gardens.

Systematics: *Menyanthes* and *Nephrophyllidium* (both monotypic) resolve as a clade that is sister to the remainder of Menyanthaceae (Figures 5.180 and 5.181). A phyletic relationship between *Menyanthes* and *Nephrophyllidium* has been evidenced repeatedly by analyses of anatomy, morphology, pollen structure, and a considerable amount of DNA sequence data (e.g., nrITS, *atpB*, *ndhF*, *rbcL*, *trnK/matK*). Some authors have recognized the North American *Menyanthes* plants as a smaller-flowered variety (*M. trifoliata* var. *minor* Raf.); however, given the extent and pattern of variability found within this species, there seems to be no compelling reason to maintain them as distinct taxonomically from Eurasian plants. The base chromosome number of *Menyanthes* is $x = 9$. *Menyanthes trifoliata* has hexaploid

($2n = 54$) and dodecaploid ($2n = 108$) cytotypes. Microsatellite genetic markers have been developed and are available for conducting population genetic studies. No hybrids involving *M. trifoliata* are known.

Comments: *Menyanthes trifoliata* occurs throughout northern North America and southward at higher elevations in the west. It has a circumboreal range that extends into Eurasia.

References: Adamczyk et al., 1990; Adamczyk-Rogozińska & Wysokińska, 1998; Addy et al., 2000; Bailey, 1882; Battersby et al., 1968; Bliss, 1958; Bohm et al., 1986; Booth, 2002; Cannings & Simaika, 2005; Champion & Clayton, 2003; Christy, 1987; Cooke, 1997; de Vos, 1958; Eimear et al., 1989; Fernald, 1929; Gustafsson et al., 1996; Han & Kim, 2006; Haraguchi, 1993, 1996, 1999; Hewett, 1964; Hildebrand, 1949; Hogg & Wein, 1988; Huang et al., 1995; Ignatowicz & Gersz, 1997; Janeczko et al., 1990; Jonsdottir et al., 2011; Junior, 1989; Kadereit, 2007; Kincaid, 1904; Knowlton, 1915; Kruse, 1998; Kuduk-Jaworska et al., 2004; Li et al., 2012; Lindholm et al., 2002; Lindsey, 1938; Lundberg & Bremer, 2003; McAtee & Bartram, 1922; McCleary, 1957; Moore, 1989; Moore et al., 1994; Nichols & Sperduto, 2012; Park et al., 2011; Racine & Walters, 1994; Salonen, 1987; Sarneel, 2013; Scholtens & Wagner, 1994; Sun et al., 2013; Thompson et al., 1998; Tippery et al., 2008; Transeau, 1903; Trudeau et al., 2013; Tunón & Bohlin, 1995; Tunón et al., 1994; Vitt & Chee, 1990.

2. *Nephrophyllidium*

Deer-cabbage; crete de coq

Etymology: from the Greek *nephros phyllas idium* ("small kidney leaf") in reference to the leaf shape

Synonyms: *Fauria*; *Menyanthes* (in part); *Villarsia* (in part)

Distribution: global: North America; **North America:** northwestern

Diversity: global: 1 species; **North America:** 1 species

Indicators (USA): OBL: *Nephrophyllidium crista-galli*

Habitat: freshwater; palustrine; **pH:** 3.6–6.9; **depth:** <1 m; **life-form(s):** emergent herb

Key morphology: shoots (to 50 cm) thick (to 2.5 cm), rhizomatous, covered by old leaf bases; leaves alternate, erect, the blades (to 14 cm) cordate or reniform, petioled (to 30 cm), the margins bluntly toothed; flowers (5–30) clustered in terminal scapose racemes or (by bifurcation) panicles, peduncles (to 50 cm) smooth, naked; flowers (to 15 mm) bisexual, distylous, radial, pedicelled (to 35 mm); corollas white, funnelform, short-tubular, 5-lobed, the lobes (to 6 mm) spreading to reflexed, their margins flanged, crisply undulate, the tip cucullate; anthers sagittate; ovary semi-inferior, styles to 5 mm (thrums); capsules (to 25 mm) conical or ovate, unilocular, 10–40-seeded, dehiscent apically by 4 valves; seeds (to 2.5 mm) light brown, ellipsoid, shiny, compressed laterally.

Life history: duration: perennial (rhizomes); **asexual reproduction:** rhizomes; **pollination:** insect; **sexual condition:** hermaphroditic (heterostylous); **fruit:** capsules (common); **local dispersal:** rhizomes, seeds; **long-distance dispersal:** seeds

Imperilment: (1) *Nephrophyllidium crista-galli* [G5]

Ecology: general: *Nephrophyllidium* is monotypic.

Nephrophyllidium crista-galli (**Menzies ex Hook.**) **Gilg** occurs in bogs, bottoms, depressions, fens, heath, marshes, meadows, moor, muskeg, seeps, slopes, swales, swamps, woodlands, and along the margins of lakes, pools, and streams at elevations of up to 2400 m. Populations are concentrated along coastal areas with the number of occurrences declining steadily inland. The plants are highly shade intolerant. They are characterized as oxylophytes, which primarily inhabit dry and wet peatlands on nitrogen-poor substrates. The substrates are acidic (pH 3.6–6.9) and are described as gleysols, gravel, Hemist (Kina), limestone, mesisols, Mor humus, peat (10–60 cm), peaty humus, saprist (Kushnea), stony, terric humisols, and typic and cumulo mesosols. Flowering occurs from late May through mid-June. The flowers are heteromorphic (distylous) and presumably self-incompatible; however, detailed studies of the reproductive biology have not been carried out. Thrum pollen is larger than pin pollen in most populations, but unlike other distylous species, is smaller in some. The gynoecium is UV reflective but the perianth is not. Pollination is carried out by bees and syrphid flies (see *Use by wildlife*), the latter sometimes resembling bees superficially. Seeds that have been dried at 25°C for 1 month germinate under a wide range of conditions. Virtually 100% germination has been obtained for fresh seeds, or for those pretreated at 0°C for 2 months, under a 30°C/25°C, 12/12 hr temperature regime. However, germination rates of unstratified seeds are reduced (~30%) under cooler temperature regimes (e.g., 20°C/10°C). Equal germination rates (99%–100%) have been obtained for seeds placed under light or dark conditions. The primary means of seed dispersal is uncertain. The plants occur primarily on landforms exposed by glacial recession 13,000 years before present. In alpine environments, both the above- and below-ground organs maintain a higher sugar content than starch, which helps to protect the plants from freezing. The leaf nitrogen content declines steadily from May to August. **Reported associates:** *Abies amabilis, A. lasiocarpa, Aconitum delphiniifolium, Actaea rubra, Alnus rubra, A. viridis, Amelanchier alnifolia, Angelica genuflexa, Aquilegia formosa, Artemisia norvegica, Aruncus dioicus, Athyrium filix-femina, Blechnum spicant, Boykinia major, Bistorta bistortoides, Calamagrostis canadensis, C. nutkaensis, Callitropsis nootkatensis, Caltha leptosepala, Campanula, Campylopus atrovirens, Cassiope mertensiana, Carex anthoxanthea, C. aquatilis, C. hoodii, C. laeviculmis, C. limosa, C. livida, C. lupulina, C. nigricans, C. obnupta, C. pauciflora, C. pluriflora, C. pyrenaica, C. rostrata, C. spectabilis, C. utriculata, Cassiope mertensiana, C. lycopodioides, Castilleja lutescens, Chamerion angustifolium, Cladina portentosa, Cladonia crispata, C. rangiferina, Claytonia sibirica, Comarum palustre, Conioselinum chinense, Coptis aspleniifolia, C. trifolia, Cornus canadensis, C. unalaschkensis, Deschampsia cespitosa, Drosera anglica, D. rotundifolia, Eleocharis quinqueflora, Elliottia pyroliflora, Empetrum nigrum, Epilobium bongardii, E. ciliatum, E. palustre, Equisetum fluviatile, Erigeron peregrinus, Eriophorum angustifolium, E. gracile, Fragaria chiloensis, Fritillaria camschatcensis, Gaultheria shallon, Gentiana douglasiana, G. sceptrum, Geranium bicknellii, G. erianthum, Geum calthifolium, Glyceria elata, Gymnocarpium dryopteris, Harrimanella stelleriana, Heracleum sphondylium, Heuchera glabra, Hylocomium splendens, Hypericum anagalloides, Hypnum subimponens, Juncus, Juniperus communis, Kalmia microphylla, K. polifolia, Kopsiopsis hookeri, Lathyrus japonicus, Leptarrhena pyrolifolia, Ligusticum scoticum, Linnaea borealis, Luetkea pectinata, Lupinus arcticus, Luzula campestris, Lycopodium annotinum, Lysichiton americanus, L. camtschatcensis, Maianthemum dilatatum, Menyanthes trifoliata, Menziesia ferruginea, Micranthes nelsoniana, M. nutkana, Microseris borealis, Mimulus guttatus, Mnium, Myrica gale, Neottia cordata, Nuphar polysepala, Oplopanax horridus, Parnassia fimbriata, Pedicularis parviflora, Phalaris arundinacea, Phyllodoce aleutica, P. empetriformis, Picea sitchensis, Pinguicula vulgaris, Pinus contorta, Platanthera dilatata, P. hyperborea, P. stricta, Pleurocladula albescens, Pleurozium schreberi, Podagrostis aequivalvis, P. thurberiana, Polypodium virginianum, Potentilla anserina, P. villosa, Primula jeffreyi, P. pauciflora, Pteridium aquilinum, Ptilium crista-castrensis, Ranunculus occidentalis, R. uncinatus, Racomitrium canescens, R. lanuginosum, Rhizomnium nudum, Rhododendron groenlandicum, Rhynchospora alba, Rhytidiadelphus loreus, Rhytidiopsis robusta, Ribes bracteosum, Rosa nutkana, Rubus chamaemorus, R. parviflorus, R. pedatus, R. pubescens, R. spectabilis, Rumex occidentalis, Sambucus racemosa, Sanguisorba officinalis, S. stipulata, Senecio triangularis, Sieversia calthifolia, Siphula ceratites, Sorbus sitchensis, Sparganium natans, Sphagnum angustifolium, S. austinii, S. capillifolium, S. centrale, S. cuspidatum, S. fuscum, S. lindbergii, S. magellanicum, S. mendocinum, S. pacificum, S. papillosum, S. recurvum, S. rubellum, S. squarrosum, S. subnitens, S. subsecundum, S. tenellum, S. teres, S. warnstorfii, Spiraea douglasii, Spiranthes romanzoffiana, Stellaria longipes, Streptopus amplexifolius, Thuja plicata, Thelypteris quelpaertensis, Triantha glutinosa, Trichophorum cespitosum, Trientalis europaea, Triglochin palustre, Tsuga heterophylla, T. mertensiana, Urtica dioica, Vaccinium cespitosum, V. deliciosum, V. ovalifolium, V. oxycoccos, V. uliginosum, V. vitis-idaea, Vahlodea atropurpurea, Veratrum viride, Viburnum edule, Viola glabella, V. langsdorffii.*

Use by wildlife: With a digestibility of 79%, *Nephrophyllidium* is an excellent forage plant and can contain as much as 24.6% protein, 13.6% fiber, 3.9% nitrogen, and 2.7% fat on a dry weight basis. It is a preferred summer food of black-tailed deer (Mammalia: Cervidae: *Odocoileus hemionus*). The flowers of *Nephrophyllidium* are visited by various insects (Insecta) including bees (Hymenoptera: Apidae: *Bombus flavifrons*), beetles (Coleoptera), butterflies (Lepidoptera: Lycaenidae: *Lycaena mariposa*), flies (Diptera: Syrphidae: *Eristalis anthophorina, Melanostoma mellinum, Myiolepta bella, Syrphus ribesii*), and wasps (Hymenoptera: Vespidae), which (except for beetles) presumably function as pollinators. The plants are hosts to rust fungi (Basidiomycota: Pucciniaceae: *Puccinia nephrophyllii*).

Economic importance: food: *Nephrophyllidium* is not edible; **medicinal:** *Nephrophyllidium* was used by the Haida people for treating respiratory illnesses. The plant extracts exhibit some inhibitory activity to a number of fungal strains; **cultivation:** *Nephrophyllidium* is sold as an ornamental pond or water garden plant; **misc. products:** none; **weeds:** none; **nonindigenous species:** none

Systematics: Phylogenetic analyses resolve *Nephrophyllidium* and *Menyanthes* (both monotypic) as a clade that is sister to the remainder of Menyanthaceae (Figures 5.180 and 5.181). Although *Nephrophyllidium* is closely related to *Menyanthes*, it has a unique flavonoid profile, which is not shared with the latter. The base chromosome number of *Nephrophyllidium* is $x = 17$. Two subspecies are recognized: the Asian tetraploid *N. crista-galli* subsp. *japonica* ($2n = 68$), and the North American hexaploid *N. crista-galli* subsp. *crista-galli* ($2n = 102$). No hybrids involving *Nephrophyllidium* are known.

Comments: *Nephrophyllidium crista-galli* occurs in Alaska, British Columbia, Oregon, and Washington with its range extending into Japan and far eastern Russia.

References: Asada et al., 2003; Boggs et al., 2010; Bohm et al., 1986; Brett et al., 2001; Bryant et al., 1983; Campbell, 1899; Ceska & Scagel, 2011; Christy, 2004; Cooper, 1939b; D'Amore & Lynn, 2002; DeVelice et al., 1999; Douglas, 1983; Gillett, 1968; Hanley & McKendrick, 1985; Kincaid, 1904; Klein, 1965; Lamb & Megill, 2003; MacKenzie & Moran, 2004; Mains, 1939; Matsuoka et al., 2012; McCutcheon, 1996; Persson, 1954; Pojar, 1974; Quickfall, 1987; Shacklette, 1961; Shibata & Nishida, 1993; Shigenobu, 1984; Shimono & Kudo, 2005; Taylor, 1932; Tippery & Les, 2008, 2009; Tippery et al., 2008; Viereck et al., 1992; Wade, 1965; Walton et al., 2014.

3. *Nymphoides*

Floating-heart, water snowflake; faux nénuphar, faux-nymphéa

Etymology: from the Greek *nymphaia oida* ("like a water lily") for its resemblance to the genus *Nymphaea*

Synonyms: *Limnanthemum*; *Schweyckerta*; *Trachysperma*; *Waldschmidia*

Distribution: global: cosmopolitan; **North America:** eastern, southern and western

Diversity: global: 50 species; **North America:** 6 species

Indicators (USA): OBL: *Nymphoides aquatica*, *N. cordata*, *N. cristata*, *N. humboldtiana*, *N. indica*, *N. peltata*

Habitat: freshwater; lacustrine, palustrine; **pH:** 4.9–7.6; **depth:** 0–2 m; **life-form(s):** floating-leaved

Key morphology: Shoots rhizomatous, congested or stolonlike; submersed leaves absent or present in juvenile plants, the blades membranous, rhomboid or deltoid, the petioles flattened; floating leaves arising from inflorescence or rhizome, ovate to orbicular, often coriaceous, the blades (to 20 cm) cordate to reniform with two basal lobes, the upper surfaces uniformly colored or variegated, the margins entire or undulate dentate, the petioles (to 1 m) slender, flexuous; inflorescence clusters 1–20, lax, umbellate, 5–40 flowered, each subtended by 1–2 floating leaves, peduncles varying in length with water depth, adventitious roots (2–5 mm thick), with or without lateral rootlets, when present arising from within flower clusters to comprise small plantlet propagules; flowers unisexual (plants dioecious or gynodioecious) or bisexual (monomorphic or distylous), pedicels (to 10 cm)

erect, corolla (to 5 cm) 4, 5, 6, or 8-merous, the lobes white or yellow, lateral and medial wings present or absent, inner surface glabrous or with dense glandular hairs, margins entire to laciniate; stamens with orange pollen (hermaphrodites, dioecious males) or lacking pollen (dioecious females), ovary superior, styles (to 7 mm) present (hemaphrodites, dioecious females) or absent (dioecious males); capsules (to 8–35 mm) globose or ellipsoid, round or strongly compressed laterally, dehiscent irregularly by breakdown of wall; seeds (4–80; to 1.8–6.0 mm) green, light/dark brown or yellowish, orbicular to ellipsoid, round to strongly compressed laterally, surface smooth or papillate, margins smooth or ringed by stiff hairs (to 2.6 mm).

Life history: duration: perennial (rhizomes); **asexual reproduction:** rhizomes, vegetative plantlets; **pollination:** insect; **sexual condition:** dioecious, gynodioecious, or hermaphroditic; **fruit:** capsules (common); **local dispersal:** seeds (water), vegetative fragments; **long-distance dispersal:** seeds (waterfowl)

Imperilment: (1) *Nymphoides aquatica* [G5]; S1 (MD, VA); S2 (MS); S3 (NC); (2) *N. cordata* [G5]; SH (LA); S1 (DE, MD, MS, NC); S2 (PA, RI, <u>NF</u>): S3 (NJ, NY, <u>QC</u>); (3) *N. cristata* [GNR]; (4) *N. humboldtiana* [unranked]; (5) *N. indica* [GNR]; (6) *N. peltata* [G5]

Ecology: general: All species within this genus are benthic pleustophytes having natant leaves, which arise from submersed stems that are anchored to the substrate. Most species occupy a zone of floating-leaved vegetation that is shallower than that of the larger "water lilies" such as *Nuphar* and *Nymphaea*. The plants display a diversity of reproductive syndromes including self-compatibility, diallelic self-incompatibility, homostylous and heterostylous hermaphroditic flowers, and unisexual flowers. Four species are dioecious, including the North American *N. aquatica* and *N. cordata*; *N. cristata* is gynodioecious. The flowers typically are outcrossed by insects (Insecta), with self-pollination occurring occasionally. The fringed corolla margins of some species increase their surface tension, making it more difficult to plunge the flower underwater; they also can trap an air bubble if immersed, thereby protecting the reproductive organs from becoming wet and nonfunctional. All of the species perennate by means of rhizomes. The thickened adventitious roots produced by some species within their inflorescences have been referred to incorrectly as tubers. The fruits mature underwater and dehisce irregularly by disintegration of their wall. The seeds sink immediately or float for a limited time. Their dispersal is mediated by water, or by external attachment to migrating waterfowl. Seeds are dormant through winter and require a period of stratification to break primary dormancy. Secondary dormancy must be overcome by exposure to light and oxygen. Germination occurs only on exposed substrates or in very shallow (<2 cm) water. The intrinsic beauty of the delicate colorful flowers and floating habit of the plants makes them prized as horticultural specimens for ornamental ponds and aquaria. However, several of the species grow aggressively, with the potential to become noxious weeds capable of shading out other species of aquatic plants and adversely impacting other aquatic organisms.

***Nymphoides aquatica* (J.F. Gmel.) Kuntze** inhabits backwaters, bogs, canals, depressions, ditches, floodplains, "houses", lagoons, lakes, marshes, ponds, pools, prairies, sinkholes, sloughs, streams, and the margins of lakes and ponds at elevations of up to 200 m. It usually occurs in still or lentic, shallow water, but can be found in depths up to 2.1 m. The plants are tolerant to a fair amount of shade. The waters primarily are acidic, soft water, and nutrient-poor with low pH (4.9–7.6; mean=6.1), alkalinity (1–26.5 mg/L $CaCO_3$; mean=9.5), calcium (0.7–23.7 mg/L; mean=9.0), chloride (4.3–25.8 mg/L; mean=14.4), total phosphorous (8–37 µg/L; mean=19), and specific conductance (28–183 µS/cm; mean=89). Despite this being a floating-leaved species, the lakes of occurrence typically are clear with relatively high Secchi disc values (0.7–4.1 m; mean=2.0). The plants also can be found on drying shores and will persist at sites that dry up annually. Substrates are described as alluvium, limestone, muck, mucky peat, mud, sand, and Shenks muck. The flowers are unisexual and the plants dioecious. Flowering proceeds from March to November. No specific pollinators have been reported but are assumed to be insects (Insecta). The seeds are produced in large numbers. Because they are eaten by waterfowl (Aves: Anatidae), presumably they also are dispersed by them. No information is available on the seed ecology. Vegetative reproduction occurs in the fall, at inflorescence sites, by the adventitious development of plantlets with large fleshy roots (to 5 mm thick) that resemble tiny bunches of bananas. These detach from the parental plants and drift on the water for extended time periods. The plantlets remain dormant until the following spring. Some unisexual populations propagate exclusively vegetatively. The plants have been cultured successfully in a 54 cm water depth using sand fortified with Osmocote (18–6–12; 228, 456, 912 g/m²) or Vigoro (6–10–4; 285 g/m²) fertilizer. The herbage has a mean C:N ratio of about 35 and N:P ratio of 40. The plants have been observed to occur more commonly in sites rooted by feral hogs (Mammalia: Suidae: *Sus scrofa*) than in nonrooted control sites. **Reported associates:** *Acer rubrum, Acrostichum danaeifolium, Aeschynomene pratensis, Agalinis, Alternanthera philoxeroides, Ampelaster carolinianus, Amphicarpum muhlenbergianum, Andropogon glomeratus, A. virginicus, Annona glabra, Axonopus furcatus, Baccharis halimifolia, Bacopa caroliniana, B. monnieri, Bidens mitis, Blechnum serrulatum, Boehmeria cylindrica, Brasenia schreberi, Cabomba caroliniana, Centella asiatica, Cephalanthus occidentalis, Ceratophyllum demersum, Cladium jamaicense, Conyza canadensis, Crinum americanum, Cyperus haspan, C. odoratus, C. polystachyos, C. retrorsus, Cyrilla racemiflora, Decodon verticillatus, Diodia virginiana, Dulichium arundinaceum, Echinochloa crusgalli, Eleocharis baldwinii, E. cellulosa, E. elongata, E. equisetoides, E. flavescens, E. geniculata, E. interstincta, Eragrostis refracta, Erechtites hieraciifolius, Eriocaulon compressum, E. decangulare, Eubotrys racemosa, Eupatorium capillifolium, E. compositifolium, E. mohrii,* *Euthamia graminifolia, Fuirena breviseta, F. scirpoidea, F. squarrosa, Funastrum clausum, Galium tinctorium, Gratiola brevifolia, Habenaria repens, Heteranthera dubia, H. reniformis, Hydrilla verticillata, Hydrocotyle ranunculoides, H. umbellata, Hymenocallis palmeri, Hypericum hypericoides, Ipomoea sagittata, Iris, Juncus megacephalus, J. scirpoides, Justicia angusta, Kosteletzkya pentacarpos, Lachnanthes caroliniana, Leersia hexandra, L. virginica, Lemna, Limnobium spongia, Limnophila, Ludwigia alata, L. arcuata, L. decurrens, L. grandiflora, L. hexapetala, L. lanceolata, L. leptocarpa, L. linifolia, L. octovalvis, L. peploides, L. pilosa, L. repens, L. sphaerocarpa, Luziola fluitans, Magnolia virginiana, Mayaca fluviatilis, Mikania scandens, Myrica cerifera, Myriophyllum heterophyllum, M. laxum, Najas, Nelumbo lutea, Nitella gracilis, Nuphar advena, Nymphaea mexicana, N. odorata, Nymphoides cordata, Nyssa biflora, N. sylvatica, Oeceoclades maculata, Oldenlandia uniflora, Orontium aquaticum, Osmunda regalis, Osmundastrum cinnamomeum, Panicum dichotomiflorum, P. hemitomon, P. verrucosum, Paspalidium geminatum, Paspalum blodgettii, Peltandra virginica, Persicaria hirsuta, P. hydropiperoides, P. punctata, Pluchea baccharis, Polygala rugelii, Pontederia cordata, Potamogeton diversifolius, P. pusillus, Proserpinaca palustris, Psilotum nudum, Pterocaulon virgatum, Rhexia cubensis, R. nashii, Rhynchospora fascicularis, R. filifolia, R. inundata, R. scirpoides, R. tracyi, Ricciocarpus natans, Saccharum giganteum, Sacciolepis striata, Sagittaria australis, S. graminea, S. lancifolia, S. latifolia, Salix caroliniana, Solidago, Sphagnum, Saururus cernuus, Taxodium ascendens, Thalia geniculata, Triadenum virginicum, T. walteri, Typha domingensis, Utricularia foliosa, U. gibba, U. juncea, U. purpurea, U. radiata, Websteria confervoides, Woodwardia virginica, Xyris ambigua, X. jupicai, X. panacea, X. smalliana.*

***Nymphoides cordata* (Elliott) Fernald** occurs in bayous, Carolina bays, depressions, dunes, lakes, mudflats, ponds, pools, seeps, sinkholes, streams (lentic), and along the margins of rivers at elevations of up to 620 m. The plants often occur in shallow water (30–89 cm; mean=65) but can grow at depths of up to 2.0 m. They also can survive along the strand of receding shorelines. Exposures range from full sun to partial shade. In the Great Lakes region, the plants have a narrow niche breadth (*T*-value of 3). There they are found mainly in oligotrophic sites with high water clarity and low turbidity, and where there is little or no human impact. In other areas, plants can occupy fairly enriched sites. The substrates are described as gravel, gravelly sand, muck, mucky sand, mud, peaty muck, sand sandy mud, and sandy peat. This is a dioecious species that flowers from May to September. Few specifics exist on the pollinators (presumably Insecta) or on other aspects of the reproductive biology. Seeds extracted from sediment cores and stored at 4°C for 3–123 days, were able to germinate on sand (under 1–2 cm water) at ambient greenhouse temperatures ranging from 16°C to 32°C over a 480-day period of analysis. They are well-represented in seed banks and comprise about 5% of the relative abundance in those sites studied. Like the closely related *Nymphoides aquatica* (earlier), asexual reproduction can occur by vegetative plantlets that are produced from the inflorescence. These bear clusters of adventitious, fusiform (to 2 mm thick), fleshy roots. Plant extracts reportedly have exhibited allelopathic activity against the growth of duckweed (Lemnaceae) and lettuce (Asteraceae: *Lactuca sativa*) under laboratory conditions. **Reported associates:** *Andropogon glomeratus, Bacopa caroliniana, Bidens beckii, Brasenia schreberi, Cabomba caroliniana, Centella asiatica, Ceratophyllum demersum, Chamaecyparis thyoides, Decodon verticillatus, Dichanthelium acuminatum, Drosera intermedia, Eleocharis acicularis, E. equisetoides, E. quadrangulata, E. flavescens, E. melanocarpa, E. robbinsii, E. tuberculosa, Eriocaulon aquaticum, E. compressum, Eupatorium leptophyllum, Euthamia graminifolia, Fimbristylis puberula, Fontinalis, Gratiola aurea, Hydrilla verticillata, Hydrocotyle umbellata, Hypericum crux-andreae, H. lissophloeus, Isoetes echinospora, Juncus militaris, J. repens, Lachnanthes caroliniana, Leersia hexandra, Lemna minor, Lobelia dortmanna, Ludwigia linearis, L. spathulata, L. suffruticosa, Luziola fluitans, Lysimachia terrestris, Mayaca fluviatilis, Myriophyllum heterophyllum, M. heterophyllum×M. laxum, M. humile, M. laxum, M. tenellum, M. verticillatum, Najas flexilis, Nitella, Nuphar microphylla, N. variegata, Nymphaea odorata, Nymphoides aquatica, Panicum hemitomon, P. verrucosum, Peltandra virginica, Pontederia cordata, Potamogeton amplifolius, P. bicupulatus, P. diversifolius, P. epihydrus, P. gramineus P. natans, P. robbinsii, Proserpinaca pectinata, Rhexia mariana, R. virginica, Rhynchospora chalarocephala, R. filifolia, R. inundata, R. perplexa, Sagittaria graminea, S. isoetiformis, S. rigida, S. teres, Schoenoplectus etuberculatus, S. pungens, Scleria reticularis, Sium suave, Sparganium angustifolium, S. fluctuans, Spirodela polyrhiza, Triadenum virginicum, Typha latifolia, Utricularia gibba, U. intermedia, U. juncea, U. macrorhiza, U. purpurea, U. radiata, U. resupinata, Vallisneria americana, Wolffia columbiana, Xyris elliottii, X. smalliana, Zizania aquatica.*

***Nymphoides cristata* (Roxb.) Kuntze** grows in canals, depressions, lakes, reservoirs, and sloughs at elevations of up to 48 m. It can occur in shallow (e.g., 20 cm) to deeper (to 2 m) water and will tolerate temperatures as low as −7°C. This is a gynodioecious species that flowers from April to November. Some seed production (but of uncertain viability) has been reported in North American populations; however, the principal means of reproduction there is asexual. Vegetative reproduction occurs by rhizome proliferation and fragmentation as well as by the formation of adventitious plantlets that bear tuberous root clusters, which can reach 0.5 m in length. Few other details on the life history of this species are known. **Reported associates (North America):** *Azolla caroliniana, Bacopa caroliniana, Cabomba caroliniana, Chara, Lemna, Najas guadalupensis, Nuphar, Pistia stratiotes, Vallisneria americana.*

***Nymphoides humboldtiana* (Kunth) Kuntze** occurs in ponds, springs, and along the margins of rivers at elevations of up to 300 m. The plants are shade tolerant and grow mainly in shallow water (0.1–0.3 m) but can occur up to 1.0 m depth. The flowers are distylous, strongly self-incompatible, and are outcrossed by bees (Hymenoptera: Apidae). The average pollen size is larger in short-style morphs (45.7 μ) than in long-style morphs (38.7 μ). Intra-morph pollinations occasionally produce from 1 to 3 seeds. Intermorph pollinations produce an average of 8.3 (long-styled morphs) to 8.6 (short-styled morphs) seeds per capsule. This species does not produce adventitious plantlets from the inflorescence region, but can propagate vegetatively by rhizome proliferation and fragmentation. The exclusive occurrence of short-styled morphs at one USA locality indicates the importance of vegetative reproduction. Because this species currently is known only from a small region of Texas, there is little life-history information available that specifically addresses the North American populations. **Reported associates (Texas):** *Shinnersia rivularis.*

***Nymphoides indica* (L.) Kuntze** grows in ponds and along the margins of lentic streams at elevations of up to 40 m. The insect-pollinated flowers are heterostylous (distylous), self-incompatible, and are produced nearly year-long (February–December). The white corollas strongly absorb UV light rather than reflect it. The morph ratios of populations can vary; however, those with equal ratios experience higher seed set (60%–87%) than biased ratio (9%–36%) or monomorphic (0%–18%) populations. Genetic markers (allozymes) also indicate a significant positive correlation between seed set and the number of multilocus genotypes (MLGs) within a population. Monomorphic populations typically contain only one MLG, most likely as a result of clonal propagation of a single founding individual. Seeds are dispersed by sinking below the parental plants or by their transport on floating plant fragments throughout a water body. In one study (in Japan), viable seeds persisted for at least 3 years in pond sediments as much as 12 m from shore in waters up to 60 cm deep. The seeds remain dormant while under water, but germinate readily when they become exposed on drying mudflats. Observed germination rates were highest for 1-year-old seeds (65%) and declined for 2-year-old (38%) and 3-year-old seeds (23%). The floating leaves produce adventitious plantlets, which have been cultured successfully in 42 cm of water (at 21°C), on sand using a 17–6–10 fertilizer (565 g/m²) formulated for an 8–9 month release, and placed 7 cm beneath the surface. The floating leaves live on average for 23 days, have a mean leaf area index of 1.9 m²/m² and can yield a seasonal biomass of 222 g/m². They decompose at a rate (mean k) of 0.0143/day. Relative growth rates are higher for seeds (54.3–64.3 mg/g day) than for vegetative propagules (41.4–57.1 mg/g day). **Reported associates (Florida):** *Orontium aquaticum, Sarracenia leucophylla, Woodwardia areolata.*

***Nymphoides peltata* (S.G. Gmel.) Kuntze** inhabits gravel pits, lakes, reservoirs, rivers, ponds, and swamps at elevations of up to 1400 m. Although the plants occur occasionally in oligotrophic sites, they are most common in alkaline (pH: 6.9–7.6; alkalinity ~3.76 meq L⁻¹), still or lentic, eutrophic waters from 0.2 to 4 m deep and rarely colonize slightly brackish sites (<125 mg Cl L⁻¹). They also can survive when exposed on wet mud, persisting at some sites even throughout the winter when temperatures can fall to −30°C. The substrates include alluvium, clay, muck, organic mud, rock, sand, silt, and silty clay. The flowering period extends from May to October. Although

Eurasian plants are heterostylous, flowers in North American populations are monomorphic with short or intermediate style lengths. Many populations are known to suffer from genetic bottlenecks and introduced populations in particular tend to have low genetic variation and to reproduce primarily by vegetative means. Clonal reproduction can result in strongly biased or monomorphic morph ratios, which interferes with sexual reproduction in these ordinarily strongly self-incompatible plants; however, homostyle genets can be self-compatible. Allozyme studies have shown that multiclonal populations are necessary in order to maintain high levels of genetic variation. A study using microsatellite markers found low clonal diversity in a population associated with high levels of inbreeding in the development of the seed bank. Capsule density can average $180\,m^{-2}$. An average yield of 26.5–29 seeds/capsule is produced in outcrossed plants; however, self-pollinated flowers (especially short-styled morphs) produce an average of 5.2 seeds. Although the flowers open and wither within 1 day, the plants produce a high flower frequency and maintain a long seasonal flowering period. Fertility is limited by dilution of compatible pollen. Insect pollinators are attracted to the large (avg. = 3.54 cm), yellow, ultraviolet reflective corollas, which are raised an average of 3.27 cm above the water surface. They also emit a faint fragrance and provide nectar. Principal pollinators in both the European and North American range include bees (Hymenoptera: Apidae) and flies (Diptera: Ephydridae; Syrphidae). The capsules mature just beneath the water surface. The seeds mature within 32–60 days after anthesis and can attain average annual densities of $3117/m^2$. They average 3 mg in mass and possess marginal hairs with a hydrophobic coating. The buoyant seeds float initially and are dispersed locally on the water surface by wind and wave action. The seeds aggregate on the water surface to form chain-like rafts. They are tolerant of desiccation (dried seeds remaining viable for up to 30 months) and are capable of long distance ectozoic dispersal as they attach externally to the head, feet, or flanks of waterfowl. The seeds are released from their vectors as they dry. They are incapable of endozoic transport by fish or water birds (see *Use by wildlife*), which digest them completely. Low wind velocities (2.4–3.0 m/s) disrupt their buoyancy, and 90% of seeds will sink after 67 h of gentle surface agitation. Sinking occurs almost immediately under simulated rain conditions. All seeds sink eventually with the onset of seasonal frost and remain dormant through the winter. The seeds require cold stratification, light, and oxygen for germination, which indicates their optimal conditions to be shallow water where light and oxygen levels are adequate. A 32% germination rate was obtained by placing seeds at 5°C for 8 weeks followed by germination on 1% agar (+ 250 mg/l gibberellic acid) under an 8/16 hr, 25°C/10°C light and temperature regime. In another experiment, stratification for 2 weeks produced high germination (63.3%) at a light intensity of 20 μmol photons m^{-2} s^{-1}. Another study obtained germination for seeds placed in aerated tap water at 20°C under a 14 h photoperiod (150 μEinsteins m^{-2} h^{-1}), following 4 weeks of cold stratification at 4°C. In North American populations, the seedlings emerge from April to July. The seeds do not germinate at sediment depths beyond 0.25 cm, but those at the water-sediment interface have high germination (74%). Most seeds float on the surface once germinated (sometimes massing in rafts), but only a small number (14%) can reestablish in shallow water (<3 cm deep). A short-term, persistent seed bank can develop having seeds densities of up to $796/m^2$. The seedlings are highly susceptible to inundation and light attenuation and survive best on bare ground exposed during periods of water-level drawdown. The plants can reach a peak August biomass (ash-free dry weight) of $372\,g/m^2$ with a net annual productivity of 1036 g m^{-2}. The herbage decomposes rapidly, reaching near completeness within 40 days. The plants reproduce vegetatively by proliferation and fragmentation of the rhizomes and stoloniferous shoots. The roots are not mycorrhizal. Microcosm competition studies indicate that the plants fare better in mixed stands of floating and submersed species, but more poorly when grown with emergent species. The average longevity of floating leaves is about 28 days. Plants grown under terrestrial conditions exhibit 89% less total biomass, 63% greater root biomass, 81% higher root:shoot ratio, and 55% greater leaf longevity than plants grown under aquatic conditions. **Reported associates (North America):** *Agrostis, Alnus rubra, Azolla caroliniana, Butomus umbellatus, Carex, Ceratophyllum, Eleocharis, Equisetum fluviatile, Eriocaulon aquaticum, Hydrocharis morsus-ranae, Iris pseudacorus, Isoetes echinospora, I. lacustris, Juncus tenuis, Leersia, Lemna minor, Lobelia dortmanna, Ludwigia palustris, Myriophyllum spicatum, M. tenellum, Najas flexilis, Nuphar variegata, Nymphaea odorata, Persicaria, Phalaris, Pistia stratiotes, Potamogeton foliosus, Rubus ulmifolius, Sagittaria graminea, Salix lucida, Sparganium, Stuckenia pectinata, Trapa natans, Typha, Vallisneria americana, Xanthium strumarium.*

Use by wildlife: *Nymphoides aquatica* is a consistent habitat element for alligators (Reptilia: Alligatoridae: *Alligator mississippiensis*). The plants are eaten by crayfish (Crustacea: Decapoda: Cambaridae: *Procambarus spiculifer, P. acutus*), Florida cooter turtles (Reptilia: Emydidae: *Pseudemys concinna*), grass carp (Osteichthyes: Cyprinidae: *Ctenopharyngodon idella*), and can comprise over 6% of the rumen contents of white-tailed deer (Mammalia: Cervidae: *Odocoileus virginianus seminolus*). The foliage is a host to several larval moths (Insecta: Lepidoptera: Crambidae: *Langessa nomophilalis, Paraponyx allionealis, P. seminealis, Synclita obliteralis*) and parasitic heterokonts (Heterokontophyta: Oomycota: Pythiaceae: *Pythium*). The seeds are eaten by waterfowl (Aves: Anatidae). *Nymphoides cordata* is a host to club fungi (Basidiomycota: Doassansiaceae: *Doassansia decipiens, Doassansiopsis limnanthemi*). There are unsubstantiated reports of *N. cristata* being eaten by comet-tailed goldfish (Osteichthyes: Cyprinidae: *Carassius auratus*). In South America (and presumably also in North America), the flowers of *N. humboldtiana* attract bees (Insecta: Hymenoptera: Apidae: *Ceratina, Florilegus*), beetles (Insecta: Coleoptera), and butterflies (Insecta: Lepidoptera). *Nymphoides indica* is used for nest construction material by the nonindigenous brown hoplo catfish (Osteichthyes: Callichthyidae: *Hoplosternum littorale*). Larval leafcutter moths (Insecta: Lepidoptera: Crambidae: *Aulacodes siennata, Parapoynx diminutalis*) have been found on the plants. *Nymphoides peltata* is eaten by red swamp crayfish (Decapoda: Cambaridae: *Procambarus clarkii*), domestic cattle (Mammalia: Bovidae: *Bos taurus*), and muskrats (Mammalia: Cricetidae: *Ondatra zibethicus*). The seeds of *N. peltata* are eaten by coots (Aves: Rallidae: *Fulica atra*) and ducks (Aves: Anatidae: *Anas platyrhynchos*) but rarely by carp (Osteichthyes: Cyprinidae: *Cyprinus carpio*). The flowers of North American populations are visited frequently by honey bees and bumble bees (Insecta: Hymenoptera: Apidae: *Apis mellifera, Bombus*), flies (Insecta: Diptera: Syrphidae), and wasps (Insecta: Hymenoptera), which function as their pollinators.

Economic importance: food: *Nymphoides cordata* plants were eaten as greens by native Canadian tribes; **medicinal:** *Nymphoides* species contain several flavonoids (kaempferol, myricetin, quercetin), which are antioxidants. *Nymphoides aquatica* is said to have weak antioxidant activity and was used as a medicinal plant by the Seminoles to treat chronic cough (turtle sickness), and as an aid in childbirth. *Nymphoides cordata* was used by the Seminoles as a cough suppressant, respiratory aid, and sedative. *Nymphoides indica* is one of nine species in a herbal formulation (recipe N040) that has high antiproliferative (anticancer) activity on human cervical adenocarcinoma (HeLa) cells. The plants also have been included in various recipes for treating epilepsy, jaundice, mental disorders, and tuberculosis. They are used by the Rama people of Nicaragua as a medicinal and to treat the side effects of snakebites. Methanolic extracts of *N. peltata* showed intermediate levels of fusion inhibition (43%) of type 1 human immunodeficiency virus (HIV-1). The plants also have been used as a remedy for snakebite; **cultivation:** *Nymphoides aquatica, N. cordata, N. cristata, N. indica,* and *N. peltata* all are cultivated as ornamental aquarium or pond and water garden plants, which sometimes are referred to as "banana plant" or "fairy water lily". *Nymphoides cordata* was being sold as a horticultural specimen as early as 1831 and several species remain readily available through mail-order commerce. *Nymphoides peltata* also is distributed under the cultivar name 'Bennettii'; **misc. products:** *Nymphoides aquatica* has been planted in wetland restoration projects. *Nymphoides peltata* has been studied for cadmium removal and other types of phytoremediation; **weeds:** *Nymphoides cristata* is highly invasive and represents a serious threat to water bodies. One account states that the plants can cover a 3.6 ha lake surface within 3–4 weeks. In South Carolina, coverage of plants on one lake burgeoned from 20 ha in 2006 to over 1500 ha in 4 years (2010). In a Florida lake, a population expanded from 0.4 ha coverage (in 2005) to 809 ha 4 years later (2009) even though herbicides were applied in 2005. The plants are recalcitrant to control efforts because herbicides cause leaf dieback, while the rhizomes remain viable. Both *N. cristata* and *N. humboltiana* are ricefield weeds in Argentina. *Nymphoides peltata* is prohibited and/or declared an invasive or noxious weed in Connecticut, Maine, Massachusetts, Oregon,

Vermont, and Washington. It also is aggressively weedy in Sweden; **nonindigenous species:** *Nymphoides aquatica* is adventive in Japan and the seeds are prohibited from import in New Zealand. *Nymphoides cristata* has been in global horticultural trade since 1982. It was introduced to Florida from Asia around 1988 as an escape from cultivated aquarium plant stocks and was introduced to Argentina as a ricefield weed. It also is reported to be a weed of deep water rice in Asia. *Nymphoides indica* was introduced to Florida as an escape from cultivation. *Nymphoides peltata* has been introduced to North America, Sweden, and New Zealand. North American plants were introduced sometime before 1882 as escapes from cultivation. Genetic data indicate that the North American plants most likely originated from European stocks.

Systematics: Phylogenetic analyses of molecular data have greatly clarified the systematics of *Nymphoides*, its circumscription, and its relationship to other Menyanthaceae. As currently circumscribed, *Nymphoides* is the most diverse genus of Menyanthaceae, where it occupies a derived phylogenetic position as the sister group of *Liparophyllum* (Figures 5.181 and 5.182). However, molecular data also reveal a complex phylogenetic history of ancestral hybridization for many species, which has resulted in discordant placements in phylogenetic trees constructed from DNA sequence data derived from maternally inherited vs. nuclear-inherited loci. However, the same phylogenetic analyses do resolve some associations of OBL North American species consistently, such as a close sister-group relationship between the temperate, dioecious *N. aquatica* and *N. cordata*, and also between the primarily neotropical *N. humboldtiana* and *N. fallax* (Figure 5.182). The trees also generally resolve *N. peltata* near the base of the *Nymphoides* clade, with *N. aquatica, N. cordata, N. cristata, N. humboldtiana,* and *N. indica* all placed within a relatively derived clade (Figure 5.182). Although long regarded as synonymous with the strikingly similar Old World *N. indica*, phylogenetic analyses have shown the New World *N. humboldtiana* to be distinct. Consequently, much of the literature is difficult to interpret because of the past tendency to report information gathered for either taxon under the name *N. indica*. The name "*N. indica*" also has been applied to yellow-petaled plants in Florida. However, those plants have been identified recently as *N. grayana* (Grisebach) Kuntze, and likely represent a natural range extension from the Carribean. This species has not been included in the present account due to the scarcity of reliable information for the North American plants. The base chromosome number of *Nymphoides* is $x = 9$. The North American representatives include diploids ($2n = 18$): *N. cristata, N. indica*; tetraploids ($2n = 36$): *N. aquatica, N. cordata, N. humboldtiana*; and a hexaploid ($2n = 54$): *N. peltata*. *Nymphoides aquatica* and *N. cordata* form natural hybrids, which are intermediate morphologically. Despite evidence of extensive past hybridization in the genus, few other natural hybrids have been reported.

Comments: *Nymphoides peltata* is found throughout the United States and eastern Canada; *N. aquatica* occurs throughout the coastal southeastern United States; *N. cordata* occurs in eastern North America and was extirpated from

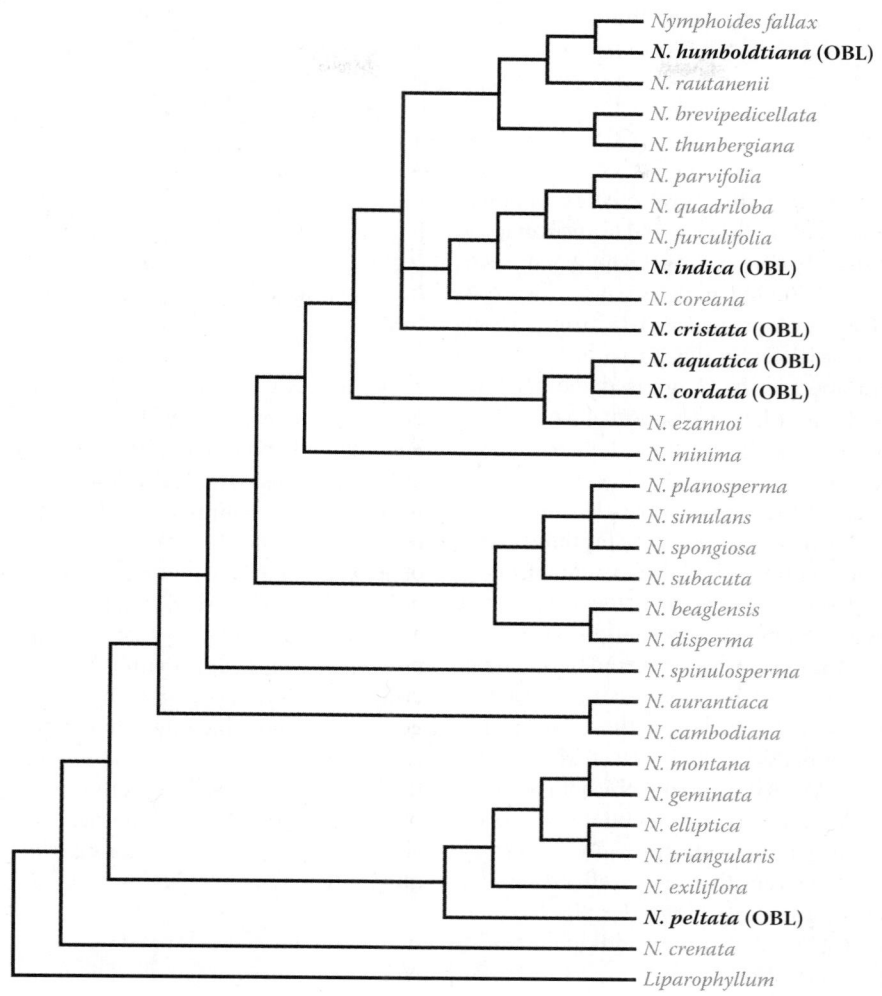

FIGURE 5.182 Interspecific relationships in *Nymphoides* as reconstructed using nrITS sequence data. The OBL North American species (bold) resolve in various portions of the tree; however, all *Nymphoides* are aquatic and would be categorized similarly if present in North America. Consequently, the genus is regarded as representing a single origin of the aquatic habit. Analyses of the same taxa using maternally inherited cpDNA sequence data (Tippery & Les, 2011; not shown) produce a tree that is similar overall, but with various discordances that reflect ancient hybridization in the history of a number of species. (Adapted from Tippery, N.P. & Les, D.H., *Syst. Bot.*, 36, 1101–1113, 2011.)

Europe during the Pleistocene. Restricted distributions currently characterize *N. cristata* (Florida, South Carolina) and *N. indica* (Florida) and *N. humboldtiana* (Texas).

References: Armstrong, 2002; Arrington et al., 1999; Austin, 2004b; Avis et al., 1997; Barrett, 1980; Baskin & Baskin, 1998; Birkenholz, 1963; Block & Rhoads, 2011; Brock, 1984; Brock et al., 1983; Buckingham & Bennett, 2001; Burks, 2002; Carr, 1831; Catling, 1992; Chang & Woo, 2003; Chao et al., 2014; Chimney & Pietro, 2006; Chuang & Ornduff, 1992; Coe, 2008; Coe & Anderson, 2005; Cook, 1990; Countryman, 1970; Croft & Chow-Fraser, 2007, 2009; Cypert, 1972; Darbyshire & Francis, 2008; DiTomaso & Healy, 2003; Eichler & Boylen, 2013; Elakovich & Wooten, 1989; Ferreira et al., 2010; Ferriter et al., 2008; Gordon & Gantz, 2011; Gumbert & Kunze, 1999; Habeck, 1996; Harms & Grodowitz, 2009; Houghton & Osibogun, 1993; Howard et al., 2011; Hoyer et al., 1996; Huang et al., 2014; Hutton, 2010; Kadono, 2004; Keddy, 1981; Kole et al., 2005; Kuhnlein & Turner, 1991; Labisky et al., 2003; Lallana, 2005; Larson, 2007; Larson & Willén, 2006; Lavid et al., 2001; Les & Mehrhoff, 1999; Lewis, 2005; Li et al., 2011; Liao et al., 2013; Madhavan et al., 2009; Main et al., 2006; Maki & Galatowitsch, 2004; Mason & van der Valk, 1992; Mazzotti et al., 2004; McCord, 2010; Milano, 1999; Moody, 1989; Mulhouse, 2004; Netherland, 2013; Nichols et al., 2001; Nico & Muench, 2004; Ornduff, 1966, 1970; Ornduff & Mosquin, 1970; Palis, 1998; Parker, 2005; Sah et al., 2010, 2012; Saunders, 2005; Shibayama & Kadono, 2003a,b, 2007a,b; Smits et al., 1989; Stoops et al., 1998; Sutton, 1993, 1994; Takagawa et al., 2005, 2006; Thomas & Jansen, 2006; Tippery & Les, 2011, 2013; Tippery et al., 2009, 2011; Tsuchiya, 1991; Uesugi et al., 2004, 2005, 2007, 2009; Van Der Velde & Van Der Heijden, 1981; Wang et al., 2005; Watts, 1971; Whiting, 1849; Wu & Yu, 2004; Zahina et al., 2007.

ORDER 9: DIPSACALES [2]

The small order Dipsacales (roughly 1000 species) contains just two families: Adoxaceae E. Meyer and Caprifoliaceae A. L. de Jussieu, which resolve as well-supported sister clades (Judd et al., 2008). Phylogenetic analyses incorporating a large amount of DNA sequence data from various loci have demonstrated consistently that the order is monophyletic (Figure 5.131) and can be diagnosed morphologically by the presence of cellular endosperm development, opposite leaves, and a 3–4-celled anther tapetum (Judd et al., 2008). However, the precise phylogenetic placement of this clade at a higher level remains somewhat unsettled, with different DNA sequence datasets yielding conflicting results. Sequences from multiple plastid loci amounting to roughly 17 kb of data resolve Dipsacales as the sister group to Paracryphiaceae with strong internal support (Tank & Donoghue, 2010). The same clade was recovered by a 17-gene phylogeny "total evidence" analysis (comprising mitochondrial, nuclear, and plastid loci); however, the clade lacked internal support in that study (Soltis et al., 2011). Moreover, a combined analysis of 18S+26S ribosomal DNA data resolved Dipsacales and Columelliaceae as a moderately supported clade with Paracryphiaceae associating instead with Apiales (Maia et al., 2014). It is apparent that more refined analytical methods will be necessary to resolve the placement of Dipsacales satisfactorily given that simply adding more data has not yielded congruent results.

Based on cytological characteristics, Benko-Iseppon and Morawetz (2000) recommended that Adoxaceae in its present circumscription be elevated to ordinal status as Viburnales, a disposition of questionable taxonomic utility.

Both Adoxaceae and Caprifoliaceae contain OBL species.

Family 1: Adoxaceae [5]

The elderberry family (Adoxaceae) contains 245 species distributed among five genera: *Adoxa* L., *Sambucus* L., *Sinadoxa* C. Y. Wu, Z. L. Wu & R. F. Huang, *Tetradoxa* C. Y. Wu, *Viburnum* L. The largest concentration of species (roughly 220) is in the genus *Viburnum* (Judd et al., 2008), which resolves as the phylogenetic sister group to the remainder of the family (Figure 5.183). Despite its relatively small size, the family is fairly diverse and contains herbaceous plants as well as woody trees and shrubs. The species characteristically have opposite, simple or compound leaves, determinate umbels, bisexual and radial flowers with reduced sepals, short styles, and pollen with a reduced exine. The fruits are blue or red drupes containing 1–5 pits (pyrenes). Ubiquitous chemical constituents include cyanogenic glycosides and iridoids. The flowers are pollinated primarily by insects such as bees and wasps (Hymenoptera) or flies (Diptera). The fruits are eaten and dispersed by birds (Aves) (Judd et al., 2008).

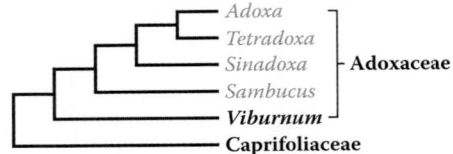

FIGURE 5.183 Phylogenetic relationships in Adoxaceae as indicated by analysis of combined *rbcL* and nrITS sequence data. North American taxa containing OBL species are shown in bold. (Adapted from Donoghue, M.J. et al., *Harvard Pap. Bot.*, 6, 459–479, 2001.)

Several Adoxaceae species have been used medicinally to treat bruises, eczema, wounds, parasitic skin infections, and snake bites. Some leaf extracts are allergenic and can cause skin rash. *Sambucus* and *Viburnum* contain hundreds of species and horticultural varieties that are planted as ornamental flowering shrubs. *Sambucus* (elderberry) fruits are used in jelly and wine making and were popularized in the 1973 song "Elderberry Wine" sung by Elton John. Some species of *Viburnum* were used by native tribes for making arrows.

Adoxaceae occur mainly in terrestrial habitats in the temperate Northern Hemisphere, but extend southward, particularly in montane environments. A few species are found facultatively in wetlands, and in some cases (e.g., *Sambucus nigra*), can be observed there quite often, despite their lack of OBL status. North America has three genera (*Adoxa*, *Sambucus*, and *Viburnum*), with a single OBL species occurring only in one:

1. ***Viburnum*** L.

1. *Viburnum*

Possumhaw, wild raisin, withe-rod; viorne

Etymology: from the Latin *viburnum*, an Etruscan name for the wayfaring-tree

Synonyms: *Lentago*; *Microtinus*; *Opulus*; *Oreinotinus*; *Solenotinus*

Distribution: global: cosmopolitan; **North America:** widespread

Diversity: global: 220 species; **North America:** 28 species

Indicators (USA): OBL; FACW: *Viburnum nudum*

Habitat: freshwater; palustrine; **pH:** 4.2–6.8; **depth:** <1 m; **life-form(s):** emergent shrub

Key morphology: stems (to 5 m) woody, arborescent or multistemmed, terminal buds naked; leaves opposite, deciduous, simple, the blades (to 15 cm) elliptic, lanceolate, lance-ovate, or obovate, margins partially revolute, entire, crenate or finely serrate, petioles (to 20 mm) winged; cymes (to 15 cm) terminal, peduncled (to 25 mm), flattened, with 4–5 main branches; flowers small, 5-merous, corolla (to 5 mm) white or cream colored; drupes pulpy, 1-seeded, crowned by persistent calyx, maturing from yellowish or pink to bluish black, surface waxy, seeds globose, obovoid.

Life history: duration: perennial (buds); **asexual reproduction:** stump sprouts; **pollination:** insect; **sexual condition:** hermaphroditic; **fruit:** drupes (common); **local dispersal:**

drupes (gravity, water); **long-distance dispersal:** drupes (birds, mammals)

Imperilment: (1) *Viburnum nudum* [G5]; SH (CT, DE, WV); S1 (IN, KY, NY, PA, RI, WI)

Ecology: general: Although only one North American *Viburnum* species is categorized as OBL, another ten (36%) have some status as wetland indicators (FACW, FAC, FACU, UPL), and often can be found growing in or near wetlands. Otherwise, the species are characteristic of more mesic forests or thicket communities. All of the species are woody shrubs or small trees with opposite leaves and fleshy, single-seeded drupes for fruits. The small flowers are arranged in showy inflorescences, which vary among species in having inflorescences consisting entirely of fertile flowers, mixtures of fertile and sterile flowers, or (in some cultivars) entirely sterile flowers. Most species are genetically self-incompatible via a sporophytic (SSI) system and require outcrossing by insects (Insecta) in order to set seed. The SSI system can be leaky, allowing for self-pollination to occur at least at low levels. The fleshy drupes are eaten by birds and small mammals, which function as their primary dispersal agents. Water might disperse drupes over short distances, but soaking results in the release of the seed, which sinks if it is viable. In many species, the seeds have "deep simple epicotyl" morphophysiological dormancy, which requires high temperatures to stimulate radicle growth and several months of cold stratification to initiate shoot development. Seeds of a few species are morphologically dormant. Many species are cultivated as ornamental shrubs for their colorful autumn foliage and showy flower and fruit displays.

Viburnum nudum L. occurs in tidal or nontidal sites including bayheads, bogs, bottomlands, cobble bars, cypress heads, depressions, ditches, drains, fens, flats, flatwoods, floodplains, gravel bars, gravel pits, hammocks, marshes, meadows, pocosins, ponds, potholes, prairies, ravines, roadsides, scrub, seeps, sloughs, swales, swamps, thickets, woodlands, and along the margins of bayous, lakes, ponds, rivers and streams at elevations of up to 1689 m. The plants are highly tolerant of flooding and exhibit a 9% increase in size index after a flood period of 6 days. They can survive in full sun, or in filtered sun or shade (0.3%–89% canopy closure). The concavity of the upper leaf surface adjusts with illumination, assuming an angle from 83° to 105° in sunlight and 180° in shade. The substrates uniformly are acidic (pH 4.2–6.8), although survival of cultivars in alkaline soil has been reported. Substrates include clay, gritty loam, loam, loamy sand, marl, muck, Paxville soil, peat, sand, sandy gravelly clay, sandy clay, sandy loam, sandy peat, sandy rock, sapric muck, serpentine, shale, and silt loam. Flowering occurs from March to mid-July once the foliage has expanded. The plants average 56–59 flowers/stem, with an open percentage of 55%–57%. The flowers are self-incompatible and are outcrossed by insects (Insecta) such as bees (Hymenoptera) and beetles (Coleoptera); however, some cultivars can be somewhat self-fertile and reduced seed set (~6%) has been observed in flowers where pollinators were excluded. The small flowers produce low quantities of nectar (<0.01 µL), which contains

about 4% sucrose. A fruit set of 28%–35% has been observed in open-pollinated plants. Pollinator limitation has been observed with pollen-supplemented flowers yielding 16%–30% higher fruit set. Reduced seed set has been observed in plants growing with the invasive *Rhamnus frangula* (Rhamnaceae), which competes for pollinators. Late secondary flowering has been observed in September. The fruits mature in mid-August and persist until late October. The seeds are morphologically dormant and do not require stratification for germination. Although some sources recommend alternating 3-month periods of warm (20°C) and cold (2°C) stratification, reduced germination rates have been observed for refrigerated seeds. Because the seeds do not mature until autumn or early winter, their natural germination is delayed until suitable temperatures return in the spring. Fresh seeds reportedly will germinate under a 26°C/18°C temperature regime. Slow germination rates of up to 18 months have been reported. The plants often occur in relatively recently burned areas. They can regenerate from stump sprouts and comprise a major feature of successional communities that develop on abandoned agricultural or logging sites. Vegetative terminal shoot cuttings can be rooted relatively easily (within 1 month), with the highest rooting percentage obtained for cuttings taken in July, and the lowest for those made in March.

Reported associates: *Abies balsamea, A. fraseri, Acer negundo, A. pensylvanicum, A. platanoides, A. rubrum, A. saccharinum, A. saccharum, A. spicatum, Agalinis, Ageratina altissima, Aesculus pavia, A. sylvatica, Agrostis hyemalis, A. perennans, Aletris farinosa, Alnus rugosa, A. serrulata, A. viridis, Amelanchier arborea, A. bartramiana, A. canadensis, Amorpha fruticosa, A. nitens, Amphicarpaea bracteata, Amsonia hubrichtii, Andropogon gerardii, A. glomeratus, A. virginicus, Anemone virginiana, Angelica triquinata, Apios americana, Aralia nudicaulis, Arctostaphylos uva-ursi, Arisaema triphyllum, Aristida palustris, A. purpurascens, A. stricta, Aristolochia serpentaria, Arnoglossum, Aronia arbutifolia, A. melanocarpa, Arundinaria gigantea, A. tecta, Asarum canadense, Asclepias rubra, Asimina parviflora, A. triloba, Asplenium platyneuron, Athyrium filix-femina, Avenella flexuosa, Axonopus fissifolius, Balduina atropurpurea, Bartonia verna, B. virginica, Bazzania trilobata, Berchemia scandens, Betula alleghaniensis, B. lenta, B. papyrifera, B. populifolia, Bidens bipinnatus, Bigelowia nudata, Bignonia capreolata, Boehmeria cylindrica, Botrychium virginianum, Boykinia aconitifolia, Brachyelytrum erectum, Burmannia capitata, Cabomba caroliniana, Calamagrostis canadensis, Callicarpa americana, Calopogon tuberosus, Caltha palustris, Calycanthus floridus, Campanula aparinoides, Campsis radicans, Cardamine bulbosa, Carex aquatilis, C. atlantica, C. baileyi, C. barrattii, C. basiantha, C. bromoides, C. brunnescens, C. canescens, C. caroliniana, C. collinsii, C. complanata, C. corrugata, C. cherokeensis, C. crinita, C. debilis, C. deflexa, C. disperma, C. echinata, C. festucacea, C. folliculata, C. frankii, C. glaucescens, C. grayi, C. gynandra, C. impressinervia, C. intumescens, C. joorii, C. lacustris, C. laevivaginata, C. leptalea, C. lonchocarpa, C. lucorum, C. lurida, C. magellanica, C. michauxiana, C. novae-angliae, C.*

oligosperma, C. pauciflora, C. pedunculata, C. pensylvanica, C. rostrata, C. scabrata, C. seorsa, C. sterilis, C. stipata, C. striata, C. striatula, C. stricta, C. torta, C. trisperma, C. turgescens, C. utriculata, C. venusta, C. verrucosa, C. wiegandii, C. woodii, Carphephorus pseudoliatris, Carpinus caroliniana, Carya aquatica, C. cordiformis, C. glabra, C. myristiciformis, C. ovata, Castilleja coccinea, Centella asiatica, Cephalanthus occidentalis, Cercis canadensis, Chamaedaphne calyculata, Chaptalia tomentosa, Chasmanthium laxum, Chelone glabra, C. lyonii, Chimaphila maculata, Chionanthus virginicus, Chrysosplenium americanum, Cicuta maculata, Cinna arundinacea, C. latifolia, Circaea alpina, C. canadensis, Cladium mariscoides, Cladonia arbuscula, C. pyxidata, C. rangiferina, C. stellaris, Cleistes, Clematis virginiana, Clethra acuminata, C. alnifolia, Cliftonia monophylla, Clintonia borealis, Collinsonia canadensis, Comarum palustre, Commelina virginica, Comptonia peregrina, Coptis trifolia, Corema conradii, Coreopsis gladiata, Cornus amomum, C. asperifolia, C. canadensis, C. florida, C. foemina, C. racemosa, C. sericea, Corydalis sempervirens, Corylus americana, C. cornuta, Crataegus flava, Cryptotaenia canadensis, Ctenium aromaticum, Cyperus haspan, Cypripedium acaule, Cyrilla racemiflora, Danthonia compressa, D. spicata, Decodon verticillatus, Decumaria barbara, Deschampsia cespitosa, Desmodium glutinosum, Dichanthelium acuminatum, D. boscii, D. clandestinum, D. dichotomum, D. ensifolium, D. polyanthes, D. scabriusculum, D. scoparium, Dicranum fuscescens, D. polysetum, D. scoparium, D. undulatum, Diervilla sessilifolia, Dioscorea villosa, Diospyros virginiana, Ditrysinia fruticosa, Doellingeria umbellata, Drosera brevifolia, D. capillaris, D. intermedia, D. rotundifolia, Dryopteris campyloptera, D. carthusiana, D. intermedia, D. ludoviciana, Dulichium arundinaceum, Egeria densa, Eleocharis acicularis, E. equisetoides, E. melanocarpa, E. quadrangulata, E. tortilis, E. tuberculosa, Elephantopus carolinianus, E. tomentosus Elymus lanceolatus, E. trachycaulus, E. virginicus, Empetrum nigrum, Epigaea repens, Epilobium coloratum, E. leptophyllum, Erigeron vernus, Eriocaulon compressum, E. decangulare, E. koernickianum, Eriophorum vaginatum, E. virginicum, Eryngium integrifolium, Eubotrys racemosa, E. recurva, Euonymus americanus, Eupatorium leucolepis, E. perfoliatum, E. pilosum, E. rotundifolium, Euphorbia purpurea, Eurybia chlorolepis, E. macrophylla, E. radula, Euthamia graminifolia, Eutrochium fistulosum, E. maculatum, Fagus grandifolia, Festuca ovina, Fraxinus americana, F. nigra, F. pennsylvanica, Galax urceolata, Galium asprellum, G. circaezans, G. kamtschaticum, G. obtusum, G. tinctorium, G. triflorum, Gaultheria hispidula, G. procumbens, Gaylussacia baccata, G. dumosa, G. frondosa, G. mosieri, G. ursina, Gelsemium rankinii, G. sempervirens, Geranium maculatum, Geum canadense, G. radiatum, G. rivale, Glyceria grandis, G. melicaria, G. striata, Goodyera pubescens, Gordonia lasianthus, Gratiola brevifolia, Gymnadeniopsis clavellata, Gymnocarpium dryopteris, Habenaria, Halesia carolina, Hamamelis virginiana, Helenium vernale, Helianthus angustifolius, H. heterophyllus,

Helonias bullata, Hexastylis arifolia, Holcus lanatus, Houstonia purpurea, H. serpyllifolia, Humbertacalia, Huperzia lucidula, Hydrocotyle americana, H. verticillata, Hylocomium splendens, Hypericum brachyphyllum, H. cruxandreae, H. densiflorum, H. denticulatum, H. fasciculatum, H. galioides, H. mutilum, H. prolificum, Hypnum imponens, Ilex collina, I. coriacea, I. decidua, I. glabra, I. montana, I. mucronata, I. myrtifolia, I. opaca, I. verticillata, Illicium floridanum, Impatiens capensis, Iris versicolor, I. virginica, Isoetes engelmannii, Itea virginica, Juglans nigra, Juncus caesariensis, J. dichotomus, J. diffusissimus, J. effusus, J. gymnocarpus, J. nodatus, J. repens, J. trigonocarpus, Juniperus communis, J. virginiana, Kalmia angustifolia, K. buxifolia, K. cuneata, K. latifolia, Lachnocaulon anceps, Larix laricina, Leersia oryzoides, L. virginica, Leucobryum glaucum, Leucothoe axillaris, L. fontanesiana, L. recurva, Ligusticum canadense, L. sinense, Lindera benzoin, Linum striatum, Liquidambar styraciflua, Linnaea borealis, Liriodendron tulipifera, Lobelia cardinalis, L. puberula, Lonicera canadensis, L. japonica, L. morrowii, L. oblongifolia, L. tatarica, Ludwigia alternifolia, L. decurrens, L. leptocarpa, L. palustris, L. pilosa, Luzula, Lycopodiella alopecuroides, Lycopodium annotinum, L. obscurum, Lycopus europaeus, L. rubellus, L. uniflorus, L. virginicus, Lysimachia ciliata, L. lanceolata, L. terrestris, L. thyrsiflora, Lyonia ligustrina, L. lucida, Lythrum lineare, Macbridea caroliniana, Macranthera flammea, Magnolia acuminata, M. virginiana, Maianthemum canadense, M. trifolium, Marshallia graminifolia, Matteuccia struthiopteris, Mayaca fluviatilis, Medeola virginiana, Melampyrum lineare, Melica mutica, Menziesia pilosa, Microstegium vimineum, Mikania scandens, Mimosa strigillosa, Mimulus alatus, Minuartia glabra, Mitchella repens, Mitella nuda, Morus rubra, Muhlenbergia expansa, Murdannia keisak, Myrica caroliniensis, M. cerifera, M. gale, M. inodora, Neottia bifolia, Nymphaea odorata, Nyssa biflora, N. sylvatica, Oclemena acuminata, O. reticulata, Oenothera fruticosa, Onoclea sensibilis, Orontium aquaticum, Orthilia secunda, Oryzopsis asperifolia, Osmunda claytoniana, O. regalis, Osmundastrum cinnamomeum, Ostrya virginiana, Oxalis montana, Oxydendrum arboreum, Oxypolis rigidior, Packera aurea, P. plattensis, P. schweinitziana, Panicum anceps, P. brachyanthum, P. virgatum, Paraleucobryum longifolium, Parnassia asarifolia, Parthenocissus quinquefolia, Paspalum laeve, Passiflora lutea, Peltandra sagittifolia, P. virginica, Penthorum sedoides, Persea palustris, Persicaria amphibia, P. arifolia, P. hydropiperoides, P. sagittata, Phalaris arundinacea, Phlox stolonifera, Phryma leptostachya, Physocarpus opulifolius, Picea glauca, P. mariana, P. rubens, Pieris floribunda, Pinckneya bracteata, Pinguicula caerulea, P. lutea, P. pumila, Pinus banksiana, P. echinata, P. elliottii, P. glabra, P. palustris, P. resinosa, P. rigida, P. serotina, P. strobus, P. taeda, Piptatherum pungens, Plagiothecium laetum, Platanthera blephariglottis, P. ciliaris, P. dilatata, P. integrilabia, P. obtusata, Platanus occidentalis, Pleurozium schreberi, Pluchea baccharis, Poa palustris, P. saltuensis, Pogonia ophioglossoides, Polemonium vanbruntiae, Polygala cymosa,

P. lutea, P. nana, P. paucifolia, P. ramosa, P. sanguine, Polypodium virginianum, Polystichum acrostichoides, Polytrichum commune, P. juniperinum, P. ohioense, P. strictum, Populus balsamifera, P. grandidentata, P. tremuloides, Proserpinaca palustris, Prunus pensylvanica, P. serotina, Pteridium aquilinum, Ptilidium ciliare, Ptilimnium nodosum, Ptilium crista-castrensis, Pycnanthemum tenuifolium, Quercus alba, Q. bicolor, Q. ellipsoidalis, Q. falcata, Q. geminata, Q. laevis, Q. laurifolia, Q. lyrata, Q. michauxii, Q. minima, Q. myrtifolia, Q. nigra, Q. pagoda, Q. phellos, Q. rubra, Q. shumardii, Q. stellata, Q. virginiana, Ranunculus recurvatus, Rhamnus caroliniana, R. frangula, Rhexia alifanus, R. lutea, R. petiolata, R. virginica, Rhizomnium appalachianum, Rhizocarpon geographicum, Rhododendron arborescens, R. calendulaceum, R. canadense, R. canescens, R. catawbiense, R. groenlandicum, R. maximum, R. minus, R. periclymenoides, R. viscosum, Rhus copallinum, Rhynchospora capitellata, R. chalarocephala, R. chapmanii, R. globularis, R. macrostachya, R. miliacea, R. plumosa, Rhytidiadelphus triquetrus, Ribes rotundifolium, R. triste, Rosa multiflora, R. palustris, Rubus alleghiensis, R. argutus, R. flagellaris, R. hispidus, R. occidentalis, R. repens, R. sachalinensis, Rudbeckia laciniata, R. scabrifolia, Rugelia nudicaulis, Sabal minor, Sabatia campanulata, S. difformis, Saccharum baldwinii, S. giganteum, Sagittaria latifolia, S. montevidensis, Salix bebbiana, S. discolor, S. humilis, S. lucida, S. nigra, S. petiolaris, S. sericea, Sambucus nigra, S. racemosa, Sanicula canadensis, S. marilandica, Sarracenia alabamensis, S. alata, S. flava, S. jonesii, S. leucophylla, S. minor, S. oreophila, S. psittacina, S. purpurea, S. rubra, Sassafras albidum, Saururus cernuus, Schizachne purpurascens, Schizachyrium scoparium, Scirpus cyperinus, S. expansus, S. lineatus, S. polyphyllus, Scleria, Scutellaria integrifolia, Sibbaldiopsis tridentata, Silphium asteriscus, Sium suave, Smilax bona-nox, S. glauca, S. hugeri, S. laurifolia, S. rotundifolia, S. tamnoides, Solidago arguta, S. caesia, S. fistulosa, S. flexicaulis, S. glomerata, S. macrophylla, S. odora, S. patula, S. puberula, S. rugosa, S. simplex, S. uliginosa, Sorbus americana, S. decora, Sorghastrum nutans, Sparganium americanum, Sphagnum affine, S. angustifolium, S. bartlettianum, S. capillifolium, S. centrale, S. compactum, S. fallax, S. fimbriatum, S. flavicomans, S. fuscum, S. girgensohnii, S. lescurii, S. magellanicum, S. palustre, S. papillosum, S. recurvum, S. rubellum, S. russowii. S. squarrosum, S. wulfianum, Spiraea alba, S. tomentosa, Sporobolus heterolepis, S. pinetorum, Steinchisma hians, Stenanthium densum, Stewartia, Streptopus lanceolatus, Styrax americanus, S. grandifolius, Symphyotrichum dumosum, S. laeve, S. lateriflorum, S. puniceum, S. racemosum, S. tradescantii, S. undulatum, Symplocarpus foetidus, Symplocos tinctoria, Taxodium ascendens, T. distichum, Taxus canadensis, Thalictrum macrostylum, T. pubescens, T. thalictroides, Thaspium trifoliatum, Thelypteris noveboracensis, T. palustris, Thuidium delicatulum, Thuja occidentalis, Tiarella cordifolia, Tilia americana, Toxicodendron radicans, T. vernix, Trautvetteria caroliniensis, Triadenum virginicum, T. walteri, Triantha racemosa, Trientalis borealis, Trillium cuneatum, T. undulatum, Tsuga canadensis, Typha latifolia, Ulmus alata, U. americana, U. rubra, Umbilicaria, Utricularia cornuta, U. gibba, U. subulata, Uvularia sessilifolia, Vaccinium angustifolium, V. corymbosum, V. elliottii, V. erythrocarpum, V. myrtilloides, V. oxycoccos, V. stamineum, V. vitis-idaea, Valerianella radiata, Veratrum viride, Vernonia noveboracensis, Viburnum acerifolium, V. dentatum, V. lantanoides, V. lentago, V. molle, V. prunifolium, V. recognitum, V. rufidulum, Viola cucullata, V. lanceolata, V. macloskeyi, V. primulifolia, V. renifolia, Vitis rotundifolia, Woodwardia areolata, W. virginica, Xanthorhiza simplicissima, Xyris ambigua, X. caroliniana, X. fimbriata, X. laxifolia, Zephyranthes atamasco, Zizania aquatica, Zizia aptera.

Use by wildlife: *Viburnum nudum* fruits are eaten by various birds (Aves) including pileated woodpeckers (Picidae: *Dryocopus pileatus*). The fruits often are not taken until they are exposed to several freeze-thaw cycles, which makes them more palatable to the avifauna. Song sparrows (Aves: Emberizidae: *Melospiza melodia*) use the plants as nesting sites. The fruits are a moderately preferred food of white-tailed deer (Mammalia: Cervidae: *Odocoileus virginianus*) and also comprise a small part of the diet of black bears (Mammalia: Ursidae: *Ursus americanus*). The plants host introduced, invasive leaf beetles (Insecta: Coleoptera: Chrysomelidae: *Pyrrhalta viburni*), citrus mealybugs (Insecta: Hemiptera: Aleyrodidae: *Aleuroplatus plumosus*; Pseudococcidae: *Planococcus citri*), and numerous moth larvae (Insecta: Lepidoptera: Drepanidae: *Oreta rosea*; Erebidae: *Euproctis chrysorrhoea, Lymantria dispar*; Geometridae: *Acasis viridata, Eupithecia ravocostaliata, E. satyrata, Operophtera brumata, Orthofidonia flavivenata, O. tinctaria, Plagodis alcoolaria, Xanthotype sospeta*; Lycaenidae: *Celastrina argiolus*; Noctuidae: *Agriopodes fallax, Autographa ampla, Eurois astricta, Harrisimemna trisignata, Lacinipolia lorea, Metaxaglaea inulta, Phlogophora periculosa, Polia purpurissata*; Notodontidae: *Schizura badia*; Saturniidae: *Hyalophora cecropia*; Sphingidae: *Darapsa choerilus, Hemaris thysbe*; Tortricidae: *Olethreutes zelleriana*; Uraniidae: *Calledapteryx dryopterata*). They also host treehoppers (Insecta: Hemiptera: Membracidae: *Enchenopa binotata*) and several sac fungi (Ascomycota: Botryosphaeriaceae: *Botryosphaeria dothidea*; Dermateaceae: *Corniculariella hystricina, Pezicula cinnamomea*; Massarinaceae: *Massaria lantanae*). The flowers attract numerous potential pollinators including ants (Insecta: Hymenoptera: Formicidae: *Camponotus pennsylvanicus, Formica aserva, Myrmica rubra*), beetles (Insecta: Coleoptera: Brentidae: *Neapion frosti*; Cerambycidae: *Callimoxys sanguinicollis, Clytus ruricola, Cosmosalia chrysocoma, Cyrtophorus verrucosus, Evodinus monticola, Leptura plebeja, Molorchus bimaculatus, Trachysida mutabilis, Trigonarthris proxima, Typocerus velutinus, Xestoleptura tibialis*; Cleridae: *Phyllobaenus humeralis*; Elateridae: *Agriotes quebecensis, A. stabilis, A. fucosus, Agrotiella bigeminata, Sericus brunneus*; Mordellidae: *Mordellistena aspersa*; Scarabaeidae: *Trichiotinus assimilis*; Tenebrionidae: *Isomira quadristriata*), bees (Insecta: Hymenoptera: Andrenidae: *Andrena miranda, A. vicina*; Halictidae: *Lasioglossum*), butterflies (Insecta: Lepidoptera), flies (Insecta: Diptera: Syrphidae: *Pterallastes thoracicus*), and wasps (Insecta: Hymenoptera: suborder: Apocrita).

Economic importance: food: The fruits of *Viburnum nudum* are edible, but are acidic and not highly acclaimed as palatable. They contain about 43% sugar and have a mean caloric value of 22.4 kJ/g (pulp) and 21.9 kJ/g (seeds). The berries and seeds were eaten as fruit by the Abnaki, Algonquin and Quebec tribes; **medicinal:** *Viburnum nudum* has long been recognized as a medicinal plant in Appalachia. The Cherokee prepared infusions from it for treating ague, fevers, and smallpox and as a preventative for recurrent convulsions. They used the root bark in tonics to induce perspiration and in infusions as a sore tongue wash; **cultivation:** *Viburnum nudum* was listed in a catalog of plants suitable for cultivation in Europe as early as 1785 and its seeds were included in trade catalogs as early as 1804. It is planted commonly as an ornamental shrub and is distributed under the cultivar names 'Angustifolium', 'BrandyWine'™, 'Bulk', 'Callaway Large Leaf', 'Callaway Small Leaf', 'Count Pulaski', 'Earth Shade', 'Longwood', 'Moonshine', 'Pink Beauty', and 'Winterthur'. It is highly recommended for planting in rain gardens and also as an alternative to invasive garden shrubs such as *Berberis bealei* (Berberidaceae) and *Ligustrum*

vulgare (Oleaceae); **misc. products:** *Viburnum nudum* seed has been used for wetland restoration projects. The twigs were twisted and made into "with" for use in binding; **weeds:** *Viburnum nudum* is not regarded as weedy; **nonindigenous species:** Despite its broad distribution as an ornamental plant, *V. nudum* has not been reported as escaping from cultivation anywhere outside of North America.

Systematics: *Viburnum* traditionally has been subdivided into 10 sections, with *V. nudum* assigned to the New World section *Lentago*. Section *Lentago* includes six other species of the Eastern United States and Mexico. Phylogenetic analyses of nrITS, *trnK*, and *psbA-trnH* IGS sequences indicate section *Viburnum* (Old World) as the sister clade to section *Lentago*; however, neither data set places *V. nudum* within either clade. Analyses of nuclear genes (nrITS, *WAXY1* and *WAXY2*), or combined nuclear and cpDNA sequences, present a weakly supported topology, where *V. nudum* resolves as the sister to a clade representing sects. *Lentago* and *Viburnum* as subclades. *Viburnum cassinoides* L. has been treated either as a variety of *V. nudum* or as a distinct species. In any case, A 10-gene sequence dataset, alone or combined with full chloroplast genome sequences, resolves *V. nudum* + *V. cassinoides*

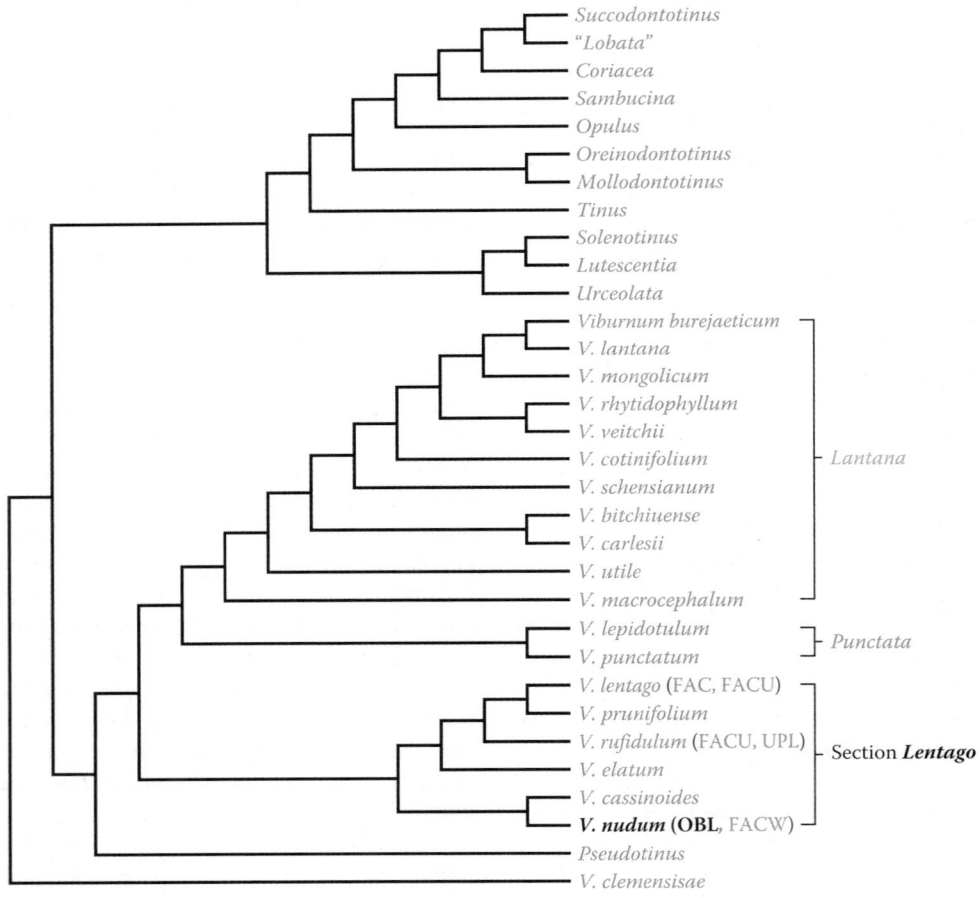

FIGURE 5.184 Phylogenetic relationships in *Viburnum* as reconstructed from multiple nuclear and chloroplast gene sequence data. Taxa are indicated either by informal clade names, or as individual *Viburnum* species. The OBL *V. nudum* (bold) and *V. cassinoides* resolve as a clade that is distinct from the remainder of section *Lentago*. Other North American wetland indicator designations are provided in parentheses. (Adapted from Clement, W.L. et al., *Amer. J. Bot.*, 101, 1029–1046, 2014.)

with high support as a highly divergent sister clade to other members of section *Lentago* (Figure 5.184). This topology represents the most compelling systematic arrangement yet derived from molecular data. The seeds of *V. nudum* also are anomalous compared to other members of section *Lentago*. Such results suggest that *V. nudum* might better be classified within an independent section. The base chromosome number of *Viburnum* is $x=8$ or $x=9$. All members of section *Lentago*, including *V. nudum*, share the diploid chromosome number of $2n=18$. No hybrids involving *V. nudum* have been reported.

Comments: *Viburnum nudum* occurs throughout eastern North America.

References: Avis et al., 1997; Becker, 1942; Bergen, 1909; Bess et al., 2014; Buc'hoz, 1785; Clement & Donoghue, 2011; Clement et al., 2014; Correll & Correll, 1975; Dehgan et al., 1989; Donoghue et al., 2004; Egolf, 1962; Forbush, 1916; Gano & McNeill, 1917; Godfrey & Wooten, 1981; Gossard, 1903; Groves, 1940; Harper, 1922; Horrell, 2013; Jacobs et al., 2008; Jernigan & Wright, 2011; Jin et al., 2010; Jones et al., 1984; Karema et al., 2010 ; Kirkaldy, 1907; Korf & Dirig, 2009; Krochmal, 1968; Landers et al., 1979; Lanicci, 2010; Lott & Knight, 1909; Lovell, 1915; Lovenshimer & Frick-Ruppert, 2013; MacRae & Rice, 2007; Majka, 2006; Majka & Lesage, 2007; M'Mahon, 1804; Moore, 2010; Moorhead et al., 2000; Nichols, 1934; Provancher, 1886; Sperduto & Nichols, 2012; Stubbs et al., 2007; Sundell & Thomas, 1988; TEPPC, 2014; Thompson, 1974; Webster et al., 2009, 2012; Winkworth & Donoghue, 2004, 2005.

Family 2: Caprifoliaceae [36]

Contemporary systematic treatments circumscribe the honeysuckle family (Caprifoliaceae) to include the former Dipsacaceae and Valerianaceae, resulting in a group comprising approximately 810 species dispersed among 36 genera (Judd et al., 2008). Honeysuckles are widespread geographically, but occur most frequently at northern latitudes in temperate climates. The monophyly of the family, as well as its sister group relationship to Adoxaceae (Figure 5.185), have been demonstrated convincingly by phylogenetic analyses incorporating morphological as well as numerous molecular data sets derived from both chloroplast and nuclear loci (Judd et al., 2008; Tank & Donoghue, 2011). Caprifoliaceae include herbaceous and woody plants with opposite leaves, but are readily distinguished by morphological synapomorphies including bilaterally symmetrical flowers (radial in Adoxaceae), elongate styles with capitate stigmas, large, spiny pollen grains, and nectar production from glandular hairs that arise from the corolla tube (Judd et al., 2008; Boyden et al., 2012). Iridoids and phenolic glycosides are found in the family (Judd et al., 2008). The flowers are showy, and are pollinated primarily by birds (Aves), bees, and wasps (Insecta: Hymenoptera), which forage for nectar (Judd et al., 2008). The fruits include achenes, berries, capsules, and drupes, which are dispersed by an assortment of abiotic and biotic vectors.

The largest genera are *Valeriana* (approx. 200 spp.) and *Lonicera* (150 spp.), which both contain OBL species, but are only remotely related (Figure 5.185). Consequently, it is evident that there have been at least two independent origins of OBL North American species within the family. A complete spectrum of wetland indicators is found in each genus with OBL taxa (Figure 5.185), suggesting that the obligate habit is more likely to arise in groups having overall high species diversity. The remaining genera (~95%) primarily are terrestrial, or occur only facultatively in wetlands.

Ornamental plants are found in many genera including *Abelia*, *Centranthus*, *Cephalaria*, *Diervilla*, *Dipelta*, *Dipsacus*, *Heptacodium*, *Knautia*, *Kolkwitzia*, *Leycesteria*, *Lonicera*, *Morina*, *Nardostachys*, *Patrinia*, *Pterocephalus*, *Succisa*, *Succisella*, *Scabiosa*, *Symphoricarpos*, *Triosteum*, *Valeriana*, *Valerianella*, *Vesalea*, *Weigela*, and *Zabelia*. Included among these is *Linnaea* L. (twinflower), the genus named for the renowned Swedish botanist Carl Linnaeus. The roots of *Valeriana* have long been used medicinally, and perfume was once made from its flowers. The amino acid name valine is derived from the genus. Other medicinal plants include *Lonicera* and *Scabiosa*. *Dipsacus* (teasel) was once used for the carding of wool fibers. Species of *Dipsacus* and *Lonicera* have become invasive in North America and elsewhere.

OBL taxa occur in two of the North American genera:

1. ***Lonicera*** L.
2. ***Valeriana*** L.

1. *Lonicera*

Honeysuckle, swamp fly honeysuckle; chèvrefeuille
Etymology: after Adam Lonitzer (1499–1569)
Synonyms: *Caprifolium*; *Chamaecerasus*; *Distegia*; *Euchylia*; *Isika*; *Metalonicera*; *Nintooa*; *Periclymenum*; *Phenianthus*; *Xylosteon*
Distribution: global: Africa, Eurasia, North America; **North America:** widespread
Diversity: global: 150 species; **North America:** 38 species
Indicators (USA): OBL: *Lonicera oblongifolia*
Habitat: freshwater; palustrine; **pH:** 6.0–7.8; **depth:** <1 m; **life-form(s):** emergent shrub
Key morphology: stems (to 2 m) erect, woody, multiple, pith white; leaves (to 8 cm) opposite, simple, short-petioled (to 5 mm), oblanceolate, oblong or ovoid, bases tapered, the apex obtuse, the margins entire; flowers sessile, paired, arising from common axillary peduncle (to 4 cm), bracts minute; corollas (to 2.5 cm) yellowish, tubular, bilabiate, upper lip 4-lobed, lower lip 1-lobed, the lips spreading; style elongate, stigma capitate; stamens and style exserted from corolla; berries (to 1 cm) paired, ovoid or subglobose, reddish to purple, distinct or partially fused, few-seeded; seeds small (to 2 mm).
Life history: duration: perennial (woody stems and buds); **asexual reproduction:** not reported; **pollination:** insect; **sexual condition:** hermaphroditic; **fruit:** berries (common); **local dispersal:** berries (gravity; animals); **long-distance dispersal:** berries (birds)
Imperilment: (1) *Lonicera oblongifolia* [G4]; SH (OH); S1 (PA); S2 (NB, SK, VT): S3 (ME, QC)

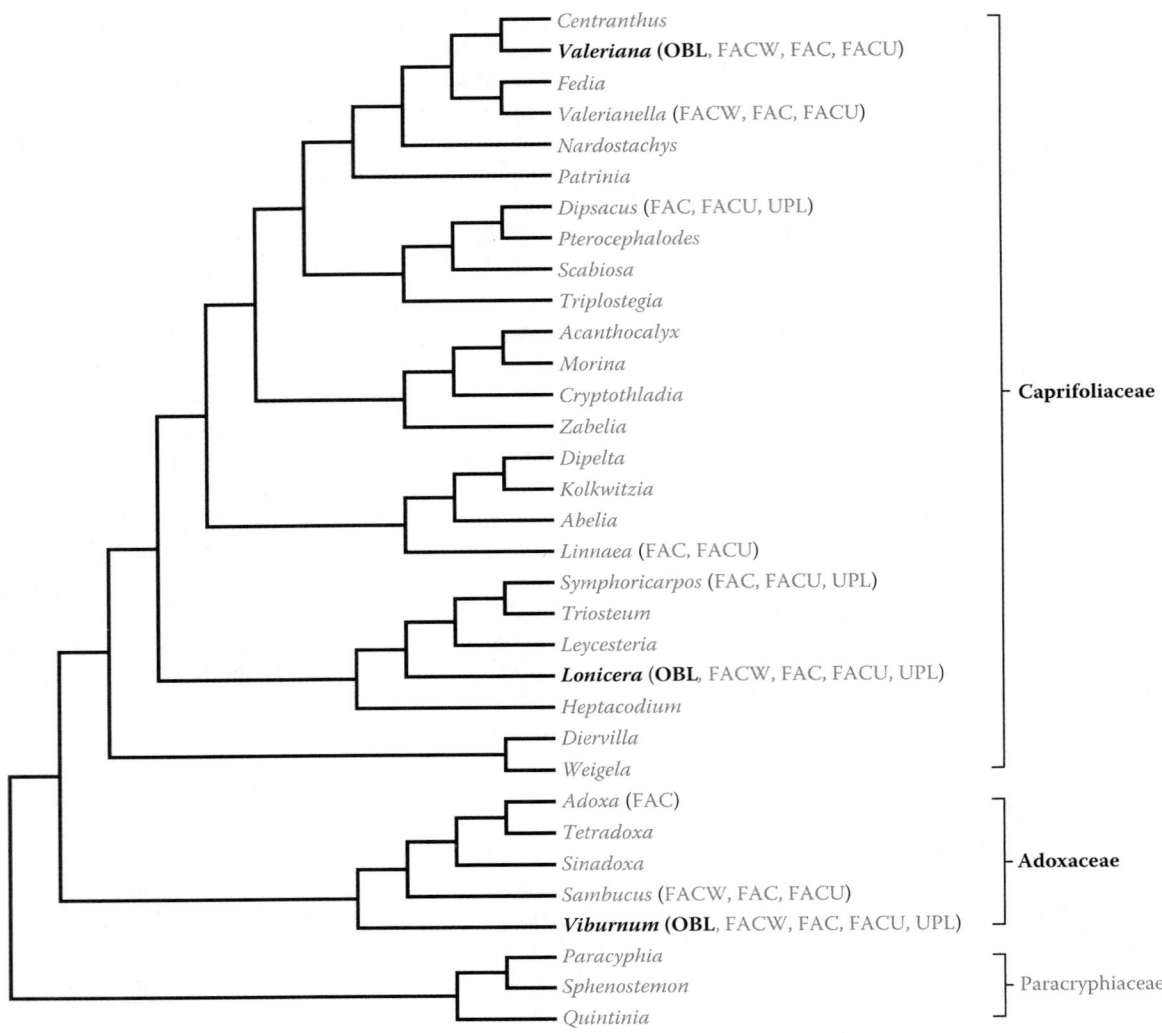

FIGURE 5.185 Phylogenetic relationships of Caprifoliaceae as reconstructed from a 10-gene cpDNA data set. In Caprifoliaceae, the North American OBL taxa (in bold) have arisen independently from within two diverse genera, which contain a spectrum of wetland indicators (shown in parentheses). (Adapted from Tank, D.C. & Donoghue, M.J., *Syst. Bot.*, 35, 425–441, 2010.)

Ecology: general: Honeysuckles include a variety of deciduous or evergreen shrubs and woody vines, which occur primarily on well-drained soils of UPL, often successional environments ranging from meadows to forests, but also in wetlands associated commonly with floodplains and riparian habitats. About 39% of the North American species have some status as wetland indicators, but only one is ranked as OBL. Details on the reproductive biology of honeysuckles are scarce, but most species appear to be pollinated mainly by bees (Insecta: Hymenoptera) and other insects, or by hummingbirds (Aves: Trochilidae). The plants can be prolific, with some species capable of producing more than 20,000 seeds per individual plant. The seeds vary from being non-dormant to requiring warm and/or cold stratification for germination. The degree to which seed banks are formed is not well understood. Dispersal is attributed mainly to fruits that are ingested and transported by birds or ungulate mammals; however, the fruits often fall in close proximity to the maternal

plants. Vegetative propagation can occur by layering and root suckering from adventitious buds on the crown, which also helps to protect the plants from serious fire damage. Some species (e.g., *Lonicera japonica*) are seriously invasive in North America.

***Lonicera oblongifolia* (Goldie) Hook.** inhabits bogs, carrs, fens, floodplains, meadows, muskegs, prairies, shores, swamps, thickets, and the margins of rivers and streams. This species is viewed as a reliable wetland indicator. Accounts vary regarding a preference for highly acidic or alkaline, calcareous sites. Substrates are circumneutral (pH 6.0–7.8) and have been described as marl or peat. Flowering occurs from May to June after the foliage develops, with fruiting occurring from June to August. The principal pollinators appear to be bees (see *Use by wildlife*); however, there have been no studies specifically conducted on the reproductive biology of this species, which would benefit from further study. The seeds are morphophysiologically dormant, and require 60–112 days of cold stratification (4°C) followed by a 30°C/20°C day/ night temperature regime to promote their germination. Germination rates can be low (<10%) and slow (21–70 days). The small-seeded (~1.9 mg) berries are disseminated by birds from August throughout the autumn. The plants have been observed to occur most often at sites where browsing by ungulate herbivores is rare. **Reported associates:** *Abies balsamea, Acer rubrum, A. spicatum, Alnus rugosa, Amelanchier arborea, A. humilis, Andromeda polifolia, Aneura pinguis, Aronia melanocarpa, A. ×prunifolia, Asclepias incarnata, Athyrium filix-femina, Aulacomnium palustre, Bazzania trilobata, Betula alleghaniensis, B. papyrifera, B. pumila, Bidens cernuus, B. trichospermus, Bromus ciliatus, Bryum pseudotriquetrum, Calamagrostis canadensis, Calliergonella cuspidata, Calopogon tuberosus, Caltha palustris, Campanula aparinoides, Campylium stellatum, Carex aquatilis, C. brunnescens, C. buxbaumii, C. chordorrhiza, C. comosa, C. diandra, C. disperma, C. eburnea, C. flava, C. folliculata, C. gracillima, C. hystericina, C. interior, C. intumescens, C. lacustris, C. lasiocarpa, C. leptalea, C. limosa, C. livida, C. magellanica, C. pauciflora, C. pellita, C. prairea, C. pseudocyperus, C. radiata, C. rugosa, C. sterilis, C. stricta, C. trisperma, C. vaginata, Chamaedaphne calyculata, Chelone glabra, Cicuta bulbifera, Cinna latifolia, Circaea alpina, Cirsium muticum, Cladium mariscoides, Clintonia borealis, Comarum palustre, Coptis trifolia, Cornus amomum, C. canadensis, C. foemina, C. sericea, Corylus americana, Cypripedium acaule, C. calceolus, C. reginae, Dasiphora floribunda, Decodon verticillatus, Doellingeria umbellata, Drosera linearis, D. rotundifolia, Dryopteris carthusiana, D. cristata, Dulichium arundinaceum, Eleocharis rostellata, Equisetum arvense, E. fluviatile, E. palustre, Erigeron nivalis, Eriophorum callitrix, E. viridicarinatum, Eupatorium perfoliatum, Euthamia graminifolia, Eutrochium maculatum, Fissidens adianthoides, Fragaria vesca, Fraxinus nigra, Galearis rotundifolia, Galium asprellum, G. boreale, G. labradoricum, G. palustre, G. triflorum, Gaultheria hispidula, G. procumbens, Gaylussacia baccata, Geum aleppicum, Glyceria striata, Gymnocarpium dryopteris, Hylocomium splendens, Hypericum kalmianum, Ilex mucronata, I. verticillata, Iris versicolor, Juncus stygius, Juniperus communis, Kalmia angustifolia, K. polifolia, Larix laricina, Lemna minor, Limprichtia revolvens, Lindera benzoin, Linnaea borealis, Lobelia kalmii, Lonicera canadensis, L. villosa, Ludwigia palustris, Lycopodium obscurum, Lycopus uniflorus, Lysimachia thyrsiflora, Maianthemum canadense, M. trifolium, Mentha arvensis, Menyanthes trifoliata, Mitchella repens, Mitella nuda, M. triflorum, Muhlenbergia glomerata, Myrica gale, M. pensylvanica, Onoclea sensibilis, Orthilia secunda Osmunda regalis, Osmundastrum cinnamomeum, Oxalis montana, Packera aurea, Parnassia glauca, Parthenocissus quinquefolia, Persicaria sagittata, Petasites frigidus, Phalaris arundinacea, Phegopteris connectilis, Picea mariana, P. rubens, Pilea pumila, Pinus strobus, Platanthera hyperborea, P. obtusata, Poa nemoralis, P. palustris, Pogonia ophioglossoides, Polemonium vanbruntiae, Populus tremuloides, Ptilium crista-castrensis, Pyrola asarifolia, Quercus bicolor, Rhamnus alnifolia, R. frangula, Rhododendron canadense, R. groenlandicum, Rhynchospora alba, R. capillacea, Rhytidiadelphus triquetrus, Ribes hirtellum, R. hudsonianum, R. triste, Rosa palustris, Rubus occidentalis, R. pubescens, R. repens, R. sachalinensis, Rumex britannica, Sagittaria latifolia, Salix bebbiana, S. candida, S. discolor, S. pedicellaris, S. petiolaris, S. pyrifolia, S. serissima, Sarracenia purpurea, Schoenoplectus acutus, Scorpidium scorpioides, Scutellaria galericulata, S. lateriflora, Sium suave, Solanum dulcamara, Solidago gigantea, S. ohioensis, S. patula, S. rugosa, S. uliginosa, Sonchus arvensis, Sphagnum angustifolium, S. contortum, S. fimbriatum, S. magellanicum, S. subtile, S. teres, S. warnstorfii, Spiraea alba, Stellaria longifolia, Symphyotrichum boreale, S. lanceolatum, S. puniceum, Symplocarpus foetidus, Taraxacum officinale, Thalictrum pubescens, Thelypteris palustris, Thuidium delicatulum, Thuja occidentalis, Tiarella cordifolia, Toxicodendron vernix, Triadenum virginicum, Triantha glutinosa, Trichocolea tomentella, Trichophorum alpinum, T. cespitosum, Trientalis borealis, Triglochin maritimum, Tsuga canadensis, Typha latifolia, Ulmus americana, Utricularia intermedia, Vaccinium angustifolium, V. corymbosum, V. oxycoccus, Valeriana uliginosa, Viburnum lentago, V. nudum, V. opulus, Viola blanda, Vitis riparia, Zizia aurea.*

Use by wildlife: *Lonicera oblongifolia* is a larval host for Baltimore checkerspot butterflies (Insecta: Lepidoptera: Nymphalidae: *Euphydryas phaeton*). It also hosts several plant pathogenic fungi including leaf blight (Basidiomycota: Platygloeaceae: *Insolibasidium deformans*), leaf spot (Ascomycota: Dothioraceae: *Kabatia lonicerae, K. mirabilis, K. periclymeni*), powdery mildew (Ascomycota: Erysiphaceae: *Microsphaera penicillata*), and other sac fungi (Ascomycota: Elsinoaceae: *Sphaceloma*; Hyponectriaceae: *Hyponectria lonicerae*; Mycosphaerellaceae: *Septoria xylostei*). The flowers attract bees (Insecta: Hymenoptera: Andrenidae: *Andrena vicina*; Anthophoridae: *Ceratina dupla, Xylocopa virginica*; Apidae: *Apis mellifera, Bombus melanopygus, B. pensylvanicus, B. ternarius, B. vagans*; Halictidae: *Agapostemon melliventris, Agapostemon splendens, Augochlorella persimilis, Lasioglossum coriaceum*; Megachilidae: *Osmia albiventris, O. atriventris*), butterflies (Insecta: Lepidoptera), and hummingbirds (Aves: Trochilidae: *Archilochus colubris*). Bees apparently are the principal pollinators, with birds and hummingbirds mainly seeking the floral nectar. Despite the common name of "fly honeysuckle", there are no reports of Diptera (Insecta) associated with the plants. The fruits are eaten by birds (Aves), which also disperse them.

Economic importance: food: *Lonicera oblongifolium* should not be eaten, as many species within the genus are at least mildly poisonous; **medicinal:** The Iroquois made a hot poultice from the bark of *Lonicera oblongifolium*, which was applied to the abdomen to ease painful urination. They also included it as an ingredient in a decoction used as a gynecological aid and in bark infusions for making sedatives to treat anxiety and depression; **cultivation:** More than 100 *Lonicera* species are in cultivation. *Lonicera oblongifolia* is sold as an ornamental native shrub and is recommended as an alternative to plant instead of invasive honeysuckle species such as *Lonicera ×bella*, *L. maackii*, *L. morrowii*, and *L. tatarica*. It sometimes is cultivated as *L. oblongifolia* f. *calyculata*; **misc. products:** none; **weeds:** *Lonicera oblongifolia* is not weedy; **nonindigenous species:** *Lonicera oblongifolia* is indigenous.

Systematics: Although no phylogenetic analysis has yet sampled all ~150 species of *Lonicera*, molecular studies including up to a third of the species have demonstrated that the genus is monophyletic. However, these studies also disagree on the precise placement of *Lonicera*, which resolves either as the sister genus to a clade comprising *Leycesteria*, *Symphoricarpos*, and *Triosteum* (10 cp genes; Figure 5.185), or (from combined nuclear and cpDNA data) is sister to a clade comprising *Leycesteria* and *Symphoricarpos*, with *Triosteum* being their next sister group. Because the topologies in both analyses exhibit comparable internal support, it is difficult to select one result as more compelling. *Lonicera oblongifolia* has thus far been excluded from phylogenetic analyses, despite an uncertain taxonomic position indicated by its placement within the monotypic subsection *Oblongifolia* of section *Isika*. Because the previously mentioned studies also show conclusively that section *Isika* is not monophyletic, the relationship of *L. oblongifolia* to its congeners presently remains unknown. The base chromosome number of *Lonicera* is $x=9$, with *L. oblongifolia* ($2n=18$) being diploid. No hybrids involving *L. oblongifolia* have been reported.

Comments: *Lonicera oblongifolia* occurs in northeastern North America.

References: Ammal & Saunders, 1952; Baskin & Baskin, 1998; Benko-Iseppon & Morawetz, 2000; Benoit, 2012; Christensen et al., 1959; Fernald, 1907; Frerker et al., 2013; Greene, 1953; Hilty, 2014; Janssen, 1967; Kost, 2001; Löve & Löve, 1982; Munger, 2005; Nichols, 1934; Rehder, 1903; Rigg, 1940; Rudenberg & Green, 1966; Stiles, 1980; Tank & Donoghue, 2010; Theis et al., 2008; Tiner, 1993b; Wheeler et al., 1983.

2. *Valeriana*

Valerian; valériane

Etymology: from the Latin *valere* ("to be healthy") for inferred medicinal properties

Synonyms: *Amplophus*; *Aretiastrum*; *Astrephia*; *Belonanthus*; *Phu*; *Phuodendron*; *Phyllactis*; *Stangea*

Distribution: global: Cosmopolitan (except Australia); **North America:** widespread

Diversity: global: 200 species; **North America:** 16 species

Indicators (USA): OBL: *Valeriana texana*, *V. uliginosa*

Habitat: freshwater; palustrine, riverine; **pH:** 7.0–8.0; **depth:** <1 m; **life-form(s):** emergent herb

Key morphology: stems (to 2.5 dm) multicipital, subscapose, and arising from taproots, or erect (to 10 dm), and arising from thick (to 7 mm) rhizomes; leaves basal and/or cauline (1–5 pairs), petiolate or subpetiolate, blades elliptic, cordate, ovate, spatulate or suborbicular, either undivided (to 35 cm) or pinnate to pinnatifid (to 21 cm), the margins dentate, entire, or undulating; inflorescence (to 12 cm) determinate, dichasial,

bracts (to 5.5 mm) reduced upward, bearing bisexual or pistillate flowers; corolla 5-lobed, rotate (to 3 mm) or funnelform (to 6 mm), white to pinkish; stamens and style conspicuously exserted, anthers 4-lobed; cypselae (to 4 mm) brownish, elliptic to lanceolate or oblong-linear, crowned by a persistent, dentate, cup-like calyx or by 11–23 plumose bristles (to 8 mm), ribbed longitudinally.

Life history: duration: perennial (rhizomes, tap roots); **asexual reproduction:** rhizomes; **pollination:** insect; **sexual condition:** dioecious, gynodioecious, hermaphroditic, or polygamodioecious; **fruit:** cypselae (common); **local dispersal:** cypselae (wind), rhizomes; **long-distance dispersal:** cypselae (wind)

Imperilment: (1) *Valeriana texana* [G3]; S2 (TX); S3 (NM); (2) *V. uliginosa* [G4]; SX (OH); S1 (IL, IN, NH, NY, VT); S2 (ME, <u>NB</u>, <u>ON</u>, WI); S3 (<u>QC</u>)

Ecology: general: *Valeriana* species occupy diverse habitats ranging from sea-level to subalpine elevations (to 3962 m) and grow from tropical to temperate latitudes, which can extend northward of 60°N. They are found in dry or moist sites that exist in or on bluffs, bogs, meadows, prairies, swamps, tundra, and woodlands. Eleven North American species (69%) can occur at least facultatively in wetlands (FACW, FAC, FACU) but only two (13%) are designated as OBL. The genus includes annual and perennial herbs that arise from tap roots (most species south of the United States border) or from rhizomes (most species north of Mexico). The flowers either are bisexual, or they are unisexual and arranged in dioecious, polygamo-dioecious, or polygamous arrays. The reproductive biology of the genus has not been studied adequately, but the bisexual flowers generally are protandrous and presumed to be primarily outcrossed. Observed pollinators include butterflies (Insecta: Lepidoptera) and small insects such as flies (Diptera: Culicidae; Tipulidae). The cypselae typically are wind dispersed, facilitated by their plumose pappus-like bristles, which are quite similar to those of many Asteraceae. Vegetative reproduction can occur by the production of rhizomes or stolons. The roots of several species have long been used medicinally as a natural sedative. Some species are cultivated as ornamentals, especially for medicinal herb gardens.

Valeriana texana **Steyerm.** occurs on cliffs or crevices in ravines, on outcrops, and in streams at elevations from 1828 to 2438 m. Occurrences often are located under the shade of trees. The substrates are coarse and have been described as boulders, granite, igneous rock, and limestone. The plants are in flower and fruit from April to July. The plants are gynodioecious with different individuals possessing either pistillate or bisexual flowers. No other information is available on the reproductive biology or seed ecology and additional study in these areas is much needed. The plants perennate from conical tap roots. **Reported associates:** *Acer grandidentatum*, *Achnatherum lobatum*, *A. robustum*, *Aquilegia chrysantha*, *Carex*, *Chaetopappa elegans*, *C. hersheyi*, *Heuchera rubescens*, *Ostrya knowltonii*, *Petrophytum caespitosum*, *Pinaropappus parvus*, *Pinus strobiformis*, *Potentilla sierraeblancae*, *Pseudotsuga menziesii*, *Ptelea trifoliata*, *Rhinotropis rimulicola*, *Selaginella pilifera*, *Viola guadalupensis*.

***Valeriana uliginosa* (Torr. & A. Gray) Rydb.** resides in bogs, fens, marshes, meadows, prairie fens, seeps, swamps, and along the margins of streams at elevations of up to 300 m. Although consistently high soil moisture levels must be maintained, the plants cannot tolerate prolonged periods of flooding. Occurences characteristically are associated with full sunlight exposures and calcareous, minerotrophic, and alkaline (pH 7.0–8.0) peat substrates. The flowers are bisexual and are produced from late May to August. The plants are in fruit from July to September. Although precise details are unavailable, the seeds reportedly germinate best (53%) after being warm dried, exposed to a period of cold stratification, and then brought to warm temperatures. The cypselae are dispersed by wind, facilitated by a persistent pappus-like calyx of plumose bristles. Although hundreds of flowers have been observed simultaneously in some sites, reproduction often occurs vegetatively from the rhizomes, especially in shaded sites. The plants grow primarily in gaps or in other open sites, which disturbance arising from fire, episodic flooding, and moderate logging can help to maintain. The maintenance of an open canopy is essential for the long-term survival of this species because flowering subsides as the canopy closes when the forests mature. The plants also are susceptible to competition from invasive flowering plant species such as *Lythrum salicaria* (Lythraceae), *Tussilago farfara* (Asteraceae), and *Phragmites australis* (Poaceae). **Reported associates:** *Abies balsamea, Acer negundo, A. rubrum, Achillea millefolium, Andromeda polifolia, Anemone canadensis, Angelica atropurpurea, Anthoxanthum nitens, Anticlea elegans, Apios americana, Apocynum cannabinum, Arisaema triphyllum, Arnoglossum plantagineum, Betula alleghaniensis, B. pumila, Bidens cernuus, Boehmeria cylindrica, Bromus ciliatus, Calamagrostis canadensis, Calopogon tuberosus, Caltha palustris, Campanula aparinoides, C. rapunculoides, Campylium stellatum, Carex aquatilis, C. aurea, C. buxbaumii, C. chordorrhiza, C. disperma, C. exilis, C. flava, C. formosa, C. gynocrates. C. lasiocarpa, C. laxiflora, C. leptalea, C. limosa, C. livida, C. schweinitzii, C. sterilis, C. stricta, C. tenuiflora, C. tetanica, C. trisperma, Chamaedaphne calyculata, Chelone glabra, Cirsium muticum, Cladium mariscoides, Clintonia borealis, Coptis trifolia, Cornus amomum, C. canadensis, C. foemina, C. racemosa, C. sericea, Corylus americana, Cypripedium acaule, C. calceolus, C. candidum, C. reginae, Dasiphora floribunda, Diervilla lonicera, Drosera rotundifolia, Eleocharis elliptica, E. obtusa, E. palustris, E. quinqueflora, E. rostellata, Elymus lanceolatus, E. trachycaulus, Equisetum arvense, E. laevigatum, Erigeron philadelphicus, Eriophorum viridicarinatum, Eurybia furcata, Eupatorium perfoliatum, Euthamia graminifolia, Eutrochium maculatum, Fragaria virginiana, Fraxinus nigra, F. pennsylvanica, Galium labradoricum, Gaultheria hispidula, Geum laciniatum, G. rivale, Gymnadeniopsis clavellata, Hylocomium splendens, Hypericum majus, Hypoxis hirsuta, Impatiens capensis, Iris versicolor, I. virginica, Juncus balticus, J. stygius, Larix laricina, Lathyrus palustris, Leersia oryzoides, Lemna minor, Liatris spicata, Lilium michiganense, L. philadelphicum, Linnaea borealis, Lobelia kalmii, Lonicera involucrata, L. morrowii, L. oblongifolia, L. villosa, Lycopus americanus, Lysimachia quadriflora, L. vulgaris, Lythrum salicaria, Maianthemum stellatum, M. trifolium, Malus pumila, Mentha piperita, Menyanthes trifoliata, Micranthes pensylvanica, Mitella nuda, Moneses uniflora, Muhlenbergia glomerata, Myosotis scorpioides, Myrica gale, M. pensylvanica, Nasturtium officinale, Orthilia secunda, Osmunda regalis, Oxypolis rigidior, Packera aurea, Parnassia caroliniana, P. glauca, Parthenocissus quinquefolia, Pedicularis lanceolata, Phalaris arundinacea, Phragmites australis, Physocarpus opulifolius, Picea mariana, Platanthera hyperborea, Pogonia ophioglossoides, Polemonium occidentale, Populus tremuloides, Prunella vulgaris, Prunus serotina, P. virginiana, Pycnanthemum virginianum, Pyrola asarifolia, Ranunculus gmelinii, Rhamnus alnifolia, R. frangula, Rhododendron groenlandicum, Rhynchospora alba, Ribes americanum, R. hirtellum, Rubus parviflorus, R. pubescens, R. sachalinensis, Rudbeckia hirta, Sagittaria montevidensis, Salix alba, S. bebbiana, S. candida, S. ×conifera, S. discolor, S. lucida, S. nigra, S. petiolaris, S. serissima, Sambucus nigra, Sanicula marilandica, Sarracenia purpurea, Schoenoplectus acutus, S. tabernaemontani, Scutellaria galericulata, Selaginella apoda, Solanum dulcamara, Solidago altissima, S. ohioensis, S. patula, S. rugosa, S. uliginosa, Sphagnum warnstorfii, Spiranthes romanzoffiana, Symphyotrichum boreale, S. laeve, S. lateriflorum, S. novae-angliae, Symplocarpus foetidus, Thalictrum dasycarpum, Thelypteris palustris, Thuja occidentalis, Toxicodendron radicans, T. vernix, Triantha glutinosa, Trichophorum alpinum, T. cespitosum, Triglochin gaspensis, T. palustre, Tussilago farfara, Typha latifolia, Ulmus americana, U. rubra, Utricularia cornuta, U. intermedia, Vaccinium macrocarpum, V. oxycoccus, Valeriana edulis, Vernonia, Viburnum dentatum, V. lentago, Viola cucullata, V. sororia, Vitis riparia, Xyris montana, Zizia aurea.*

Use by wildlife: *Valeriana uliginosa* is a host to plant pathogenic rusts (fungi: Basidiomycota: Pucciniaceae: *Puccinia commutata*) and sac fungi (Ascomycota: Mycosphaerellaceae: *Ramularia valerianae*; Pleosporaceae: *Pleospora herbarum*). Its flowers are visited by tiger swallowtail butterflies (Insecta: Lepidoptera: Papilionidae: *Papilio glaucus*).

Economic importance: food: not edible; **medicinal:** *Valeriana* roots have been used medicinally for centuries owing to their antispasmodic and sedative properties. The principal active ingredients are sesquiterpenoid valerenic acids and valepotriates, which vary widely in concentration among the different species. *Valeriana uliginosa* was used often by the Menominee tribe as an analgesic for cramps and headaches (root infusions), as a purgative, as a sedative for mental, lung, and throat disorders, and to promote the healing of cuts and wounds (pulverized root poultice); **cultivation:** The OBL *Valeriana* species are not cultivated; **misc. products:** The Menominee held a root of *V. uliginosa* in their mouth while arguing, as a charm to prevent their opponent from winning; **weeds:** Neither *V. texana* nor *V. uliginosa* is weedy; **nonindigenous species:** *Valeriana officinalis* has been introduced to North America but it is not a wetland indicator.

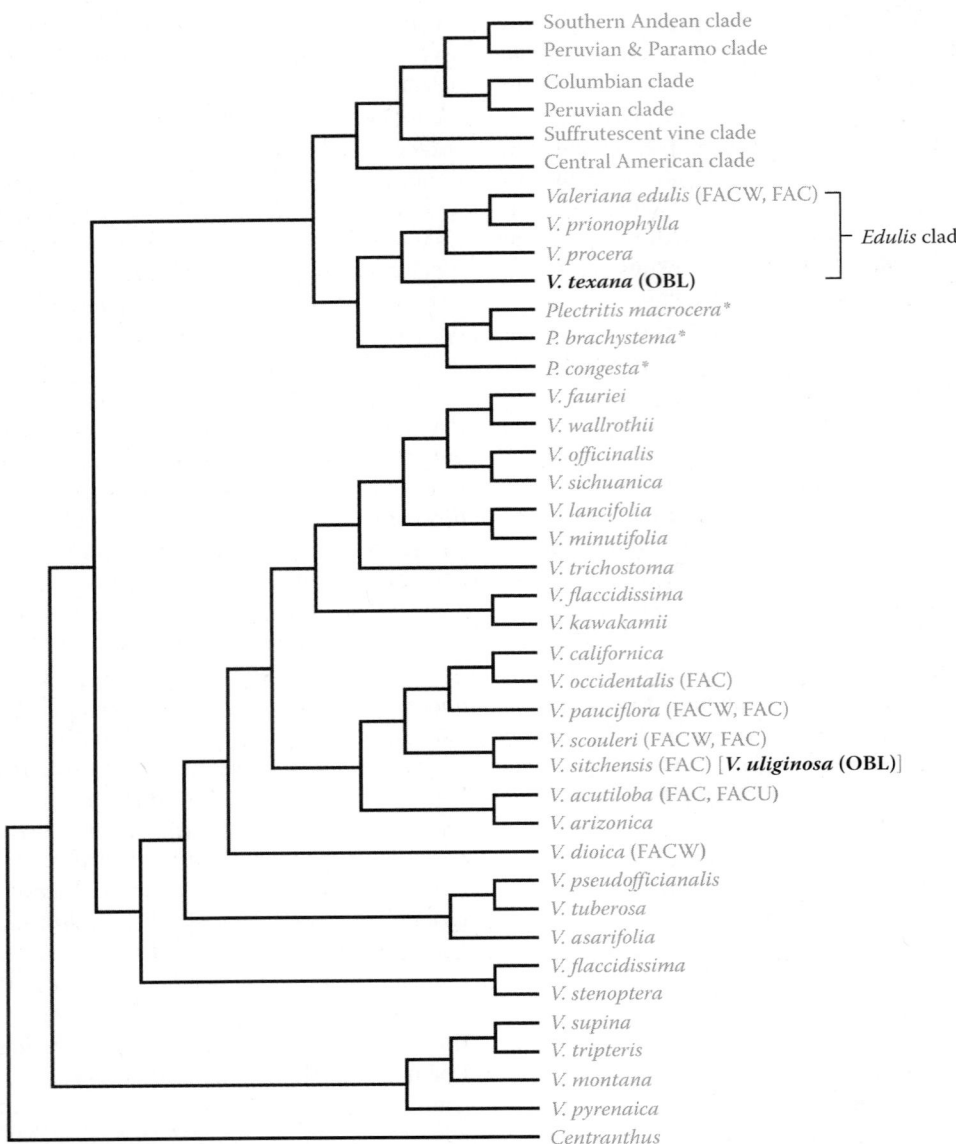

FIGURE 5.186 Phylogenetic relationships in *Valeriana* inferred by the analysis of combined nuclear and plastid DNA sequence data. The two OBL North American taxa (bold) occur in distinct clades, which indicates independent origins of the habit in the genus [note: *V. sitchensis* is used as a proxy for *V. uliginosa* (not sampled), which often is treated as a subspecies of the former]. These results show *V. scouleri* as the sister species to *V. uliginosa* (i.e., *V. sitchensis*); and resolve *V. texana* as the sister species to the remainder of the "*Edulis* clade" (equivalent to series *Edule* of Meyer, 1951). The genus *Plectritis* (asterisked), which resolves in different phylogenetic positions depending on the analysis, is sister to the *Edulis* clade in this example, which raises some doubt regarding the monophyly of *Valeriana* as currently circumscribed. (Adapted from Bell, C.D. et al., *Mol. Phylogenet. Evol.*, 63, 724–737, 2012.)

Systematics: Although the cypselae with their persistent, plumose calyx, are remarkably convergent with those found in Asteraceae, *Valeriana* (Dipsacales) actually is quite remotely related to Asteraceae (Asterales) and is not their close relative as once was thought (Figure 5.131). Rather, *Valeriana* occupies a relatively derived phylogenetic position within the family that is now broadly defined as Caprifoliaceae (Figure 5.185). The precise placement of *Valeriana* in Caprifoliaceae is more difficult to elucidate, even though a large number of taxa and genetic loci have been surveyed in phylogenetic analyses. A study of 21 *Valeriana* species incorporating combined DNA sequence data from nrITS and five plastid loci resolved the group as a clade, with the exception of one species (*V. celtica*). In that analysis *Centranthus* and *Plectritis* resolved as a sister clade to *Valeriana* with strong support. Increased sampling of *Valeriana* (69 species; nrITS+3 cpDNA loci) and other taxa yielded a topology in which *Plectritis* became embedded rather deeply within *Valeriana*, *Centranthus* became its sister clade, and *Valerianella* (plus *Fedia*) resolved as the subsequent sister group. Although the *Plectritis* subclade was characterized by a long branch, the nodal support for that conflicting topology also was very strong. In the most comprehensive analysis to date

(105 *Valeriana* species; nrITS+9 cpDNA loci), the *Plectritis* clade repositioned once again, this time as the sister group to the "*Edulis* clade" (essentially series *Edule*), which includes the OBL *V. texana* (Figure 5.186). The uncertain placement of *Plectritis* notwithstanding, the results do consistently implicate *Centranthus* as the sister group of *Valeriana* (in exclusion of *V. celtica*). There also is disagreement on whether *V. uliginosa* should be maintained as a distinct species or merged with the western *V. sitchensis* as a subspecies, which has been the convention for decades. However, *V. sitchensis* (FAC) is not an OBL plant, inhabits a discrete geographical range, and differs both ecologically and morphologically from *V. uliginosa*. Consequently, the two taxa have been retained here as distinct species until their taxonomic status can be resolved more satisfactorily. At this time only material of *V. sitchensis* has been included in phylogenetic analyses, thus providing no insight that might help to resolve this question. However, given this taxonomic dilemma, *V. uliginosa* presumably is at least closely related to *V. sitchensis*, thus allowing the latter to represent a reasonable proxy of the former in the published phylogenetic analyses. Because those studies consistently place the two taxa in different clades, two independent origins of the OBL habit within the genus are indicated (Figure 5.186). This result is consistent with the previous placement of the two

species within different taxonomic series in the genus: series *Officinales* (*V. sitchensis*/*V. uliginosa*) and series *Edule* (*V. texana*). Phylogenetic analyses resolve *V. scouleri* as the sister species to *V. uliginosa* [i.e., *V. sitchensis*], but place *V. texana* either as the sister species to *V. edulis*, or as the sister to the entire "*Edulis* clade" (Figure 5.186). The base chromosome number of *Valeriana* varies from $x=7$–9 among the different species groups. The count of $2n=96$ attributed to *V. uliginosa* would indicate that it is a dodecaploid ($x=8$); however, this count probably is based on material of *V. sitchensis* and further documentation is necessary. No count has been reported for *V. texensis*. No hybrids are reported that involve either *V. texensis* or *V. uliginosa*.

Comments: *Valeriana uliginosa* occurs in northeastern North America, whereas *V. texana* is restricted to the Guadalupe Mountains of New Mexico and Texas.

References: Bell, 2004; Bell & Donoghue, 2005; Bell et al., 2012; Blake, 1918; Cain & Slater, 1948; Dearness & House, 1921; Dugal, 1990; Farnsworth, 2004; Ferguson, 1965; Fernald, 1908; Gao & Björk, 2000; Hershey, 1940; Kennedy & Strong, 2012; Metcalf & Griscom, 1917; Mullet et al., 2008; NYNHP, 2013; Pinna et al., 2010; Powell & Wauer, 1990; Schmidt, 2003b; Soreng & Spellenberg, 1984; Steyermark, 1932; St. Hilaire, 2003; Voss, 1954; Weatherbee & Price, 2001.

References

Aarvik, L., S. Svendsen, V. Berg, K. Berggren & L. O. Hansen. 1994. Atlas of the Lepidoptera of Norway. Part 1. Gelechioidea: Oecophoridae, Agonoxenidae, Batrachedridae, Momphidae, Cosmopterigidae, Scythridae, Blastobasidae. *Insecta Norveg.* 5: 1–72.

Abad, M. J., P. Bermejo, E. Gonzales, I. Iglesias, A. Irurzun & L. Carrasco. 1999. Antiviral activity of Bolivian plant extracts. *Gen. Pharmacol.* 32: 499–503.

Abbassi, K., Z. Atay-Kadiri & S. Ghaout. 2003. Biological effects of alkaloids extracted from three plants of Moroccan arid areas on the desert locust. *Physiol. Entomol.* 28: 232–236.

Abbott, J. R. 2009. Phylogeny of the Polygalaceae and a revision of *Badiera*. PhD dissertation. University of Florida, Gainesville, FL. 291 pp.

Abbott, J. R. 2011. Notes on the disintegration of *Polygala* (Polygalaceae), with four new genera for the flora of North America. *J. Bot. Res. Inst. Texas* 5: 125–137.

Abbott, T. P., R. E. Peterson, L. W. Tjarks, D. M. Palmer & M. O. Bagby. 1990. Major extractable components in *Asclepias linaria* (Asclepiadaceae) and *Ilex verticillata* (Aquifoliaceae), two potential hydrocarbon crops. *Econ. Bot.* 44: 278–284.

Abdallah, M. A. 2012. Phytoremediation of heavy metals from aqueous solutions by two aquatic macrophytes, *Ceratophyllum demersum* and *Lemna gibba* L. *Environ. Technol.* 33: 1609–1614.

Abdel-Baset, Z. H., L. Southwick, W. G. Padolina, H. Yoshioka, T. J. Mabry & S. B. Jones Jr. 1971. Sesquiterpene lactones: a survey of 21 United States taxa from the genus *Vernonia* (Compositae). *Phytochemistry* 10: 2201–2204.

Abel, K. & J. Friesen. 1991. *Aboriginal Resource Use in Canada: Historical and Legal Aspects.* University of Manitoba Press, Winnipeg, MB. 343 pp.

Abernethy, V. J. & N. J. Willby. 1999. Changes along a disturbance gradient in the density and composition of propagule banks in floodplain aquatic habitats. *Plant Ecol.* 140: 177–190.

Abrahamson, W. G. 1984. Species responses to fire on the Florida Lake Wales ridge. *Amer. J. Bot.* 71: 35–43.

Abrahamson, W. G. & S. P. Vander Kloet. 2014. The reproduction and ecology of *Hypericum edisonianum*: an endangered Florida endemic. *Castanea* 79: 168–181.

Abrams, L. 1944. *Illustrated Flora of the Pacific States Washington, Oregon, and California. Vol. II: Polygonaceae to Krameriaceae, Buckwheats to Kramerias.* Stanford University Press, Stanford, CA. 643 pp.

Abrams, L. 1960. *Illustrated Flora of the Pacific States. Vol. IV: Bignonias to Sunflowers.* Stanford University Press, Stanford, CA. 740 pp.

Abu-Asab, M. S. & P. D. Cantino. 1994. Systematic implications of pollen morphology in subfamilies *Lamioideae* and *Pogostemonoideae* (Labiatae). *Ann. Missouri Bot. Gard.* 81: 653–686.

Abubakar, M. A., R. M. Zulkifli, W. N. A. W. Hassan, A. H. M. Shariff, N. A. N. N. Malek, Z. Zakaria & F. Ahmad. 2015. Antibacterial properties of *Persicaria minor* (Huds.) ethanolic and aqueous-ethanolic leaf extracts. *J. Appl. Pharm. Sci.* 5: S50–S56.

Abulude, F., A. Olarenwaju & M. Ogunkoya. 2010. Evaluation of boron in selected aquatic weeds as feed or food. *Electronic J. Polish Agric. Univ.* 13: #07. Available online: http://www.ejpau.media.pl/volume13/issue1/art-07.html.

Ackerly, D. D. & M. J. Donoghue. 1998. Leaf size, sapling allometry, and Corner's rules: phylogeny and correlated evolution in maples (*Acer*). *Amer. Naturalist* 152: 767–791.

Adam, P. 1990. *Saltmarsh Ecology.* Cambridge University Press, Cambridge, UK. 461 pp.

Adamczyk, U., S. A. Brown, E. G. Lewars & L. Światek. 1990. Lactones of *Menyanthes trifoliata. Plant Méd. Phytothér.* 24: 73–78.

Adamczyk-Rogozińska, U. & H. Wysokińska. 1998. Organ and plantlet regeneration of *Menyanthes trifoliata* through tissue culture. *Acta Soc. Bot. Polon.* 67: 161–166.

Adamec, L. 1995. Ecological requirements of *Aldrovanda vesiculosa*. Testing of its new potential sites in the Czech Republic. *Acta Bot. Gallica* 142: 673–680.

Adamec, L. 1997. Mineral nutrition of carnivorous plants: a review. *Bot. Rev.* 63: 273–299.

Adamec, L. 1999. Turion overwintering of aquatic carnivorous plants. *Carniv. Plant Newslett.* 28: 19–24.

Adamec, L. 2000. Rootless aquatic plant *Aldrovanda vesiculosa*: physiological polarity, mineral nutrition, and importance of carnivory. *Biol. Plant.* 43: 113–119.

Adamec, L. 2007. Investment in carnivory in *Utricularia stygia* and *U. intermedia* with dimorphic shoots. *Preslia* 79: 127–139.

Adamec, L. & J. Lev. 2002. Ecological differences between *Utricularia ochroleuca* and *U. intermedia* habitats. *Carniv. Plant Newslett.* 31: 14–18.

Adamec, L. & K. Kondo. 2002. Optimization of medium for growing the aquatic carnivorous plant *Aldrovanda vesiculosa* in vitro. *Plant Biotechnol.* 19: 283–286.

Adamec, L. & M. Kovářová. 2006. Field growth characteristics of two aquatic carnivorous plants, *Aldrovanda vesiculosa* and *Utricularia australis. Folia Geobot.* 41: 395–406.

Adams, C. C., G. P. Burns, T. L. Hankinson, B. Moore & N. Taylor. 1920. Plants and animals of Mount Marcy, New York, part II. *Ecology* 1: 204–233.

Adams, F. S. 1969. Winterbud production and function in *Brasenia schreberi. Rhodora* 71: 417–433.

Adams, J. 1927. The germination of the seeds of some plants with fleshy fruits. *Amer. J. Bot.* 14: 415–428.

Adams, L. D., S. Buchmann, A. D. Howell & J. Tsang. 2010. *A Study of Insect Pollinators Associated with DoD TER-S Flowering Plants, Including Identification of Habitat Types Where They Co-Occur by Military Installations in the Southeastern United States.* Project Number #09-391, December, 2010. Department of Defense Legacy Program, Arlington, VA. 83 pp.

Adams, S. & A. Dawson. 2009. Fountain of fountains: exploring the dynamics of seasonal wetlands at the Bouverie Preserve. *Ardeid* 2009: 1–3.

Adams, S., J. Wirka & D. Gluesenkamp. 2007. Saving Sonoma's sunshine: vernal pool restoration at Bouverie Preserve. *Ardeid* 2007: 1–3.

Adams, V. M., D. M. Marsh & J. S. Knox. 2005. Importance of the seed bank for population viability and population monitoring in a threatened wetland herb. *Biol. Conserv.* 124: 425–436.

Adamus, P. R. 2005. *Science Review and Data Analysis for Tidal Wetlands of the Oregon coast. Part 2 of a Hydrogeomorphic Guidebook.* Report to Coos Watershed Association, U. S. Environmental Protection Agency, and Oregon Department of State Lands, Salem, OR. 216 pp.

Addy, H. D., S. Hambleton & R. S. Currah. 2000. Distribution and molecular characterization of the root endophyte *Phialocephala fortinii* along an environmental gradient in the boreal forest of Alberta. *Mycol. Res.* 104: 1213–1221.

Adebayo, A., I. Watanabe & J. K. Ladha. 1989. Epiphytic occurrence of *Azorhizobium caulinodans* and other rhizobia on host and nonhost legumes. *Appl. Environ. Microbiol.* 55: 2407–2409.

Adler, L. S. 2002. Host effects on herbivory and pollination in a hemiparasitic plant. *Ecology* 83: 2700–2710.

Adler, L. S. & M. Wink. 2001. Transfer of quinolizidine alkaloids from hosts to hemiparasites in two *Castilleja-Lupinus* associations: analysis of floral and vegetative tissues. *Biochem. Syst. Ecol.* 29: 551–561.

Adrienne, B., B. Venables & E. M. Barrows. 1985. Skippers: pollinators or nectar thieves? *J. Lepid. Soc.* 39: 299–312.

Affolter, J. M. 1985. A monograph of the genus *Lilaeopsis* (Umbelliferae). *Syst. Bot. Monogr.* 6: 1–140.

Afolayan, A. T., O. J. Sharaibi & K. M. Idowu. 2013. Phytochemical analysis and in vitro antioxidant activity of *Nymphaea lotus* L. *Int. J. Pharmacol.* 9: 297–304.

Afonso, A. S. D. S. 2013. Effects of flower position on the sexual specialization within inflorescences. MS thesis. University of Coimbra, Coimbra, Portugal. 72 pp.

Agami, M. & K. R. Reddy. 1991. Interrelationships between *Eichhornia crassipes* (Mart.) Solms and *Hydrocotyle umbellata* L. *Aquat. Bot.* 39: 147–157.

Ager, T. A. & L. P. Ager. 1980. Ethnobotany of the Eskimos of Nelson Island, Alaska. *Arctic Anthropol.* 17: 26–48.

Agrawal, A. A. & M. Fishbein. 2006. Plant defense syndromes. *Ecology* 87: S132–S149.

Aguilar-Ortigoza, C. J., V. Sosa & M. Aguilar-Ortigoza. 2003. Toxic phenols in various Anacardiaceae species. *Econ. Bot.* 57: 354–364.

Aguilar-Ortigoza, C., V. Sosa & G. Angeles. 2004. Phylogenetic relationships of three genera in Anacardiaceae: *Bonetiella*, *Pseudosmodingium*, and *Smodingium*. *Brittonia* 56: 169–184.

Aguilera, X. G., V. B. Alvarez & F. Chavez-Ramirez. 2005. Nesting ecology and productivity of the Cuban sandhill crane on the Isle of Youth, Cuba. *Proc. N. Amer. Crane Workshop* 9: 225–236.

Ahedor, A. R. & W. Elisens. 2006. Genetic diversity and differentiation among populations of the *Mecardonia acuminata* (Plantaginaceae) complex inferred from inter simple sequence repeats [ISSR] banding patterns. Botany 2006 meeting (abstract).

Ahemd, S. A. & E. M. Kamel. 2013. Phenolic constituents and biological activity of the genus *Pluchea*. *Der Pharma Chem.* 5: 109–114.

Ahlgren, C. E. 1960. Some effects of rire on reproduction and growth of vegetation in northeastern Minnesota. *Ecology* 41: 431–445.

Ahmad, H. & J. D. Hitchmough. 2007. Germination and emergence of understorey and tall canopy forbs used in naturalistic sowing mixes. A comparison of performance in vitro vs the field. *Seed Sci. Technol.* 35: 624–637.

Ahti, T. & A. Henssen. 1965. New localities for *Cavernularia hultenii* in eastern and western North America. *Bryologist* 68: 85–89.

Aiken, S. G. 1981. A conspectus of *Myriophyllum* (Haloragaceae) in North America. *Brittonia* 33: 57–69.

Aiken, S. G. & K. F. Walz. 1979. Turions of *Myriophyllum exalbescens*. *Aquat. Bot.* 6: 357–363.

Aiken, S. G., M. J. Dallwitz, L. L. Consaul, C. L. McJannet, R. L. Boles, G. W. Argus, J. M. Gillett, P. J. Scott, R. Elven, M. C. LeBlanc, L. J. Gillespie, A. K. Brysting, H. Solstad & J. G. Harris. 2007. *Flora of the Canadian Arctic Archipelago: Descriptions, Illustrations, Identification, and Information Retrieval.* NRC Research Press, National Research Council of Canada, Ottawa, Canada. Available online: http://nature.ca/aaflora/data (accessed September 6, 2016).

Aikman, K., D. Bergman, J. Ebinger & D. Seigler. 1996. Variation of cyanogenesis in some plant species of the midwestern United States. *Biochem. Syst. Ecol.* 24: 637–645.

Ailstock, M. S., C. M. Norman & P. J. Bushmann. 2001. Common reed *Phragmites australis*: control and effects upon biodiversity in freshwater nontidal wetlands. *Restorat. Ecol.* 9: 49–59.

Ajima, M., S. Tsuda & H. Tsuda. 1999. A preliminary study of seed germination characteristics of *Cicuta virosa* L. *Actinia* 12: 159–166.

Akerreta, S., M. I. Calvo & R. Y. Cavero. 2010. Ethnoveterinary knowledge in Navarra (Iberian Peninsula). *J. Ethnopharmacol.* 130: 369–378.

Akinjogunla, O. J., A. A. Adegoke, I. P. Udokang & B. C. Adebayo-Tayo. 2009. Antimicrobial potential of *Nymphaea lotus* (Nymphaeaceae) against wound pathogens. *J. Med. Plants Res.* 3: 138–141.

Akman, M., A. V. Bhikharie, E. H. McLean, A. Boonman, E. J. W. Visser, M. E. Schranz & P. H. van Tienderen. 2012. Wait or escape? Contrasting submergence tolerance strategies of *Rorippa amphibia*, *Rorippa sylvestris* and their hybrid. *Ann. Bot.* 109: 1263–1276.

Alam, M. N., M. R. Islam, M. S. Biozid, M. I. A. Chowdury, M. M. U. Mazumdar, M. A. Islam & Z. Bin Anwar. 2016. Effects of methanolic extract of *Nymphaea capensis* leaves on the sedation of mice and cytotoxicity of brine shrimp. *Adv. Biol. Res.* 10: 1–9.

Alarcon-Aguilara, F. J., R. Roman-Ramos, S. Perez-Gutierrez, A. Aguilar-Contreras, C. C. Contreras-Weber & J. L. Flores-Saenz. 1998. Study of the anti-hyperglycemic effect of plants used as antidiabetics. *J. Ethnopharmacol.* 61: 101–110.

Alarcón, R., N. M. Waser & J. Ollerton. 2008. Year-to-year variation in the topology of a plant-pollinator interaction network. *Oikos* 117: 1796–1807.

Albach, D. C. & M. W. Chase. 2001. Paraphyly of *Veronica* (Veroniceae; Scrophulariaceae): evidence from the internal transcribed spacer (ITS) sequences of nuclear ribosomal DNA. *J. Plant Res.* 114: 9–18.

Albach, D. C., P. S. Soltis, D. E. Soltis & R. G. Olmstead. 2001. Phylogenetic analysis of the Asteridae s.l. based on sequences of four genes. *Ann. Missouri Bot. Gard.* 88: 163–212.

Albach, D. C., M. M. Martínez-Ortega, M. A. Fischer & M. W. Chase. 2004a. A new classification of the tribe Veroniceae: problems and a possible solution. *Taxon* 53: 429–452.

Albach, D. C., M. M. Martínez-Ortega, M. A. Fischer & M. W. Chase. 2004b. Evolution of Veroniceae: a phylogenetic perspective. *Ann. Missouri Bot. Gard.* 91: 275–320.

Albach, D. C., H. M. Meudt & B. Oxelman. 2005. Piecing together the "new" Plantaginaceae. *Amer. J. Bot.* 92: 297–315.

Al-Barwani, F. M. & E. A. Eltayeb. 2004. Antifungal compounds from induced *Conium maculatum* L. plants. *Biochem. Syst. Ecol.* 32: 1097–1108.

Albert, D. A. & A. J. Tepley. 2005. *Development of Bioassessment Procedures for Non-Forested Depressional Wetlands in Michigan.* Report Number 2005–09. Michigan Department of Environmental Quality, Lansing, MI. 108 pp.

Albert, V. A., S. E. Williams & M. W. Chase. 1992. Carnivorous plants: phylogeny and structural evolution. *Science* 257: 1491–1495.

Albrecht, M., K. M. Kneeland, E. Lindroth & J. E. Foster. 2013. Genetic diversity and relatedness of the mangrove *Rhizophora mangle* L. (Rhizophoraceae) using amplified fragment polymorphism (AFLP) among locations in Florida, USA and the Caribbean. *J. Coastal. Conserv.* 17: 483–491.

Aldrich, J. H., G. W. Knox & J. G. Norcini. 2006. Florida Coreopsis (*Coreopsis floridana*): a new fall-flowering perennial. *Proc. S. Nurs. Assoc. Res. Conf.* 51: 663–665.

Aldrich, J. H., J. G. Norcini, G. W. Knox & T. M. Batey. 2007. Container production of *Coreopsis floridana*. *Proc. S. Nurs. Assoc. Res. Conf.* 52: 468–472.

Aldrich, J. R., J. A. Bacone & M. D. Hutchison. 1981. Limestone glades of Harrison County, Indiana. *Proc. Indiana Acad. Sci.* 91: 480–485.

Aldrich, J. W. 1943. Biological survey of the bogs and swamps in northeastern Ohio. *Amer. Midl. Naturalist* 30: 346–402.

Aleric, K. M. & L. K. Kirkman. 2005. Growth and photosynthetic responses of the federally endangered shrub, *Lindera melissifolia* (Lauraceae), to varied light environments. *Amer. J. Bot.* 92: 682–689.

Alexander, D. G. & R. A. Schlising. 1998. Patterns in time and space for rare macroinvertebrates and vascular plants in vernal pool ecosystems at the Vina Plains Preserve, and implications for pool landscape management. pp. 161–168. *In*: C. W. Witham, E. T. Bauder, D. Belk, W. R. Ferren Jr. & R. Ornduff (eds.), *Ecology, Conservation, and Management of Vernal Pool Ecosystems Proceedings from a 1996 Conference*. California Native Plant Society, Sacramento, CA.

Alexander, E. B. 2007. *Serpentine Geoecology of Western North America: Geology, Soils, and Vegetation*. Oxford University Press, New York. 512 pp.

Alexander, E. J. 1938. Habitat hunting in mountains of the southeast with student gardeners. *J. New York Bot. Gard.* 39: 228–240.

Alexander, E. J. & H. K. Svenson. 1943. The field trip to the New Jersey coast and Pine Barrens Friday and Saturday, June 26–27, 1942. *Torreya* 43: 170–173.

Alexander, J. 2016. The Utah Native Plant Society rare plant list: version 2. *Calochortiana* 3: 3–247.

Alexander, S. N., L.-A. C. Hayek & A. Weeks. 2012. A subspecific revision of North American saltmarsh mallow *Kosteletzkya pentacarpos* (L.) Ledeb. (Malvaceae). *Castanea* 77: 106–122.

Alexander, W. C. 1987. Aggressive behavior of wintering diving ducks (Aythyini). *Wilson Bull.* 99: 38–49.

Alford, J. D. & L. C. Anderson. 2002. The taxonomy and morphology of *Macranthera flammea* (Orobanchaceae). *Sida* 20: 189–204.

Al-Helal, A. A. 1996. Studies on germination of *Rumex dentatus* L. seeds. *J. Arid Environ.* 33: 39–47.

Ali, M. A., J. Lee, S. Y. Kim & F. M. A. Al-Hemaid. 2012. Molecular phylogenetic study of *Cardamine amaraeformis* Nakai using nuclear and chloroplast DNA markers. *Genet. Mol. Res.* 11: 3086–3090.

Alice, L. A. & C. S. Campbell. 1999. Phylogeny of *Rubus* (Rosaceae) based on nuclear ribosomal DNA internal transcribed spacer region sequences. *Amer. J. Bot.* 86: 81–97.

Alice, L. A., T. Eriksson, B. Eriksen & C. S. Campbell. 2001. Hybridization and gene flow between distantly related species of *Rubus* (Rosaceae): evidence from nuclear ribosomal DNA internal transcribed spacer sequences. *Syst. Bot.* 26: 769–778.

Alice, L. A., D. H. Goldman, J. A. Macklin & G. Moore. 2014. 2. *Rubus* Linnaeus. pp. 28–56. *In*: N. R. Morin (convening ed.), *Flora of North America North of Mexico, Vol. 9: Magnoliophyta: Picramniaceae to Rosaceae*. Oxford University Press, New York.

Alison, R. M. 1975. Breeding biology and behavior of the oldsquaw (*Clangula hyemalis* L.). *Ornithological Monogr.* 18: 1–52.

Aliyu, A. B., A. M. Musa, M. S. Abdullahi & A. O. Oyewale. 2008. Phytochemical and antibacterial properties of *Ludwigia suffruticosa* (Wild) Iliv. Ex. O. Ktze (Onagraceae). *Int. J. Pure Appl. Sci.* 2: 1–5.

Alkire, B. H. & J. E. Simon. 1995. Response of Midwestern peppermint (*Mentha × piperita* L.) and native spearmint (*M. spicata* L.) to rate and form of nitrogen fertilizer. *Acta Hort.* 426: 537–549.

Allan, G. & J. M. Porter. 2000. Tribal delimitation and phylogenetic relationships of tribes *Loteae* and *Coronilleae* (Faboideae: Fabaceae) with special reference to *Lotus*: evidence from nuclear ribosomal ITS sequences. *Amer. J. Bot.* 87: 1871–1881.

Allard, D. J. 2001. *Pedicularis lanceolata Michx. (Swamp woodbetony) conservation and research plan*. New England Wild Flower Society, Framingham, MA. 22 pp.

Allbritton, G., J. G. Norcini & J. H. Aldrich. 2002. Natural height control of container grown *Eupatorium fistulosum*. *J. Environ. Hort.* 20: 232–235.

Allen, C. M. 1983. *Murdannia keisak* (Hassk.) Hand-Mazz. (Commelinaceae), *Bothriochloa hybrida* (Gould) Gould (Poaceae), and *Scutellaria racemosa* Pers. (Lamiaceae) new to Louisiana. *Sida* 10: 189–190.

Allen, G. A. 1984. Morphological and cytological variation in the western North American *Aster occidentalis* complex (Asteraceae). *Syst. Bot.* 9: 175–191.

Allen, J. A. 2002. *Laguncularia racemosa* (L.) C.F. Gaertn. pp. 537–539. *In*: J. Vozzo (ed.), *Tropical Tree Seed Manual: Part II, Species Descriptions*. Agricultural Handbook 712. U. S. Department of Agriculture, Washington, DC.

Allen, J. A. & K. W. Krauss. 2006. Influence of propagule flotation longevity and light availability on establishment of introduced mangrove species in Hawai'i. *Pacific. Sci.* 60: 367–376.

Allen, L., J. D. Johnson & K. Vujnovic. 2002. *Small Patch Communities of La Butte Creek Wildland Provincial Park*. Report to the Parks and Protected Areas, Alberta Community Development, Edmonton, Alberta. Alberta Natural Heritage Information Centre, Edmonton, AB. 38 pp.

Allen, R. B., R. K. Peet & W. L. Baker. 1991. Gradient analysis of latitudinal variation in southern Rocky Mountain forests. *J. Biogeogr.* 18: 123–139.

Allen, W. R. & P. M. Sheppard. 1971. Copper tolerance in some Californian populations of the monkey flower, *Mimulus guttatus*. *Proc. Roy. Soc. London. Ser. B Biol. Sci.* 177: 177–196.

Allen-Diaz, B., R. D. Jackson, J. W. Bartolome, K. W. Tate & L. G. Oates. 2004. Long-term grazing study in spring-fed wetlands reveals management tradeoffs. *Calif. Agric.* 58: 145–149.

Aller, A. R. 1956. A taxonomic and ecologic study of the flora of Monument Peak, Oregon. *Amer. Midl. Naturalist* 56: 454–472.

Allison, J. R. 2011. Synopsis of the *Hypericum denticulatum* complex (Hypericaceae). *Castanea* 76: 99–115.

Allison, L. N. 1967. Beggar-ticks cause mortality among fingerling Coho salmon. *Prog. Fish Cult.* 29: 113.

Allison, S. K. 1995. Recovery from small-scale anthropogenic disturbances by northern California salt marsh plant assemblages. *Ecol. Appl.* 5: 693–702.

Allphin, L. & K. T. Harper. 1994. Habitat requirements for *Erigeron kachinensis*, a rare endemic of the Colorado plateau. *Great Basin Naturalist* 54 193–203.

Allphin, L. & K. T. Harper. 1997. Demography and life history characteristics of the rare Kachina daisy (*Erigeron kachinensis*, Asteraceae). *Amer. Midl. Naturalist* 138: 109–120.

Allphin, L. & M. Windham. 2002. Morphological and genetic variation among populations of the rare Kachina daisy (*Erigeron kachinensis*) from southeastern Utah. *West. N. Amer. Naturalist* 64: 423–436.

Allphin, L., M. Windham & K. T. Harper. 1996. A genetic evaluation of three potential races of the rare Kachina daisy. pp. 68–76. *In*: J. Mashchinski, H. D. Hammond & L. Holter (eds.), *Southwestern Rare and Endangered Plants: Proceedings of the Second Conference*; September 11–14, 1995; Flagstaff, AZ. General Technical Report RM-GTR-283. U. S. Department of Agriculture, Forest Service, Rocky Mountain Forest and Range Experiment Station, Fort Collins, CO. 328 pp.

Allphin, L., D. Wiens & K. T. Harper. 2002. The relative effects of resources and genetics on reproductive success in the rare Kachina daisy, *Erigeron kachinensis* (Asteraceae). *Int. J. Plant Sci.* 163: 599–612.

Alm, T. 2005. *Pinguicula vulgaris* (Lentibulariaceae) and its uses in Norway. *Sida* 21: 2249–2274.

Almborn, O. 1983. Pitcher-plant, *Sarracenia purpurea*, naturalized in Sweden. *Sven. Bot. Tidskr.* 77: 209–216.

Almeida-Cortez, J. S., B. Shipley & J. T. Arnason. 1999. Do plant species with high relative growth rates have poorer chemical defences? *Funct. Ecol.* 13: 819–827.

Almquist, H. & A. J. K. Calhoun. 2003. A coastal, southern-outlier, patterned fen: Lily Fen, Swans Island, Maine. *N. E. Naturalist* 10: 119–130.

Al-Shehbaz, I. A. 1986. New wool-alien Cruciferae (Brassicaceae) in eastern North America. *Rhodora* 88: 347–355.

Al-Shehbaz, I. A. 2010a. 11. Brassicaceae Burnett. pp. 224–746. *In*: Flora of North America Editorial Committee (ed.), *Flora of North America North of Mexico: Vol. 7, Magnoliophyta: Salicaceae to Brassicaceae*. Oxford University Press, New York.

Al-Shehbaz, I. A. 2010b. 59. *Eutrema* R. Brown. pp. 555–556. *In*: Flora of North America Editorial Committee (ed.), *Flora of North America North of Mexico: Vol. 7, Magnoliophyta: Salicaceae to Brassicaceae*. Oxford University Press, New York.

Al-Shehbaz, I. A. 2010c. 43. *Nasturtium* W. T. Aiton in W. Aiton and W. T. Aiton. pp. 489–492. *In*: Flora of North America Editorial Committee (ed.), *Flora of North America North of Mexico: Vol. 7, Magnoliophyta: Salicaceae to Brassicaceae*. Oxford University Press, New York.

Al-Shehbaz, I. A. 2010d. 45. *Rorippa* Scopoli. pp. 493–505. *In*: Flora of North America Editorial Committee (ed.), *Flora of North America North of Mexico: Vol. 7, Magnoliophyta: Salicaceae to Brassicaceae*. Oxford University Press, New York.

Al-Shehbaz, I. A. 2010e. 47. *Subularia* Linnaeus. p. 509 *In*: Flora of North America Editorial Committee (ed.), *Flora of North America North of Mexico: Vol. 7, Magnoliophyta: Salicaceae to Brassicaceae*. Oxford University Press, New York.

Al-Shehbaz, I. A. 2012. A generic and tribal synopsis of the Brassicaceae (Cruciferae). *Taxon* 61: 931–954.

Al-Shehbaz, I. A. & J. F. Gaskin. 2010. 68. *Lepidium* Linnaeus. pp. 570–595. *In*: Flora of North America Editorial Committee (ed.), *Flora of North America North of Mexico: Vol. 7, Magnoliophyta: Salicaceae to Brassicaceae*. Oxford University Press, New York.

Al-Shehbaz, I. A. & R. A. Price. 1998. Delimitation of the genus *Nasturtium* (Brassicaceae). *Novon* 8: 124–126.

Al-Shehbaz, I. A. & S. I. Warwick. 2005. A synopsis of *Eutrema* (Brassicaceae). *Harvard Pap. Bot.* 10: 129–135.

Al-Shehbaz, I. A., K. Marhold & J. Lihova. 2010. 40. *Cardamine* Linnaeus. pp. 464–484. *In*: Flora of North America Editorial Committee (ed.), *Flora of North America North of Mexico: Vol. 7, Magnoliophyta: Salicaceae to Brassicaceae*. Oxford University Press, New York.

Al Shamma, A., S. D. Drake, L. F. Guagliard, L. A. Mitschar & J. K. Swazz. 1982. Antimicrobial alkaloids from *Boehmeria cylindrica*. *Phytochemistry* 21: 445–487.

Alsos, I. G., S. Spjelkavik & T. Engelskjon. 2003. Seed bank size and composition of *Betula nana*, *Vaccinium uliginosum*, and *Campanula rotundifolia* habitats in Svalbard and northern Norway. *Can. J. Bot.* 81: 220–231.

Alsos, I. G., E. Muller & P. B. Eidesen. 2013. Germinating seeds or bulbils in 87 of 113 tested Arctic species indicate potential for ex situ seed bank storage. *Polar Biol.* 36: 819–830.

Altfeld, L. F. 2003. Herbivore abundance in simple and diverse habitats: the direct and indirect effects of plant diversity and habitat structure. MS thesis. University of South Florida, Tampa, FL. 40 pp.

Altom, J. V. & D. S. Murray. 1996. Factors affecting eclipta (*Eclipta prostrata*) seed germination. *Weed Technol.* 10: 727–731.

Alverson, W. S., K. G. Karol, D. A. Baum, M. W. Chase, S. M. Swensen, R. McCourt & K. J. Sytsma. 1998. Circumscription of the Malvales and relationships to other Rosidae: evidence from *rbcL* sequence data. *Amer. J. Bot.* 85: 876–887.

Alverson, W. S., B. A. Whitlock, R. Nyffeler, C. Bayer & D. A. Baum. 1999. Phylogeny of the core Malvales: evidence from *ndhF* sequence data. *Amer. J. Bot.* 86: 1474–1486.

Alyokhin, A. V., R. H. Messing & J. J. Duan. 2001. Utilization of the exotic weed *Pluchea odorata* (Asteraceae) and related plants by the introduced biological control agent *Acinia picturata* (Diptera: Tephritidae) in Hawaii. *Biocontrol Sci. Technol.* 11: 703–710.

American Herbal Pharmacology Delegation (AHPD). 1975. *Herbal Pharmacology in the People's Republic of China*. National Academy of Sciences Press, Washington, DC. 255 pp. Available online: http://www.swsbm.com/Ephemera/China_herbs.pdf (accessed February 27, 2007).

Ammal, E. K. J. & B. Saunders. 1952. Chromosome numbers in species of *Lonicera*. *Kew Bull.* 7: 539–541.

Amon, J. P., C. A. Thompson, Q. J. Carpenter & J. Miner. 2002. Temperate zone fens of the glaciated Midwestern USA. *Wetlands* 22: 301–317.

Amsberry, K. 2001. Conservation biology of *Plagiobothrys hirtus* (Boraginaceae): evaluation of life history and population enhancement. MS thesis. Oregon State University, Corvallis, OR. 132 pp.

Amusan, A. A. & I. F. Adeniyi. 2005. Genesis, classification and heavy metal retention potential of soils in mangrove forest, Niger Delta, Nigeria. *J. Human Ecol.* 17: 255–261.

Ananthi, J., A. Prakasam & K. V. Pugalendi. 2003. Antihyperglycemic activity of *Eclipta alba* leaf on alloxan-induced diabetic rats. *Yale J. Biol. Med.* 76: 97–102.

Anderberg, A. A. 1991. Taxonomy and phylogeny of the tribe Plucheeae (Asteraceae). *Plant Syst. Evol.* 176: 145–177.

Anderberg, A. A. & X. Zhang. 2002. Phylogenetic relationships of Cyrillaceae and Clethraceae (Ericales) with special emphasis on the genus *Purdiaea* Planch. *Org. Diversity Evol.* 2: 127–137.

Anderberg, A. A., B. Stähl & M. Källersjö. 1998. Phylogenetic relationships in the *Primulales* inferred from *rbcL* sequence data. *Plant Syst. Evol.* 211: 93–102.

Anderberg, A. A., B. Stähl & M. Källersjö. 2000. Maesaceae, a new primuloid family in the order Ericales s.l. *Taxon* 49: 183–187.

Anderberg, A. A., C. Rydin & M. Källersjö. 2002. Phylogenetic relationships in the order Ericales s.l.: analyses of molecular data from five genes from the plastid and mitochondrial genomes. *Amer. J. Bot.* 89: 677–687.

Anderberg, A. A., P. Eldenas, R. J. Bayer & M. Englund. 2005. Evolutionary relationships in the Asteraceae tribe Inuleae (incl. Plucheeae) evidenced by DNA sequences of *ndhF*; with notes on the systematic positions of some aberrant genera. *Org. Diversity Evol.* 5: 135–146.

Anderberg, A. A., U. Manns & M. Källersjö. 2007. Phylogeny and floral evolution of the Lysimachieae (Ericales, Myrsinaceae): evidence from *ndhF* sequence data. *Willdenowia* 37: 407–421.

Andersen, B. A. 1996. *Desert Plants of Utah*. Utah State University Press, Logan, UT. 114 pp.

Andersen, D. C. & K. B. Armitage. 1976. Caloric content of Rocky Mountain subalpine and alpine plants. *J. Range Manag.* 29: 344–345.

Andersen, D. C., R. S. Hoffmann & K. B. Armitage. 1979. Aboveground productivity and floristic structure of a high subalpine herbaceous meadow. *Arct. Alp. Res.* 11: 467–476.

Andersen, T., F. O. Andersen & O. Pedersen. 2006. Increased CO_2 in the water around *Littorella uniflora* raises the sediment O_2 concentration. *Aquat. Bot.* 84: 294–300.

Anderson, D., J. Brandt, L. Wright & D. Davidson. 2005. Ecology and floristics of Knife Island, a gull and cormorant rookery on Lake Superior, near Two Harbors, Lake County, Minnesota. *Michigan Bot.* 44: 95–104.

Anderson, D. S. & R. B. Davis. 1998. *The Flora and Plant communities of Maine Peatlands*. Technical Bulletin 170. Maine Agricultural and Forest Experiment Station, University of Maine, Orono, ME. 98 pp.

Anderson, G. J., G. Bernardello, T. F. Stuessy & D. J. Crawford. 2001. Breeding system and pollination of selected plants endemic to Juan Fernández Islands. *Amer. J. Bot.* 88: 220–233.

Anderson, I. A. & J. W. Busch. 2006. Relaxed pollinator-mediated selection weakens floral integration in self-compatible taxa of *Leavenworthia* (Brassicaceae). *Amer. J. Bot.* 93: 860–867.

Anderson, I. B., W. H. Mullen, J. E. Meeker, S. C. Khojasteh-Bakht, S. Oishi, S. D. Nelson & P. D. Blanc. 1996. Pennyroyal toxicity: measurement of toxic metabolite levels in two cases and review of the literature. *Ann. Intern. Med.* 124: 726–734.

Anderson, J. M. 1930. Hay fever plants of Utah. *Calif. West. Med.* 33: 723–724.

Anderson, J. P. 1939. Plants used by the Eskimo of the Northern Bering Sea and Arctic regions of Alaska. *Amer. J. Bot.* 26: 714–716.

Anderson, J. P. 1940. Notes on Alaskan rust Fungi. *Bull. Torrey Bot. Club* 67: 413–416.

Anderson, K. L. & D. J. Leopold. 2002. The role of canopy gaps in maintaining vascular plant diversity at a forested wetland in New York state. *J. Torrey Bot. Soc.* 129: 238–250.

Anderson, L. C. 1987. *Boltonia apalachicolensis* (Asteraceae): a new species from Florida. *Syst. Bot.* 12: 133–138.

Anderson, L. C. 1991. *Paronychia chartacea* ssp. minima (Caryophyllaceae): a new subspecies of a rare Florida endemic. *Sida* 14: 435–441.

Anderson, L. C. 1999. Striking sexual dimorphism in *Lindera subcoriacea* (Lauraceae). *Sida* 18: 1085–1091

Anderson, L. C. 2006. 232. Arnoglossum Rafinesque. pp. 622–625. *In*: Flora of North America Editorial Committee (ed.), *Flora of North America North of Mexico: Vol. 20, Magnoliophyta: Asteridae (in part): Asteraceae, Part 2.* Oxford University Press, New York.

Anderson, M. K. 2009. *The Ozette Prairies of Olympic National Park: Their Former Indigenous Uses and Management.* Final Report to Olympic National Park, Port Angeles, Washington, DC; USDA Natural Resources Conservation Service, Davis, CA. 158 pp.

Anderson, N. O. & P. D. Ascher. 1993. Male and female fertility of loosestrife (*Lytrhrum*) cultivars. *J. Amer. Soc. Hort. Sci.* 118: 851–858.

Anderson, P. J. 1919. Index to American species of *Phyllosticta*. *Mycologia* 11: 66–79.

Anderson, R. C. & M. H. Beare. 1983. Breeding system and pollination ecology of *Trientalis borealis* (Primulaceae). *Amer. J. Bot.* 70: 408–415.

Anderson, R. C. & O. L. Loucks. 1973. Aspects of the biology of *Trientalis borealis* Raf. *Ecology* 54: 798–808.

Anderson, R. C., A. E. Liberta & L. A. Dickman. 1984. Interaction of vascular plants and vesicular-arbuscular mycorrhizal Fungi across a soil moisture-nutrient gradient. *Oecologia* 64: 111–117.

Anderson, R. C., E. A. Corbett, M. R. Anderson, G. A. Corbett & T. M. Kelley. 2001. High white-tailed deer density has negative impact on tallgrass prairie forbs. *J. Torrey Bot. Soc.* 128: 381–392.

Anderson, R. R., R. G. Brown & R. D. Rappleye. 1968. Water quality and plant distribution along the Upper Patuxent River, Maryland. *Chesapeake Sci.* 9: 145–156.

Anderson, R. S., J. Hasbargen, P. A. Koehler & E. J. Feiler. 1999. Late Wisconsin and Holocene subalpine forests of the Markagunt Plateau of Utah, southwestern Colorado Plateau, U.S.A. *Arct. Antarc. Alp. Res.* 31: 366–378.

Anderson, W. A. 1943. A fen in northwestern Iowa. *Amer. Midl. Naturalist* 29: 787–791.

Andersson, E. & C. Nilsson. 2002. Temporal variation in the drift of plant litter and propagules in a small boreal river. *Freshwater Biol.* 47: 1674–1684.

Andersson, E., C. Nilsson & M. E. Johansson. 2000. Plant dispersal in boreal rivers and its relation to the diversity of riparian flora. *J. Biogeogr.* 27: 1095–1106.

Andersson, L. & A. Antonelli. 2005. Phylogeny of the tribe Cinchoneae (Rubiaceae), its position in Cinchonoideae, and description of a new genus, *Ciliosemina. Taxon* 54: 17–28.

Andersson, L. & J. H. Rova. 1999. The *rpsl6* intron and the phylogeny of the *Rubioideae* (Rubiaceae). *Plant Syst. Evol.* 214: 161–186.

Andersson, S., L. A. Nilsson, I. Groth & G. Bergstrom. 2002. Floral scents in butterfly-pollinated plants: possible convergence in chemical composition. *Bot. J. Linn. Soc.* 140: 129–153.

Andrade-Cetto, A. & M. Heinrich. 2005. Mexican plants with hypoglycaemic effect used in the treatment of diabetes. *J. Ethnopharmacol.* 99: 325–348.

Andreas, B. K. & G. R. Bryan. 1990. The vegetation of three *Sphagnum*-dominated basin-type bogs in northeastern Ohio. *Ohio J. Sci.* 90: 54–66.

Andreasen, K. 2012. Phylogeny, hybridization, and evolution of habit and breeding system in Sidalcea and Eremalche (Malvaceae). *Int. J. Plant Sci.* 173: 532–548.

Andreasen, K. & B. G. Baldwin. 2001. Unequal evolutionary rates between annual and perennial lineages of checker mallows (*Sidalcea*, Malvaceae): evidence from 18S–26S rDNA internal and external transcribed spacers. *Mol. Biol. Evol.* 18: 936–944.

Andreasen, K. & B. G. Baldwin. 2003a. Reexamination of relationships, habital evolution, and phylogeography of checker mallows (*Sidalcea*, Malvaceae) based on molecular phylogenetic data. *Amer. J. Bot.* 90: 436–444.

Andreasen, K. & B. G. Baldwin. 2003b. Nuclear ribosomal DNA sequence polymorphism and hybridization in checker mallows (*Sidalcea*, Malvaceae). *Mol. Phylogen. Evol.* 29: 563–581.

Andreasen, K. & B. Bremer. 2000. Combined phylogenetic analysis in the Rubiaceae-Ixoroideae: morphology, nuclear and chloroplast DNA data. *Amer. J. Bot.* 87: 1731–1748.

Andres, T. C. & G. P. Nabhan. 1988. Taxonomic rank and rarity of *Cucurbita okeechobeensis*. *Cucurb. Genet. Coop.* 11: 85.

Andrews, A. C. 1958. The mints of the Greeks and Romans and their condimentary uses. *Osiris* 13: 127–149.

Andrus, R. E. 1988. Two new taxa of *Sphagnum* in section Cuspidata. *Bryologist* 91: 364–366.

Angelo, R. & D. E. Boufford. 2011. Atlas of the flora of New England: Paeoniaceae to Ericaceae. *Phytoneuron* 53: 1–18.

Angerstein, M. B. & D. E. Lemke. 1994. First records of the aquatic weed *Hygrophila polysperma* (Acanthaceae) from Texas. *Sida* 16: 365–371.

Angert, A. L. 2005. The ecology and evolution of elevation range limits in monkeyflowers (*Mimulus cardinalis* and *Mimulus lewisii*). PhD dissertation. Michigan State University, East Lansing, MI. 188 pp.

Angert, A. L. 2006. Demography of central and marginal populations of monkeyflowers (*Mimulus cardinalis* and *M. lewisii*). *Ecology* 87: 2014–2025.

Angoa-Román, M. D. J., S. H. Bullock & T. Kawashima. 2005. Composition and dynamics of the seed bank of coastal scrub in Baja California. *Madroño* 52: 11–20.

Anh, V. T., N. T. Tram, L. T. Klank, P. D. Cam & A. Dalsgaard. 2007. Faecal and protozoan parasite contamination of water spinach (*Ipomoea aquatica*) cultivated in urban wastewater in Phnom Penh, Cambodia. *Trop. Med. Int. Health* 12: 73–81.

Ankli, A., O. Sticher & M. Heinrich. 1999. Medical ethnobotany of the Yucatec Maya: healers consensus as a quantitative criterion. *Econ. Bot.* 53: 144–160.

Anon. 1879. Portraits of new and notable plants. *J. Hort. Cottage Gard.* 36: 381.

Anon. 1985. *Hydric Soils of the State of Wisconsin 1985*. United States Department of Agriculture, Soil Conservation Service, Washington, DC. 17 pp.

Antlfinger, A. E. 1981. The genetic basis of microdifferentiation in natural and experimental populations of *Borrichia frutescens* in relation to salinity. *Evolution* 35: 1056–1068.

Antlfinger, A. E. 1982. Genetic neighborhood structure of the salt marsh composite, *Borrichia frutescens*. *J. Heredity* 73: 128–132.

Antlfinger, A. E. & E. L. Dunn. 1979. Seasonal patterns of CO_2 and water vapor exchange of three salt marsh succulents. *Oecologia* 43: 249–260.

Antonelli, A. 2008. Higher level phylogeny and evolutionary trends in Campanulaceae subfam. Lobelioideae: molecular signal overshadows morphology. *Mol. Phylogen. Evol.* 46: 1–18.

APG (Angiosperm Phylogeny Group). 2003. An update of the Angiosperm Phylogeny Group classification for the orders and families of flowering plants: APG II. *Bot. J. Linn. Soc.* 141: 399–436.

APG III. 2009. An update of the Angiosperm Phylogeny Group classification for the orders and families of flowering plants: APG III. *Bot. J. Linn. Soc.* 161: 105–121.

Applequist, W. L. & R. S. Wallace. 2001. Phylogeny of the portulacaceous cohort based on *ndhF* sequence data. *Syst. Bot.* 26: 406–419.

Araki, S. 2002. Fruit and seed production of *Utricularia intermedia* Hayne in Hokkaido, Japan. *Bull. Water Plant Soc. Japan* 75: 14–17 [in Japanese with English abstract].

Araki, S. & I. Washitani. 2000. Seed dormancy/germination traits of seven *Persicaria* species and their implication in soil seedbank strategy. *Ecol. Res.* 15: 33–46.

Arasu, N. T. 1970. Self-incompatibility in *Ribes*. *Euphytica* 19: 373–378.

Arathi, H. S. & J. K. Kelly. 2004. Corolla morphology facilitates both autogamy and bumblebee pollination in *Mimulus guttatus*. *Int. J. Plant Sci.* 165: 1039–1045.

Aravanopoulos, F. A. 2000. Absence of association between heterozygosity and biomass production in *Salix exigua* Nutt. *Theor. Appl. Genet.* 100: 1203–1208.

Aravanopoulos, F. A. & L. Zsuffa. 1998. Heterozygosity and biomass production in *Salix eriocephala*. *Heredity* 81: 396–403.

Arber, A. 1920. *Water Plants: A Study of Aquatic Angiosperms*. Cambridge University Press, Cambridge, UK. 436 pp.

Arceo-Gómez, G., L. Abdala-Roberts, A. Jankowiak, C. Kohler, G. A. Meindl, C. M. Navarro-Fernández, V. Parra-Tabla, T. L. Ashman & C. Alonso. 2016. Patterns of among-and within-species variation in heterospecific pollen receipt: the importance of ecological generalization. *Amer. J. Bot.* 103: 396–407.

Archibald, J. K., M. E. Mort, D. J. Crawford & J. K. Kelly. 2005. Life history affects the evolution of reproductive isolation among species of *Coreopsis* (Asteraceae). *Evolution* 59: 2362–2369.

Archibald, K. D. 1957. Report of the forest Aphidae (Homoptera) of Nova Scotia together with species and host plant lists. *Ann. Entomol. Soc. Amer.* 50: 38–42.

Archibold, O. W. & L. Hume. 1983. A preliminary survey of seed input into fallow fields in Saskatchewan. *Can. J. Bot.* 61: 1216–1221.

Arenas, P. & G. F. Scarpa. 2007. Edible wild plants of the Chorote Indians, Gran Chaco, Argentina. *Bot. J. Linn. Soc.* 153: 73–85.

Arens, P., W. Durka, J. H. Wernke-Lenting & M. J. M. Smulders. 2004. Isolation and characterization of microsatellite loci in *Geum urbanum* (Rosaceae) and their transferability within the genus *Geum*. *Mol. Ecol. Notes* 4: 209–212.

Arft, A. M., M. D. Walker, J. Gurevitch, J. M. Alatalo, M. S. Bret-Harte, M. Dale, M. Diemer, F. Gugerli, G. H. R. Henry, M. H. Jones, R. D. Hollister, I. S. Jonsdottir, K. Laine, E. Levesque, G. M. Marion, U. Molau, P. Molgaard, U. Nordenhall, V. Raszhivin, C. H. Robinson, G. Starr, A. Stenstrom, M. Stenstrom, O. Totland, P. L. Turner, L. J. Walker, P. J. Webber, J. M. Welker & P. A. Wookey. 1999. Responses of tundra plants to experimental warming: meta-analysis of the international tundra experiment. *Ecol. Monogr.* 69: 491–511.

Argentina, J. E. 2006. *Podostemum ceratophyllum* and patterns of fish occurrence and richness in a southern Appalachian river. MS thesis. University of Georgia, Athens, GA. 109 pp.

Argentina, J. E., M. C. Freeman & B. J. Freeman. 2010. Predictors of occurrence of the aquatic macrophyte *Podostemum ceratophyllum* in a southern Appalachian river. *S. E. Naturalist (Steuben)* 9: 465–476.

Argus, G. W. 1999. Classification of *Salix* in the New World. *Bot. Electron. News* 227: 1–6. Available online: http://www.ou.edu/cas/botany-micro/ben/ben227.html.

Argus, G. W. 2010. 2. *Salix* Linnaeus. pp. 23–162. *In*: Flora of North America Editorial Committee (ed.), *Flora of North America North of Mexico: Vol. 7, Magnoliophyta: Salicaceae to Brassicaceae*. Oxford University Press, New York.

Argus, G. W., C. L. McJannet & M. J. Dallwitz. 2000. *Salicaceae of the Canadian Arctic Archipelago: Descriptions, Illustrations, Identification, and Information Retrieval*. Version: 2nd November 2000. Available online: http://www.nature.ca/aaflora/data/index.htm

Arhangelsky, K. 2005. *Non-native Plant Species of Prince of Wales Island, Alaska. Summary of Survey Findings*. Final Report for USDA Forest Service, State and Private Forestry. Turnstone Environmental Consultants, Inc., Portland, OR.

Arisawa, M., T. Hayashi, M. Shimizu, N. Morita, H. Bai, S. Kuze & Y. Ito. 1991. Isolation and cytotoxicity of two new flavonoids from *Chrysosplenium grayanum* and related flavonols. *J. Nat. Prod.* 54: 898–901.

Arisawa, M., H. Bai, S. Shimizu, S. Koshimra, M. Tanaka, T. Sasaki & N. Morita. 1992. Isolation and identification of a cytotoxic principle from *Chrysosplenium grayanum* Maxim. (Saxifragaceae) and its antitumor activities. *Chem. Pharm. Bull.* 40: 3274–3276.

Arisawa, M., T. Hatashita, Y. Numata, M. Tanaka & T. Sasaki. 1997. Cytotoxic principles from *Chrysosplenium flagelliferum*. *Int. J. Pharmacogn.* 35: 141–143.

Armbruster, W. S. & A. D. McGuire. 1991. Experimental assessment of reproductive interactions between sympatric *Aster* and *Erigeron* (Asteraceae) in interior Alaska. *Amer. J. Bot.* 78: 1449–1457.

ARMC (Agriculture & Resource Management Council of Australia & New Zealand, Australian & New Zealand Environment & Conservation Council and Forestry Ministers). 2001. *Weeds of National Significance: Pond Apple (Annona glabra) Strategic Plan*. National Weeds Strategy Executive Committee, Launceston, Australia.

Armitage, A. R., K. E. Boyer, R. R. Vance & R. F. Ambrose. 2006. Restoring assemblages of salt marsh halophytes in the presence of a rapidly colonizing dominant species. *Wetlands* 26: 667–676.

Armitage, K. B. 1979. Food selectivity by yellow-bellied marmots. *J. Mammalogy* 60: 628–629.

Armstrong, E. A. & D. J. Crawford. 1996. Origin of *Trichostema austromontanum* s.l. Lewis (Lamiaceae). *Amer. J. Bot.* 83: S137.

Armstrong, J. E. 2002. Fringe science: are the corollas of *Nymphoides* (Menyanthaceae) flowers adapted for surface tension interactions? *Amer. J. Bot.* 89: 362–365.

Árnason, S. H., A. E. T. Thórsson, B. Magnússon, M. Philipp, H. Adsersen & K. Anamthawat-Jónsson. 2014. Spatial genetic structure of the sea sandwort (*Honckenya peploides*) on Surtsey: an immigrant's journey. *Biogeosciences* 11: 6495–6507.

Aronson, J. A. 1989. *Haloph: A Data Base of Salt Tolerant Plants of the World*. Office of Arid Lands Studies, University of Arizona, Tucson, AZ. 77 pp.

Aronson, J. A. & E. E. Whitehead. 1989. *HALOPH: A Data Base of Salt Tolerant Plants of the World*. University of Arizona, Tucson, AZ. 77 pp.

Aronson, M. F. J. & S. Galatowitsch. 2008. Long-term vegetation development of restored prairie pothole wetlands. *Wetlands* 28: 883–895.

Arrighi, J. F., C. Chaintreuil, F. Cartieaux, C. Cardi, M. Rodier-Goud, S. C. Brown, M. Boursot, A. D'Hont, B. Dreyfus & E. Giraud. 2014. Radiation of the Nod-independent *Aeschynomene* relies on multiple allopolyploid speciation events. *New Phytol.* 201: 1457–1468.

Arrington, D. A., L. A. Toth & J. W. Koebel Jr. 1999. Effects of rooting by feral hogs *Sus scrofa* L. on the structure of a floodplain vegetation assemblage. *Wetlands* 19: 535–544.

Arroyo, M. T. K. 1973. Chiasma frequency evidence on the evolution of autogamy in *Limnanthes floccosa* (Limnanthaceae). *Evolution* 27: 679–688.

Arroyo, M. T. K. 1975. Electrophoretic studies of genetic variation in natural populations of allogamous *Limnanthes alba* and autogamous *Limnanthes floccosa* (Limnanthaceae). *Heredity* 35: 153–164.

Arroyo-García, R., F. Lefort, M. T. de Andrés, J. Ibáñez, J. Borrego, N. Jouve, F. Cabello & J. M. Martínez-Zapater. 2002. Chloroplast microsatellite polymorphisms in *Vitis* species. *Genome* 45: 1142–1149.

Arry, S. J. B. 1998. Managing the Sacramento Valley vernal pool landscape to sustain the native flora. pp. 136–140. *In*: C. W. Witham, E. T. Bauder, D. Belk, W. R. Ferren Jr. & R. Ornduff (eds.), *Ecology, Conservation, and Management of Vernal Pool Ecosystems—Proceedings from a 1996 Conference*. California Native Plant Society, Sacramento, CA.

Arthur, J. C. 1898. Indiana plant rusts listed in accordance with latest nomenclature. *Proc. Indiana Acad. Sci.* 8: 174–186.

Arthur, J. C. 1909. Cultures of Uredineae in 1908. *Mycologia* 1: 225–256.

Arthur, J. C. 1910. New species of Uredineae-VII. *Bull. Torrey Bot. Club* 37: 569–580.

Arts, G. H. P. & R. A. J. M. van der Heijden. 1990. Germination ecology of *Littorella uniflora* (L.) Aschers. *Aquat. Bot.* 37: 139–151.

Artz, D. R. & K. D. Waddington. 2006. The effects of neighbouring tree islands on pollinator density and diversity, and on pollination of a wet prairie species, *Asclepias lanceolata* (Apocynaceae). *J. Ecol.* 94: 597–608.

Asada, T., B. G. Warner & J. Pojar. 2003. Environmental factors responsible for shaping an open peatland-forest complex on the hypermaritime north coast of British Columbia. *Can. J. For. Res.* 33: 2380–2394.

Aschenbrenner, A.-K., S. Horakh & O. Spring. 2013. Linear glandular trichomes of *Helianthus* (Asteraceae): morphology, localization, metabolite activity and occurrence. *AoB Plants* 5: plt028. doi:10.1093/aobpla/plt028.

Ashe, D. M. 2010. Endangered and threatened wildlife and plants; 12-month finding on a petition to list *Cirsium wrightii* (Wright's marsh thistle) as endangered or threatened. *Fed. Reg.* 75: 67925–67944.

Ashe, W. W. 1897. Notes on the woody plants of the south Atlantic states. *Bot. Gaz.* 24: 373–377.

Ashman, T. L. 1992a. The relative importance of inbreeding and maternal sex in determining progeny fitness in *Sidalcea oregana* ssp. *spicata*, a gynodioecious plant. *Evolution* 46: 1862–1874.

Ashman, T. L. 1992b. Indirect costs of seed production within and between seasons in a gynodioecious species. *Oecologia* 92: 266–272.

Ashman, T. L. 1994. Reproductive allocation in hermaphrodite and female plants of *Sidalcea oregana* ssp. *spicata* (Malvaceae) using 4 currencies. *Amer. J. Bot.* 81: 433–438.

Ashman, T. L. & M. L. Stanton. 1991. Seasonal variation in pollination dynamics of the sexually dimorphic species, *Sidalcea oregana* ssp. *spicata* (Malvaceae). *Ecology* 72: 993–1003.

Ashworth, A. C. & D. J. Cantrill. 2004. Neogene vegetation of the Meyer Desert Formation (Sirius Group) Transantarctic Mountains, Antarctica. *Palaeogeogr. Palaeoclimatol. Palaeoecol.* 213: 65–82.

Askew, S. D. & J. W. Wilcut. 2002. Ladysthumb interference and seed production in cotton. *Weed Sci.* 50: 326–332.

Aslam, M. S., B. A. Choudhary, M. Uzair & A. S. Ijaz. 2012. The genus *Ranunculus*: a phytochemical and ethnopharmacological review. *Int. J. Pharm. Pharm. Sci.* 4: 15–22.

Asmussen, C. B. & A. Liston. 1998. Chloroplast DNA characters, phylogeny, and classification of *Lathyrus* (Fabaceae). *Amer. J. Bot.* 85: 387–401.

Aspinwall, N. & T. Christian. 1992a. Clonal structure, genotype diversity, and seed production in populations of *Filipendula rubra* (Rosaceae) from the northcentral United States. *Amer. J. Bot.* 79: 294–299.

Aspinwall, N. & T. Christian. 1992b. Pollination biology, seed production, and population structure in queen-of-the-prairie, *Filipendula rubra* (Rosaceae) at Botkin Fen, Missouri. *Amer. J. Bot.* 79: 488–494.

Asprey, G. F. & P. Thornton. 1953. Medicinal plants of Jamaica. Parts I & II. *West Indian Med. J.* 2: 233–252; 3: 17–41.

Astiz, V., L. A. Iriarte, A. Flemmer & L. F. Hernández. 2011. Self-compatibility in modern hybrids of sunflower (*Helianthus annuus* L.). Fruit set in open and self-pollinated (bag isolated) plants grown in two different locations. *Helia* 34: 129–138.

Atia, A., A. Debez, Z. Barhoumi, C. Abdelly & A. Smaoui. 2010. Localization and composition of seed oils of *Crithmum maritimum* L. (Apiaceae). *Afr. J. Biotechnol.* 9: 6482–6485.

Atkins, S. 2004. Verbenaceae. pp. 449–468. *In*: J. W. Kadereit (ed.), *The Families and Genera of Vascular Plants, Vol. VII. Flowering Plants, Dicotyledons: Lamiales (except Acanthaceae including Avicenniaceae).* Springer-Verlag, Berlin, Germany.

Atkinson, R. B., J. E. Perry, E. Smith & J. Cairns. 1993. Use of created wetland delineation and weighted averages as a component of assessment. *Wetlands* 13: 185–193.

Atkinson, R. B., J. E. Perry, G. B. Noe, W. L. Daniels & J. Cairns Jr. 2010. Primary productivity in 20-year old created wetlands in southwestern Virginia. *Wetlands* 30: 200–210.

Atwood, E. L. 1941. White-tailed deer foods of the United States. *J. Wildl. Manag.* 5: 314–332.

Aubé, M. & L. Caron. 2001. The mangroves of the north coast of Haiti: a preliminary assessment. *Wetlands Ecol. Manag.* 9: 271–278.

Auble, G. T., J. M. Friedman & M. L. Scott. 1994. Relating riparian vegetation to present and future streamflows. *Ecol. Appl.* 4: 544–554.

Auclair, A. N. D., A. Bouchard & J. Pajaczkowski. 1976. Productivity relations in a *Carex*-dominated ecosystem. *Oecologia* 26: 9–31.

Auf'mkolk, M., J. C. Ingbar, S. M. Amir, H. Winterhoff, H. Sourgens, R. D. Hesch & S. H. Ingbar. 1984. Inhibition by certain plant extracts of the binding and adenylate cyclase stimulatory effect of bovine thyrotropin in human thyroid membranes. *Endocrinology* 115: 527–534.

Aulbach-Smith, C. A. & S. J. de Kozlowski. 1990. *Aquatic and Wetland Plants of South Carolina*. South Carolina Water Resources Commission, Columbia, SC. 123 pp.

Aulio, K. 1986. CAM-like photosynthesis in *Littorella uniflora* (L.) Aschers.: the role of humidity. *Ann. Bot.* 58: 273–275.

Ausden, M., M. Hall, P. Pearson & T. Strudwick. 2005. The effects of cattle grazing on tall-herb fen vegetation and mollusks. *Biol. Conserv.* 122: 317–326.

Austin, D. F. 1973. The American *Erycibeae* (Convolvulaceae): *Maripa*, *Dicranostyles*, and *Lysiostyles* I. systematics. *Ann. Missouri Bot. Gard.* 60: 306–412.

Austin, D. F. 1992. Rare Convolvulaceae in the southwestern United States. *Ann. Missouri Bot. Gard.* 79: 8–16.

Austin, D. F. 2000. A revision of *Cressa* L. (Convolvulaceae). *Bot. J. Linn. Soc.* 133: 27–39.

Austin, D. F. 2004a. Convolvulaceae (morning glory family). pp. 113–115. *In*: N. Smith, S. A. Mori, A. Henderson, D. W. Stevenson & S. V. Heald (eds.), *Flowering Plants of the Neotropics*. Princeton University Press, Princeton, NJ. 594 pp.

Austin, D. F. 2004b. *Florida Ethnobotany*. CRC Press, Boca Raton, FL. 909 pp.

Austin, D. F. 2007. Water spinach (*Ipomoea aquatica*, Convolvulaceae): a food gone wild. *Ethnobot. Res. Appl.* 5: 123–146.

Austin, G. T. & P. J. Leary. 2008. Larval hostplants of butterflies in Nevada. *Holarc. Lepid.* 12: 1–134.

Austin, J. E., J. R. Keough & W. H. Pyle. 2007. Effects of habitat management treatments on plant community composition and biomass in a montane wetland. *Wetlands* 27: 570–587.

Austin, M. A., D. A. Buffett, D. J. Nicolson, G. G. E. Scudder & V. Stevens (eds.). 2008. *Taking Nature's Pulse: The Status of Biodiversity in British Columbia*. Biodiversity BC, Victoria, BC. 268 pp. Available online: http://www.biodiversitybc.org/EN/main/where/132.html.

Aviado, D. M. & H. Reutter. 1969. Antimalarial and antiarrhythmic activity of plant extracts. 2. Acid extracts of plants. *Med. Exp.* 19: 95–100.

Avis, P. (et al.). (B. Brown, K. Buscher, F. Coskun, Phil Coulling, Skye Hanford, Jon Harrod, Jay Horn, Chris Mankoff, Bob Peet, Linda Prince, Alice Stanford, Steve Seiberling, Rickie White, Ken Wurdack) 1997. Proceedings. UNC sometimes annual phytogeographical excusion to the Florida panhandle. March 8–14, 1997. Available online: http://labs.bio.unc.edu/Peet/PEL/PGE/PGE1997_report.pdf (accessed December 2, 2013).

Awad, R., J. T. Arnason, V. Trudeau, C. Bergeron, J. W. Budzinski, B. C. Foster & Z. Merali. 2003. Phytochemical and biological analysis of Skullcap (*Scutellaria lateriflora* L.): a medicinal plant with anxiolytic properties. *Phytomedicine* 10: 640–649.

Awadalla, P. & K. Ritland. 1997. Microsatellite variation and evolution in the *Mimulus guttatus* species complex with contrasting mating systems. *Mol. Biol. Evol.* 14: 1023–1034.

Ayala, F. & J. W. O'Leary. 1995. Growth and physiology of *Salicornia bigelovii* Torr. at sub-optimal salinity. *Int. J. Plant Sci.* 156: 197–205.

Ayers, T. 2012. *Lobelia*. pp. 594–595. *In*: B. G. Baldwin, D. H. Goldman, D. J. Keil, R. Patterson, T. J. Rosatti & D. H. Wilken (eds.), *The Jepson Manual (2nd ed.)*. University of California Press, Berkeley, CA.

Ayres, D., E. Fleishman & D. Zippin. 2007. Genetic structure of the endangered plant *Cordylanthus palmatus* within seasonal alkali wetlands. The Society for Conservation Biology, Annual Meeting [abstract].

Ayres, D. R. & C. M. Sloop. 2008. *Genetic Structure of Three Endangered Plants of the Santa Rosa Plain: Burke's Goldfields (Lasthenia burkei), Sonoma sunshine (Blennosperma bakeri), and Sebastopol meadowfoam (Limnanthes vinculans)*. California Department of Fish and Game, Sacramento, CA. 26 pp.

Azcárate, F. M., L. Arqueros, A. M. Sánchez & B. Peco. 2005. Seed and fruit selection by harvester ants, *Messor barbarus*, in Mediterranean grassland and scrubland. *Funct. Ecol.* 19: 273–283.

Aziz, S. & M. A. Khan. 1996. Seed bank dynamics of a semi-arid coastal shrub community in Pakistan. *J. Arid Environ.* 34: 81–87.

Aziz, T. & D. M. Sylvia. 1995. Activity and species composition of arbuscular mycorrhizal fungi following soil removal. *Ecol. Appl.* 5: 776–784.

Aziz, T., D. M. Sylvia & R. F. Doren. 1995. Activity and species composition of arbuscular mycorrhizal Fungi following soil removal. *Ecol. Appl.* 5: 776–784.

Azuma, H., L. B. Thien & S. Kawano. 1999. Molecular phylogeny of *Magnolia* (Magnoliaceae) inferred from cpDNA sequences and evolutionary divergence of floral scents. *J. Plant Res.* 112: 291–306.

Azuma, H., J. G. García-Franco, V. Rico-Gray & L. B. Thien. 2001. Molecular phylogeny of the Magnoliaceae: the biogeography of tropical and temperate disjunctions. *Amer. J. Bot.* 88: 2275–2285.

Azuma, H., R. B. Figlar, P. Del Tredici, K. Camelbeke, A. Palmarola-Bejerano & M. S. Romanov. 2011. Intraspecific sequence variation of cpDNA shows two distinct groups within *Magnolia virginiana* L. of eastern North America and Cuba. *Castanea* 76: 118–123.

Azuma, T., T. Kajita, J. Yokoyama & H. Ohashi. 2000. Phylogenetic relationships of *Salix* (Salicaceae) based on *rbcL* sequence data. *Amer. J. Bot.* 87: 67–75.

Bach, C. E. 2001. Long-term effects of insect herbivory and sand accretion on plant succession on sand dunes. *Ecology* 82: 1401–1416.

Bačić, T. 1983. A contribution to the study of spring and fall generation of the species *Bidens tripartitus* L. *Zborn. Matice Srpske Prir. Nauke* 64: 149–157.

Bacigalupo, N. M. & E. L. Cabral. 1999. Revision of the American species of the genus *Diodia* (Rubiaceae, Spermacoceae). *Darwiniana* 37: 153–165.

Bacone, J. A., L. A. Casebere & M. D. Hutchinson. 1983. Glades and barrens of Crawford and Perry Counties, Indiana. *Proc. Indiana Acad. Sci.* 92: 367–373.

Baden III, J., W. T. Batson & R. Stalter. 1975. Factors affecting the distribution of vegetation of abandoned rice fields, Georgetown Co., South Carolina. *Castanea* 40: 171–184.

Badisa, R. B., V. L. Badisa, E. H. Walker & L. M. Latinwo. 2007. Potent cytotoxic activity of *Saururus cernuus* extract on human colon and breast carcinoma cultures under normoxic conditions. *Anticancer Res.* 27: 189–193.

Bagger, J. & T. V. Madsen. 2004. Morphological acclimation of aquatic *Littorella uniflora* to sediment CO_2 concentration and wave exposure. *Funct. Ecol.* 18: 946–951.

Bai, Y., M. H. Benn, W. Majak & R. McDiarmid. 1996. Extraction and HPLC determination of ranunculin in species of the buttercup family. *J. Agric. Food Chem.* 44: 2235–2238.

Bailey, A. W. & C. E. Poulton. 1968. Plant communities and environmental interrelationship in a portion of the Tillamook burn, northwestern Oregon. *Ecology* 49: 1–13.

Bailey, C. D., M. A. Koch, M. Mayer, K. Mummenhoff, S. L. O'Kane Jr., S. I. Warwick, M. D. Windham & I. A. Al-Shehbaz. 2006a. Toward a global phylogeny of the Brassicaceae. *Mol. Biol. Evol.* 23: 2142–2160.

Bailey, D. E., J. E. Perry & D. A. DeBerry. 2006b. *Aeschynomene virginica* (Fabaceae) habitat in a tidal marsh, James City county, Virginia. *Banisteria* 27: 3–9.

Bailey, F. M. 1922. Cactus wrens' nests in southern Arizona. *Condor* 24: 163–168.

Bailey, M., S. A. Petrie & S. S. Badzinski. 2008. Diet of mute swans in lower Great Lakes coastal marshes. *J. Wildl. Manag.* 72: 726–732.

Bailey, M. F. 2002. A cost of restoration of male fertility in a gynodioecious species, *Lobelia siphilitica*. *Evolution* 56: 2178–2186.

Bailey, T. S. 2008. *Miraculum Naturae. Venus's Flytrap*. Trafford Publishing, Victoria, BC. 279 pp.

Bailey, W. W. 1882. Fall-blooming of *Menyanthes trifoliata*. *Bot. Gaz.* 7: 147–148.

Baillie, P. W. 2012. Seasonal growth and development of the subarctic plant *Honckenya peploides* subsp. *robusta* (Caryophyllaceae) on Niantic Bay, Connecticut, USA. *Rhodora* 114: 148–162.

Bain, J. F. & J. L. Golden. 2000. A phylogeny of *Packera* (Senecioneae; Asteraceae) based on internal transcribed spacer region sequence data and a broad sampling of outgroups. *Mol. Phylogen. Evol.* 16: 331–338.

Bainard, J. D., L. D. Bainard, T. A. Henry, A. J. Fazekas & S. G. Newmaster. 2012. A multivariate analysis of variation in genome size and endoreduplication in angiosperms reveals strong phylogenetic signal and association with phenotypic traits. *New Phytol.* 196: 1240–1250.

Baird, J. H. & R. Dickens. 1991. Germination and emergence of Virginia buttonweed (*Diodia virginiana*). *Weed Sci.* 39: 37–41.

Baird, J. H., R. R. Dute & R. Dickens. 1992. Ontogeny, anatomy, and reproductive biology of vegetative reproductive organs of *Diodia virginiana* L. (Rubiaceae). *Int. J. Plant Sci.* 153: 320–328.

Baird, W. V. & J. L. Riopel. 1985. Surface characteristics of root and haustorial hairs of parasitic Scrophulariaceae. *Bot. Gaz.* 146: 63–69.

Bajwa, R., A. Yaqoob & A. Javaid. 2001. Seasonal variation in VAM in wetland plants. *Pakistan J. Biol. Sci.* 4: 464–470.

Baker, A. & R. Parsons. 1997. Rapid assimilation of recently fixed N2 in root nodules of *Myrica gale*. *Physiol. Plant.* 99: 640–647.

Baker, D. L. & N. T. Hobbs. 1982. Composition and quality of elk summer diets in Colorado. *J. Wildl. Manag.* 46: 694–703.

Baker, H. G. 1953. Dimorphism and monomorphism in the Plumbaginaceae II. Pollen and stigmata in the genus *Limonium*. *Ann. Bot.* 17: 433–446.

Baker, H. G. 1959. The contribution of autecological and genecological studies to our knowledge of the past migrations of plants. *Amer. Naturalist* 93: 255–272.

Baker, H. G. 1961. The adaptation of flowering plants to nocturnal and crepuscular pollinators. *Q. Rev. Biol.* 36: 64–73.

Baker, H. G. 1975. Sugar concentrations in nectars from hummingbird flowers. *Biotropica* 7: 37–41.

Baker, H. G. & I. Baker. 1967. The cytotaxonomy of *Filipendula* (Rosaceae) and its implications. *Amer. J. Bot.* 54: 1027–1034.

Baker, H. G. & I. Baker. 1977. Intraspecific constancy of floral nectar amino acid complements. *Bot. Gaz.* 138: 183–91.

Baker, J. D. & R. W. Cruden. 1991. Thrips-mediated self-pollination of two facultatively xenogamous wetland species. *Amer. J. Bot.* 78: 959–963.

Baker, M. L. 1991. Increasing seed germination percentage of *Crataegus opaca* (mayhaw) by fermentation. *HortScience* 26: 496 [abstract].

Baker, R. H., C. C. Newman & F. Wilke. 1945. Food habits of the raccoon in eastern Texas. *J. Wildl. Manag.* 9: 45–48.

Bakerink, J. A., S. M. Gospe Jr., R. J. Dimand & M. W. Eldridge. 1996. Multiple organ failure after ingestion of pennyroyal oil from herbal tea in two infants. *Pediatrics* 98: 944–947.

Balciunas, J. K. & B. Villegas. 2007. Laboratory and realized host ranges of *Chaetorellia succinea* (Diptera: Tephritidae), an unintentionally introduced natural enemy of yellow starthistle. *Environ. Entomol.* 36: 849–857.

Baldev, B. & A. Lang. 1965. Control of flower formation by growth retardants and gibberellin in *Samolus parviflorus*, a long-day plant. *Amer. J. Bot.* 52: 408–417.

Balduf, W. V. 1958. The puparium of *Rhagoletis basiola* (O.S.): variation in length and form (Trypetidae, Diptera). *Ohio J. Sci.* 58: 7–14.

Baldwin, A. H. 2007. Vegetation and seed bank studies of saltpulsed swamps of the Nanticoke River, Chesapeake Bay. pp. 139–160. *In*: W. H. Conner, T. W. Doyle & K. W. Krauss (eds.), *Ecology of Tidal Freshwater Forested Wetlands of the Southeastern United States*. Springer, Dordrecht, The Netherlands.

Baldwin, A. H. 2013. Nitrogen and phosphorus differentially affect annual and perennial plants in tidal freshwater and oligohaline wetlands. *Estuaries Coasts* 36: 547–558.

Baldwin, A. H. & E. F. Derico. 1999. The seed bank of a restored tidal freshwater marsh in Washington, DC. *Urban Ecosyst.* 3: 5–20.

Baldwin, A. H. & I. A. Mendelssohn. 1998a. Response of two oligohaline marsh communities to lethal and nonlethal disturbance. *Oecologia* 116: 543–555.

Baldwin, A. H. & I. A. Mendelssohn. 1998b. Effects of salinity and water level on coastal marshes: an experimental test of disturbance as a catalyst for vegetation change. *Aquat.* Bot. 61: 255–268.

Baldwin, A. H. & F. N. Pendleton. 2003. Interactive effects of animal disturbance and elevation on vegetation of a tidal freshwater marsh. *Estuaries* 26: 905–915.

Baldwin, A. H. & P. J. Sharpe. 2009. *Responses of Species-Rich Tidal Freshwater Wetlands to Sea Level Rise: A Mesocosm Study*. Annual Report FY 2008. Maryland Water Resources Research Center, College Park, MD. 34 pp.

Baldwin, A. H., K. L. McKee & I. A. Mendelssohn. 1996. The influence of vegetation, salinity, and inundation on seed banks of oligohaline coastal marshes. *Amer. J. Bot.* 83: 470–479.

Baldwin, A. H., M. S. Egnotovich & E. Clarke. 2001. Hydrologic change and vegetation of tidal freshwater marshes: field, greenhouse, and seed-bank experiments. *Wetlands* 21: 519–531.

Baldwin, A. H., K. M. Kettenring & D. F. Whigham. 2010. Seed banks of *Phragmites australis*-dominated brackish wetlands: relationships to seed viability, inundation, and land cover. *Aquat. Bot.* 93: 163–169.

Baldwin, B. G. 2006. Contrasting patterns and processes of evolutionary change in the tarweed-silversword lineage: revisiting clausen, Keck, and Hiesey's findings. *Ann. Missouri Bot. Gard.* 93: 64–93.

Baldwin, B. G. 2012. Raillardella. p. 411 *In*: B. G. Baldwin, D. H. Goldman, D. J. Keil, R. Patterson, T. J. Rosatti & D. H. Wilken (eds.), *The Jepson Manual (2nd ed.)*. University of California Press, Berkeley, CA.

Baldwin, B. G. & J. L. Strother. 2006. 331. *Raillardella* (A. Gray) Bentham & Hooker f. pp. 256–257. *In*: Flora of North America Editorial Committee (ed.), *Flora of North America North of Mexico: Vol. 21, Magnoliophyta: Asteridae (in part): Asteraceae, Part 3*. Oxford University Press, New York.

Baldwin, B. G. & B. L. Wessa. 2000. Origin and relationships of the tarweed-silversword lineage (Compositae-Madiinae). *Amer. J. Bot.* 87: 1890–1908.

Baldwin, B. G., B. L. Wessa & J. L. Panero. 2002. Nuclear rDNA evidence for major lineages of helenioid Heliantheae (Compositae). *Syst. Bot.* 27: 161–198.

Baldwin, R. A. & L. C. Bender. 2009. Food and nutritional componentsof diets of black bear in Rocky Mountain National Park, Colorado. *Can. J. Zool.* 87: 1000–1008.

Ball, P. W. 2003a. 22. *Salicornia* Linnaeus. pp. 382–385. *In*: N. R. Morin (convening ed.), *Flora of North America North of Mexico, Vol. 4: Magnoliophyta: Caryophyllidae, Part 1*. Oxford University Press, New York.

Ball, P. W. 2003b. 23. *Sarcocornia* A. J. Scott. pp. 385–387. *In*: N. R. Morin (convening ed.), *Flora of North America North of Mexico, Vol. 4: Magnoliophyta: Caryophyllidae, Part 1*. Oxford University Press, New York.

Ball, P. W. 2012. *Salicornia*. Pickleweed. pp. 645–646. *In*: B. G. Baldwin, D. H. Goldman, D. J. Keil, R. Patterson, T. J. Rosatti & D. H. Wilken (eds.), *The Jepson Manual (2nd ed.)*. University of California Press, Berkeley, CA.

Ballard Jr., H. E. & K. J. Sytsma. 2000. Evolution and biogeography of the woody Hawaiian violets (*Viola*, Violaceae): Arctic origins, herbaceous ancestry and bird dispersal. *Evolution* 54: 1521–1532.

Ballard, H. E. & S. C. Gawler. 1994. Distribution, habitat and conservation of *Viola novae-angliae*. *Michigan Bot.* 33: 35–52.

Ballard Jr., H. E., K. J. Sytsma & R. R. Kowal. 1999. Shrinking the violets: phylogenetic relationships of infrageneric groups in *Viola* (Violacae) based on internal transcribed spacer DNA sequences. *Syst. Bot.* 23: 439–458.

Ballard, R. 1988. A new species of *Bidens* (Asteraceae) from San Luis Potosi, Mexico. *Syst. Bot.* 13: 184–186.

Balogh, L. 2001. Invasive alien plants threatening the natural vegetation of Őrség Landscape Protection Area (Hungary). pp. 185–197. *In*: G. Brundu, J. Brock, I. Camarda, L. Child & M. Wade (eds.), *Plant Invasions: Species Ecology and Ecosystem Management*. Backhuys Publishers, Leiden, The Netherlands.

Baltazar, J. M., L. Trierveiler-Pereira & C. Loguercio-Leite. 2009. A checklist of xylophilous Basidiomycetes (Basidiomycota) in mangroves. *Mycotaxon* 107: 221–224.

Baltensperger, A. A. 2004. Comparison of prey capturing efficiency between two species of sundew, *Drosera linearis* and *Drosera rotundifolia*. *Michigan Bot.* 43: 15–20.

Baltzer, J. L., H. L. Hewlin, E. G. Reekie, P. D. Taylor & J. S. Boates. 2002a. The impact of flower harvesting on seedling recruitment in sea lavender (*Limonium carolinianum*, Plumbaginaceae). *Rhodora* 104: 280–295.

Baltzer, J. L., E. G. Reekie, H. L. Hewlin, P. D. Taylor & J. S. Boates. 2002b. Impact of flower harvesting on the salt marsh plant *Limonium carolinianum*. *Can. J. Bot.* 80: 841–851.

Bamberg, S. A. & J. Major. 1968. Ecology of the vegetation and soils associated with calcareous parent materials in three alpine regions of Montana. *Ecol. Monogr.* 38: 127–167.

Bandaranayake, W. M. 1998. Traditional and medicinal uses of mangroves. *Mangroves Salt Marshes* 2: 133–148.

Banerjee, S., D. Kar, A. Banerjee & D. Palit. 2012. Utilization of some aquatic macrophytes in Borobandh: a lentic water body in Durgapur, West Bengal, India: implications for socio-economic upliftment of local stakeholder. *Indian J. Appl. Pure Biol.* 27: 83–92.

Banfi, E., G. Galasso & A. Soldano. 2005. Notes on systematics and taxonomy for the Italian vascular flora. 1. *Atti Soc. Ital. Sci. Nat. Mus. Civico Storia Nat. Milano* 146: 219–244.

Bank II, T. P. 1953. Ecology of prehistoric Aleutian village sites. *Ecology* 34: 246–264.

Barbour, J. H. 1897. Some country remedies and their uses. *Folklore* 8: 386–390.

Barbour, M. G. 1978. The effect of competition and salinity on the growth of a salt marsh plant species. *Oecologia* 37: 93–99.

Barbour, M. G., A. I. Solomeshch & J. J. Buck. 2007. *Classification, Ecological Characterization, and Presence of Listed Plant Taxa of Vernal Pool Associations in California*. U. S. Fish & Wildlife Service, agreement/study no. 814205G238. University of California, Davis, CA. 117 pp.

Barbour, M. G., A. I. Solomeshch, R. F. Holland, C. W. Witham, R. L. Macdonald, S. S. Cilliers, J. A. Molina, J. J. Buck & J. M. Hillman. 2005. Vernal pool vegetation of California: communities of long-inundated deep habitats. *Phytocoenologia* 35: 177–200.

Barker, N. G. & G. B. Williamson. 1988. Effects of a winter fire on *Sarracenia alata* and *S. psittacina*. *Amer. J. Bot.* 75: 138–143.

Barker, W. R. 1982. Evolution and biogeography of arid Australian Scrophulariaceae. pp. 341–350. *In*: W. R. Barker & P. J. M. Greenslade (eds.), *Evolution of the Flora and Fauna of Arid Australia*. Peacock Publications, Adelaide, SA.

Barker, W. R. 1992. New Australasian species of *Peplidium* and *Glossostigma* (Scrophulariaceae). *J. Adelaide Bot. Gard.* 15: 71–74.

Barker, W. T. 1997. 2. *Planera* J. F. Gmelin. P. 376 *In*: N. R. Morin (convening ed.), *Flora of North America North of Mexico, Vol. 3: Magnoliophyta: Magnoliidae and Hamamelidae*. Oxford University Press, New York.

Barker, W. R., G. L. Nesom, P. M. Beardsley & N. S. Fraga. 2012. A taxonomic conspectus of Phrymaceae: A narrowed circumscription for *Mimulus*, new and resurrected genera, and new names and combinations. *Phytoneuron* 39: 1–60.

Barkley, T. M. 1985. Generic boundaries in the Senecioneae. *Taxon* 34: 17–21.

Barkley, T. M. 1988. Variation among the aureoid senecios of North America: a geohistorical interpretation. *Bot. Rev.* 54: 82–106.

Barkley, T. M. 2006. 215. *Senecio* Linnaeus. pp. 544–569. *In*: Flora of North America Editorial Committee (ed.), *Flora of North America North of Mexico: Vol. 20, Magnoliophyta: Asteridae (in part): Asteraceae, Part 2*. Oxford University Press, New York.

Barkley, T. M., L. Brouillet & J. L. Strother. 2006. 187. Asteraceae. pp. 3–69. *In*: Flora of North America Editorial Committee (ed.), *Flora of North America North of Mexico, Vol. 19: Magnoliophyta: Asteridae, Part 8*. Oxford University Press, New York.

Barko, J. W. & R. M. Smart. 1981. Sediment-based nutrition of submersed macrophytes. *Aquat. Bot.* 10: 339–352.

Barkov, D. V. & E. A. Kurashov. 2011. Food composition and feeding rate of the lake Baikal invader *Gmelinoides fasciatus* (Stebbing, 1899) in Lake Ladoga. *Inland Water Biol.* 4: 346–356.

Bärlocher, F. 1982. On the ecology of Ingoldian Fungi. *BioScience* 32: 581–586.

Barneby, R. C. 1989. *Intermountain Flora, Vol. 3, Part B. Fabales*. The New York Botanical Garden, Bronx, NY. 279 pp.

Barnes, H. F. 1952. The gall midges of St. John's wort (*Hypericum* spp.), with descriptions of two new species. *Bull. Entomol. Res.* 42: 697–705.

Barnes, M. A., C. L. Jerde, D. Keller, W. L. Chadderton, J. G. Howeth & D. M. Lodge. 2013. Viability of aquatic plant fragments following desiccation. *Invas. Plant Sci. Manag.* 6: 320–325.

Barney, R. J., S. M. Clark & E. G. Riley. 2007. Annotated list of the leaf beetles (Coleoptera: Chrysomelidae) of Kentucky: subfamily Cassidinae. *J. Kentucky Acad. Sci.* 68: 132–144.

Barrat-Segretain, M.-H. & G. Bornette. 2000. Regeneration and colonization abilities of aquatic plant fragments: effect of disturbance seasonality. *Hydrobiologia* 421: 31–39.

Barrat-Segretain, M.-H., G. Bornette & A. Hering-Vilas-Bôas. 1998. Comparative abilities of vegetative regeneration among aquatic plants growing in disturbed habitats. *Aquat. Bot.* 60: 201–211.

Barrett, S. C. H. 1980. Dimorphic incompatibility and gender in *Nymphoides indica* (Menyanthaceae). *Can. J. Bot.* 58: 1938–1942.

Barrett, S. C. H. & D. E. Seaman. 1980. The weed flora of Californian rice fields. *Aquat. Bot.* 9: 351–376.

Barrett, S. C. H. & J. L. Strother. 1978. Taxonomy and natural history of *Bacopa* (Scrophulariaceae) in California. *Syst. Bot.* 3: 408–419.

Barringer, K. & W. Burger. 2000. Family #193 Scrophulariaceae. *Fieldiana Bot.* 41: 1–69.

Barrios, Y. & N. Ramírez. 2008. Outbreeding depression and reproductive biology of *Nymphaea ampla* (Salisb.) DC. (Nymphaeaceae). *Acta Bot. Venez.* 31: 539–556.

Barrows, E. M. & G. Gordh. 1974. Insect associates of the bagworm moth, *Thyridopteryx ephemeraeformis* (Lepidoptera: Psychidae), in Kansas. *J. Kansas Entomol. Soc.* 47: 156–161.

Barry, S. J. 1998. Managing the Sacramento Valley vernal pool andscape to sustain the native flora. pp. 236–240. *In*: C. W. Witham, E. T. Bauder, D. Belk, W. R. Ferren Jr. & R. Ornduff (eds.), *Ecology, Conservation, and Management of Vernal Pool Ecosystems—Proceedings from a 1996 Conference*. California Native Plant Society, Sacramento, CA.

Bart, D. & T. Davenport. 2015. The influence of legacy impacted seed banks on vegetation recovery in a post-agricultural fen complex. *Wetlands Ecol. Manag.* 23: 405–418.

Bartgis, R. 1983. Aquatic vegetation of a natural marl lake in West Virginia. *Proc. West Virginia Acad. Sci.* 55: 92–101.

Bartgis, R. L. 1997. The distribution of the endangered plant *Ptilimnium nodosum* (Rose) Mathias (Apiaceae) in the Potomac River drainage. *Castanea* 62: 55–59.

Bartgis, R. L. & G. E. Lang. 1984. Marl wetlands in eastern West Virginia: distribution, rare plant species, and recent history. *Castanea* 49: 17–25.

Barthlott, W., C. Neinhuis, R. Jetter, T. Bourauel & M. Riederer. 1996. Waterlily, poppy, or sycamore: on the systematic position of *Nelumbo*. *Flora* 191: 169–174.

Bartholomew, C. S., D. Prowell & T. Griswold. 2006. An annotated checklist of bees (Hymenoptera: Apoidea) in longleaf pine savannas of southern Louisiana and Mississippi. *J. Kansas Entomol. Soc.* 79: 184–198.

Bartkowska, M. P. & M. O. Johnston. 2012. Pollinators cause stronger selection than herbivores on floral traits in *Lobelia cardinalis* (Lobeliaceae). *New Phytol.* 193: 1039–1048.

Bartleman, A-P., K. Miyanishi, C. R. Burn & M. M. Côté. 2001. Development of vegetation communities in a retrogressive thaw slump near Mayo, Yukon Territory: a 10-year assessment. *Arctic* 54: 149–156.

Barton, L. V. 1960. Storage of seeds of *Lobelia cardinalis* L. *Contrib. Boyce Thompson Inst. Plant Res.* 20: 395–401.

Barton, M. L., I. D. Medel, K. K. Johnston & C. R. Whitcraft. 2016. Seed collection and germination strategies for common wetland and coastal sage scrub species in southern California. *Bull. South. Calif. Acad. Sci.* 115: 41–71.

Bartosik, M. B. 2010. Observations of seaside sparrow (*Ammodramus maritimus*) on Texas gulf coast. *Bull. Texas Ornithol. Soc.* 43: 11–24.

Bartow, A. 2003. Propagation protocol for production of *Myosotis laxa* Lehm. seeds; USDA NRCS—Corvallis Plant Materials Center, Corvallis. *In*: *Native Plant Network*. Forest Research Nursery, College of Natural Resources, University of Idaho, Moscow, ID. Available online: http://www.nativeplantnetwork.org (accessed April 24, 2008).

Bartow, A. 2007a. Propagation protocol for production of *Downingia elegans* Torrey seeds; Natural Resources Conservation Service—Corvallis Plant Materials Center, Corvallis. *In*: *Native Plant Network*. Forest Research Nursery, College of Natural Resources, University of Idaho, Moscow, ID. Available online: http://www.nativeplantnetwork.org (accessed September 11, 2014).

Bartow, A. 2007b. Propagation protocol for production of *Downingia yina* Applegate seeds; USDA NRCS—Corvallis Plant Materials Center, Corvallis. *In: Native Plant Network*. Forest Research Nursery, College of Natural Resources, University of Idaho, Moscow, ID. Available online: http://www.native-plantnetwork.org (accessed September 11, 2014).

Bartram, E. B. 1913. Some noteworthy plants of Bear Swamp. *Bartonia* 6: 8–16.

Basinger, M. A. 2003. Status of Fraser's Loosestrife (*Lysimchia fraseri* Duby) in Illinois. *Trans. Illinois State Acad. Sci.* 96: 1–6.

Baskaran, P. & N. Jayabalan. 2005. An efficient micropropagation system for *Eclipta alba*: a valuable medicinal herb. *In Vitro Cell. Dev. Biol. Plant* 41: 532–539.

Baskin, C. C. & J. M. Baskin. 1994. Annual dormancy cycle and influence of flooding in buried seeds of mudflat populations of the summer annual *Leucospora multifida*. *Ecoscience* 1: 47–53.

Baskin, C. C. & J. M. Baskin. 1998. *Seeds: Ecology, Biogeography, and Evolution of Dormancy and Germination*. Academic Press, San Diego, CA. 666 pp.

Baskin, C. C. & J. M. Baskin. 2002a. Propagation protocol for production of container *Pedicularis groenlandica* Retz. plants; University of Kentucky, Lexington, Kentucky. *In: Native Plant Network*. Forest Research Nursery, College of Natural Resources, University of Idaho, Moscow, ID. Available online: http://www.nativeplantnetwork.org (accessed August 24, 2007).

Baskin, C. C. & J. M. Baskin. 2002b. Propagation protocol for production of container *Jamesianthus alabamensis* Blake & Sherff plants; University of Kentucky, Lexington, Kentucky. *In: Native Plant Network*. Forest Research Nursery, College of Natural Resources, University of Idaho, Moscow, ID. Available online: http://www.nativeplantnetwork.org (accessed March 6, 2014).

Baskin, C. C. & J. M. Baskin. 2006. Symposium: the natural history of soil seed banks of arable land. *Weed Sci.* 54: 549–557.

Baskin, C. C., J. M. Baskin & E. W. Chester. 1996. Seed germination ecology of the aquatic winter annual *Hottonia inflata*. *Aquat. Bot.* 54: 51–57.

Baskin, C. C., J. M. Baskin & E. W. Chester. 1999a. Seed dormancy and germination in *Rhexia mariana* var. interior (Melastomataceae) and eco-evolutionary implications. *Can. J. Bot.* 77: 488–493.

Baskin, C. C., J. M. Baskin & E. W. Chester. 1999b. Seed dormancy in the wetland winter annual *Ptilimnium nuttallii* (Apiaceae). *Wetlands* 19: 359–364.

Baskin, C. C., J. M. Baskin & E. W. Chester. 2002. Effects of flooding and temperature on dormancy break in seeds of the summer annual mudflat species *Ammannia coccinea* and *Rotala ramosior* (Lythraceae). *Wetlands* 22: 661–668.

Baskin, C. C., J. M. Baskin & E. W. Chester. 2003. Ecological life cycle of *Trepocarpus aethusae* (Nutt.) ex DC. and comparisons with two other winter annual Apiaceae native to eastern United States. *Castanea* 68: 43–55.

Baskin, C. C., J. M. Baskin & M. A. Leck. 1993. Afterripening pattern during cold stratification of achenes of ten perennial Asteraceae from Eastern North America, and evolutionary implication. *Plant Species Biol.* 8: 61–65.

Baskin, C. C., P. Milberg, L. Andersson & J. M. Baskin. 2001. Seed dormancy-breaking and germination requirements of *Drosera anglica*, an insectivorous species of the Northern Hemisphere. *Acta Oecol.* 22: 1–8.

Baskin, J. M. & C. C. Baskin. 1971a. Germination of winter annuals in July and survival of the seedlings. *Bull. Torrey Bot. Club* 98: 272–276.

Baskin, J. M. & C. C. Baskin. 1971b. Germination ecology and adaptation to habitat in *Leavenworthia* spp. (Cruciferae). *Amer. Midl. Naturalist* 85: 22–35.

Baskin, J. M. & C. C. Baskin. 1973. Delayed germination in seeds of Phacelia dubia var. *dubia. Can. J. Bot.* 51: 2481–2486.

Baskin, J. M. & C. C. Baskin. 1974. Influence of low winter temperatures on flowering of winter annuals. *Castanea* 39: 340–345.

Baskin, J. M. & C. C. Baskin. 1977a. An undescribed cedar glade community in middle Tennessee. *Castanea* 42: 140–145.

Baskin, J. M. & C. C. Baskin. 1977b. *Leavenworthia torulosa* Gray: an endangered plant species in Kentucky. *Castanea* 42: 15–17.

Baskin, J. M. & C. C. Baskin. 1983. Seasonal changes in the germination responses of seeds of *Veronica peregrina* during burial, and ecological implications. *Can. J. Bot.* 61: 3332–3336.

Baskin, J. M. & C. C. Baskin. 1988. Endemism in rock outcrop plant communities of unglaciated eastern United States: an evaluation of the roles of the edaphic, genetic and light factors. *J. Biogeogr.* 15: 829–840.

Baskin, J. M. & C. C. Baskin. 1990. Seed germination ecology of poison hemlock, *Conium maculatum. Can. J. Bot.* 68: 2018–2024.

Baskin, J. M. & C. C. Baskin. 1996. Bessey Picklesimer's little-known quantitative study on the vegetation of a cedar glade in the central basin of Tennessee. *Castanea* 61: 25–37.

Baskin, J. M., C. C. Baskin & J. R. Aldrich. 1987. The cedar glade endemic *Viola egglestonii* Brainerd: new to Indiana and eastern Tennessee. *Castanea* 52: 68–71.

Baskin, J. M., C. C. Baskin & D. M. Spooner. 1989. Role of temperature, light and date: seeds were exhumed from soil on germination of four wetland perennials. *Aquat. Bot.* 35: 387–394.

Baskin, J. M., R. W. Tyndall, M. Chaffins & C. C. Baskin. 1998. Effect of salinity on germination and viability of nondormant seeds of the federal threatened species *Aeschynomene virginica* (Fabaceae). *J. Torrey Bot. Soc.* 125: 246–248

Baskin, Y. 1994. California's ephemeral vernal pools may be a good model for speciation. *BioScience* 44: 384–388.

Bassett, B., D. Menown & C. Gemming. 1993. *Nuisance Aquatic Plants in Missouri Ponds and Lakes*. Missouri Department of Conservation, Jefferson City, MO. 12 pp.

Bassett, I. J. & C. W. Crompton. 1978. The genus *Suaeda* (Chenopodiaceae) in Canada. *Can. J. Bot.* 56: 581–591.

Bastida, J. M., J. M. Alcántara, P. J. Rey, P. Vargas & C. M. Herrera. 2010. Extended phylogeny of *Aquilegia*: the biogeographical and ecological patterns of two simultaneous but contrasting radiations. *Plant Syst. Evol.* 284: 171–185.

Batcher, M. 2003. *Saururus cernuus (Lizard's tail) Conservation and Research Plan for New England*. New England Wild Flower Society, Framingham, MA. 34 pp.

Baten, W. D. 1936. Influence of position on structure of inflorescences of *Cicuta maculata. Biometrika* 28: 64–83.

Batra, L. R. 1988. *Monilinia gaylussaciae*, a new species pathogenic on huckleberries (*Gaylussacia*) in North America. *Mycologia* 80: 653–659.

Batt, B. D. J. (ed.). 1997. *Arctic Ecosystems in Peril: Report of the Arctic Goose Habitat Working Group*. Arctic goose joint venture special publication. U. S. Fish and Wildlife Service, Washington, DC, and Canadian Wildlife Service, Ottawa, ON. 120 pp.

Battaglia, L. L., P. R. Minchin & D. W. Pritchett. 2002. Sixteen years of old-field succession and reestablishment of a bottomland hardwood forest in the lower Mississipppi alluvial valley. *Wetlands* 22: 1–17.

Battaglia, L. L., J. S. Denslow & T. G. Hargis. 2007. Does woody species establishment alter herbaceous community composition of freshwater floating marshes? *J. Coastal Res.* 23: 1580–1587.

Battaglia, L. L., D. W. Pritchett & P. R. Minchin. 2008. Evaluating dispersal limitation in passive bottomland forest restoration. *Restorat. Ecol.* 16: 417–424.

Battersby, A. R., A. R. Burnett, G. D. Knowles & P. G. Parsons. 1968. Seco-cyclopentane glucosides from *Menyanthes trifoliata*: foliamenthin, dihydrofoliamenthin, and menthiafolin. *Chem. Commun. (London)* 21: 1277–1280.

Battjes, J., K. L. Chambers & K. Bachmann. 1994. Evolution of microsporangium numbers in *Microseris* (Asteraceae: Lactuceae). *Amer. J. Bot.* 81: 641–647.

Bauder, E. T. 1989. Drought stress and competition effects on the local distribution of *Pogogyne abramsii. Ecology* 70: 1083–1089.

Bauder, E. T. 1992. Ecological monitoring of *Downingia concolor* ssp. *brevior* (Cuyamaca Lake downingia) and *Limnanthes gracilis* ssp. *parishii* (Parish's slender meadowfoam), Cuyamaca Rancho State Park and Cuyamaca Valley, San Diego County, California. California Department of Parks and Recreation, Southern Division, San Diego, CA. 87 pp.

Bauder, E. T. 2000. Inundation effects on small-scale plant distributions in San Diego, California vernal pools. *Aquat. Ecol.* 34: 43–61.

Bauder, E. T. & S. McMillan. 1998. Current distribution and historical extent of vernal pools in southern California and northern Baja California, Mexico. pp. 56–70. *In*: C. W. Witham, E. T. Bauder, D. Belk, W. R. Ferren Jr. & R. Ornduff (eds.), *Ecology, Conservation, and Management of Vernal Pool Ecosystems— Proceedings from a 1996 Conference.* California Native Plant Society, Sacramento, CA.

Bauder, E. T., D. A. Kreager & S. C. McMillan. 1998. *Vernal Pools of Southern California Recovery Plan.* U. S. Department of the Interior, Fish and Wildlife Service, Region One, Portland, OR. 153 pp.

Bauer, H. L. 1930. Vegetation of the Tehachapi Mountains, California. *Ecology* 11: 263–280.

Bauer, P. J. 1983. Bumblebee pollination relationships on the Beartooth Plateau tundra of southern Montana. *Amer. J. Bot.* 70: 134–144.

Baum, D. A., K. J. Sytsma & P. C. Hoch. 1994. The phylogeny of *Epilobium* (Onagraceae) based on nuclear ribosomal DNA sequences. *Syst. Bot.* 19: 363–388.

Baye, P. R., P. M. Faber & B. Grewell. 2000. Tidal marsh plants of the San Francisco estuary. pp. 9–33. *In*: P. R. Olofson (ed.), *Baylands Ecosystem Species and Community Profiles: Life Histories and Environmental Requirements of Key Plants, Fish and Wildlife.* San Francisco Bay Regional Water Quality Control Board, Oakland, CA.

Bayer, C. M., F. Fay, A. Y. D. Bruijn, V. Savolainen, C. M. Morton, K. Kubitzki, W. S. Alverson & M. W. Chase. 1999. Support for an expanded family concept of Malvaceae within a recircumscribed order Malvales: a combined analysis of plastid *atpB* and *rbcL* DNA sequences. *Bot. J. Linn. Soc.* 129: 267–303.

Bayer, R. J., L. Hufford & D. E. Soltis. 1996. Phylogenetic relationships in Sarraceniaceae based on *rbcL* and ITS sequences. *Syst. Bot.* 21: 121–134.

Bayley, S. E. & R. L. Mewhort. 2004. Plant community structure and functional differences between marshes and fens in the southern boreal region of Alberta, Canada. *Wetlands* 24: 277–294.

Bazely, D. R. & R. L. Jefferies. 1986. Changes in the composition and standing crop of salt-marsh communities in response to the removal of a grazer. *J. Ecol.* 74: 693–706.

Bazzaz, F. A. 1991. Habitat selection in plants. *Amer. Naturalist* 137: S116–S130.

Beal, E. O. 1956. Taxonomic revision of the genus *Nuphar* Sm. of North America and Europe. *J. Elisha Mitchell Sci. Soc.* 72: 317–346.

Beal, E. O. 1977. *A Manual of Marsh and Aquatic Vascular Plants of North Carolina with Habitat Data.* Technical Bulletin No 247. North Carolina Agricultural Experiment Station, Raleigh, NC. 298 pp.

Beal, E. O. & R. M. Southall. 1977. The taxonomic significance of experimental selection by vernalization in *Nuphar* (Nymphaeaceae). *Syst. Bot.* 2: 49–60.

Beal, E. O. & J. W. Thieret. 1986. *Aquatic and Wetland Plants of Kentucky.* Scientific and Technical Series No 5. Kentucky Nature Preserves Commission, Frankfort, KY. 315 pp.

Beal, W. J. 1898. *Seed Dispersal.* Ginn & Co. Publishers, Boston, MA. 87 pp.

Beal, W. J. 1900. Notes on *Cabomba caroliniana* A. Gray. *Bull. Torrey Bot. Club* 27: 86–86.

Bean, L. J. & K. S. Saubel. 1972. *Temalpakh (From the Earth); Cahuilla Indian Knowledge and Usage of Plants.* Malki Museum Press, Banning, CA.

Beard, T. D. 1973. *Overwinter Drawdown: Impact on the Aquatic Vegetation in Murphy Flowage, Wisconsin.* Technical Bulletin No. 61. Department of Natural Resources, Madison, WI. 14 pp.

Beardsley, P. M. & W. R. Barker. 2005. Patterns of evolution in Australian *Mimulus* and related genera (Phrymaceae~Scrophulariaceae): a molecular phylogeny using chloroplast and nuclear sequence data. *Aust. Syst. Bot.* 18: 61–73.

Beardsley, P. M. & R. G. Olmstead. 2002. Redefining Phrymaceae: the placement of *Mimulus*, tribe Mimuleae, and *Phryma. Amer. J. Bot.* 89: 1093–1102.

Beardsley, P. M., A. Yen & R. G. Olmstead. 2003. AFLP phylogeny of *Mimulus section Erythranthe* and the evolution of hummingbird pollination. *Evolution* 57: 1397–1410.

Beardsley, P. M., S. E. Schoenig, J. B. Whittall & R. G. Olmstead. 2004. Patterns of evolution in western North American *Mimulus* (Phrymaceae). *Amer. J. Bot.* 91: 474–489.

Beas, B. J., L. M. Smith, K. R. Hickman, T. G. LaGrange & R. Stutheit. 2013. Seed bank responses to wetland restoration: do restored wetlands resemble reference conditions following sediment removal? *Aquat. Bot.* 108: 7–15.

Beatley, J. C. 1956. The winter-green herbaceous flowering plants of Ohio. *Ohio J. Sci.* 6: 349–377.

Beattie, D. J., C. F. Deneke, E. J. Holcomb & J. W. White. 1989. The effects of photoperiod and temperature on flowering of *Physostegia virginiana* 'Summer Snow' and 'Vivid' as potted plants. *Acta Hort.* 252: 227–234.

Beattie, M. H. 1995. Endangered and threatened wildlife and plants; determination of endangered status for ten plants and threatened status for two plants from serpentine habitats in the San Francisco Bay region of California. *Fed. Reg.* 60: 6671–6685.

Beauchamp, V. B., N. Ghuznavi, S. M. Koontz & R. P. Roberts. 2013. Edges, exotics and deer: the seed bank of a suburban secondary successional temperate deciduous forest. *Appl. Veg. Sci.* 16: 571–584.

Beauchamp, W. M. 1900. Iroquois women. *J. Amer. Folk.* 13: 81–91.

Beaven, G. F. & H. J. Oosting. 1939. Pocomoke swamp: a study of a cypress swamp on the eastern shore of Maryland. *Bull. Torrey Bot. Club* 66: 367–389.

Beaver, D. L., R. G. Osborn & T. W. Custer. 1980. Nest-site and colony characteristics of wading birds in selected Atlantic coast colonies. *Wilson Bull.* 92: 200–220.

Beck, J. B., G. L. Nesom, P. J. Calie, G. I. Baird, R. L. Small & E. E. Schilling. 2004. Is subtribe Solidagininae (Asteraceae) monophyletic? *Taxon* 53: 691–698.

Beck, J. B., I. A. Al-Shehbaz & B. A. Schaal. 2006. Leavenworthia (Brassicaceae) revisited: testing classic systematic and mating system hypotheses. *Syst. Bot.* 31: 151–159.

Beck, M. W., L. K. Hatch, B. Vondracek & R. D. Valley. 2010. Development of a macrophyte-based index of biotic integrity for Minnesota lakes. *Ecol. Indic.* 10: 968–979.

Becker, G. C. 1942. Notes on the pileated woodpecker in Wisconsin. *Passenger Pigeon* 4: 29–34.

Beck-Nielsen, D. & T. V. Madsen. 2001. Occurrence of vesicular-arbuscular mycorrhiza in aquatic macrophytes from lakes and streams. *Aquat. Bot.* 71: 141–148.

Beckwith, S. L. & H. J. Hosford. 1957. A report on seasonal food habits and life history notes of the Florida duck in the vicinity of Lake Okeechobee, Glades County, Florida. *Amer. Midl. Naturalist* 57: 461–473.

Bedoya, A. M. & S. Madriñán. 2015. Evolution of the aquatic habit in *Ludwigia* (Onagraceae): Morpho-anatomical adaptive strategies in the Neotropics. *Aquat. Bot.* 120: 352–362.

Begum, Z., S. S. Shaukat & I. A. Siddiqui. 2003. Suppression of *Meloidogyne javanica* by *Conyza canadensis*, *Blumea obliqua*, *Amaranthus viridis* and *Eclipta prostrata*. *Pakistan J. Plant Pathol.* 2: 174–180.

Bekker, R. M., E. J. Lammerts, A. Schutters & A. P. Grootjans. 1999. Vegetation development in dune slacks: the role of persistent seeds banks. *J. Veg. Sci.* 10: 745–754.

Bélair, G. & D. L. Benoit. 1996. Host suitability of 32 common weeds to *Meloidogyne hapla* in organic soils of southwestern Quebec. *J. Nematol.* 28(4S): 643–647.

Belal, A. E., B. Leith, J. Solway & I. Springuel (eds.). 1997. *Environmental Valuation and Management of Plants in Wadi Allaqi, Egypt*. Final Report. File 95-1005-01/02 127-01. International Development Research Centre (IDRC) Canada, Ottawa, ON.

Belden, R. C., W. B. Frankenberger, R. T. McBride & S. T. Schwikert. 1988. Panther habitat use in southern Florida. *J. Wildl. Manag.* 52: 660–663.

Belden Jr., A., A. C. Chazal & C. S. Hobson. 2003. *A Natural Heritage Inventory of Fourteen Headwater Sites in the Dragon Run Watershed*. Natural Heritage Technical Report 03–16, December 2003. Middle Peninsula Planning District Commission, Saluda, VA. 76 pp.

Belden Jr., A., G. P. Fleming, J. C. Ludwig, J. F. Townsend, N. E. Van Alstine, C. E. Stevens & T. F. Wieboldt. 2004. Virginia. *Castanea* 69: 144–153.

Bell, C. D. 2004a. Preliminary phylogeny of Valerianaceae (Dipsacales) inferred from nuclear and chloroplast DNA sequence data. *Mol. Phylogen. Evol.* 31: 340–350.

Bell, C. D. & M. J. Donoghue. 2005. Phylogeny and biogeography of Valerianaceae (Dipsacales) with special reference to the South American valerians. *Org. Diversity. Evol.* 5: 147–159.

Bell, C. D., A. Kutschker & M. T. K. Arroyo. 2012. Phylogeny and diversification of Valerianacea (Dipsacales) in the southern Andes. *Mol. Phylogen. Evol.* 63: 724–737.

Bell, C. E. 2004b. Poisonous plants in the Apiaceae. *Proc. Calif. Weed Sci. Soc.* 56: 111–113.

Bell, C. R. 1971. Breeding systems and floral biology of the Umbelliferae; or, evidence for specialization in unspecialized flowers. pp. 93–107. In: V. H. Heywood (ed.), *The Biology and Chemistry of the Umbelliferae*. Academic Press, London.

Bell, C. R. & L. Constance. 1957. Chromosome numbers in Umbelliferae. *Amer. J. Bot.* 44: 565–572.

Bell, C. R. & L. Constance. 1960. Chromosome numbers in Umbelliferae. II. *Amer. J. Bot.* 47: 24–32.

Bell, C. R. & R. Kane. 1981. An anomalous rhizomatous population of *Cicuta maculata* L. at Mountain Lake, Virginia. *Castanea* 46: 4–7.

Bell, D. T., J. A. Plummer & S. K. Taylor. 1993. Seed germination ecology in southwestern Western Australia. *Bot. Rev.* 59: 24–73.

Bell, F. W., M. Kershaw, I. Aubin, N. Thiffault, J. Dacosta & A. Wiensczyk. 2011. Ecology and traits of plant species that compete with boreal and temperate forest conifers: an overview of available information and its use in forest management in Canada. *Forest. Chron.* 87: 161–174.

Bell, J. M., J. D. Karron & R. J. Mitchell. 2005. Interspecific competition for pollination lowers seed production and outcrossing in *Mimulus ringens*. *Ecology* 86: 762–771.

Bell, N. B. & L. J. Lester. 1978. Genetic and morphological detection of introgression in a clinal population of *Sabatia* section *Campestria* (Gentianaceae). *Syst. Bot.* 3: 87–104.

Belleau, F. & G. Collin. 1993. Composition of the essential oil of *Ledum groenlandicum*. *Phytochemistry* 33: 117–121.

Bellemakers, M. J. S., M. Maessen, G. M. Verheggen & J. G. M. Roelofs. 1996. Effects of liming on shallow acidified moorland pools: a culture and a seed bank experiment. *Aquat. Bot.* 54: 37–50.

Bello, M. A., A. Bruneau, F. Forest & J. A. Hawkins. 2009. Elusive relationships within order Fabales: phylogenetic analyses using *matK* and *rbcL* sequence data. *Syst. Bot.* 34: 102–114.

Bellrose, F. C. 1941. Duck food plants of the Illinois River Valley. *Bull. Illinois. Nat. Hist. Surv.* 21: 237–280.

Belt, S. & S. Kirk. 2009. Plant fact sheet for spotted joe pye weed (*Eupatoriadelphus maculatus* (L.) King & H. Rob. var. *maculatus*). USDA-Natural Resources Conservation Service, Norman A. Berg National Plant Materials Center, Beltsville, MD.

Bender, S. R., A. L. Swinehart & J. P. Boardinan. 2013. Seventeen years of change in two *Sphagnum* bogs in Noble County, Indiana. *Proc. Indiana Acad. Sci.* 121: 110–120.

Benedict, N. B. 1983. Plant associations of subalpine meadows, Sequoia National Park, California. *Arct. Alp. Res.* 15: 383–396.

Ben Fadhel, N. & M. Boussaïd. 2004. Genetic diversity in wild Tunisian populations of *Mentha pulegium* L. (Lamiaceae). *Genet. Resour. Crop Evol.* 51: 309–321.

Benjamin, F. H. 1934. Descriptions of some native Trypetid flies with notes on their habits. *Tech. Bull. USDA*. 401: 1–95.

Benko-Iseppon, A. M. & W. Morawetz. 2000. Viburnales: cytological features and a new circumscription. *Taxon* 49: 5–16.

Benner, J. M. & R. T. Bowyer. 1988. Selection of trees for rubs by white-tailed deer in Maine. *J. Mammalogy* 69: 624–627.

Bennett, B. A. & R. S. Mulder. 2009. Natives gone wild: climate change and a history of a Yukon invasion. pp. 235–248. *In*: S. J. Darbyshire & R. Prasad (eds.), *Weeds Across Borders*. Alberta Invasive Plants Council, Lethbridge, AB.

Bennett, B. A., P. M. Catling, W. J. Cody & G. W. Argus. 2010. New records of vascular plants in the Yukon Territory VIII. *Can. Field-Naturalist* 124: 1–27.

Bennett, B. C. & R. Alarcon. 2015. Hunting and hallucinogens: the use psychoactive and other plants to improve the hunting ability of dogs. *J. Ethnopharmacol.* 171: 171–183.

Bennett, G. M. & N. A. Moran. 2013. Small, smaller, smallest: the origins and evolution of ancient dual symbioses in a phloem-feeding insect. *Genome Biol. Evol.* 5: 1675–1688.

Bennett, J. R. & S. Mathews. 2006. Phylogeny of the parasitic plant family Orobanchaceae inferred from phytochrome A. *Amer. J. Bot.* 93: 1039–1051.

Bennett, M. 2003. *Pulmonarias and the Borage Family.* Timber Press, Portland, OR. 240 pp.

Benoit, J. 2012. Clyde River Wetlands Natural Community Mapping Project. Final Report. Northwoods Stewardship Center, East Charleston, VT. 22 pp.

Benoliel, D. 2011. *Northwest Foraging: The Classic Guide to Edible Plants of the Pacific Northwest.* Skipstone, Seattle, WA. 192 pp.

Benson, L. 1940. The North American subdivisions of *Ranunculus.* *Amer. J. Bot.* 27: 799–807.

Benson, L. 1942. North American Ranunculi-IV. *Bull. Torrey Bot. Club* 69: 298–316.

Benson, L. 1948. A treatise on the North American Ranunculi. *Amer. Midl. Naturalist* 40: 1–261.

Benson, N. C. & L. L. Kurth. 1996. Vegetation establishment on rehabilitated bulldozer lines after the 1988 Red Bench fire in Glacier National Park. pp. 164–167. *In*: J. K. Brown, R. W. Mutch, C. W. Spoon & R. H. Wakimoto (eds.), *Proceedings: Symposium on Fire in Wilderness and Park Management.* Diane Publishing Co., Darby, PA.

Benvenuti, S. 2007. Weed seed movement and dispersal strategies in the agricultural environment. *Weed Biol. Manag.* 7: 141–157.

Benvenuti, S. & M. Macchia. 1997. Germination ecophysiology of bur beggarticks (*Bidens tripartita*) as affected by light and oxygen. *Weed Sci.* 45: 696–700.

Benzecry, A. 2012. Greenhouse experimental methods towards in-situ burial and restoration of contaminated sites in submerged wetlands. *Amer. J. Environ. Eng.* 2: 114–122.

Berenbaum, M. 1981. Patterns of furanocoumarin distribution and insect herbivory in the Umbelliferae: plant chemistry and community structure. *Ecology* 62: 1254–1266.

Berenbaum, M. 1983. Coumarins and caterpillars: a case for coevolution. *Evolution* 37: 163–179.

Berenbaum, M. 2009. Effects of linear furanocoumarins on an adapted specialist insect (*Papilio polyxenes*). *Ecol. Entomol.* 6: 345–351.

Berenbaum, M. R. & T. L. Harrison. 1994. *Agonopterix alstroemeriana* (Oecophoridae) and other lepidopteran associates of poison hemlock (*Conium maculatum*) in East Central Illinois. *Great Lakes Entomol.* 27: 1–5.

Berenguera, B., L. M. Sánchez, A. Quílez, M. López-Barreiro, O. de Haro, J. Gálvez & M. J. Martín. 2006. Protective and antioxidant effects of *Rhizophora mangle* L. against NSAID-induced gastric ulcers. *J. Ethnopharmacol.* 103: 194–200.

Berg, R. Y. 1975. Myrmecochorous plants in Australia and their dispersal by ants. *Aust. J. Bot.* 23: 475–508.

Bergen, J. Y. 1909. Concavity of leaves and illumination. *Bot. Gaz.* 48: 459–461.

Berger, A. J. & D. F. Parmelee. 1952. The Alder flycatcher in Washtenaw County, Michigan: breeding distribution and cowbird parasitism. *Wilson Bull.* 64: 33–38.

Berger, B. & W. Elisens. 2005. A reevaluation of the *Lindernia dubia* species complex. Botany 2005 meeting (abstract).

Bergeron, C., A. Marston, R. Gauthier & K. Hostettmann. 1996. Screening of plants used by North American Indians for antifungal, bactericidal, and molluscidal activities. *Int. J. Pharmacogn.* 34: 233–242.

Bergeron, C., A. Marston, R. Gauthier & K. Hostettmann. 1997. Iridoids and secoiridoids from *Gentiana linearis.* *Phytochemistry* 44: 633–637.

Bergerud, A. T. 1972. Food habits of Newfoundland caribou. *J. Wildl. Manag.* 36: 913–923.

Bergerud, A. T. & M. J. Nolan. 1970. Food habits of hand-reared caribou *Rangifer tarandus* L. in Newfoundland. *Oikos* 21: 348–350.

Bergerud, A. T., H. E. Butler & D. R. Miller. 1984. Antipredator tactics of calving caribou: dispersion in mountains. *Can. J. Zool.* 62: 1566–1575.

Bergman, M. 2002. Can saliva from moose, *Alces alces*, affect growth responses in the sallow, *Salix caprea*? *Oikos* 96: 164–168.

Berkowitz, J. F., S. Page & C. V. Noble. 2014. Potential disconnect between observations of hydrophytic vegetation, wetland hydrology indicators, and hydric soils in unique pitcher plant bog habitats of the southern Gulf Coast. *S. E. Naturalist* 13: 721–734.

Bermingham, L. H. 2010. Deer herbivory and habitat type influence long-term population dynamics of a rare wetland plant. *Plant Ecol.* 210: 359–378.

Bermúdez, A. & D. Velázquez. 2002. Etnobotánica médica de una comunidad campesina del estado Trujillo, Venezuela: un estudio preliminar usando técnicas cuantitativas. *Rev. Fac. Farm.* 44: 2–6.

Bern, A. L. 1997. Studies on nitrogen and phosphorus uptake by the carnivorous bladderwort, *Utricularia foliosa,* in south Florida wetlands. MS thesis. Florida International University, Miami, FL.

Bernardello, G., G. J. Anderson, T. F. Stuessy & D. J. Crawford. 2001. A survey of floral traits, breeding systems, floral visitors, and pollination systems of the angiosperms of the Juan Fernández Islands (Chile). *Bot. Rev.* 67: 255–308.

Bertin, R. I. 1982. The ruby-throated hummingbird and its major food plants: ranges, flowering phenology, and migration. *Can. J. Zool.* 60: 210–219.

Bertin, R. I. & M. A. Kerwin. 1998. Floral sex ratios and gynomonoecy in Aster (Asteraceae). *Amer. J. Bot.* 85: 235–244.

Bertin, R. I., D. B. Connors & H. M. Kleinman. 2010. Differential herbivory on disk and ray flowers of gynomonoecious asters and goldenrods (Asteraceae). *Biol. J. Linn. Soc.* 101: 544–552.

Bertness, M. D., L. Gough & S. W. Shumway. 1992. Salt tolerances and the distribution of fugitive salt marsh plants. *Ecology* 73: 1842–1851.

Bess, J. 2005. *Conservation Assessment for the Reflexed Indiangrass Leafhopper (Flexamia reflexa (Osborn and Ball)).* USDA Forest Service, Eastern Region, Milwaukee, WI. 42 pp.

Bess, J. A., R. A. Chimner & L. C. Kangas. 2014. Ditch restoration in a large northern Michigan fen: vegetation response and basic porewater chemistry. *Ecol. Restorat.* 32: 260–274.

Betz, R. F. & M. H. Cole. 1969. The Peacock Prairie—a study of a virgin Illinois mesic black-soil prairie forty years after initial study. *Trans. Illinois. Acad. Sci.* 62: 44–53.

Betz, R. F. & H. F. Lamp. 1992. Species composition of old settler savanna and sand prairie cemeteries in northern Illinois and northwestern Indiana. pp. 79–87. *In*: D. D. Smith & C. A. Jacobs (eds.), *Proceedings of the Twelfth North American Prairie Conference.* University of Northern Iowa, Cedar Falls, IA.

Beuzenberg, E. J. & J. B. Hair. 1983. Contributions to a chromosome atlas of the New Zealand flora—25 miscellaneous species. *New Zealand. J. Bot.* 21: 13–20.

Beyra-Matos, A. & M. Lavin. 1999. A monograph of *Pictetia* (Papilionoideae; Leguminosae) and a review of the tribe *Aeschynomeneae*. *Syst. Bot. Monogr.* 56: 1–93.

Bharathan, G., T. E. Goliber, C. Moore, S. Kessler, T. Pham & N. R. Sinha. 2002. Homologies in leaf form inferred from *KNOX1* gene expression during development. *Science* 296: 1858–1860.

Bhatkar, A. P. & W. H. Whitcomb. 1975. Rearing arboreal ants in glass tubing. *Florida. Entomol.* 58: 59–63.

Bhattacherjee, A., T. Ghosh, R. Sil & A. Datta. 2014. Isolation and characterisation of methanol-soluble fraction of *Alternanthera philoxeroides* (Mart.)–evaluation of their antioxidant, α-glucosidase inhibitory and antimicrobial activity in in vitro systems. *Nat. Prod. Res.* 28: 2199–2202.

Bhowmik, S. & B. K. Datta. 2013. Pollen production in relation to ecological class of some hydrophytes and marsh plants. *Amer. J. Plant Sci.* 4: 324–332.

Bhusari, N. V., D. M. Mate & K. H. Makde. 2005. Pollen of *Apis* honey from Maharashtra. *Grana* 44: 216–224.

Biabiany, M., V. Roumy, T. Hennebelle, N. François, B. Sendid, M. Pottier, E. M. Aliouat, I. Rouaud, F. L.-L. Dévéhat, H. Joseph, P. Bourgeois, S. Sahpaz & F. Bailleul. 2013. Antifungal activity of 10 Guadeloupean plants. *Phytother. Res.* 27: 1640–1645.

Bick, G. H. & J. C. Bick. 1963. Behavior and population structure of the damselfly, *Enallagma civile* (Hagen) (Odonata: Coenagriidae). *Southwest. Naturalist* 8: 57–84.

Bickel, T. O. 2012. Ecology of the submersed aquatic weed *Cabomba caroliniana* in Australia. pp. 21–24. *In*: V. Eldershaw (ed.), *Proceedings of the 18th Australasian Weeds Conference*. Weed Society of Victoria Inc., Melbourne, Australia.

Bickel, T. O. 2015. A boat hitchhiker's guide to survival: *Cabomba caroliniana* desiccation resistance and survival ability. *Hydrobiologia* 746: 123–134.

Bierner, M. W. 1974. The taxa of *Helenium* in Tennessee. *Castanea* 39: 346–349.

Bierner, M. W. 2006. 385. *Helenium* Linnaeus. pp. 426–434. *In*: Flora of North America Editorial Committee (ed.), *Flora of North America North of Mexico: Vol. 21, Magnoliophyta: Asteridae (in part): Asteraceae, Part 3*. Oxford University Press, New York.

Bills, J. W., E. H. Roalson, J. W. Busch & P. B. Eidesen. 2015. Environmental and genetic correlates of allocation to sexual reproduction in the circumpolar plant *Bistorta vivipara*. *Amer. J. Bot.* 102: 1174–1186.

Biondi, E. & S. Casavecchia. 2001. Phytosociological survey of Northern California dunes. *Plant Biosyst.* 135: 351–362.

Birkenholz, D. E. 1963. A study of the life history and ecology of the round-tailed muskrat (*Neofiber alleni* True) in north-central Florida. *Ecol. Monogr.* 33: 255–280.

Birnbaum, S. J., J. M. Poole & P. S. Williamson. 2011. Reintroduction of star cactus *Astrophytum asterias* by seed sowing and seedling transplanting, Las Estrellas Preserve, Texas, USA. *Conserv. Evid.* 8: 43–52.

Biswas, S. R. & A. U. Mallik. 2010. Disturbance effects on species diversity and functional diversity in riparian and upland plant communities. *Ecology* 91: 28–35.

Bittrich, V. 1993. Caryophyllaceae. pp. 206–236. *In*: K. Kubitzki, J. Rohwer & V. Bittrich (eds.), *The Families and Genera of Vascular Plants, Vol. II. Flowering Plants, Dicotyledons: Magnoliid, Hamamelid and Caryophyllid Families*. Springer-Verlag, Berlin, Germany.

Bjerknes, A.-L., Ø. Totland, S. J. Hegland & A. Nielsen. 2007. Do alien plant invasions really affect pollination success in native plant species? *Biol. Conserv.* 138: 1–12.

Blackman, M. W. 1922. *Mississippi Bark Beetles*. Technical Bulletin 11. Mississippi Agricultural Experiment Station, Agricultural College, MS. 130 pp.

Blackwell, L. R. 2006. *Great Basin: A Guide to Common Wildflowers of the High Deserts of Nevada, Utah, and Oregon*. Globe Pequot Press, Guilford, CT. 281 pp.

Blair, W. F. 1936. The Florida marsh rabbit. *J. Mammalogy* 17: 197–207.

Blair, W. F. 1940. A study of prairie deer-mouse populations in southern Michigan. *Amer. Midl. Naturalist* 24: 273–305.

Blake, S. F. 1918. Notes on the flora of New Brunswick. *Rhodora* 20: 101–107.

Blancaver, M. E. A., K. Itoh & K. Usui. 2002. Response of the sulfonylurea herbicide-resistant *Rotala indica* Koehne var. *uliginosa* Koehne to bispyribac sodium and imazamox. *Weed Biol. Manag.* 2: 60–63.

Blanchan, N. 1901. *Nature's Garden: An Aid to Knowledge of our Wild Flowers and their Insect Visitors*. Doubleday, Page & Co., New York. 415 pp.

Blanchard, B. 1992. Endangered and threatened wildlife and plants; determination of threatened status for the sensitive joint-vetch (*Aeschynomene virginica*). *Fed. Reg.* 57: 21569–21574.

Blanchard, B. 1993. Endangered and threatened wildlife and plants; endangered or threatened status for five Florida plants. *Fed. Reg.* 58: 37432–37443.

Blanchard Jr., O. J. 1974. Chromosome number in *Kosteletzkya* Presl (*Malvaceae*). *Rhodora* 76: 64–66.

Blanchard Jr., O. J. 2008. Innovations in *Hibiscus* and *Kosteletzkya* (*Malvaceae, Hibisceae*). *Novon* 18: 4–8.

Blanchard Jr., O. J. 2012. Chromosome numbers, phytogeography, and evolution in *Kosteletzkya* (Malvaceae). *Rhodora* 114: 37–49.

Blanchard Jr., O. J. 2013. Experimental hybridization and chromosome pairing in *Kosteletzkya* (Malvaceae, Malvoideae, Hibisceae), and possible implications for phylogeny and phytogeography in the genus. *Comp. Cytogenet.* 7: 73–101.

Blanckaert, I., K. Vancraeynest, R. L. Swennen, F. J. Espinosa-García, D. Piñero & R. Lira-Saade. 2007. Non-crop resources and the role of indigenous knowledge in semi-arid production of Mexico. *Agric. Ecosyst. Environ.* 119: 39–48.

Blanken, P. D. & W. R. Rouse. 1996. Evidence of water conservation mechanisms in several subarctic wetland species. *J. Appl. Ecol.* 33: 842–850.

Blasdale, W. C. 1919. A preliminary list of the Uredinales of California. *Univ. Calif. Publ. Bot.* 7: 101–157.

Blatchley, W. S. 1898. Some Indiana Acrididae.–IV. *Can. Entomol.* 30: 54–64.

Bledsoe, C., P. Klein & L. C. Bliss. 1990. A survey of mycorrhizal plants on Truelove Lowland, Devon Island, N.W.T., Canada. *Can. J. Bot.* 68: 1848–1856.

Bleeker, W. 2004. Genetic variation and self-incompatibility within and outside a *Rorippa* hybrid zone (Brassicaceae). *Plant Syst. Evol.* 246: 35–44.

Bleeker, W. & H. Hurka. 2001. Introgressive hybridization in *Rorippa* (Brassicaceae): gene flow and its consequences in natural and anthropogenic habitats. *Mol. Ecol.* 10: 2013–2022.

Bleeker, W., C. Weber-Sparenberg & H. Hurka. 2002. Chloroplast DNA variation and biogeography in the genus *Rorippa* Scop. (Brassicaceae). *Plant Biol.* 4: 104–111.

Blickle, R. L. 1955. Observations on the habits of Tabanidae. *Ohio J. Sci.* 55: 308–310.

Bliss, A. 1976. *Rocky Mountain Dye Plants*. Anne Bliss, Boulder, CO. 120 pp.

Bliss, L. C. 1958. Seed germination in arctic and alpine species. *Arctic* 1958: 180–188.

Bliss, L. C. 1971. Arctic and alpine plant life cycles. *Annu. Rev. Ecol. Syst.* 2: 405–438.

Bliss, L. C. & G. W. Cox. 1964. Plant community and soil variation within a northern Indiana prairie. *Amer. Midl. Naturalist* 72: 115–128.

Bliss, M. 1986. The morphology, fertility and chromosomes of *Mimulus glabratus* var. *michiganensis and M. glabratus* var. *fremontii* (Schrophulariaceae) [sic!]. *Amer. Midl. Naturalist* 116: 125–131.

Bliss, S. A. & P. H. Zedler. 1998. The germination process in vernal pools: sensitivity to environmental conditions and effects on community structure. *Oecologia* 113: 67–73.

Blits, K. C. & J. L. Gallagher. 1990a. Salinity tolerance of *Kosteletzkya virginica*. I. Shoot growth, ion and water relations. *Plant Cell Environ.* 13: 409–418.

Blits, K. C. & J. L. Gallagher. 1990b. Salinity tolerance of *Kosteletzkya virginica*. II. Root growth, lipid content, ion and water relations. *Plant Cell Environ.* 13: 419–425.

Block, T. A. & A. F. Rhoads. 2011. *Aquatic Plants of Pennsylvania: A Complete Reference Guide.* University of Pennsylvania Press, Philadelphia, PA. 320 pp.

Blondeau, M. 1989. La flore vasculaire des environs d'Akulivik, Nouveau-Québec. *Provancheria* 23: 1–80.

Blood, L. E., H. J. Pitoniak & J. H. Titus. 2010. Seed bank of a bottomland swamp in western New York. *Castanea* 75: 19–38.

Bloom, W. L. 1977.Chromosomal differentiation between *Cirsium discolor* and *C. muticum* and the origin of supernumerary chromosomes. *Syst. Bot.* 2: 1–13.

Blotnick, J. R., J. Rho & H. B. Gunner. 1980. Ecological characteristics of the rhizosphere microflora of *Myriophyllum heterophyllum. J. Environ. Qual.* 9: 207–210.

Blus, L. J. & J. A. Keahey. 1978. Variation in reproductivity with age in the brown pelican. *Auk* 95: 128–134.

Boal, C. W. & D. E. Andersen. 2005. Microhabitat characteristics of Lapland longspur, *Calcarius lapponicus*, nests at Cape Churchill, Manitoba. *Can. Field Naturalist* 119: 208–213.

Böcher, J., N. P. Kristensen, T. Pape & L. Vilhelmsen (eds.). 2015. *The Greenland Entomofauna: An Identification Manual of Insects, Spiders and Their Allies.* BRILL, Leiden, The Netherlands. 898 pp.

Boches, P. S., N. V. Bassil & L. J. Rowland. 2005. Microsatellite markers for *Vaccinium* from EST and genomic libraries. *Mol. Ecol. Notes* 5: 657–660.

Bodamer, B. L. & M. L. Ostrofsky. 2010. The use of aquatic plants by populations of the zebra mussel (*Dreissena polymorpha*) (Bivalvia: Dreissenidae) in a small glacial lake. *Nautilus* 124: 100–106.

Bodo Slotta, T. A., D. P. Horvath & M. E. Foley. 2012. Phylogeny of *Cirsium* spp. in North America: host specificity does not follow phylogeny. *Plants* 1: 61–73.

Bodkin, P. C., D. H. N. Spence & D. C. Weeks. 1980. Photoreversible control of heterophylly in *Hippuris vulgaris* L. *New Phytol.* 84: 533–542.

Boe, J. S. 1991. Breeding habitat selection by eared grebes in Minnesota. PhD dissertation. North Dakota State University, Fargo, ND. 242 pp.

Boe, J. S. 1994. Nest site selection by eared grebes in Minnesota. *Condor* 96: 19–35.

Boecklen, W. J., P. W. Price & S. Mopper. 1990. Sex and drugs and herbivores: sex-biased herbivory in arroyo willow (*Salix lasiolepis*) *Ecology* 71: 581–588.

Boedeltje, G., J. P. Bakker, R. M. Bekker, J. M. Van Groenendael & M. Soesbergen. 2003a. Plant dispersal in a lowland stream in relation to occurrence and three specific life-history traits of the species in the species pool. *J. Ecol.* 91: 855–866.

Boedeltje, G., J. P. Bakker & G. N. J. ter Heerdt. 2003b. Potential role of propagule banks in the development of aquatic vegetation in backwaters along navigation canals. *Aquat. Bot.* 77: 53–69.

Boedeltje, G., J. P. Bakker, A. T. Brinke, J. M. Van Groenendael & M. Soesbergen. 2004. Dispersal phenology of hydrochorous plants in relation to discharge, seed release time and buoyancy of seeds: the flood pulse concept supported. *J. Ecol.* 92: 786–796.

Boeger, M. R. T. & M. E. Poulson. 2003. Morphological adaptations and photosynthetic rates of amphibious *Veronica anagallis-aquatica* L. (Scrophulariaceae) under different flow regimes. *Aquat. Bot.* 75: 123–135.

Boeger, R. T. 1992. The influence of substratum and water velocity on growth of *Ranunculus aquatilis* L. (Ranunculaceae). *Aquat. Bot.* 42: 351–359.

Boggs, C. L., C. E. Holdren, I. G. Kulahci, T. C. Bonebrake, B. D. Inouye, J. P. Fay, A. McMillan, E. H. Williams & P. R. Ehrlich. 2006. Delayed population explosion of an introduced butterfly. *J. Anim. Ecol.* 75: 466–475.

Boggs, K., S. C. Klein, J. Grunblatt, T. Boucher, B. Koltun, M. Sturdy & G. P. Streveler. 2010. Alpine and subalpine vegetation chronosequences following deglaciation in coastal Alaska. *Arct. Antarc. Alp. Res.* 42: 385–395.

Bohart, G. E. & T. Griswold. 1996. A revision of the rophitine genus *Protodufourea* (Hymenoptera: Halictidae). *J. Kansas Entomol. Soc.* 69: 177–184.

Bohley, K., O. Joos, H. Hartmann, R. Sage, S. Liede-Schumann & G. Kadereit. 2015. Phylogeny of Sesuvioideae (Aizoaceae)–Biogeography, leaf anatomy and the evolution of C₄ photosynthesis. *Perspect. Plant Ecol. Evol. Syst.* 17: 116–130.

Bohlmann, F. & P. K. Mahanta. 1978. Neue eudesman-derivate aus *Pluchea foetida. Phytochemistry* 17: 1189–1190.

Bohlmann, F., L. N. Dutta, W. Knauf, H. Robinson & R. M. King. 1980. Neue sequiterpenlactone aus *Aster umbellatus. Phytochemistry* 19: 433–436.

Bohlmann, F., J. Jakupovic, A. K. Dhar, R. M. King & H. Robinson. 1981. Heliangolides and diterpenes from *Hartwrightia floridana. Phytochemistry* 20: 843–845.

Bohm, B. A. & H. M. Banek. 1987. Flavonoid variation in *Lasthenia burkei. Biochem. Syst. Ecol.* 15: 57–59.

Bohm, B. A., K. W. Nicholls & R. Ornduff. 1986. Flavonoids of the Menyanthaceae: intra-and interfamilial relationships. *Amer. J. Bot.* 73: 204–213.

Bohm, B. A., J. Y. Yang, J. E. Page & D. E. Soltis. 1999. Flavonoids, DNA and relationships of *Itea* and *Pterostemon. Biochem. Syst. Ecol.* 27: 79–83.

Boivin, G. 1999. Integrated management for carrot weevil. *Integr. Pest Manag. Rev.* 4: 21–37.

Boizard, S. D. & S. J. Mitchell. 2011. Resistance of red mangrove (*Rhizophora mangle* L.) seedlings to deflection and extraction. *Trees* 25: 371–381.

Boldt, P. E., S. S. Rosenthal & R. Srinivasan. 1998. Distribution of field bindweed and hedge bindweed in the USA. *J. Prod. Agric.* 11: 377–381.

Bolen, E. G. 1964. Plant ecology of spring-fed salt marshes in western Utah. *Ecol. Monogr.* 34: 143–166.

Bolland, H. R., J. Gutierrez & C. H. W. Flechtmann. 1998. *World Catalogue of the Spider Mite Family (Acari: Tetranychidae).* Brill, Leiden, The Netherlands. 392 pp.

Bollman, M. A., M. J. Storm, G. A. King & L. S. Watrud. 2012. Wetland and riparian plant communities at risk of invasion by transgenic herbicide-resistant *Agrostis* spp. in central Oregon. *Plant Ecol.* 213: 355–370.

Bolmgren, K. & B. Oxelman. 2004. Generic limits in *Rhamnus* s.l. (Rhamnaceae) inferred from nuclear and chloroplast DNA sequence phylogenies. *Taxon* 53: 383–390.

Bolser, R. C., M. E. Hay, N. Lindquist, W. Fenical & D. Wilson. 1998. Chemical defenses of freshwater macrophytes against crayfish herbivory. *J. Chem. Ecol.* 24: 1639–1658.

Bonamonte, D., L. Mundo, M. Daddabbo & C. Foti. 2001. Allergic contact dermatitis from *Mentha spicata* (spearmint). *Contact Dermatitis* 45: 298–298.

Bonanno, S. E., D. J. Leopold & L. R. St. Hilaire. 1998. Vegetation of a freshwater dune barrier under high and low recreational uses. *J. Torrey Bot. Soc.* 125: 40–50.

Bonasera, J., J. Lynch & M. A. Leck. 1979. Comparison of the allelo-pathic potential of four marsh species. *Bull. Torrey Bot. Club* 106: 217–222.

Bondarenko, A. S., L. A. Bakina, E. M. Kleiner, V. I. Scheichenko, M. A. Gilzin, A. S. Khoklov, M. A. Poddubnaya & T. I. Skorobogatko. 1968. Biological properties and chemical nature of the antibiotic from *Bidens cernuus* L. *Antibiotiki* 13: 167–171.

Bonde, E. K. 1965. Studies on the germination of seeds of Colorado alpine plants. *Univ. Colo. Stud., Ser. Biol.* 14: 1–16.

Bonde, E. K. 1969. Plant disseminules in wind-blown debris from a glacier in Colorado. *Arct. Alp. Res.* 1: 135–139.

Bone, T. 2007. A phylogenetic and biogeographical study of the umbellifer genus *Lilaeopsis*. MS thesis. Department of Plant Biology, University of Illinois at Urbana-Champaign, Urbana, IL. 342 pp.

Bone, T. S., S. R. Downie, J. M. Affolter & K. Spalik. 2011. A phy-logenetic and biogeographic study of the genus *Lilaeopsis* (Apiaceae tribe Oenantheae). *Syst. Bot.* 36: 789–805.

Bonilla-Barbosa, J. R. & A. Novelo Retana. 1995. *Manual de Identificatión de Plantas Acuáticas del Parque Nacional Lagunas de Zempoala, México*. Cuardenos del Instituto de Biología 26. Universidad Nacional Autónoma de México, México City. 168 pp.

Bonilla-Barbosa, J., A. Novelo, Y. H. Orozco & J. Márquez-Guzmán. 2000. Comparative seed morphology of Mexican *Nymphaea* species. *Aquat. Bot.* 68: 189–204.

Bonner, J. L., J. T. Anderson, J. S. Rentch & W. N. Grafton. 2009. Vegetative composition and community structure associated with beaver ponds in Canaan valley, West Virginia, USA. *Wetlands Ecol. Manag.* 17: 543–554.

Bonnin, I., B. Colas, C. Bacles, A.-C. Holl, F. Hendoux, B. Destiné & F. Viard. 2002. Population structure of an endan-gered species living in contrasted habitats: *Parnassia palus-tris* (Saxifragaceae). *Mol. Ecol.* 11: 979–990.

Bontrager, M., K. Webster, M. Elvin & I. M. Parker. 2014. The effects of habitat and competitive/facilitative interactions on reintroduction success of the endangered wetland herb, *Arenaria paludicola. Plant Ecol.* 215: 467–478.

Boone, D. D., G. H. Fenwick & F. Hirst. 1984. The rediscovery of *Oxypolis canbyi* on the Delmarva Peninsula. *Bartonia* 50: 21–22.

Boorman, L. A. 1968. Some aspects of the reproductive biology of *Limonium vulgare* Mill., and *Limonium humile* Mill. *Ann. Bot.* 32: 803–824.

Boose, D. L., B. R. Harmeling & R. T. Turcotte. 2005. Genetic varia-tion in eastern Washington populations of *Navarretia leuco-cephala* (Polemoniaceae) a vernal pool endemic. *Madroño* 52: 99–106.

Boot, D. H. 1914. Comparison of field and forest floras in Monona County, Iowa. *Proc. Iowa Acad. Sci.* 21: 53–58.

Booth, R. K. 2002. Testate amoebae as paleoindicators of surface-moisture changes on Michigan peatlands: modern ecology and hydrological calibration. *J. Paleolimnol.* 28: 329–348.

Borah, A., R. N. S. Yadav & B. G. Unni. 2011. In vitro antioxidant and free radical scavenging activity of *Alternanthera sessilis. Int. J. Pharm. Sci. Res.* 2: 1502–1506.

Borchardt, J. R., D. L. Wyse, C. C. Sheaffer, K. L. Kauppi, R. G. Fulcher, N. J. Ehlke, D. D. Biesboer & R. F. Bey. 2008a. Antioxidant and antimicrobial activity of seed from plants of the Mississippi river basin. *J. Med. Plants Res.* 2: 81–93.

Borchardt, J. R., D. L. Wyse, C. C. Sheaffer, K. L. Kauppi, R. G. Fulcher, N. J. Ehlke, D. D. Biesboer & R. F. Bey. 2008b. Antimicrobial activity of native and naturalized plants of Minnesota and Wisconsin. *J. Med. Plants Res.* 2: 98–110.

Borg, S. J. T., A. Janse & M. M. Kwak. 1980. Life cycle variation in *Pedicularis palustris* L. (Scrophulariaceae). *Acta Bot. Neerl.* 29: 397–405.

Borgen, L. & U.-M. Hultgård. 2003. *Parnassia palustris*: a geneti-cally diverse species in Scandinavia. *Bot. J. Linn. Soc.* 142: 347–372.

Borgias, D. 2004. *Effects of Livestock Grazing and the Development of Grazing Best Management Practices for the Vernal Pool-Mounded Prairies of the Agate Desert, Jackson County, Oregon*. U. S. Fish and Wildlife Service, Portland, OR. 55 pp.

Borkent, C. J. & L. D. Harder. 2007. Flies (Diptera) as pollinators of two dioecious plants: behaviour and implications for plant mating. *Can. Entomol.* 139: 235–246.

Borkholder, B. D., A. J. Edwards & C. A. Olson. 2002. *Biological, Physical, and Chemical Characteristics of the Cloquet River from Indian Lake to Island Lake, 2000–2001*. Technical Report #36. Fond du Lac Division of Resource Management, Cloquet, MN. 43 pp.

Borland, J. 1994. Growing indian paintbrush. *Amer. Nurseryman* 179(6): 49–53.

Borman, S., R. Korth & J. Temte. 1997. *Through the Looking Glass: A Field Guide to Aquatic Plants*. FH-207–97. Department of Natural Resources, University of Wisconsin, Stevens Points, WI. 248 pp.

Bornette, G., C. Amoros & D. Chessel. 1994. Effect of allogenic processes on successional rates in former river channels. *J. Veg. Sci.* 5: 237–246.

Bornstein, A. J. 1997. Myricaceae blume. pp. 429–435. *In*: N. R. Morin (convening ed.), *Flora of North America North of Mexico, Vol. 3: Magnoliophyta: Magnoliidae and Hamamelidae*. Oxford University Press, New York.

Boro, A. & G. C. Sarma. 2013. Ethnic uses of some wetland plants by the Bodo community in Udalgiri district of Assam, India. *Pleione* 7: 155–159.

Borsch, T. & W. Barthlott. 1994. Classification and distribution of the genus *Nelumbo* Adans. (Nelumbonaceae). *Beitr. Biol. Pflanz.* 68: 421–450.

Borsch, T., K. W. Hilu, J. H. Wiersema, C. Lohne, W. Barthlott & V. Wilde. 2007. Phylogeny of *Nymphae* (Nymphaeaceae): evidence from substitutions and microstructural changes in the chloroplast *trnT-trnF* region. *Int. J. Plant Sci.* 168: 639–671.

Borsch, T., J. H. Wiersema, C. B. Hellquist, C. Löhne & K. Govers. 2014. Speciation in North American water lilies: evidence for the hybrid origin of the newly discovered Canadian endemic *Nymphaea loriana* sp. nov. (Nymphaeaceae) in a past contact zone. *Botany* 92: 867–882.

Borthwick, H. A., M. W. Parker & S. B. Hendricks. 1950. Recent developments in the control of flowering by photoperiod. *Amer. Naturalist* 84: 117–134.

Bosch, M. & N. M. Waser. 1999. Effects of local density on pollination and reproduction in *Delphinium nuttallianum* and *Aconitum columbianum* (Ranunculaceae). *Amer. J. Bot.* 86: 871–879.

Bose, A., S. D. Ray & M. Sahoo. 2012. Evaluation of analgesic and antioxidant potential of ethanolic extract of *Nymphaea alba* rhizome. *Oxidants Antioxidants Med. Sci.* 1: 217–223.

Bosi, G., A. M. Mercuri, C. Guarnieri & M. B. Mazzanti. 2009. Luxury food and ornamental plants at the 15th century A.D. Renaissance court of the Este family (Ferrara, northern Italy). *Veg. Hist. Archaeobot.* 18: 389–402.

Bossuyt, B. & M. Hermy. 2004. Seed bank assembly follows vegetation succession in dune slacks. *J. Veg. Sci.* 15: 449–456.

Bossuyt, B., E. Cosyns & M. Hoffmann. 2007. The role of soil seed banks in the restoration of dry acidic dune grassland after burning of *Ulex europaeus* scrub. *Appl. Veg. Sci.* 10: 131–138.

Bostick, P. E. 1977. Dissemination of some Florida plants by way of commercial peat shipments. *Castanea* 42: 106–108.

Boston, H. L. & M. S. Adams. 1986. The contribution of crassulacean acid metabolism to the annual productivity of two aquatic vascular plants. *Oecologia* 68: 615–622.

Boston, H. L. & M. S. Adams. 1987. Productivity, growth and photosynthesis of two small `isoetid' plants, *Littorella uniflora* and *Isoetes macrospora. J. Ecol.* 75: 333–350.

Botts, P. S., J. M. Lawrence, B. W. Witz & C. W. Kovach. 1990. Plasticity in morphology, proximate composition, and energy content of *Hygrophila polysperma* (Roxb.) Anders. *Aquat. Bot.* 36: 207–214.

Bouchard, A., S. Hay, Y. Bergeron & A. Leduc. 1987. Phytogeographical and life-form analysis of the vascular flora of Gros Morne National Park, Newfoundland, Canada. *J. Biogeogr.* 14: 343–358.

Boudreau, A., D. M. Cheng, C. Ruiz, D. Ribnicky, L. Allain, C. R. Brassieur, D. P. Turnipseed, W. T. Cefalu & Z. E. Floyd. 2014. Screening native botanicals for bioactivity: an interdisciplinary approach. *Nutrition* 30: S11–S16.

Boufford, D. E. 1997. 26. Urticaceae Jussieu. Nettle family. pp. 400–413. *In*: N. R. Morin (convening ed.), *Flora of North America North of Mexico, Vol. 3: Magnoliophyta: Magnoliidae and Hamamelidae.* Oxford University Press, New York.

Boulanger, J. & A. G. MacHutchon. 2005. *Black Bear Inventory Plan for Haida Gwaii, B.C.* Biodiversity Branch, B.C. Ministry of Water, Land and Air protection, Victoria, BC. 55 pp.

Boule, M. E. 1979. The vegetation of Fisherman Island, Virginia. *Castanea* 44: 98–108.

Boulger, G. S. 1875. The sleep of flowers. *Nature* 12: 513–514.

Boulos, L. & R. Al-Hasan. 1986. Ten species new to the flora of Kuwait and Bahrain. *Arab Gulf J. Sci. Res.* 4: 437–447.

Boutin, C. J. & L. Harper. 1991. A comparative study of the population dynamics of five species of *Veronica* in natural habitats. *J. Ecol.* 79: 199–221.

Boutin, C., H. B. Lee, E. T. Peart, P. S. Batchelor & R. J. Maguire. 2000. Effects of the sulfonylurea herbicide metsulfuron methyl on growth and reproduction of five wetland and terrestrial plant species. *Environ. Toxicol. Chem.* 19: 2532–2541.

Bowden, W. M. 1940. Diploidy, polyploidy, and winter hardiness relationships in the flowering plants. *Amer. J. Bot.* 27: 357–371.

Bowden, W. M. 1945. A list of chromosome numbers in higher plants. II. Menispermaceae to Verbenaceae. *Amer. J. Bot.* 32: 191–201.

Bowden, W. M. 1959. Phylogenetic relationships of twenty-one species of *Lobelia* L. section *Lobelia. Bull. Torrey Bot. Club* 86: 94–108.

Bowden, W. M. 1960. Cytotaxonomy of *Lobelia* L. section *Lobelia.* II. Four narrow-leaved species and five medium-flowered species. *Can. J. Genet. Cytol.* 2: 11–27.

Bowden, W. M. 1961. Interspecific hybridization in *Lobelia* L. section *Lobelia. Can. J. Bot.* 39: 1679–1693.

Bowerman, L. & R. D. Goos. 1991. Fungi associated with living leaves of *Nymphaea odorata. Mycologia* 83: 513–516.

Bowers, M. D. 1980. Unpalatability as a defense strategy of *Euphydryas phaeton* (Lepidoptera: Nymphalidae). *Evolution* 34: 586–600.

Bowers, M. D. 1984. Iridoid glycosides and host-plant specificity in larvae of the buckeye butterfly, *Junonia coenia* (Nymphalidae). *J. Chem. Ecol.* 10: 1567–1577.

Bowers, M. D., K. Boockvar & S. K. Collinge. 1993. Iridoid glycosides of *Chelone glabra* (Scrophulariaceae) and their sequestration by larvae of a sawfly, *Tenthredo grandis* (Tenthredinidae). *J. Chem. Ecol.* 19: 815–823.

Bowles, M. L. 1991. Some aspects of the status and ecology of seven rare wetland plant species in the Chicago region of northeastern Illinois. *Erigenia* 11: 52–66.

Bowles, M. L. & S. I. Apfelbaum. 1989. Effects of land use and stochastic events on the heart-leaved plantain (*Plantago cordata* Lam.) in an Illinois stream system. *Nat. Areas J.* 9: 90–101.

Bowles, M. L., P. D. Kelsey & J. L. McBride. 2005. Relationships among environmental factors, vegetation zones, and species richness in a North American calcareous prairie fen. *Wetlands* 25: 685–696.

Bowles, M., J. McBride, N. Stoynoff & K. Johnson. 1996. Temporal changes in vegetation composition and structure in a fire-managed prairie fen. *Nat. Areas J.* 16: 275–288.

Box, T. W., J. Powell & D. L. Drawe. 1967. Influence of fire on south Texas chaparral communities. *Ecology* 48: 955–961.

Boyd, C. E. 1968. Fresh-water plants: a potential source of protein. *Econ. Bot.* 22: 359–368.

Boyd, C. E. & P. S. McGinty. 1981. Percentage digestible dry matter and crude protein in dried aquatic weeds. *Econ. Bot.* 35: 296–299.

Boyd, C. E. & W. W. Walley. 1972. Production and chemical composition of *Saururus cernuus* L. at sites of different fertility. *Ecology* 53: 927–932.

Boyd, S. 2012. Crassulaceae. pp. 665–676. *In*: B. G. Baldwin, D. H. Goldman, D. J. Keil, R. Patterson, T. J. Rosatti & D. H. Wilken (eds.), *The Jepson Manual (2nd ed.).* University of California Press, Berkeley, CA.

Boyden, G. S., M. J. Donoghue & D. G. Howarth. 2012. Duplications and expression of *RADIALIS*-like genes in Dipsacales. *Int. J. Plant Sci.* 173: 971–983.

Boyer, K. E. & A. P. Burdick. 2010. Control of *Lepidium latifolium* (perennial pepperweed) and recovery of native plants in tidal marshes of the San Francisco Estuary. *Wetlands Ecol. Manag.* 18: 731–743.

Boyer, T. & R. Carter. 2011. Community analysis of green pitcher plant (*Sarracenia oreophila*) bogs in Alabama. *Castanea* 76: 364–376.

Boyle, M. F., R. K. Peet, T. R. Wentworth & M. P. Schafale. 2007. *Natural Vegetation of the Carolinas: Classification and Description of Plant Communities of Bladen County, NC and Vicinity.* A report prepared for the Ecosystem Enhancement Program, North Carolina Department of Environment and Natural Resources, in partial fulfillments of contract D07042. University of North Carolina, Chapel Hill, North Carolina. 58 pp.

Boyle, M. F., R. K. Peet, T. R. Wentworth, M. P. Schafale & M. Lee. 2009. *Natural Vegetation of the Carolinas: Classification and Description of Plant Communities of the Upper Tar, Roanoke, Meherrin, Chowan, and Cashie Rivers*. A report prepared for the Ecosystem Enhancement Program, North Carolina Department of Environment and Natural Resources, in partial fulfillments of contract D07042. Version 1. May 19, 2009. University of North Carolina, Chapel Hill, NC. 67 pp.

Boylen, C. W., L. W. Eichler & J. D. Madsen. 1999. Loss of native aquatic plant species in a community dominated by Eurasian watermilfoil. *Hydrobiologia* 415: 207–211.

Brackley-Tolman, F. 2001. *The Alpine Garden, Mount Washington, New Hampshire*. New England Wild Flower Society, Framingham, MA. Available online: http://www.newfs.org/powerofplace/alpine.html (accessed February 8, 2006).

Bradley, K. A. & G. D. Gann. 1999. *Status Summaries of 12 Rockland Plant Taxa in Southern Florida*. Institute for Regional Conservation, Miami, FL. 80 pp.

Bradley, P. M. & E. L. Dunn. 1989. Effects of sulfide on the growth of three salt marsh halophytes of the southeastern United States. *Amer. J. Bot.* 76: 1707–1713.

Bragazza, L. & R. Gerdol. 2002. Are nutrient availability and acidity-alkalinity gradients related in *Sphagnum*-dominated peatlands? *J. Veg. Sci.* 13: 473–482.

Bram, M. R. 1998. Sex expression, sex-specific traits, and inbreeding depression in freshwater and salt marsh populations of *Amaranthus cannabinus* (L.) Sauer, a dioecious annual. PhD dissertation. Rutgers University, New Brunswick, NJ.

Bram, M. R. 2002. Effects of inbreeding in three populations of the dioecious annual *Amaranthus cannabinus* (Amaranthaceae). *J. Torrey Bot. Soc.* 129: 298–310.

Bram, M. R. & J. A. Quinn. 2000. Sex expression, sex-specific traits, and the effects of salinity on growth and reproduction of *Amaranthus cannabinus* (Amaranthaceae), a dioecious annual. *Amer. J. Bot.* 87: 1609–1618.

Brambilla, D. P. & B. C. Sutton. 1969. *Host Index of Species Deposited in the Mycological Herbarium (WINF(M)) of the Forest Research Laboratory, Winnepeg, Manitoba (February 1, 1969)*. Internal Report MS-90. Forestry Branch, Forest Research Laboratory, Winnepeg, MB. 104 pp.

Brandão, M. G. L., A. U. Krettli, L. S. R. Soares, C. G. C. Nery & H. C. Marinuzzi. 1997. Antimalarial activity of extracts and fractions from *Bidens pilosa* and other *Bidens* species (Asteraceae) correlated with the presence of acetylene and flavonoid compounds. *J. Ethnopharmacol.* 57: 131–138.

Brändel, M. 2004. The role of temperature in the regulation of dormancy and germination of two related summer-annual mudflat species. *Aquat. Bot.* 79: 15–32.

Brändel, M. 2006. Effect of temperatures on dormancy and germination in three species in the Lamiaceae occurring in northern wetlands. *Wetlands Ecol. Manag.* 14: 11–28.

Brandt, R., M. Lomonosova, K. Weising, N. Wagner & H. Freitag. 2015. Phylogeny and biogeography of *Suaeda* subg. *Brezia* (Chenopodiaceae/Amaranthaceae) in the Americas. *Plant Syst. Evol.* 301: 2351–2375.

Brantjes, N. B. M. & O. C. De Vos. 1981. The explosive release of pollen in flowers of *Hyptis* (Lamiaceae). *New Phytol.* 87: 425–430.

Bräuchler, C., H. Meimberg, T. Abele & G. Heubl. 2005. Polyphyly of the genus *Micromeria* (Lamiaceae)—evidence from cpDNA sequence data. *Taxon* 54: 639–650.

Bräuchler, C., H. Meimberg & G. Heubl. 2006. New names in Old World *Clinopodium*—the transfer of the species of *Micromeria* sect. *Pseudomelissa* to *Clinopodium*. *Taxon* 55: 977–981.

Braun, A. F. 1940. Aster and goldenrod seed-feeding species of *Coleophora* (Lepidoptera). *Can. Entomol.* 72: 178–182.

Braun, E. L. 1928. Glacial and post-glacial plant migrations indicated by relic colonies of southern Ohio. *Ecology* 9: 284–302.

Braun, E. L. 1935. The vegetation of Pine Mountain, Kentucky: an analysis of the influence of soils and slope exposure as determined by geological structure. *Amer. Midl. Naturalist* 16: 517–565.

Braun, E. L. 1936. Notes on root behavior of certain trees and shrubs of the Illinoian till plain of southwestern Ohio. *Ohio J. Sci.* 36: 141–146.

Braun, E. L. 1937. A remarkable colony of coastal plain plants on the Cumberland Plateau in Laurel County, Kentucky. *Amer. Midl. Naturalist* 18: 363–366.

Braun, U., N. Ale-Agha, A. Bolay, H. Boyle, U. Brielmaier-Liebetanz, D. Emgenbroich, J. Kruse & V. Kummer. 2013. New records of powdery mildew fungi (Erysiphaceae). *Schlechtendalia* 19: 39–46.

Bravo, H. R., S. V. Copaja & J. S. Martin. 2004. Contents of 1,4-benzoxazin-3-ones and 2-benzoxazolinone from *Stenandrium dulce* (Nees). *J. Biosci.* 59: 177–180.

Bray, J. R. 1957. Climax forest herbs in prairie. *Amer. Midl. Naturalist* 58: 434–440.

Breckpot, C. 1997. *Aldrovanda vesiculosa*: description, distribution, ecology and cultivation. *Carniv. Plant Newslett.* 26: 73–82.

Breden, F. & M. J. Wade. 1989. Selection within and between kin groups of the imported willow leaf beetle. *Amer. Naturalist* 134: 35–50.

Breden, T. F., Y. Alger, K. S. Walz & A. G. Windisch. 2001. *Classification of Vegetation Communities of New Jersey: Second Iteration*. Association for Biodiversity Information and New Jersey Natural Heritage Program, Office of Natural Lands Management, Division of Parks and Forestry, New Jersey Department of Environmental Protection, Trenton, NJ. 230 pp.

Bremer, B. 1996. Combined and separate analyses of morphological and molecular data in the plant family Rubiaceae. *Cladistics* 12: 21–40.

Bremer, B. & R. K. Jansen. 1991. Comparative restriction site mapping of chloroplast DNA implies new phylogenetic relationships within Rubiaceae. *Amer. J. Bot.* 78: 198–213.

Bremer, B. & J.-F. Manen. 2000. Phylogeny and classification of the subfamily Rubioideae (Rubiaceae). *Plant Syst. Evol.* 225: 43–72.

Bremer, B., K. Andreasen & D. Olsson. 1995. Subfamilial and tribal relationships in the Rubiaceae based on *rbcL* sequence data. *Ann. Missouri Bot. Gard.* 82: 383–397.

Bremer, B., R. K. Jansen, B. Oxelman, M. Backlund, H. Lantz & K.-J. Kim. 1999. More characters or more taxa for a robust phylogeny—case study from the coffee family (Rubiaceae). *Syst. Biol.* 48: 413–435.

Bremer, B., K. Bremer, N. Heirdari, P. Erixon, R. G. Olmstead, M. Kallersjo, A. A. Anderberg & E. Barkhordarian. 2002. Phylogenetics of asterids based on 3 coding and 3 non-coding chloroplast DNA markers and the utility of non-coding DNA at higher taxonomic levels. *Mol. Phylogen. Evol.* 24: 274–301.

Bremer, K., A. Backlund, B. Sennblad, U. Swenson, K. Andreasen, M. Hjertson, J. Lundberg, M. Backlund & B. Bremer. 2001. A phylogenetic analysis of 100+ genera and 50+ families of euasterids based on morphological and molecular data with notes on possible higher level morphological synapomorphies. *Plant Syst. Evol.* 229: 137–169.

Brenckle, J. F. 1918. North Dakota Fungi: II. *Mycologia* 10: 199–221.

Brenckle, J. F. 1941. Notes on *Polygonum* (*Avicularia*). *Bull. Torrey Bot. Club* 68: 491–495.

Brenner, F. J., R. B. Kelly & J. Kelly. 1982. Mammalian community characteristics on surface mine lands in Pennsylvania. *Environ. Manag.* 6: 241–249.

Brescacin, C. R. 2010. The role of the feral pig (*Sus scrofa*) as a disturbance agent and seed disperser in central Florida's natural lands. MS thesis. University of Central Florida, Orlando, FL. 120 pp.

Bret-Harte, M. S., G. R. Shaver, J. P. Zoerner, J. F. Johnstone, J. L. Wagner, A. S. Chavez, R. F. Gunkelman, S. C. Lippert & J. A. Laundre. 2001. Developmental plasticity allows *Betula nana* to dominate tundra subjected to an altered environment. *Ecology* 82: 18–32.

Brett, R. B., K. Klinka & H. Qian. 2001. *Classification of High-Elevation, Non-Forested Plant Communities in Coastal British Columbia*. Scientia Silvica Extension Series number 29. Forest Sciences Department, University of British Columbia, Vancouver, BC.

Breuss, O. 2016. *Byssoloma maderense* is not endemic to Macaronesia. *Evansia* 33: 54–62.

Brewbaker, J. L. 1967. The distribution and phylogenetic significance of binucleate and trinucleate pollen grains in the angiosperms. *Amer. J. Bot.* 54: 1069–1083.

Brewer, J. S. 1999a. Effects of competition, litter, and disturbance on an annual carnivorous plant (*Utricularia juncea*). *Plant Ecol.* 140: 159–165.

Brewer, J. S. 1999b. Effects of fire, competition and soil disturbances on regeneration of a carnivorous plant (*Drosera capillaris*). *Amer. Midl. Naturalist* 141: 28–42.

Brewer, J. S. 2003. Why don't carnivorous pitcher plants compete with non-carnivorous plants for nutrients? *Ecology* 84: 451–462.

Brewer, J. S. 2005. The lack of favorable responses of an endangered pitcher plant to habitat restoration. *Restorat. Ecol.* 13: 710–717.

Brewer, J. S. & J. B. Grace. 1990. Plant community structure in an oligohaline tidal marsh. *Vegetatio* 90: 93–107.

Brewer, J. S., J. M. Levine & M. D. Bertness. 1997. Effects of biomass removal and elevation on species richness in a New England salt marsh. *Oikos* 80: 333–341.

Brewer, J. S., D. J. Baker, A. S. Nero, A. L. Patterson, R. S. Roberts & L. M. Turner. 2011. Carnivory in plants as a beneficial trait in wetlands. *Aquat. Bot.* 94: 62–70.

Brewer, M. J., J. S. Armstrong, E. G. Medrano & J. F. Esquivel. 2012. Association of Verde plant bug, *Creontiades signatus* (Hemiptera: Miridae), with cotton boll rot. *J. Cotton Sci.* 16: 144–151.

Bridges, E. L. & S. L. Orzell. 1989. Longleaf pine communities of the west Gulf coastal plain. *Nat. Areas J.* 9: 246–262.

Bried, J. T. & S. Krotzer. 2005. New species records for Mississippi: an expected dragonfly and an unexpected damselfly. *J. Mississippi Acad. Sci.* 50: 233–234.

Bried, J. T., B. D. Herman & G. N. Ervin. 2007. Umbrella potential of plants and dragonflies for wetland conservation: a quantitative case study using the umbrella index. *J. Appl. Ecol.* 44: 833–842.

Briggs, D., M. Block & S. Jennings. 1989. The possibility of determining the age of colonies of clonally propagating herbaceous species from historic records: the case of *Aster novi-belgii* L. (first recorded as *A. salignus* Willd.) at Wicken Fen Nature Reserve, Cambridgeshire, England. *New Phytol.* 112: 577–584.

Briggs, M. & R. Fitzgerald. 1986. Aliens and adventives. *Downingia elegans* refound in East Sussex. *BSBI News* 44: 20–21.

Brink, D. E. & J. A. Woods. 1997. 15. *Aconitum* Linnaeus. pp. 191–195. *In*: N. R. Morin (convening ed.), *Flora of North America North of Mexico, Vol. 3: Magnoliophyta: Magnoliidae and Hamamelidae*. Oxford University Press, New York.

Britton, D. M. 1951. Cytogenetic studies on the Boraginaceae. *Brittonia* 7: 233–266.

Britton, N. L. 1887. Note on the flowers of *Populus heterophylla*, L. *Bull. Torrey Bot. Club* 14: 114–115.

Broberg, C. L., J. H. Borden & L. M. Humble. 2001. Host range, attack dynamics, and impact of *Cryptorhynchus lapathi* (Coleoptera: Curculionidae) on *Salix* (Salicaceae) spp. *Can. Entomol.* 133: 119–130.

Brochmann, C. & A. Håpnes. 2001. Reproductive strategies in some arctic *Saxifraga* (Saxifragaceae), with emphasis on the narrow endemic *S. svalbardensis* and its parental species. *Bot. J. Linn. Soc.* 137: 31–49.

Brock, M. A. & K. H. Rogers. 1998. The regeneration potential of the seed bank of an ephemeral floodplain in South Africa. *Aquat. Bot.* 61: 123–135.

Brock, S. C., D. H. Arner & D. E. Steffen. 1994. Seed yields of four moist-soil plants on Noxubee National Wildlife Refuge. *Proc. Annu. Conf. S. E. Assoc. Fish. Wildl. Agencies* 48: 38–47.

Brock, T. D. 1970. Photosynthesis by algal epiphytes of *Utricularia* in Everglades National Park. *Bull. Mar. Sci.* 20: 952–956.

Brock, T. C. M. 1984. Aspects of the decomposition of *Nymphoides peltata* (Gmel.) O. Kuntze (*Menyanthaceae*). *Aquat. Bot.* 19: 131–156.

Brock, T. C. M., G. H. P. Arts, I. L. M. Goossen & A. H. M. Rutenfrans. 1983. Structure and annual biomass production of *Nymphoides peltata* (Gmel.) O. Kuntze (Menyanthaceae). *Aquat. Bot.* 17: 167–188.

Brockett, B. L. 1970. Chromosome number of *Sium suave*. *Ohio J. Sci.* 70: 122.

Brockington, S. F., R. Alexandre, J. Ramdial, M. J. Moore, S. Crawley, A. Dhingra, K. Hilu, D. E. Soltis & P. S. Soltis. 2009. Phylogeny of the Caryophyllales sensu lato: revisiting hypotheses on pollination biology and perianth differentiation in the core Caryophyllales. *Int. J. Plant Sci.* 170: 627–643.

Brockington, S., P. Dos Santos, B. Glover & L. R. De Craene. 2013. Androecial evolution in Caryophyllales in light of a paraphyletic Molluginaceae. *Amer. J. Bot.* 100: 1757–1778.

Brodie, W. 1909. Galls found in the vicinity of Toronto—No. 3. *Can. Entomol.* 41: 157–160.

Broek, T., R. Diggelen & R. Bobbink. 2005. Variation in seed buoyancy of species in wetland ecosystems with different flooding dynamics. *J. Veg. Sci.* 16: 579–586.

Brooks, R. A. & S. S. Bell. 2005. The distribution and abundance of *Sphaeroma terebrans*, a wood-boring isopod of red mangrove (*Rhizophora mangle*) habitat within Tampa Bay. *Bull. Mar. Sci.* 76: 27–46.

Brookshire, E. N. J., J. B. Kauffman, D. Lytjen & N. Otting. 2002. Cumulative effects of wild ungulate and livestock herbivory on riparian willows. *Oecologia* 132: 559–566.

Broome, C. R. 1978. Chromosome numbers and meiosis in North and Central American species of *Centaurium* (Gentianaceae). *Syst. Bot.* 3: 299–312.

Brotherson, J. D. 1983. Species composition, distribution, and phytosociology of Kalsow Prairie, a mesic tall-grass prairie in Iowa. *Great Basin Naturalist* 43: 137–167.

Brotherson, J. D. & S. J. Barnes. 1984. Habitat relationships of *Glaux maritima* in central Utah. *Great Basin Naturalist* 44: 299–309.

Brotherson, J. D., L. A. Szyska & W. E. Evenson. 1980. Poisonous plants of Utah. *Great Basin Naturalist* 40: 229–253.

Brouillet, L. 2006a. 121. *Arctanthemum* (Tzvelev) Tzvelev. pp. 535–537. *In*: Flora of North America Editorial Committee (ed.), *Flora of North America North of Mexico: Vol. 19, Magnoliophyta: Asteridae (in part): Asteraceae, Part 1.* Oxford University Press, New York.

Brouillet, L. 2006b. 193. *Eurybia* (Cassini) Cassini. pp. 365–382. *In*: Flora of North America Editorial Committee (ed.), *Flora of North America North of Mexico: Vol. 20, Magnoliophyta: Asteridae (in part): Asteraceae, Part 2.* Oxford University Press, New York.

Brouillet, L. 2006c. 149. *Oclemena* Greene. pp. 78–81. *In*: Flora of North America Editorial Committee (ed.), *Flora of North America North of Mexico: Vol. 20, Magnoliophyta: Asteridae (in part): Asteraceae, Part 2.* Oxford University Press, New York.

Brouillet, L. 2008. The taxonomy of North American Loti (Fabaceae: *Loteae*): new names in *Acmispon* and *Hosackia*. *J. Bot. Res. Inst. Texas* 24: 387–394.

Brouillet, L. 2009. 1. *Cascadia* A. M. Johnson. pp. 48–49. *In*: N. R. Morin (convening ed.), *Flora of North America North of Mexico, Vol. 8: Magnoliophyta: Paeoniaceae to Ericaceae.* Oxford University Press, New York.

Brouillet, L. & P. E. Evander. 2009a. 2. *Micranthes* Haworth. pp. 49–70. *In*: N. R. Morin (convening ed.), *Flora of North America North of Mexico, Vol. 8: Magnoliophyta: Paeoniaceae to Ericaceae.* Oxford University Press, New York.

Brouillet, L. & P. E. Evander. 2009b. 23. *Saxifraga* Linnaeus. pp. 132–146. *In*: N. R. Morin (convening ed.), *Flora of North America North of Mexico, Vol. 8: Magnoliophyta: Paeoniaceae to Ericaceae.* Oxford University Press, New York.

Brouillet, L. & J.-P. Simon. 1981. An ecogeographical analysis of the distribution of *Aster acuminatus* Michaux and *A. nemoralis* Aiton (Asteraceae: Astereae). *Rhodora* 83: 521–550.

Brouillet, L. & J.-P. Simon. 1998. Adaptation and acclimation of higher plants at the enzyme level: thermal properties of NAD malate dehydrogenase of two species of *Aster* (Asteraceae) and their hybrid adapted to contrasting habitats. *Can. J. Bot.* 58: 1474–1481.

Brouillet, L., S. Hay, P. Turcotte & A. Bouchard. 1998. La flore vasculaire alpine du plateau Big Level, au parc national du Gros-Morne, Terre-Neuve. *Géogr. Phys. Quatern.* 52: 1–19.

Brouillet, L., J. C. Semple, G. A. Allen, K. L. Chambers & S. D. Sundberg. 2006. 214. *Symphyotrichum* Nees. pp. 465–539. *In*: Flora of North America Editorial Committee (ed.), *Flora of North America North of Mexico: Vol. 20, Magnoliophyta: Asteridae (in part): Asteraceae, Part 2.* Oxford University Press, New York.

Brouwer, E., R. Bobbink & J. G. M. Roelofs. 2002. Restoration of aquatic macrophyte vegetation in acidified and eutrophied softwater lakes: an overview. *Aquat. Bot.* 73: 405–431.

Brown, A. 1879. Ballast plants in New York city and its vicinity. *Bull. Torrey Bot. Club* 59: 353–360.

Brown, A. M. & C. Bledsoe. 1996. Spatial and temporal dynamics of mycorrhizas in *Jaumea carnosa*, a tidal saltmarsh halophyte. *J. Ecol.* 84: 703–715.

Brown, A. O. & J. N. McNeil. 2006. Fruit production in cranberry (Ericaceae: *Vaccinium macrocarpon*): a bet-hedging strategy to optimize reproductive effort. *Amer. J. Bot.* 93: 910–916.

Brown, B. J. & R. J. Mitchell. 2001. Competition for pollination: effects of pollen of an invasive plant on seed set of a native congener. *Oecologia* 129: 43–49.

Brown, B., R. Mitchell & S. Graham. 2002. Competition for pollination between an invasive species (purple loosestrife) and a native congener. *Ecology* 83: 2328–2336.

Brown, B. T. & J. W. Goertz. 1978. Reproduction and nest site selection by red-winged blackbirds in north Louisiana. *Wilson Bull.* 90: 261–270.

Brown, C. A. 1943. Vegetation and lake level correlations at Catahoula Lake, Louisiana. *Geogr. Rev.* 33: 435–445.

Brown, C. P. 1946. Food of Maine ruffed grouse by seasons and cover types. *J. Wildl. Manag.* 10: 17–28.

Brown, C. R. & S. K. Jain. 1979. Reproductive system and pattern of genetic variation in two *Limnanthes* species. *Theor. Appl. Genet.* 54: 181–190.

Brown, C. S. & R. L. Bugg. 2001. Effects of established perennial grasses on introduction of native forbs in California. *Restorat. Ecol.* 9: 38–48.

Brown, H. D. 1987. Aquatic macrophytes of Lake Mize, Florida, 1968–1980. *Bull. Torrey Bot. Club* 114: 180–182.

Brown, L. E. & S. J. Marcus. 1998. Notes on the flora of Texas with additions and other significant records. *Sida* 18: 315–324.

Brown, L. E., P. V. Roling, J. L. Aplaca & E. Keith. 2011. Notes on the flora of Texas, Arkansas, and Louisiana, with additions and other records. IV. *Phytoneuron* 2011–18: 1–12.

Brown, M. T. & R. E. Tighe. 1992. *Vegetation Composition and Cover at Sunnyhill Farm.* Special Publication SJ93-SP8. CFW-92-01. Center for Wetlands and Water Resources, University of Florida, Gainesville, FL. 38 pp.

Brown, O. P. 1878. *The Complete Herbalist.* Published by the author, Jersey City, NJ.

Brown, R. T. 1967. Influence of naturally occurring compounds on germination and growth of jack pine. *Ecology* 48: 542–546.

Brown, W. L. 1999. Evaluation of cattle grazing effects on floristic composition in eastern Washington vernal pools. MS thesis. University of Washington, Seattle, WA. 142 pp.

Broyles, P. 1987. A flora of Vina Plains Preserve, Tehama County, California. *Madroño* 34: 209–227.

Bruederle, L. P., N. Vorsa & J. R. Ballington. 1991. Population genetic structure in diploid blueberry *Vaccinium* section *Cyanococcus* (Ericaceae). *Amer. J. Bot.* 78: 230–237.

Brumback, W. E. 1989. Notes on propagation of rare New England species. *Rhodora* 91: 154–162.

Bruneau, A., F. Forest, P. S. Herendeen, B. B. Klitgaard & G. P. Lewis. 2001. Phylogenetic relationships in the Caesalpinioideae (Leguminosae) as inferred from chloroplast *trnL* intron sequences. *Syst. Bot.* 26: 487–514.

Bruneau, A., J. R. Starr & S. Joly. 2007. Phylogenetic relationships in the genus *Rosa*: new evidence from chloroplast DNA sequences and an appraisal of current knowledge. *Syst. Bot.* 32: 366–378.

Bruneau, A., M. Mercure, G. P. Lewis & P. S. Herendeen. 2008. Phylogenetic patterns and diversification in the caesalpinioid legumes. *Botany* 86: 697–718.

Bruno, J. F. & C. W. Kennedy. 2000. Patch-size dependent habitat modification and facilitation on New England cobble beaches by *Spartina alterniflora*. *Oecologia* 122: 98–108.

Bruno, M., M. Cruciata, M. L. Bondi, F. Piozzi, M. C. de la Torre, B. Rodriguez & O. Servettaz. 1998. Neo-clerodane diterpenoids from *Scutellaria lateriflora*. *Phytochemistry* 48: 687–691.

Brunsfeld, S. J. & C. K. Anttila. 2004. Complexities in the phylogeny of *Salix* (Salicaceae). Botany 2004 meeting (abstract).

Brunsfeld, S. J., D. E. Soltis & P. S. Soltis. 1991. Patterns of genetic variation in *Salix* section *Longifoliae* (Salicaceae). *Amer. J. Bot.* 78: 855–869.

Bruun, H. H., S. Österdahl, J. Moen & A. Angerbjörn. 2005. Distinct patterns in alpine vegetation around dens of the arctic fox. *Ecography* 28: 81–87.

Bryant, J. P., F. S. Chapin, III & D. R. Klein. 1983. Carbon/nutrient balance of boreal plants in relation to vertebrate herbivore. *Oikos* 40: 357–368.

Bryson, C. T., W. I. McDearman & K. L. Gordon. 1988. *Carex exilis* Dewey (Cyperaceae) in Mississippi bogs. *Sida* 13: 171–175.

Bryson, C. T., J. R. MacDonald & R. Warren. 1994. Notes on *Carex* (Cyperaceae), with *C. godfreyi* new to Alabama and *C. communis* and *C. scoparia* new to Mississippi. *Sida* 16: 355–360.

Bryson, C. T., V. L. Maddox & R. Carter. 2008. Spread of Cuban club-rush (*Oxycaryum cubense*) in the southeastern United States. *Invasive Plant Sci. Manag.* 1: 326–329.

Brysting, A. K., P. J. Scott & S. G. Aiken. 2001. *Caryophyllaceae of the Canadian Arctic Archipelago: Descriptions, Illustrations, Identification, and Information Retrieval.* Version: 29th April 2003. Available online: https://nature.ca/aaflora/data/www/ca.htm

Buchmann, S., L. D. Adams, A. D. Howell & M. Weiss. 2010. *A Study of Insect Pollinators Associated with DoD TER-S Flowering Plants, Including Identification of Habitat Types Where they Co-Occur by Military Installation in the Western United States.* Project Number 08-391. Legacy Resource Management Program, Department of Defense, Arlington, VA. 67 pp.

Buc'hoz, P.-J. 1785. Catalogue latin et françois des arbres et arbustes qu'on peut cultiveren France: et qui peuvent résister en pleine terre pendant l'hiver; Auquel on a joint la liste des Plantes nouvelles graveés & publiées tout récemment. A Londres. 101 pp.

Buchsbaum, R. N., J. Catena, E. Hutchins & M. J. James-Pirri. 2006. Changes in salt marsh vegetation, *Phragmites australis*, and nekton in response to increased tidal flushing in a New England salt marsh. *Wetlands* 26: 544–557.

Buchsbaum, R. N., L. A. Deegan, J. Horowitz, R. H. Garritt, A. E. Giblin, J. P. Ludlam & D. H. Shull. 2009. Effects of regular salt marsh haying on marsh plants, algae, invertebrates and birds at Plum Island Sound, Massachusetts. *Wetlands Ecol. Manag.* 17: 469–487.

Buck, J. J. 2004. Temporal vegetation dynamics in central and northern California vernal pools. MS thesis. University of California, Davis, CA. 49 pp.

Buckallew, R. R. 2007. Comparison of bare root vs potted plants, species selection, and caging types for restoration of a prairie wetland, and quantitative analysis and descriptive survey of plant communities and assocaiations at Lewisville Lake Environmental Learning Area (LLELA), Lewisville, TX. PhD dissertation. University of North Texas, Denton, TX. 531 pp.

Buck-Diaz, J. & J. Evens. 2011. *Carrizo Plain National Monument Vegetation Classification and Mapping Project.* California Native Plant Society, Vegetation Program, Sacramento, CA. 59 pp.

Buck-Diaz, J., S. Batiuk & J. M. Evens. 2012. *Vegetation Alliances and Associations of the Great Valley Ecoregion, California.* California Native Plant Society, Vegetation Program, Sacramento, CA. 473 pp.

Buckingham, G. R. & C. A. Bennett. 2001. Life history and laboratory host range tests of *Parapoynx seminealis* (Walker) (Crambidae: Nymphulinae) in Florida, U.S.A. *J. Lepid. Soc.* 55: 111–118.

Buckley, N. E. 2012. Mating system biology of the Florida native plant: *Illicium parviflorum*. MS thesis. University of Tennessee, Knoxville, TN. 40 pp.

Bucyanayandi, J.-D. & J.-M. Bergeron. 1990. Effects of food quality on feeding patterns of meadow voles (*Microtus pennsylvanicus*) along a community gradient. *J. Mammalogy* 71: 390–396.

Buddell II, G. F. & J. W. Thieret. 1997. 6. Saururaceae E. Meyer. pp. 37–38. *In*: N. R. Morin (convening ed.), *Flora of North America North of Mexico, Vol. 3: Magnoliophyta: Magnoliidae and Hamamelidae.* Oxford University Press, New York.

Budzianowski, J., L. Skrzypczak & K. Kukulczanka. 1993. Phenolic compounds of *Drosera intermedia* and *D. spathulata* from *in vitro* cultures. *Acta Hort.* 330: 277–280.

Buell, M. F. 1946. Jerome Bog, a peat-filled "Carolina bay." *Bull. Torrey Bot. Club* 73: 24–33.

Buell, M. F. & H. F. Buell. 1975. Moat bogs in the Itasca Park area, Minnesota. *Bull. Torrey Bot. Club* 102: 6–9.

Buell, M. F. & W. A. Wistendahl. 1955. Flood plain forests of the Raritan River. *Bull. Torrey Bot. Club* 82: 463–472.

Buhler, D. D. & R. G. Hartzler. 2001. Emergence and persistence of seed of velvetleaf, common waterhemp, woolly cupgrass, and giant foxtail. *Weed Sci.* 49: 230–235.

Bukaveckas, P. A. 1988. Effects of calcite treatment on primary producers in acidified Adirondack lakes—response of macrophyte communities. *Lake Reservoir Manag.* 4: 107–113.

Bull, L. B., C. C. J. Culvenor & A. T. Dick. 1968. Plant species containing pyrrolizidine alkaloids. pp. 234–248. *In*: L. B. Bull et al. (eds.), *The Pyrrolizidine Alkaloids.* Elsevier, New York.

Bullitta, S., G. Piluzza & L. Viegi. 2007. Plant resources used for traditional ethnoveterinary phytotherapy in Sardinia (Italy). *Genet. Resour. Crop Evol.* 54: 1447–1464.

Bunsawat, J., N. E. Elliott, K. L. Hertweck, E. Sproles & L. A. Alice. 2004. Phylogenetics of *Mentha* (Lamiaceae): evidence from chloroplast DNA sequences. *Syst. Bot.* 29: 959–964.

Burbott, A. J. & W. D. Loomis. 1967. Effects of light and temperature on the monoterpenes of peppermint. *Plant Physiol.* 42: 20–28.

Burg, M. E., D. R. Tripp & E. S. Rosenberg. 1980. Plant associations and primary productivity of the Nisqually salt marsh on southern Puget Sound, Washington. *Northwest Sci.* 54: 222–236.

Burge, M. N. & I. Isaac. 1977. Predisposition of *Aster* to *Phialophora* wilt. *Ann. Appl. Biol.* 86: 353–358.

Burger, J. 1979. Resource partitioning: nest site selection in mixed species colonies of herons, egrets and ibises. *Amer. Midl. Naturalist* 101: 191–210.

Burger, O., O. Itzhak, M. Tabak, E. I. Weiss, N. Sharon & I. Neeman. 2000. A high molecular mass constituent of cranberry juice inhibits *Helicobacter pylori* adhesion to human gastric mucus. *FEMS Immunol. Med. Microbiol.* 29: 295–301.

Burk, C. J. 1977. A four year analysis of vegetation following an oil spill in a freshwater marsh. *J. Appl. Ecol.* 14: 515–522.

Burke, H. R. 1963. New species of Texas weevils, with notes on others (Coleoptera, Curculionidae). *Southwest. Naturalist* 8: 162–172.

Burke, H. R., J. A. Jackman & M. Rose. 1994. Insects *Associated with Woody Ornamental Plants in Texas.* Texas Agricultural Extension Service, Texas A&M University, College Station, TX. 166 pp.

Burke, J. L., J. L. Golden, S. Dobing & J. F. Bain. 2012. Phylogeographic patterns in *Packera subnuda* reveal an east-west divergence along the continental divide and reduced haplotype diversity in northern populations. *Botany* 90: 473–480.

Burke, J. M., A. Sanchez, K. Kron & M. Luckow. 2010. Placing the woody tropical genera of Polygonaceae: a hypothesis of character evolution and phylogeny. *Amer. J. Bot.* 97: 1377–1390.

Burke, M. K., S. L. King, D. Gartner & M. H. Eisenbies. 2003. Vegetation, soil, and flooding relationships in a blackwater floodplain forest. *Wetlands* 23: 988–1002.

Burks, K. 1994. The effects of population size and density on the pollination biology of a threatened thistle (*Cirsium vinaceum*). MS thesis. Department of Biology, New Mexico State University, Las Cruces, NM. 99 pp.

Burks, K. C. 2002. *Nymphoides cristata* (Roxb.) Kuntze, a recent adventive expanding as a pest plant in Florida. *Castanea* 67: 206–211.

Burks, K. C. & D. F. Austin. 1999. *Ipomoea asarifolia* (Convolvulaceae), another potential exotic pest in the United States. *Sida* 14: 1267–1272.

Burn, C. R. & P. A. Friele. 1989. Geomorphology, vegetation succession, soil characteristics and permafrost in retrogressive thaw slumps near Mayo, Yukon Territory. *Arctic* 42: 31–40.

Burnett, W. C., S. B. Jones & T. J. Mabry. 1977. Evolutionary implications of sesquiterpene lactones in *Vernonia* (Compositae) and mammalian herbivores. *Taxon* 26: 203–207.

Burns, G. P. 1911. A botanical survey of the Huron River valley. VIII. Edaphic conditions in peat bogs of southern Michigan. *Bot. Gaz.* 52: 105–125.

Burns, G. W. 1942. The taxonomy and cytology of *Saxifraga pensylvanica* L. and related forms. *Amer. Midl. Naturalist* 28: 127–160.

Burns, R. M. & B. H. Honkala. 1990. *Silvics of North America: Vol. 2. Hardwoods.* Agriculture Handbook 654. U. S. Department of Agriculture, Forest Service, Washington, DC. 877 pp.

Burnside, O. C., R. G. Wilson, S. Weisberg & K. G. Hubbard. 1996. Seed longevity of 41 weed species buried 17 years in eastern and western Nebraska. *Weed Sci.* 44: 74–86.

Burr, C. A. 1979. The pollination ecology of *Sarracenia purpurea* in Cranberry Bog, Weybridge, Vermont (Addison Co.). MS thesis. Middlebury College, Middlebury, VT.

Burr, T. J. & M. N. Schroth. 1977. Occurrence of soft-rot *Erwinia* spp. in soil and plant material. *Phytopathology* 67: 1382–1387.

Burrell, C. C. 1997. *The Natural Water Garden: Pools, Ponds, Marshes and Bogs for Gardens Everywhere.* Brooklyn Botanic Garden, Brooklyn, NY. 112 pp.

Burrill, T. J. 1888. The Ustilagineæ, or smuts; with a list of Illinois species. *Proc. Amer. Soc. Microsc.* 10: 45–57.

Burrill, T. J. & F. S. Earle. 1887. Parasitic fungi of Illinois. Part II. *Illinois. Nat. Hist. Surv. Bull.* 2: 387–432.

Burrows, G. E. 1996. *Urechites lutea* (L) Britton toxicity in cattle. *Vet. Human. Toxicol.* 38: 313–314.

Burrows, G. E. & R. J. Tyrl. 2012. *Toxic Plants of North America.* Wiley-Blackwell, Ames, IA. 1390 pp.

Burton, T. M. 1977. Population estimates, feeding habits and nutrient and energy relationships of *Notophthalmus v. viridescens*, in Mirror Lake, New Hampshire. *Copeia* 1977: 139–143.

Busbee, W. S., W. H. Conner, D. M. Allen & J. D. Lanham. 2003. Composition and aboveground productivity of three seasonally flooded depressional forested wetlands in coastal South Carolina. *S. E. Naturalist (Steuben)* 2: 335–346.

Busch, D. E., W. F. Loftus & O. L. Bass Jr. 1998. Long-term hydrologic effects on marsh plant community structure in the southern Everglades. *Wetlands* 18: 230–241.

Busch, J. W. & L. Urban. 2011. Insights gained from 50 years of studying the evolution of self-compatibility in *Leavenworthia* (Brassicaceae). *Evol. Biol.* 38: 15–27.

Busey, P. & D. L. Johnston. 2006. Impact of cultural factors on weed populations in St. Augustinegrass turf. *Weed Sci.* 54: 961–967.

Bush, G. L. & M. D. Huettel. 1969. Cytogenetics and description of a new North American species of the Neotropical genus *Cecidocharella* (Diptera: Tephritidae). *Ann. Entomol. Soc. Amer.* 63: 88–91.

Bussey, III, R. O., A. A. Sy-Cordero, M. Figueroa, F. S. Carter, J. O. Falkinham, III, N. H. Oberlies & N. B. Cech. 2014. Antimycobacterial furofuran lignans from the roots of *Anemopsis californica. Plant Med.* 80: 498–501.

Bussey, III, R. O., A. Kaur, D. A. Todd, J. M. Egan, T. El-Elimat, T. N. Graf, H. A. Raja, N. H. Oberlies & N. B. Cech. 2015. Comparison of the chemistry and diversity of endophytes isolated from wild-harvested and greenhouse-cultivated yerba mansa (*Anemopsis californica*). *Phytochem. Lett.* 11: 202–208.

Butkus, V. & K. Pliszka. 1993. The highbush blueberry: a new cultivated species. *Acta Hort.* 346: 81–86.

Buys, M. H. & H. H. Hilger. 2003. Boraginaceae cymes are exclusively scorpioid and not helicoid. *Taxon* 52: 719–724.

Byers, D. L. 1995. Pollen quantity and quality as explanations for low seed set in small populations exemplified by *Eupatorium* (Asteraceae). *Amer. J. Bot.* 82: 1000–1006.

Byers, D. L. 1998. Effect of cross proximity on progeny fitness in a rare and a common species of *Eupatorium* (Asteraceae). *Amer. J. Bot.* 85: 644–653.

Byers, D. L. & T. R. Meagher. 1997. A comparison of demographic characteristics in a rare and a common species of *Eupatorium. Ecol. Appl.* 7: 519–530.

Byers, E. A., J. P. Vanderhorst & B. P. Streets. 2007. *Classification and Conservation Assessment of High Elevation Wetland Communities in the Allegheny Mountains of West Virginia.* West Virginia Natural Heritage Program, West Virginia Division of Natural Resources, Wildlife Resources Section, Elkins, WV. 192 pp.

Byrd, G. V. 1984. Vascular vegetation of Buldir Island, Aleutian Islands, Alaska, compared to another Aleutian Island. *Arctic* 37: 37–48.

Byrd, K. B. & M. Kelly. 2006. Salt marsh vegetation response to edaphic and topographic changes from upland sedimentation in a pacific estuary. *Wetlands* 26: 813–829.

Bywater, M. & G. E. Wickens. 1984. New World species of the genus *Crassula. Kew Bull.* 39: 699–728.

Caccavari, M. & G. Fagúndez. 2010. Pollen spectra of honeys from the Middle Delta of the Paraná River (Argentina) and their environmental relationship. *Span. J. Agric. Res.* 8: 42–52.

Caceres, A., A. V. Alvarez, A. E. Ovando & B. E. Samayoa. 1991. Plants used in Guatemala for the treatment of respiratory diseases. 1. Screening of 68 plants against gram-positive bacteria. *J. Ethnopharmacol.* 31: 193–208.

Caesar, L. 1916. The imported willow and poplar borer or curculio (*Cryptorrhynchus lapathi*, L.). pp. 33–40. *In: 46th Annual Report of the Entomological Society of Ontario 1915.* Toronto, ON.

Cain, S. A. 1928. Plant succession and ecological history of a central Indiana swamp. *Bot. Gaz.* 86: 384–401.

Cain, S. A. & W. T. Penfound. 1938. Aceretum rubri: the red maple swamp forest of central Long Island. *Amer. Midl. Naturalist* 19: 390–416.

Cain, S. A. & J. V. Slater. 1948. The vegetation of Sodon Lake. *Amer. Midl. Naturalist* 40: 741–762.

Caires, K. G., C. A. Bobisud, S. P. Martin & T. T. Sekioka. 1992. In vitro callus production and plant regeneration from petioles of watercress (*Nasturtium microphyllum* Boenn. ex. Reichb.). *HortScience* 27: 697.

Calcagno, M. P., J. Coll, J. Lloria, F. Faini & M. E. Alonso-Amelot. 2002. Evaluation of synergism in the feeding deterrence of some furanocoumarins on *Spodoptera littoralis*. *J. Chem. Ecol.* 28: 175–191.

Calder, J. A. & D. B. O. Savile. 1960. Studies in Saxifragaceae: III. *Saxifraga odontoloma* and *lyallii*, and North American subspecies of *S. punctata*. *Can. J. Bot.* 38: 409–435.

Calero, M. M., C. La Casa, V. Motilva, A. López & C. Alarcón de la Lastra. 1996. Healing process induced by a flavonic fraction of *Bidens aurea* on chronic gastric lesion in rat. Role of angiogenesis and neutrophil inhibition. *Z. Naturf. C. J. Biosci.* 51: 570–577.

Calhoun, J. V. 1985. An annotated list of the butterflies and skipeprs of Lawrence County, Ohio. *J. Lepid. Soc.* 39: 284–298.

Calkins, W. W. 1879. January flora of the Indian River County, Florida. *Bot. Gaz.* 4: 242.

Callaghan, D. A. 1998. *Lythrum hyssopifolium* L. *J. Ecol.* 86: 1065–1072.

Callaway, J. C. & J. B. Zedler. 1998. Interactions between a salt marsh native perennial (*Salicornia virginica*) and an exotic annual (*Polypogon monspeliensis*) under varied salinity and hydroperiod. *Wetlands Ecol. Manag.* 5: 179–194.

Callaway, J. C., G. Sullivan & J. B. Zedler. 2003. Species-rich plantings increase biomass and nitrogen accumulation in a wetland restoration experiment. *Ecol. Appl.* 13: 1626–1639.

Callaway, R. M. 1994. Facilitative and interfering effects of *Arthrocnemum subterminale* on winter annuals. *Ecology* 75: 681–686.

Callaway, R. M. & L. King. 1996. Temperature-driven variation in substrate oxygenation and the balance of competition and facilitation. *Ecology* 77: 1189–1195.

Calmes, M. A. & J. C. Zasada. 1982. Some reproductive traits of four shrub species in the black spruce forest type of Alaska. *Can. Field Naturalist* 96: 35–40.

Calviño, C. I., S. G. Martínez & S. R. Downie. 2008. The evolutionary history of *Eryngium* (Apiaceae, Saniculoideae): rapid radiations, long distance dispersals, and hybridizations. *Mol. Phylogen. Evol.* 46: 1129–1150.

Cameron, K. M., K. J. Wurdack & R. W. Jobson. 2002. *Aldrovanda* is sister to *Dionaea* (Droseraceae): molecular evidence for the common origin of snap-traps among carnivorous plants. Botany 2002 meeting (abstract).

Camp, W. H. 1945. The North American blueberries with notes on other groups of Vacciniaceae. *Brittonia* 5: 203–275.

Campbell, A. & G. E. Bradfield. 1989. Comparison of plant community-environment relations in two estuarine marshes of northern British Columbia. *Can. J. Bot.* 67: 146–155.

Campbell, C. S., C. W. Greene & T. A. Dickinson. 1991. Reproductive biology in subfam. Maloideae (Rosaceae). *Syst. Bot.* 16: 333–349.

Campbell, C. S., M. J. Donoghue, B. G. Baldwin & M. Wojciechowski. 1995. Phylogenetic relationships in Maloideae (Rosaceae): evidence from sequences of the internal transcribed spacers of the nuclear ribosomal DNA and its congruence with morphology. *Amer. J. Bot.* 82: 903–918.

Campbell, D. & J. Bergeron. 2012. Natural revegetation of winter roads on peatlands in the Hudson Bay Lowland, Canada. *Arct. Antarc. Alp. Res.* 44: 155–163.

Campbell, D. H. 1899. The northern Pacific coast. *Amer. Naturalist* 33: 391–401.

Campbell, D. R. 1987. Interpopulational variation in fruit production: the role of pollination-limitation in the Olympic Mountains. *Amer. J. Bot.* 74: 269–273.

Campbell, D. R. & L. Rochefort. 2003. Germination and seedling growth of bog plants in relation to the recolonization of milled peatlands. *Plant Ecol.* 169: 71–84.

Campbell, D. R., N. M. Waser & E. J. Melendez-Ackerman. 1997. Analyzing pollinator-mediated selection in a plant hybrid zone: hummingbird visitation patterns on three spatial scales. *Amer. Naturalist* 149: 295–315.

Campbell, D. R., L. Rochefort & C. Lavoie. 2003. Determining the immigration potential of plants colonizing disturbed environments: the case of milled peatlands in Quebec. *J. Appl. Ecol.* 40: 78–91.

Campbell, H. W. & A. B. Irvine. 1977. Feeding ecology of the West Indian manatee *Trichechus manatus* Linnaeus. *Aquaculture* 12: 249–251.

Campbell, J. E. & D. J. Gibson. 2001. The effect of seeds of exotic species transported via horse dung on vegetation along trail corridors. *Plant Ecol.* 157: 23–35.

Campbell, J. J. N. & W. R. Seymour Jr. 2011. The vegetation of Pulliam Prairie, Chickasaw County, Mississippi: a significant remnant of pre-Columbian landscape in the Black Belt. *J. Mississippi Acad. Sci.* 56: 248–263.

Campbell, J. M., W. J. Clark & R. Kosinskiz. 1982. A technique for examining microspatial distribution of Cladocera associated with shallow water macrophytes. *Hydrobiologia* 97: 225–232.

Campbell, J. M., W. C. Fenton & J. F. Majewski. 1989. *Pollinators of Rare and Endangered Plants of Presque Isle*. Mercyhurst College, Erie, PA. 30 pp.

Campbell, L. W. 1933. Nesting of the mourning warbler near Toledo, Ohio. *Auk* 50: 117–119.

Campbell, P. D. 1999. *Survival Skills of Native California*. Gibbs Smith, Layton, UT. 449 pp.

Campbell, S. P., J. L. Frair & J. P. Gibbs. 2010. *Competition and Coexistence Between the Federally-threatened Chittenango Ovate Amber Snail* (Novisuccinea chittenangoensis) *and a Non-Native Snail* (Succinea sp. *B*). Final Progress Report to the U. S. Fish and Wildlife Service—July 2010. SUNY College of Environmental Science and Forestry, Syracuse, NY. 56 pp.

Camras, S. 2007. A new conopid fly from Florida and Georgia (Diptera: Conopidae). *Insecta Mundi* 7: 1–4.

Cane, J. H., D. Schiffhauer & L. J. Kervin. 1996. Pollination, foraging, and nesting ecology of the leaf-cutting bee *Megachile* (Delomegachile) *addenda* (Hymenoptera: Megachilidae) on cranberry beds. *Ann. Entomol. Soc. Amer.* 89: 361–367.

Canne, J. M. 1984. Chromosome numbers and the taxonomy of North American *Agalinis* (Scrophulariaceae). *Can. J. Bot.* 62: 454–456.

Cannings, R. A. & J. P. Simaika. 2005. *Lestes disjunctus* and L. *forcipatus* (Odonata: Lestidae): an evaluation of status and distribution in British Columbia. *J. Entomol. Soc. B. C.* 102: 57–63.

Cannon, P. F. 2009. *Bactrodesmium betulicola*. [Descriptions of Fungi and Bacteria]. *IMI Descr. Fungi Bact.* 1814: 1–182.

Cantino, P. D. 1979. *Physostegia godfreyi* (Lamiaceae), a new species from northern Florida. *Rhodora* 81: 409–417.

Cantino, P. D. 1981. Change of status for *Physostegia virginiana* var. *ledinghamii* (Labiatae) and evidence for a hybrid origin. *Rhodora* 83: 111–118.

Cantino, P. D. 1982. A monograph of the genus *Physostegia* (Labiateae). *Contr. Gray Herb.* 211: 1–105.

Cantino, P. D. 1985. Chromosome studies in subtribe *Melittidinae* (Labiatae) and systematic implications. *Syst. Bot.* 10: 1–6.

Cantino, P. D. & S. J. Wagstaff. 1998. A reexamination of North American *Satureja* s.l. (Lamiaceae) in light of molecular evidence. *Brittonia* 50: 63–70.

Cantino, P. D., S. J. Wagstaff & R. G. Olmstead. 1998. *Caryopteris* (Lamiaceae) and the conflict between phylogenetic and pragmatic considerations in botanical nomenclature. *Syst. Bot.* 23: 369–386.

Cantrell, C. L., T. Lu, F. R. Fronczek, N. H. Fischer, L. B. Adams & S. G. Franzblau. 1996. Antimycobacterial cycloartanes from *Borrichia frutescens*. *J. Nat. Prod.* 59: 1131–1136.

Cao, Y., B. Xuan, B. Peng, C. Li, X. Chai & P. Tu. 2016. The genus *Lindera*: a source of structurally diverse molecules having pharmacological significance. *Phytochem. Rev.* 15: 869–906.

Capers, R. S. & D. H. Les. 2001. An unusual population of *Podostemum ceratophyllum* Michx. (Podostemaceae) in a tidal, Connecticut river. *Rhodora* 103: 219–223.

Capers, R. S., R. Selsky & G. J. Bugbee. 2010. The relative importance of local conditions and regional processes in structuring aquatic plant communities. *Freshwater Biol.* 55: 952–966.

Caplan, E. B. 1966. Differential feeding and niche relationships among Orthoptera. *Ecology* 47: 1074–1076.

Capperino, M. E. & E. L. Schneider. 1985. Floral biology of *Nymphaea mexicana* Zucc. (Nymphaeaceae). *Aquat. Bot.* 23: 83–93.

Cappers, R. T. J. 1993. Seed dispersal by water: a contribution to the interpretation of seed assemblages. *Veg. Hist. Archaeobot.* 2: 173–186.

Caraballo, A., B. Caraballo & A. Rodríguez-Acosta. 2004. Preliminary assessment of medicinal plants used as antimalarials in the southeastern Venezuelan Amazon. *Rev. Soc. Bras. Med. Trop.* 37: 186–188.

Caraco, N. F. & J. J. Cole. 2002. Contrasting impacts of a native and alien macrophyte on dissolved oxygen in a large river. *Ecol. Appl.* 12: 1496–1509.

Carballo, M., M. D. Mudry, I. B. Larripa, E. Villamil & M. D'Aquino. 1992. Genotoxic action of an aqueous extract of *Heliotropium curassavicum* var. *argentinum*. *Mutat. Res.* 279: 245–253.

Cardina, J., E. Regnier & K. Harrison. 1991. Long-term tillage effects on seed banks in three Ohio soils. *Weed Sci.* 39: 186–194.

Cardoso, D., R. T. Pennington, L. P. de Queiroz, J. S. Boatwright, B.-E. Van Wyk, M. F. Wojciechowski & M. Lavin. 2013. Reconstructing the deep-branching relationships of the papilionoid legumes. *S. Afr. J. Bot.* 89: 58–75.

Cariaga, K. A., J. F. Pruski, R. Oviedo, A. A. Anderberg, C. E. Lewis & J. Francisco-Ortega. 2008. Phylogeny and systematic position of *Feddea* (Asteraceae: Feddeeae): a taxonomically enigmatic and critically endangered genus endemic to Cuba. *Syst. Bot.* 33: 193–202.

Caris, P. L. & E. F. Smets. 2004. A floral ontogenetic study on the sister group relationship between the genus *Samolus* (Primulaceae) and the Theophrastaceae. *Amer. J. Bot.* 91: 627–643.

Carl III, G., P. B. Hamel, M. S. Devall & N. M. Schiff. 2004. Hermit thrush is the first observed dispersal agent for pondberry (*Lindera melissifolia*). *Castanea* 69: 1–8.

Carleial, S., A. Delgado-Salinas, C. A. Domínguez & T. Terrazas. 2015. Reflexed flowers in *Aeschynomene amorphoides* (Fabaceae: Faboideae): a mechanism promoting pollination specialization? *Bot. J. Linn. Soc.* 177: 657–666.

Carleton, T. J. 1979. Floristic variation and zonation in the boreal forest South of James Bay: a cluster seeking approach. *Vegetatio* 39: 147–160.

Carleton, W. M. 1966. Food habits of two sympatric Colorado sciurids. *J. Mammalogy* 47: 91–103.

Carlquist, S. 1981. Chance dispersal: long-distance dispersal of organisms, widely accepted as a major cause of distribution patterns, poses challenging problems of analysis. *Amer. Sci.* 69: 509–516.

Carlquist, S. & Q. Pauly. 1985. Experimental studies on epizoochorous dispersal in Californian plants. *Aliso* 11: 167–177.

Carlsen, T., W. Bleeker, H. Hurka, R. Elven & C. Brochmann. 2009. Biogeography and phylogeny of "*Cardamine*" (Brassicaceae). *Ann. Missouri Bot. Gard.* 96: 215–236.

Carlson, J. R. 1992. Selection, production, and use of riparian plant materials for the western Unites States. pp. 55–67. *In*: T. D. Landis (Tech. Coordinator), *Proceedings, Intermountain Forest Nursery Association; 1991 August 12–16; Park City, UT*. General Technical Report RM-211. U. S. Department of Agriculture, Forest Service, Rocky Mountain Forest and Range Experiment Station, Fort Collins, CO.

Carlson, M. L. & M. Shephard. 2007. Is the spread of non-native plants in Alaska accelerating? pp. 111–127. *In*: T. B. Harrington & S. H. Reichard (eds.), *Meeting the Challenge: Invasive Plants in Pacific Northwest Ecosystems*. General Technical Report PNW-GTR-694. U. S. Department of Agriculture, Forest Service, Pacific Northwest Research Station, Portland, OR.

Carlson, R. A. & J. B. Moyle. 1968. *Key to the Common Aquatic Plants of Minnesota*. Federal Aid Project FW-1-R-12. Special Publication No. 53, Minnesota Department of Conservation, Division of Game and Fish, Section of Technical Services, St. Paul, MN. 64 pp.

Carpenter, A. 1978. The microdistribution of the freshwater shrimp *Paratya curvirostris* (Decapoda: Atyidae) in Saltwater Creek, North Canterbury. *Mauri Ora* 6: 23–26.

Carpenter, J. M., B. R. Garcete-Barrett & M. G. Hermes. 2006. Catalog of the Neotropical Masarinae (Hymenoptera, Vespidae). *Rev. Bras. Entomol.* 50: 335–340.

Carpenter, S. R. & N. J. McCreary. 1985. Effects of fish nests on pattern and zonation of submersed macrophytes in a softwater lake. *Aquat. Bot.* 22: 21–32.

Carpenter, S. R. & J. E. Titus. 1984. Composition and spatial heterogeneity of submersed vegetation in a softwater lake in Wisconsin. *Vegetatio* 57: 153–165.

Carr, D. E. & M. D. Eubanks. 2002. Inbreeding alters resistance to insect herbivory and host plant quality in *Mimulus guttatus* (Scrophulariaceae). *Evolution* 56: 22–30.

Carr, M. E., W. B. Roth & M. O. Bagby. 1986. Potential resource materials from Ohio plants. *Econ. Bot.* 40: 434–441.

Carr, R. 1831. *Periodical Catalogue of American Trees, Shrubs, Plants, and Seeds, Cultivated and for Sale at the Bartram Botanic Garden, near Philadelphia*. Russell & Martien, Philadelphia, PA. 84 pp.

Carr, S. C. 2007. Floristic and environmental variation of pyrogenic pinelands in the southeastern coastal plain: description, classification, and restoration. PhD dissertation. University of Florida, Gainesville, FL. 184 pp.

Carr, S. C., K. M. Robertson & R. K. Peet. 2010. A vegetation classification of fire-dependent pinelands of Florida. *Castanea* 75: 153–189.

Carroll, B. 1982. Tissue culture of *Pinguicula*. *Carniv. Plant Newslett.* 11: 93–96.

Carroll, B. 1997. Application of genetic analyses to native plant populations, pp. 93–97. *In*: T. D. Landis & J. R. Thompson (tech. coords.), *National Proceedings, Forest and Conservation Nursery Associations*. Gen. Tech. Rep. PNW-GTR-419. U. S. Department of Agriculture, Forest Service, Pacific Northwest Research Station, Portland, OR.

Carroll, D. M. 2003. Bryophytes as indicators of water level and salinity change along the northeast Cape Fear River. MS thesis. Department of Biological Sciences, University of North Carolina at Wilmington, Wilmington, NC. 42 pp.

Carstens, B. C., R. S. Brennan, V. Chua, C. V. Duffie, M. G. Harvey, R. A. Koch, C. D. McMahan, B. J. Nelson, C. E. Newman, J. D. Satler, G. Seeholzer, K. Posbic, D. C. Tanks & J. Sullivan. 2013. Model selection as a tool for phylogeographic inference: an example from the willow *Salix melanopsis*. *Mol. Ecol.* 22: 4014–4028.

Carter, D. A. 1973. A preliminary investigation of the location and general ecology of mat plants in the Rotorua Lake District. *Tane; J. Auckland Univ. Field Club* 19: 233–242.

Carter, R., M. W. Morris & C. T. Bryson. 1990. Some rare or otherwise interesting vascular plants from the delta region of Mississippi. *Castanea* 55: 40–55.

Carter, R., T. Boyer, H. McCoy & A. Londo. 2006. Community analysis of pitcher plant bogs of the Little River Canyon National Preserve, Alabama. pp. 486–489. *In*: K. F. Connor (ed.), *Proceedings of the 13th Biennial Southern Silvicultural Research Conference.* Gen. Tech. Rep. SRS–92. U.S. Department of Agriculture, Forest Service, Southern Research Station, Asheville, NC.

Carter, R., W. W. Baker & M. W. Morris. 2009. Contributions to the flora of Georgia, U.S.A. *Vulpia* 8: 1–54.

Carter, R., J. C. Jones & R. H. Goddard. 2014. *Sphenoclea zeylanica* (Sphenocleaceae) in North America—dispersal, ecology, and morphology. *Castanea* 79: 33–50.

Carter, R. E., M. D. MacKenzie & D. H. Gjerstad. 1999. Ecological land classification in the Southern Loam hills of south Alabama. *For. Ecol. Manag.* 114: 395–404.

Carulli, J. P. & D. E. Fairbrothers. 1988. Allozyme variation in three eastern United States species of *Aeschynomene* (Fabaceae), including the rare *A. virginica*. *Syst. Bot.* 13: 559–566.

Caruso, C. M. 2012. Sexual dimorphism in floral traits of gynodioecious *Lobelia siphilitica* (Lobeliaceae) is consistent across populations. *Botany* 90: 1245–1251.

Caruso, C. M. & A. L. Case. 2007. Sex ratio variation in gynodioecious *Lobelia siphilitica*: effects of population size and geographic location. *J. Evol. Biol.* 20: 1396–1405.

Caruso, C. M., H. Maherali & R. B. Jackson. 2003. Gender-specific floral and physiological traits: implications for the maintenance of females in gynodioecious *Lobelia siphilitica*. *Oecologia* 135: 524–531.

Caruso, C. M., H. Maherali, A. Mikulyuk, K. Carlson & R. B. Jackson. 2005. Genetic variance and covariance for physiological traits in *Lobelia*: are there constraints on adaptive evolution? *Evolution* 59: 826–837.

Caruso, C. M., H. Maherali & M. Sherrard. 2006. Plasticity of physiology in *Lobelia*: testing for adaptation and constraint. *Evolution* 60: 980–990.

Caruso, C. M., S. L. Scott, J. C. Wray & C. A. Walsh. 2010. Pollinators, herbivores, and the maintenance of flower color variation: a case study with *Lobelia siphilitica*. *Int. J. Plant Sci.* 171: 1020–1028.

Casazza, M. L., P. S. Coates, M. R. Miller, C. T. Overton & D. R. Yparraguirre. 2012. Hunting influences the diel patterns in habitat selection by northern pintails *Anas acuta*. *Wildl. Biol.* 18: 1–13.

Case Jr., F. W. & R. B. Case. 1974. *Sarracenia alabamensis*, a newly recognized species from central Alabama. *Rhodora* 76: 650–665.

Casebere, L. A. & M. J. Lodato. 2011. The four-toed salamander (*Hemidactylium scutatum*) in Indiana: past and present. *Proc. Indiana Acad. Sci.* 119: 111–129.

Casper, S. J. 1962. On *Pinguicula macroceras* Link in North America. *Rhodora* 64: 212–221.

Casper, S. J. & H.-D. Krausch. 1981. *Süsswasserflora von Mitteleuropa. Band 24: Pteridophyta und Anthophyta. 2. Teil: Saururaceae bis Asteraceae.* Gustav Fischer Verlag, Stuttgart, Germany. pp. 413–943.

Cassani, J. R. 1985. Biology of *Simyra henrici* (Lepidoptera: Noctuidae) in southwest Florida. *Florida. Entomol.* 68: 645–652.

Cassell, B. A. 2010. Plant propagation protocol for *Downingia yina*. Available online: http://courses.washington.edu/esrm412/protocols/DOYI.pdf (accessed September 11, 2014).

Cassis, G. 1984. A systematic study of the subfamily Dicyphinae (Heteroptera: Miridae). PhD dissertation, Oregon State University, Corvallis, OR. 403 pp.

Castelblanco-Martínez, D. N., B. E. Morales-Vela, H. A. Hernández-Arana & J. A. Padilla-Saldivar. 2009. Diet of the manatees (*Trichechus manatus manatus*) in Chetumal Bay, Mexico. *Latin Amer. J. Aquat. Mammals.* 7: 39–46.

Castellanos, J. A. & E. Rijo. 1982. *Hydrocotyle umbellata*, a new host for *Orobanche ramosa* in Cuba. *Cienc. Tec. Agric.* 5: S99–S101.

Castelli, R. M., J. C. Chambers & R. J. Tausch. 2000. Soil-plant relations along a soil-water gradient in Great Basin. *Wetlands* 20: 251–266.

Castells, E. & M. R. Berenbaum. 2008. Resistance of the generalist moth *Trichoplusia ni* (Noctuidae) to a novel chemical defense in the invasive plant *Conium maculatum*. *Chemoecology* 18: 11–18.

Castlebury, L. A., D. F. Farr, A. Y. Rossman & W. Jaklitsch. 2003. *Diaporthe angelicae* comb. nov., a modern description and placement of *Diaporthopsis* in *Diaporthe*. *Mycoscience* 44: 203–208.

Castro-Morales, L. M., P. F. Quintana-Ascencio, J. E. Fauth, K. J. Ponzio & D. L. Hall. 2014. Environmental factors affecting germination and seedling survival of Carolina willow (*Salix caroliniana*). *Wetlands* 34: 469–478.

Catalfamo, J. L., W. B. Martin Jr. & H. Birecka. 1982. Accumulation of alkaloids and their necines in *Heliotropium curassavicum*, *H. spathulatum* and *H. indicum*. *Phytochemistry* 21: 2669–2675.

Cates, R. G. & G. H. Orians. 1975. Sucessional status and the palatability of plants to generalized herbivores. *Ecology* 56: 410–418.

Catling, H. D. 1992. *Rice in Deep Water*. The Macmillan Press, Ltd., London. 542 pp.

Catling, P. M. 1998. A synopsis of the genus *Proserpinaca* in the southeastern United States. *Castanea* 63: 408–414.

Catling, P. M., B. Freedman, C. Stewart, J. J. Kerekes & L. P. Lefkovitch. 1986. Aquatic plants of acid lakes in Kejimkujik National Park, Nova Scotia; floristic composition and relation to water chemistry. *Can. J. Bot.* 64: 724–729.

Catling, P. M., G. Mitrow, E. Haber, U. Posluszny & W. A. Charlton. 2003. The biology of Canadian weeds. 124. *Hydrocharis morsus-ranae* L. *Can. J. Plant Sci.* 83: 1001–1016.

Caton, B. P., T. C. Foin & J. E. Hill. 1997. Phenotypic plasticity of *Ammannia* spp. in competition with rice. *Weed Res.* 37: 33–38.

Cattell, M. V. & S. A. Karl. 2004. Genetics and morphology in a *Borrichia frutescens* and *B. arborescens* (Asteraceae) hybrid zone. *Amer. J. Bot.* 91: 1757–1766.

Cavalieri, A. J. & A. H. C. Huang. 1977. Effect of NaCl on the *in vitro* activity of malate dehydrogenase in salt marsh halophytes of the U.S. *Physiol. Plant.* 41: 79–84.

Cavalieri, A. J. & A. H. C. Huang. 1979. Evaluation of proline accumulation in the adaptation of diverse species of marsh halophytes to the saline environment. *Amer. J. Bot.* 66: 307–312.

Cavalli, G., T. Riis & A. Baattrup-Pedersen. 2012. Bicarbonate use in three aquatic plants. *Aquat. Bot.* 98: 57–60.

Cavieres, L. A. & M. T. K. Arroyo. 2000. Seed germination response to cold stratification period and thermal regime in *Phacelia secunda* (Hydrophyllaceae): Altitudinal variation in the Mediterranean Andes of central Chile. *Plant Ecol.* 149: 1–8.

Cayouette, J., M. Blondeau & P. M. Catling. 1997. Pollen abortion in the *Ranunculus gmelinii—hyperboreus* group (Ranunculaceae, section *Hecatonia*) and its taxonomic implications. *Rhodora* 99: 263–274.

Cázares, E., J. M. Trappe & A. Jumpponen. 2005. Mycorrhiza-plant colonization patterns on a subalpine glacier forefront as a model system of primary succession. *Mycorrhiza* 15: 405–416.

Cellot, B., F. Mouillot & C. P. Henry. 1998. Flood drift and propagule bank of aquatic macrophytes in a riverine wetland. *J. Veg. Sci.* 9: 631–640.

Cely, J. E. 1979. The ecology and distribution of banana waterlily and its utilization by Canvasback ducks. *Proc. Ann. Conf. S. E. Assoc. Game Fish Comm.* 33: 43–47.

Cerana, M. M. 2004. Flower morphology and pollination in *Mikania* (Asteraceae). *Flora* 199: 168–177.

Cerkauskas, R. F., S. T. Koike, H. R. Azad, D. T. Lowery & L. W. Stobbs. 2006. Diseases, pests, and abiotic disorders of greenhouse-grown water spinach (*Ipomoea aquatica*) in Ontario and California. *Can. J. Plant Pathol.* 28: 63–70.

Cerón-Souza, I., E. Rivera-Ocasio, E. Medina, J. A. Jiménez, W. O. McMillan & E. Bermingham. 2010. Hybridization and introgression in New World red mangroves, *Rhizophora* (Rhizophoraceae). *Amer. J. Bot.* 97: 945–957.

Cerón-Souza, M. T., E. Rivera-Ocasio, S. M. Funk & W. O. McMillan. 2006. Development of six microsatellite loci for black mangrove (*Avicennia germinans*). *Mol. Ecol. Notes* 6: 692–694.

Cervera, M. T., O. P. Rajora, V. Storme, B. Ivens, M. Van Montagu & W. Boerjan. 2003. Intraspecific and interspecific genetic and phylogenetic relationships in the genus *Populus* based on ALFP markers. (unpublished abstract).

Ceska, A. & O. Ceska. 1986. More on the techniques for collecting aquatic and marsh plants. *Ann. Missouri Bot. Gard.* 73: 825–827.

Ceska, A. & A. M. Scagel. 2011. *Indicator Plants of Coastal British Columbia*. University of British Columbia Press, Vancouver, BC. 296 pp.

Ceska, A. & P. D. Warrington. 1976. *Myriophyllum farwellii* (Haloragaceae) in British Columbia. *Rhodora* 78: 75–78.

Ceska, O., A. Ceska & P. D. Warrington. 1986. *Myriophyllum quitense* and *Myriophyllum ussuriense* (Haloragaceae) in British Columbia, Canada. *Brittonia* 38: 73–81.

Cha, D. H., C. G. Hochwender, E. M. Bosecker, R. E. Tucker, A. D. Kaufman, R. S. Fritz & R. R. Smyth. 2009. Do exotic generalist predators alter host plant preference of a native willow beetle? *Agric. For. Entomol.* 11: 175–184.

Cha, D. H., T. H. Q. Powell, J. L. Feder & C. E. Linn Jr. 2011. Identification of fruit volatiles from green hawthorn (*Crataegus viridis*) and blueberry hawthorn (*Crataegus brachyacantha*) host plants attractive to different phenotypes of *Rhagoletis pomonella* flies in the Southern United States. *J. Chem. Ecol.* 37: 974–983.

Chabreck, R. H. 1958. Beaver-forest relationships in St. Tammany Parish, Louisiana. *J. Wildl. Manag.* 22: 179–183.

Chabreck, R. H. & A. W. Palmisano. 1973. The effects of hurricane Camille on the marshes of the Mississippi River delta. *Ecology* 54: 1118–1123.

Chabreck, R. H., R. B. Thompson & A. B. Ensminger. 1977. Chronic dermatitis in nutria in Louisiana. *J. Wildl. Dis.* 13: 333–334.

Chadde, S. A. 1998. *Great Lakes Wetland Flora: A Complete, Illustrated Guide to the Aquatic and Wetland Plants of the Upper Midwest*. PocketFlora Press, Calumet, MI. 569 pp.

Chafin, L. G. 2007. *Field Guide to the Rare Plants of Georgia*. State Botanical Garden of Georgia, Athens, GA. 529 pp.

Chafin, L. G. 2008. *Hartwrightia*. Rare Plant Species Profiles. Georgia Department of Natural Resources. Wildlife Resources Division. Available online: http://www.georgiawildlife.org/node/2627 (accessed November 27, 2013).

Chakraborty, R., M. S. Mondal & S. K. Mukherjee. 2016. Ethnobotanical information on some aquatic plants of South 24 Parganas, West Bengal. *Plant Sci. Today* 3: 109–114.

Chalchat, J.-C., S. Petrovic, Z. Maksimovic & M. Gorunovic. 2009. Composition of essential oil of *Bidens cernua* L., Asteraceae from Serbia. *J. Essential. Oil Res.* 21: 41–42.

Chalmers, R. J. & C. S. Loftin. 2006. Wetland and microhabitat use by nesting four-toed salamanders in Maine. *J. Herpetol.* 40: 478–485.

Chamberlain, S. J., D. H. Wardrop, M. S. Fennessy & D. DeBerry. 2013. Hydrophytes in the Mid-Atlantic region: ecology, communities, assessment, and diversity. pp. 159–258. *In*: R. P. Brooks & D. H. Wardrop (eds.), *Mid-Atlantic Freshwater Wetlands: Advances in Wetlands Science, Management, Policy, and Practice*. Springer, New York.

Chamberlin, R. V. 1911. The ethno-botany of the Gosiute Indians of Utah. *Mem. Amer. Anthro. Assoc.* 2: 329–405.

Chambers, H. 2001. Oregon plantains: the natives are diverse and we tolerate the weeds! *Oregon Fl. Newslett.* 7: 18–20.

Chambers, H. L. & K. E. Hummer. 1994. Chromosome counts in the *Mentha* collection at the USDA: ARS National Clonal Germplasm Repository. *Taxon* 43: 423–432.

Chambers, K. L. 1963. Amphitropical species pairs in *Microseris* and *Agoseris* (Compositae: Cichorieae). *Q. Rev. Biol.* 38: 124–140.

Chambers, K. L. 2006a. 68. *Microseris* D. Don. pp. 338–346. *In*: Flora of North America Editorial Committee (ed.), *Flora of North America North of Mexico, Vol. 19: Magnoliophyta: Asteridae, Part 8*. Oxford University Press, New York.

Chambers, K. L. 2006b. 80. *Phalacroseris* A. Gray. p. 374. *In*: Flora of North America Editorial Committee (ed.), *Flora of North America North of Mexico, Vol. 19: Magnoliophyta: Asteridae, Part 8*. Oxford University Press, New York.

Chambers, K. L. 2007. Nomenclatural notes and lectotypes in the *Allocarya* section of *Plagiobothrys*. *Madroño* 54: 322–325.

Chambers, K. L. 2012. *Phalacroseris*. p. 401. *In*: B. G. Baldwin, D. H. Goldman, D. J. Keil, R. Patterson, T. J. Rosatti & D. H. Wilken (eds.), *The Jepson Manual (2nd ed.)*. University of California Press, Berkeley, CA.

Chambers, V. H. 1968. Pollens collected by species of *Andrena* (Hymenoptera: Apidae). *Proc. Roy. Entomol. Soc. London. Ser. A.* 43: 155–160.

Champion, P. D. & J. S. Clayton. 2003. The evaluation and management of aquatic weeds in New Zealand. pp. 429–434. *In*: L. Child, J. H. Brock, G. Brundu, K. Prach, K. Pyšek, P. M. Wade & M. Williamson (eds.), *Plant Invasions: Ecological Threats and Management Solutions*. Backhuys, Leiden, The Netherlands.

Champion, P. D. & C. C. Tanner. 2000. Seasonality of macrophytes and interaction with flow in a New Zealand lowland stream. *Hydrobiologia* 441: 1–12.

Chan, K. F. & M. Sun. 1997. Genetic diversity and relationships detected by isozyme and RAPD analysis of crop and wild species of *Amaranthus*. *Theor. Appl. Genet.* 95: 865–873.

Chan, R. & R. Ornduff. 2006. 354. *Lasthenia* Cassini. pp. 336–347. *In*: Flora of North America Editorial Committee (ed.), *Flora of North America North of Mexico: Vol. 21, Magnoliophyta: Asteridae (in part): Asteraceae, Part 3*. Oxford University Press, New York.

Chan, R., B. G. Baldwin & R. Ornduff. 2001. Goldfields revisited: a molecular phylogenetic perspective on the evolution of *Lasthenia* (Compositae: Heliantheae sensu lato). *Int. J. Plant Sci.* 162: 1347–1360.

Chanderbali, A. S., H. van der Werff & S. S. Renner. 2001. Phylogeny and historical biogeography of Lauraceae: evidence from the chloroplast and nuclear genomes. *Ann. Missouri Bot. Gard.* 88: 104–134.

Chandler, D. C. 1937. Fate of typical lake plankton in streams. *Ecol. Monogr.* 7: 445–479.

Chandler, G. & G. M. Plunkett. 2004. Evolution in Apiales: nuclear and chloroplast markers together in (almost) perfect harmony. *Bot. J. Linn. Soc.* 144: 123–147.

Chandler, R. F. & S. N. Hooper. 1982. Herbal remedies of the maritime indians: a preliminary screening: Part III. *J. Ethnopharmacol.* 6: 275–285.

Chandra, P. & K. Kulshreshtha. 2004. Chromium accumulation and toxicity in aquatic vascular plants. *Bot. Rev.* 70: 313–327.

Chandran, R. & K. V. Bhavanandan. 1981. Chromosome number reports LXXII. *Taxon* 30: 698.

Chang, C. I., C. C. Kuo, J. Y. Chang & Y. H. Kuo. 2004. Three new oleanane-type triterpenes from *Ludwigia octovalvis* with cytotoxic activity against two human cancer cell lines. *J. Nat. Prod. (Lloydia)* 67: 91–93.

Chang, E. R., R. L. Jefferies & T. J. Carleton. 2001. Relationship between vegetation and soil seed banks in an arctic coastal marsh. *J. Ecol.* 89: 367–384.

Chang, E. R., E. L. Zozaya, D. P. J. Kuijper & J. P. Bakker. 2005. Seed dispersal by small herbivores and tidal water: are they important filters in the assembly of salt-marsh communities? *Funct. Ecol.* 19: 665–673.

Chang, R. S., Y. He, H. D. Tabba & K. M. Smith. 1988. Inhibitor against the human immunodeficiency virus in aqueous extracts of *Alternanthera philoxeroides* (Kong Xin Xian). *Chin. Med. J.* 101: 861.

Chang, Y.-S. & E.-R. Woo. 2003. Korean medicinal plants inhibiting to Human Immunodeficiency Virus type 1 (HIV-1) fusion. *Phytother. Res.* 17: 426–429.

Channell, R. B. & C. E. Wood Jr. 1962. The Leitneriaceae in the southeastern United States. *J. Arnold Arbor.* 43: 435–438.

Chao, P.-Y., S.-Y. Lin, K.-H. Lin, Y.-F. Liu, J.-I. Hsu, C.-M. Yang & J.-Y. Lai. 2014. Antioxidant activity in extracts of 27 indigenous Taiwanese vegetables. *Nutrients* 6: 2115–2130.

Chapman, D. S. & R. M. Augé. 1994. Physiological mechanisms of drought resistance in four native ornamental perennials. *J. Amer. Soc. Hort. Sci.* 119: 299–306.

Chapman, G. W. Jr. & R. J. Horvat. 1993. Chemical compositional changes in two genetically diverse cultivars of mayhaw fruit at three maturity stages. *J. Agric. Food Chem.* 41: 1550–1552.

Chapman, J. A. & G. R. Willner. 1978. *Sylvilagus audubonii*. *Mammal. Species* 106: 1–4.

Chapman, V. J. 1936. The halophyte problem in the light of recent investigations. *Q. Rev. Biol.* 11: 209–220.

Chapman, V. J. 1947. *Suaeda maritima* (L.) Dum. *J. Ecol.* 35: 293–302.

Chapman, V. J., J. M. A. Brown, C. F. Hill & J. L. Carr. 1974. Biology of excessive weed growth in the hydro-electric lakes of the Waikato River, New Zealand. *Hydrobiologia* 44: 349–363.

Charlesworth, B. 1992. Evolutionary rates in partially self-fertilizing species. *Amer. Naturalist* 140: 126–148.

Charlesworth, D. & Z. Yang. 1998. Allozyme diversity in *Leavenworthia* populations with different inbreeding levels. *Heredity* 81: 453–461.

Chase, M. W., D. E. Soltis, R. G. Olmstead, D. Morgan, D. H. Les, B. D. Mishler, M. R. Duvall, R. A. Price, H. G. Hills, Y.-L. Qiu, K. A. Kron, J. H. Rettig, E. Conti, J. D. Palmer, J. R. Manhart, K. J. Systma, H. J. Michaels, W. J. Kress, K. G. Karol, W. D. Clark, M. Hedren, B. S. Gaut, R. K. Jansen, K.-J. Kim, C. F. Wimpee, J. F. Smith, G. R. Furnier, S. H. Strauss, Q.-Y. Xiang, G. M. Plunkett, P. S. Soltis, S. M. Swensen, S. E. Williams, P. A. Gadek, C. J. Quinn, L. E. Eguiarte, E. Golenberg, G. H. Learn Jr., S. W. Graham, S. C. H. Barrett, S. Dayanandan & V. A. Albert. 1993. Phylogenetics of seed plants: an analysis of nucleotide sequences from the plastid gene *rbcL*. Ann. Missouri Bot. Gard. 80: 528–580.

Chase, M. W., S. Zmarzty, M. D. Lledó, K. J. Wurdack, S. M. Swensen & M. F. Fay. 2002. When in doubt, put it in Flacourtiaceae: a molecular phylogenetic analysis based on plastid *rbcL* DNA sequences. *Kew Bull.* 57: 141–181.

Chase, V. A. & P. H. Raven. 1975. Evolutionary and ecological relationships between *Aquilegia formosa* and *A. pubescens* (Ranunculaceae), two perennial plants. *Evolution* 29: 474–486.

Chassot, P., S. Nemomissa, Y.-M. Yuan & P. Küpfer. 2001. High paraphyly of *Swertia* L. (Gentianaceae) in the *Gentianella*-lineage as revealed by nuclear and chloroplast DNA sequence variation. *Plant Syst. Evol.* 229: 1–21.

Chastant, J. E., R. A. Botta & D. E. Gawlik. 2015. First record of roseate spoonbills (*Platalea ajaja*) nesting on Lake Okeechobee since 1874. *Florida. Field Naturalist* 43: 114–118.

Chatrou, L. W., M. P. Escribano, M. A. Viruel, J. W. Maas, J. E. Richardson & J. I. Hormaza. 2009. Flanking regions of monomorphic microsatellite loci provide a new source of data for plant species-level phylogenetics. *Mol. Phylogen. Evol.* 53: 726–733.

Chatrou, L. W., M. D. Pirie, R. H. Erkens, T. L. Couvreur, K. M. Neubig, J. R. Abbott, J. B. Mols, J. W. Maas, R. M. Saunders & M. W. Chase. 2012. A new subfamilial and tribal classification of the pantropical flowering plant family Annonaceae informed by molecular phylogenetics. *Bot. J. Linn. Soc.* 169: 5–40.

Chaubal, M. G., R. M. Baxter & G. C. Walker. 1962. Paper chromatography of alkaloidal extracts of *Lobelia* species. *J. Pharm. Sci.* 51: 885–888.

Chaudhuri, P. K., R. Srivastava, S. Kumar & S. Kumar. 2002. Phytotoxic and antimicrobial constituents of *Bacopa monnieri* and *Holmskioldia sanguinea*. *Phytother. Res.* 18: 114–117.

Chauhan, B. S. & D. E. Johnson. 2008. Influence of environmental factors on seed germination and seedling emergence of eclipta (*Eclipta prostrata*) in a tropical environment. *Weed Sci.* 56: 383–388.

Chavasiri, W., W. Prukchareon, P. Sawasdee & S. Zungsontiporn. 2005. Allelochemicals from *Hydrocotyle umbellata* Linn. pp. 15–18. *In*: J. D. I. Harper, M. An, H. Wu & J. H. Kent (eds.), "*Establishing the Scientific Base*"; *Proceedings and Selected Papers of the Fourth World Congress on Allelopathy, August 21–26*. Charles Sturt University, Wagga Wagga, NSW, Australia.

Chaw, S. M. & M. T. Kao. 1989. *Lindernia dubia* var. *anagallidea* (Michaux) Pennel (Scrophulariaceae): a newly naturalized plant in Taiwan. *J. Taiwan Mus.* 42: 95–100.

Cheers, G. 1992. *Letts Guide to Carnivorous Plants of the World.* Charles Letts & Co. Ltd., London. 174 pp.

Chen, C. P., C. C. Lin & T. Namba. 1989. Screening of Taiwanese crude drugs for antibacterial activity against *Streptococcus mutans. J. Ethnopharmacol.* 27: 285–95.

Chen, J.-H., H. Sun, J. Wen & Y.-P. Yang. 2010. Molecular phylogeny of *Salix* L. (Salicaceae) inferred from three chloroplast datasets and its systematic implications. *Taxon* 59: 29–37.

Chen, L.-Y., S.-Y. Zhao, K.-S. Mao, D. H. Les, Q.-F. Wang & M. L. Moody. 2014. Historical biogeography of Haloragaceae: an out of Australia hypothesis with multiple intercontinental dispersals. *Mol. Phylogen. Evol.* 78: 87–95.

Chen, M., L.-L. Zhang, J. Li, X.-J. He & J.-C. Cai. 2015. Bioaccumulation and tolerance characteristics of a submerged plant (*Ceratophyllum demersum* L.) exposed to toxic metal lead. *Ecotoxicol. Environ. Saf.* 122: 313–321.

Chen, S., T. Xia, Y. Wang, J. Liu & S. Chen. 2005. Molecular systematics and biogeography of *Crawfurdia, Metagentiana* and *Tripterospermum* (Gentianaceae) based on nuclear ribosomal and plastid DNA sequences. *Ann. Bot.* 96: 413–424.

Chen, S.-H. & M.-J. Wu. 2001. Notes on two newly naturalized plants in Taiwan. *Taiwania* 46: 85–92.

Chen, X., Y. Du, J. Nan, X. Zhang, X. Qin, Y. Wang, J. Hou, Q. Wang & J. Yang. 2013. Brevilin A, a novel natural product, inhibits janus kinase activity and blocks STAT3 signaling in cancer cells. *PLoS One* 8: e63697.

Chen, Y.-P., B. Li, R. G. Olmstead, P. D. Cantino, E.-D. Liu & C.-L. Xiang. 2014. Phylogenetic placement of the enigmatic genus *Holocheila* (Lamiaceae) inferred from plastid DNA sequences. *Taxon* 63: 355–366.

Chen, Y.-Q., W.-W. Jiang, J.-R. Huang, H.-Y. Cai & L. Sun. 2009. Preliminary study on the physiological indices of heat tolerance of Aster *novi-belgii* leaves. *J. Jiangsu Forest. Sci. Technol.* 36: 15–17, 28.

Chen, Y.-Y., X.-R. Fan, Z. Li, W. Li & W.-M. Huang. 2016. Low level of genetic variation and restricted gene flow in water lily *Nymphaea tetragona* populations from the Amur River. *Aquat. Bot.* doi:10.1016/j. aquabot.2016.10.003.

Chen, Z. & J. Li. 2004. Phylogenetics and biogeography of *Alnus* (Betulaceae) inferred from sequences of nuclear ribosomal DNA ITS region. *Int. J. Plant Sci.* 165: 325–335.

Chen, Z., Z. Lei, J. Zhou & J. Chen. 2001. A preliminary study of winter seed bank of dominant submerged macrophytes in lake Liangzi. *Acta Hydrobiol. Sin.* 25: 152–158.

Chen, Z.-D., S. R. Manchester & H.-Y. Sun. 1999. Phylogeny and evolution of the Betulaceae as inferred from DNA sequences, morphology, and paleobotany. *Amer. J. Bot.* 86: 1168–1181.

Chen, Z.-Y., Z.-J. Xiong, X.-Y. Pan, S.-Q. Shen, Y.-P. Geng, C.-Y. Xu, J.-K. Chen & W.-J. Zhang. 2015. Variation of genome size and the ribosomal DNA ITS region of *Alternanthera philoxeroides* (Amaranthaceae) in Argentina, the USA, and China. *J. Syst. Evol.* 53: 82–87.

Cheney, M. & K. L. Marr. 2007. Cowbane, *Oxypolis occidentalis*, a new native plant species for the Queen Charlotte Islands, British Columbia. *Can. Field Naturalist* 121: 421–422.

Cheng, J. & L. Xie. 2014. Molecular phylogeny and historical biogeography of *Caltha* (Ranunculaceae) based on analyses of multiple nuclear and plastid sequences. *J. Syst. Evol.* 52: 51–67.

Cherry, J. A. & L. Gough. 2006. Temporary floating island formation maintains wetland plant species richness: the role of the seed bank. *Aquat. Bot.* 85: 29–36.

Chesnut, V. K. & E. V. Wilcox. 1901. *The Stock-Poisoning Plants of Montana: A Preliminary Report.* Government Printing Office, Washington, DC. 150 pp.

Chiari, A. 2005. Propagation protocol for meadow beauty *Rhexia virginica* L. (Melastomataceae). *Native Plants J.* 6: 118–120.

Chick, J. H., R. J. Cosgriff & L. S. Gittinger. 2003. Fish as potential dispersal agents for floodplain plants: first evidence in North America. *Can. J. Fish. Aquat. Sci.* 60: 1437–1439.

Childers, C. C. & H. A. Denmark. 2011. Phytoseiidae (Acari: Mesostigmata) within citrus orchards in Florida: species distribution, relative and seasonal abundance within trees, associated vines and ground cover plants. *Exp. Appl. Acarol.* 54: 331–371.

Childers, C. C., J. C. V. Rodrigues & W. C. Welbourn. 2003a. Host plants of *Brevipalpus californicus, B. obovatus*, and *B. phoenicis* (Acari: Tenuipalpidae) and their potential involvement in the spread of viral diseases vectored by these mites. *Exp. Appl. Acarol.* 30: 29–105.

Childers, D. L., R. F. Doren, R. Jones, G. B. Noe, M. Rugge & L. J. Scinto. 2003b. Decadal change in vegetation and soil phosphorus pattern across the Everglades Landscape. *J. Environ. Qual.* 32: 344–362.

Chilton, E. W. 1990. Macroinvertebrate communities associated with three aquatic macrophytes (*Ceratophyllum demersum, Myriophyllum spicatum*, and *Vallisneria americana*) in Lake Onalaska, Wisconsin. *J. Freshwater Ecol.* 5: 455–466.

Chimney, M. J. & K. C. Pietro. 2006. Decomposition of macrophyte litter in a subtropical constructed wetland in south Florida (USA). *Ecol. Eng.* 27: 301–321.

Ching, L. S. & S. Mohamed. 2001. Alpha-tocopherol content in 62 edible tropical plants. *J. Agric. Food Chem.* 49: 3101–3105.

Chinnappa, C. C. & J. K. Morton. 1984. Studies on the *Stellaria longipes* Goldie complex (Caryophyllaceae). *Syst. Bot.* 9: 60–73.

Chinnappa, C. C., G. M. Donald, R. Sasidharan & R. J. N. Emery. 2005. The biology of *Stellaria longipes* (Caryophyllaceae). *Can. J. Bot.* 83: 1367–1383.

Chippindale, H. G. & W. E. J. Milton. 1934. On the viable seeds present in the soil beneath pastures. *J. Ecol.* 22: 508–531.

Chipping, D. H. 1994. *Chorro Creek Bog Thistle Recovery Project.* Final report submitted to the California Department of Fish and Game, Natural Heritage Division, Sacramento, CA. 86 pp.

Chiscano, J. L. P. 1999. *Bacopa rotundifolia* (Mich) Wettst. (Scrophulariaceae), new for Europa. *Stud. Bot.* 18: 137.

Chistokhodova, N., C. Nguyen, T. Calvino, I. Kachirskaia, G. Cunningham & D. H. Miles. 2002. Antithrombin activity of medicinal plants from central Florida. *J. Ethnopharmacol.* 81: 277–280.

Chittka, L. 1999. Bees, white flowers, and the color hexagon – a reassessment? No, not yet. Comments on the contribution by Vorobyev et al. *Naturwissenschaften* 86: 595–597.

Chmielewski, J. G. 1999. Consequences of achene biomass, within-achene allocation patterns, and pappus on germination in ray and disc achenes of *Aster umbellatus* var. *umbellatus* (Asteraceae). *Can. J. Bot.* 77: 426–433.

Chmielewski, J. G. 2005. The effects of achene weight, orientation and storage on germination in *Eclipta prostrata* (L.) L. (Asteraceae). *Curr. Top. Plant Biol.* 6: 37–40.

Chmielewski, J. G. & S. Ruit. 2002. Interrelationships among achene weight, orientation, and germination in the asters. *Doellingeria umbellata* var. *umbellata, Symphyotrichum novae-angliae* and *S. puniceum* (Asteraceae). *Bartonia* 61: 15–26.

Chmielewski, J. G. & J. C. Semple. 2001. The biology of Canadian weeds. 113. *Symphyotrichum lanceolatum* (Willd.) Nesom [*Aster lanceolatus* Willd.] and *S. lateriflorum* (L.) Löve & Löve [*Aster lateriflorus* (L.) Britt.]. *Can. J. Plant Sci.* 81: 829–849.

Chmielewski, J. G. & S. R. Strain. 2007. Achene aerodynamics in species of *Doellingeria, Eurybia, Oclemena*, and *Symphyotrichum. J. Agric. Food Environ. Sci.* 1: 1–10.

Chmielewski, J. G., G. S. Ringius & J. C. Semple. 1987. The cytogeography of *Solidago uliginosa* (Compositae: Astereae) in the Great Lakes region. *Can. J. Bot.* 65: 1045–1046.

Chmura, G. L., P. Chase & J. Bercovitch. 1997. Climatic controls of the middle marsh zone in the Bay of Fundy. *Estuaries* 20: 689–699.

Chmura, G. L., D. M. Burdick & G. E. Moore. 2012. Recovering salt marsh ecosystem services through tidal restoration. pp. 233–251. *In*: C. T. Roman & D. M. Burdick (eds.), *Tidal Marsh Restoration: A Synthesis of Science and Practice*. Island Press, Washington, DC.

Cho, H. J. 2009. *Aquatic Plants of the Mississippi Coast*. Mississippi-Alabama Sea Grant, MASGP-09-029. 137 pp.

Cho, J. Y., P. S. Kim, J. Park, E. S. Yoo, K. U. Baik, Y.-K. Kim & M. H. Park. 2000. Inhibitor of tumor necrosis factor-alpha production in lipopolysaccharide-stimulated RAW264.7 cells from *Amorpha fruticosa. J. Ethnopharmacol.* 70: 127–133.

Choate, J. S. 1967. Factors influencing nesting success of Eiders in Penobscot Bay, Maine. *J. Wildl. Manag.* 31: 769–777.

Choesin, D. N. & R. E. J. Boerner. 2000. Vegetation and ground water alkalinity of Betsch Fen, a remnant periglacial fen in south central Ohio. *Castanea* 65: 193–206.

Choesin, D. & R. E. J. Boerner. 2002. Vegetation boundary detection: a comparison of two approaches applied to field data. *Plant Ecol.* 158: 85–96.

Choi, B., J. C. Dewey, J. A. Hatten, A. W. Ezell & Z. Fan. 2012. Changes in vegetative communities and water table dynamics following timber harvesting in small headwater streams. *For. Ecol. Manag.* 281: 1–11.

Choi, J. S. & M. K. Huh. 2007. Comparison of mating systems in Korean populations of *Oenanthe javanica. Hort. Environ. Biotechnol.* 48: 212–216.

Cholewa, A. F. 2009a. 3. *Lysimachia* L. pp. 308–318 *In*: N. R. Morin (convening ed.), *Flora of North America North of Mexico, Vol. 8: Magnoliophyta: Paeoniaceae to Ericaceae*. Oxford University Press, New York.

Cholewa, A. F. 2009b. 3. *Samolus* Linnaeus. pp. 254–256. *In*: N. R. Morin (convening ed.), *Flora of North America North of Mexico, Vol. 8: Magnoliophyta: Paeoniaceae to Ericaceae*. Oxford University Press, New York.

Cholewa, A. F. 2009c. 9. Theophrastaceae D. Don. pp. 251–256. *In*: N. R. Morin (convening ed.), *Flora of North America North of Mexico, Vol. 8: Magnoliophyta: Paeoniaceae to Ericaceae*. Oxford University Press, New York.

Cholewa, A. F. & D. M. Henderson. 1984a. *Primula alcalina* (Primulaceae): a new species from Idaho. *Brittonia* 36: 59–62.

Cholewa, A. F. & D. M. Henderson. 1984b. Biosystematics of *Sisyrinchium* section *Bermudiana* (Iridaceae) of the Rocky Mountains. *Brittonia* 36: 342–363.

Chorianopoulou, S. N., D. L. Bouranis & J. B. Drossopoulos. 2001. Oxygen transport by *Apium nodiflorum. J. Plant Physiol.* 158: 905–913.

Chowdhuri, D. K., D. Parmar, P. Kakkar, R. Shukla, P. K. Seth & R. C. Srimal. 2002. Antistress effects of bacosides of *Bacopa monnieri*: modulation of Hsp70 expression, superoxide dismutase and cytochrome P450 activity in rat brain. *Phytother. Res.* 16: 639–645.

Christapher, P. V., S. Parasuraman, J. M. Christina, M. Z. Asmawi & M. Vikneswaran. 2015. Review on *Polygonum minus*. Huds, a commonly used food additive in Southeast Asia. *Pharmacogn. Res.* 7: 1–6.

Christensen, E. M., J. J. Clausen & J. T. Curtis. 1959. Phytosociology of the lowland forests of northern Wisconsin. *Amer. Midl. Naturalist* 62: 232–247.

Christensen, K. K. & K. Sand-Jensen. 1998. Precipitated iron and manganese plaques restrict root uptake of phosphorus in *Lobelia dortmanna. Can. J. Bot.* 76: 2158–2163.

Christensen, L. P. & K. Brandt. 2006. Bioactive polyacetylenes in food plants of the Apiaceae family: occurrence, bioactivity and analysis. *J. Pharm. Biomed. Analysis.* 41: 683–693.

Christy, N. L. 1987. Distyly, pollen flow and seed set in *Menyanthes trifoliata* (Menyanthaceae). MS thesis. University of British Columbia, Vancouver, BC. 119 pp.

Christy, J. A. 1993. *Classification and Catalog of Native Plant Communities in Oregon*. Oregon Natural Heritage Program, Corvallis, OR. 77 pp.

Christy, J. A. 2004. *Native Freshwater Wetland Plant Associations of Northwestern Oregon*. Oregon Natural Heritage Information Center, Oregon State University, Corvallis, OR. 246 pp.

Christy, J. A. 2013. *Wet Meadow Plant Associations, Malheur National Wildlife Refuge, Harney, County, Oregon*. Oregon Biodiversity Information Center, Institute for Natural Resources, Portland State University, Portland, OR. 73 pp.

Christy, J. A. & T. A. Meyer. 1991. Bryophytes of algific talus slopes in Wisconsin's driftless area. *Rhodora* 93: 242–247.

Christybapita, D., M. Divyagnaneswari & R. D. Michael. 2007. Oral administration of *Eclipta alba* leaf aqueous extract enhances the non-specific immune responses and disease resistance of *Oreochromis mossambicus. Fish Shellfish Immun.* 23: 840–852.

Chrysler, M. A. 1938. The winter buds of *Brasenia. Bull. Torrey Bot. Club* 65: 277–283.

Chu, G. 2003. 12. *Suckleya* A. Gray. pp. 305–306. *In*: N. R. Morin (convening ed.), *Flora of North America North of Mexico, Vol. 4: Magnoliophyta: Caryophyllidae, Part 1*. Oxford University Press, New York.

Chu, G. L., H. C. Stutz & S. C. Sanderson. 1991. Morphology and taxonomic position of *Suckleya suckleyana* (Chenopodiaceae). *Amer. J. Bot.* 78: 63–68.

Chuang, T. I. & L. R. Heckard. 1971. Observations on root-parasitism in *Cordylanthus* (Scrophulariaceae). *Amer. J. Bot.* 58: 218–228.

Chuang, T. I. & L. R. Heckard. 1973. Taxonomy of *Cordylanthus* subgenus *Hemistegia* (Scrophulariaceae). *Brittonia* 25: 135–158.

Chuang, T. I. & L. R. Heckard. 1986. Systematics and evolution of *Cordylanthus* (Scrophulariaceae-Pedicularieae). *Syst. Bot. Monogr.* 10: 1–105.

Chuang, T. I. & L. R. Heckard. 1991. Generic realignment and synopsis of subtribe *Castillejinae* (Scrophulariaceae-tribe Pedicularieae). *Syst. Bot.* 16: 644–666.

Chuang, T. I. & L. R. Heckard. 1992a. Nomenclatural changes of some Californian *Castilleja* (Scrophulariaceae). *Novon* 2: 185–189.

Chuang, T. I. & L. R. Heckard. 1992b. Chromosome numbers of some North American Scrophulariaceae, mostly Californian. *Madroño* 39: 137–149.

Chuang, T. I. & R. Ornduff. 1992. Seed morphology and systematics of Menyanthaceae. *Amer. J. Bot.* 79: 1396–1406.

Church, S. A. 2003. Molecular phylogenetics of *Houstonia* (Rubiaceae): descending aneuploidy and breeding system evolution in the radiation of the lineage across North America. *Mol. Phylogen. Evol.* 27: 223–238.

Churchill, D. B., A. G. Berlage, D. M. Bilsland & T. M. Cooper. 1990. Conditioning wildflower seed. *Trans. Amer. Soc. Agric. Eng.* 33: 549–552.

Churchill, J. R. 1902. Some plants from Prince Edward Island. *Rhodora* 4: 31–36.

Cieslak, T., J. S. Polepalli, A. White, K. Müller, T. Borsch, W. Barthlott, J. Steiger, A. Marchant & L. Legendre. 2005. Phylogenetic analysis of *Pinguicula* (Lentibulariaceae): chloroplast DNA sequences and morphology support several geographically distinct radiations. *Amer. J. Bot.* 92: 1723–1736.

Civico, R. S. A. & K. Moody. 1979. The effect of the time and depth of submergence on growth and development of some weed species. *Philipp. J. Weed Sci.* 6: 41–49.

Clark, D. L. & M. V. Wilson. 1998. *Fire Effects on Wetland Prairie Plant Species*. Final report, Project No. 14-48-0001-96749. U.S. Fish and Wildlife Service, Western Oregon Refuges, Corvallis, OR.

Clark, D. L. & M. V. Wilson. 2001. Fire, mowing, and hand-removal of woody species in restoring a native wetland prairie in the Williamette Valley of Oregon. *Wetlands* 21: 135–144.

Clark, F. H. 2003. *Rhynchospora nitens* (Vahl) A. Gray Short-Beaked Bald-Sedge. New England Plant Conservation Program, conservation and research plan. New England Wild Flower Society, Framingham, MA. 26 pp.

Clark, J. M. 2011. Plant biomass allocation and competitive interactions in coastal wetlands of the Chesapeake Bay: experimental and observational studies. MS thesis. University of Maryland, College Park, MD. 105 pp.

Clark, J. R. 1997a. Endangered and threatened wildlife and plants; determination of endangered status for nine plants from the grasslands or mesic areas of the central coast of California. *Fed. Reg.* 62: 55791–55808.

Clark, J. R. 1997b. Endangered and threatened wildlife and plants; determination of endangered status for two tidal marsh plants—*Cirsium hydrophilum* var. *hydrophilum* (Suisun thistle) and *Cordylanthus mollis* ssp. *mollis* (soft bird's-beak) from the San Francisco Bay area of California. *Fed. Reg.* 62: 61916–61925.

Clark, J. R. 1998. Endangered and threatened wildlife and plants; determination of threatened status for Virginia sneezeweed (*Helenium virginicum*), a plant From the Shenandoah Valley of Virginia. *Fed. Reg.* 63: 59239–59244.

Clark, J. R. 2000. Endangered and threatened wildlife and plants; endangered status for the plant *Plagiobothrys hirtus* (rough popcornflower). *Fed. Reg.* 65: 3866–3875.

Clark, J. R. & V. C. Pence. 1999. *In vitro* propagation of *Lobelia boykinii*, a rare wetland species. *In Vitro Cell Dev.* Biol. 35: 64A.

Clark, M. A., J. Siegrist & P. A. Keddy. 2008. Patterns of frequency in species-rich vegetation in pine savannas: effects of soil moisture and scale. *Ecoscience* 15: 529–535.

Clark, S. M. 1983. A revision of the genus *Microrhopala* (Coleoptera: Chrysomelidae) in America north of Mexico. *Great Basin Naturalist* 43: 597–618.

Clark, W. E. 1978. Notes on the life history, and descriptions of the larva and pupa of *Neotylopterus pallidus* (LeConte) (Coleoptera: Curculionidae), a seed predator of *Forestiera acuminata* (Michx.) Poir. (Oleaceae). *Coleopt. Bull.* 32: 177–184.

Clarke, P. J. 1995. The population dynamics of the mangrove *Avicennia marina*; demographic synthesis and predictive modeling. *Hydrobiologia* 295: 83–88.

Clarke, P. J. & P. J. Myerscough. 1991a. Buoyancy of *Avicennia marina* propagules in south-eastern Australia. *Aust. J. Bot.* 39: 77–83.

Clarke, P. J. & P. J. Myerscough. 1991b. Floral biology and reproductive phenology of *Avicennia marina* in south-eastern Australia. *Aust. J. Bot.* 39: 283–293.

Clarkson, C., V. J. Maharaj, N. R. Crouch, O. M. Grace, P. Pillay, M. G. Matsabisa, N. Bhagwandin, P. J. Smith & P. I. Folb. 2004. *In vitro* antiplasmodial activity of medicinal plants native to or naturalised in South Africa. *J. Ethnopharmacol.* 92: 177–191.

Clarkson, R. B. 1958. Scotch heather in North America. *Castanea* 23: 119–130.

Clausen, J. & W. M. Hiesey. 1958. *Experimental Studies on the Nature of Species. IV. Genetic Structure of Ecological Races*. Carnegie Inst. Wash. Publ. 615, Washington, DC. 312 pp.

Clausen, J., D. D. Keck & W. M. Hiesey. 1940. *Experimental Studies on the Nature of Species. I. The Effect of Varied Environments on Western North American plants*. Carnegie Inst. Wash. Publ. 520. Washington, DC. 452 pp.

Clausen, J. C., I. M. Ortega, C. M. Glaude, R. A. Relyea, G. Garay & O. Guineo. 2006. Classification of wetlands in a Patagonian National Park, Chile. *Wetlands* 26: 217–229.

Clausen, K. E. 1966. Studies of incompatibility in Betula. pp. 48–52.*In: Joint Proceedings, Second Genetics Workshop of Society of American Foresters and Seventh Lake States Forest Tree Improvement Conference*. United States Forest Research Paper, NC-6. North Central Forest Experiment Station, St. Paul, MN.

Clausing, G. & S. S. Renner. 2001. Molecular phylogenetics of Melastomataceae and Memecyclaceae: implications for character evolution. *Amer. J. Bot.* 88: 486–498.

Clausnitzer, D. & J. H. Huddleston. 2002. Wetland determination of a southeast Oregon vernal pool and management implications. *Wetlands* 22: 677–685.

Clausnitzer, D., J. H. Huddleston, E. Horn, M. Keller & C. Leet. 2003. Hydric soils in a southeastern Oregon vernal pool. *J. Soil Sci. Soc. Amer.* 67: 951–960.

Clawson, A. B. 1933. Alpine kalmia (*Kalmia microphylla*) as a stock-poisoning plant. *Bull. USDA* 391: 1–9.

Clay, D. V., C. Nash & J. A. Bailey. 1991. An association between triazine resistance and powdery mildew resistance in *Epilobium ciliatum* and *Senecio vulgaris*. *Pestic. Sci.* 33: 189–196.

Clayton, R. & D. Orton. 2004. Contact allergy to spearmint oil in a patient with oral lichen planus. *Contact Derm.* 51: 314–315.

Clemants, S. E. 2003. 9. *Alternanthera* Forsskål. pp. 447–451. *In*: N. R. Morin (convening ed.), *Flora of North America North of Mexico, Vol. 4: Magnoliophyta: Caryophyllidae, Part 1*. Oxford University Press, New York.

Clemants, S. E. & S. L. Mosyakin. 2003. 7. *Chenopodium* Linnaeus. pp. 275–299. *In*: N. R. Morin (convening ed.), *Flora of North America North of Mexico, Vol. 4: Magnoliophyta: Caryophyllidae, Part 1*. Oxford University Press, New York.

Clement, W. L. & M. J. Donoghue. 2011. Dissolution of *Viburnum* section *Megalotinus* (Adoxaceae) of southeast Asia and its implications for morphological evolution and biogeography. *Int. J. Plant Sci.* 172: 559–573.

Clement, W. L., M. Arakaki, P. Sweeney, E. J. Edwards & M. J. Donoghue. 2014. A chloroplast tree for *Viburnum* (Adoxaceae) and its implications for phylogenetic classification and character evolution. *Amer. J. Bot.* 101: 1029–1046.

Clewell, A. F. 1985. *Guide to the Vascular Plants of the Florida Panhandle.* Florida State University Press, Tallahassee, FL. 605 pp.

Clifford, H. T. 1959. Seed dispersal by motor vehicles. *J. Ecol.* 47: 311–315.

Clokey, I. W. 1951. Flora of the Charleston Mountains, Clark County, Nevada. *Univ. Calif. Publ. Bot.* 24: 1–274.

Cloyd, R. A. & E. R. Zaborski. 2004. Fungus gnats, *Bradysia* spp. (Diptera: Sciaridae), and other arthropods in commercial bagged soilless growing media and rooted plant plugs. *J. Econ. Entomol.* 97: 503–510.

Coats, R., M. A. Showers & B. Pavlik. 1993. Management plan for an alkali sink and its endangered plant *Cordylanthus palmatus*. *Environ. Manag.* 17: 115–127.

Cobbaert, D., L. Rochefort & J. S. Price. 2004. Experimental restoration of a fen plant community after peat mining. *Appl. Veg. Sci.* 7: 209–220.

Cockel, C. P. 2010. Alien and native plants of urban river corridors: a study of riparian plant propagule dynamics along the River Brent, greater London. PhD dissertation. Queen Mary University of London, London. 283 pp.

Cody, W. J. 1954. A history of *Tillaea aquatica* (Crassulaceae) in Canada and Alaska. *Rhodora* 56: 96–101.

Cody, W. J. 2000. *Flora of the Yukon Territory, 2nd Edition.* NRC Research Press, Ottawa, ON. 669 pp.

Cody, W. J., K. L. Reading & J. M. Line. 2003. Additions and range extensions to the vascular plant flora of the continental Northwest Territories, Canada, II. *Can. Field Naturalist* 117: 448–465.

Coe, F. G. 2008. Ethnobotany of the Rama of southeastern Nicaragua and comparisons with Miskitu plant lore. *Econ. Bot.* 62: 40–59.

Coe, F. G. & G. J. Anderson. 2005. Snakebite ethnopharmacopoeia of eastern Nicaragua. *J. Ethnopharmacol.* 96: 303–323.

Coelho de Souza, G., A. P. S. Haas, G. L. von Poser, E. E. S. Schapoval & E. Elisabetsky. 2004. Ethnopharmacological studies of antimicrobial remedies in the south of Brazil. *J. Ethnopharmacol.* 90: 135–143.

Coffey, V. J. & S. B. Jones Jr. 1980. Biosystematics of *Lysimachia* section *Seleucia* (Primulaceae). *Brittonia* 32: 309–322.

Cofrancesco Jr., A. F. 1984. *Alligatorweed and its Biocontrol Agents.* Information Exchange Bulletin Vol A-84-3. Aquatic Plant Control Research Program. Department of the Army, U.S. Army Corps of Engineers, Washington, DC. 6 pp.

Cohen, S., R. Braham & F. Sanchez. 2004. Seed bank viability in disturbed longleaf pine sites. *Restorat. Ecol.* 12: 503–515.

Coin, P. 2005. *Marshallia*: beetle magnet. *New Hope Audubon Soc. Newslett.* 31(3): 2.

Cole, C. A. 1992. Wetland vegetation ecology on a reclaimed coal surface mine in southern Illinois, USA. *Wetlands Ecol. Manag.* 2: 135–142.

Cole, C. T. 1998. *Genetic Variation and Population Differentiation in* Polemonium occidentale *var.* lacustre. Report Number 9507713 to the Wisconsin Department of Natural Resources, Madison, WI.

Cole, C. T. 2003. Genetic variation in rare and common plants. *Annu. Rev. Ecol. Evol. Syst.* 34: 213–237.

Cole, D. N. 1995. Experimental trampling of vegetation. II. Predictors of resistance and resilience. *J. Appl. Ecol.* 32: 215–224.

Cole, D. N. & C. A. Monz. 2002. Trampling disturbance of high-elevation vegetation, Wind River Mountains, Wyoming, U.S.A. *Arctic Antarc. Alp. Res.* 34: 365–376.

Coleman, L. B. 2007. An evaluation of the natural and provisioned feeding rates of semi-free ranging ringtailed lemurs (*Lemur catta*) on St. Catherines Island, GA. MA thesis. Texas State University-San Marcos, San Marcos, TX. 48 pp.

Coletta, J. 2014. Evaluation of native and ornamental plant species for establishment and pollutant capture in bioretention basins. MS thesis. Michigan State University, East Lansing, MI. 133 pp.

Colin, L. J. & C. E. Jones. 1980. Pollen energetics and pollination modes. *Amer. J. Bot.* 67: 210–215.

Collet, D. M. 2010. Rearing experiment to determine the willow host range of *Rabdophaga* spp. in Alaska. *AKES Newslett.* 3: 9–11.

Collin, C. L. & J. A. Shykoff. 2003. Outcrossing rates in the gynomonoecious-gynodioecious species *Dianthus sylvestris* (Caryophyllaceae). *Amer. J. Bot.* 90: 579–585.

Collinge, S. K., C. Ray & J. T. Marty. 2013. A long-term comparison of hydrology and plant community composition in constructed versus naturally occurring vernal pools. *Restorat. Ecol.* 21: 704–712.

Collins, B. & G. Wein. 1995. Seed bank and vegetation of a constructed reservoir. *Wetlands* 15: 374–385.

Collins, D. P., W. C. Conway, C. D. Mason & J. W. Gunnels. 2013. Seed bank potential of moist-soil managed wetlands in east-central Texas. *Wetlands Ecol. Manag.* 21: 353–366.

Collins, S. L. & W. H. Blackwell Jr. 1979. *Bassia* (Chenopodiaceae) in North America. *Sida* 8: 57–64.

Colwell, A. E. L., C. J. Sheviak & P. E. Moore. 2007. A new *Platanthera* (Orchidaceae) from Yosemite National Park. *Madroño* 54: 86–93.

Combes, R. 1965. Contribution a l'étude de *Veronica anagallis-aquatica* L.: existence d'une pseudo-cléistogamie expérimentale. *Rev. Gén. Bot.* 72: 323–330.

Combroux, I. C. S., G. Bornette & C. Amoros. 2002. Plant regenerative strategies after a major disturbance: the case of a riverine wetland restoration. *Wetlands* 22: 234–246.

Combs, D. L. & R. D. Drobney. 1991. *Aquatic and Wetland Plants of Missouri.* Missouri Cooperative Fish and Wildlife Research Unit, U.S. Fish and Wildlife Service, Columbia, MO. 352 pp.

Comer, P., K. Goodin, G. Hammerson, S. Menard, M. Pyne, M. Reid, M. Robles, M. Russo, L. Sneddon, K. Snow, A. Tomaino & M. Tuffly. 2005. *Biodiversity Values of Geographically Isolated Wetlands: An Analysis of 20 U.S. States.* NatureServe, Arlington, VA. 49 pp.

Comes, R. D., V. F. Bruns & A. D. Kelley. 1978. Longevity of certain weed and crop seeds in fresh water. *Weed Sci.* 26: 336–344.

Compton, B. D. 1993. Upper North Wakashan and Southern Tsimshian ethnobotany: the knowledge and usage of plants. PhD dissertation. University of British Columbia, Victoria, BC. 530 pp.

Conard, H. S. 1905. The waterlilies: a monograph of the genus *Nymphaea*. *Publ. Carnegie Inst. Wash.* 4: 1–279.

Conard, H. S. 1924. Second survey of the vegetation of a Long Island salt marsh. *Ecology* 5: 379–388.

Conard, H. S. 1935. The plant associations of central Long Island. A study in descriptive plant sociology. *Amer. Midl. Naturalist* 16: 433–516.

Conard, H. S. & G. C. Galligar. 1929. Third survey of a Long Island salt marsh. *Ecology* 10: 326–336.

Conard, H. S. & H. Hus. 1914. *Water-lilies and How to Grow Them.* Doubleday, Page & Co., New York. 228 pp.

Conger, A. C. 1912. Some entomophilous flowers of Cedar Point, Ohio. *Ohio Naturalist* 12: 500–504.

Conn, J. S., C. A. Stockdale & J. C. Morgan. 2008. Characterizing pathways of invasive plant spread to Alaska: I. Propagules from container-grown ornamentals. *Invasive Plant Sci. Manag.* 1: 331–336.

Connelly, W. J., D. J. Orth & R. K. Smith. 1999. Habitat of the riverweed darter, *Etheostoma podostemone* Jordan, and the decline of riverweed, *Podostemum ceratophyllum*, in the tributaries of the Roanoke River, Virginia. *J. Freshwater Ecol.* 14: 93–102.

Conner, W. H., J. G. Gosselink & R. T. Parrondo. 1981. Comparison of the vegetation of three Louisiana swamp sites with different flooding regimes. *Amer. J. Bot.* 68: 320–331.

Conners, I. L. 1967. An annotated index of plant diseases in Canada and Fungi recorded on plants in Alaska, Canada and Greenland. *Rep. Res. Dept. Agri. Can.* 1251: 1–381.

Connor, K., G. Schaefer, J. Donahoo, M. Devall, E. Gardiner, T. Hawkins, D. Wilson, N. Schiff, P. Hamel & T. Leininger. 2007. Development, fatty acid composition, and storage of drupes and seeds from the endangered pondberry (*Lindera melissifolia*). *Biol. Conserv.* 137: 489–496.

Connor, K. F., G. M. Schaefer, J. B. Donahoo, M. S. Devall, E. S. Gardiner, T. D. Leininger, A. D. Wilson, N. M. Schiff, P. B. Hamel & S. J. Zarnoch. 2012. *Lindera melissifolia* seed bank study in a lower Mississippi Alluvial Valley bottomland forest. *Seed Technol.* 34: 163–172.

Constance, L. & J. M. Affolter. 2004. Apiaceae. pp. 20–22. *In*: N. Smith, S. A. Mori, A. Henderson, D. W. Stevenson & S. V. Heald (eds.), *Flowering Plants of the Neotropics*. Princeton University Press, Princeton, NJ. 594 pp.

Constantinidis, T., G. Kamari & D. Phitos. 1997. A cytological study of 28 phanerogams from the mountains of SE Sterea Ellas, Greece. *Willdenowia* 27: 121–142.

Conti, E., A. Litt, P. G. Wilson, S. A. Graham, B. G. Briggs, L. A. S. Johnson & K. J. Sytsma. 1997. Interfamilial relationships in Myrtales: molecular phylogeny and patterns of morphological evolution. *Syst. Bot.* 22: 629–647.

Conti, E., T. Eriksson, J. Schönenberger, K. J. Sytsma & D. A. Baum. 2002. Early Tertiary out-of-India dispersal of Crypteroniaceae: evidence from phylogeny and molecular dating. *Evolution* 56: 1931–1931.

Cook, C. D. K. 1966. A monographic study of *Ranunculus* subgenus *Batrachium* (DC.) A. Gray. *Mitteil. Bot. Staat. München* 6: 47–237.

Cook, C. D. K. 1973. New and noteworthy plants from the northern Italian ricefields. *Ber. Schweiz. Bot. Ges.* 83: 54–65.

Cook, C. D. K. 1978. The *Hippuris* syndrome. pp. 164–176. *In*: H. E. Street (ed.), *Essays in Plant Taxonomy*. Academic Press, London and New York.

Cook, C. D. K. 1979. A revision of the genus *Rotala* (Lythraceae). *Boissiera* 29: 1–156.

Cook, C. D. K. 1988. Wind pollination in aquatic angiosperms. *Ann. Bot. Gard.* 75: 768–777.

Cook, C. D. K. 1990. Seed dispersal of *Nymphoides peltata* (S.G. Gmelin) O. Kuntze (Menyanthaceae). *Aquat. Bot.* 37: 325–340.

Cook, C. D. K. 1996a. *Aquatic Plant Book*. SPB Academic Publishing bv, Amsterdam, The Netherlands. 228 pp.

Cook, C. D. K. 1996b. *Aquatic and Wetland Plants of India: A Reference Book and Identification Manual for the Vascular Plants Found in Permanent or Seasonal Fresh Water in the Subcontinent of India South of the Himalayas*. Oxford University Press, New York. 385 pp.

Cook, C. D. K. 2004. *Aquatic and Wetland Plants of Southern Africa. An Identification Manual for the Stoneworts (Charophytina), Liverworts (Marchantiopsida), Mosses (Bryopsida), Quillworts (Lycopodiopsida), Ferns (Polypodiopsida) and Flowering Plants (Magnoliopsida) Which Grow in Water and Wetlands of Namibia, Botswana, Swaziland, Lesotho and Republic of South Africa*. Backhuys Publishers BV, Leiden, The Netherlands. 281 pp.

Cook, C. D. K., B. J. Gut, E. M. Rix, J. Schneller & M. Seitz. 1974. *Water Plants of the World: a Manual for the Identification of the Genera of Freshwater Macrophytes*. Dr. W. Junk b.v., Publishers, The Hague, The Netherlands. 561 pp.

Cook, D. A., D. M. Decker & J. L. Gallagher. 1989. Regeneration of *Kosteletzkya virginica* (L.) Presl. Seashore Mallow from callus cultures. *Plant Cell Tissue Organ Cult.* 17: 111–119.

Cook, M. B. 1986. Hybridization between *Dodecatheon alpinum* (A. Gray) Green and *D. redolens* (Hall) H. J. Thompson (Primulaceae) in the southern Sierra Nevada. MA thesis. California State University, Fullerton, CA. 85 pp.

Cook, S. A. & M. P. Johnson. 1968. Adaptation to heterogeneous environments. I. Variation in heterophylly in *Ranunculus flammula* L. *Evolution* 22: 496–516.

Cook, W. C. 1967. *Life History, Host Plants, and Migrations of the Beet Leafhopper in the Western United States*. Technical Bulletin No. 1365. Agricultural Research Service, U.S. Department of Agriculture, Washington, DC. 122 pp.

Cooke, S. S. (ed.). 1997. *A Field Guide to the Common Wetland Plants of Western Washington & Northwestern Oregon*. Seattle Audubon Society, Seattle, WA. 417 pp.

Cooke, J. C. & M. W. Lefor. 1998. The mycorrhizal status of selected plant species from Connecticut wetlands and transition zones. *Restorat. Ecol.* 6: 214–222.

Cooke, W. B. 1969. The 1965 Illinois foray. *Mycologia* 61: 817–822.

Cooper, C. M., M. T. Moore, E. R. Bennett, S. Smith Jr., J. L. Farris, C. D. Milam & F. D. Shields Jr. 2004a. Innovative uses of vegetated drainage ditches for reducing agricultural runoff. *Water Sci. Technol.* 49: 117–123.

Cooper, D. J. 1989. *A Handbook of Wetland Plants of the Rocky Mountain Region*. EPA Region VIII. U.S. Environmental Protection Agency, Denver, CO. 125 pp.

Cooper, D. J. 1991. Additions to the peatland flora of the southern Rocky Mountains: habitat descriptions and water chemistry. *Madroño* 38: 139–141.

Cooper, D. J. 1996. Water and soil chemistry, floristics, and phytosociology of the extreme rich High Creek fen in South Park, Colorado, U.S.A. *Can. J. Bot.* 74: 1801–1811.

Cooper, E. J., I. G. Alsos, D. Hagen, F. M. Smith, S. J. Coulson & I. D. Hodkinson. 2004b. Plant recruitment in the High Arctic: seed bank and seedling emergence on Svalbard. *J. Veg. Sci.* 15: 115–124.

Cooper, D. J. & J. S. Sanderson. 1997. A montane *Kobresia myosuroides* fen community type in the southern Rocky Mountains of Colorado, U.S.A. *Arct. Alp. Res.* 29: 300–303.

Cooper, D. J. & E. C. Wolf. 2006. *Fens of the Sierra Nevada, California*. Department of Forest, Rangeland and Watershed Stewardship, Colorado State University, Fort Collins, CO. 47 pp.

Cooper, K. W. 1952. Records and flower preferences of masarid wasps. II. Polytropy or oligotropy in *Pseudomasaris*? (Hymenoptera: Vespidae). *Amer. Midl. Naturalist* 48: 103–110.

Cooper, K. W. & J. Bequaert. 1950. Records and flower preferences of masarid wasps (Hymenoptera: Vespidae). *Psyche* 57: 137–142.

Cooper, R. L., J. M. Osborn & C. T. Philbrick. 2000. Comparative pollen morphology and ultrastructure of the Callitrichaceae. *Amer. J. Bot.* 87: 161–175.

Cooper, S. V. & W. M. Jones. 2004. *A Plant Community Classification for Kootenai National Forest Peatlands*. Report to the Kootenai National Forest, Montana. Montana Natural Heritage Program, Helena, MT. 19 pp. (+ appendices).

Cooper, S. V., C. Jean & B. L. Heidel. 1999. *Plant Associations and Related Botanical Inventory of the Beaverhead Mountains Section, Montana*. United States Department of the Interior, Bureau of Land Management, Billings, MT. 245 pp.

Cooper, W. S. 1913. The climax forest of Isle Royale, Lake Superior, and its development. II. *Bot. Gaz.* 55: 115–140.

Cooper, W. S. 1939a. Additions to the flora of the Glacier Bay National Monument, Alaska, 1935–1936. *Bull. Torrey Bot. Club* 66: 453–456.

Cooper, W. S. 1939b. A fourth expedition to Glacier Bay, Alaska. *Ecology* 20: 130–155.

Cooper, W. S. 1942. Vegetation of the Prince William Sound region, Alaska; with a brief excursion into post-Pleistocene climatic history. *Ecol. Monogr.* 12: 1–22.

Cooperrider, T. S. & M. M. Galang. 1965. A *Pluchea* hybrid from the Pacific. *Amer. J. Bot.* 52: 1020–1026.

Cooperrider, T. S. & G. A. McCready. 1970. Chromosome numbers in *Chelone* (Scrophulariaceae). *Brittonia* 22: 175–183.

Copley, G. 1999. A life history study of *Hottonia inflata*. MS thesis. University of Connecticut, Storrs, CT. 113 pp.

Coppedge, B. R. & J. H. Shaw. 1997. Effects of horning and rubbing behavior by bison (*Bison bison*) on woody vegetation in a tallgrass prairie landscape. *Amer. Midl. Naturalist* 138: 189–196.

Cordazzo, C. V., V. L. Caetano & C. S. B. Costa. 2007. *Jaumea linearifolia* (Juss.) DC. (Asteraceae), primeiro registro para o Brasil. *Iheringia* 62: 99–102.

Cordes, J. M. & S. R. Downie. 2007. Ascertaining the phylogenetic position and infrageneric relationships of *Conium* (Apiaceae), the odd man out. Botany and Plant Biology 2007 joint congress (abstract).

Cordes, J. M. & S. R. Downie. 2008. Molecular systematic evidence supporting distinction of a rare Greek endemic and its southern African relatives: the genus *Conium* redefined. Botany 2008 meetings (abstract).

Cordo, H. A., C. J. Deloach & R. Ferrer. 1982. The weevils *Lixellus, Tanysphiroideus*, and *Cyrtobagous* that feed on *Hydrocotyle* and *Salvinia* in Argentina. *Coleopt. Bull.* 2: 279–286.

Córdova, W. H. P., J. Tabart, A. G. Quesdada, A. Sipel, A. L. P. Hill, C. Kevers & J. Dommes. 2010. Antioxidant capacity of three Cuban species of the genus *Pluchea* Cass. (Asteraceae). *J. Food Biochem.* 34: 249–261.

Core, E. L. 1967. Ethnobotany of the southern Appalachian aborigines. *Econ. Bot.* 21: 198–214.

Cornwell, W. K., B. L. Bedford & C. T. Chapin. 2001. Occurrence of arbuscular mycorrhizal fungi in a phosphorus-poor wetland and mycorrhizal response to phosphorus fertilization. *Amer. J. Bot.* 88: 1824–1829.

Correll, D. S. 1966. Some additions and corrections to the flora of Texas—III. *Rhodora* 68: 420–428.

Correll, D. S. & H. B. Correll. 1975. *Aquatic and Wetland Plants of Southwestern United States.* 2 volumes. Stanford University Press, Stanford, CA. 1777 pp.

COS (County of Sacramento) 2010. *Draft South Sacramento Habitat Conservation Plan.* Sacramento County, CA. unpaginated.

COSEWIC. 2004a. COSEWIC *Assessment and Update Status Report on the Victorin's Gentian* Gentianopsis procera macounii *var.* victorinii *in Canada.* Committee on the Status of Endangered Wildlife in Canada. Ottawa, ON. vii, 24 pp. Available online: http://www.registrelep-sararegistry.gc.ca/document/default_e.cfm?documentID=453.

COSEWIC. 2004b. COSEWIC *Assessment and Update Status Report on the Victorin's Water-Hemlock,* Cicuta maculata *var.* victorinii *in Canada.* Committee on the Status of Endangered Wildlife in Canada. Ottawa, ON. vii, 21 pp. Available online: http://www.registrelep-sararegistry.gc.ca/document/default_e.cfm?documentID=428.

COSEWIC. 2004c. COSEWIC *Assessment and Update Status Report on the Eastern Lilaeopsis* Lilaeopsis chinensis *in Canada.* Committee on the Status of Endangered Wildlife in Canada.

Ottawa, ON. vi, 18 pp. Available online: http://publications.gc.ca/site/archivee-archived.html?url=http://publications.gc.ca/collections/Collection/CW69-14-392-2004E.pdf.

Cossard, G., J. Sannier, H. Sauquet, C. Damerval, L. R. de Craene, F. Jabbour & S. Nadot. 2016. Subfamilial and tribal relationships of Ranunculaceae: evidence from eight molecular markers. *Plant Syst. Evol.* 302: 419–431.

Costea, M. & F. Tardif. 2003. *Koenigia islandica* (Polygonaceae) new for Utah. *Sida* 20: 1317.

Costea, M., F. J. Tardif & H. R. Hinds. 2005a. 30. *Polygonum* Linnaeus. pp. 547–571. *In*: Flora North America Editorial Committee (ed.), *Flora of North America North of Mexico, Vol. 5: Magnoliophyta: Caryophyllidae, part 2.* Oxford University Press, New York.

Costea, M., S. E. Weaver & F. J. Tardif. 2005b. The biology of invasive alien plants in Canada. 3. *Amaranthus tuberculatus* (Moq.) Sauer var. *rudis* (Sauer) Costea & Tardif. *Can. J. Plant Sci.* 85: 507–522.

Costea, M., M. A. R. Wright & S. Stefanović. 2009. Untangling the systematics of salt marsh dodders: *Cuscuta pacifica*, a new segregate species from *Cuscuta salina* (Convolvulaceae). *Syst. Bot.* 34: 787–795.

Costello, D. F. 1936. Tussock meadows in southeastern Wisconsin. *Bot. Gaz.* 97: 610–648.

Costello, D. F. 1944. Important species of the major forage types in Colorado and Wyoming. *Ecol. Monogr.* 14: 107–134.

Costelloe, B. H. 1988. Pollination ecology of *Gentiana andrewsii*. *Ohio J. Sci.* 88: 132–138.

Cosyns, E. 2004. *Ungulate Seed Dispersal: Aspects of Endozoochory in a Seminatural Landscape.* Institute of Nature Conservation, Brussels, Belgium. 178 pp.

Cosyns, E., S. Claerbout, I. Lamoot & M. Hoffmann. 2005. Endozoochorous seed dispersal by cattle and horse in a spatially heterogeneous landscape. *Plant Ecol.* 178: 149–162.

Cottam, C. 1939. *Food Habits of North American Diving Ducks.* Technical Bulletin No. 643. U.S. Department of Agriculture, Washington, DC. 139 pp.

Cottam, C. & P. Knappen. 1939. Food of some uncommon North American birds. *Auk* 56: 138–169.

Couch, R. & E. Nelson. 1988. *Myriophyllum quitense* (Haloragaceae) in the United States. *Brittonia* 40: 85–88.

Coulter, J. M. 1882. Some notes on *Physostegia virginiana. Bot. Gaz.* 7: 111–112.

Coulter, J. M. & J. N. Rose. 1887. Notes on Umbellifera of E. United States. VIII. *Bot. Gaz.* 12: 291–295.

Coulter, J. M. & J. N. Rose. 1889. Notes on North American Umbellifera. I. *Bot. Gaz.* 14: 274–284.

Coulter, M. W. 1953. Mallard nesting in Maine. *Auk* 70: 490.

Coulter, S. M. 1904. An ecological comparison of some typical swamp areas. *Missouri Bot. Gard. Ann. Rep.* 1904: 39–71.

Countryman, W. D. 1970. The history, spread and present distribution of some immigrant aquatic weeds in New England. *Hyacinth Control J.* 8: 50–52.

Cousens, M. I., D. G. Lacey & J. M. Scheller. 1988. Safe sites and the ecological life history of *Lorinseria areolata. Amer. J. Bot.* 75: 797–807.

Couvreur, M., B. Vandenberghe, K. Verheyen & M. Hermy. 2004. An experimental assessment of seed adhesivity on animal furs. *Seed Sci. Res.* 14: 147–159.

Coville, F. V. 1897. Notes on the plants used by the Klamath Indians of Oregon. *Contr. U.S. Natl. Herb.* 5: 87–108.

Coville, F. V. 1902. Wokas, a primitive food of the Klamath Indians. *Rept. U.S. Natl. Mus.* 130: 725–740.

Cowan, I. M. 1945. The ecological relationships of the food of the Columbian black-tailed deer, *Odocoileus hemionus columbianus* (Richardson), in the Coast Forest region of southern Vancouver Island, British Columbia. *Ecol. Monogr.* 15: 109–139.

Cowardin, L. M., V. Carter, F. C. Golet & E. T. LaRoe. 1979. *Classification of Wetlands and Deepwater Habitats of the United States.* FWS/OBS-79/31, Office of Biological Services, Fish & Wildlife Service. U.S. Department of the Interior, Washington, DC. 131 pp.

Cowgill, U. M. 1973. Biogeochemical cycles for the chemical elements in *Nymphaea odorata* Ait. and the aphid *Rhopalosiphum nymphaeae* (L.) living in Linsley Pond. *Sci. Total Environ.* 2: 259–303.

Cox, G. W. 2001. An inventory and analysis of the alien plant flora of New Mexico. *New Mexico Bot.* 17: 1–7.

Cox, P. B. & L. E. Urbatsch. 1994. A taxonomic revision of *Rudbeckia* subg. *Macrocline* (Asteraceae: Heliantheae: Rudbeckiinae). *Castanea* 59: 300–318.

Cozza, R., G. Galanti, M. B. Bitonti & A. M. Innocenti. 1994. Effect of storage at low-temperature on the germination of the waterchestnut (*Trapa natans* L.). *Phyton (Horn)* 34: 315–320.

Craddock, C. L. & L. F. Huenneke. 1997. Aquatic seed dispersal and its implications in *Cirsium vinaceum*, a threatened endemic thistle of New Mexico. *Amer. Midl. Naturalist* 138: 215–219.

Crain, C. M. 2008. Interactions between marsh plant species vary in direction and strength depending on environmental and consumer context. *J. Ecol.* 96: 166–173.

Crain, C. M., L. K. Albertson & M. D. Bertness. 2008. Secondary succession dynamics in estuarine marshes across landscape-scale salinity gradients. *Ecology* 89: 2889–2899.

Craine, S. I. 2002. *Rhexia mariana* L. (Maryland meadowbeauty) Conservation and Research Plan for New England. New England Wild Flower Society, Framingham, MA. Available online: http://www.newfs.org.

Craine, S. I. & C. M. Orians. 2004. Pitch pine (*Pinus rigida* Mill.) invasion of Cape Cod pond shores alters abiotic environment and inhibits indigenous herbaceous species. *Biol. Conserv.* 116: 181–189.

Crandall, R. M. & W. J. Platt. 2012. Habitat and fire heterogeneity explain the co-occurrence of congeneric resprouter and reseeder *Hypericum* spp. along a Florida pine savanna ecocline. *Plant Ecol.* 213: 1643–1654.

Crandall, R. M. & R. J. Tyrl. 2006. Vascular flora of the Pushmataha Wildlife Management Area, Pushmataha County, Oklahoma. *Castanea* 71: 65–79.

Crane, M. F. 1990. *Darlingtonia californica. In: Fire Effects Information System.* U.S. Department of Agriculture, Forest Service, Rocky Mountain Research Station, Fire Sciences Laboratory (Producer). Available online: http://www.fs.fed.us/database/feis/ (accessed August 14, 2006).

Crane, P. E. 2001. Morphology, taxonomy, and nomenclature of the *Chrysomyxa ledi* complex and related rust fungi on spruce and Ericaceae in North America and Europe. *Can. J. Bot.* 79: 957–982.

Cranfill, R. 1981. Bog clubmosses (*Lycopodiella*) in Kentucky. *Amer. Fern J.* 71: 97–100.

Craven, S. R. & R. A. Hunt. 1984a. Fall food habits of Canada geese in Wisconsin. *J. Wildl. Manag.* 48: 169–173.

Craven, S. R. & R. A. Hunt. 1984b. Food habits of Canada geese on the coast of Hudson Bay. *J. Wildl. Manag.* 48: 567–569.

Crawford, D. J. & R. L. Hartman. 1972. Chromosome numbers and taxonomic notes for Rocky Mountain Umbelliferae. *Amer. J. Bot.* 59: 386–392.

Crawford, D. J. & M. E. Mort. 2005. Phylogeny of Eastern North American *Coreopsis* (Asteraceae-Coreopsideae): insights from nuclear and plastid sequences, and comments on character evolution. *Amer. J. Bot.* 92: 330–336.

Crawford, D. J. & R. Ornduff. 1989. Enzyme electrophoresis and evolutionary relationships among three species of *Lasthenia* (Asteraceae: Heliantheae). *Amer. J. Bot.* 76: 289–296.

Crawford, D. J. & E. B. Smith. 1983. The distribution of anthochlor floral pigments in North American *Coreopsis* (Compositae): taxonomic and phyletic interpretations. *Amer. J. Bot.* 70: 355–362.

Crawford, V. 1981. *Wetland Plants of King County and the Puget Sound Lowlands.* King County Planning Division, Seattle, WA. 80 pp.

Crawley, S. S. & K. W. Hilu. 2012. Caryophyllales: Evaluating phylogenetic signal in *trnK* intron versus *matK. J. Syst. Evol.* 50: 387–410.

Creed, R. P. Jr. & S. P. Sheldon. 1993. The effect of feeding by a North American weevil, *Euhrychiopsis lecontei*, on Eurasian watermilfoil (*Myriophyllum spicatum*). *Aquat. Bot.* 45: 245–256.

Creed, R. P. & S. P. Sheldon. 1994. Aquatic weevils (Coleoptera: Curculionidae) associated with northern watermilfoil (*Myriophyllum sibiricum*) in Alberta, Canada. *Entomol. News* 105: 98–102.

Crespo, M. B., L. Serra & A. Jua. 1998. *Solenopsis* (Lobeliaceae): a genus endemic in the Mediterranean region. *Plant Syst. Evol.* 210: 211–229.

Crespo, M. B., A. R. De Leon & S. Ríos. 2002. *Borrichia* Adans. (Asteraceae, Heliantheae), a new record for the Mediterranean flora. *Isr. J. Plant Sci.* 50: 239–242.

Crête, M., J.-P. Ouellet & L. Lesage. 2001. Comparative effects on plants of caribou/reindeer, moose and white-tailed deer herbivory. *Arctic* 54: 407–417.

Cretini, K. F., J. M. Visser, K. W. Krauss & G. D. Steyer. 2012. Development and use of a floristic quality index for coastal Louisiana marshes. *Environ. Monit. Assess.* 184: 2389–2403.

Crevecœur, F. F. 1905. Additions to the list of Kansas Diptera. *Trans. Kansas. Acad. Sci.* 20: 90–96.

Cripps, C. L. & L. H. Eddington. 2005. Distribution of mycorrhizal types among alpine vascular plant families on the Beartooth Plateau, Rocky Mountains, U.S.A., in reference to large-scale patterns in arctic-alpine habitats. *Arctic Antarc. Alp. Res.* 37: 177–188.

Crispens Jr., C. G., I. O. Buss & C. F. Yocom. 1960. Food habits of the California quail in eastern Washington. *Condor* 62: 473–477.

Crocker, R. L. & J. Major. 1955. Soil development in relation to vegetation and surface age at Glacier Bay, Alaska. *J. Ecol.* 43: 427–448.

Crockett, S., R. Baur, O. Kunert, F. Belaj & E. Sigel. 2016. A new chromanone derivative isolated from *Hypericum lissophloeus* (Hypericaceae) potentiates $GABA_A$ receptor currents in a subunit specific fashion. *Bioorg. Med. Chem.* 24: 681–685.

Crockett, S. L., B. Demirçi, K. Husnu Can Baser & I. A. Khan. 2008. Volatile constituents of *Hypericum* L. section *Myriandra* (Clusiaceae): species of the *H. fasciculatum* Lam. alliance. *J. Essent. Oil Res.* 20: 244–249.

Croel, R. C. & J. M. Kneitel. 2011a. Cattle waste reduces plant diversity in vernal pool mesocosms. *Aquat. Bot.* 95: 140–145.

Croel, R. C. & J. M. Kneitel. 2011b. Ecosystem-level effects of bioturbation by the tadpole shrimp *Lepidurus packardi* in temporary pond mesocosms. *Hydrobiologia* 665: 169–181.

Croft, M. V. & P. Chow-Fraser. 2007. Use and development of the wetland macrophyte index to detect water quality in fish habitat of Great Lakes coastal marshes. *J. Great Lakes Res.* 33: 172–197.

Croft, M. V. & P. Chow-Fraser. 2009. Non-random sampling and its role in habitat conservation: a comparison of three wetland macrophyte sampling protocols. *Biodivers. Conserv.* 18: 2283–2306.

Crone, E. E. & J. L. Gehring. 1998. Population viability of *Rorippa columbiae*: multiple models and spatial trend data. *Conserv. Biol.* 12: 1054–1065.

Crone, E. E. & D. R. Taylor. 1996. Complex dynamics in experimental populations of an annual plant, *Cardamine pensylvanica*. *Ecology* 77: 289–299.

Cronin, E. H., P. Ogden, J. A. Young & W. Laycock. 1978. The ecological niches of poisonous plants in range communities. *J. Range Manag.* 5: 328–334.

Cronin, G., K. D. Wissing & D. M. Lodge. 1998. Comparative feeding selectivity of herbivorous insects on water lilies: aquatic vs. semi-terrestrial insects and submersed vs. floating leaves. *Freshwater Biol.* 39: 243–257.

Cronquist, A. 1981. *An Integrated System of Classification of Flowering Plants*. Columbia University Press, New York. 1262 pp.

Cronquist, A., A. H. Holmgren, N. H. Holmgren, J. L. Reveal & P. K. Holmgren. 1984. *Intermountain Flora. Volume 4: Subclass Asteridae (Except Asteraceae)*. The New York Botanical Garden, Bronx, NY. 573 pp.

Cross, A. 2012. Aldrovanda: *The Waterwheel Plant*. Redfern Natural History Productions, Dorset, UK. 249 pp.

Cross, A. T., L. M. Skates, L. Adamec, C. M. Hammond, P. M. Sheridan & K. W. Dixon. 2015. Population ecology of the endangered aquatic carnivorous macrophyte *Aldrovanda vesiculosa* at a naturalised site in North America. *Freshwater Biol.* 60: 1772–1783.

Crosslé, K. & M. A. Brock. 2002. How do water regime and clipping influence wetland plant establishment from seed banks and subsequent reproduction? *Aquat. Bot.* 74: 43–56.

Crosswhite, F. S. 1965. Variation in *Chelone glabra* in Wisconsin (Scrophulariaceae). *Michigan. Bot.* 4: 62–66.

Crouch, V. E. & M. S. Golden. 1997. Floristics of a bottomland forest and adjacent uplands near the Tombigbee River, Choctaw County, Alabama. *Castanea* 62: 219–238.

Crow, G. E. & C. B. Hellquist. 1983. *Aquatic Vascular Plants of New England: Part 6. Trapaceae, Haloragaceae, Hippuridaceae.* Station Bulletin 524. New Hampshire Agricultural Experiment Station, University of New Hampshire, Durham, NH. 26 pp.

Crow, G. E. & C. B. Hellquist. 1985. *Aquatic Vascular Plants of New England: Part 8. Lentibulariaceae.* Station Bulletin 528. New Hampshire Agricultural Experiment Station, University of New Hampshire, Durham, NH. 22 pp.

Crow, G. E. & C. B. Hellquist. 2000a. *Aquatic and Wetland Plants of Northeastern North America. Volume One. Pteridophytes, Gymnosperms, and Angiosperms: Dicotyledons.* The University of Wisconsin Press, Madison, WI. 480 pp.

Crow, G. E. & C. B. Hellquist. 2000b. *Aquatic and Wetland Plants of Northeastern North America. Volume Two. Angiosperms: Monocotyledons.* The University of Wisconsin Press, Madison, WI. 400 pp.

Crow, G. E., D. I. Rivera & C. Charpentier. 1987. Aquatic vascular plants of two Costa Rican ponds. *Selbeyana* 10: 31–35.

Crowden, R. K., J. B. Harborne & V. H. Heywood. 1969. Chemosystematics of the umbelliferae—a general survey. *Phytochemistry* 8: 1963–1984.

Crowder, A., J. M. Bristow, M. R. King & S. van der Kloet. 1977. The aquatic macrophytes of some lakes in southeastern Ontario. *Naturalist Can.* 104: 457–464.

Crowder, A., J. M. Bristow, M. R. King Crowe, E. A., A. J. Busacca & J. P. Reganold. 1994. Vegetation zones and soil characteristics in vernal pools in the channeled scabland of eastern Washington. *Great Basin Nat.* 54: 234–247.

Crowe, D. R. & W. H. Parker. 1981. Hybridization and agamospermy of *Bidens* in northwestern Ontario. *Taxon* 30: 749–760.

Crowe, E. & G. Kudray. 2003. *Wetland Assessment of the Whitewater Watershed.* Report to U.S. Bureau of Land Management, Malta Field Office. Montana Natural Heritage Program, Helena, MT. 34 pp.

Crowe, E. A., B. L. Kovalchik & M. J. Kerr. 2004. *Riparian and Wetland Vegetation of Central and Eastern Oregon.* Oregon State University, Portland, OR. 473 pp.

Cruden, R. W. 1972. Pollination biology of *Nemophila menziesii* (Hydrophyllaceae) with comments on the evolution of oligolectic bees. *Evolution* 26: 373–389.

Cruden, R. W. 1977. Pollen-ovule ratios: a conservative indicator of breeding systems in flowering plants. *Evolution* 31: 32–46.

Cruden, R. W. & S. M. Hermann-Parker. 1977. Temporal dioecism: an alternative to dioecism. *Evolution* 31: 863–866.

Cruden, R. W. & D. L. Lyon. 1985. Correlations among stigma depth, style length, and pollen grain size: do they reflect function or phylogeny? *Bot. Gaz.* 146: 143–149.

Crum, H., E. G. Fisher & H. C. Burtt. 1972. *Splachnum ampullaceum* in West Virginia. *Castanea* 37: 253–257.

Cruzan, M. B., P. R. Neal & M. F. Willson. 1988. Floral display in *Phyla incisa*: consequences for male and female reproductive success. *Evolution* 42: 505–515.

Cuénoud, P., M. A. Del Pero Martinez, P.-A. Loizeau, R. Spichiger, S. Andrews & J.-F. Manen. 2000. Molecular phylogeny and biogeography of the genus *Ilex* L. (Aquifoliaceae). *Ann. Bot.* 85: 111–122.

Cuénoud, P., V. Savolainen, L. W. Chatrou, M. Powell, R. J. Grayer & M. W. Chase. 2002. Molecular phylogenetics of Caryophyllales based on nuclear 18S rDNA and plastid *rbcL, atpB,* and *matK* DNA sequences. *Amer. J. Bot.* 89: 132–144.

Cullina, W. 2002. *Native Trees, Shrubs, and Vines: A Guide to Using, Growing, and Propagating North American Woody Plants.* Houghton Mifflin Co., Boston, MA. 368 pp.

Cullina, W. 2009. *Understanding Perennials: A New Look at an Old Favorite.* Houghton Mifflin Harcourt, Boston, MA. 256 pp.

Culver, D. C. & A. J. Beattie. 1978. Myrmecochory in *Viola*: dynamics of seed-ant interactions in some West Virginia species. *J. Ecol.* 66: 53–72.

Cumberland, M. S. & L. K. Kirkman. 2013. The effects of the red imported fire ant on seed fate in the longleaf pine ecosystem. *Plant Ecol.* 214: 717–724.

Cypert, E. 1961. Effects of fires in the Okefenokee Swamp in 1954 and 1955. *Amer. Midl. Naturalist* 66: 485–503.

Cypert, E. 1972. The origin of houses in the Okefenokee prairies. *Amer. Midl. Naturalist* 87: 448–458.

Czarnecka, J., B. Czarnecka & M. Garbacz. 2012. Secondary dispersal of seeds by the Magpie *Pica pica* L. in agricultural landscape. *Ann. Univ. Mariae Curie-Sklodowska, C, Biol.* 67: 13–25.

Czech, B., P. Krausman & P. K. Devers. 2000. Economic associations among causes of species endangerment in the United States. *BioScience* 50: 593–601.

Daehler, C. C. 2003. Performance comparisons of co-occurring native and alien invasive plants: implications for conservation and restoration. *Ann. Rev. Ecol. Evol. Syst.* 34: 183–211.

Dahlem, G. A. & R. F. C. Naczi. 2006. Flesh flies (Diptera: Sarcophagidae) associated with North American pitcher plants (Sarraceniaceae), with descriptions of three new species. *Ann. Entomol. Soc. Amer.* 99: 218–240.

Dahlgren, J. P. & J. Ehrlén. 2005. Distribution patterns of vascular plants in lakes: the role of metapopulation dynamics. *Ecography* 28: 49–58.

Dahlgren, K. V. D. 1922. Selbsterilitat innerhalb Klonen von *Lysimachia nummularia. Hereditas* 3: 200–210.

Dale, H. M. 1984a. Hydrostatic pressure and aquatic plant growth: a laboratory study. *Hydrobiologia* 111: 193–200.

Dale, E. 1984b. *Wetlands Forest Communities as Indicators of Flooding Potential in Backwater Areas of River Bottomlands.* Project completion report, publication no. 106, Sept. 1984. USGS Project G-829-08. Arkansas Water Resources Research Center, Fayetteville, AR. 84 pp.

Dale, J. E. & G. H. Egley. 1971. Stimulation of witchweed germination by run-off water and plant tissues. *Weed Sci.* 19: 678–681.

Dale, V. H., S. C. Beyeler & B. Jackson. 2002. Understory vegetation indicators of anthropogenic disturbance in longleaf pine forests at Fort Benning, Georgia, USA. *Ecol. Indicators.* 1: 155–170.

Dalsgaard, B., D. W. Carstensen, A. Kirkconnell, A. M. M. González, O. M. García, A. Timmermann & W. J. Sutherland. 2012. Floral traits of plants visited by the bee hummingbird (*Mellisuga helenae*). *Ornitol. Neotrop.* 23: 143–149.

Damm, C. 2001. A phytosociological study of Glacier National Park, Montana, U.S.A., with notes on the syntaxonomy of alpine vegetation in western North America. Dissertation zur Erlangung des Doktorgrades der Mathematisch-Naturwissenschaftlichen Fakultäten der Georg-August-Universität zu Göttingen, Göttingen, Germany. 297 pp.

D'Amore, D. V. & W. C. Lynn. 2002. Classification of forested histosols in southeast Alaska. *J. Soil Sci. Soc. Amer.* 66: 554–562.

Damtoft, S., S. R. Jensen, J. Thorsen, P. Mølgard & C. E. Olsen. 1994. Iridoids and verbascoside in Callitrichaceae, Hippuridaceae and Lentibulariaceae. *Phytochemistry* 36: 927–929.

Dana, M. N., W. A. Skroch & D. M. Boone. 1965. Granular herbicides for cranberry bogs. *Weeds* 13: 5–7.

Dang, T. T. & C. C. Chinnappa. 2007. The reproductive biology of *Stellaria longipes* Goldie (Caryophyllaceae) in North America. *Flora* 202: 403–407.

D'Angelo, E. M., A. D. Karathanasis, E. J. Sparks, S. A. Ritchey & S. A. Wehr-McChesney. 2005. Soil carbon and microbial communities at mitigated and late successional bottomland forest wetlands. *Wetlands* 25: 162–175.

Daniel, T. F. 1984. A revision of *Stenandrium* (Acanthaceae) in Mexico and adjacent regions. *Ann. Missouri Bot. Gard.* 71: 1028–1043.

Daniel, T. F., B. D. Parfitt & M. A. Baker. 1984. Chromosome numbers and their systematic implications in some North American Acanthaceae. *Syst. Bot.* 9: 346–355.

Darby, P. C., P. V. Darby & R. E. Bennetts. 1996. Spatial relationships of foraging and roost sites used by Snail Kites at Lake Kissimmee and Water Conservation Area 3A. Florida. *Florida. Field Naturalist* 24: 1–24.

Darbyshire, S. J. & A. Francis. 2008. The biology of invasive alien plants in Canada. 10. *Nymphoides peltata* (S. G. Gmel.) Kuntze. *Can. J. Plant Sci.* 88: 811–829.

D'Arcy, W. G. 1971. Flora of Panama. Part IX. Family 178. Plantaginaceae. *Ann. Missouri Bot. Gard.* 58: 363–369.

D'Arcy, W. G. 1979. Flora of Panama. Part IX. Family 171. Scrophulariaceae. *Ann. Missouri Bot. Gard.* 66: 173–272.

Darinot, F. & A. Morand. 2001. Management of wet meadows in the Lavours marsh implementing grazing. pp. 86–93. *In:* L. Andersson, R. Marciau, H. Paltto, B. Tardy & H. Read (eds.), *Tools in Preserving Biodiversity in Nemoral and Boreonemoral Biomes of Europe.* Nature Conservation Experience Exchange (Naconex) Textbook 1. Education and Culture Leonardo da Vinci, Töreboda Tryckeri A. B., Sweden.

Darke, R. 2005. *Amsonia* in cultivation. *Plantsman* 4: 72–75.

Darlow, N. 2000. *Behavior, Habitat Usage and Oviposition of the Mitchell's Satyr Butterfly, Neonympha mitchellii mitchellii.* Report to the U.S. Fish and Wildlife Service, Ft. Snelling, MN 43 pp.

Darokar, M. P., S. P. S. Khanuja, A. K. Shasany & S. Kumar. 2001. Low levels of genetic diversity detected by RAPD analysis in geographically distinct accessions of *Bacopa monnieri. Genet. Resour. Crop Evol.* 48: 555–558.

Darris, D. C. 2002. Ability of Pacific Northwest native shrubs to root from hardwood cuttings (with summary of propagation methods for 22 species). Plant Materials Technical Note No. 30. U.S. Dept. of Agriculture Natural Resources Conservation Service. Portland, OR. 20 pp.

Darwin, C. 1859. *On the Origin of Species by Means of Natural Selection.* John Murray, London. 460 pp.

Darwin, C. 1862. On the two forms, or dimorphic condition, in the species of *Primula,* and on their remarkable sexual relationships. *Proc. Linn. Soc. London* 6: 105–139.

Darwin, C. 1877. *The Different Forms of Flowers on Plants of the Same Species.* John Murray, London. 352 pp.

Darwin, C. 1892. *The Variation of Animals and Plants under Domestication.* Vol. II. D. Appleton & Co., New York. 495 pp.

Da Silva Bezerra Guerra, K. S., R. L. C. Silva, M. B. S. Maia & A. Schwarz. 2012. Embryo and fetal toxicity of *Mentha* ×*villosa* essential oil in Wistar rats. *Pharm. Biol.* 50: 871–877.

Dassanayake, M., J. S. Haas, H. J. Bohnert & J. M. Cheeseman. 2010. Comparative transcriptomics for mangrove species: an expanding resource. *Funct. Integr. Genomics* 10: 523–532.

Datta, S. C. & K. K. Biswas. 1979. Autecological studies on weeds of West Bengal. VIII. *Alternanthera sessilis* (L.) DC. *Bull. Bot. Soc. Bengal* 33: 5–26.

Daubenmire, R. 1990. The *Magnolia grandiflora-Quercus virginiana* forest of Florida. *Amer. Midl. Naturalist* 123: 331–347.

Davenport, L. J. 1988. A monograph of *Hydrolea* (Hydrophyllaceae). *Rhodora* 90: 169–208.

David, P. G. 1994. Wading bird use of Lake Okeechobee relative to fluctuating water levels. *Wilson Bull.* 106: 719–732.

Davidson, B. L. 2000. *Lewisias.* Timber Press, Portland, OR. 238 pp.

Davidson, R. B., C. Baker, M. McElveen & W. K. Conner. 1997. Hydroxydanaidal and the courtship of *Haploa* (Arctiidae). *J. Lepid. Soc.* 51: 288–294.

Davidson, W. M. 1909. Notes on Aphididae collected in the vicinity of Stanford University. *J. Econ. Entomol.* 2: 299–305.

Davis, A. F. 1993. Rare wetland plants and their habitats in Pennsylvania. *Proc. Acad. Nat. Sci. Philadelphia* 144: 254–262.

Davis, C. C. & M. W. Chase. 2004. Elatinaceae are sister to Malpighiaceae; Peridiscaceae belong to Saxifragales. *Amer. J. Bot.* 91: 262–273.

Davis, E. E., S. French & R. C. Venette. 2005a. Mini risk assessmentsummer fruit tortrix moth, *Adoxophyes orana* (Fischer von Roslerstamm, 1834) [Lepidoptera: Tortricidae]. University of Minnesota and USDA Forest Service, St. Paul, MN. 48 pp.

Davis, G. J. 1967. Proserpinaca: photoperiodic and chemical differentiation of leaf development and flowering. *Plant Physiol.* 42: 667–668.

Davis, J. C. 1993. The hydrology and plant community relations of Canelo Hills Cienega, an emergent wetland in southeastern Arizona. MS thesis. The University of Arizona, Tucson, AZ. 210 pp.

Davis, J. D., S. D. Hendrix, D. M. Debinski & C. J. Hemsley. 2008. Butterfly, bee and forb community composition and crosstaxon incongruence in tallgrass prairie fragments. *J. Insect Conserv.* 12: 69–79.

Davis, J. D. & S. A. Banack. 2012. Ethnobotany of the Kiluhikturmiut Inuinnait of Kugluktuk, Nunavut, Canada. *Ethnobiol. Lett.* 3: 78–90.

Davis, J. J. 1903. Third supplementary list of parasitic fungi of Wisconsin. *Trans. Wisconsin Acad. Sci.* 14: 83–106.

Davis, J. J. 1910. A list of the Aphididae of Illinois, with notes on some of the species. *J. Econ. Entomol.* 3: 482–499.

Davis, J. J. 1914. A provisional list of the parasitic Fungi of Wisconsin. *Trans. Wisconsin Acad. Sci.* 17: 846–964.

Davis, J. J. 1915. Notes on parasitic Fungi of Wisconsin–I. *Trans. Wisconsin Acad.* Sci. 18: 78–92.

Davis, J. J. 1929. Notes on parasitic Fungi in Wisconsin. XVIII. *Trans. Wisconsin Acad. Sci.* 24: 253–261.

Davis, S. M., E. E. Gaiser, W. F. Loftus & A. E. Huffman. 2005c. Southern marl prairies conceptual ecological model. *Wetlands* 25: 821–831.

Davis, S. M., D. L. Childers, J. J. Lorenz, H. R. Wanless & T. E. Hopkins. 2005b. A conceptual model of ecological interactions in the mangrove estuaries of the Florida Everglades. *Wetlands* 25: 832–842.

Davis, T. A. 1974. Enantiomorphic structures in plants. *Proc. Indian Natl. Sci. Acad. B.* 40: 424–429.

Davy, A. J., G. F. Bishop & C. S. B. Costa. 2001. *Salicornia* L. (*Salicornia pusilla* J. Woods, *S. ramosissima* J. Woods, *S. europaea* L., *S. obscura* P.W. Ball & Tutin, *S. nitens* P.W. Ball & Tutin, *S. fragilis* P.W. Ball & Tutin and *S. dolichostachya* Moss). *J. Ecol.* 89: 681–707.

Dawe, C. E. & E. G. Reekie. 2007. The effects of flooding regime on the rare Atlantic coastal plain species *Hydrocoytle umbellata*. *Can. J. Bot.* 85: 167–184.

Dawe, N. K., G. E. Bradfield, W. S. Boyd, D. E. C. Trethewey & A. N. Zolbrod. 2000. Marsh creation in a northern Pacific Estuary: is thirteen years of monitoring vegetation dynamics enough? *Conserv. Ecol.* 4: 12. Available online: http://www.consecol.org/vol4/iss2/art12/.

Dawe, N. K., W. S. Boyd, R. Buechert & A. C. Stewart. 2011. Recent, significant changes-to the native marsh vegetation of the Little Qualicum River estuary, British Columbia; a case of too many Canada Geese (*Branta canadensis*)? *British Columbia Birds* 21: 11–31.

Dawson, M. I. 1989. Contributions to a chromosome atlas of the New Zealand flora—30 miscellaneous species. *New Zealand. J. Bot.* 27: 163–165.

Day, J. W. 1975. Autecology of *Leitneria floridana*. PhD dissertation. Mississippi State University, Starkville, MS.

Day, T. A. & R. G. Wright. 1989. Positive plant spatial association with *Eriogonum ovalifolium* in primary succession on cinder cones: seed-trapping nurse plants. *Vegetatio* 80: 37–45.

De, S., D. C. Das, T. Mondal & M. Das. 2014. Investigation of the antibacterial and antifungal activity of *Cardanthera difformis* Druce whole plant extracts against some clinical pathogens. *Int. J. Bioassays* 3: 3464–3468.

De Almeida, J. D. & H. Freitas. 2006. Exotic naturalized flora of continental Portugal: A reassessment. *Bot. Complut.* 30: 117–130.

Dean, W. R. J., S. J. Milton, R. G. Ryan & C. L. Moloney. 1994. The role of disturbance in the establishment of indigenous and alien plants at Inaccessible and Nightingale Islands in the South Atlantic Ocean. *Vegetatio* 113: 13–23.

Dearness, J. 1924. New and noteworthy Fungi: III. *Mycologia* 16: 143–176.

Dearness, J. 1926. New and noteworthy Fungi: IV. *Mycologia* 18: 236–255.

Dearness, J. 1929. New and noteworthy Fungi: VI. *Mycologia* 21: 326–332.

Dearness, J. & H. D. House. 1921. New or noteworthy species of Fungi II. *New York. State Mus. Bull.* 233–234: 32–43.

Deaver, E., M. T. Moore, C. M. Cooper & S. S. Knight. 2005. Efficiency of three aquatic macrophytes in mitigating nutrient runoff. *Int. J. Ecol. Environ. Sci.* 31: 1–7.

DeBarros, N. B. 2010. Floral resource provisioning for bees in Pennsylvania and the Mid-Atlantic region. MS thesis. The Pennsylvania State University, State College, PA. 120 pp.

DeBell, D. S. & A. W. Naylor. 1972. Some factors affecting germination of swamp tupelo seeds. *Ecology* 53: 504–506.

DeBenedetti, S. H. & D. J. Parsons. 1984. Postfire succession in a Sierran subalpine meadow. *Amer. Midl. Naturalist* 111: 118–125.

DeBerry, D. A. & J. E. Perry. 2005. A drawdown flora in Virginia. *Castanea* 70: 276–286.

DeBerry, D. A. & J. E. Perry. 2007. Noteworthy collections: Virginia. *Castanea* 72: 119–120.

Debez, A., D. Saadaoui, I. Slama, B. Huchzermeyer & C. Abdelly. 2010. Responses of *Batis maritima* plants challenged with up to two-fold seawater NaCl salinity. *J. Plant Nutr. Soil Sci.* 173: 291–299.

Debinski, D. M., H. Wickham, K. Kindscher, J. C. Caruthers & M. Germino. 2010. Montane meadow change during drought varies with background hydrologic regime and plant functional group. *Ecology* 91: 1672–1681.

DeBruin, E. 1996. *Status Evaluations of Four Rare Plant Species in the Organ Mountains of New Mexico.* National Biological Service, Washington, DC. 22 pp.

De Carvalho, C. C. C. R. & M. M. R. Da Fonseca. 2006. Carvone: why and how should one bother to produce this terpene. *Food Chem.* 95: 413–422.

Decker, G. C. 1932. Biology of the bidens borer, *Epiblema Otiosana* (Clemens) (Lepidoptera, Olethreutidæ). *J. New York Entomol. Soc.* 40: 503–509.

Decker, K. 2006a. *Salix arizonica* Dorn (Arizona willow): a technical conservation assessment. USDA Forest Service, Rocky Mountain Region. Available online: http://www.fs.fed.us/r2/projects/scp/assessments/salixarizonica.pdf (accessed February 26, 2016).

Decker, K. 2006b. *Salix candida* Flueggé ex Wild. (sageleaf willow): a technical conservation assessment. USDA Forest Service, Rocky Mountain Region. Available online: http://www.fs.fed.us/r2/projects/scp/assessments/salixcandida.pdf (accessed March 4, 2016).

Decker, K., D. R. Culver & D. G. Anderson. 2006. *Kobresia simpliciuscula* (Wahlenberg) Mackenzie (simple bog sedge): a technical conservation assessment. USDA Forest Service, Rocky Mountain Region, Species Conservation Project. Colorado Natural Heritage Program, Colorado State University, Fort Collins, CO. 33 pp.

DeCoursey, R. M. 1963. The life histories of *Banasa dimidiata* and *Banasa calva* (Hemiptera: Pentatomidae). *Ann. Entomol. Soc. Amer.* 56: 687–693.

De Craene, L. P. 2005. Floral developmental evidence for the systematic position of *Batis* (Bataceae). *Amer. J. Bot.* 92: 752–760.

DeGraaf, R. M. 2002. *Trees, Shrubs, and Vines for Attracting Birds.* University Press of New England, Lebanon, NH. 224 pp.

De Grandpré, L., D. Gagnon & Y. Bergeron. 1993. Changes in the understory of Canadian southern boreal forest after fire. *J. Veg. Sci.* 4: 803–810.

De Groot, W. J., P. A. Thomas & R. W. Wein. 1997. *Betula nana* L. and *Betula glandulosa* Michx. *J. Ecol.* 85: 241–264.

De Groote, L. W., H. K. Ober, J. H. Aldrich, J. G. Norcini & G. W. Knox. 2011. Susceptibility of cultivated native wildflowers to deer damage. *S. E. Naturalist* 10: 761–771.

Degtjareva, G. V., S. J. Casper, F. H. Hellwig, A. R. Schmidt, J. Steiger & D. D. Sokoloff. 2006. Morphology and nrITS phylogeny of the genus *Pinguicula* L. (Lentibulariaceae), with special attention to embryo evolution. *Plant Biol.* 8: 778–790.

Degtjareva, G. V., T. E. Kramina, D. D. Sokoloff, T. H. Samigullin, C. M. Valiejo-Roman & A. S. Antonov. 2006. Phylogeny of the genus *Lotus* (Leguminosae, *Loteae*): evidence from nrITS sequences and morphology. *Botany* 84: 813–830.

DeHart, K. S., G. A. Meindl, D. J. Bain & T. L. Ashman. 2014. Elemental composition of serpentine plants depends on habitat affinity and organ type. *J. Plant Nutr. Soil Sci.* 177: 851–859.

Dehgan, B., M. Gooch, F. Almira & M. Kane. 1989. Vegetative propagation of Florida native plants: III. Shrubs. *Proc. Florida State Hort. Soc.* 102: 254–260.

Deka, N. & N. Devi. 2015. Aquatic angiosperm [sic!] of BTC area, Assam, with reference to their traditional uses. *Asian J. Plant Sci. Res.* 5: 9–13.

De Lange, P. J., T. J. P. de Lange & F. J. T. de Lange. 2005. New exotic plant records, and range extensions for naturalised plants in the northern North Island, New Zealand. *J. Auckland Bot. Soc.* 60: 130–147.

De la Peña, M. R. 2011. *Observaciones de Campo en la Alimentación de las Aves.* Musea Provincial de Ciencias Naturales, Santa Fe, Argentina. 88 pp.

Delatte, É. & O. Chabrerie. 2008. Performances des plantes herbacées forestières dans la dispersion de leurs graines par la fourmi *Myrmica ruginodis*. *C. R. Biol.* 331: 309–320.

De Leon, D. 1961. New false spider mites with notes on some previously described species (Acarina: Tenuipalpidae). *Florida Entomol.* 44: 167–179.

Della, A., D. Paraskeva-Hadjichambi & A. C. Hadjichambis. 2006. An ethnobotanical survey of wild edible plants of Paphos and Larnaca countryside of Cyprus. *J. Ethnobiol. Ethnomed.* 2: 34.

Dellafiore, C. M., J. B. Gallego-Fernández & S. Muñoz-Vallés. 2007. The contribution of endozoochory to the colonization and vegetation composition of recently formed sand coastal dunes. *Res. Lett. Ecol.* 2007: 1–3.

Dellavalle-Sanvictores, M. 2000. Phylogenetic analysis of the genus *Plagiobothrys* (Boraginaceae) using morpological and nutlet pericarp ultrastructural data. MS thesis. San Diego State University, San Diego, CA.

Dellinger, S. C. 1936. Baby cradles of the Ozark Bluff Dwellers. *Amer. Antiq.* 1: 197–214.

Del Moral, R. & D. M. Wood. 1993. Early primary succession on a barren volcanic plain at Mount St. Helens, Washington. *Amer. J. Bot.* 80: 981–991.

Delprete, P. G. 1996. Systematics, typification, distribution, and reproductive biology of *Pinckneya bracteata* (Rubiaceae). *Plant Syst. Evol.* 201: 243–261.

De Miranda, J. R., M. A. Thomas, D. A. Thurman & A. B. Tomsett. 1990. Metallothionein genes from the flowering plant *Mimulus guttatus*. *FEBS Lett.* 260: 277–280.

Demma, J., H. El-Seedi, E. Engidawork, T. L. Aboye, U. Göransson & B. Hellman. 2013. An *in vitro* study on the DNA damaging effects of phytochemicals partially isolated from an extract of *Glinus lotoides*. *Phytother. Res.* 27: 507–514.

Den Breeÿen, A., J. Z. Groenewald, G. J. M. Verkley & P. W. Crous. 2006. Morphological and molecular characterisation of Mycosphaerellaceae associated with the invasive weed, *Chromolaena odorata*. *Fungal Diversity* 23: 89–110.

Deng, J.-B., B. T. Drew, E. V. Mavrodiev, M. A. Gitzendanner, P. S. Soltis & D. E. Soltis. 2015. Phylogeny, divergence times, and historical biogeography of the angiosperm family Saxifragaceae. *Mol. Phylogen. Evol.* 83: 86–98.

Denmark, H. A. & M. H. Muma. 1966. Revision of the genus *Proprioseius* Chant, 1957 (Acarina: Phytoseiidae). *Florida Entomol.* 49: 253–264.

Dennis, W. M. 1982. Ecological notes on *Jamesianthus alabamensis* Blake and Sherff (Asteraceae) and an hypothesis on its endemism. *Sida* 9: 210–214.

Dennis, W. M. & B. E. Wofford. 1976. Evidence for the hybrid origin of *Proserpinaca intermedia*. *A. S. B. Bull.* 23: 54.

Dennis, W. M., T. L. Goldsby & A. L. Bates. 1977. *Selected Aquatic and Wetland Plants of the Tennessee Valley.* Water Quality and Ecology Branch, Division of Environmental Planning, Tennessee Valley Authority, Muscle Shoals, AL. 159 pp.

Dennis, W. M., A. M. Evans & B. E. Wofford. 1979. Disjunct populations of *Isoëtes macrospora* in southeastern Tennessee. *Amer. Fern J.* 69: 97–99.

Denoth, M. & J. H. Myers. 2007. Competition between *Lythrum salicaria* and a rare species: combining evidence from experiments and long-term monitoring. *Plant Ecol.* 191: 153–161.

Denslow, J. S. & L. L. Battaglia. 2002. Stand composition and structure across a changing hydrologic gradient: Jean Lafitte National Park, Louisiana, USA. *Wetlands* 22: 738–752.

Densmore, F. 1974. *How Indians Use Wild Plants for Food, Medicine and Crafts.* Dover, New York. 160 pp.

Densmore, R. V. 1997. Effect of day length on germination of seeds collected in Alaska. *Amer. J. Bot.* 84: 274–278.

Densmore, R. & J. Zasada. 1983. Seed dispersal and dormancy patterns in northern willows: ecological and evolutionary significance. *Can. J. Bot.* 61: 3207–3216.

Denton, R. 2003. Genetic variation in *Trifolium bolanderi* A. Gray, a narrow endemic, compared with *Trifolium longipes* Nutt. American Society of Plant Biologists (abstract).

DePoe, C. E. 1969. *Bacopa egensis* (Poeppig) Pennell (Scrophulariaceae) in the United States. *Sida* 3: 313–318.

DePoe, C. E. & E. O. Beal. 1969. Origin and maintenance of clinal variation in *Nuphar* (Nymphaeaceae). *Brittonia* 21: 15–28.

de Queiroz Pinto, A. C., M. C. Rocha Cordeiro, S. R. M. de Andrade, F. R. Ferreira, H. A. da Cunha Filgueiras & R. Elesbão Alves. 2002. Five important species of *Annona*. International Centre for Underutilised Crops, Southampton, UK.

Deschamp, J. A. 1977. Forage preferences of mule deer in the lodgepole pine ecosystem, Ashley National Forest, UTNI. MS thesis. Utah State University, Logan, UT. 76 pp.

Deschamp, P. A. & T. J. Cooke. 1985. Leaf dimorphism in the aquatic angiosperm *Callitriche heterophylla*. *Amer. J. Bot.* 72: 1377–1387.

De Sousa, D. P., E. V. M. Junior, F. S. Oliveira, R. N. de Almeida, X. P. Nunes & J. M. Barbosa-Filho. 2007. Antinociceptive activity of structural analogues of rotundifolone: structure-activity relationship. *Z. Naturforsch.* 62: 39–42.

Desrochers, A. M. & B. Dodge. 2003. Phylogenetic relationships in *Lasthenia* (Heliantheae: Asteraceae) based on nuclear rDNA internal transcribed spacer (ITS) sequence data. *Syst. Bot.* 28: 208–215.

Desroches, M. B., M. Lavoie & C. Lavoie. 2013. Establishing the value of a salt marsh as a potential benchmark: vegetation surveys and paleoecological analyses as assessment tools. *Botany* 91: 774–785.

Dessein, S. 2003. Systematic studies in the Spermacoceae (Rubiaceae). PhD thesis. Katholieke Universiteit Leuven, Leuven, Belgium. 407 pp.

Desserud, P., M. Wood & D. Warner. 2006. *Forage Production Survey of Riparian Areas in the Grassland and Parkland Natural Regions of Alberta: Cows and Fish*. Report no. 028. Alberta Riparian Habitat Management Society, Lethbridge, AB 128 pp.

Determann, R., L. Kirkman & H. Nourse. 1997. Plant conservation by propagation. The case for *Macranthera* and *Schwalbea*. *Tipularia* 12: 2–12.

Detmers, F. 1910a. The medicinal plants of Ohio. *Ohio Naturalist* 10: 55–60; 73–85.

Detmers, F. 1910b. A floristic survey of Orchard Island. *Ohio J. Sci.* 11: 200–210.

Devall, M. S. 2013. The endangered pondberry (*Lindera melissifolia* [Walter] Blume, Lauraceae). *Nat. Areas J.* 33: 455–465.

Devall, M., N. Schiff & D. Boyette. 2001. Ecology and reproductive biology of the endangered pondberry, *Lindera melissifolia* (Walt) Blume. *Nat. Areas J.* 21: 250–258.

de Vaulx, R. D. & M. Pitrat. 1979. Interspecific cross between *Cucurbita pepo* and *C. martinezii*. *Cucurbit. Genet. Coop. Rep.* 2: 35.

DeVelice, R. L., C. J. Hubbard, K. Boggs, S. Boudreau, M. Potkin, T. Boucher & C. Wertheim. 1999. *Plant Community Types of the Chugach National Forest: Southcentral Alaska*. Technical Publication R10-TP-76. USDA Forest Service, Chugach National Forest, Alaska, Anchorage, AK. 375 pp.

Devi, S. P., S. Kumaria, S. R. Rao & P. Tandon. 2016. Carnivorous plants as a source of potent bioactive compound: naphthoquinones. *Trop. Plant Biol.* 9: 267–279. doi:10.1007/s12042-016-9177-0.

DeVlaming, V. & V. W. Proctor. 1968. Dispersal of aquatic organisms: viability of seeds recovered from the droppings of captive killdeer and mallard ducks. *Amer. J. Bot.* 55: 20–26.

Devlin, B. 1988. The effects of stress on reproductive characters of *Lobelia cardinalis*. *Ecology* 69: 1716–1720.

Devlin, B. & A. G. Stephenson. 1984. Factors that influence the duration of the staminate and pistillate phases of *Lobelia cardinalis* flowers. *Bot. Gaz.* 145: 323–328.

Devlin, B. & A. G. Stephenson. 1985. Sex differential floral longevity, nectar secretion, and pollinator foraging in a protandrous species. *Amer. J. Bot.* 72: 303–310.

Devlin, B. & A. G. Stephenson. 1987. Sexual variations among plants of a perfect-flowered species. *Amer. Naturalist* 130: 199–218.

Devlin, B., J. B. Horton & A. G. Stephenson. 1987. Patterns of nectar production of *Lobelia cardinalis*. *Amer. Midl. Naturalist* 117: 289–295.

De Vos, A. 1958. Summer observations on moose behavior in Ontario. *J. Mammalogy* 39: 128–139.

Devy, M. S. & P. Davidar. 2003. Pollination systems of trees in Kakachi, a mid-elevation wet evergreen forest in Western Ghats, India. *Amer. J. Bot.* 90: 650–657.

Dewanji, A., S. Chanda, L. Si, S. Barik & S. Matai. 1997. Extractability and nutritional value of leaf protein from tropical aquatic plants. *Plant Foods Human Nutr.* 50: 349–357.

DeWoody, J., J. D. Nason & M. Smith. 2011. Hybridization between the threatened herb *Boltonia decurrens* (Asteraceae) and its widespread congener, *B. asteroides*. *Botany* 89: 191–201.

Dey, P., S. Chandra, P. Chatterjee & S. Bhattacharya. 2011. Neuropharmacological properties of *Mikania scandens* (L.) Willd. (Asteraceae). *J. Adv. Pharm. Technol. Res.* 2: 255–259.

Dey, P., S. Chandra, P. Chatterjee & S. Bhattacharya. 2012. Allelopathic potential of aerial parts from *Mikania scandens* (L.) Willd. *World J. Agric. Sci.* 8: 203–207.

Deyrup, M. & J. Trager. 1986. Ants of the Archbold Biological Station, Highlands County, Florida (Hymenoptera: Formi-cidae). *Florida Entomol.* 69: 206–228.

Deyrup, M., J. Edirisinghe & B. Norden. 2002. The diversity and floral hosts of bees at the Archbold Biological Station, Florida (Hymenoptera: Apoidea). *Insecta Mundi* 16: 87–120.

Deyrup, M., J. Kraus & T. Eisner. 2004. A Florida caterpillar and other arthropods inhabiting the webs of a subsocial spider (Lepidoptera: Pyralidae; Araneida: Theridiidae). *Florida Entomol.* 87: 554–558.

De-Yuan, H. 1991. A biosystematic study on *Ranunculus* subgenus *Batrachium* in S Sweden. *Nord. J. Bot.* 11: 41–59.

Dhankar, R., A. Kaushik & S. Taxak. 1998. Accumulation of organic solutes by some native plant species from semi-arid, north-western India. *Ecol. Environ. Conserv.* 4: 57–63.

Dhawan, R. S. 2007. Germination potential and growth behaviour of *Eclipta alba*. *Indian J. Weed Sci.* 39: 116–119.

D'hondt, B., L. Vansteenbrugge, K. Van Den Berge, J. Bastiaens & M. Hoffmann. 2011. Scat analysis reveals a wide set of plant species to be potentially dispersed by foxes. *Plant Ecol. Evol.* 144: 106–110.

Diamond Jr., A. R., D. R. Folkerts & R. S. Boyd. 2006. Pollination biology, seed dispersal, and recruitment in *Rudbeckia auriculata* (Perdue) Kral, a rare southeastern endemic. *Castanea* 71: 226–238.

Dias Filho, M. B. 1996. Germination and emergence of *Stachytarpheta cayennensis* and *Ipomoea asarifolia*. *Planta Daninha* 14: 118–126.

Dias Filho, M. B. 1999. Potential for seed bank formation in two weedy species from Brazilian Amazonia. *Planta Daninha* 17: 183–188.

Dias, J. F. G., S. Virtuoso, A. Davet, M. M. Cunico, M. D. Miguel, O. G. Miguel, C. G. Auer, A. Grigoletti-Júnior, A. B. Oliveira & M. L. Ferronato. 2006. Atividade antibacteriana e antifúngica de extratos etanólicos de *Aster lanceolatus* Willd., Asteraceae. *Braz. J. Pharmacogn.* 16: 83–87.

Dias, J. de F. G., G. M. Obdúlio & D. M. Marilis. 2009. Composition of essential oil and allelopathic activity of aromatic water of *Aster lanceolatus* Willd. (Asteraceae). *Braz. J. Pharm. Sci.* 45: 469–474.

Di Castri, F., A. J. Hansen & M. Debussche. 1990. *Biological Invasions in Europe and the Mediterranean Basin.* Kluwer Academic Publishers, Dordrecht, The Netherlands. 463 pp.

Dickinson, T. A., R. C. Evans & C. S. Campbell. 2000. Phylogenetic relationships between *Crataegus* and *Mespilus* (Rosaceae subf. Maloideae) based on rDNA sequence variation. *Amer. J. Bot.* 87: S122–S123.

Dicks, L. V., D. A. Showler & W. J. Sutherland. 2010. *Bee Conservation—Evidence for the Effects of Interventions.* Pelagic Publishing, Exeter, UK. 146 pp.

Diehl, J. W. & P. B. McEvoy. 1989. Impact of the cinnabar moth (*Tyria jacobaeae*) on *Senecio triangularis*, a non-target native plant in Oregon. pp. 119–126. *In*: E. S. Del Fosse (ed.), *Proceedings of the VII International Symposium on the Biological Control of Weeds.* Ministry of Agriculture and Forestry, Rome, Italy.

Dieringer, G., R. L. Cabrera & M. Mottaleb. 2014. Ecological relationship between floral thermogenesis and pollination in *Nelumbo lutea* (Nelumbonaceae). *Amer. J. Bot.* 101: 357–364.

Dieterich, R. A. & J. K. Morton. 1990. *Reindeer Health Aide Manual.* AFES Misc. Pub 90-4 CES 100H-00046. Agricultural and Forestry Experiment Station, Cooperative Extension Service, University of Alaska Fairbanks and U.S. Department of Agriculture. 77 pp.

Dietrich, W., P. H. Raven & W. L. Wagner. 1985. Revision of *Oenothera* sect. *Oenothera* subsect. *Emersonia* (Onagraceae). *Syst. Bot.* 10: 29–48.

Dietz, S. M. 1923. The role of the genus *Rhamnus* in the dissemination of crown rust. *Farmers' Bull.* USDA 1162: 1–19.

Di Gaspero, G., E. Peterlunger, R. Testolin, K. J. Edwards & G. Cipriani. 2000. Conservation of microsatellite loci within the genus *Vitis. Theor. Appl. Genet.* 101: 301–308.

Diggle, P. K., M. A. Meixner, A. B. Carroll & C. F. Aschwanden. 2002. Barriers to sexual reproduction in *Polygonum viviparum*: a comparative developmental analysis of *P. viviparum and P. bistortoides. Ann. Bot.* 89: 145–156.

Diggory, Z. E. & V. T. Parker. 2011. Seed supply and revegetation dynamics at restored tidal marshes, Napa River, California. *Restorat. Ecol.* 19: 121–130.

Dikova, B. 1989. Wild-growing hosts of the cucumber mosaic virus. *Rasteniev'dni. Nauki* 26: 57–64.

Dilcher, D. L. & H. Wang. 2009. An Early Cretaceous fruit with affinities to Ceratophyllaceae. *Amer. J. Bot.* 96: 2256–2269.

Dillingham, C. 2005. *Conservation assessment for Meesia trique-tra (L.) Aongstr. (three-ranked hump-moss) and Meesia uligi-nosa Hedwig (broad-nerved hump-moss) in California with a focus on the Sierra Nevada bioregion.* VMS Enterprise Team, Quincy, CA. 29 pp.

Dionne, J.-C. 1976. L'action glacielle dans les schores du littoral de la baie de James. *Cah. Géogr. Québec* 20: 303–326.

Dirr, M. A. & C. W. Heuser Jr. 2006. *The Reference Manual of Woody Plant Propagation: From Seed to Tissue Culture.* Varsity Press, Inc., Cary, NC. 410 pp.

Dirrigl Jr., F. J. & R. H. Mohlenbrock. 2012. Land snails in ephemeral pools at Ottine Swamp, Gonzales County, Texas. *Southwest. Naturalist* 57: 353–355.

Dirschl, H. J. 1969. Foods of lesser scaup and blue-winged teal in the Saskatchewan River delta. *J. Wildl. Manag.* 33: 77–87.

Dítě, D., J. Navrátilová, M. Hájek, M. Valachovič & D. Pukajová. 2006. Habitat variability and classification of *Utricularia* communities: comparison of peat depressions in Slovakia and the Třeboň basin. *Preslia* 78: 331–343.

DiTomaso, J. M. & E. A. Healy. 2003. *Aquatic and Riparian Weeds of the West.* Publication 3421. Agriculture and Natural Resources, University of California, Oakland, CA. 442 pp.

Dittes, J. C. & J. L. Guardino. 2002. Chapter 3: Rare plants. pp. 55–150. *In*: J. E. Vollmar (ed.), *Wildlife and Rare Plant Ecology of Eastern Merced County's Vernal Pool Grasslands.* Vollmar Consulting, Berkeley, CA.

Dittmar, L. A. & R. K. Neely. 1999. Wetland seed bank response to sedimentation varying in loading rate and texture. *Wetlands* 19: 341–351.

Dixon, M. D. & W. C. Johnson. 1999. Riparian vegetation along the Middle Snake River, Idaho, zonation, geographical trends, and historical changes. *Great Basin Naturalist* 59: 18–34.

Dmitriev, D. A. & C. H. Dietrich. 2009. Review of the species of New World Erythroneurini (Hemiptera: Cicadellidae: Typhlocybinae) III. Genus *Erythridula. Illinois Nat. Hist. Surv. Bull.* 38: 215–334.

Do, K. 2006. A Determination of phylogeny and hybridization history within *Clematis* L. (Ranunculaceae) using actin and nitrate reductase intron sequences. PhD dissertation. University of South Florida, Tampa, FL. 45 pp.

Dobeš, C. & J. Paule. 2010. A comprehensive chloroplast DNA-based phylogeny of the genus *Potentilla* (Rosaceae): implica-tions for its geographic origin, phylogeography and generic circumscription. *Mol. Phylogen. Evol.* 56: 156–175.

Dodamani, S. S., R. D. Sanakal & B. B. Kaliwal. 2012. Antidiabetic efficacy of ethanolic leaf extract of *Nymphaea odorata* in alloxan induced diabetic mice. *Int. J. Pharm. Pharm. Sci.* 4: 338–341.

Dodd, J. D. & R. T. Coupland. 1966. Vegetation of saline areas in Saskatchewan. *Ecology* 47: 958–967.

Dodd, M. B. & S. J. Orr. 1995. Seasonal growth, phosphate response, and drought tolerance of 11 perennial legume species grown in a hill-country soil. *New Zealand J. Agric. Res.* 38: 7–20.

Dodd, R. S., Z. Afzal-Rafii & A. Bousquet-Mélou. 2000. Evolutionary divergence in the pan-Atlantic mangrove *Avicennia germi-nans. New Phytol.* 145: 115–125.

Dodd, R. S., Z. Afzal-Rafii, N. Kashani & J. Budrick. 2002. Land barriers and open oceans: effects on gene diversity and popu-lation structure in *Avicennia germinans* L. (Avicenniaceae). *Mol. Ecol.* 11: 1327–1338.

Dodds, D. G. 1960. Food competition and range relationships of moose and snowshoe hare in Newfoundland. *J. Wildl. Manag.* 24: 52–60.

Doering, K. C. 1942. Host plant records of Cercopidae in North America, north of Mexico (Homoptera) (continued). *J. Kansas Entomol. Soc.* 15: 73–92.

Doffitt, C. & M. Fishbein. 2005. Phylogenetic relationships of the genus *Amsonia* (Apocynaceae) in North America based on *rpoB-TRNC* and *rpl16* sequence data. *J. Mississippi Acad. Sci.* 50: 67 (abstract).

Dolan, R. W. 1984. The effect of seed size and maternal source on individual size in a population of *Ludwigia leptocarpa* (Onagraceae). *Amer. J. Bot.* 71: 1302–1307.

Dolan, R. W. 2004. Conservation assessment for roundleaf water-hyssop (*Bacopa rotundifolia* (Michx.) Wettst.). USDA Forest Service, Eastern Region, Milwaukee, WI.

Dolan, R. W. & R. R. Sharitz. 1984. Population dynamics of *Ludwigia leptocarpa* (Onagraceae) and some factors affecting size hierarchies in a natural population. *J. Ecol.* 72: 1031–1041.

Dole, J. A. 1990. Role of corolla abscission in delayed self-pollination of *Mimulus guttatus* (Scrophulariaceae). *Amer. J. Bot.* 77: 1505–1507.

Dole, J. A. 1992. Reproductive assurance mechanisms in three taxa of the *Mimulus guttatus* complex (Scrophulariaceae). *Amer. J. Bot.* 79: 650–659.

Dole, J. A. & M. Sun. 1992. Field and genetic survey of the endangered Butte County meadowfoam—*Limnanthes floccosa* subsp. *californica* (Limnanthaceae). *Conserv. Biol.* 6: 549–558.

Dölle, M. & W. Schmidt. 2009. The relationship between soil seed bank, above-ground vegetation and disturbance intensity on old-field successional permanent plots. *Appl. Veg. Sci.* 12: 415–428.

Donahue, W. H. 1954. Some plant communities in the anthracite region of northeastern Pennsylvania. *Amer. Midl. Naturalist* 51: 203–231.

Donoghue, M. J., T. Eriksson, P. A. Reeves & R. G. Olmstead. 2001. Phylogeny and phylogenetic taxonomy of Dipsacales, with special reference to *Sinadoxa* and *Tetradoxa* (Adoxaceae). *Harvard Pap. Bot.* 6: 459–479.

Donoghue, M. J., B. G. Baldwin, J. Li & R. C. Winkworth. 2004. *Viburnum* phylogeny based on chloroplast *trnK* intron and nuclear ribosomal ITS DNA sequences. *Syst. Bot.* 29: 188–198.

Donnelly, L. M., M. M. Jenderek, J. P. Prince, P. A. Reeves, A. Brown & R. M. Hannan. 2008. Genetic diversity in the USDA *Limnanthes* germplasm collection assessed by simple sequence repeats. *Plant Genet. Resour. Characteriz. Utiliz.* 7: 33–41.

Doolittle, J. A., M. A. Hardisky & M. F. Gross. 1990. A ground-penetrating radar study of active layer thicknesses in areas of moist sedge and wet sedge tundra near Bethel, Alaska, U.S.A. *Arct. Alp. Res.* 22: 175–182.

Doran, C. W. 1943. Activities and grazing habits of sheep on summer ranges. *J. Forest.* 41: 253–258.

Dorken, M. E. & C. G. Eckert. 2001. Severely reduced sexual reproduction in northern populations of a clonal plant, *Decodon verticillatus* (Lythraceae). *J. Ecol.* 89: 339–350.

Dorman, H. J. D., M. Kosar, K. Kahlos, Y. Holm & R. Hiltunen. 2003. Antioxidant properties and composition of aqueous extracts from *Mentha* species, hybrids, varieties, and cultivars. *J. Agric. Food Chem.* 51: 4563–4569.

Dorn, N. J., G. Cronin & D. M. Lodge. 2001. Feeding preferences and performance of an aquatic lepidopteran on macrophytes: plant hosts as food and habitat. *Oecologia* 128: 406–415.

Dorn, R. D. 1970. Moose and cattle food habits in southwest Montana. *J. Wildl. Manag.* 34: 559–564.

Dorr, L. J. 1981. The pollination ecology of *Zenobia* (Ericaceae). *Amer. J. Bot.* 68: 1325–1332.

Dorr, L. J. 2009. 40. *Zenobia* D. Don. P. 506 *In*: N. R. Morin (convening ed.), *Flora of North America North of Mexico, Vol. 8: Magnoliophyta: Paeoniaceae to Ericaceae.* Oxford University Press, New York.

Dörrstock, S., R. Seine, S. Porembski & W. Barthlott. 1996. First record of *Utricularia juncea* (Lentibulariaceae) for tropical Africa. *Kew Bull.* 51: 579–583.

Dorsey, B. L., T. Haevermans, X. Aubriot, J. J. Morawetz, R. Riina, V. W. Steinmann & P. E. Berry. 2013. Phylogenetics, morphological evolution, and classification of *Euphorbia* subgenus *Euphorbia*. *Taxon* 62: 291–315.

Dosmann, M. S. 2002. Stratification improves and is likely required for germination of *Aconitum sinomontanum*. *HortTechnology* 12: 423–425.

dos Santos, A. F. & A. E. Sant'Ana. 2001. Molluscicidal properties of some species of *Annona*. *Phytomedicine* 8: 115–20.

Dötterl, S., U. Glück, A. Jürgens, J. Woodring & G. Aas. 2014. Floral reward, advertisement and attractiveness to honey bees in dioecious *Salix caprea*. *PLoS One* 9: e93421.

Doucet, C. M. & J. M. Fryxell. 1993. The effect of nutritional quality on forage preference by beavers. *Oikos* 67: 201–208.

Douglas, D. A. 1981. The balance between vegetative and sexual reproduction of *Mimulus primuloides* (Scrophulariaceae) at different altitudes in California. *J. Ecol.* 69: 295–310.

Douglas, G. W. & J. M. Illingworth. 2004. Conservation evaluation of the Pacific population of tall woolly-heads, *Psilocarphus elatior*, an endangered herb in Canada. *Can. Field Naturalist* 118: 169–173.

Douglas, G. W., J. Penny & K. Barton. 2003. Status of dwarf woolly-heads (*Psilocarphus brevissimus* var. *brevissimus*) in British Columbia. B.C. Minist. Sustainable Resour. Manag., Conservation Data Centre, and B.C. Minist. Water, Land and Air Protection, Biodiversity Branch, Victoria, BC. Wildl. Bull. No. B-109. 7 pp.

Douglas, S. 1983. Floral color patterns and pollinator attraction in a bog habitat. *Can. J. Bot.* 61: 3494–3501.

Douhan, L. I. & D. A. Johnson. 2001. Vegetative compatibility and pathogenicity of *Verticillium dahliae* from spearmint and peppermint. *Plant Dis.* 85: 297–302.

Douhovnikoff, V. R. & S. Dodd. 2003. Intra-clonal variation and a similarity threshold for identification of clones: application to *Salix exigua* using AFLP molecular markers. *Theor. Appl. Genet.* 106: 1307–1307.

Doumlele, D. G., B. K. Fowler & G. M. Silberhorn. 1985. Vegetative community structure of a tidal freshwater swamp in Virginia. *Wetlands* 4: 129–145.

Dovrat, G., A. Perevolotsky & G. Ne'eman. 2012. Wild boars as seed dispersal agents of exotic plants from agricultural lands to conservation areas. *J. Arid Environ.* 78: 49–54.

Dow, R. L., R. N. Inserra, R. P. Esser & K. R. Langdon. 1990. Distribution, hosts, and morphological characteristics of *Tylenchulus palustris* in Florida and Bermuda. *J. Nematol.* 22(4S): 724–728.

Dowhan, J. J. & R. Rozsa. 1989. Flora of Fire Island, Suffolk County, New York. *Bull. Torrey Bot. Club* 116: 265–282.

Downer, C. C. 2001. Observations on the diet and habitat of the mountain tapir (*Tapirus pinchaque*). *J. Zool.* 254: 279–291.

Downie, S. R. & D. S. Katz-Downie. 1996. A molecular phylogeny of Apiaceae subfamily Apioideae: evidence from nuclear ribosomal DNA internal transcribed spacer sequences. *Amer. J. Bot.* 83: 234–251.

Downie, S. R. & J. D. Palmer. 1992. Restriction site mapping of the chloroplast DNA inverted repeat: a molecular phylogeny of the Asteridae. *Ann. Missouri Bot. Gard.* 79: 266–283.

Downie, S. R., S. Ramanatha, D. S. Katz-Downie & E. Llanas. 1998. Molecular systematics of Apiaceae subfamily Apioideae: phylogenetic analysis of nuclear ribosomal DNA internal transcribed spacer and plastid *rpoC1* intron sequences. *Amer. J. Bot.* 85: 563–591.

Downie, S. R., D. S. Katz-Downie & M. F. Watson. 2000. A phylogeny of the flowering plant family Apiaceae based on chloroplast DNA *rpl16* and *rpoC1* intron sequences: towards a suprageneric classification of subfamily Apioideae. *Amer. J. Bot.* 87: 273–292.

Downie, S. R., G. M. Plunkett, M. F. Watson, K. Spalik, D. S. Katz-Downie, C. M. Valiejo-Roman, E. I. Terentieva, A. V. Troitsky, B.-Y. Lee, J. Lahham & A. El-Oqlah. 2001. Tribes and clades within Apiaceae subfamily *Apioideae*: the contribution of molecular data. *Edinb. J. Bot.* 58: 301–330.

Downie, S. R., R. L. Hartman, F.-J. Sun & D. S. Katz-Downie. 2002. Polyphyly of the spring-parsleys (Cymopterus): molecular and morphological evidence suggests complex relationships among the perennial endemic genera of western North American Apiaceae. *Can. J. Bot.* 80: 1295–1324.

Downie, S. R., D. S. Katz-Downie, F. J. Sun & C. S. Lee. 2008. Phylogeny and biogeography of Apiaceae tribe Oenantheae inferred from nuclear rDNA ITS and cpDNA *psbI-5'trnK*$^{(UUU)}$ sequences, with emphasis on the North American endemics clade. *Can. J. Bot.* 86: 1039–1064.

Downing, J. 1996. Native plant materials for economic development in southeast Alaska. Senior thesis (ST 2006-04), School of Agriculture and Land Resources Management, University of Alaska Fairbanks, Fairbanks, AK. 45 pp.

Downton, W. J. S. 1975. The occurrence of C4 photosynthesis among plants. *Photosynthetica* 9: 96–105.

Doyle, G. 2003. *Survey of Selected Seeps and Springs within the Bureau of Land Management's Gunnison Field Office Management Area (Gunnison and Saguache Counties, CO).* Colorado Natural Heritage Program, Colorado State University, Fort Collins, CO. 79 pp.

Doyle, J. J., D. E. Soltis & P. S. Soltis. 1985. An intergeneric hybrid in the Saxifragaceae: evidence from ribosomal RNA genes. *Amer. J. Bot.* 72: 1388–1391.

Doyle, P. J. 1972. Regional stream sediment reconnaissance and trace element content of rock, soild and plant material in eastern Yukon Territory. MS thesis. University of British Columbia, Vancouver, BC. 160 pp.

Doyle, R. D., M. D. Francis & R. M. Smart. 2003. Interference competition between *Ludwigia repens* and *Hygrophila polysperma*: two morphologically similar aquatic plant species. *Aquat. Bot.* 77: 223–234.

Drabble, E. & H. Drabble. 1927. Some flowers and their dipteran visitors. *New Phytol.* 26: 115–123.

Drawe, D. L. 1968. Mid-summer diet of deer on the Welder Wildlife Refuge. *J. Range Manag.* 21: 164–166.

Drayton, B. & R. B. Primack. 2012. Success rates for reintroductions of eight perennial plant species after 15 years. *Restorat. Ecol.* 20: 299–303.

Dress, W. J., S. J. Newell, A. J. Nastase & J. C. Ford. 1997. Analysis of amino acids in nectar from pitchers of *Sarracenia purpurea* (Sarraceniaceae). *Amer. J. Bot.* 84: 1264–1271.

Dressler, R. L., D. W. Hall, K. D. Perkins & N. H. Williams. 1987. *Identification Manual for Wetland Plant Species of Florida.* Institute of Food & Agricultural Sciences, University of Florida, Gainesville, FL. 297 pp.

Drew, J. V. & R. E. Shanks. 1965. Landscape relationships of soils and vegetation in the forest-tundra ecotone, upper Firth River Valley, Alaska-Canada. *Ecol. Monogr.* 35: 285–306.

Drewa, P. B., W. J. Platt & E. B. Moser. 2002. Community structure along elevation gradients in headwater regions of longleaf pine savannas. *Plant Ecol.* 160: 61–78.

Drezner, T. D., P. L. Fall & J. C. Stromberg. 2001. Plant distribution and dispersal mechanisms at the Hassayampa River preserve, Arizona, USA. *Glob. Ecol. Biogeogr.* 10: 205–217.

Drohan, P. J., C. N. Ross, J. T. Anderson, R. F. Fortney & J. S. Rentch. 2006. Soil and hydrological drivers of *Typha latifolia* encroachment in a marl wetland. *Wetlands Ecol. Manag.* 14: 107–122.

Druva-Lusite, I. & G. Ievinsh. 2010. Diversity of arbuscular mycorrhizal symbiosis in plants from coastal habitats. *Environ. Exp. Biol.* 8: 17–34.

Duan, Y. W., T. F. Zhang, Y. P. He & J. Q. Liu. 2009. Insect and wind pollination of an alpine biennial *Aconitum gymnandrum* (Ranunculaceae). *Plant Biol.* 11: 796–802.

DuBarry Jr., A. P. 1963. Germination of bottomland tree seed while immersed in water. *J. Forest.* 61: 225–226.

Duberstein, J. A., W. H. Conner & K. W. Krauss. 2014. Woody vegetation communities of tidal freshwater swamps in South Carolina, Georgia and Florida (US) with comparisons to similar systems in the US and South America. *J. Veg. Sci.* 25: 848–862.

Dubey, A. N. 1986. *Sphenoclea zeylanica* as organic nitrogen source for rice production. *Res. Dev. Rep.* 3: 72–75.

Ducey, T. F., J. O. Miller, M. W. Lang, A. A. Szogi, P. G. Hunt, D. E. Fenstermacher, M. C. Rabenhorst & G. W. McCarty. 2015. Soil physicochemical conditions, denitrification rates, and abundance in North Carolina Coastal Plain restored wetlands. *J. Environ. Qual.* 44: 1011–1022.

Dudash, M. R. & K. Ritland. 1991. Multiple paternity and self-fertilization in relation to floral age in *Mimulus guttatus* (Scrophulariaceae). *Amer. J. Bot.* 78: 1746–1753.

Dudle, D. A., P. Mutikainen & L. F. Delph. 2001. Genetics of sex determination in the gynodioecious species *Lobelia siphilitica*: evidence from two populations. *Heredity* 86: 265–276.

Duever, L. C. 1984. Natural communities: seepage communities. *Palmetto* 4: 1–2, 10–11.

Duever, M. J. & L. A. Riopelle. 1983. Successional sequences and rates on tree islands in the Okefenokee Swamp. *Amer. Midl. Naturalist* 110: 186–191.

Dugal, A. W. 1990. Albion Road wetlands: Part 1. *Trail & Landscape* 24: 56–78.

Dugdale, T. M., K. L. Butler, D. Clements & T. D. Hunt. 2013. Survival of cabomba (*Cabomba caroliniana*) during lake drawdown within mounds of stranded vegetation. *Lake Reserv. Manag.* 29: 61–67.

Duke, J. A. & M. J. Bogenschutz-Godwin. 2002. *Handbook of Medicinal Herbs.* CRC Press, Boca Raton, FL. 870 pp.

Duke, J. A. & V. M. Vásquez. 1994. *Amazonian Ethnobotanical Dictionary.* CRC Press, Boca Raton, FL. 215 pp.

Duke, J. A. & K. K. Wain. 1981. *Medicinal Plants of the World. Computer Index with More Than 85,000 Entries. 3 vols.* Plants Genetics and Germplasm Institute, Agriculture Research Service, Beltsville, MD. 1654 pp.

Duke, N. C. 1991. A systematic revision of the mangrove genus *Avicennia* (Avicenniaceae) in Australasia. *Aust. Syst. Bot.* 4: 299–324.

Duke, N. C., J. A. H. Benzie, J. A. Goodall & E. R. Ballment. 1998. Genetic structure and evolution of species in the mangrove genus *Avicennia* (Avicenniaceae) in the Indo-West Pacific. *Evolution* 52: 1612–1626.

Dumas, P. C. 1956. The ecological relations of sympatry in *Plethodon dunni* and *Plethodon vehiculum*. *Ecology* 37: 484–495.

Duminil, J., S. Fineschi, A. Hampe, P. Jordano, D. Salvini, G. G. Vendramin & R. J. Petit. 2007. Can population genetic structure be predicted from life-history traits? *Amer. Naturalist* 169: 662–672.

Duncan, R. S., C. P. Elliott, B. L. Fluker & B. R. Kuhajda. 2010. Habitat use of the watercress darter (*Etheostoma nuchale*): an endangered fish in an urban landscape. *Amer. Midl. Naturalist* 164: 9–21.

Duncan, W. H. 1964. New *Elatine* (Elatinaceae) populations in the southeastern United States. *Rhodora* 66: 47–53.

Dungan, J. D. & R. G. Wright. 2005. Summer diet composition of moose in Rocky Mountain National Park, Colorado. *Alces* 41: 139–146.

Dunlop, D. 2001. *Sclerolepis uniflora* (Walter) BSP. One-flowered *Sclerolepis*. New England Plant Conservation Program, Conservation and Research Plan. New England Wild Flower Society, Framingham, MA. 14 pp.

Dunn, S. T. 1905. *Alien Flora of Britain*. West, Newman & Co., London. 208 pp.

Durewicz, A. L. 2012. Inheritance of chloroplast DNA (cpDNA) in *Lobelia siphilitica*. PhD dissertation. Kent State University, Kent, OH. 62 pp.

Dvorak, B. & A. Volder. 2010. Green roof vegetation for North American ecoregions: a literature review. *Landscape Urban Plan.* 96: 197–213.

Dwire, K. A., J. B. Kauffman & J. E. Baham. 2006. Plant species distribution in relation to water-table depth and soil redox potential in montane riparian meadows. *Wetlands* 26: 131–146.

Dybas, C. L. 2016. Wild medicine. The search for cures from nature. *BioScience* 66(5): 341–349. doi:10.1093/biosci/biw031.

Dzhus, M. A. 2014. Alien species of American origin on cranberry plantation in Belarus. *Bot. Zhurn.* 99: 540–554.

Dziadyk, B. & G. K. Clambey. 1980. Floristic composition of plant communities in a western Minnesota tallgrass prairie. pp. 45–54. *In*: C. L. Kucera (ed.), *Proceedings of the 7th North American Prairie Conference*. Southwest Missouri State University, Springfield, MO.

Eakins, D. 1994. Intraspecific genetic diversity in *Blennosperma bakeri* Heiser (Asteraceae). MA thesis. Sonoma State University, Sonoma, CA.

Eallonardo Jr., A. S. & D. J. Leopold. 2014. Inland salt marshes of the northeastern United States: stress, disturbance and compositional stability. *Wetlands* 34: 155–166.

Eames, E. H. 1933. A new *Ludvigia* from New England. *Rhodora* 35: 227–230.

Easley, M. C. & W. S. Judd. 1990. Vascular flora of the southern upland property of Paynes Prairie State Preserve, Alachua County, Florida. *Castanea* 55: 142–186.

Easley, M. C. & W. S. Judd. 1993. Vascular flora of Little Talbot Island, Duval County, Florida. *Castanea* 58: 162–177.

East, E. M. 1940. The distribution of self-sterility in the flowering plants. *Proc. Amer. Philos. Soc.* 82: 449–518.

Easterly, N. W. 1957. A morphological study of *Ptilimnium*. *Brittonia* 9: 136–145.

Eastwood, B. 1856. *The Cranberry and its Culture: A Complete Manual for the Cultivation of the Cranberry with a Description of the Best Varieties*. Orange Judd & Co., New York. 120 pp.

Eberl, R. 2011. Mycorrhizal association with native and invasive cordgrass *Spartina* spp. in San Francisco Bay, California. *Aquat. Biol.* 14: 1–7.

Ebinger, J. E. & J. A. Bacone. 1981. Vegetation survey of hillside seeps at Turkey Run State Park. *Proc. Indiana Acad. Sci.* 90: 390–394.

Echols, S. L. & W. B. Zomlefer. 2010. Vascular plant flora of the remnant blackland prairies in Oaky Woods Wildlife Management Area, Houston County, Georgia. *Castanea* 75: 78–100.

Echt, C. S., D. Deemer, T. Kubisiak & C. D. Nelson. 2006. Microsatellites for *Lindera* species. *Mol. Ecol. Notes* 6: 1171–1173.

Echt, C. S., D. Demeer & D. Gustafson. 2011. Patterns of differentiation among endangered pondberry populations. *Conserv. Genet.* 12: 1015–1026.

Eck, P. & N. F. Childers (eds.). 1966. *Blueberry Culture*. Rutgers University Press, New Brunswick, NJ. 378 pp.

Eckel, P. M. 2002. *Epilobium parviflorum*, a rare European introduction along the Niagara River. *NYFA Newslett.* 13(2): 3–5.

Eckenwalder, J. E. 1996. Systematics and evolution of *Populus*. pp. 7–32. *In*: R. F. Stettler, H. D. Bradshaw Jr., P. E. Heilman & T. M. Hinckley (eds.), *Biology of Populus and its Implications for Management and Conservation*. NRC Research Press, National Research Council of Canada, Ottawa, ON.

Eckert, C. G. 2002. Effect of geographic variation in pollinator fauna on the mating system of *Decodon verticillatus* (Lythraceae). *Int. J. Plant Sci.* 163: 123–132.

Eckert, C. G. & M. Allen. 1997. Cryptic self-incompatibility in tristylous *Decodon verticillatus* (Lythraceae). *Amer. J. Bot.* 84: 1391–1397.

Eckert, C. G. & S. C. H. Barrett. 1992. Stochastic loss of style morphs from populations of tristylous *Lythrum salicaria* and *Decodon verticillatus* (Lythraceae). *Evolution* 46: 1014–1029.

Eckert, C. G. & S. C. H. Barrett. 1993a. Clonal reproduction and patterns of genotypic diversity in *Decodon verticillatus* (Lythraceae). *Amer. J. Bot.* 80: 1175–1182.

Eckert, C. G. & S. C. H. Barrett. 1993b. The inheritance of tristyly in *Decodon verticillatus* (Lythraceae). *Heredity* 71: 473–480.

Eckert, C. G. & S. C. H. Barrett. 1994a. Tristyly, self-compatibility and floral variation in *Decodon verticillatus* (Lythraceae). *Biol. J. Linn. Soc.* 53: 1–30.

Eckert, C. G. & S. C. H. Barrett. 1994b. Post-pollination mechanisms and the maintenance of outcrossing in self-compatible, tristylous, *Decodon verticillatus* (Lythraceae). *Heredity* 72: 396–411.

Eckert, C. G. & S. C. H. Barrett. 1994c. Inbreeding depression in partially self-fertilizing *Decodon verticillatus* (Lythraceae): population-genetic and experimental analysis. *Evolution* 48: 952–964.

Eckert, C. G. & S. C. H. Barrett. 1995. Style morph ratios in tristylous *Decodon verticillatus* (Lythraceae): selection vs. historical contingency. *Ecology* 76: 1051–1066.

Eckert, C. G. & A. Schaefer. 1998. Does self-pollination provide reproductive assurance in *Aquilegia canadensis* (Ranunculaceae)? *Amer. J. Bot.* 85: 919–924.

Eckert, C. G., D. Manicacci & S. C. H. Barrett. 1996. Genetic drift and founder effect in native versus introduced populations of an invading plant, *Lythrum salicaria* (Lythraceae). *Evolution* 50: 1512–1519.

Eckert, C. G., M. E. Dorken & S. A. Mitchell. 1999. Loss of sex in clonal populations of a flowering plant, *Decodon verticillatus* (Lythraceae). *Evolution* 53: 1079–1092.

Eckstein, R. L. & P. S. Karlsson. 2001. The effect of reproduction on nitrogen use-efficiency of three species of the carnivorous genus *Pinguicula*. *J. Ecol.* 89: 798–806.

Eddie, W. M. M., T. Shulkina, J. Gaskin, R. C. Haberle & R. K. Jansen. 2003. Phylogeny of Campanulaceae s. str. inferred from ITS sequences of nuclear ribosomal DNA. *Ann. Missouri Bot. Gard.* 90: 554–575.

Eddy, S. 1925. Fresh water algal succession. *Trans. Amer. Microsc. Soc.* 44: 138–147.

Edelman, A. J. 2003. *Marmota olympus. Mammal Species* 736: 1–5.

Edgar, E. 1958. Studies in New Zealand *Cotulas. Trans. R. Soc. New Zealand* 85: 357–377.

Edgin, B. 2004. Status and distribution of Illinois populations of *Stenanthium gramineum* (Ker-Gawl.) Morong, grass-leaved lily (Liliaceae): an endangered plant in Illinois. *Castanea* 69: 216–225.

Edgin, B. W., E. McClain, B. Gillespie & J. E. Ebinger. 2003. Vegetation composition and structure of Eversgerd post oak flat-woods, Clinton County, Illinois. *N. E. Naturalist* 10: 111–118.

Edmonstone, T. 1841. On the dyes of the Shetland Isles. *Trans Bot. Soc. Edinb.* 1: 123–126.

Edwards, A. L. & R. Wyatt. 1994. Population genetics of the rare *Asclepias texana* and its widespread sister species, *A. perennis. Syst. Bot.* 19: 291–307.

Edwards, A. L., R. Wyatt & R. R. Sharitz. 1994. Seed buoyancy and viability of the wetland milkweed *Asclepias perennis* and an upland milkweed, *Asclepias exaltata. Bull. Torrey Bot. Club* 121: 160–169.

Edwards, C. E., D. E. Soltis & P. S. Soltis. 2006. Molecular phylogeny of *Conradina* and other scrub mints (Lamiaceae) from the southeastern USA: evidence for hybridization in pleistocene refugia? *Syst. Bot.* 31: 193–207.

Efroymson, R. A., M. J. Peterson, N. R. Giffen, M. G. Ryon, J. G. Smith, W. W. Hargrove, W. K. Roy, C. J. Welsh, D. L. Druckenbrod & H. D. Quarles. 2008. Investigating habitat value to inform contaminant remediation options: case study. *J. Environ. Manag.* 88: 1452–1470.

Egan, A. N. & K. A. Crandall. 2008. Divergence and diversification in North American Psoraleeae (Fabaceae) due to climate change. *BMC Biol.* 6: 55. doi:10.1186/1741-7007-6-55.

Egan, T. P. & I. A. Ungar. 2000. Similarity between seed banks and above-ground vegetation along a salinity gradient. *J. Veg. Sci.* 11: 189–194.

Egger, J. M. & S. Malaby. 2015. *Castilleja collegiorum* (Orobanchaceae), a new species from the Cascade Mountains of southern Oregon, and the status of *Castilleja lassenensis Eastw. Phytoneuron* 2015-33: 1–13.

Egger, J. M., J. A. Ruygt & D. C. Tank. 2012. *Castilleja ambigua* var. *meadii* (Orobanchaceae): a new variety from Napa County, California. *Phytoneuron* 2012–68: 1–12.

Egolf, D. R. 1962. A cytological study of the genus *Viburnum. J. Arnold Arbor.* 43: 132–172.

Ehrenfeld, J. G. 2005. Vegetation of forested wetlands in urban and suburban landscapes in New Jersey. *J. Torrey Bot. Soc.* 132: 262–279.

Eicher, A. L. 1988. *Soil-Plant Correlations in Wetlands and Adjacent Uplands of the San Francisco Bay Estuary, California.* Biological Report 88 (21). U.S. Fish & Wildlife Service. 35 pp.

Eichler, L. & C. Boylen. 2009. *Finkle Brook Delta, Lake George, NY; Aquatic plant survey – 2009.* Darrin Fresh Water Institute, Bolton Landing, New York. 12 pp.

Eichler, L. & C. Boylen. 2013. *Aquatic Vegetation of Canada Lake, West Lake and Green Lake, Town of Caroga, New York.* Darrin Fresh Water Institute, Bolton Landing, New York. 27 pp.

Eichler, L. & C. Boylen. 2014. *Aquatic Vegetation of Lake Groton, Groton, Vermont.* Darrin Fresh Water Institute, Bolton Landing, New York. 11 pp.

Eilers, H. P. 1974. Plants, plant communities, net production and tide levels; the ecological biogeography of the Nehalem salt marshes, Tillamook County, Oregon. PhD dissertation. Oregon State University, Corvallis, OR. 368 pp.

Eimear, M., N. Lughadha & J. A. N. Parnell. 1989. Heterostyly and gene-flow in *Menyanthes trifoliata* L. (Menyanthaceae). *Bot. J. Linn. Soc.* 100: 337–354.

Eiseman, C. S. & A. S. Jensen. 2015. Insects feeding on sea lavender (Plumbaginaceae: *Limonium carolinianum* [Walt.] Britt) along the New England coast. *Entomol. News* 124: 364–369.

Eisner, T., M. Eisner & D. Aneshansley. 1973. Ultraviolet patterns on rear of flowers: basis of disparity of buds and blossoms. *Proc. Natl. Acad. Sci. U.S.A.* 70: 1002–1004.

Eisto, A.-K., M. Kuitunen, A. Lammi, V. Saari, J. Suhonen, S. Syrjäsuo & P. M. Tikka. 2000. Population persistence and offspring fitness in the rare bellflower *Campanula cervicaria* in relation to population size and habitat quality. *Conserv. Biol.* 14: 1413–1421.

Elakovich, S. D. & J. W. Wooten. 1987. An examination of the phytotoxicity of the water shield, *Brasenia schreberi. J. Chem. Ecol.* 13: 1935–1940.

Elakovich, S. D. & J. W. Wooten. 1989. Allelopathic potential of 16 aquatic and wetland plants. *J. Aquat. Plant Manag.* 27: 78–84.

Elam, C. E. 2007. Flora, plant communities, and soils of a significant natural area in the middle Atlantic coastal plain (Craven County, North Carolina). MS thesis. North Carolina State University, Raleigh, NC. 94 pp.

Elam, C. E. & R. W. Cantrell. 2011. *Lafayette Creek Property—Phases I and II Umbrella Regional Mitigation Plans for Florida Department of Transportation Projects.* Cardno ENTRIX Inc., Houston, TX. 57 pp.

Elam, C. E., J. M. Stucky, T. R. Wentworth & J. D. Gregory. 2009. Vascular flora, plant communities, and soils of a significant natural area in the middle Atlantic Coastal Plain (Craven County, North Carolina). *Castanea* 74: 53–77.

Elam, D. R. 1998. Population genetics of vernal pool plants: theory, data and conservation implications. pp. 180–189. *In*: C. W. Witham, E. T. Bauder, D. Belk, W. R. Ferren Jr. & R. Ornduff (eds.), *Ecology, Conservation, and Management of Vernal Pool Ecosystems—Proceedings from a 1996 Conference.* California Native Plant Society, Sacramento, CA.

Elangovan, V., S. Govindasamy, N. Ramamoorthy & K. Balasubramanian. 1995. *In vitro* studies on the anticancer activity of *Bacopa monnieri. Fitoterapia* 66: 211–215.

Elansary, H. O. M., L. Adamec & H. Storchova. 2010. Uniformity of organellar DNA in *Aldrovanda vesiculosa*, an endangered aquatic carnivorous species, distributed across four continents. *Aquat. Bot.* 92: 214–220.

Elder, C. L. 1994. Reproductive biology of the California pitcher plant (*Darlingtonia californica*). *Fremontia* 22: 29.

Elder, C. L. 1997. Reproductive biology of *Darlingtonia californica.* MS thesis. Humboldt State University, Arcata, CA. 67 pp.

Eldridge, J. 1975. Bush medicine in the Exumas and Long Island, Bahamas: a field study. *Econ. Bot.* 29: 307–332.

Eleuterius, L. N. 1972. The marshes of Mississippi. *Castanea* 37: 153–168.

Eleuterius, L. N. & J. D. Caldwell. 1984. Flowering phenology of tidal marsh plants in Mississippi. *Castanea* 49: 172–179.

Eleuterius, L. N. & S. McDaniel. 1978. The salt marsh flora of Mississippi. *Castanea* 43: 86–95.

El-Gazzar, A. & L. Watson. 1968. Labiatae: taxonomy and susceptibility to *Puccinia menthae* Pers. *New Phytol.* 67: 739–743.

Elger, A., G. Bornette, M.-H. Barrat-Segretain & C. Amoros. 2004. Disturbances as a structuring factor of plant palatability in aquatic communities. *Ecology* 85: 304–311.

Elias, T. S. 1987. *The Complete Trees of North America. Field Guide and Natural History.* Gramercy Publishing Co., New York. 948 pp.

Elias, T. S. & P. A. Dykeman. 1990. *Edible Wild Plants: A North American Field Guide.* Sterling Publishing Company, Inc., New York. 286 pp.

El-Lakany, M. H. & J. R. Dugle. 1972. DNA content in relation to phylogeny of selected boreal forest plants. *Evolution* 26: 427–434.

Ellett, C. W. 1966. Host Range of the Erysiphaceae of Ohio. *Ohio J. Sci.* 66: 570–581.

Ellett, C. W. 1970. Annotated list of the Personosporales of Ohio (I. Albuginaceae and Peronosporaceae). *Ohio J. Sci.* 70: 218–226.

Ellis, J. B. & B. M. Everhart. 1885a. North American species of *Ramularia*: with descriptions of the species. *J. Mycol.* 1: 73–83.

Ellis, J. B. & B. M. Everhart. 1885b. The North American species of *Gloeosporium. J. Mycol.* 1: 109–119.

Ellis, J. B. & B. M. Everhart. 1890. New North American fungi. *Proc. Acad. Nat. Sci. Philadelphia* 42: 219–249.

Ellis, J. R. & D. E. McCauley. 2009. Phenotypic differentiation in fitness related traits between populations of an extremely rare sunflower: conservation management of isolated populations. *Biol. Conserv.* 142: 1836–1843.

Ellis, J. R., C. H. Pashley, J. M. Burke & D. E. McCauley. 2006. High genetic diversity in a rare and endangered sunflower as compared to a common congener. *Mol. Ecol.* 15: 2345–2355.

Ellis, J. R., K. E. Bentley & D. E. McCauley. 2008. Detection of rare paternal chloroplast inheritance in controlled crosses of the endangered sunflower *Helianthus verticillatus. Heredity* 100: 574–580.

Ellison, A. M. 2001. Interspecific and intraspecific variation in seed size and germination requirements of *Sarracenia* (Sarraceniaceae). *Amer. J. Bot.* 88: 429–437.

Ellison, A. M. & E. J. Farnsworth. 2005. The cost of carnivory for *Darlingtonia californica* (Sarraceniaceae): evidence from relationships among leaf traits. *Amer. J. Bot.* 92: 1085–1093.

Ellison, A. M. & N. J. Gotelli. 2001. Evolutionary ecology of carnivorous plants. *Trends Ecol. Evol.* 16: 623–629.

Ellison, A. M. & J. N. Parker. 2002. Seed dispersal and seedling establishment of *Sarracenia purpurea* (Sarraceniaceae). *Amer. J. Bot.* 89: 1024–1026.

Ellison, A. M., E. J. Farnsworth & R. R. Twilley. 1996. Facultative mutualism between red mangroves and root-fouling sponges in Belizean mangal. *Ecology* 77: 2431–2444.

Ellison, A. M., E. J. Farnsworth & N. J. Gotelli. 2002. Ant diversity in pitcher-plant bogs of Massachusetts. *N.E. Naturalist* 9: 267–284.

Ellison, A. M., H. L. Buckley, T. E. Miller & N. J. Gotelli. 2004. Morphological variation in *Sarracenia purpurea* (Sarraceniaceae): geographic, environmental, and taxonomic correlates. *Amer. J. Bot.* 91: 1930–1935.

Ellison, A. M., C. C. Davis, P. J. Calie & R. F. C. Naczi. 2014. Pitcher plants (*Sarracenia*) provide a 21st-century perspective on infraspecific ranks and interspecific hybrids: a modest proposal* for appropriate recognition and usage. *Syst. Bot.* 39: 939–949. doi:10.1600/036364414X681473.

Ellison, L. 1949. Establishment of vegetation on depleted subalpine range as influenced by microenvironment. *Ecol. Monogr.* 19: 95–121.

Ellison, L. 1954. Subalpine vegetation of the Wasatch Plateau, Utah. *Ecol. Monogr.* 24: 89–184.

El-Lissy, O. 2007. *Treatment Program for Light Brown Apple Moth in Santa Cruz and Northern Monterey Counties, California. Environmental Assessment, September 2007.* Marketing and Regulatory Programs, Animal and Plant Health Inspection Service (APHIS), U.S. Department of Agriculture, Riverdale, MD. 79 pp.

Ellmore, G. S. 1981. Root dimorphism in *Ludwigia peploides* (Onagraceae): structure and gas content of mature roots. *Amer. J. Bot.* 68: 557–568.

Else, M. J. & D. N. Riemer. 1984. Factors affecting germination of seeds of fragrant waterlily (*Nymphaea odorata*). *J. Aquat. Plant Manag.* 22: 22–25.

Elsey, R. M., S. G. Platt & M. Shirley. 2015. An unusual beaver (*Castor canadensis*) lodge in a Louisiana coastal marsh. *S. E. Naturalist* 14: N28–N30.

Elsey-Quirk, T. & M. A. Leck. 2015. Patterns of seed bank and vegetation diversity along a tidal freshwater river. *Amer. J. Bot.* 102: 1996–2012.

Elsey-Quirk, T., B. A. Middleton & C. E. Proffitt. 2009. Seed flotation and germination of salt marsh plants: the effects of stratification, salinity, and/or inundation regime. *Aquat. Bot.* 91: 40–46.

Elster, C. & L. Perdomo. 1999. Rooting and vegetative propagation in *Laguncularia racemosa. Aquat. Bot.* 63: 83–93.

Elvin, M. A. & A. C. Sanders. 2003. A new species of *Monardella* (Lamiaceae) from Baja California, Mexico, and southern California, United States. *Novon* 13: 425–432.

El-Zaher, A., M. A. Mustafa, A. Badr, M. A. El-Galaly, A. A. Mobarak & M. G. Hassan. 2005. Genetic diversity among *Mentha* populations in Egypt as reflected by isozyme polymorphism. *Int. J. Bot.* 1: 188–195.

Elzinga, C. 1997. Habitat conservation assessment and strategy for the alkaline primrose (*Primula alcalina*). Idaho State Conservation Effort, Idaho Conservation Data Center, Boise, ID.

Emadzade, K., C. Lehnebach, P. Lockhart & E. Hörandl. 2010. A molecular phylogeny, morphology and classification of genera of Ranunculeae (Ranunculaceae). *Taxon* 59: 809–828.

Emadzade, K., M. J. Lebmann, M. H. Hoffmann, N. Tkach, F. A. Lone & E. Hörandl. 2015. Phylogenetic relationships and evolution of high mountain buttercups (*Ranunculus*) in North America and Central Asia. *Perspect. Plant Ecol. Evol. Syst.* 17: 131–141.

Emboden, W. A. 1983. The ethnobotany of the Dresden Codex with especial reference to the narcotic *Nymphaea ampla. Bot. Mus. Leafl.* 29: 87–132.

Emerson, F. W. 1921. Subterranean organs of bog plants. *Bot. Gaz.* 72: 359–374.

Emerton, J. H. 1888. Notes on *Melitaea phaeton. Psyche* 5: 54.

Emery, N. C. 2009. Ecological limits and fitness consequences of cross-gradient pollen movement in *Lasthenia fremontii. Amer. Naturalist* 174: 221–235.

Emery, N. C., K. J. Rice & M. L. Stanton. 2011. Fitness variation and local distribution limits in an annual plant population. *Evolution* 65: 1011–1020.

Emery, N. C., E. J. Forrestel, G. Jui, M. S. Park, B. G. Baldwin & D. D. Ackerly. 2012a. Niche evolution across spatial scales: climate and habitat specialization in California *Lasthenia* (Asteraceae). *Ecology* 93: S151–S166.

Emery, N. C., L. T. Martinez, E. Forrestel, B. G. Baldwin & D. D. Ackerly. 2012b. The ecology, evolution, and diversification of the vernal pool niche in *Lasthenia* (Madieae, Asteraceae). pp. 39–58. *In*: D. G. Alexander & R. A. Schlising (eds.), *Research and Recovery in Vernal Pool Landscapes*. Studies from the Herbarium, vol. 16. Chico State University, Chico, CA.

Emery, R. J. N. & C. C. Chinnappa. 1992. Natural hybridization between *Stellaria longipes* and *Stellaria borealis* (Caryophyllaceae). *Can. J. Bot.* 70: 1717–1723.

Emmel, T. C., M. C. Minno & B. A. Drummond. 1992. *Florissant Butterflies: A Guide to the Fossil and Present-Day Species of Central Colorado*. Stanford University Press, Stanford, CA. 148 pp.

Enders, R. K. 1932. Food of the muskrat in summer. *Ohio J. Sci.* 32: 21–30.

Endo, R., N. Shirasawa-Seo, Y. Adachi & K. Kanahama. 2002. *Agrobacterium*-mediated transformation of *Oenanthe javanica* (Blume) DC. plants. *Plant Biotechnol.* 19: 365–368.

Endress, M. E. 2001. Apocynaceae and Asclepiadaceae: united they stand. *Haseltonia* 8: 2–9.

Endress, M. E. & P. Bruyns. 2000. A revised classification of the Apocynaceae s.l. *Bot. Rev.* 66: 1–56.

Endress, P. K. 1992. Evolution and floral diversity: the phylogenetic surroundings of *Arabidopsis* and *Antirrhinum*. *Int. J. Plant Sci.* 153: S106–S122.

Endress, P. K. 1994. *Diversity and Evolutionary Biology of Tropical Flowers*. Cambridge University Press, Cambridge, UK. 525 pp.

Ene-OjoAtawodi, S. & G. S. Onaolapo. 2010. Comparative in vitro antioxidant potential of different parts of *Ipomoea asarifolia*, Roemer & Schultes, *Guiera senegalensis*, J. F. Gmel and *Anisopus mannii* N. E. Brown. *Brazilian J. Pharm. Sci.* 46: 245–250.

Engel, S. & S. A. Nichols. 1994. Aquatic macrophyte growth in a turbid windswept lake. *J. Freshwater Ecol.* 9: 97–109.

Engeman, R. M., A. Stevens, J. Allen, J. Dunlap, M. Daniel, D. Teague & B. Constantin. 2007. Feral swine management for conservation of an imperiled wetland habitat: Florida's vanishing seepage slopes. *Biol. Conserv.* 134: 440–446.

Enser, R. W. 2001. *Hypericum adpressum* (creeping St. John's-wort) conservation and research plan. New England Wild Flower Society, Framingham, MA. 11 pp. Available online: http://www.newfs.org.

Enser, R. W. 2004. *Sabatia stellaris* Pursh (Sea pink) conservation and research plan for New England. New England Wild Flower Society, Framingham, MA. 19 pp.

Ensign, S. H., C. R. Hupp, G. B. Noe, K. W. Krauss & C. L. Stagg. 2014. Sediment accretion in tidal freshwater forests and oligohaline marshes of the Waccamaw and Savannah rivers, USA. *Estuaries Coasts* 37: 1107–1119.

Entrup, A. 2015. Restoration of a wet longleaf pine (*Pinus palustris*) savanna in southeast Louisiana: burning toward reference conditions. MS thesis. University of New Orleans, New Orleans, LA. 57 pp.

Environmental Laboratory. 1987. *Corps of Engineers Wetlands Delineation Manual*. Technical Report Y-87-1, US Army Engineer Waterways Experiment Station, Vicksburg, MS. 117 pp.

Epling, C. & C. Jativa. 1964. Revision del genero *Satureja* en America del Sur. *Brittonia* 16: 393–416.

Epling, C. & C. Jativa. 1966. A descriptive key to the species of *Satureja* indigenous to North America. *Brittonia* 18: 244–248.

Epling, C. & H. Lewis. 1952. Increase of the adaptive range of the genus *Delphinium*. *Evolution* 6: 253–267.

Epstein, E. & E. Cohn. 1971. Biochemical changes in terminal root galls caused by an ectoparasitic nematode, *Longidorus africanus*: amino acids. *J. Nematol.* 3: 334–340.

Erbar, C. & P. Leins. 2004. Callitrichaceae. pp. 50–56. *In*: J. W. Kadereit (ed.), *The Families and Genera of Vascular Plants, Vol. VII. Flowering Plants, Dicotyledons: Lamiales (except Acanthaceae including Avicenniaceae)*. Springer-Verlag, Berlin, Germany.

Erbar, C., S. Porembski & P. Leins. 2005. Contributions to the systematic position of *Hydrolea* (Hydroleaceae) based on floral development. *Plant Syst. Evol.* 252: 71–83.

Erben, M. 1996. The significance of hybridization on the forming of species in the genus *Viola*. *Bocconea* 5: 113–118.

Erickson, T. A. & C. F. Puttock. 2006. *Hawai'i Wetland Field Guide: An Ecological and Identification Guide to Wetlands and Wetland Plants of the Hawaiian Islands*. Bess Press, Inc., Honolulu, HI. 294 pp.

Eriksen, C. H. 1968. Aspects of the limno-ecology of *Corophium spinicorne* Stimpson (Amphipoda) and *Gnorimosphaeroma oregonensis* (Dana) (Isopoda). *Crustaceana* 14: 1–12.

Eriksson, O. 1986. Survivorship, reproduction and dynamics of ramets of *Potentilla anserina* on a Baltic seashore meadow. *Plant Ecol.* 67: 17–25.

Eriksson, O. 1987. Regulation of seed-set and gender variation in the hermaphroditic plant *Potentilla anserina*. *Oikos* 49: 165–171.

Eriksson, O. 1988. Ramet behaviour and population growth in the clonal herb *Potentilla anserina*. *J. Ecol.* 76: 522–536.

Eriksson, T., M. J. Donoghue & M. S. Hibbs. 1998. Phylogenetic analysis of *Potentilla* using DNA sequences of nuclear ribosomal internal transcribed spacers (ITS), and implications for the classification of Rosoideae (Rosaceae). *Plant Syst. Evol.* 211: 155–179.

Eriksson, T., M. S. Hibbs, A. D. Yoder, C. F. Delwiche & M. J. Donoghue. 2003. The phylogeny of Rosoideae (Rosaceae) based on sequences of the internal transcribed spacers (ITS) of nuclear ribosomal DNA and the *trnL/F* region of chloroplast DNA. *Int. J. Plant Sci.* 164: 197–211.

Erlanson, E. W. 1929. Cytological conditions and evidences for hybridity in North American wild roses. *Bot. Gaz.* 87: 443–506.

Erlanson, E. W. 1938. Phylogeny and polyploidy in *Rosa*. *New Phytol.* 37: 72–81.

Erman, N. A. 1984. The use of riparian systems by aquatic insects. pp. 177–182. *In*: R. E. Warner & K. M. Hendrix (eds.), *California Riparian Gystems*. University of California Press, Berkeley, CA.

Erman, N. A. 2002. *Lessons From a Long-Term Study of Springs and Spring Invertebrates (Sierra Nevada, California, U.S.A.) and Implications for Conservation and Management. Conference Proceedings*. Spring-fed Wetlands: Important Scientific and Cultural Resources of the Intermountain Region, 2002. 13 pp. Available online: http://www.wetlands.dri.edu.

Ernst, E. 2002. Herbal medicinal products during pregnancy: are they safe? BJOG: *Int. J. Obstet. Gyn.* 109: 227–235.

Errington, P. L. 1941. Versatility in feeding and population maintenance of the muskrat. *J. Wildl. Manag.* 5: 68–89.

Ersöz, T., D. Tasdemir, I. Calis & C. M. Ireland. 2002a. Phenylethanoid glycosides from *Scutellaria galericulata*. *Turk. J. Chem.* 26: 465–471.

Ersöz, T., M. Ozalp, M. Ekizoglu & I. Callis. 2002b. Antimicrobial activities of the phenylethanoid glycosides from *Scutellaria galericulata*. *Hacettepe Univ. Eczacilik Fak. Derg. (Turkey)* 22: 1–8.

Ertter, B. 2000. Floristic surprises in North America north of Mexico. *Ann. Missouri Bot. Gard.* 87: 81–109.

Ertter, B. & J. L. Reveal. 2014. 19. *Comarum* Linnaeus. pp. 300–302. *In*: N. R. Morin (convening ed.), *Flora of North America North of Mexico, Vol. 9: Magnoliophyta: Picramniaceae to Rosaceae.* Oxford University Press, New York.

Ertter, B., R. Elven, J. L. Reveal & D. F. Murray. 2014. 8. *Potentilla* Linnaeus. pp. 121–218. *In*: N. R. Morin (convening ed.), *Flora of North America North of Mexico, Vol. 9: Magnoliophyta: Picramniaceae to Rosaceae.* Oxford University Press, New York.

Ertuğ, F. 2000. An ethnobotanical study in central Anatolia (Turkey). *Econ. Bot.* 54: 155–182.

Ervin, G. N. 2005. Spatio-temporally variable effects of a dominant macrophyte on vascular plant neighbors. *Wetlands* 25: 317–325.

Ervin, G. N. 2007. An experimental study on the facilitative effects of tussock structure among wetland plants. *Wetlands* 27: 620–630.

Eserman, L. A., G. P. Tiley, R. L. Jarret, J. H. Leebens-Mack & R. E. Miller. 2014. Phylogenetics and diversification of morning glories (tribe *Ipomoeeae*, Convolvulaceae) based on whole plastome sequences. *Amer. J. Bot.* 101: 92–103.

Eslami, S. V. 2011. Comparative germination and emergence ecology of two populations of common lambsquarters (*Chenopodium album*) from Iran and Denmark. *Weed Sci.* 59: 90–97.

Esler, A. E. 1987. The naturalisation of plants in urban Auckland, New Zealand. 3. Catalogue of naturalised species. *New Zealand J. Bot.* 25: 539–558.

Esquivel, J. F. & S. V. Esquivel. 2009. Identification of cotton fleahopper (Hemiptera: Miridae) host plants in central Texas and compendium of reported hosts in the United States. *Environ. Entomol.* 38: 766

Esser, H.-J., P. van Welzen & T. Djarwaningsih. 1997. A phylogenetic classification of the Malesian *Hippomaneae* (Euphorbiaceae). *Syst. Bot.* 22: 617–628.

Esser, L. L. 1992a. *Salix monticola*. *In*: *Fire Effects Information System.* U.S. Department of Agriculture, Forest Service, Rocky Mountain Research Station, Fire Sciences Laboratory (Producer). Available online: http://www.fs.fed.us/database/feis/ (accessed March 14, 2005).

Esser, L. L. 1992b. *Salix boothii*. *In*: *Fire Effects Information System.* U. S. Department of Agriculture, Forest Service, Rocky Mountain Research Station, Fire Sciences Laboratory (Producer). Available online: http://www.fs.fed.us/database/feis/ (accessed March 2, 2016).

Esser, L. L. 1995a. *Spiraea douglasii*. *In*: *Fire Effects Information System.* U. S. Department of Agriculture, Forest Service, Rocky Mountain Research Station, Fire Sciences Laboratory (Producer). Available online: http://www.fs.fed.us/database/feis/ (accessed February 20, 2006).

Esser, L. L. 1995b. *Heracleum lanatum*. *In*: *Fire Effects Information System.* U. S. Department of Agriculture, Forest Service, Rocky Mountain Research Station, Fire Sciences Laboratory (Producer). Available online: http://www.fs.fed.us/database/feis/ (accessed April 19, 2010).

Esser, R. P. 1983. Reproductive development of *Verutus volvingentis* (Tylenchida: Heteroderidae). *J. Nematol.* 15: 576–581.

Estes, D. & R. L. Small. 2008. Phylogenetic relationships of the monotypic genus *Amphianthus* (Plantaginaceae tribe Gratioleae) inferred from chloroplast DNA sequences. *Syst. Bot.* 33: 176–182.

Estes, J. R. 1969. Evidence for autoploid evolution in the *Artemisia ludoviciana* complex of the Pacific Northwest. *Brittonia* 21: 29–43.

Estes, J. R. & R. W. Thorp. 1974. Pollination in *Ludwigia peploides* ssp. *glabrescens* (Onagraceae). *Bull. Torrey Bot. Club* 101: 272–276.

Estevez, E. D., J. Sprinkel & R. A. Mattson. 2000. Responses of Suwannee River tidal SAV to ENSO-controlled climate variability. pp. 133–143. *In*: H. S. Greening (ed.), *Seagrass Management: It's Not Just Nutrients!* Tampa Bay Estuary Program, St. Petersburg, FL.

Evans, D. K. 1979. Floristics of the middle Mississippi River sand and mud flats. *Castanea* 44: 8–24.

Evans, E. W. 1982. Habitat differences in feeding habits and body size of the predatory stinkbug *Perillus circumcinctus* (Hemiptera: Pentatomidae). *J. New York Entomol. Soc.* 90: 129–133.

Evans, F. C. & R. Holdenried. 1943. A population study of the Beechey ground squirrel in central California. *J. Mammalgy* 24: 231–260.

Evans, G. A. 2007. *Host Plant List of the Whiteflies (Aleyrodidae) of the World.* USDA/Animal Plant Health Inspection Service (APHIS). Version. 70611; 11 June 2007. 290 pp. Available online: http://entomofaune.qc.ca/entomofaune/aleurodes/references/Evans_2007_Hosts_whiteflies.pdf.Evans, H. E. & J. E. Gillaspy. 1964. Observations on the ethology of digger wasps of the genus *Steniolia* (Hymenoptera: Sphecidae: Bembicini). *Amer. Midl. Naturalist* 72: 257–280.

Evans, J., T. Luna & D. Wick. 2001a. Propagation protocol for production of container *Parnassia fimbriata* Konig. plants (490 ml containers); Glacier National Park, West Glacier, Montana. *In*: *Native Plant Network.* Forest Research Nursery, College of Natural Resources, University of Idaho, Moscow, ID. Available online: http://www.nativeplantnetwork.org (accessed April 12, 2006).

Evans, J., T. Luna & D. Wick. 2001b. Propagation protocol for production of container *Pedicularis groenlandica* Retz. plants (172 ml containers); USDI NPS-Glacier National Park, West Glacier, Montana. *In*: *Native Plant Network.* Forest Research Nursery, College of Natural Resources, University of Idaho, Moscow, ID. Available online: http://www.nativeplantnetwork.org (accessed August 24, 2007).

Evans, M. E. K., E. Menges & D. R. Gordon. 2003. Reproductive biology of three sympatric endangered plants endemic to Florida scrub. *Biol. Conserv.* 111: 235–246.

Evans, M. E. K., D. J. Hearn, W. J. Hahn, J. M. Spangle & D. L. Venable. 2005. Climate and life-history evolution in evening primroses (*Oenothera*, Onagraceae): a phylogenetic comparative analysis. *Evolution* 59: 1914–1927.

Evans, R., C. Campbell, D. Potter, D. Morgan, T. Eriksson, L. Alice, S.-H. Oh, E. Bortiri, F. Gao, J. Smedmark & M. Arsenault. 2002a. A Rosaceae phylogeny. Botany 2002 meeting (abstract).

Evans, R. E., B. R. MacRoberts, T. C. Gibson & M. H. MacRoberts. 2002b. Mass capture of insects by the pitcher plant *Sarracenia alata* (Sarraceniaceae) in Southwest Louisiana and Southeast Texas. *Tex. J. Sci.* 54: 339–346.

Evans-Peters, G. R. 2010. Assessing biological values of wetland reserve program wetlands for wintering waterfowl. MS thesis. Oregon State University, Corvallis, OR. 89 pp.

Evans-Peters, G. R., B. D. Dugger & M. J. Petrie. 2012. Plant community composition and waterfowl food production on wetland reserve program easements compared to those on managed public lands in western Oregon and Washington. *Wetlands* 32: 391–399.

Evens, J. & S. San. 2004. *Vegetation Associations of a Serpentine Area: Coyote Ridge, Santa Clara County, California.* California Natural Heritage Program, California Department of Fish and Game, Santa Clara, CA. 165 pp.

Evens, J. & S. San. 2005. *Vegetation Alliances of the San Dieguito River Park Region, San Diego County, California.* California Native Plant Society, Sacramento, CA. 265 pp.

Evens, J. G., G. W. Page, S. A. Laymon & R. W. Stallcup. 1991. Distribution, relative abundance and status of the California black rail in western North America. *Condor* 93: 952–966.

Everett, P. C. 2012. *A Second Summary of the Horticulture and Propagation of California Native Plants at the Rancho Santa Ana Botanic Garden, 1950–1970.* Rancho Santa Ana Botanic Garden, Claremont, CA. 514 pp.

Everitt, J. H., R. S. Fletcher, H. S. Elder & C. Yang. 2008. Mapping giant salvinia with satellite imagery and image analysis. *Environ. Monit. Assess.* 139: 35–40.

Evers, D. E., C. E. Sasser, J. G. Gosselink, D. A. Fuller & J. M. Visser. 1998. The impact of vertebrate herbivores on wetland vegetation in Atchafalaya Bay, Louisiana. *Estuaries* 21: 1–13.

Evert, D. S. 1957. *Dionaea* transplants in the New Jersey Pine Barrens. *Bartonia* 29: 3–4.

Ewanchuk, P. J. & M. D. Bertness. 2003. Recovery of a northern New England salt marsh plant community from winter icing. *Oecologia* 136: 616–626.

Ewanchuk, P. J. & M. D. Bertness. 2004. Structure and organization of a northern New England salt marsh plant community. *J. Ecol.* 92: 72–85.

Ewing, B. C. 2001. *Mimulus moschatus Dougl. ex Lindl. Musk Flower.* New England Plant Conservation Program, conservation and research plan. New England Wild Flower Society, Framingham, MA. 14 pp.

Eybert, M.-C. & P. Constant. 1998. Diet of nestling Linnets (*Acanthis cannabina* L.). *J. Ornithol.* 139: 277–286.

Eyles, D. E. 1941. A phytosociological study of the *Castalia-Myriophyllum* community of Georgia coastal plain boggy ponds. *Amer. Midl. Naturalist* 26: 421–438.

Eyles, D. E. & J. L. Robertson Jr. 1963. *A Guide and Key to the Aquatic Plants of the Southeastern United States.* Circular 158, Fish & Wildlife Service, Bureau of Sport Fisheries & Wildlife. U. S. Government Printing Office, Washington, DC. 151 pp.

Ezcurra, E., R. S. Felger, A. D. Russell & M. Equihua. 1988. Freshwater islands in a desert sand sea: the hydrology, flora and phytogeography of the Gran Desierto oases of northwestern Mexico. *Desert Plants* 9: 35–63.

Faber, P. M. 1982. *Common Wetland plants of Coastal California. A Field Guide for the Layman.* Pickleweed Press, Mill Valley, CA. 111 pp.

Faber, P. M. 2004. *Design Guidelines for Tidal Wetland Restoration in San Francisco Bay.* Phillip Williams & Associates, Ltd., The Bay Institute and California State Coastal Conservancy, Oakland, CA. 83 pp.

Faber-Langendoen, D. & P. F. Maycock. 1994. A vegetation analysis of tallgrass prairie in southern Ontario. pp. 17–32. *In*: R. G. Wickett, P. D. Lewis, A. Woodliffe & P. Pratt (eds.), *Proceedings of the Thirteenth North American Prairie Conference.* Department of Parks and Recreation, Windsor, ON.

Faghir, M. B., F. Attar, A. Farazmand & S. K. Osaloo. 2014. Phylogeny of the genus *Potentilla* (Rosaceae) in Iran based on nrDNA ITS and cpDNA *trnL-F* sequences with a focus on leaf and style characters' evolution. *Turk. J. Bot.* 38: 417–429.

Fahrig, L., B. Hayden & R. Dolan. 1993. Distribution of barrier island plants in relation to overwash disturbance: a test of life history theory. *J. Coastal Res.* 9: 403–412.

Fairbarns, M. 2003. Smooth goldfields, *Lasthenia glaberrima*: a new plant for Canada. *Bot. Electron. Newslett.* 313: 1.

Fairbarns, M. 2004. *COSEWIC Assessment and Status Report on the Rosy Owl-Clover Orthocarpus bracteosus in Canada.* Committee on the Status of Endangered Wildlife in Canada, Ottawa, ON. 18 pp.

Falińska, K. 1999. Seed bank dynamics in abandoned meadows during a 20-year period in the Białowieża National Park. *J. Veg. Sci.* 87: 461–475.

Falter, C. M. & R. Naskali. 1975. *Aquatic Macrophytes of the Columbia and Snake River Drainages (United States).* U. S. Army Corps of Engineers, Walla Walla District. U. S. Government Printing Office, Washington, DC. 275 pp.

Fan, C. & Q.-Y. Xiang. 2003. Phylogenetic analyses of Cornales based on 26S rRNA and combined 26S rDNA-mat*K*-rbc*L* sequence data. *Amer. J. Bot.* 90: 1357–1372.

Fan, D.-M., J.-H. Chen, Y. Meng, J. Wen, J.-L. Huang & Y.-P. Yang. 2013. Molecular phylogeny of *Koenigia* L. (Polygonaceae: Persicarieae): implications for classification, character evolution and biogeography. *Mol. Phylogen. Evol.* 69: 1093–1100.

Fan, P. & K.-L. Wang. 2011. Evaluation of cold resistance of ornamental species for planting as urban rooftop greening. *Forest. Stud. China* 13: 239–244.

Fang, X. & J. L. McLaughlin. 1990. Ursolic acid, a cytotoxic component of the berries of *Ilex verticillata*. *Fitoterapia* 61: 176–177.

Farag, M. A. & D. A. Al-Mahdy. 2013. Comparative study of the chemical composition and biological activities of *Magnolia grandiflora* and *Magnolia virginiana* flower essential oils. *Nat. Prod. Res.* 27: 1091–1097.

Fargione, M. J., P. S. Curtis & M. E. Richmond. 1991. *Resistance of Plants to Deer Damage.* Cornell Cooperative Extension of Chemung County, Elmira, New York. 7 pp.

Farlow, W. G. & A. B. Seymour. 1888. *A Provisional Host-Index of the Fungi of the United States. Part 1. Polypetalae.* Harvard University, Cambridge, MA. 219 pp.

Farmer, A. M. 1989. *Lobelia dortmanna* L. *J. Ecol.* 77: 1161–1173.

Farmer, A. M. & D. H. N. Spence. 1987. Flowering, germination and zonation of the submerged aquatic plant *Lobelia dortmanna* L. *J. Ecol.* 75: 1065–1076.

Farmer, A. M., S. C. Maberly & G. Bowes. 1986. Activities of carboxylation enzymes in freshwater macrophytes. *J. Exp. Bot.* 37: 1568–1573.

Farmer, R. E. Jr. 1978. Propagation of a southern Appalachian population of fringed gentian. *Bull. Torrey Bot. Club* 105: 139–142.

Farmer, R. E. Jr. 1980. Germination and juvenile growth characteristics of *Parnassia asarifolia*. *Bull. Torrey Bot. Club* 107: 19–23.

Farmer, R. E. Jr. & M. Cunningham. 1981. Seed dormancy of red maple in East Tennessee. *For. Sci.* 27: 446–448.

Farnsworth, E. J. 2004. Patterns of plant invasions at sites with rare plant species throughout New England. *Rhodora* 106: 97–117.

Farnsworth, E. J. & L. A. Meyerson. 1999. Species composition and inter-annual dynamics of a freshwater tidal plant community following removal of the invasive grass, *Phragmites australis*. *Biol. Invas.* 1: 115–127.

Farnsworth, N. R., N. K. Hart, S. R. Johns, J. A. Lamberton & W. Messmer. 1969. Alkaloids of *Boehmeria cylindrica* (family Urticaceae): identification of a cytotoxic agent, highly active against Eagle's 9KB carcinoma of the nasopharynx in cell culture, as crytopleurine. *Aust. J. Chem.* 22: 1805–1807.

Farooq, M. & S. A. Siddiqui. 1965. Abnormal ovules and embryo sacs in *Utricularia vulgaris* var. *americana* A. Gray. *Naturwissenschaften* 52: 90.

Farr, D. F. & A. Y. Rossman. 2010. *Fungal Databases.* Systematic Mycology and Microbiology Laboratory, ARS, USDA. Available online: http://nt.ars-grin.gov/fungaldatabases/ (accessed May 19, 2010).

Farr, D. F. & A. Y. Rossman. 2013a. *Fungal Databases.* Systematic Mycology and Microbiology Laboratory, ARS, USDA. Available online: http://nt.ars-grin.gov/fungaldatabases/ (accessed May 30, 2013).

Farr, D. F. & A. Y. Rossman. 2013b. *Fungal Databases.* Systematic Mycology and Microbiology Laboratory, ARS, USDA. Available online: http://nt.ars-grin.gov/fungaldatabases/ (accessed May 31, 2013).

Farrell, B. & C. Mitter. 1990. Phylogenesis of insect/plant interactions: have *Phyllobrotica* leaf beetles (Chrysomelidae) and the Lamiales diversified in parallel? *Evolution* 44: 1389–1403.

Farris, M. E., G. J. Keever, J. R. Kessler & J. W. Olive. 2011. Cyclanilide promotes shoot production and flowering of *Coreopsis* and coneflower during nursery production. *J. Environ. Hort.* 29: 108–114.

Faruqi, S. A. & I. V. Holt. 1961. Studies of teratology in *Heliotropium curassavicum* L. *Proc. Oklahoma Acad. Sci.* 41: 19–22.

Fashing, N. J. 1994. A new species of *Leipothrix* (Prostigmata: Eriophyidae) from the cobra Lily, *Darlingtonia californica* (Sarraceniaceae). *Int. J. Acarol.* 20: 99–101.

Fashing, N. J. & B. M. O'Connor. 1984. *Sarraceniopus:* a new genus for histiostomatid mites inhabiting the pitchers of the Sarraceniaceae (Astigmata: Histiostomatidae). *Int. J. Acarol.* 10: 217–227.

Fassett, N. C. 1921. *Sium suave:* a new and an old form. *Rhodora* 23: 111–113.

Fassett, N. C. 1928. The vegetation of the estuaries of northeastern North America. *Proc. Boston Soc. Nat. Hist.* 39: 73–130.

Fassett, N. C. 1939. *Elatine* and other aquatics. *Rhodora* 41: 367–377.

Fassett, N. C. 1951. *Callitriche* in the New World. *Rhodora* 53: 137–155, 161–182, 185–194, 209–222.

Fassett, N. C. 1953a. North American *Ceratophyllum. Comun. Inst. Trop. Invest. Ci. Univ. El Salvador* 2: 25–45.

Fassett, N. C. 1953b. A monograph of *Cabomba. Castanea* 13: 116–128.

Fassett, N. C. 1953c. *Proserpinaca. Comun. Inst. Trop. Invest. Ci. Univ. El Salvador* 2: 139–162.

Fassett, N. C. 1957. *A Manual of Aquatic Plants.* With revision appendix by E. C. Ogden. The University of Wisconsin Press, Madison, WI. 405 pp.

Faubert, J., D. F. Bastien, M. Lapointe & C. Roy. 2012a. Cinq hépatiques nouvelles pour le Québec. *Carnets Bryol.* 2: 12–16.

Faubert, J., J. Gagnon, B. Tremblay & L. Couillard. 2012b. Mise à jour de la publication Les bryophytes rares du Québec. Espèces prioritaires pour la conservation. *Carnets Bryol.* 2: 53–56.

Faust, W. Z. 1972. A biosystematic study of the *Interiores* species group of the genus *Vernonia* (Compositae). *Brittonia* 24: 363–378.

Fay, A. M., S. M. Still & M. A. Bennett. 1993. Optimum germination temperature of *Rudbeckia fulgida. HortTechnology* 3: 433–435.

Fay, A. M., M. A. Bennett & S. M. Still. 1994. Osmotic seed priming of *Rudbeckia fulgida* improves germination and expands germination range. *HortScience* 29: 868–870.

Fay, M. F., R. G. Olmstead, J. E. Richardson, E. Santiago, G. T. Prance & M. W. Chase. 1998. Molecular data support the inclusion of *Duckeodendron cestroides* in Solanaceae. *Kew Bull.* 53: 203–212.

Fayette, K. & L. Bruederle. 2001. Morphometric analyses support specific status for *Eutrema penlandii* Rollins (Brassicaceae). Botany 2001 meeting (abstract).

Fehrmann, S., C. T. Philbrick & R. Halliburton. 2012. Intraspecific variation in *Podostemum ceratophyllum* (Podostemaceae): evidence of refugia and colonization since the last glacial maximum. *Amer. J. Bot.* 99: 145–151.

Feist, M. A. E. 2009. Clarifications concerning the nomenclature and taxonomy of *Oxypolis ternata* (Apiaceae). *J. Bot. Res. Inst. Texas.* 3: 661–666.

Feist, M. A. E. & S. R. Downie. 2008. A phylogenetic study of *Oxypolis* and *Ptilimnium* (Apiaceae) based on nuclear rDNA ITS sequences. *Syst. Bot.* 33: 447–458.

Feist, M. A. E., M. J. Morris, L. R. Phillippe, J. E. Ebinger & W. E. McClain. 2008. Sand prairie communities of Matanzas Nature Preserve, Mason County, Illinois. *Castanea* 73: 177–187.

Feist, M. A. E., S. R. Downie, A. R. Magee & M. Liu. 2012. Generic delimitations for *Oxypolis* and *Ptilimnium. Taxon* 61: 402–418.

Feldman, S. R. & J. P. Lewis. 2005. Effects of fire on the structure and diversity of a *Spartina argentinensis* tall grassland. *Appl. Veg. Sci.* 8: 77–84.

Feldman, S. R., C. Alzugaray & J. P. Lewis. 2007. Relación entre la vegetación y el banco de semillas de un espartillar de *Spartina argentinensis. Cienc. Invest. Agrar.* 34: 41–48.

Felger, R. S. & M. B. Moser. 1985. *People of the Desert and Sea.* University of Arizona Press, Tucson, AZ. 435 pp.

Fell, E. W. 1957. Plants of a northern Illinois sand deposit. *Amer. Midl. Naturalist* 58: 441–451.

Feller, I. C. & M. Sitnik (eds.). 1996. *Mangrove Ecology: A Manual for a Field Course.* Smithsonian Environmental Research Center, Smithsonian Institution, Washington, DC. 135 pp.

Fellers, G. M., P. M. Kleeman, D. A. W. Miller, B. J. Halstead & W. A. Link. 2013. Population size, survival, growth, and movements of *Rana sierrae. Herpetologica* 69: 147–162.

Fellows, M. Q. N. & J. B. Zedler. 2005. Effects of the non-native grass *Parapholis incurva* (Poaceae), on the rare and endangered hemiparasite, *Cordylanthus maritimus* subsp. *maritimus* (Scrophulariaceae). *Madroño* 52: 91–98.

Felt, E. P. 1901. Scale insects of importance and list of the species in New York state. *Bull. New York State Mus.* 46: 291–377.

Felt, E. P. 1908. Studies in Cecidomyiidae II. *Bull. New York State Mus. Nat. Hist.* 124: 307–514.

Felter, H. W. & J. U. Lloyd. 1898. *King's American Dispensatory.* 18th Ed., 3rd Revision (2 vols.). Ohio Valley Co., Cincinnatti, OH. 2172 pp. Available online: http://henriettesherbal.com/eclectic/kings/intro.html.

Feng, J., G. J. Seiler, T. J. Gulya & C. C. Jan. 2006. Development of *Sclerotinia* stem rot resistant germplasm utilizing hexaploid *Helianthus* species. *Proceedings of the 28th Sunflower Research Workshop,* Fargo, ND, January 11–12, 2006. Available online: http://www.sunflowernsa.com/research/research-workshop/documents/Feng_Sclerotinia_06.pdf.

Feng, T., S. R. Downie, Y. Yu, X. Zhang, W. Chen, X. He & S. Liu. 2009. Molecular systematics of *Angelica* and allied genera (Apiaceae) from the Hengduan Mountains of China based on nrDNA ITS sequences: phylogenetic affinities and biogeographic implications. *J. Plant Res.* 122: 403–414.

Fenner, M. 1983. Relationships between seed weight, ash content and seedling growth in twenty-four species of Compositae. *New Phytol.* 95: 697–706.

Fenster, C. B. & D. E. Carr. 1997. Genetics of sex allocation in *Mimulus* (Scrophulariaceae). *J. Evol. Biol.* 10: 641–661.

Fenton, W. N. 1986. A further note on Iroquois suicide. *Ethnohistory* 33: 448–457.

Ferdy, J. B., L. Despres & B. Godelle. 2002. Evolution of mutualism between globeflowers and their pollinating flies. *J. Theor. Biol.* 217: 219–234.

Ferguson, D. C. 1975. *Host Records for Lepidoptera Reared in Eastern North America*. Technical Bulletin No. 1521, Agricultural Research Service. United States Department of Agriculture, Washington, DC. 49 pp.

Ferguson, D. M. 1998. Phylogenetic analysis and relationships in Hydrophyllaceae based on *ndhF* sequence data. *Syst. Bot.* 23: 253–268.

Ferguson, E. & R. P. Wunderlin. 2006. A vascular plant inventory of Starkey Wilderness Preserve, Pasco County, Florida. *Sida* 22: 635–659.

Ferguson, I. K. 1965. The genera of Valerianaceae and Dipsacaceae in the southeastern United States. *J. Arnold Arbor.* 46: 218–231.

Ferlatte, W. J. 1978. Notes on two rare, endemic species from the Klamath Region of northern California, *Phacelia dalesiana* (Hydrophyllaceae) and *Raillardella pringlei* (Compositae). *Madroño* 25: 138.

Fernald, M. L. 1899. Two plants of the crowfoot family. *Rhodora* 1: 48–52.

Fernald, M. L. 1901. The vascular plants of Mt. Katahdin. *Rhodora* 3: 166–177.

Fernald, M. L. 1907. The soil preferences of certain alpine and subalpine plants. *Rhodora* 9: 149–193.

Fernald, M. L. 1908. Some northern plants possibly to be found in Vermont. *Bull. Vt. Bot. Club* 3: 29–34.

Fernald, M. L. 1911. The northern variety of *Gaylussacia dumosa*. *Rhodora* 13: 95–99.

Fernald, M. L. 1912. *Sclerolepis uniflora* in Massachusetts. *Rhodora* 14: 23–24.

Fernald, M. L. 1916. A calciphile variety of *Andromeda glaucophylla*. *Rhodora* 18: 100–102.

Fernald, M. L. 1918. The North American *Littorella*. *Rhodora* 20: 232–233.

Fernald, M. L. 1921. The Gray Herbarium expedition to Nova Scotia, 1920. *Contrib. Gray Herb.* 63: 89–111, 130–171, 184–195, 223–245, 257–278, 284–300.

Fernald, M. L. 1929. *Menyanthes trifoliata*, var. *minor*. *Rhodora* 31: 195–198.

Fernald, M. L. 1935. Midsummer vascular plants of southeastern Virginia. *Rhodora* 37: 378–413.

Fernald, M. L. 1940. A century of additions to the flora of Virginia: Part III. Phytogeographic considerations. *Rhodora* 42: 503–521.

Fernald, M. L. 1950. *Gray's Manual of Botany. 8th Ed.* D. Van Nostrand Company, New York. 1632 pp.

Fernald, M. L. & A. C. Kinsey. 1943. *Edible Wild Plants of Eastern North America*. Idlewild Press, Cornwall-On-Hudson, New York. 452 pp.

Fernández-Cuesta, Á., L. Velasco & J. M. Fernández-Martínez. 2011. Phytosterols in the seeds of wild sunflower species. *Helia* 34: 31–38.

Fernández-Illescas, F., F. J. J. Nieva, M. Á. de las Heras & A. F. Muñoz-Rodríguez. 2011. Dichogamy in *Salicornieae* species: establishment of floral sex phases and evaluation of their frequency and efficacy in four species. *Plant Syst. Evol.* 296: 255–264.

Fernando, E. S., P. A. Gadek & C. J. Quinn. 1995. Simaroubaceae, an artificial construct: evidence from *rbc*L sequence variation. *Amer. J. Bot.* 82: 92–103.

Fernando, W. & H. V. Rupasinghe. 2013. Anticancer properties of phytochemicals present in medicinal plants of North America. *InTech* 2013: 159–180. doi:10.5772/55859.

Ferreira, F. A., R. P. Mormul, G. Pedralli, V. Joana Pott & A. Pott. 2010. Estrutura da comunidade de macrófitas aquáticas em três lagoas do Parque Estadual do Rio Doce, Minas Gerais, Brasil. *Hoehnea* 37: 43–52.

Ferren Jr., W. R. 2003. 3. *Sesuvium* L. pp. 80–81. *In*: N. R. Morin (convening ed.), *Flora of North America North of Mexico, Vol. 4: Magnoliophyta: Caryophyllidae, Part 1*. Oxford University Press, New York.

Ferren Jr., W. R. & A. E. Schuyler. 1980. Intertidal vascular plants of river systems near Philadelphia. *Proc. Acad. Nat. Sci. Philadelphia* 132: 86–120.

Ferren Jr., W. R. & H. J. Schenk. 2003. 25. *Suaeda* P. Forsskål ex J. F. Gmelin. pp. 390–397. *In*: N. R. Morin (convening ed.), *Flora of North America North of Mexico, Vol. 4: Magnoliophyta: Caryophyllidae, Part 1*. Oxford University Press, New York.

Ferren Jr., W. R., D. M. Hubbard, S. Wiseman, A. K. Parikh & N. Gale. 1998. Review of ten years of vernal pool restoration and creation in Santa Barbara, CA. pp. 206–216. *In*: C. W. Witham, E. T. Bauder, D. Belk, W. R. Ferren Jr. & R. Ornduff (eds.), *Ecology, Conservation, and Management of Vernal Pool Ecosystems—Proceedings from a 1996 Conference*. California Native Plant Society, Sacramento, CA.

Ferris, J. P., C. B. Boyce, R. C. Briner, B. Douglas, J. L. Kirkpatrick & J. A. Weisbach. 1966a. Lythraceae alkaloids. Structure and stereochemistry of the major alkaloids of *Decodon* and *Heimia*. *Tetrahedron Lett.* 30: 3641–3649.

Ferris, J. P., R. C. Briner, C. B. Boyce & M. J. Wolf. 1966b. Lythraceae alkaloids. Structure and stereochemistry of the biphenyl ether alkaloids of *Decodon verticillatus*. *Tetrahedron Lett.* 42: 5125–5128.

Ferris, R. S. 1970. *Flowers of Point Reyes National Seashore*. University of California Press, Berkeley, CA. 119 pp.

Ferriter, A., B. Doren, R. Winston, D. Thayer, B. Miller, B. Thomas, M. Barrett, T. Pernas, S. Hardin, J. Lane, M. Kobza, D. Schmitz, M. Bodle, L. Toth, L. Rodgers, P. Pratt, S. Snow & C. Goodyear. 2008. Chapter 9: The status of nonindigenous species in the south Florida environment. pp. 9–1–9–101. *In*: S. Graham & W. Wagman (eds.), *2008 South Florida Environmental Report, Volume I*. South Florida Water Management District, West Palm Beach, FL.

Fertig, W. 2000. *Ecological Assessment and Monitoring Program for Northern Blackberry* (Rubus acaulis) *in Bighorn National Forest, Wyoming*. Unpublished report prepared for the Bighorn National Forest by the Wyoming Natural Diversity Database, Laramie, WY. 43 pp.

Fertig, W. & L. Welp. 1998. *Status Report on Persistent Sepal Yellowcress* (Rorippa calycina) *in Wyoming*. Unpublished report prepared for the Bureau of Land Management Wyoming State Office by the Wyoming Natural Diversity Database, Laramie, WY. 107 pp.

Fessel, K. E. & B. A. Middleton. 2000. Survivorship of woody plant seeds in bald cypress swamps in southern Illinois. Botany 2000 meeting (abstract).

Feuer, S. & A. S. Tomb. 1977. Pollen morphology and detailed structure of family Compositae, tribe Cichorieae. II. Subtribe Microseridinae. *Amer. J. Bot.* 64: 230–245.

Fiasson, J. L., K. Gluchoff-Fiasson & G. Dahlgren. 1997. Flavonoid patterns in European *Ranunculus* L. subgenus *Batrachium* (Ranunculaceae). *Biochem. Syst. Ecol.* 25: 327–333.

Fiedler, A. K. & D. A. Landis. 2007. Attractiveness of Michigan native plants to arthropod natural enemies and herbivores. *Environ. Entomol.* 36: 751–765.

Fiedler, A., K. Tuell, R. Isaacs & D. Landis. 2007a. *Attracting Beneficial Insects with Native Flowering Plants.* Extension Bulletin E-2973. Michigan State University, East Lansing, MI. 5 pp.

Fiedler, P. L., M. E. Keever, B. J. Grewell & D. J. Partridge. 2007b. Rare plants in the Golden Gate Estuary (California): the relationship between scale and understanding. *Aust. J. Bot.* 55: 206–220.

Field, R. T. & K. R. Philipp. 2000. Vegetation changes in the freshwater tidal marsh of the Delaware estuary. *Wetlands Ecol. Manag.* 8: 79–88.

Fields, J. R., T. R. Simpson, R. W. Manning & F. L. Rose. 2003. Food habits and selective foraging by the Texas river cooter (*Pseudemys texana*) in Spring Lake, Hays County, Texas. *J. Herpetol.* 37: 726–729.

Figueredo, S. M., F. P. do Nascimento, C. S. Freitas, C. H. Baggio, C. Soldi, M. G. Pizzolatti, M. D. C. C. de Ibarrola, R. L. D. de Arrua & A. R. Santos. 2011. Antinociceptive and gastroprotective actions of ethanolic extract from *Pluchea sagittalis* (Lam.) Cabrera. *J. Ethnopharmacol.* 135: 603–609.

Figueroa, M. E., J. M. Castillo, S. Redondo, T. Luque, E. M. Castellanos, F. J. Nieva, C. J. Luque, A. E. Rubio-Casal & A. J. Davy. 2003. Facilitated invasion by hybridization of *Sarcocornia* species in a salt-marsh succession. *J. Ecol.* 91: 616–626.

Figuerola, J., A. J. Green & L. Santamaría. 2003. Passive internal transport of aquatic organisms by waterfowl in Doñana, south-west Spain. *Glob. Ecol. Biogeogr.* 12: 427–436.

Fijridiyanto, I. A. & N. Murakami. 2009a. Phylogeny of *Litsea* and related genera (Laureae-Lauraceae) based on analysis of *rpb2* gene sequences. *J. Plant Res.* 122: 283–298.

Fijridiyanto, I. A. & N. Murakami. 2009b. Molecular systematics of Malesian *Litsea* Lam. (Lauraceae) and putative related genera. *Acta Phytotaxon. Geobot.* 60: 1–18.

Filatov, D. A. & D. Charlesworth. 1999. DNA polymorphism, haplotype structure and balancing selection in the *Leavenworthia PgiC* locus. *Genetics* 153: 1423–1434.

Filiz, E., E. Osma, A. Kandemir, H. Tombuloglu, S. Birbilener & M. Aydin. 2014. Assessment of genetic diversity and phylogenetic relationships of endangered endemic plant *Barbarea integrifolia* DC. (Brassicaceae) in Turkey. *Turk. J. Bot.* 38: 1169–1181.

Finke, M. D. & J. M. Scriber. 1988. Influence on larval growth of the eastern black swallowtail butterfly *Papilio polyxenes* (Lepidoptera: Papilionidae) of seasonal changes in nutritional parameters of Umbelliferae species. *Amer. Midl. Naturalist* 119: 45–62.

Finnerty, T. L., J. M. Zajicek & M. A. Hussey. 1992. Use of seed priming to bypass stratification requirements of three *Aquilegia* species. *HortScience* 27: 310–313.

Fior, S., M. Li, B. Oxelman, R. Viola, S. A. Hodges, L. Ometto & C. Varotto. 2013. Spatiotemporal reconstruction of the *Aquilegia* rapid radiation through next-generation sequencing of rapidly evolving cpDNA regions. *New Phytol.* 198: 579–592.

Fischer, E. 1997. A revision of the genus *Dopatrium* (Scrophulariaceae-Gratioloideae). *Nordic J. Bot.* 17: 527–555.

Fischer, E. 2004. Scrophulariaceae. pp. 333–432. *In*: J. W. Kadereit (ed.), *The Families and Genera of Vascular Plants, Vol. VII. Flowering Plants, Dicotyledons: Lamiales (except Acanthaceae including Avicenniaceae).* Springer-Verlag, Berlin, Germany.

Fischer, M. & D. Matthies. 1997. Mating structure and inbreeding and outbreeding depression in the rare plant *Gentianella germanica* (Gentianaceae). *Amer. J. Bot.* 84: 1685–1692.

Fischer, R. A., C. O. Martin, K. Robertson, W. R. Whitworth & M. G. Harper. 1999. *Management of Bottomland Hardwoods and Deepwater Swamps for Threatened and Endangered Species.* Technical Report SERDP-99-5. U. S. Army Engineer Research and Development Center, Vicksburg, MS. 76 pp.

Fischer, E., B. Schäferhoff & K. Müller. 2013. The phylogeny of Linderniaceae—the new genus *Linderniella*, and new combinations within *Bonnaya*, *Craterostigma*, *Lindernia*, *Micranthemum*, *Torenia* and *Vandellia*. *Willdenowia* 43: 209–238.

Fish, D. & D. W. Hall. 1978. Succession and stratification of aquatic insects inhabiting the leaves of the insectivorous pitcher plant, *Sarracenia purpurea*. *Amer. Midl. Naturalist* 99: 172–183.

Fishbein, M. & D. E. Soltis. 2004. Further resolution of the rapid radiation of Saxifragales (angiosperms, eudicots) supported by mixed-model Bayesian analysis. *Syst. Bot.* 29: 883–891.

Fishbein, M. & D. W. Stevens. 2005. Resurrection of *Seutera* Reichenbach (Apocynaceae, Asclepiadoideae). *Novon* 15: 531–533.

Fishbein, M., D. Gori & D. Meggs. 1997. Prescribed burning as a management tool for Sky Island bioregion wetlands with reference to the management of the endangered orchid *Spiranthes delitescens*. pp. 468–477. *In*: L. F. DeBano et al. (eds.), *Biodiversity and the Management of the Madrean Archipelago: The Sky Islands of Southwestern United States and Northwestern Mexico.* USDA Forest Service Technical Report RM-GTR-264, Rocky Mountain Forest and Range Experiment Station, USDA, Fort Collins, CO.

Fishbein, M., C. Hibsch-Jetter, D. E. Soltis & L. Hufford. 2001. Phylogeny of Saxifragales: patterns of floral evolution and taxonomic revision. *Syst. Biol.* 50: 817–847.

Fishbein, M., S. P. Lynch, D. Chuba, C. Ellison, D. Goyder, M. W. Chase & R. Mason-Gamer. 2006. Phylogeny, biogeography, and systematics of *Asclepias* (Apocynaceae). Botany 2006 meeting (abstract).

Fishel, F., W. Johnson, D. Peterson, M. Loux & C. Sprague. 2000. *Early Spring Weeds of No-Till Crop Production.* North Central Regional Extension Publication No. NCR 614. MU Extension, University of Missouri-Columbia, Columbia, MO. 23 pp.

Fisher, A. S., G. S. Podniesinski & D. J. Leopold. 1996. Effects of drainage ditches on vegetation patterns in abandoned agricultural peatlands in central New York. *Wetlands* 16: 397–409.

Fisher, D. D., H. Jochen Schenk, J. A. Thorsch & W. R. Ferren Jr. 1997. Leaf anatomy and subgeneric affiliations of C_3 and C_4 species of *Suaeda* (Chenopodiaceae) in North America. *Amer. J. Bot.* 84: 1198–1210.

Fisher, M. A. & P. Z. Fulé. 2004. Changes in forest vegetation and arbuscular mycorrhizae along a steep elevation gradient in Arizona. *For. Ecol. Manag.* 200: 293–311.

Fishman, L., A. J. Kelly, E. Morgan & J. H. Willis. 2001. A genetic map in the *Mimulus guttatus* species complex reveals transmission ratio distortion due to heterospecific interactions. *Genetics* 159: 1701–1716.

Fisk, W. A., O. Agbai, H. A. Lev-Tov & R. K. Sivamani. 2014. The use of botanically derived agents for hyperpigmentation: a systematic review. *J. Amer. Acad. Dermatol.* 70: 352–365.

Fitts, R. D. 1995. Reproductive ecology of the rare endemic primrose *Primula alcalina* in its varied habitat. MS thesis. College of Natural Resources, Utah State University, Logan, UT.

Fitzpatrick, J. W., M. Lammertink, M. D. Luneau Jr., T. W. Gallagher, B. R. Harrison, G. M. Sparling, K. V. Rosenberg, R. W. Rohrbaugh, E. C. H. Swarthout, P. H. Wrege, S. B. Swarthout, M. S. Dantzker, R. A. Charif, T. R. Barksdale, J. V. Remsen Jr., S. D. Simon & D. Zollner. 2005. Ivory-billed woodpecker (*Campephilus principalis*) persists in continental North America. *Science* 308: 1460–1462.

Flanagan, L. B., C. S. Cook & J. R. Ehleringer. 1997. Unusually low carbon isotope ratios in plants from hanging gardens in southern Utah. *Oecologia* 111: 481–489.

Fleck, J. & W. K. Fitt. 1999. Degrading mangrove leaves of *Rhizophora mangle* Linne provide a natural cue for settlement and metamorphosis of the upside down jellyfish *Cassiopea xamachana* Bigelow. *J. Exp. Mar. Biol. Ecol.* 234: 83–94.

Fleck, J., W. K. Fitt & M. G. Hahn. 1999. A proline-rich peptide originating from the decomposing mangrove leaves is one natural metamorphic cue of the tropical jellyfish *Cassiopea xamachana*. *Mar. Ecol. Prog. Ser.* 183: 115–124.

Fleckenstein, E. L. 2007. The influence of salinity on the germination and distribution of *Taxodium distichum* (L.) Rich, bald cypress, along the northeast Cape Fear River. MS thesis. University of North Carolina, Wilmington, NC. 72 pp.

Fleishman, E., A. E. Launer, K. R. Switky, U. Yandell, J. Heywood & D. D. Murphy. 2001. Rules and exceptions in conservation genetics: genetic assessment of the endangered plant *Cordylanthus palmatus* and its implications for management planning. *Biol. Conserv.* 98: 45–53.

Fleming, M. M. S., J. M. Stucky & C. Brownie. 2007. Effects and importance of soil wetness and neighbor vegetation on *Solidago verna* M. A. Curtis ex Torrey & A. Gray (spring-flowering goldenrod) [Asteraceae] transplant survivorship and growth. *Castanea* 72: 205–213.

Flessa, H. 1994. Plant-induced changes in the redox potential of the rhizospheres of the submerged vascular macrophytes *Myriophyllum verticillatum* L. and *Ranunculus circinatus* L. *Aquat. Bot.* 47: 119–129.

Fletcher, P., S. Gourley, J. Gove, P. Hammen, W. Hoar, J. Homer, S. Hundley, K. Kettenring, G. Loomis, S. Lussier, W. Mahaney, D. Marceau, T. Peragallo, S. Pilgrim, D. Rocque, M. Schweisberg, M. Sheehan, F. Smigelski, L. Sommer, S. Tessitore, P. Veneman & W. Wright (New England Hydric Soils Technical Committee). 1998. *Field Indicators for Identifying Hydric Soils in New England, Version 2*. New England Interstate Water Pollution Control Commission, Wilmington, MA. 76 pp.

Flinn, M. A. & J. K. Pringle. 1983. Heat tolerance of rhizomes of several understory species. *Can. J. Bot.* 61: 452–457.

Flinn, K. M., M. J. Lechowicz & M. J. Waterway. 2008. Plans species diversity and composition of wetlands within an upland forest. *Amer. J. Bot.* 95: 1216–1224.

Flint, M. B. 1882. The exogenous flora of Lincoln Co., Mississippi. *Bot. Gaz.* 7: 74–76.

Flood, M. J., II. 2010. Anti-tumor natural product research focused on plants found in the southern appalachian region. MS thesis. Western Carolina University, Cullowhee, NC. 76 pp.

Flores-Tavizón, E., M. T. Alarcón-Herrera, S. González-Elizondo & E. J. Olguín. 2003. Arsenic tolerating plants from mine sites and hot springs in the semi-arid region of Chihuahua, Mexico. *Acta Biotechnol.* 23: 113–119.

Flowers, R. W., D. G. Furth & M. C. Thomas. 1994. Notes on the distribution and biology of some Florida leaf beetles (Coleoptera: Chrysomelidae). *Coleopt. Bull.* 48: 79–89.

Flowers, S. 1934. Vegetation of the Great Salt Lake region. *Bot. Gaz.* 95: 353–418.

Flowers, T. J., M. A. Hajibagheri & N. J. W. Clipson. 1986. Halophytes. *Q. Rev. Biol.* 61: 313–337.

Floyd, J. W. 2002. Phylogenetic and biogeographic patterns in *Gaylussacia* (Ericaceae) based on morphological, nuclear DNA, and chloroplast DNA variation. *Syst. Bot.* 27: 99–115.

Floyd, R. H., S. Ferrazzano, B. W. Josey & J. R. Applegate. 2015. *Aldrovanda vesiculosa* at Fort A.P. Hill, Virginia. *Castanea* 80: 211–217.

Flynn, K. M., K. L. McKee & I. A. Mendelssohn. 1995. Recovery of freshwater marsh vegetation after a saltwater intrusion event. *Oecologia* 103: 63–72.

Földesi, D. & T. Havas. 1979. Peppermint (*Mentha piperita*) propagation by rooted stolon shoots. *Herba Hung.* 18: 63–73.

Foley, D. J. (ed.). 1974. *Herbs for Use and for Delight*. Herb Society of America: an anthology from The Herbarist. Dover Publications, New York. 324 pp.

Folk, R. A. & J. V. Freudenstein. 2014. Phylogenetic relationships and character evolution in *Heuchera* (Saxifragaceae) on the basis of multiple nuclear loci. *Amer. J. Bot.* 101: 1532–1550.

Folkerts, G. W. 1989. Facultative rhizome dimorphism in *Sarracenia psittacina* Michx. (Sarraceniaceae): an adaptation to deeping substrate. *Phytomorphology* 39: 285–289.

Folkerts, G. W. & J. D. Freeman. 1989. *Pinguicula lutea* Walt. f. *alba*, f. nov. (Lentibulariaceae), a white-flowered form of the yellow butterwort. *Castanea* 54: 40–42.

Fonda, R. W. 1974. Forest succession in relation to river terrace development in Olympic National Park, Washington. *Ecology* 55: 927–942.

Foo, L. Y., Y. Lu, A. B. Howell & N. Vorsa. 2000. The structure of cranberry proanthocyanidins which inhibit adherence of uropathogenic P-fimbriated *Escherichia coli* in vitro. *Phytochemistry* 54: 173–181.

Foote, B. A. 1961. The marsh flies of Idaho and adjoining areas (Diptera: Sciomyzidae). *Amer. Midl. Naturalist* 65: 144–167.

Foote, B. A. 1967. Biology and immature stages of fruit flies: the genus *Icterica* (Diptera: Tephritidae). *Ann. Entomol. Soc. Amer.* 60: 1295–1305.

Forbes, B. C. 1996. Plant communities of archaeological sites, abandoned dwellings, and trampled tundra in the eastern Canadian Arctic: a multivariate analysis. *Arctic* 49: 141–154.

Forbes, B. C., J. J. Ebersole & B. Strandberg. 2001. Anthropogenic disturbance and patch dynamics in circumpolar Arctic ecosystems. *Conserv. Biol.* 15: 954–969.

Forbes, M. G., H. D. Alexander & K. H. Dunton. 2008. Effects of pulsed riverine versus non-pulsed wastewater inputs of freshwater on plant community structure in a semi-arid salt marsh. *Wetlands* 28: 984–994.

Forbis, T. A. 2009. Negative associations between seedlings and adult plants in two alpine plant communities. *Arct. Antarc. Alp. Res.* 41: 301–308.

Forbis, T. A. & P. K. Diggle. 2001. Subnivean embryo development in the alpine herb *Caltha leptosepala* (Ranunculaceae). *Can. J. Bot.* 79: 635–642.

Forbush, E. H. 1916. *Food Plants to Attract Birds and Protect Fruit.* Circular no. 49. Wright & Potter Printing Co., Boston, MA. 21 pp.

Ford, B. A. 1997. 13. *Caltha* Linnaeus. pp. 187–189. *In:* N. R. Morin (convening ed.), *Flora of North America North of Mexico, Vol. 3: Magnoliophyta: Magnoliidae and Hamamelidae.* Oxford University Press, New York.

Forest, F., A. Bruneau, J. A. Hawkins, T. Kajita, J. J. Doyle & P. R. Crane. 2002. The sister of the Leguminosae revealed: phylogenetic relationships in the Fabales determined using *trnL* and *rbcL* sequences. Botany 2002 meeting (abstract).

Forest, F., M. W. Chase, C. Persson, P. R. Crane & J. A. Hawkins. 2007. The role of biotic and abiotic factors in evolution of ant dispersal in the milkwort family (Polygalaceae). *Evolution* 61: 1675–1694.

Forman, R. T. T. & R. E. Boerner. 1981. Fire frequency and the pine barrens of New Jersey. *Bull. Torrey Bot. Club* 108: 34–50.

Forys, E. & S. R. Humphrey. 1999. The importance of patch attributes and context to the management and recovery of an endangered lagomorphs. *Landscape Ecol.* 14: 177–185.

Fosberg, F. R., M.-H. Sachet & O. Royce. 1979. A geographical checklist of the Micronesian dicotyledonae. *Micronesica* 15: 1–295.

Foster, D. R. & G. A. King. 1984. Landscape features, vegetation and developmental history of a patterned fen in south-eastern Labrador, Canada. *J. Ecol.* 72: 115–143.

Foster, D. R. & P. H. Glaser. 1986. The raised bogs of south-eastern Labrador, Canada: classification, distribution, vegetation and recent dynamics. *J. Ecol.* 74: 47–71.

Foster, M., E. Carroll & V. D. Hipkins. 1997. Saw-toothed *Lewisia*: to be or not to be. *Fremontia* 25: 15–19.

Fowler, J. F., S. Overby & B. Smith. 2012. *Climate Driven Changes in Engelmann Spruce Stands at Timberline in the La Sal Mountains.* U. S. Forest Service, Rocky Mountain Research Station, Flagstaff, AZ. 21 pp.

Fox, H. 1914. Data on the orthopteran faunistics of eastern Pennsylvania and southern New Jersey. *Proc. Acad. Nat. Sci. Philadelphia* 66: 441–534.

Fox, L., I. Valiela & E. L. Kinney. 2012. Vegetation cover and elevation in long-term experimental nutrient-enrichment plots in Great Sippewissett salt marsh, Cape Cod, Massachusetts: implications for eutrophication and sea level rise. *Estuaries Coasts* 35: 445–458.

Frahm, J.-P. 1993. *Campylopus japonicus* new to North America north of Mexico. *Bryologist* 96: 142–144.

Frampton, G. T. 1995. Endangered and Threatened Wildlife and Plants; proposal to determine the least chub (*Iotichthys phlegethontis*) an endangered species with critical habitat. *Fed. Reg.* 60(189): 50518–50530.

Francini, R. B. & E. D. S. Rovati. 2011. Seasonal extrafloral nectar production by *Laguncularia racemosa* (L.) C.F. Gaertn (Combretaceae) in southeast Brazil. *Int. J. Bot.* 7: 122–125.

Francis, W. 2001. Rockclimbing is damaging cliff-dwelling plants in the Red River Gorge. *The Lady-Slipper* 16: 1–3.

Francis, W. 2011. Native plants in peril: rock climbing is damaging cliff-dwelling plants and archeological sites in the Red River Gorge, and now threatens Natural Bridge State Park. *The Lady-Slipper* 26(4): 1, 4–5.

Francko, D. A. 1986. Studies on *Nelumbo lutea* (Willd.) Pers. II. Effects of pH on photosynthetic carbon assimilation. *Aquat. Bot.* 26: 119–127.

Frank, A. A. & W. M. Reed. 1987. *Conium maculatum* (poison hemlock) toxicosis in a flock of range turkeys. *Avian Dis.* 31: 386–388.

Frank, B. S., W. B. Michelson, K. E. Panter & D. R.Gardner. 1995. Ingestion of poison hemlock (*Conium maculatum*). *West J. Med.* 163: 573–574.

Frankland, F. & T. A. Nelson. 1999. *Effects of Deer Grazing on Spring Wildflowers at Beall Woods Nature Preserve.* Final Project Report. Wildlife Preservation Fund, Large Project. 44 pp. Available online: http://www.dnr.illinois.gov/grants/documents/wpfgrantreports/1999l04w.pdf.

Franklin, M. A. 2001. Factors affecting seed production in natural populations of *Lysimachia asperulifolia* Poir. (Primulaceae), a rare, self-incompatible plant species. MS thesis. North Carolina State University, Raleigh, NC. 52 pp.

Franks, S. J. 2003. Facilitation in multiple life-history stages: evidence for nucleated succession in coastal dunes. *Plant Ecol.* 168: 1–11.

Frankton, C. & R. J. Moore. 1963. Cytotaxonomy of *Cirsium muticum, Cirsium discolor,* and *Cirsium altissimum. Can. J. Bot.* 41: 73–84.

Frantz, V. 1995. *Recovery Plan for Rough-Leaved Loosestrife (Lysimachia asperulaefolia).* U. S. Fish and Wildlife Service, Southeast Region, Atlanta, GA. 32 pp.

Franzke, A., K. Pollmann, W. Bleeker, R. Kohrt & H. Hurka. 1998. Molecular systematics of *Cardamine* and allied genera (Brassicaceae): ITS and non-coding chloroplast DNA. *Folia Geobot.* 33: 225–240.

Franzyk, H., C. E. Olsen & S. R. Jensen. 2004. Dopaol 2-keto-and 2, 3-diketoglycosides from *Chelone obliqua. J. Nat. Prod.* 67: 1052–1054.

Frappier, B., R. T. Eckert & T. D. Lee. 2003. Potential impacts of the invasive exotic shrub *Rhamnus frangula* L. (glossy buckthorn) on forests of southern New Hampshire. *N. E. Naturalist* 10: 277–296.

Fraser, D. F. 2000. Going, gone, and missing in action: the extinct, extirpated, and historic wildlife of British Columbia. pp. 19–26. *In:* L. M. Darling (ed.), *Proceedings of a Conference on the Biology and Management of Species and Habitats at Risk, Kamloops, B.C., February 15–19, 1999.* Volume 1. British Columbia Ministry of Environment, Lands and Parks, Victoria, B.C. and University College of the Cariboo, Kamloops, BC. 490 pp.

Fraser, D., D. Arthur, J. K. Morton & B. K. Thompson. 1980. Aquatic feeding by moose *Alces alces* in a Canadian lake. *Ecography Holarc. Ecol.* 3: 218–223.

Fraser, L. H. & J. P. Karnezis. 2005. A comparative assessment of seedling survival and biomass accumulation for fourteen wetland plant species grown under minor water-depth differences. *Wetlands* 25: 520–530.

Fraser, L. H. & E. B. Madson. 2008. The interacting effects of herbivore exclosures and seed addition in a wet meadow. *Oikos* 117: 1057–1063.

Fraser, L. H., K. Mulac & F. B.-G. Moore. 2014. Germination of 14 freshwater wetland plants as affected by oxygen and light. *Aquat. Bot.* 114: 29–34.

Fraser, M.-H., A. Cuerrier, P. S. Haddad, J. T. Arnason, P. L. Owen & T. Johns. 2007. Medicinal plants of Cree communities (Québec, Canada): antioxidant activity of plants used to treat type 2 diabetes symptoms. *Can. J. Physiol. Pharmacol.* 85: 1200–1214.

Fraser, W. P. 1919. Cultures of heteroecious rusts in 1918. *Mycologia* 11: 129–133.

Frasure, J. R. 1979. The effect of three grazing management systems on cattle diets on the Welder Wildlife Refuge. MS thesis in Range Science. Texas Tech University, Lubbock, TX. 93 pp.

Frederick, G. P. & R. J. Gutiérrez. 1992. Habitat use and population characteristics of the white-tailed ptarmigan in the Sierra Nevada, California. *Condor* 94: 889–902.

Free, J. B. 1963. The flower constancy of honeybees. *J. Anim. Ecol.* 32: 119–131.

Freedman, B., N. Hill, J. Svoboda & G. Henry. 1982. Seed banks and seedling occurrence in a high arctic oasis at Alexandra Fjord, Ellesmere Island, Canada. *Can. J. Bot.* 60: 2112–2118.

Freedman, B., W. Maass & P. Parfenov. 1992. The thread-leaved sundew, *Drosera filiformis*, in Nova Scotia: an assessment of risks of a proposal to mine fuel peat from its habitat. *Can. Field Naturalist* 106: 534–542.

Freeland, W. J. 1974. Vole cycles: another hypothesis. *Amer. Naturalist* 108: 238–245.

Freeman, C. C. 2009. 6. Penthoraceae Rydberg ex Britton. pp. 230–231. *In*: N. R. Morin (convening ed.), *Flora of North America North of Mexico, Vol. 8: Magnoliophyta: Paeoniaceae to Ericaceae*. Oxford University Press, New York.

Freeman, C. C. & H. R. Hinds. 2005. *Bistorta* (Linnaeus) Scopoli. pp. 594–597. *In*: Flora North America Editorial Committee (eds.), *Flora of North America North of Mexico, Vol. 5: Magnoliophyta: Caryophyllidae, part 2*. Oxford University Press, New York.

Freeman, C. C. & J. L. Reveal. 2005. 44. Polygonaceae Jussieu. Buckwheat Family. pp. 216–218. *In*: Flora North America Editorial Committee (eds.), *Flora of North America North of Mexico, Vol. 5: Magnoliophyta: Caryophyllidae, part 2*. Oxford University Press, New York.

Freeman, C. C., W. D. Kettle, K. Kindscher, R. E. Brooks, V. C. Varner & C. M. Pitcher. 1991. Vascular plants of the Kansas ecological reserves. pp. 23–47. *In*: W. D. Kettle & D. O. Whittemore (eds.), *Ecology and Hydrogeology of the Kansas Ecological Reserves and the Baker University Wetlands*. Open-file Report 91–35. Kansas Geological Survey, Lawrence, KS.

Freeman, L. 1994. Distribution of *Darlingtonia californica* on Mt. Eddy, California. MS thesis. California State University at Chico, Chico, CA.

Freeman, L. 1996. *Darlingtonia californica: Correcting errors perpetuated in the literature*. Available online: http://www.siskiyous.edu/shasta/art/smith/darcal.htm.

Freestone, A. L. 2006. Facilitation drives local abundance and regional distribution of a rare plant in a harsh environment. *Ecology* 87: 2728–2735.

Freestone, A. L. & B. D. Inouye. 2006. Dispersal limitation and environmental heterogeneity shape scale-dependent diversity patterns in plant communities. *Ecology* 87: 2425–2432 [*Ecological Archives* E087–147–A1].

Freire, S. E., A. M. Arambarri, N. D. Bayón, G. Sancho, E. Urtubey, C. Monti, M. C. Novoa & M. N. Colares. 2005. Epidermal characteristics of toxic plants for cattle from the Salado river basin (Buenos Aires, Argentina). *Bol. Soc. Argent. Bot.* 40: 241–281.

Freitas, R. F. & C. S. Costa. 2014. Germination responses to salt stress of two intertidal populations of the perennial glasswort *Sarcocornia ambigua*. *Aquat. Bot.* 117: 12–17.

Freitas, L., L. Galetto & M. Sazima. 2006. Pollination by hummingbirds and bees in eight syntopic species and a putative hybrid of Ericaceae in Southeastern Brazil. *Plant Syst. Evol.* 258: 49–61.

French, S. P., M. G. French & R. R. Knight. 1994. Grizzly bear use of army cutworm moths in the Yellowstone ecosystem. *Int. Conf. Bear Res. Manag.* 9: 389–399.

Frenkel, R. E., H. P. Eilers & C. A. Jefferson. 1981. Oregon coastal salt marsh upper limits and tidal datums. *Estuaries Coasts* 4: 198–205.

Frenot, Y., S. L. Chown, J. Whinam, P. M. Selkirk, P. Convey, M. Skotnicki & D. M. Bergstrom. 2005. Biological invasions in the Antarctic: extent, impacts and implications. *Biol. Rev.* 80: 45–72.

Frerker, K., G. Sonnier & D. M. Waller. 2013. Browsing rates and ratios provide reliable indices of ungulate impacts on forest plant communities. *For. Ecol. Manag.* 291: 55–64.

Friauf, J. J. 1953. An ecological study of the Dermaptera and Orthoptera of the Welaka area in northern Florida. *Ecol. Monogr.* 23: 79–126.

Fricke, G. & L. Steubing. 1984. Distribution of macrophytes and microphytes in hard water of the Eder Reservoir influents. *Arch. Hydrobiol.* 101: 361–372.

Fridriksson, S. & B. Johnsen. 1968. The colonization of vascular plants on Surtsey in 1967. *Surtsey Res. Prog. Rep.* 4: 31–38.

Friedman, J. M., W. R. Osterkamp & W. M. Lewis Jr. 1996. Channel narrowing and vegetation development following a Great Plains flood. *Ecology* 77: 2167–2181.

Fries, R. E. 1959. Annonaceae. pp. 1–171. *In*: A. Engler & K. Prantl (eds.), *Die natürlichen Pflanzenfamilien* (2nd ed.) 17A. Duncker & Humblot, Berlin, Germany.

Freitag, J. H. & H. P. Severin. 1945. Poison-hemlock-ringspot virus and its transmission by aphids to celery. *Hilgardia* 16: 389–410.

Frias-Torres, S. 2006. Habitat use of juvenile goliath grouper *Epinephelus itajara* in the Florida Keys, USA. *Endanger. Species Res.* 2: 1–6.

Fritsch, P. W. 1997. A Revision of *Styrax* (Styracaceae) for western Texas, Mexico, and Mesoamerica. *Ann. Missouri Bot. Gard.* 84: 705–761.

Fritsch, P. W. 2001. Phylogeny and biogeography of the flowering plant genus *Styrax* (Styracaceae) based on chloroplast DNA restriction sites and DNA sequences of the internal transcribed spacer region. *Mol. Phylogen. Evol.* 19: 387–408.

Fritsch, P. W. 2009. 15. Styracaceae de Candolle & Sprengel. pp. 339–347. *In*: N. R. Morin (convening ed.), *Flora of North America North of Mexico, Vol. 8: Magnoliophyta: Paeoniaceae to Ericaceae*. Oxford University Press, New York.

Fritsch, P. W., C. M. Morton, T. Chen & C. Meldrum. 2001. Phylogeny and biogeography of the Styracaceae. *Int. J. Plant Sci.* 162: S95–S116.

Fritsch, P. W., F. Almeda, S. S. Renner, A. B. Martins & B. C. Cruz. 2004. Phylogeny and circumscription of the near-endemic Brazilian tribe Microlicieae (Melastomataceae). *Amer. J. Bot.* 91: 1105–1114.

Fritsch, P. W., F. Almeda, A. B. Martins, B. C. Cruz & D. Estes. 2007. Rediscovery and phylogenetic placement of *Philcoxia minensis* (Plantaginaceae), with a test of carnivory. *Proc. Calif. Acad. Sci.* 58: 447–467.

Fritsch, R. 1971. Chromosomenzahlen von Pflanzen der Insel Kuba II. *Genet. Resour. Crop Evol.* 19: 305–313.

Fritz, G. B., F. J. Shaughnessy & T. J. Mulligan. 2009. Brassbuttons: an introduced species in a restored salt marsh (Oregon). *Ecol. Restorat.* 27: 389–391.

Fritz, K. M. & J. W. Feminella. 2003. Substratum stability associated with the riverine macrophyte *Justicia americana*. *Freshwater Biol.* 48: 1630–1639.

Fritz, K. M., M. M. Gangloff & J. W. Feminella. 2004a. Habitat modification by the stream macrophyte *Justicia americana* and its effects on biota. *Oecologia* 140: 388–397.

Fritz, K. M., M. A. Evans & J. W. Feminella. 2004b. Factors affecting biomass allocation in the riverine macrophyte *Justicia americana*. *Aquat. Bot.* 78: 279–288.

Fritz, R. S., C. G. Hochwender, D. A. Lewkiewicz, S. Bothwell & C. M. Orians. 2001. Seedling herbivory by slugs in a willow hybrid system: developmental changes in damage, chemical defense, and plant performance. *Oecologia* 129: 87–97.

Frodin, D. 2004. Araliaceae (Ginseng or Ivy Family). pp. 28-31. *In*: N. Smith, S. A. Mori, A. Henderson, D. W. Stevenson & S. V. Heald (eds.), *Flowering Plants of the Neotropics*. Princeton University Press, Princeton, NJ. 594 pp.

Frohlich, M. W. 1976. Appearance of vegetation in ultraviolet light: absorbing flowers, reflecting backgrounds. *Science* 194: 839–841.

Frolik, A. L. 1941. Vegetation on the peat lands of Dane County, Wisconsin. *Ecol. Monogr.* 11: 117–140.

Fruchter, J. 2005. Do large, infrequent disturbances release estuarine wetlands from coastal squeezing? MS thesis. Department of Plant Biology, Southern Illinois University, Carbondale, IL. 164 pp.

Fry, T. C. 1920. Plant migration along a partly drained lake. *Publ. Puget Sound Biol. Sta.* 2: 393–398.

Fryday, A. M. & K. A. Glew. 2003. *Stereocaulon nivale*, comb. nov., yet another crustose species in the genus. *Bryologist* 106: 565–568.

Fryxell, P. A. 1999. *Pavonia* Cavanilles (Malvaceae). *Fl. Neotrop. Monogr.* 76: 1–285.

Fuentes-Bazan, S., G. Mansion & T. Borsch. 2012. Towards a species level tree of the globally diverse genus *Chenopodium* (Chenopodiaceae). *Mol. Phylogen. Evol.* 62: 359–374.

Fujii, N. & K. Senni. 2006. Phylogeography of Japanese alpine plants. *Taxon* 55: 43–52.

Fujii, N., K. Ueda, Y. Watano & T. Shimizu. 1997. Intraspecific sequence variation of chloroplast DNA in *Pedicularis chamissonis* Steven (Scrophulariaceae) and geographic structuring of the Japanese "alpine" plants. *J. Plant Res.* 110: 195–207.

Fujii, N., K. Ueda, Y. Watano & T. Shimizu. 2001. Two genotypes of *Pedicularis chamissonis* (Scrophulariaceae) distributed at Mt. Gassan, Japan: additional genetic and morphological studies. *J. Plant Res.* 114: 133–140.

Fuke, Y., Y. Haga, H. Ono, T. Nomura & K. Ryoyama. 1998. Anti-carcinogenic activity of 6-methylsulfinylhexyl isothiocyanate, an active anti-proliferative principal of wasabi (*Eutrema wasabi* Maxim.). *Cytotechnology* 25: 197–203.

Fuller, D. A., C. E. Sasser, W. B. Johnson & J. G. Gosselink. 1985. The effects of herbivory on vegetation on islands in Atchafalaya Bay, Louisiana. *Wetlands* 4: 105–113.

Fuller, T. C. & E. M. McClintock. 1986. *Poisonous Plants of California*. University of California Press, Berkeley, CA. 433 pp.

Funk, V. A. & M. J. Fuller. 1978. A floristic survey of the seeps of Calloway County, Kentucky. *Castanea* 43: 162–172.

Furches, M. S., R. L. Small & A. Furches. 2013. Genetic diversity in three endangered pitcher plant species (*Sarracenia*; Sarraceniaceae) is lower than widespread congeners. *Amer. J. Bot.* 100: 2092–2101.

Furlong, M. & V. Pill. 1980. *Edible? Incredible! Pondlife*. Naturegraph Publishers Inc., Happy Camp, CA. 95 pp.

Furlow, J. J. 1979. The systematics of the American species of *Alnus* (Betulaceae). *Rhodora* 81: 1–121; 151–248.

Furlow, J. J. 1997. 31. Betulaceae Gray. Birch family. pp. 507–538. *In*: N. R. Morin (convening ed.), *Flora of North America North of Mexico, Vol. 3: Magnoliophyta: Magnoliidae and Hamamelidae*. Oxford University Press, New York.

Furness, N. H. & M. K. Upadhyaya. 2002. Differential susceptibility of agricultural weeds to ultraviolet-B radiation. *Can. J. Plant Sci.* 82: 789–796.

Gaascht, F., M. Dicato & M. Diederich. 2013. Venus flytrap (*Dionaea muscipula* Solander ex Ellis) contains powerful compounds that prevent and cure cancer. *Front. Oncol.* 3: Article 202, 1–18. doi:10.3389/fonc.2013.00202.

Gabel, J. D. & D. H. Les. 2001. *Neobeckia aquatica* North American lake cress. New England plant conservation program, conservation and research plan. New England Wild Flower Society, Framingham, MA. 42 pp.

Gabriel, B. C. & A. A. de la Cruz. 1974. Species composition, standing stock, and net primary production of a salt marsh community in Mississippi. *Chesapeake Sci.* 15: 72–77.

Gaddy, L. L. 1982. The floristics of three South Carolina pine savannahs. *Castanea* 47: 393–402.

Gadek, P. A., E. S. Fernando, C. J. Quinn, S. B. Hoot, T. Terrazas, M. C. Sheahan & M. W. Chase. 1996. Sapindales: molecular delimitation and infraordinal groups. *Amer. J. Bot.* 83: 802–811.

Gafner, S., C. Bergeron, L. L. Batcha, J. Reich, J. T. Arnason, J. E. Burdette, J. M. Pezzuto & C. K. Angerhofer. 2003a. Inhibition of [3H]-LSD binding to 5-HT7 receptors by flavonoids from *Scutellaria lateriflora*. *J. Nat. Prod.* 66: 535–537.

Gafner, S., S. Sudberg, É. M. Sudberg, C. Bergeron, L. L. Batcha, H. Guinaudeau, R. Gauthier & C. K. Angerhofer. 2003b. Analysis of *Scutellaria lateriflora* and its adulterants *Teucrium canadense* and *Teucrium chamaedrys* by LC-UV/MS, TLC and digital photo-microscopy. *J. AOAC Int.* 86: 453–460.

Gage, D. 1985. Chemical data and their bearing upon generic delineations in the Eupatorieae. *Taxon* 34: 61–71.

Gagné, J.-M. & G. Houle. 2002. Factors responsible for *Honckenya peploides* (Caryophyllaceae) and *Leymus mollis* (Poaceae) spatial segregation on subarctic coastal dunes. *Amer. J. Bot.* 89: 479–485.

Gai, J. P., P. Christie, G. Feng & X. L. Li. 2006. Twenty years of research on community composition and species distribution of arbuscular mycorrhizal fungi in China: a review. *Mycorrhiza* 16: 229–239.

Galatowitsch, S. M. & A. G. van der Valk. 1996a. The vegetation of restored and natural prairie wetlands. *Ecol. Appl.* 6: 102–112.

Galatowitsch, S. M. & A. G. van der Valk. 1996b. Vegetation and environmental conditions in recently restored wetlands in the prairie pothole region of the USA. *Vegetatio* 126: 89–99.

Galbany-Casals, M., S. Andrés-Sánchez, N. Garcia-Jacas, A. Susanna, E. Rico & M. M. Martínez-Ortega. 2010. How many of Cassini anagrams should there be? Molecular systematics and phylogenetic relationships in the *Filago* group (Asteraceae, Gnaphalieae), with special focus on the genus *Filago*. *Taxon* 59: 1671–1689.

Gallant, D., C. H. Bérubé, E. Tremblay & L. Vasseur. 2004. An extensive study of the foraging ecology of beavers (*Castor canadensis*) in relation to habitat quality. *Can. J. Zool.* 82: 922–933.

Galloway, L. F. 1995. Response to natural environmental heterogeneity: maternal effects and selection on life-history characters and plasticities in *Mimulus guttatus*. *Evolution* 49: 1095–1107.

Galloway, L. F., J. R. Etterson & J. L. Hamrick. 2003. Outcrossing rate and inbreeding depression in the herbaceous autotetraploid, *Campanula americana*. *Heredity* 90: 308–315.

Gálvez, M., C. Martín-Cordero, M. López-Lázaro, F. Cortés & M. J. Ayuso. 2003. Cytotoxic effect of *Plantago* spp. on cancer cell lines. *J. Ethnopharmacol.* 88: 125–130.

Ganders, F. R. 1979. The biology of heterostyly. *New Zealand J. Bot.* 17: 607–635.

Ganders, F. R., M. Berbee & M. Pirseyedi. 2000. ITS base sequence phylogeny in *Bidens* (Asteraceae): evidence for the continental relatives of Hawaiian and Marquesan *Bidens*. *Syst. Bot.* 25: 122–133.

Gandhi, S. D., A. F. Heesacker, C. A. Freeman, J. Argyris, K. Bradford & S. J. Knapp. 2005. The self-incompatibility locus (S) and quantitative trait loci for self-pollination and seed dormancy in sunflower. *Theor. Appl. Genet.* 111: 619–629.

Gandolfo, M. A., K. C. Nixon & W. L. Crepet. 2004. Cretaceous flowers of Nymphaeaceae and implications for complex insect entrapment pollination mechanisms in early Angiosperms. *Proc. Natl. Acad. Sci. U.S.A.* 101: 8056–8060.

Gano, L. 1917. A study in physiographic ecology in northern Florida. *Bot. Gaz.* 63: 337–372.

Gano, L. & J. McNeill. 1917. Evaporation records from the Gulf Coast. *Bot. Gaz.* 64: 318–329.

Ganter, B., F. Cooke & P. Mineau. 1996. Long-term vegetation changes in a Snow Goose nesting habitat. *Can. J. Zool.* 74: 965–969.

Gao, H. H., W. Li, J. Yang, Y. Wang, G. Q. Guo & G. C. Zheng. 2003. Effect of 6-benzyladenine and casein hydrolysate on micropropagation of *Amorpha fruticosa*. *Biol. Plant.* 47: 145–148.

Gao, L. M., D. Z. Li, C. Q. Zhang & J. B. Yang. 2002. Infrageneric and sectional relationships in the genus *Rhododendron* (Ericaceae) inferred from ITS sequence data. *Acta Bot. Sin.* 44: 1351–1356.

Gao, X. Q. & L. Björk. 2000. Valerenic acid derivatives and valepotriates among individuals, varieties and species of *Valeriana*. *Fitoterapia* 71: 19–24.

Gao, Y. P., G. H. Zheng & L. V. Gusta. 1998. Potassium hydroxide improves seed germination and emergence in five native plant species. *HortScience* 33: 274–276.

Garbary, D. J. & B. R. Taylor. 2007. Flowering during January in Antigonish County, Nova Scotia. *Can. Field Naturalist* 121: 76–80.

Garbary, D. J., A. G. Miller, R. Scrosati, K.-Y. Kim & W. B. Schofield. 2008. Distribution and salinity tolerance of intertidal mosses from Nova Scotian salt marshes. *Bryologist* 111: 282–291.

Garber, A. P. 1877. Botanical rambles in east Florida. *Bot. Gaz.* 2: 70–72.

Garciá, E. A. S. 2013. Germinación de seis especies de humedal de agua dulce bajo distintas condiciones de salinidad e inundación. Thesis. Universidad Veracruzana, Xalapa, Veracruz, Mexico. 55 pp.

Garcia-Barriga, H. 1974–1975. *Flora Medicinal de Colombia: Botánica Médica (3 vols.)*. Instituto de Ciencias Naturales, Universidad Nacional, Bogota, CO. 459 pp.

Garcia-Murillo, P. 1993. *Nymphaea mexicana* Zuccarini in the Iberian Peninsula. *Aquat. Bot.* 44: 407–409.

Gardner, M. & M. Macnair. 2000. Factors affecting the co-existence of the serpentine endemic *Mimulus nudatus* Curran and its presumed progenitor, *Mimulus guttatus* Fischer ex DC. *Biol. J. Linn. Soc.* 69: 443–459.

Gardner, R. L., T. P. Arbour & D. Boone. 2009. Noteworthy collection-Ohio. *Michigan Bot.* 48: 45–46.

Gargiullo, M. B. 2010. *A Guide to Native Plants of the New York City Region*. Rutgers University Press, Piscataway, NJ. 338 pp.

Gargiullo, M. B. & E. W. Stiles. 1991. Chemical and nutritional differences between two bird-dispersed fruits: *Ilex opaca* and *Ilex verticillata*. *J. Chem. Ecol.* 17: 1091–1106.

Gargiullo, M. B. & E. W. Stiles. 1993. Development of secondary metabolites in the fruit pulp of *Ilex opaca* and *Ilex verticillata*. *Bull. Torrey Bot. Club* 120: 423–430.

Garrett, A. O. 1921. Smuts and rusts of Utah: IV. *Mycologia* 13: 101–110.

Garrett, G. P., R. J. Edwards & C. Hubbs 2004. Discovery of a new population of Devils River minnow (*Dionda diaboli*), with implications for conservation of the species. *Southwest. Naturalist* 49: 435–441.

Gashwiler, J. S. 1970. Plant and mammal changes on a clearcut in west-central Oregon. *Ecology* 51: 1018–1026.

Gastony, G. J. & D. E. Soltis. 1977. Chromosome studies of *Parnassia* and *Lepuropetalon* (Saxifragaceae) from the eastern United States. A new base number for *Parnassia*. *Rhodora* 79: 573–578.

Gates, F. C. 1911. A bog in central Illinois. *Torreya* 11: 205–211.

Gates, F. C. 1929. Heat and the flowering of *Utricularia resupinata*. *Ecology* 10: 353–354.

Gates, F. C. 1940. Recent migrational trends in the distribution of weeds in Kansas. *Trans. Kansas Acad. Sci.* 43: 99–117.

Gaudet, C. L. & P. A. Keddy. 1988. A comparative approach to predicting competitive ability from plant traits. *Nature* 334: 242–243.

Gaudet, C. L. & P. A. Keddy. 1995. Competitive performance and species distribution in shoreline plant communities: a comparative approach. *Ecology* 76: 280–291.

Gaudeul, M. & I. Till-Bottraud. 2003. Low selfing in a mass-flowering, endangered perennial, *Eryngium alpinum* L. (Apiaceae). *Amer. J. Bot.* 90: 716–723.

Gaudeul, M. & I. Till-Bottraud. 2004. Reproductive ecology of the endangered alpine species *Eryngium alpinum* L. (Apiaceae): phenology, gene dispersal and reproductive success. *Ann. Bot.* 93: 711–721.

Gauthier, G. & C. Rousseau. 1973. L'écologie du *Floerkea proserpinacoides* Willd. à l'île aux Grues, Montmagny (Québec). *Natural Can.* 100: 371–383.

Ge, C. & P. Wan. 1990. Cytological study on *Eclipta prostrata* L. *China J. Chin. Mater. Med.* 15: 656–8, 702.

Gebhardt, Y., S. Witte, G. Forkmann, R. Lukačin, U. Matern & S. Martens. 2005. Molecular evolution of flavonoid dioxygenases in the family Apiaceae. *Phytochemistry* 66: 1273–1284.

Geisler, S. 1926. Soil reactions in relation to plant successions in the Cincinnati region. *Ecology* 7: 163–184.

Geißler, N., R. Schnetter & M.-L. Schnetter. 2002. The pneumathodes of *Laguncularia racemosa*: little known rootlets of surprising structure, and notes on a new fluorescent dye for lipophilic substances. *Plant Biol.* 4: 729–739.

Genesis Nursery. 2013. Compositae part deux. Available online: http://www.genesisnurseryinc.com/Up%20Ur%20C/C14%20 asters%20gai-oli.pdf (accessed April 28, 2014).

Gerber, C., L. Deeter, K. Hylton & B. Stilwill. 2011. Preliminary study of sodium chloride tolerance of *Rudbeckia fulgida* var. *speciosa* 'Goldsturm', *Heuchera americana* 'Dale's Variety' and *Aquilegia* ×*cultorum* 'Crimson Star' grown in greenhouse conditions. *J. Environ. Hort.* 29: 223–228.

Gerber, D. T. 1994. Physiological ecology of seven North American *Myriophyllum* species (Haloragaceae). PhD dissertation. University of Wisconsin-Milwaukee, Milwaukee, WI. 188 pp.

Gerber, D. T. & D. H. Les. 1994. Comparison of leaf morphology among submersed species of *Myriophyllum* (Haloragaceae) from different habitats and geographical distributions. *Amer. J. Bot.* 81: 973–979.

Gerber, D. T. & D. H. Les. 1996. Habitat differences among seven species of *Myriophyllum* (Haloragaceae) in Wisconsin and Michigan. *Michigan Bot.* 35: 75–86.

Gerber, M. A. 1985. The relationship of plant size to self-pollination in *Mertensia ciliata*. *Ecology* 66: 762–772.

Gerhardt, F. & S. K. Collinge. 2003. Exotic plant invasions of vernal pools in the Central Valley of California, USA. *J. Biogeogr.* 30: 1043–1052.

Gerloff, G. C., D. G. Moore & J. T. Curtis. 1966. Selective absorption of mineral elements by native plants of Wisconsin. *Plant Soil* 25: 393–405.

Germ, M. & A. Gaberščik. 2003. Comparison of aerial and submerged leaves in two amphibious species, *Myosotis scorpioides* and *Ranunculus trichophyllus*. *Photosynthetica* 41: 91–96.

Gerritsen, J. & H. S. Greening. 1989. Marsh seed banks of the Okefenokee Swamp: effects of hydrologic regime and nutrients. *Ecology* 70: 750–763.

Gershenzon, J., T. J. Mabry, J. D. Korpa & I. Bernala. 1984. Germacranolides from *Helianthus californicus*. *Phytochemistry* 23: 2561–2571.

Gerst, J. L. 1996. Endangered and threatened wildlife and plants; Determination of endangered status for three wetland species found in southern Arizona and northern Sonora, Mexico. *Fed. Reg.* 62: 665–689.

Gervais, J. A. & N. T. Wheelwright. 1994. Winter fruit removal in four plant species in Maine. *Maine Naturalist* 2: 15–24.

Getter, K. L. & D. B. Rowe. 2008. *Selecting Plants for Extensive Green Roofs in the United States*. Extension Bulletin E-3047. Michigan State University, Lansing, MI. 9 pp.

Gettys, L. A. 2012. Genetic control of whitef color in scarlet rosemallow (*Hibiscus coccineus* Walter). *J. Heredity* 103: 594–597.

Getz, L. L. 1961. Factors influencing the local distribution of shrews. *Amer. Midl. Naturalist* 65: 67–88.

Ghadiri, H. & M. Niazi. 2005. Effects of stratification, scarification, alternating temperature and light on seed dormancy of *Rumex dentatus*, *Amaranthus retroflexus* and *Chenopodium album*. *Iran. J. Weed Sci.* 1: 93–109.

Ghazoul, J. 2005. Pollen and seed dispersal among dispersed plants. *Biol. Rev.* 80: 413–443.

Ghedini, P. C. & C. E. Almeida. 2007. Butanolic extract of *Aster squamatus* aerial parts is the active fraction responsible to the antiulcer and gastric acid antisecretory effects. *Latin Amer. J. Pharm.* 26: 889–892.

Ghisalberti, E. L., M. Pennacchio & E. Alexander. 1998. Survey of secondary plant metaboilites with cardiovascular activity. *Pharm. Biol.* 36: 237–279.

Ghosh, T., T. K. Maity, A. Bose, G. K. Dash & M. Das. 2007. Antimicrobial activity of various fractions of ethanol extract of *Bacopa monnieri* Linn. aerial parts. *Indian J. Pharm. Sci.* 69: 312–314.

Gibbons, S., M. Oluwatuyi, N. C. Veitch & A. I. Gray. 2003. Bacterial resistance modifying agents from *Lycopus europaeus*. *Phytochemistry* 62: 83–87.

Gibbons, W., R. R. Haynes & J. L. Thomas. 1990. *Poisonous Plants and Venomous Animals of Alabama and Adjoining States*. University of Alabama Press, Tuscaloosa, AL. 345 pp.

Gibbs, J. J. 2009. New species in the *Lasioglossum petrellum* species group identified through an integrative taxonomic approach. *Can. Entomol.* 141: 371–396.

Gibbs, P. E. & S. Talavera. 2001. Breeding system studies with three species of *Anagallis* (Primulaceae): self-incompatibility and reduced female fertility in *A. monelli* L. *Ann. Bot.* 88: 139–144.

Gibson, D. J., B. A. Middleton, K. Foster, Y. A. K. Honu, E. W. Hoyer & M. Mathis. 2005. Species frequency dynamics in an old-field succession: effects of disturbance, fertilization and scale. *J. Veg. Sci.* 16: 415–422.

Gibson, K. D., A. J. Fischer & T. C. Foin. 2001. Shading and the growth and photosynthetic responses of *Ammannia coccinnea*. *Weed Res.* 41: 59–67.

Gibson, T. C. 1991a. Differential escape of insects from carnivorous plant traps. *Amer. Midl. Naturalist* 125: 55–62.

Gibson, T. C. 1991b. Competition among threadleaf sundews for limited insect resources. *Amer. Naturalist* 138: 785–789.

Gielis, C. 2003. *World Catalogue of Insects. Volume 4: Pterophoroidea & Alucitoidea (Lepidoptera)*. Apollo Books, Stenstrup, Denmark. 198 pp.

Gilbert, G. S. & W. P. Sousa. 2002. Host specialization among wood-decay polypore fungi in a Caribbean mangrove forest. *Biotropica* 34: 396–404.

Gilbert, K. G. & D. T. Cooke. 2001. Dyes from plants: past usage, present understanding and potential. *Plant Growth Regulat.* 34: 57–69.

Giles-Johnson, D. E. L., A. S. Thorpe & E. C. Gray. 2011. *Habitat Monitoring and Improvement for Cordylanthus maritimus ssp. palustris*. 2011 Progress Report. Institute for Applied Ecology for USDI Bureau of Land Management, Corvallis, OR. 25 pp.

Gill, L. S. & J. K. Morton. 1978. *Scutellaria churchilliana*-hybrid or species? *Syst. Bot.* 3: 342–348.

Gill, R. E., C. A. Babcock, C. M. Handel, W. R. Butler Jr. & D. G. Raveling. 1996. Migration, fidelity, and use of autumn staging grounds in Alaska by Cackling Canada Geese *Branta canadensis minima*. *Wildfowl* 47: 42–61.

Gillett, G. W. 1961. An experimental study of variation in the *Phacelia sericea* complex. *Amer. J. Bot.* 48: 1–7.

Gillett, J. M. 1957. A revision of the North American species of *Gentianella* Moench. *Ann. Missouri Bot. Gard.* 44: 195–269.

Gillett, J. M. 1968. The systematics of the Asiatic and American populations of *Fauria crista-galli* (Menyanthaceae). *Can. J. Bot.* 46: 92–96.

Gilliam, F. S., W. J. Platt & R. K. Peet. 2006. Natural disturbances and the physiognomy of pine savannas: a phenomenological model. *Appl. Veg. Sci.* 9: 83–96.

Gillis, E. A., S. F. Morrison, G. D. Zazula & D. S. Hik. 2005. Evidence for selective caching by arctic ground squirrels living in alpine meadows in the Yukon. *Arctic* 58: 354–360.

Gillis, W. T. 1971. The systematics and ecology of poison-ivy and the poison-oaks (*Toxicodendron*, Anacardiaceae). *Rhodora* 73: 72–159; 161–237; 370–443; 465–540.

Gilman, E. F. 1999. *Urechites lutea*. Fact Sheet FPS-595. Institute of Food and Agricultural Sciences, Cooperative Extension Service, University of Florida, Gainesville, FL. 3 pp.

Gilreath, J. P. & P. R. Gilreath. 1983. Weed control in seepage irrigated fall transplanted broccoli and cauliflower. *Proc. Florida State Hort. Soc.* 69: 77–79.

Gimingham, C. H., D. Welch, E. J. Clement & P. Lane. 2002. A large population of *Plagiobothrys scouleri* (Boraginaceae) in northeast Scotland, and notes on occurrences elsewhere in Britain. *Watsonia* 24: 159–169.

Girardin, M.-P., J. Tardif & Y. Bergeron. 2001. Gradient analysis of *Larix laricina* dominated wetlands in Canada's southeastern boreal forest. *Can. J. Bot.* 79: 444–456.

Giroux, J.-F. & J. Bedard. 1987. The effects of grazing by greater snow geese on the vegetation of tidal marshes in the St. Lawrence Estuary. *J. Appl. Ecol.* 24: 773–788.

Gladstar, R. & P. Hirsch. 2000. *Planting the Future: Saving our Medicinal Herbs.* Healing Arts Press, Rochester, VT. 310 pp.

Glaser, P. H., G. A. Wheeler, E. Gorham & H. E. Wright Jr. 1981. The patterned mires of the Red Lake peatland, northern Minnesota: vegetation, water chemistry and landforms. *J. Ecol.* 69: 575–599.

Glattstein, J. 1991. The daisies of Autumn. *Arnoldia* 51: 23–31.

Glawe, D. A. n.d. *Pacific Northwest Fungi Database.* Department of Plant Pathology, Washington State University, Puyallup, WA. Available online: http://cru23.cahe.wsu.edu/fungi/programs/aboutDatabase.asp (accessed March 14, 2005).

Glazener, W. C. 1946. Food habits of wild geese on the Gulf Coast of Texas. *J. Wildl. Manag.* 10: 322–329.

Gleason, H. A. 1917. A prairie near Ann Arbor, Michigan. *Rhodora* 19: 163–165.

Gleason, H. A. 1923. Evolution and geographical distribution of the genus *Vernonia* in North America. *Amer. J. Bot.* 10: 187–202.

Gleason, R. A., B. A. Tangen, M. K. Laubhan, R. G. Finocchiaro & J. F. Stamm. 2009. *Literature Review and Database of Relations between Salinity and Aquatic Biota: Applications to Bowdoin National Wildlife Refuge, Montana.* U. S. Geological Survey Scientific Investigations Report 2009-5098. U. S. Geological Survey, Reston, VA. 76 pp.

Gleason, S. M. 2002. Flora of the Upper Klamath River Canyon, Klamath County, Oregon. *Kalmiopsis* 9: 16–21.

Glenn, A. & M. S. Bodri. 2012. Fungal endophyte diversity in *Sarracenia.* *PLoS One* 7(3): e32980. doi:10.1371/journal.pone.0032980.

Glenn, E. P., J. O'Leary, M. Watson, T. Thompson & R. O. Kuehl. 1991. *Salicornia bigelovii* Torr.: an oilseed halophyte for seawater irrigation. *Science* 251: 1065–1067.

Glenn, E., T. L. Thompson, R. Frye, J. Riley & D. Baumgartner. 1995. Effects of salinity on growth and evapotranspiration of *Typha domingensis* Pers. *Aquat. Bot.* 52: 75–91.

Glenn, E. P., P. L. Nagler, R. C. Brusca & O. Hinojosa-Huerta. 2006. Coastal wetlands of the northern Gulf of California: inventory and conservation status. *Mar. Freshwater Ecosyst.* 16: 5–28.

Glenn, M. S. & M.-K. Woo. 1997. Spring and summer hydrology of a valley-bottom wetland, Ellesmere Island, Northwest Territories, Canada. *Wetlands* 17: 321–329.

Glitzenstein, J. S., D. R. Streng & D. D. Wade. 1998. A promising new start for a new population of *Parnassia caroliniana* Michx. pp. 44–58. *In*: J. S. Kush (ed.), *Proceedings of the Longleaf Pine Ecosystem Restoration Symposium: Ecological Restoration and Regional Conservation Strategies.* Longleaf Alliance Report No. 3. Auburn, AL.

Glitzenstein, J. S., D. R. Streng, D. D. Wade & J. Brubaker. 2001. Starting new populations of longleaf pine ground-layer plants in the outer coastal plain of South Carolina, USA. *Nat. Areas J.* 21: 89–110.

Glover, J. B. & M. A. Floyd. 2004. Larvae of the genus *Nectopsyche* (Trichoptera: Leptoceridae) in eastern North America, including a new species from North Carolina. *J. N. Amer. Benthol. Soc.* 23: 526–541.

Gluch, O. 2005. *Pinguicula* species (Lentibulariaceae) from the southeastern United States: observations of different habitats in Florida. *Acta Bot. Gallica* 152: 197–204.

Glück, H. 1924. *Biologische und morphologische Untersuchungen über Wasser-und Sumpfgewächse. Teil IV. Untergetauchte und Schwimmblattflora.* Gustav Fischer, Jena, Germany. 746 pp.

Glück, H. 1934. Novae species et varietates Generis Limosellae. *Notizbl. Königl. Bot. Gart. Berlin* 12: 71–78.

Gluesenkamp, D. & J. Wirka. 2006. Sonoma Valley vernal pools: is nitrogen pollution harming fragile ecosystems? *Ardeid* (n.v.): 8–9.

Gnanaraj, W. E., J. M. Antonisamy, K. M. Subramanian & S. Nallyan. 2011. Micropropagation of *Alternanthera sessilis* (L.) using shoot tip and nodal segments. *Iran. J. Biotechnol.* 9: 206–212.

Gobert, V., S. Moja, M. Colson & P. Taberlet. 2002. Hybridization in the section *Mentha* (Lamiaceae) inferred from AFLP markers. *Amer. J. Bot.* 89: 2017–2023.

Godefroid, S. & N. Koedam. 2004. Interspecific variation in soil compaction sensitivity among forest floor species. *Biol. Conserv.* 119: 207–217.

Godefroid, S. & N. Koedam. 2004. The impact of forest paths upon adjacent vegetation: effects of the path surfacing material on the species composition and soil compaction. *Biol. Conserv.* 119: 405–419.

Godefroid, S., D. Monbaliu & N. Koedam. 2007. The role of soil and microclimatic variables in the distribution patterns of urban wasteland flora in Brussels, Belgium. *Landscape Urban Plan.* 80: 45–55.

Godefroid, S., A. Van de Vyver & T. Vanderborght. 2010. Germination capacity and viability of threatened species collections in seed banks. *Biodivers. Conserv.* 19: 1365–1383.

Godfrey, G. A. 1975. Home range characteristics of ruffed grouse broods in Minnesota. *J. Wildl. Manag.* 39: 287–298.

Godfrey, R. K. 1961. *Plantago cordata* still grows in Georgia. *Castanea* 26: 119–120.

Godfrey, R. K. 1988. *Trees, Shrubs, and Woody Vines of Northern Florida and Adjacent Georgia and Alabama.* The University of Georgia Press, Athens, GA. 551 pp.

Godfrey, R. K. & A. F. Clewell. 1965. Polygamodioecious *Leitneria floridana* (Leitneriaceae). *Sida* 2: 172–173.

Godfrey, R. K. & R. Kral. 1958. A new species of *Vicia* (Leguminosae) in Florida. *Rhodora* 60: 256–258.

Godfrey, R. K. & H. L. Stripling. 1961. A synopsis of *Pinguicula* (Lentibulariaceae) in the southeastern United States. *Amer. Midl. Naturalist* 66: 395–409.

Godfrey, R. K. & J. W. Wooten. 1979. *Aquatic and Wetland Plants of Southeastern United States. Vol. I: Monocotyledons.* University of Georgia Press, Athens, GA. 712 pp.

Godfrey, R. K. & J. W. Wooten. 1981. *Aquatic and Wetland plants of Southeastern United States. Vol. II: Dicotyledons.* University of Georgia Press, Athens, GA. 933 pp.

Godt, M. J. W. & J. L. Hamrick. 1996a. Allozyme diversity in the endangered shrub *Lindera melissifolia* (Lauraceae) and its widespread congener *Lindera benzoin.* *Can. J. For. Res.* 26: 2080–2087.

Godt, M. J. W. & J. L. Hamrick. 1996b. Genetic structure of two endangered pitcher plants, *Sarracenia jonesii* and *Sarracenia oreophila* (Sarraceniaceae). *Amer. J. Bot.* 83: 1016–1023.

Godt, M. J. W. & J. L. Hamrick. 1998. Allozyme diversity in the endangered pitcher plant *Sarracenia rubra* ssp. *alabamensis* (Sarraceniaceae) and its close relative *S. rubra* ssp. *rubra.* *Amer. J. Bot.* 85: 802–810.

Godt, M. J. W. & J. L. Hamrick. 1999. Genetic divergence among infraspecific taxa of *Sarracenia purpurea.* *Syst. Bot.* 23: 427–438.

Goeden, R. D. & D. W. Ricker. 1982. Poison hemlock, *Conium maculatum*, in southern California-an alien weed attacked by few insects. *Ann. Entomol. Soc. Amer.* 75: 173–176.

Goeger, D. E., P. R. Cheeke, H. S. Ramsdell, S. S. Nicholson & D. R. Buhler. 1983. Comparison of the toxicities of *Senecio jacobaea*, *Senecio vulgaris* and *Senecio glabellus* in rats. *Toxicol. Lett.* 15: 19–23.

Goepfert, D. 1974. Karyotypes and DNA content in species of *Ranunculus* L. and related genera. *Bot. Not.* 127: 464–489.

Goering, K. J., R. Eslick & D. L. Brelsford. 1965. A search for high erucic acid containing oils in the Cruciferae. *Econ. Bot.* 19: 251–256.

Goertzen, L. R. & R. S. Boyd. 2007. Genetic diversity and clonality in the federally endangered plant *Clematis socialis* Kral (Ranunculaceae). *J. Torrey Bot. Soc.* 134: 433–440.

Goertzen, L. R., J. J. Cannone, R. R. Gutell & R. K. Jansen. 2003. ITS secondary structure derived from comparative analysis: implications for sequence alignment and phylogeny of the Asteraceae. *Mol. Phylogen. Evol.* 29: 216–234.

Goetsch, L., A. J. Eckert & B. D. Hall. 2005. The molecular systematics of *Rhododendron* (Ericaceae): a phylogeny based upon *RPB2* gene sequences. *Syst. Bot.* 30: 616–626.

Gohar, A. A., G. T. Maatooq, E. M. Mrawan, A. A. Zaki & Y. Takaya. 2012. Two oleananes from *Ammannia auriculata* Willd. *Naturalist Prod. Res.* 26: 1328–1333.

Going, B., J. Simpson & T. Even. 2008. The influence of light on the growth of watercress (*Nasturtium officinale* R. Br.). *Hydrobiologia* 607: 75–85.

Goldblatt, P. 1976a. New or noteworthy chromosome records in the angiosperms. *Ann. Missouri Bot. Gard.* 63: 889–895.

Goldblatt, P. 1976b. Chromosome number and its significance in *Batis maritima* (Bataceae). *J. Arnold Arbor.* 57: 526–530.

Goldblatt, P. 1979. Miscellaneous chromosome counts in angiosperms, II. including new family and generic records. *Ann. Missouri Bot. Gard.* 66: 856–861.

Goldblatt, P., P. Bernhardt, P. Vogan & J. C. Manning. 2004. Pollination by fungus gnats (Diptera: Mycetophilidae) and self-recognition sites in *Tolmiea menziesii* (Saxifragaceae). *Plant Syst. Evol.* 244: 55–67.

Golden, J. L. & J. F. Bain. 2000. Phylogeographic patterns and high levels of chloroplast DNA diversity in four *Packera* (Asteraceae) species in southwestern Alberta. *Evolution* 54: 1566–1579.

Golden, J. L., P. Achuff & J. F. Bain. 2008. Genetic divergence of *Cirsium scariosum* in eastern and western Canada. *Ecoscience* 15: 293–297.

Golden, M. S. 1979. Forest vegetation of the lower Alabama piedmont. *Ecology* 60: 770–782.

Goldenberg, D. M. & D. B. Zobel. 1997. Allocation, growth and estimated population structure of *Corydalis aquae-gelidae*, a rare riparian plant. *Northwest Sci.* 71: 196–204.

Goldenberg, D. M. & D. B. Zobel. 1999. Habitat relations of *Corydalis aquae-gelidae*, a rare riparian plant. *Northwest Sci.* 73: 94–105.

Goldmann, A., M.-L. Milat, P.-H. Ducrot, J.-Y. Lallemand, M. Maille, A. Lepingle, I. Charpin & D. Tepfer. 1990. Tropane derivatives from *Calystegia sepium*. *Phytochemistry* 29: 2125–2127.

Goldsby, T. L. & D. R. Sanders, Sr. 1977. Effects of consecutive water fluctuations on submersed vegetation of Black Lake, Louisiana. *J. Aquat. Plant Manag.* 15: 23–28.

Goliber, T. E. & L. J. Feldman. 1990. Developmental analysis of leaf plasticity in the heterophyllous aquatic plant *Hippuris vulgaris*. *Amer. J. Bot.* 77: 399–412.

Golley, F. B., G. A. Petrides & J. F. McCormick. 1965. A survey of the vegetation of the boiling springs natural area, South Carolina. *Bull. Torrey Bot. Club* 92: 355–363.

Goloboff, P. A., S. A. Catalano, J. Marcos Mirande, C. A. Szumik, J. Salvador Arias, M. Kallersjo & J. S. Farris. 2009. Phylogenetic analysis of 73 060 taxa corroborates major eukaryotic groups. *Cladistics* 25: 211–230.

Goltz, J. P. 2001. Nature news: botany ramblings. *N. B. Naturalist* 27: n.p.

Golumbia, T. 2001. Classification of plant communities in Gwaii Haanas National Park Reserve and Haida Heritage site. MS thesis. University of British Columbia, Vancouver, BC. 176 pp.

Goman, M. 2001. Statistical analysis of modern seed assemblages from the San Francisco Bay: applications for the reconstruction of paleo-salinity and paleo-tidal inundation. *J. Paleolimnol.* 24: 393–409.

Gomez, B., V. Daviero-Gomez, C. Coiffard, C. Martin-Closas & D. L. Dilcher. 2015. *Montsechia*, an ancient aquatic angiosperm. *Proc. Natl. Acad. Sci. U.S.A.* 112: 10985–10988.

Gong, L., H. S. Paris, G. Stift, M. Pachner, J. Vollmann & T. Lelley. 2013. Genetic relationships and evolution in *Cucurbita* as viewed with simple sequence repeat polymorphisms: the centrality of *C. okeechobeensis*. *Genet. Resour. Crop Evol.* 60: 1531–1546.

Gonsoulin, G. J. 1974. A revision of *Styrax* (Styracaceae) in North America, Central America, and the Caribbean. *Sida* 5: 191–258.

Gonzalez, V. & T. Griswold. 2013. Wool carder bees of the genus *Anthidium* in the Western Hemisphere (Hymenoptera: Megachilidae): diversity, host plant associations, phylogeny, and biogeography. Paper 1251. Publications from USDA-ARS/UNL Faculty. Available online: http://digitalcommons.unl.edu/usdaarsfacpub/1251.

Good, R. 1924. The germination of *Hippuris vulgaris* L. *J. Linn. Soc. Bot.* 46: 443–448.

Gooding, G. & J. R. Langford. 2004. Characteristics of tree roosts of Rafinesque's big-eared bat and southeastern bat in northeastern Louisiana. *Southwest. Naturalist* 49: 61–67.

Gooding, L. N. & L. N. Goodding. 1961. Why Sycamore Canyon in Santa Cruz County should be preserved as a nature sanctuary or natural area. *J. Arizona Acad. Sci.* 1: 113–115.

Goodman, A. M., G. G. Ganf, G. C. Dandy, H. R. Maier & M. S. Gibbs. 2010. The response of freshwater plants to salinity pulses. *Aquat. Bot.* 93: 59–67.

Goodrich, K. R. & R. A. Raguso. 2009. The olfactory component of floral display in *Asimina* and *Deeringothamnus* (Annonaceae). *New Phytol.* 183: 457–469.

Goodson, J. M., A. M. Gurnell, P. G. Angold & I. P. Morrissey. 2002. Riparian seed banks along the lower River Dove, UK: their structure and ecological implications. *Geomorphology* 47: 45–60.

Goodson, J. M., A. M. Gurnell, P. G. Angold & I. P. Morrissey. 2003. Evidence for hydrochory and the deposition of viable seeds within winter flow-deposited sediments: the River Dove, Derbyshire, UK. *River Res. Appl.* 19: 317–334.

Goodwillie, C. & W. R. Franch. 2006. An experimental study of the effects of nutrient addition and mowing on a ditched wetland plant community: results of the first year. *J. North Carolina Acad. Sci.* 122: 106–117.

Goodwin, R. H. 1937. The cyto-genetics of two species of *Solidago* and its bearing on their polymorphy in nature. *Amer. J. Bot.* 24: 425–432.

Goodwin, T. M. & W. R. Marion. 1979. Seasonal activity ranges and habitat preferences of adult alligators in a north-central Florida lake. *J. Herpetol.* 13: 157–163.

Gorai, D., S. K. Jash, R. K. Singh & A. Gangopadhyay. 2014. Chemical and pharmacological aspects of *Limnophila aromatica* (Scrophulariaceae): an overview. *Amer. J. Phytomed. Clinic. Therapeut.* 2: 348–356.

Gorchov, D. L. 1990. Pattern, adaptation, and constraint in fruiting synchrony within vertebrate-dispersed woody plants. *Oikos* 58: 169–180.

Gordon, D. R. & C. A. Gantz. 2011. Risk assessment for invasiveness differs for aquatic and terrestrial plant species. *Biol. Invas.* 13: 1829–1842.

Gordon, E. 1999. Effect of size and number of seeds on germination and seedling size in six helophyte species. *Fragm. Florist. Geobot.* 44: 429–436.

Gordon, R. B. 1933. A unique raised bog at Urbana, Ohio. *Ohio J. Sci.* 33: 453–459.

Gordon, T. (ed.). 1989. 1988 field trips. *Bartonia* 55: 63–67.

Gordon, T. (ed.). 2009. 2005–2006 field trips. *Bartonia* 64: 55–76.

Gordon, T. (ed.). 2011. 2007–2008 field trips. *Bartonia* 65: 126–147.

Gordon, T. (ed.). 2013. 2009–2011 field trips. *Bartonia* 66: 82–119.

Goremykin, V. V., K. I. Hirsch-Ernst, S. Wölfl & F. H. Hellwig. 2004. The chloroplast genome of *Nymphaea alba*: whole-genome analyses and the problem of identifying the most basal angiosperm. *Mol. Biol. Evol.* 21: 1445–1454.

Gosling, D. C. L. 1986. Ecology of the Cerambycidae (Coleoptera) of the Huron Mountains in northern Michigan. *Great Lakes Entomol.* 19: 153–162.

Gosling, L. M. & S. J. Baker. 1980. Acidity fluctuations at a broadland site in Norfolk. *J. Appl. Ecol.* 17: 479–490.

Gossard, H. A. 1903. *White Fly (Aleyrodes citri).* Bulletin no. 67. Florida Agricultural Experiment Station. E. O. Painter & Co., DeLand, FL. 666 pp.

Goswami, D. A. & B. Matfield. 1975. Cytogenetic studies in the genus *Potentilla* L. *New Phytol.* 75: 135–146.

Göthberg, A., M. Greger, K. Holm & B. E. Bengtsson. 2004. Influence of nutrient levels on uptake and effects of mercury, cadmium, and lead in water spinach. *Environ. Qual.* 33: 1247–1255.

Göthberg, A. & M. Greger. 2006. Formation of methyl mercury in an aquatic macrophyte. *Chemosphere* 65: 2096–2105.

Goto-Yamamoto, N., H. Mouri, M. Azumi & K. J. Edwards. 2006. Development of grape microsatellite markers and microsatellite analysis including Oriental cultivars. *Amer. J. Enol. Vitic.* 57: 105–108.

Gotsch, S. G. & A. M. Ellison. 1998. Seed germination of the northern pitcher plant, *Sarracenia purpurea*. *N. E. Naturalist* 5: 175–182.

Gottesfeld, L. M. J. 1994. Wet'suwet'en ethnobotany: traditional plant uses. *J. Ethnobiol.* 14: 185–210.

Gottlieb, A. M., G. C. Giberti & L. Poggio. 2005. Molecular analyses of the genus *Ilex* (Aquifoliaceae) in southern South America, evidence from AFLP and ITS sequence data. *Amer. J. Bot.* 92: 352–369.

Gotto, J. W. & B. F. Taylor. 1976. N₂ fixation associated with decaying leaves of the red mangrove (*Rhizophora mangle*). *Appl. Environ. Microbiol.* 31: 781–783.

Gottsberger, G. 1999. Pollination and evolution in neotropical Annonaceae. *Plant Species Biol.* 14: 143–152.

Gottschling, M., H. H. Hilger, M. Wolf & N. Diane. 2001. Secondary structure of the ITS1 transcript and its application in a reconstruction of the phylogeny of Boraginales. *Plant Biol.* 3: 629–636.

Goun, E. A., V. M. Petrichenko, S. U. Solodnikov, T. V. Suhinina, M. A. Kline, G. Cunningham, C. Nguyen & H. Mile. 2002. Anticancer and antithrombin activity of Russian plants. *J. Ethnopharmacol.* 81: 337–342.

Goyal, P. K. & R. R. Aggarwal. 2013. A review on phytochemical and biological investigation of plant genus *Pluchea*. *Indo Amer. J. Pharm. Res.* 3: 3373–3392.

Graae, B. J., S. Pagh & H. H. Bruun. 2004. An experimental evaluation of the Arctic fox (*Alopex lagopus*) as a seed disperser. *Arct. Antarc. Alp. Res.* 36: 468–473.

Grabas, G. P. & T. M. Laverty. 1999. The effect of purple loosestrife (*Lythrum salicaria*; Lythraceae) on the pollination and reproductive success of sympatric co-flowering wetland plants. *Ecoscience* 6: 230–242.

Grace, J. B., L. Allain & C. Allen. 2000. Vegetation associations in a rare community type—coastal tallgrass prairie. *Plant Ecol.* 147: 105–115.

Graenicher, S. 1911. Bees of northern Wisconsin. *Bull. Public Mus. Milwaukee* 1: 221–249.

Graenicher, S. 1927. Bees of the genus *Halictus* from Miami, Florida. *Psyche* 34: 203–208.

Graffis, A. 2013. The effects of a parasitic plant (*Cuscuta howelliana*) on vernal pool plant diversity. MS thesis. California State University, Sacramento, CA. 70 pp.

Graham, A., J. W. Nowicke, J. J. Skvarla, S. A. Graham, V. Patel & S. Lee. 1987. Palynology and systematics of the Lythraceae II. Genera *Haitia* through *Peplis*. *Amer. J. Bot.* 74: 829–850.

Graham, B. F. & A. L. Rebuck. 1958. The effect of drainage on the establishment and growth of pond pine (*Pinus serotina*). *Ecology* 39: 33–36.

Graham, C. J. & J. S. Kuehny. 2003. Extraction temperature alters phytochemical concentrations and quality of Mayhaw juice. *Acta Hort.* 628: 823–828.

Graham, E. E., J. F. Tooker & L. M. Hanks. 2012. Floral host plants of adult beetles in central Illinois: an historical perspective. *Ann. Entomol. Soc. Amer.* 105: 287–297.

Graham, H. W. & L. K. Henry. 1933. Plant succession at the borders of a kettle-hole lake. *Bull. Torrey Bot. Club* 60: 301–315.

Graham, S. A. 1979. The origin of *Ammannia* ×*coccinea* Rottboell. *Taxon* 28: 169–178.

Graham, S. A. 1985. A revision of *Ammannia* (Lythraceae) in the Western Hemisphere. *J. Arnold Arbor.* 66: 395–420.

Graham, S. A. 1989. Chromosome numbers in *Cuphea* (Lythraceae): new counts and a summary. *Amer. J. Bot.* 76: 1530–1540.

Graham, S. A. 1992. New chromosome counts in Lythraceae—systematic and evolutionary implications. *Acta Bot. Mexico* 17: 45–51.

Graham, S. A. & T. B. Cavalcanti. 2001. New chromosome counts in the Lythraceae and a review of chromosome numbers in the family. *Syst. Bot.* 26: 445–458.

Graham, S. A. & K. Gandhi. 2013. Nomenclatural changes resulting from the transfer of *Nesaea* and *Hionanthera* to *Ammannia* (Lythraceae). *Harvard Pap. Bot.* 18: 71–90.

Graham, S. A., J. V. Crisci & P. C. Hoch. 1993. Cladistic analysis of the Lythraceae sensu lato based on morphological characters. *Bot. J. Linn. Soc.* 113: 1–33.

Graham, S. A., M. Diazgranados & J. C. Barber. 2011. Relationships among the confounding genera *Ammannia*, *Hionanthera*, *Nesaea* and *Rotala* (Lythraceae). *Bot. J. Linn. Soc.* 166: 1–19.

Graham, S. A., J. Hall, K. Sytsma & S.-H. Shi. 2005. Phylogenetic analyses of the Lythraceae based on four gene regions and morphology. *Int. J. Plant Sci.* 166: 995–1017.

Gramling, J. M. 2010. Potential effects of laurel wilt on the flora of North America. *S. E. Naturalist* 9: 827–836.

Grand, L. F. & C. S. Vernia. 2002. A preliminary checklist of fungi in the Nags Head Woods Maritime Forest in North Carolina. *Castanea* 67: 324–328.

Grand, L. F. & C. S. Vernia. 2004. Biogeography and hosts of poroid wood decay fungi in North Carolina: species of *Phellinus* and *Schizopora*. *Mycotaxon* 89: 181–184.

Grand, L. F., C. S. Vernia & M. J. Munster. 2009. Biogeography and hosts of poroid wood decay fungi in North Carolina: species of *Trametes* and *Trichaptum*. *Mycotaxon* 106: 243–246.

Grandin, M. 1971. Dormancy of hibernacles of *Glaux maritima* L. and its adaptation to salinity. *Oecol. Plant* 6: 203–208.

Grant, A. L. 1924. A monograph of the genus *Mimulus*. *Ann. Missouri Bot. Gard.* 11: 99–388.

Grant, J. J. W. 2013. Suitability of Canadian-bred and native plant species for extensive green roofs in northern Nova Scotia. MS thesis. Dalhousie University, Halifax, NS. 81 pp.

Grant, K. A. & V. Grant. 1968. *Hummingbirds and Their Flowers*. Columbia University Press, New York. 115 pp.

Grant, V. 1949. Pollination systems as isolating mechanisms in angiosperms. *Evolution* 3: 82–97.

Grant, V. 1994. Modes and origins of mechanical and ethological isolation in angiosperms. *Proc. Natl. Acad. Sci. U.S.A.* 91: 3–10.

Grant, W. F. 1953. A cytotaxonomic study in the genus *Eupatorium*. *Amer. J. Bot.* 40: 729–742.

Grant, W. F. 1955. A cytogenetic study in the Acanthaceae. *Brittonia* 8: 121–149.

Grant, W. F. 1995. A chromosome atlas and interspecific-intergenic index for *Lotus* and *Tetragonolobus* (Fabaceae). *Can. J. Bot.* 73: 1787–1809.

Gratani, L., M. F. Crescente, G. Fabrini & L. Varone. 2008. Growth pattern of *Bidens cernua* L.: relationships between relative growth rate and its physiological and morphological components. *Photosynthetica* 46: 179–184.

Graves, J. D. & K. Taylor. 1988. A comparative study of *Geum rivale* L. and *Geum urbanum* L. to determine those factors controlling their altitudinal distribution: III. The response of germination to temperature. *New Phytol.* 110: 391–398.

Graves, W. R. & J. L. Gallagher. 2003. Resistance to salinity of *Alnus maritima* from disjunct wetlands: symptoms of salt injury, comparison to other shrubs, and effect of inundation. *Wetlands* 23: 394–405.

Greca, M. D., A. Fiorentino, P. Monaco & L. Previtera. 1994. Oleanane glycosides from *Hydrocotyle ranunculoides*. *Phytochemistry* 36: 1479–1483.

Green, A. J., J. Figuerola & M. I. Sánchez. 2002. Implications of waterbird ecology for the dispersal of aquatic organisms. *Acta Oecol.* 23: 177–189.

Green, A. J., M. Soons, A.-L. Brochet & E. Kleyheeg. 2016. Dispersal of plants by waterbirds. pp. 147–195. *In*: Ç. H. Sekercioglu, D. G. Wenny & C. J. Whelan (eds.), *Why Birds Matter: Avian Ecological Function and Ecosystem Services*. University of Chicago Press, Chicago, IL.

Green, E. K. & S. M. Galatowitsch. 2002. Effects of *Phalaris arundinacea* and nitrate-N addition on the establishment of wetland plant communities. *J. Appl. Ecol.* 39: 134–144.

Green, M. M. 1925. Notes on some mammals of Montmorency County, Michigan. *J. Mammalogy* 6: 173–178.

Green, P. S. 1962. Watercress in the new world. *Rhodora* 64: 32–43.

Green, P. S. 2004. Oleaceae. pp. 296–306. *In*: J. W. Kadereit (ed.), *The Families and Genera of Vascular Plants, Vol. VII. Flowering Plants, Dicotyledons: Lamiales (except Acanthaceae including Avicenniaceae)*. Springer-Verlag, Berlin, Germany.

Green, T. W. & G. E. Bohart. 1975. The pollination ecology of *Astragalus cibarius* and *Astragalus utahensis* (Leguminosae). *Amer. J. Bot.* 62: 379–386.

Greenberg, A. K. & M. J. Donoghue. 2011. Molecular systematics and character evolution in Caryophyllaceae. *Taxon* 60: 1637–1652.

Greene, E. 1998. *Utricularia radiata* Small, a floating bladderwort, disjunct in the southern Adirondacks. *New York Fl. Assoc. Newslett.* 9(1): 4.

Greene, H. C. 1945. Notes on Wisconsin parasitic Fungi. VII. *Amer. Midl. Naturalist* 34: 258–270.

Greene, H. C. 1948. Notes on Wisconsin parasitic Fungi. X. *Amer. Midl. Naturalist* 39: 444–456.

Greene, H. C. 1949. Notes on Wisconsin parasitic Fungi. XIII. *Amer. Midl. Naturalist* 41: 740–758.

Greene, H. C. 1953. Notes on Wisconsin parasitic Fungi. XIX. *Amer. Midl. Naturalist* 50: 501–508.

Greene, H. C. 1956. Notes on Wisconsin parasitic Fungi. XXI. *Trans. Wisconsin Acad. Sci.* 44: 29–43.

Greene, H. C. & J. T. Curtis. 1950. Germination studies of Wisconsin prairie plants. *Amer. Midl. Naturalist* 43: 186–194.

Greenfield, M. D. & M. G. Karandinos. 1979. Resource partitioning of the sex communication channel in clearwing moths (Lepidoptera: Sesiidae) of Wisconsin. *Ecol. Monogr.* 49: 403–426.

Greilhuber, J., T. Borsch, K. Müller, A. Worberg, S. Porembski & W. Barthlott. 2006. Smallest angiosperm genomes found in Lentibulariaceae, with chromosomes of bacterial size. *Plant Biol.* 8: 770–777.

Greimler, J., B. Hermanowski & C.-G. Jang. 2004. A re-evaluation of morphological characters in European *Gentianella* section *Gentianella* (Gentianaceae). *Plant Syst. Evol.* 248: 143–169.

Greller, A. M., D. C. Locke, V. Kilanowski & G. E. Lotowycz. 1990. Changes in vegetation composition and soil acidity between 1922 and 1985 at a site on the north shore of Long Island, New York. *Bull. Torrey Bot. Club* 117: 450–458.

Greulich, S. & G. Bornette. 2003. Being evergreen in an aquatic habitat with attenuated seasonal contrasts: a major competitive advantage? *Plant Ecol.* 167: 9–18.

Grevenstuk, T., S. Gonçalves, J. M. F. Nogueira, M. G. Bernardo-Gil & A. Romano. 2012a. Recovery of high purity plumbagin from *Drosera intermedia*. *Industr. Crops Prod.* 35: 257–260.

Grevenstuk, T., S. Gonçalves, T. Domingos, C. Quintas, J. J. van der Hooft, J. Vervoort & A. Romano. 2012b. Inhibitory activity of plumbagin produced by *Drosera intermedia* on food spoilage fungi. *J. Sci. Food Agric.* 92: 1638–1642.

Grewell, B. J., M. A. DaPrato, P. R. Hyde & E. Rejmankova. 2003. *Experimental Reintroduction of Endangered Soft Bird's Beak to Restored Habitat in Suisun Marsh*. Final report submitted to CALFED Ecosystem Restoration Project 99-N05. University of California, Davis, CA.

Grewell, B. J. 2004. Species diversity in northern California salt marshes: functional significance of parasitic plant interactions. PhD dissertation. University of California, Davis, CA. 143 pp.

Grewell, B. J. 2005. *Population Census and Status of the Endangered Soft Bird's-Beak* (Cordylanthus mollis ssp. mollis) *at Benicia State Recreation Area and Rush Ranch in Solano County, California*. Final Report prepared for the Solano County Water Agency, Vacaville, CA. 69 pp.

Grewell, B. J. 2008. Parasite facilitates plant species coexistence in a coastal wetland. *Ecology* 89: 1481–1488.

Gridling, M., N. Stark, S. Madlener, A. Lackner, R. Popescu, B. Benedek, R. Diaz, F. M. Tut, T. P. N. Vo, D. Huber, M. Gollinger, P. Saiko, A. Ozmen, W. Mosgoeller, R. De Martin, R. Eytner, K.-H. Wagner, M. Grusch, M. Fritzer-Szekeres, T. Szekeres, B. Kopp, R. Frisch & G. Krupitza. 2009. *In vitro* anti-cancer activity of two ethno-pharmacological healing plants from Guatemala *Pluchea odorata* and *Phlebodium decumanum*. *Int. J. Oncol.* 34: 1117–1128.

Grieve, M. 1980. *A Modern Herbal: The Medicinal, Culinary, Cosmetic and Economic Properties, Cultivation and Folklore of Herbs, Grasses, Fungi, Shrubs and Trees with all their Modern Scientific Uses.* Penguin Books, Harmondsworth, UK. 912 pp.

Griffin, D. 2009. The ethnobiology of the central Yup'ik Eskimo, southwestern Alaska. *Alaska J. Anthropol.* 7: 81–100.

Griffin, III, D. & D. A. Breil. 1982. Notes on *Frullania cobrensis* Gott. ex Steph. in Cuba & Florida. *Bryologist* 85: 438–441.

Griffith, A. B. 2014. Secondary dispersal in *Aeschynomene virginica*: do floating seeds really find a new home? *Nat. Areas J.* 34: 488–494.

Griffith, A. B. & I. N. Forseth. 2002. Primary and secondary seed dispersal of a rare, tidal wetland annual, *Aeschynomene virginica. Wetlands* 22: 696–704.

Griffith, A. B. & I. N. Forseth. 2003. Establishment and reproduction of *Aeschynomene virginica* (L.) (Fabaceae) a rare, annual, wetland species in relation to vegetation removal and water level. *Plant Ecol.* 167: 117–125.

Griffith, A. B. & I. N. Forseth. 2005. Population matrix models of *Aeschynomene virginica*, a rare annual plant: implications for conservation. *Ecol. Appl.* 15: 222–233.

Griffith, A. B. & I. N. Forseth. 2006. The role of a seed bank in establishment and persistence of *Aeschynomene virginica*, a rare wetland annual. *N. E. Naturalist* 13: 235–246.

Griffiths, G. C. D. & M. D. Piercey-Normore. 1995. A new Agromyzid (Diptera) leaf-miner of mountain holly (*Nemopanthus*, Aquifoliaceae) from the Avalon Peninsula, Newfoundland. *Can. Field Naturalist* 109: 23–26.

Griggs, R. F. 1936. The vegetation of the Katmai district. *Ecology* 17: 380–417.

Grime, J. P. & R. Hunt. 1975. Relative growth-rate: its range and adaptive significance in a local flora. *J. Ecol.* 63: 393–422.

Grime, J. P., G. Mason, A. V. Curtis, J. Rodman & S. R. Band. 1981. A comparative study of germination characteristics in a local flora. *J. Ecol.* 69: 1017–1059.

Grittinger, T. F. 1970. String bog in southern Wisconsin. *Ecology* 51: 928–930.

Grombone-Guaratini, M. T., V. N. Solferini & J. Semir. 2004. Reproductive biology in species of *Bidens* L. (Asteraceae). *Sci. Agric. (Piracicaba)* 6: 185–189.

Grondin, P. & J. Ouzilleau. 1980. Les tourbières du sud de la Jamésie, Québec. *Géogr. Phys. Quatern.* 34: 267–299.

Gronquist, M., A. Bezzerides, A. Attygalle, J. Meinwald, M. Eisner & T. Eisner. 2001. Attractive and defensive functions of the ultraviolet pigments of a flower (*Hypericum calycinum*). *Proc. Natl. Acad. Sci. U.S.A.* 98: 13745–13750.

Gross, E. M. 1999. Allelopathy in benthic and littoral areas: case studies on allelochemicals from benthic Cyanobacteria and submersed macrophytes. pp. 179–199. *In*: Inderjit, K. M. M. Dakshini & C. L. Foy (eds.), *Principles and Practices in Plant Ecology: Allelochemical Interactions.* CRC Press, Boca Raton, FL.

Gross, K. L. 1990. A comparison of methods for estimating seed numbers in the soil. *J. Ecol.* 78: 1079–1093.

Gross, R. S. & P. A. Werner 1983. Relationships among flowering phenology, insect visitors, and seed-set of individuals: experimental studies on four co-occurring species of goldenrod (Solidago: Compositae). *Ecol. Monogr.* 53: 95–117.

Grøstad, T., T. H. Melseth & R. Halvorsen. 1999. Vandreveronika *Veronica peregrina* i Norge. Tre nyfunn fra Vestfold. *Blyttia* 57: 132–137.

Groth, A. T., L. Lovett-Doust & J. Lovett-Doust. 1996. Population density and module demography in *Trapa natans* (Trapaceae), an annual, clonal aquatic macrophyte. *Amer. J. Bot.* 83: 1406–1415.

Groves, J. W. 1937. Three Dermateaceae occurring on *Nemopanthus. Mycologia* 29: 66–80.

Groves, J. W. 1940. Some *Dermatea* species and their conidial stages. *Mycologia* 32: 736–751.

Groves, J. W. 1947. *Pezicula morthieri* on *Rhamnus. Mycologia* 39: 328–333.

Grubben, G. J. H. 2004. *Ipomoea aquatica* Forssk. [Internet] Record from Protabase. G. J. H. Grubben & O. A. Denton (eds.), *PROTA (Plant Resources of Tropical Africa/Ressources végétales de l'Afrique tropicale)*, Wageningen, Netherlands. Available online: http://database.prota.org/search.htm (accessed May 13, 2008).

Grubben, G. J. H. & O. A. Denton (eds.). 2004. *Plant Resources of Tropical Africa 2. Vegetables.* PROTA Foundation, Wageningen, Netherlands/Backhuys Publishers, Leiden, Netherlands/CTA, Wageningen, The Netherlands. 668 pp.

Gruberová, H., K. Bendová & K. Prach. 2001. Seed ecology of alien *Bidens frondosa* in comparison with native species of the genus. pp. 99–104. *In*: G. Brundu, J. Brock, I. Camarda, L. Child & M. Wade (eds.), *Plant Invasions: Species Ecology and Ecosystem Management.* Backhuys Publishers, Leiden, The Netherlands.

Gruchy, J. H. B. de. 1938. *A Preliminary Study of the Larger Aquatic Plants of Oklahoma with Special Reference to their Value in Fish Culture.* Technical Bulletin No. 4. Stillwater Agricultural and Mechanical College, Agricultural Experiment Station, Stillwater, OK. 31 pp.

Grundel, R., R. P. Jean, K. J. Frohnapple, J. Gibbs, G. A. Glowacki & N. B. Pavlovic. 2011. A survey of bees (Hymenoptera: Apoidea) of the Indiana Dunes and northwest Indiana, USA. *J. Kansas Entomol. Soc.* 84: 105–138.

Gu, C., L. Wang, L. Zhang, Y. Liu, M. Yang, Z. Yuan, S. Li & Y. Han. 2013. Characterization of genes encoding granule-bound starch synthase in sacred lotus reveals phylogenetic affinity of *Nelumbo* to Proteales. *Plant Mol. Biol. Rep.* 31: 1157–1165.

Guard, B. J. 1995. *Wetland Plants of Oregon & Washington.* Lone Pine Publishing, Redmond, WA 239 pp.

Guarino, L. (ed.). 1997. *Traditional African Vegetables. Promoting the Conservation and Use of Underutilized and Neglected Crops.* 16. Proceedings of the IPGRI International Workshop on Genetic Resources of Traditional Vegetables in Africa: Conservation and Use, 29–31 August 1995, ICRAF-HQ, Nairobi, Kenya. Institute of Plant Genetics and Crop Plant Research, Gatersleben/International Plant Genetic Resources Institute, Rome, Italy.

Guarrera, P. M. 2003. Food medicine and minor nourishment in the folk traditions of Central Italy (Marche, Abruzzo and Latium). *Fitoterapia* 74: 515–544.

Gucker, C. L. 2006. *Ledum groenlandicum. In*: *Fire Effects Information System.* U. S. Department of Agriculture, Forest Service, Rocky Mountain Research Station, Fire Sciences Laboratory (Producer). Available online: http://www.fs.fed.us/database/feis/ (accessed March 14, 2006).

Guerke, W. R. 1974. A floristic study of the Hepaticae and Anthocerotae of the Florida parishes, Louisiana. *Bryologist* 77: 593–600.

Guerrant Jr., E. O. & A. Raven. 1998. Seed germination and storability studies of 69 plant taxa native to the Willamette Valley wet prairie. pp. 25–31. *In*: R. Rose & D. L. Haase (eds.), *Native Plants: Propagating and Planting.* Oregon State University, Corvallis, OR.

Guertin, S. 2013. Endangered and threatened wildlife and plants; endangered status for *Physaria globosa* (Short's bladder-pod), *Helianthus verticillatus* (whorled sunflower), and *Leavenworthia crassa* (fleshy-fruit gladecress). *Fed. Reg.* 78: 47109–47134.

Guggisberg, A., G. Mansion, S. Kelso & E. Conti. 2006. Evolution of biogeographic patterns, ploidy levels, and breeding systems in a diploid–polyploid species complex of *Primula*. *New Phytol.* 171: 617–632.

Gui, S., Z. Wu, H. Zhang, Y. Zheng, Z. Zhu, D. Liang & Y. Ding. 2016. The mitochondrial genome map of *Nelumbo nucifera* reveals ancient evolutionary features. *Sci. Rep.* 6: 30158. doi:10.1038/srep30158.

Guisande, C., C. Andrade, C. Granado-Lorencio, S. R. Duque & M. Núñez-Avellaneda. 2000. Effects of zooplankton and conductivity on tropical *Utricularia foliosa* investment in carnivory. *Aquat. Ecol.* 34: 137–142.

Gul, B. & M. A. Khan. 2003. Effect of growth regulators and osmotica in alleviating salinity effects on the germination of *Salicornia utahensis*. *Pakistan J. Bot.* 35: 877–886.

Gul, B. & D. J. Weber. 2001. Seed bank dynamics in a Great Basin salt playa. *J. Arid Environ.* 49: 785–794.

Gul, B., R. Ansari & M. A. Khan. 2009. Salt tolerance of *Salicornia utahensis* from the great basin desert. *Pakistan J. Bot.* 41: 2925–2932.

Gul, B., R. Ansari, T. J. Flowers & M. A. Khan. 2013. Germination strategies of halophyte seeds under salinity. *Environ. Exp. Bot.* 92: 4–18.

Gullion, G. W. 1962. Organization and movements of coveys of a Gambel Quail population. *Condor* 64: 402–415.

Gulya, T. J., G. J. Seiler, G. Kong & L. F. Marek. 2007. Exploration and collection of rare *Helianthus* species from Southeastern United States. *Helia* 30: 13–24.

Gumbert, A. & J. Kunze. 1999. Inflorescence height affects visitation behavior of bees: a case study of an aquatic plant community in Bolivia. *Biotropica* 31: 466–477.

Gunther, E. 1973. *Ethnobotany of Western Washington*. University of Washington Press, Seattle, WA. 74 pp.

Guntli, D., S. Burgos, Y. Moënne-Loccoz & G. Défago. 1999. Calystegine degradation capacities of microbial rhizosphere communities of *Zea mays* (calystegine-negative) and *Calystegia sepium* (calystegine-positive). *Microbiol. Ecol.* 28: 75–84.

Guo, H. & S. C. Pennings. 2012. Post-mortem ecosystem engineering by oysters creates habitat for a rare marsh plant. *Oecologia* 170: 789–798.

Guo, J. Y. & A. D. Bradshaw. 1993. The flow of nutrients and energy through a Chinese farming system. *J. Appl. Ecol.* 30: 86–94.

Guo, W. & Z. H. Hu. 2012. Effects of stolon severing on the expansion of *Alternanthera philoxeroides* from terrestrial to contaminated aquatic habitats. *Plant Species Biol.* 27: 46–52.

Guppy, H. B. 1906. *Obervations of a Naturalist in the Pacific Between 1896 and 1899. Volume II: Plant-Dispersal.* Macmillan & Co., Ltd., New York. 627 pp.

Gupta, D. P. 1959. Vascular anatomy of the flower of *Sphenoclea zeylanica* Gaertn. and some other related species. *Proc. Indian Natl. Sci. Acad. B.* 25: 55–64.

Gupta, P. L. 1992. Seed germination study of *Eclipta prostrata* Linn. *Advances Plant Sci.* 5: 187–189.

Gurnell, A., J. Goodson, K. Thompson, N. Clifford & P. Armitage. 2007a. The river-bed: a dynamic store for plant propagules? *Earth Surf. Processes Landforms* 32: 1257–1272.

Gurnell, A., J. Goodson, K. Thompson, O. Mountford & N. Clifford. 2007b. Three seedling emergence methods in soil seed bank studies: implications for interpretation of propagule deposition in riparian zones. *Seed Sci. Res.* 17: 183–199.

Gustafson, D. J., A. P. Giunta Jr. & C. S. Echt. 2013. Extensive clonal growth and biased sex ratios of an endangered dioecious shrub, *Lindera melissifolia* (Walt) Blume (Lauraceae). *J. Torrey Bot. Club Soc.* 140: 133–144.

Gustafson, F. G. 1954. A study of riboflavin, thiamine, niacin and ascorbic acid content of plants in northern Alaska. *Bull. Torrey Bot. Club* 81: 313–322.

Gustafsson, M. H. G. & K. Bremer. 1995. Morphology and phylogenetic interrelationships of the Asteraceae, Calyceraceae, Campanulaceae, Goodeniaceae, and related families (Asterales). *Amer. J. Bot.* 82: 250–265.

Gustafsson, M. H. G., A. Backlund & B. Bremer. 1996. Phylogeny of the Asterales sensu lato based on *rbcL* sequences with particular reference to the Goodeniaceae. *Plant. Syst. Evol.* 199: 217–242.

Gustafsson, M. H. G., V. Bittrich & P. F. Stevens. 2002. Phylogeny of Clusiaceae based on *rbcL* sequences. *Int. J. Plant Sci.* 163: 1045–1054.

Guthery, F. S. 1975. Food habits of sandhill cranes in southern Texas. *J. Wildl. Manag.* 39: 221–223.

Guttenberg, G. 1881. Notes on the Flora of Presquellsle, Pa. *Bull. Torrey Bot. Club* 8: 28–29.

Guttman, S. I. & L. A. Weigt. 1989. Electrophoretic evidence of relationships among *Quercus* (oaks) of eastern North America. *Can. J. Bot.* 67: 339–351.

Haack, R. A., E. Jendek, H. Liu, K. R. Marchant, T. R. Petrice, T. M. Poland & H. Ye. 2002. The emerald ash borer: a new exotic pest in North America. *Newslett. Michigan Entomol. Soc.* 47(3–4): 1–5.

Haag, K. H., T. M. Lee & D. C. Herndon. 2005. *Bathymetry and Vegetation in Isolated Marsh and Cypress Wetlands in the Northern Tampa Bay Area, 2000–2004*. U. S. Geological Survey Scientific Investigations Report 2005–5109. U. S. Geological Survey, Reston, VA. 49 pp.

Haase, E. F. 1972. Survey of floodplain vegetation along the lower Gila River in southwestern Arizona. *J. Arizona Acad. Sci.* 7: 75–81.

Habeck, D. H. 1996. *Australian Moths for Hydrilla Control*. Technical Report A-96-10. U. S. Army Corps of Engineers Waterways Experiment Station, Vicksburg, MI. 40 pp.

Haber, E. 1979. *Utricularia geminiscapa* at Mer Bleue and range extensions in eastern Canada. *Can. Field Naturalist* 93: 391–398.

Haber, E. 1999. European water chestnut-water chestnut, bull nut, Jesuit nut, water-caltrop. *Trapa natans* L. Water-nut Family – Trapaceae. Invasive Exotic Plants of Canada Fact Sheet No. 13. National Botanical Services, Ottawa, Ontario. Available online: http://24.114.142.233/nbs/ipcan/factnut.html (accessed May 19, 2005).

Haberle, R. C., A. Dang, T. Lee, C. Peñaflor, H. Cortes-Burns, A. Oestreich, L. Raubeson, N. Cellinese, E. J. Edwards, S.-T. Kim, W. M. M. Eddie & R. K. Jansen. 2009. Taxonomic and biogeographic implications of a phylogentic analysis of the Campanulaceae based on three chloroplast genes. *Taxon* 58: 715–734.

Habtemariam, S. 2008. Activity-guided isolation and identification of free radical-scavenging components from ethanolic extract of boneset (leaves of *Eupatorium perfoliatum*). *Nat. Prod. Commun.* 3: 1317–1320.

Habtemariam, S. & A. M. Macpherson. 2000. Cytotoxicity and antibacterial activity of ethanol extract from leaves of a herbal drug, Boneset (*Eupatorium perfoliatum*). *Phytother. Res.* 14: 575–577.

Hadley, E. B. & L. C. Bliss. 1964. Energy relationships of alpine plants of Mt. Washington, New Hampshire. *Ecol. Monogr.* 34: 331–357.

Haegele, E. 2007. Gardening for birds. *Hybrid* (Spring): 1, 3, 7.

Häffner, E. 2000. On the phylogeny of the subtribe Carduinae (tribe Cardueae, Compositae). *Englera* 21: 3–208.

Hagan, J. M., S. Pealer & A. A. Whitman. 2006. Do small headwater streams have a riparian zone defined by plant communities? *Can. J. For. Res.* 36: 2131–2140.

Hagemann, J. M., F. R. Earle, I. A. Wolff & A. S. Barclay. 1967. Search for new industrial oils. XIV. Seed oils of Labiatae. *Lipids* 2: 371–380.

Hagen, K. B. von & J. W. Kadereit. 2001. The phylogeny of *Gentianella* (Gentianaceae) and its colonization of the southern hemisphere as revealed by nuclear and chloroplast DNA sequence variation. *Org. Diversity Evol.* 1: 61–79.

Hagen, K. B. von & J. W. Kadereit. 2002. Phylogeny and flower evolution of the *Swertiinae* (Gentianaceae-*Gentianeae*): homoplasy and the principle of variable proportions. *Syst. Bot.* 27: 548–572.

Hager, H. A. & R. D. Vinebrooke. 2004. Positive relationships between invasive purple loosestrife (*Lythrum salicaria*) and plant species diversity and abundance in Minnesota wetlands. *Can. J. Bot.* 82: 763–773.

Hagman, M. 1971. On self-and cross-incompatibility shown by *Betula verrucosa* Ehrh. and *Betula pubescens* Ehrh. *Commun. Inst. For. Fenn.* 73: 1–125.

Hagy, H. M. & R. M. Kaminski. 2012. Apparent seed use by ducks in moist-soil wetlands of the Mississippi Alluvial Valley. *J. Wildl. Manag.* 76: 1053–1061.

Haines, A. 2000. Clarifying the taxonomy of *Salicornia sensu lato* in the northeastern United States. *Botanical Notes* 2: 1–4.

Haines, A. 2001. Identification and ecology of rare *Chenopodium* in Maine. *Botanical Notes* 5: 1–7.

Haines, A. 2003. The taxonomic status of *Bidens heterodoxa*. *Botanical Notes* 9: 1–3.

Haines, A. 2006. 158. *Euthamia* (Nuttall) Cassini. pp. 97–100. *In*: Flora of North America Editorial Committee (eds.), *Flora of North America North of Mexico: Vol. 20, Magnoliophyta: Asteridae (in part): Asteraceae, Part 2*. Oxford University Press, New York.

Haines, A., A. Cutko & D. Sperduto. 2004. Records for *Carex rostrata* (Cyperaceae) in New England. *Rhodora* 106: 287–290.

Haines, E. B. & C. L. Montague. 1979. Food sources of estuarine invertebrates analyzed using¹³C/¹²C ratios. *Ecology* 60: 48–56.

Hájek, M., L. Tichý, B. S. Schamp, D, Zelený, J. Roleček, P. Hájková, I. Apostolova & D. Dítě. 2007. Testing the species pool hypothesis for mire vegetation: exploring the influence of pH specialists and habitat history. *Oikos* 116: 1311–1322.

Hajra, A. 1987. Biochemical investigations on the protein-calorie availability in grass carp (*Ctenopharyngodon idella* Val.) from an aquatic weed (*Ceratophyllum demersum* Linn.) in the tropics. *Aquaculture* 61: 113–120.

Hakim, M. A., A. S. Juraimi, M. M. Hanafi, A. Selamat, M. R. Ismail & S. M. R. Karim. 2011. Studies on seed germination and growth in weed species of rice field under salinity stress. *J. Environ. Biol.* 32: 529–536.

Halberstein, R. A. 2005. Medicinal plants: historical and cross-cultural usage patterns. *Ann. Epidem.* 15: 686–699.

Halchak, J. L., D. M. Seliskar & J. L. Gallagher. 2011. Root system architecture of *Kosteletzkya pentacarpos* (Malvaceae) and belowground environmental influences on root and aerial growth dynamics. *Amer. J. Bot.* 98: 163–74.

Halıcı, M. G., D. L. Hawksworth, M. Candan & A. Ö. Türk. 2010. A new lichenicolous species of *Capronia* (Ascomycota, Herpotrichiellaceae), with a key to the known lichenicolous species of the genus. *Fungal Diversity* 40: 37–40.

Hall, B. R., D. J. Raynal & D. J. Leopold. 2001. Environmental influences on plant species composition in ground-water seeps in the Catskill Mountains of New York. *Wetlands* 21: 125–134.

Hall, D. W., M. Minno & J. F. Butler. 2007. *Little Metalmark, Calephelis virginiensis (Guérin-Ménéville) (Insecta: Lepidoptera: Riodinidae)*. IFAS Extension Bulletin EENY-407, University of Florida, Gainseville, FL. 4 pp.

Hall, G. B. 2010. *Minimum Levels Reevaluation: Sylvan Lake, Seminole County, Florida*. Technical Publication SJ2010-XX. St. Johns River Water Management District, Palatka, FL. 142 pp.

Hall, H. M. & F. E. Clements. 1923. The phylogenetic method in taxonomy: the North American species of *Artemisia*, *Chrysothamnus*, and *Atriplex*. *Carn. Inst. Wash. Publ. Bot.* 326: 1–355.

Hall, H. M. & F. L. Long. 1921. *Rubber-Content of North American Plants*. Publication number 313. Carnegie Institution of Washington, Washington, DC. 68 pp.

Hall, J., P. Pierce & G. Lawson. 1971. *Common Plants of the Volta Lake*. The University of Ghana, Legon. 123 pp.

Hall, J. C. 2008. Systematics of Capparaceae and Cleomaceae: an evaluation of the generic delimitations of *Capparis* and *Cleome* using plastid DNA sequence data. *Botany* 86: 682–696.

Hall, J. C., K. J. Sytsma & H. H. Iltis. 2002. Phylogeny of Capparaceae and Brassicaceae based on chloroplast sequence data. *Amer. J. Bot.* 89: 1826–1842.

Hall, T. F. 1940. The biology of *Saururus cernuus* L. *Amer. Midl. Naturalist* 24: 253–260.

Hall, T. F. & W. T. Penfound. 1934. The biology of the American lotus, *Nelumbo lutea* (Wild.) Pers. *Amer. Midl. Naturalist* 31: 744–758.

Hall, T. F. & W. T. Penfound. 1939. A phytosociological study of a cypress-gum swamp in southeastern Louisiana. *Amer. Midl. Naturalist* 21: 378–395.

Hallmark, C. T. & C. Morgan. 2010. *Calvert Mine Fieldtrip Guidebook, July 14, 2010, for the Southern Regional Soil Survey Conference*. Texas A&M University, College Station, TX (unpaginated).

Halpern, C. B. & M. E. Harmon. 1983. Early plant succession on the Muddy River mudflow, Mount St. Helens, Washington. *Amer. Midl. Naturalist* 110: 97–106.

Halpern, C. B., S. A. Evans & S. Nielson. 1999. Soil seed banks in young, closed-canopy forests of the Olympic Peninsula, Washington: potential contributions to understory reinitiation. *Can. J. Bot.* 77: 922–935.

Halsted, B. D. 1888. Iowa Peronosporeæ and a dry season. *Bot. Gaz.* 13: 52–59.

Halsted, B. D. 1890. A new white smut. *Bull. Torrey Bot. Club* 17: 95–97.

Halsted, B. D. 1893. Notes upon a new *Exobasidium*. *Bull. Torrey Bot. Club* 20: 437–440.

Halstead, J. M., J. Michaud, S. Hallas-Burt & J. P. Gibbs. 2003. Hedonic analysis of effects of a nonnative invader (*Myriophyllum heterophyllum*) on New Hampshire (USA) lakefront properties. *Environ. Manag.* 32: 391–398.

Hambäck, P. A. & P. Ekerholm. 1997. Mechanisms of apparent competition in seasonal environments: an example with vole herbivory. *Oikos* 80: 276–288.

Hambäck, P. A., J. Ågren & L. Ericson. 2000. Associational resistance: insect damage to purple loosestrife reduced in thickets of sweet gale. *Ecology* 81: 1784–1794.

Hameed, I. & G. Dastagir. 2009. Nutritional analyses of *Rumex hastatus* D. Don, *Rumex dentatus* Linn and *Rumex nepalensis* Spreng. *African J. Biotechnol.* 8: 4131–4133.

Hamer, D. & S. Herrero. 1987. Grizzly bear food and habitat in the front ranges of Banff National Park, Alberta. *Int. Conf. Bear Res. Manag.* 7: 199–213.

Hamerstrom Jr., F. N. & J. Blake. 1939a. Central Wisconsin muskrat study. *Amer. Midl. Naturalist* 21: 514–520.

Hamerstrom Jr., F. N. & J. Blake. 1939b. Winter movements and winter foods of white-tailed deer in central Wisconsin. *J. Mammalogy* 20: 206–215.

Hamilton, A. N. & F. L. Bunnell. 1987. Foraging strategies of coastal grizzly bears in the Kimsquit River valley, British Columbia. *Int. Conf. Bear Res. Manag.* 7: 187–197.

Hamilton, E. H. & L. D. Peterson. 2003. *Response of Vegetation to Burning in a Subalpine Forest Cutblock in Central British Columbia: Otter Creek Site.* Research Report 23. Research Branch, British Columbia Ministry of Forests, Victoria, BC. 60 pp.

Hamilton, G. R. & G. W. Kessinger. 1967. Self-pollination of *Dionaea muscipula*. *Trans. Missouri Acad. Sci.* 1: 10–12.

Hamilton, K. G. & D. W. Langor. 1987. Leafhopper fauna of Newfoundland and Cape Breton Islands (Rhynchota: Homoptera: Cicadellidae). *Can. Entomol.* 119: 663–695.

Hamilton, K. G. A. 2014. Extraordinary endemism: the Nearctic leafhoppers of *Gloridonus* Ball and *Ballana* DeLong (Hemiptera: Cicadellidae). *J. Kansas Entomol. Soc.* 87: 111–204.

Hamilton, M. B. 1996. Relatedness measured by oligonucleotide probe DNA fingerprints and an estimate of the mating system of sea lavender (*Limonium carolinianum*). *Theor. Appl. Genet.* 93: 249–256.

Hamilton, M. B. 1997. Genetic fingerprint-inferred population subdivision and spatial genetic tests for isolation by distance and adaptation in the coastal plant *Limonium carolinianum*. *Evolution* 51: 1457–1468.

Hamilton IV, R., J. W. Reid & R. M. Duffield. 2000. Rare copepod, *Paracyclops canadensis* (Willey), common in leaves of *Sarracenia purpurea* L. *N. E. Naturalist* 7: 17–24.

Hamilton Jr., W. J. 1938. Life history notes on the northern pine mouse. *J. Mammalogy* 19: 163–170.

Hamilton Jr., W. J. 1941. The food of small forest mammals in eastern United States. *J. Mammalogy* 22: 250–263.

Hamm, J. J., S. D. Pair & O. G. Marti, Jr. 1986. Incidence and host range of a new ascovirus isolated from fall armyworm, *Spodoptera frugiperda* (Lepidoptera: Noctuidae). *Florida Entomol.* 69: 524–531.

Hammer, R. L. 2002. *Everglades Wildflowers*. Globe Pequot Press, Guilford, CT. 256 pp.

Hammer, R. L. 2004. *Florida Keys Wildflowers: A Field Guide to Wildflowers, Trees, Shrubs, and Woody Vines of the Florida Keys*. Globe Pequot Press, Guilford, CT. 231 pp.

Hammer, U. T. & J. M. Heseltine. 1988. Aquatic macrophytes in saline lakes of the Canadian prairies. *Hydrobiologia* 158: 101–116.

Hammond, M. C. 1943. Beaver on the Lower Souris Refuge. *J. Wildl. Manag.* 7: 316–321.

Hamzeh, M. & S. Dayanandan. 2004. Phylogeny of *Populus* (Salicaceae) based on nucleotide sequences of chloroplast *trnT-trnF* region and nuclear rDNA. *Amer. J. Bot.* 91: 1398–1408.

Han, H.-Y. & B. A. McPheron. 1994. Phylogenetic study of selected tephritid flies (Insecta: Diptera: Tephritidae) using partial sequences of the nuclear 18S ribosomal DNA. *Biochem. Syst. Ecol.* 22: 447–457.

Han, H.-Y. & A. L. Norrbom. 2005. A systematic revision of the New World species of *Trypeta* Meigen (Diptera: Tephritidae). *Syst. Entomol.* 30: 208–247.

Han, M. & J. G. Kim. 2006. Water-holding capacity of a floating peat mat determines the survival and growth of *Menyanthes trifoliata* L. (bog bean) in an oligotrophic lake. *J. Plant Biol.* 49: 102–105.

Han, Y. C., C. Z. Teng, F. H. Chang, G. W. Robert, M. Q. Zhou, Z. L. Hu & Y. C. Song. 2007. Analyses of genetic relationships in *Nelumbo nucifera* using nuclear ribosomal ITS sequence data, ISSR and RAPD markers. *Aquat. Bot.* 87: 141–146.

Han, Y.-Q., Z.-M. Huang, X.-B. Yang, H.-Z. Liu & G.-X. Wu. 2008. *In vivo* and *in vitro* anti-hepatitis B virus activity of total phenolics from *Oenanthe javanica*. *J. Ethnopharmacol.* 118: 148–153.

Hanberry, P., B. B. Hanberry, S. Demarais, B. D. Leopold & J. Fleeman. 2014. Impact on plant communities by white-tailed deer in Mississippi, USA. *Plant Ecol. Diversity* 7: 541–548.

Hancock, B. L. 1942. Cytological and ecological notes on some species of *Galium* L. em. Scop. *New Phytol.* 41: 70–78.

Handa, I. T., R. Harmsen & R. L. Jefferies. 2002. Patterns of vegetation change and the recovery potential of degraded areas in a coastal marsh system of the Hudson Bay lowlands. *J. Ecol.* 90: 86–99.

Handayani, E. S. & Z. S. Nugraha. 2016. Soursop leaf extract increases neuroglia and hepatic degeneration in female rats. *Universa Med.* 34: 17–24.

Hanley, T. A. & J. D. McKendrick. 1985. Potential nutritional limitations for black-tailed deer in a spruce-hemlock forest, southeastern Alaska. *J. Wildl. Manag.* 49: 103–114.

Hanley, T. A. & R. D. Taber. 1980. Selective plant species inhibition by elk and deer in three conifer communities in Western Washington. *For. Sci.* 26: 97–107.

Hanlon, T. J., C. E. Williams & W. J. Moriarity. 1998. Species composition of soil seed banks of Allegheny Plateau riparian forests. *J. Torrey Bot. Soc.* 125: 199–215.

Hanna, W. F. 1938. The discharge of conidia in species of *Entyloma*. *Mycologia* 30: 526–536.

Hansen, B. F. & R. P. Wunderlin. 1986. *Pentalinon* Voigt, an earlier name for *Urechites* Müll. Arg. (Apocynaceae). *Taxon* 35: 166–168.

Hansen, D. R., S. G. Dastidar, Z. Cai, C. Penaflor, J. V. Kuehl, J. L. Boore & R. K. Jansen. 2007. Phylogenetic and evolutionary implications of complete chloroplast genome sequences of four early-diverging angiosperms: *Buxus* (Buxaceae), *Chloranthus* (Chloranthaceae), *Dioscorea* (Dioscoreaceae), and *Illicium* (Schisandraceae). *Mol. Phylogen. Evol.* 45: 547–563.

Hanson, H. C. 1951. Characteristics of some grassland, marsh, and other plant communities in western Alaska. *Ecol. Monogr.* 21: 317–378.

Hao, G., R. M. K. Saunders & M.-L. Chye. 2000. A phylogenetic analysis of the Illiciaceae based on sequences of internal transcribed spacers (ITS) of nuclear ribosomal DNA. *Plant Syst. Evol.* 223: 81–90.

Hao, G., Y.-M. Yuan, C.-M. Hu, X.-J. Ge & N.-X. Zhao. 2004. Molecular phylogeny of *Lysimachia* (Myrsinaceae) based on chloroplast *trnL*–F and nuclear ribosomal ITS sequences. *Mol. Phylogen. Evol.* 31: 323–339.

Hao, J. H., S. Qiang, T. Chrobock, M. van Kleunen & Q. Q. Liu. 2011. A test of baker's law: breeding systems of invasive species of Asteraceae in China. *Biol. Invas.* 13: 571–580.

Hapner, J. A. & J. A. Reinartz. 2005. Vegetation of the Ulao Swamp, a disturbed harwood-conifer swamp in southeastern Wisconsin. *Fieldstation Bull.* 31: 1–48.

Haraguchi, A. 1993. Phenotypic and phenological plasticity of an aquatic macrophyte *Menyanthes trifoliata* L. *J. Plant Res.* 106: 31–35.

Haraguchi, A. 1996. Rhizome growth of *Menyanthes trifoliata* L. in a population on a floating peat mat in Mizorogaike Pond, central Japan. *Aquat. Bot.* 53: 163–173.

Haraguchi, A. 1999. Seed production of *Menyanthes trifoliata* inside and outside a *Phragmites australis* canopy. *J. Fac. Agric. Hokkaido Univ.* 69: 27–30.

Harborne, J. B. 1975. Flavonoid sulphates: a new class of sulphur compounds in higher plants. *Phytochemistry* 14: 1147–1155.

Harborne, J. B. & H. Baxter. 1993. *Phytochemical Dictionary-A Handbook of Bioactive Compounds from Plants.* CRC Press, Boca Raton, FL. 791 pp.

Hardcastle, E. L. & D. X. Williams. 2001. A status report on Harperella, *Ptilimnium nodosum* (Rose) Mathias, in Arkansas. *J. Arkansas Acad. Sci.* 55: 177–178.

Hardee, D. D., G. D. Jones & L. C. Adams. 1999. Emergence, movement, and host plants of boll weevils (Coleoptera: Curculionidae) in the Delta of Mississippi. *J. Econ. Entomol.* 92: 130–139.

Harder, L. D. 1985. Morphology as a predictor of flower choice by bumble bees. *Ecology* 66: 198–210.

Hardie, W. J. & T. P. O'Brien. 1988. Considerations of the biological significance of some volatile constituents of grape (*Vitis* spp.). *Austral. J. Bot.* 36: 107–117.

Hardig, T. M., S. J. Brunsfeld, R. S. Fritz, M. Morgan & C. M. Orians. 2000. Morphological and molecular evidence for hybridization and introgression in a willow (*Salix*) hybrid zone. *Mol. Ecol.* 9: 9–24.

Hardig, T. M., C. K. Anttila & S. J. Brunsfeld. 2010. A phylogenetic analysis of *Salix* (Salicaceae) based on *matK* and ribosomal DNA sequence data. *J. Bot.* 2010: 1–12. doi:10.1155/2010/197696.

Hardikar, S. W. 1922. On *Rhododendron* poisoning. *J. Pharm. Exp. Therap.* 20: 17–44.

Hardin, J. W. & R. L. Beckmann. 1982. Atlas of foliar surface features in woody plants, V. *Fraxinus* (Oleaceae) of eastern North America. *Brittonia* 34: 129–140.

Hardway, T. M., K. Spalik, M. F. Watson, D. S. Katz-Downie & S. R. Downie. 2004. Circumscription of Apiaceae tribe Oenantheae. *S. Afr. J. Bot.* 70: 393–406.

Hardy, L. M. & L. R. Raymond. 1991. Observations on the activity of the pickerel frog, *Rana palustris* (Anura: Ranidae), in northern Louisiana. *J. Herpetol.* 25: 220–222.

Harley, J. L. & E. L. Harley. 1987. A check-list of mycorrhiza in the British flora. *New Phytol.* 105: S1–S102.

Harley, R. M. 1983. *Hyptis alata*, amphitropically disjunct in the Americas. Notes on new world Labiatae. V. *Kew Bull.* 38: 47–52.

Harley, R. M. & C. A. Brighton. 1977. Chromosome numbers in the genus *Mentha*. *J. Linn. Soc. Bot.* 74: 71–96.

Harley, R. M. & A. P. Granda. 2000. List of species of tropical American *Clinopodium* (*Labiatae*), with new combinations. *Kew Bull.* 55: 917–927.

Harley, R. M., S. Atkins, A. L. Budantsev, P. D. Cantino, B. J. Conn, R. Grayer, M. M. Harley, R. de Kok, T. Krestovskaja, R. Morales, A. J. Paton, O. Ryding & T. Upson. 2004. Labiatae. pp. 167–275. *In*: J. W. Kadereit (ed.), *The Families and Genera of Vascular Plants, Vol. VII. Flowering Plants, Dicotyledons: Lamiales (except Acanthaceae including Avicenniaceae).* Springer-Verlag, Berlin, Germany.

Harmaja, H. 2002. *Rhododendron subulatum*, comb. nova (Ericaceae). *Ann. Bot. Fenn.* 39: 183–184.

Harmon, J. M. & J. F. Franklin. 1995. *Seed Rain and Seed Bank of Third-and Fifth-Order Streams on the Western Slope of the Cascade Range.* Research Paper PNW-RP-480. Forest Service, United States Department of Agriculture, Pacific Northwest Research Station, Corvallis, OR. 27 pp.

Harms, N. E. & M. J. Grodowitz. 2009. Insect herbivores of aquatic and wetland plants in the United States: a checklist from literature. *J. Aquat. Plant Manag.* 47: 73–96.

Harms, S. 1999. Prey selection in three species of the carnivorous aquatic plant *Utricularia* (bladderwort). *Archiv Hydrobiol.* 146: 449–470.

Harms, V. L. 1978. The white floating marsh marigold, *Caltha natans*, in Saskatchewan. *Blue Jay* 36: 186–188.

Harms, W. R. 1990. *Fraxinus profunda* (Bush) Bush. Pumpkin Ash. pp. 256–281. *In*: R. M. Burns & B. H. Honkala (eds.), *Silvics of North America. Vol. 2, Hardwoods.* Agriculture Handbook 654. USDA, Forest Service, Washington, DC.

Harney, D. J., E. Mager, E. Cobbins & A. J. Huerta. 1998. Effects of sewage sludge on plant and soil heavy metal content. *Ohio J. Sci.* 98: A40.

Harper, F. 1920. The Florida water-rat (*Neofiber alleni*) in Okefinokee Swamp, Georgia. *J. Mammalogy* 1: 65–66.

Harper, F. 1935. Records of amphibians in the southeastern states. *Amer. Midl. Naturalist* 16: 275–310.

Harper, F. 1937. A season with Holbrook's chorus frog (*Pseudacris ornata*). *Amer. Midl. Naturalist* 18: 260–272.

Harper, K. A. & S. E. MacDonald. 2001. Structure and composition of riparian boreal forest: new methods for analyzing edge influence. *Ecology* 82: 649–659.

Harper, M. G., A.-M. Trame & M. G. Hohmann. 1998. *Management of Herbaceous Seeps and Wet Savannas for Threatened and Endangered Species.* USACERL Technical Report 98/70. U. S. Army Corps of Engineers, Construction Engineering Research Laboratories, Champaign, IL. 84 pp.

Harper, R. M. 1900. Notes on the flora of south Georgia. *Bull. Torrey Bot. Club* 27: 413–436.

Harper, R. M. 1903. Botanical explorations in Georgia during the summer of 1901.-II. Noteworthy species. *Bull. Torrey Bot. Club* 30: 319–342.

Harper, R. M. 1904. Explorations in the Coastal Plain of Georgia during the season of 1902. *Bull. Torrey Bot. Club* 31: 9–27.

Harper, R. M. 1910. A botanical and geological trip on the Warrior and Tombigbee rivers in the coastal plain of Alabama. *Bull. Torrey Bot. Club* 37: 107–126.

Harper, R. M. 1914. The aquatic vegetation of Squaw Shoals, Tuscaloosa County, Alabama. *Torreya* 14: 149–155.

Harper, R. M. 1922. Some pine-barren bogs in central Alabama. *Torreya* 22: 57–60.

Harper, R. M. 1932. *Erigenia bulbosa* and some associated and related plants in Alabama. *Torreya* 32: 141–146.

Harper, R. M. 1944. Notes on *Plantago*, with special reference to *P. cordata*. *Castanea* 9: 121–130.

Harper, R. M. 1945. *Plantago cordata*: a supplementary note. *Castanea* 10: 54.

Harrington, J. 1986. Influences on forb species' visibility. pp. 256–261. *In*: G. K. Clambey and R. H. Pemble (eds.), *Proceedings of the Ninth North American Prairie Conference.* Tri-College University Center for Environmental Studies, Fargo, ND.

Harrington, M. G., K. J. Edwards, S. A. Johnson, M. W. Chase & P. A. Gadek. 2005. Phylogenetic inference in Sapindaceae sensu lato using plastid *matK* and *rbcL* DNA sequences. *Syst. Bot.* 30: 366–382.

Harris, A. G. & J. K. Marr. 2009. *Caltha natans* Pallas (Ranunculaceae) new for Michigan and Thunder Bay District, Ontario. *Michigan Bot.* 48: 72–77.

Harris, B. D. 1968. Chromosome numbers and evolution in North American species of *Linum*. *Amer. J. Bot.* 55: 1197–1204.

Harris, C. S., A. Cuerrier, E. Lamont, P. S. Haddad, J. T. Arnason, S. A. L. Bennett & T. Johns. 2014. Investigating wild berries as a dietary approach to reducing the formation of advanced glycation endproducts: chemical correlates of in vitro antiglycation activity. *Plant Foods Human Nutr.* 69: 71–77.

Harris, S. A., S. C. Maberly & R. J. Abbott. 1992. Genetic variation within and between populations of *Myriophyllum alterniflorum* DC. *Aquat. Bot.* 44: 1–21.

Harrison Jr., A. J. & R. H. Chabreck. 1988. Duck food production in openings in forested wetlands. pp. 339–352. *In*: M. W. Weller (ed.), *Waterfowl in Winter*. University of Minnesota Press, Minneapolis, MN.

Harrison, J. W. & W. M. Knapp. 2010. *Ecological Classification of Groundwater-Fed Seepage Wetlands of the Maryland Coastal Plain*. Maryland Department of Natural Resources, Wildlife and Heritage Service, Natural Heritage Program, Annapolis, MD. 100 pp.

Harrison, S. 1999. Local and regional diversity in a patchy landscape: native, alien, and endemic herbs on serpentine. *Ecology* 80: 70–80.

Harrison, S., J. Maron & G. Huxel. 2000. Regional turnover and fluctuation in populations of five plants confined to serpentine seeps. *Conserv. Biol.* 14: 769–779.

Harrod, J. J. 1964. The distribution of invertebrates on submerged aquatic plants in a chalk stream. *J. Anim. Ecol.* 33: 335–348.

Harshberger, J. W. 1909. The vegetation of the salt marshes and of the salt and fresh water ponds of northern coastal New Jersey. *Proc. Acad. Nat. Sci. Philadelphia* 61: 373–400.

Harshberger, J. W. 1929. Preliminary notes on American snow patches and their plants. *Ecology* 10: 275–281.

Hart, K. H. & P. A. Cox. 1995. Dispersal ecology of *Nuphar luteum* (L.) Sibthorp and Smith: abiotic seed dispersal mechanisms. *Bot. J. Linn. Soc.* 119: 87–100.

Hart, T. W. & W. H. Eshbaugh. 1976. The biosystematics of *Cardamine bulbosa* (Muhlenb.) B.S.P. and *C. douglassii* Britt. *Rhodora* 78: 329–419.

Hartman, R. L. & D. J. Bogler. 2006. 198. *Arida* (R. L. Hartman) D. R. Morgan & R. L. Hartman. pp. 401–405. *In*: Flora of North America Editorial Committee (eds.), *Flora of North America North of Mexico: Vol. 20, Magnoliophyta: Asteridae (in part): Asteraceae, Part 2*. Oxford University Press, New York.

Hartman, R. L., R. K. Rabeler & F. H. Utech. 2005. 14. *Arenaria* Linnaeus. pp. 51–56. *In*: Flora North America Editorial Committee (eds.), *Flora of North America North of Mexico*, Vol. 5: Magnoliophyta: Caryophyllidae, part 2. Oxford University Press, New York.

Hartwell, J. L. 1967. Plants used against cancer. A survey. *Lloydia* 30: 379–437.

Hartwell, J. L. 1971. Plants used against cancer. A survey. *Lloydia* 34: 204–255.

Hartzler, R. G., B. A. Battles & D. Nordby. 2004. Effect of common waterhemp emergence date on growth and fecundity in soybean. *Weed Sci.* 52: 242–245.

Harvey, C. A. 2000. Windbreaks enhance seed dispersal into agricultural landscapes in Monteverde, Costa Rica. *Ecol. Appl.* 10: 155–173.

Harvey, L. R. H. 1903. An ecological excursion to Mount Ktaadn. *Rhodora* 5: 41–52.

Harwood, E. & M. Sytsma. 2003. Risk assessment for Chinese water spinach (*Ipomoea aquatica*) in Oregon. Center for Lakes and Reservoirs, Portland State University, Portland, OR. 9 pp.

Hasan, S. M. R., M. Jamila, M. M. Majumder, R. Akter, M. M. Hossain, M. E. H. Mazumder, M. A. Alam, R. Jahangir, M. S. Rana, M. Arif & S. Rahman. 2009. Analgesic and antioxidant activity of the hydromethanolic extract of *Mikania scandens* (L.) Willd. leaves. *Amer. J. Pharmacol. Toxicol.* 4: 1–7.

Hasebe, M., T. Ando & K. Iwatsuki. 1998. Intrageneric relationships of maple trees based on the chloroplast DNA restriction fragment length polymorphisms. *J. Plant Res.* 111: 441–451.

Hasegawa, G. R. 2000. Pharmacy in the American Civil War. *Amer. J. Health-System Pharm.* 57: 475–489.

Hasegawa, G. R. 2007. Quinine substitutes in the Confederate army. *Military Med.* 172: 650–655.

Hasegawa, G. R. & F. T. Hambrecht. 2003. The Confederate medical laboratories. *S. Med. J.* 96: 1221–1230.

Hashimoto, T. 2002. Molecular genetic analysis of left-right handedness in plants. *Philos. Trans. Biol. Sci.* 357: 799–808.

Haskins, M. L. & W. J. Hayden. 1987. Anatomy and affinities of *Penthorum*. *Amer. J. Bot.* 74: 162–175.

Haslam, S., C. Sinker & P. Wolseley. 1975. British water plants. *Field Studies* 4(2): 243–351.

Haslam, S. M. 1978. *River Plants: The Macrophytic Vegetation of Watercourses*. Cambridge University Press, Cambridge, UK. 396 pp.

Haslam, S. M. 1987. *River Plants of Western Europe: The Macrophytic Vegetation of Watercourses of the European Economic Community*. Cambridge University Press, Cambridge, UK. 512 pp.

Haslam, S. M. & P. A. Wolseley. 1981. *River Vegetation: Its Identification, Assessment and Management. A Field Guide to the Macrophytic Vegetation of British Watercourses*. Cambridge University Press, Cambridge, UK. 154 pp.

Hassan, N. S., J. Thiede & S. Liede-Schumann. 2005. Phylogenetic analysis of Sesuvioideae (Aizoaceae) inferred from nrDNA internal transcribed spacer (ITS) sequences and morphological data. *Plant Syst. Evol.* 255: 121–143.

Hatch, S. L. & A. T. Slack. 2008. *Lilaeopsis carolinensis* (Apiaceae) a species new to Texas and a key to *Lilaeopsis* in Texas. *J. Bot. Res. Inst. Texas.* 2: 1497–1498.

Hatcher, C. R. & A. G. Hart. 2014. Venus flytrap seedlings show growth-related prey size specificity. *Int. J. Ecol.* doi:10.1155/2014/135207.

Haukos, D. A. & L. M. Smith. 1997. *Common Flora of the Playa Lakes*. Texas Tech University Press, Lubbock, TX. 196 pp.

Haukos, D. A & L. M. Smith. 2001. Temporal emergence patterns of seedlings from playa wetlands. *Wetlands* 21: 274–280.

Hauptli, H., B. D. Webster & S. Jain. 1978. Variation in nutlet morphology of *Limnanthes*. *Amer. J. Bot.* 65: 615–624.

Hauser, E. J. P. 1964. The Rubiaceae of Ohio. *Ohio J. Sci.* 64: 27–35.

Havens, K. 1994. Clonal repeatability of in vitro pollen tube growth rates in *Oenothera organensis* (Onagraceae). *Amer. J. Bot.* 81: 161–165.

Havens, K. & L. F. Delph. 1996. Differential seed maturation uncouples fertilization and siring success in *Oenothera organensis* (Onagraceae). *Heredity* 76: 623–632.

Hawkins, T. K. & E. L. Richards. 1995. A floristic study of two bogs on Crowley's Ridge in Greene County, Arkansas. *Castanea* 60: 233–244.

Hawkins, T. S., N. M. Schiff, T. D. Leininger, E. S. Gardiner, M. S. Devall, P. B. Hamel, A. D. Wilson & K. F. Connor. 2009. Growth and intraspecific competitive abilities of the dioecious *Lindera melissifolia* (Lauraceae) in varied flooding regimes. *J. Torrey Bot. Soc.* 136: 91–101.

Hawkins, T. S., D. A. Skojac Jr., B. R. Lockhart, T. D. Leininger, M. S. Devall & N. M. Schiff. 2009a. Bottomland forests in the lower Mississippi alluvial valley associated with the endangered *Lindera melissifolia*. *Castanea* 74: 105–113.

Hawkins, T. S., D. A. Skojac Jr., N. M. Schiff & T. D. Leininger. 2010. Floristic composition and potential competitors in *Lindera melissifolia* (Lauraceae) colonies in Mississippi with reference to hydrologic regime. *J. Bot. Res. Inst. Texas.* 4: 381–390.

Hawkins, T. S., J. L. Walck & S. N. Hidayati. 2011. Seed ecology of *Lindera melissifolia* (Lauraceae) as it relates to rarity of the species. *J. Torrey Bot. Soc.* 138: 298–307.

Haworth, K. & G. R. McPherson. 1994. Effects of *Quercus emoryi* on herbaceous vegetation in a semi-arid savanna. *Vegetatio* 112: 153–159.

Hayden, E. & T. Hownsell. 2012. Eyelid tongue, stately fen beacon, and a raft floating on bladders: wild orchid safari at Little Soldier's Pond; 22 July 2012. *Sarracenia* 20: 3–7.

Hays, J. F. 2010. *Agalinis flexicaulis* sp. nov. (Orobanchaceae: Lamiales), a new species from northeast Florida. *J. Bot. Res. Inst. Texas.* 4: 1–6.

Hazlett, B. T. 1989. *The Aquatic Vegetation and Flora of Sleeping Bear Dunes National Lakeshore, Benzie and Leelanau counties, Michigan.* Tech. Rep. No. 15. University of Michigan Biological Station, Douglas Lake, MI. 66 pp.

Hazlett, D. L. 1998. *Vascular Plant Species of the Pawnee National Grassland.* General Technical Report RMRS-GTR-17. U. S. Department of Agriculture, Forest Service, Rocky Mountain Research Station. Fort Collins, CO. 26 pp.

He, Q. & Y.-J. Park. 2013. Evaluation of genetic structure of amaranth accessions from the United States. *Weed Turfgrass Sci.* 2: 230–235.

He, Z., C. Ruan, P. Qin, D. M. Seliskar & J. L. Gallagher. 2003. *Kosteletzkya virginica*, a halophytic species with potential for agroecotechnology in Jiangsu Province, China. *Ecol. Eng.* 21: 271–276.

He, Z-C., J.-Q. Li & H.-C. Wang. 2004. Karyomorphology of *Davidia involucrata* and *Camptotheca acuminata*, with special reference to their systematic positions. *Bot. J. Linn. Soc.* 144: 193–198.

He, Z. F., M. J. Mao, H. Yu, X. M. Wang & H. P. Li. 2008. First report of a strain of *Alternanthera* yellow vein virus infecting *Eclipta prostrata* (L.) L. (Compositae) in China. *J. Phytopathol.* 156: 496–498.

Healy, A. J. 1944. Some additions to the naturalised flora of New Zealand. *Trans. Proc. Roy. Soc. New Zealand* 74: 221–231.

Healy, A. J. 1946. Contributions to a knowledge of the naturalized flora of New Zealand: No. 1. *Trans. Roy. Soc. New Zealand* 75: 399–404.

Heard, S. B. 1998. Capture rates of invertebrate prey by the pitcher plant, *Sarracenia purpurea* L. *Amer. Midl. Naturalist* 139: 79–89.

Heatwole, D. W. 2004. Insect-habitat assocations in salt marshes of northern Puget Sound: implications of tidal restriction and predicted response to restoration. MS thesis. University of Washington, Seattle, WA. 115 pp.

Heaven, J. B., F. E. Gross & A. T. Gannon. 2003. Comparison of a natural and a created emergent marsh wetland. *S. E. Naturalist* 2: 195–206.

Hébert, R., C. Samson & J. Huot. 2008. Factors influencing the abundance of berry plants for black bears, *Ursus americanus*, in Quebec. *Can. Field Naturalist* 122: 212–220.

Hecht, S. S., S. G. Carmella & S. E. Murphy. 1999. Effects of watercress consumption on urinary metabolites of nicotine in smokers. *Cancer Epid. Biomark. Prev.* 8: 907–913.

Heckard, L. R. 1968. Chromosome numbers and polyploidy in *Castilleja* (Scrophulariaceae). *Brittonia* 20: 212–226.

Heckard, L. R. & T.-I. Chuang. 1977. Chromosome numbers, polyploidy, and hybridization in *Castilleja* (Scrophulariaceae) of the Great Basin and Rocky Mountains. *Brittonia* 29: 159–172.

Hedberg, O. 1969. Evolution and speciation in a tropical high mountain flora. *Biol. J. Linn. Soc.* 1: 135–148.

Hedberg, O. 1997. The genus *Koenigia* L. emend. Hedberg (Polygonaceae). *Bot. J. Linn. Soc.* 124: 295–330.

Hedeen, S. E. 1972. Food and feeding behavior of the mink frog, *Rana septentrionalis* Baird, in Minnesota. *Amer. Midl. Naturalist* 88: 291–300.

Heenan, P. B., A. D. Mitchell & M. Koch. 2002. Molecular systematics of the New Zealand *Pachycladon* (Brassicaceae) complex: generic circumscription and relationships to *Arabidopsis* s. l. and *Arabis* s. l. *New Zealand. J. Bot.* 40: 543–562.

Heenan, P. B., P. J. de Lange & P. I. Knightbridge. 2004. *Utricularia geminiscapa* (Lentibulariaceae), a naturalised aquatic bladderwort in the South Island, New Zealand. *New Zealand. J. Bot.* 42: 247–251.

Hegazy, A. K. 1994. Trade-off between sexual and vegetative reproduction of the weedy *Heliotropium curassavicum*. *J. Arid Environ.* 27: 209–220.

Hegazy, A. K., M. I. Soliman & I. A. Mashaly. 1994. Perspectives on the biology of *Heliotropium curassavicum* in the Deltaic Mediterranean coast of Egypt. *Arab Gulf J. Sci. Res.* 12: 525–545.

Hegazy, A. K., W. M. Amer & A. A. Khedr. 2001. Allelopathic effect of *Nymphaea lotus* L. on growth and yield of cultivated rice around Lake Manzala (Nile Delta). *Hydrobiologia* 464: 133–142.

Hegner, R. W. 1926. The interrelations of Protozoa and the utricles of *Utricularia*. *Biol. Bull.* 50: 239–270.

Heide, O. & Y. Gauslaa. 1999. Development strategies of *Koenigia islandica*, a high-arctic annual plant: *Ecography* 22: 637–642.

Heidel, B. L. 1983. Unusual plant assemblage in Walsh County, North Dakota. *Prairie Nat.* 15: 61–62.

Heidel, B. 2004. *Sullivantia hapemanii* var. *hapemanii* (Coult. & Fisher) Coult. (Hapeman's coolwort): a technical conservation assessment. USDAForest Service, Rocky Mountain Region. Available online: http://www.fs.fed.us/r2/projects/scp/assessments/sullivantiahapemaniivarhapemanii.pdf (accessed May 4, 2006).

Heidel, B. & G. Jones. 2006. *Botanical and Ecological Characteristics of Fens in the Medicine Bow Mountains, Medicine Bow National Forest Albany and Carbon Counties, Wyoming.* Prepared for the Medicine Bow-Routt National Forest. Wyoming Natural Diversity Database, University of Wyoming, Laramie, WY. 48 pp.

Heimbinder, E. 2001. Revegetation of a San Francisco coastal salt marsh. *Native Plant J.* 2: 54–59.

Heinken, T. & D. Raudnitschka. 2002. Do wild ungulates contribute to the dispersal of vascular plants in central European forests by epizoochory? A case study in NE Germany. *Forstwiss. Centralbl.* 121: 179–194.

Heinold, B. D., B. A. Gill & B. C. Kondratieff. 2013. Recent collection and DNA barcode of the rare coffee pot snowfly *Capnia nelsoni* (Plecoptera: Capniidae). *Illiesia* 9: 14–17.

Heinrich, B. 1975a. Bee flowers: a hypothesis on flower variety and blooming times. *Evolution* 29: 325–334.

Heinrich, B. 1975b. Energetics of pollination. *Annual Rev. Ecol. Syst.* 6: 139–170.

Heinrich, B. 1976. Resource partitioning among some eusocial insects: bumblebees. *Ecology* 57: 874–889.

Heinrich, B. 1979. "Majoring" and "minoring" by foraging bumblebees, *Bombus vagans*: an experimental analysis. *Ecology* 60: 245–255.

Heinrich, M. 2000. Ethnobotany and its role in drug development. *Phytother. Res.* 14: 479–488.

Heins, M. N. 1984. Ecology and behavior of black-billed whistling duck broods in south Texas. MS thesis. Texas Tech University, Lubbock, TX. 84 pp.

Heiser Jr., C. B. 1962. Some observations on pollination and compatibility in *Magnolia*. *Proc. Indiana Acad. Sci.* 72: 259–266.

Heiser Jr., C. B. & T. W. Whitaker. 1948. Chromosome number, polyploidy, and growth habit in California weeds. *Amer. J. Bot.* 35: 179–186.

Heiser, C. B. & D. M. Smith. 1955. New chromosome numbers in *Helianthus* and related genera (Compositae). *Proc. Indiana Acad. Sci.* 64: 250–253.

Heisey, R. M. & C. C. Delwiche. 1985. Allelopathic effects of *Trichostema lanceolatum* (Labiatae) in the California annual grassland. *J. Ecol.* 73: 729–742.

Hejda, M., P. Pyšek & V. Jarošík. 2009. Impact of invasive plants on the species richness, diversity and composition of invaded communities. *J. Ecol.* 97: 393–403.

Held, D. W. 2004. Relative susceptibility of woody landscape plants to Japanese beetle (*Coleoptera: Scarabaeidae*). *J. Arboric.* 30: 328–335.

Helenurm, K. & L. S. Parsons. 1997. Genetic variation and the reintroduction of *Cordylanthus maritimus* ssp. *maritimus* to Sweetwater Marsh, California. *Restorat. Ecol.* 5: 236–244.

Helfgott, D. M., J. Francisco-Ortega, A. Santos-Guerra, R. K. Jansen & B. B. Simpson. 2000. Biogeography and breeding system evolution of the woody *Bencomia* alliance (Rosaceae) in Macaronesia based on ITS sequence data. *Syst. Bot.* 25: 82–97.

Heller, C. A. 1953. *Edible and Poisonous Plants of Alaska.* Cooperative Agricultural Extension Service, University of Alaska, College, AK. 167 pp.

Hellquist, C. B. 1972. Range extensions of vascular aquatic plants in New England. *Rhodora* 74: 131–141.

Hellquist, C. B. 2003. *Nymphaea leibergii* Morong. Pygmy waterlily. New England plant conservation program, conservation and research plan. New England Wild Flower Society, Framingham, MA 17 pp.

Hellquist, C. B. & G. E. Crow. 2003. The vascular flora of Mud Pond peatland, Carroll County, New Hampshire. *Rhodora* 105: 153–177.

Hellquist, C. B. & J. Straub. 2001. *A Guide to Selected Invasive Non-Native Aquatic Species in Massachusetts.* Lakes & Ponds Program, Massachusetts Department of Environmental Management, Boston, MA. 19 pp.

Helm, D. 1982. Multivariate analysis of alpine snow-patch vegetation cover near Milner Pass, Rocky Mountain National Park, Colorado, U.S.A. *Arct. Alp. Res.* 14: 87–95.

Helm, S. R. & R. H. Chabreck. 2006. Notes on food habits of swamp rabbits (*Sylvilagus aquaticus*) in the Atchafalaya Basin, Louisiana. *J. Mississippi Acad. Sci.* 51: 129–133.

Henderson, N. C. 1962. A taxonomic revision of the genus *Lycopus* (Labiatae). *Amer. Midl. Naturalist* 68: 95–138.

Hendrix, S. D. 1984. Reactions of *Heracleum lanatum* to floral herbivory by *Depressaria pastinacella*. *Ecology* 65: 191–197.

Hendrix, S. D. & I-F. Sun. 1989. Inter-and intraspecific variation in seed mass in seven species of umbellifer. *New Phytol.* 112: 445–451.

Hendry, G. A. F., K. Thompson, C. J. Moss, E. Edwards & P. C. Thorpe. 1994. Seed persistence: a correlation between seed longevity in the soil and ortho-dihydroxyphenol concentration. *Funct. Ecol.* 8: 658–664.

Hengwei, L., G. M. Xiang & R. G. Zhen. 2007. A comparison of nutritive components in *Apium graveolens* and wild *Oenanthe javanica* plants. *Chinese Wild Plant Resour.* 1: 36–38.

Henrickson, J. 1989. A new species of *Leucospora* (Scrophulariaceae) from the Chichuahuan desert of Mexico. *Aliso* 12: 435–439.

Henry, D. G. 1973. Foliar nutrient concentrations of some Minnesota forest species. *Minnesota Forest. Res. Notes* 241: 1–4.

Henry, G. H. R. 1998. Environmental influences on the structure of sedge meadows in the Canadian High Arctic. *Plant Ecol.* 134: 119–129.

Henry, R. D. & A. R. Scott. 1981. Time of introduction of the alien component of the spontaneous Illinois vascular flora. *Amer. Midl. Naturalist* 106: 318–324.

Henry, T. J. 1989. *Cariniocoris*, a new phyline plant bug genus from the eastern United States, with a discussion of generic relationships (Heteroptera: Miridae). *J. New York Entomol. Soc.* 97: 87–99.

Henry, T. J. 1991. *Melanotrichus whiteheadi*, a new crucifer-feeding plant bug from the southeastern United States, with new records for the genus and a key to the species of eastern North America (Heteroptera: Miridae: Orthotylinae). *Proc. Entomol. Soc. Washington* 93: 449–456.

Henry, T. J. 2007. Synopsis of the Eastern North American species of the plant bug genus *Parthenicus*, with descriptions of three new species and a revised key (Heteroptera: Miridae: Orthotylinae). *Amer. Mus. Novit.* 3593: 1–30.

Hensel, A., M. Maas, J. Sendker, M. Lechtenberg, F. Petereit, A. Deters, T. Schmidt & T. Stark. 2011. *Eupatorium perfoliatum* L.: Phytochemistry, traditional use and current applications. *J. Ethnopharmacol.* 138: 641–651.

Heppner, J. B. & D. H. Habeck. 1976. Insects associated with *Polygonum* (Polygonaceae) in north central Florida. I. Introduction and Lepidoptera. *Florida Entomol.* 59: 231–239.

Herbert, J., M. W. Chase, M. Möller & R. J. Abbott. 2006. Nuclear and plastid DNA sequences confirm the placement of the enigmatic *Canacomyrica monticola* in Myricaceae. *Taxon* 55: 349–357.

Herbold, B. & P. B. Moyle. 1989. *The Ecology of the Sacremento-San Joaquin Delta: A Community Profile.* Biological Report 85(7.22). U. S. Department of the Interior, Fish and Wildlife Service, Research and Development. National Wetlands Research Center, Washington, DC.

Herlong, D. D. 1979. Aquatic Pyralidae (Lepidoptera: Nymphulinae) in South Carolina. *Florida Entomol.* 62: 188–193.

Hermanutz, L. A., D. J. Innes & I. M. Weis. 1989. Clonal structure of arctic dwarf birch (*Betula glandulosa*) at its northern limit. *Amer. J. Bot.* 76: 755–761.

Hernandez, J. L. 2013. Wild bee communities in grassland habitats of the Central Valley of California: Drivers of diversity and community structure. PhD dissertation. University of California, Berkeley, CA. 95 pp.

Herndon, J. A. 1968. The snake-master legend in west Kentucky. *Western Folklore* 27: 112–114.

Herring, B. & A. Davis. 2004. *Inventory of Rare and Endemic Plants and Rare Land and Riverine Vertebrates of Silver River and Silver Springs.* Special Publication SJ2004-SP36. Florida Natural Areas Inventory, Tallahassee, FL. 69 pp.

Hersch, E. I. & B. A. Roy. 2007. Context-dependent pollinator behavior: an explanation for patterns of hybridization among three species of Indian paintbrush. *Evolution* 61: 111–124.

Hershey, A. L. 1940. Notes on plants of New Mexico—II. *Leafl. West. Bot.* 2: 257–258.

Hershkovitz, M. A. & S. B. Hogan. 2003. 4. *Lewisia* Pursh. pp. 476–485. *In*: N. R. Morin (convening ed.), *Flora of North America North of Mexico, Vol. 4: Magnoliophyta: Caryophyllidae, Part 1.* Oxford University Press, New York.

Hershkovitz, M. A. & E. A. Zimmer. 2000. Ribosomal DNA evidence and disjunctions of Western American Portulacaceae. *Mol. Phylogen. Evol.* 15: 419–439.

Herz, W. & M. Bruno. 1986. Heliangolides, kauranes and other constituents of *Helianthus heterophyllus. Phytochemistry* 25: 1913–1916.

Herz, W. & M. V. Lakshmikantham. 1965. Constituents of *Helenium* species—XVII: Sesquiterpene lactones of *Helenium thurberi* Gray and the stereochemistry of bigelovin. *Tetrahedron* 21: 1711–1715.

Herz, W. & P. Santhanam. 1967. Constituents of *Helenium* species. XX. Virginolide, a new guaianolide from *Helenium virginicum. J. Organic Chem.* 32: 507–510.

Herz, W., R. B. Mitra, K. Rabindran & N. Viswanathan. 1962. Constituents of Helenium species. XI. The structure of pinnatifidin. *J. Organic Chem.* 27: 4041–4043.

Herz, W., M. Charles & P. S. Subramaniam. 1968. Sesquiterpene lactones of *Helenium alternifolium* (Spreng.) Cabrera. Structures of brevilin A, linifolin A, and alternilin. *J. Organic Chem.* 33: 2780–2784.

Heslop-Harrison, Y. 2004. Biological Flora of the British Isles, No. 237. *Pinguicula* L. *J. Ecol.* 92: 1071–1118.

Heslop-Harrison, Y. & R. B. Knox. 1971. A cytochemical study of the leaf-gland enzymes of insectivorous plants of the genus *Pinguicula. Planta* 96: 183–211.

Hess, B. R. 2009. The response of vegetation to chemical and hydrological gradients in the IMI fen, Henry County, Indiana. MS thesis. Ball State University, Muncie, IN 62 pp.

Hess, C. A. & F. C. James. 1998. Diet of the red-cockaded woodpecker in the Apalachicola National Forest. *J. Wildl. Manag.* 62: 509–517.

Hesselein, C. P. & D. W. Boyd Jr. 2003. Strawberry rootworm biology and control. *S. Nursery Assoc. Res. Conf. Proc.* 48: 174–176.

Hessl, A. & S. Spakman. 1995. *Effects of Fire on Threatened and Endangered Plants: An Annotated Bibliography.* Report 2. U. S. Department of the Interior, National Biological Service Information and Technology, Washington, DC. 55 pp.

Hestand, R. S., B. E. May, D. P. Schultz & C. R. Walker. 1973. Ecological implications of water levels on plant growth in a shallow water reservoir. *Hyacinth Control J.* 11: 54–58.

Heus, P. 2003. An historical prairie remant in Virginia. *Native Plants* (Fall): 105–106.

Heusser, C. J. 1978. Postglacial vegetation on Adak Island, Aleutian Islands, Alaska. *Bull. Torrey Bot. Club* 105: 18–23.

Heusser, C. J., R. L. Schuster & A. K. Gilkey. 1954. Geobotanical studies on the Taku glacier anomaly. *Geogr. Rev.* 44: 224–239.

Hevesi, B. T., P. J. Houghton, S. Habtemariam & Á. Kéry. 2009. Antioxidant and antiinflammatory effect of *Epilobium parviflorum* Schreb. *Phytother. Res.* 23: 719–724.

Hewett, D. G. 1964. *Menyanthes trifoliata* L. *J. Ecol.* 52: 723–735.

Hewitt, D. G. & C. T. Robbins. 1996. Estimating grizzly bear food habits from fecal analysis. *Wildl. Soc. Bull.* 24: 547–550.

Hewitt, N. & K. Miyanishi. 1997. The role of mammals in maintaining plant species richness in a floating *Typha* marsh in southern Ontario. *Biodivers. Conserv.* 6: 1085–1102.

Heyes, G. E. 1979. The effect of clipping date and height on forage yields, nutritive quality and stoked food reserves of a Ckelcotin wetland meadow. MS thesis. University of British Columbia, Vancouver, BC. 123 pp.

Heyligers, P. C. 2007. The role of currents in the dispersal of introduced seashore plants around Australia. *Cunninghamia* 10: 167–188.

Hickman, J. C. 1974. Pollination by ants: a low-energy system. *Science* 184: 1290–1292.

Hickman, J. C. (ed.). 1993. *The Jepson Manual: Higher Plants of California.* University of California Press, Berkeley, CA. 1424 pp.

Hicks, P. L. 2003. *Seed Dispersal Dynamics in a Restored Tidal Marsh: Implications for Restoration Success.* Final Report. Garden Club of America, New York. 18 pp.

Hiebert, R. D., D. A. Wilcox & N. B. Pavlovic. 1986. Vegetation patterns in and among pannes (calcareous intradunal ponds) at the Indiana Dunes National Lakeshore, Indiana. *Amer. Midl. Naturalist* 116: 276–281.

Higman, D. 1972. Emergent vascular plants of Chesapeake Bay wetlands. *Chesapeake Sci.* 13: S89–S93.

Higman, P. J. & M. R. Penskar. 1996. Special plant abstract for *Bartonia paniculata* (panicled screw-stem). Michigan Department of Natural Resources-Forest Management Division and Wildlife Division, Lansing, MI 2 pp.

Hik, D. S., R. L. Jefferies & A. R. E. Sinclair. 1992. Foraging by geese, isostatic uplift and asymmetry in the development of salt-marsh plant communities. *J. Ecol.* 80: 395–406.

Hildebrand, H. 1949. Notes on *Rana sylvatica* in the Labrador Peninsula. *Copeia* 1949: 168–172.

Hilgard, E. W. 1873. *Supplementary and Final Report of a Geological Reconnaissance of the State of Louisiana.* New Orleans, LA. 44 pp.

Hill, B. H. & J. R. Webster. 1984. Productivity of *Podostemum ceratophyllum* in the New River, Virginia. *Amer. J. Bot.* 71: 130–136.

Hill, B. H. & J. R. Webster. 2004. Aquatic macrophyte breakdown in an Appalachian river. *Hydrobiologia* 89: 53–59.

Hill, E. G. 1891. The fertilization of three native plants. *Bull. Torrey Bot. Club* 18: 111–118.

Hill, L. M. 1976. Morphological and cytological evidence for introgression in *Aster acuminatus* Michx. in the southern Appalachians. *Castanea* 4: 148–155.

Hill, N. M. & P. A. Keddy. 1992. Prediction of rarities from habitat variables: coastal plain plants on Nova Scotian lakeshores. *Ecology* 73: 1852–1859.

Hill, N. M. & S. P. Vander Kloet. 2005. Longevity of experimentally buried seed in *Vaccinium*: relationship to climate, reproductive factors and natural seed banks. *J. Ecol.* 93: 1167–1176.

Hill, N. M., P. A. Keddy & I. C. Wisheu. 1998. A hydrological model for predicting the effects of dams on the shoreline vegetation of Lakes and Reservoirs. *Environ. Manag.* 22: 723–736.

Hill, N. M., S. P. Vander Kloet & D. J. Garbary. 2012. The regeneration ecology of *Empetrum nigrum*, the black crowberry, on coastal heathland in Nova Scotia. *Botany* 90: 379–392.

Hill. P. 1987. A diversity of hollies. *Arnoldia* 47: 2–13.

Hill, R. R. & D. Harris. 1943. Food preferences of Black Hills deer. *J. Wildl. Manag.* 7: 233–235.

Hill, S. R. 1988. New plant records for Maryland with an additional note on *Nymphaea tetragona* (Nymphaeaceae) pollination. *Castanea* 53: 164–6.

Hill, S. R. 1992. Calciphiles and calcareous habitats of South Carolina. *Castanea* 57: 25–33.

Hill, S. R. 2003. Conservation assessment for twining screwstem (*Bartonia paniculata*) (Michx.) Muhl. Illinois Natural History Survey, Center for Biodiversity, Champaign, IL. 29 pp.

Hill, S. R. 2006. *Conservation Assessment for the Kidneyleaf Mud-Plantain* (Heteranthera reniformis *Ruiz & Pavon*). Technical Report 2006 (5). Illinois Natural History Survey, Center for Wildlife and Plant Ecology, Champaign, IL. 34 pp.

Hill, S. R. 2012. *Sidalcea.* pp. 887–896. *In*: B. G. Baldwin, D. H. Goldman, D. J. Keil, R. Patterson, T. J. Rosatti & D. H. Wilken (eds.), *The Jepson Manual (2nd ed.)*. University of California Press, Berkeley, CA.

Hilliard Jr., J. R. 1982. Endophytic oviposition by *Leptysma Marginicollis Marginicollis* and *Stenacris Vitreipennis* (Orthoptera: Acrididae: Leptysminae) with life history notes. *Trans. Amer. Entomol. Soc.* 108: 153–180.

Hilton, J. L. 1993. Aspects of the ecology and life history of the granite pool sprite, *Amphianthus pusillus* Torrey (Scrophulariaceae). PhD dissertation. Auburn University, Auburn, AL.

Hilton, J. L. & R. S. Boyd. 1996. Microhabitat requirements and seed/microsite limitation of the rare granite outcrop endemic *Amphianthus pusillus* (Scrophulariaceae). *Bull. Torrey Bot. Club* 123: 189–196.

Hilty, J. 2014. Insect visitors of Illinois wildflowers. World Wide Web electronic publication. Available online: http://www.illinoiswildflowers.info/flower_insects/ (accessed June 2014).

Himmelreich, S., I. Breitwieser & C. Oberprieler. 2012. Phylogeny, biogeography, and evolution of sex expression in the southern hemisphere genus *Leptinella* (Compositae, Anthemideae). *Mol. Phylogen. Evol.* 65: 464–481.

Hines, A. H. & G. M. Ruiz (eds.). 2000. *Marine Invasive Species and Biodiversity of South Central Alaska*. Regional Citizens' Advisory Council of Prince William Sound, Anchorage, AK. 75 pp.

Hinman, S. E. & J. S. Brewer. 2007. Responses of two frequently-burned wet pine savannas to an extended period without fire. *J. Torrey Bot. Soc.* 134: 512–526.

Hinsinger, D. D., J. Basak, M. Gaudeul, C. Cruaud, P. Bertolino, N. Frascaria-Lacoste & J. Bousquet. 2013. The phylogeny and biogeographic history of ashes (*Fraxinus*, Oleaceae) highlight the roles of migration and vicariance in the diversification of temperate trees. *PLoS One* 8(11): e80431.

Hirai, N., S.-I. Sakashita, T. Sano, T. Inoue, H. Ohigashi, C.-U. Premasthira, Y. Asakawa, J. Harada & Y. Fuji. 2000. Allelochemicals of the tropical weed *Sphenoclea zeylanica*. *Phytochemistry* 55: 131–140.

Hiratsuka, Y. & P. J. Maruyama. 1976. *Castilleja miniata*, a new alternate host of *Cronartium ribicola*. *Plant Dis. Reporter* 60: 241.

Hiroomi, A. 2001. Effect of shading on weed emergence in paddy fields. *J. Weed Sci. Technol.* 46: 31–36.

Hirst, F. 1983. Field report on the Delmarva flora, I. *Bartonia* 49: 59–68.

Hirthe, G. & S. Porembski. 2003. Pollination of *Nymphaea lotus* (Nymphaeaceae) by rhinoceros beetles and bees in the Northeastern Ivory Coast. *Plant Biol.* 5: 670–676.

Hiscock, S. J. & S. M. McInnis. 2003. The diversity of self-incompatibility systems in flowering plants. *Plant Biol. (Stuttgart)* 5: 23–32.

Hitchcock, C. L., A. Cronquist, M. Ownbey & J. W. Thompson. 1964. *Vascular Plants of the Pacific Northwest. Part 2. Salicaceae to Saxifragaceae*. University of Washington Press, Seattle, WA. 597 pp.

Hitchmough, J. & K. Fieldhouse (eds.). 2004. *Plant User Handbook: A Guide to Effective Specifying*. Blackwell Publishing, Oxford, UK. 400 pp.

Hitchmough, J. & M. Wagner. 2011. Slug grazing effects on seedling and adult life stages of North American Prairie plants used in designed urban landscapes. *Urban Ecosyst.* 14: 279–302.

Hitchmough, J., M. de la Fleur & C. Findlay. 2004. Establishing North American prairie vegetation in urban parks in northern England: Part 1. Effect of sowing season, sowing rate and soil type. *Landscape Urban Plan.* 66: 75–90.

Hladun, N. P., A. S. Bondarenko, S. S. Nahorna & O. V. Smyrnova. 2002. Investigation of the activity of the preparation cerbiden against *Candida* spp. *Mikrobiol. Zhurn.* 64: 57–61.

Hoagland, B. W. 2002. A classification and analysis of emergent wetland vegetation in western Oklahoma. *Proc. Oklahoma Acad. Sci.* 82: 5–14.

Hoagland, B. W. & A. K. Buthod. 2005. Vascular flora of a gypsum dominated site in Major County, Oklahoma. *Proc. Oklahoma Acad. Sci.* 85: 1–8.

Hoagland, B. W. & F. L. Johnson. 2005. Vascular flora of the Deep Fork river in Okmulgee, Creek and Okfuskee counties, Oklahoma. *Publ. Oklahoma Biol. Surv.* 6: 15–29.

Hoagland, K. E. & G. M. Davis. 1987. The succineid snail fauna of Chittenango Falls, New York: taxonomic status with comparisons to other relevant taxa. *Proc. Acad. Nat. Sci. Philadelphia* 139: 465–526.

Hoch, P. C., J. V. Crisci, H. Tobe & P. E. Berry. 1993. A cladistic analysis of the plant family Onagraceae. *Syst. Bot.* 18: 31–47.

Hodges, S. A. & M. L. Arnold. 1994a. Floral and ecological isolation between *Aquilegia formosa* and *Aquilegia pubescens*. *Proc. Natl. Acad. Sci. U.S.A.* 91: 2493–2496.

Hodges, S. A. & M. L. Arnold. 1994b. Columbines: a geographically widespread species flock. *Proc. Natl. Acad. Sci. U.S.A.* 91: 5129–5132.

Hodges, S. A., M. Fulton, J. Y. Yang & J. B. Whittall. 2003. Verne Grant and evolutionary studies of *Aquilegia*. *New Phytol.* 161: 113–120.

Hodges, S. R. & K. A. Bradley. 2006. *Distribution and Population Size of Five Candidate Plant Taxa in the Florida Keys:* Argythamnia blodgettii, Chamaecrista lineata *var.* keyensis, Indigofera mucronata *var.* keyensis, Linum arenicola, *and* Sideroxylon reclinatum *subsp.* austrofloridense. Final Report. Contract Number: 401815G011. January 23, 2006. United States Fish and Wildlife Service, South Florida Ecosystem Office, Vero Beach, FL. 79 pp.

Hodgson, B. E. 1928. *The Host Plants of the European Corn Borer in New England*. Technical Bulletin No. 77. U. S, Department of Agriculture, Washington, DC. 63 pp.

Hodkinson, D. J. & K. Thompson. 1997. Plant dispersal: the role of man. *J. Appl. Ecol.* 34: 1484–1496.

Hoffman, B. A. & C. J. Dawes. 1997. Vegetational and abiotic analysis of the salterns of mangals and salt marshes of the west coast of Florida. *J. Coastal Res.* 13: 147–154.

Hoffman, G. R. & D. L. Hazlett. 1977. Effects of aqueous *Artemisia* extracts and volatile substances on germination of selected species. *J. Range Manag.* 30: 134–137.

Hoffmann, M. H., K. B. von Hagen, E. Hörandl, M. Röser & N. V. Tkach. 2010. Sources of the Arctic flora: origins of Arctic species in *Ranunculus* and related genera. *Int. J. Plant Sci.* 171: 90–106.

Hoffmann, R. S. & R. D. Taber. 1960. Notes on *Sorex* in the northern Rocky Mountain alpine zone. *J. Mammalogy* 41: 230–234.

Hogg, E. H. & R. W. Wein. 1988. The contribution of *Typha* components to floating mat buoyancy. *Ecology* 69: 1025–1031.

Hoggard, G. D., P. J. Kores, M. Molvray & R. K. Hoggard. 2004. The phylogeny of *Gaura* (Onagraceae) based on ITS, ETS, and *trnL-F* sequence data. *Amer. J. Bot.* 91: 139–148.

Hoggard, R. K., P. J. Kores, M. Molvray, G. D. Hoggard & D. A. Broughton. 2003. Molecular systematics and biogeography of the amphibious genus *Littorella* (Plantaginaceae). *Amer. J. Bot.* 90: 429–435.

Hogsden, K. L., E. P. S. Sager & T. C. Hutchinson. 2007. The impacts of the non-native macrophyte *Cabomba caroliniana* on littoral biota of Kasshabog Lake, Ontario. *J. Great Lakes Res.* 33: 497–504.

Hogue, E. J. 1976. Seed dormancy of nodding beggarticks (*Bidens cernua* L.). *Weed Sci.* 24: 375–378.

Hokanson, K. & J. Hancock. 2000. Early-acting inbreeding depression in three species of *Vaccinium* (Ericaceae). *Sexual Plant Reprod.* 13: 145–150.

Holch, A. E., E. W. Hertel, W. O. Oakes & H. H. Whitwell. 1941. Root habits of certain plants of the foothill and alpine belts of Rocky Mountain National Park. *Ecol. Monogr.* 11: 327–345.

Holcroft Weerstra, A. C. 2001. *Preliminary Classification of Silver Sagebrush* (Artemisia cana) *Community Types.* Report prepared for Alberta Natural Heritage Information Centre, Edmonton, Alberta. Biota Consultants, Cochrane, AB. 155 pp.

Holden, C. 1999. 'Extinct' Oregon flower reappears. *Science* 284: 2083.

Holder, G. L., M. K. Johnson & J. L. Baker. 1980. Cattle grazing and management of dusky seaside sparrow habitat. *Wildl. Soc. Bull.* 8: 105–109.

Holland, R. F. 1986. *Preliminary Descriptions of the Terrestrial Natural Communities of California.* Department of Fish and Game, State of California, Sacramento, CA. 156 pp.

Holland, R. F. & J. D. Morefield. 2002. *Current Knowledge and Conservation Status of* Polyctenium williamsiae *Rollins* (Brassicaceae; including Polyctenium fremontii var. confertum *Rollins), the Williams Combleaf.* Status report prepared for the U. S. Fish and Wildlife Service, Nevada State Office, Reno, NV. 40 pp.

Hollingsworth, P. M., M. Tebbitt, K. J. Watson & R. J. Gornall. 1998. Conservation genetics of an arctic species, *Saxifraga rivularis* L., in Britain. *Bot. J. Linn. Soc.* 128: 1–14.

Hollister, R. D. & P. J. Webber. 2000. Biotic validation of small open-top chambers in a tundra ecosystem. *Glob. Change Biol.* 6: 835–842.

Holloway, P. S. & G. E. M. Matheke. 2003. Seed germination of burnet, *Sanguisorba* spp. *Native Plant J.* 4: 95–99.

Holm, Jr., G. O., E. Evers & C. E. Sasser. 2011. *The Nutria in Louisiana: A Current and Historical Perspective.* Final Report prepared for The Lake Pontchartrain Basin Foundation. Louisiana State University, Baton Rouge, LA. 58 pp.

Holm, L. & K. Holm. 1996. Four northern Mycosphaerellae. *Acta Univ. Upsal. Symb. Bot. Upsal.* 31: 285–295.

Holm, T. 1925. Hibernation and rejuvenation, exemplified by North American herbs. *Amer. Midl. Naturalist* 9: 439–476; 477–512.

Holmes, W. C. 1981. *Mikania* (Compositae) of the United States. *Sida* 9: 147–158.

Holmes, W. C. 1990. The genus *Mikania* (Compositae: Eupatorieae) in Mexico. *Sida Bot. Misc.* 5: 1–45.

Holmes, W. C. 1991. Dioecy in *Mikania* (Compositae: Eupatorieae). *Plant Syst. Evol.* 175: 87–92.

Holmes, W. C. 1993. The genus *Mikania* (Compositae: Eupatorieae) in the Greater Antilles. *Sida Bot. Misc.* 9: 1–69.

Holmes, W. C. 2006. 417. *Mikania* Willdenow. pp. 545–547. *In*: Flora of North America Editorial Committee (eds.), *Flora of North America North of Mexico: Vol. 21, Magnoliophyta: Asteridae (in part): Asteraceae, Part 3.* Oxford University Press, New York.

Holmgren, N. L. 1973. Five new species of *Castilleja* (Scrophulariaceae) from the intermountain region. *Bull. Torrey Bot. Club* 100: 83–93.

Holmgren, N. L. 2003. 2. *Nitrophila* S. Watson. P. 263 *In*: N. R. Morin (convening ed.), *Flora of North America North of Mexico, Vol. 4: Magnoliophyta: Caryophyllidae, Part 1.* Oxford University Press, New York.

Holtzman, J. A. 1990. The pollination biology of *Anemopsis californica* Hooker (Saururaceae): research in progress. *Crossosoma* 16: 1–11.

Holzinger, J. 1893. List of plants new to Florida. *Contr. U. S. Natl. Herb.* 1: 288.

Holway, J. G. & R. T. Ward. 1965. Phenology of alpine plants in northern Colorado. *Ecology* 46: 73–83.

Hölzel, N. & A. Otte. 2001. The impact of flooding regime on the soil seed bank of flood-meadows. *J. Veg. Sci.* 12: 209–218.

Home, F. 1781. *Methodus Materiae Medicae.* Impensis Jo. Bell & Gul. Creech, Edinburgh. 90 pp.

Homoya, M. A. 1983. A floristic survey of acid seep springs in Martin and Dubois Counties, Indiana. *Proc. Indiana Acad. Sci.* 93: 323–331.

Homoya, M. A. & D. B. Abrell. 2005. A natural occurrence of the Federally endangered Short's goldenrod (*Solidago shortii* T. & G.) [Asteraceae] in Indiana: its discovery, habitat, and associated flora. *Castanea* 70: 255–262.

Homoya, M. A. & C. L. Hedge. 1982. The upland sinkhole swamps and ponds of Harrison County, Indiana. *Proc. Indiana Acad. Sci.* 92: 383–388.

Homoya, M. A. & D. A. Rayner. 1987. *Carex socialis* (Cyperaceae): a new record for the Atlantic Coastal Plain. *Castanea* 52: 311–313.

Homoya, M. A., D. B. Abrell, J. R. Aldrich & T. W. Post. 1985. The natural regions of Indiana. *Proc. Indiana Acad. Sci.* 94: 245–268.

Hong, C. X. 2006. Downy mildew of *Rudbeckia fulgida* cv. Goldsturm by *Plasmopara halstedii* in Virginia. *Plant Dis.* 90: 1461.

Hong, J. C., M. T. Momol, J. B. Jones, P. S. Ji, S. M. Olson, C. Allen, A. Perez, P. Pradhanang & K. Guven. 2008. Detection of *Ralstonia solanacearum* in irrigation ponds and aquatic weeds associated with the ponds in north Florida. *Plant Dis.* 92: 1674–1682.

Hong, S.-P. & H.-K. Moon. 2003. Gynodioecy in *Lycopus maackianus* Makino (Lamiaceae) in Korea: floral dimorphism and nutlet production. *Flora* 198: 461–467.

Hook, D. & L. Brown. 1973. Root adaptations and relative flood tolerance of five hardwood species. *For. Sci.* 19: 225–229.

Hooke, J. 2006. Burn severity effects in spruce-fir forests of northwestern Wyoming. *Rx Effects* 1(6): 1, 7, 9.

Hooker, K. L. & H. M. Westbury. 1991. Development of wetland plant communities in a new reservoir. pp. 45–60. *In*: F. J. Webb Jr. (ed.), *Proceedings of the Eighteenth Annual Conference on Wetlands Restoration and Creation: May 16–17, 1991.* Plant City, FL.

Hoot, S. B. 1995. Phylogeny of the Ranunculaceae based on preliminary *atpB*, *rbcL* and 18S nuclear ribosomal DNA sequence data. *Plant Syst. Evol.* 9: S241–S251.

Hoot, S. B., J. W. Kadereit, F. R. Blattner, K. B. Jork, A. E. Schwarzbach & P. R. Crane. 1997. Data congruence and phylogeny of the Papaveraceae s.l. based on four data sets: *atpB* and *rbcL* sequences, *trnK* restriction sites, and morphological characters. *Syst. Bot.* 22: 575–590.

Hoot, S. B., S. Magallón-Puebla & P. R. Crane. 1999. Phylogeny of basal eudicots based on three molecular datasets: *atpB* and *rbcL* sequences, *trnK* restriction sites and morphological characters. *Ann. Missouri Bot. Gard.* 86: 119–131.

Hopfensperger, K. N. & A. H. Baldwin. 2009. Spatial and temporal dynamics of floating and drift-line seeds at a tidal freshwater marsh on the Potomac River, USA. *Plant Ecol.* 201: 677–686.

Hopfensperger, K. N. & K. A. M. Engelhardt. 2008. Annual species abundance in a tidal freshwater marsh: germination and survival across an elevational gradient. *Wetlands* 28: 521–526.

Hopfensperger, K. N., K. A. M. Engelhardt & T. R. Lookingbill. 2009. Vegetation and seed bank dynamics in a tidal freshwater marsh. *J. Veg. Sci.* 20: 767–778.

Hopkins, C. E. O. 1969. Vegetation of fresh-water springs of southern Illinois. *Castanea* 34: 121–145.

Hopkins, D. R. & V. T. Parker. 1984. A study of the seed bank of a salt marsh in northern San Francisco Bay. *Amer. J. Bot.* 71: 348–355.

Hoppe, R. T., L. M. Smith & D. B. Wester. 1986. Foods of wintering diving ducks in South Carolina. *J. Field Ornithology* 57: 126–134.

Horak, M. J. & T. M. Loughin. 2000. Growth analysis of four *Amaranthus* species. *Weed Sci.* 48: 347–355.

Horn, C. N. 2011. Heterophylly of *Didiplis diandra* (Nutt. ex A. DC.) Wood (Lythraceae) and a key to some rooted shallow water and shoreline herbs of the Mid-Atlantic Piedmont. *Castanea* 76: 272–278.

Horn, D. D. & P. Somers. 1981. *Neviusia alabamensis* (Rosaceae) in Tennessee. *Sida* 9: 90–92.

Hornbeck, J. H., C. H. Sieg & D. J. Reyher. 2003. *Conservation Assessment for the Autumn Willow in the Black Hills National Forest, South Dakota and Wyoming.* United States Department of Agriculture, Forest Service, Rocky Mountain Region, Black Hills National Forest, Custer, SD. 38 pp.

Hornok, L. 1992. Peppermint (*Mentha piperita* L.). pp. 187–196. *In*: L. Hornok (ed.), *Cultivation and Processing of Medicinal Plants.* Akademia Kiado, Budapest, Hungary. 338 pp.

Horr, W. H. & R. L. McGregor. 1948. A raised marsh near Muscotah, Kansas. *Trans. Kansas Acad. Sci.* 51: 197–200.

Horrell, L. B. 2013. An index for estimating forage quality for white-tailed deer across nine primary habitat types in Louisiana. MS thesis. University of Georgia, Athens, GA. 84 pp.

Horsburgh, M., J. C. Semple & P. G. Kevan. 2011. Relative pollinator effectiveness of insect floral visitors to two sympatric species of wild aster: *Symphyotrichum lanceolatum* (Willd.) Nesom and *S. lateriflorum* (L.) Löve & Löve (Asteraceae: Astereae). *Rhodora* 113: 64–86.

Horsley, S. B. 1977. Allelopathic inhibition of black cherry by fern, grass, goldenrod, and aster. *Can. J. For. Res.* 7: 205–216.

Horvat, R. J. & G. W. Chapman. 2007. Identification of volatile compounds from ripe Mayhaw fruit (*Crataegus opaca*, *C. aestzvalzs*, and *C. ruifula*. *J. Food Qual.* 14: 307–312.

Hosamani, K. M., S. S. Ganjihal & D. V. Chavadi. 2004. *Alternanthera triandra* seed oil: a moderate source of ricinoleic acid and its possible industrial utilisation. *Industr. Crops Prod.* 19: 133–136.

Hosier, E. & T. E. Eaton. 1980. The impact of vehicles on dune and grassland vegetation on a south-eastern North Carolina barrier beach. *J. Appl. Ecol.* 17: 173–182.

Hosokawa, K., M. Minami, K. Kawahara, I. Nakamura & T. Shibata. 2000. Discrimination among three species of medicinal *Scutellaria* plants using RAPD markers. *Plant Med.* 66: 270–272.

Hosokawa, K., M. Minami, I. Nakamura, A. Hishida & T. Shibata. 2005. The sequences of the plastid gene *rpl16* and the *rpl16-rpl14* spacer region allow discrimination among six species of *Scutellaria*. *J. Ethnopharmacol.* 99: 105–108.

Hossain, A. I., M. Faisal, S. Rahman, R. Jahan & M. Rahmatullah. 2014. A preliminary evaluation of antihyperglycemic and analgesic activity of *Alternanthera sessilis* aerial parts. *BMC Complem. Altern. Med.* 14: 169. doi:10.1186/1472-6882-14-169.

Hossain, M. A., J. Y. Park, J. Y. Kim, J. W. Suh & S. C. Park. 2014. Synergistic effect and antiquorum sensing activity of *Nymphaea tetragona* (water lily) extract. *BioMed Res. Int.* 2014: doi:10.1155/2014/562173.

Hotchkiss, N. & R. E. Stewart. 1947. Vegetation of the Patuxent Research Refuge, Maryland. *Amer. Midl. Naturalist* 38: 1–75.

Houghton, J. P. & L. Uhlig. 2003. A tidal habitat restoration success story—the Union Slough restoration project. pp. 1–15. *In*: T. W. Droscher & D. A. Fraser (eds.), *Proceedings of the 2003 Georgia Basin/Puget Sound Research Conference.* Olympia, WA.

Houghton, P. J. & I. M. Osibogun. 1993. Flowering plants used against snakebite. *J. Ethnopharmacol.* 39: 1–29.

Houle, G. 2002. The advantage of early flowering in the spring ephemeral annual plant *Floerkea proserpinacoides*. *New Phytol.* 154: 689–694.

Houle, G. & D. L. Phillips. 1988. The soil seed bank of granite outcrop plant communities. *Oikos* 52: 87–93.

Houle, G., M. F. McKenna & L. Lapointe. 2001. Spatiotemporal dynamics of *Floerkea proserpinacoides* (Limnanthaceae), an annual plant of the deciduous forest of eastern North America. *Amer. J. Bot.* 88: 594–607.

Hovatter, S. R., C. Dejelo, A. L. Case & C. B. Blackwood. 2011. Metacommunity organization of soil microorganisms depends on habitat defined by presence of *Lobelia siphilitica* plants. *Ecology* 92: 57–65.

Howard, D. J. & S. H. Berlocher. 1998. *Endless Forms: Species and Speciation.* Oxford University Press, New York. 496 pp.

Howard, F. W., R. W. Pemberton, G. S. Hodges, B. R. Steinberg, D. A. McLean & H. O. Liu. 2006. Host plant range of lobate lac scale, *Paratachardina lobata*, in Florida. *Proc. Florida State Hort. Soc.* 119: 398–408.

Howard, G. W. & F. W. Chege. 2007. Invasions by plants in the inland waters and wetlands of Africa. pp. 193–208. *In*: F. Gherardi (ed.), *Biological Invaders in Inland Waters: Profiles, Distribution, and Threats.* Invading Nature-Springer Series In Invasion Ecology. Volume 2. Springer, Dordrecht, The Netherlands.

Howard, H. W. & A. G. Lyon. 1952a. *Biological Flora of the British Isles.* L.C. (Ed. 11), No. 89. *Nasturtium officinale* R. Br. (*Rorippa nasturtium-aquaticum* (L.) Hayek). *J. Ecol.* 40: 228–238.

Howard, H. W. & A. G. Lyon. 1952b. *Biological Flora of the British Isles.* L.C. (ed. 11). *Nasturtium microphyllum* Boenningh. ex Rchb. (*Nasturtium uniseriatum* Howard & Manton; *Rorippa microphylla* (Boenn.) Hyl.). *J. Ecol.* 40: 239–245.

Howard, R. J. & L. Allain. 2012. *Effects of a Drawdown on Plant Communities in a Freshwater Impoundment at Lacassine National Wildlife Refuge, Louisiana.* Scientific Investigations Report 2012–5221. U. S. Department of the Interior, U. S. Geological Survey, Reston, VA. 27 pp.

Howard, R. J. & I. A. Mendelssohn. 1999. Salinity as a constraint on growth of oligohaline marsh macrophytes. I. Species variation in stress tolerance. *Amer. J. Bot.* 86: 785–794.

Howard, R. J. & C. J. Wells. 2009. Plant community establishment following drawdown of a reservoir in southern Arkansas, USA. *Wetlands Ecol. Manag.* 17: 565–583.

Howard, R. J., T. C. Michot & L. Allain. 2011. *Vegetation of Lacassine National Wildlife Refuge, Louisiana—Recent Plant Communities with Comparison to a Three-Decade-Old Survey.* U. S. Geological Survey Scientific Investigations Report 2011–5174. U. S. Geological Survey, Reston, VA. 16 pp.

Howard, W. L. 1915. An experimental study of the rest period in plants: seeds (fourth report). *Res. Bull. Missouri Agric. Exp. Stn.* 17: 3–58.

Howat, D. R. 2000. *Acceptable Salinity, Sodicity and pH Values for Boreal Forest Reclamation.* Report #ESD/LM/00-2 (on-line edition). Alberta Environment, Environmental Sciences Division, Edmonton, AB. 191 pp.

Howell, J. T. 1931. The genus *Pogogyne. Proc. Calif. Acad. Sci.* 20: 105–128.

Howell, J. T. 1970. *Marin Flora: Manual of the Flowering Plants and Ferns of Marin County, California.* University of California Press, Berkeley, CA. 366 pp.

Howell, J. W. 1937. A fossil pollen study of Kokomo bog. Howard county, Indiana. *Butler Univ. Bot. Stud.* 4: 117–127.

Hoyer, M. V., D. E. Canfield Jr., C. A. Horsburgh & K. Brown. 1996. *Florida Freshwater Plants: A Handbook of Common Aquatic Plants in Florida Lakes.* University of Florida, Institute of Food & Agricultural Sciences, Gainesville, FL. 264 pp.

Hoyo, Y. & S. Tsuyuzaki. 2015. Sexual and vegetative reproduction of the sympatric congeners *Drosera anglica* and *Drosera rotundifolia. Flora* 210: 60–65.

Hroudová, Z., P. Zákravský, L. Hrouda & I. Ostrý. 1992. *Oenanthe aquatica* (L.) Poir.: Seed reproduction, population structure, habitat conditions and distribution in Czechoslovakia. *Folia Geobot. Phytotax.* 27: 301–335.

Hsieh, C.-F. & S.-M. Chaw. 1987. *Diodia virginiana* L. (Rubiaceae) in Hsinchu: new to Taiwan. *Bot. Bull. Acad. Sin.* 28: 43–48.

Hsu, C. C., R. H. Dobberstein, A. S. Bingel, H. H. Fong, N. R. Farnsworth & J. F. Morton. 1981. Biological and phytochemical investigation of plants XVI: *Strumpfia maritima* (Rubiaceae). *J. Pharm. Sci.* 70: 682–683.

Hsu, C. L., S. C. Fang & G. C. Yen. 2013. Anti-inflammatory effects of phenolic compounds isolated from the flowers of *Nymphaea mexicana* Zucc. *Food Funct.* 4: 1216–1222.

Huang, C., H. Tunon & L. Bohlin. 1995. Anti-inflammatory compounds isolated from *Menyanthes trifoliata* L. *Acta Pharmaceut. Sin.* 30: 621–626.

Huang, D.-J., H.-J. Chen, C.-D. Lin & Y.-H. Lin. 2005. Antioxidant and antiproliferative activities of water spinach (*Ipomoea aquatica* Forsk) constituents. *Bot. Bull. Acad. Sin.* 46: 99–106.

Huang, M. J., J. V. Freudenstein & D. J. Crawford. 2000. Phylogenetic relationships of the *Caryopteris-Trichostema* complex (Lamiaceae) based on *ndhF* sequence data. *Amer. J. Bot.* 87: S174–S175.

Huang, M., D. J. Crawford, J. V. Freudenstein & P. D. Cantino. 2008. Systematics of *Trichostema* (Lamiaceae): evidence from ITS, *ndh*F, and morphology. *Syst. Bot.* 33: 437–446.

Huang, W., K. Chen, X. Shi, K. Ren & W. Li. 2014. The contribution of seeds to the recruitment of a *Nymphoides peltata* population. *Limnol.-Ecol. Manag. Inland Waters* 44: 1–8.

Huang, Y.-L. & S.-H. Shi. 2002. Phylogenetics of Lythraceae *sensu lato*: a preliminary analysis based on chloroplast *rbc*L gene, *psaA-ycf3* spacer, and nuclear rDNA internal transcribed spacer (ITS) sequences. *Int. J. Plant Sci.* 163: 215–225.

Huang, Y., P. W. Fritsch & S. Shi. 2003. A revision of the imbricate group of *Styrax* series *Cyrta* (Styracaceae) in Asia. *Ann. Missouri Bot. Gard.* 90: 491–553.

Hubbard, D. E. & A. Boe. 1988. IVDDM and chemical constituents in selected species of wetland plants. *Proc. South Dakota Acad. Sci.* 67: 44–58.

Hubbard, J. R. & W. S. Judd. 2013. Floristics of Silver River State Park, Marion County, Florida. *Rhodora* 115: 250–280.

Huber, A. M. 2005. Moisture requirements for the germination of early seedling survival of *Cirsium loncholepis*: a thesis. MS thesis. California Polytechnic State University, San Luis Obispo, CA.

Huenneke, L. F. 1983. Understory response to gaps caused by the death of *Ulmus americana* in central New York. *Bull. Torrey Bot. Club* 110: 170–175.

Huenneke, L. F. 1985. Spatial distribution of genetic individuals in thickets of *Alnus incana* ssp. *rugosa*, a clonal shrub. *Amer. J. Bot.* 72: 152–158.

Huenneke, L. F. & R. R. Sharitz. 1986. Microsite abundance and distribution of woody seedlings in a South Carolina cypress-tupelo swamp. *Amer. Midl. Naturalist* 115: 328–335.

Huenneke, L. F. & J. K. Thomson. 1995. Potential interference between a threatened endemic thistle and an invasive nonnative plant. *Conserv. Biol.* 9: 416–425.

Huff, S. H., R. L. Harkess, B. S. Baldwin, G. R. Bachman & E. K. Blythe. n.d. Optimization of select native seed propagation. Unpublished manuscript. Available online: http://ipps-srna.org/pdf/2011Papers/24-%20Huff.pdf.

Huffman, J. M. & W. S. Judd. 1998. Vascular flora of Myakka River State Park, Sarasota and Manatee counties, Florida. *Castanea* 63: 25–50.

Huffman, R. T. & R. I. Lonard. 1983. Successional patterns on floating vegetation mats in a southwestern Arkansas bald cypress swamp. *Castanea* 48: 73–78.

Huffman, T. 2004. *Investigation of the Presence and Geographic Extent of Wetlands on Terrace Point and Younger Lagoon Reserve, University of California, Santa Cruz.* The Huffman-Broadway Group, Inc., Larkspur, CA. 35 pp.

Hufford, L. 1995. Seed morphology of Hydrangeaceae and its phylogenetic implications. *Int. J. Plant Sci.* 156: 555–580.

Hufford, L. 1997. A phylogenetic analysis of Hydrangeaceae based on morphological data. *Int. J. Plant Sci.* 158: 652–672.

Hufford, L. & M. McMahon. 2004. Morphological evolution and systematics of *Synthyris* and *Besseya* (Veronicaceae): a phylogenetic analysis. *Syst. Bot.* 29: 716–736.

Hufford, L., M. L. Moody & D. E. Soltis. 2001. A phylogenetic analysis of Hydrangeaceae based on sequences of the plastid gene *matK* and their combination with *rbcL* and morphological data. *Int. J. Plant Sci.* 162: 835–846.

Hughes, J. W. & W. B. Cass. 1997. Pattern and process of a floodplain forest, Vermont, USA: predicted responses of vegetation to perturbation. *J. Appl. Ecol.* 34: 594–612.

Hughes, M., J. A. Smith, A. E. Mayfield III, M. C. Minno & K. Shin. 2014. First report of laurel wilt disease caused by *Raffaelea lauricola* on pondspice in Florida. *Plant Dis.* 98: 379–383.

Hughes, R., K. Bachmann, N. Smirnoff & M. R. Macnair. 2001. The role of drought tolerance in serpentine tolerance in the *Mimulus guttatus* Fischer ex DC. complex *S. Afr. J. Sci.* 97: 581–586.

Huguet, V., M. Gouy, P. Normand, J. F. Zimpfer & M. P. Fernandez. 2005. Molecular phylogeny of Myricaceae: a reexamination of host-symbiont specifcity. *Mol. Phylogen. Evol.* 34: 557–568.

Huh, M. K. 1999. Genetic diversity and population structure of Korean alder (*Alnus japonica*; Betulaceae). *Can. J. For. Res.* 29: 1311–1316.

Huh, M. K., J. S. Choi, S. G. Moon & H. W. Huh. 2002. Genetic diversity of natural and cultivated populations of *Oenanthe javanica* in Korea. *J. Plant Biol.* 45: 83–89.

Huhta, A.-P., K. Hellström, P. Rautio & J. Tuomi. 2003. Grazing tolerance of *Gentianella amarella* and other monocarpic herbs: why is tolerance highest at low damage levels? *Plant Ecol.* 166: 49–61.

Humbert, L., D. Gagnon, D. Kneeshaw & C. Messier. 2007. A shade tolerance index for common understory species of northeastern North America. *Ecol. Indicators* 7: 195–207.

Humeera, N., A. N. Kamili, S. A.Bandh, B. A. Lone & N. Gousia. 2013. Antimicrobial and antioxidant activities of alcoholic extracts of *Rumex dentatus* L. *Microbial Pathog.* 57: 17–20.

Humphreys, H. N. 1857. *River Gardens; Being an Account of the Best Methods of Cultivating Fresh-Water Plants in Aquaria, in Such a Manner as to Afford Suitable Abodes to Ornamental Fish, and Many Interesting Kinds of Aquatic Animals.* Sampson Low, Son & Co., London. 108 pp.

Hung, K. H., B. A. Schaal, T. W. Hsu, Y. C. Chiang, C. I. Peng & T. Y. Chiang. 2009. Phylogenetic relationships of diploid and polyploid species in *Ludwigia* sect. *Isnardia* (Onagraceae) based on chloroplast and nuclear DNAs. *Taxon* 58: 1216–1226.

Hunt, J. 1992. Feeding ecology of valley pocket gophers (*Thomomys bottae sanctidiegi*) on a California coastal grassland. *Amer. Midl. Naturalist* 127: 41–51.

Hunt, J., B. Anderson, B. Phillips, R. Tjeerdema, B. Largay, M. Beretti & A. Bern. 2008. Use of toxicity identification evaluations to determine the pesticide mitigation effectiveness of onfarm vegetated treatment systems. *Environ. Pollut., A.* 156: 348–358.

Hunt, K. W. 1943. Floating mats on a southeastern coastal plain reservoir. *Bull. Torrey Bot. Club* 70: 481–488.

Hunt, W. R. 1927. Miscellaneous collections of North American rusts. *Mycologia* 19: 286–288.

Hunter, J. E. 1986. Some responses of *Sidalcea calycosa* (Malvaceae) to fire. *Madroño* 33: 305–307.

Hunt-Joshi, T. R., R. B. Root & B. Blossey. 2005. Disruption of weed biological control by an opportunistic Mirid predator. *Ecol. Appl.* 15: 861–870.

Huntley, M. J. W. 2012. High-resolution Late Holocene climate change and human impacts on a hypermaritime peatland on Haida Gwaii, BC. MS thesis. Simon Fraser University, Vancouver, BC. 76 pp.

Huntzinger, M., R. Karban & J. L. Maron. 2011. Small mammals cause non-trophic effects on habitat and associated snails in a native system. *Oecologia* 167: 1085–1091.

Hurd Jr., P. D. & E. G. Linsley. 1976. The bee family Oxaeidae with a revision of the North American species (Hymenoptera: Apoidea). *Smithsonian Contr. Zool.* 220: 1–75.

Hurd Jr., P. D., W. E. LaBerge & E. G. Linsley. 1980. *Principal Sunflower Bees of North America with Emphasis on the Southwestern United States (Hymenoptera: Apoidea).* Smithsonian Contributions to Zoology, number 310. Smithsonian Institution Press, Washington, DC. 158 pp.

Hurd, T. M., D. J. Raynal & C. R. Schwintzer. 2001. Symbiotic N2 fixation of *Alnus incana ssp. rugosa* in shrub wetlands of the Adirondack Mountains, New York, USA. *Oecologia* 126: 94–103.

Hurst, C. A. 2001. Sinker cypress: treasures of a lost landscape. MA thesis. The Department of Geography and Anthropology, Louisiana State University, Baton Rouge, LA. 117 pp.

Hurst, C. C. 1928. Differential polyploidy in the genus *Rosa* L. *Z. Indukt. Abstammungs-Vererbungsl.* 46: S46–S47.

Hussain, I. & H. Kahn. 2010. Investigation of heavy metals content in medicinal plant, *Eclipta alba* L. *J. Chem. Soc. Pakistan* 32: 28–33.

Hussein, A. A., B. Rodriguez, M. P. Martinez-Alcazar & F. H. Cano. 1999. Diterpenoids from *Lycopus europaeus* and *Nepeta septemcrenata*: revised structures and new isopimarane derivatives. *Tetrahedron* 55: 7375–7388.

Hussner, A. 2012. Alien aquatic plant species in European countries. *Weed Res.* 52: 297–306.

Hussner, A. & R. Lösch. 2007. Growth and photosynthesis of *Hydrocotyle ranunculoides* L. fil. in Central Europe. *Flora* 202: 653–660.

Hussner, A. & C. Meyer. 2009. The influence of water level on the growth and photosynthesis of *Hydrocotyle ranunculoides* L.fil. *Flora* 204: 755–761.

Hutchens Jr., J. J. & J. O. Luken. 2015. Prey capture success by established and introduced populations of the Venus flytrap (*Dionaea muscipula*). *Ecol. Restorat.* 33: 171–177.

Hutchens Jr., J. J., J. B. Wallace & E. D. Romaniszyn. 2004. Role of *Podostemum ceratophyllum* Michx. in structuring benthic macroinvertebrate assemblages in a southern Appalachian river. *J. N. Amer. Benthol. Soc.* 23: 713–727.

Hutchings, S. S. 1932. Light in relation to the seed germination of *Mimulus ringens* L. *Amer. J. Bot.* 19: 632–643.

Hutchinson, G. E. 1970. The chemical ecology of three species of *Myriophyllum* (Angiospermae, Haloragaceae). *Limnol. Oceanogr.* 15: 1–5.

Hutchinson, G. E. 1975. *A Treatise on Limnology. Volume 3: Limnological Botany.* John Wiley & Sons, New York. 660 pp.

Hutchinson, R. A., J. H. Viers & J. F. Quinn. 2007. *Soil Seed Bank Analysis from the Cosumnes River Preserve Lepidium Control Experiment.* Project report ERP02D-P66. Department of Environmental Science & Policy, University of California, Davis, CA. 91 pp.

Hutton, E. E. 1974. A large vegetational formation of Cretaceous and Tertiary origin in West Virginia. *Castanea* 39: 71–76.

Hutton, E. E. 1976. Dissemination of perigynia in *Carex pauciflora*. *Castanea* 41: 346–348.

Huxtable, R. J. 1980. Herbal teas and toxins: novel aspects of pyrrolizidine poisoning in the United States. *Perspect. Biol. Med.* 24: 1–14.

Huxtable, R. J. 1989. Human health implications of pyrrolizidine alkaloids and herbs containing them. pp. 41–86. *In*: P. R. Cheeke (ed.), *Toxicants of Plant Origin.* CRC Press, Boca Raton, FL.

Iamonico, D. & I. Sánchez-Del Pino. 2016. Taxonomic revision of the genus *Alternanthera* (Amaranthaceae) in Italy. *Plant Biosyst.* 150: 333–342.

Igersheim, A. 1993. The character states of the Caribbean monotypic endemic *Strumpfia* (Rubiaceae). *Nordic J. Bot.* 13: 545–559.

Ignatowicz, S. & M. Gersz. 1997. Extracts of medical herbs as repellents and attractants for the dry bean weevil, *Acanthoscelides obtectus* Say (Coleoptera: Bruchidae). *Polskie Pismo Entomol.* 66: 151–159.

Ikeda, H. & K. Itoh. 2001. Germination and water dispersal of seeds from a threatened plant species *Penthorum chinense*. *Ecol. Res.* 16: 99–106.

Ikeda, E. & R. Miura. 1994. A note on the production of cleistogamous flowers in *Lindernia dubia* (L.) Pennel (Scrophulariaceae) in a paddy field. *Weed Res. (Japan)* 39: 177–179.

Ikusima, I. & J. G. Gentil. 1996. Evaluation of the faster initial decomposition of tropical floating leaves of *Nymphaea elegans* Hook. *Ecol. Res.* 11: 201–206.

Imai, H., K. Osawa, H. Yasuda, H. Hamashima, T. Arai & M. Sasatsu. 2001. Inhibition by the essential oils of peppermint and spearmint of the growth of pathogenic bacteria. *Microbios* 106: S31–S39.

Imbert, F. M. & J. H. Richards. 1993. Protandry, incompatibility, and secondary pollen presentation in *Cephalanthus occidentalis* (Rubiaceae). *Amer. J. Bot.* 80: 395–404.

Imbruce, V. 2007. Bringing southeast Asia to the southeast United States: new forms of alternative agriculture in Homestead, Florida. *Agric. Human Values* 24: 41–59.

Imbs, A. B. & L. Q. Pham. 1995. Lipid composition of ten edible seed species from North Vietnam. *J. Amer. Oil Chem. Soc.* 72: 957–961.

Imtiaz, B., A. Waheed, A. Rehman, H. Ullah, H. Iqbal, A Wahab, M Almas & I. Ahmad. 2012. Antimicrobial activity of *Malva neglecta* and *Nasturtium microphyllum*. *Int. J. Res. Ayurveda Pharm.* 3: 808–810.

Ingham, E. R. & M. V. Wilson. 1999. The mycorrhizal colonization of six wetland plant species at sites differing in land use history. *Mycorrhiza* 9: 233–235.

Ingles, L. G. 1952. The ecology of the mountain pocket gopher, *Thomomoys monticola*. *Ecology* 33: 87–95.

Ingles, L. G. 1961. Home range and habitats of the wandering shrew. *J. Mammalogy* 42: 455–462.

Ingolia, M., T. P. Young & E. G. Sutter. 2008. Germination ecology of *Rorippa subumbellata* (Tahoe yellow cress), an endangered, endemic species of Lake Tahoe. *Seed Sci. Technol.* 36: 621–632.

Ingram, D. L., P. G. Webb & R. H. Biggs. 1986. Interactions of exposure time and temperature on thermostability and protein content of excised *Illicium parviflorum* roots. *Plant Soil* 96: 69–76.

Ingrouille, M. J., M. W. Chase, M. F. Fay, D. Bowman, M. van der Bank & A. D. E. Bruijn. 2002. Systematics of Vitaceae from the viewpoint of plastid *rbc*L sequence data. *Bot. J. Linn. Soc.* 138: 421–432.

Inouye, B. D & N. Underwood. 2004. Persistence of a fugitive species: *Mimulus angustatus* on gopher mounds. Annual Meeting of the Ecological Society of America, Portland, OR (poster).

Inouye, D. W., B. Barr, K. B. Armitage & B. D. Inouye. 2000. Climate change is affecting altitudinal migrants and hibernating species. *Proc. Natl. Acad. Sci. U.S.A* 97: 1630–1633.

Inouye, D. W., B. M. H. Larson, A. Ssymank & P. G. Kevan. 2015. Flies and flowers III. Ecology of foraging and pollination. *J. Pollination Ecol.* 16: 115–133.

Inserra, R. N., J. H. O'Bannon & L. W. Duncan. 1990a. Native hosts of *Pratylenchus coffeae* in Florida. Nematology circular. Florida Department of Agriculture and Consumer Services, Division of Plant Industry, Gainesville, FL. 3 pp.

Inserra, R. N., N. Vovlas, A. P. Nyczepir, E. J. Wehunt & A. M. Golden. 1990b. *Tylenchulus palustris* parasitizing peach trees in the United States. *J. Nematol.* 22: 45–55.

Ionta, G. M., N. H. Williams, W. M. Whitten & W. S. Judd. 2004. Relationships within a temperate genus of Melastomataceae: *Rhexia*, the meadow beauties. Botany 2004 meeting (abstract).

Ionta, G. M., W. S. Judd, N. H. Williams & W. M. Whitten. 2007. Phylogenetic relationships in *Rhexia* (Melastomataceae): evidence from DNA sequence data and morphology. *Int. J. Plant Sci.* 168: 1055–1066.

Irvine, F. R. & R. S. Trickett. 1953. Waterlilies as food. *Kew Bull.* 8: 363–370.

Isaacs, R., J. Tuell, A. Fiedler, M. Gardiner & D. Landis. 2009. Maximizing arthropod-mediated ecosystem services in agricultural landscapes: the role of native plants. *Front. Ecol. Environ.* 7: 196–203.

Isaak, D., W. H. Marshall & F. M. Buell. 1959. A record of reverse plant succession in a tamarack bog. *Ecology* 40: 317–320.

Isely, D. 1990. *Vascular Flora of the Southeastern United States. Volume 3, Part 2. Leguminosae (Fabaceae)*. University of North Carolina Press, Chapel Hill, NC. 258 pp.

Isely, D. & W. H. Wright. 1951. Noxious weeds and their seeds: miscellaneous species. *Proc. Assoc. Off. Seed Analysts* 41: 139–144.

Islam, M. N., C. A. Wilson & T. R. Watkins. 1982. Nutritional analysis of seashore mallow seed, *Kosteletzkya virginica*. *J. Agric. Food Chem.* 30: 1195–1198.

Islam, M. N., F. Downey & C. K. Y. Ng. 2011. Comparative analysis of bioactive phytochemicals from *Scutellaria baicalensis, Scutellaria lateriflora, Scutellaria racemosa, Scutellaria tomentosa* and *Scutellaria wrightii* by LC-DAD-MS. *Metabolomics* 7: 446–453.

Ismailoglu, U. B., I. Saracoglu, U. S. Harput & I. Sahin-Erdemli. 2002. Effects of phenylpropanoid and iridoid glycosides on free radical-induced impairment of endothelium-dependent relaxation in rat aortic rings. *J. Ethnopharmacol.* 79: 193–197.

Ito, M., K. Watanabe, Y. Kita, T. Kawahara, D. J. Crawford & T. Yahara. 2000. Phylogeny and phytogeography of *Eupatorium* (Eupatorieae, Asteraceae): insights from sequence data of the nrDNA ITS regions and cpDNA RFLP. *J. Plant Res.* 113: 79–89.

Ito, T. M. & T. U. Esugi. 2004. Invasive alien species in Japan: the status quo and the new regulation for prevention of their adverse effects. *Glob. Environ. Res.* 8: 171–191.

Itoh, K. 1994. Weed ecology and its control in south-east tropical countries. *Jap. J. Trop. Agric.* 38: 369–373.

Ives, S. A. 1923. Maturation and germination of seeds of *Ilex opaca*. *Bot. Gaz.* 76: 60–77.

Ivey, C. T. & R. Wyatt. 1999. Family outcrossing rates and neighborhood floral density in natural populations of swamp milkweed (*Asclepias incarnata*): potential statistical artifacts. *Theor. Appl. Genet.* 98: 1063–1071.

Ivey, C. T., S. R. Lipow & R. Wyatt. 1999. Mating systems and interfertility of swamp milkweed (*Asclepias incarnata* ssp. *incarnata* and ssp. *pulchra*). *Heredity* 82: 25–35.

Ivey, C. T., P. Martinez & R. Wyatt. 2003. Variation in pollinator effectiveness in swamp milkweed, *Asclepias incarnata* (Apocynaceae). *Amer. J. Bot.* 90: 214–225.

Ivie, R. A. & W. H. Watson. 1974. Cuauhtemone. *Acta Cryst.* B30: 2891.

Iwamoto, A., R. Izumidate & L. P. R. De Craene. 2015. Floral anatomy and vegetative development in *Ceratophyllum demersum*: a morphological picture of an "unsolved" plant. *Amer. J. Bot.* 102: 1578–1589.

Izhaki, I. 2002. Emodin: a secondary metabolite with multiple ecological functions in higher plants. *New Phytol.* 155: 205–217.

Jabbour, F. & S. S. Renner. 2012. A phylogeny of Delphinieae (Ranunculaceae) shows that *Aconitum* is nested within *Delphinium* and that Late Miocene transitions to long life cycles in the Himalayas and Southwest China coincide with bursts in diversification. *Mol. Phylogen. Evol.* 62: 928–942.

Jackson, H. S. 1915. The Ustilaginales of Indiana. *Proc. Indiana Acad. Sci.* 25: 429–476.

Jackson, H. S. 1918. The Uredinales of Oregon. *Brooklyn Bot. Gard. Mem.*1: 198–297.

Jackson, H. S. 1921. The Ustilaginales of Indiana II. *Proc. Indiana Acad. Sci.* 1920: 157–164.

Jackson, H. S. 1922. New or noteworthy rusts on Carduaceae. *Mycologia* 14: 104–120.

Jackson, L. E. & L. C. Bliss. 1982. Distribution of ephemeral herbaceous plants near treeline in the Sierra Nevada, California, U.S.A. *Arct. Alp. Res.* 14: 33–43.

Jacobs, B., M. J. Donoghue, F. Bouman, S. Huysmans & E. Smets. 2008. Evolution and phylogenetic importance of endocarp and seed characters in *Viburnum* (Adoxaceae). *Int. J. Plant Sci.* 169: 409–431.

Jacobs, M. J. & H. J. Macisaac. 2009. Modelling spread of the invasive macrophyte *Cabomba caroliniana. Freshwater Biol.* 54: 296–305.

Jacobs, S. W. L, E. Perrett, M. Brock, K. H. Bowmer, G. McCorkelle, J. Rawlings, J. Stricker & G. Rr. Sainty. 1993. *Ludwigia peruviana*-description and biology. *Proceedings I, 10th Australian Weeds Conference and 14th Asian Pacific Weed Science Society Conference* 6–10 Sept 1993, Brisbane, Australia.

Jacono, C. C. & V. V. Vandiver Jr. 2007. *Rotala rotundifolia*, Purple Loosestrife of the South? *Aquatics* 29: 4–9.

Jacquemart, A-L. 1997. *Vaccinium oxycoccos* L. (*Oxycoccus palustris* Pers.) and *Vaccinium microcarpum* (Turcz. ex Rupr.) Schmalh. (*Oxycoccus microcarpus* Turcz. ex Rupr.). *J. Ecol.* 85: 381–396.

Jacquemart, A-L. 1998. *Andromeda polifolia* L. *J. Ecol.* 86: 527–541.

Jacques, D. R. 1973. Reconnaissance botany of alpine ecosystems of Prince of Wales Island, southeast Alaska. MS thesis. Oregon State University, Corvallis, OR. 133 pp.

Jaeger, C. H. & R. K. Monson. 1992. Adaptive significance of nitrogen storage in *Bistorta bistortoides*, an alpine herb. *Oecologia* 92: 578–585.

Jain, A., M. Sundriyal, S. Roshnibala, R. Kotoky, P. B. Kanjilal, H. B. Singh & R. C. Sundriyal. 2011. Dietary use and conservation concern of edible wetland plants at Indo-Burma hotspot: a case study from northeast India. *J. Ethnobiol. Ethnomed.* 7: 29.

Jain, R. K., M. R. Khan & V. Kumar. 2012. Rice root-knot nematode (*Meloidogyne graminicola*) infestation in rice. *Arch. Phytopathol. Plant Protect.* 45: 635–645.

Jain, S. K. 1978. Local dispersal of *Limnanthes* nutlets: an experiment with artificial vernal pools. *Can. J. Bot.* 56: 1995–1997.

Jaiswal, N., V. Bhatia, S. P. Srivastava, A. K. Srivastava & A. K. Tamrakara. 2012. Antidiabetic effect of *Eclipta alba* associated with the inhibition of alpha-glucosidase and aldose reductase. *Nat. Prod. Res.* 26: 2363–2367.

Jalalpure, S. S., N. Agrawal, M. B. Patil, R. Chimkode & A. Tripathi. 2008. Antimicrobial and wound healing activities of leaves of *Alternanthera sessilis* Linn. *Int. J. Green Pharm.* 2: 141–144.

James, M L. & J. B. Zedler. 2000. Dynamics of wetland and upland subshrubs at the salt marsh-coastal sage scrub ecotone. *Amer. Midl. Naturalist* 143: 298–311.

James, M. T. 1960. The soldier flies or Stratiomyidae of California. *Bull. Calif. Insect Surv.* 6: 79–122.

James, R. L. 1948. Some hummingbird flowers east of the Mississippi. *Castanea* 13: 97–109.

Janeczko, Z., J. Sendra, K. Kmieć & C. H. Brieskorna. 1990. A triterpenoid glycoside from *Menyanthes trifoliata. Phytochemistry* 29: 3885–3887.

Jang, C.-G., A. N. Müllner & J. Greimler. 2005. Conflicting patterns of genetic and morphological variation in European *Gentianella* section *Gentianella. Bot. J. Linn. Soc.* 148: 175–187.

Jankovsky-Jones, M. 1999. *Conservation Strategy for Wetlands in East-Central Idaho.* Natural Resource Policy Bureau, Idaho Department of Fish and Game. Boise, ID. 68 pp.

Janousek, C. N. & C. L. Folger. 2012. Patterns of distribution and environmental correlates of macroalgal assemblages and sediment chlorophyll A in Oregon tideal wetlands. *J. Phycol.* 48: 1448–1457.

Jansen, R. K., C. Kaittanis, S.-B. Lee, C. Saski, J. Tomkins, A. J. Alverson & H. Daniell. 2006. Phylogenetic analyses of *Vitis* (Vitaceae) based on complete chloroplast genome sequences: effects of taxon sampling and phylogenetic methods on resolving relationships among rosids. *BMC Evol. Biol.* 6: 32. doi:10.1186/1471-2148-6-32.

Jansen, R. K. M. F. Wojciechowski, E. Sanniyasi, S.-B. Lee & H. Daniell. 2008. Complete plastid genome sequence of the chickpea (*Cicer arietinum*) and the phylogenetic distribution of *rps12* and *clpP* intron losses among legumes (Leguminosae). *Mol. Phylogen. Evol.* 48: 1204–1217.

Jansen, S., T. Watanabe & E. Smets. 2002. Aluminium accumulation in leaves of 127 species in Melastomataceae, with comments on the order Myrtales. *Ann. Bot.* 90: 53–64.

Janssen, C. R. 1967. A floristic study of forests and bog vegetation, northwestern Minnesota. *Ecology* 48: 751–765.

Janssens, J. A. & T. A. Johnson. 2001. *Status Report on the Vegetation of the Sioux Nation Fen Complex, Yellow Medicine County, Minnesota: Year 2000.* March 27, 2001. Report to the Minnesota Department of Natural Resources, Division of Waters, St. Paul, MN. 32 pp.

Jansson, R., U. Zinko & D. M. Merritt, C. Nilsson. 2005. Hydrochory increases riparian plant species richness: a comparison between a free-flowing and a regulated river. *J. Ecol.* 93: 1094–1103.

Jantrarotai, P. 1991. Water hyacinth (*Eichhornia crassipes*) and water pennywort (*Hydrocotyle ranunculoides*) use in treatment of poultry wastewater. *Diss. Abstr. Int. B.* 51: 89.

Janzen, D. H. 1984. Dispersal of small seeds by big herbivores: foliage is the fruit. *Amer. Naturalist* 123: 338–353.

Jacques, D. R. 1973. Reconnaissance botany of alpine ecosystems of Prince of Wales Island, southeast Alaska. MS thesis. Oregon State University, Corvallis, OR. 133 pp.

Jarolímová V. 1998. Chromosome counts of *Rorippa sylvestris* in the Czech Republic. *Preslia* 70: 69–73.

Jarolímová V. 2005. Experimental hybridization between species of the genus *Rorippa. Preslia* 77: 277–296.

Jaroszewicz, B. & E. Pirożnikow. 2008. Diversity of plant species eaten and dispersed by the European bison *Bison bonasus* in Białowieża Forest. *Eur. Bison Conserv. Newslett.* 1: 14–29.

Järvinen, P., A. Palmé, L. O. Morales, M. Lännenpää, M. Keinänen, T. Sopanen & M. Lascoux. 2004. Phylogenetic relationships of *Betula* species (Betulaceae) based on nuclear *ADH* and chloroplast *mat*K sequences. *Amer. J. Bot.* 91: 1834–1845.

Jarzen, D. M. & J. A. Hogsette. 2008. Pollen from the exoskeletons of stable flies, *Stomoxys calcitrans* (Linnaeus 1758), in Gainesville, Florida, USA. *Palynology* 32: 77–81.

Jauzein, P. 1991. *Eclipta prostrata* (L.) L. a weed of rice fields in the Camargue. *Monde Plant* 86: 15–16.

Javed, S. M. & L. E. Urbatsch. 2014. *Tetragonia tetragonioides* (Aizoaceae) discovered in Louisiana, USA. *J. Bot. Res. Inst. Texas.* 8: 661–662.

Jayachandran, K. & J. Fisher. 2008. Arbuscular mycorrhizae and their role in plant restoration in native ecosystems. pp. 195–209. *In*: Z. A. Siddiqui, M. S. Akhtar & K. Futai (eds.), *Mycorrhizae: Sustainable Agriculture and Forestry*. Springer Science, Dordrecht, The Netherlands.

Jayathirthaa, M. G. & S. H. Mishraa. 2004. Preliminary immuno-modulatory activities of methanol extracts of *Eclipta alba* and *Centella asiatica*. *Phytomedicine* 11: 361–365.

Jaynes, M. L. & S. R. Carpenter. 1986. Effects of vascular and non-vascular macrophytes on sediment redox and solute dynamics. *Ecology* 67: 875–882.

Jaynes, R. A. 1968. Interspecific crosses in *Kalmia*. *Amer. J. Bot.* 55: 1120–1125.

Jaynes, R. A. 1969. Chromosome counts of *Kalmia* species and revaluation of *K. polifolia* var. *microphylla*. *Rhodora* 71: 280–284.

Jaynes, R. A. 1988. *Kalmia: The Laurel Book II*. Timber Press, Portland, OR. 220 pp.

Jean, M. & A. Bouchard. 1993. Riverine wetland vegetation: importance of small-scale and large-scale environmental variation. *J. Veg. Sci.* 4: 609–620.

Jeandroz, S., A. Roy & J. Bousquet. 1997. Phylogeny and phylogeography of the circumpolar genus *Fraxinus* (Oleaceae) based on internal transcribed spacer sequences of nuclear ribosomal DNA. *Mol. Phylogen. Evol.* 7: 241–251.

Jedlička, J. & K. Prach. 2006. A comparison of two North-American asters invading in central Europe. *Flora* 201: 652–657.

Jefferies, R. L. 1977. The vegetation of salt marshes at some coastal sites in Arctic North America. *J. Ecol.* 65: 661–672.

Jefferies, R. L. & R. F. Rockwell. 2002. Foraging geese, vegetation loss and soil degradation in an Arctic salt marsh. *Appl. Veg. Sci.* 5: 7–16.

Jefferies, R. L., A. Jensen & K. F. Abraham. 1979. Vegetational development and the effect of geese on vegetation at La Perouse Bay, Manitoba. *Can. J. Bot.* 57: 1439–1450.

Jefferson, C. A. 1974. Plant communities and succession in Oregon coastal salt marshes. PhD dissertation. Oregon State University. Corvallis, OR. 192 pp.

Jegede, I. A., F. C. Nwinyi, J. Ibrahim, G. Ugbabe, S. Dzarma & O. F. Kunle. 2009. Investigation of phytochemical, anti-inflammatory and anti-nociceptive properties of *Ipomoea asarifolia* leaves. *J. Med. Plant Res.* 3: 160–165.

Jeglum, J. K. 1971. Plant indicators of pH and water level in peatlands at Candle Lake, Saskatchewan. *Can. J. Bot.* 49: 1661–1676.

Jehl Jr., J. 1973. Breeding biology and systematic relationships of the stilt sandpiper. *Wilson Bull.* 85: 115–147.

Jenkins, K. J. & E. E. Starkey. 1991. Food habits of Roosevelt elk. *Rangelands* 13: 261–265.

Jenkins, K. J. & E. E. Starkey. 1993. Winter forages and diets of elk in old-growth and regenerating coniferous forests in western Washington. *Amer. Midl. Naturalist* 130: 299–313.

Jennings, D. 1941. Fall food habits of the bobwhite quail in eastern Kansas. *Trans. Kansas Acad. Sci.* 44: 420–426.

Jennings, O. E. 1942. The verticillate eupatoriums of western Pennsylvania. *Castanea* 7: 43–48.

Jensch, D. & P. Poschlod. 2008. Germination ecology of two closely related taxa in the genus *Oenanthe*: fine tuning for the habitat? *Aquat. Bot.* 89: 345–351.

Jensen, K. 2004. Dormancy patterns, germination ecology, and seed-bank types of twenty temperate fen grassland species. *Wetlands* 24: 152–166.

Jensen, K. & C. Meyer. 2001. Effects of light competition and litter on the performance of *Viola palustris* and on species composition and diversity of an abandoned fen grassland. *Plant Ecol.* 155: 169–181.

Jensen, M. K., J. K. Vogt, S. Bressendorff, A. Seguin-Orlando, M. Petersen, T. Sicheritz-Pontén & J. Mundy. 2015. Transcriptome and genome size analysis of the Venus flytrap. *PLoS One* 10(4): e0123887. doi:10.1371/journal.pone.0123887.

Jensen, N. J. 2007. *The Habitat of* Astragalus pycnostachyus *var.* lanosissimus *(Ventura Marsh Milk-Vetch) and an Assessment of Potential Future Planting Sites*. California Native Plant Society, Channel Islands Chapter. Ojai, CA. 95 pp.

Jensen, S. R. & B. J. Nielsen. 1985. Hygrophiloside, an iridoid glucoside from *Hygrophila difformis* (Acanthaceae). *Phytochemistry* 24: 602–603.

Jensen, S. R. & J. Schripsema. 2002. Chemotaxonomy and pharmacology of Gentianaceae. pp. 573–631. *In*: L. Struwe & V. A. Albert (eds.), *Gentianaceae: Systematics and natural history*. Cambridge University Press, Cambridge, UK.

Jensen, S. R., H. Franzyk & E. Wallander. 2002. Chemotaxonomy of the Oleaceae: iridoids as taxonomic markers. *Phytochemistry* 60: 213–231.

Jensen, U., S. B. Hoot, J. T. Johansson & K. Kosuge. 1995. Systematics and phylogeny of the Ranunculaceae-a revised family concept on the basis of molecular data. *Plant Syst. Evol.* 9: S273–S280.

Jerling, L. 1988a. Population dynamics of *Glaux maritima* (L.) along a distributional cline. *Plant Ecol.* 74: 161–170.

Jerling, L. 1988b. Clone dynamics, population dynamics and vegetation pattern of *Glaux maritima* on a Baltic sea shore meadow. *Plant Ecol.* 74: 171–185.

Jernigan, K. J. & A. N. Wright. 2011. Effect of repeated short interval flooding events on root and shoot growth of four landscape shrub taxa. *J. Environ. Hort.* 29: 220–222.

Jervis, M. A., N. A. C. Kidd, M. G. Fitton, T. Huddleston & H. A. Dawah. 1993. Flower-visiting by hymenopteran parasitoids. *J. Nat. Hist.* 27: 67–105.

Jeschke, M. R., P. J. Tranel & A. L. Rayburn. 2003. DNA content analysis of smooth pigweed (*Amaranthus hybridus*) and tall waterhemp (*A. tuberculatus*): implications for hybrid detection. *Weed Sci.* 51: 1–3.

Jeske, C. W., H. F. Percival & J. E. Thul. 1993. Food habits of ring-necked ducks wintering in Florida. *Proc. Ann. Conf. S. E. Assoc. Fish Wildl. Agencies* 47: 130–137.

Jesurun, J., S. Jagadeesh, S. Ganesan, V. Rao & M. Eerike. 2013. Anti-inflammatory activity of ethanolic extract of *Nymphaea alba* flower in Swiss albino mice. *Int. J. Med. Res. Health Sci.* 2: 474–478.

Jha, A. K., K. Prasad, V. Kumar & K. Prasad. 2009. Biosynthesis of silver nanoparticles using *Eclipta* leaf. *Biotechnol. Prog.* 25: 1476–1479.

Jiang, N., W. B. Yu, H. Z. Li & K. Y. Guan. 2010. Floral traits, pollination ecology and breeding system of three *Clematis* species (Ranunculaceae) in Yunnan province, southwestern China. *Austral. J. Bot.* 58: 115–123.

Jiménez-Osornio, F. M. V. Z. J., J. Kumamoto & C. Wasser. 1996. Allelopathic activity of *Chenopodium ambrosioides* L. *Biochem. Syst. Ecol.* 24: 195–205.

Jiménez-Pérez, N. D. C. & F. G. Lorea-Hernández. 2009. Identity and delimitation of the American species of *Litsea* Lam. (Lauraceae): a morphological approach. *Plant Syst. Evol.* 283: 19–32.

Jin, B., L. Wang, J. Wang, N.-J. Teng, X.-D. He, X.-J. Mu & Y.-L. Wang. 2010. The structure and roles of sterile flowers in *Viburnum macrocephalum* f. *keteleeri* (Adoxaceae). *Plant Biol.* 12: 853–862.

Jing, S.-R., Y.-F. Lin, T.-W. Wang & D.-Y. Lee. 2002. Microcosm wetlands for wastewater treatment with different hydraulic loading rates and macrophytes. *J. Environ. Qual.* 31: 690–696.

Jin-Gui, M. A. [et al.] 2012. Research of salt resistance of *Hibiscus grandiflorus*. *J. Anhui Agric. Sci.* 21: 13.

Jiwajinda, S., V. Santisopasri, A. Murakami, O.-K. Kim, H. W. Kim & H. Ohigashi. 2002. Suppressive fffects of edible Thai plants on superoxide and nitric oxide generation. *Asian Pacific J. Cancer Prev.* 3: 215–223.

Jobes, D. V., D. L. Hurley & L. B. Thien. 1998. Molecular systematics of the Magnoliaceae: a review of recent molecular studies. pp. 59–70. *In*: D. Hunt (ed.), *Magnolias and Their Allies*. D. Hunt Publisher, Sherborne, UK.

Jobson, R. W. & V. A. Albert. 2002. Molecular rates parallel diversification contrasts between carnivorous plant sister lineages. *Cladistics* 18: 127–136.

Jobson, R. W., J. Playford, K. M. Cameron & V. A. Albert. 2003. Molecular phylogenetics of Lentibulariaceae inferred from plastid *rps*16 intron and *trn*L-F DNA sequences: implications for character evolution and biogeography. *Syst. Bot.* 28: 157–171.

Joel, D. M., B. E. Juniper & A. Dafni. 1985. Ultraviolet patterns in the traps of carnivorous plants. *New Phytol.* 101: 585–593.

Jog, S. K., J. T. Kartesz, J. R. Johansen & G. J. Wilder. 2005. Floristic study of Highland Heights Community Park, Cuyahoga County, Ohio. *Castanea* 70: 136–145.

Johansson, J. T. 1995. A revised chloroplast DNA phylogeny of the Ranunculaceae. *Plant Syst. Evol.* 9: S253–S261.

Johansson, J. T. 1998. Chloroplast DNA restriction site mapping and the phylogeny of *Ranunculus* (Ranunculaceae). *Plant Syst. Evol.* 213: 1–19.

Johansson, J. T. & R. K. Jansen. 1993. Chloroplast DNA variation and phylogeny of the Ranunculaceae. *Plant Syst. Evol.* 187: 29–49.

Johansson, M. E. & P. A. Keddy. 1991. Intensity and asymmetry of competition between plant pairs of different degrees of similarity: an experimental study on two guilds of wetland plants. *Oikos* 60: 27–34.

John-Africa, L. B., M. S. Idris-Usman, B. Adzu & K. S. Gamaniel. 2012. Protective effects of the aqueous extract of *Nymphaea lotus* L.(Nymphaeaceae) against ethanol-induced gastric ulcers. *Int. J. Biol. Chem. Sci.* 6: 1917–1925.

Johnson, A. M. 1927. The status of *Saxifraga nuttallii*. *Amer. J. Bot.* 14: 38–43.

Johnson, A. M. & D. J. Leopold. 1994. Vascular plant species richness and rarity across a minerotrophic gradient in wetlands of St. Lawrence County, New York, USA. *Biodivers. Conservation* 3: 606–627.

Johnson, D. A. 1942. Chromosomes of *Jamesianthus*. *Rhodora* 44: 280.

Johnson, D. A., L. M. Carris & J. D. Rogers. 1997. Morphological and molecular characterization of *Colletotrichum nymphaeae* and *C. nupharicola* sp. nov. on water-lilies (*Nymphaea* and *Nuphar*). *Mycol. Res.* 101: 641–649.

Johnson, D. A., G. Pimentel & F. M. Dugan. 2008. *Cladosporium herbarum* causes a leaf spot on *Caltha leptosepala* (marsh-marigold) in western North America. *Plant Health Prog.* doi:10.1094/PHP-2008-1121-01-RS.

Johnson, D. M. 2003. Phylogenetic significance of spiral and distichous architecture in the Annonaceae. *Syst. Bot.* 28: 503–511.

Johnson, D. S. 1935. The development of the shoot, male flower and seedling of *Batis maritima* L. *Bull. Torrey Bot. Club* 62: 19–31.

Johnson, J. B. & D. A. Steingraeber. 2003. The vegetation and ecological gradients of calcareous mires in the South Park valley, Colorado. *Can. J. Bot.* 81: 201–219.

Johnson, K. S. 1999. Comparative detoxification of plant (*Magnolia virginiana*) allelochemicals by generalist and specialist Saturniid silkmoths. *J. Chem. Ecol.* 25: 253–269.

Johnson, K. W. & N. A. Altman. 1999. Canonical correspondence analysis as an approximation to Gaussian ordination. *Environmetrics* 10: 39–52.

Johnson, L. A., J. L. Schultz, D. E. Soltis & P. S. Soltis. 1996. Monophyly and generic relationships of Polemoniaceae based on *matK* sequences. *Amer. J. Bot.* 83: 1207–1224.

Johnson, L. M. K. & L. F. Galloway. 2008. From horticultural plantings into wild populations: movement of pollen and genes in *Lobelia cardinalis*. *Plant Ecol.* 197: 55–67.

Johnson, M. F. 1974. Eupatorieae (Asteraceae) in Virginia: *Eupatorium* L. *Castanea* 39: 205–228.

Johnson, M. P. & S. A. Cook. 1968. "Clutch size" in buttercups. *Amer. Naturalist* 102: 405–411.

Johnson, M. T. & C. J. Rothfels. 2001. The establishment and proliferation of the rare exotic plant, *Lythrum hyssopifolia*, Hyssop-leaved Loosestrife, at a pond in Guelph, Ontario. *Can. Field-Naturalist* 115: 229–233.

Johnson, P. L. & W. D. Billings. 1962. The alpine vegetation of the Beartooth Plateau in relation to cryopedogenic processes and patterns. *Ecol. Monogr.* 32: 105–135.

Johnson, P. N. 1982. Naturalised plants in south-west South Island, New Zealand. *New Zealand J. Bot.* 20: 131–142.

Johnson, P. N. & P. A. Brooke. 1998. *Wetland Plants in New Zealand*. Manaaki Whenua Press, Lincoln, New Zealand. 320 pp.

Johnson, R. G. & R. C. Anderson. 1986. The seed bank of a tallgrass prairie in Illinois. *Amer. Midl. Naturalist* 115: 123–130.

Johnson, R. L. 1990a. *Populus heterophylla* L. Swamp cottonwood. pp. 551–554. *In*: R. M. Burns & B. H. Honkala (eds.), *Silvics of North America. Vol. 2, Hardwoods*. Agriculture Handbook 654. USDA, Forest Service, Washington, DC.

Johnson, R. L. 1990b. *Nyssa aquatica* L. Water Tupelo. pp. 474–478. *In*: R. M. Burns & B. H. Honkala (eds.), *Silvics of North America. Vol. 2, Hardwoods*. Agriculture Handbook 654. USDA, Forest Service, Washington, DC.

Johnson, R. L., E. M. Gross & N. G. Hairston Jr. 1997. Decline of the invasive submersed macrophyte *Myriophyllum spicatum* (Haloragaceae) associated with herbivory by larvae of *Acentria ephemerella* (Lepidoptera). *Aquat. Ecol.* 31: 273–282.

Johnson, S. 2004. Effects of water level and phosphorus enrichment on seedling emergence from marsh seed banks collected from northern Belize. *Aquat. Bot.* 79: 311–323.

Johnson, T. 1999. *CRC Ethnobotany Desk Reference*. CRC Press, Boca Raton, FL. 1211 pp.

Johnston, A. & E. Reekie. 2008. Regardless of whether rising atmospheric carbon dioxide levels increase air temperature, flowering phenology will be affected. *Int. J. Plant Sci.* 169: 1210–1218.

Johnston, A., L. M. Bezeau & S. Smoliak. 1968. Chemical composition and in vitro digestibility of alpine tundra plants. *J. Wildl. Manag.* 32: 773–777.

Johnston, C. A. 2003. Shrub species as indicators of wetland sedimentation. *Wetlands* 23: 911–920.

Johnston, C. A., B. L. Bedford, M. Bourdaghs, T. Brown, C. Frieswyk, M. Tulbure, L. Vaccaro & J. B. Zedler. 2007. Plant species indicators of physical environment in Great Lakes coastal wetlands. *J. Great Lakes Res.* 33: 106–124 (special issue 3).

Johnston, M. O. 1991. Pollen limitation of female reproduction in *Lobelia cardinalis* and *L. siphilitica*. *Ecology* 72: 1500–1503.

Johnston, M. O. 1992. Effects of cross and self-fertilization on progeny fitness in *Lobelia cardinalis* and *L. siphilitica*. *Evolution* 46: 688–702.

Johnstone, I. M. 1982. Yellow waterlily (*Nymphaea mexicana*) in Lake Ohakuri, North Island, New Zealand. *New Zealand J. Bot.* 20: 387–389.

Jokerst, J. D. 1992. *Pogogyne floribunda* (Lamiaceae), a new species from the great basin in northeastern California. *Aliso* 13: 347–353.

Joly, S., J. R. Starr, W. H. Lewis & A. Bruneau. 2006. Polyploid and hybrid evolution in roses east of the Rocky Mountains. *Amer. J. Bot.* 93: 412–425.

Jones, A. G. 1978. Observations on reproduction and phenology in some perennial asters. *Amer. Midl. Naturalist* 99: 184–197.

Jones, A. G. & D. A. Young. 1983. Generic concepts of *Aster* (Asteraceae): a comparison of cladistic, phenetic, and cytological approaches. *Syst. Bot.* 8: 71–84.

Jones, C. C. & R. Del Moral. 2005. Effects of microsite conditions on seedling establishment on the foreland of Coleman Glacier, Washington. *J. Veg. Sci.* 16: 293–300.

Jones, C. D. (ed.). 2006. *Ontario Lepidoptera 2003–2004*. Toronto Entomologists' Association, Toronto, ON. 87 pp.

Jones, C. H. 1941. Studies in Ohio floristics. I. Vegetation of Ohio bogs. *Amer. Midl. Naturalist* 26: 674–689.

Jones, D. E. 2007. *Poison arrows: North American Indian Hunting and Warfare*. University of Texas Press, Austin TX. 144 pp.

Jones, F. R. 1924. A mycorrhizal fungus in the roots of legumes and some other plants. *J. Agric. Res.* 29: 459–470.

Jones, G. 2005. *Vascular Flora of the Fort Whyte Centre, Winnipeg, Manitoba: 2002–2004*. Report No. 2005-02. Habitat Management and Ecosystem Monitoring Section, Wildlife and Ecosystem Protection Branch, Manitoba Conservation, Winnipeg, MB. 37 pp.

Jones, G. D. & K. C. Allen. 2013. Pollen analyses of tarnished plant bugs. *Palynology* 37: 170–176.

Jones, G. N. 1937. New records of vascular plants in Washington. *Madroño* 4: 34–37.

Jones, G. P. & W. Fertig. 1999a. *Ecological Evaluation of the Potential Pat O'Hara Mountain Research Natural Area Within the Shoshone National Forest, Park County, Wyoming*. Report to the Shoshone National Forest, U. S. Department of Agriculture Forest Service. University of Wyoming, Laramie, WY. 50 pp.

Jones, G. P. & W. Fertig. 1999b. *Ecological evaluation of the Potential Beartooth Butte Research Natural Area Within the Shoshone National Forest, Park County, Wyoming*. Report to the Shoshone National Forest, U. S. Department of Agriculture Forest Service. University of Wyoming, Laramie, WY. 56 pp.

Jones, J. 2013. DRAFT Recovery strategy for the Willowleaf Aster (*Symphyotrichum praealtum*) in Ontario. Ontario Recovery Strategy Series. Prepared for the Ontario Ministry of Natural Resources, Peterborough, ON. 29 pp.

Jones, J. R. J. L. 1944. Some food plants of Lepidopterous larvae. List No. 10. *Proc. Entomol. Soc. British Columbia* 41: 4.

Jones, K., A. A. Anderberg, L. P. Ronse De Craene & L. Wanntorp. 2012. Origin, diversification, and evolution of *Samolus valerandi* (Samolaceae, Ericales). *Plant Syst. Evol.* 298: 1523–1531.

Jones, K. N. 2000. *Trollius laxus* Salisb. Spreading globeflower. New England Plant Conservation Program, conservation and research plan. New England Wild Flower Society, Framingham, MA. 17 pp.

Jones, K. N. & S. M. Klemetti. 2012. Managing marginal populations of the rare wetland plant *Trollius laxus* Salisbury (spreading globeflower): consideration of light levels, herbivory, and pollination. *N. E. Naturalist* 19: 267–278.

Jones Jr., M. P. 2001a. Endangered and threatened wildlife and plants; final rule for endangered status for *Astragalus pycnostachyus* var. *lanosissimus* (Ventura marsh milkvetch). *Fed. Reg.* 66: 27901–27908.

Jones Jr., M. P. 2001b. Endangered and threatened wildlife and plants; final designation of critical habitat for *Sidalcea oregana* var. *calva* (Wenatchee Mountains checker-mallow). *Fed. Reg.* 66: 46536–46548.

Jones, P. & Y. Arocha. 2006. A natural infection of *Hebe* is associated with an isolate of 'Candidatus Phytoplasma asteris' causing a yellowing and little-leaf disease in the UK. *Plant Pathol.* 55: 821–821.

Jones, R. H. & K. W. McLeod. 1990. Growth and photosynthetic responses to a range of light environments in Chinese tallowtree and Carolina ash seedlings. *For. Sci.* 36: 851–862.

Jones, R. L. 1994. The status of *Helianthus eggertii* Small in the southeastern United States. *Castanea* 59: 319–330.

Jones, S. M., D. H. Van Lear & S. K. Cox. 1984. A vegetation-landform classification of forest sites within the upper Coastal Plain of South Carolina. *Bull. Torrey Bot. Club* 111: 349–360.

Jones, W. P., T. Lobo-Echeverri, Q. Mi, H. Chai, D. Lee, D. D. Soejarto, G. A. Cordell, J. M. Pezzuto, S. M. Swanson & A. D. Kinghorn. 2005. Antitumour activity of 3-chlorodeoxylapachol, a naphthoquinone from *Avicennia germinans* collected from an experimental plot in southern Florida. *J. Pharm. Pharmacol.* 57: 1101–1108.

Jones Jr., S. B. 1966. Experimental hybridizations in *Vernonia* (Compositae). *Brittonia* 18: 39–44.

Jones Jr., S. B. 1968. Chromosome numbers in southeastern United States Compositae, I. *Bull. Torrey Bot. Club* 95: 393–395.

Jones Jr., S. B. 1970. Chromosome numbers in Compositae. *Bull. Torrey Bot. Club* 97: 168–171.

Jones Jr., S. B. 1972. A systematic study of the *Fasciculatae* group of *Vernonia* (Compositae). *Brittonia* 24: 28–45.

Jones Jr., S. B. 1976. Cytogenetics and affinities of *Vernonia* (Compositae) from the Mexican Highlands and eastern North America. *Evolution* 30: 455–462.

Jonsdottir, G., S. Omarsdottir, A. Vikingsson, I. Hardardottir & J. Freysdottir. 2011. Aqueous extracts from *Menyanthes trifoliata* and *Achillea millefolium* affect maturation of human dendritic cells and their activation of allogeneic CD4+ T cells in vitro. *J. Ethnopharmacol.* 136: 88–93.

Jonsell, B. 1968. *Studies in the North-West European Species of Rorippa s. str.* Symbolae Botanicae Upsalienses XIX:2. A.-B. Lundequistska Bokhandeln, Uppsala. 221 pp. + 11 plates.

Jorgensen, M.-H., R. Elven, A. Tribsch, T. M. Gabrielsen, B. Stedje & C. Brochmann. 2000. Taxonomy and evolutionary relationships in the *Saxifraga rivularis* complex. *Syst. Bot.* 31: 702–729.

Jorgenson, M. T. 2000. Hierarchical organization of ecosystems at multiple spatial scales on the Yukon-Kuskokwim delta, Alaska, U.S.A. *Arct. Antarc. Alp. Res.* 32: 221–239.

Jost, M. A., J. Hamr, I. Filion & F. F. Mallory. 1999. Forage selection by elk in habitats common to the French River-Burwash region of Ontario. *Can. J. Zool.* 77: 1429–1438.

Joyner, J. M. & E. W. Chester. 1994. The vascular flora of Cross Creeks National Wildlife Refuge, Stewart County, Tennessee. *Castanea* 59: 117–145.

Judd, F. W. & R. I. Lonard. 2002. Species richness and diversity of brackish and salt marshes in the Rio Grande delta. *J. Coastal Res.* 18: 751–759.

Judd, F. W. & R. I. Lonard. 2004. Community ecology of freshwater, brackish and salt marshes of the Rio Grande delta. *Tex. J. Sci.* 56: 103–122.

Judd, F. W., R. I. Lonard & S. L. Sides. 1977. Vegetation of South Padre Island, Texas in relation to topography. *Southwest. Naturalist* 22: 31–48.

Judd, W. S. 1982. The taxonomic status of *Oxypolis greenmanii* (Apiaceae). *Rhodora* 84: 265–279.

Judd, W. S. 1984. A taxonomic revision of *Agarista* (Ericaceae). *J. Arnold Arbor.* 65: 255–342.

Judd, W. S. & R. G. Olmstead. 2004. A survey of tricolpate (eudicot) phylogenetic relationships. *Amer. J. Bot.* 91: 1627–1644.

Judd, W. S. & K. Waselkov. 2006. A phylogenetic analysis of *Leucothoe* s.l. (Ericaceae; Gaultherieae) based on morphological characters. Botany 2006 meeting (abstract).

Judd, W. S., R. W. Sanders & M. J. Donoghue. 1994. Angiosperm family pairs: preliminary cladistic analyses. *Harvard Pap. Bot.* 5: 1–51.

Judd, W. S., C. S. Campbell, E. A. Kellogg, P. F. Stevens & M. J. Donoghue. 2002. *Plant Systematics: A Phylogenetic Approach. 2nd Ed.* Sinauer Associates, Inc., Sunderland, MA. 576 pp.

Judd, W. S., C. S. Campbell, E. A. Kellogg, P. F. Stevens & M. J. Donoghue. 2008. *Plant Systematics: A Phylogenetic Approach. 3rd Ed.* Sinauer Associates, Inc., Sunderland, MA. 611 pp.

Judd, W. S., C. S. Campbell, E. A. Kellogg, P. F. Stevens & M. J. Donoghue. 2016. *Plant Systematics: A Phylogenetic Approach. 4th Ed.* Sinauer Associates, Inc., Sunderland, MA. 677 pp.

Judd, W. W. 1964a. Insects associated with flowering marsh marigold, *Caltha palustris* L., at London, Ontario. *Can. Entomol.* 96: 1472–1476.

Judd, W. W. 1964b. A study of the population of insects emerging as adults from Saunders Pond at London, Ontario. *Amer. Midl. Naturalist* 71: 402–414.

Judziewicz, E. J. & J. C. Nekola. 1997. Recent Wisconsin records for some interesting vascular plants in the western Great Lakes region. *Michigan Bot.* 36: 91–118.

Jumpponen, A. & J. M. Trappe. 1998. Dark septate endophytes: a review of facultative biotrophic root-colonizing Fungi. *New Phytol.* 140: 295–310.

Jumpponen, A., K. Mattson, J. M. Trappe & R. Ohtonen. 1998. Effects of established willows on primary succession on Lyman Glacier forefront, North Cascade Range, Washington, U.S.A.: evidence for simultaneous canopy inhibition and soil facilitation. *Arct. Antarc. Alp. Res.* 30: 31–39.

Jumpponen, A., H. Väre, K. G. Mattson, R. Ohtonen & J. M. Trappe. 1999. Characterization of 'safe sites' for pioneers in primary succession on recently deglaciated terrain. *J. Ecol.* 87: 98–105.

Junior, P. 1989. Further investigations regarding distribution and structure of the bitter principles from *Menyanthes trifoliata. Plant Med.* 55: 83–87.

Juniper, B. E., R. J. Robins & D. M. Joel. 1989. *The Carnivorous Plants.* Academic Press, London. 353 pp.

Jurado, E., J. Navar, H. Villalón & M. Pando. 2001. Germination associated with season and sunlight for Tamaulipan thornscrub plants in north-eastern Mexico. *J. Arid. Environ.* 49: 833–841.

Juraimi, A. S., M. S. Ahmad-Hamdani, A. R. Anuar, M. Azmi, M. P. Anwar & M. K. Uddin. 2012. Effect of water regimes on germination of weed seeds in a Malaysian rice field. *Austral. J. Crop Sci.* 6: 598–605.

Jürgens, A. & S. Dötterl. 2004. Chemical composition of anther volatiles in Ranunculaceae: genera-specific profiles in *Anemone, Aquilegia, Caltha, Pulsatilla, Ranunculus,* and *Trollius* species. *Amer. J. Bot.* 91: 1969–1980.

Jürgens, A., A. M. El-Sayed & D. M. Suckling. 2009. Do carnivorous plants use volatiles for attracting prey insects? *Funct. Ecol.* 23: 875–887.

Jürgens, A., U. Glüc, G. Aas & S. Dötterl. 2014. Diel fragrance pattern correlates with olfactory preferences of diurnal and nocturnal flower visitors in *Salix caprea* (Salicaceae). *Bot. J. Linn. Soc.* 175: 624–640.

Jurik, T. W., S.-C. Wang & A. G. van der Valk. 1994. Effects of sediment load on seedling emergence from wetland seed banks. *Wetlands* 14: 159–165.

Justin, S. H. F. W. & W. Armstrong. 1987. The anatomical characteristics of roots and plant response to soil flooding. *New Phytol.* 106: 465–495.

Jutila b. Erkkilä, H. M. 1998. Seed banks of grazed and ungrazed Baltic seashore meadows. *J. Veg. Sci.* 9: 395–408.

Jutila, H. M. 2002. Seed banks of river delta meadows on the west coast of Finland. *Ann. Bot. Fenn.* 39: 49–61.

Kaczynski, K. M., D. J. Cooper & W. R. Jacobi. 2014. Interactions of sapsuckers and *Cytospora* canker can facilitate decline of riparian willows. *Botany* 92: 485–493.

Kadereit, G. 2007. Menyanthaceae. pp. 599–604. *In*: J. W. Kadereit & C. Jeffrey, (eds.), *The Families and Genera of Vascular Plants, Vol. VIII, Flowering plants, Eudicots: Asterales.* Springer-Verlag, Berlin, Germany.

Kadereit, G. & H. Freitag. 2011. Molecular phylogeny of Camphorosmeae (Camphorosmoideae, Chenopodiaceae): implications for biogeography, evolution of C_4-photosynthesis and taxonomy. *Taxon* 60: 51–78.

Kadereit, G., T. Borsch, K. Weising & H. Freitag. 2003. Phylogeny of Amaranthaceae and Chenopodiaceae and the evolution of C_4 photosynthesis. *Int. J. Plant Sci.* 164: 959–986.

Kadereit, G., L. Mucina & H. Freitag. 2006. Phylogeny of Salicornioideae (Chenopodiaceae): diversification, biogeography, and evolutionary trends in leaf and flower morphology. *Taxon* 55: 617–642.

Kadereit, G., P. Ball, S. Beer, L. Mucina, D. Sokoloff, P. Teege, A. E. Yaprak & H. Freitag. 2007. A taxonomic nightmare comes true: phylogeny and biogeography of glassworts (*Salicornia* L., Chenopodiaceae). *Taxon* 56: 1143–1170.

Kadereit, J. W. & K. B. von Hagen. 2003. The evolution of flower morphology in Gentianaceae-Swertiinae and the roles of key innovations and niche width for the diversification of *Gentianella* and *Halenia* in South America. *Int. J. Plant Sci.* 164: S441–S452.

Kadereit, J. W., M. Repplinger, N. Schmalz, C. H. Uhink & A. Wörz. 2008. The phylogeny and biogeography of Apiaceae subf. Saniculoideae tribe Saniculeae: from south to north and south again. *Taxon* 57: 365–382.

Kado, T., A. Fujimoto, L.-H. Giang, M. Tuan, P. N. Hong, K. Harada & H. Tachida. 2004. Genetic structures of natural populations of three mangrove species, *Avicennia marina, Kandelia candel* and *Lumnitzera racemosa,* in Vietnam revealed by maturase sequences of plastid DNA. *Plant Species Biol.* 19: 91–99.

Kadono, Y. 2004. Alien aquatic plants naturalized in Japan: history and present status. *Glob. Environ. Res.* 8: 163–169.

Kadono, Y. & E. L. Schneider. 1986. Floral biology of *Trapa natans* var. *japonica*. *Bot. Mag. (Tokyo)* 99: 435–499.

Kahn, M. A. & D. J. Weber. 1986. Factors influencing seed germination in *Salicornia pacifica* var. *utahensis*. *Amer. J. Bot.* 73: 1163–1167.

Kai, W. & Z. Zhiwei. 2006. Occurrence of arbuscular mycorrhizas and dark septate endophytes in hydrophytes from lakes and streams in southwest China. *Int. Rev. Hydrobiol.* 91: 29–37.

Kainradl, E. 1927. Beiträge zur Biologie von *Hydrolea spinosa* L. mit besonderer Berücksichtigung von Fruchtwand und Samenentwicklung. *Sitzungsber. Akad. Wiss. Wien, Math.-Naturwiss. Kl., Abt. 1* 136: 167–194.

Kajita, T., H. Ohashi, Y. Tateishi, C. D. Bailey & J. J. Doyle. 2001. *rbcL* and legume phylogeny, with particular reference to Phaseoleae, Millettieae, and allies. *Syst. Bot.* 26: 515–536.

Kalk, H. J. 2011. The role of coastal plant community response to climate change: implications for restoring ecosystem resiliency. Theses. Paper 742. Southern Illinois University Carbondale, Carbondale, IL. 116 pp.

Källersjö, M. & B. Ståhl. 2003. Phylogeny of Theophrastaceae (Ericales s. lat.). *Int. J. Plant Sci.* 164: 579–591.

Källersjö, M., G. Bergqvist & A. A. Anderberg. 2000. Generic realignment in primuloid families of the Ericales s.l.: a phylogenetic analysis based on DNA sequences from three chloroplast genes and morphology. *Amer. J. Bot.* 87: 1325–1341.

Kalmbacher, R. S., K. R. Long & F. G. Martin. 1984. Seasonal mineral concentration in diets of esophageally fistulated steers on three range areas. *J. Range Manag.* 37: 36–39.

Kalmbacher, R. S., N. Cellinese & F. Martin. 2005. Seeds obtained by vacuuming the soil surface after fire compared with soil seedbank in a flatwoods plant community. *Native Plant J.* 6: 233–241.

Kameyama, Y. & M. Ohara. 2006. Genetic structure in aquatic bladderworts: clonal propagation and hybrid perpetuation. *Ann. Bot.* 98: 1017–1024.

Kampny, C. M. & N. G. Dengler. 1997. Evolution of flower shape in Veroniceae (Scrophulariaceae). *Plant Syst. Evol.* 205: 1–25.

Kanazawa, A., S. Watanabe, T. Nakamoto, N. Tsutsumi & A. Hirai. 1998. Phylogenetic relationships in the genus *Nelumbo* based on polymorphism and quantitative variations in mitochondrial DNA. *Genes Genet. Syst.* 73: 39–44.

Kandalepas, D., K. J. Stevens, G. P. Shaffer & W. J. Platt. 2010. How abundant are root-colonizing Fungi in southeastern Louisiana's degraded marshes? *Wetlands* 30: 189–199.

Kane, M. E. & L. S. Albert. 1982. Environmental and growth regulator effects on heterophylly and growth of *Proserpinaca intermedia* (Haloragaceae). *Aquat. Bot.* 13: 73–85.

Kane, M. E. & L. S. Albert. 1987. Integrative regulation of leaf morphogenesis by gibberellic acid and abscisic acids in the aquatic angiosperm *Proserpinaca palustris* L. *Aquat. Bot.* 28: 89–96.

Kangas, P. C. & G. L. Hannan. 1985. Vegetation on muskrat mounds in a Michigan marsh. *Amer. Midl. Naturalist* 113: 392–396.

Kanouse, S. C. 2003. Nekton use and growth in three brackish marsh pond microhabitats. MS thesis. School of Renewable Natural Resources, Louisiana State University, Baton Rouge, LA. 68 pp.

Kapadia, G. J., A. Ramdass & F. Bada. 1990. Pyrrolizidine alkaloids of *Senecio glabellus*. *Int. J. Crude Drug Res.* 28: 67–71.

Kaplan, D. R. 1967. Floral morphology, organogenesis and interpretation of the inferior ovary in *Downingia bacigalupii*. *Amer. J. Bot.* 54: 1274–1290.

Kaplan, D. R. 1970. Comparative development and morphological interpretation of 'rachis-leaves' in Umbelliferae. *Bot. J. Linn. Soc.* 63: S1101–S126.

Kaplan, S. M. & D. L. Mulcahy. 1971. Mode of pollination and floral sexuality in *Thalictrum*. *Evolution* 25: 659–668.

Kapoor, B. M. 1978. Supernumerary chromosomes of some species of *Solidago* and a related taxon. *Caryologia* 31: 315–330.

Kapralov, M. V., H. Akhani, E. V. Voznesenskaya, G. Edwards, V. Franceschi & E. H. Roalson. 2006. Phylogenetic relationships in the Salicornioideae/Suaedoideae/Salsoloideae s.l. (Chenopodiaceae) clade and a clarification of the phylogenetic position of *Bienertia* and *Alexandra* using multiple DNA sequence datasets. *Syst. Bot.* 31: 571–585.

Kar, A., S. Panda & S. Bharti. 2002. Relative efficacy of three medicinal plant extracts in the alteration of thyroid hormone concentrations in male mice. *J. Ethnopharmacol.* 81: 281–285.

Karacetin, E. 2007. Biotic barriers to colonizing new hosts by the cinnabar moth *Tyria jacobaeae* (L.) (Lepidoptera: Arctiidae). PhD dissertation. Oregon State University, Corvallis, OR. 158 pp.

Karagatzides, J. D. & A. M. Ellison. 2009. Construction costs, payback times, and the leaf economics of carnivorous plants. *Amer. J. Bot.* 96: 1612–1619.

Karale, S. S., S. A. Jadhav, N. B. Chougule, S. S. Awati & A. A. Patil. 2013. Evaluation of analgesic, antipyretic and anti-inflammatory activities of *Ceratophyllum demersum* Linn. in albino rats. *Curr. Pharm. Res.* 3: 1027–1030.

Karaman-Castro, V. & L. E. Urbatsch. 2006. 189. *Boltonia* L'Héritier. pp. 353–357. *In*: Flora of North America Editorial Committee (eds.), *Flora of North America North of Mexico: Vol. 20, Magnoliophyta: Asteridae (in part): Asteraceae, Part 2*. Oxford University Press, New York.

Karaman-Castro, V. & L. E. Urbatsch. 2009. Phylogeny of *Hinterhubera* group and related genera (Hinterhuberinae: Astereae) based on the nrDNA ITS and ETS sequences. *Syst. Bot.* 34: 805–817.

Karatas, M., M. Aasim & M. Çiftçioğlu. 2014. Adventitious shoot regeneration of roundleaf toothcup-*Rotala rotundifolia* (Buch-Ham. ex Roxb) Koehne. *J. Anim. Plant Sci.* 24: 838–842.

Karberg, J. M. & M. R. Gale. 2006. Genetic diversity and distribution of *Sarracenia purpurea* (Sarraceniaceae) in the western Lake Superior basin. *Can. J. Bot.* 84: 235–242.

Kardošová, A., M. Matulová & A. Malovíková. 1998. (4-O-Methyl-α-D-glucurono)-D-xylan from *Rudbeckia fulgida*, var. *sullivantii* (Boynton et Beadle) Cronq. *Carbohydr. Res.* 308: 99–106.

Kårehed, J. 2001. Multiple origin of the tropical forest tree family Icacinaceae. *Amer. J. Bot.* 88: 2259–2274.

Karema, J., S. A. Woods, F. Drummond & C. Stubbs. 2010. The relationships between Apocrita wasp populations and flowering plants in Maine's wild lowbush blueberry agroecosystems. *Biocontrol Sci. Technol.* 20: 257–274.

Kärkönen, A., L. K. Simola & T. Koponen. 1999. Micropropagation of several Japanese woody plants for horticultural purposes. *Ann. Bot. Fenn.* 36: 21–31.

Karlin, E. F. 1978. Major environmental influences on the pattern of *Ledum groenlandicum* in mire systems. PhD dissertation. University of Alberta, Edmonton, AB. 140 pp.

Karlin, E. F. & L. C. Bliss. 1983. Germination ecology of *Ledum groenlandicum* and *Ledum palustre* spp. *decumbens*. *Arct. Alp. Res.* 15: 397–404.

Karlin, E. F. & L. M. Lynn. 1988. Dwarf-shrub bogs of the southern Catskill mountain region of New York state: geographic changes in the flora of peatlands in northern New Jersey and southern New York. *Bull. Torrey Bot. Club* 115: 209–217.

Karling, J. S. 1943. The life history of Anisolpidium ectocarpii gen. nov. et sp. nov., and a synopsis and classification of other Fungi with anteriorly uniflagellate zoospores. *Amer. J. Bot.* 30: 637–648.

Karling, J. S. 1955a. Prosori in *Synchytrium*. *Bull. Torrey Bot. Club* 82: 218–236.

Karling, J. S. 1955b. *Synchytrium ranunculi* Cook. *Mycologia* 47: 130–139.

Karling, J. S. 1956. Unrecorded hosts and species of *Physoderma*. *Bull. Torrey Bot. Club* 83: 292–299.

Karlson, D. T., Q.-Y. Xiang, V. E. Stirm, A. M. Shirazi & E. N. Ashworth. 2004. Phylogenetic analyses in *Cornus* substantiate ancestry of xylem supercooling freezing behavior and reveal lineage of desiccation related proteins. *Plant Physiol.* 135: 1654–1665.

Karlsson, P. S. 1988. Seasonal patterns of nitrogen, phosphorus and potassium utilization by three *Pinguicula* species. *Funct. Ecol.* 2: 203–209.

Karlsson, P. S. & B. Carlsson. 1984. Why does *Pinguicula vulgaris* L. trap insects? *New Phytol.* 97: 25–30.

Karlsson, P. S., K. O. Nordell, S. Eirefelt & A. Svensson. 1987. Trapping efficiency of three carnivorous *Pinguicula* species. *Oecologia* 73: 518–521.

Karlsson, P. S., B. M. Svensson, B. Å. Carlsson & K. O. Nordell. 1990. Resource investment in reproduction and its consequences in three *Pinguicula* species. *Oikos* 59: 393–398.

Karlsson, P. S., K. O. Nordell, B. Å. Carlsson & B. M. Svensson. 1991. The effect of soil nutrient status on prey utilization in four carnivorous plants. *Oecologia* 86: 1–7.

Karlsson, P. S., L. M. Thorén & H. M. Hanslin. 1994. Prey capture by three *Pinguicula* species in a subarctic environment. *Oecologia* 99: 188–193.

Karp, K., M. Mänd, M. Starast & T. Paal. 2004. Nectar production of *Rubus arcticus. Agron. Res.* 2: 57–61.

Karrenberg, S., P. J. Edwards & J. Kollmann. 2002. The life history of Salicaceae living in the active zone of floodplains. *Freshwater Biol.* 47: 733–748.

Karron, J. D. 1987. A comparison of levels of genetic polymorphism and self-compatibility in geographically restricted and widespread plant congeners. *Evol. Ecol.* 1: 47–58.

Karron, J. D. 1991. Patterns of genetic variation and breeding systems in rare plant species. pp. 87–98. *In*: D. A. I. Falk & K. E. Holsinger (eds.), *Genetics and Conservation of Rare Plants*. Oxford University Press, New York. 304 pp.

Karron, J. D., N. N. Thumser, R. Tucker & A. J. Hessenauer. 1995a. The influence of population density on outcrossing rates in *Mimulus ringens. Heredity* 75: 175–180.

Karron, J. D., R. Tucker, N. N. Thumser & J. A. Reinartz. 1995b. Comparison of pollinator flight movements and gene dispersal patterns in *Mimulus ringens. Heredity* 75: 612–617.

Karron, J. D., R. T. Jackson, N. N. Thumser & S. L. Schlicht. 1997. Outcrossing rates of individual *Mimulus ringens* genets are correlated with anther-stigma separation. *Heredity* 79: 365–370.

Karron, J. D., R. J. Mitchell, K. G. Holmquist, J. M. Bell & B. Funk. 2004. The influence of floral display size on selfing rates in *Mimulus ringens. Heredity* 92: 242–248.

Karthikumar, S., K. Vigneswari & J. Jegatheesan. 2007. Screening of antibacterial and antioxidant activities of leaves of *Eclipta prostrata* (L). *Sci. Res. Essay* 2: 101–104.

Kashiwada, Y., A. Aoshima, Y. Ikeshiro, Y. P. Chen, H. Furukawa, M. Itoigawa, T. Fujioka, K. Mihashi, L. M. Cosentino, S. L. Morris-Natschke & K. H. Lee. 2005. Anti-HIV benzylisoquinoline alkaloids and flavonoids from the leaves of *Nelumbo nucifera*, and structure–activity correlations with related alkaloids. *Bioorg. Med. Chem.* 13: 443–448.

Kasselmann, C. 1995. *Aquarienpflanzen*. Verlag Eugen Ulmer, Stuttgart, Germany. 472 pp.

Kasselmann, C. 2003. *Aquarium Plants*. Krieger Publishing Company, Malabar, FL. 518 pp.

Kathiresan, K. & B. L. Bingham. 2001. Biology of mangroves and mangrove ecosystems. *Advances Mar. Biol.* 40: 81–251.

Katsar, C. S., W. B. Hunter & X. H. Sinisterra. 2007. Phytoreovirus-like sequences isolated from salivary glands of the glassy-winged sharpshooter *Homalodisca vitripennis* (Hemiptera: Cicadellidae). *Florida Entomol.* 90: 196–203.

Katz-Downie, D. S., C. M. Valiejo-Roman, E. I. Terentieva, A. V. Troitsky, M. G. Pimenov, B. Lee & S. R. Downie. 1999. Towards a molecular phylogeny of Apiaceae subfamily Apioideae: additional information from nuclear ribosomal DNA ITS sequences. *Plant Syst. Evol.* 216: 167–195.

Kaufmann, T. 1970. Studies on the biology and ecology of *Pyrrhalta nymphaea* (Col. Chrysomelidae) in Alaska with special reference to population dynamics. *Amer. Midl. Naturalist* 83: 496–509.

Kaul, M. L. H. 1975. Cytogenetical studies on ecological races of *Mecardonia dianthera* (Sw.) Pennell. III. Induced polyploidy. *Cytologia* 40: 97–111.

Kaur, R. 2011. Assessment of phenotypic and genotypic diversity in *Eclipta alba* (L.) Hassk: an important medicinal plant. PhD dissertation. Department of Botany, Punjabi University Patiala, India. 236 pp.

Kautsky, L. 1988. Life strategies of aquatic soft bottom macrophytes. *Oikos* 53: 126–135.

Kavimani, S., K. T. Manisenthilkumar, R. Ilango & G. Krishnamoorthy. 1999. Effect of the methanolic extract of *Glinus lotoides* on Dalton's ascitic lymphoma. *Biol. Pharm. Bull.* 22: 1251–1252.

Kawahara, A. Y. 2006. Biology of the snout butterflies (Nymphalidae, Libytheinae) Part 2: *Libytheana* Michener. *Trans. Lepid. Soc. Jpn.* 57: 265–277.

Kawakubo, N. 1990. Dioecism of the genus *Callicarpa* (Verbenaceae) in the Bonin (Ogasawara) Islands. *J. Plant Res.* 103: 57–66.

Kawamura, H. 2004. Symbolic and political ecology among contemporary Nez Perce Indians in Idaho, USA: functions and meanings of hunting, fishing, and gathering practices. *Agric. Human Values* 21: 157–169.

Kawiak, A., A. Królicka & E. Lojkowska. 2011. In vitro cultures of *Drosera aliciae* as a source of a cytotoxic naphthoquinone: ramentaceone. *Biotechnol. Lett.* 33: 2309–2316.

Kay, M. A. 1996. *Healing with Plants in the American and Mexican West*. University of Arizona Press, Tucson, AZ. 318 pp.

Kaye, T., W. Messinger, S. Massey & R. Meinke. 1990. *Gratiola heterosepala: Inventory and Breeding System Evaluation*. Unpublished VI-19 report to the U. S. Bureau of Land Management, Lakeview, OR. 22 pp.

Kaye, T. N. 2001. A morphometric evaluation of *Corydalis caseana* and its subspecies with special attention to *C. aquae-gelidae*. Institute for Applied Ecology and USDA Forest Service, Willamette National Forest, Corvallis, OR. 16 pp.

Kaye, T. N., M. Gisler & R. Fiegener (eds.). 2003. *Proceedings of a Conference on Native Plant Restoration and Management on Public Lands in the Pacific Northwest: Rare Plants, Invasive Species and Ecosystem Management*. Institute for Applied Ecology, Corvallis, OR. 51 pp.

Kaye, T. N., R. T. Massatti, A. S. Thorpe, I. S. Silvernail & R. J. Meinke. 2011. Developing recovery protocols for the federally-listed species *Lomatium cookii*. Institute for Applied Ecology, Corvallis, OR. 43 pp.

Kazuo, S. 1994. Pollinator restriction in the narrow-tube flower type of *Mertensia ciliata* (James) G. Don (Boraginaceae). *Plant Species Biol.* 9: 69–73.

Keammerer, H. B. & S. D. Hacker. 2013. Negative and neutral marsh plant interactions dominate in early life stages and across physical gradients in an Oregon estuary. *Plant Ecol.* 214: 303–315.

Kearney, T. H. & R. H. Peebles. 1960. *Arizona Flora* (with supplement). University of California Press, Berkeley, CA. 1085 pp.

Kearns, C. A. & D. W. Inouye. 1994. Fly pollination of *Linum lewisii* (Linaceae). *Amer. J. Bot.* 81: 1091–1095.

Keating, R. C., P. C. Hoch & P. H. Raven. 1982. Perennation in *Epilobium* (Onagraceae) and its relation to classification and ecology. *Syst. Bot.* 7: 379–404.

Keddy, P. A. 1981. Vegetation with Atlantic Coastal Plain affinities in Axe Lake, near Georgian Bay, Ontario. *Can. Field-Naturalist* 95: 241–248.

Keddy, P. A. 1983. Shoreline vegetation in Axe Lake, Ontario: effects of exposure on zonation patterns. *Ecology* 64: 331–344.

Keddy, P. A. 1984. Plant zonation on lakeshores in Nova Scotia: a test of the resource specialization hypothesis. *J. Ecol.* 72: 797–808.

Keddy, P. A. 1985. Lakeshore plants in the Tusket River Valley, Nova Scotia: the distribution and status of some rare species including *Coreopsis rosea* and *Sabatia kennedyana*. *Rhodora* 87: 309–320.

Keddy, P. A. 1989. Effects of competition from shrubs on herbaceous wetland plants: a 4-year field experiment. *Can. J. Bot.* 67: 708–716.

Keddy, P. A. & P. Constabel. 1986. Germination of ten shoreline plants in relation to seed size, soil particle size and water level: an experimental study. *J. Ecol.* 74: 133–141.

Keddy, P. A. & T. H. Ellis. 1985. Seedling recruitment of 11 wetland plant species along a water level gradient: shared or distinct responses? *Can. J. Bot.* 63: 1876–1879.

Keddy, P. A. & A. A. Reznicek. 1982. The role of seed banks in the persistence of Ontario, Canada's coastal plain flora. *Amer. J. Bot.* 69: 13–22.

Keddy, P. A. & I. C. Wisheu. 1989. Ecology, biogeography, and conservation of coastal plain plants: some general principles from the study of Nova Scotian wetlands. *Rhodora* 91: 72–94.

Keddy, P., L. H. Fraser & I. C. Wisheu. 1998. A comparative approach to examine competitive response of 48 wetland plant species. *J. Veg. Sci.* 9: 777–786.

Keddy, P., C. Gaudet & L. H. Fraser. 2000. Effects of low and high nutrients on the competitive hierarchy of 26 shoreline plants. *J. Ecol.* 88: 413–423.

Keddy, P. A., L. Twolan-Strutt & I. C. Wisheu. 1994. Competitive effect and response rankings in 20 wetland plants: are they consistent across three environments? *J. Ecol.* 82: 635–643.

Keddy, P. A., L. Smith, D. R. Campbell, M. Clark & G. Montz. 2006. Patterns of herbaceous plant diversity in southeastern Louisiana pine savannas. *Appl. Veg. Sci.* 9: 17–26.

Keeland, B. D. & L. E. Gorham. 2009. Delayed tree mortality in the Atchafalaya Basin of southern Louisiana following hurricane Andrew. *Wetlands* 29: 101–111.

Keeland, B. D., R. O. Draugelis-Dale & J. W. McCoy. 2010. Tree growth and mortality during 20 years of managing a green-tree reservoir in Arkansas, USA. *Wetlands* 30: 345–357.

Keeler, K. H. 1978. Intra-population differentiation in annual plants. II. Electrophoretic variation in *Veronica peregrina*. *Evolution* 32: 638–645.

Keeler, R. F. & L. D. Balls. 1978. Teratogenic effects in cattle of *Conium maculatum* and *Conium* alkaloids and analogs. *Clin. Toxicol.* 12: 49–64.

Keeley, J. E. 1998. CAM photosynthesis in submerged aquatic plants. *Bot. Rev.* 64: 121–175.

Keeley, J. E. 1999. Photosynthetic pathway diversity in a seasonal pool community. *Funct. Ecol.* 13: 106–118.

Keeley, J. E. & C. J. Fotheringham. 1998. Smoke-induced seed germination in California chaparral. *Ecology* 79: 2320–2336.

Keeley, J. E. & D. R. Sandquist. 1991. Diurnal photosynthesis cycle in CAM and non-CAM seasonal-pool aquatic macrophytes. *Ecology* 72: 716–727.

Keeley, J. E. & P. H. Zedler. 1998. Characterization and global distribution of vernal pools. pp. 1–14. *In:* C. W. Witham, E. T. Bauder, D. Belk, W. R. Ferren Jr. & R. Ornduff (eds.), *Ecology, Conservation, and Management of Vernal Pool Ecosystems—Proceedings from a 1996 Conference.* California Native Plant Society, Sacramento, CA.

Keeley, S. C. & R. K. Jansen. 1991. Evidence from chloroplast DNA for the recognition of a new tribe, the Tarchonantheae, and the tribal placement of *Pluchea* (Asteraceae). *Syst. Bot.* 16: 173–181.

Keeley, S. C., Z. H. Forsman & R. Chan. 2007. A phylogeny of the "evil tribe" (Vernonieae: Compositae) reveals Old/New World long distance dispersal: support from separate and combined congruent datasets (*trnL-F, ndhF*, ITS). *Mol. Phylogen. Evol.* 44: 89–103.

Keener, C. S. 1976. Studies in the Ranunculaceae of the southeastern United States II. *Thalictrum* L. *Rhodora* 78: 457–472.

Keener, P. D. & J. E. Hampton. 1968. A rust on *Anemopsis californica*. *Madroño* 19: 218–220.

Keigley, R. B., M. R. Frisina & C. Fager. 2003. A method for determining the onset year of intense browsing. *J. Range Manag.* 56: 33–38.

Keil, D. J. 2006. 13. *Cirsium* Miller. pp. 95–164. *In:* Flora of North America Editorial Committee (eds.), *Flora of North America North of Mexico, Vol. 19: Magnoliophyta: Asteridae, Part 8.* Oxford University Press, New York.

Keil, D. J. & M. A. Elvin. 2010. *Helianthus inexpectatus* (Asteraceae), a tetraploid perennial new Species from southern California. *Aliso* 28: 59–62.

Keil, D. J., M. A. Luckow & D. J. Pinkava. 1988. Chromosome studies in Asteraceae from the United States, Mexico, the West Indies, and South America. *Amer. J. Bot.* 75: 652–668.

Keiper, J. B., P. L. Brutsche & B. A. Foote. 1998. Acalyptrate Diptera associated with water willow, *Justicia americana* (Acanthaceae). *Proc. Entomol. Soc. Washington* 100: 576–587.

Keith, L. B. 1961. A study of waterfowl ecology on small impoundments in southeastern Alberta. *Wildl. Monogr.* 6: 3–88.

Kelch, D. G. & B. G. Baldwin. 2003. Phylogeny and ecological radiation of New World thistles (*Cirsium*. Cardueae – Compositae) based on ITS and ETS rDNA sequence data. *Mol. Ecol.* 12: 141–151.

Kelch, D. G. & A. Murdock. 2012. Flora of the Carquinez Strait Region, Contra Costa and Solano Counties, California. *Madroño* 59: 47–108.

Keller, B. E. M. 2000. Plant diversity in *Lythrum, Phragmites*, and *Typha* marshes, Massachusetts, U.S.A. *Wetlands Ecol. Manag.* 8: 391–401.

Keller, S. & S. Armbruster. 1989. Pollination of *Hyptis capitata* by Eumenid wasps in Panama. *Biotropica* 21: 190–192.

Kellerman, W. A. 1908. Index to North American mycology. *J. Mycol.* 14: 75–85.

Kellerman, W. A. & M. A. Carleton. 1885. Second list of Kansas parasitic Fungi, together with their host-plants. *Trans. Ann. Meet. Kansas Acad. Sci.* 10: 88–99.

Kelley, A. D. & V. F. Bruns. 1975. Dissemination of weed seeds by irrigation water. *Weed Sci.* 23: 486–493.

Kelley, C. J., J. R. Mahajan, L. C. Brooks, L. A. Neubert, W. R. Breneman & M. Carmack. 1975. Polyphenolic acids of *Lithospermum ruderale* Dougl. ex Lehm. (Boraginaceae). 1. Isolation and structure determination of lithospermic acid. *J. Org. Chem.* 40: 1804–1815.

Kellman, M. 1974. Preliminary seed budgets for two plant communities in coastal British Columbia. *J. Biogeogr.* 1: 123–133.

Kellman, M. 1978. Microdistribution of viable weed seed in two tropical soils. *J. Biogeogr.* 5: 291–300.

Kellogg, C. H. & S. D. Bridgham. 2002. Colonization during early succession of restored freshwater marshes. *Can. J. Bot.* 80: 176–185.

Kellogg, C. H., S. D. Bridgham & S. A. Leicht. 2003. Effects of water level, shade and time on germination and growth of freshwater marsh plants along a simulated successional gradient. *J. Ecol.* 91: 274–282.

Kelly, A. J. & J. H. Willis. 1998. Polymorphic microsatellite loci in *Mimulus guttatus* and related species. *Mol. Ecol.* 7: 769–774.

Kelly, D. 1984. Seeds per fruit as a function of fruits per plant in 'depauperate' annuals and biennials. *New Phytol.* 96: 103–114.

Kelly, D. 1989a. Demography of short-lived plants in chalk grassland. I. Life cycle variation in annuals and strict biennials. *J. Ecol.* 77: 747–769.

Kelly, D. 1989b. Demography of short-lived plants in chalk grassland. II. Control of mortality and fecundity. *J. Ecol.* 77: 770–784.

Kelly, L. M. 1997. A cladistic analysis of *Asarum* (Aristolochiaceae) and implications for the evolution of herkogamy. *Amer. J. Bot.* 84: 1752–1765.

Kelly, L. M. 1998. Phylogenetic relationships in *Asarum* (Aristolochiaceae) based on morphology and ITS sequences. *Amer. J. Bot.* 85: 1454–1467.

Kelly, L. M. 2001. Taxonomy of *Asarum* section *Asarum* (Aristolochiaceae). *Syst. Bot.* 26: 17–53.

Kelly, L. M. & F. González. 2003. Phylogenetic relationships in Aristolochiaceae. *Syst. Bot.* 28: 236–249.

Kelsey, F. D. 1889. Study of Montana Erysipheae. *Bot. Gaz.* 14: 285–288.

Kemppainen, E. & R. Lampinen. 1987. Idännukki (*Androsace filiformis*) vakiintuneena Janakkalassa. [*Androsace filiformis* as an established weed in Janakkala, S. Finland]. *Lutukka* 3: 77–80.

Kenaley, S. C., L. B. Smart & G. W. Hudler. 2014. Genetic evidence for three discrete taxa of *Melampsora* (Pucciniales) affecting willows (*Salix* spp.) in New York State. *Fungal Biol.* 18: 704–720.

Kendall, R. O. 1964. Larval foodplants for twenty-six species of Rhopalocera (Papilionoidea) from Texas. *J. Lepid. Soc.* 18: 129–157.

Kennedy, C. W. & J. F. Bruno. 2000. Restriction of the upper distribution of New England cobble beach plants by wave-related disturbance. *J. Ecol.* 88: 856–868.

Kennedy Jr., H. E. 1983. Water tupelo in the Atchafalaya Basin does not benefit from thinning. Research Note SO-298. U. S. Department of Agriculture, Forest Service, Southern Forest Experiment Station. New Orleans, LA. 3 pp.

Kennedy, K. L. & A. Strong. 2012. *Seedbanking the Rare Plants of Texas. Phase I.* Project #E-112–1, Final Report. Texas Parks & Wildlife, Austin, TX. 133 pp.

Kenow, K. P. & J. E. Lyon. 2009. Composition of the seed bank in drawdown areas of navigation pool 8 of the upper Mississippi River. *River Res. Appl.* 25: 194–207.

Kent, G. H. 1908. *The Ware Collection of Blaschka Glass Flower Models. A Short Desciption of Their Makers and Their Making.* Caustic-Claflin Co., Cambridge, MA. 42 pp.

Kent, R. J. & D. E. Johnson. 2001. Influence of flood depth and duration on growth of lowland rice weeds, Cote d'Ivoire. *Crop Protect.* 20: 691–694.

Kephart, S. R. 1981. Breeding systems in *Asclepias incarnata* L., *A. syriaca* L., and *A. verticillata* L. *Amer. J. Bot.* 68: 226–232.

Kephart, S. R. 1983. The partitioning of pollinators among three species of *Asclepias*. *Ecology* 64: 120–133.

Kerbes, R. H., P. M. Kotanen & R. L. Jefferies. 1990. Destruction of wetland habitats by lesser snow geese: a keystone species on the west coast of Hudson Bay. *J. Appl. Ecol.* 27: 242–258.

Kercher, S. M. & J. B. Zedler. 2004. Flood tolerance in wetland angiosperms: a comparison of invasive and noninvasive species. *Aquat. Bot.* 80: 89–102.

Kercher, S. M., A. Herr-Turoff & J. B. Zedler. 2007. Understanding invasion as a process: the case of *Phalaris arundinacea* in wet prairies. *Biol. Invas.* 9: 657–665.

Kern, F. D. 1919. North American rusts on *Cyperus* and *Eleocharis*. *Mycologia* 11: 134–147.

Kern, R. A., W. G. Shriver, J. L. Bowman, L. R. Mitchell & D. L. Bounds. 2012. Seaside sparrow reproductive success in relation to prescribed fire. *J. Wildl. Manag.* 76: 932–939.

Kerner von Marilaun, A. 1895. *The Natural History of Plants: Their Forms, Growth, Reproduction, and Distribution.* Half-volume III (translated into English by F. W. Oliver). Henry Holt & Co., New York. 496 pp.

Kerr, M. S. 2004. A phylogenetic and biogeographic analysis of Sanguisorbeae (Rosaceae) with emphasis on the Pleistocene radiation of the high Andean genus *Polylepis*. PhD dissertation. University of Maryland, College Park, MD. 191 pp.

Kerwin, J. A. 1972. Distribution of the salt marsh snail (*Melampus bidentatus* Say) in relation to marsh plants in the Poropotank River area, Virginia. *Chesapeake Sci.* 13: 150–153.

Kerwin, J. A. & R. A. Pedigo. 1971. Synecology of a Virginia salt marsh. *Chesapeake Sci.* 12: 125–130.

Kesseli, R. & S. K. Jain. 1984. An ecological genetic study of gynodioecy in *Limnanthes douglasii* (Limnanthaceae). *Amer. J. Bot.* 71: 775–786.

Kesseli, R. & S. K. Jain. 1985. Breeding systems and population structure in *Limnanthes*. *Theor. Appl. Genet.* 71: 292–299.

Kessler Jr., J. R., J. L. Sibley, B. K. Behe, D. M. Quinn & J. S. Bannon. 2000. Herbaceous perennial trials in central Alabama, 1996–97. *HortTechnology* 10: 222–228.

Kessler, P. J. A. 1993. Annonaceae. pp. 93–129. *In*: K. Kubitzki, J. Rohwer & V. Bittrich (eds.), *The Families and Genera of Vascular Plants, Vol. II. Flowering Plants, Dicotyledons: Magnoliid, Hamamelid and Caryophyllid Families.* Springer-Verlag, Berlin, Germany.

Kesting, J. R., I.-L. Tolderlund, A. F. Pedersen, M. Witt, J. W. Jaroszewski & D. Staerk. 2009. Piperidine and tetrahydropyridine alkaloids from *Lobelia siphilitica* and *Hippobroma longiflora*. *J. Nat. Prod.* 72: 312–315.

Kether, F. B. H., M. A. Mahjoub, S. Ammar, K. Majouli & Z. Mighri. 2012. Two new alkaloids and a new polyphenolic compound from *Cotula coronopifolia*. *Chem. Nat. Compounds* 47: 955–958.

Kevan, P. G. 1973. Parasitoid wasps as flower visitors in the Canadian high arctic. *Anz. Schädlingsk. Pflanzen Umweltschutz.* 46: 3–7.

Khaing, A. A., K. T. Moe, J.-W. Chung, H.-J. Baek & Y.-J. Park. 2013. Genetic diversity and population structure of the selected core set in *Amaranthus* using SSR markers. *Plant Breed.* 132: 165–173.

Khalik, K. A., L. J. G. van der Maesen, W. J. M. Koopman & R. G. van den Berg. 2002. Numerical taxonomic study of some tribes of Brassicaceae from Egypt. *Plant Syst. Evol.* 233: 207–221.

Khamwan, K., A. Akaracharanya, S. Chareonpornwattana, Y.-E. Choi, T. Nakamura, Y. Yamaguchi, H. Sano & A. Shinmyo. 2003. Genetic transformation of water spinach (*Ipomoea aquatica*). *Plant Biotechnol.* 20: 335–338.

Khan, H. A. & G. S. Brush. 1994. Nutrient and metal accumulation in a freshwater tidal marsh. *Estuaries* 17: 345–360.

Khan, M. A. & B. Gul. 1999. Seed bank strategies of coastal populations at the Arabian Sea coast. pp. 227–230. *In*: *Proceedings of the Symposium on Shrubland Ecotones*. USDA Forest Service, Rocky Mountain Research Station, Ogden, UT.

Khan, M. A. & D. J. Weber. 2006. *Ecophysiology of High Salinity Tolerant Plants*. Springer, New York. 399 pp.

Khan, M. A., B. Gul & D. J. Weber. 2000. Germination responses of *Salicornia rubra* to temperature and salinity. *J. Arid Environ.* 45: 207–214.

Khan, M. A., B. Gul & D. J. Weber. 2001. Effect of salinity on the growth and ion content of *Salicornia rubra*. *Commun. Soil Sci. Plant Analysis* 32: 2965–2977.

Khan, M. M., K. Thompson & M. Usman. 2002. Correlation of seed unsaturated fatty acids with seed persistence in the soil. *Pakistan J. Agric. Sci.* 39: 300–306.

Khanduri, P., R. Tandon, P. L. Uniyal, V. Bhat & A. K. Pandey. 2015. Comparative morphology and molecular systematics of Indian Podostemaceae. *Plant Syst. Evol.* 301: 861–882.

Khang, H. V., M. Hatayama & C. Inoue. 2012. Arsenic accumulation by aquatic macrophyte coontail (*Ceratophyllum demersum* L.) exposed to arsenite, and the effect of iron on the uptake of arsenite and arsenate. *Environ. Exp. Bot.* 83: 47–52.

Khanh, T. D., M. I. Chung, T. D. Xuan & S. Tawata. 2005. The exploitation of crop allelopathy in sustainable agricultural production. *J. Agron. Crop Sci.* 191: 172–184.

Khanna, V. G. & K. Kannabiran. 2007. Larvicidal effect of *Hemidesmus indicus*, *Gymnema sylvestre*, and *Eclipta prostrata* against *Culex qinquifaciatus* mosquito larvae. *African J. Biotechnol.* 6: 307–311.

Khanuja, S. P. S., A. K. Shasany, A. Srivastava & S. Kumar. 2000. Assessment of genetic relationships in *Mentha* species. *Euphytica* 111: 121–125.

Khare, C. P. 2007. *Indian Medicinal Plants: An Illustrated Dictionary*. Springer, New York. 900 pp.

Khayyal, M., M. El-Ghazaly, S. Kenawy, M. Seif-el-Nasr, L. Mahran, Y. Kafafi & S. Okpanyi. 2001. Antiulcerogenic effect of some gastrointestinally acting plant extracts and their combination. *Arzneim. Forsch.* 51: S545–S553.

Khedr, A. H. & A. K. Hegazy. 1998. Ecology of the rampant weed *Nymphaea lotus* L. Willdenow in natural and ricefield habitats of the Nile delta, Egypt. *Hydrobiologia* 386: 119–129.

Khoury, C. K., S. Greene, J. Wiersema, N. Maxted, A. Jarvis & P. C. Struik. 2013. An inventory of crop wild relatives of the United States. *Crop Sci.* 53: 1496–1508.

Khuroo, A. A., I. Rashid, Z. Reshi, G. H. Dar & B. A. Wafai. 2007. The alien flora of Kashmir Himalaya. *Biol. Invas.* 9: 269–292.

Kiang, Y. T. 1971. Pollination study in a natural population of *Mimulus guttatus*. *Evolution* 26: 308–310.

Kie, J. G. & S. A. Myler. 1987. Use of fertilization and grazing exclusion in mitigating lost meadow production in the Sierra Nevada, California, USA. *Environ. Manag.* 11: 641–648.

Kiehn, M. 1995. Chromosome survey of the Rubiaceae. *Ann. Missouri Bot. Gard.* 82: 398–408.

Kiill, L. H. P. & N. T. Ranga. 2003. Pollination ecology of *Ipomoea asarifolia* (Ders.) Roem. & Schult. (Convolvulaceae) in the semi-arid area of Pernambuco. *Acta Bot. Brasilica* 17: 355–362.

Kil, J.-S., Y. Son, Y.-K. Cheong, N.-H. Kim, H. J. Jeong, J.-W. Kwon, E.-J. Lee, T.-O. Kwon, H.-T. Chung & H.-O. Pae. 2012. Okanin, a chalcone found in the genus *Bidens*, and 3-penten-2-one inhibit inducible nitric oxide synthase expression via heme oxygenase-1 induction in RAW264.7 macrophages activated with lipopolysaccharide. *J. Clin. Biochem. Nutr.* 50: 53–58.

Kim, B. J., J. H. Kim, H. P. Kim & M. Y. Heo. 1997. Biological screening of 100 plant extracts for cosmetic use (II): antioxidative activity and free radical scavenging activity. *Int. J. Cosmetic Sci.* 19: 299–307.

Kim, C., H. R. Na & H.-K. Choi. 2010. Molecular genotyping of *Trapa bispinosa* and *T. japonica* (Trapaceae) based on nuclear AP2 and chloroplast DNA *trnL-F* region. *Amer. J. Bot.* 97: e149–e152.

Kim, C., J. Jung, H. R. Na, S. W. Kim, W. Li, Y. Kadono, H. Shin & H.-K. Choi. 2012. Population genetic structure of the endangered *Brasenia schreberi* in South Korea based on nuclear ribosomal spacer and chloroplast DNA sequences. *J. Plant Biol.* 55: 81–91.

Kim, C., T. Deng, M. Chase, D.-G. Zhang, Z.-L. Nie & H. Sun. 2015. Phylogeny and character evolution of Urticeae. *Taxon* 64: 65–78.

Kim, J. Y., K.-H. Kim, Y. J. Lee, S. H. Lee, J. C. Park & D. H. Nam. 2009. *Oenanthe javanica* extract accelerates ethanol metabolism in ethanol-treated animals. *BMB Rep.* 42: 482–485.

Kim, K. D. 2002. Plant invasion and management in turf-dominated waste landfills in South Korea. *Land Degrad. Dev.* 13: 257–267.

Kim, S. & Y. Suh. 2013. Phylogeny of Magnoliaceae based on ten chloroplast DNA regions. *J. Plant Biol.* 56: 290–305.

Kim, S., C-W. Park, Y-D. Kim & Y. Suh. 2001. Phylogenetic relationships in family Magnoliaceae inferred from *ndhF* sequences. *Amer. J. Bot.* 88: 717–728.

Kim, S., D. E. Soltis, P. S. Soltis, M. J. Zanis & Y. Suh. 2004. Phylogenetic relationships among early-diverging eudicots based on four genes: were the eudicots ancestrally woody? *Mol. Phylogen. Evol.* 31: 16–30.

Kim, S.-C., D. J. Crawford, M. Tadesse, M. Berbee, F. R. Ganders, M. Pirseyedi & E. J. Esselman. 1999. ITS sequences and phylogenetic relationships in *Bidens* and *Coreopsis* (Asteraceae). *Syst. Bot.* 24: 480–493.

Kim, S.-T. & M. J. Donoghue. 2008a. Molecular phylogeny of *Persicaria* (Persicarieae, Polygonaceae). *Syst. Bot.* 33: 77–86.

Kim, S.-T. & M. J. Donoghue. 2008b. Incongruence between cpDNA and nrITS trees indicates extensive hybridization within *Eupersicaria* (Polygonaceae). *Amer. J. Bot.* 95: 1122–1135.

Kim, S.-T., S. E. Sultan & M. J. Donoghue. 2003. Phylogeny and evolution of *Polygonum* sect. *Persicaria* (Polygonaceae) in eastern North America. Botany 2003 meeting (abstract).

Kim, S. T., S. E. Sultan & M. J. Donoghue. 2008. Allopolyploid speciation in *Persicaria* (Polygonaceae): insights from a low-copy nuclear region. *Proc. Natl. Acad. Sci. U.S.A.* 105: 12370–12375.

Kimball, K. D. & D. M. Weihrauch. 2000. Alpine vegetation communities and the alpine-treeline ecotone boundary in New England as biomonitors for climate change. RMRS-P-15-VOL-3. *USDA For. Serv. Proc.* 3: 93–101.

Kimball, R. T. & D. J. Crawford. 2004. Phylogeny of Coreopsideae (Asteraceae) using ITS sequences suggests lability in reproductive characters. *Mol. Phylogen. Evol.* 33: 127–139.

Kincaid, T. 1904. The insects of Alaska: Introduction. pp. 1–34. *In*: C. H. Merriam (ed.), *Alaska. Vol. VIII. Insects. Part 1*. Doubleday, Page & Co., New York.

Kindscher, K. 1994. Rockefeller Prairie: A case study on the use of plant guild classification of a tallgrass prairie. pp. 123–140. *In*: R. G. Wickett, P. D. Lewis, A. Woodliffe & P. Pratt (eds.), *Proceedings of the Thirteenth North American Prairie Conference. Proceedings of the Thirteenth North American Prairie Conference, 6–9 August, 1992*. Department of Parks and Recreation, Windsor, ON.

Kindscher, K. & D. P. Hurlburt. 1998. Huron Smith's ethnobotany of the Hocąk (Winnebago). *Econ. Bot.* 52: 352–372.

Kindscher, K. & L. L. Tieszen. 1998. Floristic and soil organic matter changes after five and thirty-five years of native tallgrass prairie restoration. *Restorat. Ecol.* 6: 181–196.

King, B. C. 2003. The ethnobotany of the Miami Tribe: traditional plant use from historical texts. Honors thesis. Miami University, Miami, OH. 188 pp.

King, G. 1992. Geomorphology of Piedmont vernal pool basins, California. *California Geogr.* 32: 19–38.

King, J. R. & R. Bamford. 1937. The chromosome number in *Ipomoea* and related genera. *J. Heredity* 28: 279–282.

King, R. M. & H. Robinson. 1970. *Eupatorium*, a composite genus of Arcto-Tertiary distribution. *Taxon* 19: 769–774.

King, R. M., D. W. Kyhos, A. M. Powell, P. H. Raven & H. Robinson. 1976. Chromosome numbers in Compositae. XIII. Eupatorieae. *Ann. Missouri Bot. Gard.* 63: 862–888.

King, S. L. & T. J. Antrobus. 2001. Canopy disturbance patterns in a bottomland hardwood forest in northeast Arkansas, USA. *Wetlands* 21: 543–553.

King, S. L. & T. J. Antrobus. 2005. Relationships between gap makers and gap fillers in an Arkansas floodplain forest. *J. Veg. Sci.* 16: 471–478.

King-Jones, S. 2001. Revision of *Pluchea* Cass. (Compositae, Plucheeae) in the Old World. *Englera* 23: 3–136.

Kirkaldy, G. W. 1907. *A Catalogue of the Hemipterous Family Aleyrodidae*. Bulletin no. 2. Paradise of the Pacific Print, Honolulu, HI. 85 pp.

Kirkbride Jr., J. H. 1985. Manipulus Rubiacearum IV. *Kerianthera* (Rubiaceae), a new genus from Amazonian Brazil. *Brittonia* 37: 109–116.

Kirkman, L. K. & R. R. Sharitz. 1994. Vegetation disturbance and maintenance of diversity in intermittently flooded Carolina bays in South Carolina. *Ecol. Appl.* 4: 177–188.

Kirkman, L. K., P. C. Goebel, L. West, M. B. Drew & B. J. Palik. 2000. Depressional wetland vegetation types: a question of plant community development. *Wetlands* 20: 373–385.

Kirkpatrick, H. E. & J. W. S. Barnes. 2000. Moss mediation of site occupation in two monkeyflowers. Ecological Society of America, annual meeting (abstract).

Kirsten, R. & D. W. Blinn. 2003. Herbivory on a chemically defended plant as a predation deterrent in *Hyalella azteca*. *Freshwater Biol.* 48: 247–254.

Kishore, V. K. 2002. Mapping quantitative trait loci underlying genome-wide recombination rate and mating system differences in meadowfoam. PhD dissertation. Oregon State University, Corvallis, OR. 187 pp.

Kistchinski, A. A. & V. E. Flint. 1974. On the biology of the spectacled eider. *Wildfowl* 25: 5–15.

Kita, Y. & M. Kato. 2001. Infrafamilial phylogeny of the aquatic angiosperm Podostemaceae inferred from the nucleotide sequences of the *matK* gene. *Plant Biol.* 3: 156–163.

Kita, Y., Y. Hirayama, R. Rutishauser, K. A. Huber & M. Kato. 2012. Molecular phylogenetic analysis of Podostemaceae: implications for taxonomy of major groups. *Bot. J. Linn. Soc.* 169: 461–492.

Kitamura, S. 1991. *Utricularia inflata* naturalized in Japan. *Acta Phytotax. Geobot.* 42: 158.

Klak, C., A. Khunou, G. Reeves & T. Hedderson. 2003. A phylogenetic hypothesis for the Aizoaceae (Caryophyllales) based on four plastid DNA regions. *Amer. J. Bot.* 90: 1433–1445.

Klein, A., J. Crawford, J. Evens, T. Keeler-Wolf & D. Hickson. 2007. *Classification of the Vegetation Alliances and Associations of the Northern Sierra Nevada Foothills, California*. Report prepared for California Department of Fish and Game. California Native Plant Society, Sacramento, CA.

Klein, D. R. 1965. Ecology of deer range in Alaska. *Ecol. Monogr.* 35: 259–284.

Klein, D. R. 1982. Fire, lichens, and caribou. *J. Range Manag.* 35: 390–395.

Klekowski Jr., E. J., R. Lowenfeld & P. K. Hepler. 1994. Mangrove genetics. II. Outcrossing and lower spontaneous mutation rates in Puerto Rican *Rhizophora*. *Int. J. Plant Sci.* 155: 373–381.

Kleyheeg, E., M. Klaassen & M. B. Soons. 2016. Seed dispersal potential by wild mallard duck as estimated from digestive tract analysis. *Freshwater Biol.* 61: 1746–1758.

Klimeš, L. & J. Klimešová. 1994. Biomass allocation in a clonal vine: effects of intraspecific competition and nutrient availability. *Folia Geobot.* 29: 237–244.

Klinka, K., V. J. Krajina, A. Ceska & A. M. Skagel. 1995. *Indicator Plants of Coastal British Columbia*. UBC Press, Vancouver, BC. 286 pp.

Klinkenberg, B. (ed.). 2004. E-Flora BC: Electronic atlas of the plants of British Columbia [www.eflora.bc.ca]. Lab for Advanced Spatial Analysis, Department of Geography, University of British Columbia, Vancouver, BC.

Klips, R. A. 1995. Genetic affinity of the rare eastern Texas endemic *Hibiscus dasycalyx* (Malvaceae). *Amer. J. Bot.* 82: 1463–1472.

Klips, R. A. 1999. Pollen competition as a reproductive isolating mechanism in *Hibiscus laevis* (Malvaceae). *Amer. J. Bot.* 86: 269–272.

Klips, R. A. 2003. Vegetation of Claridon Railroad Prairie, a remnant of the Sandusky Plains of central Ohio. *Castanea* 68: 135–142.

Klips, R. A. & A. A. Snow. 1997. Delayed autonomous self-pollination in *Hibiscus laevis* (Malvaceae). *Amer. J. Bot.* 84: 48.

Kloot, P. M. 1984. The introduced elements of the flora of Southern Australia. *J. Biogeogr.* 11: 63–78.

Klussmann, W. G. & J. T. Davis. 1988. *Common Aquatic Plants of Texas*. B-1018 (revised edition). Texas Agricultural Extension Service, The Texas A&M University System, College Station, TX. 8 pp.

Klymko, J., C. S. Blaney & D. G. Anderson. 2012. The first record of Dorcas Copper (*Lycaena dorcas*) from Nova Scotia. *J. Acad. Entomol. Soc.* 8: 41–42.

Klyver, F. D. 1931. Major plant communities in a transect of the Sierra Nevada mountains of California. *Ecology* 12: 1–17.

Knapp, W. M., R. F. C. Naczi, W. D. Longbottom, C. A. Davis, W. A. McAvoy, C. T. Frye, J. W. Harrison & P. Stango, III. 2011. Floristic discoveries in Delaware, Maryland, and Virginia. *Phytoneuron* 2011 64: 1–26.

Knapik, L. J., G. W. Scotter & W. W. Pettapiece. 1973. Alpine soil and plant community relationships of the Sunshine area, Banff National Park. *Arct. Alp. Res.* 5: A161–A170.

Knepper, D. A., D. M. Johnson & L. J. Musselman. 2002. *Marsilea mutica* in Virginia. *Amer. Fern J.* 92: 243–244.

Knight, A. P. & R. G. Walter. 2001. *A Guide to Plant Poisoning of Animals in North America*. Teton New Media, Jackson, WY. 367 pp.

Knight, S. E. & T. M. Frost. 1991. Bladder control in *Utricularia macrorhiza*: lake-specific variation in plant investment in carnivory. *Ecology* 72: 728–734.

Knight, T. M. & T. E. Miller. 2004. Local adaptation within a population of *Hydrocotyle bonariensis*. *Evol. Ecol. Res.* 6: 103–114.

Knisley, C. B. & H. A. Denmark. 1978. New Phytoseiid mites from successional and climax plant communities in New Jersey. *Florida Entomol.* 61: 5–17.

Knotts, K. & R. Sacher. 2000. *Provisional Checklist of Names/ Epithets of* Nymphaea *L.* International Waterlily and Water Gardening Society, Cocoa beach, FL. 84 pp.

Knowlton, C. H. 1915. Plants and plant societies at Roque Bluffs, Maine. *Rhodora* 17: 145–155.

Knox, E. B. 2014. The dynamic history of plastid genomes in the Campanulaceae sensu lato is unique among angiosperms. *Proc. Natl. Acad. Sci. U.S.A.* 111: 11097–11102.

Knox, E. B. & J. D. Palmer. 1999. The chloroplast genome arrangement of *Lobelia thuliniana* (Lobeliaceae): expansion of the inverted repeat in an ancestor of the Campanulales. *Plant Syst. Evol.* 214: 49–64.

Knox, J. S. 1987. An experimental garden test of characters used to distinguish *Helenium virginicum* Blake from *H. autumnale* L. *Castanea* 52: 52–58.

Knox, J. S. 1997. A nine year demographic study of *Helenium virginicum* (Asteraceae), a narrow endemic seasonal wetland plant. *J. Torrey Bot. Soc.* 124: 236–245.

Knox, J. S., M. J. Gutowski, D. C. Marshall & O. G. Rand. 1995. Tests of the genetic bases of character differences between *Helenium virginicum* and *H. autumnale* (Asteraceae) using common gardens and transplant studies. *Syst. Bot.* 20: 120–131.

Knuth, P. 1906. *Handbook of Flower Pollination Based upon Hermann Müller's Work 'The Fertilisation of Flowers by Insects.' Volume I. Introduction and Literature.* Oxford at the Clarendon Press, Oxford, UK. 382 pp.

Knuth, P. 1908. *Handbook of Flower Pollination Based upon Hermann Müller's Work 'The Fertilisation of Flowers by Insects.' Volume II. Observations of Flower Pollination Made in Europe and the Arctic Regions on Species Belonging to the Natural Orders: Ranunculaceae to Stylidieae.* Clarendon Press, Oxford, UK. 703 pp.

Knuth, P. 1909. *Handbook of Flower Pollination Based upon Hermann Müller's Work 'The Fertilisation of Flowers by Insects.' Volume III. Observations on Flower Pollination Made in Europe and the Arctic Regions on Species Belonging to the Natural Orders: Goodenovieae to Cycadeae.* Clarendon Press, Oxford, UK. 644 pp.

Ko, R. K., M.-C. Kang, B.-S. Kim, J.-H. Han, G.-O. Kim & N. H. Lee. 2009. A new phloroglucinol glycoside from *Aster subulatus* Michx. *Bull. Korean Chem. Soc.* 30: 1167–1169.

Koch, F. H. & W. D. Smith. 2008. Spatio-temporal analysis of *Xyleborus glabratus* (Coleoptera: Circulionidae: Scolytinae) invasion in eastern US forests. *Environ. Entomol.* 37: 442–452.

Koch, K., M. Bennemann, H. F. Bohn, D. C. Albach & W. Barthlott. 2013. Surface microstructures of daisy florets (Asteraceae) and characterization of their anisotropic wetting. *Bioinspir. Biomim.* 8: 1–15. doi:10.1088/1748-3182/8/3/036005.

Koch, M. A., B. Haubold & T. Mitchell-Olds. 2000. Comparative evolutionary analysis of chalcone synthase and alcohol dehydrogenase loci in *Arabidopsis*, *Arabis*, and related genera (Brassicaceae). *Mol. Biol. Evol.* 17: 1483–1498.

Koch, M., B. Haubold & T. Mitchell-Olds. 2001. Molecular systematics of the Brassicaceae: evidence from coding plastidic *matK* and nuclear *Chs* sequences. *Amer. J. Bot.* 88: 534–544.

Kock, O. 1987. New medicinal plant: *Amsonia tabernaemontana*. *Kert. Szölészet* 36: 6.

Kocyan, A., L.-B. Zhang, H. Schaefer & S. S. Renner. 2007. A multi-locus chloroplast phylogeny for the Cucurbitaceae and its implications for character evolution and classification. *Mol. Phylogen. Evol.* 44: 553–577.

Koehl, V., L. B. Thien, B., E. G. Heij & T. L. Sage. 2004. The causes of self-sterility in natural populations of the relictual angiosperm, *Illicium floridanum* (Illiciaceae). *Ann. Bot.* 94: 43–50.

Koeman-Kwak, M. 1973. The pollination of *Pedicularis palustris* by nectar thieves (short-tongued bumblebees). *Acta Bot. Neerl.* 22: 608–615.

Koevenig, J. L. 1976. Effect of climate, soil physiography and seed germination on the distribution of river birch (*Betula nigra*). *Rhodora* 78: 420–437.

Kohl, L. M. 2008. Population dynamics and dispersal gradient of *Aphelenchoides fragariae* in the woody ornamental *Lantana camera*. MS thesis. North Carolina State University, Raleigh, NC. 96 pp.

Kohl, L. M. 2011. Astronauts of the nematode world: an aerial view of foliar nematode biology, epidemiology, and host range. *APSnet Features*. doi:10.1094/APSnetFeature-2011-0111.

Kohout, P., Z. Sýkorová, M. Čtvrtlíková, J. Rydlová, J. Suda, M. Vohník & R. Sudová. 2012. Surprising spectra of root-associated fungi in submerged aquatic plants. *FEMS Microbiol. Ecol.* 80: 216–235.

Koi, S. 2008. Nectar sources for *Eumaeus atala* (Lepidoptera: Lycaenidae: Theclinae). *Florida Entomol.* 91: 118–120.

Koizumi, H., T. Kibe, T. Nakadai, Y. Yazaki, M. Adachi, M. Inatomi, M. Kondo & T. Ohtsuka. 2004. Effect of free-air CO_2 enrichment on structures of weed communities and CO_2 exchange at the flood-water surface in a rice paddy field. pp. 473-485. *In*: M. Shiyomi, H. Kawahata, H. Koizumi, A. Tsuda & Y. Awaya (eds.), *Global Environmental Change in the Ocean and on Land.* Terrapub, Tokyo, Japan.

Kokkinou, A., K. Yannakopoulou, I. M. Mavridis & D. Mentzafos. 2001. Structure of the complex of β-cyclodextrin with β-naphthyloxyacetic acid in the solid state and in aqueous solution. *Carbohyd. Res.* 332: 85–94.

Kole, P. L., H. R. Jadhav, P. Thakurdesai & A. N. Nagappa. 2005. Cosmetic potential of herbal extracts. *Nat. Prod. Radiance* 4: 315–321.

Kondo, K. 1969. Chromosome numbers of carnivorous plants. *Bull. Torrey Bot. Club* 96: 322–328.

Kondo, K. 1971. Germination and developmental morphology of seeds in *Utricularia cornuta* Michx. and *Utricularia juncea* Vahl. *Rhodora* 73: 541–547.

Kondo, K. 1972. A comparison of variability in *Utricularia cornuta* and *Utricularia juncea*. *Amer. J. Bot.* 59. 23–37.

Kondo, K. 1976. A cytotaxonomic study in some species of *Drosera*. *Rhodora* 78: 532–541.

Kondo, K., M. Segawa & K. Nehira. 1978a. Anatomical studies on seeds and seedlings of some *Utricularia* (Lentibulariaceae). *Brittonia* 30: 89–95.

Kondo, K., L. J. Musselman & W. F. Mann Jr. 1978b. Karyomorphological studies in some parasitic species of the Scrophulariaceae, I. *Brittonia* 30: 345–354.

Kong, L. D., Y. Cai, W. W. Huang, C. H. Cheng & R. X. Tan. 2000. Inhibition of xanthine oxidase by some Chinese medicinal plants used to treat gout. *J. Ethnopharmacol.* 73: 199–207.

Koning, C. O. 2005. Vegetation patterns resulting from spatial and temporal variability in hydrology, soils, and trampling in an isolated basin marsh, New Hampshire, USA. *Wetlands* 25: 239–251.

Konoshima, T. & K.-H. Lee. 1986. Antitumor agents, 85. Cicutoxin, an antileukemic principle from *Cicuta maculata*, and the cytotoxicity of the related derivatives. *J. Nat. Prod.* 49: 1117–1121.

Konuma, A. & R. Terauchi. 2001. Population genetic structure of the self-compatible annual herb *Polygonum thunbergii* (Polygonaceae) detected by multilocus DNA fingerprinting. *Amer. Midl. Naturalist* 146: 122–127.

Konuma, A. & T. Yahara. 1997. Temporally changing male reproductive success and resource allocation strategy in protandrous *Heracleum lanatum* (Apiaceae). *J. Plant Res.* 110: 227–234.

Koo, K. A., S. H. Sung, J. H. Park, S. H. Kim, K. Y. Lee & Y. C. Kim. 2005. *In vitro* neuroprotective activities of phenylethanoid glycosides from *Callicarpa dichotoma*. *Plant Med.* 71: 778–780.

Koontz, J. A., P. S. Soltis & D. E. Soltis. 2004. Using phylogeny reconstruction to test hypotheses of hybrid origin in *Delphinium* section *Diedropetala* (Ranunculaceae). *Syst. Bot.* 29: 345–357.

Koopman, M. M. & D. A. Baum. 2005. A phylogenetic analysis of the *Hibisceae* (Malvaceae) radiation on Madagascar. Botany 2005 meeting (abstract).

Köppler, M.-R., I. Kowarik, N. Kühn & M. von der Lippe. 2014. Enhancing wasteland vegetation by adding ornamentals: opportunities and constraints for establishing steppe and prairie species on urban demolition sites. *Landscape Urban Plann* 126: 1–9.

Koprowski, J. L. 1994. *Sciurus niger. Mamm. Species* 479: 1–9.

Koptur, S., J. O'Brien & J. R. Snyder. 2000. Community patterns of seedling recruitment after summer fire in two pine rocklands. Greater Everglades Ecosystem Restoration (G.E.E.R.) Science Conference (abstract).

Korf, R. P. & R. Dirig. 2009. Discomycetes exsiccati—fascicles 5 and 6. *Mycotaxon* 107: 25–34.

Kornhall, P. & B. Bremer. 2004. New circumscription of the tribe Limoselleae (Scrophulariaceae) that includes the taxa of the tribe Manuleeae. *Bot. J. Linn. Soc.* 146: 453–467.

Korschgen, L. J. 1962. Food habits of greater prairie chickens in Missouri. *Amer. Midl. Naturalist* 68: 307–318.

Koryak, M. & R. J. Reilly. 1984. Vascular riffle flora of Appalachian streams: the ecology and effects of acid mine drainage on *Justicia americana* (L.) Vahl. *Proc. Pennsylvania Acad. Sci.* 58: 55–60.

Kosiba, P. 1992. Studies on the ecology of *Utricularia vulgaris* L. II. Physical, chemical and biotic factors and the growth of *Utricularia vulgaris* L. in cultures *in vitro*. *Ekol. Polska* 40: 193–212.

Kossuth, S. & R. L. Scheer. 1990. *Nyssa ogeche* Bartr. ex Marsh. Ogeechee Tupelo. pp. 479–481. *In*: R. M. Burns & B. H. Honkala (eds.), *Silvics of North America. Vol. 2, Hardwoods.* Agriculture Handbook 654. USDA, Forest Service, Washington, DC.

Kost, M. A. 2001. *Potential Indicators for Assessing Biological Integrity of Forested, Depressional Wetlands in Southern Michigan.* Michigan Department of Environmental Quality, Report 2001–10. Michigan Natural Features Inventory, Lansing, MI. 69 pp.

Kost, M. A. & D. De Steven. 2000. Plant community responses to prescribed burning in Wisconsin sedge meadows. *Nat. Areas J.* 20: 36–45.

Kost, M. A., D. A. Albert, J. G. Cohen, B. S. Slaughter, R. K. Schillo, C. R. Weber & K. A. Chapman. 2007. *Natural Communities of Michigan: Classification and Description.* Michigan Natural Features Inventory, Michigan Department of Natural Resources, Lansing, MI. 317 pp.

Krafft, C. C. & S. N. Handel. 1991. The role of carnivory in the growth and reproduction of *Drosera filiformis* and *D. rotundifolia. Bull. Torrey Bot. Club* 118: 12–19.

Kral, R. 1966a. *Xyris* (Xyridaceae) of the continental United States and Canada. *Sida* 2: 177–260.

Kral, R. 1966b. Observations on the flora of the southeastern United States with special reference to northern Louisiana. *Sida* 2: 395–408.

Kral, R. 1981. Some distributional reports of weedy or naturalized foreign species of vascular plants for the southern states, particularly Alabama and middle Tennessee. *Castanea* 46: 334–339.

Kral, R. 1997. 2. Annonaceae Jussieu. pp. 11–20. *In*: N. R. Morin (convening ed.), *Flora of North America North of Mexico, Vol. 3: Magnoliophyta: Magnoliidae and Hamamelidae.* Oxford University Press, New York.

Kral, R. & P. E. Bostick. 1969. The genus *Rhexia* (Melastomataceae). *Sida* 3: 387–440.

Krasnoff, S. B. & D. E. Dussourd. 1989. Dihydropyrrolizine attractants for arctiid moths that visit plants containing pyrrolizidine alkaloids. *J. Chem. Ecol.* 15: 47–60.

Krasny, M. E., J. C. Zasada & K. A. Vogt. 1988. Adventitious rooting of four Salicaceae species in response to a flooding event. *Can. J. Bot.* 66: 2597–2598.

Kraus, T. E. C., R. J. Zasoski & R. A. Dahlgren. 2004. Fertility and pH effects on polyphenol and condensed tannin concentrations in foliage and roots. *Plant Soil* 262: 95–109.

Krauss, P. 1987. *Old Field Succession in Everglades National Park.* South Florida Research Center Report. SFRC-87/03. Everglades National Park, South Florida Research Center, Homestead, FL.

Krebs, S. L. & J. F. Hancock.1991. Embryonic genetic load in the highbush blueberry, *Vaccinium corymbosum* (Ericaceae). *Amer. J. Bot.* 78: 1427–1437.

Krefting, L. W. & E. Roe. 1949. The role of some birds and mammals in seed germination. *Ecol. Monogr.* 19: 284–286.

Kress, W. J., G. D. Maddox & C. S. Roesel. 1994. Genetic variation and protection priorities in *Ptilimnium nodosum* (Apiaceae), an endangered plant of the eastern United States. *Conserv. Biol.* 8: 271–276.

Kreuzwieser, J., U. Scheerer, J. Kruse, T. Burzlaff, A. Honsel, S. Alfarraj, P. Georgiev, J. P. Schnitzler, A. Ghirardo, I. Kreuzer & R. Hedrich. 2014. The Venus flytrap attracts insects by the release of volatile organic compounds. *J. Exp. Bot.* 65: 755–766.

Krewer, G. 2000. *Experiments and Observations on Growing Mayhaws as a Crop in South Georgia and North Florida.* Departmental Fact Sheet #H-00–053. Cooperative Extension Service, College of Agricultural and Environmental Sciences, The University of Georgia, Athen, GA.

Krieger, J. R. & P. S. Kourtev. 2012. Bacterial diversity in three distinct sub-habitats within the pitchers of the northern pitcher plant, *Sarracenia purpurea. FEMS Microbiol. Ecol.* 79: 555–567.

Krings, A. 2005. Neotypification of *Ceropegia palustris* and *Lyonia maritima* (Apocynaceae: Asclepiadoideae). *Sida* 21: 1507–1513.

Krings, A. & J. C. Neal. 2001a. A *Scutellaria* (Lamiaceae) new to North Carolina and a key to the small-flowered Carolina congeners. *Sida* 19: 735–739.

Krings, A. & J. C. Neal. 2001b. South American skullcap (*Scutellaria racemosa*: Lamiaceae) in the southeastern United States. *Sida* 19: 1171–1179.

Krischik, V. A. & M. R. Zbinden. n.d. *Conservation of Beneficial Insects and Nutrient Abatement with the Use of Native Plants in Urban Landscapes.* Department of Entomology, University of Minnesota, St. Paul, MN. 38 pp.

Krishanu, S. 2012. A details [sic!] study on *Hygrophila difformis.* *Int. J. Pharmaceut. Chem. Biol. Sci.* 2: 494–499.

Krishnan, K., V. G. Khanna & S. Hameed. 2010. Antiviral activity of dasyscyphin C extracted from *Eclipta prostrata* against fish nodavirus. *J. Antivirals Antiretrovirals* 1: 29–32.

Kristiansen, K., C. W. Hansen & K. Brandt. 1997. Flower induction in seedlings of *Aster novi-belgii* and selection before and after vegetative propagation. *Euphytica* 93: 361–367.

Krmpotic, E., N. R. Farnsworth & W. M. Messmer. 1972. Cryptopleurine, an active antiviral alkaloid from *Boehmeria cylindrica* (L.) Sw. (Urticaceae). *J. Pharm. Sci.* 61: 1508–1509.

Krochmal, A. 1968. Medicinal plants and Appalachia. *Econ. Bot.* 22: 332–337.

Kron, K. A. 1997. Phylogenetic relationships of Rhododendroideae (Ericaceae). *Amer. J. Bot.* 84: 973–980.

Kron, K. A. & W. S. Judd. 1990. Phylogenetic relationships within the Rhodoreae (Ericaceae) with specific comments on the placement of *Ledum.* *Syst. Bot.* 15: 57–68.

Kron, K. A. & W. S. Judd. 1997. Systematics of the *Lyonia* group (Andromedeae, Ericaceae) and the use of species as terminals in higher-level cladistic analyses. *Syst. Bot.* 22: 479–492.

Kron, K. A. & J. M. King. 1996. Cladistic relationships of *Kalmia, Leiophyllum,* and *Loiseleuria* (Phyllodoceae, Ericaceae) based on *rbcL* and nrITS data. *Syst. Bot.* 21: 17–29.

Kron, K. A., W. S. Judd & D. M. Crayn. 1999. Phylogenetic analyses of Andromedeae (Ericaceae subfam. Vaccinioideae). *Amer. J. Bot.* 86: 1290–1300.

Kron, K. A., W. S. Judd, P. F. Stevens, D. M. Crayn, A. A. Anderberg, P. A. Gadek, C. J. Quinn & J. L. Luteyn. 2002a. A phylogenetic classification of Ericaceae: molecular and morphological evidence. *Bot. Rev.* 68: 335–423.

Kron, K. A., E. A. Powell & J. L. Luteyn. 2002b. Phylogenetic relationships within the blueberry tribe (Vaccinieae, Ericaceae) based on sequence data from *matK* and nuclear ribosomal ITS regions, with comments on the placement of *Satyria.* *Amer. J. Bot.* 89: 327–336.

Kruchek, B. L. 2004. Use of tidal marsh and upland habitats by the marsh rice rat (*Oryzomys palustris*). *J. Mammalogy* 85: 569–575.

Kruckeberg, A. R. 1957. A study of the perennial species of *Sidalcea,* Part II: Chromosome numbers and interspecific hybridizations. *Univ. Washington Publ. Biol.* 18: 83–93.

Kruckeberg, A. R. & F. L. Hedglin. 1963. Natural and artificial hybrids of *Besseya* and *Synthyris* (Scrophulariaceae). *Madroño* 17: 109–115.

Krueger, W. C. 1972. Evaluating animal forage preference. *J. Range Manag.* 25: 471–475.

Krull, J. N. 1970. Aquatic plant-macroinvertebrate associations and waterfowl. *J. Wildl. Manag.* 34: 707–718.

Kruse, J. J. 1998. New Wisconsin records for a *Hemileuca* (Lepidoptera: Saturniidae) using *Menyanthes trifoliata* (Solanales: Menyanthaceae) and *Betula pumila* (Betulaceae). *Great Lakes Entomol.* 31: 109–112.

Krychak-Furtado, S., A. L. P. Silva, O. G. Miguel, J. de F. G. Dias, M. D. Miguel, S. S. Costa & R. R. B. Negrelle. 2011. Effectiveness of Asteraceae extracts on Trichostrongylidae eggs development in sheep. *Rev. Brasil. Parasitol. Veterin.* 20: 215–218.

Ku, Y. C., K. H. Park & Y. J. Oh. 1993. Changes of weed flora under direct seeded rice cultivation in dry paddy field. *Korean J. Weed Sci.* 13: 159–163.

Kubanek, J., M. E. Hay, P. J. Brown, N. Lindquist & W. Fenical. 2001. Lignoid chemical defenses in the freshwater macrophyte *Saururus cernuus.* *Chemoecology* 11: 1–8.

Kubo, N., M. Hirai, A. Kaneko, D. Tanaka & K. Kasumi. 2009. Development and characterization of simple sequence repeat (SSR) markers in the water lotus (*Nelumbo nucifera*). *Aquat. Bot.* 90: 191–184.

Kudo, G. 1992. Performance and phenology of alpine herbs along a snow-melting gradient. *Ecol. Res.* 7: 297–304.

Kudoh, H. & D. F. Whigham. 1997. Microgeographic genetic structure and gene flow in *Hibiscus moscheutos* (Malvaceae) populations. *Amer. J. Bot.* 84: 1285–1293.

Kudoh, H. & D. F. Whigham. 1998. The effect of petal-size manipulation on pollinator/seed predator mediated female reproductive success of *Hibiscus moscheutos* (Malvaceae) populations. *Oecologia* 117: 70–79.

Kudoh, H. & D. F. Whigham. 2001. A genetic analysis of hydrologically dispersed seeds of *Hibiscus moscheutos* (Malvaceae). *Amer. J. Bot.* 88: 588–593.

Kuduk-Jaworska, J., J. Szpunar, K. Gasiorowski & B. Brokos. 2004. Immunomodulating polysaccharide fractions of *Menyanthes trifoliata* L. *Z. Naturforsch.* 59: 485–493.

Kuebel, K. R. & A. O. Tucker. 1988. Vietnamese culinary herbs in the United States. *Econ. Bot.* 42: 413–419.

Kufeld, R. C. 1973. Foods eaten by the Rocky Mountain elk. *J. Range Manag.* 26: 106–113.

Kuhn, T. J., H. D. Safford, B. E. Jones & K. W. Tate. 2011. Aspen (*Populus tremuloides*) stands and their contribution to plant diversity in a semiarid coniferous landscape. *Plant Ecol.* 212: 1451–1463.

Kühn, U. 1993. Chenopodiaceae. pp. 253–281. *In*: K. Kubitzki, J. Rohwer & V. Bittrich (eds.), *The Families and Genera of Vascular Plants, Vol. II. Flowering Plants, Dicotyledons: Magnoliid, Hamamelid and Caryophyllid Families.* Springer-Verlag, Berlin, Germany.

Kuhnlein, H. V. & N. J. Turner. 1991. *Traditional Plant Foods of Canadian Indigenous Peoples: Nutrition, Botany and Use.* Gordon & Breach Science Publishers, Amsteldijk, The Netherlands. 635 pp.

Kuhnlein, H. V., N. J. Turner & P. D. Kluckner. 1982. Nutritional significance of two important root foods (springbank clover and Pacific silverweed) used by Native People of the coast of British Columbia. *Ecol. Food Nutr.* 12: 89–95.

Kuijper, D. P., J. P. Bakker, E. J. Cooper, R. Ubels, I. S. Jónsdóttir & M. J. Loonen. 2006. Intensive grazing by Barnacle geese depletes High Arctic seed bank. *Botany* 84: 995–1004.

Kuitunen, T. & T. Lahtonen. 1994. Greenhouse weeds in a market garden in Jyväskylä, central Finland, in 1984–1992. *Lutukka* 10: 21–28.

Kujawski, J. 2001. Propagation protocol for poison sumac (*Toxicodendron vernix*). *Native Plant J.* 2: 112–113.

Kukongviriyapan, U., S. Luangaram, K. Leekhaosoong, V. Kukongviriyapan & S. Preeprame. 2007. Antioxidant and vascular protective activities of *Cratoxylum formosum, Syzygium gratum* and *Limnophila aromatica.* *Biol. Pharm. Bull.* 30: 661–666.

Kulbaba, M. W. 2004. Investigating epizoochorous adaptations to mammalian furs. Honours thesis. Department of Biology, The University of Winnipeg, Winnipeg, MB. 39 pp.

Kumar, M., S. K. Prasad & S. Hemalatha. 2014. A current update on the phytopharmacological aspects of *Houttuynia cordata* Thunb. *Pharmacogn. Rev.* 8: 22–35.

Kumaria, C. S., S. Govindasamya & E. Sukumarb. 2006. Lipid lowering activity of *Eclipta prostrata* in experimental hyperlipidemia. *J. Ethnopharmacol.* 105: 332–335.

Kunii, H. 1988. Longevity and germinability of buried seed in *Trapa* sp. *Mem. Fac. Sci. Shimane Univ.* 22: 83–91.

Kunii, H. 1993. Rhizome longevity in two floating-leaved aquatic macrophytes, *Nymphaea tetragona* and *Brasenia schreberi*. *J. Aquat. Plant Manag.* 31: 94–97.

Kunii, H. & M. Aramaki. 1992. Annual net production and life span of floating leaves in *Nymphaea tetragona* Georgi: a comparison with other floating-leaved macrophytes. *Hydrobiologia* 242: 185–193.

Kunkel, G. 1984. Plants for human consumption. Koeltz Scientific Books, Koenigsten, Germany. 393 pp.

Kunza, A. E. & S. C. Pennings. 2008. Patterns of plant diversity in Georgia and Texas salt marshes. *Estuaries Coasts* 31: 673–681.

Kupchan, S. M. & M. I. Suffness. 1967. Tumor inhibitors XXII. Senecionine and senecionine N-oxide, the active principles of *Senecio triangularis*. *J. Pharm. Sci.* 56: 541–543.

Küpeli, E., U. S. Harput, M. Varel, E. Yesilada & I. Saracoglu. 2005. Bioassay-guided isolation of iridoid glucosides with antinociceptive and anti-inflammatory activities from *Veronica anagallis-aquatica* L. *J. Ethnopharmacol.* 102: 170–176.

Kurashige, Y., J.-I. Etoh, T. Handa, K. Takayanagi & T. Yukawa. 2001. Sectional relationships in the genus *Rhododendron* (Ericaceae): evidence from *matK* and *trnK* intron sequences. *Plant Syst. Evol.* 228: 1–14.

Kurihara, H., J. Kawabata & M. Hatano. 1993. Geraniin, a hydrolyzable tannin from *Nymphaea tetragona* Georgi (Nymphaeaceae). *Biosci. Biotechnol. Biochem.* 57: 1570–1571.

Kurka, R. 1990. *Lindernia dubia*: nový zavlečený druh v Československu [*Lindernia dubia*: a new introduced species in Czechoslovakia]. *Zprávy Českoslov. Bot. Společn.* 25: 47–48.

Kurtz, A. M., J. M. Bahr, Q. J. Carpenter & R. J. Hunt. 2007. The importance of subsurface geology for water source and vegetation communities in Cherokee Marsh, Wisconsin. *Wetlands* 27: 189–202.

Kurup, R., A. J. Johnson, S. Sankar, A. A. Hussain, C. S. Kumar & B. Sabulal. 2013. Fluorescent prey traps in carnivorous plants. *Plant Biol.* 15: 611–615.

Kutcher, T. E., K. B. Raposa & F. Golet. 2004. *Habitat Classification and Inventory for the Narragansett Bay National Estuarine Research Reserve*. Technical Report Series 2004:2. NBNERR, Department of Natural Resources Science, University of Rhode Island, Kingston, RI. 60 pp.

Kuwabara, A., K. Ikegami, T. Koshiba & T. Nagata. 2003. Effects of ethylene and abscisic acid upon heterophylly in *Ludwigia arcuata* (Onagraceae). *Planta* 217: 880–887.

Kyhos, D. W., G. D. Carr & B. G. Baldwin. 1990. Biodiversity and cytogenetics of the tarweeds (Asteraceae: Heliantheae-Madiinae). *Ann. Missouri Bot. Gard.* 77: 84–95.

Laakkonen, L., R. W. Jobson & V. A. Albert. 2006. A new model for the evolution of carnivory in the bladderwort plant (*Utricularia*): adaptive changes in cytochrome *c* oxidase (COX) provide respiratory power. *Plant Biol.* 8: 758–764.

Laanei, M. M., B. E. Croff & R. Wahlstrøm. 1983. Cytotype distribution in the *Campanula rotundifolia* complex in Norway, and cyto-morphological characteristics of diploid and tetraploid groups. *Hereditas* 99: 21–48.

Laberge, W. E. & D. W. Ribble. 1975. A revision of the bees of the genus *Andrena* of the Western Hemisphere. Part VII. *Subgenus Euandrena. Trans. Amer. Entomol. Soc.* 101: 371–446.

Labisky, R. F., C. C. Hurd, M. K. Oli & R. S. Barwick. 2003. Foods of white-tailed deer in the Florida Everglades: the significance of *Crinum. S. E. Naturalist* 2: 261–270.

Labrecque, M. & T. I. Teodorescu. 2005. Field performance and biomass production of 12 willow and poplar clones in short-rotation coppice in southern Quebec (Canada). *Biomass Bioenergy* 29: 1–9.

Lacey, E. P., A. Royo, R. Bates & D. Herr. 2001. The role of population dynamic models in biogeographic studies: an illustration from a study of *Lobelia boykinii*, a rare species endemic to the Carolina bays. *Castanea* 66: 115–125.

Lachance, D. & C. Lavoie. 2002. Reconstructing the biological invasion of european water-horehound, *Lycopus europaeus* (Labiatae), along the St. Lawrence River, Québec. *Rhodora* 104: 151–160.

Lachance, M.-A., C. A. Rosa, W. T. Starmer, B. Schlag-Edler, J. S. F. Barker & J. M. Bowles. 1998. *Metschnikowia continentalis* var. *borealis*, *Metschnikowia continentalis* var. *continentalis*, and *Metschnikowia hibisci*, new heterothallic haploid yeasts from ephemeral flowers and associated insects. *Can. J. Microbiol.* 44: 279–288.

Lack, A. J. 1982. The ecology of flowers of chalk grassland and their insect pollinators. *J. Ecol.* 70: 773–790.

Lackney, V. K. 1981. The parasitism of *Pedicularis lanceolata* Michx., a root hemiparasite. *Bull. Torrey Bot. Club* 108: 422–429.

Lacoul, P. & B. Freedman. 2006a. Environmental influences on aquatic plants in freshwater ecosystems. *Environ. Rev.* 14: 89–136.

Lacoul, P. & B. Freedman. 2006b. Recent observation of a proliferation of *Ranunculus trichophyllus* Chaix. in high-altitude lakes of the Mount Everest region. *Arct. Antarc. Alp. Res.* 38: 394–398.

Lacoursière, E., P. Pontbriand & J.-P. Dumas. 1976. Première etapé de l'évolution écologique de l'Île aux terns, Québec. *Naturaliste Can.* 103: 169–189.

LaDeau, S. L. & A. M. Ellison. 1999. Seed bank composition of a northeastern U. S. tussock swamp. *Wetlands* 19: 255–261.

Laessle, A. M. & O. E. Frye Jr. 1956. A food study of the Florida bobwhite *Colinus virginianus floridanus (Coues). J. Wildl. Manag.* 20: 125–131.

Lahlou, S., R. F. L. Carneiro-Leão & J. H. Leal-Cardoso. 2002. Cardiovascular effects of the essential oil of *Mentha ×villosa* in DOCA-salt-hypertensive rats. *Phytomedicine* 9: 715–720.

Lahring, H. 2003. *Water and Wetland Plants of the Prairie Provinces*. Canadian Plains Research Center, University of Regina, Regina, SK. 327 pp.

Lai, M. N., S. M. Wang, P. C. Chen, Y. Y. Chen & J. D. Wang. 2010. Population-based case–control study of Chinese herbal products containing aristolochic acid and urinary tract cancer risk. *J. Natl. Cancer Inst.* 102: 179–186.

Lai, W.-L., S.-Q. Wang, C.-L. Peng & Z.-H. Chen. 2011. Root features related to plant growth and nutrient removal of 35 wetland plants. *Water Res.* 45: 3941–3950.

Laidig, K. J. & R. A. Zampella. 1996. *Stream Vegetation Data for Twenty Long-Term Study Sites in the New Jersey Pinelands*. Pinelands Commission, New Lisbon, NJ. 78 pp.

Lake, M. S., P. Gunther, D. C. Grossi & H. L. Bauer. 2000. Seed bank ecology of Butterfield Creek watershed along Old Plank Road trail. *Trans. Illinois Acad. Sci.* 93: 261–270.

Lako, J., V. C. Trenerry, M. Wahlqvist, N. Wattanapenpaiboon, S. Sotheeswaran & R. Premier. 2007. Phytochemical flavonols, carotenoids and the antioxidant properties of a wide selection of Fijian fruit, vegetables and other readily available foods. *Food Chem.* 101: 1727–1741.

Lallana, V. H. 1990. Dispersal units in aquatic environments of the middle Paraná River and its tributary, the Saladillo River. pp. 151–159. *In: Proceedings of the EWRS Symposium on Aquatic Weeds.* Uppsala, Sweden.

Lallana, V. H. 2005. Lista de malezas del cultivo de arroz en Entre Ríos, Argentina. *Ecosistemas* 14: 162–167.

Lam, D. R. E. 1998. Population genetics of vernal pool plants: theory, data and conservation implications. pp. 180–189. *In:* C. W. Witham, E. T. Bauder, D. Belk, W. R. Ferren Jr. & R. Ornduff (eds.), *Ecology, Conservation, and Management of Vernal Pool Ecosystems-Proceedings from a 1996 Conference.* California Native Plant Society, Sacramento, CA.

Lamaison, J. L., C. Petitjean-Freytet & A. Carnat. 1991. Medicinal Lamiaceae with antioxidant properties, a potential source of rosmarinic acid. *Pharm. Acta Helv.* 66: 185–188.

Lamb, E. G. & A. U. Mallik. 2003. Plant species traits across a riparian-zone/forest ecotone. *J. Veg. Sci.* 14: 853–858. (internet supplement: http://www.opuluspress.se/pub/archives/JVS/J014/J014-024A.pdf).

Lamb, E. G. & W. Megill. 2003. The shoreline fringe forest and adjacent peatlands of the southern central British Columbia coast. *Can. Field Naturalist* 117: 209–217.

Lamb, E. G., A. U. Mallik & R. W. Mackereth. 2003. The early impact of adjacent clearcutting and forest fire on riparian zone vegetation in northwestern Ontario. *For. Ecol. Manag.* 177: 529–538.

Lamb, R. 1991. *Pinguicula villosa*: the northern butterwort. *Carniv. Plant Newslett.* 20: 73–77.

Lamb-Frye, A. S. & K. A. Kron. 2003. *rbcL* phylogeny and character evolution in Polygonaceae. *Syst. Bot.* 28: 326–332.

Lammers, T. G. 1993. Chromosome numbers of Campanulaceae. III. Review and integration of data for subfamily Lobelioideae. *Amer. J. Bot.* 80: 660–675.

Lammers, T. G. 2003. A new combination in *Gentianopsis virgata* (Rafinesque) Holub (Gentianaceae). *Michigan Bot.* 42: 163.

Lammers, T. G. 2011. Revision of the infrageneric classification of *Lobelia* L. (Campanulaceae: Lobelioideae). *Ann. Missouri Bot. Gard.* 98: 37–62.

Lammers, W. T. 1958. A study of certain environmental and physiological factors influencing the adaptation of three granite outcrop endemics: *Amphianthus pusillus* Torr., *Isoetes melanospora* Engelm., and *Diamorpha cymosa* (Nutt.) Briiton, PhD dissertation. Emory University, Atlanta, GA. 85 pp.

Lamondia, J. A. 1995. Response of perennial herbaceous ornamentals to *Meloidogyne hapla. J. Nematol.* 27(4S): 645–648.

Lamont, E. E. 1988. Status of *Schizaea pusilla* in New York, with notes on some early collections. *Amer. Fern J.* 88: 158–164.

Lamont, E. E. 2006. 401. *Sclerolepis* Cassini. pp. 488–489. *In:* Flora of North America Editorial Committee (eds.), *Flora of North America North of Mexico: Vol. 21, Magnoliophyta: Asteridae (in part): Asteraceae, Part 3.* Oxford University Press, New York.

Lamont, E. E. & J. M. Fitzgerald. 2001. Noteworthy plants reported from the Torrey range—2000. *J. Torrey Bot. Soc.* 128: 409–414.

Lamont, E. E. & S. M. Young. 2004. Noteworthy plants reported from the Torrey Range—2002 and 2003. *J. Torrey Bot. Soc.* 131: 394–402.

Lamont, E. E. & S. M. Young. 2005. *Juncus diffusissimus*, an addition to the flora of New York, with notes on its recent spread in the United States. *J. Torrey Bot. Soc.* 132: 635–643.

Lamont, E. E., R. Sivertsen, C. Doyle & L. Adamec. 2013. Extant populations of *Aldrovanda vesiculosa* (Droseraceae) in the New World. *J. Torrey Bot. Soc.* 140: 517–522.

Landenberger, R. E. & J. B. McGraw. 2004. Seed-bank characteristics in mixed-mesophytic forest clearcuts and edges: does "edge effect" extend to the seed bank? *Can. J. Bot.* 82: 992–1000.

Landers, J. L., T. T. Fendley & A. S. Johnson. 1977. Feeding ecology of wood ducks in South Carolina. *J. Wildl. Manag.* 41: 118–127.

Landers, J. L., R. J. Hamilton, A. S. Johnson & R. L. Marchinton. 1979. Foods and habitat of black bears in southeastern North Carolina. *J. Wildl. Manag.* 43: 143–153.

Landman, G. B. & E. S. Menges. 1999. Dynamics of woody bayhead invasion into seasonal ponds in south central Florida. *Castanea* 64: 130–137.

Landman, G. B., R. K. Kolka & R. R. Sharitz. 2007. Soil seed bank analysis of planted and naturally revegetating thermally-disturbed riparian wetland forests. *Wetlands* 27: 211–223.

Land Protection. 2004. Pond apple *Annona glabra.* Fact sheet PP58, QNRM01273. Department of Natural Resources and Mines, Queensland, Australia. 3 pp.

Landry, C. L. 2013. Pollinator-mediated competition between two co-flowering Neotropical mangrove species, *Avicennia germinans* (Avicenniaceae) and *Laguncularia racemosa* (Combretaceae). *Ann. Bot.* 111: 207–214.

Landry, C. L. & B. J. Rathcke. 2007. Do inbreeding depression and relative male fitness explain the maintenance of androdioecy in white mangrove, *Laguncularia racemosa* (Combretaceae)? *New Phytol.* 176: 891–901.

Landry, C. L., B. J. Rathcke & L. B. Kass. 2009. Distribution of androdioecious and hermaphroditic populations of the mangrove *Laguncularia racemosa* (Combretaceae) in Florida and the Bahamas. *J. Trop. Ecol.* 25: 75–83.

Landry, M. & L. C. Cwynar. 2005. History of the endangered thread-leaved sundew (*Drosera filiformis*) in southern Nova Scotia. *Can. J. Bot.* 83: 14–21.

Lane, A. L. & J. Kubanek. 2006. Structure-activity relationship of chemical defenses from the freshwater plant *Micranthemum umbrosum. Phytochemistry* 67: 1224–1231.

Lane, J. J. & W. A. Mitchell. 1997. *Species profile: Alligator Snapping Turtle* (Macroclemys temminckii) *on Military Installations in the Southeastern United States.* Technical Report SERDP-97–9, U. S. Army Engineer Waterways Experiment Station, Vicksburg, MS. 17 pp.

Lang, N. L. 2006. The soil seed bank of an Oregon montane meadow: consequences of conifer encroachment and implications for restoration. MS thesis. University of Washington, Seattle, WA. 32 pp.

Langel, D., D. Ober & P. B. Pelser. 2011. The evolution of pyrrolizidine alkaloid biosynthesis and diversity in the Senecioneae. *Phytochem. Rev.* 10: 3–74.

Långström, E. & M. W. Chase. 2002. Tribes of Boraginoideae (Boraginaceae) and placement of *Antiphytum, Echiochilon, Ogastemma* and *Sericostoma*: a phylogenetic analysis based on *atpB* plastid DNA sequence data. *Plant Syst. Evol.* 234: 137–153.

Lanicci, R. 2010. *Garden Secrets for Attracting Birds.* Moseley Road, Inc., Irvington, New York. 160 pp.

Lanoue, K. Z., P. G. Wolf, S. Browning & E. E. Hood. 1996. Phylogenetic analysis of restriction-site variation in wild and cultivated *Amaranthus* species (Amaranthaceae). *Theor. Appl. Genet.* 93: 722–732.

Lans, C., N. Turner, G. Brauer, G. Lourenco & K. Georges. 2006. Ethnoveterinary medicines used for horses in Trinidad and in British Columbia, Canada. *J. Ethnobiol. Ethnomed.* 7: 2–31.

Lansdown, R. V. 2009. Nomenclatural Notes on *Callitriche* (Callitrichaceae) in North America. *Novon* 19: 364–369.

Laphitz, R. M. L., Y. Ma & J. C. Semple. 2011. A multivariate study of *Solidago* subsect. *Junceae* and a new species in South America (Asteraceae: Astereae). *Novon* 21: 219–225.

Largent, D. L., N. Sugihara & C. Wishner. 1980. Occurrence of mycorrhizae on ericaceous and pyrolaceous shrubs and subshrubs in northern California. *Can. J. Bot.* 58: 2274–2279.

La Roi, G. H. & R. J. Hnatiuk. 1980. The *Pinus contorta* forests of Banff and Jasper National Parks: a study in comparative synecology and syntaxonomy. *Ecol. Monogr.* 50: 1–29.

LaRosa, R. J., D. A. Rogers, T. P. Rooney & D. M. Waller. 2004. Does steeplebush (*Spiraea tomentosa*) facilitate pollination of Virginia meadow beauty (*Rhexia virginica*)? *Michigan Bot.* 43: 57–63.

Larsen, H. S. 1963. Effects of soaking in water on acorn germination of four southern oaks. *For. Sci.* 9: 236–241.

Larsen, J. A. 1929. Fires and forest succession in the Bitterroot Mountains of northern Idaho. *Ecology* 10: 67–76.

Larsen, J. A. 1965. The vegetation of the Ennadai Lake area, N.W.T.: studies in subarctic and arctic bioclimatology. *Ecol. Monogr.* 35: 37–59.

Larsen, R. C., K. S. Kim & H. A. Scott. 1991. Properties and cytopathology of diodia vein chlorosis virus: a new whitefly-transmitted virus. *Phytopathology* 81: 227–232.

Larson, B. M. H. & S. C. H. Barrett. 1999. The pollination ecology of buzz-pollinated *Rhexia virginica* (Melastomataceae). *Amer. J. Bot.* 86: 502–511.

Larson, D. 2007. Reproduction strategies in introduced *Nymphoides peltata* populations revealed by genetic markers. *Aquat. Bot.* 86: 402–406.

Larson, D. & E. Willén. 2006. Non-indigenous and invasive water plants in Sweden. *Svensk Bot. Tidskr.* 100: 5–15.

Larson, J. L. 1992. Seed germination requirements of four species co-occurring in a Wisconsin sedge weadow. *Univ. Wisconsin-Milwaukee Field Sta. Bull.* 25: 21–26.

La Rue, C. D. 1943. Regeneration in *Radicula aquatica*. *Michigan Acad.* 28: 51–61.

La Starza, S. R., C. Damiano, A. Frattarelli, P. Ferrazza & A. Figliolia. 2000. Cadmium detoxification by aquatic plant *Hygrophilia corymbosa* "stricta", in vitro growth. *In*: First Scientific Workshop, Phytoremediation 2000: State of the Art in Europe (an Intercontinental Comparison). COST Action 837, Plant biotechnology for the removal of organic pollutants and toxic metals from wastewaters and contaminated sites, held in Herisonissos, Crete, Greece, April 6–8. Available online: http://lbewww.epfl.ch/COST837/.

Latham, R. Earl, J. E. Thompson, S. A. Riley & A. W. Wibiralske. 1996. The Pocono till barrens: shrub savanna persisting on soils favoring forest. *Bull. Torrey Bot. Club* 123: 330–349.

Latimer, J. & J. Freeborn. 2009. New uses of PGRs in ornamentals: configure (6-BA)increases branching of herbaceous perennials. *Proc. Plant Growth Regulat. Soc. Amer.* 36: 88–93.

Latta, R. & K. Ritland. 1994. The relationship between inbreeding depression and prior inbreeding among populations of four *Mimulus* taxa. *Evolution* 48: 806–817.

Laureto, P. J. & T. J. Barkman. 2011. Nuclear and chloroplast DNA suggest a complex single origin for the threatened allopolyploid *Solidago houghtonii* (Asteraceae) involving reticulate evolution and introgression. *Syst. Bot.* 36: 209–226.

Lauridsen, T. L. & D. M. Lodge. 1996. Avoidance by *Daphnia magna* of fish and macrophytes: chemical cues and predator-mediated use of macrophyte habitat. *Limnol. Oceanogr.* 41: 794–798.

Lauron-Moreau, A., F. E. Pitre, L. Brouillet & M. Labrecque. 2013. Microsatellite markers of willow species and characterization of 11 polymorphic microsatellites for *Salix eriocephala* (Salicaceae), a potential native species for biomass production in Canada. *Plants* 2: 203–210.

Lavid, N., A. Schwartz, E. Lewinsohn & E. Tel-Or. 2001. Phenols and phenol oxidases are involved in cadmium accumulation in the water plants *Nymphoides peltata* (Menyanthaceae) and *Nymphaea* (Nymphaeaceae). *Planta* 214: 189–195.

Lavin, M., R. T. Pennington, B. Klitgaard, J. Sprent, H. C. De Lima & P. Gasson. 2001. The dalbergioid legumes (Fabaceae): delimitation of a pantropical monophyletic clade. *Amer. J. Bot.* 88: 503–533.

Lavin, M., M. F. Wojciechowski, P. Gasson, C. Hughes & E. Wheeler. 2003. Phylogeny of robinioid legumes (Fabaceae) revisited: *Coursetia* and *Gliricidia* recircumscribed, and a biogeographical appraisal of the Caribbean endemics. *Syst. Bot.* 28: 387–409.

Lawal, U., H. Ibrahim, A. Agunu & Y. Abdulahi. 2010. Anti-inflammatory and analgesic activity of water extract from *Ipomoea asarifolia* Desr (Convolvulaceae). *Afr. J. Biotechnol.* 9: 8877–8880.

Lawrence, B. A. 2005. Studies to facilitate reintroduction of golden paintbrush (*Castilleja levisecta*) to the Willamette Valley, Oregon. MS thesis. Department of Botany and Plant Pathology, Oregon State University, Corvallis, OR.

Lawrence, B. A. & T. N. Kaye. 2003. *Vegetation Survey of the North Fork Silver Creek RNA, Medford District, Bureau of Land Management*. Institute for Applied Ecology, Corvallis, OR. 17 pp.

Lawrence, B. A. & T. N. Kaye. 2005. Growing *Castilleja* for restoration and the garden. *Rock Gard. Q.* 63: 128–134.

Lawrence, B. M. 1993. Labiatae oils-mother nature's chemical factory. pp. 188–206. *In*: *Essential Oils (1988–1991)*. Allured Publishing Corporation, Carol Stream, IL.

Lawrence, B. M. (ed.). 2006. *Mint: The Genus* Mentha. CRC Press, Boca Raton, FL. 556 pp.

Lawrence, D. B. 1939. Some features of the vegetation of the Columbia River Gorge with special reference to asymmetry in forest trees. *Ecol. Monogr.* 9: 217–257.

Lawrence, G. & H. Biswell. 1972. Effect of forest manipulation on deer habitat in Giant Sequoia. *J. Wildl. Manag.* 36: 595–605.

Lawrence, W. H. 1905. Notes on the Erysiphaceae of Washington. *J. Mycol.* 11: 106–108.

Lawton, L.-A. & D. Gravatt. 2006. Composition of insect prey of the pitcher plant *Sarracenia alata* from an East Texas bog. Botany 2006 meeting (abstract).

Laymon, S. A. 1998. Yellow-billed Cuckoo (*Coccycus americanus*). *In*: *The Riparian Bird Conservation Plan: A Strategy for Reversing the Decline of Riparian-Associated Birds in California*. California Partners in Flight. Available online: http://www.prbo.org/calpif/htmldocs/riparian_v-2.html.

Lazar, K. A. 2006. Characterization of rare plant species in the vernal pools of California. MS thesis. University of California, Davis, CA. 344 pp.

Lazarine, P. 1980. *Common Wetland Plants of Southeast Texas*. U. S. Army Corps of Engineers, Galveston District, Galveston, TX. 154 pp.

LeBlanc, C. M. & D. J. Leopold. 1992. Demography and age structure of a central New York shrub-carr 94 years after fire. *Bull. Torrey Bot. Club* 119: 50–64.

LeBlond, R. J., E. E. Schilling, R. D. Porcher, B. A. Sorrie, J. F. Townsend, P. D. McMillan & A. S. Weakley. 2007. *Eupatorium paludicola* (Asteraceae): a new species from the coastal plain of North and South Carolina. *Rhodora* 109: 137–144.

LeBlond, R. J., S. M. Tessel & D. B. Poindexter. 2015. *Scleria bellii* (Cyperaceae), a distinctive and uncommon nutsedge from the southern U.S., Cuba, and Mexico. *J. Bot. Res. Inst. Tex.* 9: 31–41.

Leck, M. A. 1996. Germination of macrophytes from a Delaware River tidal freshwater wetland. *Bull. Torrey Bot. Club* 123: 48–67.

Leck, M. A. 2003. Seed-bank and vegetation development in a created tidal freshwater wetland on the Delaware River, Trenton, New Jersey, USA. *Wetlands* 23: 310–343.

Leck, M. A. 2012. Dispersal potential of a tidal river and colonization of a created tidal freshwater marsh. *AoB Plants* 5: 10.1093/aobpla/pls050.

Leck, M. A. & M. A. Brock. 2000. Ecological and evolutionary trends in wetlands: evidence from seeds and seed banks in New South Wales, Australia and New Jersey, USA. *Plant Species Biol.* 15: 97–112.

Leck, M. A. & K. J. Graveline. 1979. The seed bank of a freshwater tidal marsh. *Amer. J. Bot.* 66: 1006–1015.

Leck, M. A. & C. F. Leck. 1998. A ten-year seed bank study of old field succession in central New Jersey. *J. Torrey Bot. Soc.* 125: 11–32.

Leck, M. A. & C. F. Leck. 2005. Vascular plants of a Delaware River tidal freshwater wetland and adjacent terrestrial areas: seed bank and vegetation comparisons of reference and constructed marshes and annotated species list. *J. Torrey Bot. Soc.* 132: 323–354.

Leck, M. A. & R. L. Simpson. 1987. Seed bank of a freshwater tidal wetland: turnover and relationship to vegetation change. *Amer. J. Bot.* 74: 360–370.

Leck, M. A. & R. L. Simpson. 1992. Effect of oil on recruitment from the seed bank of two tidal freshwater wetlands. *Wetlands Ecol. Manag.* 1: 223–231.

Leck, M. A. & R. L. Simpson. 1993. Seeds and seedlings of the Hamilton Marshes, a Delaware River tidal freshwater wetland. *Proc. Acad. Nat. Sci. Philadelphia* 144: 267–281.

Leck, M. A. & R. L. Simpson. 1995. Ten-year seed bank and vegetation dynamics of a tidal freshwater marsh. *Amer. J. Bot.* 82: 1547–1557.

Leck, M. A., C. C. Baskin & J. M. Baskin. 1994. Ecology of *Bidens laevis* (Asteraceae) from a tidal freshwater wetland. *Bull. Torrey Bot. Club* 121: 230–239.

Leck, M. A., V. T. Parker & R. L. Simpson (eds.). 2008. *Seedling Ecology and Evolution.* Cambridge University Press, New York. 513 pp.

Leck, M. A., A. H. Baldwin, V. T. Parker, L. Schile & D. F. Whigham. 2009. Chapter 5: Plant communities of tidal freshwater wetlands of the continental USA and Canada. pp. 41–58. *In*: A. Barendregt, D. F. Whigham & A. H. Baldwin (eds.), *Tidal Freshwater Wetlands.* Backhuys Publishers, Leiden, The Netherlands.

Leckie, S., M. Vellend, G. Bell, M. J. Waterway & M. J. Lechowicz. 2000. The seed bank in an old-growth, temperate deciduous forest. *Can. J. Bot.* 78: 181–192.

Leclerc-Potvin, C. & K. Ritland. 1994. Modes of self-fertilization in *Mimulus guttatus* (Scrophulariaceae): a field experiment. *Amer. J. Bot.* 81: 199–205.

Lederhouse, R. C., S. G. Codella, D. W. Grossmueller & A. D. Maccarone. 1992. Host plant-based territoriality in the white peacock butterfly, *Anartia jatrophae* (Lepidoptera: Nymphalidae). *J. Insect Behav.* 5: 721–728.

Lee, C.-S. & S. R. Downie. 2006. Phylogenetic relationships within *Cicuta* (Apiaceae tribe Oenantheae) inferred from nrDNA ITS and cpDNA sequence data. *Can. J. Bot.* 84: 453–468.

Lee, D. S. 1972. The thread-leaf sundew, *Drosera filiformis*, on the coastal plain of Maryland. *Castanea* 37: 302.

Lee, H. K. & K. Moody. 1988. Seed viability and growth characteristics of *Eclipta prostrata* (L.) L. *Korean J. Weed Sci.* 8: 309–316.

Lee, J., B. G. Baldwin & L. D.Gottlieb. 2003. Phylogenetic relationships among the primarily North American genera of Cichorieae (Compositae) based on analysis of 18S-26S nuclear rDNA ITS and ETS sequences. *Syst. Bot.* 28: 616–626.

Lee, J. G., M. A. Kost & D. L. Cuthrell. 2006. *A Characterization of Hine's Emerald Dragonfly (*Somatochlora hineana *Williamson) Habitat in Michigan.* Report Number 2006–01. U. S. Fish and Wildlife Service, Twin Cities, MN. 54 pp.

Lee, J.-Y., K. Mummenhoff & J. L. Bowman. 2002. Allopolyploidization and evolution of species with reduced floral structures in *Lepidium* L. (Brassicaceae). *Proc. Natl. Acad. Sci. U.S.A.* 99: 16835–16840.

Lee, K. & M. R. Berenbaum. 1992. Ecological aspects of antioxidant enzymes and glutathione-S-transferases in three *Papilio* species. *Biochem. Syst. Ecol.* 20: 197–207.

Lee, K., C.-C. Shen, C.-F. Lin, S.-Y. Li & Y.-L. Huang. 2012. A phenolic derivative and two diacetylenes from *Symphyotrichum subulatum*. *Plant Med.* 78: 1780–1783.

Lee, K.-H. & T. O. Soine. 2006. Coumarins IX: coumarins of *Sphenosciadium capitellatum* (A. Gray). *J. Pharmaceut. Sci.* 58: 675–681.

Lee, K. Y., E. J. Jeong, H. S. Lee & Y. C. Kim. 2006. Acteoside of *Callicarpa dichotoma* attenuates scopolamine-induced memory impairments. *Biol. Pharm. Bull.* 29: 71–74.

Lee, N. S., T. Sang, D. J. Crawford, S. H. Yeau & S.-C. Kim. 1996. Molecular divergence between disjunct taxa in eastern Asia and eastern North America. *Amer. J. Bot.* 83: 1373–1378.

Lee, P. 2004. The impact of burn intensity from wildfires on seed and vegetative banks, and emergent understory in aspendominated boreal forests. *Can. J. Bot.* 82: 1468–1480.

Lee, P. & D. Scott. 2007. *East Anglian Wetland Bees and Wasps.* Hymettus Ltd., West Sussex, UK. 23 pp.

Legare, M. L. & W. R. Eddleman. 2001. Home range size, nest-site selection and nesting success of black rails in Florida. *J. Field Ornithol.* 72: 170–177.

Légaré, S., Y. Bergeron, A. Leduc & D. Paré. 2001. Comparison of the understory vegetation in boreal forest types of southwest Quebec. *Can. J. Bot.* 79: 1019–1027.

Legault, J., T. Perron, V. Mshvildadze, K. Girard-Lalancette, S. Perron, C. Laprise, P. Sirois & A. Pichette. 2011. Antioxidant and anti-inflammatory activities of quercetin 7-O-β-D-glucopyranoside from the leaves of *Brasenia schreberi*. *J. Med. Food* 14: 1127–1134.

Legendre, L. 2000. The genus *Pinguicula* L. (Lentibulariaceae): an overview. *Acta Bot. Gallica* 147: 77–95.

Lehman, R. L., R. O'Brien & T. White. 2009. *Plants of the Texas Coastal Bend.* Texas A&M University Press, College Station, TX. 368 pp.

Lehmann, W. & H. Jage. 2004. Phytoparasitic fungi in the city of Magdeburg (Saxony-Anhalt). *Boletus* 27: 125–144.

Lehnebach, C. A., A. Cano, C. Monsalve, P. McLenachan, E. Hörandl & P. Lockhart. 2007. Phylogenetic relationships of the monotypic Peruvian genus *Laccopetalum* (Ranunculaceae). *Plant Syst. Evol.* 264: 109–116.

Leininger, T. D., A. D. Wilson & D. G. Lester. 1997. Hurricane Andrew damage in relation to wood decay Fungi and insects in bottomland hardwoods of the Atchafalaya Basin, Louisiana. *J. Coastal Res.* 13: 1290–1293.

Leins, P. & C. Erbar. 2004. Hippuridaceae. pp. 163–166. *In*: J. W. Kadereit (ed.), *The Families and Genera of Vascular Plants, Vol. VII. Flowering Plants, Dicotyledons: Lamiales (except Acanthaceae including Avicenniaceae).* Springer-Verlag, Berlin, Germany.

Lelej, A. S., M. Y. Proshchalykin, A. N. Kupianskaya, M. V. Berezin & E. Y. Tkacheva. 2012. Trophical links of bumble bees (Hymenoptera, Apidae: *Bombus* Latreille, 1802) on North-West Pacific islands, the Russian Far East. *Euro-Asian Entomol. J.* 11: 261–269.

Lemly, J. M. 2007. Fens of Yellowstone National Park, USA: Regional and local controls over plant species distribution. MS thesis. Colorado State University, Fort Collins, CO. 134 pp.

Lempe, J., K. J. Stevens & R. L. Peterson. 2001. Shoot responses of six Lythraceae species to flooding. *Plant Biol. (Stuttgart)* 3: 186–193.

Lenaghan, S. C., K. Serpersu, L. Xia, W. He & M Zhang. 2011. A naturally occurring nanomaterial from the sundew (*Drosera*) for tissue engineering. *Bioinspiration Biomimetics* 6: 046009–10.1088/1748–3182/6/4/046009.

Lendemer, J. C. & R. C. Harris. 2014. Studies in lichens and lichenicolous fungi–No. 19: Further notes on species from the Coastal Plain of southeastern North America. *Opusc. Philolichenum* 13: 155–176.

Lenski, H. 1986. *Downingia elegans* (Dougl.) Torr.: Eine mit Grassaatgut eingeschleppte Adventivpflanze. *Gottinger Flor. Rundbr.* 19: 75–77.

Lenski, H. 1987. Nachtrag zu *Downingia elegans* (Douglas) Torr. *Flor. Rundbr.* 21: 24.

Lenssen, J. P. M., G. E. ten Dolle & C. W. P. M. Blom. 1998. The effect of flooding on the recruitment of reed marsh and tall forb plant species. *Plant Ecol.* 139: 13–23.

Lenssen, J. P. M., F. B. J. Menting, W. H. van der Putten & C. W. P. M. Blom. 1999. Effects of sediment type and water level on biomass production of wetland plant species. *Aquat. Bot.* 64: 151–165.

Lenssen, J. P. M., F. B. J. Menting, W. H. Van der Putten & C. W. P. M. Blom. 2000. Vegetative reproduction by species with different adaptations to shallow-flooded habitats. *New Phytol.* 145: 61–70.

Lentini, F. & F. Venza. 2007. Wild food plants of popular use in Sicily. *J. Ethnobiol Ethnomed.* 3: 15.

Lenz, M. J. 2004. Noteworthy collections. California. *Parnassia cirrata. Madroño* 51: 331.

Leon, R. G. & M. D. K. Owen. 2003. Regulation of weed seed dormancy through light and temperature interactions. *Weed Sci.* 51: 752–758.

Leonard, M. D. 1936. Additions to the New York State list of aphids with notes on other New York species. *J. New York Entomol. Soc.* 44: 177–185.

Leonard, M. D. 1959. A preliminary list of the aphids of Missouri. *J. Kansas Entomol. Soc.* 32: 9–18.

Leonard, M. D. 1964. Additional records of New Jersey aphids. *J. New York Entomol. Soc.* 72: 79–101.

Leonard, M. D. 1967. Further records of New Jersey aphids (Homoptera: Aphididae). *J. New York Entomol. Soc.* 75: 77–92.

Leong, J. M. & E. L. Bailey. 2000. The incidence of a generalist thrips herbivore among natural and translocated patches of an endangered vernal pool plant, *Blennosperma bakeri. Restorat. Ecol.* 8: 127–134.

Le Page, C. & P. A. Keddy 1998. Reserves of buried seeds in beaver ponds. *Wetlands* 18: 242–248.

Leppik, E. E. 1967. Some viewpoints on the phylogeny of rust Fungi. VI. biogenic radiation. *Mycologia* 59: 568–579.

LeResche, R. E. & J. L. Davis. 1973. Importance of nonbrowse foods to moose on the Kenai Peninsula, Alaska. *J. Wildl. Manag.* 37: 279–287.

Lerner, J. M. 1997. Mycorrhizal interactions of selected species of endangered New England flora. MS thesis. University of Massachusetts, Amherst, MA. 209 pp.

Leroux, G. D., D.-L. Benoît & S. Banville. 1996. Effect of crop rotations on weed control, *Bidens cernua* and *Erigeron canadensis* populations, and carrot yields in organic soils. *Crop Protect.* 15: 171–178.

Lertzman, K. P. 1981. Pollen transfer: processes and consequences. MS thesis. University of British Columbia, Vancouver, BC. 154 pp.

Les, D. H. 1985. The taxonomic significance of plumule morphology in *Ceratophyllum* (Ceratophyllaceae). *Syst. Bot.* 10: 338–346.

Les, D. H. 1986a. The phytogeography of *Ceratophyllum demersum* and *C. echinatum* (Ceratophyllaceae) in glaciated North America. *Can. J. Bot.* 64: 498–509.

Les, D. H. 1986b. The evolution of achene morphology in *Ceratophyllum* (Ceratophyllaceae), I. Fruit-spine variation and relationships of *C. demersum, C. submersum,* and *C. apiculatum. Syst. Bot.* 11: 549–558.

Les, 1986c. Systematics and evolution of *Ceratophyllum* L. (Ceratophyllaceae): a monograph. PhD dissertation, The Ohio State University, Columbus, OH. 418 pp.

Les, D. H. 1988a. The origin and affinities of the Ceratophyllaceae. *Taxon* 37: 326–345.

Les, D. H. 1988b. Breeding systems, population structure, and evolution in hydrophilous angiosperms. *Ann. Missouri Bot. Gard.* 75: 819–835.

Les, D. H. 1988c. The evolution of achene morphology in *Ceratophyllum* (Ceratophyll-aceae), III. Relationships of the "facially-spined" group. *Syst. Bot.* 13: 509–518.

Les, D. H. 1988d. The evolution of achene morphology in *Ceratophyllum* (Ceratophyll-aceae), II. Fruit variation and systematics of the "spiny-margined" group. *Syst. Bot.* 13: 73–86.

Les, D. H. 1989. The evolution of achene morphology in *Ceratophyllum* (Ceratophyllaceae), IV. Summary of proposed relationships and evolutionary trends. *Syst. Bot.* 14: 254–262.

Les, D. H. 1991. Genetic diversity in the monoecious hydrophile *Ceratophyllum* (Ceratophyllaceae). *Amer. J. Bot.* 78: 1070–1082.

Les, D. H. 1993. Ceratophyllaceae. pp. 246–250. *In:* K. Kubitzki, J. Rohwer & V. Bittrich (eds.), The Families and Genera of Vascular Plants, Vol. II. Flowering Plants, Dicotyledons: Magnoliid, Hamamelid and Caryophyllid Families. Springer-Verlag, Berlin, Germany.

Les, D. H. 1994. Molecular systematics and taxonomy of lake cress (*Neobeckia aquatica*; Brassicaceae), an imperiled aquatic mustard. *Aquat. Bot.* 49: 149–165.

Les, D. H. 1997. 14. Ceratophyllaceae Gray. pp. 81–84. *In:* N. R. Morin (convening ed.), *Flora of North America North of Mexico, Vol. 3: Magnoliophyta: Magnoliidae and Hamamelidae.* Oxford University Press, New York.

Les, D. H. 2002. Nonindigenous aquatic plants: a garden of earthly delight. *Lakeline* 22(1): 20–24.

Les, D. H. 2004. Cabombaceae. pp. 72–73. *In:* N. Smith, S. A. Mori, A. Henderson, D. W. Stevenson & S. V. Heald (eds.), *Flowering Plants of the Neotropics.* Princeton University Press, Princeton, NJ. 594 pp.

Les, D. H. 2015. Water from the rock: Ancient aquatic angiosperms flow from the fossil record. *Proc. Natl. Acad. Sci. U.S.A.* 112: 10825–10826.

Les, D. H. & M. A. Miller. 1979. A biometric analysis of lenticel development in Red-Osier Dogwood (*Cornus stolonifera* Michx.) in relation to soil moisture content. *Michigan Acad.* 12: 217–219.

Les, D. H. & R. P. Wunderlin. 1980. *Hygrophila polysperma* (Acanthaceae) in Florida. *Florida Sci.* 44: 189–192.

Les, D. H. & R. L. Stuckey. 1985. The introduction and spread of *Veronica beccabunga* (Scrophulariaceae) in eastern North America. *Rhodora* 87: 503–515.

Les, D. H. & D. T. Gerber. 1991. *Laboratory Growth Experiments for Selected Aquatic Plants. Final Report (July, 1990–October, 1991).* Wisconsin DNR, Bureau of Water Resources Management and Bureau of Research, Milwaukee, WI.

Les, D. H. & L. J. Mehrhoff. 1999. Introduction of nonindigenous aquatic vascular plants in southern New England: a historical perspective. *Biol. Invas.* 1: 281–300.

Les, D. H., D. K. Garvin & C. F. Wimpee. 1991. Molecular evolutionary history of ancient aquatic angiosperms. *Proc. Natl. Acad. Sci. U.S.A.* 88: 10119–10123.

Les, D. H., C. T. Philbrick & A. Novelo R. 1997. The phylogenetic position of river-weeds (Podostemaceae): insights from *rbc*L sequence data. *Aquat. Bot.* 57: 5–27.

Les, D. H., E. L. Schneider, D. J. Padgett, P. S. Soltis, D. E. Soltis & M. Zanis. 1999. Phylogeny, classification and floral evolution of water lilies (Nymphaeales): A synthesis of non-molecular, *rbc*L, *mat*K and 18S rDNA data. *Syst. Bot.* 24: 28–46.

Les, D. H., D. J. Crawford, R. T. Kimball, M. L. Moody & E. Landolt. 2003. Biogeography of discontinuously distributed hydrophytes: a molecular appraisal of intercontinental disjunctions. *Int. J. Plant Sci.* 164: 917–932.

Les, D. H., M. L. Moody, A. Doran & W. L. Phillips. 2004. A genetically confirmed intersubgeneric hybrid in *Nymphaea* L. (Nymphaeaceae Salisb.). *HortScience* 39: 219–222.

Les, D. H., R. S. Capers & N. Tippery. 2006. Introduction of *Glossostigma* (Phrymaceae) to North America: a taxonomic and ecological overview. *Amer. J. Bot.* 93: 927–939.

LeSage, L. 1984. Egg, larva, and pupa of *Lexiphanes saponatus* (Coleoptera: Chrysomelidae: Cryptocephalinae). *Can. Entomol.* 116: 537–548.

LeSage, L. 1986. A taxonomic monograph of the Nearctic galerucine genus *Ophraella* Wilcox (Coleoptera: Chrysomelidae). *Mem. Entomol. Soc. Canada* 118: 3–75.

Lesica, P. 1992. Autecology of the endangered plant *Howellia aquatilis*; implications for management and reserve design. *Ecol. Appl.* 2: 411–421.

Lesica, P. 1997. Spread of *Phalaris arundinacea* adversely impacts the endangered plant *Howellia aquatilis. Great Basin Naturalist* 57: 366–368.

Lesica, P. 2003. *Conserving Globally Rare Plants on Lands Administered by the Dillon Office of the Bureau of Land Management.* Report to the USDI Bureau of Land Management, Dillon Office. Montana Natural Heritage Program, Helena, MT. 22 pp.

Lesica, P. & B. McCune. 2004. Decline of arctic-alpine plants at the southern margin of their range following a decade of climatic warming. *J. Veg. Sci.* 15: 679–690.

Lesica, P., R. F. Leary, F. W. Allendorf & D. E. Bilderback. 1988. Lack of genic diversity within and among populations of an endangered plant, *Howellia aquatilis. Conserv. Biol.* 2: 275–282.

Lesuisse, D. J. Berjonneau, C. Ciot, P. Devaux, B. Doucet, J. F. Gourvest, B. Khemis, C. Lang, R. Legrand, M. Lowinski, P. Maquin, A. Parent, B. Schoot, G. Teutsch, H. Chodounská & A. Kasal. 1996. Determination of Oenothein B as the active 5-α-reductase-inhibiting principle of the folk medicine *Epilobium parviflorum. J. Nat. Prod.* 59: 490–492.

Levin, D. A. 1988. The paternity pools of plants. *Amer. Naturalist* 132: 309–317.

Levin, D. A. & H. W. Kerster. 1969. The dependence of bee-mediated pollen and gene dispersal upon plant density. *Evolution* 23: 560–571.

Levin, D. A., K. Ritter & N. C. Ellstrand. 1979. Protein polymorphism in the narrow endemic *Oenothera organensis. Evolution* 33: 534–542.

Levin, R. A., W. L. Wagner, P. C. Hoch, M. Nepokroeff, J. C. Pires, E. A. Zimmer & K. J. Sytsma. 2003. Family-level relationships of Onagraceae based on chloroplast *rbc*L and *ndh*F data. *Amer. J. Bot.* 90: 107–115.

Levin, R. A., W. L. Wagner, P. C. Hoch, W. J. Hahn, A. Rodriguez, D. A. Baum, L. Katinas, E. A. Zimmer & K. J. Sytsma. 2004. Paraphyly in tribe Onagreae: Insights into phylogenetic relationships of Onagraceae based on nuclear and chloroplast sequence data. *Syst. Bot.* 29: 147–164.

Levine, J. M. 2000. Complex interactions in a streamside plant community. *Ecology* 81: 3431–3444.

Levri, M. A. 2000. A measure of the various modes of inbreeding in *Kalmia latifolia. Ann. Bot.* 86: 415–420.

Levy, F. 1988. Effects of pollen source and time of pollination on seed production and seed weight in *Phacelia dubia* and *P. maculata* (Hydrophyllaceae). *Amer. Midl. Naturalist* 119: 193–198.

Levy, F. & C. L. Neal. 1999. Spatial and temporal genetic structure in chloroplast and allozyme markers in *Phacelia dubia* implicate genetic drift. *Heredity* 82: 422–431.

Lewis, C. G. 2005. Linkages among vegetative substrate quality, biomass production, and decomposition in maintaining Everglades ridge and slough vegetative communities. MS thesis. University of Florida, Gainesville, FL. 58 pp.

Lewis, D. H. & D. C. Smith. 1967. Sugar alcohols (polyols) in fungi and green plants. I. distribution, physiology and metabolism. *New Phytol.* 66: 143–184.

Lewis, D. Q. 2000. A revision of the New World species of *Lindernia* (Scrophulariaceae). *Castanea* 65: 93–122.

Lewis, F. J., E. S. Dowding & E. H. Moss. 1928. The vegetation of Alberta: II. the swamp, moor and bog forest vegetation of central Alberta. *J. Ecol.* 16: 19–70.

Lewis, H. 1945. A revision of the genus *Trichostema. Brittonia* 5: 276–303.

Lewis, H. 1960. Chromosome numbers and phylogeny of *Trichostema. Brittonia* 12: 93–97.

Lewis, K. P. 1980. Vegetative reproduction in populations of *Justicia americana* in Ohio and Alabama. *Ohio J. Sci.* 80: 134–137.

Lewis, W. H. 1961. Chromosome numbers for five American species of *Callicarpa, Lantana,* and *Phyla* (Verbenaceae). *Southwest. Naturalist* 6: 47–48.

Lewis, W. H. 1962a. Chromosome numbers in North American Rubiaceae. *Brittonia* 14: 285–290.

Lewis, W. H. 1962b. Phylogenetic study of *Hedyotis* (Rubiaceae) in North America. *Amer. J. Bot.* 49: 855–865.

Lewis, W. H. 1965. Cytopalynological study of African Hedyotideae (Rubiaceae). *Ann. Missouri Bot. Gard.* 52: 182–211.

Lewis, W. H. & R. L. Oliver. 1961. Cytogeography and phylogeny of the North American species of *Verbena. Amer. J. Bot.* 48: 638–643.

Lewis, W. H. & R. L. Oliver. 1965. Realignment of *Calystegia* and *Convolvulus* (Convolvulaceae). *Ann. Missouri Bot. Gard.* 52: 217–222.

Lewis, W. H., G. L. Stripling & R. G. Ross. 1962. Chromosome numbers for some angiosperms of the southern United States and Mexico. *Rhodora* 64: 147–161.

Lewis, W. H. & M. P. F. Elvin-Lewis. 2003. *Medical Botany: Plants Affecting Human Health (2nd Edition)*. John Wiley & Sons, Hoboken, NJ. 812 pp.

Lewis, W. H., B. Ertter & A. Bruneau. 2015. 7. *Rosa* Linnaeus. pp. 75–119. *In*: N. R. Morin (convening ed.), *Flora of North America North of Mexico, Vol. 9: Magnoliophyta: Picramniaceae to Rosaceae*. Oxford University Press, New York.

Ley, E. L., S. Buchmann, G. Kauffman & K. McGuire. n.d. Selecting Plants for Pollinators: A Regional Guide for Farmers, Land Managers, and Gardeners in the Ecological Region of the Outer Coastal Plain Mixed Province Including the States of: Delaware, Florida, Georgia, Louisiana, North Carolina, South Carolina and Parts of Alabama, Maryland, Mississippi, Texas, and Virginia. The Pollinator Partnership™/North American Pollinator Protection Campaign, San Francisco, CA.

Leyer, I. 2006. Dispersal, diversity and distribution patterns in pioneer vegetation: The role of river-floodplain connectivity. *J. Veg. Sci.* 17: 407–416.

Li, H.-L. 1957. Chromosome studies in the azaleas of eastern North America. *Amer. J. Bot.* 44: 8–14.

Li, H.-T., H.-M. Wu, H.-L. Chen, C.-M. Liu & C.-Y. Chen. 2013. The pharmacological activities of (–)-anonaine. *Molecules* 18: 8257–8263.

Li, J. & D. C. Christophel. 2000. Systematic relationships within the *Litsea* complex (Lauraceae): a cladistic analysis on the basis of morphological and leaf cuticle data. *Aust. Syst. Bot.* 13: 1–13.

Li, J. & J. G. Conran. 2003. Phylogenetic relationships in Magnoliaceae subfam. Magnolioideae: a morphological cladistic analysis. *Plant Syst. Evol.* 242: 33–47.

Li, J., D. C. Christophel, J. G. Conran & H. W. Li. 2004. Phylogenetic relationships within the 'core'Laureae (*Litsea* complex, Lauraceae) inferred from sequences of the chloroplast gene *matK* and nuclear ribosomal DNA ITS regions. *Plant Syst. Evol.* 246: 19–34.

Li, J., S. Shoup & Z. Chen. 2005. Phylogenetics of *Betula* (Betulaceae) inferred from sequences of nuclear ribosomal DNA. *Rhodora* 107: 69–86.

Li, J., S. Shoup & Z. Chen. 2007. Phylogenetic relationships of diploid species of *Betula* (Betulaceae) inferred from DNA sequences of nuclear nitrate reductase. *Syst. Bot.* 32: 357–365.

Li, J., Y. Liu, J. Luo, P. Liu & C. Zhang. 2012. Excellent lubricating behavior of *Brasenia schreberi* mucilage. *Langmuir* 28: 7797–7802.

Li, J. K. & S. Q. Huang. 2009. Effective pollinators of Asian sacred lotus (*Nelumbo nucifera*): contemporary pollinators may not reflect the historical pollination syndrome. *Ann. Bot.* 104: 845–851.

Li, L. P., H. K. Wang, J. J. Chang, A. T. McPhail, D. R. McPhail, H. Terada, T. Konoshima, M. Kokumai, M. Kozuka, J. R. Estes & K. H. Lee. 1993. Antitumor agents 138. Rotenoids and iso-flavones as cytotoxic constituents from *Amorpha fruticosa*. *J. Nat. Prod.* 56: 690.

Li, S. M., W. D. Ke, Y. Y. Ye & L. C. Huang. 2011. Studies on pollination biology and hybridization of *Brasenia schreberi*. *China Veget.* 12: 023.

Li, S. M., W. D. Ke & H. L. Zhu. 2013. Effect of different treatments on seed germination of water shield (*Brasenia schreberi* JF Gmel). *China Veget.* 10: 018.

Li, W. & F. Recknagel. 2002. *In situ* removal of dissolved phosphorus in irrigation drainage water by planted floats: preliminary results from growth chamber experiment. *Agric. Eco-Syst. Environ.* 90: 9–15.

Li, W.-J., X.-L. Pu & Z.-Y. Chen. 2012. Study on in vitro culture and plant regeneration of *Menyanthes trifoliata*. *Northern Hort.* 23: 44.

Li, Y. & F. R. Stermitz. 1988. Pyrrolizidine alkaloids from *Mertensia* species of Colorado. *J. Nat. Prod.* 51: 1289–1290.

Li, Y., P. Svetlana, J. Yao & C. Li. 2014. A review on the taxonomic, evolutionary and phytogeographic studies of the lotus plant (Nelumbonaceae: *Nelumbo*). *Acta Geol. Sinica* 88: 1252–1261.

Li, Z., Z. Wu, H. Liu, X. Hao, C. Zhang & Y. Wu. 2010. *Spiraea salicifolia*: a new plant host of "*Candidatus Phytoplasma ziziphi*"-related phytoplasma. *J. Gen. Plant Pathol.* 76: 299–301.

Li, Z., D. Yu & J. Xu. 2011. Adaptation to water level variation: responses of a floating-leaved macrophyte *Nymphoides peltata* to terrestrial habitats. *Ann. Limnol. Int. J. Limnol.* 47: 97–102.

Liao, Y.-Y., X.-L. Yue, Y.-H. Guo, W. R. Gituru, Q.-F. Wang & J.-M. Chen. 2013. Genotypic diversity and genetic structure of populations of the distylous aquatic plant *Nymphoides peltata* (Menyanthaceae) in China. *J. Syst. Evol.* 51: 536–544.

Libert, B. & V. R. Franceschi. 1987. Oxalate in crop plants. *J. Agric. Food Chem.* 35: 926–938.

Lichthardt, J. & R. K. Moseley. 2000. Ecological assessment of *Howellia aquatilis* habitat at the Harvard-Palouse River flood plain site, Idaho. Conservation Data Center, Idaho Department of Fish and Game, Natural Resource Policy Bureau. Boise, ID. 32 pp.

Lichthardt, J. & K. Gray. 2002. Monitoring of *Howellia aquatilis* (water howellia) and its habitat at the Harvard-Palouse River flood plain site, Idaho: third year results. Conservation Data Center, Idaho Department of Fish and Game, Natural Resource Policy Bureau. Boise, ID. 12 pp.

Lichvar, R. W., N. C. Melvin, M. L. Butterwick & W. N. Kirchner. 2012. *National Wetland Plant List Indicator Rating Definitions*. ERDC/CRREL TN-12-1. Wetland Regulatory Assistance Program, U. S. Army Corps of Engineers, Washington, DC. 7 pp.

Lidén, M., T. Fukuhara & T. Axberg. 1995. Phylogeny of *Corydalis*, ITS and morphology. *Plant Syst. Evol. Suppl.* 9: 183–188.

Liede, S. & A. Täuber. 2000. *Sarcostemma* R.Br. (Apocynaceae-Asclepiadoideae)-a controversial generic circumscription reconsidered: Evidence from *trn*L-F spacers. *Plant Syst. Evol.* 225: 133–140.

Liede, S. & A. Täuber. 2002. Circumscription of the genus *Cynanchum* (Apocynaceae-Asclepiadoideae). *Syst. Bot.* 27: 789–800.

Liede-Schumann, S., A. Rapini, D. J. Goyder & M. W. Chase. 2005. Phylogenetics of the New World subtribes of Asclepiadeae (Apocynaceae–Asclepiadoideae): Metastelmatinae, Oxypetalinae and Gonolobinae. *Syst. Bot.* 30: 184–195.

Lieneman, C. 1929. A host index to the North American species of the genus *Cercospora*. *Ann. Missouri Bot. Gard.* 16: 1–52.

Light, H. M., M. R. Darst, M. T. MacLaughlin & S. W. Sprecher. 1993. *Hydrology, Vegetation, and Soils of Four North Florida River Flood Plains with an Evaluation of State and Federal Wetland Determinations*. U. S. Geological Survey, Water-Resources Investigation Report 93-4033. U. S. Geological Survey, Earth Science Information Center, Denver, CO. 94 pp.

Light, H. M., M. R. Darst, L. J. Lewis & D. A. Howell. 2002. *Hydrology, Vegetation, and Soils of Riverine and Tidal Floodplain Forests of the Lower Suwannee River, Florida, and Potential Impacts of Flow Reductions.* U. S. Geological Survey Professional Paper 1656A. U. S. Geological Survey, Branch of Information Services, Denver, CO. 124 pp.

Lihua, M. & Y. Liu. 2009. Inhibitory effects of *Hygrophila difformis* on the growth of *Microcystis aeruginosa. Environ. Pollut. Control* 1: 135–245.

Likar, M. & M. Regvar. 2013. Isolates of dark septate endophytes reduce metal uptake and improve physiology of *Salix caprea* L. *Plant Soil* 370: 593–604.

Lin, J., R. J. Smith Jr. & R. H. Dilday. 1992. Allelopathic activity of rice germplasm on weeds. *Proc. South. Weed Sci. Soc.* 45: 99.

Lin, J., J. P. Gibbs & L. B. Smart. 2009. Population genetic structure of native versus naturalized sympatric shrub willows (*Salix*; Salicaceae). *Amer. J. Bot.* 96:771–785.

Lin, S. C., Y. H. Lin, S. J. Shyuu & C. C. Lin. 1994. Hepatoprotective effects of Taiwan folk medicine: *Alternanthera sessilis* on liver damage induced by various hepatotoxins. *Phytother. Res.* 8: 391–398.

Lin, Y., G.-X. Wang, W. Li & M. Ito. 2004. Secondary structure prediction of acetolactate synthase protein in sulfonylurea herbicide resistant *Limnophila sessiliflora. J. Pestic. Sci.* 29: 1–5.

Lin, Z. Q. & N. Terry. 1998. Managing High Selenium in Agricultural Drainage Water by Agroforestry Systems: Role of Selenium Volatilization. DWR B80665. Department of Water Resources, Sacramento, CA. 67 pp.

Lincicome, D. A. 1998. *The Rare Perennial* Balduina atropurpurea *(Asteraceae) at Fort Stewart, Georgia.* Technical Report 98/75. U. S. Army Corps of Engineers, Construction Engineering Research Laboratories (USACERL), Champaign, IL. 133 pp.

Lind, C. T. & G. Cottam. 1969. The submerged aquatics of University Bay: a study in eutrophication. *Amer. Midl. Naturalist* 81: 353–369.

Linder, D. H. 1938. New California Fungi. *Mycologia* 30: 664–671.

Lindgren, C. J. & R. T. Clay. 1993. Fertility of "Morden Pink" *Lythrum virgatum* L. transplanted into wild stands of *L. salicaria* in Manitoba. *HortScience* 28: 954.

Lindholm, P., J. Gullbo, P. Claeson, U. Goransson, S. Johansson, A. Backlund, R. Larsson & L. Bohlin. 2002. Selective cytotoxicity evaluation in anticancer drug screening of fractionated plant extracts. *J. Biomol. Screening* 7: 333–340.

Lindig-Cisneros, R. & J. B. Zedler. 2002. *Phalaris arundinacea* seedling establishment: effects of canopy complexity in fen, mesocosm, and restoration experiments. *Can. J. Bot.* 80: 617–624.

Lindquist, E. D., D. K. Foster, S. P. Wilcock & J. S. Erikson. 2013. Rapid assessment tools for conserving woodland vernal pools in the northern Blue Ridge Mountains. *N. E. Naturalist* 20: 397–418.

Lindqvist, C. & V. A. Albert. 2002. Origin of the Hawaiian endemic mints within North American *Stachys* (Lamiaceae). *Amer. J. Bot.* 89: 1709–1724.

Lindqvist-Kreuze, H., H. Koponen & J. P. T Valkonen. 2003. Genetic diversity of arctic bramble (*Rubus arcticus* L. subsp. *arcticus*) as measured by amplified fragment length polymorphism. *Can. J. Bot.* 81: 805–813.

Lindsey, A. A. 1938. Anatomical evidence for the Menyanthaceae. *Amer. J. Bot.* 25: 480–485.

Lindsey, A. A. 1948. Terron vegetation in New Mexico. *Ecology* 29: 470–478.

Lindsey, A. A. 1953. Notes on some plant communities in the Northern Mackenzie Basin, Canada. *Bot. Gaz.* 115: 44–55.

Lindsey, A. A., R. O. Petty, D. K. Sterling & W. Van Asdall. 1961. Vegetation and environment along the Wabash and Tippecanoe rivers. *Ecol. Monogr.* 31: 105–156.

Linhart, Y. B. 1974. Intra-population differentiation in annual plants I. *Veronica peregrina* L. raised under non-competitive conditions. *Evolution* 28: 232–243.

Linhart, Y. B. 1976. Density-dependent seed germination strategies in colonizing versus non-colonizing plant species. *J. Ecol.* 64: 375–380.

Linke, J., S. E. Franklin, F. Huettmann & G. B. Stenhouse. 2005. Seismic cutlines, changing landscape metrics and grizzly bear landscape use in Alberta. *Landscape Ecol.* 20: 811–826.

Linke, W. R. 1963. *Drosera filiformis* in Connecticut. *Rhodora* 65: 273–273.

Linzon, S. N., W. D. McIlveen & P. J. Temple. 1973. Sulphur dioxide injury to vegetation in the vicinity of a sulphite pulp and paper mill. *Water Air Soil Pollut.* 2: 129–134.

Liogier, H. A. 1985. Descriptive Flora of Puerto Rico and Adjacent Islands. Vol. 1. Spermatophyta. Universidad de Puerto Rico, Río Piedras, PR. 352 pp.

Liouane, K., K. B. H. Salah, H. Ben Abdelkader, M. A. Mahjoub, M. Aouni, K. Said & Z. Mighri. 2012. Antibacterial and antifungal activities of *Cotula coronopifolia. Afr. J. Microbiol. Res.* 6: 4662–4666.

Lipow, S. R. & R. Wyatt. 1999. Diallel crosses reveal patterns of variation in fruit-set, seed mass, and seed number in *Asclepias incarnata. Heredity* 83: 310–318.

Lipow, S. R. & R. Wyatt. 2000. Towards an understanding of the mixed breeding system of swamp milkweed (*Asclepias incarnata*). *J. Torrey Bot. Soc.* 127: 193–199.

Lippert, B. E. & D. L. Jameson. 1964. Plant succession in temporary ponds of the Willamette Valley, Oregon. *Amer. Midl. Naturalist* 71: 181–197.

Lippok, B., A. A. Gardine, P. S. Williamson & S. S. Renner. 2000. Pollination by flies, bees, and beetles of *Nuphar ozarkana* and *N. advena* (Nymphaeaceae). *Amer. J. Bot.* 87: 898–902.

Lira, R. & J. Caballero. 2002. Ethnobotany of the wild Mexican Cucurbitaceae. *Econ. Bot.* 56: 380–398.

Lira-Medeiros, C. F., C. Parisod, R. A. Fernandes, C. S. Mata, M. A. Cardoso & P. C. Ferreira. 2010. Epigenetic variation in mangrove plants occurring in contrasting natural environment. *PLoS One* 5(4):e10326.

Lira-Salazar, G., E. Marines-Montiel, J. Torres-Monzón, F. Hernández-Hernández & J. S. Salas-Benito. 2006. Effects of homeopathic medications *Eupatorium perfoliatum* and *Arsenicum album* on parasitemia of *Plasmodium berghei*-infected mice. *Homeopathy* 95: 223–228.

Lis, R. 2014. 44. *Spiraea* Linnaeus. pp. 398–411. *In*: N. R. Morin (convening ed.), *Flora of North America North of Mexico, Vol. 9: Magnoliophyta: Picramniaceae to Rosaceae.* Oxford University Press, New York.

Lisberg, A. E. & D. K. Young. 2003. An annotated checklist of Wisconsin Mordellidae (Coleoptera). *Insecta Mundi* 17: 195–202.

Liston, A. & T. Thorsen. 2000. Conservation genetics of *Corydalis aquae-gelidae* (Fumariaceae). Oregon State University, Corvallis, OR.

Liston, S. E. & J. C. Trexler. 2005. Spatiotemporal patterns in community structure of macroinvertebrates inhabiting calcareous periphyton mats. *J. N. Amer. Benthol. Soc.* 24: 832–844.

Lite, S. J. & J. C. Stromberg. 2003. Groundwater and surface water thresholds for maintaining *Populus-Salix* forests, San Pedro River, southern Arizona. pp. 17–28. *In: Saltcedar and Water Resources in the West Symposium (Conference Proceedings).* San Angelo, TX.

Little, E. C. S. 1979. *Handbook of Utilization of Aquatic Plants.* FAO Fisheries Technical Paper No 187. Food and Agricultural Organization of the United Nations, Rome, Italy. 176 pp.

Little Jr., E. L. 1938. The vegetation of Muskogee County, Oklahoma. *Amer. Midl. Naturalist* 19: 559–572.

Little Jr., E. L. 1939. The vegetation of the Caddo County canyons, Oklahoma. *Ecology* 20: 1–10.

Little Jr., E. L. 1944. Layering after a heavy snow storm in Maryland. *Ecology* 25: 112–113.

Littlefield, J. L. 1991. Parasitism of *Rhinocyllus conicus* Froelich (Coleoptera: Curculionidae) in Wyoming. *Can. Entomol.* 123: 929–932.

Liu, G.-H., J. Zhou, W. Li and Y. Cheng. 2005. The seed bank in a subtropical freshwater marsh: implications for wetland restoration. *Aquat. Bot.* 81: 1–11.

Liu, G.-H., W. Li, J. Zhou, W.-Z. Liu, D. Yang & A. J. Davy. 2006. How does the propagule bank contribute to cyclic vegetation change in a lakeshore marsh with seasonal drawdown? *Aquat. Bot.* 84: 137–143.

Liu, H. & T. P. Spira. 2001. Influence of seed age and inbreeding on germination and seedling growth in *Hibiscus moscheutos* (Malvaceae). *J. Torrey Bot. Soc.* 128: 16–24.

Liu, J., A. S. Davis & P. J. Tranel. 2012. Pollen biology and dispersal dynamics in waterhemp (*Amaranthus tuberculatus*). *Weed Sci.* 60: 416–422.

Liu, S., K. E. Denford, J. E. Ebinger, J. G. Packer & G. C. Tucker. 2009. 28. *Kalmia* Linnaeus. pp. 480–485. *In*: N. R. Morin (convening ed.), *Flora of North America North of Mexico, Vol. 8: Magnoliophyta: Paeoniaceae to Ericaceae.* Oxford University Press, New York.

Liu, S. Y., F. Sporer, M. Wink, J. Jourdane, R. Henning, Y. L. Li & A. Ruppel. 1997. Anthraquinones in *Rheum palmatum* and *Rumex dentatus* (Polygonaceae), and phorbol esters in *Jatropha curcas* (Euphorbiaceae) with molluscicidal activity against the schistosome vector snails *Oncomelania, Biomphalaria,* and *Bulinus. Trop. Med. Int. Health* 2: 179–188.

Livshultz, T. 2010. The phylogenetic position of milkweeds (Apocynaceae subfamilies Secamonoideae and Asclepiadoideae): evidence from the nucleus and chloroplast. *Taxon* 59: 1016–1030.

Lledó, M. D., M. B. Crespo, K. M. Cameron, M. F. Fay & M. W. Chase. 1998. Systematics of Plumbaginaceae based on cladistic analysis of *rbcL* sequence data. *Syst. Bot.* 23: 21–29.

Lledó, M. D., M. Erben & M. B. Crespo. 2003. *Myriolepis,* a new genus segregated from *Limonium* (Plumbaginaceae). *Taxon* 52: 67–73.

Llewellyn, D. W. & G. P. Shaffer. 1993. Marsh restoration in the presence of intense herbivory: The role of *Justicia lanceolata* (Chapm.) Small. *Wetlands* 13: 176–184.

Lloyd, D. G. 1965. Evolution of self-compatibility and racial differentiation in *Leavenworthia* (Cruciferae). *Contr. Gray Herb.* 195: 3–134.

Lloyd, D. G. 1972. Breeding systems in *Cotula* L. (Compositae, Anthemideae): I. The array of monoclinous and diclinous systems. *New Phytol.* 71: 1181–1194.

Lloyd, D. G. & C. J. Webb. 1987. The reinstatement of *Leptinella* at generic rank, and the status of the 'Cotuleae' (Asteraceae, Anthemideae). *New Zealand J. Bot.* 25: 99–105.

Lloyd, F. E. 1976. *The Carnivorous Plants.* Dover Publications, Inc., New York. 352 pp.

Lo, E. Y. Y., S. Stefanović & T. A. Dickinson. 2007. Molecular reappraisal of relationships between *Crataegus* and *Mespilus* (Rosaceae, Pyreae)—two genera or one? *Syst. Bot.* 32: 596–616.

Lo, E. Y. Y., S. Stefanović, K. I. Christensen & T. A. Dickinson. 2009. Evidence for genetic association between East Asian and western North American *Crataegus* L.(Rosaceae) and rapid divergence of the eastern North American lineages based on multiple DNA sequences. *Mol. Phylogen. Evol.* 51: 157–168.

Lobello, G., M. Fambrini, R. Baraldi, B. Lercari & C. Pugliesi. 2000. Hormonal influence on photocontrol of the protandry in the genus *Helianthus. J. Exp. Bot.* 51: 1403–1412.

Lockhart, B. E. L., J. Fetzer & J. Westendrop. 2002. Previously unreported viral diseases of *Aster, Heuchera, Lobelia, Pulmonaria* and *Physostegia* in the USA. *Acta Hort.* 568: 221–224.

Lockhart, B. R., E. S. Gardiner, T. Stautz & T. D. Leininger. 2012. Development and plasticity of endangered shrub *Lindera melissifolia* (Lauraceae) seedlings under contrasting light regimes. *Plant Species Biol.* 27: 30–45.

Lockhart, B. R., E. S. Gardiner, T. D. Leininger, P. B. Hamel, K. F. Connor, M. S. Devall, N. M. Schiff & A. D. Wilson. 2013. *Lindera melissifolia* responses to flood durations and light regimes suggest strategies for recovery and conservation. *Plant Ecol.* 214: 893–905.

Lockhart, B. R., E. S. Gardiner & T. D. Leininger. 2015. Initial response of pondberry released from heavy shade. pp. 218–223. In: A. G. Holley, K. F. Connor & J. D. Haywood (eds.). *Proceedings of the 17th Biennial Southern Silvicultural Research Conference.* e-Gen. Tech. Rep. SRS–203. U. S. Department of Agriculture, Forest Service, Southern Research Station, Asheville, NC.

Lockwood, J. L., K. H. Fenn, T. L. Warren, R. Hirsch-Jacobson, A. VanHolt & A. Fargue. 1999. Defining nest site microhabitat and preferences to aid in the recovery of the Cape Sable seaside sparrow. Available online: http://www.nicholas.duke.edu/people/faculty/pimm/cssp/cssspdf/microhab.pdf (accessed December 4, 2013).

Lockwood, M. A., C. P. Griffin, M. E. Morrow, C. J. Randel & N. J. Silvy. 2005. Survival, movements, and reproduction of released captive-reared Attwater's prairie-chicken. *J. Wildl. Manag.* 69: 1251–1258.

Locky, D. A., S. E. Bayley & D. H. Vitt. 2005. The vegetational ecology of black spruce swamps, fens, and bogs in southern boreal Manitoba, Canada. *Wetlands* 25: 564–582.

Loconte, H., L. Campbell & D. W. Stevenson, 2012. Ordinal and family relationships of Ranunculid genera. pp. 99–118. *In*: U. Jensen & J. W. Kadereit (eds.). *Systematics and Evolution of the Ranunculiflorae.* Springer Verlag, Wien, Austria. 367 pp.

Lodha, R. & A. Bagga. 2000. Traditional Indian systems of medicine. *Ann. Acad. Med. Singapore* 29: 37–41.

Loeffler, C. 2013. *Eurybia radula* life history information (personal communication).

Loeffler, C. C. 1994. Natural history of leaf-folding caterpillars, *Dichomeris* spp. (Gelechiidae), on goldenrods and asters. *J. New York Entomol. Soc.* 102: 405–428.

Loehrlein, M. & S. Siqueira. 2005. Self-incompatibility in pink tickseed, *Coreopsis rosea* Nutt. *HortScience* 40: 1001–1002.

Logacheva, M. D., C. M. Valiejo-Roman & M. G. Pimenov. 2008. ITS phylogeny of West Asian *Heracleum* species and related taxa of Umbelliferae–Tordylieae W.D.J.Koch, with notes on evolution of their *psbA-trnH* sequences. *Plant Syst. Evol.* 270: 139–157.

Logan, J. M. 2003. *A Vascular Plant Inventory of Springs and Seeps of Buffalo National River.* Technical report NPS/HTLN/P637001B001. Heartland Network, Inventory and Monitoring Program, National Park Service, Republic, MO. 18 pp.

Logan, M. H. 1973. Digestive disorders and plant medicinals in highlands Guatemala. *Anthropos* 68: 538–547.

Logarzo, G. A., M. A. Casalinuovo, R. V. Piccinali, K. Braun & E. Hasson. 2010. Geographic host use variability and host range evolutionary dynamics in the phytophagous insect *Apagomerella versicolor* (Cerambycidae). *Oecologia* 165: 387–402.

Lohne, C., J. H. Wiersema & T. Borsch. 2009. The unusual *Ondinea*, actually just another Australian water-lily of *Nymphaea* subg. Anecphya (Nymphaeaceae). *Willdenowia* 39: 55–58.

Loizeau, P.-A. & R. Spichiger. 2004. Aqifoliaceae (Holly family). pp. 26–27. *In*: N. Smith, S. A. Mori, A. Henderson, D. W. Stevenson & S. V. Heald (eds.), *Flowering Plants of the Neotropics*. Princeton University Press, Princeton, NJ. 594 pp.

Loizeau, P.-A., V. Savolainen, S. Andrews & R. Spichiger. (in press). Aquifoliaceae. *In*: K. Kubitzki (ed.), *The Families and Genera of Vascular Plants*. Springer-Verlag, Berlin, Germany.

Lollar, A. Q., D. C. Coleman & C. E. Boyd. 1971. Carnivorous pathway of phosphorus uptake in *Utricularia inflata*. *Arch. Hydrobiol.* 69: 400–404.

Lonard, R. I., F. W. Judd, R. Stalter & J. J. Brown. 2016. A review of the North American halophyte *Suaeda linearis* (Ell.) Moq. pp. 73–84. *In*: M. A. Khan, B. Boër, M. Öztürk, M. Clüsener-Godt, B. Gul & S.-W. Breckle (eds.), *Sabkha Ecosystems Volume V: The Americas*. Springer International Publishing AG, Basel, Switzerland.

Londo, A. J. & R. C. Carter. 2002. *Soil pH and Tree Species Suitability in Mississippi*. Publication 2311. Extension Service of Mississippi State University, Mississippi State, MS. 2 pp.

Long, J. W. & A. L. Medina. 2007. Title: Geologic associations of Arizona willow in the White Mountains, Arizona. pp. 49–58. *In*: P. Barlow-Irick, J. Anderson & C. McDonald (Technical Eds.), *Southwestern Rare and Endangered Plants: Proceedings of the Fourth Conference; March 22–26, 2004; Las Cruces, New Mexico*. RMRS-P-48CD. United Sattes Department of Agriculture, Forest Service, Rocky Mountain Research Station. Fort Collins, CO.

Long, J. W., A. Tecle & B. M. Burnette. 2003. Marsh development at restoration sites on the White Mountain Apache reservation, Arizona. *J. Amer. Water Res. Assoc.* 39: 1345–1359.

Long, R. W. 1966. Biosystematics of the *Helianthus nuttallii* complex (Compositae). *Brittonia* 18: 64–79.

Long, R. W. & O. Lakela. 1971. *A Flora of Tropical Florida*. University of Miami Press, Coral Gables, FL. 962 pp.

Looney, P. B. & D. J. Gibson. 1995. The relationship between the soil seed bank and above-ground vegetation of a coastal barrier island. *J. Veg. Sci.* 6: 825–836.

Loos, P. 2014. Ornamental garden plants native to the Four Corners Region. pp. 19–41. *In*: Proceedings of Symposium 2014 Texarkana: "Northeast Texas: A Diversity of Ecosystems", October 16–19, 2014. Texarkana, TX.

Loose, J. L., F. A. Drummond, C. Stubbs, S. Woods & S. Hoffman. 2005. *Conservation and Management of Native Bees in Cranberry*. Technical Bulletin 191. Maine Agricultural and Forest Experiment Station, The University of Maine, Orono, ME. 27 pp.

López, T. A., M. S. Cid & M. L. Bianchini. 1999. Biochemistry of hemlock (*Conium maculatum* L.) alkaloids and their acute and chronic toxicity in livestock. A review. *Toxicon* 37: 841–865.

LoPresti, E. F., I. S. Pearse & G. K. Charles. 2015. The siren song of a sticky plant: columbines provision mutualist arthropods by attracting and killing passerby insects. *Ecology* 96: 2862–2869.

Losada, J. M. 2014. *Magnolia virginiana*: Ephemeral courting for millions of years. *Arnoldia* 71(3): 19–27.

Lott, R. H. & N. Knight. 1909. The fruit of *Viburnum nudum*. *Proc. Iowa Acad. Sci.* 16: 145–150.

Louda, S. & J. E. Rodman. 1996. Insect herbivory as a major factor in the shade distribution of a native crucifer (*Cardamine cordifolia* A. Gray, bittercress). *J. Ecol.* 84: 229–237.

Löve, Á. & P. Sarkar. 1957. Chromosomes and relationships of *Koenigia islandica*. *Can. J. Bot.* 35: 507–514.

Löve, Á. & D. Löve. 1954. Vegetation of a prairie marsh. *Bull. Torrey Bot. Club* 81: 16–34.

Löve, Á. & D. Löve. 1982. IOPB chromosome number reports LXXV. *Taxon* 31: 344–360.

Loveless, C. M. 1959. A study of the vegetation in the Florida Everglades. *Ecology* 40: 1–9.

Lovell, J. H. 1898. The insect-visitors of flowers. *Bull. Torrey Bot. Club* 25: 382–390.

Lovell, J. H. 1899. The colors of northern monocotyledonous flowers. *Amer. Naturalist* 33: 493–504.

Lovell, J. H. 1903. The colors of northern gamopetalous flowers. *Amer. Naturalist* 37: 365–384.

Lovell, J. H. 1907. The Colletidae of southern Maine. *Can. Entomol.* 39: 363–365.

Lovell, J. H. 1908. The Halictidae of southern Maine. *Psyche* 15: 32–40.

Lovell, J. H. 1912. The color sense of the honey-bee: the pollination of green flowers. *Amer. Naturalist* 46: 83–107.

Lovell, J. H. 1913. A vernal bee (*Colletes inæqualis* Say). *Psyche* 20: 147–148.

Lovell, J. H. 1915. A preliminary list of the anthophilous Coleoptera of New England. *Psyche* 22: 109–116.

Lovell, J. H. & H. B. Lovell. 1935. Pollination of the Ericaceae: *Chamaedaphne* and *Xolisma*. *Rhodora* 37: 157–61.

Lovell, H. B. & J. H. Lovell. 1936. Pollination of the Ericaceae: IV. *Ledum* and *Pyrola*. *Rhodora* 38: 90–94.

Lovenshimer, J. & J. Frick-Ruppert. 2013. Nutritional values of six native and introduced fall-ripening fruit species in western North Carolina. *Bios* 84: 218–226.

Lowcock, L. A. & R. W Murphy. 1990. Seed dispersal by amphibian vectors: passive transport of bur-marigold, *Bidens cernua*, achenes by migrating salamanders, genus *Ambystoma*. *Can. Field-Naturalist* 104: 298–300.

Lowden, R. M. 1969. A vascular flora of Winous Point, Ottawa and Sandusky Counties, Ohio. *Ohio J. Sci.* 69: 257–284.

Lowden, R. M. 1981. Distribution of *Ceratophyllum muricatum* Chamisso in the West Indies. *Aquat. Bot.* 10: 85–87.

Lowenfeld, R. & E. J. Klekowski. 1992. Mangrove genetics. I. Mating system and mutation rates of *Rhizophora mangle* in Florida and San Salvador Island, Bahamas. *Int. J. Plant Sci.* 153: 394–399.

LSA. 2010. Contra Costa Goldfields in Solano County, California: Results of Contra Costa Goldfields Monitoring, 2006–2009, Solano County, California. LSA Project No. SCD430, SCD0601, SWG0701, SWG0801, and SWG0901. LSA Associates, Inc., Point Richmond, CA. 83 pp.

Lu, K. L. 1982. Pollination biology of *Asarum caudatum* (Aristolochiaceae) in northern California. *Syst. Bot.* 7: 150–157.

Lu, Z.-J., L.-F. Li, M.-X. Jiang, H.-D. Huang & D.-C. Bao. 2010. Can the soil seed bank contribute to revegetation of the drawdown zone in the Three Gorges Reservoir region? *Plant Ecol.* 209: 153–165.

Lubelczyk, C. B., S. P. Elias, P.W. Rand, M. S. Holman, E. H. Lacombe & R. P. Smith. 2004. Habitat associations of *Ixodes scapularis* (Acari: Ixodidae) in Maine. *Environ. Entomol.* 33: 900–906.

Luc, M., J. G. Baldwin & H. Bell. 1986. *Pratylenchus morettoi* n. sp. (Nemata: Pratylenchidae). *Rev. Nématol.* 9: 119–123.

Lucas, F. A. 1893. The food of hummingbirds. *Auk* 10: 311–315.

Ludwig, J. C., K. McCarthy, A. Rome & R. W. Tyndall. 1986. *Management Plans for Significant Plant and Wildlife Habitat Areas of Maryland's Eastern Shore: Worcester County.* Maryland Natural Heritage Program, Annapolis, MD. 75 pp.

Luken, J. O. 2005. *Dionaea muscipula* (Venus flytrap) establishment, release, and response of associated species in mowed patches on the rims of Carolina bays. *Restorat. Ecol.* 13: 678–684.

Luken, J. O. 2012. Long-term outcomes of Venus flytrap (*Dionaea muscipula*) establishment. *Restorat. Ecol.* 20: 669–670.

Luken, J. O. & J. W. Thieret. 2001. Floristic relationships of mud flats and shorelines at Cave Run Lake, Kentucky. *Castanea* 66: 336–351.

Luker, M. A., S. A. Graham, J. V. Freudenstein & C. Bult. 1999. An emerging phylogeny of the genus *Cuphea* (Lythraceae) based on nuclear ribosomal ITS sequences, p. 409 (abstract). XVI International Botanical Congress, St. Louis, MO.

Lumbreras, A., G. Navarro, C. Pardo & J. A. Molina. 2011. Aquatic *Ranunculus* communities in the northern hemisphere: A global review. *Plant Biosyst.* 145(suppl.): 118–122.

Lumbreras, A., J. A. Molina, A. Benavent, A. Marticorena & C. Pardo. 2014. Disentangling the taxonomy and ecology of South American *Ranunculus* subgen. *Batrachium. Aquat. Bot.* 114: 21–28.

Luo, S.-X., S.-M. Chaw, D. Zhang & S. S. Renner. 2010. Flower heating following anthesis and the evolution of gall midge pollination in Schisandraceae. *Amer. J. Bot.* 97: 1220–1228.

Luna, T. 2005. Propagation protocol for Indian paintbrush *Castilleja* species. *Native Plant J.* 6: 62–68.

Lundberg, J. & K. Bremer. 2003. A phylogenetic study of the order Asterales using one morphological and three molecular data sets. *Int. J. Plant Sci.* 164: 553–578.

Lundqvist, A. 1992. The self-incompatibility system in *Caltha palustris* (Ranunculaceae). *Hereditas* 117: 145–51.

Lunsford, D. E. 1938. Studies in the life cycle of *Amphianthus pusillus* Torr. AA Thesis. Emory University, Atlanta, GA. 88 pp.

Lusk, J. M & E. G. Reekie. 2007. The effect of growing season length and water level fluctuations on growth and survival of two rare and at risk Atlantic Coastal Plain flora species, *Coreopsis rosea* and *Hydrocotyle umbellata. Can. J. Bot.* 85: 119–131.

Luteyn, J. L. 1976. Revision of *Limonium* (Plumbaginaceae) in Eastern North America. *Brittonia* 28: 303–317.

Luteyn, J. L. 2004. Ericaceae (Heath family). pp. 140–143. *In:* N. Smith, S. A. Mori, A. Henderson, D. W. Stevenson & S. V. Heald (eds.), *Flowering Plants of the Neotropics.* Princeton University Press, Princeton, NJ. 594 pp.

Lutz, H. J. 1958. Observation on 'diamond willow,' with particular reference to its occurrence in Alaska. *Amer. Midl. Naturalist* 60: 176–185.

Lyder, J. 2009. Endangered and threatened wildlife and plants; proposed designation of critical habitat for *Limnanthes floccosa* ssp. *grandiflora* (large-flowered woolly meadowfoam) and *Lomatium cookii* (Cook's lomatium). *Fed. Reg.* 74: 37314–37392.

Lynch, S. P., L. E. Watson, P. Meyer, B. Farrell & R. Olmstead. 2001. Molecular phylogeny of North American *Asclepias* (Asclepiadaceae). Botany 2001 meeting (abstract).

Lynn, L. M. 1984. The vegetation of Little Cedar Bog, southeastern New York. *Bull. Torrey Bot. Club* 111: 90–95.

Lynn, L. M. & E. F. Karlin. 1985. The vegetation of the low-shrub bogs of northern New Jersey and adjacent New York: ecosystems at their southern limit. *Bull. Torrey Bot. Club* 112: 436–444.

Ma, C. J., K. Y. Lee, E. J. Jeong, S. H. Kim, J. Park, Y. H. Choi, Y. C. Kim & S. H. Sung. 2010. Persicarin from water dropwort (*Oenanthe javanica*) protects primary cultured rat cortical cells from glutamate-induced neurotoxicity. *Phytother. Res.* 24: 913–918.

Maas, D. 1989. Germination characteristics of some plant species from calcareous fens in southern Germany and their implications for the seed bank. *Ecography* 12: 337–344.

Maas, M., A. M. Deters & A. Hensel. 2011a. Anti-inflammatory activity of *Eupatorium perfoliatum* L. extracts, eupafolin, and dimeric guaianolide via iNOS inhibitory activity and modulation of inflammation-related cytokines and chemokines. *J. Ethnopharmacol.* 137: 371–381.

Maas, M., A. Hensel, F. Batista da Costa, R. Brun, M. Kaiser & T. J. Schmidt. 2011b. An unusual dimeric guaianolide with antiprotozoal activity and further sesquiterpene lactones from *Eupatorium perfoliatum. Phytochemistry* 72: 635–644.

Mabbayad, M. O. & A. K. Watson. 1995. Biological control of gooseweed (*Sphenoclea zeylanica* Gaertn.) with an *Alternaria* sp. *Crop Protect.* 14: 429–433.

Mabbott, D. C. 1920. *Food Habits of Seven Species of American Shoal-Water Ducks.* Bulletin No 862, U. S. Department of Agriculture. Government Printing Office, Washington, DC. 68 pp.

Maberly, S. C. & T. V. Madsen. 2002. Use of bicarbonate ions as a source of carbon in photosynthesis by *Callitriche hermaphroditica. Aquat. Bot.* 73: 1–7.

Macanawai, A. R., M. D. Day, T. Tumaneng-Diete & S.W. Adkins. 2012. The impact of rainfall upon pollination and reproduction of *Mikania micrantha* in Viti Levu, Fiji. *Pakistan J. Weed Sci. Res.* 18: 367–375.

MacCracken, J. G. 2002. Response of forest floor vertebrates to riparian hardwood conversion along the Bear River, southwest Washington. *Forest Sci.* 48: 299–308.

MacCracken, J. G., V. Van Ballenberghe & J. M. Peek. 1997. Habitat relationships of moose on the Copper River Delta in coastal south-central Alaska. *Wildl. Monogr.* 136: 3–52.

MacDonagh, P. 2010. *Native Seed Mix Design for Roadsides.* Final Report #2010–20. Minnesota Dpeartment of Transportation, Research Services, Office of Policy Analysis, Research and Innovation. St. Paul, MN. 83 pp.

MacDonald, I. D. 2003. *A Rare Plant Survey of the Wainwright Dunes Ecological Reserve: Supplementary Data.* Alberta Sustainable Resource Development, Resource Data Branch, Public Lands Division, Edmonton, AB. 283 pp.

MacDonald, K. B. & M. G. Barbour. 1974. Beach and salt marsh vegetation of the North American Pacific coast. pp. 175–234. *In:* R. J. Reimold & W. H. Queen (eds.), *Ecology of Halophytes.* Academic Press, New York.

MacDonald, M. A. & P. B. Cavers. 1991. The biology of Canadian weeds. 97. *Barbarea vulgaris* R.Br. *Can. J. Plant Sci.* 71: 149–166.

Macdonald, S. E., C. C. Chinnappa, D. M. Reid & B. G. Purdy. 1987. Population differentiation of the *Stellaria longipes* complex within Saskatchewan's Athabasca sand dunes. *Can. J. Bot.* 65: 1726–1732.

MacDonald, S. E., C. C. Chinnappa & D. M. Reid. 1988. Evolution of phenotypic plasticity in the *Stellaria longipes* complex: comparisons among cytotypes and habitats. *Evolution* 42: 1036–1046.

Mace, R. D. & C. J. Jonkel. 1986. Local food habits of the grizzly bear in Montana. *Int. Conf. Bear Res. Manag.* 6: 105–110.

Mac Elwee, A. 1900. The flora of the Edgehill Ridge near Willow Grove and its ecology. *Proc. Acad. Nat. Sci. Philadelphia* 52: 482–490.

MacHutchon, A. G. 1999. *Black Bear Inventory, Clayoquot Sound, B.C. – Volume I: Habitat Inventory.* British Columbia Ministry of Environment, Lands and Parks, Vancouver Island Region. Nanaimo, BC. 93 pp.

MacHutchon, A. G., S. Himmer & C. A. Bryden. 1993. *Khutzeymateen Valley Grizzly Bear Study: Final Report.* British Columbia Ministry of Environment, Lands and Parks, Wildlife Report R-25, Wildlife Habitat Restoration Report. WHR-31. British Columbia Ministry of Forests, Vancouver, BC. 107 pp.

Macior, L. W. 1967. Pollen-foraging behavior of *Bombus* in relation to pollination of nototribic flowers. *Amer. J. Bot.* 54: 359–364.

Macior, L. W. 1968. Pollination adaptation in *Pedicularis groenlandica*. *Amer. J. Bot.* 55: 927–932.

Macior, L. W. 1969. Pollination adaptation in *Pedicularis lanceolata*. *Amer. J. Bot.* 56: 853–859.

Macior, L. W. 1970. The pollination ecology of *Pedicularis* in Colorado. *Amer. J. Bot.* 57: 716–728.

Macior, L. W. 1973. The pollination ecology of *Pedicularis* on Mount Rainier. *Amer. J. Bot.* 60: 863–871.

Macior, L. W. 1977. The pollination ecology of *Pedicularis* (Scrophulariaceae) in the Sierra Nevada of California. *Bull. Torrey Bot. Club* 104: 148–154.

Macior, L. W. 1983. The pollination dynamics of sympatric species of *Pedicularis* (Scrophulariaceae). *Amer. J. Bot.* 70: 844–853.

Macior, L. W. 1993. Pollination ecology of *Pedicularis palustris* L. (Scrophulariaceae) in North America. *Plant Species Biol.* 8: 35–44.

Mack, J. J. & R. E. J. Boerner. 2004. At the tip of the prairie peninsula: vegetation of Daughmer Savannah, Crawford County, Ohio. *Castanea* 69: 309–323.

MacKay, P. 2003. Mojave Desert Wildflowers: A Field Guide to Wildflowers, Trees, and Shrubs. Globe Pequot Press, Guilford, CT. 352 pp.

MacKenzie, W., D. Remington & J. Shaw. 2000. *Estuaries on the North Coast of British Columbia: A Reconnaissance Survey of Selected Sites.* British Columbia Ministry of Forests, Research Branch, Smithers, BC. 38 pp.

MacKenzie, W. H. & J. R Moran. 2004. *Wetlands of British Columbia: A Guide to Identification.* Land Management Handbook No. 52. Research Branch, The Ministry of Forests, Victoria, BC. 287 pp.

Mackey, D., S. Gregory, W. Matthews, J. Claar & I. Ball. 1987. Impacts of Water Levels on Breeding Canada Geese and Methods for Mitigation and Management in the Southern Flathead Valley, Montana. Project No. 1983-00200. BPA Report DOE/BP-10062-3. Bonneville Power Administration, Division of Fish and Wildlife, Portland, OR. 171 pp.

Mackun, I. R., D. J. Leopold & D. J. Raynal. 1994. Short-term responses of wetland vegetation after liming of an Adirondack watershed. *Ecol. Appl.* 4: 535–543.

MacMillan, C. 1891. Notes on Fungi affecting leaves of *Sarracenia purpurea* in Minnesota. *Bull. Torrey Bot. Club* 18: 214–215.

Mac Nally, R. C., 1995. *Ecological versatility and community ecology.* Cambridge University Press, Cambridge, UK. 435 pp.

Macoun, J. M. 1899. XXIII.-A list of the plants of the Pribilof Islands, Bering Sea: with notes on their distribution. pp. 559–587. *In*: D. S. Jordan (ed.), *The Fur Seals and Fur-seal Islands of the North Pacific Ocean, Part 3.* Government Printing Office, Washington, DC.

MacRae, T. C. & G. H. Nelson. 2003. Distributional and biological notes on Buprestidae (Coleoptera) in North and Central America and the West Indies, with validation of one species. *Coleopt. Bull.* 57: 57–70.

MacRae, T. C. & M. E. Rice. 2007. Biological and distributional observations on North American Cerambycidae (Coleoptera). *Coleopt. Bull.* 61: 227–263.

MacRoberts, B. R., M. H. MacRoberts & L. S. Jackson. 2001. Floristics and management of pitcher plant bogs in northern Natchitoches and Winn Parishes, Louisiana. *Proc. Louisiana Acad. Sci.* 64: 14–21.

MacRoberts, B. R., M. H. MacRoberts, D. C. Rudolph & D. W. Peterson. 2014. Floristics of ephemeral ponds in east-central Texas. *S. E. Naturalist (Steuben)* 13: 15–25.

MacRoberts, M. H. & B. R. MacRoberts. 2004. *Sarracenia purpurea* (Sarraceniaceae) in Louisiana. *Sida* 21: 1149–1152.

Maddox, D. & R. Bartgis. 1990. Harperella (*Ptilimnium nodosum* (Rose) Mathias) recovery plan. U. S. Fish and Wildlife Service, Newton Corner, MA. 56 pp.

Madhavan, V., S. Arora, A. Murali & S. N. Yoganarasimhan. 2009. Anti-convulsant activity of aqueous and alcohol extracts of roots and rhizomes of *Nymphoides indica* (L.) Kuntze in swiss albino mice. *J. Nat. Remedies* 9: 68–73.

Madsen, T. P. 2006. A Vascular Plant Inventory and Vegetation Analysis of the Johnson County Heritage Trust's Hora Woods in Johnson County, Iowa. University of Iowa, Iowa City, IA. 16 pp.

Madsen, T. V. & M. Soendergaard. 1983. The effects of current velocity on the photosynthesis of *Callitriche stagnalis* Scop. *Aquat. Bot.* 15: 187–193.

Magdych, W. P. 1979. The microdistribution of mayflies (Ephemeroptera) in *Myriophyllum* beds in Pennington Creek, Johnston County, Oklahoma. *Hydrobiologia* 66: 161–175.

Magee, T. K., P. L. Ringold & M. A. Bollman. 2008. Alien species importance in native vegetation along wadeable streams, John Day River basin, Oregon, USA. *Plant Ecol.* 195: 287–307.

Maguire, T. L., P. Saenger, P. Baverstock & R. Henry. 2000. Microsatellite analysis of genetic structure in the mangrove species *Avicennia marina* (Forsk.) Vierh. (Avicenniaceae). *Mol. Ecol.* 9: 1853–1862.

Mahler, M. J. 1980. *Limnophila*, a new exotic pest. *Aquatics* 2: 4–7.

Mahy, G., L. P. Bruederle, B. Connors, M. Van Hofwegen & N. Vorsa. 2000. Allozyme evidence for genetic autopolyploidy and high genetic diversity in tetraploid cranberry, *Vaccinium oxycoccos* (Ericaceae). *Amer. J. Bot.* 87: 1882–1889.

Maia, V. H., M. A. Gitzendanner, P. S. Soltis, G. K.-S. Wong & D. E. Soltis. 2014. Angiosperm phylogeny based on 18S/26S rDNA sequence data: constructing a large data set using next-generation sequence data. *Int. J. Plant Sci.* 175: 613–650.

Maier, C. T. 2003. Distribution, hosts, abundance, and seasonal flight activity of the exotic leafroller, *Archips fuscocupreanus* Walsingham (Lepidoptera: Tortricidae), in the Northeastern United States. *Ann. Entomol. Soc. Amer.* 96: 660–666.

Maier, C. T. 2011. First state records of flower flies (Diptera: Syrphidae): *Copestylum vittatum* Thompson in Connecticut and *Mixogaster johnsoni* Hull in Rhode Island. *Proc. Entomol. Soc. Washington* 113: 218–221.

Maillette, L. 1992. Plasticity of modular reiteration in *Potentilla anserina*. *J. Ecol.* 80: 231–239.

Maillette, L., R. J. N. Emery, C. C. Chinnappa & N. K. Kimm. 2000. Module demography does not mirror differentiation among populations of *Stellaria longipes* along an elevational gradient. *Plant Ecol.* 149: 143–156.

Main, M. B., G. M. Allen & K. A. Langeland. 2006. *Creating Wildlife Habitat with Native Florida Freshwater Wetland Plants.* Circular 912. Florida Cooperative Extension Service, Institute of Food and Agricultural Sciences, University of Florida, Gainesville, FL. 11 pp.

Mains. E. B. 1921. Unusual rusts on *Nyssa* and *Urticastrum. Amer. J. Bot.* 8: 442–451.

Mains. E. B. 1939. New and unusual species of Uredinales. *Bull. Torrey Bot. Club* 66: 617–621.

Maity, T. & A. Ahmad. 2012. Protective effect of *Mikania scandens* (L.) Willd. against isoniazid induced hepatotoxicity in rats. *Int. J. Pharm. Pharm. Sci., (Supp. 3)* 4: 466–469.

Maity, T., A. Ahmad & N. Pahari. 2012. Evaluation of hepatotherapeutic effects of *Mikania scandens* (L.) Willd. on alcohol induced hepatotoxicity in rats. *Int. J. Pharm. Pharm. Sci.* 4: 490–494.

Majak, W., B. M. Brooke & R. T. Ogilvie. 2008. *Stock-poisoning Plants of western Canada.* 93 pp. Available online: http://www1.foragebeef.ca/$foragebeef/frgebeef.nsf/all/frg33/$FILE/rangepoisonousplants.pdf (accessed March 17, 2014).

Majka, C. G. 2006. The checkered beetles (Coleoptera: Cleridae) of the Maritime Provinces of Canada. *Zootaxa* 1385: 31–46.

Majka, C. G. & A. R. Cline. 2006. Nitidulidae and Kateretidae (Coleoptera: Cucujoidea) of the Maritime provinces of Canada. I. New records from Nova Scotia and Prince Edward Island. *Can. Entomol.* 138: 314–332.

Majka, C. G. & L. Lesage. 2007. Introduced leaf beetles of the maritime provinces, 3: The viburnum leaf beetle, *Pyrrhalta viburni* (Paykull) (Coleoptera: Chrysomelidae). *Proc. Entomol. Soc. Washington* 109: 454–462.

Majka, C. G., R. S. Anderson, D. F. McAlpine & R. P. Webster. 2007. The weevils (Coleoptera: Curculionoidea) of the maritime provinces of Canada, I: new records from New Brunswick. *Can. Entomol.* 139: 378–396.

Majka, C. G., R. Webster & A. Cline. 2008. New records of Nitidulidae and Kateretidae (Coleoptera) from New Brunswick, Canada. *ZooKeys* 2: 337–356.

Maki, K. & S. Galatowitsch. 2004. Movement of invasive aquatic plants into Minnesota (USA) through horticultural trade. *Biol. Conserv.* 118: 389–396.

Makings, E. 2013. Flora and vegetation of the Saint David and Lewis Springs cienegas, Cochise County, Arizona. pp. 71–76. *In*: G. J. Gottfried, P. F. Folliott, B. S.Gebow, L. G. Eskew & L. C. Collins (eds.), *Merging Science and Management in a Rrapidly Changing World: Biodiversity and Management of the Madrean Archipelago III and 7th Conference on Research and Resource Management in the Southwestern Deserts; 2012 May 1–5; Tucson, AZ.* Proceedings. RMRS-P-67. U. S. Department of Agriculture, Forest Service, Rocky Mountain Research Station, Fort Collins, CO.

Malaipan, S. 1983. Studies of insect pollinators of water convovulae and yard long bean. *Res. Rep. Kasetsart Univ. Bangkok* 1983: 82–83.

Malalavidhane, T. S., S. M. D. Nalinie Wickramasinghe & E. R. Jansz. 2000. Oral hypoglycaemic activity of *Ipomoea aquatica. J. Ethnopharmacol.* 72: 293–298.

Malcom, J. W. & W. R. Radke. 2008. Livestock trampling and *Lilaeopsis schaffneriana* var. *recurva* (Brassicaceae). *Madroño* 55: 81–81.

Maldonado San Martín, A. P., L. Adamec, J. Suda, T. H. M. Mes & H. Štorchová. 2003. Genetic variation within the endangered species *Aldrovanda vesiculosa* (Droseraceae) as revealed by RAPD analysis. *Aquat. Bot.* 75: 159–172.

Maliakal, S. K., E. S. Menges & J. S. Denslow. 2000. Community composition and regeneration of Lake Wales Ridge wiregrass flatwoods in relation to time-since-fire. *J. Torrey Bot. Soc.* 127: 125–138.

Malling, H. 1957. The chromosome number of *Honckenya peploides* (L.) Ehrh., with a note on its mode of sex determination. *Hereditas* 43: 517–524.

Malloch, D. & B. Malloch. 1982. The mycorrhizal status of boreal plants: additional species from northeastern Ontario. *Can. J. Bot.* 60: 1035–1040.

Maloney, D. C. & G. A. Lamberti. 1995. Rapid decomposition of summer-input leaves in a northern Michigan stream. *Amer. Midl. Naturalist* 133: 184–195.

Mamedov, N., Z. Gardner & L. E. Craker. 2005. Medicinal plants used in Russia and central Asia for the treatment of selected skin conditions. *J. Herbs Spices Med. Plant* 11: 191–222.

Mamolos, A. P. & D. S. Veresoglou. 2000. Patterns of root activity and responses of species to nutrients in vegetation of fertile alluvial soil. *Plant Ecol.* 148: 245–253.

Mancuso, M. 1996. Vegetation Monitoring at the Nature Conservancy's Flat Ranch Preserve: 1996 Results. Idaho Department of Fish and Game, Boise, ID. 8 pp.

Mancuso, M. 2001. Monitoring Tobias' saxifrage (*Saxifraga bryophora* var. *tobiasiae*) on the Payette National Forest: first year results. Idaho Department of Fish and Game, Boise, ID. 36 pp.

Mandáková, T., A. D. Gloss, N. K. Whiteman & M. A. Lysak. 2016. How diploidization turned a tetraploid into a pseudotriploid. *Amer. J. Bot.* 103: 1–10.

Mandeel, Q. & A. Taha. 2005. Assessment of in vitro antifungal activities of various extracts of indigenous Bahraini medicinal plants. *Pharmaceut. Biol.* 43: 340–348.

Mandel, J. R. 2010. Clonal diversity, spatial dynamics, and small genetic population size in the rare sunflower, *Helianthus verticillatus. Conserv. Genet.* 11: 2055–2059.

Mandossian, A. J. 1965. Plant associates of *Sarracenia purpurea* (pitcher plant) in acid and alkaline habitats. *Michigan Bot.* 4: 107–114.

Mandossian, A. & R. P. McIntosh. 1960. Vegetation zonation on the shore of a small lake. *Amer. Midl. Naturalist* 64: 301–308.

Manen, J.-F. 2000. Relaxation of evolutionary constraints in promoters of the plastid gene *atpB* in a particular Rubiaceae lineage. *Plant Syst. Evol.* 224: 235–241.

Manen, J.-F., A. Natali & F. Ehrendorfer. 1994. Phylogeny of *Rubiaceae-Rubieae* inferred from the sequence of a cpDNA intergene region. *Plant Syst. Evol.* 190: 195–211.

Manen, J.-F. M. C. Boulter & Y. Naciri-Graven. 2002. The complex history of the genus *Ilex* L. (Aquifoliaceae): evidence from the comparison of plastid and nuclear DNA sequences and from fossil data. *Plant Syst. Evol.* 235: 79–98.

Mann Jr., W. F. & L. J. Musselman. 1981. Autotrophic growth of southern root parasites. *Amer. Midl. Naturalist* 106: 203–205.

Manning, K., E. Petrunak, M. Lebo, A. González-Sarrías, N. P. Seeram & G. E. Henry. 2011. Acylphloroglucinol and xanthones from *Hypericum ellipticum. Phytochemistry* 72: 662–667.

Manns, U. & A. A. Anderberg. 2005. Molecular phylogeny of *Anagallis* (Myrsinaceae) based on ITS, *trn*L-F, and *ndh*F sequence data. *Int. J. Plant Sci.* 166: 1019–1028.

Manos, P. S. & D. E. Stone. 2001. Evolution, phylogeny, and systematics of the Juglandaceae. *Ann. Missouri Bot. Gard.* 88: 231–269.

Manos, P. S., J. J. Doyle & K. C. Nixon. 1999. Phylogeny, biogeography, and processes of molecular differentiation in *Quercus* subgenus *Quercus* (Fagaceae). *Mol. Phylogen. Evol.* 12: 333–349.

Manos, P. S., R. E. Miller & P. Wilkin. 2001a. Phylogenetic analysis of *Ipomoea*, *Argyreia*, *Stictocardia*, and *Turbina* suggests a generalized model of morphological evolution in morning glories. *Syst. Bot.* 26: 585–602.

Manos, P. S., Z.-K. Zhou & C. H. Cannon. 2001b. Systematics of Fagaceae: phylogenetic tests of reproductive trait evolution. *Int. J. Plant Sci.* 162: 1361–1379.

Mansberg, L. & T. R. Wentworth. 1984. Vegetation and soils of a serpentine barren in western North Carolina. *Bull. Torrey Bot. Club* 111: 273–286.

Mansion, G. & L. Struwe. 2004. Generic delimitation and phylogenetic relationships within the subtribe Chironiinae (Chironieae: Gentianaceae), with special reference to *Centaurium*: evidence from nrDNA and cpDNA sequences. *Mol. Phylogen. Evol.* 32: 951–977.

Mansion, G. & L. Zeltner. 2004. Phylogenetic relationships within the New World endemic *Zeltnera* (Gentianaceae-Chironiinae) inferred from molecular and karyological data. *Amer. J. Bot.* 91: 2069–2086.

Mansion, G., G. Parolly, A. A. Crowl, E. Mavrodiev, N. Cellinese, M. Oganesian, K. Fraunhofer, G. Kamari, D. Phitos, R. Haberle, G. Akaydin, N. Ikinci, Thomas Raus & T. Borsch. 2012. How to handle speciose clades? Mass taxon-sampling as a strategy towards illuminating the natural history of *Campanula* (Campanuloideae). *PLoS One*, 7(11): e50076.

Manske, R. H. F. 1938. Lobinaline, an alkaloid from *Lobelia cardinalis* L. *Can. J. Res.* 16: 445–448.

Manson, C. 2003. Endangered and Threatened Wildlife and Plants; Final designation of critical habitat for four vernal pool crustaceans and eleven vernal pool plants in California and southern Oregon. *Fed. Reg.* 68: 46684–46867.

Manson, C. 2005. Endangered and Threatened Wildlife and Plants; Final Designation of Critical Habitat for Four Vernal Pool Crustaceans and Eleven Vernal Pool Plants in California and Southern Oregon; Evaluation of Economic Exclusions From August 2003 Final Designation; Final Rule. *Fed. Reg.* 70: 46923–46999.

Marble, J. C., T. L. Corley & M. H. Conklin. 1999. Representative plant and algal uptake of metals near Globe, Arizona. pp. 239–245. *In*: D. W. Morganwalp & H. T. Buxton (eds.), *Contamination from Hardrock Mining*: U. S. Geological Survey Water-Resources Investigations Report 99-4018A. Proceedings of the Technical Meeting, Charleston, South Carolina, March 8–12, 1999 (Volume 1 of 3). Toxic Substances Hydrology Program, U. S. Geological Survey, Reston, VA.

Marcinko, S. E. & J. L. Randall. 2008. Protandry, mating systems, and sex expression in the federally endangered *Ptilimnium nodosum* (Apiaceae). *J. Torrey Bot. Soc.* 135: 178–188.

Marco, J. L. 1985. Iridoid glucosides of *Pinguicula vulgaris*. *J. Nat. Prod.* 48: 338.

Marcussen, T., K. S. Jakobsen, J. Danihelka, H. E. Ballard, K. Blaxland, A. K. Brysting & B. Oxelman. 2012. Inferring species networks from gene trees in high-polyploid North American and Hawaiian violets (*Viola*, Violaceae). *Syst. Biol.* 61: 107–126.

Marhold, K. (ed.) 2012. IAPT/IOPB chromosome data 14. *Taxon* 61: 1336–1345.

Marhold, K., J. Lihová, M. Perný & W. Bleeker. 2004. Comparative ITS and AFLP analysis of diploid *Cardamine* (Brassicaceae) taxa from closely related polyploid complexes. *Ann. Bot.* 93: 507–520.

Marie-Victorin, F. 1930. Le genre *Rorippa* dans le Québec. *Contr. Inst. Bot. Univ. Montréal* 17: 1–17.

Marino, R. 2008. Homeopathy and collective health: The case of Dengue epidemics. *Int. J. High Dilution Res.* 7: 179–185.

Markle, M. S. 1915. The phytecology of peat bogs near Richmond, Indiana. *Proc. Ind. Acad. Sci.* 25: 359–375.

Markert, A., N. Steffan, K. Ploss, S. Hellwig, U. Steiner, C. Drewke, S.-M. Li, W. Boland & E. Leistner. 2008. Biosynthesis and accumulation of ergoline alkaloids in a mutualistic association between *Ipomoea asarifolia* (Convolvulaceae) and a clavicipitalean fungus. *Plant Physiol.* 147: 296–305.

Markon, C. J. & D. V. Derksen. 1994. Identification of tundra land cover near Teshekpuk Lake, Alaska using SPOT satellite data. *Arctic* 47: 222–231.

Markow, S. 2004. Survey for *Stephanomeria fluminea* on the Bridger-Teton National Forest. Wyoming Natural Diversity Database, Laramie, WY. 23 pp.

Marrotte, R. R., G. L. Chmura & P. A. Stone. 2012. The utility of Nymphaeaceae sclereids in paleoenvironmental research. *Rev. Palaeobot. Palynol.* 169: 29–37.

Marquardt, E. S. & S. C. Pennings. 2010. Constraints on host use by a parasitic plant. *Oecologia* 164: 177–184.

Marques, M. R., C. Stüker, N. Kichik, T. Tarragó, E. Giralt, A. F. Morel & I. I. Dalcol. 2010. Flavonoids with prolyl oligopeptidase inhibitory activity isolated from *Scutellaria racemosa* Pers. *Fitoterapia* 81: 552–556.

Marr, J. K., M. R. Penskar & D. A. Albert. 2009. Rare plant species and plant community types of Manitou Island and Gull Rock, Keweenaw County, Michigan. *Michigan Bot.* 48: 97–120.

Marr, K. L., R. J. Hebda & W. H. MacKenzie. 2012. New alpine plant records for British Columbia and a previously unrecognized biogeographical element in western North America. *Botany* 90: 445–55.

Marrero, H. J., J. P. Torretta & G. Pompozzi. 2013. Triple interaction network among flowers, flower visitors and crab spiders in a grassland ecosystem. *Stud. Neotrop. Fauna Environ.* 48: 153–164.

Marrs-Smith, G. E. 1983. Vegetation and flora of the San Bernadino Ranch, Cochise County, Arizona. MS thesis. Arizona State University, Phoenix, AZ. 95 pp.

Martin, A. C. & F. M. Uhler. 1951. *Food of Game Ducks in the United States and Canada*. Research Report 30, Fish & Wildlife Service, U. S. Department of the Interior, Washington, DC. 308 pp.

Martin, A. C., R. C. Erickson & J. H. Steenis. 1957. *Improving Duck Marshes by Weed Control. Circular 19—Revised*. Fish and Wildlife Service, Department of the Interior, Washington, DC. 60 pp.

Martin, B. D. & E. W. Lathrop. 1986. Niche partitioning in *Downingia bella* and *D. cuspidata* (Campanulaceae) in the vernal pools of the Santa Rosa Plateau Preserve, California. *Madroño* 33: 284–299.

Martin, C. E., A. E. Lubbers & J. A. Teeri. 1982. Variability in crassulacean acid metabolism: a survey of North Carolina succulent species. *Bot. Gaz.* 143: 491–497.

Martin, G. 1887. Enumeration and description of the Septorias of North America. *J. Mycol.* 3: 73–82.

Martin, J. 1988. Different feeding strategies of two sympatric hummingbird species. *Condor* 90: 233–236.

Martin, K. L. 2006. Restoration of fire-dependent depression wetlands. PhD dissertation, University of Georgia, Athens, GA. 71 pp.

Martin, M. A. 1981. A unique natural area in Dane County, Wisconsin. *Ohio Biol. Surv. Biol. Notes* 15: 187–189.

Martinez, M. A. 2006. Habitat usage by the Page springsnail, *Pyrgulopsis morrisoni* (Gastropoda: Hydrobiidae), from central Arizona. *The Veliger* 48: 8–16.

Martins, A. M., F. M. Abilio, P. G. de Oliveira, R. P. Feltrin, F. S. A. de Lima, P. de O. Antonelli, D. Teixeira Vilela, C. G. Smith III, C. Tidwell, Paul Hamela, M. Devall, K. Connor, T. Leininger, N. Schiff & A. D. Wilson. 2015. Pondberry (*Lindera melissifolia*, Lauraceae) seed and seedling dispersers and predators. *Global Ecol. Conserv.* 4: 358–368.

Martins, L., C. Oberprieler & F. H. Hellwig. 2003. A phylogenetic analysis of Primulaceae s.l. based on internal transcribed spacer (ITS) DNA sequence data. *Plant Syst. Evol.* 237: 75–85.

Marushia, R. G. & E. B. Allen. 2011. Control of exotic annual grasses to restore native forbs in abandoned agricultural land. *Restorat. Ecol.* 19: 45–54.

Marx, H. E., N. O'Leary, Y.-W. Yuan, P. Lu-Irving, D. C. Tank, M. E. Múlgura & R. G. Olmstead. 2010. A molecular phylogeny and classification of Verbenaceae. *Amer. J. Bot.* 97: 1647–1663.

Masangkay, R. F., M. O. Mabbayad, T. C. Paulitz & A. K. Watson. 1999. Host range of *Alternaria alternata* f.sp. *sphenocleae* causing leaf blight of *Sphenoclea zeylanica*. *Can. J. Bot.* 77: 103–112.

Maschinski, J. 2001. Impacts of ungulate herbivores on a rare willow at the southern edge of its range. *Biol. Conservation* 101: 119–130.

Mason, C. T. 1998. Gentianaceae. Gentian Family. *J. Arizona-Nevada Acad. Sci.* 30: 84–95.

Mason, D. & A. G. van der Valk. 1992. Growth responses of *Nymphoides indica* seedlings and vegetative propagules along a water depth gradient. *Aquat. Bot.* 42: 339–350.

Mason, H. L. 1955. New species of *Elatine* in California. *Madroño* 13: 239–240.

Mason, H. L. 1957. *A Flora of the Marshes of California*. University of California Press, Berkeley, CA. 878 pp.

Massey, B. W., R. Zembal & P. D. Jorgensen. 1984. Nesting habitat of the light-footed clapper rail in southern California. *J. Field Ornithology* 55: 67–80.

Mast, A. R. & J. L. Reveal. 2007. Transfer of *Dodecatheon* to *Primula* (Primulaceae). *Brittonia* 59: 79–82.

Mast, A. R., D. M. S. Feller, S. Kelso & E. Conti. 2004. Buzz-pollinated *Dodecatheon* originated from within the heterostylous *Primula* subgenus *Auriculastrum* (Primulaceae): A 7-region cpDNA phylogeny and its implications for floral evolution. *Amer. J. Bot.* 91: 926–942.

Mast, A. R., S. Kelso & E. Conti. 2006. Are any primroses (*Primula*) primitively monomorphic? *New Phytol.* 171: 605–616.

Mast, A. R., S. Kelso, A. J. Richards, D. J. Lang, D. M. S. Feller & E. Conti. 2001. Phylogenetic relationships in *Primula* L. and related genera (Primulaceae) based on noncoding chloroplast DNA. *Int. J. Plant Sci.* 162:1381–1400.

Masters, C. O. 1974. *Encyclopedia of the Water-Lily*. T.F.H. Publications Inc., Ltd., Neptune City, NJ. 512 pp.

Masuda, Y., T. Yukawa & K. Kondo. 2009. Molecular phylogenetic analysis of members of *Chrysanthemum* and its related genera in the tribe Anthemideae, the Asteraceae in East Asia on the basis of the internal transcribed spacer (ITS) region and the external transcribed spacer (ETS) region of nrDNA. *Chromosome Bot.* 4: 25–36.

Máthé Jr., I., Á. Vadász, T. Keve, I. Máthé Sen & Á. Máthé. 1983. Time-dependent variations in the alkaloid production of *Amsonia* and *Rhazya* species. *Acta Hort.* 132:101–110.

Mathews, K. G., N. Dunne, E. York & L. Struwe. 2009. A phylogenetic analysis and taxonomic revision of *Bartonia* (Gentianaceae: Gentianeae), based on molecular and morphological evidence. *Syst. Bot.* 34: 162–172.

Mathews, S. & M. J. Donoghue. 1999. The root of angiosperm phylogeny inferred from duplicate phytochrome genes. *Science* 286: 947–950.

Mathews, S. & M. Lavin. 1998. A biosystematic study of *Castilleja crista-galli* (Scrophulariaceae): an allopolyploid origin reexamined. *Syst. Bot.* 23: 213–230.

Mathias, M. E. 1994. Magic, myth and medicine. *Econ. Bot.* 48: 3–7.

Mathias, M. E. & L. Constance. 1941a. A synopsis of the North American species of *Eryngium*. *Amer. Midl. Naturalist* 25: 361–387.

Mathias, M. E. & L. Constance. 1941b. *Limnosciadium*, a new genus of Umbelliferae. *Amer. J. Bot.* 28: 162–163.

Mathias, M. & L. Constance. 1961. Umbelliferae. pp. 263–329. *In*: C. L. Lundell (ed.), *Flora of Texas, Vol. 3*. Texas Research Foundation, Renner, TX.

Mathies, P. S., W. C. Holmes & A. S. Allen. 1983. The vascular flora of Cunningham Brake, a cypress-gum swamp in Natchitoches Parish, Louisiana. *Castanea* 48: 24–31.

Mathur, S. & S. Kumar. 2001. Reproductive biology of *Bacopa monnieri*. *J. Genetics Breed.* 55: 101–109.

Matlack, G. R., D. J. Gibson & R. E. Good. 1993. Regeneration of the shrub *Gaylussacia baccata* and associated species after low-intensity fire in an Atlantic coastal plain forest. *Amer. J. Bot.* 80: 119–126.

Matsuoka, S. M., J. A. Johnson & D. A. Dellasala. 2012. Succession of bird communities in young temperate rainforests following thinning. *J. Wildl. Manag.* 76: 919–931.

Matthews, D. L., D. H. Habeck & D. W. Hall. 1990. Annotated checklist of the Pterophoridae (Lepidoptera) of Florida including larval food plant records. *Florida Entomol.* 73: 613–621.

Matthews, J. F. & W. H. Murdy. 1969. A study of *Isoetes* common to the granite outcrops of the southeastern piedmont, United States. *Bot. Gaz.* 130: 53–61.

Matthews, J. F., J. R. Allison, R. T. Ware, Sr. & C. Nordman. 2002. *Helianthus verticillatus* Small (Asteraceae) rediscovered and redescribed. *Castanea* 67: 13–24.

Matthews, M. L. & P. K. Endress. 2005. Comparative floral structure and systematics in Celastrales (Celastraceae, Parnassiaceae, Lepidobotryaceae). *Bot. J. Linn. Soc.* 149: 129–194.

Matthews, P. D. & G. J. Haas. 1993. Antimicrobial activity of some edible plants: lotus (*Nelumbo nucifera*), coffee, and others. *J. Food Prot.* 56: 66–68.

Matthies, D. 1997. Parasite-host interactions in *Castilleja* and *Orthocarpus*. *Can. J. Bot.* 75: 1252–1260.

Mattrick, C. 2001. *Rotala ramosior* (L.) Koehne (Toothcup). Conservation and research plan. New England Wild Flower Society, Framingham, MA. 27 pp.

Mattson, D. J., B. M. Blanchard & R. R. Knight. 1991. Food habits of Yellowstone grizzly bears, 1977–1987. *Can. J. Zool.* 69: 1619–1629.

Matus, G., R. Verhagen, R. M. Bekker & A. P. Grootjans. 2003. Restoration of the *Cirsio dissecti-Molinietum* in The Netherlands: Can we rely on soil seed banks? *Appl. Veg. Sci.* 6: 73–84.

Maun, M. A. 1998. Adaptations of plants to burial in coastal sand dunes. *Can. J. Bot.* 76: 713–738.

Maurin, O., M. W. Chase, M. Jordaan & M. Van der Bank. 2010. Phylogenetic relationships of Combretaceae inferred from nuclear and plastid DNA sequence data: implications for generic classification. *Bot. J. Linn. Soc.* 162: 453–467.

Maurya, R., S. Srivastava, D. K. Kulshreshta & C. M. Gupta. 2004. Traditional remedies for fertility regulation. *Curr. Med. Chem.* 11: 1431–1450.

Mawdsley, J. R. 1999. Redescription and notes on the biology of *Amecocerus senilis* (Leconte) (Coleoptera: Melyridae: Dasytinae). *J. New York Entomol. Soc.* 107: 68–72.

Mawdsley, J. R. 2002. Ecological notes on species of Cleridae (Insecta: Coleoptera) associated with the prairie flora of central North America. *Great Lakes Entomol.* 35: 15–22.

Mawdsley, J. R. 2003. The importance of species of Dasytinae (Coleoptera: Melyridae) as pollinators in Western North America. *Coleopterists Bull.* 57: 154–160.

Mayer, M. 2004. Phylogeny of the *Thelypodieae* (Brassicaceae), based on ITS and *ndhF* sequence analysis. Botany 2004 meeting (abstract).

Mayfield, H. F. 1972. Bird bones identified from Indian sites at western end of Lake Erie. *Condor* 74: 344–347.

Mayol, M. & J. A. Rossello. 2001. Why nuclear ribosomal DNA spacers (ITS) tell different stories in *Quercus. Mol. Phylogen. Evol.* 19: 167–176.

Mazer, S. J. & D. E. Lowry. 2003. Environmental, genetic, and seed mass effects on winged seed production in the heteromorphic *Spergularia marina* (Caryophyllaceae). *Funct. Ecol.* 17: 637–650.

Mazzotti, F. J., L. A. Brandt, K. G. Rice, R. R. Borkhataria, G. Martin & K. Minkowski. 2004. *Ecological Characterization of Aquatic Refugia in the Arthur R. Marshall Loxahatchee National Wildlife Refuge. Final Report.* U. S. Department of Interior, U. S. National Park Service, Critical Ecosystems Studies Initiative, Homestead, FL. 29 pp.

McArthur, E. D., A. C. Blauer, S. B. Monsen & S. C. Sanderson. 1995. Plant inventory, succession, and reclamation alternatives on disturbed lands in Grand Teton National Park. pp. 343–358. *In*: B. A. Roundy, E. D. McArthur, J. S. Hayley & D. K. Mann (eds.), *Proceedings: Wild Land Shrub and Arid Land Restoration Symposium.* General Technical Report INT-GTR-315. U. S. Department of Agriculture, Forest Service, Intermountain Research Station, Ogden, UT.

McAtee, W. L. 1918. *Food Habits of the Mallard ducks of the United States.* U. S. Department of Agriculture Bulletin No 720. Government Printing Office, Washington, DC. 36 pp.

McAtee, W. L. 1920. Notes on the jack pine plains of Michigan. *Bull. Torrey Bot. Club* 47: 187–190.

McAtee, W. L. 1925. Notes on drift, vegetable balls, and aquatic insects as a food product of inland waters. *Ecology* 6: 288–302.

McAtee, W. L. 1939. *Wildfowl Food Plants.* Collegiate Press, Inc., Ames, IA. 141 pp.

McAtee, W. L. 1947. Distribution of seeds by birds. *Amer. Midl. Naturalist* 38: 214–223.

McAtee, W. L. & E. B. Bartram. 1922. Additions to the flora of the Pribilof Islands. *Torreya* 22: 67–68.

McAvoy, B. 1931. Ecological survey of the Bella Coola region. *Bot. Gaz.* 92: 141–171.

McAvoy, W. A. and R. M. Wilson. 2014. Rediscovery of *Lobelia boykinii* (Campanulaceae) in Delaware. *Phytoneuron* 2014–23: 1–4.

McAvoy, W. A., T. S. Patrick & L. M. Kruse. 2015. Rediscovery of *Dichanthelium hirstii* (Poaceae) in Georgia. *Phytoneuron* 2015–7: 1–8.

McBain & Trush, Inc. 2004. A Vegetation Description of the Lower Rocky Gulch Restoration Project Area, Humboldt County, California. McBain & Trush, Inc., Arcata, CA. 19 pp.

McCabe, T. L. 1985. The natural history of *Oncocnemis piffardi* (Walker) (Lepidoptera: Noctuidae). *J. New York Entomol. Soc.* 93: 1027–1031.

McCaffrey, C. A. & R. D. Dueser. 1990. Plant associations on the Virginia barrier islands. *Viginia J. Sci.* 41: 282–299.

McCain, C. & J. A. Christy. 2005. *Field Guide to Riparian Plant Communities in Northwestern Oregon.* Technical Paper R6-NR-ECOL-TP-01-05. United States Department of Agriculture, Forest Service, Pacific Northwest Region, Portland, OR. 357 pp.

McCanny, S. J., P. A. Keddy, T. J. Arnason, C. L. Gaudet, D. R. J. Moore & B. Shipley. 1990. Fertility and the food quality of wetland plants: A test of the resource availability hypothesis. *Oikos* 59: 373–381.

McCarron, J. K., K.W. McLeod & W.H. Conner. 1998. Flood and salinity stress of wetland woody species, buttonbush (*Cephalanthus occidentalis*) and swamp tupelo (*Nyssa sylvatica* var. *biflora*). *Wetlands* 18: 165–175.

McCarter, S. M. & J. A. Payne. 1993. Fire blight caused by *Erwinia amylovora* on mayhaw in Georgia. *Plant Dis.* 77: 1262.

McCarthy, B. C., C. J. Small & D. L. Rubino. 2001. Composition, structure and dynamics of Dysart Woods, an old-growth mixed mesophytic forest of southeastern Ohio. *Forest Ecol. Manag.* 140: 193–213.

McCarthy, S. & J. P. Evans. 2000. Population dynamics of overcup oak (*Quercus lyrata*) in a seasonally flooded karst depression. *J. Torrey Bot. Club* 127: 9–18.

McCarty, L. B., J. W. Everest, D. W. Hall, T. R. Murphy & F. Yelverton. 2001. *Color Atlas of Turfgrass Weeds.* Ann Arbor Press, Chelsea, MI. 269 pp.

McCleary, J. A. 1957. Notes on Arizona plants. *Southwest. Naturalist* 2: 152–154.

McClellan, M. H., T. Brock & J. F. Baichtal. 2003. *Calcareous Fens in Southeast Alaska.* Pacific Northwest Research Station Research Note (PNW-RN-536). United States Department of Agriculture, Forest Service, Portland, OR. 10 pp.

McClelland, M. K. & I. A. Ungar. 1970. The influence of edaphic factors on *Betula nigra* L. distribution in southeastern Ohio. *Castanea* 35: 99–117.

McConnell, J. & R. Muniappan. 1991. Introduced ornamental plants that have become weeds on Guam. *Micronesia* 3: S47-S49.

McCord, L. 2010. Santee Cooper—crested floating heart update. *Newslett. S. Carolina Aquatic Plant Manag. Soc.* 31: 4–5.

McCormac, J. S. 1993. First record of *Leersia lenticularis* Michx. in the Great Lakes drainage. *Castanea* 58: 54–58.

McCormick, P. V. & R. E. Gibble. 2014. Effects of soil chemistry on plant germination and growth in a northern Everglades peatland. *Wetlands* 34: 979–988.

McCormick, M. A., M. A. Cubeta & L. F. Grand. 2013. Geography and hosts of the wood decay fungi *Fomes fasciatus* and *Fomes fomentarius* in the United States. *N. Amer. Fungi* 8(2): 1–53.

McCormick, S. & B. Bohm. 1986. Methylated flavonoids from *Artemisia lindleyana. J. Nat. Prod.* 49: 167–186.

McCoy, J. W. & F. R. Stermitz. 1983. Alkaloids from *Castilleja miniata* and *Penstemon whippleanus*, two host species for the plume moth, *Amblyptilia* (*Platyptilia*) *pica. J. Nat. Prod.* 46: 902–907.

McCoy, R. W. & D. C. Hales. 1974. A survey of eight streams in eastern South Dakota: physical and chemical characteristics, vascular plants, insects and fishes. *Proc. South Dakota Acad. Sci.* 53: 202–219.

McCracken, A., J. D. Bainard, M. C. Miller & B. C. Husband. 2013. Pathways of introduction of the invasive aquatic plant *Cabomba caroliniana. Ecol Evol.* 3: 1427–1439.

McCullough, H. A. 1948. Plant succession on fallen logs in a virgin spruce-fir forest. *Ecology* 29: 508–513.

McCune, L. M. 2012. Traditional medicinal plants of indigenous peoples of Canada and their antioxidant activity in relation to treatment of diabetes. pp. 221–234. *In*: R. R. Watson & V. R. Preedy (eds.), *Bioactive Food as Dietary Interventions for Diabetes*. Elsevier, Boston, MA.

McCutcheon, A. R. 1996. Ethnopharmacology of western North American plants with special focus on the genus *Artemisia* L. PhD dissertation, University of British Columbia, Vancouver, BC. 437 pp.

McCutcheon, A. R., R. W. Stokes, L. M. Thorson, S. M. Ellis, R. E. W. Hancock & G. H. N. Towers. 1997. Anti-mycobacterial screening of British Columbian medicinal plants. *Pharmaceut. Biol.* 35: 77–83.

McDade, L. A., T. F. Daniel, C. A. Kiel & K. Vollesen. 2005. Phylogenetic relationships among Acantheae (Acanthaceae): major lineages present contrasting patterns of molecular evolution and morphological differentiation. *Syst. Bot.* 30: 834–862.

McDade, L. A. & J. A. Weeks. 2004. Nectar in hummingbird-pollinated Neotropical plants I: Patterns of production and variability in 12 species. *Biotropica* 36: 196–215.

McDade, L. A., S. E. Masta, M. L. Moody & E. Waters. 2000. Phylogenetic relationships among Acanthaceae: evidence from two genomes. *Syst. Bot.* 25: 106–121.

McDaniel, B. 1971a. The armored scale insects of Texas (Homoptera: Coccoidea: Diaspididae): Part IV. The tribe Diaspidini. *Southwest. Naturalist* 15: 275–308.

McDaniel, S. 1971b. The genus *Sarracenia* (Sarraceniaceae). *Bull. Tall Timbers Res. Stn.* 9: 1–36.

McDearman, W. W. 1984 *Arnoglossum sulcatum* (Asteraceae) in Mississippi. *Sida* 10: 258–259.

McDill, J. M. 2009. Molecular phylogenetic studies in the Linaceae and *Linum*, with implications for their systematics and historical biogeography. PhD dissertation. University of Texas, Austin, TX. 309 pp.

McDill, J., M. Repplinger, B. B. Simpson & J. W. Kadereit. 2009. The phylogeny of *Linum* and Linaceae subfamily Linoideae, with implications for their systematics, biogeography, and evolution of heterostyly. *Syst. Bot.* 34: 386–405.

McDill, J. R. & B. B. Simpson. 2011. Molecular phylogenetics of Linaceae with complete generic sampling and data from two plastid genes. *Bot. J. Linn. Soc.* 165: 64–83.

McDonald, A. W., J. P. Bakker & K. Vegelin. 1996. Seed bank classification and its importance for the restoration of species-rich flood-meadows. *J. Veg. Sci.* 7: 157–164.

McDonald, G. I., B. A. Richardson, P. J. Zambino, N. B. Klopfenstein & M.-S. Kim. 2006. *Pedicularis* and *Castilleja* are natural hosts of *Cronartium ribicola* in North America: a first report. *Forest Pathol.* 36: 73–82.

McDonald, I. & S. Andrews. 1981. Genetic interaction of *Cronartium ribicola* and *Ribes hudsonianum* var. *petiolare. Forest Sci.* 27: 758–763.

McDonald, J. N. 1976. Blue winged warbler: midseason diet and feeding behavior. *Ohio J. Sci.* 76: 16–17.

McDougall, K. L., J. W. Morgan, N. G. Walsh & R. J. Williams. 2005. Plant invasions in treeless vegetation of the Australian Alps. *Perspect. Plant Ecol. Evol. Syst.* 7: 159–171.

McDougall, W. B. & O. E. Glasgow. 1929. Mycorhizas of the Compositae. *Amer. J. Bot.* 16: 225–228.

McEvoy, P. B. & C. S. Cox. 1987. Wind dispersal distances in dimorphic achenes of ragwort, *Senecio jacobea. Ecology* 68: 2006–2015.

McFarland, D. G. & S. J. Rogers. 1998. The aquatic macrophyte seed bank in Lake Onalaska, Wisconsin. *J. Aquat. Plant Manag.* 36: 33–39.

McGaha, Y. J. 1954. Contribution to the biology of some Lepidoptera which feed on certain aquatic flowering plants. *Trans. Amer. Microscop. Soc.* 73: 167–177.

McGill, N. R. 1973. A comparison of the vascular flora of two lakes in northern Champaign County, Ohio. MS thesis. The Ohio State University, Columbus, OH. 59 pp.

McGilvrey, F. B. 1966. Fall food habits of wood ducks from Lake Marion, South Carolina. *J. Wildl. Manag.* 30: 193–195.

McGregor, M. 2008. Saxifrages: A Definitive Guide to the 2000 Species, Hybrids & Cultivars. Timber Press, Portland, OR. 384 pp.

McGregor, M. A., D. R. Bayne, J. G. Steeger, E. C. Webber & E. Reutebuch. 1996. The potential for biological control of water primrose (*Ludwigia grandiflora*) by the water primrose flea beetle (*Lysathia ludoviciana*) in the southeastern United States. *J. Aquat. Plant Manag.* 34: 74–76.

McGregor, R. L. 1948. First year invasion of plants on an exposed lake bed. *Trans. Kansas Acad. Sci.* 51: 324–327.

McGregor, R. L. 1960. Ferns and allies in Kansas. *Amer. Fern J.* 50: 62–66.

McIntyre, A. P., R. A. Schmitz & C. M. Crisafulli. 2006. Associations of the Van Dyke's salamander (*Plethodon vandykei*) with geomorphic conditions in headwall seeps of the Cascade range, Washington state. *J. Herpetol.* 40: 309–322.

McJannet, C. L., P. A. Keddy & F. R. Pick. 1995. Nitrogen and phosphorus tissue concentrations in 41 wetland plants: a comparison across habitats and functional groups. *Funct. Ecol.* 9: 231–238.

McKain, M. R., M. A. Chapman & A. L. Ingram. 2010. Confirmation of the hybrid origin of *Eupatorium* ×*truncatum* (Asteraceae) using nuclear and plastid markers. *Castanea* 75: 381–387.

McKee, K. L. 1995. Interspecific variation in growth, biomass partitioning and defensive characteristics of neotropical mangrove seedlings: response to light and nutrient availability. *Amer. J. Bot.* 82: 299–307.

McKee, K. L., I. A. Mendelssohn & M. D. Materne. 2004. Acute salt marsh dieback in the Mississippi River deltaic plain: a drought-induced phenomenon? *Glob. Ecol. Biogeogr.* 13: 65–73.

McKellar, M. A., C. W.Deren & K. H. Quesenberry. 1991. Outcrossing in *Aeschynomene. Crop Sci.* 31: 476–478.

McKelvey, R. W., M. C. Dennington & D. Mossop. 1983. The status and distribution of trumpeter swans (*Cygnus buccinator*) in the Yukon. *Arctic* 36: 76–81.

McKendrick, J. D. 1987. Plant succession on disturbed sites, North Slope, Alaska, U.S.A. *Arct. Alp. Res.* 19: 554–565.

McKenna, D. D., A. R. Zangerl & M. R. Berenbaum. 2001. A native hymenopteran predator of *Agonopteirx alstroemeriana* (Lepidoptera: Oecophoridae) in East-Central Illinois. *Great Lakes Entomol.* 34: 71–75.

McKenna, M. F. & G. Houle. 2000. Under-saturated distribution of *Floerkea proserpinacoides* Willd. (Limnanthaceae) at the northern limit of its distribution. *EcoScience* 7: 466–473.

McKenzie, P. M., T. Smith & C. T. Witsell. 2003. *Carex conoidea* (Cyperaceae) new to Arkansas and notes on its occurrence in Arkansas and Missouri. *Sida* 20: 1727–1730.

McKenzie, P. M., C. T. Witsell, L. R. Phillippe, C. S. Reid, M. A. Homoya, S. B. Rolfsmeier & C. A. Morse. 2009. Status assessment of *Eleocharis wolfii* (Cyperaceae) in the United States. *J. Bot. Res. Inst. Texas* 3: 831–854.

McKone, M. J., R. Ostertag, J. T. Rauscher, D. A. Heiser & F. L. Russell. 1995. An exception to Darwin's syndrome: floral position, protogyny, and insect visitation in *Besseya bullii* (Scrophulariaceae). *Oecologia* 101: 68–74.

McLain, D. K. 1986. Niche differentiation and the evolution of ethological isolation in a soldier beetle hybrid zone. *Oikos* 47: 159–167.

McLaughlin, E. G. 1974. Autecological studies of three species of *Callitriche* native in California. *Ecol. Monogr.* 44: 1–16.

McLaughlin, W. T. 1932. Atlantic coastal plain plants in the sand barrens of northwestern Wisconsin. *Ecol. Monogr.* 2: 335–383.

McLeod, K. W. & T. G. Ciravolo.1997. Differential sensitivity of *Nyssa aquatica* and *Taxodium distichum* seedlings grown in fly ash amended sand. *Wetlands* 17: 330–335.

McLeod, K. W., M. R. Reed & E. A. Nelson. 2001. Influence of a willow canopy on tree seedling establishment for wetland restoration. *Wetlands* 21: 395–402.

McMahon, M. & L. Hufford. 2004a. Floral simplification in the *Amorpheae* (Papilionoideae). Botany 2004 meeting (abstract).

McMahon, M. & L. Hufford. 2004b. Phylogeny of *Amorpheae* (Fabaceae: *Papilionoideae*). *Amer. J. Bot.* 91: 1219–1230.

McMaster, R. T. 1994. Ecology, reproductive biology and population genetics of *Ophioglossum vulgatum* (Ophioglossaceae) in Massachusetts. *Rhodora* 96: 259–286.

McMaster, R. T. 2001. The population biology of *Liparis loeselii*, Loesel's twayblade, in a Massachusetts wetland. *N. E. Naturalist* 8: 163–178.

McMaster, R. T. 2003. *Populus heterophylla* (Swamp cottonwood). Conservation and research plan for New England. New England Wild Flower Society, Framingham, MA. 28 pp.

McMillan, C. 1975. Interaction of soil texture with salinity tolerances of black mangrove (*Avicennia*) and white mangrove (*Laguncularia*) from North America. pp. 561–566. *In*: G. Walsh, S. Snedaker & H. Teas (eds.), *Proceedings of the International Symposium on Biology and Management of Mangroves*. East-West Center, Honolulu, HI.

McMillan, P. D. & R. D. Porcher. 2005. Noteworthy collections: South Carolina. *Castanea* 70: 237–240.

McMillan, S. C. 1995. A morphometric and systematic study of the southern California species in the genus *Pogogyne* (Lamiaceae). *Amer. J. Bot.* 82: S149–S150.

McMinn, R. G. 1951. The vegetation of a burn near Blaney Lake, British Columbia. *Ecology* 32: 135–140.

McMullen, C. K. 1987. Breeding systems of selected Galapagos Islands angiosperms. *Amer. J. Bot.* 74: 1694–1705.

McNair, M. 1992. Preliminary studies on the genetics and evolution of the serpentine endemic *Mimulus nudatus* Curran. pp. 409–419. *In*: A. J. M. Baker, J. Proctor & R. D. Reeves (eds.), *The Vegetation of Ultramafic (Serpentine) Soils*. Intercept Limited, Andover, UK.

McNeill, J. 1975. A generic revision of Portulacaceae tribe *Montieae* using techniques of numerical taxonomy. *Can. J. Bot.* 53: 789–809.

McPherson, J. E. & R. M. McPherson. 2000. *Stink Bugs of Economic Importance in America North of Mexico*. CRC Press, Boca Raton, FL. 255 pp.

McVaugh, R. 1936. Studies in the taxonomy of eastern North American species of *Lobelia*. *Rhodora* 38: 241–263; 273–298; 305–329; 316–362.

McVaugh, R. 1941. A monograph on the genus *Downingia*. *Mem. Torrey Bot. Club* 19: 1–57.

McVaugh, R. 1943a. The vegetation of the granitic flat-rocks of the southeastern United States. *Ecol. Monogr.* 13: 119–166.

McVaugh, R. 1943b. Campanulales. Campanulaceae. Lobelioideae. *N. Amer. Fl.* 32A: 1–134.

McVaugh, R. 1957. Establishment of vegetation on sand-flats along the Hudson River, New York—II. The period 1945–1955. *Ecology* 38: 23–29.

McVeigh, A., J. E. Carey & T. C. G. Rich. 2005. Chiltern Gentian, *Gentianella germanica* (Gentianaceae) in Britain: distribution. *Watsonia* 25: 339–367.

Mead, B. R. 1998. *Phytomass in Southeast Alaska*. Research Paper PNW-RP-505. U. S. Department of Agriculture, Forest Service, Pacific Northwest Research Station. Portland, OR. 48 pp.

Mead, B. R. 2000. *Phytomass in Southwest Alaska*. Research Paper PNW-RP-523. U. S. Department of Agriculture, Forest Service, Pacific Northwest Research Station. Portland, OR. 164 pp.

Mead, B. R. 2002. *Constancy and Cover of Plants in the Petersburg and Wrangell Districts, Tongass National Forest and Associated Private and Other Public Lands, Southeast Alaska*. Research Paper PNW-RP-540. U. S. Department of Agriculture, Forest Service, Pacific Northwest Research Station. Portland, OR. 112 pp.

Meagher, T. R., J. Antonovics & R. Primack. 1978. Experimental ecological genetics in *Plantago*. III. Genetic variation and demography in relation to survival of *Plantago cordata*, a rare species. *Biol. Conserv.* 14: 243–257.

Mealey, S. P. 1980. The natural food habits of grizzly bears in Yellowstone National Park, 1973–74. pp. 281–292. *In*: C. J. Martinka & K. L. McArthur (eds.), *Bears: their Biology and Management, Vol. 4*. A selection of papers from the Fourth International Conference on Bear Research and Management, February 1977. International Association of Bear Research and Management, Kalispell, MT.

Meanley, B. 1965. The roosting behavior of the Red-winged Black-bird in the southern United States. *Wilson Bull.* 77: 217–228.

Medeiros, R. B. & J. J. Steiner. 2002. Influência de sistemas de rotação de sementes de gramíneas forrageiras temperadas na composição do banco de sementes invasoras no solo. *Rev. Brasil. Sementes* 24: 118–128.

Medina, E., E. Cuevas & A. Lugo. 2007. Nutrient and salt relations of *Pterocarpus officinalis* L. in coastal wetlands of the Caribbean: assessment through leaf and soil analyses. *Trees* 21: 321–327.

Meeuse, B. J. & E. L. Schneider. 1979. *Nymphaea* revisited: a preliminary communication. *Israel J. Bot.* 28: 65–79.

Mehringer Jr., P. J., S. F. Arno & K. L. Petersen. 1977. Postglacial history of lost trail pass bog, Bitterroot Mountains, Montana. *Arct. Alp. Res.* 9: 345–368.

Meiman, P. J., M. S. Thorne, Q. D. Skinner, M. A. Smith & J. L. Dodd. 2009. Wild ungulate herbivory of willow on two National Forest allotments in Wyoming. *Rangeland Ecol. Manag.* 62: 460–469.

Meiners, S. J. & E. W. Stiles. 1997. Selective predation on the seeds of woody plants. *J. Torrey Bot. Soc.* 124: 67–70.

Meinke, R. J. 1982. *Threatened and Endangered Vascular Plants of Oregon: An Illustrated Guide*. U. S. Fish and Wildlife Service (Region 1), Portland, OR. 326 pp.

Meinke, R. J. 1995a. Assessment of the genus *Mimulus* (Scrophulariaceae) within the interior Columbia River basin of Oregon and Washington. Department of Botany and Plant Pathology, Oregon State University, Corvallis, OR. 79 pp.

Meinke, R. J. 1995b. *Mimulus evanescens* (Scrophulariaceae): a new annual species from the northern Great Basin. *Great Basin Naturalist* 55: 249–257.

Meinke, R. J. 2001. *Comparative Studies of Germination and Reproductive Ecology in Four Rare Species of* Mimulus *(Scrophulariaceae).* Research Report No. HLP001040. U. S. Bureau of Land Management, Oregon District Office, Lakeview, OR. 25 pp.

Meira, M., E. P. da Silva, J. M. David & J. P. David. 2012. Review of the genus *Ipomoea*: traditional uses, chemistry and biological activities. *Rev. Brasileira Farmacognosia* 22: 682–713.

Meisel, J., N. Trushenski & E. Weiher. 2002. A gradient analysis of oak savanna community composition in western Wisconsin. *J. Torrey Bot. Soc.* 129: 115–124.

Meitzen, K. M. 2009. Lateral channel migration effects on riparian forest structure and composition, Congaree River, South Carolina, USA. *Wetlands* 29: 465–475.

Melcher, I. M., F. Bouman & A. M. Cleef. 2000. Seed dispersal in Páramo plants: epizoochorous and hydrochorous taxa. *Plant Biol.* 2: 40–52.

Melchert, T. E. 2010. Chromosome counts of *Bidens*, *Cosmos* and *Thelesperms* species (Asteraceae, Coreopsidinae). *Phytologia* 92: 312–333.

Meléndez-Ackerman, E. J., C. Cortés, J. Sustache, S. Aragón, M. Morales-Vargas, M. García-Bermúdez & D. S. Fernández. 2008. Diet of feral goats in Mona Island reserve, Puerto Rico. *Caribb. J. Sci.* 44: 199–205.

Mellichamp, T. L. 2009. 1. *Darlingtonia* Torrey. pp. 349–350. *In*: N. R. Morin (convening ed.), *Flora of North America North of Mexico, Vol. 8: Magnoliophyta: Paeoniaceae to Ericaceae.* Oxford University Press, New York.

Mellichamp, T. L. 2015. Droseraceae Salisbury. Sundew family. pp. 418–425. *In*: Flora North America Editorial Committee (eds.), *Flora of North America North of Mexico, Vol. 6: Magnoliophyta: Cucurbitaceae to Droseraceae.* Oxford University Press, New York.

Mellichamp, T. L. & F. W. Case. 2009. 2. *Sarracenia* Linnaeus. pp. 350–362. *In*: N. R. Morin (convening ed.), *Flora of North America North of Mexico, Vol. 8: Magnoliophyta: Paeoniaceae to Ericaceae.* Oxford University Press, New York.

Mellmann-Brown, S. 2004. *Botanical and Ecological Inventory of Selected Peatland Sites on the Shoshone National Forest.* Report prepared for Shoshone National forest by Wyoming Natural Diversity Database, Laramie, WY. 39 pp.

Melo, P. A., M. C. Do Nascimento, W. B. Mors & G. Suarez-Kurtz. 1994. Inhibition of the myotoxic and hemorrhagic activities of crotalid venoms by *Eclipta prostrata* (Asteraceae) extracts and constituents. *Toxicon* 32: 595–603.

Melouk, H. A., J. P. Damicone & K. E. Jackson. 1992. *Eclipta prostrata*, a new weed host for *Sclerotinia minor*. *Plant Dis.* 76: 101.

Mendall, H. L. 1949. Food habits in relation to black duck management in Maine. *J. Wildl. Manag.* 13: 64–101.

Méndez, M. & P. S. Karlsson. 2004. Between-population variation in size-dependent reproduction and reproductive allocation in *Pinguicula vulgaris* (Lentibulariaceae) and its environmental correlates. *Oikos* 104: 59–70.

Méndez, M. & P. S. Karlsson. 2005. Nutrient stoichiometry in *Pinguicula vulgaris*: nutrient availability, plant size, and reproductive status. *Ecology* 86: 982–991.

Menegus, F., L. Cattaruzza, L. Scaglioni & E. Ragg. 1992. Effects of oxygen level on metabolism and development of seedlings of *Trapa natans* and two ecologically related species. *Plant Physiol.* 86: 168–172.

Meng, S.-W. 2004. Analysis of phylogeny: A case study on Saururaceae. *Pract. Bioinf.* 2004: 245–268.

Meng, S.-W., Z.-D. Chen, D.-Z. Li & H.-X. Liang. 2002. Phylogeny of Saururaceae based on mitochondrial *matR* gene sequence data. *J. Plant Res.* 115: 71–76.

Meng, S.-W., A. W. Douglas, D.-Z. Li, Z.-D. Chen, H.-X. Liang & J.-B. Yang. 2003. Phylogeny of Saururaceae based on morphology and five regions from three plant genomes. *Ann. Missouri Bot. Gard.* 90: 592–602.

Menges, E. S. & D. M. Waller. 1983. Plant strategies in relation to elevation and light in floodplain herbs. *Amer. Naturalist* 122: 454–473.

Mengesha, A. E. & B.-B. C. Youan. 2010. Anticancer activity and nutritional value of extracts of the seed of *Glinus lotoides*. *J. Nutr. Sci. Vitaminol.* 56: 311–318.

Merritt, D. M. & D. J. Cooper. 2000. Riparian vegetation and channel change in response to river regulation: a comparative study of regulated and unregulated streams in the Green River Basin, USA. *Regulat. Rivers Res. Manag.* 16: 543–564.

Merritt, D. M. & E. E. Wohl. 2006. Plant dispersal along rivers fragmented by dams. *River Res. Appl.* 22: 1–26.

Merritt, J. F. & J. M. Merritt. 1978. Population ecology and energy relationships of *Clethrionomys gapperi* in a Colorado subalpine forest. *J. Mammalogy* 59: 576–598.

Meseguer, A. S., I. Sanmartín, T. Marcussen & B. E. Pfeil. 2014. Utility of low-copy nuclear markers in phylogenetic reconstruction of *Hypericum* L.(Hypericaceae). *Plant Syst. Evol.* 300: 1503–1514.

Messinger, W., K. Hummer & A. Liston. 1999. *Ribes* (Grossulariaceae) phylogeny as indicated by restriction-site polymorphisms of PCR-amplified chloroplast DNA. *Plant Syst. Evol.* 217: 185–195.

Messmore, N. A. & J. S. Knox. 1997. The breeding system of the narrow endemic, *Helenium virginicum* (Asteraceae). *J. Torrey Bot. Soc.* 124: 318–321.

Metcalf, F. P. & L. Griscom. 1917. Notes on rare New York state plants. *Rhodora* 19: 28–37.

Methé, B. A., R. J. Soracco, J. D. Madsen & C. W. Boylen. 1993. Seed production and growth of waterchestnut as influenced by cutting. *J. Aquat. Plant Manag.* 31: 154–157.

Meyer, C. K., S. G. Baer & M. R. Whiles. 2008. Ecosystem recovery across a chronosequence of restored wetlands in the Platte River Valley. *Ecosystems* 11: 193–208.

Meyer, F. G. 1951. *Valeriana* in North America and the West Indies (Valerianaceae). *Ann. Missouri Bot. Gard.* 38: 377–503.

Meyer, F. G. 1997. 1. Magnoliaceae Jussieu. pp. 3–10. *In*: N. R. Morin (convening ed.), *Flora of North America North of Mexico, Vol. 3: Magnoliophyta: Magnoliidae and Hamamelidae.* Oxford University Press, New York.

Meyer, J.-Y. & C. Lavergne. 2004. *Beautés fatales*: Acanthaceae species as invasive alien plants on tropical Indo-Pacific Islands. *Diversity Distrib.* 10: 333–347.

Meyer, S. E. & S. L. Carlson. 2004. Comparative seed germination biology and seed propagation of eight intermountain species of indian paintbrush. pp. 125–130. *In*: A. L. Hild, N. L. Shaw, S. E. Meyer, T. D. Booth & E. D. McArthur (compilers), *Proceedings: Seed and Soil Dynamics in Shrubland Ecosystems*; 2002 Aug 12–16; Laramie, WY. Proceedings. RMRS-P-31. USDA Forest Service, Rocky Mountain Research Station, Ogden, UT.

Meyers, D. G. & J. R. Strickler. 1979. Capture enhancement in a carnivorous aquatic plant: function of antennae and bristles in *Utricularia vulgaris*. *Science* 203: 1022–1025.

Meyers, S. C., A. Liston & R. Meinke. 2010. A molecular phylogeny of *Limnanthes* (Limnanthaceae) and investigation of an anomalous *Limnanthes* population from California, USA. *Syst. Bot.* 35: 552–558.

Meyers, S. C., A. Liston & R. Meinke. 2012. An evaluation of putative sympatric speciation within *Limnanthes* (Limnanthaceae). *PLoS One* 7: e36480. doi:10.1371/journal.pone.0036480.

Meyers-Rice, B. 1998. *Darlingtonia californica* 'Othello'. *Carniv. Plant Newslett.* 27: 40–42.

Meyers-Rice, B. A. 2000. New cultivars (*Dionaea* 'Dentate Traps', *Dionaea* 'Sawtooth', *Dionaea* Dentate Traps cultivar group). *Carniv. Plant Newslett.* 29: 14–21.

Michel, F. A. & K. Henein. 2007. *Natural Re-Vegetation of Arsenic-Bearing Alkaline Tailings at Cobalt, Ontario.* Mining and the Environment IV Conference, October 19–27, 2007. Sudbury, ON. 6 pp.

Michels, M. G., L. C. T. Bertolini, A. F. Esteves, P. Moreira & S. C. Franca. 2011. Anticoccidial effects of coumestans from *Eclipta alba* for sustainable control of *Eimeria tenella* parasitosis in poultry production. *Veterin. Parasitol.* 177: 55–60.

Michener, C. D. 1936a. Western bees of the genus *Ceratina*, subgenus *Zaodontomerus*. *Amer. Mus. Novit.* 844: 1–13.

Michener, C. D. 1936b. Some North American Osmiinae (Hymenoptera, Apoidea). *Amer. Mus. Novit.* 875: 1–30.

Michener, C. D. 1939. A revision of the genus *Ashmeadiella* (Hymen., Megachilidae). *Amer. Midl. Naturalist* 22: 1–84.

Michez, D. & S. Patiny. 2005. World revision of the oil-collecting bee genus *Macropis* Panzer 1809 (Hymenoptera: Apoidea: Melittidae) with a description of a new species from Laos. *Ann. Soc. Entomol. France* 41: 15–28.

Middleton, B. A. 1990. Effect of water depth and clipping frequency on the growth and survival of four wetland plant species. *Aquat. Bot.* 37: 189–196.

Middleton, B. A. 1995a. *The Role of Flooding in Seed Dispersal: Restoration of Cypress Swamps along the Cache River, Illinois.* Research Report 220. Water Resources Center, University of Illinois at Urbana-Champaign. Champaign, IL. 97 pp.

Middleton, B. A. 1995b. Sampling devices for a measurement of seed rain and hydrochory in rivers. *J. Torrey Bot. Soc.* 122: 152–155.

Middleton, B. 2000. Hydrochory, seed banks, and regeneration dynamics along the landscape boundaries of a forested wetland. *Plant Ecol.* 146: 169–184.

Middleton, B. 2002a. Winter burning and the reduction of *Cornus sericea* in sedge meadows in southern Wisconsin. *Restorat. Ecol.* 10: 723–730.

Middleton, B. 2002b. Nonequilibrium dynamics of sedge meadows grazed by cattle in southern Wisconsin. *Plant Ecol.* 161: 89–110.

Middleton, B. 2003. Soil seed banks and the potential restoration of forested wetlands after farming. *J. Appl. Ecol.* 40: 1025–1034.

Middleton, B. A. 2009. Regeneration potential of *Taxodium distichum* swamps and climate change. *Plant Ecol.* 202: 257–274.

Middleton, B. A. & D. H. Mason. 1992. Seed herbivory by nilgai, feral cattle, and wild boar in the Keoladeo National Park, India. *Biotropica* 1: 538–543.

Middleton, B. A. & K. L. McKee. 2011. Soil warming alters seed-bank responses across the geographic range of freshwater *Taxodium distichum* (Cupressaceae) swamps. *Amer. J. Bot.* 98: 1943–1955.

Middleton, D. J. & C. C. Wilcox. 1990. Chromosome counts in *Gaultheria* and related genera. *Edinburgh J. Bot.* 47: 303–313.

Miedzobrodski, K. 1960. *Oenanthe aquatica* poisoning in heifers. *Medycyna Weterynaryjna* 16: 608–609.

Miklovic, S. & S. M. Galatowitsch. 2005. Effect of NaCl and *Typha angustifolia* L. on marsh community establishment: A greenhouse study. *Wetlands* 25: 420–429.

Mikulic-Petkovsek, M., V. Schmitzer, A. Slatnar, F. Stampar & R. Veberic. 2012. Composition of sugars, organic acids, and total phenolics in 25 wild or cultivated berry species. *J. Food Sci.* 77: C1064–C1070.

Milano, G. R. 1999. Restoration of coastal wetlands in southeastern Florida. *Wetland J.* 11: 15–24.

Milberg, P. & B. Stridh. 1994. Seed bank of annual mudflat species at Lake Vikarsjoen, central Sweden. *Svensk Bot. Tidskr.* 88: 237–240.

Miles, D. H. & U. Kokpol. 1976. Tumor inhibitors II: constituents and antitumor activity of *Sarracenia flava*. *J. Pharm. Sci.* 65: 284–285.

Miles, D. H., U. Kokpol, L. H. Zalkow, S. J. Steindel & J. B. Nabors. 1974. Tumor inhibitors. I. antitumor activity of *Sarracenia flava*. *J. Pharm. Sci.* 63: 613–615.

Miles, P. D. & W. B. Smith. 2009. *Specific Gravity and Other Properties of Wood and Bark for 156 Tree Species Found in North America.* Research Note NRS-38. U. S. Department of Agriculture, Forest Service. Newtown Square, PA. 35 pp.

Milius, S. 2002. Are they really extinct? Searches for plants and animals so rare that they may not be there at all. *Sci. News* 161: 168.

Miller, A. J., D. A. Young & J. Wen. 2001. Phylogeny and biogeography of *Rhus* (Anacardiaceae) based on ITS sequence data. *Int. J. Plant Sci.* 162: 1401–1407.

Miller, F. A. 1914. The propagation of medicinal plants. *Bull. Torrey Bot. Club* 41: 105–129.

Miller, G. S. & P. C. Standley. 1912. The North American species of *Nymphaea* [*Nuphar*]. *Contr. U. S. Natl. Herb.* 16: 1–109.

Miller, J. M. 2003a. 3. *Claytonia* Linnaeus. pp. 465–475. *In*: N. R. Morin (convening ed.), *Flora of North America North of Mexico, Vol. 4: Magnoliophyta: Caryophyllidae, Part 1.* Oxford University Press, New York.

Miller, J. M. 2003b. 5. *Montia* Linnaeus. pp. 485–488. *In*: N. R. Morin (convening ed.), *Flora of North America North of Mexico, Vol. 4: Magnoliophyta: Caryophyllidae, Part 1.* Oxford University Press, New York.

Miller, J. M. & B. A. Bohm. 1980. Flavonoid variation in some North Americna *Saxifraga* species. *Biochem. Syst. Ecol.* 8: 279–284.

Miller, J. M., K. L. Chambers & C. E. Fellows. 1984. Cytogeographic patterns and relationships in the *Claytonia sibirica* complex (Portulacaceae). *Syst. Bot.* 9: 266–271.

Miller, J. S. 1988. A revised treatment of Boraginaceae for Panama. *Ann. Missouri Bot. Gard.* 75: 456–521.

Miller, J. S. & J. L. Stanton-Geddes. 2007. Gynodioecy in *Lobelia siphilitica* and *L. spicata* (Lobeliaceae) from western Massachusetts. *J. Torrey Bot. Soc.* 134: 349–361.

Miller, M. R. 1983. Foraging dives by post-breeding Northern Pintails. *Wilson Bull.* 95: 294–296.

Miller, R. E., T. R. Buckley & P. S. Manos. 2002. An examination of the monophyly of morning glory taxa using Bayesian phylogenetic inference. *Syst. Biol.* 51: 740–753.

Miller, R. E., M. D. Rausher & P. S. Manos. 1999. Phylogenetic systematics of *Ipomoea* (Convolvulaceae) based on ITS and Waxy sequences. *Syst. Bot.* 24: 209–227.

Miller, R. L. 1931. A contribution to the life history and habits of the celery leaf tyer *Phlyctaenia rubigalis* Guenee in Florida. *Florida Entomol.* 15: 28–34.

Miller, S. G., S. P. Bratton & J. Hadidian. 1992. Impacts of white-tailed deer on endangered and threatened vascular plants. *Nat. Areas J.* 12: 67–74.

Millett, J., B. M. Svensson, J. Newton & H. Rydin. 2012. Reliance on prey-derived nitrogen by the carnivorous plant *Drosera rotundifolia* decreases with increasing nitrogen deposition. *New Phytol.* 195: 182–188.

Milligan, H. 2008. Aquatic and terrestrial foraging by a subarctic herbivore: the beaver. MS thesis. Department of Natural Resource Sciences, McGill University, Macdonald Campus, Montreal, QC.

Mills, E. L., D. L. Strayer, M. D. Scheuerell & J. T. Carlton. 1996. Exotic species in the Hudson River basin: A history of invasions and introductions. *Estuaries* 19: 814–823.

Mills, J. E., J. A. Reinartz, G. A. Meyer & E. B. Young. 2009. Exotic shrub invasion in an undisturbed wetland has little community-level effect over a 15-year period. *Biol. Invas.* 11: 1803–1820.

Milton, W. E. J. 1939. The occurrence of buried viable seeds in soils at different elevations and on a salt marsh. *J. Ecol.* 27: 149–159.

Mimica-Dukić, N, B. Božin, M. Soković, B. Mihajlović & M. Matavulj. 2003. Antimicrobial and antioxidant activities of three *Mentha* species essential oils. *Plant Med.* 69: 413–419.

Minchinton, T. E. 2006. Rafting on wrack as a mode of dispersal for plants in coastal marshes. *Aquat. Bot.* 84: 372–376.

Ming, R., R. Van Buren, Y. Liu, M. Yang, Y. Han, L.-T. Li, Q. Zhang, M.-J. Kim, M. C. Schatz, M. Campbell, J. Li, J. E. Bowers, H. Tang, E. Lyons, A. A Ferguson, G. Narzisi, D. R. Nelson, C. E Blaby-Haas, A. R Gschwend, Y. Jiao, J. P Der, F. Zeng, J. Han, X. J. Min, K. A Hudson, R. Singh, A. K. Grennan, S. J. Karpowicz, J. R. Watling, K. Ito, S. A. Robinson, M. E. Hudson, Q. Yu, T. C. Mockler, A. Carroll, Y. Zheng, R. Sunkar, R. Jia, N. Chen, J. Arro, C. M. Wai, E. Wafula, A. Spence, Y. Han, L. Xu, J. Zhang, R. Peery, M. J Haus, W. Xiong, J. A. Walsh, J. Wu, M.-L. Wang, Y. J Zhu, R. E. Paull, A. B. Britt, C. Du, S. R. Downie, M. A. Schuler, T. P. Michael, S. P. Long, D. R. Ort, J. W. Schopf, D. R. Gang, N. Jiang, M. Yandell, C. W. dePamphilis, S. S. Merchant, A. H. Paterson, B. B. Buchanan, S. Li & J. Shen-Miller. 2013. Genome of the long-living sacred lotus (*Nelumbo nucifera* Gaertn.). *Genome Biol.* 14: R41. doi:10.1186/gb-2013-14-5-r41.

Minno, M. C. & C. Slaughter. 2003. New record of the endangered lakeside sunflower, *Helianthus carnosus* (Asteraceae), from Putnam County, Florida. *Florida Sci.* 66: 291–293.

Mitchell, A. D. & P. B. Heenan. 2000. Systematic relationships of New Zealand endemic Brassicaceae inferred from nrDNA ITS sequence data. *Syst. Bot.* 25: 98–105.

Mitchell, C. C. & W. A. Niering. 1993. Vegetation change in a topogenic bog following beaver flooding. *Bull. Torrey Bot. Club* 120: 136–147.

Mitchell, D. K. 2011. Wet flatwoods restoration after decades of fire suppresion. MS thesis. University of Florida, Gainesville, FL. 77 pp.

Mitchell, D. S. (ed.). 1974. *Aquatic Vegetation and its Use and Control.* UNESCO, Paris, France. 135 pp.

Mitchell, E. 1926. Germination of seeds of plants native to Dutchess County, New York. *Bot. Gaz.* 81: 108–112.

Mitchell, R. J., J. D. Karron, K. G. Holmquist & J. M. Bell. 2004. The influence of *Mimulus ringens* floral display size on pollinator visitation patterns. *Funct. Ecol.* 18: 116–124.

Mitchell, R. J., J. D. Karron, K. G. Holmquist & J. M. Bell. 2005. Patterns of multiple paternity in fruits of *Mimulus ringens* (Phrymaceae). *Amer. J. Bot.* 92: 885–890.

Mitchell, R. S. 1970. A re-evaluation of *Polygonum meisnerianum* in North America. *Rhodora* 72: 182–188.

Mitchell, R. S. 1994. Discovery of *Utricularia inflata* Walt. A new native species for New York state. *New York Fl. Assoc. Newslett.* 5(2): 1–2.

Mitchell, R. S., T. E. Maenza-Gmelch & J. G. Barbour. 1994. *Utricularia inflata* Walt. (Lentibulariaceae), new to New York State. *Bull. Torrey Bot. Club* 121: 295–297.

Mitchell, S. A. & M. H. Ahmad. 2006. A Review of medicinal plant research at the University of the West Indies, Jamaica, 1948–2001. *W. Indian Med. J.* 55: 243–269.

Mitich, L. W. 1994. Beggarticks. *Weed Technol.* 8: 172–175.

Mitich, L. W. 1998. Poison-hemlock (*Conium maculatum* L.). *Weed Technol.* 12: 194–197.

Mito, T. & T. Uesugi. 2004. Invasive alien species in Japan: the status quo and the new regulation for prevention of their adverse effects. *Global Environ. Res.* 8: 171–191.

Mitra, S. S. 1999. Ecology and behavior of yellow warblers breeding in Rhode Island's great swamp. *N. E. Naturalist* 6: 249–262.

Mitscher, L. A., Y. H. Park, A. Al-Shamma, P. B. Hudson & T. Haas. 1981. Amorfrutin A and B, bibenzyl antimicrobial agents from *Amorpha fruticosa. Phytochemistry* 20: 781–785.

Miura, R., M. Hosotani & M. Ito. 2007. An awnless variant of *Bidens tripartita. J. Weed Sci. Technol.* 52: 130–136.

Miwa, K. & L. J. Meinke. 2015. Developmental biology and effects of adult diet on consumption, longevity, and fecundity of *Colaspis crinicornis* (Coleoptera: Chrysomelidae). *J. Insect Sci.* 15: 78; doi: 10.1093/jisesa/iev062.

Mix, A. J. 1954. Additions and emendations to a monograph of the genus *Taphrina. Trans. Kansas Acad. Sci.* 57: 55–65.

Miyajima, D. 1995. Causes of low double-flowered seed production in breeding Zinnia. *J. Amer. Soc. Hort. Sci.* 120: 759–764.

M'Mahon, B. 1804. *A Catalogue of American Seeds Sold by Bernard M'Mahon, Seedsman, Philadelphia.* Bartholomew Graves, Philadelphia, PA. 30 pp.

Mody, N. V., R. Henson, P. A. Hedin, U. Kokpol & D. H. Miles. 1976. Isolation of the insect paralyzing agent coniine from *Sarracenia flava. Cell. Mol. Life Sci.* 32: 829–830.

Moeller, R. E. 1978a. Carbon uptake by the submersed hydrophyte *Utricularia purpurea. Aquat. Bot.* 12: 209–216.

Moeller, R. E. 1978b. Seasonal changes in biomass, tissue chemistry, and net production of the evergreen hydrophyte, *Lobelia dortmanna. Can. J. Bot.* 56: 1425–1433.

Moeller, R. E. 1980. The temperature-determined growing season of a submerged hydrophyte: tissue chemistry and biomass turnover of *Utricularia purpurea. Freshwater Biol.* 10: 391–400.

Moerman, D. E. 1996. An analysis of the food plants and drug plants of native North America. *J. Ethnopharmacol.* 52: 1–22.

Moerman, D. E. 1998. *Native American Ethnobotany.* Timber Press, Portland, OR. 927 pp.

Mohamed, S. M., E. M. Hassan & S. A. El Toumy. 2011. Phytochemical and biological studies on *Aster novi-belgii. Planta Med.* 77: PG3.

Mohan Ram, H. Y. & S. Rao. 1982. *In vitro* induction of aerial leaves and of precocious flowering in submerged shoots of *Limnophila indica* by abscisic acid. *Planta* 155: 521–523.

Mohlenbrock, R. H. 1959a. A floristic study of a southern Illinois swampy area. *Ohio J. Sci.* 9: 89–100.

Mohlenbrock, R. H. 1959b. Plant communities in Jackson County, Illinois. *Bull. Torrey Bot. Club* 86: 109–119.

Mohlenbrock, R. H., G. B. Dillard & T. S. Abney. 1961. A survey of southern Illinois aquatic vascular plants. *Ohio J. Sci.* 61: 262–273.

Mohr, C. 1901. Plant life of Alabama. *Contr. U. S. Natl. Herb.* 6: 1–921.

Molanoh-Flores, B. 2001. Reproductive biology of *Eryngium yuccifolium* (Apiaceae), a prairie species. *J. Torrey Bot. Soc.* 128: 1–6.

Molau, U. 1992. On the occurrence of sexual reproduction in *Saxifraga cernua* and *Saxifraga foliolosa* (Saxifragaceae). *Nordic J. Bot.* 12: 197–203.

Molau, U. 1993. Reproductive ecology of the 3 Nordic *Pinguicula* species (Lentibulariaceae). *Nordic J. Bot.* 13: 149–157.

Molau, U. & H. C. Prentice. 1992. Reproductive system and population structure in three arctic *Saxifraga* spp. *J. Ecol.* 80: 149–161.

Mold, R. J. 2012. *Ecology of Halophytes.* Academic Press, New York. 620 pp.

Moldenke, H. N. 1958. Hybridity in the Verbenaceae. *Amer. Midl. Naturalist* 59: 333–370.

Moldowan, P. D., M. G. Keevil, P. B. Mills, R. J. Brooks & J. D. Litzgus. 2016. Diet and feeding behaviour of snapping turtles (*Chelydra serpentina*) and midland painted turtles (*Chrysemys picta marginata*) in Algonquin Provincial Park, Ontario. *Can. Field-Nat.* 129: 403–408.

Moline, P. 2001. *Podostemum* and *Crenias* (Podostemaceae) – American riverweeds – infrageneric systematic relationships using molecular and morphological methods. MS thesis. Institut für Systematische Botanik, Universität Zürich, Zürich, Switzerland.

Moll, D. 1976. Food and feeding strategies of the Ouachita map turtle (*Graptemys pseudogeographica ouachitensis*). *Amer. Midl. Naturalist* 96: 478–482.

Møller, C. L. & K. Sand-Jensen. 2011. High sensitivity of *Lobelia dortmanna* to sediment oxygen depletion following organic enrichment. *New Phytol.* 190: 320–331.

Molnár V., A., J. P. Tóth, G. Sramkó, O. Horváth, A. Popiela, A. Mesterházy & B. A. Lukács. 2015. Flood induced phenotypic plasticity in amphibious genus *Elatine* (Elatinaceae). *PeerJ.* 3: e1473.

Molofsky, J., C. M. Danforth & E. E. Crone. 2014. Nutrient enrichment alters dynamics in experimental plant populations. *Populat. Ecol.* 56: 97–107.

Molyneux, R. J. & L. F. James. 1982. Loco intoxication: indolizidine alkaloids of spotted locoweed (*Astragalus lentiginosus*). *Science* 216: 190–191.

Moncada, K., N. J. Ehlke, G. Muehlbauer, C. Sheaffer & D. Wyse. 2005. *Assessment of AFLP-Based Genetic Variation in Three Native Plant Species across the State of Minnesota.* Final Report MN/RC-2005–46. Minnesota Department of Transportation, Research Services Section, St. Paul, MN. 78 pp.

Monda, M. J. & J. T. Ratti. 1988. Niche overlap and habitat use by sympatric duck broods in eastern Washington. *J. Wildl. Manag.* 52: 95–103.

Monette, D. & S. H. Markwith. 2012. Hydrochory in the Florida Everglades: Temporal and spatial variation in seed dispersal phenology, hydrology, and restoration of wetland structure *Ecol. Restorat.* 30: 180–191.

Mongelli, E., C. Desmarchelier, J. Coussio & G. Ciccia.1997. The potential effects of allelopathic mechanisms on plant species diversity and distribution determined by the wheat rootlet growth inhibition bioassay in South American plants. *Rev. Chilena Hist. Nat.* 70: 83–89.

Monk, C. D. 1966. An ecological study of hardwood swamps in north-central Florida. *Ecology* 47: 649–654.

Monk, C. D. & T. W. Brown. 1965. Ecological consideration of cypress heads in northcentral Florida. *Amer. Midl. Naturalist* 74: 126–140.

Monro, A. K. 2006. The revision of species-rich genera: a phylogenetic framework for the strategic revision of *Pilea* (Urticaceae) based on cpDNA, nrDNA, and morphology. *Amer. J. Bot.* 93: 426–441.

Monson, R. K. & C. H. Jaeger. 1991. Photosynthetic characteristics of C_3-C_4 intermediate *Flaveria floridana* (Asteraceae) in natural habitats: evidence of advantages to C_3-C_4 photosynthesis at high leaf temperatures. *Amer. J. Bot.* 78: 795–800.

Montes, L., J. C. Cerono & M. C. Nuciari. 1993. *Bidens aurea* (Asteraceae: Heliantheae): adventive cytotypes in Argentina. *Phytologia* 75: 192–198.

Monty, A.-M. & R. E. Emerson. 2003. 19. Eastern wood rat. *Neotoma floridana* and allies. pp. 381–393. *In*: G. A. Feldhamer, B. C. Thompson & J. A. Chapman (eds.), *Wild Mammals of North America: Biology, Management, and Conservation.*, 2nd ed. The Johns Hopkins University Press, Baltimore, MD.

Monty, A., C. Stainier, F. Lebeau, N. Piereti & G. Mahy. 2008. Seed rain patern of the invasive weed *Senecio inaequidens* (Asteraceae). *Belg. J. Bot.* 141: 51–63.

Montz, G. N. 1977. *Wetland Plants of the New Orleans District.* U. S. Army Corps of Engineers, Regulatory Functions Branch, New Orleans, LA. 62 pp.

Montz, G. N. 1978. The submerged vegetation of Lake Pontchartrain, Louisiana. *Castanea* 43: 115–128.

Moody, K. 1981. *Major Weeds of Rice in South and Southeast Asia.* International Rice Research Institute, Manila, Philippines. 79 pp.

Moody, K. 1989. *Weeds Reported in Rice in South and Southeast Asia.* International Rice Research Institute, Manila, Philippines. 442 pp.

Moody, M. L. 2004. Systematics of the angiosperm family Haloragaceae R. Br. emphasizing the aquatic genus *Myriophyllum*: phylogeny, hybridization and character evolution. PhD dissertation, University of Connecticut, Storrs, CT. 246 pp.

Moody, M. L. & D. H. Les. 2002. Evidence of hybridity in invasive watermilfoil (*Myriophyllum*) populations. *Proc. Natl. Acad. Sci. U.S.A.* 99: 14867–14871.

Moody, M. L. & D. H. Les. 2007. Phylogenetic systematics and character evolution in the angiosperm family Haloragaceae. *Amer. J. Bot.* 94: 2005–2025.

Moon, D. C., A. M. Rossi & P. Stiling. 2000. The effects of abiotically induced changes in host plant quality (and morphology) on a salt marsh planthopper and its parasitoid. *Ecol. Entomol.* 25: 325–331.

Moon, D. C., A. Rossi, K. Stokes & J. Moon. 2008. Effects of the pitcher plant mining moth *Exyra semicrocea* on the hooded pitcher plant *Sarracenia minor*. *Amer. Midl. Naturalist* 159: 321–326.

Moon, D. C., A. M. Rossi, J. Depaz, L. McKelvey, S. Elias, E. Wheeler & J. Moon. 2010. Ants provide nutritional and defensive benefits to the carnivorous plant *Sarracenia minor*. *Oecologia* 164: 185–192.

Moon, H. K. & S. P. Hong. 2003. Pollen morphology of the genus *Lycopus* (Lamiaceae). *Ann. Bot. Fenn.* 40: 191–198.

Moon, H. K. & S. P. Hong. 2006. Nutlet morphology and anatomy of the genus *Lycopus* (Lamiaceae: Mentheae). *J. Plant Res.* 119: 633–644.

Moon, M., M. R. Rattray, F. E. Putz & G. Bowes. 1993. Acclimatization to flooding of the herbaceous vine, *Mikania scandens*. *Funct. Ecol.* 7: 610–615.

Mooney, H. A. 1980. Photosynthetic plasticity of populations of *Heliotropium curassavicum* L. originating from differing thermal regimes. *Oecologia* 45: 372–376.

Mooney, J. 1891. The sacred formulas of the Cherokees. *Rep. (Annual) Bur. Amer. Ethnol.* 7: 302–397.

Moore, A. J. & L. Bohs. 2007. Phylogeny of *Balsamorhiza* and *Wyethia* (Asteraceae: Heliantheae) using ITS, ETS, and *trnK* sequence data. *Syst. Bot.* 32: 682–691.

Moore, A. W. 1930. Six Utah mammal records. *J. Mammalogy* 11: 87–88.

Moore, B. C., J. E. Laferb & W. H. Funkc. 1994. Influence of aquatic macrophytes on phosphorus and sediment porewater chemistry in a freshwater wetland. *Aquat. Bot.* 49: 137–148.

Moore, C., M. Bastian & H. Hunt. 2001. *Long Term Vegetation and Faunal Succession in an Artificial Northern California Vernal Pool System*. Report No. FHWA/CA/TL-2001/36. California Department of Transportation, Sacramento, CA. 49 pp.

Moore, C., M. Bastian & H. Hunt. 2003. *Long Term Evaluation of Characteristics in an Artificial Northern California Vernal Pool System*. California Department of Transportation, Sacremento, CA. 77 pp.

Moore, D. L. 1976. Changes in the marsh and aquatic vascular flora of East Harbor State Park, Ottawa County, Ohio, since 1895. *Ohio J. Sci.* 76: 78–86.

Moore, G. 2010. International registration of cultivar names for unassigned woody genera July 2006-December 2009. *HortScience* 45: 332.

Moore, G. E., C. R. Peter, D. M. Burdick & D. R. Keirstead. 2009. Status of the eastern grasswort, *Lilaeopsis chinensis* (Apiaceae), in the Great Bay Estuary, New Hampshire, U.S.A. *Rhodora* 111: 171–188.

Moore, J. W. & R. M. Tryon Jr. 1946. A new record for *Isoëtes melanopoda*. *Amer. Fern J.* 36: 89–91.

Moore, K. K., L. Fisher & D. L. Sutton. 2006. Nursery production techniques for obligate wetland species. *J. Aquat. Plant Manag.* 44: 37–40.

Moore, M. 1989. *Medicinal Plants of the Desert and Canyon West*. Museum of New Mexico Press, Santa Fe, NM. 368 pp.

Moore, M. J., P. S. Soltis, C. D. Bell, J. G. Burleigh & D. E. Soltis. 2010. Phylogenetic analysis of 83 plastid genes further resolves the early diversification of eudicots. *Proc. Nat. Acad. Sci. U.S.A.* 107: 4623–4628.

Moore, P. L., K. D. Holl & D. M. Wood. 2011. Strategies for restoring native riparian understory plants along the Sacramento River: timing, shade, non-native control, and planting method. *San Francisco Estuary Watershed Sci.* 9: 1–15.

Moore, R. J. (ed.). 1973. Index to plant chromosome numbers 1967–1971. *Regnum Veg.* 90: 1–539.

Moore, T. R. 1989. Plant production, decomposition, and carbon efflux in a subarctic patterned fen. *Arct. Alp. Res.* 21: 156–162.

Moorhead, K. K. & K. R. Reddy. 1988. Oxygen transport through selected aquatic macrophytes. *J. Environ. Qual.* 17: 138–142.

Moorhead, K. K. & K. R. Reddy. 1990. Carbon and nitrogen transformations in wastewater during treatment with *Hydrocotyle umbellata* L. *Aquat. Bot.* 37: 153–161.

Moorhead, K. K., R. E. Moynihan & S. L. Simpson. 2000. Soil characteristics of four southern Appalachian fens in North Carolina, USA. *Wetlands* 20: 560–564.

Moral, R. del & C. Jones. 2002. Vegetation development on pumice at Mount St. Helens, USA. *Plant Ecol.* 162: 9–22.

Morales Valverde, R. 1993. Synopsis and distribution of the genus *Micromeria*. *Bot. Complut.* 18: 157–168.

Moran, J., B. van Rijswijk, V. Traicevski, E.W. Kitajima, A. M. Mackenzie & A. J. Gibbs. 2002. Potyviruses, novel and known, in cultivated and wild species of the family *Apiaceae* in Australia. *Arch. Virol.* 147: 1855–1867.

Moran, N. 1984. The genus *Uroleucon* (Homoptera: Aphididae) in Michigan: key, host records, biological notes, and descriptions of three new species. *J. Kansas Entomol. Soc.* 57: 596–616.

Moran, R. C. 1981. Prairie fens in northeastern Illinois: floristic composition and disturbance. pp. 164–168. *In*: R. L. Stuckey & K. J. Reese (eds.), *Proceedings Sixth North American Prairie Conference. Ohio Biological Survey Notes. No. 15*. The Ohio State University, Columbus, OH.

Moran, R. V. 2009. 1. *Crassula* Linnaeus. pp. 150–155. *In*: N. R. Morin (convening ed.), *Flora of North America North of Mexico, Vol. 8: Magnoliophyta: Paeoniaceae to Ericaceae*. Oxford University Press, New York.

Moravcová, L., P. Pyšek, V. Jarošík, V. Havlíčková & P. Zákravský. 2010. Reproductive characteristics of neophytes in the Czech Republic: traits of invasive and non-invasive species. *Preslia* 82: 365–390.

Morefield, J. D. 1988. Floristic habitats of the White Mountains, California and Nevada: a local approach to plant communities. pp. 1–17. *In*: C. A. Hall and V. Doyle-Jones (eds.), *Plant Biology of Eastern California*. White Mountain Research Station, University of California, Los Angeles, CA.

Morefield, J. D. 1992a. Evolution and systematics of *Stylocline* (Asteraceae: Inuleae). PhD dissertation. Claremont Graduate School, Claremont, CA.

Morefield, J. D. 1992b. Resurrection and revision of *Hesperevax* (Asteraceae: Inuleae). *Syst. Bot.* 17: 293–310.

Morefield, J. D. 1992c. Notes on the status of *Psilocarphus berteri* (Asteraceae: Inuleae). *Madroño* 39: 155–157.

Morefield, J. D. 2006a. 105. *Hesperevax* (A. Gray) A. Gray. pp. 467–470. *In*: Flora of North America Editorial Committee (eds.), *Flora of North America North of Mexico, Vol. 19: Magnoliophyta: Asteridae, Part 8*. Oxford University Press, New York.

Morefield, J. D. 2006b. 101. *Psilocarphus* Nuttall. pp. 456–459. *In*: Flora of North America Editorial Committee (eds.), *Flora of North America North of Mexico, Vol. 19: Magnoliophyta: Asteridae (in part): Asteraceae, Part 3*. Oxford University Press, New York.

Morefield, J. D. 2012. *Hesperevax*. pp. 347–349. *In*: B. G. Baldwin, D. H. Goldman, D. J. Keil, R. Patterson, T. J. Rosatti and D. H. Wilken (eds.), *The Jepson Manual (2nd ed.)*. University of California Press, Berkeley, CA.

Morgan, D. R. 2003. nrDNA external transcribed spacer (ETS) sequence data, reticulate evolution, and the systematics of *Machaeranthera* (Asteraceae). *Syst. Bot.* 28: 179–190.

Morgan, D. R. & R. L. Hartman. 2003. A synopsis of *Machaeranthera* (Asteraceae: Astereae), with recognition of segregate genera. *Sida* 20: 1837–1416.

Morgan, D. R. & B. Holland. 2012. Systematics of Symphyotrichinae (Asteraceae: Astereae): disagreements between two nuclear regions suggest a complex evolutionary history. *Syst. Bot.* 37: 818–832.

Morgan, D. R. & B. B. Simpson. 1992. A systematic study of *Machaeranthera* (Asteraceae) and related groups using restriction site analysis of chloroplast DNA. *Syst. Bot.* 17: 511–531.

Morgan, D. R., D. E. Soltis & K. R. Robertson. 1994. Systematic and evolutionary implications of *rbc*L sequence variation in Rosaceae. *Amer. J. Bot.* 81: 890–903.

Morgan, J. M. A. 2008. Comparison of environmental substrate gradients and calcium selectivity in plant species of calcareous fens in Massachusetts. MS thesis. University of Massachusetts, Amherst, MA. 198 pp.

Morgan, S. 1983. *Lindera melissifolium:* A rare southeastern shrub. *Nat. Areas J.* 3: 62–67.

Morhardt, S. & E. Morhardt. 2004. *California Desert Flowers: An Introduction to Families, Genera, and Species.* University of California Press, Berkeley, CA. 295 pp.

Morin, N. R. 2005. 45. Plumbaginaceae Jussieu. Leadwort family. pp. 602–603. *In*: Flora North America Editorial Committee (eds.), *Flora of North America North of Mexico*, Vol. 5: Magnoliophyta: Caryophyllidae, part 2. Oxford University Press, New York.

Morin, N. R. 2009. 3. Grossulariaceae de Candolle. pp. 8–42. *In*: N. R. Morin (convening ed.), *Flora of North America North of Mexico, Vol. 8: Magnoliophyta: Paeoniaceae to Ericaceae.* Oxford University Press, New York.

Morin, N. R. 2010. 2. *Limnanthes* R. Brown. pp. 179–183. *In*: Flora of North America Editorial Committee (eds.), *Flora of North America North of Mexico: Vol. 7, Magnoliophyta: Salicaceae to Brassicaceae.* Oxford University Press, New York.

Morin, N. R. 2012a. Campanulaceae. *Hesperevax*. pp. 588–598. *In:* B. G. Baldwin, D. H. Goldman, D. J. Keil, R. Patterson, T. J. Rosatti and D. H. Wilken (eds.), *The Jepson Manual (2nd ed.).* University of California Press, Berkeley, CA.

Morin, N. R. 2012b. Campanulaceae. *Howellia.* pp. 594 *in* B. G. Baldwin, D. H. Goldman, D. J. Keil, R. Patterson, T. J. Rosatti and D. H. Wilken (eds.), *The Jepson Manual (2nd ed.).* University of California Press, Berkeley, CA.

Morin, N. R. 2012c. Campanulaceae. *Legenere.* pp. 594 *in* B. G. Baldwin, D. H. Goldman, D. J. Keil, R. Patterson, T. J. Rosatti and D. H. Wilken (eds.), *The Jepson Manual (2nd ed.).* University of California Press, Berkeley, CA.

Morin, N. R. 2012d. Campanulaceae. *Porterella.* pp. 598 *in* B. G. Baldwin, D. H. Goldman, D. J. Keil, R. Patterson, T. J. Rosatti and D. H. Wilken (eds.), *The Jepson Manual (2nd ed.).* University of California Press, Berkeley, CA.

Mørk, E. K., K. Henriksen, H. Brinch-Pedersen, K. Kristiansen & K. K. Petersen. 2012. An efficient protocol for regeneration and transformation of *Symphyotrichum novi-belgii. Plant Cell Tiss. Organ Cult.* 108: 501–512.

Morozowska, M., A. Czarna, I. Jędrzejczyk & J. Bocianowski. 2015. Genome size, leaf, fruit and seed traits–taxonomic tools for species identification in the genus *Nasturtium* R. Br. *Acta Biol. Cracov., Ser. Bot.* 57: 114–124.

Morris, A. B., C. D. Bell, J. W. Clayton, W. S. Judd, D. E. Soltis & P. S. Soltis. 2007. Phylogeny and divergence time estimation in *Illicium* with implications for New World biogeography. *Syst. Bot.* 32: 236–249.

Morris, J. A. 2007. A molecular phylogeny of the Lythraceae and inference of the evolution of heterostyly. PhD dissertation. Kent State University. Kent, OH. 107 pp.

Morris, J. A., S. A. Graham & A. E. Schwarzbach. 2005a. Placement of the monotypic genus *Didiplis* within the Lythraceae. Botany 2005 meeting (abstract).

Morris, J. A., S. A. Graham & A. E. Schwarzbach. 2005b. A preliminary phylogeny of the genus *Lythrum.* Botany 2005 meeting (abstract).

Morris, J. B., S. M. Yang & L. Wilson. 1983. Reaction of *Helianthus* species to *Alternaria helianthi. Plant Dis.* 67: 539–540.

Morris, J. C. 2007. The vascular flora of Boiling Spring Lakes Preserve, Brunswick County, North Carolina. MS thesis. Department of Biology and Marine Biology, University of North Carolina Wilmington, Wilmington, NC. 66 pp.

Morris, M.W. 2013. The genus *Platanthera* (Orchidaceae) in Mississippi. *J. Bot. Res. Inst. Texas* 7: 323–339.

Morrison, B. Y. 1929. Azaleas and Rhododendrons from seed. *Circ. U.S.D.A.* 68: 1–8.

Morrison, P. H. & H. M. Smith IV. 2007. Rare Plant and Vegetation Survey of Bottle Beach, Grayland Beach, Twin Harbors, Westhaven and Westport Light State Parks. Pacific Biodiversity Institute, Winthrop, WA. 149 pp.

Morrison, W. E. & M. E. Hay. 2011. Feeding and growth of native, invasive and non-invasive alien apple snails (Ampullariidae) in the United States: invasives eat more and grow more. *Biol. Invas.* 13: 945–955.

Mors, W. B., J. P. Parente, M. H. da Silva, P. A. Melo & G. Suarez-Kurtz. 1989. Neutralization of lethal and myotoxic activities of South American rattlesnake venom by extracts and constituents of the plant *Eclipta prostrata* (Asteraceae). *Toxicon* 27: 1003–1009.

Morse, D. H. 2005. Initial responses to substrates by naïve spiderlings: single and simultaneous choices. *Anim. Behav.* 70: 319–328.

Morse, J. C. & D. R. Lenat. 2005. A new species of *Ceraclea* (Trichoptera: Leptoceridae) preying on snails. *J. N. Amer. Benthol. Soc.* 24: 872–879.

Morse, K. A. 2008. Vascular plant inventory of Deer Creek Center property in Selma, Oregon. MS thesis. Department of Biology, Southern Oregon University, Ashland, OR. 124 pp.

Mort, M. E., D. E. Soltis, P. S. Soltis, J. Francisco-Ortega & A. Santos-Guerra. 2001. Phylogenetic relationships and evolution of Crassulaceae inferred from *matK* sequence data. *Amer. J. Bot.* 88: 76–91.

Mort, M. E., C. P. Randle, R. T. Kimball, M. Tadesse & D. J. Crawford. 2008. Phylogeny of Coreopsideae (Asteraceae) inferred from nuclear and plastid DNA sequences. *Taxon* 57: 109–120.

Mort, M. E., C. P. Randle, P. Burgoyne, G. Smith, E. Jaarsveld & S. D. Hopper. 2009. Analyses of cpDNA *matK* sequence data place *Tillaea* (Crassulaceae) within *Crassula. Plant Syst. Evol.* 283: 211–217.

Morteau, B., G. Triffault-Bouchet, L. Martel, R. Galvez & S. Leroueil. 2009. Treatment of salted road runoffs using *Typha latifolia, Spergularia canadensis*, and *Atriplex patula*: A comparison of their salt removal potential. *J. ASTM Int.* 6: 1–7.

Morton, C. M. 2011. Newly sequenced nuclear gene (*Xdh*) for inferring angiosperm phylogeny. *Ann. Missouri Bot. Gard.* 98: 63–89.

Morton, J. F. 1964. Honeybee plants of south Florida. *Proc. Florida State Hort. Soc.* 77: 415–436.

Morton, J. F. & G. H. Snyder. 1978. Trial of water celery as an aquatic flavoring herb for Everglades farmlands. *Proc. Florida State Hort. Soc.* 91: 301–305.

Morton, J. K. 1979. Obervations on Houghton's goldenrod (*Solidago houghtonii*). *Michigan Bot.* 18: 31–35.

Morton, J. K. 2005. 22. *Stellaria* Linnaeus. pp. 96–114. *In*: Flora North America Editorial Committee (eds.), *Flora of North America North of Mexico*, Vol. 5: Magnoliophyta: Caryophyllidae, part 2. Oxford University Press, New York.

Morzaria-Luna, H. N. & J. B. Zedler. 2007. Does seed availability limit plant establishment during salt marsh restoration? *Estuaries Coasts* 30: 12–25.

Moseley, R. K. 1991. A field investigation of park milkvetch (*Astragalus leptaleus*) in Idaho. Idaho Department of Fish and Game, Boise, ID. 15 pp.

Moseley, R. K. 1992. *Ecological and Floristic Inventory of Birch Creek fen, Lemhi and Clark Counties, Idaho.* Idaho Department of Fish and Game, Boise, ID. 29 pp.

Moseley, R. K. 1995. *The Ecology of Geothermal Springs in South-Central Idaho*. Agreement No. 0414-CCS-95-010. Conservation Data Center, Idaho Department of Fish and Game, Boise, ID. 94 pp.

Moseley, R. K. & A. Pitner. 1996. *Rare Bryophytes and Lichens in Idaho: Status of Our Knowledge*. Conservation Data Center, Idaho Department of Fish and Game, Boise, ID. 50 pp.

Moser, B. R., B. S. Dien, D. M. Seliskar & J. L. Gallagher. 2013. Seashore mallow (*Kosteletzkya pentacarpos*) as a salt-tolerant feedstock for production of biodiesel and ethanol. *Renew. Energy* 50: 833–839.

Moss, E. H. 1953. Marsh and bog vegetation in northwestern Alberta. *Can. J. Bot.* 31: 448–470.

Mosseler, A. 1990. Hybrid performance and species crossability relationships in willows (*Salix*) *Can. J. Bot.* 68: 2329–2338.

Mosseler, A. & C. S. Papadopol. 1989. Seasonal isolation as a reproductive barrier among sympatric *Salix* species. *Can. J. Bot.* 67: 2563–2570.

Mossman, R. E. 2009. Seed dispersal and reproduction patterns among Everglades plants. PhD dissertation, Florida International University, Miami, FL. 125 pp.

Mosyakin, S. L. 2003. 16. *Bassia* Allioni. pp. 309–310. *In*: N. R. Morin (convening ed.), *Flora of North America North of Mexico, Vol. 4: Magnoliophyta: Caryophyllidae, Part 1*. Oxford University Press, New York.

Mosyakin, S. L. 2005. 26. *Rumex* Linnaeus. pp. 489–533. *In*: Flora North America Editorial Committee (eds.), *Flora of North America North of Mexico, Vol. 5: Magnoliophyta: Caryophyllidae, part 2*. Oxford University Press, New York.

Mosyakin, S. L. & K. R. Robertson. 2003. 3. *Amaranthus* Linnaeus. pp. 410–435. *In*: N. R. Morin (convening ed.), *Flora of North America North of Mexico, Vol. 4: Magnoliophyta: Caryophyllidae, Part 1*. Oxford University Press, New York.

Motley, T. J. & L. Raz. 2004. Origins, evolution, and systematics of Hawaiian and Pacific *Chamaesyce* (Euphorbiaceae). Botany 2004 meeting (abstract).

Motley, T. J., K. J. Wurdack & P. G. Delprete. 2005. Molecular systematics of the Catesbaeeae-Chiococceae complex (Rubiaceae): flower and fruit evolution and biogeographic implications. *Amer. J. Bot.* 92: 316–329.

Mottl, L. M. C. M. Mabry & D. R. Farrar. 2006. Seven-year survival of perennial herbaceous transplants in temperate woodland restoration. *Restorat. Ecol.* 14: 330–338.

Mouissie, A. M. 2004. Seed dispersal by large herbivores: implications for the restoration of plant biodiversity. Doctoral dissertation, Rijksuniversiteit Groningen, Groningen, The Netherlands. 120 pp.

Moyer, R. D. & E. L. Bridges. 2015. *Xyris chapmanii*, an overlooked *Xyris* species of the New Jersey pine barrens. *Bartonia* 67: 58–74.

Moylan, E. C., J. R. Bennett, M. A. Carine, R. G. Olmstead & R. W. Scotland. 2004. Phylogenetic relationships among *Strobilanthes* s.l. (Acanthaceae): evidence from ITS nrDNA, *trnL-F* cpDNA, and morphology. *Amer. J. Bot.* 91: 724–735.

Moyle, J. B. 1945. Some chemical factors influencing the distribution of aquatic plants in Minnesota. *Amer. Midl. Naturalist* 34: 402–420.

Mueller, M. H. & A. G. van der Valk. 2002. The potential role of ducks in wetland seed dispersal. *Wetlands* 22: 170–178.

Muenscher, W. C. 1936. Storage and germination of seeds of aquatic plants. Agricultural Experiment Station, Cornell University, Ithaca, New York. 17 pp.

Muenscher, W. C. 1940. Fruits and seedlings of *Ceratophyllum*. *Amer. J. Bot.* 27: 231–233.

Muenscher, W. C. 1944. *Aquatic Plants of the United States*. Comstock Publishing Associates, Ithaca, New York. 374 pp.

Muenscher, W. C. 1955. *Weeds. 2nd Ed.* The Macmillan Company, New York. 560 pp.

Mühlberg, H. 1982. *The Complete Guide to Water Plants*. EP Publishing Limited, Leipzig, Germany. 392 pp.

Muir, P. S. & R. K. Moseley. 1994. Responses of *Primula alcalina*, a threatened species of alkaline seeps, to site and grazing. *Nat. Areas J.* 14: 269–279.

Mukherjee, P. K., S. N. Kakali Saha Giri, M. Pal & B. P. Saha. 1995. Antifungal screening of *Nelumbo nucifera* (Nymphaeaceae) rhizome extract. *Indian J. Microb.* 35: 327–330.

Mukherjee, P. K., D. Mukherjee, A. K. Maji, S. Rai & M. Heinrich. 2009. The sacred lotus (*Nelumbo nucifera*)–phytochemical and therapeutic profile. *J. Pharm. Pharmacol.* 61: 407–422.

Mukhin, V. A. & A. A. Betekhtina. 2006. Adaptive significance of endomycorrhizas for herbaceous plants. *Russ. J. Ecol.* 37: 1–6.

Muldavin, E., P. Durkin, M. Bradley, M. Stuever & P. Mehlhop. 2000. *Handbook of Wetland Vegetation Communities of New Mexico. Volume I: Classification and Community Descriptions*. New Mexico Natural Heritage Program, University of New Mexico, Albuquerque, NM. 172 pp.

Mulder, C. P. H., R. W. Ruess & J. S. Sedinger. 1996. Effects of environmental manipulations on *Triglochin palustris*: implications for the role of goose herbivory in controlling its distribution. *J. Ecol.* 84: 267–278.

Mulé, L. P. 1983. The Chicago Ridge Prairie—a floral summary. pp. 112–117. *In*: R. Brewer, (ed.), *Proceedings of the Eighth North American Prairie Conference: 1–4 August 1982*. Western Michigan University, Kalamazoo, MI.

Mulhouse, J. M. 2004. Vegetation change in herbaceous Carolina bays of the upper coastal plain: dynamics during drought. MS thesis. University of Georgia, Athens, GA. 108 pp.

Mulhouse, J. M., L. E. Burbage & R. R. Sharitz. 2005. Seed bank-vegetation relationships in herbaceous Carolina bays: Responses to climatic variability. *Wetlands* 25: 738–747.

Muller, C. H. 1953. The association of desert annuals with shrubs. *Amer. J. Bot.* 40: 53–60.

Muller, J. O. 1979. Fourth addition to the supplemental list of Macrolepidoptera of New Jersey. *J Lepid. Soc.* 33: 174–178.

Müller, J. 2006. Systematics of *Baccharis* (Compositae-Astereae) in Bolivia, including an overview of the genus. *Syst. Bot. Monogr.* 76: 1–341.

Müller, K. & T. Borsch. 2005. Phylogenetics of *Utricularia* (Lentibulariaceae) and molecular evolution of the *trnK* intron in a lineage with high substitution rates. *Plant Syst. Evol.* 250: 39–67.

Müller, K., T. Borsch, L. Legendre, S. Porembski, I. Theisen & W. Barthlott. 2004. Evolution of carnivory in Lentibulariaceae and the Lamiales. *Plant Biol.* 6: 477–490.

Müller, L. 1933. Über den Bau und die Entwicklung des Bewegungsmechanismus von *Physostegia virginiana*. *Planta* 18: 651–663.

Mullet, T. C., F. Armstrong, B. Zank & C. M. Ritzi. 2008. Predicting *Viola guadalupensis* (Violaceae) habitat in the Guadalupe Mountains using GIS: Evidence of a new isolated population. *J. Bot. Res. Inst. Texas* 2: 677–684.

Mulligan, G. A. 1964. Chromosome numbers of the family Cruciferae I. *Can. J. Bot.* 42: 1509–1520.

Mulligan, G. A. 2008. *Common Plants of the Northern United States and Canada Reported to Have Caused Poisonings, A Dermatitis or Hay Fever in Humans*. Canadian Weed Science Society. Available online: http://www.cwss-scm.ca/Weeds/poisonous_weeds.htm (accessed April 8, 2008).

Mulligan, G. A. & J. A. Calder. 1964. The genus *Subularia* (Cruciferae). *Rhodora* 66: 127–135.

Mulligan, G. B. & D. B. Munro. 1981. The biology of Canadian weeds. 48. *Cicuta maculata* L., *C. douglasii* (DC.) Coult. & Rose and *C. virosa* L. *Can. J. Plant Sci.* 61: 93–105.

Mulligan, G. A. & D. B. Munro. 1983. The status of *Stachys palustris* (Labiatae) in North America. *Can. J. Bot.* 61: 679–682.

Mulligan G. A. & D. B. Munro. 1989. Taxonomy of species of North American *Stachys* (Labiatae) found north of Mexico. *Naturaliste Can.* 116: 35–51.

Mulligan, G. A. & A. E. Porsild. 1966. *Rorippa calycina* in the Northwest Territories. *Can. J. Bot.* 44: 1105–1106.

Mullin, S. J. 1995. Estuarine fish populations among red mangrove prop roots of small overwash islands. *Wetlands* 15: 324–329.

Mullin, W. J., S. Peacock, D. C. Loewen & N. J. Turner. 1997. Macronutrients content of Yellow Glacier Lily and Balsamroot; root vegetables used by indigenous peoples of northwestern North America. *Food Res. Int.* 30: 769–775.

Mullins, W. H. and E. G. Bizeau. 1978. Summer foods of sandhill cranes in Idaho. *Auk* 95: 175–178.

Muminović, J., A. E. Melchinger & T. Lübberstedt. 2004. Prospects for celeriac (*Apium graveolens* var. *rapaceum*) improvement by using genetic resources of *Apium*, as determined by AFLP markers and morphological characterization. *Plant Genet. Resour.* 2: 189–198.

Mummenhoff, K., H. Brueggemann & J. L. Bowman. 2001. Chloroplast DNA phylogeny and biogeography of *Lepidium* (Brassicaceae). *Amer. J. Bot.* 88: 2051–2063.

Mummenhoff, K., P. Linder, N. Friesen, J. Bowman, J.-Y. Lee & A. Franzke. 2004. Molecular evidence for bicontinental hybridogenous genomic constitution in *Lepidium* sensu stricto (Brassicaceae) species from Australia and New Zealand. *Amer. J. Bot.* 91: 254–261.

Mummenhoff, K., A. Polster, A. Mühlhausen & G. Theißen. 2009. *Lepidium* as a model system for studying the evolution of fruit development in Brassicaceae. *J. Exp. Bot.* 60: 1503–1513.

Mundry, K. W. & H. Priess. 1971. Structural elements of viral ribonucleic acid and their variation: II.^{32}P-oligonucleotide maps of large G-lacking segments of RNA of tobacco mosaic virus wild strains. *Virology* 46: 86–97.

Munger, G. T. 2002. *Lythrum salicaria. In: Fire Effects Information System*. U. S. Department of Agriculture, Forest Service, Rocky Mountain Research Station, Fire Sciences Laboratory (Producer). Available online: http://www.fs.fed.us/database/feis/ (accessed May 9, 2005).

Munger, G. T. 2005. *Lonicera* spp. *In: Fire Effects Information System*. U. S. Department of Agriculture, Forest Service, Rocky Mountain Research Station, Fire Sciences Laboratory (Producer). Available online: http://www.fs.fed.us/database/feis/ (accessed October 21, 2014).

Munson, E. S. 1992. Influence of nest cover on habitat selection in clay-colored sparrows. *Wilson Bull.* 104: 525–529.

Munz, P. A. 1946. The cultivated and wild columbines. *Gentes Herb.* 7: 1–150.

Munz, P. A. & I. M. Johnston. 1922. Miscellaneous notes on plants of southern California-II. *Bull. Torrey Bot. Club* 49: 349–359.

Munz, P. A. & D. D. Keck. 1973. *A California Flora with Supplement.* University of California Press, Berkeley, CA. 1905 pp.

Murashige, T. 1974. Plant propagation through tissue culture. *Ann. Rev. Plant Physiol.* 25: 135–166.

Murdock, N. & D. Rayner. 1990. Recovery plan for Canby's dropwort (*Oxypolis canbyi* [Coulter and Rose] Fernald). U. S. Fish and Wildlife Service, Atlanta, GA. 40 pp.

Murdock, N. & A. Weakley. 1991. Recovery plan for small-anthered bittercress (*Cardamine micranthera* Rollins). U. S. Fish and Wildlife Service, Atlanta, GA. 22 pp.

Murie, A. 1934. The moose of Isle Royale. *Univ. Michigan Mus. Zool. Misc. Publ.* 25: 1–44.

Murie, A. 1961. Some food habits of the marten. *J. Mammalogy* 42: 516–521.

Murillo, H., D. W. A. Hunt & S. L. VanLaerhoven. 2013. First records of *Chrysodeixis chalcites* (Lepidoptera: Noctuidae: Plusiinae) for east-central Canada. *Can. Entomol.* 145: 338–342.

Murphy, J. E., K. B. Beckmen, J. K. Johnson, R. B. Cope, T. Lawmaster & V. R. Beasley. 2002. Toxic and feeding deterrent effects of native aquatic macrophytes on exotic grass carp (*Ctenopharyngodon idella*). *Ecotoxicology* 11: 243–254.

Murphy, K. J. 2002. Plant communities and plant diversity in softwater lakes of northern Europe. *Aquat. Bot.* 73: 287–324.

Murphy, M. L. 2009. Examination of a California coastal sage scrub seed bank that has been invaded by exotic annual species. *UCR Undergraduate Res. J.* 3: 29–34.

Murphy, M. L. & E. B. Allen. 2009. Examination of a California coastal sage scrub seed bank that has been invaded by exotic annual species. *UCR Undergraduate. Res. J.* 3: 29–34.

Murphy, P. B. & R. S. Boyd. 1999. Population status and habitat characterization of the endangered plant, *Sarracenia rubra* subspecies *alabamensis. Castanea* 64: 101–113.

Murphy, S. M., B. Kessel & L. J. Vining. 1984. Waterfowl populations and limnologic characteristics of taiga ponds. *J. Wildl. Manag.* 48: 1156–1163.

Murray, G., P. C. Boxall & R. W. Wein. 2005. Distribution, abundance, and utilization of wild berries by the Gwich'in people in the Mackenzie River delta region. *Econ. Bot.* 59: 174–184.

Murray, M. J. 1960. The genetic basis for a third ketone group in *Mentha spicata* L. *Genetics* 45: 931–937.

Murray-Gulde, C. L., G. M. Huddleston III, K. V. Garber & J. H. Rodgers Jr. 2005. Contributions of *Schoenoplectus californicus* in a constructed wetland system receiving copper contaminated wastewater. *Water Air Soil Pollut.* 163: 355–378.

Murrill, W. A. 1940a. Ecologic notes on the violets of Alachua County, Florida. *Ecology.* 21: 282–284.

Murrill, W. A. 1940b. Alachua County, Florida, soils and violets. *Ecology* 21: 512–513.

Murza, G. L. & A. R. Davis. 2003. Comparative flower structure of three species of sundew (*Drosera anglica, Drosera linearis,* and *Drosera rotundifolia*) in relation to breeding system. *Can. J. Bot.* 81: 1129–1142.

Murza, G. L., J. R. Heaver & A. R. Davis. 2006. Minor pollinator–prey conflict in the carnivorous plant, *Drosera anglica. Plant Ecol.* 184: 43–52.

Musselman, L. J. 1972. Root parasitism in *Macranthera flammea* and *Tomanthera auriculata* (Scrophulariaceae). *J. Elisha Mitchell Sci. Soc.* 88: 5860.

Musselman, L. J. & W. F. Mann Jr. 1977. Host plants of some Rhinanthoideae (Scrophulariaceae) of eastern North America. *Plant Syst. Evol.* 127: 45–53.

Musselman, L. J. & W. F. Mann Jr. 1979. Haustorial frequency of some root parasites in culture. *New Phytol.* 83: 479–483.

Musselman, L. J. & H. J. Wiggins. 2013. *The Quick Guide to Wild Edible Plants: Easy to Pick, Easy to Prepare.* The Johns Hopkins University Press, Baltimore, MD. 144 pp.

Musselman, L. J., C. S. Harris & W. F. Mann Jr. 1978. *Agalinis purpurea*: a parasitic weed on sycamore, sweetgum, and loblolly pine. *Tree Planters' Notes* 29: 24–25.

Musselman, L. J., R. D. Bray & D. A. Knepper. 1995. *Isoetes ×bruntonii (Isoetes engelmannii × I. hyemalis)*, a new hybrid quillwort from Virginia. *Amer. Fern J.* 86: 8–15.

Musselman, L. J., W. C. Taylor & R. D. Bray. 2001. *Isoetes mattaponica* (Isoetaceae), a new diploid quillwort from freshwater tidal marshes of Virginia. *Novon* 11: 200–204.

Mutikainen, P. & L. F. Delph. 1998. Inbreeding depression in gynodioecious *Lobelia siphilitica*: among-family differences override between-morph differences. *Evolution* 52: 1572–1582.

Myers, J. A., M. Vellend, S. Gardescu & P. L. Marks. 2004. Seed dispersal by white-tailed deer: implications for long-distance dispersal, invasion, and migration of plants in eastern North America. *Oecologia* 139: 35–44.

Mymudes, M. S. 1991. Morphological and genetic variability in *Plantago cordata* Lam. (Plantaginaceae), a rare aquatic plant. MS thesis. University of Wisconsin-Milwaukee, Milwaukee, WI. 83 pp.

Mymudes, M. S. & D. H. Les. 1993. Morphological and genetic variability in *Plantago cordata* (Plantaginaceae), a threatened aquatic plant. *Amer. J. Bot.* 80: 351–359.

Nachlinger, J. L. 1988. *Soil-Vegetation Correlations in Riparian and Emergent Wetlands, Lyon County, Nevada.* U. S. Fish and Wildlife Service Biological Report 88(17). Washington, DC. 39 pp.

Nachtrieb, J. G., M. J. Grodowitz & R. M. Smart. 2011. Impact of invertebrates on three aquatic macrophytes: American pondweed, Illinois pondweed, and Mexican water lily. *Nordic J. Bot.* 11: 179–203.

Naczi, R. F. C., E. M. Soper, F. W. Case Jr. & R. B. Case. 1999. *Sarracenia rosea* (Sarraceniaceae), a new species of pitcher plant from the southeastern United States. *Sida* 18: 1183–1206.

Nadja, D., H. Förther & H. H. Hilger. 2002. A systematic analysis of *Heliotropium, Tournefortia*, and allied taxa of the Heliotropiaceae (Boraginales) based on ITS1 sequences and morphological data. *Amer. J. Bot.* 89: 287–295.

Nagata, K. M. 1971. Hawaiian medicinal plants. *Econ. Bot.* 25: 245–254.

Nagata, K. M. 1985. Early plant introductions in Hawai'i. *Hawaiian J. Hist.* 19: 35–61.

Nagel, J. M. & K. L. Griffin. 2001. Construction cost and invasive potential: comparing *Lythrum salicaria* (Lythraceae) with co-occurring native species along pond banks. *Amer. J. Bot.* 88: 2252–2258.

Nagorsen, D. W. 1987. *Marmota vancouverensis*. Mamm. *Species* 270: 1–5.

Nazaire, M. & L. Hufford. 2012. A broad phylogenetic analysis of Boraginaceae: implications for the relationships of *Mertensia*. *Syst. Bot.* 37: 758–783.

Nazaire, M. & L. Hufford. 2014. Phylogenetic systematics of the genus *Mertensia* (Boraginaceae). *Syst. Bot.* 39: 268–303.

Nahrstedt, A., M. Lechtenberg, A. Brinker, D. S. Seigler & R. Hegnauer. 1993. 4-hydroxymandelonitrile glucosides, dhurrin in *Suckleya suckleyana* and taxiphyllin in *Girgensohnia oppositiflora* (Chenopodiaceae). *Phytochemistry* 33: 847–859.

Naiman, R. J. 1979. Preliminary food studies of *Cyprinodon macularius* and *Cyprinodon nevadensis* (Cyprinodontidae). *Southwest. Naturalist* 24: 538–541.

Nakamura, G. & J. K. Nelson (eds.). 2001. *Illustrated Field Guide to Selected Rare Plants of Northern California.* Publication 3395. Agriculture and Natural Resources, University of California, Oakland, CA. 370 pp.

Nakanishi, H. 2002. Splash seed dispersal by raindrops. *Ecol. Res.* 17: 663–671.

Nakano, C. & I. Washitani. 2003. Variability and specialization of plant–pollinator systems in a northern maritime grassland. *Ecol. Res.* 18: 221–246.

Nakasone, K. K. 2006. *Dendrothele griseocana* (Corticiaceae) and related taxa with hyphal pegs. *Nova Hedwigia* 83: 99–108.

Nakatani, K. & T. Kusanagi. 1991. Effect of photoperiod and temperature on growth characteristics, especially heading or flower bud appearance of upland weeds. *Weed Res. (Tokyo)* 36: 74–81.

Nakayama, H., K. Fukushima, T. Fukuda, J. Yokoyama & S. Kimura. 2014. Molecular phylogeny determined using chloroplast DNA inferred a new phylogenetic relationship of *Rorippa aquatica* (Eaton) EJ Palmer & Steyermark (Brassicaceae)—lake cress. *Amer. J. Plant Sci.* 5: 48–54.

Namestnik, S. A., J. R. Thomas & B. S. Slaughter. 2012. Two recent plant discoveries in Missouri: *Cladium mariscus* subsp. *jamaicense* (Cyperaceae) and *Utricularia minor* (Lentibulariaceae). *Phytoneuron* 2012–92: 1–6.

Naruhashi, N. & Y. Iwatsubo. 1998. Chromosome numbers and distributions of *Oenanthe javanica* (Umbelliferae) in Japan. *J. Phytogeogr. Taxon.* 46: 161–166.

Nass, R. & H. Rimpler. 1996. Distribution of iridoids in different populations of *Physostegia virginiana* and some remarks on iridoids from *Avicennia officinalis* and *Scrophularia ningpoensis*. *Phytochemistry* 41: 489–498.

Natali, A., J.-F. Manen & F. Ehrendorfer. 1995. Phylogeny of the Rubiaceae-Rubioideae, in particular the tribe Rubieae: evidence from a non-coding chloroplast DNA Sequence. *Ann. Missouri Bot. Gard.* 82: 428–439.

National Academy of Sciences. 1976. *Making Aquatic Weeds Useful: Some Perspectives for Developing Countries.* National Academy of Sciences Press, Washington, DC. 174 pp.

NatureServe. 2003. *NatureServe Explorer: An Online Encyclopedia of Life* [web application]. Version 1.8. NatureServe, Arlington, Virginia. Available online: http://www.natureserve.org/explorer.

Naudain, E. H. 1885. *Pinckneya pubens*, Michaux. (Georgia Bark.). *Amer. J. Pharm.* 57: (no pagination).

Nauman, C. E. 1981. A re-examination of *Mikania* Willd. (Compositae) in Florida. *Bull. Torrey Bot. Club* 108: 467–471.

Navarro, E., J. Bousquet, A. Moiroud, A. Munive, D. Piou & P. Normand. 2003. Molecular phylogeny of *Alnus* (Betulaceae), inferred from nuclear ribosomal DNA ITS sequences. *Plant Soil* 254: 207–217.

Nawwar, M. A., N. A. Youb, M. A. El-Raey, S. S. Zaghloul, A. M. Hashem, E. S. Mostafa, O. Eldahshan, V. Werner, A. Becker, B. Haertel, U. Lindequist & M. W. Linscheid. 2014. Polyphenols in *Ammania auriculata*: Structures, antioxidative activity and cytotoxicity. *Pharmazie* 69: 860–864.

Neal, W. A. & C. Norquist. 1992. Alabama canebrake pitcher plant (*Sarracenia rubra* ssp. *alabamensis*) Recovery Plan. U. S. Fish & Wildlife Service, Jackson, MS.

Neck, R. W. 1976. Lepidopteran foodplant records from Texas. *J. Res. Lepid.* 15: 75–82.

Nedorostova, L., P. Kloucek, L. Kokoska, M. Stolcova & J. Pulkrabek. 2009. Antimicrobial properties of selected essential oils in vapour phase against foodborne bacteria. *Food Control* 20: 157–160.

Neel, M. C. 2002. Conservation implications of the reproductive ecology of *Agalinis acuta* (Scrophulariaceae). *Amer. J. Bot.* 89: 972–980.

Neel, M. C. & M. P. Cummings. 2004. Section-level relationships of North American *Agalinis* (Orobanchaceae) based on DNA sequence analysis of three chloroplast gene regions. *BMC Evo Bio* 4: 15. doi:10.1186/1471-2148-4-15.

Neff, K. P. & A. H. Baldwin. 2005. Seed dispersal into wetlands: techniques and results for a restored tidal freshwater marsh. *Wetlands* 25: 392–404.

Neff, K. P. & J. L. Vankat. 1982. Survey of the vegetation and flora of a wetland in Kiser Lake State Park, Champaign County, Ohio. *Ohio J. Sci.* 82: 252–259.

Neff, K. P., K. Rusello & A. H. Baldwin. 2009. Rapid seed bank development in restored tidal freshwater wetlands. *Restorat. Ecol.* 17: 539–548.

Negrón-Ortiz, V. & L. E. Watson. 2003. Hypotheses for the colonization of the Caribbean basin by two genera of the Rubiaceae: *Erithalis* and *Ernodea*. *Syst. Bot.* 28: 442–451.

Neid, S. L. 2006. *Utricularia minor* L. (lesser bladderwort): a technical conservation assessment. USDA Forest Service, Rocky Mountain Region. Available online: http://www.fs.fed.us/r2/projects/scp/assessments/utriculariaminor.pdf (accessed June 7, 2007).

Neiland, B. J. 1958. Forest and adjacent burn in the Tillamook burn Area of northwestern Oregon. *Ecology* 39: 660–671.

Neinhuis, C., S. Wanke, K. W. Hilu, K. Müller & T. Borsch. 2005. Phylogeny of Aristolochiaceae based on parsimony, likelihood, and Bayesian analyses of *trnL-trnF* sequences. *Plant Syst. Evol.* 250: 7–26.

Nekola, J. C. 2004. Vascular plant compositional gradients within and between Iowa fens. *J. Veg. Sci.* 15: 771–780.

Nekola, J. C. & T. G. Lammers. 1989. Vascular flora of Brayton-Horsley prairie: a remnant prairie and spring fen complex in eastern Iowa. *Castanea* 54: 238–254.

Nelson, A. 1912. New plants from Idaho. *Bot. Gaz.* 54: 404–418.

Nelson, A. D., W. J. Elisens & D. Benesh. 1998. Notes on chromosome numbers in *Chelone* (Scrophulariaceae). *Castanea* 63: 183–187.

Nelson, A. D. & W. J. Elisens. 1999. Polyploid evolution and biogeography in *Chelone* (Scrophulariaceae): morphological and isozyme evidence. *Amer. J. Bot.* 86: 1487–1501.

Nelson, A. D., J. R. Goetze, I. G. Negrete, V. E. French, M. P. Johnson & L. M. Macke. 2000. Vegetational analysis and floristics of four communities in the Big Ball Hill region of Padre Island National Seashore. *Southwest. Naturalist* 45: 431–442.

Nelson, A. R. & K. Kashima. 1993. Diatom zonation in southern Oregon tidal marshes relative to vascular plants, Foraminifera, and sea level. *J. Coastal Res.* 9: 673–697.

Nelson, D. C. & R. C. Anderson 1983. Factors related to the distribution of prairie plants along a moisture gradient. *Amer. Midl. Naturalist* 109: 367–375.

Nelson, E. N. & R. W. Couch. 1985. *Aquatic Plants of Oklahoma I: Submersed, Floating-leaved, and Selected Emergent Macrophytes*. Nelson & Couch, Oral Roberts University, Tulsa, OK. 113 pp.

Nelson, G. 2006. Atlantic Coastal Plain Wildflowers: A Guide to Common Wildflowers of the Coastal Regions of Virginia, North Carolina, South Carolina, Georgia, and Northeastern Florida. The Globe Pequot Press, Guilford, CT. 272 pp.

Nelson, J. B. 1986. *The Natural Communities of South Carolina: Initial Classification and Description*. South Carolina Wildlife and Marine Resources Department, Charleston, SC. 64 pp.

Nelson, J. K. & W. E. Harmon. 1993. Discovery of *Subularia aquatica* L. in Colorado and the extension of its range. *Rhodora* 95: 155–157.

Nelson, L. S., A. B. Stewart & K. D. Getsinger. 2002. Fluridone effects on fanwort and water marigold. *J. Aquat. Plant Manag.* 40: 58–63.

Nepokroeff, M., W. L. Wagner, E. A. Zimmer, S. G. Weller, A. K. Sakai & R. K. Rabeler. 2001. Origin of the Hawaiian subfam. *Alsinoideae* and preliminary relationships in Caryophyllaceae inferred from *mat*K and *trn*L C-F sequence data. Botany 2001 meeting (abstract).

Nepokroeff, M., W. L. Wagner, S. G. Weller, P. S. Soltis, E. A. Zimmer, A. K. Sakai & D. E. Soltis. 2002. Origin and diversification of the endemic Hawaiian genus *Schiedea* (Caryophyllaceae subfamily *Alsinoideae*) inferred from combined molecular and morphological data. Botany 2002 meeting (abstract).

Nesom, G. L. 1994. Review of the taxonomy of *Aster* sensu lato (Asteraceae), emphasizing the New World species. *Phytologia* 77: 141–297.

Nesom, G. L. 2000. Generic Conspectus of the Tribe Astereae (Asteraceae) in North America and Central America, the Antilles, and Hawaii. BRIT Press, Ft. Worth, TX. 96 pp.

Nesom, G. L. 2001. An anomalous population of *Aster* (Asteraceae: Astereae) sensu lato in Michigan. *Sida* 19: 625–632.

Nesom, G. L. 2005. Taxonomy of the *Symphyotrichum* (*Aster*) *subulatum* group and *Symphyotrichum* (*Aster*) *tenuifolium* (Asteraceae: Astereae). *Sida* 21: 2125–2140.

Nesom, G. L. 2006a. 411. *Carphephorus* Cassini. pp. 535–538 *In*: Flora of North America Editorial Committee (eds.), *Flora of North America North of Mexico: Vol. 21, Magnoliophyta: Asteridae (in part): Asteraceae, Part 3*. Oxford University Press, New York.

Nesom, G. L. 2006b. 186. *Erigeron* Linnaeus. pp. 256–348. *In*: Flora of North America Editorial Committee (eds.), *Flora of North America North of Mexico: Vol. 20, Magnoliophyta: Asteridae (in part): Asteraceae, Part 3*. Oxford University Press, New York.

Nesom, G. L. 2006c. 92. *Gnaphalium* Linnaeus. pp. 428–430. *In*: Flora of North America Editorial Committee (eds.), *Flora of North America North of Mexico: Vol. 19, Magnoliophyta: Asteridae (in part): Asteraceae, Part 3*. Oxford University Press, New York.

Nesom, G. L. 2006d. 413. *Hartwrightia* A. Gray ex S. Watson. p. 540. *In*: Flora of North America Editorial Committee (eds.), *Flora of North America North of Mexico: Vol. 21, Magnoliophyta: Asteridae (in part): Asteraceae, Part 3*. Oxford University Press, New York.

Nesom, G. L. 2006e. 111. *Pluchea* Cassini. pp. 478–483. *In*: Flora of North America Editorial Committee (eds.), *Flora of North America North of Mexico: Vol. 19, Magnoliophyta: Asteridae (in part): Asteraceae, Part 3*. Oxford University Press, New York.

Nesom, G. L. 2006f. 400. *Shinnersia* R. M. King & H. Robinson. p. 488. *In*: Flora of North America Editorial Committee (eds.), *Flora of North America North of Mexico: Vol. 21, Magnoliophyta: Asteridae (in part): Asteraceae, Part 3*. Oxford University Press, New York.

Nesom, G. L. 2008. Classification of subtribe Conyzinae (Asteraceae: Astereae). *Lundellia* 11: 8–38.

Nesom, G. L. 2012. Infrageneric classification of *Rhexia* (Melastomataceae). *Phytoneuron* 2012–15: 1–9.

Nesom, G. L. 2014. Phylogeny of *Fraxinus* sect. *Melioides* (Oleaceae): Review and an alternative hypothesis. *Phytoneuron* 2014-95: 1–9.

Nesom, G. L. & R. J. O'Kennon. 2008. Major plant communities of Lake Meredith National Recreation Area and Alibates Flint Quarries National Monument. *Phytologia* 90: 391–405.

Netherland, M. D. 2013. Biology and management of invasive aquatic plants in the southeastern United States. pp. 19–27. *In*: R. Johansen, L. D. Estes, S. W. Hamilton & A. N. Barrass (eds.), *Proceedings of the 14th Symposium on the Natural History of Lower Tennessee and Cumberland River Valleys.* The Center of Excellence for Field Biology, Austin Peay State University, Clarksville, TN.

Nettel, A., F. Rafii & R. S. Dodd. 2005. Characterization of microsatellite markers for the mangrove tree *Avicennia germinans* L. (Avicenniaceae). *Mol. Ecol. Notes* 5: 103–105.

Neubig, K. M., O. J. Blanchard Jr., W. M. Whitten & S. F. McDaniel. 2015. Molecular phylogenetics of *Kosteletzkya* (Malvaceae, Hibisceae) reveals multiple independent and successive polyploid speciation events. *Bot. J. Linn. Soc.* 179: 421–435.

Newberry, J. S. 1887. *Food and Fiber Plants of the North American Indians.* D. Appleton and Company, New York. 16 pp.

Newcombe, F. C. 1922. Significance of the behavior of sensitive stigmas. *Amer. J. Bot.* 9: 99–120.

Newell, D. L. & A. B. Morris. 2010. Clonal structure of wild populations and origins of horticultural stocks of *Illicium parviflorum* (Illiciaceae). *Amer. J. Bot.* 97: 1574–1578.

Newell, S. J. & A. J. Nastase. 1998. Efficiency of insect capture by *Sarracenia purpurea* (Sarraceniaceae), the northern pitcher plant. *Amer. J. Bot.* 85: 88–91.

Newman, E. A., M. E. Harte, N. Lowell, M. Wilber & J. Harte. 2014. Empirical tests of within-and across-species energetics in a diverse plant community. *Ecology* 95: 2815–2825.

Newman, L. F. 1948. Some notes on the pharmacology and therapeutic value of folk-medicines, II. *Folklore* 59: 145–156.

Newman, R. M., Z. Hanscom & W. C. Kerfoot. 1992. The watercress glucosinolate myrosinase system: a feeding deterrent to caddisflies, snails and amphipods. *Oecologia* 92: 1–7.

Newmaster, S. G., A. G. Harris & L. J. Kershaw. 1997. *Wetland Plants of Ontario.* Lone Pine Publishing, Redmond, WA. 240 pp.

Newmaster, S., A. Fazekas, R. Subramanyam, R. Steeves, C. LaCroix & J. Maloles. 2012. *Population Genetics of Symphyotrichum praealtum (Poir.) G.L. Nesom (Synonym = Aster praealtus Poir.) in Southern Ontario.* Unpublished report to OMNR and manuscript in press. 37 pp.

Neyland, R. 2007. The effects of Hurricane Rita on the aquatic vascular flora in a large fresh-water marsh in Cameron Parish, Louisiana. *Castanea* 72: 1–7.

Neyland, R. & M. Merchant. 2006. Systematic relationships of Sarraceniaceae inferred from nuclear ribosomal DNA sequences. *Madroño* 53: 223–232.

Neyland, R. & H. A. Meyer. 1997. Species diversity of Louisiana chenier woody vegetation remnants. *J. Torrey Bot. Soc.* 124: 254–261.

Neyland, R., M. Hennigan & L. E. Urbatsch. 2004. A vascular flora survey of emergent creek bed microhabitats of Kisatchie Bayou tributaries in Natchitoches Parish, Louisiana. *Sida* 21: 1141–1147.

Ngai, J. T. & R. L. Jefferies. 2004. Nutrient limitation of plant growth and forage quality in Arctic coastal marshes. *J. Ecol.* 92: 1001–1010.

Nguyen, A. Q. 2006. *Into the Vietnamese Kitchen: Treasured Foodways, Modern Flavors.* Ten Speed Press, Berkeley, CA. 344 pp.

Nguyen, L. P., J. Hamr & G. H. Parker. 2004. Wild Turkey, *Meleagris gallopavo silvestris,* behavior in central Ontario during winter. *Can. Field-Naturalist* 118: 251–255.

Nichols, G. E. 1915. The vegetation of Connecticut. IV. Plant societies in lowlands. *Bull. Torrey Bot. Club* 42:169–217.

Nichols, G. E. 1920. The vegetation of Connecticut. VII. The associations of depositing areas along the seacoast. *Bull. Torrey Bot. Club* 47: 511–548.

Nichols, G. E. 1934. The influence of exposure to winter temperatures upon seed germination in various native American plants. *Ecology* 15: 364–373.

Nichols, H. C. 1931. Seed germination in "*Utricularia gibba*", "*U. cleistogama*", and "*U. geminiscapa.*" MA thesis. George Washington University, Washington, DC.

Nichols, S. A. 1975. The impact of overwinter drawdown on the aquatic vegetation of the Chippewa flowage, Wisconsin. *Trans. Wisconsin Acad. Sci.* 63: 176–186.

Nichols, S. A. 1999a. *Distribution and Habitat Descriptions of Wisconsin Lake Plants.* Bulletin 96. Wisconsin Geological & Natural History Survey, Madison, WI. 268 pp.

Nichols, S. A. 1999b. Floristic quality assessment of Wisconsin lake plant communities with example applications. *Lake Reservoir Manag.* 15: 133–141.

Nichols, S. A. & L. A. J. Buchan. 1997. Use of native macrophytes as indicators of suitable Eurasian watermilfoil habitat in Wisconsin lakes. *J. Aquat. Plant Manag.* 35: 21–24.

Nichols, W. F. & V. C. Nichols. 2008. The land use history, flora, and natural communities of the Isles Of Shoals, Rye, New Hampshire and Kittery, Maine. *Rhodora* 110: 245–295.

Nichols, W. F. & D. Sperduto. 2012. A circumneutral patterned fen in northern New Hampshire. *Rhodora* 114: 202–208.

Nichols, W. F., J. M. Hoy & D. Sperduto. 2001. *Open Riparian Communities and Riparian Complexes in New Hampshire.* New Hampshire Natural Heritage Inventory, DRED Division of Forests & Lands and The Nature Conservancy, Concord, NH. 101 pp.

Nichols, W. F., G. E. Moore, N. P. Ritter & C. R. Peter. 2013. A globally rare coastal salt pond marsh system at Odiorne Point State Park, Rye, New Hampshire. *Rhodora* 115: 1–27.

Nicholson, B. J. 1995. The wetlands of Elk Island National Park: Vegetation classification, water chemistry, and hydrotopographic relationships. *Wetlands* 15: 119–133.

Nickell, L. G. 1959. Antimicrobial activity of vascular plants. *Econ. Bot.* 13: 281–318.

Nickell, W. P. 1951. Studies of habitats, territory, and nests of the eastern goldfinch. *Auk* 68: 447–470.

Nickerson, J. & G. Drouin. 2004. The sequence of the largest subunit of RNA polymerase II is a useful marker for inferring seed plant phylogeny. *Mol. Phylogen. Evol.* 31: 403–415.

Nico, L. G. & A. M. Muench. 2004. Nests and nest habitats of the invasive catfish *Hoplosternum littorale* in Lake Tohopekaliga, Florida: A novel association with non-native *Hydrilla verticillata.* *S. E. Naturalist* 3: 451–466.

Nicol, J., S. Muston, P. D'Santos, B. McCarthy & S. Zukowski. 2007. Impact of sheep grazing on the soil seed bank of a managed ephemeral wetland: implications for management. *Austral. J. Bot.* 55: 103–109.

Nicolas, A. N. & G. M. Plunkett. 2009. The demise of subfamily Hydrocotyloideae (Apiaceae) and the re-alignment of its genera across the entire order Apiales. *Mol. Phylogen. Evol.* 53: 134–151.

Nie, Z. L., J. Wen & H. Sun. 2007. Phylogeny and biogeography of *Sassafras* (Lauraceae) disjunct between eastern Asia and eastern North America. *Plant Syst. Evol.* 267: 191–203.

Nie, Z. L., J. Wen, H. Azuma, Y. L. Qiu, H. Sun, Y. Meng, W. B. Sun & E. A. Zimmer. 2008. Phylogenetic and biogeographic complexity of Magnoliaceae in the Northern Hemisphere inferred from three nuclear data sets. *Mol. Phylogen. Evol.* 48: 1027–1040.

Nielsen, D. W. 1990. Arthropod communities associated with *Darlingtonia californica.* *Ann. Entomol. Soc. Amer.* 83: 189–200.

Nielsen, E. L. & J. B. Moyle. 1941. Forest invasion and succession on the basins of two catastrophically drained lakes in northern Minnesota. *Amer. Midl. Naturalist* 25: 564–579.

Nielsen, K. B., R. Kjøller, P. A. Olsson, P. F. Schweiger, F. Ø. Andersen & S. Rosendahl. 2004. Colonization intensity and molecular diversity of arbuscular mycorrhizal fungi in the aquatic plants *Littorella uniflora* and *Lobelia dortmanna* in Southern Sweden. *Mycol. Res.* 108: 616–625.

Nielsen, S. L. & K. Sand-Jensen. 1989. Regulation of photosynthetic rates of submerged rooted macrophytes. *Oecologia* 81: 364–368.

Nielsen, S. L., E. Gacia & K. Sand-Jensen. 1991. Land plants of amphibious *Littorella uniflora* (L.) Aschers. maintain utilization of CO_2 from the sediment. *Oecologia* 88: 258–262.

Niering, W. A. & R. H. Goodwin. 1973. *Inland Wetland Plants of Connecticut.* Bulletin No 19, The Connecticut Arboretum, Connecticut College, New London, CT. 24 pp.

Nieuwland, J. A. 1914. Notes on cleistogamous flowers of violets. II. *Amer. Midl. Naturalist* 3: 198–200.

Nieuwland, J. A. 1916. Habits of waterlily seedlings. *Amer. Midl. Naturalist* 4: 291–297.

Nijland, W., R. de Jong, S. M. de Jong, M. A. Wulder, C. W. Bater & N. C. Coops. 2014. Monitoring plant condition and phenology using infrared sensitive consumer grade digital cameras. *Agric. Forest Meteorol.* 184: 98–106.

Nikolin, E. G. & V. V. Petrovskii. 1988. *Saxifraga lyallii* (Saxifragaceae): novyi vid dlya flory SSSR. *Bot. Zhurn.* 73: 1026–1027.

Nikolova, M., L. Evstatieva & T. D. Nguyen. 2011. Screening of plant extracts for antioxidant properties. *Bot. Serbica* 35: 43–48.

Nilsson, S. G. & I. N. Nilsson. 1978. Species richness and dispersal of vascular plants to islands in Lake Mockeln, southern Sweden. *Ecology* 59: 473–480.

Nishihiro, J., S. Miyawaki, N. Fujiwara & I. Washitani. 2004. Regeneration failure of lakeshore plants under an artificially altered water regime. *Ecol. Res.* 19: 613–623.

Nishihiro, J., M. A. Nishihiro & I. Washitani. 2006. Assessing the potential for recovery of lakeshore vegetation: species richness of sediment propagule banks. *Ecol. Res.* 21: 436–445.

Nitao, J. K. 1987. Test for toxicity of coniine to a polyphagous herbivore, *Heliothis zea* (Lepidoptera: Noctuidae). *Environ. Entomol.* 16: 656–659.

Nitao, J. K., M. G. Nair, D. L. Thorogood, K. S. Johnson & J. M. Scriber. 1991. Bioactive neolignans from the leaves of *Magnolia virginiana. Phytochemistry* 30: 2193–2195.

Nitao, J. K., K. S. Johnson, J. M. Scriber & M. G. Nair. 1992. *Magnolia virginiana* neolignan compounds as chemical barriers to swallowtail butterfly host use. *J. Chem. Ecol.* 18: 1661–1671.

Nixon, C. M. & J. Ely. 1969. Foods eaten by a beaver colony in southeast Ohio. *Ohio J. Sci.* 69: 313–319.

Nixon, E. S. & J. R. Ward. 1981. Distribution of *Schoenolirion wrightii* (Liliaceae) and *Bartonia texana* (Gentianaceae). *Sida* 9: 64–69.

Nixon. K. C. 1997. 5. *Quercus* Linnaeus. pp. 445–506. *In:* N. R. Morin (convening ed.), *Flora of North America North of Mexico, Vol. 3: Magnoliophyta: Magnoliidae and Hamamelidae.* Oxford University Press, New York.

Noe, G. B. & J. B. Zedler. 2000. Differential effects of four abiotic factors on the germination of salt marsh annuals. *Amer. J. Bot.* 87: 1679–1692.

Noe, G. B. & J. B. Zedler. 2001. Spatio-temporal variation of salt marsh seedling establishment in relation to the abiotic and biotic environment. *J. Veg. Sci.* 12: 61–74.

Noe, G. B., L. J. Scinto, J. Taylor, D. L. Childers & R. D. Jones. 2003. Phosphorus cycling and partitioning in an oligotrophic Everglades wetland ecosystem: a radioisotope tracing study. *Freshwater Biol.* 48: 1993–2008.

Noel, F., M. C. Boisselier-Dubayle, J. Lambourdiere, N. Machon, J. Moret & S. Samadi. 2005. Characterization of seven polymorphic microsatellites for the study of two Ranunculaceae: *Ranunculus nodiflorus* L., a rare endangered species and *Ranunculus flammula* L., a common closely related species. *Mol. Ecol. Notes* 5: 827–829.

Nolfo-Clements, L. E. 2006. Vegetative survey of wetland habitats at Jean Lafitte National Historical Park and Preserve in southeastern Louisiana. *S. E. Naturalist (Steuben)* 5: 499–514.

Nolin, D. B. & J. R. Runkle. 1985. Prairies and fens of Bath Township, Greene County, Ohio: 1802 and 1984. *Ohio J. Sci.* 85: 125–130.

Norcini, J. 2010. *Wildflower Survey, 2010: Big Bend.* Final Report, October 25, 2010. Available online: http://flawildflowers.org/resources/pdfs/pdf10/BigBendWfSurvey-2010-FinalRpt-low-rez.pdf (accessed August 6, 2014).

Norcini, J. & G. Nelson. 2010. *Wildflower Survey, 2010: Panhandle.* Final Report, November 22, 2010. Available online: http://www.flawildflowers.org/resources/pdfs/FWF%20Research/PanhandleWfSurvey-2010-FinalRpt-200DPI.pdf (accessed September 25, 2014).

Norcini, J. G. 2002. *Common Native Wildflowers of North Florida.* Circular 1246. Environmental Horticulture Department, Florida Cooperative Extension Service, Institute of Food and Agricultural Sciences, University of Florida, Gainesville, FL. 10 pp.

Norcini, J. G. & J. H. Aldrich. 2007. Storage effects on dormancy and germination of native tickseed species. *HortTechnol.* 17: 505–512.

Norcini, J. G. & J. H. Aldrich. 2008. Assessing viability of prevariety germplasm of native *Coreopsis* species. *HortScience* 43: 1870–1881.

Norcini, J. G., J. H. Aldrich & G. Allbritton. 2012. *Native Wildflowers: Container Production of Joe-Pye Weed from Seed.* 3 pp. Document ENH876, Florida Cooperative Extension Service, Institute of Food and Agricultural Sciences, University of Florida, Gainesville, FL.

Nordman, C. 2004. Vascular Plant Community Classification for Stones River National Battlefield. NatureServe, Durham, NC. 157 pp.

North, D. S. & R. B. Nelson. 1985. Anticholinergic agents in cicutoxin poisoning. *W. J. Med.* 143: 250.

Norton, H. H. 1981. Plant use in Kaigani Haida culture: correction of an ethnohistorical oversight. *Econ. Bot.* 35: 434–449.

Nosál'ová, G., A. Kardosová & S. Franová. 2000. Antitussive activity of a glucuronoxylan from *Rudbeckia fulgida* compared to the potency of two polysaccharide complexes from the same herb. *Die Pharm.* 55: 65–68.

Noss, R. F. 2013. *Forgotten Grasslands of the South: Natural History and Conservation.* Island Press, Washington, DC. 320 pp.

Noureddin, M. I., T. Furumoto, Y. Ishida & H. Fukui. 2004. Absorption and metabolism of bisphenol A, a possible endocrine disruptor, in the aquatic edible plant, water convolvulus (*Ipomoea aquatica*). *Biosci. Biotech. Biochem.* 68: 1398–1402.

Novak, J. A. & B. A. Foote. 1968. Biology and immature stages of fruit flies: *Paroxyna albiceps* (Diptera: Tephritidae). *J. Kansas Entomol. Soc.* 41: 108–119.

Nowak, D. J. & T. D. Sydnor. 1992. *Popularity of Tree Species and Cultivars in the United States*. General Technical Report NE-166. Forest Service, Northeastern Forest Experiment Station, U. S. Department of Agriculture, Radnor, PA. 44 pp.

Nowak, J. 2003. The effect of nutrient solution concentration and growth retardants on growth and flowering of bidens (*Bidens aurea* (Ait.) Scheriff). *Zesz. Probl. Postepow Nauk Roln.* 491: 181–185.

Nowicke, J. W., S. G. Shetler & N. Morin. 1992. Exine structure of pantoporate *Campanula* (Campanulaceae) species. *Ann. Missouri Bot. Gard.* 79: 65–80.

Noyce, K. V. & P. L. Coy. 1990. Abundance and productivity of bear food species in different forest types of northcentral Minnesota. *Int. Conf. Bear Res. Manag.* 8: 169–181.

Noyes, R. D. 2000a. Diplospory and parthenogenesis in sexual × agamospermous (apomictic) *Erigeron* (Asteraceae) hybrids. *Int. J. Plant Sci.* 161: 1–12.

Noyes, R. D. 2000b. Biogeographical and evolutionary insights on *Erigeron* and allies (Asteraceae) from ITS sequence data. *Plant Syst. Evol.* 220: 93–114.

Noyes, R. D. 2007. Apomixis in the Asteraceae: diamonds in the rough. *Funct. Plant Sci. Biotechnol.* 1: 207–222.

Noyes, R. D. & J. R. Allison. 2005. Cytology, ovule development, and pollen quality in sexual *Erigeron strigosus* (Asteraceae). *Int. J. Plant Sci.* 166: 49–59.

Noyes, R. D. & L. H. Rieseberg. 1999. ITS sequence data support a single origin for North American Astereae (Asteraceae) and reflect deep geographical divisions in *Aster* s.l. *Amer. J. Bot.* 86: 398–412.

Noyes, R. D., D. E. Soltis & P. S. Soltis. 1995. Genetic and cytological investigations in sexual *Erigeron compositus* (Asteraceae). *Syst. Bot.* 20: 132–146.

Nürk, N. M., S. Madriñán, M. A. Carine, M. W. Chase & F. R. Blattner. 2013. Molecular phylogenetics and morphological evolution of St. John's wort (*Hypericum*; Hypericaceae). *Mol. Phylogen. Evol.* 66: 1–16.

Nuzzo, V. 1978. Propagation and planting of prairie forbs and grasses in southern Wisconsin. pp. 182–189. *In*: D. C. Glenn-Lewin & R. Q. Landers Jr. (eds.), *Fifth Midwest Prairie Conference Proceedings: Iowa State University, Ames, August 22–24, 1976*. Iowa State University, Ames, IA.

Nybom, H., G. Werlemark, D. G. Esselink & B. Vosman. 2005. Sexual preferences linked to rose taxonomy and cytology. *Acta Hort.* 690: 21–28.

Nyffeler, R. & U. Eggli. 2010. Disintegrating Portulacaceae: a new familial classification of the suborder Portulacineae (Caryophyllales) based on molecular and morphological data. *Taxon* 59: 227–240.

Nyman, J. A. 2011. Integrating successional ecology and the delta lobe cycle in wetland research and restoration. *Estuaries Coasts* 37: 1490. doi:10.1007/s12237-013-9747-4.

NYNHP, 2013. NYNHP conservation guide-marsh voalerian (*Valeriana uliginosa*). New York Natural Heritage Program, Albany, NY. 6 pp.

Nyoka, S. E. & C. Ferguson. 1999. Pollinators of *Darlingtonia californica* Torr., the California pitcher plant. *Nat. Areas J.* 19: 386–391.

Oberholser, H. C. 1925. The relations of vegetation to bird life in Texas. *Amer. Midl. Naturalist* 9: 564–594.

O'Brien, C. W. 1981. The larger (4.5⁺ mm.) *Listronotus* of America, north of Mexico (Cylindrorhininae, Curculionidae, Coleoptera). *Trans. Amer. Entomol. Soc.* 107: 69–123.

O'Brien, E. L. & J. B. Zedler. 2006. Accelerating the restoration of vegetation in a southern California salt marsh. *Wetlands Ecol. Manag.* 14: 269–286.

O'Brien, M. H. 1980. The pollination biology of a pavement plain: pollinator visitation patterns. *Oecologia* 47: 213–218.

Odion, D. C. 2000. Seed banks of long-unburned stands of maritime chaparral: composition, germination behavior, and survival with fire. *Madroño* 47: 195–203.

Odion, D. C., T. L. Dudley & C. M. D'Antonio. 1988. Cattle grazing in southeastern Sierran meadows: ecosystem change and prospects for recovery. pp. 277–292. *In*: C. A. Hall and V. Doyle-Jones (eds.), *Plant Biology of Eastern California*. Mary DeDecker Symposium, White Mountain Research Station, University of California, Los Angeles, CA.

Odion, D. C., R. M. Callaway, W. R. Ferren Jr. & F. W. Davis. 1992. Vegetation of Fish Slough, an Owens Valley wetland ecosystem. pp. 173–197. *In*: C. A. Hall, V. Doyle-Jones & B. Widawski (eds.), *History of Water: Eastern Sierra Nevada, Owens Valley, White-Inyo Mountains*. University of California Press, Berkeley, CA.

Odum, E. P. 1942. Annual cycle of the black-capped chickadee. *Auk* 59: 499–531.

Offord, H. R., C. R. Quick & V. D. Moss. 1944. Self-incompatibility in several species of *Ribes* in the western states. *J. Agric. Res.* 68: 65–71.

Often, A., T. Berg & O. Stabbetorp. 2003. Nurseries are stepping-stones for expanding weeds. *Blyttia* 61: 37–47.

Often, A., O. Stabbetorp & B. Økland. 2006. The role of imported pulpwood for the influx of exotic plants to Norway. *Norsk Geogr. Tidsskr.* 60: 295–302.

Ogburn, R. M. & E. J. Edwards. 2015. Life history lability underlies rapid climate niche evolution in the angiosperm clade Montiaceae. *Mol. Phylogen. Evol.* 92: 181–192.

Ogden, E. C., J. K. Dean, C. W. Boylen & R. B. Sheldon. 1976. *Field Guide to the Aquatic Plants of Lake George, New York*. Bulletin No 426, New York State Museum, Albany, NY. 65 pp.

Ogden, J. 1978. Variation in *Calystegia* R.Br. (Convolvulaceae) in New Zealand. *New Zealand J. Bot.* 16: 123–140.

Ogihara, H., F. Endou, S. Furukawa, H. Matsufuji, K. Suzuki & H. Anzai. 2013. Antimicrobial activity of the carnivorous plant *Dionaea muscipula* against food-related pathogenic and putrefactive bacteria. *Biocontrol Sci.* 18: 151–155.

Oginuma, K., P. H. Raven & H. Tobe. 1990. Karyomorphology and relationships of Celtidaceae and Ulmaceae (Urticales). *Bot. Mag. (Tokyo)* 103: 113–131.

Ogle, D. W. 1989. Barns Chapel swamp: An unusual arbor-vitae (*Thuja occidentalis* L.) site in Washington county, Virginia. *Castanea* 54: 200–202.

Ogren, T. L. 2003. *Safe Sex in the Garden: and Other Propositions for an Allergy-Free World*. Ten Speed Press, Berkeley, CA. 208 pp.

Ogunwenmo, K. O. 2006. Variation in fruit and seed morphology, germination and seedling behaviour of some taxa of *Ipomoea* L. (Convolvulaceae). *Feddes Repert.* 117: 207–216.

Oh, I-C., T. Denk & E. M. Friis. 2003. Evolution of *Illicium* (Illiciaceae): Mapping morphological characters on the molecular tree. *Plant Syst. Evol.* 240: 175–209.

Oh, J., J. J. Bowling, J. F. Carroll, B. Demirci, K. H. Başer, T. D. Leininger, U. R. Bernier & M. T. Hamann. 2012. Natural product studies of US endangered plants: Volatile components of *Lindera melissifolia* (Lauraceae) repel mosquitoes and ticks. *Phytochemistry* 80: 28–36.

Ohga, I. 1923. On the longevity of seeds of *Nelumbo nucifera*. *Bot. Mag. (Tokyo)* 37: 87–95.

Ohtonen, R., H. Fritze, T. Pennanen, A. Jumpponen & J. Trappe. 1999. Ecosystem properties and microbial community changes in primary succession on a glacier forefront. *Oecologia* 119: 239–246.

O'Kane Jr., S. L. & I. A. Al-Shehbaz. 2003. Phylogenetic limits of *Arabidopsis* (Brassicaceae) based on sequences of nuclear ribosomal DNA. *Ann. Missouri Bot. Gard.* 90: 603–612.

Okeniyi, J. O., C. A. Loto & A. P. I. Popopla. 2014. Corrosion inhibition performance of *Rhizophora mangle* L bark-extract on concrete steel-reinforcement in industrial/microbial simulating-environment. *Int. J. Electrochem. Sci.* 9: 4205–4216.

Okuyama, Y., M. Kato & N. Murakami. 2004. Pollination by fungus gnats in four species of the genus *Mitella* (Saxifragaceae). *Bot. J. Linn. Soc.* 144: 449–460.

Okuyama, Y., N. Fujii, M. Wakabayashi, A. Kawakita, M. Ito, M. Watanabe, N. Murakami & M. Kato. 2005. Nonuniform concerted evolution and chloroplast capture: heterogeneity of observed introgression patterns in three molecular data partition phylogenies of Asian *Mitella* (Saxifragaceae). *Mol. Biol. Evol.* 22: 285–296.

Okuyama, Y., O. Pellmyr & M. Kato. 2008. Parallel floral adaptations to pollination by fungus gnats within the genus *Mitella* (Saxifragaceae). *Mol. Phylogen. Evol.* 46: 560–575.

Okuyama, Y., A. S. Akifumi & M. Kato. 2012. Entangling ancient allotetraploidization in Asian *Mitella*: an integrated approach for multilocus combinations. *Mol. Biol. Evol.* 29: 429–439.

Oldham, M. J. 1983. Halberd-leaved rose mallow (*Hibiscus laevis* All.: Malvaceae): an overlooked element of the Canadian flora. *Plant Press (Mississauga)* 1: 78–79.

Olesen, J. M. & E. Warncke. 1989a. Flowering and seasonal changes in flower sex ratio and frequency of flower visitors in a population of *Saxifraga hirculus*. *Holarc. Ecol.* 12: 21–30.

Olesen, J. M. & E. Warncke. 1989b. Predation and transfer of pollen in a population of *Saxifraga hirculus* L. *Holarc. Ecol.* 12: 87–95.

Olesen, J. M. & E. Warncke. 1989c. Temporal changes in pollen flow and neighbourhood structure in a population of *Saxifraga hirculus* L. *Oecologia* 79: 205–211.

Olive, L. S. 1948. Taxonomic notes on Louisiana Fungi: I. *Mycologia* 40: 6–20.

Oliveira, R. S., M. Vosátka, J. C. Dodd & P. M. L. Castro. 2005. Studies on the diversity of arbuscular mycorrhizal fungi and the efficacy of two native isolates in a highly alkaline anthropogenic sediment. *Mycorrhiza* 16: 23–31.

Oliver, C., P. M. Hollingsworth & R. J. Gornall. 2006. Chloroplast DNA phylogeography of the arctic-montane species *Saxifraga hirculus* (Saxifragaceae). *Heredity* 96: 222–231.

Olmstead, R. G. 1989. Phylogeny, phenotypic evolution, and biogeography of the *Scutellaria angustifolia* complex (Lamiaceae): inference from morphological and molecular data. *Syst. Bot.* 14: 320–338.

Olmstead, R. G. & L. Bohs. 2007. A summary of molecular systematic research in Solanaceae: 1982–2006. *Acta Hort.* 745: 255–268.

Olmstead, R. G., H. J. Michaels, K. M. Scott & J. D. Palmer. 1992. Monophyly of the Asteridae and identification of their major lineages inferred from DNA sequences of *rbcL*. *Ann. Missouri Bot. Gard.* 79: 249–265.

Olmstead, R. G., B. Bremer, K. M. Scott & J. D. Palmer. 1993. A parsimony analysis of the Asteridae sensu lato based on *rbcL* sequences. *Ann. Missouri Bot. Gard.* 80: 700–722.

Olmstead, R. G., K.-J. Kim, R. K. Jansen & S. J. Wagstaff. 2000. The phylogeny of the Asteridae sensu lato based on chloroplast *ndhF* gene sequences. *Mol. Phylogen. Evol.* 16: 96–112.

Olmstead, R. G., C. W. dePamphilis, A. D. Wolfe, N. D. Young, W. J. Elisens & P. A. Reeves. 2001. Disintegration of the Scrophulariaceae. *Amer. J. Bot.* 88: 348–361.

Olmstead, R., Y. Yuan, H. Marx, I. E. Peralta & M. Múlgara. 2007. Verbenaceae: A phylogenetic travelogue through its Argentine center of diversity. Botany 2007 meeting (abstract).

Olofsdotter, M., B. E. Valverde & K. H. Madsen. 2000. Herbicide resistant rice (*Oryza sativa* L.): Global implications for weedy rice and weed management. *Ann. Appl. Biol.* 137: 279–295.

Olofson, P. R. (ed.). 2000. Goals Project 2000. Baylands Ecosystem Species and Community Profiles: Life Histories and Environmental Requirements of Key Plants, Fish and Wildlife. San Francisco Bay Regional Water Quality Control Board, Oakland, CA. 408 pp.

Olsen, J. D., G. D. Manners & S. W. Pelletier. 1990. Poisonous properties of larkspur (*Delphinium* spp.). *Collect. Bot. (Barcelona)* 19: 141–151.

Olsen, R. T. & J. M. Ruter. 2001. Preliminary study shows that cold, moist stratification increases germination of 2 native *Illicium* species. *Native Plant J.* 2: 79–83.

Olson, P. E. & J. S. Fletcher. 2000. Ecological recovery of vegetation at a former industrial sludge basin and its implications to phytoremediation. *Environ. Sci. Pollut. Res.* 7: 195–204.

Onaindia M., B. G. de Bikuña & I. Benito. 1996. Aquatic plants in relation to environmental factors in northern Spain. *J. Environ. Manag.* 47: 123–137.

O'Neill, K. M., J. E. Fultz & M. A. Ivie. 2008. Distribution of adult Cerambycidae and Buprestidae (Coleoptera) in a subalpine forest under shelterwood management. *Coleopterists Bull.* 62: 27–36.

O'Neill, M. J. 1986. Initial colonization of periphyton on natural and artificial apices of *Myriophyllum heterophyllum* Michx. *Freshwater Biol.* 16: 685–694.

Oosting, D. P. & D. K. Parshall. 1978. Ecological notes on the butterflies of the Churchill Region of Northern Manitoba. *J. Res. Lepid.* 17: 188–203.

Oosting, H. J. 1945. Tolerance to salt spary of plants of coastal dunes. *Ecology*: 26: 85–89.

Oosting, H. J. & L. E. Anderson. 1937. The vegetation of a barefaced cliff in western North Carolina. *Ecology* 18: 280–292.

Oosting, H. J. & W. D. Billings. 1943. The red fir forest of the Sierra Nevada: Abietum Magnificae. *Ecol. Monogr.* 13: 259–274.

Oppel, C. B., D. E. Dussourd & U. Garimella. 2009. Visualizing a plant defense and insect counterploy: alkaloid distribution in *Lobelia* leaves trenched by a Plusiine caterpillar. *J. Chem. Ecol.* 35: 625–634.

Opsahl, S. P., S. W. Golladay, L. L. Smith & S. E. Allums. 2010. Resource-consumer relationships and baseline stable isotopic signatures of food webs in isolated wetlands. *Wetlands* 30: 1213–1224.

O'Quinn, R. & L. Hufford. 2002. Phylogenetic relationships in *Montieae* (Portulacaceae) based on DNA sequences from nrITS. Botany 2002 meeting (abstract).

Orban, I. & J. Bouharmont. 1995. Reproductive biology of *Nymphaea capensis* Thunb. var. *zanzibariensis* (Casp.) Verdc. (Nymphaeaceae). *Bot. J. Linn. Soc.* 119: 35–43.

Orchard, A. E. 1975. Taxonomic revisions in the family Haloragaceae. The genera *Haloragis*, *Haloragodendron*, *Glischrocaryon*, *Meziella* & *Gonocarpus*. *Bull. Auckland Inst. Mus.* 10: 1–299.

Orchard, A. E. 1980. *Myriophyllum* (Haloragaceae) in Australasia. I. New Zealand: a revision of the genus and a synopsis of the family. *Brunonia* 2: 247–287.

Orchard, A. E. 1981. A revision of South American *Myriophyllum* (Haloragaceae) and its repercussions on some Australian and North American species. *Brunonia* 4: 27–65.

Orchard, A. E. 1986. *Myriophyllum* (Haloragaceae) in Australasia. II. The Australian species. *Brunonia* 8: 173–291.

Orchard, A. E. & E. W. Cross. 2013. A revision of the Australian species of *Eclipta* (Asteraceae: Ecliptinae), with discussion of extra-Australian taxa. *Nuytsia* 23: 43–62.

Orcutt, C. R. 1883. Aquatic plants of San Diego. *Science* 5: 441.

Ørgaard, M. 1990. The genus *Cabomba* (Cabombaceae): a taxonomic study. *Nordic J. Bot.* 11: 179–203.

Ørgaard, M., H. W. E. van Bruggen & P. J. van der Vlugt. 1992. Die Familie Cabombaceae (*Cabomba* und *Brasenia*). Aqua-Planta, Sonderheft No 3. VDA-Arbeitskreis Wasserpflanzen, Berlin, Germany. 43 pp.

Orians C. M. & T. Floyd. 1997. The susceptibility of parental and hybrid willows to plant enemies under contrasting soil nutrient conditions. *Oecologia* 109: 407–413.

Orians, C. M., C. H. Huang, A. Wild, K. A. Dorfman, P. Zee, M. T. T. Dao & R. S. Fritz. 1997. Willow hybridization differentially affects preference and performance of herbivorous beetles. *Entomol. Exp. Appl.* 83: 285–294.

Orians C. M., D. I. Bolnick, B. M. Roche, R. S. Fritz & T. Floyd. 1999. Water availability alters the relative performance of *Salix sericea*, *Salix eriocephala*, and their F$_1$ hybrids. *Can. J. Bot.* 77: 514–522.

Orendovici, T., J. M. Skelly, J. A. Ferdinand, J. E. Savage, M.-J. Sanz & G. C. Smith. 2003. Response of native plants of northeastern United States and southern Spain to ozone exposures; determining exposure/response relationships. *Environ. Pollut.* 125: 31–40.

Orłowski, G. & J. Czarnecka. 2009. Granivory of birds and seed dispersal: viable seeds of *Amaranthus retroflexus* L. recovered from the droppings of the grey partridge *Perdix perdix* L. *Polish J. Ecol.* 57: 191–196.

Ornduff, R. 1963. Experimental studies in two genera of Helenieae (Compositae): *Blennosperma* and *Lasthenia*. *Q. Rev. Biol.* 38: 141–150.

Ornduff, R. 1964. Biosystematics of *Blennosperma* (Compositae). *Brittonia* 16: 289–295.

Ornduff, R. 1966. The origin of dioecism from heterostyly in *Nymphoides* (Menyanthaceae). *Evolution* 20: 309–314.

Ornduff, R. 1969a. Reproductive biology in relation to systematics. *Taxon* 18: 121–133.

Ornduff, R. 1969b. The origin and relationships of *Lasthenia burkei* (Compositae). *Amer. J. Bot.* 56: 1042–1047.

Ornduff, R. 1969c. *Limnanthes vinculans*, a new California endemic. *Brittonia* 21: 11–14.

Ornduff, R. 1970. Cytogeography of *Nymphoides* (Menyanthaceae). *Taxon* 19: 715–719.

Ornduff, R. 1978. Features of pollen flow in dimorphic species of *Lythrum* section *Euhyssopifolia*. *Amer. J. Bot.* 65: 1077–1083.

Ornduff, R. 1988. Distyly and monomorphism in *Villarsia* (Menyanthaceae): some evolutionary considerations. *Ann. Missouri Bot. Gard.* 75: 761–767.

Ornduff, R. & T. J. Crovello. 1968. Numerical taxonomy of Limnanthaceae. *Amer. J. Bot.* 55: 173–182.

Ornduff, R. & T. Mosquin. 1970. Variation in the spectral qualities of flowers in the *Nymphoides indica* complex (Menyanthaceae) and its possible adaptive significance. *Can. J. Bot.* 48: 603–605.

Ornduff, R., N. A. M. Saleh & B. A. Bohm. 1973. The flavonoids and affinities of *Blennosperma* and *Crocidium* (Compositae). *Taxon* 22: 407–412.

Orr, C. C. & O. J. Dickerson. 1966. Nematodes in true prairie soils of Kansas. *Trans. Kansas Acad. Sci.* 69: 317–334.

Orrell Elliston, L. C. 2006. The natural history, genetics and population biology of *Sabatia kennedyana* (Plymouth Gentian): an endangered plant of atlantic coastal pondshores. PhD thesis. University of Massachusetts-Boston, Boston, MA. 147 pp.

Orsenigo, J. R. & T. A. Zitter. 1971. Vegetable virus problems in south Florida as related to weed science. *Proc. Florida State Hort. Soc.* 84: 168–171.

Ortega, C.A., A. O. M. María & J. C. Gianello. 2000. Chemical components and biological activity of *Bidens subalternans*, *B. aurea* (Astereaceae) and *Zuccagnia puntacta* (Fabaceae). *Molecules* 5: 465–467.

Ortwine-Boes, C. & J. Silbernagel. 2003. Bumblebee conservation in and around cranberry marshes. *In*: *School Proceedings 2003, Wisconsin Cranberry Crop Management Library*. 10 pp. Available online: http://www.hort.wisc.edu/cran.

Orzell, S. L. & E. Bridges. 1993. *Eriocaulon nigrobracteatum* (Eriocaulaceae), a new species from the Florida panhandle, with a characterization of its poor fen habitat. *Phytologia* 74: 104–124.

Orzell, S. L. & E. Bridges. 2006. Floristic composition and species richness of subtropical seasonally wet *Muhlenbergia sericea* prairies in portions of central and south Florida. pp. 136–150. *In*: R. F. Noss (ed.), *Land of Fire and Water: The Florida Dry Prairie Ecosystem*. Avon Park Air Force Range and Department of Defense, Avon Park, FL.

Orzell, S. L. and D. R. Kurz. 1986. Floristic analysis of prairie fens in the southeastern Missouri Ozarks. pp. 50–58. *In*: G. K. Clambey and R. H. Pemble (eds.), *Proceedings of the Ninth North American Prairie Conference*. Tri-College University Center for Environmental Studies, Fargo, ND.

Osaloo, S. K., A. A. Maassoumi & N. Murakami. 2003. Molecular systematics of the genus *Astragalus* L. (Fabaceae): phylogenetic analyses of nuclear ribosomal DNA internal transcribed spacers and chloroplast gene *ndhF* sequences. *Plant Syst. Evol.* 242: 1–32.

Osborn, J. M. & E. L. Schneider. 1988. Morphological studies of the Nymphaeaceae *sensu lato*. XVI. The floral biology of *Brasenia schreberi*. *Ann. Missouri Bot. Gard.* 75: 778–794.

Osborn, J. M., T. N. Taylor & E. L. Schneider. 1991. Pollen morphology and ultrastructure of the Cabombaceae: Correlations with pollination biology. *Amer. J. Bot.* 78: 1367–1378.

Osborne, W. P. & W. H Lewis. 1962. Chromosome numbers of *Linum* from the southern United States and Mexico. *Sida* 1: 63–68.

Osmond, C. B., N. Valaane, S. M. Haslam, P. Uotila & Z. Roksandic. 1981. Comparisons of $\delta^{13}C$ values in leaves of aquatic macrophytes from different habitats in Britain and Finland; some implications for photosynthetic processes in aquatic plants. *Oecologia* 50: 117–124.

Osorio, R. 2009. Mock bishop's weed—An overlooked native. Available online: http://www.rufino.info/articles/ptilimnium_capillaceum.pdf.

Ostaff, D. P., A. Mosseler, R. C. Johns, S. Javorek, J. Klymko & J. S. Ascher. 2015a. Willows (*Salix* spp.) as pollen and nectar sources for sustaining fruit and berry pollinating insects. *Can. J. Plant Sci.* 95: 505–516.

Ostaff, D. P., J. S. Ascher, S. Javorek & A. Mosseler. 2015b. New records of *Andrena* (Hymenoptera: Andrenidae) in New Brunswick, Canada. *J. Acad. Entomol. Soc.* 11: 5–8.

Ostenfeld, C. H. 1925. Vegetation of north Greenland. *Bot. Gaz.* 80: 213–218.

Osuna, M. D., A. J. Fischer & R. De Prado. 2003. Herbicide resistance in *Aster squamatus* conferred by a less sensitive form of acetolactate synthase. *Pest Manag. Sci.* 59: 1210–1216.

Osvald, H. 1935. A bog at Hartford, Michigan. *Ecology* 16: 520–528.

Oswald, W. W., E. D. Doughty, G. Ne'eman, R. Ne'eman & A. M. Ellison. 2011. Pollen morphology and its relationship to taxonomy of the genus *Sarracenia* (Sarraceniaceae). *Rhodora* 955: 235–251.

Ott, J. R. 1991. The biology of *Acanthoscelides alboscutellatus* (Coleoptera: Bruchidae) on its host plant, *Ludwigia alternifolia* (L.) (Onagraceae). *Proc. Entomol. Soc. Washington* 93: 641–651.

Otte, D. 1975. Plant preference and plant succession. A consideration of evolution of plant preference in *Schistocerca*. *Oecologia* 18: 129–144.

Ottenbreit, K. A. & R. J. Staniforth. 1992. Life cycle and age structure of ramets in an expanding population of *Salix exigua* (sandbar willow). *Can. J. Bot.* 70: 1141–1146.

Ottenbreit, K. A. & R. J. Staniforth. 1994. Crossability of naturalized and cultivated *Lythrum* taxa. *Can. J. Bot.* 72:337–341.

Otto, C. & B. S. Svensson. 1981. How do macrophytes growing in or close to water reduce their consumption by aquatic herbivores? *Hydrobiologia* 78: 107–112.

Ouren T. 1978. The impact of shipping on the invasion of alien plants to Norway. *GeoJournal* 2: 123–132.

Outcalt, K. W. 1990. *Nyssa sylvatica* var. *biflora* (Walt.) Sara. Swamp Tupelo. pp. 482–489. *In*: R. M. Burns & B. H. Honkala (eds.), *Silvics of North America. Vol. 2, Hardwoods.* Agriculture Handbook 654. USDA, Forest Service, Washington, DC.

Overlease, W. & E. Overlease. 2011. A note on the host species of mistletoe (*Phoradendron leucarpum*) in the eastern United States. *Bartonia* 65: 105–111.

Ovesna, Z., A. Vachalkova & K. Horvathova. 2004. Taraxasterol and beta-sitosterol: new naturally compounds with chemoprotective/chemopreventive effects. *Neoplasma* 51: 407–414.

Owens, C. S., J. D. Madsen, R. M. Smart & R. M. Stewart. 2001. Dispersal of native and nonnative aquatic plant species in the San Marcos River, Texas. *J. Aquat. Plant Manag.* 39: 75–79.

Owens, N. L. & G. N. Cole. 2003. 25 years of vegetational changes in a glacial drift hill prairie community in east-central Illinois. *Trans. Illinois State Acad. Sci.* 96: 265–269.

Owens, S. J. & J. L. Ubera-Jiménez. 1992. Breeding systems in Labiatae. pp. 257–280. *In*: R. M. Harley & T. Reynolds (eds.), *Advances in Labiatae science.* Royal Botanic Gardens, Kew, UK.

Ownbey, G. B. 1951. Natural hybridization in the genus *Cirsium*-I. *C. discolor* (Muhl. Ex Willd.) Spreng. × *C. muticum* Michx. *Bull. Torrey Bot. Club* 78: 233–253.

Ownbey, G. B. 1968. Cytotaxonomic notes on eleven species of *Cirsium* native to Mexico. *Brittonia* 20: 336–342.

Ownbey, G. B., P. H. Raven & D. W. Kyhos. 1975. Chromosome numbers in some North American species of the genus *Cirsium*. III. Western United States, Mexico, and Guatemala. *Brittonia* 27: 297–304.

Oxelman, B., P. Kornhall, R. G. Olmstead & B. Bremer. 2005. Further disintegration of Scrophulariaceae. *Taxon* 54: 411–425.

Oyedeji, A. A. & J. F. N. Abowei. 2012. The classification, distribution, control and economic importance of aquatic plants. *Int. J. Fish. Aquat. Sci.* 1: 118–128.

Oyedeji, O., M. Oziegbe & F. O. Taiwo. 2011. Antibacterial, antifungal and phytochemical analysis of crude extracts from the leaves of *Ludwigia abyssinica* A. Rich. and *Ludwigia decurrens* Walter. *J. Med. Plant Res.* 5: 1192–1199.

Oziegbe, M., J. O. Faluyi & A. Oluwaranti. 2010. Effect of seed age and soil texture on the germination of some *Ludwigia* species (Onagraceae) in Nigeria. *Acta Bot. Croat.* 69: 249–257.

Öztürk, A. & M. A. Fischer. 1982. Karyosystematics of *Veronica* sect. *Beccabunga* (Scrophulariaceae) with special reference to the taxa in Turkey. *Plant Syst. Evol.* 140: 307–319.

Pabst, R. J. & T. A. Spies. 1998. Distribution of herbs and shrubs in relation to landform and canopy cover in riparian forests of coastal Oregon. *Can. J. Bot.* 76: 298–315.

Packer, J. G. 1963. The taxonomy of some North American species of *Chrysosplenium* L., section *Alternifolia* Franchet. *Can. J. Bot.* 41: 85–103.

Packer, J. G. 2003. 40. Portulacaceae Adanson. pp. 457–504. *In*: N. R. Morin (convening ed.), *Flora of North America North of Mexico, Vol. 4: Magnoliophyta: Caryophyllidae, Part 1.* Oxford University Press, New York.

Packer, J. G. & C. C. Freeman. 2005. 35. *Koenigia*. pp. 600–601. *In*: Flora North America Editorial Committee (eds.), *Flora of North America North of Mexico, Vol. 5: Magnoliophyta: Caryophyllidae, part 2.* Oxford University Press, New York.

Packer, L. 1987. The triungulin larva of *Nemognatha* (*Pauronemognatha*) *punctulata* Le Conte (Coleoptera: Meloidae) with a description of the nest of its host-*Megachile brevis pseudobrevis* Say (Hymenoptera: Megachilidae). *J. Kansas Entomol. Soc.* 60: 280–287.

Padgett, D. J. 1998. Phenetic distinction between the dwarf yellow water-lilies: *Nuphar microphylla* and *N. pumila* (Nymphaeaceae). *Can. J. Bot.* 76: 1755–1762.

Padgett, D. J. 1999. Nomenclatural novelties in *Nuphar* (Nymphaeaceae). *Sida* 18: 823–826.

Padgett, D. J. 2003. Phenetic studies in *Nuphar* Sm. (Nymphaeaceae): variation in sect. *Nuphar*. *Plant Syst. Evol.* 239: 187–197.

Padgett, D. J. 2007. A monograph of *Nuphar* (Nymphaeaceae). *Rhodora* 109: 1–95.

Padgett, D. J. & D. H. Les. 2004. Nymphaeaceae. pp. 271–273. *In*: N. Smith, S. A. Mori, A. Henderson, D. W. Stevenson & S. V. Heald (eds.), *Flowering Plants of the Neotropics.* Princeton University Press, Princeton, NJ. 594 pp.

Padgett, D. J., D. H. Les & G. E. Crow. 1998. Evidence for the hybrid origin of *Nuphar* ×*rubrodisca* (Nymphaeaceae). *Amer. J. Bot.* 85: 1468–1476.

Padgett, D. J., D. H. Les & G. E. Crow. 1999. Phylogenetic relationships in *Nuphar* (Nymphaeaceae): evidence from morphology, chloroplast DNA, and nuclear ribosomal DNA. *Amer. J. Bot.* 86: 1316–1324.

Padgett, D. J., L. Cook, L. Horky, J. Noris & K. Vale. 2004. Seed production and germination in Long's Bittercress (*Cardamine longii*) of Massachusetts. *N. E. Naturalist* 11: 49–56.

Padgett, D. J., J. J. Carboni & D. J. Schepis. 2010. The dietary composition of *Chrysemys picta picta* (eastern painted turtles) with special reference to the seeds of aquatic macrophytes. *N. E. Naturalist* 17: 305–312.

Pagano, A. M. & J. E. Titus. 2004. Submersed macrophyte growth at low pH: contrasting responses of three species to dissolved inorganic carbon enrichment and sediment type. *Aquat. Bot.* 79: 65–74.

Page, H. M. 1997. Importance of vascular plant and algal production tomacro-invertebrate consumers in a southern California salt marsh. *Estuarine Coastal Shelf Sci.* 45: 823–834.

Paine, L. K. & C. A. Ribic. 2002. Comparison of riparian plant communities under four land management systems in southwestern Wisconsin. *Agric. Ecosyst. Environ.* 92: 93–105.

Pak, J.-H., J.-K. Park & S. S. Whang. 2001. Systematic implications of fruit wall anatomy and surface sculpturing of *Microseris* (Asteraceae, Lactuceae) and relatives. *Int. J. Plant Sci.* 162: 209–220.

Pakeman, R. J. & J. L. Small. 2005. The role of the seed bank, seed rain and the timing of disturbance in gap regeneration. *J. Veg. Sci.* 16: 121–130.

Pakeman, R. J., J. Engelen & J. P. Attwood. 1999. Rabbit endozoochroy and seedbank build-up in an acidic grassland. *Plant Ecol.* 145: 83–90.

Pakeman, R. J., G. Digneffe & J. L. Small. 2002. Ecological correlates of endozoochory by herbivores. *Funct. Ecol.* 16: 296–304.

Pal, D. & K. Samanta. 2011. CNS activities of ethanol extract of aerial parts of *Hygrophila difformis* in mice. *Acta Pol. Pharm.* 68: 75–81.

Pal, D. K., K. Samanta & P. Maity. 2010. Evaluation of antioxidant activity of aerial parts of *Hygrophila difformis*. *Asian J. Chem.* 22: 2459–2461.

Palis, J. G. 1998. Breeding biology of the gopher frog, *Rana capito*, in western Florida. *J. Herpetol.* 32: 217–223.

Palmé, A. 2003. Evolutionary history and chloroplast DNA variation in three plant genera: *Betula*, *Corylus* and *Salix*.: The impact of post-glacial colonisation and hybridisation. Universitetstryckeriet Ekonomikum, Uppsala, Sweden. 59 pp.

Palmeiro, N. M. S., C. E. Almeida, P. C. Ghedini, L. S. Goulart & B. Baldisserotto. 2002. Analgesic and anti-inflammatory properties of *Plantago australis* hydroalcoholic extract. *Acta Farm. Bonaerense* 21: 89–92.

Palmeiro, N. M. S., C. E. Almeida, P. C. Ghedini, L. S. Goulart, M. C. F. Pereira, S. Huber, J. E. P. da Silva & S. Lopes. 2003. Oral subchronic toxicity of aqueous crude extract of *Plantago australis* leaves. *J. Ethnopharmacol.* 88: 15–18.

Palmer, E. J. 1919. Texas Pteridophyta: I. *Amer. Fern J.* 9: 17–22.

Palmer, I. E., T. G. Ranney, N. P. Lynch & R. E. Bir. 2009. Crossability, cytogenetics, and reproductive pathways in *Rudbeckia* subgenus *Rudbeckia*. *HortScience* 44: 44–48.

Palmer, M. A., S. L. Bell & I. Butterfield. 1992. A botanical classification of standing waters in Britain: Applications for conservation and monitoring. *Aquat. Conserv.* 2: 125–143.

Palomino, S. S., M. J. Abad, L. M. Bedoya, J. García, E. Gonzales, X. Chiriboga, P. Bermejo & J. Alcami. 2002. Screening of South American plants against human immunodeficiency virus: preliminary fractionation of aqueous extract from *Baccharis trinervis*. *Biol. Pharm. Bull.* 25: 1147–1150.

Pammel, L. H. 1908. Flora northern Iowa peat bogs. *Iowa Geol. Surv. Ann. Rep.* 19: 735–778.

Panasahatham, S. 2000. Biology, ecology and management of *Scaptomyza apicalis* Hardy (Diptera: Drosophilidae) on meadowfoam, *Limnanthes alba* Benth. in western Oregon. PhD dissertation. Oregon State University, Corvallis, OR. 96 pp.

Panchaphong, P. 1987. Preliminary study on water convolvulus (*Ipomoea aquatica* Forsk.) improvement II. Floral biology and breeding behaviors of 2x and 4x. Kasetsart University, Bangkok, Thailand. 14 pp.

Pandey, A. K., U. T. Palni & N. N. Tripathi. 2006. Repellent activity of some essential oils against two stored product beetles *Callosobruchus chinensis* L. and *C. maculatus* F. (Coleoptera: Bruchidae) with reference to *Chenopodium ambrosioides* L. oil for the safety of pigeon pea seeds. *J. Food Sci. Technol.* 51: 4066–4071.

Pandey, V. N. & A. K. Srivastava. 1989. *Veronica anagallis-aquatica* L., a potential source of leaf protein. *Aquat. Bot.* 34: 385–388.

Panero, J. L. & V. A. Funk. 2008. The value of sampling anomalous taxa in phylogenetic studies: major clades of the Asteraceae revealed. *Mol. Phylogen. Evol.* 47: 757–782.

Panero, J. L., R. K. Jansen & J. A. Clevinger. 1999. phylogenetic relationships of subtribe Ecliptinae (Asteraceae: Heliantheae) based on chloroplast DNA restriction site data. *Amer. J. Bot.* 86: 413–427.

Panetta, F. D. 1985. Population studies on pennyroyal mint (*Mentha pulegium* L.) II. Seed banks. *Weed Res.* 25: 311–315.

Pannell, J. R. 2002. The evolution and maintenance of androdioecy. *Ann. Rev. Ecol. Syst.* 33: 397–425.

Panter, K. E., D. C. Baker & P. O. Kechele. 1996. Water hemlock (*Cicuta douglasii*) toxicoses in sheep: pathologic description and prevention of lesions and death. *J. Veterin. Diagn. Invest.* 8: 474–480.

Papp, C. S. 1959. Discussion of synonymy and first illustrations of larva and pupa of *Coreopsomela elegans* (Olivier 1807) from California (Notes on North American Coleoptera, No. 6). *J. Kansas Entomol. Soc.* 32: 137–141.

Pappers, S. M., G. van der Velde, N. J. Ouborg & J. M. van Groenendael. 2002. Genetically based polymorphisms in morphology and life history associated with putative host races of the water lily leaf beetle, *Galerucella nymphaeae*. *Evolution* 56: 1610–1621.

Parachnowitsch, A. L. & E. Elle. 2005. Insect visitation to wildflowers in the endangered garry oak, *Quercus garryana*, ecosystem of British Columbia. *Can. Field Naturalist* 119: 245–253.

Parachnowitsch, A. L., C. M. Caruso, S. A. Campbell & A. Kessler. 2012. *Lobelia siphilitica* plants that escape herbivory in time also have reduced latex production. *PLoS One* 7(5): e37745. doi:10.1371/journal.pone.0037745.

Paratley, R. D. & T. J. Fahey. 1986. Vegetation-environment relations in a conifer swamp in central New York. *Bull. Torrey Bot. Club* 113: 357–371.

Parfitt, B. D. 1997. 14. *Trollius* Linnaeus. pp. 189–190. *In*: N. R. Morin (convening ed.), *Flora of North America North of Mexico, Vol. 3: Magnoliophyta: Magnoliidae and Hamamelidae*. Oxford University Press, New York.

Parisod, C., C. Trippi & N. Galland. 2005. Genetic variability and founder effect in the pitcher plant *Sarracenia purpurea* (Sarraceniaceae) in populations introduced into Switzerland: from inbreeding to invasion. *Ann. Bot.* 95: 277–286.

Park, C. H., T. Tanaka, J. H. Kim, E. J. Cho, J. C. Park, N. Shibahara & T. Yokozawa. 2011. Hepato-protective effects of loganin, iridoid glycoside from Corni Fructus, against hyperglycemia-activated signaling pathway in liver of type 2 diabetic db/db mice. *Toxicology* 290: 14–21.

Park, J.-C., Y.-B. Yu, J.-H. Lee, M. Hattori, C.-K. Lee & J.-W. Choi. 1996. Protective effect of *Oenanthe javanica* on the hepatic lipid peroxidation in bromobenzene-treated rats and its bioactive component. *Plant Med.* 62: 488–490.

Park, K.-R. & A. Backlund. 2002. Origin of the cyathium-bearing Euphorbieae (Euphorbiaceae): phylogenetic study based on morphological characters. *Bot. Bull. Acad. Sin.* 43: 57–62.

Park, K.-R. & W. J. Elisens. 2000. A phylogenetic study of tribe Euphorbieae (Euphorbiaceae). *Int. J. Plant Sci.* 161: 425–434.

Park, M. M. & D. Festerling Jr. 1997. 22. *Thalictrum* Linnaeus. pp. 258–271. *In*: N. R. Morin (convening ed.), *Flora of North America North of Mexico, Vol. 3: Magnoliophyta: Magnoliidae and Hamamelidae*. Oxford University Press, New York.

Parker, E. D., M. F. Hirshfield & J. W. Gibbons. 1973. Ecological comparisons of thermally affected aquatic environments. *J. Water Pollut. Control Fed.* 45: 726–733.

Parker, I. M., R. R. Nakamura & D. W. Schemske. 1995. Reproductive allocation and the fitness consequences of selfing in two sympatric species of *Epilobium* (Onagraceae) with contrasting mating systems. *Amer. J. Bot.* 82: 1007–1016.

Parker, J. D. 2005. Plant-herbivore interactions: consequences for the structure of freshwater communities and exotic plant invasions. PhD dissertation, Georgia Institute of Technology, Atlanta, GA. 150 pp.

Parker, J. D. & M. E. Hay. 2005. Biotic resistance to plant invasions? Native herbivores prefer non-native plants. *Ecol. Lett.* 8: 959–967.

Parker, J. D., D. O. Collins, J. Kubanek, M. Cameron Sullards, D. Bostwick & M. E. Hay. 2006. Chemical defenses promote persistence of the aquatic plant *Micranthemum umbrosum*. *J. Chem. Ecol.* 32: 815–833.

Parker, J. D., D. E. Burkepile, D. O. Collins, J. Kubanek & M. E. Hay. 2007a. Stream mosses as chemically-defended refugia for freshwater macroinvertebrates. *Oikos* 116: 302–312.

Parker, J. D., C. C. Caudill & M. E. Hay. 2007b. Beaver herbivory on aquatic plants. *Oecologia* 151: 616–625.

Parker, V. T. & M. A. Leck. 1985. Relationships of seed banks to plant distribution patterns in a freshwater tidal wetland. *Amer. J. Bot.* 72: 161–174.

Parks, G. E., M. A. Dietrich & K. S. Schumaker. 2002. Increased vacuolar Na^+/H^+ exchange activity in *Salicornia bigelovii* Torr. in response to NaCl. *J. Exp. Bot.* 371: 1055–1065.

Parks, M. S. & P. E. Elvander. 2012. Saxifragaceae. Saxifrage family. pp. 1234–1244. *In*: B. G. Baldwin, D. H. Goldman, D. J. Keil, R. Patterson, T. J. Rosatti and D. H. Wilken (eds.), *The Jepson Manual (2nd ed.)*. University of California Press, Berkeley, CA.

Parmelee, J. A. & D. B. O. Savile. 1954. Life history and relationship of the rusts of *Sparganium* and *Acorus*. *Mycologia* 46: 823–836.

Parmenter, R. R. 1980. Effects of food availability and water temperature on the feeding ecology of pond sliders (*Chrysemys s. scripta*). *Copeia* 1980: 503–514.

Parolin, P. 2006. Ombrohydrochory: Rain-operated seed dispersal in plants–With special regard to jet-action dispersal in Aizoaceae. *Flora* 201: 511–518.

Parrella, G., P. Gognalons, K. Gebre-Selassiè, C. Vovlas & G. Marchoux. 2003. An update of the host range of tomato spotted wilt virus. *J. Plant Pathol.* 85: 227–264.

Parrish, J. A. D. & F. A. Bazzaz. 1979. Difference in pollination niche relationships in early and late successional plant communities. *Ecology* 60: 597–610.

Parsons L. S. & J. B. Zedler. 1997. Factors affecting reestablishment of an endangered annual plant at a California salt marsh. *Ecol. Appl.* 7: 253–267.

Parsons, W. T. 1973. *Noxious Weeds of Victoria*. Inkata Press, Ltd., Melbourne, Australia. 300 pp.

Partridge, J. W. 2001. Biological flora of the British Isles: *Persicaria amphibia* (L.) Gray (*Polygonum amphibium* L.). *J. Ecol.* 89: 487–501.

Partridge, T. R. & J. B. Wilson. 1987. Germination in relation to salinity in some plants of salt marshes in Otago, New Zealand. *New Zealand J. Bot.* 25: 255–261.

Passmore, M. F. 1981. Population biology of the common ground-dove and ecological relationships with mourning and white-winged doves in south Texas. PhD dissertation, Texas A & M University, College Station, TX. 96 pp.

Pasqualetto, P. L. & P. H. Dunn. 1990. Propagation of *Cirsium douglasii* and *Cirsium andrewsii* by tissue culture for use as test plants in biological control of weeds research. pp. 191–193. *In*: E. S. Delfosse (ed.), *Proceedings of the VII International Symposium on Biological Control of Weeds*. Instituto Sperimentale per la Patologia Vegetale (MAF), Rome, Italy.

Pasternack, G. B., W. B. Hilgartner & G. S. Brush. 2000. Biogeomorphology of an upper Chesapeake Bay river-mouth tidal freshwater marsh. *Wetlands* 20: 520–537.

Pastore, J. F. B. & J. R. Abbott. 2012. Taxonomic notes and new combinations for *Asemeia* (Polygalaceae). *Kew Bull.* 67: 801–813.

Paszota, P., M. Escalante-Perez, L. R. Thomsen, M. W. Risør, A. Dembski, L. Sanglas, T. A. Nielsen, H. Karring, I. B. Thøgersen, R. Hedrich, J. J. Enghild, I. Kreuzer & K. W. Sanggaard. 2014. Secreted major Venus flytrap chitinase enables digestion of Arthropod prey. *Biochim. Biophys. Acta (BBA) Proteins Proteomics* 1844: 374–383.

Patchell, M. J. 2013. *Vegetation and Rare Plant Survey of Lois Hole Centennial Provincial Park, Summer 2013*. Big Lake Environment Support Society, St. Albert, AB. 56 pp.

Paterson, I. G. & M. Snyder. 1999. Genetic evidence supporting the taxonomy of *Geum peckii* (Rosaceae) and *G. radiatum* as separate species. *Rhodora* 101: 325–340.

Pates, A. L. & G. C. Madsen. 1955. Occurrence of antimicrobial substances in chlorophyllose plants growing in Florida. II. *Bot. Gaz.* 116: 250–261.

Patiño, S. & J. Grace. 2002. The cooling of convolvulaceous flowers in a tropical environment. *Plant Cell Environ.* 25: 41–51.

Patnaik, S. 1976. Autecology of *Ipomoea aquatica* Forsk. *J. Int. Fish. Soc. India.* 8: 77–82.

Paton, A. J., D. Springate, S. Suddee, D. Otieno, R. J. Grayer, M. M. Harley, F. Willis, M. S. J. Simmonds, M. P. Powell & V. Savolainen. 2004. Phylogeny and evolution of basils and allies (Ocimeae, Labiatae) based on three plastid DNA regions. *Mol. Phylogen. Evol.* 31: 277–299.

Patrick, T. S., J. R. Allison & G. A. Krakow. 1995. *Protected Plants of Georgia: An Information Manual on Plants Designated by the State of Georgia as Endangered, Threatened, Rare, or Unusual*. Georgia Natural Heritage Program, Georgia Department of Natural Resources, Social Circle, GA. 246 pp.

Patten Jr., B. C. 1956. Notes on the biology of *Myriophyllum spicatum* L. in a New Jersey lake. *Bull. Torrey Bot. Club* 83: 5–18.

Patton, J. E. & W. S. Judd. 1988. A phenological study of 20 vascular plant species occurring on the Paynes Prairie basin, Alachua County, Florida. *Castanea* 53: 149–163.

Patzelt, A., U. Wild & J. Pfadenhauer. 2001. Restoration of wet fen meadows by topsoil removal: vegetation development and germination biology of fen species. *Restorat. Ecol.* 9: 127–136.

Paulsen, E., P. S. Skov & K. E. Andersen. 1998. Immediate skin and mucosal symptoms from pot plants and vegetables in gardeners and greenhouse workers. *Contact Dermatitis* 39: 166–170.

Pavek, D. S. 1992. *Asclepias incarnata. In: Fire Effects Information System*. U. S. Department of Agriculture, Forest Service, Rocky Mountain Research Station, Fire Sciences Laboratory (Producer). Available online: http://www.fs.fed.us/database/feis/ (accessed November 10, 2006).

Pavek, D. S. 1993. *Chamaedaphne calyculata. In: Fire Effects Information System*. U. S. Department of Agriculture, Forest Service, Rocky Mountain Research Station, Fire Sciences Laboratory (Producer). Available online: http://www.fs.fed.us/database/feis/ (accessed May 26, 2006).

Pavela, R. 2009. Larvicidal effects of some Euro-Asiatic plants against *Culex quinquefasciatus* Say larvae (Diptera: Culicidae). *Parasitol. Res.* 105: 887–892.

Pavlik, B., D. Murphy et al. 2002. Draft conservation strategy for Tahoe yellow cress (*Rorippa subumbellata*). Tahoe Regional Planning Agency. Lake Tahoe, NV. 107 pp.

Pavlovič, A., M. Krausko, M. Libiaková & L. Adamec. 2014. Feeding on prey increases photosynthetic efficiency in the carnivorous sundew *Drosera capensis*. *Ann. Bot.* 113: 69–78.

Pavol, E. Jr., M. Hájek & P. Hájková. 2009. A European warm waters neophyte *Shinnersia rivularis*—new alien species to the Slovak flora. *Biologia* 64: 684–686.

Payne, D. 2010. A survey of the vascular flora of Beaufort County, South Carolina. MS thesis. Clemson University, Clemson, SC. 252 pp.

Payne, J. A. & G. W. Krewer. 1990. Mayhaw: a new fruit crop for the South. pp. 317–321. *In*: J. Janick & J. E. Simon (eds). *Advances in New Crops*. Timber Press, Portland, OR.

Payne, J. A., G. W. Krewer & R. R. Fitenmiller. 1990. Mayhaws: trees of pomological and ornamental interest. *HortScience* 25: 246, 375.

Payson, E. B. 1918. The North American species of *Aquilegia*. *Contr. U. S. Natl. Herb.* 20: 133–157.

Peach, M. & J. B. Zedler. 2006. How tussocks structure sedge meadow vegetation. *Wetlands* 26: 322–335.

Peacock, J. W. 2000. Immature stages of the marbled underwing, *Catocala marmorata* (Noctuidae). *J. Lepid. Soc.* 54: 107–110.

Pearsall, W. H. 1917. The aquatic and marsh vegetation of Esthwaite water. *J. Ecol.* 5: 180–202.

Pearson, J. A. & M. J. Leoschke. 1992. Floristic composition and conservation status of fens in Iowa. *J. Iowa Acad. Sci.* 99: 41–52.

Pearson, P. G. 1954. Mammals of Gulf Hammock, Levy County, Florida. *Amer. Midl. Naturalist* 51: 468–480.

Pedersen, O. 1993. Long-distance water transport in aquatic plants. *Plant Physiol.* 103: 1369–1375.

Pedersen O. & K. Sand-Jensen. 1992. Adaptations of submerged *Lobelia dortmanna* to aerial life form: morphology, carbon sources and oxygen dynamics. *Oikos* 65: 89–96.

Pedersen, O., T. Andersen, K. Ikejima, M. D. Z. Hossain & F. Ø. Andersen. 2006. A multidisciplinary approach to understanding the recent and historical occurrence of the freshwater plant, *Littorella uniflora*. *Freshwater Biol.* 51: 865–877.

Pegtel, D. M. 1998. Rare vascular plant species at risk: recovery by seeding? *Appl. Veg. Sci.* 1: 67–74.

Peinado, M., J. L. Aguirre & M. de la Cruz. 1998. A phytosociological survey of the boreal forest (*Vaccinio-Piceetea*) in North America. *Plant Ecol.* 137: 151–202.

Peinado, M., J. L. Aguirre, J. Delgadillo & J. M. Martínez-Parras. 2005a. A phytosociological survey of the chionophilous communities of western North America. Part I: temperate and Mediterranean associations. *Plant Ecol.* 180: 187–241.

Peinado, M., J. L. Aguirre, J. Delgadillo, J. González & J. M. Martínez-Parras. 2005b. A phytosociological survey of the chionophilous communities of western North America. Part II: boreal associations. *Plant Ecol.* 180: 243–256.

Pell, S. K. & L. E. Urbatsch. 2001. Tribal relationships and character evolution in the cashew family (Anacardiaceae): inferences from three regions of the chloroplast genome. Botany 2001 meeting (abstract).

Pellerin, S., J. Huot & S. D. Côté. 2006. Long term effects of deer browsing and trampling on the vegetation of peatlands. *Biol. Conserv.* 128: 316–326.

Pellett, F. C. 1920. *American Honey Plants, Together with Those Which are of Special Value to the Beekeeper as Sources of Pollen*. American Bee Journal, Hamilton, IL. 207 pp.

Pellett, F. C. 1920. *American Honey Plants, Together with Those Which are of Special Value to the Beekeeper as Sources of Pollen*. American Bee Journal, Hamilton, IL. 207 pp.

Pellicer, J., L. J. Kelly, C. Magdalena & I. J. Leitch. 2013. Insights into the dynamics of genome size and chromosome evolution in the early diverging angiosperm lineage Nymphaeales (water lilies). *Genome* 56: 1–13.

Pellmyr, O. 1987. Multiple sex expressions in *Cimicifuga simplex*: dichogamy destabilizes hermaphroditism. *Biol. J. Linn. Soc.* 31: 161–174.

Pelser, P. B., B. Gravendeel & R. van der Meijden. 2002. Tackling speciose genera: species composition and phylogenetic position of *Senecio* sect. *Jacobaea* (Asteraceae) based on plastid and nrDNA sequences. *Amer. J. Bot.* 89: 929–939.

Pelser, P. B., B. Nordenstam, J. W. Kadereit & L. E. Watson. 2007. An ITS phylogeny of tribe Senecioneae (Asteraceae) and a new delimitation of *Senecio* L. *Taxon* 56: 1077–1104.

Pelser, P. B., A. H. Kennedy, E. J. Tepe, J. B. Shidler, B. Nordenstam, J. W. Kadereit & L. E. Watson. 2010. Patterns and causes of incongruence between plastid and nuclear Senecioneae (Asteraceae) phylogenies. *Amer. J. Bot.* 97: 856–873.

Pelton, J. 1961. An investigation of the ecology of *Mertensia ciliata* in Colorado. *Ecology* 42: 38–52.

Pemberton, R. W. 2000. Predictable risk to native plants in weed biological control. *Oecologia* 125: 489–494.

Pence, V. C. & J. R. Clark. 2005. Desiccation, cryopreservation and germination of seeds of the rare wetland species, *Plantago cordata* Lam. *Seed Sci. Technol.* 33: 767–770.

Pence, V. C., E. O. Guerrant & A. N. Raven. 2006. Cytokinin stimulation of seed germination in *Rorippa subumbellata* Rollins. *Seed Sci. Technol.* 34: 241–245.

Penfound, W. T. 1940a. The biology of *Achyranthes philoxeroides* (Mart.) Standley. *Amer. Midl. Naturalist* 24: 248–252.

Penfound, W. T. 1940b. The biology of *Dianthera americana* L. *The Amer. Midl. Naturalist* 24: 242–247.

Penfound, W. T. 1953. Plant communities of Oklahoma lakes. *Ecology* 34: 561–583.

Penfound, W. T. & A. G. Watkins. 1937. Phytosociological studies in the pinelands of southeastern Louisiana. *Amer. Midl. Naturalist* 18: 661–682.

Penfound, W. T. & E. S. Hathaway. 1938. Plant communities in the marshlands of southeastern Louisiana. *Ecol. Monogr.* 8: 1–56.

Penfound, W. T. & T. T. Earle. 1948. The biology of the water hyacinth. *Ecol. Monogr.* 18: 447–472.

Penfound, W. T., B. G. Efron & J. J. Morrison. 1930. A survey of the herbaceous plants in New Orleans in relation to allergy. *J. Allergy* 1: 369–374.

Penfound, W. T., T. F. Hall & D. Hess. 1945. The spring phenology of plants in and around the reservoirs in north Alabama with particular reference to malaria control. *Ecology* 26: 332–352.

Peng, C.-I. 1984. *Ludwigia ravenii* (Onagraceae), a new species from the coastal plain of the southeastern United States. *Syst. Bot.* 9: 129–132.

Peng, C.-I. 1988. The biosytematics of *Ludwigia* sect. *Microcarpium* (Onagraceae). *Ann. Missouri Bot. Gard.* 75: 970–1003.

Peng C.-I. 1989. The systematics and evolution of *Ludwigia* sect. *Microcarpium* (Onagraceae). *Ann. Missouri Bot. Gard.* 76: 221–302.

Peng, C.-I., C.-H. Chen, W.-P. Leu & H.-F. Yen. 1998. *Pluchea* Cass. (Asteraceae: Inuleae) in Taiwan. *Bot. Bull. Acad. Sin.* 39: 287–297.

Peng, C.-I., C. L. Schmidt, P. C. Hoch & P. H. Raven. 2005. Systematics and evolution of *Ludwigia* section *Dantia* (Onagraceae). *Ann. Missouri Bot. Gard.* 92: 307–359.

Peng, H., C. Qu, Z. Yang & Y. Liu. 1997. Studies on antiviral effect of extract from *Alternanthera philoxeroides* Griseb. on epidemic hemorrhagic fever virus in vivo. *Antiviral Res.* 34: 90.

Pennell, F. W. 1910. Flora of the Conowingo barrens of southeastern Pennsylvania. *Proc. Acad. Nat. Sci. Philadelphia* 62: 541–584.

Pennell, F. W. 1919. Scrophulariaceæ of the Southeastern United States. *Proc. Acad. Nat. Sci. Philadelphia* 71: 224–291.

Pennell, F. W. 1929. *Agalinis* and allies in North America: II. *Proc. Acad. Nat. Sci. Philadelphia* 81: 111–249.

Pennell, F. W. 1933. A revision of *Synthyris* and *Besseya*. *Proc. Acad. Nat. Sci. Philadelphia* 85: 77–106.

Pennell, F. W. 1935. The Scrophulariaceae of eastern temperate North America. *Monogr. Acad. Nat. Sci. Philadelphia* 1: 1–650.

Pennings, S. C. & R. M. Callaway. 2000. The advantages of clonal integration under different ecological conditions: a community-wide test. *Ecology* 81:709–716.

Pennings, S. C. & D. J. Moore. 2001. Zonation of shrubs in western Atlantic salt marshes. *Oecologia* 126: 587–594.

Pennings, S. C. & C. L. Richards. 1998. Effects of wrack burial in salt-stressed habitats: *Batis maritima* in a southwest Atlantic salt marsh. *Ecography* 21: 630–638.

Pennings, S. C., E. L. Siska & M. D. Bertness. 2001. Latitudinal differences in plant palatability in Atlantic coast salt marshes. *Ecology* 82: 1344–1359.

Pennings, S. C., E. R. Selig, L. T. Houser & M. D. Bertness. 2003. Geographic variation in positive and negative interactions among salt marsh plants. *Ecology* 84: 1527–1538.

Penskar, M. R. 1997. Recovery plan for Houghton's goldenrod (*Solidago houghtonii* A. Gray). U. S. Fish and Wildlife Service, Ft. Snelling, MN. 58 pp.

Percifield, R. J., J. S. Hawkins, J.-A. McCoy, M. P. Widrlechner & J. F. Wendel. 2007. Genetic diversity in *Hypericum* and AFLP markers for species-specific identification of *H. perforatum* L. *Plant Med.* 73: 1614–1621.

Percival, M. S. 1955. The presentation of pollen in certain angiosperms and its collection by *Apis mellifera*. *New Phytol.* 54: 353–368.

Percy, D. M., A. Rung & M. S. Hoddle. 2012. An annotated checklist of the psyllids of California (Hemiptera: Psylloidea). *Zootaxa* 3193: 1–27.

Peredery, O. & M. A. Persinger. 2004. Herbal treatment following post-seizure induction in rat by lithium pilocarpine: *Scutellaria lateriflora* (Skullcap), *Gelsemium sempervirens* (Gelsemium) and *Datura stramonium* (Jimson Weed) may prevent development of spontaneous seizures. *Phytother. Res.* 18: 700–705.

Pereira, A. M. S., B. W. Bertoni, A. Menezes Jr., P. S. Pereira & S. C. Franca. 1998. Soil pH and production of biomass and wedelolactone in field grown *Eclipta alba*. *J. Herbs Spices Med. Plant* 6: 43–48.

Perez, C. J., P. J. Zwank & D. W. Smith. 1996. Survival, movements and habitat use of Aplomado falcons released in southern Texas. *J. Raptor Res.* 30:175–182.

Pérez, D. J., M. L. Menone, E. L. Camadro & V. J. Moreno. 2008. Genotoxicity evaluation of the insecticide endosulfan in the wetland macrophyte *Bidens laevis* L. *Environ. Pollut.* 153: 695–698.

Pérez, D. J., G. Lukaszewicz, M. L. Menone & E. L. Camadro. 2011. Sensitivity of *Bidens laevis* L. to mutagenic compounds. Use of chromosomal aberrations as biomarkers of genotoxicity. *Environ. Pollut.* 159: 281–286.

Perez, R. M. 2003. Antiviral activity of compounds isolated from plants. *Pharm. Biol.* 41: 107–157.

Pérez-García, F., E. Marín, S. Cañigueral & T. Adzet. 1996. Anti-inflammatory action of *Pluchea sagittalis*: Involvement of an antioxidant mechanism. *Life Sci.* 59: 2033–2040.

Pérez-García, F., E. Marín, T. Adzet & S. Cañigueral. 2001. Activity of plant extracts on the respiratory burst and the stress protein synthesis. *Phytomedicine* 8: 31–38.

Perleberg, D. & N. Brown. 2004. Aquatic Vegetation of North Union Lake (DOW 21-0095-00), Stony Lake (DOW 21-0101-00), Lottie Lake (DOW 21-0105-00), Douglas County, Minnesota, August 9–12, 2004. MNDNR Ecological Services Division, Brainerd, MN. 17 pp.

Peroni, P. A. 1994. Seed size and dispersal potential of *Acer rubrum* (Aceraceae) samaras produced by populations in early and late successional environments. *Amer. J. Bot.* 81: 1428–1434.

Perry, J. D. 1971. Biosystematic studies in the North American genus *Sabatia* (Gentianaceae). *Rhodora* 73: 309–369.

Perry, J. E. & R. B. Atkinson. 1997. Plant diversity along a salinity gradient of four marshes on the York and Pamunkey Rivers in Virginia. *Castanea* 62: 112–118.

Perry, J. E. & C. H. Hershner. 1999.Temporal changes in the vegetation pattern in a tidal freshwater marsh. *Wetlands* 19: 90–99.

Perry, L. M. 1933. A revision of the North American species of *Verbena*. *Ann. Missouri Bot. Gard.* 20: 239–356; 358–362.

Perry, M. C. & F. M. Uhler. 1981. Asiatic clam (*Corbicula manilensis*) and other foods used by waterfowl in the James River, Virginia. *Estuaries* 4: 229–233.

Perry, M. C. & F. M. Uhler. 1982. Food habits of diving ducks in the Carolinas. *Proc. Ann. Conf. S. E. Assoc. Fish. Wildl. Agencies* 36: 492–504.

Persson, C. 2001. Phylogenetic relationships in Polygalaceae based on plastid DNA sequences from the *trnL-F* region. *Taxon* 50: 763–779.

Persson, H. 1952. Critical or otherwise interesting bryophytes from Alaska-Yukon. *Bryologist* 55: 1–25.

Persson, H. 1954. Mosses of Alaska-Yukon. *Bryologist* 57: 189–217.

Persson, H. & H. T. Shacklette. 1959. *Drepanocladus trichophyllus* found in North America. *Bryologist* 62: 251–254.

Perveen, A. 1999. Contributions to the pollen morphology of the family Compositae. *Turkish J. Biol.* 23: 523–535.

Pessin, L. J. 1938.The effect of vegetation on the growth of longleaf pine seedlings. *Ecol. Monogr.* 8: 115–149.

Petersen, F. P. & D. E. Fairbrothers. 1983. A serotaxonomic appraisal of *Amphipterygium* and *Leitneria*—two amentiferous taxa of Rutiflorae (Rosidae). *Syst. Bot.* 8: 134–148.

Petersen, G., O. Seberg & S. Larsen. 2002. The phylogenetic and taxonomic position of *Lilaeopsis* (Apiaceae), with notes on the applicability of ITS sequence data for phylogenetic reconstruction. *Austral. Syst. Bot.* 15: 181–191.

Peterson, J. E. & A. H. Baldwin. 2004. Variation in wetland seed banks across a tidal freshwater landscape. *Amer. J. Bot.* 91: 1251–1259.

Peterson, L. A. 1977. *A Field Guide to Edible Wild Plants: Eastern and Central North America*. Houghton Mifflin, New York. 330 pp.

Petrović, J., N. Stavretović, S. Ćurčić, I. Jelić & B. Mijović. 2013. Invasive plant species and ground beetles and ants as potential of the biological control: a case of the Bojčin forest nature monument (Vojvodina Province, Serbia). *Šumarski List* 137: 61–69.

Petrunak, E., A. C. Kester, Y. Liu, C. S. Bowen-Forbes, M. G. Nair & G. E. Henry. 2009. New benzophenone O-glucoside from *Hypericum ellipticum*. *Nat. Prod. Commun.* 4: 507–510.

Pettengill, J. B. & M. C. Neel. 2008. Phylogenetic patterns and conservation among North American members of the genus *Agalinis* (Orobanchaceae). *BMC Evol. Biol.* 8(1): 264. doi: 10.1186/1471-2148-8-264.

Pettengill, J. B. & M. C. Neel. 2011. A sequential approach using genetic and morphological analyses to test species status: the case of United States federally endangered *Agalinis acuta* (Orobanchaceae). *Amer. J. Bot.* 98: 859–871.

Pettingill Jr., O. S. 1939. Additional information on the food of the American woodcock. *Wilson Bull.* 51: 78–82.

Petty, R. O. and A. M. Petty. 2005. *Wild Plants in Flower– Wetlands and Quiet Waters of the Midwest*. Quarry Books, Bloomington, IN. 100 pp.

Pfauth, M. & M. Sytsma. 2005. *Alaska Aquatic Plant Survey Report 2005*. Center for Lakes and Reservoirs Publications and Presentations. Paper 12. 20 pp. Available online: http://pdxscholar.library.pdx.edu/centerforlakes_pub/12.

Pfeil, B. E., C. L. Brubaker, L. A. Craven & M. D. Crisp. 2002. Phylogeny of *Hibiscus* and the tribe *Hibisceae* (Malvaceae) using chloroplast DNA sequences of *ndhF* and the *rpl16* intron. *Syst. Bot.* 27: 333–350.

Pfeil, B. E., C. L. Brubaker, L. A. Craven & M. D. Crisp. 2004. Paralogy and orthology in the Malvaceae *rpb2* gene family: Investigation of gene duplication in *Hibiscus*. *Mol. Biol. Evol.* 21: 1428–1437.

Philbrick, C. T. 1984a. Aspects of floral biology, breeding system, and seed and seedling biology in *Podostemum ceratophyllum* (Podostemaceae). *Syst. Bot.* 9: 166–174.

Philbrick, C. T. 1984b. Pollen tube growth within vegetative tissues of *Callitriche* (Callitrichaceae). *Amer. J. Bot.* 71: 882–886.

Philbrick, C. T. 1992. Isozyme variation and population structure in *Podostemum ceratophyllum* Michx. (Podostemaceae): implications for colonization of glaciated North America. *Aquat. Bot.* 43: 311–325.

Philbrick, C. T. 1993. Underwater cross-pollination in *Callitriche hermaphroditica* (Callitrichaceae): Evidence from random amplified polymorphic DNA markers. *Amer. J. Bot.* 80: 391–394.

Philbrick, C. T. & G. J. Anderson. 1992. Pollination biology in the Callitrichaceae. *Syst. Bot.* 17: 282–292.

Philbrick, C. T. & L. M. Bernardello. 1992. Taxonomic and geographic distribution of internal geitonogamy in New World *Callitriche* (Callitrichaceae). *Amer. J. Bot.* 79: 887–890.

Philbrick, C. T. & D. H. Les. 1996. Evolution of aquatic angiosperm reproductive systems. *BioScience* 46: 813–826.

Philbrick, C. T. & D. H. Les. 2000. Phylogenetic studies in *Callitriche* (Callitrichaceae): implications for interpretation of ecological, karyological and pollination system evolution. *Aquat. Bot.* 68: 123–141.

Philbrick, C. T. & A. Novelo R. 2004. Monograph of *Podostemum* (Podostemaceae). *Syst. Bot. Monogr.* 70: 1–106.

Philbrick, C. T., R. A. Aakjar & R. L. Stuckey. 1998. Invasion and spread of *Callitriche stagnalis* (Callitrichaceae) in North America. *Rhodora* 100: 25–38.

Philbrick, C. T., M. Vomela & A. R. Novelo. 2006. Preanthesis cleistogamy in the genus *Podostemum* (Podostemaceae). *Rhodora* 108: 195–202.

Philbrick, C. T., P. K. B. Philbrick & B. M. Lester. 2015. Root fragments as dispersal propagules in the aquatic angiosperm *Podostemum ceratophyllum* Michx. (hornleaf riverweed, Podostemaceae). *N. E. Naturalist* 22: 643–647.

Philcox, D. 1970. A taxonomic revision of the genus *Limnophila* R. Br. (Scrophulariaceae). *Kew Bull.* 24: 101–170.

Philcox, D. 1990. Scrophulariaceae. *In*: G. V. Pope & E. Launert (eds.). *Flora Zambesiaca, volume 8, part 2*. Royal Botanic Gardens, Kew, UK. 179 pp.

Philipp, M. 1980. Reproductive biology of *Stellaria longipes* Goldie as revealed by a cultivation experiment. *New Phytol.* 85: 557–569.

Phillippe, L. R., W. C. Handel, S. L. Horn, F. M. Harty & J. E. Ebinger. 2003. Vascular flora of Momence wetlands, Kankakee County, Illinois. *Trans. Illinois State Acad. Sci.* 96: 271–294.

Phillips, R. B. 1982. Systematics of *Parnassia* (Parnassiaceae): generic overview and revision of North American taxa. PhD dissertation. University of California, Berkeley, CA.

Phillips, R. L., N. K. McDougald & J. Sullins. 1996. Plant preference of sheep grazing in the Mojave Desert. *Rangelands* 18: 141–144.

Philomena P. A. & C. K. Shah. 1985. Unusual germination and seedling development in two monocotyledonous dicotyledons. *Proc. Indian Acad. Sci.* 95: 221–225.

Phimmasan, H., S. Kongvongxay, C. Ty & T. R. Preston. 2004. Water spinach (*Ipomoea aquatica*) and Stylo 184 (*Stylosanthes guianensis* CIAT 184) as basal diets for growing rabbits. *Livestock Res. Rural Dev.* 16: 46–59.

Phipps, J. B. 1988. *Crataegus* (Maloideae, Rosaceae) of the Southeastern United States, I. Introduction and series Aestivales. *J. Arnold Arbor.* 69: 401–431.

Phipps, J. B. 1998. Synopsis of *Crataegus* series *Apiifoliae, Cordatae, Microcarpae*, and *Brevispinae* (Rosaceae subfam. Maloideae). *Ann. Missouri Bot. Gard.* 85: 475–491.

Phipps, J. B. 1999. The relationships of the American black fruited hawthorns *Crataegus erythropoda*, *C. rivularis*, *C. saligna* and *C. brachyacantha* to *C.* ser. *Douglasianae* (Rosaceae). *Sida* 18: 647–660.

Phipps, J. B. 2014. 64. *Crataegus* Linnaeus. pp. 491–643. *In*: N. R. Morin (convening ed.), *Flora of North America North of Mexico, Vol. 9: Magnoliophyta: Picramniaceae to Rosaceae*. Oxford University Press, New York.

Picchioni, G. A. & C. J. Graham. 2001. Salinity, growth, and ion uptake selectivity of container-grown *Crataegus opaca*. *Sci. Hort.* 90: 151–166.

Pickart, A. 2006. Vegetation of Diked Herbaceous Wetlands of Humboldt Bay National Wildlife Refuge: Classification, Description, and Ecology. U. S. Fish and Wildlife Service, Humboldt Bay National Wildlife Refuge, Arcata, CA. 81 pp.

Pickens, A. L. 1927. Unique method of pollination by the ruby-throat. *Auk* 44: 14–17.

Pickens, A. L. 1930. Favorite colors of hummingbirds. *Auk* 47: 346–352.

Pickford, G. D. & E. H. Reid. 1943. Competition of elk and domestic livestock for summer range forage. *J. Wildl. Manag.* 7: 328–332.

Picking, D. J. & P. L. M. Veneman. 2004. Vegetation patterns in a calcareous sloping fen of southwestern Massachusetts, USA. *Wetlands* 24: 514–528.

Pickwell, G. & E. Smith. 1938. The Texas nighthawk in its summer home. *Condor* 40: 193–215.

Picó, B. & F. Nuez. 2000. Minor crops of Mesoamerica in early sources (II). Herbs used as condiments. *Genet. Resour. Crop Evol.* 47: 541–552.

Piechura, J. E. & D. E. Fairbrothers. 1983. The use of protein-serological characters in the systematics of the family Oleaceae. *Amer. J. Bot.* 70: 780–789.

Piehl, M. A. 1965. Studies of root parasitism in *Pedicularis lanceolata*. *Michigan Bot.* 4: 75–81.

Pierce, R. J. 1977. *Wetland Plants of the Eastern United States*. NADP 200–1–1 (includes supplement 1, 1979.). U. S. Army Corps of Engineers, North Atlantic Division, New York. 328 pp.

Pieroni, A. & L. L. Price. 2006. *Eating and Healing: Traditional Food as Medicine*. Haworth Press Inc., Binghamton, New York. 406 pp.

Pieroni, A., S. Nebel, R. F. Santoro & M. Heinrich. 2005. Food for two seasons: Culinary uses of non-cultivated local vegetables and mushrooms in a south Italian village. *Int. J. Food Sci. Nutr.* 56: 245–272.

Pieroni, A., V. Janiak, C. M. Dürr, S. Lüdeke, E. Trachsel & M. Heinrich. 2002. *In vitro* antioxidant activity of non-cultivated vegetables of ethnic Albanians in southern Italy. *Phytother. Res.* 16: 467–473.

Pietropaolo, J. & P. Pietropaolo. 1993. *Carnivorous Plants of the World*. Timber Press, Inc., Portland, OR. 206 pp.

Pigliucci, M. 2004. Natural selection and its limits: where ecology meets evolution. pp. 29–34. *In*: R. Casagrandi & P. Melià (eds.) *Ecologia. Atti del XIII Congresso Nazionale della Società Italiana di Ecologia (Como, 8–10 settembre 2003)*. Aracne, Rome, Italy.

Pigliucci, M. & C. D. Schlichting. 1995. Ontogenetic reaction norms in *Lobelia siphilitica* (Lobeliaceae): response to shading. *Ecology* 76: 2134–2144.

Pigliucci, M., P. Diiorio & C. D. Schlichting. 1997. Phenotypic plasticity of growth trajectories in two species of *Lobelia* in response to nutrient availability. *J. Ecol.* 85: 265–276.

Pill, W. G., R. H. Bender, A. C. Pie, J. K. Marvel & E. E. Veacock. 2000. Responses of six wildflower species to seed matric priming. *J. Environ. Hort.* 18: 160–165.

Pinder, J. E., T. G. Hinton & F. W. Whicker. 2006. Foliar uptake of cesium from the water column by aquatic macrophytes. *J. Environ. Radioact.* 85: 23–47.

Pinna, S., H. Varady-Szabo & M. Côté. 2010. Les espèces à statut précaire associées à la forêt gaspésienne. Consortium en foresterie Gaspésie-Les-Îles, Gaspé, QC. 31 pp.

Pino, J. & E. de Roa. 2007. Population biology of *Kosteletzkya pentacarpos* (Malvaceae) in the Llobregat delta (Catalonia, NE of Spain). *Plant Ecol.* 188: 1–16.

Piovano, M. A. & L. M. Bernardello. 1991. Chromosome numbers in Argentinean Acanthaceae. *Syst. Bot.* 16: 89–97.

Piqueras, J. 1999. Herbivory and ramet performance in the clonal herb *Trientalis europaea* L. *J. Ecol.* 87: 450–460.

Pisula, N. L. & S. J. Meiners. 2010. Allelopathic effects of goldenrod species on turnover in successional communities. *Amer. Midl. Naturalist* 163: 161–172.

Pitts-Singer, T. L., J. L. Hanula & J. L. Walker. 2002. Insect pollinators of three rare plants in a Florida longleaf pine forest. *Florida Entomol.* 85: 308–316.

Płachno, B. J., L. Adamec, I. K. Lichtscheidl, M. Peroutka, W. Adlassnig & J. Vrba. 2006. Fluorescence labelling of phosphatase activity in digestive glands of carnivorous plants. *Plant Biol.* 8: 813–820.

Plante, S. 2000. Documentation chromosomique. Contribution no. 3. *Ludoviciana* 29: 81–82.

Platenkamp, G. A. J. 1998. Patterns of vernal pool biodiversity at Beale Air Force Base. pp. 151–160. *In:* C.W. Witham, E. T. Bauder, D. Belk, W. R. Ferren Jr. & R. Ornduff (eds), *Ecology, Conservation, and Management of Vernal Pool Ecosystems-Proceedings from a 1996 Conference*. California Native Plant Society, Sacramento, CA.

Platt, S. G., R. M. Elsey, H. Liu, T. R. Rainwater, J. C. Nifong, A. E. Rosenblatt, M. R. Heithaus & F. J. Mazzotti. 2013. Frugivory and seed dispersal by crocodilians: an overlooked form of saurochory? *J. Zool.* 291: 87–99.

Platt, W. J., S. M. Carr, M. Reilly & J. Fahr. 2006. Pine savanna overstorey influences on ground-cover biodiversity. *Appl. Veg. Sci.* 9: 37–50.

Plitmann, U. 2002. Agamospermy is much more common than conceived: A hypothesis. *Israel J. Plant Sci.* 50: S111–S117.

Plunkett, G. M. 2001. Relationship of the order Apiales to subclass *Asteridae*: a re-evaluation of morphological characters based on insights from molecular data. *Edinburgh J. Bot.* 58: 183–200.

Plunkett, G. M. & G. W. Hall. 1995. The vascular flora and vegetation of western Isle of Wight County, Virginia. *Castanea* 60: 30–59.

Plunkett, G. M. & P. P. Lowry II. 2001. Relationships among 'ancient araliads' and their significance for the systematics of Apiales. *Mol. Phylogen. Evol.* 19: 259–276.

Plunkett, G. M., D. E. Soltis & P. S. Soltis 1996a. Higher level relationships of Apiales (Apiaceae and Araliaceae) based on phylogenetic analysis of *rbcL* sequences. *Amer. J. Bot.* 83: 499–515.

Plunkett, G. M., D. E. Soltis & P. S. Soltis 1996b. Evolutionary patterns in Apiaceae: inferences based on *matK* sequence data. *Syst. Bot.* 21: 477–495.

Plunkett, G. M., D. E. Soltis & P. S. Soltis. 1997. Clarification of the relationship between Apiaceae and Araliaceae based on *matK* and *rbcL* sequence data. *Amer. J. Bot.* 84: 565–580.

Plunkett, G. M., J. Wen & P. P. Lowry. 2004a. Infrafamilial classifications and characters in Araliaceae: Insights from the phylogenetic analysis of nuclear (ITS) and plastid (*trnL-trnF*) sequence data. *Plant Syst. Evol.* 245: 1–39.

Plunkett, G. M., G. T. Chandler, P. P. Lowry, S. M. Pinney & T. S. Sprenkle. 2004b. Recent advances in understanding Apiales and a revised classification. *S. Afr. J. Bot.* 70: 371–381.

Podolak, I., A. Galanty & Z. Janeczko. 2005. Cytotoxic activity of embelin from *Lysimachia punctata*. *Fitoterapia* 76: 333–335.

Podoplelova, Y. & G. Ryzhakov. 2005. Phylogenetic analysis of the order Nymphaeales based on the nucleotide sequences of the chloroplast ITS2–4 region. *Plant Sci.* 169: 606–611.

Pogue, M. G. 2005. The Plusiinae (Lepidoptera: Noctuidae) of Great Smoky Mountains National Park. *Zootaxa* 1032: 1–28.

Poiani, K. A. & P. M. Dixon. 1995. Seed banks of Carolina bays: potential contributions from surrounding landscape vegetation. *Amer. Midl. Naturalist* 134: 140–154.

Poiani, K. A. & W. C. Johnson. 1988. Evaluation of the emergence method in estimating seed bank composition of prairie wetlands. *Aquat. Bot.* 32: 91–97.

Poindexter, D. B. & J. B. Nelson. 2011. A new hedge-nettle (*Stachys*: Lamiaceae) from the southern Appalachian mountains. *J. Bot. Res. Inst. Texas* 5: 405–414.

Pojar, J. 1973. Pollination of typically anemophilous salt marsh plants by bumble bees, *Bombus terricola occidentalis* Grne. *Amer. Midl. Naturalist* 89: 448–451.

Pojar, J. 1974. Reproductive dynamics of four plant communities of southwestern British Columbia. *Can. J. Bot.* 52: 1819–1834.

Pojar, J. 1975. Hummingbird flowers of British Columbia. *Syesis* 8: 25–28.

Pojar, J. 1991. Chapter 19: Non-tidal wetlands. pp. 275–280. *In: D.* Meidinger & J. Pojar (eds.), *Ecosystems of British Columbia*. British Columbia Ministry of Forests, Victoria, BC.

Polania, J. 1990. Physiological adaptations in some species of mangroves. *Acta Biol. Columb.* 2: 23–36.

Poljakoff-Mayber, A., G. F. Somers, E. Werker & J. L. Gallagher. 1992. Seeds of *Kosteletzkya virginica* (Malvaceae): Their structure, germination and salt tolerance. I. Seed structure and germination. *Amer. J. Bot.* 79: 249–256.

Poljakoff-Mayber, A., G. F. Somers, E. Werker & J. L. Gallagher. 1994. Seeds of *Kosteletzkya virginica* (Malvaceae): Their structure, germination and salt tolerance. II. Germination and salt tolerance. *Amer. J. Bot.* 81: 54–59.

Polley, H. W. & L. L. Wallace. 1986. The relationship of plant species heterogeneity to soil variation in buffalo wallows. *Southwest. Naturalist* 31: 493–501.

Pollock, M. M., R. J. Naiman & T. A. Hanley. 1998. Plant species richness in riparian wetlands—a test of biodiversity theory. *Ecology* 79: 94–105.

Pontieri, V. & T. L. Sage. 1999. Evidence of stigmatic self-incompatibility, pollination induced ovule enlargement and transmitting tissue exudates in the paleoherb, *Saururus cernuus* L. (Saururaceae). *Ann. Bot.* 84: 507–519.

Ponzio, K. J., S. J. Miller & M. Ann Lee. 2004. Long-term effects of prescribed fire on *Cladium jamaicense* Crantz and *Typha domingensis* Pers. densities. *Wetlands Ecol. Manag.* 12: 123–133.

Pool, R. J. 1910. The Erysiphaceae of Nebraska. *Univ. Nebraska Stud.* 10: 59–84.

Poole, J. & D. E. Bowles. 1999. Habitat characterization of Texas wild-rice (*Zizania texana* Hitchcock), an endangered aquatic macrophyte from the San Marcos River, TX, USA. *Aquatic Conservation: Mar. Freshwater Ecosyst.* 9: 291–302.

Poole, J. M., W. R. Carr, D. M. Price & J. R. Singhurst. 2007. *Rare Plants of Texas: A Field Guide.* Texas A & M University Press, College Station, TX. 640 pp.

Popov, S. V., R. G. Ovodova, G. Y. Popova, I. R. Nikitina & Y. S. Ovodov. 2005. Adhesion of human neutrophils to fibronectin is inhibited by comaruman, pectin of marsh cinquefoil *Comarum palustre* L., and by its fragments. *Biochemistry (Moscow)* 70: 108–112.

Porcher, F. P. 1863. *Resources of the Southern Fields and Forests, Medical, Economical, and Agricultural. Being also a Medical Botany of the Confederate States; with Practical Information on the Useful Properties of the Trees, Plants, and Shrubs.* Evans & Cogswell, Charleston, SC. 601 pp.

Porcher, R. D. 1981. The vascular flora of the Francis Beidler Forest in Four Holes Swamp, Berkeley and Dorchester counties, South Carolina. *Castanea* 46: 248–280.

Porcher, R. D. & D. A. Rayner. 2001. *A Guide to the Wildflowers of South Carolina.* University of South Carolina Press, Columbia, SC. 496 pp.

Porembski, S. & M. Koch. 1999. Inulin occurrence in *Sphenoclea* (Sphenocleaceae). *Plant Biol.* 1: 288–289.

Porter, C. C. 1983. Ecological notes on lower Río Grande Valley *Augochloropsis* and *Agapostemon* (Hymenoptera: Halictidae). *Florida Entomol.* 66: 344–353.

Porter, J. M. & N. Fraga. 2005. A quantitative analysis of pollen variation in two southern California perennial *Helianthus* (Heliantheae: Asteraceae). *Crossosoma* 31: 1–12.

Porter Jr., C. L. 1967. Composition and productivity of a subtropical prairie. *Ecology* 48: 937–942.

Porter-Utley, K. 1997. Santeria: an ethnobotanical study in Miami, Florida, USA. MS thesis. University of Florida, Gainesville, FL. 786 pp.

Portinga, R. L. & R. A. Moen. 2015. A novel method of performing moose browse surveys. *Alces* 51: 107–22.

Posluszny, U., M. J. Sharp & P. A. Keddy. 1984. Vegetative propagation in *Rhexia virginica* (Melastomataceae): some morphological and ecological considerations. *Can. J. Bot.* 62: 2118–2121.

Pospelova, M. L. & O. D. Barnaulov. 2000. The antihypoxant and antioxidant effects of medicinal plants as the basis for their use in destructive diseases of the brain. *Human Physiol.* 26: 86–97.

Posto, A. L. & L. A. Prather. 2003. The evolutionary and taxonomic implications of RAPD data on the genetic relationships of *Mimulus michiganensis* (comb. et stat. nov.: Scrophulariaceae). *Syst. Bot.* 28: 172–178.

Potgieter, K. & V. A. Albert. 2001. Phylogenetic relationships within Apocynaceae s.l. based on *trnL* intron and *trnL-F* spacer sequences and propagule characters. *Ann. Missouri Bot. Gard.* 88: 523–549.

Potter, D., F. Gao, P. E. Bortiri, S.-H. Oh & S. Baggett. 2002. Phylogenetic relationships in Rosaceae inferred from chloroplast *matK* and *trnL-trnF* nucleotide sequence data. *Plant Syst. Evol.* 231: 77–89.

Potter, D., T. Eriksson, R. C. Evans, S. Oh, J. E. E. Smedmark, D. R. Morgan, M. Kerr, K. R. Robertson, M. Arsenault, T. A. Dickinson & C. S. Campbell. 2007a. Phylogeny and classification of Rosaceae. *Plant Syst. Evol.* 266: 5–43.

Potter, D., S. M. Still, T. Grebenc, D. Ballian, G. Božič, J. Franjiæ & H. Kraigher. 2007b. Phylogenetic relationships in tribe Spiraeeae (Rosaceae) inferred from nucleotide sequence data. *Plant Syst. Evol.* 266: 105–118.

Potts, L. & J. J. Krupa. 2016. Does the dwarf sundew (*Drosera brevifolia*) attract prey? *Amer. Midl. Naturalist* 175: 233–241.

Powell, A. M. & B. Wauer. 1990. A new species of *Viola* (Violaceae) from the Guadalupe Mountains, Trans-Pecos Texas. *Sida* 14: 1–6.

Powell, A. M., D. W. Kyhos & P. H. Raven. 1974. Chromosome numbers in Compositae. X. *Amer. J. Bot.* 61: 909–913.

Powell, A. N. 1993. Nesting habitat of Belding's Savannah sparrows in coastal salt marshes. *Wetlands* 13: 219–223.

Powell, E. A. & K. A. Kron. 2001. An analysis of the phylogenetic relationships in the wintergreen group (*Diplycosia, Gaultheria, Pernettya, Tepuia*; Ericaceae). *Syst. Bot.* 26: 808–817.

Powell, G. S. 2000. Charred, non-maize seed concentrations in the American bottom area: examples from the Westpark Site (11-MO-96), Monroe County, Illinois. *Midcont. J. Archaeol.* 25: 27–48.

Powell, K. I. & T. M. Knight. 2009. Effects of nutrient addition and competition on biomass of five *Cirsium* species (Asteraceae), including a serpentine endemic. *Int. J. Plant Sci.* 170: 918–925.

Powell, K. I., K. N. Krakos & T. M. Knight. 2011. Comparing the reproductive success and pollination biology of an invasive plant to its rare and common native congeners: a case study in the genus *Cirsium* (Asteraceae). *Biol. Invas.* 13: 905–917.

Powell, M., V. Savolainen, P. Cuénoud, J.-F. Manen & S. Andrews. 2000. The mountain holly (*Nemopanthus mucronatus*, Aquifoliaceae) revisited with molecular data. *Kew Bull.* 55: 341–347.

Powers, K. D., R. E. Noble & R. H. Chabreck. 1978. Seed distribution by waterfowl in southwestern Louisiana. *J. Wildl. Manag.* 42: 598–605.

Pozharitskaya, O. N., A. N. Shikov, M. N. Makarova, V. M. Kosman, N. M. Faustova, S. V. Tesakova, V. G. Makarov & B. Galambosi. 2010. Anti-inflammatory activity of a HPLC-fingerprinted aqueous infusion of aerial part of *Bidens tripartita* L. *Phytomedicine* 17: 463–468.

Prabhu, N., J. P. Innocent, P. Chinnaswamy, K. Natarajaseenivasan & L. Sarayu. 2008. *In vitro* evaluation of *Eclipta alba* against serogroups of *Leptospira interrogans*. *Indian J. Pharm. Sci.* 70: 788–791.

Pradhanang, P. M. & M. T. Momol. 2001. Survival of *Ralstonia solanacearum* in soil under irrigated rice culture and aquatic weeds. *J. Phytopathol.* 149: 707–711.

Prankevicius, A. B. & D. M.Cameron. 1991. Bacterial dinitrogen fixation in the leaf of the northern pitcher plant (*Sarracenia purpurea*). *Can. J. Bot.* 69: 2296–2298.

Pranty, B. & M. D. Scheuerell. 1997. First summer record of the Henslow's Sparrow in Florida. *Florida Field Naturalist* 25: 64–66.

Prasad, J. S. & Y. S. Rao. 1979. Nematicidal properties of the weed *Eclipta alba* Hassk (Compositae). *Rivista Parassitol.* 40: 87–90.

Prasad, K. V., K. Bharathi & K. K. Srinivasan. 1994. Evaluation of *Ammannia baccifera* Linn. for antiurolithic activity in albino rats. *Indian J. Exp. Biol.* 32: 311–313.

Prather, L. A., C. J. Ferguson & R. K. Jansen. 2000. Polemoniaceae phylogeny and classification: implications of sequence data from the chloroplast gene *ndhF. Amer. J. Bot.* 87: 1300–1308.

Prather, L. A., A. K. Monfils, A. L. Posto & R. A. Williams. 2002. Monophyly and phylogeny of *Monarda* (Lamiaceae): evidence from the internal transcribed spacer (ITS) region of nuclear ribosomal DNA. *Syst. Bot.* 27: 127–137.

Pratt, D. B. & L. G. Clark. 2001. *Amaranthus rudis* and *A. tuberculatus*-one species or two? *J. Torrey Bot. Soc.* 128: 282–296.

Prayoonrat, P. 2005. Biodiversity of medicinal weeds in Chonburi region, Thailand. *ISHS Acta Hort.* 675: 23–29.

Prena, J. 2008. Review of *Odontocorynus* Schönherr (Coleoptera: Curculionidae: Baridinae) with descriptions of four new species. *Coleopterists Bull.* 62: 243–277.

Prendusi, T. E., D. U. Atwood, B. R. Palmer & R. Rodriguez. 1996. Interagency conservation biology program for Arizona willow (*Salix arizonica* Dorn). pp. 224–230. *In*: J. Maschinski, H. D. Hammond & L. Holter (eds.), *Southwestern Rare and Endangered Plants: Proceedings of the Second Conference.* General Technical Report RM-GTR 283. United States Department of Agriculture, Rocky Mountain Forest and Range Experiment Station, Fort Collins, CO.

Prenger, J. 2005. *John C. and Mariana Jones/Hungryland WEA Plant Community Type Mapping. Analysis Results.* Florida Fish & Wildlife Conservation Commission, Florida Wildlife Research Institute & Division of Habitat and Species Conservation, Florida Natural Areas Inventory. Available online: http://myfwc.com/media/121434/Hungryland_Community_Mapping_Report.pdf (accessed December 5, 2013).

Prescott, G. W. 1953. Preliminary notes on the ecology of freshwater algae in the Arctic slope, Alaska, with descriptions of some new species. *Amer. Midl. Naturalist* 50: 463–473.

Prescott, G. W. 1980. *How to Know the Aquatic Plants. 2nd Edition.* The pictured key nature series. Wm. C. Brown Company Publishers, Dubuque, IA. 158 pp.

Preston, C. D. & J. M. Croft. 1997. *Aquatic Plants in Britain and Ireland.* Harley Books, Colchester, Essex, UK. 365 pp.

Preston, R. E. 1986. Pollen-ovule ratios in the Cruciferae. *Amer. J. Bot.* 73: 1732–1740.

Preston, R. E. 2011. *Brodiaea matsonii* (Asparagaceae: Brodiaeoideae) a new species from Shasta County, California. *Madroño* 57: 261–267.

Prevete, K. J., R. T. Fernandez & W. B. Miller. 2000. Drought response of three ornamental herbaceous perennials. *J. Amer. Soc. Hort. Sci.* 125: 310–317.

Price, E. T. 1960. Root digging in the Appalachians: the geography of botanical drugs. *Geogr. Rev.* 50: 1–20.

Price, H. J. & K. Bachmann. 1975. DNA content and evolution in the Microseridinae. *Amer. J. Bot.* 62: 262–267.

Price, H. J. & M. Baranova. 1976. Evolution of DNA content in higher plants. *Bot. Rev.* 42: 27–52.

Price, J. 2001. DBS spring field trip: Walker Ridge walkabout. *Lasthenia* 18: 6.

Price, P. W. & M. F. Wilson. 1979. Abundance of herbivores on six milkweed species in Illinois. *Amer. Midl. Naturalist* 101: 76–86.

Price, P. W., G. L. Waring, R. Julkunen-Tiitto, J. Tahvanainen, H. Mooney & T. P. Craig. 1989. Carbon-nutrient balance hypothesis in within-species phytochemical variation of *Salix lasiolepis. J. Chem. Ecol.* 15: 1117–1131.

Price, W. C. 1940. Comparative host ranges of six plant viruses. *Amer. J. Bot.* 27: 530–541.

Priester, D. S. 1990. *Magnolia virginiana* L. Sweet Bay. pp. 449–454. *In:* R. M. Burns & B. H. Honkala (eds.), *Silvics of North America. Vol. 2, Hardwoods.* Agriculture Handbook 654. USDA, Forest Service, Washington, DC.

Primack, R. B. 1978a. Regulation of seed yield in *Plantago. J. Ecol.* 66: 835–847.

Primack, R. B. 1978b. Evolutionary aspects of wind pollination in the genus *Plantago* (Plantaginaceae). *New Phytol.* 81: 449–458.

Primack, R. B. 1979. Reproductive effort in annual and perennial species of *Plantago* (Plantaginaceae). *Amer. Naturalist* 114: 51–62.

Primack, R. B. 1982. Ultraviolet patterns in flowers, or flowers as viewed by insects. *Arnoldia* 42: 139–146.

Primack, R. B. & C. McCall. 1986. Gender variation in a red maple population (*Acer rubrum*; Aceraceae): a seven-year study of a "polygamodioecious" species. *Amer. J. Bot.* 73: 1239–1248.

Pringle, J. S. 1967. Taxonomy of *Gentiana*, section Pneumonanthae, in eastern North America. *Brittonia* 19: 1–32.

Pringle, J. S. 1968. The status and distribution of *Gentiana linearis* and *G. rubricaulis* in the Upper Great Lakes region. *Michigan Bot.* 7: 99–112.

Pringle, J. S. 1977. *Gentiana linearis* (Gentianaceae) in the southern Appalachians. *Castanea* 42: 1–8.

Pringle, J. S. 1982. The distribution of *Solidago ohioensis. Michigan Bot.* 21: 51–57.

Pringle, J. S. 1997. 6. *Clematis* Linnaeus subg. *Viorna.* pp. 167–176. *In*: N. R. Morin (convening ed.), *Flora of North America North of Mexico, Vol. 3: Magnoliophyta: Magnoliidae and Hamamelidae.* Oxford University Press, New York.

Pringle, J. S. 2004. History and eponymy of the genus name *Amsonia* (Apocynaceae). *Sida* 21: 379–387.

Prins, R. 1968. Comparative ecology of the crayfishes *Orconectes rusticus rusticus* and *Cambarus tenebrosus* in Doe Run, Meade County, Kentucky. *Int. Rev. Ges. Hydrobiol. Hydrograph.* 53: 667–714.

Probert, R. J., J. B. Dickie & M. R. Hart. 1989. Analysis of the effect of cold stratification on the germination response to light and alternating temperatures using selected seed populations of *Ranunculus sceleratus* L. *J. Exp. Bot.* 40: 293–301.

Probert, R. J., S. V. Bogh, A. J. Smith & G. E. Wechsberg. 1991. The effects of priming on seed longevity in *Ranunculus sceleratus* L. *Seed Sci. Res.* 4: 243–249.

Proell, J. 2009. Population sex ratio and size affect pollination, reproductive success, and seed germination in gynodioecious *Lobelia siphilitica*: evidence using experimental populations and microsatellite genotypes. MS thesis. Kent State University, Kent, OH. 96 pp.

Proeseler, G., A. Stanarius, H. G. Kontzog & H. Barth. 1990. Virus infections of aquatic plants. *Arch. Phytopathol. Pflanzenschutz* 26: 19–24.

Proffitt, C. E., R. L. Chiasson, A. B. Owens, K. R. Edwards & S. E. Travis. 2005. *Spartina alterniflora* genotype influences facilitation and suppression of high marsh species colonizing an early successional salt marsh. *J. Ecol.* 93: 404–416.

Proffitt, C. E., E. C. Milbrandt & S. E. Travis. 2006. Red mangrove (*Rhizophora mangle*) reproduction and seedling colonization after Hurricane Charley: comparisons of Charlotte Harbor and Tampa Bay. *Estuaries Coasts* 29: 972–978.

Provancher, L. 1886. Petite Faune Entomologique du Canada et Particulièrement de la Province de Québec: Vol. III, Cinquième Ordre, les Hémiptères. C. Darveau, Quebec. 205 pp.

Prusak, A. C., J. O'Neal & J. Kubanek. 2005. Prevalence of chemical defenses among freshwater plants. *J. Chem. Ecol.* 31: 1145–1160.

Pruski, J. F. & R. L. Hartman. 2012. Synopsis of *Leucosyris*, including synonymous *Arida* (Compositae: Astereae). *Phytoneuron* 98: 1–15.

Pucci, J. R. 2007. The effects of invasive plants on the endangered sunflower, *Pentachaeta lyonii*. MS thesis. California State University, Northridge, CA. 52 pp.

Puff, C. 1976. The *Galium trifidum* group (*Galium* sect. *Aparinoides*, Rubiaceae). *Can. J. Bot.* 54: 1911–1925.

Puff, C. 1977. The *Galium obtusum* group (*Galium* sect. *Aparinoides*, Rubiaceae). *Bull. Torrey Bot. Club* 104: 117–125.

Puff, C., A. Igersheim, R. Buchner & U. Rohrhofer. 1995. The united stamens of Rubiaceae. Morphology, anatomy; their role in pollination ecology. *Ann. Missouri Bot. Gard.* 82: 357–382.

Puijalon, S. & G. Bornette. 2004. Morphological variation of two taxonomically distant plant species along a natural flow velocity gradient. *New Phytol.* 163: 651–660.

Puijalon, S., J.-P. Lena & G. Bornette. 2007. Interactive effects of nutrient and mechanical stresses on plant morphology. *Ann. Bot. (Oxford)* 100: 1297–1305.

Puijalon, S., F. Piola & G. Bornette. 2008. Abiotic stresses increase plant regeneration ability. *Evol. Ecol.* 22: 493–506.

Pulido, C., D. J. H. Keijsers, E. C. H. E. T. Lucassen, O. Pedersen & J. G. M. Roelofs. 2012. Elevated alkalinity and sulfate adversely affect the aquatic macrophyte *Lobelia dortmanna*. *Aquat. Ecol.* 46: 283–295.

Punzalan, D., F. H. Rodd & L. Rowe. 2008. Contemporary sexual selection on sexually dimorphic traits in the ambush bug *Phymata americana*. *Behav. Ecol.* 19: 860–870.

Purcell, N. J. 1976. *Epilobium parviflorum* Schreb. (Onagraceae) established in North America. *Rhodora* 78: 785–787.

Purcifull, D. E. & T. A. Zitter. 1971. Virus diseases affecting lettuce and endive in Florida. *Proc. Florida State Hort. Soc.* 84: 165–168.

Purdy, B. G., R. J. Bayer & S. E. Macdonald. 1994. Genetic variation, breeding system evolution, and conservation of the narrow sand dune endemic *Stellaria arenicola* and the widespread *S. longipes* (Caryophyllaceae). *Amer. J. Bot.* 81: 904–911.

Purdy, B. G., S. E. Macdonald & V. J. Lieffers. 2005. Naturally saline boreal communities as models for reclamation of saline oil sand tailings. *Restorat. Ecol.* 13: 667–677.

Purer, E. A. 1939. Ecological study of vernal pools, San Diego county. *Ecology* 20: 217–229.

Putz, N. & K. H. A. Schmidt. 1999. 'Underground plant mobility' and 'dispersal of diaspores.' Two exemplary case studies for useful examinations of functional morphology. *Syst. Geogr. Plant* 68: 39–50.

Pyke, G. H. 1982. Local geographic distributions of bumblebees near Crested Butte, Colorado: competition and community structure. *Ecology* 63: 555–573.

Pyke, G. H., D. W. Inouye & J. D. Thomson. 2012. Local geographic distributions of bumble bees near Crested Butte, Colorado: competition and community structure revisited. *Environ. Entomol.* 41: 1332–1349.

Pyšek, P., J. Sádlo & B. Mandák. 2002. Catalogue of alien plants of the Czech Republic. *Preslia* 74: 97–186.

Qasair, R., A. D. Headley & D. P. M. Comber. 2003. The distribution and ecology of the arctic plant Iceland purslane (*Koenigia islandica*) in Scotland. *Pakistan J. Biol. Sci.* 6: 252–254.

Qiu, Y-L., M. W. Chase, D. H. Les & C. R. Parks. 1993. Molecular phylogenetics of the Magnoliidae: cladistic analyses of nucleotide sequences of ther plastid gene *rbcL*. *Ann. Missouri Bot. Gard.* 80: 587–606.

Qiu, Y-L., M. W. Chase & C. R. Parks. 1995. A chloroplast DNA phylogenetic study of the eastern Asia-eastern North America disjunct section *Rytidospermum* of *Magnolia* (Magnoliaceae). *Amer. J. Bot.* 82: 1582–1588.

Qiu, Y-L., J. Lee, F. Bernasconi-Quadroni, D. E. Soltis, P. S. Soltis, M. Zanis, E. A. Zimmer, Z. Chen, V. Savolainen & M. W. Chase. 2000. Phylogeny of basal angiosperms: analyses of five genes from three genomes. *Int. J. Plant Sci.* 161: S3–S27.

Qiu, X. & L. Li. 2000. Contact dermatitis owing to *Oenanthe javanica* Blume DC: a case report. *J. Clin. Dermatol.* 29: 233.

Qiyan, L., A. Ye, C. Xu & G. Liu. 2006. Study on methods of breaking seed dormancy in water dropwort, *Oenanthe javanica* (BL.) DC. *Seed* 25: 34–37.

Quanyu, D., C. Shuwei & Z. Xiuying. 1998. Purification of gold-containing wastewater by *Oenanthe javanica* and accumulation of gold in it. *Chinese J. Appl. Ecol.* 9: 107–109.

Quarterman, E. 1957. Early plant succession on abandoned cropland in the central basin of Tennessee. *Ecology* 38: 300–309.

Quattrocchi, U. 2000. *CRC World Dictionary of Plant Names: Common Names, Scientific Names, Eponyms, Synonyms, and Etymology*. 4 vols. CRC Press, Boca Raton, FL. 2896 pp.

Quattrocchi, U. 2012. *CRC World Dictionary of Medicinal and Poisonous Plants: Common Names, Scientific Names, Eponyms, Synonyms, and Etymology*. 5 vols. CRC Press, Boca Raton, FL. 3960 pp.

Quickfall, G. S. 1987. Paludification and climate on the Queen Charlotte Islands during the past 8,000 years. MS thesis. Simon Fraser University, Burnaby, BC. 99 pp.

Quilliam, R. S. & D. L. Jones. 2010. Fungal root endophytes of the carnivorous plant *Drosera rotundifolia*. *Mycorrhiza* 20: 341–348.

Quimby, D. C. 1951. The life history and ecology of the jumping mouse, *Zapus Hudsonius*. *Ecol. Monogr.* 21: 61–95.

Quinby, P. A. 2000. First-year impacts of shelterwood logging on understory vegetation in an old-growth pine stand in central Ontario, Canada. *Environ. Conserv.* 27: 229–241.

Quinlan, A., M. R. T. Dale & C. C. Gates. 2003. Effects of prescribed burning on herbaceous and woody vegetation in northern lowland meadows. *Restorat. Ecol.* 11: 343–350.

Quinn, L. D., M. Kolipinski, V. R. Coelho, B. Davis, J.-M. Vianney, O. Batjargal, M. Alas & S. Ghosh. 2008. Germination of invasive plant seeds after digestion by horses in California. *Nat. Areas. J.* 28: 356–362.

Raabová J., G. Hans, A.-M. Risterucci, A.-L. Jacquemart & O. Raspé. 2010. Development and multiplexing of microsatellite markers in the polyploid perennial herb, *Menyanthes trifoliata* (Menyanthaceae). *Amer. J. Bot.* 97: e31–e33.

Rabb, R. L. 1960. Biological studies of *Polistes* in North Carolina (Hymenoptera: Vespidae). *Ann. Entomol. Soc. Amer.* 53: 111–121.

Rabeler, R. K. 1993. The occurrence of anther smut, *Ustilago violacea* s.l., on *Stellaria borealis* (Caryophyllaceae) in North America. *Contr. Univ. Michigan Herb.* 19: 165–169.

Rabeler, R. K. & R. L. Hartman. 2005a. 43. Caryophyllaceae Jussieu. pp. 5–8. *In*: Flora North America Editorial Committee (eds.), *Flora of North America North of Mexico*, Vol. 5: Magnoliophyta: Caryophyllidae, part 2. Oxford University Press, New York.

Rabeler, R. K. & R. L. Hartman. 2005b. 3. *Spergularia* (Persoon) J. Presl & C. Presl. pp. 16–23. *In*: Flora North America Editorial Committee (eds.), *Flora of North America North of Mexico*, Vol. 5: Magnoliophyta: Caryophyllidae, part 2. Oxford University Press, New York.

Rabinowitz, D. 1978. Dispersal properties of mangrove propagules. *Biotropica* 10: 47–57.

Rabinowitz, D. 1981. Buried viable seeds in a North American tall-grass prairie: the resemblance of their abundance and composition to dispersing seeds. *Oikos* 36: 191–195.

Rabinowitz, D. & J. K. Rapp. 1985. Colonization and establishment of Missouri prairie plants on artificial soil disturbances. III. Species abundance distributions, survivorship, and rarity. *Amer. J. Bot.* 72: 1635–1640.

Rabinowitz, D., J. K. Rapp, V. L. Sork, B. J. Rathcke, G. A. Reese & J. C. Weaver. 1981. Phenological properties of wind-and insect-pollinated prairie plants. *Ecology* 62: 49–56.

Racine, C. H. & J. C. Walters. 1994. Groundwater-discharge fens in the Tanana lowlands, interior Alaska, U.S.A. *Arct. Alp. Res.* 26: 418–426.

Racine, C. H., J. C. Walters & M. T. Jorgenson. 1998a. Airboat use and disturbance of floating mat fen wetlands in interior Alaska, U.S.A. *Arctic* 51: 371–377.

Racine, C. H., M. T. Jorgenson & J. C. Walters. 1998b. Thermokarst vegetation in lowland birch forests on the Tanana Flats, interior Alaska, U.S.A. pp. 927–933 *in* A.G. Lewkowicz & M. Allard (eds.), *Proceedings of the 7th International Conference on Permafrost. Yellowknife, NT, 23–27 June, 1998*. Collection Nordicana No.55. Centre d'études nordiques, Université Laval, Quebec City, Quebec.

Raffauf, R. F. 1996. *Plant Alkaloids: A Guide to their Discovery and Distribution*. Haworth Press, Inc., New York. 279 pp.

Raffauf, R. F. & A. Higurashi. 1988. Notes on the toxicity of *Sphenoclea zeylanica*. *Rev. Gen.* 16: 99–105.

Ragupathy, S. & A. Mahadevan. 1993. Distribution of vesicular-arbuscular mycorrhizae in the plants and rhizosphere soils of the tropical plains, Tamil Nadu, India. *Mycorrhiza* 3: 123–136.

Rahimi, S. & S. Khatoon. 2004. *In situ* studies on the plants exposed to industrial and agricultural pollution in the vicinity of Karachi. I: Effects on meiosis. *Int. J. Biol. Biotechnol.* 1: 293–300.

Rahman, M. M., M. I. I. Wahed, M. Helal, U. Biswas, M. G. Sadik & M. E. Haque. 2001. *In vitro* antibacterial activity of the compounds of *Trapa bispinosa* Roxb. *Sciences* 1: 214–216.

Rahmanzadeh, R., K. Müller, E. Fischer, D. Bartels & T. Borsch. 2005. Linderniaceae and Gratiolaceae are further lineages distinct from Scrophulariaceae (Lamiales). *Plant Biol.* 7: 67–78.

Rahn, K. 1996. A phylogenetic study of the Plantaginaceae. *Bot. J. Linn. Soc.* 120: 145–198.

Rai, U. N. & S. Sinha. 2001. Distribution of metals in aquatic edible plants: *Trapa natans* (Roxb.) Makino and *Ipomoea aquatica* Forsk. *Environ. Monit. Assessm.* 70: 241–252.

Raible, B. 2004. White Rock fen. *Kalmiopsis* 11: 30–35.

Raitviir, A., J. Haines & E. Müller. 1991. A re-evaluation of the ascomycetous genus *Solenopezia*. *Sydowia* 43: 219–227.

Raja, H. A., H. A. Violi & C. A. Shearer. 2010. Freshwater ascomycetes: *Alascospora evergladensis*, a new genus and species from the Florida Everglades. *Mycologia* 102: 33–38.

Rajakaruna, N. 2003. Edaphic differentiation in *Lasthenia*: a model for studies in evolutionary ecology. *Madroño* 50: 34–40.

Rajakaruna, N., M. Y. Siddiqi, J. Whitton, B. A. Bohm & A. D. M. Glass. 2003. Differential responses to Na^+/K^+ and Ca^{2+}/Mg^{2+} in two edaphic races of the *Lasthenia californica* (Asteraceae) complex: A case for parallel evolution of physiological traits. *New Phytol.* 157: 93–103.

Rajakumar, G. & A. A. Rahuman. 2011. Larvicidal activity of synthesized silver nanoparticles using *Eclipta prostrata* leaf extract against filariasis and malaria vectors. *Acta Trop.* 118: 196–203.

Raju, M. V. S. 1969. Development of floral organs in the sites of leaf primordia in *Pinguicula vulgaris*. *Amer. J. Bot.* 56: 507–514.

Ramaley, F. 1919. Vegetation of undrained depressions on the Sacramento plains. *Bot. Gaz.* 68: 380–387.

Ramaley, F. 1934. Influence of supplemental light on blooming. *Bot. Gaz.* 96: 165–174.

Ramsey, J., H. D. Bradshaw Jr. & D. W. Schemske. 2003. Components of reproductive isolation between the monkey-flowers *Mimulus lewisii* and *M. cardinalis* (Phrymaceae). *Evolution* 57: 1520–1534.

Ramstetter, J. & J. Mott-White. 2001. *Ludwigia polycarpa* (Many-fruited false-loosestrife) conservation and research plan. New England Plant Conservation Program, Framingham, MA. 18 pp.

Rand, E. L. 1899. *Subularia aquatica* on Mt. Desert Island. *Rhodora* 1: 155–156.

Rand, T. A. 2000. Seed dispersal, habitat suitability and the distribution of halophytes across a salt marsh tidal gradient. *J. Ecol.* 88: 608–621.

Rands, R. L. 1953. The waterlily in Maya art: a complex of alleged Asiatic origin. *Bull. Bur. Amer. Ethol.* 151: 75–153.

Randhawa, A. S. & K. I. Beamish. 1970. Observations on the morphology, anatomy, classification, and reproductive cycle of *Saxifraga ferruginea*. *Can. J. Bot.* 48: 299–312.

Randhawa, A. S. & K. I. Beamish. 1972. The distribution of *Saxifraga ferruginea* and the problem of refugia in northwestern North America. *Can. J. Bot.* 50: 79–87.

Rangineni, V., D. Sharada & S. Saxena. 2007. Diuretic, hypotensive, and hypocholesterolemic effects of *Eclipta alba* in mild hypertensive subjects: a pilot study. *J. Med. Food* 10: 143–148.

Ranjan, R., A. Marczewski, T. Chojnacki, J. Hertel & E. Swiezewska. 2001. Search for polyprenols in leaves of evergreen and deciduous Ericaceae plants. *Acta Biochim. Polon.* 48: 579–584.

Ranney, T. & D. Gillooly. 2014. New insights into breeding and propagating magnolias. pp. 441–449. *In*: *Proceedings of the 2014 Annual Meeting of the International Plant Propagators Society 1085*. Bellefonte, PA.

Ranua, V. A. & C. Weinig. 2010. Mixed-mating strategies and their sensitivity to abiotic variation in *Viola lanceolata* L. (Violaceae). *Open Ecol. J.* 3: 83–94.

Rao, J. V., K. S. Aithal & K. K. Srinivasan. 1989. Antimicrobial activity of the essential oil of *Limnophila gratissima*. *Fitoterapia* 60: 376–377.

Rao, S. & H. Y. Mohan Ram. 1981. Regeneration of whole plants from cultured root tips of *Limnophila indica*. *Can. J. Bot.* 59: 969–973.

Rapp, J. M., M. Fugler, C. K. Armbruster & N. S. Clark. 2001. *Atchafalaya Sediment Delivery.* Progress Report No. 1. Monitoring Series No. AT-02-MSPR-0599-1. National Wetlands Research Center, Lafayette, LA. 38 pp.

Rasran, L. & K. Vogt. 2008. Diversity of plant propagules deposited in drift lines at the Voss River (W Norway). *Mitt. Arbeitsgem. Geobot. Schleswig-Holstein & Hamburg* 65: 363–374.

Rasser, M. K., N. L. Fowler & K. H. Dunton. 2013. Elevation and plant community distribution in a microtidal salt marsh of the western Gulf of Mexico. *Wetlands* 33: 575–583.

Rathcke, B. 1988. Flowering phenologies in a shrub community: competition and constraints. *J. Ecol.* 76: 975–994.

Rathcke, B. J. 2000. Birds, pollination reliability, and green flowers in an endemic island shrub, *Pavonia bahamensis* (Malvaceae). *Rhodora* 102: 392–414.

Ratliff, R. D. 1985. *Meadows in the Sierra Nevada of California: State of Knowledge.* General Technical Report PSW-84. Pacific Southwest Forest and Range Experiment Station. Berkeley, CA. 52 pp.

Ratliff, R. D. & R. G. Denton. 1993, Bolander's clover in the central Sierra Nevada: a sensitive species? *Madroño* 40: 166–173.

Raubeson, L. A., R. Peery, T. W Chumley, C. Dziubek, H. M. Fourcade, J. L. Boore & R. K. Jansen. 2007. Comparative chloroplast genomics: analyses including new sequences from the angiosperms *Nuphar advena* and *Ranunculus macranthus. BMC Genomics* 8:174. doi:10.1186/1471-2164-8-174.

Raulings, E., K. Morris, R. Thompson & R. Mac Nally. 2011. Do birds of a feather disperse plants together? *Freshwater Biol.* 56: 1390–1402.

Rave, D. P. & G. A. Baldassarre. 1989. Activity budget of greenwinged teal wintering in coastal wetlands of Louisiana. *J. Wildl. Manag.* 53: 753–759.

Raven, P. H. 1979. A survey of reproductive biology in Onagraceae. *New Zealand J. Bot.* 17: 575–593.

Raven, P. H. & W. Tai. 1979. Observations of chromosomes in *Ludwigia* (Onagraceae). *Ann. Missouri Bot. Gard.* 66: 862–879.

Raven, P. H., W. Dietrich & W. Stubbe. 1979. An outline of the systematics of *Oenothera* subsect. *Euoenothera* (Onagraceae). *Syst. Bot.* 4: 242–252.

Rawinski, T. 1982. The ecology and management of purple loosestrife (*Lythrum salicaria* L.) in central New York. MS thesis. Cornell University, Ithaca, New York.

Rawinski, T. J. & J. C. Ludwig. 1992. *Critical Natural Areas, Exemplary Wetlands, and Endangered Species Habitats in Southeastern Virginia: Results of the 1991 Inventory Encompassing Prince George County, Surry County, Isle of Wight County, Chesapeake City, Suffolk City, and Virginia Beach City.* Natural Heritage Tech. Rep. 92–14. Virginia Department of Conservation and Recreation, Division of Natural Heratage, Richmond, VA. 87 pp.

Ray, A. C., H. J. Williams & J. C. Reagor. 1987. Pyrrolizidine alkaloids from *Senecio longilobus* and *Senecio glabellus. Phytochemistry* 26: 2431–2433.

Ray, A. M., A. J. Rebertus & H. L. Ray. 2001. Macrophyte succession in Minnesota beaver ponds. *Can. J. Bot.* 79: 487–499.

Ray, J. D. Jr. 1956. The genus *Lysimachia* in the New World. *Illinois Biol. Monogr.* 24: 1–160.

Raynal D. J. & F. A. Bazzaz. 1973. Establishment of early successional plant populations on forest and prairie soil. *Ecology* 54: 1335–1341.

Rayner, M. R. 1978. Some aspects of the ecology and management of *Chenopodium rubrum* L. in the Delta Marsh, Manitoba. MS thesis. Department of Botany, University of Manitoba, Winnipeg, MB. 86 pp.

Raynor, L. A. 1952. Cytotaxonomic studies of *Geum. Amer. J. Bot.* 39: 713–719.

Razafimandimbison, S. G. & B. Bremer. 2002. Phylogeny and classification of Naucleeae s.l. (Rubiaceae) inferred from molecular (ITS, *rBCL*, and *tRNT-F*) and morphological data. *Amer. J. Bot.* 89: 1027–1041.

Razafimandimbison, S. G., J. Moog, H. Lantz, U. Maschwitz & B. Bremer. 2005. Re-assessment of monophyly, evolution of myrmecophytism, and rapid radiation in *Neonauclea* s.s. (Rubiaceae). *Mol. Phylogen. Evol.* 34: 334–354.

Razi, B. A. 1950. A contribution towards the study of the dispersal mechanisms in flowering plants of Mysore (south India). *Ecology* 31: 282–286.

Razifard, H. 2016. Systematics of *Elatine* L. (Elatinaceae). PhD dissertation, The University of Connecticut. Storrs, CT. 103 pp.

Razifard, H., G. C. Tucker & D. H. Les. 2016. *Elatine* L. pp. 349–353. *In:* Flora North America Editorial Committee (eds.), *Flora of North America North of Mexico*, Vol. 12: Magnoliophyta: Vitaceae to Garryaceae. Oxford University Press, New York.

Read, E. 1999. New cultivars (*Dionaea* 'Red Piranha'). *Carniv. Plant Newslett.* 28: 99.

Readel, K. E., D. S. Seigler & D. A. Young. 2003. 5-methoxycanthin-6-one from *Leitneria floridana* (Simaroubaceae). *Biochem. Syst. Ecol.* 31:167–170.

Reader, R. J. 1975. Competative relationships of some bog ericads for major insect pollinators. *Can. J. Bot.* 53: 1300–1305.

Reader, R. J. 1977. Bog ericad flowers: self-compatibility and relative attractiveness to bees. *Can. J. Bot.* 55: 2279–2287.

Reader, R. J. 1978. Contribution of overwintering leaves to the growth of three broad-leaved evergreen shrubs belonging to the Ericaceae family. *Can. J. Bot.* 56: 1248–1261.

Reader, R. J. 1979. Flower cold hardiness: a potential determinant of the flowering sequence exhibited by bog ericads. *Can. J. Bot.* 57: 997–999.

Reader, R. J. 1982. Geographic variation in the shoot productivity of bog shrubs and some environmental correlates. *Can. J. Bot.* 60: 340–348.

Reader, R. J. 1983. Modelling geographic variation in the timing of shoot extension by ericaceous shrubs. *Can. J. Bot.* 61: 2032–2037.

Reader, R. J. 1987. Loss of species from deciduous forest understorey immediately following selective tree harvesting. *Biol. Conserv.* 42: 231–244.

Reams Jr., W. M. 1953. The occurrence and ontogeny of hydathodes in *Hygrophila polysperma* T. Anders. *New Phytol.* 52: 8–13.

Rechinger, K. H. 1937. The North American species of *Rumex. Publ. Field Mus. Nat. Hist., Bot. Ser.* 17: 1–151.

Record, S. 2011. Plant species associated with a regionally rare hemiparasitic plant, *Pedicularis lanceolata* (Orobanchaceae), throughout its geographic range. *Rhodora* 113: 125–159.

Redbo-Torstensson, P. & A. Telenius. 1995. Primary and secondary seed dispersal by wind and water in *Spergularia salina. Ecography* 18: 230–237.

Reddoch, J. M. & A. H. Reddoch. 2005. Consequences of beaver, *Castor canadensis*, flooding on a small shore fen in southwestern Quebec. *Can. Field Naturalist* 119: 385–394.

Reddy, K. R. & W. F. DeBusk. 1984. Growth characteristics of aquatic macrophytes cultured in nutrient-enriched water: I. water hyacinth, water lettuce, and pennywort. *Econ. Bot.* 38: 229–239.

Reddy, K. R. & J. C. Tucker. 1985. Growth and nutrient uptake of pennywort (*Hydrocotyle umbellata* L.), as influenced by the nitrogen concentration of the water. *J. Aquat. Plant Manag.* 23: 35–40.

Redmond, K. 2007. Browsing the bog. *Field Sta. Bull. Univ. Wisconsin-Milwaukee* 32: 1–56.

Ree, R. H. 2005. Phylogeny and the evolution of floral diversity in *Pedicularis* (Orobanchaceae). *Int. J. Plant Sci.* 166: 595–613.

Reed, G. M. 1913. The powdery mildews: Erysiphaceæ. *Trans. Amer. Microscop. Soc.* 32: 219–258.

Reed, G. M. 1936. Physiological specialization of parasitic fungi. *Brooklyn Bot. Gard. Mem.* 1: 348–409.

Reed, P. B. 1986. *Wetland Plants of the State of Wisconsin 1986.* WELUT-86/W12.49. United States Department of Interior, Fish & Wildlife Service, Washington, DC. 26 pp.

Reed, P. B. 1988. *National List of Plant Species that Occur in Wetlands.* U. S. Fish and Wildlife Service, Washington, DC.

Reed, W. R. 1993. *Salix gooddingii. In: Fire Effects Information System.* U. S. Department of Agriculture, Forest Service, Rocky Mountain Research Station, Fire Sciences Laboratory (Producer). Available online: http://www.fs.fed.us/database/feis/ (accessed March 11, 2005).

Reese, M. C. & K. S. Lubinski. 1983. A survey and annotated checklist of late summer aquatic and floodplain vascular flora, middle and lower pool 26, Mississippi and Illinois Rivers. *Castanea* 48: 305–316.

Reese, N. L. & R. R. Haynes. 2002. Noteworthy collections: Alabama. *Castanea* 67: 216.

Reese, W. D. 1976. Two tropical mosses new to Louisiana. *Castanea* 41: 213–215.

Reeves, D. M. & W. W. Woessner. 2004. Hydrologic controls on the survival of Water Howellia (*Howellia aquatilis*) and implications of land management. *J. Hydrol.* 287: 1–18.

Reeves, R. D., R. M. Macfarlane & R. R. Brooks. 1983. Accumulation of nickel and zinc by western North American genera containing serpentine-tolerant species. *Amer. J. Bot.* 70: 1297–1303.

Reeves, T. 1977. The genus *Botrychium* (Ophioglossaceae) in Arizona. *Amer. Fern J.* 67: 33–39.

Refulio-Rodriguez, N. F. & R. G. Olmstead. 2014. Phylogeny of Lamiidae. *Amer. J. Bot.* 101: 287–299.

Reginatto, F. H., M. L. Athayde, G. Gosmann & E. P. Schenkel. 1999. Methylxanthines accumulation in *Ilex* species-caffeine and theobromine in erva-mate (*Ilex paraguariensis*) and other *Ilex* species. *J. Brazil. Chem. Soc.* 10: 443–446.

Rehder, A. 1903. Synopsis of the genus *Lonicera. Missouri Bot. Gard. Ann. Rep.* 1903: 27–232.

Reichardt, P. B., J. P. Bryant, B. J. Anderson, D. Phillips, T. P. Clausen, M. Meyer & K. Frisby. 1990. Germacrone defends Labrador tea from browsing by snowshoe hares. *J. Chem. Ecol.* 16: 1961–1970.

Reid, B. 2001. *Ludwigia sphaerocarpa* (globe-fruited false-loosestrife) conservation and research plan. New England Plant Conservation Program, Framingham, MA.

Reid, C. S. & L. Urbatsch. 2012. Noteworthy plant records from Louisiana. *J. Bot. Res. Inst. Texas* 6: 273–278.

Reinartz, J. A. and D. H. Les. 1994. Bottleneck-induced dissolution of self-incompatibility and breeding system consequences in *Aster furcatus* (Asteraceae). *Amer. J. Bot.* 81: 446–455.

Reinert, G. W. & R. K. Godfrey. 1962. Reappraisal of *Utricularia inflata* and *U. radiata* (Lentibulariaceae). *Amer. J. Bot.* 49: 213–220.

Rejmánková, E. 1992. Ecology of creeping macrophytes with special reference to *Ludwigia peploides* (HBK) Raven. *Aquat. Bot.* 43: 282–299.

Remington, J. P., H. C. Wood, S. P. Sadtler, C. H. LaWall, H. Draemer & J. F. Anderson (eds.). 1918. *The Dispensatory of the United States of America (20th Ed.).* J. B. Lippincott, Philadelphia, PA. 2010 pp.

Renda, M. T. & H. L. Rodgers.1995. Restoration of tidal wetlands along the Indian River Lagoon. *Bull. Mar. Sci.* 57: 283–285.

Renne, I. J. & B. F. Tracy. 2007. Disturbance persistence in managed grasslands: shifts in aboveground community structure and the weed seed bank. *Plant Ecol.* 190: 71–80.

Renner, S. S. 1989. A survey of reproductive biology in Neotropical Melastomataceae and Memecylaceae. *Ann. Missouri Bot. Gard.* 76: 496–518.

Renner, S. S. 1999. Circumscription and phylogeny of Laurales: evidence from molecular and morphological data. *Amer. J. Bot.* 86: 1301–1315.

Renner, S. S. 2004. Melastomataceae (Black Mouth Family). pp. 240–243. *In:* N. Smith, S. A. Mori, A. Henderson, D. W. Stevenson & S. V. Heald (eds.), *Flowering Plants of the Neotropics.* Princeton University Press, Princeton, NJ. 594 pp.

Renner, T. & C. D. Specht. 2011. A sticky situation: assessing adaptations for plant carnivory in the Caryophyllales by means of stochastic character mapping. *Int. J. Plant Sci.* 172: 889–901.

Renner, T. & C. D. Specht. 2012. Molecular and functional evolution of class I chitinases for plant carnivory in the Caryophyllales. *Mol. Biol. Evol.* 29: 2971–2985.

Rentch, J. S. & R. H. Fortney. 1997. The vegetation of West Virginia grass bald communities. *Castanea* 62: 147–160.

Renz, M. J., S. J. Steinmaus, D. S. Gilmer & J. M. DiTomaso. 2012. Spread dynamics of perennial pepperweed (*Lepidium latifolium*) in two seasonal wetland areas. *Invas. Plant Sci. Manag.* 5: 57–68.

Resch, J. F., D. F. Rosberger, J. Meinwald & J. W. Appling. 1982. Biologically active pyrrolizidine alkaloids from the true forget-me-not, *Myosotis scorpioides. J. Nat. Prod.* 45: 358–362.

Reveal, J. L. 1978. *Status Report on* Nitrophila mohavensis. U. S. Fish and Wildlife Service, Portland, OR.

Reveal, J. L. 2003. Demographics and ecology of the Amargosa niterwort (*Nitrophila mohavensis*) and Ash Meadows gumplant (*Grendelia fraxino-pratensis*) of the Carson Slough area. San Diego State University, San Diego, California. Available online: http://www.serg.sdsu.edu/SERG/restorationproj/mojave%20desert/deathvalleyfinal.htm.

Reveal, J. L. 2006. Revision of *Dodecatheon* (Primulaceae). Submission copy: 10 May 2006; revised 1 Aug 2006. Available online: http://www.life.umd.edu/emeritus/reveal/pbio/fna/dodecatheon.html#redo.

Reveal, J.L., C. R. Broome & J. C. Beatley. 1973. A new *Centaurium* (Gentianaceae) from the Death Valley region of Nevada and California. *Bull. Torrey Bot. Club* 100: 353–356.

Reyes, G. P., D. Kneeshaw, L. De Grandpré & A. Leduc. 2010. Changes in woody vegetation abundance and diversity after natural disturbances causing different levels of mortality. *J. Veg. Sci.* 21: 406–417.

Reynolds, D. N. 1984. Populational dynamics of three annual species of alpine plants in the Rocky Mountains. *Oecologia* 62: 250–255.

Reynolds, H. W., R. M. Hansen & D. G. Peden. 1978. Diets of the Slave River lowland bison herd, Northwest Territories, Canada. *J. Wildl. Manag.* 42: 581–590.

Reynolds, L. K. & K. E. Boyer. 2010. Perennial pepperweed (*Lepidium latifolium*): properties of invaded tidal marshes. *Invas. Plant Sci. Manag.* 3: 130–138.

Reynolds, T. 2005. Hemlock alkaloids from Socrates to poison aloes. *Phytochemistry* 66: 1399–1406.

Reznicek, A. A. & P. F. Maycock. 1983. Composition of an isolated prairie in central Ontario. *Can. J. Bot.* 61: 3107–3116.

Rheinhardt, R. D. & M. M. Brinson. 2002. *An Evaluation of North Carolina Department of Transportation Wetland Mitigation Sites: Selected Case Studies-Phase II Report.* Report No.

FHWA/NC/2002-009, March 2002. U. S. Department of Transportation, Research and Special Programs Administration, Washington, DC.

Rho, J. & H. B. Gunner. 1978. Microfloral response to aquatic weed decomposition. *Water Res.* 12: 165–170.

Rho, Y. D. & M. H. Lee. 2004. Germination characteristics of *Bidens tripartita* and *Bidens frondosa* occuring in paddy fields. *Korean J. Weed Sci.* 24: 299–307.

Ribble, D. W. 1968. A new subgenus, *Belandrena*, of the genus *Andrena* (Hymenoptera: Apoidea). *J. Kansas Entomol. Soc.* 41: 220–236.

Rice, B. A. 2002. Noteworthy collections. California. *Drosera aliciae* Hamet (Droseraceae); *Drosera capensis* L. (Droseraceae). *Madroño* 49: 193–194.

Rich, S. M., M. Ludwig & T. D. Colmer. 2012. Aquatic adventitious root development in partially and completely submerged wetland plants *Cotula coronopifolia* and *Meionectes brownii*. *Ann. Bot.* 110: 405–414.

Richards, A. J. 1986. *Plant Breeding Systems*. George Allen & Unwin, London, UK. 529 pp.

Richards, C. L., J. L. Hamrick, L. A. Donovan & R. Mauricio. 2004. Unexpectedly high clonal diversity of two salt marsh perennials across a severe environmental gradient. *Ecol. Lett.* 7: 1155–1162.

Richards, J. H. 2001. Bladder function in *Utricularia purpurea* (Lentibulariaceae): is carnivory important? *Amer. J. Bot.* 88: 170–176.

Richards, J. H. & B. Fry. 1996. Nitrogen acquisition by the aquatic bladderworts, *Utricularia foliosa* L. and *U. purpurea* Walt. *Amer. J. Bot.* 83: S75.

Richards, J. H. & D. N. Kuhn. 1996. Genetic diversity and autogamy in *Utricularia foliosa* L. *Amer. J. Bot.* 83: S48.

Richards, J. H., T. G. Troxler, D. W. Lee & M. S. Zimmerman. 2011. Experimental determination of effects of water depth on *Nymphaea odorata* growth, morphology and biomass allocation. *Aquat. Bot.* 95: 9–16.

Richards, W. R. 1969. *Cepegillettea viridis* a new aphid from Ontario, with a review of the genus (Homoptera: Aphididae). *Can. Entomol.* 101: 963–970.

Richards, W. R. 1972. Review of the *Solidago*-inhabiting aphids in Canada with descriptions of three new species (Homoptera: Aphididae). *Can. Entomol.* 104: 1–34.

Richards, W. R. 1976. A host index for species of Aphidoidea described during 1935 to 1969. *Can. Entomol.* 108: 499–550.

Richardson, A. & K. King. 2008. *Neptunia plena* (Fabaceae: Mimosoideae) rediscovered in Texas. *J. Bot. Res. Inst. Texas* 2: 1491–1493.

Richardson, F. 1967. Black tern nest and egg moving experiments. *Murrelet* 48: 52–56.

Richardson, J. E., M. F. Fay, Q. C. B. Cronk, D. Bowman & M. W. Chase. 2000. A phylogenetic analysis of Rhamnaceae using *rbcL* and *trnL-F* plastid sequences. *Amer. J. Bot.* 87: 1309–1324.

Richardson, K., H. Griffiths, M. L. Reed, J. A. Raven & N. M. Griffiths. 1984. Inorganic carbon assimilation in the isoetids, *Isoetes lacustris* L. and *Lobelia dortmanna* L. *Oecologia* 61: 115–121.

Richburg, J. A., W. A. Patterson & F. Lowenstein. 2001. Effects of road salt and *Phragmites australis* invasion on the vegetation of a western Massachusetts calcareous lake-basin fen. *Wetlands* 21: 247–255.

Richerson, S. 1997. *Interim Management Guide for* Lewisia cantelovii *and* L. serrata. Eldorado, Plumas, Shasta-Trinity and Tahoe National Forests. USDA Forest Service. Washington, DC. 81 pp.

Richter, R. & J. C. Stromberg. 2005. Soil seed banks of two montane riparian areas: implications for restoration. *Biodivers. Conserv.* 14: 993–1016.

Ridder, F. & A. A. Dhondt. 1992. The demography of a clonal herbaceous perennial plant, the longleaved sundew *Drosera intermedia*, in different heathland habitats. *Ecography* 15: 129–143.

Riefner Jr., R. E. & S. R. Hill. 1983. Notes on infrequent and threatened plants of Maryland including new state records. *Castanea* 48: 117–137.

Riemer, D. N. 1985. Seed germiantion in spatterdock (*Nuphar advena* Ait.). *J. Aquat. Plant Manag.* 23: 46–47.

Riemer, D. N. & R. D. Ilnicki. 1968. Reproduction and overwintering of *Cabomba* in New Jersey. *Weed Sci.* 16: 101–102.

Rigat, M., M. À. Bonet, S. Garcia, T. Garnatje and J. Vallès. 2007. Studies on pharmaceutical ethnobotany in the high river Ter valley (Pyrenees, Catalonia, Iberian Peninsula). *J. Ethnopharmacol.* 113: 267–277.

Rigg, G. B. 1922. A bog forest. *Ecology* 3: 207–213.

Rigg, G. B. 1925. Some sphagnum bogs of the north Pacific coast of America. *Ecology* 6:. 260–278.

Rigg, G. B. 1937. Some raised bogs of southeastern Alaska with notes on flat bogs and muskegs. *Amer. J. Bot.* 24: 194–198.

Rigg, G. B. 1940. Comparisons of the development of some sphagnum bogs of the Atlantic coast, the interior, and the Pacific coast. *Amer. J. Bot.* 27: 1–14.

Rigg, G. B. 1942. A raised cattail-tule bog in Yellowstone National Park. *Amer. Midl. Naturalist* 27: 766–771.

Rightmyer, M. G. 2008. A Review of the Cleptoparasitic Bee Genus Triepeolus (Hymenoptera: Apidae).—Part I (Zootaxa 1710). Magnolia Press, Auckland, New Zealand. 170 pp.

Rigney, C. L. 2013. Habitat characterization and biology of the threatened Dakota skipper (*Hesperia dacotae*) in Manitoba. MS thesis. Department of Biology, University of Winnipeg, Winnipeg, MB. 242 pp.

Rimer, R. L. & J. W. Summers. 2006. Range and ecology of *Helenium virginicum* in the Missouri Ozarks. *S. E. Naturalist* 5: 515–522.

Rindge, F. H. 1952. Taxonomic and life history notes on North American *Eupithecia* (Lepidoptera, Geometridae). *Amer. Mus. Novit.* 1569: 1–27.

Ristich, S. S., S. W. Fredrick & E. H. Buckley. 1976. Transplantation of *Typha* and the distribution of vegetation and algae in a reclaimed estuarine marsh. *Bull. Torrey Bot. Club* 103: 157–164.

Ritchie, J. C. 1957. The vegetation of northern Manitoba: II. A prisere on the Hudson Bay lowlands. *Ecology* 38: 429–435.

Ritchie, J. C. 1977. The modern and late Quaternary vegetation of the Campbell-dolomite uplands, near Inuvik, N.W.T., Canada. *Ecol. Monogr.* 47: 401–423.

Ritchie, J. C. 1982. The modern and late-Quaternary vegetation of the Doll Creek area, North Yukon, Canada. *New Phytol.* 90: 563–603.

Ritch-Krc, E. M., S. Thomas, N. J. Turner & G. H. N. Towers. 1996. Carrier herbal medicine: traditional and contemporary plant use. *J. Ethnopharmacol.* 52: 85–94.

Ritland, C. & K. Ritland. 1989. Variation of sex allocation among eight taxa of the *Mimulus guttatus* species complex (Scrophulariaceae). *Amer. J. Bot.* 76: 1731–1740.

Ritland, K. 1989. Correlated matings in the partial selfer *Mimulus guttatus*. *Evolution* 43: 848–859.

Ritland, K. & S. Jain. 1984. The comparative life histories of two annual *Limnanthes* species in a temporally variable environment. *Amer. Naturalist* 124: 656–679.

Ritland, K. & M. Leblanc. 2004. Mating system of four inbreeding monkeyflower (*Mimulus*) species revealed using 'progeny-pair' analysis of highly informative microsatellite markers. *Plant Species Biol.* 19: 149–157.

Ritter, N. P. & G. E. Crow. 1998. *Myriophyllum quitense* Kunth (Haloragaceae) in Bolivia: A terrestrial growth-form with bisexual flowers. *Aquat. Bot.* 60: 389–395.

Rivadavia, F., K. Kondo, M. Kato & M. Hasebe. 2003. Phylogeny of the sundews, *Drosera* (Droseraceae), based on chloroplast *rbcL* and nuclear 18S ribosomal DNA sequences. *Amer. J. Bot.* 90: 123–130.

Rivas-Martínez, S., D. Sánchez-Mata & M. Costa. 1999. North American boreal and western temperate forest vegetation (Syntaxonomical synopsis of the potential natural plant communities of North America, II). *Itin. Geobot.* 12: 5–316.

Rivera, D., C. Obon, C. Inocencio, M. Heinrich, A. Verde, J. Fajardo & R. Llorach. 2005. The ethnobotanical study of local Mediterranean food plants as medicinal resources in southern Spain. *J. Physiol. Pharmacol.* 56: S97–S114.

Rizza, A., G. Campobasso, P. H. Dunn & M. Stazi. 1988. *Cheilosia corydon* (Diptera: Syrphidae), a candidate for the biological control of musk thistle in North America. *Ann. Entomol. Soc. Amer.* 81: 225–232.

Ro, K.-E. & B. A. McPheron. 1997. Molecular phylogeny of the *Aquilegia* group (Ranunculaceae) based on internal transcribed spacers and the 5.8S nuclear ribosomal DNA. *Biochem. Syst. Ecol.* 25: 445–461.

Ro, K.-E., C. S. Keener & B. A. McPheron. 1997. Molecular phylogenetic study of the Ranunculaceae: utility of the nuclear 26S ribosomal DNA in inferring intrafamilial relationships. *Mol. Phylogen. Evol.* 8: 117–127.

Robart, B. W. 2000. The systematics of *Pedicularis bracteosa*: morphometrics, development, pollination ecology, and molecular phylogenetics. PhD dissertation. Illinois State University, Normal, IL.

Robart, B. W. 2005. Morphological diversification and taxonomy among the varieties of *Pedicularis bracteosa* Benth. (Orobanchaceae). *Syst. Bot.* 30: 644–656.

Robart, B. W., C. Gladys, T. Frank & S. Kilpatrick. 2015. Phylogeny and biogeography of North American and Asian *Pedicularis* (Orobanchaceae). *Syst. Bot.* 40: 229–258.

Robbins, G. T. 1944. North American species of *Androsace*. *Amer. Midl. Naturalist* 32: 137–163.

Robbins, W. 1940. *Alien Plants Growing Without Cultivation in California*. Agricultural Experiment Station. Bulletin 637. University of California, Berkeley, CA. 128 pp.

Robbins, W. W. 1918. Successions of vegetation in Boulder Park, Colorado. *Bot. Gaz.* 65: 493–525.

Robe, W. E. & H. Griffiths. 1998. Adaptations for an amphibious life: changes in leaf morphology, growth rate, carbon and nitrogen investment, and reproduction during adjustment to emersion by the freshwater macrophyte *Littorella uniflora*. *New Phytol.* 140: 9–23.

Robert, M., B. Jobin, F. Shaffer, L. Robillard & B. Gagnon. 2004. Yellow rail distribution and numbers in southern James Bay, Québec, Canada. *Waterbirds* 27: 282–288.

Roberts, A. 1984. *Guide to Wetland Ecosystems of the Sub Boreal Spruce Subzone, Caribou Forest Region, British Columbia*. Research Branch, British Columbia Ministry of Forests, Vancouver, BC. 53 pp.

Roberts, D. A., R. Singer & C. W. Boylen. 1985. Submersed macrophyte communities of Adirondack lakes (New York, U.S.A.) of varying degrees of acidity. *Aquat. Bot.* 21: 219–235.

Roberts, H. A. 1979. Periodicity of seedling emergence and seed survival in some Umbelliferae. *J. Appl. Ecol.* 16: 195–201.

Roberts, H. A. 1986. Seed persistence in soil and seasonal emergence in plant species from different habitats. *J. Appl. Ecol.* 23: 639–656.

Roberts, M. L. 1972. *Wolffia* in the bladders of *Utricularia*: an "herbivorous" plant? *Michigan Bot.* 11: 67–69.

Roberts, M. L. 1985. The cytology, biology and systematics of *Megalodonta beckii* (Compositae). *Aquat. Bot.* 21: 99–110.

Roberts, M. L. & R. R. Haynes. 1983. Ballistic seed dispersal in *Illicium* (Illiciaceae). *Plant Syst. Evol.* 143: 227–232.

Roberts, T. H. & D. H. Arner. 1984. Food habits of beaver in east-central Mississippi. *J. Wildl. Manag.* 48: 1414–1419.

Roberts, T. L. & J. L. Vankat. 1991. Floristics of a chronosequence corresponding to old field-deciduous forest succession in southwestern Ohio. II. Seed banks. *Bull. Torrey Bot. Club* 118: 377–384.

Robertson, A. W., C. Mountjoy, B. E. Faulkner, M. V. Roberts & M. R. Macnair. 1999. Bumble bee selection of *Mimulus guttatus* flowers: the effects of pollen quality and reward depletion. *Ecology* 80: 2594–2606.

Robertson, C. 1887. Insect relations of certain Asclepiads. II. *Bot. Gaz.* 12: 244–250.

Robertson, C. 1888. Proterogynous umbelliferæ. *Bot. Gaz.* 13: 193.

Robertson, C. 1889a. Synopsis of North American species of the genus *Oxybelus*. *Trans. Amer. Entomol. Soc.* 16: 77–85.

Robertson, C. 1889b. Flowers and insects. II. *Bot. Gaz.* 14: 172–178.

Robertson, C. 1891. Descriptions of new species of North American bees. *Trans. Amer. Entomol. Soc.* 18: 49–66.

Robertson, C. 1894. Flowers and insects: Rosaceae and Compositae. *Trans. Acad. Sci. St. Louis* 6: 435–480.

Robertson, C. 1922. Synopsis of Panurgidæ (Hymenoptera). *Psyche* 29: 159–173.

Robertson, C. 1924a. Flowers and insects. XXIII. *Bot. Gaz.* 78: 68–84.

Robertson, C. 1924b. Flower visits of insects II. *Psyche* 31: 93–111.

Robertson, C. 1928. *Flowers and Insects; Lists of Visitors of Four Hundred and Fifty-Three Flowers*. The Science Press Printing Co., Lancaster, PA. 221 pp.

Robertson, C. 1929. *Flowers and Insects; Lists of Visitors of Four Hundred and Fifty-Three Flowers*. The Science Press Printing Co., Lancaster, PA. 221 pp.

Robertson, H. A. & K. R. James. 2007. Plant establishment from the seed bank of a degraded floodplain wetland: a comparison of two alternative management scenarios. *Plant Ecol.* 188: 145–164.

Robertson, K. R. 2006. Distributions of tree species along point bars of 10 rivers in the south-eastern US Coastal Plain. *J. Biogeogr.* 33: 121–132.

Robertson, K. R. & S. E. Clemants. 2003. 39. Amaranthaceae Jussieu. pp. 405–406. *In*: N. R. Morin (convening ed.), *Flora of North America North of Mexico, Vol. 4: Magnoliophyta: Caryophyllidae, Part 1*. Oxford University Press, New York.

Robertson, K. R., J. B. Phipps, J. R. Rohrer & P. G. Smith. 1991. A synopsis of genera in Maloideae (Rosaceae). *Syst. Bot.* 16: 376–394.

Robertson, P. A., G. T. Weaver & J. A. Cavanaugh. 1978. Vegetation and tree species patterns near the northern terminus of the southern floodplain forest. *Ecol. Monogr.* 48: 249–267.

Robinson, A. F. 1978. Possible impacts of silvicultural activities on proposed endangered and threatened plant species of pine flatwoods. pp. 336–342. *In*: W. E. Balmer (ed.), *Proceedings: Soil Moisture, Site Productivity Symposium, Myrtle Beach, SC, Nov. 1–3, 1977*. Department of Agriculture, Forest Service, Southeastern Area, State and Private Forestry, Atlanta, GA.

Robinson, A. G. & G. A. Bradley. 1968. A revised list of the aphids of Manitoba. *Manitoba Entomol.* 2: 60–65.

Robinson, A. T. 2005. *Aquatic Plant Survey of Willow Creek Reservoir, Prescott, Arizona, September 7, 2005.* Report to city of Prescott Parks, Recreation and Library Department. Arizona Game and Fish Department, Research Branch, Phoenix, AZ. 9 pp.

Robinson, B. H., M. Marchetti, C. Moni, L. Schroeter, C. van den Dijssel, G. Milne, N. S. Bolan & S. Mahimairaja. 2006. Arsenic accumulation by aquatic and terrestrial plants: Taupo Volcanic Zone, New Zealand. pp. 235–247. *In*: R. Naidu, E. Smith, G. Owens, P. Bhattacharya & P. Nadebaum (eds.), *Managing Arsenic in the Environment: From Soil to Human Health.* CSIRO Publishing, Adelaide, Australia.

Robinson, G. S., P. R. Ackery, I. J. Kitching, G. W. Beccaloni & L. M. Hernandez. 2002. Hostplants of the moth and butterfly caterpillars of America north of Mexico. *Mem. Amer. Entomol. Inst.* 69: 1–824.

Robinson, H., A. M. Powell, G. D. Carr, R. M. King & J. F. Weedin. 1989. Chromosome numbers in Compositae, XVI: Eupatorieae II. *Ann. Missouri Bot. Gard.* 76: 1004–1011.

Robinson, R. W. & J. T. Puchalski. 1980. Synonymy of *Cucurbita martinezii* and *C. okeechobeensis. Cucurbit Genet. Coop. Rep.* 3: 45–46.

Robson, N. K. B. & P. Adams. 1968. Chromosome numbers in *Hypericum* and related genera. *Brittonia* 20: 95–106.

Robuck, O. W. 1985. *The Common Plants of the Muskegs of Southeast Alaska.* Miscellaneous Publication. United States Department of Agriculture, Pacific Northwest Forest and Range Experiment Station, Portland, OR. 131 pp.

Rocha, A. C., E. C. Canal, E. Campostrini, F. O. Reis & G. R. F. Cuzzuol. 2009. Influence of chromium in *Laguncularia racemosa* (L). Gaertn f. physiology. *Brazil. J. Plant Physiol.* 21: 87–94.

Roche, B. M. & R. S. Fritz. 1997. Genetics of resistance of *Salix sericea* to a diverse community of herbivores. *Evolution* 51: 1490–1498.

Rock, H. F. L. 1957. A revision of the vernal species of *Helenium* (Compositae). *Rhodora* 59: 128–158; 203–216.

Rockwell, S. M. & J. L. Stephens. 2014. *Song Sparrow Habitat Selection on the Trinity River.* Report No. KBO-2014-0010. Klamath Bird Observatory, Ashland, OR. 29 pp.

Rodger, E. A. 1933. Wound healing in submerged plants. *Amer. Midl. Naturalist* 14: 704–713.

Rodger, L. 1998. *Tallgrass Communities of Southern Ontario: A Recovery Plan.* World Wildlife Fund Canada and the Ontario Ministry of Natural Resources. Toronto, ON. 66 pp.

Rodman, J. E., P. S. Soltis, D. E. Soltis, K. J. Sytsma & K. G. Karol. 1998. Parallel evolution of glucosinolate biosynthesis inferred from congruent nuclear and plastid gene phylogenies. *Amer. J. Bot.* 85: 997–1006.

Rodrigues, B. F. & A. N. Naik. 2009. Arbuscular mycorrhizal fungi of the "Khazan land" agro-ecosystem. pp. 141–150. *In*: K. R. Sridhar (ed.), *Frontiers in Fungal Ecology, Diversity and Metabolites.* I. K. International, New Delhi, India.

Rodrigues, C. F. B., H. H. Gaeta, M. N. Belchor, M. J. P. Ferreira, M. V. T. Pinho, D. D. O. Toyama & M. H. Toyama. 2015. Evaluation of potential thrombin inhibitors from the white mangrove (*Laguncularia racemosa* (L.) C.F. Gaertn.). *Mar. Drugs* 13: 4505–4519.

Rodriguez, B., M. C. de la Torre, B. Rodriguez & P. Gomez-Serranillos. 1996. Neo-clerodane diterpenoids from *Scutellaria galericulata. Phytochemistry* 41: 247–253.

Rodríguez, B., M. C. de la Torre, B. Rodríguez, M. Bruno, F. Piozzi, G. Savona, M. S. J. Simmonds, W. M. Blaney & A. Perales. 1993. *Neo*-clerodane insect antifeedants from *Scutellaria galericulata. Phytochemistry* 33: 309–315.

Rodriguez, S., A. Marston, J.-L. Wolfender & K. Hostettmann. 1998. Iridoids and secoiridoids in the Gentianaceae. *Curr. Organic Chem.* 2: 627–648.

Rodríguez-Riano, T. & A. Dafni. 2007. Pollen-stigma interference in two gynodioecious species of Lamiaceae with intermediate individuals. *Ann. Bot.* 100: 423–431.

Rodriguez-Sahagun, A., D. Rojas-Bravo, G. J. Acevedo-Hernandez, M. I. Torres-Moran, F. Zurita-Martinez, M. Gutierrez-Lomeli, C. L. Del Toro-Sanchez & O. Castellanos-Hernandez. 2012. Efficient protocols for *in vitro* axillary bud proliferation and somatic embryogenesis of the medicinal plant *Anemopsis californica. J. Med. Plant Res.* 6: 3859–3864.

Rogers, C. E. & J. C. Garrison. 1975. Seed destruction in indigobush amorpha by a seed beetle. *J. Range Manag.* 28: 241–242.

Rogers, C. E., T. E. Thompson & M. J. Wellik. 1980. Survival of *Bothynus gibbosus* (Coleoptera: Scarabaeidae) on *Helianthus* species. *J. Kansas Entomol. Soc.* 53: 490–494.

Rogers, C. M. 1963. Yellow flowered species of *Linum* in eastern North America. *Brittonia* 15: 97–122.

Rogers, C. M. 1972. The taxonomic significance of the fatty acid content of seeds of *Linum. Brittonia* 24: 415–419.

Rogers, D. L., C. I. Millar & R. D. Westfall. 1996. Genetic diversity within species. pp. 759–838. *In*: *Status of the Sierra Nevada.* Final Report to Congress. Wildlands Resources Center Report #93. University of California, Davis, CA.

Rogers, G., S. Walker, M. Tubbs & J. Henderson. 2002. Ecology and conservation status of three "spring annual" herbs in dryland ecosystems of New Zealand. *New Zealand J. Bot.* 40: 649–669.

Rogers, G. K. 2005. The genera of Rubiaceae in the southeastern United States, part II. Subfamily Rubioideae, and subfamily Cinchonoideae revisited (*Chiococca, Erithalis*, and *Guettarda*). *Harvard Pap. Bot.* 10: 1–45.

Rogers, J. G. 1996. Endangered and threatened species; notice of reclassification of 96 candidate taxa. *Fed. Reg.* 61 (40): 7457–7463.

Rogers, J. G. 1997. Endangered and threatened wildlife and plants; endangered status for four plants from vernal pools and mesic areas in northern California. *Fed. Reg.* 62 (117): 33029–33038.

Rogers, W. L., J. M. Cruse-Sanders, R. Determann & R. L. Malmberg. 2010. Development and characterization of microsatellite markers in *Sarracenia* L. (pitcher plant) species. *Conserv. Genet. Resour.* 2: 75–79.

Rohrer, J. R. 2014. 4. *Geum* Linnaeus. pp. 58–70. *In*: N. R. Morin (convening ed.), *Flora of North America North of Mexico, Vol. 9: Magnoliophyta: Picramniaceae to Rosaceae.* Oxford University Press, New York.

Rohwer, J. G. 1993. Lauraceae. pp. 366–391. *In*: K. Kubitzki, J. Rohwer, and V. Bittrich (eds.), *The Families and Genera of Vascular Plants, Vol. II. Flowering Plants, Dicotyledons: Magnoliid, Hamamelid and Caryophyllid Families.* Springer-Verlag, Berlin, Germany.

Rohwer, J. G. 2000. Toward a phylogenetic classification of the Lauraceae: evidence from *matK* sequences. *Syst. Bot.* 25: 60–71.

Roitman, J. N. 1983. The pyrrolizidine alkaloids of *Senecio triangularis. Austral. J. Chem.* 36: 1203–1213.

Rojas, L. B. & A. Usubillaga. 2000. Composition of the essential oil of *Satureja brownei* (SW.) Briq. from Venezuela. *Flav. Fragr. J.* 15: 21–22.

Roley, S. 2005. *Lewis Lake Environmental Assessment.* CAP Report 084. Center for Urban and Regional Affairs (CURA), University of Minnesota, Minneapolis, MN. 31 pp.

Rolfsmeier, S. B., R. F. Steinauer & D. M. Sutherland. 1999. New floristic records for Nebraska-5. *Trans. Nebraska Acad. Sci.* 25: 15–22.

Rollins, R. C. 1993. *The Cruciferae of Continental North America.* Stanford University Press, Stanford, CA. 976 pp.

Roman, C. T., N. E. Barrett & J. W. Portnoy. 2001. Aquatic vegetation and trophic condition of Cape Cod (Massachusetts, U.S.A.) kettle ponds. *Hydrobiologia* 443: 31–42.

Römermann, C., O. Tackenberg & P. Poschlod. 2005. How to predict attachment potential of seeds to sheep and cattle coat from simple morphological seed traits. *Oikos* 110: 219–230.

Rondelaud, D., G. Dreyfuss, B. Bouteille & M. L. Dardé. 2000. Changes in human fasciolosis in a temperate area: about some observations over a 28-year period in central France. *Parasit. Res.* 86: 1432–1955.

Rønsted, N., E. Gøbel, H. Franzyk, S. R. Jensen & C. E. Olsen. 2000. Chemotaxonomy of *Plantago.* Iridoid glucosides and caffeoyl phenylethanoid glycosides. *Phytochemistry* 55: 337–348.

Rønsted, N., M. W. Chase, D. C. Albach & M. A. Bello. 2002. Phylogenetic relationships within *Plantago* (Plantaginaceae): evidence from nuclear ribosomal ITS and plastid *trnL-F* sequence data. *Bot. J. Linn. Soc.* 139: 323–338.

Roodenrys, S., D. Booth, S. Bulzomi, A. Phipps, C. Micallef & J. Smoker. 2002. Chronic effects of Brahmi (*Bacopa monnieri*) on human memory. *Neuropsychopharmacology* 27: 279–281.

Rooke, J. B. 1984. The invertebrate fauna of four macrophytes in a lotic system. *Freshwater Biol.* 14: 507–513.

Roquet, C., L. Sáez, J. J. Aldasoro, A. Susanna, M. L. Alarcón & N. Garcia-Jacas. 2008. Natural delineation, molecular phylogeny and floral evolution in *Campanula. Syst. Bot.* 33: 203–217.

Rosa, B. A., L. Malek & W. Qin. 2009. The development of the pitcher plant *Sarracenia purpurea* into a potentially valuable recombinant protein production system. *Biotechnol. Mol. Biol. Rev.* 3: 105–110.

Rose, J. N. 1911. Two new species of *Harperella. Contr. U. S. Natl. Herb.* 13: 289–290.

Rosen, D. J. & S. Zamirpour. 2014. *Avicennia germinans* (Acanthaceae) gaining ground in southeast Texas? *Phytoneuron* 2014–74: 1–4.

Rosen, D. J., A. D. Caskey, W. C. Conway & D. A. Haukos. 2013. Vascular flora of saline lakes in the southern high plains of Texas and eastern New Mexico. *J. Bot. Res. Inst. Texas* 7: 595–602.

Rosenblatt, A. E., S. Zona, M. R. Heithaus & F. J. Mazzotti. 2014. Are seeds consumed by crocodilians viable? A test of the crocodilian saurochory hypothesis. *S. E. Naturalist* 13: N26–N29.

Rosenfeld, K. M. 2004. Ecology of Bird Island, North Carolina: an uninhabited, undeveloped barrier island. MS thesis. North Carolina State University, Raleigh, NC. 158 pp.

Roshon, R. D., G. R. Stephenson & R. F. Horton. 1996. Comparison of five media for the axenic culture of *Myriophyllum sibiricum* Komarov. *Hydrobiologia* 340: 17–22.

Rosman, A. J., H. Razifard, G. C. Tucker & D. H. Les. 2016. New records of *Elatine ambigua* (Elatinaceae), a species nonindigenous to North America. *Rhodora* 118: 235–242.

Ross, M. S. & P. L Ruiz. 1998. *A Study of the Distribution of Several South Florida endemic Plants in the Florida Keys.* Report to the U. S. Fish and Wildlife Service. Southeast Environmental Research Program, Florida International University, Miami, FL.

Ross, M. S., J. F. Meeder, J. P. Sah, P. L. Ruiz & G. J. Telesnicki. 2000. The southeast saline Everglades revisited: 50 Years of coastal vegetation change. *J. Veg. Sci.* 11: 101–112.

Ross, M. S., P. L. Ruiz, J. Sah, S. Stofella, N. Timilsina & E. Hanan. 2006. *Marl Prairie/Slough Gradients; Patterns and Trends in Shark Slough and Adjacent Marl Prairies (CERP Monitoring Activity 3.1.3.5), First Annual Report (2005).* SERC Research Reports. Paper 84. Available online: http://digitalcommons.fiu.edu/sercrp/84.

Rossato, L. V., S. B. Tedesco, H. D. Laughinghouse, J. G. Farias & F. T. Nicoloso. 2010. Alterations in the mitotic index of *Allium cepa* induced by infusions of *Pluchea sagittalis* submitted to three different cultivation systems. *Anais Acad. Brasil. Ci.* 82: 857–860.

Rossato, L. V., F. T. Nicoloso, J. G. Farias, D. Cargnelluti, L. A. Tabaldi, F. G. Antes, V. L. Dressler, V. M. Morsch & M. R. C. Schetinger. 2012. Effects of lead on the growth, lead accumulation and physiological responses of *Pluchea sagittalis. Ecotoxicology* 21: 111–123.

Rosatti, T. J. 1986. The genera of Sphenocleaceae and Campanulaceae in the southeastern United States. *J. Arnold Arbor.* 67: 1–64.

Rossetto, M., B. R. Jackes, K. D. Scott & R. J. Henry. 2002. Is the genus *Cissus* (Vitaceae) monophyletic? Evidence from plastid and nuclear ribosomal DNA. *Syst. Bot.* 27: 522–533.

Rossbach, G. B. 1963. Distributional and taxonomic notes on some plants collected in West Virginia and nearby states. *Castanea* 28: 10–38.

Rossbach, R. P. 1940. *Spergularia* in North America and South America. *Rhodora* 42: 57–83; 105–143; 158–193; 203–213.

Rossell, I. M., K. K. Moorhead, H. Alvarado & R. J. Warren, II. 2008. Succession of a southern Appalachian mountain wetland six years following hydrologic and microtopographic restoration. *Restorat. Ecol.* 17: 205–214.

Rossi, A. M., B. V. Brodbeck & D. R. Strong. 1996. Response of xylem-feeding leafhopper to host plant species and plant quality. *J. Chem. Ecol.* 22: 653–671.

Rossow, R. A. 1987. Revisión del género *Mecardonia* (Scrophulariaceae). *Candollea* 42: 431–474.

Rouse, C. H. 1941. Notes on winter foraging habits of antelopes in Oklahoma. *J. Mammlogy* 22: 57–60.

Rousi, A. 1965. Biosystematic studies on the species aggregate *Potentilla anserina* L. *Ann. Bot. Fenn.* 2: 47–112.

Rowlee, 1894. The aëration of organs and tissues in *Mikania* and other phanerogams. *Proc. Amer. Microscop. Soc.* 14: 143–166.

Rova, J. H. E., P. G. Delprete, L. Andersson & V. A. Albert. 2002. A *trnL-F* cpDNA sequence study of the Condamineeae-Rondeletieae-Sipaneeae complex with implications on the phylogeny of the Rubiaceae. *Amer. J. Bot.* 89: 145–159.

Roy, J. & H. A. Mooney. 1982. Physiological adaptation and plasticity to water stress of coastal and desert populations of *Heliotropium curassavicum* L. *Oecologia* 52: 370–375.

Roy, R. K., M. Thakur & V. K. Dixit. 2008. Hair growth promoting activity of *Eclipta alba* in male albino rats. *Arch. Dermatol. Res.* 300: 357–364.

Roy, S., R. Ihantola & O. Haenninen. 1992. Peroxidase activity in lake macrophytes and its relation to pollution tolerance. *Environ. Exp. Bot.* 32: 457–464.

Royo, A. A., R. Bates & E. P. Lacey. 2008. Demographic constraints in three populations of *Lobelia boykinii*: a rare wetland endemic. *J. Torrey Bot. Soc.* 135: 189–199.

Rozefelds, A. C. F., L. Cave, D. I. Morris & A. M. Buchanan. 1999. The weed invasion in Tasmania since 1970. *Austral. J. Bot.* 47: 23–48.

Rozema, J. & I. Riphagen. 1977. Physiology and ecologic relevance of salt secretion by the salt gland of *Glaux maritima* L. *Oecologia* 29: 349–357.

Rozema, J., W. Arp, J. Van Diggelen, M. Van Esbroek & R. Broekman. 1986. Occurrence and ecological significance of vesicular arbuscular mycorrhiza in the salt marsh environment. *Acta Bot. Neerl.* 35: 457–467.

Rozenfeld, S. B. & I. S. Sheremetiev. 2014. Barnacle Goose (*Branta leucopsis*) feeding ecology and trophic relationships on Kolguev Island: The usage patterns of nutritional resources in tundra and seashore habitats. *Biol. Bull.* 41: 645–656.

Ruan, C. J., P. Qin & Z. X. He. 2004. Delayed autonomous selfing in *Kosteletzkya virginica* (Malvaceae). *S. African J. Bot.* 70: 639–644.

Ruan, C. J., J. A. da Silva & P. Qin. 2010. Style curvature and its adaptive significance in the Malvaceae. *Plant Syst. Evol.* 288: 13–23.

Ruas, P. M., C. F. Ruas, E. M.D. Maffei, M. A. Marin-Morales & M. L.R. Aguiar-Perecin. 2000. Chromosome studies in the genus *Mikania* (Asteraceae). *Genet. Mol. Biol.* 23: 979–984.

Rubino, D. L., C. E. Williams & W. J. Moriarity. 2002. Herbaceous layer contrast and alien plant occurrence in utility corridors and riparian forests of the Allegheny high plateau. *J. Torrey Bot. Soc.* 129: 125–135.

Ruch, D. G., A. Schoultz & K. S. Badger. 1998. The flora and vegetation of Ginn Woods, Ball State University, Delaware County, Indiana. *Proc. Indiana Acad. Sci.* 107: 17–60.

Ruch, D. G., B. G. Torke, B. R. Hess, K. S. Badger & P. E. Rothrock. 2008. The vascular flora and vegetational communities of the wetland complex on the IMI property in Henry County, near Luray, Indiana. *Proc. Indiana Acad. Sci.* 117: 142–158.

Ruch, D. G., B. G. Torke, B. R. Hess, K. S. Badger & P. E. Rothrock. 2009. The vascular flora and plant communities of the Bennett wetland complex in Henry County, Indiana. *Proc. Indiana Acad. Sci.* 118: 39–54.

Rudd, V. 1955. American species of *Aeschynomene*. *Contr. U. S. Natl. Herb.* 32: 1–172.

Rudenberg, L. & P. S. Green. 1966. A karyological survey of *Lonicera*, I. *J. Arnold Arbor.* 47: 222–247.

Rüdiger, W. 1974. *Ihr Name ist Apis: Kleine Kulturgeschichte der Bienen.* Heinrich Mack, Illertissen, Germany.

Rudolph, D. C., C. A. Ely, R. R. Schaefer, J. H. Williamson & R. E. Thill. 2006. The Diana Fritillary (*Speyeria diana*) and Great Spangled Fritillary (*S. cybele*) dependence on fire in the Ouachita Mountains of Arkansas. *J. Lepid. Soc.* 60: 218–226.

Rueda-Puente, E., T. Castellanos, E. Troyo-Diéguez, J. L. Díaz de León-Alvarez & B. Murillo-Amador. 2002. Effects of a nitrogen-fixing indigenous bacterium (*Klebsiella pneumoniae*) on the growth and development of the halophyte *Salicornia bigelovii* as a new crop for saline environments. *J. Agron. Crop Sci.* 189: 323–332.

Rüeger, H. & M. H. Benn. 1983. The alkaloids of *Senecio triangularis* Hook. *Can. J. Chem.* 61: 2526–2529.

Ruehle, J. L. 1971. Nematodes parasitic on forest trees: III. Reproduction on selected hardwoods. *J. Nematol.* 3: 170–173.

Rufatto, L. C., A. Gower, J. Schwambach & S. Moura. 2012. Genus *Mikania*: chemical composition and phytotherapeutical activity. *Rev. Brasileira Farmacogn.* 22: 1384–1403.

Ruhfel, B. R., M. A. Gitzendanner, P. S. Soltis, D. E. Soltis & J. G. Burleigh. 2014. From algae to angiosperms-inferring the phylogeny of green plants (Viridiplantae) from 360 plastid genomes. *BMC Evol. Biol.* 14: 23. doi:10.1186/1471-2148-14-23.

Ruiz Avila, R. J. & V. V. Klemm. 1996. Management of *Hydrocotyle ranunculoides* L. f., an aquatic invasive weed of urban waterways in Western Australia. *Hydrobiologia* 340: 187–190.

Rundell, H. & M. Woods. 2001. The vascular flora of Ech Lake, Alabama. *Castanea* 66: 352–362.

Rundle, W. D. & L. H. Fredrickson. 1981. Managing seasonally flooded impoundments for migrant rails and shorebirds. *Wildl. Soc. Bull.* 9: 80–87.

Rundle, W. D. & M. W. Sayre. 1983. Feeding ecology of migrant soras in southeastern Missouri. *J. Wildl. Manag.* 47: 1153–1159.

Runkel, S. T. & D. M. Roosa. 1999. *Wildflowers and Other Plants of Iowa Wetlands.* Iowa State University Press, Ames, IA. 372 pp.

Runkle, E. S., R. D. Heins, A. C. Cameron & W. H. Carlson. 1999. Photoperiod and cold treatment regulate flowering of *Rudbeckia fulgida* 'Goldsturm'. *HortScience* 34: 55–58.

Runquist, R. D. 2012. Pollinator-mediated competition between two congeners, *Limnanthes douglasii* subsp. *rosea* and *L. alba* (Limnanthaceae). *Amer. J. Bot.* 99: 1125–1132.

Runquist, R. D. 2013. Community phenology and its consequences for plant-pollinator interactions and pollen limitation in a vernal pool plant. *Int. J. Plant Sci.* 174: 853–862.

Ruotsalainen, A. L. & S. Aikio. 2004. Mycorrhizal inoculum and performance of nonmycorrhizal *Carex bigelowii* and mycorrhizal *Trientalis europaea*. *Can. J. Bot.* 82: 443–449.

Rusk, H. M. 1942. Field trips of the club. *Torreya* 42: 196–198.

Russell, N. H. 1954. Three field studies of hybridization in the stemless white violets. *Amer. J. Bot.* 41: 679–686.

Russell, N. H. 1955. Local introgression between *Viola cucullata* Ait. and *V. septentrionalis* Green. *Evolution* 9: 436–440.

Russell, R. J. 1942. Flotant. *Geogr. Rev.* 32: 74–98.

Russell, W. H., J. R. McBride & K. Carnell. 2003. Influence of environmental factors on the regeneration of hardwood species on three streams in the Sierra Nevada. *Madroño* 50: 21–27.

Russo, L., N. DeBarros, S. Yang, K. Shea & D. Mortensen. 2013. Supporting crop pollinators with floral resources: network-based phenological matching. *Ecol. Evol.* 3: 3125–3140.

Rust, S. K. 2002. *Canada Lynx Habitat Inventory-Pine Creek, Coeur D'Alene basin, Idaho.* Conservation Data Center, Idaho Department of Fish and Game, Boise, ID. 24 pp.

Rustaiyan, A., S. Masoudi, N. Ameri, K. Samiee & A. Monfared. 2006. Volatile constituents of *Ballota aucheri* Boiss., *Stachys benthamiana* Boiss. and *Perovskia abrotanoides* Karel. growing wild in Iran. *J. Essent. Oil Res.* 18: 218–221.

Ruthven, A. G. 1911. *A Biological Survey of the Sand Dune Region on the South Shore of Saginaw Bay, Michigan.* Michigan Geological and Biological Survey. Publication 4. Biological series 2. Lansing, MI. 347 pp.

Rutishauser, R. & B. Isler. 2001. Developmental genetics and morphological evolution of flowering plants, especially bladderworts (*Utricularia*): fuzzy Arberian morphology complements classical morphology. *Ann. Bot.* 88: 1173–1202.

Ruygt, J. 1994. Ecological Studies and Demographic Monitoring of Soft Bird's Beak (*Cordylanthus mollis* ssp. *mollis*), a California Rare Plant Species, and Habitat Recommendations. Report to the Endangered Plant Program, California Department of Fish and Game, Sacramento, CA. 173 pp.

Ryan, A. B. & K. E. Boyer. 2012. Nitrogen further promotes a dominant salt marsh plant in an increasingly saline environment. *J. Plant Ecol.* 5: 429–441.

Rydberg, A. 1920. Notes on Rosaceae-XII. *Bull. Torrey Bot. Club* 47: 45–66.

Rymer, T. 2011. Jewels of the Pacific Northwest. *Rock Gard.* 126: 1–15.

Ryynanen, A. 1973. *Rubus arcticus* and its cultivation. *Ann. Agric. Fenniae* 12: 1–76.

Saarela, J. M., H. S. Rai, J. A. Doyle, P. K. Endress, S. Mathews, A. D. Marchant, B. G. Briggs & S. W. Graham. 2007. Hydatellaceae identified as a new branch near the base of the angiosperm phylogenetic tree. *Nature* 446: 312–315.

Saarela, J. M., P. C. Sokoloff, L. J. Gillespie, L. L. Consaul & R. D. Bull. 2013. DNA barcoding the Canadian Arctic flora: core plastid barcodes (*rbcL* + *matK*) for 490 vascular plant species. *PLoS One* 8: e77982.

Saba, F. 1974. Life history and population dynamics of *Tetranychus tumidus* in Florida (Acarina: Tetranychidae). *Florida Entomol.* 57: 47–63.

Sabancilar, E., F. Aydin, Y. Bek, M. G. Ozden, M. Ozcan, N. Senturk, T. Canturk & A. Y. Turanli. 2011. Treatment of melasma with a depigmentation cream determined with colorimetry. *J. Cosmet. Laser Therapy* 13: 255–259.

Sabourin, A. 2006. Nouvelles des comités. Comité flore québécoise. *Fl. Quebeca* 11: 2–3.

Sacchi, C. F. & P. W. Price. 1988. Pollination of the arroyo willow, *Salix lasiolepis*: role of insects and wind. *Amer. J. Bot.* 75: 1387–1393.

Sacchi, C. F. & P. W. Price. 1992. The relative roles of abiotic and biotic factors in seedling demography of arroyo willow (*Salix lasiolepis*: Salicaceae). *Amer. J. Bot.* 79: 395–405.

Sacher, R. 2005. *Hybridizing Waterlilies: State of the Art.* IWGS, Bradenton, FL. 31 pp.

Sadeghian, S., S. Zarre, R. K. Rabeler & G. Heubl. 2015. Molecular phylogenetic analysis of Arenaria (Caryophyllaceae: tribe Arenarieae) and its allies inferred from nuclear DNA internal transcribed spacer and plastid DNA rps16 sequences. *Bot. J. Linn. Soc.* 178: 648–669.

Saeidi Mehrvarz, S. & A. Kharabian. 2005. Chromosome counts of some *Veronica* L. (Scrophulariaceae) species from Iran. *Turk. J. Bot.* 29: 263–267

Saetersdal, M. & H. J. B. Birks. 1997. A comparative ecological study of Norwegian mountain plants in relation to possible future climatic change. *J. Biogeogr.* 24: 127–152.

Safford, W. E. 1888. Botanizing in the Strait of Magellan. *Bull. Torrey Bot. Club* 15: 15–20.

Saggoo, M. I. S., R. C. Gupta & R. Kaur. 2010. Seasonal variation in chiasma frequency among three morphotypes of *Eclipta alba*. *Chromosome Bot.* 5: 33–36.

Sah, J. P., M. S. Ross, P. L. Ruiz, D. T. Jones, R. Travieso, S. Stoffella, N. Timilsina, E. Hanan, H. Cooley, J. R. Snyder & B. Barrios. 2007. *Effect of Hydrologic Restoration on the Habitat of the Cape Sable Seaside Sparrow.* Annual Report of 2005–2006. Everglades National Park, Homestead, FL. 45 pp.

Sah, J. P., M. S. Ross & S. Stoffella. 2010. *Developing a Data-Driven Classification of South Florida Plant Communities.* National Park Service: South Florida Caribbean Network (NPS/SFCN). Cooperative agreement # H5000 06 0104. Southeast Environmental Research Center, Florida International University, Miami, FL. 109 pp.

Sah, J. P., M. S. Ross, P. L. Ruiz & S. Subedi. 2012. *Monitoring of Tree Island Condition in the Southern Everglades.* Annual Report – 2011 (Cooperative Agreement #: W912HZ-09-2-0019 Modification No.: P00001). Florida International University, Miami, FL. 68 pp.

Sah, J. P., M. S. Ross & P. L. Ruiz. 2013. Landscape Pattern – Marl Prairie/Slough Gradient: Vegetation Composition Along the Gradient and Decadal Vegetation Change Pattern in Shark Slough. Paper 101. Annual Report 2012 (2013). SERC Research Reports. Smithsonian Environmental Research Center, Edgewater, MD. 55 pp.

Saha, A. K., L. D. S. L. O. Sternberg & F. Miralles-Wilhelm. 2009. Linking water sources with foliar nutrient status in upland plant communities in the Everglades National Park, USA. *Ecohydrology* 2: 42–54.

Sahid, I., Z. N. Faezah & N. K. Ho. 1995. Weed populations and their buried seeds in rice fields of the Muda area, Kedah, Malaysia. *Pertanika J. Trop. Agric. Sci.* 18: 21–28.

Saidak, W. J. & S. H. Nelson. 1962. Weed control in ornamental nurseries. *Weeds* 10: 311–315.

Sainty, G. R. & S. W. L. Jacobs. 1981. *Waterplants of New South Wales.* Water Resources Commission, New South Wales, Australia. 550 pp.

Saikkonen, K., S. Koivunen, T. Vuorisalo & P. Mutikainen. 1998. Interactive effects of pollination and heavy metals on resource allocation in *Potentilla anserina* L. *Ecology* 79: 1620–1629.

St. Hilaire, L. 2002. *Ranunculus lapponicus* L. Lapland Buttercup. Conservation and Research Plan for New England. New England Wild Flower Society, Framingham, MA.

St. Hilaire, L. 2003. *Valeriana uliginosa* (Torr. & Gray) Rydb. Marsh valerian. Conservation and Research Plan for New England. New England Wild Flower Society, Framingham, MA. 27 pp.

St. John, H. 1942. The water lily *Nymphaea odorata*, a cultivated plant in the state of Washington. *Leafl. W. Bot.* 3: 142–144.

St. John, H. & W. D. Courtney. 1924. The flora of Epsom lake. *Amer. J. Bot.* 11: 100–107.

St. Omer, L. 2004. Small-scale resource heterogeneity among halophytic plant species in an upper salt marsh community. *Aquat. Bot.* 78: 337–448.

St. Omer, L. & W. H. Schlesinger. 1980. Regulation of NaCl in *Jaumea carnosa* (Asteraceae), a salt marsh species, and its effect on leaf succulence. *Amer. J. Bot.* 67: 1448–1454.

Sakagami, Y, H. Murata, T. Nakanishi, Y. Inatomi, K. Watabe, M. Iinuma, T. Tanaka, J. Murata & F. A. Lang. 2001. Inhibitory effect of plant extracts on production of verotoxin by enterohemorrhagic *Escherichia coli* O157: H7. *J. Health Sci.* 47: 473–477.

Sakai, A. & K. Otsuka.1970. Freezing resistance of alpine plants. *Ecology* 51: 665–671.

Sakhanokho, H. F. 2009. Sulfuric acid and hot water treatments enhance ex vitro and in vitro germination of *Hibiscus* seed. *African J. Biotechnol.* 8: 6185–6190.

Saleem, M., A. Alam & S. Sultana. 2000. Attenuation of benzoyl peroxide-mediated cutaneous oxidative stress and hyperproliferative response by the prophylactic treatment of mice with spearmint (*Mentha spicata*). *Food Chem. Toxicol.* 38: 939–948.

Salick, J. & E. Pfeffer. 1999. The interplay of hybridization and clonal reproduction in the evolution of willows: experiments with hybrids of *S. eriocephala* [R] & *S. exigua* [X] and *S. eriocephala* & *S. petiolaris* [P]. *Plant Ecol.* 141: 163–178.

Salisbury, E. J. 1967. The reproduction and germination of *Limosella aquatica*. *Ann. Bot.* 31: 147–162.

Salisbury, E. 1970. The pioneer vegetation of exposed muds and its biological features. *Philos. Trans. Ser. B.* 259: 207–255.

Salisbury, E. 1978. A note on seed production and frequency. *Proc. Roy. Soc. London, Ser. B Biol. Sci.* 200: 485–487.

Salsbury, G. A. 1984. The weevil genus *Apion* in Kansas. *Trans. Kansas Acad. Sci.* 87: 41–52.

Salonen, J., T. Hyvönen & H. Jalli. 2001. Weeds in spring cereal fields in Finland-a third survey. *Agric. Food Sci. Finland* 10: 347–364.

Salonen, V. 1987. Relationship between the seed rain and the establishment of vegetation in two areas abandoned after peat harvesting. *Holarc. Ecol.* 10: 171–174.

Salvucci, M. E. & G. Bowes. 1981. Induction of reduced photorespiratory activity in submersed and amphibious aquatic macrophytes. *Plant Physiol.* 67: 335–340.

Samanta, K., E. Hossain & D. K. Pal. 2012. Anthelmintic activity of *Hygrophila difformis* Blume. *J. Buffalo Sci.* 1: 35–38.

Samarth, R. M., M. Panwar, M. Kumar & A. Kumar. 2006. Protective effects of *Mentha piperita* Linn on benzo[*a*]pyrene-induced lung carcinogenicity and mutagenicity in Swiss albino mice. *Mutagenesis* 21: 61–66.

Samuel, R., W. Pinsker, S. Balasubhramaniam & W. Morawetz. 1991. Allozyme diversity and systematics in *Annonaceae* – a pilot project. *Plant Syst. Evol.* 178: 125–134.

Sanabria-Aranda, L., A. González-Bermúdez, N. N. Torres, C. Guisande, A. Manjarrés-Hernández, V. Valoyes-Valois, J. Díaz-Olarte, C. Andrade-Sossa & S. R. Duque. 2006. Predation by the tropical plant *Utricularia foliosa*. *Freshwater Biol.* 51: 1999–2008.

Sanchez, A. & W. K. Smith. 2015. No evidence for photoinhibition of photosynthesis in alpine *Caltha leptosepala* DC. *Alpine Bot.* 125: 41–50.

Sánchez, M. P. O. 1999. *Plantas Medicinales del Estado de Chihuahua I.* Universidad Autónoma de Ciudad Juárez, Chihuahua, México. 127 pp.

Sánchez-Del Pino, I., T. J. Motley & T. Borsch. 2012. Molecular phylogenetics of *Alternanthera* (Gomphrenoideae, Amaranthaceae): resolving a complex taxonomic history caused by different interpretations of morphological characters in a lineage with C_4 and C_3–C_4 intermediate species. *Bot. J. Linn. Soc.* 169: 493–517.

Sanchez-Vilas J. & R. Retuerto. 2012. Response of the sexes of the subdioecious plant *Honckenya peploides* to nutrients under different salt spray conditions. *Ecol. Res.* 27: 163–171.

Sánchez-Vilas, J., R. Bermúdez & R. Retuerto. 2012. Soil water content and patterns of allocation to below-and above-ground biomass in the sexes of the subdioecious plant *Honckenya peploides*. *Ann. Bot.* 110: 839–848.

Sanchez-Vilas, J., M. Philipp & R. Retuerto. 2010. Unexpectedly high genetic variation in large unisexual clumps of the subdioecious plant *Honckenya peploides* (Caryophyllaceae). *Plant Biol.* 12: 518–525.

Sanders, D. W., P. Weatherwax & L. S. McClung. 1945. Antibacterial substances from plants collected in Indiana. *J. Bacteriol.* 49: 611–615.

Sanders, R. W. 1987. Taxonomic significance of chromosome observations in Caribbean species of *Lantana* (Verbenaceae). *Amer. J. Bot.* 74: 914–920.

Sanders, R. W. & P. D. Cantino. 1984. Nomenclature of the subdivisions of the Lamiaceae. *Taxon* 33: 64–72.

Sanderson, M. A., S. C. Goslee, K. D. Klement & K. J. Soder. 2007. Soil seed bank composition in pastures of diverse mixtures of temperate forages. *Agron. J.* 99: 1514–1520.

Sand-Jensen, K. 1983. Photosynthetic carbon sources of stream macrophytes. *J. Exp. Bot.* 34: 198–210.

Sand-Jensen, K., C. Prahl & H. Stokholm. 1982. Oxygen release from roots of submerged aquatic macrophytes. *Oikos* 38: 349–354.

Sandvik, S. M. & Ø. Totland. 2003. Quantitative importance of staminodes for female reproductive success in *Parnassia palustris* under contrasting environmental conditions. *Can. J. Bot.* 81: 49–56.

Sanford, M. T. 1988. *Beekeeping: Florida Bee Botany.* Circular 686. Entomology and Nematology Department, Florida Cooperative Extension Service, Institute of Food and Agricultural Sciences, University of Florida, Gainesville, FL. 5 pp.

Sanger, J. E. & E. Gorham. 1973. A comparison of the abundance and diversity of fossil pigments in wetland peats and woodland humus layers. *Ecology* 54: 605–611.

Sanjur, O. I., D. R. Piperno, T. C. Andres & L. Wessel-Beaver. 2002. Phylogenetic relationships among domesticated and wild species of *Cucurbita* (Cucurbitaceae) inferred from a mitochondrial gene: Implications for crop plant evolution and areas of origin. *Proc. Natl. Acad. Sci. U.S.A.* 99: 535–540.

Sanz, M., R. Vilatersana, O. Hidalgo, N. Garcia-Jacas, A. Susanna, G. M. Schneeweiss & J. Vallès. 2008. Molecular phylogeny and evolution of floral characters of *Artemisia* and allies (Anthemideae, Asteraceae): evidence from nrDNA ETS and ITS sequences. *Taxon* 57: 66–78.

Sargent, R. D. & S. P. Otto. 2004. A phylogenetic analysis of pollination mode and the evolution of dichogamy in angiosperms. *Evol. Ecol. Res.* 6: 1183–1199.

Sargent, R. D., S. W. Kembel, N. C. Emery, E. J. Forrestel & D. D. Ackerly. 2011. Effect of local community phylogenetic structure on pollen limitation in an obligately insect-pollinated plant. *Amer. J. Bot.* 98: 283–289.

Sarker, S. D. & L. Nahar. 2004. Natural medicine: the genus *Angelica*. *Curr. Med. Chem.* 11: 1479–1500.

Sarker, S. D., J.-P. Girault, R. Lafont & L. N. Dinan. 1997. Ecdysteroid xylosides from *Limnanthes douglasii*. *Phytochemistry* 44: 513–521.

Sarneel, J. M. 2013. The dispersal capacity of vegetative propagules of riparian fen species. *Hydrobiologia* 710: 219–225.

Sarracino, J. M. & N. Vorsa. 1991. Self and cross fertility in cranberry. *Euphytica* 58: 129–136.

Sasser, C. E., J. G. Gosselink, E. M. Swenson & D. E. Evers. 1995. Hydrologic, vegetation, and substrate characteristics of floating marshes in sediment-rich wetlands of the Mississippi river delta plain, Louisiana, USA. *Wetlands Ecol.* 3: 171–187.

Sasser, C. E., J. G. Gosselink, E. M. Swenson, C. M. Swarzenski & N. C. Leibowitz. 1996. Vegetation, substrate and hydrology in floating marshes in the Mississippi river delta plain wetlands, USA. *Vegetatio* 122: 129–142.

Saunders, C. F. 1900. The pine barrens of New Jersey. *Proc. Acad. Nat. Sci. Philadelphia* 52: 544–549.

Saunders, C. F. 2011. *Edible and Useful Wild Plants of the United States and Canada.* Dover Publications, Mineola, New York. 320 pp.

Saunders, K. 2005. First record of *Nymphoides indica* (Menyanthaceae) in Texas. *Sida* 21: 2441–2443.

Saunders Jr., J. K. 1955. Food habits and range use of the Rocky Mountain goat in the Crazy Mountains, Montana. *J. Wildl. Manag.* 19: 429–437.

Savard, L., M. Michaud & J. Bousquet. 1993. Genetic diversity and phylogenetic relationships between birches and alders using ITS, 18S rRNA, and *rbc*L gene sequences. *Mol. Phylogen. Evol.* 2: 112–118.

Savile, D. B. O. 1953. Splash-cup dispersal mechanism in *Chrysosplenium* and *Mitella*. *Science* 117: 250–251.

Savile, D. B. O. 1956. Known dispersal rates and migratory potentials as clues to the origin of the North American biota. *Amer. Midl. Naturalist* 56: 434–453.

Savile, D. B. O. 1964. General ecology and vascular plants of the Hazen Camp area. *Arctic* 17: 237–258.

Savile, D. B. O. 1975. Evolution and biogeography of Saxifragaceae with guidance from their rust parasites. *Ann. Missouri Bot. Gard.* 62: 354–361.

Savile, D. B. O. & B. D. O. Savile. 1953. Short-season adaptations in the rust Fungi. *Mycologia* 45: 75–87.

Savka, M. A. & J. O. Dawson. 1985. *Culture of Ovules Containing Immature Embryos of* Populus heterophylla *L. In Vitro.* Forestry Research Report No. 85-1. Department of Forestry, Agricultural Experiment Station, University of Illinois, Urbana-Champaign, IL. 4 pp.

Savolainen, V., M. F. Fay, D. C. Albach, A. Backlund, M. van der Bank, K. M. Cameron, S. A. Johnson, M. D. Lledó, J.-C. Pintaud, M. Powell, M. C. Sheahan, D. E. Soltis, P. S. Soltis, P. Weston, W. M. Whitten, K. J. Wurdack & M. W. Chase. 2000. Phylogeny of the eudicots: a nearly complete familial analysis based on *rbcL* gene sequences. *Kew Bull.* 55: 257–309.

Sawant, M., J. C. Isaac & S. Narayanan. 2004. Analgesic studies on total alkaloids and alcohol extracts of *Eclipta alba* (Linn.) Hassk. *Phytother. Res.* 18: 111–113.

Sayre, C. 2001. A new species of *Bidens* (Asteraceae: Heliantheae) from Starbuck Island provides evidence for a second colonization of Pacific Islands by the genus. MS thesis. The University of British Columbia, Vancouver, BC. 61 pp.

Sax, K. 1930. Chromosome stability in the genus *Rhododendron*. *Amer. J. Bot.* 17: 247–251.

Scanga, S. E. 2011. Effects of light intensity and groundwater level on the growth of a globally rare fen plant. *Wetlands* 31: 773–781.

Scanga, S. E. & D. J. Leopold. 2010. Population vigor of a rare, wetland, understory herb in relation to light and hydrology. *J. Torrey Bot. Soc.* 137: 297–311.

Scatizzi, A., A. Di Maggio, D. Rizzi, A. M. Sebastio & C. Basile. 1993. Acute renal failure due to tubular necrosis caused by wildfowl-mediated hemlock poisoning. *Renal Failure* 15: 93–96.

Schaefer, H. M. & G. D. Ruxton. 2014. Fenestration: a window of opportunity for carnivorous plants. *Biol Lett.* 10(4): 20140134. doi:10.1098/rsbl.2014.0134.

Schafale, M. P. & A. S. Weakley. 1990. *Classification of the Natural Communities of North Carolina, Third approximation.* North Carolina Natural Heritage Program, Division of Parks and Recreation, Department of Environment and Natural Resources. Raleigh, NC. 321 pp.

Schaffer, B. 1998. Flooding responses and water-use ffficiency of subtropical and tropical fruit trees in an environmentally-sensitive wetland. *Ann. Bot.* 81: 475–481.

Schaffner, J. H. 1902. The self-pruning of woody plants. *Ohio J. Sci.* 2: 171–174.

Schaffner, J. H. 1904. Poisonous and other injurious plants of Ohio. *Ohio Naturalist* 4: 69–73.

Schamel, D. 1977. Breeding of the common eider (*Somateria mollissima*) on the Beaufort Sea coast of Alaska. *Condor* 79: 478–485.

Schanzer, I. A. 1994. Taxonomic revision of the genus *Filipendula* Mill. (Rosaceae). *J. Jap. Bot.* 69: 290–319.

Schanzer, I. A. 2014. 1. *Filipendula* Miller. pp. 23–27. *In*: N. R. Morin (convening ed.), *Flora of North America North of Mexico, Vol. 9: Magnoliophyta: Picramniaceae to Rosaceae.* Oxford University Press, New York.

Schat, H. & M. Scholten. 1986. Effects of salinity on growth, survival and life history of four short-lived pioneers from brackish dune slacks. *Acta Oecol.* 7: 221–231.

Scheen, A.-C. & V. A. Albert. 2004. Molecular phylogenetics of the *Physostegia* group (Lamioideae: Lamiaceae). Botany 2004 meeting (abstract).

Scheen, A.-C., C. Lindqvist, C. G. Fossdal & V. A. Albert. 2008. Molecular phylogenetics of tribe Synandreae, a North American lineage of lamioid mints (Lamiaceae). *Cladistics* 24: 299–314.

Schemske, D. W. & H. D. Bradshaw Jr. 1999. Pollinator preference and the evolution of floral traits in monkeyflowers (*Mimulus*). *Proc. Natl. Acad. Sci. U.S.A.* 96: 11910–11915.

Schender, D., K. Katz & M. W. Gates. 2014. Review of *Hyperimerus* (Pteromalidae: Asaphinae) in North America, with redescription of *Hyperimerus corvus* (Girault). *Proc. Entomol. Soc. Washington* 116: 408–420.

Schenk, M. F., C. N. Thienpont, W. J. Koopman, L. J. Gilissen & M. J. Smulders. 2008. Phylogenetic relationships in *Betula* (Betulaceae) based on AFLP markers. *Tree Genet. Genomes* 4: 911–924.

Schep, L. J., R. J. Slaughter, G. Becket & D. M. G. Beasley. 2009. Poisoning due to water hemlock. *Clin. Toxicol.* 47: 270–278.

Scherm, H. & A. T. Savelle. 2003. Epidemic development of hawthorn leaf blight (*Monilinia johnsonii*) on mayhaw (*Crataegus aestivalis* and *C. opaca*) in Georgia. *Plant Dis.* 87: 539–543.

Schery, R. W. 1972. *Plants for Man, 2nd Edition.* Prentice Hall, Inc., Englewood Cliffs, NJ. 657 pp.

Schiefer, T. L. 1999. First records of interspecific hybrids between two *Limenitis* sp. in Mississippi. *News Lepid. Soc.* 41: 99.

Schierenbeck, K. A. & F. Phipps. 2010. Population genetics of *Howellia aquatilis* (Campanulaceae) in disjunct locations throughout the Pacific Northwest. *Genetica* 138: 1161–1169.

Schiller, J. R., P. H. Zedler & C. H. Black. 2000. The effect of density-dependent insect visits, flowering phenology, and plant size on seed set of the endangered vernal pool plant *Pogogyne abramsii* (Lamiaceae) in natural compared to created vernal pools. *Wetlands* 20: 386–396.

Schilling, E. E. 2001. Phylogeny of *Helianthus* and related genera. *Oléagineaux* 8: 22–25.

Schilling, E. E. 2006. 299. *Helianthus* Linnaeus. pp. 141–169. *In*: Flora of North America Editorial Committee (eds.), *Flora of North America North of Mexico: Vol. 21, Magnoliophyta: Asteridae (in part): Asteraceae, Part 3.* Oxford University Press, New York.

Schilling, E. E. 2011a. Hybrid genera in Liatrinae (Asteraceae: Eupatorieae). *Mol. Phylogen. Evol.* 59: 158–167.

Schilling, E. E. 2011b. Systematics of the *Eupatorium album* complex (Asteraceae) from eastern North America. *Syst. Bot.* 36: 1088–1100.

Schilling, E. E. & A. Floden. 2012. Barcoding the Asteraceae of Tennessee, tribes Gnaphalieae and Inuleae. *Phytoneuron* 2012–99: 1–6.

Schilling, E. E. & A. Floden. 2013. Barcoding the Asteraceae of Tennessee, tribes Helenieae and Polymnieae. *Phytoneuron* 2013–81: 1–6.

Schilling, E. E., J. L. Panero & P. B. Cox. 1999. Chloroplast DNA restriction site data support a narrowed interpretation of *Eupatorium* (Asteraceae). *Plant Syst. Evol.* 219: 209–223.

Schilling, E. E., C. R. Linder, R. D. Noyes & L. H. Rieseberg. 1998. Phylogenetic relationships in *Helianthus* (Asteraceae) based on nuclear ribosomal DNA internal transcribed spacer region sequence data. *Syst. Bot.* 23: 177–187.

Schilling, E. E., J. B. Beck, P. J. Calie & R. L. Small. 2008. Molecular analysis of *Solidaster* cv. Lemore, a hybrid goldenrod (Asteraceae). *J. Bot. Res. Inst. Texas* 2008: 7–18.

Schilling, N. 1976. Distribution of l-(+)-bornesitol in the Gentianaceae and Menyanthaceae. *Phytochemistry* 15: 824–826.

Schimpf, D. J. 1977. Seed weight of *Amaranthus retroflexus* in relation to moisture and length of growing season. *Ecology* 58: 450–453.

Schleidlinger, C. R. 1981. Population studies in *Pogogyne abramsii*. pp. 223–231. *In*: S. Jain and P. Moyle (eds.), *Vernal Pools and Intermittent Streams*. Institute of Ecology, University of California. Davis, CA.

Schlesinger, W. H. 1978. On the relative dominance of shrubs in Okefenokee Swamp. *Amer. Naturalist* 112: 949–954.

Schlessman, M. A. & F. R. Barrie. 2004. Protogyny in Apiaceae, subfamily Apioideae: systematic and geographic distributions, associated traits, and evolutionary hypotheses. *S. African J. Bot.* 70: 475–487.

Schlichting, C. D. & B. Devlin. 1992. Pollen and ovule sources affect seed production of *Lobelia cardinalis* (Lobeliaceae). *Amer. J. Bot.* 79: 891–898.

Schlising, R. A. & E. L. Sanders. 1982. Quantitative analysis of vegetation at the Richvale vernal pools, California. *Amer. J. Bot.* 69: 734–742.

Schlising, R. A. & E. L. Sanders. 1983. Vascular plants of Richvale vernal pools, Butte County, California. *Madroño* 30: S19-S30.

Schloesser, D. W. 1986. *A Field Guide to Valuable Underwater Aquatic Plants of the Great Lakes*. Extension Bulletin E-1902, Cooperative Extension Service, Michigan State University, Lansing, MI. 32 pp.

Schlosser, K. 2009. North Carolina imperiled plants: *Pinguicula lutea*, *P. pumila*, butterwort. *Friends of NC Plant Conserv. Field Notes* 1: 11–14.

Schmalzer, P. A. 1995. Biodiversity of saline and brackish marshes of the Indian River lagoon: Historic and current patterns. *Bull. Mar. Sci.* 57: 37–48.

Schmid, B. & F. A. Bazzaz. 1994. Crown construction, leaf dynamics, and carbon gain in owo perennials with contrasting architecture. *Ecol. Monogr.* 64: 177–203.

Schmidt, A. C. 2005. A vascular plant inventory and description of the twelve plant community types found in the University of South Florida Ecological Research Area, Hillsborough County, Florida. MS thesis. Department of Biology, University of South Florida, Tampa, FL. 118 pp.

Schmidt, B. L. & W. F. Millington. 1968. Regulation of leaf shape in *Proserpinaca palustris*. *Bull. Torrey Bot. Club* 95: 264–286.

Schmidt, G. J. & E. E. Schilling. 2000. Phylogeny and biogeography of *Eupatorium* (Asteraceae: Eupatorieae) based on nuclear ITS sequence data. *Amer. J. Bot.* 87: 716–726.

Schmidt Jr., J. A. 1957. Comparative morphology of the *Zinnia*, *Phlox* and cucurbit powdery mildews. *Amer. J. Bot.* 44: 120–125.

Schmidt, K. and K. Jensen. 2000. Genetic structure and AFLP variation of remnant populations in the rare plant *Pedicularis palustris* (Scrophulariaceae) and its relation to population size and reproductive components. *Amer. J. Bot.* 87: 678–689.

Schmidt, L. J. 2003a. Conservation assessment for *Polemonium occidentale* v. *lacustre* western Jacob's ladder. Conservation assessment for the Eastern Region of the Forest Service, Milwaukee, WI. 16 pp.

Schmidt, L. J. 2003b. Conservation assessment for *Valeriana uliginosa* marsh valerian. Conservation assessment for the Eastern Region of the Forest Service, Milwaukee, Wisconsin. 16 pp.

Schmidt, M., K. Sommer, K. Wolf-Ulrich, H. Ellenberg & G. von Oheimb. 2004. Dispersal of vascular plants by game in northern Germany. Part I: Roe deer (*Capreolus capreolus*) and wild boar (*Sus scrofa*). *Eur. J. Forest Res.* 123: 167–176.

Schmidt, M. G. 1980. *Growing California Native Plants*. University of California Press, Berkeley, CA. 400 pp.

Schmidt, S. K. & K. M. Scow. 1986. Mycorrhizal Fungi on the Galapagos Islands. *Biotropica* 18: 236–240.

Schmidt, T. J. 1999. Novel seco-prezizaane sesquiterpenes from North American *Illicium* species. *J. Nat. Prod.* 62: 684–687.

Schmidt, W. 1989. Plant dispersal by motor cars. *Vegetatio* 80: 147–152.

Schmitt, J. 1980. Pollinator foraging behavior and gene dispersal in *Senecio* (Compositae). *Evolution* 34: 934–943.

Schmitt, J. 1983. Flowering plant density and pollinator visitation in *Senecio*. *Oecologia* 60: 97–102.

Schmitt Jr., J. A. 1955. The host specialization of *Erysiphe cichoracearum* from *Zinnia*, *Phlox* and cucurbits. *Mycologia* 47: 688–701.

Schmitz, D. C. & L. E. Nall. 1984. Status of *Hygrophila polysperma* in Florida. *Aquatics* 6(3): 11–14.

Schnabel, A. & J. F. Wendel. 1998. Cladistic biogeography of *Gleditsia* (Leguminosae) based on *ndhF* and *rpl16* chloroplast gene sequences. *Amer. J. Bot.* 85: 1753–1765.

Schnabel, A., P. E. McDonel & J. F. Wendel. 2003. Phylogenetic relationships in *Gleditsia* (Leguminosae) based on ITS sequences. *Amer. J. Bot.* 90: 310–320.

Schneider, A. C., W. A. Freyman, C. M. Guilliams, Y. P. Springer & B. G. Baldwin. 2016. Pleistocene radiation of the serpentine-adapted genus *Hesperolinon* and other divergence times in Linaceae (Malpighiales). *Amer. J. Bot.* 103: 1–12.

Schneider, E. L. 1982. Notes on the floral biology of *Nymphaea elegans* (Nymphaeaceae) in Texas. *Aquat. Bot.* 12: 197–200.

Schneider, E. L. & J. D. Buchanan. 1980. Morphological studies of the Nymphaeaceae. XI. The floral biology of *Nelumbo pentapetala*. *Amer. J. Bot.* 67: 182–193.

Schneider, E. L. & S. Carlquist. 1996a. Vessels in *Nelumbo* (Nelumbonaceae). *Amer. J. Bot.* 83: 1101–1106.

Schneider, E. L. & S. Carlquist. 1996b. Conductive tissue in *Ceratophyllum demersum* (Ceratophyllaceae). *Sida* 17: 437–443.

Schneider, E. L. & T. Chaney. 1981. The floral biology of *Nymphaea odorata* (Nymphaeaceae). *Southwest. Naturalist* 26: 159–165.

Schneider, E. L. & J. M. Jeter. 1982. Morphological studies of the Nymphaeaceae. XII. The floral biology of *Cabomba caroliniana*. *Amer. J. Bot.* 69: 1410–1419.

Schneider, S. & A. Melzer. 2003. The trophic index of macrophytes (TIM): a new tool for indicating the trophic state of running waters. *Int. Rev. Hydrobiol.* 88: 49–67.

Schneider, E. L. & L. A. Moore. 1977. Morphological studies of the Nymphaeaceae. VII. The floral biology of *Nuphar lutea subsp. macrophylla*. *Brittonia* 29: 88–99.

Schneider, R. L. & R. R. Sharitz. 1986. Seed bank dynamics in a southeastern riverine swamp. *Amer. J. Bot.* 73: 1022–1030.

Schneider, R. L. & R. R. Sharitz. 1988. Hydrochory and regeneration in a bald cypress-water tupelo swamp forest. *Ecology* 69: 1055–1063.

Schneider, M. J. & F. R. Stermitz. 1990. Uptake of host plant alkaloids by root parasitic *Pedicularis* spp. *Phytochemistry* 29: 1811–1814.

Schneider, E. L., P. S. Williamson & D. C. Whitenberg. 1990. Hot sex in water lilies. *Water Gard. J.* 6(4): 41–51.

Schnell, D. E. 1977. Infraspecific variation in *Sarracenia rubra* Walt.: some observations. *Castanea* 42: 149–170.

Schnell, D. E. 1980. Notes on *Utricularia simulans* Pilger (Lentibulariaceae) in Southern Florida. *Castanea* 45: 270–276.

Schnell, D. E. 1982. Notes on *Drosera linearis* Goldie in northeastern lower Michigan. *Castanea* 47: 313–328.

Schnell, D. E. 1983. Notes on the pollination of *Sarracenia flava* L. (Sarraceniaceae) in the Piedmont province of North Carolina. *Rhodora* 85: 405–420.

Schnell, D. E. 1993. *Sarracenia purpurea* L. ssp. *venosa* (Raf.) Wherry var. *burkii* Schnell (Sarraceniaceae)—A new variety of the Gulf coastal plain. *Rhodora* 95: 6–10.

Schnell, D. E. 2002. *Carnivorous Plants of the United States and Canada. 2nd Edition.* Timber Press, Inc., Portland, OR. 468 pp.

Schnepf, E. & J. Busch. 1976. Morphology and kinetics of slime secretion in glands of *Mimulus tilingii. Z. Pflanzenphysiol.* 79: 62–71.

Schoennagel, T., D. M. Waller, M. G.Turner & W. H. Romme. 2004. The effect of fire interval on post-fire understorey communities in Yellowstone National Park. *J. Veg. Sci.* 15: 797–806.

Scholtens, B. G. 1991. Host plants and habitats of the Baltimore checkerspot butterfly, *Euphydryas phaeton* (Lepidoptera: Nymphalidae), in the Great Lakes region. *Great Lakes Entomol.* 24: 207–217.

Scholtens, B. G. & W. H. Wagner Jr. 1994. Biology of the genus *Hemileuca* (Lepidoptera: Saturniidae) in Michigan. *Great Lakes Entomol.* 27: 197–207.

Schönenberger, J. & E. Conti. 2003. Molecular phylogeny and floral evolution of Penaeaceae, Oliniaceae, Rhynchocalycaceae, and Alzateaceae (Myrtales). *Amer. J. Bot.* 90: 293–309.

Schönenberger, J., A. A. Anderberg & K. J. Sytsma. 2005. Molecular phylogenetics and patterns of floral evolution in the Ericales. *Int. J. Plant Sci.* 166: 265–288.

Schoof-van Pelt, M. M. 1973. *Littorelletea: A Study of the Vegetation of Some Amphiphytic Communities of Western Europe.* Stichting Studentenpers, Nijmegen, the Netherlands. 216 pp.

Schooler, S. S. 2008. Shade as a management tool for the invasive submerged macrophyte, *Cabomba caroliniana. J. Aquat. Plant Manag.* 46: 168–171.

Schooler, S. S. 2012. *Alternanthera philoxeroides* (Martius) Grisebach (alligator weed). pp. 25–27. *In*: R. A. Francis (ed.), *A Handbook of Global Freshwater Invasive Species.* Earthscan, New York.

Schooler, S., W. Cabrera-Walsh & M. Julien. 2009. 6. *Cabomba caroliniana* Gray (Cabombaceae). pp. 87–107. *In*: R. Muniappan, G. V. P. Reddy & A. Raman (eds.), *Biological Control of Tropical Weeds Using Arthropods.* Cambridge University Press, Cambridge, UK.

Schoolmaster Jr., D. R. 2005. *Impatiens capensis* (Balsaminaceae) Meerb. is a necessary nurse host for the parasitic plant *Cuscuta gronovii* (Cuscutaceae) Willd. ex J.A. Schultes in southeastern Michigan wetlands. *Amer. Midl. Naturalist* 153: 33–40.

Schotsman, H. D. 1954. A taxonomic spectrum of the section *Eucallitriche* in the Netherlands. *Acta Bot. Neerl.* 3: 313–384.

Schotsman, H. D. 1967. *Les Callitriches. Flore de France—Vol. I.* P. Lechevalier, Paris, France. 152 pp.

Schornherst, R. O. 1943. Phytogeographic studies of the mosses of northern Florida. *Amer. Midl. Naturalist* 29: 509–532.

Schrader, J. A. & W. R. Graves. 2002. Infraspecific systematics of *Alnus maritima* (Betulaceae) from three widely disjunct provenances. *Castanea* 67: 380–401.

Schrader, J. A. & W. R. Graves. 2004. Systematics of *Alnus maritima* (seaside alder) resolved by ISSR polymorphisms and morphological characters. *J. Amer. Soc. Hort. Sci.* 129: 231–236.

Schrader, J. A. & W. R. Graves. 2011. Taxonomy of *Leitneria* (Simaroubaceae) resolved by ISSR, ITS, and morphometric characterization. *Castanea* 76: 313–338.

Schroeder, M. S. & S. G. Weller. 1997. Self-incompatibility and clonal growth in *Anemopsis californica. Plant Species Biol.* 12: 55–59.

Schroth, A. B. 1996. An ethnographic review of grinding, pounding, pulverizing, and smoothing with stones. *Pacif. Coast Archaeol. Soc. Q.* 32: 55–75.

Schubauer, J. P. & R. R. Parmenter. 1981. Winter feeding by aquatic turtles in a southeastern reservoir. *J. Herpetol.* 15: 444–447.

Schubert, K. & U. Braun. 2005. Taxonomic revision of the genus *Cladosporium* s. lat. 1. Species reallocated to *Fusicladium, Parastenella, Passalora, Pseudocercospora* and *Stenella. Mycol. Progr.* 4: 101–109.

Schuettpelz, E. & S. B. Hoot. 2004. Phylogeny and biogeography of *Caltha* (Ranunculaceae) based on chloroplast and nuclear DNA sequences. *Amer. J. Bot.* 91: 247–253.

Schuh, R. T. & M. D. Schwartz. 2005. Review of North American *Chlamydatus* Curtis species, with new synonymy and the description of two new species (Heteroptera: Miridae: Phylinae). *Amer. Mus. Novit.* 3471: 1–55.

Schultheis, L. M. 2001. Systematics of *Downingia* (Campanulaceae) based on molecular sequence data: Implications for floral and chromosome evolution. *Syst. Bot.* 26: 603–621.

Schultheis, L. M. 2010. Morphologically cryptic species within *Downingia yina* (Campanulaceae). *Madroño* 57: 20–41.

Schultheis, L. M. 2012. *Downingia.* pp. 590–591. *In*: B. G. Baldwin, D. H. Goldman, D. J. Keil, R. Patterson, T. J. Rosatti and D. H. Wilken (eds.), *The Jepson Manual (2nd ed.).* University of California Press, Berkeley, CA.

Schultheis, L. M. & M. J. Donoghue. 2004. Molecular phylogeny and biogeography of *Ribes* (Grossulariaceae), with an emphasis on gooseberries (subg. *Grossularia). Syst. Bot.* 29: 77–96.

Schultz, J., P. Beyer & J. Williams. 2001. Propagation protocol for production of container *Ledum groenlandicum* Oeder plants; Hiawatha National Forest, Marquette, Michigan. *In*: Native Plant Network. Moscow (ID): University of Idaho, College of Natural Resources, Forest Research Nursery. Available online: http://www.nativeplantnetwork.org (accessed June 7, 2006).

Schultz, J. H. 1946. The introduction of *Viola lanceolata* into the Pacific Northwest. *Madroño* 8: 191–193.

Schulz, B., J. Döring & G. Gottsberger. 1991. Apparatus for measuring the fall velocity of anemochorous diaspores, with results from two plant communities. *Oecologia* 86: 454–456.

Schulze, D. M., J. L. Walker & T. P. Spira. 2002. Germination and seed bank studies *of Macbridea alba* (Lamiaceae), a Federally threatened plant. *Castanea* 67: 280–289.

Schumacher, G. J. 1956. A qualitative and quantitative study of the plankton algae in southwestern Georgia. *Amer. Midl. Naturalist* 56: 88–115.

Schuster, T. M., J. L. Reveal & K. A. Kron. 2011. Phylogeny of Polygoneae (Polygonaceae: Polygonoideae). *Taxon* 60: 1653–1666.

Schuster, T. M., J. L. Reveal, M. J. Bayly & K. A. Kron. 2015. An updated molecular phylogeny of Polygonoideae (Polygonaceae): Relationships of *Oxygonum, Pteroxygonum,* and *Rumex,* and a new circumscription of *Koenigia. Taxon* 64: 1188–1208.

Schuyler, A. E. 1989. Intertidal variants of *Bacopa rotundifolia* and *B. innominata* in the Chesapeake drainage. *Bartonia* 55: 18–22.

Schuyler, A. E., S. B. Andersen & V. J. Kolaga. 1993. Plant zonation changes in the tidal portion of the Delaware River. *Proc. Acad. Nat. Sci. Philadelphia* 144: 263–266.

Schwaegerle, K. E. 1983. Population growth of the pitcher plant, *Sarracenia purpurea* L., at Cranberry Bog, Licking County, Ohio. *Ohio J. Sci.* 83: 19–22.

Schwartz, A. 1954. Observations on the Big Pine Key cotton rat. *J. Mammalogy* 35: 260–263.

Schwartz, M. D. 1984. A revision of the black grass bug genus *Irbisia* Reuter (Heteroptera: Miridae). *J. New York Entomol. Soc.* 92: 193–306.

Schwartz, O. A. 1985. Lack of protein polymorphism is the endemic relict *Chrysosplenium iowense* (Saxifragaceae). *Can. J. Bot.* 63: 2031–2034.

Schwartz, M. D. & G. G. E. Scudder. 2000. Miridae (Heteroptera) new to Canada, with some taxonomic changes. *J. New York Entomol. Soc.* 108: 248–267.

Schwarz, H. F. 1927. Notes on some anthidiine bees of Montana and California. *Amer. Mus. Novitates* 277: 1–8.

Schwarzbach, A. E. & L. A. McDade. 2002. Phylogenetic relationships of the mangrove family Avicenniaceae based on chloroplast and nuclear ribosomal DNA sequences. *Syst. Bot.* 27: 84–98.

Schwarzbach, A. E. & R. E. Ricklefs. 2000. Systematic affinities of Rhizophoraceae and Anisophylleaceae, and intergeneric relationships within Rhizophoraceae, based on chloroplast DNA, nuclear ribosomal DNA, and morphology. *Amer. J. Bot.* 87: 547–564.

Schwarzbach, A. E. & R. E. Ricklefs. 2001. The use of molecular data in mangrove plant research. *Wetlands Ecol. Manag.* 9: 195–201.

Schweitzer, D. F. 2007. *Enigmogramma baigera* (Noctuidae, Plusiinae) as a specialized transient pest of *Lobelia* in New Jersey. *J. Lepid. Soc.* 61: 55–56.

Schwintzer, C. R. 1978. Nutrient and water levels in a small Michigan bog with high tree mortality. *Amer. Midl. Naturalist* 100: 441–451.

Schwintzer, C. R. & A. Ostrofsky. 1989. Factors affecting germination of *Myrica gale* seeds. *Can. J. Forest Res.* 19: 1105–1109.

Scianna, J. D. & J. Lapp. 2005. *Effects of Erosion Control Blanket on Germination and Germinant Survival of Six Native Species and Potential Management Implications.* Plant Materials Technical Note No. MT-47. United States Department of Agriculture, Natural Resources Conservation Service, Bozeman, MT. 4 pp.

Scifres, C. J., J. W. McAtee & D. L. Drawe. 1980. Botanical, edaphic, and water relationships of Gulf cordgrass (*Spartina spartinae* [Trin.] Hitchc.) and associated communities. *Southwest. Naturalist* 25: 397–409.

Scocco, C., P. Corvi Mora & C. Corti. 1998. Introduction of *Amsonia tabernaemontana* Walt. in hilly area: germination test and first searches about the results in rutin. *Acta Hort.* 457: 357–362.

Scogin, R. & K. Zakar. 1976. Anthochlor pigments and floral UV patterns in the genus *Bidens*. *Biochem. Syst. Ecol.* 4: 165–167.

Scotland, R. W., J. A. Sweere, P. A. Reeves & R. G. Olmstead. 1995. Higher-level systematics of Acanthaceae determined by chloroplast DNA sequences. *Amer. J. Bot.* 82: 266–275.

Scott, J. A. 1978. The identity of the Rocky Mountain *Lycaena dorcas-helloides* complex (Lycaenidae). *J. Res. Lepid.* 17: 40–50.

Scott, J. A. 1986. Larval hostplant records for butterflies and skippers (mainly from western U. S.), with notes on their natural history. *Papilio* 4: 1–37.

Scott, J. A. 2014. *Lepidoptera of North America. 13. Flower Visitation by Colorado Butterflies (40, 615 Records) with a Review of the Literature on Pollination of Colorado Plants and Butterfly Attraction (Lepidoptera: Hesperioidea and Papilionoidea).* C. P. Gillette Museum of Arthropod Diversity, Colorado State University, Fort Collins, CO. 190 pp.

Scott, L. & B. Molano-Flores. 2007. Reproductive ecology of *Rudbeckia fulgida* Ait. var. *sullivantii* (C. L. Boynt and Beadle) Cronq. (Asteraceae) in northeastern Illinois. *J. Torrey Bot. Soc.* 134: 362–368.

Scott, L., B. Molano-Flores & J. A. Koontz. 2007. Comparisons of genetic variation and outcrossing potential between the sensitive sSpecies *Rudbeckia fulgida* var. *sullivantii* (Asteraceae) and its cultivar. *Trans. Illinois State Acad. Sci.* 100: 129–144.

Scott, R., D. Goodrich, D. Williams, J. Stromberg & J. Leenhouts. 2001. *San Pedro Riparian National Conservation Area (SPRNCA) Water Needs Study.* Year 2001 Progress Report. Upper San Pedro Partnership. Available online: http://www.usppartnership.com/docs/USPP-WN%20Project%20Report%202001%20with%20Figures.pdf (accessed December 5, 2013).

Scotter, G. W. 1975. Permafrost profiles in the Continental Divide region of Alberta and British Columbia. *Arct. Alp. Res.* 7: 93–95.

Scribailo, R. W. & M. S. Alix. 2002. First reports of *Ceratophyllum echinatum* A. Gray from Indiana with notes on the distribution, ecology and phytosociology of the species. *J. Torrey Bot. Soc.* 129: 164–171.

Scribailo, R. W. & M. S. Alix. 2006. *Myriophyllum tenellum* (Haloragaceae): an addition to the aquatic plant flora of Indiana. *Rhodora* 108: 76–79.

Sculthorpe, C. D. 1967. *The Biology of Aquatic Vascular Plants.* Edward Arnold (Publishers) Ltd., London, UK. 610 pp.

Seabloom, E. W., A. G. van der Valk & K. A. Moloney. 1998. The role of water depth and soil temperature in determining initial composition of prairie wetland coenoclines. *Plant Ecol.* 138: 203–216.

Sealy, S. G. 1989. Incidental "egg dumping" by the House Wren in a Yellow Warbler nest. *Wilson Bull.* 101: 491–493.

Seaman, D. E. & W. A. Porterfield. 1964. Control of aquatic weeds by the snail *Marisa cornuarietis*. *Weeds* 12:87–92.

Searcy, K. B. & R. Ascher. 2001. The first record of *Populus heterophylla* (swamp cottonwood, Salicaceae) in Massachusetts. *Rhodora* 103: 224–226.

Seaver, F. J. 1909. The Hypocreales of North America: I. *Mycologia* 1: 41–76.

Seavey, F. & J. Seavey. 2012. *Caloplaca lecanorae* (Teloschistaceae), a new lichenicolous lichen and several additions to the North American lichenized mycota from Everglades National Park. *Bryologist* 115: 322–328.

Seavey, S. R. & P. H. Raven. 1977a. Chromosomal evolution in *Epilobium* sect. *Epilobium* (Onagraceae). *Plant Syst. Evol.* 127: 107–119.

Seavey, S. R. & P. H. Raven. 1977b. Chromosomal evolution in *Epilobium* sect. *Epilobium* (Onagraceae), II. *Plant Syst. Evol.* 128: 195–200.

Seddon, B. 1972. Aquatic macrophytes as limnological indicators. *Freshwater Biol.* 2: 107–130.

Sedinger, J. S. 1986. Biases in comparison of proventricular and esophageal food samples from cackling Canada geese. *J. Wildl. Manag.* 50: 221–222.

Sedinger, J. S. & D. G. Raveling. 1984. Dietary selectivity in relation to availability and quality of food for goslings of cackling geese. *Auk* 101: 295–306.

Segadas-Vianna, F. 1951. A phytosociological and ecological study of cattail stands in Oakland County, Michigan. *J. Ecol.* 39: 316–329.

Segal, S. 1967. Some notes on the ecology of *Ranunculus hederaceus* L. *Vegetatio* 15: 1–26.

Sei, M. & A. H. Porter. 2003. Microhabitat-specific early-larval survival of the maritime ringlet (*Coenonympha tullia nipisiquit*). *Anim. Conserv.* 6: 55–61.

Seibert, A. C. & R. B. Pearce. 1993. Growth analysis of weed and crop species with reference to seed weight. *Weed Sci.* 41: 52–56.

Seigler, D. S. 1976. Plants of the northeastern United States that produce cyanogenic compounds. *Econ. Bot.* 30: 395–407.

Seigler, D. S. 1999. *Plant Secondary Metabolism.* Springer, New York. 759 pp.

Seiler, G. J., T. J. Gulya & G. Kong. 2010. Oil concentration and fatty acid profile of wild *Helianthus* species from the southeastern United States. *Industr. Crops Prod.* 31: 527–533.

Sekine, T., M. Sugano, A. Majid & Y. Fujii. 2007. Antifungal effects of volatile compounds from black zira (*Bunium persicum*) and other spices and herbs. *J. Chem. Ecol.* 33: 2123–2132.

Self, C. A., R. H. Chabreck & T. Joanen. 1974. Food preferences of deer in Louisiana coastal marshes. pp. 548–556. *In*: *Proceedings of the Twenty-Eighth Annual Conference, Southeastern Association of Game and Fish Commissioners.* White Sulphur Springs, WV.

Seliskar, D. M. 1985. Morphometric variations of five tidal marsh halophytes along environmental gradients. *Amer. J. Bot.* 72: 1340–1352.

Sellers, R. 1979. *Waterbird Use of and Management Considerations for Cook Inlet State Game Refuges.* Alaska Department of Fish and Game, Juneau, AK. 42 pp.

Selliah, S. 2009. La phylogénie moléculaire du genre nord-américain *Eurybia* (Asteraceae: Astereae) et ses proches parents (*Oreostemma, Herrickia, Triniteurybia*). MS thesis. Département de sciences biologiques, Institut de recherche en biologie végétale (IRBV), Faculté des Arts et Sciences, Université de Montréal, Montréal, Quebec. 179 pp.

Selliah, S. & L. Brouillet. 2008. Molecular phylogeny of the North American eurybioid asters (Asteraceae, Astereae) based on the nuclear ribosomal internal and external transcribed spacers. *Botany* 86: 901–915.

Semple, J. C. 1977. Chromosome numbers and karyotypes in *Borrichia* (Compositae). *Syst. Bot.* 2: 287–291.

Semple, J. C. 1978. A revision of the genus *Borrichia* Adans. (Compositae). *Ann. Missouri Bot. Gard.* 65: 681–693.

Semple, J. C. 1982. Observations on morphology and cytology of *Aster hemisphaericus, A. paludosus,* and *A. chapmanii* (Asteraceae) with comments on chromosomal base number and phylogeny of *Aster* subg. *Aster* sect. *Heleastrum. Syst. Bot.* 7: 60–70.

Semple, J. C. 2006. 211. *Ampelaster* G. L. Nesom. p. 460. *In*: Flora of North America Editorial Committee (eds.), *Flora of North America North of Mexico: Vol. 20, Magnoliophyta: Asteridae (in part): Asteraceae, Part 2.* Oxford University Press, New York.

Semple, J. C. 2012. Typification of *Solidago gracillima* (Asteraceae: Astereae) and application of the name. *Phytoneuron* 2012–107: 1–10.

Semple, J. C. 2013. Application of the names *Solidago stricta* and *S. virgata* (Asteraceae: Astereae). *Phytoneuron* 2013–42: 1–3.

Semple, J. C. & R. A. Brammall. 1982. Wild *Aster lanceolatus* × *lateriflorus* hybrids in Ontario and comments on the origin of *A. ontarionis* (Compositae-Astereae). *Can. J. Bot.* 60: 1895–1906.

Semple, J. C. & L. Brouillet. 1980. A synopsis of North American asters: the subgenera, sections and subsections of *Aster* and *Lasallea. Amer. J. Bot.* 67: 1010–1026.

Semple, J. C. & J. G. Chmielewski. 2006. 144. *Doellingeria* Nees. pp. 43–45. *In*: Flora of North America Editorial Committee (eds.), *Flora of North America North of Mexico: Vol. 20, Magnoliophyta: Asteridae (in part): Asteraceae, Part 2.* Oxford University Press, New York.

Semple, J. C. & R. E. Cook. 2006. 163. *Solidago* Linnaeus. pp. 107–166. *In*: Flora of North America Editorial Committee (eds.), *Flora of North America North of Mexico: Vol. 20, Magnoliophyta: Asteridae (in part): Asteraceae, Part 2.* Oxford University Press, New York.

Semple, J. C. & K. S. Semple. 1977. *Borrichia* ×*cubana* (*B. frutescens* × *arborescens*): interspecific hybridization in the Florida keys. *Syst. Bot.* 2: 292–301.

Semple, J. C., J. G. Chmielewski & C. C. Chinnappa. 1983. Chromosome number determinations in *Aster* L. (Compositae) with comments on cytogeography, phylogeny and chromosome morphology. *Amer. J. Bot.* 70: 1432–1443.

Semple, J. C., G. S. Ringius, C. Leeder & G. Morton. 1984. Chromosome numbers of goldenrods, *Euthamia* and *Solidago* (Compositae: Astereae). II. Additional counts with comments on cytogeography. *Brittonia* 36: 280–292.

Semple, J. C., L. Tong & P. Pastolero. 2012. Neotypification of *Solidago salicina* (Asteraceae: Astereae) and a multivariate comparison with *S. patula. Phytoneuron* 2012–56: 1–6.

Senanayake, Y. D. A. & R. S. Bringhurst. 1967. Origin of *Fragaria* polyploids. I. Cytological analysis. *Amer. J. Bot.* 54: 221–228.

Sengupta, R., B. Middleton, C. Yan, M. Zuro & H. Hartman. 2005. Propagule deposition and landscape characteristics of source forests of *Rhizophora mangle* in coastal landscapes in Florida. *Landscape Ecol.* 20: 63–72.

Senters, A. E. & D. E. Soltis. 2003. Phylogenetic relationships in *Ribes* (Grossulariaceae) inferred from ITS sequence data. *Taxon* 52: 51–66.

Serra, L. & M. B. Crespo. 1997. An outline revision of the subtribe Siphocampylinae (Lobeliaceae). *Lagascalia* 19: 881–888.

Setoguchi, H. & I. Watanabe. 2000. Intersectional gene flow between insular endemics of *Ilex* (Aquifoliaceae) on the Bonin Islands and the Ryukyu Islands. *Amer. J. Bot.* 87: 793–810.

Setter, S. D., M. J. Setter, M. F. Graham & J. S. Vitelli. 2008. Buoyancy and germination of pond apple (*Annona glabra* L.) propagules in fresh and salt water. *Weed Manag.* 2008: 140–142.

Severns, P. M., L. Boldt & S. Villegas. 2006. Conserving a wetland butterfly: quantifying early lifestage survival through seasonal flooding, adult nectar, and habitat preference. *J. Insect Conserv.* 10: 361–370.

Seybold, C. A., W. Mersie, J. Huang & C. McNamee. 2002. Soil redox, pH, temperature, and water-table patterns of a freshwater tidal wetland. *Wetlands* 22: 149–158.

Seymour, R. S. & P. Schultz-Motel. 1998. Thermoregulating lotus flowers. *Nature* 383: 305.

Seymour, S. D. 2011. Vegetation of non-alluvial wetlands of the Southeastern Piedmont. MS thesis. University of North Carolina, Chapel Hill, NC. 153 pp.

SFWO. 2009. *Eryngium constancei* (Loch Lomond coyote-thistle) 5-year review: summary and evaluation. U. S. Fish and Wildlife Service, Sacramento Fish and Wildlife Office, Sacramento, CA.

Shacklette, H. T. 1961. Substrate relationships of some bryophyte communities on Latouche Island, Alaska. *Bryologist* 64: 1–16.

Shad, A. A., H. U. Shah & J. Bakht. 2013. Ethnobotanical assessment and nutritive potential of wild food plants. *J. Anim. Plant Sci.* 23: 92–97.

Shaffer, G. P., C. E. Sasser, J. G. Gosselink & M. Rejmanek. 1992. Vegetation dynamics in the emerging Atchafalaya delta, Louisiana, USA. *J. Ecol.* 80: 677–687.

Shah, U., W. N. Baba, M. Ahmad, A. Shah, A. Gani, F. A. Masoodi, A. Gani & B. A. Ashwar. 2014. In vitro antioxidant and anti-proliferative activities of seed extracts of *Nymphaea mexicana* in different solvents and GC-MS analysis. *Int. J. Drug Dev. Res.* 6: 68–79.

Shamey, A. M. 2011. Spatial relations of weed management practices and agroeconomically dominant weed species with organic dairy farms in southwestern Wisconsin. MS thesis. The Pennsylvania State University, State College, PA. 104 pp.

Shannon, E. L. 1953. The production of root hairs by aquatic plants. *Amer. Midl. Naturalist* 50: 474–479.

Shao, S.-G., L. Zhang, Y.-Q. Zhao & X. Zheng. 2011. The effect of seawater stress on seed germination of *Eclipta prostrata*. *N. Hort.* 19: 158–160.

Shapiro, A. M. 1974a. Butterflies of the Suisun Marsh, California. *J. Res. Lepid.* 13: 191–206.

Shapiro, A. M. 1974b. The butterfly fauna of the Sacramento Valley, California. *J. Res. Lepid.* 13: 73–82.

Shapiro, A. M. 1990. Ball Mountain revisited: anomalous species richness of a montane barrier zone. *J. Res. Lepid.* 29: 143–156.

Sharitz, R. R. 2003. Carolina bay wetlands: unique habitats of the southeastern United States. *Wetlands* 23: 550–562.

Sharma, A. 2012. Role of uneven-aged silviculture and the soil seed bank in restoration of longleaf pine-slash pine (*Pinus palustris-Pinus elliottii*) ecosystems. PhD dissertation. University of Florida, Gainseville, FL. 156 pp.

Sharma, J. & W. R. Graves. 2004a. Germination of *Leitneria floridana* seeds from disjunct populations. *HortScience* 39: 1695–1699.

Sharma, J. & W. R. Graves. 2004b. Midwinter cold-tolerance of *Leitneria floridana* (Leitneriaceae) from three provenances. *J. Environ. Hort.* 22: 88–92.

Sharma, J. & W. R. Graves. 2005. Propagation of *Rhamnus alnifolia* and *Rhamnus lanceolata* by seeds and cuttings. *J. Environ. Hort.* 23: 86–90.

Sharma, N. 2005. Micropropagation of *Bacopa monneiri* L. Penn.-an important medicinal plant. MS thesis. Department of Biotechnology and Environmental Sciences, Thapar Institute of Engineering and Technology, Patiala, India. 70 pp.

Sharp, J. L., R. S. Sojda, M. Greenwood, D. O. Rosenberry & J. M. Warren. 2013. Statistical classification of vegetation and water depths in montane wetlands. *Ecohydrology* 6: 173–181.

Sharp, M. J. & P. A. Keddy. 1985. Biomass accumulation by *Rhexia virginica* and *Triadenum fraseri* along two lakeshore gradients: a field experiment. *Can. J. Bot.* 63: 1806–1810.

Sharpe, P. J. & A. H. Baldwin. 2009. Patterns of wetland plant species richness across estuarine gradients of Chesapeake Bay. *Wetlands* 29: 225–235.

Sharpe, P. J. & A. H. Baldwin. 2012. Tidal marsh plant community response to sea-level rise: A mesocosm study. *Aquat. Bot.* 101: 34–40.

Sharratt, B., M. Zhang & S. Sparrow. 2006. Twenty years of tillage research in subarctic Alaska I. Impact on soil strength, aggregation, roughness, and residue cover. *Soil Tillage Res.* 91: 75–81.

Shasany, A. K., M. P. Darokar, S. Dhawan, A. K. Gupta, S. Gupta, A. K. Shukla, N. K. Patra & S. P. S. Khanuja. 2005a. Use of RAPD and AFLP markers to identify inter-and intraspecific hybrids of *Mentha*. *J. Heredity* 96: 542–549.

Shasany, A. K., A. K. Shukla, S. Gupta, S. Rajkumar & S. P. S. Khanuja. 2005b. AFLP analysis for genetic relationships among *Mentha* species. *Plant Genet. Resour. Newslett.* 144: 14–19.

Shaver, G. R., J. A. Laundre, A. E. Giblin & K. J. Nadelhoffer. 1996. Changes in live plant biomass, primary production, and species composition along a riverside toposequence in arctic Alaska, U.S.A. *Arct. Alp. Res.* 28: 363–379.

Shaw, C. G. 1951. New species of the Peronosporaceae. *Mycologia* 43: 445–455.

Shaw, J. M. H. 2008. What ever happened to *Cacalia* L. (Asteraceae)? *Hanburyana* 3: 10–16.

Shaw, W. T. 1930. The lemming mouse in North America and its occurrence in the state of Washington. *Murrelet* 11: 7–10.

Shay, G. (ed.). 1990. *Saline Agriculture: Salt-Tolerant Plants for Developing Countries*. Report of a panel of the Board on Science and Technology for International Development, National Research Council. National Academy Press, Washington, DC. 152 pp.

Shay, J. M. 1986. Vegetation dynamics in the Delta Marsh, Manitoba. pp. 65–70. *In*: G. K. Clambey & R. H. Pemble (eds.), *Proceedings of the Ninth North American Prairie Conference, 1984*. Tri-College University, Fargo, ND.

Shea, M. M., P. M. Dixon & R. R. Sharitz. 1993. Size differences, sex ratio, and spatial distribution of male and female water tupelo, *Nyssa aquatica* (Nyssaceae). *Amer. J. Bot.* 80: 26–30.

Sheela, K., K. G. Nath, D. Vijayalakshmi, G. M. Yankanchi & R. B. Patil. 2004. Proximate composition of underutilized green leafy vegetables in Southern Karnataka. *J. Human Ecol.* 15: 227–229.

Sheidai, M., S. Mosaferi, M. Keshavarzi, Z. Noormohammadi & S. Ghasemzadeh-Baraki. 2016. Genetic diversity in different populations of *Persicaria minor* (Polygonaceae), a medicinal plant. *Nucleus* 59: 115–121.

Sheldon, J. C. & F. M. Burrows.1973. The dispersal effectiveness of the achene-pappus units of selected Compositae in steady winds with convection. *New Phytol.* 72: 665–675.

Sheldon, R. B. & C. W. Boylen. 1977. Maximum depth inhabited by aquatic vascular plants. *Amer. Midl. Naturalist* 97: 248–254.

Sheldon, S. P. & K. N. Jones. 2001. Restricted gene flow according to host plant in an herbivore feeding on native and exotic watermilfoils (*Myriophyllum*: Haloragaceae). *Int. J. Plant Sci.* 162: 793–799.

Shellhammer, H. 2012. Mammals of China Camp State Park and Rush Ranch open space preserve. *San Francisco Estuary and Watershed Sci.* 10: 1–10.

Shelly, J. S. & R. K. Moseley. 1988. *Report on the Conservation Status of* Howellia aquatilis, *a Candidate Threatened Species*. Natural Heritage Section, Nongame Wildlife/Endangered Species Program, Idaho Department of Fish and Game. Boise, ID. 43 pp.

Shemluck, M. 1982. Medicinal and other uses of the Compositae by Indians in the United States and Canada. *J. Ethnopharmacol.* 5: 303–358.

Sheng, L., K. Ji & L. Yu. 2014. Karyotype analysis on 11 species of the genus *Clematis*. *Brazil. J. Bot.* 37: 601–608.

Shen-Miller, J., M. B. Mudgett, J. W. Schopf, S. Clarke & R. Berger. 1995. Exceptional seed longevity and robust growth: ancient sacred lotus from China. *Amer. J. Bot.* 82: 1367–1380.

Shennan, I., A. J. Long, M. M. Rutherford, J. B. Innes, F. M. Green & K. J. Walker. 1998. Tidal marsh stratigraphy, sea-level change and large earthquakes—II: Submergence events during the last 3500 years at Netarts Bay, Oregon, USA. *Quatern. Sci. Rev.* 17: 365–393.

Sherff, E. E. 1912. The vegetation of Skokie Marsh, with special reference to subterranean organs and their interrelationships. *Bot. Gaz.* 53: 415–435.

Sherff, E. E. 1940. A new genus of Compositae from northwestern Alabama. *Publ. Field Mus. Nat. Hist., Bot. Ser.* 22: 399–403.

Sheridan, C. D. & T. A. Spies. 2005. Vegetation-environment relationships in zero-order basins in coastal Oregon. *Can. J. For. Res.* 35: 340–355.

Sheridan, D. J. 1990. Aquatic macrophyte community dynamics of Wallace Lake, Washington Co., Wisconsin. MS thesis. University of Wisconsin-Milwaukee, Milwaukee, WI. 157 pp.

Sheridan, G. E. C., J. R. Claxton, J. M. Clarkson & D. Blakesley. 2001. Genetic diversity within commercial populations of watercress (*Rorippa nasturtium-aquaticum*), and between allied Brassicaceae inferred from RAPD-PCR. *Euphytica* 122: 319–325.

Sheridan, P. M. & D. N. Karowe. 2000. Inbreeding, outbreeding, and heterosis in the yellow pitcher plant, *Sarracenia flava* (Sarraceniaceae), in Virginia. *Amer. J. Bot.* 87: 1628–1633.

Sheridan, P. M. & R. R. Mills. 1998. Genetics of anthocyanin deficiency in *Sarracenia* L. *HortScience* 33: 930–944.

Sheridan, P. M., S. L. Orzell & E. L. Bridges. 1997. Powerline easements as refugia for state rare seepage and pineland plant taxa. pp. 451–460. *In*: J. R. Williams, J. W. Goodrich-Mahoney, J. R. Wisniewski & J. Wisniewski (eds.), *The Sixth International Symposium on Environmental Concerns in Rights-of-Way Management*. Elsevier Science, Oxford, UK.

Shetler, S. G. & N. R. Morin. 1986. Seed morphology in North American Campanulaceae. *Ann. Missouri Bot. Gard.* 73: 653–688.

Sheviak, C. J. 1989. A new *Spiranthes* (Orchidaceae) from Ash Meadows, Nevada. *Rhodora* 91: 225–234.

Shi, H. Z., N. N. Gao, Y. Z. Li, J. G. Yu, Q. C. Fan & G. E. Bai. 2002. Effects of active fractions from *Lycopus lucidus* L. F04 on erythrocyte rheology. *Space Med. Med. Engi. (Beijing)* 15: 331–334.

Shi, S., Y. Huang, F. Tan, X. He & D. E. Boufford. 2000. Analysis of the Sonneratiaceae and its relationship to Lythraceae based on ITS sequences of nrDNA. *Int. J. Plant Sci.* 113: 253–258.

Shi, X., G. Chang & F. Xu. 1992. Biological characteristics and morphological changes during seed germination of *Nymphaea tetragona* Georgi. *J. Nanjing Normal Univ.* 2 (1992): 011.

Shibata, O. & T. Nishida. 1993. Seasonal changes in sugar and starch content of the alpine snowbed plants, *Primula cuneifolia* ssp. *hakusanensis* and *Fauria crista-galli*, in Japan. *Arct. Alp. Res.* 25: 207–210.

Shibayama, Y. & Y. Kadono. 2003a. Heterostyly in *Nymphoides indica* (Menyanthaceae) in Japan. *Acta Phytotax. Geobot.* 54: 77–80.

Shibayama, Y. & Y. Kadono. 2003b. Floral morph composition and pollen limitation in the seed set of *Nymphoides indica* populations. *Ecol. Res.* 18: 725–737.

Shibayama, Y. & Y. Kadono. 2007a. Reproductive success and genetic structure of populations of the heterostylous aquatic plant *Nymphoides indica* (L.) Kuntze (Menyanthaceae). *Aquat. Bot.* 86: 1–8.

Shibayama, Y. & Y. Kadono. 2007b. The effect of water-level fluctuations on seedling recruitment in an aquatic macrophyte *Nymphoides indica* (L.) Kuntze (Menyanthaceae). *Aquat. Bot.* 87: 320–324.

Shields, J. W. 1966. Preliminary reports on the flora of Wisconsin No. 58. Hydrophyllaceae–waterleaf family. *Trans. Wisconsin Acad. Sci. Arts Lett.* 55: 255–259.

Shigenobu, Y. 1984. Karyomorphological studies in some genera of Gentianaceae II. *Gentiana* and its allied four genera. *Bull. Coll. Child Dev. Kochi Womens Univ.* 8: 55–104.

Shimada, T., T. Matsushita & M. Otani. 1997. Plant regeneration from leaf explants of *Primula cuneifolia* var. *hakusanensis*, "Hakusan-kozakura." *Plant Biotechnol.* 14: 47–50.

Shimai, H., Y. Masuda, C. M. P. Valdés & K. Kondo. 2007. Phylogenetic analysis of Cuban *Pinguicula* (Lentibulariaceae) based on internal transcribed spacer (ITS) region. *Chromosome Bot.* 2: 151–158.

Shimamura, R., N. Kachi, H. Kudoh & D. F. Whigham. 2005. Visitation of a specialist pollen feeder *Althaeus hibisci* Oliver (Coleoptera: Bruchidae) to flowers of *Hibiscus moscheutos* L. (Malvaceae). *J. Torrey Bot. Club.* 132: 197–203.

Shimizu, T. & T. Uchida. 1993. Hybridization between North American *Acer rubrum* L. and Japanese *A. pycnanthum* K. Koch (Aceraceae). *J. Phytogeogr. Taxon.* 41: 63–69.

Shimono, Y. & G. Kudo. 2005. Comparisons of germination traits of alpine plants between fellfield and snowbed habitats. *Ecol. Res.* 20: 189–197.

Shipley, B. 1989. The use of above-ground maximum relative growth rate as an accurate predictor of whole-plant maximum relative growth rate. *Funct. Ecol.* 3: 771–775.

Shipley, B. & M. Parent. 1991. Germination responses of 64 wetland species in relation to seed size, minimum time to reproduction and seedling growth rate. *Funct. Ecol.* 5: 111–118.

Shipley, B. & R. H. Peters. 1990. A test of the Tilman model of plant strategies: relative growth rate and biomass partitioning. *Amer. Naturalist* 136: 139–153.

Shipley, B. & T.-T. Vu. 2002. Dry matter content as a measure of dry matter concentration in plants and their parts. *New Phytol.* 153: 359–364.

Shipley, B., P. A. Keddy, C. Gaudet & D. R. J. Moore. 1991. A model of species density in shoreline vegetation. *Ecology* 72: 1658–1667.

Shneyer, V. S., N. G. Kutyavina & M. G. Pimenov. 2003. Systematic relationships within and between *Peucedanum* and *Angelica* (Umbelliferae-Peucedaneae) inferred from immunological studies of seed proteins. *Plant Syst. Evol.* 236: 175–194.

Short, P. S. 2002. A new species of *Glinus* L.(Molluginaceae) from the Northern Territory, Australia. *Telopea* 9: 761–763.

Showers, M. A. 2010. *Seed Collection and Banking of 50 Plant Species of Critical Conservation Concern*. Final Performance Report. Grant E-2-P-31, California Department of Fish and Game, Sacramento, CA.

Shrader, J. A. & W. R. Graves. 2002. Infraspecific systematics of *Alnus maritima* (Betulaceae) from three widely disjunct provenances. *Castanea* 67: 380–401.

Shrestha, G. & P. D. Stahl. 2008. Carbon accumulation and storage in semi-arid sagebrush steppe: Effects of long-term grazing exclusion. *Agric. Ecosyst. Environ.* 125: 173–181.

Shu, Z. Z. 1994. *Callicarpa* L. pp. 4–15. *In*: Flora of China Editorial Committee (eds.), *Flora of China. Vol. 17 (Verbenaceae through Solanaceae)*. Science Press, Beijing, and Missouri Botanical Garden Press, St. Louis, MO. 342 pp.

Shull, G. H. 1903. Geographic distribution of *Isoetes saccharata*. *Bot. Gaz.* 36: 187–202.

Shull, G. H. 1905. Stages in the development of *Sium cicutaefolium*. Publication 30. Carnegie Institution of Washington, Washington, DC. 28 pp.

Shull, G. H. 1914. The longevity of submerged seeds. *Plant World* 17: 329–437.

Shultz, L. M. 2006. 119. *Artemisia* L. pp. 503–534. *In*: Flora of North America Editorial Committee (eds.), *Flora of North America North of Mexico, Vol. 19: Magnoliophyta: Asteridae, Part 8*. Oxford University Press, New York.

Shyamala, M. & A. Arulanantham. 2009. *Eclipta alba* as corrosion pickling inhibitor on mild steel in hydrochloric acid. *J. Mater. Sci. Technol.* 25: 633–636.

Siekaniec, G. E. 2010. Endangered and threatened wildlife and plants; 12-month finding on a petition to delist *Cirsium vinaceum* (Sacramento Mountains thistle). *Fed. Reg.* 75: 30757–30769.

Sieren, D. J. 1981. The taxonomy of the genus *Euthamia*. *Rhodora* 83: 551–579.

Sigafoos, R. S. 1951. Soil instability in tundra vegetation. *Ohio J. Sci.* 51: 281–298.

Sigler, L. & C. F. C. Gibas. 2005. Utility of a cultural method for identification of the ericoid mycobiont *Oidiodendron maius* confirmed by ITS sequence analysis. *Stud. Mycol.* 53: 63–74.

Sigmon, L., S. Hoopes, M. Booker, C. Waters, K. Salpeter & B. Touchette. 2013. Breaking dormancy during flood and drought: sublethal growth and physiological responses of three emergent wetland herbs used in bioretention basins. *Wetlands Ecol. Manag.* 21: 45–54.

Sikes, K., D. Roach-McIntosh & D. Stout. 2012. *Vegetation Assessment and Ranking of Fen and Wet Meadow Sites of the Shasta-Trinity National Forest, California*. California Native Plant Society, Sacramento, CA. 84 pp.

Sikes, K., D. Cooper, S. Weis, T. Keeler-Wolf, M. Barbour, D. Ikeda, D. Stout & J. Evens. 2013. *Fen Conservation and Vegetation Assessment in the National Forests of the Sierra Nevada and Adjacent Mountains, California*. (Revised, public version 2). Unpublished report to the United States Forest Service, Region 5, Vallejo, CA. 314 pp.

Silberhorn, G. M. 1976. *Tidal Wetland Plants of Virginia*. Educational Series No 19. Virginia Institute of Marine Science, Gloucester Point, VA. 86 pp.

Silberhorn, G. M. 1982. *Common Plants of the Mid-Atlantic Coast: A Field Guide*. The Johns Hopkins University Press, Baltimore, MD. 256 pp.

Silva, M. R. O., A. C. Almeida, F. V. F. Arruda & N. Gusmão. 2011. Endophytic fungi from Brazilian mangrove plant *Laguncularia racemosa* (L.) Gaertn. (Combretaceae): their antimicrobial potential. *Commun. Curr. Res. Technol. Advances* 2: 1260–1266.

Silvani, V. A., S. Fracchia, L. Fernández, M. Pérgola & A. Godeas. 2008. A simple method to obtain endophytic microorganisms from field-collected roots. *Soil Biol. Biochem.* 40: 1259–1263.

Silveira, J. G. 2000a. Vernal pools and relict duneland at Arena Plains. *Fremontia* 27/28: 38–47.

Silveira, J. G. 2000b. Alkali vernal pools at Sacramento National Wildlife Refuge. *Fremontia* 27/28: 10–18.

Silvertown, J. & M. Tremlett. 1989. Interactive effects of disturbance and shade upon colonization of grassland: An experiment with *Anthriscus sylvestris* (L.) Hoffm., *Conium maculatum* L., *Daucus carota* L. and *Heracleum sphondylium* L. *Funct. Ecol.* 3: 229–235.

Simak M. 1982. Germination and storage of *Salix caprea* L. and *Populus tremula* L. seeds. pp. 142–160. *In*: B.S.P. Wang & J. A. Pitel (eds.), *Proceedings of the International Symposium of Forest Tree Seed Storage, Petawawa/Canada*. Environment Canada, Canada Forestry Service, Ottawa, ON.

Simmonds, N. W. 1945. *Polygonum* L. em. Gaertn. *J. Ecol.* 33: 117–120.

Simmons, R. & M. Strong. 2002. Fall line magnolia bogs of the mid-Atlantic region. *Marilandica* 11(1): 19–20.

Simmons, R. H., M. T. Strong & J. M. Parrish. 2008. Noteworthy collections: Virginia. *Castanea* 73: 328–332.

Simonnet, X. & N. Delabays. 1993. The rate of cross-pollination in *Epilobium parviflorum* Schreb. *Acta Hort.* 330: 197–202.

Simms, E. L. 1985. Growth response to clipping and nutrient addition in *Lyonia lucida* and *Zenobia pulverulenta*. *Amer. Midl. Naturalist* 114: 44–50.

Simms, E. L. 1987. The effect of nitrogen and phosphorus addition on the growth, reproduction, and nutrient dynamics of two ericaceous shrubs. *Oecologia* 71: 541–547.

Simon, J. P. 1970 Comparative serology of the order Nymphaeales. I. Preliminary survey on the relationships of *Nelumbo*. *Aliso* 7: 243–261.

Simon, N. P. P. & F. E. Schwab. 2005. The response of conifer and broad-leaved trees and shrubs to wildfire and clearcut logging in the boreal forests of central Labrador. *N. J. Appl. Forest.* 22: 35–41.

Simonot, D., J. McColl & D. Thorne. 2002. Tyrosinase inhibitors: activity of a *Rumex* extract in combination with kojic acid and arbutin. *Cosmet. Toiletr. Mag.* 117(3): 52–56.

Simpson, B. B., J. L. Neff & D. S. Siegler. 1983. Floral biology and floral rewards of *Lysimachia* (Primulaceae). *Amer. Midl. Naturalist* 110: 249–256.

Simpson, M. J. A., D. F. MacIntosh, J. B. Cloughley & A. E. Stuart. 1996. Past, present and future utilisation of *Myrica gale* (Myricaceae). *Econ. Bot.* 50: 122–129.

Simpson, R. L., R. E. Good, M. A. Leck & D. F. Whigham. 1983. The ecology of freshwater tidal wetlands. *BioScience* 33: 255–259.

Sims, L. E. & H. J. Price. 1985. Nuclear DNA content variation in *Helianthus* (Asteraceae). *Amer. J. Bot.* 72: 1213–1219.

Sims, S. R. 1980. Diapause dynamics and host plant suitability of *Papilio zelicaon* (Lepidoptera: Papilionidae). *Amer. Midl. Naturalist* 103: 375–384.

Sims, S. R. 2007. Diapause dynamics, seasonal phenology, and upal color dimorphism of *Papilio polyxenes* in southern Florida, USA. *Entomol. Exp. Appl.* 123: 239–245.

Simurda, M. C. & J. S. Knox. 2000. ITS sequence evidence for the disjunct distribution between Virginia and Missouri of the narrow endemic *Helenium virginicum*. *J. Torrey Bot. Soc.* 127: 316–323.

Simurda, M. C., D. C. Marshall & J. S. Knox. 2005. Phylogeography of the narrow endemic, *Helenium virginicum* (Asteraceae), based upon ITS sequence comparisons. *Syst. Bot.* 30: 887–898.

Singer, M. S. & J. O. Stireman III. 2001. How foraging tactics determine host-plant use by a polyphagous caterpillar. *Oecologia* 129: 98–105.

Singer, R., D. A. Roberts & C. W. Boylen. 1983. The macrophytic community of an acidic lake in Adirondack (New York, U.S.A.): a new depth record for aquatic angiosperms. *Aquat. Bot.* 16: 49–57.

Singh, J., N. P. Mishra, G. Joshi, S. C. Singh, A. Sharma & S. P. S. Khanuja. 2005. Traditional uses of *Bacopa monnieri* (Brahmi). *J. Med. Aromatic Plant Sci* 27: 122–124.

Singhurst, J. R. & E. L. Bridges. 2007. Two additions to the flora of Oklahoma and notes on *Xyris jupicai* (Xyridaceae) in Oklahoma. *Phytologia* 89: 211–218.

Singhurst, J. R. & W. C. Holmes. 2012. *Sanguisorba minor* (Rosaceae) adventive in Texas. *Phytoneuron* 93: 1–3.

Singhurst, J. R., B. A. Sorrie & W. C. Holmes. 2012. *Andropogon glaucopsis* (Poaceae) in Texas. *Phytoneuron* 2012–16: 1–3.

Singhurst, J. R., M. White, J. N. Mink & W. C. Holmes. 2011a. *Castilleja coccinea* (Orobanchaceae): New to Texas. *Phytoneuron* 2011–32: 1–3.

Singhurst, J. R., A. E. Rushing, C. K. Hanks & W. C. Holmes. 2011b. *Isoetes texana* (Isoetaceae): A new species from the Texas Coastal Bend. *Phytoneuron* 2011–22: 1–6.

Singhurst, J. R., N. Shackelford, W. Newman, J. N. Mink & W. C. Holmes. 2014. The ecology and abundance of *Hymenoxys texana* (Asteraceae). *Phytoneuron* 2014–19: 1–19.

Singhurst, J. R., L. L. Sanchez, D. Frels Jr., T. W. Schwertner, M. Mitchell, S. Moren & W. C. Holmes. 2007. The vascular flora of Mason Mountain Wildlife Management Area, Mason County, Texas. *S. E. Naturalist* 6: 683–692.

Singleton, J. R. 1951. Production and utilization of waterfowl food plants on the east Texas gulf coast. *J. Wildl. Manag.* 15: 46–56.

Sinha, A. R. P. 1987. Report of B chromosome in *Mecardonia procumbens* (Miller) Small. *Cytologia* 52: 373–375.

Sinha, S. 1999. Accumulation of Cu, Cd, Cr, Mn and Pb from artificially contaminated soil by *Bacopa monnieri*. *Environ. Monit. Assessm.* 57: 253–264.

Sinha, S. & P. Chandra. 1990. Removal of Cu and Cd from water by *Bacopa monnieri* L. *Water Air Soil Pollut.* 51: 271–276.

Sinn, B. T., L. M. Kelly & J. V. Freudenstein. 2015. Putative floral brood-site mimicry, loss of autonomous selfing, and reduced vegetative growth are significantly correlated with increased diversification in *Asarum* (Aristolochiaceae). *Mol. Phylogen. Evol.* 89: 194–204.

Sinnott, E. W. 1912. The pond flora of Cape Cod. *Rhodora* 14: 25–34.

Sinnott, Q. P. 1985. A revision of *Ribes* L. subg. *Grossularia* (Mill.) Pers. sect. *Grossularia* (Mill.) Nutt. (Grossulariaceae) in North America. *Rhodora* 87: 189–286.

Sipple, W. S. & W. A. Klockner 1980. A unique wetland in Maryland. *Castanea* 45: 60–69.

Siragusa, A. J., J. E. Swenson & D. A. Casamatta. 2007. Culturable bacteria present in the fluid of the hooded-pitcher plant *Sarracenia minor* based on 16S rDNA gene sequence data. *Microb. Ecol.* 54: 324–331.

Sîrbu, C. & A. Oprea. 2010. New and rare plants from the flora of Moldavia (Romania). *Cercet. Agron. Moldova* 43: 31–42.

Siripun, K. C. & E. E. Schilling. 2006. 392. *Eupatorium* Linnaeus. pp. 462–471. *In*: Flora of North America Editorial Committee (eds.), *Flora of North America North of Mexico: Vol. 21, Magnoliophyta: Asteridae (in part): Asteraceae, Part 3.* Oxford University Press, New York.

Sites, R. W. & J. E. McPherson. 1981. A list of the butterflies (Lepidoptera: Papilionoidea) of the La Rue-Pine Hills Ecological Area. *Great Lakes Entomol.* 14: 81–85.

Sivinski, B. 2011. *Agalinis calycina* (Leoncita false-foxglove): A Conservation Status Assessment. Unpublished report prepared for NM Energy, Minerals and Natural Resources Department and USDI-Fish & Wildlife Service, Region 2. Available online: http://www.npsnm.org/wp-content/uploads/2012/09/S-6-Agalinis-calycina.pdf.

Sivinski, R. & K. Lightfoot. 1993. Sacramento Mountains thistle (*Cirsium vinaceum*) recovery plan. U. S. Fish & Wildlife Service, Albuquerque, NM. 23 pp.

Skawinski, P. M. 2010. *Aquatic Plants of Wisconsin. A Photographic Field Guide to Submerged and Floating-Leaf Aquatic Plants.* Published by the author. 150 pp.

Skeate, S. T. 1987. Interactions between birds and fruits in a northern Florida hammock community. *Ecology* 68: 297–309.

Skene, K. R., J. I. Sprent, J. A. Raven & L. Herdman. 2000. *Myrica gale* L. *J. Ecol.* 88: 1079–1094.

Skinner, M. P. 1928. The elk situation. *J. Mammalogy* 9: 309–317.

Skinner, M. W. & B. A. Sorrie. 2002. Conservation and ecology of *Lilium pyrophilum*, a new species of Liliaceae from the sandhills region of the Carolinas and Virginia, U.S.A. *Novon* 12: 94–105.

Skinner, W. R. & E. S. Telfer. 1974. Spring, summer, and fall foods of deer in New Brunswick. *J. Wildl. Manag.* 38: 210–214.

Sklenář, P. 2016. Seasonal variation of freezing resistance mechanisms in north-temperate alpine plants. *Alpine Bot.* doi: 10.1007/s00035-016-0174-6.

Skogerboe, J. G., A. G. Poovey, K. D. Getsinger & G. Kudray. 2003. *Invasion of Eurasian Watermilfoil in Lakes of the Western Upper Peninsula, Michigan.* Final Report ERDC/EL TR-03-10 prepared for the U. S. Army Corps of Engineers, Washington, DC.

Skroch, W. A. & M. N. Dana. 1965. Sources of weed infestation in cranberry fields. *Weeds* 13: 263–267.

Skutch, A. F. 1929. Early stages of plant succession following forest fires. *Ecology* 10: 177–190.

Skvarla, J. J. & B. L. Turner. 1966. Pollen wall ultrastructure and its bearing on the systematic position of *Blennosperma* and *Crocidium* (Compositae). *Amer. J. Bot.* 53: 555–563.

Slack, A. 1981. *Carnivorous Plants.* The MIT Press, Cambridge, MA. 240 pp.

Sladen, F. W. L. 1916. Honey sources of Canada. *Amer. Bee J.* 56: 376–377.

Slater, J. A. 1952. A contribution to the biology of the subfamily Cyminae (Heteroptera: Lygaeidae). *Ann. Entomol. Soc. Amer.* 45: 315–326.

Slater, J. B. 2008. *Minimum Levels Determination: Lake Hiawassee, Orange County, Florida.* Technical Publication SJ2008-X. St. Johns River Water Management District, Palatka, FL. 123 pp.

Sletten, K. K. & G. E. Larson. 1984. Possible relationships between surface water chemistry and aquatic plants in the northern Great Plains. *Proc. South Dakota Acad. Sci.* 63: 70–76.

Slocum, M. G. & I. A. Mendelssohn. 2008. Effects of three stressors on vegetation in an oligohaline marsh. *Freshwater Biol.* 53: 1783–1796.

Slocum, P. D., P. Robinson & F. Perry. 1996. *Water Gardening: Water Lilies and Lotuses.* Timber Press, Portland, OR. 322 pp.

Slomba, J. M., F. B. Essig & R. G. James. 2001. A re-evaluation of the infrageneric classification of *Clematis* (Ranunculaceae) using chloroplast DNA sequences. Botany 2001 meeting (abstract).

Slomba, J. M., J. R. Garey & F. B. Essig. 2004. The actin I intron—a phylogenetically informative DNA region in *Clematis* (Ranunculaceae). *Sida* 28: 879–886.

Sloop, C. M., C. Pickens & S. P. Gordon. 2011. Conservation genetics of Butte County meadowfoam (*Limnanthes floccosa* ssp. *californica* Arroyo), an endangered vernal pool endemic. *Conserv. Genet.* 12: 311–323.

Sloop, C. M., K. Gilmore, H. Brown & N. E. Rank. 2012a. *An Investigation of the Ecology and Seed Bank Dynamics of Burke's Goldfields (Lasthenia burkei), Sonoma Sunshine (Blennosperma bakeri), and Sebastopol Meadowfoam (Limnanthes vinculans) in Natural and Constructed Vernal Pools.* Final Project Report E-2-P-35. California Department of Fish and Game, Habitat Conservation Division, Sacramento, CA. 50 pp.

Sloop, C. M., R. Eberl & D. R. Ayres. 2012b. Genetic diversity and structure in the annual vernal pool endemic *Limnanthes vinculans* Ornduff (Limnanthaceae): implications of breeding system and restoration practices. *Conservation Genet.* 13: 1365–1379.

Sluis, W. & J. Tandarich. 2004. Siltation and hydrologic regime determine species composition in herbaceous floodplain communities. *Plant Ecol.* 173: 115–124.

Slusher, C. E., M. J. Vepraskas & S. W. Broome. 2014. Evaluating responses of four wetland plant species to different hydroperiods. *J. Environ. Qual.* 43: 723–731.

Small, E. 1976. Insect pollinators of the Mer Bleue peat bog of Ottawa. *Can. Field- Naturalist* 90: 22–28.

Small, E. & P. M. Catling. 2000. Poorly known economic plants of Canada – 26. Labrador Tea, *Ledum palustre* sensu lato (*Rhododendron tomentosum*). *Bull. Can. Bot. Assoc.* 33: 31–36.

Small, J. K. 1896. Studies in the botany of the southeastern United States.-VI. *Bull. Torrey Bot. Club* 23: 295–301.

Small, J. K. 1900. Notes and descriptions of North American plants II. *Bull. Torrey Bot. Club* 27: 275–281.

Small, R. L. 2004. Phylogeny of *Hibiscus* sect. *Muenchhusia* (Malvaceae) based on chloroplast *rpL16* and *ndhF*, and nuclear ITS and GBSSI sequences. *Syst. Bot.* 29: 385–392.

Smaoui, A., J. Jouini, M. Rabhi, G. Bouzaien, A. Albouchi & C. Abdelly. 2011. Physiological and anatomical adaptations induced by flooding in *Cotula coronopifolia*. *Acta Biol. Hung.* 62: 182–193.

Smedmark, J. E. E. & T. Eriksson. 2002. Phylogenetic relationships of *Geum* (Rosaceae) and relatives inferred from the nrITS and *trnL-trnF* regions. *Syst. Bot.* 27: 303–317.

Smeins, F. E. & D. D. Diamond. 1983. Remnant grasslands of the Fayette Prairie, Texas. *Amer. Midl. Naturalist* 110: 1–13.

Smeins, F. E. & D. E. Olsen. 1970. Species composition and production of a native northwestern Minnesota tall grass prairie. *Amer. Midl. Naturalist* 84: 398–410.

Smiley, F. J. 1915. The alpine and subalpine vegetation of the Lake Tahoe region. *Bot. Gaz.* 59: 265–286.

Smiley, J. T. & N. E. Rank. 1986. Predator protection versus rapid growth in a montane leaf beetle. *Oecologia* 70: 106–112.

Smirnov, V. V., L. A. Bakina, A. S. Bondarenko, G. T. Petrenko & O. V. Yevseyenko. 1995. A new sesquiterpene phenol from *Bidens cernua* L. with antimicrobial activity. *Rastitel'n. Resursy.* 31: 31–36.

Smissen, R. D., J. C. Clement, P. J. Garnock-Jones & G. K. Chambers. 2002. Subfamilial relationships within Caryophyllaceae as inferred from 5' *ndhF* sequences. *Amer. J. Bot.* 89: 1336–1341.

Smissen, R. D., M. Galbany-Casals & I. Breitwieser. 2011. Ancient allopolyploidy in the everlasting daisies (Asteraceae: Gnaphalieae): Complex relationships among extant clades. *Taxon* 60: 649–662.

Smith, A. H. 1867. On colonies of plants observed near Philadelphia. *Proc. Acad. Nat. Sci. Philadelphia* 19: 15–24.

Smith, B. A., R. L. Brown, W. Laberge & T. Griswold. 2012. A faunistic survey of bees (Hymenoptera: Apoidea) in the Black Belt Prairie of Mississippi. *J. Kansas Entomol. Soc.* 85: 32–47.

Smith, B. N. & B. L. Turner. 1975. Distribution of Kranz syndrome among Asteraceae. *Amer. J. Bot.* 62: 541–545.

Smith, C. C. 1940. Notes on the food and parasites of the rabbits of a lowland area in Oklahoma. *J. Wildl. Manag.* 4: 429–431.

Smith III, C. G., P. B. Hamel, M. S. Devall & N. M. Schiff. 2004. Hermit thrush is the first observed dispersal agent for pondberry. *Castanea* 69: 1–8.

Smith, E. B. 1975. The chromosome numbers of North American *Coreopsis* with phyletic interpretations. *Bot. Gaz.* 136: 78–86.

Smith, E. B. 1976. A biosystematic survey of *Coreopsis* in eastern United States and Canada. *Sida* 6: 123–215.

Smith, E. B. 1982. Phyletic trends in section *Coreopsis* of the genus *Coreopsis* (Compositae). *Bot. Gaz.* 143: 121–124.

Smith, E. B. 1983. Phyletic trends in sections *Eublepharis* and *Calliopsis* of the genus *Coreopsis* (Compositae). *Amer. J. Bot.* 70: 549–554.

Smith, F. F., R. E. Webb, G. W. Argus, J. A. Dickerson & H. W. Everett. 1983. Willow beaked-gall midge, *Mayetiola rigidae* (Osten Sacken), (Diptera: Cecidomyiidae): differential susceptibility of willows. *Environ. Entomol.* 12: 1175–1184.

Smith, G. L., L. C. Anderson & W. S. Flory. 2001. A new species of *Hymenocallis* (Amaryllidaceae) in the lower central Florida panhandle. *Novon* 11: 233–240.

Smith, G. W. 1973. Arctic pharmacognosia. *Arctic* 26: 324–333.

Smith, H. H. 1928. Ethnobotany of the Meskwaki Indians. *Bull. Milw. Pub. Mus.* 4: 175–326.

Smith, H. I. 1929. Materia medica of the Bella Coola and neighbouring tribes of British Columbia. *Bull. Natl. Mus. Canada* 59: 47–68.

Smith, J. A., R. A. Blanchette & G. Newcombe. 2004. Molecular and morphological characterization of the willow rust fungus, *Melampsora epitea*, from arctic and temperate hosts in North America. *Mycologia* 96: 1330–1338.

Smith, L. 1996. *The Rare and Sensitive Natural Wetland Plant Communities of Interior Louisiana.* Louisiana Natural Heritage Program, Louisiana Department of Wildlife & Fisheries, Baton Rouge, LA. 40 pp.

Smith, L. M. & J. A. Kadlec. 1983. Seed banks and their role during drawdown of a North American marsh. *J. Appl. Ecol.* 20: 673–684.

Smith, M., T. Keevin, P. Mettler-McClure & R. Barkau. 1998. Effect of the flood of 1993 on *Boltonia decurrens*, a rare floodplain plant. *Regulat. Rivers Res. Manag.* 14: 191–202.

Smith, M. S., J. D. Fridley, M. Goebel & T. L. Bauerle. 2014. Links between belowground and aboveground resource-related traits reveal species growth strategies that promote invasive advantages. *PLoS One* 9(8): e104189. doi:10.1371/journal.pone.0104189.

Smith Jr., R. J. 1988. Weed thresholds in southern U. S. rice, *Oryza sativa.* *Weed Technol.* 2: 232–241.

Smith, R. N. 1991. Endangered and threatened wildlife and plants; determination of endangered status for three plants, *Blennosperma bakeri* (Sonoma sunshine or Baker's stickyseed), *Lasthenia burkei* (Burke's goldfields), and *Limnanthes vinculans* (Sebastopol meadowfoam). *Fed. Reg.* 56: 61174–61182.

Smith, R. N. 1993. Endangered and threatened wildlife and plants; the plant *Eutrema penlandii* (Penland alpine fen mustard) determined to be a threatened species. *Fed. Reg.* 58: 40539–40547.

Smith, R. R. & D. B. Ward. 1976. Taxonomy of the genus *Polygala* series Descurrentes (Polygalaceae). *Sida* 6: 284–310.

Smith, S. B., K. H. McPherson, J. M. Backer, B. J. Pierce, D. W. Podlesak & S. R. McWilliams. 2007. Fruit quality and consumption by songbirds during autumn migration. *Wilson J. Ornithol.* 119: 419–428.

Smith, S. M. & M. C. Tyrrell. 2012. Effects of mud fiddler crabs (*Uca pugnax*) on the recruitment of halophyte seedlings in salt marsh dieback areas of Cape Cod (Massachusetts, USA). *Ecol. Res.* 27: 233–237.

Smith, S. M., P. V. McCormick, J. A. Leeds & P. B. Garrett. 2002. Constraints of seed bank species composition and water depth for restoring vegetation in the Florida Everglades, U.S.A. *Restorat. Ecol.* 10: 138–145.

Smith, S. M., C. T. Roman, M.-J. James-Pirri, K. Chapman, J. Portnoy & E. Gwilliam. 2008. Responses of plant communities to incremental hydrologic restoration of a tide-restricted salt marsh in southern New England (Massachusetts, U.S.A.). *Restorat. Ecol.* 17: 606–618.

Smith, T. E. 2003. Observations on the experimental planting of *Lindera melissifolia* (Walter) Blume in southeastern Missouri after ten years. *Castanea* 68: 75–80.

Smith, W. K. 1981a. Temperature and water relation patterns in subalpine understory plants. *Oecologia* 48: 353–359.

Smith, W. R. 1981b. Status report: *Chrysosplenium iowense* Rydb. Minnesota Heritage Program, St. Paul, MN. 4 pp.

Smits, A. J. M., R. Van Ruremonde & G. Van der Velde. 1989. Seed dispersal of three nymphaeid macrophytes. *Aquat. Bot.* 35: 167–180.

Smolders, A., C. den Hartog & J. G. M. Roelofs. 1995. A pseudoviviparous specimen of *Lobelia dortmanna* L. in Lake Haptatjørn (S.W. Norway). *Aquat. Bot.* 49: 269–271.

Smyth, B. B. 1903. Preliminary list of medicinal and economic Kansas plants, with their reputed therapeutic properties. *Trans. Kansas Acad. Sci.* 18: 191–209.

Snedden, G. A. & G. D. Steyer. 2013. Predictive occurrence models for coastal wetland plant communities: Delineating hydrologic response surfaces with multinomial logistic regression. *Estuar. Coastal Shelf Sci.* 118: 11–23.

Sneddon, B. V. 1977. A biosystematic study of *Microseris* subgenus *Monermos* (Compositae: Cichorieae). PhD dissertation, Victoria University, Wellington, New Zealand. 53 pp.

Snodgrass, G. L. 2003. Role of reproductive diapause in the adaption of the tarnished plant gug (Heteroptera: Miridae) to its winter habitat in the Mississippi River delta. *Environ. Entomol.* 32: 945–952.

Snodgrass, G. L., W. P. Scott & J. W. Smith. 1984a. Host plants of *Taylorilygus pallidulus* and *Polymerus basalis* (Hemiptera: Miridae) in the delta of Arkansas, Louisiana, and Mississippi. *Florida Entomol.* 67: 402–408.

Snodgrass, G. L., T. J. Henry & W. P. Scott. 1984b. An annotated list of the Miridae (Heteroptera) found in the Yazoo-Mississippi delta and associated areas in Arkansas and Louisiana. *Proc. Entomol. Soc. Washington* 86: 845–860.

Snow, A. A. & T. P. Spira. 1993. Individual variation in the vigor of self pollen and selfed progeny in *Hibiscus moscheutos* (Malvaceae). *Amer. J. Bot.* 80: 160–164.

Snow, A. A. & T. P. Spira. 1996. Pollen-tube competition and male fitness in *Hibiscus moscheutos*. *Evolution* 50: 1866–1870.

Snow, A. A., T. P. Spira, R. Simpson & R. A. Klips. 1996. The ecology of geitonogamous pollination. pp. 191–216. *In*: D. G. Lloyd & S. C. H. Barrett (eds.), *Floral Biology*. Chapman & Hall, New York.

Snow, A. A., T. P. Spira & H. Liu. 2000. Effects of sequential pollination on the success of 'fast' and 'slow' pollen donors in *Hibiscus moscheutos* (Malvaceae). *Amer. J. Bot.* 87: 1656–1659.

Snowden, R. E. D. & B. D. Wheeler. 1993. Iron toxicity to fen plant species. *J. Ecol.* 81: 35–46.

Snyder, D. 1996. The genus *Rhexia* in New Jersey. *Bartonia* 59: 55–70.

Snyder, E., A. Francis & S. J. Darbyshire. 2016. Biology of invasive alien plants in Canada. 13. *Stratiotes aloides* L. *Can. J. Plant Sci.* 96: 225–242.

Snyder, S. A. 1991. *Cephalanthus occidentalis*. *In*: *Fire Effects Information System*. U. S. Department of Agriculture, Forest Service, Rocky Mountain Research Station, Fire Sciences Laboratory (Producer). Available online: http://www.fs.fed.us/database/feis/ (accessed December 28, 2006).

Sobota, A. E. 1984. Inhibition of bacterial adherence by cranberry juice: potential use for the treatment of urinary tract infections. *J. Urol.* 131: 1013–1016.

Sobral-Leite, M., J. A. de Siqueira Filho, C. Erbar & I. C. Machado. 2011. Anthecology and reproductive system of *Mourera fluviatilis* (Podostemaceae): pollination by bees and xenogamy in a predominantly anemophilous and autogamous family? *Aquat. Bot.* 95: 77–87.

Sobrero, R., V. E. Campos, S. M. Giannoni & L. A. Ebensperger. 2010. *Octomys mimax* (Rodentia: Octodontidae). *Mammal. Species* 42: 49–57.

Sockman, K. W. 2008. Ovulation order mediates a trade-off between pre-hatching and post-hatching viability in an altricial bird. *PLoS One* 3(3): e1785. doi:10.1371/journal.pone.0001785.

Soejima, A. & J. Wen. 2006. Phylogenetic analysis of the grape family (Vitaceae) based on three chloroplast markers. *Amer. J. Bot.* 93: 278–287.

Soerjani, M., A. J. G. H. Kostermans & G. Tjitrosoepomo (eds.). 1987. *Weeds of Rice in Indonesia*. Balai Pustaka, Jakarta, Indonesia. 716 pp.

Sohmer, S. H. & D. F. Sefton. 1978. The reproductive biology of *Nelumbo pentapetala* (Nelumbonaceae) on the Upper Mississippi River. II. The insects associated with the transfer of pollen. *Brittonia* 30: 355–364.

Sokoloff, D. D., G. V. Degtjareva, P. K. Endress, M. V. Remizowa, T. H. Samigullin & C. M. Valiejo-Roman. 2007. Inflorescence and early flower development in *Loteae* (Leguminosae) in a phylogenetic and taxonomic context. *Int. J. Plant Sci.* 168: 801–833.

Solarz, S. L. & R. M. Newman. 1996. Oviposition specificity and behavior of the watermilfoil specialist *Euhrychiopsis lecontei*. *Oecologia* 106: 337–344.

Solbreck, C. & I. Pehrson. 1979. Relations between environment, migration and reproduction in a seed bug, *Neacoryphus bicrucis* (Say) (Heteroptera: Lygaeidae). *Oecologia* 43: 51–62.

Solbrig, O. T., D. W. Kyhos, M. Powell & P. H. Raven. 1972. Chromosome numbers in Compositae VIII: Heliantheae. *Amer. J. Bot.* 59: 869–878.

Solbrig, O. T., W. F. Curtis, D. T. Kincaid & S. J. Newell. 1988. Studies on the population biology of the genus *Viola*. VI. The demography of *V. fimbriatula* and *V. lanceolata*. *J. Ecol.* 76: 301–319.

Solis-Garza, G. & P. Jenkins. 1998. Riparian vegetation on the Rio Saota Cruz, Sonora. pp. 100–118. *In*: G. J. Gottfried, C. B. Edminster & M. C. Dillon (compilers), *Cross Border Waters: Fragile Treasures for the 21st Century*. Ninth U.S./Mexico Border States Conference on Recreation, Parks, and Wildlife. June 3–6, 1998. U. S. Department of Agriculture, Forest Service, Rocky Mountain Research Station, USDA Forest Service Proceedings RMRS-P-5. Fort Collins, CO.

Soltis, D. E. 1982. Allozymic variability in *Sullivantia* (Saxifragaceae). *Syst. Bot.* 7: 26–34.

Soltis, D. E. 1984. Autopolyploidy in *Tolmiea menziesii* (Saxifragaceae). *Amer. J. Bot.* 71: 1171–1174.

Soltis, D. E. 1987. Karyotypes and relationships among *Bolandra*, *Boykinia*, *Peltoboykinia*, and *Suksdorfia* (Saxifragaceae: Saxifrageae). *Syst. Bot.* 12: 14–20.

Soltis, D. E. 1988. Karyotypes of *Bensoniella*, *Conimitella*, *Lithophragma*, and *Mitella*, and relationships in Saxifrageae (Saxifragaceae). *Syst. Bot.* 13: 64–72.

Soltis, D. E. 1991. A revision of *Sullivantia* (Saxifragaceae). *Brittonia* 43: 27–53.

Soltis, D. E. 2009. 17. *Sullivantia* Torrey & A. Gray. pp. 121–124. *In*: N. R. Morin (convening ed.), *Flora of North America North of Mexico, Vol. 8: Magnoliophyta: Paeoniaceae to Ericaceae*. Oxford University Press, New York.

Soltis, D. E. & B. A. Bohm. 1982. Flavonoids of *Penthorum sedoides*. *Biochem. Syst. Ecol.* 10: 221–224.

Soltis, D. E. & L. H. Rieseberg. 1986. Autopolyploidy in *Tolmiea menziesii* (Saxifragaceae): genetic insights from enzyme electrophoresis. *Amer. J. Bot.* 73: 310–318.

Soltis, D. E. & P. D. Soltis. 1997. Phylogenetic relationships in Saxifragaceae sensu lato: a comparison of topologies based on 18S rDNA and *rbcL* sequences. *Amer. J. Bot.* 84: 504–522.

Soltis, D. E., P. S. Soltis, T. A. Ranker & B. D. Ness. 1989. Chloroplast DNA variation in a wild plant, *Tolmiea menziesii*. *Genetics* 121: 819–826.

Soltis, D. E., P. S. Soltis & B. D. Ness. 1990. Maternal inheritance of the chloroplast genome in *Heuchera* and *Tolmiea* (Saxifragaceae). *J. Heredity* 81: 168–170.

Soltis, D. E., P. S. Soltis, T. G. Collier & M. L. Edgerton. 1991. Chloroplast DNA variation within and among genera of the *Heuchera* group (Saxifragaceae): evidence for chloroplast transfer and paraphyly. *Amer. J. Bot.* 78: 1091–1112.

Soltis, D. E., L. A. Johnson & C. Looney. 1996a. Discordance between ITS and chloroplast topologies in the *Boykinia* group (Saxifragaceae). *Syst. Bot.* 21: 169–185.

Soltis, D. E., R. K. Kuzoff, E. Conti, R. Gornall & K. Ferguson. 1996b. *matK* and *rbcL* gene sequence data indicate that *Saxifraga* (Saxifragaceae) is polyphyletic. *Amer. J. Bot.* 83: 371–382.

Soltis, D. E., P. S. Soltis, M. W. Chase, M. E. Mort, D. C. Albach, M. Zanis, V. Savolainen, W. H. Hahn, S. B. Hoot, M. F. Fay, M. Axtell, S. M. Swensen, L. M. Prince, W. J. Kress, K. C. Nixon & J. S. Farris. 2000. Angiosperm phylogeny inferred from 18S rDNA, *rbcL*, and *atpB* sequences. *Bot. J. Linn. Soc.* 133: 381–461.

Soltis, D. E., R. K. Kuzoff, M. E. Mort, M. Zanis, M. Fishbein, L. Hufford, J. Koontz & M. K. Arroyo. 2001a. Elucidating deep-level phylogenetic relationships in Saxifragaceae using sequences for six chloroplastic and nuclear DNA regions. *Ann. Missouri Bot. Gard.* 88: 669–693.

Soltis, D. E., M. Tago-Nakazawa, Q.-Y. Xiang, S. Kawano, J. Murata, M. Wakabayashi & C. Hibsch-Jetter. 2001b. Phylogenetic relationships and evolution in *Chrysosplenium* (Saxifragaceae) based on *matK* sequence data. *Amer. J. Bot.* 88: 883–893.

Soltis, D. E., P. S. Soltis, P. K. Endress & M. W. Chase. 2005. *Phylogeny and Evolution of Angiosperms.* Sinauer Associates, Inc., Sunderland, MA. 370 pp.

Soltis, D. E., W. S. Judd, P. S. Soltis & P. E. Elvander. 2009. 11. *Tolmiea* Torrey & A. Gray. pp. 107–108. *In*: N. R. Morin (convening ed.), *Flora of North America North of Mexico, Vol. 8: Magnoliophyta: Paeoniaceae to Ericaceae.* Oxford University Press, New York.

Soltis, D. E., S. A. Smith, N. Cellinese, K. J. Wurdack, D. C. Tank, S. F. Brockington, N. F. Refulio-Rodriguez, J. B. Walker, M. J. Moore, B. S. Carlsward, C. D. Bell, M. Latvis, S. Crawley, C. Black, D. Diouf, Z. Xi, C. A. Rushworth, M. A. Gitzendanner, K. J. Sytsma, Y.-L. Qiu, K. W. Hilu, C. C. Davis, M. J. Sanderson, R. S. Beaman, R. G. Olmstead, W. S. Judd, M. J. Donoghue & P. S. Soltis. 2011. Angiosperm phylogeny: 17 genes, 640 taxa. *Amer. J. Bot.* 98: 704–730.

Soper, L. R. & N. F. Payne. 1997. Relationship of introduced mink, an island race of muskrat, and marginal habitat. *Ann. Zool. Fennici* 34: 251–258.

Soreng, R. J. & R. W. Spellenberg. 1984. An unusual new *Chaetopappa* (Asteraceae-Astereae) from New Mexico. *Syst. Bot.* 9: 1–5.

Sorensen, J. T. & D. J. Holden. 1974. Germination of native prairie forb seeds. *J. Range Manag.* 27: 123–126.

Sorenson, D. R. & W. T. Jackson. 1968. The utilization of paramecia by the carnivorous plant *Utricularia gibba*. *Planta* 83: 166–170.

Sorrie, B. 1990. *Myosurus minimus* (Ranunculaceae) in New England with notes on flower morphology. *Rhodora* 92: 103–104.

Sorrie, B. A. 1997. Notes on *Lycopus cokeri* (Lamiaceae). *Castanea* 62: 119–126.

Sorrie, B. A. 2005. Alien vascular plants in Massachusetts. *Rhodora* 107: 284–329.

Sorrie, B.A. 2014. Noteworthy records from Dare and Tyrrell counties, North Carolina. *Phytoneuron* 2014–47: 1–15.

Sorrie, B. A. & S. W. Leonard. 1999. Noteworthy records of Mississippi vascular plants. *Sida* 18: 889–908.

Sorrie, B. A. & A. S. Weakley. 2001. Coastal plain vascular plant endemics: phytogeographic patterns. *Castanea* 66: 50–82.

Sorrie, B. A. & A. S. Weakley. 2007. Notes on the *Gaylussacia dumosa* complex (Ericaceae). *J. Bot. Res. Inst. Texas* 1: 333–344.

Sorrie, B. A., M. H. MacRoberts, B. R. MacRoberts & W. S. Birmingham. 2003. *Oxypolis ternata* (Apiaceae) deleted from the Texas flora. *Sida* 20: 1323–1324.

Sorrie, B. A., J. B. Gray & P. J. Crutchfield. 2006. The vascular flora of the longleaf pine ecosystem of Fort Bragg and Weymouth Woods, North Carolina. *Castanea* 71: 129–161.

Sorrie, B. A., A. S. Weakley & G. C. Tucker. 2009. 46. *Gaylussacia* Kunth. pp. 530–535. *In*: N. R. Morin (convening ed.), *Flora of North America North of Mexico, Vol. 8: Magnoliophyta: Paeoniaceae to Ericaceae.* Oxford University Press, New York.

Sosa, M. M. & G. Seijo. 2002. Chromosome studies in Argentinian species of *Stemodia* L. (Scrophulariaceae). *Cytologia* 67: 261–266.

Sousa, W. P., S. P. Quek & B. J. Mitchell. 2003. Regeneration of *Rhizophora mangle* in a Caribbean mangrove forest: interacting effects of canopy disturbance and a stem-boring beetle. *Oecologia* 137: 436–445.

South Florida Water Management District (SFWMD). 2003. *Southern Golden Gate Estates Watershed Planning Assistance Cooperative Study.* Final Report. South Florida Water Management District, Big Cypress Basin & United States Department of Agriculture, Natural Resources Conservation Service. Available online: http://www.sfwmd.gov/org/bcb/reports/sgoldengaterpt/appendix_a.pdf (accessed February 23, 2007).

Soza, V. L., K. Haworth & V. S. Di Stilio. 2013. Timing and consequences of recurrent polyploidy in meadow-rues (*Thalictrum*, Ranunculaceae). *Mol. Biol. Evol.* 30: 1940–1954.

Soza, V. L., J. Brunet, A. Liston, P. S. Smith & V. S. Di Stilio. 2012. Phylogenetic insights into the correlates of dioecy in meadow-rues (*Thalictrum*, Ranunculaceae). *Mol. Phylogen. Evol.* 63: 180–192.

Spaeth, D. F., R. T. Bowyer, T. R. Stephenson, P. S. Barboza & V. Van Ballenberghe. 2002. Nutritional quality of willows for moose: effects of twig age and diameter. *Alces* 38: 143–154.

Spahn, S. A. & T. W. Sherry. 1999. Cadmium and lead exposure associated with reduced growth rates, poorer fledging success of little blue heron chicks (*Egretta caerulea*) in south Louisiana wetlands. *Arch. Environ. Contam. Toxicol.* 37: 377–384.

Spalik, K. & S. R. Downie. 2006. The evolutionary history of *Sium* sensu lato (Apiaceae): dispersal, vicariance, and domestication as inferred from ITS rDNA phylogeny. *Amer. J. Bot.* 93: 747–761.

Spalik, K., J.-P. Reduron & S. R. Downie. 2004. The phylogenetic position of *Peucedanum* sensu lato and allied genera and their placement in tribe *Selineae* (Apiaceae, subfamily Apioideae). *Plant Syst. Evol.* 243: 189–210.

Spalik, K., S. R. Downie & M. R. Watson. 2009. Generic delimitations within the *Sium* alliance (Apiaceae tribe *Oenantheae*) inferred from cpDNA rps16–5′trnK(UUU) and nrDNA ITS sequences. *Taxon* 58: 735–748.

Sparrow, F. K. 1957. Observations on chytridiaceous parasites of phanerogams. VII. A *Physoderma* on *Lycopus americanus*. *Amer. J. Bot.* 44: 661–665.

Sparrow, F. K. 1975. Observations on chytridiaceous parasites of phanerogams. XXIII. Notes on *Physoderma*. *Mycologia* 67: 552–568.

Sparrow, F. K. 1976. Observations on chytridiaceous parasites of phanerogams. XXV. *Physoderma johnsii* Sparrow, a parasite of *Caltha palustris* L. *Amer. J. Bot.* 63: 602–607.

Speck, F. G. 1941. A list of plant curatives obtained from the Houma Indians of Louisiana. *Primitive Man* 14: 49–73.

Speichert, G. & S. Speichert. 2004. *Encyclopedia of Water Garden Plants*. Timber Press, Portland, OR. 386 pp.

Spence, J. R. 2005. Notes on significant collections and additions to the flora of Glen Canyon National Recreation Area, Utah and Arizona, between 1992 and 2004. *W. N. Amer. Naturalist* 65: 103–111.

Spence, J. R. & N. R. Henderson. 1993. Tinaja and hanging garden vegetation of Capitol Reef National Park, southern Utah, U.S.A. *J. Arid Environ.* 24: 21–36.

Spence, J. R. & J. R. Shevock. 2012. *Ptychostomum pacificum* (Bryaceae), a new fen species from California, Oregon, and western Nevada, USA. *Madroño* 59: 156–162.

Spencer, D. F. & G. G. Ksander. 1998. Using videotaped transects to estimate submersed plant abundance in Fall River, California. *J. Aquat. Plant Manag.* 36: 130–137.

Spencer, K. A. 1981. *A Revisionary Study of the Leaf-Mining Flies (Agromyzidae) of California*. University of California, Division of Agricultural Sciences, Special Publication 3273. University of California Press, Berkeley, CA. 489 pp.

Spencer, L. J. & S. G. Bousquin. 2014. Interim responses of floodplain wetland vegetation to phase I of the Kissimmee River restoration project: Comparisons of vegetation maps from five periods in the river's history. *Restorat. Ecol.* 22: 397–408.

Spencer, S. C. & J. M. Porter. 1997. Evolutionary diversification and adaptation to novel environments in *Navarretia* (Polemoniaceae). *Syst. Bot.* 22: 649–668.

Spencer, S. C. & L. H. Rieseberg. 1998. Evolution of amphibious vernal pool specialist annuals: putative vernal pool adaptive traits in *Navarretia* (Polemoniaceae). pp. 76–85. *In:* C. W. Witham, E. T. Bauder, D. Belk, W. R. Ferren Jr. & R. Ornduff (eds.), *Ecology, Conservation, and Management of Vernal Pool Ecosystems—Proceedings from a 1996 Conference*. California Native Plant Society, Sacramento, CA.

Spencer, W. & G. Bowes. 1985. *Limnophila* and *Hygrophila*: a review and physiological assessment of their weed potential in Florida. *J. Aquat. Plant Manag.* 23: 7–16.

Sperduto, D. D. 1996. *Scleria reticularis* (Cyperaceae) new to New Hampshire. *Rhodora* 98: 99–102.

Sperduto, D. D. 2011. *Natural Community Systems of New Hampshire*. New Hampshire Natural Heritage Bureau, Concord, NH. 134 pp.

Sperduto, D. D. & W. F. Nichols. 2004. *Natural Communities of New Hampshire*. New Hampshire Natural Heritage Bureau, Concord, New Hampshire. University of New Hampshire Cooperative Extension, Durham, NH. 229 pp.

Sperduto, D. D. & W. F. Nichols. 2012. *Natural Communities of New Hampshire*. 2nd ed. New Hampshire Natural Heritage Bureau, Concord, New Hampshire. University of New Hampshire Cooperative Extension, Durham, NH. 266 pp.

Sperduto, D. D., W. F. Nichols & N. Cleavitt. 2000. *Bogs and Fens of New Hampshire*. A report submitted to the U. S. Environmental Protection Agency. New Hampshire Natural Heritage Inventory, DRED Division of Forests and Lands, Concord, NH. 38 pp.

Sperry, T. S. 1993. Inbreeding depression in the highly self-fertilizing annual herb, *Mimulus floribundus*. Honor's thesis. University of Oregon, Eugene, OR. 15 pp.

Spetzman, L. A. 1959. *Vegetation of the Arctic Slope of Alaska*. Geological Survey Professional Paper 302-B. United States Government Printing Office, Washington, DC. 58 pp.

Spieles, J. B., P. J. Comer, D. A. Albert & M. A. Kost. 1999. *Natural Community Abstract for Prairie Fen*. Michigan Natural Features Inventory, Lansing, MI. 4 pp.

Spinner, G. P. & J. S. Bishop. 1950. Chemical analysis of some wild-life foods in Connecticut. *J. Wildl. Manag.* 14: 175–180.

Spira, T. P. 1980. Floral parameters, breeding system and pollinator type in *Trichostema* (Labiatae). *Amer. J. Bot.* 67: 278–284.

Spira, T. P. & O. D. Pollak. 1986. Comparative reproductive biology of alpine biennial and perennial gentians (*Gentiana*: Gentianaceae) in California. *Amer. J. Bot.* 73: 39–47.

Spira, T. P., A. A. Snow, D. F. Whigham & J. Leak. 1992. Flower visitation, pollen deposition, and pollen-tube competition in *Hibiscus moscheutos* (Malvaceae). *Amer. J. Bot.* 79: 428–433.

Spira, T. P., A. A. Snow & M. N. Puterbaugh. 1996. The timing and effectiveness of sequential pollinations in *Hibiscus moscheutos*. *Oecologia* 105: 230–235.

Spirling, L. I. & I. R. Daniels. 2001. Peppermint: more than just an after-dinner mint. *J. Roy. Soc. Promot. Health* 121: 62–63.

Spongberg, S. A. 1976. Some old and new interspecific *Magnolia* hybrids. *Arnoldia* 36: 129–145.

Spooner, D. M. 1984. Infraspecific variation in *Gratiola viscidula* Pennell (Scrophulariaceae). *Rhodora* 86: 79–87.

Sprinkles, C. & G. R. Bachman. 1999. Germination of woody plants using coir as a peat alternative. *Proc. South. Nursery Assoc. Res. Conf.* 44: 343–345.

Squires, M. M. & L. F. W. Lesack. 2003. The relation between sediment nutrient content and macrophyte biomass and community structure along a water transparency gradient among lakes of the Mackenzie Delta. *Can. J. Fisheries Aquat. Sci.* 60: 333–343.

Squitier, J. M. & J. L. Capinera. 2002. Host selection by grasshoppers (Orthoptera: Acrididae) inhabiting semi-aquatic environments. *Florida Entomol.* 85: 336–340.

Šraj-Kržič, N., P. Pongrac, M. Klemenc, A. Kladnik, M. Regvar & A. Gaberščik. 2006. Mycorrhizal colonisation in plants from intermittent aquatic habitats. *Aquat. Bot.* 85: 331–336.

Srivastava, P. & K. Shanker. 2012. *Pluchea lanceolata* (Rasana): Chemical and biological potential of Rasayana herb used in traditional system of medicine. *Fitoterapia* 83: 1371–1385.

Srivastava, R. C. & Nyishi Community. 2010. Traditional knowledge of *Nyishi* (*Daffla*) tribe of Arunachal Pradesh. *Indian J. Tradit. Knowl.* 9: 26–37.

Srivastava, S., U. C. Lavania, N. K. Misra & S. Basu. 2002. Chromosome behaviour in diploid [sic] and its bearing on tetraploid meiosis in *Bacopa monnieri* (L) Pennell. *Nucleus* 45: 57–60.

Stace, H. M. & S. H. James. 1996. Another perspective on cyto-evolution in Lobelioideae (Campanulaceae). *Amer. J. Bot.* 83: 1356–1364.

Ståhl, B. 2004a. Cyrillaceae (*Cyrilla* family). pp. 125–126. *In:* N. Smith, S. A. Mori, A. Henderson, D. W. Stevenson & S. V. Heald (eds.), *Flowering Plants of the Neotropics*. Princeton University Press, Princeton, NJ. 594 pp.

Ståhl, B. 2004b. Samolaceae. pp. 387–389. *In:* K. Kubitzki, J. Rohwer, and V. Bittrich (eds.), *The Families and Genera of Vascular Plants, Vol. VI. Flowering Plants, Dicotyledons: Celastrales, Oxalidales, Rosales, Ericales*. Springer-Verlag, Berlin, Germany. 489 pp.

Ståhl, B. 2004c. Theophrastaceae (*Theophrasta* family). pp. 371–372. *In:* N. Smith, S. A. Mori, A. Henderson, D. W. Stevenson & S. V. Heald (eds.), *Flowering Plants of the Neotropics*. Princeton University Press, Princeton, NJ. 594 pp.

Stallins, J. A. 2005. Stability domains in barrier island dune systems. *Ecol. Complexity* 2: 410–430.

Stalter, R. 1973. Factors influencing the distribution of vegetation of the Cooper River estuary. *Castanea* 38: 18–24.

Stalter, R. 2004. The flora on the High Line, New York City, New York. *J. Torrey Bot. Club* 131: 387–393.

Stalter, R. & J. Baden. 1994. A twenty year comparison of vegetation of three abandoned rice fields, Georgetown County, South Carolina. *Castanea* 59: 69–77.

Stalter, R. & W. T. Batson. 1969. Transplantation of salt marsh vegetation, Georgetown, South Carolina. *Ecology* 50: 1087–1089.

Stalter, R. & W. T. Batson. 1973. Seed viability in salt marsh taxa, Georgetown County, South Carolina. *Castanea* 38: 109–110.

Stalter, R. & E. E. Lamont. 1990. The vascular flora of Assateague Island, Virginia. *Bull. Torrey Bot. Club* 117: 48–56.

Stamp, N. E. 1984. Effect of defoliation by checkerspot caterpillars (*Euphydryas phaeton*) and sawfly larvae (*Macrophya nigra* and *Tenthredo grandis*) on their host plants (*Chelone* spp.). *Oecologia* 63: 275–280.

Stamp, N. E. 1987. Availability of resources for predators of *Chelone* seeds and their parasitoids. *Amer. Midl. Naturalist* 117: 265–279.

Standley, P. C. 1914. The genus *Arthrocnemum* in North America. *J. Washington Acad. Sci.* 4: 398–399.

Standley, P. C. 1916. Fungi of New Mexico. *Mycologia* 8: 142–177.

Standley, P. C. 1919. A new locality for *Senecio crawfordii*. *Rhodora* 21: 117–120.

Staniforth, R. J. & P. B. Cavers. 1976. An experimental study of water dispersal in *Polygonum* spp. *Can. J. Bot.* 54: 2587–2596.

Staniforth, R. J., N. Griller & C. Lajzerowicz. 1998. Soil seed banks from coastal subarctic ecosystems of Bird Cove, Hudson Bay. *Écoscience* 5: 241–249.

Stanley, R. A. 1970. Studies on nutrition, photosynthesis, and respiration in *Myriophyllum spicatum* L. PhD dissertation, Duke University. University Microfilms, Ann Arbor, MI. 128 pp.

Stansly, P. A., D. J. Schuster & T. X. Liu. 1997. Apparent parasitism of *Bemisia argentifolii* (Homoptera: Aleyrodidae) by aphelinidae (Hymenoptera) on vegetable crops and associated weeds in south Florida. *Biol. Control* 9: 49–57.

Stanton, A. E. & B. M. Pavlik. 2009. *Implementation of the Conservation Strategy for Tahoe Yellow Cress* (Rorippa subumbellata). 2008 Annual Report. BMP Ecosciences, San Francisco, CA. 87 pp.

Stanton, F. W. & H. C. Smith. 1957. *Key to Some Important Aquatic Plants of Oregon*. Miscellaneous Wildlife Publication No. 2. Oregon State Game Commission, Portland, OR. 14 pp.

Staples, G. & M. S. Kristiansen. 1999. *Ethnic Culinary Herbs: A Guide to Identification and Cultivation in Hawaii*. University of Hawai'i Press, Honolulu, HI. 136 pp.

Starr, J. R. & A. Bruneau, 2002. Phylogeny of *Rosa* L. (Rosaceae) based on *trn*L-F intron and spacer sequences. Botany 2002 meeting (abstract).

Start, A. N. & T. Handasyde. 2002. Using photographs to document environmental change: the effects of dams on the riparian environment of the lower Ord River. *Austral. J. Bot.* 50: 465–480.

Stauffer, R. C. 1987. *Charles Darwin's Natural Selection*. Cambridge University Press, Cambridge, UK. 704 pp.

Stebbins, G. L. & K. Daly. 1961. Changes in the variation pattern of a hybrid population of *Helianthus* over an eight-year period. *Evolution* 15: 60–71.

Steck, G. J. 1984. *Chaetostomella undosa* (Diptera: Tephritidae): biology, ecology, and larval description. *Ann. Entomol. Soc. Amer.* 77: 669–678.

Steele, K. P. & R. Vilgalys. 1994. Phylogenetic analysis of Polemoniaceae using nucleotide sequences of the plastid gene *matK*. *Syst. Bot.* 19: 126–142.

Steenkamp, V., M. C. Gouws, M. Gulumian, E. E. Elgorashi & J. van Staden. 2006. Studies on antibacterial, anti-inflammatory and antioxidant activity of herbal remedies used in the treatment of benign prostatic hyperplasia and prostatitis. *J. Ethnopharmacol.* 103: 71–75.

Steers, R. J. & E. B. Allen. 2012. Impact of recurrent fire on annual plants: a case study from the western edge of the Colorado Desert. *Madroño* 59: 14–24.

Stefanović, S., L. Krueger & R. G. Olmstead. 2002. Monophyly of the Convolvulaceae and circumscription of their major lineages based on DNA sequences of multiple chloroplast loci. *Amer. J. Bot.* 89: 1510–1522.

Stefanović, S., D. F. Austin & R. G. Olmstead. 2003. Classification of Convolvulaceae: a phylogenetic approach. *Syst. Bot.* 28: 791–806.

Steffen, J. F. 1997. Seed treatment and propagation methods. pp. 151–162. *In:* S. Packard & C. F. Mutel (eds.), *The Tallgrass Restoration Handbook for Prairies, Savannas, and Woodlands*. Island Press, Washington, DC.

Steffen, S., P. Ball, L. Mucina & G. Kadereit. 2015. Phylogeny, biogeography and ecological diversification of *Sarcocornia* (Salicornioideae, Amaranthaceae). *Ann. Bot.* 115: 353–368.

Stegmaier Jr., C. E. 1966. Host plants and parasites of *Liriomyza munda* in Florida (Diptera: Agromyzidae). *Florida Entomol.* 49: 81–86.

Stegmaier Jr., C. E. 1967a. Notes on a seed-feeding Tephritidae, *Paracantha forficula*, (Diptera) in Florida. *Florida Entomol.* 50: 157–160.

Stegmaier Jr., C. E. 1967b. Some new host plant records and parasites of *Phytobia* (*Amauromyza*) *maculosa* in Florida (Diptera: Agromyzidae). *Florida Entomol.* 50: 99–101.

Stegmaier Jr., C. E. 1967c. *Pluchea odorata*, a new host record for *Acinia picturata* (Diptera, Tephritidae). *Florida Entomol.* 50: 53–55.

Stegmaier Jr., C. E. 1968a. Notes on the biology of *Trupanea actinobola* (Diptera: Tephritidae). *Florida Entomol.* 51: 95–99.

Stegmaier Jr., C. E. 1968b. *Erigeron*, a host plant genus of Tephritids (Diptera). *Florida Entomol.* 51: 45–50.

Stegmaier Jr., C. E. 1972. Parasitic Hymenoptera bred from the family Agromyzidae (Diptera) with special reference to south Florida. *Florida Entomol.* 55: 273–282.

Stegmaier Jr., C. E. 1973. Some insects associated with the Joe-Pye Weed, *Eupatorium coelestinum* (Compositae), from south Florida. *Florida Entomol.* 56: 61–65.

Stehn, R. A., C. P. Dau, B. Conant & W. I. Butler Jr. 1993. Decline of spectacled eiders nesting in western Alaska. *Arctic* 46: 264–277.

Stein, J., D. Binion & R. Acciavatti. 2003. *Field Guide to Native Oak Species of Eastern North America*. FHTET-2003–1. Forest Service, United States Department of Agriculture, Morgantown, WV. 161 pp.

Stein, S., P. W. Price, W. G. Abrahamson & C. F. Sacchi. 1992. The effect of fire on stimulating willow regrowth and subsequent attack by grasshoppers and elk. *Oikos* 65: 190–196.

Steinauer, G. & S. Rolfkmeier. 2003. *Terrestrial Natural Communities of Nebraska (Version 111)*. Nebraska Natural Heritage Program, Nebraska Game and Parks Commission, Lincoln, NE. 162 pp.

Steinauer, G., S. Rolfkmeier & J. P. Hardy. 1996. Inventory and floristics of sandhills fens in Cherry County, Nebraska. *Trans. Nebraska Acad. Sci.* 23: 9–21.

Steinbach, K. & G. Gottsberger. 1994. Phenology and pollination biology of five *Ranunculus* species in Giessen, central Germany. *Phyton* 34: 203–218.

Steinbauer, G. P. & B. Grigsby. 1960. Dormancy and germination of the docks (*Rumex* spp.). *Proc. Assoc. Off. Seed Analysts* 50: 112–117.

Steinberg, B. 1980. Vegetation of the Atlantic coastal ridge of Broward County, Florida based on 1940 imagery. *Florida Sci.* 43: 7–12.

Steiner, U., S. Leibner, C. L. Schardl, A. Leuchtmann & E. Leistner. 2011. *Periglandula*, a new fungal genus within the Clavicipitaceae and its association with Convolvulaceae. *Mycologia* 103: 1133–1145.

Stelzer, R. J., N. E. Raine, K. D. Schmitt & L. Chittka. 2010. Effects of aposematic coloration on predation risk in bumblebees? A comparison between differently coloured populations, with consideration of the ultraviolet. *J. Zool.* 282: 75–83.

Stenberg, L. 1995. Floristic observations from the province of Norrbotten, northern Sweden, 1994. *Svensk Bot. Tidskr.* 89: 45–50.

Stensvold, M. C., D. R. Farrar & C. Johnson-Groh. 2002. Two new species of moonworts (*Botrychium* subg. *Botrychium*) from Alaska. *Amer. Fern J.* 92: 150–160.

Stephen, A., R. Suresh, M. Balaji, S. Sarathkumar, S. Parthiban & S. Vinothraj. 2015. Diversity of vegetables from the markets of Chennai, Tamil Nadu, India. *J. Acad. Industr. Res.* 3: 536–541.

Stephens, K. M. & R. M. Dowling. 2001. *Wetland Plants of Queensland. A Field Guide*. CSIRO Publishing, Collingwood, VIC, Australia. 146 pp.

Stephenson, A. G., S. V. Good & D. W. Vogler. 2000. Interrelationships among inbreeding depression, plasticity in the self-incompatibility system, and the breeding system of *Campanula rapunculoides* L. (Campanulaceae). *Ann. Bot.* 85: S211-S219, S2000 I A.

Sterk, A. A. 1969. Biosystematic studies on *Spergularia*. *Acta Bot. Neerl.* 18: 639–650.

Stern, K. R. 1997. 3. *Corydalis* de Candolle. pp. 348–355. *In*: N. R. Morin (convening ed.), *Flora of North America North of Mexico, Vol. 3: Magnoliophyta: Magnoliidae and Hamamelidae*. Oxford University Press, New York.

Stern, M. A., J. F. Morawski & G. A. Rosenberg. 1993. Rediscovery and status of a disjunct population of breeding yellow rails in southern Oregon. *Condor* 95: 1024–1027.

Steubing, L., C. Ramírez & M. Alberdi. 1980. Energy content of water- and bog-plant associations in the region of Valdivia (Chile). *Vegetatio* 43: 153–161.

Steury, B. W. & C. A. Davis, 2003. Vascular flora of Piscataway and Fort Washington National Parks, Prince Georges and Charles Counties, Maryland. *Castanea* 68: 271–299.

Steven, D. D. & W. Franke. 1990. Germination and growth of *Ranunculus cymbalaria*, an endangered wetland plant. *Michigan Bot.* 29: 83–87.

Stevens, K. J., S.-Y. Kim, S. Adhikari, V. Vadapalli & B. J. Venables. 2009. Effects of triclosan on seed germination and seedling development of three wetland plants: *Sesbania herbacea*, *Eclipta prostrata*, and *Bidens frondosa*. *Environ. Toxicol. Chem.* 28: 2598–2609.

Stevens, K. J., M. R. Wellner & M. F. Acevedo. 2010. Dark septate endophyte and arbuscular mycorrhizal status of vegetation colonizing a bottomland hardwood forest after a 100 year flood. *Aquat. Bot.* 92: 105–111.

Stevens, K. L., S. C. Witt & C. E. Turner. 1990. Polyacetylenes in related thistles of the subtribes Centaureinae and Carduinae. *Biochem. Syst. Ecol.* 18: 229–232.

Stevens, O. A. 1924. Some effects of the first fall freeze. *Amer. Midl. Naturalist* 9: 14–17.

Stevens, O. A. 1932. The number and weight of seeds produced by weeds. *Amer. J. Bot.* 19: 784–794.

Stevens, O. A. 1957. Weights of seeds and numbers per plant. *Weeds* 5: 46–55.

Stevens, P. F. 2001. *Angiosperm Phylogeny Website*. Version 8, June 2007. Available online: http://www.mobot.org/MOBOT/research/APweb/ (accessed January 6, 2011).

Stevens, R., K. R. Jorgensen, S. A. Young & S. B. Monsen. 1996. *Forb and Shrub Seed Production Guide for Utah*. AG 501, December 1996. Cooperative Extension Service, Utah State University, Logan, UT. 52 pp.

Steward, A. N., L. R. J. Dennis & H. M. Gilkey. 1963. *Aquatic Plants of the Pacific Northwest with Vegetative Keys. 2nd Edition*. Oregon State Monographs, Studies in Botany No 11. Oregon State University Press, Corvallis, OR. 261 pp.

Stewart, C. D., S. K. Braman & A. F. Pendley. 2002. Functional response of the azalea plant bug (Heteroptera: Miridae) and a green lacewing *Chrysoperla rufilabris* (Neuroptera: Chrysopidae), two predators of the azalea lace bug (Heteroptera: Tingidae). *Environ. Entomol.* 31: 1184–1190.

Stewart Jr., C. N. & E. T. Nilsen. 1993. Association of edaphic factors and vegetation in several isolated Appalachian peat bogs. *Bull. Torrey Bot. Club* 120: 128–135.

Stewart, F. & J. Grace. 1984. An experimental study of hybridization between *H. mategazzianum* Somm. et Levier and *H. sphondylium* L. ssp. *sphondylium*. *Watsonia* 15: 75–83.

Stewart, G. R., J. A. Lee & T. O. Orebamjo. 1973. Nitrogen metabolism of halophytes. II. Nitrate availability and utilization. *New Phytol.* 72: 539–546.

Stewart, H. M. & J. M. Canne-Hilliker. 1998. Floral development of *Agalinis neoscotica*, *Agalinis paupercula* var. *borealis*, and *Agalinis purpurea* (Scrophulariaceae): implications for taxonomy and mating system. *Int. J. Plant Sci.* 159: 418–439.

Stewart, R. E. 1953. A life history study of the yellow-throat. *Wilson Bull.* 65: 99–115.

Stewart, R. E. & H. A. Kantrud. 1972. *Vegetation of Prairie Potholes, North Dakota, in Relation to Quality of Water and Other Environmental Factors*. Geological Survey Professional Paper 585-D. U. S. Bureau of Sport Fisheries and Wildlife, Northern Prairie Wildlife Research Center, Jamestown, ND. 35 pp.

Steyermark, J. A. 1931. A study of plant distribution in relation to the acidity of various soils in Missouri. *Ann. Missouri Bot. Gard.* 18: 41–55.

Steyermark, J. A. 1932. Some new spermatophytes from Texas. *Ann. Missouri Bot. Gard.* 19: 389–395.

Steyermark, J. A. 1949. *Lindera melissifolia*. *Rhodora* 51: 153–162.

Stieglitz, W. O. 1972. Food habits of the Florida duck. *J. Wildl. Manag.* 36: 422–428.

Stiles, E. W. 1980. Patterns of fruit presentation and seed dispersal in bird-disseminated woody plants in the eastern deciduous forest. *Amer. Naturalist* 116: 670–688.

Stiling, P., A. M. Rossi, D. R. Strong & D. M. Johnson. 1992. Life history and parasites of *Asphondylia borrichiae* (Diptera: Cecidomyiidae), a gall maker on *Borrichia frutescens*. *Florida Entomol.* 75: 130–137.

Still, S. & D. Potter. 2005. Phylogenetic relationships in tribe Spiraeeae (Rosaceae) inferred from nucleotide sequence data. Botany 2005 meeting (abstract).

Stireman III, J. O., H. Devlin, T. G. Carr & P. Abbot. 2010. Evolutionary diversification of the gall midge genus *Asteromyia* (Cecidomyiidae) in a multitrophic ecological context. *Mol. Phylogen. Evol.* 54: 194–210.

Stoetzel, M. B., A. S. Jensen & G. L. Miller. 1999. Reevaluation of the genus *Hyalomyzus* Richards with the description of two new species (Homoptera: Aphididae). *Ann. Entomol. Soc. Amer.* 92: 488–513.

Stoffolano, Jr., J. G., M. Rice & W. L. Murphy. 2015. *Sepedon fuscipennis* Loew (Diptera: Sciomyzidae): elucidation of external morphology by use of SEM of the head, legs, and postabdomen of adults. *Proc. Entomol. Soc. Washington* 117: 209–225.

Stokstad, E. 2016. How the Venus flytrap acquired its taste for meat. *Science* 352: 756.

Stollberg, B. P. 1950. Food habits of shoal-water ducks on Horicon Marsh, Wisconsin. *J. Wildl. Manag.* 14: 214–217.

Stone, D. E. 1959. A unique balanced breeding system in the vernal pool mousetails. *Evolution* 13: 151–174.

Stone, D. E. 1997. 1. *Carya* Nuttall. pp. 417–425. *In*: N. R. Morin (convening ed.), *Flora of North America North of Mexico, Vol. 3: Magnoliophyta: Magnoliidae and Hamamelidae*. Oxford University Press, New York.

Stone, D. E. & J. L. Freeman. 1968. Cytotaxonomy of *Illicium floridanum* and *I. parviflorum* (Illiciaceae*)*. *J. Arnold Arb.* 49: 41–51.

Stone, D. E., G. A. Adrouny & R. H. Flake. 1969. New World Juglandaceae. II. Hickory nut oils, phenetic similarities, and evolutionary implications in the genus *Carya. Amer. J. Bot.* 56: 928–935.

Stone, J. L. & B. A. Drummond. 2006. Rare estuary monkeyflower in Merrymeeting Bay is genetically distinct. *N. E. Naturalist* 13: 179–190.

Stone, R. D., W. B. Davilla & D. W. Taylor. 1988. *Status Survey of the Grass Tribe* Orcuttieae *and* Chamaesyce hooveri *(Euphorbiaceae) in the Central Valley of California*. Technical Report Prepared for Office of Endangered Species, USFWS, Contract #14-16-0001-85115. Sacramento, CA. 248 pp.

Stoner, J. G. & J. E. Rasmussen. 1983. Plant dermatitis. *J. Amer. Acad. Dermatol.* 9: 1–15.

Stoops, C. A., P. H. Adler & J. W. McCreadie. 1998. Ecology of aquatic Lepidoptera (Crambidae: Nymphulinae) in South Carolina, USA. *Hydrobiologia* 379: 33–40.

Stotz, N. G. 2000. *Historic Reconstruction of the Ecology of the Rio Grande/Río Bravo Channel and Floodplain in the Chihuahuan Desert*. World Wildlife Fund, Washington, DC. 152 pp.

Stoudt, J. H. 1944. Food preferences of mallards on the Chippewa National Forest, Minnesota. *J. Wildl. Manag.* 8: 100–112.

Stoughton, J. A. & W. A. Marcus. 2000. Persistent impacts of trace metals from mining on floodplain grass communities along Soda Butte Creek, Yellowstone National Park. *Environ. Manag.* 25: 305–320.

Stout, D. 2012. Fens: A remarkable habitat in the Sierra Nevada. *Fremontia* 40: 36–40.

Stoynoff, N. A. 1993. A quantitative analysis of the vegetation of Bluff Spring Fen nature preserve. *Trans. Illinois State Acad. Sci.* 86: 93–110.

Straka, J. R. & B. M. Starzomski. 2015. Fruitful factors: what limits seed production of flowering plants in the alpine? *Oecologia* 178: 249–260.

Strakosh, T. R., J. L. Eitzmann, K. B. Gido & C. S. Guy. 2005. The response of water willow *Justicia americana* to different water inundation and desiccation regimes. *N. Amer. J. Fish. Manag.* 25: 1476–1485.

Stralberg, D., M. P. Herzog, N. Nur, K. A. Tuxen & M. Kelly. 2010. Predicting avian abundance within and across tidal marshes uUsing fine-scale vegetation and geomorphic metrics. *Wetlands* 30: 475–487.

Strand, A. E. & B. G. Milligan. 1996. Genetics and conservation biology: Assessing historical gene flow in *Aquilegia* populations of the southwest. pp. 138–145. *In*: J. Maschinski, H. D. Hammond & L. Holter (eds.), *Southwestern Rare and Endangered Plants: Proceedings of the Second Conference*. General Technical Report RM GTR-283. United States Department of Agriculture Forest Service, Fort Collins, CO. 330 pp.

Strand, A. E., B. G. Milligan & C. M. Pruitt. 1996. Are populations islands? Analysis of chloroplast DNA variation in *Aquilegia*. *Evolution* 50: 1822–1829.

Stratman, M. R. & M. R. Pelton. 1999. Feeding ecology of black bears in northwest Florida. *Florida Field Naturalist* 27: 95–102.

Straub, S. C. K, M. J. Moore, P. S. Soltis, D. E. Soltis, A. Liston & T. Livshultz. 2014. Phylogenetic signal detection from an ancient rapid radiation: Effects of noise reduction, long-branch attraction, and model selection in crown clade Apocynaceae. *Mol. Phylogen. Evol.* 80: 169–185.

Strayer, D. L., C. Lutz, H. M. Malcom, K. Munger & W. H. Shaw. 2003. Invertebrate communities associated with a native (*Vallisneria americana*) and an alien (*Trapa natans*) macrophyte in a large river. *Freshwater Biol.* 48: 1938–1949.

Strecker, A. L., R. Miller & V. Morgan. 2014. *OSMB Final Report, Task 4: Oregon Lake Watch (2014)*. Paper 34. Center for Lakes and Reservoirs Publications and Presentations. Available online: http://pdxscholar.library.pdx.edu/centerforlakes_pub/34.

Strefeler, M. S., E. Darmo, R. L. Becker & E. J. Katovich. 1996a. Isozyme variation in cultivars of purple loosestrife (*Lythrum* sp.). *HortScience* 31: 279–282.

Strefeler, M. S., E. Darmo, R. L. Becker & E. J. Katovich. 1996b. Isozyme characterization of genetic diversity in Minnesota populations of purple loosestrife, *Lythrum salicaria* (Lythraceae). *Amer. J. Bot.* 83: 265–273.

Strickler, K., V. L. Scott & R. L. Fischer. 1996. Comparative nesting ecology of two sympatric leafcutting bees that differ in body size (Hymenoptera: Megachilidae). *J. Kansas Entomol. Soc.* 69: 26–44.

Strimaitis, A. M. & S. P. Sheldon. 2011. A comparison of macroinvertebrate and epiphyte density and diversity on native and exotic complex macrophytes in three Vermont lakes. *N. E. Naturalist* 18: 149–160.

Stringer, B. 2013. The sunflowers of South Carolina. *J. South Carolina Native Plant Soc.* Summer, 2013: 1, 4–10.

Stromberg, J. C., K. J. Bagstad, J. M. Leenhouts, S. J. Lite & E. Makings. 2005. Effects of stream flow intermittency on riparian vegetation of a semiarid region river (San Pedro River, Arizona). *River Res. Appl.* 21: 925–938.

Stromberg, J. C., A. F. Hazelton & M. S. White. 2009. Plant species richness in ephemeral and perennial reaches of a dryland river. *Biodivers. Conserv.* 18: 663–677.

Stromberg, J. C., L. Butler, A. F. Hazelton & J. A. Boudell. 2011. Seed size, sediment, and spatial heterogeneity: post-flood species coexistence in dryland riparian ecosystems. *Wetlands* 31: 1187–1197.

Stromberg-Wilkins, J. C. 1984. Autecological studies of *Drosera linearis*, a threatened sundew species. *Univ. Wisconsin-Milwaukee Field Sta. Bull.* 17: 1–16.

Strong, M. T. & C. L. Kelloff. 1994. Intertidal vascular plants of Brent Marsh, Potomac River, Stafford County, Virginia. *Castanea* 59: 354–366.

Strong, W. L. 2000. Vegetation development on reclaimed lands in the Coal Valley Mine of western Alberta, Canada. *Can. J. Bot.* 78: 110–118.

Strother, J. L. 2006a. 242. *Blennosperma* Lessing. pp. 640–641. *In*: Flora of North America Editorial Committee (eds.), *Flora of North America North of Mexico: Vol. 20, Magnoliophyta: Asteridae (in part): Asteraceae, Part 2*. Oxford University Press, New York.

Strother, J. L. 2006b. 308. *Coreopsis* Linnaeus. pp. 185–198. *In*: Flora of North America Editorial Committee (eds.), *Flora of North America North of Mexico: Vol. 21, Magnoliophyta: Asteridae (in part): Asteraceae, Part 3*. Oxford University Press, New York.

Strother, J. L. 2006c. 289. *Eclipta* Linnaeus. pp. 128–129. *In*: Flora of North America Editorial Committee (eds.), *Flora of North America North of Mexico: Vol. 21, Magnoliophyta: Asteridae (in part): Asteraceae, Part 3*. Oxford University Press, New York.

Strother, J. L. 2006d. 362. *Jamesianthus* S. F. Blake & Sherff. p. 377. *In*: Flora of North America Editorial Committee (eds.), *Flora of North America North of Mexico: Vol. 21, Magnoliophyta: Asteridae (in part): Asteraceae, Part 3*. Oxford University Press, New York.

Strother, J. L. 2006e. 330. *Jaumea* Persoon. pp. 253–254. *In*: Flora of North America Editorial Committee (eds.), *Flora of North America North of Mexico: Vol. 21, Magnoliophyta: Asteridae (in part): Asteraceae, Part 3*. Oxford University Press, New York.

Strother, J. L. 2006f. 33. *Vernonia* Schreber. pp. 206–213. *In*: Flora of North America Editorial Committee (eds.), *Flora of North America North of Mexico: Vol. 21, Magnoliophyta: Asteridae (in part): Asteraceae, Part 3*. Oxford University Press, New York.

Strother, J. L. & R. R. Weedon. 2006. 312. *Bidens* Linnaeus. pp. 205–218. *In*: Flora of North America Editorial Committee (eds.), *Flora of North America North of Mexico: Vol. 21, Magnoliophyta: Asteridae (in part): Asteraceae, Part 3*. Oxford University Press, New York.

Strother, J. L., R. L. Moe & G. L. Nesom. 2012. *Gamochaeta argyrinea* (Asteraceae) naturalized in California. *Phytoneuron* 2012–52: 1–5.

Stroud, J. L. & R. N. Collins. 2014. Improved detection of coastal acid sulfate soil hotspots through biomonitoring of metal (loid) accumulation in water lilies (*Nymphaea capensis*). *Sci. Total Environ.* 487: 500–505.

Struwe, L. & V. A. Albert (eds.). 2002. *Gentianaceae: Systematics and Natural History.* Cambridge University Press, Cambridge, UK. 652 pp.

Struwe, L., J. W. Kadereit, J. Klackenberg, S. Nilsson, M. Thiv, K. B. von Hagen & V. A. Albert. 2002. Systematics, character evolution, and biogeography of Gentianaceae, including a new tribal and subtribal classification. pp. 21–309. *In*: L. Struwe & V. A. Albert (eds.), *Gentianaceae: Systematics and Natural History.* Cambridge University Press, Cambridge, MA.

Stuart, A. E. & C. L. E. Stuart. 1998. A microscope slide test for the evaluation of insect repellents as used with *Culicoides impunctatus. Entomol. Exper. Appl.* 89: 277–280.

Stuart, R. J. & S. Polavarapu. 2000. Egg-mass variability and differential parasitism of *Choristoneura parallela* (Lepidoptera: Tortricidae) by endemic *Trichogramma minutum* (Hymenoptera: Trichogrammatidae). *Ann. Entomol. Soc. Amer.* 93: 1076–1084.

Stubbs, C. S., H. A. Jacobson, E. A. Osgood & F. A. Drummond. 1992. *Alternative Forage Plants for Native (Wild) Bees Associated with Lowbush Blueberry,* Vaccinium *spp., in Maine.* Technical Bulletin 148. Maine Agricultural Experiment Station, University of Maine Orono, ME. 54 pp.

Stubbs, C. S., F. Drummond & H. Ginsberg. 2007. *Effects of Invasive Plant Species on Pollinator Service and Reproduction in Native Plants at Acadia National Park.* Technical Report NPS/NER/NRTR--2007/096. National Park Service. Boston, MA. 85 pp.

Stuber, O. S. 2013. The relationship between land use and the ecological integrity of isolated wetlands in the Dougherty Plain, Georgia, USA. MS thesis. University of Georgia, Athens, GA. 166 pp.

Stuckey, I. H. & L. L. Gould. 2000. *Coastal Plants from Cape Cod to Cape Canaveral.* The University of North Carolina Press, Chapel Hill, NC. 305 pp.

Stuckey, R. L. 1966a. The distribution of *Rorippa sylvestris* (Cruciferae) in North America. *Sida* 2: 361–376.

Stuckey, R. L. 1966b. Differences in habitats and associates of the varieties of *Rorippa islandica* in the Douglas Lake region of Michigan. *Michigan Bot.* 5: 99–108.

Stuckey, R. L. 1969. The introduction and spread of *Lycopus asper* (western water-horehound) in the western Lake Erie and Lake St. Clair region. *Michigan Bot.* 8: 111–120.

Stuckey, R. L. 1970. Distributional History of *Epilobium hirsutum* (great hairy willow-herb) in North America. *Rhodora* 72: 164–181.

Stuckey, R. L. 1971. Changes of vascular aquatic flowering plants during 70 years in Put-In-Bay Harbor, Lake Erie, Ohio. *Ohio J. Sci.* 71: 321–342.

Stuckey, R. L. 1972. Taxonomy and distribution of the genus *Rorippa* (Cruciferae) in North America. *Sida* 4: 279–430.

Stuckey, R. L., J. R. Wehrmeister & R. J. Bartolotta. 1978. Submersed aquatic vascular plants in ice-covered ponds of central Ohio. *Rhodora* 80: 575–580.

Stucky, J. M. & R. Coxe. 1999. The loss of a unique wetland in the Piedmont, North Carolina. *Castanea* 64: 287–298.

Stumpf, C. F. & P. L. Lambdin. 2000. Distribution and known host records for *Planchonia stentae* (Hemiptera: Coccoidea: Asterolecaniidae). *Florida Entomol.* 83: 368–369.

Stutzenbaker, C. D. 1999. *Aquatic and Wetland Plants of the Western Gulf Coast.* Texas Parks & Wildlife Press/University of Texas Press, Austin, TX. 465 pp.

Subhashini, T., B. Krishnaveni & R. C. Srinivas. 2010. Anti-inflammatory activity of leaf extracts of *Alternanthera sessilis. Hygeia* 2: 54–56.

Subramanian, P. S., S. Mohanraj, P. A. Cockrum, C. C. J. Culvenor, J. A. Edgar, J. L. Frahn & L. W. Smith. 1980. The alkaloids of *Heliotropium curassavicum. Austral. J. Chem.* 33: 1357–1363.

Suh, Y., K. Heo & C.-W. Park. 2000. Phylogenetic relationships of maples (*Acer* L.; Aceraceae) implied by nuclear ribosomal ITS sequences. *J. Plant Res.* 113: 193–202.

Sukhorukov, A. P. & M. V. Nilova. 2016. A new species of *Arthrocnemum* (Salicornioideae: Chenopodiaceae-Amaranthaceae) from West Africa, with a revised characterization of the genus. *Bot. Lett.* doi:10.1080/23818107.2016.1185033.

Suksri, S., S. Premcharoen, C. Thawatphan & S. Sangthongprow. 2005. Ethnobotany in Bung Khong Long non-hunting area, northeast Thailand. *Kasetsart J.* 39: 519–533.

Sullivan, G. & J. B. Zedler. 1999. Functional redundancy among tidal marsh halophytes: a test. *Oikos* 84: 246–260.

Sullivan, J. 2001. A note on the death of Socrates. *Class. Q.* 51: 608–610.

Sullivan, V. I. 1975. Pollen and pollination in the genus *Eupatorium* (Compositae). *Can. J. Bot.* 53: 582–589.

Sultan, S. E. & F. A. Bazzaz. 1993a. Phenotypic plasticity in *Polygonum persicaria*. I. Diversity and uniformity in genotypic norms of reaction to light. *Evolution* 47: 1009–1031.

Sultan, S. E. & F. A. Bazzaz. 1993b. Phenotypic plasticity in *Polygonum persicaria*. II. Norms of reaction to soil moisture and the maintenance of genetic diversity. *Evolution* 47: 1032–1049.

Sultan, S. E. & F. A. Bazzaz. 1993c. Phenotypic plasticity in *Polygonum persicaria*. III. The evolution of ecological breadth for nutrient environment. *Evolution* 47: 1050–1071.

Sultana, S. & M. Saleem. 2004. *Salix caprea* inhibits skin carcinogenesis in murine skin: inhibition of oxidative stress, ornithine decarboxylase activity and DNA synthesis. *J. Ethnopharmacol.* 91: 267–276.

Sun, F.-J. & S. R. Downie. 2004. A molecular systematic investigation of *Cymopterus* and its allies (Apiaceae) based on phylogenetic analyses of nuclear (ITS) and plastid (*rps16* intron) DNA sequences. *S. Afr. J. Bot.* 70: 407–416.

Sun, F.-J., S. R. Downie & R. L. Hartman. 2004. An ITS-based phylogenetic analysis of the perennial, endemic Apiaceae subfamily Apioideae of western North America. *Syst. Bot.* 29: 419–431.

Sun, H, L. Li, A. Zhang, N. Zhang, H. Lv, W. Sun & X. Wang. 2013. Protective effects of sweroside on human MG-63 cells and rat osteoblasts. *Fitoterapia* 84: 174–179.

Sun, M. & F. R. Ganders. 1988. Mixed mating systems in Hawaiian *Bidens* (Asteraceae). *Evolution* 42: 516–527.

Sun, S. S., P. F. Gugger, Q. F. Wang & J. M. Chen. 2016. Identification of a R2R3-MYB gene regulating anthocyanin biosynthesis and relationships between its variation and flower color difference in lotus (*Nelumbo* Adans.). *PeerJ* 4: e2369.

Sundell, E. & R. D. Thomas. 1988. Four new records of *Cyperus* (Cyperaceae) in Arkansas. *Sida* 13: 259–261.

Sundue, M. 2005. Field trip reports. *J. Torrey Bot. Soc.* 132: 368–374.

Sundue, M. 2006. Addendum to field trip reports. *J. Torrey Bot. Soc.* 133: 210–211.

Sundue, M. 2007. Field trip reports. *J. Torrey Bot. Soc.* 134: 144–152.

Sungkajantranon, O. 1991. Seed development and seed storability of water convolvulus (*Ipomoea aquatica* Forsk.). Kasetsart University, Bangkok, Thailand. 92 pp.

Sustaita, D., P. F. Quickert, L. Patterson, L. Barthman-Thompson & S. Estrella. 2011. Salt marsh harvest mouse demography and habitat use in the Suisun Marsh, California. *J. Wildl. Manag.* 75: 1498–1507.

Sutherland, S. & R. K. Vickery Jr. 1988. Trade-offs between sexual and asexual reproduction in the genus *Mimulus. Oecologia* 76: 330–335.

Sutter, R. D. & M. Boyer. 1994. *The Seed Bank of Three Rare Species in Southeastern Pond Cypress Savanna:* Rhexia aristosa, Lobelia boykinii, *and* Oxypolis canbyii. Unpublished report. Southeast Regional Office of The Nature Conservancy, Chapel Hill, NC.

Sutton, B. D. & G. J. Steck. 2005. An annotated checklist of the Tephritidae (Diptera) of Florida. *Insecta Mundi* 19: 227–245.

Sutton, D. L. 1993. Culture of banana-lilies. *Proc. Florida State Hort. Soc.* 106: 332–335.

Sutton, D. L. 1994. Culture of water-snowflake. *Proc. Florida State Hort. Soc.* 107: 409–413.

Sutton, D. L. 1995. *Hygrophila* is replacing *Hydrilla* in south Florida. *Aquatics* 17(3): 4–10.

Svejcar, T. & G. M. Riegel. 1998. Spatial pattern of gas exchange for montane moist meadow species. *J. Veg. Sci.* 9: 85–94.

Svendsen, G. E. 1974. Behavioral and environmental factors in the spatial distribution and popualtion dynamics of a yellow-bellied marmot population. *Ecology* 55: 760–771.

Svenson, H. K. 1935. Plants from the estuary of the Hudson River. *Torreya* 35: 117–125.

Svensson, B. M., B. A. Carlsson, P. S. Karlsson & K. O. Nordell. 1993. Comparative long-term demography of three species of *Pinguicula. J. Ecol.* 81: 635–645.

Swales, D. E. 1979. Nectaries of certain Arctic and sub-Arctic plants with notes on pollination. *Rhodora* 81: 363–407.

Swan, D. C. 2010. The North American lotus (*Nelumbo lutea* Willd Pers.)-sacred food of the Osage people. *Ethnobot. Res. Appl.* 8: 249–253.

Swanson, C. P. & R. Nelson. 1942. Spindle abnormalities in *Mentha. Bot. Gaz.* 104: 273–280.

Swanson, D. R. 2012. An updated synopsis of the Pentatomoidea (Heteroptera) of Michigan. *Great Lakes Entomol.* 45: 263–311.

Sweeney, P. W. & R. A. Price. 2000. Polyphyly of the genus *Dentaria* (Brassicaceae): evidence from *trnL* intron and *ndhF* sequence data. *Syst. Bot.* 25: 468–478.

Sweetman, H. L. 1928. Notes on insects inhabiting the roots of weeds. *Ann. Entomol. Soc. Amer.* 21: 594–600.

Sweetman, H. L. 1944. Selection of woody plants as winter food by the cottontail rabbit. *Ecology* 25: 467–472.

Sweigart, A. L. & J. H. Willis. 2003. Patterns of nucleotide diversity in two species of *Mimulus* are affected by mating system and asymmetric introgression. *Evolution* 57: 2490–2506.

Sweigart, A. L., N. H. Martin & J. H. Willis. 2008. Patterns of nucleotide variation and reproductive isolation between a *Mimulus* allotetraploid and its progenitor species. *Mol. Ecol.* 17: 2089–2100.

Swengel, A. B. 1997. Habitat associations of sympatric violet-feeding fritillaries (*Euptoieta, Speyeria, Boloria*) (Lepidoptera: Nymphalidae) in tallgrass prairie. *Great Lakes Entomol.* 30: 1–18.

Swengel, A. B. & S. R. Swengel. 2010. The butterfly fauna of Wisconsin bogs: lessons for conservation. *Biodivers. Conserv.* 19: 3565–3581.

Swenson, U. & K. Bremer. 1997. Patterns of floral evolution of four Asteraceae genera (Senecioneae, Blennospermatinae) and the origin of white flowers in New Zealand. *Syst. Biol.* 46: 407–425.

Swenson, U. & K. Bremer. 1999. On the circumscription of the Blennospermatinae (Asteraceae, Senecioneae) based on *ndhF* sequence data. *Taxon* 48: 7–14.

Swift, I. 2008. Ecological and biogeographical observations on Cerambycidae (Coleoptera) from California, USA. *Insecta Mundi* 26: 1–7.

Swindale, D. N. & J. T. Curtis. 1957. Phytosociology of the larger submerged plants in Wisconsin lakes. *Ecology* 38: 397–407.

Swindells, P. 1983. *Waterlilies.* Timber Press, Portland, OR. 159 pp.

Swingle, W. T. 1889. First addition to the list of Kansas Peronosporaceae. *Trans. Kansas Acad. Sci.* 12: 129–134.

Swink, F. A. 1952. A phenological study of the flora of the Chicago region. *Amer. Midl. Naturalist* 48: 758–768.

Switzer, P. V. & G. F. Grether. 2000. Characteristics and possible functions of traditional night roosting aggregations in rubyspot damselflies. *Behaviour* 137: 401–416.

Sykes, W. R. 1981. Checklist of dicotyledons naturalised in New Zealand. 7. Scrophulariales. *New Zealand J. Bot.* 19: 53–57.

Sytsma, M. D. & L. W. J. Anderson. 1993. Nutrient limitation in *Myriophyllum aquaticum*. *J. Freshwater Ecol.* 8: 165–176.

Sytsma, K. J., J. Morawetz, C. Pires, M. Nepokroeff, E. Conti, M. Zjhra, J. C. Hall & M. W. Chase. 2002. Urticalean rosids: circumscription, rosid ancestry, and phylogenetics based on *rbcL*, *trnL-F*, and *ndhF* sequences. *Amer. J. Bot.* 89: 1531–1546.

Szczepaniec, A. & M. J. Raupp. 2007. Residual toxicity of imidacloprid to hawthorn lace bug, *Corythuca cydoniae*, feeding on cotoneasters in landscapes and containers. *J. Environ. Hort.* 25: 43–46.

Szpitter, A., M. Narajczyk, M. Maciag-Dorszynska, G. Wegrzyn, E. Lojkowska & A. Krolicka. 2014. Effect of *Dionaea muscipula* extract and plumbagin on maceration of potato tissue by *Pectobacterium atrosepticum*. *Ann. Appl. Biol.* 164: 404–414.

Taboada, M. A., G. Rubio & R. S. Lavado. 1998. The deterioration of tall wheatgrass pastures on saline sodic soils. *J. Range Manag.* 51: 241–246.

Tadesse, M., D. J. Crawford & E. B. Smith. 1996. Generic concepts in *Bidens* and *Coreopsis* (Compositae): an overview. pp. 493–498. *In*: L. J. V. van Maesen et al. (eds.), *The Biodiversdity of African Plants*. Kluwer Academic Publishers, Dordrecht, The Netherlands.

Tadesse, M., D. J. Crawford & S-C. Kim. 2001. A cladistic analysis of morphological features in *Bidens* L. and *Coreopsis* L. (Asteraceae-Heliantheae) with notes on generic delimitation and systematics. *Biol. Skr.* 54: 85–102.

Takagawa, S., J. Nishihiro & I. Washitani. 2005. Safe sites for establishment of *Nymphoides peltata* seedlings for recovering the population from the soil seed bank. *Ecol. Res.* 20: 661–667.

Takagawa, S., I. Washitani, R. Uesugi & Y. Tsumura. 2006. Influence of inbreeding depression on a lake population of *Nymphoides peltata* after restoration from the soil seed bank. *Conserv. Genet.* 7: 705–716.

Takahashi, H. 1988. Ontogenetic development of pollen tetrads of *Drosera capensis* L. *Bot. Gaz.* 149: 275–282.

Takahashi, H. 2001. Infraspecific variation of the floral morphs of *Primula cuneifolia* Ledeb. (Primulaceae) in the Kurils and Hokkaido. International Symposium on Kuril Island Biodiversity, Sapporo, Japan, 19 May, 2001 (abstract).

Takano, A. & Y. Kadono. 2005. Allozyme variations and classification of *Trapa* (Trapaceae) in Japan. *Aquat. Bot.* 83: 108–118.

Takashi, Y. 1999. *Bacopa procumbens* (Mill.) Greenm. naturalized in the Bonin Islands. *J. Jap. Bot.* 74: 256–257.

Takhtajan, A. 1997. *Diversity and Classification of Flowering Plants*. Columbia University Press, New York. 620 pp.

Takos, M. J. 1947. A semi-quantitative study of muskrat food habits. *J. Wildl. Manag.* 11: 331–339.

Talavera, M., P. L. Ortiz, M. Arista, R. Berjano & E. Imbert. 2010. Disentangling sources of maternal effects in the heterocarpic species *Rumex bucephalophorus*. *Perspect. Plant Ecol. Evol. Syst.* 12: 295–304.

Talavera, M., L. Navarro-Sampedro, P. L. Ortiz & M. Arista. 2013. Phylogeography and seed dispersal in islands: the case of *Rumex bucephalophorus* subsp. *canariensis* (Polygonaceae). *Ann. Bot.* 111: 249–260.

Talbot, S. L., S. J. Looman & S. L. Welsh. 1988. *Cicuta bulbifera* L. (Umbelliferae) in Alaska. *Great Basin Naturalist* 48: 382.

Talbot, S. S. & S. L. Talbot. 1994. Numerical classification of the coastal vegetation of Attu Island, Aleutian Islands, Alaska. *J. Veg. Sci.* 5: 867–876.

Talbot, S. S., S. L. Talbot & W. B. Schofield. 1995. Contribution toward an understanding of *Polystichum aleuticum* C. Chr. on Adak Island, Alaska. *Amer. Fern J.* 85: 83–88.

Talbot, S. S., S. L. Talbot, J. W. Thomson & W. B. Schofielde. 2001. Lichens from St. Matthew and St. Paul Islands, Bering Sea, Alaska. *Bryologist* 104: 47–58.

Talbot, S. S., S. L. Talbot, J. W. Thomson, F. J. A. Daniëls & W. B. Schofielde. 2002. Lichens from Simeonof Wilderness, Shumagin Islands, southwestern Alaska. *Bryologist* 105: 111–121.

Talley, T. S. & L. A. Levin. 1999. Macrofaunal succession and community structure in *Salicornia* marshes of southern California. *Estuar. Coast. Shelf Sci.* 49: 713–731.

Tamang, B. 2005. Vegetation and soil quality changes associated with reclaiming phosphate-mine clay settling areas with fast-growing trees. MS thesis. University of Florida, Gainesville, FL. 134 pp.

Tan, B. H. & W. S. Judd. 1995. A floristic inventory of O'Leno State Park and Northeast River Rise State Preserve, Alachua and Columbia Counties, Florida. *Castanea* 60: 141–165.

Tanahara, A. & M. Maki. 2010. Genetic diversity and population genetic differentiation in the endangered annual weed, *Bidens cernua* (Compositae), and two common congeners in Japan. *Weed Biol. Manag.* 10: 113–119.

Tank, D. C. 2002. Subtribe *Castillejinae* revisited (tribe *Rhinantheae*: Orobanchaceae): a molecular phylogenetic analysis. *Bot. Electronic Newslett.* 300.

Tank, D. C. 2006. Molecular phylogenetics of *Castilleja* and subtribe *Castillejinae* (Orobanchaceae). PhD dissertation, University of Washington, Seattle, WA. 166 pp.

Tank, D. C. & M. J. Donoghue. 2010. Phylogeny and phylogenetic nomenclature of the Campanulidae based on an expanded sample of genes and taxa. *Syst. Bot.* 35: 425–441.

Tank, D. C. & R. G. Olmstead. 2008. From annuals to perennials: phylogeny of subtribe Castillejinae (Orobanchaceae). *Amer. J. Bot.* 95: 608–625.

Tank, D. C., P. M. Beardsley, S. A., Kelchner & R. G. Olmstead. 2006. Review of the systematics of Scrophulariaceae s.l. and their current disposition. *Austral. Syst. Bot.* 19: 289–307.

Tank, D. C., J. M. Egger & R. G. Olmstead. 2009. Phylogenetic classification of subtribe Castillejinae (Orobanchaceae). *Syst. Bot.* 34: 182–197.

Tanner, V. M. 1958. Life history notes on *Calligrapha multipunctata multipunctata* (Say) (Coleoptera, Chrysomelidae). *Great Basin Naturalist* 18: 101–103.

Tantaquidgeon, G. 1932. Notes on the origin and uses of plants of the Lake St. John Montagnais. *J. Amer. Folklore* 45: 265–267.

Taranhalli, A. D., A. M. Kadam, S. S. Karale & Y. B. Warke. 2011. Evaluation of antidiarrhoeal and wound healing potentials of *Ceratophyllum demersum* Linn. *Latin Amer. J. Pharm.* 30: 297–303.

Tarbell, D. (ed.). 2004. *Riparian Vegetation and Wetlands Technical Report, Sacramento Municipal Utility District Upper American River Project (FERC project no. 2101) and Pacific Gas and Electric Company Chili Bar Project (FERC project no. 2155)*. Sacramento Municipal Utility District and Pacific Gas & Electric Company, Sacramento, CA. 220 pp.

Tardío, J., M. Molina, L. Aceituno-Mata, M. Pardo-de-Santayana, R. Morales, V. Fernández-Ruiz, P. Morales, P. García, M. Cámara & M. C. Sánchez-Mata. 2011. *Montia fontana* L.

(Portulacaceae), an interesting wild vegetable traditionally consumed in the Iberian Peninsula. *Genet. Resour. Crop Evol.* 58: 1105–1118.

Tarver, D. P., J. A. Rodgers, M. J. Mahler & R. L. Lazor. 1978. *Aquatic and Wetland Plants of Florida.* Bureau of Aquatic Plant Research and Control, Florida Department of Natural Resources, Tallahassee, FL. 127 pp.

Tatic, B. & W. Zukowski. 1973. *Bidens vulgata* Greene in Yugoslavia. *Bull. Inst. Jard. Bot. Univ. Beograd* 8: 125–128.

Taylor, B. R. & S. Raney. 2013. Correlation between ATV tracks and density of a rare plant (*Drosera filiformis*) in a Nova Scotia bog. *Rhodora* 115: 158–169.

Taylor, D., J. MacGregor, V. Bishop & B. Knowles. 1995. *Inventory of Habitat Guilds and Dependant or Strongly Associated Species for PETS Flora & Fauna, Nongame, and Game.* AMS Supplement #5. Daniel Boone National Forest, Winchester, KY. 34 pp.

Taylor, D. S. 2012. Removing the sands (sins?) of our past: dredge spoil removal and saltmarsh restoration along the Indian River Lagoon, Florida (USA). *Wetlands Ecol. Manag.* 20: 213–218.

Taylor, D. W. 2008. Phylogenetic analysis of Cabombaceae and Nymphaeaceae based on vegetative and leaf architectural characters. *Taxon* 57: 1082–1095.

Taylor, H. 1945. Cyto-taxonomy and phylogeny of the Oleaceae. *Brittonia* 5: 337–367.

Taylor, J. 1977. *A Catalog of Vascular Aquatic and Wetland Plants that Grow in Oklahoma.* Publication No 1. Herbarium, Southeastern Oklahoma State University, Durant, OK. 75 pp.

Taylor, K., D. C. Havill, J. Pearson & J. Woodall. 2002. Biological Flora of the British Isles. No. 222, List Br. Vasc. Pl. (1958), no. 371, 1 *Trientalis europaea* L. *J. Ecol.* 90: 404–418.

Taylor, K. L. & J. B. Grace. 1995. The effects of vertebrate herbivory on plant community structure in the coastal marshes of the Pearl River, Louisiana, USA. *Wetlands* 15: 68–73.

Taylor, K. L., J. B. Grace, G. R. Guntenspergen & A. L. Foote. 1994. The interactive effects of herbivory and fire on an oligohaline marsh, Little Lake, Louisiana, USA. *Wetlands* 14: 82–87.

Taylor, N. 2004. Foster flat. *Kalmiopsis* 11: 17–22.

Taylor, P. 1989. *The Genus* Utricularia – *A Taxonomic Monograph.* Royal Botanic Gardens, Kew, London, UK. 724 pp.

Taylor, R. F. 1932. The successional trend and its relation to second-growth forests in southeastern Alaska. *Ecology* 13: 381–391.

Taylor, R. J. 1981. Shoreline vegetation of the Arctic Alaska coast. *Arctic* 34: 37–42.

Taylor, V. L., K. T. Harper & L. L. Mead. 1996. Stem growth and longevity dynamics for *Salix arizonica* Dorn. *Great Basin Naturalist* 56: 294–299.

Teeri, J. A. 1976. Phytotron analysis of a photoperiodic response in a high Arctic plant species. *Ecology* 57: 374–379.

Tehon, L. R., C. C. Morrill & R. Graham. 1946. *Illinois Plants Poisonous to Livestock.* Circular 599. University of Illinois, College of Agriculture, Extension Service in Agriculture and Home Economics, Urbana, IL. 103 pp.

Teketay, D. 1998. The joint role of alternating temperatures and light quality in the germination of *Veronica anagallis-aquatica* and *V. javanica. Trop. Ecol.* 39: 179–184.

Telenius, A. & P. Torstensson. 1989. The seed dimorphism of *Spergularia marina* in relation to dispersal by wind and water. *Oecologia* 80: 206–210.

Telfer, E. S. 1974. Vertical distribution of cervid and snowshoe hare browsing. *J. Wildl. Manag.* 38: 944–946.

Teng, N.-J., Y.-L. Wang, C.-Q. Sun, W.-M. Fang & F.-D. Chen. 2012. Factors influencing fecundity in experimental crosses of water lotus (*Nelumbo nucifera* Gaertn.) cultivars. *BMC Plant Biol.* 12: 82. doi:10.1186/1471-2229-12-82.

Tenorio, R. C. & T. D. Drezner. 2006. Native and invasive vegetation of karst springs in Wisconsin's Driftless area. *Hydrobiologia* 568: 499–505.

Tepfer, D., A. Goldmann, N. Pamboukdjian, M. Maille, A. Lepingle, D. Chevalier, J. Dénarié & C. Rosenberg. 1988. A plasmid of *Rhizobium meliloti* 41 encodes catabolism of two compounds from root exudate of *Calystegium sepium. J. Bacteriol.* 170: 1153–1161.

TEPPC. 2014. Tennessee's native plant alternatives to exotic invasives: a garden and landscape guide. Tennessee Exotic Pest Plant Council. Available online: https://mastergardener.tennessee.edu/2009%20WS/Native_Substitutes.pdf (accessed October 15, 2014).

Teppner, H., F. Ehrendorfer & C. Puff. 1976. Karyosystematic notes on the *Galium palustre*-group (Rubiaceae). *Taxon* 25: 95–97.

Ter Heerdt, G. N. J., G. L. Verweij, R. M. Bekker & J. P. Bakker. 1996. An improved method for seed-bank analysis: seedling emergence after removing the soil by sieving. *Funct. Ecol.* 10: 144–151.

Terrel, T. L. 1972. The swamp rabbit (*Sylvilagus aquaticus*) in Indiana. *Amer. Midl. Naturalist* 87: 283–295.

Terrell, E. E. & W. H. Lewis. 1990. *Oldenlandiopsis* (Rubiaceae), a new genus from the Caribbean basin, based on *Oldenlandia callitrichoides* Grisebach. *Brittonia* 42: 185–190.

Tesky, J. L. 1992. *Salix nigra. In: Fire Effects Information System.* U. S. Department of Agriculture, Forest Service, Rocky Mountain Research Station, Fire Sciences Laboratory (Producer). Available online: http://www.fs.fed.us/database/feis/ (accessed March 15, 2005).

Tessene, M. F. 1969. Systematic and ecological studies on *Plantago cordata. Michigan Bot.* 8: 72–104.

Tester, J. R. 1996. Effects of fire frequency on plant species in oak savanna in east-central Minnesota. *Bull. Torrey Bot. Club* 123: 304–308.

Tewksbury, J. J., D. J. Levey, N. M. Haddad, S. Sargent, J. L. Orrock, A. Weldon, B. J. Danielson, J. Brinkerhoff, E. I. Damschen & P. Townsend. 2002. Corridors affect plants, animals, and their interactions in fragmented landscapes. *Proc. Natl. Acad. Sci. U.S.A.* 99: 12923–12926.

Tewtrakul, S., S. Subhadhirasakul, S. Cheenpracha & C. Karalai. 2007. HIV-1 protease and HIV-1 integrase inhibitory substances from *Eclipta prostrata. Phytother. Res.* 21: 1092–1095.

Thabrew, W. V. de. 1981. *Popular Tropical Aquarium Plants.* Thornhill Press, Cheltenham, UK. 200 pp.

Thakur, V. D. & S. A. Mengi. 2005. Neuropharmacological profile of *Eclipta alba* (Linn.) Hassk. *J. Ethnopharmacol.* 102: 23–31.

Thebeau, L. C. & B. R. Chapman. 1984. Laughing gull nest placement on Little Pelican Island, Galveston Bay. *Southwest. Naturalist* 29: 247–256.

Theis, N., M. J. Donoghue & J. Li. 2008. Phylogenetics of the Caprifolieae and *Lonicera* (Dipsacales) based on nuclear and chloroplast DNA sequences. *Syst. Bot.* 33: 776–783.

Theodose, T. A. & J. B. Roths. 1999. Variation in nutrient availability and plant species diversity across forb and graminoid zones of a Northern New England high salt marsh. *Plant Ecol.* 143: 219–228.

Thien, L. B. 1974. Floral biology of *Magnolia. Amer. J. Bot.* 61: 1037–1045.

Thien, L. B., E. G. Ellagaard, M. S. Devall, S. E. Ellgaard & P. F. Ramp. 1994. Population structure and reproductive biology of *Saururus cernuus* L. (Saururaceae). *Plant Species Biol.* 9: 47–55.

Thien, L. B., H. Azuma & S. Kawano. 2000. New perspectives on the pollination biology of basal angiosperms. *Int. J. Plant Sci.* 161: S225–S235.

Thieret, J. W. 1966. Seeds of some United States Phytolaccaceae and Aizoaceae. *Sida* 2: 352–360.

Thieret, J.W. 1969. *Rumex obovatus* and *Rumex paraguayensis* (Polygonaceae) in Louisiana: new to North America. *Sida* 3: 445.

Thieret, J. W. 1970. *Bacopa repens* (Scrophulariaceae) in the conterminous United States. *Castanea* 35: 132–136.

Thieret, J. W. 1971. Quadrat study of a bottomland forest in St. Martin Parish, Louisiana. *Castanea* 36: 174–181.

Thieret, J. W. 1972. *Rotala indica* (Lythraceae) in Louisiana. *Sida* 5: 45.

Thill, R. E. 1984. Deer and cattle diets on Louisiana pine-hardwood sites. *J. Wildl. Managr.* 48: 788–798.

Thomas, B. & D. Vince-Prue. 1997. *Photoperiodism in Plants. 2nd Ed.* Academic Press, San Diego, CA. 428 pp.

Thomas, D. D. 2002. Propagation protocol for North American pitcher plants (*Sarracenia* L.). *Native Plant J.* 3: 50–53.

Thomas, E. S. 1933. *Neoconocephalus lyristes* (Rehn and Hebard) in the middle west (Tettigoniidae, Orthopt.). *Ann. Entomol. Soc. Amer.* 26: 303–308.

Thomas, E. S. & R. D. Alexander. 1962. Systematic and behavioral studies on the meadow grasshoppers of the *Orchelimum concinnum* group (Orthoptera: Tettigoniidae). *Occ. Pap. Mus. Zool. Univ. Michigan* 626: 1–31.

Thomas, R. B & K. P. Jansen. 2006. *Pseudemys floridana* – Florida Cooter. *Chelonian Res. Monogr.* 3: 338–347.

Thomas, R. D. 1985. A newly discovered habitat for *Isoëtes melanopoda* in Louisiana. *Amer. Fern J.* 75: 77–79.

Thomas, R. D., W. H. Wagner Jr. & M. R. Mesler. 1973. Log fern (*Dryopteris celsa*) and related species in Louisiana. *Castanea* 38: 269–274.

Thommen, G. H. & D. F. Westlake. 1981. Factors affecting the distribution of populations of *Apium nodiflorum* and *Nasturtium officinale* in small chalk streams. *Aquat. Bot.* 11: 21–36.

Thompson, B. C. & R. D. Slack. 1982. Physical aspects of colony selection by least terns on the Texas coast. *Colon. Waterbirds* 5: 161–168.

Thompson, D. Q., R. L. Stuckey & E. B. Thompson. 1987. *Spread, Impact, and Control of Purple Loosestrife (Lythrum salicaria) in North American Wetlands.* Research Report No. 2. USDI, Fish and Wildlife Service, Washington, DC. 55 pp.

Thompson, F. C. 1974. The genus *Pterallastes* Loew (Diptera: Syrphidae). *J. New York Entomol. Soc.* 82: 15–29.

Thompson, F. L., L. A. Hermanutz & D. J. Innes. 1998. The reproductive ecology of island populations of distylous *Menyanthes trifoliata* (Menyanthaceae). *Can. J. Bot.* 76: 818–828.

Thompson, J. T., R. Van Buren & K. T. Harper. 2003. Genetic analysis of the rare species *Salix arizonica* (Salicaceae) and associated willows in Arizona and Utah. *W. N. Amer. Naturalist* 63: 273–282.

Thompson, K. & J. P. Grime. 1979. Seasonal variation in the seed banks of herbaceous species in ten contrasting habitats. *J. Ecol.* 67: 893–921.

Thompson, K. & J. P. Grime. 1983. A comparative study of germination responses to diurnally-fluctuating temperatures. *J. Appl. Ecol.* 20: 141–156.

Thompson, K.A., D. M. Sora, K. S. Cross, J. M. St. Germain & K. Cottenie. 2014. Mucilage reduces leaf herbivory in Schreber's watershield, *Brasenia schreberi* JF Gmel. (Cabombaceae). *Botany* 92: 412–416.

Thompson, P. A. 1969a. Germination of *Lycopus europaeus* L. in response to fluctuating temperatures and light. *J. Exp. Bot.* 20: 1–11.

Thompson, P. A. 1969b. Germination of species of Labiatae in response to gibberellins *Physiol. Plant (Copenhagen)* 22: 575–586.

Thompson, R. L. 2001. *Botanical Survey of Myrtle Island Research Natural Area, Oregon.* Gen. Tech. Rep. PNW-GTR-507. U. S. Department of Agriculture, Forest Service, Pacific Northwest Research Station, Portland, OR. 27 pp.

Thompson, R. L. & S. R. Green. 2010. Vascular plants of an abandoned limestone quarry in Garrard County, Kentucky. *Castanea* 75: 245–258.

Thompson, R. L. & L. E. McKinney. 2006. Vascular flora and plant habitats of an abandoned limestone quarry at Center Hill Dam, DeKalb County, Tennessee. *Castanea* 71: 54–64.

Thompson, R. L., R. L. Jones, J. R. Abbott & W. N. Denton. 2000. *Botanical Survey of Rock Creek Research Natural Area, Kentucky.* General Technical Report NE-272. USDA Forest Service, Newtown Square, PA. 23 pp.

Thompson, V. & N. Mohd-Saleh. 1995. Spittle maggots: studies on *Cladochaeta* fly larvae living in association with *Clastoptera* spittlebug nymphs. *Amer. Midl. Naturalist* 134: 215–225.

Thomsen, C. D., W. A. Williams, M. Vayssiéres, F. L. Bell & M. George. 1993. Controlled grazing on annual grassland decreases yellow starthistle. *California Agric.* 47: 36–40.

Thomson, J. D. 1980. Skewed flowering distributions and pollinator attraction. *Ecology* 61: 572–579.

Thomson, J. D. 1981a. Spatial and temporal components of resource assessment by flower-feeding insects. *J. Anim. Ecol.* 50: 49–59.

Thomson, J. D. 1981b. Field measures of flower constancy in bumblebees. *Amer. Midl. Naturalist* 105: 377–380.

Thomson, J. D. 1982. Patterns of visitation by animal pollinators. *Oikos* 39: 241–250.

Thor, G. 1988. The genus *Utricularia* in the Nordic countries, with special emphasis on *U. stygia* and *U. ochroleuca*. *Nordic J. Bot.* 8: 213–225.

Thorén, L. M., J. Tuomi, T. Kämäräinen & K. Laine. 2003. Resource availability affects investment in carnivory in *Drosera rotundifolia*. *New Phytol.* 159: 507–511.

Thorp, F. Jr & A. W. Deem. 1939. *Suckleya suckleyana* a poisonous plant. *J. Amer. Vet. Med. Assoc.* 94: 192–197.

Thorp, R. W. 1979. Structural, behavioral, and physiological adaptations of bees (Apoidea) for collecting pollen. *Ann. Missouri Bot. Gard.* 66: 788–812.

Thorp, R. W. 2014. Vernal pool flowers and their specialist bee pollinators. California vernal pools. Available online: http://www.vernalpools.org/Thorp/ (accessed March 11, 2014).

Thorp, R. W. & W. E. LaBerge. 2005. A revision of the bees of the genus *Andrena* of the Western Hemisphere. Part XV—Subgenus *Hesperandrena*. *Illinois Nat. Hist. Surv. Bull.* 37: 65–94.

Thorp, R. W. & J. M. Leong. 1995. *Determining Effective Mitigation Techniques for Vernal Pool Wetlands: Effect of Host-Specific Pollinators on Vernal Pool Plants.* California Department of Transportation, Sacramento, CA. 114 pp.

Thorp, R. W. & J. M. Leong. 1998. Specialist bee pollinators of showy vernal pool flowers. pp. 169–179. *In:* C.W. Witham, E. T. Bauder, D. Belk, W. R. Ferren Jr. & R. Orduff (eds.), *Ecology, Conservation, and Management of Vernal Pool Ecosystems-Proceedings from a 1996 Conference.* California Native Plant Society, Sacramento, CA.

Thorp, R. W., D. S. Horning & L. L. Dunning. 1983. Bumble bees and cuckoo bumble bees of California (Hymenoptera, Apidae). *Bull. Calif. Insect Surv.* 23: 1–79.

Thorsen, M. J., K. J. M. Dickinson & P. J. Seddon. 2009. Seed dispersal systems in the New Zealand flora. *Perspect. Plant Ecol. Evol. Syst.* 11: 285–309.

Thulin, M. & B. Bremer. 2004. Studies in the tribe Spermacoceae (Rubiaceae-Rubioideae): the circumscriptions of *Amphiasma* and *Pentanopsis* and the affinities of *Phylohydrax*. *Plant Syst. Evol.* 247: 233–239.

Thum, M. 1986. The significance of opportunistic predators for the sympatric carnivorous plant species *Drosera intermedia* and *Drosera rotundifolia*. *Oecologia* 81: 397–400.

Thunhorst, G. A. 1995. The seed bank as a buffer to change in population size, distribution, and genetic composition in the rare plant *Rhexia aristosa* at Antioch Church Bay with implications to conservation. MA thesis. University of North Carolina at Chapel Hill, Chapel Hill, NC. 85pp.

Tidestrom, I. 1913. *Sphenoclea zeylanica* and *Caperonia palustris* in the southern United States. *Contr. U. S. Natl. Herb.* 16: 305–307.

Tikhodeev, O. N. & M. Y. Tikhodeeva. 2001. Variability of the flower structure in European starflower (*Trientalis europaea* L.) in natural populations. *Russ. J. Ecol. Vol.* 32: 206–210.

Tilman, D. 1997. Community invasibility, recruitment limitation, and grassland biodiversity. *Ecology* 78: 81–92.

Tilly, L. J. 1968. The structure and dynamics of Cone Spring. *Ecol. Monogr.* 38: 169–197.

Timberlake, P. H. 1937. New anthophorid bees from California. *Amer. Mus. Novitat.* 958: 1–17.

Timberlake, P. H. 1939. New species of bees of the genus *Dufourea* from California (Hymenoptera, Apoidea). *Ann. Entomol. Soc. Amer.* 32: 395–414.

Timberlake, P. H. 1941. Three new dufoureine bees from California (Hymenoptera, Apoidea). *Ann. Entomol. Soc. Amer.* 34: 38–42.

Timberlake, P. H. 1951. New species of *Anthophora* from the western United States (Hymenoptera, Apoidea). *J. New York Entomol. Soc.* 59: 51–62.

Timme, R. E. & I. De Geofroy. 2001. A molecular phylogeny of the genus *Polemonium* (Polemoniaceae) using ITS sequence[s]. Botany 2001 meeting (abstract).

Timme, R. E. & R. Patterson. 2001. The molecular phylogeny of *Polemonium* (Polemoniaceae). Thesis. San Francisco State University, San Francisco, CA.

Timme, R. E., B. B. Simpson & C. R. Linder. 2007. High-resolution phylogeny for *Helianthus* (Asteraceae) using the 18S-26S ribosomal DNA external transcribed spacer. *Amer. J. Bot.* 94: 1837–1852.

Timoney, K. P. 2001. String and net-patterned salt marshes: rare landscape elements of boreal Canada. *Can. Field Naturalist* 115: 406–412.

Tiner, R. W., Jr. 1987. *A Field Guide to Coastal Wetland Plants of the Northeastern United States*. The University of Massachusetts Press, Amherst, MA. 286 pp.

Tiner, R. W. 1993a. *Field Guide to Coastal Wetland Plants of the Southeastern United States*. The University of Massachusetts Press, Amherst, MA. 328 pp.

Tiner, R. W. 1993b. Using plants as indicators of wetland. *Proc. Acad. Nat. Sci. Philadelphia* 144: 240–253.

Tiner, R. W. 1999. *Wetland Indicators: A Guide to Wetland Identification, Delineation, Classification, and Mapping*. Lewis Publishers, Boca Raton, FL. 392 pp.

Tiner, R. W. 2009. *Field Guide to Tidal Wetland plants of the Northeastern United States and Neighboring Canada: Vegetation of Beaches, Tidal Flats, Rocky Shores, Marshes, Swamps, and Coastal Ponds*. University of Massachusetts Press, Amherst, MA. 459 pp.

Tinsley, B. 2012. The ecological roles of *Podostemum ceratophyllum* and *Cladophora* in the habitat and dietary preferences of the riverine caddisfly *Hydropsyche simulans*. Honors College Capstone Experience/Thesis Projects. Paper 359. Western Kentucky University, Bowling Green, KY. 43 pp.

Tippery, N. P. & D. H. Les. 2008. Phylogenetic analysis of the internal transcribed spacer (ITS) region in Menyanthaceae using predicted secondary structure. *Mol. Phylogen. Evol.* 49: 526–537.

Tippery, N. P. & D. H. Les. 2009. A new genus and new combinations in Australian *Villarsia* (Menyanthaceae). *Novon* 19: 404–411.

Tippery, N. P. & D. H. Les. 2011. Phylogenetic relationships and morphological evolution in *Nymphoides* (Menyanthaceae). *Syst. Bot.* 36: 1101–1113.

Tippery, N. P. & D. H. Les. 2013. Hybridization and systematics of dioecious North American *Nymphoides* (*N. aquatica* and *N. cordata*; Menyanthaceae). *Aquat. Bot.* 104: 127–137.

Tippery, N. P., D. H. Les, D. J. Padgett & S. W. L. Jacobs. 2008. Generic circumscription in Menyanthaceae: a phylogenetic approach. *Syst. Bot.* 33: 598–612.

Tippery, N. P., D. H. Les, J. C. Regalado Jr., L. V. Averyanov, V. N. Long & P. H. Raven. 2009. Transfer of *Villarsia cambodiana* to *Nymphoides* (Menyanthaceae). *Syst. Bot.* 34: 818–823.

Tippery, N. P., D. H. Les & C. R. Williams. 2011. *Nymphoides humboldtiana* (Menyanthaceae) in Uvalde County, Texas: a new record for the U.S.A. *J. Bot. Res. Inst. Texas* 5: 889–890.

Tippery, N. P., E. E. Schilling, J. L. Panero, D. H. Les & C. S. Williams. 2014. Independent origins of aquatic Eupatorieae (Asteraceae). *Syst. Bot.* 39: 1217–1225.

Tippery, N. P., D. H. Les & E. Peredo. 2015. *Nymphoides grayana* (Menyanthaceae) in Florida verified by DNA and morphological data. *J. Torrey Bot. Soc.* 142: 325–330.

Tipping, P. W. & G. Campobasso. 1997. Impact of *Tyta luctuosa* (Lepidoptera: Noctuidae) on hedge bindweed (*Calystegia sepium*) in corn (*Zea mays*). *Weed Technol.* 11: 731–733.

Tirmenstein, D. 1987a. *Potentilla glandulosa. In*: Fire Effects Information System. U. S. Department of Agriculture, Forest Service, Rocky Mountain Research Station, Fire Sciences Laboratory (Producer). Available online: http://www.fs.fed.us/database/feis/ (accessed February 10, 2006).

Tirmenstein, D. A. 1987b. *Potentilla newberryi. In*: Fire Effects Information System. U. S. Department of Agriculture, Forest Service, Rocky Mountain Research Station, Fire Sciences Laboratory (Producer). Available online: http://www.fs.fed.us/database/feis/ (accessed February 10, 2006).

Titus, B. D., S. S. Sidhu & A. U. Mallik. 1995. *A Summary of Some Studies on Kalmia angustifolia L.: A Problem Species in Newfoundland Forestry*. Information Report N-X-296. Natural Resources Canada, Canadian Forest Service, Newfoundland & Labrador Region, St. John's, Newfoundland. 68 pp.

Titus, J. E. & P. G. Sullivan. 2001. Heterophylly in the yellow waterlily, *Nuphar variegata* (Nymphaeaceae): effects of [CO_2], natural sediment type, and water depth. *Amer. J. Bot.* 88: 1469–1478.

Titus, J. H. 1990. Microtopography and woody plant regeneration in a hardwood floodplain swamp in Florida. *Bull. Torrey Bot. Club* 117: 429–437.

Titus J. H. 1991. Seed bank of a hardwood floodplain swamp in Florida. *Castanea* 56: 117–127.

Titus J. H. & P. J. Titus. 2008a. Assessing the reintroduction potential of the endangered Huachuca water umbel in southeastern Arizona. *Ecol. Restorat. N. Amer.* 26: 311–320.

Titus J. H. & P. J. Titus. 2008b. Ecological monitoring of the endangered Huachuca water umbel (*Lilaeopsis schaffneriana* ssp. *recurva*: Apiaceae). *Southwest. Naturalist* 53: 458–465.

Titus J. H. & P. J. Titus. 2008c. Seedbank of Bingham Cienega, a spring-fed marsh in southeastern Arizona. *Southwest. Naturalist* 53: 393–399.

Tluczek, M. 2012. Diet, nutrients, and free water requirements of pronghorn antelope on Perry Mesa, Arizona. MS thesis, Arizona State University, Tempe, AZ. 118 pp.

Tkach, N., M. Röser & M. H. Hoffmann. 2015. Molecular phylogenetics, character evolution and systematics of the genus *Micranthes* (Saxifragaceae). *Bot. J. Linn. Soc.* 178: 47–66.

Tobe, H. 2011. Embryological evidence supports the transfer of *Leitneria floridana* to the family Simaroubaceae. *Ann. Missouri Bot. Gard.* 98: 277–293.

Tobe, H., P. H. Raven & S. A. Graham. 1986. Chromosome counts for some Lythraceae sens. str. (Myrtales), and the base number of the family. *Taxon* 35: 13–20.

Tobe, J. D., K. C. Burks, R. W. Cantrell, M. A. Garland, M. E. Sweeley, D. W. Hall, P. Wallace, G. Anglin, G. Nelson, J. R. Cooper, D. Bickner, K. Gilbert, N. Aymond, K. Greenwood & N. Raymond. 1998. *Florida Wetland Plants: An Identification Manual*. University of Florida, Institute of Food & Agricultural Sciences, Gainesville, FL. 598 pp.

Todd, J. B. 1927. Winter food of cottontail rabbits. *J. Mammalogy* 8: 222–228.

Tofern, B., P. Mann, M. Kaloga, K. Jenett-Siems, L. Witte & E. Eich. 1999. Aliphatic pyrrolidine amides from two tropical convolvulaceous species. *Phytochemistry* 52: 1437–1441.

Toft, J., J. Cordell & C. Simenstad. 1999. More non-indigenous species? First records of one amphipod and two isopods in the delta. *IEP Newslett.* 12(4): 35–37.

Tokuoka T. 2008. Molecular phylogenetic analysis of Violaceae (Malpighiales) based on plastid and nuclear DNA sequences. *J. Plant Res.* 121: 253–260.

Tokuoka, T. & H. Tobe. 2006. Phylogenetic analyses of Malpighiales using plastid and nuclear DNA sequences, with particular reference to the embryology of Euphorbiaceae sens. str. *J. Plant Res.* 119: 599–616.

Tölken, H. R. 1967. The species of *Arthrocnemum* and *Salicornia* (Chenopodiaceae) in Southern Africa. *Bothalia* 9: 255–307.

Tollsten, L., J. T. Knudsen & L. G. Bergström. 1994. Floral scent in generalistic *Angelica* (Apiaceae): an adaptive character? *Biochem. Syst. Ecol.* 22: 161–169.

Tolstead, W. L. 1942. A note on unusual plants in the flora of northwestern Nebraska. *Amer. Midl. Naturalist* 28: 475–481.

Tomar, S. K. & R. L. Sharma. 2002. Fodders and feeding practices of cattle and sheep in Kashmir (India). *Trop. Agric. Res. Extens.* 5: 48–52.

Tomczykowa, M., M. Tomczyk, P. Jakoniuk & E. Tryniszewska. 2008. Antimicrobial and antifungal activities of the extracts and essential oils of *Bidens tripartita*. *Folia Histochem. Cytobiol.* 46: 389–393.

Tomczykowa, M., K. Leszczyńska, M. Tomczyk, E. Tryniszewska & D. Kalemba. 2011. Composition of the essential oil of *Bidens tripartita* L. roots and its antibacterial and antifungal activities. *J. Med.* 14: 428–433.

Tomlinson, P. B. 1986. *The Botany of Mangroves*. Cambridge University Press, Cambridge, MA. 441 pp.

Tompkins, R. D., W. C. Stringer, K. Richardson, E. A. Mikhailova & W. C. Bridges Jr. 2010. A newly documented and significant Piedmont prairie site with a *Helianthus schweinitzii* Torrey & A. Gray (Schweinitz's sunflower) population. *J. Torrey Bot. Soc.* 137: 120–129.

Toneatto, F., T. P. Hauser, J. K. Nielsen & M. Ørgaard. 2012. Genetic diversity and similarity in the *Barbarea vulgaris* complex (Brassicaceae). *Nordic J. Bot.* 30: 506–512.

Tooker, J. F. & L. M. Hanks. 2000. Flowering plant hosts of adult hymenopteran parasitoids of central Illinois. *Ann. Entomol. Soc. Amer.* 93: 580–588.

Tooker, J. F., P. F. Reagel & L. M. Hanks. 2002. Nectar sources of day-flying Lepidoptera of central Illinois. *Ann. Entomol. Soc. Amer.* 95: 84–96.

Tooker, J. F., M. Hauser & L. M. Hanks. 2006. Floral host plants of Syrphidae and Tachinidae (Diptera) of central Illinois. *Ann. Entomol. Soc. Amer.* 99: 96–112.

Tooley, P. W. & M. Browning. 2009. Susceptibility to *Phytophthora ramorum* and inoculum production potential of some common Eastern forest understory plant species. *Plant Dis.* 93: 249–256.

Topić, J. & L. Ilijanić. 2003. *Veronica peregrina* L. and *Veronica scardica* Griseb. (Scrophulariaceae), new species in Croatian flora. *Natura Croatica* 12: 253–258.

Topinka, J. R., A. J. Donovan & B. May. 2004. Characterization of microsatellite loci in Kearney's bluestar (*Amsonia kearneyana*) and cross-amplification in other *Amsonia* species. *Mol. Ecol. Notes* 4: 710.

Torimaru, T. & N. Tomaru. 2005. Fine-scale clonal structure and diversity within patches of a clone-forming dioecious shrub, *Ilex leucoclada* (Aquifoliaceae). *Ann. Bot.* 95: 295–304.

Toriyama, K., K. L. Heong & B. Hardy (eds.). 2005. *Rice is Life: Scientific Perspectives for the 21st Century*. International Rice Research Institute, Manila, Philippines. 590 pp.

Torres, C. & L. Galetto. 2002. Are nectar sugar composition and corolla tube length related to the diversity of insects that visit Asteraceae flowers? *Plant Biol.* 4: 360–366.

Torrey, J. 1853. On the *Darlingtonia californica*, a new pitcher-plant, from northern California. *Smithsonian Contr. Knowl.* 6: 1–8.

Tóth, B. H., B. Blazics & Á. Kéry. 2009. Polyphenol composition and antioxidant capacity of *Epilobium* species. *J. Pharm. Biomed. Analysis* 49: 26–31.

Tóth, K., D. Bogyó & O. Valkó. 2016. Endozoochorous seed dispersal potential of grey geese *Anser* spp. in Hortobágy National Park, Hungary. *Plant Ecol.* 217: 1015–1024.

Toth, L. A. 2005. Plant community structure and temporal variability in a channelized subtropical floodplain. *S. E. Naturalist (Humboldt)* 4: 393–408.

Townsend, C. C. 1993. Amaranthaceae. pp. 70–91. In: K. Kubitzki, J. Rohwer, and V. Bittrich (eds.), *The Families and Genera of Vascular Plants, Vol. II. Flowering Plants, Dicotyledons: Magnoliid, Hamamelid and Caryophyllid Families*. Springer-Verlag, Berlin, Germany.

Townsend, P. A. 2001. Relationships between vegetation patterns and hydroperiod on the Roanoke River floodplain, North Carolina. *Plant Ecol.* 156: 43–58.

Tracy, B. F. & M. A. Sanderson. 2000. Seedbank diversity in grazing lands of the northeast United States. *J. Range Manag.* 53: 114–118.

Tracy, S. M. & F. S. Earle. 1896. New species of Fungi from Mississippi. *Bull. Torrey Bot. Club* 23: 205–211.

Tracy, S. M. & F. S. Earle. 1899. New Fungi from Mississippi. *Bull. Torrey Bot. Club* 26: 493–495.

Transeau, E. N. 1903. On the geographic distribution and ecological relations of the bog plant societies of northern North America. *Bot. Gaz.* 36: 401–420.

Transeau, E. N. 1905. The bogs and bog flora of the Huron River valley (continued). *Bot. Gaz.* 40: 418–448.

Trappey, A. F., C. E. Johnson & P. W. Wilson. 2007. Consumer acceptance of mayhaw (*Crataegus opaca* Hook, and Arn.) juice blended with muscadine grape (*Vitis rotundifolia* Michx.) juice. *Int. J. Fruit Sci.* 6: 53–65.

Traveset, A., M. F. Willson & M. Verdú. 2004. Characteristics of fleshy fruits in southeast Alaska: phylogenetic comparison with fruits from Illinois. *Ecography* 27: 41–48.

Tremlett, M., J. W. Silvertown & C. Tucker. 1984. An analysis of spatial and temporal variation in seedling survival of a monocarpic perennial, *Conium maculatum. Oikos* 43: 41–45.

Tremolieres, M., R. Carbiener, A. Ortscheit & J.-P. Klein. 1994. Changes in aquatic vegetation in Rhine floodplain streams in Alsace in relation to disturbance. *J. Veg. Sci.* 5: 169–178.

Trefry, S. A. & D. S. Hik. 2009. Eavesdropping on the neighbourhood: collared pika (*Ochotona collaris*) responses to playback calls of conspecifics and heterospecifics. *Ethology* 115: 928–938.

Trial, H. & J. B. Dimond. 1979. Emodin in buckthorn: a feeding deterrent to phytophagous insects. *Can. Entomol.* 111: 207–212.

Trift, I., M. Källersjo & A. Anderberg. 2002. The monophyly of *Primula* (Primulaceae) evaluated by analysis of sequences from the chloroplast gene *rbcL. Syst. Bot.* 27: 396–407.

Tripp, E. A. 2007. Ruellieae (Acanthaceae). Version 17 February 2007. *In: The Tree of Life Web Project*. Available online: http://tolweb.org/Ruellieae/.

Trock, D. K. 2006. 216. *Packera* Á. Löve & D. Löve. pp. 570–602. *In*: Flora of North America Editorial Committee (eds.), *Flora of North America North of Mexico: Vol. 20, Magnoliophyta: Asteridae (in part): Asteraceae, Part 2*. Oxford University Press, New York.

Tröndle, D., S. Schröder, H.-H. Kassemeyer, C. Kiefer, M. A. Koch & P. Nick. 2010. Molecular phylogeny of the genus *Vitis* (Vitaceae) based on plastid markers. *Amer. J. Bot.* 97: 1168–1178.

Trowbridge, W. B. 2007. The role of stochasticity and priority effects in floodplain restoration. *Ecol. Appl.* 17: 1312–1324.

Troy, M. R. & D. E. Wimber. 1968. Evidence for a constancy of the DNA synthetic period between diploid-polyploid groups in plants. *Exp. Cell Res.* 53: 145–154.

Trucco, F., M. R. Jeschke, A. L. Rayburn & P. J. Tranel. 2005. *Amaranthus hybridus* can be pollinated frequently by *A. tuberculatus* under field conditions. *Heredity* 94: 64–70.

Trudeau, N. C., M. Garneau & L. Pelletier. 2013. Methane fluxes from a patterned fen of the northeastern part of the La Grande river watershed, James Bay, Canada. *Biogeochemistry* 113: 409–422.

Trumble, J. T., W. Dercks, C. F. Quiros & R. C. Beier. 1990. Host plant resistance and linear furanocoumarin content of *Apium* accessions. *J. Econ. Entomol.* 83: 519–525.

Truscott, A-M., C. Soulsby, S. C. F. Palmer, L. Newell & P. E. Hulme. 2006. The dispersal characteristics of the invasive plant *Mimulus guttatus* and the ecological significance of increased occurrence of high-flow events. *J. Ecol.* 94: 1080–1091.

Tryon, C. A. & N. W. Easterly. 1975. Plant communities of the Irwin Prairie and adjacent wooded areas. *Castanea* 40: 201–213.

Trzcinski, M. K., S. J. Walde & P. D. Taylor. 2003. Colonisation of pitcher plant leaves at several spatial scales. *Ecol. Entomol.* 28: 482–489.

Turesson, G. 1914. Slope exposure as a factor in the distribution of *Pseudotsuga taxifolia* in arid parts of Washington. *Bull. Torrey Bot. Club* 41: 337–345.

Turner, J. A. 1928. Relation of the distribution of certain Compositae to the hydrogen-ion concentration of the soil. *Bull. Torrey Bot. Club* 55: 199–213.

Tsang, A. C. W. & R. T. Corlett. 2005. Reproductive biology of the *Ilex* species (Aquifoliaceae) in Hong Kong, China. *Can. J. Bot.* 83: 1645–1654.

Tseng, C. F., S. Iwakami, A. Mikajiri, M. Shibuya, F. Hanaoka, Y. Ebizuka, K. Padmawinata & U. Sankawa. 1992. Inhibition of in vitro prostaglandin and leukotriene biosyntheses by cinnamoyl-beta-phenethylamine and N-acyldopamine derivatives. *Chem. Pharm. Bull. (Tokyo)* 40: 396–400.

Tsuchiya, T. 1991. Leaf life span of floating-leaved plants. *Vegetatio* 97: 149–160.

Tsuda, Y. & L. Marion. 1963. The alkaloids of *Eupatorium maculatum* L. *Can. J. Chem.* 41: 1919–1923.

Tsukaya, H., T. Fukuda & J. Yokoyama. 2003. Hybridization and introgression between *Callicarpa japonica* and *C. mollis* (Verbenaceae) in central Japan, as inferred from nuclear and chloroplast DNA sequences. *Mol. Ecol.* 12: 3003–3011.

Tsuyuzaki, S. & C. Miyoshi. 2009. Effects of smoke, heat, darkness and cold stratification on seed germination of 40 species in a cool temperate zone in northern Japan. *Plant Biol.* 11: 369–378.

Tsuyuzaki, S., J. H. Titus & R. del Moral. 1997. Seedling establishment patterns on the pumice plain, Mount St. Helens, Washington. *J. Veg. Sci.* 8: 727–734.

Tucker, A., M. Maciarello & D. McCrory. 1992. The essential oil of *Micromeria brownei* (Swartz) Benth. var. *pilosiuscula* Gray. *J. Essential Oil Res.* 4: 301–302.

Tucker, A. O. & H. L. Chambers. 2002. *Mentha canadensis* L. (Lamiaceae): A relict amphidiploid from the lower Tertiary. *Taxon* 51: 703–718.

Tucker, A. O. & N. H. Dill. 1989. Nomenclature and distribution of *Eupatorium* ×*truncatum*, with comments on the status of *E. resinosum* var. *kentuckiense* (Asteraceae). *Castanea* 54: 43–48.

Tucker, A. O. & D. E. Fairbrothers. 1981. A euploid series in an F_1 interspecific hybrid progeny of *Mentha* (Lamiaceae). *Bull. Torrey Bot. Club* 108: 51–53.

Tucker, A. O. & D. E. Fairbrothers. 1990. The origin of *Mentha* ×*gracilis* (Lamiaceae). I. Chromosome numbers, fertility, and three morphological characters. *Econ. Bot.* 44: 183–213.

Tucker, A. O., N. H. Dill, T. D. Pizzolato & R. D. Kral. 1983. Nomenclature, distribution, chromosome numbers, and fruit morphology of *Oxypolis canbyi* and *O. filiformis* (Apiaceae). *Syst. Bot.* 8: 299–304.

Tucker, G. C. 1986. The genera of Elatinaceae in the southeastern United States. *J. Arnold Arbor.* 67: 471–483.

Tucker, G. C. 2010. 5. Limnanthaceae R. Brown. p. 172. *In*: Flora of North America Editorial Committee (eds.), *Flora of North America North of Mexico: Vol. 7, Magnoliophyta: Salicaceae to Brassicaceae*. Oxford University Press, New York.

Tucker, G. C. 2016a. Elatinaceae Dumortier. Waterwort family. p. 348. *In*: Flora of North America Editorial Committee (eds.), *Flora of North America North of Mexico*: Vol. 12: Magnoliophyta: Vitaceae to Garryaceae. Oxford University Press, New York.

Tucker, G. C. 2016b. *Bergia* Linaneus. p. 349. *In*: Flora of North America Editorial Committee (eds.), *Flora of North America North of Mexico*: Vol. 12: Magnoliophyta: Vitaceae to Garryaceae. Oxford University Press, New York.

Tuell, J. K., A. K. Fiedler, D. Landis & R. Isaacs. 2008. Visitation by wild and managed bees (Hymenoptera: Apoidea) to eastern U. S. native plants for use in conservation programs. *Environ. Entomol.* 37: 707–718.

Tullock, J. H. 2006. *Freshwater Aquarium Models: Recipes for Creating Beautiful Aquariums That Thrive.* Wiley Publishing, Inc., Hoboken, NJ. 280 pp.

Tumlison, B. & B. Serviss. 2007. Discovery of a second record of seaside heliotrope. (*Heliotropium curassavicum* L.) (Boraginaceae) in Arkansas. *J. Arkansas Acad. Sci.* 61: 137–139.

Tumosa, C. S. 1977. A lectin in *Nemopanthus mucronatus* to papain treated porcine erythrocytes. *Cell Mol. Life Sci.* 33: 1531.

Tunnell, J. W. Jr., B. Hardegree & D. W. Hicks. 1995. Environmental impact and recovery of a high marsh pipeline oil spill and burn site, Upper Copano Bay, TX. pp. 133–138. *In: Proceedings of the 1995 Oil Spill Conference.* Long Beach, CA.

Tunón, H. & L. Bohlin. 1995. Anti-inflammatory studies on *Menyanthes trifoliata* related to the effect shown against renal failure in rats. *Phytomedicine* 2: 103–112.

Tunón, H., L. Bohlin & G. Öjteg. 1994. The effect of *Menyanthes trifoliata* L. on acute renal failure might be due to PAF-inhibition. *Phytomedicine* 1: 39–45.

Turesson, G. 1916. *Lysichiton camtschatcense* (L) Schott, and its behavior in sphagnum bogs. *Amer. J. Bot.* 3: 189–209.

Turki, Z. & M. Sheded. 2002. Some observations on the weed flora of rice fields in the Nile Delta, Egypt. *Feddes Repert.* 113 394–403.

Turnbull, C. L., A. J. Beattie & F. M. Hanzawa. 1983. Seed dispersal by ants in the Rocky Mountains. *Southwest. Naturalist* 28: 289–293.

Turner, B. L. 2014. Taxonomic overview of *Eustoma* (Gentianaceae). *Phytologia* 96: 7–11.

Turner, B. L. & C. C. Cowan. 1993. Taxonomic overview of *Stemodia* (Scrophulariaceae) for North America and the West Indies. *Phytologia* 74: 61–103.

Turner, B. L. & M. W. Turner. 2003. *Triglochin concinna* (Juncaginaceae), a new family, genus, and species for Texas. *Sida* 20: 1721–1722.

Turner, C. E., R. W. Pemberton & S. S. Rosenthal. 1987a. Host range and new host records for the plume moth *Platyptilia carduidactyla* (Lepidoptera: Pterophoridae) from California thistles (Asteraceae). *Proc. Entomol. Soc. Washington* 89: 132–136.

Turner, C. E., R. W. Pemberton & S. S. Rosenthal. 1987b. Host utilization of native *Cirsium* thistles (Asteraceae) by the introduced weevil *Rhinocyllus conicus* (Coleoptera: Curculionidae) in California. *Environ. Entomol.* 16: 111–115.

Turner, J. A. 1928. Relation of the distribution of certain Compositae to the hydrogen-ion concentration of the soil. *Bull. Torrey Bot. Club* 55: 199–213.

Turner, L. M. 1934. Grassland in the floodplain of Illinois rivers. *Amer. Midl. Naturalist* 15: 770–780.

Turner, L. M. 1936. Ecological studies in the lower Illinois river valley. *Bot. Gaz.* 97: 689–727.

Turner, M. W. 1996. Systematic study of the genus *Brazoria* (*Lamiaceae*), and *Warnockia* (*Lamiaceae*), a new genus from Texas. *Plant Syst. Evol.* 203: 65–82.

Turner, N. J. 1995. *Food Plants of Coastal First Peoples.* University of British Columbia Press, Vancouver, BC. 164 pp.

Turner, N. J. 1998. *Plant Technology of First Peoples in British Columbia.* University of British Columbia Press, Vancouver, BC. 288 pp.

Turner, N. J. & M. A. M. Bell. 1971. The ethnobotany of the Coast Salish Indians of Vancouver Island, I and II. *Econ. Bot.* 25: 63–104; 335–339.

Turner, N. J. & H. V. Kuhnlein. 1982. Two important "root" foods of the Northwest Coast Indians: springbank clover (*Trifolium wormskioldii*) and Pacific silverweed (*Potentilla anserina* ssp. *pacifica*). *Econ. Bot.* 36: 411–432.

Turner, N. J., J. Thomas, B. Carlson & R. Ogilvie. 1983. *Ethnobotany of the Nitinaht Indians of Vancouver Island.* Occasional Papers No. 24. British Columbia Provincial Museum, Victoria, BC.

Turner, N. J., I. J. Davidson-Hunt & M. O'Flaherty. 2003. Living on the edge: ecological and cultural edges as sources of diversity for social-ecological resilience. *Human Ecol.* 31: 439–461.

Turner, S. D., J. P. Amon, R. M. Schneble & C. F. Friese. 2000. Mycorrhizal fungi associated with plan ts in ground-water fed wetlands. *Wetlands* 20: 200–204.

Tuskan, G. A., L. E. Gunter, Z. K. Yang, T-M. Yin, M. M. Sewell & S. P. DiFazio. 2004. Characterization of microsatellites revealed by genomic sequencing of *Populus trichocarpa*. *Can. J. Forest Res.* 34: 85–93.

Tutin, T. G., V. H. Heywood, N. A. Burges, D. H. Valentine, S. M. Walters & D. A. Webb. 1964. *Flora Europaea. Volume I: Lycopodiaceae to Platanaceae.* Cambridge University Press, Cambridge, UK. 496 pp.

Tuttle, D. M. 1954. Notes on the bionomics of six species of *Apion* (Curculionidae, Coleoptera). *Ann. Entomol. Soc. Amer.* 47: 301–307.

Tuttle, S. E. & D. M. Carroll. 2005. Movements and behavior of hatchling wood turtles (*Glyptemys insculpta*). *N. E. Naturalist* 12: 331–348.

Tyler-Julian, K., O. Demirozer, R. Figlar, T. Skarlinsky, J. Funderburk & G. Knox. 2012. New host records for *Caliothrips striatus* (Thysanoptera: Thripidae) on *Magnolia* spp. with the description of the second instar. *Florida Entomol.* 95: 485–488.

Tymon, L. S., D. A. Glawe & D. A. Johnson. 2015. *Puccinia areolata*, *P. treleasiana*, and *P. gemella* on marshmarigold (*Caltha* spp.) in subalpine habitats in Northwestern United States. *N. Amer. Fungi* 10: 1–10.

Tyndall, R. W., K. A. McCarthy, J. C. Ludwig & A. Rome. 1990. Vegetation of six Carolina bays in Maryland. *Castanea* 55: 1–21.

Tyrrell, L. E. 1987. A floristic survey of buttonbush swamps in Gahanna Woods State Nature Preserve Franklin County, Ohio. *Michigan Bot.* 26: 29–37.

Tzonev, R. 2007. *Eclipta prostrata* (Asteraceae): a new alien species for the Bulgarian flora. *Phytol. Balcan.* 13: 79–80.

Uchytal, R. J. 1989a. *Alnus incana* ssp. *tenuifolia*. *In: Fire Effects Information System.* U.S. Department of Agriculture, Forest Service, Rocky Mountain Research Station, Fire Sciences Laboratory (Producer). Available online: http://www.fs.fed. us/database/feis/ (accessed June 9, 2004].

Uchytil, R. J. 1989b. *Salix exigua*. *In: Fire Effects Information System.* U. S. Department of Agriculture, Forest Service, Rocky Mountain Research Station, Fire Sciences Laboratory (Producer). Available online: http://www.fs.fed.us/database/ feis/ (accessed March 11, 2005).

Uchytil, R. J. 1989c. *Salix lasiandra*. *In: Fire Effects Information System.* U. S. Department of Agriculture, Forest Service, Rocky Mountain Research Station, Fire Sciences Laboratory (Producer). Available online: http://www.fs.fed.us/database/ feis / (accessed March 11, 2005).

Uchytil, R. J. 1989d. *Salix lemmonii*. *In: Fire Effects Information System.* U. S. Department of Agriculture, Forest Service, Rocky Mountain Research Station, Fire Sciences Laboratory (Producer). Available online: http://www.fs.fed.us/database/ feis/ (accessed March 11, 2005).

Uchytil, R. J. 1989e. *Salix lutea*. *In: Fire Effects Information System.* U. S. Department of Agriculture, Forest Service, Rocky Mountain Research Station, Fire Sciences Laboratory (Producer). Available online: http://www.fs.fed.us/database/

feis/ (accessed March 11, 2005).

Uchytil, R. J. 1991a. *Salix drummondiana. In: Fire Effects Information System.* U. S. Department of Agriculture, Forest Service, Rocky Mountain Research Station, Fire Sciences Laboratory (Producer). Available online: http://www.fs.fed.us/database/feis/ (accessed March 11, 2005).

Uchytil, R. J. 1991b. *Salix planifolia. In: Fire Effects Information System.* U. S. Department of Agriculture, Forest Service, Rocky Mountain Research Station, Fire Sciences Laboratory (Producer). Available online: http://www.fs.fed.us/database/feis/ (accessed March 17, 2005).

Uchytil, R. J. 1992. *Salix myrtillifolia. In: Fire Effects Information System.* U. S. Department of Agriculture, Forest Service, Rocky Mountain Research Station, Fire Sciences Laboratory (Producer). Available online: http://www.fs.fed.us/database/feis/ (accessed March 14, 2005).

Uchytil, R. J. 1993. *Vaccinium corymbosum. In: Fire Effects Information System.* U. S. Department of Agriculture, Forest Service, Rocky Mountain Research Station, Fire Sciences Laboratory (Producer). Available online: http://www.fs.fed.us/database/feis/ (accessed June 23, 2006).

Ueda, K., K. Kosuge & H. Tobe. 1997. A molecular phylogeny of Celtidaceae and Ulmaceae (Urticales) based on *rbcL* nucleotide sequences. *J. Plant Res.* 110: 171–178.

Ueda, Y. & S. Akimoto. 2001. Cross-and self-compatibility in various species of the genus *Rosa. J. Hort. Sci. Biotechnol.* 76: 392–395.

Ueno, S. & Y. Kadono. 2001. Monoecious plants of *Myriophyllum ussuriense* (Regel) Maxim. in Japan. *J. Plant Res.* 114: 375–376.

Uesugi, R., K. Goka, J. Nishihiro & I. Washitani. 2004. Allozyme polymorphism and conservation of the Lake Kasumigaura population of *Nymphoides peltata. Aquat. Bot.* 79: 203–210.

Uesugi, R., N. Tani, K. Goka, J. Nishihiro, Y. Tsumura & I. Washitani. 2005. Isolation and characterization of highly polymorphic microsatellites in the aquatic plant, *Nymphoides peltata* (Menyanthaceae). *Mol. Ecol. Notes* 5: 343–345.

Uesugi, R., J. Nishihiro, Y. Tsumura & I. Washitani. 2007. Restoration of genetic diversity from soil seed banks in a threatened aquatic plant, *Nymphoides peltata. Conserv. Genet.* 8: 111–121.

Uesugi, R., J. Nishihiro & I. Washitani. 2009. Population status and genetic diversity of *Nymphoides peltata* in Japan. *Jap. J. Conserv. Ecol.* 14: 13–24.

Ulanowicz, R. 1997. *Ecology, the Ascending Perspective.* Columbia University Press, New York. 224 pp.

Umar, K. J., L. G. Hassan, S. M. Dangoggo & M. J. Ladan. 2007. Nutritional composition of water spinach (*Ipomoea aquatica* Forsk.) leaves. *J. Appl. Sci.* 7: 803–809.

Umemoto, S. & H. Koyama. 2007. A new species of *Eclipta* (Compositae: Heliantheae) and its allies in eastern Asia. *Thai Forest Bull., Bot.* 35: 108–118.

Umemoto, S. & H. Yamaguchi. 1999. Immigration mode of *Eclipta* into Japan. *Sci. Rep. Coll. Agric. Osaka Pref. Univ.* 51: 25–31.

Umemoto, S., H. Kobayashi & K. Ueki. 1987. The genecological variation of the photoperiodism in *Eclipta prostrata* (L.) S. L. from Japan and Taiwan. *Proc. 11th Asian Pacific Weed Sci. Soc. Conf.* 1: 39–44.

Underwood, L. M. 1893. List of Cryptogams at present known to inhabit the State of Indiana. *Proc. Indiana Acad. Sci.* 1893: 30–67.

Ungar, I. A. 1962. Influence of salinity on seed germination in succulent halophytes. *Ecology* 43: 763–764.

Ungar, I. A. 1964. A phytosociological analysis of the Big Salt Marsh, Stafford County, Kansas. *Trans. Kansas Acad. Sci.* 67: 50–64.

Ungar, I. A. 1973. Salinity tolerance of inland halophytic vegetation of North America. *Bull. Soc. Bot. France* 120: 217–222.

Ungar, I. A. 1978. Halophyte seed germination. *Bot. Rev.* 44: 233–264.

Ungar, I. A. 1984. Alleviation of seed dormancy in *Spergularia marina. Bot. Gaz.* 145: 33–36.

Ungar, I. A. 1987. Population ecology of halophyte seeds. *Bot. Rev.* 53: 301–334.

Ungar, I. A. 1988. Brief note: A significant seed bank for *Spergularia marina* (Caryophyllaceae). *Ohio J. Sci.* 88: 200–202.

Ungar, I. A., W. Hogan & M. McClelland. 1969. Plant communities of saline soils at Lincoln, Nebraska. *Amer. Midl. Naturalist* 82: 564–577.

Unks, R. R., T. H. Shear, A. Krings & R. R. Braham. 2014. Environmental controls of reproduction and early growth of *Lindera melissifolia* (Lauraceae). *Castanea* 79: 266–277.

Uphof, J. C. T. 1922. Ecological relations of plants in southeastern Missouri. *Amer. J. Bot.* 9: 1–17.

Uprety, Y., A. Lacasse & H. Asselin. 2015. Traditional uses of medicinal plants from the Canadian boreal forest for the management of chronic pain syndromes. *Pain Practice* 16: 459–466.

Urban, L. & C. D. Bailey. 2013. Phylogeny of *Leavenworthia* and *Selenia* (Brassicaceae). *Syst. Bot.* 38: 723–736.

Urban, R. A., J. E. Titus & W.-X. Zhu. 2006. An invasive macrophyte alters sediment chemistry due to suppression of a native isoetid. *Ecosyst. Ecol.* 148: 455–463.

Urbatsch, L. E. 2013. Plants new and noteworthy for Louisiana and Mississippi. *Phytoneuron* 2013–14: 1–7.

Urbatsch, L. E. & P. B. Cox. 2006. 261. *Rudbeckia* Linnaeus. pp. 44–60. *In:* Flora of North America Editorial Committee (eds.), *Flora of North America North of Mexico: Vol. 21, Magnoliophyta: Asteridae (in part): Asteraceae, Part 3.* Oxford University Press, New York.

Urbatsch, L. E. & R. K. Jansen. 1995. Phylogenetic affinities among and within the coneflower genera (Asteraceae, Heliantheae), a chloroplast DNA analysis. *Syst. Bot.* 20: 28–39.

Urbatsch, L. E., J. D. Bacon, R. L. Hartman, M. C. Johnston, T. J. Watson Jr. & G. L. Webster. 1975. Chromosome numbers for North American Euphorbiaceae. *Amer. J. Bot.* 62: 494–500.

Urbatsch, L. E., B. G. Baldwin & M. J. Donoghue. 2000. Phylogeny of the coneflowers and relatives (Heliantheae: Asteraceae) based on nuclear rDNA internal transcribed spacer (ITS) sequences and chlorplast DNA restriction site data. *Syst. Bot.* 25: 539–565.

Urbatsch, L. E., R. P. Roberts & V. Karaman. 2003. Phylogenetic evaluation of *Xylothamia, Gundlachia,* and related genera (Asteraceae, Astereae) based on ETS and ITS nrDNA sequence data. *Amer. J. Bot.* 90: 634–649.

USACE. 2016. NWPL-*National Wetland Plant List.* U. S. Army Corps of Engineers. Available online: http://rsgisias.crrel.usace.army.mil/NWPL/ (accessed October 24, 2016)

USDA. 1948. *Woody-plant Seed Manual.* Miscellaneous Publication No. 654. United States Government Printing Office, Washington, DC. 416 pp.

USDA. 1984. *Soils and Vegetation of the Apalachicola National Forest.* United States Department of Agriculture Forest Service, Southern Region, Atlanta, GA. 165 pp.

USDA. 1998. Have you seen this plant? It's mud mat. APHIS 81–35–001 Pest alert. United States Department of Agriculture, Riverdale, MD.

USDA. 2002. Conservation assessment for Arctic raspberry (*Rubus acaulis*) Michx. USDA Forest Service, Eastern Region, Milwaukee, WI. 27 pp.

USDA, NRCS. 2004. *The PLANTS Database, Version 3.5.* National Plant Data Center, Baton Rouge, LA 70874–4490 USA. Available online: http://plants.usda.gov.

USDA. 2005. *Field Guide to Riparian Plant Communities in Northwestern Oregon.* Technical Paper R6-NR-ECOL-TP-01–05. United States Department of Agriculture, Forest Service, Pacific Northwest Region, Portland, OR. 114 pp.

USDA. 2007. *Soil Survey of the Tahoe Basin Area, California and Nevada.* United States Department of Agriculture, Natural Resources Conservation Service. Available online: http://soils.usda.gov/survey/printed_surveys/.

USEPA. 1995. *Bioindicators for Assessing Ecological Integrity of Prairie Wetlands.* Report # EPA/600/R-96/082. U. S. Environmental Protection Agency, Washington, DC.

USFWS. 1988. Endangered and threatened wildlife and plants; determination of *Ptilimnium nodosum* to be an endangered species. *Fed. Reg.* 50: 37978–37982.

USFWS. 1993a. Recovery plan for three granite outcrop plant species. U. S. Fish and Wildlife Service, Jackson, MI. 41 pp.

USFWS. 1993b. U. S. Fish and Wildlife Service. Sacramento Mountains Thistle (*Cirsium vinaceum*) recovery plan. U. S. Fish and Wildlife Service, Albuquerque, NM. 23 pp.

USFWS. 1994. Recovery plan for four plants of the lower Apalachicola region, Florida: *Euphorbia telephioides* (telephus spurge), *Macbridea alba* (white birds-in-a-nest), *Pinguicula ionantha* (Godfrey's butterwort), and *Scutellaria floridana* (Florida skullcap). U. S. Fish and Wildlife Service, Atlanta, GA. 32 pp.

USFWS. 1995a. Okeechobee gourd (*Cucurbita okeechobeensis* ssp. *okeechobeensis*). Endangered and threatened species of the southeastern United States (the red book). U. S. Fish and Wildlife Service, Region 4, Asheville, NC. Available online: http://endangered.fws.gov/i/q/saqad.html.

USFWS. 1995b. Sensitive joint-vetch (*Aeschynomene virginica*). Endangered and threatened species of the southeastern United States (the red book). U. S. Fish and Wildlife Service, Region 4, Asheville, NC. Available online: http://endangered.fws.gov/i/q/saqad.html.

USFWS. 1998a. Recovery plan for marsh sandwort (*Arenaria paludicola*) and Gambel's watercress (*Rorippa gambelii*). U. S. Fish and Wildlife Service, Portland, OR. 50 pp.

USFWS. 1998b. Recovery plan for the pedate checkermallow (*Sidalcea pedata*) and the slender-petaled mustard (*Thelypodium stenopetalum*). U. S. Fish and Wildlife Service, Portland, OR. 68 pp.

USFWS. 1998c. *Vernal Pools of Southern California: Recovery Plan.* U. S. Fish and Wildlife Service, Portland, OR. 113 pp.

USFWS. 1999. *South Florida Multi-Species Recovery Plan.* U. S. Fish and Wildlife Service, Atlanta, GA. 2179 pp.

USFWS. 2002. Draft recovery plan for the rough popcorn flower (*Plagiobothrys hirtus*). U. S. Fish and Wildlife Service, Portland, OR. 48 pp.

USFWS. 2003. Recovery plan for the rough popcornflower (*Plagiobothrys hirtus*). U. S. Fish and Wildlife Service, Portland, OR. 60 pp.

USFWS. 2004. Draft recovery plan for vernal pool ecosystems of California and southern Oregon. Region 1. U. S. Fish and Wildlife Service, Portland, OR. [Pagination irregular].

USFWS. 2006a. Recovery plan for the Carson wandering skipper (*Pseudocopaeodes eunus obscurus*). U. S. Fish and Wildlife Service, Sacramento, CA. 94 pp.

USFWS. 2006b. Draft recovery plan for listed species of the Rogue Valley vernal pool and Illinois Valley wet meadow ecosystems. U. S. Fish and Wildlife Service, Portland, OR. 136 pp.

USFWS. 2007. Chorro Creek bog thistle (*Cirsium fontinale* var. *obispoense*). 5-Year Review: summary and evaluation. Ventura Field Office, U. S. Fish and Wildlife Service, Ventura, CA. 16 pp.

USFWS. 2009. Draft recovery plan for tidal marsh ecosystems of northern and central California. U. S. Fish and Wildlife Service, Sacramento, CA. 636 pp.

USFWS. 2010. Sacramento Mountains thistle (*Cirsium vinaceum*). 5-year review: summary and evaluation. U. S. Fish & Wildlife Service, Albuquerque, NM. 45 pp.

USFWS. 2013a. Cutthroat grass communities. Multi-Species Recovery Plan for South Florida. Available online: http://www.fws.gov/verobeach/ListedSpeciesMSRP.html.

USFWS. 2013b. U. S. Fish and Wildlife Service. Recovery plan for tidal marsh ecosystems of northern and central California. Sacramento, CA. 605 pp.

USFWS. 2016a. U. S. Fish and Wildlife Service. Recovery Plan for the Santa Rosa Plain: *Blennosperma bakeri* (Sonoma sunshine); *Lasthenia burkei* (Burke's goldfields); *Limnanthes vinculans* (Sebastopol meadowfoam); California Tiger Salamander Sonoma County Distinct Population Segment (*Ambystoma californiense*). U. S. Fish and Wildlife Service, Pacific Southwest Region, Sacramento, CA. 128 pp.

USFWS. 2016b. Mesic temperate hammock. Multi-Species Recovery Plan for South Florida Available online: https://www.fws.gov/southeast/vbpdfs/commun/mth.pdf.

Usher, G. A. 1974. *Dictionary of Plants Used by Man.* Constable, London, UK. 619 pp.

Ushimaru, A. & K. Kikuzawa. 1999. Variation of breeding system, floral rewards, and reproductive success in clonal *Calystegia* species (Convolvulaceae). *Amer. J. Bot.* 86: 436–446.

Usinger, R. L. 1956. *Aquatic Insects of California: With Keys to North American Genera and California species.* University of California Press, Berkeley, CA. 508 pp.

Utelli, A. B. & B. A. Roy. 2000. Pollinator abundance and behavior on *Aconitum lycoctonum* (Ranunculaceae): an analysis of the quantity and quality components of pollination. *Oikos* 89: 461–470.

Uttal, L. J. 1987. The genus *Vaccinium* L. (Ericaceae) in Virginia. *Castanea* 52: 231–255.

Vaccaro, L. E., B. L. Bedford & C. A. Johnston. 2009. Litter accumulation promotes dominance of invasive species of cattails (*Typha* spp.) in Lake Ontario wetlands. *Wetlands* 29: 1036–1048.

Vaezi, J. 2008. Origin of *Symphyotricltum anticostense* (Asteraceae: Astereae), an endemic species of the Gulf of St. Lawrence. PhD dissertation, Université de Montréal, Montreal, QC. 132 pp.

Vaezi, J. & L. Brouillet. 2009. Phylogenetic relationships among diploid species of *Symphyotrichum* (Asteraceae: Astereae) based on two nuclear markers, ITS and GAPDH. *Mol. Phylogen. Evol.* 51: 540–553.

Vaghti, M. & T. Keeler-Wolf. 2004. *Suisun Marsh Vegetation Mapping Change Detection 2003: A report to the California Department of Water Resources.* Wildlife Habitat Data Analysis Branch, California Department of Fish and Game, Sacramento, CA. 43 pp.

Vaidhayakarn, U. 1987. Preliminary study on water convolvulus (*Ipomoea aquatica*) improvement I. Morphological characters of 2x and 4x plants. Kasetsart University, Bangkok, Thailand. 14 pp.

Vaithiyanathan, P. & C. J. Richardson. 1999. Macrophyte species changes in the Everglades: examination along a eutrophication gradient. *J. Environ. Qual.* 28: 1347–1358.

Valiejo-Roman, C. M., E. I. Terentieva, T. H. Samigullin & M. G. Pimenov. 2002. nrDNA ITS sequences and affinities of Sino-Himalayan Apioideae (Umbelliferae). *Taxon* 51: 685–701.

Valkó, O., B. Tóthmérész, A. Kelemen, E. Simon, T. Miglécz, B. A. Lukács & P. Török. 2014. Environmental factors driving seed bank diversity in alkali grasslands. *Agric. Ecosyst. Environ.* 182: 80–87.

Vallejo-Marín, M. & S. J. Hiscock. 2016. Hybridization and hybrid speciation under global change. *New Phytol.* doi:10.1111/nph.14004.

Vallès, J. & E. D. McArthur. 2001. *Artemisia* systematics and phylogeny: cytogenetic and molecular insights. USDA Forest Service Proceedings RMRS-P-21. pp. 67–74. *In*: E. D. McArthur & D. J. Fairbanks (eds.), *Proceedings: Shrubland Ecosystem Genetics and Biodiversity; 2000 June 13–15; Provo, UT. Ogden.* U. S. Department of Agriculture Forest Service, Rocky Mountain Research Station, Fort Collins, CO.

Valley, R. D. & R. M. Newman. 1998. Competitive interactions between Eurasian watermilfoil and northern watermilfoil in experimental tanks. *J. Aquat. Plant Manag.* 36: 121–126.

Van, T. K. & P. T. Madeira. 1998. Random amplified polymorphic DNA analysis of water spinach (*Ipomoea aquatica*) in Florida. *J. Aquat. Plant Manag.* 36: 107–111.

Van Alstine, N. E. 2000. Virginia sneezeweed (*Helenium virginicum*) recovery plan. Technical/Agency Draft. U. S. Fish & Wildlife Service, Hadley, MA. 54 pp.

Van Alstine, N. E., A. C. Chazal & K. M. McCoy. 2001. *A Biological Survey of the Coastal Plain Depression Ponds (Sinkholes) of Colonial National Historical Park, Yorktown, Virginia.* Natural Heritage Technical Report 01-9. Virginia Department of Conservation and Recreation, Division of Natural Heritage, Richmond, VA. 56 pp.

Van Auken, O. W., M. Grunstra & S. C. Brown. 2007. Composition and structure of a west Texas salt marsh. *Madroño* 54: 138–147.

Van Bodegom, P. M., M. de Kanter, C. Bakker & R. Aerts. 2005. Radial oxygen loss, a plastic property of dune slack plant species. *Plant Soil* 271: 351–364.

Van Damme, E. J. M., A. Barre, P. Rougé & W. J. Peumans. 2004. Cytoplasmic/nuclear plant lectins: a new story *Trends Plant Sci.* 9: 484–489.

Van Deelen, T. R. 1991. *Alnus rugosa. In*: *Fire Effects Information System.* U. S. Department of Agriculture, Forest Service, Rocky Mountain Research Station, Fire Sciences Laboratory (Producer). Available online: http://www.fs.fed.us/database/feis/ (accessed June 9, 2004).

Van den Broek. T., R. van Diggelen & R. Bobbink. 2005. Variation in seed buoyancy of species in wetland ecosystems with different flooding dynamics. *J. Veg. Sci.* 16: 579–586.

Van de Kerckhove, G. A., M. L. Smither-Kopperl & H. C. Kistler. 2002. First report of *Sphaeropsis tumefaciens* on an endangered St. John's-Wort in Florida. *Dis. Notes* 86: 1177.

Van der Laan, F. M. & J. C. Arends. 1985. Cytotaxonomy of the Apocynaceae. *Genetica* 68: 3–35.

Van der Pijl, L. 1969. *Principles of Dispersal in Higher Plants.* Springer-Verlag, Berlin, Germany. 153 pp.

Van der Sman A. J. M., O. F. R. Van Tongeren & C. W. P. M. Blom. 1988. Growth and reproduction of *Rumex maritimus* and *Chenopodium rubrum* under different waterlogging regimes. *Acta Bot. Neerl.* 37: 439–450.

Van der Sman, A. J. M., C. W. P. M. Blom & H. M. Van de Steeg 1992. Phenology and seed production in *Chenopodium rubrum, Rumex maritimus* and *Rumex palustris* as related to photoperiod in river flood plains. *Can. J. Bot.* 70: 392–400.

Van der Toorn, J. 1980. On the ecology of *Cotula coronopifolia* L. and *Ranunculus sceleratus* L. I. Geographic distribution, habitat, and field observation. *Acta Bot. Neerl.* 29: 385–396.

Van der Toorn, J. & H. J. ten Hove. 1982. On the ecology of *Cotula coronopifolia* L. and *Ranunculus sceleratus* L. II.–Experiment on germination, seed longevity, and seedling survival. *Acta Oecol.* 3: 409–418.

Van der Valk, A. G. 2013. Seed banks of drained floodplain, drained palustrine, and undrained wetlands in Iowa, USA. *Wetlands* 33: 183–190.

Van der Valk, A. G. & C. B. Davis. 1978. The role of seed banks in the vegetation dynamics of prairie glacial marshes. *Ecology* 59: 322–335.

Van der Valk, A. G. & T. R. Rosburg. 1997. Seed bank composition along a phosphorous gradient in the northern Florida Everglades. *Wetlands* 17: 228–236.

Van der Valk, A. G. & J. T. A. Verhoeven. 1988. Potential role of seed banks and understory species in restoring quaking fens from floating forests. *Vegetatio* 76: 3–13.

Van der Valk, A. G., P. Wetzel, E. Cline & F. H. Sklar. 2008. Restoring tree islands in the Everglades: experimental studies of tree seedling survival and growth. *Restorat. Ecol.* 16: 281–289.

Van Der Velde, G. & L.A. Van Der Heijden. 1981. The floral biology and seed production of *Nymphoides peltata* (GMEL.) O. Kuntze (Menyanthaceae). *Aquat. Bot.* 10: 261–293.

Van der Werf, H. 1997. 5. Lauraceae Jussieu. pp. 26–36. *In*: N. R. Morin (convening ed.), *Flora of North America North of Mexico, Vol. 3: Magnoliophyta: Magnoliidae and Hamamelidae.* Oxford University Press, New York.

Van De Wiel, C. C. M., J. Van Der Schoot, J. L. C. H. Van Valkenburg, H. Duistermaat & M. J. M. Smulders. 2009. DNA barcoding discriminates the noxious invasive plant species, floating pennywort (*Hydrocotyle ranunculoides* L.f.), from non-invasive relatives. *Mol. Ecol. Resour.* 9: 1086–1091.

Van Houtte, L. (ed.). 1845. *Flore des Serres et des Jardins de l'Europe.* Volume 1. Ghent, Belgium. 510 pp.

Van Geel, B., A. Protopopov, I. Bull, E. Duijm, F. Gill, Y. Lammers, A. Nieman, N. Rudaya, S. Trofimova, A. N. Tikhonov & R. Vos. 2014. Multiproxy diet analysis of the last meal of an early Holocene Yakutian bison. *J. Quatern. Sci.* 29: 261–268.

Vander Kloet, S. P. 1980. The taxonomy of the highbush blueberry, *Vaccinium corymbosum. Can. J. Bot.* 58: 1187–1201.

Vander Kloet, S. P. 1992. On the etymology of *Vaccinium* L. *Rhodora* 94: 371–373.

Vander Kloet, S. P. 2009. 45. *Vaccinium* Linnaeus. pp. 515–530. *In*: N. R. Morin (convening ed.), *Flora of North America North of Mexico, Vol. 8: Magnoliophyta: Paeoniaceae to Ericaceae.* Oxford University Press, New York.

Vander Kloet, S. P. & P. J. Austin-Smith. 1986. Energetics, patterns and timing of seed dispersal in *Vaccinium* section *Cyanococcus. Amer. Midl. Naturalist* 115: 386–396.

Vander Kloet, S. P., T. S. Avery, P. J. Vander Kloet & G. R. Milton. 2012. Restoration ecology: aiding and abetting secondary succession on abandoned peat mines in Nova Scotia and New Brunswick, Canada. *Mires Peat* 10: 1–20.

Vanderwier, J. M. & J. C. Newman. 1984. Observations of haustoria and host preference in *Cordylanthus maritimus* subsp. *maritimus* (Scrophulariaceae) at Mugu Lagoon. *Madroño* 31: 185–186.

Van Deusen, M. & D. W. Kaufman. 1977. Distribution of *Peromyscus leucopus* within prairie woods. *Trans. Kansas Acad. Sci.* 80: 151–154.

Vandevender, J. 2008. Propagation protocol for production of field-grown *Alnus serrulata* (Aiton) Willd. 'Panbowl' plants (Bareroot); USDA NRCS-Appalachian Plant Materials Center, Alderson, West Virginia. *In: Native Plant Network*. Forest Research Nursery, College of Natural Resources, University of Idaho, Moscow, ID. Available online: http://www.native-plantnetwork.org (accessed March 31, 2016).

Van Dijk, G. M., D. D. Thayer & W. T. Haller. 1986. Growth of *Hygrophila* in flowing water. *J. Aquat. Plant Manag.* 24: 85–87.

Van Dijk, W. F., J. van Ruijven, F. Berendse & G. R. de Snoo. 2014. The effectiveness of ditch banks as dispersal corridor for plants in agricultural landscapes depends on species' dispersal traits. *Biol. Conserv.* 171: 91–98.

Vandiver Jr., V. V. 1980. "Hygrophila." *Aquatics* 2(4): 4–11.

Van Dorp, D., W. P. M. van den Hoek & C. Daleboudt. 1996. Seed dispersal capacity of six perennial grassland species measured in a wind tunnel at varying wind speed and height. *Can. J. Bot.* 74: 1956–1963.

Van Ham, R. C. H. J. & H. T. Hart. 1998. Phylogenetic relationships in the Crassulaceae inferred from chloroplast DNA restriction-site variation. *Amer. J. Bot.* 85: 123–134.

Van Handel, E., J. S. Haeger & C. W. Hansen. 1972. The sugars of some Florida nectars. *Amer. J. Bot.* 59: 1030–1032.

Vankar, P. S., R. Shanker & J. Srivastava. 2007. Ultrasonic dyeing of cotton fabric with aqueous extract of *Eclipta alba*. *Dyes Pigments* 72: 33–37.

Van Laerhoven, S. L., D. R. Gillespie & B. D. Roitberg. 2006. Patch retention time in an omnivore, *Dicyphus hesperus* is dependent on both host plant and prey type. *J. Insect Behav.* 19: 613–621.

Van Leeuwen, C. H., J. M. Sarneel, J. Paassen W. J. Rip & E. S. Bakker. 2014. Hydrology, shore morphology and species traits affect seed dispersal, germination and community assembly in shoreline plant communities. *J. Ecol.* 102: 998–1007.

Vannimwegen, R. E. & D. M. Debinski. 2004. *Nest Success of Yellow Warblers in Willow Habitats: The Role of Surface Water and Snakes*. University of Wyoming National Park Service Research Center Annual Report: Vol. 28, Article 9. Available online: http://repository.uwyo.edu/uwnpsrc_reports/vol28/iss1/9.

Van Riemsdijk, I. 2014. Unravelling the impact of substitution rates and time on branch lengths in phylogenetic trees. MS thesis. Wageningen University and Research Center, Wageningen, The Netherlands. 124 pp.

Van Strien, A. J., J. Van Der Linden, T. C. P. Melman & M. A. W. Noordervliet. 1989. Factors affecting the vegetation of ditch banks in peat areas in the western Netherlands. *J. Appl. Ecol.* 26: 989–1004.

Varopoulos, A. 1979. Breeding systems in *Myosotis scorpioides* L. 1. Self-incompatibility. *Heredity* 42: 149–157.

Varshney, S. P. & B. D. Sharma. 1979. Responses of saline and non-saline populations of *Eclipta alba* to salinity. *Can. J. Plant Sci.* 59: 539–540.

Vartapetian, B. B. & I. N. Andreeva. 1986. Mitochondrial ultrastructure of three hygrophyte species at anoxia and in anoxic glucose-supplemented medium. *J. Exp. Bot.* 37: 685–692.

Vasas, A., O. Orbán-Gyapai & J. Hohmann. 2015. The genus *Rumex*: review of traditional uses, phytochemistry and pharmacology. *J. Ethnopharmacol.* 175: 198–228.

Vasconcelos, T., M. Tavares & N. Gaspar. 1999. Aquatic plants in the rice fields of the Tagus Valley, Portugal. *Hydrobiologia* 415: 59–65.

Vasey, M. C., V. T. Parker, J. C. Callaway, E. R. Herbert & L. M. Schile. 2012. Tidal wetland vegetation in the San Francisco Bay-Delta estuary. *San Francisco Estuary Watershed Sci.* 10: 1–16.

Veerasethakul, S. & J. S. Lassetter. 1981. Karyotype relationships of native New World *Vicia* species (Leguminosae). *Rhodora* 83: 595–606.

Velasco, L. & F. D. Goffman. 1999. Tocopherol and fatty acid composition of twenty-five species of Onagraceae Juss. *Bot. J. Linn. Soc.* 129: 359–366.

Velázquez-Velázquez, E. & J. J. Schmitter-Soto. 2004. Conservation status of the San Cristóbal pupfish *Profundulus hildebrandi* Miller (Teleostei: Profundulidae) in the face of urban growth in Chiapas, Mexico. *Aquat. Conserv.* 14: 201–209.

Venable, N. J. 1986. *Aquatic Plants: Guide to Aquatic and Wetland plants of West Virginia*. Series 803. Cooperative Extension Service, West Virginia University, Morgantown, WV. 88 pp.

Venables, B. A. B. & E. M. Barrows. 1985. Skippers: pollinators or nectar thieves? *J. Lepid. Soc.* 39: 299–312.

Venkata Rao, J., K. S. Aithal & K. K. Srinivasan. 1989. Antimicrobial activity of the essential oil of *Limnophila gratissima*. *Fitoterapia* 60: 376–377.

Venkatesan, S. & R. Ravi. 2004. Antifungal activity of *Eclipta alba*. *Indian J. Pharm. Sci.* 66: 97–98.

Venkateswarlu, V. 1984. Algae and aquatic plants in the treatment of industrial waste effluents. *J. Environ. Biol.* 5: 15–21.

Vera, N., R. Misico, M. G. Sierra, Y. Asakawa & A. Bardón. 2008. Eudesmanes from *Pluchea sagittalis*. Their antifeedant activity on *Spodoptera frugiperda*. *Phytochemistry* 69: 1689–1694.

Verbeek, N. A. M. 1967. Breeding biology and ecology of the horned lark in alpine tundra. *Wilson Bull.* 79: 208–218.

Verhey, D. M. 2007. Endangered and threatened wildlife and plants; designation of critical habitat for *Cirsium hydrophilum* var. *hydrophilum* (Suisun thistle) and *Cordylanthus mollis* ssp. *mollis* (soft bird's-beak). *Fed. Reg.* 72: 18518–18553.

Verloove, F. 2003. Graanadventieven nieuw voor de Belgische flora, hoofdzakeüjk in 1999 en 2000. *Dumortiera* 80: 45–53.

Vestal, A. G. 1914. A black-soil prairie station in northeastern Illinois. *Bull. Torrey Bot. Club* 41: 351–363.

Vibrans, H. 1998. Native maize field weed communities in south-central Mexico. *Weed Res.* 38: 153–166.

Vickery, B. S. J. & P. D. Vickery. 1983. Note on the status of *Agalinis maritima* (Raf.) Raf. in Maine. *Rhodora* 85: 267–269.

Vickery, R. 1995. *A Dictionary of Plant Lore*. Oxford University Press, Oxford, UK. 437 pp.

Vickery Jr., R. K. 1964. Barriers to gene exchange between members of the *Mimulus guttatus* complex (Scrophulariaceae). *Evolution* 18: 52–69.

Vickery Jr., R. K. 1990a. Pollination experiments in the *Mimulus cardinalis-M. lewisii* complex. *Great Basin Naturalist* 50: 155–159.

Vickery Jr., R. K. 1990b. Close correspondence of allozyme groups to geographic races in the *Mimulus glabratus* complex (Scrophulariaceae). *Syst. Bot.* 15: 481–496.

Vickery Jr., R. K. 1991. Crossing relationships of *Mimulus glabratus* var. *michigansis* (Scrophulariaceae). *Amer. Midl. Naturalist* 125: 368–371.

Vickery Jr., R. K. 1999. Remarkable waxing, waning, and wandering of populations of *Mimulus guttatus*: An unexpected example of global warming. *Great Basin Naturalist* 59: 112–126.

Vickery Jr., R. K., D. R. Phillips & P. R. Wonsavage. 1986. Seed dispersal in *Mimulus guttatus* by wind and deer. *Amer. Midl. Naturalist* 116: 206–208.

Vickery Jr., R. K., J. W. Ajioka, E. S. C. Lee & K. D. Johnson. 1989. Allozyme-based relationships of the populations and taxa of section *Erythranthe* (*Mimulus*). *Amer. Midl. Naturalist* 121: 232–244.

Viegi, L., A. Pieroni, P. M. Guarrera & R. Vangelisti. 2003. A review of plants used in folk veterinary medicine in Italy as basis for a databank. *J. Ethnopharmacol.* 89: 221–244.

Viereck, L. 2010. *Alaska Trees and Shrubs.* University of Alaska Press, Fairbanks, AK. 370 pp.

Viereck, L. A., C. T. Dyrness, A. R. Batten & K. J. Wenzlick. 1992. *The Alaska Vegetation Classification.* Gen. Tech. Rep. PNW-GTR-286. U. S. Department of Agriculture, Forest Service, Pacific Northwest Research Station, Portland, OR. 278 pp.

Viglasky, J., I. Andrejcak, J. Huska & J. Suchomel. 2009. Amaranth (*Amarantus* L.) is a potential source of raw material for biofuels production. *Agron. Res.* 7: 865–873.

Vignolio, O. R. & O. N. Fernández. 2006. Seed dispersal in hare (*Lepus europaeus*) faecal pellets in flooding Pampa grasslands. *Rev. Argent. Prod. Anim.* 26: 31–38.

Vijaya, A. S. 2010. Comparative account of soil physico chemical characteristics of wet land and wet land converted to rubber plantation-a case study in Nalloor village, Marthandam, Kanyakumari Dist, Tamilnadu, India. *J. Basic Appl. Biol.* 4: 213–216.

Vijayakumar, M., R. Govindarajan, A. Shirwaikar, V. Kumar, A. K. S. Rawat, S. Mehrotra & P. Pushpangadan. 2005. Free radical scavenging and lipid peroxidation inhibition potential of *Hygrophila auriculata. Nat. Prod. Sci.* 11: 22–26.

Vijverberg, C. A. 2001. Adaptive radiation of Australian and New Zealand *Microseris* (Asteraceae). PhD dissertation, University of Amsterdam, Amsterdam, The Netherlands. 183 pp.

Vijverberg, K. & K. Bachmann. 1999. Molecular evolution of a tandemly repeated *trnF* (GAA) gene in the chloroplast genomes of *Microseris* (Asteraceae) and the use of structural mutations in phylogenetic analyses. *Mol. Biol. Evol.* 16: 1329–1340.

Vijverberg, K., T. H. M. Mes & K. Bachmann. 1999. Chloroplast DNA evidence for the evolution of *Microseris* (Asteraceae) in Australia and New Zealand after long-distance dispersal from western North America. *Amer. J. Bot.* 86: 1448–1463.

Villani, P. J. & S. A. Etnier. 2008. Natural history of heterophylly in *Nymphaea odorata* ssp. *tuberosa* (Nymphaeaceae). *N. E. Naturalist* 15: 177–188.

Villanueva, V. R., L. K. Simola & M. Mardon. 1985. Polyamines in turions and young plants of *Hydrocharis morsus-ranae* and *Utricularia intermedia. Phytochemistry* 24: 171–172.

Villarreal Q., J. Á., M. Á. Carranza P., E. Estrada C. & A. Rodríguez G. 2006. Flora riparia de los Rios Sabinas y San Rodrigo, Coahuila, México. *Acta Bot. Mex.* 75: 1–20.

Villarreal, S., R. D. Hollister, D. R. Johnson, M. J. Lara, P. J. Webber & C. E. Tweedie. 2012. Tundra vegetation change near Barrow, Alaska (1972–2010). *Environ. Res. Lett.* 7: 015508.

Villaseñor, J. L. & F. J. Espinosa-Garcia. 2004. The alien flowering plants of Mexico. *Diversity Distrib.* 10: 113–123.

Vimala S., S. Rohana, A. A. Rashih & M. Juliza. 2012. Antioxidant evaluation in Malaysian medicinal plant: *Persicaria minor* (Huds.) leaf. *Sci. J. Med. Clinical Trials* 2012: ISSN: 2276–7487.

Vince, S. W. & A. A. Snow. 1984. Plant zonation in an Alaskan salt marsh: I. Distribution, abundance and environmental factors. *J. Ecol.* 72: 651–667.

Vincent, K. A. L. T. Aucoin & N. S. Clark. 2003. *Freshwater Bayou Wetlands.* ME-04 (XME-21), Second Priority List Hydrologic Restoration Project of the Coastal Wetlands Planning, Protection, and Restoration Act (Public Law 101–646). Coastal Restoration Division, Louisiana Department of Natural Resources, Baton Rouge, LA. 36 pp.

Vincent M. A. 1997. 9. Illiciaceae (de Candolle) A. C. Smith. Star-anise family. pp. 59–61 *In*: N. R. Morin (convening ed.), *Flora of North America North of Mexico, Vol. 3: Magnoliophyta: Magnoliidae and Hamamelidae.* Oxford University Press, New York.

Vincent, M. A. 2003. 2. *Glinus* Linnaeus. pp. 509–512. *In*: N. R. Morin (convening ed.), *Flora of North America North of Mexico, Vol. 4: Magnoliophyta: Caryophyllidae, Part 1.* Oxford University Press, New York.

Vincent, R. E. 1958. The larger plants of Little Kitoi Lake. *Amer. Midl. Naturalist* 60: 212–218.

Vincieri, F. F., S. A. Coran, V. Giannellini & M. Bambagiotti-Alberti. 1985. Oxygenated C15 polyacetylenes from *Oenanthe aquatica* fruits. *Plant Med.* 2: 107–110.

Vinha, D., L. F. Alves, L. B. P. Zaidan & M. T. Grombone-Guaratini. 2011. The potential of the soil seed bank for the regeneration of a tropical urban forest dominated by bamboo. *Landscape Urban Plan.* 99: 178–185.

Visser, J. M. & C. E. Sasser. 2009. The effect of environmental factors on floating fresh marsh end-of-season biomass. *Aquat. Bot.* 91: 205–212.

Visser, J. M., C. E. Sasser, R. H. Chabreck & R. G. Linscombe. 1998. Marsh vegetation types of the Mississippi River deltaic plain. *Estuaries* 21: 818–828.

Visser, J. M., C. E. Sasser, R. H. Chabreck & R. G. Linscombe. 1999. Long-term vegetation change in Louisiana tidal marshes. *Wetlands* 19: 168–175.

Visser, J. M., C. E. Sasser, R. G. Linscombe & R. H. Chabreck. 2000. Marsh vegetation types of the Chenier plain, Louisiana, USA. *Estuaries* 23: 318–327.

Visser, J. M., C. E. Sasser, R. H. Chabreck & R. G. Linscombe. 2002. The impact of a severe drought on the vegetation of a subtropical estuary. *Estuaries* 25: 1184–1195.

Vitt, D. H. & W.-L. Chee. 1990. The relationships of vegetation to surface water chemistry and peat chemistry in fens of Alberta, Canada. *Vegetatio* 89: 87–106.

Vitt, D. H. & D. G. Horton. 1990. Rich fen bryophytes in Missouri: ecological comments and three state records. *Bryologist* 93: 62–65.

Vittoz, P. & R. Engler. 2007. Seed dispersal distances: a typology based on dispersal modes and plant traits. *Bot. Helv.* 117: 109–124.

Vivian-Smith, G. & E. W. Stiles. 1994. Dispersal of salt marsh seeds on the feet and feathers of waterfowl. *Wetlands* 14: 316–319.

Vivrette, N. J., J. E. Bleck & W. R. Ferren Jr. 2003. 36. Aizoaceae Martinov. pp. 75–76. *In*: N. R. Morin (convening ed.), *Flora of North America North of Mexico, Vol. 4: Magnoliophyta: Caryophyllidae, Part 1.* Oxford University Press, New York.

Vizgirdas, R. S. & E. M. Rey-Vizgirdas. 2005. *Wild Plants of the Sierra Nevada.* University of Nevada Press, Reno, NV. 384 pp.

Vlyssides, A., E. M. Baramoouti & S. Mai. 2005. Heavy metal removal from water resources using the aquatic plant *Apium nodiflorum. Commun. Soil Sci. Plant Analysis* 36: 1075–1081.

Voegtlin, D. J., R. J. O'Neil & W. R. Graves. 2004. Tests of suitability of overwintering hosts of *Aphis glycines*: Identification of a new host association with *Rhamnus alnifolia* L'Heritier. *Ann. Entomol. Soc. Amer.* 97: 233–234.

Voeks, R. A. 1997. *Sacred Leaves of Candomblé: African Magic, Medicine, and Religion in Brazil.* University of Texas Press, Austin, TX. 256 pp.

Vogler, D. W. & A. G. Stephenson. 2001. The potential for mixed mating in a self-incompatible plant. *Int. J. Plant Sci.* 162: 801–805.

Vogt, K., L. Rasran & K. Jensen. 2004. Water-borne seed transport and seed deposition during flooding in a small river-valley in Northern Germany. *Flora* 199: 377–388.

Vogt, K., L. Rasran & K. Jensen. 2006. Seed deposition in drift lines during an extreme flooding event-Evidence for hydrochorous dispersal? *Basic Appl. Ecol.* 7: 422–432.

Vogt, K., L. Rasran & K. Jensen. 2007. Seed deposition in drift lines: Opportunity or hazard for species establishment? *Aquat. Bot.* 86: 385–392.

Vohora, S. B., T. Khanna, M. Athar & B. Ahmad. 1997. Analgesic activity of bacosine, a new triterpene isolated from *Bacopa monnieri*. *Fitoterapia* 68: 361–365.

Voigt, J. W. & R. H. Mohlenbrock. 1964. *Plant Communities of Southern Illinois*. Southern Illinois University Press, Carbondale, IL. 202 pp.

Voirin, B., C. Bayet, O. Faure & F. Jullien.1999. Free flavonoid aglycones as markers of parentage in *Mentha aquatica*, *M. citrata*, *M. spicata* and *M. ×piperita*. *Phytochemistry* 50: 1189–1193.

Volkova, P. A., P. Trávníček & C. Brochmann. 2010. Evolutionary dynamics across discontinuous freshwater systems: Rapid expansions and repeated allopolyploid origins in the Palearctic white water-lilies (*Nymphaea*). *Taxon* 59: 483–494.

Vollmar, A. W. Schäfera & H. Wagner. 1986. Immunologically active polysaccharides of *Eupatorium cannabinum* and *Eupatorium perfoliatum*. *Phytochemistry* 25: 377–381.

Vonhoff, C. & H. Winterhoff. 2006. Kardiale Effekte von *Lycopus europaeus* L. im Tierexperiment. *Z. Phytother.* 27: 110–119.

Vonhoff, C., A. Baumgartner, M. Hegger, B. Korte, A. Biller & H. Winterhoff. 2006. Extract of *Lycopus europaeus* L. reduces cardiac signs of hyperthyroidism in rats. *Life Sci.* 78:1063–1070.

Von Kesseler, E. 1932. Observations on chromosome number in *Althaea rosea*, *Callirhoe involucrata*, and *Hibiscus coccineus*. *Amer. J. Bot.* 19: 128–130.

Von Oheimb, G., M. Schmidt, W.-U. Kriebitzsch & H. Ellenberg. 2005. Dispersal of vascular plants by game in northern Germany. Part II: Red deer (*Cervus elaphus*). *Eur. J. Forest Res.* 124: 55–65.

Vorsa, N. 1997. On a wing: The genetics and taxonomy of *Vaccinium* species from a pollination perspective. *Acta Hort.* 446: 59–66.

Voss, E. G. 1950. Observations on the Michigan flora. III, the flora of Green Island (Mackinac County). *Ohio J. Sci.* 50: 182–190.

Voss, E. G. 1954. The butterflies of Emmet and Cheboygan Counties, Michigan with other notes on northern Michigan butterflies. *Amer. Midl. Naturalist* 51: 87–104.

Voss, E. G. 1985. *Michigan Flora. Part II: Dicots (Saururaceae-Cornaceae)*. Bulletin 59. Cranbrook Institute of Science, Bloomfield Hills, MI. 724 pp.

Voss, E. G. 1996. *Michigan Flora. Part III: Dicots (Pyrolaceae-Compositae)*. Bulletin 61. Cranbrook Institute of Science, Bloomfield Hills, MI. 622 pp.

Vossler, F. G. 2012. Flower visits, nesting and nest defence behaviour of stingless bees (Apidae: Meliponini): suitability of the bee species for meliponiculture in the Argentinean Chaco region. *Apidologie* 43: 139. doi:10.1007/s13592-011-0097-6.

Vroege, P. W. & P. Stelleman. 1990. Insect and wind pollination in *Salix repens* L. and *Salix caprea* L. *Israel J. Bot.* 39: 125–132.

Vujnovic, K., L. Allen, J. D. Johnson & D. Vujnovic. 2005. *Survey of Rare Vascular Plants in La Butte Creek, Wildland Provincial Park*. Alberta Natural Heritage Information Centre, Parks and Protected Areas, Alberta Community Development, Edmonton, AB. 73 pp.

Wacker, M. & N. M. Kelly. 2004. Changes in vernal pool edaphic settings through mitigation at the project and landscape scale. *Wetlands Ecol. Manag.* 12: 165–178.

Waddington, K. D. 1979. Divergence in inflorescence height: an evolutionary response to pollinator fidelity. *Oecologia* 40: 43–50.

Wade, G. L. & M. T. Mengak. 2010. Deer-tolerant ornamental plants. Circular 985. Cooperative Extension, University of Georgia, Athens, GA. 8 pp.

Wade, L. K. 1965. Vegetation and history of the sphagnum bogs of the Tofino area, Vancouver Island. MS thesis. University of British Columbia, Vancouver, BC. 125 pp.

Wagenaar, E. B. 1969. End-to-end chromosome attachments in mitotic interphase and their possible significance to meiotic chromosome pairing. *Chromosoma* 26: 410–426.

Wagner, D. L., D. Adamski & R. L. Brown. 2004. A new species of *Mompha* Hübner (Lepidoptera: Coleophoridae: Momphinae) from buttonbush (*Cephalanthus occidentalis* L.) with descriptions of the early stages. *Proc. Entomol. Soc. Washington* 106: 1–18.

Wagner, H. & B. Fessler. 1986. *In vitro* 5-lipoxygenase inhibition by *Eclipta alba* extracts and the coumestan derivative wedelolactone. *Plant Med. (Stuttgart)* 52: 374–377.

Wagner, H., A. Proksch, I. Riess-Maurer, A. Vollmar, S. Odenthal, H. Stuppner, K. Jurcic, M. Le Turdu & J. N. Fang. 1985. Immunostimulating polysaccharides (heteroglycans) of higher plants. *Arzneimittelforschung* 35: 1069–1075.

Wagner, H., B. Geyer, Y. Kiso, H. Hikino & G. S. Rao. 1986. Coumestans as the main active principles of the liver drugs *Eclipta alba* and *Wedelia calendulacea*. *Plant Med. (Stuttgart)* 52: 370–374.

Wagner, I. & A. M. Simons. 2008. Intraspecific divergence in seed germination traits between high-and low-latitude populations of the Arctic-alpine annual *Koenigia islandica*. *Arctic Antarc. Alpine Res.* 40: 233–239.

Wagner, I. & A. M. Simons. 2009a. Divergence among arctic and alpine populations of the annual *Koenigia islandica*: morphology, life-history, and phenology. *Ecography* 32: 114–122.

Wagner, I. & A. M. Simons. 2009b. Divergence in germination traits among arctic and alpine populations of *Koenigia islandica*: light requirements. *Plant Ecol.* 204: 145–153.

Wagner, M., P. Poschlod & R. P. Setchfield. 2003. Soil seed bank in managed and abandoned semi-natural meadows in Soomaa National Park, Estonia. *Ann. Bot. Fenn.* 40: 87–100.

Wagner, W. L. 2005a. Systematics of *Oenothera* sections *Contortae*, *Eremia*, and *Ravenia* (Onagraceae). *Syst. Bot.* 30: 332–355.

Wagner, W. L. 2005b. 26. *Honckenya* Ehrhart. pp. 137–140. *In*: Flora North America Editorial Committee (eds.), *Flora of North America North of Mexico*, Vol. 5: Magnoliophyta: Caryophyllidae, part 2. Oxford University Press, New York.

Wagner, W. L., P. C. Hoch & P. H. Raven. 2007. Revised classification of the Onagraceae. *Syst. Bot. Monogr.* 83: 1–240.

Wagstaff, S. J. & F. Hennion. 2007. Evolution and biogeography of *Lyallia* and *Hectorella* (Portulacaceae), geographically isolated sisters from the Southern Hemisphere. *Antarc. Sci.* 19: 417–426.

Wagstaff, S. J. & R. G. Olmstead. 1997. Phylogeny of Labiatae and Verbenaceae inferred from *rbcL* sequences. *Syst. Bot.* 22: 165–179.

Wagstaff, S. J., R. G. Olmstead & P. D. Cantino. 1995. Parsimony analysis of cpDNA restriction site variation in subfamily Nepetoideae (Labiatae). *Amer. J. Bot.* 82: 886–892.

Wagstaff, S. J., L. Hickerson, R. Spangler, P. A, Reeves & R. G. Olmstead. 1998. Phylogeny in *Labiatae* s. l., inferred from cpDNA sequences. *Plant Syst. Evol.* 209: 265–274.

Wagstaff, S. J., I. Breitwieser & U. Swenson. 2006. Origin and relationships of the Austral genus *Abrotanella* (Asteraceae) inferred from DNA sequences. *Taxon* 55: 95–106.

Wahl, H.A. 1954. A preliminary study of the genus *Chenopodium* in North America. *Bartonia* 27: 1–46.

Wahlberg, N. 2001. The phylogenetics and biochemistry of host-plant specialization in Melitaeine butterflies (Lepidoptera: Nymphalidae). *Evolution* 55: 522–537.

Wahlert, G. A., T. Marcussen, J. de Paula-Souza, M. Fen & H. E. Ballard Jr. 2014. A phylogeny of the Violaceae (Malpighiales) inferred from plastid DNA sequences: implications for generic diversity and intrafamilial classification. *Syst. Bot.* 39: 239–252.

Wain, R. P., W. T. Haller & D. F. Martin. 1983. Genetic relationship among three forms of *Cabomba*. *J. Aquat. Plant Manag.* 21: 96–98.

Wainwright, C. M. 1978. Hymenopteran territoriality and its influences on the pollination ecology of *Lupinus arizonicus*. *S. W. Naturalist* 23: 605–615.

Wainwright, C. E., E. M. Wolkovich & E. E. Cleland. 2012. Seasonal priority effects: implications for invasion and restoration in a semi-arid system. *J. Appl. Ecol.* 49: 234–241.

Wakefield, A. E., N. J. Gotelli, S. E. Wittman & A. M. Ellison. 2005. Prey addition alters nutrient stoichiometry of the carnivorous plant *Sarracenia purpurea*. *Ecology* 86: 1737–1743.

Walck, J. L., J. M. Baskin & C. C. Baskin. 1996: An ecologically and evolutionarily meaningful definition of a persistent seed bank in *Solidago*. *Amer. J. Bot.* 83: S78-S79.

Walck, J. L., J. M. Baskin & C. C. Baskin. 1998. A comparative study of the seed germination biology of a narrow endemic and two geographically-widespread species of *Solidago* (Asteraceae). 6. Seed bank. *Seed Sci. Res.* 8: 65–74.

Walck, J. L., C. C. Baskin & J. M. Baskin. 1999. Seeds of *Thalictrum mirabile* (Ranunculaceae) require cold stratification for loss of nondeep simple morphophysiological dormancy. *Can. J. Bot.* 77: 1769–1776.

Walden, G. K., L. M. Garrison, G. S. Spicer, F. W. Cipriano & R. Patterson. 2014. Phylogenies and chromosome evolution of *Phacelia* (Boraginaceae: Hydrophylloideae) inferred from nuclear ribosomal and chloroplast sequence data. *Madroño* 61: 16–47.

Waldron, G., M. Gartshore & K. Colthurst. 1996. Pumpkin ash, *Fraxinus profunda*, in southwestern Ontario. *Can. Field-Naturalist* 110: 615–619.

Walford, G., G. Jones, W. Fertig, S. Mellman-Brown & K. E. Houston. 2001. *Riparian and Wetland Plant Community Types of the Shoshone National Forest*. Gen. Tech. Rept. RMRS-GTR-85. U. S. Dept. of Agriculture, Forest Service, Rocky Mountain Research Station, Ogden, UT. 122 pp.

Walker, B. H. & C. F. Wehrhahn. 1971. Relationships between derived vegetation gradients and measured environmental variables in Saskatchewan wetlands. *Ecology* 52: 85–95.

Walker, D. A. & K. R. Everett. 1991. Loess ecosystems of northern Alaska: regional gradient and toposequence at Prudhoe Bay. *Ecol. Monogr.* 61: 437–464.

Walker, J. & R. K. Peet. 1984. Composition and species diversity of pine-wiregrass savannas of the Green Swamp, North Carolina. *Vegetatio* 55: 163–179.

Walker, J., M. S. Lindberg, M. C. MacCluskie, M. J. Petrula & J. S. Sedinger. 2005. Nest survival of scaup and other ducks in the boreal forest of Alaska. *J. Wildl. Manag.* 69: 582–591.

Walker, J. B. & K. J. Sytsma. 2007. Staminal evolution in the genus *Salvia* (Lamiaceae): molecular phylogenetic evidence for multiple origins of the staminal lever. *Ann. Bot.* 100: 375–391.

Walker, J. L. & A. M. Silletti. 2005. A three-year demographic study of Harper's beauty (*Harperocallis flava* McDaniel), an endangered Florida endemic. *J. Torrey Bot. Soc.* 132: 551–560.

Walker, M. D., D. A. Walker & N. A. Auerbach. 1994. Plant communities of a tussock tundra landscape in the Brooks Range foothills, Alaska. *J. Veg. Sci.* 5: 843–866.

Walker, X. J., J. F. Basinger & S. G. W. Kaminskyj. 2010. Endorhizal Fungi in *Ranunculus* from western and Arctic Canada: predominance of fine endophytes at high latitudes. *Open Mycol. J.* 4: 1–9.

Walkinshaw, L. H. 1935. Studies of the short-billed marsh wren (*Cistothorus stellaris*) in Michigan. *Auk* 52: 362–369.

Walkinshaw, L. H. 1966. Summer biology of Traill's flycatcher. *Wilson Bull.* 78: 31–46.

Walkinshaw, L. H. & C. J. Henry. 1957. Yellow-bellied flycatcher nesting in Michigan. *Auk* 74: 293–304.

Wall, M. A., M. Timmerman-Erskine & R. S. Boyd. 2003. Conservation impact of climatic variability on pollination of the Federally endangered plant, *Clematis socialis* (Ranunculaceae). *S. E. Naturalist (Steuben)* 2: 11–24.

Wall, W. A. 2007. Characterization of a North Carolina lower Coastal Plain wet pine savanna. MS thesis. North Carolina State University, Raleigh, NC. 77 pp.

Wall, W. A., M. G. Hohmann, A. S. Walker & J. B. Gray. 2013. Sex ratios and population persistence in the rare shrub *Lindera subcoriacea* Wofford. *Plant Ecol.* 214: 1105–1114.

Wallace, P. M., D. M. Kent & E. R. Rich. 1996. Responses of wetland tree species to hydrology and soils. *Restorat. Ecol.* 4: 33–41.

Wallace, R. L. 1978. Substrate selection by larvae of the sessile rotifer *Ptygura beauchampi*. *Ecology* 59: 221–227.

Wallace, R. S. & R. K. Jansen. 1990. Systematic implications of chloroplast DNA variation in the genus *Microseris* (Asteraceae: Lactuceae). *Syst. Bot.* 15: 606–616.

Wallace, C. S. & S. R. Szarek. 1981. Ecophysiological studies of Sonoran desert plants. VII. Photosynthetic gas exchange of winter ephemerals from sun and shade environments. *Oecologia* 51: 57–61.

Wallander, E. 2008. Systematics of *Fraxinus* (Oleaceae) and evolution of dioecy. *Plant Syst. Evol.* 273: 25–49.

Wallander, E. & V. A. Albert. 2000. Phylogeny and classification of Oleaceae based on *rps16* and *trnL-F* sequence data. *Amer. J. Bot.* 87: 1827–1841.

Wallenstein, A. & L. S. Albert. 1963. Plant morphology: Its control in *Proserpinaca* by photoperiod, temperature, and gibberellic acid. *Science* 140: 998–1000.

Wallentinus, I. 2002. Introduced marine algae and vascular plants in European aquatic environments. pp. 27–52. *In*: E. Leppäkowski et al. (eds.), *Invasive Aquatic Species of Europe*. Kluwer Academic Publishers, Dordrecht, The Netherlands.

Walley, R. C. 2007. Environmental factors affecting the distribution of native and invasive aquatic plants in the Atchafalaya River basin, Louisiana, U.S.A. MS thesis. School of Renewable Natural Resources, Louisiana State University, Baton Rouge, LA. 106 pp.

Wallis, C. 2001. *Proposed Protocols for Inventories of Rare Plants of the Grassland Natural Region*. Alberta Species at Risk Report No. 21. Alberta Environment, Natural Resources Service. Edmonton, AB. 305 pp.

Wallmo, O. C., W. L. Regelin & D. W. Reichert. 1972. Forage use by mule deer relative to logging in Colorado. *J. Wildl. Manag.* 36: 1025–1033.

Walters, E. L., E. H. Miller & P. E. Lowther. 2002. Red breasted Sapsucker, Red naped Sapsucker. pp. 1–32. *In*: A. Poole & F, Gill (eds.), *The Birds of North America, No. 663*. Cornell Laboratory of Ornithology and The Academy of Natural Sciences, Ithaca, New York.

Walters, T. W. & D. S. Decker-Walters. 1993. Systematics of the endangered Okeechobee gourd (*Cucurbita okeechobeensis*: Cucurbitaceae). *Syst. Bot.* 18: 175–187.

Walters, T. W. & R. Wyatt. 1982. The vascular flora of granite outcrops in the central mineral region of Texas. *Bull. Torrey Bot. Club* 109: 344–364.

Walters, T. W., D. S. Decker-Walters & S. Katz. 1992. Seeking the elusive Okeechobee gourd. *Fairchild Trop. Gard. Bull.* 47: 23–30.

Walti, A. 1945. Determination of the nature of the volatile base from the rhizome of the pitcher plant *Sarracenia purpurea*. *J. Amer. Chem. Soc.* 67: 2271.

Walton, G. B. 1994. *Report for Field Season 1994 Status Survey for* Caltha natans *and* Sparganium glomeratum *in Minnesota*. Conservation Biology Research, Natural Heritage Program, Minnesota Department of Natural Resources, Minneapolis, MN. 8 pp.

Walton, G. B. 1995. *Report for the 1994–1995 Status Survey for* Sparganium glomeratum *in Minnesota*. Natural Heritage & Nongame Research Program, Division of Ecological Services, Minnesota Department of Natural Resources, Minneapolis, MN. 43 pp.

Walton, J. 1922. A Spitsbergen salt marsh: with observations on the ecological phenomena attendant on the emergence of land from the sea. *J. Ecol.* 10: 109–121.

Walton, K., T. Gotthardt, T. Nawrocki & J. Reimer. 2014. *Prince of Wales Island Amphibian Survey, 2013*. Preliminary results. Alaska Natural Heritage Program, University of Alaska Anchorage, Anchorage, AK. 29 pp.

Walton, G. B. & O. Lakela. 1995. Report for the 1994-1995 *Status Survey for Sparganium glomeratum in Minnesota*. Minnesota Department of Natural Resources, Natural Heritage & Nongame Research Program, Division of Ecological Services. Minneapolis, MN. 43 pp.

Wan-Ibrahima, W. I., K. Sidikb & U. R. Kuppusamya. 2010. A high antioxidant level in edible plants is associated with genotoxic properties. *Food Chem.* 122: 1139–1144.

Wang, B., W. Li & J. Wang. 2005. Genetic diversity of *Alternanthera philoxeroides* in China. *Aquat. Bot.* 81: 277–283.

Wang, G. G. & K. J. Kemball. 2005. Effects of fire severity on early development of understory vegetation. *Can. J. Forest Res.* 35: 254–262.

Wang, G. W., W. T. Hu, B. K. Huang & L. P. Qin. 2011. *Illicium verum*: A review on its botany, traditional use, chemistry and pharmacology. *J. Ethnopharmacol.* 136: 10–20.

Wang, G.-X., H. Watanabe, A. Uchino & K. Itoh. 2000. Response of a sulfonylurea (SU)-resistant biotype of *Limnophila sessiliflora* to selected SU and alternative herbicides. *Pestic. Biochem. Physiol.* 68: 59–66.

Wang, H. & D.-Z. Li. 2005. Pollination biology of four *Pedicularis* species (Scrophulariaceae) in northwestern Yunnan, China. *Ann. Missouri Bot. Gard.* 92: 127–138.

Wang, H., H.-J. He, J.-Q. Chen & L. Lu. 2010a. Palynological data on Illiciaceae and Schisandraceae confirm phylogenetic relationships within these two basally-branching angiosperm families. *Flora* 205: 221–228.

Wang, W., H. Hu, X.-G. Xiang, S.-X. Yu & Z.-D. Chen. 2010b. Phylogenetic placements of *Calathodes* and *Megaleranthis* (Ranunculaceae): Evidence from molecular and morphological data. *Taxon* 59: 1712–1720.

Wang, W., A.-M. Lu, Y. Ren, M. E. Endress & Z.-D. Chen. 2009. Phylogeny and classification of Ranunculales: Evidence from four molecular loci and morphological data. *Perspect. Plant Ecol. Evol. Syst.* 11: 81–110.

Wang, Y., Q.-F. Wang, Y.-H. Guo & S. C. H. Barrett. 2005. Reproductive consequences of interactions between clonal growth and sexual reproduction in *Nymphoides peltata*: a distylous aquatic plant. *New Phytol.* 165: 329–336.

Wang, Y.-J., X.-J. Li, G. Hao & J.-Q. Liu. 2004a. Molecular phylogeny and biogeography of *Androsace* (Primulaceae) and the convergent evolution of cushion morphology. *Acta Phytotax. Sin.* 42: 481–499.

Wang, Z.-F., J. L. Hamrick & M. J. W. Godt. 2004b. High genetic diversity in *Sarracenia leucophylla* (Sarraceniaceae), a carnivorous wetland herb. *J. Heredity* 95: 234–243.

Warashina, T. & T. Noro. 2000. Steroidal glycosides from the aerial part of *Asclepias incarnata* L. II. *Chem. Pharm. Bull.* 48: 99–107.

Ward, D. B. 1977. Keys to the flora of Florida. 4. *Nymphaea* (Nymphaeaceae). *Phytologia* 37: 443–448.

Ward, D. B. & A. F. Clewell. 1989. Atlantic white cedar (*Chamaecyparis thyoides*) in the southern states. *Florida Sci.* 52: 8–47.

Ward, D. B. & M. C. Minno. 2002. Rediscovery of the endangered Okeechobee gourd (*Cucurbita okeechobeensis*) along the St. Johns River, Florida, where last reported by William Bartram in 1774. *Castanea* 67: 201–206.

Ward, L. A,. C. Wilson, C. Saenz, L. K. Harrell, G. J. Steck & R. Wharton. 2013. New host plant and distribution records of Tephritidae (Diptera) from Texas, with notes on parasitism of Tephritidae by Opiinae (Hymenoptera: Braconidae). *Proc. Entomol. Soc. Washington* 115: 96–102.

Ward, P. 1968. Fire in relation to waterfowl habitat of the delta marshes. pp. 255–267. *In*: *Proceedings, Annual Tall Timbers Fire Ecology Conference; 1968 March 14–15; Tallahassee, Florida*. No. 8. Tall Timbers Research Station, Tallahassee, FL.

Wardlow, I. F., M. W. Moncur & C. J. Totterdell. 1989a. The growth and development of *Caltha introloba* F. Muell. I. The pattern and control of flowering. *Austral. J. Bot.* 37: 275–289.

Wardlow, I. F., M. W. Moncur & C. J. Totterdell. 1989b. The growth and development of *Caltha introloba* F. Muell. II. The regulation of germination, growth and photosynthesis by temperature. *Austral. J. Bot.* 37: 291–303.

Ware, C., D. M. Bergstrom, E. Müller & I. Greve Alsos. 2012. Humans introduce viable seeds to the Arctic on footwear. *Biol. Invas.* 14: 567–577.

Ware, R. A. 1915. Josselyn Botanical Society of Maine. *Bull. Josselyn Bot. Soc. Maine* 5: 1–4.

Warner, R. E. & K. M. Hendrix (eds.). 1984. *California Riparian Systems: Ecology, Conservation and Productive Management*. University of California Press, Berkeley, CA. 1035 pp.

Warner, S. R. 1926. Distribution of native plants and weeds on certain soil types in eastern Texas. *Bot. Gaz.* 82: 345–372.

Warners, D. P. & D. C. Laughlin. 1999. Evidence for a species-level distinction of two co-occurring asters: *Aster puniceus* L. and *Aster firmus* Nees. *Michigan Bot.* 38: 19–31.

Warnock, M. J. 1997. 16. *Delphinium* Linnaeus. pp. 196–240. *In*: N. R. Morin (convening ed.), *Flora of North America North of Mexico, Vol. 3: Magnoliophyta: Magnoliidae and Hamamelidae*. Oxford University Press, New York.

Warren, K. A. 2001. Habitat use, nest success, and management recommendations for grassland birds of the Canaan Valley National Wildlife Refuge, West Virginia. MS thesis. West Virginia University, Morgantown, WV. 147 pp.

Warren, P. H. 1993. Insect herbivory on water mint: you can't get there from here? *Ecography* 16: 11–15.

Warrington, P. D. 1986. The pH tolerance of the aquatic plants of British Columbia, Part I. Literature survey of the pH limits of aquatic plants of the world. 157 pp. Available online: http://www.env.gov.bc.ca/wat/wq/plants/plantph.pdf.

Warwick, S. I. 1991. Herbicide resistance in weedy plants: physiology and population biology. *Ann. Rev. Ecol. Syst.* 22: 95–114.

Warwick, S. I., I. A. Al-Shehbaz & C. A. Sauder. 2006. Phylogenetic position of *Arabis arenicola* and generic limits of *Eutrema* and *Aphragmus* (Brassicaceae) based on sequences of nuclear ribosomal DNA. *Can. J. Bot.* 84: 269–281.

Warwick, S. I., K. Mummenhoff, C. A. Sauder, M. A. Koch & I. A. Al-Shehbaz. 2010. Closing the gaps: phylogenetic relationships in the Brassicaceae based on DNA sequence data of nuclear ribosomal ITS region. *Plant Syst. Evol.* 285: 209–232.

Waselkov, K. 2013. Population genetics and phylogenetic context of weed evolution in the genus *Amaranthus* (Amaranthaceae). PhD dissertation. Washington University, St. Louis, MO. 252 pp.

Waselkov, K. & K. M. Olsen. 2014. Population genetics and origin of the native North American agricultural weed waterhemp (*Amaranthus tuberculatus*; Amaranthaceae). *Amer. J. Bot.* 101: 1726–1736.

Waser, N. M., R. K. Vickery Jr. & M. V. Price. 1982. Patterns of seed dispersal and population differentiation in *Mimulus guttatus*. *Evolution* 36: 753–761.

Wasshausen, D. 2004. Acanthaceae. pp. 3–7. *In:* N. Smith, S. A. Mori, A. Henderson, D. W. Stevenson & S. V. Heald (eds.), *Flowering Plants of the Neotropics.* Princeton University Press, Princeton, NJ. 594 pp.

Wassink, E. & C. M. Caruso. 2013. Effect of coflowering *Mimulus ringens* on phenotypic selection on floral traits of gynodioecious *Lobelia siphilitica*. *Botany* 91: 745–751.

Wasule, D. D. 2011. Hair dyeing activity of *Eclipta* species. *J. Pharm. Res. Opin.* 1: 159–160.

Watanabe, K., R. M. King, T. Yahara, M. Ito, J. Yokoyama, T. Suzuki & D. J. Crawford. 1995. Chromosomal cytology and evolution in Eupatorieae (Asteraceae). *Ann. Missouri Bot. Gard.* 82: 581–592.

Watanabe, K., T. Yahara, G. Hashimoto, Y. Nagatani, A. Soejima, T. Kawahara & M. Nakazawa. 2007. Chromosome numbers and karyotypes in Asteraceae. *Ann. Missouri Bot. Gard.* 94: 643–654.

Watrous, K. M. & J. H. Cane. 2011. Breeding biology of the threadstalk milkvetch, *Astragalus filipes* (Fabaceae), with a review of the genus. *Amer. Midl. Naturalist* 165: 225–240.

Watson, E. E. 1928. Contributions to a monograph of the genus *Helianthus*. *Pap. Michigan Acad. Sci.* 9: 305–476.

Watson, J. W., K. R. McAllister, D. J. Pierce & A. Alvarado. 2000. *Ecology of a Remnant Population of Oregon Spotted Frogs (Rana pretiosa) in Thurston County, Washington.* Final Report. Washington Department of Fish and Wildlife, Olympia, WA. 85 pp.

Watson, L. E. 2006a. 126. *Cotula* Linnaeus. pp. 543–544. *In:* Flora of North America Editorial Committee (eds.), *Flora of North America North of Mexico: Vol. 19, Magnoliophyta: Asteridae (in part): Asteraceae, Part 1.* Oxford University Press, New York.

Watson, L. E. 2006b. 391. *Marshallia* Schreber. pp. 456–458. *In:* Flora of North America Editorial Committee (eds.), *Flora of North America North of Mexico: Vol. 21, Magnoliophyta: Asteridae (in part): Asteraceae, Part 3.* Oxford University Press, New York.

Watson, L. E. & J. R. Estes. 1990. Biosystematic and phenetic analysis of *Marshallia*. *Syst. Bot.* 15: 403–414.

Watson, L. E., W. J. Elisens & J. R. Estes.1994a. Genetic variation within and among populations of the *Marshallia graminifolia* complex (Asteraceae). *Biochem. Syst. Ecol.* 22: 577–582.

Watson, L. E., G. E. Uno, N. A. McCarty & A. B. Kornkven. 1994b. Conservation biology of a rare plant species, *Eriocaulon kornickianum* (Eriocaulaceae). *Amer. J. Bot.* 81: 980–986.

Watson, L. E., P. L. Bates, T. M. Evans, M. M. Unwin & J. R. Estes. 2002. Molecular phylogeny of subtribe Artemisiinae (Asteraceae), including *Artemisia* and its allied and segregate genera. *BMC Evol. Biol.* 2: 17. doi:10.1186/1471-2148-2-17.

Watts, W. A. 1971. The identity of *Menyanthes microsperma* n. sp. foss. from the Gort Interglacial, Ireland. *New Phytol.* 70: 435–436.

Waldbauer, G. P. & A. W. Ghent. 1984. Flower associations and mating behavior or its absence at blossoms by *Spilomyia* spp. (Diptera, Syrphidae). *Great Lakes Entomol.* 17: 13–16.

Weakley, A. S. 2005. The ongoing task of describing our region's flora. *NCBG Newslett.* January-February: 9.

Weakley, A. S. 2015. 25. *Sanguisorba* Linnaeus. pp. 321–323. *In:* N. R. Morin (convening ed.), *Flora of North America North of Mexico, Vol. 9: Magnoliophyta: Picramniaceae to Rosaceae.* Oxford University Press, New York.

Weakley, A. S. & G. L. Nesom. 2004. A new species of *Ptilimnium* (Apiaceae) from the Atlantic coast. *Sida* 21: 743–752.

Weakley, A. S. & M. P. Schafale. 1994. Non-alluvial wetlands of the Southern Blue Ridge—Diversity in a threatened ecosystem. *Water Air Soil Pollut.* 77: 359–383.

Weatherbee, E. E. & D. Price. 2001. Middleville Fen, Barry County plant list. Available online: http://tracmvl.webs.com/MvlFen.pdf (accessed October 23, 2014).

Weaver, J. E. 1960. Flood plain vegetation of the central Missouri Valley and contacts of woodland with prairie. *Ecol. Monogr.* 30: 37–64.

Weaver Jr., R. E. & P. J. Anderson. 2007. Botany section. *Tri-ology* 46: 1–5.

Weaver, V. & R. Adams. 1996. Horses as vectors in the dispersal of weeds into native vegetation. *Proc. 11th Austral. Weeds Conf.* 30: 383–397.

Webb, C. J. 1998. The selection of pollen and seed dispersal in plants. *Plant Species Biol.* 13: 57–67.

Webb, C. J. & E. J. Beuzenberg. 1987. Contributions to a chromosome atlas of the New Zealand flora-corrections and additions to number 21 Umbelliferae (*Hydrocotyle*). *New Zealand J. Bot.* 25: 371–372.

Webber, J. M. & P. W. Ball. 1980. Introgression in Canadian populations of *Lycopus americanus* Muhl. and *L. europaeus* L. (Labaitae). *Rhodora* 82: 281–304.

Weber, J. & S. Rooney. 1994. Josselyn Botanical Society's annual meeting, Unity College, July 1993. *Maine Naturalist* 2: 50–52.

Weber, J. A. 1972. The importance of turions in the propagation of *Myriophyllum exalbescens* (Haloragidaceae) in Douglas Lake, Michigan. *Michigan Bot.* 11: 115–121.

Weber, J. A. & L. D. Noodén. 1974. Turion formation in *Myriophyllum verticillatum*: phenology and its interpretation. *Michigan Bot.* 13: 151–158.

Weber, J. A. & L. D. Noodén. 1976a. Environmental and hormonal control of turion formation in *Myriophyllum verticillatum*. *Plant Cell Physiol.* 17: 721–731.

Weber, J. A. & L. D. Noodén. 1976b. Environmental and hormonal control of turion germination in *Myriophyllum verticillatum*. *Amer. J. Bot.* 63: 936–944.

Weber, R. M. 1979. Species biology of *Chrysosplenium iowense*. MA thesis. University of Northern Iowa, Cedar Falls, IA. 94 pp.

Webster, D., P. Taschereau, T. D. G. Less & T. Jurgens. 2006. Immunostimulant properties of *Heracleum maximum* Bartr. *J. Ethnopharmacol.* 106: 360–363.

Webster, D., P. Taschereau, R. J. Belland, C. Sand & R. P. Rennie. 2008. Antifungal activity of medicinal plant extracts; preliminary screening studies. *J. Ethnopharmacol.* 115: 140–146.

Webster, G. L., W. V. Brown & B. N. Smith. 1975. Systematics of photosynthetic carbon fixation pathways in *Euphorbia*. *Taxon* 24: 27–33.

Webster, R. P., D. B. McCorquodale & C. G. Majka. 2009. New records of Cerambycidae (Coleoptera) for New Brunswick, Nova Scotia, and Prince Edward Island, Canada. *ZooKeys* 22: 285–308.

Webster, R. P., R. S. Anderson, J. D. Sweeney & I. DeMerchant. 2012. New Coleoptera records from New Brunswick, Canada: Anthribidae, Brentidae, Dryophthoridae, Brachyceridae, and Curculionidae, with additions to the fauna of Quebec, Nova Scotia and Prince Edward Island. *ZooKeys* 179: 349–406.

Webster, S. D. 1991. A chromatographic investigation of the flavonoids of *Ranunculus* L. subgenus *Batrachium* (DC.) A. Gray (water buttercups) and selected species in subgenus *Ranunculus*. *Aquat. Bot.* 40: 11–26.

Weeks, K. 2009. Population ecology of the floodplain herb *Macbridea caroliniana* (Lamiaceae) with investigations on the species' habitat, breeding system and genetic diversity. All Dissertations. Paper 437. Available online: http://tigerprints. clemson.edu/all_dissertations/437.

Weeks, K. & J. Walker. 2006. Microhabitat indicators for *Macbridea caroliniana*, a rare mint of floodplain communities. Poster, Ecological Society of America, Annual Meeting (2006).

Weems Jr., H. V. 1953. Notes on collecting syrphid flies (Diptera: Syrphidae). *Florida Entomol.* 36: 91–98.

Wehtje, G. R., C. H. Gilliam, T. L. Grey & E. K. Blythe. 2006. Potential for halosulfuron to control *Eclipta* (*Eclipta prostrata*) in container-grown landscape plants and its sorption to container rooting substrate. *Weed Technol.* 20: 361–367.

Wei, P.-H., W.-P. Chen & R.-R. Chen. 1994. Study on the karyotype analysis of Nymphaeaceae and its taxonomic position. *Acta Phytotax. Sin.* 32: 293–300.

Wei, C. H., A. Kun & S. H. Yen. 2007. A preliminary analyses of the phylogeny of the Ethmiinae moths (Lepidoptera: Gelechioidea) with special reference to the evolutionary patterns of host use. *Acta Zool. Acad. Scient. Hungaricae* 53: 61–100.

Weigend, M., M. Gottschling, F. Selvi & H. H. Hilger. 2010. Fossil and extant western hemisphere Boragineae, and the polyphyly of "Trigonotideae" Riedl (Boraginaceae: Boraginoideae). *Syst. Bot.* 35: 409–419.

Weigend, M., F. Luebert, F. Selvi, G. Brokamp & H. H. Hilger. 2013. Multiple origins for Hound's tongues (*Cynoglossum* L.) and Navel seeds (*Omphalodes* Mill.)—the phylogeny of the borage family (Boraginaceae s.str.). *Mol. Phylog. Evol.* 68: 604–618.

Weigend, M., O. Mohr & T. Motley. 2002. Phylogeny and classification of the genus *Ribes* (Grossulariaceae) based on 5S-NTS sequences and morphological and anatomical data. *Bot. Jahrb. Syst.* 124: 163–182.

Weigend, M., F. Luebert, M. Gottschling, T. L. Couvreur, H. H. Hilger & J. S. Miller. 2014. From capsules to nutlets—phylogenetic relationships in the Boraginales. *Cladistics* 30: 508–518.

Weiher, E. & P. A. Keddy. 1995. The assembly of experimental wetland plant communities. *Oikos* 73: 323–335.

Weiher, E. R., C. W. Boylen & P. A. Bukaveckas. 1994. Alterations in aquatic plant community structure following liming of an acidic Adirondack lake. *Can. J. Fisheries Aquat. Sci.* 51: 20–24.

Weiher, E., I. C. Wisheu, P. A. Keddy & D. R. Moore. 1996. Establishment, persistence, and management implications of experimental wetland plant communities. *Wetlands* 16: 208–218.

Wein, E. E. & M. M. R. Freeman. 1995. Frequency of traditional food use by three Yukon First Nations living in four communities. *Arctic* 48: 161–171.

Weinmann, F., M. Boulé, K. Brunner, J. Malek & V. Yoshino. 1984. *Wetland Plants of the Pacific Northwest*. U. S. Army Corps of Engineers, Seattle District, Seattle, WA. 85 pp.

Weis, I. M. & L. A. Hermanutz. 1993. Pollination dynamics of arctic dwarf birch (*Betula glandulosa*; Betulaceae) and its role in the loss of seed production. *Amer. J. Bot.* 80: 1021–1027.

Weishampel, P. A. & B. L. Bedford. 2006. Wetland dicots and monocots differ in colonization by arbuscular mycorrhizal fungi and dark septate endophytes. *Mycorrhiza* 16: 495–502.

Weiss, M. R. 1995. Floral color change: a widespread functional convergence. *Amer. J. Bot.* 82: 167–185.

Welch, B. A., C. B. Davis & R. J. Gates. 2006. Dominant environmental factors in wetland plant communities invaded by *Phragmites australis* in East Harbor, Ohio, USA. *Wetlands Ecol. Manag.* 14: 511–525.

Welling, C. H. & R. L. Becker. 1990. Seed bank dynamics of *Lythrum salicaria* L.: implications for control of this species in North America. *Aquat. Bot.* 38: 303–309.

Wells, A. F. 2006. Deep Canyon and Subalpine Riparian and Wetland Plant Associations of the Malheur, Umatilla, and Wallowa-Whitman National Forests. General Technical Report PNW-GTR-682. U. S. Department of Agriculture, Forest Service, Pacific Northwest Research Station, Portland, OR. 277 pp.

Wells, B. W. 1928. Plant communities of the coastal plain of North Carolina and their successional relations. *Ecology* 9: 230–242.

Wells, C. & M. Pigliucci. 2000. Heterophylly in aquatic plants: considering the evidence for adaptive plasticity. *Perspect. Plant Ecol. Evol. Syst.* 3: 1–18.

Wells, E. D. 1996. Classification of peatland vegetation in Atlantic Canada. *J. Veg. Sci.* 7: 847–878.

Wells, E. F. & P. E. Elvander. 2009. 4. Saxifragaceae Jussieu. pp. 43–146. *In*: N. R. Morin (convening ed.), *Flora of North America North of Mexico, Vol. 8: Magnoliophyta: Paeoniaceae to Ericaceae*. Oxford University Press, New York.

Wells, F. H., W. K. Lauenroth & J. B. Bradford. 2012. Recreational trails as corridors for alien plants in the Rocky Mountains, USA. *W. N. Amer. Naturalist* 72: 507–533.

Welsh, S. L. 1974. *Anderson's Flora of Alaska and Adjacent Parts of Canada*. Brigham Young University Press, Provo, UT. 724 pp.

Welsh, S. L. 1989. On the distribution of Utah's hanging gardens. *Great Basin Naturalist* 49: 1–30.

Welsh, S. L. 2003. 20. *Atriplex* L. pp. 322–381. *In*: N. R. Morin (convening ed.), *Flora of North America North of Mexico, Vol. 4: Magnoliophyta: Caryophyllidae, Part 1*. Oxford University Press, New York.

Welsh, S. L. & C. A. Toft. 1981. Biotic communities of hanging gardens in southeastern Utah. *Res. Rep. Natl. Geogr. Soc.* 13: 663–681.

Wen, J. & T. F. Stuessy. 1993. The phylogeny and biogeography of *Nyssa* (Cornaceae). *Syst. Bot.* 18: 68–79.

Wen, J,. Z.-L. Nie & S. M. Ickert-Bond. 2016. Intercontinental disjunctions between eastern Asia and western North America in vascular plants highlight the biogeographic importance of the Bering land bridge from late Cretaceous to Neogene. *J. Syst. Evol.* 54: 469–490.

Weniger, B., M. Haag-Berrurier & R. Anton. 1982. Plants of Haiti used as antifertility agents. *J. Ethnopharmacol.* 6: 67–84.

Went, F. W. 1958. *Synthyris ranunculina. Ann. Missouri Bot. Gard.* 45: 305–312.

Went, F. W., G. Juhren & M. C. Juhren. 1952. Fire and biotic factors affecting germination. *Ecology* 33: 351–364.

Werier, D. A. & R. F. C. Naczi. 2012. *Carex secalina* (Cyperaceae), an introduced sedge new to North America. *Rhodora* 114: 349–365.

Wernegreen, J. J., E. E. Harding & M. A. Riley. 1997. *Rhizobium* gone native: Unexpected plasmid stability of indigenous *Rhizobium leguminosarum*. *Proc. Natl. Acad. Sci. U.S.A.* 94: 5483–5488.

Werner, K. J. & J. B. Zedler. 2002. How sege meadow soils, microtopography, and vegetation respond to sedimentation. *Wetlands* 22: 451–466.

Wersal, R. M., J. D. Madsen & M. L. Tagert. 2008. *Littoral Zone Aquatic Plant Community Assessment of the Ross Barnett Reservoir, MS for 2007*. An Annual Report to the Pearl River Valley Water Supply District. GeoResources Institute Report 5027. Mississippi Water Resources Research Institute, Mississippi State University, MS. 19 pp.

West, N. E. & G. A. Reese. 1991. Comparison of some methods for collecting and analyzing data on aboveground net production and diversity of herbaceous vegetation in a northern Utah subalpine context. *Vegetatio* 96: 145–163.

Westergaard, K. B., M. H. Jørgensen, T. M. Gabrielsen, I. G. Alsos & C. Brochmann. 2010. The extreme Beringian/Atlantic disjunction in *Saxifraga rivularis* (Saxifragaceae) has formed at least twice. *J. Biogeogr.* 37: 1262–1276.

Westervelt, K., E. Largay, R. Coxe, W. McAvoy, S. Perles, G. Podniesinski, L. Sneddon & K. Walz. 2006. *A Guide to the Natural Communities of the Delaware Estuary: Version I*. NatureServe, Arlington, VA. 336 pp.

Westman, W. E. 1975. Edaphic climax pattern of the pygmy forest region of California. *Ecol. Monogr.* 45: 109–135.

Wetter, M. A. 1983. Micromorphological characters and generic delimitation of some New World Senecioneae (Asteraceae). *Brittonia* 35: 1–22.

Wetzel, P. R., A. G. van der Valk & L. A. Toth. 2001. Restoration of wetland vegetation on the Kissimmee River floodplain: potential role of seed banks. *Wetlands* 21: 189–198.

Wetzel, P. R., W. M. Kitchens, J. M. Brush & M. L. Dusek. 2004. Use of a reciprocal transplant study to measure the rate of plant community change in a tidal marsh along a salinity gradient. *Wetlands* 24: 879–890.

Whalen, M. A. 1987. Systematics of *Frankenia* (Frankeniaceae) in North and South America. *Syst. Bot. Monogr.* 17: 1–93.

Wheatley, M. & J. Bentz. 2002. *A Preliminary Classification of Plant Communities in the Central Parkland Natural Sub-Region of Alberta*. Geowest Environmental Consultants, Ltd., Edmonton, AB. 87 pp.

Wheeler, A. G. 1988. *Diabrotica cristata*, a chrysomelid (Coleoptera) of relict Midwestern prairies discovered in eastern serpentine barrens. *Entomol. News* 99: 134–142.

Wheeler Jr., A. G. 1982. *Clanoneurum americanum* (Diptera: Ephydridae), a leafminer of the littoral chenopod *Suaeda linearis*. *Proc. Entomol. Soc. Washington* 84: 297–300.

Wheeler Jr., A. G. 1983. The small milkweed bug, *Lygaeus kalmii* (Hemiptera: Lygaeidae): milkweed specialist or opportunist? *J. New York Entomol. Soc.* 91: 57–62.

Wheeler Jr., A. G. 1987. Hedge bindweed, *Calystegia sepium* (Convolvulaceae), an adventitious host of the Chrysanthemum lace bug, *Corythucha marmorata* (Heteroptera: Tingidae). *Proc. Entomol. Soc. Washington* 89: 200.

Wheeler Jr., A. G. 1988. "Violent deaths" of soldier beetles (Coleoptera: Cantharidae) revisited: new records of the fungal pathogen *Eryniopsis lampyridarum* (Zygomycetes: Entomophthoraceae). *Coleopterists Bull.* 42: 233–236.

Wheeler Jr., A. G. & E. R. Hoebeke. 1982. Host plants and nymphal descriptions of *Acanalonia pumila* and *Cyarda* sp. near *acutissima* (Homoptera, Fulgoroidea: Acanaloniidae and Flatidae). *Florida Entomol.* 65: 340–349.

Wheeler, G. A., P. H. Glaser, E. Gorham, C. M. Wetmore, F. D. Bowers & J. A. Janssens. 1983. Contributions to the flora of the Red Lake peatland, northern Minnesota, with special attention to *Carex*. *Amer. Midl. Naturalist* 110: 62–96.

Wheeler, J. F. G. 1942. The discovery of the Nemertean *Gorgonorhynchus* and its bearing on evolutionary theory. *Amer. Naturalist* 76: 470–493.

Wheelwright, N. T. 1986. The diet of American robins: an analysis of U. S. Biological Survey records. *Auk* 103: 710–725.

Wherry, E. T. 1920. Soil tests of Ericaceae and other reaction-sensitive families in northern Vermont and New Hampshire. *Rhodora* 22: 33–49.

Wherry, E. T. 1927. Divergent soil reaction preferences of related plants. *Ecology* 8: 197–206.

Wherry, E. T. 1929. Acidity relations of the sarracenias. *J. Washington Acad. Sci.* 19: 379–390.

Whicker, F. W., J. E. Pinder III, J. W. Bowling, J. J. Alberts & I. L. Brisbin Jr. 1990. Distribution of long-lived radionuclides in an abandoned reactor cooling reservoir. *Ecol. Monogr.* 60: 471–496.

Whitaker, J. D., J. W. McCord, P. P. Maier, A. L. Segars, M. L. Rekow, N. Shea, J. Ayers & R. Browder. 2004. *An Ecological Characterization of Coastal Hammock Islands in South Carolina*. Marine Resources Division, South Carolina Department of Natural Resources Charleston, SC. 115 pp.

Whitaker Jr., J. O. 1963. Food, habitat and parasites of the woodland jumping mouse in central New York. *J. Mammalogy* 44: 316–321.

Whitcraft, C. R., B. J. Grewell & P. R. Baye. 2011. Estuarine vegetation at Rush Ranch open space preserve, San Franciso Bay National Estuarine Research Reserve, California. *San Francisco Estuary Watershed Sci.* 9: 1–29.

White, A. L., C. Boutin, R. L Dalton, B. Henkelman & D. Carpenter. 2007. Germination requirements for 29 terrestrial and wetland wild plant species appropriate for phytotoxicity testing. *Pest Manag. Sci.* 65: 19–26.

White, D. 1998. Rare plants of the bluegrass. *Naturally Kentucky* 26: 1, 4.

White, D. A. 1983. Plant communities of the lower Pearl River basin, Louisiana. *Amer. Midl. Naturalist* 110: 381–396.

White, D. A. & M. J. Simmons. 1988. Productivity of the marshes at the mouth of the Pearl River, Louisiana. *Castanea* 53: 215–224.

White, D. A. & S. A. Skojac. 2002. Remnant bottomland forests near the terminus of the Mississippi River in southeastern Louisiana. *Castanea* 67: 134–145.

White, D. A. & L. B. Thien. 1985. The pollination of *Illicium parviflorum* Mich. ex Vent. (Illiciaceae). *J. Elisha Mitchell Sci. Soc.* 101: 115–118.

White, D. H., C. A. Mitchell & E. Cromartie. 1982. Nesting ecology of roseate spoonbills at Nueces Bay, Texas. *Auk* 99: 275–284.

White, D. L. & N. C. Drozda. 2006. Status of *Solidago albopilosa* Braun (white-haired goldenrod) [Asteraceae], a Kentucky endemic. *Castanea* 71: 124–128.

White, J. J. 1885. *Cranberry Culture*. Orange Judd Co., New York. 131 pp.

White, K. L. 1965. Shrub-carrs of southeastern Wisconsin. *Ecology* 46: 286–304.

White, M. & S. W. Harris. 1966. Winter occurrence, foods, and habitat use of snipe in northwest California. *J. Wildl. Manag.* 30: 23–34.

Whitehouse, C. 2007. *Eupatorium*: Joe Pye weeds and their ornamental relatives. *Plantsman* 6: 242–247.

Whitehouse, E. 1933. Plant succession on central Texas granite. *Ecology* 14: 391–405.

Whitehouse, H. E. & S. E. Bayley. 2005. Vegetation patterns and biodiversity of peatland plant communities surrounding mid-boreal wetland ponds in Alberta, Canada. *Can. J. Bot.* 83: 621–637.

Whiteside, J. E. 1887. Antidote for *Cicuta* poisoning. *Bot. Gaz.* 12: 113.

Whiting, J. B. 1849. *Manual of Flower Gardening for Ladies: With Directions for the Propagation & Management of the Plants Usually Cultivated in the Flower Garden.* David Bogue, London. 144 pp.

Whiting, P., T. Savchenko, S. D. Sarker, H. H. Rees & L. Dinan. 1998. Phytoecdysteroids in the genus *Limonium* (Plumbaginaceae). *Biochem. Syst. Ecol.* 26: 695–698.

Whiting, P. W., A. Clouston & P. Kerlin. 2002. Black cohosh and other herbal remedies associated with acute hepatitis. *Med. J. Australia* 177: 440–443.

Whitley, J. R., B. Bassett, J. G. Dillard & R. A. Haefner. 1990. *Water Plants for Missouri Ponds.* Missouri Department of Conservation, Jefferson City, MO. 151 pp.

Whitlow, T. H. & L. B. Anella. 1999. Response of *Acer rubrum* L. (red maple) genotypes and cultivars to soil anaerobiosis. *Acta Hort.* 496: 393–400.

Whitman, R. L., S. E. Byers, D. A. Shively, D. M. Ferguson & M. Byappanahalli. 2005. Occurrence and growth characteristics of *Escherichia coli* and enterococci within the accumulated fluid of the northern pitcher plant (*Sarracenia purpurea* L.). *Can. J. Microbiol.* 51: 1027–1037.

Whitney, G. G. & R. E. Moeller. 1982. An analysis of the vegetation of Mt. Cardigan, New Hampshire: a rocky, subalpine New England summit. *Bull. Torrey Bot. Club* 109: 177–188.

Whitson, A. R. 1905. A report on cranberry investigations. *Bull. Wisconsin Agr. Exp. Sta.* 119: 1–77.

Whittall, J. B., M. L Carlson, P. M. Beardsley, R. J. Meinke & A. Liston. 2006. The *Mimulus moschatus* alliance (Phrymaceae): molecular and morphological phylogenetics and their conservation implications. *Syst. Bot.* 31: 380–397.

Whittemore, A. T. 1997a. 21. *Aquilegia* L. pp. 249–258. *In*: N. R. Morin (convening ed.), *Flora of North America North of Mexico, Vol. 3: Magnoliophyta: Magnoliidae and Hamamelidae.* Oxford University Press, New York.

Whittemore, A. T. 1997b. 3. *Myosurus* L. pp. 135–138. *In*: N. R. Morin (convening ed.), *Flora of North America North of Mexico, Vol. 3: Magnoliophyta: Magnoliidae and Hamamelidae.* Oxford University Press, New York.

Whittemore, A. T. 1997c. 2. *Ranunculus* L. pp. 88–135. *In*: N. R. Morin (convening ed.), *Flora of North America North of Mexico, Vol. 3: Magnoliophyta: Magnoliidae and Hamamelidae.* Oxford University Press, New York.

Whittemore, A. T., M. R. Mesler & K. L. Lu. 1997. 2. *Asarum* L. pp. 50–53. *In*: N. R. Morin (convening ed.), *Flora of North America North of Mexico, Vol. 3: Magnoliophyta: Magnoliidae and Hamamelidae.* Oxford University Press, New York.

Whyte, R. J. & B. W. Cain. 1981. Wildlife habitat on grazed or ungrazed small pond shorelines in south Texas. *J. Range Manag.* 34: 64–68.

Wichman, B., R. K. Peet & T. R. Wentworth. 2007. *Natural Vegetation of the Carolinas: Classification and Description of Montane Non-Alluvial Wetlands of the Southern Appalachian Region.* Ecosystem Enhancement Program, North Carolina Department of Environment and Natural Resources, Raleigh, NC. 34 pp.

Widgren, Å. 2003. *Cotula coronopifolia*-spreading in Sweden? *Svensk Bot. Tidskr.* 97: 130–133.

Wiedenfeld, H., G. Hösch, E. Roeder & T. Dingermann. 2009. Lycopsamine and cumambrin B from *Eupatorium maculatum*. *Pharmazie* 64: 415–416.

Wiedenmann, R. N. 2005. Non-target feeding by *Galerucella calmariensis* on sandbar willow (*Salix interior*) in Illinois. *Great Lakes Entomol.* 38: 100–103.

Wieder, R. K., A. M. McCormick & G. E. Lang. 1981. Vegetational analysis of Big Run Bog, a nonglaciated *Sphagnum* bog in West Virginia. *Castanea* 46: 16–29.

Wieder, R. K., C. A. Bennett & G. E. Lang. 1984. Flowering phenology at Big Run Bog, West Virginia. *Amer. J. Bot.* 71: 203–209.

Wieder, R. K., J. B. Yavitt, G. E. Lang & C. A. Bennett. 1989. Aboveground net primary production at Big Run Bog, West Virginia. *Castanea* 54: 209–216.

Wiefenbach, D. 1993. An investigation into the origins of *Eupatorium leucolepis* var. *novae-angliae*. MS thesis. University of Southwestern Louisiana, Lafayette, LA.

Wiegand, K. M. 1897. *Galium trifidum* and its North American allies. *Bull. Torrey Bot. Club* 24: 389–403.

Wiegrefe, S. J., K. J. Sytsma & R. P. Guries. 1998. The Ulmaceae, one family or two? Evidence from chloroplast DNA restriction site mapping. *Plant Syst. Evol.* 210: 249–270.

Wieland, R. G. 2000. Ecology and vegetation of LeFleur's Bluff State Park, Jackson, Mississippi. *J. Mississippi Acad. Sci.* 45: 150–185.

Wiersema, J. H. 1982. Distributional records for *Nymphaea lotos* (Nymphaeaceae) in the Western Hemisphere. *Sida* 9: 230–234.

Wiersema, J. H. 1987. A monograph of *Nymphaea* subgenus *Hydrocallis* (Nymphaeaceae). *Syst. Bot. Monogr.* 16: 1–112.

Wiersema, J. H. 1988. Reproductive biology of *Nymphaea* (Nymphaeaceae). *Ann. Missouri Bot. Gard.* 75: 795–804.

Wiersema, J. H. 1996. *Nymphaea tetragona* and *Nymphaea leibergii* (Nymphaeaceae): two species of diminutive waterlilies in North America. *Brittonia* 48: 520–531.

Wiersema, J. H. 1997a. 2. *Nymphaea* Linnaeus. pp. 71–77. *In*: N. R. Morin (convening ed.), *Flora of North America North of Mexico, Vol. 3: Magnoliophyta: Magnoliidae and Hamamelidae.* Oxford University Press, New York.

Wiersema, J. H. 1997b. 11. Nelumbonaceae Dumortier. pp. 64–65. *In*: N. R. Morin (convening ed.), *Flora of North America North of Mexico, Vol. 3: Magnoliophyta: Magnoliidae and Hamamelidae.* Oxford University Press, New York.

Wiersema, J. H. & C. B. Hellquist. 1994. Nomenclatural notes in Nymphaeaceae for the North American flora. *Rhodora* 96: 170–178.

Wiersema, J. H. & C. B. Hellquist. 1997. 12. Nymphaeaceae. Salisbury. pp. 66–77. *In*: N. R. Morin (convening ed.), *Flora of North America North of Mexico, Vol. 3: Magnoliophyta: Magnoliidae and Hamamelidae.* Oxford University Press, New York.

Wiersema, J. H., A. Novelo R. & J. R. Bonilla-Barbosa. 2008. Taxonomy and typification of *Nymphaea ampla* (Salisb.) DC. sensu lato (Nymphaeaceae). *Taxon* 57: 967–974.

Wiese, J. L., J. F. Meadow & J. A. Lapp. 2012. Seed weights for northern Rocky Mountain native plants with an emphasis on Glacier National Park. *Native Plant J.* 13: 39–49.

Wigand, C., B. Carlisle, J. Smith, M. Carullo, D. Fillis, M. Charpentier, R. McKinney, R. Johnson & J. Heltshe. 2011. Development and validation of rapid assessment indices of condition for coastal tidal wetlands in southern New England, USA. *Environ. Monit. Assess.* 182: 31–46.

Wiggins, G. J. 2009. Non-target host utilization of thistle species by introduced biological control agents and spatial prediction of non-target feeding habitats. PhD dissertation, The University of Tennessee, Knoxville, TN. 160 pp.

Wilbur, H. M. 1976. Life history evolution in seven milkweeds of the genus *Asclepias*. *J. Ecol.* 64: 223–240.

Wilbur, R. L. 1955. A revision of the North American genus *Sabatia* (Gentianaceae). *Rhodora* 57: 87–91.

Wilbur, R. L. 1975. A revision of the North American genus *Amorpha* (Leguminosae-Psoraleae). *Rhodora* 77: 337–409.

Wilbur, R. L. 1994. The Myricaceae of the United States and Canada: genera, subgenera, and series. *Sida* 16: 93–107.

Wilcock, C. C. 1974. Population variation, differentiation and reproductive biology of *Stachys palustris* L. *S. sylvatica* L. and *S. ×ambigua* Sm. *New Phytol.* 73: 1233–1241.

Wilcox, C., N. F. Smith & J. M. Lessmann. 2009. Fiddler crab burrowing affects growth and production of the white mangrove (*Laguncularia racemosa*) in a restored Florida coastal marsh. *Mar. Biol.* 156: 2255–2266.

Wilcox, D. A. & J. E. Meeker. 1981. Disturbance effects on aquatic vegetation in regulated and unregulated lakes in northern Minnesota. *Can. J. Bot.* 69: 1542–1551.

Wilde, S. B., T. M. Murphy, C. P. Hope, S. K. Habrun, J. Kempton, A. Birrenkott, F. Wiley, W. W. Bowerman & A. J. Lewitus. 2005. Avian vacuolar myelinopathy linked to exotic aquatic plants and a novel Cyanobacterial species. *Environ. Toxicol.* 20: 348–353.

Wildsmith, W. 1884. The flower garden. The "mixed "style. pp. 348–354. *In*: D. T. Fish (ed.), *Cassell's Popular Gardening*. Volume 1. Cassell and Company, London.

Wilken, D. and T. Wardlaw. 2001. Ecological and life history characteristics of Ventura marsh milk-vetch (*Astragalus pycnostachyus* var. *lanosissimus*) and their implications for recovery. Department of Fish and Game, South Coast Region. San Diego, CA. 55 pp.

Wilken, G. C. 1970. The ecology of gathering in a Mexican farming region. *Econ. Bot.* 24: 286–295.

Wilkins, B. T. 1957. Range use, food habits, and agricultural relationships of the mule deer, Bridger Mountains, Montana. *J. Wildl. Manag.* 21: 159–169.

Wilkinson, S. M. & M. H. Beck. 1994. Allergic contact dermatitis from menthol in peppermint. *Contact Dermatitis* 30: 42–43.

Willaman, J. J. & B. G. Schubert. 1961. *Alkaloid-Bearing Plants and their Contained Alkaloids*. Technical Bulletin 1234, Agricultural Research Service, U. S. Department of Agriculture, Washington, DC. 287 pp.

Willemstein, S. C. 1987. *An Evolutionary Basis for Pollination Ecology*. Brill Archive, Leiden, The Netherlands. 425 pp.

Williams, A. H. 1997. New food plants and first Wisconsin records of *Publilia modesta* var. *brunnea* (Hemiptera: Membracidae). *Great Lakes Entomol.* 30: 61–63.

Williams, A. H. 2000. Wisconsin Cydnidae (Hemiptera: Heteroptera). *Great Lakes Entomol.* 33: 161–164.

Williams, A. H. 2003. *Oxypolis rigidior*, a new larval food plant record for *Papilio polyxenes* (Papilionidae). *J. Lepid. Soc.* 57: 149–150.

Williams, A. H. 2015a. Feeding records of true bugs (Hemiptera: Heteroptera) from Wisconsin. *Great Lakes Entomol.* 48: 192–198.

Williams, C. E. 1990a. New host plants for adult *Systena hudsonias* (Coleoptera: Chrysomelidae) from southwestern Virginia. *Great Lakes Entomol.* 23: 149–150.

Williams, C. E. 2015b. The salamander species assemblage and environment of forested seeps of the Allegheny High Plateau, northwestern Pennsylvania, USA. *Herpetol. Notes* 8: 99–106.

Williams, B. R. M., T. C. Mitchell, J. R. I. Wood, D. J. Harris, R. W. Scotland & M. A. Carine. 2014. Integrating DNA barcode data in a monographic study of *Convolvulus*. *Taxon* 63: 1287–1306.

Williams, C. E., W. J. Moriarity & G. L. Walters. 1999a. Overstory and herbaceous layer of a riparian savanna in northwestern Pennsylvania. *Castanea* 64: 90–97.

Williams, C. E., W. J. Moriarity, G. L. Walters & L. Hill. 1999b. Influence of inundation potential and forest overstory on the ground-layer vegetation of Allegheny Plateau riparian forests. *Amer. Midl. Naturalist* 141: 323–338.

Williams, C. E., E. V. Mosbacher & W. J. Moriarity. 2000. Use of turtlehead (*Chelone glabra* L.) and other herbaceous plants to assess intensity of white-tailed deer browsing on Allegheny Plateau riparian forests, USA. *Biol. Conservation* 92: 207–215.

Williams, C. F., J. Ruvinsky, P. E. Scott & D. K. Hews. 2001. Pollination, breeding system, and genetic structure in two sympatric *Delphinium* (Ranunculaceae) species. *Amer. J. Bot.* 88: 1623–1633.

Williams, C. S. & W. H. Marshall. 1938. Duck nesting studies, Bear River Migratory Bird Refuge, Utah, 1937. *J. Wildl. Manag.* 2: 29–48.

Williams, D. D. 2006. *The Biology of Temporary Waters*. Oxford University Press, New York. 348 pp.

Williams, D. F. 1988. *Ecology and Management of the Riparian Brush Rabbit in Caswell Memorial State Park*. Interagency Agreement 4–305–6108. California Department of Parks and Recreation, Lodi, California and California State University, Stanislaus, Turlock, CA. 31 pp.

Williams, J. 2005. A Morphometric analysis of the *Amsonia tabernaemontana* Walt. complex (Apocynaceae). Botany 2005 meeting (abstract).

Williams, L. A. D. 1999. *Rhizophora mangle* (Rhizophoraceae) triterpenoids with insecticidal activity. *Naturwissenschaften* 86: 450–452.

Williams, L. O. 1937. A monograph of the genus *Mertensia* in North America. *Ann. Missouri Bot. Gard.* 24: 17–159.

Williams, L. O. 1972. Tropical American plants, XII. *Fieldiana, Bot.* 34: 101–132.

Williams, P. J., J. R. Robb & D. R. Karns. 2012. Habitat selection by Crawfish Frogs (*Lithobates areolatus*) in a large mixed grassland/forest habitat. *J. Herpetol.* 46: 682–688.

Williams, S. C., J. S. Warda & U. Ramakrishnan. 2008. Endozoochory by white-tailed deer (*Odocoileus virginianus*) across a suburban/woodland interface. *Forest Ecol. Manag.* 255: 940–947.

Williams, T. A. 1891. *Host-Plant List of North American Aphididae*. Special Bulletin Number 1, Department of Entomology, University of Nebraska, Lincoln, NE. 28 pp.

Williams, T. Y. 1990b. *Salix serissima*. *In*: Fire Effects Information System. U. S. Department of Agriculture, Forest Service, Rocky Mountain Research Station, Fire Sciences Laboratory (Producer). Available online: http://www.fs.fed.us/database/feis/ (accessed March 18, 2005).

Williams, T. Y. 1990c. *Pinguicula vulgaris*. *In*: Fire Effects Information System. U. S. Department of Agriculture, Forest Service, Rocky Mountain Research Station, Fire Sciences Laboratory (Producer). Available online: http://www.fs.fed.us/database/feis/ (accessed May 9, 2007).

Williams, T. Y. 1990d. *Utricularia intermedia*. *In*: Fire Effects Information System. U. S. Department of Agriculture, Forest Service, Rocky Mountain Research Station, Fire Sciences Laboratory (Producer). Available online: http://www.fs.fed.us/database/feis/ (accessed June 5, 2007).

Williamson, G. B. & E. M. Black. 1981. Mimicry in hummingbird-pollinated plants? *Ecology* 62: 494–496.

Williamson, P. S. & E. L. Schneider. 1993. Nelumbonaceae. pp. 470–472. In: K. Kubitzki, J. Rohwer, and V. Bittrich (eds.), *The Families and Genera of Vascular Plants, Vol. II. Flowering Plants, Dicotyledons: Magnoliid, Hamamelid and Caryophyllid Families*. Springer-Verlag, Berlin, Germany.

Willis, J. H. 1970. More about mud-mats, *Glossostigma* spp. *Vict. Naturalist* 87: 249–250.

Willis, J. H. 1993. Effects of different levels of inbreeding on fitness components in *Mimulus guttatus*. *Evolution* 47: 864–876.

Willmer, P. 2011. *Pollination and Floral Ecology*. Princeton University Press, Princeton, NJ. 828 pp.

Willson, M. F. 1993. Mammals as seed-dispersal mutualists in North America. *Oikos* 67: 159–176.

Wilm, B. W. & J. B. Taft. 1998. *Trepocarpus aethusae* Nutt. (Apiaceae) in Illinois. *Trans. Illinois State Acad. Sci.* 91: 53–56.

Wilson, A. M. 2011. A comparison of vegetation in artificially isolated wetlands on West Galveston Island. MS thesis. Texas A&M University, College Station, TX. 73 pp.

Wilson, C. E., S. J. Darbyshire & R. Jones. 2007. The biology of invasive alien plants in Canada. 7. *Cabomba caroliniana* A. Gray. *Can. J. Plant Sci.* 87: 615–638.

Wilson, G. W. 1908. Studies in North American Peronosporales-IV. Host index. *Bull. Torrey Bot. Club* 35: 543–554.

Wilson, J. S., L. E. Wilson, L. D. Loftis & T. Griswold. 2010. The montane bee fauna of north central Washington, USA, with floral associations. *W. N. Amer. Naturalist* 70: 198–207.

Wilson L. F. 1968. Life history and habits of the pine cone willow gall midge, Rhabdophaga strobiloides (Diptera: Cecidomyiidae), in Michigan. *Can. Entomol.* 100: 430–433.

Wilson, L. R. 1935. Lake development and plant succession in Vilas County, Wisconsin. *Ecol. Monogr.* 5: 207–247.

Wilson, M. R. 1978. Notes on ethnobotany in Inuktitut. *W. Can. J. Anthropol.* 8: 180–196.

Wilson, M. V., K. P. Connelly & L. E. Lantz. 1993. *Plant Species, Habitat, and Site Information for Fern Ridge Reservoir. A Component of the Project to Develop Management Guidelines for Native Wetland Communities*. Army Corps of Engineers, Waterways Experiment Station. Vicksburg, MS. 59 pp.

Wilson, M. V., C. A. Ingersoll, M. G. Wilson & D. L. Clark. 2004. Why pest plant control and native plant establishment failed: a restoration autopsy. *Nat. Areas J.* 24: 23–31.

Wilson, P. 1994. The east-facing flowers of *Drosera tracyi*. *Amer. Midl. Naturalist* 131: 366–369.

Wilson, P. 1995. Variation in the intensity of pollination in *Drosera tracyi*: selection is strongest when resources are intermediate. *Evol. Ecol.* 9: 382–396.

Wilson, R. S. 1948. The summer bird life of Attu. *Condor* 50: 124–129.

Wilson, S. B. & P. J. Stoffella. 2006. Using compost for container production of ornamental wetland and flatwood species native to Florida. *Native Plant* 7: 293–300.

Wilson, S. D. 1985. The growth of *Drosera intermedia* in nutrient-rich habitats: the role of insectivory and interspecific competition. *Can. J. Bot.* 63: 2468–2469.

Wilson, S. D. & P. A. Keddy. 1985. Plant zonation on a shoreline gradient: physiological response curves of component species. *J. Ecol.* 73: 851–860.

Wilson, S. D. & P. A. Keddy. 1988. Species richness, survivorship, and biomass accumulation along an environmental gradient. *Oikos* 53: 375–380.

Wilson, S. D., D. R. J. Moore & P. A. Keddy. 1993. Relationships of marsh seed banks to vegetation patterns along environmental gradients. *Freshwater Biol.* 29: 361–370.

Wilson, S. W. & J. E. McPherson. 1981. Life history of *Megamelus davisi* with descriptions of immature stages. *Ann. Entomol. Soc. Amer.* 74: 345–350.

Wilzer, K. A., F. R. Fronczek, L. E. Urbatsch & N. H. Fischer. 1989. Coumarins from *Aster praealtus*. *Phytochemistry* 28: 1729–1735.

Wimmer, F. E. 1948. Vorarbeiten zur Monographie der Campanulaceae-Lobelioideae: II. Trib. Lobelieae. *Ann. Naturhist. Mus. Wien* 56: 317–374.

Windell, J. T., B. E. Willard, D. J. Cooper, S. Q. Foster, F. Knud-Hansen, L. P. Rink & G. N. Kiladis. 1986. *An Ecological Characterization of Rocky Mountain Montane and Subalpine Wetlands*. Biological Report 86(11). United States Department of the Interior, Fish & Wildlife Service, Washington, DC. 299 pp.

Windeløv, H. 1998. *Tropica Aquarium Plants*. Tropica Aquarium Plants, Egå, Denmark. 96 pp.

Windler, D. R. 1965. Evidence of natural hybridization between *Mimulus ringens* and *Mimulus alatus* (Scrophulariaceae). *Rhodora* 78: 641–649.

Winfield, M. O., P. J. Wilson, M. Labra & J. S. Parker. 2003. A brief evolutionary excursion comes to an end: the genetic relationship of British species of *Gentianella* sect. *Gentianella* (Gentianaceae). *Plant Syst. Evol.* 237: 137–151.

Winkworth, R. C. & M. J. Donoghue. 2004. *Viburnum* phylogeny: evidence from the duplicated nuclear gene *GBSSI*. *Mol. Phylogen. Evol.* 33: 109–126.

Winkworth, R. C. & M. J. Donoghue. 2005. *Viburnum* phylogeny based on combined molecular data: implications for taxonomy and biogeography. *Amer. J. Bot.* 92: 653–666.

Winkworth, R. C., J. Grau, A. W. Robertson & P. J. Lockhart. 2002. The origins and evolution of the genus *Myosotis* L. (Boraginaceae). *Mol. Phylogen. Evol.* 24: 180–193.

Winsor, J. 1983. Persistence by habitat dominance in the annual *Impatiens capensis* (Balsaminaceae). *J. Ecol.* 71: 451–466.

Winstead, N. A. & S. L. King. 2006. Least bittern nesting sites at Reelfoot Lake, Tennessee. *S. E. Naturalist (Steuben)* 5: 317–320.

Winston, R. D. & P. R. Gorham. 1979a. Turions and dormancy states in *Utricularia vulgaris*. *Can. J. Bot.* 57: 2740–2749.

Winston, R. D. & P. R. Gorham. 1979b. Roles of endogenous and exogenous growth regulators in dormancy of *Utricularia vulgaris*. *Can. J. Bot.* 57: 2750–2759.

Winter, C., S. Lehmann & M. Diekmann. 2008. Determinants of reproductive success: a comparative study of five endangered river corridor plants in fragmented habitats. *Biol. Conserv.* 141: 1095–1104.

Winter, K. 1981. C_4 plants of high biomass in arid regions of Asia: occurrence of C_4 photosynthesis in Chenopodiaceae and Polygonaceae from the Middle East and USSR. *Oecologia* 48: 100–106.

Winterhoff, H., H. G. Gumbinger & H. Sourgens. 1988. On the antigonadotropic activity of *Lithospermum* and *Lycopus* species and some of their phenolic constituents. *Planta Med.* 54: 101–106.

Winterringer, G. S. & A. C. Lopinot. 1977. *Aquatic Plants of Illinois*. Illinois State Museum, Popular Science Series, vol. 6. Illinois State Museum, Springfield, IL. 142 pp.

Wise, D. A. & M. Y. Menzel. 1971. Genetic affinities of the North American species of *Hibiscus* sect. *Trionum*. *Brittonia* 23: 425–437.

Wiser, S. K. 1998. Comparison of Southern Appalachian high-elevation outcrop plant communities with their Northern Appalachian counterparts. *J. Biogeogr.* 25: 501–513.

Wiser, S. K., R. K. Peet & P. S. White. 1996. High-elevation rock outcrop vegetation of the Southern Appalachian Mountains. *J. Veg. Sci.* 7: 703–722.

Wisheu, I. C. & P. A. Keddy. 1989. The conservation and management of a threatened coastal plain plant community in Eastern North America (Nova Scotia, Canada). *Biol. Conserv.* 48: 229–238.

Wisheu, I. C. & P. A. Keddy. 1991. Seed banks of a rare wetland plant community: distribution patterns and effects of human-induced disturbance. *J. Veg. Sci.* 2: 181–188.

Wissemann, V. & C. M. Ritz. 2005. The genus *Rosa* (Rosoideae, Rosaceae) revisited: molecular analysis of nrITS-1 and *atp*B-*rbc*L intergenic spacer (IGS) versus conventional taxonomy. *Bot. J. Linn. Soc.* 147: 275–290.

Wistrom, C. & A. H. Purcell. 2005. The fate of *Xylella fastidiosa* in vineyard weeds and other alternate hosts in California. *Plant Dis.* 89: 994–999.

Wit, H.C.D. De. 1964. *Aquarium Plants.* Blandford Press, London, UK. 255 pp.

Witham, C. W. 1992. The role of vernal pools in the 1992 mass dispersal of *Vanessa cardui* (Nymphalidae) with new larval host-plant records. *J. Res. Lepid.* 30: 302–304.

Witham, C. W. 2006. *Field Guide to the Vernal Pools of Mather Field, Sacramento County.* California Native Plant Society, Sacramento Valley Chapter, Sacramento, CA. 48 pp.

Witham, C. W. & G. A. Kareofelas. 1994. *Botanical Resources Inventory at Calhoun Cut Ecological Reserve Following California's Recent Drought.* Final Report. California Department of Fish and Game, Natural Heritage Division, Sacramento, CA. 33 pp.

Witmer, G., N. Snow, L. Humberg & T. Salmon. 2009. Vole problems, management options, and research needs in the United States. pp. 235–249. *In*: J. R. Boulanger (ed.), *Proceedings of the 13th Wildlife Damage Management Conferences.* Paper 140. Available online: http://digitalcommons.unl.edu/icwdm_wdmconfproc/140.

Witmer, M. C. 1996. Annual diet of cedar waxwings based on U. S. Biological Survey Records (1885–1950) compared to diet of American robins: contrasts in dietary patterns and natural history. *Auk* 113: 414–430.

Witsell, T. & B. Baker. 2006. The vascular flora of the South Fork native plant preserve, Van Buren County, Arkansas. *J. Arkansas Acad. Sci.* 60: 144–164.

Wittenberger, J. F. 1978. The breeding biology of an isolated bobolink population in Oregon. *Condor* 80: 355–371.

Wittstock, U., K.-H. Lichtnow & E. Teuscher. 1997. Effects of cicutoxin and related polyacetylenes from *Cicuta virosa* on neuronal action potentials: A comparative study on the mechanism of the convulsive action. *Plant Med.* 63: 120–124.

Wittstock, U., F. Hadacek, G. Wurz, E. Teuscher & H. Greger. 1995. Polyacetylenes from water hemlock, *Cicuta virosa. Plant Med.* 61: 439–445.

Wium-Andersen, S. 1971. Photosynthetic uptake of free CO_2, by the roots of *Lobelia dortmanna. Physiol. Plant* 25: 245–248.

Wofford, B. E. 1997. 1. *Lindera* Thunberg. pp. 27–29. *In*: N. R. Morin (convening ed.), *Flora of North America North of Mexico, Vol. 3: Magnoliophyta: Magnoliidae and Hamamelidae.* Oxford University Press, New York.

Wohl, D. L. & J. V. McArthur. 1998. Actinomycete-flora associated with submersed freshwater macrophytes. *FEMS Microbiol. Ecol.* 26: 135–140.

Wojciechowski, M. F. 2003. Reconstructing the phylogeny of legumes (Leguminosae): an early 21st century perspective. pp. 5–35. *In*: B. B. Klitgaard & A. Bruneau (eds.), *Advances in Legume Systematics, Part 10, Higher Level Systematics.* Royal Botanic Gardens, Kew, UK.

Wojciechowski, M. F., M. J. Sanderson & J.-M. Hu. 1999. Evidence on the monophyly of *Astragalus* (Fabaceae) and its major subgroups based on nuclear ribosomal DNA ITS and chloroplast DNA *trnL* intron data. *Syst. Bot.* 24: 409–437.

Wojciechowski, M. F., M. J. Sanderson, K. P. Steele & A. Liston. 2000. Molecular phylogeny of the "temperate herbaceous tribes" of Papilionoid legumes: a supertree approach. pp. 277–298. *In*: P. S. Herendeen and A. Bruneau (eds.), *Advances in Legume Systematics 9.* Royal Botanic Gardens, Kew, UK.

Wojciechowski, M. F., M. Lavin & M. J. Sanderson. 2004. A phylogeny of legumes (Leguminosae) based on analysis of the plastid *matK* gene resolves many well-supported subclades within the family. *Amer. J. Bot.* 91: 1845–1861.

Wolcott, G. N. 1937. An animal census of two pastures and a meadow in northern New York. *Ecol. Monogr.* 7: 1–90.

Wolfe, A. D., W. J. Elisens, L. E. Watson & C. W. Depamphilis. 1997. Using restriction-site variation of PCR-amplified cpDNA genes for phylogenetic analysis of tribe Cheloneae (Scrophulariaceae). *Amer. J. Bot.* 84: 555–564.

Wolfe, A. D., S. L. Datwyler & C. P. Randle. 2002. A phylogenetic and biogeographic analysis of the Cheloneae (Scrophulariaceae) based on ITS and matK sequence data. *Syst. Bot.* 27: 138–148.

Wolfe, A. D., C. P. Randle, S. L. Datwyler, J. J. Morawetz, N. Arguedas & J. Diaz. 2006a. Phylogeny, taxonomic affinities, and biogeography of *Penstemon* (Plantaginaceae) based on ITS and cpDNA sequence data. *Amer. J. Bot.* 93: 1699–1713.

Wolfe, B. E., P. A. Weishampel & J. N. Klironomos. 2006b. Arbuscular mycorrhizal fungi and water table affect wetland plant community composition. *J. Ecol.* 94: 905–914.

Wolfe, C. B. Jr. & J. D. Pittillo. 1977. Some ecological factors influencing the distribution of *Betula nigra* L. in western North Carolina. *Castanea* 42: 18–30.

Wolff, F. E., D. T. McKay Jr. & D. K. Norman. 2005. *Inactive and Abandoned Mine Lands—Van Stone mine, Northport Mining District, Stevens County, Washington.* Information Circular 100. Washington Department of Natural Resources, Division of Geology and Earth Resources, Olympia, WA. 18 pp.

Wolff, S. L. & R. L. Jeffries. 1987. Taxonomic status of diploid *Salicornia europaea* (s.l.) (Chenopodiaceae) in northeastern North America. *Can. J. Bot.* 65: 1420–1426.

Wolfson, P. & D. L. Hoffmann. 2003. An investigation into the efficacy of *Scutellaria lateriflora* in healthy volunteers. *Altern. Therapies* 9: 74–78.

Wolken, P. M., C. H. Sieg & S. E. Williams. 2001. Quantifying suitable habitat of the threatened western prairie fringed orchid. *J. Range Manag.* 54: 611–616.

Wolniak, M., M. Tomczykowa, M.Tomczyk, J. Gudej & I. Wawer. 2007. Antioxidant activity of extracts and flavonoids from *Bidens tripartita. Acta Polon. Pharm.* 64: 441–447.

Wolters, M. & J. P. Bakker. 2002. Soil seed bank and driftline composition along a successional gradient on a temperate salt marsh. *Appl. Veg. Sci.* 5: 55–62.

Wolverton, B. C. & R. C. McDonald. 1981. Energy from vascular plant wastewater treatment systems. *Econ. Bot.* 35: 224–232.

Woo, I., J. Y. Takekawa, A. R. Westhoff, K. L. Turner & G. T. Downard. 2004. *The Tubbs Setback Restoration Project: 2004 Annual Report.* Unpublished Report, U. S. Geological Survey, Western Ecological Research Center, San Francisco Bay Estuary Field Station, Vallejo, CA. 44 pp.

Woo, I., J. Y. Takekawa, A. Rowan, L. Dembosz & R. Gardiner. 2008. *The Benicia-Martinez (BenMar) Restoration Project: 2007 Open-File Report*. U. S. Geological Survey, Western Ecological Research Center, San Francisco Bay Estuary Field Station, Vallejo, CA. 46 pp.

Wood Jr., C. E. 1961. A study of hybridization in *Downingia* (Campanulaceae). *J. Arnold Arbor.* 42: 219–262.

Wood, G. B. & F. Bache. 1854. *The Dispensatory of the United States of America, 10th Ed.* Lippincott, Grambo & Co., Philadelphia, PA. 1480 pp.

Wood, H. C., C. H. LaWall, H. W. Youngken, A. Osol, I. Griffith & L. Gershenfeld. 1940. *The Dispensatory of the United States of America, 22nd Ed (with Supplement).* J. B. Lippincott Company, Philadelphia, PA. 1894 [+ 76] pp.

Woodard, C. A. 2008. Poison hemlock (*Conium maculatum* L.): Biology, implications for pastures and responses to herbicides. MS thesis. University of Missouri, Columbia, MO. 71 pp.

Woodcock, E. F. 1925. Observations on the poisonous plants of Michigan. *Amer. J. Bot.* 12: 116–131.

Woodcock, T. S. & L. J. Pekkola. 2014. Development of a pollination service measurement (PSM) method using potted plant phytometry. *Environ. Monitor. Assess.* 186: 5041–5057.

Woodcock, T., L. Pekkola, R. Wildfong, K. Fellows & P. Kevan. 2013. *Expansion of the Pollination Service Measurement (PSM) Concept in Southern Ontario.* Environment Canada-CESI Program, Gatineau, QC. 26 pp.

Woodland, I. A. 1984. Flower records for anthophilous Cerambycidae in a southwestern Michigan woodland (Coleoptera). *Great Lakes Entomol.* 17: 79–82.

Woods, K., K. W. Hilu, J. H. Wiersema & T. Borsch. 2005. Pattern of variation and systematics of *Nymphaea odorata*: I. Evidence from morphology and inter-simple sequence repeats (ISSRs). *Syst. Bot.* 30: 471–480.

Woods, S. W. & D. J. Cooper. 2005. Hydrologic factors affecting initial willow seedling establishment along a subalpine stream, Colorado, U.S.A. *Arctic Antarc. Alpine Res.* 37: 636–643.

Woodson Jr., R. E. 1928. Studies in the Apocynaceae. III. A monograph of the genus *Amsonia*. *Ann. Missouri Bot. Gard.* 15: 379–434.

Woodson Jr., R. E. 1936. Studies in the Apocynaceae. IV. The American genera of Echitoideae. *Ann. Missouri Bot. Gard.* 23: 169–438.

Woodson Jr., R. E. 1954. The North American species of *Asclepias* L. *Ann. Missouri Bot. Gard.* 41: 1–211.

Woodson Jr., R. E., R. W. Schery & J. W. Nowicke. 1969. Flora of Panama. Part IX. Family 167. Boraginaceae. *Ann. Missouri Bot. Gard.* 56: 33–69.

Woodson Jr., R. E., R. W. Schery & D. F. Austin. 1975. Flora of Panama. Part IX. Family 164. Convolvulaceae. *Ann. Missouri Bot. Gard.* 62: 157–224.

Woodson Jr., R. E., R. W. Schery & W. G. D'Arcy. 1979. Flora of Panama. Part IX. Family 171. Scrophulariaceae. *Ann. Missouri Bot. Gard.* 66: 173–274.

Woolhouse, S. 2012. The biology and ecology of six rare plants from Plumas National Forest, northern California, USA. MS thesis. San Jose State University, San Jose, CA. 88 pp.

Woradulayapinij, W., N. Soonthornchareonnon & C. Wiwat. 2005. *In vitro* HIV type 1 reverse transcriptase inhibitory activities of Thai medicinal plants and *Canna indica* L. rhizomes. *J. Ethnopharmacol.* 101: 84–89.

Worley, A. C. & L. D. Harder. 1996. Size-dependent resource allocation and costs of reproduction in *Pinguicula vulgaris* (Lentibulariaceae). *J. Ecol.* 84: 195–206.

Worley, I. A. 1969. *Haplomitrium hookeri* from western North America. *Bryologist* 72: 225–232.

Wright, A. H. & A. A. Wright. 1932. The habitats and composition of the vegetation of Okefinokee swamp, Georgia. *Ecol. Monogr.* 2: 109–232.

Wright, D. H. 1988. Temporal changes in nectar availability and *Bombus appositus* (Hymenoptera: Apidae) foraging profits *Southwest. Naturalist* 33: 219–227.

Wright Jr., H. E. & A. M. Bent. 1968. Vegetation bands around Dead Man Lake, Chuska Mountain, New Mexico. *Amer. Midl. Naturalist* 79: 8–30.

Wright, J. C. & I. K. Mills. 1967. Productivity studies on the Madison River, Yellowstone National Park. *Limnol. Oceanogr.* 12: 568–577.

Wright, J. P., A. S. Flecker & C. G. Jones. 2003. Local vs. landscape controls on plant species richness in beaver meadows. *Ecology* 84: 3162–3173.

Wright, R. D. 1989. Reproduction of *Lindera melissifolia* in Arkansas. *Proc. Arkansas Acad. Sci.* 43: 69–70.

Wright, R. D. 1990. Photosynthetic competence of an endangered shrub, *Lindera melissifolia*. *Proc. Arkansas Acad. Sci.* 44: 118–120.

Wright, R. D. & A. Conway. 1994. Sex ratio and success, an assessment of *Lindera melissifolia* in Arkansas. *Proc. Arkansas Acad. Sci.* 48: 230–233.

Wu, A. 2012. Can spiders (*Argiope aurantia*) indirectly affect the fitness of orange coneflowers (*Rudbeckia fulgida*) by limiting pollinator visitation? MS thesis. The University of Akron, Akron, OH. 64 pp.

Wu, C. A., D. B. Lowry, A. M. Cooley, K. M. Wright, Y. W. Lee & J. H. Willis. 2008. *Mimulus* is an emerging model system for the integration of ecological and genomic studies. *Heredity* 100: 220–230.

Wu, C.-H., H.-T. Hsieh, J.-A. Lin & G.-C. Yen. 2013. *Alternanthera paronychioides* protects pancreatic β-cells from glucotoxicity by its antioxidant, antiapoptotic and insulin secretagogue actions. *Food Chem.* 139: 362–370.

Wu, D., H. Wang, D.-Z. Li & S. Blackmore. 2005a. Pollen morphology of *Parnassia* L. (Parnassiaceae) and its systematic implications. *J. Integrative Plant Biol.* 47: 2–12.

Wu, D., H. Wang, J.-M. Lu & D.-Z Li. 2005b. Comparative morphology of leaf epidermis in *Parnassia* (Parnassiaceae) from China. *Acta Phytotax. Sin.* 43: 210–224.

Wu, L.-Y. 1962. *Paratylenchus brevihastus* n. sp. (Criconematidae: Nematoda). *Can. J. Zool.* 40: 391–393.

Wu, S.-H., C.-F. Hsieh & M. Rejmánek. 2004. Catalogue of the naturalized flora of Taiwan. *Taiwania* 49: 16–31.

Wu, S.-H., T. Y. A. Yang, Y.-C. Teng, C.-Y. Chang, K.-C. Yang & C.-F. Hsieh. 2010. Insights of the latest naturalized flora of Taiwan: change in the past eight years. *Taiwania* 55: 139–159.

Wu, Z. & D. Yu. 2004. The effects of competition on growth and biomass allocation in *Nymphoides peltata* (Gmel.) O. Kuntze growing in microcosm. *Hydrobiologia* 527: 241–250.

Wu, Z., S. Gui, Z. Quan, L. Pan, S. Wang, W. Ke, D. Liang and Y. Ding. 2014a. A precise chloroplast genome of *Nelumbo nucifera* (Nelumbonaceae) evaluated with Sanger, Illumina MiSeq, and PacBio RS II sequencing platforms: insight into the plastid evolution of basal eudicots. *BMC Plant Biol.* 14: 289. doi:10.1186/s12870-014-0289-0.

Wu, Z., S. Gui, S. Wang and Y. Ding. 2014b. Molecular evolution and functional characterisation of an ancient phenylalanine ammonia-lyase gene (*NnPAL1*) from *Nelumbo nucifera*: novel insight into the evolution of the PAL family in angiosperms. *BMC Evol. Biol.* 14: 100. doi:10.1186/1471-2148-14-100.

Wunderlin, R. P. & B. F. Hensen. 2001. Seven new combinations in the Florida flora. *Novon* 11: 366–369.

Wunderlin, R. P. & D. H. Les. 1980. *Nymphaea ampla* (Nymphaeaceae), a waterlily new to Florida. *Phytologia* 45: 82–84.

Wunderlin, R. P., B. F. Hansen & D. W. Hall. 1988. The vascular flora of central Florida: taxonomic and nomenclatural changes, additional taxa, II. *Sida* 13: 83–91.

Wunderlin, R. P., B. F. Hansen & L. C. Anderson. 2002. Plants new to the United States and Florida. *Sida* 20: 813–817.

Wurdack, J. J. & R. Kral. 1982. The genera of Melastomataceae in the southeastern United States. *J. Arnold Arbor.* 63: 429–439.

Wurdack, K. J. & C. C. Davis. 2009. Malpighiales phylogenetics: Gaining ground on one of the most recalcitrant clades in the angiosperm tree of life. *Amer. J. Bot.* 96: 1551–1570.

Wurdack, K. J., P. Hoffmann & M. W. Chase. 2005. Molecular phylogenetic analysis of uniovulate Euphorbiaceae (Euphorbiaceae sensu stricto) using plastid *rbcL* and *trnL-F* DNA sequences. *Amer. J. Bot.* 92: 1397–1420.

Wurzburger, N. & C. S. Bledsoe. 2001. Comparison of ericoid and ectomycorrhizal colonization and ectomycorrhizal morphotypes in mixed conifer and pygmy forests on the northern California coast. *Can. J. Bot.* 79: 1202–1210.

Wyatt, R. 1986. Ecology and evolution of self-pollination in *Arenaria uniflora* (Caryophyllaceae). *J. Ecol.* 74: 403–418.

Wyatt, R. & S. B. Broyles. 1992. Hybridization in North American *Asclepias*. III. isozyme evidence. *Syst. Bot.* 17: 640–648.

Wyatt, R. & N. Fowler. 1977. The vascular flora and vegetation of the North Carolina granite outcrops. *Bull. Torrey Bot. Club* 104: 245–253.

Wyatt, R., A. L. Edwards, S. R. Lipow & C. T. Ivey. 1998. The rare *Asclepias texana* and its widespread sister species, *A. perennis*, are self-incompatible and interfertile. *Syst. Bot.* 23: 151–156.

Wyatt, R., S. B. Broyles & S. R. Lipow. 2000. Pollen-ovule Ratios in milkweeds (Asclepiadaceae): an exception that probes the rule. *Syst. Bot.* 25: 171–180.

Xavier, K. S. & C. M. Rogers. 1963. Pollen morphology as a taxonomic tool in *Linum*. *Rhodora* 65: 137–145.

Xi, Z., B. R. Ruhfel, H. Schaefer, A. M. Amorim, M. Sugumaran, K. J. Wurdack, P. K. Endress, M. L. Matthews, P. F. Stevens, S. Mathews & C. C. Davis. 2012. Phylogenomics and a posteriori data partitioning resolve the Cretaceous angiosperm radiation Malpighiales. *Proc. Natl. Acad. Sci. U.S.A.* 109: 17519–17524.

Xiang, Q.-Y., D. T. Thomas & Q. P. Xiang. 2011. Resolving and dating the phylogeny of Cornales – Effects of taxon sampling, data partitions, and fossil calibrations. *Mol. Phylogen. Evol.* 59: 123–138.

Xie, L., J. Wen & L.-Q. Li. 2011. Phylogenetic analyses of *Clematis* (Ranunculaceae) based on sequences of nuclear ribosomal ITS and three plastid regions. *Syst. Bot.* 36: 907–921.

Xiuwen, Z., Z. Xiaorui & J. Xiaobai. 1997. *Flowering Lotus of China*. Jindun Publishing House, Beijing, China. 128 pp.

Xu, Z., F. R. Chang, H. K. Wang, Y. Kashiwada, A. T. McPhail, K. F. Bastow, Y. Tachibana, M. Cosentino & K. H. Lee. 2000. Anti-HIV agents 45[1] and antitumor agents 205[2]. Two new sesquiterpenes, leitneridanins A and B, and the cytotoxic and anti-HIV principles from *Leitneria floridana*. *J. Nat. Prod.* 63: 1712–1715.

Xue, X.-F., H. Sadeghi, X.-Y. Hong & S. Sinaie. 2011. Nine eriophyoid mite species from Iran (Acari, Eriophyidae). *Zookeys* 143: 23–45.

Ya, V. B. 2015. Aquatic plants of the Far East of Russia: a review on their use in medicine, pharmacological activity. *Bangladesh J. Med. Sci.* 14: 9–13.

Yaacob, O. & H. D. Tindall. 1995. *Mangosteen Cultivation*. FAO Plant Production and Protection Paper 129. Food and Agriculture Organization of the United Nations, Rome, Italy. 100 pp.

Yager, L. Y. & S. W. Leonard. 2005. Rare plant species on Camp Shelby training site, MS. pp. 145–150. *In*: J. S., Kush (ed.), *Longleaf Pine: Making Dollar$ and Sense, Proceedings of the Fifth Longleaf Alliance Regional Conference; 2004 October 12–15*. Longleaf Alliance Report No. 8. Hattiesburg, MS.

Yahara, T., T. Kawahara, D. J. Crawford, M. Ito and K. Watanabe. 1989. Extensive gene duplications in diploid *Eupatorium* (Asteraceae). *Amer. J. Bot.* 76: 1247–1253.

Yamada, S., S. Okubo, Y. Kitagawa & K. Takeuchi. 2007. Restoration of weed communities in abandoned rice paddy fields in the Tama Hills, central Japan. *Agric. Eco-Syst. Environ.* 119: 88–102.

Yamaguchi, T., E. Mukai & Y. Nakayama. 2005. [A genetic evaluation on *Lindernia* species emerging from the seed-bank in a wetland restoration in Azamenose site, the Matsuura River, Saga prefecture, Japan.] *Sci. Rep. Graduate School Agric. Biol. Sci. Osaka Pref. Univ.* 57: 25–32 (in Japanese with English abstract).

Yan, F., W. Wang, L. Liu, X. Li & G. Liang. 2006. Cytological studies on *Hibiscus esculentus* L. and *Hibiscus coccineus* (Medicus) Walt. *S. W. Hort.* 1: 5.

Yang, L.-P. 2009. Study on the germination of *Oenanthe javanica* (Blume) DC seeds. *Agric. Technol. Serv.* 26: 163–164.

Yang, M., Y. Han, L. Xu, J. Zhao & Y. Liu. 2012. Comparative analysis of genetic diversity of lotus (*Nelumbo*) using SSR and SRAP markers. *Sci. Hort.* 142: 185–195.

Yang, M., F. Liu, Y. Han, L. Xu, N. Juntawong & Y. Liu. 2013. Genetic diversity and structure in populations of *Nelumbo* from America, Thailand and China: Implications for conservation and breeding. *Aquat. Bot.* 107: 1–7.

Yang, Q.-H., W.-H. Ye, X. Deng, H.-L. Cao, Y. Zhang & K.-Y. Xu. 2005. Seed germination eco-physiology of *Mikania micrantha* H.B.K. *Bot. Bull. Acad. Sin.* 46: 293–299.

Yang, R.-Y., S. C. S. Tsou, T.-C. Lee, W.-J. Wu, P. M. Hanson, G. Kuo, L. M. Engle & P.-Y. Lai. 2006. Distribution of 127 edible plant species for antioxidant activities by two assays. *J. Sci. Food Agric.* 86: 2395–2403.

Yang, R.-Y., S. Lin & G. Kuo. 2008. Content and distribution of flavonoids among 91 edible plant species. *Asian Pacific J. Clin. Nutr.* 17: S275-S279.

Yang, X. B., Z. M. Huang, W. B. Cao, M. Zheng, H. Y. Chen & J. Z. Zhang. 2000. Antidiabetic effect of *Oenanthe javanica* flavone. *Acta Pharm. Sin.* 21: 239–242.

Yang, Y. & P. E. Berry. 2011. Phylogenetics of the *Chamaesyce* clade (*Euphorbia*, Euphorbiaceae): Reticulate evolution and long-distance dispersal in a prominent C_4 lineage. *Amer. J. Bot.* 98: 1486–1503.

Yang, Y., R. Riina, J. J. Morawetz, T. Haevermans, X. Aubriot & P. E. Berry. 2012. Molecular phylogenetics and classification of *Euphorbia* subgenus *Chamaesyce*: a group with high diversity in photosynthetic systems and growth forms. *Taxon* 61: 764–789.

Yang, Y., M. J. Moore, S. F. Brockington, D. E. Soltis, G. K.-S. Wong, E. J. Carpenter, Y. Zhang, L. Chen, Z. Yan, Y. Xie, R. F. Sage, S. Covshoff, J. M. Hibberd, M. N. Nelson & S. A. Smith. 2015. Dissecting molecular evolution in the highly diverse plant clade Caryophyllales using transcriptome sequencing. *Mol. Biol. Evol.* 32: 2001–2014.

Yang, Y.-P. & S.-H. Yen. 1997. Notes on *Limnophila* (Scrophulariaceae) of Taiwan. *Bot. Bull. Acad. Sin.* 38: 285–295.

Yang, Y.-W., K.-N. Lai, P.-Y. Tai, D.-P. Ma & W.-H. Li. 1999. Molecular phylogenetic studies of *Brassica, Rorippa, Arabidopsis*, and allied genera based on the internal transcribed spacer region of 18S-25S rDNA. *Mol. Phylogen. Evol.* 13: 455–462.

Yano, S. 1997. Silique burst of *Cardamine scutata* (Cruciferae) as a physical inducible defense against seed predatory caterpillars. *Res. Pop. Ecol.* 39: 95–100.

Yanovsky, E. 1936. *Food Plants of the North American Indians*. Miscellaneous Publication Number 237. U. S. Department of Agriculture, Washington, DC. 83 pp.

Yaowakhan, P., M. Kruatrachue, P. Pokethitiyook & V. Soonthornsarathool. 2005. Removal of lead using some aquatic macrophytes. *Bull. Environ. Contam. Toxicol.* 75: 723–730.

Yarbrough, J. A. 1936. The foliar embryos of *Tolmiea menziesii*. *Amer. J. Bot.* 23: 16–20.

Yasuda, K. & H. Yamaguchi. 2005. Genetic diversity of vegetable water pepper (*Persicaria hydropiper* (L.) Spach) as revealed by RAPD markers. *Breeding Sci.* 55: 7–14.

Yatskievych, G. & J. A. Raveill. 2001. Notes on the increasing proportion of non-native angiosperms in the Missouri flora, with reports of three new genera for the state. *Sida* 19: 701–709.

Yatsu, Y., N. Kachi & H. Kudoh. 2003. Ecological distribution and phenology of an invasive species, *Cardamine hirsuta* L., and its native counterpart, *Cardamine flexuosa* With., in central Japan. *Plant Species Biol.* 18: 35–42.

Yavorska, O. G. 2009. The North American species of the non-native flora of the Kyiv urban area (Ukraine): a checklist and analysis. *Biodivers. Res. Conserv.* 13: 25–30.

Ye, W. H., J. Li, H. L. Cao & X. J. Ge. 2003. Genetic uniformity of *Alternanthera philoxeroides* in South China. *Weed Res.* 43: 297–302.

Yi, T., A. J. Miller & J. Wen. 2004a. Phylogenetic and biogeographic diversification of *Rhus* (Anacardiaceae) in the Northern Hemisphere. *Mol. Phylogen. Evol.* 33: 861–879.

Yi, T., P. P. Lowry II., G. M. Plunkett & J. Wen. 2004b. Chromosomal evolution in Araliaceae and close relatives. *Taxon* 53: 987–1005.

Yi, Y. L., Y. Lei, Y. B. Yin, H. Y. Zhang & G. X. Wang. 2012. The antialgal activity of 40 medicinal plants against *Microcystis aeruginosa*. *J. Appl. Phycol.* 24: 847–856.

Yin, L., B. P. Colman, B. M. McGill, J. P. Wright & E. S. Bernhardt. 2012. Effects of silver nanoparticle exposure on germination and early growth of eleven wetland plants. *PLoS One* 7(10): e47674. doi:10.1371/journal.pone.0047674.

Yocum, C. F. 1951. *Waterfowl and Their Food Plants in Washington*. University of Washington Press, Seattle, WA. 272 pp.

Yoo, M.-J. & C.-W. Park. 2001. Molecular phylogeny of *Polygonum* sect. *Echinocaulon* (Polygonaceae) based on chloroplast and nuclear DNA sequences. Botany 2001 meeting (abstract).

Yoon, J., X. Cao, Q. Zhou & L. Q. Ma. 2006. Accumulation of Pb, Cu, and Zn in native plants growing on a contaminated Florida site. *Sci. Total Environ.* 368: 456–464.

Yoon, J. S., H. M. Kim, A. K. Yadunandam, N. H. Kim, H. A. Jung, J. S. Choi, C. Y. Kim & G. D. Kim. 2013. Neferine isolated from *Nelumbo nucifera* enhances anti-cancer activities in Hep3B cells: molecular mechanisms of cell cycle arrest, ER stress induced apoptosis and anti-angiogenic response. *Phytomedicine* 20: 1013–1022.

York, D. 1997. A fire ecology study of a Sierra Nevada foothill basaltic mesa grassland. *Madroño* 44: 374–383.

York, H. H. 1905. The hibernacula of Ohio water plants. *Ohio Naturalist* 5: 291–293.

Yoshie, F. 2008. Dormancy of alpine and subalpine perennial forbs. *Ecol. Res.* 23: 35–40.

Yoshino, N., G.-X. Wang, M. Ito, B. Auld, H. Kohara & T. Enomoto. 2006. Naturalization and dissemination of two subspecies of *Lindernia dubia* (Scrophulariaceae) in Japan. *Weed Biol. Manag.* 6: 174–176.

Young, B. 2001. Propagation protocol for production of container *Oenanthe sarmentosa* Presl. plants. Golden Gate National Parks, San Francisco, CA.

Young, J. A. & C. D. Clements. 2003. Seed germination of willow species from a desert riparian ecosystem. *J. Range Manag.* 56: 496–500.

Young, N. D., K. E. Steiner & C. W. dePamphilis. 1999. The evolution of parasitism in Scrophulariaceae/Orobanchaceae: plastid gene sequences refute an evolutionary transition series. *Ann. Missouri Bot. Gard.* 86: 876–893.

Young, O. P. 1986. Host plants of the tarnished plant bug, *Lygus lineolaris* (Heteroptera: Miridae). *Ann. Entomol. Soc. Amer.* 79: 747–762.

Young, S. M. & T. W. Weldy. 2004. *New York Natural Heritage Program Rare Plant Status List*. New York Natural Heritage Program, Albany, NY. 105 pp.

Young, V. A. 1936. Certain sociological aspects associated with plant competition between native and foreign species in a saline area. *Ecology* 17: 133–142.

Yu, F., Y. Chen & M. Dong. 2001. Clonal integration enhances survival and performance of *Potentilla anserina*, suffering from partial sand burial on Ordos plateau, China. *Evol. Ecol.* 15: 303–318.

Yu, T.-W., M. Xu & R. H. Dashwood. 2004. Antimutagenic activity of spearmint. *Environ. Mol. Mutagen.* 44: 387–393.

Yu, Y. 2015. Research of dyeing performance of natural plant dye *Alternanthera paronychioides* on wool fabric. *Wool Textile J.* 9: 15.

Yuan, M., W. H. Carlson, R. D. Heins & A. C. Cameron. 1998a. Effect of forcing temperature on time to flower of *Coreopsis grandiflora*, *Gaillardia* × *grandiflora*, *Leucanthemum* × *superbum*, and *Rudbeckia fulgida*. *HortScience* 33: 663–667.

Yuan, M., W. H. Carlson, R. D. Heins & A. C. Cameron. 1998b. Determining the duration of the juvenile phase of *Coreopsis grandiflora* (Hogg ex Sweet.), *Gaillardia* ×*grandiflora* (Van Houtte), *Heuchera sanguinea* (Engelm.) and *Rudbeckia fulgida* (Ait.). *Sci. Hort.* 72: 135–150.

Yuan, X., L. Morano, R. Bromley, S. Spring-Pearson, R. Stouthamer & L. Nunney. 2010. Multilocus sequence typing of *Xylella fastidiosa* causing Pierce's disease and oleander leaf scorch in the United States. *Phytopathology* 100: 601–611.

Yuan, Y.-M., P. Küpfer & J. J. Doyle. 1996. Infrageneric phylogeny of the genus *Gentiana* (Gentianaceae) inferred from nucleotide sequences of the internal transcribed spacers (ITS) of nuclear ribosomal DNA. *Amer. J. Bot.* 83: 641–652.

Yuan, Y.-M., P. Küpfer & L. Zeltner. 1998. Chromosomal evolution of *Gentiana* and *Jaeschkea (Gentianaceae),* with further documentation of chromosome data for 35 species from western China. *Plant Syst. Evol.* 210: 231–247.

Yukio, H. 1965. Chromosome analysis in the tribe Astereae. *Jap. J. Genet.* 40: 63–71.

Yun, H. Y., A. M. Minnis, L. J. Dixon, L. A. Castlebury & S. M. Douglas. 2010. First report of *Uromyces acuminatus* on *Honckenya peploides*, the endangered seabeach sandwort. *Plant Dis.* 94: 279.

Zagrebel'nyi, S. V. 2003. Den ecology of the arctic fox *Alopex lagopus beringensis* (Carnivora, Canidae) on Bering Island, Commander Islands. *Russ. J. Ecol.* 34: 114–121.

Zahina, J. G., W. P. Said, R. Grein & M. Duever. 2007. *Pre-Development Vegetation Communities of Southern Florida.* Technical Publication HESM-02. South Florida Water Management District, Miami, FL. 136 pp.

Zald, H. S. J. 2010. Patterns of tree establishment and vegetation composition in relation to climate and topography of a subalpine meadow landscape, Jefferson Park, Oregon, USA. PhD dissertation, Oregon State University, Corvallis, OR. 180 pp.

Zalewski, M., B. Bis, M. Łapińska, P. Frankiewicz & W. Puchalski. 1998. The importance of the riparian ecotone and river hydraulics for sustainable basin-scale restoration scenarios. *Aquatic Conservation* 8: 287–307.

Zalewska-Gałosz, J., M. Jopek & T. Ilnicki. 2014. Hybridization in *Batrachium* group: Controversial delimitation between heterophyllous *Ranunculus penicillatus* and the hybrid *Ranunculus fluitans* × *R. peltatus. Aquat. Bot.* 120: 160–168.

Zambino, P., G. McDonald & B. Richardson. 2005. White pine blister rust: pathologists identify new hosts. *Nutcracker Notes* 8: 11.

Zammit, C. & P. H. Zedler. 1990. Seed yield, seed size and germination behaviour in the annual *Pogogyne abramsii. Oecologia* 84: 24–28.

Zammit, C. & P. H. Zedler. 1994. Organisation of the soil seed bank in mixed chaparral. *Plant Ecol.* 111: 1–16.

Zampella, R. A. & K. J. Laidig. 1997. Effect of watershed disturbance on pinelands stream vegetation. *J. Torrey Bot. Soc.* 124: 52–66.

Zampella, R. A., J. F. Bunnell, K. J. Laidig & C. L. Dow. 2001. The Mullica River Basin: A Report to the Pinelands Commission on the Status of the Landscape and Selected Aquatic and Wetland Resources. The New Jersey Pinelands Commission, New Lisbon, NJ. 371 pp.

Zampella, R. A., J. F. Bunnell, K. J. Laidig & N. A. Procopio. 2006. Monitoring the Ecological Integrity of Pinelands Wetlands: A Comparison of Wetland Landscapes, Hydrology, and Stream Communities in Pinelands Watersheds Draining Active-Cranberry Bogs, Abandoned-Cranberry Bogs, and Forest Land. Final report submitted to the U. S. Environmental protection Agency, February, 2006. The New Jersey Pinelands Commission, New Lisbon, NJ. 135 pp.

Zardini, E. & P. H. Raven. 1992. A new section of *Ludwigia* (Onagraceae) with a key to the sections of the genus. *Syst. Bot.* 17: 481–485.

Zardini, E. M., H. Gu & P. H. Raven. 1991. On the separation of two species within the *Ludwigia uruguayensis* complex (Onagraceae). *Syst. Bot.* 16: 242–244.

Zaremba, R. E. 2004. *Sabatia campanulata* (L.) Torrey, slender marsh-pink. New England Plant Conservation Program, conservation and research plan. New England Wild Flower Society, Framingham, MA. 21 pp.

Zartman, C. E. & J. D. Pittillo. 1998. Spray cliff communities of the Chattooga basin. *Castanea* 63: 217–240.

Zavitkovski, J. 1976. Ground vegetation biomass, production, and efficiency of energy utilization in some northern Wisconsin forest ecosystems. *Ecology* 57: 694–706.

Zazula, G. D., D. G. Froese, S. A. Elias, S. Kuzmina & R. W. Mathewes. 2011. Early Wisconsinan (MIS 4) Arctic ground squirrel middens and a squirrel-eye-view of the mammoth-steppe. *Quatern. Sci. Rev.* 30: 2220–2237.

Zebell, R. K. & P. L. Fiedler. 1996. *Restoration and Recovery of Mason's* Lilaeopsis: *Phase II.* Final Report. San Francisco State University, San Francisco, CA. 37 pp.

Zedler, J. B. 2000. *Handbook for Restoring Tidal Wetlands.* CRC Press, Boca Raton, FL. 464 pp.

Zedler, J. B., J. C. Callaway, J. S. Desmond, G. Vivian-Smith, G. D. Williams, G. Sullivan, A. E. Brewster & B. K. Bradshaw. 1999. Californian salt-marsh vegetation: an improved model of spatial pattern. *Ecosystems* 2: 19–35.

Zedler, P. H. 1987. The ecology of southern California vernal pools: a community profile. *U. S. Fish Wildl. Serv. Biol. Rep.* 35: 1–135.

Zedler, P. H. 2003. Vernal pools and the concept of "isolated wetlands." *Wetlands* 23: 597–607.

Zedler, P. H. & C. Black. 1992. Seed dispersal by a generalized herbivore: rabbits as dispersal vectors in a semiarid California vernal pool landscape. *Amer. Midl. Naturalist* 128: 1–10.

Zedler, P. H. & C. Black. 2004. Exotic plant invasions in an endemic-rich habitat: The spread of an introduced Australian grass, *Agrostis avenacea* J. F. Gmel., in California vernal pools. *Austral Ecol.* 29: 537–546.

Zeng, L., Q. Zhang, R. Sun, H. Kong, N. Zhang & H. Ma. 2014. Resolution of deep angiosperm phylogeny using conserved nuclear genes and estimates of early divergence times. *Nat. Commun.* 5: doi:10.1038/ncomms5956.

Zerbe, S., I.-K. Choi & I. Kowarik. 2004. Characteristics and habitats of non-native plant species in the city of Chonju, southern Korea. *Ecol. Res.* 19: 91–98.

Zettler, F. W. & T. E. Freeman. 1972. Plant pathogens as biocontrols of aquatic weeds. *Ann. Rev. Phytopathology* 10: 455–470.

Zettler, L. W., N. S. Ahuja & T. M. McInnis Jr. 1996. Insect pollination of the endangered monkey-face orchid (*Platanthera integrilabia*) in McMinn County, Tennessee: One last glimpse of a once common spectacle. *Castanea* 61: 14–24.

Zhang, C., X. Chen, B. Wei, Z. Li, S. Liu & Q. Li. 2009. Study on removal efficiency of nitrogen and phosphorus from sewage by aquatic macrophtes under two cultivation modes. *Southwest China J. Agric. Sci.* 22: 786–790.

Zhang, L.-B. & M. P. Simmons. 2006. Phylogeny and delimitation of the Celastrales inferred from nuclear and plastid genes. *Syst. Bot.* 31: 122–137.

Zhang, L.-B., M. P. Simmons, A. Kocyan & S. S. Renner. 2006. Phylogeny of the Cucurbitales based on DNA sequences from nine loci from three genomes. *Mol. Phyogen. Evol.* 39: 305–322.

Zhang, L. J., S. F. Yeh, Y. T. Yu, L. M. Y. Kuo & Y. H. Kuo. 2011. Antioxidative flavonol glucuronides and AntiHBsAg flavonol from *Rotala rotundifolia. J. Tradit. Complem. Med.* 1: 57–63.

Zhang, X. & J. A. Mosjidis. 1998. Rapid prediction of mating system of *Vicia* species. *Crop Sci.* 38: 872–875.

Zhang, Z., S. Li & S. Zhang. 2005. Six new triterpenoid saponins from the root and stem bark of *Cephalanthus occidentalis. Nat. Prod. Chem.* 71: 355–361.

Zhao, H.-B., F.-D. Chen, S.-M. Chen, G.-S. Wu & W.-M. Guo. 2010. Molecular phylogeny of *Chrysanthemum, Ajania* and its allies (Anthemideae, Asteraceae) as inferred from nuclear ribosomal ITS and chloroplast *trn*L-F IGS sequences. *Plant Syst. Evol.* 284: 153–169.

Zhen, W.-L., W. Wang, D.-S.Kong, Y.-F. Zhan, L.-P. Ding & Y.-F. Teng. 2009. Salt tolerance of four kinds of herbaceous perennial flowers. *J. N. E. Forest. Univ.* 37: 61–63.

Zheng, C. & D. Sankoff. 2014. Practical halving; the *Nelumbo nucifera* evidence on early eudicot evolution. *Computat. Biol. Chem.* 50: 75–81.

Zheng, J., H. Hintelmann, B. Dimock & M. S. Dzurko. 2003. Speciation of arsenic in water, sediment, and plants of the Moira watershed, Canada, using HPLC coupled to high resolution ICP-MS. *Analysis Bioanal. Chem.* 377: 14–24.

Zhong, J. & E. A. Kellogg. 2014. Duplication and expression of *CYC2*-like genes in the origin and maintenance of corolla zygomorphy in Lamiales. *New Phytol.* 205: 852–868.

Zhou, J., H. Peng, S. R. Downie, Z.-W. Liu & X. Gong. 2008. A molecular phylogeny of Chinese Apiaceae subfamily Apioideae inferred from nuclear ribosomal DNA internal transcribed spacer sequences. *Taxon* 57: 402–416.

Zhou, J.-S., F.-G. Wang & F.-W. Xing. 2010. *Pluchea sagittalis*, a naturalized medical plant in mainland China. *Guihaia* 4: 10.

Zhou, X., J. Wang, L. Xue, X. Xu & L. Yang. 2005. N and P removal characteristics of a eutrophic water body under floating plants. *J. Appl. Ecol.* 16: 2199–2203.

Zhu, N., X. Li, G. Liu, X. Shi, M. Gui, C. Sun & Y. Jin. 2009. Constituents from aerial parts of *Bidens ceruna* L. and their DPPH radical scavenging activity. *Chem. Res. Chinese Univ.* 25: 328–331.

Zielinski, Q. B. 1953. Chromosome numbers and meiotic studies in *Ribes. Bot. Gaz.* 114: 265–274.

Ziemkiewicz, P. F. & E. H. Cronin. 1981. Germination of seed of three varieties of spotted locoweed. *J. Range Manag.* 34: 94–97.

Zika, P. F. 1988. Contributions to the flora of the Lake Champlain valley, New York and Vermont, II. *Bull. Torrey Bot. Club* 115: 218–220.

Zika, P. F. 1996. *Pilularia americana* A. Braun in Klamath County, Oregon. *Amer. Fern J.* 86: 26.

Zika, P. F. 2003. Notes on the provenance of some eastern wetland species disjunct in western North America. *J. Torrey Bot. Soc.* 130: 43–46.

Zimmerman, P. W. & A. E. Hitchcock. 1929. Vegetative propagation of holly. *Amer. J. Bot.* 16: 556–570.

Zimmerman, T. G. & S. H. Reichard. 2005. Factors affecting persistence of Wenatchee Mountains checker-mallow: an exploratory look at a rare endemic. *N. W. Sci.* 79: 172–178.

Zobel, A. M. & R. E. March. 1993. Autofluorescence reveals dfferent histological localizations of furanocoumarins in fruit of some Umbelliferae and Leguminosae. *Ann. Bot.* 71: 251–255.

Zohary, M. & D. Heller. 1984. The genus *Trifolium*. The Israel Academy of Sciences & Humanities, Jerusalem. 606 pp.

Zomlefer, W. B. & D. E. Giannasi. 2005. Floristic Survey of Castillo de San Marcos National Monument, St. Augustine, Florida. *Castanea* 70: 222–236.

Zomlefer, W. B., D. E. Giannasi & W. S. Judd. 2007. A floristic survey of National Park Service areas of Timucuan ccological and historic preserve (including Fort Caroline National Memorial), Duval County, Florida. *J. Bot. Res. Inst. Texas* 1: 1157–1178.

Zomlefer, W. B., D. E. Giannasi, K. A. Bettinger, S. L. Echols & L. M. Kruse. 2008. Vascular plant survey of Cumberland Island National Seashore, Camden County, Georgia. *Castanea* 73: 251–282.

Zungsontiporn, S. 2002. Global invasive plants in Thailand and its [sic!] status and a case study of *Hydrocotyle umbellata* L. Available online: http://www.fftc.agnet.org/activities/sw/2006/589543823/paper-615244292.pdf.

Zutter, C. 2009. Paleoethnobotanical contributions to 18th-century Inuit economy: an example from Uivak, Labrador. *J. N. Atlantic* 2: 23–32 [Special Volume 1].

Zwank, P. J., T. H. White Jr. & F. G. Kimmel. 1988. Female turkey habitat use in Mississippi River batture. *J. Wildl. Manag.* 52: 253–260.

Zygadlo, J. A., R. E. Morero, R. E. Abburra & C. A. Guzman. 1994. Fatty acid composition in seed oils of some Onagraceae. *J. Amer. Oil Chem. Soc.* 71: 915–916.

Index